U0215406

A Dictionary of Seed Plant Names

Vol. 1 In Latin, Chinese and English (A-D)

种子植物名称

卷1 拉汉英名称 (A-D)

(1-139738)

尚衍重　编著

中国林业出版社

图书在版编目(CIP)数据

种子植物名称. 卷1, 拉汉英名称. A－D / 尚衍重编著.
—北京: 中国林业出版社, 2012.6

ISBN 978－7－5038－6655－5

Ⅰ. ①种… Ⅱ. ①尚… Ⅲ. ①种子植物－专有名称－拉丁语、汉语、英语 Ⅳ. ①Q949.4－61

中国版本图书馆 CIP 数据核字(2012)第139138号

中国林业出版社·自然保护图书出版中心

出 版 人: 金 旻
策划编辑: 温 晋
责任编辑: 刘家玲 温 晋 周军见 李 敏

出版 中国林业出版社(100009 北京市西城区刘海胡同7号)
网址 http://lycb.forestry.gov.cn
E-mail wildlife_cfph@163.com 电话 010－83225836
发行 中国林业出版社
营销电话: (010)83284650 83227566
印刷 北京中科印刷有限公司
版次 2012年6月第1版
印次 2012年6月第1次
开本 889mm×1194mm 1/16
印张 126.5
字数 10455千字
印数 1～2000册
定价 768.00元

内容简介

这是国内外第一部植物学名标准化、多功能的大型工具书。首批出版5卷:《卷1　拉汉英名称(A－D)》、《卷2　拉汉英名称(E－O)》、《卷3　拉汉英名称(P－Z)》、《卷4　中文名称索引》、《卷5　英日俄名称索引》。本书共收录拉丁名418831条，包括全部科名和属名，命名人均依据《国际植物命名法规》的要求标准化。5种文字的名称总计967723条，可以互相检索，故本书相当于20部单项词典。书中还用符号给出了非国产植物、木本植物、草本植物、中国特有种子植物属及国家确定的濒危植物等信息。它不仅对植物学及相关学科的高级研究人员有重要参考价值，而且也是农业、林业、医药、环境保护、生物工程、植物检疫、花卉园艺、新闻出版、外贸旅游等领域的教师、学生和工作人员不可或缺的工具书。

About this Book

This is the first, in China or abroad, large-scale, multifunctional reference book having standardized scientific names of plants. Its first publication comprises five volumes: *Vol.* 1 *In Latin, Chinese and English(A－D)*, *Vol.* 2 *In Latin, Chinese and English(E－O)*, *Vol.* 3 *In Latin, Chinese and English(P－Z)*, *Vol.* 4 *Chinese Index* and *Vol.* 5 *English, Japanese and Russian Indices.* This book contains a total of 418, 831 Latin entries, including all the names of families and genera. The scientific names, including the author names, are standardized according to the *International Code of Botanical Nomenclature*. The plant names in 5 languages compiled in this book amount to a total of 967, 723 and they can be cross-referenced, making it equivalent to 20 bilingual dictionaries. Also, symbols are used to indicate those non-native to China, woody or herbaceous, seed plant genera unique to China and the plants listed as "endangered" by relevant government bodies. This dictionary serves as a highly valuable source of reference not only for scientists and academics working in plant sciences and related fields, but also for teachers, students and workers in the areas of agriculture, forestry, medicine, environmental protection and conservation, biotechnology, plant quarantine, horticulture, press and publication, foreign trade and tourism industries.

序　一

尚衍重教授编著的《种子植物名称》付梓之际，请我写序，欣然命笔。因为与已出版的几本同类书相比，本书更有特色。

首先是本书所收植物拉丁名的标准化。植物名称的标准化是植物研究、交流和利用的必要条件。国际植物学会每隔六年修订一次《国际植物命名法规》，目的也在于此。为纠正拉丁名命名人缩写方面的混乱，东京法规还指定《Authors of Plant Names》(Brummitt & Powell，1992)一书作为命名人缩写的蓝本，其后所有版本的《国际植物命名法规》引证的所有命名人均依据《Authors of Plant Names》一书标准化。但当今国内外出版物中拉丁名不规范的现象十分严重。尚衍重教授在规范拉丁名方面做了大量工作。本书收录的约42万条拉丁名命名人缩写全部依据《Authors of Plant Names》一书标准化；词尾不符合拉丁文语法者，加词中带有连字符“-”者，自动名中带有命名人者，命名人中漏缺或误写者等，尚衍重教授亦参照有关资料做了订正。对如此大量的拉丁名标准化，本书在国内外还是第一部。它对规范植物名称，特别是中国种子植物名称将起到重要作用。

其次，本书收录拉丁文、中文、日文、俄文、英文5种文字名称。仅从语种方面，本书即相当于20部单项词典，应用面更广，对植物学方面的国际交流意义更大。

第三，信息量大。本书收入5种文字名称共约100万条，并包含了全部中国种子植物。这是迄今为止收录最全的一部书。书中还标出了木本植物、草本植物，科、属的有关数据等资料，一书多用，为读者提供了很多方便。

简言之，本书不仅对植物分类学者有重要参考价值，而且对农业、林业、医药、园林等部门的工作者来说，都是一部不可多得的好书。

中国科学院　院士
第三世界科学院　院士
植物分类学报　原主编
中国科学院植物研究所　研究员
洪德元

Preface 1

Prof. Yanzhong Shang asked me to write a few opening words in the time when his book *A dictionary of Seed Plant Names* is being sent to printing and I'm delighted to oblige. Compare to other books of similar kind, his book has very distinctive features.

Firstly, the book has standardized the Latin names of plants. Standardization of plant names is the prerequisite to research, exchange of information and utilization of plants. That is why the International Association for Plant Taxonomy revises the *International Code of Botanical Nomenclature* every six years. To resolve the confusion in abbreviations of author names in plant Latin names, the *Tokyo Code* (1994) designated the *Authors of Plant Names* (Brummitt & Powell, 1992) as the blueprint for author name abbreviations and all the author names cited in the later editions of the *International Code of Botanical Nomenclature* have been standardized according to the *Authors of Plant Names*. At present, however, a long-standing and widespread problem is that plant Latin names in various publications worldwide do not always follow the standard format sanctioned by the *Code*. Prof. Yanzhong Shang's achievement in standardization of plant Latin names is phenomenal. For the ca 420,000 Latin names compiled in this dictionary, all abbreviations of author names are standardized according to the *Authors of Plant Names*. Never before, has standardization of plant Latin names been achieved in such a grand scale. Moreover, incorrect terminations, misuse of the hyphen "–", and missing or mistaken author names in scientific names have been corrected. I am sure that this book will play a hugely important role in standardization of plant names, especially of the seed plant names of China.

Secondly, this book has compiled plant names in five different languages, Latin, Chinese, Japanese, Russian and English. Just for the variety of languages alone, this book can be used as 20 bilingual dictionaries and will greatly benefit international communication and exchange of knowledge in plant science.

Thirdly, this book contains immensely rich information. It has compiled a total of a million plant names in five languages, including the names of all seed plants in China, making it the most comprehensive book of plant names ever published. Furthermore, this book is multifunctional, providing readers the added benefits of the information on plant families, genera and whether woody or herbaceous etc.

In short, this is an exceptionally valuable reference book not only for plant scientists, but also for workers and students in Agriculture, Forestry, Medicine, Horticulture and other relevant fields.

The Chinese Academy of Sciences **Academician**
The Third World Academy of Sciences **Academician**
Acta Phytotaxonomica Sinica Honorary **Editor in Chief** (former)
Institute of Botany, The Chinese Academy of Sciences **Professor**

Deyuan Hong

序 二

尚衍重教授编著的《种子植物名称》即将付梓。作为尚衍重教授的师兄，我很高兴写几句话。

这样一部传世巨著，尚衍重教授能以一人之力完成，确令世人匪夷所思。且不讲一人查阅5种文字的资料，也不讲微机输入和校对的工作量，仅规范80万人次的命名人缩写和42万条拉丁名这两项工作，组织一个专家班子，10年也未必能完成。但细想起来又非偶然。尚衍重教授的聪明和刻苦在我们头几届研究生中是公认的。尚衍重教授的导师邵力平教授是国内外著名的森林病理学家和真菌学家，其渊博的知识和严谨的学风闻名于世。尚衍重教授秉承其师之衣钵，可谓顺理成章。

这部著作的出版，不仅对规范植物名称具有重要意义，而且为需要查阅植物名称的诸多领域的科学家和工作人员将带来极大的方便，节省很多时间；同时也是母校的光荣和骄傲。在祝贺和感谢尚衍重教授之际，我也希望东林学子能发扬尚衍重教授的这种博学、严谨、拼搏之精神，多出几部好书，为中国林业的发展多做点贡献。

东北林业大学　原校长

中国工程院　院　士

Preface 2

As *A dictionary of Seed Plant Names* compiled by Prof. Yanzhong Shang is being sent to the press, I am very pleased to write a few opening words.

It is beyond anyone's imagination that Prof. Yanzhong Shang has single-handedly written such a *magnum opus*. The standardization of ca 800,000 author name abbreviations and of 420,000 plant Latin names alone would have taken at least 10 years for a group of specialists. This is not even counting the time and efforts spent in going through the vast amount of the literature, typing and proof reading of the manuscript. Looking back, Prof. Yanzhong Shang's such an outstanding achievement may not come as a complete surprise. Some 30 years ago, when Prof. Shang was doing his postgraduate studies in our university, his intelligence and hard-working attitude were well recognized among us postgraduate students. Prof. Liping Shao, the mentor of Prof. Yanzhong Shang, is the world-famous forest pathologist and mycologist. Prof. Shao is renowned for his profound knowledge and rigorous approach to science and Prof. Yanzhong Shang has certainly carried well forward the standard set by his mentor.

The publication of *A dictionary of Seed Plant Names* is an enormous contribution to the standardization of plant names and will greatly benefit scientists and workers in need of search of plant names across relevant scientific fields. It also brings honor and pride to our *alma mater*. With congratulations and gratitude to Prof. Yanzhong Shang, I sincerely hope that students from our university can learn from his extraordinary erudition, wholehearted devotion and rigorous approach to science and will write many more excellent books to contribute to the development of forestry in China.

Northeast Forestry University **President** (former)
The Chinese Academy of Engineering **Academician**
Jian Li

序 三

学弟尚衍重教授编著的《种子植物名称》杀青之际，请我作序，乐于从命。

我之所以称尚衍重教授为学弟是有原因的。恢复招考研究生的第一年(1978年)，我们二人同时报考王云章教授的研究生。学弟考试成绩优良，王先生已准备录取，但由于应试的外文语种是日语而未能被中国科学院微生物研究所认可，遂从师于东北林业大学邵力平教授。学弟研究生毕业后到内蒙古林学院(今内蒙古农业大学)任教30多年来，始终与我保持联系，互相探讨学问，可以称之为关系密切的师兄弟。

学弟治学严谨，工作勤奋。且不讲全书的编撰，仅标准化80万人次学名命名人，不知要花费多大的精力！其中只有极少数人未能查到出处，但他却特意加了"?"以提醒读者，学风之严谨，可见一斑。

学弟积累资料30余年，伏案耕作十几年，编著出这部如此丰富的名汇，令人钦佩。我是一个菌物学工作者，编辑《菌物学报》20余年，工作中经常为遇到大量的不规范的植物名称而深感头痛。我编著《中国真菌志 锈菌目》时，曾想把寄主植物名称的命名人缩写按照《Authors of Plant Names》(Brummitt & Powell，1992)的标准加以规范，但因名称出处难以稽考而未能如愿。迄今为止国内外还没有一本标准化的植物名称工具书，这是因为规范这些名称需要花费大量的心血与时间，更需要深厚的学术功底和齐全的文献资料，一般人很难具备这些条件。现在学弟对40多万条学名包括中国的全部种子植物认真进行了考证，按照《国际植物命名法规》的规定将不规范的书写方式予以标准化，意义深远。

这是一部国内外前所未有的标准化的大型植物名称工具书。我相信，不仅植物学工作者，而且所有与植物名称有关联的其他学科和部门的工作者，都会铭记和感谢尚衍重教授的这一非凡劳动。

《菌物学报》原主编
中国科学院微生物研究所 研究员

Preface 3

As *A dictionary of Seed Plant Names* compiled by my old friend and colleague Prof. Yanzhong Shang is ready to be sent to the press, I' m very pleased to write a few words as a preface on his request.

In 1978, when postgraduate study programs in China were first resumed after the Cultural Revolution, Yanzhong and I both took the same entrance examination to study under Prof. Yunzhang Wang. He achieved excellent results in the examination and Prof. Wang was ready to take him on board. Unfortunately, because the foreign language he opted for was Japanese, he was not accepted by the Institute of Microbiology, Academia Sinica. Thereafter, he pursued his postgraduate studies under the guidance of Prof. Liping Shao at the Northeast Forestry University. After achieving his master degree, he taught at Inner Mongolia Forestry College (now Inner Mongolia Agricultural University) for more than 30 years. Throughout our professional career, we have always kept in touch, discussing various issues we both are interested in, and remained as close friends and colleagues.

Yanzhong is known for his rigorous approach and extremely hard-working attitude toward science. Simply writing such a monumental piece of scientific literature is beyond anyone' s reach. Yet, Yanzhong has corrected and standardized the author names which appear in some 800, 000 places. It is hard to imagine the amount of time and efforts he has dedicated in just standardizing the author names. In the dictionary, only a very small number of author names cannot be verified and, in each case, a question mark "?" is added to remind readers. Such details surely reflect the extent of his rigorous attitude to science.

Yanzhong has spent more 30 years to gather an astronomical amount of information for the book and devoted more than 12 years of his career to complete this *magnum opus*. As a mycologist and being the editor of the *Mycosystema* for more than 20 years, I deeply felt the headaches caused by countless non-standardized plant names. While I was compiling the *Uredinales* for the *Flora Fungorum Sinicorum*, I intended to standardize abbreviations of author names of host plants according to the *Authors of plant names* (Brummitt & Powell, 1992), but failed to do so because it was very difficult to find the sources of those names. To date, no reference books of plant names in the world have yet standardized plant names as it requires endless and painstaking efforts and, more importantly, in-depth knowledge of the science and comprehensive search of the literature. Writing this dictionary, Yanzhong has meticulously checked more than 400, 000 scientific names of plants, including all seed plants in China, and standardized all non-standard names according to the *International Code of Botanical Nomenclature*. Publication of this book has far-reaching implications because this is the first grand-scale reference book of standardized plant names ever published in China or abroad. I believe that workers in plant sciences and in many other fields where plant names are used will always remember and appreciate this extraordinary contribution by Prof. Yanzhong Shang.

Mycosystema **Editor in Chief** (former)
Institute of Microbiology, Academia Sinica **Professor** **Jianyun Zhuang**

序　四

序，也称叙、绪、引等。它是用来说明书籍著述、出版意旨、编排体例和作者情况的。学术著作的序一般由业内权威人士撰写。这些内容的诠释、说明在我主编的《现代科学技术写作大词典》中都有详细的阐述，所以对于写序的重要性和写序人的身份地位我不会是不清楚的。尚衍重教授是我国著名菌物学家、中国菌物学会名誉理事长邵力平先生的研究生开门弟子，邵先生对他的做人和学问曾有耳提面命、言传身教之恩义，对《种子植物名称》和即将出版的《中国树病》的编撰，更是给予了极大的关注和指导。结果邵先生还是谢绝了为本书作序，理由是“包子有肉不在褶儿上”。

“包子有肉不在褶儿上”，我细细咀嚼和玩味邵先生这句话，感慨颇多。首先，把书的内容和水平比作包子的馅儿，把序比作包子的褶儿，太妥帖、太生动了。我以前曾惊讶老舍先生把三鲜馅包子的褶儿比作人脸上的皱纹，其生动和形象，已令人永世难忘。而把序比作包子上的褶儿，惟有像邵先生这样的学问大家，才能修炼到对事物信手拈来、随心所欲的境界。其次，我感到了邵先生品质人格的震撼力。现在有些“学术界”，恕我直言，真不敢恭维。按常理，手把手教出来的弟子，精心指导出来的成果又是一部价可传世的巨著，邵先生作序本应是当仁不让或者是欣然应诺的事情，可他来一句“包子有肉不在褶儿上”，轻飘飘的，淡淡的，似纱，像雾，但却重重地压得我喘不上气来。因为尚衍重教授早就有话，邵先生若不写，这篇序我是逃不掉的。

序既然是包子的褶儿，没褶儿也不行，而且外观感觉和包装效果自然不可小觑。再者，作为彼此敬重的朋友，他多年超乎常人的辛苦我历历在目。现在大功告成，我也应该写点东西表示祝贺。如果“褶儿”捏得不雅致，大家尽可以尝尝馅儿。谁吃包子，大概一门心思都会在馅上。

读者大概很难想象，这部巨著《种子植物名称》竟是由《中国树病》逼出来的副产品。尚衍重教授早在30多年前读硕士时，就萌发了编写《中国树病》的想法，之后在教学科研工作中不断积攒资料，聚沙成塔，集腋成裘。可当《中国树病》雏形初现时，一种困惑却使尚教授无法将《中国树病》定稿，原因就是中外文献中的树名十分混乱。树病树病，树之病也，树名都不准确，树病如何无误？正如人们常说的皮之不存，毛将焉附？无奈之下，尚教授一咬牙一跺脚，决定先搞一本《种子植物名称》出来！

没想到这一义愤之举，竟花费了他十多年的光阴。

首先，中国的种子植物到底有多少种，多少年来一直是笔糊涂账。如木本植物，原中国林业科学研究院院长郑万钧院士主编、全国46家研究院所和高等院校参编的《中国树木志》及全国林业院校统编教材《树木学》均估计为8000种。为了准确地算清这笔糊涂账，尚教授在资料收集上采取了竭泽而渔的做法，将国内外有关植物分类、树木志、植物志等资料几乎彻底地过滤一遍，查阅资料达10亿多字，第一次将中国的木本植物定位于1679属19668种(含种下单元)。他把这一前无古人的创举，谦虚地归功于计算机的现代科学手段。

其次，本书几乎收尽了中国种子植物5种文字的名称，但它并不是简单地汇总。尚教授要对这些名称进行考证，并且按有关法规，诸如拉丁学名命名人缩写、同物异名、同名异物、不规范名称等逐一订正、规范。如此繁复枯燥但又极具科学价值的工作，计算机就无能为力了。这得靠尚教授几十年的学业功力、大量藏书、敬业精神和十多年几乎每天15个小时的伏案劳作，来将疑难一一化解。尚教授的工具书和专业书堪称“富甲一方”；此外，国家图书馆、中国科学院图书馆、中国科学院植物研究所图书馆、中国林业科学研究院图书馆、中国农业科学院图书馆、北京师范大学图书馆以及内蒙古几所大学图书馆的工作人员都被他疯狂的工作精神所感动，纷纷大开绿灯，才使得他能以一人之力、十多年之时、半生之功完成这部划时代的学术巨著，给后人留下一部得心应手、排疑解惑的工具书。

第三，本书内容全面。它不仅将中国的木本植物、草本植物“一网打尽”，考虑到读者的实用性，还将中国特有种子植物属、濒危植物、外国重要植物、全球种子植物所有科和属的重要资料等尽收其中。拳拳之心，可见一斑。

还要强调的是本书的编排特点。学术界有句名言“科学是无国界的”，但语言文字是有界限的。随着全球学术界的交流、融合逐步增强，语言障碍也日渐显著。一些小型词典倒是有两三种文字的，但把5种文字的100万条植物名称和多种信息汇编于一书，可以互查，编排简洁新颖，查找方便迅速，那简直是匪夷所思！然而，尚衍重教授的这部书就做到了！而且不必翻译即可发行全球。这样，凡懂拉、汉、日、俄、英5种语言之一的读者，想查某个植物名称，你就能从书中立刻查到它的准确学名、其他几种语言的名称与写法。

一部书，4000万字，5种语言，100万条名称；一部书，第一次将种子植物学名标准化，而且多达42万条，并包括全部科名和属名；一部书，将全球种子植物的全部科和属给出中文名称和相关资料！然而谁能想到，所有这些奇迹，竟然是由尚教授一人13年完成！透过这部厚厚的传奇巨著，有谁能够不为尚衍重教授深厚的学术功底、崇高的学术境界、严谨的治学态度和顽强的拼搏精神而叹服！至于它给读者带来的便捷，它对中国乃至世界学术界的影响，桃李不言，下自成蹊，我相信历史自会评说。

我写的序，或捏的褶儿，能不能为这部书锦上添花已不重要。包子一熟，打开笼屉，包子里精制美味的馅儿自然会散发出诱人的馨香，弥漫在读者群，飘香于学术界。

内蒙古农业大学 教授

Preface 4

It is commonly known that a preface is written to explain the purpose and structural arrangement of the book and to say something about the author. Usually, a preface of an academic work should be written by an authority in its field. As the author of *A Dictionary of Modern Science and Technology Writing*, I could not have been unaware of the importance and the status of the person who writes the preface. Prof. Yanzhong Shang was the first graduate student of Prof. Liping Shao, the famous mycologist and the Honorary Director of the Mycological Society of China. Throughout the preparation of *A dictionary of Seed Plant Names* and the preparation of the *Tree Diseases of China*, Prof. Shao has offered valuable guidance and unreserved support. However, Prof. Liping Shao declined the request to write a preface for this dictionary, giving the reason by using a Chinese saying "the deliciousness of Chinese Baozi couldn't be judged from the folds on its cover".

I thought over Mr. Shao's words and felt a lot. First, how vivid a metaphor it is to compare the content of the book to deliciousness of Chinese Baozi, and the preface to the folds on the cover of Baozi. I was once surprised and much impressed by the comparison of the wrinkles on the human face to the folds on the cover of Baozi by the late Chinese literary master Lao She, and now I come to realize that only such a master scholar as Prof. Shao has the talent to use common idiomatic saying to express his profound meaning. Secondly, I was deeply impressed by the greatness of Prof. Shao's character. Since Prof. Shang was the first graduate student tutored by him and the work itself is an invaluable one, it is quite natural that Prof. Shao should have been delighted to write a preface for his student. However, to the surprise of most people, he declined the request with the reply that "the deliciousness of Chinese Baozi couldn't be judged from the folds on its cover". Though his words were brief and plain, I still felt the great pressure on me as Prof. Shang had asked that I should write him the preface if Prof. Shao would not.

Since the preface is compared to the folds of Baozi, it will not be done without the "folds". What's more, as a friend with mutual respect, I have witnessed the extraordinary, "super human" efforts he has forwarded in writing this dictionary. The dictionary is now completed and I am compelled to write a few words to express my heartiest congratulations.

It is hard to believe that *A dictionary of Seed Plant Names* is actually a byproduct of another book, *Tree Diseases of China*. When he was studying for his master degree some 30 years ago, Prof. Yanzhong Shang had an idea of compiling the book *Tree Diseases of China*. In the years of teaching and conducting research, he took every opportunity to gather necessary information in preparation for his book. However, he was unable to finalize the *Tree Diseases of China* because the names of trees in the literature, either from China or abroad, were in a state of confusion. It was impossible to complete a book on diseases of trees without knowing correct names of the trees. Just as the old Chinese saying goes, "With the skin gone, what can the hair adhere to?" No other way round, Prof. Yanzhong Shang decided, once and for all, to compile *A dictionary of Seed Plant Names* first. It was never expected that this decision would take him more than 10 years of long and unimaginably hard journey to succeed.

Firstly, there has never been any reliable figure on the number of seed plants native to China. For example, there are about 8,000 woody plants in China according to the estimate from *Sylva Sinica*, jointly compiled by scholars from 46 Chinese institutions with Mr. Wanjun Zheng, the president of Chinese Academy of Forest Sciences, as editor in chief, and from the *Dendrology*, a standard textbook compiled and used by forest universities and colleges nationwide. In order to find accurate answers, Prof. Yanzhong Shang has looked into almost all the literature on plant taxonomy, tree flora and plant flora in the world, going through more than a billion words during the process. His work has, for the first time, provided conclusive statistics that a total of 19, 668 species (including infraspecific taxa) belonging to 1, 679 genera are native to China. He modestly credited this remarkable achievement to the work of the modern gadget-computer.

Secondly, this book has compiled almost all the names of seed plants in China in five languages. Rather than simply putting plant names in a book, Prof. Yanzhong Shang has verified and corrected a huge number of names by standardizing author name abbreviations in Latin names, synonyms, homonyms, and non-standard names. The work has been painstaking, extremely laborious and more often mundane, but the end result is a book of great scientific value, which computer alone would never achieve. Such an achievement can only attribute to his profound knowledge built up over decades, a huge collection of reference books, his wholehearted devotion to science and, above all, working 15 hours almost every day for more than 10 years. I was very much impressed with his personal library which contains the most comprehensive collection of reference books on the subject. Also, touched by his passionate dedication to work, librarians from the National Library of China and the libraries of the Chinese Academy of Sciences, Institute of Botany, Chinese Academy of Forest Sciences, Chinese Academy of Agricultural Sciences and of several universities in Inner Mongolia have tried their very best to help him, making it possible for him to complete such a monumental piece of work on his own in more than 10 years.

Thirdly, this book is the most comprehensive of its kind. Not only it covers all the woody plants and herbaceous plants native to China, but also it provides information on the genera of seed plants unique to China, endangered plants, important non-native plants and the literature on all the families and genera of seed plants in the world.

Another feature of this dictionary is the way it is compiled. A famous saying in the academic world goes: "Science has no national boundaries." Languages, however, do have boundaries. With global integration of science becoming ever more important, language barriers present an increasingly prominent problem. There are some small dictionaries which contain plant names in two or three languages. By sheer contrast, this dictionary contains about a million plant names in five languages plus other relevant information, making it easily accessible from any of the five languages. Thus, any reader who knows one of the five languages can easily find the plant names in other four languages.

In brief, this is a book containing 40 million words, five languages and a million plant names, a book which has, for the first time in history, standardized scientific names of seed plants, amounting to 418, 831 in total, and the names of all the families and genera, and a book which gives Chinese names and the sources of reference to all the families and genera and provides interpretations of scientific names for all the genera. Nobody can believe that such a great piece of scientific literature has been miraculously accomplished by Prof. Shang single handedly through 13 years of wholehearted devotion and unimaginably hard work! From this *magnum opus*, one can only admire Prof. Yanzhong Shang's profound and in-depth knowledge, outstanding academic standard, rigorous approach to science and unrelenting perseverance. As for the usefulness and the benefit of this dictionary to the scientific communities in China and abroad, I believe that the future will tell in due course.

In fact, it is not important whether the preface I wrote, or the folds I made, can add any shine to the work, for we all know that when the Chinese Baozi is ready, the attractive smell will come out of it as soon as the steamer is opened. I am confident that, like the delicious Chinese Baozi, the scent of the dictionary of Prof. Yanzhong Shang will always linger in the academic world.

Inner Mongolia Agricultural University **Professor** **Yuming Ma**

前 言

一、编撰本书的缘由

植物是人类及其他一切生物赖以生存的物质基础。认识、研究、利用植物，是全人类永恒的课题。准确鉴定植物，科学命名植物，规范使用名称，乃是研究、利用植物的首要条件。植物学之父林奈(Carl von Linne)有句名言可谓一语中的："不知道事物的名称，就不会认识事物"。对此笔者深有体会。早在1974年就曾将日文版《图说树病新讲》翻译完毕，但其中的一些树木名称至今仍空缺着，译稿当然也不可能出版。后来在教学与科研中经常查阅大量文献，同样对植物名称特别恼火。例如要查找一个日文或俄文名称的拉丁名和中文名称，常常忙了几天而一无所获。有时一种植物又查到几个名称，难定取舍，令人苦不堪言。

1995年，笔者开始整理撰写《中国树病》，又遇到这个问题，且情况更为糟糕！植物名称在各种文献中用得太混乱了！例如汉语中叫接骨草(接骨丹、接骨藤、接骨药等)的植物达242种。Orchis mascula (L.) L.(强壮红门兰)英文名称达199个。读者若细翻本书就可发现，同物异名和同名异物(包括学名和各种语言的普通名称)在30个以上的植物不在少数，超过100个者亦不鲜见。笔者的《中国树病》拟将囊括中国所有木本植物病害并适当收录外国树病，首要任务是搞准确植物名称特别是树木名称，迫切需要一部全面、准确、规范的拉、英、汉、日、俄等方面的植物名称词典，可惜如此全面的工具书世上还没有。笔者不想在自己的著作中出现那么多错误(尽管是寄主植物名称方面——责任不应该由我负)贻误后人，只能经常翻腾大堆的资料卡片，有时还要去图书馆查证，而且常常查不到！烦躁的心情与耗费的时间都令人难以忍受，一怒之下愤然将写到半途的《中国树病》书稿放下，索性集中全力整理编写此书。虽说磨刀不误砍柴工，但欲砍柴先打刀，也着实令人唏嘘。

笔者最初计划编写的是《拉汉日树木名称》，在一些专家的建议下，先后加入了俄文名称和英文名称。在《拉汉日俄英树木名称》基本完稿时，收到庄剑云博士寄来的《Authors of Plant Names》，此书是《国际植物命名法规》(东京法规，1994)指定的标准化命名人缩写的蓝本。本来笔者所收入的拉丁名，已经全部与《中国植物志》、《中国树木志》、《苏联植物志》、《日本植物志》、《台湾植物志》、《欧洲植物志》、《北美植物志》等数百册中外权威文献核对过，但与《Authors of Plant Names》一对照，令人大吃一惊，误用和不规范的命名人缩写多得令人咋舌！笔者曾将当时在国内影响最大的两本工具书的前10页做了一个统计，仅仅学名，平均每100条中，不规范和漏误之处分别为62处和102处！植物学名不规范的现象，笔者在编写《中国树病》和本书的过程中已深有感触，但在国内外的权威书刊中达到如此严重的地步，是万万没有料到的。作为工具书，全面实用当然十分重要，但准确规范更是根本。一部工具书若给人以不准确或不规范的名称，谬种流传，害的是一批人、几代人，而且这个混乱传播在各类书刊中，后人根本无法纠正。在这种情况下，笔者是不可能再将《拉汉日俄英树木名称》原稿付印了，只能把规范学名的工作提到日程，并干脆连同草本植物和药用植物一起搞。逼上梁山，今知其义矣。

二、规范学名的几点说明

植物的拉丁名称为科学名称(scientific name)，简称学名。其他所有语言的名称均称为普通名(common name)，或称俗名或地方名(vernacular name)。但是，不管普通名还是学名，现在都存在很多问题。

本书收进的植物名称，笔者依据有关法规，做了力所能及的订正与规范。当然，重点是学名，因为学名不仅全世界通用，而且有《国际植物命名法规》作依据。由于迄今为止全世界还没有一部标准化的植物学名工具书，加上需要标准化的数量又是如此巨大，使得此项工作特别艰难，耗费了笔者大量的精力和时间。

要想规范学名，首要条件是搞准确学名。这里面有三层意义：一是这种植物的准确学名是什么？二是它的合理分类地位在何处？三是这个名称的命名人到底是谁？前两条虽然工作量极大，还算好办，本书收入的同物异名就反映了这些观点。最后一条是异常难做的。命名人写全名的还好标准化，但绝大多数不规范的缩写并非如此，令人很难确认。例如中国命名人中，张、王两姓的命名人在本书中出现9000余处，而且很多名称中仅仅是个姓；而这两个姓中仅仅写姓者全部都是不标准写法！要想把这些不标准的书写方式全部订正过来，真可谓“难于上青天”啊。

为了搞准确学名，也为了尽量收全中文和其他几种语言名称，笔者查阅了巨量文献，现在已经无法详列；书后所附参考文献仅是一部分。这些文献中的学名常常有矛盾。有疑问的学名再与邱园索引(1～2卷)和邱园索引补编(1～26卷)核对。然后，笔者又将本书收录的学名与《国际植物名称索引》的1000000余条相关学名编排在一起核对。

但是，这也仅仅起到搞准确学名的作用，而并非标准化，因为上述所有文献的学名全未标准化，就连《国际植物名称索引》和《邱园索引》中的学名也是如此，笔误、命名人缩写错误和其他漏误屡见不鲜。

笔者从6个方面对学名进行了规范：①命名人缩写不标准者；②加词词尾错误者；③加词中带有连字符“-”者；④命名人漏缺或误用者；⑤违反《国际植物命名法规》和《国际栽培植物命名法规》的名称；⑥自动名带命名人者。

(一)命名人缩写的标准化

《国际植物命名法规》历次版本对命名人缩写均有规定，1994年的东京法规和其后的法规均以《Authors of Plant Names》(Brummitt et Powell，1992)一书作为命名人标准化的蓝本。

为标准化命名人缩写，笔者将本书中出现的命名人逐一与《Authors of Plant Names》核对。《Authors of Plant Names》出版后新涌现出的作者，笔者则根据《Authors of Plant Names》采用的原则加以处理。

尽管笔者作了如此努力，仍有个别命名人未能查到出处。这些人名在书中均加了“?”，以提醒读者注意。即便如此，仍有漏误之可能。规范学名不是一个人的事情，不是一代人的事情，也不是一个国家的事情，诚恳欢迎读者指正。

(二)加词词尾的规范

学名中的属名是有性别的。种加词和种下加词，形容词应与属名的性保持一致；同位语则不考虑性。导致加词词尾错误的原因主要有3点：其一是不懂上述二者的区别；其二是对属名性别的误判；其三是对加词词性的误判。

源于人名和地名的加词也常常出现与属名性别不一致的错误。

(三)带有连字符“-”的加词

学名中的种加词、种下加词只能用一个词。如果是复合词，两个词干是拉丁语者用“i”连接，希腊语用“o”连接；同种语言用“i”连接，不同语言用“o”连接。两种情况用连字符“-”：两个词独立存在，或者连字符“-”两边字母相同。

(四)命名人漏缺或误用

在国内外文献中，命名人误用和漏缺者并不罕见，常常出错的如ex、in、apud、et、f. 等，ex与et误用者更多些。笔者发现的均予以订正。

（五）违反《国际植物命名法规》和《国际栽培植物命名法规》的学名

《国际栽培植物命名法规》自1953年首版问世至今，已经出版了7个版本，时间也超过了半个世纪，第5、6、7版都有中文版出版。但是，真正了解并遵循它的学者并不多。栽培植物学名的最大问题是"品种"的学名，依法规规定应使用单引号，但在已查文献中，大量使用双引号、cv. 等错误写法。本书中的栽培植物学名，全部依据法规做了订正。

最近几版的《国际植物命名法规》都附有保留科名、保留属名、废弃属名名录。对违反法规者，笔者均做了订正。

（六）自动名

自动名是不允许带命名人的。笔者曾纠正了一批此类错误。但是后来考虑到节省篇幅问题，则把自动名全部删除了。这里提出，加以提醒。

目前，国内外很多人对学名的标准化重视不足，甚至不以为然。但标准化毕竟是必由之路，也代表了治学态度和学术水平。规范植物学名是全世界所有国家植物学者义不容辞的、艰巨的，然而又是必须完成的任务。现在国内外有些核心学术期刊已经要求所投稿件中的学名包括命名人缩写必须规范，否则不予刊登，显示了对学术和后人的强烈责任感，应该引起所有学者的震撼！

三、关于属名的说明

在生物分类系统的7个主要等级中，"属"是很重要的。因为种和种下名称的第一个词都是属名，科的词干源于属名，多数目的词干也源于属名。

属的学名是不能随意改动的，除非通过法规"保留"的方式。

全世界的种子植物属，约13000～17000个。本书收入属名58746个，其中"好属"16436个，同物异名42310个。属名均带有标准化的命名人。为了不同读者查询的方便，本书采用了"小属"观点（虽然笔者并不全部赞同这些观点），同时又给出"小属"应该隶属的"大属"。"好属"都给出了中文名称。中国植物的属名全部取自现有文献。国外植物的属名，有些采用于文献包括网络文献，其中刘冰和刘夙拟订的名称约有2000条。翻译不当的属名作为异名收入，笔者另行拟订，例如 Kingiella，有人译为肯基拉兰属，笔者订正为金氏兰属，因为该属名源于英国植物学者的名字 G. King。笔者拟订将近10000个，主要根据是拉丁名含义或形态特征、分布等。

植物的属名是有性别的，它决定了形容词性加词的性别。属名性别误判的情况在文献中很常见。主要原因是，尽管所有属名不管来源于何种文字都作为拉丁文对待，但是由于法规的规定，它们的性别又并非与拉丁文语法完全一致。法规规定，"属名保留植物学传统赋予的性，与经典用法或作者的原始用法无关"。"复合属名的性取决于复合词中最后一词主格的性。然而，如果该词尾被改变，其性别亦随之改变"。"性不明显的随意构成的属名，或用作属名的地名或形容词，其性别由原作者指定。如果该原作者没有指定性别，可由下一个作者指定"。"无论原作者的指定如何，以-anthes，-oides或-odes 结尾的属名被视为阴性。以-ites 结尾的属名被视为阳性"。"当一个属分为两个属或多个属时，新属的名称应与被保留的属的性一致"。

在属名条目下，给出了发表年份。保留属名和废弃属名也都依据维也纳法规加了标注。

特别强调一点，国内外所有较大型的植物志、专著和工具书中，几乎都存在违反《国际植物命名法规》的属名：①不用法规的保留属名而误用了其他名称；②法规废弃的属名还在作为正确名称使用。笔者依据2006年的维也纳法规，全部做了订正。还请读者注意，保留属名和废弃属名不是固定不变的，要及时留意新法规。

四、关于科名的说明

所有科名的词干都来源于模式属名，但法规确定有9个科有互用名称，这9对互用名称中，每对都有一个名称来自模式属名。

本书收入科名4124个，其中"好科"1061个，异名3063个。"好科"都给出了中文名称。全部科名都有标准化命名人。同属名一样，对于"好科"和异名，学者们是仁者见仁，智者见智。笔者如此处理，仅仅是为了读者的不同需要。在科名条目下，笔者给出了中文名称和外文名称。维也纳法规的保留科名也加了标注。笔者新拟订的中文科名，参照《国际植物命名法规》的原则，均取自模式属名。

法规规定：括号内的作者引证不适用于属级以上的名称(不包括属级)。因此，科的命名人不允许引证括号内的人名。这是文献中常见的一个错误，本书均做了订正。

科和属的某些内容，将另成一卷出版。

五、中文名称的说明

本书给出中文名称的条目176817条，共收入中文名称280213条。

由于中国幅员辽阔，历史悠久，所以植物的同物异名、同名异物也特别多。有的植物中文异名竟达200余个。

我国一直没有一部与《国际植物命名法规》相类似的植物中文名称命名法规。尽管笔者感到很多植物的现有名称应该规范，但在没有法规的情况下，不想再增添新的名称，所以本书中的中国植物的汉语正名均从《中国植物志》、《中国树木志》、《拉汉英种子植物名称》、《中国种子植物科属词典》等文献中挑选。笔者只对极少数名称做了订正。对只有一个名称的同名异物，笔者则根据拉丁文含义或形态特征、产地、命名人等新拟了一些中文名称。

种下的分类单元，通常原种名称应给予自动名，新发现的种类另起新名。笔者对不符合此例的有些种下名称做了订正，新拟了中文名称。例如，《中国植物志》中将太平山冬青的新变种 Ilex sugeroki Maxim. var. brevipedunculata (Maxim.)S. Y. Hu 称为太平山冬青，本书将其改为短梗太平山冬青。

文献中属于笔误的中文名称，笔者直接做了订正，不再收录，如“苹芨”为“苹茇”的笔误；只有个别收了进来，如“山样子”虽为“山様子”的笔误，但好几部重要书都这样写，考虑再三还是作为异名收进来，否则读者将无法查到它的正确名称。

非国产植物的中文名称，笔者充分尊重文献已有的名称，但是也订正了一批翻译不妥者。例如，把印多尼亚属(Indoneesiella)改为印度爵床属等。笔者拟订和订正的名称中难免有不尽人意者，有待后人完善。

六、外文普通名的说明

书中收入了日、俄、英3种外文名称22万条。收入名称的原则是必须有拉丁名为依据。这些名称中，同物异名和同名异物很常见。因为没有法规，无法确定正名和异名。笔者也极少增添新的外文名称。

日文名称中，为了读者查找老文献的方便，古老写法的名称也收了进来。

七、词典的编撰

作为一名大量使用此类工具书的工作者，又查阅了巨量的文献，对不同编排的利弊感受很深。本书如何编排才能既使读者查找方便、节省时间，又降低印刷成本？笔者颇费了番脑筋。最后采用了现在这种二级检索的方式。正文按拉丁文字母顺序排列，同时每一个条目又给予编号。

书中收入的5种语言名称可以互相检索。从语种方面，本书即具有20种功能，换言之相当于20部单项词典。

本书收入多数读者常用的科、属、种、亚种、变种、变型、亚变种、亚变型、杂交种、品种等10类常用名称。

对中国特有属、外国植物、木本植物、草本植物、濒危植物等信息，本书用符号做了标注，直接放在了条目中，未做成附录。这样处理，既节省了很大的篇幅，同时也会令读者在查找名称时顺便即可得到一些相关信息。

八、结束语

2001年暑期，内蒙古农业大学的一位教师奉日本导师之命查阅一种树木的资料。她在图书馆查找未得结果，后来经图书管理员推荐找到我家，结果仅用了1分钟，问题便迎刃而解。能为规范植物名称奠定一块基石，为学者节省一些时间，为有关学科的发展与国际交流起些作用，余愿足矣。

历史的误会造就了本书。本书也留下了一桩笑谈。

Foreword

1. Background

Life on Earth depends on plants. To mankind, a never-ending quest is to recognize, study and utilize plants and accurate identification is a prerequisite to subsequent activities. The father of Botany C. Linnaeus had a famous saying: "You won't know a thing until you know its name." I learned such a fact through my personal experience. Early back in 1974, I translated the *New Illustrated Tree Diseases* from Japanese to Chinese but was unable to publish the translated version because some of the tree names were, and still are until today, missing. In later days, I was particularly troubled with plant names while I was doing extensive literature search for teaching and research at the university. Sometimes, for example, finding the Latin and Chinese names for a Japanese or Russian name could take several days with no avail.

In 1995, I started to write the *Tree Disease of China* and came across the same old problem. Much worse than I had realized, plant names from various sources of the literature were extremely confusing. For example, the Chinese name 接骨草 (bone-healing herb) and similar names, i. e. 接骨丹(-herbal extract)、接骨藤 (-vine)、接骨药 (-medicine) etc, refer to 242 plant species while *Orchis mascula* (L.) L. has up to 199 different English names. Readers can find in this book that it is not rare that a plant has more than 30, or even 100, synonyms and homonyms (including the scientific name and the common names in other languages). At that time, I had planned to cover all the diseases of woody plants in China and some non-native tree diseases in the *Tree Diseases of China*. I found in the process of gathering the information that the first and foremost important task was to know the correct names of plants, the names of trees in particular. What I urgently needed was a complete, accurate and standardized dictionary of plant names in Latin, English, Chinese, Japanese and Russian languages, but unfortunately there was no such a dictionary available anywhere in the world. I felt very uneasy about the idea of having so many mistakes in my work (even with the names of host plants) and misleading readers. So that I had to resort to searching through huge volumes of references, often without much success. This became almost unbearable because of the sheer amount of time and efforts spent or, more often, wasted. I had no other option but to put aside the original plan of writing the book of the *Tree Diseases of China* and to concentrate all my efforts on writing this dictionary first.

At the very beginning, I intended to write *A Dictionary of Woody Plant Names in Latin, Chinese, and Japanese* and later I added Russian and English names after I sought advice from a number of experts. When the book titled *A Dictionary of Seed Plant Names in Latin, Chinese, Japanese, Russian and English* was almost finished, I received the book *Authors of Plant Names* (Brummitt & Powell, 1992) sent by my colleague Dr. Jianyun Zhuang. This book is the blueprint of the standardized abbreviations of author names designated by the *International Code of Botanical Nomenclature* (*Tokyo Code*, 1994). Actually, I had already checked all the Latin names gathered in my dictionary referring hundreds of authoritative reference books from China and abroad, such as *Flora Republicae Popularis Sinicae* (*Flora of China*), *Sylva Sinica* (*Trees of China*), *Flora of USSR*, *Flora of Japan*, *Flora of Taiwan*, *Flora Europaea* and *Flora of North America*. To my horror, compared with the *Authors of Plant Names*, I discovered that there were an amazing amount of misused and non-standardized abbreviations of author names in my book. Then I checked the first 10 pages of two most influential reference dictionaries in China and found that, only scientific names alone, there were respectively 62 and 102 non-standardized, missing or misused author names in every 100 entries. Although I was deeply aware of the issue of non-standardized plant names when I was writing the *Tree Diseases of China* and was preparing to compile this dictionary, I had never expected that the problem was so serious, even in the most authoritative literature. For a reference book, although comprehensiveness and usefulness are important, the basic requirement is to be accurate and formally acceptable. If a reference book having inaccurate or non-standard names is used for a long time, misuse of plant names will spread widely, misleading people for generations and, eventually, making it difficult to correct the mistakes. Because of this, I was unable to send the original version of *A Dictionary of Woody Plant Names in Latin*, *Chinese*, *Japanese*, *Russian and English* to the press, and had to take on the task of standardizing scientific names. Furthermore, I decided to include the names of herbaceous plants and medicinal plants. I hope that, apart from serving today's practical needs, it could provide a useful book of reference for later generations as well. Now I can understand the true meaning of "forced to rebel".

2. Standardization of Scientific Names

Latin names of plants are referred as scientific names. On the other hand, names in all other languages are referred as common names or vernacular names. In reality, there are many problems with either the scientific names or the common names.

In writing the book, I tried, as far as possible, to correct and standardize all the plant names gathered according to relevant codes and rules. Obviously, the scientific name was the main focus, because scientific names are used universally and sanctioned by the *International Code of Botanical Nomenclature*. When I was writing this book, no dictionary of standardized scientific names of plants was, and still is to date, available anywhere in the world and the standardization of plant names was seemingly a never-ending task. With no other option, I had to spend a vast amount of my time and efforts to standardize scientific names.

A prerequisite for standardization of scientific names is to make sure whether a scientific name is accurate. Here three questions need to be answered. Firstly, what is the correct scientific name of the plant? Secondly, what is its appropriate taxonomic rank? Thirdly, who is (are) the legitimate author(s) for this name? Although huge volumes of work could be involved, the first two questions can be answered with relative ease. Answering the last question can be extremely difficult. Some plant names having full authority names are easy to standardize, but it is not the case with the great majority of the plant names having non-standardized abbreviations of author names. For example, surnames Zhang and Wang appear in more than 9000 cases in the dictionary and, when I started writing this book, only their surnames were given in numerous cases. According to the *Code*, however, it is incorrect to use only surnames of authors alone. Correcting all the non-standard names proved to be so difficult and, to me, was truly a "mission impossible".

To identify accurate scientific names and to gather plant names in other languages, I had to wade through the vast ocean of literature. The references listed at the end of this book are merely a fraction of what I came across. Scientific names from various sources of the literature very often contradict each other. These scientific names in question were checked using *the Index Kewensis* (Volumes 1-2) and *the Index Kewensis-Supplementum* (Volumes 1-24). Following that, I checked all the scientific names included in this dictionary by comparing with over 1, 000, 000 relevant scientific names listed in *The International Plant Names Index*. However, this was merely a step to find the correct scientific names and the standardization of scientific names was yet to come. Scientific names in the literature, even in *The International Plant Names Index*, the *Index Kewensis* and the *Index Kewensis-Supplementum*, were not standardized and it was commonplace to find misspelling, non-standardized abbreviations of author names and other errors.

Writing this dictionary, I standardized scientific names in the following six cases: ①abbreviations of author names are not standardized; ② termination of the name does not match with Latin grammar; ③there is a hyphen " – " in added words; ④author names are missing or misused; ⑤names do not follow the rules of the *International Code of Botanical Nomenclature* or the *International Code of the Nomenclature for Cultivated Plants*, ⑥ there is an author name(s) in an autonym.

(1) Standardization of abbreviations of author names

Each edition of the *International Code of Botanical Nomenclature* stipulates the writing of author name abbreviations and the *Tokyo Code* (1994) and others followed subsequently use the *Authors of Plant Names* as the blueprint for standardization of author name abbreviations.

To standardize author name abbreviations, I checked all the author names in this book and verified using the *Authors of Plant Names*. For the authors who are not included in the *Authors of Plant Names*, I applied the same principles as described in the book.

Despite my best effort to clarify the authorship, on some occasions, author names are still missing. These cases are marked with "?" to remind readers. Yet, it is still possible that missing or misused named can be found in this book, especially in the cases of the Chinese authors having only surnames in the literature. Standardization of scientific names is beyond the task for an individual, a generation or a country. Therefore, I strongly encourage readers to find out mistakes and to correct them.

(2) Termination of epithets

A generic name has its own gender. Specific epithet and infraspecific epithet should be consistent with the gender of genus. In contrast, appositives have no such requirement. Incorrect termination is mainly due to that (a) distinction between generic names and epithets is ignored, (b) gender of the generic name is mistaken and (c) the gender of the epithet is inconsistent with the generic name. The error of inconsistency between generic name and epithet also occurs often in the epithets originated from the names of people or locations.

(3) A hyphen " – " in epithet

Only one word of epithet or infraspecific epithet can be used in scientific names. In a compound word, two words are combined together. If two etyma are from Latin language, they are connected with "i" and, if from Greek language, connected with "o". In two cases, (a) if an epithet is formed of words that usually stand independently, or (b) if the letters before and after the hyphen are the same, a hyphen " – " is used to link the two words.

(4) Missing or misused author names

In the literature, it is not uncommon that author names are misused or missing, often in the use of ex, in, apud, et, f., especially ex and et. Wherever possible, these mistakes are corrected.

(5) Names contrary to *the International Code of Botanical Nomenclature* and the *International Code of Nomenclature for Cultivated Plants*.

Since 1953, for more than half a century, a total of seven editions of the *International Code of Nomenclature for Cultivated Plants* have been published. The 5th, 6th and 7th editions of the *Code* were translated into Chinese but relatively few workers followed the rules defined in the *Code*. In the literature, especially that from China, quotation marks " " (correct ones are ' ') and the abbreviation cv. were very often used incorrectly for cultivars. These mistakes are corrected by the author.

Recent editions of the *International Code of Botanical Nomenclature* listed the names of families and genera to be conserved or to be rejected. Names not in line with the names "to be conserved" in the *Code* are treated accordingly in this book.

(6) Autonym

Autonym is not permitted to have author name(s). Readers should note that all autonyms are omitted in this book to save space.

At present, many workers in China or abroad do not pay enough attention to the standardization of scientific names or do not take it seriously. In terms of the advancement of science, standardization of scientific names is the only way forward. It also reflects the attitude towards learning and the academic standard of a professional or a country. Therefore, standardization of plant scientific names is the responsibility of plant scientists worldwide and is a tough challenge we must meet. Nowadays, some core scientific journals require that scientific names in submitted papers, including the author name abbreviations, must be standardized, or else not be accepted. This reflects strong obligations we need to fulfill for the sake of the advancement of science and for future generations and, at the same time, it gives us a wake-up call as well!

3. Generic names

Of the seven major taxonomic ranks in classification of living organisms, the "genus" rank is the most important. This is because the first word of the name of a species is generic name and family names and the majority of the names of the rank "order" stem from generic names. Also, generic names cannot be changed unless sanctioned by the *Codes*.

There are 13, 000 – 17, 000 genera of seed plants worldwide. A total of 58, 746 generic names are included in this book and, of these, 16, 436 are regards as "good" genera and 42, 310 as synonyms. In this book, all generic names have standardized author names. Also, "narrow sense" generic names are adopted (although I do not necessarily agree with such treatments) and, for the convenience of cross-reference, "broad sense" generic names are given to the "narrow sense" generic names. All the "good" genera are given Chinese names. For plants native to China, all the generic names are adopted from existing names in the literature. For plants non-native to China, some generic names are adopted from the literature, of which approximately 2, 000 were proposed by Bing Liu and Su Liu. For non-native genera, inappropriate Chinese translations are treated as synonyms. In this dictionary, nearly 10, 000 Chinese names are newly given by the author mainly according to the meaning of the Latin name, morphological characteristics and distribution.

Generic names of plants have gender. A specific epithet or a subspecific epithet, treated as an adjective, must agree in gender with the generic name. In the literature, the gender of generic names is very often mistaken. It is mainly due to that, although all genus names are treated as Latin regardless of their derivation, according to the rules of the *Code*, the gender of generic names does not exactly follow the rules of Latin grammar. The *Code* states "A generic name retains the gender assigned by botanical tradition, irrespective of classical usage or the author's original usage." "Compound generic names take the gender of the last word in the nominative case in the compound. If the termination is altered, however, the gender is altered accordingly." "Arbitrarily formed generic names or vernacular names or adjectives used as generic names, whose gender is not apparent, take the gender assigned to them by their authors. If the original author failed to indicate the gender, the next subsequent author may choose a gender." "Generic names ending in -anthes, -oides or -odes are treated as feminine and those ending in -ites as masculine, irrespective of the gender assigned to them by the original author." "When a genus is divided into two or more genera, the gender of the new generic name or names should be that of the generic name that is retained." In this book, the gender is given to all the genus names in the interpretations of generic names.

Under generic names, the year of publication is given and conserved generic names and rejected generic names are also indicated according to the *Vienna Code* (2006) in this book.

It is extremely important to point out that generic names which do not conform with the *International Code of Botanical Nomenclature* appear in almost all major publications of flora, monographs and reference books, either from China or abroad. Very often, instead of using legitimate generic names retained by the *Code*, other generic names are applied while the names designated as obsolete according to the *Code* are used as correct names. In this book, mistakes in generic names are corrected according the rules of the *Vienna Code* (2006). Also, I would like to remind readers that the "conserved" or "rejected" names can change and therefore one needs to pay attention to the update of the *Code*.

4. Family names

Family name stems from the name of the type genus. However, the *Code* allows that nine families to have alternative names and one of the names in each pair originated from the type genus. Therefore, readers can simply refer to the type genus for interpretation of the family name.

Of a total of 4, 124 family names included in this book, 1, 061 were treated as "good" families and other 3, 063 as synonyms. Chinese names are given to all the "good" families and the author names are standardized.

Descriptions under family names also include Chinese names and the names in other languages. Family names conserved according to the *Vienna Code* are indicated in the book. All the newly-proposed Chinese names for the families are based on the names of type genera according to the rules of the *International Code of Botanical Nomenclature*.

According to the *Code*, author names in brackets are not permitted in the names of the ranks above the genus and, therefore, such common errors are corrected by the author.

Further descriptions of families and genera will be published in another volume in the future.

5. Chinese names

In this dictionary, 176, 817 entries have Chinese names and a total of 280, 213 Chinese names are complied. Probably due to its vast expanse, great natural diversity and long history of civilization, there are an astonishing amount of synonyms and homonyms of plants in Chinese, some plants even having up to 200 different names.

To date, there are no rules of a code similar to the *International Code of Botanical Nomenclature* governing how to name plants in Chinese. Although I feel that many current plant names should be changed according to certain rules, before such rules are available, I did not attempt to add new Chinese names to the existing ones and, instead, Chinese names in this book are adopted from major references such as the *Flora Republicae Popularis Sinicae*, *Sylva Sinica*, *Dictionary of Seed-plants Names*: *Latin-Chinese-English* and *A Dictionary of the Families and Genera of Chinese Seed Plants*. Only a very small number of names are modified by the author. For those having only one homonym, new Chinese names are proposed by the author based on the meaning of the Latin name, morphological characteristics, natural distribution and author's name etc.

With regard to the infraspecific taxa of plants, usually the original species name should be endorsed as an autonym and a newly discovered taxon be given another name. The author modified some names of infraspecific taxa which did not tally with this tradition and gave new Chinese names. For example, in *Flora Republicae Popularis Sinicae*, the new variety of *Ilex sugeroki* Maxim. var. *brevipedunculata* (Maxim.) S. Y. Hu is called 太平山冬青(Taipingshandongqing) in Chinese, whereas in this book it is called 短柄太平山冬青(Duanbingtaipingshandongqing).

In this dictionary, some Chinese names having obvious typographical errors were corrected directly without including the original ones with error. For example, 荜茇 is not included because it is the typographical error of 荜茇. Only occasionally, typographical errors such as "山样子", are included in the dictionary because, although "山样子" is an error of "山檨子", it has widely been used in several important reference books. Inclusion of such errors may help readers find their correct names.

For Chinese names of the plants not native to China, existing names are retained wherever possible. However, the names which are regarded to be inappropriate translations are modified by the author. In total, over 100, 000 Chinese names are either proposed or modified by the author. Some of these names may be less desirable and I wish that better suited names can be given by readers in the future.

6. Vernacular names in other languages

A total of 220, 000 plant names in three other languages, Japanese, Russian and English, are complied in this book. All the names included in this book have Latin names which can be referred to. Many of these names are synonyms and homonyms but, as no rules on how to give vernacular names are available, it was not possible to distinguish the "good" names from others. In any case, very rarely new names in foreign languages are given by the author in this book.

For Japanese names, the names in old Japanese are also included to help readers in search of old references.

7. Compilation of the dictionary

Searching through a vast amount of dictionaries and reference books year in and year out, I deeply felt the importance of how reference books similar to this are structured. It took me some time to decide the structure of the dictionary in order to make it easy to use, to save reader's time and to lower the printing costs. In the end, the dictionary is indexed in two ways; the entries in the text are arranged in an alphabetical order of Latin while each entry is given a number.

In this dictionary, plant names can be cross-referenced in five different languages. Thus, the dictionary has 20 language-converting functions; in another word, it can be used as 20 individual bilingual dictionaries.

The content of this book covers the great majority of commonly used names of families, genera, species, subspecies, varieties, subvarieties, forms, subforms, hybrids and cultivars.

In this dictionary, symbols are used to provide information on plants unique to China, woody or herbaceous and endangered plants etc. Such arrangement substantially saves space and, at the same time, instantly provides readers relevant information.

8. Concluding remarks

I recall that, in the summer of 2001, a teacher from Inner Mongolia Agriculture University was asked by her Japanese supervisor to gather some information on a tree species. After spending some time in the library without success, she turned up on my doorstep on the suggestions by the librarian and it took her only a minute to solve the problem. It would be more than satisfactory if this dictionary can provide a stone of foundation for standardization of plant names, can offer the means of time-saving and the convenience for readers in search for plant names and related information and can make a humble contribution to the advancement of science and international exchange of knowledge in relevant fields.

A wrong twist of history created this dictionary. Thus this book also leaves behind something to wonder about.

致　谢

本书能够以笔者比较满意的面貌出版，不敢忘记恩师邵力平先生的言传身教及对本书的指导性意见。师兄庄剑云博士提供了许多帮助，对本书的出版特别是标准化起到了重要作用。内蒙古大学植物标本馆是由中国植物分类学界的先辈、《内蒙古植物志》主编马毓泉教授所创办，该馆植物分类学文献之丰富在全国也是佼佼者，其现任负责人曹瑞教授为本书的编撰与出版提供了全方位支持。张海清硕士专门为本书编制了多个程序，对保证质量和加快进度起到了重要作用。乔才元硕士、段立清博士、徐国联博士、王传璧教授、刘文霞硕士、马思聪硕士、马思睿硕士、吕全博士、田呈明博士、刘伟成博士、贺伟教授、张和平博士、王林和博士、姜伊敏教授、杨冬梅硕士、林丽萍硕士、范毅教授、特木钦高级工程师、李冬梅教授、张梅雨教授、王玉涛博士、邹晓林硕士、邹立杰高级工程师、李荣亮高级工程师、姜俊清编审、刘冰博士、杨婷博士等都帮助做了一些工作，或从国内外提供资料。原内蒙古林学院图书馆、内蒙古农业大学图书馆、内蒙古大学图书馆、中国科学院植物研究所图书馆、中国科学院图书馆、中国林业科学研究院图书馆、中国农业科学院图书馆、北京林业大学图书馆、国家图书馆、北京师范大学图书馆等都曾提供过很多方便。老友马玉明教授著作等身，写作经验丰富，对本书的编写提出了很好的建议，并为本书作序。朱宗元高级工程师在中国植物学史方面的知识与资料极为丰富，给笔者不少帮助。师弟裴明浩博士在英国科研工作十分繁忙，但是仍然花费很多时间与精力推敲辅文的英文文稿；刘萍和范晓彬教授为辅文的翻译亦做了不少工作。中国林业出版社原总编辑陈利、邵权熙副总编辑和温晋主任对本书一见钟情，当场拍板出版，显现出其睿智的目光和对科学专著的深情；温晋主任还亲自出任责任编辑。责任编辑刘家玲、周军见和李敏为本书的完善和出版付出了大量心血；他们的敬业精神、深厚功底、严谨学风和娴熟技术都令人钦佩。还要感谢夫人张玉华和孩子刘佳、尚雨、林云威、尚宁、邢凤羽为此书成稿给予的全方位支持。

感谢聂绍荃教授、马毓泉教授、王庆礼博士、马恩伟教授、黄普华教授、李秉韬教授、张玉钧教授、马其云教授、夏振岱教授等在审稿中提出的有益意见。

十分感谢洪德元院士为本书热情作序；感谢母校东北林业大学校长李坚院士的真诚鼓励。

特别要感谢国家出版基金委员会，把本书作为国家重点项目给予资助。

Acknowledgement

As this book is being published to my reasonable satisfaction, I am deeply indebted to my beloved teacher Prof. Liping Shao for mentoring me and for offering me guidance in writing the book. My senior colleague Dr. Jianyun Zhuang helped me a great deal, especially in the standardization of plant scientific names. The Herbarium of the Inner Mongolia University, founded by Prof. Yuquan Ma, one of the forefathers of Plant Taxonomy in China and the Editor-in-Chief of the *Flora of Inner Mongolia*, is one of the best in China in terms of the collection of the literature on Plant Taxonomy. Prof. Rui Cao, the current head of the herbarium, offered every possible help in writing and publication of this book. I thank Haiqing Zhang for writing several computer programs specifically for this book, which helped greatly to ensure the quality and speedy completion of the book. My thanks are also due to Caiyuan Qiao, Dr. Liqing Duan, Dr. Guolian Xu, Prof. Chuanbi Wang, Wenxia Liu, Sicong Ma, Sirui Ma, Dr. Quan Lü, Dr. Chengming Tian, Dr. Weicheng Liu, Prof. Wei He, Dr. Heping Zhang, Dr. Linhe Wang, Prof. Yimin Jiang , Dongmei Yang, Liping Lin, Prof. Yi Fan, Senior Engineer (SE) Temuqin, Prof. Dongmei Li, Prof. Meiyu Zhang, Dr. Yutao Wang, Xiaolin Zou, SE Lijie Zou, SE Rongliang Li, Editor Junqing Jiang, Dr. Bing Liu, and Dr. Ting Yang for their help and for their assistance in search of the literature. Tremendous support was received from the Library of the previous Inner Mongolia Forestry College, Library of Inner Mongolia Agricultural University, Library of Inner Mongolia University, Library of the Institute of Botany, Chinese Academy of Sciences, Library of the Chinese Academy of Sciences, Library of the Chinese Academy of Forestry, Library of the Chinese Academy of Agricultural Sciences, Library of Beijing Forestry University, Library of Beijing Normal University, and the National Library of China. My old friend Prof. Yuming Ma, a well-known expert in science writing, offered me extremely useful advice and wrote a preface for this book. SE Zongyuan Zhu also helped me enormously with his exceptional knowledge of the history of Chinese Botany and with his extensive collection of the references. My close friend and colleague Dr. Ming-Hao Pei helped translate the introductory parts of the book into English despite his busy research schedules in the UK. Ping Liu and Prof. Xiaobin Fan also offered much help in English translation. The former Editor-in-Chief of the China Forestry Publishing House Li Chen, Deputy Editor-in-Chief Quanxi Shao and the Head of Sales Office Jin Wen all "fell in love at first sight" with this book. I am grateful to them for their wisdom, vision and their affection for monographic works of science. Jin Wen even took on the role of one of the editors in charge of this book. Editors responsible for this book, Jialing Liu, Jujian Zhou and Min Li devoted enormous efforts to refining and finalizing this book for publication. My admiration goes to them for their dedication, great expertise, rigorous approach to science and extraordinary skills. I thank my wife Yuhua Zhang and my children, Jia Liu, Yu Shang, Yunwei Lin, Ning Shang and Fengyu Xing for their sacrifice, love and support throughout writing this book.

I would like to thank Prof. Shaoquan Nie, Prof. Yuquan Ma, Dr. Qingli Wang, Prof. Enwei Ma, Prof. Puhua Huang, Prof. Bingtao Li, Prof. Yujun Zhang, Prof. Qiyun Ma, Prof. Zhendai Xia for their constructive comments and suggestions in reviewing the manuscript.

I am truly grateful to Academician Deyuan Hong for kindly writing a preface for this book and to Dr Jian Li, the President of my alma mater, Northeast Forestry University, and academician of the Chinese Academy of Engineering (CAE), for his enthusiastic support.

My special thanks go to the National Publication Foundation of China for listing this book as a national key project and providing financial support.

使用指南

一、本书现出版5卷:《卷1　拉汉英名称(A-D)》、《卷2　拉汉英名称(E-O)》、《卷3　拉汉英名称(P-Z)》、《卷4　中文名称索引》、《卷5　英日俄名称索引》。

二、5卷中的序号是统一的，因此各卷之间可以互相检索。

三、拉汉英名称(卷1~3)

1. 收录科、属、种、亚种、变种、变型、亚变种、亚变型、杂交种和品种等10类名称。

2. 3卷条目按照拉丁名字母顺序排列。

3. 每词条按拉丁文、中文、英文依次排列，中间用";"隔开。同种文字中不同的名称之间用","隔开。

4. 中文名称中，正名紧跟学名，异名则放在正名后面的括号内。

5. 杂交种学名，"×"的前后各留一空格；其种下单元的等级用"nothovar."、"nothof."、"nothosubsp."等表示。

6. 英文名称中单独实词的首字母均大写。连字符"-"后面的实词，专有名词首字母大写，其他词类一律小写。

7. 书中标注的(保留科名)、(保留属名)、(废弃属名)均依据《国际植物命名法规》2006年维也纳法规。

8. 学名作者依据《国际植物命名法规》建议的《Authors of Plant Names》标准化。《Authors of Plant Names》未收入的作者，笔者则根据《Authors of Plant Names》采用的原则加以处理。笔者未能核准的作者及缺如的作者，均加了"?"以提醒。

9. 书中标号代表：●为木本植物，■为草本植物，★为中国特有植物属，☆为外国植物，◇为国家有关部门确定的濒危植物。

10. 栽培植物学名中的品种加词标志，全部采用《国际栽培植物命名法规》规定的单引号。

四、中文名称索引(卷4)

1. 名称按汉语拼音字母顺序排列。

2. 序号为黑体字表示该名称为正名，序号为白体字表示该名称为异名。

五、英日俄名称索引(卷5)

1. 英文名称索引按字母顺序排列。

2. 日文名称索引中包含了两个索引：一个索引按照序号排列，另一个索引则按照假名顺序排列。日文名称采用片假名。

3. 俄文名称索引中包含了两个索引：一个索引按照序号排列，另一个索引则按照字母顺序排列。字母的大小写遵从该文习惯。

User's Guide

1. The dictionary comprises 5 volumes: *Vol. 1 In Latin, Chinese and English (A – D)*, *Vol. 2 In Latin, Chinese and English (E – O)*, *Vol. 3 In Latin, Chinese and English (P – Z)*, *Vol. 4 Chinese Index* and *Vol. 5 English, Japanese, Russian and Indices.*

2. The serial numbers are unified in the 5 volumes and enabling readers to cross reference between the volumes.

3. Latin, Chinese and English Names (Volume 1 – 3)

1) The names of the following taxonomic ranks or categories are included: family, genus, species, subspecies, variety, subvariety, form, subform, hybrid and cultivar.

2) Entries in the volumes are listed in the alphabetical order of Latin names.

3) Plant names in each entry are arranged in the order of Latin, Chinese and English, separated by ";". Different names in the same languages are separated by ",".

4) For Chinese names, the correct name follows immediately after the scientific name and synonyms are placed in brackets after the correct Chinese name.

5) For hybrid names, there is a single space before and after " × ". At infraspecific level, "nothovar.", "nothof." and "nothosubf." are used to refer to hybrid variants.

6) For English names, a capital letter is used for the first letter of a content word. After the hyphen " – ", a small letter is used for the first letter of a content word and a capital letter is used for the first letter of a proper noun.

7) The notes in brackets, i. e. (conserved family name), (conserved generic name) and (rejected generic name), are given according to the *International Code of Botanical Nomenclature* (Vienna, 2006).

8) Author names are standardized according to the *Authors of Plant Names*, as endorsed by the *International Code of Botanical Nomenclature.* For the authors who are not included in the *Authors of Plant Names*, the same principles as described in the book are applied. The question mark "?" is added to show that the author name is missing or, if not, cannot be verified.

9) Symbols in the book stand for: ● woody plant, ■ herbaceous plant, ★ genus unique to China, ☆ non – native plant, ◇ plant listed as "endangered" by relevant government bodies.

10) For cultivated plants, single quotation marks ' ' are used for cultivar names according the *International Code of Nomenclature for Cultivated Plants.*

4. The Chinese Index (Volume 4)

1) The names are listed in the alphabetical order of Chinese Pinyin.

2) Correct names are printed in boldface and synonyms in "Kai Ti".

5. The English, Japanese and Russian Indices (Volume 5)

1) English names are in the alphabetical order of English.

2) Russian names are indexed in two ways: (a) in the serial number and (b) in the alphabetical order. The letter cases of Russian names follow the traditions of the language.

3) Japanese names are indexed in two ways: (a) in the serial number and (b) in the order of Japanese Kana. They are printed in Katakana.

目　录

1 Aa Rchb. f. (1854);阿兰属■☆
2 Aa paleacea (Kunth) Schltr.;阿兰■☆
3 Aakesia Baill. = Akesia Tussac ●☆
4 Aakesia Baill. = Blighia K. König ●☆
5 Aalium Lam. ex Kuntze = Sauropus Blume ●■
6 Aalius Kuntze = Sauropus Blume ●■
7 Aalius Lam. = Breynia J. R. Forst. et G. Forst. (保留属名)●
8 Aalius Lam. ex Kuntze = Sauropus Blume ●■
9 Aalius Rumph. = Breynia J. R. Forst. et G. Forst. (保留属名)●
10 Aalius Rumph. ex Kuntze = Sauropus Blume ●■
11 Aalius Rumph. ex Lam. = Breynia J. R. Forst. et G. Forst. (保留属名)●
12 Aalius Rumph. ex Lam. = Sauropus Blume ●■
13 Aalius trinervius Kuntze = Sauropus trinervius (Wall.) Hook. f. et Thomson ex Müll. Arg. ●
14 Aama B. D. Jacks. = Aamia Hassk. ●
15 Aama B. D. Jacks. = Adamia Wall. ●
16 Aama B. D. Jacks. = Dichroa Lour. ●
17 Aamia Hassk. = Adamia Wall. ●
18 Aamia Hassk. = Dichroa Lour. ●
19 Aapaca Metzdorff = Uapaca Baill. ■☆
20 Aaronsohnia Warb. et Eig(1927);阿氏菊属(肋脂菊属)■☆
21 Aaronsohnia factorovskyi Warb. et Eig;阿氏菊■☆
22 Aaronsohnia pubescens (Desf.) K. Bremer et Humphries;毛阿氏菊■☆
23 Aaronsohnia pubescens (Desf.) K. Bremer et Humphries subsp. maroccana (Ball) Förther et Podlech;马罗阿氏菊■☆
24 Aasa Houtt. = Tetracera L. ●
25 Ababella Comm. ex Moewes = Turraea L. ●
26 Abacosa Alef. = Vicia L. ■
27 Abalemis Raf. = Anemone L. (保留属名)■
28 Abalon Adans. = Helonias L. ■☆
29 Abalum Adans. = Abalon Adans. ■☆
30 Abalum Adans. = Helonias L. ■☆
31 Abama Adans. = Narthecium Huds. (保留属名)■
32 Abama americana (Ker Gawl.) Morong = Narthecium americanum Ker Gawl. ■☆
33 Abama californica (Baker) A. Heller = Narthecium californicum Baker ■☆
34 Abama montana Small = Narthecium americanum Ker Gawl. ■☆
35 Abama occidentalis (A. Gray) A. Heller = Narthecium californicum Baker ■☆
36 Abamaceae J. Agardh = Melianthaceae Horan. (保留科名)●☆
37 Abaminaceae J. Agardh = Liliaceae Juss. (保留科名)■●
38 Abamineae J. Agardh = Liliaceae Juss. (保留科名)■●
39 Abandion Adans. = Bulbocodium L. ■☆
40 Abandium Adans. = Abandion Adans ■☆
41 Abandium Adans. = Bulbocodium L. ■☆
42 Abaphus Raf. = Abapus Adans ■☆
43 Abapus Adans. = Gethyllis L. ■☆
44 Abapus Adans. = Papiria Thunb. ■☆
45 Abarema Pittier(1927);阿巴豆属●☆
46 Abarema angulata (Benth.) Kosterm. = Archidendron clypearia (Jack) I. C. Nielsen ●
47 Abarema angulata (Benth.) Kosterm. = Pithecellobium clypearia (Jack) Benth. ●
48 Abarema clypearia (Jack) Kosterm. = Archidendron clypearia (Jack) I. C. Nielsen ●
49 Abarema clypearia (Jack) Kosterm. = Pithecellobium clypearia (Jack) Benth. ●
50 Abarema dalatensis Kosterm. = Cylindrokelupha dalatensis (Kosterm.) T. L. Wu ●
51 Abarema kerrii (Gagnep.) Kosterm. = Archidendron kerrii (Gagnep.) I. C. Nielsen ●
52 Abarema kerrii (Gagnep.) Kosterm. = Cylindrokelupha kerrii (Gagnep.) T. L. Wu ●
53 Abarema lucida (Benth.) Kosterm. = Archidendron lucidum (Benth.) I. C. Nielsen ●
54 Abarema lucida (Benth.) Kosterm. = Pithecellobium lucidum Benth. ●
55 Abarema robinsonii (Gagnep.) Kosterm. = Archidendron robinsonii (Gagnep.) I. C. Nielsen ●
56 Abarema robinsonii (Gagnep.) Kosterm. = Cylindrokelupha robinsonii (Gagnep.) Kosterm. ●
57 Abarema utilis (Chun et F. C. How) Kosterm. = Archidendron utile (Chun et F. C. How) I. C. Nielsen ●
58 Abarema utilis (Chun et F. C. How) Kosterm. = Pithecellobium utile Chun et F. C. How ●
59 Abarema yunnanensis Kosterm. = Archidendron kerrii (Gagnep.) I. C. Nielsen ●
60 Abarema yunnanensis Kosterm. = Cylindrokelupha yunnanensis (Kosterm.) T. L. Wu ●
61 Abasaloa Benth. et Hook. f. = Abasoloa La Llave ■☆
62 Abasicarpon (Andrz. ex Rchb.) Rchb. = Arabis L. ●■
63 Abasicarpon (Andrz. ex Rchb.) Rchb. = Cheiranthus L. ●■
64 Abasicarpon Andrz. ex Rchb. = Arabis L. ●■
65 Abasoloa La Llave = Eclipta L. (保留属名)■
66 Abasoloa La Llave ex Lex. = Eclipta L. (保留属名)■
67 Abasoloa La Llave(1824);阿瓦菊属■☆
68 Abasoloa taboada La Llave;阿巴菊■☆
69 Abatia Ruiz et Pav. (1794);阿巴特木属(阿巴特属)●☆
70 Abatia parviflora Ruiz et Pav.;阿巴特木(阿巴特)●☆
71 Abatia rugosa Ruiz et Pav.;染料阿巴特木●☆
72 Abauria Becc. = Koompassia Maingay ex Benth. ●☆
73 Abavo Risler = Adansonia L. ●
74 Abaxianthus M. A. Clem. et D. L. Jones(2002);远轴兰属■☆
75 Abazicarpus Andrz. ex DC. = Abasicarpon (Andrz. ex Rchb.) Rchb. ●■
76 Abazicarpus Andrz. ex DC. = Cheiranthus L. ●■
77 Abbevillea O. Berg = Campomanesia Ruiz et Pav. ●☆
78 Abbotia Raf. = Triglochin L. ■
79 Abbottia F. Muell. (1875);北澳茜草属●☆
80 Abbottia F. Muell. = Timonius DC. (保留属名)●
81 Abbottia singularis F. Muell.;北澳茜草●☆
82 Abdominea J. J. Sm. (1914);阿道米尼兰属;Abdominea ■☆
83 Abdominea minimiflora (Hook. f.) J. J. Sm.;阿道米尼兰;Leastflower Abdominea ■☆
84 Abdra Greene = Draba L. ■
85 Abdulmajidia Whitmore(1974);马来亚玉蕊属●☆
86 Abdulmajidia chaniana Whitmore;马来亚玉蕊●☆
87 Abebaia Baehni = Manilkara Adans. (保留属名)●
88 Abelemis Britton = Abalemis Raf. ■
89 Abelemis Britton = Anemone L. (保留属名)■
90 Abelemis Raf. ex Britton = Abalemis Raf. ■

91 Abelia R. Br. (1818);六道木属(六条木属,糯米条属);Abelia ●
92 Abelia angustifolia Bureau ex Franch.;巴塘六道木(狭叶六道木)●
93 Abelia angustifolia Bureau ex Franch. = Zabelia triflora (R. Br. ex Wall.) Makino ●
94 Abelia anhweiensis Nakai = Abelia dielsii (Graebn.) Rehder ●
95 Abelia anhweiensis Nakai = Zabelia biflora (Turcz.) Makino ●
96 Abelia aschersoniana (Graebn.) Rehder = Abelia chinensis R. Br. ●
97 Abelia azanderii (Graebn.) Rehder = Abelia dielsii (Graebn.) Rehder ●
98 Abelia biflora Turcz. = Zabelia biflora (Turcz.) Makino ●
99 Abelia biflora Turcz. f. minor (Nakai) C. F. Fang = Zabelia biflora (Turcz.) Makino ●
100 Abelia biflora Turcz. var. alpina A. I. Baranov et Skvortsov = Abelia biflora Turcz. ●
101 Abelia biflora Turcz. var. coreana (Nakai) C. F. Fang = Zabelia biflora (Turcz.) Makino ●
102 Abelia biflora Turcz. var. minor Nakai = Zabelia biflora (Turcz.) Makino ●
103 Abelia brachystemon (Diels) Rehder = Zabelia biflora (Turcz.) Makino ●
104 Abelia brachystemon Herb. ex Rehder = Abelia dielsii (Graebn.) Rehder ●
105 Abelia buddleioides W. W. Sm.;醉鱼草状六道木;Tutterflybuch-like Abelia ●
106 Abelia buddleioides W. W. Sm. = Zabelia triflora (R. Br. ex Wall.) Makino ●
107 Abelia buddleioides W. W. Sm. var. divergens W. W. Sm. = Abelia buddleioides W. W. Sm. ●
108 Abelia buddleioides W. W. Sm. var. divergens W. W. Sm. = Zabelia triflora (R. Br. ex Wall.) Makino ●
109 Abelia buddleioides W. W. Sm. var. intercedens Hand.-Mazz. = Abelia buddleioides W. W. Sm. ●
110 Abelia buddleioides W. W. Sm. var. intercedens Hand.-Mazz. = Zabelia triflora (R. Br. ex Wall.) Makino ●
111 Abelia buddleioides W. W. Sm. var. stenantha Hand.-Mazz. = Abelia buddleioides W. W. Sm. ●
112 Abelia buddleioides W. W. Sm. var. stenantha Hand.-Mazz. = Zabelia triflora (R. Br. ex Wall.) Makino ●
113 Abelia chinensis R. Br.;糯米条(白花树,茶条树,大叶白马骨,华北六条木,山柳树,台湾糯米条);China Abelia,Chinese Abelia,Taiwan Abelia ●
114 Abelia chinensis R. Br. var. ionandra (Hayata) Masam.;台湾糯米条●
115 Abelia chinensis R. Br. var. ionandra (Hayata) Masam. = Abelia chinensis R. Br. ●
116 Abelia chowii G. Hoo = Abelia parvifolia Hemsl. ●
117 Abelia coreana Nakai;朝鲜六道木;Korean Abelia ●☆
118 Abelia coreana Nakai = Abelia biflora Turcz. ●
119 Abelia corymbosa Regel et Schmalh.;伞房花状六道木;Corymbose Abelia ●☆
120 Abelia davidii Hance = Abelia biflora Turcz. ●
121 Abelia davidii Hance = Zabelia biflora (Turcz.) Makino ●
122 Abelia dielsii (Graebn.) Rehder;南方六道木(太白六道木);Diels Abelia,Southern Abelia ●
123 Abelia engleriana (Graebn.) Rehder;短枝六道木(梗花六道木,通梗花,莲梗花,紫荆丫);Engler Abelia ●
124 Abelia engleriana (Graebn.) Rehder = Abelia uniflora R. Br. ●
125 Abelia floribunda Decne.;多花六道木;Abelia Multiflower ●
126 Abelia floribunda Zabel = Abelia grandiflora (Rovelli ex André) Rehder ●
127 Abelia forrestii (Diels) W. W. Sm.;细瘦六道木;Forrest Abelia,Thin Abelia ●
128 Abelia gracilenta W. W. Sm. = Abelia forrestii (Diels) W. W. Sm. ●
129 Abelia gracilenta W. W. Sm. var. microphylla W. W. Sm. = Abelia forrestii (Diels) W. W. Sm. ●
130 Abelia graebneriana (André) Rehder = Abelia engleriana (Graebn.) Rehder ●
131 Abelia graebneriana Rehder = Abelia uniflora R. Br. ●
132 Abelia grandiflora (André) Rehder 'Prostrata';俯卧大花六道木;Prostrate Abelia ●☆
133 Abelia grandiflora (André) Rehder 'Sunrise';日出大花六道木;Sunrise Abelia ●☆
134 Abelia grandiflora (André) Rehder = Abelia grandiflora (Rovelli ex André) Rehder ●
135 Abelia grandiflora (Rovelli ex André) Rehder;大花六道木(二翅六道木);Glossy Abelia,Large-flowered Abelia ●
136 Abelia hanceana M. Martens ex Hance = Abelia chinensis R. Br. ●
137 Abelia hanceana Mart. ex Hance = Abelia chinensis R. Br. ●
138 Abelia hersii Nakai = Abelia dielsii (Graebn.) Rehder ●
139 Abelia hersii Nakai = Zabelia biflora (Turcz.) Makino ●
140 Abelia howii G. Hoo = Abelia parvifolia Hemsl. ●
141 Abelia integrifolia Koidz. = Zabelia integrifolia (Koidz.) Makino ex Ikuse et S. Kuros. ●☆
142 Abelia ionandra Hayata = Abelia chinensis R. Br. var. ionandra (Hayata) Masam. ●
143 Abelia ionandra Hayata = Abelia chinensis R. Br. ●
144 Abelia longituba Rehder = Abelia parvifolia Hemsl. ●
145 Abelia longituba Rehder = Abelia schumannii (Graebn.) Rehder ●
146 Abelia longituba Rehder = Abelia uniflora R. Br. ●
147 Abelia macrotera (Graebn. et Buchw.) Rehder;二翅六道木(假拉药藤,空心树,紫荆桠);Twowing Abelia,Two-winged Abelia ●
148 Abelia macrotera (Graebn. et Buchw.) Rehder = Abelia uniflora R. Br. ●
149 Abelia mairei H. Lév. = Abelia parvifolia Hemsl. ●
150 Abelia mairei H. Lév. = Abelia uniflora R. Br. ●
151 Abelia microphylla (W. W. Sm.) Golubk. = Abelia forrestii (Diels) W. W. Sm. ●
152 Abelia mosanensis T. H. Chung ex Nakai;芳香六道木;Fragrant Abelia,Sweet Abelia ●☆
153 Abelia multiflora Zabel = Abelia grandiflora (Rovelli ex André) Rehder ●
154 Abelia myrtilloides Rehder = Abelia parvifolia Hemsl. ●
155 Abelia myrtilloides Rehder = Abelia uniflora R. Br. ●
156 Abelia onkocarpa (Graebn.) Rehder = Zabelia biflora (Turcz.) Makino ●
157 Abelia onkocarpa Rehder = Abelia dielsii (Graebn.) Rehder ●
158 Abelia parvifolia Hemsl. = Abelia uniflora R. Br. ●
159 Abelia rupestris L. hybrida Rovelli ex Schaedtler = Abelia grandiflora (Rovelli ex André) Rehder ●
160 Abelia rupestris L. Späth = Abelia grandiflora (Rovelli ex André) Rehder ●
161 Abelia rupestris Lindl. = Abelia chinensis R. Br. ●
162 Abelia rupestris Lindl. f. grandiflora Rovelli ex André = Abelia grandiflora (Rovelli ex André) Rehder ●

163 Abelia schischkinii Golubk. = Abelia parvifolia Hemsl. ●
164 Abelia schischkinii Golubk. = Abelia uniflora R. Br. ●
165 Abelia schumannii (Graebn.) Rehder = Abelia parvifolia Hemsl. ●
166 Abelia schumannii (Graebn.) Rehder = Abelia uniflora R. Br. ●
167 Abelia serrata Siebold et Zucc.;黄花六道木(齿叶六道木,日本小六道木);Serrate Abelia ●
168 Abelia serrata Siebold et Zucc. f. gymnocarpa (Graebn.) Sugim.;裸果齿叶六道木●☆
169 Abelia serrata Siebold et Zucc. f. obspathulata (Koidz.) Sugim.;倒匙苞齿叶六道木●☆
170 Abelia serrata Siebold et Zucc. f. sanguinea (Sugim.) Sugim.;红色齿叶六道木●☆
171 Abelia serrata Siebold et Zucc. f. tomentosa (Koidz.) Sugim. = Abelia serrata Siebold et Zucc. var. tomentosa (Koidz.) Nakai ●☆
172 Abelia serrata Siebold et Zucc. var. buchwaldii (Graebn.) Nakai = Abelia serrata Siebold et Zucc. ●
173 Abelia serrata Siebold et Zucc. var. gymnocarpa (Graebn.) Nakai = Abelia serrata Siebold et Zucc. f. gymnocarpa (Graebn.) Sugim. ●☆
174 Abelia serrata Siebold et Zucc. var. integerrima Nakai = Zabelia integrifolia (Koidz.) Makino ex Ikuse et S. Kuros. ●☆
175 Abelia serrata Siebold et Zucc. var. obspathulata Koidz. = Abelia serrata Siebold et Zucc. f. obspathulata (Koidz.) Sugim. ●☆
176 Abelia serrata Siebold et Zucc. var. tomentosa (Koidz.) Nakai;毛齿叶六道木●☆
177 Abelia spathulata Siebold et Zucc.;日本六道木(温州六道木);Japanese Abelia ●
178 Abelia spathulata Siebold et Zucc. f. colorata (H. Hara et S. Kuros.) H. Hara et S. Kuros.;多色日本六道木●☆
179 Abelia spathulata Siebold et Zucc. f. duplexa H. Ohba;重瓣日本六道木●☆
180 Abelia spathulata Siebold et Zucc. f. flavescens Makino;黄花日本六道木;Yellowflower Japane Abelia ●☆
181 Abelia spathulata Siebold et Zucc. f. lucida Makino;光日本六道木;Smooth Japanese Abelia ●☆
182 Abelia spathulata Siebold et Zucc. f. pilosa Nakai;毛日本六道木;Pilose Japanese Abelia ●☆
183 Abelia spathulata Siebold et Zucc. var. colorata H. Hara et S. Kuros. = Abelia spathulata Siebold et Zucc. f. colorata (H. Hara et S. Kuros.) H. Hara et S. Kuros. ●☆
184 Abelia spathulata Siebold et Zucc. var. micrantha Nakai;小花日本六道木;Small-flowered Japanese Abelia ●☆
185 Abelia spathulata Siebold et Zucc. var. sanguinea Makino;红色日本六道木●☆
186 Abelia spathulata Siebold et Zucc. var. stenophylla Honda;狭叶日本六道木●☆
187 Abelia spathulata Siebold et Zucc. var. tetrasepala Koidz. = Abelia tetrasepala (Koidz.) H. Hara et S. Kuros. ●☆
188 Abelia splendens Hort. ex K. Koch;纤细六道木●☆
189 Abelia tereticalyx (Graebn.) Rehder = Abelia parvifolia Hemsl. ●
190 Abelia tereticalyx (Graebn.) Rehder = Abelia uniflora R. Br. ●
191 Abelia tetrasepala (Koidz.) H. Hara et S. Kuros.;四萼日本六道木●☆
192 Abelia tomentosa Koidz. = Abelia serrata Siebold et Zucc. var. tomentosa (Koidz.) Nakai ●☆
193 Abelia triflora R. Br. = Zabelia triflora (R. Br. ex Wall.) Makino ●
194 Abelia triflora R. Br. ex Wall. = Zabelia triflora (R. Br. ex Wall.) Makino ●
195 Abelia triflora R. Br. ex Wall. subsp. parvifolia (C. B. Clarke) Wendelbo;小叶三花六道木●☆
196 Abelia triflora R. Br. subsp. parvifolia (C. B. Clarke) Wendelbo = Abelia triflora R. Br. ex Wall. subsp. parvifolia (C. B. Clarke) Wendelbo ●☆
197 Abelia trifolia R. Br. ex Wall.;三叶六道木●
198 Abelia umbellata (Graebn. et Buchw.) Rehder;伞花六道木;Umbellaflower Abelia, Umbella-flowered Abelia ●
199 Abelia uniflora Hort. = Abelia grandiflora (Rovelli ex André) Rehder ●
200 Abelia uniflora R. Br.;小叶六道木(白花莲花梗,单花六道木,独花六道木,二翅六道木,福建六道木,鸡肚子,鸡壳肚花,棵棵兜,舒曼六道木,泰宁六道木);Littleleaf Abelia, Little-leaved Abelia, Parvifoliate Abelia, Schumann's Abelia ●
201 Abelia verticillata H. Lév. = Abelia parvifolia Hemsl. ●
202 Abelia verticillata H. Lév. = Abelia uniflora R. Br. ●
203 Abelia zanderi (Graebn.) Rehder = Abelia dielsii (Graebn.) Rehder ●
204 Abelia zanderi (Graebn.) Rehder = Zabelia biflora (Turcz.) Makino ●
205 Abelicea Baill. = Zelkova Spach(保留属名)●
206 Abelicea Rchb. = Zelkova Spach(保留属名)●
207 Abelicea hirta C. K. Schneid. = Zelkova serrata (Thunb.) Makino ●
208 Abelicia Kuntze = Abelicea Rchb. ●
209 Abeliophyllum Nakai(1919);六道木叶属(白花连翘属,翅果连翘属);Forsythia ●☆
210 Abeliophyllum distichum Nakai;六道木叶(白花连翘,白连翘);Korean Abelia, White Forsythia ●☆
211 Abeliophyllum distichum Nakai 'Roseum';粉红白连翘;Pink 'Forsythia' ●☆
212 Abelmoschus Medik. (1787);秋葵属(黄葵属);Abelmoschus ●■
213 Abelmoschus caillei (A. Chev.) Stevels;卡耶秋葵■☆
214 Abelmoschus cancellatus (Roxb.) Voigt = Abelmoschus crinitus Wall. ■
215 Abelmoschus coccineus S. Y. Hu = Abelmoschus sagittifolius (Kurz) Merr. ■
216 Abelmoschus coccineus S. Y. Hu var. aceraifolius S. Y. Hu = Abelmoschus sagittifolius (Kurz) Merr. ■
217 Abelmoschus crinitus Wall.;长毛黄葵(白背木,黄花马宁,黄葵,黄茄花,假芙蓉,山芙蓉,山尖茶,纹叶木槿,野芙蓉,野棉花);Crinite Abelmoschus ■
218 Abelmoschus cruentus Walp. = Hibiscus sabdariffa L. ●■
219 Abelmoschus esculentus (L.) Moench;咖啡黄葵(潺茄,糊麻,黄秋葵,毛茄,秋葵,山茄,食用秋葵,羊角豆,越南芝麻);Bandakai, Bankokai, Edible Abelmoschus, Gobbo, Gobo, Gombo, Gumbo, Lady's Fingers, Lady's-finger, Okra ■
220 Abelmoschus esculentus (L.) Moench var. praecox (Forssk.) A. Chev.;早咖啡黄葵■☆
221 Abelmoschus esculentus (L.) Moench var. textilis A. Chev.;织叶咖啡黄葵■☆
222 Abelmoschus esquirolii (H. Lév.) S. Y. Hu = Abelmoschus sagittifolius (Kurz) Merr. ■
223 Abelmoschus ficulneus (L.) Wight et Arn. = Abelmoschus ficulneus (L.) Wight et Arn. ex Wight ■☆
224 Abelmoschus ficulneus (L.) Wight et Arn. ex Wight;无花果黄葵■☆
225 Abelmoschus glutinotextilis Kagawa;黏黄葵■☆
226 Abelmoschus grewioides Baker f.;非洲秋葵■☆

227 Abelmoschus hainanensis S. Y. Hu = Abelmoschus crinitus Wall. ■
228 Abelmoschus indicus G. Don;印度黄葵;Indian Abelmoschus ■☆
229 Abelmoschus manihot (L.) Medik.;黄蜀葵(霸天伞,豹子眼睛花,侧金盏,啜脓兰,大麻,花药草,黄芙蓉,黄花莲,黄葵,黄秋葵,黄色葵,黄稀饭花,火炮药,鸡爪兰,假棉花,假阳桃,金花捷根,疽疮药,弥勒佛掌,棉花蒿,棉花葵,荞面花,秋葵,水芙蓉,辛麻,野芙蓉,野棉花,追风药);Aibika, Chinese Musk-mallow, Queen-of-the-summer, Sunset Abelmoschus, Sunset Hibiscus ■
230 Abelmoschus manihot (L.) Medik. subsp. pungens (Roxb.) Hochr. = Abelmoschus manihot (L.) Medik. var. pungens (Roxb.) Hochr. ■
231 Abelmoschus manihot (L.) Medik. subsp. tetraphyllus (Hornem.) Borss. Waalk. var. pungens (Roxb.) Hochr. = Abelmoschus manihot (L.) Medik. var. pungens (Roxb.) Hochr. ■
232 Abelmoschus manihot (L.) Medik. var. betulifolius (Mast.) Hochr.;桦叶黄蜀葵■☆
233 Abelmoschus manihot (L.) Medik. var. pungens (Roxb.) Hochr.;刚毛黄蜀葵(豹子眼睛花,大黏药,大苏子,刚毛黄秋葵,黄芙蓉,黄火麻,黄麻,黄秋葵,火炮药,桐麻,辛麻,眼睛花,野棉花,竹芙蓉);Setose Abelmoschus ■
234 Abelmoschus manihot (L.) Medik. var. zenkeri (Gürke) A. Chev. = Hibiscus zenkeri Gürke ●☆
235 Abelmoschus moschatus (L.) Medik.;黄葵(芙蓉麻,鬼布,假芙蓉,假棉花,假棉桃,假三稔,假山稔,假杨桃,碌毒草,毛夹,毛荚,鸟笼胶,三脚鳖,三脚破,山芙蓉,山胶播,山油麻,麝香秋葵,麝香球葵,水芙蓉,香黄葵,香葵,药虎,野芙蓉,野棉,野棉花,野油麻);Ambrette Seed, Musk Mallow, Musk Okra, Muskmallow, Silk Flower ■
236 Abelmoschus moschatus (L.) Medik. var. betulifolius (Mast.) Hochr.;桦叶黄葵■☆
237 Abelmoschus moschatus (L.) Medik. var. tuberosus (Span) Borss. Waalk. = Abelmoschus sagittifolius (Kurz) Merr. ■
238 Abelmoschus muliensis K. M. Feng;木里秋葵;Muli Abelmoschus ■
239 Abelmoschus mutabilis (L.) Wall. ex Hassk. = Hibiscus mutabilis L. ●
240 Abelmoschus pungens (Roxb.) Voigt = Abelmoschus manihot (L.) Medik. var. pungens (Roxb.) Hochr. ■
241 Abelmoschus rostellatus Walp. = Hibiscus surattensis L. ■
242 Abelmoschus sagittifolius (Kurz) Merr.;箭叶秋葵(红花马宁,剑叶秋葵,秋葵,水芙蓉,铜皮,五指山参,小红芙蓉,岩酸,梓桐花);Arrowleaf Abelmoschus ■
243 Abelmoschus strictus Voigt;灌木黄葵●☆
244 Abelmoschus venustus (Blume) Walp. = Hibiscus indicus (Burm. f.) Hochr. ●
245 Abelmoschus verrucosus (Guillaumin et Perr.) Walp. = Hibiscus cannabinus L. ■
246 Abelmoschus verrucosus Walp. = Hibiscus cannabinus L. ■
247 Abels Lindl. = Caucaea Schltr. ■☆
248 Abena (Schauer) Hitchc. = Stachytarpheta Vahl(保留属名)●■
249 Abena Neck. = Stachytarpheta Vahl(保留属名)●■
250 Abena Neck. ex Hitchc. = Stachytarpheta Vahl(保留属名)●■
251 Aberemoa Aubl.(废弃属名) = Guatteria Ruiz et Pav.(保留属名)●☆
252 Aberia Hochst. = Dovyalis E. Mey. ex Arn. ●
253 Aberia abyssinica (A. Rich.) Clos = Dovyalis abyssinica (A. Rich.) Warb. ●☆
254 Aberia caffra Hook. f. et Harv. = Dovyalis caffra (Hook. f. et Harv.) Hook. f. ●☆
255 Aberia edulis T. Anderson = Dovyalis caffra (Hook. f. et Harv.) Hook. f. ●☆
256 Aberia gardnerii Clos = Dovyalis hebecarpa (Gardner) Warb. ●
257 Aberia hebecarpa (Gardner) Kuntze = Dovyalis hebecarpa (Gardner) Warb. ●
258 Aberia longispina Harv. = Dovyalis longispina (Harv.) Warb. ●☆
259 Aberia macrocalyx Oliv. = Dovyalis macrocalyx (Oliv.) Warb. ●☆
260 Aberia mollis Oliv. = Dovyalis mollis (Oliv.) Warb. ●☆
261 Aberia tristis Sond. = Dovyalis zeyheri (Sond.) Warb. ●☆
262 Aberia verrucosa Hochst. = Dovyalis verrucosa (Hochst.) Warb. ●☆
263 Aberia zeyheri Sond. = Dovyalis zeyheri (Sond.) Warb. ●☆
264 Aberrantia (Luer) Luer = Pleurothallis R. Br. ■☆
265 Aberrantia (Luer) Luer(2004);阿拜兰属■☆
266 Aberrantia Luer = Aberrantia (Luer) Luer ■☆
267 Abesina Neck. = Verbesina L.(保留属名)●■☆
268 Abies D. Don = Abies Mill. ●
269 Abies Mill.(1754);冷杉属;Fir, Silver Fir, Whitewood ●
270 Abies ajanensis F. Schmidt = Picea jezoensis (Siebold et Zucc.) Carrière var. microsperma (Lindl.) W. C. Cheng et L. K. Fu ●
271 Abies ajanensis F. Schmidt var. microsperma (Lindl.) Mast. = Picea jezoensis (Siebold et Zucc.) Carrière var. microsperma (Lindl.) W. C. Cheng et L. K. Fu ●
272 Abies alba Mill.;欧洲冷杉(白枞,欧洲白冷杉,欧洲银冷杉,银枞);Christmas Tree, Edeltanne, European Fir, European Silver Fir, Red Fir, Silver Fir, Strasburg Turpentine, Weisstanne, Whitewood ●☆
273 Abies alba Mill. var. columnaris (Carrière) Rehder;柱状银枞;Columnar Silver Fir ●☆
274 Abies alba Mill. var. compacta (Pers.) Rehder;矮银枞;Dwarf Silver Fir ●☆
275 Abies alba Mill. var. pendula (Carrière) Asch. et Graebn.;垂枝银枞;Weeping Silver Fir ●☆
276 Abies alba Mill. var. pyramidalis (Carrière) Voss;塔形银枞;Sentinel Silver Fir ●☆
277 Abies alcockiana Lindl. ex K. Koch = Picea alcoquiana (Veitch ex Lindl.) Carrière ●☆
278 Abies amabilis (Douglas ex Loudon) J. Forbes 'Spreading Star';匍匐星温哥华冷杉●☆
279 Abies amabilis (Douglas ex Loudon) J. Forbes = Abies amabilis Douglas ex J. Forbes ●☆
280 Abies amabilis Douglas ex J. Forbes;温哥华冷杉(太平洋冷杉,太平洋银枞);Amalilis Fir, Beautiful Fir, Bratiful Fir, Cascades Fir, Lovely Fir, Pacific Fir, Pacific Silver Fir, Red Silver Fir, Silver Fir, White Fir ●☆
281 Abies arizonica Merriam = Abies lasiocarpa (Hook.) Nutt. var. arizonica (Merr.) Lemmon ●☆
282 Abies balsamea (L.) Mill.;香脂冷杉(胶枞,胶冷杉,树胶冷杉);American Silver Fir, Balm of Gilead Fir, Balm of Gilead, Balsam Fir, Canada Balsam, Canadapitch, Dwarf Balsam Fir, Eastern Fir ●☆
283 Abies balsamea (L.) Mill. f. hudsonia Bosc ex Carrière;哈得逊香脂冷杉(哈得逊胶枞,哈得逊冷杉);Hudson Fir ●☆
284 Abies balsamea (L.) Mill. f. phanerolepis (Fernald) Rehder = Abies balsamea (L.) Mill. subsp. phanerolepis (Fernald) A. E. Murray ●☆
285 Abies balsamea (L.) Mill. subsp. fraseri (Pursh) A. E. Murray;弗氏香脂冷杉(弗氏胶枞);Fraser Balsam Fir ●☆
286 Abies balsamea (L.) Mill. subsp. lasiocarpa (Hook.) B. Boivin;

毛鳞香脂冷杉●☆

287 Abies balsamea (L.) Mill. subsp. phanerolepis (Fernald) A. E. Murray;显鳞香脂冷杉(显胶枞,显鳞胶枞);Bracted Balsam Fir ●☆

288 Abies balsamea (L.) Mill. var. fallax ?;亚高山香脂冷杉;Subalpine Fir ●☆

289 Abies balsamea (L.) Mill. var. hudsonia (Jacq.) Sarg. = Abies balsamea (L.) Mill. f. hudsonia Bosc ex Carrière ●☆

290 Abies balsamea (L.) Mill. var. macrocarpa Kent.;大果香脂冷杉(大果胶枞,大球果胶枞,球果胶枞);Bigcone Balsam Fir ●☆

291 Abies balsamea (L.) Mill. var. nana (Nelson) Carrière;矮香脂冷杉(矮胶枞);Dwarf Balsam Fir ●☆

292 Abies balsamea (L.) Mill. var. phanerolepis Fernald = Abies balsamea (L.) Mill. subsp. phanerolepis (Fernald) A. E. Murray ●☆

293 Abies beshanzuensis M. H. Wu;百山祖冷杉;Baishanzu Fir ●◇

294 Abies beshanzuensis M. H. Wu var. ziyuanensis (L. K. Fu et S. L. Mo) L. K. Fu et Nan Li;资源冷杉;Ziyuan Fir ●◇

295 Abies bicolor Maxim. = Picea alcoquiana (Veitch ex Lindl.) Carrière ●☆

296 Abies bifida Siebold et Zucc. = Abies firma Siebold et Zucc. ●

297 Abies bifolia A. Murray = Abies lasiocarpa (Hook.) Nutt. ●☆

298 Abies borisii-regis Mattf.;鲍里斯冷杉;King Boris' Fir ●☆

299 Abies bornmuelleriana Mattf.;鲍氏冷杉;Bornmueller Fir ●☆

300 Abies brachyphylla Maxim. = Abies homolepis Siebold et Zucc. ●☆

301 Abies brachytyla Franch. = Picea brachytyla (Franch.) E. Pritz. ●◇

302 Abies bracteata (D. Don) Nutt. = Abies venusta (Douglas) K. Koch ●

303 Abies bracteata (D. Don) Poit. = Abies venusta (Douglas) K. Koch ●

304 Abies brunoniana (Wall.) Lindl. = Tsuga dumosa (D. Don) Eichler ●

305 Abies canadensis Mill. = Picea glauca (Moench) Voss ●

306 Abies cephalonica Loudon;希腊冷杉(塞法罗尼亚冷杉);Apollo Fir, Bristlecone Fir, Grecian Fir, Greece Fir, Greek Fir, Shaanxi Fir, Shensi Fir ●☆

307 Abies cephalonica Loudon 'Meyer's Dwarf';迈尔矮生希腊冷杉●☆

308 Abies cephalonica Loudon var. apollinis (Link) Beissn.;阿波罗冷杉;Apollo Fir ●☆

309 Abies chayuensis W. C. Cheng et L. K. Fu;察隅冷杉;Chayu Fir ●

310 Abies chengii Rushforth = Abies forrestii Coltm. -Rog. ●

311 Abies chensiensis Tiegh.;秦岭冷杉(陕西冷杉);Chinling Fir, Chinling Mountains Fir, Qinling Fir, Shaanxi Fir ●◇

312 Abies chensiensis Tiegh. var. ernestii (Rehder) Liu = Abies ernestii Rehder ●

313 Abies chinensis Franch. = Tsuga chinensis (Franch.) Pritz. ex Diels ●

314 Abies cilicica (Antoine et Kotschy) Carrière;西里西亚冷杉(土耳其冷杉,西里西卡冷杉);Cilicia Fir, Cilician Fir, Cilicica Abies, Cilicica Fir ●☆

315 Abies concolor (Gordon et Glend.) Hildebrandt = Abies concolor (Gordon et Glend.) Lindl. ex Hildebrandt ●☆

316 Abies concolor (Gordon et Glend.) Hildebrandt var. lowiana (Gordon) Lemmon = Abies lowiana A. Murray ●☆

317 Abies concolor (Gordon et Glend.) Lindl. ex Hildebrandt;白冷杉(灰叶冷杉,科罗拉多白冷杉,科罗拉多冷杉,银冷杉);Blue Fir, California White Fir, Colorado Fir, Colorado White Fir, Concolor Fir, Rocky Mountain White Fir, Silver Fir, Western Balsam Fir, White Balsam Fir, White Colorado Fir, White Fir ●☆

318 Abies concolor (Gordon et Glend.) Lindl. ex Hildebrandt 'Argentea';银叶白冷杉●☆

319 Abies concolor (Gordon et Glend.) Lindl. ex Hildebrandt 'Compactea';致密白冷杉;Compact White Fir ●☆

320 Abies concolor (Gordon et Glend.) Lindl. ex Hildebrandt 'Masonic Broom';马松白冷杉(马松银冷杉)●☆

321 Abies concolor (Gordon et Glend.) Lindl. ex Hildebrandt = Abies lowiana A. Murray ●☆

322 Abies concolor (Gordon et Glend.) Lindl. ex Hildebrandt var. aurea Beissn.;金叶白冷杉;Golden White Fir ●☆

323 Abies concolor (Gordon et Glend.) Lindl. ex Hildebrandt var. brevifolia Beissn.;短叶白冷杉;Shortleaf White Fir ●☆

324 Abies concolor (Gordon et Glend.) Lindl. ex Hildebrandt var. conica Slavin;圆锥白冷杉(圆锥异色冷杉);Conical White Fir ●☆

325 Abies concolor (Gordon et Glend.) Lindl. ex Hildebrandt var. globosa Niemetz;球形白冷杉;Globe White Fir ●☆

326 Abies concolor (Gordon et Glend.) Lindl. ex Hildebrandt var. lowiana (Gordon) Lemmon = Abies lowiana A. Murray ●☆

327 Abies concolor (Gordon et Glend.) Lindl. ex Hildebrandt var. pendula Beissn.;垂枝白冷杉;Weeping White Fir ●☆

328 Abies davidiana Franch. = Keteleeria davidiana (Bertrand) Beissn. ●

329 Abies davidiana Franch. var. nukiangensis (W. C. Cheng et L. K. Fu) Farjon et Silba = Abies nukiangensis W. C. Cheng et L. K. Fu ●

330 Abies davidiana Franch. var. smithii (Vig. et Gaussen) T. S. Liu = Abies smithii (Vig. et Gaussen) Flous ●

331 Abies dayuanensis Q. X. Liu;大院冷杉;Dayuan Fir ●

332 Abies dayuanensis Q. X. Liu = Abies beshanzuensis M. H. Wu var. ziyuanensis (L. K. Fu et S. L. Mo) L. K. Fu et Nan Li ●◇

333 Abies delavayi Franch.;苍山冷杉(大理冷杉,高山枞,冷杉,塔杉,唐则);Delavay Fir, Delavay Silver Fir, Delavay's Fir ●

334 Abies delavayi Franch. var. faberi (Mast.) D. R. Hunt = Abies faberi (Mast.) Craib ●

335 Abies delavayi Franch. var. faxoniana (Rehder et E. H. Wilson) A. B. Jacks. = Abies faxoniana Rehder et E. H. Wilson ●

336 Abies delavayi Franch. var. forrestii (Coltm. -Rog.) A. B. Jacks. = Abies forrestii Coltm. -Rog. ●

337 Abies delavayi Franch. var. georgei (Orr) Melville = Abies georgei Orr ●◇

338 Abies delavayi Franch. var. motuoensis W. C. Cheng et L. K. Fu;墨脱冷杉;Motuo Fir ●

339 Abies delavayi Franch. var. nukiangensis (W. C. Cheng et L. K. Fu) Farjon et Silba = Abies nukiangensis W. C. Cheng et L. K. Fu ●

340 Abies delavayi Franch. var. smithii (Vig. et Gaussen) T. S. Liu = Abies smithii (Vig. et Gaussen) Flous ●

341 Abies densa Griff. ex Parker;锡金冷杉(亚东冷杉);Sikkim Fir ●

342 Abies diversifolia Maxim. = Tsuga diversiflora (Maxim.) Mast. ●☆

343 Abies douglasii Lindl. = Pseudotsuga menziesii (Mirb.) Franco ●

344 Abies douglasii Lindl. var. macrocarpa Torr. = Pseudotsuga macrocarpa (Torr.) Mayr ●◇

345 Abies dumosa Loudon = Tsuga dumosa (D. Don) Eichler ●

346 Abies ernestii Rehder;黄果冷杉;Ernest Fir ●

347 Abies ernestii Rehder var. salouenensis (Borderes et Gaussen) W. C. Cheng et L. K. Fu;云南黄果冷杉(大果黄果冷杉,澜沧冷杉);Salouen Fir ●

348 Abies excelsa Poir.;挪威冷杉;Norway Fir ●

349 Abies faberi (Mast.) Craib;冷杉;Faber Fir ●

350 Abies faberi (Mast.) Craib var. beshanzuensis (M. H. Wu) Silba =

Abies beshanzuensis M. H. Wu ●◇
351 Abies faberi (Mast.) Craib var. minensis (Borderes et Gaussen) Silba = Abies faxoniana Rehder et E. H. Wilson ●
352 Abies faberi (Mast.) Craib var. ziyuanensis (L. K. Fu et S. L. Mo) Silba = Abies beshanzuensis M. H. Wu var. ziyuanensis (L. K. Fu et S. L. Mo) L. K. Fu et Nan Li ●◇
353 Abies falcata Raf. = Picea sitchensis (Bong.) Carrière ●☆
354 Abies fanjingshanensis W. L. Huang, Y. L. Tu et S. Z. Fang;梵净山冷杉(梵净冷杉);Fanjingshan Fir ●◇
355 Abies fansipanensis Q. P. Xiang, L. K. Fu et Nan Li;越南冷杉;Vietnam Fir ●
356 Abies fargesii Franch.;巴山冷杉(川枞,鄂西冷杉,华枞,冷杉,蒲松,朴松,四川冷杉,松墨,太白冷杉,洮河冷杉);Farges Fir, Sutchuen Fir ●
357 Abies fargesii Franch. var. fanjingshanensis (W. L. Huang, Y. L. Tu et S. Z. Fang) Silba = Abies fanjingshanensis W. L. Huang, Y. L. Tu et S. Z. Fang ●◇
358 Abies fargesii Franch. var. faxoniana (Rehder et E. H. Wilson) Tang S. Liu = Abies faxoniana Rehder et E. H. Wilson ●
359 Abies fargesii Franch. var. hupehensis Silba = Abies fargesii Franch. ●
360 Abies fargesii Franch. var. sutchuenensis Franch. = Abies fargesii Franch. ●
361 Abies fargesii Franch. var. tieghemi Borderes et Gaussen = Abies fargesii Franch. ●
362 Abies faxoniana Rehder et E. H. Wilson;岷江冷杉(柔毛枞,柔毛冷杉);Faxon Fir, Faxons Fir ●
363 Abies ferreana Borderes et Gaussen;中甸冷杉;Ferre Fir ●
364 Abies ferreana Borderes et Gaussen var. longibracteata L. K. Fu et Nan Li;长苞中甸冷杉(长白中甸冷杉);Longibracteate Ferre Fir ●
365 Abies firma Siebold et Zucc.;日本冷杉(枞,枞,日本枞);Jananese Fir, Japan Fir, Japanese Silver Fir, Momi Fir, Strong White Fir ●
366 Abies firma Siebold et Zucc. f. pendula Honda;垂枝日本冷杉●☆
367 Abies firma Siebold et Zucc. var. bifida (Siebold et Zucc.) Mast. = Abies firma Siebold et Zucc. ●
368 Abies firma Siebold et Zucc. var. momi (Siebold) Mast. = Abies firma Siebold et Zucc. ●
369 Abies forrestii Coltm. -Rog.;川滇冷杉(毛枝冷杉,云南枞);Forrest Fir ●
370 Abies forrestii Coltm. -Rog. var. chayuensis (W. C. Cheng et L. K. Fu) Silba = Abies chayuensis W. C. Cheng et L. K. Fu ●
371 Abies forrestii Coltm. -Rog. var. chengii (Rushforth) Silba = Abies chayuensis W. C. Cheng et L. K. Fu ●
372 Abies forrestii Coltm. -Rog. var. ferreana (Borderes et Gaussen) Farjon et Silba = Abies ferreana Borderes et Gaussen ●
373 Abies forrestii Coltm. -Rog. var. georgei (Orr) Farjon = Abies georgei Orr ●◇
374 Abies forrestii Coltm. -Rog. var. smithii Vig. et Gaussen = Abies georgei Orr var. smithii (Vig. et Gaussen) W. C. Cheng et L. K. Fu ●
375 Abies fortunei A. Murray = Keteleeria fortunei (A. Murray) Carrière ●
376 Abies fraseri (Pursh) Poir. = Abies fraseri Lindl. ●☆
377 Abies fraseri Lindl.;弗氏冷杉(福莱胶枞,美东香脂冷杉,南部香脂冷杉);Alleghanies, Balsam, Fraser Balsam Fir, Fraser Fir, Fraser's Balsam Fir, Fraser's Fir, She Balsam, Shebalsam, She-balsam, Southern Balsam Fir, Southern Fir ●☆
378 Abies fraseri Lindl. var. prostrata Rehder;平卧福氏冷杉(平卧福莱胶枞,平卧福氏胶枞);Prostrate Fraser Balsam Fir ●☆
379 Abies georgei Orr;长苞冷杉(西康冷杉);George's Fir, Longbract Fir ●◇
380 Abies georgei Orr var. smithii (Vig. et Gaussen) W. C. Cheng et L. K. Fu = Abies smithii (Vig. et Gaussen) Flous ●
381 Abies glehnii F. Schmidt = Picea glehnii (F. Schmidt) Mast. ●☆
382 Abies gmelinii Rupr. = Larix gmelinii (Rupr.) Rupr. ●
383 Abies gracilis Kom.;勘察加冷杉;Kamtschatka Fir ●☆
384 Abies grandis (Douglas ex D. Don) Lindl.;北美冷杉(大冷杉,巨冷杉,巨美冷杉);Giant Fir, Grand Fir, Grand White Fir, Great Fir, Lowland Fir, Lowland White Fir, Norway Fir, Oregon Fir, Vancouver Fir, White Fir ●
385 Abies griffithiana Lindl. et Gordon = Larix griffithiana (Lindl. et Gordon) Carrière ●
386 Abies guatemalensis Rehder;危地马拉冷杉;Guatemala Fir, Guatemalan Fir ●☆
387 Abies heterophylla Raf. = Tsuga heterophylla (Raf.) Sarg. ●☆
388 Abies holophylla Maxim.;辽东冷杉(白松,东北杉木,沙松,杉松,杉松冷杉,针枞);Manchurian Fir, Needle Fir, Nikko Fir ●
389 Abies homolepis Siebold et Zucc.;日光冷杉;Kawakami Fir, Nikko Fir, Nikko Silver Fir ●☆
390 Abies homolepis Siebold et Zucc. var. umbellata (Mayr) E. H. Wilson = Abies umbellata Mayr ●☆
391 Abies hookeriana A. Murray bis = Tsuga mertensiana (Bong.) Carrière ●☆
392 Abies jezoensis Siebold et Zucc. = Picea jezoensis (Siebold et Zucc.) Carrière ●☆
393 Abies kaempferi Lindl. = Larix kaempferi (Lamb.) Carrière ●
394 Abies kaempferi Lindl. = Pseudolarix amabilis (J. Nelson) Rehder ●◇
395 Abies kansouensis Borderes et Gaussen = Abies fargesii Franch. ●
396 Abies kawakamii (Hayata) T. Ito;台湾冷杉(白松,柏松);Formosan Fir, Kawakami Fir, Korean Fir, Taiwan Fir ●◇
397 Abies khutrow Loudon = Picea smithiana (Wall.) Boiss. ●◇
398 Abies koreana E. H. Wilson;朝鲜冷杉;Korean Fir ●☆
399 Abies koreana E. H. Wilson 'Aurea';金色朝鲜冷杉;Golden Korean Fir ●☆
400 Abies koreana E. H. Wilson 'Compact Dwarf';矮密朝鲜冷杉●☆
401 Abies koreana E. H. Wilson 'Horstmann's Silberlocke';霍斯曼矮生朝鲜冷杉●☆
402 Abies lasiocarpa (Hook.) Nutt.;落基山冷杉(落山矶冷杉,落杉矶冷杉,毛果冷杉,亚利桑那州冷杉);Alpine Fir, Cork Bark Fir, Cork Fir, Corkbark Fir, Rocky Mountain Alpine Fir, Rocky Mountain Fir, Rocky Mountain Subalpine Fir, Subalpine Fir, Sub-alpine Fir ●☆
403 Abies lasiocarpa (Hook.) Nutt. var. arizonica (Merr.) Lemmon;栓皮冷杉(亚利桑那落基山冷杉);Arizona Cork Fir, Cork Fir, Corkbark Fir, Cork-barked Fir ●☆
404 Abies ledebourii Rupr. = Larix sibirica (Münchh.) Ledeb. ●
405 Abies leptolepis Siebold et Zucc. = Larix kaempferi (Lamb.) Carrière ●
406 Abies likiangensis Franch. = Picea likiangensis (Franch.) E. Pritz. ●
407 Abies lowiana A. Murray;太平洋冷杉(娄氏白冷杉,娄氏冷杉,太平洋白冷杉);Californian Fir, Low's White Fir, Pacific White Fir, Sierra White Fir ●☆
408 Abies macrocarpa Vasey = Pseudotsuga macrocarpa (Vasey) Mayr ●◇
409 Abies magnifica A. Murray;红果冷杉(加利福尼亚州冷杉,加州红冷杉,沙斯塔冷杉,沙斯塔山冷杉);California Red Fir, Californian Red Fir, Golden Fir, Red Fir, Shasta Fir, Shasta Red Fir, Silvertip ●☆

410 Abies magnifica A. Murray var. argentea Beissn.；银叶红果冷杉；Silverleaf Red Fir ●☆

411 Abies magnifica A. Murray var. glauca Beissn.；粉绿叶红果冷杉；Azure Red Fir ●☆

412 Abies magnifica A. Murray var. shastensis (Lemmon) Lemmon；沙斯塔红果冷杉(沙斯塔红冷杉,夏斯塔红果冷杉)；Shasta Fir, Shasta Red Fir ●☆

413 Abies mariana Mill. = Picea mariana (Mill.) Britton, Sterns et Poggenb. ●☆

414 Abies mariesii Mast.；八甲田山冷杉(大白叶冷杉,马氏冷杉)；Maries Fir, Marie's Fir, Todomatsu Fir ●☆

415 Abies mariesii Mast. f. hayachinensis Hayashi = Abies mariesii Mast. ●☆

416 Abies mariesii Mast. var. kawakamii Hayata = Abies kawakamii (Hayata) T. Ito ●◇

417 Abies marocana Trab.；摩洛哥冷杉●☆

418 Abies mayriana (Miyabe et Kudo) Miyabe et Kudo = Abies sachalinensis (F. Schmidt) Mast. var. mayriana Miyabe et Kudo ●☆

419 Abies menziesii (Douglas ex D. Don) Lindl. = Picea sitchensis (Bong.) Carrière ●☆

420 Abies menziesii Mirb. = Pseudotsuga menziesii (Mirb.) Franco ●

421 Abies microsperma Lindl. = Picea jezoensis (Siebold et Zucc.) Carrière var. microsperma (Lindl.) W. C. Cheng et L. K. Fu ●

422 Abies minensis Borderes et Gaussen = Abies fargesii Franch. var. faxoniana (Rehder et E. H. Wilson) Tang S. Liu ●

423 Abies morinda Loudon = Picea smithiana (Wall.) Boiss. ●◇

424 Abies mucronata Raf. = Pseudotsuga menziesii (Mirb.) Franco ●

425 Abies nebrodensis (Lojac.) Mattei = Abies nebrodensis Mattei ●☆

426 Abies nebrodensis Mattei；西西里冷杉；Sicilian Fir ●☆

427 Abies nephrolepis (Turcz.) Maxim.；臭冷杉(白枞,白果枞,白松,臭枞,臭松,东陵冷杉,华北冷杉,冷杉)；East Siberian Fir, Khingan Fir, Manchurian Fir, Siberian White Fir ●

428 Abies nobilis Lindl. = Abies procera Rehder ●☆

429 Abies nordmanniana (Steven) Spach；高加索冷杉(黑海冷杉)；Caucasia Fir, Caucasian Fir, Nordmann Christmas Tree, Nordmann Fir ●☆

430 Abies nordmanniana (Steven) Spach 'Golden Spreader'；金叶阔冠高加索冷杉●☆

431 Abies nordmanniana (Steven) Spach var. aurea Beissn.；金叶高加索冷杉；Golden Nordmann Fir ●☆

432 Abies nordmanniana (Steven) Spach var. tortifolia Rehder；扭叶高加索冷杉；Twistleaf Nordmann Fir ●☆

433 Abies nukiangensis W. C. Cheng et L. K. Fu；怒江冷杉；Nujiang Fir, Nukiang Fir ●

434 Abies numidica Carrière = Abies numidica de Lannoy ex Carrière ●☆

435 Abies numidica de Lannoy ex Carrière；阿尔及利亚冷杉(双米地冷杉)；Algeria Fir, Algerian Fir ●☆

436 Abies numidica de Lannoy ex Carrière var. glauca Beissn.；粉绿叶阿尔及利亚冷杉；Blue Algerian Fir ●☆

437 Abies pattoniana A. Murray bis = Tsuga mertensiana (Bong.) Carrière ●☆

438 Abies pichta Forbes = Abies sibirica Ledeb. ●

439 Abies pindrow (Royle) Spach；长叶冷杉(西喜马拉雅冷杉,西喜马拉雅银冷杉,印度冷杉)；Himalayan Silver Fir, Pindrow Fir, West Himalayan Fir, West Himalayan Silver Fir ●☆

440 Abies pindrow (Royle) Spach var. brevifolia Dallim. et Jacks.；短叶冷杉；Shortleaf Pindrow Fir, Short-needle Pindrow Fir ●☆

441 Abies pindrow Royle = Abies pindrow (Royle) Spach ●☆

442 Abies pinsapo Boiss.；西班牙冷杉；Hedgehog Fir, Spain Fir, Spanish Fir, Spanish Silver Spruce ●☆

443 Abies pinsapo Boiss. 'Glauca'；粉绿西班牙冷杉(粉绿叶西班牙冷杉,蓝灰西班牙冷杉)；Blue Noble Fir, Blue Spanish Fir ●☆

444 Abies pinsapo Boiss. 'Kelleris'；矮壮西班牙冷杉●☆

445 Abies pinsapo Boiss. subsp. hispanica (Chambray) Maire = Abies pinsapo Boiss. ●☆

446 Abies pinsapo Boiss. subsp. marocana (Trab.) Emb. et Maire = Abies marocana Trab. ●☆

447 Abies pinsapo Boiss. var. glauca Carrière = Abies pinsapo Boiss. 'Glauca' ●☆

448 Abies pinsapo Boiss. var. marocana (Trab.) Ceballos et Martín Bol. = Abies marocana Trab. ●☆

449 Abies pinsapo Boiss. var. pendula Beissn.；垂枝西班牙冷杉(西班牙悬垂冷杉)；Spanish Pendulous Fir, Weeping Spanish Fir ●☆

450 Abies pinsapo Boiss. var. tazaotana (Villar) Govaerts = Abies marocana Trab. ●☆

451 Abies polita Siebold et Zucc. = Picea polita (Siebold et Zucc.) Carrière ●

452 Abies polita Siebold et Zucc. = Picea torano (Siebold ex K. Koch) Koehne ●

453 Abies procera Rehder；高大冷杉(大冷杉,壮丽红冷杉,壮丽冷杉)；Bracted Fir, Noble Fir, Red Fir ●☆

454 Abies procera Rehder var. glauca (Ravenscr.) L. H. Bailey；粉绿叶壮丽冷杉；Bluleaf Noble Fir ●☆

455 Abies recurvata Mast.；紫果冷杉；Min Fir, Purplecone Fir ●

456 Abies recurvata Mast. var. ernestii (Redher) C. T. Kuan = Abies ernestii Rehder ●

457 Abies religiosa (Kunth) Schltdl. = Abies religiosa Lindl. ●☆

458 Abies religiosa Lindl.；墨西哥冷杉(神圣冷杉)；Guatemala Fir, Mexican Fir, Sacred Fir ●☆

459 Abies rolii Borderes et Gaussen = Abies ferreana Borderes et Gaussen ●

460 Abies sachalinensis (F. Schmidt) Mast.；库页冷杉(萨哈林冷杉)；Sachalin Fir, Sakhalin Fir ●☆

461 Abies sachalinensis (F. Schmidt) Mast. f. Corticosa (Tatew.) Hayashi = Abies sachalinensis (F. Schmidt) Mast. ●☆

462 Abies sachalinensis (F. Schmidt) Mast. var. corticosa Tatew. = Abies sachalinensis (F. Schmidt) Mast. ●☆

463 Abies sachalinensis (F. Schmidt) Mast. var. mayriana Miyabe et Kudo；迈氏库页冷杉；Sachalin Fir, Sakhalin Fir ●☆

464 Abies sachalinensis (F. Schmidt) Mast. var. nemorensis Mayr；林中库页冷杉；Wilson Fir, Wilson Sachalin Fir ●☆

465 Abies sachalinensis (F. Schmidt) Mast. var. nemorensis Mayr = Abies sachalinensis (F. Schmidt) Mast. ●☆

466 Abies sachalinensis (F. Schmidt) Mast. var. nemorensis Mayr = Abies wilsonii Miyabe et Kudo ●☆

467 Abies sachalinensis (F. Schmidt) Mast. var. sikokiana Kusaka；四国冷杉；Sikoku Sachalin Fir ●☆

468 Abies sacra Franch. = Keteleeria davidiana (Bertrand) Beissn. ●

469 Abies salouenensis Borderes et Gaussen = Abies ernestii Rehder var. salouenensis (Borderes et Gaussen) W. C. Cheng et L. K. Fu ●

470 Abies schrenkiana Lindl. et Gordon = Picea schrenkiana Fisch. et C. A. Mey. ●

471]Abies semenovii Fedtsch.；谢明诺夫冷杉；Semenov Fir ●☆

472 Abies shensiensis Pritz. = Abies chensiensis Tiegh. ●◇

473 Abies shikokiana Nakai = Abies veitchii Lindl. var. reflexa Koidz. ●☆
474 Abies shikokiana Nakai = Abies veitchii Lindl. var. sikokiana (Nakai) Kusaka ●☆
475 Abies sibiico-nephrolepis Taken. et S. S. Chien = Abies nephrolepis (Turcz.) Maxim. ●
476 Abies sibirica Ledeb.;西伯里亚冷杉(新疆冷杉);Siberian Fir, Xinjiang Fir ●
477 Abies sibirica Ledeb. var. nephrolepis Trautv. = Abies nephrolepis (Turcz.) Maxim. ●
478 Abies smithiana Lindl. = Picea smithiana (Wall.) Boiss. ●◇
479 Abies smithii (Vig. et Gaussen) Flous;急尖长苞冷杉(乌蒙冷杉);Smith Fir ●
480 Abies spectabilis (D. Don) Spach;喜马拉雅冷杉(西藏冷杉);East Himalayan Fir, Himalayan Fir, Indian Silver Fir, Pindrow Fir ●
481 Abies spectabilis (D. Don) Spach var. brevifolia (Henry) Rehder;短叶喜马拉雅冷杉(光皮喜马拉雅冷杉);Smoothbark Himalayan Fir ●
482 Abies spectabilis (D. Don) Spach var. densa (Griff.) Silba = Abies densa Griff. ex Parker ●
483 Abies spectabilis (D. Don) Spach var. vrevifolia (Henry) Rehder;光皮喜马拉雅冷杉●☆
484 Abies spinulosa Griff. = Picea spinulosa (Griff.) Henry ●
485 Abies squamata Mast.;鳞皮冷杉;Flaky Fir ●
486 Abies squamata Mast. = Abies fargesii Franch. ●
487 Abies subalpina Engelm. = Abies bifolia A. Murray ●☆
488 Abies sutchuenensis (Franch.) Rehder et E. H. Wilson = Abies fargesii Franch. ●
489 Abies taxifolia Lamb. = Pseudotsuga menziesii (Mirb.) Franco ●
490 Abies taxifolia Poir. = Pseudotsuga menziesii (Mirb.) Franco ●
491 Abies tazaotana Villar = Abies marocana Trab. ●☆
492 Abies theisha David = Tsuga chinensis (Franch.) Pritz. ex Diels ●
493 Abies torano Siebold ex K. Koch = Picea torano (Siebold ex K. Koch) Koehne ●
494 Abies tsuga Siebold et Zucc. = Tsuga sieboldii Carrière ●☆
495 Abies umbellata Mayr;伞冷杉●☆
496 Abies veitchii Lindl.;富士山冷杉(白叶冷杉,日本白冷杉,日本灰白冷杉);Veitch Fir, Veitch's Silver Fir ●☆
497 Abies veitchii Lindl. f. olivacea (Shiras.) Kusaka;卵形富士山冷杉(绿果白叶冷杉,绿果富士山冷杉);Olivacea Veitch Fir ●☆
498 Abies veitchii Lindl. var. komagatakensis Hayashi;渡岛富士山冷杉●☆
499 Abies veitchii Lindl. var. nikoensis Mayr;小果富士山冷杉(小果白叶冷杉);Littlecone Veitch Fir ●
500 Abies veitchii Lindl. var. nikoensis Mayr = Abies veitchii Lindl. ●☆
501 Abies veitchii Lindl. var. olivacea Shiras. = Abies veitchii Lindl. f. olivacea (Shiras.) Kusaka ●☆
502 Abies veitchii Lindl. var. reflexa Koidz.;小白叶冷杉(四国冷杉);Sikoku Fir ●☆
503 Abies veitchii Lindl. var. reflexa Koidz. = Abies veitchii Lindl. ●☆
504 Abies veitchii Lindl. var. reflexa Koidz. f. viridis Kusaka;绿果小白叶冷杉●☆
505 Abies veitchii Lindl. var. sachalinensis F. Schmidt = Abies sachalinensis (F. Schmidt) Mast. ●☆
506 Abies veitchii Lindl. var. shikokiana (Nakai) Kusaka = Abies veitchii Lindl. var. reflexa Koidz. ●☆
507 Abies venusta (Douglas) K. Koch;亮果冷杉(硬苞冷杉);Bristle Cone Fir, Bristlecone Fir, Santa Lucia Fir, Santalucia Fir ●
508 Abies webbiana (Wall.) Lindl. = Abies spectabilis (D. Don) Spach ●
509 Abies webbiana (Wall.) Lindl. var. pindrow Royle = Abies pindrow (Royle) Spach ●☆
510 Abies wilsonii Miyabe et Kudo;威氏冷杉;Wilson Fir, Wilson Sakhalin Fir ●☆
511 Abies yoneyame Satao = Abies holophylla Maxim. ●
512 Abies yuana Borderes et Gaussen = Abies ferreana Borderes et Gaussen ●
513 Abies yuanpaoshanensis Y. J. Lü et L. K. Fu;元宝山冷杉;Yuanbao Mountain Fir, Yuanbaoshan Fir ●◇
514 Abies yunnanensis Franch. = Tsuga dumosa (D. Don) Eichler ●
515 Abies ziyuanensis L. K. Fu et S. L. Mo = Abies beshanzuensis M. H. Wu var. ziyuanensis (L. K. Fu et S. L. Mo) L. K. Fu et Nan Li ●◇
516 Abietaceae Bercht. et J. Presl = Abietaceae Gray(保留科名)●
517 Abietaceae Bercht. et J. Presl = Pinaceae Spreng. ex F. Rudolphi(保留科名)●
518 Abietaceae Gray(1822)(保留科名);冷杉科●
519 Abietaceae Gray(保留科名) = Pinaceae Spreng. ex F. Rudolphi ●
520 Abietia Kent = Keteleeria Carrière ●
521 Abietia Kent = Pseudotsuga Carrière + Keteleeria Carrière ●
522 Abietia Kent = Pseudotsuga Carrière ●
523 Abietia doglasii (Lindl.) Kent = Pseudotsuga menziesii (Mirb.) Franco ●
524 Abietia doglasii (Lindl.) Kent var. macrocarpa (Torr.) Kent = Pseudotsuga macrocarpa (Torr.) Mayr ●◇
525 Abietia fortunei (A. Murray) Kent = Keteleeria fortunei (A. Murray) Carrière ●
526 Abiga St. -Lag. = Ajuga L. ●■
527 Abildgaardia Vahl = Fimbristylis Vahl(保留属名)■
528 Abildgaardia Vahl(1805);阿氏莎草属■☆
529 Abildgaardia abbreviata Lye;减短阿氏莎草■☆
530 Abildgaardia abortiva (Steud.) Lye;败育阿氏莎草■☆
531 Abildgaardia acutespicata Lye;尖穗阿氏莎草■☆
532 Abildgaardia afroorientalis Lye;东非阿氏莎草■☆
533 Abildgaardia angustespicata Lye;窄穗阿氏莎草■☆
534 Abildgaardia argentobrunnea (C. B. Clarke) Lye;银褐阿氏莎草(银褐莎草)■☆
535 Abildgaardia barbata P. Beauv. = Bulbostylis filamentosa (Vahl) C. B. Clarke var. barbata C. B. Clarke ■☆
536 Abildgaardia boeckeleriana (Schweinf.) Lye;伯克阿氏莎草■☆
537 Abildgaardia boeckeleriana (Schweinf.) Lye var. transiens (K. Schum.) Lye;横伯克阿氏莎草■☆
538 Abildgaardia buchananii (C. B. Clarke) Lye;布坎南阿氏莎草■☆
539 Abildgaardia burchellii (Ficalho et Hiern) Lye;布尔阿氏莎草■☆
540 Abildgaardia capillaris (L.) Lye;发状阿氏莎草■☆
541 Abildgaardia capitata Lye;头状阿氏莎草■☆
542 Abildgaardia cardiocarpoides (Cherm.) Lye;心果阿氏莎草■☆
543 Abildgaardia clarkeana (Hutch. ex M. Bodard) Lye;克拉尔阿氏莎草■☆
544 Abildgaardia coleotricha (Hochst. ex A. Rich.) Lye;毛鞘莎草■☆
545 Abildgaardia coleotricha (Hochst. ex A. Rich.) Lye var. miegei (Bodard) Lye;米氏毛鞘莎草■☆
546 Abildgaardia collina (Ridl.) Lye;丘陵莎草■☆
547 Abildgaardia congolensis (De Wild.) Lye;刚果阿氏莎草■☆
548 Abildgaardia contexta (Nees) Lye;黏着阿氏莎草(黏着莎草)■☆
549 Abildgaardia cruciformis Lye;十字阿氏莎草(十字莎草)■☆

550 Abildgaardia densa (Wall.) Lye;密集阿氏莎草■☆
551 Abildgaardia densa (Wall.) Lye subsp. afromontana Lye;非洲山生阿氏莎草■☆
552 Abildgaardia densecaespitosa Lye;密丛阿氏莎草■☆
553 Abildgaardia densiflora Lye;密花阿氏莎草■☆
554 Abildgaardia elegantissima Lye;优雅阿氏莎草■☆
555 Abildgaardia eragrostis Nees et Meyen = Fimbristylis eragrostis (Nees et Meyen) Hance ■
556 Abildgaardia erratica (Hook. f.) Lye;移动阿氏莎草■☆
557 Abildgaardia erratica (Hook. f.) Lye subsp. schoenoides (Kunth) Lye = Bulbostylis schoenoides (Kunth) C. B. Clarke ■☆
558 Abildgaardia filamentosa (Vahl) Lye;丝状阿氏莎草■☆
559 Abildgaardia filamentosa (Vahl) Lye var. holubii (C. B. Clarke) Lye = Abildgaardia collina (Ridl.) Lye ■☆
560 Abildgaardia filamentosa (Vahl) Lye var. metralis (Cherm.) Lye = Abildgaardia filamentosa (Vahl) Lye ■☆
561 Abildgaardia fimbristyloides F. Muell. = Fimbristylis fimbristyloides (F. Muell.) Druce ■
562 Abildgaardia fusca Nees = Fimbristylis fusca (Nees) Benth. ■
563 Abildgaardia glaberrima (Kük.) Lye;光秃阿氏莎草■☆
564 Abildgaardia hensii (C. B. Clarke) Lye = Abildgaardia hispidula (Vahl) Lye subsp. brachyphylla (Cherm.) Lye ■☆
565 Abildgaardia hispidula (Vahl) Lye;粗毛阿氏莎草■☆
566 Abildgaardia hispidula (Vahl) Lye subsp. brachyphylla (Cherm.) Lye;短叶粗毛阿氏莎草■☆
567 Abildgaardia hispidula (Vahl) Lye subsp. filiformis (C. B. Clarke) Lye;线形粗毛阿氏莎草■☆
568 Abildgaardia hispidula (Vahl) Lye subsp. halophila Lye;喜盐粗毛阿氏莎草■☆
569 Abildgaardia hispidula (Vahl) Lye subsp. intermedia Lye;全叶粗毛阿氏莎草■☆
570 Abildgaardia hispidula (Vahl) Lye subsp. pyriformis (Lye) Lye;梨形粗毛阿氏莎草■☆
571 Abildgaardia hispidula (Vahl) Lye subsp. senegalensis (Cherm.) J. -P. Lebrun et Stork;塞内加尔粗毛阿氏莎草■☆
572 Abildgaardia hispidula (Vahl) Lye var. oligostachys (Hochst. ex A. Rich.) Lye = Abildgaardia oligostachys (Hochst. ex A. Rich.) Lye ■☆
573 Abildgaardia hispidula (Vahl) Lye var. pyriformis Lye = Abildgaardia hispidula (Vahl) Lye subsp. pyriformis (Lye) Lye ■☆
574 Abildgaardia hygrophila (Gordon-Gray) Lye;喜水阿氏莎草■☆
575 Abildgaardia igneotonsa (Raymond) Kornas;火焰阿氏莎草■☆
576 Abildgaardia johnstonii (C. B. Clarke) Lye;约翰斯顿阿氏莎草■☆
577 Abildgaardia lacunosa Lye;具腔阿氏莎草■☆
578 Abildgaardia lanifera (Boeck.) Lye;绵毛阿氏莎草■☆
579 Abildgaardia laxespicata Lye;疏穗阿氏莎草■☆
580 Abildgaardia leiolepis (Kük.) Lye;光鳞阿氏莎草■☆
581 Abildgaardia longespicata Lye;长穗阿氏莎草■☆
582 Abildgaardia macra (Ridl.) Lye;瘦弱阿氏莎草■☆
583 Abildgaardia macroanthela Lye;大羽阿氏莎草■☆
584 Abildgaardia macrostachya Lye;大穗阿氏莎草■☆
585 Abildgaardia malawiensis Lye;马拉维阿氏莎草■☆
586 Abildgaardia megastachys (Ridl.) Lye;热非大穗阿氏莎草■☆
587 Abildgaardia metalliphila Lye;光泽阿氏莎草■☆
588 Abildgaardia microcarpa Lye;小果阿氏莎草■☆
589 Abildgaardia microcephala Lye;小头阿氏莎草■☆
590 Abildgaardia microelegans Lye;小雅阿氏莎草■☆
591 Abildgaardia microrotundata Lye;小圆阿氏莎草■☆
592 Abildgaardia monostachya (L.) Vahl = Abildgaardia ovata (Burm. f.) Král ■
593 Abildgaardia monostachya (L.) Vahl = Fimbristylis ovata (Burm. f.) J. Kern ■
594 Abildgaardia monostachya L. = Fimbristylis monostachya (L.) Hassk. ■
595 Abildgaardia monostachya Vahl = Abildgaardia ovata (Burm. f.) Král ■
596 Abildgaardia natalensis Gand.;纳塔尔阿氏莎草■☆
597 Abildgaardia nudiuscula Lye;裸阿氏莎草■☆
598 Abildgaardia oligostachys (Hochst. ex A. Rich.) Lye;少穗阿氏莎草■☆
599 Abildgaardia oritrephes (Ridl.) Lye;山地阿氏莎草■☆
600 Abildgaardia oritrephes (Ridl.) Lye var. major Meneses = Abildgaardia oritrephes (Ridl.) Lye ■☆
601 Abildgaardia ovata (Burm. f.) Král = Fimbristylis ovata (Burm. f.) J. Kern. ■
602 Abildgaardia pallescens Lye;苍白阿氏莎草■☆
603 Abildgaardia parva (Ridl.) Lye = Abildgaardia pusilla (Hochst. ex A. Rich.) Lye ■☆
604 Abildgaardia parvinux (C. B. Clarke) Lye = Bulbostylis parvinux C. B. Clarke ■☆
605 Abildgaardia pilosa (Willd.) Nees;疏毛阿氏莎草■☆
606 Abildgaardia pluricephala Lye;多头阿氏莎草■☆
607 Abildgaardia pusilla (Hochst. ex A. Rich.) Lye;弱小阿氏莎草■☆
608 Abildgaardia pusilla (Hochst. ex A. Rich.) Lye subsp. congolensis (De Wild.) Lye = Abildgaardia congolensis (De Wild.) Lye ■☆
609 Abildgaardia pusilla (Hochst. ex A. Rich.) Lye subsp. yalingensis (Cherm.) Lye = Abildgaardia pusilla (Hochst. ex A. Rich.) Lye ■☆
610 Abildgaardia rhizomatosa Lye;根茎阿氏莎草■☆
611 Abildgaardia rotundata (Kük.) Lye;近圆阿氏莎草■☆
612 Abildgaardia schimperiana (A. Rich.) Lye;欣珀阿氏莎草■☆
613 Abildgaardia schlechteri (C. B. Clarke) Lye = Bulbostylis schlechteri C. B. Clarke ■☆
614 Abildgaardia scleropus (C. B. Clarke) Lye;硬梗阿氏莎草■☆
615 Abildgaardia scrobiculata Lye;蜂窝状阿氏莎草■☆
616 Abildgaardia setifolia (Hochst. ex A. Rich.) Lye;刚毛阿氏莎草■☆
617 Abildgaardia sphaerocarpa (Boeck.) Lye;球果阿氏莎草■☆
618 Abildgaardia squarrosa Lye;粗糙阿氏莎草■☆
619 Abildgaardia striatella (C. B. Clarke) Lye;细纹阿氏莎草■☆
620 Abildgaardia subumbellata Lye;近伞形阿氏莎草■☆
621 Abildgaardia tanzaniae Lye;坦桑尼亚阿氏莎草■☆
622 Abildgaardia taylorii (K. Schum.) Lye;泰勒阿氏莎草■☆
623 Abildgaardia trabeculata (C. B. Clarke) Lye;横条阿氏莎草■☆
624 Abildgaardia trabeculata (C. B. Clarke) Lye var. microglumis Lye;小壳横条阿氏莎草■☆
625 Abildgaardia triflora (L.) Abeyw.;三花阿氏莎草■☆
626 Abildgaardia tristachya Vahl = Abildgaardia triflora (L.) Abeyw. ■☆
627 Abildgaardia ugandensis Lye;乌干达阿氏莎草■☆
628 Abildgaardia vanderystii (Cherm.) Lye;范德阿氏莎草■☆
629 Abildgaardia variegata (Gordon-Gray) Lye = Fimbristylis variegata Gordon-Gray ■☆
630 Abildgaardia wallichiana (Schult.) Lye;瓦里阿氏莎草■☆
631 Abildgaardia willdenowii (Kunth) Lye;威尔阿氏莎草■☆
632 Abildgaardia wombaliensis (De Wild.) Lye;沃姆阿氏莎草■☆
633 Abildgaardia yalingensis (Cherm.) Lye = Abildgaardia pusilla (Hochst. ex A. Rich.) Lye ■☆

634 Abildgardia Rchb. = Abildgaardia Vahl ■☆
635 Abildgardia Rchb. = Fimbristylis Vahl(保留属名)■
636 Abilgaardia Poir. = Abildgaardia Vahl ■☆
637 Abioton Raf. = Capnophyllum Gaertn. ■☆
638 Ablania Aubl. = Sloanea L. ●
639 Abobra Naudin(1862);阿波瓜属;Craneberry Gourd ■☆
640 Abobra tenuifolia Naudin = Abobra viridiflora Naudin ■☆
641 Abobra viridiflora Naudin;阿波瓜;Craneberry Gourd ■☆
642 Aboita Adans. = Cinna L. ■
643 Abola Adans. = Cinna L. ■
644 Abola Lindl. = Caucaea Schltr. ■☆
645 Abolaria Neck. = Globularia L. ●☆
646 Abolboda Bonpl. = Abolboda Humb. ■☆
647 Abolboda Humb. (1813);三棱黄眼草属(阿波黄眼草属)■☆
648 Abolboda Humb. et Bonpl. = Abolboda Humb. ■☆
649 Abolboda acaulis Maguire;无茎三棱黄眼草■☆
650 Abolboda americana (Aubl.) Lanj.;美洲三棱黄眼草■☆
651 Abolboda brasiliensis Kunth;巴西三棱黄眼草■☆
652 Abolboda gracilis Huber;细三棱黄眼草■☆
653 Abolboda grandis Griseb.;大三棱黄眼草■☆
654 Abolboda macrostachya Spruce;大穗三棱黄眼草■☆
655 Abolboda minima Maguire;小三棱黄眼草■☆
656 Abolboda paniculata Maguire;圆锥三棱黄眼草■☆
657 Abolboda pulchella Humb. et Bonpl.;美丽三棱黄眼草■☆
658 Abolboda rigida (Malme) Steyerm.;硬三棱黄眼草■☆
659 Abolboda uniflora Maguire;单花三棱黄眼草■☆
660 Abolbodaceae (Suess. et Beyerle) Nakai = Xyridaceae C. Agardh (保留科名)■
661 Abolbodaceae (Suess. et Beyerle) Nakai;三棱黄眼草科■☆
662 Abolbodaceae Nakai = Abolbodaceae (Suess. et Beyerle) Nakai ■☆
663 Abolbodaceae Nakai = Xyridaceae C. Agardh(保留科名)■
664 Aborchis Steud. = Disa P. J. Bergius ■☆
665 Aboriella Bennet = Pilea Lindl. (保留属名)■
666 Aboriella Bennet(1981);长穗冷水花属■
667 Aboriella myriantha (Dunn) Bennet;长穗冷水花;Longspike Clearweed,Longspike Coldwaterflower ■
668 Aboriella myriantha (Dunn) Bennet = Pilea myriantha (Dunn) C. J. Chen ■
669 Abortopetalum O. Deg. = Abutilon Mill. ●■
670 Abramsia Gillespie = Airosperma K. Schum. et Lauterb. ■☆
671 Abrochis Neck. = Disa P. J. Bergius ■☆
672 Abroma Jacq. (1776);昂天莲属(水麻属);Abroma,Ambroma ●
673 Abroma Jacq. = Ambroma L. f. ●
674 Abroma Mart. = Theobroma L. ●
675 Abroma augusta (L.) L. f.;昂天莲(假芙蓉,水麻);Ambroma, Common Ambroma, Cotton Abroma, Cotton Ambroma, Devil's Bit Cotton,Devil's Cotton ●
676 Abroma fastuosa Jacq. = Abroma augusta (L.) L. f. ●
677 Abroma fastuosa Jacq. = Ambroma augusta (L.) L. f. ●
678 Abromeitia Mez = Fitchia Hook. f. ●☆
679 Abromeitia Mez = Fittingia Mez ●☆
680 Abromeitia Mez(1922);翅果紫金牛属●☆
681 Abromeitia pterocarpa Mez;翅果紫金牛●☆
682 Abromeitiella Mez = Deuterocohnia Mez ■☆
683 Abromeitiella Mez(1927);阿根廷菠萝属■☆
684 Abromeitiella brevifolia (Griseb.) A. Cast.;短叶阿根廷菠萝;Bromeliad ■☆
685 Abromeitiella brevifolia A. Cast. = Abromeitiella brevifolia (Griseb.) A. Cast. ■☆
686 Abromeitiella pulvinata Mez;阿根廷菠萝■☆
687 Abronia Juss. (1789);沙马鞭属(匍匐美女樱属,沙地马鞭草属,叶子草属);Abronia,Sand Verbena,Sand-verbena ■☆
688 Abronia alba Eastw.;白沙马鞭(白沙地马鞭草)■☆
689 Abronia alpina Brandegee;山生沙马鞭(山生沙地马鞭草)■☆
690 Abronia ameliae Lundell;阿米沙马鞭(阿米沙地马鞭草);Amelia's Sand Verbena,Heart's Delight ■☆
691 Abronia angustifolia Greene;狭叶沙马鞭(狭叶沙地马鞭草);Narrow-leaf Sand Verbena,Purple Sand Verbena ■☆
692 Abronia angustifolia Greene var. arizonica (Standl.) Kearney et Peebles = Abronia angustifolia Greene ■☆
693 Abronia argillosa S. L. Welsh et Goodrich;克莱沙马鞭(克莱沙地马鞭草);Clay Sand-verbena ■☆
694 Abronia aurita Abrams = Abronia villosa S. Watson var. aurita (Abrams) Jeps. ■☆
695 Abronia bigelovii Heimerl;毕氏沙马鞭(毕氏沙地马鞭草)■☆
696 Abronia breviflora Standl. = Abronia umbellata Lam. var. breviflora (Standl.) L. A. Galloway ■☆
697 Abronia carnea Greene = Tripterocalyx wootonii Standl. ■☆
698 Abronia crux-maltae Kellogg = Tripterocalyx crux-maltae (Kellogg) Standl. ■☆
699 Abronia elliptica A. Nelson;椭圆沙马鞭(椭圆沙地马鞭草)■☆
700 Abronia fragrans Nutt. ex Hook.;白花沙马鞭(白花沙地马鞭草,白沙地马鞭草);Fragrant Sand Verbena, Fragrant White Sand-verbena,White Sand Verbena ■☆
701 Abronia fragrans Nutt. ex Hook. var. elliptica (A. Nelson) M. E. Jones = Abronia elliptica A. Nelson ■☆
702 Abronia latifolia Eschsch.;黄花沙马鞭(黄花匍匐美女樱,黄花沙地马鞭草);Yellow Sand Verbena,Yellow Sand-verbena ■☆
703 Abronia macrocarpa L. A. Galloway;大果沙马鞭(大果沙地马鞭草);Large-fruited Sand-verbena ■☆
704 Abronia maritima Nutt. ex S. Watson;红花沙马鞭(红花沙地马鞭草,红沙地马鞭草);Beach Pancake,Red Sand Verbena,Red Sand-verbena ■☆
705 Abronia micrantha Torr. = Tripterocalyx micranthus (Torr.) Hook. ■☆
706 Abronia nana S. Watson;小沙马鞭(小沙地马鞭草)■☆
707 Abronia nana S. Watson var. harrisii S. L. Welsh = Abronia elliptica A. Nelson ■☆
708 Abronia pogonantha Heimerl;摩札沙马鞭(摩札沙地马鞭草);Mojave Sand Verbena ■☆
709 Abronia torreyi Standl. = Abronia angustifolia Greene ■☆
710 Abronia umbellata Lam.;粉红沙马鞭(粉红沙地马鞭草,匍匐美女樱);Beach Sand Verbena,Pink Sand Verbena,Pink Sand-verbena ■☆
711 Abronia umbellata Lam. subsp. alba (Eastw.) A. E. Murray = Abronia alba Eastw. ■☆
712 Abronia umbellata Lam. subsp. breviflora (Standl.) Munz = Abronia umbellata Lam. var. breviflora (Standl.) L. A. Galloway ■☆
713 Abronia umbellata Lam. var. breviflora (Standl.) L. A. Galloway;短花粉红沙马鞭(短花粉红沙地马鞭草)■☆
714 Abronia villosa S. Watson;长毛沙马鞭(长毛沙地马鞭草);Desert Sand Verbena,Desert Sand-verbena ■☆
715 Abronia villosa S. Watson var. aurita (Abrams) Jeps.;黄沙马鞭(黄沙地马鞭草)■☆

716 Abronia wootonii (Standl.) Tidestr. = Tripterocalyx wootonii Standl. ■☆
717 Abrophaes Raf. = Leonicenia Scop.(废弃属名)●☆
718 Abrophaes Raf. = Miconia Ruiz et Pav.(保留属名)●☆
719 Abrophyllaceae Nakai = Carpodetaceae Fenzl ●☆
720 Abrophyllaceae Nakai = Escalloniaceae R. Br. ex Dumort.(保留科名)●
721 Abrophyllaceae Nakai = Rousseaceae DC. ●☆
722 Abrophyllaceae Nakai;东澳木科(王冠果科)●☆
723 Abrophyllum Hook. f. = Abrophyllum Hook. f. ex Benth. ●☆
724 Abrophyllum Hook. f. ex Benth.(1864);东澳木属●☆
725 Abrophyllum microcarpum (Bailey) Domin;小果东澳木●☆
726 Abrophyllum ornans Hook. f.;东澳木●☆
727 Abrotanella Cass.(1825);垫菊属■☆
728 Abrotanella crassipes Skottsb.;粗梗垫菊■☆
729 Abrotanella filiformis Petrie;线形垫菊■☆
730 Abrotanella linearifolia A. Gray;线叶垫菊■☆
731 Abrotanella purpurea Swenson;紫垫菊■☆
732 Abrotanum Duhamel = Artemisia L. ●■
733 Abrotanum L. = Artemisia L. ●■
734 Abrotanum Mill. = Artemisia L. ●■
735 Abrotanum alpestre Jord. et Fourr.;高山状垫菊■☆
736 Abrotanum frutescens Gilib.;灌状垫菊■☆
737 Abrotanum xerophilum Jord. et Fourr.;沙垫菊■☆
738 Abrus Adans.(1763);相思子属(鸡母珠属,相思豆属);Abrus, Lucky Bean, Rosary Pea, Rosarypea, Rosary-pea ●■
739 Abrus acutifolius R. Vig. = Abrus diversifoliolatus Breteler ●☆
740 Abrus aureus R. Vig.;黄相思子■☆
741 Abrus baladensis Thulin;巴拉德相思子■☆
742 Abrus canescens Welw. ex Baker;灰相思子■☆
743 Abrus cantoniensis Hance = Abrus fruticulosus Wall. ex Wight et Arn. ●■
744 Abrus cyaneus R. Vig. = Abrus aureus R. Vig. ■☆
745 Abrus cyaneus R. Vig. = Abrus precatorius L. subsp. africanus Verdc. ●■☆
746 Abrus diversifoliolatus Breteler;异叶相思子●☆
747 Abrus fruticulosus Wall. ex Wight et Arn.;广东相思子(大黄草,地香根,广州相思子,红母鸡草,黄石草,黄食草,黄头草,黄仔蕹,鸡骨草,假牛甘子,山弯豆,石门坎,土甘草,细叶鸡骨草,细叶龙鳞草,相思子,小叶龙鳞草,猪腰草); Canton Abrus, Guangdong Abrus, Guangzhou Abrus ●■
748 Abrus fruticulosus Wall. ex Wight et Arn. = Abrus cantoniensis Hance ●■
749 Abrus fruticulosus Wall. ex Wight et Arn. = Abrus melanospermus Hassk. ●■☆
750 Abrus fruticulosus Wall. ex Wight et Arn. = Abrus mollis Hance ●■
751 Abrus fruticulosus Wall. ex Wight et Arn. = Abrus pulchellus Wall. ex Thwaites ●■
752 Abrus gawenensis Thulin;加瓦尼相思子●■☆
753 Abrus gorsei Berhaut = Abrus melanospermus Hassk. ●■☆
754 Abrus gracilis Pires de Lima = Abrus melanospermus Hassk. subsp. tenuiflorus (Benth.) D. Harder ●■☆
755 Abrus grandiflorus R. Vig. = Abrus aureus R. Vig. ■☆
756 Abrus laevigatus E. Mey.;平滑相思子●■☆
757 Abrus madagascariensis R. Vig.;马岛相思子●■☆
758 Abrus madagascariensis R. Vig. var. dumensis R. Vig. = Abrus aureus R. Vig. ■☆
759 Abrus madagascariensis R. Vig. var. parvifolius R. Vig. = Abrus parvifolius (R. Vig.) Verdc. ●■☆
760 Abrus melanospermus Hassk.;黑籽相思子●■☆
761 Abrus melanospermus Hassk. subsp. suffruticosus (Boutique) D. Harder;灌木黑籽相思子●☆
762 Abrus melanospermus Hassk. subsp. tenuiflorus (Benth.) D. Harder;细花黑籽马岛相思子●■☆
763 Abrus mollis Hance;毛相思子(金不换,芒尾蛇,毛鸡骨草,牛甘藤,蜻蜓藤,油甘草,油甘藤); Hairy Abrus, Hairy Pea, Hairy Rosarypea ●■
764 Abrus parvifolius (R. Vig.) Verdc.;小叶相思子●■☆
765 Abrus precatorius L.;相思子(八重山珊瑚,赤小豆,毒红豆,观音子,鬼眼子,海红豆,黑头小鸡,红豆,红公卯,红黑豆,红漆豆,红珠木,猴子眼,鸡丹真珠,鸡母珠,鸡眼子,郎君豆,郎君子,马料豆,美人豆,南丹真珠,南国红豆,山甘草,珊瑚豆,唐小豆,土甘草,土甘草豆,畏差草,相思豆,相思藤,相槵豆,相槵子,小人草,蟹眼豆,鸳鸯豆,云南豆子); Bead Vine, Black-eyed Susan, Coral Pea, Coral-bead Plant, Coralhead Plant, Crab's Eye, Crab's Eyes, Crab's Stone, Crab's-eye Vine, Guinea Pea, Indian Licorice, Indian Liquorice, Jamaican Licorice, Jequerity Seeds, Jequirity Bean, Jequirity Rosary Pea, Jequirity Rosarypea, Jequirity Rosary-pea, Jequirity, Jumble Beads, Ladybird Beads, Licorice Vine, Love Bean, Love Nut, Love Pea, Lucky Bean, Lucky Beans, Minnie Minnies, Paternoster Bean, Paternoster Beans, Prayer Beads, Prayer Beans, Precatory Bean, Precatory, Precatory-pea, Rati, Red Beadvine, Redhead Vine, Red-headed Vine, Rosary Pea, Rosarypea Tree, Rosary-pea Tree, Rosarypea, Rosary-pea, Weather Plant, Weatherplant, Weathervine, Wild Liquorice ●■
766 Abrus precatorius L. subsp. africanus Verdc.;非洲相思子●■☆
767 Abrus pulchellus Wall. = Abrus melanospermus Hassk. ●■☆
768 Abrus pulchellus Wall. ex Thwaites;美丽相思子(鸡骨草,土甘草); Beautiful Abrus, Beautiful Rosarypea ●■
769 Abrus pulchellus Wall. ex Thwaites = Abrus melanospermus Hassk. ●■☆
770 Abrus pulchellus Wall. ex Thwaites subsp. cantoniensis (Hance) Verdc. = Abrus cantoniensis Hance ●■
771 Abrus pulchellus Wall. ex Thwaites subsp. mollis (Hance) Verdc. = Abrus mollis Hance ●■
772 Abrus pulchellus Wall. ex Thwaites subsp. suffruticosus (Boutique) Verdc. = Abrus melanospermus Hassk. subsp. suffruticosus (Boutique) D. Harder ●☆
773 Abrus pulchellus Wall. ex Thwaites subsp. tenuiflorus (Benth.) Verdc. = Abrus melanospermus Hassk. subsp. tenuiflorus (Benth.) D. Harder ●■☆
774 Abrus pulchellus Wall. subsp. cantoniensis (Hance) Verdc. = Abrus cantoniensis Hance ●■
775 Abrus sambiranensis R. Vig.;桑比相思子■☆
776 Abrus sandwithiana Krukoff et Tharneby;桑威相思子■☆
777 Abrus schimperi Hochst. ex Baker;欣珀相思子■☆
778 Abrus schimperi Hochst. ex Baker subsp. africanus (Vatke) Verdc.;非洲施氏相思子■☆
779 Abrus schimperi Hochst. ex Baker subsp. oblongus Verdc.;矩圆欣珀相思子■☆
780 Abrus somalensis Taub.;索马里相思子■☆
781 Abrus squamulosus E. Mey. = Abrus precatorius L. subsp. africanus Verdc. ●■☆
782 Abrus stictospermus Berhaut = Abrus pulchellus Wall. ex Thwaites ●■

783 Abrus suffruticosus Boutique = Abrus melanospermus Hassk. subsp. suffruticosus (Boutique) D. Harder ●☆

784 Abrus tenuiflorus Benth. = Abrus melanospermus Hassk. subsp. tenuiflorus (Benth.) D. Harder ●■☆

785 Abrus tunguensis Pires de Lima = Abrus precatorius L. subsp. africanus Verdc. ●■☆

786 Abryanthemum Neck. = Carpobrotus N. E. Br. ●☆

787 Abryanthemum Neck. ex Rothm. = Carpobrotus N. E. Br. ●☆

788 Abryanthemum edule (L.) Rothm. = Carpobrotus edulis (L.) N. E. Br. ●☆

789 Absinthium L. = Artemisia L. ●■

790 Absinthium Mill. = Artemisia L. ●■

791 Absinthium Tourn. ex L. = Artemisia L. ●■

792 Absinthium bipedale Gilib. = Artemisia absinthium L. ■

793 Absinthium divaricatum (Weber ex Stechm.) Fisch. ex Besser = Artemisia anethifolia Weber ex Stechm. ■

794 Absinthium divaricatum Fisch. ex Besser = Artemisia anethifolia Weber ex Stechm. ■

795 Absinthium frigidum (Willd.) Besser var. fischerianum Besser = Artemisia frigida Willd. ■

796 Absinthium frigidum (Willd.) Besser var. willdenowianum Besser = Artemisia frigida Willd. ■

797 Absinthium frigidum Besser = Artemisia frigida Willd. ■

798 Absinthium grandiflorum Besser = Artemisia sericea Weber ex Stechm. ■

799 Absinthium lagocephalum Besser = Artemisia lagocephala (Fisch. ex Besser) Fisch. ex DC. ●■

800 Absinthium lagocephalum Fisch. ex Besser = Artemisia lagocephala (Fisch. ex Besser) DC. ●■

801 Absinthium nitens Besser = Artemisia sericea Weber ex Stechm. ■

802 Absinthium nitens Steven ex Besser = Artemisia sericea (Besser) Weber ●

803 Absinthium rupestre (L.) Besser = Artemisia rupestris L. ■

804 Absinthium rupestre (L.) Schrank = Artemisia rupestris L. ■

805 Absinthium rupestre (L.) Schrank var. oelandicum Besser = Artemisia rupestris L. ■

806 Absinthium rupestre (L.) Schrank var. thurigiacum Besser = Artemisia rupestris L. ■

807 Absinthium sericeum Besser = Artemisia sericea Weber ex Stechm. ■

808 Absinthium sieversianum (Ehrh. ex Willd.) Besser = Artemisia sieversiana Ehrh. ex Willd. ■

809 Absinthium sieversianum Besser = Artemisia sieversiana Ehrh. ex Willd. ■

810 Absinthium viride Besser = Artemisia rupestris L. ■

811 Absinthium viridiflorum Besser = Artemisia rupestris L. ■

812 Absinthium viridifolium (Ledeb.) Besser var. rupestre (L.) Besser = Artemisia rupestris L. ■

813 Absinthium viridifolium Besser = Artemisia rupestris L. ■

814 Absinthium vulgare Lam. = Artemisia absinthium L. ■

815 Absinthium vulgare Lam. = Artemisia vulgaris L. ■

816 Absintion Adans. = Absinthium Mill. ●■

817 Absolmsia Kuntze(1891);滑藤属;Absolmsia ■

818 Absolmsia oligophylla Tsiang = Tylophora oligophylla (Tsiang) M. G. Gilbert, W. D. Stevens et P. T. Li ●

819 Absynthium G. Gaertn., B. Mey. et Scherb. = Absinthium L. ●■

820 Abulfali Adans. = Thymbra L. ●☆

821 Abumon Adans.(废弃属名) = Agapanthus L'Hér.(保留属名)■☆

822 Abuta Aubl.(1775);脱皮藤属(阿布塔草属,阿布藤属)●☆

823 Abuta acutifolia Miers;尖叶脱皮藤●☆

824 Abuta barbata Miers;髯毛脱皮藤●☆

825 Abuta boliviana Rusby;玻利维亚脱皮藤●☆

826 Abuta brevifolia Krukoff et Moldenke;短叶脱皮藤●☆

827 Abuta candicans Rich.;灰白脱皮藤●☆

828 Abuta grisebachii Triana et Planch.;格氏脱皮藤(格氏阿布藤)●☆

829 Abuta obovata Diels;倒卵脱皮藤(倒卵阿布藤)●☆

830 Abuta platyphylla Mart. ex Eichler;宽叶脱皮藤●☆

831 Abuta rufescens Aubl.;红棕脱皮藤(红棕阿布藤);Cmianas, White Pareira Rroot ●☆

832 Abuta splendida Krukoff et Moldenke;闪光脱皮藤(阿布塔草)●☆

833 Abuta trinervis (Rusby) Moldenke;三脉脱皮藤●☆

834 Abutilaea F. Muell. = Abutilon Mill. ●■

835 Abutilodes Kuntze = Modiola Moench ■☆

836 Abutilodes Siegel = Modiola Moench ■☆

837 Abutilon Mill.(1754);苘麻属(风铃花属);Abutilon, Chinese Hemp, Chinese Jute, Chinese Lantern, Chingma, Flowering Malpe, Flowering-malpe, Indian Mallow, Manchurian Jute, Velvetleaf ●■

838 Abutilon Tourn. ex Adans. = Abutilon Mill. ●■

839 Abutilon abutilon (L.) Huth = Abutilon theophrastii Medik. ●■

840 Abutilon abutilon (L.) Rusby = Abutilon theophrastii Medik. ●■

841 Abutilon agnesae Borzí = Abutilon indicum (L.) Sweet subsp. guineense (Schumach.) Borss. Waalk. ■

842 Abutilon albescens Miq. = Abutilon indicum (L.) Sweet subsp. albescens (Miq.) Borss. Waalk. ●■☆

843 Abutilon albidum (Willd.) Sweet;白苘麻■☆

844 Abutilon albidum (Willd.) Sweet subsp. fruticosum (Guillaumin et Perr.) Maire = Abutilon fruticosum Guillaumin et Perr. ●☆

845 Abutilon americanum (L.) Sweet;美洲苘麻;Twelve-o'clock ●☆

846 Abutilon anglosomaliae Cufod. ex Thulin;安索苘麻■☆

847 Abutilon angulatum (Guillaumin et Perr.) Mast.;角苘麻■☆

848 Abutilon angulatum (Guillaumin et Perr.) Mast. var. macrophyllum (Baker f.) Hochr.;大叶角苘麻■☆

849 Abutilon asiaticum (L.) Sweet = Abutilon indicum (L.) Sweet var. guineense (Schumach.) K. M. Feng ■

850 Abutilon asiaticum (L.) Sweet var. lobulatum Robyns et Lawalrée;小裂片苘麻■☆

851 Abutilon asperifolium Ulbr.;糙叶苘麻■☆

852 Abutilon auritum (Link) Sweet = Abutilon auritum (Wall. ex Link) Sweet ●☆

853 Abutilon auritum (Wall. ex Link) Sweet;长耳苘麻;Asian Indian Mallow ●☆

854 Abutilon austro-africanum Hochr.;南非苘麻■☆

855 Abutilon avicennae (L.) Gaertn. = Abutilon theophrastii Medik. ●■

856 Abutilon avicennae Gaertn. = Abutilon theophrastii Medik. ●■

857 Abutilon avicennae Gaertn. var. chinense Skvortsov = Abutilon theophrastii Medik. ●■

858 Abutilon avicennae Gaertn. var. chinensenigrum Skvortsov = Abutilon theophrastii Medik. ●■

859 Abutilon avicennae Gaertn. var. genuina Skvortsov = Abutilon theophrastii Medik. ●■

860 Abutilon badium Husain et Baquar = Abutilon indicum (L.) Sweet ●■

861 Abutilon bakerianum Rose = Abutilon theophrastii Medik. ●■

862 Abutilon benadirense Mattei = Abutilon pannosum (G. Forst.) Schltdl. ■☆

863 Abutilon betschuanicum Ulbr. = Abutilon matopense Gibbs ■☆
864 Abutilon bidentatum A. Rich.;双齿苘麻■☆
865 Abutilon bidentatum Hochst. ex A. Rich. var. forrestii (S. Y. Hu) Abedin = Abutilon guineense (Schumach.) Baker f. et Exell var. forrestii (S. Y. Hu) Y. Tang ■
866 Abutilon blepharocarpum Mattei = Abutilon indicum (L.) Sweet subsp. guineense (Schumach.) Borss. Waalk. ■
867 Abutilon braunii Baker f.;布劳恩苘麻■☆
868 Abutilon bussei Gürke ex Ulbr.;布瑟苘麻■☆
869 Abutilon cabrae De Wild. et T. Durand = Abutilon mauritianum (Jacq.) Medik. ■☆
870 Abutilon cavaleriei H. Lév. = Abutilon indicum (L.) Sweet ●■
871 Abutilon cecilii N. E. Br.;塞西尔苘麻■☆
872 Abutilon cornutum (Humb. et Bonpl. ex Willd.) Sweet = Abutilon pakistanicum Jafri et Ali ●☆
873 Abutilon crassinervium Hochst. ex Mattei = Abutilon longicuspe Hochst. ex A. Rich. ■☆
874 Abutilon crispum (L.) Medik. = Herissantia crispa (L.) Brizicky ■
875 Abutilon cysticarpum Hance = Abutilon indicum (L.) Sweet ●■
876 Abutilon darwinii Hook. f.;橙红苘麻●☆
877 Abutilon densevillosum Mattei = Abutilon indicum (L.) Sweet subsp. guineense (Schumach.) Borss. Waalk. ■
878 Abutilon denticulatum (Fresen.) Webb = Abutilon fruticosum Guillaumin et Perr. ●☆
879 Abutilon dinteri Ulbr.;丁特苘麻■☆
880 Abutilon dubium Mattei = Abutilon fruticosum Guillaumin et Perr. ●☆
881 Abutilon eetveldeanum De Wild. et T. Durand = Abutilon angulatum (Guillaumin et Perr.) Mast. ■☆
882 Abutilon elaeocarpoides Webb = Abutilon ramosum (Cav.) Guillaumin et Perr. ■☆
883 Abutilon endlichii Ulbr.;恩德苘麻■☆
884 Abutilon engleranum Ulbr.;恩格勒苘麻■☆
885 Abutilon erythraeum Mattei;淡红苘麻■☆
886 Abutilon esquirolii H. Lév. = Urena repanda Roxb. ■
887 Abutilon eufigarii Chiov.;欧氏苘麻●
888 Abutilon figarianum Webb;菲加里苘麻■☆
889 Abutilon flanaganii A. Meeuse;弗拉纳根苘麻■☆
890 Abutilon flavum Ulbr. = Abutilon pycnodon Hochr. ■☆
891 Abutilon forrestii S. Y. Hu = Abutilon bidentatum Hochst. ex A. Rich. var. forrestii (S. Y. Hu) Abedin ■
892 Abutilon forrestii S. Y. Hu = Abutilon guineense (Schumach.) Baker f. et Exell var. forrestii (S. Y. Hu) Y. Tang ■
893 Abutilon forrestii S. Y. Hu = Abutilon indicum (L.) Sweet var. forrestii (S. Y. Hu) K. M. Feng ■
894 Abutilon fruticosum Guillaumin et Perr.;灌木苘麻●☆
895 Abutilon galpinii A. Meeuse;盖尔苘麻■☆
896 Abutilon gebauerianum Hand.-Mazz.;滇西苘麻;Gebauer Abutilon,West Yunnan Abutilon ●
897 Abutilon glaucum (Cav.) Sweet = Abutilon pannosum (G. Forst.) Schltdl. ■☆
898 Abutilon glaucum Webb = Abutilon pannosum (G. Forst.) Schltdl. ■☆
899 Abutilon grandiflorum G. Don;大花苘麻■☆
900 Abutilon grandifolium (Willd.) Sweet;大叶苘麻;Hairy Abutilon, Hairy Indian Mallow ●☆
901 Abutilon grantii A. Meeuse;格兰特苘麻■☆
902 Abutilon graveolens (Roxb. ex Hornem.) Wight et Arn. ex Wight = Abutilon hirtum (Lam.) Sweet ■
903 Abutilon graveolens (Roxb. ex Hornem.) Wight et Arn. ex Wight var. hirtum (Lam.) Mast. = Abutilon hirtum (Lam.) Sweet ■
904 Abutilon graveolens (Roxb.) Wight et Arn. = Abutilon hirtum (Lam.) Sweet ■
905 Abutilon guineense (Schumach.) Baker f. et Exell;几内亚磨盘草(几内冬葵子,台磨盘草,台湾冬葵子,台湾苘麻,亚洲黄花稔);Guinea Abutilon,Taiwan Abutilon ■
906 Abutilon guineense (Schumach.) Baker f. et Exell = Abutilon indicum (L.) Sweet var. guineense (Schumach.) K. M. Feng ■
907 Abutilon guineense (Schumach.) Baker f. et Exell var. forrestii (S. Y. Hu) Y. Tang;小花磨盘草;Forrest Abutilon ■
908 Abutilon harmsianum Ulbr. = Abutilon ramosum (Cav.) Guillaumin et Perr. ■☆
909 Abutilon heterotrichum Hochst. ex Mattei = Abutilon hirtum (Lam.) Sweet ■
910 Abutilon hirtum (Lam.) Sweet;恶味苘麻(黄花磨盘草);Hirsute Abutilon ■
911 Abutilon hirtum (Lam.) Sweet var. heterotrichum (Hochst. ex Mattei) Cufod.;异毛恶味苘麻■☆
912 Abutilon hirtum (Lam.) Sweet var. yuanmouense K. M. Feng;元谋恶味苘麻;Yuanmou Hirsute Abutilon ■
913 Abutilon hybridum Voss;观赏苘麻(杂种苘麻);Chinese Bellflower, Chinese Lantern, Flowering Malpe, Flowering-malpe, Garden Abutilon,Hybrid Abutilon,Variegated Flowering Maple ●☆
914 Abutilon hybridum Voss 'Apricot';杏黄苘麻●☆
915 Abutilon hybridum Voss 'Ashford Red';爱须福红苘麻●☆
916 Abutilon hybridum Voss 'Boule de Neige';保勒耐格苘麻●☆
917 Abutilon hybridum Voss 'Canary Bird';金丝鸟苘麻●☆
918 Abutilon hybridum Voss 'Cannington Carol';坎宁顿苘麻●☆
919 Abutilon hybridum Voss 'Caprice';狂想曲苘麻●
920 Abutilon hybridum Voss 'Cerise Queen';樱桃皇后苘麻●☆
921 Abutilon hybridum Voss 'Fire Ball';火球苘麻●
922 Abutilon hybridum Voss 'Fuchsiarum';矮金钟苘麻●
923 Abutilon hybridum Voss 'Golden Fleece';金发苘麻●
924 Abutilon hybridum Voss 'Maxima';大观赏苘麻●
925 Abutilon hybridum Voss 'Moonchimes';月谐苘麻●☆
926 Abutilon hybridum Voss 'Nabob';娜波苘麻●☆
927 Abutilon hybridum Voss 'Savitzer';白纹苘麻●
928 Abutilon hybridum Voss 'Snglant';赤色苘麻●
929 Abutilon hybridum Voss 'Snow Ball';雪球苘麻●
930 Abutilon hybridum Voss 'Souvenir Bonn';波恩纪念苘麻●☆
931 Abutilon impressum Hochst. ex Mattei = Abutilon figarianum Webb ■☆
932 Abutilon incanum (Link) Sweet;印度苘麻;Indian Mallow ●☆
933 Abutilon indicum (L.) G. Don = Abutilon indicum (L.) Sweet ●■
934 Abutilon indicum (L.) Sweet;磨盘草(白麻,倒绑茶,冬葵子,耳响草,金花草,帽笼草,帽笼子,帽子盾,米兰草,米篮草,磨挡草,磨谷子,磨龙子,磨砻草,磨笼草,磨笼子,磨盘子,磨片果,磨牙草,磨仔草,磨子盾草,磨子树,牛牯子麻,牛响草,石磨子,四米草,台磨草,土砻盾,牙仔草,研子盾草,印度苘麻);India Abutilon,Indian Abutilon,Kanska ●■
935 Abutilon indicum (L.) Sweet subsp. albescens (Miq.) Borss. Waalk.;白磨盘草●■☆
936 Abutilon indicum (L.) Sweet subsp. guineense (Schumach.) Borss. Waalk. = Abutilon indicum (L.) Sweet var. guineense (Schumach.) K. M. Feng ■

937 Abutilon indicum (L.) Sweet var. forrestii (S. Y. Hu) K. M. Feng = Abutilon guineense (Schumach.) Baker f. et Exell var. forrestii (S. Y. Hu) Y. Tang ■

938 Abutilon indicum (L.) Sweet var. guineense (Schumach.) K. M. Feng = Abutilon guineense (Schumach.) Baker f. et Exell ■

939 Abutilon indicum (L.) Sweet var. hirtum (Lam.) Griseb. = Abutilon hirtum (Lam.) Sweet ■

940 Abutilon indicum (L.) Sweet var. microphyllum Hochr. = Abutilon indicum (L.) Sweet ●■

941 Abutilon indicum (L.) Sweet var. populifolium (Lam.) Wight et Arn. = Abutilon indicum (L.) Sweet ●■

942 Abutilon indicum (L.) Sweet var. welwitschii Baker f. = Abutilon mauritianum (Jacq.) Medik. ■☆

943 Abutilon indicum (L.) Sweet var. wittei Hochr. = Abutilon mauritianum (Jacq.) Medik. ■☆

944 Abutilon intermedium Hochst. ex Garcke = Abutilon angulatum (Guillaumin et Perr.) Mast. ■☆

945 Abutilon intermedium Hochst. ex Garcke var. macrophyllum Baker f. = Abutilon angulatum (Guillaumin et Perr.) Mast. var. macrophyllum (Baker f.) Hochr. ■☆

946 Abutilon kauaiense Hochr. = Abutilon grandifolium (Willd.) Sweet ●☆

947 Abutilon kotschyi Hochst. ex Webb = Abutilon fruticosum Guillaumin et Perr. ●☆

948 Abutilon laxiflorum Guillaumin et Perr. = Wissadula rostrata (Schumach.) Hook. f. ●☆

949 Abutilon longicuspe Hochst. ex A. Rich.;长尖苘麻■☆

950 Abutilon longicuspe Hochst. ex A. Rich. var. epilosum Exell = Abutilon longicuspe Hochst. ex A. Rich. ■☆

951 Abutilon longipes Mattei = Abutilon mauritianum (Jacq.) Medik. ■☆

952 Abutilon lugardii Hochr. et Schinz = Abutilon hirtum (Lam.) Sweet ■

953 Abutilon macropodum Guillaumin et Perr.;大齿苘麻■☆

954 Abutilon marlothii Ulbr. = Abutilon pycnodon Hochr. ■☆

955 Abutilon matopense Gibbs;马托苘麻■☆

956 Abutilon mauritianum (Jacq.) Medik.;长梗苘麻■☆

957 Abutilon megapotamicum (Spreng.) A. St. -Hil. et Naudin;大河苘麻(古巴苘麻,美佳苘麻);Brazilian Abutilon, Chinese Lantern, Flowering Maple, Trailing Abutilon ●☆

958 Abutilon membranifolium Baker f. = Abutilon engleranum Ulbr. ■☆

959 Abutilon mendoncae Baker f.;门东萨苘麻■☆

960 Abutilon messinicum Burtt Davy = Abutilon matopense Gibbs ■☆

961 Abutilon microcarpum Mattei = Abutilon bidentatum A. Rich. ■☆

962 Abutilon microphyllum A. Rich. = Abutilon fruticosum Guillaumin et Perr. ●☆

963 Abutilon molle Baker = Abutilon grandifolium (Willd.) Sweet ●☆

964 Abutilon muticum (Delile ex DC.) Sweet = Abutilon pannosum (G. Forst.) Schltdl. ■☆

965 Abutilon muticum (Delile) Webb = Abutilon pannosum (G. Forst.) Schltdl. ■☆

966 Abutilon pakistanicum Jafri et Ali;巴基斯坦苘麻●☆

967 Abutilon palmeri A. Gray;帕氏苘麻;Indian Mallow, Palmer's Indian Mallow, Superstition Mallow ●☆

968 Abutilon paniculatum Hand. -Mazz.;圆锥苘麻;Paniculate Abutilon ●

969 Abutilon pannosum (G. Forst.) Schltdl.;毡毛苘麻■☆

970 Abutilon periplocifolium (L.) G. Don = Wissadula periplocifolia (L.) Thwaites ●

971 Abutilon periplocifolium (L.) Sweet = Sida periplocifolia L. ●

972 Abutilon periplocifolium (L.) Sweet = Wissadula periplocifolia (L.) C. Presl ex Thwaites ●

973 Abutilon periplocifolium (L.) Sweet var. zeylanica DC. = Wissadula periplocifolia (L.) C. Presl ex Thwaites ●

974 Abutilon persicum (Burm. f.) Merr.;波斯苘麻■☆

975 Abutilon pictum (Gillies ex Hook.) Walp.;缟花苘麻(金铃花,中国灯笼花);Chinese Lantern, Variegated Chinese Lantern ●☆

976 Abutilon piloso-cinereum A. Meeuse;灰毛苘麻■☆

977 Abutilon populifolium Sweet = Abutilon indicum (L.) Sweet ●■

978 Abutilon pritchardii Exell et Hillc.;普理查德苘麻■☆

979 Abutilon pseudangulatum Hochr. = Abutilon angulatum (Guillaumin et Perr.) Mast. var. macrophyllum (Baker f.) Hochr. ■☆

980 Abutilon pycnodon Hochr.;密齿苘麻■☆

981 Abutilon ramosum (Cav.) Guillaumin et Perr.;多枝苘麻■☆

982 Abutilon rehmannii Baker f.;拉赫曼苘麻■☆

983 Abutilon rigidum G. Don = Sida rigida (G. Don) D. Dietr. ●☆

984 Abutilon roseum Hand. -Mazz.;红花苘麻;Redflower Abutilon ■

985 Abutilon rotundifolium Mattei;圆叶苘麻■☆

986 Abutilon salmoneum Ulbr. = Abutilon pycnodon Hochr. ■☆

987 Abutilon schinzii Ulbr.;斑花苘麻;Variegated Flowering Maple ■☆

988 Abutilon seineri Ulbr. = Abutilon rehmannii Baker f. ■☆

989 Abutilon sinense Oliv.;华苘麻(豹子眼睛花,老熊花,野苘麻);China Abutilon, Chinese Abutilon ●

990 Abutilon sinense Oliv. var. edentatum K. M. Feng;无齿华苘麻;Edentate Abutilon ●

991 Abutilon sinense Oliv. var. gebauerianum Hand. -Mazz. = Abutilon gebauerianum Hand. -Mazz. ●

992 Abutilon sinense Oliv. var. typica Hochr. = Abutilon sinense Oliv. ●

993 Abutilon sinense Oliv. var. yunnanensis Hochr. = Abutilon gebauerianum Hand. -Mazz. ●

994 Abutilon smenospermum Pic. Serm. = Abutilon cecilii N. E. Br. ■☆

995 Abutilon somalense Mattei;索马里苘麻●☆

996 Abutilon sonneratianum (Cav.) Sweet;索南里苘麻●☆

997 Abutilon striatum G. F. Dicks.;金铃花(灯笼花,风铃花,金棒花,条纹苘麻,纹瓣悬铃花,猩猩花);Redvein Abutilon, Striped Abutilon ●

998 Abutilon striatum G. F. Dicks. ex Lindl. var. thompsonii Veitch;汤普森风铃花(汤姆逊风铃花)●☆

999 Abutilon taiwanensis S. Y. Hu = Abutilon indicum (L.) Sweet var. guineense (Schumach.) K. M. Feng ■

1000 Abutilon theophrastii Medik.;苘麻(八角乌,白麻,车轮草,椿麻,冬葵,空麻,孔麻,磨盘草,磨盘树,青麻,苘,手巾花,塘麻,桐麻,野棉花,野苎麻);America Jute, Bellflower, Butter Print, Butterprint, Butter-print, China Jute, Chinese Hemp, Chingma Abutilon, Indian Mallow, Manchurian Jute, Pie Marker, Pie Print, Piemarker, Pie-Marker, Pie-print, Velvet Leaf, Velvet Weed, Velvetleaf, Velvet-leaf ●■

1001 Abutilon theophrastii Medik. var. chinense (Skvortsov) S. Y. Hu = Abutilon theophrastii Medik. ●■

1002 Abutilon theophrastii Medik. var. nigrum (Skvortsov) S. Y. Hu = Abutilon theophrastii Medik. ●■

1003 Abutilon tortuosum Guillaumin et Perr. = Abutilon grandifolium (Willd.) Sweet ●☆

1004 Abutilon tridens Standl. et Steyerm.;三齿苘麻●☆

1005 Abutilon trisulcatum (Jacq.) Urb.;三沟苘麻●☆

1006 Abutilon umtaliense Baker f. = Abutilon sonneratianum (Cav.) Sweet ●☆

1007 Abutilon velutinum G. Don;短绒毛苘麻■☆
1008 Abutilon vexillarium E. Morren = Abutilon megapotamicum (Spreng.) A. St. -Hil. et Naudin ●☆
1009 Abutilon vitifolium (Cav.) G. Don;葡萄叶苘麻●☆
1010 Abutilon webbianum Mattei = Abutilon figarianum Webb ■☆
1011 Abutilon wituense Baker f. = Pavonia procumbens (Wall. ex Wight et Arn.) Walp. ●☆
1012 Abutilon zanzibaricum Bojer ex Mast. = Abutilon mauritianum (Jacq.) Medik. ■☆
1013 Abutilothamnus Ulbr. (1915);南美麻属●☆
1014 Abutilothamnus grewiifolius Ulbr.;南美麻●☆
1015 Abutilothamnus yaracuyensis Fryxell;亚拉圭南美麻●☆
1016 Abutua Batsch = Abuta Aubl. ●☆
1017 Abutua Lour. = Gnetum L. ●
1018 Acacallis Lindl. (1853);美唇兰属(阿卡卡里兰属,阿卡兰属);Acacallis ■☆
1019 Acacallis Lindl. = Aganisia Lindl. ■☆
1020 Acacallis cyanea Lindl.;美唇兰(阿卡卡里兰,阿卡兰);Blue Acacallis ■☆
1021 Acachmena H. P. Fuchs(废弃属名) = Cuspidaria (DC.) Besser (废弃属名)●☆
1022 Acachmena H. P. Fuchs(废弃属名) = Erysimum L. ●■
1023 Acacia Hill. = Acacia Mill. (保留属名)●■
1024 Acacia Mill. (1754)(保留属名);金合欢属(相思树属);Acacia, Aden Gum, Barbary Gum, E. Indian Gum, Gum-tree, Indian Gum, Morocco Gum, Wattle ●■
1025 Acacia Willd. = Acacia Mill. (保留属名)●■
1026 Acacia abyssinica Hochst. ex Benth.;平顶金合欢;Flat-top Acacia ●☆
1027 Acacia abyssinica Hochst. ex Benth. subsp. calophylla Brenan;美叶平顶金合欢;Flat-top Acacia ●☆
1028 Acacia acicularis Humb. et Bonpl. ex Willd. = Acacia minuta (M. E. Jones) R. M. Beauch. ●☆
1029 Acacia acinacea Lindl.;拉特氏金合欢(金尘金合欢,拉特氏相思树);Gold-dust Wattle, Gold-dust, Latrobe Acacia ●☆
1030 Acacia acuminata Benth.;红木相思;Raspberry Acacia, Raspberry Jam Tree, Raspberry Jam ●☆
1031 Acacia adansonii Guillaumin et Perr. = Acacia nilotica (L.) Willd. ex Delile subsp. adstringens (Schumach. et Thonn.) Roberty ●☆
1032 Acacia adenocalyx Brenan et Exell;腺萼金合欢●☆
1033 Acacia adstringens (Schumach. et Thonn.) Berhaut = Acacia nilotica (L.) Willd. ex Delile subsp. adstringens (Schumach. et Thonn.) Roberty ●☆
1034 Acacia adunca A. Cunn. ex G. Don;瓦兰加拉金合欢;Wallangara Wattle ●☆
1035 Acacia albida Delile;微白金合欢(白楞);Ana Tree, Apple Ring Acacia, Apple-ring Acacia, Winter Thorn, Winterthorn ●
1036 Acacia albida Delile = Faidherbia albida (Delile) A. Chev. ■☆
1037 Acacia alpina F. Muell.;高山金合欢;Alpine Wattle, Ana Tree, Apple Ring Acacia ●☆
1038 Acacia ambigua Hoffmanns.;可疑金合欢●☆
1039 Acacia amblygona A. Cunn. ex Benth.;扇金合欢;Fan Wattle ●☆
1040 Acacia amboensis Schinz = Acacia sieberiana DC. var. woodii (Burtt Davy) Keay et Brenan ●☆
1041 Acacia amentacea DC. = Acacia rigidula Benth. ●☆
1042 Acacia amoena H. L. Wendl.;可爱相思;Boomerang Wattle ●☆
1043 Acacia ampliceps Maslin;大花相思●☆
1044 Acacia amythethophylla Steud. ex A. Rich.;密叶相思●☆
1045 Acacia ancistrocarpa Maiden;钩果金合欢(钩荚相思)●☆
1046 Acacia ancistroclada Brenan;钩枝金合欢●☆
1047 Acacia andongensis Welw. ex Hiern;安东金合欢●☆
1048 Acacia aneura F. Muell. = Acacia aneura F. Muell. ex Benth. ●☆
1049 Acacia aneura F. Muell. ex Benth.;无脉相思树(缺脉相思,无脉相思);Mulga Acacia, Mulga, Mulgamulga ●☆
1050 Acacia angustisiliqua (Lam.) Desf. = Desmanthus virgatus (L.) Willd. ●■
1051 Acacia angustissima (Mill.) Kuntze;极狭金合欢;Fern Acacia, Prairie Acacia, Timbre, Whiteball Acacia ●☆
1052 Acacia antunesii Harms;安图内思金合欢●☆
1053 Acacia aphylla Maslin;无叶金合欢;Twisted Desert Wattle ●☆
1054 Acacia arabica (Lam.) Willd. = Acacia nilotica (L.) Willd. ex Delile ●
1055 Acacia arabica (Lam.) Willd. var. adansoniana Dubard = Acacia nilotica (L.) Willd. ex Delile subsp. adstringens (Schumach. et Thonn.) Roberty ●☆
1056 Acacia arabica (Lam.) Willd. var. indica Benth. = Acacia nilotica (L.) Willd. ex Delile subsp. indica (Benth.) Brenan ●☆
1057 Acacia arabica (Lam.) Willd. var. kraussiana Benth. = Acacia nilotica (L.) Willd. ex Delile subsp. kraussiana (Benth.) Brenan ●☆
1058 Acacia arenaria Schinz;沙生金合欢●☆
1059 Acacia armata (Willd.) Batt. = Astragalus armatus Willd. ■☆
1060 Acacia armata R. Br.;刺相思树;Kangaroo Thorn ●☆
1061 Acacia arrophula D. Don = Acacia pennata (L.) Willd. ●■
1062 Acacia arrophula D. Don ex Wall. = Acacia megaladena Desv. ●
1063 Acacia asak (Forssk.) Willd.;阿萨克相思●☆
1064 Acacia atacorensis Aubrév. et Pellegr. = Acacia hockii De Wild. ●☆
1065 Acacia atomiphylla Burch. = Acacia haematoxylon Willd. ●☆
1066 Acacia aulacocarpa A. Cunn. ex Benth.;纹荚相思(沟果荆)●☆
1067 Acacia auriculiformis A. Cunn. ex Benth.;大叶相思(阿列克楞,大叶相思树,耳荚相思树,耳形金合欢,耳叶相思,耳状金合欢);Auriculate Acacia, Earleaf Acacia, Ear-leaved Acacia, Ear-pod Wattle, Northern Black Wattle ●
1068 Acacia aurisparsa Drake = Albizia aurisparsa (Drake) R. Vig. ●☆
1069 Acacia baileyana F. Muell.;贝利氏相思树(贝利氏金合欢,贝利相思树,灰叶相思树,考塔孟德相思树);Bailey Acacia, Bailey's Mimosa, Cootamundra Wattle, Golden Mimosa ●☆
1070 Acacia baileyana F. Muell. 'Purpurea';紫叶贝利氏相思树;Cootamundra Wattle ●☆
1071 Acacia bancroftiana Bertero ex Colla;班氏金合欢;Bancroft's Wattle ●☆
1072 Acacia barbertonensis Schweick. = Acacia borleae Burtt Davy ●☆
1073 Acacia baronii Villiers et Du Puy;巴龙金合欢●☆
1074 Acacia beauverdiana Ewart et Sharman;博维金合欢●☆
1075 Acacia bellula Drake;可爱合欢●☆
1076 Acacia benadirensis (Chiov.) Chiov. = Acacia horrida (L.) Willd. subsp. benadirensis (Chiov.) Hillc. et Brenan ●☆
1077 Acacia benthamii Meisn.;本瑟姆金合欢(本氏金合欢)●☆
1078 Acacia benthamii Rochebr. = Acacia nilotica (L.) Willd. ex Delile subsp. kraussiana (Benth.) Brenan ●☆
1079 Acacia bequaertii De Wild. = Acacia goetzei Harms ●☆
1080 Acacia berlandieri Benth.;伯氏金合欢;Berlandier's Acacia, Guajillo, Plains Acacia ●☆
1081 Acacia bimucronata DC. = Mimosa bimucronata (DC.) Kuntze ■

1082 Acacia binervata DC.;双脉相思树(双脉栲,双生脉相思树);Coast Myall ●☆

1083 Acacia binervia (J. C. Wendl.) J. F. Macbr.;双生脉相思树;Coastal Myall,Sally Wattle ●☆

1084 Acacia binervia J. F. Macbr. = Acacia binervia (J. C. Wendl.) J. F. Macbr. ●☆

1085 Acacia blommaertii De Wild = Acacia sieberiana DC. ●☆

1086 Acacia boboensis Aubrév. = Acacia hockii De Wild. ●☆

1087 Acacia boivinii (E. Fourn.) Baill. = Albizia boivinii E. Fourn. ●☆

1088 Acacia boormanii Maiden;布尔曼相思树;Snowy River Wattle ●☆

1089 Acacia borleae Burtt Davy;波尔相思树●☆

1090 Acacia brachybotrya Benth.;银叶相思树;Gray Mulga ●☆

1091 Acacia brevispica Harms;短穗金合欢;Wait-a-bit Thorn ●☆

1092 Acacia brevispica Harms subsp. dregeana (Benth.) Brenan;德雷相思树●☆

1093 Acacia brevispica Harms var. dregeana (Benth.) J. H. Ross et Gordon-Gray = Acacia brevispica Harms subsp. dregeana (Benth.) Brenan ●☆

1094 Acacia bricchettiana Chiov. = Acacia edgeworthii T. Anderson ●☆

1095 Acacia brosigii Harms = Acacia nigrescens Oliv. ●☆

1096 Acacia brunioides A. Cunn. ex G. Don;澳东金合欢●☆

1097 Acacia buchananii Harms = Acacia amythethophylla Steud. ex A. Rich. ●☆

1098 Acacia bullockii Brenan;布洛克金合欢●☆

1099 Acacia burkei Benth.;布尔克金合欢●☆

1100 Acacia burkittii F. Muell. ex Benth.;针叶金合欢;Pinbush Wattle ●☆

1101 Acacia burttii Baker f.;伯特金合欢●☆

1102 Acacia bussei Harms ex Y. Sjostedt;布瑟金合欢(小金合欢);Busse Acacia ●☆

1103 Acacia bussei Harms ex Y. Sjostedt var. benadirensis Chiov. = Acacia horrida (L.) Willd. subsp. benadirensis (Chiov.) Hillc. et Brenan ●☆

1104 Acacia buxifolia A. Cunn.;黄杨叶金合欢;Box-leaf Wattle ●☆

1105 Acacia bynoeana Benth.;白诺金合欢;Dwarf Nealie ●☆

1106 Acacia caesia (L.) Roxb. ex Wall. = Acacia caesia (L.) Walld. ●

1107 Acacia caesia (L.) Walld.;尖叶相思(尖叶印度相思,尖叶印度相思树,藤相思树,相思树);Caesious Acacia, Sharp-leaved Acacia ●

1108 Acacia caesia (L.) Walld. var. subnuda (Craib) I. C. Nielsen = Acacia caesia (L.) Willd. ●

1109 Acacia caffra (Thunb.) Willd.;开菲尔金合欢●☆

1110 Acacia caffra (Thunb.) Willd. var. campylacantha (Hochst. ex A. Rich.) Aubrév. = Acacia polyacantha Willd. subsp. campylacantha (Hochst. ex A. Rich.) Brenan ●☆

1111 Acacia caffra (Thunb.) Willd. var. longa Glover = Acacia caffra (Thunb.) Willd. ●☆

1112 Acacia caffra (Thunb.) Willd. var. namaquensis Eckl. et Zeyh. = Acacia caffra (Thunb.) Willd. ●☆

1113 Acacia caffra (Thunb.) Willd. var. pechuelii Kuntze = Acacia erubescens Welw. ex Oliv. ●☆

1114 Acacia caffra (Thunb.) Willd. var. tomentosa Glover = Acacia caffra (Thunb.) Willd. ●☆

1115 Acacia caffra (Thunb.) Willd. var. transvaalensis Glover = Acacia caffra (Thunb.) Willd. ●☆

1116 Acacia calamifolia Sieber ex Steud.;藤叶相思树(藤相思树);Broom Wattle,Wallowa ●☆

1117 Acacia campylacantha Hochst. ex A. Rich.;弯刺金合欢;Hook Thorn,Hook Thorn Tree ●☆

1118 Acacia campylacantha Hochst. ex A. Rich. = Acacia polyacantha Willd. subsp. campylacantha (Hochst. ex A. Rich.) Brenan ●☆

1119 Acacia cardiophylla A. Cunn. ex Benth.;心叶金合欢;Wyalong Wattle ●☆

1120 Acacia catechu (L. f.) Willd.;儿茶(阿仙药,阿仙药树,百药煎,粉口儿茶,孩儿茶,黑儿茶,乌爹泥,乌丁,乌丁泥,乌垒泥);Black Catechu,Black Cutch,Catch,Catechu,Catechu Acacia,Cutch-tree,Khair,Khair Gum,Khair Tree,Khair-tree,Khayer,Wadalee-gum,Wadaleetree ●

1121 Acacia catechu (L. f.) Willd. var. baumii Roberty = Acacia fleckii Schinz ●☆

1122 Acacia catechu (L. f.) Willd. var. campylacantha (Hochst. ex A. Rich.) Roberty = Acacia polyacantha Willd. subsp. campylacantha (Hochst. ex A. Rich.) Brenan ●☆

1123 Acacia catechu (L. f.) Willd. var. erythrantha (Steud. ex A. Rich.) Roberty = Acacia polyacantha Willd. subsp. campylacantha (Hochst. ex A. Rich.) Brenan ●☆

1124 Acacia catechu (L. f.) Willd. var. hecatophylla (Steud. ex A. Rich.) Roberty = Acacia hecatophylla Steud. ex A. Rich. ●☆

1125 Acacia catechu (L. f.) Willd. var. wallichiana (DC.) P. C. Huang;无刺儿茶;Wallich Catch ●

1126 Acacia catechu (L. f.) Willd. var. wallichiana (DC.) P. C. Huang = Acacia catechu (L. f.) Willd. ●

1127 Acacia cavaleriei H. Lév. = Acacia delavayi Franch. ●

1128 Acacia cavenia Hook. et Arn.;加芬相思树(南美相思树);Caven,Cavenia,Espino,Espino-caven ●☆

1129 Acacia cavenia Molina = Acacia cavenia Hook. et Arn. ●☆

1130 Acacia cernua Thulin et Hassan;俯垂金合欢●☆

1131 Acacia chariensis A. Chev. = Acacia hockii De Wild. ●☆

1132 Acacia cheilanthifolia Chiov.;碎米蕨叶金合欢●☆

1133 Acacia cheilanthifolia Chiov. var. hirtella ? = Acacia cheilanthifolia Chiov. ●☆

1134 Acacia chinchillensis Tindale;秦奇金合欢;Chinese Wattle ●☆

1135 Acacia chrysothrix Taub. = Acacia rovumae Oliv. ●☆

1136 Acacia cinerea Schinz = Acacia fleckii Schinz ●☆

1137 Acacia circummarginata Chiov.;索马里金合欢●☆

1138 Acacia clavigera E. Mey. = Acacia robusta Burch. subsp. clavigera (E. Mey.) Brenan ●☆

1139 Acacia cognata Maiden;窄叶金合欢;Bower Wattle, Narrow-leaf Bower Wattle,River Wattle ●☆

1140 Acacia colei Maslin et L. A. J. Thomson;鞘尾相思●☆

1141 Acacia comorensis Baill. = Albizia glaberrima (Schumach. et Thonn.) Benth. ●☆

1142 Acacia concinna (Willd.) DC.;藤金合欢(金合欢,小金合欢);Glory Acacia,Sinuate Acacia,Soap Pod ●■

1143 Acacia concinna (Willd.) DC. = Acacia sinuata (Lour.) Merr. ●■

1144 Acacia conferta A. Cunn. ex Benth.;密叶金合欢(密枝相思);Crowded-leaf Wattle ●☆

1145 Acacia confusa Merr.;台湾相思(番子树,松丝,台湾柳,台湾相思树,相思,相思树,相思仔,香丝树);Formosan Koa, Small Philippine Acacia,Small Philippine Wattle,Taiwan Acacia ●

1146 Acacia confusa Merr. = Acacia richii A. Gray ●☆

1147 Acacia confusa Merr. var. inamurae Hayata = Acacia confusa Merr. ●

1148 Acacia constricta Benth.;白金合欢;Largancillo,Mescat Acacia, White Thorn,Whitethorn,Whitethorn Acacia ●☆

1149 Acacia continua Benth.;尖刺金合欢;Thorn Wattle ●☆
1150 Acacia cornigera Willd.;牛角相思树(牛角楟);Bull Horn Thorn, Bull-horn Acacia, Bull-horn Thorn, Bullhorn Wattle, Cockspur, Swollen-thorn Acacia ●☆
1151 Acacia covenyi Tindale;蓝金合欢;Blue Bush ●☆
1152 Acacia cowleana Tate;寇氏相思●☆
1153 Acacia craspedocarpa F. Muell.;蜡叶金合欢;Leatherleaf Acacia, Waxleaf Acacia ●☆
1154 Acacia crassa Pedley;肥大金合欢●☆
1155 Acacia cufodontii Chiov. = Acacia senegal (L.) Willd. ●
1156 Acacia cultriformis A. Cunn. ex G. Don;刀叶相思树(刀状相思树);Knife Acacia, Knife Leaf Wattle, Knife-leaf Acacia, Knife-leaf Waggle, Knife-leaf Wattle, Plough Share Wattle ●☆
1157 Acacia cultriformis A. Cunn. ex G. Don 'Austraflora Cascade';瀑布金合欢;Knife Acacia, Knife Leaf Wattle ●☆
1158 Acacia cuthbertsonii Luehm.;卡伯金合欢●☆
1159 Acacia cyanophylla Lindl.;蓝叶金合欢(澳洲金合欢);Blueleaf Acacia, Blueleaf Wattle, Blue-leaved Wattle, Golden Wattle, Golden Willow, Orange Wattle, Port Jackson Willow ●
1160 Acacia cyanophylla Lindl. = Acacia saligna (Labill.) H. L. Wendl. ●☆
1161 Acacia cyclopis A. Cunn. ex G. Don;巨相思树;Coastal Wattle, Cyclops Acacia ●☆
1162 Acacia cyclopis A. Cunn. ex Loudon = Acacia cyclopis A. Cunn. ex G. Don ●☆
1163 Acacia dalzielii Craib = Acacia amythethophylla Steud. ex A. Rich. ●☆
1164 Acacia davyi N. E. Br.;戴维银荆树●☆
1165 Acacia dealbata Link;银荆树(阿卡锡,澳洲白色金合欢,白粉金合欢,德尔楟,圣诞树,银儿茶,银合欢,银荆,银楟,银楟皮树,鱼骨槐,鱼骨松);Blue Wattle, Mimosa, Silver Acacia, Silver Wattle, Silvergreen-wattle Acacia, Silver-green-wattle Acacia ●
1166 Acacia deanei (R. T. Baker) R. T. Baker ex Welch, Coombs et McGlynn;迪氏荆●☆
1167 Acacia decora Rchb.;美丽相思树(美好相思树);Graceful Wattle, Showy Wattle, Western Golden Wattle ●☆
1168 Acacia decurrens (J. C. Wendl.) Willd.;线叶金合欢(阿卡锡,澳洲细叶金合欢,楟皮树,绿儿茶,绿荆,绿荆树,青楟,圣诞树,下延金合欢,下延相思树,鱼骨槐,鱼骨松);Black Wattle, Early Black Wattle, Green Wattle, Green Wattle Acacia, Green-wattle Acacia, Syndery Green Wattle ●
1169 Acacia decurrens (J. C. Wendl.) Willd. var. mollis Lindl.;澳洲金合欢●
1170 Acacia decurrens (J. C. Wendl.) Willd. var. mollis Lindl. = Acacia dealbata Link ●
1171 Acacia decurrens (J. C. Wendl.) Willd. var. mollissima Link = Acacia mearnsii De Wild. ●
1172 Acacia decurrens Willd. var. dealbata (Link) Maiden = Acacia dealbata Link ●
1173 Acacia decurrens Willd. var. mollis Lindl. = Acacia dealbata Link ●
1174 Acacia dekindtiana A. Chev. = Acacia karroo Hayne ●☆
1175 Acacia delagoensis Harms = Acacia welwitschii Oliv. subsp. delagoensis (Harms) J. H. Ross et Brenan ●☆
1176 Acacia delavayi Franch.;光叶金合欢(阔叶金合欢,阔叶相思树,老虎刺,老虎棘,丽江金合欢,酸格,酸格刺);Delavay Acacia ●■
1177 Acacia delavayi Franch. var. kunmingensis C. Chen et H. Sun;昆明金合欢;Kunming Acacia ●■
1178 Acacia densispina Thulin;密刺金合欢●☆
1179 Acacia detinens Burch. = Acacia mellifera (Vahl) Benth. subsp. detinens (Burch.) Brenan ●☆
1180 Acacia dewevrei De Wild. et T. Durand = Acacia lujae De Wild. ●☆
1181 Acacia dietrichiana F. Muell.;昆士兰相思树;Queensland Acacia ●☆
1182 Acacia dolichocephala Harms;长头金合欢●☆
1183 Acacia dolichocephala Harms = Acacia sphaerocephala Cham. et Schltdl. ●☆
1184 Acacia doratoxylon Meisn.;枪木相思;Brigalow, Currawong Acacia, Spear-wood ●☆
1185 Acacia drakei R. Vig. = Acacia sakalava Drake ●☆
1186 Acacia drepanolobium Harms ex Y. Sjostedt;肯尼亚金合欢;E. Africa Gum, Whistling Thorn ●☆
1187 Acacia drummondii Lindl.;德拉蒙德金合欢;Drummond Wattle ●☆
1188 Acacia dudgeonii Craib ex Holland;达吉金合欢●☆
1189 Acacia dulcis Marloth et Engl. = Acacia erubescens Welw. ex Oliv. ●☆
1190 Acacia dyeri P. P. Sw.;戴尔金合欢●☆
1191 Acacia eburnea (L. f.) Willd.;象牙白金合欢;Cockspur Thorn ●☆
1192 Acacia edgeworthii T. Anderson;埃奇沃斯金合欢●☆
1193 Acacia eggelingii Baker f. = Acacia persiciflora Pax ●☆
1194 Acacia ehrenbergiana Hayne;埃伦金合欢●☆
1195 Acacia elata A. Cunn. ex Benth.;高耸金合欢;Cedar Wattle ●☆
1196 Acacia elatior Brenan;较高金合欢●☆
1197 Acacia elephantina Burch. = Elephantorrhiza elephantina (Burch.) Skeels ●☆
1198 Acacia elephantorrhiza DC. = Elephantorrhiza elephantina (Burch.) Skeels ●☆
1199 Acacia elongata Sieber ex DC.;长荚相思树;Swamp Wattle ●☆
1200 Acacia emoryana Benth. = Acacia berlandieri Benth. ●☆
1201 Acacia engleri Schinz = Dichrostachys cinerea (L.) Wight et Arn. var. africana Brenan et Brummitt ●☆
1202 Acacia enterocarpa R. V. Sm.;肠果金合欢;Jumping Wattle ●☆
1203 Acacia erentiniana A. Chev.;埃伦丁金合欢●☆
1204 Acacia eriocarpa Brenan;毛果金合欢●☆
1205 Acacia erioloba E. Mey. = Acacia erioloba Edgew. ●☆
1206 Acacia erioloba Edgew.;骆驼刺金合欢;Camel Thorn, Giraffe Thorn ●☆
1207 Acacia erubescens Welw. ex Oliv.;渐红金合欢●☆
1208 Acacia erythraea Chiov. = Acacia edgeworthii T. Anderson ●☆
1209 Acacia erythrantha Steud. ex A. Rich. = Acacia polyacantha Willd. subsp. campylacantha (Hochst. ex A. Rich.) Brenan ●☆
1210 Acacia erythrocalyx Brenan;红萼金合欢●☆
1211 Acacia erythrophloea Brenan;红皮金合欢●☆
1212 Acacia esculenta DC. = Leucaena esculenta (DC.) Benth. ●☆
1213 Acacia estrophiolata F. Muell.;暗蛇小叶金合欢;Ironwood ●☆
1214 Acacia etbaica Schweinf. subsp. platycarpa Brenan;阔果金合欢●☆
1215 Acacia etbaica Schweinf. subsp. uncinata Brenan;具钩金合欢●☆
1216 Acacia exuvialis I. Verd.;剥落金合欢●☆
1217 Acacia falcata Willd.;镰刀金合欢(镰刀状金合欢,镰叶楟,扭叶楟);Falcate Acacia ●☆
1218 Acacia fallax E. Mey. = Acacia caffra (Thunb.) Willd. ●☆
1219 Acacia farnensiana (L.) Willd.;金合欢(刺球花,莿球花,莿球,番苏木,牛角花,绒祖刺,天黄豆树,消息花,鸭皂树,洋梅花刺,楹树,猪牙皂树);Cassie, Cassie Flower, Cassie-oil-plant, Castle,

Huisache, Huisache Dulce, Mimosa Bush, Opopanax, Poninac, Popinac, Sponge Tree, Spongetree, Sweet Acacia, Sweet Briar, West Indian Blackthorn ●

1220 Acacia fasciculata Guillaumin et Perr. = Acacia tortilis (Forssk.) Hayne var. raddiana (Savi) Brenan ●☆

1221 Acacia fasciculata Guillaumin et Perr. var. pubescens (A. Chev.) A. Chev. = Acacia tortilis (Forssk.) Hayne var. pubescens A. Chev. ●☆

1222 Acacia ferox Benth. = Acacia burkei Benth. ●☆

1223 Acacia ferruginea DC.;锈色金合欢●☆

1224 Acacia fimbriata A. Cunn. ex G. Don;流苏金合欢;Fringed Wattle ●☆

1225 Acacia fischeri Harms;菲舍尔金合欢●☆

1226 Acacia fistula Schweinf. = Acacia seyal Delile var. fistula (Schweinf.) Oliv. ●☆

1227 Acacia flagellaris Thulin;鞭状金合欢●☆

1228 Acacia flava (Forssk.) Schweinf. f. chariensis (A. Chev.) Roberty = Acacia hockii De Wild. ●☆

1229 Acacia flava (Forssk.) Schweinf. f. ehrenbergiana ? = Acacia ehrenbergiana Hayne ●☆

1230 Acacia flava (Forssk.) Schweinf. f. fistula (Schweinf.) Roberty = Acacia seyal Delile var. fistula (Schweinf.) Oliv. ●☆

1231 Acacia flava (Forssk.) Schweinf. var. atacorensis (Aubrév. et Pellegr.) Aubrév. = Acacia hockii De Wild. ●☆

1232 Acacia flava (Forssk.) Schweinf. var. seyal (Delile) Roberty = Acacia seyal Delile ●☆

1233 Acacia fleckii Schinz;福来氏金合欢●☆

1234 Acacia flexicaulis Benth. = Pithecellobium flexicaule (Benth.) J. M. Coult. ●

1235 Acacia flexifolia A. Cunn. ex Benth.;弯叶金合欢;Bean-leaf Wattle ●☆

1236 Acacia floribunda (Vent.) Willd. = Acacia retinodes Schltdl. ●☆

1237 Acacia formicarum Harms = Acacia drepanolobium Harms ex Y. Sjostedt ●☆

1238 Acacia galpinii Burtt Davy;南非相思树;Apiesdoring, Galpin Acacia ●☆

1239 Acacia gandalensis Thulin;甘达尔金合欢●☆

1240 Acacia gansbergensis Schinz = Acacia hereroensis Engl. ●☆

1241 Acacia gerrardii Benth.;盖氏金合欢(元兰榜)●☆

1242 Acacia gerrardii Benth. var. calvescens Brenan;光秃金合欢●☆

1243 Acacia gerrardii Benth. var. latisiliqua Brenan;宽荚盖氏金合欢●☆

1244 Acacia gillettiae Burtt Davy = Acacia luederitzii Engl. var. retinens (Sim) J. H. Ross et Brenan ●☆

1245 Acacia giraffae Burch.;吉氏相思树(吉拉夫氏相思树);Camel Thorn ●☆

1246 Acacia giraffae Burch. = Acacia erioloba Edgew. ●☆

1247 Acacia giraffae Willd. var. espinosa Kuntze = Acacia erioloba E. Mey. ●☆

1248 Acacia glandulifera Schinz = Acacia nebrownii Burtt Davy ●☆

1249 Acacia glauca (L.) Moench;苏门答腊金合欢(灰金合欢,苏门答腊合欢);Glaucous Acacia, Grey-blue Acacia ●

1250 Acacia glauca (L.) Moench = Leucaena leucocephala (Lam.) de Wit ●

1251 Acacia glauca (L.) Willd. = Leucaena leucocephala (Lam.) de Wit ●

1252 Acacia glauca Willd. = Leucaena leucocephala (Lam.) de Wit ●

1253 Acacia glaucescens Willd. = Acacia binervia (J. C. Wendl.) J. F. Macbr. ●☆

1254 Acacia glaucophylla Steud. ex A. Rich. = Acacia asak (Forssk.) Willd. ●☆

1255 Acacia glaucoptera Benth.;灰绿翅金合欢;Clay Wattle, Claybush Wattle, Clay-bush Wattle, Queen Wattle ●☆

1256 Acacia gloveri Gilliland = Acacia edgeworthii T. Anderson ●☆

1257 Acacia goeringii Schinz = Acacia luederitzii Engl. ●☆

1258 Acacia goetzei Harms;戈策金合欢●☆

1259 Acacia goetzei Harms subsp. microphylla Brenan;小叶戈策金合欢●☆

1260 Acacia gorinii Chiov. = Acacia oerfota (Forssk.) Schweinf. ●☆

1261 Acacia gossweileri Baker f. = Acacia goetzei Harms subsp. microphylla Brenan ●☆

1262 Acacia gourmaensis A. Chev.;古尔马金合欢●☆

1263 Acacia grandicornuta Gerstner;大角金合欢●☆

1264 Acacia greggii A. Gray;格雷格金合欢(猫爪相思树);Cat Claw Acacia, Catclaw, Catclaw Acacia, Cat' s Claws, Cat' s-claws, Devil' s Claws, Devilsclaw, Devil' s-claw, Gregg Catclaw, Gregg Catdaw, Tear-blanket, Texas Mimosa, Wait-a-minute ●☆

1265 Acacia greveana Baill. = Albizia greveana (Baill.) Baron ●☆

1266 Acacia gummifera Willd.;摩洛哥橡胶树;Barbary Gum, Mogador Acacia, Morocco Gum ●☆

1267 Acacia haematoxylon Willd.;血材金合欢●☆

1268 Acacia hainanensis Hayata;海南金合欢;Hainan Acacia ●

1269 Acacia hainanensis Hayata = Acacia pennata (L.) Willd. subsp. hainanensis (Hayata) I. C. Nielsen ●

1270 Acacia hainanensis Hayata = Acacia pennata (L.) Willd. ●■

1271 Acacia hakeoides A. Cunn. ex Benth.;哈克金合欢;Western Black Wattle ●☆

1272 Acacia hamulosa Benth.;袋鼠刺;Kangaron Thorn, Paradox Acacia, Thorn Tree ●☆

1273 Acacia harmsiana Dinter = Acacia kirkii Oliv. ●☆

1274 Acacia harpophylla F. Muell. ex Benth.;镰叶金合欢;Brigalow ●☆

1275 Acacia havilandiorum Maiden;细叶金合欢;Needle Wattle ●☆

1276 Acacia hebeclada DC.;柔毛枝金合欢●☆

1277 Acacia hebeclada DC. subsp. chobiensis (O. B. Mill.) A. Schreib.;乔贝金合欢●☆

1278 Acacia hebeclada DC. subsp. tristis (Welw. ex Oliv.) A. Schreib.;暗枝金合欢●☆

1279 Acacia hebeclada DC. var. stolonifera (Burch.) Dinter = Acacia hebeclada DC. ●☆

1280 Acacia hebecladoides Harms = Acacia gerrardii Benth. ●☆

1281 Acacia hecatophylla Steud. ex A. Rich.;多叶金合欢●☆

1282 Acacia hereroensis Engl.;百叶枝金合欢●☆

1283 Acacia hermannii Baker f. = Acacia arenaria Schinz ●☆

1284 Acacia heteracantha Burch. = Acacia tortilis (Forssk.) Hayne subsp. heteracantha (Burch.) Brenan ●☆

1285 Acacia heterophylla Willd.;异叶金合欢●☆

1286 Acacia hildebrandtii (Vatke) Drake;希尔德金合欢●☆

1287 Acacia hindsii Benth.;亨兹金合欢●☆

1288 Acacia hirtella E. Mey. = Acacia karroo Hayne ●☆

1289 Acacia hirtella Willd. var. inermis Walp. = Acacia karroo Hayne ●☆

1290 Acacia hockii De Wild.;胡克金合欢;White Thorn ●☆

1291 Acacia holosericea A. Cunn. ex G. Don;厚叶相思树(绢毛相思,绒毛相思树,丝毛相思,细丝毛金合欢);Thick-leaved Acacia ●☆

1292 Acacia homalophylla A. Cunn. ex Benth.;矛叶相思树(平展叶相思树);Fragrant Myall, Gidgee, Gidgee Myall, Mayll, Myallwood,

Spearwood, Violet Wood, Yaran, Yarran ●☆
1293 Acacia homalophylla Benth. = Acacia homalophylla A. Cunn. ex Benth. ●☆
1294 Acacia hookii de Wild.;霍克金合欢●☆
1295 Acacia horrida (L.) Willd. subsp. benadirensis (Chiov.) Hillc. et Brenan;贝纳迪尔金合欢●☆
1296 Acacia horrida (L.) Willd. var. transvaalensis Burtt Davy = Acacia karroo Hayne ●☆
1297 Acacia horrida Willd.;非洲相思树(刺金合欢,非洲橡胶树);Allthom Acacia, Cape Gum, Dornbloom ●☆
1298 Acacia hova Drake = Albizia boivinii E. Fourn. ●☆
1299 Acacia howittii F. Muell.;何威特金合欢(黏柄金合欢);Howitt's Wattle, Sticky Wattle ●☆
1300 Acacia humifusa Chiov. = Acacia edgeworthii T. Anderson ●☆
1301 Acacia impervia Gilliland = Acacia zizyphispina Chiov. ●☆
1302 Acacia implexa Benth.;轻木相思树(轻木相思,旋荚相思树,纵错金合欢);Australian Hickory, Lightwood, Lightwood Hickory, Screw-pod Wattle ●☆
1303 Acacia inaequilatera Domin;不等边金合欢●☆
1304 Acacia inconflagrabilis Gerstner = Acacia karroo Hayne ●☆
1305 Acacia inermis Marloth = Albizia anthelmintica Brongn. ●
1306 Acacia intermedia A. Cunn. = Angophora floribunda (Sm.) Sweet ●☆
1307 Acacia intermedia A. Cunn. ex Hook. = Angophora floribunda (Sm.) Sweet ●☆
1308 Acacia intsia (L.) Willd.;印度金合欢(臭菜藤,藤相思树,印度相思树);Climbing Acacia, Indian Wattle, Soap-bark ●
1309 Acacia intsia (L.) Willd. var. caesia (L.) Wight et Arn. = Acacia caesia (L.) Roxb. ex Wall. ●
1310 Acacia intsia (L.) Willd. var. oxyphylla Geahan ex Baker = Acacia caesia (L.) Willd. ●
1311 Acacia intsia Willd. var. caesia (L.) Wight et Arn. ex Baker = Acacia caesia (L.) Willd. ●
1312 Acacia irrorata Sieber ex Spreng.;露珠金合欢●☆
1313 Acacia isenbergiana Schimp. ex A. Rich. = Albizia isenbergiana (A. Rich.) E. Fourn. ●☆
1314 Acacia iteaphylla Benth.;鼠刺叶金合欢;Flinders Range Wattle ●☆
1315 Acacia ixiophylla Benth.;艾克叶金合欢●☆
1316 Acacia jaubertiana (Fourn.) Baill. = Albizia jaubertiana Fourn. ●☆
1317 Acacia joachimii Harms = Acacia goetzei Harms subsp. microphylla Brenan ●☆
1318 Acacia julibrissin Durazz. var. mollis (Wall.) Benth. = Albizia mollis (Wall.) Boivin ●
1319 Acacia julibrissin Willd. = Albizia julibrissin (Willd.) Durazz. ●
1320 Acacia julibrissin Willd. var. mollis Wall. = Albizia mollis (Wall.) Boivin ●
1321 Acacia juniperina Willd.;柏叶相思树(杜松相思树);Prickly Moses, Prickly Wattle ●☆
1322 Acacia kalachariensis Schinz = Dichrostachys cinerea (L.) Wight et Arn. var. setulosa (Welw. ex Oliv.) Brenan et Brummitt ●☆
1323 Acacia kamerunensis Gand.;喀麦隆金合欢●☆
1324 Acacia karroo Hayne;卡路金合欢(甜刺金合欢);Cape Gum, Karroo Thorn, Mimosa Thorn, Sweet Thorn ●☆
1325 Acacia karroo Hayne var. transvaalensis (Burtt Davy) Burtt Davy = Acacia karroo Hayne ●☆
1326 Acacia katangensis De Wild. = Acacia sieberiana DC. var. woodii (Burtt Davy) Keay et Brenan ●☆
1327 Acacia kempeana F. Muell.;澳大利亚金合欢;Witchetty Bush ●☆
1328 Acacia kettlewelliae Maiden;凯特威氏相思树;Buffalo Wattle ●☆
1329 Acacia kinionge De Wild. = Acacia goetzei Harms subsp. microphylla Brenan ●☆
1330 Acacia kirkii Oliv.;柯克金合欢●☆
1331 Acacia kirkii Oliv. subsp. mildbraedii (Harms) Brenan;米尔金合欢●☆
1332 Acacia kirkii Oliv. var. intermedia Brenan = Acacia kirkii Oliv. ●☆
1333 Acacia kirkii Oliv. var. sublaevis Brenan;近平滑金合欢●☆
1334 Acacia koa A. Gray;夏威夷金合欢(柯阿金合欢);Koa, Koa Acacia ●
1335 Acacia kosiensis P. P. Sw.;柯西金合欢●☆
1336 Acacia kraussiana Meisn. ex Benth. = Acacia nilotica (L.) Willd. ex Delile subsp. kraussiana (Meisn. ex Benth.) Brenan ●☆
1337 Acacia kwebensis N. E. Br. = Acacia erubescens Welw. ex Oliv. ●☆
1338 Acacia laeta R. Br. ex Benth.;愉快金合欢●☆
1339 Acacia lahai Steud. et Hochst. ex Benth.;红金合欢;Red Thorn ●☆
1340 Acacia lanigera A. Cunn.;绒毛金合欢;Woolly Wattle ●☆
1341 Acacia lasiopetala Oliv.;毛瓣金合欢●☆
1342 Acacia lathouwersii Staner = Acacia drepanolobium Harms ex Y. Sjostedt ●☆
1343 Acacia latistipulata Harms;宽托叶金合欢●☆
1344 Acacia latronum (L. f.) Willd. subsp. benadirensis (Chiov.) Brenan = Acacia horrida (L.) Willd. subsp. benadirensis (Chiov.) Hillc. et Brenan ●☆
1345 Acacia lebbeck (L.) Willd. = Albizia lebbeck (L.) Benth. ●
1346 Acacia lebbekoides DC. = Albizia lebbekoides (DC.) Benth. ●☆
1347 Acacia lenticularis Buch. -Ham.;大叶楞皮树●
1348 Acacia leprosa Sieber ex DC.;皮屑相思树(皮屑金合欢);Cinnamon Wattle ●☆
1349 Acacia leptocarpa A. Cunn. ex Benth.;细果金合欢(细果荆)●☆
1350 Acacia leucacantha Vatke = Acacia zanzibarica (S. Moore) Taub. ●☆
1351 Acacia leucophloea (Roxb.) Willd.;白韧金合欢(白韧相思树);Whitebark Acacia, Whitephloem Acacia ●
1352 Acacia leucophloea Willd. = Acacia leucophloea (Roxb.) Willd. ●
1353 Acacia leucospira Brenan;白刺金合欢●☆
1354 Acacia ligulata Aiton ex Steud.;具舌金合欢(具舌相思)●☆
1355 Acacia linifolia Willd.;亚麻叶金合欢;Flax Wattle ●☆
1356 Acacia litakunensis Burch. = Acacia tortilis (Forssk.) Hayne subsp. heteracantha (Burch.) Brenan ●☆
1357 Acacia longepetiolata Schinz = Acacia erubescens Welw. ex Oliv. ●☆
1358 Acacia longifolia (Andréws) Willd.;长叶相思树(长叶相思,悉尼相思树);Sydney Acacia, Sydney Golden Wattle, Sydney Wattle, White Sallow ●☆
1359 Acacia longifolia (Andréws) Willd. var. floribunda (Vent.) F. Muell. ex Benth.;多花长叶相思树●☆
1360 Acacia longifolia (Andréws) Willd. var. sophorea (Labill.) F. Muell. ex Benth.;槐相思●☆
1361 Acacia lophantha Willd. = Paraserianthes lophantha (Willd.) I. C. Nielsen ●☆
1362 Acacia luederitzii Engl.;吕德金合欢●☆
1363 Acacia luederitzii Engl. var. retinens (Sim) J. H. Ross et Brenan;密集金合欢●☆
1364 Acacia lujae De Wild.;卢亚金合欢●☆
1365 Acacia lysiphloia F. Muell.;松皮金合欢●☆
1366 Acacia macalusoi Mattei = Acacia rovumae Oliv. ●☆

1367 Acacia macradenia Benth. ;大腺相思●☆

1368 Acacia macrophylla Bunge = Albizia kalkora (Roxb.) Prain ●

1369 Acacia macrophylla Bunge = Albizia macrophylla (Bunge) P. C. Huang ●

1370 Acacia macrostachya Rchb. ex DC. ;大穗相思●☆

1371 Acacia macrostachya Rchb. ex DC. var. spinosissima A. Chev. = Acacia macrostachya Rchb. ex DC. ●☆

1372 Acacia macrothyrsa Harms = Acacia amythethophylla Steud. ex A. Rich. ●☆

1373 Acacia maidenii F. Muell. ;迈氏金合欢;Black Wattle,Maiden's Wattle ●☆

1374 Acacia malacocephala Harms;软头金合欢●☆

1375 Acacia mangium Willd. ;曼荆金合欢(阔叶相思树,马占相思,直干相思树);Forest Mangrove,Mangium,Wattle ●☆

1376 Acacia manubensis J. H. Ross;马努金合欢●☆

1377 Acacia maras Engl. = Acacia tortilis (Forssk.) Hayne subsp. heteracantha (Burch.) Brenan ●☆

1378 Acacia marlothii Engl. = Albizia anthelmintica Brongn. ●

1379 Acacia mbuluensis Brenan;姆布卢金合欢●☆

1380 Acacia mearnsii De Wild. ;黑荆树(澳洲金合欢,黑儿茶,黑荆,黑栲皮树,栲皮树,圣诞树);Black Wattle,Black Wattle Acacia,Late Black Wattle,Mearns Acacia,Wattle ●

1381 Acacia megaladena Desv. ;钝叶金合欢(臭菜藤,臭香椿,黑儿茶,黑栲,柔毛金合欢);Obtuseleaf Acacia,Obtuse-leaved Acacia ●

1382 Acacia megaladena Desv. = Acacia pennata (L.) Willd. ●■

1383 Acacia megaladena Desv. var. garrettii I. C. Nielsen;盘腺金合欢;Garrett Obtuseleaf Acacia ●

1384 Acacia melanoxylon R. Br. ;黑木金合欢(澳洲黑檀,黑荆树,黑木合欢,黑木相思树,迷拉纳栲,丝栲,乌木相思);Australia Blackwood,Australian Acacia,Australian Black Wood,Australian Blackwood,Black Acacia,Black Wattle,Black Wood,Blackwood,Blackwood Acacia,E. Australian Blackwood,Green Wattle,Swart Hoat ●

1385 Acacia mellei I. Verd. = Acacia hereroensis Engl. ●☆

1386 Acacia mellifera (Vahl) Benth. ;具蜜金合欢;Honeyed Acacia,Wait-a-bit Thorn ●

1387 Acacia mellifera (Vahl) Benth. subsp. detinens (Burch.) Brenan;代亭金合欢●☆

1388 Acacia mellifera Benth. = Acacia mellifera (Vahl) Benth. ●

1389 Acacia menabeensis Villiers et Du Puy;微花金合欢●☆

1390 Acacia meridionalis Villiers et Du Puy;南方金合欢●☆

1391 Acacia merkeri Harms = Acacia oerfota (Forssk.) Schweinf. ●☆

1392 Acacia mildbraedii Harms = Acacia kirkii Oliv. subsp. mildbraedii (Harms) Brenan ●☆

1393 Acacia milleriana Standl. ;米勒金合欢●☆

1394 Acacia millifolia S. Watson;蕨叶金合欢;Fernleaf Acacia ●☆

1395 Acacia minuta (M. E. Jones) R. M. Beauch. ;极小金合欢;Mimosa Bush,Sweet Wattle ●☆

1396 Acacia minutifolia Drake = Acacia menabeensis Villiers et Du Puy ●☆

1397 Acacia misera Vatke = Acacia reficiens Wawra subsp. misera (Vatke) Brenan ●☆

1398 Acacia modesta Wall. ;适度金合欢●☆

1399 Acacia moggii Thulin et Tardelli;莫格金合欢●☆

1400 Acacia mollis Wall. = Albizia mollis (Wall.) Boivin ●

1401 Acacia mollissima Willd. ;柔毛金合欢(黑栲,黑栲皮树,栲皮树);Black Wattle ●

1402 Acacia mollissima Willd. = Acacia mearnsii De Wild. ●

1403 Acacia mollissima Willd. = Acacia megaladena Desv. ●

1404 Acacia monga De Wild. = Acacia sieberiana DC. var. woodii (Burtt Davy) Keay et Brenan ●☆

1405 Acacia montana P. P. Sw. = Acacia theronii P. P. Sw. ●☆

1406 Acacia monticola Brenan et Exell = Acacia montigena Brenan ●☆

1407 Acacia montigena Brenan;山地金合欢●☆

1408 Acacia morondavensis Drake = Acacia rovumae Oliv. ●☆

1409 Acacia mossambicensis Bolle = Faidherbia albida (Delile) A. Chev. ■☆

1410 Acacia muelleriana Maiden et R. T. Baker;缪勒金合欢;Mueller's Wattle ●☆

1411 Acacia multijuga Meisn. = Acacia caffra (Thunb.) Willd. ●☆

1412 Acacia myrmecophila R. Vig. ;蚂蚁金合欢●☆

1413 Acacia myrtifolia Willd. ;香桃木金合欢(夹竹桃叶相思);Myrtle Wattle,Myrtle-leaved Mimosa ●☆

1414 Acacia natalitia E. Mey. ;纳塔利特金合欢●☆

1415 Acacia nebneb Adans. = Acacia nilotica (L.) Willd. ex Delile subsp. tomentosa (Benth.) Brenan ●☆

1416 Acacia neboueb Baill. = Acacia nilotica (L.) Willd. ex Delile subsp. tomentosa (Benth.) Brenan ●☆

1417 Acacia nebrownii Burtt Davy;内布金合欢●☆

1418 Acacia nefasia (Hochst. ex A. Rich.) Schweinf. = Acacia sieberiana DC. var. woodii (Burtt Davy) Keay et Brenan ●☆

1419 Acacia negrii Pic. Serm. ;内格里金合欢●☆

1420 Acacia nemu Willd. = Albizia julibrissin Durazz. ●

1421 Acacia neovernicosa Isely;黏金合欢;Viscid Acacia ●☆

1422 Acacia neriifolia A. Cunn. ex Benth. ;夹竹桃叶金合欢;Bald Acacia,Oleander Wattle,Silver Wattle ●☆

1423 Acacia nervosula Chiov. = Albizia obbadiensis (Chiov.) Brenan ●☆

1424 Acacia nigrescens Oliv. ;瘤金合欢;Knob Thorn,Knob-thorn ●☆

1425 Acacia nigrescens Oliv. var. pallens Benth. = Acacia nigrescens Oliv. ●☆

1426 Acacia nilotica (L.) Delile = Acacia nilotica (L.) Willd. ex Delile ●

1427 Acacia nilotica (L.) Willd. ex Delile;阿拉伯金合欢(阿拉伯胶金合欢,阿拉伯胶树,阿拉伯相思树,阿联金合欢,埃及相思树,胶树,西非金合欢);Arabian Acacia,Babul Acacia,Babul Bark,Babul Tree,Egyptian Mimosa,Egyptian Thorn,Gambia Pods,Gum Arabic Tree,Gum-arabic Tree,Motse,Sant Pods,Santwood,Snut,Thorny Acacia ●

1428 Acacia nilotica (L.) Willd. ex Delile subsp. adansonii (Guillaumin et Perr.) Brenan = Acacia nilotica (L.) Willd. ex Delile subsp. adstringens (Schumach. et Thonn.) Roberty ●☆

1429 Acacia nilotica (L.) Willd. ex Delile subsp. adstringens (Schumach. et Thonn.) Roberty;收敛金合欢●☆

1430 Acacia nilotica (L.) Willd. ex Delile subsp. indica (Benth.) Brenan;印度阿拉伯金合欢●☆

1431 Acacia nilotica (L.) Willd. ex Delile subsp. kraussiana (Benth.) Brenan = Acacia nilotica (L.) Willd. ex Delile subsp. kraussiana (Meisn. ex Benth.) Brenan ●☆

1432 Acacia nilotica (L.) Willd. ex Delile subsp. kraussiana (Meisn. ex Benth.) Brenan;克劳斯金合欢●☆

1433 Acacia nilotica (L.) Willd. ex Delile subsp. leiocarpa Brenan;光果金合欢●☆

1434 Acacia nilotica (L.) Willd. ex Delile subsp. subalata (Vatke) Brenan;苏巴金合欢●☆

1435 Acacia nilotica (L.) Willd. ex Delile subsp. tomentosa (Benth.)

Brenan;毛阿拉伯金合欢●☆
1436 Acacia nilotica (L.) Willd. ex Delile var. adansoniana (Dubard) A. F. Hill = Acacia nilotica (L.) Willd. ex Delile subsp. adstringens (Schumach. et Thonn.) Roberty ●☆
1437 Acacia nilotica (L.) Willd. ex Delile var. adansonii (Guillaumin et Perr.) Kuntze = Acacia nilotica (L.) Willd. ex Delile subsp. adstringens (Schumach. et Thonn.) Roberty ●☆
1438 Acacia nilotica (L.) Willd. ex Delile var. adstringens Roberty = Acacia nilotica (L.) Willd. ex Delile subsp. adstringens (Schumach. et Thonn.) Roberty ●☆
1439 Acacia nilotica (L.) Willd. ex Delile var. indica (Benth.) A. F. Hill = Acacia nilotica (L.) Willd. ex Delile subsp. indica (Benth.) Brenan ●☆
1440 Acacia nilotica (L.) Willd. ex Delile var. kirkii (Oliv.) Roberty = Acacia kirkii Oliv. ●☆
1441 Acacia nilotica (L.) Willd. ex Delile var. kraussiana (Benth.) A. F. Hill = Acacia nilotica (L.) Willd. ex Delile subsp. kraussiana (Benth.) Brenan ●☆
1442 Acacia nilotica (L.) Willd. ex Delile var. vera Roberty = Acacia nilotica (L.) Willd. ex Delile ●
1443 Acacia nossibiensis Drake = Acacia hildebrandtii (Vatke) Drake ●☆
1444 Acacia notabilis F. Muell.;显著金合欢;Notable Wattle ●☆
1445 Acacia nubica Benth. = Acacia oerfota (Forssk.) Schweinf. ●☆
1446 Acacia obbiadensis Chiov. = Albizia obbadiensis (Chiov.) Brenan ●☆
1447 Acacia occidentalis Rose;西部金合欢;Sonoran Tree Catclaw ●☆
1448 Acacia ochracea Thulin et Hassan;淡黄褐金合欢●☆
1449 Acacia oerfota (Forssk.) Schweinf.;高林金合欢●☆
1450 Acacia oerfota (Forssk.) Schweinf. var. brevifolia Boulos;短叶高林金合欢●☆
1451 Acacia ogadensis Chiov.;欧加金合欢●☆
1452 Acacia oliveri Vatke = Acacia senegal (L.) Willd. ●
1453 Acacia ormocarpoides P. J. H. Hurter;链荚木金合欢●☆
1454 Acacia ornithophora Sweet = Acacia paradoxa Chiov. ●☆
1455 Acacia oxycedrus Sieber ex DC.;穗花金合欢;Spike Wattle ●☆
1456 Acacia oxysprion Chiov.;尖籽金合欢●☆
1457 Acacia pallens (Benth.) Rolfe = Acacia nigrescens Oliv. ●☆
1458 Acacia paniculata J. F. Macbr.;圆锥花序金合欢(圆锥相思树);Sunshine Wattle ●☆
1459 Acacia paolii Chiov.;保尔金合欢●☆
1460 Acacia paolii Chiov. subsp. paucijuga Brenan;少轭金合欢●☆
1461 Acacia pappii Gand. = Acacia tortilis (Forssk.) Hayne var. spirocarpa (Hochst. ex A. Rich.) Brenan ●☆
1462 Acacia paradoxa Chiov. = Acacia hamulosa Benth. ●☆
1463 Acacia parramattensis Tindale;南威尔士金合欢;South Wales Wattle ●☆
1464 Acacia passargei Harms = Acacia nigrescens Oliv. ●☆
1465 Acacia pendula A. Cunn. ex G. Don;垂枝相思树;Boree, Myall, Myall-wood, Se Bolt, Weeping Acacia, Weeping Myall, Weeping Wattle ●☆
1466 Acacia pennata (L.) Willd.;羽叶金合欢(倒钩藤,红皮毒鱼藤,红藤,加力酸藤,加烈酸藤,龙骨刺,南蛇簕,南蛇簕藤,蛇藤,细南蛇);Plume Acacia ●■
1467 Acacia pennata (L.) Willd. = Acacia pendula A. Cunn. ex G. Don ●☆
1468 Acacia pennata (L.) Willd. subsp. hainanensis (Hayata) I. C. Nielsen;海南羽叶金合欢●
1469 Acacia pennata (L.) Willd. var. arrophula (D. Don ex Wall.) Baker = Acacia megaladena Desv. ●
1470 Acacia pennata (L.) Willd. var. arrophula (D. Don) Baker = Acacia megaladena Desv. ●
1471 Acacia pennata (L.) Willd. var. arrophula (D. Don) Baker = Acacia pennata (L.) Willd. ●■
1472 Acacia pennata (L.) Willd. var. dregeana Benth. = Acacia brevispica Harms subsp. dregeana (Benth.) Brenan ●☆
1473 Acacia pennatula (Schltdl. et Cham.) Benth.;羽金合欢;Fernleaf Acacia, Huizache, Sierra Madre Acacia ●☆
1474 Acacia penninervis Sieber ex DC.;羽脉相思树;Blackwood, Mountain Hickory ●☆
1475 Acacia pentagona (Schumach. et Thonn.) Hook. f.;五角金合欢●☆
1476 Acacia pentaptera Welw. = Acacia pentagona (Schumach. et Thonn.) Hook. f. ●☆
1477 Acacia permixta Burtt Davy;混乱金合欢●☆
1478 Acacia permixta Burtt Davy var. glabra ? = Acacia tenuispina I. Verd. ●☆
1479 Acacia perrieri Drake = Albizia perrieri (Drake) R. Vig. ex Capuron ●☆
1480 Acacia perrottii Warb. = Acacia nigrescens Oliv. ●☆
1481 Acacia persiciflora Pax;桃叶金合欢●☆
1482 Acacia pervillei Benth.;佩里耶金合欢●☆
1483 Acacia petersiana Bolle = Acacia tortilis (Forssk.) Hayne var. spirocarpa (Hochst. ex A. Rich.) Brenan ●☆
1484 Acacia petrensis Thulin;皮特拉金合欢●☆
1485 Acacia peuce F. Muell.;棒状金合欢;Waddy Wood ●☆
1486 Acacia pilispina Pic. Serm.;毛刺金合欢●☆
1487 Acacia podalyriifolia A. Cunn. = Acacia podalyriifolia A. Cunn. ex G. Don ●☆
1488 Acacia podalyriifolia A. Cunn. ex G. Don;珍珠相思树(真珠相思树);Mount Morgan Wattle, Mountain Morgan Wattle, Mt. Morgan Wattle, Pearl Acacia, Pearl Wattle, Queensland Silver Wattle, Queensland Wattle ●☆
1489 Acacia polhillii Villiers et Du Puy;普尔金合欢●☆
1490 Acacia polyacantha Willd.;多花相思树(多刺相思树);Catechu Tree, Falcon's Claw Acacia ●☆
1491 Acacia polyacantha Willd. subsp. campylacantha (Hochst. ex A. Rich.) Brenan = Acacia campylacantha Hochst. ex A. Rich. ●☆
1492 Acacia polystachya A. Cunn. ex Benth.;多穗金合欢●☆
1493 Acacia prasinata Hunde;草绿金合欢●☆
1494 Acacia pravissima F. Muell.;极弯相思树(三角叶相思树);Ovens Acacia, Ovens Wattle, Wedge-leaved Wattle, Winged Acacia ●☆
1495 Acacia procera (Roxb.) Willd. = Albizia procera (Roxb.) Benth. ●
1496 Acacia prominens A. Cunn. ex G. Don;凸起金合欢(凸起相思树);Golden Rain Wattle, Golden-rain Wattle, Gosford Wattle ●☆
1497 Acacia prorsispinula Stapf = Acacia amythethophylla Steud. ex A. Rich. ●☆
1498 Acacia provincialis A. Camus = Acacia retinodes Schltdl. ●☆
1499 Acacia pruinescens Kurz;粉背金合欢(粉被金合欢);Glaucousleaf Wattle, Pruinose Acacia ●
1500 Acacia pruinescens Kurz var. luchunensis C. Chen et H. Sun;阔叶粉背金合欢;Luchun Glaucousleaf Wattle ●
1501 Acacia pruinescens Kurz var. luchunensis C. Chen et H. Sun = Acacia pruinescens Kurz ●
1502 Acacia pruinosa A. Cunn. ex Benth.;粉茎相思树;Frosty Wattle

●☆
1503 Acacia pseudofistula Harms;假管金合欢●☆
1504 Acacia pseudoglauca Chiov.;灰绿金合欢●☆
1505 Acacia pseudonigrescens Brenan et J. H. Ross;假变黑金合欢●☆
1506 Acacia pseudosocotrana Chiov. = Acacia edgeworthii T. Anderson ●☆
1507 Acacia psoralea DC. = Mimosa psoralea (DC.) Benth. ●☆
1508 Acacia pubescens (Vent.) R. Br.;毛金合欢;Downy Wattle, Hairy Wattle ●☆
1509 Acacia pubescens W. T. Aiton = Acacia pubescens (Vent.) R. Br. ●☆
1510 Acacia puccioniana Chiov. = Dichrostachys kirkii Benth. ●☆
1511 Acacia pulchella R. Br.;艳美相思树;Western Prickly Moses ●☆
1512 Acacia pulverulenta Schltdl. = Leucaena pulverulenta (Schltdl.) Benth. ●☆
1513 Acacia purpurascens Vatke = Acacia sieberiana DC. ●☆
1514 Acacia purpurea Bolle;紫金合欢●☆
1515 Acacia pustula Maiden et Blakely;普斯土拉金合欢●☆
1516 Acacia pycnantha Benth.;密花金合欢(金儿茶,金榜);Broadleaf Wattle, Denseflower Acacia, Golden Wattle ●
1517 Acacia quintanilhae Torre;金塔合欢●☆
1518 Acacia raddiana Savi;拉德合欢;Shittim Wood ●☆
1519 Acacia raddiana Savi = Acacia tortilis (Forssk.) Hayne var. raddiana (Savi) Brenan ●☆
1520 Acacia raddiana Savi var. pubescens (A. Chev.) A. F. Hill = Acacia tortilis (Forssk.) Hayne var. pubescens A. Chev. ●☆
1521 Acacia redacta J. H. Ross = Calliandra redacta (J. H. Ross) Thulin et Asfaw ●☆
1522 Acacia redolens Maslin;沙地金合欢;Bank Catclaw, Desert Carpet, Prostrate Acacia ●☆
1523 Acacia reficiens Wawra;圆叶金合欢;Round-leaf Wattle, Wavy-leaf Wattle ●☆
1524 Acacia reficiens Wawra subsp. misera (Vatke) Brenan;瘦弱沙地金合欢●☆
1525 Acacia rehmanniana Schinz;丝绒金合欢;Silky Acacia ●☆
1526 Acacia reticulata (L.) Willd. = Acacia karroo Hayne ●☆
1527 Acacia retinens Sim = Acacia luederitzii Engl. var. retinens (Sim) J. H. Ross et Brenan ●☆
1528 Acacia retinodes Schltdl.;树胶状相思树(多花金合欢,多花楞,多花相思);Cooba, Everblooming Acacia, Flowery Acacia, Ovens Wattle, Sally Wattle, Water Wattle, White Sallow Wattle, Wirilda ●☆
1529 Acacia rhetinocarpa J. M. Black;脂金合欢;Wattle ●☆
1530 Acacia richii A. Gray;理氏金合欢(台湾相思)●☆
1531 Acacia rigens A. Cunn. ex G. Don;刚硬金合欢;Needle Wattle, Needle-bush Wattle ●☆
1532 Acacia rigidula Benth.;黑果金合欢;Blackbrush, Blackbrush Acacia, Chaparro Prieto ●☆
1533 Acacia robbertsii P. P. Sw.;罗伯茨金合欢●☆
1534 Acacia robecchii Pirotta;罗贝克金合欢●☆
1535 Acacia robusta Burch.;粗壮金合欢●☆
1536 Acacia robusta Burch. subsp. clavigera (E. Mey.) Brenan;棍棒粗壮金合欢●☆
1537 Acacia robusta Burch. subsp. usambarensis (Taub.) Brenan;乌桑巴拉金合欢●☆
1538 Acacia robynsiana Merxm. et A. Schreib.;罗默金合欢;Roemer Acacia, Roemer Catclaw, Roemer's Acacia ●☆
1539 Acacia roemeriana Scheele = Acacia robynsiana Merxm. et A. Schreib. ●☆
1540 Acacia rogersii Burtt Davy = Acacia nebrownii Burtt Davy ●☆
1541 Acacia rostellifera Benth.;臭金合欢;Skunk Tree, Summer Scented Wattle ●☆
1542 Acacia rostrata Sim = Acacia senegal (L.) Willd. var. rostrata Brenan ●☆
1543 Acacia rovumae Oliv.;洛夫金合欢●☆
1544 Acacia rubida A. Cunn.;变红金合欢;Red-leaved Wattle, Red-stem Acacia, Red-stemmed Acacia ●☆
1545 Acacia rufobrunnea N. E. Br. = Acacia arenaria Schinz ●☆
1546 Acacia rugata (Lam.) Buch. -Ham. ex Voigt = Acacia concinna (Willd.) DC. ●■
1547 Acacia rugata (Lam.) Buch. -Ham. ex Voigt = Acacia sinuata (Lour.) Merr. ●■
1548 Acacia rupestris Stocks ex Boiss. = Acacia senegal (L.) Willd. ●
1549 Acacia rupicola F. Muell.;岩生金合欢●☆
1550 Acacia sacleuxii A. Chev. = Acacia robusta Burch. subsp. usambarensis (Taub.) Brenan ●☆
1551 Acacia sakalava Drake;萨卡相思树●☆
1552 Acacia salicina Lindl.;柳相思树;Australian Willow, Cooba, Willow Acacia, Willow Wattle ●☆
1553 Acacia saligna (Lindl.) H. L. Wendl.;金环相思树(垂柳相思,柳叶金合欢);Blue Leaf Wattle, Cooba, Golden-wreath Wattle, Orange Wattle, Port Jackson, Port Jackson Willow, Weeping Wattle, Western Australian Golden Wattle ●☆
1554 Acacia salinarum DC.;盐地金合欢●☆
1555 Acacia saltiana Steud. = Acacia lahai Steud. et Hochst. ex Benth. ●☆
1556 Acacia sambesiaca Schinz = Acacia robusta Burch. subsp. clavigera (E. Mey.) Brenan ●☆
1557 Acacia samoryana A. Chev. = Acacia dudgeonii Craib ex Holland ●☆
1558 Acacia sanguinea Hochst. ex A. Rich. = Acacia venosa Hochst. ex Benth. ●☆
1559 Acacia sarcophylla Chiov. = Acacia oerfota (Forssk.) Schweinf. var. brevifolia Boulos ●☆
1560 Acacia schaffneri (S. Watson) F. J. Herm.;扭曲金合欢;Huizache Chino, Twisted Acacia ●☆
1561 Acacia schinoides Benth.;拟欣兹金合欢●☆
1562 Acacia schlechteri Harms;施莱金合欢●☆
1563 Acacia schliebenii Harms = Acacia nigrescens Oliv. ●☆
1564 Acacia schweinfurthii Brenan et Exell;施韦金合欢●☆
1565 Acacia sclerophylla Lindl.;硬木相思●☆
1566 Acacia scorpioides (L.) W. Wight = Acacia nilotica (L.) Willd. ex Delile ●
1567 Acacia scorpioides (L.) W. Wight subsp. adstringens (Schumach. et Thonn.) Quézel et Santa = Acacia nilotica (L.) Willd. ex Delile subsp. adstringens (Schumach. et Thonn.) Roberty ●☆
1568 Acacia scorpioides (L.) W. Wight subsp. nilotica (L.) A. Chev. = Acacia nilotica (L.) Willd. ex Delile ●
1569 Acacia scorpioides (L.) W. Wight var. pubescens A. Chev. = Acacia nilotica (L.) Willd. ex Delile subsp. tomentosa (Benth.) Brenan ●☆
1570 Acacia sekhukhuniensis P. J. H. Hurter;塞库金合欢●☆
1571 Acacia semlikiensis De Wild. = Acacia kirkii Oliv. subsp. mildbraedii (Harms) Brenan ●☆
1572 Acacia senegal (L.) Willd.;阿拉伯胶树(刺合欢);Arab Gumtree, Arabic Acacia, Arabic Gum, Arabic Gum Tree, Arabic Gumtree, Arabic Tree, Gum Acacia, Gum Arabic, Gum Senegal,

Gumarabic Acacia, Gum-arabic Acacia, Gum-arabic Tree, Kolhol Gum, Senegal Gum, Sudan Gum Arabic Tree, Sudan Gum-arabic, Tree-horned Acacia ●

1573 Acacia senegal (L.) Willd. subsp. mellifera (Vahl) Roberty = Acacia mellifera (Vahl) Benth. ●

1574 Acacia senegal (L.) Willd. var. kerensis Schweinf.;克尔阿拉伯胶树●☆

1575 Acacia senegal (L.) Willd. var. laeta (R. Br. ex Benth.) Roberty = Acacia laeta R. Br. ex Benth. ●☆

1576 Acacia senegal (L.) Willd. var. leiorhachis Brenan;平滑轴阿拉伯胶树●☆

1577 Acacia senegal (L.) Willd. var. leiorhachis Brenan = Acacia circummarginata Chiov. ●☆

1578 Acacia senegal (L.) Willd. var. platysprion Chiov. = Acacia senegal (L.) Willd. ●

1579 Acacia senegal (L.) Willd. var. pseudoglaucophylla Chiov. = Acacia senegal (L.) Willd. ●

1580 Acacia senegal (L.) Willd. var. rostrata Brenan;具喙阿拉伯胶树●☆

1581 Acacia senegal (L.) Willd. var. rupestris (Stocks ex Boiss.) Roberty = Acacia senegal (L.) Willd. ●

1582 Acacia senegal (L.) Willd. var. samoryana (A. Chev.) Roberty = Acacia dudgeonii Craib ex Holland ●☆

1583 Acacia senegal (L.) Willd. var. sanguinea (Hochst. ex A. Rich.) Roberty = Acacia venosa Hochst. ex Benth. ●☆

1584 Acacia senegal (L.) Willd. var. venosa (Hochst. ex Benth.) Roberty = Acacia venosa Hochst. ex Benth. ●☆

1585 Acacia senegal (L.) Willd. var. verek (Guillaumin et Perr.) Roberty = Acacia senegal (L.) Willd. ●

1586 Acacia sennii Chiov. = Acacia zanzibarica (S. Moore) Taub. ●☆

1587 Acacia sericocephala Fenzl = Albizia amara (Roxb.) Boivin subsp. sericocephala (Benth.) Brenan ●☆

1588 Acacia seyal Delile;多刺金合欢(塞伊尔金合欢,塞伊尔相思树);Gum-arabic Tree, Shittim Wood, Tahl Gum, Thirty Thorn, Whistling Tree ●☆

1589 Acacia seyal Delile var. fistula (Schweinf.) Oliv.;管状多刺金合欢●☆

1590 Acacia seyal Delile var. multijuga Schweinf. ex Baker f. = Acacia hockii De Wild. ●☆

1591 Acacia seyal Delile. f. fistula (Schweinf.) Cufod. = Acacia seyal Delile var. fistula (Schweinf.) Oliv. ●☆

1592 Acacia sieberiana DC.;西伯尔金合欢●☆

1593 Acacia sieberiana DC. f. eusieberiana Roberty = Acacia sieberiana DC. ●☆

1594 Acacia sieberiana DC. f. nefasia (Hochst. ex A. Rich.) Roberty = Acacia sieberiana DC. var. woodii (Burtt Davy) Keay et Brenan ●☆

1595 Acacia sieberiana DC. var. kagerensis Troupin = Acacia sieberiana DC. var. woodii (Burtt Davy) Keay et Brenan ●☆

1596 Acacia sieberiana DC. var. orientalis Troupin = Acacia sieberiana DC. var. woodii (Burtt Davy) Keay et Brenan ●☆

1597 Acacia sieberiana DC. var. sing (Guillaumin et Perr.) Roberty = Acacia sieberiana DC. ●☆

1598 Acacia sieberiana DC. var. vermoesenii (De Wild.) Keay et Brenan = Acacia sieberiana DC. var. woodii (Burtt Davy) Keay et Brenan ●☆

1599 Acacia sieberiana DC. var. vermoesenii (De Wild.) Troupin = Acacia sieberiana DC. var. woodii (Burtt Davy) Keay et Brenan ●☆

1600 Acacia sieberiana DC. var. villosa A. Chev.;长柔毛西伯尔金合欢●☆

1601 Acacia sieberiana DC. var. woodii (Burtt Davy) Keay et Brenan;伍德金合欢(屋得金合欢);Paperbark Thorn, Paper-bark Thorn ●☆

1602 Acacia sieberiana Tausch;喜氏相思;Umbrella Thorn ●☆

1603 Acacia silvestris Tindale;林金合欢●☆

1604 Acacia silvicola G. C. C. Gilbert et Boutique = Acacia pentagona (Schumach. et Thonn.) Hook. f. ●☆

1605 Acacia sing Guillaumin et Perr. = Acacia sieberiana DC. ●☆

1606 Acacia sinuata (Lour.) Merr.;缺损金合欢(藤金合欢)●☆

1607 Acacia socotrana Balf. f. = Acacia edgeworthii T. Anderson ●☆

1608 Acacia solenota Pedley;昆士兰金合欢●☆

1609 Acacia songwensis Harms = Acacia xanthophloea Benth. ●☆

1610 Acacia sophorae R. Br.;海滨金合欢;Coast Wattle, Coastal Wattle ●☆

1611 Acacia spadicigera Cham. et Schltdl.;牛角金合欢;Bull's Horn Acacia ●☆

1612 Acacia spectabilis A. Cunn. ex Benth.;显著相思树;Glory Wattle, Mudgee Wattle ●☆

1613 Acacia sphaerocephala Cham. et Schltdl.;圆头金合欢(蚁荆);Bee Wattle, Bull's-horn Acacia ●☆

1614 Acacia spinosa E. Mey. = Dichrostachys cinerea (L.) Wight et Arn. var. africana Brenan et Brummitt ●☆

1615 Acacia spirocarpa Hochst. ex A. Rich.;螺果金合欢;Coliledfruit Acacia, Spiral-fruit Gum ●

1616 Acacia spirocarpa Hochst. ex A. Rich. = Acacia tortilis (Forssk.) Hayne var. spirocarpa (Hochst. ex A. Rich.) Brenan ●☆

1617 Acacia spirocarpoides Engl. = Acacia tortilis (Forssk.) Hayne subsp. heteracantha (Burch.) Brenan ●☆

1618 Acacia spirorbis Labill.;螺旋相思树●☆

1619 Acacia stefaninii Chiov. = Acacia reficiens Wawra subsp. misera (Vatke) Brenan ●☆

1620 Acacia stellata (Forssk.) Willd. = Pterolobium stellatum (Forssk.) Brenan ●☆

1621 Acacia stenocarpa Hochst. ex A. Rich. var. boboensis Aubrév. = Acacia hockii De Wild. ●☆

1622 Acacia stenocarpa Hochst. ex A. Rich. var. chariensis (A. Chev.) Aubrév. = Acacia hockii De Wild. ●☆

1623 Acacia stenophylla A. Cunn. ex Benth.;狭叶相思树;Eumong, River Cooba Wattle, Shoestring Acacia ●☆

1624 Acacia stipulacea Royle = Albizia chinensis (Osbeck) Merr. ●

1625 Acacia stipulata DC. = Albizia chinensis (Osbeck) Merr. ●

1626 Acacia stolonifera Burch. = Acacia hebeclada DC. ●☆

1627 Acacia stolonifera Burch. var. chobiensis O. B. Mill. = Acacia hebeclada DC. subsp. chobiensis (O. B. Mill.) A. Schreib. ●☆

1628 Acacia stuhlmannii Taub.;斯图尔曼金合欢●☆

1629 Acacia suaresensis Baill. = Albizia polyphylla E. Fourn. ●☆

1630 Acacia suaveolens Willd.;香甜相思树(香甜相思);Sweet Acacia, Sweet Wattle, Sweet-scented Wattle ●☆

1631 Acacia subalata Vatke = Acacia nilotica (L.) Willd. ex Delile subsp. subalata (Vatke) Brenan ●☆

1632 Acacia subporosa F. Muell. = Acacia cognata Maiden ●☆

1633 Acacia subrhombea Baill. = Albizia viridis E. Fourn. var. zygioides (Baill.) Villiers ●☆

1634 Acacia subtomentosa De Wild. = Acacia gerrardii Benth. ●☆

1635 Acacia sultani Chiov. = Acacia edgeworthii T. Anderson ●☆

1636 Acacia suma Kurz ex Brandis.;苏马金合欢●☆

1637 Acacia swazica Burtt Davy;斯威士金合欢●☆

1638 Acacia taitensis Vatke = Acacia nilotica (L.) Willd. ex Delile

subsp. subalata (Vatke) Brenan ●☆
1639 Acacia tanganyikensis Brenan;坦噶尼喀金合欢●☆
1640 Acacia taylori Brenan et Exell;泰勒金合欢●☆
1641 Acacia tenax Marloth = Acacia mellifera (Vahl) Benth. subsp. detinens (Burch.) Brenan ●☆
1642 Acacia teniana Harms;无刺金合欢(盐丰金合欢);Spineless Acacia,Ten Wattle,Yanfen Acacia ●
1643 Acacia tenuispina I. Verd.;细刺金合欢●☆
1644 Acacia tephrodermis Brenan;灰皮叶金合欢●☆
1645 Acacia tephrophylla Thulin;灰叶金合欢●☆
1646 Acacia terminalis (Salisb.) J. F. Macbr.;顶生金合欢;Cedar Wattle,Peppermint-tree Wattle,Sunshine Wattle ●☆
1647 Acacia tetragonophylla F. Muell.;四棱金合欢(四棱叶金合欢);Dead Finish,Prickly Wattle ●☆
1648 Acacia theronii P. P. Sw.;太龙金合欢●☆
1649 Acacia thomasii Harms;托马斯金合欢●☆
1650 Acacia tindaleae Pedley;金顶金合欢;Golden-top Wattle ●☆
1651 Acacia tonkinensis I. C. Nielsen;滇南金合欢(双腺金合欢);Tonkin Acacia ●
1652 Acacia torrei Brenan;托雷金合欢(托勒金合欢)●☆
1653 Acacia tortilis (Forssk.) Hayne;突尼斯相思树(扭旋金合欢,伞形金合欢);Tunisian Gum Acacia,Umbrella Thorn ●☆
1654 Acacia tortilis (Forssk.) Hayne subsp. heteracantha (Burch.) Brenan;异刺托勒金合欢●☆
1655 Acacia tortilis (Forssk.) Hayne subsp. raddiana (Savi) Brenan = Acacia tortilis (Forssk.) Hayne var. raddiana (Savi) Brenan ●☆
1656 Acacia tortilis (Forssk.) Hayne var. crinita Chiov.;长软毛金合欢●☆
1657 Acacia tortilis (Forssk.) Hayne var. flava Nongon.;黄托勒金合欢●☆
1658 Acacia tortilis (Forssk.) Hayne var. lenticellosa Chiov. = Acacia tortilis (Forssk.) Hayne var. raddiana (Savi) Brenan ●☆
1659 Acacia tortilis (Forssk.) Hayne var. pubescens A. Chev.;短柔毛托勒金合欢●☆
1660 Acacia tortilis (Forssk.) Hayne var. raddiana (Savi) Brenan;拉德金合欢●☆
1661 Acacia tortilis (Forssk.) Hayne var. spirocarpa (Hochst. ex A. Rich.) Brenan;螺果托勒金合欢●☆
1662 Acacia tortilis (Forssk.) Hayne. f. raddiana (Savi) Roberty = Acacia tortilis (Forssk.) Hayne var. raddiana (Savi) Brenan ●☆
1663 Acacia tortilis (Forssk.) Hayne. f. spirocarpa (Hochst. ex A. Rich.) Roberty = Acacia tortilis (Forssk.) Hayne var. spirocarpa (Hochst. ex A. Rich.) Brenan ●☆
1664 Acacia tortilis Hayne = Acacia tortilis (Forssk.) Hayne ●☆
1665 Acacia trachycarpa Pritz.;糙果金合欢(粗果金合欢);Minnieritchie ●☆
1666 Acacia translucens A. Cunn.;半透明金合欢●☆
1667 Acacia trentiniani A. Chev. = Acacia laeta R. Br. ex Benth. ●☆
1668 Acacia trichopetala Drake = Albizia viridis E. Fourn. var. zygioides (Baill.) Villiers ●☆
1669 Acacia triptera Benth.;翅金合欢;Spurwing Wattle, Spur-wing Wattle ●☆
1670 Acacia trispinosa Marloth et Engl. = Acacia senegal (L.) Willd. var. rostrata Brenan ●☆
1671 Acacia tristis Welw. ex Oliv. = Acacia hebeclada DC. subsp. tristis (Welw. ex Oliv.) A. Schreib. ●☆
1672 Acacia tumida F. Muell. ex Benth.;厚荚相思(膨厚荚相思)●☆
1673 Acacia turnbulliana Brenan;特恩布尔金合欢●☆
1674 Acacia ulicifolia (Salisb.) Court = Acacia juniperina Willd. ●☆
1675 Acacia ulugurensis Taub. ex Harms = Acacia goetzei Harms subsp. microphylla Brenan ●☆
1676 Acacia uncinata Engl. = Acacia reficiens Wawra ●☆
1677 Acacia usambarensis Taub. = Acacia robusta Burch. subsp. usambarensis (Taub.) Brenan ●☆
1678 Acacia vanmeelii G. C. C. Gilbert et Boutique = Acacia goetzei Harms subsp. microphylla Brenan ●☆
1679 Acacia venosa Hochst. ex Benth.;多脉金合欢●☆
1680 Acacia vera Garsault;埃及金合欢;Egyptian Thorn ●☆
1681 Acacia veracruzensis Schenck = Acacia sphaerocephala Cham. et Schltdl. ●☆
1682 Acacia verek Guillaumin et Perr. = Acacia senegal (L.) Willd. ●
1683 Acacia vermoesenii De Wild. = Acacia sieberiana DC. var. woodii (Burtt Davy) Keay et Brenan ●☆
1684 Acacia verniciflua A. Cunn.;亮叶相思●☆
1685 Acacia verrucifera Harms = Acacia kirkii Oliv. ●☆
1686 Acacia verticillata (L'Hér.) Willd.;轮生相思树(轮生相思,轮生叶金合欢,针叶相思树);Prickly Moses,Prickly-leaved Wattle,Star Acacia,Whorl-leaved Acacia ●☆
1687 Acacia verugera Schweinf. = Acacia sieberiana DC. ●☆
1688 Acacia verugera Schweinf. var. latisiliqua Harms = Acacia sieberiana DC. var. woodii (Burtt Davy) Keay et Brenan ●☆
1689 Acacia vestita Ker Gawl.;被覆金合欢;Hairy Wattle,Weeping Boree ●☆
1690 Acacia victoriae Benth.;胜利金合欢;Bramble Acacia,Bramble Wattle,Gundabluey ●☆
1691 Acacia vietnamensis I. C. Nielsen;越南金合欢(藤金合欢);Vietnam Acacia ●
1692 Acacia viguieri Villiers et Du Puy;维基耶金合欢●☆
1693 Acacia virchowiana Vatke et Hildebrandt = Acacia oerfota (Forssk.) Schweinf. ●☆
1694 Acacia virgata (L.) Gaertn. = Desmanthus virgatus (L.) Willd. ●■
1695 Acacia viridiramis Burch. = Xerocladia viridiramis (Burch.) Taub. ●☆
1696 Acacia viridis (E. Fourn.) Baill. = Albizia viridis E. Fourn. ●☆
1697 Acacia visco Lorentz ex Griseb.;秘斯卡榜;Visco Acacia ●☆
1698 Acacia volkii Suess. = Acacia senegal (L.) Willd. var. rostrata Brenan ●☆
1699 Acacia wallichiana DC. = Acacia catechu (L. f.) Willd. ●
1700 Acacia walteri Suess. = Acacia nebrownii Burtt Davy ●☆
1701 Acacia walwalensis Gilliland;瓦尔金合欢●☆
1702 Acacia welwitschii Oliv.;韦尔金合欢●☆
1703 Acacia welwitschii Oliv. subsp. delagoensis (Harms) J. H. Ross et Brenan;迪拉果韦尔金合欢●☆
1704 Acacia wilhelmiana F. Muell.;威氏金合欢;Dwarf Nealing,Palo Blanco,Wilhelm's Wattle,Willard's Acacia ●☆
1705 Acacia willardiana Rose;威拉尔德金合欢;Palo Blanco,Willard's Acacia ●☆
1706 Acacia woodii Burtt Davy = Acacia sieberiana DC. var. woodii (Burtt Davy) Keay et Brenan ●☆
1707 Acacia wrightii Benth. ex A. Gray;赖特金合欢;Texas Catclaw,Wright Acacia,Wright Catclaw ●☆
1708 Acacia xanthophloea Benth.;黄皮金合欢;Fever Tree,Naivansha Thorn ●☆
1709 Acacia xiphocarpa Hochst. ex Benth. = Acacia abyssinica Hochst.

ex Benth. ●☆
1710 Acacia xiphoclada Baker = Acacia heterophylla Willd. ●☆
1711 Acacia yunnanensis Franch.;云南金合欢(滇金合欢,合欢树,云南相思,云南相思树);Yunnan Acacia ●
1712 Acacia zanzibarica (S. Moore) Taub.;海岸金合欢;Coast Whistling Thorn ●☆
1713 Acacia zanzibarica (S. Moore) Taub. var. microphylla Brenan;小叶海岸金合欢●☆
1714 Acacia zanzibarica Taub. = Acacia zanzibarica (S. Moore) Taub. ●☆
1715 Acacia zizyphispina Chiov.;枣刺金合欢●☆
1716 Acacia zygioides Baill. = Albizia viridis E. Fourn. var. zygioides (Baill.) Villiers ●☆
1717 Acaciaceae E. Mey. = Fabaceae Lindl.(保留科名)●■
1718 Acaciaceae E. Mey. = Leguminosae Juss.(保留科名)●■
1719 Acaciaceae Schimp.;金合欢科●
1720 Acaciaceae Schimp. = Fabaceae Lindl.(保留科名)●■
1721 Acaciaceae Schimp. = Leguminosae Juss.(保留科名)●■
1722 Acaciella Britton et Rose = Acacia Mill.(保留属名)●■
1723 Acaciella Britton et Rose(1928);小金合欢属●■
1724 Acaciella angustissima (Mill.) Britton et Rose;窄小金合欢●☆
1725 Acaciella bicolor Britton et Rose;二色小金合欢●☆
1726 Acaciella ciliata Britton et Rose;睫毛小金合欢●☆
1727 Acaciella hirta (Nutt.) Britton et Rose = Acacia angustissima (Mill.) Kuntze ●☆
1728 Acaciella laevis Britton et Rose;平滑小金合欢●☆
1729 Acaciella villosa (Sw.) Britton et Rose;小金合欢●
1730 Acaciella villosa (Sw.) Britton et Rose = Acacia glauca (L.) Moench ●
1731 Acaciella villosa (Sw.) Willd. = Acacia glauca (L.) Moench ●
1732 Acaciopsis Britton et Rose = Acacia Mill.(保留属名)●■
1733 Acaciopsis Britton et Rose(1928);拟金合欢属●☆
1734 Acaciopsis pringlei (Rose) Britton et Rose;拟金合欢●☆
1735 Acacium Steud. = Aracium Neck. ■
1736 Acacium Steud. = Crepis L. ■
1737 Acadella Britton et Rose = Acacia Mill.(保留属名)●■
1738 Acaena L. = Acaena Mutis ex L. ●■☆
1739 Acaena Mutis ex L.(1771);芒刺果属(刺果薇属,红刺头属,猬莓属);Acaena, Bidgee-widgee, Bidi-bidi, New Zeyland Bur, Pirripirri-bur, Sheep-bur ●■☆
1740 Acaena anserinifolia Druce = Acaena novae-zelandiae Kirk ●☆
1741 Acaena argentea Ruiz et Pav.;银色芒刺果●☆
1742 Acaena boliviana Gand.;玻利维亚芒刺果●☆
1743 Acaena buchananii Hook. f.;绿猬莓●☆
1744 Acaena caesiglauca (Bitter) Bergmans;灰蓝芒刺果(天蓝猬莓);Gray Sheepbur, Gray-glaucous Acaena, New Zealand Bur, Sheer Burr ●☆
1745 Acaena cylindristachya Ruiz et Pav.;柱穗芒刺果●☆
1746 Acaena elongata L.;伸长芒刺果●☆
1747 Acaena eupatoria Bitter;泽兰芒刺果●☆
1748 Acaena inermis Hook. f. = Acaena microphylla Hook. f. ●☆
1749 Acaena latebrosa Aiton;横卧芒刺果●☆
1750 Acaena macrorhiza Hook. f. = Acaena cylindristachya Ruiz et Pav. ●☆
1751 Acaena microphylla Hook. f.;小叶红刺头(无刺芒刺果,小叶猬莓);Sheep Burr, Small-leaf Acaena, Spineless Acaena ●☆
1752 Acaena novae-zelandiae Kirk;新西兰芒刺果(藜叶芒刺果,新西兰猬莓);Biddy-biddy, Bidgee-widgee, Bidibid, Bidi-bidi, Bronze Pirri-pirri-bur, New Zealand Bur, Pirri-pirri Bur, Pirri-pirri-bur ●☆
1753 Acaena ovalifolia Ruiz et Pav.;卵叶芒刺果;Two-spined Acaena ●☆
1754 Acaena ovalifolia Ruiz et Pav. var. subserrata Bitter = Acaena ovalifolia Ruiz et Pav. ●☆
1755 Acaena ovalis Pers. = Acaena ovalifolia Ruiz et Pav. ●☆
1756 Acaena pallida (Kirk) Allan;苍白芒刺果;Pale Biddy-biddy ●☆
1757 Acaena pinnata Citerne;羽状芒刺果●☆
1758 Acaena pusilla (Bitter) Allan = Acaena novae-zelandiae Kirk ●☆
1759 Acaena stricta Griseb.;直立芒刺果●☆
1760 Acaena subincisa Wedd.;小刺芒刺果●☆
1761 Acaenops (Schrad.) Schrad. ex Fourr. = Virga Hill ■
1762 Acaenops Schrad. ex Fourr. = Virga Hill ■
1763 Acaenops Schrad. ex Steud. = Dipsacus L. ■
1764 Acajou Mill. = Anacardium L. ●
1765 Acajou Tourn. ex Adans. = Anacardium L. ●
1766 Acajuba Gaertn. = Acajou Mill. ●☆
1767 Acajuba Gaertn. = Anacardium L. ●
1768 Acakia Raf. = Acacia Mill.(保留属名)●■
1769 Acalypha L.(1753);铁苋菜属(铁苋属);Copper Leaf, Copperleaf, Copper-leaf, Threeseed Mercury ●■
1770 Acalypha abortiva Hochst. ex Baill. = Acalypha crenata Hochst. ex A. Rich. ■☆
1771 Acalypha acmophylla Hemsl.;尾叶铁苋菜;Caudate-leaved Copper-leaf, Tailleaf Copperleaf ●
1772 Acalypha acrogyna Pax;大柱铁苋菜●☆
1773 Acalypha acuta Thunb. = Adenocline acuta (Thunb.) Baill. ■☆
1774 Acalypha adenotricha A. Rich. = Acalypha ornata Hochst. ex A. Rich. ■☆
1775 Acalypha akoensis Hayata;屏东铁苋菜(屏东铁苋);Pingdong Copperleaf, Pingdong Copper-leaf, Pingtung Copperleaf ●
1776 Acalypha allenii Hutch.;阿伦铁苋菜●☆
1777 Acalypha alopecuroidea Jacq.;狐尾铁苋菜●☆
1778 Acalypha ambigua Pax;可疑铁苋菜■☆
1779 Acalypha amentacea Roxb.;柔荑铁苋菜;Catch Me if You Can ●
1780 Acalypha amentacea Roxb. subsp. wilkesiana (Müll. Arg.) Fosberg;威氏柔荑铁苋菜;Wilkes' Acalypha ●
1781 Acalypha angatensis Blanco;台湾铁苋菜(台湾铁苋);Taiwan Copperleaf, Taiwan Copper-leaf ●
1782 Acalypha angolensis Müll. Arg. = Acalypha wilkesiana Müll. Arg. ●
1783 Acalypha angustata Sond.;细铁苋菜■☆
1784 Acalypha angustata Sond. var. glabra ? = Acalypha angustata Sond. ■☆
1785 Acalypha angustifolia Sw.;窄叶铁苋菜■☆
1786 Acalypha angustissima Pax;极细铁苋菜■☆
1787 Acalypha annobonae Pax et K. Hoffm.;安诺本铁苋菜■☆
1788 Acalypha aristata Kunth;具芒铁苋菜;Field Copperleaf ■☆
1789 Acalypha arvensis Poepp. et Endl.;田野铁苋菜●☆
1790 Acalypha australis L.;铁苋菜(半边珠,蚌壳草,布袋口,藏珠草,草蚌含珠,撮斗撮金珠,撮斗里装珍珠,撮斗珍珠,大青草,灯盏窝,粪斗草,凤眼草,海蚌含珠,海底藏珍珠,含珠草,寒热草,金畚斗,金盘野苋菜,老鼠耳朵草,痢疾草,六合草,麻子草,猫眼菜,猫眼草,鸟屎麻,喷水草,皮撮珍珠,瓢里藏珠,瓢里珍珠,七盏灯,人苋,肉草,沙罐草,沙罐头,山黄麻,田螺草,铁灯碗,筒筒草,下合花,蚬草,小耳朵草,血布袋,血见愁,野黄麻,野苦麻,野六麻,野络麻,野麻草,野棉花,叶里藏珠,叶里存珠,叶里含珠,叶里仙桃,叶下双桃,萤火虫草,玉碗捧真珠,珍珠草);Australian Acalypha, Copperleaf, Virginia Copperleaf ■

1791 Acalypha australis L. f. glareosa (Rupr.) H. Hara = Acalypha australis L. ■
1792 Acalypha australis L. f. velutina (Honda) Ohwi = Acalypha australis L. ■
1793 Acalypha australis L. var. glareosa ? = Acalypha australis L. ■
1794 Acalypha australis L. var. lanceolata Hayata; 细叶榎草; Thin-leaved Virginia Copperleaf ■
1795 Acalypha australis L. var. lanceolata Hayata = Acalypha australis L. ■
1796 Acalypha australis L. var. velutina Honda = Acalypha australis L. ■
1797 Acalypha bailloniana Müll. Arg. = Acalypha indica L. var. bailloniana (Müll. Arg.) Hutch. ■☆
1798 Acalypha benguelensis Müll. Arg.;本格拉铁苋菜☆
1799 Acalypha bequaertii Staner = Acalypha psilostachya Hochst. ex A. Rich. var. glandulosa Hutch. ■☆
1800 Acalypha betulina E. Mey. = Acalypha glabrata Thunb. ■☆
1801 Acalypha betulina E. Mey. var. latifolia Sond. = Acalypha glabrata Thunb. ■☆
1802 Acalypha bipartita Müll. Arg.;二深裂铁苋菜■☆
1803 Acalypha boehmerioides Miq. = Acalypha lanceolata Willd. ■
1804 Acalypha boiviniana Baill.;博伊文铁苋菜■☆
1805 Acalypha brachiata C. Krauss;塞内加尔铁苋菜■☆
1806 Acalypha brachystachya Hornem.;裂苞铁苋菜(短穗铁苋菜,短序铁苋菜,鸡眼草);Splitbract Copperleaf ■
1807 Acalypha brachystachya Hornem. = Acalypha supera Forssk. ■☆
1808 Acalypha bussei Hutch.;布瑟铁苋菜■☆
1809 Acalypha capensis (L. f.) Prain et Hutch.;好望角铁苋菜■☆
1810 Acalypha caperonioides Baill.;皱折铁苋菜■☆
1811 Acalypha caturus Blume;尖尾铁苋菜(兰屿铁苋,绿岛铁苋);Caudate Copper-leaf, Lutao Copperleaf, Sharptail Copperleaf ●
1812 Acalypha ceraceopunctata Pax;蜡斑铁苋菜■☆
1813 Acalypha chamaedrifolia (Lam.) Müll. Arg. = Acalypha reptans Sw. ●☆
1814 Acalypha chariensis Beille = Acalypha brachiata C. Krauss ■☆
1815 Acalypha chinensis Roxb. = Acalypha australis L. ■
1816 Acalypha chirindica S. Moore;奇林达铁苋菜■☆
1817 Acalypha chrysadenia Suess. et Friedrich = Acalypha fruticosa Forssk. ●■☆
1818 Acalypha ciliata Forssk.;缘毛铁苋菜■☆
1819 Acalypha clutioides Radcl. -Sm.;油芦子铁苋菜■☆
1820 Acalypha compacta Guilf. ex C. T. White = Acalypha amentacea Roxb. subsp. wilkesiana (Müll. Arg.) Fosberg ●
1821 Acalypha crassa Buchinger ex Meisn. = Acalypha peduncularis E. Mey. ex Meisn. var. crassa (Buchinger ex Meisn.) Müll. Arg. ■☆
1822 Acalypha crenata Hochst. ex A. Rich.;圆齿铁苋菜■☆
1823 Acalypha cristata Radcl. -Sm.;冠状铁苋菜■☆
1824 Acalypha crotonoides Pax = Acalypha polymorpha Hutch. ex Müll. Arg. ■☆
1825 Acalypha crotonoides Pax var. caudata Hutch. = Acalypha polymorpha Hutch. ex Müll. Arg. ■☆
1826 Acalypha crotonoides Pax var. cordata Hutch. = Acalypha polymorpha Hutch. ex Müll. Arg. ■☆
1827 Acalypha cupricola Robyns = Acalypha cupricola Robyns ex G. A. Levin ■☆
1828 Acalypha cupricola Robyns ex G. A. Levin;喜铜铁苋菜■☆
1829 Acalypha deamii (Weath.) H. E. Ahles; 大籽铁苋菜; Large-seeded Mercury ☆
1830 Acalypha decumbens Thunb. = Acalypha capensis (L. f.) Prain et Hutch. ■☆
1831 Acalypha decumbens Thunb. var. villosa (Thunb.) Müll. Arg. = Acalypha capensis (L. f.) Prain et Hutch. ■☆
1832 Acalypha deltoidea Robyns et Lawalrée;三角铁苋菜■☆
1833 Acalypha dentata Schumach. et Thonn. = Mallotus oppositifolius (Geiseler) Müll. Arg. ●☆
1834 Acalypha depressinerva (Kuntze) K. Schum.;凹脉铁苋菜■☆
1835 Acalypha dewevrei Pax = Acalypha neptunica Müll. Arg. var. pubescens (Pax) Hutch. ■☆
1836 Acalypha dikuluwensis P. A. Duvign. et Dewit;迪库铁苋菜■☆
1837 Acalypha dregei Gand. = Acalypha peduncularis E. Mey. ex Meisn. ■☆
1838 Acalypha dumertorum Pax = Acalypha ambigua Pax ■☆
1839 Acalypha dumetorum Müll. Arg.;灌丛铁苋菜■☆
1840 Acalypha echinus Pax et K. Hoffm.;刺铁苋菜■☆
1841 Acalypha ecklonii Baill.;埃氏铁苋菜■☆
1842 Acalypha elegantula Hochst. ex A. Rich. = Acalypha supera Forssk. ■☆
1843 Acalypha elskensi De Wild. = Acalypha ambigua Pax ■☆
1844 Acalypha engleri Pax;恩格勒铁苋菜■☆
1845 Acalypha entumenica Prain = Acalypha glandulifolia Buchinger ex Meisn. ■☆
1846 Acalypha eriophylla Hutch.;绵毛叶铁苋菜■☆
1847 Acalypha eriophylloides S. Moore;拟绵毛叶铁苋菜■☆
1848 Acalypha evrardii Gagnep. = Acalypha siamensis Oliv. ex Gagnep. ●
1849 Acalypha fallax Müll. Arg. = Acalypha lanceolata Willd. ■
1850 Acalypha fimbriata Schumach. et Thonn.;流苏铁苋菜■☆
1851 Acalypha formosa Hayata = Acalypha angatensis Blanco ●
1852 Acalypha fruticosa Forssk.;灌木铁苋菜●■☆
1853 Acalypha fruticosa Forssk. var. eglandulosa Radcl. -Sm.;无腺铁苋菜■☆
1854 Acalypha fruticosa Forssk. var. villosa Hutch.;长柔毛铁苋菜■☆
1855 Acalypha fuscescens Müll. Arg.;浅棕铁苋菜■☆
1856 Acalypha gagnepainii Merr. = Acalypha kerrii Craib ●
1857 Acalypha gemina (Lour.) Spreng. = Acalypha segetalis Müll. Arg. ■☆
1858 Acalypha gemina (Lour.) Spreng. var. brevibracteata (Müll. Arg.) Müll. Arg. = Acalypha segetalis Müll. Arg. ■☆
1859 Acalypha gemina (Lour.) Spreng. var. exserta (Müll. Arg.) Müll. Arg. = Acalypha segetalis Müll. Arg. ■☆
1860 Acalypha gemina Spreng. var. genuina Müll. Arg. = Acalypha australis L. ■
1861 Acalypha gillmanii Radcl. -Sm.;双生铁苋菜■☆
1862 Acalypha glabrata Thunb.;光铁苋菜■☆
1863 Acalypha glabrata Thunb. var. genuina Müll. Arg. = Acalypha glabrata Thunb. ■☆
1864 Acalypha glabrata Thunb. var. latifolia (Sond.) Müll. Arg. = Acalypha glabrata Thunb. ■☆
1865 Acalypha glabrata Thunb. var. pilosa Pax;疏毛铁苋菜■☆
1866 Acalypha glabrata Thunb. var. pilosior (Kuntze) Prain = Acalypha glabrata Thunb. var. pilosa Pax ■☆
1867 Acalypha glandulifolia Buchinger ex Meisn.;腺叶铁苋菜■☆
1868 Acalypha glomerata Hutch. = Acalypha lanceolata Willd. var. glandulosa (Müll. Arg.) Radcl. -Sm. ■☆
1869 Acalypha godseffiana Mast.;黄纹铁苋菜(黄纹铁苋);Copper-leaf ●
1870 Acalypha godseffiana Mast. = Acalypha amentacea Roxb. subsp.

wilkesiana (Müll. Arg.) Fosberg ●
1871 Acalypha godseffiana Mast. = Acalypha wilkesiana Müll. Arg. 'Marginata' ●
1872 Acalypha godseffiana Mast. var. heterophylla Hort.;细叶铁苋●
1873 Acalypha godseffiana Mast. var. marginata (J. J. Sm.) Pax et K. Hoffm. = Acalypha amentacea Roxb. subsp. wilkesiana (Müll. Arg.) Fosberg ●
1874 Acalypha goetzei Pax et K. Hoffm. = Acalypha polymorpha Hutch. ex Müll. Arg. ■☆
1875 Acalypha gossweileri S. Moore;戈斯铁苋●
1876 Acalypha gracilens A. Gray;短柄铁苋菜;Short-stalk Copper-leaf, Slender Three-seeded-mercury ■☆
1877 Acalypha gracilens A. Gray var. delzii Lill. W. Mill. = Acalypha gracilens A. Gray ■☆
1878 Acalypha gracilens A. Gray var. fraseri (Müll. Arg.) Weath. = Acalypha gracilens A. Gray ■☆
1879 Acalypha gracilescens A. Gray var. monococca Engelm. ex A. Gray = Acalypha monococca (Engelm. ex A. Gray) Lill. W. Mill. et Gandhi ■☆
1880 Acalypha grandidentata Müll. Arg.;大齿铁苋菜■☆
1881 Acalypha grandis Benth. var. akoensis (Hayata) Hurus. = Acalypha akoensis Hayata ●
1882 Acalypha grandis Benth. var. formosana (Hayata) Hurus. = Acalypha angatensis Blanco ●
1883 Acalypha grandis Benth. var. kotoensis (Hayata) Hurus. = Acalypha caturus Blume ●
1884 Acalypha grandis Benth. var. longi-acuminata (Hayata) Hurus. = Acalypha caturus Blume ●
1885 Acalypha grantii Baker et Hutch. = Acalypha ornata Hochst. ex A. Rich. ■☆
1886 Acalypha guatemalensis Pax et K. Hoffm.;危地马拉铁苋菜■☆
1887 Acalypha guineensis J. K. Morton et G. A. Levin;几内亚铁苋菜■☆
1888 Acalypha hainanensis Merr. et Chun; 海南铁苋菜; Hainan Copperleaf, Hainan Copper-leaf ●
1889 Acalypha hamiltoniana Bruant;变叶铁苋菜(变叶铁苋)●
1890 Acalypha haplostyla Pax = Acalypha brachiata C. Krauss ■☆
1891 Acalypha helenae Buscal. et Muschl.;海伦娜铁苋菜■☆
1892 Acalypha heterostachya Gagnep. = Acalypha kerrii Craib ●
1893 Acalypha hirsuta Hochst. ex A. Rich. = Acalypha brachiata C. Krauss ■☆
1894 Acalypha hispida Burm. f.;红穗铁苋菜(长穗铁苋,刚毛铁苋菜,狗尾红,红花铁苋,红穗铁苋,猫尾红,铁苋菜);Chenille Copperleaf, Chenille Copper-leaf, Chenille Plant, Hispid Copperleaf, Red Hot Cattail, Red Hot Copper-leaf, Red-hot Cat Tail, Red-hot Cat's Tail, Redhot Cat's Tails, Red-hot Cat's-tail, Red-hot Cattail, Redhot Cat-tail, Red-hot Cat-tail, Redspike Copperleaf ●
1895 Acalypha hochstetteriana Müll. Arg.;霍赫铁苋菜■☆
1896 Acalypha holtzii Pax et K. Hoffm.;霍尔茨铁苋菜■☆
1897 Acalypha homblei De Wild. = Acalypha ambigua Pax ■☆
1898 Acalypha hontauyuensis H. Keng = Acalypha suirenbiensis Yamam. ●
1899 Acalypha huillensis Pax et K. Hoffm.;威拉铁苋菜■☆
1900 Acalypha indica L.;热带铁苋菜(印度铁苋,印度铁苋菜);India Copperleaf ■
1901 Acalypha indica L. var. abortiva (Hochst. ex Baill.) Müll. Arg. = Acalypha crenata Hochst. ex A. Rich. ■☆
1902 Acalypha indica L. var. bailloniana (Müll. Arg.) Hutch.;巴永铁苋菜■☆
1903 Acalypha indica L. var. minima (H. Keng) S. F. Huang et T. C. Huang = Acalypha australis L. ■
1904 Acalypha indica L. var. minima (H. Keng) S. F. Huang et T. C. Huang = Acalypha minima H. Keng ■
1905 Acalypha intermedia De Wild. = Acalypha ambigua Pax ■☆
1906 Acalypha johnstoni Pax = Acalypha psilostachya Hochst. ex A. Rich. ■☆
1907 Acalypha kerrii Craib;卵叶铁苋菜;Kerr Copperleaf, Ovateleaf Copperleaf ●
1908 Acalypha kilimandscharica Pax et K. Hoffm. = Acalypha fruticosa Forssk. var. eglandulosa Radcl. -Sm. ■☆
1909 Acalypha koraensis Radcl. -Sm.;朝鲜铁苋菜■☆
1910 Acalypha kotoensis Hayata = Acalypha caturus Blume ●
1911 Acalypha lamiifolia Scheele = Acalypha capensis (L. f.) Prain et Hutch. ■☆
1912 Acalypha lanceolata Willd.;麻叶铁苋菜;Hempleaf Copperleaf ■
1913 Acalypha lanceolata Willd. var. glandulosa (Müll. Arg.) Radcl. -Sm.;腺点麻叶铁苋菜■☆
1914 Acalypha leonensis Benth. = Mareya micrantha (Benth.) Müll. Arg. ■☆
1915 Acalypha livingstoniana Müll. Arg. = Acalypha ornata Hochst. ex A. Rich. ■☆
1916 Acalypha longi-acuminata Hayata;长柄铁苋(尖尾铁苋);Long-acuminateleaf Copperleaf ●
1917 Acalypha longi-acuminata Hayata = Acalypha caturus Blume ●
1918 Acalypha macrostachya Griseb.;大穗铁苋菜■☆
1919 Acalypha mairei (H. Lév.) C. K. Schneid.;毛叶铁苋菜;Hairleaf Copperleaf, Maire Copperleaf ●
1920 Acalypha malawiensis Radcl. -Sm.;马拉维铁苋菜■☆
1921 Acalypha manniana Müll. Arg.;曼氏铁苋菜■☆
1922 Acalypha marginata J. J. Sm. = Acalypha amentacea Roxb. subsp. wilkesiana (Müll. Arg.) Fosberg ●
1923 Acalypha matsudae Hayata;恒春铁苋菜(恒春铁苋);Hengchun Copperleaf, Matsuda Copperleaf ●
1924 Acalypha mentiens Gand. = Acalypha peduncularis E. Mey. ex Meisn. ■☆
1925 Acalypha micrantha Benth. = Mareya micrantha (Benth.) Müll. Arg. ■☆
1926 Acalypha mildbraediana Pax var. pubescens ? = Acalypha neptunica Müll. Arg. var. pubescens (Pax) Hutch. ■☆
1927 Acalypha minima H. Keng;小铁苋菜(小叶铁苋)■
1928 Acalypha minima H. Keng = Acalypha australis L. ■
1929 Acalypha moggii Compton = Acalypha ornata Hochst. ex A. Rich. ■☆
1930 Acalypha monococca (Engelm. ex A. Gray) Lill. W. Mill. et Gandhi;单籽铁苋菜;One-seeded Mercury ■☆
1931 Acalypha monostachya Benth.;单穗铁苋菜;Copper Leaf ■☆
1932 Acalypha multifida N. E. Br.;多裂铁苋菜■☆
1933 Acalypha neptunica Müll. Arg.;内普丘恩铁苋菜■☆
1934 Acalypha neptunica Müll. Arg. var. glabrescens (Pax) Pax et K. Hoffm. = Acalypha neptunica Müll. Arg. ■☆
1935 Acalypha neptunica Müll. Arg. var. pubescens (Pax) Hutch.;短柔毛铁苋菜■☆
1936 Acalypha neptunica Müll. Arg. var. vestita Pax et K. Hoffm. = Acalypha neptunica Müll. Arg. var. pubescens (Pax) Hutch. ■☆
1937 Acalypha nigritiana Müll. Arg. = Acalypha ornata Hochst. ex A. Rich. ■☆
1938 Acalypha nyasica Hutch.;尼亚斯铁苋菜■☆

1939 Acalypha obtusa Thunb. = Leidesia procumbens (L.) Prain ■☆
1940 Acalypha obtusata Spreng. ex Steud. = Adenocline violifolia (Kunze) Prain ■☆
1941 Acalypha ornata Hochst. ex A. Rich.;装饰铁苋菜■☆
1942 Acalypha ornata Hochst. ex A. Rich. var. bracteosa Müll. Arg. = Acalypha ornata Hochst. ex A. Rich. ■☆
1943 Acalypha ornata Hochst. ex A. Rich. var. glandulosa Müll. Arg. = Acalypha ornata Hochst. ex A. Rich. ■☆
1944 Acalypha ornata Hochst. ex A. Rich. var. pilosa Müll. Arg. = Acalypha ornata Hochst. ex A. Rich. ■☆
1945 Acalypha ostryifolia Riddell;糙梗铁苋菜;Roughpod Copperleaf ■☆
1946 Acalypha paniculata Miq. = Acalypha racemosa Baill. ■☆
1947 Acalypha pauciflora Hornem. = Acalypha australis L. ■
1948 Acalypha pauciflora Hornem. var. glareosa ? = Acalypha australis L. ■
1949 Acalypha paucifolia Baker et Hutch.;少叶铁苋菜■☆
1950 Acalypha peduncularis E. Mey. ex Meisn.;花梗铁苋菜■☆
1951 Acalypha peduncularis E. Mey. ex Meisn. var. crassa (Buchinger ex Meisn.) Müll. Arg.;粗花梗铁苋菜■☆
1952 Acalypha peduncularis Pax = Acalypha caperonioides Baill. ■☆
1953 Acalypha peduncularis Pax var. caperonioides (Baill.) Müll. Arg. = Acalypha caperonioides Baill. ■☆
1954 Acalypha peduncularis Pax var. glabtata Sond. = Acalypha caperonioides Baill. ■☆
1955 Acalypha pendula C. Wright ex Griseb.;火尾铁苋菜;Firetail ☆
1956 Acalypha petiolaris Hochst. ex C. Krauss = Acalypha brachiata C. Krauss ■☆
1957 Acalypha petiolaris Sond. = Acalypha sonderiana Müll. Arg. ●
1958 Acalypha phleoides Cav.;梯牧草铁苋菜■☆
1959 Acalypha polymorpha Hutch. ex Müll. Arg.;多形铁苋菜■☆
1960 Acalypha polymorpha Hutch. ex Müll. Arg. var. angustifolia Müll. Arg. = Acalypha ambigua Pax ■☆
1961 Acalypha polymorpha Hutch. ex Müll. Arg. var. depauretana Müll. Arg. = Acalypha ambigua Pax ■☆
1962 Acalypha psilostachya Hochst. ex A. Rich.;裸穗铁苋菜■☆
1963 Acalypha psilostachya Hochst. ex A. Rich. var. glandulosa Hutch.;具腺铁苋菜■☆
1964 Acalypha psilostachyoides Pax = Acalypha chirindica S. Moore ■☆
1965 Acalypha pubiflora Baill.;毛花铁苋菜■☆
1966 Acalypha punctata Meisn.;斑点铁苋菜■☆
1967 Acalypha punctata Meisn. var. rogersii Prain;罗杰斯铁苋菜■☆
1968 Acalypha racemosa Baill.;总花铁苋菜■☆
1969 Acalypha rehmanni Pax = Acalypha villicaulis Hochst. ex A. Rich. ■☆
1970 Acalypha reptans Sw.;红尾铁苋菜(猫尾红);Red Cat's Tail, Red Cat-tails ●☆
1971 Acalypha rhomboidea Raf.;菱形铁苋菜;Rhombic Copperleaf, Rhombic Copper-leaf, Three-seed Mercury, Three-seeded Mercury, Three-seeded-mercury ●☆
1972 Acalypha sanderi N. E. Br. = Acalypha hispida Burm. f. ●
1973 Acalypha schinzii Pax = Acalypha depressinerva (Kuntze) K. Schum. ■☆
1974 Acalypha schlechteri Gand. = Acalypha peduncularis E. Mey. ex Meisn. var. crassa (Buchinger ex Meisn.) Müll. Arg. ■☆
1975 Acalypha schneideriana Pax et K. Hoffm.;丽江铁苋菜;Lijiang Copperleaf, Schneider Copperleaf ●
1976 Acalypha scleropumila A. Chev.;硬侏儒铁苋菜■☆
1977 Acalypha segetalis Müll. Arg.;谷地铁苋菜■☆
1978 Acalypha senegalensis Pax et K. Hoffm. = Acalypha brachiata C. Krauss ■☆
1979 Acalypha senensis Klotzsch = Acalypha brachiata C. Krauss ■☆
1980 Acalypha sessilis Poir. var. brevibracteata Müll. Arg. = Acalypha segetalis Müll. Arg. ■☆
1981 Acalypha sessilis Poir. var. excerta Müll. Arg. = Acalypha segetalis Müll. Arg. ■☆
1982 Acalypha shirensis Hutch. ex Pax = Acalypha polymorpha Hutch. ex Müll. Arg. ■☆
1983 Acalypha siamensis Oliv. ex Gagnep.;菱叶铁苋菜;Rhombleaf Copperleaf, Siam Copperleaf ●
1984 Acalypha siamensis Oliv. ex Gagnep. = Acalypha kerrii Craib ●
1985 Acalypha sidifolia A. Rich. = Acalypha brachiata C. Krauss ■☆
1986 Acalypha sigensis Pax et K. Hoffm. = Acalypha engleri Pax ■☆
1987 Acalypha somalensis Pax = Acalypha indica L. ■
1988 Acalypha somalium Müll. Arg. = Acalypha indica L. ■
1989 Acalypha sonderi Gand. = Acalypha peduncularis E. Mey. ex Meisn. ■☆
1990 Acalypha sonderiana Müll. Arg.;森诺铁苋菜●
1991 Acalypha spiciflora Burm. f.;穗序铁苋菜●
1992 Acalypha stuhlmannii Pax = Acalypha polymorpha Hutch. ex Müll. Arg. ■☆
1993 Acalypha subsessilis Hutch. = Acalypha neptunica Müll. Arg. ■☆
1994 Acalypha subsessilis Hutch. var. glabra Pax et K. Hoffm. = Acalypha neptunica Müll. Arg. ■☆
1995 Acalypha subsessilis Hutch. var. mollis ? = Acalypha neptunica Müll. Arg. ■☆
1996 Acalypha subsessilis Pax = Acalypha neptunica Müll. Arg. ■☆
1997 Acalypha suirebiensis Yamam.;花莲铁苋菜(红头铁苋,红头屿铁苋,花莲铁苋,台东铁苋,台东苋);Hontauyu Copperleaf, Hualian Copperleaf, Taitung Copper-leaf, Taiug Acalypha ●
1998 Acalypha supera Forssk.;上升铁苋菜■☆
1999 Acalypha swynnertonii S. Moore = Acalypha ornata Hochst. ex A. Rich. ■☆
2000 Acalypha szechuanensis Hutch. = Acalypha acmophylla Hemsl. ●
2001 Acalypha tenuis Müll. Arg. = Acalypha villicaulis Hochst. ex A. Rich. ■☆
2002 Acalypha teusczii Pax = Acalypha benguelensis Müll. Arg. ☆
2003 Acalypha transvaalensis Gand. = Acalypha caperonioides Baill. ■☆
2004 Acalypha tricolor Veitch = Acalypha wilkesiana Müll. Arg. ●
2005 Acalypha vahliana Müll. Arg. = Acalypha ciliata Forssk. ■☆
2006 Acalypha vahliana Oliv. = Acalypha crenata Hochst. ex A. Rich. ■☆
2007 Acalypha villicaulis Hochst. ex A. Rich. = Acalypha brachiata C. Krauss ■☆
2008 Acalypha virginica L.;弗吉尼亚铁苋菜(维吉尼亚铁苋菜);Virginia Copper-leaf, Virginia Three-seeded-mercury ●☆
2009 Acalypha virginica L. var. rhomboidea (Raf.) Cooperr. = Acalypha rhomboidea Raf. ●☆
2010 Acalypha volkensii Pax;福尔铁苋菜●☆
2011 Acalypha wilkesiana Müll. Arg.;红桑(红叶铁苋菜,金边莲,金边桑,木本金线莲,威氏铁苋);Beef Steak Plant, Beefsteak Plant, Copper Leaf, Copperleaf, Copper-leaf, Jacob's Coat, Match-me-if-you-can, Painted Copperleaf, Painted Copper-leaf, Red Copperleaf, Virginia Copperleaf, Wilkes Copperleaf ●
2012 Acalypha wilkesiana Müll. Arg. 'Hoffmannii';镶边旋叶铁苋●
2013 Acalypha wilkesiana Müll. Arg. 'Marginata Mini Red';红边小红

桑●

2014 Acalypha wilkesiana Müll. Arg. 'Marginata';金边红桑(红边铁苋,金边莲,金边桑,木本金线莲,皱边红桑)●

2015 Acalypha wilkesiana Müll. Arg. = Acalypha amentacea Roxb. subsp. wilkesiana (Müll. Arg.) Fosberg ●

2016 Acalypha wilkesiana Müll. Arg. var. macafeana W. Mill.;长卡灰铁苋●

2017 Acalypha wilkesiana Müll. Arg. var. macrophylla Hort.;大叶铁苋;Big-leaved Wilkes Copperleaf ●

2018 Acalypha wilkesiana Müll. Arg. var. marginata W. Mill. = Acalypha wilkesiana Müll. Arg. 'Marginata' ●

2019 Acalypha wilkesiana Müll. Arg. var. musaica W. Mill.;彩叶铁苋●

2020 Acalypha wilmsii Pax ex Prain et Hutch.;维尔姆斯铁苋菜■☆

2021 Acalypha wui H. S. Kiu;印禅铁苋菜;Wu's Copperleaf ●

2022 Acalypha wythei Hutch. = Acalypha wilkesiana Müll. Arg. ●

2023 Acalypha zambesica Müll. Arg. = Acalypha brachiata C. Krauss ■☆

2024 Acalypha zeyheri Baill. = Acalypha peduncularis E. Mey. ex Meisn. ■☆

2025 Acalyphaceae J. Agardh = Euphorbiaceae Juss. (保留科名)●■

2026 Acalyphaceae J. Agardh;铁苋菜科●■

2027 Acalyphaceae Juss. ex Menge = Euphorbiaceae Juss. (保留科名)●■

2028 Acalyphes Hassk. = Acalypha L. ●■

2029 Acalyphopsis Pax et K. Hoffm. (1924);拟铁苋菜属■☆

2030 Acalyphopsis Pax et K. Hoffm. = Acalypha L. ●■

2031 Acalyphopsis celebica Pax et K. Hoffm.;拟铁苋菜■☆

2032 Acampe Lindl. (1853)(保留属名);脆兰属(脆兰花属);Acampe,Fragileorchis ■

2033 Acampe dentata Lindl.;齿脆兰■☆

2034 Acampe hayatae Szlach. = Sarcophyton taiwanianum (Hayata) Garay ■

2035 Acampe longifolia (Lindl.) Lindl. = Acampe rigida (Buch.-Ham. ex Sm.) P. F. Hunt ■

2036 Acampe mombasensis Rendle = Acampe pachyglossa Rchb. f. ■☆

2037 Acampe multiflora Lindl. = Acampe rigida (Buch.-Ham. ex Sm.) P. F. Hunt ■

2038 Acampe nyassana Schltr. = Acampe pachyglossa Rchb. f. ■☆

2039 Acampe ochracea (Lindl.) Hochr.;窄果脆兰;Narrowfruit Fragileorchis ■

2040 Acampe pachyglossa Rchb. f.;厚舌脆兰■☆

2041 Acampe pachyglossa Rchb. f. subsp. renschiana (Rchb. f.) Senghas = Acampe pachyglossa Rchb. f. ■☆

2042 Acampe papillosa (Lindl.) Lindl.;短序脆兰(乳突脆兰);Papillose Fragileorchis ■

2043 Acampe papillosa Lindl. = Acampe papillosa (Lindl.) Lindl. ■

2044 Acampe praemorsa (Roxb.) Blatt. et McCann;啮齿脆兰■☆

2045 Acampe renschiana Rchb. f. = Acampe pachyglossa Rchb. f. ■☆

2046 Acampe rigida (Buch.-Ham. ex Sm.) P. F. Hunt;多花脆兰(长叶脆兰,长叶假万代兰,脆兰,黑山蔗,焦兰,乌猿蔗,香蕉兰);Flowery Fragileorchis ■

2047 Acampe taiwaniana S. S. Ying = Acampe rigida (Buch.-Ham. ex Sm.) P. F. Hunt ■

2048 Acamptoclados Nash = Eragrostis Wolf ■

2049 Acamptopappus (A. Gray) A. Gray (1873);直冠菊属(美洲菊属);Goldenhead ■☆

2050 Acamptopappus A. Gray = Acamptopappus (A. Gray) A. Gray ■☆

2051 Acamptopappus microcephalus M. E. Jones;小头直冠菊■☆

2052 Acamptopappus shockleyi A. Gray;直冠菊■☆

2053 Acamptopappus sphaerocephalus A. Gray;球头直冠菊■☆

2054 Acamptopappus sphaerocephalus A. Gray var. hirtellus S. F. Blake;毛球头直冠菊■☆

2055 Acana Durand = Acanos Adans ■

2056 Acanos Adans. = Onopordum L. ■

2057 Acanos spina Scop. = Onopordum acanthium L. ■

2058 Acantacaryx Arruda ex Kost. = Caryocar F. Allam. ex L. ●☆

2059 Acanthacaryx Steud. = Acantacaryx Arruda ex Kost. ●☆

2060 Acanthaceae Juss. (1789)(保留科名);爵床科;Acanthus Family,Bear's-breech Family ●■

2061 Acanthambrosia Rydb. = Ambrosia L. ●■

2062 Acanthanthus Y. Ito = Acantholobivia Backeb. ●

2063 Acanthanthus Y. Ito = Echinopsis Zucc. ●

2064 Acantharia Rojas (1879);阿根廷豆属■☆

2065 Acantharia melanocarpa Rojas;阿根廷豆■☆

2066 Acanthella Hook. f. (1867);小老鼠簕属☆

2067 Acanthella montana Gleason;山地小老鼠簕☆

2068 Acanthella pulchra Gleason;美丽小老鼠簕☆

2069 Acanthella sprucei Hook. f.;小老鼠簕☆

2070 Acanthephippium Blume = Acanthephippium Blume ex Endl. ■

2071 Acanthephippium Blume ex Endl. (1825);坛花兰属;Acanthophippium,Saddleorchis ■

2072 Acanthephippium bicolora Lindl.;二色坛花兰;Twocolored Acanthophippium ■

2073 Acanthephippium chrysoglossum Schltr.;黄坛花兰■☆

2074 Acanthephippium javanicum Blume;爪哇坛花兰;Java Acanthophippium ■

2075 Acanthephippium mantinianum Lindl. et Cogn.;菲律宾坛花兰;Philippine Acanthophippium ■☆

2076 Acanthephippium odoratum Aver.;芳香坛花兰■☆

2077 Acanthephippium pictum Fukuy.;着色坛花兰(延龄钟馗兰);Painted Acanthophippium ■

2078 Acanthephippium pictum Fukuy. = Acanthephippium sylhetense Lindl. ■

2079 Acanthephippium sinense Rolfe;中华坛花兰(花兰,坛花兰);China Saddleorchis,Chinese Acanthophippium ■

2080 Acanthephippium striatum Lindl.;锥囊坛花兰(台湾坛花兰,条斑坛花兰,一叶坛花兰,一叶钟馗兰);Striate Acanthophippium,Striate Saddleorchis ■

2081 Acanthephippium sylhetense Lindl.;坛花兰(台湾坛花兰,台湾钟馗兰,罈花兰,香坛花兰,钟馗兰);Common Saddleorchis,Fragrant Acanthophippium ■

2082 Acanthephippium sylhetense Lindl. var. pictum (Fukuy.) T. Hashim. = Acanthephippium pictum Fukuy. ■

2083 Acanthephippium thailandicum Seidenf.;泰国坛花兰■☆

2084 Acanthephippium unguiculatum (Hayata) Fukuy.;具爪坛花兰(一叶钟馗兰);Clawy Acanthophippium ■

2085 Acanthephippium unguiculatum (Hayata) Fukuy. = Acanthephippium striatum Lindl. ■

2086 Acanthephippium yamamotoi Hayata = Acanthephippium sylhetense Lindl. ■

2087 Acanthinophyllum Allemão = Clarisia Ruiz et Pav. (保留属名)●☆

2088 Acanthinophyllum Burger = Clarisia Ruiz et Pav. (保留属名)●☆

2089 Acanthium Fabr. = Onopordum L. ■

2090 Acanthium Haller = Onopordum L. ■

2091 Acanthium Heist. ex Fabr. = Onopordum L. ■

2092 Acanthobotrya Eckl. et Zeyh. = Lebeckia Thunb. ■☆

2093 Acanthobotrya armata (Thunb.) Eckl. et Zeyh. = Wiborgia mucronata (L. f.) Druce ■☆
2094 Acanthobotrya sessilifolia Eckl. et Zeyh. = Lebeckia sessilifolia (Eckl. et Zeyh.) Benth. ■☆
2095 Acanthobotrys Clem. = Acanthobotrya Eckl. et Zeyh. ■☆
2096 Acanthocalycium Backeb. (1936);刺萼掌属(南美球形仙人掌属,有刺萼属)●■☆
2097 Acanthocalycium Backeb. = Echinopsis Zucc. ●
2098 Acanthocalycium aurantiacum Rausch;橘花刺萼掌(橘色花有刺萼)■☆
2099 Acanthocalycium brevispinum F. Ritter;短刺萼掌■☆
2100 Acanthocalycium glaucum F. Ritter;灰刺萼掌(苍白有刺萼,灰刺萼)■☆
2101 Acanthocalycium klimpeliana (Weidlich et Werderm.) Backeb.;凤冠球(凤冠丸)■☆
2102 Acanthocalycium klimpeliana (Weidlich et Werderm.) Backeb. = Acanthocalycium spiniflorum (K. Schum.) Backeb. ■☆
2103 Acanthocalycium peitscherianum Backeb.;樱春球(樱春丸)■☆
2104 Acanthocalycium peitscherianum Backeb. = Acanthocalycium spiniflorum (K. Schum.) Backeb. ■☆
2105 Acanthocalycium spiniflorum (K. Schum.) Backeb.;花冠球(花冠丸)■☆
2106 Acanthocalycium thionanthum (Speg.) Backeb.;硫花刺萼掌(硫磺色花有刺萼)■☆
2107 Acanthocalycium violaceum (Werderm.) Backeb.;紫盛球(蓝紫色有刺萼,紫盛丸)■☆
2108 Acanthocalyx (DC.) M. J. Cannon = Acanthocalyx (DC.) Tiegh. ■
2109 Acanthocalyx (DC.) Tiegh. (1909);刺萼属(刺参属,刺续断属)■
2110 Acanthocalyx (DC.) Tiegh. = Morina L. ■
2111 Acanthocalyx alba (Hand.-Mazz.) M. J. Cannon;白刺萼(白花刺续断,刺萼);Whiteflower Morina ■
2112 Acanthocalyx delavayi (Franch.) M. J. Cannon = Acanthocalyx nepalensis (D. Don) M. J. Cannon subsp. delavayi (Franch.) D. Y. Hong et F. Barrie ■
2113 Acanthocalyx delavayi (Franch.) M. J. Cannon = Morina nepalensis D. Don ■
2114 Acanthocalyx nepalensis (D. Don) D. Don = Acanthocalyx nepalensis (D. Don) M. J. Cannon subsp. delavayi (Franch.) D. Y. Hong et F. Barrie ■
2115 Acanthocalyx nepalensis (D. Don) M. J. Cannon = Acanthocalyx nepalensis (D. Don) M. J. Cannon subsp. delavayi (Franch.) D. Y. Hong et F. Barrie ■
2116 Acanthocalyx nepalensis (D. Don) M. J. Cannon = Morina nepalensis D. Don ■
2117 Acanthocalyx nepalensis (D. Don) M. J. Cannon subsp. delavayi (Franch.) D. Y. Hong et F. Barrie;德氏刺续断■
2118 Acanthocardamum Thell. (1906);刺碎米荠属■☆
2119 Acanthocardamum erinaceum Thell.;刺碎米荠■☆
2120 Acanthocarpaea Dalla Torre et Harms = Acanthocarpea Klotzsch ●■☆
2121 Acanthocarpaea scabrida Klotzsch = Limeum sulcatum (Klotzsch) Hutch. var. scabridum (Klotzsch) Friedrich ■☆
2122 Acanthocarpaea sulcata Klotzsch = Limeum sulcatum (Klotzsch) Hutch. ■☆
2123 Acanthocarpea Klotzsch = Limeum L. ●■☆
2124 Acanthocarpus Kuntze = Acanthocarpea Klotzsch ●■☆
2125 Acanthocarpus Lehm. (1848);刺果蕉属■☆
2126 Acanthocarpus preissii Lehm.;刺果蕉■☆
2127 Acanthocarya Arruda ex Endl. = Acantacaryx Arruda ex Kost. ●☆
2128 Acanthocarya Arruda ex Endl. = Caryocar F. Allam. ex L. ●☆
2129 Acanthocaulon Klotzsch = Platygyna P. Mercier☆
2130 Acanthocaulon Klotzsch ex Endl. = Platygyna P. Mercier☆
2131 Acanthocephala Backeb. = Brasilicactus Backeb. ■☆
2132 Acanthocephala Backeb. = Parodia Speg. (保留属名)●
2133 Acanthocephalus Kar. et Kir. (1842);棘头花属(刺头花属,刺头菊属);Acanthocephalus ■
2134 Acanthocephalus amplexifolius Kar. et Kir.;棘头花(刺头花);Amplexifolious Acanthocephalus ■
2135 Acanthocephalus benthamianus Regel;本瑟姆棘头花■☆
2136 Acanthocereus (A. Berger) Britton et Rose = Acanthocereus (Engelm. ex A. Berger) Britton et Rose ●☆
2137 Acanthocereus (Engelm. ex A. Berger) Britton et Rose(1909);刺萼柱属(柱形仙人掌属)●☆
2138 Acanthocereus Britton et Rose = Acanthocereus (Engelm. ex A. Berger) Britton et Rose ●☆
2139 Acanthocereus acutangulus (Sweet) A. Berger = Acanthocereus tetragonus (L.) Hummelinck ■
2140 Acanthocereus brasiliensis Britton et Rose = Acanthocereus tetragonus (L.) Hummelinck ■
2141 Acanthocereus brasiliensis Britton et Rose = Pseudoacanthocereus brasiliensis (Britton et Rose) F. Ritter ■☆
2142 Acanthocereus columbianus Britton et Rose = Acanthocereus tetragonus (L.) Hummelinck ■
2143 Acanthocereus floridanus Small ex Britton et Rose = Acanthocereus tetragonus (L.) Hummelinck ■
2144 Acanthocereus pentagonus (L.) Britton et Rose = Acanthocereus tetragonus (L.) Hummelinck ■
2145 Acanthocereus pitajaya (Jacq.) Dugand ex Croizat = Acanthocereus tetragonus (L.) Hummelinck ■
2146 Acanthocereus princeps (Pfeiff.) Backeb. = Acanthocereus tetragonus (L.) Hummelinck ■
2147 Acanthocereus tetragonus (L.) Hummelinck;待宵仙人柱;Barbed Wire Cereus,Pitahaya Anaranjada ■
2148 Acanthochiton Torr. (1853);墨西哥刺被苋属(刺被苋属,美洲苋属,墨西哥苋属)■☆
2149 Acanthochiton Torr. = Amaranthus L. ■
2150 Acanthochiton wrightii Torr.;墨西哥刺被苋(刺被苋,墨西哥苋)■☆
2151 Acanthochiton wrightii Torr. = Amaranthus acanthochiton J. D. Sauer ■☆
2152 Acanthochlamydaceae (S. L. Chen) P. C. Kao;芒苞草科;Acanthochlamys Family ■
2153 Acanthochlamydaceae (S. L. Chen) P. C. Kao = Anthericaceae J. Agardh ●■☆
2154 Acanthochlamydaceae P. C. Kao = Acanthochlamydaceae (S. L. Chen) P. C. Kao ●■☆
2155 Acanthochlamydaceae P. C. Kao = Anthericaceae J. Agardh ●■☆
2156 Acanthochlamydaceae P. C. Kao = Velloziaceae J. Agardh(保留科名)■
2157 Acanthochlamys P. C. Kao (1980);芒苞草属;Acanthochlamys,Brecteate Acanthochlamys ■★
2158 Acanthochlamys bracteata P. C. Kao;芒苞草;Bracteate Acanthochlamys ■
2159 Acanthocladium F. Muell. (1861);刺枝菊属■☆

2160 Acanthocladium F. Muell. = Helichrysum Mill.(保留属名)●■
2161 Acanthocladium dockeri F. Muell.;刺枝菊■☆
2162 Acanthocladus Klotzsch ex Hassk.(1864);枝刺远志属●☆
2163 Acanthocladus Klotzsch ex Hassk. = Polygala L. ●■
2164 Acanthocladus albicans A. W. Benn.;白枝刺远志●☆
2165 Acanthocladus brasiliensis Klotzsch = Acanthocladus brasiliensis Klotzsch ex Hassk. ●☆
2166 Acanthocladus brasiliensis Klotzsch ex Hassk.;巴西枝刺远志●☆
2167 Acanthocladus microphylla Griseb.;小叶枝刺远志●☆
2168 Acanthocladus scleroxylon (Ducke) B. Eriksen et B. Stahl;硬木枝刺远志●☆
2169 Acanthococos Barb. Rodr.(1900);南美椰子属●☆
2170 Acanthococos Barb. Rodr. = Acrocomia Mart. ●☆
2171 Acanthococos hassleri Barb. Rodr.;南美椰子●☆
2172 Acanthodesmos C. D. Adams et duQuesnay(1971);刺链菊属(贾梅卡菊属)■☆
2173 Acanthodesmos distichus C. D. Adams et du Quesnay;刺链菊(贾梅卡菊)■☆
2174 Acanthodion Lem. = Acanthodium Delile ●■
2175 Acanthodium Delile = Blepharis Juss. ●■
2176 Acanthodium angustum Nees = Blepharis angusta (Nees) T. Anderson ■☆
2177 Acanthodium capense (L. f.) Nees var. inermis Nees = Blepharis inermis (Nees) C. B. Clarke ■☆
2178 Acanthodium capense (L. f.) Nees var. villosum Nees = Blepharis capensis (L. f.) Pers. ■☆
2179 Acanthodium capense Nees = Blepharis capensis (L. f.) Pers. ■☆
2180 Acanthodium diversispinum Nees = Blepharis diversispina (Nees) C. B. Clarke ■☆
2181 Acanthodium diversispinum Nees = Blepharis serrulata (Nees) Ficalho et Hiern ■☆
2182 Acanthodium furcatum (L. f.) Nees = Blepharis furcata (L. f.) Pers. ■☆
2183 Acanthodium glabrum Nees = Acanthopsis carduifolia (L. f.) Schinz ■☆
2184 Acanthodium glaucum Nees = Acanthopsis glauca (Nees) Schinz ■☆
2185 Acanthodium grossum Nees = Blepharis grossa (Nees) T. Anderson ■☆
2186 Acanthodium hirtinervium Nees = Blepharis hirtinervia (Nees) T. Anderson ■☆
2187 Acanthodium hirtum Hochst. ex Nees = Blepharis linariifolia Pers. ■☆
2188 Acanthodium hoffmannseggianum Nees = Acanthopsis hoffmannseggiana (Nees) C. B. Clarke ■☆
2189 Acanthodium marginatum Nees = Blepharis marginata (Nees) C. B. Clarke ■☆
2190 Acanthodium procumbens (L. f.) Nees = Blepharis procumbens (L. f.) Pers. ■☆
2191 Acanthodium serrulatum Nees = Blepharis serrulata (Nees) Ficalho et Hiern ■☆
2192 Acanthodium sinuatum Nees = Blepharis sinuata (Nees) C. B. Clarke ■☆
2193 Acanthodium spathulare Nees = Acanthopsis spathularis (Nees) Schinz ■☆
2194 Acanthodium spicatum Delile = Blepharis ciliaris (L.) B. L. Burtt ■☆
2195 Acanthodium squarrosum Nees = Blepharis squarrosa (Nees) T. Anderson ■☆
2196 Acanthodus Raf. = Acanthus L. ●■
2197 Acanthogilia A. G. Day et Moran(1986);北美花荵属●☆
2198 Acanthogilia gloriosa (Brandegee) A. G. Day et Moran;北美花荵●☆
2199 Acanthoglossum Blume = Pholidota Lindl. ex Hook. ■
2200 Acanthogonum Torr.(1857);美国刺花蓼属●■☆
2201 Acanthogonum Torr. = Chorizanthe R. Br. ex Benth. ●■☆
2202 Acanthogonum corrugatum Torr. = Chorizanthe corrugata (Torr.) Torr. et A. Gray ■☆
2203 Acanthogonum polygonoides (Torr. et A. Gray) Goodman = Chorizanthe polygonoides Torr. et A. Gray ■☆
2204 Acanthogonum polygonoides (Torr. et A. Gray) Goodman var. longispinum Goodman = Chorizanthe polygonoides Torr. et A. Gray var. longispina (Goodman) Munz ■☆
2205 Acanthogonum rigidum Torr. = Chorizanthe rigida (Torr.) Torr. et A. Gray ■☆
2206 Acantholepis Less.(1831);棘苞菊属(刺苞菊属);Acantholepis ■
2207 Acantholepis orientalis Less.;棘苞菊;Oriental Acantholepis ■
2208 Acantholimon Boiss.(1846)(保留属名);彩花属(刺矶松属,刺雪属);Prickly Thrift,Pricklythrift,Prickly-thrift ●
2209 Acantholimon acerosum (Willd.) Boiss.;针叶彩花(针状彩花);Acicular Pricklythrift ●
2210 Acantholimon alaicum Czerniak.;中亚彩花●☆
2211 Acantholimon alatavicum Bunge;刺叶彩花(彩花,刺矶松);Common Pricklythrift,Common Prickly-thrift ●
2212 Acantholimon alatavicum Bunge var. laevigatum T. X. Peng = Acantholimon laevigatum (T. X. Peng) Kamelin ●
2213 Acantholimon albertii Regel;阿伯特彩花;Albert Pricklythrift ●
2214 Acantholimon alexandri Fed.;阿赖彩花●☆
2215 Acantholimon androsaceum Boiss.;点地梅状彩花●☆
2216 Acantholimon armenum Boiss. et A. Huet;亚美尼亚彩花;Armenia Pricklythrift ●
2217 Acantholimon aulieatense Czerniak.;奥里爱坦彩花;Aulieat Pricklythrift ●
2218 Acantholimon avenaceum Bunge;燕麦状彩花;Oat-like Pricklythrift ●
2219 Acantholimon blandum Czerniak.;光滑彩花●☆
2220 Acantholimon borodinii Krasn.;细叶彩花;Borodin Pricklythrift, Borodin Prickly-thrift ●
2221 Acantholimon bracteatum (Girard) Boiss.;有苞彩花(大苞彩花);Bracteate Pricklythrift ●
2222 Acantholimon bromifolium Boiss. ex Bunge;雀麦叶彩花●☆
2223 Acantholimon calocephalum Aitch. et Hemsl.;美头彩花●☆
2224 Acantholimon caryophyllaceum Boiss.;石竹状彩花;Pink Pricklythrift ●
2225 Acantholimon compactum Korovin;紧密彩花●☆
2226 Acantholimon desertorum Regel;荒漠彩花●☆
2227 Acantholimon diapensioides Boiss.;小叶彩花(刺矶松);Diapensia Pricklythrift, Diapensia-like Prickly-thrift, Little-leaf Pricklythrift ●
2228 Acantholimon diapensioides Boiss. var. longifolia O. Fetsch. = Acantholimon hedinii Ostenf. ●
2229 Acantholimon echinus (L.) Bunge;刺彩花;Spiny Pricklythrift ●
2230 Acantholimon ekatherinae (B. Fedtsch.) Czerniak.;叶卡氏彩花●☆
2231 Acantholimon erinaceum (Jaub. et Spach) Lincz.;小刺彩花●☆
2232 Acantholimon erythraeum Bunge;淡红彩花●☆
2233 Acantholimon faustii Trautv.;福斯特彩花●☆
2234 Acantholimon fetisowi Regel;费氏彩花●☆

2235 Acantholimon fominii Kusn.;福明氏彩花●☆
2236 Acantholimon gaudanense Czerniak.;戈丹彩花●☆
2237 Acantholimon glumaceum (Jaub. et Spach) Boiss.;颖状彩花;Glume-like Pricklythrift,Prickly Thrift ●
2238 Acantholimon gontscharovii Czerniak.;高恩恰洛夫彩花●☆
2239 Acantholimon hedinii Ostenf.;海丁彩花;Hedin Pricklythrift,Hedin Prickly-thrift ●
2240 Acantholimon hissaricum Lincz.;希萨尔彩花●☆
2241 Acantholimon hohenacheiri (Jaub. et Spach) Boiss.;呵海纳契彩花;Hohenacheir Pricklythrift ●
2242 Acantholimon incomptum Boiss. et Buhse;装饰彩花●☆
2243 Acantholimon karatavicum Pavlov;卡拉塔夫彩花●☆
2244 Acantholimon karelinii (Shcheg.) Bunge;花柴彩花;Karelin Pricklythrift ●
2245 Acantholimon karelinii Bunge;卡氏彩花●
2246 Acantholimon kaschgaricum Lincz.;喀什彩花●
2247 Acantholimon khorassanicum Czerniak.;呼罗珊彩花●☆
2248 Acantholimon knorringianum Lincz.;克诺林彩花●☆
2249 Acantholimon kokandense Bunge;浩罕彩花;Haohan Prickly-thrift,Kokanic Pricklythrift ●
2250 Acantholimon korolkovii (Regel) Korovin;考氏彩花●☆
2251 Acantholimon laevigatum (T. X. Peng) Kamelin;光萼彩花;Glabrous-callyx Common Pricklythrift,Smoothcalyx Pricklythrift ●
2252 Acantholimon laxum Czerniak;松散彩花●☆
2253 Acantholimon leptostachyum Aitch. et Hemsl.;纤穗彩花●☆
2254 Acantholimon lepturoides (Jaub. et Spach) Boiss.;细穗彩花;Lepturus Pricklythrift ●
2255 Acantholimon longiflorum Boiss.;长花彩花●☆
2256 Acantholimon longiscapum Bokhari;长葶彩花●☆
2257 Acantholimon lycopodioides (Girard) Boiss.;石松彩花●
2258 Acantholimon lycopodioides Boiss. = Acantholimon lycopodioides (Girard) Boiss. ●
2259 Acantholimon majewianum O. Fedtsch. et B. Fedtsch.;马耶彩花●☆
2260 Acantholimon margaritae Korovin;马尔彩花●☆
2261 Acantholimon melananthum Boiss.;黑萼彩花●☆
2262 Acantholimon mikeschinii Lincz.;米氏彩花●☆
2263 Acantholimon minshelkense Pavlov;科马罗夫彩花●☆
2264 Acantholimon mirandum Lincz.;奇异彩花●☆
2265 Acantholimon mirum Lincz.;中亚奇异彩花●☆
2266 Acantholimon munroanum Aitch. et Hemsl.;短尖彩花●☆
2267 Acantholimon nikitinii Lincz.;尼氏彩花●☆
2268 Acantholimon nuratavicum Zakirov;努拉套彩花●☆
2269 Acantholimon nuristanicum Kitam.;努里斯坦彩花●☆
2270 Acantholimon pamiricum Czerniak.;帕米尔彩花●
2271 Acantholimon parviflorum Regel;小花彩花●☆
2272 Acantholimon polystachyum Boiss.;多穗彩花●☆
2273 Acantholimon popovii Czerniak.;乌恰彩花;Papov Pricklythrift,Papov Prickly-thrift,Xingjiang Pricklythrift ●
2274 Acantholimon procumbens Czerniak.;平铺彩花●☆
2275 Acantholimon pskemense Lincz.;普斯克姆彩花●☆
2276 Acantholimon pterostegium Bunge;翅盖彩花●☆
2277 Acantholimon pulchellum Korovin;玲珑彩花;Pretty Pricklythrift,Pretty Prickly-thrift ●
2278 Acantholimon purpureum Korovin;紫彩花●☆
2279 Acantholimon quinquelobum Bunge;五裂彩花●☆
2280 Acantholimon raddeanum Czerniak.;拉德彩花●
2281 Acantholimon roborowskii Czerniak.;新疆彩花;Roborowsk Prickly-thrift ●
2282 Acantholimon ruprechtii Bunge;鲁氏彩花●
2283 Acantholimon sackenii Bunge;萨肯彩花●☆
2284 Acantholimon sahendicum Boiss. et Buhse;萨亨迪彩花●☆
2285 Acantholimon schemachense Grossh.;晒马彩花●☆
2286 Acantholimon scirpinum Bunge;藨草彩花●☆
2287 Acantholimon scorpius Boiss.;蝎尾彩花●☆
2288 Acantholimon squarrosum Pavlov;粗鳞彩花●☆
2289 Acantholimon stocksii Boiss.;斯托克斯彩花●☆
2290 Acantholimon strictum Czerniak.;条纹彩花●☆
2291 Acantholimon tarbagataicum Gamajun.;塔尔巴哈彩花;Tarbagatai Pricklythrift,Tarbagatai Prickly-thrift ●
2292 Acantholimon tataricum Boiss.;鞑靼彩花;Tatarian Pricklythrift,Tatarian Prickly-thrift ●
2293 Acantholimon tenuiflorum Boiss.;细花彩花●
2294 Acantholimon tianschanicum Czerniak.;天山彩花;Tianshan Pricklythrift,Tianshan Prickly-thrift ●
2295 Acantholimon titovii Lincz.;梯氏彩花●
2296 Acantholimon tragacanthinum Boiss.;羊角刺彩花●☆
2297 Acantholimon truncatum Bunge;平截彩花●☆
2298 Acantholimon varivtzevae Czerniak.;瓦里夫彩花●☆
2299 Acantholimon velutinum Czerniak.;短绒毛彩花●☆
2300 Acantholimon venustum Boiss.;大花彩花(雅致彩花);Largeflower Pricklythrift,Lovely Pricklythrift ●
2301 Acantholimon virens Czerniak.;绿彩花●☆
2302 Acantholinum K. Koch = Acantholimon Boiss.(保留属名)●
2303 Acantholippia Griseb.(1874);刺甜舌草属●☆
2304 Acantholippia deserticola (Phil.) Moldenke;沙地刺甜舌草●☆
2305 Acantholippia salsoloides Griseb.;刺甜舌草●☆
2306 Acantholippia trifida (Gay) Moldenke;三裂刺甜舌草●☆
2307 Acantholobivia Backeb. = Echinopsis Zucc. ●
2308 Acantholobivia Backeb. = Lobivia Britton et Rose ■
2309 Acantholobivia Y. Ito = Acanthanthus Y. Ito ●
2310 Acantholobivia Y. Ito = Rebutia K. Schum. ●
2311 Acantholobivia Y. Ito(1957);橙饰球属●☆
2312 Acantholobivia euanthema (Backeb.) Y. Ito = Rebutia aureiflora Backeb. ●☆
2313 Acantholobivia tegeleriana (Backeb.) Backeb.;橙饰球(橙饰丸)●
2314 Acantholobivia tegeleriana (Backeb.) Backeb. = Echinopsis tegeleriana (Backeb.) D. R. Hunt ●☆
2315 Acantholoma Gaudich. ex Baill. = Pachystroma Müll. Arg. ☆
2316 Acanthomintha (A. Gray) A. Gray = Acanthomintha (A. Gray) Benth. et Hook. f. ■☆
2317 Acanthomintha (A. Gray) Benth. et Hook. f.(1876);刺薄荷属■☆
2318 Acanthomintha A. Gray = Acanthomintha (A. Gray) Benth. et Hook. f. ■☆
2319 Acanthomintha duttonii (Abrams) Jokerst;杜顿刺薄荷■☆
2320 Acanthomintha ilicifolia A. Gray;刺薄荷■☆
2321 Acanthomintha obovata Jeps.;倒卵刺薄荷■☆
2322 Acanthonema Hook. f.(1862)(保留属名);刺丝草属(刺林草属);Acanthonema ■☆
2323 Acanthonema J. Agardh(废弃属名) = Acanthonema Hook. f.(保留属名)■☆
2324 Acanthonema diandrum (Engl.) B. L. Burtt;刺丝草■☆
2325 Acanthonema strigosum Hook. f.;粗毛刺丝草(刺林草);Strigose Acanthonema ■☆
2326 Acanthonotus Benth. = Indigofera L. ●■

2327 Acanthonotus echinata (Willd.) Benth. = Indigofera nummulariifolia (L.) Livera ex Alston ■
2328 Acanthonotus echinatus Benth. = Indigofera nummularifolia (L.) Livera ex Alston ■
2329 Acanthonychia (DC.) Rohrb. = Cardionema DC. ■☆
2330 Acanthonychia (DC.) Rohrb. = Pentacaena Bartl. ■☆
2331 Acanthonychia Rohrb. = Cardionema DC. ■☆
2332 Acanthopale C. B. Clarke(1899);刺林草属(刺苞花属)●■☆
2333 Acanthopale albosetulosa C. B. Clarke;白刺林草●☆
2334 Acanthopale azaleoides C. B. Clarke;杜鹃刺林草●☆
2335 Acanthopale buchholzii (Lindau) C. B. Clarke = Dischistocalyx grandifolius C. B. Clarke ●☆
2336 Acanthopale cameronica Bremek. = Acanthopale decempedalis C. B. Clarke ●☆
2337 Acanthopale confertiflora (Lindau) C. B. Clarke;密花刺林草●☆
2338 Acanthopale dalzielii W. W. Sm. = Pteroptychia dalziellii (Sm.) H. S. Lo ●■
2339 Acanthopale debilis (Hemsl.) C. B. Clarke ex S. Moore = Championella tetrasperma (Champ. ex Benth.) Bremek. ●
2340 Acanthopale decempedalis C. B. Clarke;疏花刺林草●☆
2341 Acanthopale divaricata (Wall.) C. B. Clarke = Diflugossa divaricata (Nees) Bremek. ■
2342 Acanthopale japonica (Thunb.) C. B. Clarke ex S. Moore = Championella japonica (Thunb.) Bremek. ●■
2343 Acanthopale labordei (H. Lév.) Hand. -Mazz. = Championella labordei (H. Lév.) E. Hossain ●■
2344 Acanthopale laxiflora (Lindau) C. B. Clarke = Acanthopale decempedalis C. B. Clarke ●☆
2345 Acanthopale longipilosa Mildbr. ex Bremek.;长毛刺林草■☆
2346 Acanthopale madagascariensis (Baker) C. B. Clarke ex Bremek.; 马岛刺林草■☆
2347 Acanthopale oligantha (Miq.) C. B. Clarke ex S. Moore = Championella oligantha (Miq.) Bremek. ●
2348 Acanthopale oligantha (Miq.) C. B. Clarke ex S. Moore = Strobilanthes oligantha Miq. ■
2349 Acanthopale pubescens (Lindau) C. B. Clarke;毛刺林草■☆
2350 Acanthopale radicans (T. Anderson ex Benth.) C. B. Clarke ex Benth. = Championella tetrasperma (Champ. ex Benth.) Bremek. ●
2351 Acanthopale tetrasperma (Champ. ex Benth.) Hand. -Mazz. = Championella tetrasperma (Champ. ex Benth.) Bremek. ●
2352 Acanthopanax (Decne. et Planch.) Miq. = Eleutherococcus Maxim. ●
2353 Acanthopanax (Decne. et Planch.) Withe = Eleutherococcus Maxim. ●
2354 Acanthopanax Miq. = Eleutherococcus Maxim. ●
2355 Acanthopanax acerifolius Schelle;尖叶五加●☆
2356 Acanthopanax aculeatus (Aiton) H. Witte = Acanthopanax trifoliatus (L.) Merr. ●
2357 Acanthopanax aculeatus Seem.;南亚五加●
2358 Acanthopanax asperatus Franch. et Sav. = Eleutherococcus divaricatus (Siebold et Zucc.) S. Y. Hu ●☆
2359 Acanthopanax baoxinensis X. P. Fang et C. K. Hsieh = Eleutherococcus baoxinensis (X. P. Fang et C. K. Hsieh) P. S. Hsu et S. L. Pan ●
2360 Acanthopanax bockii R. Vig. = Nothopanax davidii (Franch.) Harms ex Diels ●
2361 Acanthopanax bodinieri H. Lév. = Brassaiopsis bodinieri (H. Lév.) J. Wen et Lowry ●
2362 Acanthopanax bodinieri H. Lév. = Brassaiopsis ciliata Seem. ●
2363 Acanthopanax bodinieri H. Lév. = Euaraliopsis ciliata (Dunn) Hutch. ●
2364 Acanthopanax brachypus Harms = Eleutherococcus brachypus (Harms) Nakai ●
2365 Acanthopanax cissifolius (Griff. ex C. B. Clarke) Harms = Eleutherococcus cissifolius (Griff. ex C. B. Clarke) Nakai ●
2366 Acanthopanax cissifolius (Griff. ex C. B. Clarke) Nakai var. glaber Y. R. Li = Eleutherococcus cissifolius (Griff. ex C. B. Clarke) Nakai ●
2367 Acanthopanax cissifolius (Griff.) Harms = Eleutherococcus cissifolius (Griff. ex C. B. Clarke) Nakai ●
2368 Acanthopanax cissifolius (Griff.) Harms var. glaber Y. R. Li = Eleutherococcus cissifolius (Griff. ex C. B. Clarke) Nakai var. glaber (Y. R. Li) P. S. Hsu et S. L. Pan ●
2369 Acanthopanax connatistylus S. C. Li et X. M. Liu = Eleutherococcus henryi Oliv. var. faberi (Harms) S. Y. Hu ●
2370 Acanthopanax cuspidatus G. Hoo; 尾叶五加; Cuspidateleaf Acanthopanax, Cuspidate-leaved Acanthopanax, Tailleaf Acanthopanax ●
2371 Acanthopanax cuspidatus G. Hoo = Eleutherococcus leucorrhizus Oliv. ●
2372 Acanthopanax cuspidatus G. Hoo var. tienchuanensis G. Hoo;天全尾叶五加;Tianquan Cuspidateleaf Acanthopanax ●
2373 Acanthopanax cuspidatus G. Hoo var. tienchuanensis G. Hoo = Acanthopanax cuspidatus G. Hoo ●
2374 Acanthopanax cuspidatus G. Hoo var. tienchuanensis G. Hoo = Eleutherococcus senticosus (Rupr. et Maxim.) Maxim. ●
2375 Acanthopanax davidii R. Vig. = Nothopanax davidii (Franch.) Harms ex Diels ●
2376 Acanthopanax delavayi R. Vig. = Nothopanax davidii (Franch.) Harms ex Diels ●
2377 Acanthopanax divaricatus (Siebold et Zucc.) Seem. = Eleutherococcus divaricatus (Siebold et Zucc.) S. Y. Hu ●☆
2378 Acanthopanax divaricatus (Siebold et Zucc.) Seem. f. inermis (Nakai) H. Hara = Eleutherococcus divaricatus (Siebold et Zucc.) S. Y. Hu f. inermis (Nakai) H. Ohashi ●☆
2379 Acanthopanax divaricatus (Siebold et Zucc.) Seem. var. inerme Makino = Eleutherococcus divaricatus (Siebold et Zucc.) S. Y. Hu f. inermis (Nakai) H. Ohashi ●☆
2380 Acanthopanax diversifolius Hemsl. = Nothopanax davidii (Franch.) Harms ex Diels ●
2381 Acanthopanax ehongensis Z. T. Zhu = Gamblea ciliata C. B. Clarke var. evodiifolia (Franch.) C. B. Shang et al. ●
2382 Acanthopanax eleutheristylus G. Hoo = Eleutherococcus eleutheristylus (G. Hoo) H. Ohashi ●
2383 Acanthopanax eleutheristylus G. Hoo var. glaucus K. M. Feng;粉绿离柱五加;Glaucous Separatestyle Acanthopanax ●
2384 Acanthopanax eleutheristylus G. Hoo var. simplex G. Hoo = Eleutherococcus eleutheristylus (G. Hoo) H. Ohashi ●
2385 Acanthopanax eleutheristylus G. Hoo var. simplex G. Hoo = Eleutherococcus eleutheristylus (G. Hoo) H. Ohashi var. simplex (G. Hoo) H. Ohashi ●
2386 Acanthopanax esquirolii H. Lév. = Brassaiopsis glomerulata (Blume) Regel ●
2387 Acanthopanax evodiifolius Franch.;吴茱萸叶五加(树三加,树五加,吴茱叶五加,吴茱萸五加,萸叶五加); Evodialeaf

Acanthopanax, Evodia-leaved Acanthopanax ●

2388 Acanthopanax evodiifolius Franch. var. ferrugineus W. W. Sm.; 锈毛吴茱萸叶五加(乔木五加,锈毛吴茱萸五加); Ferrugineous Evodialeaf Acanthopanax ●

2389 Acanthopanax evodiifolius Franch. var. ferrugineus W. W. Sm. = Gamblea ciliata C. B. Clarke ●

2390 Acanthopanax evodiifolius Franch. var. glaucus K. M. Feng = Gamblea ciliata C. B. Clarke ●

2391 Acanthopanax evodiifolius Franch. var. gracilis W. W. Sm.; 细梗吴茱萸叶五加(短梗吴茱萸叶五加); Gracilepeduncle Evodialeaf Acanthopanax, Slenderstalk Acanthopanax ●

2392 Acanthopanax evodiifolius Franch. var. gracilis W. W. Sm. = Gamblea ciliata C. B. Clarke ●

2393 Acanthopanax evodiifolius Franch. var. pseudo-evodiifolius K. M. Feng; 滇南吴茱萸叶五加(假吴茱萸叶五加); South Yunnan Acanthopanax ●

2394 Acanthopanax evodiifolius Franch. var. pseudo-evodiifolius K. M. Feng = Gamblea pseudoevodiifolia (K. M. Feng) C. B. Shang, Lowry et Frodin ●

2395 Acanthopanax fargesii (Franch.) C. B. Shang; 中华五加; China Acanthopanax, Chinese Acanthopanax ●

2396 Acanthopanax fargesii (Franch.) C. B. Shang = Chengiopanax fargesii (Franch.) C. B. Shang et J. Y. Huang ●

2397 Acanthopanax giraldii Harms = Eleutherococcus giraldii (Harms) Nakai ●

2398 Acanthopanax giraldii Harms var. hispidus G. Hoo = Acanthopanax giraldii Harms ●

2399 Acanthopanax giraldii Harms var. hispidus G. Hoo = Eleutherococcus giraldii (Harms) Nakai f. hispidus (G. Hoo) H. Ohashi ●

2400 Acanthopanax giraldii Harms var. hispidus G. Hoo = Eleutherococcus giraldii (Harms) Nakai ●

2401 Acanthopanax giraldii Harms var. inermis Harms et Rehder = Eleutherococcus giraldii (Harms) Nakai ●

2402 Acanthopanax giraldii Harms var. pilosulus Rehder = Eleutherococcus wilsonii (Harms) Nakai var. pilosulus (Rehder) P. S. Hsu et S. L. Pan ●

2403 Acanthopanax gracilistylus W. W. Sm. = Eleutherococcus nodiflorus (Dunn) S. Y. Hu ●

2404 Acanthopanax gracilistylus W. W. Sm. var. major G. Hoo = Eleutherococcus nodiflorus (Dunn) S. Y. Hu ●

2405 Acanthopanax gracilistylus W. W. Sm. var. nodiflorus (Dunn) H. L. Li = Eleutherococcus nodiflorus (Dunn) S. Y. Hu ●

2406 Acanthopanax gracilistylus W. W. Sm. var. pubescens (Pamp.) H. L. Li; 短毛五加; Pubescent Slenderstyle Acanthopanax, Pubescent-leaf Slenderstyle Acanthopanax ●

2407 Acanthopanax gracilistylus W. W. Sm. var. pubescens (Pamp.) H. L. Li = Eleutherococcus nodiflorus (Dunn) S. Y. Hu ●

2408 Acanthopanax gracilistylus W. W. Sm. var. trifoliolatus C. B. Shang; 三叶五加(三叶细柱五加); Threeleaf Slenderstyle Acanthopanax ●

2409 Acanthopanax gracilistylus W. W. Sm. var. trifoliolatus C. B. Shang = Eleutherococcus nodiflorus (Dunn) S. Y. Hu ●

2410 Acanthopanax gracilistylus W. W. Sm. var. villosulus (Harms) H. L. Li; 柔毛五加; Villoseleaf Slenderstyle Acanthopanax ●

2411 Acanthopanax gracilistylus W. W. Sm. var. villosulus (Harms) H. L. Li = Eleutherococcus nodiflorus (Dunn) S. Y. Hu ●

2412 Acanthopanax henryi (Oliv.) Harms; 亨利五加(糙叶五加,刺五加); Henry Acanthopanax ●

2413 Acanthopanax henryi (Oliv.) Harms = Eleutherococcus henryi Oliv. ●

2414 Acanthopanax henryi (Oliv.) Harms var. faberi Harms; 费伯糙叶五加(毛梗糙叶五加); Pubescentpedicel Henry Acanthopanax ●

2415 Acanthopanax henryi (Oliv.) Harms var. faberi Harms = Eleutherococcus henryi Oliv. var. faberi (Harms) S. Y. Hu ●

2416 Acanthopanax higoensis Hatus. = Eleutherococcus higoensis (Hatus.) H. Ohba ●☆

2417 Acanthopanax hondae Matsuda = Acanthopanax gracilistylus W. W. Sm. ●

2418 Acanthopanax hondae Matsuda = Eleutherococcus nodiflorus (Dunn) S. Y. Hu ●

2419 Acanthopanax humillimus Y. S. Lian et Xue L. Chen = Eleutherococcus giraldii (Harms) Nakai ●

2420 Acanthopanax hypoleucus Makino = Eleutherococcus hypoleucus (Makino) Nakai ●☆

2421 Acanthopanax innovans (Siebold et Zucc.) Franch. et Sav.; 新生五加(新生刺楸)●☆

2422 Acanthopanax innovans (Siebold et Zucc.) Franch. et Sav. = Gamblea innovans (Siebold et Zucc.) C. B. Shang, Lowry et Frodin ●☆

2423 Acanthopanax japonicus Franch. et Sav. = Acanthopanax nipponicus Makino ●☆

2424 Acanthopanax japonicus Franch. et Sav. = Eleutherococcus spinosus (L. f.) S. Y. Hu var. japonicus (Franch. et Sav.) H. Ohba ●☆

2425 Acanthopanax japonicus Franch. et Sav. f. ionanthus Nakai = Eleutherococcus spinosus (L. f.) S. Y. Hu var. japonicus (Franch. et Sav.) H. Ohba f. ionanthus (Nakai) H. Ohba ●☆

2426 Acanthopanax japonicus Franch. et Sav. f. kiusianus (Nakai) H. Hara = Eleutherococcus spinosus (L. f.) S. Y. Hu var. japonicus (Franch. et Sav.) H. Ohba f. kiusianus (Nakai) H. Ohba ●☆

2427 Acanthopanax japonicus Franch. et Sav. var. nikaianus (Koidz. ex Nakai) H. Hara = Eleutherococcus spinosus (L. f.) S. Y. Hu var. nikaianus (Koidz. ex Nakai) H. Ohba ●☆

2428 Acanthopanax lasiogyne Harms = Eleutherococcus lasiogyne (Harms) S. Y. Hu ●

2429 Acanthopanax lasiogyne Harms var. ferrugineus Y. R. Li; 锈毛康定五加; Rusty-hair Kangding Acanthopanax ●

2430 Acanthopanax lasiogyne Harms var. ferrugineus Y. R. Li = Eleutherococcus lasiogyne (Harms) S. Y. Hu ●

2431 Acanthopanax leucorrhizus (Oliv.) Harms = Eleutherococcus leucorrhizus Oliv. ●

2432 Acanthopanax leucorrhizus (Oliv.) Harms f. angustifoliatus G. Hoo = Eleutherococcus leucorrhizus Oliv. ●

2433 Acanthopanax leucorrhizus (Oliv.) Harms var. angustifoliolatus (G. Hoo) C. B. Shang; 窄叶叶藤五加(藤五加狭叶变型); Narrowleaf Acanthopanax ●

2434 Acanthopanax leucorrhizus (Oliv.) Harms var. angustifoliolatus (G. Hoo) C. B. Shang = Eleutherococcus leucorrhizus Oliv. ●

2435 Acanthopanax leucorrhizus (Oliv.) Harms var. angustifoliolatus G. Hoo = Acanthopanax leucorrhizus (Oliv.) Harms var. angustifoliolatus (G. Hoo) C. B. Shang ●

2436 Acanthopanax leucorrhizus (Oliv.) Harms var. axilaritomentosus G. Hoo; 腋毛藤五加; Axillarytomentose Whiteroot Acanthopanax ●

2437 Acanthopanax leucorrhizus (Oliv.) Harms var. axillaritomentosus G. Hoo = Eleutherococcus leucorrhizus Oliv. ●

2438 Acanthopanax leucorrhizus (Oliv.) Harms var. brevipedunculatus Y. R. Ling;短梗藤五加;Short-stalk Cane Acanthopanax ●

2439 Acanthopanax leucorrhizus (Oliv.) Harms var. fulvescens Harms et Rehder;糙叶藤五加(毛叶藤五加);Muricateleaf Whiteroot Acanthopanax ●

2440 Acanthopanax leucorrhizus (Oliv.) Harms var. fulvescens Harms et Rehder = Eleutherococcus leucorrhizus Oliv. var. fulvescens (Harms et Rehder) Nakai ●

2441 Acanthopanax leucorrhizus (Oliv.) Harms var. scaberulus Harms = Eleutherococcus leucorrhizus Oliv. var. scaberulus (Harms et Rehder) Nakai ●

2442 Acanthopanax leucorrhizus Oliv. var. scaberulus Harms et Rehder = Eleutherococcus leucorrhizus Oliv. var. scaberulus (Harms et Rehder) Nakai ●

2443 Acanthopanax longipes Hand.-Mazz. = Acanthopanax leucorrhizus (Oliv.) Harms var. fulvescens Harms et Rehder ●

2444 Acanthopanax malayanus M. R. Hend.;马来五加●

2445 Acanthopanax nanpingensis X. P. Fang et C. K. Hsieh;南坪五加;Nanping Acanthopanax ●

2446 Acanthopanax nanpingensis X. P. Fang et C. K. Hsieh = Eleutherococcus wilsonii (Harms) Nakai ●

2447 Acanthopanax nikaianus Koidz. ex Nakai = Eleutherococcus spinosus (L. f.) S. Y. Hu var. nikaianus (Koidz. ex Nakai) H. Ohba ●☆

2448 Acanthopanax nipponicus Makino = Eleutherococcus spinosus (L. f.) S. Y. Hu var. japonicus (Franch. et Sav.) H. Ohba ●☆

2449 Acanthopanax nodiflorus Dunn = Acanthopanax gracilistylus W. W. Sm. var. nodiflorus (Dunn) H. L. Li ●

2450 Acanthopanax nodiflorus Dunn = Eleutherococcus nodiflorus (Dunn) S. Y. Hu ●

2451 Acanthopanax obovatus G. Hoo;倒卵叶五加;Obovateleaf Acanthopanax, Obovate-leaved Acanthopanax ●

2452 Acanthopanax obovatus G. Hoo = Eleutherococcus brachypus (Harms) Nakai ●

2453 Acanthopanax pentaphyllus (Siebold et Zucc.) Marchal = Acanthopanax sieboldianus Makino ●

2454 Acanthopanax phanerophlebius Merr. et Chun = Brassaiopsis phanerophlebia (Merr. et Chun) C. N. Ho ●

2455 Acanthopanax phanerophlebius Merr. et Chun = Brassaiopsis tripteris (H. Lév.) Rehder ●

2456 Acanthopanax rehderianus Harms = Eleutherococcus rehderianus (Harms) Nakai ●

2457 Acanthopanax rehderianus Harms var. longipedunculatus G. Hoo = Eleutherococcus rehderianus (Harms) Nakai var. longipedunculatus (G. Hoo) H. Ohashi ●

2458 Acanthopanax rehderianus Harms var. longipedunculatus G. Hoo = Eleutherococcus rehderianus (Harms) Nakai ●

2459 Acanthopanax ricinifolius (Siebold et Zucc.) Seem. = Kalopanax septemlobus (Thunb.) Koidz. ●

2460 Acanthopanax ricinifolius Seem. = Kalopanax septemlobus (Thunb.) Nakai ●

2461 Acanthopanax ricinifolius Seem. var. maximowiczii Schneid. = Kalopanax septemlobus (Thunb.) Koidz. f. maximowiczii (Van Houtte) H. Ohashi ●

2462 Acanthopanax rosthornii (Harms) R. Vig. = Macropanax rosthornii (Harms) C. Y. Wu ex G. Hoo ●

2463 Acanthopanax rosthornii R. Vig. = Macropanax rosthornii (Harms) C. Y. Wu ex G. Hoo ●

2464 Acanthopanax scandens G. Hoo = Eleutherococcus scandens (G. Hoo) H. Ohashi ●

2465 Acanthopanax scandens Poir. = Eleutherococcus nodiflorus (Dunn) S. Y. Hu ●

2466 Acanthopanax sciadophylloides Franch. et Sav. = Chengiopanax sciadophylloides (Franch. et Sav.) C. B. Shang et J. Y. Huang ●☆

2467 Acanthopanax senticosus (Rupr. et Maxim.) Harms = Eleutherococcus senticosus (Rupr. et Maxim.) Maxim. ●

2468 Acanthopanax senticosus (Rupr. et Maxim.) Harms f. inermis (Kom.) Harms = Eleutherococcus divaricatus (Siebold et Zucc.) S. Y. Hu f. inermis (Nakai) H. Ohashi ●☆

2469 Acanthopanax senticosus (Rupr. et Maxim.) Harms f. subinermis (Regel) Harms = Eleutherococcus senticosus (Rupr. et Maxim.) Maxim. ●

2470 Acanthopanax senticosus (Rupr. et Maxim.) Harms f. subinermis (Regel) H. L. Li = Eleutherococcus senticosus (Rupr. et Maxim.) Maxim. ●

2471 Acanthopanax senticosus (Rupr. et Maxim.) Harms var. brevistamineus S. F. Gu;短蕊刺五加;Short-stamen Manyprickle Acanthopanax ●

2472 Acanthopanax senticosus (Rupr. et Maxim.) Harms var. brevistamineus S. F. Gu = Eleutherococcus senticosus (Rupr. et Maxim.) Maxim. ●

2473 Acanthopanax senticosus (Rupr. et Maxim.) Harms var. subinermis (Regel) Kitag. = Eleutherococcus divaricatus (Siebold et Zucc.) S. Y. Hu f. inermis (Nakai) H. Ohashi ●☆

2474 Acanthopanax senticosus (Rupr. et Maxim.) Harms var. subinermis (Regel) Kitag. = Eleutherococcus senticosus (Rupr. et Maxim.) Maxim. ●

2475 Acanthopanax sepium Seem. = Acanthopanax trifoliatus (L.) Merr. ●

2476 Acanthopanax sepium Seem. = Eleutherococcus trifoliatus (L.) S. Y. Hu ●

2477 Acanthopanax septemlobus (Thunb.) Koidz. ex Rehder = Kalopanax septemlobus (Thunb.) Koidz. ●

2478 Acanthopanax septemlobus (Thunb.) Koidz. ex Rehder var. magnificus (Zabel) W. C. Cheng = Kalopanax septemlobus (Thunb.) Koidz. ●

2479 Acanthopanax septemlobus (Thunb.) Koidz. ex Rehder var. maximowiczii (Van Houtte) W. C. Cheng = Kalopanax septemlobus (Thunb.) Koidz. ●

2480 Acanthopanax septemlobus (Thunb.) Koidz. var. magnificus W. C. Cheng = Kalopanax septemlobus (Thunb.) Koidz. f. maximowiczii (Van Houtte) H. Ohashi ●

2481 Acanthopanax septemlobus Koidz. ex Rehder = Kalopanax septemlobus (Thunb.) Nakai ●

2482 Acanthopanax septemlobus Koidz. ex Rehder var. maximowiczii W. C. Cheng = Kalopanax septemlobus (Thunb.) Koidz. f. maximowiczii (Van Houtte) H. Ohashi ●

2483 Acanthopanax sessiliflorus (Rupr. et Maxim.) Seem. = Eleutherococcus sessiliflorus (Rupr. et Maxim.) S. Y. Hu ●

2484 Acanthopanax sessiliflorus (Rupr. et Maxim.) Seem. var. parviceps Rehder = Eleutherococcus sessiliflorus (Rupr. et Maxim.) S. Y. Hu var. parviceps (Rehder) S. Y. Hu ●

2485 Acanthopanax sessiliflorus (Rupr. et Maxim.) Seem. var. parviceps Rehder = Eleutherococcus sessiliflorus (Rupr. et Maxim.) S. Y. Hu ●

2486 Acanthopanax setchuenensis Harms = Eleutherococcus leucorrhizus Oliv. var. setchuenensis (Harms ex Diels) C. B. Shang et J. Y. Huang ●

2487 Acanthopanax setchuenensis Harms ex Diels = Eleutherococcus leucorrhizus Oliv. var. setchuenensis (Harms ex Diels) C. B. Shang et J. Y. Huang ●

2488 Acanthopanax setchuenensis Harms ex Diels var. latifoliatus G. Hoo;阔叶蜀五加;Broad-leaf Sichuan Acanthopanax ●

2489 Acanthopanax setchuenensis Harms ex Diels var. latifoliatus G. Hoo = Eleutherococcus leucorrhizus Oliv. var. setchuenensis (Harms ex Diels) C. B. Shang et J. Y. Huang ●

2490 Acanthopanax setchuenensis Harms Harms var. latifoliatus G. Hoo = Eleutherococcus leucorrhizus Oliv. var. setchuenensis (Harms ex Diels) C. B. Shang et J. Y. Huang ●

2491 Acanthopanax setosus (H. L. Li) C. B. Shang = Eleutherococcus setosus (H. L. Li) Y. R. Ling ●

2492 Acanthopanax setulosus Franch. = Eleutherococcus setulosus (Franch.) S. Y. Hu ●

2493 Acanthopanax sieboldianus Makino = Eleutherococcus sieboldianus (Makino) Koidz. ●

2494 Acanthopanax sieboldianus Makino f. variegatus (Rehder) Rehder = Eleutherococcus sieboldianus (Makino) Koidz. f. variegatus (Rehder) S. Y. Hu ●

2495 Acanthopanax simonii Simon-Louis ex C. K. Schneid.;刚毛五加(白五加,雷五加,西门五加,西蒙五加);Hispid Acanthopanax ●

2496 Acanthopanax simonii Simon-Louis ex C. K. Schneid. var. longipedicellatus G. Hoo;长梗刚毛五加;Longpedicellate Acanthopanax ●

2497 Acanthopanax simonii Simon-Louis ex Mouill. = Eleutherococcus leucorrhizus Oliv. var. scaberulus (Harms et Rehder) Nakai ●

2498 Acanthopanax simonii Simon-Louis ex Mouill. var. longipedicellatus G. Hoo = Eleutherococcus leucorrhizus Oliv. var. scaberulus (Harms et Rehder) Nakai ●

2499 Acanthopanax sinensis G. Hoo = Acanthopanax fargesii (Franch.) C. B. Shang ●

2500 Acanthopanax sinensis G. Hoo = Chengiopanax fargesii (Franch.) C. B. Shang et J. Y. Huang ●

2501 Acanthopanax spinosus (L. f.) Miq. = Eleutherococcus spinosus (L. f.) S. Y. Hu ●☆

2502 Acanthopanax spinosus (L. f.) Miq. f. variegatus Makino et Nemoto;白边疏刺五加●☆

2503 Acanthopanax spinosus (L. f.) Miq. var. pubescens Pamp. = Acanthopanax gracilistylus W. W. Sm. var. pubescens (Pamp.) H. L. Li ●

2504 Acanthopanax spinosus (L. f.) Miq. var. pubescens Pamp. = Eleutherococcus nodiflorus (Dunn) S. Y. Hu ●

2505 Acanthopanax stenophyllus Harms;太白五加(太白山五加,狭叶五加);Narrowleaf Acanthopanax, Stenophyllous Acanthopanax, Taibaishan Acanthopanax ●

2506 Acanthopanax stenophyllus Harms = Eleutherococcus wilsonii (Harms) Nakai ●

2507 Acanthopanax stenophyllus Harms f. angustissimus Rehder = Eleutherococcus wilsonii (Harms) Nakai ●

2508 Acanthopanax stenophyllus Harms f. dilatatus Rehder = Eleutherococcus wilsonii (Harms) Nakai ●

2509 Acanthopanax ternatus Rehder = Eleutherococcus lasiogyne (Harms) S. Y. Hu ●

2510 Acanthopanax trichodon Franch. et Sav. = Eleutherococcus trichodon (Franch. et Sav.) H. Ohashi ●☆

2511 Acanthopanax trifoliatus (L.) Merr. = Eleutherococcus trifoliatus (L.) S. Y. Hu ●

2512 Acanthopanax trifoliatus (L.) Merr. var. setosus H. L. Li = Acanthopanax setosus (H. L. Li) C. B. Shang ●

2513 Acanthopanax trifoliatus (L.) Merr. var. setosus H. L. Li = Eleutherococcus setosus (H. L. Li) Y. R. Ling ●

2514 Acanthopanax verticillatus G. Hoo = Eleutherococcus verticillatus (G. Hoo) H. Ohashi ●

2515 Acanthopanax villosulus Harms = Acanthopanax gracilistylus W. W. Sm. var. villosulus (Harms) H. L. Li ●

2516 Acanthopanax villosulus Harms = Eleutherococcus nodiflorus (Dunn) S. Y. Hu ●

2517 Acanthopanax wardii W. W. Sm. = Acanthopanax lasiogyne Harms ●

2518 Acanthopanax wardii W. W. Sm. = Eleutherococcus lasiogyne (Harms) S. Y. Hu ●

2519 Acanthopanax wilsonii (Harms) Nakai var. pilosulus (Rehder) X. P. Fang et C. K. Hsieh = Eleutherococcus wilsonii (Harms) Nakai var. pilosulus (Rehder) P. S. Hsu et S. L. Pan ●

2520 Acanthopanax wilsonii Harms = Eleutherococcus wilsonii (Harms) Nakai ●

2521 Acanthopanax xizangensis Y. R. Li;西藏五加;Xizang Acanthopanax ●

2522 Acanthopanax yui H. L. Li;云南五加(德钦五加);Yu Acanthopanax ●

2523 Acanthopanax yui H. L. Li = Acanthopanax giraldii Harms ●

2524 Acanthopanax yui H. L. Li = Eleutherococcus giraldii (Harms) Nakai ●

2525 Acanthopanax yui H. L. Li var. longipedunculatus G. Hoo = Acanthopanax verticillatus G. Hoo ●

2526 Acanthopanax yui H. L. Li var. longipedunculatus G. Hoo = Eleutherococcus giraldii (Harms) Nakai ●

2527 Acanthopanax yui H. L. Li var. parvispinosus G. Hoo = Acanthopanax yui H. L. Li ●

2528 Acanthopanax yui H. L. Li var. parvispinosus G. Hoo = Eleutherococcus giraldii (Harms) Nakai ●

2529 Acanthopanax yui H. L. Li var. villosus Y. R. Li = Acanthopanax giraldii Harms ●

2530 Acanthopanax yui H. L. Li var. villosus Y. R. Li = Eleutherococcus giraldii (Harms) Nakai ●

2531 Acanthopanax zhejiangensis X. J. Xue et S. T. Fang;浙江五加;Zhejiang Acanthopanax ●

2532 Acanthopanax zhejiangensis X. J. Xue et S. T. Fang = Eleutherococcus setulosus (Franch.) S. Y. Hu ●

2533 Acanthopetalus Y. Ito = Arthrocereus A. Berger(保留属名)●☆

2534 Acanthopetalus Y. Ito = Echinopsis Zucc. ●

2535 Acanthopetalus Y. Ito = Setiechinopsis (Backeb.) de Hass ■☆

2536 Acanthopetalus Y. Ito(1957);刺瓣掌属●☆

2537 Acanthopetalus mirabilis (Speg.) Y. Ito;刺瓣掌■☆

2538 Acanthophaca Nevski = Astragalus L. ●■

2539 Acanthophippium Blume = Acanthephippium Blume ■

2540 Acanthophippium bicolor Lindl. = Acanthephippium bicolora Lindl. ■

2541 Acanthophippium javanicum Blume = Acanthephippium javanicum Blume ■

2542 Acanthophippium mantinianum Lindl. et Cogn. = Acanthephippium mantinianum Lindl. et Cogn. ■☆
2543 Acanthophippium pictum Fukuy. = Acanthephippium pictum Fukuy. ■
2544 Acanthophippium sinense Rolfe = Acanthephippium sinense Rolfe ■
2545 Acanthophippium striatum Lindl. = Acanthephippium striatum Lindl. ■
2546 Acanthophippium sylhetens Lindl. = Acanthephippium sylhetense Lindl. ■
2547 Acanthophippium unguiculatum (Hayata) Fukuy. = Acanthephippium unguiculatum (Hayata) Fukuy. ■
2548 Acanthophippium unguiculatum (Hayata) Fukuy. = Acanthophippium striatum Lindl. ■
2549 Acanthophippium yamamotoi Hayata = Acanthephippium sylhetense Lindl. ■
2550 Acanthophoenix H. Wendl. (1867);刺棕榈属(茨根榈属,刺根榈属,刺海枣属,刺椰属,刺椰子属,刺棕属,红脉椰属,棘榈属);Barbel Palm, Spine Palm, Spine-Areca ●☆
2551 Acanthophoenix crinita H. Hendl.;长毛刺棕榈;Wooly Spine Palm ●☆
2552 Acanthophoenix nobilis Blatt. = Deckenia nobilis H. Wendl. ●☆
2553 Acanthophoenix rubra H. Hendl.;红刺棕榈;Barbel Palm, Red Spine Palm ●☆
2554 Acanthophora Merr. = Aralia L. ●■
2555 Acanthophyllum C. A. Mey. (1831);刺叶属(刺叶石竹属);Spinepink ■
2556 Acanthophyllum Hook. et Arn. = Nassauvia Comm. ex Juss. ●☆
2557 Acanthophyllum Less. = Cichorium L. ■
2558 Acanthophyllum albidum Schischk.;白刺叶■☆
2559 Acanthophyllum crassifolium Boiss.;密叶刺叶■☆
2560 Acanthophyllum glaucum Bunge ex Boiss.;灰蓝刺叶■☆
2561 Acanthophyllum grandiflorum Stocks;大花刺叶■☆
2562 Acanthophyllum latifolium Lipsky;宽叶刺叶■☆
2563 Acanthophyllum laxiflorum Boiss.;疏花刺叶■☆
2564 Acanthophyllum microcephalum Boiss.;小头刺叶■☆
2565 Acanthophyllum pungens (Bunge) Boiss.;刺叶(刺叶石竹);Rockgarden Spinepink ■
2566 Acanthophyllum spinosum C. A. Mey. = Acanthophyllum pungens (Bunge) Boiss. ■
2567 Acanthophyton Less. = Cichorium L. ■
2568 Acanthophyton Sch. Bip. = Acanthophyllum Less. ■
2569 Acanthophytum Less. = Cichorium L. ■
2570 Acanthoplana C. Koch = Polylophium Boiss. ☆
2571 Acanthoprasium (Benth.) Spenn. = Ballota L. ●■☆
2572 Acanthoprasium (Benth.) Spenn. = Elbunis Raf. ●■☆
2573 Acanthoprasium Spenn. = Ballota L. ●■☆
2574 Acanthopsis Harv. (1842);拟老鼠簕属■☆
2575 Acanthopsis carduifolia (L. f.) Schinz;刺叶拟老鼠簕■☆
2576 Acanthopsis carduifolia (L. f.) Schinz var. glabra (Nees) C. B. Clarke = Acanthopsis carduifolia (L. f.) Schinz ■☆
2577 Acanthopsis carduifolia (L. f.) Schinz var. longearistata Schinz = Acanthopsis disperma Nees ■☆
2578 Acanthopsis disperma Nees;双籽拟老鼠簕■☆
2579 Acanthopsis glauca (Nees) Schinz;灰拟老鼠簕■☆
2580 Acanthopsis hoffmannseggiana (Nees) C. B. Clarke;豪夫拟老鼠簕■☆
2581 Acanthopsis horrida Nees;多刺拟老鼠簕■☆
2582 Acanthopsis scullyi (S. Moore) Oberm.;斯库里拟老鼠簕■☆
2583 Acanthopsis spathularis (Nees) Schinz;匙叶拟老鼠簕■☆
2584 Acanthopsis trispina C. B. Clarke;三刺拟老鼠簕■☆
2585 Acanthopteron Britton = Mimosa L. ●■
2586 Acanthopteron Britton(1928);刺翼豆属●☆
2587 Acanthopteron laceratum (Rose) Britton;刺翼豆●☆
2588 Acanthopyxis Miq. ex Lanj. = Caperonia A. St. -Hil. ■☆
2589 Acanthorhipsalis (K. Schum.) Britton et Rose = Lepismium Pfeiff. ●☆
2590 Acanthorhipsalis (K. Schum.) Britton et Rose(1923);刺丝苇属(有刺丝苇属)●☆
2591 Acanthorhipsalis Britton et Rose = Lepismium Pfeiff. ●☆
2592 Acanthorhipsalis Britton et Rose = Rhipsalis Gaertn. (保留属名)●
2593 Acanthorhipsalis Kimnach = Lepismium Pfeiff. ●☆
2594 Acanthorhiza H. Wendl. = Cryosophila Blume ●☆
2595 Acanthorrhinum Rothm. (1943);针玄参属●☆
2596 Acanthorrhinum ramosissimum (Coss. et Durieu) Rothm.;针玄参●☆
2597 Acanthorrhiza H. Wendl. = Cryosophila Blume ●☆
2598 Acanthosabal Prosch. = Acoelorrhaphe H. Wendl. ●☆
2599 Acanthoscyphus Small = Oxytheca Nutt. ■☆
2600 Acanthosicyos Welw. = Acanthosicyos Welw. ex Hook. f. ●☆
2601 Acanthosicyos Welw. ex Hook. f. (1867);南非葫芦树属●☆
2602 Acanthosicyos horridus Welw. ex Hook. f.;沙丘南非葫芦树;Narra, Narras ●☆
2603 Acanthosicyos naudinianus (Sond.) C. Jeffrey;南非葫芦树●☆
2604 Acanthosicyus Post et Kuntze = Acanthosicyos Welw. ex Hook. f. ●☆
2605 Acanthosonchus (Sch. Bip.) Kirp. = Atalanthus D. Don ■☆
2606 Acanthosonchus Don ex Hoffm. = Sonchus L. ■
2607 Acanthosperma Vell. = Acicarpha Juss. ■☆
2608 Acanthospermum Schrank(1820)(保留属名);刺苞果属(刺苞菊属);Starbur ■
2609 Acanthospermum australe (Loefl.) Kuntze;刺苞果(刺苞菊);Paraguay Starbur ■
2610 Acanthospermum brasilum Schrank = Acanthospermum australe (Loefl.) Kuntze ■
2611 Acanthospermum glabratum (DC.) Wild;光刺苞果菊■☆
2612 Acanthospermum hispidum DC.;硬毛刺苞果(刚毛刺苞菊);Hispid Starbur ■
2613 Acanthospermum humile DC.;矮刺苞果;Low Starburr ■☆
2614 Acanthospermum xanthoides (Kunth) DC. var. glabratum DC. = Acanthospermum glabratum (DC.) Wild ■☆
2615 Acanthospermum xanthoides (Kunth) DC. var. obtusifolium DC. = Acanthospermum australe (Loefl.) Kuntze ■
2616 Acanthosphaera Lemmerm. (1899);刺球桑属●☆
2617 Acanthosphaera Warb. = Acanthosphaera Lemmerm. ●☆
2618 Acanthosphaera Warb. = Naucleopsis Miq. ●☆
2619 Acanthosphaera Warb. = Uleodendron Rauschert ●☆
2620 Acanthosphaera ulei Warb.;刺球桑●☆
2621 Acanthospora Spreng. = Bonapartea Ruiz et Pav. ■☆
2622 Acanthospora Spreng. = Tillandsia L. ■☆
2623 Acanthostachys Klotzsch(1840);松球凤梨属(刺穗凤梨属,松球属)■☆
2624 Acanthostachys Link = Acanthostachys Klotzsch ■☆
2625 Acanthostachys strobilacea (Schult. f.) Link, Klotzsch et Otto;松球凤梨(松果凤梨,松球菠萝)■☆
2626 Acanthostachyum Benth. et Hook. f. = Acanthostachys Klotzsch ■☆

2627 Acanthostelma Bidgood et Brummitt = Hoya R. Br. ●
2628 Acanthostelma Bidgood et Brummitt(1985);刺冠爵床属●☆
2629 Acanthostelma thymifolium (Chiov.) Bidgood et Brummitt = Crabbea thymifolia (Chiov.) Thulin ■☆
2630 Acanthostelma thymifolium (Chiov.) Bidgood et Brummitt = Neuracanthus thymifolius Chiov. ■☆
2631 Acanthostemma (Blume) Blume = Hoya R. Br. ●
2632 Acanthostemma Blume = Hoya R. Br. ●
2633 Acanthostyles R. M. King et H. Rob. (1971);刺柱菊属☆
2634 Acanthostyles buniifolius (Hook. et Arn.) R. M. King et H. Rob.;刺柱菊■☆
2635 Acanthosyris (Eichler) Griseb. (1879);刺沙针属●☆
2636 Acanthosyris Griseb. = Acanthosyris (Eichler) Griseb. ●☆
2637 Acanthosyris glabrata (Stapf) Stauffer;无刺沙针●☆
2638 Acanthosyris spinescens Griseb.;刺沙针●☆
2639 Acanthothamnus Brandegee(1909);刺灌卫矛属(刺灌属)●☆
2640 Acanthothamnus aphyllus Standl.;无叶刺灌卫矛●☆
2641 Acanthothamnus viridis Brandegee;刺灌卫矛●☆
2642 Acanthothapsus Gand. = Verbascum L. ■●
2643 Acanthotheca DC. = Dimorphotheca Vaill.(保留属名)■●☆
2644 Acanthotheca dentata DC. = Dimorphotheca sinuata DC. ■☆
2645 Acanthotheca integrifolia DC. = Dimorphotheca sinuata DC. ■☆
2646 Acanthotheca pinnatisecta DC. = Osteospermum pinnatum (Thunb.) Norl. ■☆
2647 Acanthotreculia Engl. (1908);非洲桑属●☆
2648 Acanthotreculia Engl. = Treculia Decne. ex Trécul ●☆
2649 Acanthotreculia winkleri Engl. = Treculia obovoidea N. E. Br. ●☆
2650 Acanthotrichilia (Urb.) O. F. Cook et G. N. Collins = Trichilia P. Browne(保留属名)●
2651 Acanthoxanthium (DC.) Fourr. = Xanthium L. ■
2652 Acanthoxanthium Fourr. = Xanthium L. ■
2653 Acanthoxanthium spinosum (L.) Fourr. = Xanthium spinosum L. ■
2654 Acanthura Lindau(1901);棘尾爵床属■☆
2655 Acanthura mattogrossensis Lindau;棘尾爵床■☆
2656 Acanthus (Tourn.) ex L. = Acanthus L. ●■
2657 Acanthus L. (1753);老鼠簕属(老鸦企属,叶蓟属);Acanthus, Bear's Breech, Bear's Breeches, Bear's-breech ●■
2658 Acanthus Tourn. ex L. = Acanthus L. ●■
2659 Acanthus arboreus C. B. Clarke = Acanthus pubescens (Thomson ex Oliv.) Engl. ●☆
2660 Acanthus arboreus Forssk. f. albiflorus Fiori = Acanthus polystachyus Delile ●☆
2661 Acanthus arboreus Forssk. var. pubescens Thomson ex Oliv. = Acanthus polystachyus Delile ●☆
2662 Acanthus arboreus Forssk. var. ruber Engl. = Acanthus sennii Chiov. ●☆
2663 Acanthus austromontanus Vollesen;南方山地老鼠簕■☆
2664 Acanthus balcanicus Heywood et I. Richardson;巴尔干老鼠簕(长叶老鼠簕);Balkan Acanthus, Bear's Breech ●☆
2665 Acanthus balcanicus Heywood et I. Richardson = Acanthus hungaricus (Borbás) Baen. ●☆
2666 Acanthus barteri T. Anderson = Acanthus montanus (Nees) T. Anderson ●☆
2667 Acanthus capensis L. f. = Blepharis capensis (L. f.) Pers. ■☆
2668 Acanthus carduifolius L. f. = Acanthopsis carduifolia (L. f.) Schinz ■☆
2669 Acanthus caudatus Lindau;尾状老鼠簕■☆
2670 Acanthus ciliaris Burm. f. = Blepharis maderaspatensis (L.) B. Heyne ex Roth ■☆
2671 Acanthus dioscoridis L.;薯蓣状老鼠簕■☆
2672 Acanthus ebracteatus Vahl;小花老鼠簕;Holly-leaved Mangrove, Jeruju, Sea Holly, Small-flowered Acanthus ●
2673 Acanthus ebracteatus Vahl var. xiamenensis (R. T. Zhang) C. Y. Wu et C. C. Hu;厦门老鼠簕;Xiamen Acanthus ●
2674 Acanthus ebracteatus Vahl var. xiamenensis (R. T. Zhang) C. Y. Wu et C. C. Hu = Acanthus xiamenensis R. T. Zhang ●
2675 Acanthus edulis Forssk. = Blepharis ciliaris (L.) B. L. Burtt ■☆
2676 Acanthus eminens C. B. Clarke;显著老鼠簕●☆
2677 Acanthus flamandii De Wild. = Acanthus polystachyus Delile ●☆
2678 Acanthus furcatus L. f. = Blepharis furcata (L. f.) Pers. ■☆
2679 Acanthus glaber E. Mey. = Acanthopsis carduifolia (L. f.) Schinz ■☆
2680 Acanthus glaucus E. Mey. = Acanthopsis glauca (Nees) Schinz ■☆
2681 Acanthus glomeratus Lam. = Blepharis procumbens (L. f.) Pers. ■☆
2682 Acanthus guineensis Heine et P. Taylor;几内亚老鼠簕●☆
2683 Acanthus hungaricus (Borbás) Baen.;匈牙利老鼠簕;Hungarian Acanthus ●☆
2684 Acanthus ilicifolius L.;老鼠簕(刺苞花,莨芀花,蚧爪簕,老鼠艻,老鼠竻,老鼠勒,老鼠怕,老鸦企,木老鼠簕,软骨牡丹,水老鼠簕);Hollyleaf Acanthus, Holly-leaved Acanthus ●
2685 Acanthus integrifolius L. f. = Blepharis integrifolia (L. f.) E. Mey. ex Schinz ■☆
2686 Acanthus kirkii T. Anderson = Sclerochiton kirkii (T. Anderson) C. B. Clarke ●☆
2687 Acanthus kulalensis Vollesen;库拉尔老鼠簕●☆
2688 Acanthus latisepalus C. B. Clarke;宽瓣老鼠簕●☆
2689 Acanthus leucostachyus Wall. = Acanthus leucostachyus Wall. ex Nees ●
2690 Acanthus leucostachyus Wall. ex Nees;刺苞老鼠簕(白穗蛤蟆花);Whitespike Acanthus, White-spiked Acanthus ●
2691 Acanthus longifolius Poir.;长叶老鼠簕;Longleaf Acanthus ●
2692 Acanthus longifolius Poir. = Acanthus hungaricus (Borbás) Baen. ●☆
2693 Acanthus lusitanicus Hort. = Acanthus mollis L. ●
2694 Acanthus maderaspatensis L. = Blepharis maderaspatensis (L.) B. Heyne ex Roth ■☆
2695 Acanthus mollis L.;蛤蟆花(莨力花,金蝉脱壳,柔毛老鼠簕);Bear's Breech, Bear's Breeches, Bear's-breech, Brank Ursine, Brank-Ursin, Cutberdill, Cutberdole, Sedocke, Soft Acanthus, Softleaved Bear's Breech ●
2696 Acanthus mollis L. subsp. platyphyllus Murb.;宽叶蛤蟆花●☆
2697 Acanthus montanus (Nees) T. Anderson;山地老鼠簕(八角簕)●☆
2698 Acanthus montanus T. Anderson = Acanthus montanus (Nees) T. Anderson ●☆
2699 Acanthus niger Mill.;黑老鼠簕;Balck Acanthus ●☆
2700 Acanthus nitidus S. Moore = Sclerochiton nitidus (S. Moore) C. B. Clarke ●☆
2701 Acanthus polycotomus (Lindau) C. B. Clarke = Crossandrella dusenii (Lindau) S. Moore ■☆
2702 Acanthus polycotomus Chiov. var. lanatus Cufod.;绵毛老鼠簕●☆
2703 Acanthus polystachyus Delile;多穗老鼠簕●☆
2704 Acanthus polystachyus Delile var. pseudopubescens Cufod. = Acanthus polystachyus Delile ●☆
2705 Acanthus procumbens L. f. = Blepharis procumbens (L. f.)

Pers. ■☆

2706 Acanthus pubescens (Thomson ex Oliv.) Engl.;短柔毛老鼠簕●☆

2707 Acanthus repens Vahl = Blepharis integrifolia (L. f.) E. Mey. ex Schinz ■☆

2708 Acanthus sennii Chiov.;森恩老鼠簕●☆

2709 Acanthus seretii De Wild.;赛雷老鼠簕●☆

2710 Acanthus spathularis E. Mey. = Acanthopsis spathularis (Nees) Schinz ■☆

2711 Acanthus spinosus L.;刺老鼠簕;Bear's Breeches,Oyster Plant, Spiny Acanthus,Spiny Bear's Breech,Spiny Bear's-breech ●☆

2712 Acanthus spinulosus Host = Acanthus mollis L. ●

2713 Acanthus tetragonus R. Br.;四角老鼠簕●☆

2714 Acanthus ueleensis De Wild.;韦莱老鼠簕●☆

2715 Acanthus ueleensis De Wild. subsp. mahaliensis Napper = Acanthus ueleensis De Wild. ●☆

2716 Acanthus ugandensis C. B. Clarke = Acanthus polystachyus Delile ●☆

2717 Acanthus vandermeireni De Wild. = Acanthus ueleensis De Wild. ●☆

2718 Acanthus vandermeirenii De Wild. var. violaceo-punctatus ? = Acanthus ueleensis De Wild. ●☆

2719 Acanthus villaeanus De Wild.;比利亚老鼠簕●☆

2720 Acanthus xiamenensis R. T. Zhang = Acanthus ebracteatus Vahl var. xiamenensis (R. T. Zhang) C. Y. Wu et C. C. Hu ●

2721 Acanthyllis Pomel = Anthyllis L. ■☆

2722 Acanthyllis numidica (Coss. et Durieu) Pomel = Astragalus armatus Willd. subsp. numidicus (Murb.) Tietz ■☆

2723 Acanthyllis tragacanthoides (Desf.) Pomel = Astragalus armatus Willd. ■☆

2724 Acareosperma Gagnep. (1919);东南亚葡萄属●☆

2725 Acareosperma spireanum Gagnep.;东南亚葡萄●☆

2726 Acarna Boehm. = Atractylis L. ■☆

2727 Acarna Hill = Cirsium Mill. + Tyrimnus (Cass.) Cass. ■☆

2728 Acarna Hill = Picnomon Adans. ■☆

2729 Acarna chinensis Bunge = Atractylodes lancea (Thunb.) DC. ■

2730 Acarnaceae Link = Asteraceae Bercht. et J. Presl(保留科名)●■

2731 Acarnaceae Link = Compositae Giseke(保留科名)●■

2732 Acarpha Griseb. (1854);南美萼角花属■☆

2733 Acarpha R. Br. ex Benth. = Latrobea Meisn. ■☆

2734 Acarpha agglomerata (Miers) Stapf;南美萼角花■☆

2735 Acarpha agglomerata Stapf = Acarpha agglomerata (Miers) Stapf ■☆

2736 Acarpha australis Griseb.;澳洲萼角花■☆

2737 Acarpha laciniata (Ball) Stapf;裂叶南美萼角花■☆

2738 Acarpha laciniata Stapf = Acarpha laciniata (Ball) Stapf ■☆

2739 Acarphaea Harv. et Gray ex A. Gray = Chaenactis DC. ●■☆

2740 Acarphaea artemisiifolia Harv. et A. Gray = Chaenactis artemisiifolia (Harv. et A. Gray) A. Gray ■☆

2741 Acaste Salisb. = Babiana Ker Gawl. ex Sims(保留属名)■☆

2742 Acaste pulchra Salisb. = Babiana angustifolia Sweet ■☆

2743 Acaulimalva Krapov. (1974);无茎锦葵属■☆

2744 Acaulimalva acaulis (Cav.) Krapov. = Acaulimalva acaulis (Dombey ex Cav.) Krapov. ■☆

2745 Acaulimalva acaulis (Dombey ex Cav.) Krapov.;无茎锦葵■☆

2746 Acaulimalva sulphurea Krapov.;硫色无茎锦葵■☆

2747 Acaulon N. E. Br. = Aistocaulon Poelln. ■☆

2748 Acaulon N. E. Br. = Aloinopsis Schwantes ■☆

2749 Acaulon rosulatum (Kensit) N. E. Br. = Aloinopsis rosulata (Kensit) Schwantes ■☆

2750 Acca O. Berg(1856);阿卡果属(菲油果属,凤榴属,南美棯属)●☆

2751 Acca sellowiana (O. Berg) Burret;凤梨阿卡果;Pineapple Guava ●☆

2752 Acca sellowiana (O. Berg) Burret = Feijoa sellowiana O. Berg ●

2753 Accara Landrum(1990);阿卡拉木属●☆

2754 Accara elegans (DC.) Landrum;巴西桃金娘●☆

2755 Accia A. St. -Hil. = Fragariopsis A. St. -Hil. ●☆

2756 Accoromba Baill. = Accovombona Endl. ■

2757 Accorombona Endl. = Callotropis G. Don ■

2758 Accovombona Endl. = Galega L. ■

2759 Acedilanthus Benth. et Hook. f. = Acelidanthus Trautv. et C. A. Mey. ■☆

2760 Acedilanthus Benth. et Hook. f. = Veratrum L. ■

2761 Acelica Rizzini = Justicia L. ●■

2762 Acelica Rizzini(1949);南美鸭嘴花属●■

2763 Acelidanthus Trautv. et C. A. Mey. (1856);肖藜芦属■☆

2764 Acelidanthus Trautv. et C. A. Mey. = Veratrum L. ■

2765 Acelidanthus anticleoides Trautv. et C. A. Mey.;肖藜芦■☆

2766 Acentra Phil. (1870);无距堇属■☆

2767 Acentra Phil. = Hybanthus Jacq. (保留属名)●■

2768 Acentra serrata Phil.;无距堇■☆

2769 Acer L. (1753);槭属;Acer,Maple,Papapsco ●

2770 Acer acuminatum Wall. ex D. Don;齿裂槭;Cut-leaved Maple ●☆

2771 Acer acutum W. P. Fang;锐角槭;Acute Maple ●

2772 Acer acutum W. P. Fang var. quinquefidum W. P. Fang et P. L. Chiu;五裂锐角槭;Fivelobed Acute Maple ●

2773 Acer acutum W. P. Fang var. tietungense W. P. Fang et M. Y. Fang;天童锐角槭;Tiantong Acute Maple,Tiantong Maple ●

2774 Acer aidzuensis (Franch.) Nakai = Acer ginnala Maxim. var. aidzuensis (Franch.) K. Ogata ●☆

2775 Acer albo-purpurescens Hayata;紫白槭(长叶槭,飞蛾子树,樟叶槭);Flying Moth Tree,Palepurple Maple,Pale-purple Maple,White-purple Maple ●

2776 Acer amamiense T. Yamaz.;天见槭●☆

2777 Acer ambiguum Dippel = Acer pictum Thunb. ●☆

2778 Acer amoenum Carrière;日本秀丽槭●☆

2779 Acer amoenum Carrière f. latilobatum (Koidz.) K. Ogata = Acer amoenum Carrière ●☆

2780 Acer amoenum Carrière f. palmatipartitum (Koidz.) K. Ogata;掌裂秀丽槭●☆

2781 Acer amoenum Carrière var. matsumurae (Koidz.) K. Ogata;松村氏紫白槭●☆

2782 Acer amoenum Carrière var. nambuanum (Koidz.) K. Ogata;南氏槭●☆

2783 Acer amplum Rehder;阔叶槭(高大槭,黄枝槭);Broadleaf Maple,Broad-leaved Maple ●

2784 Acer amplum Rehder var. covexum (W. P. Fang) W. P. Fang;凸果阔叶槭(马蹄槭,凸果贵阳槭);Protruded Fruit Broadleaf Maple ●

2785 Acer amplum Rehder var. jianshuiense W. P. Fang;建水阔叶槭(建水槭);Jianshui Maple ●

2786 Acer amplum Rehder var. tientaiense (C. K. Schneid.) Rehder;天台阔叶槭(天台高大槭,天台黄枝槭);Tiantai Broadleaf Maple, Tiantai Maple ●

2787 Acer angustilobum Hu = Acer wilsonii Rehder ●

2788 Acer angustilobum Hu f. longicaudatum W. P. Fang;长尾窄裂槭;Longcaudate Maple ●

2789 Acer angustilobum Hu f. longicaudatum W. P. Fang = Acer wilsonii

Rehder var. longicaudatum (W. P. Fang) W. P. Fang ●

2790 Acer angustilobum Hu var. sichouense W. P. Fang et M. Y. Fang = Acer sichouense (W. P. Fang et M. Y. Fang) W. P. Fang ●

2791 Acer anhweiense W. P. Fang et M. Y. Fang;安徽槭;Anhui Maple ●

2792 Acer anhweiense W. P. Fang et M. Y. Fang var. brachypterum W. P. Fang et P. L. Chiu;短翅安徽槭;Shortwing Anhui Maple ●

2793 Acer argutum Maxim.;尖齿槭(麻叶槭,锐齿槭);Sharptoothed Maple ●

2794 Acer argutum Maxim. f. latialatum Sugim. = Acer argutum Maxim. ●

2795 Acer argutum Maxim. f. minus Sugim. = Acer argutum Maxim. ●

2796 Acer australe (Momot.) Ohwi et Momot.;澳洲槭(南方褐毛日本槭,南方毛脉槭)●☆

2797 Acer barbatum Michx.;簇毛槭(佛罗里达槭,佛罗里达糖槭);Florida Maple, Florida Sugar Maple, Hammock Maple, Southern Sugar Maple ●☆

2798 Acer barbinerve Maxim. ex Miq.;髯脉槭(簇毛槭,辽吉槭,辽吉槭树,毛脉槭,髯毛槭,髭毛脉槭,髯毛脉槭);Barbate-veined Maple, Barbedvein Maple, Ussuri Maple ●

2799 Acer barbinerve Maxim. ex Miq. var. chanbaischanense S. L. Tung;长白簇毛槭;Cangbaishan Barbedveiin Maple ●

2800 Acer betulifolium Maxim. = Acer tetramerum Pax var. betulifolium (Maxim.) Rehder ●

2801 Acer bicolor F. Chun;两色槭;Twocolor Maple, Two-colored Maple ●

2802 Acer bicolor F. Chun var. serratifolium (W. P. Fang) W. P. Fang;粗齿两色槭;Serrate Twocolor Maple ●

2803 Acer bicolor F. Chun var. serrulatum (F. P. Metcalf) W. P. Fang;圆齿两色槭;Serrulate Twocolor Maple ●

2804 Acer bodinieri H. Lév. = Acer mono Maxim. var. tricuspis (Rehder) Rehder ●

2805 Acer bodinieri H. Lév. var. covexum W. P. Fang = Acer amplum Rehder var. covexum (W. P. Fang) W. P. Fang ●

2806 Acer bornmuelleri Borbás;波姆雷槭;Bornmueller Maple ●☆

2807 Acer boscii Spach;波士槭;Bose Maple ●☆

2808 Acer brachystephyanum T. Z. Hsu;短瓣槭;Shortbract Maple ●

2809 Acer brevilobum Hesse ex Rehder = Acer nipponicum H. Hara ●☆

2810 Acer brevipes Gagnep.;短梗槭●☆

2811 Acer buergerianum Miq.;三角槭(三角枫);Buerger Maple, Trident Maple ●

2812 Acer buergerianum Miq. f. integrifolium (Makino) H. Hara;全缘叶三角槭●☆

2813 Acer buergerianum Miq. subsp. formosanum (Hayata ex Koidz.) A. E. Murray = Acer buergerianum Miq. var. formosanum (Hayata ex Koidz.) Sasaki ●

2814 Acer buergerianum Miq. subsp. yui (W. P. Fang) A. E. Murray = Acer yui W. P. Fang ●

2815 Acer buergerianum Miq. var. formosanum (Hayata ex Koidz.) Sasaki;台湾三角槭(三角枫,台湾三角枫);Taiwan Buerger Maple, Taiwan Maple, Taiwan Trident Maple ●

2816 Acer buergerianum Miq. var. horizontale F. P. Metcalf;平翅三角槭(平翅三角枫);Flat Buerger Maple, Flat Trident Maple ●

2817 Acer buergerianum Miq. var. integrifolium (Makino) Nemoto = Acer buergerianum Miq. f. integrifolium (Makino) H. Hara ●☆

2818 Acer buergerianum Miq. var. jiujiangense Z. X. Yu;九江三角槭;Jiujiang Maple ●

2819 Acer buergerianum Miq. var. kaiscianense (Pamp.) W. P. Fang;界山三角槭;Jieshan Maple, Jieshan Trident Maple ●

2820 Acer buergerianum Miq. var. ningpoense (Hance) A. E. Murray;宁波三角槭;Ningbo Buerger Maple, Ningbo Maple, Ningbo Trident Maple, Trident Maple ●

2821 Acer buergerianum Miq. var. ningpoense (Hance) Rehder = Acer buergerianum Miq. var. ningpoense (Hance) A. E. Murray ●

2822 Acer buergerianum Miq. var. trinerve (Dippel) Rehder;三脉三角枫;Threevein Buerger Maple, Threevein Trident Maple ●

2823 Acer buergerianum Miq. var. trinerve (Dippel) Rehder = Acer buergerianum Miq. ●

2824 Acer buergerianum Miq. var. yetangense W. P. Fang et M. Y. Fang;雁荡三角槭(雁荡山三角枫);Yandang Buerger Maple, Yandang Maple, Yandang Trident Maple ●

2825 Acer buzimpala Buch. -Ham. ex D. Don = Acer oblongum Wall. ex DC. ●

2826 Acer caesium Wall. ex Brandis;深灰槭(粉白槭);Bluegray Maple, Blue-gray Maple ●

2827 Acer caesium Wall. ex Brandis subsp. giraldii (Pax) A. E. Murray;太白深灰槭(古拉德槭,纪氏槭);Girald Bluegray Maple, Taibai Maple ●

2828 Acer californicum Torr. et A. Gray;加州槭●☆

2829 Acer caloneurum C. Y. Wu et T. Z. Hsu;美脉槭;Beautiful-nerve Maple, Beautiful-nerved Maple ●

2830 Acer campbellii Hook. f. et Thomson ex Brand. subsp. chekiangense (W. P. Fang) A. E. Murray = Acer pubinerve Rehder ●

2831 Acer campbellii Hook. f. et Thomson ex Brand. subsp. flabellatum (Rehder) A. E. Murray = Acer flabellatum Rehder ●

2832 Acer campbellii Hook. f. et Thomson ex Brand. subsp. oliveranum (Pax) A. E. Murray = Acer olivaceum W. P. Fang et P. L. Chiu ●

2833 Acer campbellii Hook. f. et Thomson ex Brand. subsp. robustum (Pax) A. E. Murray = Acer robustum Pax ●

2834 Acer campbellii Hook. f. et Thomson ex Brand. subsp. schneideranum (Pax et K. Hoffm.) A. E. Murray = Acer schneiderianum Pax et K. Hoffm. ●

2835 Acer campbellii Hook. f. et Thomson ex Brand. var. serratifolia ? = Acer campbellii Hook. f. et Thomson ex Brand. ●

2836 Acer campbellii Hook. f. et Thomson ex Hiern;藏南槭(喜马拉雅槭);Campbell Maple ●

2837 Acer campestre L.;栓皮槭(篱笆槭,桐状槭);Bird's Tongue, Boats, Box-of-matches, Cat Oak, Cats-and-keys, Common Dogoak Maple, Common Maple, Dog Oak, English Maple, Field Maple, Haskett, Hatchets-and-billhooks, Hedge Maple, Hooks-and-hatchets, Ketty Keys, Ketty-keys, Keys, Kit Keys, Kitty Keys, Lady's Locket, Locks-and-keys, Maple, Maplin, Maser, Maser Tree, Masertree, Mauple, Maypole-ing Tree, Mazer, Money-in-both-pockets, Oak, Propellers, Shacklers, Small-leaved Maple, Spinning Jenny, Ulmer Pipes, Whistlewood, Whitty-bush, Zigzag ●

2838 Acer campestre L. f. albo-variegatum Hayne;白斑叶栓皮槭;Variegated Hedge Maple ●☆

2839 Acer campestre L. f. postelense Lauche;金黄叶栓皮槭;Goldenleaf Hedge Maple ●☆

2840 Acer campestre L. f. pulverulentum Kirchn.;粉叶栓皮槭;Powdered Hedge Maple ●☆

2841 Acer campestre L. f. schwerinii Hesse;紫叶栓皮槭;Purpleleaf Hedge Maple ●☆

2842 Acer campestre L. f. tauricum Kirchn.;金牛栓皮槭;Taurus Hedge Maple ●☆

2843 Acer campestre L. var. austriacum DC.;奥地利栓皮槭;Austrian Hedge Maple ●☆

2844 Acer campestre L. var. hebecarpum DC.;野栓皮槭;Downy-fruited Field Maple, Wild Hedge Maple ●☆

2845 Acer campestre L. var. leiocarpum Tausch;光果栓皮槭;Smoothfruit Hedge Maple ●☆

2846 Acer capillipes Maxim.;细柄槭(毛柄槭);Japanese Maple, Japanese Strippedbark Maple, Red Snakebark Maple, Slender-stalked Maple, Snake-bark Maple ●☆

2847 Acer capillipes Maxim. ex Miq. var. morifolium Hatus.;桑叶细柄槭●☆

2848 Acer capillipes Maxim. var. morifolium (Koidz.) Hatus. = Acer morifolium Koidz. ●☆

2849 Acer cappadocicum Gled.;青皮槭;Cappadocian Maple, Caucasian Maple, Coliseum Maple ●

2850 Acer cappadocicum Gled. 'Aureum';黄叶青皮槭(金黄青皮槭) ●☆

2851 Acer cappadocicum Gled. f. rubrum (Kirchn.) Rehder;红青皮槭;Red Coliseum Maple ●☆

2852 Acer cappadocicum Gled. f. tricaudatum (Rehder ex Veitch) Rehder = Acer cappadocicum Gled. var. tricaudatum (Rehder ex Veitch) Rehder ●

2853 Acer cappadocicum Gled. f. tricolor (Carrière) Rehder;三色青皮槭;Tricolor Coliseum Maple ●☆

2854 Acer cappadocicum Gled. subsp. amplum (Rehder) A. E. Murray = Acer amplum Rehder ●

2855 Acer cappadocicum Gled. subsp. catalpifolium (Rehder) A. E. Murray = Acer catalpifolium Rehder ●◇

2856 Acer cappadocicum Gled. subsp. mono (Maxim.) A. E. Murray = Acer mono Maxim. ●

2857 Acer cappadocicum Gled. subsp. sinicum (Rehder) Hand. -Mazz. = Acer cappadocicum Gled. var. sinicum Rehder ●

2858 Acer cappadocicum Gled. subsp. sinicum Rehder var. tricaudatum (Rehder ex Veitch) W. P. Fang = Acer cappadocicum Gled. var. tricaudatum (Rehder ex Veitch) Rehder ●

2859 Acer cappadocicum Gled. subsp. truncatum (Bunge) A. E. Murray = Acer truncatum Bunge ●

2860 Acer cappadocicum Gled. var. aureum ? = Acer cappadocicum Gled. 'Aureum' ●☆

2861 Acer cappadocicum Gled. var. brevialatum W. P. Fang;短翅青皮槭;Shortwing Coliseum Maple ●

2862 Acer cappadocicum Gled. var. cultratum (Wall.) W. P. Fang = Acer cappadocicum Gled. var. indicum (Pax) Rehder ●

2863 Acer cappadocicum Gled. var. cultratum (Wall.) W. P. Fang = Acer cappadocicum Gled. ●

2864 Acer cappadocicum Gled. var. indicum (Pax) Rehder = Acer cappadocicum Gled. ●

2865 Acer cappadocicum Gled. var. rotundilobum A. E. Murray = Acer tenellum Pax ●

2866 Acer cappadocicum Gled. var. sinicum Rehder;小叶青皮槭;Littleleaf Coliseum Maple ●

2867 Acer cappadocicum Gled. var. tricaudatum (Rehder ex Veitch) Rehder;三尾青皮槭(裂叶青皮槭);Tricaudate Coliseum Maple ●

2868 Acer carpinifolium Siebold et Zucc.;鹅耳枥槭(鹅耳枥叶槭);Hornbeam, Hornbeam Maple, Hornbeam-leaved Maple, Japanese Maple ●☆

2869 Acer carpinifolium Siebold et Zucc. f. magnificum Sugim.;华丽鹅耳枥槭●☆

2870 Acer catalpifolium Rehder;梓叶槭;Catalpa Maple, Catalpaleaf Maple, Catalpa-leaved Maple, Caudute Maple ●◇

2871 Acer catalpifolium Rehder subsp. xinganense W. P. Fang;兴安梓叶槭(兴安槭);Xing'an Maple ●

2872 Acer caudatifolium Hayata;尖尾槭;Caudate-leaved Maple ●☆

2873 Acer caudatum sensu Brandis = Acer acuminatum Wall. ex D. Don ●☆

2874 Acer caudatum Wall.;长尾槭;Caudate Maple, Longtail Maple ●

2875 Acer caudatum Wall. subsp. georgei (Diels) A. E. Murray = Acer caudatum Wall. var. prattii Rehder ●

2876 Acer caudatum Wall. subsp. ukuranduense (Trautv. et C. A. Mey.) A. E. Murray = Acer ukurunduense Trautv. et C. A. Mey ●

2877 Acer caudatum Wall. var. erosum (Pax) Rehder = Acer caudatum Wall. var. multiserratum (Maxim.) Rehder ●

2878 Acer caudatum Wall. var. multiserratum (Maxim.) Rehder;多齿长尾槭(光叶长尾槭,陕甘长尾槭);Multiserrate Caudate Maple ●

2879 Acer caudatum Wall. var. prattii Rehder;川滇长尾槭(川康长尾槭,康藏长尾槭);Pratt Caudate Maple ●

2880 Acer caudatum Wall. var. ukurunduense (Trautv. et C. A. Mey.) A. E. Murray = Acer ukurunduense Trautv. et C. A. Mey. ●

2881 Acer caudatum Wall. var. ukurunduense (Trautv. et C. A. Mey.) Rehder = Acer ukurunduense Trautv. et C. A. Mey. ●

2882 Acer cavaleriei H. Lév. = Acer davidii Franch. ●

2883 Acer ceriferum Rehder;蜡枝槭;Ceriferous Maple, Wax Maple ●

2884 Acer changhuaense (W. P. Fang et M. Y. Fang) W. P. Fang et P. L. Chiu;昌化槭(昌化稀花槭);Changhua Maple ●

2885 Acer chekiangensis Hu et W. P. Fang = Aesculus chinensis Bunge var. chekiangensis (Hu et W. P. Fang) W. P. Fang ●

2886 Acer chienii Hu et W. C. Cheng;怒江槭(雨农槭);Chien's Maple, Nujiang Maple ●

2887 Acer chingii Hu;黔桂槭(桂北槭,罗城槭,苗山槭,秦氏槭,子农槭);Ching Maple, Nujiang Maple ●

2888 Acer chunii W. P. Fang;乳源槭;Chun Maple ●

2889 Acer chunii W. P. Fang subsp. dimorphyllum W. P. Fang;两型叶乳源槭(两型乳源槭);Heteroleaved Chun Maple ●

2890 Acer cinerascens Boiss.;西亚槭;W. Asian Maple, West Asian Maple ●

2891 Acer cinnamomifolium Hayata;樟叶槭(桂叶槭);Cinnamonleaf Maple ●

2892 Acer cinnamomifolium Hayata var. microphyllum W. P. Fang et S. Ye Liang;小叶樟叶槭;Small-leaf Cinnamonleaf Maple ●

2893 Acer circinatum Pursh;藤槭(旋卷槭,旋槭,圆叶槭);Round-leaved Maple, Vine Maple ●

2894 Acer circumlobatum Maxim. var. pseudosieboldianum Pax = Acer pseudosieboldianum (Pax) Kom. ●

2895 Acer cissifolium (Siebold et Zucc.) C. Koch = Acer cissifolium (Siebold et Zucc.) K. Koch ●

2896 Acer cissifolium (Siebold et Zucc.) K. Koch;蔹叶槭(白粉藤叶槭,蔹莓槭,日本三叶槭,三叶槭);Ivyleaved Maple, Treebine Leaf Maple, Treebineleaf Maple, Vine-leaf Maple ●

2897 Acer cissifolium (Siebold et Zucc.) K. Koch subsp. henryi (Pax) A. E. Murray = Acer henryi Pax ●

2898 Acer cissifolium K. Koch = Acer cissifolium (Siebold et Zucc.) K. Koch ●

2899 Acer confertifolium Merr. et F. P. Metcalf;密叶槭(细齿水杨槭);Denseleaved Maple, Dense-leaved Maple ●

2900 Acer confertifolium Merr. et F. P. Metcalf var. serrulatum (Dunn) W. P. Fang;细齿密叶槭;Serrulate Denseleaved Maple ●

2901 Acer cordatum Pax;紫果槭(紫槭);Cordate-leaed Maple, Cordateleaf Maple ●
2902 Acer cordatum Pax var. jinggangshanense Z. X. Yu;井冈山紫果槭;Jinggangshan Cordateleaf Maple ●
2903 Acer cordatum Pax var. microcordatum F. P. Metcalf;小紫果槭;Microcordateleaf Maple ●
2904 Acer cordatum Pax var. subtrinervium (F. P. Metcalf) W. P. Fang;长柄紫果槭;Longstalk Cordateleaf Maple ●
2905 Acer coriaceifolium H. Lév.;革叶槭;Coriaceousleaf Maple, Coriaceous-leaved Maple ●
2906 Acer coriaceifolium H. Lév. subsp. obscurilobum A. E. Murray = Acer sycopseoides Chun ●
2907 Acer coriaceifolium H. Lév. var. microcarpum W. P. Fang et S. S. Chang;小果革叶槭;Littlefruit Coriaceousleaf Maple, Small Fruit Coriaceousleaf Maple, Smallfruit Coriaceousleaf Maple ●
2908 Acer coriaceum Bosc ex Tausch;杂种革叶槭(革叶槭);Hybrid Leathery-leaf Maple ●☆
2909 Acer crassum Hu et W. C. Cheng;厚叶槭;Thickleaf Maple, Thick-leaved Maple ●
2910 Acer crataegifolium Siebold et Zucc.;山楂叶槭(瓜叶槭,山楂槭,楂叶槭);Hawthorn Maple, Hawthornleaf Maple ●
2911 Acer crataegifolium Siebold et Zucc. 'Veitchii' = Acer crataegifolium Siebold et Zucc. var. veitchii Nicholson ●☆
2912 Acer crataegifolium Siebold et Zucc. f. macrophyllum (H. Hara) Hayashi;大叶山楂叶槭●☆
2913 Acer crataegifolium Siebold et Zucc. f. veitchii (G. Nicholson) Schwer.;维奇山楂叶槭(维茨山楂叶槭);Veitch Hawthornleaf Maple ●☆
2914 Acer crataegifolium Siebold et Zucc. var. macrophyllum H. Hara = Acer crataegifolium Siebold et Zucc. f. macrophyllum (H. Hara) Hayashi ●☆
2915 Acer crataegifolium Siebold et Zucc. var. veitchii Nicholson = Acer crataegifolium Siebold et Zucc. f. veitchii (G. Nicholson) Schwer. ●☆
2916 Acer creticum L. = Acer orientale L. ●
2917 Acer cultratum Wall. = Acer cappadocicum Gled. var. indicum (Pax) Rehder ●
2918 Acer cultratum Wall. = Acer cappadocicum Gled. ●
2919 Acer dasycarpum Ehrh. = Acer saccharinum L. ●
2920 Acer davidii Franch.;青榨槭(大卫槭,青虾蟆,青榨子,青柞);David Maple, David's Maple, Father David's Maple, Pere David's Maple, Snake-bark Maple ●
2921 Acer davidii Franch. var. acuminatifolium W. P. Fang = Acer davidii Franch. ●
2922 Acer davidii Franch. var. glabrescens Pax = Acer davidii Franch. ●
2923 Acer davidii Franch. var. grandifolium S. Ye Liang et Y. Q. Huang;大叶青榨槭;Bigleaf David Maple ●
2924 Acer davidii Franch. var. horizontale Pax = Acer grosseri Pax ●
2925 Acer davidii Franch. var. tomentellum Schwer. = Acer davidii Franch. ●
2926 Acer decandrum Merr.;十蕊槭(海南槭,阔翅槭);Tenstamen Maple, Ten-stamen Maple ●
2927 Acer dedyle Maxim. = Acer ukurunduense Trautv. et C. A. Mey ●
2928 Acer diabolicum Blume ex K. Koch;鬼槭(恶魔槭,红鬼槭,魔鬼槭);Devil Maple, Horned Maple, Red Devil Maple ●
2929 Acer diabolicum Blume ex K. Koch subsp. barbinerve (Maxim. ex Miq.) Wesm. = Acer barbinerve Maxim. ex Miq. ●
2930 Acer diabolicum Blume ex K. Koch subsp. sinopurpurascens (W. C. Cheng) A. E. Murray = Acer sinopurpurascens W. C. Cheng ●
2931 Acer diabolicum Blume ex K. Koch var. purpurascens (Franch. et Sav.) Rehder;红鬼槭●
2932 Acer diabolicum Blume ex K. Koch var. purpurascens (Franch. et Sav.) Rehder = Acer diabolicum Blume ex K. Koch ●
2933 Acer dieckii Pax;戴克槭;Dieck Maple ●☆
2934 Acer dielsii H. Lév. = Dipteronia sinensis Oliv. ●
2935 Acer dimorphifolium F. P. Metcalf = Acer reticulatum Champ. ex Benth. var. dimorphifolium (F. P. Metcalf) W. P. Fang et W. K. Hu ●
2936 Acer discolor Maxim.;异色槭;Diversecolor Maple, Diverse-colored Maple ●
2937 Acer discolor Rehder = Acer oblongum Wall. ex DC. ●
2938 Acer dissectum Thunb. var. tenuifolium (Koidz.) Koidz. = Acer tenuifolium (Koidz.) Koidz. ●☆
2939 Acer distylum Siebold et Zucc.;二柱槭(椴叶槭);Lime-leaved Maple ●☆
2940 Acer divergens K. Koch et Pax = Acer divergens K. Koch et Pax ex Pax ●☆
2941 Acer divergens K. Koch et Pax ex Pax;叉状槭●☆
2942 Acer duplicato-serratum Hayata;重齿槭(台湾掌叶槭);Double-teeth Maple, Double-toothed Maple, Hairy Japanese Maple ●
2943 Acer earpinifolium Siebold et Zucc.;枥叶槭;Hornbean Maple ●☆
2944 Acer elegantulum W. P. Fang et P. L. Chiu;秀丽槭(五角枫,丫角枫);Elegant Maple ●
2945 Acer elegantulum W. P. Fang et P. L. Chiu var. macrurum W. P. Fang et P. L. Chiu;长尾秀丽槭;Longtail Elegant Maple ●
2946 Acer emeiense T. Z. Hsu;峨眉槭(峨眉秀丽槭);Emei Maple ●
2947 Acer erianthum Schwer.;毛花槭(阔翅槭);Hairyflower Maple, Hairy-flowered Maple ●
2948 Acer eriocarpum Michx. = Acer saccharinum L. ●
2949 Acer erosum Pax = Acer caudatum Wall. var. multiserratum (Maxim.) Rehder ●
2950 Acer eucalyptoides W. P. Fang et Y. T. Wu;桉状槭(桉叶槭);Eucalyptusleaved Maple, Eucalyptuslike Maple, Eucalyptus-like Maple ●
2951 Acer fabri Hance;罗浮槭(红翅槭,红蝴蝶,蝴蝶果,蝴蝶花);Faber Maple, Luofu Maple ●
2952 Acer fabri Hance f. rubrocarpum (F. P. Metcalf) Rehder = Acer fabri Hance var. rubrocarpum F. P. Metcalf ●
2953 Acer fabri Hance var. dolichophyllum W. P. Fang et S. Ye Liang;长叶罗浮槭;Longleaf Faber Maple ●
2954 Acer fabri Hance var. gracillimum W. P. Fang;特瘦罗浮槭;Thin Faber Maple ●
2955 Acer fabri Hance var. megalocarpum Hu et W. C. Cheng;大果罗浮槭(大果红翅槭);Bigfruit Faber Maple ●
2956 Acer fabri Hance var. rubrocarpum F. P. Metcalf;红果罗浮槭;Red Fruit Faber Maple ●
2957 Acer fabri Hance var. tongguense Z. X. Yu;铜鼓罗浮槭(铜鼓槭);Tonggu Maple ●
2958 Acer fabri Hance var. virescens W. P. Fang;毛梗罗浮槭(绿果罗浮槭,毛梗红翅槭);Hairystalk Faber Maple ●
2959 Acer fargesi Franch. ex Rehder = Acer fabri Hance var. rubrocarpum F. P. Metcalf ●
2960 Acer fauriei H. Lév. et Vaniot = Acer negundo L. ●
2961 Acer fengii A. E. Murray = Acer kwangnanense Hu et W. C. Cheng ●
2962 Acer fenzelianum Hand. -Mazz.;河口槭;Fenzel Maple ●
2963 Acer firmianioides W. C. Cheng;梧桐槭;Phoenix Maple,

Phoenixtree Maple ●
2964 Acer flabellatum Rehder;扇叶槭(七裂槭);Fan Maple,Flabellateleaf Maple,Flabellate-leaved Maple ●
2965 Acer flabellatum Rehder var. yunnanense (Rehder) W. P. Fang;云南扇叶槭(云南槭树);Yunnan Flabellateleaf Maple ●
2966 Acer floridanum (Chapm.) Pax;佛罗里达槭;Florida Maple,Florida Sugar Maple ●☆
2967 Acer floridanum (Chapm.) Pax = Acer saccharum Marshall ●
2968 Acer floridanum (Chapm.) Pax f. villipes (Rehder) Fernald = Acer floridanum (Chapm.) Pax var. villipes Rehder ●☆
2969 Acer floridanum (Chapm.) Pax var. villipes Rehder;毛佛罗里达槭;Hairy Florida Maple ●☆
2970 Acer forrestii Diels;丽江槭(和氏槭);Forrest Maple ●
2971 Acer forrestii Diels f. caudatilobum Rehder = Acer pectinatum Wall. ex Nicholson f. caudatilobum (Rehder) W. P. Fang ●
2972 Acer franchetii Pax;房县槭(富氏槭,山枫香树);Franchet Maple ●
2973 Acer franchetii Pax var. acuminatilobum W. P. Fang et H. F. Chow = Acer kungshanense W. P. Fang et C. Yu Chang var. acuminatilobum (W. P. Fang et H. F. Chow) W. P. Fang ●
2974 Acer franchetii Pax var. megalocarpum W. P. Fang et W. K. Hu;大果房县槭;Bigfruit Franchet Maple ●
2975 Acer franchetii Pax var. schoenermarkiae (Pax) W. P. Fang et H. F. Chow = Acer franchetii Pax ●
2976 Acer freemanii E. Murray;弗里曼槭;Freeman's Maple,Maple ●☆
2977 Acer fulvescens Rehder;黄毛槭;Yellowhair Maple,Yellow-haired Maple ●
2978 Acer fulvescens Rehder subsp. danbaense W. P. Fang;丹巴黄毛槭;Danba Yellowhair Maple ●
2979 Acer fulvescens Rehder subsp. fupingense (W. P. Fang et W. K. Hu) W. P. Fang et W. K. Hu;陕甘黄毛槭;Fuping Yellowhair Maple ●
2980 Acer fulvescens Rehder subsp. fuscescens W. P. Fang;褐脉黄毛槭;Brownvein Yellowhair Maple ●
2981 Acer fulvescens Rehder subsp. pentalobum (W. P. Fang et Soong) W. P. Fang et Soong;五裂黄毛槭;Pentalobe Yellowhair Maple ●
2982 Acer fulvescens Rehder var. fupingense W. P. Fang et W. K. Hu = Acer fulvescens Rehder subsp. fupingense (W. P. Fang et W. K. Hu) W. P. Fang et W. K. Hu ●
2983 Acer fulvescens Rehder var. pentalobum W. P. Fang et Soong = Acer fulvescens Rehder subsp. fuscescens W. P. Fang ●
2984 Acer ginnala Maxim.;茶条槭(茶条,黑枫,华北茶条槭,女儿红,青桑头);Amur Maple,Crimsonleaved Maple,Russian Maple ●
2985 Acer ginnala Maxim. subsp. euginnala Pax = Acer ginnala Maxim. ●
2986 Acer ginnala Maxim. subsp. theiferum (W. P. Fang) W. P. Fang;苦茶槭(茶条,茶条槭,鸡骨枫,苦津茶,女儿红,青桑,青桑头,桑芽,银桑叶);Bitter Amur Maple ●
2987 Acer ginnala Maxim. var. aidzuensis (Franch.) K. Ogata;会津槭●☆
2988 Acer ginnala Maxim. var. euginnala Pax = Acer ginnala Maxim. ●
2989 Acer ginnala Maxim. var. semenovii (Regel et Herder) Pax = Acer semenovii Regel et Herder ●
2990 Acer giraldii Pax = Acer caesium Wall. ex Brandis subsp. giraldii (Pax) A. E. Murray ●
2991 Acer glabrum Torr.;光槭(光叶槭,落基山槭,三叶光槭);Dwarf Maple,Mountain Maple,Rock Maple,Rocky Mountain Maple ●
2992 Acer glabrum Torr. f. trisectum Sarg.;三裂光槭;Trilobed Rocky Mountain Maple ●☆
2993 Acer glabrum Torr. var. douglasii (Hook.) Dippel;道格拉斯光槭;Douglas Rocky Mountain Maple ●☆
2994 Acer glabrum Torr. var. rhodocarpum Schwer.;蔷薇果光槭;Rose-fruit Rocky Mountain Maple ●☆
2995 Acer glabrum Torr. var. tripartitum (Nutt. ex Torr. et A. Gray) Pax;三叶光槭;Trileaf Rocky Mountain Maple ●☆
2996 Acer gracile W. P. Fang et M. Y. Fang;纤瘦槭;Gracile Maple ●
2997 Acer gracilifolium W. P. Fang et C. C. Fu;长叶槭;Long-leaf Maple ●
2998 Acer granatense Boiss.;格拉槭●☆
2999 Acer granatense Boiss. subsp. xauense (Pau et Font Quer) Dobignard;达乌槭●☆
3000 Acer grandidentatum Nutt. ex Torr. et Gray;北美大齿槭(北美槭,大齿糖槭);Big Tooth Maple,Bigtooth Maple,Big-tooth Maple,Canyon Maple ●
3001 Acer grandidentatum Nutt. var. brachypterum (Wooton et Standl.) E. J. Palmer;北美西南大齿槭;Southwestern Big-tooth Maple ●☆
3002 Acer grandidentatum Nutt. var. sinuosum (Rehder) Littlew.;波状北美大齿槭(波状大齿槭);Uvalde Big-tooth Maple ●☆
3003 Acer griseum (Franch.) Pax;血皮槭(马梨光);Bloodbark Maple,Bloody-bark Maple,Chinese Paperbark Maple,Fullmoon Maple,Paper Bark Maple,Paperbark Maple,Paper-bark Maple ●
3004 Acer grosseri Pax;葛罗槭(葛萝槭);Grosser Maple,Snake-skin Maple ●
3005 Acer grosseri Pax var. forrestii (Diels) Hand.-Mazz. = Acer forrestii Diels ●
3006 Acer grosseri Pax var. hersii (Rehder) Rehder;长裂葛罗槭;Hers Maple ●
3007 Acer grosseri Pax var. hersii (Rehder) Rehder = Acer grosseri Pax ●
3008 Acer guanense W. P. Fang;灌县槭;Guanxian Maple ●◇
3009 Acer guizhouense Y. K. Li;贵州槭;Guizhou Maple ●
3010 Acer hainanense F. Chun et W. P. Fang;海南槭;Hainan Maple ●
3011 Acer hayatae H. Lév. et Vaniot = Acer miyabei Maxim. ●☆
3012 Acer hayatae H. Lév. et Vaniot var. glabrum ? = Acer mono Maxim. ●
3013 Acer heldreichii Boiss. et Heldr.;巴尔干槭(希腊槭,紫巴尔干槭);Balkan Maple,Balkans Maple,Greek Maple,Heldreich's Maple ●☆
3014 Acer heldreichii Boiss. et Heldr. f. purpuratum Schwer.;紫巴尔干槭;Purple Balkan Maple ●☆
3015 Acer henryi Pax;建始槭(亨利槭,亨利槭树,亨氏槭,三叶槭);Henry's Maple ●
3016 Acer henryi Pax f. intermedium W. P. Fang = Acer henryi Pax ●
3017 Acer heptalobum Diels;七裂槭;Sevenlobe Maple,Seven-lobe Maple ●
3018 Acer hersii Rehder;赫氏槭●☆
3019 Acer hilaense Hu et W. C. Cheng;海拉槭(昌宁槭,顺宁槭);Haila Maple,Hila Maple ●
3020 Acer hookeri Miq.;锐齿槭;Hooker Maple ●
3021 Acer hookeri Miq. var. normale Schwer. = Acer hookeri Miq. ●
3022 Acer hookeri Miq. var. orbiculae W. P. Fang et Y. T. Wu;圆叶锐齿槭;Round-leaved Hooker Maple ●
3023 Acer huangpingense T. Z. Hsu;黄平槭;Huangping Maple ●
3024 Acer huianum W. P. Fang et C. F. Hsieh;胡氏槭(步曾槭,勐海槭);Hu Maple,Menghai Maple ●
3025 Acer hybridum Bosc;杂种槭;Hybrid Maple ●☆
3026 Acer hypoleucum Hayata;灰毛槭(里白槭);Grayhair Maple,Gray-haired Maple,Pale-leaf Maple ●
3027 Acer hypoleucum Hayata = Acer albo-purpurescens Hayata ●
3028 Acer hypotrichum Franch. ex W. P. Fang = Acer longipes Franch.

ex Rehder ●
3029 Acer hyrcanum Fisch. et C. A. Mey.；五指槭(海卡槭)；Balkan Maple，Hyrcanian Maple ●☆
3030 Acer ibericum M. Bieb.；西班牙槭(乔治亚槭)●☆
3031 Acer insulare Makino；海岛槭(川上氏槭，大屯尖叶槭)；Kawakami Maple ●
3032 Acer insulare Makino var. caudatifolium ?；川上氏槭●
3033 Acer interius Britton = Acer negundo L. var. interius (Britton) Sarg. ●☆
3034 Acer japonicum Thunb.；日本槭(团扇槭，羽扇槭)；Downy Japanese Maple，Full Moon Maple，Fullmoon Maple，Full-moon Maple，Japanese Maple ●
3035 Acer japonicum Thunb. 'Aconitifolium'；扇叶日本槭(乌头叶日本槭，乌头叶羽扇槭)；Cut-leaf Full-moon Maple，Fanleaf Maple，Fernleaf Maple ●☆
3036 Acer japonicum Thunb. 'Aureum'；金叶日本槭(黄叶羽扇槭，金黄日本槭)；Golden Fullmoon Maple ●☆
3037 Acer japonicum Thunb. 'Microphyllum'；小叶日本槭；Smallleaf Fullmoon Maple ●☆
3038 Acer japonicum Thunb. 'Parsonsii'；帕森斯日本槭●☆
3039 Acer japonicum Thunb. 'Vitifolium'；葡萄叶日本槭(葡萄叶羽扇槭)●☆
3040 Acer japonicum Thunb. f. aconitifolium (Meehan) Rehder = Acer japonicum Thunb. 'Aconitifolium' ●☆
3041 Acer japonicum Thunb. f. aureum Schwer. = Acer japonicum Thunb. 'Aureum' ●☆
3042 Acer japonicum Thunb. f. microphyllum (Koidz.) Rehder = Acer japonicum Thunb. 'Microphyllum' ●☆
3043 Acer japonicum Thunb. f. villosum (Koidz.) H. Hara；毛叶日本槭；Hairyleaf Fullmoon Maple ●☆
3044 Acer japonicum Thunb. f. viscosum Hayashi = Acer japonicum Thunb. ●
3045 Acer japonicum Thunb. var. aconitifolium Meehan = Acer japonicum Thunb. 'Aconitifolium' ●☆
3046 Acer japonicum Thunb. var. circumlobatum (Maxim.) Koidz.；圆裂日本槭●☆
3047 Acer japonicum Thunb. var. insulare (Pax) Ohwi = Acer japonicum Thunb. var. circumlobatum (Maxim.) Koidz. ●☆
3048 Acer japonicum Thunb. var. kasado Koidz.；十三裂日本槭●☆
3049 Acer japonicum Thunb. var. kobakoense (Nakai) H. Hara = Acer japonicum Thunb. ●
3050 Acer japonicum Thunb. var. macrophyllum Koidz.；大叶日本槭；Bigleaf Fullmoon Maple ●☆
3051 Acer japonicum Thunb. var. microphyllum Koidz. = Acer japonicum Thunb. 'Microphyllum' ●☆
3052 Acer japonicum Thunb. var. sayoshigure Koidz.；小夜雨日本槭●☆
3053 Acer japonicum Thunb. var. stenolobum Hara = Acer japonicum Thunb. ●
3054 Acer japonicum Thunb. var. villosum Hara = Acer japonicum Thunb. f. villosum (Koidz.) H. Hara ●☆
3055 Acer jingdongense T. Z. Hsu；景东槭；Jingdong Maple ●
3056 Acer johnedwardianum F. P. Metcalf = Acer confertifolium Merr. et F. P. Metcalf var. serrulatum (Dunn) W. P. Fang ●
3057 Acer kansuense W. P. Fang et C. W. Chang = Acer mandshuricum Maxim. subsp. kansuense (W. P. Fang et C. W. Chang) W. P. Fang ●
3058 Acer kawakamii Koidz.；尖叶槭(大屯尖叶槭)；Kawakami Maple ●
3059 Acer kawakamii Koidz. = Acer caudatifolium Hayata ●☆
3060 Acer kawakamii Koidz. var. taitonmontanum (Hayata) H. L. Li；大屯尖叶槭●
3061 Acer kawakamii Koidz. var. taitonmontanum (Hayata) H. L. Li = Acer kawakamii Koidz. ●
3062 Acer kiangxiense W. P. Fang et M. Y. Fang；江西槭(韩槭)；Jiangxi Maple ●
3063 Acer kiukiangense Hu et W. C. Cheng；俅江槭；Kiukiang Maple，Qiujiang Maple ●
3064 Acer komarovii Pojark.；小楷槭；Komarov Maple ●
3065 Acer kungshanense W. P. Fang et C. Yu Chang；贡山槭；Gongshan Maple ●
3066 Acer kungshanense W. P. Fang et C. Yu Chang var. acuminatilobum (W. P. Fang et H. F. Chow) W. P. Fang；尖裂贡山槭(尖裂房县槭)；Acute Gongshan Maple ●
3067 Acer kuomeii W. P. Fang et M. Y. Fang；密果槭(国楣槭)；Feng Maple，Kuomei Maple ●
3068 Acer kwangnanense Hu et W. C. Cheng；广南槭；Guangnan Maple，Kwangnan Maple ●
3069 Acer kwangsiense W. P. Fang et M. Y. Fang = Acer tonkinense Lecomte subsp. kwangsiense (W. P. Fang et M. Y. Fang) W. P. Fang ●
3070 Acer kweilinense W. P. Fang et M. Y. Fang；桂林槭(三角枫)；Guilin Maple，Kweilin Maple ●
3071 Acer laetum C. A. Mey. = Acer cappadocicum Gled. ●
3072 Acer laetum C. A. Mey. var. cultratum (Wall.) Pax = Acer cappadocicum Gled. var. indicum (Pax) Rehder ●
3073 Acer laetum C. A. Mey. var. cultratum (Wall.) Pax = Acer cappadocicum Gled. ●
3074 Acer laetum C. A. Mey. var. tomentosulum Rehder = Acer longipes Franch. ex Rehder ●
3075 Acer laevigatum Wall.；光叶槭；Smoothleaf Maple，Smooth-leaved Maple ●
3076 Acer laevigatum Wall. var. angustum Pax = Acer laevigatum Wall. ●
3077 Acer laevigatum Wall. var. fargesii (Rehder) Veitch = Acer fabri Hance var. rubrocarpum F. P. Metcalf ●
3078 Acer laevigatum Wall. var. salweenense (W. W. Sm.) Cowan ex W. P. Fang；怒江光叶槭；Nujiang Smoothleaf Maple ●
3079 Acer laevigatum Wall. var. typicum Pax = Acer laevigatum Wall. ●
3080 Acer laikuanii Y. Ling；将乐槭(来官槭，闽西槭)；Laiguan Maple，Laikwan Maple ●
3081 Acer laisuense W. P. Fang et W. K. Hu；来苏槭；Laisu Maple ●
3082 Acer lanceolatum Molliard；剑叶槭(披针叶槭)；Lanceolate Maple ●
3083 Acer lanpingense W. P. Fang et M. Y. Fang；兰坪槭；Lanping Maple ●
3084 Acer lasiocarpum H. Lév. et Vaniot = Acer ukurunduense Trautv. et C. A. Mey. ●
3085 Acer latilobum (Koidz.) Koidz. = Acer pictum Thunb. subsp. glaucum (Koidz.) H. Ohashi ●☆
3086 Acer laurifolium D. Don = Acer oblongum Wall. ex DC. ●
3087 Acer laurinum Hassk.；月桂叶槭●☆
3088 Acer laurinum Hassk. subsp. decandrum (Merr.) A. E. Murray = Acer decandrum Merr. ●
3089 Acer lauyuense W. P. Fang ex C. C. Fu；涝峪槭；Laoyu Maple，Lauyu Maple ●
3090 Acer laxiflorum Pax；疏花槭(川康槭)；Laxflower Maple，Laxiflowered Maple ●
3091 Acer laxiflorum Pax var. dolichophyllum W. P. Fang；长叶疏花槭；Longleaf Laxflower Maple，Longleaved Laxflower Maple ●

3092 Acer laxiflorum Pax var. genuinum Pax = Acer laxiflorum Pax ●

3093 Acer laxiflorum Pax var. integrifolium W. P. Fang = Acer davidii Franch. ●

3094 Acer laxiflorum Pax var. ningpoense Pax = Acer davidii Franch. ●

3095 Acer legonsanicum Y. K. Li;雷公山槭;Leigongshan Maple ●

3096 Acer leiopodum (Hand. -Mazz.) W. P. Fang et H. F. Chow;秃梗槭; Glabrousstalk Maple, Hairless Stalk Maple, Hairless-stalked Maple ●

3097 Acer leipoense W. P. Fang et Soong;雷波槭;Leibo Maple ●

3098 Acer leipoense W. P. Fang et Soong subsp. leucotrichum W. P. Fang;白毛雷波槭;White Hair Leibo Maple,Whitehair Leibo Maple ●

3099 Acer leptophyllum W. P. Fang;细叶槭;Littleleaf Maple,Thinleaf Maple,Thin-leaved Maple ●

3100 Acer leucoderme Small;白皮槭(灰白皮槭);Calk Maple,Chalk Maple,White-bark Maple,Whitebarked Maple ●☆

3101 Acer lichuanense C. D. Chu et G. G. Tang;利川槭;Lichuan Maple ●

3102 Acer linganense W. P. Fang et P. L. Chiu;临安槭;Lin'an Maple ●

3103 Acer lingii W. P. Fang;福州槭(君范槭);Fuzhou Maple,Ling Maple ●

3104 Acer liquidambarifolium Hu et W. C. Cheng = Acer tonkinense Lecomte subsp. liquidambarifolium (Hu et W. C. Cheng) W. P. Fang ●

3105 Acer litseifolium Hayata;木姜叶槭(长叶槭);Litseleaf Maple, Litse-leaved Maple ●

3106 Acer litseifolium Hayata = Acer albo-purpurescens Hayata ●

3107 Acer lobelii Ten.;罗伯利槭(蛋白石槭,意大利青皮槭);Italian Maple,Lobel Maple,Lobel's Maple ●☆

3108 Acer lobelii Ten. = Acer truncatum Bunge ●

3109 Acer lobelii Ten. subsp. pictum (Thunb.) Wesm. = Acer mono Maxim. ●

3110 Acer lobelii Ten. var. indicum Pax = Acer cappadocicum Gled. var. indicum (Pax) Rehder ●

3111 Acer lobelii Ten. var. indicum Pax = Acer cappadocicum Gled. ●

3112 Acer lobulatum Nakai = Acer truncatum Bunge ●

3113 Acer lobulatum Nakai var. rubripes Nakai = Acer truncatum Bunge ●

3114 Acer longicarpum Hu et W. C. Cheng;长翅槭;Longwing Maple, Long-winged Maple ●

3115 Acer longipedicellatum C. Y. Wu; 川南槭; Longipedicellate Maple,Long-stalk Maple ●

3116 Acer longipes Franch. ex Rehder;长柄槭(长序槭);Longstalk Maple,Long-stalked Maple ●

3117 Acer longipes Franch. ex Rehder var. chengbuense W. P. Fang;城步长柄槭;Chengbu Longstalk Maple ●

3118 Acer longipes Franch. ex Rehder var. hunanense W. P. Fang et W. K. Hu = Acer nayongense W. P. Fang var. hunanense (W. P. Fang et W. K. Hu) W. P. Fang et W. K. Hu ●

3119 Acer longipes Franch. ex Rehder var. nanchuanense W. P. Fang;南川长柄槭;Nanchuan Longstalk Maple ●

3120 Acer longipes Franch. ex Rehder var. pubigerum (W. P. Fang) W. P. Fang;卷毛长柄槭(卷长柄槭);Twist Hair Longstalk Maple, Twisthair Longstalk Maple ●

3121 Acer longipes Franch. ex Rehder var. tientaiense C. K. Schneid. = Acer amplum Rehder var. tientaiense (C. K. Schneid.) Rehder ●

3122 Acer longipes Franch. ex Rehder var. typicum C. K. Schneid. = Acer amplum Rehder var. tientaiense (C. K. Schneid.) Rehder ●

3123 Acer longipes Franch. ex Rehder var. weixiense W. P. Fang;维西长柄槭;Weixi Longstalk Maple ●

3124 Acer lucidum F. P. Metcalf;亮叶槭(红翅槭,蝴蝶花,蝴蝶槭); Lucidleaf Maple,Lucid-leaved Maple ●

3125 Acer lungshengense W. P. Fang et L. C. Hu;龙胜槭;Longsheng Maple,Lungsheng Maple ●

3126 Acer machilifolium Hu et W. C. Cheng;楠叶槭;Machilus-leaved Maple ●

3127 Acer macrophyllum Pursh;大叶槭(白枫,俄勒冈枫,枫木,阔叶枫,太平洋海岸枫,西部枫);Bigleaf Maple,Big-leaf Maple, Broadleaf Maple, Broad-leaf Maple, Japanese Maple, Large-leaved Maple, Maple, Oregon Maple, Pacific Coast Maple, Pacific Maple, Western Maple,White Maple ●

3128 Acer mairei H. Lév. = Pterocarya stenoptera C. DC. ●

3129 Acer mandshuricum Maxim.;东北槭(白牛槭,关东槭,满洲槭); Manshurian Maple,NE. China Maple ●

3130 Acer mandshuricum Maxim. subsp. kansuense (W. P. Fang et C. W. Chang) W. P. Fang;甘肃槭;Gansu Maple,Kansu Maple ●

3131 Acer mapienense W. P. Fang;马边槭(纤瘦槭);Mabian Maple, Mapien Maple ●

3132 Acer mapienense W. P. Fang formosanum (Koidz.) Sasaki = Acer duplicato-serratum Hayata ●

3133 Acer martinii Jord. = Acer monspessulanum L. subsp. martinii (Jord.) P. Fourn. ●☆

3134 Acer matsumurae (Koidz.) Koidz. = Acer amoenum Carrière var. matsumurae (Koidz.) K. Ogata ●☆

3135 Acer maximowiczianum Miq. = Acer nikoense Maxim. ●

3136 Acer maximowiczianum Miq. subsp. megalocarpum (Rehder) A. E. Murray = Acer nikoense Maxim. ●

3137 Acer maximowiczii Pax;五尖槭(马氏槭,日光槭,重齿槭); Maximowicz Maple,Nikko Maple ●

3138 Acer maximowiczii Pax subsp. porphyrophyllum W. P. Fang;紫叶五尖槭;Purpleleaf Maximowicz Maple ●

3139 Acer maximowiczii Pax var. minor W. W. Sm. = Acer forrestii Diels ●

3140 Acer mayrii Schwer.;迈尔槭;Mayrs Maple ●☆

3141 Acer mayrii Schwer. = Acer pictum Thunb. subsp. mayrii (Schwer.) H. Ohashi ●☆

3142 Acer megalodum W. P. Fang et H. Y. Su;大齿槭; Largetooth Maple,Large-toothed Maple ●

3143 Acer metcalfii Rehder;南岭槭;F. P. Metcalf Maple,Metcalf Maple ●

3144 Acer miaoshanicum W. P. Fang;苗山槭;Miaoshan Maple ●

3145 Acer miaotaiense P. C. Tsoong;庙台槭(留坝槭);Miaotai Maple ●◇

3146 Acer miaotaiense P. C. Tsoong var. glabrum M. C. Wang;光果庙台槭;Smoothfruit Miaotai Maple ●

3147 Acer micranthum Siebold et Zucc.;长梗槭(塔槭,小花槭); Pagoda Maple ●

3148 Acer mirabile Hand. -Mazz. = Acer wardii W. W. Sm. ●

3149 Acer miyabei Maxim.;宫部氏槭(米亚贝槭,田高槭);Miyabe Maple ●☆

3150 Acer miyabei Maxim. f. shibatae (Nakai) K. Ogata;柴田槭●☆

3151 Acer miyabei Maxim. subsp. miaotaiense (P. C. Tsoong) A. E. Murray = Acer miaotaiense P. C. Tsoong ●◇

3152 Acer miyabei Maxim. var. shibatae (Nakai) H. Hara = Acer miyabei Maxim. f. shibatae (Nakai) K. Ogata ●☆

3153 Acer molle sensu Pax = Acer caesium Wall. ex Brandis ●

3154 Acer mono Maxim.;色木槭(地锦槭,槭木,色木,色树,水色树,五角枫,五角槭,五龙皮);Mono Maple,Painted Maple ●

3155 Acer mono Maxim. = Acer pictum Thunb. subsp. mono (Maxim.) H. Ohashi ●

3156 Acer mono Maxim. f. ambiguum (Pax) Rehder = Acer pictum

Thunb. f. ambiguum (Pax) H. Ohashi ●☆

3157 Acer mono Maxim. f. connivens (G. Nicholson) Rehder;合翅五角枫;Connivent Wing Mono Maple ●

3158 Acer mono Maxim. f. heterophyllum Nakai = Acer mono Maxim. var. marmoratum (G. Nicholson) H. Hara f. dissectum (Wesm.) Rehder ●

3159 Acer mono Maxim. f. heterophyllum Nakai = Acer pictum Thunb. f. dissectum (Wesm.) H. Ohashi ●

3160 Acer mono Maxim. f. marmoratum (G. Nicholson) Rehder;彩纹五角枫;Painted Mono Maple ●

3161 Acer mono Maxim. f. marmoratum (G. Nicholson) Rehder = Acer pictum Thunb. ●☆

3162 Acer mono Maxim. f. marmoratum (G. Nicholson) Rehder = Acer pictum Thunb. f. ambiguum (Pax) H. Ohashi ●☆

3163 Acer mono Maxim. f. septemlobum W. P. Fang et Soong;七裂地锦槭;Sevenlobed Maple ●

3164 Acer mono Maxim. f. septemlobum W. P. Fang et Soong = Acer tenellum Pax var. septemlobum (W. P. Fang et Soong) W. P. Fang et Soong ●

3165 Acer mono Maxim. f. tashiroi (Hisauti) H. Hara = Acer mono Maxim. var. marmoratum (G. Nicholson) H. Hara f. dissectum (Wesm.) Rehder ●

3166 Acer mono Maxim. f. tashiroi (Hisauti) H. Hara = Acer pictum Thunb. subsp. dissectum (Wesm.) H. Ohashi ●☆

3167 Acer mono Maxim. f. tricuspis (Rehder) W. P. Fang = Acer mono Maxim. var. tricuspis (Rehder) Rehder ●

3168 Acer mono Maxim. subsp. ambiguum (Pax) Kitam. = Acer pictum Thunb. f. ambiguum (Pax) H. Ohashi ●☆

3169 Acer mono Maxim. subsp. glaucum (Koidz.) Kitam. = Acer pictum Thunb. subsp. glaucum (Koidz.) H. Ohashi ●☆

3170 Acer mono Maxim. subsp. mayrii (Schwer.) Kitam. = Acer mayrii Schwer. ●☆

3171 Acer mono Maxim. subsp. mayrii (Schwer.) Kitam. = Acer pictum Thunb. subsp. mayrii (Schwer.) H. Ohashi ●☆

3172 Acer mono Maxim. subsp. savatieri (Pax) Kitam. = Acer pictum Thunb. subsp. savatieri (Pax) H. Ohashi ●☆

3173 Acer mono Maxim. subsp. taishakuense (K. Ogata) Kitam. = Acer pictum Thunb. subsp. taishakuense (K. Ogata) H. Ohashi ●☆

3174 Acer mono Maxim. var. ambiguum (Pax) Rehder;疏毛五角枫;Ambiguous Mono Maple ●

3175 Acer mono Maxim. var. ambiguum (Pax) Rehder = Acer pictum Thunb. f. ambiguum (Pax) H. Ohashi ●☆

3176 Acer mono Maxim. var. connivens (G. Nicholson) H. Hara = Acer pictum Thunb. subsp. dissectum (Wesm.) H. Ohashi f. connivens (G. Nicholson) H. Ohashi ●☆

3177 Acer mono Maxim. var. connivens (G. Nicholson) H. Hara f. puberulum K. Ogata = Acer pictum Thunb. subsp. dissectum (Wesm.) H. Ohashi f. puberulum (K. Ogata) H. Ohashi ●☆

3178 Acer mono Maxim. var. connivens (G. Nicholson) H. Hara f. subtrifidum (Makino) Rehder = Acer pictum Thunb. subsp. dissectum (Wesm.) H. Ohashi f. connivens (G. Nicholson) H. Ohashi ●☆

3179 Acer mono Maxim. var. dissectum (Pax) Honda;深裂五角枫;Cutleaf Mono Maple ●

3180 Acer mono Maxim. var. dissectum (Wesm.) H. Hara f. connivens (G. Nicholson) Rehder = Acer pictum Thunb. subsp. dissectum (Wesm.) H. Ohashi f. connivens (G. Nicholson) H. Ohashi ●☆

3181 Acer mono Maxim. var. dissectum (Wesm.) Honda = Acer mono Maxim. var. marmoratum (G. Nicholson) H. Hara f. dissectum (Wesm.) Rehder ●

3182 Acer mono Maxim. var. dissectum (Wesm.) Honda = Acer pictum Thunb. subsp. dissectum (Wesm.) H. Ohashi ●☆

3183 Acer mono Maxim. var. glabrum (H. Lév. et Vaniot) H. Hara = Acer mono Maxim. ●

3184 Acer mono Maxim. var. glabrum (H. Lév. et Vaniot) H. Hara = Acer pictum Thunb. subsp. mono (Maxim.) H. Ohashi ●

3185 Acer mono Maxim. var. glabrum (H. Lév. et Vaniot) H. Hara f. latialatum (H. Hara) H. Hara = Acer pictum Thunb. subsp. mono (Maxim.) H. Ohashi ●

3186 Acer mono Maxim. var. glabrum (H. Lév. et Vaniot) H. Hara f. latialatum (H. Hara) H. Hara = Acer mono Maxim. ●

3187 Acer mono Maxim. var. glabrum (H. Lév. et Vaniot) H. Hara f. magnificum (H. Hara) H. Hara = Acer pictum Thunb. subsp. mono (Maxim.) H. Ohashi f. magnificum (H. Hara) H. Ohashi ●☆

3188 Acer mono Maxim. var. glaucum (Koidz.) Honda = Acer mono Maxim. ●

3189 Acer mono Maxim. var. glaucum (Koidz.) Honda = Acer pictum Thunb. subsp. glaucum (Koidz.) H. Ohashi ●☆

3190 Acer mono Maxim. var. incurvatum W. P. Fang et P. L. Chiu;弯翅色木槭;Wingcurved Mono Maple ●

3191 Acer mono Maxim. var. macropterum W. P. Fang;大翅色木槭;Bigwing Mono Maple ●

3192 Acer mono Maxim. var. magnificum H. Hara = Acer pictum Thunb. subsp. mono (Maxim.) H. Ohashi f. magnificum (H. Hara) H. Ohashi ●☆

3193 Acer mono Maxim. var. marmoratum (G. Nicholson) H. Hara f. connivens (G. Nicholson) Rehder = Acer pictum Thunb. subsp. dissectum (Wesm.) H. Ohashi f. connivens (G. Nicholson) H. Ohashi ●☆

3194 Acer mono Maxim. var. marmoratum (G. Nicholson) H. Hara f. dissectum (Wesm.) Rehder = Acer pictum Thunb. subsp. dissectum (Wesm.) H. Ohashi ●☆

3195 Acer mono Maxim. var. marmoratum (G. Nicholson) H. Hara f. piliferum K. Ogata = Acer pictum Thunb. subsp. dissectum (Wesm.) H. Ohashi f. piliferum (K. Ogata) H. Ohashi ●☆

3196 Acer mono Maxim. var. marmoratum (G. Nicholson) H. HaraAcer mono Maxim. var. mayrii (Schwer.) Murai = Acer mayrii Schwer. ●☆

3197 Acer mono Maxim. var. mayrii (Schwer.) Murai = Acer pictum Thunb. subsp. mayrii (Schwer.) H. Ohashi ●☆

3198 Acer mono Maxim. var. mayrii (Schwer.) Nakai;迈尔色木槭;Mayr Mono Maple ●

3199 Acer mono Maxim. var. minshanicum W. P. Fang;岷山色木槭;Minshan Mono Maple ●

3200 Acer mono Maxim. var. pubigerum (W. P. Fang) W. P. Fang = Acer longipes Franch. ex Rehder var. pubigerum (W. P. Fang) W. P. Fang ●

3201 Acer mono Maxim. var. savatieri (Pax) Nakai = Acer pictum Thunb. subsp. savatieri (Pax) H. Ohashi ●☆

3202 Acer mono Maxim. var. taishakuense K. Ogata = Acer pictum Thunb. subsp. taishakuense (K. Ogata) H. Ohashi ●☆

3203 Acer mono Maxim. var. trichobasis Nakai = Acer pictum Thunb. subsp. savatieri (Pax) H. Ohashi ●☆

3204 Acer mono Maxim. var. tricuspis (Rehder) Rehder;三尖色木槭(三裂叶色木槭);Trilobed Mono Maple ●

3205 Acer monocarpon Nakai;虾夷单果槭;One-fruited Maple ●☆

3206 Acer monspessulanum L.;三裂槭(蒙比利埃槭);Montpellier Maple ●☆

3207 Acer monspessulanum L. subsp. martinii (Jord.) P. Fourn.;马丁三裂槭●☆

3208 Acer morifolium Koidz.;桑叶槭●☆

3209 Acer morrisonense Hayata;红色槭(台湾红榨槭);Blush-red Maple, Reddening Maple, Snakebark Maple ●

3210 Acer morrisonense Hayata = Acer caudatifolium Hayata ●☆

3211 Acer muliense W. P. Fang et W. K. Hu;木里槭;Muli Maple ●

3212 Acer muliense W. P. Fang et W. K. Hu = Acer stachyophyllum Hiern var. pentaneurum (W. P. Fang et W. K. Hu) W. P. Fang ●

3213 Acer muliense W. P. Fang et W. K. Hu var. pentaneurum W. P. Fang et W. K. Hu;五脉木里槭;Five-nerve Muli Maple ●

3214 Acer muliense W. P. Fang et W. K. Hu var. pentaneurum W. P. Fang et W. K. Hu = Acer stachyophyllum Hiern var. pentaneurum (W. P. Fang et W. K. Hu) W. P. Fang ●

3215 Acer multiserratum Maxim. = Acer caudatum Wall. var. multiserratum (Maxim.) Rehder ●

3216 Acer nayongense W. P. Fang;纳雍槭;Nayong Maple ●

3217 Acer nayongense W. P. Fang var. hunanense (W. P. Fang et W. K. Hu) W. P. Fang et W. K. Hu;湖南槭;Hunan Maple ●

3218 Acer negundo L.;梣叶槭(白蜡槭,复叶槭,美国枫树,美国槭,糖槭);Ashleaf Maple, Ash-leaved Maple, Ash-leaved Negundo, Box Elder, Box Elder Maple, Boxelder, Boxelder Maple, Box-elder Maple, Manitoba Maple, Negundo, Variegated Box Elder ●

3219 Acer negundo L. 'Aureo-variegatum';黄斑梣叶槭(黄斑叶梣叶槭,金叶梣叶槭);Golden Variegated Boxelder Maple ●☆

3220 Acer negundo L. 'Elegans;金边阔梣叶槭●☆

3221 Acer negundo L. 'Flamingo';火烈鸟梣叶槭;Flamingo Boxelder ●☆

3222 Acer negundo L. 'Variegatum' = Acer negundo L. var. variegatum Jacq. ●☆

3223 Acer negundo L. f. auratum Spüth;金星梣叶槭;Golden Star Boxelder Maple ●☆

3224 Acer negundo L. f. aureo-variegatum Wesm. = Acer negundo L. 'Aureo-variegatum' ●☆

3225 Acer negundo L. var. arizonicum Sarg.;亚利桑那梣叶槭;Arizona Boxelder Maple ●☆

3226 Acer negundo L. var. aureo-marginatum Schwer.;金边梣叶槭;Golden Margin Boxelder Maple ●☆

3227 Acer negundo L. var. californicum (Torr. et A. Gray) Sarg.;加州梣叶槭;California Boxelder Maple ●☆

3228 Acer negundo L. var. interius (Britton) Sarg.;内地梣叶槭;Ash-leaved Maple, Box Elder, Inland Boxelder Maple ●☆

3229 Acer negundo L. var. pseudocalifornicum Schwer.;假加州梣叶槭●☆

3230 Acer negundo L. var. texanum Pax;得州梣叶槭;Texas Boxelder Maple ●☆

3231 Acer negundo L. var. texanum Pax f. latifolium Sarg.;宽叶得州梣叶槭;Broadleaved Texas Boxelder Maple ●☆

3232 Acer negundo L. var. variegatum Jacq.;花叶梣叶槭(斑纹梣叶槭);Variegated Boxelder Maple ●☆

3233 Acer negundo L. var. violaceum (K. Koch) Dippel;堇色槭(紫梣叶槭);Box Elder, Violet Boxelder Maple ●☆

3234 Acer negundo L. var. violaceum (Kirchn.) H. Jaeger = Acer negundo L. var. violaceum (K. Koch) Dippel ●☆

3235 Acer nepalense Pax = Acer oblongum Wall. ex DC. ●

3236 Acer nigrum F. Michx.;黑槭;Black Maple, Black Sugar Maple, Black Sugar-maple, Hard Maple, Rock Maple ●

3237 Acer nigrum F. Michx. = Acer saccharum Marshall ●

3238 Acer nigrum F. Michx. subsp. saccharophorum (K. Koch) R. T. Clausen = Acer saccharum Marshall ●

3239 Acer nigrum F. Michx. var. glaucum (F. Schmidt) Fosberg = Acer saccharum Marshall ●

3240 Acer nigrum F. Michx. var. palmeri Sarg.;掌叶黑槭;Palmer Black Maple ●☆

3241 Acer nigrum F. Michx. var. palmeri Sarg. = Acer nigrum F. Michx. ●

3242 Acer nigrum F. Michx. var. saccharophorum (K. Koch) R. T. Clausen = Acer saccharum Marshall ●

3243 Acer nikoense Maxim.;毛果槭(马氏槭,毛黑槭,日本槭,日光槭);Nikko Maple ●

3244 Acer nikoense Maxim. subsp. megalocarpum (Rehder) A. E. Murray = Acer nikoense Maxim. ●

3245 Acer nikoense Maxim. var. griseum Franch. = Acer griseum (Franch.) Pax ●

3246 Acer nikoense Maxim. var. megalocarpum Rehder = Acer nikoense Maxim. ●

3247 Acer ningpoense Hance = Acer buergerianum Miq. var. ningpoense (Hance) Rehder ●

3248 Acer nipponicum H. Hara;褐毛日本槭(日本长序槭,日本槭);Nippon Maple ●☆

3249 Acer nipponicum H. Hara subsp. orientale T. Yamaz.;东方褐毛日本槭●☆

3250 Acer nipponicum H. Hara subsp. orientale T. Yamaz. var. koshinense T. Yamaz.;高志槭●☆

3251 Acer nipponicum H. Hara var. australe T. Yamaz.;南方褐毛日本槭●☆

3252 Acer nipponicum H. Hara var. australe T. Yamaz. = Acer australe (Momot.) Ohwi et Momot. ●☆

3253 Acer nipponicum H. Hara var. orientale T. Yamaz. = Acer nipponicum H. Hara subsp. orientale T. Yamaz. ●☆

3254 Acer oblongifolium Dippel = Acer oblongum Wall. ex DC. ●

3255 Acer oblongum Wall. ex DC.;飞蛾槭(飞蛾子树,鸡火树,见风干);Evergreen Maple, Himalayan Maple, Oblongleaf Maple, Oblong-leaved Maple ●

3256 Acer oblongum Wall. ex DC. subsp. itoanum (Hayata) Hatus. ex Shimabuku;伊藤飞蛾槭●

3257 Acer oblongum Wall. ex DC. var. biauritum W. W. Sm. = Acer paxii Franch. ●

3258 Acer oblongum Wall. ex DC. var. concolor Pax;绿叶飞蛾槭;Green Oblongleaf Maple ●

3259 Acer oblongum Wall. ex DC. var. erythrocarpum H. Lév. = Acer paxii Franch. ●

3260 Acer oblongum Wall. ex DC. var. glaucum Schwer. = Acer oblongum Wall. ex DC. ●

3261 Acer oblongum Wall. ex DC. var. itoanum Hayata = Acer oblongum Wall. ex DC. subsp. itoanum (Hayata) Hatus. ex Shimabuku ●

3262 Acer oblongum Wall. ex DC. var. itoanum Hayata = Acer oblongum Wall. ex DC. ●

3263 Acer oblongum Wall. ex DC. var. laevigatum (Wall.) Wesm. = Acer laevigatum Wall. ●

3264 Acer oblongum Wall. ex DC. var. latialatum Pax;宽翅飞蛾槭(鄂西飞蛾槭);Broadwing Oblongleaf Maple ●

3265 Acer oblongum Wall. ex DC. var. macrocarpum Hu = Acer

cinnamomifolium Hayata ●

3266 Acer oblongum Wall. ex DC. var. omeiense W. P. Fang et Soong;峨眉飞蛾槭;Emei Oblongleaf Maple ●

3267 Acer oblongum Wall. ex DC. var. pachyphyllum W. P. Fang et Y. C. Wu;厚叶飞蛾槭;Thickleaf Oblongleaf Maple ●

3268 Acer oblongum Wall. ex DC. var. trilobum Henry;三裂飞蛾槭;Trilobed Oblongleaf Maple ●

3269 Acer oblongum Wall. ex DC. var. wenxianse L. C. Wang;文县飞蛾槭;Wenxian Oblongleaf Maple ●

3270 Acer obtusatum Willd.;钝头槭●☆

3271 Acer obtusatum Willd. subsp. africanum Pax = Acer opalus Mill. var. africanum (Pax) Murray ●☆

3272 Acer obtusatum Willd. var. erythrocarpum Batt. = Acer obtusatum Willd. ●☆

3273 Acer okamotoanum Nakai;朝鲜五角槭(朝鲜五角枫);Okamoto Maple ●☆

3274 Acer oligocarpum W. P. Fang et L. C. Hu;少果槭;Fewfruit Maple,Fewfruited Maple,Oligocarpous Maple ●

3275 Acer olivaceum W. P. Fang et P. L. Chiu;橄榄槭;Olivaceous Maple ●

3276 Acer olivaceum W. P. Fang et P. L. Chiu = Acer elegantulum W. P. Fang et P. L. Chiu ●

3277 Acer oliverianum Pax;五裂槭;Oliver Maple ●

3278 Acer oliverianum Pax subsp. formosanum (Koidz.) A. E. Murray;台湾五裂槭;Taiwan Oliver Maple ●

3279 Acer oliverianum Pax var. microcarpum Hayata = Acer oliverianum Pax subsp. formosanum (Koidz.) A. E. Murray ●

3280 Acer oliverianum Pax var. nakaharae Hayata = Acer oliverianum Pax subsp. formosanum (Koidz.) A. E. Murray ●

3281 Acer oliverianum Pax var. nakaharae Hayata subvar. formosanum Koidz. = Acer oliverianum Pax subsp. formosanum (Koidz.) A. E. Murray ●

3282 Acer oliverianum Pax var. nakaharae Hayata subvar. longistamineum Hayata = Acer oliverianum Pax subsp. formosanum (Koidz.) A. E. Murray ●

3283 Acer oliverianum Pax var. nakaharae Hayata subvar. trilobum Koidz. = Acer tutcheri Duthie ●

3284 Acer oliverianum Pax var. serrulatum (Dunn) Rehder = Acer serrulatum Hayata ●

3285 Acer oliverianum Pax var. tutcheri (Duthie) F. P. Metcalf ex Krussm. = Acer tutcheri Duthie ●

3286 Acer opalus Mill.;意大利槭;Italian Maple,Nikko Maple ●☆

3287 Acer opalus Mill. subsp. granatense (Boiss.) Willk. = Acer granatense Boiss. ●☆

3288 Acer opalus Mill. var. africanum (Pax) Murray;非洲槭●☆

3289 Acer opalus Mill. var. obtusatum (Kit.) Henry;锐尖意大利槭;Obtuse Italian Maple ●☆

3290 Acer opalus Mill. var. tomentosum (Tausch) Rehder;绒毛意大利槭;Tomentose Italian Maple ●☆

3291 Acer opalus Mill. var. xauense Pau et Font Quer = Acer granatense Boiss. subsp. xauense (Pau et Font Quer) Dobignard ●☆

3292 Acer opulifolium Chaix = Acer obtusatum Willd. ●☆

3293 Acer opulifolium Chaix var. obtusatum (Willd.) Batt. = Acer obtusatum Willd. ●☆

3294 Acer orientale L.;东方槭;Oriental Maple ●

3295 Acer ovatifolium Koidz. = Acer caudatifolium Hayata ●☆

3296 Acer oxyodon Franch. ex W. P. Fang = Acer erianthum Schwer. ●

3297 Acer paihengii W. P. Fang;富宁槭(伯衡槭,丽槭);Funing Maple,Paiheng Maple ●

3298 Acer palmatum Raf ‘Sango Kaku’ = Acer palmatum Thunb. ‘Sango-kaku’ ●☆

3299 Acer palmatum Raf. = Acer macrophyllum Pursh ●

3300 Acer palmatum Thunb.;鸡爪槭(鸡爪枫,鸡爪树,槭树);Greenleaf Japanese Maple,Japan Maple,Japanese Cultivar Maple,Japanese Maple,Red Japanese Maple ●

3301 Acer palmatum Thunb. ‘Atrolineare’;深纹鸡爪槭(条绿鸡爪槭);Dark Linear Japanese Maple ●☆

3302 Acer palmatum Thunb. ‘Atropurpureum’;红枫(红槭,深紫鸡爪槭,紫红鸡爪槭,紫叶鸡爪槭,紫叶槭);Japanese Maple,Purpule Japanese Maple ●

3303 Acer palmatum Thunb. ‘Bloodgood’;红果鸡爪槭(血红鸡爪槭);Bloodgood Japanese Maple ●☆

3304 Acer palmatum Thunb. ‘Butterfly’;蝴蝶鸡爪槭●☆

3305 Acer palmatum Thunb. ‘Chishio’;企喜鸡爪槭●☆

3306 Acer palmatum Thunb. ‘Chitoseyama’;辉红鸡爪槭●☆

3307 Acer palmatum Thunb. ‘Corallinum’;珊瑚鸡爪槭●☆

3308 Acer palmatum Thunb. ‘Dissectum Nigrum’;黑羽毛槭●☆

3309 Acer palmatum Thunb. ‘Dissectum’;羽毛槭(常红鸡爪槭,细叶鸡爪槭,羽毛枫);Cutleaved Japanese Maple,Everred Japanese Maple ●

3310 Acer palmatum Thunb. ‘Ever Red’ = Acer palmatum Thunb. ‘Dissectum Nigrum’ ●☆

3311 Acer palmatum Thunb. ‘Garnet’;石榴红鸡爪槭●☆

3312 Acer palmatum Thunb. ‘Green Lace’;翠绿花边鸡爪槭●☆

3313 Acer palmatum Thunb. ‘Heptalobum Rubrum’;七裂红鸡爪槭●☆

3314 Acer palmatum Thunb. ‘Higasayama’;夕佳鸡爪槭●☆

3315 Acer palmatum Thunb. ‘Katsura’;卡苏鸡爪槭;The Wig Japanese Maple ●☆

3316 Acer palmatum Thunb. ‘Linearilobium Rubrum’ = Acer palmatum Thunb. ‘Atrolineare’ ●☆

3317 Acer palmatum Thunb. ‘Margaret Bee’;玛格蜂鸡爪槭●☆

3318 Acer palmatum Thunb. ‘Moonfire’;月火鸡爪槭●☆

3319 Acer palmatum Thunb. ‘Nicholsonii’;尼科尔森鸡爪槭●☆

3320 Acer palmatum Thunb. ‘Nigrum’;紫墨鸡爪槭●☆

3321 Acer palmatum Thunb. ‘Orengeola’;赤橙鸡爪槭●☆

3322 Acer palmatum Thunb. ‘Ornatum’;深红细叶●☆

3323 Acer palmatum Thunb. ‘Osakazuki’;大叶鸡爪槭●☆

3324 Acer palmatum Thunb. ‘Oshu-beni’;深红奥舒鸡爪槭●☆

3325 Acer palmatum Thunb. ‘Red Dragon’;红龙鸡爪槭;Red Laceleaf Maple ●☆

3326 Acer palmatum Thunb. ‘Red Filigree Lace’;紫红细叶鸡爪槭●☆

3327 Acer palmatum Thunb. ‘Reticulatum’;网纹鸡爪槭●☆

3328 Acer palmatum Thunb. ‘Sango-kaku’;赤干鸡爪槭(红皮鸡爪槭);Coral Bark Japanese Maple,Coral-bark Maple ●☆

3329 Acer palmatum Thunb. ‘Scolopendrifolium’;荷叶蕨鸡爪槭●☆

3330 Acer palmatum Thunb. ‘Scolopendrifolium’ = Acer palmatum Thunb. ‘Atrolineare’ ●☆

3331 Acer palmatum Thunb. ‘Seiyu’;直立绿羽鸡爪槭;Upright Laceleaf Maple ●☆

3332 Acer palmatum Thunb. ‘Senkaki’ = Acer palmatum Thunb. ‘Sango-kaku’ ●☆

3333 Acer palmatum Thunb. ‘Shin-deshojo’;春艳鸡爪槭●☆

3334 Acer palmatum Thunb. ‘Shishigashira’;狮子头鸡爪槭;Lion´s Mane Maple ●☆

3335 Acer palmatum Thunb. 'Suminagashi'；高紫鸡爪槭；Maroon-leaved Japanese Maple ●☆
3336 Acer palmatum Thunb. 'Trompenburg'；特朋堡鸡爪槭；Trompenburg Maple ●☆
3337 Acer palmatum Thunb. 'Villa Taranto'；塔冉妥别墅鸡爪槭●☆
3338 Acer palmatum Thunb. 'Waterf.'；瀑布鸡爪槭●☆
3339 Acer palmatum Thunb. = Acer amoenum Carrière ●☆
3340 Acer palmatum Thunb. f. atrolineare Schwer. = Acer palmatum Thunb. 'Atrolineare' ●☆
3341 Acer palmatum Thunb. f. atropurpureum (Van Houtte) Schwer. = Acer palmatum Thunb. 'Atropurpureum' ●
3342 Acer palmatum Thunb. f. ornatum (Carrière) André；红纹鸡爪槭；Red Laceleaf Japanese Maple ●☆
3343 Acer palmatum Thunb. f. reticulatum André；网脉鸡爪槭；Reticulate Japanese Maple ●☆
3344 Acer palmatum Thunb. f. roseo-marginatum (Van Houtte) G. Nicholson；红晕鸡爪槭；Rose Margine Japanese Maple ●☆
3345 Acer palmatum Thunb. f. rubrum Schwer.；红叶鸡爪槭；Redleaf Japanese Maple ●☆
3346 Acer palmatum Thunb. f. sanguineum Lem.；血红鸡爪槭；Sanguine Japanese Maple ●☆
3347 Acer palmatum Thunb. f. versicolor (Van Houtte) Schwer.；春槭；Versicolor Japanese Maple ●☆
3348 Acer palmatum Thunb. subsp. amoenum (Carrière) H. Hara = Acer amoenum Carrière ●☆
3349 Acer palmatum Thunb. subsp. matsumurae Koidz. = Acer amoenum Carrière var. matsumurae (Koidz.) K. Ogata ●☆
3350 Acer palmatum Thunb. var. amoenum (Carrière) Ohwi = Acer amoenum Carrière ●☆
3351 Acer palmatum Thunb. var. amoenum (Carrière) Ohwi f. latilobatum (Koidz.) Ohwi ex H. Hara = Acer amoenum Carrière ●☆
3352 Acer palmatum Thunb. var. amoenum (Carrière) Ohwi subvar. palmatipartitum Koidz. = Acer amoenum Carrière f. palmatipartitum (Koidz.) K. Ogata ●☆
3353 Acer palmatum Thunb. var. atropurpureum Van Houtte = Acer palmatum Thunb. 'Atropurpureum' ●
3354 Acer palmatum Thunb. var. dissectum ? = Acer palmatum Thunb. 'Dissectum' ●
3355 Acer palmatum Thunb. var. heptalobum Rehder；七爪槭；Sevenclaw Japanese Maple ●
3356 Acer palmatum Thunb. var. linearilobum Miq.；线裂鸡爪槭；Linearlobe Japanese Maple ●☆
3357 Acer palmatum Thunb. var. matsumurae (Koidz.) Makino ex W. T. Lee = Acer amoenum Carrière var. matsumurae (Koidz.) K. Ogata ●☆
3358 Acer palmatum Thunb. var. matumurae (Koidz.) Makino f. latialatum (Nakai) H. Hara = Acer amoenum Carrière var. matsumurae (Koidz.) K. Ogata ●☆
3359 Acer palmatum Thunb. var. matumurae (Koidz.) Makino f. miyajimense (Nakai) H. Hara = Acer amoenum Carrière var. matsumurae (Koidz.) K. Ogata ●☆
3360 Acer palmatum Thunb. var. nambuanum (Koidz.) H. Hara = Acer amoenum Carrière var. nambuanum (Koidz.) K. Ogata ●☆
3361 Acer palmatum Thunb. var. pubescens H. L. Li；台湾掌叶槭●
3362 Acer palmatum Thunb. var. pubescens H. L. Li = Acer duplicato-serratum Hayata ●
3363 Acer palmatum Thunb. var. septalobum ?；七裂鸡爪槭；Sevenlobe Japanese Maple ●☆
3364 Acer palmatum Thunb. var. sessilifolium Maxim.；无柄鸡爪槭；Sessileleaf Japanese Maple ●☆
3365 Acer palmatum Thunb. var. subtrilobum K. Koch = Acer buergerianum Miq. ●
3366 Acer palmatum Thunb. var. thunbergii Pax；小鸡爪槭(蓑衣槭)；Thunberg Japanese Maple ●
3367 Acer palmifolium Borkh.；掌叶槭●☆
3368 Acer papilio King = Acer caudatum Wall. ●
3369 Acer parviflorum Franch. et Sav. = Acer nipponicum H. Hara ●☆
3370 Acer pashanicum W. P. Fang et Soong；巴山槭；Bashan Maple, Pashan Maple ●
3371 Acer pauciflorum W. P. Fang；稀花槭(蜡枝槭,稀毛槭)；Few-flowered Maple ●
3372 Acer pauciflorum W. P. Fang et Soong var. changhuaense W. P. Fang et M. Y. Fang = Acer changhuaense (W. P. Fang et M. Y. Fang) W. P. Fang et P. L. Chiu ●
3373 Acer pavolinii Pamp. = Acer grosseri Pax ●
3374 Acer paxii Franch.；金沙江槭(川滇三角枫,金河槭,金江槭,金沙槭,三角枫)；Pax Maple ●
3375 Acer paxii Franch. var. genuinum Pax = Acer paxii Franch. ●
3376 Acer paxii Franch. var. integrifolium H. Lév. = Acer oblongum Wall. ex DC. ●
3377 Acer paxii Franch. var. ningpoense (Hance) Pax = Acer buergerianum Miq. var. ningpoense (Hance) A. E. Murray ●
3378 Acer paxii Franch. var. semilunatum W. P. Fang；半圆叶金江槭(半圆金沙槭,半圆叶金沙槭)；Crescent Pax Maple ●
3379 Acer pectinatum Wall. ex Nicholson；篦齿槭；Pectinated Maple ●
3380 Acer pectinatum Wall. ex Nicholson f. caudatilobum (Rehder) W. P. Fang；尖尾篦齿槭；Caudate Pectinated Maple ●
3381 Acer pectinatum Wall. ex Nicholson subsp. formosanum A. E. Murray = Acer caudatifolium Hayata ●☆
3382 Acer pectinatum Wall. ex Nicholson subsp. maximowiczii (Pax) A. E. Murray = Acer maximowiczii Pax ●
3383 Acer pectinatum Wall. ex Nicholson subsp. taronense (Hand.-Mazz.) A. E. Murray = Acer taronense Hand.-Mazz. ●
3384 Acer pectinatum Wall. ex Nicholson var. caudatilobum (Rehder) A. E. Murray = Acer pectinatum Wall. ex Nicholson f. caudatilobum (Rehder) W. P. Fang ●
3385 Acer pectinatum Wall. subsp. formosanum A. E. Murray = Acer caudatifolium Hayata ●☆
3386 Acer pectinatum Wall. subsp. forrestii (Diels) A. E. Murray = Acer forrestii Diels ●
3387 Acer pectinatum Wall. subsp. laxiflorum (Pax) A. E. Murray = Acer laxiflorum Pax ●
3388 Acer pectinatum Wall. subsp. maximowiczii (Pax) A. E. Murray = Acer maximowiczii Pax ●
3389 Acer pectinatum Wall. subsp. taronense (Hand.-Mazz.) A. E. Murray = Acer taronense Hand.-Mazz. ●
3390 Acer pectinatum Wall. var. caudatilobum (Rehder) A. E. Murray = Acer pectinatum Wall. ex Nicholson f. caudatilobum (Rehder) W. P. Fang ●
3391 Acer pedunculatum K. S. Hao = Acer griseum (Franch.) Pax ●
3392 Acer pehpeiense W. P. Fang et H. Y. Su；北碚槭；Beibei Maple, Pehpei Maple ●
3393 Acer pensylvanicum L.；条纹槭(宾夕法尼亚槭,宾州槭)；Goosefoot, Goosefoot Maple Stripe Bark, Maple, Moose Maple,

Moosewood, Snake-bark Maple, Striped Dogwood, Striped Maple ●
3394 Acer pensylvanicum L. f. erythrocladum Spüth;红枝条纹槭;Red-branched Striped Maple ●
3395 Acer pensylvanicum L. var. tegmentosum (Maxim.) Wesm. = Acer tegmentosum Maxim. ●
3396 Acer pentaphyllum Diels;五叶槭(五小叶槭);Fiveleaf Maple, Pentaleaved Maple ●◇
3397 Acer peronai Schwer.;伯罗那槭;Perona Maple ●☆
3398 Acer pictum Thunb.;花纹槭;Painted Mono Maple ●☆
3399 Acer pictum Thunb. = Kalopanax septemlobus (Thunb.) Koidz. ●
3400 Acer pictum Thunb. f. ambiguum (Pax) H. Ohashi;可疑花纹槭●☆
3401 Acer pictum Thunb. f. dissectum (Wesm.) H. Ohashi = Acer pictum Thunb. subsp. dissectum (Wesm.) H. Ohashi ●☆
3402 Acer pictum Thunb. f. mono (Maxim.) H. Ohashi = Acer mono Maxim. ●
3403 Acer pictum Thunb. f. tricuspis Rehder = Acer mono Maxim. var. tricuspis (Rehder) Rehder ●
3404 Acer pictum Thunb. subsp. dissectum (Wesm.) H. Ohashi;多裂花纹槭●☆
3405 Acer pictum Thunb. subsp. dissectum (Wesm.) H. Ohashi f. connivens (G. Nicholson) H. Ohashi;密脉花纹槭●☆
3406 Acer pictum Thunb. subsp. dissectum (Wesm.) H. Ohashi f. piliferum (K. Ogata) H. Ohashi;软毛花纹槭●☆
3407 Acer pictum Thunb. subsp. dissectum (Wesm.) H. Ohashi f. puberulum (K. Ogata) H. Ohashi;短柔毛多裂花纹槭●☆
3408 Acer pictum Thunb. subsp. glaucum (Koidz.) H. Ohashi;灰花纹槭●☆
3409 Acer pictum Thunb. subsp. mayrii (Schwer.) H. Ohashi;梅理花纹槭●☆
3410 Acer pictum Thunb. subsp. mono (Maxim.) H. Ohashi = Acer mono Maxim. ●
3411 Acer pictum Thunb. subsp. mono (Maxim.) H. Ohashi f. magnificum (H. Hara) H. Ohashi;华丽槭●☆
3412 Acer pictum Thunb. subsp. savatieri (Pax) H. Ohashi;萨氏槭●☆
3413 Acer pictum Thunb. subsp. taishakuense (K. Ogata) H. Ohashi;帝释山槭●☆
3414 Acer pictum Thunb. var. connivens G. Nicholson = Acer pictum Thunb. subsp. dissectum (Wesm.) H. Ohashi f. connivens (G. Nicholson) H. Ohashi ●☆
3415 Acer pictum Thunb. var. dissectum Wesm. = Acer mono Maxim. var. marmoratum (G. Nicholson) H. Hara f. dissectum (Wesm.) Rehder ●
3416 Acer pictum Thunb. var. dissectum Wesm. = Acer pictum Thunb. subsp. dissectum (Wesm.) H. Ohashi ●☆
3417 Acer pictum Thunb. var. glaucum Koidz. = Acer pictum Thunb. subsp. glaucum (Koidz.) H. Ohashi ●☆
3418 Acer pictum Thunb. var. mayrii (Schwer.) Henry = Acer mayrii Schwer. ●☆
3419 Acer pictum Thunb. var. mayrii (Schwer.) Henry = Acer pictum Thunb. subsp. mayrii (Schwer.) H. Ohashi ●☆
3420 Acer pictum Thunb. var. mono (Maxim.) Maxim. ex Franch. = Acer pictum Thunb. subsp. mono (Maxim.) H. Ohashi ●
3421 Acer pictum Thunb. var. parviflorum (Regel) C. K. Schneid. = Acer mono Maxim. ●
3422 Acer pictum Thunb. var. savatieri Pax = Acer pictum Thunb. subsp. savatieri (Pax) H. Ohashi ●☆
3423 Acer pilosum Maxim.;疏毛槭(陇秦槭,秦陇槭);Fewhair Maple, Pilose Maple ●
3424 Acer pilosum Maxim. var. stenolobum (Rehder) W. P. Fang = Acer stenolobum Rehder ●
3425 Acer platanoides L.;挪威槭(尖叶槭,桐状槭);Bosnian Maple, Norway Maple, Plane Maple ●☆
3426 Acer platanoides L. 'Columnare';柱冠挪威槭●☆
3427 Acer platanoides L. 'Crimson Column';鲜红柱挪威槭●☆
3428 Acer platanoides L. 'Crimson King';红王挪威槭(绯红王挪威槭)●☆
3429 Acer platanoides L. 'Culcullatum';兜状挪威槭●☆
3430 Acer platanoides L. 'Drummondii';德拉蒙德挪威槭(詹孟德花叶挪威槭);Variegata Norway Maple ●☆
3431 Acer platanoides L. 'Emerald Queen';绿宝石皇后挪威槭(绿后挪威槭)●☆
3432 Acer platanoides L. 'Erectum';直立挪威槭;Erect Norway Maple ●☆
3433 Acer platanoides L. 'Globosum';圆冠挪威槭●☆
3434 Acer platanoides L. 'Goldworth Purple';高氏紫挪威槭●☆
3435 Acer platanoides L. 'Laciniatum';鹰爪挪威槭;Eagle-claw Norway Maple, Eagle's-claw Maple ●☆
3436 Acer platanoides L. 'Lorbergii';罗伯格挪威槭;Loberg Norway Maple ●
3437 Acer platanoides L. 'Palmatifidum';掌裂挪威槭(细裂挪威槭,掌叶挪威槭);Palmate Norway Maple ●☆
3438 Acer platanoides L. 'Pond';池塘挪威槭●☆
3439 Acer platanoides L. 'Pyramidale Nanum';矮小金字塔挪威槭●☆
3440 Acer platanoides L. 'Royal Red';品红挪威槭●☆
3441 Acer platanoides L. 'Rubrum';红色挪威槭(红挪威槭);Red Norway Maple ●☆
3442 Acer platanoides L. 'Schwedleri';雪卫德挪威槭(深绿挪威槭,施维德挪威槭);Darkgreen Norway Maple ●☆
3443 Acer platanoides L. 'Sumershade';暗绿挪威槭●☆
3444 Acer platanoides L. 'Undulatum';波状挪威槭●☆
3445 Acer platanoides L. 'Walderseii';瓦尔德塞挪威槭;Waldersii Maple ●☆
3446 Acer platanoides L. f. columnare Carrière;圆柱挪威槭;Columnar Norway Maple ●☆
3447 Acer platanoides L. f. erectum A. D. Slavin = Acer platanoides L. 'Erectum' ●☆
3448 Acer platanoides L. f. stollii Spüth;史多尔挪威槭;Stoll Norway Maple ●☆
3449 Acer platanoides L. var. globosum Nicholson;球状挪威槭;Globose Norway Maple ●☆
3450 Acer platanoides L. var. laciniatum Hovey = Acer platanoides L. 'Laciniatum' ●☆
3451 Acer platanoides L. var. lorbergii Van Houtte = Acer platanoides L. 'Lorbergii' ●
3452 Acer platanoides L. var. palmatifidum Tausch = Acer platanoides L. 'Palmatifidum' ●☆
3453 Acer platanoides L. var. rubrum Herder = Acer platanoides L. 'Rubrum' ●☆
3454 Acer platanoides L. var. schwedleri Nicholson = Acer platanoides L. ●☆
3455 Acer platanoides L. var. schwedleri Nicholson = Acer platanoides L. 'Schwedleri' ●☆
3456 Acer platanoides L. var. variegatum West.;斑叶挪威槭;Silver-variegared Platanus-leaved Maple ●☆

3457 Acer poliophyllum W. P. Fang et Y. T. Wu;灰叶槭;Greyleaf Maple, Grey-leaved Maple ●◇
3458 Acer polymorphum Thunb. ex A. Murray;多形槭●☆
3459 Acer polymorphum Thunb. ex A. Murray = Acer palmatum Thunb. ●
3460 Acer prainii H. Lév. = Acer fabri Hance var. rubrocarpum F. P. Metcalf ●
3461 Acer prolificum W. P. Fang et M. Y. Fang;多果槭;Fruitful Maple, Manyfruit Maple, Multifruited Maple ●
3462 Acer pseudoplatanus L.;欧亚槭(大枫木,假悬铃木,扣子木,染色槭,桐叶槭,无花果,悬铃木槭,洋桐槭);Aeroplanes, Angels, Birds, Birds' Wings, Butterfly, Buttonwood, Cats-and-keys, Chats, Dicky Birds, Dragonflies, Faddy-tree, Fly-angels, Flyaways, Flying Dutchman, Flying Dutchmen, Flying-angels, Great Maple, Hairwood, Harewood, Horse Shoes, Horseshoes, Kays, Keys, Knives-and-forks, Lady's Keys, Large-leaved Maple, Locks-and-keys, Maple, May, May-tree, Mock Plane, Peweep-tree, Pigeons, Plane, Planetree Maple, Plane-tree Maple, Propellers, Sacymore, Scots Plane, Scottish Maple, Secymore, Seggy, Segumber, Share, Succamore, Summer Tree, Sycamore, Sycamore Maple, Tulip Tree, Whistle Tree, Whistlewood, Wild Fig-tree, Wings, Zigzag ●
3463 Acer pseudoplatanus L.'Atropurpureum';紫叶欧亚槭;Spaeth Maple ●☆
3464 Acer pseudoplatanus L.'Brilliantissimum';极美欧亚槭(辉煌欧亚槭);Variegata Sycamore Maple ●☆
3465 Acer pseudoplatanus L.'Leopoldii';雷波得欧亚槭;Leopold Sycamore Maple, Variegata Sycamore Maple ●☆
3466 Acer pseudoplatanus L.'Negenia';内良尼欧亚槭●☆
3467 Acer pseudoplatanus L.'Prinz Handjery';帕林兹汉爵欧亚槭●☆
3468 Acer pseudoplatanus L.'Simmon Louis Freves';斑叶欧亚槭●☆
3469 Acer pseudoplatanus L.'Worleei';红柄欧亚槭;Worle Sycamore Maple ●☆
3470 Acer pseudoplatanus L. f. corstorphinense Schwer.;鲜黄欧亚槭;Bright Yellow-leaved Sycamore Maple ●☆
3471 Acer pseudoplatanus L. f. euchlorum Spüth;大果欧亚槭;Bigkey Sycamore Maple ●☆
3472 Acer pseudoplatanus L. f. flavo-variegatum Hayne;黄斑欧亚槭(黄斑叶欧亚槭);Yellow-variegated Sycamore Maple ●☆
3473 Acer pseudoplatanus L. f. leopoldii Lem. = Acer pseudoplatanus L.'Leopoldii' ●☆
3474 Acer pseudoplatanus L. f. purpureum Loudon;紫欧亚槭;Purple-leaved Planetree Maple ●☆
3475 Acer pseudoplatanus L. f. tomentosum Tausch;毛欧亚槭;Woolly Planetree Maple ●☆
3476 Acer pseudoplatanus L. f. variegatum West.;斑点欧亚槭;Variegated Planetree Maple ●☆
3477 Acer pseudoplatanus L. f. worleei K. Rosenthal = Acer pseudoplatanus L.'Worleei' ●☆
3478 Acer pseudoplatanus L. var. erythrocarpum Carrière;红果欧亚槭;Red-fruit Sycamore Maple ●☆
3479 Acer pseudosieboldianum (Pax) Kom.;紫花槭(丹枫,假色槭);Purplebloom Maple, Purple-flowered Maple ●
3480 Acer pseudosieboldianum (Pax) Kom. var. koreanum Nakai;小果紫花槭;Korean Purplebloom Maple, Littlefruit Purplebloom Maple ●
3481 Acer pubescens Franch.;短毛槭●☆
3482 Acer pubinerve Rehder;毛脉槭;Hairyvein Maple, Hairy-veined Maple ●
3483 Acer pubinerve Rehder var. apiferum W. P. Fang et P. L. Chiu;细果毛脉槭;Little-fruit Hairyvein Maple ●
3484 Acer pubinerve Rehder var. kwangtungense (Chun) W. P. Fang;广东毛脉槭;Guangdong Hairyvein Maple, Kwangtung Maple ●
3485 Acer pubipalmatum W. P. Fang;毛脉鸡爪槭(毛鸡爪槭);Haircockclaw Maple, Hairy-palmate Maple ●
3486 Acer pubipalmatum W. P. Fang var. pulcherrimum W. P. Fang et P. L. Chiu;美丽毛脉鸡爪槭(美丽毛鸡爪槭);Beautiful Hairy-palmate Maple ●
3487 Acer pubipetiolatum Hu et W. C. Cheng;毛柄槭;Hairy-petiolate Maple, Hairy-petiole Maple ●
3488 Acer pubipetiolatum Hu et W. C. Cheng var. pingpienense W. P. Fang et W. K. Hu;屏边毛柄槭;Pingbian Hairy-petiole Maple, Pingpien Hairy-petiole Maple ●
3489 Acer purpurascens Franch. et Sav. = Acer diabolicum Blume ex K. Koch ●
3490 Acer pusillum Schwer.;灌木槭;Shrub Maple ●☆
3491 Acer pycnanthum C. Koch = Acer pycnanthum K. Koch ●☆
3492 Acer pycnanthum K. Koch;密花槭;Dense-flowered Maple ●☆
3493 Acer ramosum Schwer.;多枝槭;Dense-branched Maple ●☆
3494 Acer regelii Pax;雷格尔槭●☆
3495 Acer reticulatum Champ. ex Benth.;网脉槭;Net-veined Maple ●
3496 Acer reticulatum Champ. ex Benth. var. dimorphifolium (F. P. Metcalf) W. P. Fang et W. K. Hu;两型叶网脉槭;Dimorphicleaf Net-veined Maple ●
3497 Acer robustum Pax;杈叶槭(杈杈叶,红色槭);Forkleaf Maple, Fork-leaved Maple ●
3498 Acer robustum Pax var. honanense W. P. Fang;河南杈叶槭(河南花叶槭);Henan Forkleaf Maple, Honan Forkleaf Maple ●
3499 Acer robustum Pax var. minus W. P. Fang;小杈叶槭(小花叶槭);Small Forkleaf Maple ●
3500 Acer rotundilobum Lam.;浅裂槭;Lobed Maple ●☆
3501 Acer rubescens Hayata 'Summer Snow';夏日红色槭●☆
3502 Acer rubescens Hayata 'Summer Surprise';夏日惊奇红色槭●☆
3503 Acer rubescens Hayata = Acer morrisonense Hayata ●
3504 Acer rubronervium Y. K. Li;红脉槭;Rednerve Maple ●
3505 Acer rubrum L.;红花槭(北美红枫,红槭,红糖槭,软枫,水枫,沼泽枫);Canadian Maple, Curled Maple, Red Maple, Redbud, Scarlet Canada Maple, Scarlet Canadian Maple, Scarlet Maple, Soft Maple, Squirters, Swamp Maple, Water Maple ●
3506 Acer rubrum L.'Arrowhead';箭头红花槭●☆
3507 Acer rubrum L.'Autumn Blaze';秋焰红花槭●☆
3508 Acer rubrum L.'Autumn Flame';秋之火红花槭;Autumn Flame Red Maple ●☆
3509 Acer rubrum L.'Autumn Glory';秋之荣耀红花槭●☆
3510 Acer rubrum L.'Bowhall';堡豪红花槭●☆
3511 Acer rubrum L.'Colummnare';柱冠红花槭(柱冠红槭)●☆
3512 Acer rubrum L.'Gerling';格灵红花槭●☆
3513 Acer rubrum L.'October Glory';壮丽十月红花槭(十月光红槭);October Glory Maple ●☆
3514 Acer rubrum L.'Red Sunset';晚霞红花槭(晚霞红槭)●☆
3515 Acer rubrum L.'Scalon';思卡伦红花槭(狭冠红槭)●☆
3516 Acer rubrum L.'Schlesinferi';施莱辛格红花槭(施莱辛格红槭)●☆
3517 Acer rubrum L.'Sunshine';阳光红花槭●☆
3518 Acer rubrum L. f. pallidiflorum (K. Koch ex Pax) Fernald = Acer rubrum L. var. pallidiflorum K. Koch ex Pax ●☆
3519 Acer rubrum L. f. tomentosum (Desf.) = Acer rubrum L. ●

3520 Acer rubrum L. var. columnare Rehder;柱形红花槭;Columnar Red Maple ●☆

3521 Acer rubrum L. var. drummondii (Hook. et Arn.) Sarg.;鲜红果红花槭(詹氏红花槭);Drummond Red Maple ●☆

3522 Acer rubrum L. var. drummondii (Hook. et Arn.) Sarg. f. rotundatum Sarg.;浅裂鲜红果红花槭;Lobed Drummond Red Maple ●☆

3523 Acer rubrum L. var. globosum Rehder;球形红花槭;Globose Red Maple ●☆

3524 Acer rubrum L. var. pallidiflorum K. Koch ex Pax;淡红花槭;Pale-flowered Red Maple ●☆

3525 Acer rubrum L. var. pycnanthum (K. Koch) Makino = Acer pycnanthum K. Koch ●☆

3526 Acer rubrum L. var. tomentosum Desf. ex Kirchhoff;毛红花槭;Hairy Red Maple ●☆

3527 Acer rubrum L. var. tomentosum Tausch = Acer rubrum L. ●

3528 Acer rubrum L. var. trilobum K. Koch;三裂红花槭;Trilobed Red Maple ●☆

3529 Acer rufinerve Siebold et Zucc.;褐脉槭(褐叶槭,红脉槭);Crimson Foliage Maple, Gray-budded Snakebark Maple, Redvein Maple, Red-vein Maple, Red-veined Maple, Snakebark Maple, Snake-bark Maple ●☆

3530 Acer rufinerve Siebold et Zucc. f. angustifolium Kitam.;狭叶褐脉槭●☆

3531 Acer rufinerve Siebold et Zucc. var. albo-limbatum Hook.;银边褐脉槭;Silver-margined Redvein Maple ●☆

3532 Acer saccharinum L.;银白槭(糖白槭,银槭);River Maple, Silver Maple, Silverleaf Maple, Soft Maple, White Maple ●

3533 Acer saccharinum L. 'Albo-variegatum';白花叶银白槭;Whiteflower Silver Maple ●

3534 Acer saccharinum L. 'Beebe Cutleaf Weeping';垂枝银白槭;Pendular Silver Maple ●☆

3535 Acer saccharinum L. 'Longifolium';长叶银槭;Longleaf Silver Maple ●

3536 Acer saccharinum L. 'Lutescens';黄叶银白槭;Luteous Silver Maple ●☆

3537 Acer saccharinum L. 'Pyramidale';塔形银白槭;Pyramid Silver Maple ●☆

3538 Acer saccharinum L. 'Silver Queen';银皇后银白槭●☆

3539 Acer saccharinum L. 'Skinneri';司内力银白槭●☆

3540 Acer saccharinum L. f. lutescens (Spüth) Pax = Acer saccharinum L. 'Lutescens' ●☆

3541 Acer saccharinum L. f. pendulum (Nicholson) Pax = Acer saccharinum L. 'Beebe Cutleaf Weeping' ●☆

3542 Acer saccharinum L. f. pyramidale (Spüth) Pax = Acer saccharinum L. 'Pyramidale' ●☆

3543 Acer saccharinum L. f. tripartitum (Schwer.) Pax;三裂银白槭;Trilobular Silver Maple ●☆

3544 Acer saccharinum L. var. laciniatum Pax;条裂银白槭;Wiers Weeping Maple ●☆

3545 Acer saccharinum L. var. laciniatum Pax = Acer saccharinum L. ●

3546 Acer saccharinum L. var. wieri Rehder = Acer saccharinum L. ●

3547 Acer saccharinum Wangenh. var. glaucum F. Schmidt = Acer saccharum Marshall ●

3548 Acer saccharophorum K. Koch = Acer saccharum Marshall ●

3549 Acer saccharum Marshall;糖槭(黑枫,石枫,糖枫,甜枫,硬枫);Bird's Eye Maple, Black Maple, Hard Maple, Rock Maple, Striped Maple, Sugar Maple, Sweet Maple ●

3550 Acer saccharum Marshall 'Columnar Selection';圆柱糖槭;Columnar Sugar Maple ●☆

3551 Acer saccharum Marshall 'Flax Hill Majesty';佛西糖槭●☆

3552 Acer saccharum Marshall 'Green Mountain';青山糖槭(绿山糖槭)●☆

3553 Acer saccharum Marshall 'Legacy';传代糖槭●☆

3554 Acer saccharum Marshall 'Seneca Chief';塞内卡糖槭●☆

3555 Acer saccharum Marshall 'Temple's Upright';高耸糖槭(柱冠糖槭)●☆

3556 Acer saccharum Marshall f. conicum Fernald;圆锥糖槭;Conic Sugar Maple ●☆

3557 Acer saccharum Marshall f. glaucum (F. Schmidt) Pax = Acer saccharum Marshall ●

3558 Acer saccharum Marshall f. glaucum (Schmidt) Pax = Acer saccharum Marshall ●

3559 Acer saccharum Marshall f. rugelii (Pax) E. J. Palmer et Steyerm. = Acer saccharum Marshall ●

3560 Acer saccharum Marshall f. schneckii (Rehder) Deam = Acer saccharum Marshall ●

3561 Acer saccharum Marshall f. schneckii (Rehder) Deam = Acer saccharum Marshall var. schneckii Rehder ●☆

3562 Acer saccharum Marshall subsp. floridanum (Chapm.) Desmarais;多花糖槭;Florida Sugar Maple ●☆

3563 Acer saccharum Marshall subsp. grandidentatum (Nutt. ex Torr. et Gray) Desmarais = Acer grandidentatum Nutt. ex Torr. et Gray ●

3564 Acer saccharum Marshall subsp. leucoderme (Small) Desmarais;白皮糖槭●☆

3565 Acer saccharum Marshall subsp. nigrum (F. Michx.) Desmarais;黑糖槭;Black Maple ●☆

3566 Acer saccharum Marshall subsp. nigrum (F. Michx.) Desmarais = Acer nigrum F. Michx. ●

3567 Acer saccharum Marshall subsp. ozarkense E. Murray = Acer saccharum Marshall var. schneckii Rehder ●☆

3568 Acer saccharum Marshall subsp. schneckii (Rehder) Desmarais = Acer saccharum Marshall var. schneckii Rehder ●☆

3569 Acer saccharum Marshall var. glaucum (F. Schmidt) Sarg. = Acer saccharum Marshall ●

3570 Acer saccharum Marshall var. glaucum (Pax) Sarg.;蓝粉糖槭;Blue Sugar Maple ●☆

3571 Acer saccharum Marshall var. grandidentatum (Nutt. ex Torr. et Gray) Sudw. = Acer grandidentatum Nutt. ex Torr. et Gray ●

3572 Acer saccharum Marshall var. monumentale (Temple) Rehder;柱形糖槭;Sentry Sugar Maple ●☆

3573 Acer saccharum Marshall var. nigrum (F. Michx.) Britton = Acer nigrum F. Michx. ●

3574 Acer saccharum Marshall var. rugelii (Pax) Rehder;罗杰糖槭;Rugel Sugar Maple ●☆

3575 Acer saccharum Marshall var. rugellii (Pax) Rehder = Acer saccharum Marshall var. schneckii Rehder ●☆

3576 Acer saccharum Marshall var. schneckii Rehder;长柔毛糖槭;Hard Maple, Schneick Sugar Maple, Sugar Maple ●☆

3577 Acer saccharum Marshall var. sinuosum (Rehder) Sarg.;北方糖槭;Northern Sugar Maple ●☆

3578 Acer saccharum Marshall var. vanvolxemii ?;沃氏糖槭;Van Volxem's Maple ●☆

3579 Acer saccharum Marshall var. viride (Schmidt) E. Murray = Acer nigrum F. Michx. ●

3580 Acer salweenense W. W. Sm. = Acer laevigatum Wall. var. salweenense (W. W. Sm.) Cowan ex W. P. Fang ●

3581 Acer schneiderianum Pax et K. Hoffm.;盐源槭(宁远槭);Schneider Maple, Yanyuan Maple ●

3582 Acer schneiderianum Pax et K. Hoffm. var. pubescens W. P. Fang et Y. T. Wu;柔毛盐源槭;Pubescent Yanyuan Maple ●

3583 Acer schoenermarkiae Pax = Acer franchetii Pax ●

3584 Acer schoenermarkiae Pax var. oxycolpum Hand. -Mazz. = Acer franchetii Pax ●

3585 Acer schwedleri Hort. ex K. Koch;施威德槭●☆

3586 Acer schwerinii Pax;施威令槭;Sehwerin Maple ●☆

3587 Acer semenovii Regel et Herder;天山槭(多节翅茶条槭);Semenov Amur Maple, Tianshan Maple, Turkestan Shrub Maple ●

3588 Acer sempervirens L.;常绿槭(克里特槭);Cretan Maple ●☆

3589 Acer septemlobum Thunb. = Kalopanax septemlobus (Thunb.) Nakai ●

3590 Acer sericeum Schwer.;丝毛槭;Silky Hair Maple ●☆

3591 Acer serrulatum Hayata;青枫(中原氏掌叶槭);Green Maple ●

3592 Acer serrulatum Hayata = Acer oliverianum Pax subsp. formosanum (Koidz.) A. E. Murray ●

3593 Acer shangszeense W. P. Fang et Soong;上思槭;Shangsi Maple ●

3594 Acer shangszeense W. P. Fang et Soong var. anfuense W. P. Fang et Soong;安福槭;Anfu Maple ●

3595 Acer shenkanense W. P. Fang ex C. C. Fu;陕甘槭;Shaanxi-Gansu Maple ●

3596 Acer shensiense W. P. Fang et L. C. Hu;陕西槭;Shaanxi Maple ●

3597 Acer shibatae Nakai = Acer miyabei Maxim. f. shibatae (Nakai) K. Ogata ●☆

3598 Acer shihweii F. Chun et W. P. Fang;平坝槭(世纬槭);Pingba Maple, Shihwe Maple ●

3599 Acer shirasawanum Koidz.;钝翅槭;Shirasawa Maple ●☆

3600 Acer shirasawanum Koidz. 'Aureum';金色钝翅槭;Golden Fullmoon Maple ●☆

3601 Acer shirasawanum Koidz. 'Autumn Moon';秋月钝翅槭●☆

3602 Acer shirasawanum Koidz. var. tenuifolium Koidz. = Acer tenuifolium (Koidz.) Koidz. ●☆

3603 Acer sichouense (W. P. Fang et M. Y. Fang) W. P. Fang;西畴槭;Xichou Maple ●

3604 Acer sieboldianum Miq.;西氏槭(塞波德槭,深裂槭,席氏槭);Siebold Maple, Siebold's Maple ●☆

3605 Acer sieboldianum Miq. f. dissectum Baba;深裂西氏槭●☆

3606 Acer sieboldianum Miq. f. microphyllum (Maxim.) H. Hara = Acer sieboldianum Miq. var. microphyllum Maxim. ●☆

3607 Acer sieboldianum Miq. var. kasatoriyama Koidz.;笠取山西氏槭(笠取山席氏槭)●☆

3608 Acer sieboldianum Miq. var. microphyllum Maxim.;小叶西氏槭(小叶塞波德槭,小叶席氏槭);Littleleaf Siebold Maple ●☆

3609 Acer sieboldianum Miq. var. tsusimense Koidz.;对岛西氏槭(对岛席氏槭)●☆

3610 Acer sikkimense Miq.;锡金槭;Sikkim Maple ●

3611 Acer sikkimense Miq. subsp. davidii (Franch.) Wesm. = Acer davidii Franch. ●

3612 Acer sikkimense Miq. subsp. hookeri (Miq.) Wesm. = Acer hookeri Miq. ●

3613 Acer sikkimense Miq. var. serrulatum Pax;细齿锡金槭;Serrulate Sikkim Maple ●

3614 Acer sikkimense Miq. var. subintegrum Schwer. = Acer sikkimense Miq. ●

3615 Acer sinense Pax;中华槭(华槭,丫角槭,丫角树);China Maple, Chinese Maple ●

3616 Acer sinense Pax subsp. chekiangense (W. P. Fang) A. E. Murray = Acer pubinerve Rehder ●

3617 Acer sinense Pax subsp. chingii (Hu) A. E. Murray = Acer chingii Hu ●

3618 Acer sinense Pax var. brevilobum W. P. Fang = Acer prolificum W. P. Fang et M. Y. Fang ●

3619 Acer sinense Pax var. concolor Pax;绿叶中华槭;Greencolored Chinese Maple ●

3620 Acer sinense Pax var. kwangtungense Chun = Acer pubinerve Rehder var. kwangtungense (Chun) W. P. Fang ●

3621 Acer sinense Pax var. longilobum W. P. Fang;深裂中华槭;Longlobed Chinese Maple ●

3622 Acer sinense Pax var. microcarbum F. P. Metcalf;小果中华槭;Littlefruit Chinense Maple ●

3623 Acer sinense Pax var. pubinerve (Rehder) W. P. Fang = Acer pubinerve Rehder ●

3624 Acer sinense Pax var. tatrophifolium Diels = Acer bicolor F. Chun ●

3625 Acer sinense Pax var. typicum Pax = Acer sinense Pax ●

3626 Acer sinense Pax var. undulatum W. P. Fang et Y. T. Wu;波缘中华槭;Undulate Chinense Maple ●

3627 Acer sino-ohlongum F. P. Metcalf;滨海槭(罗浮槭);Luofu Maple ●

3628 Acer sinopurpurascens W. C. Cheng;天目槭;Tianmu Maple, Tianmushan Maple ●

3629 Acer spicatum Lam.;穗花槭(穗状槭);Moose Maple, Moosewood, Mountain Maple ●

3630 Acer spicatum Lam. var. ukurunduense (Trautv. et C. A. Mey.) Maxim. = Acer ukurunduense Trautv. et C. A. Mey. ●

3631 Acer stachyanthum Franch. ex W. P. Fang = Acer erianthum Schwer. ●

3632 Acer stachyophyllum Hiern;毛叶槭;Hairyleaf Maple, Hairy-leaved Maple ●

3633 Acer stachyophyllum Hiern subsp. betulifolium (Maxim.) P. C. de Jong = Acer tetramerum Pax var. betulifolium (Maxim.) Rehder ●

3634 Acer stachyophyllum Hiern var. pentaneurum (W. P. Fang et W. K. Hu) W. P. Fang;五脉毛叶槭;Five-veined Hairyleaf Maple ●

3635 Acer stenobotrys Franch. ex W. P. Fang = Acer henryi Pax ●

3636 Acer stenocarpum Britton = Acer rubrum L. ●

3637 Acer stenolobum Rehder;细裂槭;Slenderlobe Maple, Slender-lobed Maple ●

3638 Acer stenolobum Rehder var. megalophyllum W. P. Fang et Y. T. Wu;大叶细裂槭;Bigleaf Slenderlobe Maple, Bigleaved Slenderlobe Maple ●

3639 Acer stenolobum Rehder var. monochlamdea S. C. Cui et J. X. Yu;单被槭●

3640 Acer sterculiaceum Griff. = Acer pictum Thunb. ●☆

3641 Acer sterculiaceum Wall.;苹婆槭;Sterculia Maple ●

3642 Acer sterculiaceum Wall. subsp. franchetii (Pax) A. E. Murray = Acer franchetii Pax ●

3643 Acer sterculiaceum Wall. subsp. thomsonii (Miq.) A. E. Murray = Acer thomsonii Miq. ●

3644 Acer sterculiaceum Wall. var. tomentosum A. E. Murray;毛苹婆槭●☆

3645 Acer stevenii Pojark.;司梯文氏槭;Steven Maple ●☆

3646 Acer subtrinervium F. P. Metcalf = Acer cordatum Pax var. subtrinervium (F. P. Metcalf) W. P. Fang ●

3647 Acer sunyiense W. P. Fang;信宜槭;Sunyi Maple,Xinyi Maple ●

3648 Acer sutchuenense Franch.;四川槭(川槭);Sichuan Maple, Szechwan Maple ●

3649 Acer sutchuenense Franch. subsp. tienchunanense (W. P. Fang et Soong) W. P. Fang;天全槭;Tianquan Maple,Tienchuan Maple ●

3650 Acer sycopseoides Chun;角叶槭(丝栗槭,丝叶槭);Hornleaf Maple,Horn-leaved Maple ●

3651 Acer syriacum Boiss. et Gaill.;叙利亚槭;Syrian Maple ●☆

3652 Acer taipuense W. P. Fang;大埔槭;Tapu Maple ●

3653 Acer taiton-montanum Hayata = Acer caudatifolium Hayata ●☆

3654 Acer taiwanense Yamam.;台湾槭;Taiwan Maple ●

3655 Acer taronense Hand. -Mazz.;独龙槭;Dulong Maple, Tulung Maple ●

3656 Acer tataricum L.;鞑靼槭;Tatar Maple,Tatarian Maple ●

3657 Acer tataricum L. f. acutipterum Jovan.;南斯拉夫槭●☆

3658 Acer tataricum L. f. rubrum Schwer.;红鞑靼槭;Red Tatarian Maple ●☆

3659 Acer tataricum L. subsp. ginnala (Maxim.) Maxim. var. euginnala Wesm. = Acer ginnala Maxim. ●

3660 Acer tataricum L. subsp. ginnala (Maxim.) Maxim. var. semenovii (Regel et Herder) Wesm. = Acer semenovii Regel et Herder ●

3661 Acer tataricum L. var. ginnala (Maxim.) Maxim. = Acer ginnala Maxim. ●

3662 Acer tataricum L. var. ginnala (Maxim.) Maxim. = Acer ginnala Maxim. subsp. theiferum (W. P. Fang) W. P. Fang ●

3663 Acer tataricum L. var. laciniatum Regel = Acer ginnala Maxim. ●

3664 Acer tataricum L. var. semenovii (Regel et Herder) Regel = Acer semenovii Regel et Herder ●

3665 Acer tegmentosum Maxim.;青楷槭(白枫,辽东槭,绿皮槭,青楷子);Manchustriple Maple,Tegmentose Maple ●

3666 Acer tegmentosum Maxim. f. rufinerve A. E. Murray = Acer chienii Hu et W. C. Cheng ●

3667 Acer tegmentosum Maxim. subsp. glaucorufinerve A. E. Murray = Acer chienii Hu et W. C. Cheng ●

3668 Acer tegmentosum Maxim. subsp. grosseri (Pax) A. E. Murray = Acer grosseri Pax ●

3669 Acer tegmentosum Maxim. subsp. grosseri (Pax) A. E. Murray var. pavolinii (Pamp.) A. E. Murray = Acer grosseri Pax ●

3670 Acer tegmentosum Maxim. subsp. hersii (Rehder) A. E. Murray = Acer hersii Rehder ●☆

3671 Acer tegmentosum Maxim. subsp. rufinerve A. E. Murray = Acer chienii Hu et W. C. Cheng ●

3672 Acer tegmentosum Maxim. var. hersii (Rehder) A. E. Murray = Acer hersii Rehder ●☆

3673 Acer tegmentosum Maxim. var. pavolinii (Pamp.) A. E. Murray = Acer grosseri Pax ●

3674 Acer tenellum Pax;薄叶槭;Tenuity-leaved Maple, Thinleaf Maple,Thin-leaved Maple ●

3675 Acer tenellum Pax var. septemlobum (W. P. Fang et Soong) W. P. Fang et Soong;七裂薄叶槭;Sevelobes Tenuity-leaved Maple ●

3676 Acer tenuifolium (Koidz.) Koidz.;小叶槭●☆

3677 Acer tetramerum Pax;四蕊槭(红色木,红色槭);Fourstamen Maple,Tetrastamen Maple ●

3678 Acer tetramerum Pax var. betulifolium (Maxim.) Rehder;桦叶四蕊槭(菱叶红色木);Birch-leaved Fourstamen Maple ●

3679 Acer tetramerum Pax var. betulifolium (Maxim.) Rehder f. latialatum Rehder = Acer tetramerum Pax var. betulifolium (Maxim.) Rehder ●

3680 Acer tetramerum Pax var. dolichurum W. P. Fang et Y. T. Wu;长尾四蕊槭;Long-tailed Fourstamen Maple ●

3681 Acer tetramerum Pax var. elobulatum Rehder = Acer stachyophyllum Hiern ●

3682 Acer tetramerum Pax var. elobulatum Rehder f. longeracemosum Rehder = Acer stachyophyllum Hiern ●

3683 Acer tetramerum Pax var. elobulatum Rehder f. mapienense W. P. Fang = Acer stachyophyllum Hiern ●

3684 Acer tetramerum Pax var. elobulatum Rehder f. viridicarpum W. P. Fang = Acer stachyophyllum Hiern ●

3685 Acer tetramerum Pax var. haopingense W. P. Fang;蒿苹四蕊槭;Haoping Fourstamen Maple ●

3686 Acer tetramerum Pax var. lobulatum Rehder;裂叶红色木(裂叶四蕊槭);Lobulate Fourstamen Maple ●

3687 Acer tetramerum Pax var. lobulatum Rehder = Acer tetramerum Pax ●

3688 Acer tetramerum Pax var. tiliifolium Rehder = Acer stachyophyllum Hiern ●

3689 Acer theiferum W. P. Fang = Acer ginnala Maxim. subsp. theiferum (W. P. Fang) W. P. Fang ●

3690 Acer thomsonii Miq.;巨果槭;Thomson Maple ●

3691 Acer thomsonii Miq. subsp. catalpifolium A. E. Murray;印度巨果槭●

3692 Acer tibetense W. P. Fang;察隅槭(西藏槭);Tibetan Maple, Tsayul Maple,Xizang Maple ●

3693 Acer tienchunanense W. P. Fang et Soong = Acer sutchuenense Franch. supsp. tienchunanense (W. P. Fang et Soong) W. P. Fang ●

3694 Acer tonkinense Lecomte;粗柄槭;Tonkin Maple, Vietnamese Maple ●

3695 Acer tonkinense Lecomte subsp. fenzelianum (Hand. -Mazz.) A. E. Murray = Acer fenzelianum Hand. -Mazz. ●

3696 Acer tonkinense Lecomte subsp. kwangsiense (W. P. Fang et M. Y. Fang) W. P. Fang;广西槭;Guangxi Tonkin Maple, Guanxi Maple,Kwangsi Maple ●

3697 Acer tonkinense Lecomte subsp. liquidambarifolium (Hu et W. C. Cheng) W. P. Fang;枫香叶槭(枫叶槭);Sweetgum-leaved Tonkin Maple ●

3698 Acer torreyi Greene;北美槭●☆

3699 Acer trautvetteri Medw. ex Trautv.;红芽槭;Redbud Maple ●

3700 Acer trialatum L. L. Deng,K. Y. Wei et G. S. Fan;三翅槭●

3701 Acer tricaudatum W. P. Fang et C. C. Fu;三尾槭;Tricaudate Maple,Tricaudated Maple ●

3702 Acer trifidum Thunb. = Dendropanax trifidus (Thunb.) Makino ex H. Hara ●

3703 Acer trifidum Thunb. ex A. Murray = Acer buergerianum Miq. ●

3704 Acer trifidum Thunb. ex A. Murray = Dendropanax trifidus (Thunb.) Makino ex H. Hara ●

3705 Acer trifidum Thunb. f. buergerianum (Miq.) Schwer. = Acer buergerianum Miq. ●

3706 Acer trifidum Thunb. f. formosanum Hayata ex Koidz. = Acer buergerianum Miq. var. formosanum (Hayata ex Koidz.) Sasaki ●

3707 Acer trifidum Thunb. f. kaiscianense Pamp. = Acer buergerianum Miq. var. kaiscianense (Pamp.) W. P. Fang ●

3708 Acer trifidum Thunb. f. ningpoense (Hance) Schwer. = Acer buergerianum Miq. var. ningpoense (Hance) A. E. Murray ●

3709 Acer trifidum Thunb. var. ningpoense (Hance) Schwer. = Acer buergerianum Miq. var. ningpoense (Hance) A. E. Murray ●

3710 Acer triflorum Kom.;三花槭(柠筋槭,伞花槭);Threeflower Maple,Threeflowered Maple,Triflower Maple,Triflowered Maple ●

3711 Acer triflorum Kom. subsp. leiopodum (Hand. -Mazz.) A. E. Murray = Acer leiopodum (Hand. -Mazz.) W. P. Fang et H. F. Chow ●

3712 Acer triflorum Kom. var. leiopodum Hand. -Mazz. = Acer leiopodum (Hand. -Mazz.) W. P. Fang et H. F. Chow ●

3713 Acer triflorum Kom. var. subcoriacea Kom.;革叶三花槭;Leatherleaved Triflower Maple ●

3714 Acer truncatum Bunge;元宝槭(瓜子叉,华北五角槭,平基槭,槭,五角枫,五脚树,元宝树);Purple Blow Maple,Purpleblow Maple,Shantung Maple,Truncate Maple,Truncate-leaved Maple ●

3715 Acer truncatum Bunge f. cordatum S. L. Tung;心叶元宝槭;Cordate Truncate-leaved Maple ●

3716 Acer truncatum Bunge subsp. mono (Maxim.) A. E. Murray = Acer mono Maxim. ●

3717 Acer truncatum Bunge var. beipiao S. L. Tung;北票元宝槭;Beipiao Truncate-leaved Maple ●

3718 Acer tschonoskii Maxim.;须川氏槭(毛脉槭);Tschonoski Maple ●☆

3719 Acer tschonoskii Maxim. subsp. australe (Momot.) Kitam. et Murata = Acer australe (Momot.) Ohwi et Momot. ●☆

3720 Acer tschonoskii Maxim. subsp. rubripes (Kom.) Kitam. et Murata = Acer komarovii Pojark. ●

3721 Acer tschonoskii Maxim. var. australe Momot. = Acer australe (Momot.) Ohwi et Momot. ●☆

3722 Acer tschonoskii Maxim. var. macrophyllum Nakai;大叶须川氏槭 ●☆

3723 Acer tschonoskii Maxim. var. rubripes Kom. = Acer komarovii Pojark. ●

3724 Acer tsinglingense W. P. Fang et C. F. Hsieh;秦岭槭;Chinling Maple,Qinling Maple ●

3725 Acer turcomanicum Pojark.;土库曼槭●☆

3726 Acer turkestanicum Pax;土耳其斯坦槭●☆

3727 Acer tutcheri Duthie;岭南槭;Tutcher Maple ●

3728 Acer tutcheri Duthie subsp. confertifolium (Merr. et F. P. Metcalf) A. E. Murray = Acer confertifolium Merr. et F. P. Metcalf ●

3729 Acer tutcheri Duthie subsp. formosanum A. E. Murray = Acer tutcheri Duthie var. shimadai Hayata ●

3730 Acer tutcheri Duthie var. serratifolium W. P. Fang = Acer bicolor F. Chun var. serratifolium (W. P. Fang) W. P. Fang ●

3731 Acer tutcheri Duthie var. shimadai Hayata;小果岭南槭(岛田氏三裂槭,台湾岭南槭);Littlefruit Tutcher Maple,Shimada Tutcher Maple ●

3732 Acer ukurunduense Trautv. et C. A. Mey.;花楷槭(查条,东北长尾槭,黄槭,辽宁槭);Northeast Caudate Maple,Ukurundu Maple ●

3733 Acer ukurunduense Trautv. et C. A. Mey. f. pilosum (Nakai) Nakai ex H. Hara;毛花楷槭●☆

3734 Acer urophyllum Maxim. = Acer maximowiczii Pax ●

3735 Acer veitchii Schwer.;槭;Veitch Maple ●☆

3736 Acer velutinum Boiss.;绒毛槭(大叶槭,毡毛槭,毡毛壮槭);Asiatic Maple,Velvet Maple,Velvety Maple ●☆

3737 Acer velutinum Boiss. var. glabrescens (Boiss. et Buhse) Rehder;波斯毡毛槭;Persian Velvet Maple ●☆

3738 Acer velutinum Boiss. var. vanvolxemi (Mast.) Rehder;粉背叶毡毛槭(范沃克毡毛槭);Vanvolsem Velvet Maple ●☆

3739 Acer velutinum Boiss. var. wolfii (Schwer.) Rehder;伏尔夫毡毛槭;Wolf Velvet Maple ●☆

3740 Acer villosum Wall.;长毛槭;Woolly-leaved Nepal Maple ●

3741 Acer villosum Wall. = Acer sterculiaceum Wall. ●

3742 Acer villosum Wall. f. euvillosum Schwer. = Acer sterculiaceum Wall. ●

3743 Acer villosum Wall. f. sterculiaceum (Wall.) Schwer. = Acer sterculiaceum Wall. ●

3744 Acer villosum Wall. var. thomsonii (Miq.) Hiern = Acer thomsonii Miq. ●

3745 Acer voveolatum C. Y. Wu;路边槭;Voveolate Maple ●

3746 Acer wangchii W. P. Fang;天峨槭(黄志槭,天蛾槭);Wangchi Maple ●

3747 Acer wangchii W. P. Fang subsp. tsinyunense W. P. Fang;缙云槭;Jinyun Maple ●

3748 Acer wardii W. W. Sm.;滇藏槭;Ward Maple ●

3749 Acer wilsonii Rehder;三峡槭(武陵槭);E. H. Wilson Maple,Wilson Maple ●

3750 Acer wilsonii Rehder var. chekiangense W. P. Fang = Acer pubinerve Rehder ●

3751 Acer wilsonii Rehder var. kwangtungense (Chun) W. P. Fang = Acer pubinerve Rehder var. kwangtungense (Chun) W. P. Fang ●

3752 Acer wilsonii Rehder var. longicaudatum (W. P. Fang) W. P. Fang;长尾三峡槭;Longtail E. H. Wilson Maple ●

3753 Acer wilsonii Rehder var. obtusum W. P. Fang et Y. T. Wu;钝角三峡槭;Obtuse E. H. Wilson Maple ●

3754 Acer wilsonii Rehder var. serrulatum Dunn = Acer confertifolium Merr. et F. P. Metcalf var. serrulatum (Dunn) W. P. Fang ●

3755 Acer wuyishanicum W. P. Fang et C. M. Tan;武夷槭;Wuyishan Maple ●

3756 Acer wuyuanense W. P. Fang et Y. T. Wu;婺源槭;Wuyuan Maple ●

3757 Acer wuyuanense W. P. Fang et Y. T. Wu var. trichopodum W. P. Fang et Y. T. Wu;毛柄婺源槭;Hairypetiole Wuyuan Maple ●

3758 Acer yangjuechi W. P. Fang et P. L. Chiu;羊角槭;Sheephorn Maple,Yangjue Maple ●◇

3759 Acer yaoshanicum W. P. Fang;瑶山槭;Yaoshan Maple ●

3760 Acer yinkunii W. P. Fang;都安槭(荫昆槭);Yinkun Maple ●

3761 Acer yui W. P. Fang;川甘槭(季川槭,小甘槭);Yu Maple ●

3762 Acer yui W. P. Fang var. leptocarpum W. P. Fang et Y. T. Wu;瘦果川甘槭(细果川甘槭);Leptocarp Yu Maple,Slender-fruit Yu Maple,Thin-fruit Yu Maple ●

3763 Acer zhongtiaoense W. P. Fang et B. L. Li;中条槭;Zhongtiao Maple,Zhongtiaoshan Maple ●

3764 Acer zoeschense Pax;平翅槭;Horizontalwing Maple ●☆

3765 Aceraceae Juss. (1789)(保留科名);槭树科;Maple Family ●

3766 Aceraceae Link = Aceraceae Juss. (保留科名)●

3767 Aceraceae Link = Sapindaceae Juss. (保留科名)●■

3768 Aceranthes Rchb. = Aceranthus C. Morren et Decne. ■

3769 Aceranthus C. Morren et Decne. = Epimedium L. ■

3770 Aceranthus diphyllus C. Morren et Decne. = Epimedium diphyllum (C. Morren et Decne.) Lodd. ■☆

3771 Aceranthus macrophyllus Blume ex K. Koch = Epimedium sagittatum (Siebold et Zucc.) Maxim. ■

3772 Aceranthus sagittatus Siebold et Zucc. = Epimedium sagittatum (Siebold et Zucc.) Maxim. ■

3773 Aceranthus triphyllus K. Koch = Epimedium sagittatum (Siebold et Zucc.) Maxim. ■

3774 Aceras R. Br. (1813);人唇兰属;Man Orchid ■☆

3775 Aceras angustifolium Lindl. = Herminium lanceum (Thunb. ex Sw.) Vuijk ■

3776 Aceras angustifolium Lindl. var. longicruris (C. Wright ex A. Gray) Miq. = Herminium lanceum (Thunb. ex Sw.) Vuijk ■

3777 Aceras anthropophorum R. Br.;人唇兰;Green Man Orchid, Man Orchid ■☆

3778 Aceras hircina (L.) Lindl. = Himantoglossum hircinum (L.) Spreng. ■☆

3779 Aceras intacta Link = Neotinea maculata (Desf.) Stearn ■☆

3780 Aceras lanceum (Thunb. ex Sw.) Steud. = Herminium lanceum (Thunb. ex Sw.) Vuijk ■

3781 Aceras longibracteatum (Biv.) Rchb. f. = Himantoglossum robertianum (Loisel.) P. Delforge ■☆

3782 Aceras longibracteatum Rchb. f.;长苞人唇兰;Longbract Man Orchid ■☆

3783 Aceras longicruris C. Wright ex A. Gray = Herminium lanceum (Thunb. ex Sw.) Vuijk ■

3784 Aceras pyramidalis (L.) Rchb. = Anacamptis pyramidalis (L.) Rich. ■☆

3785 Acerates Elliott = Asclepias L. ■

3786 Acerates Elliott(1817);北美萝藦属■☆

3787 Acerates angustifolia Decne.;窄叶北美萝藦■☆

3788 Acerates auriculata Engelm. = Acerates auriculata Engelm. ex Torr. ■☆

3789 Acerates auriculata Engelm. ex Torr.;耳状北美萝藦■☆

3790 Acerates bifida Rusby ex A. Gray;双裂北美萝藦■☆

3791 Acerates floridana Hitchc.;佛罗里达北美萝藦■☆

3792 Acerates hirtella Pennell = Asclepias hirtella (Pennell) Woodson ■☆

3793 Acerates lanuginosa (Nutt.) Decne. = Asclepias lanuginosa Nutt. ■☆

3794 Acerates longifolia Elliott;长叶北美萝藦■☆

3795 Acerates monocephala Lapham ex A. Gray = Asclepias lanuginosa Nutt. ■☆

3796 Acerates viridiflora (Raf.) Pursh ex Eaton = Asclepias viridiflora Raf. ■☆

3797 Acerates viridiflora (Raf.) Pursh ex Eaton var. ivesii Britton = Asclepias viridiflora Raf. ■☆

3798 Acerates viridiflora (Raf.) Pursh ex Eaton var. linearis A. Gray = Asclepias viridiflora Raf. ■☆

3799 Acerates viridiflora Elliott;绿花北美萝藦■☆

3800 Aceratium DC. (1824);无距杜英属●☆

3801 Aceratium angustifolium A. C. Sm.;窄叶无距杜英●☆

3802 Aceratium breviflorum Schltr.;短花无距杜英●☆

3803 Aceratium ferrugineum C. T. White;锈色无距杜英●☆

3804 Aceratium megalospermum (F. Muell.) Baigooy;大籽无距杜英●☆

3805 Aceratium molle Schltr.;绢毛无距杜英●☆

3806 Aceratium oppositifolium DC.;对叶无距杜英●☆

3807 Aceratium pachypetalum Schltr. = Aceratium pachypetalum Schltr. et A. C. Sm. ●☆

3808 Aceratium pachypetalum Schltr. et A. C. Sm.;毛瓣无距杜英●☆

3809 Aceratium parvifolium Schltr. et A. C. Sm.;小叶无距杜英●☆

3810 Aceratium tomentosum Coode;绒毛无距杜英●☆

3811 Aceratorchis Schltr. (1922);无距兰属■☆

3812 Aceratorchis Schltr. = Orchis L. ■

3813 Aceratorchis albiflora Schltr.;白花无距兰■☆

3814 Aceratorchis albiflora Schltr. = Galearis tschiliensis (Schltr.) S. C. Chen, P. J. Cribb et S. W. Gale ■

3815 Aceratorchis albiflora Schltr. = Orchis tschiliensis (Schltr.) Soó ■

3816 Aceratorchis tschiliensis Schltr. = Galearis tschiliensis (Schltr.) S. C. Chen, P. J. Cribb et S. W. Gale ■

3817 Aceratorchis tschiliensis Schltr. = Orchis tschiliensis (Schltr.) Soó ■

3818 Aceriphyllum Engl. (1891);丹顶草属■

3819 Aceriphyllum Engl. = Mukdenia Koidz. ■

3820 Aceriphyllum bossii Engl.;丹顶草■☆

3821 Aceriphyllum rossii (Oliv.) Engl. = Mukdenia rossii (Oliv.) Koidz. ■

3822 Acerophyllum Post et Kuntze = Aceriphyllum Engl. ■☆

3823 Acerotis Raf. = Asclepias L. ■

3824 Acetosa Mill. = Rumex L. ●■

3825 Acetosa Tourn. ex Mill. = Rumex L. ●■

3826 Acetosa abyssinica (Jacq.) Á. Löve et B. M. Kapoor = Rumex abyssinica Jacq. ■☆

3827 Acetosa acetosella (L.) Mill. = Rumex acetosella L. ■

3828 Acetosa alpestris (Jacq.) Á. Löve subsp. lapponica (Hiitonen) Á. Löve = Rumex lapponica (Hiitonen) Czernov ■☆

3829 Acetosa angiocarpa (Murb.) Holub = Rumex acetosella L. subsp. angiocarpa (Murb.) Murb. ■

3830 Acetosa aristidis (Coss.) Á. Löve et B. M. Kapoor = Rumex aristidis Coss. ■☆

3831 Acetosa bucephalophora (L.) Fourr. = Rumex bucephalophora L. ■☆

3832 Acetosa cypria (Murb.) Á. Löve et B. M. Kapoor = Rumex cyprius Murb. ■☆

3833 Acetosa ellenbeckii (Damm.) Á. Löve et B. M. Kapoor = Rumex ellenbeckii Dammer ■☆

3834 Acetosa gracilescens (Rech. f.) Á. Löve et Evenson = Rumex paucifolia Nutt. ■☆

3835 Acetosa hastata Moench = Rumex acetosella L. ■

3836 Acetosa hastatula (Baldwin) Á. Löve = Rumex hastatula Baldwin ex Elliott ●■

3837 Acetosa indurata (Boiss. et Reut.) Holub = Rumex indurata Boiss. et Reut. ■☆

3838 Acetosa intermedia (DC.) Fourr. = Rumex intermedia DC. ■☆

3839 Acetosa lapponica (Hiitonen) Holub = Rumex lapponica (Hiitonen) Czernov ■☆

3840 Acetosa lunaria (L.) Mill. = Rumex lunaria L. ■☆

3841 Acetosa maderensis (Lowe) Á. Löve et B. M. Kapoor = Rumex maderensis Lowe ■☆

3842 Acetosa nervosa (Vahl) Á. Löve et B. M. Kapoor = Rumex nervosa Vahl ■

3843 Acetosa obtusifolia (L.) M. Gómez = Rumex obtusifolia L. ■

3844 Acetosa papilio (Coss.) Á. Löve et B. M. Kapoor = Rumex papilio Coss. et Balansa ■☆

3845 Acetosa patientia (L.) M. Gómez = Rumex patientia L. ■

3846 Acetosa paucifolia (Nutt.) Á. Löve = Rumex paucifolia Nutt. ■☆

3847 Acetosa picta (Forssk.) Á. Löve et B. M. Kapoor = Rumex picta Forssk. ■☆

3848 Acetosa pratensis Mill. = Rumex acetosa L. ■

3849 Acetosa pratensis Mill. = Rumex acetosella L. ■

3850 Acetosa rosea (L.) Mill. = Rumex rosea L. ■☆

3851 Acetosa sagittata (Thunb.) L. A. S. Johnson et B. G. Briggs = Rumex sagittata Thunb. ■☆

3852 Acetosa simpliciflora (Murb.) Á. Löve et B. M. Kapoor = Rumex simpliciflora Murb. ■☆
3853 Acetosa thyrsiflora (Fingerh.) Á. Löve = Rumex thyrsiflora Fingerh. ■
3854 Acetosa thyrsiflora (Fingerh.) Á. Löve et D. Löve = Rumex thyrsiflora Fingerh. ■
3855 Acetosa thyrsoides (Desf.) Á. Löve et B. M. Kapoor = Rumex thyrsoides Desf. ■☆
3856 Acetosa tuberosa (L.) Chaz. = Rumex tuberosa L. ■☆
3857 Acetosa vesicaria (L.) Á. Löve = Rumex vesicaria L. ■☆
3858 Acetosella (Meisn.) Fourr. = Rumex L. ●■
3859 Acetosella Kuntze = Oxalis L. ●■
3860 Acetosella acetosella (L.) Small = Rumex acetosella L. ■
3861 Acetosella beringensis (Jurtzev et V. V. Petrovsky) Á. Löve et D. Löve = Rumex beringensis Jurtzev et V. V. Petrovsky ■☆
3862 Acetosella chinensis (Haw. ex G. Don) Kuntze = Oxalis stricta L. ■
3863 Acetosella chinensis Kuntze = Oxalis stricta L. ■
3864 Acetosella corniculata (L.) Kuntze = Oxalis corniculata L. ■
3865 Acetosella gracilescens (Rech. f.) Á. Löve = Rumex paucifolia Nutt. ■☆
3866 Acetosella graminifolia (Rudolph ex Lamb.) Á. Löve = Rumex graminifolia Rudolph ex Lamb. ■☆
3867 Acetosella griffithii Kuntze = Oxalis acetosella L. subsp. griffithii (Edgew. et Hook. f.) H. Hara ■
3868 Acetosella krausei (Jurtzev et V. V. Petrovsky) Á. Löve et D. Löve = Rumex krausei Jurtzev et V. V. Petrovsky ■☆
3869 Acetosella obtriangulata (Maxim.) Kuntze = Oxalis acetosella L. subsp. japonica (Franch. et Sav.) H. Hara ■
3870 Acetosella obtriangulata (Maxim.) Kuntze = Oxalis obtriangulata Maxim. ■
3871 Acetosella paucifolia (Nutt.) Á. Löve = Rumex paucifolia Nutt. ■☆
3872 Acetosella vulgaris (Koch) Fourr. = Rumex acetosella L. ■
3873 Acetosella vulgaris Fourr. = Rumex acetosella L. subsp. pyrenaica (Pourr. et Lapeyr.) Akeroyd ■☆
3874 Acetosella vulgaris Fourr. = Rumex acetosella L. ■
3875 Achaemenes St. -Lag. = Achimenes Pers. (保留属名) ■☆
3876 Achaenipodium Brandegee = Verbesina L. (保留属名) ●■☆
3877 Achaeta E. Fourn. = Calamagrostis Adans. + Koeleria Pers. ■
3878 Achaeta E. Fourn. = Calamagrostis Adans. ■
3879 Achaete Hack. = Achaeta E. Fourn. ■
3880 Achaetogeron A. Gray = Erigeron L. ●■
3881 Achaetogeron chihuahuensis Larsen = Erigeron versicolor (Greenm.) G. L. Nesom ■☆
3882 Achaetogeron pringlei Larsen = Erigeron oreophilus Greenm. ■☆
3883 Achaetogeron versicolor Greenm. = Erigeron versicolor (Greenm.) G. L. Nesom ■☆
3884 Achania Sw. = Malvaviscus Fabr. ●
3885 Achania malvaviscus (L.) Sw. = Malvaviscus arboreus Cav. ●
3886 Achantia A. Chev. = Mansonia J. R. Drumm. ex Prain ●☆
3887 Achantia altissima A. Chev. = Mansonia altissima (A. Chev.) A. Chev. ●☆
3888 Acharagma (N. P. Taylor) A. D. Zimmerman ex Glass = Escobaria Britton et Rose ●☆
3889 Acharagma (N. P. Taylor) A. D. Zimmerman ex Glass(1997);墨松笠属●☆
3890 Acharagma (N. P. Taylor) Glass = Acharagma (N. P. Taylor) A. D. Zimmerman ex Glass ●☆
3891 Acharagma (N. P. Taylor) Glass = Escobaria Britton et Rose ●☆
3892 Acharia Thunb. (1794);脊脐子属(柄果木属,宿冠花属)●☆
3893 Acharia tragodes Thunb.;脊脐子(柄果木,宿冠花)●☆
3894 Achariaceae Harms(1897)(保留科名);脊脐子科(柄果木科,宿冠花科,钟花科)●■☆
3895 Acharitea Benth. = Nesogenes A. DC. ●■☆
3896 Achariterium Bluff et Fingerh. = Filago L. (保留属名) ■
3897 Achariterium Bluff et Fingerh. = Oglifa (Cass.) Cass. ■
3898 Achasma Griff. = Etlingera Roxb. ■
3899 Achasma megalocheilos Griff. = Etlingera littoralis Gieseke ■
3900 Achasma yunnanense T. L. Wu et S. J. Chen = Etlingera yunnanensis (T. L. Wu et S. J. Chen) R. M. Sm. ■
3901 Achatocarpaceae Heimerl(1934)(保留科名);玛瑙果科(透镜籽科)●☆
3902 Achatocarpus Triana(1858);玛瑙果属(透镜籽属)●☆
3903 Achatocarpus nigricans Triana;玛瑙果●☆
3904 Achemora Raf. = Archemora DC. ■☆
3905 Achemora Raf. = Oxypolis Raf. ■☆
3906 Achetaria Cham. et Schltdl. (1827);蝉玄参属■☆
3907 Achetaria alpicola Rydb.;高山蝉玄参■☆
3908 Achetaria angustissima Rydb.;窄蝉玄参■☆
3909 Achetaria arenicola A. Heller;沙地蝉玄参■☆
3910 Achetaria borealis Bong.;北方蝉玄参■☆
3911 Achetaria floribunda Kuntze;多花蝉玄参■☆
3912 Achillaea L. = Achetaria Cham. et Schltdl. ■☆
3913 Achillea L. (1753);蓍属(锯叶草属,蓍草属);Moonshine Yarrow, Yarrow ■
3914 Achillea acuminata (Ledeb.) Sch. Bip.;齿叶蓍(单叶蓍,蜈蚣草);Toothedleaf Yarrow ■
3915 Achillea acuminata (Ledeb.) Sch. Bip. = Achillea ptarmica L. var. acuminata (Ledeb.) Heimerl ■
3916 Achillea aegyptiaca Steud.;埃及蓍草;Egyptian Yarrow ■☆
3917 Achillea agerata L.;香蓍草(常春蓍草);Sweet Yarrow, Sweet-nancy ■☆
3918 Achillea ageratifolia (Sibth. et Sm.) Boiss.;银毛蓍草;Greek Yarrow ■☆
3919 Achillea albana Stev.;阿尔巴蓍■☆
3920 Achillea allptica DC.;阿列蓍■☆
3921 Achillea alpina L.;高山蓍(飞天蜈蚣,高山蓍草,锯草,锯齿草,乱头发,千条蜈蚣,黔一枝蒿,蓍,蓍草,蓍草叶,蜈蚣草,蜈蚣蒿,一枝蒿,蚰蜒草,羽衣草);Alpine Yarrow ■
3922 Achillea alpina L. subsp. camtschatica (Heimerl) Kitam.;勘察加高山蓍■☆
3923 Achillea alpina L. subsp. japonica (Heimerl) Kitam.;日本高山蓍■☆
3924 Achillea alpina L. subsp. pulchra (Koidz.) Kitam.;美丽高山蓍■☆
3925 Achillea alpina L. subsp. subcartilaginea (Heimerl) Kitam.;软骨高山蓍■☆
3926 Achillea alpina L. var. angustifolia (H. Hara) Kitam. = Achillea alpina L. subsp. japonica (Heimerl) Kitam. ■☆
3927 Achillea alpina L. var. brevidens (Makino) Kitam. ex Akasawa = Achillea alpina L. subsp. subcartilaginea (Heimerl) Kitam. ■☆
3928 Achillea alpina L. var. brevidens (Makino) Kitam. ex Akasawa = Achillea alpina L. subsp. pulchra (Koidz.) Kitam. ■☆
3929 Achillea alpina L. var. camtschatica (Heimerl) Kitam. = Achillea alpina L. subsp. camtschatica (Heimerl) Kitam. ■☆

3930 Achillea alpina L. var. discoidea (Regel) Kitam.;盘状高山蓍■☆

3931 Achillea alpina L. var. longiligulata H. Hara = Achillea sibirica Ledeb. ■

3932 Achillea alpina L. var. pulchra (Koidz.) Kitam. = Achillea alpina L. subsp. pulchra (Koidz.) Kitam. ■☆

3933 Achillea angustissima Rydb. = Achillea millefolia L. subsp. lanulosa (Nutt.) Piper ■☆

3934 Achillea asiatica Serg.;亚洲蓍;Asia Yarrow, Asiatic Yarrow ■

3935 Achillea atrata L.;黑蓍草;Chamomile-leaved Milfoil ■☆

3936 Achillea aurea Lam.;金花蓍草■☆

3937 Achillea biebersteinii Afan.;比伯蓍■☆

3938 Achillea biserrata M. Bieb.;双齿蓍■☆

3939 Achillea borealis Bong.;北方蓍■☆

3940 Achillea camtschatica (Heimerl) Botsch. = Achillea alpina L. subsp. camtschatica (Heimerl) Kitam. ■☆

3941 Achillea cartilaginea Ledeb.;软骨质蓍草■☆

3942 Achillea clavennae L.;银叶蓍草;Silvery Milfoil, Silvery Yarrow, Siverweed Yarrow ■☆

3943 Achillea coarctata Poir.;密集蓍草■☆

3944 Achillea collina Becker;山地蓍■☆

3945 Achillea cuneatiloba Boiss. et Buhse;楔形蓍■☆

3946 Achillea distans Waldst. et Kit. ex Willd.;分离蓍;Alps Yarrow, Tall Yarrow ■☆

3947 Achillea eradiata Piper = Achillea millefolia L. subsp. lanulosa (Nutt.) Piper ■☆

3948 Achillea filipendulina Lam.;凤尾蓍草(黄花蓍草);Fernleaf Yarrow, Yellow Yarrow ■☆

3949 Achillea filipendulina Lam. 'Gold Plate';金盘黄花蓍草■☆

3950 Achillea fragrantissima (Forssk.) Sch. Bip.;极香蓍草■☆

3951 Achillea glaberrima Klokov;光蓍■☆

3952 Achillea gracilis Raf. = Achillea millefolia L. subsp. lanulosa (Nutt.) Piper ■☆

3953 Achillea impatiens L.;褐苞蓍;Brownbract Yarrow ■

3954 Achillea impatiens L. subsp. euimpatiens Heimerl = Achillea impatiens L. ■

3955 Achillea impatiens L. subsp. ledebouri (Heimerl) Heimerl = Achillea ledebouri Heimerl ■

3956 Achillea kermanica Gand.;凯尔马蓍■☆

3957 Achillea lanulosa Nutt.;西部毛蓍;Western Yarrow, Woolly Yarrow ■☆

3958 Achillea lanulosa Nutt. = Achillea millefolia L. subsp. lanulosa (Nutt.) Piper ■☆

3959 Achillea lanulosa Nutt. = Achillea millefolia L. ■

3960 Achillea lanulosa Nutt. f. peroutkyi F. Seym. = Achillea millefolia L. subsp. lanulosa (Nutt.) Piper ■☆

3961 Achillea lanulosa Nutt. f. rubicunda Farw. = Achillea millefolia L. subsp. lanulosa (Nutt.) Piper ■☆

3962 Achillea lanulosa Nutt. subsp. typica D. D. Keck = Achillea millefolia L. subsp. lanulosa (Nutt.) Piper ■☆

3963 Achillea lanulosa Nutt. var. eradiata (Piper) M. Peck = Achillea millefolia L. subsp. lanulosa (Nutt.) Piper ■☆

3964 Achillea latiloba Ledeb. ex Nordm.;宽裂蓍■☆

3965 Achillea laxiflora Pollard et Cockerell = Achillea millefolia L. subsp. lanulosa (Nutt.) Piper ■☆

3966 Achillea ledebouri Heimerl;阿尔泰蓍;Altai Yarrow, Ledebour Yarrow ■

3967 Achillea leptophylla M. Bieb.;细叶蓍■☆

3968 Achillea leptophylla M. Bieb. subsp. spithamaea (Coss. et Durieu) Maire;细距蓍■☆

3969 Achillea leptophylla M. Bieb. var. major (Batt.) Maire = Achillea leptophylla M. Bieb. subsp. spithamaea (Coss. et Durieu) Maire ■☆

3970 Achillea ligustica All.;女贞蓍;Ligurian Yarrow, Southern Yarrow ■☆

3971 Achillea ligustica All. var. foliosa Ball = Achillea ligustica All. ■☆

3972 Achillea macrocephala Rupr. = Achillea ptarmica L. subsp. macrocephala (Rupr.) Heimerl ■☆

3973 Achillea magna L.;大蓍草;Brownnedge Yarrow ■☆

3974 Achillea maritima (L.) Ehrend. et Y. P. Guo;滨海蓍■☆

3975 Achillea maura Humbert;莫尔蓍■☆

3976 Achillea micrantha M. Bieb.;小花蓍;Smallflower Yarrow ■☆

3977 Achillea micranthoides Klokov;假小花蓍■☆

3978 Achillea millefolia L.;蓍(多叶蓍,锯草,苗蒿,欧蓍,欧蓍草,千叶蓍,蜈蚣蒿,洋蓍草,一苗蒿,一枝蒿,一枝箭);Angel Flower, Arrowroot, Bad Man's Plaything, Bloodwort, Bunch-o'-daisies, Camel, Cammil, Cammock, Carpenter Grass, Carpenter's Herb, Carpenter's Weed, Common Milfoil, Common Yarrow, Death-flower, Devil's Nettle, Devil's Plaything, Devil's Rattle, Dog Daisy, Eerie, Errie, Fever Plant, Field Hop, Goose Tongue, Goose-tongue, Gordolaba, Green Arrow, Hemming-and-sewing, Herb-of-the-seven-cures, Hundred-leaved Grass, Knight's Milfoil, Lady's Lace, Melancholy, Milfoil, Milfoil Yarrow, Mille Flower, Moleery Tea, Moleery-tea, Mother-die, Mother-of-thousands, Musk Milfoil, Nosebleed, Old Man's Mustard, Old Man's Pepper, Plumajillo, Sanguinary, Seven Years' love, Seven-years'-love, Snake's Grass, Sneezewort, Sneezlngs, Soldier's Woundwort, Staunch-girs, Staunch-grass, Sweet Maudlin, Sweet Nut, Sweet Nuts, Tansy, Thousand-leaf, Thousand-leaf Grass, Thousand-leaved Clover, Thousand-seal, Thousand-weed, Traveller's Ease, Venus Tree, Wild Pepper, Woundwort, Yallow, Yarra-grass, Yarrel, Yarroway ■

3979 Achillea millefolia L. 'Fire King';火皇蓍■☆

3980 Achillea millefolia L. 'Rosea';蔷薇蓍;Rose Yarrow ■☆

3981 Achillea millefolia L. subsp. lanulosa (Nutt.) Piper;绵毛蓍;Common Yarrow, Milfoil ■☆

3982 Achillea millefolia L. subsp. lanulosa (Nutt.) Piper f. rubicunda Farw. = Achillea millefolia L. subsp. lanulosa (Nutt.) Piper ■☆

3983 Achillea millefolia L. subsp. occidentalis (DC.) Hyl. = Achillea millefolia L. subsp. lanulosa (Nutt.) Piper ■☆

3984 Achillea millefolia L. subsp. pallidotegula B. Boivin = Achillea millefolia L. subsp. lanulosa (Nutt.) Piper ■☆

3985 Achillea millefolia L. var. aspleniifolia (Vent.) Farw. = Achillea millefolia L. subsp. lanulosa (Nutt.) Piper ■☆

3986 Achillea millefolia L. var. gracilis Raf. ex DC. = Achillea millefolia L. subsp. lanulosa (Nutt.) Piper ■☆

3987 Achillea millefolia L. var. lanulosa (Nutt.) Piper = Achillea millefolia L. subsp. lanulosa (Nutt.) Piper ■☆

3988 Achillea millefolia L. var. lanulosa (Nutt.) Piper et Beattie = Achillea millefolia L. ■

3989 Achillea millefolia L. var. mandshurica Kitag. = Achillea asiatica Serg. ■

3990 Achillea millefolia L. var. occidentalis DC.;西方毛蓍;Woolly Yarrow ■☆

3991 Achillea millefolia L. var. occidentalis DC. = Achillea millefolia L. subsp. lanulosa (Nutt.) Piper ■☆

3992 Achillea millefolia L. var. rosea (Desf.) Torr. et A. Gray =

Achillea millefolia L. subsp. lanulosa (Nutt.) Piper ■☆
3993 Achillea millefolia L. var. russeolata B. Boivin = Achillea millefolia L. subsp. lanulosa (Nutt.) Piper ■☆
3994 Achillea mongolica Fisch. ex Spreng. = Achillea alpina L. ■
3995 Achillea moschata Jacq.;麝香蓍草;Musk Milfoil, Musky Yarrow ■☆
3996 Achillea multiflora Hook.;多花蓍■☆
3997 Achillea nana L.;矮蓍草;Dwarf Milfoil, Dwarf Yarrow ■☆
3998 Achillea nobilis L.;富贵蓍;Noble Yarrow ■☆
3999 Achillea occidentalis (DC.) Raf. ex Rydb. = Achillea millefolia L. subsp. lanulosa (Nutt.) Piper ■☆
4000 Achillea ochroleuca Ehrh.;黄白蓍■☆
4001 Achillea odorata L.;异香蓍■☆
4002 Achillea odorata L. subsp. pectinata (Lam.) Briq. = Achillea odorata L. ■☆
4003 Achillea odorata L. var. microphylla (Willd.) Willk. = Achillea odorata L. ■☆
4004 Achillea pannonica Scheele;潘城蓍■☆
4005 Achillea ptarmica L.;蛐蜓蓍(单叶蓍,珠蓍);Adder's Tongue, Bachelor's Buttons, Bastard Pellitory, Fair Maid of France, Fair Maids of France, False Pellitory, Goose Tongue, Goose-tongue, Hardhead, Mnleery Tea, Moleery-tea, Neesewort, Nosewort, Old Man's Pepper, Old Man's Pepper-box, Pepper-girse, Seven Years' Love, Seven-years'-love, Shirt Buttons, Sholgirse, Sjolgirse, Sneezeweed, Sneezewort, Sneezewort Yarrow, Sneez-wort, While Tansy, Whiteweed, Wild Fire, Wild Pellitory, Wild-fire ■
4006 Achillea ptarmica L. 'The Pearl';珍珠蛐蜓蓍(珍珠蓍草)■☆
4007 Achillea ptarmica L. subsp. euptarmica Heimerl var. acuminata (Ledeb.) Heimerl = Achillea acuminata (Ledeb.) Sch. Bip. ■
4008 Achillea ptarmica L. subsp. macrocephala (Rupr.) Heimerl;大头蛐蜓蓍■☆
4009 Achillea ptarmica L. subsp. macrocephala (Rupr.) Heimerl var. yezoensis Kitam.;北海道蛐蜓蓍■☆
4010 Achillea ptarmica L. subsp. macrocephala Heimerl var. angustifolia Heimerl = Achillea acuminata (Ledeb.) Sch. Bip. ■
4011 Achillea ptarmica L. var. acuminata (Ledeb.) Heimerl = Achillea acuminata (Ledeb.) Sch. Bip. ■
4012 Achillea ptarmica L. var. macrocephala (Rupr.) Masam. = Achillea ptarmica L. subsp. macrocephala (Rupr.) Heimerl ■☆
4013 Achillea ptarmica L. var. speciosa (DC.) Herder = Achillea ptarmica L. subsp. macrocephala (Rupr.) Heimerl ■☆
4014 Achillea ptarmicoides Maxim.;短瓣蓍(白蒿古花);Shortpetal Yarrow ■
4015 Achillea ptarmicoides Maxim. = Achillea alpina L. var. discoidea (Regel) Kitam. ■☆
4016 Achillea pulchra Koidz. = Achillea alpina L. subsp. pulchra (Koidz.) Kitam. ■☆
4017 Achillea pulchra Koidz. var. angustifolia H. Hara = Achillea alpina L. subsp. japonica (Heimerl) Kitam. ■☆
4018 Achillea rosea Desf. = Achillea millefolia L. subsp. lanulosa (Nutt.) Piper ■☆
4019 Achillea rupestris Huter ex Becker;岩生蓍草;Rocky Yarrow ■☆
4020 Achillea sachokiana Sosn.;萨奇科蓍■☆
4021 Achillea salicifolia Besser;柳叶蓍;Willowleaf Yarrow ■
4022 Achillea santolina L.;檀香蓍■☆
4023 Achillea santolinoides Lag.;银香菊蓍■☆
4024 Achillea schischkinii Sosn. ex Grossh.;希施蓍■☆
4025 Achillea schurii Sch. Bip.;舒拉蓍■☆
4026 Achillea septentrionalis (Serg.) Botsch.;北方腺蓍■☆
4027 Achillea setacea Waldst. et Kit.;丝叶蓍;Setaceous Yarrow ■
4028 Achillea sibirica Ledeb.;西伯利亚蓍(黄芪,蓍,蓍草);Siberia Yarrow ■
4029 Achillea sibirica Ledeb. = Achillea alpina L. var. longiligulata H. Hara ■
4030 Achillea sibirica Ledeb. = Achillea alpina L. ■
4031 Achillea sibirica Ledeb. = Achillea ptarmicoides Maxim. ■
4032 Achillea sibirica Ledeb. subsp. camtschatica Heimerl = Achillea alpina L. subsp. camtschatica (Heimerl) Kitam. ■☆
4033 Achillea sibirica Ledeb. subsp. japonica Heimerl = Achillea alpina L. subsp. japonica (Heimerl) Kitam. ■☆
4034 Achillea sibirica Ledeb. subsp. mongolica (Fisch. ex Spreng.) Heimerl = Achillea alpina L. ■
4035 Achillea sibirica Ledeb. subsp. ptarmicoides (Maxim.) Heimarl = Achillea alpina L. var. discoidea (Regel) Kitam. ■☆
4036 Achillea sibirica Ledeb. subsp. ptarmicoides (Maxim.) Heimerl = Achillea ptarmicoides Maxim. ■
4037 Achillea sibirica Ledeb. subsp. pulchra (Koidz.) Kitam. = Achillea alpina L. subsp. pulchra (Koidz.) Kitam. ■☆
4038 Achillea sibirica Ledeb. subsp. subcartilaginea Heimerl = Achillea alpina L. subsp. subcartilaginea (Heimerl) Kitam. ■☆
4039 Achillea sibirica Ledeb. subsp. wilsoniana Heimerl ex Hand.-Mazz. = Achillea wilsoniana (Heimerl ex Hand.-Mazz.) Heimerl ■
4040 Achillea sibirica Ledeb. var. angustifolia (H. Hara) Ohwi = Achillea alpina L. subsp. japonica (Heimerl) Kitam. ■☆
4041 Achillea sibirica Ledeb. var. brevidens (Makino) Ohwi = Achillea alpina L. subsp. subcartilaginea (Heimerl) Kitam. ■☆
4042 Achillea sibirica Ledeb. var. camtschatica (Heimerl) Ohwi = Achillea alpina L. subsp. camtschatica (Heimerl) Kitam. ■☆
4043 Achillea sibirica Ledeb. var. discoidea Regel = Achillea alpina L. var. discoidea (Regel) Kitam. ■☆
4044 Achillea sibirica Ledeb. var. discoidea Regel = Achillea ptarmicoides Maxim. ■
4045 Achillea sibirica Ledeb. var. japonica (Heimerl) Ohwi = Achillea alpina L. subsp. japonica (Heimerl) Kitam. ■☆
4046 Achillea sibirica Ledeb. var. ptarmicoides (Maxim.) Makino = Achillea ptarmicoides Maxim. ■
4047 Achillea sibirica Ledeb. var. ptarmicoides (Maxim.) Makino = Achillea alpina L. var. discoidea (Regel) Kitam. ■☆
4048 Achillea sibirica Ledeb. var. pulchra (Koidz.) Ohwi = Achillea alpina L. subsp. pulchra (Koidz.) Kitam. ■☆
4049 Achillea speciosa DC. = Achillea alpina L. subsp. camtschatica (Heimerl) Kitam. ■☆
4050 Achillea spithamea Coss. = Achillea leptophylla M. Bieb. subsp. spithamaea (Coss. et Durieu) Maire ■☆
4051 Achillea spithamea Coss. var. major Batt. = Achillea leptophylla M. Bieb. subsp. spithamaea (Coss. et Durieu) Maire ■☆
4052 Achillea stricta (W. D. J. Koch) Schleich. ex Gremli;锯蓍■☆
4053 Achillea subcartilaginea (Heimerl) Heimerl = Achillea alpina L. subsp. subcartilaginea (Heimerl) Kitam. ■☆
4054 Achillea tanacetifolia All.;艾菊叶蓍■☆
4055 Achillea tenuifolia Lam.;窄叶蓍草■☆
4056 Achillea tomentosa L.;绒毛蓍草(小锯叶草);Solwherf, Woolly Yarrow, Yellow Yarrow ■☆
4057 Achillea tomentosa Pursh = Achillea millefolia L. subsp. lanulosa (Nutt.) Piper ■☆

4058 Achillea trichophylla Schrenk ex Fisch. et C. A. Mey. = Handelia trichophylla (Schrenk ex Fisch. et C. A. Mey.) Heimerl ■

4059 Achillea umbellata Sibth. et Sm.;伞形蓍草;Umbel Yarrow ■☆

4060 Achillea vermicularis Trin.;蠕虫蓍■☆

4061 Achillea wilhelmsii K. Koch;威尔蓍■☆

4062 Achillea wilsoniana (Heimerl ex Hand. -Mazz.) Heimerl;云南蓍(白花一枝蒿,刀口伤皮,刀口药,飞天蜈蚣,蒿子跌打,乱头发,马茴香,茅草一枝蒿,千叶蓍,蓍草,四乱蒿,土一枝蒿,蜈蚣草,西南蓍,西南蓍草,细杨柳,野一枝蒿,一支蒿,一枝蒿);E. H. Wilson Yarrow,Wilson's Yarrow ■

4063 Achillea wilsoniana Heimerl ex Hand. -Mazz. = Achillea wilsoniana (Heimerl ex Hand. -Mazz.) Heimerl ■

4064 Achillea wilsoniana Heimerl ex Hand. -Mazz. f. obconica Heimerl = Achillea wilsoniana (Heimerl ex Hand. -Mazz.) Heimerl ■

4065 Achilleopsis Turcz. = Rulingia R. Br. (保留属名)●☆

4066 Achillios St. -Lag. = Achillea L. ■

4067 Achilus Hemsl. = Globba L. ■

4068 Achimenes P. Browne(废弃属名) = Achimenes Pers. (保留属名)■☆

4069 Achimenes P. Browne(废弃属名) = Columnea L. ●■☆

4070 Achimenes Pers. (1806)(保留属名);猴面蝴蝶草属(长筒花属,耐寒苣苔属);Achimenes,Cupid's Bower,Cupid's-bower,Hot Water Plant,Hot-water Plant,Monkeyfaced Pansy,Mother's Tears,Orchid Pansy ■☆

4071 Achimenes Vahl = Artanema D. Don(保留属名)■☆

4072 Achimenes Vahl = Bahel Adans. (废弃属名)■☆

4073 Achimenes antirrhina (DC.) C. V. Morton;金鱼草长筒花■☆

4074 Achimenes candida Lindl.;白花猴面蝴蝶草■☆

4075 Achimenes coccinea Pers. = Achimenes erecta (Lam.) H. P. Fuchs ■☆

4076 Achimenes erecta (Lam.) H. P. Fuchs;直立猴面蝴蝶草(轮叶长筒花);Erect Achimenes ■☆

4077 Achimenes ghiesbrechtii Lindl. = Achimenes heterophylla (Mart.) DC. ■☆

4078 Achimenes grandiflora DC.;大花猴面蝴蝶草(大长筒花,大花耐寒苣苔);Big Purple Achimenes,Hot Water Plant ■☆

4079 Achimenes heterophylla (Mart.) DC.;异叶猴面蝴蝶草;Diverse-leaf Achimenes ■☆

4080 Achimenes heterophylla DC. = Achimenes heterophylla (Mart.) DC. ■☆

4081 Achimenes ignescens Lem. = Achimenes heterophylla (Mart.) DC. ■☆

4082 Achimenes longiflora DC.;长花猴面蝴蝶草(长筒花);Magic Flower,Monkey-faced Pansy,Trumpet Achimenes ■☆

4083 Achimenes mexicana Benth. et Hook.;墨西哥猴面蝴蝶草■☆

4084 Achimenes misera Lindl.;细弱猴面蝴蝶草■☆

4085 Achimenes ocellata Hook.;眼斑猴面蝴蝶草;Eye-spotted Achimenes ■☆

4086 Achimenes pedunculata Benth.;序柄猴面蝴蝶草;Pedunculate Achimenes ■☆

4087 Achimenes pulchella Hitchc. = Achimenes erecta (Lam.) H. P. Fuchs ■☆

4088 Achimenes rosea Lindl. = Achimenes erecta (Lam.) H. P. Fuchs ■☆

4089 Achimenes scheerii Hemsl. = Achimenes mexicana Benth. et Hook. ■☆

4090 Achimenes skinneri Lindl.;斯氏猴面蝴蝶草■☆

4091 Achimenes warszewicziana Otto ex Regel = Achimenes misera Lindl. ■☆

4092 Achimus Poir. = Achymus Vahl ex Juss. ●

4093 Achimus Poir. = Streblus Lour. ●

4094 Achiranthes P. Browne = Achyranthes L. (保留属名)■

4095 Achirida Horan. = Canna L. ■

4096 Achiridia Baker = Achirida Horan. ■

4097 Achironia Steud. = Achyronia L. ●☆

4098 Achlaena Griseb. = Arthropogon Nees ■☆

4099 Achlydosa M. A. Clem. et D. L. Jones = Lyperanthus R. Br. ■☆

4100 Achlydosa M. A. Clem. et D. L. Jones(2002);多腺蕨叶梅属■☆

4101 Achlyphila Maguire et Wurdack(1960);荫地黄眼草属■☆

4102 Achlyphila disticha Maguire et Wurdack;荫地黄眼草■☆

4103 Achlys DC. (1821);裸花草属(阿葎属);Deer-foot,Vanilla-leaf ■☆

4104 Achlys japonica Maxim.;裸花草(日本裸花草);Japanese Vanilla-leaf ■☆

4105 Achlys triphylla (Sm.) DC.;三叶裸花草;Deerfoot,Deer-foot,Vanilla Leaf,Vanilla-leaf ■☆

4106 Achlys triphylla (Sm.) DC. subsp. japonica (Maxim.) Kitam. = Achlys japonica Maxim. ■☆

4107 Achlys triphylla DC. subsp. japonica (Maxim.) Kitam. = Achlys japonica Maxim. ■☆

4108 Achmaea Steud. = Aechmea Ruiz et Pav. (保留属名)■☆

4109 Achmandra Arn. = Bryonia L. ■☆

4110 Achmandra Wight = Aechmandra Arn. ■☆

4111 Achmandra Wight = Kedrostis Medik. ■☆

4112 Achmea Poir. = Aechmea Ruiz et Pav. (保留属名)■☆

4113 Achnatherum P. Beauv. (1812);芨芨草属;Achnatherum,Jijigrass,Speargrass ■

4114 Achnatherum P. Beauv. = Stipa L. ■

4115 Achnatherum argenteum P. Beauv.;银色芨芨草■

4116 Achnatherum avinoides (Honda) Y. L. Chang = Achnatherum sibiricum (L.) Keng ■

4117 Achnatherum brandisii (Mez) Z. L. Wu;展序芨芨草■

4118 Achnatherum breviaristatum Keng et P. C. Kuo;短芒芨芨草;Shortaristate Jijigrass,Shortawn Speargrass,Shortbristled Needlegrass ■

4119 Achnatherum bromoides (L.) P. Beauv.;燕麦芨芨草■☆

4120 Achnatherum calamagrostis (L.) P. Beauv.;拂子茅芨芨草;Calamagrostis Speargrass ■

4121 Achnatherum calamagrostis (L.) P. Beauv. subsp. mesatlanticum (Quézel) Dobignard;梅萨芨芨草■☆

4122 Achnatherum capense P. Beauv.;好望角芨芨草;Cape Ricegrass ■☆

4123 Achnatherum caraganum (Trin. et Rupr.) Nevski;小芨芨草(锦鸡儿芨芨);Little Speargrass,Small Jijigrass ■

4124 Achnatherum caudatum (Trin.) S. W. L. Jacobs et J. Everett;智利芨芨草;Chilean Ricegrass ■☆

4125 Achnatherum chinense (Hitchc.) Tzvelev;中华芨芨草■

4126 Achnatherum chingii (Hitchc.) Keng = Achnatherum chingii (Hitchc.) Keng ex P. C. Kuo ■

4127 Achnatherum chingii (Hitchc.) Keng ex P. C. Kuo;细叶芨芨草(秦氏芨芨草);Ching Jijigrass,Ching Speargrass ■

4128 Achnatherum chingii (Hitchc.) Keng ex P. C. Kuo var. laxum S. L. Lu;林荫芨芨草;Loose Speargrass ■

4129 Achnatherum clandestinum (Hack.) Barkworth;墨西哥芨芨草;Mexican Ricegrass ■☆

4130 Achnatherum coreanum (Honda) Ohwi;朝鲜直芒草(大叶直芒草)■

4131 Achnatherum duthiei (Hook. f.) P. C. Kuo et S. L. Lu;藏芨芨草;Duthie Speargrass, Xizang Jijigrass ■
4132 Achnatherum effusum (Maxim.) Y. L. Chang = Achnatherum extremiorientale (H. Hara) Keng ex P. C. Kuo ■
4133 Achnatherum extremiorientale (H. Hara) Keng = Achnatherum extremiorientale (H. Hara) Keng ex P. C. Kuo ■
4134 Achnatherum extremiorientale (H. Hara) Keng ex P. C. Kuo;远东芨芨草(极东针茅,展穗芨芨草);Extremeeast Feathergrass, Extremioriental Jijigrass, Extremioriental Speargrass ■
4135 Achnatherum extremiorientale (H. Hara) Keng ex P. C. Kuo = Stipa pekinensis Hance ■
4136 Achnatherum henryi (Rendle) S. M. Phillips et Z. L. Wu;湖北芨芨草■
4137 Achnatherum henryi (Rendle) S. M. Phillips et Z. L. Wu var. acutum (L. Liou ex Z. L. Wu) S. M. Phillips et Z. L. Wu;尖颖湖北芨芨草(尖颖芨芨草)■
4138 Achnatherum hookeri (Stapf) Keng = Trikeraia hookeri (Stapf) Bor ■
4139 Achnatherum hymenoides (Roem. et Schult.) Barkworth;印度芨芨草;Indian Rice Grass, Indian Ricegrass ■☆
4140 Achnatherum inaequiglume Keng = Achnatherum inaequiglume Keng ex P. C. Kuo ■
4141 Achnatherum inaequiglume Keng ex P. C. Kuo;异颖芨芨草;Unequalglume Jijigrass, Unequalglume Speargrass ■
4142 Achnatherum inebrians (Hance) Keng;醉马草(阿尔善醉马草,米米蒿,药草,药老,醉针茅);Inebriate Jijigrass, Inebriate Speargrass ■
4143 Achnatherum jacquemontii (Jaub. et Spach) P. C. Kuo et S. L. Lu;干生芨芨草;Jacquemont Speargrass ■
4144 Achnatherum lanceolatum P. Beauv.;披针叶芨芨草■☆
4145 Achnatherum longiaristatum (Boiss. et Hausskn.) Keng et P. C. Kuo;长芒芨芨草■☆
4146 Achnatherum mesatlanticum (Quézel) Inb Tattou = Achnatherum calamagrostis (L.) P. Beauv. subsp. mesatlanticum (Quézel) Dobignard ■☆
4147 Achnatherum mongholicum (Turcz. ex Trin.) Ohwi = Ptilagrostis mongholica (Turcz. ex Trin.) Griseb. ■
4148 Achnatherum nakaii (Honda) Tateoka;朝阳芨芨草(中井芨芨草);Nakai Jijigrass, Nakai Speargrass ■
4149 Achnatherum pappiformis (Keng) Keng = Trikeraia pappiformis (Keng) P. C. Kuo et S. L. Lu ■
4150 Achnatherum pekinense (Hance) Ohwi;京芒草(京羽茅);Beijing Jijigrass, Beijing Speargrass ■
4151 Achnatherum pekinense (Hance) Ohwi = Stipa pekinensis Hance ■
4152 Achnatherum pekinense (Hance) Ohwi subsp. effusum (Maxim.) T. Koyama = Stipa pekinensis Hance ■
4153 Achnatherum psilantherum Keng ex P. C. Keng;光药芨芨草;Nakedanther Jijigrass, Nakedanther Speargrass ■
4154 Achnatherum pubicalyx (Ohwi) Keng = Achnatherum pubicalyx (Ohwi) Keng ex P. C. Kuo ■
4155 Achnatherum pubicalyx (Ohwi) Keng ex P. C. Kuo;毛颖芨芨草;Hairyglume Jijigrass, Hairyglume Speargrass ■
4156 Achnatherum purpurascens (Hack.) Keng = Stipa regeliana Hack. ■
4157 Achnatherum purpurascens (Hitchc.) Keng;紫花芨芨草;Purpleflower Speargrass ■
4158 Achnatherum regelianum (Hack.) P. C. Keng = Achnatherum purpurascens (Hitchc.) Keng ■
4159 Achnatherum saposhnikovii (Roshev.) Nevski;钝基芨芨草(钝基草)■
4160 Achnatherum sibiricum (L.) Keng;羽茅(光颖芨芨草,燕麦芨芨草,醉马草);Siberian Feathergrass, Siberian Jijigrass, Siberian Speargrass ■
4161 Achnatherum sibiricum (L.) Keng = Stipa sibirica (L.) Lam. ■
4162 Achnatherum sibiricum (L.) Keng var. qinghaiense Y. J. Wang;青海芨芨草;Qinghai Speargrass ■
4163 Achnatherum splendens (Trin.) Nevski;芨芨草(积机草,席萁草,枳机草,枳芨草);Jijigrass, Lovely Achnatherum, Lovely Jijigrass, Shining Speargrass ■
4164 Achnatherum tenuifolium P. Beauv. = Agrostis canina L. ■
4165 Achneria Benth. = Pentaschistis (Nees) Spach ■☆
4166 Achneria Munro = Afrachneria Sprague ■☆
4167 Achneria Munro = Pentaschistis (Nees) Spach ■☆
4168 Achneria Munro ex Benth. et Hook. f. = Afrachneria Sprague ■☆
4169 Achneria P. Beauv. = Eriachne R. Br. ■
4170 Achneria ampla (Nees) T. Durand et Schinz = Pentaschistis ampla (Nees) McClean ■☆
4171 Achneria assimilis (Steud.) T. Durand et Schinz = Pentaschistis ecklonii (Nees) McClean ■☆
4172 Achneria aurea (Steud.) T. Durand et Schinz = Pentaschistis aurea (Steud.) McClean ■☆
4173 Achneria capensis (Steud.) T. Durand et Schinz = Pentaschistis malouinensis (Steud.) Clayton ■☆
4174 Achneria capillaris (Thunb.) Stapf = Pentaschistis capillaris (Thunb.) McClean ■☆
4175 Achneria curvifolia (Hack.) Stapf = Pentaschistis ecklonii (Nees) McClean ■☆
4176 Achneria ecklonii (Nees) T. Durand et Schinz = Pentaschistis ecklonii (Nees) McClean ■☆
4177 Achneria fasciculata Peter = Poa kilimanjarica (Hedberg) Markgr.-Dann. ■☆
4178 Achneria galpinii Stapf = Pentaschistis galpinii (Stapf) McClean ■☆
4179 Achneria hirsuta (Nees) Stapf = Pentaschistis aurea (Steud.) McClean subsp. pilosogluma (McClean) H. P. Linder ■☆
4180 Achneria microphylla (Nees) T. Durand et Schinz = Pentaschistis microphylla (Nees) McClean ■☆
4181 Achneria pallida (Nees) T. Durand et Schinz = Pentaschistis ampla (Nees) McClean ■☆
4182 Achneria setifolia (Thunb.) Stapf = Pentaschistis setifolia (Thunb.) McClean ■☆
4183 Achneria tuberculata (Nees) T. Durand et Schinz = Pentaschistis setifolia (Thunb.) McClean ■☆
4184 Achnodon Link = Phleum L. ■
4185 Achnodonton P. Beauv. = Phleum L. ■
4186 Achnophora F. Muell. (1883);鞘莲菀属(澳洲菊属)■☆
4187 Achnophora tatei F. Muell.;鞘莲菀■☆
4188 Achnopogon Maguire, Steyerm. et Wurdack (1957);鳞菊木属●☆
4189 Achnopogon virgatus Maguire, Steyerm. et Wurdack;鳞菊木●☆
4190 Achoriphragma Soják = Leiospora (C. A. Mey.) Dvorák ■
4191 Achoriphragma Soják = Parrya R. Br. ●■
4192 Achoriphragma Soják (1982);不离膜芥属■
4193 Achoriphragma ajanense (N. Busch) Soják = Parrya nudicaulis (L.) Regel ■
4194 Achoriphragma beketovii (Krasn.) Soják = Parrya beketovii Krasn. ■

4195 Achoriphragma lancifolium (Popov) Soják = Parrya lancifolia Popov ■
4196 Achoriphragma nudicaule (L.) Soják = Parrya nudicaulis (L.) Regel ■
4197 Achoriphragma pinnatifidum (Kar. et Kir.) Soják = Parrya pinnatifida Kar. et Kir. ■
4198 Achoriphragma stenocarpum (Kar. et Kir.) Soják = Parrya pinnatifida Kar. et Kir. ■
4199 Achraceae Roberty = Sapotaceae Juss.(保留科名)●
4200 Achradaceae Vest = Sapotaceae Juss.(保留科名)●
4201 Achradelpha O. F. Cook = Calospermum Pierre ●
4202 Achradelpha O. F. Cook = Pouteria Aubl. ●
4203 Achradotypus Baill. = Pycnandra Benth. ●☆
4204 Achrantes Pfeiff. = Achranthes Dumort. ■
4205 Achranthes Dumort. = Achyranthes L.(保留属名)■
4206 Achras L.(废弃属名) = Manilkara Adans.(保留属名)●
4207 Achras amammosa L. = Manilkara zapota (L.) P. Royen ●
4208 Achras australis R. Br. = Planchonella australis (R. Br.) Pierre ●☆
4209 Achras australis R. Br. = Sersalisia australis (R. Br.) Domin ●☆
4210 Achras sericea Schumach. = Chrysophyllum albidum G. Don ●☆
4211 Achras zapota L. = Manilkara zapota (L.) P. Royen ●
4212 Achras zapotilla (Jacq.) Coville = Manilkara zapota (L.) P. Royen ●
4213 Achratinis Kuntze = Arachnitis Phil.(保留属名)■☆
4214 Achratinitaceae Barkley = Corsiaceae Becc.(保留科名)■
4215 Achroanthes Raf.(废弃属名) = Malaxis Sol. ex Sw. ■
4216 Achroanthes Raf.(废弃属名) = Microstylis (Nutt.) Eaton(保留属名)■☆
4217 Achroanthes corymbosa (S. Watson) Greene = Malaxis corymbosa (S. Watson) Kuntze ■☆
4218 Achroanthes floridana (Chapm.) Greene = Malaxis spicata Sw. ■☆
4219 Achroanthes monophylla (L.) Greene = Malaxis monophylla (L.) Sw. ■
4220 Achroanthes montana (Rothr.) Greene = Malaxis soulei L. O. Williams ■☆
4221 Achroanthes unifolia (Michx.) Raf. = Malaxis unifolia Michx. ■☆
4222 Achrochloa B. D. Jacks. = Achrochloa Griseb. ■
4223 Achrochloa B. D. Jacks. = Airochloa Link ■
4224 Achrochloa B. D. Jacks. = Koeleria Pers. ■
4225 Achrochloa Griseb. = Airochloa Link ■
4226 Achromochlaena Post et Kuntze = Achromolaena Cass. ●☆
4227 Achromolaena Cass. = Cassinia R. Br.(保留属名)●☆
4228 Achroostachys Benth. = Athroostachys Benth. ex Benth. et Hook. f. ●☆
4229 Achrouteria Eyma = Chrysophyllum L. ●
4230 Achrysum A. Gray = Calocephalus R. Br. ●■☆
4231 Achuaria Gereau(1990);阿丘芸香属●☆
4232 Achuaria hirsuta Gereau;阿丘芸香●☆
4233 Achudemia Blume = Pilea Lindl.(保留属名)■
4234 Achudemia Blume(1856);山美豆属■☆
4235 Achudemia insignis Migo = Pilea japonica (Maxim.) Hand.-Mazz. ■
4236 Achudemia japonica Maxim.;日本山美豆■☆
4237 Achudemia japonica Maxim. = Pilea japonica (Maxim.) Hand.-Mazz. ■
4238 Achudemia javanica Blume;山美豆■☆
4239 Achudenia Benth. et Hook. f. = Achudemia Blume ■☆
4240 Achupalla Humb. = Puya Molina ■☆
4241 Achymenes Batsch = Achimenes Pers.(保留属名)■☆
4242 Achymenes Batsch = Artanema D. Don(保留属名)■☆
4243 Achymus Vahl ex Juss. = Streblus Lour. ●
4244 Achyrachaena Schauer(1838);拂妻菊属■☆
4245 Achyrachaena mollis Schauer;拂妻菊■☆
4246 Achyracharna Walp. = Achyrachaena Schauer ■☆
4247 Achyrantes L. = Achyranthes L.(保留属名)■
4248 Achyranthaceae Raf. = Amaranthaceae Juss.(保留科名)●■
4249 Achyranthes L.(1753)(保留属名);牛膝属;Achyranthes, Chaff-flower ■
4250 Achyranthes alba Eckl. et Zeyh. ex Moq. = Achyropsis leptostachya (E. Mey. ex Meisn.) Baker et C. B. Clarke ■☆
4251 Achyranthes alternifolia L. = Digera muricata (L.) Mart. ■☆
4252 Achyranthes amaranthoides Lam. = Cladostachys frutescens D. Don ●
4253 Achyranthes amaranthoides Lam. = Deeringia amaranthoides (Lam.) Merr. ●■
4254 Achyranthes angustifolia (Vahl) Lopr. = Pandiaka angustifolia (Vahl) Hepper ■☆
4255 Achyranthes annua Dinter = Achyranthes aspera L. var. sicula L. ■☆
4256 Achyranthes aquatica R. Br. = Centrostachys aquatica (R. Br.) Wall. ■☆
4257 Achyranthes argentea Lam. = Achyranthes aspera L. var. sicula L. ■☆
4258 Achyranthes argentea Thwaites = Achyranthes aspera L. var. argentea (Thwaites) Hook. f. ■
4259 Achyranthes aspera L.;土牛膝(白基牛七,白牛七,粗毛牛膝,撮鼻草,倒刺草,倒梗草,倒钩草,倒挂草,倒扣草,倒扣簕,倒勒草,倒捋草,倒吞吞,对节草,鹅膝,虎鞭草,鸡撮鼻,鸡骨草,鸡骨黄,鸡骨癀,南蛇牙草,牛七风,牛舌大黄,牛舌头,牛獭鼻,牛藤,牛膝,破布粘,铁马鞭,土牛七,鸭脚节,印度牛膝,鱼鳞菜,粘身草);Common Achyranthes, Devil's Horsewhip, Hold-me-tight, Prickly Chaff Flower ■
4260 Achyranthes aspera L. f. robustoides Suess. = Achyranthes aspera L. var. late-ovata Boerl. ■☆
4261 Achyranthes aspera L. var. argentea (Lam.) Boiss. = Achyranthes aspera L. var. sicula L. ■☆
4262 Achyranthes aspera L. var. argentea (Thwaites) Hook. f.;银毛土牛膝;Silverhair Achyranthes ■
4263 Achyranthes aspera L. var. indica L.;钝叶土牛膝(倒刺草,钝头牛膝,钝叶牛膝,土牛膝,印度牛膝);Hold-me-tight, Native Bloodleaf, Obtuseleaf Achyranthes ■
4264 Achyranthes aspera L. var. indica L. = Achyranthes aspera L. ■
4265 Achyranthes aspera L. var. late-ovata Boerl.;宽卵形土牛膝■☆
4266 Achyranthes aspera L. var. nigro-olivacea Suess. = Achyranthes aspera L. var. pubescens (Moq.) C. C. Towns. ■☆
4267 Achyranthes aspera L. var. obtusifolia (Lam.) Suess. = Achyranthes aspera L. ■
4268 Achyranthes aspera L. var. pubescens (Moq.) C. C. Towns.;毛土牛膝;Devil's Horsewhip ■☆
4269 Achyranthes aspera L. var. rubrofusca (Wight) Hook. f.;褐叶土牛膝(禾叶土牛膝,红褐粗毛牛膝,台湾牛膝,土牛膝,云牛膝,紫茎牛膝);Brownleaf Achyranthes ■
4270 Achyranthes aspera L. var. sicula L.;西西里土牛膝■☆
4271 Achyranthes atropurpurea Lam. = Pupalia lappacea (L.) A. Juss. ■☆
4272 Achyranthes benthamii Lopr. = Pandiaka angustifolia (Vahl) Hepper ■☆

4273 Achyranthes bidentata Blume;牛膝(白牛膝,白牛膝草,百倍,杜牛膝,对节菜,对节草,红牛膝,红叶牛膝,怀牛膝,怀夕,怀膝,淮牛膝,鸡胶骨,脚斯蹬,接骨草,马麦草,牛茎,牛磕膝,牛髁膝,牛克膝,牛夕,牛郄,山牛膝,山苋,山苋菜,四季花,通天柱杖,透骨草,土牛膝);Achyranthes,Twotooth Achyranthes ■

4274 Achyranthes bidentata Blume var. hachijoensis (Honda) H. Hara;八丈岛牛膝;Japanese Chaff Flower ■☆

4275 Achyranthes bidentata Blume var. japonica Miq.;少毛牛膝(日本牛膝);Japanese Achyranthes, Japanese Chaff Flower, Japanese Twotooth Achyranthes ■

4276 Achyranthes bidentata Blume var. longifolia Makino = Achyranthes longifolia (Makino) Makino ■

4277 Achyranthes bidentata Blume var. tomentosa (Honda) H. Hara = Achyranthes fauriei H. Lév. et Vaniot ■

4278 Achyranthes brachiata L. = Nothosaerva brachiata (L.) Wight ■☆

4279 Achyranthes breviflora Baker = Centemopsis kirkii (Hook. f.) Schinz ■☆

4280 Achyranthes carsonii Baker = Pandiaka carsonii (Baker) C. B. Clarke ■☆

4281 Achyranthes conferta Schinz = Centemopsis conferta (Schinz) Suess. ■☆

4282 Achyranthes corymbosa L. = Polycarpaea corymbosa (L.) Lam. ■

4283 Achyranthes cylindrica Bojer = Cyathula cylindrica Moq. ■☆

4284 Achyranthes elegantissima Schinz = Pandiaka elegantissima (Schinz) Dandy ■☆

4285 Achyranthes fasciculata (Suess.) C. C. Towns.;簇生牛膝■☆

4286 Achyranthes fauriei H. Lév. et Vaniot;日本牛膝■

4287 Achyranthes fauriei H. Lév. et Vaniot f. rotundifolia Ohwi;圆叶日本牛膝■☆

4288 Achyranthes fauriei H. Lév. et Vaniot var. japonica (Miq.) Hiyama = Achyranthes bidentata Blume var. japonica Miq. ■

4289 Achyranthes ferruginea Roxb. = Psilotrichum ferrugineum (Roxb.) Moq. ■

4290 Achyranthes fruticosa Lam. var. pubescens Moq. = Achyranthes aspera L. var. pubescens (Moq.) C. C. Towns. ■☆

4291 Achyranthes geminata Thonn. = Cyathula achyranthoides (Kunth) Moq. ■☆

4292 Achyranthes heudelotii Moq. = Pandiaka angustifolia (Vahl) Hepper ■☆

4293 Achyranthes indica (L.) Mill. = Achyranthes aspera L. ■

4294 Achyranthes involucrata Moq. = Pandiaka involucrata (Moq.) B. D. Jacks. ■☆

4295 Achyranthes japonica (Miq.) Nakai = Achyranthes bidentata Blume var. japonica Miq. ■

4296 Achyranthes japonica (Miq.) Nakai var. hachijoensis Honda = Achyranthes bidentata Blume var. hachijoensis (Honda) H. Hara ■☆

4297 Achyranthes javanica (Burm. f.) Pers. = Aerva javanica (Burm. f.) Juss. ex Schult. ■☆

4298 Achyranthes lanata L. = Aerva lanata (L.) Juss. ex Schult. ■☆

4299 Achyranthes lanuginosa Nutt. = Tidestromia lanuginosa (Nutt.) Standl. ■☆

4300 Achyranthes lanuginosa Schinz = Pandiaka lanuginosa (Schinz) Schinz ■☆

4301 Achyranthes lappacea L. = Pupalia lappacea (L.) A. Juss. ■☆

4302 Achyranthes leiantha (Seub.) Standl. = Alternanthera pungens Kunth ■

4303 Achyranthes leptostachya E. Mey. ex Meisn. = Achyropsis leptostachya (E. Mey. ex Meisn.) Baker et C. B. Clarke ■☆

4304 Achyranthes longifolia (Makino) Makino;柳叶牛膝(长叶牛膝,粗毛牛膝,倒扣簕,杜牛膝,拐牛膝,红牛七,红牛膝,鸡掇鼻,剪刀牛膝,老鸹窝,荔支红,牛克膝,山牛膝,苏木红,透血红,土牛夕,土牛膝,尾膝,未膝,味牛膝,狭叶牛膝,野牛膝);Willowleaf Achyranthes ■

4305 Achyranthes longifolia (Makino) Makino f. rubra C. N. Ho;红柳叶牛膝;Red Willowleaf Achyranthes ■

4306 Achyranthes maritima (Mart.) Standl. = Alternanthera maritima (Mart.) A. St.-Hil. ■☆

4307 Achyranthes mauritiana Moq. = Achyranthes bidentata Blume ■

4308 Achyranthes megaphylla Y. H. Li;大叶牛膝;Bigleaf Achyranthes ■

4309 Achyranthes mollis Thonn. = Pupalia lappacea (L.) A. Juss. var. velutina (Moq.) Hook. f. ■☆

4310 Achyranthes monsoniae (L. f.) Pers. = Trichuriella monsoniae (L. f.) Bennet ■

4311 Achyranthes monsoniae Pers. = Trichurus monsoniae (L. f.) C. C. Towns. ■

4312 Achyranthes muricata L. = Digera muricata (L.) Mart. ■☆

4313 Achyranthes nivea Aiton = Polycarpaea nivea (Aiton) Webb ■☆

4314 Achyranthes nodosa Vahl;结节牛膝■☆

4315 Achyranthes oblanceolata Schinz = Pandiaka elegantissima (Schinz) Dandy ■☆

4316 Achyranthes obovata Peter = Achyranthes aspera L. ■

4317 Achyranthes obtusifolia Lam. = Achyranthes aspera L. var. indica L. ■

4318 Achyranthes obtusifolia Lam. = Achyranthes aspera L. ■

4319 Achyranthes ogatai Yamam.;小叶牛膝(南天牛膝);Ogata Achyranthes ■

4320 Achyranthes pedicellata Lopr.;梗花牛膝■☆

4321 Achyranthes philoxeroides (C. Mart.) Standl. = Alternanthera philoxeroides (Mart.) Griseb. ■

4322 Achyranthes polystachia Forssk. = Celosia polystachia (Forssk.) C. C. Towns. ■☆

4323 Achyranthes prostrata L. = Cyathula prostrata (L.) Blume ■

4324 Achyranthes radicans Cav. = Alternanthera sessilis (L.) R. Br. ex DC. ■

4325 Achyranthes ramosissima (Mart.) Standl. = Alternanthera flavescens Kunth ■☆

4326 Achyranthes repens L. = Alternanthera pungens Humb. ■

4327 Achyranthes robusta C. H. Wright = Achyranthes aspera L. ■

4328 Achyranthes rubrofusca Wight = Achyranthes aspera L. var. rubrofusca (Wight) Hook. f. ■

4329 Achyranthes rubrolutea Lopr. = Pandiaka rubro-lutea (Lopr.) C. C. Towns. ■☆

4330 Achyranthes sanguinolenta L. = Aerva sanguinolenta (L.) Blume ●

4331 Achyranthes scandens Roxb. = Aerva sanguinolenta (L.) Blume ●

4332 Achyranthes schinzii (Standl.) Cufod. = Pandiaka lanuginosa (Schinz) Schinz ■☆

4333 Achyranthes schweinfurthii Schinz = Pandiaka welwitschii (Schinz) Hiern ■☆

4334 Achyranthes sericea Jos. König ex Roxb. = Psilotrichum sericeum (Jos. König ex Roxb.) Dalzell ●☆

4335 Achyranthes setacea Roth = Trichuriella monsoniae (L. f.) Bennet ■

4336 Achyranthes sicula (L.) All. = Achyranthes aspera L. var. sicula L. ■☆

4337 Achyranthes stellata Willd. = Polycarpaea stellata (Willd.) DC. ■☆

4338 Achyranthes talbotii Hutch. et Dalziel;塔尔博特牛膝■☆
4339 Achyranthes tenuifolia Willd. = Polycarpaea tenuifolia (Willd.) DC. ■☆
4340 Achyranthes thonningii Schumach. = Pupalia lappacea (L.) A. Juss. var. velutina (Moq.) Hook. f. ■☆
4341 Achyranthes tomentosa Roth = Cyathula tomentosa (Roth) Moq. ■
4342 Achyranthes uncinulata Schrad. = Cyathula uncinulata (Schrad.) Schinz ■☆
4343 Achyranthes verticillata Thunb.;轮生牛膝■☆
4344 Achyranthes villosa Forssk. = Aerva lanata (L.) Juss. ex Schult. ■☆
4345 Achyranthes viridis Lopr.;绿牛膝■☆
4346 Achyranthes welwitschii Schinz = Pandiaka welwitschii (Schinz) Hiern ■☆
4347 Achyranthes winteri Peter;温特牛膝■☆
4348 Achyranthus Neck. = Achyranthes L.(保留属名)■
4349 Achyrastrum Neck. = Hyoseris L. ■☆
4350 Achyrobaccharis Sch. Bip. = Baccharis L.(保留属名)●■☆
4351 Achyrobaccharis Sch. Bip. ex Walp. = Baccharis L.(保留属名)●■☆
4352 Achyrocalyx Benoist(1930);糠萼爵床属■☆
4353 Achyrocalyx decaryi Benoist;糠萼爵床■☆
4354 Achyrocalyx pungens Benoist;锐尖糠萼爵床■☆
4355 Achyrocalyx vicinus Benoist = Achyrocalyx decaryi Benoist ■☆
4356 Achyrochoma B. D. Jacks. = Achyrocoma Cass. ●■
4357 Achyrochoma B. D. Jacks. = Vernonia Schreb.(保留属名)●■
4358 Achyrocline (Less.) DC.(1838);多头金绒草属■☆
4359 Achyrocline (Less.) DC. = Helichrysum Mill.(保留属名)●■
4360 Achyrocline Less. = Achyrocline (Less.) DC. ■☆
4361 Achyrocline adoensis Sch. Bip. ex A. Rich. = Helichrysum schimperi (Sch. Bip. ex A. Rich.) Moeser ●☆
4362 Achyrocline batocana Oliv. et Hiern = Helichrysum kraussii Sch. Bip. ■☆
4363 Achyrocline glumacea (DC.) Oliv. et Hiern = Helichrysum glumaceum DC. ■☆
4364 Achyrocline hochstetteri Sch. Bip. ex A. Rich. = Helichrysum stenopterum DC. ●☆
4365 Achyrocline insularis Humbert = Helichrysum globosum A. Rich. ■☆
4366 Achyrocline luzuloides Sch. Bip. ex Vatke = Helichrysum glumaceum DC. ■☆
4367 Achyrocline luzuloides Sch. Bip. ex Vatke var. alpina Mattei = Helichrysum glumaceum DC. ■☆
4368 Achyrocline pumila Klatt = Helichrysum somalense Baker f. ●☆
4369 Achyrocline schimperi Sch. Bip. ex A. Rich. = Helichrysum schimperi (Sch. Bip. ex A. Rich.) Moeser ●☆
4370 Achyrocline sclerochlaena Vatke = Helichrysum sclerochlaenum (Vatke) Moeser ●☆
4371 Achyrocline steetzii Vatke = Helichrysum kraussii Sch. Bip. ■☆
4372 Achyrocline stenoptera (DC.) Hilliard et B. L. Burtt = Helichrysum stenopterum DC. ●☆
4373 Achyrocoma Cass. = Vernonia Schreb.(保留属名)●■
4374 Achyrocoma Post et Kuntze = Achyrocome Schrank ●☆
4375 Achyrocome Schrank = Elytropappus Cass. ●☆
4376 Achyrodes Boehm ex Kuntze = Lamarckia Moench(保留属名)■☆
4377 Achyrodes Boehm(废弃属名) = Lamarckia Moench(保留属名)■☆
4378 Achyroma J. C. Wendl. = Achyrocoma Cass. ●■
4379 Achyroma J. C. Wendl. = Vernonia Schreb.(保留属名)●■
4380 Achyronia Boehm. = Aspalathus L. ●☆
4381 Achyronia J. C. Wendl. = Priestleya DC. ■☆
4382 Achyronia L. = Aspalathus L. ●☆
4383 Achyronia Royen ex L. = Aspalathus L. ●☆
4384 Achyronia acuminata (Lam.) Kuntze = Aspalathus acuminata Lam. ●☆
4385 Achyronia anthyllodes (L.) Kuntze = Aspalathus aspalathoides (L.) R. Dahlgren ●☆
4386 Achyronia arida (E. Mey.) Kuntze = Aspalathus arida E. Mey. ●☆
4387 Achyronia asparagoides (L. f.) Kuntze = Aspalathus asparagoides L. f. ●☆
4388 Achyronia calcarata (Harv.) Kuntze = Aspalathus calcarata Harv. ●☆
4389 Achyronia callosa (L.) Kuntze = Aspalathus callosa L. ●☆
4390 Achyronia capitata (L.) Kuntze = Aspalathus capitata L. ●☆
4391 Achyronia carnosa (P. J. Bergius) Kuntze = Aspalathus carnosa P. J. Bergius ●☆
4392 Achyronia chenopoda (L.) Kuntze = Aspalathus chenopoda L. ●☆
4393 Achyronia ciliaris (L.) Kuntze = Aspalathus ciliaris L. ●☆
4394 Achyronia collina (Eckl. et Zeyh.) Kuntze = Aspalathus collina Eckl. et Zeyh. ●☆
4395 Achyronia erythrodes (Eckl. et Zeyh.) Kuntze = Aspalathus erythrodes Eckl. et Zeyh. ●☆
4396 Achyronia ferox (Harv.) Kuntze = Aspalathus ferox Harv. ●☆
4397 Achyronia latibracteata Kuntze = Aspalathus latibracteata (Kuntze) K. Schum. ●☆
4398 Achyronia leucophaea (Harv.) Kuntze = Aspalathus ciliaris L. ●☆
4399 Achyronia longipes (Harv.) Kuntze = Aspalathus longipes Harv. ●☆
4400 Achyronia marginalis (Eckl. et Zeyh.) Kuntze = Aspalathus marginalis Eckl. et Zeyh. ●☆
4401 Achyronia opaca (Eckl. et Zeyh.) Kuntze = Aspalathus opaca Eckl. et Zeyh. ●☆
4402 Achyronia pachyloba (Benth.) Kuntze = Aspalathus pachyloba Benth. ●☆
4403 Achyronia pallescens (Eckl. et Zeyh.) Kuntze = Aspalathus pallescens Eckl. et Zeyh. ●☆
4404 Achyronia pappeana (Harv.) Kuntze = Aspalathus opaca Eckl. et Zeyh. subsp. pappeana (Harv.) R. Dahlgren ●☆
4405 Achyronia prostrata (Eckl. et Zeyh.) Kuntze = Aspalathus prostrata Eckl. et Zeyh. ●☆
4406 Achyronia recurva (Benth.) Kuntze = Aspalathus recurva Benth. ●☆
4407 Achyronia rubens (Thunb.) Kuntze = Aspalathus rubens Thunb. ●☆
4408 Achyronia rubro-fusca (Eckl. et Zeyh.) Kuntze = Aspalathus asparagoides L. f. subsp. rubro-fusca (Eckl. et Zeyh.) R. Dahlgren ●☆
4409 Achyronia sanguinea (Thunb.) Kuntze = Aspalathus sanguinea Thunb. ●☆
4410 Achyronia spinescens (Thunb.) Kuntze = Aspalathus spinescens Thunb. ●☆
4411 Achyronia subulata (Thunb.) Kuntze = Aspalathus subulata Thunb. ●☆
4412 Achyronia ulicina (Eckl. et Zeyh.) Kuntze = Aspalathus ulicina Eckl. et Zeyh. ●☆
4413 Achyronia villosa J. C. Wendl. = Liparia angustifolia (Eckl. et Zeyh.) A. L. Schutte ■☆
4414 Achyronychia Torr. et A. Gray(1868);霜垫花属;Frost-mat, Onyx Flower ■☆
4415 Achyronychia parryi Hemsl.;霜垫花■☆
4416 Achyronychia rixfordii Brandegee = Scopulophila rixfordii (Brandegee)

Munz et I. M. Johnst. ■☆
4417 Achyropappus Kunth = Bahia Lag. + Schkuhria Roth(保留属名)■☆
4418 Achyropappus Kunth(1818);秕冠菊属■☆
4419 Achyropappus M. Bieb. ex Fisch. = Tricholepis DC. ■
4420 Achyropappus anthemoides Kunth;秕冠菊■☆
4421 Achyropappus woodhousei A. Gray = Picradeniopsis woodhousei (A. Gray) Rydb. ■☆
4422 Achyrophorus Adans. = Hypochaeris L. ■
4423 Achyrophorus Guett. = Hypochaeris L. ■
4424 Achyrophorus Scop. = Hypochaeris L. ■
4425 Achyrophorus aurantiacus DC. = Hypochaeris ciliata (Thunb.) Makino ■
4426 Achyrophorus chillensis (Kunth) Sch. Bip. = Hypochaeris chillensis (Kunth) Britton ■☆
4427 Achyrophorus ciliatus (Thunb.) Sch. Bip. = Hypochaeris ciliata (Thunb.) Makino ■
4428 Achyrophorus ciliatus (Thunb.) Schultz = Hypochaeris ciliata (Thunb.) Makino ■
4429 Achyrophorus crepidioides (Miyabe et Kudo) Kitag. = Hypochaeris crepidioides (Miyabe et Kudo) Tatew. et Kitam. ■☆
4430 Achyrophorus grandiflorus (Ledeb.) Ledeb. = Hypochaeris ciliata (Thunb.) Makino ■
4431 Achyrophorus maculatus (L.) Scop. = Hypochaeris maculata L. ■
4432 Achyrophorus microcephalus Sch. Bip. = Hypochaeris microcephala (Sch. Bip.) Cabrera ■☆
4433 Achyrophorus roseus Less. = Pinaropappus roseus (Less.) Less. ■☆
4434 Achyrophorus uniflorus F. W. Schmidt = Hypochaeris uniflora Vill. ■☆
4435 Achyropsis (Moq.) Benth. et Hook. f. (1880);尖被苋属■☆
4436 Achyropsis (Moq.) Hook. f. = Achyropsis (Moq.) Benth. et Hook. f. ■☆
4437 Achyropsis Benth. et Hook. f. = Achyropsis (Moq.) Benth. et Hook. f. ■☆
4438 Achyropsis alba Eckl. et Zeyh. ex Moq. = Achyropsis leptostachya (E. Mey. ex Meisn.) Baker et C. B. Clarke ■☆
4439 Achyropsis avicularis (E. Mey.) Hook. f.;鸟状尖被苋■☆
4440 Achyropsis conferta (Schinz) Schinz = Centemopsis conferta (Schinz) Suess. ■☆
4441 Achyropsis filifolia C. C. Towns.;线叶尖被苋■☆
4442 Achyropsis fischeri R. E. Fr. = Achyropsis fruticulosa C. B. Clarke ■☆
4443 Achyropsis fruticulosa C. B. Clarke;多枝尖被苋■☆
4444 Achyropsis gracilis C. C. Towns.;纤细尖被苋■☆
4445 Achyropsis graminea Suess. et Overkott = Centemopsis graminea (Suess. et Overkott) C. C. Towns. ■☆
4446 Achyropsis greenwayi Suess. = Achyropsis fruticulosa C. B. Clarke ■☆
4447 Achyropsis laniceps C. B. Clarke;毛梗尖被苋■☆
4448 Achyropsis laniceps C. B. Clarke f. robynsii (Schinz) Cavaco = Achyropsis laniceps C. B. Clarke ■☆
4449 Achyropsis laricifolia Peter = Centemopsis gracilenta (Hiern) Schinz ■☆
4450 Achyropsis leptostachya (E. Mey. ex Meisn.) Baker et C. B. Clarke;细穗尖被苋■☆
4451 Achyropsis longipedunculata (Peter) Suess. = Centemopsis longipedunculata (Peter) C. C. Towns. ■☆
4452 Achyropsis minutissima Lambinon = Centemopsis longipedunculata (Peter) C. C. Towns. ■☆
4453 Achyropsis oxyuris Suess. et Overkott = Centemopsis gracilenta (Hiern) Schinz ■☆
4454 Achyropsis robynsii Schinz = Achyropsis laniceps C. B. Clarke ■☆
4455 Achyroseris Sch. Bip. = Scorzonera L. ■
4456 Achyrospermum Blume(1826);鳞果草属;Achyrospermum ●■
4457 Achyrospermum aethiopicum Welw.;埃塞俄比亚鳞果草■☆
4458 Achyrospermum africanum Hook. f. ex Baker;非洲鳞果草■☆
4459 Achyrospermum axillare E. A. Bruce;腋花鳞果草■☆
4460 Achyrospermum carvalhoi Gürke;卡瓦略鳞果草■☆
4461 Achyrospermum ciliatum Gürke;缘毛鳞果草■☆
4462 Achyrospermum cryptanthum Baker;隐花鳞果草■☆
4463 Achyrospermum dasytrichum Perkins;多毛鳞果草■☆
4464 Achyrospermum densiflorum Blume;鳞果草;Philippine Achyrospermum ■
4465 Achyrospermum densiflorum Perkins = Achyrospermum oblongifolium Baker ■☆
4466 Achyrospermum erythrobotrys Perkins;淡红穗鳞果草■☆
4467 Achyrospermum fruticosum Benth.;灌木鳞果草●☆
4468 Achyrospermum laterale Baker;侧生鳞果草■☆
4469 Achyrospermum mearnsii Standl. = Achyrospermum schimperi (Hochst. ex Briq.) Perkins ex Mildbr. ■☆
4470 Achyrospermum micranthum Perkins;非洲小花鳞果草■☆
4471 Achyrospermum mildbraedii Perkins = Achyrospermum parviflorum S. Moore ■☆
4472 Achyrospermum nyasanum Baker = Achyrospermum cryptanthum Baker ■☆
4473 Achyrospermum oblongifolium Baker;矩圆鳞果草■☆
4474 Achyrospermum parviflorum S. Moore;小花鳞果草■☆
4475 Achyrospermum peulhorum E. A. Bruce = Achyrospermum africanum Hook. f. ex Baker ■☆
4476 Achyrospermum philippinense Benth. = Achyrospermump densiflorum Blume ■
4477 Achyrospermum phlomoides Blume = Achyrospermum densiflorum Blume ■
4478 Achyrospermum purpureum Phillipson;紫鳞果草■☆
4479 Achyrospermum radicans Gürke;辐射鳞果草■☆
4480 Achyrospermum radicans Gürke var. grandibracteatum E. A. Bruce;大苞鳞果草■☆
4481 Achyrospermum schimperi (Hochst. ex Briq.) Perkins ex Mildbr.;欣珀鳞果草■☆
4482 Achyrospermum schlechteri Gürke;施莱鳞果草■☆
4483 Achyrospermum serratum E. A. Bruce;具齿鳞果草■☆
4484 Achyrospermum squamosum Chikuni;多鳞鳞果草■☆
4485 Achyrospermum swina Perkins = Achyrospermum cryptanthum Baker ■☆
4486 Achyrospermum tisserantii Letouzey;蒂斯朗特鳞果草■☆
4487 Achyrospermum urens Baker;螯毛鳞果草■☆
4488 Achyrospermum wallichianum Benth. = Achyrospermum densiflorum Blume ■
4489 Achyrospermum wallichianum Benth. ex Hook. f.;西藏鳞果草;Wallich Achyrospermum, Xiazang Achyrospermum ■
4490 Achyrospermum wallichianum Benth. ex Hook. f. = Achyrospermum densiflorum Blume ■
4491 Achyrothalamus O. Hoffm. = Erythrocephalum Benth. ●■☆
4492 Achyrothalamus marginatus O. Hoffm. = Erythrocephalum marginatum (O. Hoffm.) S. Ortiz et Cout. ■☆

4493 Achyrothalamus teitensis O. Hoffm. = Erythrocephalum marginatum (O. Hoffm.) S. Ortiz et Cout. ■☆
4494 Acia Schreb. = Acioa Aubl. + Couepia Aubl. ●☆
4495 Acia Schreb. = Acioa Aubl. ●☆
4496 Aciachna Post et Kuntze = Aciachne Benth. ■☆
4497 Aciachne Benth. (1881);南美针茅属■☆
4498 Aciachne pulvinata Benth.;南美针茅■☆
4499 Aciachne uniflora Baill.;单花南美针茅■☆
4500 Acialyptus B. D. Jacks. = Acicalyptus A. Gray ●
4501 Acianthella D. L. Jones et M. A. Clem. (2004);小钻花兰属■☆
4502 Acianthella aegeridantennata (N. Hallé) D. L. Jones et M. A. Clem.;小钻花兰■☆
4503 Acianthella amplexicaulis (F. M. Bailey) D. L. Jones et M. A. Clem.;抱茎小钻花兰■☆
4504 Acianthera Post et Kuntze = Acisanthera P. Browne ●■☆
4505 Acianthera Scheidw. = Pleurothallis R. Br. ■☆
4506 Acianthopsis M. A. Clem. et D. L. Jones(2002);类钻花兰属■☆
4507 Acianthopsis Szlach. = Acianthopsis M. A. Clem. et D. L. Jones ■☆
4508 Acianthopsis bracteata (Rendle) M. A. Clem. et D. L. Jones;具苞类钻花兰■☆
4509 Acianthopsis grandiflora (Schltr.) M. A. Clem. et D. L. Jones;大花类钻花兰■☆
4510 Acianthopsis oxyglossa (Schltr.) M. A. Clem. et D. L. Jones;尖舌类钻花兰■☆
4511 Acianthus R. Br. (1810);钻花兰属(蚊兰属);Gnat Orchid, Mosquito Orchid, Pixie Caps, Acianthus ■☆
4512 Acianthus exsertus R. Br.;钻花兰;Exsert Acianthus ■☆
4513 Acianthus fornicatus R. Br.;拱形钻花兰(蚊兰);Gnat Orchid, Pixie Orchid ■☆
4514 Acianthus petiolatus D. Don = Liparis petiolata (D. Don) P. F. Hunt et Summerh. ■
4515 Acianthus reniformis (R. Br.) Schltr.;肾叶钻花兰(肾叶蚊兰);Reniformleaf Acianthus ■☆
4516 Acianthus sinclairii Hook. f.;辛氏钻花兰;Sinclair Acianthus ■☆
4517 Acicalyptus A. Gray = Cleistocalyx Blume ●
4518 Acicalyptus A. Gray = Syzygium R. Br. ex Gaertn. (保留属名)●
4519 Acicarpa R. Br. = Acicarpha Juss. ■☆
4520 Acicarpa Raddi = Digitaria Haller(保留属名)■
4521 Acicarpa Raddi = Panicum L. ■
4522 Acicarpa Raddi = Trichachne Nees ■☆
4523 Acicarpha Juss. (1803);热美萼角花属;Acidanthera ■☆
4524 Acicarpha tribuloides Juss.;热美萼角花;Madam Gorgon ■☆
4525 Acicarphaea Walp. = Acarphaea Harv. et Gray ex A. Gray ●■☆
4526 Acicarphaea Walp. = Chaenactis DC. ●■☆
4527 Acicarpus Post et Kuntze = Acicarpa Raddi ■
4528 Acicarpus Post et Kuntze = Panicum L. ■
4529 Aciclinium Torr. et A. Gray = Bigelowia DC. (保留属名)●☆
4530 Acidandra Mart. ex Spreng. = Zollernia Wied-Neuw. et Nees ●☆
4531 Acidanthera Hochst. = Gladiolus L. ■
4532 Acidanthera aequinoctialis (Herb.) Baker = Gladiolus aequinoctialis Herb. ■☆
4533 Acidanthera amoena (A. Chev.) Jacq.-Fél. = Gladiolus chevalieranus Marais ■☆
4534 Acidanthera bicolor Hochst.;菖蒲鸢尾(二色菖蒲鸢尾);Abyssinian Gladiolus, Fragrant Gladiolus, Magpie Gladiolus, Peacock Flower, Sword Lily, Twocolour Acidanthera ■☆
4535 Acidanthera bicolor Hochst. = Gladiolus callianthus Marais ■☆
4536 Acidanthera bicolor Hochst. = Gladiolus murielae Kelway ■☆
4537 Acidanthera brachystachys Baker = Babiana brachystachys (Baker) G. J. Lewis ■☆
4538 Acidanthera brevicaulis Baker = Hesperantha brevicaulis (Baker) G. J. Lewis ■☆
4539 Acidanthera brevicollis Baker = Gladiolus gueinzii Kunze ■☆
4540 Acidanthera candida Rendle = Gladiolus candidus (Rendle) Goldblatt ■☆
4541 Acidanthera capensis (Houtt.) Benth. ex Baker = Tritonia flabellifolia (D. Delaroche) G. J. Lewis ■☆
4542 Acidanthera divina Vaupel = Gladiolus aequinoctialis Herb. var. divina (Vaupel) Marais ■☆
4543 Acidanthera euryphylla (Harms) Diels = Savannosiphon euryphyllus (Harms) Goldblatt et Marais ■☆
4544 Acidanthera flabellifolia (D. Delaroche) N. E. Br. = Tritonia flabellifolia (D. Delaroche) G. J. Lewis ■☆
4545 Acidanthera flexuosa (L. f.) Baker = Tritoniopsis flexuosa (L. f.) G. J. Lewis ■☆
4546 Acidanthera forsythiana Baker = Gladiolus floribundus Jacq. ■☆
4547 Acidanthera fourcadei L. Bolus = Geissorhiza fourcadei (L. Bolus) G. J. Lewis ■☆
4548 Acidanthera goetzei Harms = Gladiolus curtifolius Marais ■☆
4549 Acidanthera gracilis Pax = Gladiolus candidus (Rendle) Goldblatt ■☆
4550 Acidanthera graminifolia Baker = Gladiolus floribundus Jacq. ■☆
4551 Acidanthera gunnisii Rendle = Gladiolus gunnisii (Rendle) Marais ■☆
4552 Acidanthera holostachya (Baker) N. E. Br. = Radinosiphon leptostachyis (Baker) N. E. Br. ■☆
4553 Acidanthera huttonii Baker = Hesperantha huttonii (Baker) Hilliard et B. L. Burtt ■☆
4554 Acidanthera iroensis (A. Chev.) A. Chev. = Gladiolus iroensis (A. Chev.) Marais ■☆
4555 Acidanthera ixioides Baker = Gladiolus ixioides (Baker) G. J. Lewis ■☆
4556 Acidanthera laxifolia Baker = Gladiolus candidus (Rendle) Goldblatt ■☆
4557 Acidanthera leptostachya (Baker) N. E. Br. = Radinosiphon leptostachyis (Baker) N. E. Br. ■☆
4558 Acidanthera lomatensis N. E. Br. = Radinosiphon lomatensis (N. E. Br.) N. E. Br. ■☆
4559 Acidanthera muirii L. Bolus = Hesperantha muirii (L. Bolus) G. J. Lewis ■☆
4560 Acidanthera murielae Hoog;缪里菖蒲鸢尾■☆
4561 Acidanthera murielae Hoog = Gladiolus murielae Kelway ■☆
4562 Acidanthera nelloi Chiov. = Gladiolus gunnisii (Rendle) Marais ■☆
4563 Acidanthera pauciflora (Baker) Benth. = Gladiolus floribundus Jacq. ■☆
4564 Acidanthera platypetala Baker = Gladiolus longicollis Baker subsp. platypetalus (Baker) Goldblatt et J. C. Manning ■☆
4565 Acidanthera rosea Schinz = Geissorhiza tenella Goldblatt ■☆
4566 Acidanthera roseoalba G. J. Lewis = Geissorhiza roseoalba (G. J. Lewis) Goldblatt ■☆
4567 Acidanthera sabulosa Schltr. = Geissorhiza tenella Goldblatt ■☆
4568 Acidanthera schimperi (Asch. et Klatt) Foster = Lapeirousia schimperi (Asch. et Klatt) Milne-Redh. ■☆
4569 Acidanthera schinzii Baker = Geissorhiza schinzii (Baker)

Goldblatt ■☆

4570 Acidanthera tubulosa (Houtt.) Baker = Geissorhiza exscapa (Thunb.) Goldblatt ■☆

4571 Acidanthera tysonii Baker = Hesperantha grandiflora G. J. Lewis ■☆

4572 Acidanthera ukambanensis Baker = Gladiolus candidus (Rendle) Goldblatt ■☆

4573 Acidanthera unicolor Hochst. ex Baker = Lapeirousia schimperi (Asch. et Klatt) Milne-Redh. ■☆

4574 Acidanthera zanzibarica Baker = Gladiolus candidus (Rendle) Goldblatt ■☆

4575 Acidanthus Clem. = Acianthus R. Br. ■☆

4576 Acidocroton Griseb. (1859);酸巴豆属(酸豆戟属)●☆

4577 Acidocroton Griseb. = Acidocroton P. Browne ●☆

4578 Acidocroton P. Browne = Flueggea Willd. ●

4579 Acidocroton montanus Urb. et Ekman;山地酸巴豆●☆

4580 Acidocroton pilosulus Urb.;毛酸巴豆●☆

4581 Acidocroton trichophyllus Urb.;毛叶酸巴豆●☆

4582 Acidodendron Kuntze = Acidodendrum Kuntze ●☆

4583 Acidodendrum Kuntze = Acinodendron Raf. ●☆

4584 Acidodendrum Kuntze = Miconia Ruiz et Pav. (保留属名)●☆

4585 Acidolepis Clem. = Acilepis D. Don ■

4586 Acidolepis Clem. = Vernonia Schreb. (保留属名)●■

4587 Acidonia L. A. S. Johnson et B. G. Briggs(1975);澳西南龙眼属●☆

4588 Acidonia microcarpa (R. Br.) L. A. S. Johnson et B. G. Briggs;澳西南龙眼●☆

4589 Acidosasa C. D. Chu et C. S. Chao = Acidosasa C. D. Chu et C. S. Chao ex P. C. Keng ●★

4590 Acidosasa C. D. Chu et C. S. Chao ex P. C. Keng(1979);酸竹属;Sour Bamboo, Sourbamboo, Sour-bamboo ●★

4591 Acidosasa bilamina W. T. Lin et Z. M. Wu = Oligostachyum spongiosum (C. D. Chu et C. S. Chao) G. H. Ye et Z. P. Wang ●

4592 Acidosasa breviclavata W. T. Lin;小叶酸竹●

4593 Acidosasa brilletii (A. Camus) C. S. Chao et Renvoize;越南酸竹;Vietnam Sour Bamboo ●

4594 Acidosasa chienouensis (T. H. Wen) C. S. Chao et T. H. Wen;粉酸竹(建瓯酸竹);Glaucous Sour Bamboo, Jian'ou Sour-bamboo, Powder Sourbamboo ●

4595 Acidosasa chinensis C. D. Chu et C. S. Chao;酸竹;Sour Bamboo, Sour-bamboo ●◇

4596 Acidosasa dayongensis C. S. Chao et H. Y. Zou;大庸酸竹;Dayong Bamboo ●

4597 Acidosasa dayongensis T. P. Yi = Acidosasa hirtiflora Z. P. Wang et G. H. Ye ●

4598 Acidosasa denigrata W. T. Lin = Pseudosasa hindsii (Munro) S. L. Chen et G. Y. Sheng ex T. G. Liang ●

4599 Acidosasa edulis (T. H. Wen) T. H. Wen;黄甜竹;Edible Chinacane, Edible Sour Bamboo ●

4600 Acidosasa fujianensis C. S. Chao et H. Y. Zhou = Acidosasa longiligula (T. H. Wen) C. S. Chao et C. D. Chu ●

4601 Acidosasa gigantea (T. H. Wen) Q. Z. Xie et W. Y. Zhang = Indosasa gigantea (T. H. Wen) T. H. Wen ●

4602 Acidosasa gigantea (T. H. Wen) Q. Z. Xie et W. Y. Zhang = Sinobambusa gigantea T. H. Wen ●

4603 Acidosasa glauca B. M. Yang = Acidosasa chienouensis (T. H. Wen) C. S. Chao et T. H. Wen ●

4604 Acidosasa gracilis W. T. Lin et X. B. Ye;小酸竹;Little Sour Bamboo ●

4605 Acidosasa guangxiensis Q. H. Dai et C. F. Huang;广西酸竹●

4606 Acidosasa heterolodicula (W. T. Lin et Z. J. Feng) W. T. Lin = Oligostachyum scabriflorum (McClure) Z. P. Wang et G. H. Ye ●

4607 Acidosasa hirtiflora Z. P. Wang et G. H. Ye;毛花酸竹;Hairy-flower Sour Bamboo, Hairy-flowered Sour-bamboo ●

4608 Acidosasa lentiginosa W. T. Lin et Z. J. Feng = Arundinaria oleosa (T. H. Wen) C. S. Chao et G. Y. Yang ●

4609 Acidosasa lentiginosa W. T. Lin et Z. J. Feng = Pleioblastus oleosus T. H. Wen ●

4610 Acidosasa lingchuanensis (C. D. Chu et C. S. Chao) Q. Z. Xie et X. Y. Chen;灵川酸竹●

4611 Acidosasa longiligula (T. H. Wen) C. S. Chao et C. D. Chu;福建酸竹;Fujian Sour Bamboo, Fujian Sour-bamboo ●

4612 Acidosasa macula W. T. Lin et Z. M. Wu = Oligostachyum scabriflorum (McClure) Z. P. Wang et G. H. Ye ●

4613 Acidosasa nanunica (McClure) C. S. Chao et G. Y. Yang;长舌酸竹(白环异枝竹,长舌茶秆竹,高舌茶秆竹,清远青篱竹,异枝竹);Longtongue Sour Bamboo ●

4614 Acidosasa notata (Z. P. Wang et G. H. Ye) S. S. You;斑箨酸竹●

4615 Acidosasa paucifolia W. T. Lin = Acidosasa nanunica (McClure) C. S. Chao et G. Y. Yang ●

4616 Acidosasa paucifolia W. T. Lin = Pseudosasa pubiflora (Keng) P. C. Keng ex D. Z. Li et L. M. Gao ●

4617 Acidosasa purpurea (J. R. Xue et T. P. Yi) P. C. Keng = Acidosasa hirtiflora Z. P. Wang et G. H. Ye ●

4618 Acidosasa venusta (McClure) Z. P. Wang et G. H. Ye ex C. S. Chao et C. D. Chu;黎竹(篱竹,坭竹);Elegant Sourbamboo, Show Sour-bamboo ●

4619 Acidosasa xiushanensis T. P. Yi;秆子草;Xiushan Sour Bamboo ●

4620 Acidosperma Clem. = Acispermum Neck. ●■

4621 Acidosperma Clem. = Coreopsis L. ●■

4622 Acidoton P. Browne(废弃属名) = Acidoton Sw. (保留属名)●☆

4623 Acidoton P. Browne(废弃属名) = Flueggea Willd. ●

4624 Acidoton P. Browne(废弃属名) = Securinega Comm. ex Juss. (保留属名)●☆

4625 Acidoton Sw. (1788)(保留属名);尖大戟属●☆

4626 Acidoton Sw. (保留属名) = Flueggea Willd. ●

4627 Acidoton ellipticus Kuntze;椭圆尖大戟●☆

4628 Acidoton flueggeodes Kuntze = Flueggea suffruticosa (Pall.) Baill. ●

4629 Acidoton leucopyrus Kuntze = Flueggea leucopyrus Willd. ●

4630 Acidoton microphyllus Urb.;小叶尖大戟●☆

4631 Acidoton obovatus (Willd.) Kuntze = Flueggea virosa (Roxb. ex Willd.) Voigt ●

4632 Acidoton ramiflorus Kuntze = Flueggea suffruticosa (Pall.) Baill. ●

4633 Acidoton virosus Kuntze = Flueggea virosa (Roxb. ex Willd.) Voigt ●

4634 Aciella Tiegh. = Amylotheca Tiegh. ●☆

4635 Acilepidopsis H. Rob. (1989);少花尖鸠菊属●■

4636 Acilepidopsis H. Rob. = Vernonia Schreb. (保留属名)●■

4637 Acilepidopsis echitifolia (Mart. ex DC.) H. Rob.;少花尖鸠菊■☆

4638 Acilepis D. Don = Vernonia Schreb. (保留属名)●■

4639 Acilepis D. Don(1825);尖鸠菊属■

4640 Acilepis aspera (Buch. -Ham.) H. Rob.;糙叶尖鸠菊■

4641 Acilepis clivorum (Hance) H. Rob.;山岗尖鸠菊■

4642 Acilepis nantcianensis (Pamp.) H. Rob.;南漳尖鸠菊■

4643 Acilepis spirei (Gand.) H. Rob.;折苞尖鸠菊■

4644 Acilepis squarrosa D. Don;尖鸠菊■

4645 Acilepis squarrosa D. Don = Vernonia squarrosa (D. Don) Less. ■
4646 Acinax Raf. = Amomum L. ■
4647 Acinax Raf. = Amomum Roxb. (保留属名)■
4648 Acineta Lindl. (1843);葡萄兰属;Acineta ■☆
4649 Acineta barkeri Lindl.;巴氏葡萄兰■☆
4650 Acineta chrysantha Lindl. et Paxton;黄花葡萄兰;Yellowflower Acineta ■☆
4651 Acineta densa Lindl.;密花葡萄兰■☆
4652 Acineta humboldtii Lindl. = Acineta superba Rchb. ■☆
4653 Acineta sulcata Rchb. f.;葡萄兰;Haughty Acineta ■☆
4654 Acineta superba Rchb.;华丽葡萄兰■☆
4655 Acinodendron Kuntze = Acinodendrum Kuntze ●☆
4656 Acinodendron Raf. = Miconia Ruiz et Pav. (保留属名)●☆
4657 Acinodendrum Kuntze = Acinodendron Raf. ●☆
4658 Acinolis Raf. = Miconia Ruiz et Pav. (保留属名)●☆
4659 Acinopetala Luer = Masdevallia Ruiz et Pav. ■☆
4660 Acinos Mill. = Clinopodium L. ●■
4661 Acinos Moench(1794);酸唇草属;Basil Thyme ■☆
4662 Acinos alpinus (L.) Moench;高山酸唇草(高山香草);Alpine Savory ■☆
4663 Acinos alpinus (L.) Moench subsp. meridionalis (Nyman) P. W. Ball;酸唇草■☆
4664 Acinos alpinus Moench = Calamintha alpina Lam. ■☆
4665 Acinos alpinus Moench subsp. meridionalis Ball. = Acinos alpinus (L.) Moench subsp. meridionalis (Nyman) P. W. Ball ■☆
4666 Acinos arvense (Lam.) Dandy;野生酸唇草;Basil Balm, Basil Thyme, Basil-thyme, Corn Mint, Mother of Thyme, Mother Thyme, Mother-of-thyme, Motherthyme, Rough Basil, Stone Basil, Wild Basil ■☆
4667 Acinos arvensis (Lam.) Dandy = Clinopodium acinos Kuntze ■☆
4668 Acinos fominii Des. -Shost.;福明酸唇草■☆
4669 Acinos graveoleus (M. Bieb.) Link;香酸唇草■☆
4670 Acinos multiflorus K. Koch ex Boiss.;多花酸唇草■☆
4671 Acinos rotundifolius Pers.;圆叶酸唇草■☆
4672 Acinos thymoides (L.) Moench = Calamintha arvensis Lam. ■☆
4673 Acinotum (DC.) Rchb. = Matthiola W. T. Aiton(保留属名)●■
4674 Acinotum Rchb. = Matthiola W. T. Aiton(保留属名)●■
4675 Acinotus Baill. = Acinotum (DC.) Rchb. ●■
4676 Acioa Aubl. (1775);热美金壳果属●☆
4677 Acioa barteri (Hook. f. ex Oliv.) Engl. = Dactyladenia barteri (Hook. f. ex Oliv.) Prance et F. White ●☆
4678 Acioa bellayana Baill. = Dactyladenia bellayana (Baill.) Prance et F. White ●☆
4679 Acioa brazzaea De Wild. = Dactyladenia dewevrei (De Wild. et T. Durand) Prance et F. White ●☆
4680 Acioa buchneri Engl. = Dactyladenia buchneri (Engl.) Prance et Sothers ●☆
4681 Acioa campestris Engl. = Dactyladenia campestris (Engl.) Prance et F. White ●☆
4682 Acioa chevalieri De Wild. = Dactyladenia chevalieri (De Wild.) Prance et F. White ●☆
4683 Acioa cinerea Engl. ex De Wild. = Dactyladenia cinerea (Engl. ex De Wild.) Prance et F. White ●☆
4684 Acioa dawei Mendes = Dactyladenia campestris (Engl.) Prance et F. White ●☆
4685 Acioa dewevrei De Wild. et T. Durand = Dactyladenia dewevrei (De Wild. et T. Durand) Prance et F. White ●☆
4686 Acioa dewevrei De Wild. et T. Durand var. reygaertii (De Wild.) Hauman = Dactyladenia dewevrei (De Wild. et T. Durand) Prance et F. White ●☆
4687 Acioa dewevrei De Wild. et T. Durand var. seretii (De Wild.) Hauman = Dactyladenia dewevrei (De Wild. et T. Durand) Prance et F. White ●☆
4688 Acioa dewevrei De Wild. et T. Durand var. vanhouttei (De Wild.) Hauman = Dactyladenia dewevrei (De Wild. et T. Durand) Prance et F. White ●☆
4689 Acioa dichotoma De Wild. = Dactyladenia dichotoma (De Wild.) Prance et F. White ●☆
4690 Acioa dinklagei Engl. = Dactyladenia dinklagei (Engl.) Prance et F. White ●☆
4691 Acioa edulis Prance;可食热美金壳果●☆
4692 Acioa eketensis De Wild. = Dactyladenia eketensis (De Wild.) Prance et F. White ●☆
4693 Acioa floribunda (Welw.) Exell;多花热美金壳果●☆
4694 Acioa floribunda (Welw.) Exell = Dactyladenia floribunda Welw. ●☆
4695 Acioa floribunda Exell = Acioa floribunda (Welw.) Exell ●☆
4696 Acioa gilletii De Wild. = Dactyladenia gilletii (De Wild.) Prance et F. White ●☆
4697 Acioa goetzeana Engl. = Hirtella zanzibarica Oliv. ●☆
4698 Acioa gossweileri Cavaco = Dactyladenia buchneri (Engl.) Prance et Sothers ●☆
4699 Acioa guianensis Aubl.;热美金壳果●☆
4700 Acioa hirsuta A. Chev. ex De Wild. = Dactyladenia hirsuta (A. Chev. ex De Wild.) Prance et F. White ●☆
4701 Acioa icondere Baill. var. welwitschii De Wild. = Dactyladenia floribunda Welw. ●☆
4702 Acioa johnstonei Hoyle = Dactyladenia johnstonei (Hoyle) Prance et F. White ●☆
4703 Acioa klaineana Pierre ex De Wild. = Dactyladenia campestris (Engl.) Prance et F. White ●☆
4704 Acioa laevis Pierre ex De Wild. = Dactyladenia laevis (Pierre ex De Wild.) Prance et F. White ●☆
4705 Acioa lanceolata Engl. = Dactyladenia barteri (Hook. f. ex Oliv.) Prance et F. White ●☆
4706 Acioa lehmbachii Engl. = Dactyladenia lehmbachii (Engl.) Prance et F. White ●☆
4707 Acioa letestui Letouzey = Dactyladenia letestui (Letouzey) Prance et F. White ●☆
4708 Acioa librevillensis Letouzey = Dactyladenia librevillensis (Letouzey) Prance et F. White ●☆
4709 Acioa lujae De Wild. = Dactyladenia buchneri (Engl.) Prance et Sothers ●☆
4710 Acioa mannii (Oliv.) Engl. = Dactyladenia mannii (Oliv.) Prance et F. White ●☆
4711 Acioa pallescens Baill. = Dactyladenia pallescens (Baill.) Prance et F. White ●☆
4712 Acioa parvifolia Engl. = Dactyladenia smeathmannii (Baill.) Prance et F. White ●☆
4713 Acioa pierrei De Wild. = Dactyladenia pierrei (De Wild.) Prance et F. White ●☆
4714 Acioa reygaertii De Wild. = Dactyladenia dewevrei (De Wild. et T. Durand) Prance et F. White ●☆
4715 Acioa rudatisii Engl. ex De Wild. = Dactyladenia lehmbachii

(Engl.) Prance et F. White ●☆
4716 Acioa sapinii De Wild. = Dactyladenia sapinii (De Wild.) Prance et F. White ●☆
4717 Acioa scabrifolia Hua = Dactyladenia scabrifolia (Hua) Prance et F. White ●☆
4718 Acioa seretii De Wild. = Dactyladenia dewevrei (De Wild. et T. Durand) Prance et F. White ●☆
4719 Acioa smeathmannii Baill. = Dactyladenia smeathmannii (Baill.) Prance et F. White ●☆
4720 Acioa stapfiana De Wild. = Dactyladenia whytei (Stapf) Prance et F. White ●☆
4721 Acioa staudtii Engl. = Dactyladenia staudtii (Engl.) Prance et F. White ●☆
4722 Acioa talbotii Baker f. = Dactyladenia staudtii (Engl.) Prance et F. White ●☆
4723 Acioa tenuiflora Dinkl. et Engl. = Dactyladenia barteri (Hook. f. ex Oliv.) Prance et F. White ●☆
4724 Acioa tessmannii Engl. = Dactyladenia letestui (Letouzey) Prance et F. White ●☆
4725 Acioa thollonii De Wild. = Dactyladenia pallescens (Baill.) Prance et F. White ●☆
4726 Acioa trillesiana Pierre ex De Wild. = Dactyladenia barteri (Hook. f. ex Oliv.) Prance et F. White ●☆
4727 Acioa unwinii De Wild. = Dactyladenia smeathmannii (Baill.) Prance et F. White ●☆
4728 Acioa vanhouttei De Wild. = Dactyladenia dewevrei (De Wild. et T. Durand) Prance et F. White ●☆
4729 Acioa whytei Stapf = Dactyladenia whytei (Stapf) Prance et F. White ●☆
4730 Acioja Gmel. = Acioa Aubl. ●☆
4731 Acion B. G. Briggs et L. A. S. Johnson(1998);多花帚灯草属■☆
4732 Acion hookeri (D. I. Morris) B. G. Briggs et L. A. S. Johnson;多花帚灯草■☆
4733 Acion monocephalum (R. Br.) B. G. Briggs et L. A. S. Johnson;单头多花帚灯草■☆
4734 Aciotis D. Don(1823);尖耳野牡丹属☆
4735 Aciotis acuminifolia Triana;尖叶尖耳野牡丹☆
4736 Aciotis annua Triana;一年尖耳野牡丹☆
4737 Aciotis brachybotria Triana;短穗尖耳野牡丹☆
4738 Aciotis discolor D. Don;异色尖耳野牡丹☆
4739 Aciotis ferruginea Triana;锈色尖耳野牡丹☆
4740 Aciotis purpurascens Triana;紫色尖耳野牡丹☆
4741 Acipetalum Turcz. = Cambessedesia DC. + Pyramia Cham. ●■☆
4742 Aciphylla J. R. Forst. et G. Forst. (1775);针叶芹属(刺刀草属)■☆
4743 Aciphylla aurea W. R. B. Oliv.;黄针叶芹;Golden Spaniard ■☆
4744 Aciphylla glaucescens W. R. B. Oliv.;渐灰针叶芹■☆
4745 Aciphylla hookeri Kirk;胡克针叶芹■☆
4746 Aciphylla montana Armstr.;山地针叶芹■☆
4747 Aciphylla scott-thomsonii Cockayne et Allan;大针叶芹(大叶针叶芹);Bayonet Plant, Giant Spaniard ■☆
4748 Aciphylla squarrosa J. R. Forst. et G. Forst.;粗糙针叶芹;Bayonet Plant, Spaniard, Speargrass, Spear-grass ■☆
4749 Aciphylla trifoliolata Petrie;三小叶针叶芹■☆
4750 Aciphylla verticillata W. R. B. Oliv.;轮生针叶芹■☆
4751 Aciphyllaea A. Gray = Hymenatherum Cass. ■☆
4752 Aciphyllum Steud. = Chorizema Labill. ●■☆
4753 Acis Salisb. = Leucojum L. ●■
4754 Acis autumnalis (L.) Herb. = Leucojum autumnale L. ■☆
4755 Acis autumnalis (L.) Schur var. oporantha (Jord. et Fourr.) Lledò et A. P. Davis et M. B. Crespo = Acis autumnalis (L.) Herb. ■☆
4756 Acis autumnalis (L.) Schur var. pulchella (Jord. et Fourr.) Lledò et A. P. Davis et M. B. Crespo = Acis autumnalis (L.) Herb. ■☆
4757 Acis oporantha Jord. et Fourr. = Acis autumnalis (L.) Schur var. oporantha (Jord. et Fourr.) Lledò et A. P. Davis et M. B. Crespo ■☆
4758 Acis pulchella Jord. et Fourr. = Acis autumnalis (L.) Schur var. pulchella (Jord. et Fourr.) Lledò et A. P. Davis et M. B. Crespo ■☆
4759 Acis tingitana (Baker) Lledò et A. P. Davis et Crespo = Leucojum tingitanum Baker ■☆
4760 Acis trichophylla (Schousb.) Sweet = Leucojum trichophyllum Schousb. ■☆
4761 Acis trichophylla (Schousb.) Sweet var. micrantha (Gatt. et Maire) Lledò et et Crespo = Acis trichophylla (Schousb.) Sweet ■☆
4762 Acisanthera P. Browne(1756);针药野牡丹属●■☆
4763 Acisanthera crassipes (Naudin) Wurdack;粗梗针药野牡丹●☆
4764 Acisanthera glandulifera Jenn.;腺点针药野牡丹●☆
4765 Acisanthera tetraptera (Cogn.) Gleason;四瓣针药野牡丹●☆
4766 Acisanthera tetraptera Gleason = Acisanthera tetraptera (Cogn.) Gleason ●☆
4767 Acisanthera uniflora (Vahl) Gleason;单花针药野牡丹●☆
4768 Acispermum Neck. = Coreopsis L. ●■
4769 Acistoma Zipp. ex Span. = Woodfordia Salisb. ●
4770 Ackama A. Cunn. (1839);澳桤木属●☆
4771 Ackama A. Cunn. = Caldcluvia D. Don ●☆
4772 Ackama australiensis (Schltr.) C. T. White;褐澳桤木;Australian Brown Alder ●☆
4773 Acladodea Ruiz et Pav. = Talisia Aubl. ●☆
4774 Acladodia Dalla Torre et Harms = Acladodea Ruiz et Pav. ●☆
4775 Acleanthus Clem. = Acleisanthes A. Gray ●■☆
4776 Acleia DC. = Senecio L. ●■
4777 Acleisanthes A. Gray(1853);喇叭茉莉属(无苞花属);Trumpets ●■☆
4778 Acleisanthes angustifolius (Torr.) R. A. Levin;窄叶无苞花■☆
4779 Acleisanthes longiflora A. Gray;长花无苞花;Angel Trumpets ■☆
4780 Acleisanthes longiflora A. Gray subsp. hirtella Standl.;毛长花无苞花■☆
4781 Acleisanthes nana I. M. Johnst.;无苞花■☆
4782 Acleisanthes obtusa Standl.;粗壮无苞花■☆
4783 Acleisanthes parvifolius (Torr.) R. A. Levin;小叶无苞花■☆
4784 Acleisanthes somalensis (Chiov.) R. A. Levin;索马里喇叭茉莉■☆
4785 Acleja Post et Kuntze = Acleia DC. ●■
4786 Acleja Post et Kuntze = Senecio L. ●■
4787 Aclema Post et Kuntze = Aklema Raf. ●■
4788 Aclema Post et Kuntze = Euphorbia L. ●■
4789 Aclinia Griff. = Dendrobium Sw. (保留属名)■
4790 Aclisanthes Post et Kuntze = Acleisanthes A. Gray ●■☆
4791 Aclisia E. Mey. = Pollia Thunb. ■
4792 Aclisia E. Mey. ex C. Presl = Pollia Thunb. ■
4793 Aclisia Hassk. = Pollia Thunb. ■
4794 Aclisia condensata (C. B. Clarke) Brückn. = Pollia condensata C. B. Clarke ■☆
4795 Aclisia gigantea Hassk. = Pollia secundiflora (Blume) Bakh. f. ■
4796 Aclisia indica Wight = Pollia secundiflora (Blume) Bakh. f. ■
4797 Aclisia secundiflora (Blume) Bakh. f. = Pollia secundiflora (Blume) Bakh. f. ■

4798 Aclisia sorzogonensis E. Mey. = Pollia secundiflora (Blume) Bakh. f. ■
4799 Aclisia subumbellata C. B. Clarke = Pollia subumbellata C. B. Clarke ■
4800 Acmadenia Bartl. et H. L. Wendl. (1824);尖腺芸香属●☆
4801 Acmadenia alternifolia Cham.;互叶尖腺芸香●☆
4802 Acmadenia apetala Dümmer = Diosma apetala (Dümmer) I. Williams ●☆
4803 Acmadenia argillophila I. Williams;白土尖腺芸香●☆
4804 Acmadenia assimilis Sond. = Euchaetis laevigata Turcz. ●☆
4805 Acmadenia baileyensis I. Williams;贝利尖腺芸香●☆
4806 Acmadenia barosmoides Dümmer = Phyllosma barosmoides (Dümmer) I. Williams ●☆
4807 Acmadenia bodkinii (Schltr.) Strid;包德尖腺芸香●☆
4808 Acmadenia bodkinii (Schltr.) Strid = Adenandra bodkinii Schltr. ●☆
4809 Acmadenia burchellii Dümmer;伯切尔尖腺芸香●☆
4810 Acmadenia candida I. Williams;纯白尖腺芸香●☆
4811 Acmadenia cassiopoides Turcz. = Macrostylis cassiopoides (Turcz.) I. Williams ●☆
4812 Acmadenia cucullata E. Mey. ex Sond. = Acmadenia mundiana Eckl. et Zeyh. ●☆
4813 Acmadenia densifolia Sond.;密花尖腺芸香●☆
4814 Acmadenia diosmoides Schltr. = Euchaetis diosmoides (Schltr.) I. Williams ●☆
4815 Acmadenia flaccida Eckl. et Zeyh.;柔软尖腺芸香●☆
4816 Acmadenia fruticosa I. Williams;灌丛尖腺芸香●☆
4817 Acmadenia gracilis Dümmer;纤细尖腺芸香●☆
4818 Acmadenia heterophylla P. E. Glover;异叶尖腺芸香●☆
4819 Acmadenia juniperina Bartl. et H. L. Wendl. = Acmadenia obtusata (Thunb.) Bartl. et H. L. Wendl. ●☆
4820 Acmadenia kiwanensis I. Williams;基温尖腺芸香●☆
4821 Acmadenia laevigata Sond. = Euchaetis laevigata Turcz. ●☆
4822 Acmadenia latifolia I. Williams;宽叶尖腺芸香●☆
4823 Acmadenia laxa I. Williams;疏松尖腺芸香●☆
4824 Acmadenia macradenia (Sond.) Dümmer;大腺尖腺芸香●☆
4825 Acmadenia macropetala (P. E. Glover) Compton;大瓣尖腺芸香●☆
4826 Acmadenia maculata I. Williams;斑点尖腺芸香●☆
4827 Acmadenia marlothii Dümmer = Agathosma rudolphii I. Williams ●☆
4828 Acmadenia matroosbergensis E. Phillips;马特卢尖腺芸香●☆
4829 Acmadenia mundiana Eckl. et Zeyh.;蒙德尖腺芸香●☆
4830 Acmadenia muraltioides Eckl. et Zeyh. = Acmadenia obtusata (Thunb.) Bartl. et H. L. Wendl. ●☆
4831 Acmadenia nivea I. Williams;雪白尖腺芸香●☆
4832 Acmadenia nivenii Sond.;尼文尖腺芸香●☆
4833 Acmadenia obtusata (Thunb.) Bartl. et H. L. Wendl.;钝尖腺芸香●☆
4834 Acmadenia obtusata Bartl. et H. L. Wendl. var. macropetala Glover = Acmadenia macropetala (P. E. Glover) Compton ●☆
4835 Acmadenia patentifolia I. Williams;展叶尖腺芸香●☆
4836 Acmadenia psilopetala Sond. = Acmadenia trigona (Eckl. et Zeyh.) Druce ●☆
4837 Acmadenia pungens Bartl. et H. L. Wendl. = Euchaetis pungens (Bartl. et H. L. Wendl.) I. Williams ●☆
4838 Acmadenia rosmarinifolia Bartl. = Agathosma rosmarinifolia (Bartl.) I. Williams ●☆
4839 Acmadenia rourkeana I. Williams;鲁尔克尖腺芸香●☆
4840 Acmadenia rupicola I. Williams;岩生尖腺芸香●
4841 Acmadenia strobilina E. Mey. = Acmadenia tetragona (L. f.) Bartl. et H. L. Wendl. ●☆
4842 Acmadenia tenax I. Williams;黏尖腺芸香●☆
4843 Acmadenia teretifolia (Link) E. Phillips;柱叶尖腺芸香●☆
4844 Acmadenia tetracarpellata I. Williams;四果尖腺芸香●☆
4845 Acmadenia tetragona (L. f.) Bartl. et H. L. Wendl.;四棱尖腺芸香●☆
4846 Acmadenia trigona (Eckl. et Zeyh.) Druce;三棱尖腺芸香●☆
4847 Acmadenia uniflora (E. Phillips) E. Phillips = Acmadenia obtusata (Thunb.) Bartl. et H. L. Wendl. ●☆
4848 Acmadenia wittebergensis (Compton) I. Williams;维特贝格尖腺芸香●☆
4849 Acmanthera (A. Juss.) Griseb. (1858);南美金虎尾属●☆
4850 Acmanthera Griesb. = Acmanthera (A. Juss.) Griseb. ●☆
4851 Acmanthera cowanii W. R. Anderson;考万南美金虎尾●☆
4852 Acmanthera duckei W. R. Anderson;杜克南美金虎尾●☆
4853 Acmanthera latifolia Griseb.;宽叶南美金虎尾●☆
4854 Acmanthera longifolia Nied.;长叶南美金虎尾●☆
4855 Acmanthera minima W. R. Anderson;小南美金虎尾●☆
4856 Acmanthera parviflora W. R. Anderson;小叶南美金虎尾●☆
4857 Acmella Rich. = Acmella Rich. ex Pers. ■
4858 Acmella Rich. ex Pers. (1807);斑花菊属(千日菊属,金钮扣属);Spotflower ■
4859 Acmella Rich. ex Pers. = Spilanthes Jacq. ■
4860 Acmella alba (L'Hér.) R. K. Jansen;白斑花菊(白千日菊)■☆
4861 Acmella calva (DC.) R. K. Jansen;美丽斑花菊(美形金钮扣)■
4862 Acmella caulirhiza Delile;茎根斑花菊(茎根千日菊)■☆
4863 Acmella ciliata (Kunth) Cass.;缘毛斑花菊(缘毛千日菊)■☆
4864 Acmella decumbens (Sm.) R. K. Jansen;俯卧斑花菊(俯卧千日菊);Creeping Spotflower ■☆
4865 Acmella mauritiana A. Rich. ex Pers. = Spilanthes mauritiana (A. Rich. ex Pers.) DC. ■☆
4866 Acmella oleracea (L.) R. K. Jansen;千日菊(桂圆花,桂圆菊,金钮扣,六神花,蔬食花菊,铁拳头,印度金钮扣);Brazil Cress, Brazilian Cress, Para Cress, Paracress Spotflower, Para-cress Spotflower, Peek-a-boo Plant, Spot Flower ■
4867 Acmella oleracea (L.) R. K. Jansen = Spilanthes oleracea L. ■
4868 Acmella oppositifolia (Lam.) R. K. Jansen;对叶千日菊■☆
4869 Acmella oppositifolia (Lam.) R. K. Jansen var. repens (Walter) R. K. Jansen;匍匐对叶千日菊■☆
4870 Acmella paniculata (Wall. ex DC.) R. K. Jansen;金钮扣(遍地红,大黄花,过海龙,黑节关,红铜水草,红细水草,黄花草,黄花苦草,苦草,拟千日菊,散血草,山骨皮,山天文草,天文草,铁拳头,小铜锤,雨伞草);Goldenbutton, Panicullate Spotflower, Toothache Plant ■
4871 Acmella paniculata (Wall. ex DC.) R. K. Jansen = Spilanthes paniculata Wall. ex DC. ■
4872 Acmella parvifolia Raf.;小叶白斑花菊■☆
4873 Acmella pilosa R. K. Jansen;毛千日菊;Hairy Spotflower ■☆
4874 Acmella pusilla (Hook. et Arn.) R. K. Jansen;侏儒千日菊;Dwarf Spotflower ■☆
4875 Acmella radicans (Jacq.) R. K. Jansen;辐射千日菊■☆
4876 Acmella uliginosa (Sw.) Cass.;沼生千日菊■☆
4877 Acmena DC. (1828);肖蒲桃属(裂胚木属,赛赤楠属);Acmena ●☆
4878 Acmena DC. = Syzygium R. Br. ex Gaertn. (保留属名)●
4879 Acmena acuminatissima (Blume) Merr. et L. M. Perry;肖蒲桃(火

炭木，荔枝母，锐叶赤楠，赛赤楠）；Acuminate Acmena，Sharpleaf Acmena，Sharp-leaved Acmena，Willow-leaf Eugenia ●
4880 Acmena acuminatissima（Blume）Merr. et L. M. Perry = Syzygium acuminatissimum（Blume）DC. ●
4881 Acmena championii Benth. = Syzygium championii（Benth.）Merr. et L. M. Perry ●
4882 Acmena gerrardii Harv. = Syzygium gerrardii（Harv.）Burtt Davy ●
4883 Acmena hemilampra（F. Muell. ex F. M. Bailey）Merr. et L. M. Perry；大叶肖蒲桃；Broad-leaved Lillypilly ●☆
4884 Acmena ingens（F. Muell. ex C. Moore）Guymer et B. Hyland；巨大肖蒲桃；Red Apple ●☆
4885 Acmena smithii（Poir.）Merr. et L. M. Perry；斯密斯肖蒲桃；Lilly Pilly，Lillypilly，Lilly-pilly，Lilly-pilly Tree ●☆
4886 Acmenospenna Kausel = Syzygium R. Br. ex Gaertn.（保留属名）●
4887 Acmenosperma claviflorum（Roxb.）Kausel = Syzygium claviflorum（Roxb.）Wall. ex A. M. Cowan et Cowan ●
4888 Acmispon Raf.（1832）；钩足豆属■☆
4889 Acmispon Raf. = Lotus L. ■
4890 Acmispon americanum（Nutt.）Rydb. = Lotus unifoliolata（Hook.）Benth. ■☆
4891 Acmispon americanus Rydb.；美洲钩足豆■☆
4892 Acmispon brachycarpus（Benth.）D. D. Sokoloff；短果钩足豆■☆
4893 Acmispon denticulatus（Drew）D. D. Sokoloff；密齿钩足豆■☆
4894 Acmispon floribundus A. Heller；繁花钩足豆■☆
4895 Acmispon glabratus A. Heller；光钩足豆■☆
4896 Acmispon gracilis A. Heller；纤细钩足豆■☆
4897 Acmispon mollis（Nutt.）A. Heller；绢毛钩足豆■☆
4898 Acmispon multiflorum Raf.；多花钩足豆■☆
4899 Acmispon parviflorus（Benth.）D. D. Sokoloff；小花钩足豆■☆
4900 Acmispon pilosus A. Heller；毛钩足豆■☆
4901 Acmispon roudairei（Bonnet）Lassen；钩足豆■☆
4902 Acmispon roudairei（Bonnet）Lassen = Lotus roudairei Bonnet ■☆
4903 Acmispon rubriflorus（H. Sharsm.）D. D. Sokoloff；红花钩足豆■☆
4904 Acmopylaceae Melikian et A. V. Bobrov = Podocarpaceae Endl.（保留科名）●
4905 Acmopylaceae Pilg. = Podocarpaceae Endl.（保留科名）●
4906 Acmopyle Pilg.（1903）；铁门杉属●☆
4907 Acmopyle alba J. Buchholz；白铁门杉●☆
4908 Acmopyle pancheri Pilg.；铁门杉●☆
4909 Acmopyleaceae Melikyan et A. V. Bobrov = Podocarpaceae Endl.（保留科名）●
4910 Acmopyleaceae Melikyan et A. V. Bobrov；铁门杉科●☆
4911 Acmostemon Pilg. = Ipomoea L.（保留属名）●■
4912 Acmostemon angolensis Pilg. = Ipomoea verbascoidea Choisy ■☆
4913 Acmostigma Post et Kuntze = Acmostima Raf. ●
4914 Acmostigma Raf. = Pavetta L. ●
4915 Acmostima Raf. = Palicourea Aubl. ●☆
4916 Acnadena Raf. = Cordia L.（保留属名）●
4917 Acnida L. = Amaranthus L. ■
4918 Acnida alabamensis Standl. = Amaranthus australis（A. Gray）J. D. Sauer ■☆
4919 Acnida altissima Riddell ex Moq. = Amaranthus tuberculatus（Moq.）J. D. Sauer ■☆
4920 Acnida altissima Riddell ex Moq. var. prostrata（Uline et W. L. Bray）Fernald = Amaranthus tuberculatus（Moq.）J. D. Sauer ■☆
4921 Acnida altissima Riddell ex Moq. var. subnuda（S. Watson）Fernald = Amaranthus tuberculatus（Moq.）J. D. Sauer ■☆
4922 Acnida australis A. Gray = Amaranthus australis（A. Gray）J. D. Sauer ■☆
4923 Acnida cannabina L. = Amaranthus cannabinus（L.）J. D. Sauer ■☆
4924 Acnida cannabina L. var. australis（A. Gray）Uline et W. L. Bray = Amaranthus australis（A. Gray）J. D. Sauer ■☆
4925 Acnida concatenata（Moq.）Small = Amaranthus tuberculatus（Moq.）J. D. Sauer ■☆
4926 Acnida cuspidata Bertero ex Spreng. = Amaranthus australis（A. Gray）J. D. Sauer ■☆
4927 Acnida floridana S. Watson = Amaranthus floridanus（S. Watson）J. D. Sauer ■☆
4928 Acnida subnuda（S. Watson）Standl. = Amaranthus tuberculatus（Moq.）J. D. Sauer ■☆
4929 Acnida tamariscina（Nutt.）A. W. Wood = Amaranthus tuberculatus（Moq.）J. D. Sauer ■☆
4930 Acnida tamariscina（Nutt.）A. W. Wood var. concatenata（Moq.）Uline et W. L. Bray = Amaranthus tuberculatus（Moq.）J. D. Sauer ■☆
4931 Acnida tamariscina（Nutt.）A. W. Wood var. prostrata Uline et W. L. Bray = Amaranthus tuberculatus（Moq.）J. D. Sauer ■☆
4932 Acnida tamariscina（Nutt.）A. W. Wood var. tuberculata（Moq.）Uline et W. L. Bray = Amaranthus tuberculatus（Moq.）J. D. Sauer ■☆
4933 Acnida tuberculata Moq. = Amaranthus tuberculatus（Moq.）J. D. Sauer ■☆
4934 Acnida tuberculata Moq. var. prostrata（Uline et W. L. Bray）B. L. Rob. = Amaranthus tuberculatus（Moq.）J. D. Sauer ■☆
4935 Acnida tuberculata Moq. var. subnuda S. Watson = Amaranthus tuberculatus（Moq.）J. D. Sauer ■☆
4936 Acnide Mitch. = Acnida L. ■
4937 Acnide Mitch. = Amaranthus L. ■
4938 Acnista Durand = Acnida L. ■
4939 Acnista Durand = Amaranthus L. ■
4940 Acnistus Schott ex L. = Acnistus Schott ●☆
4941 Acnistus Schott（1829）；阿克尼茄树属（阿克尼茄属）；Wild Tobacco ●☆
4942 Acnistus arborescens（L.）Schltdl.；阿克尼茄树（阿克尼茄）●☆
4943 Acnistus arborescens Schltdl. = Acnistus arborescens（L.）Schltdl. ●☆
4944 Acocanthera G. Don = Acokanthera G. Don ●☆
4945 Acocanthera Post et Kuntze = Acokanthera G. Don ●☆
4946 Acoeloraphe Post et Kuntze = Acoeloraphis Durand ●☆
4947 Acoeloraphis Durand = Acoelorrhaphe H. Wendl. ●☆
4948 Acoelorhaphe H. Wendl. = Acoelorrhaphe H. Wendl. ●☆
4949 Acoelorrhaphe H. Wendl.（1879）；沼地棕属（阿斯罗榈属，常湿地棕榈属，丛立刺榈属，沼泽棕属）；Everglades Palm ●☆
4950 Acoelorrhaphe arborescens Becc. = Acoelorrhaphe wrightii Becc. ●☆
4951 Acoelorrhaphe wrightii（Griseb. et H. Wendl.）Becc.；常湿地棕榈（丛生棕榈，沼地棕）；Everglades Palm，Madeira Palm，Paurotis Palm，Saw Cabbage Palm，Silver Palm，Silver Saw Palm，Silver Saw Palmetto ●☆
4952 Acoelorrhaphe wrightii Becc. = Acoelorrhaphe wrightii（Griseb. et H. Wendl.）Becc. ●☆
4953 Acoidium Lindl. = Trichocentrum Poepp. et Endl. ■☆
4954 Acokanthera G. Don（1837）；尖药木属（长药花属，毒夹竹桃属，非洲简明毒树属）；Bushman's Poison，Bushman's-poison，Poison Bush，Poison Tree，Winter-sweet ●☆
4955 Acokanthera abyssinica K. Schum.；阿比西尼亚尖药木（阿比西尼亚毒夹竹桃，埃塞尖药木）●☆

4956 Acokanthera abyssinica K. Schum. = Acokanthera schimperi (A. DC.) Schweinf. ●☆
4957 Acokanthera deflersii Schweinf. ex Lewin = Acokanthera schimperi (A. DC.) Schweinf. ●☆
4958 Acokanthera friesiorum Markgr. = Acokanthera schimperi (A. DC.) Schweinf. ●☆
4959 Acokanthera laevigata Kupicha;平滑尖药木(平滑毒夹竹桃)●☆
4960 Acokanthera lamarkii G. Don = Acokanthera oppositifolia (Lam.) Codd ●
4961 Acokanthera longiflora Stapf;大花尖药木(长花尖药木,长叶毒夹竹桃,长叶尖药木);Longflower Bushman's Poison, Longflower Bushman's-poison ●☆
4962 Acokanthera longiflora Stapf = Acokanthera oppositifolia (Lam.) Codd ●
4963 Acokanthera lycioides (Roem. et Schult.) G. Don;澳非尖药木(毒夹竹桃);Wintersweet ●☆
4964 Acokanthera lycioides G. Don = Acokanthera lycioides (Roem. et Schult.) G. Don ●☆
4965 Acokanthera oblongifolia (Hochst.) Codd;长圆叶尖药木(美丽假虎刺,美丽尖药木);African Wintersweet, Dune Poison Bush, Wintersweet, Winter-sweet ●☆
4966 Acokanthera oblongifolia Benth. et Hook. f.;矩圆叶尖药木(矩圆叶毒夹竹桃);Bushman's Poison, Poison Bush ●☆
4967 Acokanthera oblongifolia Codd = Acokanthera oblongifolia (Hochst.) Codd ●☆
4968 Acokanthera oppositifolia (Lam.) Codd;对叶尖药木(长药花);Bushman's Poison, Common Poison Bush, Oppositeleaf Bushman's Poison ●
4969 Acokanthera oppositifolia Codd = Acokanthera oppositifolia (Lam.) Codd ●
4970 Acokanthera ouabaio Cathelineau ex Lewin;尖药木●☆
4971 Acokanthera ouabaio Lewin = Acokanthera schimperi (A. DC.) Schweinf. ●☆
4972 Acokanthera pubescens (Roem. et Schult.) G. Don;柔毛尖药木●☆
4973 Acokanthera rhodesica Merxm. = Acokanthera oppositifolia (Lam.) Codd ●
4974 Acokanthera rotundata (Codd) Kupicha;圆尖药木●☆
4975 Acokanthera schimperi (A. DC.) Benth. et Hook. f. ex Schweinf.;施氏尖药木●☆
4976 Acokanthera schimperi (A. DC.) Schweinf. = Acokanthera schimperi (A. DC.) Benth. et Hook. f. ex Schweinf. ●☆
4977 Acokanthera schimperi (A. DC.) Schweinf. var. rotundata Codd = Acokanthera rotundata (Codd) Kupicha ●☆
4978 Acokanthera schimperi Benth. et Hook. f. = Acokanthera schimperi (A. DC.) Benth. et Hook. f. ex Schweinf. ●☆
4979 Acokanthera spectabilis (Sond.) Hook. f. = Acokanthera oblongifolia (Hochst.) Codd ●☆
4980 Acokanthera spectabilis G. Don = Acokanthera oblongifolia (Hochst.) Codd ●☆
4981 Acokanthera venenata G. Don;毒尖药木(铁枣);Bushman's Poison ●☆
4982 Acoma Adans. = Homalium Jacq. ●
4983 Acoma Benth. = Coreocarpus Benth. ■☆
4984 Acomastylis Greene = Geum L. ■
4985 Acomastylis Greene et F. Bolle = Acomastylis Greene ■
4986 Acomastylis Greene(1906);羽叶花属;Acomastylis, Pinnaflower ■
4987 Acomastylis calthifolia (Sm.) F. Bolle var. nipponica (F. Bolle) H. Hara = Geum calthifolium Menzies ex Sm. ■☆
4988 Acomastylis elata (Royle) F. Bolle;羽叶花(狭叶路边青);Tall Acomastylis, Tall Pinnaflower ■
4989 Acomastylis elata (Royle) F. Bolle = Geum elatum Wall. ex Hook. f. ■
4990 Acomastylis elata (Royle) F. Bolle var. humilis (Royle) F. Bolle;矮生羽叶花;Dwarf Acomastylis, Dwarf Pinnaflower ■
4991 Acomastylis elata (Royle) F. Bolle var. leiocarpa (Evans) F. Bolle;光果羽叶花(秦岭无尾果);Smoothfruit Acomastylis, Smoothfruit Pinnaflower ■
4992 Acomastylis elata (Royle) F. Bolle var. leiocarpa (Evans) F. Bolle = Acomastylis elata (Royle) F. Bolle ■
4993 Acomastylis macrantha (Kearney) F. Bolle;大花羽叶花■☆
4994 Acomastylis macrosepala (Ludlow) Te T. Yu et C. L. Li;大萼羽叶花(大萼路边青);Largesepal Acomastylis, Largesepal Pinnaflower ■
4995 Acomastylis rossii (R. Br.) Greene;罗斯羽叶花■☆
4996 Acomastylis sikkimensis (Prain) F. Bolle;锡金羽叶花■☆
4997 Acome Baker = Cleome L. ●■
4998 Acomides Sol. = Xanthorrhoea Sm. ●■☆
4999 Acomis F. Muell. (1864);棕鼠麹属■☆
5000 Acomis acoma (F. Muell.) Druce;棕鼠麹■☆
5001 Acomis acoma Druce = Acomis acoma (F. Muell.) Druce ■☆
5002 Acomosperma K. Schum. ex Ule(1908);无毛果属☆
5003 Acomosperma K. Sckum. = Acomosperma K. Schum. ex Ule☆
5004 Aconceveibum Miq. = Mallotus Lour. ●
5005 Aconitaceae Bercht. et J. Presl = Ranunculaceae Juss. (保留科名) ●■
5006 Aconitella Spach = Delphinium L. ■
5007 Aconitella Spach(1839);小乌头属■☆
5008 Aconitella delphinioides Spach;小乌头■☆
5009 Aconitopsis Kem. -Nath. (1940);类乌头属■☆
5010 Aconitopsis Kem. -Nath. = Aconitella Spach ■☆
5011 Aconitopsis Kem. -Nath. = Aconitum L. ■
5012 Aconitopsis barbata (Bunge) Kem. -Nath.;毛类乌头■☆
5013 Aconitopsis malayana Ridl.;类乌头■☆
5014 Aconitum L. (1753);乌头属;Aconite, Aconitum, Monkshood, Monk's-hood, Wolf's Bane, Wolfsbane ■
5015 Aconitum abietetorum W. T. Wang et L. Q. Li;冷杉林乌头;Firforest Monkshood ■
5016 Aconitum acutiusculum H. R. Fletcher et Lauener;尖萼乌头;Acutate Monkshood, Acutatesepal Monkshood ■
5017 Aconitum acutiusculum H. R. Fletcher et Lauener var. aureopilosum W. T. Wang;展毛尖萼乌头;Yellowpilose Acutatesepal Monkshood ■
5018 Aconitum aggregatifolium C. C. Chang ex W. T. Wang = Aconitum cavaleriei H. Lév. et Vaniot var. aggregatifolium (C. C. Chang ex W. T. Wang) W. T. Wang ■
5019 Aconitum alatavicum Vorosch.;细叶乌头(乌头,阿拉套乌头);Alatai Monkshood ■
5020 Aconitum alboflavidum W. T. Wang = Aconitum pendulicarpum C. C. Chang ex W. T. Wang ■
5021 Aconitum alboviolaceum Kom.;两色乌头;Twocolored Monkshood ■
5022 Aconitum alboviolaceum Kom. f. albiflorum S. H. Li et Y. Huei Huang = Aconitum alboviolaceum Kom. var. albiflorum (S. H. Li et Y. Huei Huang) S. H. Li ■
5023 Aconitum alboviolaceum Kom. f. purpurascens (Nakai) Kitag. = Aconitum alboviolaceum Kom. var. purpurascens Nakai ■

5024 Aconitum albovioIaceum Kom. var. albiflorum (S. H. Li et Y. Huei Huang) S. H. Li; 白花两色乌头; Whiteflower Twocolored Monkshood ■

5025 Aconitum alboviolaceum Kom. var. albiflorum (S. H. Li et Y. Huei Huang) S. H. Li = Aconitum alboviolaceum Kom. ■

5026 Aconitum alboviolaceum Kom. var. erectum W. T. Wang;直立两色乌头;Erect Twocolored Monkshood ■

5027 Aconitum alboviolaceum Kom. var. purpurascens Nakai; 紫花乌头;Purpleflower Twocolored Monkshood ■

5028 Aconitum alboviolaceum Kom. var. purpurascens Nakai = Aconitum alboviolaceum Kom. ■

5029 Aconitum alboviolaceum Kom. var. typicum Nakai = Aconitum alboviolaceum Kom. ■

5030 Aconitum alpino-nepalense Tamura; 高峰乌头; Nepal Alpine Monkshood ■

5031 Aconitum altaicum Steinb.;阿尔泰乌头;Altai Monkshood ■☆

5032 Aconitum ambiguum Rchb.;兴安乌头;Khing'an Monkshood, Xing'an Monkshood ■

5033 Aconitum ambiguum Rchb. f. multisectum S. H. Li et Y. Huei Huang = Aconitum ambiguum Rchb. ■

5034 Aconitum amurense Nakai = Aconitum villosum Rchb. var. amurense (Nakai) S. H. Li et Y. Huei Huang ■

5035 Aconitum anglicum Stapf = Aconitum napellus L. ■☆

5036 Aconitum angusticassidatum Steinb.; 狭盔乌头; Narrowhelmet Monkshood ■☆

5037 Aconitum angustisegmentum W. T. Wang = Aconitum refracticarpum C. C. Chang ex W. T. Wang ■

5038 Aconitum angustius (W. T. Wang) W. T. Wang = Aconitum sinomontanum Nakai var. angustius W. T. Wang ■

5039 Aconitum anthora L.;南欧乌头(安索乌头,黄花乌头,铁棒槌); Pyrenees Monkshood, Pyrenees Monk's-hood, Yellow Monk's-hood ■☆

5040 Aconitum anthora L. var. gilvum Maxim. = Aconitum pendulum Busch ■

5041 Aconitum anthoroideum DC.;拟黄花乌头(乌头,新疆乌头); False Yellowflower Monkshood ■

5042 Aconitum apetalum (Huth) B. Fedtsch.;空茎乌头;Apetalous Monkshood, Petalless Monkshood ■

5043 Aconitum arcuatum Maxim.;拱形乌头;Arcuate Monkshood ■

5044 Aconitum arcuatum Maxim. = Aconitum fischeri Rchb. var. arcuatum (Maxim.) Regel ■

5045 Aconitum atropurpureum Hand.-Mazz. = Aconitum hemsleyanum E. Pritz. ex Diels var. atropurpureum (Hand.-Mazz.) W. T. Wang ■

5046 Aconitum austroyunnanense W. T. Wang;滇南草乌(草乌,大草乌,大黑牛,滇南乌头,七星草乌,树乌头,铜皮,小黑牛); S. Yunnan Monkshood, South Yunnan Monkshood ■

5047 Aconitum austroyunnanense W. T. Wang = Aconitum hemsleyanum E. Pritz. ex Diels ■

5048 Aconitum baicalense Turcz. ex Steud.; 贝加尔乌头; Baical Monkshood ■

5049 Aconitum bailangense Y. Z. Zhao;白狼乌头;Bailang Monkshood ■

5050 Aconitum bakeri Greene;巴凯尔乌头■☆

5051 Aconitum balfourii Stapf;西藏乌头(藏草乌,西藏草乌,亚东乌头)■

5052 Aconitum barbatum Patrin ex Pers.;细叶黄乌头(扁毒,扁特,便特,牛扁,曲芍,髯毛乌头,乌头);Slenderleaf Monkshood ■

5053 Aconitum barbatum Patrin ex Pers. subsp. pekinense (Vorosch.) Gubanov = Aconitum barbatum Patrin ex Pers. var. puberulum Ledeb. ■

5054 Aconitum barbatum Patrin ex Pers. var. gmelinii (Rchb.) Ledeb. = Aconitum barbatum Patrin ex Pers. ■

5055 Aconitum barbatum Patrin ex Pers. var. hispidum DC.;西伯利亚乌头(瓣子艽,朝天恒,黑大艽,黑秦艽,黑尾大艽,黄花乌头,马尾大艽,牛扁,细叶黄乌头);Siberian Monkshood ■

5056 Aconitum barbatum Patrin ex Pers. var. parviflorum Reverdin et Polozhij;小花髯毛乌头■☆

5057 Aconitum barbatum Patrin ex Pers. var. puberulum Ledeb.;牛扁(北方乌头,扁毒,扁桃叶根,扁特,牛扁乌头,曲芍);Puberulent Monkshood ■

5058 Aconitum bartletii Yamam. var. formosanum (Tamura) T. S. Liu et C. F. Hsieh = Aconitum formosanum Tamura ■

5059 Aconitum bartlettii Yamam.;奇莱乌头■

5060 Aconitum bartlettii Yamam. = Aconitum fukutomei Hayata ■

5061 Aconitum bartlettii Yamam. var. fukutomei (Hayata) T. S. Liu et C. F. Hsieh = Aconitum fukutomei Hayata ■

5062 Aconitum benzilanense T. L. Ming; 奔子栏乌头; Benzilan Monkshood ■

5063 Aconitum benzilanense T. L. Ming = Aconitum acutiusculum H. R. Fletcher et Lauener var. aureopilosum W. T. Wang ■

5064 Aconitum benzilanense T. L. Ming = Aconitum ouvrardianum Hand.-Mazz. ■

5065 Aconitum biflorum Fisch. ex DC.; 二花乌头; Twoflower Monkshood ■☆

5066 Aconitum birobidshanicum Vorosch.; 带岭乌头; Dailing Monkshood, Tailing Monkshood ■

5067 Aconitum bisma (Buch.-Ham.) Rapaics;掌裂乌头;Palmateleaf Monkshood ■

5068 Aconitum bisma (Buch.-Ham.) Rapaics var. taronense Hand.-Mazz. = Aconitum taronense (Hand.-Mazz.) H. R. Fletcher et Lauener ■

5069 Aconitum brachypodum Diels;短柄乌头(磨三转,三转半,生根子,搜山虎,铁棒槌,雪上一枝蒿,一枝蒿);Shortstalk Monkshood ■

5070 Aconitum brachypodum Diels var. crispulum W. T. Wang;曲毛短柄乌头(伏毛短柄乌头);Curvedhair Shortstalk Monkshood ■

5071 Aconitum brachypodum Diels var. laxiflorum H. R. Fletcher et Lauener;展毛短柄乌头(疏花短柄乌头,雪上一枝蒿); Looseflower Shortstalk Monkshood ■

5072 Aconitum bracteolatum Lauener;宽苞乌头;Broadbract Monkshood ■

5073 Aconitum bracteolosum W. T. Wang; 显苞乌头; Bracteose Monkshood ■

5074 Aconitum bracteolosum W. T. Wang = Aconitum tongolense Ulbr. ■

5075 Aconitum brevicalcaratum (Finet et Gagnep.) Diels;短距乌头(大草乌,短距牛扁);Shortspur Monkshood ■

5076 Aconitum brevicalcaratum (Finet et Gagnep.) Diels f. bracteatum (Finet et Gagnep.) Hand.-Mazz. = Aconitum brevicalcaratum (Finet et Gagnep.) Diels ■

5077 Aconitum brevicalcaratum (Finet et Gagnep.) Diels var. lauenerianum (Finet et Gagnep.) Diels (H. R. Fletcher) W. T. Wang;弯短距乌头;Shortbentspur Monkshood ■

5078 Aconitum brevicalcaratum (Finet et Gagnep.) Diels var. lauenerianum (Finet et Gagnep.) Diels (H. R. Fletcher) W. T. Wang = Aconitum brevicalcaratum (Finet et Gagnep.) Diels ■

5079 Aconitum brevicalcaratum (Finet et Gagnep.) Diels var. parviflorum F. H. Chen et S. Liu;小花短距乌头(短距乌头)■

5080 Aconitum brevicalcaratum (Finet et Gagnep.) Diels var. pauciflorum F. H. Chen et S. Liu = Aconitum brevicalcaratum (Finet

et Gagnep.) Diels ■
5081 Aconitum brevilimbum Lauener; 短唇乌头; Brevilimbate Monkshood, Shortlimb Monkshood ■
5082 Aconitum brevipetalum W. T. Wang; 短瓣乌头; Brevipetalous Monkshood, Shortpetal Monkshood ■
5083 Aconitum brevipetalum W. T. Wang = Aconitum ouvrardianum Hand. -Mazz. ■
5084 Aconitum brunneum Hand. -Mazz. ;褐紫乌头;Brown Monkshood ■
5085 Aconitum bulbiferum Howell;北美珠芽乌头■☆
5086 Aconitum bulbiferum Rchb. = Aconitum variegatum L. ■☆
5087 Aconitum bulbilliferum Hand. -Mazz. ; 珠芽乌头; Bulbil Monkshood, Bulbilliferous Monkshood ■
5088 Aconitum bullatifolium H. Lév. ;白弩箭药(黄腊一支蒿,小白撑,雪山一支蒿);Bullatiform Monkshood ■
5089 Aconitum bullatifolium H. Lév. = Aconitum nagarum Stapf var. heterotrichum H. R. Fletcher et Lauener ■
5090 Aconitum bullatifolium H. Lév. var. dielsianum (Airy Shaw) H. R. Fletcher et Lauener = Aconitum nagarum Stapf var. heterotrichum H. R. Fletcher et Lauener f. dielsianum Airy Shaw ■
5091 Aconitum bullatifolium H. Lév. var. dielsianum (Airy Shaw) H. R. Fletcher et Lauener = Aconitum nagarum Stapf var. heterotrichum H. R. Fletcher et Lauener ■
5092 Aconitum bullatifolium H. Lév. var. homaotrichum W. T. Wang = Aconitum nagarum Stapf ■
5093 Aconitum bullatifolium H. Lév. var. leiocarpum (W. T. Wang) W. T. Wang = Aconitum nagarum Stapf var. acaule (Finet et Gagnep.) Q. E. Yang ■
5094 Aconitum bulleyanum Diels;滇西乌头;W. Yunnan Monkshood, West Yunnan Monkshood ■
5095 Aconitum cammarum Jacq. ;有毒乌头■☆
5096 Aconitum campylorrhynchum Hand. -Mazz. ;弯喙乌头;Curvebeak Monkshood ■
5097 Aconitum campylorrhynchum Hand. -Mazz. var. patentipilum W. T. Wang;展毛弯喙乌头;Patentpilose Curvebeak Monkshood ■
5098 Aconitum campylorrhynchum Hand. -Mazz. var. tenuipes W. T. Wang;细梗弯喙乌头;Thinstalk Curvebeak Monkshood ■
5099 Aconitum cannabifolium Franch. ex Finet et Gagnep. ;大麻叶乌头(羊角七);Hempleaf Monkshood ■
5100 Aconitum carmichaelii Debeaux;乌头(白乌,草乌,侧子,莔,莔子,川乌,川乌头,大草乌,大川乌,鹅儿花,附子,虎掌,即子,金乌,九子不离母,九子不离娘,漏篮,漏篮子,木鳖子,十二元脚,铁花,乌药,五毒,小草乌,盐乌头);Carmichael's Monkshood, Common Monkshood, Monkshood ■
5101 Aconitum carmichaelii Debeaux 'Arendsii';阿伦氏乌头■☆
5102 Aconitum carmichaelii Debeaux var. albovillosum F. H. Chen et Y. Liu = Aconitum forrestii Stapf var. albovillosum (F. H. Chen et S. Liu) W. T. Wang ■
5103 Aconitum carmichaelii Debeaux var. angustius W. T. Wang et P. K. Hsiao;狭菱裂乌头;Narrowsplit Monkshood ■
5104 Aconitum carmichaelii Debeaux var. fortunei (Hemsl.) W. T. Wang et P. K. Hsiao = Aconitum carmichaelii Debeaux var. truppelianum (Ulbr.) W. T. Wang et P. K. Hsiao ■
5105 Aconitum carmichaelii Debeaux var. hwangshanicum W. T. Wang;黄山乌头(吓虎打);Huangshan Monkshood, Huangshan Mountain Monkshood ■
5106 Aconitum carmichaelii Debeaux var. pubscens W. T. Wang et P. K. Hsiao;毛叶乌头(草乌,大乌药,密毛乌头,乌药);Hairyleaf Monkshood ■
5107 Aconitum carmichaelii Debeaux var. tripartitum W. T. Wang;深裂乌头;Tripartite Monkshood ■
5108 Aconitum carmichaelii Debeaux var. truppelianum (Ulbr.) W. T. Wang et P. K. Hsiao;展毛乌头(草乌);Patenthairy Monkshood ■
5109 Aconitum carmichaelii Debeaux var. truppelianum (Ulbr.) W. T. Wang et P. K. Hsiao = Aconitum liaodungense Nakai ■
5110 Aconitum cavaleriei H. Lév. et Vaniot;黔川乌头(水八角莲);Cavalerie Monkshood ■
5111 Aconitum cavaleriei H. Lév. et Vaniot = Aconitum scaposum Franch. ■
5112 Aconitum cavaleriei H. Lév. et Vaniot var. aggregatifolium (C. C. Chang ex W. T. Wang) W. T. Wang;聚叶黔川乌头;Aggregateleaf Monkshood ■
5113 Aconitum cavaleriei H. Lév. et Vaniot var. aggregatifolium (C. C. Chang ex W. T. Wang) W. T. Wang = Aconitum scaposum Franch. ■
5114 Aconitum changianum W. T. Wang;察瓦龙乌头(铁罗汉,张氏乌头);Chang Monkshood, Chawalong Monkshood ■
5115 Aconitum chasmanthum Stapf ex Holmes;展花乌头;Patentflower Monkshood ■
5116 Aconitum chayuense W. T. Wang;察隅乌头;Chayu Monkshood ■
5117 Aconitum chenianum W. T. Wang;陈氏乌头;Chen Monkshood ■
5118 Aconitum chenianum W. T. Wang = Aconitum tongolense Ulbr. ■
5119 Aconitum chiachaense W. T. Wang; 加查乌头; Chiacha Monkshood, Gyaca Monkshood, Jiacha Monkshood ■
5120 Aconitum chiachaense W. T. Wang var. glandulosum W. T. Wang;腺毛加查乌头;Glandular Hair Monkshood ■
5121 Aconitum chienningense W. T. Wang; 乾宁乌头; Chienning Monkshood, Qianning Monkshood ■
5122 Aconitum chienningense W. T. Wang var. lasiocarpum W. T. Wang;毛果乾宁乌头;Hairyfruit Chienning Monkshood, Hairyfruit Qianning Monkshood ■
5123 Aconitum chilienshanicum W. T. Wang;祁连山乌头(铁棒槌);Chilien Moutain Monkshood, Qilianshan Monkshood ■
5124 Aconitum chinense Pax = Aconitum carmichaelii Debeaux var. truppelianum (Ulbr.) W. T. Wang et P. K. Hsiao ■
5125 Aconitum chinense Pax var. hwangshanicum W. T. Wang et P. K. Hsiao = Aconitum carmichaelii Debeaux var. hwangshanicum W. T. Wang ■
5126 Aconitum chinense Siebold ex Paxton = Aconitum liaodungense Nakai ■
5127 Aconitum chingtungense W. T. Wang; 景东乌头; Jingdong Monkshood ■
5128 Aconitum chingtungense W. T. Wang = Aconitum hemsleyanum E. Pritz. ex Diels var. chingtungense (W. T. Wang) W. T. Wang ■
5129 Aconitum chingtungense W. T. Wang = Aconitum hemsleyanum E. Pritz. ex Diels ■
5130 Aconitum chloranthum Hand. -Mazz. = Aconitum scaposum Franch. var. hupehanum Rapaics ■
5131 Aconitum chrysotrichum W. T. Wang; 黄毛乌头; Yellowhair Monkshood ■
5132 Aconitum chuianum W. T. Wang;拟哈巴乌头(朱氏乌头);Chu Monkshood ■
5133 Aconitum chuianum W. T. Wang = Aconitum habaense W. T. Wang ■
5134 Aconitum chuosjiaense W. T. Wang;绰斯甲乌头(绰斯乌头);Chuosijia Monkshood ■
5135 Aconitum ciliare DC. = Aconitum volubile Pall. ex Koelle var.

pubescens Regel ■

5136 Aconitum columbianum Nutt.;哥伦比亚乌头;Columbian Monk's-hood,Western Monkshood,Western Monk's-hood ■☆

5137 Aconitum columbianum Nutt. subsp. pallidum Piper;苍白哥伦比亚乌头■☆

5138 Aconitum columbianum Nutt. subsp. pallidum Piper = Aconitum columbianum Nutt. ■☆

5139 Aconitum columbianum Nutt. var. bakeri (Greene) H. D. Harr.;巴氏乌头■☆

5140 Aconitum columbianum Nutt. var. bakeri (Greene) H. D. Harr. = Aconitum columbianum Nutt. ■☆

5141 Aconitum columbianum Nutt. var. ochroleucum A. Nelson = Aconitum columbianum Nutt. ■☆

5142 Aconitum contortum Finet et Gagnep.;苍山乌头(七星草乌,五虎下西川);Cangshan Monkshood ■

5143 Aconitum contortum Finet et Gagnep. var. villosulipes W. T. Wang;紫苍山乌头(紫乌头);Violet Korea Monkshood ■

5144 Aconitum coreanum (H. Lév.) Rapaics;黄花乌头(白附子,白花子,百步草,关白附,黄乌拉花,黄乌拉藤,两头菜,山喇叭花,鼠尾草,乌拉花,小喇叭花,药虱子草,竹节白附);Korea Monkshood,Korean Monkshood ■

5145 Aconitum coriaceifolium W. T. Wang;革叶乌头;Letherleaf Monkshood ■

5146 Aconitum coriaceum H. Lév. = Aconitum racemulosum Franch. ■

5147 Aconitum coriophyllum Hand.-Mazz. = Aconitum nagarum Stapf var. acaule (Finet et Gagnep.) Q. E. Yang ■

5148 Aconitum crassicaule W. T. Wang;粗茎乌头(紫乌头);Crassicauliferous Monkshood,Thickstem Monkshood ■

5149 Aconitum crassicaule W. T. Wang = Aconitum hemsleyanum E. Pritz. ex Diels ■

5150 Aconitum crassiflorum Hand.-Mazz.;粗花乌头;Thickflower Monkshood ■

5151 Aconitum crassifolium Steinb.;厚叶乌头■☆

5152 Aconitum creagromorphum Lauener;叉苞乌头;Forkbract Monkshood,Furcatebract Monkshood ■

5153 Aconitum cymbulatum (Schmalh.) Lipsky;舟形乌头;Cymbaeform Monkshood ■☆

5154 Aconitum czekanovskyi Steinb.;契卡乌头■☆

5155 Aconitum daxinganlingense Y. Z. Zhao;大兴安岭乌头;Daxing'anling Monkshood ■

5156 Aconitum delavayi Franch.;马耳山乌头;Delavay Monkshood,Maershan Monkshood ■

5157 Aconitum delavayi Franch. var. coreanum H. Lév. = Aconitum coreanum (H. Lév.) Rapaics ■

5158 Aconitum delavayi Franch. var. leiocarpum Finet et Gagnep. = Aconitum episcopale H. Lév. ■

5159 Aconitum delphiniifolium DC. = Aconitum napellus L. ■☆

5160 Aconitum desoulavyi Kom.;代苏乌头(德氏乌头)■☆

5161 Aconitum dielsianum Airy Shaw = Aconitum nagarum Stapf var. acaule (Finet et Gagnep.) Q. E. Yang ■

5162 Aconitum diqingense Q. E. Yang et Z. D. Fang;迪庆乌头;Diqing Monkshood ■

5163 Aconitum dissectum D. Don = Aconitum gammiei Stapf ■

5164 Aconitum divaricatum Finet et Gagnep. = Aconitum tatsienense Finet et Gagnep. var. divaricatum (Finet et Gagnep.) W. T. Wang ■

5165 Aconitum dolichorhynchum W. T. Wang;长柱乌头;Longstigma Monkshood,Longstyle Monkshood ■

5166 Aconitum dolichorhynchum W. T. Wang = Aconitum rockii H. R. Fletcher et Lauener var. fengii (W. T. Wang) W. T. Wang ■

5167 Aconitum dolichostachyum W. T. Wang;长序乌头;Long Monkshood,Longstachys Monkshood ■

5168 Aconitum drientale Mill.;东方乌头■

5169 Aconitum duclouxii H. Lév.;宾川乌头;Ducloux Monkshood ■

5170 Aconitum duclouxii H. Lév. var. ecalcaratum H. R. Fletcher et lauener;无距宾川乌头(白药);Spurless Ducloux Monkshood ■

5171 Aconitum dunhuaense S. H. Li;敦化乌头;Dunhua Monkshood ■

5172 Aconitum elliotii Lauener;墨脱乌头;Elliott Monkshood,Modog Monkshood,Motuo Monkshood ■

5173 Aconitum elliotii Lauener var. doshongense (Lauener) W. T. Wang;短梗墨脱乌头;Shortpedicel Elliott Monkshood ■

5174 Aconitum elliotii Lauener var. glabrescens W. T. Wang et L. Q. Li;光梗墨脱乌头;Smoothpedicel Elliott Monkshood ■

5175 Aconitum elliotii Lauener var. pilopetalum W. T. Wang et L. Q. Li;毛瓣墨脱乌头;Hairbract Elliott Monkshood ■

5176 Aconitum elwesii Stapf;藏南乌头(藏南藤乌,藏南藤乌头);Elwes Monkshood ■

5177 Aconitum episcopale H. Lév.;西南乌头(草乌,藤乌,亚东乌头,野弩箭药,紫草乌,紫乌头);Purple Monkshood ■

5178 Aconitum episcopale H. Lév. var. villosulipes W. T. Wang;紫西南乌头(紫乌头)■

5179 Aconitum episcopale H. Lév. var. villosum W. T. Wang = Aconitum delavayi Franch. ■

5180 Aconitum euryanthum Hand.-Mazz. = Aconitum stylosum Stapf ■

5181 Aconitum excelsum Rchb. = Aconitum septentrionale Koelle ■

5182 Aconitum falciforme Hand.-Mazz.;镰形乌头(绿盔乌头);Falcate Monkshood ■

5183 Aconitum fangianum W. T. Wang;刷经寺乌头(方氏乌头);Fang Monkshood,Shuajingsi Monkshood ■

5184 Aconitum fanjingshanicum W. T. Wang;梵净山乌头;Fanjingshan Monkshood ■

5185 Aconitum fauriei H. Lév. et Vaniot;蛇岛乌头;Chinese Monkshood,Faurie Monkshood ■

5186 Aconitum fengii W. T. Wang = Aconitum rockii H. R. Fletcher et Lauener var. fengii (W. T. Wang) W. T. Wang ■

5187 Aconitum fengii W. T. Wang var. crispulum Q. E. Yang;曲毛石膏山乌头(曲毛冯氏乌头);Benthair Feng Monkshood ■

5188 Aconitum ferox Wall.;多刺乌头(印度乌头);Indian Aconite,Nepal Aconite ■☆

5189 Aconitum ferox Wall. var. atrox ? = Aconitum balfourii Stapf ■

5190 Aconitum ferox Wall. var. naviculare Brühl = Aconitum naviculare (Brühl) Stapf ■

5191 Aconitum ferox Wall. var. spicata ? = Aconitum spicatum Stapf ■

5192 Aconitum finetianum Hand.-Mazz.;赣皖乌头(牛虱鞭);Finet Monkshood ■

5193 Aconitum fischeri Rchb.;薄叶乌头(侧子,川乌,附子,光乌,天雄,乌毒,乌独,乌头);Azure Monkshood,Fischer Monkshood ■

5194 Aconitum fischeri Rchb. f. pilocarpum S. H. Li et Y. Huei Huang = Aconitum fischeri Rchb. var. arcuatum (Maxim.) Regel ■

5195 Aconitum fischeri Rchb. var. arcuatum (Maxim.) Regel;弯枝乌头;Curvibranch Monkshood ■

5196 Aconitum fischeri Rchb. var. arcuatum (Maxim.) Regel f. pilocarpum S. H. Li et Y. Huei Huang = Aconitum fischeri Rchb. var. arcuatum (Maxim.) Regel ■

5197 Aconitum flaccidum Rchb.;柔软乌头■☆

5198 Aconitum flagellate (F. Schmidt) Steinb.;鞭乌头■☆

5199 Aconitum flavidum W. T. Wang;淡黄乌头;Paleyellow Monkshood ■

5200 Aconitum flavum Hand. -Mazz.;伏毛铁棒槌(断肠草,伏毛铁棒棒,伏毛铁棒锤,两头尖,磨三转,铁棒槌,乌药,小草乌,一枝蒿);Yellow Monkshood ■

5201 Aconitum flavum Hand. -Mazz. var. galeatum W. T. Wang = Aconitum yinschanicum Y. Z. Zhao ■

5202 Aconitum flerovii Steinb.;和氏乌头■☆

5203 Aconitum fletcherianum G. Taylor;独花乌头(多花乌头);Oneflower Monkshood ■

5204 Aconitum formosanum Tamura;台湾乌头(蔓乌头);Taiwan Monkshood ■

5205 Aconitum forrestii Stapf;丽江乌头(黄草乌);Forrest Monkshood ■

5206 Aconitum forrestii Stapf var. albovillosum (F. H. Chen et Tang S. Liu) W. T. Wang;毛果丽江乌头;Hairyfruit Forrest Monkshood ■

5207 Aconitum fortunei Hemsl. = Aconitum carmichaelii Debeaux var. truppelianum (Ulbr.) W. T. Wang et P. K. Hsiao ■

5208 Aconitum franchetii Finet et Gagnep.;大渡乌头;Dadu Monkshood, Franchet Monkshood ■

5209 Aconitum franchetii Finet et Gagnep. var. geniculatum W. T. Wang et P. K. Hsiao;膝瓣大渡乌头;Geniculate Franchet Monkshood ■

5210 Aconitum franchetii Finet et Gagnep. var. glabrescens W. T. Wang;光序大渡乌头;Smooth Franchet Monkshood ■

5211 Aconitum franchetii Finet et Gagnep. var. lasiocalyx W. T. Wang et P. K. Hsiao;毛萼大渡乌头;Hairysepal Franchet Monkshood ■

5212 Aconitum franchetii Finet et Gagnep. var. subnaviculare W. T. Wang;低盔大渡乌头;Subnaviculare Monkshood ■

5213 Aconitum franchetii Finet et Gagnep. var. villosulum W. T. Wang;展毛大渡乌头;Villose Franchet Monkshood ■

5214 Aconitum fukutomei Hayata;梨山乌头(蔓乌头,奇莱乌头,台湾乌头);Bartlett Monkshood, Fukutome Monkshood, Lishan Monkshood ■

5215 Aconitum fukutomei Hayata var. formosanum (Tamura) Yen C. Yang et T. C. Huang = Aconitum formosanum Tamura ■

5216 Aconitum funckii Rchb.;富克乌头■☆

5217 Aconitum fusungense S. H. Li et Y. Huei Huang;抚松乌头;Fusong Monkshood ■

5218 Aconitum gammiei Stapf;错那乌头;Nakao Monkshood ■

5219 Aconitum geniculatum H. R. Fletcher et Lauener;膝瓣乌头(大草乌);Geniculate Monkshood, Geniculatepetal Monkshood ■

5220 Aconitum geniculatum H. R. Fletcher et Lauener var. humilis W. T. Wang;低盔膝瓣乌头;Low Geniculatepetal Monkshood ■

5221 Aconitum geniculatum H. R. Fletcher et Lauener var. humilis W. T. Wang = Aconitum geniculatum H. R. Fletcher et Lauener ■

5222 Aconitum geniculatum H. R. Fletcher et Lauener var. longicalcaratum M. Li;长距膝瓣乌头■

5223 Aconitum geniculatum H. R. Fletcher et Lauener var. unguiculatum W. T. Wang;盔膝瓣乌头(东川乌头,爪盔膝瓣乌头);Unguicular Geniculatepetal Monkshood ■

5224 Aconitum geniculatum H. R. Fletcher et Lauener var. unguiculatum W. T. Wang = Aconitum geniculatum H. R. Fletcher et Lauener ■

5225 Aconitum georgei Comber;长喙乌头;George Monkshood, Longbeak Monkshood ■

5226 Aconitum geraniifolium Host;老鹳草叶乌头■☆

5227 Aconitum geranioides Greene;假老鹳草叶乌头■☆

5228 Aconitum gezaense W. T. Wang et L. Q. Li;格咱乌头;Gezan Monkshood ■

5229 Aconitum gezaense W. T. Wang et L. Q. Li = Aconitum tongolense Ulbr. ■

5230 Aconitum gibbiferum Rchb. = Aconitum kusnezoffii Rchb. var. gibbiferum (Rchb.) Regel ■

5231 Aconitum gigas H. Lév. et Vaniot;巨乌头■☆

5232 Aconitum gilvum (Maxim.) Hand. -Mazz. = Aconitum flavum Hand. -Mazz. ■

5233 Aconitum glabrisepalum W. T. Wang;无毛乌头;Glabrous Monkshood ■

5234 Aconitum gmelinii Rchb. = Aconitum barbatum Patrin ex Pers. ■

5235 Aconitum gracile Rchb. ex Gáyer;纤细乌头■☆

5236 Aconitum grandiflorum Hegetschw.;大花乌头■☆

5237 Aconitum gymnandrum Maxim.;露蕊乌头(泽兰);Nakedstamen Monkshood ■

5238 Aconitum gymnandrum Maxim. f. leucanthum W. T. Wang;白花露蕊乌头;Whiteflower Nakedstamen Monkshood ■

5239 Aconitum gymnandrum Maxim. f. leucanthum W. T. Wang = Aconitum gymnandrum Maxim. ■

5240 Aconitum habaense W. T. Wang;哈巴乌头;Haba Monkshood ■

5241 Aconitum halleri Rchb.;哈勒乌头■☆

5242 Aconitum hamatipetalum W. T. Wang;钩瓣乌头;Hamatipetalous Monkshood, Hookpetal Monkshood ■

5243 Aconitum handelianum Comber;剑川乌头;Handel Monkshood, Jianchuan Monkshood ■

5244 Aconitum handelianum Comber = Aconitum pulchellum Hand. -Mazz. ■

5245 Aconitum handelianum Comber var. laxipilosum Hand. -Mazz.;疏毛剑川乌头;Laxpilose Handel Monkshood ■

5246 Aconitum helleri Greene;赫勒乌头■☆

5247 Aconitum hemsleyanum E. Pritz. ex Diels;瓜叶乌头(白乌头,草乌,见血封喉,老汉背姥姥,老汉背娃娃,蔓乌头,藤草乌,藤儿乌,藤乌,藤乌头,乌毒,血乌,盐附子,羊角七,鱼夫子);Hemsley Monkshood ■

5248 Aconitum hemsleyanum E. Pritz. ex Diels var. atropurpureum (Hand. -Mazz.) W. T. Wang;展毛瓜叶乌头;Darkpurple Hemsley Monkshood ■

5249 Aconitum hemsleyanum E. Pritz. ex Diels var. chingtungense (W. T. Wang) W. T. Wang;截基瓜叶乌头;Truncatebase Hemsley Monkshood ■

5250 Aconitum hemsleyanum E. Pritz. ex Diels var. chingtungense (W. T. Wang) W. T. Wang = Aconitum hemsleyanum E. Pritz. ex Diels ■

5251 Aconitum hemsleyanum E. Pritz. ex Diels var. circinatum W. T. Wang;拳距瓜叶乌头(草乌,藤乌,铁头和尚,血乌,掌距瓜叶乌头);Circinate Hemsley Monkshood ■

5252 Aconitum hemsleyanum E. Pritz. ex Diels var. circinatum W. T. Wang = Aconitum hemsleyanum E. Pritz. ex Diels ■

5253 Aconitum hemsleyanum E. Pritz. ex Diels var. elongatum W. T. Wang;长距瓜叶乌头;Elongatespur Hemsley Monkshood ■

5254 Aconitum hemsleyanum E. Pritz. ex Diels var. elongatum W. T. Wang = Aconitum hemsleyanum E. Pritz. ex Diels ■

5255 Aconitum hemsleyanum E. Pritz. ex Diels var. hsiae (W. T. Wang) W. T. Wang;珠芽瓜叶乌头;Bulbil Hemsley Monkshood ■

5256 Aconitum hemsleyanum E. Pritz. ex Diels var. hsiae (W. T. Wang) W. T. Wang = Aconitum hemsleyanum E. Pritz. ex Diels ■

5257 Aconitum hemsleyanum E. Pritz. ex Diels var. lasianthum W. T. Wang et L. Q. Li;毛萼瓜叶乌头;Haircalyx Hemsley Monkshood ■

5258 Aconitum hemsleyanum E. Pritz. ex Diels var. lasianthum W. T.

Wang et L. Q. Li = Aconitum hemsleyanum E. Pritz. ex Diels ■
5259 Aconitum hemsleyanum E. Pritz. ex Diels var. leucantum P. Guo et M. R. Jia;白花瓜叶乌头;Whiteflower Hemsley Monkshood ■
5260 Aconitum hemsleyanum E. Pritz. ex Diels var. leucantum P. Guo et M. R. Jia = Aconitum hemsleyanum E. Pritz. ex Diels ■
5261 Aconitum hemsleyanum E. Pritz. ex Diels var. pilopetalum W. T. Wang et L. Q. Li;毛瓣瓜叶乌头;Hairsepal Hemsley Monkshood ■
5262 Aconitum hemsleyanum E. Pritz. ex Diels var. pilopetalum W. T. Wang et L. Q. Li = Aconitum hemsleyanum E. Pritz. ex Diels ■
5263 Aconitum hemsleyanum E. Pritz. ex Diels var. puberulum W. T. Wang et L. Q. Li;毛枝瓜叶乌头;Hairbranch Hemsley Monkshood ■
5264 Aconitum hemsleyanum E. Pritz. ex Diels var. puberulum W. T. Wang = Aconitum hemsleyanum E. Pritz. ex Diels ■
5265 Aconitum hemsleyanum E. Pritz. ex Diels var. unguiculatum W. T. Wang;爪盔瓜叶乌头;Unguiculate Hemsley Monkshood ■
5266 Aconitum hemsleyanum E. Pritz. ex Diels var. unguiculatum W. T. Wang = Aconitum hemsleyanum E. Pritz. ex Diels ■
5267 Aconitum hemsleyanum E. Pritz. ex Diels var. xizangense W. T. Wang et L. Q. Li;西藏瓜叶乌头;Xizang Hemsley Monkshood ■
5268 Aconitum henryi E. Pritz. ;川鄂乌头;Henry Monkshood ■
5269 Aconitum henryi E. Pritz. var. compositum Hand. -Mazz. ;细裂川鄂乌头;Composite Henry Monkshood ■
5270 Aconitum henryi E. Pritz. var. pilocarpum W. T. Wang et L. Q. Li;毛果川鄂乌头;Hairfruit Henry Monkshood ■
5271 Aconitum henryi E. Pritz. var. villosum W. T. Wang;展毛川鄂乌头;Villous Henry Monkshood ■
5272 Aconitum heterophyllum Wall. ex Royle;异叶乌头;Atis Root ■☆
5273 Aconitum hicksii Lauener;同嘎乌头;Hicks Monkshood,Tongga Monkshood ■
5274 Aconitum hispidum DC. = Aconitum barbatum Patrin ex Pers. var. hispidum DC. ■
5275 Aconitum hondoense Nakai;本州乌头■☆
5276 Aconitum hookeri Stapf;胡克乌头■☆
5277 Aconitum hopeiense (W. T. Wang) Vorosch. = Aconitum leucostomum Vorosch. var. hopeiense W. T. Wang ■
5278 Aconitum hoppeanum Rchb. ;霍氏乌头■☆
5279 Aconitum hoppii Rchb. ;霍普乌头■☆
5280 Aconitum hortense Hoppe ex Rchb. ;田园乌头■☆
5281 Aconitum howellii A. Nelson et J. F. Macbr. ;豪氏乌头■☆
5282 Aconitum hsiae W. T. Wang = Aconitum hemsleyanum E. Pritz. ex Diels ■
5283 Aconitum huiliense Hand. -Mazz. ;会理乌头;Huili Monkshood ■
5284 Aconitum huizenense T. L. Ming;会泽乌头;Huize Monkshood ■
5285 Aconitum huizenense T. L. Ming = Aconitum brachypodum Diels ■
5286 Aconitum huizenense T. L. Ming = Aconitum gammiei Stapf ■
5287 Aconitum ichangense (Finet et Gagnep.) Hand. -Mazz. ;巴东乌头;Badong Monkshood,Ichang Monkshood ■
5288 Aconitum incisofidum W. T. Wang;缺刻乌头;Incised Monkshood ■
5289 Aconitum infectum Greene = Aconitum columbianum Nutt. ■☆
5290 Aconitum iochanicum Ulbr. ;滇北乌头;N. Yunan Monkshood, North Yunan Monkshood ■
5291 Aconitum iochanicum Ulbr. var. robustum F. H. Chen et S. Liu = Aconitum tanguticum (Maxim.) Stapf ■
5292 Aconitum jaluense Kom. ;鸭绿乌头;Yalu Monkshood ■
5293 Aconitum jaluense Kom. f. glabrescens (Nakai) Kitag. = Aconitum jaluense Kom. var. glabrescens Nakai ■
5294 Aconitum jaluense Kom. var. glabrescens Nakai;光梗鸭绿乌头(东北乌头,靰鞡花,靰鞡花蓝);Glabrous Yalu Monkshood ■
5295 Aconitum jaluense Kom. var. paniculigerum (Nakai) S. X. Li = Aconitum paniculigerum Nakai ■
5296 Aconitum jaluense Kom. var. triphyllum (Nakai) U. C. La;三叶鸭绿乌头■
5297 Aconitum jaluense Kom. var. truncatum S. H. Li et Y. Huei Huang;截基鸭绿乌头;Truncate Yalu Monkshood ■
5298 Aconitum japonicum Thunb. ;日本乌头;Japan Monkshood ■☆
5299 Aconitum japonicum Thunb. subsp. maritimum (Nakai ex Tamura et Namba) Kadota;海滨日本乌头■☆
5300 Aconitum japonicum Thunb. subsp. napiforme (H. Lév. et Vaniot) Kadota;萝卜乌头■
5301 Aconitum japonicum Thunb. var. truppelianum Ulbr. = Aconitum carmichaelii Debeaux var. truppelianum (Ulbr.) W. T. Wang et P. K. Hsiao ■
5302 Aconitum jeholense Nakai et Kitag. ;热河乌头(矮华北乌头,低矮华北乌头,华北乌头);Dwarf North China Monkshood ■
5303 Aconitum jeholense Thunb. var. angustius (W. T. Wang) Y. Z. Zhao;华北乌头(狭裂准噶尔乌头);N. China Monkshood ■
5304 Aconitum jilongense W. T. Wang et L. Q. Li;吉隆乌头;Jilong Monkshood ■
5305 Aconitum jinchengensis L. C. Wang et Silba = Aconitum sinomontanum Nakai ■
5306 Aconitum jinyangense W. T. Wang;金阳乌头;Jinyang Monkshood ■
5307 Aconitum jiulongense W. T. Wang;九龙乌头;Jiulong Monkshood ■
5308 Aconitum jiulongense W. T. Wang = Aconitum carmichaelii Debeaux ■
5309 Aconitum jucundum Diels = Aconitum scaposum Franch. var. hupehanum Rapaics ■
5310 Aconitum jucundum Diels var. chloranthum (Hand. -Mazz.) Hand. -Mazz. = Aconitum scaposum Franch. var. hupehanum Rapaics ■
5311 Aconitum kagerpuense W. T. Wang;卡卡波乌头;Kakabo Monkshood ■
5312 Aconitum kagerpuense W. T. Wang = Aconitum ouvrardianum Hand. -Mazz. ■
5313 Aconitum kamtschaticum Willd. ;勘察加乌头;Kamtschatka Monkshood ■☆
5314 Aconitum karakolicum Rapaics;多根乌头(草乌);Manyroot Monkshood ■
5315 Aconitum karakolicum Rapaics var. patentipilum W. T. Wang;展毛多根乌头;Patentpillose Manyroot Monkshood ■
5316 Aconitum kialaense W. T. Wang;卡拉乌头;Kala Monkshood ■
5317 Aconitum kialaense W. T. Wang = Aconitum crassiflorum Hand. -Mazz. ■
5318 Aconitum kirinense Nakai;吉林乌头;Jilin Monkshood,Kirin Monkshood ■
5319 Aconitum kirinense Nakai var. australe W. T. Wang;毛果吉林乌头;Hairyfruit Jilin Monkshood,Hairyfruit Kirin Monkshood ■
5320 Aconitum kirinense Nakai var. heterophyllum W. T. Wang;异裂吉林乌头;Heteroleaf Jilin Monkshood ■
5321 Aconitum kitagawai Nakai = Aconitum carmichaelii Debeaux var. truppelianum (Ulbr.) W. T. Wang et P. K. Hsiao ■
5322 Aconitum kiusianum Nakai;九州乌头■☆
5323 Aconitum kojimae Ohwi;锐裂乌头;Kojima Monkshood, Sharpdissected Monkshood ■
5324 Aconitum kojimae Ohwi var. lasiocarpum ?;粗果锐裂乌头(粗果乌头)■☆

5325 Aconitum kojimae Ohwi var. ramosum Tamura;分枝锐裂乌头;Ramose Sharpdissected Monkshood ■

5326 Aconitum komarovii Steinb.;科马罗夫乌头(构氏乌头)■☆

5327 Aconitum kongboense Lauener;工布乌头(雪山一支蒿);Gongbu Monkshood,Kongpo Monkshood ■

5328 Aconitum kongboense Lauener var. polycarpum W. T. Wang et L. Q. Li;多果工布乌头;Manyfruit Gongbu Monkshood ■

5329 Aconitum kongboense Lauener var. polycarpum W. T. Wang et L. Q. Li = Aconitum tongolense Ulbr. ■

5330 Aconitum kongboense Lauener var. villosum W. T. Wang;展毛工布乌头;Villous Gongbu Monkshood ■

5331 Aconitum koreanum Rapaics;朝鲜乌头■

5332 Aconitum krylovii Steinb.;克雷乌头■

5333 Aconitum kungshanense W. T. Wang;贡山乌头;Gongshan Monkshood,Kungshan Monkshood ■

5334 Aconitum kungshanense W. T. Wang = Aconitum taronense (Hand. -Mazz.) H. R. Fletcher et Lauener ■

5335 Aconitum kusnezoffii Rchb.;北乌头(百步草,草乌,草乌头,大草乌,帝秋,毒公,独白草,断肠草,多根乌头,莨,耿子,果负,蒿叶乌头,淮乌头,黄山乌头,芨,鸡毒,鸡头草,金鸦,堇,堇草,蓝附子,蓝花草,蓝乌拉花,勒革拉花,两头尖,千秋,僧鞋菊,双鸾菊,土附子,乌喙,乌头,五毒根,靰鞡花,奚毒,细叶草乌,显柱乌头,小黑牛,小叶芦,小叶鸦儿芦,穴种,鸦头,亚东乌头,药,鹦哥菊,鸳鸯菊,圆锥序乌头,直喙乌头,竹节乌头,准噶尔草乌,紫草乌);Chinese Aconite,Kusnezoff Monkshood ■

5336 Aconitum kusnezoffii Rchb. subsp. birobidshanicum (Vorosch.) Luferov = Aconitum birobidshanicum Vorosch. ■

5337 Aconitum kusnezoffii Rchb. var. bodinieri (H. Lév. et Vaniot) Finet et Gagnep. = Aconitum carmichaelii Debeaux ■

5338 Aconitum kusnezoffii Rchb. var. crispulum W. T. Wang;伏毛北乌头(伏毛草乌头);Prostratehair Kusnezoff Monkshood ■

5339 Aconitum kusnezoffii Rchb. var. gibbiferum (Rchb.) Regel;宽裂北乌头;Broadlobed Kusnezoff Monkshood ■

5340 Aconitum kusnezoffii Rchb. var. pilosum Y. Z. Zhao;疏毛草乌头■

5341 Aconitum kusnezoffii Rchb. var. wulingense (Nakai) W. T. Wang = Aconitum paniculigerum Nakai var. wulingense (Nakai) W. T. Wang ■

5342 Aconitum laevicaule W. T. Wang;光茎乌头■

5343 Aconitum laevicaule W. T. Wang = Aconitum rockii H. R. Fletcher et Lauener var. fengii (W. T. Wang) W. T. Wang ■

5344 Aconitum lamarckii Rchb. = Aconitum vulparia Rchb. ■☆

5345 Aconitum lasianthum Simonk.;毛花乌头■☆

5346 Aconitum lasiocarpum Rchb.;毛果乌头■☆

5347 Aconitum lasiostomum Rchb.;毛喉乌头;Hairystomate Monkshood ■☆

5348 Aconitum lauenerianum H. R. Fletcher = Aconitum brevicalcaratum (Finet et Gagnep.) Diels var. lauenerianum (H. R. Fletcher) W. T. Wang ■

5349 Aconitum lauenerianum H. R. Fletcher = Aconitum brevicalcaratum (Finet et Gagnep.) Diels ■

5350 Aconitum laxiflorum Schleich.;疏花乌头■☆

5351 Aconitum legendrei Hand. -Mazz.;冕宁乌头;Legendre Monkshood,Mianning Monkshood ■

5352 Aconitum leiostachyum W. T. Wang;光序乌头;Glabrousinflorescence Monkshood,W. China Monkshood ■

5353 Aconitum leiwuqiense W. T. Wang;类乌齐乌头;Leiwuqi Monkshood ■

5354 Aconitum lengyelii Gáyer;伦杰尔乌头■☆

5355 Aconitum lengyelii Gáyer nothosubsp. walasii Mitka;瓦拉斯乌头■☆

5356 Aconitum leptanthum Rchb.;细花乌头■☆

5357 Aconitum leptophyllum Rchb. ex Gáyer;线叶乌头■☆

5358 Aconitum leucanthum Rchb.;白花乌头■☆

5359 Aconitum leucostomum Vorosch.;白喉乌头(麻布七);Whitethroat Monkshood ■

5360 Aconitum leucostomum Vorosch. var. hopeiense W. T. Wang;河北白喉乌头;Hebei Whitethroat Monkshood ■

5361 Aconitum lhasaense Lauener = Aconitum kongboense Lauener ■

5362 Aconitum liangshanicum W. T. Wang;凉山乌头(草乌,雪乌);Liangshan Monkshood ■

5363 Aconitum liaotungense Nakai;辽东乌头;Liaodong Monkshood ■

5364 Aconitum liaotungense Nakai = Aconitum carmichaelii Debeaux var. truppelianum (Ulbr.) W. T. Wang et P. K. Hsiao ■

5365 Aconitum lihsienense W. T. Wang;理县乌头;Lihsien Monkshood,Lixian Monkshood ■

5366 Aconitum likiangense F. H. Chen et S. Liu = Aconitum forrestii Stapf ■

5367 Aconitum liljestrandii Hand. -Mazz.;贡嘎乌头;Gonggashan Monkshood,Konka Monkshood ■

5368 Aconitum liljestrandii Hand. -Mazz. var. falcatum W. T. Wang;马尔康乌头;Falcate Gonggashan Monkshood,Liljestrand Monkshood ■

5369 Aconitum liljestrandii Hand. -Mazz. var. fangianum (W. T. Wang) Y. Luo et Q. E. Yang;方氏乌头■

5370 Aconitum lioui W. T. Wang;秦岭乌头(刘氏乌头);Liou Monkshood,Qinling Monkshood ■

5371 Aconitum lobulatum W. T. Wang;浅裂乌头;Lobed Monkshood ■

5372 Aconitum loczyanum Rapaics;丽人草■☆

5373 Aconitum lonchodontum Hand. -Mazz.;长齿乌头;Longtooth Monkshood ■

5374 Aconitum longe-cassidatum Nakai;高帽乌头(长盔乌头);Tallhelmet Monkshood ■

5375 Aconitum longilobum W. T. Wang;长裂乌头;Longbreach Monkshood,Longilobate Monkshood ■

5376 Aconitum longipedicellatum Lauener;长梗乌头;Longipedicel Monkshood,Longipedicellate Monkshood ■

5377 Aconitum longipetiolatum Lauener;长柄乌头;Longipetiolate Monkshood,Longstalk Monkshood ■

5378 Aconitum longiramosum W. T. Wang;长枝乌头;Longbranch Monkshood,Longshoot Monkshood ■

5379 Aconitum longtouense T. L. Ming = Aconitum franchetii Finet et Gagnep. ■

5380 Aconitum longtouense T. L. Ming = Aconitum georgei Comber ■

5381 Aconitum lucidusculam Nakai;光泽乌头■☆

5382 Aconitum ludlowii Exell;江孜乌头(雪上一枝蒿);Jiangzi Monkshood,Ludlow Monkshood ■

5383 Aconitum luningense W. T. Wang;芦宁乌头;Luning Monkshood ■

5384 Aconitum luridum Hook. f. et Thomson = Aconitum novoluridum Munz ■

5385 Aconitum lushanense Migo = Aconitum carmichaelii Debeaux ■

5386 Aconitum luteum H. Lév. et Vaniot = Aconitum barbatum Patrin ex Pers. ■

5387 Aconitum lycoctonifolium W. T. Wang et L. Q. Li;牛扁叶乌头;Lycoctonifoliate Monkshood ■

5388 Aconitum lycoctonum Besser = Aconitum orientale Mill. ■☆

5389 Aconitum lycoctonum L.;欧美乌头;Badger's Bane,Badger's-bane, Wolfsbane ■☆

5390 Aconitum lycoctonum L. = Aconitum vulparia Rchb. ■☆

5391 Aconitum lycoctonum L. subsp. moldavicum (Hacq.) Jalas = Aconitum lycoctonum L. ■☆

5392 Aconitum lycoctonum L. subsp. neapolitanum (Ten.) Nyman = Aconitum vulparia Rchb. subsp. neapolitanum (Ten.) Munoz Garm. ■☆

5393 Aconitum lycoctonum L. var. atlanticum (Coss.) Batt. et Trab. = Aconitum vulparia Rchb. subsp. neapolitanum (Ten.) Munoz Garm. ■☆

5394 Aconitum lycoctonum L. var. barbatum (Pers.) Finet et Gagnep. = Aconitum barbatum Patrin ex Pers. ■

5395 Aconitum lycoctonum L. var. brevicalcaratum Finet et Gagnep. = Aconitum brevicalcaratum (Finet et Gagnep.) Diels ■

5396 Aconitum lycoctonum L. var. brevicalcaratum Finet et Gagnep. f. bracteatum Finet et Gagnep. = Aconitum brevicalcaratum (Finet et Gagnep.) Diels var. parviflorum F. H. Chen et S. Liu ■

5397 Aconitum lycoctonum L. var. brevicalcaratum Finet et Gagnep. f. bracteatum Finet et Gagnep. = Aconitum brevicalcaratum (Finet et Gagnep.) Diels ■

5398 Aconitum lycoctonum L. var. circinatum H. Lév. = Aconitum scaposum Franch. var. hupehanum Rapaics ■

5399 Aconitum lycoctonum L. var. efoliatum Rapaics = Aconitum scaposum Franch. ■

5400 Aconitum lycoctonum L. var. ranunculoides Finet et Gagnep. = Aconitum scaposum Franch. ■

5401 Aconitum lycoctonum L. var. rerayense Litard. et Maire = Aconitum vulparia Rchb. subsp. neapolitanum (Ten.) Munoz Garm. ■☆

5402 Aconitum lycoctonum L. var. vulparium (Rchb.) Regel = Aconitum brevicalcaratum (Finet et Gagnep.) Diels ■

5403 Aconitum macrorhynchum Turcz.; 大嘴乌头(细叶乌头); Bigbeak Monkshood ■

5404 Aconitum macrorhynchum Turcz. f. tenuissimum (Nakai et Kitag.) S. H. Li et Y. Huei Huang; 伷枝乌头; Tenuous Bigbeak Monkshood ■

5405 Aconitum macrorhynchum Turcz. f. tenuissimum (Nakai et Kitag.) S. H. Li et Y. Huei Huang = Aconitum macrorhynchum Turcz. ■

5406 Aconitum macrorhynchum Turcz. var. octocarpum P. K. Chang et S. H. Li et Y. Huei Huang = Aconitum macrorhynchum Turcz. ■

5407 Aconitum macrorhynchum Turcz. var. tenuissimum (Nakai et Kitag.) S. H. Li et Y. Huei Huang = Aconitum macrorhynchum Turcz. ■

5408 Aconitum macrorhynchum Turcz. var. viviparum P. K. Chang et B. Y. Wang = Aconitum macrorhynchum Turcz. f. tenuissimum (Nakai et Kitag.) S. H. Li et Y. Huei Huang ■

5409 Aconitum magnibracteolatum W. T. Wang; 巨苞乌头; Giantbract Monkshood ■

5410 Aconitum mairei H. Lév. = Aconitum vilmorinianum Kom. ■

5411 Aconitum manshuricum Nakai = Aconitum jaluense Kom. var. glabrescens Nakai ■

5412 Aconitum maowenense W. T. Wang; 茂汶乌头; Maowen Monkshood ■

5413 Aconitum maximum Pall. ex DC.; 大乌头(马氏乌头); Big Monkshood ■☆

5414 Aconitum micranthum Nakai; 小花嘴乌头■☆

5415 Aconitum microphyllum Gaudin ex Steud.; 小叶嘴乌头■☆

5416 Aconitum microstachyum Rchb.; 小嘴乌头■☆

5417 Aconitum milinense W. T. Wang; 米林乌头; Milin Monkshood ■

5418 Aconitum mitakense Nakai; 御岳乌头■☆

5419 Aconitum moldavicum Hacq. ex Rchb. f. pilocarpum (W. T. Wang) Tamura et Lauener = Aconitum sinomontanum Nakai var. pilocarpum W. T. Wang ■

5420 Aconitum moldavicum Hacq. ex Rchb. var. sinomontanum (Nakai) Tamura et Lauener = Aconitum sinomontanum Nakai ■

5421 Aconitum monanthum Nakai; 高山乌头; Alpine Monkshood ■

5422 Aconitum monticola Steinb.; 山地乌头; Hill Monkshood, Montane Monkshood ■

5423 Aconitum multifidum Royle = Aconitum violaceum Jacquem. ex Stapf ■☆

5424 Aconitum nagarum Stapf; 保山乌头(保山附片, 草乌, 兰花草乌, 七星草乌, 山乌头, 水乌头, 小黑牛); Baoshan Monkshood, Paoshan Monkshood ■

5425 Aconitum nagarum Stapf f. dielsianum (Airy Shaw) W. T. Wang = Aconitum nagarum Stapf var. heterotrichum H. R. Fletcher et Lauener ■

5426 Aconitum nagarum Stapf f. ecalcaratum (Airy Shaw) W. T. Wang; 无距保山乌头; Spurless Baoshan Monkshood, Spurless Monkshood, Spurless Paoshan Monkshood ■

5427 Aconitum nagarum Stapf f. ecalcaratum (Airy Shaw) W. T. Wang = Aconitum nagarum Stapf ■

5428 Aconitum nagarum Stapf f. leiocarpum (W. T. Wang) W. T. Wang = Aconitum nagarum Stapf var. heterotrichum H. R. Fletcher et Lauener ■

5429 Aconitum nagarum Stapf var. acaule (Finet et Gagnep.) Q. E. Yang; 小白撑(厚叶乌头, 黄蜡一枝蒿, 泡叶乌头, 雪上一枝蒿, 皱叶乌头); Thickleaf Monkshood, Unequalhair Baoshan Monkshood, Unequalhair Paoshan Monkshood ■

5430 Aconitum nagarum Stapf var. acaule (Finet et Gagnep.) Q. E. Yang = Aconitum duclouxii H. Lév. ■

5431 Aconitum nagarum Stapf var. ecalcaratum Airy Shaw = Aconitum nagarum Stapf ■

5432 Aconitum nagarum Stapf var. heterotrichum H. R. Fletcher et Lauener f. leiocarpum (W. T. Wang) W. T. Wang; 光果小白撑(光果白弩箭药, 树乌, 小黑牛) ■

5433 Aconitum nagarum Stapf var. heterotrichum H. R. Fletcher et Lauener f. dielsianum Airy Shaw = Aconitum nagarum Stapf var. heterotrichum H. R. Fletcher et Lauener ■

5434 Aconitum nagarum Stapf var. heterotrichum H. R. Fletcher et Lauener f. dielsianum (Airy Shaw) W. T. Wang = Aconitum nagarum Stapf var. heterotrichum H. R. Fletcher et Lauener ■

5435 Aconitum nagarum Stapf var. heterotrichum H. R. Fletcher et Lauener f. leiocarpum (W. T. Wang) W. T. Wang = Aconitum nagarum Stapf var. heterotrichum H. R. Fletcher et Lauener ■

5436 Aconitum nagarum Stapf var. heterotrichum H. R. Fletcher et Lauener; 无距小白撑(小黑牛) ■

5437 Aconitum nagarum Stapf var. heterotrichum H. R. Fletcher et Lauener = Aconitum nagarum Stapf var. acaule (Finet et Gagnep.) Q. E. Yang ■

5438 Aconitum nagarum Stapf var. lasiandrum W. T. Wang; 宣威乌头(草乌, 宣威一枝蒿, 雪上一枝蒿, 一枝蒿); Haoryanther Baoshan Monkshood, Haoryanther Paoshan Monkshood ■

5439 Aconitum nakaoi Tamura = Aconitum gammiei Stapf ■

5440 Aconitum namlaense W. T. Wang; 纳木拉乌头; Namula Monkshood ■

5441 Aconitum napelloides Hand.-Mazz. = Aconitum sinonapelloides W. T. Wang ■

5442 Aconitum napellus L.;欧洲乌头(堵喇,附子,萝卜状乌头,欧乌头,舟形乌头); Aaron's Beard, Aconite, Aconite Monkshood, Aconite Monks-hood, Adam-and-eve, Bear's Foot, Bee-in-a-bush, Bird of Paradise, Blue Peter, Blue Rocket, Bluebottle, Boots-and-shoes, Cat's Tail, Cat's Tails, Chariot-and-horses, Coach-and-horses, Cuckold's Cap, Cuckoo's Cap, Cuekold's Cap, Devil's Nightcap, Dobbin-in-the-ark, Dove Flower, Doves-in-the-ark, Drawn By Doves, European Monkshood, Face-in-hood, Fox's Mouth, Friar's Cap, Garden Aconite, Garden Monkshood, Grandmother's Bonnet, Grandmother's Bonnets, Grandmother's Nightcap, Granny-jump-out-of-bed, Granny's Bonnet, Granny's Bonnets, Granny's Nightcap, Granny's Shoes, Granny's Slipper, Granny's Slippers, Helmet Flower, Helmet-flower, Jacob's Chariot, Jacob's Ladder, Ladies-in-a-ship, Lady Dove And Her Coach And Pair, Lady Lavinia's Dove Carriage, Lady's Slipper, Libbardine, Libbard's Bane, Luckie's Mutch, Monk's Hood, Monkeys Hood, Monk's Cowl, Monkshood, Monk's-hood, Mother's Nightcap, Noah's Ark, Official Monkshead, Officinal Acotite, Old Wife Hood, Old Wives' Mutch, Old Wives' Mutches, Old Woman's Nightcap, Parson-in-the-pulpit, Policeman's Helmet, Poliman's Helmet, Priest's Pintle, Soldier's Cap, True Monkshood, Turk's Cap, Turtle Doves, Venus' Chariot, Venus'-chariot, Wolf's Bane ■☆

5443 Aconitum napellus L. var. acaule Finet et Gagnep. = Aconitum duclouxii H. Lév. ■

5444 Aconitum napellus L. var. acaule Finet et Gagnep. = Aconitum nagarum Stapf var. acaule (Finet et Gagnep.) Q. E. Yang ■

5445 Aconitum napellus L. var. polyanthum Finet et Gagnep. = Aconitum polyanthum (Finet et Gagnep.) Hand. -Mazz. ■

5446 Aconitum napellus L. var. refractum Finet et Gagnep. = Aconitum refractum (Finet et Gagnep.) Hand. -Mazz. ■

5447 Aconitum napellus L. var. sessiliflorum Finet et Gagnep. = Aconitum sessiliflorum (Finet et Gagnep.) Hand. -Mazz. ■

5448 Aconitum napellus L. var. turkestanicum B. Fedtsch. = Aconitum karakolicum Rapaics ■

5449 Aconitum napiforme H. Lév. et Vaniot = Aconitum japonicum Thunb. subsp. napiforme (H. Lév. et Vaniot) Kadota ■

5450 Aconitum napiforme H. Lév. et Vaniot var. albiflorum Y. N. Lee;白花萝卜乌头■

5451 Aconitum nasutum Fisch. ex Rchb.;鼻乌头■☆

5452 Aconitum naviculare (Brühl) Stapf;船盔乌头(船形乌头,雪乌);Scaphoidhelmet Monkshood ■

5453 Aconitum nemorum Popov;林地乌头;Woodland Monkshood ■

5454 Aconitum nielamuense W. T. Wang;聂拉木乌头;Nielamu Monkshood ■

5455 Aconitum ningwuense W. T. Wang;宁武乌头;Ningwu Monkshood ■

5456 Aconitum noveboracense A. Gray ex Coville = Aconitum columbianum Nutt. ■☆

5457 Aconitum noveboracense A. Gray ex Coville var. quasiciliatum Fassett = Aconitum columbianum Nutt. ■☆

5458 Aconitum novoluridum Munz;展喙乌头;Patentbeak Monkshood ■

5459 Aconitum nutantiflorum C. C. Chang ex W. T. Wang;垂花乌头;Nutantflower Monkshood, Nutantiflorous Monkshood ■

5460 Aconitum ochranthum C. A. Mey. = Aconitum barbatum Patrin ex Pers. var. puberulum Ledeb. ■

5461 Aconitum orientale Mill.;俄罗斯乌头;Russian Aconite ■☆

5462 Aconitum ouvrardianum Hand. -Mazz.;德钦乌头;Dechen Monkshood, Deqin Monkshood ■

5463 Aconitum ouvrardianum Hand. -Mazz. var. acutiusculum (H. R. Fletcher et Lauener) Q. E. Yang et Y. Luo = Aconitum acutiusculum H. R. Fletcher et Lauener ■

5464 Aconitum ouvrardianum Hand. -Mazz. var. pilopes W. T. Wang et L. Q. Li;毛爪德钦乌头;Pilose Deqin Monkshood ■

5465 Aconitum ouvrardianum Hand. -Mazz. var. pilopes W. T. Wang et L. Q. Li = Aconitum ouvrardianum Hand. -Mazz. ■

5466 Aconitum paishanense Kitag.;草地乌头;Grassland Monkshood, Meadow Monkshood ■

5467 Aconitum palmatum D. Don;掌叶乌头■☆

5468 Aconitum paniculatum Lam.;欧洲圆锥乌头;Panicled Monkshood, Panicled Monk's-hood ■☆

5469 Aconitum paniculigerum Nakai;圆锥乌头(草乌);Panicled Monkshood ■

5470 Aconitum paniculigerum Nakai var. leiocarpum Nakai ex Kitag. = Aconitum paniculigerum Nakai ■

5471 Aconitum paniculigerum Nakai var. leiocarpum Nakai ex Kitag. f. glabrescens Nakai = Aconitum paniculigerum Nakai ■

5472 Aconitum paniculigerum Nakai var. wulingense (Nakai) W. T. Wang;疏毛圆锥乌头(雾灵乌头);Wuling Monkshood, Wuling Panicled Monkshood ■

5473 Aconitum paniculigerum Nakai var. wulingense (Nakai) W. T. Wang = Aconitum wulingense Nakai ■

5474 Aconitum parabrachypodum Lauener = Aconitum gammiei Stapf ■

5475 Aconitum paradoxum Rchb.;奇异乌头■☆

5476 Aconitum parcifolium Q. E. Yang et Z. D. Fang;疏叶乌头;Parcifoliate Monkshood ■

5477 Aconitum parvifolium Host;小叶乌头■☆

5478 Aconitum pauciflorum Bertol.;少花乌头■☆

5479 Aconitum pekinense Vorosch. = Aconitum barbatum Patrin ex Pers. var. puberulum Ledeb. ■

5480 Aconitum pendulicarpum C. C. Chang ex W. T. Wang;垂果乌头(淡黄乌头);Droopfruit Monkshood, Pendulicarpous Monkshood ■

5481 Aconitum pendulicarpum C. C. Chang ex W. T. Wang var. circinatum W. T. Wang;长距垂果乌头;Longspur Droopfruit Monkshood ■

5482 Aconitum pendulicarpum C. C. Chang ex W. T. Wang var. circinatum W. T. Wang = Aconitum pendulicarpum C. C. Chang ex W. T. Wang ■

5483 Aconitum pendulum Busch;铁棒锤(八百棒,三转半,铁牛七,雪上一枝蒿,一枝箭);Pendulous Monkshood ■

5484 Aconitum phyllostegium Hand. -Mazz.;木里乌头;Muli Monkshood ■

5485 Aconitum phyllostegium Hand. -Mazz. var. pilosum H. R. Fletcher et Lauener;伏毛木里乌头;Pillose Muli Monkshood ■

5486 Aconitum piepunense Hand. -Mazz.;中甸乌头;Zhongdian Monkshood ■

5487 Aconitum piepunense Hand. -Mazz. var. pilosum Comber;疏毛中甸乌头;Pilous Zhongdian Monkshood ■

5488 Aconitum pilopetalum W. T. Wang et L. Q. Li;毛瓣乌头;Hairpetal Monkshood ■

5489 Aconitum platysepalum Greene;宽瓣乌头■☆

5490 Aconitum polyanthum (Finet et Gagnep.) Hand. -Mazz.;多花乌头(独花乌头,多头乌头);Manyflower Monkshood, Multicapital Monkshood ■

5491 Aconitum polyanthum (Finet et Gagnep.) Hand. -Mazz. var. puberulum W. T. Wang;毛萼多花乌头;Hairycalyx Manyflower

Monkshood ■
5492 Aconitum polycarpum C. C. Chang ex W. T. Wang；多果乌头；Manyfruit Monkshood, Multicarpous Monkshood ■
5493 Aconitum polyschistum Hand. -Mazz.；多裂乌头；Manysplitted Monkshood, Multifid Monkshood ■
5494 Aconitum pomeense W. T. Wang；波密乌头；Bomi Monkshood, Pome Monkshood ■
5495 Aconitum popovii Steinb. et Schischk. ex Sipliv.；波氏乌头■☆
5496 Aconitum potaninii Kom. et Hand. -Mazz.；密花乌头；Denseflower Monkshood, Potanin Monkshood ■
5497 Aconitum prominens Lauener；露瓣乌头；Prominent Monkshood ■
5498 Aconitum pseudobrunneum W. T. Wang；小花乌头；Smallflower Monkshood ■
5499 Aconitum pseudodivaricatum W. T. Wang；全裂乌头；Falsedivaricate Monkshood, Trisect Monkshood ■
5500 Aconitum pseudogeniculatum W. T. Wang；拟膝瓣乌头；Falsegrniculate Monkshood, Geniculatelike Monkshood ■
5501 Aconitum pseudogeniculatum W. T. Wang var. pubipes W. T. Wang；黄毛梗乌头；Yellow Falsegrniculate Monkshood ■
5502 Aconitum pseudohuiliense C. C. Chang ex W. T. Wang；雷波乌头；Leibo Monkshood ■
5503 Aconitum pseudokongboense W. T. Wang et L. Q. Li；拟工布乌头；Gongbu-like Monkshood ■
5504 Aconitum pseudostapfianum W. T. Wang；拟玉龙乌头(黑心解)；False Stapf Monkshood, Stapflike Monkshood ■
5505 Aconitum pseudostapfianum W. T. Wang = Aconitum stapfianum Hand. -Mazz. ■
5506 Aconitum pubiceps (Rupr.) Trautv.；毛头乌头■☆
5507 Aconitum pukeense W. T. Wang；普格乌头；Puge Monkshood, Pukeen Monkshood ■
5508 Aconitum pukeense W. T. Wang = Aconitum geniculatum H. R. Fletcher et Lauener ■
5509 Aconitum pukeense W. T. Wang var. brevipes W. T. Wang；短梗普格乌头；Short-stalked Puge Monkshood ■
5510 Aconitum pulchellum Hand. -Mazz.；美丽乌头；Lovely Monkshood ■
5511 Aconitum pulchellum Hand. -Mazz. var. hispidum Lauener；毛瓣美丽乌头；Hairypetal Lovely Monkshood ■
5512 Aconitum pulchellum Hand. -Mazz. var. racemosum W. T. Wang；长序美丽乌头；Longracemose Lovely Monkshood ■
5513 Aconitum pulchellum Hand. -Mazz. var. racemosum W. T. Wang = Aconitum pulchellum Hand. -Mazz. ■
5514 Aconitum pulcherrimum Nakai = Aconitum kusnezoffii Rchb. ■
5515 Aconitum pulcherrimum Nakai var. dissectum (Regel) Nakai = Aconitum kusnezoffii Rchb. ■
5516 Aconitum pulcherrimum Nakai var. tenuisectum (Regel) Nakai = Aconitum kusnezoffii Rchb. ■
5517 Aconitum pycnanthum W. T. Wang；密序乌头；Denseraceme Monkshood ■
5518 Aconitum pyrenaicum L. = Aconitum barbatum Patrin ex Pers. ■
5519 Aconitum qinghaiense Kadota；青海乌头；Qinghai Monkshood ■
5520 Aconitum racemulosum Franch.；岩乌头(雪上一枝蒿，崖乌头草，岩乌，岩乌子，一枝蒿)；Cliff Monkshood ■
5521 Aconitum racemulosum Franch. var. austrokoreense (Koidz.) Y. N. Lee；南韩乌头■☆
5522 Aconitum racemulosum Franch. var. grandibracteolatum W. T. Wang；巨苞岩乌头；Grandibracted Cliff Monkshood ■
5523 Aconitum racemulosum Franch. var. pengzhouense W. J. Zhang et G. H. Chen；彭州乌头；Pengzhou Monkshood ■
5524 Aconitum racemulosum Franch. var. pengzhouense W. J. Zhang et G. H. Chen = Aconitum racemulosum Franch. ■
5525 Aconitum raddeanum Regel；大苞乌头；Bigbract Monkshood ■
5526 Aconitum ramulosum W. T. Wang；多枝乌头；Manybranch Monkshood, Ramose Monkshood ■
5527 Aconitum ramulosum W. T. Wang = Aconitum piepunense Hand. -Mazz. ■
5528 Aconitum ranunculoides Turcz.；毛茛叶乌头；Ranunculifoliate Monkshood ■
5529 Aconitum refracticarpum C. C. Chang ex W. T. Wang；弯果乌头；Refracted Monkshood, Refractedfruit Monkshood ■
5530 Aconitum refracticarpum C. C. Chang ex W. T. Wang = Aconitum tsaii W. T. Wang ■
5531 Aconitum refractum (Finet et Gagnep.) Hand. -Mazz.；狭裂乌头；Narrowlobe Monkshood ■
5532 Aconitum rhombifolium F. H. Chen；菱叶乌头；Rhombiclesf Monkshood, Rhomboidleaf Monkshood ■
5533 Aconitum rhombifolium F. H. Chen var. leiocarpum W. T. Wang；光果菱叶乌头；Smoothfruit Rhomboidleaf Monkshood ■
5534 Aconitum richardsonianum Lauener；直序乌头；Richardson Monkshood ■
5535 Aconitum richardsonianum Lauener var. crispulum W. T. Wang = Aconitum richardsonianum Lauener var. pseudosessiliflorum (Lauener) W. T. Wang ■
5536 Aconitum richardsonianum Lauener var. pseudosessiliflorum (Lauener) W. T. Wang；伏毛直序乌头；Falsesessileflower Monkshood ■
5537 Aconitum rilongense Kadota；邛崃山乌头■
5538 Aconitum rockii H. R. Fletcher et Lauener；拟康定乌头；Rock Monkshood ■
5539 Aconitum rockii H. R. Fletcher et Lauener var. fengii (W. T. Wang) W. T. Wang；石膏山乌头(冯氏乌头)；Feng Monkshood, Feng Rock Monkshood ■
5540 Aconitum rockii H. R. Fletcher et Lauener var. ramosum W. T. Wang = Aconitum piepunense Hand. -Mazz. var. pilosum Comber ■
5541 Aconitum rockii H. R. Fletcher et Lauener var. ramosum W. T. Wang = Aconitum rockii H. R. Fletcher et Lauener ■
5542 Aconitum rongchuense Lauener = Aconitum kongboense Lauener var. villosum W. T. Wang ■
5543 Aconitum rotundifolium Kar. et Kir.；圆叶乌头(乌头)；Rotundleaf Monkshood ■
5544 Aconitum rotundifolium Kar. et Kir. var. sessiliflorum (Finet et Gagnep.) Rapaics = Aconitum sessiliflorum (Finet et Gagnep.) Hand. -Mazz. ■
5545 Aconitum rotundifolium Kar. et Kir. var. tanguticum Maxim. = Aconitum tanguticum (Maxim.) Stapf ■
5546 Aconitum sachalinense F. Schmidt；库页乌头；Sachalin Monkshood ■☆
5547 Aconitum sachalinense F. Schmidt subsp. yezoense (Nakai) Kadota = Aconitum yezoense Nakai ■☆
5548 Aconitum saposhnikovii B. Fedtsch.；萨波乌头■☆
5549 Aconitum scaposum Franch.；花葶乌头(活血莲，土莎莲，一口血)；Scape Monkshood ■
5550 Aconitum scaposum Franch. var. chloranthum (Hand. -Mazz.) Lauener et Tamura = Aconitum scaposum Franch. var. hupehanum Rapaics ■

5551 Aconitum scaposum Franch. var. efolianum Rapaics = Aconitum scaposum Franch. ■

5552 Aconitum scaposum Franch. var. hupehanum Rapaics;等叶花葶乌头(碎骨还阳);Hupeh Monkshood ■

5553 Aconitum scaposum Franch. var. patentipilum W. T. Wang = Aconitum scaposum Franch. ■

5554 Aconitum scaposum Franch. var. pseudovaginatum Rapaics = Aconitum scaposum Franch. var. vaginatum (Pritz.) Rapaics ■

5555 Aconitum scaposum Franch. var. pyramidale Franch. = Aconitum scaposum Franch. ■

5556 Aconitum scaposum Franch. var. vaginatum (Pritz.) Rapaics;聚叶花葶乌头(独儿七,活血连,活血莲,墨七,鞘柄乌头,笄尖七,土沙莲,土莎连,血三七);Sheathed Monkshood ■

5557 Aconitum sczukini Turcz. subsp. subalpinum (Baran.) Vorosch.;亚高山宽叶蔓乌头■☆

5558 Aconitum sczukinii Turcz.;宽叶蔓乌头(藤乌头);Sczukin Monkshood ■

5559 Aconitum sczukinii Turcz. var. hemsleyanum Rapaics = Aconitum hemsleyanum E. Pritz. ex Diels ■

5560 Aconitum sczukinii Turcz. var. pauciflorum Rapaics = Aconitum racemulosum Franch. ■

5561 Aconitum secundiflorum W. T. Wang;侧花乌头;Secundiflower Monkshood ■

5562 Aconitum semigaleatum Pall. ex Rchb. var. ichangense Finet et Gagnep. = Aconitum ichangense (Finet et Gagnep.) Hand. -Mazz. ■

5563 Aconitum septentrionale Koelle;北方乌头(高乌头,紫花高乌头);Purpleflower High Monkshood ■

5564 Aconitum sessiliflorum (Finet et Gagnep.) Hand. -Mazz.;缩梗乌头(铁棒七);Sessileflower Monkshood,Subsessile Monkshood ■

5565 Aconitum setosum Grint.;刚毛乌头■☆

5566 Aconitum shensiense W. T. Wang;陕西乌头;Shaanxi Monkshood, Shensi Monkshood ■

5567 Aconitum sherriffii Lauener;谢里夫乌头■☆

5568 Aconitum shimianense W. T. Wang;石棉乌头;Shimian Monkshood ■

5569 Aconitum sibiricum Poir. = Aconitum barbatum Patrin ex Pers. var. hispidum DC. ■

5570 Aconitum sichotense Kom.;锡浩特乌头■☆

5571 Aconitum sikangense Hand. -Mazz. = Aconitum tatsienense Finet et Gagnep. ■

5572 Aconitum sinchiangense W. T. Wang;新疆乌头;Sinchiang Monkshood,Xinjiang Monkshood ■

5573 Aconitum sinoaxillare W. T. Wang;腋花乌头;Axillaryflower Monkshood ■

5574 Aconitum sinomontanum Nakai;高乌头(背网子,辫子七,穿心莲,穿心莲牛扁,穿心莲乌头,花花七,九连环,口袋七,龙骨七,龙蹄叶,龙膝,麻布袋,麻布口袋,麻布七,麻布芪,破骨七,七连环,碎骨还阳,蓑衣七,通天袋,统天袋,统仙袋,网子七);Tall Monkshood ■

5575 Aconitum sinomontanum Nakai var. angustius W. T. Wang;狭盔高乌头(裂叶牛扁,野草乌);Narrowhelmet Tall Monkshood ■

5576 Aconitum sinomontanum Nakai var. pilocarpum W. T. Wang;毛果高乌头;Hairyfruit Tall Monkshood ■

5577 Aconitum sinonapelloides W. T. Wang;拟缺刻乌头;False Incised Monkshood ■

5578 Aconitum sinonapelloides W. T. Wang var. subulatum W. T. Wang et P. K. Hsiao;钻苞拟缺刻乌头;Subulate False Incised Monkshood ■

5579 Aconitum sinonapelloides W. T. Wang var. weisiense W. T. Wang;展毛拟缺刻乌头;Weisi False Incised Monkshood, Weixi False Incised Monkshood ■

5580 Aconitum sinonapelloides W. T. Wang var. weisiense W. T. Wang = Aconitum ouvrardianum Hand. -Mazz. ■

5581 Aconitum smirnovii Steinb.;斯氏乌头(阿尔泰乌头);Altai Monkshood ■

5582 Aconitum smithii Hand. -Mazz.;山西乌头;Shansi Monkshood, Shanxi Monkshood ■

5583 Aconitum smithii Hand. -Mazz. var. tenuilobum W. T. Wang;狭裂山西乌头;Narrowlobed Shanxi Monkshood, Thinlobate Shanxi Monkshood ■

5584 Aconitum smithii Hand. -Mazz. var. tenuilobum W. T. Wang = Aconitum smithii Hand. -Mazz. ■

5585 Aconitum soongaricum (Regel) Stapf;准噶尔乌头(草乌,乌头);Dzungar Monkshood,Dzungaria Monkshood ■

5586 Aconitum soongaricum Stapf var. angustius W. T. Wang = Aconitum jeholense Thunb. var. angustius (W. T. Wang) Y. Z. Zhao ■

5587 Aconitum soongaricum Stapf var. jeholense (Nakai et Kitag.) W. T. Wang = Aconitum jeholense Nakai et Kitag. ■

5588 Aconitum soongaricum Stapf var. pubescens Steinb.;毛序准噶尔乌头;Pubescent Dzungaria Monkshood ■

5589 Aconitum souliei Finet et Gagnep.;茨开乌头;Cikai Monkshood, Soulie Monkshood ■

5590 Aconitum souliei Finet et Gagnep. var. glabrum Comber = Aconitum phyllostegium Hand. -Mazz. ■

5591 Aconitum souliei Finet et Gagnep. var. pumilum Finet et Gagnep. = Aconitum souliei Finet et Gagnep. ■

5592 Aconitum spathulatum W. T. Wang;匙苞乌头(匙形乌头);Spathulate Monkshood,Spoonbract Monkshood ■

5593 Aconitum spicatum Stapf;亚东乌头;Spiked Monkshood, Yadong Monkshood ■

5594 Aconitum spiripetalum Hand. -Mazz.;螺瓣乌头;Spiralpetal Monkshood,Spiripetalous Monkshood ■

5595 Aconitum sprengelii Rchb.;施普顿乌头■☆

5596 Aconitum squarrosum L. ex DC. = Aconitum barbatum Patrin ex Pers. ■

5597 Aconitum staintonii Lauener;斯坦顿乌头■☆

5598 Aconitum stapfianum Hand. -Mazz.;玉龙乌头(黑心解,藤子草乌);Stapf Monkshood ■

5599 Aconitum stapfianum Hand. -Mazz. var. pubipes W. T. Wang;毛梗玉龙乌头;Hairstalk Stapf Monkshood ■

5600 Aconitum stoerkianum Rchb.;斯托尔克乌头■☆

5601 Aconitum stoloniferum Vorosch.;匍匐乌头■☆

5602 Aconitum stramineiflorum C. C. Chang ex W. T. Wang;草黄乌头(草黄花乌头);Stramineous Monkshood ■

5603 Aconitum stramineiflorum C. C. Chang ex W. T. Wang = Aconitum rockii H. R. Fletcher et Lauener var. fengii (W. T. Wang) W. T. Wang ■

5604 Aconitum stylosoides W. T. Wang;拟显柱乌头;Distinctstylelike Monkshood,Styloselike Monkshood ■

5605 Aconitum stylosum Stapf;显柱乌头(草乌,大草乌);Distinctstyle Monkshood ■

5606 Aconitum stylosum Stapf f. albidum F. H. Chen et S. Liu = Aconitum stylosum Stapf ■

5607 Aconitum stylosum Stapf var. doshongense Lauener = Aconitum elliotii Lauener var. doshongense (Lauener) W. T. Wang ■

5608 Aconitum stylosum Stapf var. geniculatum H. R. Fletcher et Lauener;膝爪显柱乌头(膝距显柱乌头);Geniculateclaw Distinctstyle Monkshood ■

5609 Aconitum stylosum Stapf var. geniculatum H. R. Fletcher et Lauener = Aconitum stylosum Stapf ■

5610 Aconitum subalpinum A. I. Baranov = Aconitum paniculigerum Nakai ■

5611 Aconitum sukaczevii Steinb.;苏氏乌头■☆

5612 Aconitum sungpanense Hand.-Mazz.;松潘乌头(草乌,火烟子,火焰子,金牛七,蔓乌药,千锤打,藤乌,羊角七);Songpan Monkshood,Sungpan Monkshood ■

5613 Aconitum sungpanense Hand.-Mazz. var. leucanthum W. T. Wang;白花松潘乌头;Whiteflower Songpan Monkshood,Whiteflower Sungpan Monkshood ■

5614 Aconitum sungpanense Hand.-Mazz. var. villosum W. T. Wang;展毛松潘乌头■

5615 Aconitum szechenyianum Gáyer = Aconitum pendulum Busch ■

5616 Aconitum taipeicum Hand.-Mazz.;太白乌头(金牛七,千锤打,乌头);Taibai Monkshood,Taipai Monkshood ■

5617 Aconitum takahashii Kitag. = Aconitum carmichaelii Debeaux var. truppelianum (Ulbr.) W. T. Wang et P. K. Hsiao ■

5618 Aconitum talassicum Popov;塔拉乌头■☆

5619 Aconitum talassicum Popov var. villosulum W. T. Wang;伊犁乌头;Yili Monkshood ■

5620 Aconitum tangense Marquand et Airy Shaw;堆拉乌头;Duila Monkshood,Tang Monkshood ■

5621 Aconitum tanguticum (Maxim.) Stapf;甘青乌头(辣辣草,山附子,唐古特乌头,翁阿鲁,翁格尔,雪乌);Tangut Monkshood ■

5622 Aconitum tanguticum (Maxim.) Stapf f. viridulum W. T. Wang = Aconitum tanguticum (Maxim.) Stapf ■

5623 Aconitum tanguticum (Maxim.) Stapf var. trichocarpum Hand.-Mazz.;毛果甘青乌头(毛果唐古特乌头);Hairyfruit Tangut Monkshood ■

5624 Aconitum tanguticum (Maxim.) Stapf var. trichocarpum Hand.-Mazz. f. robustum (F. H. Chen et S. Liu) W. T. Wang;粗壮毛果甘青乌头■

5625 Aconitum taronense (Hand.-Mazz.) H. R. Fletcher et Lauener;独龙乌头;Dulong Monkshood,Taron Monkshood ■

5626 Aconitum tatsienense Finet et Gagnep.;康定乌头;Kangding Monkshood ■

5627 Aconitum tatsienense Finet et Gagnep. var. divaricatum (Finet et Gagnep.) W. T. Wang;展枝康定乌头;Divaricatebranch Kangding Monkshood ■

5628 Aconitum tenuicaule W. T. Wang;细茎乌头;Tenuicauliferous Monkshood,Tenuousstem Monkshood ■

5629 Aconitum tenuicaule W. T. Wang = Aconitum ouvrardianum Hand.-Mazz. ■

5630 Aconitum tenuissimum Nakai = Aconitum macrorhynchum Turcz. ■

5631 Aconitum tianschanicum (Adolf) Lipsch.;天山乌头■

5632 Aconitum tokii Nakai = Aconitum paniculigerum Nakai var. wulingense (Nakai) W. T. Wang ■

5633 Aconitum tongolense Ulbr.;东俄洛乌头(新都桥乌头);Hsintucho Monkshood,Xinduqiao Monkshood ■

5634 Aconitum tongolense Ulbr. var. patentipilum Q. E. Yang et Z. D. Fang = Aconitum ouvrardianum Hand.-Mazz. ■

5635 Aconitum transsectum Diels;直缘乌头(大草乌,黑草乌,小黑牛);Transsect Monkshood,Transsected Monkshood ■

5636 Aconitum tranzschelii Steinb.;特拉氏乌头■☆

5637 Aconitum tripartitum H. R. Fletcher et Lauener = Aconitum episcopale H. Lév. ■

5638 Aconitum triphylloides Nakai = Aconitum kusnezoffii Rchb. ■

5639 Aconitum triphyllum Nakai;三叶乌头■☆

5640 Aconitum triphyllum Nakai var. manshuricum Nakai = Aconitum jaluense Kom. ■

5641 Aconitum truppelianum (Ulbr.) Nakai = Aconitum carmichaelii Debeaux var. truppelianum (Ulbr.) W. T. Wang et P. K. Hsiao ■

5642 Aconitum tsaii W. T. Wang;碧江乌头(蔡氏乌头);Tsai Monkshood ■

5643 Aconitum tsaii W. T. Wang f. purpureum W. T. Wang;紫花碧江乌头;Purpleflower Tsai Monkshood ■

5644 Aconitum tsaii W. T. Wang f. purpureum W. T. Wang = Aconitum tsaii W. T. Wang ■

5645 Aconitum tsaii W. T. Wang var. geniculatum W. T. Wang = Aconitum tsaii W. T. Wang ■

5646 Aconitum tsaii W. T. Wang var. puberulum W. T. Wang;毛茎碧江乌头;Puberulous Tsai Monkshood ■

5647 Aconitum tsaii W. T. Wang var. puberulum W. T. Wang = Aconitum tsaii W. T. Wang ■

5648 Aconitum tsangpoense Lauener = Aconitum kongboense Lauener ■

5649 Aconitum tschangbaischanense S. H. Li et T. Huei Huang;长白乌头;Changbai Monkshood,Tschangbai Monkshood ■

5650 Aconitum tuguancunense Q. E. Yang;土官村乌头;Tuguancun Monkshood ■

5651 Aconitum tuguancunense Q. E. Yang = Aconitum pseudostapfianum W. T. Wang ■

5652 Aconitum turczaninowii Vorosch. = Aconitum jeholense Thunb. var. angustius (W. T. Wang) Y. Z. Zhao ■

5653 Aconitum umbrosum (Korsh.) Kom. = Aconitum paishanense Kitag. ■

5654 Aconitum uncinatum L.;钩状乌头;Monkshood,Wild Monkshood,Wild Monk's-hood ■☆

5655 Aconitum uncinatum L. subsp. noveboracense (A. Gray ex Coville) Hardin;新布洛斯乌头■☆

5656 Aconitum vaginatum E. Pritz. ex Diels = Aconitum scaposum Franch. var. vaginatum (Pritz.) Rapaics ■

5657 Aconitum vaginatum E. Pritz. ex Diels var. xanthanthum Hand.-Mazz. = Aconitum scaposum Franch. var. vaginatum (Pritz.) Rapaics ■

5658 Aconitum validinerve W. T. Wang;显脉乌头;Distinctvein Monkshood ■

5659 Aconitum variegatum L.;银色乌头(杂色乌头);Manchurian Monkshood,Manchurian Monk's-hood,Variegated Monkshood ■☆

5660 Aconitum variegatum L. = Aconitum novoluridum Munz ■

5661 Aconitum venatorium Diels = Aconitum nagarum Stapf ■

5662 Aconitum venatorium Diels var. ecalcaratum Airy Shaw = Aconitum nagarum Stapf ■

5663 Aconitum villosum Rchb.;白毛乌头;Whitevillous Monkshood ■

5664 Aconitum villosum Rchb. subsp. tschangbaischanense (S. X. Li et Y. H. Huang) S. X. Li = Aconitum tschangbaischanense S. H. Li et T. Huei Huang ■

5665 Aconitum villosum Rchb. var. amurense (Nakai) S. H. Li et Y. Huei Huang;缠绕白毛乌头;Amur Whitevillous Monkshood ■

5666 Aconitum villosum Rchb. var. daxinganlinense (Y. Z. Zhao) S. X. Li = Aconitum daxinganlingense Y. Z. Zhao ■

5667 Aconitum vilmorinianum Kom.;黄草乌(草乌,大草乌,滇草乌,黄乌头,昆明堵喇,昆明乌头);Vilmorin Monkshood ■
5668 Aconitum vilmorinianum Kom. var. altifidum W. T. Wang;深裂黄草乌(藤乌,西南乌头);Deeplobe Vilmorin Monkshood ■
5669 Aconitum vilmorinianum Kom. var. altifidum W. T. Wang = Aconitum episcopale H. Lév. ■
5670 Aconitum vilmorinianum Kom. var. patentipilum W. T. Wang;展毛黄草乌;Patenehairy Vilmorin Monkshood ■
5671 Aconitum violaceum Jacquem. ex Stapf;堇色乌头■☆
5672 Aconitum viridiflorum Lauener = Aconitum kongboense Lauener ■
5673 Aconitum virosum D. Don = Aconitum ferox Wall. ■☆
5674 Aconitum volubile Pall. ex Koelle;蔓乌头(鸡头草,细茎蔓乌头,狭叶蔓乌头,狭叶乌头);Twining Monkshood ■
5675 Aconitum volubile Pall. ex Koelle var. pubescens Regel;卷毛蔓乌头;Pubescent Twining Monkshood ■
5676 Aconitum vulparia Rchb.; 狼毒乌头; Badger's Bane, Badgersbane, Blue Rocket, Captain-over-the-garden, Chariot-and-horses, Cuckoo's Cap, Foxbane, Monkeys Hood, Noah's Ark, Parson-in-the-pulpit, Pope's Ode, Soldier's Cap, Turk's Cap, Wolf's Bane, Wolfsbane, Wolf's-bane, Wolfsbane Monkshood, Yellow Aconite ■☆
5677 Aconitum vulparia Rchb. subsp. neapolitanum (Ten.) Munoz Garm.;北非乌头■☆
5678 Aconitum wangii Q. E. Yang = Aconitum episcopale H. Lév. ■
5679 Aconitum wangyedianense Y. Z. Zhao;旺业甸乌头;Wangyedian Monkshood ■
5680 Aconitum wardii H. R. Fletcher et Lauener;滇川乌头(滇川牛扁);Ward Monkshood ■
5681 Aconitum wardii H. R. Fletcher et Lauener = Aconitum crassiflorum Hand. -Mazz. ■
5682 Aconitum wardii H. R. Fletcher et Lauener f. flavidum H. R. Fletcher et Lauener = Aconitum wardii H. R. Fletcher et Lauener ■
5683 Aconitum wardii H. R. Fletcher et Lauener var. hopeiense (W. T. Wang) Tamura et Lauener = Aconitum leucostomum Vorosch. var. hopeiense W. T. Wang ■
5684 Aconitum wardii H. R. Fletcher et Lauener var. trisectum W. T. Wang et L. Q. Li;全裂滇川乌头;Trisected Ward Monkshood ■
5685 Aconitum wardii H. R. Fletcher et Lauener var. trisectum W. T. Wang et L. Q. Li = Aconitum crassiflorum Hand. -Mazz. ■
5686 Aconitum weileri Gilli;韦勒乌头■☆
5687 Aconitum weixiense W. T. Wang;维西乌头;Weixi Monkshood ■
5688 Aconitum weixiense W. T. Wang = Aconitum hemsleyanum E. Pritz. ex Diels ■
5689 Aconitum willdenowii Rchb. ex Gáyer;威尔乌头■☆
5690 Aconitum williamsii Lauener;威廉斯乌头■☆
5691 Aconitum wilsonii Stapf ex Veitch = Aconitum carmichaelii Debeaux ■
5692 Aconitum winkleri Rapaics;温克勒乌头■☆
5693 Aconitum wolongense W. T. Wang;卧龙乌头■
5694 Aconitum wolongense W. T. Wang = Aconitum franchetii Finet et Gagnep. var. villosulum W. T. Wang ■
5695 Aconitum wolongense W. T. Wang = Aconitum tanguticum (Maxim.) Stapf var. trichocarpum Hand. -Mazz. ■
5696 Aconitum woroschilovii Luferov;沃氏乌头■☆
5697 Aconitum wuchagouense Y. Z. Zhao;阿尔山乌头(五叉沟乌头);Wuchagou Monkshood ■
5698 Aconitum wulfenianum Rchb. ex Steud.;武尔芬乌头■☆
5699 Aconitum wulingense Nakai = Aconitum paniculigerum Nakai var. wulingense (Nakai) W. T. Wang ■
5700 Aconitum xiangchengense W. T. Wang; 乡城乌头; Xiangcheng Monkshood ■
5701 Aconitum yachiangense W. T. Wang; 雅江乌头; Yachiang Monkshood, Yajiang Monkshood ■
5702 Aconitum yamamotoanum Ohwi;雪山乌头;Yamamoto Monkshood ■
5703 Aconitum yamatsutae Nakai = Aconitum kusnezoffii Rchb. ■
5704 Aconitum yangii W. T. Wang et L. Q. Li; 竞生乌头; Yang Monkshood ■
5705 Aconitum yangii W. T. Wang et L. Q. Li var. villosulum W. T. Wang et L. Q. Li;展毛竞生乌头;Villose Yang Monkshood ■
5706 Aconitum yanyuanense W. T. Wang et L. Q. Li; 盐源乌头; Yanyuan Monkshood ■
5707 Aconitum yezoense Nakai;虾夷乌头■☆
5708 Aconitum yinschanicum Y. Z. Zhao; 阴山乌头; Galeate Monkshood, Yinshan Monkshood ■
5709 Aconitum yunlingense Q. E. Yang et Z. D. Fang; 云岭乌头; Yunling Monkshood ■
5710 Aconitum zhaojiueense W. T. Wang et P. K. Hsiao;昭觉乌头; Zhaojue Monkshood ■
5711 Aconium Engl. = Aeonium Webb et Berthel. ●■☆
5712 Aconogonon (Meisn.) Rchb. = Aconogonum (Meisn.) Rchb. ■☆
5713 Aconogonon (Meisn.) Rchb. = Persicaria (L.) Mill. ■
5714 Aconogonon Rchb. = Persicaria (L.) Mill. ■
5715 Aconogonon ajanense (Regel et Tiling) H. Hara = Polygonum ajanense (Regel et Tiling) Grig. ■
5716 Aconogonon alpinum (All.) Schur = Polygonum alpinum All. ■
5717 Aconogonon angustifolium (Pall.) H. Hara = Polygonum angustifolium Pall. ■
5718 Aconogonon bucharicum (Grig.) Holub = Polygonum coriarium Grig. ■
5719 Aconogonon campanulatum (Hook. f.) H. Hara = Polygonum campanulatum Hook. f. ■
5720 Aconogonon campanulatum (Hook. f.) H. Hara var. fulvidum (Hook. f.) H. Hara = Polygonum campanulatum Hook. f. var. fulvidum Hook. f. ■
5721 Aconogonon campanulatum (Hook. f.) H. Hara var. oblongum (Meisn.) H. Hara = Polygonum campanulatum Hook. f. ■
5722 Aconogonon coriarium (Grig.) Soják = Polygonum coriarium Grig. ■
5723 Aconogonon coriarium (Grig.) Soják subsp. bucharicum (Grig.) Soják = Polygonum coriarium Grig. ■
5724 Aconogonon divaricatum (L.) Nakai = Polygonum divaricatum L. ■
5725 Aconogonon divaricatum (L.) Nakai ex T. Mori = Polygonum divaricatum L. ■
5726 Aconogonon hookeri (Meisn.) H. Hara = Polygonum hookeri Meisn. ■
5727 Aconogonon hypoleucum (Nakai ex Ohwi) Soják = Fallopia multiflora (Thunb.) Haraldson ■
5728 Aconogonon laxmannii (Lepech.) Á. Löve et D. Löve = Polygonum ochreatum L. ■
5729 Aconogonon lichiangense (W. W. Sm.) Soják = Polygonum lichiangense W. W. Sm. ■
5730 Aconogonon limosum (Kom.) H. Hara = Polygonum limosum Kom. ■
5731 Aconogonon molle (D. Don) H. Hara = Polygonum molle D. Don ●■
5732 Aconogonon molle (D. Don) H. Hara var. frondosum (Meisn.) H.

Hara = Polygonum molle D. Don var. frondosum (Meisn.) A. J. Li ●■
5733 Aconogonon molle (D. Don) H. Hara var. paniculatum (Blume) Yonek. et H. Ohashi = Polygonum molle D. Don var. frondosum (Meisn.) A. J. Li ●■
5734 Aconogonon molle (D. Don) H. Hara var. rude (Meisn.) H. Hara = Polygonum molle D. Don var. rude (Meisn.) A. J. Li ■
5735 Aconogonon nakaii (H. Hara) H. Hara = Polygonum nakaii (H. Hara) Ohwi ■☆
5736 Aconogonon ochreatum (L.) H. Hara = Polygonum ochreatum L. ■
5737 Aconogonon ocreatum (L.) H. Hara = Polygonum ochreatum L. ■
5738 Aconogonon ocreatum (L.) H. Hara var. laxmannii (Lepech.) Tzvelev = Polygonum ochreatum L. ■
5739 Aconogonon pamiricum (Korsh.) H. Hara = Polygonum sibiricum Laxm. var. thomsonii Meisn. ex Stewart ■
5740 Aconogonon paniculatum (Blume) Haraldson = Polygonum molle D. Don var. frondosum (Meisn.) A. J. Li ●■
5741 Aconogonon platyphyllum (S. X. Li et Y. L. Chang) Holub = Polygonum platyphyllum S. X. Li et Y. L. Chang ■
5742 Aconogonon polystachyum (Wall. ex Meisn.) Král = Persicaria wallichii Greuter et Burdet ■☆
5743 Aconogonon polystachyum (Wall. ex Meisn.) Král = Polygonum polystachyum Wall. ex Meisn. ●■
5744 Aconogonon rhombitepalum S. P. Hong;滇蓼■
5745 Aconogonon savatieri (Nakai) Tzvelev = Aconogonon weyrichii (F. Schmidt) H. Hara var. alpinum (Maxim.) H. Hara ■☆
5746 Aconogonon sibiricum (Laxm.) H. Hara = Polygonum sibiricum Laxm. ■
5747 Aconogonon sibiricum (Laxm.) H. Hara subsp. thomsonii (Meisn.) Soják = Polygonum sibiricum Laxm. var. thomsonii Meisn. ex Stewart ■
5748 Aconogonon tibeticum (Hemsl.) Soják = Polygonum tibeticum Hemsl. ●■
5749 Aconogonon tortuosum (D. Don) H. Hara = Polygonum tortuosum D. Don ■
5750 Aconogonon tortuosum (D. Don) H. Hara var. glabrifolium S. P. Hong = Polygonum tibeticum Hemsl. ●■
5751 Aconogonon tortuosum (D. Don) H. Hara var. tibetanum (Meisn.) S. P. Hong = Polygonum tortuosum D. Don ■
5752 Aconogonon weyrichii (F. Schmidt) H. Hara = Polygonum weyrichii F. Schmidt ■☆
5753 Aconogonon weyrichii (F. Schmidt) H. Hara var. alpinum (Maxim.) H. Hara = Polygonum weyrichii F. Schmidt var. alpinum Maxim. ■☆
5754 Aconogonum (D. Don) H. Hara = Polygonum molle D. Don ●■
5755 Aconogonum (Meisn.) Rchb. (1837);肖蓼属(虎杖属)■☆
5756 Aconogonum (Meisn.) Rchb. = Persicaria (L.) Mill. ■
5757 Aconogonum Rchb. = Persicaria (L.) Mill. ■
5758 Aconogonum campanulatum (Hook. f.) H. Hara = Polygonum campanulatum Hook. f. ■
5759 Aconogonum hookeri (Meisn.) H. Hara = Polygonum hookeri Meisn. ■
5760 Aconogonum molle D. Don var. frondosum (Meisn.) H. Hara = Polygonum molle D. Don var. frondosum (Meisn.) A. J. Li ●■
5761 Aconogonum molle D. Don var. rude (Meisn.) H. Hara = Polygonum molle D. Don var. rude (Meisn.) A. J. Li ■
5762 Aconogonum sibiricum (Laxm.) H. Hara = Polygonum sibiricum Laxm. ■
5763 Aconogonum sibiricum (Laxm.) H. Hara subsp. thomsonii (Meisn.) Soják = Polygonum sibiricum Laxm. var. thomsonii Meisn. ex Stewart ■
5764 Aconogonum tortuosum (D. Don) H. Hara = Polygonum tortuosum D. Don ■
5765 Acontias Schott = Xanthosoma Schott ■
5766 Acopanea Steyerm. = Bonnetia Mart. (保留属名)●☆
5767 Acophorum Gaudich. ex Steud. (1840);针梗禾属■☆
5768 Acophorum Steud. = Acophorum Gaudich. ex Steud. ■☆
5769 Acophorum caerulescens Gaudich.;针梗禾■☆
5770 Acoraceae C. Agardh = Araceae Juss. (保留科名)●■
5771 Acoraceae Martinov. (1820);菖蒲科■
5772 Acorellus Palla = Cyperus L. ■
5773 Acorellus Palla ex Kneuck. = Cyperus L. ■
5774 Acorellus Palla ex Kneuck. = Juncellus (Griseb.) C. B. Clarke ■
5775 Acorellus distachyus (All.) Palla = Cyperus distans L. f. ■
5776 Acorellus laevigatus (L.) Palla = Cyperus laevigatus L. ■☆
5777 Acorellus pannonicus (Jacq.) Palla = Juncellus pannonicus (Jacq.) C. B. Clarke ■
5778 Acoridium Nees et Meyen = Dendrochilum Blume ■
5779 Acoridium Nees et Meyen(1843);无孔兰属■☆
5780 Acoridium angustifolium Ames;窄叶无孔兰■☆
5781 Acoridium cobbianum Rolfe = Dendrochilum cobbianum Rchb. f. ■☆
5782 Acoridium filiforme Rolfe = Dendrochilum filiforme Lindl. ■☆
5783 Acoridium glumaceum Rolfe = Dendrochilum glumaceum Lindl. ■☆
5784 Acoridium graminifolium Ames;禾叶无孔兰■☆
5785 Acoridium latifolium Rolfe;宽叶无孔兰■☆
5786 Acoridium longifolium Rolfe;长叶无孔兰■☆
5787 Acoroides Sol. = Xanthorrhoea Sm. ●■☆
5788 Acorus L. (1753);菖蒲属;Sweet Flag, Sweet-flag ■
5789 Acorus americanus (Raf.) Raf.;美洲菖蒲;Sweet-flag ■☆
5790 Acorus americanus Raf. = Acorus calamus L. ■
5791 Acorus asiaticus Nakai = Acorus calamus L. ■
5792 Acorus brevispathus K. M. Liu = Acorus tatarinowii Schott ■
5793 Acorus calamus L.;菖蒲(白菖,白菖蒲,昌阳,臭草,臭菖,臭菖蒲,臭蒲,大菖蒲,大叶菖蒲,地心,家菖蒲,剑菖蒲,剑叶菖蒲,茎蒲,兰荪,凌水挡,泥昌,泥菖,泥菖蒲,蒲剑,山菖蒲,十香和,石菖蒲,水八角草,水昌,水菖,水菖蒲,水剑草,水宿,土菖蒲,溪菖蒲,溪荪,香蒲,尧韭,野菖蒲,野枇杷);Acorus, Bastard Flag, Bee Wort, Calamus Root, Calamus Sweetflag, Cinnamon Iris, Cinnamon Sedge, Drug Sweet Flag, Drug Sweetflag, Flag Root, Flagroot, Gladden, Gladdon, Grass, Grass Myrtle, Myrtle Flag, Myrtle Grass, Myrtle Sedge, Sweet Calomel, Sweet Cane, Sweet Flag, Sweet Myrtle, Sweet Rush, Sweet Sedge, Sweet Seg, Sweetflag, Sweet-flag, Sweetgrass, Vegetable Calomel ■
5794 Acorus calamus L. 'Argenteostriatus';黄条菖蒲(银纹菖蒲);Myrtle Flag, Sweet Flag, Sweet-flag, Variegated Sweet Flag ■☆
5795 Acorus calamus L. 'Variegatus' = Acorus calamus L. 'Argenteostriatus' ■☆
5796 Acorus calamus L. var. americanus (Raf.) H. Wulff = Acorus americanus (Raf.) Raf. ■☆
5797 Acorus calamus L. var. americanus Raf. = Acorus americanus (Raf.) Raf. ■☆
5798 Acorus calamus L. var. angustatus Besser;窄菖蒲(日本菖蒲)■☆
5799 Acorus calamus L. var. asiaticus Pers. = Acorus calamus L. ■
5800 Acorus calamus L. var. macrospadiceus Yamam.;大穗石菖蒲;Bigspike Drug Sweetflag ■

5801 Acorus calamus L. var. verus L.;细根菖蒲■

5802 Acorus calamus L. var. vulgaris L. = Acorus calamus L. ■

5803 Acorus gramineus Sol. ex Aiton;金菖蒲(昌阳,菖蒲,建菖蒲,金钱蒲,九节菖蒲,凌水档,钱蒲,十香和,石菖蒲,石上菖蒲,水剑草,随手香,鲜菖蒲,小石菖蒲,尧韭);Dwarf Sweetflag,Grassleaf Sweetflag,Grass-leaf Sweetflag,Grassleaved Sweetflag,Grassy-leaved Sweet Flag,Japanese Rush,Japanese Sweet Flag,Japanese Sweetflag,Japanese Sweet-flag,Slender Sweet-flag ■

5804 Acorus gramineus Sol. ex Aiton 'Pusillus' = Acorus gramineus Sol. ex Aiton var. pusillus (Siebold) Engl. ■

5805 Acorus gramineus Sol. ex Aiton 'Variegatus';条纹菖蒲(花叶石菖蒲);Dwf Varieg Sweet Flag,Variegate Grassleaf Sweetflag ■

5806 Acorus gramineus Sol. ex Aiton var. crassispadix ? = Acorus tatarinowii Schott ■

5807 Acorus gramineus Sol. ex Aiton var. flavomarginatus K. M. Liu = Acorus gramineus Sol. ex Aiton ■

5808 Acorus gramineus Sol. ex Aiton var. macrospadiceus Yamam. = Acorus gramineus Sol. ex Aiton ■

5809 Acorus gramineus Sol. ex Aiton var. pusillus (Siebold ex Schott) Engl.;细叶菖蒲(金钱蒲,钱菖蒲,钱蒲,随手香);Thinleaf Grassleaf Sweetflag ■

5810 Acorus gramineus Sol. ex Aiton var. pusillus (Siebold ex Schott) Engl. = Acorus gramineus Sol. ex Aiton ■

5811 Acorus gramineus Sol. ex Aiton var. pusillus (Siebold) Engl. = Acorus gramineus Sol. ex Aiton var. pusillus (Siebold ex Schott) Engl. ■

5812 Acorus gramineus Sol. ex Aiton var. pusillus (Siebold) Engl. = Acorus gramineus Sol. ex Aiton ■

5813 Acorus gramineus Sol. ex Aiton var. variegatus ? = Acorus gramineus Sol. ex Aiton 'Variegatus' ■

5814 Acorus japonicus ? = Acorus gramineus Sol. ex Aiton ■

5815 Acorus latifolius Z. Y. Zhu;宽叶菖蒲;Broadleaf Sweetflag ■

5816 Acorus latifolius Z. Y. Zhu = Acorus tatarinowii Schott ■

5817 Acorus macrospadiceus (Yamam.) F. N. Wei et Y. K. Li;茴香菖蒲■

5818 Acorus pusillus Siebold;京石菖蒲;Beijing Sweet Flag ■

5819 Acorus pusillus Siebold = Acorus gramineus Sol. ex Aiton ■

5820 Acorus rumphianus S. Y. Hu;长苞菖蒲;Rumph Sweet Flag ■

5821 Acorus tatarinowii Schott;石菖蒲(薄菖蒲,昌本,昌羊,昌阳,菖蒲,臭菖,粉菖,回手香,剑草,剑叶菖蒲,九节菖蒲,苦菖蒲,木蜡,山艾,山菖蒲,石蜈蚣,水菖蒲,水剑草,水蜈蚣,随手香,望见消,溪菖,香草,香菖蒲,小石菖蒲,岩菖蒲,阳春雪,尧韭,尧时韭,药菖蒲,野韭菜,夜晚香,紫耳)■

5822 Acorus xiangyeus Z. Y. Zhu;香叶菖蒲;Xiangye Sweet Flag ■

5823 Acorus xiangyeus Z. Y. Zhu = Acorus gramineus Sol. ex Aiton ■

5824 Acosmia Benth. = Gypsophila L. ●■

5825 Acosmia Benth. ex G. Don = Gypsophila L. ●■

5826 Acosmium Schott = Sweetia Spreng. (保留属名)●☆

5827 Acosmium Schott(1827);无饰豆属(埃可豆属)●☆

5828 Acosmium brachystachyum (Benth.) Yakovlev;短穗无饰豆●☆

5829 Acosmium dasycarpum (Vogel) Yakovlev;毛果无饰豆●☆

5830 Acosmium parvifolium (Harms) Yakovlev;小叶无饰豆●☆

5831 Acosmium stipulare (Harms) Yakovlev = Dicraeopetalum stipulare Harms ●☆

5832 Acosmium tenuifolium (Vogel) Yakovlev;细叶无饰豆●☆

5833 Acosmus Desv. = Aspicarpa Rich. ●☆

5834 Acosta Adans. = Centaurea L. (保留属名)●■

5835 Acosta DC. = Spiracantha Kunth ■☆

5836 Acosta Lour. = Vaccinium L. ●

5837 Acosta Ruiz et Pav. = Moutabea Aubl. ●☆

5838 Acosta boissieri (DC.) Holub = Centaurea boissieri DC. ■☆

5839 Acosta boissieri (DC.) Holub subsp. atlantica (Font Quer) Fern. Casas et Susanna = Centaurea boissieri DC. subsp. atlantica (Font Quer) Blanca ■☆

5840 Acosta boissieri (DC.) Holub subsp. calvescens (Maire) Fern. Casas et Susanna = Centaurea boissieri DC. ■☆

5841 Acosta diffusa (Lam.) Soják = Centaurea diffusa Lam. ■

5842 Acosta monticola (DC.) Holub = Centaurea monticola DC. ■☆

5843 Acosta pomeliana (Batt.) Holub = Centaurea pomeliana Batt. ■☆

5844 Acosta resupinata (Coss.) Holub = Centaurea resupinata Coss. ■☆

5845 Acosta resupinata (Coss.) Holub subsp. degenii (Sennen) Fern. Casas et Susanna = Centaurea resupinata Coss. subsp. degenii (Sennen) Fern. Casas et Susanna ■☆

5846 Acosta resupinata (Coss.) Holub subsp. spachii (Willk.) Fern. Casas et Susanna = Centaurea resupinata Coss. subsp. spachii (Willk.) Fern. Casas et Susanna ■☆

5847 Acosta spicata Lour. = Vaccinium bracteatum Thunb. ●

5848 Acostaea Schltr. (1923);无脉兰属■☆

5849 Acostaea bicornis Luer;双角无脉兰■☆

5850 Acostaea trilobata Luer;三裂无脉兰■☆

5851 Acostaea unicornis Luer;单角无脉兰■☆

5852 Acostia Swallen(1968);细无脊草属■☆

5853 Acostia gracilis Swallen;细无脊草■☆

5854 Acouba Aubl. (废弃属名) = Dalbergia L. f. (保留属名)●

5855 Acourea Scop. = Acouroa Aubl. (废弃属名)●

5856 Acouroa Aubl. (废弃属名) = Dalbergia L. f. (保留属名)●

5857 Acouroua Taub. = Acouroa Aubl. (废弃属名)●

5858 Acourtia D. Don = Perezia Lag. ■☆

5859 Acourtia D. Don(1830);沙牡丹属;Desertpeony ●☆

5860 Acourtia runcinata (D. Don) B. L. Turner = Acourtia runcinata (Lag. ex D. Don) B. L. Turner ■☆

5861 Acourtia runcinata (Lag. ex D. Don) B. L. Turner;沙牡丹;Peonia ■☆

5862 Acourtia wrightii (A. Gray) Reveal et R. M. King;赖氏沙牡丹;Desert Holly ■☆

5863 Acquartia Endl. = Aquartia Jacq. ●■

5864 Acquartia Endl. = Solanum L. ●■

5865 Acrachne Chiov. = Acrachne Wight et Arn. ex Chiov. ■

5866 Acrachne Wight et Arn. ex Chiov. (1907);尖稃草属(假龙爪茅属);Acrachne ■

5867 Acrachne Wight et Arn. ex Lindl. = Acrachne Wight et Arn. ex Chiov. ■

5868 Acrachne Wight et Arn. ex Lindl. et Chiov. = Acrachne Wight et Arn. ex Chiov. ■

5869 Acrachne eleusinoides Wight et Arn. ex Steud. = Acrachne racemosa (B. Heyne ex Roth et Schult.) Ohwi ■

5870 Acrachne perrieri (A. Camus) S. M. Phillips;佩里耶尖稃草■☆

5871 Acrachne racemosa (B. Heyne ex Roth et Schult.) Ohwi;尖稃草(微药假龙爪茅);Goosegrass,Raceme Acrachne ■

5872 Acrachne racemosa (B. Heyne ex Roth) Ohwi = Acrachne racemosa (B. Heyne ex Roth et Schult.) Ohwi ■

5873 Acrachne racemosa (Roth et Schult.) Ohwi = Acrachne racemosa (B. Heyne ex Roth et Schult.) Ohwi ■

5874 Acrachne verticillata (Roxb.) Chiov. = Acrachne racemosa (B.

Heyne ex Roth et Schult.) Ohwi ■
5875 Acrachne verticillata (Roxb.) Lindl. ex Chiov. = Acrachne racemosa (B. Heyne ex Roth et Schult.) Ohwi ■
5876 Acradenia Kippist(1853);白木属(塔斯马尼亚芸木属)●☆
5877 Acradenia frankliniae Kippist;白木(塔斯马尼亚芸木);Whitey Wood ●☆
5878 Acraea Lindl. = Pterichis Lindl. ■☆
5879 Acrandra O. Berg = Campomanesia Ruiz et Pav. ●☆
5880 Acranthemum Tiegh. = Agelanthus Tiegh. ●☆
5881 Acranthemum Tiegh. = Tapinanthus (Blume) Rchb.(保留属名)●☆
5882 Acranthemum natalitium (Meisn.) Tiegh. = Agelanthus natalitius (Meisn.) Polhill et Wiens ●☆
5883 Acranthemum zeyheri (Harv.) Tiegh. = Agelanthus natalitius (Meisn.) Polhill et Wiens subsp. zeyheri (Harv.) Polhill et Wiens ●☆
5884 Acranthera Arn. ex Meisn.(1838)(保留属名);尖药花属;Acranthera, Tineanther ●
5885 Acranthera sinensis C. Y. Wu;中华尖药花(华尖药花,尖药花);China Tineanther, Chinese Acranthera ●
5886 Acranthus Clem. = Acrosanthes Eckl. et Zeyh. ■☆
5887 Acranthus Hook. f. = Aeranthes Lindl. ■☆
5888 Acratherum Hochst. ex Rich. = Acratherum Link ■
5889 Acratherum Link = Arundinella Raddi ■
5890 Acratherum miliaceum Link = Arundinella nepalensis Trin. ■
5891 Acratherum pumilum Hochst. ex A. Rich. = Arundinella pumila (Hochst. ex A. Rich.) Steud. ■☆
5892 Acreugenia Kausel = Myrcianthes O. Berg ●☆
5893 Acridocarpus Guill. et Perr.(1831)(保留属名);虫果金虎尾属●☆
5894 Acridocarpus alopecurus Sprague;狐狸虫果金虎尾●☆
5895 Acridocarpus alternifolius (Schumach. et Thonn.) Nied.;互叶虫果金虎尾●☆
5896 Acridocarpus angolensis A. Juss. = Sphedamnocarpus pruriens (A. Juss.) Szyszyl. ●☆
5897 Acridocarpus ballyi Launert;博利虫果金虎尾●☆
5898 Acridocarpus brevipetiolatus Engl. = Acridocarpus longifolius (G. Don) Hook. f. ●☆
5899 Acridocarpus camerunensis Nied.;喀麦隆虫果金虎尾●☆
5900 Acridocarpus chevalieri Sprague;舍瓦利耶金虎尾●☆
5901 Acridocarpus chloropterus Oliv.;绿翅虫果金虎尾●☆
5902 Acridocarpus congestus Launert;密集虫果金虎尾●☆
5903 Acridocarpus congolensis Sprague;刚果虫果金虎尾●☆
5904 Acridocarpus corymbosus Hook. f. = Acridocarpus alternifolius (Schumach. et Thonn.) Nied. ●☆
5905 Acridocarpus ferrugineus Engl. = Acridocarpus glaucescens Engl. var. ferrugineus (Engl.) Launert ●☆
5906 Acridocarpus galphimiifolius A. Juss. = Sphedamnocarpus pruriens (A. Juss.) Szyszyl. subsp. galphimiifolius (A. Juss.) P. D. de Villiers et D. J. Botha ●☆
5907 Acridocarpus glaucescens Engl.;灰绿虫果金虎尾●☆
5908 Acridocarpus glaucescens Engl. var. ferrugineus (Engl.) Launert;锈色虫果金虎尾●☆
5909 Acridocarpus glaucescens Engl. var. graniticus Fiori;花岗岩虫果金虎尾●☆
5910 Acridocarpus goossensii De Wild. = Acridocarpus smeathmannii (DC.) Guillaumin et Perr. ●☆
5911 Acridocarpus hemicyclopterus Sprague = Acridocarpus spectabilis (Nied.) Doorn-Hoekm. ●☆
5912 Acridocarpus humbertii Arènes;亨伯特虫果金虎尾●☆
5913 Acridocarpus katangensis De Wild.;加丹加虫果金虎尾●☆
5914 Acridocarpus kerstingii Engl. = Acridocarpus spectabilis (Nied.) Doorn-Hoekm. ●☆
5915 Acridocarpus laurentii De Wild. = Acridocarpus congolensis Sprague ●☆
5916 Acridocarpus ledermannii Engl. = Acridocarpus longifolius (G. Don) Hook. f. ●☆
5917 Acridocarpus longifolius (G. Don) Hook. f.;长叶虫果金虎尾●☆
5918 Acridocarpus longifolius (G. Don) Hook. f. f. brevialata R. Wilczek;短翅虫果金虎尾●☆
5919 Acridocarpus macrocalyx Engl.;大萼虫果金虎尾●☆
5920 Acridocarpus mayumbensis Gonc. et Launert;马永巴虫果金虎尾●☆
5921 Acridocarpus monodii Arènes et Jaeger = Acridocarpus monodii Arènes et Jaeger ex Birnbaum et Florence ●☆
5922 Acridocarpus monodii Arènes et Jaeger ex Birnbaum et Florence;莫诺虫果金虎尾●☆
5923 Acridocarpus natalitius A. Juss.;纳塔尔虫果金虎尾●☆
5924 Acridocarpus natalitius A. Juss. var. acuminatus Nied. = Acridocarpus natalitius A. Juss. ●☆
5925 Acridocarpus natalitius A. Juss. var. linearifolius Launert;线叶纳塔尔虫果金虎尾●☆
5926 Acridocarpus natalitius A. Juss. var. obtusus Nied. = Acridocarpus natalitius A. Juss. ●☆
5927 Acridocarpus orientalis A. Juss.;东方虫果金虎尾●☆
5928 Acridocarpus pauciglandulosus Launert;少腺虫果金虎尾●☆
5929 Acridocarpus pondoensis Engl. ex Nied. = Acridocarpus natalitius A. Juss. ●☆
5930 Acridocarpus prasinus Exell;草绿虫果金虎尾●☆
5931 Acridocarpus pruriens A. Juss. = Sphedamnocarpus pruriens (A. Juss.) Szyszyl. ●☆
5932 Acridocarpus pruriens A. Juss. var. laevigatus Sond. = Sphedamnocarpus pruriens (A. Juss.) Szyszyl. subsp. galphimiifolius (A. Juss.) P. D. de Villiers et D. J. Botha ●☆
5933 Acridocarpus reticulatus Burtt Davy = Acridocarpus natalitius A. Juss. ●☆
5934 Acridocarpus rudis De Wild. et T. Durand = Acridocarpus longifolius (G. Don) Hook. f. ●☆
5935 Acridocarpus rufescens Hutch. = Acridocarpus katangensis De Wild. ●☆
5936 Acridocarpus scheffleri Engl. = Acridocarpus ugandensis Sprague ●☆
5937 Acridocarpus smeathmannii (DC.) Guillaumin et Perr.;斯米虫果金虎尾●☆
5938 Acridocarpus smeathmannii (DC.) Guillaumin et Perr. var. staudtii Engl. = Acridocarpus staudtii (Engl.) Engl. ●☆
5939 Acridocarpus spectabilis (Nied.) Doorn-Hoekm.;壮观虫果金虎尾●☆
5940 Acridocarpus staudtii (Engl.) Engl.;施陶虫果金虎尾●☆
5941 Acridocarpus ugandensis Sprague;乌干达虫果金虎尾●☆
5942 Acridocarpus vanderystii R. Wilczek;范德虫果金虎尾●☆
5943 Acridocarpus zanzibaricus A. Juss.;桑给巴尔虫果金虎尾●☆
5944 Acridocarpus zanzibaricus A. Juss. var. brachyphyllus Chiov.;短叶桑给巴尔虫果金虎尾●☆
5945 Acrilia Griseb. = Trichilia P. Browne(保留属名)●
5946 Acrilla C. DC. = Acrilia Griseb. ●
5947 Acriopsis Blume(1825);合萼兰属(阿瑞奥普兰属,合柱兰属);

Acriopsis ■
5948 Acriopsis Reinw. ex Blume = Acriopsis Blume ■
5949 Acriopsis indica Wight;合萼兰;Acriopsis ■
5950 Acriopsis javanica Reinw. ex Blume;爪哇合萼兰(爪哇阿瑞奥普兰);Java Acriopsis ■☆
5951 Acriopsis latifolia Rolfe;大叶合萼兰(大叶阿瑞奥普兰);Broadleaf Acriopsis ■☆
5952 Acriopsis ridleyi Hook. f.;阿瑞合萼兰(阿瑞奥普兰);Rigley Acriopsis ■☆
5953 Acrisione B. Nord. (1985);箭药千里光属●☆
5954 Acrisione cymosa (Remy) B. Nord.;箭药千里光●☆
5955 Acrisione denticulata (Hook. et Arn.) B. Nord.;齿叶箭药千里光●☆
5956 Acrista O. F. Cook = Euterpe Gaertn. (废弃属名)●☆
5957 Acrista O. F. Cook = Euterpe Mart. (保留属名)●☆
5958 Acrista O. F. Cook = Prestoea Hook. f. (保留属名)●☆
5959 Acristaceae O. F. Cook = Arecaceae Bercht. et J. Presl(保留科名)●
5960 Acristaceae O. F. Cook = Palmae Juss. (保留科名)●
5961 Acritochaete Pilg. (1902);乱毛颖草属■
5962 Acritochaete volkensii Pilg.;乱毛颖草■☆
5963 Acritopappus R. M. King et H. Rob. (1972);短柔毛菊属(短冠菊属)■☆
5964 Acritopappus heterolepis (Baker) R. M. King et H. Rob.;异鳞短柔毛菊■☆
5965 Acritopappus micropappus (Baker) R. M. King et H. Rob.;短柔毛菊■☆
5966 Acriulus Ridl. = Scleria P. J. Bergius ■
5967 Acriulus griegiifolius Ridl. = Scleria griegiifolia (Ridl.) C. B. Clarke ■☆
5968 Acriulus leopoldianus C. B. Clarke = Scleria griegiifolia (Ridl.) C. B. Clarke ■☆
5969 Acriulus madagascariensis Ridl. = Scleria griegiifolia (Ridl.) C. B. Clarke ■☆
5970 Acriulus titan C. B. Clarke = Scleria griegiifolia (Ridl.) C. B. Clarke ■☆
5971 Acriviola Mill. = Tropaeolum L. ■
5972 Acroanthes Raf. = Malaxis Sol. ex Sw. ■
5973 Acroblastum Sol. = Balanophora J. R. Forst. et G. Forst. ■
5974 Acroblastum Sol. ex Setchell = Balanophora J. R. Forst. et G. Forst. ■
5975 Acroblastum Sol. ex Setchell = Polyplethia (Griff.) Tiegh. ■
5976 Acrobotrys K. Schum. et K. Krause(1908);顶穗茜属☆
5977 Acrobotrys discolor K. Schum. et Krause;顶穗茜☆
5978 Acrocarpidium Miq. = Peperomia Ruiz et Pav. ■
5979 Acrocarpus Nees = Cryptangium Schrad. ex Nees ■☆
5980 Acrocarpus Wight ex Arn. (1838);尖果苏木属(梣叶豆属,顶果木属,顶果树属,顶果苏木属);Acrocarpus ●
5981 Acrocarpus fraxinifolius Arn. = Acrocarpus fraxinifolius Wight et Arn. ●◇
5982 Acrocarpus fraxinifolius Arn. var. guangxiensis S. L. Mo et Y. Wei = Acrocarpus fraxinifolius Wight ex Arn. ●
5983 Acrocarpus fraxinifolius Wight et Arn.;尖果苏木(白蜡尖果苏木,顶果木,顶果树,顶果苏木,广西顶果木,泡椿);Kuranjan, Pink Cedar, Red Cedar, Shingle Tree ●◇
5984 Acrocarpus fraxinifolius Wight et Arn. var. guangxiensis X. L. Mo et Y. Wei;广西顶果木;Guangxi Acrocarpus ●
5985 Acrocarpus fraxinifolius Wight et Arn. var. guangxiensis X. L. Mo et Y. Wei = Acrocarpus fraxinifolius Wight et Arn. ●◇
5986 Acrocentron Cass. = Centaurea L. (保留属名)●■
5987 Acrocentrum Post et Kuntze = Acrocentron Cass. ●■
5988 Acrocephalium Hassk. = Acrocephalus Benth. ■
5989 Acrocephalus Benth. (1829);尖头花属(顶头花属);Acrocephalus ■
5990 Acrocephalus Benth. = Haumaniastrum P. A. Duvign. et Plancke ●■☆
5991 Acrocephalus Benth. = Platostoma P. Beauv. ■☆
5992 Acrocephalus abyssinicus Hochst. ex Chiov. = Haumaniastrum villosum (Benth.) A. J. Paton ●☆
5993 Acrocephalus alboviridis Hutch. = Haumaniastrum alboviride (Hutch.) P. A. Duvign. et Plancke ●☆
5994 Acrocephalus angolensis Gürke = Haumaniastrum villosum (Benth.) A. J. Paton ●☆
5995 Acrocephalus axillaris Benth.;腋生尖头花;Axillary Acrocephalus ■☆
5996 Acrocephalus barakaensis De Wild. = Haumaniastrum caeruleum (Oliv.) P. A. Duvign. et Plancke ●☆
5997 Acrocephalus barbatus Robyns et Lebrun = Haumaniastrum dissitifolium (Baker) A. J. Paton ●☆
5998 Acrocephalus bequaertii De Wild. = Haumaniastrum caeruleum (Oliv.) P. A. Duvign. et Plancke ●☆
5999 Acrocephalus buddleioides S. Moore = Haumaniastrum praealtum (Briq.) P. A. Duvign. et Plancke ●☆
6000 Acrocephalus buettneri Gürke = Haumaniastrum buettneri (Gürke) J. K. Morton ●☆
6001 Acrocephalus caeruleus Oliv. = Haumaniastrum caeruleum (Oliv.) P. A. Duvign. et Plancke ●☆
6002 Acrocephalus callianthus Briq. = Haumaniastrum villosum (Benth.) A. J. Paton ●☆
6003 Acrocephalus campicola Briq. = Haumaniastrum caeruleum (Oliv.) P. A. Duvign. et Plancke ●☆
6004 Acrocephalus canonensis G. Taylor = Haumaniastrum minor (Briq.) A. J. Paton ●☆
6005 Acrocephalus capitatus Benth. = Acrocephalus indicus (Burm. f.) Kuntze ■
6006 Acrocephalus capitellatus (L. f.) Druce;小头尖头花■☆
6007 Acrocephalus centratheroides Baker = Haumaniastrum caeruleum (Oliv.) P. A. Duvign. et Plancke ●☆
6008 Acrocephalus chevalieri Briq. = Haumaniastrum villosum (Benth.) A. J. Paton ●☆
6009 Acrocephalus chirindensis S. Moore = Haumaniastrum dissitifolium (Baker) A. J. Paton ●☆
6010 Acrocephalus claessensii Robyns et Lebrun = Haumaniastrum caeruleum (Oliv.) P. A. Duvign. et Plancke ●☆
6011 Acrocephalus coeruleus Oliv. var. genuinus Briq. = Haumaniastrum caeruleum (Oliv.) P. A. Duvign. et Plancke ●☆
6012 Acrocephalus coriaceus Robyns et Lebrun = Haumaniastrum coriaceum (Robyns et Lebrun) A. J. Paton ●☆
6013 Acrocephalus crinitus Briq. = Haumaniastrum caeruleum (Oliv.) P. A. Duvign. et Plancke ●☆
6014 Acrocephalus cubanquensis R. D. Good = Haumaniastrum cubanquense (R. D. Good) A. J. Paton ●☆
6015 Acrocephalus cyaneo-bracteatus De Wild. = Haumaniastrum katangense (S. Moore) P. A. Duvign. et Plancke ●☆
6016 Acrocephalus cylindraceus Oliv. = Haumaniastrum villosum (Benth.) A. J. Paton ●☆

6017 Acrocephalus debeerstii Briq. ex T. Durand et De Wild. = Haumaniastrum villosum (Benth.) A. J. Paton ●☆

6018 Acrocephalus degiorgii Robyns et Lebrun = Haumaniastrum caeruleum (Oliv.) P. A. Duvign. et Plancke ●☆

6019 Acrocephalus demeusei Briq. = Haumaniastrum caeruleum (Oliv.) P. A. Duvign. et Plancke ●☆

6020 Acrocephalus descampsii Briq. = Haumaniastrum caeruleum (Oliv.) P. A. Duvign. et Plancke ●☆

6021 Acrocephalus dewevrei Briq. ex De Wild. = Haumaniastrum villosum (Benth.) A. J. Paton ●☆

6022 Acrocephalus dissitifolius Baker = Haumaniastrum dissitifolium (Baker) A. J. Paton ●☆

6023 Acrocephalus divaricatus Briq. = Haumaniastrum villosum (Benth.) A. J. Paton ●☆

6024 Acrocephalus doloensis De Wild. = Haumaniastrum caeruleum (Oliv.) P. A. Duvign. et Plancke ●☆

6025 Acrocephalus elongatus Briq. = Haumaniastrum villosum (Benth.) A. J. Paton ●☆

6026 Acrocephalus elskensii Robyns et Lebrun = Haumaniastrum villosum (Benth.) A. J. Paton ●☆

6027 Acrocephalus erectifolius N. E. Br. = Haumaniastrum venosum (Baker) Agnew ●☆

6028 Acrocephalus fischeri Gürke = Haumaniastrum villosum (Benth.) A. J. Paton ●☆

6029 Acrocephalus fruticosus Dunn = Elsholtzia capituligera C. Y. Wu ●

6030 Acrocephalus galeopsifolius Baker = Haumaniastrum villosum (Benth.) A. J. Paton ●☆

6031 Acrocephalus glaucescens Robyns et Lebrun = Haumaniastrum caeruleum (Oliv.) P. A. Duvign. et Plancke ●☆

6032 Acrocephalus goetzei Gürke = Haumaniastrum venosum (Baker) Agnew ●☆

6033 Acrocephalus gracilis Briq. = Haumaniastrum caeruleum (Oliv.) P. A. Duvign. et Plancke ●☆

6034 Acrocephalus graminifolius Robyns = Haumaniastrum graminifolium (Robyns) A. J. Paton ●☆

6035 Acrocephalus hensii Briq. = Haumaniastrum caeruleum (Oliv.) P. A. Duvign. et Plancke ●☆

6036 Acrocephalus heudelotii Briq. = Haumaniastrum buettneri (Gürke) J. K. Morton ●☆

6037 Acrocephalus homblei De Wild. = Haumaniastrum praealtum (Briq.) P. A. Duvign. et Plancke var. homblei (De Wild.) A. J. Paton ●☆

6038 Acrocephalus hyptoides Baker = Haumaniastrum minor (Briq.) A. J. Paton ●☆

6039 Acrocephalus indicus (Burm. f.) Kuntze;尖头花(顶头花,水薄荷,头状尖头花,团花草,鱼香草);Capitate Acrocephalus, Indian Acrocephalus ■

6040 Acrocephalus iododermis Briq. = Haumaniastrum caeruleum (Oliv.) P. A. Duvign. et Plancke ●☆

6041 Acrocephalus kaessneri S. Moore = Haumaniastrum kaessneri (S. Moore) P. A. Duvign. et Plancke ●☆

6042 Acrocephalus kambovianus P. A. Duvign. et Plancke = Haumaniastrum timpermannii (P. A. Duvign. et Plancke) P. A. Duvign. et Plancke ●☆

6043 Acrocephalus katangensis S. Moore = Haumaniastrum katangense (S. Moore) P. A. Duvign. et Plancke ●☆

6044 Acrocephalus kipilaensis Robyns = Haumaniastrum praealtum (Briq.) P. A. Duvign. et Plancke var. homblei (De Wild.) A. J. Paton ●☆

6045 Acrocephalus kundelungense De Wild. = Haumaniastrum lantanoides (S. Moore) P. A. Duvign. et Plancke ●☆

6046 Acrocephalus lagoensis Baker = Haumaniastrum caeruleum (Oliv.) P. A. Duvign. et Plancke ●☆

6047 Acrocephalus lantanoides S. Moore = Haumaniastrum lantanoides (S. Moore) P. A. Duvign. et Plancke ●☆

6048 Acrocephalus laurentii Briq. = Haumaniastrum caeruleum (Oliv.) P. A. Duvign. et Plancke ●☆

6049 Acrocephalus lescrauwaetii Robyns et Lebrun = Haumaniastrum caeruleum (Oliv.) P. A. Duvign. et Plancke ●☆

6050 Acrocephalus lilacinoides De Wild. = Haumaniastrum caeruleum (Oliv.) P. A. Duvign. et Plancke ●☆

6051 Acrocephalus lilacinus Oliv. = Haumaniastrum caeruleum (Oliv.) P. A. Duvign. et Plancke ●☆

6052 Acrocephalus linearifolius De Wild. = Haumaniastrum linearifolium (De Wild.) P. A. Duvign. et Plancke ●☆

6053 Acrocephalus lippioides Baker = Haumaniastrum praealtum (Briq.) P. A. Duvign. et Plancke ●☆

6054 Acrocephalus longecuspidatus Robyns et Lebrun = Haumaniastrum caeruleum (Oliv.) P. A. Duvign. et Plancke ●☆

6055 Acrocephalus martreti A. Chev. = Haumaniastrum caeruleum (Oliv.) P. A. Duvign. et Plancke ●☆

6056 Acrocephalus masuianus Briq. = Haumaniastrum caeruleum (Oliv.) P. A. Duvign. et Plancke ●☆

6057 Acrocephalus mechowianus Briq. = Haumaniastrum praealtum (Briq.) P. A. Duvign. et Plancke ●☆

6058 Acrocephalus mildbraedii Perkins = Haumaniastrum caeruleum (Oliv.) P. A. Duvign. et Plancke ●☆

6059 Acrocephalus minor Briq. = Haumaniastrum minor (Briq.) A. J. Paton ●☆

6060 Acrocephalus monocephalus Baker = Haumaniastrum caeruleum (Oliv.) P. A. Duvign. et Plancke ●☆

6061 Acrocephalus obovatifolius Robyns et Lebrun = Haumaniastrum dissitifolium (Baker) A. J. Paton ●☆

6062 Acrocephalus paniculatus Briq. = Haumaniastrum paniculatum (Briq.) A. J. Paton ●☆

6063 Acrocephalus picturatus S. Moore = Haumaniastrum villosum (Benth.) A. J. Paton ●☆

6064 Acrocephalus poggeanus Briq. = Haumaniastrum villosum (Benth.) A. J. Paton ●☆

6065 Acrocephalus polyneurus S. Moore = Haumaniastrum polyneurum (S. Moore) P. A. Duvign. et Plancke ●☆

6066 Acrocephalus polytrichus Baker = Haumaniastrum caeruleum (Oliv.) P. A. Duvign. et Plancke ●☆

6067 Acrocephalus porphyrophyllus Baker = Haumaniastrum caeruleum (Oliv.) P. A. Duvign. et Plancke ●☆

6068 Acrocephalus praealtus Briq. = Haumaniastrum praealtum (Briq.) P. A. Duvign. et Plancke ●☆

6069 Acrocephalus pseudosericeus G. Taylor = Haumaniastrum caeruleum (Oliv.) P. A. Duvign. et Plancke ●☆

6070 Acrocephalus quarrei Robyns et Lebrun = Haumaniastrum caeruleum (Oliv.) P. A. Duvign. et Plancke ●☆

6071 Acrocephalus ramosissimus A. Chev. = Haumaniastrum villosum (Benth.) A. J. Paton ●☆

6072 Acrocephalus reticulatus Briq. = Haumaniastrum praealtum

(Briq.) P. A. Duvign. et Plancke ●☆

6073 Acrocephalus ringoetii De Wild. = Haumaniastrum rupestre (R. E. Fr.) A. J. Paton ●☆

6074 Acrocephalus robertii Robyns = Haumaniastrum robertii (Robyns) P. A. Duvign. et Plancke ●☆

6075 Acrocephalus rosulatus De Wild. = Haumaniastrum rosulatum (De Wild.) P. A. Duvign. et Plancke ●☆

6076 Acrocephalus rupestris R. E. Fr. = Haumaniastrum rupestre (R. E. Fr.) A. J. Paton ●☆

6077 Acrocephalus sapinii Robyns et Lebrun = Haumaniastrum caeruleum (Oliv.) P. A. Duvign. et Plancke ●☆

6078 Acrocephalus schweinfurthii Briq. = Haumaniastrum caeruleum (Oliv.) P. A. Duvign. et Plancke ●☆

6079 Acrocephalus semilignosus P. A. Duvign. et Plancke = Haumaniastrum semilignosum (P. A. Duvign. et Plancke) P. A. Duvign. et Plancke ●☆

6080 Acrocephalus seretii De Wild. = Haumaniastrum caeruleum (Oliv.) P. A. Duvign. et Plancke ●☆

6081 Acrocephalus sericeus Briq. = Haumaniastrum sericeum (Briq.) A. J. Paton ●☆

6082 Acrocephalus sordidus Briq. = Haumaniastrum caeruleum (Oliv.) P. A. Duvign. et Plancke ●☆

6083 Acrocephalus speciosus E. A. Bruce = Haumaniastrum speciosum (E. A. Bruce) A. J. Paton ●☆

6084 Acrocephalus stormsii Briq. = Haumaniastrum villosum (Benth.) A. J. Paton ●☆

6085 Acrocephalus suberosus Robyns et Lebrun = Haumaniastrum suberosum (Robyns et Lebrun) P. A. Duvign. et Plancke ●☆

6086 Acrocephalus succisifolius Baker = Haumaniastrum praealtum (Briq.) P. A. Duvign. et Plancke var. succisifolium (Baker) A. J. Paton ●☆

6087 Acrocephalus termiticola Robyns = Haumaniastrum rosulatum (De Wild.) P. A. Duvign. et Plancke ●☆

6088 Acrocephalus timpermannii P. A. Duvign. et Plancke = Haumaniastrum timpermannii (P. A. Duvign. et Plancke) P. A. Duvign. et Plancke ●☆

6089 Acrocephalus triramosus N. E. Br. = Haumaniastrum triramosum (N. E. Br.) A. J. Paton ●☆

6090 Acrocephalus upembensis Robyns = Haumaniastrum rosulatum (De Wild.) P. A. Duvign. et Plancke ●☆

6091 Acrocephalus vandenbrandei P. A. Duvign. et Plancke = Haumaniastrum vandenbrandei (P. A. Duvign. et Plancke) P. A. Duvign. et Plancke ●☆

6092 Acrocephalus vanderystii De Wild. = Haumaniastrum caeruleum (Oliv.) P. A. Duvign. et Plancke ●☆

6093 Acrocephalus venosus Baker = Haumaniastrum venosum (Baker) Agnew ●☆

6094 Acrocephalus verbenaceus Vatke = Haumaniastrum villosum (Benth.) A. J. Paton ●☆

6095 Acrocephalus villosus Benth. = Haumaniastrum villosum (Benth.) A. J. Paton ●☆

6096 Acrocephalus viridulus Robyns = Haumaniastrum rosulatum (De Wild.) P. A. Duvign. et Plancke ●☆

6097 Acrocephalus welwitschii Briq. = Haumaniastrum minor (Briq.) A. J. Paton ●☆

6098 Acrocephalus zambesiacus Baker ex Gürke = Haumaniastrum villosum (Benth.) A. J. Paton ●☆

6099 Acroceras Stapf(1920);凤头黍属;Acroceras ■

6100 Acroceras amplectens Stapf;环抱凤头黍■☆

6101 Acroceras attenuatum Renvoize;狭变凤头黍■☆

6102 Acroceras basicladum Stapf = Acroceras amplectens Stapf ■☆

6103 Acroceras crassiapiculatum (Merr.) Alston = Acroceras munroanum (Balansa) Henrard ■

6104 Acroceras diffusum L. C. Chia = Setiacis diffusa (L. C. Chia) S. L. Chen et Y. X. Jin ■

6105 Acroceras gabunense (Hack.) Clayton;加蓬凤头黍■☆

6106 Acroceras hubbardii (A. Camus) Clayton;哈伯德黍■☆

6107 Acroceras ivohibense A. Camus;伊武希贝凤头黍■☆

6108 Acroceras macrum Stapf;大凤头黍;Nile Grass ■☆

6109 Acroceras munroanum (Balansa) Henrard;凤头黍(门氏凤头黍);Common Acroceras ■

6110 Acroceras oryzoides Stapf;稻状凤头黍(拟菰凤头黍);Oryza-like Acroceras ■☆

6111 Acroceras oryzoides Stapf = Acroceras zizanioides (Kunth) Dandy ■☆

6112 Acroceras pilgerianum Schweick. = Panicum pilgerianum (Schweick.) Clayton ■☆

6113 Acroceras tonkinense (Balansa) C. E. Hubb. ex Bor = Neohusnotia tonkinensis (Balansa) A. Camus ■

6114 Acroceras zizanioides (Kunth) Dandy;类菰凤头黍■☆

6115 Acrochaene Lindl. = Monomeria Lindl. ■

6116 Acrochaene rimannii Rchb. f. = Sunipia rimannii (Rchb. f.) Seidenf. ■

6117 Acrochaete Peter = Setaria P. Beauv. (保留属名)■

6118 Acrochaete Peter = Tansaniochloa Rauschert ■

6119 Acrochaete pseudaristata Peter = Setaria pseudaristata (Peter) Pilg. ■☆

6120 Acrochloa Griseb. = Airochloa Link ■

6121 Acrochloa Griseb. = Koeleria Pers. ■

6122 Acroclasia C. Presl = Mentzelia L. ●■☆

6123 Acroclinium A. Gray = Helipterum DC. ex Lindl. ■☆

6124 Acroclinium A. Gray = Rhodanthe Lindl. ●■☆

6125 Acroclinium roseum Hook. = Helipterum roseum (Hook.) Benth. ■☆

6126 Acrocoelium Baill. = Leptaulus Benth. ●☆

6127 Acrocoelium congolanum Baill. = Leptaulus zenkeri Engl. ●☆

6128 Acrocomia Mart. (1824);格鲁棕属(垂花椆属,刺干椰属,刺茎椰子属,刺茎棕属,大刺可可椰子属,顶束毛椆属,可雅棕属);Acrocomia, Grugru Palm, Gru-gru Palm ●☆

6129 Acrocomia aculeata (Jacq.) Lodd. ex Mart.;皮刺格鲁棕;Grugru Acrocomia, Mecaw Palm ●☆

6130 Acrocomia armentalis L. H. Bailey;亚美尼亚格鲁棕●☆

6131 Acrocomia crispa (Kunth) C. F. Baker ex Becc. = Gastrococos crispa (Kunth) H. E. Moore ●☆

6132 Acrocomia lasiospatha Mart.;毛格鲁棕●☆

6133 Acrocomia media O. F. Cook;波多黎各格鲁棕;Puerto Rico Acrocomia ●☆

6134 Acrocomia mexicana Karw. ex Mart.;墨西哥格鲁棕;Coyoli, Coyoli Palm ●☆

6135 Acrocomia mexicana Karw. ex Mart. = Acrocomia aculeata (Jacq.) Lodd. ex Mart. ●☆

6136 Acrocomia sclerocarpa Mart.;核果格鲁棕;Corozo, Gru-gru, Grugru Nut, Gru-gru Palm, Macaw Palm, Mbocaya, Mucaia Acrocomia, Mucaja Acrocomia, Palm Hearts, Paraguay Palm ●☆

6137 Acrocomia totai Mart.;多太格鲁棕;Gru-gru,Grugru Palm,Mbocarya,Paraguay Palm,Totai Palm ●☆
6138 Acrocorion Adans. = Galanthus L. ■☆
6139 Acrocorium Post et Kuntze = Acrocorion Adans. ■☆
6140 Acrocoryna Post et Kuntze = Acrocoryne Turcz ●☆
6141 Acrocoryne Turcz. = Metastelma R. Br. ●☆
6142 Acrodiclidium Nees = Licaria Aubl. ●☆
6143 Acrodiclidium Nees et Mart. = Licaria Aubl. ●☆
6144 Acrodiclidium triandrum (Sw.) Lundell = Licaria triandra (Sw.) Kosterm. ●☆
6145 Acrodon N. E. Br.(1927);斗鱼草属■☆
6146 Acrodon bellidiflorus (L.) N. E. Br.;斗鱼草■☆
6147 Acrodon bellidiflorus (L.) N. E. Br. var. striatus (Haw.) N. E. Br. = Acrodon bellidiflorus (L.) N. E. Br. ■☆
6148 Acrodon bellidiflorus (L.) N. E. Br. var. viridis (Haw.) N. E. Br. = Acrodon bellidiflorus (L.) N. E. Br. ■☆
6149 Acrodon deminutus Klak;简缩斗鱼草■☆
6150 Acrodon duplessiae (L. Bolus) Glen = Acrodon bellidiflorus (L.) N. E. Br. ■☆
6151 Acrodon leptophyllus (L. Bolus) Glen = Acrodon subulatus (Mill.) N. E. Br. ■☆
6152 Acrodon parvifolius du Plessis;小叶斗鱼草■☆
6153 Acrodon purpureostylus (L. Bolus) Burgoyne;紫柱斗鱼草■☆
6154 Acrodon subulatus (Mill.) N. E. Br.;细叶斗鱼草■☆
6155 Acrodryon Spreng. = Cephalanthus L. ●
6156 Acrodrys Clem. = Acrodryon Spreng. ●
6157 Acroelytrum Steud. = Lophatherum Brongn. ■
6158 Acroelytrum japonicum Steud. = Lophatherum gracile Brongn. ■
6159 Acroglochia Gerard. = Acroglochin Schrad. ■
6160 Acroglochin Schrad.(1822);千针苋属;Acroglochin ■
6161 Acroglochin Schrad. ex Schult. = Acroglochin Schrad. ■
6162 Acroglochin chenopodioides Schrad. ex Schult. f. = Acroglochin persicarioides (Poir.) Moq. ■
6163 Acroglochin obtusifolia Blom = Acroglochin persicarioides (Poir.) Moq. ■
6164 Acroglochin persicarioides (Poir.) Moq.;千针苋;Common Acroglochin ■
6165 Acroglochin persicarioides (Poir.) Moq. var. muliensis Soong;木里千针苋;Muli Acroglochin ■
6166 Acroglochin persicarioides (Poir.) Moq. var. multiflora Soong;多花千针苋;Manyflower Acroglochin ■
6167 Acroglyphe E. Mey. = Annesorhiza Cham. et Schltdl. ■☆
6168 Acrolasia C. Presl = Mentzelia L. ●■☆
6169 Acrolepis Schrad. = Ficinia Schrad.(保留属名)■☆
6170 Acrolepis ferruginea Boeck. = Ficinia ferruginea (Boeck.) C. B. Clarke ●☆
6171 Acrolepis ramosissima (Kunth) Boeck. = Ficinia ramosissima Kunth ■☆
6172 Acrolepis trichodes Schrad. = Ficinia trichodes (Schrad.) Benth. et Hook. f. ●☆
6173 Acrolinium Engl. = Acroclinium A. Gray ■☆
6174 Acrolinium Engl. = Helipterum DC. ex Lindl. ■☆
6175 Acrolobus Klotzsch = Heisteria Jacq.(保留属名)●☆
6176 Acrolophia Pfitzer(1887);冠顶兰属■☆
6177 Acrolophia barbata (Thunb.) H. P. Linder;髯毛冠顶兰■☆
6178 Acrolophia bolusii Rolfe;博卢斯冠顶兰■☆
6179 Acrolophia capensis (P. J. Bergius) Fourc.;好望角冠顶兰■☆
6180 Acrolophia capensis (P. J. Bergius) Fourc. var. lamellata (Lindl.) Schelpe = Acrolophia lamellata (Lindl.) Schltr. et Bolus ■☆
6181 Acrolophia cochlearis (Lindl.) Schltr. et Bolus;螺状冠顶兰■☆
6182 Acrolophia comosa (Sond.) Schltr. et Bolus = Acrolophia capensis (P. J. Bergius) Fourc. ■☆
6183 Acrolophia fimbriata Schltr. = Acrolophia micrantha (Lindl.) Pfitzer ■☆
6184 Acrolophia lamellata (Lindl.) Schltr. et Bolus;片状冠顶兰■☆
6185 Acrolophia lunata (Schltr.) Schltr. et Bolus = Acrolophia barbata (Thunb.) H. P. Linder ■☆
6186 Acrolophia micrantha (Lindl.) Pfitzer;小花冠顶兰■☆
6187 Acrolophia micrantha (Lindl.) Schltr. et Bolus = Acrolophia micrantha (Lindl.) Pfitzer ■☆
6188 Acrolophia paniculata P. J. Cribb = Eulophia callichroma Rchb. f. ■☆
6189 Acrolophia parvula Schltr. = Acrolophia ustulata (Bolus) Schltr. et Bolus ■☆
6190 Acrolophia sphaerocarpa (Sond.) Schltr. et Bolus = Acrolophia capensis (P. J. Bergius) Fourc. ■☆
6191 Acrolophia triste (L. f.) Schltr. et Bolus = Acrolophia capensis (P. J. Bergius) Fourc. ■☆
6192 Acrolophia ustulata (Bolus) Schltr. et Bolus;暗褐冠顶兰■☆
6193 Acrolophus Cass. = Centaurea L.(保留属名)●■
6194 Acrolophus squarrosus (Willd.) Nevski = Centaurea squarrosa Willd. ■
6195 Acronema Edgew. = Acronema Falc. ex Edgew. ■
6196 Acronema Falc. ex Edgew.(1845);丝瓣芹属;Acronema ■
6197 Acronema alpinum S. L. Liou et R. H. Shan;高山丝瓣芹;Alpine Acronema ■
6198 Acronema astrantiifolium (H. Wolff) M. Hiroe.;星叶丝瓣芹(石滩丝瓣芹);Starryleaf Acronema ■
6199 Acronema astrantiifolium H. Wolff = Acronema astrantiifolium (H. Wolff) M. Hiroe. ■
6200 Acronema brevipedicellatum Z. H. Pan et M. F. Watson;短柄丝瓣芹■
6201 Acronema chienii R. H. Shan = Acronema chienii R. H. Shan et S. L. Liou ■
6202 Acronema chienii R. H. Shan et S. L. Liou;条叶丝瓣芹;Chie Acronema ■
6203 Acronema chienii R. H. Shan et S. L. Liou var. dissectum R. H. Shan;细裂条叶丝瓣芹(细裂丝瓣芹);Thinlobed Chie Acronema ■
6204 Acronema chienii R. H. Shan var. dissectum R. H. Shan = Acronema chienii R. H. Shan et S. L. Liou var. dissectum R. H. Shan ■
6205 Acronema chinense H. Wolff;尖瓣芹;China Acronema ■
6206 Acronema chinense H. Wolff var. humile S. L. Liou et R. H. Shan;矮尖瓣芹;Dwraf China Acronema ■
6207 Acronema commutatum H. Wolff;多变丝瓣芹(变化丝瓣芹);Changed Acronema ■
6208 Acronema edosmioides (H. Boissieu) Pimenov et Kljuykov = Cyclorhiza peucedanifolia (Franch.) Constance ■
6209 Acronema forrestii H. Wolff;中甸丝瓣芹(疏齿丝瓣芹);Forrest Acronema ■
6210 Acronema gracile S. L. Liou et R. H. Shan;细梗丝瓣芹;Slenderstalk Acronema ■
6211 Acronema graminifolium (H. Wolff) S. L. Liou et R. H. Shan;禾叶丝瓣芹;Grassleaf Acronema ■
6212 Acronema handelii H. Wolff;块根丝瓣芹(中甸丝瓣芹);Hendl Acronema ■

6213 Acronema hookeri (C. B. Clarke) H. Wolff;锡金丝瓣芹(丝瓣芹);Hooker Acronema ■

6214 Acronema hookeri (C. B. Clarke) H. Wolff var. graminifolium (W. W. Sm.) H. Wolff. = Acronema graminifolium (H. Wolff) S. L. Liou et R. H. Shan ■

6215 Acronema hookeri (C. B. Clarke) H. Wolff var. graminifolium W. W. Sm. = Acronema graminifolium (H. Wolff) S. L. Liou et R. H. Shan ■

6216 Acronema johrianum Babu;单羽丝瓣芹■

6217 Acronema minus (M. F. Watson) M. F. Watson et Z. H. Pan;矮小丝瓣芹■

6218 Acronema muscicola (Hand. -Mazz.) Hand. -Mazz.;苔间丝瓣芹(苔生丝瓣芹);Moss-living Acronema ■

6219 Acronema nervosum H. Wolff;羽轴丝瓣芹;Nervedrachis Acronema ■

6220 Acronema paniculatum (Franch.) H. Wolff;圆锥丝瓣芹(锥序丝瓣芹);Paniculate Acronema ■

6221 Acronema radiatum (W. W. Sm.) H. Wolff;环辐丝瓣芹;Radiate Acronema ■

6222 Acronema schneideri H. Wolff;丽江丝瓣芹(线叶丝瓣芹);Lijiang Acronema ■

6223 Acronema sichuanense S. L. Liou et R. H. Shan;四川丝瓣芹;Sichuan Acronema ■

6224 Acronema tenerum (Wall.) Edgew.;丝瓣芹;Common Acronema ■

6225 Acronema wolffianum Fedde ex H. Wolff = Sinocarum schizopetalum (Franch.) H. Wolff ex R. H. Shan et F. T. Pu var. bijiangense (S. L. Liou) X. T. Liu ■

6226 Acronema xizangense S. L. Liou et R. H. Shan;西藏丝瓣芹;Xizang Acronema ■

6227 Acronema yadongense S. L. Liou;亚东丝瓣芹;Yadong Acronema ■

6228 Acronia C. Presl = Pleurothallis R. Br. ■☆

6229 Acronoda Hassk. = Acronodia Blume ●

6230 Acronodia Blume = Elaeocarpus L. ●

6231 Acronozus Steud. = Acronodia Blume ●

6232 Acronozus Steud. = Acrozus Spreng. ●

6233 Acronychia J. R. Forst. et G. Forst. (1775)(保留属名);山油柑属(降真香属);Acronychia ●

6234 Acronychia acuminata T. G. Hartley;渐尖山油柑●☆

6235 Acronychia albiflora Rechinger;白花山油柑●☆

6236 Acronychia arborea Blume;树山油柑●☆

6237 Acronychia baueri Schott;鲍尔山油柑(澳山油柑);Bauer Acronychia ●☆

6238 Acronychia eriocarpa Pancher ex Guillaumin;毛果山油柑●☆

6239 Acronychia esquirolii H. Lév. = Alstonia yunnanensis Diels ●

6240 Acronychia laevis J. R. Forst. et G. Forst.;平滑山油柑●☆

6241 Acronychia laevis J. R. Forst. et G. Forst. var. leucocarpa F. M. Bailey;白果平滑山油柑●☆

6242 Acronychia laevis J. R. Forst. et G. Forst. var. purpurea F. M. Bailey;紫果平滑山油柑●☆

6243 Acronychia laurifolia Blume;桂叶山油柑●☆

6244 Acronychia laurifolia Blume = Acronychia pedunculata (L.) Miq. ●

6245 Acronychia leiocarpa P. S. Green;光果山油柑●☆

6246 Acronychia minahassae (Teijsm. et Binn.) Miq. = Melicope triphylla (Lam.) Merr. ●

6247 Acronychia oblongifolia Endl. ex Heynh.;矩圆叶山油柑●☆

6248 Acronychia obovata Merr.;倒卵叶山油柑●☆

6249 Acronychia oligophlebia Merr.;贡甲(白山柑);Fewveined Acronychia, Few-veined Acronychia, Gongjia Acronychia ●

6250 Acronychia ovalifolia Pancher ex Guillaumin;卵叶山油柑●☆

6251 Acronychia pedunculata (L.) Miq.;山油柑(长柄山油柑,降真香,沙柑木,沙塘木,砂糖木,山柑,山橘,石苓舅,水浓叶);Acronychia, Pedunculate Acronychia ●

6252 Acropera Lindl. = Gongora Ruiz et Pav. ■☆

6253 Acropetalum A. Juss. = Dombeya Cav. (保留属名)●☆

6254 Acropetalum A. Juss. = Xeropetalum Delile ●☆

6255 Acropetalum Delile ex A. Juss. = Dombeya Cav. (保留属名)●☆

6256 Acropetalum Delile ex A. Juss. = Xeropetalum Delile ●☆

6257 Acrophyllum Benth. (1838);尖叶火把树属●☆

6258 Acrophyllum E. Mey. = Pappea Eckl. et Zeyh. ●☆

6259 Acrophyllum australe (A. Cunn.) Hoogland;尖叶火把树●☆

6260 Acroplanes K. Schum. = Donax Lour. + Schumannianthus Gagnep. ■☆

6261 Acroplanes K. Schum. = Ilythuria Raf. ■

6262 Acropodium Desv. = Aspalathus L. ●☆

6263 Acropogon Schltr. (1906);顶须桐属●☆

6264 Acropogon fatsioides Schltr.;顶须桐●☆

6265 Acropogon macrocarpus Morat et Chalopin;大果顶须桐●☆

6266 Acropogon megaphyllus (Bureau et Poiss. ex Guillaumin) Morat;大叶顶须桐●☆

6267 Acropogon pilosus Morat et Chalopin;毛顶须桐●☆

6268 Acropselion Spach = Acrospelion Besser ex Roem. et Schult. ■

6269 Acropselion Spach = Trisetum Pers. ■

6270 Acropsis Asch. et Graebn. = Airopsis Desv. ■☆

6271 Acroptilion Endl. = Acroptilon Cass. ■

6272 Acroptilon Cass. (1827);顶羽菊属;Acroptilon, Russian Knapweed ■

6273 Acroptilon Cass. = Rhaponticum Adans. ■

6274 Acroptilon angustifolium Cass. = Acroptilon repens (L.) DC. ■

6275 Acroptilon australe Iljin = Acroptilon repens (L.) DC. ■

6276 Acroptilon obtusifolium Cass. = Acroptilon repens (L.) DC. ■

6277 Acroptilon picris (Pall. ex Willd.) C. A. Mey. = Acroptilon repens (L.) DC. ■

6278 Acroptilon repens (L.) DC.;顶羽菊(灰叫驴,苦蒿,毛连矢车菊,匍匐顶须桐,匍匐矢车菊,窄叶顶须桐);Creeping Acroptilon, Hardhead, Hardheads, Russian Knapweed ■

6279 Acroptilon repens (L.) DC. = Centaurea repens L. ■

6280 Acroptilon repens (L.) DC. subsp. australe (Iljin) Gubanov;南方匍匐顶羽菊●☆

6281 Acroptilon repens DC. = Acroptilon repens (L.) DC. ■

6282 Acroptilon serratum Cass. = Acroptilon repens (L.) DC. ■

6283 Acroptilon subdentatum Cass. = Acroptilon repens (L.) DC. ■

6284 Acrorchis Dressier(1990);顶花兰属■☆

6285 Acrorchis roseola Dressier;顶花兰■☆

6286 Acrosanthes Eckl. et Zeyh. (1837);干裂番杏属■☆

6287 Acrosanthes Engl. = Acrossanthes C. Presl ●☆

6288 Acrosanthes Engl. = Vismia Vand. (保留属名)●☆

6289 Acrosanthes anceps (Thunb.) Sond.;干裂番杏■☆

6290 Acrosanthes angustifolia Eckl. et Zeyh.;窄叶干裂番杏■☆

6291 Acrosanthes decandra Fenzl = Acrosanthes humifusa (Thunb.) Sond. ■☆

6292 Acrosanthes fistulosa Eckl. et Zeyh. = Acrosanthes anceps (Thunb.) Sond. ■☆

6293 Acrosanthes humifusa (Thunb.) Sond.;矮小干裂番杏■☆

6294 Acrosanthes microphylla Adamson;小叶干裂番杏■☆

6295 Acrosanthes teretifolia Eckl. et Zeyh.;四叶干裂番杏■☆
6296 Acrosanthus Clem. = Acrosanthes Eckl. et Zeyh. ■☆
6297 Acroschizocarpus Gombocz = Christolea Cambess. ■
6298 Acroschizocarpus Gombocz = Smelowskia C. A. Mey. ex Ledebour (保留属名)■
6299 Acrosepalum Pierre = Ancistrocarpus Oliv. (保留属名)●☆
6300 Acrospelion Besser = Trisetum Pers. ■
6301 Acrospelion Besser ex Roem. et Schult. = Trisetum Pers. ■
6302 Acrospelion Besser ex Trin. = Trisetum Pers. ■
6303 Acrospelion Steud. = Trisetum Pers. ■
6304 Acrospelion Wittst. = Acrospelion Besser ex Roem. et Schult. ■
6305 Acrospira Welw. ex Baker = Debesia Kuntze ■☆
6306 Acrospira asphodeloides Welw. ex Baker = Chlorophytum stolzii (K. Krause) Kativu ■☆
6307 Acrospira lilioides A. Chev. = Chlorophytum tuberosum (Roxb.) Baker ■☆
6308 Acrossanthes C. Presl = Vismia Vand. (保留属名)●☆
6309 Acrossanthus Baill. = Acrossanthes C. Presl ●☆
6310 Acrossanthus C. Presl = Vismia Vand. (保留属名)●☆
6311 Acrostachys (Benth.) Tiegh. = Helixanthera Lour. ●
6312 Acrostachys Tiegh. = Helixanthera Lour. ●
6313 Acrostachys kirkii (Oliv.) Tiegh. = Helixanthera kirkii (Oliv.) Danser ●☆
6314 Acrostachys sandersonii Tiegh. = Helixanthera woodii (Schltr. et K. Krause) Danser ●☆
6315 Acrostemon Klotzsch = Erica L. ●☆
6316 Acrostemon Klotzsch(1838);尖蕊杜鹃属●☆
6317 Acrostemon barkerae Compton = Erica eriocephala Lam. ●☆
6318 Acrostemon concinnus N. E. Br. = Erica russakiana E. G. H. Oliv. ●☆
6319 Acrostemon equisetoides Klotzsch = Erica eriocephala Lam. ●☆
6320 Acrostemon eriocephalus (Klotzsch) N. E. Br. = Erica pilosiflora E. G. H. Oliv. ●☆
6321 Acrostemon fourcadei L. Guthrie = Erica angulosa E. G. H. Oliv. ●☆
6322 Acrostemon glandulosus Rach = Erica eriocephala Lam. ●☆
6323 Acrostemon hirsutus (Thunb.) Klotzsch = Erica eriocephala Lam. ●☆
6324 Acrostemon incanus Klotzsch = Erica eriocephala Lam. ●☆
6325 Acrostemon incurvus (Klotzsch) Benth. = Erica eriocephala Lam. ●☆
6326 Acrostemon schlechteri N. E. Br. = Erica radicans (L. Guthrie) E. G. H. Oliv. subsp. schlechteri (N. E. Br.) E. G. H. Oliv. ●☆
6327 Acrostemon stokoei L. Guthrie = Erica eriocephala Lam. ●☆
6328 Acrostemon thunbergii (G. Don) Alm et T. C. E. Fr. = Erica eriocephala Lam. ●☆
6329 Acrostemon viscidus N. E. Br. = Erica arachnocalyx E. G. H. Oliv. ●☆
6330 Acrostemon xeranthemifolius (Salisb.) E. G. H. Oliv. = Erica xeranthemifolia Salisb. ●☆
6331 Acrostephanus Tiegh. = Tapinanthus (Blume) Rchb. (保留属名) ●☆
6332 Acrostephanus buchneri (Engl.) Tiegh. = Tapinanthus buchneri (Engl.) Danser ●☆
6333 Acrostephanus coronatus Tiegh. = Tapinanthus coronatus (Tiegh.) Danser ●☆
6334 Acrostephanus dependens (Engl.) Tiegh. = Tapinanthus dependens (Engl.) Danser ●☆
6335 Acrostephanus ogowensis (Engl.) Tiegh. = Tapinanthus ogowensis (Engl.) Danser ●☆
6336 Acrostephanus poggei (Engl.) Tiegh. = Tapinanthus constrictiflorus (Engl.) Danser ●☆
6337 Acrostephanus syringifolius Tiegh. = Tapinanthus constrictiflorus (Engl.) Danser ●☆
6338 Acrostephanus truncatus (Engl.) Tiegh. = Tapinanthus belvisii (DC.) Danser ●☆
6339 Acrostephanus tschintschochensis (Engl.) Tiegh. = Tapinanthus buchneri (Engl.) Danser ●☆
6340 Acrostiche Dietr. = Acrotriche R. Br. ●☆
6341 Acrostigma O. F. Cook et Doyle = Catoblastus H. Wendl. ●☆
6342 Acrostigma Post et Kuntze = Acmostima Raf. ●
6343 Acrostigma Post et Kuntze = Palicourea Aubl. ●☆
6344 Acrostoma Didr. = Macrocnemum P. Browne ●☆
6345 Acrostoma Didr. = Remijia DC. ●☆
6346 Acrostylia Frapp. ex Cordem. = Cynorkis Thouars ■☆
6347 Acrostylis Post et Kuntze = Acrostylia Frapp. ex Cordem. ■☆
6348 Acrosynanthus Urb. = Remijia DC. ●☆
6349 Acrotaphros Steud. ex Hochst. = Ormocarpum P. Beauv. (保留属名)●
6350 Acrotaphros bibracteata Steud. ex A. Rich. = Ormocarpum pubescens (Hochst.) Cufod. ●☆
6351 Acrothamnus Quinn = Leucopogon R. Br. (保留属名)●☆
6352 Acrothamnus Quinn(2005);昆氏尖苞木属●☆
6353 Acrothrix Clem. = Acrotriche R. Br. ●☆
6354 Acrothrix Clem. ex Airy Shaw = Acrotriche R. Br. ●☆
6355 Acrotiche Poir. = Acrothrix Clem. ●☆
6356 Acrotoma Post et Kuntze = Acrotome Benth. ex Endl. ●■☆
6357 Acrotome Benth. = Acrotome Benth. ex Endl. ●■☆
6358 Acrotome Benth. ex Endl. (1838);顶片草属●■☆
6359 Acrotome amboensis Briq. = Acrotome inflata Benth. ■☆
6360 Acrotome angustifolia G. Taylor;窄叶顶片草■☆
6361 Acrotome belckii Gürke = Acrotome fleckii (Gürke) Launert ■☆
6362 Acrotome fleckii (Gürke) Launert;福来顶片草■☆
6363 Acrotome hispida Benth.;毛顶片草■☆
6364 Acrotome inflata Benth.;顶片草■☆
6365 Acrotome lancifolia Bremek. et Oberm. = Acrotome angustifolia G. Taylor ■☆
6366 Acrotome mozambiquensis G. Taylor;莫桑比克顶片草■☆
6367 Acrotome pallescens Benth.;苍白顶片草■☆
6368 Acrotome tenuis G. Taylor;细顶片草■☆
6369 Acrotome thorncroftii V. Naray.;托尔顶片草■☆
6370 Acrotrema Jack(1820);顶孔五桠果属●■☆
6371 Acrotrema lanceolatum Hook.;剑叶顶孔五桠果●☆
6372 Acrotrema sylvaticum Thwaites;林地顶孔五桠果●☆
6373 Acrotrema uniflorum Hook.;单花顶孔五桠果●☆
6374 Acrotrema wightianum Wight et Arn.;赖特顶孔五桠果●☆
6375 Acrotriche R. Br. (1810);顶毛石南属●☆
6376 Acrotriche affinis DC.;近缘顶毛石南●☆
6377 Acrotriche aggregata R. Br.;团集顶毛石南●☆
6378 Acrotriche lancifolia Hislop;披针叶顶毛石南●☆
6379 Acrotriche latifolia A. Cunn. ex DC.;宽叶顶毛石南●☆
6380 Acrotriche leucocarpa P. C. Jobson et Whiffin;白果顶毛石南●☆
6381 Acrotriche ovalifolia R. Br.;卵叶顶毛石南●☆
6382 Acrotriche parviflora (Stschegl.) Hislop;小花顶毛石南●☆
6383 Acrotriche rigida B. R. Paterson;硬顶毛石南●☆

6384 Acrotriche serrulata R. Br.;小齿顶毛石南●☆
6385 Acrotriche subcordata DC.;亚心形顶毛石南●☆
6386 Acroxis Steud. = Muhlenbergia Schreb. ■
6387 Acroxis Trin. ex Steud. = Muhlenbergia Schreb. ■
6388 Acrozus Spreng. = Acronodia Blume ●
6389 Acrozus Spreng. = Elaeocarpus L. ●
6390 Acrumen Gallesio = Citrus L. ●
6391 Acrymia Prain(1908);无霜草属●☆
6392 Acrymia ajugiflora Prain;无霜草●☆
6393 Acryphyllum Lindl. = Arcyphyllum Elliott ●■
6394 Acryphyllum Lindl. = Tephrosia Pers.(保留属名)●■
6395 Acsmithia Hoogland = Spiraeanthemum A. Gray ●☆
6396 Acsmithia Hoogland ex W. C. Dickison = Acsmithia Hoogland ●☆
6397 Acsmithia Hoogland(1979);多叶螺花树属●☆
6398 Acsmithia densiflora (Brongn. et Gris) Hoogland;密花多叶螺花树●☆
6399 Acsmithia grandiflora R. J. Carp. et A. M. Buchanan;大花多叶螺花树●☆
6400 Acsmithia integrifolia (Pulle) Hoogland;全缘多叶螺花树●☆
6401 Acsmithia laxiflora Hoogland;疏花多叶螺花树●☆
6402 Acsmithia parvifolia (Schltr.) Hoogland;小叶多叶螺花树●☆
6403 Acsmithia pubescens (Pamp.) Hoogland;毛多叶螺花树●☆
6404 Acsmithia reticulata Hoogland ex W. C. Dickison;网脉多叶螺花树●☆
6405 Actaea L. (1753);类叶升麻属(绿豆升麻属);Baneberry, Coralberry, Doll's-eyes, Snakeberry ■
6406 Actaea Lour. = Tetracera L. ●
6407 Actaea acerina Prantl = Cimicifuga japonica (Thunb.) Spreng. ■
6408 Actaea acuminata Wall.;尖类叶升麻■☆
6409 Actaea acuminata Wall. subsp. asiatica (H. Hara) Luferov = Actaea asiatica H. Hara ■
6410 Actaea alba Mill. = Actaea pachypoda Elliott ■
6411 Actaea arguta Nutt. = Actaea rubra (Aiton) Willd. ■☆
6412 Actaea arguta Nutt. ex Torr. et A. Gray;锐齿类叶升麻;Western Baneberry ■☆
6413 Actaea arguta Nutt. var. viridiflora (Greene) Tidestr. = Actaea rubra (Aiton) Willd. ■☆
6414 Actaea asiatica H. Hara;类叶升麻(茶七,金丝三七,开喉箭,绿豆升麻,马尾升麻,米升麻,升麻,五角莲);Asian Baneberry ■
6415 Actaea aspera Lour. = Tetracera asiatica (Lour.) Hoogland ●
6416 Actaea brachycarpa (P. K. Hsiao) J. Compton = Cimicifuga brachycarpa P. K. Hsiao ■
6417 Actaea cimicifuga L. = Cimicifuga foetida L. ■
6418 Actaea cimicifuga L. var. simplex DC. = Cimicifuga simplex Wormsk. ex DC. ■
6419 Actaea dahurica (Turcz. ex Fisch. et C. A. Mey.) Turcz. ex Fisch. et C. A. Mey. = Cimicifuga dahurica (Turcz. ex Fisch. et C. A. Mey.) Maxim. ■
6420 Actaea dahurica Turcz. ex Fisch. et C. A. Mey. = Cimicifuga dahurica (Turcz. ex Fisch. et C. A. Mey.) Maxim. ■
6421 Actaea eburnea Rydb. = Actaea rubra (Aiton) Willd. ■☆
6422 Actaea erythrocarpa Fisch. ex Fisch. et C. A. Mey.;红果类叶升麻;Redfruit Baneberry ■
6423 Actaea frigida (Royle) Prantl = Cimicifuga foetida L. ■
6424 Actaea frigida Wall. = Cimicifuga foetida L. ■
6425 Actaea heracleifolia (Kom.) J. Compton = Cimicifuga heracleifolia Kom. ■
6426 Actaea japonica Thunb. = Cimicifuga japonica (Thunb.) Spreng. ■
6427 Actaea ludovici B. Boivin;杂种类叶升麻;Hybrid Actaea ■☆
6428 Actaea mairei (H. Lév.) J. Compton = Cimicifuga foetida L. ■
6429 Actaea mairei (H. Lév.) J. Compton var. foliolosa (P. K. Hsiao) J. Compton = Cimicifuga foetida L. var. foliolosa P. K. Hsiao ■
6430 Actaea monogyna Walter = Cimicifuga racemosa (L.) Nutt. ■☆
6431 Actaea neglecta Gillman = Actaea rubra (Aiton) Willd. ■☆
6432 Actaea odorata ?;芳香类叶升麻■☆
6433 Actaea pachypoda Elliott;白果类叶升麻(粗柄类叶升麻);Baneberry, Doll's Eyes, Doll's-eyes, Necklace Weed, Snakeroot, Thick-pedicled Baneberry, Toadroot, White Baneberry, White Cohosh ■
6434 Actaea pachypoda Elliott f. microcarpa (DC.) Fassett = Actaea pachypoda Elliott ■
6435 Actaea pachypoda Elliott f. rubrocarpa (Killip) Fernald = Actaea pachypoda Elliott ■
6436 Actaea pterosperma Turcz. ex Fisch. et C. A. Mey. = Cimicifuga dahurica (Turcz. ex Fisch. et C. A. Mey.) Maxim. ■
6437 Actaea purpurea (P. K. Hsiao) J. Compton = Cimicifuga japonica (Thunb.) Spreng. ■
6438 Actaea racemosa L.;美丽类叶升麻;Black Cohosh ■☆
6439 Actaea racemosa L. = Cimicifuga racemosa (L.) Nutt. ■☆
6440 Actaea richardsonii ?;理查逊类叶升麻■☆
6441 Actaea rubra (Aiton) Willd.;美国类叶升麻(红果类叶升麻,红类叶升麻);Baneberry, Black Cohosh, Dolls-eyes, Red Baneberry, Red Bane-berry ■☆
6442 Actaea rubra (Aiton) Willd. f. neglecta (Gillman) B. L. Rob. = Actaea rubra (Aiton) Willd. ■☆
6443 Actaea rubra (Aiton) Willd. subsp. arguta (Nutt.) Hultén = Actaea rubra (Aiton) Willd. ■☆
6444 Actaea rubra (Aiton) Willd. var. arguta (Nutt.) Lawson = Actaea rubra (Aiton) Willd. ■☆
6445 Actaea rubra (Aiton) Willd. var. dissecta Britton = Actaea rubra (Aiton) Willd. ■☆
6446 Actaea rubra (Aiton) Willd. var. gigantea R. R. Gates = Actaea rubra (Aiton) Willd. ■☆
6447 Actaea simplex (DC.) Wormsk. ex Fisch. et C. A. Mey. = Cimicifuga simplex Wormsk. ex DC. ■
6448 Actaea spicata L.;穗花类叶升麻;Baneberry, Black Baneberry, Black Cohosh, Bugbane, Doll's Eyes, Grapewort, Herb Christopher, Toadroot ■☆
6449 Actaea spicata L. subsp. rubra (Aiton) Hultén = Actaea rubra (Aiton) Willd. ■☆
6450 Actaea spicata L. var. arguta (Nutt.) Torr. = Actaea rubra (Aiton) Willd. ■☆
6451 Actaea spicata L. var. asiatica (Hara) S. H. Li et Y. Huei Huang = Tetracera asiatica (Lour.) Hoogland ●
6452 Actaea spicata L. var. erythrocarpa (Fisch. ex Fisch. et C. A. Mey.) Turcz. = Actaea erythrocarpa Fisch. ex Fisch. et C. A. Mey. ■
6453 Actaea spicata L. var. rubra Aiton = Actaea rubra (Aiton) Willd. ■☆
6454 Actaea vaginata (Maxim.) J. Compton = Souliea vaginata (Maxim.) Franch. ■
6455 Actaea viridiflora Greene = Actaea rubra (Aiton) Willd. ■☆
6456 Actaea yunnanensis (P. K. Hsiao) J. Compton = Cimicifuga yunnanensis P. K. Hsiao ■
6457 Actaeaceae Bercht. et J. Presl = Ranunculaceae Juss.(保留科名)●■

6458 Actaeaceae Raf. = Ranunculaceae Juss.(保留科名)●■

6459 Actaeogeton Rchb. = Actegeton Blume ●

6460 Actaeogeton Rchb. = Azima Lam. ●

6461 Actaeogeton Steud. = Scirpus L.(保留属名)■

6462 Actartife Raf. = Boltonia L' Hér. ■☆

6463 Actea Raf. = Actaea L. ■

6464 Actegeton Blume = Azima Lam. ●

6465 Actegiton Endl. = Actegeton Blume ●

6466 Actephila Blume(1826);喜光花属(滨木属);Actephila ●

6467 Actephila africana Pax = Pentabrachion reticulatum Müll. Arg. ●☆

6468 Actephila dolichantha Croizat = Actephila excelsa (Dalzell) Müll. Arg. ●

6469 Actephila excelsa (Dalzell) Müll. Arg.;毛喜光花(长花喜光花,喜光花);Long-flowered Actephila ●

6470 Actephila grandifolia Baill.;大叶喜光花●☆

6471 Actephila inopinata Croizat = Actephila merrilliana Chun ●

6472 Actephila javanica Miq.;爪哇喜光花●☆

6473 Actephila latifolia Benth.;宽叶喜光花●☆

6474 Actephila longipedicellata P. I. Forst.;长梗喜光花●☆

6475 Actephila macrantha Gagnep.;大花喜光花●☆

6476 Actephila merrilliana Chun;喜光花(海南喜光花);Merrill Actephila ●

6477 Actephila ovalis Gage;卵形喜光花●☆

6478 Actephila rectinervis Kurz;网脉喜光花●☆

6479 Actephila reticulata (Müll. Arg.) Pax = Pentabrachion reticulatum Müll. Arg. ●☆

6480 Actephila sessilifolia Benth.;无柄喜光花●☆

6481 Actephila subessilis Pierre ex Gagnep.;短柄喜光花;Shortstalk Actephila, Shortstalked Actephila ●

6482 Actephila trichogyna Airy Shaw;毛蕊喜光花●☆

6483 Actephilopsis Ridl.(1923);类喜光花属●

6484 Actephilopsis Ridl. = Trigonostemon Blume(保留属名)●

6485 Actephilopsis Ridl. = Tylosepalum Kurz ex Teijsm. et Binn. ●

6486 Actephilopsis malayana Ridl.;类喜光花●

6487 Acticarnopus Raf. = Actinocarpus Raf. ■☆

6488 Acticarnopus Raf. = Chaetopappa DC. ■☆

6489 Actimeris Raf. = Actinomeris Nutt.(保留属名)■☆

6490 Actinanthella Balle(1954);水芹状寄生属●☆

6491 Actinanthella menyharthii (Engl. et Schinz ex Schinz) Balle;水芹状寄生●☆

6492 Actinanthella wyliei (Sprague) Wiens;非洲水芹状寄生●☆

6493 Actinanthus Ehrenb. = Oenanthe L. ■

6494 Actinea Juss. = Helenium L. ■

6495 Actinea subintegra (Cockerell) S. F. Blake = Hymenoxys subintegra Cockerell ■☆

6496 Actinella Juss. ex Nutt. = Actinea Juss. ■

6497 Actinella Juss. ex Nutt. = Gaillardia Foug. ■

6498 Actinella Pers. = Actinea Juss. ■

6499 Actinella argentea A. Gray = Tetraneuris argentea (A. Gray) Greene ■☆

6500 Actinella bigelovii A. Gray = Hymenoxys bigelowii (A. Gray) K. L. Parker ■☆

6501 Actinella bigelovii A. Gray = Macdougalia bigelovii A. Heller ■☆

6502 Actinella brandegeei Porter ex A. Gray = Hymenoxys brandegeei (Porter ex A. Gray) K. L. Parker ■☆

6503 Actinella cooperi A. Gray = Hymenoxys cooperi (A. Gray) Cockerell ■☆

6504 Actinella grandiflora Torr. et A. Gray = Hymenoxys grandiflora (Torr. et A. Gray) K. L. Parker ■☆

6505 Actinella lanata Pursh = Eriophyllum lanatum (Pursh) J. Forbes ■☆

6506 Actinella palmeri A. Gray = Plateilema palmeri (A. Gray) Cockerell ■☆

6507 Actinella richardsonii (Hook.) Nutt. var. floribunda A. Gray = Hymenoxys richardsonii (Hook.) Cockerell var. floribunda (A. Gray) K. L. Parker ■☆

6508 Actinella rusbyi A. Gray = Hymenoxys rusbyi (A. Gray) Cockerell ■☆

6509 Actinella texana J. M. Coult. et Rose = Hymenoxys texana (J. M. Coult. et Rose) Cockerell ■☆

6510 Actinella torreyana Nutt. = Tetraneuris torreyana (Nutt.) Greene ■☆

6511 Actinella vaseyi A. Gray = Hymenoxys vaseyi (A. Gray) Cockerell ■☆

6512 Actinia Griff. = Dendrobium Sw.(保留属名)■

6513 Actinidia Lindl. (1836);猕猴桃属;Actinidia, Chinese Gooseberry, Kiwi, Kiwi Fruit, Kiwifruit ●

6514 Actinidia albicalyx R. G. Li et M. Y. Liang;白萼猕猴桃;Whitecalyx Kiwifruit ●

6515 Actinidia arguta (Siebold et Zucc.) Planch. ex Miq.;软枣猕猴桃(毛梨子,猕猴梨,猕猴李,软枣子,藤瓜,藤梨,阳桃,洋桃梨,洋桃藤,圆枣子);Bower Actinidia, Bower Kiwifruit, Hardy Kiwi, Tara Actinidia, Tara Vine ●

6516 Actinidia arguta (Siebold et Zucc.) Planch. ex Miq. f. platyphylla (A. Gray ex Miq.) H. Ohba;宽叶软枣猕猴桃●☆

6517 Actinidia arguta (Siebold et Zucc.) Planch. ex Miq. f. rufinervis (Nakai) Sugim.;红脉软枣猕猴桃●☆

6518 Actinidia arguta (Siebold et Zucc.) Planch. ex Miq. var. cordifolia (Miq.) Bean;心叶猕猴桃;Heartleaf Actinidia, Heartleaf Kiwifruit ●

6519 Actinidia arguta (Siebold et Zucc.) Planch. ex Miq. var. giraldii (Diels) Vorosch.;陕西猕猴桃;Girald Actinidia, Girald Kiwifruit ●

6520 Actinidia arguta (Siebold et Zucc.) Planch. ex Miq. var. hypoleuca (Nakai) Kitam.;里白软枣猕猴桃●☆

6521 Actinidia arguta (Siebold et Zucc.) Planch. ex Miq. var. nervosa C. F. Liang;凸脉猕猴桃;Convexvein Actinidia, Convexvein Kiwifruit ●

6522 Actinidia arguta (Siebold et Zucc.) Planch. ex Miq. var. nervosa C. F. Liang = Actinidia arguta (Siebold et Zucc.) Planch. ex Miq. var. giraldii (Diels) Vorosch. ●

6523 Actinidia arguta (Siebold et Zucc.) Planch. ex Miq. var. platyphylla (A. Gray) Nakai;日本猕猴桃;Japanese Actinidia ●☆

6524 Actinidia arguta (Siebold et Zucc.) Planch. ex Miq. var. platyphylla (A. Gray ex Miq.) Nakai = Actinidia arguta (Siebold et Zucc.) Planch. ex Miq. f. platyphylla (A. Gray ex Miq.) H. Ohba ●☆

6525 Actinidia arguta (Siebold et Zucc.) Planch. ex Miq. var. purpurea (Rehder) C. F. Liang;紫果猕猴桃(牛奶果,牛奶奶,小羊桃,羊奶奶,羊奶子);Purplefruit Actinidia, Purplefruit Kiwifruit ●

6526 Actinidia arguta (Siebold et Zucc.) Planch. ex Miq. var. rufinervis Nakai = Actinidia arguta (Siebold et Zucc.) Planch. ex Miq. f. rufinervis (Nakai) Sugim. ●☆

6527 Actinidia arisanensis Hayata;阿里山猕猴桃●

6528 Actinidia arisanensis Hayata = Actinidia callosa Lindl. ●

6529 Actinidia asymmetrica F. Chun = Actinidia glaucophylla F. Chun var. asymmetrica (F. Chun) C. F. Liang ●

6530 Actinidia callosa Lindl.;硬齿猕猴桃(阿里山猕猴桃,京梨,台湾猕猴桃);Alishan Actinidia, Alishan Kiwifruit, Arishan Actinidia, Callose Actinidia, Callose Kiwifruit, Formosan Actinidia, Taiwan

Actinidia ●

6531 Actinidia callosa Lindl. var. acuminata C. F. Liang;尖叶猕猴桃;Sharpleaf Actinidia,Sharpleaf Kiwifruit ●

6532 Actinidia callosa Lindl. var. coriacea Finet et Gagnep. = Actinidia rubricaulis Dunn var. coriacea (Finet et Gagnep.) C. F. Liang ●

6533 Actinidia callosa Lindl. var. discolor C. F. Liang;异色猕猴桃(鸡考果,京梨猕猴桃);Discolored Actinidia,Discolored Kiwifruit ●

6534 Actinidia callosa Lindl. var. ephippioidea C. F. Liang;驼齿猕猴桃;Camel-tooth Actinidia ●

6535 Actinidia callosa Lindl. var. formosana Finet et Gagnep.;台湾猕猴桃●

6536 Actinidia callosa Lindl. var. formosana Finet et Gagnep. = Actinidia callosa Lindl. ●

6537 Actinidia callosa Lindl. var. henryi Maxim.;京梨猕猴桃(鸡考果,水梨儿藤,水梨藤,异色猕猴桃);Henry Actinidia,Henry Kiwifruit ●

6538 Actinidia callosa Lindl. var. indochinensis (Merr.) H. L. Li = Actinidia indochinensis Merr. ●

6539 Actinidia callosa Lindl. var. pilosula Finet et Gagnep. = Actinidia pilosula (Finet et Gagnep.) Stapf ex Hand. -Mazz. ●◇

6540 Actinidia callosa Lindl. var. pubiramura C. Y. Wu;毛枝秤花藤;Hairbranch Callose Actinidia ●

6541 Actinidia callosa Lindl. var. pubiramura C. Y. Wu = Actinidia callosa Lindl. ●

6542 Actinidia callosa Lindl. var. sabiifolia Dunn = Actinidia sabiifolia Dunn ●

6543 Actinidia callosa Lindl. var. strigilosa C. F. Liang;毛叶硬齿猕猴桃(秤砣梨);Henryleaf Callose Actinidia,Henryleaf Callose Kiwifruit ●

6544 Actinidia callosa Lindl. var. trichogyna Finet et Gagnep. = Actinidia trichogyna Franch. ●

6545 Actinidia carnosifolia C. Y. Wu;肉叶猕猴桃;Fleshyleaf Actinidia,Fleshyleaf Kiwifruit,Fleshy-leaved Actinidia ●

6546 Actinidia carnosifolia C. Y. Wu var. glaucescens C. F. Liang;奶果猕猴桃(小果冬藤);Milk-fruit Actinidia,Milk-fruit Kiwifruit ●

6547 Actinidia championii Benth. = Actinidia latifolia (Gardner et Champ.) Merr. ●

6548 Actinidia championii Benth. var. mollis Dunn = Actinidia latifolia (Gardner et Champ.) Merr. var. mollis (Dunn) Hand. -Mazz. ●

6549 Actinidia changii P. S. Hsu;章氏猕猴桃;Zhang's Actinidia,Zhang's Kiwifruit ●

6550 Actinidia changii P. S. Hsu = Actinidia melanandra Franch. ●

6551 Actinidia chartacea Hu = Actinidia arguta (Siebold et Zucc.) Planch. ex Miq. var. purpurea (Rehder) C. F. Liang ●

6552 Actinidia chengkouensis C. Yu Chang;城口猕猴桃;Chengkou Actinidia,Chengkou Kiwifruit ●

6553 Actinidia chinensis Planch.;中华猕猴桃(白毛桃,大红袍,大零核,鬼桃,猴仔梨,猴子梨,狐狸桃,江藤梨,金梨,毛梨子,毛栗树,毛叶猕猴桃,猕猴梨,猕猴桃,木子,山洋桃,绳梨,藤梨,羊桃,阳桃,杨桃,洋桃,野梨);China Gooseberry,Chinese Actinidia,Chinese Gooseberry,Hupeh Vine,Kiwi,Kiwi Fruit,Yangtao,Yangtao Actinidia,Yangtao Kiwifruit ●

6554 Actinidia chinensis Planch. f. chlorocarpa C. F. Liang = Actinidia chinensis Planch. var. deliciosa (A. Chev.) A. Chev. ●

6555 Actinidia chinensis Planch. f. jinggangshanensis C. F. Liang = Actinidia chinensis Planch. var. jinggangshanensis (C. F. Liang) C. F. Liang et A. R. Ferguson ●

6556 Actinidia chinensis Planch. f. longipila C. F. Liang et R. Z. Wang = Actinidia chinensis Planch. var. deliciosa (A. Chev.) A. Chev. ●

6557 Actinidia chinensis Planch. f. rufopulpa C. F. Liang et R. H. Huang = Actinidia chinensis Planch. var. rufopulpa (C. F. Liang et R. H. Huang) C. F. Liang et A. R. Ferguson ●

6558 Actinidia chinensis Planch. var. deliciosa (A. Chev.) A. Chev.;美味猕猴桃;Chinese Gooseberry,Delicious Actinidia,Delicious Kiwifruit,Hispid Actinidia,Kiwi,Kiwi Fruit,Yangtao ●

6559 Actinidia chinensis Planch. var. hispida C. F. Liang;硬毛猕猴桃(毛阳桃);Hispid Actinidia,Hispid Kiwifruit ●

6560 Actinidia chinensis Planch. var. hispida C. F. Liang = Actinidia chinensis Planch. var. deliciosa (A. Chev.) A. Chev. ●

6561 Actinidia chinensis Planch. var. hispida C. F. Liang f. chlorocarpa C. F. Liang et R. Z. Wang;绿果猕猴桃●

6562 Actinidia chinensis Planch. var. hispida C. F. Liang f. longipila C. F. Liang et R. Z. Wang;长毛猕猴桃●

6563 Actinidia chinensis Planch. var. jinggangshanensis (C. F. Liang) C. F. Liang et A. R. Ferguson;井冈山猕猴桃;Jinggangshan Actinidia,Jinggangshan Kiwifruit ●

6564 Actinidia chinensis Planch. var. lageniformis S. Y. Wang et C. F. Chen;葫果猕猴桃;Lageniforme Actinidia,Lageniforme Kiwifruit ●

6565 Actinidia chinensis Planch. var. nephrocarpa S. Y. Wang et C. F. Chen;肾果猕猴桃;Kidnyfruit Actinidia ●

6566 Actinidia chinensis Planch. var. rufopulpa (C. F. Liang et R. H. Huang) C. F. Liang et A. R. Ferguson;红肉猕猴桃;Redfleshy Yangtao Actinidia ●

6567 Actinidia chinensis Planch. var. setosa H. L. Li;刺毛猕猴桃(台湾羊桃);Setulose Actinidia,Setulose Kiwifruit,Taiwan Actinidia ●

6568 Actinidia chrysantha C. F. Liang;金花猕猴桃;Goldenflower Actinidia,Goldenflower Kiwifruit,Golden-flowered Actinidia ●◇

6569 Actinidia cinerascens C. F. Liang = Actinidia fulvicoma Hance var. cinerascens (C. F. Liang) J. Q. Li et Soejarto ●

6570 Actinidia cinerascens C. F. Liang var. longipetiolata C. F. Liang;长叶柄猕猴桃;Longstalk Actinidia,Longstalk Kiwifruit ●

6571 Actinidia cinerascens C. F. Liang var. longipetiolata C. F. Liang = Actinidia fulvicoma Hance var. cinerascens (C. F. Liang) J. Q. Li et Soejarto ●

6572 Actinidia cinerascens C. F. Liang var. tenuifolia C. F. Liang;菲叶猕猴桃;Tenuousleaf Actinidia,Tenuousleaf Kiwifruit ●

6573 Actinidia cinerascens C. F. Liang var. tenuifolia C. F. Liang. = Actinidia fulvicoma Hance var. cinerascens (C. F. Liang) J. Q. Li et Soejarto ●

6574 Actinidia cordifolia Miq. = Actinidia arguta (Siebold et Zucc.) Planch. ex Miq. var. cordifolia (Miq.) Bean ●

6575 Actinidia coriacea (Finet et Gagnep.) Dunn = Actinidia rubricaulis Dunn var. coriacea (Finet et Gagnep.) C. F. Liang ●

6576 Actinidia curvidens Dunn = Actinidia callosa Lindl. var. henryi Maxim. ●

6577 Actinidia cylindrica C. F. Liang;柱果猕猴桃;Column Fruit Actinidia,Cylindric Kiwifruit,Cylindrical Actinidia ●

6578 Actinidia cylindrica C. F. Liang f. obtusifolia C. F. Liang;钝叶猕猴桃;Obtuse-leaf Actinidia,Obtuse-leaf Kiwifruit ●

6579 Actinidia cylindrica C. F. Liang var. reticulata C. F. Liang;网脉猕猴桃;Net-veined Actinidia,Net-veined Kiwifruit ●

6580 Actinidia davidii Franch. = Actinidia eriantha Benth. ●

6581 Actinidia deliciosa (A. Chev.) C. F. Liang et A. R. Ferguson = Actinidia chinensis Planch. ●

6582 Actinidia deliciosa (A. Chev.) C. F. Liang et A. R. Ferguson = Actinidia chinensis Planch. var. deliciosa (A. Chev.) A. Chev. ●

6583 Actinidia deliciosa (A. Chev.) C. F. Liang et A. R. Ferguson var. chlorocarpa (C. F. Liang) C. F. Liang et A. R. Ferguson = Actinidia chinensis Planch. var. deliciosa (A. Chev.) A. Chev. ●

6584 Actinidia deliciosa (A. Chev.) C. F. Liang et A. R. Ferguson var. coloris T. H. Lin et X. Y. Xiong;彩色猕猴桃;Colorus Actinidia, Colorus Kiwifruit ●

6585 Actinidia deliciosa (A. Chev.) C. F. Liang et A. R. Ferguson var. coloris T. H. Lin et X. Y. Xiong = Actinidia chinensis Planch. var. deliciosa (A. Chev.) A. Chev. ●

6586 Actinidia deliciosa (A. Chev.) C. F. Liang et A. R. Ferguson var. longipila (C. F. Liang et R. Z. Wang) C. F. Liang et A. R. Ferguson = Actinidia chinensis Planch. var. deliciosa (A. Chev.) A. Chev. ●

6587 Actinidia diversicolona R. G. Li;二色猕猴桃;Bicolor Kiwifruit ●

6588 Actinidia eriantha Benth.;毛花猕猴桃(白毛卵,白毛桃,白葡萄,白藤梨,白洋桃,毛冬瓜,毛狗卵,毛花阳桃,毛花杨桃,毛卵,毛藤里公,山蒲桃,生毛藤梨);Hairyflower Actinidia, Hairyflower Kiwifruit, Hairy-flowered Actinidia ●

6589 Actinidia eriantha Benth. f. alba C. F. Gan;白色毛花猕猴桃;Whitle Hairyflower Actinidia, Whitle Hairyflower Kiwifruit ●

6590 Actinidia eriantha Benth. var. brunnea C. F. Liang;棕毛毛花猕猴桃;Brownhairy Hairyflower Actinidia ●

6591 Actinidia eriantha Benth. var. calvescens C. F. Liang;秃果毛花猕猴桃;Barefruit Hairyflower Actinidia, Barefruit Hairyflower Kiwifruit ●

6592 Actinidia fanjingshanensis S. D. Shi et Q. B. Wang;梵净山猕猴桃;Fanjingshan Actinidia, Fanjingshan Kiwifruit ●

6593 Actinidia farinosa C. F. Liang;粉毛猕猴桃;Farinose Kiwifruit, Powdery Hair Actinidia, Powdery-haired Actinidia ●

6594 Actinidia fasciculoides C. F. Liang;簇花猕猴桃;Clustered Flower Actinidia, Clustered-flowered Actinidia, Fascicledflower Kiwifruit ●

6595 Actinidia fasciculoides C. F. Liang var. cuneata C. F. Liang;楔叶猕猴桃;Cuneate-leaf Actinidia, Cuneate-leaf Kiwifruit ●

6596 Actinidia fasciculoides C. F. Liang var. orbiculata C. F. Liang;圆叶猕猴桃;Round Leaf Actinidia ●

6597 Actinidia formosana Hayata = Actinidia callosa Lindl. var. formosana Finet et Gagnep. ●

6598 Actinidia fortunatii Finet et Gagnep.;光萼猕猴桃(粉绿猕猴桃,条叶猕猴桃);Beltleaf Actinidia, Fortunat Actinidia, Fortunat Kiwifruit ●

6599 Actinidia fulvicoma Hance;黄毛猕猴桃(棕毛猕猴桃);Yellow-haired Actinidia, Yellowhairy Actinidia, Yellowhairy Kiwifruit ●

6600 Actinidia fulvicoma Hance f. arachnoidea C. F. Liang = Actinidia fulvicoma Hance var. lanata (Hemsl.) C. F. Liang f. arachnoidea C. F. Liang ●

6601 Actinidia fulvicoma Hance f. hirsuta (Finet et Gagnep.) C. F. Liang = Actinidia fulvicoma Hance var. lanata (Hemsl.) C. F. Liang f. hirsuta (Finet et Gagnep.) C. F. Liang ●

6602 Actinidia fulvicoma Hance f. hirsuta (Finet et Gagnep.) C. F. Liang = Actinidia fulvicoma Hance var. hirsuta Finet et Gagnep. ●

6603 Actinidia fulvicoma Hance f. lanata Hemsl. = Actinidia fulvicoma Hance var. hirsuta Finet et Gagnep. ●

6604 Actinidia fulvicoma Hance var. cinerascens (C. F. Liang) J. Q. Li et Soejarto;灰毛猕猴桃;Gray-hair Actinidia, Gray-hair Kiwifruit, Gray-haired Actinidia ●

6605 Actinidia fulvicoma Hance var. hirsuta Finet et Gagnep.;粗毛猕猴桃(糙毛猕猴桃)●

6606 Actinidia fulvicoma Hance var. hirsuta Finet et Gagnep. = Actinidia fulvicoma Hance var. lanata (Hemsl.) C. F. Liang f. hirsuta (Finet et Gagnep.) C. F. Liang ●

6607 Actinidia fulvicoma Hance var. lanata (Hemsl.) C. F. Liang;绵毛猕猴桃;Woolly Actinidia ●

6608 Actinidia fulvicoma Hance var. lanata (Hemsl.) C. F. Liang f. arachnoidea C. F. Liang;丝毛猕猴桃●

6609 Actinidia fulvicoma Hance var. lanata (Hemsl.) C. F. Liang f. hirsuta (Finet et Gagnep.) C. F. Liang = Actinidia fulvicoma Hance var. hirsuta Finet et Gagnep. ●

6610 Actinidia fulvicoma Hance var. pachyphylla (Dunn) H. L. Li;厚叶猕猴桃;Thickleaf Actinidia, Thickleaf Kiwifruit ●

6611 Actinidia gagnepaini Nakai = Actinidia kolomicta (Rupr. et Maxim.) Maxim. ●

6612 Actinidia giraldii Diels = Actinidia arguta (Siebold et Zucc.) Planch. ex Miq. var. giraldii (Diels) Vorosch. ●

6613 Actinidia glabra H. L. Li. = Actinidia indochinensis Merr. ●

6614 Actinidia glabra L.;光叶猕猴桃;Glabrous Leaf Actinidia ●☆

6615 Actinidia glauco-callosa C. Y. Wu;粉叶猕猴桃;Farinoseleaf Kiwifruit, Powdery-leaf Actinidia, Powdery-leaved Actinidia ●

6616 Actinidia glaucophylla F. Chun;华南猕猴桃(猴子果,羊奶奶);Greyleaf Actinidia, Grey-leaved Actinidia, S. China Kiwifruit ●

6617 Actinidia glaucophylla F. Chun var. asymmetrica (F. Chun) C. F. Liang;耳叶猕猴桃(歪叶猕猴桃);Asymmetrical Leaf Actinidia, Asymmetrical Leaf Kiwifruit, Auriculate Leaf Actinidia, Auriculate Leaf Kiwifruit ●

6618 Actinidia glaucophylla F. Chun var. robusa C. F. Liang;粗叶猕猴桃;Thickleaf Actinidia, Thickleaf Kiwifruit ●

6619 Actinidia glaucophylla F. Chun var. rotunda C. F. Liang;团叶猕猴桃;Roundleaf Actinidia, Roundleaf Kiwifruit ●

6620 Actinidia globasa C. F. Liang;圆果猕猴桃;Circular Actinidia Ballfruit Kiwifruit, Circular Fruit Actinidia ●

6621 Actinidia gnaphalocarpa Hayata = Actinidia latifolia (Gardner et Champ.) Merr. ●

6622 Actinidia gracilis C. F. Liang;纤小猕猴桃;Tenuous Vine Actinidia, Tenuous-vined Actinidia, Weak Kiwifruit ●

6623 Actinidia grandiflora C. F. Liang;大花猕猴桃;Bigflower Kiwifruit, Big-flowered Actinidia, Large Flower Actinidia ●◇

6624 Actinidia guilinensis C. F. Liang;桂林猕猴桃;Guilin Kiwifruit ●

6625 Actinidia hemsleyana Dunn;长叶猕猴桃;Hemsley Actinidia, Hemsley Kiwifruit ●

6626 Actinidia hemsleyana Dunn var. kengiana (F. P. Metcalf) C. F. Liang;粗齿猕猴桃;Thick Tooth Actinidia, Thick Tooth Kiwifruit ●

6627 Actinidia henanensis C. F. Liang;河南猕猴桃;Henan Actinidia, Henan Kiwifruit, Honan Actinidia ●◇

6628 Actinidia henry Dunn;蒙自猕猴桃;Henry Actinidia, Henry Kiwifruit ●

6629 Actinidia henry Dunn var. glabricaulis (C. Y. Wu) C. F. Liang;光茎蒙自猕猴桃(光茎猕猴桃);Glabrous-stem Actinidia ●

6630 Actinidia henry Dunn var. polyodonta Hand. -Mazz.;多齿猕猴桃;Manytoothed Actinidia ●

6631 Actinidia holotricha Finet et Gagnep.;全毛猕猴桃;Holo-haired Actinidia, Holohairy Actinidia, Holohairy Kiwifruit ●

6632 Actinidia hubeiensis H. M. Sun et R. H. Huang;湖北猕猴桃;Hubei Actinidia, Hubei Kiwifruit ●

6633 Actinidia hypoleuca Nakai;白背猕猴桃;Whiteback Actinidia, Whiteback Kiwifruit ●

6634 Actinidia hypoleuca Nakai = Actinidia arguta (Siebold et Zucc.) Planch. ex Miq. var. hypoleuca (Nakai) Kitam. ●☆

6635 Actinidia indochinensis Merr.;中越猕猴桃;Indochina Actinidia, Indo-China Actinidia, Indochina Kiwifruit, Indochinese Actinidia ●

6636 Actinidia indochinensis Merr. var. ovatifolia R. G. Li, X. G. Wang et L. Mo;卵圆叶猕猴桃●

6637 Actinidia japonica Nakai = Actinidia arguta (Siebold et Zucc.) Planch. ex Miq. var. platyphylla (A. Gray) Nakai ●☆

6638 Actinidia jiangkouensis S. D. Shi et Z. S. Zhang;江口猕猴桃;Jiangkou Actinidia, Jiangkou Kiwifruit ●

6639 Actinidia kolomicta (Rupr. et Maxim.) Maxim.;狗枣猕猴桃(狗枣猕猴梨,狗枣子,深山木天蓼);Kiwi Vine, Kiwi, Kolomikta Actinidia, Kolomikta Kiwifruit, Kolomikta Vine, Kolomikta-vine Actinidia, Kolomikta-vine, Kolomikta-vined Actinidia, Manchurian Gooseberry ●

6640 Actinidia kolomicta (Rupr. et Maxim.) Maxim. var. gagnepainii (Nakai) H. L. Li = Actinidia kolomicta (Rupr. et Maxim.) Maxim. ●

6641 Actinidia kungshanensis C. Y. Wu et S. K. Chen = Actinidia pilosula (Finet et Gagnep.) Stapf ex Hand. -Mazz. ●◇

6642 Actinidia kwangsiensis H. L. Li = Actinidia arguta (Siebold et Zucc.) Planch. ex Miq. var. giraldii (Diels) Vorosch. ●

6643 Actinidia kwangsiensis H. L. Li = Actinidia melanandra Franch. var. kwangsiensis (H. L. Li) C. F. Liang ●

6644 Actinidia laevissima C. F. Liang;滑叶猕猴桃;Smoothleaf Actinidia, Smoothleaf Kiwifruit, Smooth-leaved Actinidia ●

6645 Actinidia laevissima C. F. Liang var. floscula S. D. Shi;小花猕猴桃;Small-flower Smoothleaf Kiwifruit ●

6646 Actinidia lanata Hemsl. = Actinidia fulvicoma Hance var. lanata (Hemsl.) C. F. Liang ●

6647 Actinidia lanceolata Dunn;小叶猕猴桃;Lanceolate Actinidia, Lanceolate Kiwifruit ●

6648 Actinidia latifolia (Gardner et Champ.) Merr.;阔叶猕猴桃(多果猕猴桃,多花猕猴桃,宽叶猕猴桃);Broadleaf Actinidia, Broadleaf Kiwifruit, Broad-leaved Actinidia ●

6649 Actinidia latifolia (Gardner et Champ.) Merr. var. deliciosa A. Chev. = Actinidia chinensis Planch. var. deliciosa (A. Chev.) A. Chev. ●

6650 Actinidia latifolia (Gardner et Champ.) Merr. var. indichinensis (H. L. Li) H. L. Li = Actinidia latifolia (Gardner et Champ.) Merr. ●

6651 Actinidia latifolia (Gardner et Champ.) Merr. var. mollis (Dunn) Hand. -Mazz.;长绒猕猴桃;Loongwoolly Actinidia, Loongwoolly Kiwifruit ●

6652 Actinidia lecomtei Nakai = Actinidia polygama (Siebold et Zucc.) Maxim. ●

6653 Actinidia leptophylla C. Y. Wu;薄叶猕猴桃;Thin-leaf Actinidia, Thin-leaf Kiwifruit, Thin-leaved Actinidia ●

6654 Actinidia liangkwangensis C. F. Liang;两广猕猴桃(鱼网藤);Guangdong-Guangxi Actinidia, Guangdong-Guangxi Kiwifruit, Kwangtung-kwangsi Actinidia, Kwangtung-kwangsi Kiwifruit ●

6655 Actinidia lijiangensis C. F. Liang et Y. X. Lu;漓江猕猴桃;Lijiang Kiwifruit ●

6656 Actinidia linguiensis R. G. Li et L. Mo;临桂猕猴桃;Lingui Kiwifruit ●

6657 Actinidia longicarpa R. G. Li et M. Y. Liang;长果猕猴桃;Longfruited Kiwifruit ●

6658 Actinidia longicauda F. Chun = Actinidia glaucophylla F. Chun ●

6659 Actinidia macrosperma C. F. Liang;大籽猕猴桃;Big-seeded Actinidia, Largeseed Actinidia, Largeseed Kiwifruit ●

6660 Actinidia macrosperma C. F. Liang var. mumoides C. F. Liang;梅叶猕猴桃;Plumleaf Actinidia, Plumleaf Kiwifruit ●

6661 Actinidia maloides H. L. Li;海棠猕猴桃(四川猕猴桃);Crabapple-like Actinidia, Crabapple-like Kiwifruit ●

6662 Actinidia maloides H. L. Li = Actinidia kolomicta (Rupr. et Maxim.) Maxim. ●

6663 Actinidia maloides H. L. Li f. cordata C. F. Liang;心叶海棠猕猴桃;Heartleaf Crabapple-like Actinidia ●

6664 Actinidia maloides H. L. Li f. cordata C. F. Liang = Actinidia kolomicta (Rupr. et Maxim.) Maxim. ●

6665 Actinidia megalocarpa Nakai = Actinidia arguta (Siebold et Zucc.) Planch. ex Miq. ●

6666 Actinidia melanandra Franch.;黑蕊猕猴桃(黑蕊羊桃);Balackstamen Actinidia, Balackstamen Kiwifruit, Balack-stamened Actinidia, Kiwi ●

6667 Actinidia melanandra Franch. var. cretacea C. F. Liang;垩叶猕猴桃;Cretaceousleaf Actinidia, Cretaceousleaf Kiwifruit ●

6668 Actinidia melanandra Franch. var. glabrescens C. F. Liang;无髯猕猴桃;Beardless Actinidia, Beardless Kiwifruit ●

6669 Actinidia melanandra Franch. var. kwangsiensis (H. L. Li) C. F. Liang;广西猕猴桃;Guangxi Actinidia, Guangxi Kiwifruit, Kwangsi Actinidia ●

6670 Actinidia melanandra Franch. var. kwangsiensis (H. L. Li) C. F. Liang. = Actinidia arguta (Siebold et Zucc.) Planch. ex Miq. var. giraldii (Diels) Vorosch. ●

6671 Actinidia melanandra Franch. var. latifolia Pritz. ex Diels = Actinidia arguta (Siebold et Zucc.) Planch. ex Miq. var. purpurea (Rehder) C. F. Liang ●

6672 Actinidia melanandra Franch. var. subconcolor C. F. Liang;褪粉猕猴桃;Concolor Actinidia, Concolor Kiwifruit ●

6673 Actinidia melliana Hand. -Mazz.;美丽猕猴桃(红毛藤,红网藤,两广猕猴桃);Mell Actinidia, Mell Kiwifruit ●

6674 Actinidia mumoides C. F. Liang = Actinidia macrosperma C. F. Liang var. mumoides C. F. Liang ●

6675 Actinidia obovata Chun ex C. F. Liang;倒卵叶猕猴桃;Obovateleaf Actinidia, Obovateleaf Kiwifruit, Obovate-leaved Actinidia ●

6676 Actinidia pachyphylla Dunn = Actinidia fulvicoma Hance var. pachyphylla (Dunn) H. L. Li ●

6677 Actinidia pentapetala R. G. Li et J. W. Li;五瓣猕猴桃●

6678 Actinidia persicina R. H. Huang et S. M. Wang;繁花猕猴桃(桃花猕猴桃);Manyflower Actinidia, Manyflower Kiwifruit ●

6679 Actinidia petelotii Diels;沙巴猕猴桃(佩氏猕猴桃);Petelot Actinidia, Petelot Kiwifruit ●

6680 Actinidia pilosula (Finet et Gagnep.) Stapf ex Hand. -Mazz.;贡山猕猴桃;Gongshan Actinidia, Gongshan Kiwifruit, Pilose Actinidia ●◇

6681 Actinidia platyphylla A. Gray = Actinidia arguta (Siebold et Zucc.) Planch. ex Miq. var. cordifolia (Miq.) Bean ●

6682 Actinidia polygama (Siebold et Zucc.) Maxim.;葛枣猕猴桃(葛枣,葛枣子,含水藤,获留,金莲枝,马枣子,木蓼,木天蓼,南扶留,楠扶留,蓬莱,藤蓼,藤天蓼,天蓼,天蓼木,天木蓼,小天蓼);Silver Vine, Silvervine Actinidia, Silver-vine Actinidia, Silvervine Kiwifruit, Silver-vine, Silver-vined Actinidia ●

6683 Actinidia polygama (Siebold et Zucc.) Maxim. var. lecomtei (Nakai) H. L. Li = Actinidia polygama (Siebold et Zucc.) Maxim. ●

6684 Actinidia polygama (Siebold et Zucc.) Maxim. var. puberula C. Yu Chang;柔毛木天蓼;Pubescent Silvervine Kiwifrui ●

6685 Actinidia pubescens H. L. Li;柔毛猕猴桃(毛猕猴桃);Supple-hair Actinidia ●

6686 Actinidia purpurea Rehder = Actinidia arguta (Siebold et Zucc.) Planch. ex Miq. var. purpurea (Rehder) C. F. Liang ●

6687 Actinidia rankanensis Hayata = Actinidia callosa Lindl. ●

6688 Actinidia remoganensis Hayata = Actinidia callosa Lindl. ●

6689 Actinidia renryi var. glabricaulis (C. Y. Wu) C. F. Liang = Actinidia rudis Dunn var. glabricaulis C. Y. Wu ●

6690 Actinidia rongshuiensis R. G. Li et X. G. Wang;融水猕猴桃;Rongshui Kiwifruit ●

6691 Actinidia rubrafimenta R. G. Li et J. W. Li;红丝猕猴桃●

6692 Actinidia rubricaulis Dunn;红茎猕猴桃;Redstem Actinidia, Redstem Kiwifruit, Red-stemmed Actinidia ●

6693 Actinidia rubricaulis Dunn var. coriacea (Finet et Gagnep.) C. F. Liang;革叶猕猴桃(秤砣梨,马奶藤,铁甲藤);Chinese Egg-gooseberry, Coriaceousleaf Actinidia, Coriaceousleaf Kiwifruit ●

6694 Actinidia rubus H. Lév.;昭通猕猴桃(藨叶猕猴桃,藨状猕猴桃);Chaotung Actinidia, Zhaotong Actinidia, Zhaotong Kiwifruit ●

6695 Actinidia rudis Dunn;糙叶猕猴桃;Rough Leaf Actinidia, Rough-leaved Actinidia, Rustylea Kiwifruit ●

6696 Actinidia rudis Dunn var. glabricaulis C. Y. Wu;光茎猕猴桃;Lucidstem Actinidia, Lucidstem Kiwifruit, Smooth-stem Actinidia, Smooth-stem Kiwifruit ●

6697 Actinidia rufa (Siebold et Zucc.) Planch. ex Miq.;喜马拉雅猕猴桃(红猕猴桃,山梨猕猴桃,腺齿猕猴桃);Himalayan Actinidia ●

6698 Actinidia rufotricha C. Y. Wu;红毛猕猴桃;Redhair Actinidia, Redhair Kiwifruit, Red-haired Actinidia ●

6699 Actinidia rufotricha C. Y. Wu var. glomerata C. F. Liang;密花猕猴桃;Denseflower Actinidia ●

6700 Actinidia sabiifolia Dunn;清风藤猕猴桃(钱叶猕猴桃);Sabialeaf Actinidia, Sabialeaf Kiwifruit, Sabia-leaved Actinidia ●

6701 Actinidia setosa (H. L. Li) C. F. Liang et A. R. Ferguson = Actinidia chinensis Planch. var. setosa H. L. Li ●

6702 Actinidia sorbifolia C. F. Liang;花楸猕猴桃;Mountainash Actinidia, Mountainash Kiwifruit, Mountain-ash-leaved Actinidia ●

6703 Actinidia stellato-pilosa C. Yu Chang;星毛猕猴桃;Starhair Kiwifruit, Stellate Actinidia, Stellate-haired Actinidia ●◇

6704 Actinidia strigosa Hook. f. et Thomson;糙毛猕猴桃;Strigose Actinidia, Strigose Kiwifruit ●

6705 Actinidia styracifolia C. F. Liang;安息香猕猴桃;Snowbellleaf Actinidia, Snowbellleaf Kiwifruit, Snowbell-leaved Actinidia ●

6706 Actinidia suberifolia C. Y. Wu;栓叶猕猴桃;Cork-leaved Actinidia, Corkyleaf Actinidia, Suberleaf Kiwifruit ●◇

6707 Actinidia subglaucifolia Metcalf = Actinidia hemsleyana Dunn ●

6708 Actinidia tetramera Maxim.;四萼猕猴桃;Fourcalyx Actinidia, Fourcalyx Kiwifruit, Tetrasepalous Actinidia ●

6709 Actinidia tetramera Maxim. var. badongensis C. F. Liang;巴东猕猴桃;Badong Actinidia, Badong Kiwifruit, Patung Actinidia ●

6710 Actinidia tetramera Maxim. var. maloides (H. L. Li) C. Y. Wu = Actinidia maloides H. L. Li ●

6711 Actinidia tonkinensis H. L. Li = Actinidia latifolia (Gardner et Champ.) Merr. ●

6712 Actinidia trichogyna Franch.;毛蕊猕猴桃;Hairypistil Actinidia, Hairypistil Kiwifruit, Hairy-pistiled Actinidia ●

6713 Actinidia truncatifolia C. Yu Chang et P. S. Liu;截叶猕猴桃;Trncate-leaf Actinidia, Trncate-leaf Kiwifruit ●

6714 Actinidia ulmifolia C. F. Liang;榆叶猕猴桃;Elm Leaf Actinidia, Elmleaf Kiwifruit, Elm-leaved Actinidia ●

6715 Actinidia umbelloides C. F. Liang;伞花猕猴桃;Umbellike Actinidia, Umbrellaflower Kiwifruit ●

6716 Actinidia umbelloides C. F. Liang var. flabellifolia C. F. Liang;扇叶猕猴桃;Fan Leaf Actinidia ●

6717 Actinidia valvata Dunn;对萼猕猴桃(猫气藤,猫人参,镊合猕猴桃,糯米饭藤,沙梨藤,痈草);Valvate Actinidia, Valvate Kiwifruit ●

6718 Actinidia valvata Dunn var. boehmeriifolia C. F. Liang;麻叶猕猴桃;Falsenettleleaf Actinidia ●

6719 Actinidia valvata Dunn var. longipedicellata L. L. Yu;长柄对萼猕猴桃;Longstalk Valvate Kiwifruit ●

6720 Actinidia venosa Rehder;显脉猕猴桃;Veiny Actinidia, Veiny Kiwifruit ●

6721 Actinidia venosa Rehder f. pubescens H. L. Li = Actinidia pubescens H. L. Li ●

6722 Actinidia venosa Rehder var. pubescens (H. L. Li) C. Y. Wu = Actinidia venosa Rehder f. pubescens H. L. Li ●

6723 Actinidia viridiflava P. S. Hsu = Actinidia melanandra Franch. ●

6724 Actinidia vitifolia C. Y. Wu;葡萄叶猕猴桃;Crapeleaf Kiwifruit, Grapeleaf Actinidia, Grape-leaved Actinidia ●

6725 Actinidia wantianensis R. G. Li et L. Mo;宛田猕猴桃;Wantian Kiwifruit ●

6726 Actinidia zhejiangensis C. F. Liang;浙江猕猴桃;Zhejiang Actinidia, Zhejiang Kiwifruit ●

6727 Actinidiaceae Engl. et Gilg = Actinidiaceae Gilg et Werderm.(保留科名)●

6728 Actinidiaceae Gilg et Werderm.(1925)(保留科名);猕猴桃科;Actinidia Family, Kiwifruit Family ●

6729 Actinidiaceae Hutch. = Actinidiaceae Gilg et Werderm.(保留科名)●

6730 Actinidiaceae Tiegh. = Actinidiaceae Gilg et Werderm.(保留科名)●

6731 Actinobole Endl. = Actinobole Fenzl ex Endl. ■☆

6732 Actinobole Endl. = Gnaphalodes A. Gray ■☆

6733 Actinobole Fenzl ex Endl.(1843);羽冠鼠麹草属■☆

6734 Actinobole drummondiana P. S. Short;德拉蒙德羽冠鼠麹草■☆

6735 Actinobole oldfieldiana P. S. Short;奥尔羽冠鼠麹草■☆

6736 Actinobole uliginosum (A. Gray) H. Eichler;羽冠鼠麹草■☆

6737 Actinocarpus R. Br. = Damasonium Mill. ■

6738 Actinocarpus Raf. = Chaetopappa DC. ■☆

6739 Actinocarya Benth.(1876);锚刺果属(星果紫草属);Actinocarya ■

6740 Actinocarya bhutanica T. Yamaz. = Microula bhutanica (T. Yamaz.) H. Hara ■

6741 Actinocarya bhutanica Yamaz. = Microula bhutanica (T. Yamaz.) H. Hara ■

6742 Actinocarya kansuensis (W. T. Wang) W. T. Wang = Actinocarya tibetica Benth. ■

6743 Actinocarya tibetica Benth.;锚刺果;Xizang Actinocarya ■

6744 Actinocaryum Post et Kuntze = Actinocarya Benth. ■

6745 Actinocephalus (Körn.) Sano = Paepalanthus Kunth(保留属名)■☆

6746 Actinocephalus (Körn.) Sano(2004);星头谷精草属■☆

6747 Actinocheita F. A. Barkley(1937);墨西哥漆属●☆

6748 Actinocheita filicina (DC.) F. A. Barkley;墨西哥漆●☆

6749 Actinochloa Roem. et Schult. = Chondrosum Desv. ■☆

6750 Actinochloa Willd. ex P. Beauv. = Bouteloua Lag.(保留属名)■
6751 Actinochloa Willd. ex Roem. et Schult. = Bouteloua Lag.(保留属名)■
6752 Actinochloa Willd. ex Roem. et Schult. = Chondrosum Desv. ■☆
6753 Actinochloa gracilis (Kunth) Willd. ex Roem. et Schult. = Bouteloua gracilis (Kunth) Lag. ex Steud. ■
6754 Actinochloris Panzer = Chloris Sw. ●■
6755 Actinochloris Steud. = Chloris Sw. ●■
6756 Actinocladum McClure ex Soderstr.(1981);射枝竹属●☆
6757 Actinocladum verticillatum (Nees) McClure ex Soderstr.;射枝竹●☆
6758 Actinocladus E. Mey. = Capnophyllum Gaertn. ■☆
6759 Actinocyclus Klotzsch = Orthilia Raf. ■
6760 Actinocyclus secundus Klotzsch = Orthilia secunda (L.) House ■
6761 Actinodaphne Nees(1831);黄肉楠属;Actinodaphne ●
6762 Actinodaphne acuminata (Blume) Meisn.;南投黄肉楠(长叶黄肉楠,长叶木姜子,臭屎楠,细叶楠,竹叶楠);Acuminate Litse, Longleaf Actinodaphne, Nanto Actinodaphne, Nantou Actinodaphne ●
6763 Actinodaphne acuminata (Blume) Meisn. = Litsea acuminata (Blume) Sa. Kurata ●
6764 Actinodaphne acutivena (Hayata) Nakai = Litsea acutivena Hayata ●
6765 Actinodaphne akoensis (Hayata) Tang S. Liu et J. C. Liao = Litsea akoensis Hayata ●
6766 Actinodaphne chinensis Blume = Litsea rotundifolia (Nees) Hemsl. var. oblongifolia (Nees) Allen ●
6767 Actinodaphne chinensis Blume var. oblongifolia Nees = Litsea rotundifolia (Nees) Hemsl. var. oblongifolia (Nees) Allen ●
6768 Actinodaphne chinensis Blume var. rotundifolia Nees = Litsea rotundifolia (Nees) Hemsl. ●
6769 Actinodaphne chinensis Nees = Litsea rotundifolia (Nees) Hemsl. var. oblongifolia (Nees) Allen ●
6770 Actinodaphne chinensis Nees var. rotundifolia (Nees) Nees = Litsea rotundifolia (Nees) Hemsl. ●
6771 Actinodaphne citrata Hayata;柠檬黄肉楠;Citrate Actinodaphne ●☆
6772 Actinodaphne citrata Hayata = Litsea cubeba (Lour.) Pers. ●
6773 Actinodaphne cochinchinensis Meisn. = Actinodaphne pilosa (Lour.) Merr. ●
6774 Actinodaphne confertiflora Meisn.;密花黄肉楠;Denseflower Actinodaphne ●
6775 Actinodaphne confertiflora Meisn. = Parasassafras confertiflorum (Meisn.) D. G. Long ●
6776 Actinodaphne confertifolia (Hemsl.) Gamble = Neolitsea confertifolia (Hemsl.) Merr. ●
6777 Actinodaphne crassa Hand.-Mazz. = Lindera megaphylla Hemsl. ●
6778 Actinodaphne cupularis (Hemsl.) Gamble;红果黄肉楠(红果楠,凉药,凉药红树,小楠木);Redfruit Actinodaphne, Red-fruited Actinodaphne ●
6779 Actinodaphne forrestii (C. K. Allen) Kosterm.;毛尖黄肉楠(毛尖树);Forrest Actinodaphne ●
6780 Actinodaphne glaucina C. K. Allen;白背黄肉楠;Glaucousback Actinodaphne, Lightgreen Leaf Actinodaphne, Light-green-leaved Actinodaphne ●
6781 Actinodaphne henryi Gamble;思茅黄肉楠(麦硬);Henry Actinodaphne ●
6782 Actinodaphne hongkongensis Chun = Neolitsea cambodiana Lecomte var. glabra Allen ●
6783 Actinodaphne hypoleucophylla Hayata = Litsea rotundifolia (Nees) Hemsl. var. oblongifolia (Nees) Allen ●
6784 Actinodaphne koshepangii Chun ex Hung T. Chang;广东黄肉楠(高氏黄肉楠);Guangdong Actinodaphne ●
6785 Actinodaphne kweichowensis Yen C. Yang et P. H. Huang;黔桂黄肉楠;Guizhou Actinodaphne ●◇
6786 Actinodaphne lancifolia (Siebold et Zucc.) Meisn. = Litsea coreana H. Lév. ●
6787 Actinodaphne lancifolia (Siebold et Zucc.) Meisn. var. sinensis C. K. Allen = Litsea coreana H. Lév. var. sinensis (Allen) Yen C. Yang et P. H. Huang ●
6788 Actinodaphne lancifolia (Siebold et Zucc.) Meisn. var. sinensis C. K. Allen = Litsea coreana H. Lév. ●
6789 Actinodaphne lecomtei C. K. Allen;柳叶黄肉楠(山桂花);Willowleaf Actinodaphne, Willow-leaved Actinodaphne ●
6790 Actinodaphne litseifolia C. K. Allen = Litsea litseifolia (Allen) Yen C. Yang et P. H. Huang ●
6791 Actinodaphne longifolia (Blume) Nakai = Actinodaphne acuminata (Blume) Meisn. ●
6792 Actinodaphne longifolia (Blume) Nakai = Litsea acuminata (Blume) Sa. Kurata ●
6793 Actinodaphne madraspatana Bedd. ex Hook. f.;马特拉斯黄肉楠●☆
6794 Actinodaphne magniflora C. K. Allen = Machilus velutina Champ. ex Benth. ●
6795 Actinodaphne menghaiensis J. Li;勐海黄肉楠●
6796 Actinodaphne monantha (Yen C. Yang et P. H. Huang) H. P. Tsui = Dodecadenia grandiflora Nees ●
6797 Actinodaphne morrisonensis (Hayata) Hayata = Litsea morrisonensis Hayata ●
6798 Actinodaphne morrisonensis (Hayata) Hayata var. nantoensis (Hayata) Yamam. = Actinodaphne acuminata (Blume) Meisn. ●
6799 Actinodaphne morrisonensis (Hayata) Hayata var. nantoensis (Hayata) Yamam. = Actinodaphne nantoensis (Hayata) Hayata ●
6800 Actinodaphne mushaensis (Hayata) Hayata;雾社黄肉楠(台中黄肉楠,雾社木姜子);Musha Actinodaphne, Mushan Actinodaphne ●
6801 Actinodaphne nakaii (Hayata) T. S. Liu et J. C. Liao = Litsea acutivena Hayata ●
6802 Actinodaphne nantoensis (Hayata) Hayata = Actinodaphne acuminata (Blume) Meisn. ●
6803 Actinodaphne oblongifolia Nees = Litsea rotundifolia (Nees) Hemsl. var. oblongifolia (Nees) Allen ●
6804 Actinodaphne obovata (Nees) Blume;倒卵叶黄肉楠(倒卵叶六驳,假蓑衣叶,七叶一把伞);Obovateleaf Actinodaphne, Obovate-leaved Actinodaphne ●
6805 Actinodaphne obscurinervia Yen C. Yang et P. H. Huang.;隐脉黄肉楠;Obscurevein Actinodaphne, Obscure-veined Actinodaphne ●
6806 Actinodaphne omeiensis (H. Liu) C. K. Allen;峨眉黄肉楠(山桂花);Emei Actinodaphne, Omei Actinodaphne, Omei Mountain Actinodaphne ●
6807 Actinodaphne paotingensis Yen C. Yang et P. H. Huang;保亭黄肉楠;Baoting Actinodaphne ●
6808 Actinodaphne pedicellata Hayata = Litsea hypophaea Hayata ●
6809 Actinodaphne pedicellata Hayata ex Matsum. et Hayata = Litsea hypophaea Hayata ●
6810 Actinodaphne pedicellata Hayata ex Matsum. et Hayata = Litsea taiwaniana Kamik. ●
6811 Actinodaphne pilosa (Lour.) Merr.;毛黄肉楠(茶胶树,胶木,老人木,毛樟,牛耳胶,刨花,刨花木,瓢花木,山枇杷,香胶,香胶木);Pilose Actinodaphne, Pilose Leaf Actinodaphne, Pilose-leaved

Actinodaphne ●

6812 Actinodaphne reticulata Meisn. var. forrestii Allen = Actinodaphne forrestii (C. K. Allen) Kosterm. ●

6813 Actinodaphne reticulata Meisn. var. omeiensis H. Liu = Actinodaphne omeiensis (H. Liu) C. K. Allen ●

6814 Actinodaphne rotundifolia (Nees) Merr. = Litsea rotundifolia (Nees) Hemsl. ●

6815 Actinodaphne rotundifolia Nees = Litsea rotundifolia (Nees) Hemsl. ●

6816 Actinodaphne sasakii (Kamik.) Tang S. Liu et J. C. Liao = Litsea akoensis Hayata var. sasakii (Kamik.) J. C. Liao ●

6817 Actinodaphne sasakii (Kamik.) Tang S. Liu et J. C. Liao = Litsea sasakii Kamik. ●

6818 Actinodaphne sessilifructa C. J. Qi et K. W. Liu;无梗黄肉楠●

6819 Actinodaphne setchuenensis (Gamble) C. K. Allen = Lindera setchuenensis Gamble ●

6820 Actinodaphne trichocarpa C. K. Allen;毛果黄肉楠;Hairyfruit Actinodaphne,Hairy-fruited Actinodaphne ●

6821 Actinodaphne tsaii Hu;马关黄肉楠(蔡氏六驳);Maguan Actinodaphne,Tsai's Actinodaphne ●

6822 Actinodium Schauer ex Schltdl. = Actinodium Schauer ●☆

6823 Actinodium Schauer(1836);辐射桃金娘属●☆

6824 Actinodium cunninghami Schau ex Lindl.;辐射桃金娘●☆

6825 Actinodium proliferum Turcz.;多育辐射桃金娘●☆

6826 Actinokentia Dammer(1906);叉叶椰属(辐堪蒂桐属,广射椰子属,玫瑰椰属)●☆

6827 Actinokentia divaricata Dammer;叉叶椰●☆

6828 Actinokentia huerlimannii H. E. Moore;许尔里曼叉叶椰●☆

6829 Actinokentia schlechteri Dammer;施莱叉叶椰●☆

6830 Actinolema Fenzl(1842);射皮芹属■☆

6831 Actinolema eryngioides Fenzl;射皮芹■☆

6832 Actinolema macrolema Boiss.;大稃射皮芹■☆

6833 Actinolepis DC. (1836);星鳞菊属;Crown Beard ●■☆

6834 Actinolepis DC. = Eriophyllum Lag. ●■☆

6835 Actinolepis lemmonii A. Gray = Syntrichopappus lemmonii (A. Gray) A. Gray ■☆

6836 Actinolepis multicaulis DC. = Eriophyllum multicaule (DC.) A. Gray ■☆

6837 Actinomeris Nutt. (1818)(保留属名);射须菊属●■☆

6838 Actinomeris Nutt. (保留属名) = Verbesina L. (保留属名)●■☆

6839 Actinomeris alternifolia (L.) DC. = Verbesina alternifolia (L.) Britton ex Kearney ■☆

6840 Actinomeris heterophylla Chapm. = Verbesina heterophylla (Chapm.) A. Gray ■☆

6841 Actinomeris longifolia A. Gray = Verbesina longifolia (A. Gray) A. Gray ■☆

6842 Actinomeris squarrosa Nutt. = Verbesina alternifolia (L.) Britton ex Kearney ■☆

6843 Actinomorphe (Miq.) Miq. = Heptapleurum Gaertn. ●■

6844 Actinomorphe (Miq.) Miq. = Schefflera J. R. Forst. et G. Forst. (保留属名)●

6845 Actinomorphe Kuntze = Actinodaphne Nees ●

6846 Actinomorphe Miq. = Heptapleurum Gaertn. ●■

6847 Actinomorphe Miq. = Schefflera J. R. Forst. et G. Forst. (保留属名)●

6848 Actinopappus Hook. f. ex A. Gray = Rutidosis DC. ■☆

6849 Actinophloeus (Becc.) Becc. (1885);射叶椰子属(辐弗鲁桐属,辐叶椰子属,海桃椰属,海桃椰子属,箭叉椰子属,皱子棕属);Clusterpalm ●

6850 Actinophloeus (Becc.) Becc. = Ptychosperma Labill. ●☆

6851 Actinophloeus Becc. = Actinophloeus (Becc.) Becc. ●

6852 Actinophloeus Becc. ex K. Schum. et Hollr. = Actinophloeus (Becc.) Becc. ●

6853 Actinophloeus angustifolia Bailey;细射叶椰子;Narrow-leaf Clusterpalm ●

6854 Actinophloeus macarthurii Becc. ex Raderm.;马氏射叶椰子;Macarthur Clusterpalm ●

6855 Actinophora A. Juss. = Actinospora Turcz. ●■

6856 Actinophora A. Juss. = Cimicifuga L. ●■

6857 Actinophora Wall. = Schoutenia Korth. ●☆

6858 Actinophora Wall. ex R. Br. = Schoutenia Korth. ●☆

6859 Actinophyllum Ruiz et Pav. = Schefflera J. R. Forst. et G. Forst. (保留属名)●

6860 Actinophyllum Ruiz et Pav. = Sciadophyllum P. Browne ●■

6861 Actinorhytis H. Wendl. et Drude(1875);星喙棕属(辐瑞提桐属,拱叶椰属,马来槟榔属,马来椰属)●

6862 Actinorhytis calapparia (Blume) H. Wendl. et Drude ex Scheff.;星喙棕●

6863 Actinorhytis calapparia (Blume) Scheff. = Actinorhytis calapparia (Blume) H. Wendl. et Drude ex Scheff. ●

6864 Actinoschoenus Benth. (1881);星穗草属(星莎属)■

6865 Actinoschoenus chinensis Benth.;华星穗草(华飘拂草,星莎);China Fluttergrass,Chinese Fimbristylis ■

6866 Actinoschoenus chinensis Benth. = Fimbristylis chinensis (Benth.) Ts. Tang et F. T. Wang ■

6867 Actinoschoenus erinaceus (Ridl.) Raymond = Sphaerocyperus erinaceus (Ridl.) Lye ■☆

6868 Actinoschoenus filiformis (Thwaites) Benth.;线星莎■☆

6869 Actinoschoenus humbertii Cherm. = Actinoschoenus filiformis (Thwaites) Benth. ■☆

6870 Actinoschoenus repens J. Raynal;匍匐星莎■☆

6871 Actinoschoenus thouarsii Benth.;马岛星莎■☆

6872 Actinoschoenus yunnanensis (C. B. Clarke) S. Yun Liang;滇星穗草■

6873 Actinoscirpus (Ohwi) R. W. Haines et Lye(1971);大藨草属(星藨属)■

6874 Actinoscirpus grossus (L. f.) Goetgh. et D. A. Simpson;大藨草(硕大藨草)■

6875 Actinoscirpus grossus (L. f.) Goetgh. et D. A. Simpson = Schoenoplectus grossus (L. f.) Palla ■

6876 Actinoseris (Endl.) Cabrera(1970);辐射苣属■☆

6877 Actinoseris Cabrera = Actinoseris (Endl.) Cabrera ■☆

6878 Actinoseris angustifolia (Gardner) Cabrera;窄叶辐射苣■☆

6879 Actinoseris polymorpha (Less.) Cabrera;多形辐射苣■☆

6880 Actinoseris polyphylla (Baker) Cabrera;多叶辐射苣■☆

6881 Actinospermum Elliott = Balduina Nutt. (保留属名)■☆

6882 Actinospermum angustifolium (Pursh) Torr. et A. Gray = Balduina angustifolia (Pursh) B. L. Rob. ■☆

6883 Actinospora Turcz. = Cimicifuga L. ●■

6884 Actinospora Turcz. ex Fisch. et C. A. Mey. = Cimicifuga L. ●■

6885 Actinospora dahurica Turcz. ex Fisch. et C. A. Mey. = Cimicifuga dahurica (Turcz. ex Fisch. et C. A. Mey.) Maxim. ■

6886 Actinospora frigida (Royle) Fisch. et C. A. Mey. = Cimicifuga foetida L. ■

6887 Actinostema Lindl. = Actinostemon Mart. ex Klotzsch ■☆
6888 Actinostemma Griff. (1845);盒子草属(合子草属);Actinostemma,Boxweed ■
6889 Actinostemma Lindl. = Actinostema Lindl. ■☆
6890 Actinostemma Lindl. = Actinostemon Mart. ex Klotzsch ■☆
6891 Actinostemma biglandulosum Hemsl. = Bolbostemma biglandulosum (Hemsl.) Franquet ■
6892 Actinostemma japonicum Miq. = Actinostemma tenerum Griff. ■
6893 Actinostemma lobatum (Maxim.) Maxim. ex Franch. et Sav. = Actinostemma tenerum Griff. ■
6894 Actinostemma lobatum (Maxim.) Maxim. ex Franch. et Sav. f. longiloba Kom. = Actinostemma tenerum Griff. ■
6895 Actinostemma lobatum (Maxim.) Maxim. ex Franch. et Sav. f. semilobatum ? = Actinostemma tenerum Griff. ■
6896 Actinostemma lobatum (Maxim.) Maxim. ex Franch. et Sav. f. subintegrum (Kom.) Kitag. = Actinostemma tenerum Griff. ■
6897 Actinostemma lobatum (Maxim.) Maxim. ex Franch. et Sav. var. genuinum Cogn. = Actinostemma tenerum Griff. ■
6898 Actinostemma lobatum (Maxim.) Maxim. ex Franch. et Sav. var. japonicum Maxim. = Actinostemma tenerum Griff. ■
6899 Actinostemma lobatum (Maxim.) Maxim. ex Franch. et Sav. var. palmatum Makino = Actinostemma tenerum Griff. ■
6900 Actinostemma lobatum (Maxim.) Maxim. ex Franch. et Sav. var. racemosum (Maxim.) Makino = Actinostemma tenerum Griff. ■
6901 Actinostemma lobatum (Maxim.) Maxim. ex Franch. et Sav. var. semilobatum ? = Actinostemma tenerum Griff. ■
6902 Actinostemma multilobum Harms = Bolbostemma paniculatum (Maxim.) Franquet ■
6903 Actinostemma palmatum (Makino) Makino = Actinostemma tenerum Griff. ■
6904 Actinostemma paniculatum Maxim. ex Cogn. = Bolbostemma paniculatum (Maxim.) Franquet ■
6905 Actinostemma parvifolium Cogn. = Actinostemma tenerum Griff. ■
6906 Actinostemma racemosum Maxim. ex Cogn. = Actinostemma tenerum Griff. ■
6907 Actinostemma tenerum Griff.;盒子草(瓣状盒子草,打破碗子藤,龟儿草,合子草,合子藤,盒儿藤,葫篓棵子,黄丝藤,马瓞儿,水荔枝,天球草,无白草,小盒子草,野瓜藤,鸳鸯木鳖);Common Boxweed,Lobed Actinostemma ■
6908 Actinostemma tenerum Griff. var. yunnanensis A. M. Lu et Zhi Y. Zhang;云南盒子草;Yunnan Actinostemma ■
6909 Actinostemon Mart. ex Klotzsch(1841);星蕊大戟属■☆
6910 Actinostemon acuminatus Klotzsch ex Baill.;尖星蕊大戟■☆
6911 Actinostemon angustifolius Klotzsch ex Regel;窄叶星蕊大戟■☆
6912 Actinostemon brachypodum (Griseb.) Urb.;短梗星蕊大戟■☆
6913 Actinostemon brachypodum Urb. = Actinostemon brachypodum (Griseb.) Urb. ■☆
6914 Actinostemon brasiliensis Pax;巴西星蕊大戟■☆
6915 Actinostemon communis Pax;普通星蕊大戟■☆
6916 Actinostemon concolor Müll. Arg.;单色星蕊大戟■☆
6917 Actinostemon echinatus Müll. Arg.;刺星蕊大戟■☆
6918 Actinostemon glabrescens Pax et K. Hoffm.;变光星蕊大戟■☆
6919 Actinostemon grandifolius Klotzsch ex Baill.;大叶星蕊大戟■☆
6920 Actinostemon lasiocarpus Baill.;毛果星蕊大戟■☆
6921 Actinostemon macrocarpus Müll. Arg.;大果星蕊大戟■☆
6922 Actinostemon multiflorus Müll. Arg.;多花星蕊大戟■☆
6923 Actinostemon parvifolius Pittier;小叶星蕊大戟■☆
6924 Actinostemon verticillatus Baill.;轮生星蕊大戟■☆
6925 Actinostigma Turcz. = Asterolasia F. Muell. ●☆
6926 Actinostigma Turcz. = Seringia J. Gay(保留属名)●☆
6927 Actinostigma Welw. (1859);星头藤黄属●☆
6928 Actinostigma Welw. = Symphonia L. f. ●☆
6929 Actinostigma speciosum Welw.;星头藤黄●☆
6930 Actinostrobaceae Lotsy = Cupressaceae Gray(保留科名)●
6931 Actinostrobus Miq. = Actinostrobus Miq. ex Lehm. ●☆
6932 Actinostrobus Miq. ex Lehm. (1845);辐球柏属(沼生柏属)●☆
6933 Actinostrobus pyramidalis Miq.;塔状辐球柏;Swan River Cypress ●☆
6934 Actinotaceae A. I. Konstant. et Melikyan = Apiaceae Lindl.(保留科名)●■
6935 Actinotaceae A. I. Konstant. et Melikyan = Umbelliferae Juss.(保留科名)●■
6936 Actinotinus Oliv. = Viburnum L. ●
6937 Actinotinus sinense Oliv. = Aesculus wilsonii Rehder ●
6938 Actinotus Labill. (1805);轮射芹属(辐射芹属);Flannel Flower ●■☆
6939 Actinotus helianthi Labill.;法绒花;Australian Flannel Flower,Flannel Flower ●☆
6940 Actinotus minor (Sm.) DC.;小轮射芹;Dwarf Flannel Flower ●☆
6941 Actipsis Raf. = Solidago L. ■
6942 Actispermum Raf. = Actinospermum Elliott ■☆
6943 Actispermum Raf. = Baldwinia Raf. ●■
6944 Actites Lander = Sonchus L. ■
6945 Actogeton Clem. = Actogiton Blume ●
6946 Actogiton Blume = Actegeton Blume ●
6947 Actogiton Blume = Azima Lam. ●
6948 Actophila Post et Kuntze = Actephila Blume ●
6949 Actoplanes K. Schum. = Donax Lour. ■
6950 Actoplanes canniformis (G. Forst.) K. Schum. = Donax canniformis (G. Forst.) K. Schum. ■
6951 Actynophloeus Becc. = Actinophloeus (Becc.) Becc. ●
6952 Acuan Medik.(废弃属名) = Desmanthus Willd.(保留属名)●■
6953 Acuan illinoense (Michx.) Kuntze = Desmanthus illinoensis (Michx.) MacMill. ex B. L. Rob. et Fernald ■☆
6954 Acuan virgatum (L.) Medik. = Desmanthus virgatus (L.) Willd. ●■
6955 Acuania Kuntze = Acuan Medik.(废弃属名)●■
6956 Acuania arborescens (Bojer ex Benth.) Kuntze = Dichrostachys arborescens (Bojer ex Benth.) Villiers ●☆
6957 Acuba Link = Aucuba Thunb. ●
6958 Acubalus Neck. = Cucubalus L. ■
6959 Acularia Raf. = Myrrhoides Heist. ex Fabr. ■☆
6960 Acularia Raf. = Scandix L. ■
6961 Acuna Endl. = Acunna Ruiz et Pav. ●☆
6962 Acunaeanthus Borhidi,Komlodi et Moncada(1981);非楔花属●☆
6963 Acunaeanthus tinifolius (Griseb.) Borhidi;非楔花●☆
6964 Acunna Ruiz et Pav. = Bejaria Mutis(保留属名)●☆
6965 Acura Hill = Scabiosa L. ●■
6966 Acuroa J. F. Gmel. = Acouroa Aubl.(废弃属名)●
6967 Acuroa J. F. Gmel. = Dalbergia L. f.(保留属名)●
6968 Acustelma Baill. = Cryptolepis R. Br. ●
6969 Acustelma Baill. = Pentopetia Decne. ■☆
6970 Acustelma grandidieri Baill. = Pentopetia cotoneaster Decne. ■☆
6971 Acuston Raf. = Fibigia Medik. ■☆

6972 Acylopsis Post et Kuntze = Akylopsis Lehm. ■
6973 Acylopsis Post et Kuntze = Matricaria L. ■
6974 Acynos Pers. = Acinos Mill. ■
6975 Acynos Pers. = Calamintha Mill. ■
6976 Acyntha Medik. (废弃属名) = Sansevieria Thunb. (保留属名) ■
6977 Acyntha abyssinica (N. E. Br.) Chiov. var. sublaevigata Chiov. = Sansevieria forskaoliana (Schult. f.) Hepper et J. R. I. Wood ■☆
6978 Acyntha conspicua Chiov. = Sansevieria forskaoliana (Schult. f.) Hepper et J. R. I. Wood ■☆
6979 Acyntha elliptica Chiov. = Sansevieria forskaoliana (Schult. f.) Hepper et J. R. I. Wood ■☆
6980 Acyntha massae Chiov. = Sansevieria nilotica Baker ■☆
6981 Acyntha patens (N. E. Br.) Chiov. = Sansevieria patens N. E. Br. ■☆
6982 Acyntha polyrhitis Chiov. = Sansevieria volkensii Gürke ■☆
6983 Acyntha powellii (N. E. Br.) Chiov. = Sansevieria powellii N. E. Br. ■☆
6984 Acyntha robusta (N. E. Br.) Chiov. = Sansevieria robusta N. E. Br. ■☆
6985 Acyntha rorida (Lanza) Chiov. = Sansevieria rorida (Lanza) N. E. Br. ■☆
6986 Acyntha stuckyi (God. -Leb.) Chiov. = Sansevieria stuckyi God. -Leb. ■☆
6987 Acyphilla Poir. = Aciphylla J. R. Forst. et G. Forst. ■☆
6988 Ada Lindl. (1816); 阿达兰属(阿达属,爱达兰属); Ada ■☆
6989 Ada allenii (L. O. Williams ex C. Schweinf.) N. H. Williams; 阿伦阿达兰■☆
6990 Ada aurantiaca Lindl.; 阿达兰(阿达,爱达兰) ■☆
6991 Ada brachypus (Rchb. f.) Pupulin; 短足阿达兰■☆
6992 Ada lehmannii Rolfe; 莱曼阿达兰(莱曼阿达) ■☆
6993 Adactylus Rolfe = Apostasia Blume ■
6994 Adamanthus Szlach. (2006); 爱达花属■☆
6995 Adamanthus Szlach. = Camaridium Lindl. ■☆
6996 Adamanthus Szlach. = Maxillaria Ruiz et Pav. ■☆
6997 Adamantinia Van den Berg et C. N. Gonç. (2004); 阿地兰属■☆
6998 Adamantogeton Schrad. ex Nees = Lagenocarpus Nees ■☆
6999 Adamantogiton Post et Kuntze = Adamantogeton Schrad. ex Nees ■☆
7000 Adamaram Adans. (废弃属名) = Terminalia L. (保留属名) ●
7001 Adambea Lam. = Adamboe Adans. ●
7002 Adambea Lam. = Catu-Adamboe Adans. ●
7003 Adamboe Adans. = Lagerstroemia L. ●
7004 Adamboe Raf. = Stictocardia Hallier f. ●■
7005 Adamea Jacq. -Fél. = Feliciadamia Bullock ●☆
7006 Adamea stenocarpa (Jacq. -Fél.) Jacq. -Fél. = Feliciadamia stenocarpa (Jacq. -Fél.) Bullock ●☆
7007 Adamia Jacq. -Fél. = Adamea Jacq. -Fél. ●☆
7008 Adamia Jacq. -Fél. = Feliciadamia Bullock ●☆
7009 Adamia Wall. = Dichroa Lour. ●
7010 Adamia chinensis Gardner et Champ. = Dichroa febrifuga Lour. ●
7011 Adamia cyanea Wall. = Dichroa febrifuga Lour. ●
7012 Adamia stenocarpa Jacq. -Fél. = Feliciadamia stenocarpa (Jacq. -Fél.) Bullock ●☆
7013 Adamia sylvatica Meisn. = Dichroa febrifuga Lour. ●
7014 Adamia versicolor Fortune = Dichroa febrifuga Lour. ●
7015 Adamsia Fisch. ex Steud. = Geum L. ■
7016 Adamsia Willd. = Puschkinia Adams ■☆
7017 Adamsia scilloides Willd. = Puschkinia scilloides Adams ■☆
7018 Adansonia L. (1753); 猴面包树属(猴面包属,猢狲面包属,猢狲木属); Adansonia, Baobab, Baobabtree, Baobab-tree ●
7019 Adansonia alba Jum. et H. Perrier = Adansonia za Baill. ●☆
7020 Adansonia bernieri Baill. ex Poiss. = Adansonia madagascariensis Baill. ●☆
7021 Adansonia bozy Jum. et H. Perrier = Adansonia za Baill. ●☆
7022 Adansonia digitata L.; 猴面包树(猢猴面树,猢狲木,指叶猴饼树); Baobab Cream of Tartar Tree, Baobab, Baobabtree, Baobab-tree, Egyptian Sour Bread, Ethiopian Sour Bread, Judas' Bag, Monkey Bread Tree, Monkey Bread, Monkey-bread Tree, Monkeybread, Monkey-bread, Sour Gourd ●
7023 Adansonia digitata L. var. congolensis A. Chev. = Adansonia digitata L. ●
7024 Adansonia fony Baill. ex H. Perrier = Adansonia rubrostipa Jum. et H. Perrier ●☆
7025 Adansonia fony Baill. ex H. Perrier var. rubrostipa (Jum. et H. Perrier) H. Perrier = Adansonia rubrostipa Jum. et H. Perrier ●☆
7026 Adansonia gibbosa (A. Cunn.) Guymer ex D. A. Baum = Adansonia gregorii F. Muell. ●☆
7027 Adansonia grandidieri Baill.; 大猴面包树●☆
7028 Adansonia gregorii F. Muell.; 澳洲猴面包树(格氏猴面包树); Australian Baobab, Boab, Bottle Tree, Cream Nut, Gourd-tree, Goury Tree ●☆
7029 Adansonia madagascariensis Baill.; 马达加斯加猴面包树; Monkey Bread Tree ●☆
7030 Adansonia perrieri Capuron; 佩里耶猴面包树●☆
7031 Adansonia rubrostipa Jum. et H. Perrier; 红梗猴面包树●☆
7032 Adansonia situla (Lour.) Spreng. = Adansonia digitata L. ●
7033 Adansonia somalensis Chiov. = Adansonia digitata L. ●
7034 Adansonia sphaerocarpa A. Chev. = Adansonia digitata L. ●
7035 Adansonia suarezensis H. Perrier; 苏亚雷斯猴面包树●☆
7036 Adansonia sulcata A. Chev. = Adansonia digitata L. ●
7037 Adansonia za Baill.; 杂猴面包树●☆
7038 Adansonia za Baill. var. boinensis H. Perrier = Adansonia za Baill. ●☆
7039 Adansonia za Baill. var. bozy (Jum. et H. Perrier) H. Perrier = Adansonia za Baill. ●☆
7040 Adansoniaceae Vest; 猴面包树科●☆
7041 Adaphus Neck. = Laurus L. ●
7042 Adarianta Knoche = Pimpinella L. ■
7043 Adarianta Knoche = Spiroceratium H. Wolff ■
7044 Adatoda Adans. = Adhatoda Mill. ●
7045 Adatoda Raf. = Justicia L. ●■
7046 Addisonia Rnsby = Helogyne Nutt. ●☆
7047 Adelanthus Endl. = Pyrenacantha Wight(保留属名) ●
7048 Adelaster Lindl. = Fittonia Coem. (保留属名) ■☆
7049 Adelaster Lindl. ex Veitch(废弃属名) = Fittonia Coem. (保留属名) ■☆
7050 Adelaster Veitch = ? Pseuderanthemum Radlk. ●■
7051 Adelaster Veitch = Fittonia Coem. (保留属名) ■☆
7052 Adelbertia Meisn. = Meriania Sw. (保留属名) ●☆
7053 Adelia L. (1759)(保留属名); 隐匿大戟属(阿德尔大戟属) ●☆
7054 Adelia P. Browne(废弃属名) = Adelia L. (保留属名) ●☆
7055 Adelia P. Browne(废弃属名) = Forestiera Poir. (保留属名) ●☆
7056 Adelia anomala Juss. ex Poir. = Erythrococca anomala (Juss. ex Poir.) Prain ●☆

7057 Adelia castanicarpa Roxb. = Chaetocarpus castanocarpus (Roxb.) Thwaites ●
7058 Adelia microphylla A. Rich.;小叶隐匿大戟●☆
7059 Adelia pubescens (Nutt.) Kuntze;毛隐匿大戟●☆
7060 Adeliodes Post et Kuntze = Adelioides R. Br. ex Benth. ●
7061 Adelioides R. Br. ex Benth. = Adeliopsis Benth. ●
7062 Adeliopsis Benth. = Hypserpa Miers ●
7063 Adelmannia Rchb. = Borrichia Adans. ●■☆
7064 Adelmeria Ridl. = Alpinia Roxb.(保留属名)■
7065 Adelobotrys DC.(1828);隐果野牡丹属●☆
7066 Adelobotrys ciliata Triana;睫毛隐果野牡丹●☆
7067 Adelobotrys laxiflora Triana;疏花隐果野牡丹●☆
7068 Adelobotrys macrantha Gleason;大花隐果野牡丹●☆
7069 Adelobotrys macrophylla Pilg.;大叶隐果野牡丹●☆
7070 Adelobotrys monticola Gleason;山地隐果野牡丹●☆
7071 Adelocaryum Brand = Cynoglossum L. + Lindelofia Lehm. ■
7072 Adelocaryum Brand = Lindelofia Lehm. ■
7073 Adelocaryum Brand(1915);隐果紫草属■☆
7074 Adelocaryum erythraeum Brand = Paracaryum erythraeum Schweinf. ex Brand ●☆
7075 Adelocaryum schlagintweitii Brand = Cynoglossum schlagintweitii (Brand) Kazmi ■
7076 Adeloda Raf. = Dicliptera Juss.(保留属名)■
7077 Adelodypsis Becc. = Dypsis Noronha ex Mart. ●☆
7078 Adelodypsis boiviniana (Baill.) Becc. = Dypsis paludosa J. Dransf. ●☆
7079 Adelodypsis gracilis (Bory ex Mart.) Becc. = Dypsis pinnatifrons Mart. ●☆
7080 Adelodypsis sambiranensis (Jum. et H. Perrier) H. P. Guérin = Dypsis pinnatifrons Mart. ●☆
7081 Adelonema Schott = Homalomena Schott ■
7082 Adelonenga (Becc.) Hook. f. = Hydriastele H. Wendl. et Drude ●
7083 Adelonenga Hook. f. = Hydriastele H. Wendl. et Drude ●
7084 Adelopetalum Fitzg. = Bulbophyllum Thouars(保留属名)■
7085 Adelosa Blume(1850);隐匿马鞭草属☆
7086 Adelosa microphylla Blume;隐匿马鞭草☆
7087 Adelostemma Hook. f. (1883);乳突果属(高冠藤属,无冠藤属);Adelostemma ■
7088 Adelostemma gracillimum (Wall. ex Wight) Hook. f.;乳突果(无冠藤);Common Adelostemma ■
7089 Adelostemma gracillimum (Wall. ex Wight) Hook. f. et Tsiang = Adelostemma gracillimum (Wall. ex Wight) Hook. f. ■
7090 Adelostemma gracillimum Hook. f. = Adelostemma gracillimum (Wall. ex Wight) Hook. f. ■
7091 Adelostemma mairei Hand.-Mazz. = Biondia yunnanensis (H. Lév.) Tsiang ●
7092 Adelostemma microcentrum Tsiang = Biondia microcentra (Tsiang) P. T. Li ●
7093 Adelostigma Steetz(1864);隐柱菊属■☆
7094 Adelostigma athrixiodes Steetz;隐柱菊■☆
7095 Adelostigma senegalense Benth.;塞内加尔隐柱菊■☆
7096 Adelphia W. R. Anderson = Triopterys L.(保留属名)●☆
7097 Adelphia W. R. Anderson(2006);牙买加三翅藤属●☆
7098 Adeltia Mirb. = Adelia L.(保留属名)●☆
7099 Ademo Post et Kuntze = Euphorbia L. ●■
7100 Adenacantha B. D. Jacks. = Adenachaena DC. ●☆
7101 Adenacantha B. D. Jacks. = Phymaspermum Less. ●☆
7102 Adenacanthus Nees = Strobilanthes Blume ●■
7103 Adenacanthus Nees(1832);腺背蓝属■
7104 Adenacanthus acuminatus Nees;腺背蓝■☆
7105 Adenacanthus latifolia Nees;宽叶腺背蓝■☆
7106 Adenacanthus longispicus H. P. Tsui;长穗腺背蓝■
7107 Adenacanthus repandus (Blume) Bremek.;匍匐腺背蓝■☆
7108 Adenacanthus rubroglandulosus (Craib) Bremek.;红腺背蓝■☆
7109 Adenaceae Dulac = Droseraceae Salisb.(保留科名)■☆
7110 Adenachaena DC. = Phymaspermum Less. ●☆
7111 Adenachaena leptophylla DC. = Phymaspermum leptophyllum (DC.) Benth. et Hook. ex B. D. Jacks. ●☆
7112 Adenachaena parvifolia DC. = Phymaspermum parvifolium (DC.) Benth. et Hook. ex B. D. Jacks. ●☆
7113 Adenandra Willd. (1809)(保留属名);阿登芸香属;Adenandra ■☆
7114 Adenandra acuta Schltr.;尖阿登芸香■☆
7115 Adenandra amoena (Lodd.) Bartl. et H. L. Wendl. = Adenandra villosa (P. J. Bergius) Licht. ex Roem. et Schult. subsp. orbicularis Strid ■☆
7116 Adenandra bodkinii Schltr. = Acmadenia bodkinii (Schltr.) Strid ●☆
7117 Adenandra brachyphylla Schltdl.;短叶阿登芸香■☆
7118 Adenandra brachyphylla Schltdl. var. glandulosa Sond. = Adenandra villosa (P. J. Bergius) Licht. ex Roem. et Schult. subsp. sonderi (Dümmer) Strid ■☆
7119 Adenandra brachyphylla Schltdl. var. heterophylla Sond. = Adenandra brachyphylla Schltdl. ■☆
7120 Adenandra brachyphylla Schltdl. var. isophylla Sond. = Adenandra brachyphylla Schltdl. ■☆
7121 Adenandra calycina (Tausch) Steud. = Adenandra villosa (P. J. Bergius) Licht. ex Roem. et Schult. subsp. sonderi (Dümmer) Strid ■☆
7122 Adenandra ciliata Sond. = Adenandra villosa (P. J. Bergius) Licht. ex Roem. et Schult. subsp. sonderi (Dümmer) Strid ■☆
7123 Adenandra coriacea Licht. ex Roem. et Schult.;革叶阿登芸香■☆
7124 Adenandra coriacea Licht. ex Roem. et Schult. var. oblongifolia Sond. = Adenandra coriacea Licht. ex Roem. et Schult. ■☆
7125 Adenandra cuspidata E. Mey. ex Bartl. et H. L. Wendl. var. glabra Sond. = Adenandra villosa (P. J. Bergius) Licht. ex Roem. et Schult. ■☆
7126 Adenandra cuspidata E. Mey. ex Bartl. et H. L. Wendl. var. villosa (P. J. Bergius) Sond. = Adenandra villosa (P. J. Bergius) Licht. ex Roem. et Schult. ■☆
7127 Adenandra dahlgrenii Strid;达尔阿登芸香■☆
7128 Adenandra fragrans (Sims) Roem. et Schult.;阿登芸香■☆
7129 Adenandra fragrans Roem. et Schult. = Adenandra fragrans (Sims) Roem. et Schult. ■☆
7130 Adenandra glandulosa Roem. et Schult. = Adenandra villosa (P. J. Bergius) Licht. ex Roem. et Schult. subsp. umbellata (J. C. Wendl.) Strid ■☆
7131 Adenandra gracilis Eckl. et Zeyh.;纤细阿登芸香■☆
7132 Adenandra gummifera Strid;产胶阿登芸香■☆
7133 Adenandra humilis Eckl. et Zeyh. = Adenandra marginata (L. f.) Roem. et Schult. subsp. humilis (Eckl. et Zeyh.) Strid ■☆
7134 Adenandra humilis Eckl. et Zeyh. var. glabra Sond. = Adenandra marginata (L. f.) Roem. et Schult. subsp. humilis (Eckl. et Zeyh.) Strid ■☆
7135 Adenandra humilis Eckl. et Zeyh. var. imbricata Sond. =

Adenandra marginata (L. f.) Roem. et Schult. subsp. humilis (Eckl. et Zeyh.) Strid ■☆

7136 Adenandra lasiantha Sond.;毛花阿登芸香■☆

7137 Adenandra linifolia Bartl. = Adenandra marginata (L. f.) Roem. et Schult. subsp. humilis (Eckl. et Zeyh.) Strid ■☆

7138 Adenandra macradenia Sond. = Acmadenia macradenia (Sond.) Dümmer ●☆

7139 Adenandra marginata (L. f.) Roem. et Schult.;具边阿登芸香■☆

7140 Adenandra marginata (L. f.) Roem. et Schult. subsp. humilis (Eckl. et Zeyh.) Strid;低矮具边阿登芸香■☆

7141 Adenandra marginata (L. f.) Roem. et Schult. subsp. mucronata Strid;短尖具边阿登芸香■☆

7142 Adenandra marginata (L. f.) Roem. et Schult. subsp. serpyllacea (Bartl.) Strid;百里香阿登芸香■☆

7143 Adenandra marginata (L. f.) Roem. et Schult. var. angustata Sond. = Adenandra marginata (L. f.) Roem. et Schult. ■☆

7144 Adenandra multiflora Strid;多花阿登芸香■☆

7145 Adenandra obtusata Sond.;钝阿登芸香■☆

7146 Adenandra odoratissima Strid;大齿阿登芸香■☆

7147 Adenandra odoratissima Strid subsp. tenuis Strid;细大齿阿登芸香■☆

7148 Adenandra ovata Thunb. = Adenandra villosa (P. J. Bergius) Licht. ex Roem. et Schult. subsp. sonderi (Dümmer) Strid ■☆

7149 Adenandra pubescens Sond. = Adenandra villosa (P. J. Bergius) Licht. ex Roem. et Schult. subsp. orbicularis Strid ■☆

7150 Adenandra rotundifolia Eckl. et Zeyh.;圆叶阿登芸香■☆

7151 Adenandra schlechteri Dümmer;施莱阿登芸香■☆

7152 Adenandra serpyllacea Bartl. = Adenandra marginata (L. f.) Roem. et Schult. subsp. serpyllacea (Bartl.) Strid ■☆

7153 Adenandra sonderi Dümmer = Adenandra villosa (P. J. Bergius) Licht. ex Roem. et Schult. subsp. sonderi (Dümmer) Strid ■☆

7154 Adenandra sub-pubescens Sond. = Adenandra villosa (P. J. Bergius) Licht. ex Roem. et Schult. subsp. umbellata (J. C. Wendl.) Strid ■☆

7155 Adenandra umbellata (J. C. Wendl.) Willd. = Adenandra villosa (P. J. Bergius) Licht. ex Roem. et Schult. subsp. umbellata (J. C. Wendl.) Strid ■☆

7156 Adenandra umbellata (J. C. Wendl.) Willd. var. glandulosa (Roem. et Schult.) Bartl. et H. L. Wendl. = Adenandra villosa (P. J. Bergius) Licht. ex Roem. et Schult. subsp. umbellata (J. C. Wendl.) Strid ■☆

7157 Adenandra umbellata (J. C. Wendl.) Willd. var. speciosa (Sims) Bartl. et H. L. Wendl. = Adenandra villosa (P. J. Bergius) Licht. ex Roem. et Schult. subsp. umbellata (J. C. Wendl.) Strid ■☆

7158 Adenandra uniflora (L.) Willd.;单花阿登芸香■☆

7159 Adenandra uniflora (L.) Willd. var. linearis (Thunb.) Sond. = Adenandra uniflora (L.) Willd. ■☆

7160 Adenandra uniflora (L.) Willd. var. pubescens Sond. = Adenandra uniflora (L.) Willd. ■☆

7161 Adenandra villosa (P. J. Bergius) Licht. ex Roem. et Schult.;长柔毛阿登芸香■☆

7162 Adenandra villosa (P. J. Bergius) Licht. ex Roem. et Schult. subsp. apiculata Strid;锐尖长柔毛阿登芸香■☆

7163 Adenandra villosa (P. J. Bergius) Licht. ex Roem. et Schult. subsp. imbricata Strid;覆瓦长柔毛阿登芸香■☆

7164 Adenandra villosa (P. J. Bergius) Licht. ex Roem. et Schult. subsp. orbicularis Strid;圆长柔毛阿登芸香■☆

7165 Adenandra villosa (P. J. Bergius) Licht. ex Roem. et Schult. subsp. pedicellata Strid;花梗长柔毛阿登芸香■☆

7166 Adenandra villosa (P. J. Bergius) Licht. ex Roem. et Schult. subsp. robusta Strid;粗壮长柔毛阿登芸香■☆

7167 Adenandra villosa (P. J. Bergius) Licht. ex Roem. et Schult. subsp. sonderi (Dümmer) Strid;桑德长柔毛阿登芸香■☆

7168 Adenandra villosa (P. J. Bergius) Licht. ex Roem. et Schult. subsp. umbellata (J. C. Wendl.) Strid;伞形长柔毛阿登芸香■☆

7169 Adenandra viscida Eckl. et Zeyh.;黏质阿登芸香■☆

7170 Adenanthe Maguire, Steyerm. et Wurdack = Tyleria Gleason ●☆

7171 Adenanthe Maguire, Steyerm. et Wurdack(1961);腺花金莲木属●☆

7172 Adenanthe bicarpellata Maguire, Steyerm. et Wurdack;腺花金莲木●☆

7173 Adenanthe ciliata Sastre;睫毛腺花金莲木●☆

7174 Adenanthellum B. Nord. (1979);腺羽菊属(腺菊属)■☆

7175 Adenanthellum osmitoides (Harv.) B. Nord.;腺羽菊(腺菊)■☆

7176 Adenanthemum B. Nord. = Adenanthellum B. Nord. ■☆

7177 Adenanthemum osmitoides (Harv.) B. Nord. = Adenanthellum osmitoides (Harv.) B. Nord. ■☆

7178 Adenanthera L. (1753);海红豆属(孔雀豆属);Bead Tree, Beadtree, Bead-tree ●

7179 Adenanthera bicolor Moon;二色海红豆●☆

7180 Adenanthera chrysostachys Benth. = Entada chrysostachys (Benth.) Drake ●☆

7181 Adenanthera falcataria L. = Albizia falcataria (L.) Fosberg ●

7182 Adenanthera falcataria L. = Falcataria moluccana (Miq.) Barneby et J. W. Grimes ●

7183 Adenanthera falcataria L. = Paraserianthes falcataria (L.) I. C. Nielsen ●

7184 Adenanthera gillettii De Wild. = Pseudoprosopis gilletii (De Wild.) Villiers ●☆

7185 Adenanthera gogo Blanco = Entada rheedei Spreng. ●

7186 Adenanthera klainei Pierre ex Baker f. = Pseudoprosopis gilletii (De Wild.) Villiers ●☆

7187 Adenanthera microsperma Teijsm. et Binn.;小果海红豆(海红豆,红豆,孔雀豆,相思格,小实孔雀豆,小籽孔雀豆);Ladycoot Beadtree, Little-seeds Beadtree, Smaollseed Coral Peadtree ●

7188 Adenanthera microsperma Teijsm. et Binn. = Adenanthera pavonina L. var. microsperma (Teijsm. et Binn.) I. C. Nielsen ●

7189 Adenanthera microsperma Teijsm. et Binn. var. luteosemiralis G. A. Fu et Y. K. Yang = Adenanthera microsperma Teijsm. et Binn. ●

7190 Adenanthera pavonina L.;海红豆(大眼星木,红豆,红金豆,红木,红球田螺意,孔雀豆,双栖树,五彩海红豆,西施格树,相思格,相思树,银珠);Aandal Beadtree, Barbados-pride, Circassian Bean, Coral Pea, Coral Peadtree, Coralwood, Jumble Beans, Peacock Flower-fence, Red Bead Tree, Red Sandal Wood, Red Sandal-wood Tree, Red Sandalwood, Red-wood, Sandal Bead Tree, Sandal Beadtree, Sandal Bean Tree, Sandal-bead Tree, Sandalwood Tree ●

7191 Adenanthera pavonina L. var. microsperma (Teijsm. et Binn.) I. C. Nielsen = Adenanthera microsperma Teijsm. et Binn. ●

7192 Adenanthera tamarindifolia Pierre = Adenanthera pavonina L. var. microsperma (Teijsm. et Binn.) I. C. Nielsen ●

7193 Adenanthera tetraptera Schumach. et Thonn. = Tetrapleura tetraptera (Schumach. et Thonn.) Taub. ■☆

7194 Adenanthera trigona Eckl. et Zeyh. = Acmadenia trigona (Eckl. et Zeyh.) Druce ●☆

7195 Adenanthera triphysa Dennst. = Ailanthus triphysa (Dennst.)

Alston ●
7196 Adenanthes Knight = Adenanthos Labill. ●☆
7197 Adenanthos Labill. (1805);壶状花属;Woolly Bush ●☆
7198 Adenanthos cuneata Labill.;滨海壶状花;Coastal Jugflower ●☆
7199 Adenanthos cunninghamii Meisn.;坎宁氏壶状花;Albany Woollybush ●☆
7200 Adenanthos cygnorum Diels;银叶壶状花;Common Woollybush ●☆
7201 Adenanthos detmoldii F. Muell.;黄壶状花;Scott Jugflower, Yellow Jugflower ●☆
7202 Adenanthos obovatus Labill.;倒卵壶状花;Basket Flower ●☆
7203 Adenanthos sericeus Labill.;绢毛壶状花;Albany Woollybush, Coastal Woollybush ●☆
7204 Adenanthus Room. et Schult. = Adenanthos Labill. ●☆
7205 Adenarake Maguire et Wurdack(1961);委内瑞拉金莲木属●☆
7206 Adenarake macrocarpa Sastre;大果委内瑞拉金莲木●☆
7207 Adenarake muriculata Maguire et Wurdack;委内瑞拉金莲木●☆
7208 Adenaria Kunth(1823);墨西哥千屈菜属■☆
7209 Adenaria Pfeiff. = Adnaria Raf. ●
7210 Adenaria Pfeiff. = Gaylussacia Kunth(保留属名)●☆
7211 Adenaria floribunda Kunth;墨西哥千屈菜■☆
7212 Adenaria purpurata Kunth;紫墨西哥千屈菜■☆
7213 Adenarium Raf. = Honckenya Ehrh. ■☆
7214 Adeneleuterophora Barb. Rodr. = Elleanthus C. Presl ■☆
7215 Adeneleuthera Kuntze = Adeneleuterophora Barb. Rodr. ■☆
7216 Adeneleutherophara Dalla Torre et Harms = Elleanthus C. Presl ■☆
7217 Adenema G. Don = Enicostema Blume. (保留属名)■☆
7218 Adenema hyssopifolia (Willd.) G. Don = Enicostema axillare (Lam.) A. Raynal ■☆
7219 Adenesma Griseb. = Adenema G. Don ■☆
7220 Adenia Forssk. (1775);蒴莲属(阿丹藤属,假西番莲属,三瓢果属);Adenia ●
7221 Adenia Torr. = Pilea Lindl. (保留属名)■
7222 Adenia aculeata (Oliv.) Engl.;皮刺蒴莲●☆
7223 Adenia aculeata (Oliv.) Engl. subsp. inermis W. J. de Wilde = Adenia inermis (W. J. de Wilde) W. J. de Wilde ●☆
7224 Adenia adenifera Engl. = Adenia cissampeloides (Planch. ex Hook.) Harms ●☆
7225 Adenia angustisecta Burtt Davy = Adenia digitata Burtt Davy ●☆
7226 Adenia apiculata (De Wild. et T. Durand) Engl. = Adenia poggei (Engl.) Engl. ●☆
7227 Adenia aspidophylla Harms = Adenia staudtii Harms ●☆
7228 Adenia ballyi Verdc.;博利蒴莲●☆
7229 Adenia bequaertii Robyns et Lawalrée;贝卡尔蒴莲●☆
7230 Adenia bequaertii Robyns et Lawalrée subsp. macranthera W. J. de Wilde;大药蒴莲●☆
7231 Adenia bequaertii Robyns et Lawalrée subsp. occidentalis W. J. de Wilde;西方蒴莲●☆
7232 Adenia buchananii Harms ex Engl. = Adenia digitata Burtt Davy ●☆
7233 Adenia cardiophylla (Mast.) Engl.;三开瓢(红牛白皮,假瓜蒌,肉杜仲,三瓢果,心叶蒴莲);Heartleaf Adenia, Heart-leaved Adenia ●
7234 Adenia chevalieri Gagnep.;蒴莲(过山参,软骨青藤,土白芍,土党参,云龙党,云龙党参,猪笼藤);Chevalier Adenia ●
7235 Adenia cissampeloides (Planch. ex Hook.) Harms;几内亚蒴莲●☆
7236 Adenia cynanchifolia (Benth.) Harms;鹅绒藤蒴莲●☆
7237 Adenia dewevrei (De Wild. et T. Durand) Engl. = Adenia poggei (Engl.) Engl. ●☆
7238 Adenia digitata (Harv.) Engl.;野蒴莲;Wild Granadilla ●☆
7239 Adenia digitata Burtt Davy = Adenia digitata (Harv.) Engl. ●☆
7240 Adenia dinklagei Hutch. et Dalziel;丁克蒴莲●☆
7241 Adenia dolichosiphon Harms;长管蒴莲●☆
7242 Adenia ellenbeckii Harms;埃伦蒴莲●☆
7243 Adenia erecta W. J. de Wilde;直立蒴莲●☆
7244 Adenia fernandesiana A. Robyns = Adenia lobata (Jacq.) Engl. ●☆
7245 Adenia formosana Hayata;假秋海棠(假西番莲,台湾拟西番莲);Taiwan Adenia ●
7246 Adenia fruticosa Burtt Davy;灌丛蒴莲●☆
7247 Adenia fruticosa Burtt Davy subsp. simplicifolia W. J. de Wilde;单叶灌丛蒴莲●☆
7248 Adenia fruticosa Burtt Davy subsp. trifoliata W. J. de Wilde;三小叶灌丛蒴莲●☆
7249 Adenia gedoensis W. J. de Wilde;盖托蒴莲●☆
7250 Adenia glauca Schinz;灰蓝蒴莲●☆
7251 Adenia globosa Engl.;球形蒴莲(徐福之酒瓮)●☆
7252 Adenia globosa Engl. subsp. curvata (Verdc.) W. J. de Wilde;弯曲球形蒴莲●☆
7253 Adenia globosa Engl. subsp. pseudoglobosa (Verdc.) W. J. de Wilde;假球形蒴莲●☆
7254 Adenia goetzei Harms;格茨蒴莲●☆
7255 Adenia gracilis Harms = Adenia cissampeloides (Planch. ex Hook.) Harms ●☆
7256 Adenia gracilis Harms subsp. pinnata W. J. de Wilde = Adenia cissampeloides (Planch. ex Hook.) Harms ●☆
7257 Adenia guineensis W. J. de Wilde = Adenia cissampeloides (Planch. ex Hook.) Harms ●☆
7258 Adenia gummifera (Harv.) Harms;产胶蒴莲●☆
7259 Adenia gummifera (Harv.) Harms = Adenia cissampeloides (Planch. ex Hook.) Harms ●☆
7260 Adenia gummifera (Harv.) Harms var. cerifera W. J. de Wilde = Adenia cissampeloides (Planch. ex Hook.) Harms ●☆
7261 Adenia hastata (Harv.) Schinz;戟叶蒴莲●☆
7262 Adenia hastata (Harv.) Schinz var. glandulifera W. J. de Wilde;腺体蒴莲●☆
7263 Adenia heterophylla (Blume) Koord.;异叶蒴莲;Diverseleaf Adenia ●
7264 Adenia huillensis (Welw.) A. Fern. et R. Fern.;威拉蒴莲●☆
7265 Adenia inermis (W. J. de Wilde) W. J. de Wilde;无刺蒴莲●☆
7266 Adenia karibaensis W. J. de Wilde;卡里巴蒴莲●☆
7267 Adenia kirkii (Mast.) Engl.;柯克蒴莲●☆
7268 Adenia lanceolata Engl.;剑叶蒴莲●☆
7269 Adenia lanceolata Engl. subsp. scheffleri (Engl. et Harms) W. J. de Wilde;谢夫勒蒴莲●☆
7270 Adenia latepetala W. J. de Wilde;宽瓣蒴莲●☆
7271 Adenia letouzeyi W. J. de Wilde = Adenia lobata (Jacq.) Engl. ●☆
7272 Adenia lewallei A. Robyns;勒瓦莱蒴莲●☆
7273 Adenia lindiensis Harms;林迪蒴莲●☆
7274 Adenia lobata (Jacq.) Engl.;浅裂蒴莲●☆
7275 Adenia lobata (Jacq.) Engl. subsp. rumicifolia (Engl. et Harms) Lye;酸模叶蒴莲●☆
7276 Adenia lobata (Jacq.) Engl. subsp. schweinfurthii (Engl.) Lye;施韦蒴莲●☆
7277 Adenia lobulata Engl. = Adenia cissampeloides (Planch. ex Hook.) Harms ●☆
7278 Adenia maclurei Merr. = Adenia chevalieri Gagnep. ●
7279 Adenia malangeana Harms;马兰加蒴莲●☆

7280 Adenia mannii (Mast.) Engl. = Adenia lobata (Jacq.) Engl. ●☆
7281 Adenia miegei Aké Assi = Adenia lobata (Jacq.) Engl. ●☆
7282 Adenia mossambicensis W. J. de Wilde;莫桑比克蒴莲●☆
7283 Adenia mukengensis Harms = Adenia cynanchifolia (Benth.) Harms ●☆
7284 Adenia multiflora Pott = Adenia digitata Burtt Davy ●☆
7285 Adenia natalensis W. J. de Wilde;纳塔尔蒴莲●☆
7286 Adenia nicobarica (Kurz.) King = Adenia penangiana (Wall. ex G. Don) J. J. de Wilde ●
7287 Adenia ovata W. J. de Wilde;卵叶蒴莲●☆
7288 Adenia panduriformis Engl.;琴形蒴莲●☆
7289 Adenia parviflora (Blanco) Cusset;大花蒴莲●☆
7290 Adenia parvifolia Gagnep. = Adenia penangiana (Wall. ex G. Don) J. J. de Wilde ●
7291 Adenia pechuelii (Engl.) Harms;埃氏蒴莲;Elephant's Foot ●☆
7292 Adenia pechuelii Harms = Adenia pechuelii (Engl.) Harms ●☆
7293 Adenia penangiana (Wall. ex G. Don) J. J. de Wilde;滇南蒴莲;South Yunnan Adenia ●
7294 Adenia poggei (Engl.) Engl.;波格蒴莲●☆
7295 Adenia pseudoglobosa Verdc.;拟球形蒴莲●☆
7296 Adenia pseudoglobosa Verdc. = Adenia globosa Engl. subsp. pseudoglobosa (Verdc.) W. J. de Wilde ●☆
7297 Adenia pseudoglobosa Verdc. subsp. curvata ? = Adenia globosa Engl. subsp. curvata (Verdc.) W. J. de Wilde ●☆
7298 Adenia pulchra M. G. Gilbert et W. J. de Wilde;美丽蒴莲●☆
7299 Adenia racemosa W. J. de Wilde;总花蒴莲●☆
7300 Adenia repanda (Burch.) Engl.;浅波蒴莲●☆
7301 Adenia reticulata (De Wild. et T. Durand) Engl. = Adenia cissampeloides (Planch. ex Hook.) Harms ●☆
7302 Adenia reticulata (De Wild. et T. Durand) Engl. var. cinerea W. J. de Wilde = Adenia cissampeloides (Planch. ex Hook.) Harms ●☆
7303 Adenia rumicifolia Engl. et Harms = Adenia lobata (Jacq.) Engl. subsp. rumicifolia (Engl. et Harms) Lye ●☆
7304 Adenia rumicifolia Engl. et Harms var. miegei (Aké Assi) W. J. de Wilde = Adenia lobata (Jacq.) Engl. ●☆
7305 Adenia scheffleri Engl. et Harms = Adenia lanceolata Engl. subsp. scheffleri (Engl. et Harms) W. J. de Wilde ●☆
7306 Adenia schliebenii Harms;施利本蒴莲●☆
7307 Adenia schweinfurthii Engl. = Adenia lobata (Jacq.) Engl. subsp. schweinfurthii (Engl.) Lye ●☆
7308 Adenia senensis Engl. = Adenia digitata Burtt Davy ●☆
7309 Adenia spinosa Burtt Davy;具刺蒴莲●☆
7310 Adenia staudtii Harms;施陶蒴莲●☆
7311 Adenia stenodactyla Harms;狭指蒴莲●☆
7312 Adenia stenophylla Harms = Adenia digitata Burtt Davy ●☆
7313 Adenia stolzii Harms;斯托尔兹蒴莲●☆
7314 Adenia stricta (Mast.) Engl.;刚直蒴莲●☆
7315 Adenia tenuispira (Stapf) Engl. = Adenia lobata (Jacq.) Engl. ●☆
7316 Adenia tisserantii A. Fern. et R. Fern.;蒂斯朗特蒴莲●☆
7317 Adenia toxicaria Harms = Adenia ellenbeckii Harms ●☆
7318 Adenia tricostata W. J. de Wilde;三脉蒴莲●☆
7319 Adenia trisecta (Mast.) Engl.;三深裂蒴莲●☆
7320 Adenia tuberifera R. E. Fr.;块茎蒴莲●☆
7321 Adenia venenata Forssk.;毒蒴莲●☆
7322 Adenia vitifolia Hutch. et Bruce = Adenia ellenbeckii Harms ●☆
7323 Adenia volkensii Harms;伏氏蒴莲;Volkens Adenia ●☆
7324 Adenia welwitschii (Mast.) Engl.;韦尔蒴莲●☆
7325 Adenia wightiana (Wall. ex Wight et Arn.) M. Roem.;赖特蒴莲●☆
7326 Adenia wightiana (Wall. ex Wight et Arn.) M. Roem. subsp. africana W. J. de Wilde;非洲赖特蒴莲●☆
7327 Adenia wilmsii Harms;维尔姆斯蒴莲●☆
7328 Adenia zambesiensis R. Fern. et A. Fern.;赞比西蒴莲●☆
7329 Adenileima Rchb. = Adenilema Blume ●
7330 Adenilema Blume = Neillia D. Don ●
7331 Adenilemma Hassk. = Adenilema Blume ●
7332 Adenimesa Nieuwl. = Mesadenia Raf. ■☆
7333 Adenium Roem. et Schult. (1819);沙漠蔷薇属(箭毒胶属,沙漠玫瑰属,天宝花属,腺叶属);Adenium, Desert Rose ●■☆
7334 Adenium arboreum Ehrenb. = Adenium obesum (Forssk.) Roem. et Schult. ●☆
7335 Adenium boehmianum Schinz;贝姆沙漠蔷薇●☆
7336 Adenium boehmianum Schinz var. swazicum (Stapf) Rowley = Adenium swazicum Stapf ●☆
7337 Adenium coetaneum Stapf = Adenium obesum (Forssk.) Roem. et Schult. ●☆
7338 Adenium honghel A. DC. = Adenium obesum (Forssk.) Roem. et Schult. ●☆
7339 Adenium lugardii N. E. Br. = Adenium oleifolium Stapf ●☆
7340 Adenium multiflorum Klotzsch;多花沙漠蔷薇;Impala Lily, Star of the Sabi ●☆
7341 Adenium namaquanum Wyley ex Harv. = Pachypodium namaquanum (Laxm. ex Harv.) Welw. ●☆
7342 Adenium namaquarium G. Hensl. = Hoodia currorii (Hook.) Decne. ■☆
7343 Adenium obesum (Forssk.) Roem. et Schult.;沙漠蔷薇(沙漠玫瑰,沙蔷薇,天宝花,壮箭毒胶);Desert Rose, Impalalily, Sabi Star ●☆
7344 Adenium obesum (Forssk.) Roem. et Schult. subsp. boehmianum (Schinz) Rowley = Adenium boehmianum Schinz ●☆
7345 Adenium obesum (Forssk.) Roem. et Schult. subsp. multiflorum (Klotzsch) Rowley = Adenium multiflorum Klotzsch ●☆
7346 Adenium obesum (Forssk.) Roem. et Schult. subsp. socotranum (Vierh.) Lavranos = Adenium obesum (Forssk.) Roem. et Schult. ●☆
7347 Adenium obesum (Forssk.) Roem. et Schult. subsp. somalense (Balf. f.) Rowley = Adenium obesum (Forssk.) Roem. et Schult. ●☆
7348 Adenium obesum (Forssk.) Roem. et Schult. subsp. swazicum (Stapf) Rowley = Adenium swazicum Stapf ●☆
7349 Adenium obesum (Forssk.) Roem. et Schult. var. multiflorum (Klotzsch) Codd = Adenium multiflorum Klotzsch ●☆
7350 Adenium obesum Roem. et Schult. subsp. oleifolium (Stapf) G. D. Rowley;奥雷叶沙漠蔷薇(奥雷叶沙漠玫瑰,油叶箭毒胶)●☆
7351 Adenium obesum Roem. et Schult. subsp. swazicum (Stapf) G. D. Rowley;斯瓦兹库沙漠蔷薇(斯瓦兹库沙漠玫瑰)●☆
7352 Adenium oleifolium Stapf = Adenium obesum Roem. et Schult. subsp. oleifolium (Stapf) G. D. Rowley ●☆
7353 Adenium socotranum Vierh.;索科特拉沙漠蔷薇;Desert Rose ●☆
7354 Adenium socotranum Vierh. = Adenium obesum (Forssk.) Roem. et Schult. ●☆
7355 Adenium somalense Balf. f.;索马里沙漠蔷薇●☆
7356 Adenium somalense Balf. f. = Adenium obesum (Forssk.) Roem. et Schult. ●☆
7357 Adenium somalense Balf. f. var. caudatipetalum Chiov. = Adenium obesum (Forssk.) Roem. et Schult. ●☆

7358 Adenium somalense Balf. f. var. crispum Chiov. = Adenium obesum (Forssk.) Roem. et Schult. ●☆
7359 Adenium speciosum Fenzl = Adenium obesum (Forssk.) Roem. et Schult. ●☆
7360 Adenium swazicum Stapf;斯瓦沙漠蔷薇●☆
7361 Adenium tricholepis Chiov. = Adenium obesum (Forssk.) Roem. et Schult. ●☆
7362 Adenleima Rchb. = Adenilema Blume ●
7363 Adenleima Rchb. = Neillia D. Don ●
7364 Adenoa Arbo(1977);腺体时钟花属●☆
7365 Adenoa cubensis (Britton et E. Wilson) Arbo;腺体时钟花●☆
7366 Adenobasium C. Presl = Sloanea L. ●
7367 Adenobium Steud. = Adenobasium C. Presl ●
7368 Adenocalymma Mart. ex Meisn. (1840)(保留属名);腺头葳属●☆
7369 Adenocalymma alliaceum Miers;葱味腺头葳●☆
7370 Adenocalymma marginatum (Cham.) A. DC.;腺头葳●☆
7371 Adenocalymma punctifolium S. F. Blake;斑叶腺头葳●☆
7372 Adenocalymna Mart. = Adenocalymma Mart. ex Meisn. (保留属名)●☆
7373 Adenocalyx Bertero ex Kunth(1823);腺萼豆属●☆
7374 Adenocalyx racemosus Bertero ex Kunth;腺萼豆●☆
7375 Adenocarpum D. Don ex Hook. et Arn. (1841);腺果菊属■☆
7376 Adenocarpum D. Don ex Hook. et Arn. = Chrysanthellum Rich. ex Pers. ■☆
7377 Adenocarpum tuberculatum D. Don ex Hook. et Arn.;腺果菊■☆
7378 Adenocarpum tuberculatum D. Don ex Hook. et Arn. = Chrysanthellum indicum DC. ■☆
7379 Adenocarpus DC. (1815);腺荚果属(腺果豆属);Silver Broom ●☆
7380 Adenocarpus Post et Kuntze = Adenocarpum D. Don ex Hook. et Arn. ■☆
7381 Adenocarpus Post et Kuntze = Chrysanthellum Rich. ex Pers. ■☆
7382 Adenocarpus anagyrifolius Coss. et Balansa;臭红豆叶腺荚果●☆
7383 Adenocarpus anagyrifolius Coss. et Balansa var. leiocarpus Maire;光果臭红豆叶腺荚果●☆
7384 Adenocarpus anagyrifolius Coss. et Balansa var. leiocarpus Maire = Adenocarpus anagyrifolius Coss. et Balansa ●☆
7385 Adenocarpus artemisifolius Jahand. et al.;蒿叶腺荚果●☆
7386 Adenocarpus battandieri (Maire) Talavera = Argyrocytisus battandieri (Maire) Raynaud ●☆
7387 Adenocarpus benguellensis Welw. ex Baker = Adenocarpus mannii (Hook. f.) Hook. f. ●☆
7388 Adenocarpus boudyi Batt. et Maire;布迪腺荚果●☆
7389 Adenocarpus boudyi Batt. et Maire var. pilosus Maire = Adenocarpus boudyi Batt. et Maire ●☆
7390 Adenocarpus boudyi Batt. et Maire var. semiglaber Maire = Adenocarpus boudyi Batt. et Maire ●☆
7391 Adenocarpus bracteatus Font Quer et Pau = Adenocarpus complicatus (L.) J. Gay subsp. bracteatus (Pau et Font Quer) Talavera et Gibbs ●☆
7392 Adenocarpus cincinnatus (Ball) Maire;卷毛腺荚果●☆
7393 Adenocarpus commutatus Guss. = Adenocarpus complicatus (L.) J. Gay subsp. commutatus (Guss.) Maire ●☆
7394 Adenocarpus complicatus (L.) J. Gay;折叠腺荚果●☆
7395 Adenocarpus complicatus (L.) J. Gay subsp. bracteatus (Pau et Font Quer) Talavera et Gibbs;苞片折叠腺荚果●☆
7396 Adenocarpus complicatus (L.) J. Gay subsp. commutatus (Guss.) Maire;普通折叠腺荚果●☆
7397 Adenocarpus complicatus (L.) J. Gay subsp. nainii (Maire) P. E. Gibbs;奈恩腺荚果●☆
7398 Adenocarpus complicatus (L.) J. Gay var. barbarus Maire = Adenocarpus complicatus (L.) J. Gay ●☆
7399 Adenocarpus complicatus (L.) J. Gay var. bracteatus (Pau et Font Quer) Maire = Adenocarpus complicatus (L.) J. Gay subsp. bracteatus (Pau et Font Quer) Talavera et Gibbs ●☆
7400 Adenocarpus complicatus (L.) J. Gay var. longivillosus Maire = Adenocarpus complicatus (L.) J. Gay ●☆
7401 Adenocarpus decorticans Boiss.;脱皮腺荚果●☆
7402 Adenocarpus decorticans Boiss. var. planifolius Maire = Adenocarpus decorticans Boiss. ●☆
7403 Adenocarpus decorticans Boiss. var. speciosus (Pomel) Batt. = Adenocarpus decorticans Boiss. ●☆
7404 Adenocarpus faurei Maire;福雷腺荚果●☆
7405 Adenocarpus foliolosus (Aiton) DC.;多叶腺荚果●☆
7406 Adenocarpus foliolosus (Aiton) DC. var. villosus Webb et Berthel. = Adenocarpus foliolosus (Aiton) DC. ●☆
7407 Adenocarpus foliolosus DC. = Adenocarpus foliolosus (Aiton) DC. ●☆
7408 Adenocarpus intermedius DC. = Adenocarpus complicatus (L.) J. Gay ●☆
7409 Adenocarpus intermedius DC. var. bracteatus (Pau et Font Quer) Maire = Adenocarpus complicatus (L.) J. Gay subsp. bracteatus (Pau et Font Quer) Talavera et Gibbs ●☆
7410 Adenocarpus intermedius DC. var. nainii Maire = Adenocarpus complicatus (L.) J. Gay subsp. nainii (Maire) P. E. Gibbs ●☆
7411 Adenocarpus intermedius DC. var. tazzekanus Humbert et Maire = Adenocarpus complicatus (L.) J. Gay ●☆
7412 Adenocarpus mannii (Hook. f.) Hook. f.;曼氏腺荚果●☆
7413 Adenocarpus mannii (Hook. f.) Hook. f. var. laevicarpa Verdc.;光果曼氏腺荚果●☆
7414 Adenocarpus nainii Maire = Adenocarpus complicatus (L.) J. Gay subsp. nainii (Maire) P. E. Gibbs ●☆
7415 Adenocarpus ombriosus Ceballos et Ortega;翁布尔腺荚果●☆
7416 Adenocarpus segonnei Maire = Teline segonnei (Maire) Raynaud ●☆
7417 Adenocarpus speciosus Pomel = Adenocarpus decorticans Boiss. ●☆
7418 Adenocarpus telonensis (Loisel.) DC.;太隆腺荚果●☆
7419 Adenocarpus telonensis (Loisel.) DC. var. grandiflorus (Boiss.) Maire = Adenocarpus telonensis (Loisel.) DC. ●☆
7420 Adenocarpus telonensis (Loisel.) DC. var. rodriguezii Sennen = Adenocarpus telonensis (Loisel.) DC. ●☆
7421 Adenocarpus telonensis (Loisel.) DC. var. transiens A. Reyn. = Adenocarpus telonensis (Loisel.) DC. ●☆
7422 Adenocarpus umbellatus Batt.;小伞腺荚果●☆
7423 Adenocarpus viscosus (Willd.) Webb et Berthel.;黏腺荚果(腺果豆)●☆
7424 Adenocarpus viscosus (Willd.) Webb et Berthel. var. spartioides Webb et Berthel. = Adenocarpus viscosus (Willd.) Webb et Berthel. ●☆
7425 Adenocarpus viscosus Webb et Berthel. = Adenocarpus viscosus (Willd.) Webb et Berthel. ●☆
7426 Adenocaullon Hook. = Adenocaulon Hook. ■
7427 Adenocaulon Hook. (1829);和尚菜属(腺梗菜属);Adenocaulon ■
7428 Adenocaulon adhaerescens Maxim. = Adenocaulon himalaicum Edgew. ■
7429 Adenocaulon bicolor Hook.;北美和尚菜;American Trail-plant ■☆

7430 Adenocaulon bicolor Hook. var. adhaerescens (Maxim.) Makino = Adenocaulon himalaicum Edgew. ■
7431 Adenocaulon bicolor Hook. var. adhaerescens Makino = Adenocaulon himalaicum Edgew. ■
7432 Adenocaulon chilense Less.;智利和尚菜■☆
7433 Adenocaulon himalaicum Edgew.;和尚菜(葫芦叶,火绒叶花,水葫芦,水马蹄草,土冬花,腺梗菜);Himalayan Adenocaulon, Himalayas Adenocaulon ■
7434 Adenocaulum Clem. = Adenocaulon Hook. ■
7435 Adenocaulus Clem. = Adenocaulon Hook. ■
7436 Adenoceras Rchb. f. et Zoll. ex Baill. = Macaranga Thouars ●
7437 Adenochaena DC. = Adenachaena DC. ●☆
7438 Adenochaena DC. = Phymaspermum Less. ●☆
7439 Adenochaena Steud. = Adenachaena DC. ■☆
7440 Adenochaeton Endl. = Adenocheton Fenzl ●
7441 Adenocheton Fenzl = Cocculus DC.(保留属名)●
7442 Adenochetus Baill. = Adenocheton Fenzl ●
7443 Adenochilus Hook. f. (1853);腺唇兰属■☆
7444 Adenochilus gracilis Hook. f.;腺唇兰■☆
7445 Adenochilus nortonii Fitzg.;诺顿腺唇兰■☆
7446 Adenochlaena Boiss. ex Baill. (1858);腺蓬属■☆
7447 Adenochlaena Boiss. ex Baill. = Cephalocroton Hochst. ●☆
7448 Adenochlaena indica Bedd. = Adenochlaena indica Bedd. ex Hook. f. ●☆
7449 Adenochlaena indica Bedd. ex Hook. f.;印度腺蓬●☆
7450 Adenochlaena leucocephala Baill.;腺蓬●☆
7451 Adenochlaena siamensis Ridl.;泰国腺蓬■☆
7452 Adenochlaena zeylanica Thwaites;斯里兰卡腺蓬●☆
7453 Adenoclina Post et Kuntze = Adenocline Turcz. ■☆
7454 Adenocline Turcz. (1843);腺床大戟属■☆
7455 Adenocline acuta (Thunb.) Baill.;尖腺床大戟■☆
7456 Adenocline bupleuroides (Meisn.) Prain = Adenocline pauciflora Turcz. ■☆
7457 Adenocline bupleuroides (Meisn.) Prain var. peglerae Prain = Adenocline pauciflora Turcz. ■☆
7458 Adenocline humilis Turcz. = Adenocline pauciflora Turcz. ■☆
7459 Adenocline mercurialis Turcz. = Adenocline acuta (Thunb.) Baill. ■☆
7460 Adenocline ovalifolia Turcz. = Adenocline pauciflora Turcz. ■☆
7461 Adenocline ovalifolia Turcz. var. rotundifolia Prain = Adenocline pauciflora Turcz. ■☆
7462 Adenocline pauciflora Turcz.;腺床大戟■☆
7463 Adenocline procumbens Benth. ex Pax = Adenocline violifolia (Kunze) Prain ■☆
7464 Adenocline serrata (Meisn.) Turcz. = Adenocline pauciflora Turcz. ■☆
7465 Adenocline sessiliflora Baill. = Adenocline pauciflora Turcz. ■☆
7466 Adenocline sessilifolia Turcz. = Adenocline pauciflora Turcz. ■☆
7467 Adenocline stricta Prain = Adenocline pauciflora Turcz. ■☆
7468 Adenocline violifolia (Kunze) Prain;堇叶腺床大戟■☆
7469 Adenocrepis Blume = Baccaurea Lour. ●
7470 Adenocritonia R. M. King et H. Rob. (1976);密腺亮泽兰属■☆
7471 Adenocritonia R. M. King et H. Rob. = Eupatorium L. ●■
7472 Adenocritonia adamsii R. M. King et H. Rob.;密腺亮泽兰■☆
7473 Adenocyclus Less. = Pollalesta Kunth ●☆
7474 Adenodaphne S. Moore = Litsea Lam.(保留属名)●
7475 Adenodiscus Turcz. = Belotia A. Rich. ●☆
7476 Adenodolichos Harms(1902);非洲长腺豆属(非洲镰扁豆属)■☆
7477 Adenodolichos acutifoliolatus Verdc.;尖托叶非洲长腺豆■☆
7478 Adenodolichos adenophorus (Harms) Harms = Adenodolichos punctatus (Micheli) Harms subsp. bussei (Harms) Verdc. ■☆
7479 Adenodolichos anchietae (Hiern) Harms = Adenodolichos rhomboideus (O. Hoffm.) Harms ■☆
7480 Adenodolichos baumii Harms;鲍姆非洲长腺豆■☆
7481 Adenodolichos bequaertii De Wild.;贝卡尔非洲长腺豆■☆
7482 Adenodolichos bequaertii De Wild. var. purpureus ? = Adenodolichos punctatus (Micheli) Harms ■☆
7483 Adenodolichos brevipetiolatus R. Wilczek;短瓣非洲长腺豆■☆
7484 Adenodolichos bussei Harms = Adenodolichos punctatus (Micheli) Harms subsp. bussei (Harms) Verdc. ■☆
7485 Adenodolichos bussei Harms var. moxicensis Torre;莫希克非洲长腺豆■☆
7486 Adenodolichos caeruleus R. Wilczek;天蓝非洲长腺豆■☆
7487 Adenodolichos dinklagei (Harms) Roberty = Dolichos dinklagei Harms ■☆
7488 Adenodolichos euryphyllus Harms = Adenodolichos rhomboideus (O. Hoffm.) Harms ■☆
7489 Adenodolichos exellii Torre;埃克塞尔非洲长腺豆■☆
7490 Adenodolichos grandifoliolatus De Wild.;大小叶非洲长腺豆■☆
7491 Adenodolichos harmsianus De Wild.;哈姆斯非洲长腺豆■☆
7492 Adenodolichos harmsianus De Wild. var. acutifoliatus Baker f. = Adenodolichos punctatus (Micheli) Harms subsp. bussei (Harms) Verdc. ■☆
7493 Adenodolichos helenae Buscal. et Muschl.;海伦娜非洲长腺豆■☆
7494 Adenodolichos huillensis Torre;威拉非洲长腺豆■☆
7495 Adenodolichos huillensis Torre var. kawambwaensis Verdc.;卡万布瓦非洲长腺豆■☆
7496 Adenodolichos kaessneri Harms;卡斯纳非洲长腺豆■☆
7497 Adenodolichos katangensis R. Wilczek;加丹加非洲长腺豆■☆
7498 Adenodolichos macrothyrsus (Harms) Harms;大序非洲长腺豆■☆
7499 Adenodolichos mendesii Torre;门代斯非洲长腺豆■☆
7500 Adenodolichos nanus R. E. Fr. = Adenodolichos rhomboideus (O. Hoffm.) Harms ■☆
7501 Adenodolichos oblongifoliolatus R. Wilczek;矩圆叶非洲长腺豆■☆
7502 Adenodolichos obtusifolius R. E. Fr. = Adenodolichos punctatus (Micheli) Harms ■☆
7503 Adenodolichos pachyrhizus De Wild. = Adenodolichos rhomboideus (O. Hoffm.) Harms ■☆
7504 Adenodolichos paniculatus (Hua) Hutch. et Dalziel;圆锥非洲长腺豆■☆
7505 Adenodolichos punctatus (Micheli) Harms;斑点非洲长腺豆■☆
7506 Adenodolichos punctatus (Micheli) Harms subsp. bussei (Harms) Verdc.;布瑟斑点非洲长腺豆■☆
7507 Adenodolichos punctatus (Micheli) Harms subsp. decumbens (Verdc.) Verdc.;横卧斑点非洲长腺豆■☆
7508 Adenodolichos punctatus (Micheli) Harms var. decumbens Verdc. = Adenodolichos punctatus (Micheli) Harms subsp. decumbens (Verdc.) Verdc. ■☆
7509 Adenodolichos rhomboideus (O. Hoffm.) Harms;菱形非洲长腺豆■☆
7510 Adenodolichos rhomboideus (O. Hoffm.) Harms var. kundelungensis R. Wilczek;昆德龙非洲长腺豆■☆
7511 Adenodolichos rhomboideus (O. Hoffm.) Harms var. lanceolatus R. Wilczek;剑叶非洲长腺豆■☆

7512 Adenodolichos rupestris Verdc.;岩生非洲长腺豆■☆
7513 Adenodolichos salvifoliolatus R. Wilczek;鼠尾草小叶非洲长腺豆■☆
7514 Adenodolichos upembaensis R. Wilczek;乌彭巴非洲长腺豆■☆
7515 Adenodolichus Post et Kuntze = Adenodolichos Harms ■☆
7516 Adenodus Lour. = Elaeocarpus L. ●
7517 Adenodus sylvestris Lour. = Elaeocarpus sylvestris (Lour.) Poir. ●
7518 Adenodus sylvestris Lour. var. ellipticus ? = Elaeocarpus sylvestris (Lour.) Poir. ●
7519 Adenoglossa B. Nord. (1976);腺舌菊属■☆
7520 Adenoglossa decurrens (Hutch.) B. Nord.;腺舌菊■☆
7521 Adenogonum Welw. ex Hiern = Engleria O. Hoffm. ●■☆
7522 Adenogonum decumbens Welw. ex Hiern = Engleria decumbens (Welw. ex Hiern) Hiern ■☆
7523 Adenogramma Rchb. (1828);坚果粟草属■☆
7524 Adenogramma asparagoides Adamson = Adenogramma teretifolia (Thunb.) Adamson ■☆
7525 Adenogramma capillaris (Eckl. et Zeyh.) Druce;发状坚果粟草■☆
7526 Adenogramma congesta Adamson;密集坚果粟草■☆
7527 Adenogramma diffusa Fenzl = Adenogramma lichtensteiniana (Schult.) Druce ■☆
7528 Adenogramma dregeana Gand. = Adenogramma physocalyx Fenzl ■☆
7529 Adenogramma galioides Fenzl = Adenogramma glomerata (L. f.) Druce ■☆
7530 Adenogramma glomerata (L. f.) Druce;团集坚果粟草■☆
7531 Adenogramma lampocarpa E. Mey. ex Fenzl = Adenogramma capillaris (Eckl. et Zeyh.) Druce ■☆
7532 Adenogramma lichtensteiniana (Schult.) Druce;利希坚果粟草■☆
7533 Adenogramma littoralis Adamson;滨海坚果粟草■☆
7534 Adenogramma physocalyx Fenzl;囊萼坚果粟草■☆
7535 Adenogramma rigida (Bartl.) Sond.;硬坚果粟草■☆
7536 Adenogramma sylvatica (Eckl. et Zeyh.) Fenzl;林地坚果粟草■☆
7537 Adenogramma teretifolia (Thunb.) Adamson;柱叶坚果粟草■☆
7538 Adenogrammaceae (Fenzl) Nakai = Molluginaceae Bartl.(保留科名)■
7539 Adenogrammaceae Nakai = Molluginaceae Bartl.(保留科名)■
7540 Adenogrammataceae (Fenzl) Nakai = Molluginaceae Bartl.(保留科名)■
7541 Adenogrammataceae Nakai = Molluginaceae Bartl.(保留科名)■
7542 Adenogyna Post et Kuntze = Adenogyne Klotzsch ●
7543 Adenogyna Post et Kuntze = Sebastiania Spreng. ●
7544 Adenogyna Raf. = Saxifraga L. ■
7545 Adenogyna Raf. = Sekika Medik. ■
7546 Adenogyne B. D. Jacks. = Adenogyna Raf. ■
7547 Adenogyne Klotzsch = Sebastiania Spreng. ●
7548 Adenogynum Rchb. f. et Zoll. = Cladogynos Zipp. ex Span. ●
7549 Adenogyras Durand = Adenogyrus Klotzsch ●
7550 Adenogyrus Klotzsch = Scolopia Schreb.(保留属名)●
7551 Adenoia Raf. = Ludwigia L. ●■
7552 Adenolepis Less. = Cosmos Cav. ■
7553 Adenolepis Sch. Bip. = Bidens L. ●■
7554 Adenolinum Rchb. = Linum L. ●■
7555 Adenolisianthus (Progel) Gilg = Irlbachia Mart. ■☆
7556 Adenolisianthus Gilg = Irlbachia Mart. ■☆
7557 Adenolisianthus Gilg(1895);巴西腺龙胆属■☆
7558 Adenolisianthus arboreus Gilg;树状巴西腺龙胆■☆
7559 Adenolisianthus virgatus Gilg;巴西腺龙胆■☆
7560 Adenolobus (Benth.) Torr. et Hillc. = Adenolobus (Harv. ex Benth. et Hook. f.) Torre et Hillc. ■☆
7561 Adenolobus (Harv. ex Benth. et Hook. f.) Torre et Hillc. (1956);腺羊蹄甲属■☆
7562 Adenolobus (Harv. ex Benth.) Torre et Hillc. = Adenolobus (Harv. ex Benth. et Hook. f.) Torre et Hillc. ■☆
7563 Adenolobus (Harv.) Torre et Hillc. = Adenolobus (Harv. ex Benth. et Hook. f.) Torre et Hillc. ■☆
7564 Adenolobus garipensis (E. Mey.) Torre et Hillc.;加里普腺羊蹄甲■☆
7565 Adenolobus mossamedensis Torre et Hillc. = Adenolobus pechuelii (Kuntze) Torre et Hillc. subsp. mossamedensis (Torre et Hillc.) Brummitt et J. H. Ross ■☆
7566 Adenolobus pechuelii (Kuntze) Torre et Hillc.;佩氏腺羊蹄甲■☆
7567 Adenolobus pechuelii (Kuntze) Torre et Hillc. subsp. mossamedensis (Torre et Hillc.) Brummitt et J. H. Ross;安哥拉腺羊蹄甲■☆
7568 Adenolobus rufescens (Lam.) A. Schmitz = Bauhinia rufescens Lam. ●☆
7569 Adenoncos Blume(1825);腺瘤兰属■☆
7570 Adenoncos macranthus Schltr.;大蛤腺瘤兰■☆
7571 Adenoncos major Ridl.;大腺瘤兰■☆
7572 Adenoncos parviflora Ridl.;小花腺瘤兰■☆
7573 Adenoncos triloba Carr;三裂腺瘤兰■☆
7574 Adenoncos uniflora J. J. Sm.;单花腺瘤兰■☆
7575 Adenoncos virens Blume;绿花腺瘤兰■☆
7576 Adenonema Bunge = Stellaria L. ■
7577 Adenonema petraeum (Bunge) Bunge var. alpinum Bunge = Stellaria petraea Bunge ■
7578 Adenoon Dalzell(1850);无冠糙毛菊属■☆
7579 Adenoon indicum Dalzell;无冠糙毛菊■☆
7580 Adenopa Raf. = Drosera L. ■
7581 Adenopappus Benth. (1840);腺毛菊属■☆
7582 Adenopappus Benth. = Tagetes L. ●■
7583 Adenopappus persicifolius Benth.;腺毛菊■☆
7584 Adenopeltis Bertero = Adenopeltis Bertero ex A. Juss. ●☆
7585 Adenopeltis Bertero ex A. Juss. (1832);腺盾大戟属●☆
7586 Adenopeltis colliguayana Bertero;腺盾大戟●☆
7587 Adenopeltis serrata (Aiton) G. L. Webster;齿叶腺盾大戟●☆
7588 Adenopetalum Klotzsch et Garcke = Euphorbia L. ●■
7589 Adenopetalum Turcz. = Vitis L. ●
7590 Adenophaedra (Müll. Arg.) Müll. Arg. (1874);亮腺大戟属●☆
7591 Adenophaedra Müll. Arg. = Adenophaedra (Müll. Arg.) Müll. Arg. ●☆
7592 Adenophaedra megalophylla Müll. Arg.;大叶亮腺大戟●☆
7593 Adenophaedra minor Ducke;大亮腺大戟●☆
7594 Adenophaedra prealta (Croizat) Croizat;亮腺大戟●☆
7595 Adenophaedra woodsoniana (Croizat) Croizat;伍氏亮腺大戟●☆
7596 Adenophora Fisch. (1823);沙参属;Gland Bellflower, Gland Bell-flower, Lady Bell, Ladybell, Ladybells, Lady's Bell, Lady's-Bell ●■
7597 Adenophora albescens C. Y. Wu = Adenophora khasiana (Hook. f. et Thomson) Collett et Hemsl. ■
7598 Adenophora alpina Nannf. = Adenophora himalayana Feer subsp. alpina (Nannf.) D. Y. Hong ■
7599 Adenophora amurica C. X. Fu et M. Y. Liou;阿穆尔沙参;Amur Ladybell ■
7600 Adenophora argyi H. Lév. = Adenophora stricta Miq. ■

7601 Adenophora atuntzensis C. Y. Wu;阿墩沙参;Adun Ladybell ■
7602 Adenophora atuntzensis C. Y. Wu = Adenophora jasionifolia Franch. ■
7603 Adenophora aurita Franch. = Adenophora stricta Miq. subsp. aurita (Franch.) D. Y. Hong et S. Ge ■
7604 Adenophora axilliflora Borbás = Adenophora stricta Miq. ■
7605 Adenophora biformifolia Y. Z. Zhao;二型叶沙参(二型沙参);Biform Ladybell ■
7606 Adenophora biformifolia Y. Z. Zhao = Adenophora potaninii Korsh. subsp. wawreana (Zahlbr.) S. Ge et D. Y. Hong ■
7607 Adenophora biloba Y. Z. Zhao = Adenophora gmelinii (Spreng.) Fisch. ■
7608 Adenophora bockiana Diels = Adenophora potaninii Korsh. ■
7609 Adenophora borealis D. Y. Hong et Y. Z. Zhao;北方沙参;Northern Ladybell ■
7610 Adenophora borealis D. Y. Hong et Y. Z. Zhao = Adenophora gmelinii (Spreng.) Fisch. ■
7611 Adenophora borealis D. Y. Hong et Y. Z. Zhao var. oreophilla Y. Z. Zhao = Adenophora gmelinii (Spreng.) Fisch. ■
7612 Adenophora borealis D. Y. Hong et Y. Z. Zhao var. oreophylla Y. Z. Zhao;山沙参;Mountain Northern Ladybell ■
7613 Adenophora brevidiscifera D. Y. Hong;短花盘沙参;Shortdiscid Ladybell,Shortdisck Ladybell ■
7614 Adenophora bulleyana Diels = Adenophora khasiana (Hook. f. et Thomson) Collett et Hemsl. ■
7615 Adenophora bulleyana Diels var. alba C. Y. Wu = Adenophora coelestis Diels ■
7616 Adenophora capillaris Hemsl.;丝裂沙参(龙胆草,毛鸡脚,泡参,泡参草,线齿沙参);Threadlobe Ladybell ■
7617 Adenophora capillaris Hemsl. subsp. leptosepala (Diels) D. Y. Hong;细萼沙参(壶花沙参);Thinsepal Ladybell,Urceolate Ladybell ■
7618 Adenophora capillaris Hemsl. subsp. paniculata (Nannf.) D. Y. Hong et S. Ge;细叶沙参(圆锥沙参,紫沙参);Paniculate Ladybell ■
7619 Adenophora chionantha C. Y. Wu = Adenophora khasiana (Hook. f. et Thomson) Collett et Hemsl. ■
7620 Adenophora coelestis Diels;天蓝沙参(滇川沙参,富民沙参,两型沙参,萝卜根沙参);Skyblue Ladybell ■
7621 Adenophora coelestis Diels var. stenophylla Diels ex C. Y. Wu = Adenophora coelestis Diels ■
7622 Adenophora coelestis Diels var. uehatae (Yamam.) Masam. = Adenophora morrisonensis Hayata subsp. uehatae (Yamam.) Lammers ■
7623 Adenophora collina Kitag. = Adenophora stenanthina (Ledeb.) Kitag. ■
7624 Adenophora communis Fisch. = Adenophora liliifolia (L.) Besser ■
7625 Adenophora communis Fisch. var. lamarkii Trautv. = Adenophora lamarkii Fisch. ■
7626 Adenophora communis Fisch. var. latifolia Trautv. = Adenophora pereskiifolia (Fisch. ex Roem. et Schult.) Fisch. ex G. Don ■
7627 Adenophora confusa Nannf. = Adenophora stricta Miq. subsp. confusa (Nannf.) D. Y. Hong ■
7628 Adenophora contracta (Kitag.) J. Z. Qiu et D. Y. Hong;缢花沙参;Contracted Ladybell ■
7629 Adenophora cordifolia D. Y. Hong;心叶沙参;Cordateleaf Ladybell,Heartleaf Ladybell ■
7630 Adenophora coronata A. DC. = Adenophora stenanthina (Ledeb.) Kitag. ■
7631 Adenophora coronopifolia (Fisch. ex Roem. et Schult.) Fisch. = Adenophora gmelinii (Spreng.) Fisch. ■
7632 Adenophora coronopifolia Fisch. = Adenophora gmelinii (Spreng.) Fisch. ■
7633 Adenophora crispata (Korsh.) Kitag. = Adenophora stenanthina (Ledeb.) Kitag. ■
7634 Adenophora crispata Turcz. ex Ledeb. = Adenophora stenanthina (Ledeb.) Kitag. ■
7635 Adenophora curvidens Nakai = Adenophora pereskiifolia (Fisch. ex Roem. et Schult.) Fisch. ex G. Don ■
7636 Adenophora denticulata Fisch. = Adenophora tricuspidata (Fisch. ex Roem. et Schult.) A. DC. ■
7637 Adenophora dimorphophylla C. Y. Wu = Adenophora khasiana (Hook. f. et Thomson) Collett et Hemsl. ■
7638 Adenophora diplodonta Diels = Adenophora khasiana (Hook. f. et Thomson) Collett et Hemsl. ■
7639 Adenophora divaricata Franch. et Sav.;展枝沙参;Spreadingbranch Ladybell ■
7640 Adenophora divaricata Franch. et Sav. f. albiflora Hayashi;白花展枝沙参■☆
7641 Adenophora divaricata Franch. et Sav. f. angustifolia Hiyama = Adenophora divaricata Franch. et Sav. ■
7642 Adenophora divaricata Franch. et Sav. var. manshurica Kitag. = Adenophora divaricata Franch. et Sav. ■
7643 Adenophora elata Nannf.;狭长花沙参(沙参);Tall Ladybell ■
7644 Adenophora elata Nannf. f. verticillata Kitag. = Adenophora wulingshanica D. Y. Hong ■
7645 Adenophora erysimoides (Vest ex Roem. et Schult.) Nakai ex Kitam. = Adenophora gmelinii (Spreng.) Fisch. ■
7646 Adenophora forrestii Diels;滇北沙参■
7647 Adenophora forrestii Diels = Adenophora jasionifolia Franch. ■
7648 Adenophora forrestii Diels var. handeliana Nannf. = Adenophora jasionifolia Franch. ■
7649 Adenophora gmelinii (Spreng.) Fisch.;狭叶沙参(厚叶沙参,柳叶沙参);Gmelin Ladybell ■
7650 Adenophora gmelinii (Spreng.) Fisch. subsp. hailinensis J. Z. Qiu et D. Y. Hong;海林沙参■
7651 Adenophora gmelinii (Spreng.) Fisch. subsp. nystroemii J. Z. Qiu et D. Y. Hong;山西沙参■
7652 Adenophora gmelinii (Spreng.) Fisch. var. coronopifolia (Fisch.) Y. Z. Zhao;柳叶沙参■
7653 Adenophora gmelinii (Spreng.) Fisch. var. coronopifolia (Fisch.) Y. Z. Zhao = Adenophora gmelinii (Spreng.) Fisch. ■
7654 Adenophora gmelinii (Spreng.) Fisch. var. pachyphylla (Kitag.) Y. Z. Zhao;厚叶沙参■
7655 Adenophora gmelinii (Spreng.) Fisch. var. pachyphylla (Kitag.) Y. Z. Zhao = Adenophora gmelinii (Spreng.) Fisch. ■
7656 Adenophora gmelinii (Spreng.) Fisch. var. stylosa A. DC. = Adenophora gmelinii (Spreng.) Fisch. ■
7657 Adenophora golubinzevaeana Reverdin;高氏沙参■☆
7658 Adenophora gracilis Nannf. = Adenophora liliifolioides Pax et K. Hoffm. ■
7659 Adenophora grandiflora Nakai;大花沙参■☆
7660 Adenophora hatsushimae Kitam.;初岛沙参;Hatsushima Ladybell ■☆
7661 Adenophora himalayana Feer;喜马拉雅沙参;Himalayan

Ladybell, Himalayas Ladybell ■
7662 Adenophora himalayana Feer subsp. alpina (Nannf.) D. Y. Hong; 高山沙参; Alpine Himalayas Ladybell, Alpine Ladybell ■
7663 Adenophora huangae C. Y. Wu; 富民沙参; Fumin Ladybell ■
7664 Adenophora huangae C. Y. Wu = Adenophora coelestis Diels ■
7665 Adenophora hubeiensis D. Y. Hong; 鄂西沙参; Hubei Ladybell ■
7666 Adenophora hunanensis Nannf. = Adenophora petiolata Pax et K. Hoffm. subsp. hunanensis (Nannf.) D. Y. Hong et S. Ge ■
7667 Adenophora hunanensis Nannf. subsp. huadungensis D. Y. Hong = Adenophora petiolata Pax et K. Hoffm. subsp. huadungensis (D. Y. Hong) D. Y. Hong et S. Ge ■
7668 Adenophora insolens Reverdin = Adenophora triphylla (Thunb.) A. DC. var. japonica (Regel) H. Hara ■☆
7669 Adenophora isabellae Hemsl. = Adenophora trachelioides Maxim. ■
7670 Adenophora ishiyamae Miyabe et Tatewaki = Adenophora pereskiifolia (Fisch. ex Roem. et Schult.) Fisch. ex G. Don ■
7671 Adenophora izuensis H. Ohba et S. Watan.; 伊豆沙参■☆
7672 Adenophora jacutica Fed.; 雅库特沙参■☆
7673 Adenophora jasionifolia Franch.; 甘孜沙参(阿墩沙参,保科参,小钟沙参); Ganzi Ladybell, Jasioneleaf Ladybell, Smallbell Ladybell ■
7674 Adenophora khasiana (Hook. f. et Thomson) Collett et Hemsl.; 云南沙参(变白沙参,丽江沙参,两型沙参,两型叶沙参,玫花沙参,泡参,雪花沙参,重齿沙参); Khasia Ladybell ■
7675 Adenophora khasiana Feer = Adenophora khasiana (Hook. f. et Thomson) Collett et Hemsl. ■
7676 Adenophora kulunensis Y. Z. Zhao = Adenophora contracta (Kitag.) J. Z. Qiu et D. Y. Hong ■
7677 Adenophora kurilensis Nakai = Adenophora triphylla (Thunb.) A. DC. var. japonica (Regel) H. Hara ■☆
7678 Adenophora lamarkii Fisch.; 天山沙参; Lamark Ladybell ■
7679 Adenophora latifolia Fisch. = Adenophora pereskiifolia (Fisch. ex Roem. et Schult.) Fisch. ex G. Don ■
7680 Adenophora leptosepala Diels = Adenophora capillaris Hemsl. subsp. leptosepala (Diels) D. Y. Hong ■
7681 Adenophora likiangensis C. Y. Wu = Adenophora khasiana (Hook. f. et Thomson) Collett et Hemsl. ■
7682 Adenophora liliifolia (L.) Besser; 新疆沙参(亚麻叶风铃草,窄叶沙参); Lady Bells, Lilyleaf Ladybell, Lily-leaved Lady Bell, Lily-leaved Ladybell ■
7683 Adenophora liliifolia (L.) Ledeb. ex A. DC. = Adenophora liliifolia (L.) Besser ■
7684 Adenophora liliifolia Fisch. = Adenophora liliifolia (L.) Besser ■
7685 Adenophora liliifolioides Pax et K. Hoffm.; 川藏沙参; Lilyleaf-like Ladybell ■
7686 Adenophora lobophylla D. Y. Hong; 裂叶沙参; Lobateleaf Ladybell, Splitleaf Ladybell ■
7687 Adenophora longipedicellata D. Y. Hong; 湖北沙参(长梗沙参); Longpedicel Ladybell ■
7688 Adenophora longisepala P. C. Tsoong = Adenophora capillaris Hemsl. ■
7689 Adenophora manshurica Nakai = Adenophora divaricata Franch. et Sav. ■
7690 Adenophora marsupiiflora (Spreng.) Fisch. = Adenophora stenanthina (Ledeb.) Kitag. ■
7691 Adenophora marsupiiflora (Spreng.) Fisch. f. crispata Korsh. = Adenophora stenanthina (Ledeb.) Kitag. ■
7692 Adenophora marsupiiflora Fisch. = Adenophora stenanthina (Ledeb.) Kitag. ■
7693 Adenophora marsupiiflora Fisch. f. crispata Korsh. = Adenophora stenanthina (Ledeb.) Kitag. ■
7694 Adenophora marsupiiflora Fisch. var. crispata (Turcz. ex Ledeb.) Kitag. = Adenophora stenanthina (Ledeb.) Kitag. ■
7695 Adenophora maximowicziana Makino; 马氏沙参; Maximowicz Ladybell ■☆
7696 Adenophora megalantha Diels = Adenophora coelestis Diels ■
7697 Adenophora micrantha D. Y. Hong; 小花沙参; Littleflower Ladybell ■
7698 Adenophora microcodon C. Y. Wu = Adenophora jasionifolia Franch. ■
7699 Adenophora moiwana Nakai = Adenophora pereskiifolia (Fisch. ex Roem. et Schult.) Fisch. ex G. Don ■
7700 Adenophora moiwana Nakai var. heterotricha ? = Adenophora pereskiifolia (Fisch. ex Roem. et Schult.) Fisch. ex G. Don ■
7701 Adenophora mongolica A. I. Baranov = Adenophora stenanthina (Ledeb.) Kitag. ■
7702 Adenophora morrisonensis Hayata; 台湾沙参(高山沙参,玉山沙参); Morrison Ladybell, Taiwan Ladybell ■
7703 Adenophora morrisonensis Hayata subsp. uehatae (Yamam.) Lammers; 玉山沙参■
7704 Adenophora morrisonensis Hayata subsp. uehatae (Yamam.) Lammers = Adenophora morrisonensis Hayata ■
7705 Adenophora morrisonensis Hayata subsp. uehatae (Yamam.) Lammers = Adenophora uehatae Yamam. ■
7706 Adenophora nikoensis Franch. et Sav.; 姬沙参■☆
7707 Adenophora nikoensis Franch. et Sav. f. globiflora Hiyama; 球花姬沙参■☆
7708 Adenophora nikoensis Franch. et Sav. f. hispidula T. Shimizu; 毛姬沙参■☆
7709 Adenophora nikoensis Franch. et Sav. f. linearifolia Takeda = Adenophora nikoensis Franch. et Sav. ■☆
7710 Adenophora nikoensis Franch. et Sav. f. nipponica (Kitam.) H. Hara = Adenophora nikoensis Franch. et Sav. ■☆
7711 Adenophora nikoensis Franch. et Sav. f. stenophylla (Kitam.) H. Hara = Adenophora nikoensis Franch. et Sav. ■☆
7712 Adenophora nikoensis Franch. et Sav. var. persicaria Ohwi ex T. Shimizu et Okazaki; 桃形姬沙参■☆
7713 Adenophora nikoensis Franch. et Sav. var. petrophila (H. Hara) H. Hara; 喜岩姬沙参■☆
7714 Adenophora nikoensis Franch. et Sav. var. stenophylla (Kitam.) Ohwi = Adenophora nikoensis Franch. et Sav. ■☆
7715 Adenophora nikoensis Franch. et Sav. var. teramotoi (Hurus. ex T. Yamaz.) Okazaki et T. Shimizu; 寺本姬沙参■☆
7716 Adenophora ningxianica D. Y. Hong; 宁夏沙参; Ningsia Ladybell, Ningxia Ladybell ■
7717 Adenophora nipponica Kitam. = Adenophora nikoensis Franch. et Sav. ■☆
7718 Adenophora nipponica Kitam. = Adenophora nikoensis Franch. et Sav. f. nipponica (Kitam.) H. Hara ■☆
7719 Adenophora nystroemii Nannf. = Adenophora gmelinii (Spreng.) Fisch. ■
7720 Adenophora obtusifolia Merr. = Adenophora tetraphylla (Thunb.) A. DC. ■
7721 Adenophora omeiensis Z. Y. Zhu = Campanula omeiensis (Z. Y. Zhu) D. Y. Hong et Z. Y. Li ■

7722 Adenophora onoi Tatew. et Kitam. = Adenophora pereskiifolia (Fisch. ex Roem. et Schult.) Fisch. ex G. Don ■

7723 Adenophora onoi Tatew. et Kitam. = Adenophora triphylla (Thunb.) A. DC. var. japonica (Regel) H. Hara ■☆

7724 Adenophora ornata Diels = Adenophora coelestis Diels ■

7725 Adenophora ornata Diels var. alba C. Y. Wu;白花装饰沙参■

7726 Adenophora pachyphylla Kitag. = Adenophora gmelinii (Spreng.) Fisch. ■

7727 Adenophora pachyrhiza Diels = Adenophora coelestis Diels ■

7728 Adenophora palustris Kom.;沼沙参(沼生沙参);Marshy Ladybell ■

7729 Adenophora palustris Kom. f. nivea T. Koyama et Asai;雪白沼沙参■☆

7730 Adenophora paniculata Nannf. = Adenophora capillaris Hemsl. subsp. paniculata (Nannf.) D. Y. Hong et S. Ge ■

7731 Adenophora paniculata Nannf. var. dentata Y. Z. Zhao;齿叶紫沙参;Dentate Paniculate Ladybell ■

7732 Adenophora paniculata Nannf. var. dentata Y. Z. Zhao = Adenophora capillaris Hemsl. subsp. paniculata (Nannf.) D. Y. Hong et S. Ge ■

7733 Adenophora paniculata Nannf. var. petiolata Y. Z. Zhao;有柄紫沙参;Petiolate Paniculate Ladybell ■

7734 Adenophora paniculata Nannf. var. petiolata Y. Z. Zhao = Adenophora capillaris Hemsl. subsp. paniculata (Nannf.) D. Y. Hong et S. Ge ■

7735 Adenophora paniculata Nannf. var. pilosa Kitag. = Adenophora capillaris Hemsl. subsp. paniculata (Nannf.) D. Y. Hong et S. Ge ■

7736 Adenophora paniculata Nannf. var. pilosa Kitag. = Adenophora paniculata Nannf. ■

7737 Adenophora paniculata Nannf. var. psilosa Kitag. = Adenophora capillaris Hemsl. subsp. paniculata (Nannf.) D. Y. Hong et S. Ge ■

7738 Adenophora paniculata Nannf. var. psilosa Kitag. = Adenophora paniculata Nannf. ■

7739 Adenophora pereskiifolia (Fisch. ex Roem. et Schult.) Fisch. ex G. Don;长白沙参(阔叶沙参);Broad-leaved Harebell, Changbaishan Ladybell, Changpai Mountains Ladybell ■

7740 Adenophora pereskiifolia (Fisch. ex Roem. et Schult.) Fisch. ex G. Don f. puberula Kitag. = Adenophora pereskiifolia (Fisch. ex Roem. et Schult.) Fisch. ex G. Don ■

7741 Adenophora pereskiifolia (Fisch. ex Roem. et Schult.) Fisch. ex G. Don subsp. alternifolia (P. Y. Fu ex Y. Z. Zhao) C. X. Fu et M. Y. Liu = Adenophora pereskiifolia (Fisch. ex Roem. et Schult.) Fisch. ex G. Don var. alternifolia P. Y. Fu ex Y. Z. Zhao ■

7742 Adenophora pereskiifolia (Fisch. ex Roem. et Schult.) Fisch. ex G. Don subsp. subalpina Baran. = Adenophora pereskiifolia (Fisch. ex Roem. et Schult.) Fisch. ex G. Don ■

7743 Adenophora pereskiifolia (Fisch. ex Roem. et Schult.) Fisch. ex G. Don subsp. subalpina Baran. f. linearifolia Baran. = Adenophora pereskiifolia (Fisch. ex Roem. et Schult.) Fisch. ex G. Don ■

7744 Adenophora pereskiifolia (Fisch. ex Roem. et Schult.) Fisch. ex G. Don var. alternifolia P. Y. Fu ex Y. Z. Zhao;兴安沙参(长叶沙参);Xing'an Ladybell ■

7745 Adenophora pereskiifolia (Fisch. ex Roem. et Schult.) Fisch. ex G. Don subsp. alternifolia (P. Y. Fu ex Y. Z. Zhao) C. X. Fu et M. Y. Liu = Adenophora pereskiifolia (Fisch. ex Roem. et Schult.) Fisch. ex G. Don ■

7746 Adenophora pereskiifolia (Fisch. ex Roem. et Schult.) Fisch. ex G. Don subsp. subalpina f. linearifolia Baranov = Adenophora pereskiifolia (Fisch. ex Roem. et Schult.) Fisch. ex G. Don ■

7747 Adenophora pereskiifolia (Fisch. ex Roem. et Schult.) Fisch. ex G. Don var. alternifolia P. Y. Fu ex Y. Z. Zhao = Adenophora pereskiifolia (Fisch. ex Roem. et Schult.) Fisch. ex G. Don ■

7748 Adenophora pereskiifolia (Fisch. ex Roem. et Schult.) Fisch. ex G. Don var. angustifolia Y. Z. Zhao;狭叶长白沙参;Narrowleaf Changbaishan Ladybell ■

7749 Adenophora pereskiifolia (Fisch. ex Roem. et Schult.) Fisch. ex G. Don var. curvidens (Nakai) Kitag. = Adenophora pereskiifolia (Fisch. ex Roem. et Schult.) Fisch. ex G. Don ■

7750 Adenophora pereskiifolia (Fisch. ex Roem. et Schult.) Fisch. ex G. Don var. heterotricha (Nakai ex H. Hara) H. Hara = Adenophora pereskiifolia (Fisch. ex Roem. et Schult.) Fisch. ex G. Don ■

7751 Adenophora pereskiifolia (Fisch. ex Roem. et Schult.) Fisch. ex G. Don var. japonica ? = Adenophora triphylla (Thunb.) A. DC. var. japonica (Regel) H. Hara ■☆

7752 Adenophora pereskiifolia (Fisch. ex Roem. et Schult.) Fisch. ex G. Don var. moiwana (Nakai) H. Hara = Adenophora pereskiifolia (Fisch. ex Roem. et Schult.) Fisch. ex G. Don ■

7753 Adenophora pereskiifolia (Fisch. ex Roem. et Schult.) Fisch. ex G. Don var. moiwana (Nakai) H. Hara f. petrophila (H. Hara) T. Shimizu = Adenophora pereskiifolia (Fisch. ex Roem. et Schult.) Fisch. ex G. Don ■

7754 Adenophora pereskiifolia (Fisch. ex Roem. et Schult.) Fisch. ex G. Don var. petrophila (H. Hara) T. Shimizu = Adenophora nikoensis Franch. et Sav. var. petrophila (H. Hara) H. Hara ■☆

7755 Adenophora pereskiifolia (Fisch. ex Roem. et Schult.) Fisch. ex G. Don var. petrophila (H. Hara) T. Shimizu = Adenophora pereskiifolia (Fisch. ex Roem. et Schult.) Fisch. ex G. Don ■

7756 Adenophora pereskiifolia (Fisch. ex Roem. et Schult.) Fisch. ex G. Don var. yamadae ? = Adenophora pereskiifolia (Fisch. ex Roem. et Schult.) Fisch. ex G. Don ■

7757 Adenophora pereskiifolia (Fisch. ex Roem. et Schult.) Fisch. ex G. Don var. uryuensis (Miyabe et Tatew.) Toyok. et Nosaka = Adenophora pereskiifolia (Fisch. ex Roem. et Schult.) Fisch. ex G. Don ■

7758 Adenophora pereskiifolia (Fisch. ex Roem. et Schult.) Fisch. ex Loudon = Adenophora pereskiifolia (Fisch. ex Roem. et Schult.) Fisch. ex G. Don ■

7759 Adenophora pereskiifolia (Fisch. ex Roem. et Schult.) G. Don = Adenophora pereskiifolia (Fisch. ex Roem. et Schult.) Fisch. ex G. Don ■

7760 Adenophora pereskiifolia G. Don = Adenophora pereskiifolia (Fisch. ex Roem. et Schult.) Fisch. ex G. Don ■

7761 Adenophora petiolata Pax et K. Hoffm.;秦岭沙参;Chinling Ladybell, Qinling Ladybell ■

7762 Adenophora petiolata Pax et K. Hoffm. subsp. huadungensis (D. Y. Hong) D. Y. Hong et S. Ge;华东杏叶沙参(华东沙参);E. China Ladybell, East China Ladybell ■

7763 Adenophora petiolata Pax et K. Hoffm. subsp. hunanensis (Nannf.) D. Y. Hong et S. Ge;杏叶沙参(宽裂沙参);Hunan Ladybell ■

7764 Adenophora petrophila (H. Hara) H. Hara = Adenophora nikoensis Franch. et Sav. var. petrophila (H. Hara) H. Hara ■☆

7765 Adenophora pinifolia Kitag.;松叶沙参(辽宁沙参);Pineleaf Ladybell ■

7766 Adenophora polyantha f. densipila Kitag. = Adenophora polyantha

Nakai ■
7767 Adenophora polyantha f. eriocaulis Kitag. = Adenophora polyantha Nakai ■
7768 Adenophora polyantha Nakai;石沙参(糙萼沙参,互叶沙参);Manyflower Ladybell ■
7769 Adenophora polyantha Nakai subsp. scabricalyx (Kitag.) J. Z. Qiu et D. Y. Hong;糙萼石沙参■
7770 Adenophora polyantha Nakai var. contracta Kitag. = Adenophora contracta (Kitag.) J. Z. Qiu et D. Y. Hong ■
7771 Adenophora polyantha Nakai var. contracta Kitag. = Adenophora polyantha Nakai ■
7772 Adenophora polyantha Nakai var. glabricalyx Kitag.;光萼松叶沙参(光萼沙参)■☆
7773 Adenophora polyantha Nakai var. glabricalyx Kitag. = Adenophora polyantha Nakai ■
7774 Adenophora polyantha Nakai var. glabricalyx Kitag. f. eriocaulis Kitag. = Adenophora polyantha Nakai ■
7775 Adenophora polyantha Nakai var. media Nakai et Kitag. = Adenophora polyantha Nakai ■
7776 Adenophora polyantha Nakai var. media Nakai et Kitag. f. densipila Kitag. = Adenophora polyantha Nakai ■
7777 Adenophora polyantha Nakai var. rhombica Y. Z. Zhao;菱叶石沙参;Rhombic-leaved Manyflower Ladybell ■
7778 Adenophora polyantha Nakai var. scabricalyx Kitag. = Adenophora polyantha Nakai ■
7779 Adenophora polyantha Nakai var. scabricalyx Kitag. = Adenophora polyantha Nakai subsp. scabricalyx (Kitag.) J. Z. Qiu et D. Y. Hong ■
7780 Adenophora polydentata P. F. Tu et G. J. Xu = Adenophora potaninii Korsh. ■
7781 Adenophora polymorpha Ledeb.;多型沙参■☆
7782 Adenophora polymorpha Ledeb. = Adenophora nikoensis Franch. et Sav. ■☆
7783 Adenophora polymorpha Ledeb. var. coronopifolia Trautv. ex Herder = Adenophora gmelinii (Spreng.) Fisch. ■
7784 Adenophora polymorpha Ledeb. var. lamarkii Herder = Adenophora lamarkii Fisch. ■
7785 Adenophora polymorpha Ledeb. var. latifolia Herder = Adenophora pereskiifolia (Fisch. ex Roem. et Schult.) Fisch. ex G. Don ■
7786 Adenophora polymorpha Ledeb. var. stricta ? = Adenophora stricta Miq. ■
7787 Adenophora polymorpha Ledeb. var. verticillata Fisch. et Sav. = Adenophora pereskiifolia (Fisch. ex Roem. et Schult.) Fisch. ex G. Don ■
7788 Adenophora polymorpha Ledeb. var. verticillata Franch. et Sav. = Adenophora tetraphylla (Thunb.) Fisch. ■
7789 Adenophora potaninii Korsh.;泡沙参(长叶沙参,灯花草,灯笼花,奶腥菜花,泡参,山沙参);Bush Ladybell,Ladybells,Longleaf Ladybell,Potanin Ladybell ■
7790 Adenophora potaninii Korsh. subsp. wawreana (Zahlbr.) S. Ge et D. Y. Hong;多歧泡沙参■
7791 Adenophora potaninii Korsh. var. bockiana (Diels) S. W. Liu = Adenophora potaninii Korsh. ■
7792 Adenophora pratensis Y. Z. Zhao;草原沙参;Prairie Ladybell ■
7793 Adenophora pubescens Hemsl. = Adenophora rupincola Hemsl. ■
7794 Adenophora pumila P. C. Tsoong = Adenophora jasionifolia Franch. ■
7795 Adenophora radiatifolia Nakai = Adenophora tetraphylla (Thunb.) Fisch. ■
7796 Adenophora raphanorrhiza C. Y. Wu;萝卜根沙参■
7797 Adenophora raphanorrhiza C. Y. Wu = Adenophora coelestis Diels ■
7798 Adenophora remotiflora (Siebold et Zucc.) Miq.;薄叶沙参(薄叶荠苨,莀苨,地参,荠苨);Scatteredflower Ladybell ■
7799 Adenophora remotiflora (Siebold et Zucc.) Miq. f. angustifola Makino;细薄叶沙参(细薄叶荠苨)■☆
7800 Adenophora remotiflora (Siebold et Zucc.) Miq. f. hirsuta (Honda) Sugim.;毛薄叶沙参(毛薄叶荠苨)■☆
7801 Adenophora remotiflora (Siebold et Zucc.) Miq. f. leucantha Honda;白花薄叶沙参(白花薄叶荠苨)■☆
7802 Adenophora remotiflora (Siebold et Zucc.) Miq. f. longifolia Kom. = Adenophora remotiflora (Siebold et Zucc.) Miq. ■
7803 Adenophora remotiflora (Siebold et Zucc.) Miq. var. hirsuta Honda = Adenophora remotiflora (Siebold et Zucc.) Miq. f. hirsuta (Honda) Sugim. ■☆
7804 Adenophora roseiflora C. Y. Wu = Adenophora khasiana (Hook. f. et Thomson) Collett et Hemsl. ■
7805 Adenophora rotundifolia H. Lév. = Adenophora stricta Miq. ■
7806 Adenophora rupestris Reverdin;岩地沙参■☆
7807 Adenophora rupincola Hemsl.;多毛沙参;Manyhairy Ladybell ■
7808 Adenophora scabridula Nannf. = Adenophora polyantha Nakai subsp. scabricalyx (Kitag.) J. Z. Qiu et D. Y. Hong ■
7809 Adenophora scabridula Nannf. var. viscida P. C. Tsoong = Adenophora polyantha Nakai subsp. scabricalyx (Kitag.) J. Z. Qiu et D. Y. Hong ■
7810 Adenophora scahridula Nannf. = Adenophora polyantha Nakai ■
7811 Adenophora sinensis A. DC.;中华沙参(华沙参);China Ladybell,Chinese Ladybell ■
7812 Adenophora sinensis A. DC. var. pilosa A. DC. = Adenophora stricta Miq. ■
7813 Adenophora smithii Nannf.;川北沙参■
7814 Adenophora smithii Nannf. = Adenophora himalayana Feer ■
7815 Adenophora smithii Nannf. f. crispa Nannf. = Adenophora himalayana Feer ■
7816 Adenophora stenanthina (Ledeb.) Kitag.;长柱沙参;Longstyle Ladybell ■
7817 Adenophora stenanthina (Ledeb.) Kitag. f. crispata (Korsh.) Kitag. = Adenophora stenanthina (Ledeb.) Kitag. ■
7818 Adenophora stenanthina (Ledeb.) Kitag. f. crispata (Korsh.) Kitag. = Adenophora stenanthina (Ledeb.) Kitag. var. crispata (Korsh.) Y. Z. Zhao ■
7819 Adenophora stenanthina (Ledeb.) Kitag. f. crispata Kitag. = Adenophora stenanthina (Ledeb.) Kitag. var. crispata (Korsh.) Y. Z. Zhao ■
7820 Adenophora stenanthina (Ledeb.) Kitag. f. crispata Kitag. = Adenophora stenanthina (Ledeb.) Kitag. ■
7821 Adenophora stenanthina (Ledeb.) Kitag. subsp. sylvatica D. Y. Hong;林沙参(林下沙参);Woodland Ladybell ■
7822 Adenophora stenanthina (Ledeb.) Kitag. var. angustilanceifolia Y. Z. Zhao = Adenophora stenanthina (Ledeb.) Kitag. ■
7823 Adenophora stenanthina (Ledeb.) Kitag. var. angustilanceofolia Y. Z. Zhao;锡林沙参■
7824 Adenophora stenanthina (Ledeb.) Kitag. var. collina (Kitag.) Y. Z. Zhao;丘沙参■
7825 Adenophora stenanthina (Ledeb.) Kitag. var. collina (Kitag.) Y. Z. Zhao = Adenophora stenanthina (Ledeb.) Kitag. ■
7826 Adenophora stenanthina (Ledeb.) Kitag. var. crispata (Korsh.)

Y. Z. Zhao;皱叶沙参■

7827 Adenophora stenanthina (Ledeb.) Kitag. var. crispata (Korsh.) Y. Z. Zhao = Adenophora stenanthina (Ledeb.) Kitag. ■

7828 Adenophora stenophylla Hemsl.;扫帚沙参(蒙古沙参,细叶沙参);Broom Ladybell,Narrowleaf Ladybell ■

7829 Adenophora stenophylla Hemsl. var. denudata Kitag. = Adenophora stenophylla Hemsl. ■

7830 Adenophora stricta Miq.;沙参(白参,白沙参,保牙参,风箱灵子,虎须,桔参,苦桑头,苦心,龙须沙参,面杆杖,南沙参,泡参,泡沙参,沙癞子,识美,土人参,文虎,文希,稳牙参,杏叶沙参,羊婆奶,知母,志取,钻天老);Ladybell,Upright Ladybell ■

7831 Adenophora stricta Miq. f. albiflora Migo;白花沙参■☆

7832 Adenophora stricta Miq. subsp. aurita (Franch.) D. Y. Hong et S. Ge;川西沙参;Longeared Ladybell ■

7833 Adenophora stricta Miq. subsp. confusa (Nannf.) D. Y. Hong;昆明沙参(罗兰参);Kunming Ladybell,Lady Bells ■

7834 Adenophora stricta Miq. subsp. henanica P. F. Tu et G. J. Xu = Adenophora stricta Miq. subsp. sessilifolia D. Y. Hong ■

7835 Adenophora stricta Miq. subsp. sessilifolia D. Y. Hong;无柄沙参(泡参);Sessile Ladybell ■

7836 Adenophora stricta Miq. var. lancifolia Honda;剑叶沙参■☆

7837 Adenophora stricta Miq. var. nanjingensis P. F. Tu et G. J. Xu;南京沙参■

7838 Adenophora stricta Miq. var. nanjingensis P. F. Tu et G. J. Xu = Adenophora stricta Miq. ■

7839 Adenophora stricta Miq. var. qinglongshanica P. F. Tu et G. J. Xu;青龙山沙参■

7840 Adenophora stricta Miq. var. qinglongshanica P. F. Tu et G. J. Xu = Adenophora stricta Miq. ■

7841 Adenophora stylosa Fisch. = Adenophora liliifolia (L.) Besser ■

7842 Adenophora suolunensis P. F. Tu et X. F. Zhao = Adenophora micrantha D. Y. Hong ■

7843 Adenophora takedae Makino;武氏沙参■☆

7844 Adenophora takedae Makino var. howozana (Takeda) Sugim. ex Okazaki;凤凰沙参■☆

7845 Adenophora tashiroi (Makino et Nakai) Makino et Nakai;田代氏沙参■☆

7846 Adenophora taurica (W. N. Sukaczev) Juz.;克里木沙参■☆

7847 Adenophora teramotoi Hurus. et T. Yamaz. var. hispidula T. Shimizu;细毛■☆

7848 Adenophora teramotoi Hurus. ex T. Yamaz. = Adenophora nikoensis Franch. et Sav. var. teramotoi (Hurus. ex T. Yamaz.) Okazaki et T. Shimizu ■☆

7849 Adenophora tetraphylla (Thunb.) A. DC. = Adenophora triphylla (Thunb.) Fisch. ■

7850 Adenophora tetraphylla (Thunb.) Fisch.;轮叶沙参(白参,白沙参,保牙参,虎须,桔参,苦心,铃儿草,面杆杖,南沙参,泡参,泡沙参,沙参,识美,四叶沙参,土人参,文虎,文希,稳牙参,细叶沙参,羊婆奶,知母,志取);Dahurian Ladybell,Fourleaf Ladybell ■

7851 Adenophora tetraphylla (Thunb.) Fisch. var. integrifolia Y. Z. Zhao;全缘轮叶沙参;Entire-leaved Fourleaf Ladybell ■

7852 Adenophora thunbergiana Kudo = Adenophora triphylla (Thunb.) A. DC. var. japonica (Regel) H. Hara ■☆

7853 Adenophora trachelioides Maxim.;荠苨(白麦根,白面根,蒡,长叶沙参,臭苏,菧苨,空沙参,老母鸡肉,梅参,苨,芪苨,甜桔梗,土桔梗,心叶沙参,杏参,杏叶菜,杏叶沙参,隐忍,隐忍草);Apricotleaf Ladybell ■

7854 Adenophora trachelioides Maxim. subsp. giangsuensis D. Y. Hong;苏南荠苨(苏南宽萼沙参);South Jiangsu Ladybell,South Kiangsu Ladybell ■

7855 Adenophora tricuspidata (Fisch. ex Roem. et Schult.) A. DC.;锯齿沙参(锯叶沙参);Serrete Ladybell ■

7856 Adenophora triphylla (Thunb.) A. DC. = Adenophora tetraphylla (Thunb.) Fisch. ■

7857 Adenophora triphylla (Thunb.) A. DC. f. alba Hatus.;白花轮叶沙参■☆

7858 Adenophora triphylla (Thunb.) A. DC. subsp. aperticampanulata Kitam. = Adenophora triphylla (Thunb.) A. DC. var. japonica (Regel) H. Hara ■☆

7859 Adenophora triphylla (Thunb.) A. DC. var. angustifolia (Regel) Kitam. = Adenophora verticillata Fisch. ■☆

7860 Adenophora triphylla (Thunb.) A. DC. var. angustifolia (Regel) Kitam. = Adenophora triphylla (Thunb.) A. DC. ■

7861 Adenophora triphylla (Thunb.) A. DC. var. hakusanensis (Nakai) Kitam. = Adenophora triphylla (Thunb.) A. DC. var. japonica (Regel) H. Hara f. violacea (H. Hara) T. Shimizu ■☆

7862 Adenophora triphylla (Thunb.) A. DC. var. insularis (Kitam.) Kitam. = Adenophora triphylla (Thunb.) A. DC. ■

7863 Adenophora triphylla (Thunb.) A. DC. var. japonica (Regel) H. Hara;日本轮叶沙参(日本三叶沙参)■☆

7864 Adenophora triphylla (Thunb.) A. DC. var. japonica (Regel) H. Hara f. albiflora (Tatew.) H. Hara;白花日本轮叶沙参■☆

7865 Adenophora triphylla (Thunb.) A. DC. var. japonica (Regel) H. Hara f. canescens (Franch. et Sav.) Kitam.;灰日本轮叶沙参■☆

7866 Adenophora triphylla (Thunb.) A. DC. var. japonica (Regel) H. Hara f. leucantha H. Hara = Adenophora triphylla (Thunb.) A. DC. var. japonica (Regel) H. Hara f. albiflora (Tatew.) H. Hara ■☆

7867 Adenophora triphylla (Thunb.) A. DC. var. japonica (Regel) H. Hara f. procumbens T. Shimizu;匍匐日本轮叶沙参■☆

7868 Adenophora triphylla (Thunb.) A. DC. var. japonica (Regel) H. Hara f. rotundifolia Hiyama;圆叶日本轮叶沙参■☆

7869 Adenophora triphylla (Thunb.) A. DC. var. japonica (Regel) H. Hara f. violacea (H. Hara) T. Shimizu;淡紫色日本轮叶沙参■☆

7870 Adenophora triphylla (Thunb.) A. DC. var. kurilensis (Nakai) Kitam. = Adenophora triphylla (Thunb.) A. DC. var. japonica (Regel) H. Hara ■☆

7871 Adenophora triphylla (Thunb.) A. DC. var. puellaris (Honda) H. Hara;美丽轮叶沙参■☆

7872 Adenophora triphylla (Thunb.) Fisch. = Adenophora triphylla (Thunb.) A. DC. ■

7873 Adenophora tsinlingensis Pax et K. Hoffm. = Adenophora himalayana Feer subsp. alpina (Nannf.) D. Y. Hong ■

7874 Adenophora uehatae Yamam. = Adenophora morrisonensis Hayata subsp. uehatae (Yamam.) Lammers ■

7875 Adenophora uehatae Yamam. = Adenophora morrisonensis Hayata ■

7876 Adenophora urceolata C. Y. Wu = Adenophora capillaris Hemsl. subsp. leptosepala (Diels) D. Y. Hong ■

7877 Adenophora urceolata Y. Z. Zhao = Adenophora contracta (Kitag.) J. Z. Qiu et D. Y. Hong ■

7878 Adenophora uryuensis Miyabe et Tatew.;瓜生沙参■☆

7879 Adenophora verticillata Fisch.;维奇沙参■☆

7880 Adenophora verticillata Fisch. = Adenophora tetraphylla (Thunb.) Fisch. ■

7881 Adenophora verticillata Fisch. = Adenophora triphylla (Thunb.)

A. DC. var. japonica (Regel) H. Hara ■☆

7882 Adenophora verticillata var. hirsuta Fisch. = Adenophora triphylla (Thunb.) A. DC. var. japonica (Regel) H. Hara ■☆

7883 Adenophora watsonii W. W. Sm. = Adenophora aurita Franch. ■

7884 Adenophora wawreana Zahlbr.;多歧沙参(多枝沙参,南沙参);Manyfork Ladybell,Wawre Ladybell ■

7885 Adenophora wawreana Zahlbr. = Adenophora potaninii Korsh. subsp. wawreana (Zahlbr.) S. Ge et D. Y. Hong ■

7886 Adenophora wawreana Zahlbr. f. oligotricha Kitag. = Adenophora potaninii Korsh. subsp. wawreana (Zahlbr.) S. Ge et D. Y. Hong ■

7887 Adenophora wawreana Zahlbr. f. oligotricha Kitag. = Adenophora wawreana Zahlbr. ■

7888 Adenophora wawreana Zahlbr. f. polytricha Kitag.;密毛多歧沙参■

7889 Adenophora wawreana Zahlbr. f. polytricha Kitag. = Adenophora potaninii Korsh. subsp. wawreana (Zahlbr.) S. Ge et D. Y. Hong ■

7890 Adenophora wawreana Zahlbr. f. polytricha Kitag. = Adenophora wawreana Zahlbr. ■

7891 Adenophora wawreana Zahlbr. var. lanceifolia Y. Z. Zhao;阴山沙参;Yinshan Ladybell ■

7892 Adenophora wawreana Zahlbr. var. lancifolia Y. Z. Zhao = Adenophora potaninii Korsh. subsp. wawreana (Zahlbr.) S. Ge et D. Y. Hong ■

7893 Adenophora wilsonii Nannf.;聚叶沙参(妇奶参,聚沙参);E. H. Wilson Ladybell ■

7894 Adenophora wulingshanica D. Y. Hong;雾灵沙参;Wuling Mountains Ladybell,Wulingshan Ladybell ■

7895 Adenophora wulingshanica D. Y. Hong var. alterna Y. Z. Zhao;互叶雾灵沙参;Alternate Wulingshan Ladybell ■

7896 Adenophora wutaiensis Hurus. = Adenophora elata Nannf. ■

7897 Adenophora yokoyamae Miyabe et Tatewaki = Adenophora pereskiifolia (Fisch. ex Roem. et Schult.) Fisch. ex G. Don ■

7898 Adenophyllum Pers. (1807);腺叶菊属●■☆

7899 Adenophyllum Pers. = Schlechtendalia Willd. (废弃属名)■☆

7900 Adenophyllum Thouars ex Baill. = Hecatea Thouars ■☆

7901 Adenophyllum Thouars ex Baill. = Omphalea L. (保留属名)■☆

7902 Adenophyllum anomalum (Canby et Rose) Strother;异常腺叶菊●☆

7903 Adenophyllum coccineum Pers.;腺叶菊●☆

7904 Adenoplea Radlk. = Buddleja L. ●■

7905 Adenoplea baccata Radlk. = Buddleja acuminata Poir. ●☆

7906 Adenoplea baroniana (Oliv.) Petit = Buddleja acuminata Poir. ●☆

7907 Adenoplea lindleyana (Fortune) Small = Buddleja lindleyana Fortune ●

7908 Adenoplea madagascariensis (Lam.) Eastw. = Buddleja madagascariensis Lam. ●

7909 Adenoplea sinuata (Willd. ex Roem. et Schult.) Radlk. = Buddleja acuminata Poir. ●☆

7910 Adenoplusia Radlk. = Buddleja L. ●■

7911 Adenoplusia axillaris Radlk. = Buddleja axillaris Willd. ex Roem. et Schult. ●☆

7912 Adenoplusia ulugurensis Melch. = Buddleja axillaris Willd. ex Roem. et Schult. ●☆

7913 Adenoplusia willdenowii Radlk. = Buddleja axillaris Willd. ex Roem. et Schult. ●☆

7914 Adenopodia C. Presl = Entada Adans. (保留属名)●

7915 Adenopodia C. Presl(1851);腺柄豆属■☆

7916 Adenopodia rotundifolia (Harms) Brenan;圆叶腺柄豆■☆

7917 Adenopodia scelerata (A. Chev.) Brenan;刺腺柄豆■☆

7918 Adenopodia schlechteri (Harms) Brenan;施莱腺柄豆■☆

7919 Adenopodia spicata (E. Mey.) C. Presl;腺柄豆■☆

7920 Adenopogon Welw. = Swertia L. ■

7921 Adenopogon stellarioides Welw. = Swertia welwitschii Engl. ■☆

7922 Adenoporces Small = Tetrapterys Cav. (保留属名)●☆

7923 Adenopus Benth. (1849);肖葫芦属■☆

7924 Adenopus Benth. = Lagenaria Ser. ■

7925 Adenopus abyssinicus Hook. f. = Lagenaria abyssinica (Hook. f.) C. Jeffrey ■☆

7926 Adenopus breviflorus Benth. = Lagenaria breviflora (Benth.) Roberty ■☆

7927 Adenopus chariensis A. Chev.;沙里肖葫芦■☆

7928 Adenopus cienkowskii Schweinf. = Peponium cienkowskii (Schweinf.) Engl. ■☆

7929 Adenopus diversifolius Cogn. = Eureiandra formosa Hook. f. ■☆

7930 Adenopus eglandulosus Hook. f. = Ruthalicia eglandulosa (Hook. f.) C. Jeffrey ■☆

7931 Adenopus guineensis (G. Don) Exell = Lagenaria guineensis (G. Don) C. Jeffrey ■☆

7932 Adenopus ledermannii Harms = Lagenaria breviflora (Benth.) Roberty ■☆

7933 Adenopus longiflorus Benth. = Lagenaria guineensis (G. Don) C. Jeffrey ■☆

7934 Adenopus multiflorus Cogn. = Lagenaria breviflora (Benth.) Roberty ■☆

7935 Adenopus noctiflorus Gilg = Lagenaria breviflora (Benth.) Roberty ■☆

7936 Adenopus pynaertii De Wild. = Lagenaria guineensis (G. Don) C. Jeffrey ■☆

7937 Adenopus reticulatus Gilg = Lagenaria abyssinica (Hook. f.) C. Jeffrey ■☆

7938 Adenopus rufus Gilg = Lagenaria rufa (Gilg) C. Jeffrey ■☆

7939 Adenorachis (DC.) Nieuwl. (1915);肖石楠属●☆

7940 Adenorachis (DC.) Nieuwl. = Aronia Medik. (保留属名)●☆

7941 Adenorachis (DC.) Nieuwl. = Photinia Lindl. ●

7942 Adenorachis Nieuwl. = Adenorachis (DC.) Nieuwl. ●☆

7943 Adenorandia Vermoesen = Gardenia Ellis(保留属名)●

7944 Adenorandia kalbreyeri (Hiern) Robbr. et Bridson;肖石楠●☆

7945 Adenorhopium Rchb. = Adenoropium Pohl ●■

7946 Adenorhopium Rchb. = Jatropha L. (保留属名)●■

7947 Adenorima Raf. = Euphorbia L. ●■

7948 Adenoropium Pohl = Jatropha L. (保留属名)●■

7949 Adenoropium elegans Pohl = Jatropha gossypiifolia L. var. elegans (Pohl) Müll. Arg. ●

7950 Adenoropium multifidum Pohl = Jatropha multifida L. ●

7951 Adenorrhopium Wittst. = Adenoropium Pohl ●■

7952 Adenosachma A. Juss. = Mycetia Reinw. ●

7953 Adenosachma Wall. = Mycetia Reinw. ●

7954 Adenosacma Post et Kuntze = Mycetia Reinw. ●

7955 Adenosacme Wall. = Mycetia Reinw. ●

7956 Adenosacme Wall. ex Endl. = Mycetia Reinw. ●

7957 Adenosacme coriacea Dunn = Mycetia coriacea (Dunn) Merr. ●◇

7958 Adenosacme coriacea Dunn = Mycetia sinensis (Hemsl.) Craib ●

7959 Adenosacme fasciculata Miq. = Mycetia obovata Kuntze ●☆

7960 Adenosacme holotricha Miq. = Mycetia holotricha Kuntze ●☆

7961 Adenosacme lanceolata Miq. = Mycetia lanceolata Kuntze ●☆

7962 Adenosacme longiflora (Wall.) Kuntze var. sinensis Hemsl. =

Mycetia sinensis (Hemsl.) Craib ●
7963 Adenosacme longifolia (Wall.) Hook. f. = Mycetia longifolia (Wall.) Kuntze ●
7964 Adenosacme macrostachya Hook. f. = Mycetia macrostachya (Hook. f.) Kuntze ●
7965 Adenosacme nepalensis Wall. = Mycetia nepalensis H. Hara ●
7966 Adenosacme stipulata Hook. f. = Mycetia stipulata Kuntze ●☆
7967 Adenosciadium H. Wolff. (1927);腺伞芹属■☆
7968 Adenosciadium arabicum H. Wolff;腺伞芹■☆
7969 Adenoscilla Gren. et Godr. = Scilla L. ■
7970 Adenoselen Spach = Adenosolen DC. ■☆
7971 Adenoselen Spach = Marasmodes DC. ■☆
7972 Adenosepalum Fourr. = Hypericum L. ●■
7973 Adenosma Nees = Synnema Benth. ●■☆
7974 Adenosma R. Br. (1810);毛麝香属;Adenosma ■
7975 Adenosma affinis Griff. = Adenosma glutinosa (L.) Druce ■
7976 Adenosma africana T. Anderson = Hygrophila africana (T. Anderson) Heine ■☆
7977 Adenosma bracteosa Bonati;多苞毛麝香■☆
7978 Adenosma buchneroides Bonati;勐腊毛麝香■
7979 Adenosma caerulea R. Br.;天蓝毛麝香;Skyblue Adenosma ■☆
7980 Adenosma camphorata Hook. f.;樟脑味毛麝香;Camphorsmell Adenosma ■☆
7981 Adenosma capitata Benth. ex Hance = Adenosma indiana (Lour.) Merr. ■
7982 Adenosma glutinosa (L.) Druce;毛麝香(饼草,解菜,酒子草,辣鸡,辣蓟,蓝花草,蓝花毛麝香,凉草,毛老虎,毛麝香草,麝香草,五郎草,五凉草,香草);Sticky Adenosma ■
7983 Adenosma glutinosa (L.) Druce var. caerulea (R. Br.) Tsoong = Adenosma caerulea R. Br. ■☆
7984 Adenosma glutinosa (L.) Merr. = Adenosma glutinosa (L.) Druce ■
7985 Adenosma indiana (Lour.) Merr.;球花毛麝香(大头陈,地松茶,黑头草,假薄荷,千锤草,神曲草,石棘,石辣,头状毛麝香,土夏枯草,乌头风);Capitate Adenosma, India Adenosma, Indian Adenosma ■
7986 Adenosma javanica(Blume) Merr.;卵萼毛麝香;Javan Adenosma ■
7987 Adenosma macrophylla Benth. ex Wall.;大叶毛麝香(毛麝香);Largeleaf Adenosma ■☆
7988 Adenosma malabarica Hook. f.;马拉巴毛麝香;Malabar Adenosma ■☆
7989 Adenosma microcephala Hook. f.;小头毛麝香;Smallhead Adenosma ■
7990 Adenosma ovata Benth.;卵叶毛麝香;Ovateleaf Adenosma ■☆
7991 Adenosma retusiloba P. C. Tsoong et T. L. Chin;凹裂毛麝香(毛麝香,山薄荷);Retuselobe Adenosma ■
7992 Adenosma subrepens Benth. ex Hook. f.;匍匐毛麝香;Subrepent Adenosma ■☆
7993 Adenosolen DC. = Marasmodes DC. ■☆
7994 Adenospermum Hook. et Arn. = Chrysanthellum Rich. ex Pers. ■☆
7995 Adenostachya Bremek. (1944);腺花爵床属■☆
7996 Adenostachya Bremek. = Strobilanthes Blume ●■
7997 Adenostachya moschifera (Blume) Bremek.;腺花爵床■☆
7998 Adenostachya parvifolia Bremek.;小叶腺花爵床■☆
7999 Adenostegia Benth. (废弃属名) = Cordylanthus Nutt. ex Benth. (保留属名) ■☆
8000 Adenostema Desport. = Adenostemma J. R. Forst. et G. Forst. ■
8001 Adenostemma Forst. = Adenostemma J. R. Forst. et G. Forst. ■
8002 Adenostemma Hook. f. = Arenaria L. ■
8003 Adenostemma Hook. f. = Odontostemma Benth. ■
8004 Adenostemma J. R. Forst. et G. Forst. (1775);下田菊属(猪耳朵属);Adenostemma ■
8005 Adenostemma angustifolium Arn.;狭叶下田菊■☆
8006 Adenostemma caffrum DC.;开菲尔下田菊■☆
8007 Adenostemma caffrum DC. var. asperum Brenan;粗糙下田菊■☆
8008 Adenostemma caffrum DC. var. longifolium (Chiov.) S. A. L. Sm.;长叶下田菊■☆
8009 Adenostemma dregei DC. = Adenostemma viscosum J. R. Forst. et G. Forst. ■☆
8010 Adenostemma fasciulatum DC.;簇生下田菊;Chamise, Chamise Greasewood ■☆
8011 Adenostemma latifolium D. Don = Adenostemma lavenia (L.) Kuntze var. latifolium (D. Don) Hand. -Mazz. ■
8012 Adenostemma lavenia (L.) Kuntze;下田菊(白龙须,风气草,汗苏麻,回筋草,胖婆娘,仁皂刺,乳痈药,水胡椒,猪耳朵叶);Common Adenostemma ■
8013 Adenostemma lavenia (L.) Kuntze var. latifolium (D. Don) Hand. -Mazz.;宽叶下田菊(阔叶下田菊,重皮冲);Broadleaf Adenostemma ■
8014 Adenostemma lavenia (L.) Kuntze var. longifolium Chiov. = Adenostemma caffrum DC. var. longifolium (Chiov.) S. A. L. Sm. ■☆
8015 Adenostemma lavenia (L.) Kuntze var. madurense (DC.) Panigrahi = Adenostemma madurense DC. ■☆
8016 Adenostemma lavenia (L.) Kuntze var. parviflorum (Blume) Hochr.;小花下田菊;Smallflower Broadleaf Adenostemma ■
8017 Adenostemma lavenia (L.) Kuntze var. parviflorum (Blume) Hochr. = Adenostemma parviflorum (Blume) DC. ■
8018 Adenostemma madurense DC.;马都拉下田菊■☆
8019 Adenostemma mauritianum DC.;毛里求斯岛下田菊■☆
8020 Adenostemma natalense DC. = Adenostemma viscosum J. R. Forst. et G. Forst. ■☆
8021 Adenostemma parviflorum (Blume) DC. = Adenostemma lavenia (L.) Kuntze var. parviflorum (Blume) Hochr. ■
8022 Adenostemma parviflorum Blume = Adenostemma lavenia (L.) Kuntze var. parviflorum (Blume) Hochr. ■
8023 Adenostemma schimperi Sch. Bip. ex A. Rich. = Adenostemma caffrum DC. var. asperum Brenan ■☆
8024 Adenostemma tinctorium (Lour.) Cass. = Adenostemma lavenia (L.) Kuntze ■
8025 Adenostemma viscosum J. R. Forst. et G. Forst. = Adenostemma lavenia (L.) Kuntze ■
8026 Adenostemma viscosum J. R. Forst. et G. Forst. var. parviflorum (Blume) Hook. f. = Adenostemma lavenia (L.) Kuntze var. parviflorum (Blume) Hochr. ■
8027 Adenostemon Spreng. = Adenostemum Pers. ●☆
8028 Adenostemum Pers. = Gomortega Ruiz et Pav. ●☆
8029 Adenostephanes Lindl. = Adenostephanus Klotzsch ●☆
8030 Adenostephanus Klotzsch = Euplassa Salisb. ex Knight ●☆
8031 Adenostoma Hook. et Arn. (1832);腺口花属(红皮木属);Chamise ●☆
8032 Adenostoma fasciculatum Hook. et Arn.;簇生腺口花(簇叶红皮木);Chamise, Greasewood ●☆
8033 Adenostoma sparsifolium Torr.;疏叶腺口花(疏叶红皮木);Red Shanks, Redshanks, Ribbon Bush, Ribbonwood ●☆

8034 Adenostyles Benth. et Hook. f. (1883);欧蟹甲属(腺柱菊属)■☆
8035 Adenostyles Cass. = Adenostyles Benth. et Hook. f. ■☆
8036 Adenostyles Cass. = Cacalia L. ●■
8037 Adenostyles Endl. = Adenostylis Blume ■☆
8038 Adenostyles alliariae (Gouan) Kern.;葱芥欧蟹甲(腺柱菊)■☆
8039 Adenostyles alpina (L.) Bluff et Fingerh.;欧蟹甲■☆
8040 Adenostyles leucophylla (Willd.) Rchb.;白叶欧蟹甲■☆
8041 Adenostyles viridis Cass.;绿欧蟹甲■☆
8042 Adenostylidaceae Bercht. et J. Presl;欧蟹甲科■☆
8043 Adenostylis Blume = Zeuxine Lindl. (保留属名)■
8044 Adenostylis Blume(1825);腺柱兰属■☆
8045 Adenostylis Engl. = Zeuxine Lindl. (保留属名)■
8046 Adenostylis Post et Kuntze = Adenostyles Cass. ■☆
8047 Adenostylis arisanensis (Hayata) Hayata = Zeuxine affinis (Lindl.) Benth. ex Hook. f. ■
8048 Adenostylis arisanensis Hayata = Zeuxine affinis (Lindl.) Benth. ex Hook. f. ■
8049 Adenostylis benguetensis Ames = Zeuxine parviflora (Ridl.) Seidenf. ■
8050 Adenostylis emarginata Blume = Pecteilis susannae (L.) Raf. ■
8051 Adenostylis formosana (Rolfe) Hayata = Zeuxine nervosa (Wall. ex Lindl.) Trimen ■
8052 Adenostylis formosana (Rolfe) Hayata = Zeuxine nervosa (Wall. ex Lindl.) Benth. ex C. B. Clarke ■
8053 Adenostylis integerrima Blume = Pecteilis susannae (L.) Raf. ■
8054 Adenostylis philippinensis Ames = Zeuxine philippinensis (Ames) Ames ■
8055 Adenostylis rhombifolia ?;菱叶腺柱兰■☆
8056 Adenostylis strateumatica (L.) Ames = Pecteilis susannae (L.) Raf. ■
8057 Adenostylis sulcata (Roxb.) Hayata = Pecteilis susannae (L.) Raf. ■
8058 Adenostylis tabiyahanensis Hayata = Zeuxine tabiyahanensis (Hayata) Hayata ■
8059 Adenostylis zamboangensis Ames = Zeuxine nervosa (Wall. ex Lindl.) Trimen ■
8060 Adenostylium Rchb. = Adenostyles Benth. et Hook. f. ■☆
8061 Adenostylium Rchb. = Adenostyles Cass. ■☆
8062 Adenothamnus D. D. Keck(1935);星木菊属●☆
8063 Adenothamnus validus (Brandegee) D. D. Keck;星木菊●☆
8064 Adenotheca Welw. ex Baker = Schizobasis Baker ■☆
8065 Adenothola Lem. = Manettia Mutis ex L. (保留属名)●■☆
8066 Adenotrachelium Nees ex Meisn. = Ocotea Aubl. ●☆
8067 Adenotrias Jaub. et Spach = Hypericum L. ●■
8068 Adenotrichia Lindl. = Senecio L. ●■
8069 Adenum G. Don = Adenium Roem. et Schult. ●■☆
8070 Adesia Eaton = Adicea Raf. ■
8071 Adesia Eaton = Pilea Lindl. (保留属名)■
8072 Adesmia DC. (1825)(保留属名);无带豆属(艾兹豆属)■☆
8073 Adesmia emarginata Clos;微缺无带豆■☆
8074 Adhatoda Mill. (1754);肖鸭嘴花属(鸭嘴花属)●
8075 Adhatoda Mill. = Justicia L. ●■
8076 Adhatoda acuminata Nees = Justicia palustris (Hochst.) T. Anderson ■☆
8077 Adhatoda anagalloides Nees = Justicia anagalloides (Nees) T. Anderson ■☆
8078 Adhatoda andromeda (Lindau) C. B. Clarke;肖鸭嘴花■☆
8079 Adhatoda anselliana Nees = Justicia anselliana (Nees) T. Anderson ■☆
8080 Adhatoda auriculata S. Moore = Adhatoda tristis Nees ■☆
8081 Adhatoda bagshawei S. Moore = Justicia francoiseana Brummitt ■☆
8082 Adhatoda betonica (L.) Nees = Justicia betonica L. ■☆
8083 Adhatoda bojeriana Nees = Justicia bojeriana (Nees) Baron ■☆
8084 Adhatoda bolomboensis (De Wild.) Heine = Justicia bolomboensis De Wild. ■☆
8085 Adhatoda buchholzii (Lindau) S. Moore;布赫肖鸭嘴花■☆
8086 Adhatoda camerunensis Heine;喀麦隆肖鸭嘴花■☆
8087 Adhatoda candicans Nees = Justicia candicans (Nees) L. D. Benson ●☆
8088 Adhatoda capensis (Thunb.) Nees = Justicia capensis Thunb. ■☆
8089 Adhatoda cheiranthifolia Nees = Justicia betonica L. ■☆
8090 Adhatoda chevalieri (Lindau) Heine = Duvernoia chevalieri Lindau ●☆
8091 Adhatoda chinensis Benth. = Calophanoides chinensis (Benth.) C. Y. Wu et H. S. Lo ●
8092 Adhatoda chinensis Champ. = Calophanoides chinensis (Champ.) C. Y. Wu et H. S. Lo ex Y. C. Tang ●
8093 Adhatoda claessensii (De Wild.) Heine = Justicia claessensii De Wild. ■☆
8094 Adhatoda cuneata (Vahl) Nees = Justicia cuneata Vahl ■☆
8095 Adhatoda densiflora (Hochst.) J. C. Manning;密花肖鸭嘴花■☆
8096 Adhatoda diffusa Benth. = Justicia insularis T. Anderson ■☆
8097 Adhatoda diosmophylla (Nees) Nees = Justicia orchioides L. f. ■☆
8098 Adhatoda divaricata Nees = Monechma divaricatum (Nees) C. B. Clarke ■☆
8099 Adhatoda duvernoia (Nees) C. B. Clarke = Duvernoia adhatodoides E. Mey. ex Nees ●☆
8100 Adhatoda engleriana (Lindau) C. B. Clarke = Justicia engleriana Lindau ■☆
8101 Adhatoda eylesii S. Moore = Isoglossa eylesii (S. Moore) Brummitt ■☆
8102 Adhatoda fasciata Nees = Justicia flava (Vahl) Vahl ■☆
8103 Adhatoda flava (Vahl) Nees = Justicia flava (Vahl) Vahl ■☆
8104 Adhatoda formosissima Klotzsch = Anisotes formosissimus (Klotzsch) Milne-Redh. ●☆
8105 Adhatoda guineensis Heine;几内亚肖鸭嘴花■☆
8106 Adhatoda hypericum Solms = Justicia odora (Forssk.) Vahl ■☆
8107 Adhatoda hyssopifolia (Nees) Nees = Justicia cuneata Vahl ■☆
8108 Adhatoda incana (Nees) Nees = Monechma incanum (Nees) C. B. Clarke ■☆
8109 Adhatoda kotschyi (Hochst.) Nees = Justicia ladanoides Lam. ■☆
8110 Adhatoda latibracteata (De Wild.) Benoist = Adhatoda buchholzii (Lindau) S. Moore ■☆
8111 Adhatoda leptantha (Nees) Nees = Siphonoglossa leptantha (Nees) Immelman ●☆
8112 Adhatoda leptostachya Nees = Justicia calyculata Deflers ■☆
8113 Adhatoda letestui (Benoist) Heine;莱泰斯图肖鸭嘴花■☆
8114 Adhatoda lupulina Nees = Justicia betonica L. ■☆
8115 Adhatoda maculata (T. Anderson) C. B. Clarke = Justicia biokoensis V. A. W. Graham ■☆
8116 Adhatoda major Nees = Justicia flava (Vahl) Vahl ■☆
8117 Adhatoda matammensis Schweinf. = Justicia matammensis (Schweinf.) Oliv. ■☆
8118 Adhatoda microphylla Klotzsch = Justicia microphylla (Klotzsch)

Lindau ●☆
8119 Adhatoda minor Nees = Justicia flava (Vahl) Vahl ■☆
8120 Adhatoda mollissima Nees = Monechma mollissimum (Nees) P. G. Mey. ■☆
8121 Adhatoda mossambicensis Klotzsch = Justicia mossambicensis (Klotzsch) Lindau ■☆
8122 Adhatoda natalensis Nees = Adhatoda densiflora (Hochst.) J. C. Manning ■☆
8123 Adhatoda nilgherrensis Nees = Justicia nilgherrensis (Nees) C. B. Clarke ■☆
8124 Adhatoda nuda Nees = Justicia tigrina Heine ■☆
8125 Adhatoda odora (Forssk.) Nees = Justicia odora (Forssk.) Vahl ■☆
8126 Adhatoda orbicularis (Lindau) C. B. Clarke = Justicia orbicularis (Lindau) V. A. W. Graham ■☆
8127 Adhatoda orchioides (L. f.) Nees var. angustifolia Nees = Monechma divaricatum (Nees) C. B. Clarke ■☆
8128 Adhatoda orchioides (L. f.) Nees var. latifolia Nees = Justicia orchioides L. f. ■☆
8129 Adhatoda palustris (Hochst.) Nees = Justicia palustris (Hochst.) T. Anderson ■☆
8130 Adhatoda paniculata Benth. = Justicia laxa T. Anderson ●☆
8131 Adhatoda patula (Nees) Nees = Justicia orchioides L. f. ■☆
8132 Adhatoda petiolaris Nees = Justicia petiolaris (Nees) T. Anderson ■☆
8133 Adhatoda plicata Nees = Justicia flava (Vahl) Vahl ■☆
8134 Adhatoda protracta (Nees) Nees = Justicia protracta (Nees) T. Anderson ■☆
8135 Adhatoda pygmaea (Nees) Nees = Justicia orchioides L. f. ■☆
8136 Adhatoda quadrifaria Nees = Calophanoides quadrifaria (Wall.) Ridl. ●
8137 Adhatoda robusta C. B. Clarke = Justicia baronii V. A. W. Graham ■☆
8138 Adhatoda rostellaria Nees = Justicia ladanoides Lam. ■☆
8139 Adhatoda rostellaria Nees var. humilis ? = Asystasia mysurensis (Roth) T. Anderson ●☆
8140 Adhatoda rostrata Hochst. ex Oliv. = Asystasia mysurensis (Roth) T. Anderson ●☆
8141 Adhatoda rotundifolia Nees = Justicia protracta (Nees) T. Anderson ■☆
8142 Adhatoda schimperiana Hochst. ex Nees = Justicia schimperiana (Hochst. ex Nees) T. Anderson ■☆
8143 Adhatoda spicata Nees = Justicia spicata (Nees) Baron ■☆
8144 Adhatoda striata Klotzsch = Justicia striata (Klotzsch) Bullock ■☆
8145 Adhatoda suaveolens Nees = Justicia flava (Vahl) Vahl ■☆
8146 Adhatoda sulcata (Vahl) Nees = Justicia flava (Vahl) Vahl ■☆
8147 Adhatoda thymifolia Nees = Justicia thymifolia (Nees) C. B. Clarke ■☆
8148 Adhatoda trinervia (Vahl) Nees = Justicia betonica L. ■☆
8149 Adhatoda tristis Nees;暗淡肖鸭嘴花■☆
8150 Adhatoda tubulosa Nees = Siphonoglossa leptantha (Nees) Immelman ●☆
8151 Adhatoda variegata Nees = Justicia betonica L. ■☆
8152 Adhatoda variegata Nees var. pallidior ? = Justicia betonica L. ■☆
8153 Adhatoda vasculosa Nees = Mananthes vasculosa (Nees) Bremek. ●
8154 Adhatoda vasica Nees = Justicia adhatoda L. ●
8155 Adhatoda ventricosa (Wall. ex Sims) Nees = Gendarussa ventricosa (Wall. ex Sims) Nees ●■
8156 Adhatoda ventricosa (Wall.) Nees = Gendarussa ventricosa (Wall. ex Sims) Nees ●■
8157 Adhatoda zeylanica Medik. = Justicia zollingeriana C. B. Clarke ●☆
8158 Adhatoda zollingeriana Nees = Calophanoides quadrifaria (Wall.) Ridl. ●
8159 Adianthum Burm. = Acacia Mill. (保留属名)●■
8160 Adicea Karin. = Adike Raf. ■
8161 Adicea Raf. = Adicea Raf. ex Britton et A. Br. ■
8162 Adicea Raf. ex Britton et A. Br. = Pilea Lindl. (保留属名)■
8163 Adicea deamii Lunell = Pilea pumila (L.) A. Gray ■
8164 Adicea fontana Lunell = Pilea fontana (Lunell) Rydb. ■☆
8165 Adicea mooreana Hiern = Laportea mooreana (Hiern) Chew ●☆
8166 Adicea pumila (L.) Raf. = Pilea pumila (L.) A. Gray ■
8167 Adicea tetraphylla Kuntze var. angolensis Hiern = Pilea angolensis (Hiern) Rendle ■☆
8168 Adike Raf. = Pilea Lindl. (保留属名)■
8169 Adike Raf. = Urtica L. ■
8170 Adina Salisb. (1808);水团花属(水冬瓜属);Adina ●
8171 Adina affinis F. C. How = Pertusadina hainanensis (F. C. How) Ridsdale ●
8172 Adina asperula Hand. -Mazz. = Sinoadina racemosa (Siebold et Zucc.) Ridsdale ●
8173 Adina cordifolia (Roxb.) Hook. f. ex Brandis = Haldina cordifolia (Roxb.) Ridsdale ●
8174 Adina galpinii Oliv. = Breonadia salicina (Vahl) Hepper et J. R. I. Wood ●☆
8175 Adina globifera Salisb. var. tonkinensis Pit. = Adina pilurifera (Lam.) Franch. ex Drake var. tonkinensis (Pit.) Merr. ex H. L. Li ●☆
8176 Adina globiflora Salisb.;球花水团花●
8177 Adina globiflora Salisb. = Adina pilulifera (Lam.) Franch. ex Drake ●
8178 Adina griffithii Hook. f. = Neonauclea griffithii (Hook. f.) Merr. ●
8179 Adina hainanensis F. C. How = Pertusadina hainanensis (F. C. How) Ridsdale ●
8180 Adina inermis (Willd.) Roberty = Mitragyna inermis (Willd.) K. Schum. ●☆
8181 Adina lasiantha K. Schum. = Breonadia salicina (Vahl) Hepper et J. R. I. Wood ●☆
8182 Adina lasiantha K. Schum. var. parviflora Hochr. = Breonadia salicina (Vahl) Hepper et J. R. I. Wood ●☆
8183 Adina ledermannii K. Krause = Hallea ledermannii (K. Krause) Verdc. ●☆
8184 Adina metcalfii Merr. ex H. L. Li = Pertusadina hainanensis (F. C. How) Ridsdale ●
8185 Adina metcalfii Merr. ex H. L. Li = Pertusadina metcalfii (Merr. ex H. L. Li) Y. F. Deng et C. M. Hu ●
8186 Adina microcephala (Delile) Hiern;小头水团花;Microcephaloid Adina, Mugonga, Redwood, Wild Oleander ●
8187 Adina microcephala (Delile) Hiern = Breonadia salicina (Vahl) Hepper et J. R. I. Wood ●☆
8188 Adina microcephala (Delile) Hiern var. galpinii (Oliv.) Hiern = Breonadia salicina (Vahl) Hepper et J. R. I. Wood ●☆
8189 Adina microcephala Hiern = Adina microcephala (Delile) Hiern ●
8190 Adina mollifolia Hutch. = Sinoadina racemosa (Siebold et Zucc.) Ridsdale ●
8191 Adina oligocephala Havil. = Khasiaclunea oligocephala (Havil.)

Ridsdale ●

8192 Adina pilulifera (Lam.) Franch. ex Drake;水团花(穿鱼柳,假马烟树,假杨梅,满山香,青龙珠,球花水团花,球花水杨梅,水黄凿,水加榴,水蓼花,水石榴,水杨柳,水杨梅,溪棉条);Globe Flower Adina,Pilular Adina ●

8193 Adina pilurifera (Lam.) Franch. ex Drake var. tonkinensis (Pit.) Merr. ex H. L. Li;北越水杨梅;Tonkin Pilular Adina ●☆

8194 Adina polycephala Benth. = Metadina trichotoma (Zoll. et Moritzi) Bakh. f. ●

8195 Adina polycephala Benth. var. glabra F. C. How = Pertusadina hainanensis (F. C. How) Ridsdale ●

8196 Adina pubicostata Merr.;毛脉水团花;Hairy-nerved Adina ●

8197 Adina pubicostata Merr. = Adina pilulifera (Lam.) Franch. ex Drake ●

8198 Adina racemosa (Siebold et Zucc.) Miq. = Sinoadina racemosa (Siebold et Zucc.) Ridsdale ●

8199 Adina rubella Hance;细叶水团花(白消木,穿鱼草,穿鱼串,串鱼木,钉木树,沙金子,水毕鸡,水红桃,水金口,水金铃,水晶,水泡木,水石榴,水杨柳,水杨梅,小叶水团花,小叶水杨梅,小叶团花,小叶杨柳,绣球花,绣球柳,杨柳渣子,鱼串鳃);Chinese Buttonbush,Glossy Adina,Thinleaf Adina,Thin-leaved Adina ●

8200 Adina rubescens Hemsl.;红变水团花木●☆

8201 Adina rubrostipulata K. Schum. = Hallea rubrostipulata (K. Schum.) Leroy ●☆

8202 Adina rubrostipulata K. Schum. var. discolor Chiov. = Hallea rubrostipulata (K. Schum.) Leroy ●☆

8203 Adina sessilifolia (Roxb.) Hook. f. = Neonauclea sessilifolia (Roxb.) Merr. ●

8204 Adina sessilifolia (Roxb.) Hook. f. ex Brandis = Neonauclea sessilifolia (Roxb.) Merr. ●

8205 Adinandra Jack(1822);黄瑞木属(红淡属,杨桐属);Adinandra ●

8206 Adinandra acutifolia Hand. -Mazz. = Adinandra bockiana Pritz. ex Diels var. acutifolia (Hand. -Mazz.) Kobuski ●

8207 Adinandra angustifolia (S. H. Chun ex H. G. Ye) B. M. Barthol. et T. L. Ming;狭叶杨桐●

8208 Adinandra auriformis L. K. Ling et S. X. Liang;耳基叶杨桐;Auriculate Adinandra ●

8209 Adinandra bockiana E. Pritz. ex Diels;川黄瑞木(川杨桐,四川红淡,瑶人茶);Bock Adinandra ●

8210 Adinandra bockiana E. Pritz. ex Diels var. acutifolia (Hand.-Mazz.) Kobuski;尖叶川黄瑞木(湖南杨桐,尖叶川杨桐,尖叶杨桐);Sharp-leaf Adinandra ●

8211 Adinandra bracteata H. L. Li = Ternstroemia insignis Y. C. Wu ●

8212 Adinandra caudata Gagnep.;尾尖叶黄瑞木;Caudate Adinandra ●

8213 Adinandra chinensis Merr. et F. P. Metcalf = Adinandra glischroloma Hand. -Mazz. ●

8214 Adinandra chinensis Merr. ex F. P. Metcalf;中华黄瑞木(华红淡);China Adinandra,Chinese Adinandra ●

8215 Adinandra chingii F. P. Metcalf = Cleyera japonica Thunb. ●

8216 Adinandra drakeana Franch. = Adinandra millettii (Hook. et Arn.) Benth. et Hook. f. ex Hance ●

8217 Adinandra dumosa Jack;丛生杨桐●☆

8218 Adinandra elegans F. C. How et W. C. Ko ex Hung T. Chang;长梗黄瑞木(长梗杨桐,狭叶杨桐);Long-stalk Adinandra,Long-stalked Adinandra ●◇

8219 Adinandra epunctata Merr. et Chun;无腺黄瑞木(无腺杨桐);Dotless Adinandra,Glanduleless Adinandra ●

8220 Adinandra filipes Merr. ex Kobuski;细梗黄瑞木(细柄杨桐,细梗杨桐);Slender-stalk Adinandra,Slender-stalked Adinandra ●◇

8221 Adinandra formosana Hayata;台湾杨桐(红淡,红淡比,绿背杨桐,牛屎茶,台湾红淡,台湾黄瑞木,秃萼红淡,秃萼台湾杨桐,尾叶红淡,尾叶台湾杨桐,硬茶仔);Caudate Taiwan Adinandra, Green-back Adinandra, Smooth-calyx Taiwan Adinandra, Taiwan Adinandra ●

8222 Adinandra formosana Hayata = Adinandra millettii Adinandra millettii (Hook. et Arn.) Benth. et Hook. f. ex Hance var. formosana (Hayata) Kobuski ●

8223 Adinandra formosana Hayata f. glabristyla Keng;秃柱台湾杨桐●

8224 Adinandra formosana Hayata var. caudata Keng = Adinandra formosana Hayata ●

8225 Adinandra formosana Hayata var. hypochlora (Hayata) Yamam. ex Keng;秃萼台湾杨桐●

8226 Adinandra formosana Hayata var. hypochlora (Hayata) Yamam. ex Keng = Adinandra formosana Hayata ●

8227 Adinandra formosana Hayata var. longipedicellata Keng;长梗台湾杨桐●

8228 Adinandra formosana Hayata var. obtusissima (Hayata) Keng;钝叶台湾杨桐(钝叶红淡,钝叶杨桐);Obtuse Taiwan Adinandra, Obtuse-leaf Adinandra ●

8229 Adinandra glischroloma Hand. -Mazz.;两广黄瑞木(睫毛杨桐,两广杨桐,亮叶杨桐,毛杨桐);Guangdong-Guangxi Adinandra, Kwangtung-kwangsi Adinandra ●

8230 Adinandra glischroloma Hand. -Mazz. var. hirta (Gagnep.) Kobuski = Adinandra hirta Gagnep. ●

8231 Adinandra glischroloma Hand. -Mazz. var. jubata (H. L. Li) Kobuski;长毛杨桐(长毛黄瑞木,美毛两广黄瑞木);Long-haired Adinandra,Maned Adinandra ●

8232 Adinandra glischroloma Hand. -Mazz. var. macrosepala (F. P. Metcalf) Kobuski;大萼黄瑞木(大萼红淡,大萼两广黄瑞木,大萼杨桐,华红淡);Largesepal Adinandra ●

8233 Adinandra grandis L. K. Ling;大黄瑞木(大杨桐);Big Adinandra,Grand Adinandra ●

8234 Adinandra greenwayii Verdc. = Melchiora schliebenii (Melch.) Kobuski var. greenwayi (Verdc.) Kobuski ●☆

8235 Adinandra hainanensis Hayata;赤点黄瑞木(赤点红淡,海南黄瑞木,海南杨桐,山稔子,油楠);Hainan Adinandra ●

8236 Adinandra hemsleyi Hand. -Mazz. ex Metcalf = Adinandra millettii (Hook. et Arn.) Benth. et Hook. f. ex Hance ●

8237 Adinandra hirta Gagnep.;粗毛黄瑞木(粗毛杨桐,硬毛亮叶杨桐);Hirsute Adinandra,Hairy Adinandra ●

8238 Adinandra hirta Gagnep. var. macrobracteata (L. K. Ling) L. K. Ling;大苞粗毛黄瑞木(大苞粗毛杨桐,大苞黄瑞木,大萼粗毛杨桐);Bigbract Adinandra,Big-bracted Adinandra ●

8239 Adinandra howii Merr. et Chun;保亭黄瑞木(保亭杨桐,琼中杨桐);Baoting Adinandra,How Adinandra ●

8240 Adinandra hypochlora Hayata = Adinandra formosana Hayata ●

8241 Adinandra incornuta (Y. C. Wu) T. L. Ming = Cleyera incornuta Y. C. Wu ●

8242 Adinandra integerrima T. Anderson ex Dyer;全缘叶杨桐(全缘杨桐);Entire-leaved Adinandra ●

8243 Adinandra intermedia Boutique et Troupin = Melchiora schliebenii (Melch.) Kobuski var. intermedia (Boutique et Troupin) Kobuski ●☆

8244 Adinandra japonica (Thunb.) T. L. Ming;日本黄瑞木;Japanese Adinandra ●

8245 Adinandra jubata H. L. Li = Adinandra glischroloma Hand. -Mazz. var. jubata (H. L. Li) Kobuski ●

8246 Adinandra lancipetala L. K. Ling;狭瓣黄瑞木(狭瓣杨桐);Lancipetal Adinandra,Lanci-petaled Adinandra ●

8247 Adinandra lasiostyla Hayata;毛柱黄瑞木(阿里山红淡,阿里山杨桐,毛柱红淡,毛柱杨桐,柱毛黄瑞木);Lasiostyle Adinandra,Woolly-styled Adinandra ●

8248 Adinandra latifolia L. K. Ling;阔叶黄瑞木(阔叶杨桐);Bigleaf Adinandra,Broad-leaved Adinandra ●

8249 Adinandra lutescens Craib = Adinandra integrrima T. Anderson ex Dyer ●

8250 Adinandra maclurei Merr. = Adinandra hainanensis Hayata ●

8251 Adinandra macrobracteata L. K. Ling = Adinandra hirta Gagnep. var. macrobracteata (L. K. Ling) L. K. Ling ●

8252 Adinandra macrocarpa H. L. Li = Ternstroemia insignis Y. C. Wu ●

8253 Adinandra macrosepala F. P. Metcalf = Adinandra glischroloma Hand. -Mazz. var. macrosepala (F. P. Metcalf) Kobuski ●

8254 Adinandra mannii Oliv. = Melchiora mannii (Oliv.) Kobuski ●☆

8255 Adinandra megaphylla Hu;大叶黄瑞木(大叶红淡,大叶杨桐,黄心果);Largeleaf Adinandra,Megaphyllous Adinandra ●

8256 Adinandra millettii (Hook. et Arn.) Benth. et Hook. f. ex Hance;毛药黄瑞木(黄板叉本,黄板叉木,黄瑞木,鸡仔茶,鸡子茶,毛药红淡,乌珠子,杨桐);Millett Adinandra ●

8257 Adinandra millettii (Hook. et Arn.) Benth. et Hook. f. ex Hance var. formosana (Hayata) Kobuski = Adinandra formosana Hayata ●

8258 Adinandra millettii (Hook. et Arn.) Benth. et Hook. f. ex Hance var. hypochlora (Hayata) A. M. Lu et Liu = Adinandra formosana Hayata ●

8259 Adinandra millettii (Hook. et Arn.) Benth. et Hook. f. ex Hance var. obtusissima (Hayata) Kobuski = Adinandra formosana Hayata var. obtusissima (Hayata) Keng ●

8260 Adinandra nigroglandulosa L. K. Ling;腺叶杨桐;Glandular Adinandra,Glandular-leaved Adinandra ●

8261 Adinandra nitida Merr. ex H. L. Li;亮叶黄瑞木(亮叶红淡,亮叶杨桐);Shiningleaf Adinandra,Shiny-leaved Adinandra ●

8262 Adinandra obscurinervia Merr. et Chun = Cleyera obscurinervis (Merr. et Chun) Hung T. Chang ●

8263 Adinandra obtusissima Hayata = Adinandra formosana Hayata var. obtusissima (Hayata) Keng ●

8264 Adinandra pedunculata Hayata = Adinandra formosana Hayata ●

8265 Adinandra petelotii Gagnep. = Adinandra megaphylla Hu ●

8266 Adinandra phlebophylla Hance = Adinandra integrrima T. Anderson ex Dyer ●

8267 Adinandra pingbianensis L. K. Ling;屏边杨桐;Pingbian Adinandra ●

8268 Adinandra retusa D. Fang et D. H. Qin;凹萼杨桐●

8269 Adinandra rubropunctata Merr. et Chun = Adinandra hainanensis Hayata ●

8270 Adinandra ryukyuensis Masam.;琉球杨桐●

8271 Adinandra schliebenii Melch. = Melchiora schliebenii (Melch.) Kobuski ●☆

8272 Adinandra schliebenii Melch. var. glabra Verdc. = Melchiora schliebenii (Melch.) Kobuski var. glabra (Verdc.) Kobuski ●☆

8273 Adinandra schliebenii Melch. var. greenwayi (Verdc.) Verdc. = Melchiora schliebenii (Melch.) Kobuski var. greenwayi (Verdc.) Kobuski ●☆

8274 Adinandra schliebenii Melch. var. intermedia (Boutique et Troupin) Verdc. = Melchiora schliebenii (Melch.) Kobuski var. intermedia (Boutique et Troupin) Kobuski ●☆

8275 Adinandra serrulata H. L. Li = Adinandra megaphylla Hu ●

8276 Adinandra stenosepala Hu = Xantolis stenosepala (Hu) Royle ●

8277 Adinandra wangii Hu;滇南黄瑞木(滇南杨桐);Wang Adinandra ●

8278 Adinandra yaeyamensis Ohwi;八重山黄瑞木●☆

8279 Adinandrella Exell = Ternstroemia Mutis ex L. f.(保留属名)●

8280 Adinandrella congolense Exell = Ternstroemia africana Melch. ●☆

8281 Adinandropsis Pitt-Schenkel = Melchiora Kobuski ●☆

8282 Adinauclea Ridsdale(1979);密乌檀属(山毛榉状茜属)●☆

8283 Adinauclea fagifolia (Havil.) Ridsd.;密乌檀(山毛榉状茜)●☆

8284 Adinobotrys Dunn = Callerya Endl. ●■

8285 Adinobotrys Dunn = Whitfordiodendron Elmer ●

8286 Adinobotrys filipes Dunn = Afgekia filipes (Dunn) R. Geesink ●◇

8287 Adinobotrys filipes Dunn = Whitfordiodendron filipes (Dunn) Dunn ●◇

8288 Adipe Raf. = Bifrenaria Lindl. ■☆

8289 Adipera Raf. = Cassia L.(保留属名)●■

8290 Adisa Steud. = Sumbaviopsis J. J. Sm. ●

8291 Adisca Blume = Mallotus Lour. + Sumbaviopsis J. J. Sm. ●

8292 Adisca Blume = Sumbaviopsis J. J. Sm. ●

8293 Adiscanthus Ducke(1922);无盘花属●☆

8294 Adiscanthus fuscifiorus Ducke;无盘花●☆

8295 Adlera Post et Kuntze = Adleria Neck. ●☆

8296 Adleria Neck. = Eperua Aubl. ●☆

8297 Adlumia Raf. = Adlumia Raf. ex DC.(保留属名)■

8298 Adlumia Raf. ex DC.(1821)(保留属名);荷包藤属(合瓣花属,藤荷包牡丹属);Adlumia,Mountainfringe,Pouchvine ■

8299 Adlumia asiatica Ohwi;荷包藤(合瓣花,藤荷包牡丹);Asia Pouchvine,Asian Mountainfringe ■

8300 Adlumia cirrhosa Raf. = Adlumia fungosa (Aiton) Greene ex Britton,Sterns et Poggenb. ■☆

8301 Adlumia fungosa (Aiton) Britton,Sterns et Poggenb. = Adlumia fungosa (Aiton) Greene ex Brittons,Stern et Poggenb. ■☆

8302 Adlumia fungosa (Aiton) Greene ex Brittons,Stern et Poggenb.;蕈状荷包藤(北美荷包藤,蕈状山缘草);Allegheny Vine,Allegheny-vine,Climbing Fumitory,Mountain Fringe ■☆

8303 Adlumia fungosa (Aiton) Greene ex Brittons,Stern et Poggenb. = Adlumia asiatica Ohwi ■

8304 Admarium Raf. = Adenarium Raf. ■☆

8305 Admarium Raf. = Honkenya Ehrh. ■☆

8306 Admirabilis Nieuwl. = Mirabilis L. ■

8307 Adnaria Raf. = Styrax L. ●

8308 Adnula Raf. = Pelexia Poit. ex Lindl.(保留属名)■

8309 Adoceton Raf. = Alternanthera Forssk. ■

8310 Adodendron DC. = Adodendrum Neck. ●☆

8311 Adodendrum Neck. = Rhodothamnus Rchb.(保留属名)●☆

8312 Adodendrum Neck. ex Kuntze = Rhodothamnus Rchb.(保留属名)●☆

8313 Adoketon Raf. = Adoceton Raf. ■

8314 Adoketon Raf. = Alternanthera Forssk. ■

8315 Adolia Lam.(废弃属名)= Scutia (Comm. ex DC.) Brongn.(保留属名)●

8316 Adolphia Meisn.(1837);阿多鼠李属(阿多路非木属,南美鼠李属)●☆

8317 Adolphia californica S. Watson;加州阿多鼠李●☆

8318 Adolphia infesta Meisn.;阿多鼠李●☆

8319 Adonanthe Spach = Adonis L.(保留属名)■

8320 Adonastrum Dalla Torre et Harms = Adoniastrum Schur ■

8321 Adoniastrum Schur = Adonis L.(保留属名)■

8322 Adonidia Becc. = Veitchia H. Wendl.(保留属名)●☆

8323 Adonigeron Fourr. = Senecio L. ●■

8324 Adonis L.(1753)(保留属名);侧金盏花属;Adonis,Pheasant's Eye,Pheasant's-eye ■

8325 Adonis aestivalis L.;夏侧金盏花(福寿草,夏福寿草);Flos Adonis,Pheasant's Eye,Pheasant's-eye,Pheasant's-eye Adonis,Summer Adonis,Summer Pheasant's Eye,Summer Pheasant's Eyes ■

8326 Adonis aestivalis L. subsp. provincialis (DC.) Steinb. = Adonis aestivalis L. subsp. squarrosa (Steven) Nyman ■☆

8327 Adonis aestivalis L. subsp. squarrosa (Steven) Nyman;粗鳞夏侧金盏花■☆

8328 Adonis aestivalis L. var. parviflora M. Bieb. = Adonis parviflora (M. Bieb.) Fisch. ex DC. ■

8329 Adonis aestivalis L. var. provincialis (DC.) Hochr. = Adonis aestivalis L. subsp. squarrosa (Steven) Nyman ■☆

8330 Adonis aethiopica Thunb. = Knowltonia filia (L. f.) T. Durand et Schinz ■☆

8331 Adonis amurensis Regel et Radde;侧金盏花(冰郎花,冰里花,冰凉花,冰了花,冰溜花,长春菊,顶冰花,福寿草,岁菊,献岁菊,雪莲,雪莲花,元日草);Amur Adonis ■

8332 Adonis annua L.;献岁花(欧侧金盏花,秋侧金盏花,秋福寿草);Adonis Flower,Autumn Adonis,Autumn Flos Adonis,Blood of Adonis,Blooddrops,Corn Pheasant's Eye,Fall Adonis,False Hellebore,Flos Adonis,Jack-in-the-green,Love-lies-bleeding,Pheasant's Eye,Pheasant's Eyes,Pheasant's-eye,Pheasant's-eye Adonis,Purple Camomile,Red Camomile,Red Mathes,Red Mathet,Red Maydweed,Red Maythe,Red Morocco,Rose-a-ruby ■

8333 Adonis annua L. subsp. autumnalis (L.) Maire et Weiller = Adonis annua L. ■

8334 Adonis annua L. subsp. baetica (Coss.) Nyman = Adonis annua L. ■

8335 Adonis annua L. var. atrorubens ? = Adonis annua L. ■

8336 Adonis annua L. var. baetica (Coss.) Maire = Adonis annua L. ■

8337 Adonis annua L. var. coccinea Maire et Wilczek = Adonis annua L. ■

8338 Adonis annua L. var. preslii (Tod.) Fiori = Adonis annua L. subsp. baetica (Coss.) Nyman ■

8339 Adonis autumnalis L. = Adonis annua L. ■

8340 Adonis bobroviana Simonov.;甘青侧金盏花(青甘侧金盏花);Bobrov Adonis ■

8341 Adonis brevistyla Franch. = Adonis davidii Franch. ■

8342 Adonis capensis L. = Knowltonia capensis (L.) Huth ■☆

8343 Adonis chrysocyatha Hook. f. et Thomson ex Hook. f.;金黄侧金盏花(福寿草,金色冰凉花);Golden Adonis ■

8344 Adonis coerulea Maxim.;蓝侧金盏花(豆月老);Skyblue Adonis ■

8345 Adonis coerulea Maxim. f. integra W. T. Wang;高蓝侧金盏花■

8346 Adonis coerulea Maxim. f. integra W. T. Wang = Adonis coerulea Maxim. ■

8347 Adonis coerulea Maxim. f. integra W. T. Wang = Adonis integra (W. T. Wang) W. T. Wang ■

8348 Adonis coerulea Maxim. f. puberula W. T. Wang = Adonis coerulea Maxim. var. puberula W. T. Wang ■

8349 Adonis coerulea Maxim. var. puberula W. T. Wang;毛蓝侧金盏花(无盖侧金盏);Hairy Skyblue Adonis ■

8350 Adonis coerulea Maxim. var. puberula W. T. Wang = Adonis coerulea Maxim. ■

8351 Adonis cyllenea Boiss.,Heldr. et Orph.;希腊侧金盏花■☆

8352 Adonis cyllenea Boiss.,Heldr. et Orph. var. paryadrica Boiss.;土耳其侧金盏花■☆

8353 Adonis davidii Franch.;短柱侧金盏花(宝兴侧金盏花,短柱福寿草,水黄连);David Adonis,Shortstyle Adonis ■

8354 Adonis delavayi Franck. = Adonis davidii Franch. ■

8355 Adonis dentata Delile;具齿侧金盏花■☆

8356 Adonis dentata Delile subsp. microcarpa (DC.) Riedl = Adonis microcarpa DC. ■☆

8357 Adonis dentata Delile var. intermedia Webb et Berthel. = Adonis microcarpa DC. subsp. intermedia (Webb et Berthel.) Valdés ■☆

8358 Adonis dentata Delile var. microcarpa (DC.) Hochst. = Adonis microcarpa DC. ■☆

8359 Adonis dentata Delile var. orientalis DC. = Adonis dentata Delile ■☆

8360 Adonis dentata Delile var. pseudoflammea Maire et Sennen = Adonis microcarpa DC. ■☆

8361 Adonis distorta Ten.;意大利侧金盏花■☆

8362 Adonis flammea Jacq.;火焰侧金盏花(火焰金盏花);Flame Adonis,Flamecoloured Adonis,Large Pheasant's Eye,Large Pheasant's Eyes ■☆

8363 Adonis integra (W. T. Wang) W. T. Wang;全缘侧金盏花(高蓝侧金盏花);Entire-leaved Adonis,Tall Skyblue Adonis ■

8364 Adonis intermedia Webb et Berthel. = Adonis microcarpa DC. subsp. intermedia (Webb et Berthel.) Valdés ■☆

8365 Adonis intermedia Webb et Berthel. var. flaviflora Webb = Adonis microcarpa DC. subsp. intermedia (Webb et Berthel.) Valdés ■☆

8366 Adonis intermedia Webb et Berthel. var. phoenicea Webb = Adonis microcarpa DC. subsp. intermedia (Webb et Berthel.) Valdés ■☆

8367 Adonis microcarpa DC.;小果侧金盏花■☆

8368 Adonis microcarpa DC. subsp. intermedia (Webb et Berthel.) Valdés;中间小果侧金盏花■☆

8369 Adonis microcarpa DC. var. dentata (Delile) Coss. et Kralik = Adonis aestivalis L. subsp. squarrosa (Steven) Nyman ■☆

8370 Adonis microcarpa DC. var. intermedia (Webb et Berthel.) Boiss. = Adonis microcarpa DC. subsp. intermedia (Webb et Berthel.) Valdés ■☆

8371 Adonis mongolica Simonov.;蒙古侧金盏花;Mongol Adonis ■

8372 Adonis multiflora Nishikawa et Koji Ito = Adonis ramosa Franch. ■

8373 Adonis nepalensis Simonov.;尼泊尔侧金盏花;Nepal Adonis ■☆

8374 Adonis parviflora (M. Bieb.) Fisch. ex DC.;小侧金盏花;Smallflower Adonis ■

8375 Adonis parviflora Fisch. ex DC. = Adonis aestivalis L. var. parviflora M. Bieb. ■

8376 Adonis pseudoamurensis W. T. Wang = Adonis ramosa Franch. ■

8377 Adonis pyrenaica DC.;比利牛斯侧金盏花;Pyrenean Pheasant's Eye ■☆

8378 Adonis ramosa Franch.;辽吉侧金盏花;False Amur Adonis ■

8379 Adonis ramosa Franch. subsp. fupingensis W. T. Wang;阜平侧金盏花;Fuping Adonis ■

8380 Adonis ramosa Franch. subsp. fupingensis W. T. Wang = Adonis ramosa Franch. ■

8381 Adonis sibirica Patrin ex Ledeb.;北侧金盏花(福寿草);Siberia Adonis ■

8382 Adonis sutchuenensis Franch.;蜀侧金盏花(毛黄连,毛连,四川侧金盏花);Sichuan Adonis,Szechwan Adonis ■

8383 Adonis tianschanica (Adolf) Lipsch.;天山侧金盏花(天山福寿

草);Tianshan Adonis,Tianshan Mountain Adonis ■
8384 Adonis transsilvanica Simonov.;匈罗侧金盏花■☆
8385 Adonis turkestanica (Korsh.) Adolf;中亚侧金盏花(土耳其斯坦侧金盏花,中亚福寿草);Turkestan Adonis ■☆
8386 Adonis turkestanica (Korsh.) Adolf var. tianschanica Adolf = Adonis tianschanica (Adolf) Lipsch. ■
8387 Adonis vernalis L.;春侧金盏花(春福寿草);Adonis, False Hellebore, Ox Eyes, Ox-eye, Spring Adonis, Spring Pheasant's Eye, Yellow Adonis, Yellow Pheasant's Eye, Yellow Pheasant's Eyes, Yellow Pheasant's-eye ■☆
8388 Adonis vernalis L. var. amurensis Finet et Gagnep. = Adonis amurensis Regel et Radde ■
8389 Adonis villosa Ledeb.;密毛侧金盏花(长毛侧金盏花,毛侧金盏花);Villose Adonis ■
8390 Adonis volgensis Steven ex DC.;伏尔加侧金盏花;Volga Adonis ■☆
8391 Adonostylis strateumatica (L.) Ames = Zeuxine strateumatica (L.) Schltr. ■
8392 Adopogon Neck. = Krigia Schreb.(保留属名)■☆
8393 Adopogon virginicum (L.) Kuntze = Krigia biflora (Walter) S. F. Blake ■☆
8394 Adorioon Raf. = Adorium Raf. ■☆
8395 Adorium Raf. = Musineon Raf. ex DC. ■☆
8396 Adoxa L.(1753);五福花属;Moschatel, Muskroot ■
8397 Adoxa inodora (Falc. ex C. B. Clarke) Nepomn.;无味五福花■
8398 Adoxa insularis Nepomn. = Adoxa moschatellina L. var. insularis (Nepomn.) S. Y. Li et Z. H. Ning ■☆
8399 Adoxa moschatelliana L.;五福花;Fairy's Clock, Five-faced Bishop, Gloriless, Good Friday, Good Friday Flower, Holewort, Hollow-root, Hollow-wort, Infirmary Clock, Lady's Mantle, Moschatel, Musk Crow Flower, Musk Crowflower, Musk Wood Crawfoot, Muskroot, Musk-root, Muskweed, Town Clock, Town-hall Clock, Townhan Clock, Whiskers, Wild Shamrock, Wood-alone ■
8400 Adoxa moschatellina L. f. japonica (H. Hara) H. Hara = Adoxa moscatelliana L. ■
8401 Adoxa moschatellina L. var. inodora Falc. ex C. B. Clarke = Adoxa moschatellina L. ■
8402 Adoxa moschatellina L. var. insularis (Nepomn.) S. Y. Li et Z. H. Ning;海岛五福花■☆
8403 Adoxa moschatellina L. var. japonica H. Hara = Adoxa moschatellina L. ■
8404 Adoxa omeiensis H. Hara = Tetradoxa omeiensis (H. Hara) C. Y. Wu ■
8405 Adoxa xizangensis G. Yao;西藏五福花;Xizang Muskroot ■
8406 Adoxaceae E. Mey.(1839)(保留科名);五福花科;Moschatel Family, Muskroot Family ●■
8407 Adoxaceae Trautv. = Adoxaceae E. Mey.(保留科名)●■
8408 Adrastaea DC. = Hibbertia Andréws ●☆
8409 Adrastea Spreng. = Adrastaea DC. ●☆
8410 Adriana Endl. = Adriana Gaudich. ●☆
8411 Adriana Gaudich.(1825);苦大戟属;Bitter Bush ●☆
8412 Adriana acerifolia Hook.;尖叶苦大戟●☆
8413 Adriana glabrata Gaudich.;光苦大戟●☆
8414 Adriana hookeri (F. Muell.) Müll. Arg.;胡克苦大戟●☆
8415 Adriana hookeri Müll. Arg. = Adriana hookeri (F. Muell.) Müll. Arg. ●☆
8416 Adriana tomentosa Gaudich.;毛苦大戟●☆
8417 Adromischus Lem.(1852);短梗景天属(天锦章属,天章属)●■☆
8418 Adromischus alstonii (Schönland et Baker f.) C. A. Sm.;阿尔短梗景天;Bulbees ■☆
8419 Adromischus alveolatus Hutchison = Adromischus marianiae (Marloth) A. Berger var. immaculatus Uitewaal ■☆
8420 Adromischus bicolor Hutchison;二色短梗景天■☆
8421 Adromischus blosianus Hutchison = Adromischus marianiae (Marloth) A. Berger var. kubusensis (Uitewaal) Toelken ■☆
8422 Adromischus bolusii (Schönland) A. Berger = Adromischus caryophyllaceus (Burm. f.) Lem. ■☆
8423 Adromischus caryophyllaceus (Burm. f.) Lem.;石竹状短梗景天■☆
8424 Adromischus casmithianus Poelln. = Adromischus marianiae (Marloth) A. Berger var. hallii (Hutchison) Toelken ■☆
8425 Adromischus clavifolius (Haw.) Lem. = Adromischus cooperi A. Berger ■☆
8426 Adromischus clavifolius (Haw.) Lem. = Adromischus cristatus (Haw.) Lem. var. clavifolius (Haw.) Toelken ■☆
8427 Adromischus cooperi (Baker) A. Berger;棒叶短梗景天(边圆瓶草,古伯天章,锦铃殿);Knuppelplakkie, Plover Eggs ■
8428 Adromischus cooperi A. Berger = Adromischus cooperi (Baker) A. Berger ■
8429 Adromischus cristatus (Haw.) Lem.;皱叶景天(天章);Crinkle Leaf Plant ■☆
8430 Adromischus cristatus (Haw.) Lem. var. clavifolius (Haw.) Toelken;棒状皱叶景天■☆
8431 Adromischus cristatus (Haw.) Lem. var. mzimvubuensis Van Jaarsv.;姆津短梗景天■☆
8432 Adromischus cristatus (Haw.) Lem. var. schonlandii (E. Phillips) Toelken;绍氏短梗景天■☆
8433 Adromischus cristatus (Haw.) Lem. var. zeyheri (Harv.) Toelken;蔡氏短梗景天■☆
8434 Adromischus cuneatus (Thunb.) Lem. = Cotyledon cuneata Thunb. ●☆
8435 Adromischus cuneatus Poelln. = Adromischus cooperi (Baker) A. Berger ■
8436 Adromischus diabolicus Toelken;魔鬼短梗景天■☆
8437 Adromischus fallax Toelken;疑惑短梗景天■☆
8438 Adromischus festivus C. A. Sm. = Adromischus cooperi (Baker) A. Berger ■
8439 Adromischus filicaulis (Eckl. et Zeyh.) C. A. Sm.;丝茎短梗景天■☆
8440 Adromischus filicaulis (Eckl. et Zeyh.) C. A. Sm. subsp. marlothii (Schönland) Toelken;马尔丝茎短梗景天■☆
8441 Adromischus fragilis Hutchison = Adromischus filicaulis (Eckl. et Zeyh.) C. A. Sm. ■☆
8442 Adromischus fragilis Hutchison var. numeesensis ? = Adromischus filicaulis (Eckl. et Zeyh.) C. A. Sm. ■☆
8443 Adromischus fusiformis (Rolfe) A. Berger = Adromischus filicaulis (Eckl. et Zeyh.) C. A. Sm. ■☆
8444 Adromischus geyeri Hutchison = Adromischus marianiae (Marloth) A. Berger var. kubusensis (Uitewaal) Toelken ■☆
8445 Adromischus grandiflorus Uitewaal = Adromischus caryophyllaceus (Burm. f.) Lem. ■☆
8446 Adromischus halesowensis Uitewaal = Adromischus cooperi (Baker) A. Berger ■
8447 Adromischus hallii Hutchison = Adromischus marianiae (Marloth) A. Berger var. hallii (Hutchison) Toelken ■☆

8448 Adromischus hemisphaericus (L.) Lem.;半球短梗景天■☆
8449 Adromischus herrei (W. F. Barker) Poelln. = Adromischus marianiae (Marloth) A. Berger var. immaculatus Uitewaal ■☆
8450 Adromischus hoerleinianus (Dinter) Poelln. = Tylecodon schaeferianus (Dinter) Toelken ●☆
8451 Adromischus hoerleinianus (Dinter) Poelln. var. schaeferi Dinter = Tylecodon schaeferianus (Dinter) Toelken ●☆
8452 Adromischus humilis (Marloth) Poelln.;低矮短梗景天■☆
8453 Adromischus inamoenus Toelken;丑短梗景天■☆
8454 Adromischus jasminiflorus (Salm-Dyck) Lem. = Adromischus caryophyllaceus (Burm. f.) Lem. ■☆
8455 Adromischus juttae Poelln.;犹他短梗景天■☆
8456 Adromischus keilhackii Werderm. = Tylecodon schaeferianus (Dinter) Toelken ●☆
8457 Adromischus kesselringianus Poelln. = Adromischus cristatus (Haw.) Lem. var. clavifolius (Haw.) Toelken ■☆
8458 Adromischus kleinioides C. A. Sm. = Adromischus filicaulis (Eckl. et Zeyh.) C. A. Sm. ■☆
8459 Adromischus kubusensis Uitewaal = Adromischus marianiae (Marloth) A. Berger var. kubusensis (Uitewaal) Toelken ■☆
8460 Adromischus leucophyllus Uitewaal;白叶短梗景天■☆
8461 Adromischus leucothrix C. A. Sm. = Tylecodon leucothrix (C. A. Sm.) Toelken ●☆
8462 Adromischus liebenbergii Hutchison;利本短梗景天■☆
8463 Adromischus liebenbergii Hutchison subsp. orientalis Van Jaarsv.; 东方短梗景天■☆
8464 Adromischus maculatus (Salm-Dyck) Lem.;斑天章(御所锦)■☆
8465 Adromischus maculatus Lem. = Adromischus maculatus (Salm-Dyck) Lem. ■☆
8466 Adromischus mammillaris (L. f.) Lem.;花豆瓣■
8467 Adromischus mammillaris (L. f.) Lem. var. filicaulis (Eckl. et Zeyh.) H. Jacobsen = Adromischus filicaulis (Eckl. et Zeyh.) C. A. Sm. ■☆
8468 Adromischus mammillaris (L. f.) Lem. var. fusiformis (Rolfe) H. Jacobsen = Adromischus filicaulis (Eckl. et Zeyh.) C. A. Sm. ■☆
8469 Adromischus mammillaris (L. f.) Lem. var. marlothii (Schönland) H. Jacobsen = Adromischus filicaulis (Eckl. et Zeyh.) C. A. Sm. subsp. marlothii (Schönland) Toelken ■☆
8470 Adromischus mammillaris (L. f.) Lem. var. rubra Poelln. = Adromischus filicaulis (Eckl. et Zeyh.) C. A. Sm. ■☆
8471 Adromischus marianiae (Marloth) A. Berger;马氏短梗景天;Brosplakkie ■☆
8472 Adromischus marianiae (Marloth) A. Berger var. antidorcadum (Poelln.) Pilbeam = Adromischus marianiae (Marloth) A. Berger var. immaculatus Uitewaal ■☆
8473 Adromischus marianiae (Marloth) A. Berger var. hallii (Hutchison) Toelken;霍尔短梗景天■☆
8474 Adromischus marianiae (Marloth) A. Berger var. immaculatus Uitewaal;无斑马氏短梗景天■☆
8475 Adromischus marianiae (Marloth) A. Berger var. kubusensis (Uitewaal) Toelken;库地马氏短梗景天■☆
8476 Adromischus marianiae (Marloth) A. Berger. f. alveoluatus (Hutchison) Pilbeam = Adromischus marianiae (Marloth) A. Berger var. immaculatus Uitewaal ■☆
8477 Adromischus marianiae (Marloth) A. Berger. f. herrei (W. F. Barker) Pilbeam = Adromischus marianiae (Marloth) A. Berger var. immaculatus Uitewaal ■☆
8478 Adromischus marianiae (Marloth) A. Berger. f. multicolor Pilbeam = Adromischus marianiae (Marloth) A. Berger var. immaculatus Uitewaal ■☆
8479 Adromischus marlothii (Schönland) A. Berger = Adromischus filicaulis (Eckl. et Zeyh.) C. A. Sm. subsp. marlothii (Schönland) Toelken ■☆
8480 Adromischus maximus Hutchison;大短梗景天■☆
8481 Adromischus mucronatus (Lam.) Lem. = Cotyledon orbiculata L. ●☆
8482 Adromischus nanus (N. E. Br.) Poelln.;小短梗景天■☆
8483 Adromischus nussbaumerianus (Poelln.) Poelln. = Adromischus cristatus (Haw.) Lem. var. clavifolius (Haw.) Toelken ■☆
8484 Adromischus pachylophus C. A. Sm. = Adromischus cooperi (Baker) A. Berger ■
8485 Adromischus pauciflorus Hutchison = Adromischus nanus (N. E. Br.) Poelln. ■☆
8486 Adromischus phillipsiae (Marloth) Poelln.;菲舍尔短梗景天■☆
8487 Adromischus poellnitzianus Werderm.;长叶熊掌草■
8488 Adromischus poellnitzianus Werderm. = Adromischus cristatus (Haw.) Lem. var. clavifolius (Haw.) Toelken ■☆
8489 Adromischus procurvus (N. E. Br.) C. A. Sm. = Adromischus triflorus (L. f.) A. Berger ■☆
8490 Adromischus pulchellus Hutchison = Adromischus alstonii (Schönland et Baker f.) C. A. Sm. ■☆
8491 Adromischus rhombifolius (Haw.) Lem.;菱形短梗景天■☆
8492 Adromischus rhombifolius Haw. var. bakeri Poelln. = Adromischus sphenophyllus C. A. Sm. ■☆
8493 Adromischus robustus Lem.;粗壮短梗景天■☆
8494 Adromischus rodinii Hutchison = Adromischus marianiae (Marloth) A. Berger var. kubusensis (Uitewaal) Toelken ■☆
8495 Adromischus rotundifolius (Haw.) C. A. Sm. = Adromischus hemisphaericus (L.) Lem. ■☆
8496 Adromischus rupicola C. A. Sm. = Adromischus trigynus (Burch.) Poelln. ■☆
8497 Adromischus saxicola C. A. Sm. = Adromischus umbraticola C. A. Sm. ■☆
8498 Adromischus schaeferianus (Dinter) A. Berger = Tylecodon schaeferianus (Dinter) Toelken ●☆
8499 Adromischus schaeferianus (Dinter) A. Berger var. keilhackii (Werderm.) Poelln. = Tylecodon schaeferianus (Dinter) Toelken ●☆
8500 Adromischus schonlandii (E. Phillips) Poelln. = Adromischus cristatus (Haw.) Lem. var. schonlandii (E. Phillips) Toelken ■☆
8501 Adromischus schuldtianus (Poelln.) Poelln. subsp. brandbergensis B. Nord. et Van Jaarsv.;布兰德山短梗景天■☆
8502 Adromischus schuldtianus (Poelln.) Poelln. subsp. juttae (Poelln.) Toelken = Adromischus juttae Poelln. ■☆
8503 Adromischus sphenophyllus C. A. Sm.;楔叶短梗景天■☆
8504 Adromischus subcompressus Poelln. = Adromischus triflorus (L. f.) A. Berger ■☆
8505 Adromischus subdistichus Makin ex Bruyns;二列短梗景天■☆
8506 Adromischus subpetiolatus Poelln. = Adromischus triflorus (L. f.) A. Berger ■☆
8507 Adromischus subrubellus Poelln. = Adromischus alstonii (Schönland et Baker f.) C. A. Sm. ■☆
8508 Adromischus subviridis Toelken;浅绿短梗景天■☆
8509 Adromischus tricolor C. A. Sm. = Adromischus filicaulis (Eckl. et Zeyh.) C. A. Sm. subsp. marlothii (Schönland) Toelken ■☆

8510 Adromischus triebneri Poelln. = Adromischus alstonii（Schönland et Baker f.）C. A. Sm. ■☆

8511 Adromischus triflorus（L. f.）A. Berger;三花短梗景天■☆

8512 Adromischus trigynus（Burch.）Poelln.;三蕊短梗景天;Calico Hearts ■☆

8513 Adromischus umbraticola C. A. Sm.;岩地短梗景天■☆

8514 Adromischus umbraticola C. A. Sm. subsp. ramosus Toelken;多枝短梗景天■☆

8515 Adromischus zeyheri（Harv.）Poelln. = Adromischus cristatus（Haw.）Lem. var. zeyheri（Harv.）Toelken ■☆

8516 Adrorhizon Hook. f.（1898）;短根兰属■☆

8517 Adrorhizon purpurascens Hook. f.;短根兰■☆

8518 Adulpa Bosc. = Mariscus Gaertn. ■

8519 Adulpa Endl. = Adulpa Bosc. ■

8520 Adupla Bosc = Cyperus L. ■

8521 Adupla Bosc ex Juss. = Mariscus Gaertn. ■

8522 Adupla Bosc ex Juss. = Schoenus L. ■

8523 Aduseta Dalla Torre et Harms = Aduseton Scop. ■

8524 Aduseton Adans.（废弃属名）= Lobularia Desv.（保留属名）■

8525 Aduseton Scop. = Adyseton Adans. ■

8526 Adventina Raf. = Galinsoga Ruiz et Pav. ●■

8527 Adventina ciliata Raf. = Galinsoga ciliata（Raf.）S. F. Blake ■

8528 Adventina ciliata Raf. = Galinsoga qnadriradiata Ruiz et Pav. ■

8529 Adyseton Adans. = Alyssum L. + Lobularia Desv. + Draba L. ■

8530 Adysetum Link = Alyssum L. + Lobularia Desv. + Draba L. ■

8531 Aeceoclades Duchartre = Aeceoclades Duchartre ex B. D. Jacks. ■☆

8532 Aeceoclades Duchartre ex B. D. Jacks. = Oeceoclades Lindl. ■☆

8533 Aeceoclades Duchartre ex B. D. Jacks. = Saccolabium Blume（保留属名）■

8534 Aechma C. Agardh = Aechmea Ruiz et Pav.（保留属名）■☆

8535 Aechmaea Brongn. = Aechmandra Arn. ■☆

8536 Aechmandra Arn. = Kedrostis Medik. ■☆

8537 Aechmandra conocarpa Dalzell et Gibson = Corallocarpus conocarpus（Dalzell et Gibson）Hook. f. ex Clarke ■☆

8538 Aechmanthera Nees（1832）;尖药草属（尖蕊花属,尖药花属,十三年花属）;Aechmanthera ●

8539 Aechmanthera gossypina（Nees）Nees;绵毛尖药草（棉毛尖药草,棉毛尖药花）;Woolly Aechmanthera ●

8540 Aechmanthera tomentosa（Wall.）Nees;尖药草（尖蕊花,尖药花,蓝花草,十三年花）;Aechmanthera, Common Aechmanthera ●

8541 Aechmanthera tomentosa（Wall.）Nees = Aechmanthera gossypina（Nees）Nees ●

8542 Aechmanthera tomentosa（Wall.）Nees var. wallichii ? = Aechmanthera gossypina（Nees）Nees ●

8543 Aechmanthera wallichii Nees = Aechmanthera gossypina（Nees）Nees ●

8544 Aechmea Ruiz et Pav.（1794）（保留属名）;光萼荷属（大萼凤梨属,附生凤梨属,光萼凤梨属,尖萼凤梨属,尖萼荷属,亮叶光萼荷属,蜻蜓凤梨属,珊瑚凤梨属,珊瑚属）;Aechmea ■☆

8545 Aechmea 'Fostex's Favorite' = Aechmea fosteriana L. B. Sm. ■☆

8546 Aechmea amazonica Ule;黑纹凤梨（黑纹菠萝）;Black Chaetinii ■

8547 Aechmea angustifolia Poepp. et Endl.;狭叶光萼荷;Narrowleaf Aechmea ■☆

8548 Aechmea bracteata（Sw.）Griseb.;红苞光萼荷■☆

8549 Aechmea bromeliifolia（Rudge）Baker;禾叶光萼荷■☆

8550 Aechmea calyculata Baker;副萼光萼荷;Calyculate Aechmea ■☆

8551 Aechmea candida E. Morren;白毛光萼荷■☆

8552 Aechmea caudata Lindm.;尾萼光萼荷;Caudute Aechmea ■☆

8553 Aechmea caudata Lindm. 'Variegata';斑纹尾萼光萼荷（变叶尾萼光萼荷）■☆

8554 Aechmea chantinii（Carrière）Baker;光萼荷（斑马菠萝,斑马凤梨,黑纹凤梨,尖萼荷）;Amazonian Zebra Plant, Chantin Aechmea, Queen of the Bromeliads ■

8555 Aechmea coelestis E. Morren;天蓝光萼荷;Aechmea ■☆

8556 Aechmea cylindrata Lindm.;圆头凤梨（圆头菠萝）■

8557 Aechmea distichantha Lem.;二列花光萼荷（二列花尖萼荷）■☆

8558 Aechmea fasciata（Lindl.）Baker;美叶光萼荷（光萼荷,横缟尖萼荷,美叶尖萼荷,蜻蜓菠萝,蜻蜓凤梨）;Air Pine, Bromeliad, Fasciate Aechmea, Silver Vase, Silver Vase Plant, Silver Vase-plant, Urn Plant ■☆

8559 Aechmea fasciata Baker = Aechmea fasciata（Lindl.）Baker ■☆

8560 Aechmea fillicaulis（Griseb.）Mez;翡翠凤梨（翡翠菠萝）■

8561 Aechmea fosteriana L. B. Sm.;福德光萼荷;Foster Aechmea, Lacquered Wine Cup ■☆

8562 Aechmea fulgens Brongn.;光亮叶萼荷（亮叶尖萼荷,珊瑚菠萝,珊瑚凤梨）;Coral Berry, Coralberry ■

8563 Aechmea fulgens Brongn. var. discolor（Hook.）Brongn.;斑纹珊瑚凤梨（斑纹珊瑚菠萝）■

8564 Aechmea gigantea Baker;大光萼荷;Giant Aechmea ■☆

8565 Aechmea gigantea Baker = Aechmea sphaerocephala（Gaudich.）Baker ■☆

8566 Aechmea lagenaria Mez = Aechmea lamarchei Mez ■☆

8567 Aechmea lamarchei Mez;豹纹凤梨（豹纹菠萝）■☆

8568 Aechmea lindenii E. Morren ex K. Koch;林登光萼荷■☆

8569 Aechmea lueddemanniana（K. Koch）Brongn. ex Mez;青花凤梨（青花菠萝）■

8570 Aechmea lueddemanniana Brongn. = Aechmea lueddemanniana（K. Koch）Brongn. ex Mez ■

8571 Aechmea magdalenae André ex Baker;纤维凤梨（纤维菠萝）;Pita Fibre ■

8572 Aechmea maginali Hort.;赤苞珊瑚凤梨（赤苞珊瑚菠萝）■☆

8573 Aechmea mariae-reginae H. Wendl.;白花光萼荷;Queen Aechmea ■☆

8574 Aechmea marmorata Mez;大理石光萼荷■☆

8575 Aechmea mertensii Schult. f.;迈尔光萼荷■☆

8576 Aechmea mexicana Baker;墨西哥光萼荷;Mexican Aechmea ■☆

8577 Aechmea miniata Baker;深红光萼荷;Red Aechmea ■☆

8578 Aechmea minima Baker var. discolor Beer;异色深红光萼荷■☆

8579 Aechmea nudicaulis Griseb.;裸茎光萼荷（裸茎尖萼荷）■☆

8580 Aechmea nudicaulis Griseb. var. aureo-rosea（Antoine）L. B. Sm.;金玫瑰裸茎光萼荷（裸茎尖萼荷）■☆

8581 Aechmea organensis Wawra;红刺光萼荷（红刺菠萝,红刺凤梨）;Redspiny Aechmea ■☆

8582 Aechmea orlandiana L. B. Sm.;大蜻蜓光萼荷（大蜻蜓凤梨）;Finger of God ■☆

8583 Aechmea ornata Baker;华美光萼荷■☆

8584 Aechmea pectinata Baker;栉齿光萼荷■☆

8585 Aechmea pineliana（Brongn.）Baker;鼓槌凤梨（鼓槌菠萝）■☆

8586 Aechmea pubescens Baker;软毛光萼荷■☆

8587 Aechmea purpurea Baker;紫色光萼荷■☆

8588 Aechmea racinae L. B. Sm.;拉氏光萼荷■☆

8589 Aechmea ramosa Mart. ex Schult. f.;多枝光萼荷■☆

8590 Aechmea recurvata（Klotzsch）L. B. Sm.;弯光萼荷（曲叶尖萼荷）;Recurved Aechmea ■☆

8591 Aechmea setigera Mart. = Aechmea setigera Mart. ex Schult. f. ■☆
8592 Aechmea setigera Mart. ex Schult. f.;刚毛光萼荷■☆
8593 Aechmea sphaerocephala (Gaudich.) Baker;球头光萼荷■☆
8594 Aechmea tessmannii Harms;泰氏光萼荷■☆
8595 Aechmea tillandsioides (Mart. ex Schult. f.) Baker;紫凤光萼荷;Tilandsialike Aechmea ■☆
8596 Aechmea triangularis L. B. Sm.;三棱光萼荷■☆
8597 Aechmea victoriana L. B. Sm.;维多利亚光萼荷■☆
8598 Aechmea weilbachii F. Dietr.;豹纹光萼荷;Weilbach Aechmea ■☆
8599 Aechmea weilbachii F. Dietr. var. leodiensis André;胜常光萼荷■☆
8600 Aechmolepis Decne. = Tacazzea Decne. ●☆
8601 Aechmolepis rosmarinifolia Decne. = Tacazzea rosmarinifolia (Decne.) N. E. Br. ●☆
8602 Aechmophora Spreng. ex Steud. = Bromus L. (保留属名)■
8603 Aechmophora Steud. = Bromus L. (保留属名)■
8604 Aectyson Raf. = Sedum L. ●■
8605 Aedemone Kotschy = Aeschynomene L. ●■
8606 Aedemone Kotschy = Herminiera Guill. et Perr. ●■
8607 Aedesia O. Hoffm. (1897);叶苞糙毛菊属■☆
8608 Aedesia baumannii O. Hoffm. = Aedesia glabra (Klatt) O. Hoffm. ■☆
8609 Aedesia engleriana Mattf.;恩格勒叶苞糙毛菊■☆
8610 Aedesia glabra (Klatt) O. Hoffm.;光叶苞糙毛菊■☆
8611 Aedesia spectabilis Mattf.;叶苞糙毛菊■☆
8612 Aedia Post et Kuntze = Aidia Lour. ●
8613 Aedia Post et Kuntze = Randia L. ●
8614 Aedmannia Spach = Oedmannia Thunb. ■☆
8615 Aedmannia Spach = Rafnia Thunb. ■☆
8616 Aedula Noronha = Orophea Blume ●
8617 Aeegiphila Sw. = Aegiphila Jacq. ●■☆
8618 Aeevidium Salisb. = Aerides Lour. ■
8619 Aegelatis Roxb. = Aegialitis R. Br. ●☆
8620 Aegenetia Roxb. = Aeginetia L. ■
8621 Aegeria Endl. = Ageria Adans. ●
8622 Aegeria Endl. = Ilex L. + Myrsine L. ●
8623 Aegialea Klotzsch = Pieris D. Don ●
8624 Aegialina Schult. = Koeleria Pers. ■
8625 Aegialina Schult. = Rostraria Trin. ■☆
8626 Aegialinites C. Presl = Aegialitis R. Br. ●☆
8627 Aegialinitis Benth. et Hook. f. = Aegialitis R. Br. ●☆
8628 Aegialitidaceae Lincz.;叉枝补血草科(紫条木科)●■
8629 Aegialitidaceae Lincz. = Plumbaginaceae Juss. (保留科名)●■
8630 Aegialitis R. Br. (1810);叉枝补血草属(紫条木属)●☆
8631 Aegialitis Trin. = Aegialina Schult. ■☆
8632 Aegialitis Trin. = Koeleria Pers. ■
8633 Aegialitis Trin. = Rostraria Trin. ■☆
8634 Aegialitis annulata R. Br.;环状叉枝补血草(紫条木)●☆
8635 Aegialitis rotundifolia Roxb.;圆叶叉枝补血草(圆叶紫条木)●☆
8636 Aegialitis tenuis Trin.;纤细叉枝补血草●☆
8637 Aegialophila Boiss. et Heldr. (1849);滨海菊属■☆
8638 Aegialophila Boiss. et Heldr. = Centaurea L. (保留属名)●■
8639 Aegialophila cretica Boiss. et Heldr.;滨海菊■☆
8640 Aegialophila longispina Candargy;长刺滨海菊■☆
8641 Aegialophila pumila (L.) Boiss. = Centaurea pumilio L. ■☆
8642 Aegianilites C. B. Clarke = Aegialinites C. Presl ●☆
8643 Aegianilites C. B. Clarke = Aegialitis R. Br. ●☆
8644 Aegiatilis Griff. = Aegialitis R. Br. ●☆
8645 Aegiceras Gaertn. (1788);桐花树属(蜡烛果属);Aegiceras, Candlefruit ●
8646 Aegiceras corniculatum (L.) Blanco;桐花树(蜡烛果);Corniculate Aegiceras, Corniculate Candlefruit, River Mangrove ●
8647 Aegiceras fragrans Koen. = Aegiceras corniculatum (L.) Blanco ●
8648 Aegiceras majus Gaertn. = Aegiceras corniculatum (L.) Blanco ●
8649 Aegiceras minus Gaertn. = Rourea minor (Gaertn.) Leenh. ●
8650 Aegicerataceae Blume = Myrsinaceae R. Br. (保留科名)●
8651 Aegicerataceae Blume;桐花树科(蜡烛果科)●■☆
8652 Aegicon Adans. = Aegilops L. (保留属名)■
8653 Aegilemma Á. Löve = Aegilops L. (保留属名)■
8654 Aegilemma kotschyi (Boiss.) Á. Löve = Aegilops kotschyi Boiss. ■☆
8655 Aegilonearum Á. Löve = Aegilops L. (保留属名)■
8656 Aegilopaceae Martinov = Gramineae Juss. (保留科名)●■
8657 Aegilopaceae Martinov = Poaceae Barnhart(保留科名)●■
8658 Aegilopodes Á. Löve = Aegilops L. (保留属名)■
8659 Aegilopodes triuncialis (L.) Á. Löve = Aegilops triuncialis L. ■
8660 Aegilops L. (1753)(保留属名);山羊草属(山羊麦属);Aegilops, Goat Grass, Goatgrass, Goat-grass ■
8661 Aegilops algeriensis Gand. = Aegilops neglecta Bertol. ■☆
8662 Aegilops aucheri Boiss.;东方山羊草■☆
8663 Aegilops bicornis (Forssk.) Jaub. et Spach;二角山羊草(两角山羊草);Two-awn Goatgrass ■☆
8664 Aegilops bicornis (Forssk.) Jaub. et Spach var. anathera Eig = Aegilops bicornis (Forssk.) Jaub. et Spach ■☆
8665 Aegilops biuncialis Vis.;两芒山羊草(两芒山羊麦,欧山羊草);Twooawn Goatgrass ■
8666 Aegilops brachyathera Pomel = Aegilops geniculata Roth ■☆
8667 Aegilops caudata L.;尾状山羊草■
8668 Aegilops caudata L. = Aegilops cylindrica Host ■
8669 Aegilops columnaris Zhuk.;小亚山羊草(扭芒山羊草);Twistedawn Goatgrass ■☆
8670 Aegilops comosa Sibth. et Sm.;顶芒山羊草(种毛山羊草);Hairy Goatgrass ■☆
8671 Aegilops crassa Boiss.;粗厚山羊草(粗山羊草,肥羊草);Persian Goatgrass, Thick Goatgrass ■☆
8672 Aegilops cylindrica Host;圆柱山羊草(具节山羊草,圆柱山羊麦,柱穗山羊草);Bearded Goat Grass, Jointed Goatgrass ■
8673 Aegilops exaltata L. = Ophiuros exaltatus (L.) Kuntze ■
8674 Aegilops geniculata Roth;膝曲山羊草;Bent Goatgrass, Ovate Goatgrass ■☆
8675 Aegilops geniculata Roth subsp. africana (Eig) Scholz;非洲膝曲山羊草■☆
8676 Aegilops geniculata Roth subsp. gibberosa (Zhuk.) Hammer;北非山羊草■☆
8677 Aegilops geniculata Roth var. africana (Eig) Hammer = Aegilops geniculata Roth subsp. africana (Eig) Scholz ■☆
8678 Aegilops geniculata Roth var. eventricosa (Eig) Hammer = Aegilops geniculata Roth ■☆
8679 Aegilops geniculata Roth var. latiaristata (Lange) Hammer = Aegilops geniculata Roth ■☆
8680 Aegilops heldreichii Holzm. ex Nyman;粗齿山羊草■☆
8681 Aegilops incurva L. = Parapholis incurva (L.) C. E. Hubb. ■
8682 Aegilops incurvata L. = Parapholis incurva (L.) C. E. Hubb. ■
8683 Aegilops juvenalis (Thell.) Eig;壮山羊草(少壮山羊草);Youthful Goatgrass ■☆
8684 Aegilops kotschyi Boiss.;黏果山羊草(柯奇山羊草);Kotschy

Goatgrass ■☆
8685 Aegilops kotschyi Boiss. var. palaestina Eig = Aegilops kotschyi Boiss. ■☆
8686 Aegilops ligustica (Savign.) Coss.;里古斯山山羊草■☆
8687 Aegilops longissima Schweinf. et Muschl. = Aegilops longissima Schweinf., Muschl. et Eig ■☆
8688 Aegilops longissima Schweinf., Muschl. et Eig;高大山羊草(长山羊草);Verylong Goatgrass ■☆
8689 Aegilops lorentii Hochst.;洛雷山羊草■☆
8690 Aegilops muricata Retz. = Eremochloa muricata (Retz.) Hack. ■
8691 Aegilops mutica Boiss.;无芒山羊草;Awnless Goatgrass ■☆
8692 Aegilops neglecta Bertol. = Aegilops neglecta Req. ex Bertol. ■☆
8693 Aegilops neglecta Bertol. subsp. recta (Zhuk.) Hammer;直立忽视山羊草■☆
8694 Aegilops neglecta Req. ex Bertol.;忽视山羊草;Neglected Goatgrass ■☆
8695 Aegilops ovata L.;卵穗山羊草(南欧山羊草,椭圆山羊麦);Ovate Goatgrass ■
8696 Aegilops ovata L. = Aegilops geniculata Roth ■☆
8697 Aegilops ovata L. subsp. atlantica Eig = Aegilops geniculata Roth subsp. gibberosa (Zhuk.) Hammer ■☆
8698 Aegilops ovata L. subsp. brachyathera (Pomel) Trab. = Aegilops geniculata Roth ■☆
8699 Aegilops ovata L. subsp. gibberosa Zhuk. = Aegilops geniculata Roth subsp. gibberosa (Zhuk.) Hammer ■☆
8700 Aegilops ovata L. subsp. triaristata (Willd.) Rouy = Aegilops neglecta Bertol. ■☆
8701 Aegilops ovata L. subsp. triticoides (Req.) Trab. = Aegilops geniculata Roth ■☆
8702 Aegilops ovata L. var. africana Eig = Aegilops geniculata Roth subsp. africana (Eig) Scholz ■☆
8703 Aegilops ovata L. var. eigiana Maire et Weiller = Aegilops geniculata Roth ■☆
8704 Aegilops ovata L. var. eventricosa Eig = Aegilops geniculata Roth ■☆
8705 Aegilops ovata L. var. latiaristata Lange = Aegilops geniculata Roth ■☆
8706 Aegilops ovata L. var. procera (Jord. et Fourr.) Rouy = Aegilops geniculata Roth ■☆
8707 Aegilops ovata L. var. pubiglumis (Jord. et Fourr.) Rouy = Aegilops geniculata Roth ■☆
8708 Aegilops ovata L. var. triaristata (Willd.) Coss. et Durieu = Aegilops neglecta Bertol. ■☆
8709 Aegilops ovata L. var. trispiculata Hack. = Aegilops neglecta Bertol. ■☆
8710 Aegilops peregrina (Hack.) Maire et Weiller;外来山羊草■☆
8711 Aegilops peregrina (Hack.) Maire et Weiller subsp. cylindrostachys (Eig et Feinbrun) Hammer;柱穗外来山羊草■☆
8712 Aegilops peregrina (Hack.) Maire et Weiller subsp. variabilis (Eig et Feinbrun) Maire = Aegilops peregrina (Hack.) Maire et Weiller var. variabilis (Eig et Feinbrun) Hammer ■☆
8713 Aegilops peregrina (Hack.) Maire et Weiller var. aristata Eig et Feinbrun = Aegilops peregrina (Hack.) Maire et Weiller ■☆
8714 Aegilops peregrina (Hack.) Maire et Weiller var. brachyathera (Boiss.) Maire et Weiller = Aegilops peregrina (Hack.) Maire et Weiller ■☆
8715 Aegilops peregrina (Hack.) Maire et Weiller var. multiaristata (Eig et Feinbrun) Hammer = Aegilops peregrina (Hack.) Maire et Weiller ■☆
8716 Aegilops peregrina (Hack.) Maire et Weiller var. mutica (Eig et Feinbrun) Hammer = Aegilops peregrina (Hack.) Maire et Weiller ■☆
8717 Aegilops peregrina (Hack.) Maire et Weiller var. variabilis (Eig et Feinbrun) Hammer;多变外来山羊草■☆
8718 Aegilops peregrina (Hack.) Melderis = Aegilops peregrina (Hack.) Maire et Weiller ■☆
8719 Aegilops procera Jord. et Fourr. = Aegilops geniculata Roth ■☆
8720 Aegilops pubiglumis Jord. et Fourr. = Aegilops peregrina (Hack.) Maire et Weiller ■☆
8721 Aegilops recta (Zhuk.) Chennav.;直山羊草■☆
8722 Aegilops searsii Feldman et Kislev;西尔斯山羊草■☆
8723 Aegilops sharonensis Eig;沙龙山羊草■☆
8724 Aegilops speltoides Tausch;拟斯卑脱山羊草(斯佩特状山羊草)■☆
8725 Aegilops squarrosa L. = Aegilops tauschii Coss. ■
8726 Aegilops subulata Pomel;钻形山羊草■☆
8727 Aegilops tauschii Coss.;节节麦(山羊草);Goatgrass, Tausch Goatgrass, Tausch's Goatgrass ■
8728 Aegilops triaristata Willd.;短穗山羊草;Shortspike Goatgrass ■
8729 Aegilops triaristata Willd. = Aegilops neglecta Bertol. ■☆
8730 Aegilops triaristata Willd. subsp. recta Zhuk. = Aegilops neglecta Bertol. subsp. recta (Zhuk.) Hammer ■☆
8731 Aegilops triaristata Willd. var. trispiculata Hack. = Aegilops neglecta Bertol. ■☆
8732 Aegilops triuncialis L.;三芒山羊草(钩刺山羊草,离果山羊草,三芒山羊麦);Barbed Goatgrass, Bard Goatgrass ■
8733 Aegilops triuncialis L. subsp. atlantica (Eig) Quézel et Santa = Aegilops geniculata Roth ■☆
8734 Aegilops triuncialis L. subsp. ovata (Eig) Quézel et Santa = Aegilops geniculata Roth ■☆
8735 Aegilops triuncialis L. subsp. triaristata (Willd.) Quézel et Santa = Aegilops neglecta Bertol. ■☆
8736 Aegilops triuncialis L. var. brachyathera Boiss. = Aegilops peregrina (Hack.) Maire et Weiller var. brachyathera (Boiss.) Maire et Weiller ■☆
8737 Aegilops umbellulata Zhuk.;伞穗山羊草(小伞山羊草);Umbel Goatgrass, Umbelspike Goatgrass ■
8738 Aegilops uniarista Vis.;单芒山羊草;Oneawn Goatgrass ■☆
8739 Aegilops variabilis Eig;易变山羊草■☆
8740 Aegilops variabilis Eig et Feinbrun = Aegilops peregrina (Hack.) Maire et Weiller var. variabilis (Eig et Feinbrun) Hammer ■☆
8741 Aegilops variabilis Eig et Feinbrun subsp. cylindrostachys ? = Aegilops peregrina (Hack.) Maire et Weiller subsp. cylindrostachys (Eig et Feinbrun) Hammer ■☆
8742 Aegilops variabilis Eig et Feinbrun var. multiaristata ? = Aegilops peregrina (Hack.) Maire et Weiller ■☆
8743 Aegilops variabilis Eig et Feinbrun var. mutica ? = Aegilops peregrina (Hack.) Maire et Weiller var. mutica (Eig et Feinbrun) Hammer ■☆
8744 Aegilops variabilis Eig et Feinbrun var. peregrina (Hack.) Eig et Feinbrun = Aegilops peregrina (Hack.) Maire et Weiller ■☆
8745 Aegilops vavilovii (Zhuk.) Chennav.;瓦维洛夫山羊草;Vavilov Goatgrass ■☆
8746 Aegilops ventricosa Tausch;偏凸山羊草;Bentricose Goatgrass, Ventricose Goatgrass ■
8747 Aegilops ventricosa Tausch subvar. comosa Coss. et Durieu = Aegilops ventricosa Tausch var. comosa (Coss. et Durieu) Eig ■

8748 Aegilops ventricosa Tausch subvar. truncata Coss. et Durieu = Aegilops ventricosa Tausch var. truncata (Coss. et Durieu) Eig ■
8749 Aegilops ventricosa Tausch var. comosa (Coss. et Durieu) Eig = Aegilops ventricosa Tausch ■
8750 Aegilops ventricosa Tausch var. prostrata Sennen et Mauricio = Aegilops ventricosa Tausch ■
8751 Aegilops ventricosa Tausch var. subulata (Pomel) Maire et Weiller = Aegilops subulata Pomel ■☆
8752 Aegilops ventricosa Tausch var. truncata (Coss. et Durieu) Eig = Aegilops ventricosa Tausch ■
8753 Aegilops ventricosa Tausch var. ventricosa ? = Aegilops ventricosa Tausch ■
8754 Aegilops ventricosa Tausch var. vulgaris Eig = Aegilops ventricosa Tausch ■
8755 Aeginetia Cav. = Bouvardia Salisb. ●■☆
8756 Aeginetia L. (1753);野菰属(蘼寄生属);Aeginetia ■
8757 Aeginetia acaulis (Roxb.) Walp.;短梗野菰;Stemless Aeginetia ■
8758 Aeginetia boninensis Nakai = Aeginetia indica Roxb. ■
8759 Aeginetia indica L. var. sekimotoana (Makino) Makino = Aeginetia indica Roxb. ■
8760 Aeginetia indica Roxb.;野菰(白茅花,蘼寄生,茶匙黄,赤膊花,官巾红,管真花,灌草菰,金锁匙,金钥匙,马口含珠,僧帽花,烧不死,蛇箭草,铁雨伞,土灵芝草,鸭脚板,烟斗花,烟管头草,芋菰草);India Aeginetia,Indian Aeginetia ■
8761 Aeginetia japonica Siebold et Zucc. = Aeginetia indica Roxb. ■
8762 Aeginetia japonica Siebold et Zucc. = Aeginetia sinensis G. Becker ■
8763 Aeginetia orientalis L.;东方野菰(东野菰)■
8764 Aeginetia pedunculata (Roxb.) Wall. = Aeginetia acaulis (Roxb.) Walp. ■
8765 Aeginetia pedunculata Wall. = Aeginetia acaulis (Roxb.) Walp. ■
8766 Aeginetia sekimotoana Makino = Aeginetia indica Roxb. ■
8767 Aeginetia sinensis G. Becker;中国野菰(草寄生,横杯草,箭杆七);China Aeginetia,Chinese Aeginetia ■
8768 Aeginetia sinensis G. Becker f. albiflora K. Asano;白花中国野菰■☆
8769 Aeginetiaceae Livera = Orobanchaceae Vent.(保留科名)●■
8770 Aeginetiaceae Livera;野菰科■
8771 Aegiphila Jacq. (1767);羊族草属●■☆
8772 Aegiphila annomala Pitt;奇羊族草●☆
8773 Aegiphila elata Sw.;高羊族草●☆
8774 Aegiphila elata Sw. var. macrophylla (Kunth) López-Pal.;大叶高羊族草●☆
8775 Aegiphila integrifolia Jacq.;全缘叶羊族草;Entireleaf Spiritweed ●☆
8776 Aegiphila laevigata Juss. = Parameria laevigata (Juss.) Moldenke ●
8777 Aegiphila peruviana Turcz.;羊族草●☆
8778 Aegiphilaceae Raf. = Labiatae Juss.(保留科名)●■
8779 Aegiphilaceae Raf. = Lamiaceae Martinov(保留科名)●■
8780 Aegiphilaceae Raf. = Verbenaceae J. St. -Hil.(保留科名)●■
8781 Aegiphyla Steud. = Aegiphila Jacq. ●■☆
8782 Aegle Corrêa ex Koenig = Aegle Corrêa(保留属名)●
8783 Aegle Corrêa(1800)(保留属名);木橘属(木桔属,印度枳属);Sepiaria ●
8784 Aegle Dulac = Aglae Dulac ■
8785 Aegle Dulac = Posidonia K. D. König(保留属名)■
8786 Aegle marmelos (L.) Corrêa = Aegle marmelos (L.) Corrêa ex Roxb. ●
8787 Aegle marmelos (L.) Corrêa ex Roxb.;木橘(孟加拉苹果,三叶木橘,印度枸橘,印度橘,印度榅桲,印度枳,硬皮橘);Bael Fruit,Baeltree,Bel,Bel Bael,Bell,Bel-tree,Bengal Quince,Golden Apple,Golden-apple,Japanese Bitter Orange,Okshit,Sepiaria ●
8788 Aegle sepiaria DC. = Citrus trifoliata L. ●
8789 Aegle sepiaria DC. = Poncirus trifoliata (L.) Raf. ●
8790 Aeglopsis Swingle(1912);西非橘属(西非枳属)●☆
8791 Aeglopsis alexandrae Chiov. = Vepris eugeniifolia (Engl.) I. Verd. ●☆
8792 Aeglopsis beguei A. Chev.;贝格西非橘●☆
8793 Aeglopsis chevalieri Swingle;西非橘●☆
8794 Aeglopsis chevalieri Swingle var. tanakae (Swingle et M. Kellerm.) A. Chev. = Citropsis gabunensis (Engl.) Swingle et M. Kellerm. ●☆
8795 Aeglopsis eggelingii M. Taylor;埃格西非橘●☆
8796 Aeglopsis mangenotii A. Chev.;曼氏西非橘●☆
8797 Aegoceras Post et Kuntze = Aegiceras Gaertn. ●
8798 Aegochloa Benth. = Navarretia Ruiz et Pav. ■☆
8799 Aegokeras Raf. (1840);山羊角芹属■☆
8800 Aegokeras Raf. = Seseli L. ■
8801 Aegokeras caespitosa (Sibth. et Sm.) Raf.;山羊角芹■☆
8802 Aegomarathrum Steud. = Cachrys L. ■
8803 Aegonychion Endl. = Aegonychon Gray ■
8804 Aegonychon Gray = Lithospermum L. ■
8805 Aegonychon purpurocaeruleum (L.) Holub = Lithospermum purpurocaeruleum L. ■☆
8806 Aegophila Post et Kuntze = Aegiphila Jacq. ●■☆
8807 Aegopicron Giseke = Maprounea Aubl. ■☆
8808 Aegopieron Giseke = Aegopricum L. ■☆
8809 Aegopodion St. -Lag. = Aegopodium L. ■
8810 Aegopodium L. (1753);羊角芹属;Bishop's Weed,Bishops Weed,Goatweed,Gout Weed,Ground Elder,Ground-elder ■
8811 Aegopodium alpestre Ledeb.;东北羊角芹(小叶芹);Alpine Goatweed ■
8812 Aegopodium alpestre Ledeb. f. scabrum Kitag.;毛脉东北羊角芹■
8813 Aegopodium alpestre Ledeb. f. scabrum Kitag. = Aegopodium alpestre Ledeb. ■
8814 Aegopodium alpestre Ledeb. f. tenerum Hara = Aegopodium alpestre Ledeb. ■
8815 Aegopodium alpestre Ledeb. f. tenuisectum Kitag. = Aegopodium alpestre Ledeb. ■
8816 Aegopodium alpestre Ledeb. var. daucifolium Gorovoj = Aegopodium alpestre Ledeb. ■
8817 Aegopodium alpestre Ledeb. var. daucifolium Gorovoj = Aegopodium alpestre Ledeb. var. tenuisectum Kitag. ■
8818 Aegopodium alpestre Ledeb. var. tenuisectum Kitag.;小叶羊角芹(细叶东北羊角芹,细叶小叶芹)■
8819 Aegopodium brachycarpum (Kom.) Schischk. = Pimpinella brachycarpa (Kom.) Nakai ■
8820 Aegopodium burttii Nasir;布尔羊角芹■☆
8821 Aegopodium handelii H. Wolff;湘桂羊角芹;Handel Goatweed ■
8822 Aegopodium henryi Diels;巴东羊角芹;Henry Goatweed ■
8823 Aegopodium kashmirica (R. R. Stewart ex Dunn) M. G. Pimenov;克什米尔羊角芹■☆
8824 Aegopodium latifolium Turcz.;宽叶羊角芹;Broadleaf Goatweed ■
8825 Aegopodium podagraria L. var. variegatum L. H. Bailey = Aegopodium podograria L. ■☆
8826 Aegopodium podograria L.;羊角芹(节竹菜);Bishop Weed,

Bishop's Cap Weed, Bishop's Goutweed, Bishop's Gout-weed, Bishop's Weed, Eltrot, English Masterwort, Farmer's Plague, Garden's Plague, Goat's Foot, Goat's Herb, Goatweed, Gout Weed, Goutweed, Gout-weed, Goutwort, Ground Ash, Ground Elder, Ground Eller, Ground-elder, Grunty Ash, Herb Gerard, Jack-jump-about, Jump-about, Kesh, Pigweed, Pot Ash, Pot-ash, Setfoil, Snow-on-the-mountain, White Ash, Wild Elder, Wild Masterwort ■☆

8827 Aegopodium podograria L. 'Variegatum';花叶羊角芹;Variegated Bishop's Weed, Variegated Gout Weed, Variegated Goutweed ■☆

8828 Aegopodium tadshikorum Schischk.;塔什克羊角芹;Tashkent Goatweed ■

8829 Aegopogon Humb. et Bonpl. ex Willd. (1806);羊须草属■☆

8830 Aegopogon P. Beanv. = Amphipogon R. Br. ■☆

8831 Aegopogon gracilis Peter = Scleria melanotricha Hochst. ex A. Rich. ■☆

8832 Aegopogon strictus (R. Br.) P. Beauv.;羊须草■☆

8833 Aegopogon strictus P. Beauv. = Aegopogon strictus (R. Br.) P. Beauv. ■☆

8834 Aegopordon Boiss. (1846);羊屁菊属■☆

8835 Aegopordon Boiss. = Jurinea Cass. ●■

8836 Aegopricon L. f. = Aegopricum L. ■☆

8837 Aegopricon L. f. = Maprounea Aubl. ■☆

8838 Aegopricum L. = Maprounea Aubl. ■☆

8839 Aegoseris Steud. = Crepis L. ■

8840 Aegotoxicon Molina = Aextoxicon Ruiz et Pav. ●☆

8841 Aegotoxicum Endl. = Aextoxicon Ruiz et Pav. ●☆

8842 Aegtoxicon Molina = Aextoxicon Ruiz et Pav. ●☆

8843 Aegylops Honck. = Aegilops L. (保留属名)■

8844 Aeiphanes Spreng. = Aiphanes Willd. ●☆

8845 Aeiphanes Spreng. = Martinezia Ruiz et Pav. (废弃属名)●☆

8846 Aelbroeckia De Moor = Aeluropus Trin. ■

8847 Aellenia Ulbr. = Halothamnus Jaub. et Spach ●■

8848 Aellenia Ulbr. et Aellen(1934);爱伦藜属(新疆藜属);Aellenia ●■

8849 Aellenia auricula (Moq.) Ulbr.;耳状爱伦藜■☆

8850 Aellenia glauca (M. Bieb.) Aellen;新疆爱伦藜(新疆藜);Sinkiang Aellenia, Xinjiang Aellenia ●

8851 Aellenia glauca (M. Bieb.) Aellen = Halothamnus glaucus (M. Bieb.) Botsch. ●

8852 Aellenia iliensis (Lipsky) Aellen;伊犁爱伦藜■☆

8853 Aellenia lancifolia (Boiss.) Ulbr.;披针叶爱伦藜■☆

8854 Aellenia turcomanica (Aellen) De Moor;土库曼爱伦藜■☆

8855 Aeluropus Trin. (1820);獐茅属(稔草属,獐毛属,樟毛属);Aeluropus ■

8856 Aeluropus arabicus var. distans Bornm. = Aeluropus macrostachyus Hack. ■☆

8857 Aeluropus brevifolius (Jos. König ex Willd.) Nees ex Steud. = Aeluropus lagopoides (L.) Trin. ex Thwaites ■☆

8858 Aeluropus erythraeus (A. Terracc.) Mattei = Aeluropus lagopoides (L.) Trin. ex Thwaites ■☆

8859 Aeluropus laevis Trin. = Aeluropus lagopoides (L.) Trin. ex Thwaites ■☆

8860 Aeluropus lagopoides (L.) Trin. ex Thwaites;平滑獐茅(鬼足獐毛,平滑獐毛,小獐毛);Smooth Aeluropus ■☆

8861 Aeluropus lagopoides (L.) Trin. ex Thwaites var. brevifolius (Jos. König ex Willd.) Chiov. = Aeluropus lagopoides (L.) Trin. ex Thwaites ■☆

8862 Aeluropus littoralis (Gouan) Parl.;疏穗獐茅(疏穗獐毛,小獐毛);Sea-shore Aeluropus ■☆

8863 Aeluropus littoralis (Gouan) Parl. subsp. pungens (M. Bieb.) Tzvelev = Aeluropus pungens (M. Bieb.) K. Koch ■

8864 Aeluropus littoralis (Gouan) Parl. subsp. repens (Desf.) Trab. = Aeluropus lagopoides (L.) Trin. ex Thwaites ■☆

8865 Aeluropus littoralis (Gouan) Parl. subsp. vulgaris (Coss.) Maire = Aeluropus littoralis (Gouan) Parl. ■☆

8866 Aeluropus littoralis (Gouan) Parl. var. intermedius Coss. et Durieu = Aeluropus littoralis (Gouan) Parl. ■☆

8867 Aeluropus littoralis (Gouan) Parl. var. pilosus H. L. Yang = Aeluropus pilosus (X. L. Yang) S. L. Chen et X. L. Yang ■

8868 Aeluropus littoralis (Gouan) Parl. var. repens (Desf.) Coss. et Durieu = Aeluropus lagopoides (L.) Trin. ex Thwaites ■☆

8869 Aeluropus littoralis (Gouan) Parl. var. sinensis Debeaux = Aeluropus sinensis (Debeaux) Tzvelev ■

8870 Aeluropus littoralis (Gouan) Parl. var. sinkiangensis K. L. Chang;新疆獐茅(新疆獐毛);Xinjiang Aeluropus ■

8871 Aeluropus littoralis (Gouan) Parl. var. vulgaris Coss. = Aeluropus littoralis (Gouan) Parl. ■☆

8872 Aeluropus macrostachyus Hack.;大穗獐茅■☆

8873 Aeluropus massauensis (Fresen.) Mattei = Aeluropus lagopoides (L.) Trin. ex Thwaites ■☆

8874 Aeluropus micrantherus Tzvelev;微药獐茅(密穗小獐毛,微药獐毛);Microanther Aeluropus ■

8875 Aeluropus mucronatus (Forssk.) Asch. = Odyssea mucronata (Forssk.) Stapf ■☆

8876 Aeluropus mucronatus (Forssk.) Asch. var. erythraeus A. Terracc. = Aeluropus lagopoides (L.) Trin. ex Thwaites ■☆

8877 Aeluropus pilosus (X. L. Yang) S. L. Chen et X. L. Yang;毛叶獐茅(毛叶獐毛);Hairleaf Aeluropus ■

8878 Aeluropus pungens (M. Bieb.) K. Koch;小獐茅(小獐毛);Small Aeluropus ■

8879 Aeluropus pungens (M. Bieb.) K. Koch var. hirtulus S. L. Chen et X. L. Yang;刺叶獐茅(刺叶獐毛)■

8880 Aeluropus pungens K. Koch = Aeluropus pungens (M. Bieb.) K. Koch ■

8881 Aeluropus repens (Desf.) Parl.;匍匐獐毛■☆

8882 Aeluropus repens (Desf.) Parl. = Aeluropus lagopoides (L.) Trin. ex Thwaites ■☆

8883 Aeluropus repens Parl. = Aeluropus repens (Desf.) Parl. ■☆

8884 Aeluropus sinensis (Debeaux) Tzvelev;獐茅(马绊草,马牙头,虾须草,小叶芦,獐毛);China Aeluropus, Chinese Aeluropus ■

8885 Aeluropus sinensis (Debeaux) Tzvelev = Aeluropus littoralis (Gouan) Parl. ■☆

8886 Aeluropus villosus Trin. ex C. A. Mey. = Aeluropus lagopoides (L.) Trin. ex Thwaites ■☆

8887 Aeluroschia Post et Kuntze = Ailuroschia Steven ●■

8888 Aeluroschia Post et Kuntze = Astragalus L. ●■

8889 Aembilla Adans. (废弃属名) = Scolopia Schreb. (保留属名)●

8890 Aenanthe Raf. = Oenanthe L. ■

8891 Aenhenrya Gopalan(1994);塔米尔兰属■☆

8892 Aenictophyton A. T. Lee(1973);谜木豆属■☆

8893 Aenictophyton reconditum A. T. Lee;谜木豆■☆

8894 Aenida Scop. = Acnida L. ■

8895 Aenida Scop. = Amaranthus L. ■

8896 Aenigmatanthera W. R. Anderson(2006);谜药木属●☆

8897 Aenothera Lam. = Oenothera L. ●■
8898 Aeolanthus Mart. = Aeollanthus Mart. ex Spreng. ■☆
8899 Aeollanthes Spreng. = Aeollanthus Mart. ex Spreng. ■☆
8900 Aeollanthus Mart. = Aeollanthus Mart. ex Spreng. ■☆
8901 Aeollanthus Mart. ex Spreng. (1825);柔花属■☆
8902 Aeollanthus Spreng. = Aeollanthus Mart. ex Spreng. ■☆
8903 Aeollanthus abyssinicus Hochst. ex Benth.;阿比西尼亚柔花■☆
8904 Aeollanthus abyssinicus Hochst. ex Benth. var. angustifolius Sacleux = Aeollanthus abyssinicus Hochst. ex Benth. ■☆
8905 Aeollanthus adenotrichus Gürke = Aeollanthus subacaulis (Baker) Hua et Briq. var. linearis (Burkill) Ryding ■☆
8906 Aeollanthus affinis De Wild. = Aeollanthus suaveolens Mart. ex Spreng. ■☆
8907 Aeollanthus alternatus Ryding;互生柔花■☆
8908 Aeollanthus ambustus Oliv.;条纹柔花■☆
8909 Aeollanthus angolensis Ryding;安哥拉柔花■☆
8910 Aeollanthus angustifolius Ryding;窄叶柔花■☆
8911 Aeollanthus bequaertii De Wild. = Aeollanthus repens Oliv. ■☆
8912 Aeollanthus biformifolius De Wild. = Aeollanthus subacaulis (Baker) Hua et Briq. var. linearis (Burkill) Ryding ■☆
8913 Aeollanthus breviflorus De Wild.;短花柔花■☆
8914 Aeollanthus buchnerianus Briq.;布赫纳柔花■☆
8915 Aeollanthus buettneri Gürke = Aeollanthus pubescens Benth. ■☆
8916 Aeollanthus butaguensis De Wild. = Aeollanthus holstii Gürke ■☆
8917 Aeollanthus calvus Briq. = Aeollanthus pubescens Benth. ■☆
8918 Aeollanthus cameronii Burkill = Aeollanthus ukamensis Gürke ■☆
8919 Aeollanthus candelabrum Briq.;楔叶柔花■☆
8920 Aeollanthus canescens Gürke = Aeollanthus buchnerianus Briq. ■☆
8921 Aeollanthus cassawa G. Taylor = Aeollanthus suaveolens Mart. ex Spreng. ■☆
8922 Aeollanthus caudatus Ryding;尾状柔花■☆
8923 Aeollanthus chevalieri Briq. = Aeollanthus pubescens Benth. ■☆
8924 Aeollanthus claessensi De Wild. = Aeollanthus suaveolens Mart. ex Spreng. ■☆
8925 Aeollanthus conglomeratus Baker = Aeollanthus engleri Briq. ■☆
8926 Aeollanthus crenatus S. Moore = Aeollanthus rehmannii Gürke ■☆
8927 Aeollanthus cryptanthus Baker = Aeollanthus engleri Briq. ■☆
8928 Aeollanthus cucullatus Ryding;僧帽状柔花■☆
8929 Aeollanthus cuneifolius Baker = Aeollanthus candelabrum Briq. ■☆
8930 Aeollanthus densiflorus Ryding;密花柔花■☆
8931 Aeollanthus edlingeri Gürke = Aeollanthus suaveolens Mart. ex Spreng. ■☆
8932 Aeollanthus elongatus Briq. = Aeollanthus pubescens Benth. ■☆
8933 Aeollanthus elongatus De Wild. = Aeollanthus subacaulis (Baker) Hua et Briq. var. linearis (Burkill) Ryding ■☆
8934 Aeollanthus elsholzioides Briq.;香薷柔花■☆
8935 Aeollanthus engleri Briq.;恩格勒柔花■☆
8936 Aeollanthus ericoides De Wild. = Aeollanthus subacaulis (Baker) Hua et Briq. var. ericoides (De Wild.) Ryding ■☆
8937 Aeollanthus floribundus Briq. = Aeollanthus engleri Briq. ■☆
8938 Aeollanthus fouta-djalonensis (De Wild.) De Wild. = Aeollanthus paradoxus (Hua) Hua et Briq. ■☆
8939 Aeollanthus fruticosus Gürke;灌丛柔花■☆
8940 Aeollanthus gamwelliae G. Taylor = Aeollanthus myrianthus Baker subsp. gamwelliae (G. Taylor) Ryding ■☆
8941 Aeollanthus glabrifolius De Wild. = Aeollanthus suaveolens Mart. ex Spreng. ■☆
8942 Aeollanthus glandulosus Gürke = Aeollanthus fruticosus Gürke ■☆
8943 Aeollanthus goetzei Gürke = Aeollanthus engleri Briq. ■☆
8944 Aeollanthus grandifolium Gilli = Aeollanthus subacaulis (Baker) Hua et Briq. ■☆
8945 Aeollanthus haumannii Van Jaarsv.;豪曼柔花■☆
8946 Aeollanthus heliotropioides Oliv. = Aeollanthus suaveolens Mart. ex Spreng. ■☆
8947 Aeollanthus holstii Gürke;霍尔柔花■☆
8948 Aeollanthus homblei De Wild.;洪布柔花■☆
8949 Aeollanthus lamborayi De Wild. = Aeollanthus suaveolens Mart. ex Spreng. ■☆
8950 Aeollanthus linearis (Burkill) Hua et Briq. = Aeollanthus subacaulis (Baker) Hua et Briq. var. linearis (Burkill) Ryding ■☆
8951 Aeollanthus lisowskii Ryding;利索柔花■☆
8952 Aeollanthus livingstonei K. Koch;利文斯通柔花■☆
8953 Aeollanthus lobatus N. E. Br.;浅裂柔花■☆
8954 Aeollanthus lujai De Wild. = Aeollanthus pubescens Benth. ■☆
8955 Aeollanthus lythroides R. E. Fr. = Aeollanthus engleri Briq. ■☆
8956 Aeollanthus medusa Baker = Aeollanthus myrianthus Baker subsp. gamwelliae (G. Taylor) Ryding ■☆
8957 Aeollanthus myrianthus Baker;多花柔花;Ninde ■☆
8958 Aeollanthus myrianthus Baker subsp. gamwelliae (G. Taylor) Ryding;加姆柔花■☆
8959 Aeollanthus namibiensis Ryding;纳米布柔花■☆
8960 Aeollanthus ndorensis Schweinf.;恩多罗柔花■☆
8961 Aeollanthus neglectus (Dinter) Launert;忽视柔花■☆
8962 Aeollanthus njassae Gürke = Aeollanthus buchnerianus Briq. ■☆
8963 Aeollanthus nodosus Hiern = Aeollanthus candelabrum Briq. ■☆
8964 Aeollanthus nyikensis Baker = Aeollanthus buchnerianus Briq. ■☆
8965 Aeollanthus obtusifolius Briq. = Aeollanthus engleri Briq. ■☆
8966 Aeollanthus paludosus Gürke = Aeollanthus engleri Briq. ■☆
8967 Aeollanthus panganiensis Gürke = Aeollanthus zanzibaricus S. Moore ■☆
8968 Aeollanthus paradoxus (Hua) Hua et Briq.;奇异柔花■☆
8969 Aeollanthus parvifolius Benth.;小花柔花■☆
8970 Aeollanthus petasatus Briq. = Aeollanthus pubescens Benth. ■☆
8971 Aeollanthus petiolatus Ryding;柄叶柔花■☆
8972 Aeollanthus pinnatifidus Hochst. ex Benth.;羽裂柔花■☆
8973 Aeollanthus pinnatifidus Hochst. ex Benth. var. tenuis Vatke = Aeollanthus pinnatifidus Hochst. ex Benth. ■☆
8974 Aeollanthus plicatus Ryding;折扇柔花■☆
8975 Aeollanthus poggei Gürke = Aeollanthus engleri Briq. ■☆
8976 Aeollanthus prittwitzianus Gürke = Aeollanthus repens Oliv. ■☆
8977 Aeollanthus pubescens Benth.;毛柔花■☆
8978 Aeollanthus pubescens Benth. var. nuda A. Chev. = Aeollanthus pubescens Benth. ■☆
8979 Aeollanthus purpureo-pilosus Wernham = Aeollanthus pubescens Benth. ■☆
8980 Aeollanthus quarrei De Wild. = Aeollanthus breviflorus De Wild. ■☆
8981 Aeollanthus rehmannii Gürke;拉赫曼柔花■☆
8982 Aeollanthus repens Oliv.;匍匐柔花■☆
8983 Aeollanthus rivularis Hiern;溪畔柔花■☆
8984 Aeollanthus robynsii De Wild. = Aeollanthus pubescens Benth. ■☆
8985 Aeollanthus rosulifolium P. A. Duvign. et Denaeyer = Aeollanthus homblei De Wild. ■☆
8986 Aeollanthus rubescens Gürke = Aeollanthus subacaulis (Baker) Hua et Briq. ■☆

8987 Aeollanthus rydingianus Van Jaarsv. et A. E. van Wyk;吕丁柔花■☆

8988 Aeollanthus salicifolius Baker = Aeollanthus subacaulis (Baker) Hua et Briq. ■☆

8989 Aeollanthus saxatilis P. A. Duvign. et Denaeyer;岩生柔花■☆

8990 Aeollanthus schliebenii Mansf. ex Brooks = Aeollanthus suaveolens Mart. ex Spreng. ■☆

8991 Aeollanthus sedoides Hiern;景天柔花■☆

8992 Aeollanthus semicylindricus De Wild. = Aeollanthus repens Oliv. ■☆

8993 Aeollanthus serpiculoides Baker;仙草柔花■☆

8994 Aeollanthus stefaninii Chiov. = Aeollanthus zanzibaricus S. Moore ■☆

8995 Aeollanthus stormsii Gürke = Aeollanthus pubescens Benth. ■☆

8996 Aeollanthus stormsioides Stopp = Aeollanthus pubescens Benth. ■☆

8997 Aeollanthus stuhlmannii Gürke;斯图尔曼柔花■☆

8998 Aeollanthus suaveolens Mart. ex Spreng.;香柔花;Macassa ■☆

8999 Aeollanthus suavis Mart. = Aeollanthus suaveolens Mart. ex Spreng. ■☆

9000 Aeollanthus subacaulis (Baker) Hua et Briq.;无茎柔花■☆

9001 Aeollanthus subacaulis (Baker) Hua et Briq. var. ericoides (De Wild.) Ryding;石南柔花■☆

9002 Aeollanthus subacaulis (Baker) Hua et Briq. var. linearis (Burkill) Ryding;线形无茎柔花■☆

9003 Aeollanthus trifidus Ryding;三裂柔花■☆

9004 Aeollanthus tuberculatus De Wild. = Aeollanthus subacaulis (Baker) Hua et Briq. var. ericoides (De Wild.) Ryding ■☆

9005 Aeollanthus tuberosus Gürke = Aeollanthus subacaulis (Baker) Hua et Briq. ■☆

9006 Aeollanthus tuberosus Hiern;块根柔花■☆

9007 Aeollanthus ukamensis Gürke;乌卡姆柔花■☆

9008 Aeollanthus uliginosus Gürke = Aeollanthus engleri Briq. ■☆

9009 Aeollanthus usambarenis Gürke = Aeollanthus repens Oliv. ■☆

9010 Aeollanthus violaceus De Wild. = Aeollanthus engleri Briq. ■☆

9011 Aeollanthus virgatus Gürke = Aeollanthus ambustus Oliv. ■☆

9012 Aeollanthus virgatus Gürke var. foliosus Stopp = Aeollanthus ambustus Oliv. ■☆

9013 Aeollanthus viscosus Ryding;黏柔花■☆

9014 Aeollanthus welwitschii Briq. = Aeollanthus candelabrum Briq. ■☆

9015 Aeollanthus xerophytiens Lebrun ex Brooks et al. = Aeollanthus myrianthus Baker ■☆

9016 Aeollanthus zanzibaricus S. Moore;桑给巴尔柔花■☆

9017 Aeolotheca Post et Kuntze = Aiolotheca DC. ■☆

9018 Aeolotheca Post et Kuntze = Zaluzania Pers. ■☆

9019 Aeonia Lindl. = Oeonia Lindl. (保留属名)■☆

9020 Aeoniopsis Rech. f. = Bukiniczia Lincz. ■☆

9021 Aeonium Webb et Berthel. (1840);莲花掌属(鳞甲草属,树莲花属);Aeonium ●■☆

9022 Aeonium aizoon (Bolle) T. Mes;长生草莲花掌●☆

9023 Aeonium arboreum (L.) Webb et Berthel.;树形莲花掌(莲花掌,木鳞甲草,艳姿);Aeonium,Tree Aeonium ●☆

9024 Aeonium arboreum Webb et Berthel. 'Atropurpurum';紫叶树形莲花掌●☆

9025 Aeonium arboreum Webb et Berthel. 'Schwarzkopf' = Aeonium arboreum Webb et Berthel. 'Zwartkop ●☆

9026 Aeonium arboreum Webb et Berthel. 'Variegatum';花叶树形莲花掌●☆

9027 Aeonium arboreum Webb et Berthel. 'Zwartkop';黑头树形莲花掌(紫叶莲花掌)●☆

9028 Aeonium arboreum Webb et Berthel. = Aeonium arboreum (L.) Webb et Berthel. ●☆

9029 Aeonium aureum (Hornem.) T. Mes;黄莲花掌;Green Rose Buds ●☆

9030 Aeonium balsamiferum Webb et Berthel.;香脂莲花掌●☆

9031 Aeonium barbatum Webb et Berthel.;髯毛莲花掌●☆

9032 Aeonium burchardii (Praeger) Praeger;伯查德莲花掌●☆

9033 Aeonium caespitosum (C. Sm.) Webb et Berthel.;丛生莲花掌●☆

9034 Aeonium canariense (L.) Webb et Berthel.;加那利莲花掌●☆

9035 Aeonium ciliatum Webb et Berthel.;纤毛莲花掌●☆

9036 Aeonium cruentum (Webb et Berthel.) Webb et Berthel.;血红莲花掌●☆

9037 Aeonium cuneatum Webb et Berthel.;楔形莲花掌●☆

9038 Aeonium decorum Bolle = Aeonium arboreum Webb et Berthel. 'Luteovariegatum' ●☆

9039 Aeonium decorum Bolle = Aeonium haworthii Webb et Berthel. 'Variegated' ●☆

9040 Aeonium diplocyclum (Bolle) T. Mes;双圆莲花掌●☆

9041 Aeonium dodrantale (Willd.) T. Mes;指距莲花掌●☆

9042 Aeonium glandulosum (Aiton) Webb et Berthel.;腺体莲花掌●☆

9043 Aeonium glutinosum (Aiton) Webb et Berthel.;黏性莲花掌●☆

9044 Aeonium gomerense (Praeger) Praeger;戈梅拉莲花掌●☆

9045 Aeonium goochiae (Webb et Berthel.) Webb et Berthel.;古氏莲花掌●☆

9046 Aeonium gorgoneum J. A. Schmidt;凶恶莲花掌●☆

9047 Aeonium haworthii (Webb et Berthel.) Webb et Berthel.;红缘莲花掌(豪氏鳞甲草,红姬);Haworth's Aeonium,Pinwheel Aeonium ●☆

9048 Aeonium haworthii (Webb et Berthel.) Webb et Berthel. 'Variegated';斑叶红缘莲花掌;Haworth's Aeonium ●☆

9049 Aeonium haworthii Webb et Berthel. = Aeonium haworthii (Webb et Berthel.) Webb et Berthel. ●☆

9050 Aeonium hierrense (Murray) Pit. et Proust;耶罗莲花掌●☆

9051 Aeonium holochrysum Webb et Berthel.;君美丽●☆

9052 Aeonium korneliuslemsii H. Y. Liu = Aeonium arboreum (L.) Webb et Berthel. ●☆

9053 Aeonium lancerottense (Praeger) Praeger;兰瑟莲花掌●☆

9054 Aeonium leucoblepharum Webb ex A. Rich.;白睫毛莲花掌●☆

9055 Aeonium leucoblepharum Webb ex A. Rich. var. glandulosum (Chiov.) Cufod. = Aeonium leucoblepharum Webb ex A. Rich. ●☆

9056 Aeonium lindleyi Webb et Berthel.;林德利莲花掌●☆

9057 Aeonium mascaense Bramwell;马斯卡莲花掌●☆

9058 Aeonium nobile (Praeger) Praeger;镜狮子●☆

9059 Aeonium palmense H. Christ;帕尔马莲花掌●☆

9060 Aeonium percarneum (Murray) Pit. et Proust;紫缘莲花掌●☆

9061 Aeonium percarneum (Murray) Pit. et Proust var. guiaense G. Kunkel = Aeonium percarneum (Murray) Pit. et Proust ●☆

9062 Aeonium pulcher ?;美丽莲花掌;Lipstick Plant ●☆

9063 Aeonium rubrolineatum Svent.;红条纹莲花掌●☆

9064 Aeonium saundersii Bolle;桑德斯莲花掌●☆

9065 Aeonium sedifolium (Bolle) Pit. et Proust;景天莲花掌●☆

9066 Aeonium simsii (Sweet) Stearn;毛叶莲花掌;Hairyleaf Aeonium ●☆

9067 Aeonium smithii (Sims) Webb et Berthel.;史密斯莲花掌●☆

9068 Aeonium spathulatum (Hornem.) Praeger;匙叶莲花掌;Spatulate Aeonium ●☆

9069 Aeonium spathulatum (Hornem.) Praeger var. cruentum (Webb et Berthel.) Praeger = Aeonium spathulatum (Hornem.) Praeger ●☆

9070 Aeonium stuessyi H. Y. Liu;斯图莲花掌●☆

9071 Aeonium subplanum Praeger;近平莲花掌●☆

9072 Aeonium tabulaeforme (Haw.) Webb et Berthel.;平叶莲花掌(明镜,盘状莲花掌);Saucer Plant,Tabletforme Aeonium ●☆

9073 Aeonium tortuosum Aiton = Aichryson tortuosum (Aiton) Webb et Berthel. ■☆

9074 Aeonium undulatum Webb et Berthel.;托盘莲花掌;Saucer Plant, Saucer-plant,Undulate Aeonium ●☆

9075 Aeonium undulatum Webb et Berthel. 'Pseudotabuliforme';拟平叶莲花掌;Saucer Plant ●☆

9076 Aeonium urbicum (Hornem.) Webb et Berthel.;大叶莲花掌;Largeleaf Aeonium ●☆

9077 Aeonium urbicum Webb et Berthel. = Aeonium urbicum (Hornem.) Webb et Berthel. ●☆

9078 Aeonium valverdense (Praeger) Praeger;瓦尔韦德莲花掌●☆

9079 Aeonium vestitum Svent.;包被莲花掌●☆

9080 Aeonium virgineum H. Christ;纯白莲花掌●☆

9081 Aeonium viscatum Bolle;胶莲花掌●☆

9082 Aeonium webbii Bolle = Aeonium gorgoneum J. A. Schmidt ●☆

9083 Aepyanthus Post et Kuntze = Aipyanthus Steven ●■

9084 Aepyanthus Post et Kuntze = Arnebia Forssk. ●■

9085 Aepyanthus Post et Kuntze = Macrotomia DC. ●☆

9086 Aequatorium B. Nord. (1978);赤道菊属●☆

9087 Aequatorium asterotrichum B. Nord.;赤道菊●☆

9088 Aera Asch. = Aira L. (保留属名)■

9089 Aerachne Hook. f. = Acrachne Wight et Arn. ex Chiov. ■

9090 Aerachne Hook. f. = Eleusine Gaertn. ■

9091 Aerangis Rchb. f. (1865);空船兰属(艾兰吉斯兰属,船形兰属)■☆

9092 Aerangis alata H. Perrier = Aerangis ellisii (B. S. Williams) Schltr. ■☆

9093 Aerangis alata H. Perrier = Angraecum ellisii Rchb. f. ■☆

9094 Aerangis alcicornis (Rchb. f.) Garay;尖角空船兰■☆

9095 Aerangis appendiculata (De Wild.) Schltr.;附物空船兰■☆

9096 Aerangis arachnopus (Rchb. f.) Schltr.;蛛网空船兰■☆

9097 Aerangis articulata (Rchb. f.) Schltr.;关节空船兰■☆

9098 Aerangis avicularia (Schltr.) Schltr. = Aerangis rostellaris (Rchb. f.) H. Perrier ■☆

9099 Aerangis batesii (Rolfe) Schltr. = Aerangis arachnopus (Rchb. f.) Schltr. ■☆

9100 Aerangis biloba (Lindl.) Schltr.;二裂空船兰■☆

9101 Aerangis biloba (Lindl.) Schltr. var. kirkii (Rchb. f.) Hawkes = Aerangis kirkii (Rchb. f.) Schltr. ■☆

9102 Aerangis biloboides (De Wild.) Schltr. = Aerangis arachnopus (Rchb. f.) Schltr. ■☆

9103 Aerangis bouarensis Chiron;布阿尔空船兰■☆

9104 Aerangis brachycarpa (A. Rich.) T. Durand et Schinz;短果空船兰■☆

9105 Aerangis brachyceras Summerh. = Cribbia brachyceras (Summerh.) Senghas ■☆

9106 Aerangis buchlohii Senghas = Aerangis rostellaris (Rchb. f.) H. Perrier ■☆

9107 Aerangis calantha (Schltr.) Schltr.;美花空船兰■☆

9108 Aerangis calligera (Rchb. f.) Garay = Aerangis articulata (Rchb. f.) Schltr. ■☆

9109 Aerangis calodictyon Summerh. = Aerangis alcicornis (Rchb. f.) Garay ■☆

9110 Aerangis carnea J. Stewart;肉色空船兰■☆

9111 Aerangis carusiana (Severino) Garay = Aerangis brachycarpa (A. Rich.) T. Durand et Schinz ■☆

9112 Aerangis caulescens Schltr. = Aerangis ellisii (B. S. Williams) Schltr. ■☆

9113 Aerangis caulescens Schltr. = Angraecum ellisii Rchb. f. ■☆

9114 Aerangis citrata (Thouars) Schltr.;柑橘空船兰■☆

9115 Aerangis clavigera H. Perrier = Aerangis macrocentra (Schltr.) Schltr. ■☆

9116 Aerangis concavipetala H. Perrier;凹瓣空船兰■☆

9117 Aerangis confusa J. Stewart;混乱空船兰■☆

9118 Aerangis coriacea Summerh.;革质空船兰■☆

9119 Aerangis crassipes Schltr. = Aerangis modesta (Hook. f.) Schltr. ■☆

9120 Aerangis cryptodon (Rchb. f.) Schltr.;隐齿空船兰■☆

9121 Aerangis decaryana H. Perrier;德卡里空船兰■☆

9122 Aerangis distincta J. Stewart et la Croix;离生空船兰■☆

9123 Aerangis elegans (Rolfe) Dandy = Aerangis flexuosa (Ridl.) Schltr. ■☆

9124 Aerangis ellisii (B. S. Williams) Schltr. = Angraecum ellisii Rchb. f. ■☆

9125 Aerangis ellisii Schltr. = Angraecum ellisii Rchb. f. ■☆

9126 Aerangis ellisii Schltr. var. grandiflora J. L. Stewart = Aerangis ellisii Schltr. ■☆

9127 Aerangis ellisii Schltr. var. grandiflora J. L. Stewart = Angraecum ellisii Rchb. f. ■☆

9128 Aerangis englerianum (Kraenzl.) Schltr. = Rangaeris muscicola (Rchb. f.) Summerh. ■☆

9129 Aerangis erythrura (Kraenzl.) Garay = Aerangis verdickii (De Wild.) Schltr. ■☆

9130 Aerangis falcifolia Schltr. = Rangaeris muscicola (Rchb. f.) Summerh. ■☆

9131 Aerangis fastuosa (Rchb. f.) Schltr.;异美空船兰■☆

9132 Aerangis flabellifolia Rchb. f. = Aerangis brachycarpa (A. Rich.) T. Durand et Schinz ■☆

9133 Aerangis flexuosa (Ridl.) Schltr.;曲折空船兰■☆

9134 Aerangis floribunda (Rolfe) Summerh. = Rangaeris muscicola (Rchb. f.) Summerh. ■☆

9135 Aerangis friesiorum Schltr. = Aerangis thomsonii (Rolfe) Schltr. ■☆

9136 Aerangis fuscata (Rchb. f.) Schltr.;暗褐空船兰■☆

9137 Aerangis gracillima(Kraenzl.) Arends et J. Stewart;纤细空船兰■☆

9138 Aerangis grantii (Baker) Schltr. = Aerangis kotschyana (Rchb. f.) Schltr. ■☆

9139 Aerangis gravenreuthii (Kraenzl.) Schltr.;格氏空船兰■☆

9140 Aerangis hologlottis (Schltr.) Schltr.;全舌空船兰■☆

9141 Aerangis hyaloides (Rchb. f.) Schltr.;无色空船兰■☆

9142 Aerangis ikopana Schltr. = Aerangis mooreana (Rolfe ex Sander) P. J. Cribb et J. L. Stewart ■☆

9143 Aerangis jacksonii J. Stewart;杰克逊空船兰■☆

9144 Aerangis kirkii (Rchb. f.) Schltr.;柯克空船兰■☆

9145 Aerangis kotschyana (Rchb. f.) Schltr.;科奇空船兰■☆

9146 Aerangis kotschyi (Rchb. f.) Rchb. f. = Aerangis kotschyana (Rchb. f.) Schltr. ■☆

9147 Aerangis laurentii (De Wild.) Schltr. = Summerhayesia laurentii (De Wild.) P. J. Cribb ■☆

9148 Aerangis lutambae Mansf. = Aerangis alcicornis (Rchb. f.) Garay ■☆

9149 Aerangis luteoalba (Kraenzl.) Schltr.;黄白空船兰■☆

9150 Aerangis luteoalba (Kraenzl.) Schltr. var. rhodosticta (Kraenzl.)

J. Stewart;红斑空船兰■☆
9151 Aerangis macrocentra (Schltr.) Schltr.;大距空船兰■☆
9152 Aerangis maireae la Croix et J. Stewart;迈雷空船兰■☆
9153 Aerangis malmquistiana Schltr. = Aerangis cryptodon (Rchb. f.) Schltr. ■☆
9154 Aerangis megaphylla Summerh.;大叶空船兰■☆
9155 Aerangis moandensis (De Wild.) Schltr. = Angraecum moandense De Wild. ■☆
9156 Aerangis modesta (Hook. f.) Schltr.;适度空船兰■☆
9157 Aerangis montana J. Stewart;山地空船兰■☆
9158 Aerangis mooreana (Rolfe ex Sander) P. J. Cribb et J. L. Stewart;穆尔空船兰■☆
9159 Aerangis muscicola (Rchb. f.) Schltr. = Rangaeris muscicola (Rchb. f.) Summerh. ■☆
9160 Aerangis mystacidii (Rchb. f.) Schltr.;粗尾空船兰■☆
9161 Aerangis mystacidioides Schltr. = Aerangis mystacidii (Rchb. f.) Schltr. ■☆
9162 Aerangis oligantha Schltr.;寡花空船兰■☆
9163 Aerangis pachyura (Rolfe) Schltr. = Aerangis mystacidii (Rchb. f.) Schltr. ■☆
9164 Aerangis pallida (W. Watson) Garay;苍白空船兰■☆
9165 Aerangis pallidiflora H. Perrier;苍白花空船兰■☆
9166 Aerangis parvula Schltr. = Aerangis calantha (Schltr.) Schltr. ■☆
9167 Aerangis phalaenopsis Schltr. ex Mildbr. = Aerangis megaphylla Summerh. ■☆
9168 Aerangis platyphylla Schltr. = Aerangis ellisii (B. S. Williams) Schltr. ■☆
9169 Aerangis platyphylla Schltr. = Angraecum ellisii Rchb. f. ■☆
9170 Aerangis potamophila (Schltr.) Schltr. = Angraecum potamophilum Schltr. ■☆
9171 Aerangis primulina (Rolfe) H. Perrier;报春空船兰■☆
9172 Aerangis pulchella (Schltr.) Schltr.;美丽空船兰■☆
9173 Aerangis pumilio Schltr. = Aerangis hyaloides (Rchb. f.) Schltr. ■☆
9174 Aerangis punctata J. L. Stewart;斑点空船兰■☆
9175 Aerangis rhodosticta (Kraenzl.) Schltr. = Aerangis luteoalba (Kraenzl.) Schltr. var. rhodosticta (Kraenzl.) J. Stewart ■☆
9176 Aerangis rohlfsiana (Kraenzl.) Schltr. = Aerangis brachycarpa (A. Rich.) T. Durand et Schinz ■☆
9177 Aerangis rostellaris (Rchb. f.) H. Perrier;喙空船兰■☆
9178 Aerangis schliebenii Mansf. = Aerangis verdickii (De Wild.) Schltr. ■☆
9179 Aerangis solheidii (De Wild.) Schltr. = Rangaeris muscicola (Rchb. f.) Summerh. ■☆
9180 Aerangis somalensis (Schltr.) Schltr.;索马里空船兰■☆
9181 Aerangis spiculata (Finet) Senghas;细刺空船兰■☆
9182 Aerangis splendida J. Stewart et la Croix;闪光空船兰■☆
9183 Aerangis stelligera Summerh.;星状空船兰■☆
9184 Aerangis stylosa (Rolfe) Schltr.;宿柱空船兰■☆
9185 Aerangis thomsonii (Rolfe) Schltr.;托马森空船兰■☆
9186 Aerangis ugandensis Summerh.;乌干达空船兰■☆
9187 Aerangis umbonata (Finet) Schltr. = Aerangis fuscata (Rchb. f.) Schltr. ■☆
9188 Aerangis venusta Schltr. = Aerangis articulata (Rchb. f.) Schltr. ■☆
9189 Aerangis verdickii (De Wild.) Schltr.;韦尔空船兰■☆
9190 Aerangis verdickii (De Wild.) Schltr. var. rusituensis (Fibeck et Dare) la Croix et P. J. Cribb;鲁西图空船兰■☆
9191 Aeranthes Lindl. (1824);气花兰属■☆
9192 Aeranthes adenopoda H. Perrier;腺梗气花兰■☆
9193 Aeranthes aemula Schltr.;匹敌气花兰■☆
9194 Aeranthes africana J. Stewart;非洲气花兰■☆
9195 Aeranthes albidiflora Toill. -Gen., Ursch et Bosser;白花气花兰■☆
9196 Aeranthes ambrensis Toill. -Gen., Ursch et Bosser;昂布尔气花兰■☆
9197 Aeranthes angustidens H. Perrier;窄齿气花兰■☆
9198 Aeranthes bathieana Schltr.;巴西气花兰■☆
9199 Aeranthes biauriculata H. Perrier = Aeranthes aemula Schltr. ■☆
9200 Aeranthes brachycentron Regel = Aeranthes grandiflora Lindl. ■☆
9201 Aeranthes brachystachyus Bojer = Beclardia macrostachya (Thouars) A. Rich. ■☆
9202 Aeranthes brevivaginans H. Perrier = Aeranthes ramosa Rolfe ■☆
9203 Aeranthes campbelliae Hermans et Bosser;弯气花兰■☆
9204 Aeranthes carnosa Toill. -Gen., Ursch et Bosser;肉质气花兰■☆
9205 Aeranthes caudata Rolfe;尾状气花兰■☆
9206 Aeranthes crassifolia Schltr.;厚叶气花兰■☆
9207 Aeranthes curnowianus Rchb. f. = Angraecum curnowianum (Rchb. f.) T. Durand et Schinz ■☆
9208 Aeranthes denticulata Toill. -Gen., Ursch et Bosser;细齿气花兰■☆
9209 Aeranthes ecalcarata H. Perrier;无距气花兰■☆
9210 Aeranthes englerianus Kraenzl. = Angraecum kranzlinianum H. Perrier ■☆
9211 Aeranthes erectiflora Senghas;直花气花兰■☆
9212 Aeranthes filipes Schltr.;丝梗气花兰■☆
9213 Aeranthes gladiifolius (Thouars) Rchb. f. = Angraecum mauritianum (Poir.) Frapp. ■☆
9214 Aeranthes gracilis Schltr. = Aeranthes subramosa Garay ■☆
9215 Aeranthes grandidieranus Rchb. f. = Neobathiea grandidierana (Rchb. f.) Garay ■☆
9216 Aeranthes grandiflora Lindl.;大花气花兰■☆
9217 Aeranthes gravenreuthii Kraenzl. = Aerangis gravenreuthii (Kraenzl.) Schltr. ■☆
9218 Aeranthes henricii Schltr. = Erasanthe henrici (Schltr.) P. J. Cribb, Hermans et D. L. Roberts ■☆
9219 Aeranthes imerinensis H. Perrier = Aeranthes caudata Rolfe ■☆
9220 Aeranthes laxiflora Schltr.;疏花气花兰■☆
9221 Aeranthes leandriana Bosser;利安气花兰■☆
9222 Aeranthes leonis Rchb. f. = Angraecum leonis (Rchb. f.) André ■☆
9223 Aeranthes longipes Schltr.;长梗气花兰■☆
9224 Aeranthes macrostachyus (Thouars) Rchb. f. = Beclardia macrostachya (Thouars) A. Rich. ■☆
9225 Aeranthes moratii Bosser;莫拉特气花兰■☆
9226 Aeranthes multinodis Bosser;多节气花兰■☆
9227 Aeranthes neoperrieri Toill. -Gen., Ursch et Bosser;佩里耶气花兰■☆
9228 Aeranthes nidus Schltr.;巢状气花兰■☆
9229 Aeranthes orophila Toill. -Gen.;喜山气花兰■☆
9230 Aeranthes orthopoda Toill. -Gen., Ursch et Bosser;直足气花兰■☆
9231 Aeranthes parkesii G. Will.;帕克斯气花兰■☆
9232 Aeranthes parvula Schltr.;较小气花兰■☆
9233 Aeranthes pectinatus (Thouars) Rchb. f. = Angraecum pectinatum Thouars ■☆
9234 Aeranthes perrieri Schltr. = Neobathiea perrieri Schltr. ■☆
9235 Aeranthes peyrotii Bosser;佩罗气花兰■☆
9236 Aeranthes polyanthemus Ridl.;多花气花兰■☆
9237 Aeranthes pseudonidus H. Perrier = Aeranthes nidus Schltr. ■☆

9238 Aeranthes pusilla Schltr. = Aeranthes setiformis Garay ■☆
9239 Aeranthes ramosa Rolfe;多枝气花兰■☆
9240 Aeranthes ramosus Cogn. = Aeranthes ramosa Rolfe ■☆
9241 Aeranthes rigidula Schltr. = Aeranthes longipes Schltr. ■☆
9242 Aeranthes robusta Senghas;粗壮气花兰■☆
9243 Aeranthes sambiranoensis Schltr.;桑比拉诺风兰■☆
9244 Aeranthes schlechteri Bosser;施莱气花兰■☆
9245 Aeranthes sesquipedalis (Thouars) Lindl. = Angraecum sesquipedale Thouars ■☆
9246 Aeranthes setiformis Garay;刚毛气花兰■☆
9247 Aeranthes setipes Schltr.;毛梗气花兰■☆
9248 Aeranthes subramosa Garay;略分枝气花兰■☆
9249 Aeranthes thouarsii S. Moore = Angraecum filicornu Thouars ■☆
9250 Aeranthes tricalcarata H. Perrier;三距气花兰■☆
9251 Aeranthes trichoplectron Rchb. f. = Angraecum trichoplectron (Rchb. f.) Schltr. ■☆
9252 Aeranthus Bartl. = Aeranthes Lindl. ■☆
9253 Aeranthus Rchb. f. = Aeranthes Lindl. + Macroplectrum Pfitzer ■☆
9254 Aeranthus Spreng. = Aeranthes Lindl. ■☆
9255 Aeranthus calceolus (Thouars) S. Moore = Angraecum calceolus Thouars ■☆
9256 Aeranthus deistelianus Kraenzl. = Tridactyle tridactylites (Rolfe) Schltr. ■☆
9257 Aeranthus erythropollinius Rchb. f. = Diaphananthe xanthopollinia (Rchb. f.) Summerh. ■☆
9258 Aeranthus filicornis (Lindl.) Rchb. f. = Mystacidium capense (L. f.) Schltr. ■☆
9259 Aeranthus gerrardii Rchb. f. = Diaphananthe xanthopollinia (Rchb. f.) Summerh. ■☆
9260 Aeranthus gracilis (Harv.) Rchb. f. = Mystacidium gracile Harv. ■☆
9261 Aeranthus lindenii (Lindl.) Rchb. f. = Dendrophylax lindenii (Lindl.) Benth. ex Rolfe ■☆
9262 Aeranthus muscicola Rchb. f. = Rangaeris muscicola (Rchb. f.) Summerh. ■☆
9263 Aeranthus pachyrrhizus Rchb. f. = Campylocentrum pachyrrhizum (Rchb. f.) Rolfe ■☆
9264 Aeranthus porrectus Rchb. f. = Harrisella porrecta (Rchb. f.) Fawc. et Rendle ■☆
9265 Aeranthus pusillus (Harv.) Rchb. f. = Mystacidium pusillum Harv. ■☆
9266 Aeranthus rutilus Rchb. f. = Rhipidoglossum rutilum (Rchb. f.) Schltr. ■☆
9267 Aeranthus spathaceus Griseb. = Campylocentrum pachyrrhizum (Rchb. f.) Rolfe ■☆
9268 Aeranthus trifurcus Rchb. f. = Angraecopsis trifurca (Rchb. f.) Schltr. ■☆
9269 Aeranthus xanthopollinius Rchb. f. = Diaphananthe xanthopollinia (Rchb. f.) Summerh. ■☆
9270 Aeria O. F. Cook = Gaussia H. Wendl. ●☆
9271 Aerides Lour. (1790);指甲兰属;Aerides, Cat's-tail Orchid, Fox Brush Orchid, Fox-tail Orchid, Fox-tail Orchids, Fox-taoil Orchis, Nailorchis ■
9272 Aerides affine Wall. = Aerides multiflorum Roxb. ■☆
9273 Aerides ampullaceum Roxb. = Ascocentrum ampullaceum (Roxb.) Schltr. ■
9274 Aerides arachnites (Blume) Lindl. = Thrixspermum centipeda Lour. ■
9275 Aerides biswasianum Ghose et Mukerjee = Papilionanthe biswasiana (Ghose et Mukerjee) Garay ■
9276 Aerides brookei Bateman = Aerides crispum Lindl. ■☆
9277 Aerides brookei Bateman ex Lindl. = Aerides crispum Lindl. ■☆
9278 Aerides calceolare Buch. -Ham. ex Sm. = Gastrochilus calceolaris (Buch. -Ham. ex Sm.) D. Don ■
9279 Aerides cornutum Roxb. = Aerides odoratum Lour. ■
9280 Aerides crassifolium Parl. et Rchb. f.;厚叶指甲兰;Thickleaf Fox-tail Orchid ■☆
9281 Aerides crispum Lindl.;皱唇指甲兰(皱指甲兰);Crisp Fox-tail Orchid ■☆
9282 Aerides cristatum (Lindl.) Wall. ex Hook. f. = Vanda cristata Lindl. ■
9283 Aerides densiflorum (Lindl.) Wall. ex Hook. f. = Robiquetia spatulata (Blume) J. J. Sm. ■
9284 Aerides difforme Wall. ex Lindl. = Ornithochilus difformis (Lindl.) Schltr. ■
9285 Aerides expansum Rchb. f. = Aerides crassifolium Parl. et Rchb. f. ■☆
9286 Aerides falcatum Lindl. et Paxton;指甲兰(短距指甲兰,镰形指甲兰);Falcate Fox-tail Orchid, Sickle Nailorchis ■
9287 Aerides fieldingii Lodd. = Aerides fieldingii Lodd. ex E. Morren ■
9288 Aerides fieldingii Lodd. ex E. Morren;费氏指甲兰(狐尾指甲兰);Fielding Fox-tail Orchid, Fox-brush Orchid ■
9289 Aerides flabellatum Rolfe ex Downie;扇唇指甲兰(扇叶指甲兰);Fanlip Nailorchis ■
9290 Aerides flavescens Schltr. = Holcoglossum flavescens (Schltr.) Z. H. Tsi ■
9291 Aerides japonicum Lindenb. et Rchb. f. = Sedirea japonica (Lindenb. et Rchb. f.) Garay et H. R. Sweet ■
9292 Aerides japonicum Rchb. f. = Sedirea japonica (Linden et Rchb. f.) Garay et H. R. Sweet ■
9293 Aerides larpentae Rchb. f. = Aerides falcatum Lindl. et Paxton ■
9294 Aerides lasiopetalum Willd = Dendrolirium lasiopetalum (Willd.) S. C. Chen et J. J. Wood ■
9295 Aerides lasiopetalum Willd. = Eria lasiopetala (Willd.) Ormerod ■
9296 Aerides latifolium Thwaites = Phalaenopsis deliciosa Rchb. f. ■
9297 Aerides lawrenceae Rchb. f.;菲岛指甲兰■
9298 Aerides lindleyanum Wight = Aerides crispum Lindl. ■☆
9299 Aerides macrostachyum (Thouars) A. Spreng. = Beclardia macrostachya (Thouars) A. Rich. ■☆
9300 Aerides mitratum Rchb. f.;尖帽指甲兰■☆
9301 Aerides multiflorum Roxb.;繁花指甲兰;Manyflower Fox-tail Orchid ■☆
9302 Aerides odoratum Lour.;香花指甲兰(香指甲兰,指甲兰);Fox-tail Orchid, Fragrant Nailorchis ■
9303 Aerides orthocentrum Hand. -Mazz. = Vanda coerulea Griff. ex Lindl. ■
9304 Aerides paniculatum Ker Gawl. = Cleisostoma paniculatum (Ker Gawl.) Garay ■
9305 Aerides quinquevulnerum Lindl.;五痕指甲兰■☆
9306 Aerides quinquevulnerum Lindl. var. purpuratum Valmayor et D. Tiu;紫花指甲兰■
9307 Aerides racemiferum (Lindl.) Wall. ex Hook. f. = Cleisostoma racemiferum (Lindl.) Garay ■
9308 Aerides retusum (L.) Sw. = Rhynchostylis retusa (L.) Blume ■
9309 Aerides rigidum Buch. -Ham. ex Sm. = Acampe rigida (Buch. -

Ham. ex Sm.) P. F. Hunt ■
9310 Aerides roseum Lodd. ex Lindl. et Paxton;多花指甲兰;Flowery Nailorchis ■
9311 Aerides savaganum Sand. ex Veitch;狐尾兰■
9312 Aerides subulatum (Blume) Lindl. = Thrixspermum subulatum Rchb. f. ■
9313 Aerides taeniale Lindl. = Kingidium taeniale (Lindl.) P. F. Hunt ■
9314 Aerides taeniale Lindl. = Phalaenopsis taenialis (Lindl.) Christenson et Pradhan ■
9315 Aerides vandarum Rchb. f. ;棒叶指甲兰(万代指甲兰,万带指甲兰);Clubleaf Fox-tail Orchid ■☆
9316 Aerides vandarum Rchb. f. = Papilionanthe vendarum (Rchb. f.) Garay ■
9317 Aerides virens Lindl. = Aerides odoratum Lour. ■
9318 Aerides warneri Hook. f. = Aerides crispum Lindl. ■☆
9319 Aerides williamsii Warn. = Aerides fieldingii Lodd. ex E. Morren ■
9320 Aeridium Pfeiff. = Aeevidium Salisb. ■
9321 Aeridium Pfeiff. = Aerides Lour. ■
9322 Aeridium Post et Kuntze = Airidium Steud. ■
9323 Aeridium Post et Kuntze = Deschampsia P. Beauv. ■
9324 Aeridium Salisb. = Aerides Lour. ■
9325 Aeridostachya (Hook. f.) Brieger = Eria Lindl. (保留属名)■
9326 Aeridostachya reptans (Kuntze) Rauschert = Conchidium japonicum (Maxim.) S. C. Chen et J. J. Wood ■
9327 Aeriphracta Rchb. = Aperiphracta Nees ex Meisn. ●☆
9328 Aeriphracta Rchb. = Ocotea Aubl. ●☆
9329 Aerisilvaea Radcl. -Sm. (1990);马拉维大戟属●☆
9330 Aerisilvaea serrata Radcl. -Sm. = Maytenus undata (Thunb.) Blakelock ●☆
9331 Aerisilvaea sylvestris Radcl. -Sm. ;马拉维大戟●☆
9332 Aerisilvaea sylvestris Radcl. -Sm. = Lingelsheimia sylvestris (Radcl. -Sm.) Radcl. -Sm. ●☆
9333 Aeritochaeta Post et Kuntze = Acritochaete Pilg. ■
9334 Aeriulus Ridl. = Scleria P. J. Bergius ■
9335 Aerobion Kaempfer ex Spreng. = Angraecum Bory ■
9336 Aerobion Spreng. = Angraecum Bory ■
9337 Aerobion Spreng. = Eulophia R. Br. (保留属名)■
9338 Aerobion Spreng. = Eulophidium Pfitzer ■☆
9339 Aerobion Spreng. = Jumellea Schltr. ■☆
9340 Aerobion Spreng. = Solenangis Schltr. ■☆
9341 Aerobion calceolus (Thouars) Spreng. = Angraecum calceolus Thouars ■☆
9342 Aerobion caulescens (Thouars) Spreng. = Angraecum caulescens Thouars ■☆
9343 Aerobion citratum (Thouars) Spreng. = Aerangis citrata (Thouars) Schltr. ■☆
9344 Aerobion crassum (Thouars) Spreng. = Angraecum crassum Thouars ■☆
9345 Aerobion filicornu (Thouars) Spreng. = Angraecum filicornu Thouars ■☆
9346 Aerobion gladiifolium (Thouars) Spreng. = Angraecum mauritianum (Poir.) Frapp. ■☆
9347 Aerobion gracile (Thouars) Spreng. = Chamaeangis gracilis Schltr. ■☆
9348 Aerobion implicatum (Thouars) Spreng. = Angraecum implicatum Thouars ■☆
9349 Aerobion maculatum (Lindl.) Spreng. = Oeceoclades maculata (Lindl.) Lindl. ■☆
9350 Aerobion multiflorum (Thouars) Spreng. = Angraecum multiflorum Thouars ■☆
9351 Aerobion parviflorum (Thouars) Spreng. = Angraecopsis parviflora (Thouars) Schltr. ■☆
9352 Aerobion pectinatum (Thouars) Spreng. = Angraecum pectinatum Thouars ■☆
9353 Aerokorion Scop. = Acrocorion Adans. ■☆
9354 Aerokorion Scop. = Galanthus L. ■☆
9355 Aeronia Lindl. = Oeonia Lindl. (保留属名)■☆
9356 Aerope (Endl.) Rchb. = Rhizophora L. ●
9357 Aeropsis Asch. et Graebner = Airopsis Desv. ■☆
9358 Aerosperma Post et Kuntze = Airosperma K. Schum. et Lauterb. ■☆
9359 Aerua A. Cunn. ex Juss. = Aerva Forssk. (保留属名)●■
9360 Aerua Juss. = Aerva Forssk. (保留属名)●■
9361 Aerva Forssk. (1775)(保留属名);白花苋属(绢毛苋属);Aerva ●■
9362 Aerva ambigua Moq. = Aerva leucura Moq. ■☆
9363 Aerva arachnoidea Gand. = Aerva lanata (L.) Juss. ex Schult. ■☆
9364 Aerva bovei (Webb) Edgew. = Aerva javanica (Burm. f.) Juss. ex Schult. var. bovei Webb ■☆
9365 Aerva brachiata (L.) Mart. = Nothosaerva brachiata (L.) Wight ■☆
9366 Aerva coriacea Schinz;革质白花苋■☆
9367 Aerva desertorum Engl. = Arthraerua leubnitziae (Kuntze) Schinz ■☆
9368 Aerva edulis Suess. = Aerva leucura Moq. ■☆
9369 Aerva elegans Moq. = Aerva lanata (L.) Juss. ex Schult. ■☆
9370 Aerva glabrata Hook. f. ;少毛白花苋;Glabrescent Aerva ●
9371 Aerva hainanensis F. C. How;海南白花苋;Hainan Aerva ●
9372 Aerva hainanensis F. C. How = Psilotrichopsis curtisii (Oliv.) C. C. Towns. var. hainanensis (F. C. How) H. S. Kiu ■
9373 Aerva humbertii Cavaco;亨伯特白花苋■☆
9374 Aerva incana Suess. = Aerva lanata (L.) Juss. ex Schult. ■☆
9375 Aerva javanica (Burm. f.) Juss. = Aerva javanica (Burm. f.) Juss. ex Schult. ■☆
9376 Aerva javanica (Burm. f.) Juss. ex Schult. ;爪哇白花苋■☆
9377 Aerva javanica (Burm. f.) Juss. ex Schult. var. bovei Webb;博韦白花苋■☆
9378 Aerva javanica Juss. = Aerva javanica (Burm. f.) Juss. ex Schult. ■☆
9379 Aerva lanata (L.) Juss. ex Schult. ;弹毛白花苋■☆
9380 Aerva lanata Juss. ex Schult. = Aerva lanata (L.) Juss. ex Schult. ■☆
9381 Aerva leubnitziae Kuntze = Arthraerua leubnitziae (Kuntze) Schinz ■☆
9382 Aerva leucura Moq. ;白尾白花苋■☆
9383 Aerva leucura Moq. var. lanatoides Suess. = Aerva leucura Moq. ■☆
9384 Aerva madagassica Suess. ;马岛白花苋●☆
9385 Aerva monsoniae (L. f.) C. Mart. = Trichuriella monsoniae (L. f.) Bennet ■
9386 Aerva monsoniae Mart. = Trichurus monsoniae (L. f.) C. C. Towns. ■
9387 Aerva mozambicensis Gand. = Aerva lanata (L.) Juss. ex Schult. ■☆
9388 Aerva pechuelii Kuntze = Calicorema capitata (Moq.) Hook. f. ■☆
9389 Aerva persica (Burm. f.) Merr. = Aerva javanica (Burm. f.)

Juss. ex Schult. ■☆

9390 Aerva persica (Burm. f.) Merr. var. bovei (Webb) Chiov. = Aerva javanica (Burm. f.) Juss. ex Schult. var. bovei Webb ■☆

9391 Aerva persica (Burm. f.) Merr. var. bovei Webb = Aerva javanica (Burm. f.) Juss. ex Schult. ■☆

9392 Aerva persica (Burm. f.) Merr. var. latifolia (Vahl) Cufod. = Aerva javanica (Burm. f.) Juss. ex Schult. ■☆

9393 Aerva persica (Burm. f.) Merr. var. latifolia Vahl = Aerva javanica (Burm. f.) Juss. ex Schult. ■☆

9394 Aerva pseudotomentosa Blatt. et Hallb. = Aerva javanica (Burm. f.) Juss. ex Schult. var. bovei Webb ■☆

9395 Aerva ruspolii Lopr. = Aerva javanica (Burm. f.) Juss. ex Schult. ■☆

9396 Aerva sanguinolenta (L.) Blume;白花苋(白牛膝,绢毛苋,烂脚浩,蔓鸡冠);Common Aerva ●

9397 Aerva sansibarica Suess. = Aerva lanata (L.) Juss. ex Schult. ■☆

9398 Aerva scandens (Roxb.) Wall. = Aerva sanguinolenta (L.) Blume ●

9399 Aerva timorensis Moq. = Aerva sanguinolenta (L.) Blume ●

9400 Aerva tomentosa Forssk. = Aerva javanica (Burm. f.) Juss. ex Schult. ■☆

9401 Aerva tomentosa Lam.;绒毛白花苋■☆

9402 Aerva transvaalensis Gand.;德兰士瓦白花苋■☆

9403 Aerva triangularifolia Cavaco;三角叶白花苋■☆

9404 Aerva velutina Moq. = Aerva sanguinolenta (L.) Blume ●

9405 Aerva wallichii Moq. = Aerva javanica (Burm. f.) Juss. ex Schult. ■☆

9406 Aesandra Pierre = Aisandra Airy Shaw ●

9407 Aesandra Pierre = Diploknema Pierre ●

9408 Aesandra Pierre ex L. Planch. = Aisandra Airy Shaw ●

9409 Aesandra Pierre ex L. Planch. = Diploknema Pierre ●

9410 Aeschinanthus Endl. = Aeschynanthus Jack(保留属名)●■

9411 Aeschinomene Nocca = Aeschynomene L. ●■

9412 Aeschrion Vell. (1829);爱舍苦木属●☆

9413 Aeschrion Vell. = Picrasma Blume ●

9414 Aeschrion crenata Vell.;爱舍苦木●☆

9415 Aeschrion excelsa Kuntze = Picrasma excelsa (Sw.) Planch. ●☆

9416 Aeschryon Pfeiff. = Aeschrion Vell. ●☆

9417 Aeschynanthus Jack(1823)(保留属名);芒毛苣苔属(口红花属);Basket Plant,Basket Vine,Basketvine,Blushwort ●■

9418 Aeschynanthus acuminatissimus W. T. Wang;长尖芒毛苣苔(长果藤);Acutest Basketvine,Longacuminate Basketvine,Long-toothed Basketvine ●

9419 Aeschynanthus acuminatus Wall. ex A. DC.;芒毛苣苔(白背风,长果藤,大叶榕藤,大叶石榕,牛奶树,石壁风,石榕);Acuminate Basketvine ●

9420 Aeschynanthus acuminatus Wall. ex A. DC. var. chinensis (Gardner et Champ.) C. B. Clarke = Aeschynanthus acuminatus Wall. ex A. DC. ●

9421 Aeschynanthus andersonii C. B. Clarke;轮叶芒毛苣苔(软叶芒毛苣苔);Verticillateleaf Basketvine,Whorlleaf Basketvine ●

9422 Aeschynanthus angustioblongus W. T. Wang;狭矩叶芒毛苣苔(狭矩芒毛苣苔);Narrow-oblong Basketvine ●

9423 Aeschynanthus angustissimus (W. T. Wang) W. T. Wang;狭叶芒毛苣苔;Narrowleaf Basketvine,Narrow-leaved Basketvine ●

9424 Aeschynanthus apicidens Hance = Lysionotus pauciflorus Maxim. ●

9425 Aeschynanthus austroyunnanensis W. T. Wang;滇南芒毛苣苔;S. Yunnan Basketvine,South Yunnan Basketvine,Yunnan Basketvine ●

9426 Aeschynanthus austroyunnanensis W. T. Wang var. guangxiensis (Chun) W. T. Wang;广西芒毛苣苔;Guangxi Basketvine ●

9427 Aeschynanthus bracteatus Wall. ex A. DC.;显苞芒毛苣苔;Bracteate Basketvine ●

9428 Aeschynanthus bracteatus Wall. ex A. DC. var. orientalis W. T. Wang;黄棕芒毛苣苔;Oroental Basketvine ●

9429 Aeschynanthus bracteatus Wall. ex A. DC. var. orientalis W. T. Wang = Aeschynanthus bracteatus Wall. ex A. DC. ●

9430 Aeschynanthus bracteatus Wall. ex A. DC. var. peelii (Hook. f. et Thomson) C. B. Clarke = Aeschynanthus bracteatus Wall. ex A. DC. ●

9431 Aeschynanthus buxifolius Hemsl.;黄杨叶芒毛苣苔(上树蜈蚣);Boxleaf Basketvine,Box-leaved Basketvine ●

9432 Aeschynanthus chinensis Gardner et Champ. = Aeschynanthus acuminatus Wall. ex A. DC. ●

9433 Aeschynanthus denticuliger W. T. Wang;小齿芒毛苣苔;Little-tooth Basketvine,Smalltooth Basketvine ●

9434 Aeschynanthus dolichanthus W. T. Wang;长花芒毛苣苔;Longflower Basketvine,Long-flowered Basketvine ●

9435 Aeschynanthus dunnii H. Lév. = Phlogacanthus pubinervius T. Anderson ●

9436 Aeschynanthus esquirolii H. Lév. = Exbucklandia populnea (R. Br. ex Griff.) R. W. Br. ●

9437 Aeschynanthus gracilis Parish ex C. B. Clarke;细芒毛苣苔;Small Basketvine,Thin Basketvine ●

9438 Aeschynanthus guangxiensis Chun ex W. T. Wang et K. Y. Pan = Aeschynanthus austroyunnanensis W. T. Wang var. guangxiensis (Chun) W. T. Wang ●

9439 Aeschynanthus hildebrandii Hemsl. ex Hook. f.;唇冠芒毛苣苔●

9440 Aeschynanthus hookeri C. B. Clarke;束花芒毛苣苔;Hooker Basketvine ●

9441 Aeschynanthus humilis Hemsl.;矮芒毛苣苔;Dwarf Basketvine ●

9442 Aeschynanthus lancilimbus W. T. Wang;披针叶芒毛苣苔(披针芒毛苣苔);Lance-leaved Basketvine,Lancelimbate Basketvine,Lanceolateleaf Basketvine ●

9443 Aeschynanthus lasianthus W. T. Wang;毛花芒毛苣苔;Cottonyflower Basketvine,Hairy-flower Basketvine,Hairy-flowered Basketvine ●

9444 Aeschynanthus lasiocalyx W. T. Wang;毛萼芒毛苣苔;Cottonycalyx Basketvine,Hairy-calyx Basketvine,Hairy-calyxed Basketvine ●

9445 Aeschynanthus levipes C. B. Clarke = Lysionotus levipes (C. B. Clarke) B. L. Burtt ●

9446 Aeschynanthus linearifolius C. E. C. Fisch.;条叶芒毛苣苔;Linear-leaved Basketvine,Line-leaf Basketvine ●

9447 Aeschynanthus linearifolius C. E. C. Fisch. var. angustissimus W. T. Wang = Aeschynanthus angustissimus (W. T. Wang) W. T. Wang ●

9448 Aeschynanthus linearifolius C. E. C. Fisch. var. oblonceolatus (Anthony) W. T. Wang;倒披针芒毛苣苔;Oblanceolate Linear-leaved Basketvine ●

9449 Aeschynanthus linearifolius C. E. C. Fisch. var. oblonceolatus (Anthony) W. T. Wang = Aeschynanthus linearifolius C. E. C. Fisch. ●

9450 Aeschynanthus lineatus Craib;线条芒毛苣苔;Linear Basketvine,Lineate Basketvine ●

9451 Aeschynanthus lobianus Hook.;洛布芒毛苣苔(毛萼口红花);Lipstick Plant ●☆

9452 Aeschynanthus longicalyx H. W. Li = Aeschynanthus sinolongicalyx

W. T. Wang ●

9453 Aeschynanthus longicalyx W. T. Wang;长萼芒毛苣苔;Long-calyx Basketvine ●

9454 Aeschynanthus longicaulis Wall. ex R. Br.;长茎芒毛苣苔;Long-stem Basketvine, Long-stemmed Basketvine ●

9455 Aeschynanthus macranthus (Merr.) Pellegr.;伞花芒毛苣苔;Big-flower Basketvine, Umbellate-flowered Basketvine, Umbrellaflower Basketvine ●

9456 Aeschynanthus maculatus Lindl.;具斑芒毛苣苔;Maculate Basketvine, Spoted Basketvine ●

9457 Aeschynanthus maculatus Lindl. var. stenophyllus C. B. Clarke = Aeschynanthus maculatus Lindl. ●

9458 Aeschynanthus marmoratus F. W. Moore;斑纹口红花(花叶口红花);Zebra Basket-plant ●☆

9459 Aeschynanthus marmoratus F. W. Moore = Aeschynanthus longicaulis Wall. ex R. Br. ●

9460 Aeschynanthus medogensis W. T. Wang;墨脱芒毛苣苔;Medog Basketvine, Motuo Basketvine ●

9461 Aeschynanthus mengxingensis W. T. Wang;勐醒芒毛苣苔;Mengxing Basketvine ●

9462 Aeschynanthus mimetes B. L. Burtt;大花芒毛苣苔;Largeflower Basketvine, Large-flower Basketvine ●

9463 Aeschynanthus moningeriae (Merr.) Chun;海南芒毛苣苔(红花芒毛苣苔);Hainan Basketvine, Red-flowered Basketvine ●

9464 Aeschynanthus novogracilis W. T. Wang = Aeschynanthus gracilis Parins ex C. B. Clarke ●

9465 Aeschynanthus oblanceolatus (Anthony) C. E. C. Fisch. = Aeschynanthus linearifolius C. E. C. Fisch. ●

9466 Aeschynanthus oblongifolius (Roxb.) G. Don = Chirita oblongifolia (Roxb.) J. Sinclair ■

9467 Aeschynanthus oblongifolius G. Don = Chirita oblongifolia (Roxb.) J. Sinclair ■

9468 Aeschynanthus pachytrichus W. T. Wang;粗毛芒毛苣苔;Hairy Basketvine, Shaggy Basketvine ●

9469 Aeschynanthus parviflora Spreng.;小花芒毛苣苔●☆

9470 Aeschynanthus parvifolia R. Br.;小叶芒毛苣苔●☆

9471 Aeschynanthus peelii Hook. f. et Thomson = Aeschynanthus bracteatus Wall. ex A. DC. ●

9472 Aeschynanthus peelii Hook. f. et Thomson var. oblanceolatus Anthony = Aeschynanthus linearifolius C. E. C. Fisch. ●

9473 Aeschynanthus planipetiolatus H. W. Li;扁柄芒毛苣苔;Flatpetiolate Basketvine, Plat-petiole Basketvine, Plat-stalked Basketvine ●

9474 Aeschynanthus poilanei Pellegr.;药用芒毛苣苔;Medicinal Basketvine, Poilane Basketvine ●

9475 Aeschynanthus pulchr (Blume) G. Don;口红花(大红芒毛苣苔,花蔓草);Lipstick Plant, Red-bugle Vine, Royal Red Bugler, Royal-red Bugler, Scarlet Basketvine ●

9476 Aeschynanthus radicans Jack;毛萼口红花(芒毛苣苔);Lipstick Plant ●☆

9477 Aeschynanthus ramosissima Wall. = Aeschynanthus parviflora Spreng. ●☆

9478 Aeschynanthus sinolongicalyx W. T. Wang;中国长萼芒毛苣苔(长萼芒毛苣苔);Long-calyx Basketvine, Long-calyxed Basketvine ●

9479 Aeschynanthus speciosus Hook.;翠锦口红花●

9480 Aeschynanthus splendens Lindl. et Paxton = Aeschynanthus speciosus Hook. ●

9481 Aeschynanthus stenosepalus Anthony;尾叶芒毛苣苔;Narrow-sepal Basketvine, Narrow-sepaled Basketvine, Tailleaf Basketvine ●

9482 Aeschynanthus superbus C. B. Clarke;华丽芒毛苣苔;Magnificent Basketvine ●

9483 Aeschynanthus tengchungensis W. T. Wang;腾冲芒毛苣苔;Tengchong Basketvine ●

9484 Aeschynanthus tenuis Hand. -Mazz. = Aeschynanthus stenosepalus Anthony ●

9485 Aeschynanthus tubulosus Anthony;筒花芒毛苣苔;Tube-flower Basketvine, Tuberous Basketvine ●

9486 Aeschynanthus tubulosus Anthony var. angustilobus Anthony;狭萼片芒毛苣苔;Narrowsepal Tube-flower Basketvine ●

9487 Aeschynanthus wardii Merr.;狭花芒毛苣苔;Narrowflower Basketvine, Ward Basketvine ●

9488 Aeschynanthus zebrinus Lem. = Aeschynanthus marmoratus F. W. Moore ●☆

9489 Aeschynomene L.(1753);合萌属(田皂荚属,田皂角属);Aeschynomene, Consprout, Joint Vetch, Joint-vetch, Pith Plant ●■

9490 Aeschynomene abyssinica (A. Rich.) Vatke;阿比西尼亚田皂角●☆

9491 Aeschynomene aculeata Schreb. = Sesbania aculeata Pers. ■

9492 Aeschynomene aculeata Schreb. = Sesbania bispinosa (Jacq.) W. Wight ■

9493 Aeschynomene acutangula Welw. ex Baker;尖角田皂角●☆

9494 Aeschynomene aegyptiaca (Pers.) Steud. = Sesbania sesban (L.) Merr. ●

9495 Aeschynomene americana L.;美洲合萌(美洲田皂角,敏感合萌);American Aeschynomene, American Jointvetch ●

9496 Aeschynomene angolensis Rossberg;安哥拉田皂角●☆

9497 Aeschynomene aphylla Wild;无叶田皂角●☆

9498 Aeschynomene arbuscula Baker f. = Aeschynomene fulgida Welw. ex Baker ■☆

9499 Aeschynomene aspera L.;粗毛合萌(粗田皂角);Shola Pith, Sola Pith ●☆

9500 Aeschynomene batekensis Troch. et Koechlin;巴泰凯田皂角■☆

9501 Aeschynomene baumii Harms;鲍姆田皂角■☆

9502 Aeschynomene baumii Harms var. kassneri (Harms) Verdc.;卡斯纳田皂角■☆

9503 Aeschynomene bella Harms;雅致田皂角●☆

9504 Aeschynomene benguellensis Torre;本格拉田皂角■☆

9505 Aeschynomene bequaertii De Wild. = Aeschynomene dimidiata Welw. ex Baker subsp. bequaertii (De Wild.) J. Léonard ■☆

9506 Aeschynomene bispinosa Jacq. = Sesbania bispinosa (Jacq.) W. Wight ■

9507 Aeschynomene bracteosa Baker;苞片田皂角■☆

9508 Aeschynomene bracteosa Baker var. delicatula (Baker f.) Verdc.;姣美田皂角■☆

9509 Aeschynomene bracteosa Baker var. major Verdc.;大苞片田皂角■☆

9510 Aeschynomene brevifolia L. ex Poir.;短叶田皂角■☆

9511 Aeschynomene brevifolia Poir. = Aeschynomene brevifolia L. ex Poir. ■☆

9512 Aeschynomene bullockii J. Léonard;布洛克田皂角●☆

9513 Aeschynomene bullockii J. Léonard var. volubilis = Aeschynomene bullockii J. Léonard ●☆

9514 Aeschynomene burttii Baker f.;伯特田皂角●☆

9515 Aeschynomene butayei De Wild. = Aeschynomene baumii Harms ■☆

9516 Aeschynomene campicola Taub. = Aeschynomene baumii Harms ■☆

9517 Aeschynomene cannabina Retz. = Sesbania cannabina (Retz.) Poir. ■
9518 Aeschynomene chimanimaniensis Verdc.;奇马尼马尼田皂角■☆
9519 Aeschynomene claessensii De Wild. = Aeschynomene abyssinica (A. Rich.) Vatke ●☆
9520 Aeschynomene crassicaulis Harms;粗茎毛田皂角■☆
9521 Aeschynomene cristata Vatke;冠毛田皂角■☆
9522 Aeschynomene cristata Vatke var. pubescens J. Léonard;短冠毛田皂角■☆
9523 Aeschynomene debilis Welw. ex Baker;弱小毛田皂角■☆
9524 Aeschynomene deightonii Hepper;戴顿田皂角■☆
9525 Aeschynomene delicatula Baker f. = Aeschynomene bracteosa Baker var. delicatula (Baker f.) Verdc. ■☆
9526 Aeschynomene dewevrei De Wild. et T. Durand = Aeschynomene cristata Vatke ■☆
9527 Aeschynomene dimidiata Welw. ex Baker;对开田皂角■☆
9528 Aeschynomene dimidiata Welw. ex Baker subsp. bequaertii (De Wild.) J. Léonard;贝卡尔田皂角■☆
9529 Aeschynomene dissitiflora Baker;离花田皂角■☆
9530 Aeschynomene djalonensis A. Chev. = Aeschynomene deightonii Hepper ■☆
9531 Aeschynomene elaphroxylon (Guillaumin et Perr.) Taub.;轻田皂角;Ambush,Ambutch ■☆
9532 Aeschynomene elisabethvilleana De Wild. = Aeschynomene bracteosa Baker ■☆
9533 Aeschynomene elongata Salisb. = Sesbania sesban (L.) Merr. ●
9534 Aeschynomene erubescens E. Mey. = Smithia erubescens (E. Mey.) Baker f. ●☆
9535 Aeschynomene evenia Wright;灌木田皂角;Shrubby Jointvetch ●☆
9536 Aeschynomene falcata (Poir.) DC.;镰形田皂角●☆
9537 Aeschynomene filipes Baill.;丝梗田皂角●☆
9538 Aeschynomene fluitans Peter;漂浮田皂角●☆
9539 Aeschynomene fulgida Welw. ex Baker;光亮田皂角■☆
9540 Aeschynomene gazensis Baker f.;加兹田皂角■☆
9541 Aeschynomene gilletii De Wild. = Aeschynomene schimperi Hochst. ex A. Rich. ●☆
9542 Aeschynomene glabrescens Welw. ex Baker;无毛田皂角■☆
9543 Aeschynomene glabrescens Welw. ex Baker var. pubescens J. Léonard;短柔毛田皂角■☆
9544 Aeschynomene glandulosa De Wild. = Aeschynomene baumii Harms ■☆
9545 Aeschynomene glauca R. E. Fr.;灰绿田皂角■☆
9546 Aeschynomene glutinosa Schinz = Aeschynomene schinzii Suess. ●☆
9547 Aeschynomene glutinosa Taub. = Aeschynomene abyssinica (A. Rich.) Vatke ●☆
9548 Aeschynomene goetzei Harms = Aeschynomene trigonocarpa Taub. ex Baker f. ●☆
9549 Aeschynomene gracilipes Taub.;细梗田皂角●☆
9550 Aeschynomene gracilipes Taub. var. brevistipitata Verdc.;短柄田皂角■☆
9551 Aeschynomene grandiflora L. = Sesbania grandiflora (L.) Pers. ●
9552 Aeschynomene grandistipulata Harms;大托叶田皂角■☆
9553 Aeschynomene harmsiana De Wild. = Aeschynomene baumii Harms ■☆
9554 Aeschynomene heurckeana Baker = Aeschynomene dissitiflora Baker ■☆
9555 Aeschynomene heurckeana Baker = Aeschynomene filipes Baill. ●☆
9556 Aeschynomene histrix Poir. var. incana (Vogel) Benth.;灰白毛田皂角;Porcupine Jointvetch ■☆
9557 Aeschynomene hockii De Wild. = Aeschynomene pygmaea Welw. ex Baker var. hebecarpa J. Léonard ■☆
9558 Aeschynomene homblei De Wild. = Aeschynomene pygmaea Welw. ex Baker var. hebecarpa J. Léonard ■☆
9559 Aeschynomene humilis N. E. Br. = Aeschynomene pygmaea Welw. ex Baker var. hebecarpa J. Léonard ■☆
9560 Aeschynomene indica L.;合萌(白梗通,大样夜合草,独木根,梗通草,海柳,禾镰草,禾镰树子,合明草,拉田草,连根拔,木稗,木排豆,气通草,梳子树,水茸角,水通草,水皂角,田皂角,蜈蚣杨柳,野豆箕,野含羞草,野寒豆,野槐树,野兰,野绿豆,野通草,野鸭树草,野皂角,夜关门,自梗通);Common Aeschynomene,Consprout,Indian Jointvetch,Shola Pith ●
9561 Aeschynomene inyangensis Willd.;伊尼扬加田皂角■☆
9562 Aeschynomene kapiriensis De Wild. = Aeschynomene abyssinica (A. Rich.) Vatke ●☆
9563 Aeschynomene kassneri Harms = Aeschynomene baumii Harms var. kassneri (Harms) Verdc. ■☆
9564 Aeschynomene katangensis De Wild.;加丹加田皂角■☆
9565 Aeschynomene katangensis De Wild. subsp. sublignosa (De Wild.) J. Léonard;近木质田皂角■☆
9566 Aeschynomene kerstingii Harms;克斯廷田皂角●☆
9567 Aeschynomene kilimandscharica Taub. ex Engl. = Aeschynomene abyssinica (A. Rich.) Vatke ●☆
9568 Aeschynomene lateritia Harms = Bakerophyton lateritium (Harms) Hutch. ex Maheshw. ●☆
9569 Aeschynomene lateriticola Verdc.;砖红田皂角●☆
9570 Aeschynomene laxiflora Bojer ex Baker;疏花田皂角●☆
9571 Aeschynomene leptobotrya Harms ex Baker f. = Aeschynomene rehmannii Schinz var. leptobotrya (Harms ex Baker f.) J. B. Gillett ●☆
9572 Aeschynomene leptophylla Harms;细叶田皂角●☆
9573 Aeschynomene leptophylla Harms subsp. magnifoliolata J. Léonard;大小叶田皂角●☆
9574 Aeschynomene leptophylla Harms var. crassituberculata Verdc.;粗疣田皂角●☆
9575 Aeschynomene mazangayana Baill. = Aeschynomene cristata Vatke ■☆
9576 Aeschynomene mearnsii De Wild. = Aeschynomene schimperi Hochst. ex A. Rich. ●☆
9577 Aeschynomene mediocris Verdc.;中位田皂角●☆
9578 Aeschynomene megalophylla Harms;大叶田皂角●☆
9579 Aeschynomene micrantha (Poir.) DC. = Aeschynomene brevifolia L. ex Poir. ■☆
9580 Aeschynomene micrantha DC.;小花田皂角■☆
9581 Aeschynomene mimosifolia Vatke;多花田皂角●☆
9582 Aeschynomene minutiflora Taub. = Aeschynomene mimosifolia Vatke ●☆
9583 Aeschynomene minutiflora Taub. subsp. grandiflora Verdc.;大花田皂角●☆
9584 Aeschynomene morumbensis Baker f. = Aeschynomene nilotica Taub. ■☆
9585 Aeschynomene mossambicensis Verdc.;莫桑比克田皂角●☆
9586 Aeschynomene mossambicensis Verdc. subsp. longestipitata (Verdc.) Vollesen;长柄莫桑比克田皂角●☆
9587 Aeschynomene mossambicensis Verdc. var. longestipitata (Verdc.) Vollesen = Aeschynomene mossambicensis Verdc. subsp.

longestipitata (Verdc.) Vollesen ●☆
9588 Aeschynomene mossoensis J. Léonard;莫斯田皂角■☆
9589 Aeschynomene mossoensis J. Léonard var. parvifolia Verdc.;小叶莫桑比克田皂角●☆
9590 Aeschynomene mossoensis J. Léonard var. pubescens ?;短柔毛莫斯田皂角●☆
9591 Aeschynomene mukuluensis De Wild. = Aeschynomene schimperi Hochst. ex A. Rich. ●☆
9592 Aeschynomene multicaulis Harms;多茎田皂角●☆
9593 Aeschynomene nambalensis Harms = Aeschynomene bracteosa Baker ■☆
9594 Aeschynomene neglecta Hepper = Bakerophyton neglectum (Hepper) Maheshw. ●☆
9595 Aeschynomene nematopoda Harms;线梗田皂角■☆
9596 Aeschynomene newtonii Schinz = Aeschynomene tenuirama Baker ●☆
9597 Aeschynomene nilotica Taub.;尼罗河田皂角■☆
9598 Aeschynomene nodulosa (Baker) Baker f.;多节田皂角■☆
9599 Aeschynomene nodulosa (Baker) Baker f. var. glabrescens J. B. Gillett;光滑田皂角●☆
9600 Aeschynomene nyassana Taub.;尼亚萨田皂角●☆
9601 Aeschynomene nyikensis Baker;尼卡田皂角■☆
9602 Aeschynomene nyikensis Baker var. gracilis Suess. = Aeschynomene mimosifolia Vatke ●☆
9603 Aeschynomene nyikensis Baker var. mossambicensis Baker f. = Aeschynomene schliebenii Harms var. mossambicensis (Baker f.) Verdc. ●☆
9604 Aeschynomene obovalis Baill. = Aeschynomene brevifolia L. ex Poir. ■☆
9605 Aeschynomene oligantha Welw. ex Baker = Aeschynomene indica L. ●
9606 Aeschynomene oligophylla Harms;寡叶田皂角■☆
9607 Aeschynomene paludicola Harms = Aeschynomene schimperi Hochst. ex A. Rich. ●☆
9608 Aeschynomene paludosa Roxb. = Sesbania javanica Miq. ■
9609 Aeschynomene paniculata Willd. ex Vogel;圆锥田皂角;Pannicle Jointvetch ●☆
9610 Aeschynomene papulosa Welw. ex Baker = Aeschynomene uniflora E. Mey. ●☆
9611 Aeschynomene pararubrofarinacea J. Léonard;假红粉田皂角■☆
9612 Aeschynomene patula Poir.;伸展田皂角■☆
9613 Aeschynomene pawekiae Verdc.;帕维基田皂角■☆
9614 Aeschynomene praticola Baker f. = Aeschynomene baumii Harms ■☆
9615 Aeschynomene pseudoglabrescens Verdc.;假光滑田皂角■☆
9616 Aeschynomene pulchella Planch. ex Baker = Bakerophyton pulchellum (Planch. ex Baker) Maheshw. ●☆
9617 Aeschynomene pulchra Vatke = Rhynchosia pulchra (Vatke) Harms ■☆
9618 Aeschynomene pygmaea Welw. ex Baker;矮小田皂角■☆
9619 Aeschynomene pygmaea Welw. ex Baker var. hebecarpa J. Léonard;柔毛矮小田皂角■☆
9620 Aeschynomene racemosa De Wild. = Aeschynomene katangensis De Wild. subsp. sublignosa (De Wild.) J. Léonard ■☆
9621 Aeschynomene recta N. E. Br. = Aeschynomene pygmaea Welw. ex Baker var. hebecarpa J. Léonard ■☆
9622 Aeschynomene rehmannii Schinz;拉赫曼田皂角●☆
9623 Aeschynomene rehmannii Schinz var. leptobotrya (Harms ex Baker f.) J. B. Gillett;细穗拉赫曼田皂角●☆
9624 Aeschynomene remota Poir. = Desmodium repandum (Vahl) DC. ●☆
9625 Aeschynomene rhodesiaca Harms;罗得西亚田皂角●☆
9626 Aeschynomene rogersii N. E. Br. = Aeschynomene katangensis De Wild. subsp. sublignosa (De Wild.) J. Léonard ■☆
9627 Aeschynomene rubrofarinacea (Taub.) F. White;红粉田皂角●☆
9628 Aeschynomene rubroviolacea J. Léonard;红堇色田皂角●☆
9629 Aeschynomene rudis Benth.;关节田皂角;Joint Vetch ●☆
9630 Aeschynomene rueppelii Baker = Aeschynomene abyssinica (A. Rich.) Vatke ●☆
9631 Aeschynomene ruspoliana Taub. ex Harms;鲁斯波利田皂角●☆
9632 Aeschynomene sansibarica Taub.;桑给巴尔田皂角●☆
9633 Aeschynomene saxicola Taub. = Bakerophyton pulchellum (Planch. ex Baker) Maheshw. ●☆
9634 Aeschynomene schimperi A. Rich. = Aeschynomene schimperi Hochst. ex A. Rich. ●☆
9635 Aeschynomene schimperi Hochst. ex A. Rich.;欣珀田皂角●☆
9636 Aeschynomene schinzii Suess.;欣兹田皂角●☆
9637 Aeschynomene schlechteri Harms ex Baker f. = Aeschynomene fluitans Peter ●☆
9638 Aeschynomene schliebenii Harms;施利本田皂角●☆
9639 Aeschynomene schliebenii Harms var. mossambicensis (Baker f.) Verdc.;莫萨田皂角●☆
9640 Aeschynomene semilunaris Hutch.;新月田皂角●☆
9641 Aeschynomene sensitiva Sw.;敏感合萌;Sensitive Jointvetch ●☆
9642 Aeschynomene sesban L. = Sesbania sesban (L.) Merr. ●
9643 Aeschynomene shirensis Taub. = Aeschynomene nodulosa (Baker) Baker f. ■☆
9644 Aeschynomene sparsiflora Baker;散花田皂角●☆
9645 Aeschynomene stellaris (Afzel. ex Baker) Roberty = Cyclocarpa stellaris Afzel. ex Baker ■☆
9646 Aeschynomene stipitata Burtt Davy;具柄田皂角●☆
9647 Aeschynomene stipulosa Verdc.;托叶田皂角●☆
9648 Aeschynomene stolzii Harms;斯托尔兹田皂角●☆
9649 Aeschynomene striata De Wild. = Aeschynomene rehmannii Schinz var. leptobotrya (Harms ex Baker f.) J. B. Gillett ●☆
9650 Aeschynomene subaphylla De Wild. = Aeschynomene tenuirama Baker ●☆
9651 Aeschynomene sublignosa De Wild. = Aeschynomene katangensis De Wild. subsp. sublignosa (De Wild.) J. Léonard ■☆
9652 Aeschynomene tambacoundensis Berhaut;坦巴昆达田皂角●☆
9653 Aeschynomene telekii Schweinf. = Aeschynomene schimperi Hochst. ex A. Rich. ●☆
9654 Aeschynomene tenuirama Baker;细枝田皂角●☆
9655 Aeschynomene tenuirama Baker var. hebecarpa Verdc.;柔毛细枝田皂角●☆
9656 Aeschynomene tenuirama Baker var. parviflora Torre;小花细枝田皂角●☆
9657 Aeschynomene trigonocarpa Taub. ex Baker f.;三棱果田皂角●☆
9658 Aeschynomene tsaratanensis Du Puy et Labat;察拉塔纳田皂角●☆
9659 Aeschynomene uniflora E. Mey.;单花田皂角●☆
9660 Aeschynomene uniflora E. Mey. var. grandiflora Verdc.;大单花田皂角●☆
9661 Aeschynomene upembensis J. Léonard;乌彭贝田皂角●☆
9662 Aeschynomene venulosa Verdc.;细脉田皂角●☆
9663 Aeschynomene venulosa Verdc. var. grandis ?;大田皂角●☆
9664 Aeschynomene virgata Cav. = Sesbania virgata (Cav.) Pers. ■☆

9665 Aeschynomene virginica (L.) Britton, Sterns et Poggenb.;弗吉尼亚田皂角●☆
9666 Aeschynomene walteri Harms = Aeschynomene mimosifolia Vatke ●☆
9667 Aeschynomene wittei Baker f. = Aeschynomene multicaulis Harms ●☆
9668 Aeschynomene youngii Baker f. = Aeschynomene pygmaea Welw. ex Baker var. hebecarpa J. Léonard ■☆
9669 Aeschynomene zigzag De Wild. = Aeschynomene bracteosa Baker ■☆
9670 Aesculaceae Bercht. et J. Presl = Hippocastanaceae A. Rich.(保留科名)●
9671 Aesculaceae Burnett = Hippocastanaceae A. Rich.(保留科名)●
9672 Aesculaceae Burnett = Sapindaceae Juss.(保留科名)●■
9673 Aesculaceae Lindl. = Hippocastanaceae A. Rich.(保留科名)●
9674 Aesculaceae Lindl. = Sapindaceae Juss.(保留科名)●■
9675 Aesculus L. (1753);七叶树属;Buck Eye, Buckeye, Horse Chestnut, Horsechestnut, Horse-chestnut ●
9676 Aesculus 'Autumn Splendor';彩秋七叶树;Autumn Splendor Buckeye ●☆
9677 Aesculus arguta Buckley;得克萨斯七叶树;Texas Buckeye, White Buckeye ●☆
9678 Aesculus assamica Griff.;长柄七叶树(滇缅七叶树);Assam Buckeye, Assam Horse-chestnut ●
9679 Aesculus californica (Spach) Nutt.;加州七叶树;American Horse Chestnut, California Buckeye, Californian Buckeye, Horse Chestnut, Horsechestnut ●
9680 Aesculus carnea S. Watson;红七叶树(红马栗,肉红七叶树);Flesh-coloured Buckeye, Red Horse Chestnut, Red Horsechestnut, Red Horse-chestnut ●
9681 Aesculus carnea S. Watson 'Briotii';布里奥特肉红七叶树●☆
9682 Aesculus chekiangensis Hu et W. P. Fang = Aesculus chinensis Bunge var. chekiangensis (Hu et W. P. Fang) W. P. Fang ●
9683 Aesculus chinensis Bunge;七叶树(草椤,狗板栗,开心果,七叶枫树,七叶莲,莎婆子,苏罗子,娑罗树,娑罗子,桫椤,桫椤树,梭椤子,索罗果,天师栗,武吉);China Buckeye, Chinese Buckeye, Chinese Horse Chestnut, Chinese Horsechestnut, Chinese Horse-chestnut ●
9684 Aesculus chinensis Bunge var. chekiangensis (Hu et W. P. Fang) W. P. Fang;浙江七叶树(开心果,莎婆子,苏罗子,梭椤子,索罗果,武吉);Zhejiang Buckeye ●
9685 Aesculus chinensis Bunge var. wilsonii (Rehder) Turland et N. H. Xia = Aesculus wilsonii Rehder ●
9686 Aesculus chinensis C. K. Schneid. = Aesculus turbinata Blume ●
9687 Aesculus chinpinensis W. P. Fang = Brassaiopsis glomerulata (Blume) Regel ●
9688 Aesculus chuniana Hu et W. P. Fang;大果七叶树(焕镛七叶树);Big-fruited Buckeye, Chun Buckeye, Chun Horse-chestnut ●
9689 Aesculus coreiacifolia W. P. Fang = Aesculus assamica Griff. ●
9690 Aesculus dissimilis Blume;虾夷七叶树●☆
9691 Aesculus dissimilis Blume = Aesculus turbinata Blume ●
9692 Aesculus flava Aiton;北美黄花七叶树(黄花七叶树,黄七叶木,七叶木);Big Buckeye, Buckeye, Sweet Buckeye, Sweet Horsechestnut, Yellow Buckeye ●☆
9693 Aesculus georgiana Sarg.;林生七叶树;Adarf Buckeye, Georgia Buckeye, Painted Buckeye ●☆
9694 Aesculus glabra Willd.;光叶七叶树(北美马栗);American Buckeye, Fetid Buckeye, Ohio Buckeye, Ohio Suckeye, Stinking Buckeye ●
9695 Aesculus glabra Willd. f. pallida (Willd.) Fernald = Aesculus glabra Willd. ●
9696 Aesculus glabra Willd. var. leucodermis Sarg. = Aesculus glabra Willd. ●
9697 Aesculus glabra Willd. var. micrantha Sarg. = Aesculus glabra Willd. ●
9698 Aesculus glabra Willd. var. monticola Sarg. = Aesculus glabra Willd. ●
9699 Aesculus glabra Willd. var. pallida (Willd.) Kirchn. = Aesculus glabra Willd. ●
9700 Aesculus glabra Willd. var. sargentii Rehder = Aesculus glabra Willd. ●
9701 Aesculus hippocastanum L.;欧洲七叶树(马栗,欧马栗);Bird Tree, Bongay, Buckeye, Buckeye Horse Chestnut, Bull's Eyes, Bur, Candles, Chestnut, Christmas Candles, Christmas Tree, Common Horse Chestnut, Common Horsechestnut, Common Horse-chestnut, Conker Tree, Conkers, Conks, Conquerors, Double-flowered Horsechestnut, European Buckeye, European Horse Chestnut, European Horse-chestnut, Fish Bones, Hobbly Honker, Hobbly-flower, Hobbly-honkers, Hoblionker, Hoblionkers, Horse Chestnut, Horsechestnut, Horse-chestnut, Horse-nut Tree, Knuckle-bleeders, Lambs, Oblionker, Robber's Lanterns, Roman Candles, Sticky-buds, Tow Tree ●
9702 Aesculus hippocastanum L. 'Baumannii';鲍曼欧洲七叶树●☆
9703 Aesculus hippocastanum L. 'Pyramidalis';金字塔欧洲七叶树●☆
9704 Aesculus hippocastanum L. 'Variegata';斑叶欧洲七叶树;Variegated Horsechestn ●☆
9705 Aesculus indica (Colebr. ex Cambess.) Hook. 'Syndey Pearce';亮叶印度七叶树●☆
9706 Aesculus indica (Colebr. ex Cambess.) Hook. f.;印度七叶树;Indian Buckeye, Indian Horse Chestnut, Indian Horse-chestnut ●☆
9707 Aesculus indica (Wall. ex Cambess.) Hook. f. = Aesculus indica (Colebr. ex Cambess.) Hook. f. ●☆
9708 Aesculus indica (Wall. ex Cambess.) Hook. f. var. concolor Browicz = Aesculus indica (Colebr. ex Cambess.) Hook. f. ●☆
9709 Aesculus japonica C. K. Schneid. = Aesculus turbinata Blume ●
9710 Aesculus khassyana C. R. Das et Majumdar;滇缅七叶树●
9711 Aesculus kwangsiensis W. P. Fang = Schefflera octophylla (Lour.) Harms ●
9712 Aesculus lantsangensis Hu et W. P. Fang;澜沧七叶树;Lancang Buckeye ●
9713 Aesculus lutea Wangenh. = Aesculus flava Aiton ●☆
9714 Aesculus macrostachya Michx. = Aesculus parviflora Walter ●☆
9715 Aesculus megaphylla Hu et W. P. Fang;大叶七叶树;Big-leaved Buckeye, Largeleaf Buckeye, Largeleaf Horsechestnut ●
9716 Aesculus neglecta Lindl.;彩花七叶树;Painted Buckeye ●☆
9717 Aesculus octandra Marshall = Aesculus flava Aiton ●☆
9718 Aesculus pallida Willd. = Aesculus glabra Willd. ●
9719 Aesculus parviflora Walter;小花七叶树;Bottlebrush Buckeye, Bottle-brush Buckeye, Buckeye, Buckeye Bottlebrush, Dwarf Buckeye, Dwarf Horsechestnut, Dwarf Horse-chestnut, White Buckeye ●☆
9720 Aesculus pavia L.;红花七叶树(北美红花七叶树,小七叶树);Firecracker Plant, Firecracker-plant, Red Buckeye, Red Horsechestnut, Red Horse-chestnut, Scarlet Buckeye ●
9721 Aesculus pavia L. 'Atrosanguinea';暗红小七叶树●☆
9722 Aesculus pavia L. 'Humilis';矮红花七叶树;Dwarf Red Buckeye ●☆

9723 Aesculus pavia L. = Aesculus turbinata Blume ●
9724 Aesculus polyneura Hu et W. P. Fang;多脉七叶树;Manyvein Buckeye,Multinervous Buckeye ●
9725 Aesculus polyneura Hu et W. P. Fang var. dongchuanensis X. W. Li et W. Y. Yin;东川七叶树;Dongchuan Manyvein Buckeye ●
9726 Aesculus rubicunda Lodd.;稍红七叶树;Red Horse-chestnut ●☆
9727 Aesculus rupicola Hu et W. P. Fang = Aesculus wangii Hu var. rupicola (Hu et W. P. Fang) W. P. Fang ●
9728 Aesculus sinensis Bean = Aesculus turbinata Blume ●
9729 Aesculus splendens Sarg.;美丽七叶树;Flame Buckeye ●☆
9730 Aesculus sylvatica Bartram = Aesculus georgiana Sarg. ●☆
9731 Aesculus tsiangii Hu et W. P. Fang;小果七叶树(菊川七叶树);Tsiang Buckeye ●
9732 Aesculus turbinata Blume;日本七叶树;Japan Buckeye,Japan Horsechestnut,Japanese Buckeye,Japanese Horse Chestnut,Japanese Horsechestnut,Japanese Horse-chestnut ●
9733 Aesculus turbinata Blume f. pubescens (Rehder) Ohwi ex Y. Endo;毛日本七叶树●☆
9734 Aesculus turbinata Blume var. pubescens Rehder = Aesculus turbinata Blume f. pubescens (Rehder) Ohwi ex Y. Endo ●☆
9735 Aesculus turbinata Blume var. pubescens Rehder = Aesculus turbinata Blume ●
9736 Aesculus wangii Hu;云南七叶树(水茄子,阴阳果);Wang Buckeye,Yunnan Buckeye ●◇
9737 Aesculus wangii Hu var. rupicola (Hu et W. P. Fang) W. P. Fang;石生七叶树;Rockliving Buckeye,Rocky Wang Buckeye ●
9738 Aesculus wilsonii Rehder;天师栗(猴板栗,开心果,莎婆子,苏罗子,娑罗果,娑罗子,梭罗树,梭椤子,索罗果,武吉);E. H. Wilson Buckeye,Wilson Buckeye,E. H. Wilson Horsechestnut ●
9739 Aetanthus (Eichler) Engl. (1889);鹰花寄生属●☆
9740 Aetanthus Engl. = Aetanthus (Eichler) Engl. ●☆
9741 Aetanthus cauliflorus Ule;茎花鹰花寄生●☆
9742 Aetanthus engelsii Engl.;恩氏鹰花寄生●☆
9743 Aetanthus holtonii Engl.;豪顿鹰花寄生●☆
9744 Aetanthus ovalis Rusby;卵鹰花寄生●☆
9745 Aethales Post et Kuntze = Sedum L. ●■
9746 Aetheilema R. Br. = Phaulopsis Willd. (保留属名) ■
9747 Aetheilema glutinosum Steud. = Phaulopsis talbotii S. Moore ■☆
9748 Aetheilema imbricatum (Forssk.) Spreng. = Phaulopsis imbricata (Forssk.) Sweet ■
9749 Aetheilema longifolium (Sims) Spreng. = Phaulopsis barteri T. Anderson ■☆
9750 Aetheilema micranthum Benth. = Phaulopsis micrantha (Benth.) C. B. Clarke ■☆
9751 Aetheilema reniforme Nees = Phaulopsis oppositifolius (J. C. Wendl.) Lindau ■
9752 Aetheilema reniforme Nees var. hispidosa ? = Phaulopsis imbricata (Forssk.) Sweet ■
9753 Aetheocephalus Gagnep. = Athroisma DC. ●■☆
9754 Aetheochlaena Post et Kuntze = Aetheolaena Cass. ●☆
9755 Aetheolaena Cass. (1827);柄叶绵头菊属●☆
9756 Aetheolaena Cass. = Lasiocephalus Schltdl. ■☆
9757 Aetheolaena Cass. = Senecio L. ●■
9758 Aetheolaena heterophylla (Turcz.) B. Nord.;互叶柄叶绵头菊●☆
9759 Aetheolaena hypoleuca (Turcz.) B. Nord.;里白柄叶绵头菊●☆
9760 Aetheolirion Forman (1962);线果吉祥草属(泰国鸭跖草属) ■☆
9761 Aetheolirion stenolobium Forman;线果吉祥草■☆
9762 Aetheonema Bubani et Penzig = Aethionema R. Br. ■☆
9763 Aetheonema R. Br. = Iberis L. ●■
9764 Aetheonema Rchb. = Gaertnera Lam. ●
9765 Aetheopappus Cass. (1827);亮毛菊属●☆
9766 Aetheopappus Cass. = Centaurea L. (保留属名) ●■
9767 Aetheopappus Cass. = Psephellus Cass. ●■☆
9768 Aetheopappus caucasicus Sosn.;高加索亮毛菊●☆
9769 Aetheopappus pulcherrimus (Willd.) Cass.;亮毛菊●☆
9770 Aetheopappus vvedenskii (Sosn.) Sosn.;韦坚斯基亮毛菊●☆
9771 Aetheorhiza Cass. (1827);梭果苣属;Tuberous Hawk's-beard ■☆
9772 Aetheorhiza Cass. = Crepis L. ■
9773 Aetheorhiza bulbosa (L.) Cass.;梭果苣;Tuberous Hawk's-beard ■☆
9774 Aetheorhiza bulbosa (L.) Cass. = Sonchus bulbosus (L.) N. Kilian et Greuter ■☆
9775 Aetheorhyncha Dressler = Chondrorhyncha (Rchb. f.) Garay ■☆
9776 Aetheorhyncha Dressler = Stenia Lindl. ■☆
9777 Aetheorhyncha Dressler (2005);亮喙兰属■☆
9778 Aetheorrhiza Rchb. = Aetheorhiza Cass. ■
9779 Aethephyllum N. E. Br. (1928);雅琴花属■☆
9780 Aethephyllum pinnatifidum (L. f.) N. E. Br.;雅琴花;Jagged-leaved Fig Marigold ■☆
9781 Aetheria Blume ex Endl. = Stenorrhynchos Rich. ex Spreng. ■☆
9782 Aetheria Endl. = Hetaeria Blume (保留属名) ■
9783 Aethiocarpa Vollesen (1986);亮果梧桐属●☆
9784 Aethiocarpa lepidota Vollesen = Harmsia lepidota (Vollesen) M. Jenny ●☆
9785 Aethionema R. Br. (1812);岩芥菜属(赤线属,小蜂室花属);Candy Mustard,Stone Cress,Stone-Cress ■☆
9786 Aethionema W. T. Aiton = Aethionema R. Br. ■☆
9787 Aethionema arabicum (L.) Andrz. ex DC.;阿拉伯岩芥菜■☆
9788 Aethionema armenum Boiss.;亚美尼亚岩芥菜■☆
9789 Aethionema caespitosum Boiss.;丛生岩芥菜■☆
9790 Aethionema cardiophyllum Boiss. et Heldr.;心叶岩芥菜■☆
9791 Aethionema carneum (Banks et Sol.) B. Fedtsch.;肉色岩芥菜■☆
9792 Aethionema carneum (Sol.) B. Fedtsch. = Aethionema carneum (Banks et Sol.) B. Fedtsch. ■☆
9793 Aethionema cordatum (Desf.) Boiss.;心形岩芥菜■☆
9794 Aethionema cristatum (Desf.) Boiss. = Aethionema carneum (Banks et Sol.) B. Fedtsch. ■☆
9795 Aethionema diastrophis Bunge;卷曲岩芥菜■☆
9796 Aethionema edentulum N. Busch;无齿岩芥菜■☆
9797 Aethionema elongatum Boiss.;长岩芥菜■☆
9798 Aethionema grandiflorum Boiss. et Hohen.;大花岩芥菜(岩芥菜);Persian Stone Cress,Persian Stone-cress ●■☆
9799 Aethionema heterocarpum Gay;异果岩芥菜■☆
9800 Aethionema heterophyllum Boiss.;异叶岩芥菜■☆
9801 Aethionema iberideum Boiss.;伊伯利亚岩芥菜■☆
9802 Aethionema levandovskyi N. Busch;赖氏岩芥菜■☆
9803 Aethionema lipskyi N. Busch;利普斯基岩芥菜■☆
9804 Aethionema marginatum (Lapeyr.) Montemurro;具边岩芥菜■☆
9805 Aethionema marginatum (Lapeyr.) Montemurro subsp. latifolium (H. Lindb.) Dobignard;宽叶具边岩芥菜■☆
9806 Aethionema membranaceum DC.;膜质岩芥菜■☆
9807 Aethionema pulchellum Boiss. et Huet;美丽岩芥菜■☆
9808 Aethionema pulchellum Boiss. et Huet = Aethionema grandiflorum Boiss. et Hohen. ●■☆

9809 Aethionema salmasium Boiss.;萨尔曼岩芥菜■☆

9810 Aethionema saxatile (L.) R. Br.;非洲岩芥菜■☆

9811 Aethionema saxatile (L.) R. Br. subsp. latifolium H. Lindb. = Aethionema marginatum (Lapeyr.) Montemurro subsp. latifolium (H. Lindb.) Dobignard ■☆

9812 Aethionema saxatile (L.) R. Br. var. anodontonema Maire = Aethionema saxatile (L.) R. Br. ■☆

9813 Aethionema saxatile (L.) R. Br. var. odontonema Maire = Aethionema saxatile (L.) R. Br. ■☆

9814 Aethionema saxatile (L.) R. Br. var. thomasianum (J. Gay) Thell. = Aethionema thomasianum J. Gay ■☆

9815 Aethionema saxatile R. Br. = Aurinia saxatilis (L.) Desv. ■

9816 Aethionema schistosum Boiss. et Kotschy;开裂岩芥菜;Stonecress ■☆

9817 Aethionema spinosum (Boiss.) N. Busch;具刺岩芥菜■☆

9818 Aethionema szovitsii Boiss.;绍氏岩芥菜■☆

9819 Aethionema thomasianum J. Gay;托马斯岩芥菜■☆

9820 Aethionema transhyrcanum (Czerniak.) N. Busch;外吉尔康岩芥菜■☆

9821 Aethionema trinervium (DC.) Boiss.;三脉岩芥菜■☆

9822 Aethionema voronovii Schischk.;沃氏岩芥菜■☆

9823 Aethionema warleyense C. K. Schneid. ex Boom;沃利岩芥菜;Stonecress ■☆

9824 Aethiopis (Benth.) Opiz = Salvia L. ●■

9825 Aethiopis Fourr. = Salvia L. ●■

9826 Aethiopsis Engl. = Aethiopis (Benth.) Opiz ●■

9827 Aethonia D. Don = Tolpis Adans. ●■☆

9828 Aethonopogon Hack. ex Kuntze = Polytrias Hack. ■

9829 Aethonopogon Kuntze = Polytrias Hack. ■

9830 Aethulla A. Gray = Ethulia L. f. ■

9831 Aethusa L. (1753);欧洲毒芹属(拟芫荽属,欧毒芹属,欧芹属);Aethusa,Fool's Parsley ■☆

9832 Aethusa cynapium L.;欧洲毒芹(毒欧芹);Ass Parsley, Ass-parsley,Cow Parsley, Devil's Wand, Dog Poison, Dog's Parsley, False Parsley,Fool's Cicely,Fool's Parsley,Fool's-parsley,Fool's-parsley Aethusa, Gypsy Flower, Kelk, Kicks, Lace Curtains, Lady's Lace,Lesser Hemlock,Pig Dock,Smaller Hemlock ■☆

9833 Aethusa leptophylla (Pers.) Spreng. = Cyclospermum leptophyllum (Pers.) Sprague ex Britton et P. Wilson ■

9834 Aethusa leptophylla Spreng. = Aethusa leptophylla (Pers.) Spreng. ■

9835 Aethusa leptophylla Spreng. = Apium leptophyllum (Pers.) F. Muell. ex Benth. ■

9836 Aethusa leptophylla Spreng. = Cyclospermum leptophyllum (Pers.) Sprague ■

9837 Aethyopys (Benth.) Opiz = Salvia L. ●■

9838 Aetia Adans. = Combretum Loefl. (保留属名)●

9839 Aetoxicon Endl. = Aextoxicon Ruiz et Pav. ●☆

9840 Aetoxylon (Airy Shaw) Airy Shaw(1950);鹰瑞香属●☆

9841 Aetoxylon sympetalum (Steenis et Domke) Airy Shaw;鹰瑞香●☆

9842 Aextoxicaceae Engl. et Gilg(1920)(保留科名);毒羊树科●☆

9843 Aextoxicon Ruiz et Pav. (1794);毒羊树属(智利大戟属,毒鹰木属,鳞枝树属,毒戟属)●☆

9844 Aextoxicon punctatum Ruiz et Pav.;毒羊树(毒戟木);Olivillo ●☆

9845 Aextoxicum Post et Kuntze = Aextoxicon Ruiz et Pav. ●☆

9846 Afarca Raf. = Sageretia Brongn. ●

9847 Affonsea A. St. -Hil. (1833);巴西豆属●■☆

9848 Affonsea A. St. -Hil. = Inga Mill. ●■☆

9849 Affonsea juglandifolia A. St. -Hil.;巴西豆■☆

9850 Affonsoa Post et Kuntze = Affonsea A. St. -Hil. ●■☆

9851 Afgekia Craib(1927);泰豆属(猪腰豆属)●

9852 Afgekia filipes (Dunn) R. Geesink;猪腰豆(冲天子,大荚藤,瓦叶藤,细梗大荚藤,细梗惠特木,细梗密束花,小血藤,猪腰耳,猪腰子);Kidney-shaped Whitfordiodendron,Porkkidneybean ●◇

9853 Afgekia filipes (Dunn) R. Geesink = Adinobotrys filipes Dunn ●

9854 Afgekia filipes (Dunn) R. Geesink var. tomentosa (Z. Wei) Y. F. Deng et H. N. Qin;毛叶猪腰豆;Tomentose Kidney-shaped Whitfordiodendron,Tomentose Porkkidneybean ●

9855 Afgekia mahidolae B. L. Burtt et Chermsir.;泰豆●

9856 Aflatunia Vassilcz. (1955);榆叶蔷薇属●☆

9857 Aflatunia Vassilcz. = Louiseania Carrière ●

9858 Aflatunia ulmifolia (Franch.) Vassilcz.;榆叶蔷薇●☆

9859 Aflatunia ulmifolia (Franch.) Vassilcz. = Prunus ulmifolia Franch. ●☆

9860 Afrachneria Sprague = Pentaschistis (Nees) Spach ■☆

9861 Afraegle (Swingle) Engl. (1915);非洲木橘属●☆

9862 Afraegle Engl. = Afraegle (Swingle) Engl. ●☆

9863 Afraegle asso Engl.;喀麦隆木橘●☆

9864 Afraegle gabonensis (Swingle) Engl.;加蓬非洲木橘●☆

9865 Afraegle mildbraedii Engl.;米尔非洲木橘●☆

9866 Afraegle paniculata (Schumach.) Engl.;非洲木橘(西非埃乐果)●☆

9867 Afraegle paniculata Engl. = Afraegle paniculata (Schumach.) Engl. ●☆

9868 Afrafzelia Pierre = Afzelia Sm. (保留属名)●

9869 Afrafzelia quanzensis (Welw.) Pierre = Afzelia quanzensis Welw. ●☆

9870 Aframmi C. Norman(1929);非洲阿米芹属■☆

9871 Aframmi angolense (C. Norman) C. Norman;安哥拉阿米芹■☆

9872 Aframmi longiradiatum (H. Wolff) Cannon;非洲阿米芹■☆

9873 Aframomum K. Schum. (1904);非洲豆蔻属(非砂仁属,非洲砂仁属);Cardamom ■☆

9874 Aframomum albiflorum Lock;白花非洲豆蔻(白花非洲砂仁)■☆

9875 Aframomum alboviolaceum (Ridl.) K. Schum.;浅堇色非洲豆蔻(浅堇色非洲砂仁)■☆

9876 Aframomum alpinum (Gagnep.) K. Schum.;高山非洲豆蔻(高山非洲砂仁)■☆

9877 Aframomum amaniense Loes.;阿马尼非洲豆蔻■☆

9878 Aframomum angustifolium (Sonn.) K. Schum.;狭叶非洲豆蔻(狭叶豆蔻,狭叶非洲砂仁);Madagascar Cardamom ■☆

9879 Aframomum arundinaceum (Oliv. et D. Hanb.) K. Schum.;苇状非洲豆蔻(苇状非洲砂仁)■☆

9880 Aframomum aulacocarpos Pellegr. ex Koechlin;沟果非洲豆蔻(沟果非砂仁)■☆

9881 Aframomum baumannii K. Schum. = Aframomum angustifolium (Sonn.) K. Schum. ■☆

9882 Aframomum biauriculatum K. Schum. = Aframomum alboviolaceum (Ridl.) K. Schum. ■☆

9883 Aframomum candidum Gagnep. = Aframomum alboviolaceum (Ridl.) K. Schum. ■☆

9884 Aframomum cereum (Hook. f.) K. Schum. = Aframomum sceptrum (Oliv. et D. Hanb.) K. Schum. ■☆

9885 Aframomum chlamydanthum Loes. et Mildbr. = Aframomum zambesiacum (Baker) K. Schum. ■☆

9886 Aframomum chrysanthum Lock;金花非洲砂仁■☆
9887 Aframomum citratum (J. Pereira ex Oliv. et D. Hanb.) K. Schum.;柠檬非洲豆蔻(柠檬非洲砂仁)■☆
9888 Aframomum coraninum ?;非洲砂仁(非砂仁)■☆
9889 Aframomum cordifolium Lock et J. B. Hall;心叶非洲豆蔻(心叶非洲砂仁)■☆
9890 Aframomum corrorima (A. Braun) P. C. M. Jansen;科拉非洲豆蔻(科拉砂仁);Korarima Cardamom ■☆
9891 Aframomum crassilabium (K. Schum.) K. Schum.;厚荚非洲豆蔻(厚荚非洲砂仁)■☆
9892 Aframomum cuspidatum (Gagnep.) K. Schum. = Aframomum exscapum (Sims) Hepper ■☆
9893 Aframomum dalzielii Hutch. = Aframomum leptolepis (K. Schum.) K. Schum. ■☆
9894 Aframomum daniellii (Hook. f.) K. Schum.;多尼非洲豆蔻(多尼非砂仁,非洲砂仁)■☆
9895 Aframomum daniellii K. Schum. = Aframomum daniellii (Hook. f.) K. Schum. ■☆
9896 Aframomum elegans Lock;雅致非洲豆蔻(雅致非洲砂仁)■☆
9897 Aframomum elliottii (Baker) K. Schum.;埃利非洲豆蔻(埃利砂仁)■☆
9898 Aframomum erythrostachyum Gagnep. = Aframomum strobilaceum (Sm.) Hepper ■☆
9899 Aframomum exscapum (Sims) Hepper = Amomum grana-paradisi L. ■☆
9900 Aframomum flavum Lock;黄非洲豆蔻(黄非洲砂仁)■☆
9901 Aframomum geocarpum Lock et J. B. Hall;地果非洲豆蔻■☆
9902 Aframomum giganteum (Oliv. et D. Hanb.) K. Schum.;巨非洲豆蔻(巨非洲砂仁)■☆
9903 Aframomum glaucophyllum (K. Schum.) K. Schum. = Aframomum subsericeum (Oliv. et D. Hanb.) K. Schum. subsp. glaucophyllum (K. Schum.) Lock ■☆
9904 Aframomum hanburyi K. Schum.;喀麦隆非洲豆蔻(喀麦隆非砂仁,喀麦隆非洲砂仁);Cameroon Cardamom, Camerouns Cardamom ■☆
9905 Aframomum kayserianum (K. Schum.) K. Schum.;凯泽非洲豆蔻(凯泽砂仁)■☆
9906 Aframomum keniense R. E. Fr. = Aframomum zambesiacum (Baker) K. Schum. ■☆
9907 Aframomum korarima (J. Pereira) K. Schum. ex Engl. = Aframomum corrorima (A. Braun) P. C. M. Jansen ■☆
9908 Aframomum latifolium K. Schum. = Aframomum alboviolaceum (Ridl.) K. Schum. ■☆
9909 Aframomum laurentii (De Wild. et T. Durand) K. Schum.;洛朗非洲豆蔻(洛朗非洲砂仁)■☆
9910 Aframomum laxiflorum Lock;疏花非洲豆蔻(疏花非洲砂仁)■☆
9911 Aframomum leonense K. Schum. = Aframomum exscapum (Sims) Hepper ■☆
9912 Aframomum leptolepis (K. Schum.) K. Schum.;细鳞非洲豆蔻(细鳞非洲砂仁)■☆
9913 Aframomum letestuanum Gagnep.;莱泰斯图非洲豆蔻(莱泰斯图非洲砂仁)■☆
9914 Aframomum limbatum (Oliv. et D. Hanb.) K. Schum.;具边非洲豆蔻(具边非洲砂仁)■☆
9915 Aframomum longiligulatum Koechlin;长舌非洲豆蔻(长舌非洲砂仁)■☆
9916 Aframomum longipetiolatum Koechlin;长叶柄非洲豆蔻(长叶柄非洲砂仁)■☆
9917 Aframomum longiscapum (Hook. f.) K. Schum.;长花茎非洲豆蔻(长花茎非洲砂仁)■☆
9918 Aframomum luteoalbum (K. Schum.) K. Schum.;黄白非洲豆蔻(黄白非洲砂仁)■☆
9919 Aframomum makandensis Dhetchuvi;马坎多非洲豆蔻■☆
9920 Aframomum mala (K. Schum.) K. Schum.;马拉非洲豆蔻(马拉非洲砂仁)■☆
9921 Aframomum mannii (Oliv. et D. Hanb.) K. Schum.;曼氏非洲豆蔻(曼氏非洲砂仁)■☆
9922 Aframomum masuianum (De Wild. et T. Durand) K. Schum. = Aframomum sceptrum (Oliv. et D. Hanb.) K. Schum. ■☆
9923 Aframomum melegueta (Roscoe) K. Schum.;斑点非洲豆蔻(斑点非洲砂仁,斑非砂仁);Alligator Pepper, Grains of Paradise, Grains-of-paradise, Greins, Greins of Paris, Guinea Grains, Guinea Pepper, Malaguetta Pepper, Malegueta Pepper, Melegueta Pepper ■☆
9924 Aframomum meleguetella K. Schum. = Aframomum melegueta (Roscoe) K. Schum. ■☆
9925 Aframomum mildbraedii Loes.;米尔德非洲豆蔻(米尔德非洲砂仁)■☆
9926 Aframomum oleraceum A. Chev.;蔬菜非洲豆蔻(蔬菜砂仁)■☆
9927 Aframomum orientale Lock;东方非洲豆蔻(东方非洲砂仁)■☆
9928 Aframomum pilosum (Oliv. et D. Hanb.) K. Schum.;疏毛非洲豆蔻(疏毛非洲砂仁)■☆
9929 Aframomum polyanthum (K. Schum.) K. Schum.;多花非洲豆蔻(多花非洲砂仁)■☆
9930 Aframomum pruinosum Gagnep.;白粉非洲豆蔻(白粉非洲砂仁)■☆
9931 Aframomum pseudostipulare Loes. et Mildbr. ex Koechlin;假托叶非洲豆蔻(假托叶砂仁)■☆
9932 Aframomum rostratum K. Schum.;喙非洲豆蔻(非洲豆蔻)■☆
9933 Aframomum sanguineum (K. Schum.) K. Schum. = Aframomum angustifolium (Sonn.) K. Schum. ■☆
9934 Aframomum sceleratum A. Chev. = Aframomum angustifolium (Sonn.) K. Schum. ■☆
9935 Aframomum sceptrum (Oliv. et D. Hanb.) K. Schum.;王杖非洲豆蔻(王杖砂仁)■☆
9936 Aframomum simiarum (A. Chev.) A. Chev. = Aframomum strobilaceum (Sm.) Hepper ■☆
9937 Aframomum singulariflorum Dhetchuvi;单花非洲豆蔻(单花砂仁)■☆
9938 Aframomum spiroligulatum A. D. Poulsen et Lock;刺舌非洲豆蔻(刺舌砂仁)■☆
9939 Aframomum stanfieldii Hepper;斯坦菲尔德非洲豆蔻(斯坦菲尔德砂仁)■☆
9940 Aframomum stipulatum (Gagnep.) K. Schum.;托叶非洲豆蔻(托叶非砂仁,托叶非洲砂仁)■☆
9941 Aframomum stipulatum (Gagnep.) K. Schum. = Aframomum alboviolaceum (Ridl.) K. Schum. ■☆
9942 Aframomum strobilaceum (Sm.) Hepper;球果非洲豆蔻(球果砂仁)■☆
9943 Aframomum subsericeum (Oliv. et D. Hanb.) K. Schum.;亚绢毛非洲豆蔻(亚绢毛砂仁)■☆
9944 Aframomum subsericeum (Oliv. et D. Hanb.) K. Schum. subsp. glaucophyllum (K. Schum.) Lock;灰绿非洲豆蔻(灰绿砂仁)■☆
9945 Aframomum sulcatum (Oliv. et D. Hanb. ex Baker) K. Schum.;纵沟非洲豆蔻(纵沟非洲砂仁)■☆

9946 Aframomum tectorum K. Schum.;屋顶非洲豆蔻(屋顶非洲砂仁)■☆
9947 Aframomum thonneri De Wild.;托内非洲豆蔻(托内非洲砂仁)■☆
9948 Aframomum uniflorum Lock et A. D. Poulsen;独花非洲豆蔻(单花非洲砂仁)■☆
9949 Aframomum usambarense Lock;乌桑巴拉非洲豆蔻(乌桑巴拉非洲砂仁)■☆
9950 Aframomum verrucosum Lock;多疣非洲豆蔻(多疣砂仁)■☆
9951 Aframomum wuerthii Dhetchuvi et Eb. Fisch.;维特非洲豆蔻■☆
9952 Aframomum zambesiacum (Baker) K. Schum.;赞比西非洲豆蔻(赞比西砂仁)■☆
9953 Aframomum zambesiacum (Baker) K. Schum. subsp. puberulum Lock;短柔毛非洲豆蔻(短柔毛砂仁)■☆
9954 Aframomum zimmermannii K. Schum. = Aframomum alpinum (Gagnep.) K. Schum. ■☆
9955 Afrardisia Mez = Ardisia Sw. (保留属名)●■
9956 Afrardisia Mez(1902);非洲紫金牛属●☆
9957 Afrardisia bequaertii De Wild. = Ardisia staudtii Gilg ●☆
9958 Afrardisia bracteata (Baker) Mez = Ardisia bracteata Baker ●☆
9959 Afrardisia brunneo-purpurea (Gilg) Mez = Ardisia staudtii Gilg ●☆
9960 Afrardisia buesgenii Gilg et G. Schellenb. = Ardisia buesgenii (Gilg et G. Schellenb.) Taton ●☆
9961 Afrardisia comosa de Wit = Ardisia comosa (de Wit) Taton ●☆
9962 Afrardisia conraui (Gilg) Mez = Ardisia conraui Gilg ●☆
9963 Afrardisia cymosa (Baker) Mez = Ardisia staudtii Gilg ●☆
9964 Afrardisia dentata Gilg et G. Schellenb. = Ardisia kivuensis Taton ●☆
9965 Afrardisia didymopora H. Perrier = Ardisia didymopora (H. Perrier) Capuron ●☆
9966 Afrardisia haemantha (Gilg) Mez = Ardisia staudtii Gilg ●☆
9967 Afrardisia hylophila Gilg et G. Schellenb. = Ardisia zenkeri Gilg ●☆
9968 Afrardisia ledermannii Gilg et G. Schellenb. = Ardisia conraui Gilg ●☆
9969 Afrardisia leucantha Gilg et G. Schellenb. = Ardisia batangaensis Taton ●☆
9970 Afrardisia mayumbensis R. D. Good = Ardisia mayumbensis (R. D. Good) Taton ●☆
9971 Afrardisia mildbraedii Gilg et G. Schellenb. = Ardisia mildbraedii (Gilg et G. Schellenb.) Taton ●☆
9972 Afrardisia oligantha Gilg et G. Schellenb. = Ardisia oligantha (Gilg et G. Schellenb.) Taton ●☆
9973 Afrardisia platyphylla Gilg et G. Schellenb. = Ardisia platyphylla (Gilg et G. Schellenb.) Taton ●☆
9974 Afrardisia polyadenia (Gilg) Mez = Ardisia polyadenia Gilg ●☆
9975 Afrardisia rosacea Gilg et G. Schellenb. = Ardisia conraui Gilg ●☆
9976 Afrardisia sadebeckiana (Gilg) Mez = Ardisia sadebeckiana Gilg ●☆
9977 Afrardisia schlechteri (Gilg) Mez = Ardisia schlechteri Gilg ●☆
9978 Afrardisia staudtii (Gilg) Mez = Ardisia staudtii Gilg ●☆
9979 Afrardisia zenkeri (Gilg) Mez = Ardisia zenkeri Gilg ●☆
9980 Afraurantium A. Chev. (1949);塞内加尔橘属●☆
9981 Afraurantium senegalensis A. Chev.;塞内加尔橘●☆
9982 Afrazelia Pierre = Afzelia Sm. (保留属名)●
9983 Afridia Duthie = Nepeta L. ●■
9984 Afrobrunnichia Hutch. et Dalziel = Brunnichia Banks ex Gaertn. ●☆
9985 Afrobrunnichia Hutch. et Dalziel(1927);西非蓼属■☆
9986 Afrobrunnichia africana (Welw.) Hutch. et Dalziel;西非蓼■☆
9987 Afrobrunnichia erecta (Asch.) Hutch. et Dalziel;直立西非蓼■☆
9988 Afrocalathea K. Schum. (1902);西非竹芋属■☆
9989 Afrocalathea rhizantha (K. Schum.) K. Schum.;西非竹芋■☆
9990 Afrocanthium (Bridson) Lantz et B. Bremer(2004);非洲鱼骨木属●☆
9991 Afrocarpus (Buchholz et N. E. Gray) C. N. Page(1989);非洲罗汉松属(阿佛罗汉松属)●☆
9992 Afrocarpus Gaussen = Afrocarpus (Buchholz et N. E. Gray) C. N. Page ●☆
9993 Afrocarpus falcata (Thunb.) Gaussen;镰非洲罗汉松(镰阿佛罗汉松);Outeniqua Yellowwood ●☆
9994 Afrocarpus gracilior (Pilg.) Gaussen;细非洲罗汉松(细阿佛罗汉松);Fern Pine,Musengera ●☆
9995 Afrocarpus usambarensis (Pilg.) Gaussen;乌桑巴拉山非洲罗汉松(乌桑巴拉山阿佛罗汉松);East African Yellow Wood ●☆
9996 Afrocarum Rauschert(1982);非洲葛缕子属■☆
9997 Afrocarum imbricatum (Schinz) Rauschert;非洲葛缕子■☆
9998 Afrocrania (Harms) Hutch. (1942);阿夫萸属(阿夫山茱萸属)●☆
9999 Afrocrania (Harms) Hutch. = Cornus L. ●
10000 Afrocrania volkensii (Harms) Hutch.;阿夫萸(阿夫山茱萸)●☆
10001 Afrocrocus J. C. Manning et Goldblatt(2008);非洲番红花属(藏红花属)■☆
10002 Afrodaphne Stapf = Beilschmiedia Nees ●
10003 Afrodaphne Stapf(1905);非洲樟属●☆
10004 Afrodaphne caudata Stapf = Beilschmiedia caudata (Stapf) A. Chev. ●☆
10005 Afrodaphne elata (Scott-Elliot) Stapf = Beilschmiedia mannii (Meisn.) Benth. et Hook. f. ●☆
10006 Afrodaphne euryneura Stapf = Beilschmiedia caudata (Stapf) A. Chev. ●☆
10007 Afrodaphne fruticosa (Engl.) Stapf = Beilschmiedia fruticosa Engl. ●☆
10008 Afrodaphne gaboonensis (Meisn.) Stapf = Beilschmiedia gaboonensis (Meisn.) Benth. et Hook. f. ●☆
10009 Afrodaphne grandifolia Stapf = Beilschmiedia grandifolia (Stapf) Robyns et R. Wilczek ●☆
10010 Afrodaphne mannii (Meisn.) Stapf = Beilschmiedia mannii (Meisn.) Benth. et Hook. f. ●☆
10011 Afrodaphne minutiflora (Meisn.) Stapf = Beilschmiedia minutiflora (Meisn.) Benth. et Hook. f. ●☆
10012 Afrodaphne nitida (Engl.) Stapf = Beilschmiedia nitida Engl. ●☆
10013 Afrodaphne obscura Stapf = Beilschmiedia obscura (Stapf) Engl. ex A. Chev. ●☆
10014 Afrodaphne preussii (Engl.) Stapf = Beilschmiedia preussii Engl. ●☆
10015 Afrodaphne sessilifolia Stapf = Beilschmiedia sessilifolia (Stapf) Engl. ex Fouilloy ●☆
10016 Afrodaphne zenkeri (Engl.) Stapf = Beilschmiedia zenkeri Engl. ●☆
10017 Afrofittonia Lindau(1913);西非银网叶属(西非爵床属)■☆
10018 Afrofittonia silvestris Lindau;西非银网叶(西非爵床)■☆
10019 Afroguatteria Boutique(1951);非洲硬蕊花属(非洲番荔枝属)●☆
10020 Afroguatteria bequaertii (De Wild.) Boutique;非洲硬蕊花(非洲番荔枝)●☆
10021 Afroguatteria globosa Paiva;球形非洲番荔枝●☆
10022 Afrohamelia Wernham = Atractogyne Pierre ●☆

10023 Afrohamelia bracteata Wernham = Atractogyne bracteata (Wernham) Hutch. et Dalziel ●☆

10024 Afroknoxia Verdc. (1981);非洲红芽大戟属■☆

10025 Afroknoxia Verdc. = Knoxia L. ■

10026 Afroknoxia manika Verdc.;非洲红芽大戟■☆

10027 Afrolicania Mildbr. = Licania Aubl. ●☆

10028 Afroligusticum C. Norman(1927);非洲藁本属■☆

10029 Afroligusticum chaerophylloides C. Norman = Afroligusticum elliotii (Engl.) C. Norman ■☆

10030 Afroligusticum elliotii (Engl.) C. Norman;非洲藁本■☆

10031 Afrolimon Lincz. (1979);南非补血草属●■☆

10032 Afrolimon Lincz. = Limonium Mill. (保留属名)●■

10033 Afrolimon amoenum (C. H. Wright) Lincz.;秀丽南非补血草●☆

10034 Afrolimon capense (L. Bolus) Lincz.;好望角补血草■☆

10035 Afrolimon longifolium (Thunb.) Lincz.;长花南非补血草■☆

10036 Afrolimon namaquanum (L. Bolus) Lincz.;纳马夸南非补血草■☆

10037 Afrolimon peregrinum (P. J. Bergius) Lincz.;紫南非补血草■☆

10038 Afrolimon purpuratum (L.) Lincz. = Afrolimon peregrinum (P. J. Bergius) Lincz. ■☆

10039 Afrolimon teretifolium (L. Bolus) Lincz.;四叶南非补血草■☆

10040 Afromendoncia Gilg = Mendoncia Vell. ex Vand. ●☆

10041 Afromendoncia Gilg ex Lindau = Mendoncia Vell. ex Vand. ●☆

10042 Afromendoncia floribunda (Pierre) Burkill = Mendoncia lindaviana (Gilg) Benoist ●☆

10043 Afromendoncia gilgiana Lindau = Mendoncia gilgiana (Lindau) Benoist ●☆

10044 Afromendoncia lindaviana Gilg = Mendoncia lindaviana (Gilg) Benoist ●☆

10045 Afromendoncia phytocrenoides Gilg = Mendoncia phytocrenoides (Gilg) Benoist ●☆

10046 Afromendonica ioidioides S. Moore = Mendoncia phytocrenoides (Gilg) Benoist var. ioides (S. Moore) Heine ●☆

10047 Afrorchis Szlach. (2006);异非洲兰属■☆

10048 Afrorhaphidophora Engl. = Rhaphidophora Hassk. ●■

10049 Afrorhaphidophora africana (N. E. Br.) Engl. = Rhaphidophora africana N. E. Br. ●☆

10050 Afrormosia Harms = Pericopsis Thwaites ●☆

10051 Afrormosia Harms(1906);非洲红豆树属(非洲红豆属);Afrormosia ●

10052 Afrormosia angolensis (Baker) De Wild. var. subtomentosa (De Wild.) Louis = Pericopsis angolensis (Baker) Meeuwen ●☆

10053 Afrormosia angolensis Harms;安哥拉红豆树(刚果非洲红豆树);Angola Afrormosia ●☆

10054 Afrormosia bequaertii De Wild. = Pericopsis angolensis (Baker) Meeuwen ●☆

10055 Afrormosia elata Harms = Pericopsis elata (Harms) Meeuwen ●

10056 Afrormosia laxiflora (Benth. ex Baker) Harms;疏花非洲红豆树 ●☆

10057 Afrormosia laxiflora (Benth.) Harms = Pericopsis laxiflora (Benth.) Meeuwen ●☆

10058 Afrormosia schliebenii Harms = Pericopsis angolensis (Baker) Meeuwen f. brasseuriana (De Wild.) Brummitt ●☆

10059 Afrosciadium P. J. D. Winter(2008);非洲伞芹属■☆

10060 Afrosersalisia A. Chev. (1943);非洲桃榄属●☆

10061 Afrosersalisia A. Chev. = Synsepalum (A. DC.) Daniell ●☆

10062 Afrosersalisia afzelii (Engl.) A. Chev. = Synsepalum afzelii (Engl.) T. D. Penn. ●☆

10063 Afrosersalisia afzelii (Engl.) A. Chev. var. ligulata (Baehni) A. Chev. = Synsepalum passargei (Engl.) T. D. Penn. ●☆

10064 Afrosersalisia cerasifera (Welw.) Aubrév. ex Heine = Synsepalum cerasiferum (Welw.) T. D. Penn. ●☆

10065 Afrosersalisia chevalieri (Engl.) Aubrév. = Synsepalum cerasiferum (Welw.) T. D. Penn. ●☆

10066 Afrosersalisia disaco (Hiern) Aubrév. = Synsepalum cerasiferum (Welw.) T. D. Penn. ●☆

10067 Afrosersalisia kaessneri (Engl.) J. H. Hemsl. = Synsepalum kaessneri (Engl.) T. D. Penn. ●☆

10068 Afrosersalisia micrantha (A. Chev.) A. Chev. = Synsepalum afzelii (Engl.) T. D. Penn. ●☆

10069 Afrosersalisia rwandensis (Troupin) Liben;卢旺达非洲桃榄●☆

10070 Afrosersalisia usambarensis (Engl.) Aubrév. = Synsepalum cerasiferum (Welw.) T. D. Penn. ●☆

10071 Afrosison H. Wolff(1912);非洲水柴胡属■☆

10072 Afrosison djurense H. Wolff;非洲水柴胡■☆

10073 Afrosison gallabatense H. Wolff;加拉巴特非洲水柴胡■☆

10074 Afrosison schweinfurthii H. Wolff;施氏非洲水柴胡■☆

10075 Afrostyrax Perkins et Gilg(1909);非洲蒜树属(非洲安息香属)●☆

10076 Afrostyrax kamerunensis Perkins et Gilg;非洲蒜树●☆

10077 Afrostyrax lepidophyllus Mildbr.;鳞叶非洲蒜树●☆

10078 Afrostyrax macranthus Mildbr.;大花非洲蒜树●☆

10079 Afrothismia (Engl.) Schltr. (1906);非洲水玉簪属■☆

10080 Afrothismia Schltr. = Afrothismia (Engl.) Schltr. ■☆

10081 Afrothismia amietii Cheek;阿米非洲水玉簪■☆

10082 Afrothismia gesnerioides H. Maas;苦苣苔非洲水玉簪■☆

10083 Afrothismia insignis Cowley;显著非洲水玉簪■☆

10084 Afrothismia korupensis Sainge et Franke;科鲁普非洲水玉簪■☆

10085 Afrothismia pachyantha Schltr.;粗花非洲水玉簪■☆

10086 Afrothismia saingei T. Franke;萨因吉非洲水玉簪■☆

10087 Afrothismia winkleri (Engl.) Schltr.;温克勒非洲水玉簪■☆

10088 Afrothismia winkleri (Engl.) Schltr. var. budongensis Cowley;布东戈非洲水玉簪■☆

10089 Afrotrewia Pax et K. Hoffm. (1914);非洲滑桃树属●☆

10090 Afrotrewia kamerunica Pax et K. Hoffm.;非洲滑桃树■☆

10091 Afrotrichloris Chiov. (1915);非洲三花禾属(非洲禾属,非洲虎尾草属)■☆

10092 Afrotrichloris hyaloptera Clayton;非洲三花禾■☆

10093 Afrotrichloris martinii Chiov.;非洲禾■☆

10094 Afrotrilepis (Gilly) J. Raynal(1963);非洲三鳞莎草属■☆

10095 Afrotrilepis jaegeri J. Raynal;非洲三鳞莎草■☆

10096 Afrotrilepis pilosa (Boeck.) J. Raynal;毛非洲三鳞莎草■☆

10097 Afrotysonia Rauschert(1982);非洲紫草属●☆

10098 Afrotysonia africana (Bolus) Rauschert;非洲紫草●☆

10099 Afrotysonia glochidiata (R. R. Mill) R. R. Mill;钩毛非洲紫草●☆

10100 Afrotysonia pilosicaulis R. R. Mill;毛茎非洲紫草☆

10101 Afrovivella A. Berger = Rosularia (DC.) Stapf ■

10102 Afrovivella simensis (Britten) A. Berger = Rosularia semiensis (J. Gay ex A. Rich.) H. Ohba ■☆

10103 Afzelia J. F. Gmel. (废弃属名) = Afzelia Sm. (保留属名)●

10104 Afzelia J. F. Gmel. (废弃属名) = Seymeria Pursh(保留属名)■☆

10105 Afzelia Sm. (1792)(保留属名);缅茄属;Afzelia,Pahudia ●

10106 Afzelia africana Sm.;缅茄(非洲缅茄);African Afzelia,Afzelia, Common Afzelia,Mahogany Bean ●☆

10107 Afzelia africana Sm. ex Pers.;非洲缅茄●☆

10108 Afzelia attenuata Klotzsch = Afzelia quanzensis Welw. ●☆

10109 Afzelia bakeri Prain;泰国缅茄(泰国阿芙苏木);Baker Afzelia, Lumpho ●

10110 Afzelia bella Harms;雅洁缅茄(艳阿芙豆,艳阿芙苏木);Beautiful Afzelia ●☆

10111 Afzelia bella Harms var. glabra Aubrév.;光雅洁缅茄●☆

10112 Afzelia bella Harms var. gracilior Keay;细雅洁缅茄●☆

10113 Afzelia bequaertii De Wild. = Afzelia bipindensis Harms ●☆

10114 Afzelia bijuga (Colebr.) A. Gray = Intsia bijuga (Colebr.) Kuntze ●☆

10115 Afzelia bijuga (Colebr.) A. Gray f. sambiranensis R. Vig. = Intsia bijuga (Colebr.) Kuntze ●☆

10116 Afzelia bipindensis Harms;比平迪缅茄●☆

10117 Afzelia bracteata Vogel ex Benth. = Afzelia parviflora (Vahl) Hepper ●☆

10118 Afzelia brieyi De Wild. = Afzelia pachyloba Harms ●☆

10119 Afzelia caudata Hoyle = Afzelia bipindensis Harms ●☆

10120 Afzelia cochinchinensis (Pierre) J. Léonard = Afzelia xylocarpa (Kurz) Craib ●

10121 Afzelia macrophylla (Nutt.) Kuntze;大叶缅茄;Mullein Foxglove, Mullein-foxglove ●☆

10122 Afzelia macrophylla (Nutt.) Kuntze = Dasistoma macrophylla (Nutt.) Raf. ●☆

10123 Afzelia microcarpa A. Chev. = Afzelia bella Harms var. gracilior Keay ●☆

10124 Afzelia pachyloba Harms;粗荚缅茄●☆

10125 Afzelia parviflora (Vahl) Hepper;小花缅茄●☆

10126 Afzelia petersiana Klotzsch = Afzelia quanzensis Welw. ●☆

10127 Afzelia quanzensis Welw.;安哥拉缅茄;Angola Afzelia, Chamfuta, Lucky Bean, Lucky Bean Tree, Mahogany Bean, Red Mahogany, Rhodesian Mahogany ●☆

10128 Afzelia rhomboidea S. Vidal;田达罗树;Malacca Teak ●

10129 Afzelia xylocarpa (Kurz) Craib;木果缅茄(泻茄,缅茄,木果阿芙苏木,木茄,细茄,印支阿芙苏木,越南缅茄);African Afzelia, Cochinchina Afzelia, Makha Huakham, Malkamong, Woodyfruit Afzelia, Woody-fruited Afzelia ●

10130 Afzelia zenkeri Harms = Afzelia pachyloba Harms ●☆

10131 Afzeliella Gilg = Guyonia Naudin ☆

10132 Afzeliella bolivari Brenan et Guinea = Guyonia ciliata Hook. f. ■☆

10133 Afzeliella ciliata (Hook. f.) Gilg = Guyonia ciliata Hook. f. ■☆

10134 Afzeliella intermedia (Cogn.) Gilg = Guyonia ciliata Hook. f. ■☆

10135 Agaisia Garay et Sweet = Aganisia Lindl. ■☆

10136 Agalina Hort. = Agalma Miq. ●

10137 Agalinis Raf. (1837)(保留属名);假毛地黄属■☆

10138 Agalinis Raf. (保留属名) = Gerardia L. (废弃属名)■☆

10139 Agalinis aspera (Douglas ex Benth.) Britton;糙假毛地黄;Rough Agalinis, Rough False Foxglove, Rough Gerardia, Tall False Foxglove ■☆

10140 Agalinis auriculata (Michx.) S. F. Blake;耳状假毛地黄;Auriculate False Foxglove, Eared False Foxglove, Ear-leaved Gerardia ■☆

10141 Agalinis besseyana (Britton) Britton = Agalinis tenuifolia (Vahl) Raf. ■☆

10142 Agalinis fasciculata (Elliott) Raf.;簇生假毛地黄;Gerardia ■☆

10143 Agalinis flava (L.) B. Boivin;金栎;Fern-leaf False Foxglove, Smooth False Foxglove, Yellow False Foxglove ■☆

10144 Agalinis gattingeri (Small) Small;糙茎假毛地黄(圆茎假毛地黄);Rough-stemmed Gerardia, Round-stem Foxglove, Round-stemmed False Foxglove ■☆

10145 Agalinis heterophylla Small ex Britton;异叶假毛地黄■☆

10146 Agalinis laevigata (Raf.) S. F. Blake;平滑假毛地黄(平滑杰勒草);Entire-leaved False Foxglove, Smooth False Foxglove ■☆

10147 Agalinis maritima Raf.;海滨假毛地黄(海滨杰勒草);Seaside Gerardia ■☆

10148 Agalinis paupercula (A. Gray) Britton;小花假毛地黄;Small-flowered False Foxglove, Small-flowered Gerardla, Smooth False Foxglove ■☆

10149 Agalinis paupercula (A. Gray) Britton var. borealis Pennell;光小花假毛地黄;Small-flowered False Foxglove, Smooth False Foxglove ■☆

10150 Agalinis pedicularia (L.) S. F. Blake = Aureolaria pedicularia (L.) Raf. ■☆

10151 Agalinis pedicularia (L.) S. F. Blake var. ambigens (Fernald) S. F. Blake = Aureolaria pedicularia (L.) Raf. var. ambigens (Fernald) Farw. ■☆

10152 Agalinis pedicularia (L.) S. F. Blake var. caesariensis (Pennell) S. F. Blake;凯萨里假毛地黄■☆

10153 Agalinis purpurea (L.) Pennell;紫假毛地黄(紫杰勒草);Purple False Foxglove, Purple Gerardia, Smooth Agalinis ■☆

10154 Agalinis purpurea (L.) Pennell var. carteri Pennell = Agalinis purpurea (L.) Pennell ■☆

10155 Agalinis purpurea (L.) Pennell var. parviflora (Benth.) B. Boivin = Agalinis paupercula (A. Gray) Britton var. borealis Pennell ■☆

10156 Agalinis skinneriana (A. W. Wood) Britton;斯氏假毛地黄;Pale False Foxglove, Skinner's False Foxglove ■☆

10157 Agalinis tenuifolia (Vahl) Raf.;普通假毛地黄;Common Agalinis, Common False Foxglove ■☆

10158 Agalinis tenuifolia (Vahl) Raf. var. macrophylla (Benth.) S. F. Blake = Agalinis tenuifolia (Vahl) Raf. ■☆

10159 Agalinis tenuifolia (Vahl) Raf. var. parviflora (Nutt.) Pennell = Agalinis tenuifolia (Vahl) Raf. ■☆

10160 Agallis Phil. (1864);智利虹膜花属■☆

10161 Agallis Phil. = Tropidocarpum Hook. ■☆

10162 Agallis montana Phil.;智利虹膜花■☆

10163 Agallochum Lam. (废弃属名) = Aquilaria Lam. (保留属名)●

10164 Agallochum sinense (Lour.) Kuntze = Aquilaria sinensis (Lour.) Spreng. ●◇

10165 Agallostachys Beer = Bromelia L. ■☆

10166 Agalma Miq. = Schefflera J. R. Forst. et G. Forst. (保留属名)●

10167 Agalma Steud. = Sonchus L. ■

10168 Agalma delavayi (Franch.) Hutch. = Schefflera delavayi (Franch.) Harms ●

10169 Agalma discolor (Merr.) Hutch. = Schefflera delavayi (Franch.) Harms ●

10170 Agalma diversifoliolatum (H. L. Li) Hutch. = Schefflera diversifoliolata H. L. Li ●

10171 Agalma dumicola (W. W. Sm.) Hutch. = Schefflera hoi (Dunn) R. Vig. ●

10172 Agalma elatum (Buch. -Ham.) Seem. = Schefflera elata (C. B. Clarke) Harms ●

10173 Agalma elatum Seem. = Schefflera elata (C. B. Clarke) Harms ●

10174 Agalma glaucum Seem. = Schefflera rhododendrifolia (Griff.) Frodin ●

10175 Agalma hainanense (Merr. et Chun) Hutch. = Schefflera

hainanensis Merr. et Chun ●
10176 Agalma hoi (Dunn) Hutch. = Schefflera hoi (Dunn) R. Vig. ●
10177 Agalma lutchuense Nakai = Schefflera bodinieri (H. Lév.) Rehder ●
10178 Agalma lutchuense Nakai = Schefflera heptaphylla (L.) Frodin ●
10179 Agalma multinervium (H. L. Li) Hutch. = Schefflera multinervia H. L. Li ●
10180 Agalma octophyllum (Lour.) Seem. = Schefflera bodinieri (H. Lév.) Rehder ●
10181 Agalma octophyllum Seem. = Schefflera heptaphylla (L.) Frodin ●
10182 Agalma schweliense (W. W. Sm.) Hutch. = Schefflera schweliensis W. W. Sm. ●
10183 Agalma taiwanianum Nakai = Schefflera taiwaniana (Nakai) Kaneh. ●
10184 Agalma tomentosum (Buch.-Ham.) Seem. = Schefflera rhododendrifolia (Griff.) Frodin ●
10185 Agalma wardii (C. Marquand et Airy Shaw) Hutch. = Schefflera wardii C. Marquand et Airy Shaw ●
10186 Agalmanthus (Endl.) Hombr. et Jacquinot = Metrosideros Banks ex Gaertn. (保留属名)●☆
10187 Agalmanthus Hombr. et Jacquinot ex Decne. = Metrosideros Banks ex Gaertn. (保留属名)●☆
10188 Agalmyla Blume(1826);菊叶苣苔属(根花属);Agalmyla ●☆
10189 Agalmyla bicolor Hilliard et B. L. Burtt;双色菊叶苣苔●☆
10190 Agalmyla biflora(Elmer) Hilliard et B. L. Burtt;双花菊叶苣苔●☆
10191 Agalmyla parasitica Kuntze;红根花;Scarlet Root-blosson ●☆
10192 Agalmyla staminea Blume;雄蕊根花;Stamen Scarlet Root-blosson ●☆
10193 Agaloma Raf. = Euphorbia L. ●■
10194 Aganippa Baill. = Aganippea Moc. et Sessé ex DC. ■☆
10195 Aganippea Moc. et Sessé ex DC. = Jaegeria Kunth ■☆
10196 Aganisia Lindl. (1839);雅兰属■☆
10197 Aganisia alba Ridl.;白雅兰■☆
10198 Aganisia coerulea Rchb. f.;蓝雅兰■☆
10199 Aganisia lepida Linden et Rchb. f.;鳞雅兰■☆
10200 Aganon Raf. = Callicarpa L. ●
10201 Aganonerion Pierre = Aganonerion Pierre ex Spire ●☆
10202 Aganonerion Pierre et Spire = Aganonerion Pierre ex Spire ●☆
10203 Aganonerion Pierre ex Spire(1906);越南夹竹桃属●☆
10204 Aganonerion polymorphum Pierre ex Spire;越南夹竹桃●☆
10205 Aganope Miq. (1855);双束鱼藤属●
10206 Aganope Miq. = Ostryocarpus Hook. f. ■☆
10207 Aganope dinghuensis (P. Y. Chen) T. C. Chen et Pedley;鼎湖鱼藤;Dinghu Fishvine, Dinghu Jewelvine ●
10208 Aganope gabonica (Baill.) Polhill;加蓬双束鱼藤■☆
10209 Aganope impressa (Dunn) Polhill;双束鱼藤■☆
10210 Aganope latifolia (Prain) T. C. Chen et Pedley;大叶鱼藤;Broadleaf Fishvine, Broadleaf Jewelvine, Broad-leaf Jewelvine ●
10211 Aganope leucobotrya (Dunn) Polhill;白穗双束鱼藤■☆
10212 Aganope lucida (Welw. ex Baker) Polhill;光亮双束鱼藤■☆
10213 Aganope thyrsiflora (Benth.) Polhill;密锥花鱼藤●
10214 Aganosma (Blume) G. Don (1837);香花藤属(阿根藤属);Aganosma ●
10215 Aganosma G. Don = Aganosma (Blume) G. Don ●
10216 Aganosma acuminata (Roxb.) G. Don = Aganosma cymosa (Roxb.) G. Don ●
10217 Aganosma acuminata (Roxb.) G. Don = Aganosma marginata (Roxb.) G. Don ●
10218 Aganosma breviloba Kerr;贵州香花藤●
10219 Aganosma calycina A. DC.;印度香花藤(香花藤)●☆
10220 Aganosma cymosa (Roxb.) G. Don;云南香花藤(黄毛香花藤,老鼠牛角);Harmand Aganosma, Yunnan Aganosma ●
10221 Aganosma cymosa (Roxb.) G. Don var. fulva Craib = Aganosma cymosa (Roxb.) G. Don ●
10222 Aganosma cymosa (Roxb.) G. Don var. goabra DC. = Aganosma cymosa (Roxb.) G. Don ●
10223 Aganosma cymosa (Roxb.) G. Don var. lanceolata Hook. f. = Aganosma cymosa (Roxb.) G. Don ●
10224 Aganosma edithae Hance = Cryptolepis sinensis (Lour.) Merr. ●
10225 Aganosma harmandiana Pierre = Aganosma cymosa (Roxb.) G. Don ●
10226 Aganosma kwangsiensis Tsiang = Aganosma siamensis Craib ●
10227 Aganosma kwangsiensis Tsiang var. longilobata Y. Wan ex C. Z. Gao;大花香花藤;Bigflower Guangxi Aganosma ●
10228 Aganosma laevis Champ. ex Benth. = Anodendron affine (Hook. et Arn.) Druce ●
10229 Aganosma marginata (Roxb.) G. Don;香花藤;Common Aganosma ●
10230 Aganosma montana Kerr = Aganosma schlechteriana H. Lév. ●
10231 Aganosma navaillei (H. Lév.) Tsiang = Aganosma schlechteriana H. Lév. ●
10232 Aganosma odora Tsiang = Aganosma schlechteriana H. Lév. ●
10233 Aganosma radiata Merr. = Aganosma schlechteriana H. Lév. ●
10234 Aganosma schlechteriana H. Lév.;海南香花藤(短瓣香花藤,贵州香花藤);Guizhou Aganosma, Hainan Aganosma ●
10235 Aganosma schlechteriana H. Lév. var. breviloba Tsiang;短瓣香花藤;Shortlobe Aganosma, Short-petal Aganosma ●
10236 Aganosma schlechteriana H. Lév. var. breviloba Tsiang = Aganosma schlechteriana H. Lév. ●
10237 Aganosma schlechteriana H. Lév. var. leptantha Tsiang;柔花香花藤;Slenderflower Aganosma, Thin-flower Aganosma ●
10238 Aganosma schlechteriana H. Lév. var. leptantha Tsiang = Aganosma schlechteriana H. Lév. ●
10239 Aganosma siamensis Craib;广西香花藤(廖刀竹,石上羊奶树);Guangxi Aganosma ●
10240 Agaosizia Spach = Camissonia Link ■☆
10241 Agapanthaceae F. Voigt = Alliaceae Borkh. (保留科名)■
10242 Agapanthaceae F. Voigt;百子莲科■☆
10243 Agapanthaceae Lotsy = Alliaceae Borkh. (保留科名)■
10244 Agapanthus L'Hér. (1789)(保留属名);百子莲属;African Lily, Africanlily, Agapanthus, Lily-of-the-nile ■☆
10245 Agapanthus africanus (L.) Hoffmanns.;百子莲(非洲百子莲);African Agapanthus, African Lily, Agapanthus, Blue African Lily, Blue Lily, Lily of The Nile, Lily-of-the-nile ■☆
10246 Agapanthus africanus (L.) Hoffmanns. var. albidus Hort.;白花百子莲■☆
10247 Agapanthus africanus (L.) Hoffmanns. var. atrocaeruleus Hort.;深蓝百子莲■☆
10248 Agapanthus africanus (L.) Hoffmanns. var. giganteus Hort.;巨大百子莲■☆
10249 Agapanthus africanus (L.) Hoffmanns. var. globosus Hort.;球花百子莲■☆
10250 Agapanthus africanus (L.) Hoffmanns. var. multiflorus Hort.;多花百子莲■☆

10251 Agapanthus campanulatus F. M. Leight.;钟花百子莲(吊钟百子莲)■☆

10252 Agapanthus campanulatus F. M. Leight. subsp. patens (F. M. Leight.) F. M. Leight.;铺展钟花百子莲■☆

10253 Agapanthus caulescens Spreng.;无茎百子莲(紫花小百子莲)■☆

10254 Agapanthus caulescens Spreng. subsp. angustifolius F. M. Leight.;窄叶无茎百子莲■☆

10255 Agapanthus caulescens Spreng. subsp. gracilis (F. M. Leight.) F. M. Leight.;纤细无茎百子莲■☆

10256 Agapanthus coddii F. M. Leight.;科德百子莲■☆

10257 Agapanthus comptonii F. M. Leight.;康普顿百子莲■☆

10258 Agapanthus comptonii F. M. Leight. subsp. comptonii ? = Agapanthus praecox Willd. subsp. minimus (Lindl.) F. M. Leight. ■☆

10259 Agapanthus comptonii F. M. Leight. subsp. longitubus ? = Agapanthus praecox Willd. subsp. minimus (Lindl.) F. M. Leight. ■☆

10260 Agapanthus dyeri F. M. Leight. = Agapanthus inapertus P. Beauv. subsp. intermedius F. M. Leight. ■☆

10261 Agapanthus gracilis F. M. Leight. = Agapanthus caulescens Spreng. subsp. gracilis (F. M. Leight.) F. M. Leight. ■☆

10262 Agapanthus hollandii F. M. Leight. = Agapanthus inapertus P. Beauv. subsp. hollandii (F. M. Leight.) F. M. Leight. ■☆

10263 Agapanthus inapertus P. Beauv.;垂花百子莲■☆

10264 Agapanthus inapertus P. Beauv. subsp. hollandii (F. M. Leight.) F. M. Leight.;霍兰垂花百子莲■☆

10265 Agapanthus inapertus P. Beauv. subsp. intermedius F. M. Leight.;间型垂花百子莲■☆

10266 Agapanthus inapertus P. Beauv. subsp. parviflorus F. M. Leight.;小垂花百子莲■☆

10267 Agapanthus inapertus P. Beauv. subsp. pendulus (L. Bolus) F. M. Leight.;下垂百子莲■☆

10268 Agapanthus longispathus F. M. Leight. = Agapanthus praecox Willd. subsp. minimus (Lindl.) F. M. Leight. ■☆

10269 Agapanthus minor Lodd. = Agapanthus africanus (L.) Hoffmanns. ■☆

10270 Agapanthus nutans F. M. Leight. = Agapanthus caulescens Spreng. subsp. gracilis (F. M. Leight.) F. M. Leight. ■☆

10271 Agapanthus orientalis F. M. Leight.;东方百子莲;African Blue Lily,Agapanthus,Lily of the Nile ■☆

10272 Agapanthus orientalis F. M. Leight. = Agapanthus praecox Willd. ■☆

10273 Agapanthus pendulus L. Bolus = Agapanthus inapertus P. Beauv. subsp. pendulus (L. Bolus) F. M. Leight. ■☆

10274 Agapanthus praecox Willd.;早花百子莲■☆

10275 Agapanthus praecox Willd. = Agapanthus africanus (L.) Hoffmanns. ■☆

10276 Agapanthus praecox Willd. = Agapanthus orientalis F. M. Leight. ■☆

10277 Agapanthus praecox Willd. subsp. minimus (Lindl.) F. M. Leight.;小早花百子莲■☆

10278 Agapanthus praecox Willd. subsp. orientalis (F. M. Leight.) F. M. Leight. = Agapanthus orientalis F. M. Leight. ■☆

10279 Agapanthus umbellatus L'Hér. = Agapanthus africanus (L.) Hoffmanns. ■☆

10280 Agapanthus umbellatus L'Hér. var. maximus Edwards = Agapanthus praecox Willd. subsp. orientalis (F. M. Leight.) F. M. Leight. ■☆

10281 Agapanthus umbellatus L'Hér. var. minimus Lindl. = Agapanthus praecox Willd. subsp. minimus (Lindl.) F. M. Leight. ■☆

10282 Agapanthus walshii L. Bolus;瓦尔什百子莲■☆

10283 Agapatea Steud. = Distichia Nees et Meyen ■☆

10284 Agapetes D. Don ex G. Don(1834);树萝卜属(爱花属,岩桃属);Agapetes ●

10285 Agapetes G. Don = Agapetes D. Don ex G. Don ●

10286 Agapetes aborensis Airy Shaw;阿波树萝卜;Abor Agapetes ●

10287 Agapetes angulata (Griff.) Benth. et Hook. f.;棱枝树萝卜;Angled Agapetes,Angular Agapetes,Ridgyshoot Agapetes ●

10288 Agapetes anonyma Airy Shaw;锈毛树萝卜;Rusthair Agapetes,Rusty-haired Agapetes ●

10289 Agapetes brachypoda Airy Shaw;短柄树萝卜;Shortstalk Agapetes,Short-stalked Agapetes ●

10290 Agapetes brachypoda Airy Shaw var. gracilis Airy Shaw;纤细短柄树萝卜;Slender Shortstalk Agapetes ●

10291 Agapetes brandisiana W. E. Evans;环萼树萝卜;Brandis Agapetes,Ring-calyxed Agapetes,Ringed Calyx Agapetes ●

10292 Agapetes bullata Dop = Vaccinium bullatum (Dop) Sleumer ●

10293 Agapetes bulleyana Diels = Vaccinium bulleyanum (Diels) Sleumer ●

10294 Agapetes bulleyana Diels var. tenuifolia J. Anthony = Vaccinium bulleyanum (Diels) Sleumer ●

10295 Agapetes burmanica W. E. Evans;缅甸树萝卜;Burma Agapetes ●

10296 Agapetes buxifolia Nutt. ex Hook. f.;黄杨叶树萝卜;Boxleaf Agapetes,Box-leaved Agapetes ●

10297 Agapetes camelliifolia S. H. Huang;茶叶树萝卜;Tealeaf Agapetes,Tea-leaved Agapetes ●

10298 Agapetes chapaensis Dop = Vaccinium brevipedicellatum C. Y. Wu ex W. P. Fang et Z. H. Pan ●

10299 Agapetes chapaensis Dop var. oblonga Dop = Vaccinium brevipedicellatum C. Y. Wu ex W. P. Fang et Z. H. Pan ●

10300 Agapetes ciliata S. H. Huang;纤毛叶树萝卜;Ciliate Agapetes,Delicatehair Agapetes,Hairy Agapetes ●

10301 Agapetes corallina Cowan = Agapetes lobbii C. B. Clarke ●

10302 Agapetes desmogyne King et Prain = Agapetes neriifolia (King et Prain) Airy Shaw ●

10303 Agapetes discolor C. B. Clarke;异色树萝卜;Discolor Agapetes,Discoloured Agapetes ●

10304 Agapetes dulongensis S. H. Huang;独龙树萝卜;Dulong Agapetes ●

10305 Agapetes dulongensis S. H. Huang = Agapetes pensilis Airy Shaw ●

10306 Agapetes emarginata (Hayata) Nakai = Vaccinium emarginatum Hayata ●

10307 Agapetes epacridea Airy Shaw;尖叶树萝卜;Sharpleaf Agapetes,Sharp-leaved Agapetes ●

10308 Agapetes flava (Hook. f.) Sleumer;黄花树萝卜;Yellow Agapetes,Yellow-flower Agapetes,Yellow-flowered Agapetes ●

10309 Agapetes forrestii W. E. Evans;伞花树萝卜;Forrest Agapetes,Umbrellaflower Agapetes ●

10310 Agapetes glandulosissima (C. Y. Wu ex R. C. Fang et Z. H. Pan) S. H. Huang = Agapetes inopinata Airy Shaw ●

10311 Agapetes graciliflora R. C. Fang;细花树萝卜;Smallflower Agapetes,Small-flowered Agapetes,Thin-flower Agapetes ●

10312 Agapetes griffithii C. B. Clarke;尾叶树萝卜;Griffith Agapetes,Tailleaf Agapetes ●

10313 Agapetes guangxiensis D. Fang;广西树萝卜;Guangxi Agapetes ●

10314 Agapetes hookeri (C. B. Clarke) Sleumer;胡克树萝卜●

10315 Agapetes hosseana Diels;红花树萝卜;Redflower Agapetes ●

10316 Agapetes hyalocheilos Airy Shaw;透明边树萝卜;Pellucidmargin Agapetes,Transparent Agapetes ●
10317 Agapetes incurvata (Griff.) Sleumer;皱叶树萝卜(内曲树萝卜);Wrinkle Agapetes, Wrinkledleaf Agapetes, Wrinkled-leaved Agapetes ●
10318 Agapetes inopinata Airy Shaw;沧源树萝卜;Cangyuan Agapetes ●
10319 Agapetes interdicta (Hand.-Mazz.) Sleumer;中型树萝卜;Intermediate Agapetes,Middling Agapetes ●
10320 Agapetes interdicta (Hand.-Mazz.) Sleumer var. stenoloba (W. E. Evans) Sleumer;狭萼中型树萝卜;Stenolobed Middling Agapetes ●
10321 Agapetes interdicta (Hand.-Mazz.) Sleumer var. stenoloba (W. E. Evans) Sleumer = Agapetes interdicta (Hand.-Mazz.) Sleumer ●
10322 Agapetes lacei Craib;灯笼花(柳叶树萝卜,深红树萝卜,树萝卜,小叶爱楠,岩龙香);Lace Agapetes, Lantern Agapetes, Willowleaf Agapetes ●◇
10323 Agapetes lacei Craib var. glaberrima Airy Shaw;无毛灯笼花;Glabrous Willowleaf Agapetes ●
10324 Agapetes lacei Craib var. tomentella Airy Shaw;绒毛灯笼花;Tomentose Willowleaf Agapetes ●
10325 Agapetes leiocarpa S. H. Huang;光果树萝卜;Smooth-fruit Agapetes,Smooth-fruited Agapetes ●
10326 Agapetes leptantha Airy Shaw = Agapetes graciliflora R. C. Fang ●
10327 Agapetes leucocarpa S. H. Huang;白果树萝卜;White-fruit Agapetes,White-fruited Agapetes ●
10328 Agapetes linearifolia C. B. Clarke;线叶树萝卜;Linear-leaved Agapetes,Line-leaf Agapetes,Threadleaf Agapetes ●
10329 Agapetes listeri (King ex C. B. Clarke) Sleumer;短锥花树萝卜;Lister Agapetes,Short-panicled Agapetes ●
10330 Agapetes lobbii C. B. Clarke;深裂树萝卜;Deeplobe Agapetes, Lobe Agapetes ●
10331 Agapetes macrantha Benth. et Hook.;大花树萝卜●☆
10332 Agapetes macrophylla C. B. Clarke;大叶树萝卜;Bigleaf Agapetes,Big-leaved Agapetes ●
10333 Agapetes malipoensis S. H. Huang;麻栗坡树萝卜;Malipo Agapetes ●
10334 Agapetes mannii Hemsl.;白花树萝卜(猴子板凳,葫芦暗消,陆次,树萝卜,小叶爱楠,瘿袋花);White Flower Agapetes, Whiteflower Agapetes,White-flowered Agapetes ●
10335 Agapetes marginata Dunn;边脉树萝卜;Marginate Agapetes, Medog Agapetes ●
10336 Agapetes medogensis S. H. Huang;墨脱树萝卜;Motuo Agapetes ●
10337 Agapetes megacarpa W. W. Sm.;大果树萝卜;Bigfruit Agapetes, Big-fruited Agapetes ●
10338 Agapetes meiniana F. Muell.;攀缘树萝卜●☆
10339 Agapetes merrilliana (Hayata) Nakai = Vaccinium delavayi Franch. subsp. merrillianum (Hayata) R. C. Fang ●
10340 Agapetes merrilliana (Hayata) Nakai = Vaccinium merrillianum Hayata ●
10341 Agapetes miniata (Griff.) Benth. et Hook. f.;朱红树萝卜;Cinnabar Agapetes,Deepred Agapetes ●
10342 Agapetes miranda Airy Shaw;坛花树萝卜;Pitcherflower Agapetes,Urceolar-flowered Agapetes ●
10343 Agapetes mitrarioides Hook. f. ex C. B. Clarke;亮红树萝卜;Bright-red Agapetes,Crimson Agapetes ●
10344 Agapetes moorei Hemsl.;树萝卜;Common Agapetes, Moore Agapetes ●
10345 Agapetes neriifolia (King et Prain) Airy Shaw;夹竹桃叶树萝卜(石萝卜,树萝卜,叶上花);Oleander Agapetes, Oleander-leaved Agapetes ●
10346 Agapetes neriifolia (King et Prain) Airy Shaw var. maxima Airy Shaw;大花夹竹桃叶树萝卜(大花树萝卜,石萝卜,树萝卜,叶上花);Big-flower Oleander Agapetes, Big-flowered Oleander-leaved Agapetes ●
10347 Agapetes nutans Dunn;垂花树萝卜;Droping Agapetes, Pendulousflower Agapetes ●
10348 Agapetes oblonga Craib;长圆叶树萝卜(长圆树萝卜);Oblong-leaf Agapetes,Oblong-leaved Agapetes ●
10349 Agapetes oblonga Craib var. longipes Airy Shaw;长梗树萝卜;Long-stalk Oblong-leaf Agapetes ●
10350 Agapetes oblonga Craib var. longipes Airy Shaw = Agapetes oblonga Craib ●
10351 Agapetes obovata (Wight) Benth. et Hook. f.;倒卵叶树萝卜;Obovate Agapetes,Obovateleaf Agapetes ●
10352 Agapetes parviflora Dunn = Vaccinium petelotii Merr. ●
10353 Agapetes pensilis Airy Shaw;倒挂树萝卜;Pendent Agapetes, Pendent Branches Agapetes,Pensil Agapetes ●
10354 Agapetes pilifera Hook. f. ex C. B. Clarke;钟花树萝卜;Bellflower Agapetes,Piliferous Agapetes ●
10355 Agapetes poilanei Dop = Vaccinium papillatum P. F. Stevens ●
10356 Agapetes praeclara C. Marquand;藏布江树萝卜(岩生树萝卜);Saxicolous Agapetes,Yaluzangbu Agapetes ●
10357 Agapetes praestigiosa Airy Shaw;听邦树萝卜;Tingbang Agapetes ●
10358 Agapetes pseudogriffithii Airy Shaw;杯梗树萝卜;Cupstalk Agapetes,False Griffith Agapetes ●
10359 Agapetes pubiflora Airy Shaw;毛花树萝卜;Hairy Flower Agapetes,Hairyflower Agapetes,Hairy-flowered Agapetes ●
10360 Agapetes pyrolifolia Airy Shaw;鹿蹄草叶树萝卜;Pyrolaleaf Agapetes,Pyrola-leaved Agapetes ●
10361 Agapetes racemosa Watt ex Kanjilal et Das = Agapetes lobbii C. B. Clarke ●
10362 Agapetes refracta Airy Shaw;折瓣树萝卜;Refracted Agapetes, Refrectpetal Agapetes ●
10363 Agapetes rubrobracteata R. C. Fang et S. H. Huang;红苞树萝卜(三齿越橘,沙俱越橘);Chapa Blueberry,Redbract Agapetes,Red-bracted Agapetes ●
10364 Agapetes rugosus (Hook. f.) Harid. et R. R. Rao;褶皱树萝卜●☆
10365 Agapetes salicifolia C. B. Clarke;柳叶树萝卜;Willow-leaf Agapetes,Willow-leaved Agapetes ●
10366 Agapetes saligna (Hook. f.) Hook. f. var. cordifolia C. B. Clarke = Agapetes hyalocheilos Airy Shaw ●
10367 Agapetes serpens (Wight) Sleumer = Pentapterygium serpens (Wight) Klotzsch ●
10368 Agapetes serrata G. Don = Vaccinium vacciniaceum (Roxb.) Sleumer ●
10369 Agapetes smithiana Sleumer;史密斯树萝卜●☆
10370 Agapetes speciosa Hemsl.;美丽树萝卜●☆
10371 Agapetes spissa Airy Shaw;丛生树萝卜;Clustered Agapetes, Dense Agapetes,Fascicular Agapetes ●
10372 Agapetes stenantha Rehder = Agapetes lobbii C. B. Clarke ●
10373 Agapetes subsessilifolia S. H. Huang, H. Sun et Z. K. Zhou;近无柄树萝卜●
10374 Agapetes vaccinacea Dunal = Vaccinium vacciniaceum (Roxb.) Sleumer ●
10375 Agapetes vacciniacea (Roxb.) Dunal = Vaccinium vacciniaceum

(Roxb.) Sleumer ●

10376 Agapetes vaccinioidea H. Lév. = Vaccinium japonicum Miq. var. sinicum (Nakai) Rehder ●

10377 Agapetes vaccinioides Dunn = Vaccinium dunalianum Wight ●

10378 Agapetes vaccinioides Dunn = Vaccinium dunnianum Sleumer ●

10379 Agapetes xizangensis S. H. Huang;西藏树萝卜;Xizang Agapetes ●

10380 Agapetes yunnanensis Franch. = Agapetes mannii Hemsl. ●

10381 Agardhia Spreng. = Qualea Aubl. ●☆

10382 Agarista D. Don = Agarista D. Don ex G. Don ●☆

10383 Agarista D. Don = Leucothoe D. Don ex G. Don + Agauria (DC.) Hook. f. ●☆

10384 Agarista D. Don ex G. Don(1834);阿加鹃属(绊足花属)●☆

10385 Agarista DC. = Coreopsis L. ●■

10386 Agarista calliopsidea DC. = Coreopsis calliopsidea (DC.) A. Gray ■☆

10387 Agarista mexicana (Hemsl.) Judd;阿加鹃(墨西哥阿加鹃)●☆

10388 Agarista populifolia (Lam.) Judd = Leucothoe axillaris (Lam.) D. Don ●☆

10389 Agarista salicifolia (Comm. ex Lam.) G. Don = Agauria salicifolia (Comm. ex Lam.) Hook. f. ex Oliv. ●☆

10390 Agasillis Spreng. = Agasyllis Hoffm. ■☆

10391 Agasillis Spreng. = Agasyllis Spreng. ■☆

10392 Agassizia A. Gray et Engelm. = Gaillardia Foug. ■

10393 Agassizia Chav. = Galvezia Dombey ex Juss. ●☆

10394 Agassizia Spach = Camissonia Link ■☆

10395 Agassizia suavis A. Gray et Engelm. = Gaillardia suavis (A. Gray et Engelm.) Britton et Rusby ■☆

10396 Agassyllis Lag. = Agasyllis Spreng. ■☆

10397 Agasta Miers = Barringtonia J. R. Forst. et G. Forst. (保留属名)●

10398 Agasta indica Miers = Barringtonia asiatica (L.) Kurz ●

10399 Agastache Clayt. = Agastache J. Clayton ex Gronov. ■

10400 Agastache Clayt. ex Gronov. = Agastache J. Clayton ex Gronov. ■

10401 Agastache Gronov. = Agastache J. Clayton ex Gronov. ■

10402 Agastache J. Clayton ex Gronov. (1762);藿香属;Anise Hyssop, Giant Hyssop, Gianthyssop, Hyssop, Hyssop of the Bible, Mexican Giant Hyssop, Mexican Giant-hyssop ■

10403 Agastache anethiodora (Nutt.) Britton = Agastache foeniculum (Pursh) Kuntze ■☆

10404 Agastache barberi (B. L. Rob.) Epling = Agastache pallida (Lindl.) Cory ■☆

10405 Agastache cana (Hook.) Wooton et Standl.;灰色藿香;Bubble Gum Mint, Double Bubble Mint, Mosquito Plant ■☆

10406 Agastache cana Wooton et Standl. = Agastache cana (Hook.) Wooton et Standl. ■☆

10407 Agastache foeniculum (Pursh) Kuntze;茴萝藿香;Anise Gianthyssop, Anise Hyssop, Anise-hyssop, Blue Giant Hyssop, Fennel Giant-hyssop, Fragrant Giant Hyssop, Giant Hyssop, Hyssop, Lavender Giant Hyssop, Licorice Mint ■☆

10408 Agastache formosana (Hayata) Hayata;台湾藿香;Taiwan Gianthyssop ■

10409 Agastache mexicana (Kunth) Lint et Epling;墨西哥藿香;Lion's Tail, Lion's Tails, Mexican Giant Hyssop, Mexican Giant-hyssop ■☆

10410 Agastache nepetoides (L.) Kuntze;黄藿香;Catnip Giant Hyssop, Catnip Giant-hyssop, Giant Hyssop, Yellow Giant Hyssop, Yellow Giant-hyssop ■☆

10411 Agastache pallida (Lindl.) Cory;苍白藿香;Pale Giant Hyssop ■☆

10412 Agastache pallidiflora A. Heller et Rydb.;苍白花藿香;Bill Williams Mountain Giant Hyssop ■☆

10413 Agastache pringlei (Briq.) Lint et Epling var. verticillata (Wooton et Standl.) R. W. Sanders;轮生藿香;Organ Mountain Giant Hyssop ■☆

10414 Agastache rugosa (Fisch. et C. A. Mey.) Kuntze;藿香(八蒿,把蒿,白薄荷,白荷,薄荷,苍告,大薄荷,大叶薄荷,兜类婆草,兜娄婆香,合香,红花小茴香,鸡苏,家茴香,拉拉香,猫巴蒿,猫巴虎,猫尾巴香,排香草,青茎薄荷,仁丹草,山薄荷,山灰香,山茴香,山猫巴,水麻叶,苏藿香,土藿香,香荆芥花,香薷,小薄荷,杏仁花,野薄荷,野藿香,野苏子,叶藿香,鱼香,鱼子苏,紫苏草);Chinese Giant Hyssop, Wrinkled Gianthyssop ■

10415 Agastache rugosa (Fisch. et C. A. Mey.) Kuntze f. albiflora Kawano;白花藿香■☆

10416 Agastache rugosa (Fisch. et C. A. Mey.) Kuntze f. hypoleuca (Kudo) H. Hara = Agastache rugosa (Fisch. et C. A. Mey.) Kuntze ■

10417 Agastache rugosa (Fisch. et C. A. Mey.) Kuntze f. lanceolata Kudo = Agastache rugosa (Fisch. et C. A. Mey.) Kuntze ■

10418 Agastache rugosa (Fisch. et C. A. Mey.) Kuntze var. hypoleuca Kudo = Agastache rugosa (Fisch. et C. A. Mey.) Kuntze ■

10419 Agastache rupestris Standl.;岩生藿香;Threadleaf Giant Hyssop ■☆

10420 Agastache scrophulariifolia (Willd.) Kuntze;玄参叶藿香;Figwort Giant Hyssop, Purple Giant Hyssop, Purple Giant-hyssop ■☆

10421 Agastache scrophulariifolia (Willd.) Kuntze var. mollis (Fernald) A. Heller;绢毛玄参叶藿香■☆

10422 Agastache urticifolia Kuntze;荨麻叶藿香;Giant Hyssop, Horse Mint, Horse Nettle, Nettle-leaf Hyssop ■☆

10423 Agastache wrightii Wooton et Standl.;赖氏藿香;Wright's Giant Hyssop ■☆

10424 Agastachis Poir. = Agastachys R. Br. ●☆

10425 Agastachys Ehrh. = Carex L. ■

10426 Agastachys R. Br. (1810);多穗山龙眼属●☆

10427 Agastachys odorata R. Br.;多穗山龙眼●☆

10428 Agasthiyamalaia S. Rajkumar et Janarth. (2007);印度稀花藤黄属●☆

10429 Agasthiyamalaia S. Rajkumar et Janarth. = Poeciloneuron Bedd. ●

10430 Agasthosma Brongn. = Agathosma Willd. (保留属名)●☆

10431 Agastianis Raf. = Broussonetia Ortega(废弃属名)●■

10432 Agastianis Raf. = Sophora L. ●■

10433 Agasulis Raf. = Ferula L. ■

10434 Agasyllis Hoffm. = Agasyllis Spreng. ■☆

10435 Agasyllis Spreng. (1813);高加索草属■☆

10436 Agasyllis caucasica Spreng.;高加索草■☆

10437 Agasyllis galbanum (L.) Spreng. = Peucedanum galbanum (L.) Drude ■☆

10438 Agasyllis latifolia Boiss.;宽叶高加索草☆

10439 Agatea A. Gray(1852);宿爿堇属■☆

10440 Agatea W. Rich ex A. Gray = Agatea A. Gray ■☆

10441 Agatea W. Rich ex A. Gray = Crossostylis J. R. Forst. et G. Forst. ●☆

10442 Agatea W. Rich ex A. Gray = Haplopetalon A. Gray ●☆

10443 Agatea violaris A. Gray;宿爿堇■☆

10444 Agathaea Cass. = Aster L. + Felicia Cass. (保留属名)●■

10445 Agathaea Cass. = Felicia Cass. (保留属名)●■

10446 Agathaea abyssinica Hochst. ex A. Rich. = Felicia dentata (A. Rich.) Dandy ■☆

10447 Agathaea amelloides (L.) DC. = Felicia amelloides (L.) Voss ■☆

10448 Agathaea amelloides (L.) DC. var. kraussii (Sch. Bip.) Sch. Bip. = Felicia aethiopica (Burm. f.) Bolus et Wolley-Dod ex Adamson et T. M. Salter ■☆

10449 Agathaea amoena Sch. Bip. = Felicia amoena (Sch. Bip.) Levyns ■☆

10450 Agathaea barbata DC. = Felicia ovata (Thunb.) Compton ●☆

10451 Agathaea barbata DC. var. subalbescens = Felicia ovata (Thunb.) Compton ●☆

10452 Agathaea bergeriana (Spreng.) DC. = Felicia bergeriana (Spreng.) O. Hoffm. ■☆

10453 Agathaea brevifolia DC. = Felicia brevifolia (DC.) Grau ■☆

10454 Agathaea coelestis Cass. = Felicia amelloides (L.) Voss ■☆

10455 Agathaea corymbosa Turcz. = Gymnostephium corymbosum (Turcz.) Harv. ■☆

10456 Agathaea dentata A. Rich. = Felicia dentata (A. Rich.) Dandy ■☆

10457 Agathaea diffusa DC. = Felicia diffusa (DC.) Grau ■☆

10458 Agathaea ecklonis (Less.) Nees = Felicia aethiopica (Burm. f.) Bolus et Wolley-Dod ex Adamson et T. M. Salter subsp. ecklonis (Less.) Grau ■☆

10459 Agathaea elongata (Thunb.) Nees = Felicia elongata (Thunb.) O. Hoffm. ■☆

10460 Agathaea hirta DC. = Felicia aculeata Grau ■☆

10461 Agathaea hispida DC. = Felicia hispida (DC.) Grau ●☆

10462 Agathaea hispida DC. var. barbigera ? = Felicia hispida (DC.) Grau ●☆

10463 Agathaea kraussii Sch. Bip. = Felicia aethiopica (Burm. f.) Bolus et Wolley-Dod ex Adamson et T. M. Salter ■☆

10464 Agathaea leiocarpa DC. = Aster leiocarpus (DC.) Harv. ■☆

10465 Agathaea linifolia (L.) G. Don ex Loudon = Euryops linifolius (L.) DC. ■☆

10466 Agathaea microphylla Cass. = Felicia aethiopica (Burm. f.) Bolus et Wolley-Dod ex Adamson et T. M. Salter ■☆

10467 Agathaea natalensis Sch. Bip. = Felicia rosulata Yeo ●☆

10468 Agathaea rotundifolia (Thunb.) Nees = Felicia amelloides (L.) Voss ■☆

10469 Agathaea scabrida DC. = Felicia scabrida (DC.) Range ●☆

10470 Agathaea serrata (Thunb.) Nees = Felicia serrata (Thunb.) Grau ●☆

10471 Agathaea stricta DC. = Felicia amoena (Sch. Bip.) Levyns subsp. stricta (DC.) Grau ■☆

10472 Agathaea strigosa (A. Spreng.) Nees = Felicia ovata (Thunb.) Compton ●☆

10473 Agathaea tenera DC. = Felicia tenera (DC.) Grau ●☆

10474 Agathaea tricolor Nees = Felicia elongata (Thunb.) O. Hoffm. ■☆

10475 Agathaea zeyheri Nees = Felicia linifolia (Harv.) Grau ■☆

10476 Agathea Endl. = Agathaea Cass. ●■

10477 Agathelepis Reichb. = Agathelpis Choisy ■☆

10478 Agathelpis Choisy = Microdon Choisy ●☆

10479 Agathelpis Choisy(1824);澳非玄参属■☆

10480 Agathelpis adunca E. Mey. = Microdon dubius (L.) Hilliard ●☆

10481 Agathelpis angustifolia Choisy = Microdon dubius (L.) Hilliard ●☆

10482 Agathelpis brevifolia E. Mey. = Microdon dubius (L.) Hilliard ●☆

10483 Agathelpis dubia (L.) Hutch. ex Wijnands = Microdon dubius (L.) Hilliard ●☆

10484 Agathelpis mucronata E. Mey. = Microdon dubius (L.) Hilliard ●☆

10485 Agathelpis nitida E. Mey. = Microdon nitidus (E. Mey.) Hilliard ●☆

10486 Agathelpis parviflora (P. J. Bergius) Choisy = Microdon parviflorus (P. J. Bergius) Hilliard ●☆

10487 Agathidaceae Baum. -Bod. = Araucariaceae Henkel et W. Hochst. (保留科名)●

10488 Agathidaceae Baum. -Bod. ex A. V. Bobrov et Melikyan = Araucariaceae Henkel et W. Hochst. (保留科名)●

10489 Agathidaceae Nakai = Araucariaceae Henkel et W. Hochst. (保留科名)●

10490 Agathidanthes Hassk. = Agathisanthes Blume ●

10491 Agathidanthes Hassk. = Nyssa Gronov. ex L. ●

10492 Agathis Salisb. (1807)(保留属名);贝壳杉属;Damar Pine, Damara Tree, Damar-pine, Dammar Pine, Dammar-pine, Kauri, Kauri Pine, New Zealand Kauri ●

10493 Agathis alba Jeffrey = Agathis dammara (Lamb.) Rich. et A. Rich. ●

10494 Agathis australis (D. Don) Steud.;新西兰贝壳杉(澳大利亚贝壳杉);Cowrie Pine, Dammar Pine, Kauri, Kauri Dammar Pine, Kauri Dammar-pine, Kauri Pine, Kauripine, New Zealand Kauri ●

10495 Agathis borneensis Warb. = Agathis dammara (Lamb.) Rich. et A. Rich. ●

10496 Agathis dammara (Lamb.) Rich. et A. Rich.;贝壳杉;Amboina Pine, Amboina Pitch Tree, Bendang, Bindang, Damar Minyak, Dammar Pine, E. India Copal, Manila Copal ●

10497 Agathis lanceolata (Pancher) Warb.;披针叶贝壳杉;Lanceolate Kauri, Lanceolate Kauri Pine ●☆

10498 Agathis latifolia Meijer Drees;宽叶贝壳杉;Broadleaf Kauri, Broadleaf Kauri Pine ●☆

10499 Agathis loranthifolia Salisb. = Agathis dammara (Lamb.) Rich. et A. Rich. ●

10500 Agathis macrophylla (Lindl.) Mast.;大叶贝壳杉;Dakua, Fijian Kauri, Largeleaf Kauri, Largeleaf Kauri Pine, Pacific Island Kauri, Vanikoro Kauri ●☆

10501 Agathis microstachya J. F. Bailey et C. T. White;细穗贝壳杉;Black Kauri, Black Kauri Pine ●☆

10502 Agathis ovata Warb.;新几内亚贝壳杉;New Guinea Kauri, New Guinea Kauri Pine ●☆

10503 Agathis philippinensis Warb.;菲律宾贝壳杉;Philippine Kauri, Philippine Kauri Pine ●☆

10504 Agathis robusta (C. Moore) F. Muell. = Agathis robusta (F. Muell.) F. M. Bailey ●☆

10505 Agathis robusta (F. Muell.) F. M. Bailey;大果贝壳杉(昆士兰贝壳杉);Australian, Big Dammar Pine, Big-fruited Dammar Pine, Dundathu Pine, Queensland Kauri, Queensland Kauri Pine, Smooth-bark Kauri, South Queensland Kauri, South Queensland Kauri Pine ●☆

10506 Agathis robusta F. M. Bailey = Agathis robusta (F. Muell.) F. M. Bailey ●☆

10507 Agathis robusta F. Muell. = Agathis robusta (F. Muell.) F. M. Bailey ●☆

10508 Agathis robusta F. Muell. subsp. nesophila Whitmore;岛生昆士兰贝壳杉●☆

10509 Agathis vitiensis (Seem.) Benth. et Hook. f. ex Drake;斐济贝壳杉;Dakua Tree, Fiji Kauri, Fijian Kauri Pine ●☆

10510 Agathis vitiensis (Seem.) Benth. et Hook. f. ex Drake = Agathis macrophylla (Lindl.) Mast. ●☆

10511 Agathisanthemum Klotzsch = Oldenlandia L. ●■

10512 Agathisanthemum Klotzsch(1861);团花茜属■☆

10513 Agathisanthemum angolense Bremek. = Agathisanthemum bojeri Klotzsch subsp. angolense (Bremek.) Verdc. ■☆

10514 Agathisanthemum assimile Bremek.;相似团花茜■☆

10515 Agathisanthemum bojeri Klotzsch;博耶尔团花茜■☆

10516 Agathisanthemum bojeri Klotzsch subsp. angolense (Bremek.) Verdc.;安哥拉团花茜■☆

10517 Agathisanthemum bojeri Klotzsch subsp. australe Bremek. = Agathisanthemum bojeri Klotzsch ■☆

10518 Agathisanthemum bojeri Klotzsch var. angolense (Bremek.) Verdc. = Agathisanthemum bojeri Klotzsch subsp. angolense (Bremek.) Verdc. ■☆

10519 Agathisanthemum bojeri Klotzsch var. glabriflorum Bremek. = Agathisanthemum bojeri Klotzsch ■☆

10520 Agathisanthemum bojeri Klotzsch var. linearifolia Verdc.;线叶博耶尔团花茜■☆

10521 Agathisanthemum chlorophyllum (Hochst.) Bremek.;绿团花茜■☆

10522 Agathisanthemum chlorophyllum (Hochst.) Bremek. var. pubescens Bremek.;毛绿团花茜■☆

10523 Agathisanthemum globosum (Hochst. ex A. Rich.) Bremek.;球团花茜■☆

10524 Agathisanthemum globosum (Hochst. ex A. Rich.) Bremek. var. subglabrum Bremek. = Agathisanthemum globosum (Hochst. ex A. Rich.) Bremek. ■☆

10525 Agathisanthemum petersii Klotzsch = Agathisanthemum globosum (Hochst. ex A. Rich.) Bremek. ■☆

10526 Agathisanthemum quadricostatum Bremek. = Agathisanthemum globosum (Hochst. ex A. Rich.) Bremek. ■☆

10527 Agathisanthemum quadricostatum Bremek. var. pubescens ? = Agathisanthemum bojeri Klotzsch var. angolense (Bremek.) Verdc. ■☆

10528 Agathisanthes Blume = Nyssa Gronov. ex L. ●

10529 Agathisanthes javanica Blume = Nyssa javanica (Blume) Wangerin ●

10530 Agathodes Rchb. = Agathotes D. Don ■

10531 Agathodes Rchb. = Swertia L. ■

10532 Agathomeria Baill. = Agathomeris Delaun. ●☆

10533 Agathomeris Deaun. = Calomeria Vent. ●■☆

10534 Agathomeris Delaun. = Humea Sm. ●☆

10535 Agathomeris Delaun. ex DC. = Humea Sm. ●☆

10536 Agathomoris Durand = Agathomeris Delaun. ●☆

10537 Agathophora (Fenzl) Bunge = Halogeton C. A. Mey. ●■

10538 Agathophora Bunge = Halogeton C. A. Mey. ●■

10539 Agathophora alopecuroides (Delile) Bunge = Halogeton alopecuroides (Delile) Moq. ■☆

10540 Agathophora alopecuroides (Delile) Bunge var. papillosa (Maire) Boulos = Agathophora alopecuroides (Delile) Bunge ■☆

10541 Agathophora alopecuroides (Delile) Bunge var. papillosa (Maire) Boulos = Halogeton alopecuroides (Delile) Moq. ■☆

10542 Agathophora galalensis Botsch. = Agathophora alopecuroides (Delile) Bunge ■☆

10543 Agathophora galalensis Botsch. = Halogeton alopecuroides (Delile) Moq. ■☆

10544 Agathophora postii (Eig) Botsch. = Agathophora alopecuroides (Delile) Bunge ■☆

10545 Agathophora postii (Eig) Botsch. = Halogeton alopecuroides (Delile) Moq. ■☆

10546 Agathophyllum Blume = Ocotea Aubl. ●☆

10547 Agathophyllum Juss. (1789);佳叶樟属(芳香厚壳桂属,拉文萨拉属)●☆

10548 Agathophyllum Juss. = Cryptocarya R. Br. (保留属名)●

10549 Agathophyllum Juss. = Ravensara Sonn. (废弃属名)●

10550 Agathophyllum aromaticum Willd.;芳香佳叶樟(芳香厚壳桂,马岛丁香);Madagascar Clove, Madagascar Cloves, Madagascar Nutmeg ●☆

10551 Agathophyllum parvifolium (Scott-Elliot) Palacky = Aspidostemon parvifolium (Scott-Elliot) van der Werff ●☆

10552 Agathophyton Moq. = Agathophytum Moq. ●■

10553 Agathophytum Moq. = Anserina Dumort. ●■

10554 Agathophytum Moq. = Chenopodium L. ●■

10555 Agathorhiza Raf. = Archangelica Hoffm. ■

10556 Agathorhiza Raf. = Archangelica Wolf ■

10557 Agathosma Willd. (1809)(保留属名);香芸木属(布楚属,布枯属,线球香属);Buchu ●☆

10558 Agathosma abrupta Pillans;平截香芸木●☆

10559 Agathosma acerosa Eckl. et Zeyh. = Agathosma bisulca (Thunb.) Bartl. et H. L. Wendl. ●☆

10560 Agathosma acocksii Pillans;阿氏香芸木●☆

10561 Agathosma acutifolia Sond. = Agathosma recurvifolia Sond. ●☆

10562 Agathosma acutissima Dümmer;尖香芸木●☆

10563 Agathosma adenocaulis Kuntze = Agathosma hirta (Lam.) Bartl. et H. L. Wendl. ●☆

10564 Agathosma adnata Pillans;贴生香芸木●☆

10565 Agathosma aemula Schltr.;匹敌香芸木●☆

10566 Agathosma affinis Sond.;近缘香芸木●☆

10567 Agathosma alpina Schltr.;高山香芸木●☆

10568 Agathosma alticola Schltr. ex Dümmer;高原香芸木●☆

10569 Agathosma anomala E. Mey. ex Sond.;异常香芸木●☆

10570 Agathosma apiculata G. Mey.;细尖香芸木●☆

10571 Agathosma asperifolia Eckl. et Zeyh.;糙叶香芸木●☆

10572 Agathosma barosmifolia Eckl. et Zeyh. = Agathosma bisulca (Thunb.) Bartl. et H. L. Wendl. ●☆

10573 Agathosma barosmoides Sond. = Agathosma puberula (Steud.) Fourc. ●☆

10574 Agathosma bathii (Dümmer) Pillans;巴斯香芸木●☆

10575 Agathosma betulina (P. J. Bergius) Pillans;圆叶香芸木(短叶布枯,圆布枯,圆海布枯,圆叶布楚);Buchu,Round-leaved Buchu ●☆

10576 Agathosma bicolor Dümmer;二色香芸木●☆

10577 Agathosma bicornuta R. A. Dyer;二角香芸木●☆

10578 Agathosma bifida (Jacq.) Bartl. et H. L. Wendl.;二裂香芸木●☆

10579 Agathosma bisulca (Thunb.) Bartl. et H. L. Wendl.;二沟香芸木●☆

10580 Agathosma blaerioides Cham. et Schltdl.;欧石南香芸木●☆

10581 Agathosma bodkinii Dümmer;博德金香芸木●☆

10582 Agathosma burchellii Dümmer = Agathosma foetidissima (Bartl. et H. L. Wendl.) Steud ●☆

10583 Agathosma capensis (L.) Dümmer;好望角香芸木●☆

10584 Agathosma capitata Sond.;头状香芸木●☆

10585 Agathosma cerefolia (Vent.) Bartl. et H. L. Wendl.;蜡叶香芸木●☆

10586 Agathosma chortophila Eckl. et Zeyh. = Agathosma capensis (L.) Dümmer ●☆

10587 Agathosma ciliaris (L.) Druce;缘毛香芸木●☆

10588 Agathosma ciliata (L.) Link;睫毛香芸木●☆

10589 Agathosma clavisepala R. A. Dyer;棒萼香芸木●☆

10590 Agathosma collina Eckl. et Zeyh.;山丘香芸木●☆
10591 Agathosma commutata Sond. = Agathosma virgata (Lam.) Bartl. et H. L. Wendl. ●☆
10592 Agathosma concava Pillans;凹香芸木●☆
10593 Agathosma conferta Pillans;密集香芸木●☆
10594 Agathosma cordifolia Pillans;心叶香芸木●☆
10595 Agathosma corymbosa (Montin) G. Don;伞序香芸木●☆
10596 Agathosma craspedota Sond.;有用香芸木●☆
10597 Agathosma crassifolia Sond.;厚叶香芸木●☆
10598 Agathosma crenulata (L.) Pillans;锯齿香芸木(锯齿布楚,卵布枯,卵海不枯,卵叶布枯);Buchu ●☆
10599 Agathosma cuspidata (J. C. Wendl.) Bartl. et H. L. Wendl. = Agathosma serpyllacea Licht. ex Roem. et Schult. ●☆
10600 Agathosma decora Dümmer = Agathosma pentachotoma E. Mey. ex Sond. ●☆
10601 Agathosma decumbens Eckl. et Zeyh. = Agathosma capensis (L.) Dümmer ●☆
10602 Agathosma decurrens Pillans;下延香芸木●☆
10603 Agathosma delicatula Compton = Agathosma capensis (L.) Dümmer ●☆
10604 Agathosma dentata Pillans;齿香芸木●☆
10605 Agathosma denticulata Dümmer = Agathosma craspedota Sond. ●☆
10606 Agathosma dielsiana Schltr. ex Dümmer;迪尔斯香芸木●☆
10607 Agathosma distans Pillans;远离香芸木●☆
10608 Agathosma divaricata Pillans;叉开香芸木●☆
10609 Agathosma dodii Dümmer = Agathosma serpyllacea Licht. ex Roem. et Schult. ●☆
10610 Agathosma dregeana Sond.;德雷香芸木●☆
10611 Agathosma eckloniana Schltdl. = Agathosma serpyllacea Licht. ex Roem. et Schult. ●☆
10612 Agathosma elata Sond.;高香芸木●☆
10613 Agathosma elegans Cham. et Schltdl.;雅致香芸木●☆
10614 Agathosma erecta (J. C. Wendl.) Bartl. et H. L. Wendl. = Agathosma capensis (L.) Dümmer ●☆
10615 Agathosma eriantha (Steud.) Steud.;毛花香芸木●☆
10616 Agathosma ericoides Schltdl. = Agathosma capensis (L.) Dümmer ●☆
10617 Agathosma esterhuyseniae Pillans;埃斯特香芸木●☆
10618 Agathosma fastigiata Eckl. et Zeyh. = Agathosma capensis (L.) Dümmer ●☆
10619 Agathosma filamentosa Schltr. = Agathosma serpyllacea Licht. ex Roem. et Schult. ●☆
10620 Agathosma filipetala Eckl. et Zeyh. = Agathosma stenopetala (Steud.) Steud. ●☆
10621 Agathosma florida Sond.;佛罗里达香芸木●☆
10622 Agathosma florulenta Sond.;多花香芸木●☆
10623 Agathosma foetidissima (Bartl. et H. L. Wendl.) Steud.;臭香芸木●☆
10624 Agathosma foliosa Sond. = Agathosma marifolia Eckl. et Zeyh. ●☆
10625 Agathosma froemblingii Dümmer = Agathosma spinescens Dümmer ●☆
10626 Agathosma geniculata Pillans;膝曲香芸木●☆
10627 Agathosma gibbosa Dümmer = Agathosma serpyllacea Licht. ex Roem. et Schult. ●☆
10628 Agathosma giftbergensis E. Phillips;吉夫特香芸木●☆
10629 Agathosma gillivrayi Sond. = Agathosma capensis (L.) Dümmer ●☆
10630 Agathosma glabrata Bartl. et H. L. Wendl.;光滑香芸木●☆
10631 Agathosma glandulosa (Thunb.) Sond.;腺香芸木●☆
10632 Agathosma gnidiiflora Dümmer;格尼瑞香香芸木●☆
10633 Agathosma gnidioides Schltdl. = Agathosma puberula (Steud.) Fourc. ●☆
10634 Agathosma gracilicaulis Sond. = Agathosma bifida (Jacq.) Bartl. et H. L. Wendl. ●☆
10635 Agathosma hirsuta Pillans;粗毛香芸木●☆
10636 Agathosma hirta (Lam.) Bartl. et H. L. Wendl.;多毛香芸木●☆
10637 Agathosma hirtella Sond. = Agathosma capensis (L.) Dümmer ●☆
10638 Agathosma hispida (Thunb.) Bartl. et H. L. Wendl.;毛香芸木●☆
10639 Agathosma hookeri Sond.;胡克香芸木●☆
10640 Agathosma humilis Sond.;低矮香芸木●☆
10641 Agathosma imbricata (L.) Willd.;覆瓦香芸木●☆
10642 Agathosma insignis (Compton) Pillans;显著香芸木●☆
10643 Agathosma involucrata Eckl. et Zeyh.;总苞香芸木●☆
10644 Agathosma joubertiana Schltdl.;朱伯特香芸木●☆
10645 Agathosma juniperifolia Bartl.;刺柏叶香芸木●☆
10646 Agathosma kougaense Pillans;科加香芸木●☆
10647 Agathosma krakadouwensis Dümmer;克拉卡多香芸木●☆
10648 Agathosma lactea Schltr. = Agathosma capensis (L.) Dümmer ●☆
10649 Agathosma lanceolata (L.) Engl.;剑叶香芸木●☆
10650 Agathosma lancifolia Eckl. et Zeyh.;披针叶香芸木●☆
10651 Agathosma latipetala Sond.;宽瓣香芸木●☆
10652 Agathosma lediformis Eckl. et Zeyh. = Agathosma bifida (Jacq.) Bartl. et H. L. Wendl. ●☆
10653 Agathosma leptospermoides Sond.;细籽香芸木●☆
10654 Agathosma linifolia (Roem. et Schult.) Licht. ex Bartl. et H. L. Wendl.;亚麻叶香芸木●☆
10655 Agathosma longicornu Pillans;长角香芸木●☆
10656 Agathosma lycopodioides Bartl. et H. L. Wendl. = Agathosma imbricata (L.) Willd. ●☆
10657 Agathosma marifolia Eckl. et Zeyh.;芋叶香芸木●☆
10658 Agathosma marlothii Dümmer;马洛斯香芸木●☆
10659 Agathosma melaleucoides Sond. = Agathosma cerefolium (Vent.) Bartl. et H. L. Wendl. ●☆
10660 Agathosma microcalyx Dümmer;小萼香芸木●☆
10661 Agathosma microcarpa (Sond.) Pillans;小果香芸木●☆
10662 Agathosma microphylla G. Mey. ex Bartl. et H. L. Wendl. = Agathosma cerefolium (Vent.) Bartl. et H. L. Wendl. ●☆
10663 Agathosma minuta Schltdl.;微小香芸木●☆
10664 Agathosma mirabilis Pillans;奇异香芸木●☆
10665 Agathosma mixta Dümmer = Agathosma peglerae Dümmer ●☆
10666 Agathosma montana Schltdl. = Agathosma bifida (Jacq.) Bartl. et H. L. Wendl. ●☆
10667 Agathosma monticola Sond. = Agathosma bifida (Jacq.) Bartl. et H. L. Wendl. ●☆
10668 Agathosma mucronulata Sond.;微凸香芸木●☆
10669 Agathosma mucronulata Sond. var. rudolphii Dümmer = Agathosma mucronulata Sond. ●☆
10670 Agathosma muirii E. Phillips;缪里香芸木●☆
10671 Agathosma muizenbergensis Dümmer = Agathosma capensis (L.) Dümmer ●☆
10672 Agathosma mundtii Cham. et Schltdl.;蒙特香芸木●☆
10673 Agathosma namaquensis Pillans;纳马夸香芸木●☆
10674 Agathosma neglecta Dümmer = Agathosma capensis (L.) Dümmer ●☆

10675 Agathosma nigra Eckl. et Zeyh. = Agathosma capensis (L.) Dümmer ●☆
10676 Agathosma nigromontana Eckl. et Zeyh. = Agathosma bifida (Jacq.) Bartl. et H. L. Wendl. ●☆
10677 Agathosma nivenii Sond. = Agathosma mundtii Cham. et Schltdl. ●☆
10678 Agathosma odoratissima (Montin) Pillans;极香香芸木●☆
10679 Agathosma orbicularis (Thunb.) Bartl. et H. L. Wendl.;圆形香芸木●☆
10680 Agathosma ovalifolia Pillans;卵叶香芸木●☆
10681 Agathosma ovata (Thunb.) Pillans;椭圆香芸木●☆
10682 Agathosma pallens Pillans;变苍白香芸木●☆
10683 Agathosma parviflora (Roem. et Schult.) Bartl. et H. L. Wendl. = Agathosma virgata (Lam.) Bartl. et H. L. Wendl. ●☆
10684 Agathosma patula G. Mey. = Agathosma capensis (L.) Dümmer ●☆
10685 Agathosma peglerae Dümmer;佩格拉香芸木●☆
10686 Agathosma pentachotoma E. Mey. ex Sond.;五叉香芸木●☆
10687 Agathosma perdita Hutch. = Agathosma lanceolata (L.) Engl. ●☆
10688 Agathosma phillipsii Dümmer;菲利香芸木●☆
10689 Agathosma pilifera Schltdl.;纤毛香芸木●☆
10690 Agathosma planifolia Sond.;平叶香芸木●☆
10691 Agathosma platypetala Eckl. et Zeyh. = Agathosma capensis (L.) Dümmer ●☆
10692 Agathosma propinqua Sond.;邻近香芸木●☆
10693 Agathosma puberula (Steud.) Fourc.;短柔毛香芸木●☆
10694 Agathosma pubescens Sond. = Agathosma elegans Cham. et Schltdl. ●☆
10695 Agathosma pubigera Sond.;短毛香芸木●☆
10696 Agathosma pulchella (L.) Link;香芸木(茴香布枯)●☆
10697 Agathosma pulchella Link = Agathosma pulchella (L.) Link ●☆
10698 Agathosma punctata Sond. = Agathosma ovalifolia Pillans ●☆
10699 Agathosma pungens (E. Mey. ex Sond.) Pillans;刺香芸木●☆
10700 Agathosma purpurea Pillans;紫芸木●☆
10701 Agathosma recurvifolia Sond.;折叶香芸木●☆
10702 Agathosma rehmanniana Dümmer;雷曼香芸木●☆
10703 Agathosma riversdalensis Dümmer;里弗斯代尔香芸木●☆
10704 Agathosma robusta Eckl. et Zeyh.;粗壮香芸木●☆
10705 Agathosma roodebergensis Compton;鲁德伯格香芸木●☆
10706 Agathosma rosmarinifolia (Bartl.) I. Williams;迷迭香叶香芸木●☆
10707 Agathosma rubra Willd. et Licht. ex Bartl. et H. L. Wendl. = Agathosma serpyllacea Licht. ex Roem. et Schult. ●☆
10708 Agathosma rubricaulis Dümmer;红茎香芸木●☆
10709 Agathosma rudolphii I. Williams;鲁道夫香芸木●☆
10710 Agathosma rugosa (Thunb.) Link = Agathosma ciliaris (L.) Druce ●☆
10711 Agathosma sabulosa Sond.;砂地香芸木●☆
10712 Agathosma salina Eckl. et Zeyh.;盐地香芸木●☆
10713 Agathosma scaberula Dümmer;粗糙香芸木●☆
10714 Agathosma schlechtendalii Sond. = Agathosma eriantha (Steud.) Steud. ●☆
10715 Agathosma serpyllacea Licht. ex Roem. et Schult.;百里香香芸木●☆
10716 Agathosma serratifolia (Curtis) Spreeth;长叶香芸木(长叶布枯)●☆
10717 Agathosma sladeniana P. E. Glover;斯莱登香芸木●☆
10718 Agathosma sonderiana Dümmer = Agathosma mundtii Cham. et Schltdl. ●☆
10719 Agathosma spicata Lichtst. ex G. Don;穗状香芸木(小穗线球香)●☆
10720 Agathosma spinescens Dümmer;细刺香芸木●☆
10721 Agathosma spinosa Sond.;具刺香芸木●☆
10722 Agathosma squamosa (Roem. et Schult.) Bartl. et H. L. Wendl.;多鳞香芸木●☆
10723 Agathosma stenopetala (Steud.) Steud.;窄瓣香芸木●☆
10724 Agathosma stenosepala Pillans;窄萼香芸木●☆
10725 Agathosma stilbeoides Dümmer;槌状香芸木●☆
10726 Agathosma stipitata Pillans;具柄香芸木●☆
10727 Agathosma stokoei Pillans;斯托克香芸木●☆
10728 Agathosma struthioloides Dümmer = Agathosma bifida (Jacq.) Bartl. et H. L. Wendl. ●☆
10729 Agathosma subteretifolia Pillans;圆柱叶香芸木●☆
10730 Agathosma tabularis Sond.;扁平香芸木●☆
10731 Agathosma tenuis Sond. = Agathosma capensis (L.) Dümmer ●☆
10732 Agathosma thymifolia Schltdl.;百里香叶香芸木●☆
10733 Agathosma trichocarpa Holmes;毛果香芸木●☆
10734 Agathosma tulbaghensis Dümmer;塔尔巴赫香芸木●☆
10735 Agathosma umbellata (Thunb.) Sond. = Agathosma bifida (Jacq.) Bartl. et H. L. Wendl. ●☆
10736 Agathosma umbonata Pillans;脐突香芸木●☆
10737 Agathosma unicarpellata (Fourc.) Pillans;单果香芸木●☆
10738 Agathosma utilis Dümmer = Agathosma craspedota Sond. ●☆
10739 Agathosma variabilis Sond. = Agathosma capensis (L.) Dümmer ●☆
10740 Agathosma ventenatiana (Roem. et Schult.) Bartl. et H. L. Wendl. = Agathosma corymbosa (Montin) G. Don ●☆
10741 Agathosma venusta (Eckl. et Zeyh.) Pillans;雅丽香芸木●☆
10742 Agathosma villosa (Willd.) Willd. = Agathosma glabrata Bartl. et H. L. Wendl. ●☆
10743 Agathosma virgata (Lam.) Bartl. et H. L. Wendl.;小花香芸木●☆
10744 Agathosma viscida Dümmer = Agathosma eriantha (Steud.) Steud. ●☆
10745 Agathosma wrightii MacOwan = Agathosma lanceolata (L.) Engl. ●☆
10746 Agathosma zwartbergense Pillans;茨瓦特伯格香芸木●☆
10747 Agathosoma N. T. Burb. = Agathosma Willd.(保留属名)●☆
10748 Agathotes D. Don = Swertia L. ■
10749 Agathotes nervosa G. Don = Swertia nervosa (G. Don) Wall. ex C. B. Clarke ■
10750 Agathotes nervosa Wall. ex G. Don = Swertia nervosa (G. Don) Wall. ex C. B. Clarke ■
10751 Agathyrsus D. Don = Lactuca L. ■
10752 Agathyrus B. D. Jacks. = Agathyrsus D. Don ■
10753 Agathyrus Raf. = ? Lactuca L. ■
10754 Agathyrus tataricus (L.) D. Don = Mulgedium tataricum (L.) DC. ■
10755 Agathyrus tataricus D. Don = Mulgedium tataricum (L.) DC. ■
10756 Agati Adans.(废弃属名) = Sesbania Scop.(保留属名)●■
10757 Agati grandiflora (L.) Desv. = Sesbania grandiflora (L.) Pers. ●
10758 Agatia Reichb. = Sesbania Scop.(保留属名)●■
10759 Agation Brongn. = Agatea A. Gray ■☆
10760 Agatophyllum Comm. ex Thouars = Agathophyllum Juss. ●☆
10761 Agatophyllum Comm. ex Thouars = Ravensara Sonn.(废弃属名)●

10762 Agatophyton Fourr. = Agathophytum Moq. ●■

10763 Agauria (DC.) Benth. = Agarista D. Don ex G. Don ●☆

10764 Agauria (DC.) Benth. et Hook. f. (1876);绊足花属●☆

10765 Agauria (DC.) Benth. et Hook. f. = Agarista D. Don ex G. Don ●☆

10766 Agauria (DC.) Hook. f. = Agarista D. Don ex G. Don ●☆

10767 Agauria Benth. et Hook. f. = Agarista D. Don ex G. Don ●☆

10768 Agauria goetzei Engl. = Agauria salicifolia (Comm. ex Lam.) Hook. f. ex Oliv. ●☆

10769 Agauria salicifolia (Comm. ex Lam.) Hook. f. ex Oliv.;绊足花●☆

10770 Agauria salicifolia (Comm. ex Lam.) Hook. f. ex Oliv. f. adenantha Sleumer = Agauria salicifolia (Comm. ex Lam.) Hook. f. ex Oliv. ●☆

10771 Agauria salicifolia (Comm. ex Lam.) Hook. f. ex Oliv. f. glandulosa Sleumer = Agauria salicifolia (Comm. ex Lam.) Hook. f. ex Oliv. ●☆

10772 Agauria salicifolia (Comm. ex Lam.) Hook. f. ex Oliv. var. intercedens Sleumer = Agauria salicifolia (Comm. ex Lam.) Hook. f. ex Oliv. ●☆

10773 Agauria salicifolia (Comm. ex Lam.) Hook. f. ex Oliv. var. latissima Engl. = Agauria salicifolia (Comm. ex Lam.) Hook. f. ex Oliv. ●☆

10774 Agauria salicifolia (Comm. ex Lam.) Hook. f. ex Oliv. var. pyrifolia (Pers.) Oliv. = Agauria salicifolia (Comm. ex Lam.) Hook. f. ex Oliv. ●☆

10775 Agavaceae Dumort. (1829)(保留科名);龙舌兰科;Agave Family,Genturyplant Family ●■

10776 Agavaceae Dumort.(保留科名) = Haemodoraceae R. Br.(保留科名)■☆

10777 Agavaceae Endl. = Agavaceae Dumort.(保留科名)●■

10778 Agave L. (1753);龙舌兰属;Agave, Century Plant, Centuryplant, Ixtle Fibre, Ixtli Fibre, Keratto, Maguey, Tampico Fibre,Tequila ■

10779 Agave abrupta Trel. = Agave americana L. var. expansa (Jacobi) Gentry ■☆

10780 Agave acuispina Trel. = Agave cantula Roxb. ex Salm-Dyck var. acuispina (Trel.) Gentry ■☆

10781 Agave albicans Jacobi = Agave mitis Mart. var. albidior (Salm-Dyck) B. Ullrich ■☆

10782 Agave albomarginata Gentry;白边龙舌兰;White-margined Agave ■☆

10783 Agave aloina C. Koch = Agave pendula Schnitts. ■☆

10784 Agave aloina K. Koch = Agave pendula Schnitts. ■☆

10785 Agave altissima Zumagl. = Agave americana L. ■

10786 Agave amaniensis Trel. et Nowell;蓝剑麻■☆

10787 Agave americana L.;龙舌兰(菠萝麻,番麻,剑兰,石莲花,世纪树,洋棕);American Agave, American Aloe, Century Plant, Century Plant Agave, Centuryplant, Centuryplant Agave, Maguey, Mescal,Sisal Plant,South American Agave ■

10788 Agave americana L. 'Marginata';黄心龙舌兰(割舌兰,黄斑龙舌兰,黄边龙舌兰,黄绿龙舌兰,金边百年兰,金边菠萝,金边菠萝麻,金边假菠萝,金边兰,金边莲,金边龙舌兰,金心龙舌兰,龙舌兰,一片消,银边龙舌兰);Am Aloe, Century Plant, Golden Margin Agave,Striped-edge Agave,Yellow-margin Agave ■

10789 Agave americana L. 'Striata' = Agave americana L. var. striata Trel. ■☆

10790 Agave americana L. var. latifolia Torr. = Agave parryi Engelm. ■☆

10791 Agave americana L. var. marginata Trel. = Agave americana L. ■

10792 Agave americana L. var. mediopicta Trel. = Agave americana L. 'Marginata' ■

10793 Agave americana L. var. oaxacensis Gentry;墨西哥龙舌兰;Century Plant ■☆

10794 Agave americana L. var. striata Trel.;白心龙舌兰(条斑龙舌兰)■☆

10795 Agave americana L. var. theometel (Zucc.) A. Terracc. = Agave americana L. ■

10796 Agave americana L. var. variegata Hibberd. = Agave americana L. 'Marginata' ■

10797 Agave angustifolia Haw.;狭叶龙舌兰(短叶龙舌兰,假菠萝麻);Maguey Lechugilla,Narrowleaf Agave,Narrow-leaf Agave ■

10798 Agave angustifolia Haw. = Agave vivipara L. ■☆

10799 Agave angustifolia Haw. var. marginata Trel.;白缘龙舌兰(金边狭叶龙舌兰)■☆

10800 Agave angustissima Engelm.;线叶龙舌兰(白丝)■☆

10801 Agave angustissima Engelm. var. woodrowii Hort.;日闪光■☆

10802 Agave arizonica Gentry et J. H. Weber;亚利桑那龙舌兰;Arizona Agave ■☆

10803 Agave aspessima Jacobi.;墨西哥糙叶龙舌兰;Rough Leaved Agave ■☆

10804 Agave atrovirens Karw. = Agave salmiana Otto ■☆

10805 Agave atrovirens Karw. var. sigmatophylla Berger = Agave salmiana Otto ■☆

10806 Agave attenuata Salm-Dyck;翠绿龙舌兰(狐尾龙舌兰);Agave, Fox Tail Agave ■☆

10807 Agave aurea Brandegee;黄龙舌兰■☆

10808 Agave bakeri H. Ross = Agave gilberti A. Berger ■☆

10809 Agave bollii A. Terracc. var. celsiana A. Terracc. = Agave mitis Mart. ■☆

10810 Agave bollii A. Terracc. var. celsiana A. Terracc. = Agave moranii Gentry ■☆

10811 Agave botterii Baker = Agave mitis Mart. ■☆

10812 Agave bouchei Jacobi = Agave mitis Mart. ■☆

10813 Agave bovicoruta S. Watson ex Engelm.;棱舌兰;Cow Horn Agave,Lechuguilla Verde ■☆

10814 Agave breedlovei Gentry = Agave angustifolia Haw. ■

10815 Agave cantula Roxb. = Agave cantula Roxb. ex Salm-Dyck ■

10816 Agave cantula Roxb. ex Salm-Dyck;马盖麻(菲律宾马盖麻,狭叶龙舌兰,狭叶龙舌兰麻,亚洲马盖麻);Bombay Aloe, Cantala, Cantala Fibre,Maguey,Manila Maguey,Philippine Agave ■

10817 Agave cantula Roxb. ex Salm-Dyck var. acuispina (Trel.) Gentry;小刺马盖麻■☆

10818 Agave celsiana Jacob = Agave mitis Mart. ■☆

10819 Agave celsii Hook. = Agave mitis Mart. ■☆

10820 Agave celsii Hook. = Agave moranii Gentry ■☆

10821 Agave celsii Hook. var. albicans (Jacobi) Gentry = Agave mitis Mart. var. albidior (Salm-Dyck) B. Ullrich ■☆

10822 Agave cernua A. Berger = Agave attenuata Salm-Dyck ■☆

10823 Agave chihuahuana Trel. = Agave parryi Engelm. ■☆

10824 Agave chrysantha Peebles;金花龙舌兰;Golden-flowered Agave ■☆

10825 Agave coarctata Jacobi = Agave salmiana Otto ■☆

10826 Agave coccinea Roezl ex Jacobi = Agave salmiana Otto ■☆

10827 Agave cochlearis Jacobi = Agave salmiana Otto ■☆

10828 Agave coespitosa Tod. = Agave pendula Schnitts. ■☆

10829 Agave collina Greenm.;鬼子母神■☆

10830 Agave colorata Gentry;着色龙舌兰;Mescal Ceniza ■☆

10831 Agave complicata Trel. ex Ochot. = Agave americana L. ■
10832 Agave complicata Trel. ex Ochot. = Agave salmiana Otto ■☆
10833 Agave concinna Angl. ex Baker = Agave mitis Mart. var. albidior (Salm-Dyck) B. Ullrich ■☆
10834 Agave costaricana Gentry = Agave angustifolia Haw. ■
10835 Agave crassispina Trel. = Agave salmiana Otto ■☆
10836 Agave crenata Jacobi;波缘龙舌兰■☆
10837 Agave cubensis Jacq. = Furcraea cubensis Vent. ■☆
10838 Agave cucullata Lem. ex Jacobi;圆帽龙舌兰■☆
10839 Agave dealbata Lem. ex Jacobi;苍白龙舌兰■☆
10840 Agave debaryana Jacobi = Agave attenuata Salm-Dyck ■☆
10841 Agave decipiens Baker;三角龙舌兰;False Sisal ■
10842 Agave densiflora Hook. = Agave mitis Mart. ■☆
10843 Agave deserti Engelm.;荒漠龙舌兰;Desert Agave, Mescal ■☆
10844 Agave echinoides Jacobi;龙须兰(龙舌兰)■
10845 Agave elongata Jacobi = Agave angustifolia Haw. ■
10846 Agave endlichiana Trel. = Agave angustifolia Haw. ■
10847 Agave excelsa Jacobi = Agave angustifolia Haw. ■
10848 Agave falcata Engelm.;镰状龙舌兰■☆
10849 Agave felina Trel. = Agave americana L. ■
10850 Agave ferdinandi-regis Berger;流星龙舌兰(笹吹雪)■
10851 Agave ferox K. Koch;巨刺龙舌兰■☆
10852 Agave filifera Salm-Dyck;丝龙舌兰(乱雪,丝状龙舌兰);Agave, Thread Agave, Thread-leaf Agave ■☆
10853 Agave foetida L. = Furcraea foetida (L.) Haw. ■☆
10854 Agave fourcroydes Lem.;黄条龙舌兰;Henequen, Henequen Agave, Yucatan ■☆
10855 Agave franzosini P. Sewell;立叶龙舌兰(细齿龙舌)■☆
10856 Agave funifera Lem. = Hesperaloe funifera (K. Koch) Trel. ■☆
10857 Agave funkiana K. Koch et C. D. Bouché;芬克龙舌兰;Ixtle De Jaumave, Jaumarve Istle, Jaumave Fibre ■☆
10858 Agave geminiflora (Togliani) Ker Gawl.;双花龙舌兰■☆
10859 Agave geminiflora Ker Gawl. = Agave geminiflora (Togliani) Ker Gawl. ■☆
10860 Agave gilberti A. Berger;吉伯特龙舌兰■☆
10861 Agave glaucescens Hook. = Agave attenuata Salm-Dyck ■☆
10862 Agave glomeruliflora ?;团花龙舌兰;Chisos Agave ■☆
10863 Agave gracilipes Trel.;细柄龙舌兰;Slimfoot Century Plant ■☆
10864 Agave gracilispina (Rol.-Goss.) Engelm. ex Trel. = Agave americana L. ■
10865 Agave haseloffii Jacobi = Agave mitis Mart. ■☆
10866 Agave havardiana Trel.;阿瓦尔龙舌兰;Chisos Agave, Havard Agave ■☆
10867 Agave heteracantha Jacobi;异刺龙舌兰■☆
10868 Agave horrida Lem. = Agave horrida Lem. ex Jacobi ■☆
10869 Agave horrida Lem. ex Jacobi;魔鬼龙舌兰■☆
10870 Agave houlletii Jacobi = Agave sisalana Perr. ex Engelm. ■
10871 Agave huachucensis Baker;美洲龙舌兰■☆
10872 Agave inaequidens K. Koch;不等齿龙舌兰■☆
10873 Agave ingens A. Berger;巨大龙舌兰■☆
10874 Agave ingens A. Berger f. picta Hort.;彩色巨大龙舌兰(白覆轮龙舌)■☆
10875 Agave ixtli Karw. = Agave angustifolia Haw. ■
10876 Agave jacobiana Salm-Dyck = Agave salmiana Otto ■☆
10877 Agave jacquiniana Schult. f. = Agave angustifolia Haw. ■
10878 Agave kellocki Jacobi = Agave attenuata Salm-Dyck ■☆
10879 Agave latissima Jacobi = Agave salmiana Otto ■☆
10880 Agave laxifolia Baker = Agave decipiens Baker ■
10881 Agave lecheguilla Torr.;白肋龙舌兰(墨西哥龙舌兰);Lechuguilla, Shindagger, Tula Ixtle ■☆
10882 Agave lehmanni Jacobi = Agave salmiana Otto ■☆
10883 Agave letonae F. W. Taylor ex Trel.;萨尔瓦多龙舌兰;Salvador Henequen, Salvador Sisal ■☆
10884 Agave letonae Trel. = Agave letonae F. W. Taylor ex Trel. ■☆
10885 Agave letonae Trel. var. marginata ? = Agave angustifolia Haw. ■
10886 Agave longiflora (Rose) G. D. Rowley = Manfreda longiflora (Rose) Verh.-Will. ■☆
10887 Agave lophantha Schiede;白磁炉■☆
10888 Agave lophantha Schiede var. latifolia (Karw.) Berger;广叶龙舌兰;Thorncrest Century Plant, Thorn-crested Agave ■☆
10889 Agave lurida Aiton = Agave vera-cruz Mill. ■☆
10890 Agave macroacantha Zucc.;大刺龙舌兰■☆
10891 Agave macroculmis Tod.;大龙舌兰;Maguey Verde, Mountain Agave ■☆
10892 Agave maculosa Hook.;斑叶龙舌兰;Texas Tuberose ■☆
10893 Agave maculosa Hook. = Manfreda maculosa Rose ■☆
10894 Agave maculosa Hook. var. brevituba Engelm. = Manfreda maculosa Rose ■☆
10895 Agave maculosa Hook. var. minor Jacobi = Manfreda maculosa Rose ■☆
10896 Agave margaritae Brandegee = Agave angustifolia Haw. ■
10897 Agave marmorata Roezl = Agave marmorata Roezl ex E. Ortiz ■☆
10898 Agave marmorata Roezl ex E. Ortiz;大理石龙舌兰■☆
10899 Agave maximiliana Saunders ex Baker;马氏龙舌兰;Maguey Tecolote ■☆
10900 Agave melliflua Trel. = Agave americana L. ■
10901 Agave micracantha Salm-Dyck = Agave mitis Mart. ■☆
10902 Agave micracantha Salm-Dyck = Agave moranii Gentry ■☆
10903 Agave micrantha Salm-Dyck var. albicans (Jacobi) A. Terracc. = Agave mitis Mart. var. albidior (Salm-Dyck) B. Ullrich ■☆
10904 Agave mitis Mart.;柔软龙舌兰■☆
10905 Agave mitis Mart. var. albidior (Salm-Dyck) B. Ullrich;白柔软龙舌兰■☆
10906 Agave mitriformis Jacobi = Agave salmiana Otto ■☆
10907 Agave moranii Gentry;莫兰龙舌兰■☆
10908 Agave multifilifera Gentry;多丝龙舌兰;Chahuiqui ■☆
10909 Agave multiflora Tod.;多花龙舌兰;Many-flowered Agave ■☆
10910 Agave murpheyi F. Gibson;墨菲龙舌兰;Hohokam Agave, Murphy's Agave ■☆
10911 Agave neglecta Small;小侧出齿龙舌兰;Wild Century Plant ■☆
10912 Agave neomexicana Wooton et Standl. = Agave parryi Engelm. var. neomexicana (Wooton et Standl.) McKechnie ■☆
10913 Agave oblongata Jacobi = Agave mitis Mart. ■☆
10914 Agave ornithobroma Gentry;鸟龙舌兰;Maguey Pajarito ■☆
10915 Agave owenii I. M. Johnst. = Agave vivipara L. ■☆
10916 Agave pacifica Trel. = Agave vivipara L. ■☆
10917 Agave palmeri Engelm.;帕尔龙舌兰;Palmer's Agave ■☆
10918 Agave parrasana A. Berger;赖公■☆
10919 Agave parryi Engelm.;巴利龙舌兰;Century Plant, Parry Agave, Parry's Agave, Parry's Centuryplant ■☆
10920 Agave parryi Engelm. var. neomexicana (Wooton et Standl.) McKechnie;新墨西哥龙舌兰;New Mexico Agave ■☆
10921 Agave parryi Engelm. var. truncata Gentry;平截龙舌兰;Mescal Agave ■☆

10922 Agave parviflora Torr.;小花龙舌兰;Smallflower Century Plant ■☆
10923 Agave pendula Schnitts.;悬垂龙舌兰■☆
10924 Agave polianthes Thiede et Eggli;多药龙舌兰;Tuberose ■☆
10925 Agave potatorum Zucc.;雷神■☆
10926 Agave potatorum Zucc. var. verschaffeltii (Lem. ex Jacobi) Berger;戟叶龙舌兰(风雷神,雷神)■☆
10927 Agave pumila De Smet ex Baker;小龙舌兰■☆
10928 Agave rasconensis Trel. = Agave americana L. ■
10929 Agave rigida Mill.;野裂颜琼麻(野烈颜琼麻)■☆
10930 Agave rigida Mill. var. sisalana (Perrine) Engelm. = Agave sisalana (Engelm.) Perr. ex Engelm. ■
10931 Agave rubescens Salm-Dyck;赤叶龙舌兰■☆
10932 Agave salmiana Otto = Agave salmiana Otto ex Salm-Dyck ■☆
10933 Agave salmiana Otto ex Salm-Dyck;墨绿龙舌兰(酒龙舌);Century Plant,Maguey,Maguey Agave,Pulque Agave ■☆
10934 Agave salmiana Otto ex Salm-Dyck var. gracilispina Rol.-Goss. = Agave americana L. ■
10935 Agave scabra Salm-Dyck;粗糙龙舌兰;Rough Agave ■☆
10936 Agave schidigera Lem.;老翁龙舌兰;Schidiger Agave ■
10937 Agave schottii Engelm.;斯氏龙舌兰;Beargrass,Schott's Agave,Schott's Century Plant,Shindagger ■☆
10938 Agave shawii Engelm.;肖氏龙舌兰;Coastal Century Plant ■☆
10939 Agave sisalana (Engelm.) Perr. = Agave sisalana (Engelm.) Perr. ex Engelm. ■
10940 Agave sisalana (Engelm.) Perr. ex Engelm.;剑麻(菠萝麻,衿麻,琼麻,西沙尔琼麻);Bahama Hemp,Hemp-plant,Sisal,Sisal Agave,Sisal Hemp ■
10941 Agave sisalana Perr. ex Engelm. = Agave sisalana (Engelm.) Perr. ex Engelm. ■
10942 Agave striata Salm-Dyck;长序龙舌兰;Espadin,Hedgehog Agave ■☆
10943 Agave stricta Salm-Dyck;刚直龙舌兰(吹上)■☆
10944 Agave subzonata Trel. = Agave americana L. ■
10945 Agave tequilana Weber;韦伯龙舌兰;Agave,Agave Azul,Weber Blue Agave ■☆
10946 Agave toumeyana Trel.;杜迈龙舌兰;Toumey's Century Plant ■☆
10947 Agave tuberosa Mill. = Furcraea tuberosa Aiton ■☆
10948 Agave utahensis Engelm.;犹他龙舌兰(青磁炉);Maguey,Mescal,Utah Agave ■☆
10949 Agave utahensis Engelm. var. nevadensis Engelm. ex Greenm. et Roush;内华达龙舌兰(尤塔龙舌兰);Nevada Agave ■☆
10950 Agave variegata Jacobi = Manfreda variegata (Jacobi) Rose ■☆
10951 Agave vera-cruz J. R. Drumm. et Prain = Agave americana L. ■
10952 Agave vera-cruz Mill. = Agave americana L. ■
10953 Agave vestita S. Watson;白发龙舌兰■☆
10954 Agave vexans Trel.;邪魔龙舌兰■☆
10955 Agave victoriae-reginae T. Moore;皇后龙舌兰(鬼脚掌,厚叶龙舌兰,积雪,积雪龙舌兰,龙角);Queen Agave,Queen Victiria Centuryplant,Queen Victoria Agave,Royal Agave ■☆
10956 Agave vilmoriniana A. Berger;维氏龙舌兰;Octopus Agave ■☆
10957 Agave virginica L. = Manfreda virginica (L.) Salisb. ex Rose ■☆
10958 Agave virginica L. f. tigrina (Engelm.) E. J. Palmer et Steyerm. = Manfreda virginica (L.) Salisb. ex Rose ■☆
10959 Agave vivipara L.;太平洋龙舌兰■☆
10960 Agave weberi Cels ex Poiss.;韦伯世纪龙舌兰;Weber Agave,Weber's Century Plant ■☆
10961 Agave wightii J. R. Drumm. et Prain;威氏龙舌兰■☆
10962 Agave wocomahi Gentry;沃克龙舌兰;Wocomahi Agave ■☆
10963 Agave woodrowii W. Watson;伍德罗龙舌兰■☆
10964 Agave xylonacantha Salm-Dyck;木刺龙舌兰■☆
10965 Agave yaquiana Trel. = Agave vivipara L. ■☆
10966 Agave zapupe Trel.;长叶龙舌兰■☆
10967 Agave zonata Trel. = Agave americana L. ■
10968 Agdestidaceae Nakai = Phytolaccaceae R. Br.(保留科名)●■
10969 Agdestidaceae Nakai(1942);萝卜藤科(毛商陆科)●☆
10970 Agdestis Moc. et Sessé ex DC. = Agdestis Sessé et Moc. ex DC. ●☆
10971 Agdestis Sessé et Moc. ex DC.(1817);萝卜藤属(爱特史迪斯属,毛商陆属)●☆
10972 Agdestis clematidea Moc. et Sessé = Agdestis clematidea Sessé et Moc. ex DC. ●☆
10973 Agdestis clematidea Sessé et Moc. ex DC.;萝卜藤(毛商陆,铁线莲爱特史迪斯);Clematis Agdestis,Rockroot ●☆
10974 Agelaea Lour. = Agelaea Sol. ex Planch. ●
10975 Agelaea Sol. ex Planch.(1850);栗豆藤属(栗豆属);Agelaea ●
10976 Agelaea annobonensis G. Schellenb. = Agelaea pentagyna (Lam.) Baill. ●☆
10977 Agelaea australis G. Schellenb.;南方栗豆藤●☆
10978 Agelaea australis G. Schellenb. = Agelaea pentagyna (Lam.) Baill. ●☆
10979 Agelaea baronii G. Schellenb. = Agelaea pentagyna (Lam.) Baill. ●☆
10980 Agelaea brevipaniculata Cummins;短序栗豆藤●☆
10981 Agelaea brevipaniculata Cummins = Agelaea paradoxa Gilg var. microcarpa Jongkind ●☆
10982 Agelaea cambodiana Pierre = Agelaea trinervis (Llanos) Merr. ●
10983 Agelaea claessensii De Wild. = Agelaea pentagyna (Lam.) Baill. ●☆
10984 Agelaea coccinea Exell;绯红栗豆藤●☆
10985 Agelaea conraui G. Schellenb. = Agelaea pentagyna (Lam.) Baill. ●☆
10986 Agelaea cordata G. Schellenb.;心叶栗豆藤●☆
10987 Agelaea cordata G. Schellenb. = Agelaea pentagyna (Lam.) Baill. ●☆
10988 Agelaea demeusei De Wild. et T. Durand = Agelaea pentagyna (Lam.) Baill. ●☆
10989 Agelaea densiflora Pierre;密花栗豆藤●☆
10990 Agelaea dewevrei De Wild. et T. Durand = Agelaea pentagyna (Lam.) Baill. ●☆
10991 Agelaea duchesnei De Wild. = Agelaea pentagyna (Lam.) Baill. ●☆
10992 Agelaea elegans G. Schellenb.;雅致栗豆藤●☆
10993 Agelaea elegans G. Schellenb. = Agelaea pentagyna (Lam.) Baill. ●☆
10994 Agelaea emetica Baill.;呕吐栗豆藤●☆
10995 Agelaea floccosa G. Schellenb. = Agelaea pentagyna (Lam.) Baill. ●☆
10996 Agelaea fragrans Gilg;香栗豆藤●☆
10997 Agelaea fragrans Gilg = Agelaea paradoxa Gilg var. microcarpa Jongkind ●☆
10998 Agelaea gabonensis Jongkind;加蓬栗豆藤●☆
10999 Agelaea glabrifolia Hance;光叶栗豆藤●☆
11000 Agelaea glandulosissima Gilg = Agelaea pentagyna (Lam.) Baill. ●☆
11001 Agelaea gracilis G. Schellenb.;纤细栗豆藤●☆

11002 Agelaea gracilis G. Schellenb. = Agelaea rubiginosa Gilg ●☆
11003 Agelaea grisea G. Schellenb. = Agelaea pentagyna (Lam.) Baill. ●☆
11004 Agelaea heterophylla Gilg;异叶栗豆藤●☆
11005 Agelaea heterophylla Gilg = Agelaea pentagyna (Lam.) Baill. ●☆
11006 Agelaea hirsuta De Wild.;毛叶栗豆藤●☆
11007 Agelaea hirsuta De Wild. = Agelaea pentagyna (Lam.) Baill. ●☆
11008 Agelaea katangensis Troupin = Agelaea pentagyna (Lam.) Baill. ●☆
11009 Agelaea kivuensis Troupin = Agelaea rubiginosa Gilg ●☆
11010 Agelaea lamarckii Planch.;拉马栗豆藤●☆
11011 Agelaea lamarckii Planch. = Agelaea pentagyna (Lam.) Baill. ●☆
11012 Agelaea laurentii De Wild. = Agelaea rubiginosa Gilg ●☆
11013 Agelaea leopoldvilleana De Wild. = Agelaea pentagyna (Lam.) Baill. ●☆
11014 Agelaea lescrauwaetii De Wild. = Agelaea pentagyna (Lam.) Baill. ●☆
11015 Agelaea longecalyculata G. Schellenb. = Agelaea pentagyna (Lam.) Baill. ●☆
11016 Agelaea longifoliata G. Schellenb.;长小叶栗豆藤(长叶栗豆藤)●☆
11017 Agelaea longifoliata G. Schellenb. = Agelaea pentagyna (Lam.) Baill. ●☆
11018 Agelaea lucida G. Schellenb. = Agelaea pentagyna (Lam.) Baill. ●☆
11019 Agelaea macrocarpa G. Schellenb.;大果栗豆藤●☆
11020 Agelaea macrocarpa G. Schellenb. = Agelaea pentagyna (Lam.) Baill. ●☆
11021 Agelaea macrophylla (Zoll.) Leenh.;大叶栗豆藤●☆
11022 Agelaea macrophysa Gilg ex G. Schellenb. = Agelaea rubiginosa Gilg ●☆
11023 Agelaea marginata G. Schellenb. = Agelaea pentagyna (Lam.) Baill. ●☆
11024 Agelaea mayottensis G. Schellenb. = Agelaea pentagyna (Lam.) Baill. ●☆
11025 Agelaea mildbraedii Gilg = Agelaea pentagyna (Lam.) Baill. ●☆
11026 Agelaea neglecta G. Schellenb. = Agelaea pentagyna (Lam.) Baill. ●☆
11027 Agelaea nitida Sol. ex Planch.;光亮栗豆藤●☆
11028 Agelaea nitida Sol. ex Planch. = Agelaea pentagyna (Lam.) Baill. ●☆
11029 Agelaea obliqua (P. Beauv.) Baill. = Agelaea pentagyna (Lam.) Baill. ●☆
11030 Agelaea obliqua (P. Beauv.) Baill. var. cordata (G. Schellenb.) Exell = Agelaea pentagyna (Lam.) Baill. ●☆
11031 Agelaea obliqua (P. Beauv.) Baill. var. usambarensis Gilg = Agelaea pentagyna (Lam.) Baill. ●☆
11032 Agelaea obovata G. Schellenb. = Agelaea pentagyna (Lam.) Baill. ●☆
11033 Agelaea oligantha Gilg ex G. Schellenb. = Agelaea pentagyna (Lam.) Baill. ●☆
11034 Agelaea ovalis G. Schellenb. = Agelaea pentagyna (Lam.) Baill. ●☆
11035 Agelaea palmata Jongkind;掌裂栗豆藤●☆
11036 Agelaea paradoxa Gilg;奇异栗豆藤●☆
11037 Agelaea paradoxa Gilg var. microcarpa Jongkind;小果奇异栗豆藤●☆
11038 Agelaea pentagyna (Lam.) Baill.;五蕊栗豆藤●☆
11039 Agelaea phaeocarpa Exell = Agelaea pentagyna (Lam.) Baill. ●☆
11040 Agelaea phaseolifolia Gilg ex G. Schellenb. = Agelaea pentagyna (Lam.) Baill. ●☆
11041 Agelaea pilosa G. Schellenb. = Agelaea pentagyna (Lam.) Baill. ●☆
11042 Agelaea poggeana Gilg;波格栗豆藤●☆
11043 Agelaea preussii Gilg = Agelaea pentagyna (Lam.) Baill. ●☆
11044 Agelaea principensis Exell = Agelaea rubiginosa Gilg ●☆
11045 Agelaea pseudobliqua G. Schellenb. = Agelaea pentagyna (Lam.) Baill. ●☆
11046 Agelaea punctulata (Hiern) G. Schellenb. = Agelaea pentagyna (Lam.) Baill. ●☆
11047 Agelaea pynaertii De Wild. = Agelaea pentagyna (Lam.) Baill. ●☆
11048 Agelaea reticulata Exell;网脉栗豆藤●☆
11049 Agelaea reticulata Exell = Agelaea pentagyna (Lam.) Baill. ●☆
11050 Agelaea rubiginosa Gilg;锈色栗豆藤●☆
11051 Agelaea schweinfurthii Gilg = Agelaea rubiginosa Gilg ●☆
11052 Agelaea setulosa G. Schellenb. = Agelaea pentagyna (Lam.) Baill. ●☆
11053 Agelaea sublanata De Wild. = Agelaea pentagyna (Lam.) Baill. ●☆
11054 Agelaea tenuinervis G. Schellenb.;细脉栗豆藤●☆
11055 Agelaea thouarsiana Baill. = Agelaea pentagyna (Lam.) Baill. ●☆
11056 Agelaea tricuspidata Gilg ex G. Schellenb. = Agelaea pentagyna (Lam.) Baill. ●☆
11057 Agelaea trifolia (Lam.) Baill. = Agelaea pentagyna (Lam.) Baill. ●☆
11058 Agelaea trifolia (Lam.) Gilg ex G. Schellenb.;三叶栗豆藤;Threeleaf Agelaea ●
11059 Agelaea trinervis (Llanos) Merr.;栗豆藤;Trinerve Agelaea, Trinerved Agelaea ●
11060 Agelaea ugandensis G. Schellenb. = Agelaea pentagyna (Lam.) Baill. ●☆
11061 Agelaea ustulata G. Schellenb. = Agelaea pentagyna (Lam.) Baill. ●☆
11062 Agelaea vanderystii G. Schellenb. = Agelaea rubiginosa Gilg ●☆
11063 Agelaea villosa Sol. ex Planch.;毛栗豆藤●☆
11064 Agelaea villosa Sol. ex Planch. = Agelaea pentagyna (Lam.) Baill. ●☆
11065 Agelaea villosiflora G. Schellenb.;长毛花栗豆藤●☆
11066 Agelaea wallichii Hook. f. = Agelaea trinervis (Llanos) Merr. ●
11067 Agelaea woodii Merr.;伍氏栗豆藤●☆
11068 Agelaea zenkeri G. Schellenb.;岑克尔栗豆藤●☆
11069 Agelaea zenkeri G. Schellenb. = Agelaea pentagyna (Lam.) Baill. ●☆
11070 Agelandra Engl. et Pax = Angelandra Endl. ●
11071 Agelandra Engl. et Pax = Croton L. ●
11072 Agelanthus Tiegh. (1895);群花寄生属●☆
11073 Agelanthus Tiegh. = Tapinanthus (Blume) Rchb. (保留属名)●☆
11074 Agelanthus atrocoronatus Polhill et Wiens;暗冠群花寄生●☆
11075 Agelanthus bipartitus Balle ex Polhill et Wiens;二深裂群花寄生●☆
11076 Agelanthus brunneus (Engl.) Balle et N. Hallé;褐群花寄生●☆
11077 Agelanthus combreticola (Lebrun et L. Touss.) Polhill et Wiens;风车藤●☆
11078 Agelanthus copaiferae (Sprague) Polhill et Wiens;古巴香脂树

群花寄生●☆

11079 Agelanthus crassifolius (Wiens) Polhill et Wiens;厚叶群花寄生●☆

11080 Agelanthus deltae (Baker et Sprague) Polhill et Wiens;德尔塔群花寄生●☆

11081 Agelanthus dichrous (Danser) Polhill et Wiens;二色群花寄生●☆

11082 Agelanthus discolor (Schinz) Balle;异色群花寄生●☆

11083 Agelanthus djurensis (Engl.) Polhill et Wiens;非洲群花寄生●☆

11084 Agelanthus dodoneifolius (DC.) Polhill et Wiens;车桑子叶群花寄生●☆

11085 Agelanthus elegantulus (Engl.) Polhill et Wiens;雅致群花寄生●☆

11086 Agelanthus entebbensis (Sprague) Polhill et Wiens;恩德培群花寄生●☆

11087 Agelanthus falcifolius (Sprague) Polhill et Wiens;镰叶群花寄生●☆

11088 Agelanthus flamignii (De Wild.) Balle = Agelanthus unyorensis (Sprague) Polhill et Wiens ●☆

11089 Agelanthus flammeus Polhill et Wiens;焰红群花寄生●☆

11090 Agelanthus fuellebornii (Engl.) Polhill et Wiens;菲勒群花寄生●☆

11091 Agelanthus glaucoviridis (Engl.) Polhill et Wiens;灰绿群花寄生●☆

11092 Agelanthus glomeratus (Engl.) Polhill et Wiens;团集群花寄生●☆

11093 Agelanthus gracilis (Toelken et Wiens) Polhill et Wiens;纤细群花寄生●☆

11094 Agelanthus guineensis Polhill et Wiens;几内亚群花寄生●☆

11095 Agelanthus heteromorphus (A. Rich.) Polhill et Wiens;异形群花寄生●☆

11096 Agelanthus igneus (Danser) Polhill et Wiens;火焰群花寄生●☆

11097 Agelanthus irangensis (Engl.) Polhill et Wiens;伊朗群花寄生●☆

11098 Agelanthus kayseri (Engl.) Polhill et Wiens;凯泽群花寄生●☆

11099 Agelanthus keilii (Engl. et K. Krause) Polhill et Wiens;凯尔群花寄生●☆

11100 Agelanthus krausei (Engl.) Polhill et Wiens;克劳斯群花寄生●☆

11101 Agelanthus kraussianus (Meisn.) Polhill et Wiens;克氏群花寄生●☆

11102 Agelanthus lancifolius Polhill et Wiens;剑叶群花寄生●☆

11103 Agelanthus longipes (Baker et Sprague) Polhill et Wiens;长梗群花寄生●☆

11104 Agelanthus lugardii (N. E. Br.) Polhill et Wiens;卢格德群花寄生●☆

11105 Agelanthus microphyllus Polhill et Wiens;小叶群花寄生●☆

11106 Agelanthus molleri (Engl.) Polhill et Wiens;默勒群花寄生●☆

11107 Agelanthus musozensis (Rendle) Polhill et Wiens;穆索济群花寄生●☆

11108 Agelanthus myrsinifolius (Engl. et K. Krause) Polhill et Wiens;铁仔叶群花寄生●☆

11109 Agelanthus natalitius (Meisn.) Polhill et Wiens;南非群花寄生●☆

11110 Agelanthus natalitius (Meisn.) Polhill et Wiens subsp. zeyheri (Harv.) Polhill et Wiens;蔡南非群花寄生●☆

11111 Agelanthus nyasicus (Baker et Sprague) Polhill et Wiens;尼亚萨群花寄生●☆

11112 Agelanthus oehleri (Engl.) Polhill et Wiens;奥勒群花寄生●☆

11113 Agelanthus pennatulus (Sprague) Polhill et Wiens;羽状群花寄生●☆

11114 Agelanthus pilosus Polhill et Wiens;疏毛群花寄生●☆

11115 Agelanthus platyphyllus (Hochst. ex A. Rich.) Balle;宽叶群花寄生●☆

11116 Agelanthus polygonifolius (Engl.) Polhill et Wiens;多角叶群花寄生●☆

11117 Agelanthus prunifolius (E. Mey. ex Harv.) Polhill et Wiens;粉叶群花寄生●☆

11118 Agelanthus pungu (De Wild.) Polhill et Wiens;刺群花寄生●☆

11119 Agelanthus rondensis (Engl.) Polhill et Wiens;龙达群花寄生●☆

11120 Agelanthus sakarensis (Engl.) Polhill et Wiens = Agelanthus tanganyikae (Engl.) Polhill et Wiens ●☆

11121 Agelanthus sambesiacus (Engl. et Schinz) Polhill et Wiens;桑比群花寄生●☆

11122 Agelanthus sansibarensis (Engl.) Polhill et Wiens;桑给巴尔群花寄生●☆

11123 Agelanthus sansibarensis (Engl.) Polhill et Wiens subsp. montanus Polhill et Wiens;山地桑给巴尔群花寄生●☆

11124 Agelanthus scassellatii (Chiov.) Polhill et Wiens;斯卡群花寄生●☆

11125 Agelanthus schweinfurthii (Engl.) Polhill et Wiens;施韦群花寄生●☆

11126 Agelanthus songeensis Balle ex Polhill et Wiens;松吉群花寄生●☆

11127 Agelanthus subulatus (Engl.) Polhill et Wiens;钻形群花寄生●☆

11128 Agelanthus tanganyikae (Engl.) Polhill et Wiens;坦噶尼喀群花寄生●☆

11129 Agelanthus terminaliae (Engl. et Gilg) Polhill et Wiens;泰米群花寄生●☆

11130 Agelanthus toroensis (Sprague) Polhill et Wiens;托罗群花寄生●☆

11131 Agelanthus transvaalensis (Sprague) Polhill et Wiens;德兰士瓦群花寄生●☆

11132 Agelanthus uhehensis (Engl.) Polhill et Wiens;乌赫群花寄生●☆

11133 Agelanthus unyorensis (Sprague) Polhill et Wiens;乌尼群花寄生●☆

11134 Agelanthus validus Polhill et Wiens;刚直群花寄生●☆

11135 Agelanthus villosiflorus (Engl.) Polhill et Wiens;长毛群花寄生●☆

11136 Agelanthus zizyphifolius (Engl.) Polhill et Wiens;枣叶群花寄生●☆

11137 Agelanthus zizyphifolius (Engl.) Polhill et Wiens subsp. vittatus (Engl.) Polhill et Wiens;粗线群花寄生●☆

11138 Agenium Nees et Pilg. = Agenium Nees ■☆

11139 Agenium Nees(1836);童颜草属■☆

11140 Agenium goyazense (Hack.) Clayton;童颜草■☆

11141 Agenium leptocladum (Hack.) Clayton;细枝童颜草■☆

11142 Agenium majus Pilg.;大童颜草■☆

11143 Agenium villosum (Nees) Pilg.;长柔毛童颜草■☆

11144 Agenora D. Don = Hypochaeris L. ■

11145 Agenora D. Don = Seriola L. ■

11146 Ageomoron Raf. = Caucalis L. ■☆

11147 Ageratella A. Gray ex S. Watson = Ageratella A. Gray ■☆

11148 Ageratella A. Gray(1887);小藿香蓟属■☆

11149 Ageratella microphylla (Sch. Bip.) A. Gray ex S. Watson;小藿香蓟■☆

11150 Ageratina O. Hoffm. = Ageratinastrum Mattf. ●☆

11151 Ageratina Spach = Eupatorium L. ●■

11152 Ageratina Spach(1841);假藿香蓟属(破坏草属,紫茎泽兰属);Snakeroot,White Snakeroot ●■

11153 Ageratina adenophora (Spreng.) R. M. King et H. Rob.;假藿香蓟(破坏草);Crofton Weed,Mexican Devil,Sticky Snakeroot ●■

11154 Ageratina altissima (L.) R. M. King et H. Rob.;皱叶泽兰(荨麻叶泽兰,团聚泽兰);Common White Snakeroot,Mistflower,White Snakeroot ■☆

11155 Ageratina altissima (L.) R. M. King et H. Rob. = Eupatorium rugosum Houtt. ■☆

11156 Ageratina altissima (L.) R. M. King et H. Rob. var. angustatum (A. Gray) S. F. Blake = Eupatorium rugosum Houtt. ■☆

11157 Ageratina altissima (L.) R. M. King et H. Rob. var. roanensis (Small) Clewell et Wooten;南非紫茎泽兰;Appalachian White Snakeroot ■☆

11158 Ageratina altissima L. var. angustata (A. Gray) Clewell et Wooten = Ageratina altissima (L.) R. M. King et H. Rob. ■☆

11159 Ageratina aromatica (L.) Spach;紫茎泽兰;Small-leaved White Snakeroot ■☆

11160 Ageratina aromatica (L.) Spach = Eupatorium aromaticum L. ■☆

11161 Ageratina fruticosa O. Hoffm. = Gutenbergia fruticosa (O. Hoffm.) C. Jeffrey ■☆

11162 Ageratina goetzeana O. Hoffm. = Ageratinastrum polyphyllum (Baker) Mattf. ■☆

11163 Ageratina havanensis (Kunth) R. M. King et H. Rob.;黑文紫茎泽兰;Havana Snakeroot ■☆

11164 Ageratina herbacea (A. Gray) R. M. King et H. Rob.;草本紫茎泽兰;Fragrant Snakeroot ■☆

11165 Ageratina jucunda (Greene) Clewell et Wooten;愉悦紫茎泽兰■☆

11166 Ageratina lemmonii (B. L. Rob.) R. M. King et H. Rob.;莱蒙紫茎泽兰;Lemmon's Snakeroot ■☆

11167 Ageratina luciae-brauniae (Fernald) R. M. King et H. Rob.;露西娅紫茎泽兰;Rockhouse White Snakeroot ■☆

11168 Ageratina occidentalis (Hook.) R. M. King et H. Rob.;西方紫茎泽兰;Western Snakeroot ■☆

11169 Ageratina paupercula (A. Gray) R. M. King et H. Rob.;贫乏紫茎泽兰;Santa Rita Snakeroot ■☆

11170 Ageratina polyphylla (Baker) O. Hoffm. = Ageratinastrum polyphyllum (Baker) Mattf. ■☆

11171 Ageratina riparia (Regel) R. M. King et H. Rob.;河岸破坏草(河岸泽兰);Hamakua Pamakani,Mist Flower,River Eupatorium,Riverside Bogorchid,Spreading Snakeroot,Spreading Snake-root ■☆

11172 Ageratina rothrockii (A. Gray) R. M. King et H. Rob.;罗氏紫茎泽兰;Rothrock's Snakeroot ■☆

11173 Ageratina shastensis (D. W. Taylor et Stebbins) R. M. King et H. Rob.;沙斯塔紫茎泽兰;Mt. Shasta Snakeroot ■☆

11174 Ageratina thyrsiflora (Greene) R. M. King et H. Rob.;聚伞泽兰;Congested Snakeroot ■☆

11175 Ageratina wrightii (A. Gray) R. M. King et H. Rob.;赖氏紫茎泽兰;Wright's Snakeroot ■☆

11176 Ageratinastrum Mattf. (1932);轮叶瘦片菊属(小破坏草属)●☆

11177 Ageratinastrum fruticosum (O. Hoffm.) Mattf. = Gutenbergia fruticosa (O. Hoffm.) C. Jeffrey ■☆

11178 Ageratinastrum glomeratum Mattf. = Brachythrix glomerata (Mattf.) C. Jeffrey ■☆

11179 Ageratinastrum goetzeanum (O. Hoffm.) Mattf. = Ageratinastrum polyphyllum (Baker) Mattf. ■☆

11180 Ageratinastrum katangense Lisowski;加丹加轮叶瘦片菊■☆

11181 Ageratinastrum lejolyanum (Adamska et Lisowski) Kalanda;勒若利轮叶瘦片菊■☆

11182 Ageratinastrum palustre Wild et G. V. Pope;沼泽轮叶瘦片菊■☆

11183 Ageratinastrum polyphyllum (Baker) Mattf.;多叶轮叶瘦片菊■☆

11184 Ageratinastrum polyphyllum (Baker) Mattf. var. marungense Lisowski;马龙加轮叶瘦片菊■☆

11185 Ageratinastrum virgatum Mattf. = Ageratinastrum polyphyllum (Baker) Mattf. ■☆

11186 Ageratiopsis Sch. Bip. ex Benth. et Hook. f. (1873);假紫茎泽兰属■☆

11187 Ageratiopsis Sch. Bip. ex Benth. et Hook. f. = Eupatorium L. ●■

11188 Ageratiopsis apiculata Sch. Bip. ex Baker;假紫茎泽兰■☆

11189 Ageratium Rchb. = Aceratium DC. ●☆

11190 Ageratium Steud. = Ageraton Adans. ●■

11191 Ageraton Adans. = Ageratum L. ■●

11192 Ageraton Adans. = Ageratum Mill. ■●

11193 Ageraton Adans. = Erinus L. ■☆

11194 Ageratum L. (1753);藿香蓟属(胜红蓟属);Ageratum,Bastard Agrimony,Floss Flower,Flossflower ■●

11195 Ageratum Mill. = Erinus L. ■☆

11196 Ageratum Tourn. ex Adans. = Erinus L. ■☆

11197 Ageratum altissimum L. = Ageratina altissima (L.) R. M. King et H. Rob. ■☆

11198 Ageratum altissimum L. = Eupatorium rugosum Houtt. ■☆

11199 Ageratum conyzoides L.;藿香蓟(白花草,白花臭草,白花藿香蓟,白花香草,白毛苦,臭草,臭垆草,狗扯尾,广马草,路遇香,绿升麻,猫尿草,毛射香,毛麝香,毛屎草,柠檬菊,脓泡草,七星菊,胜红蓟,胜红药,水丁药,咸虾草,咸虾花,消炎草,野棉花,油贴贴果,鱼眼草,重阳草,紫花毛草);Ageratum,Bastard Agrimony,Billy-goat Weed,Goat's Weed,Goatweed,Maile-hohono,Tropic Ageratum,Tropical Whiteweed ■

11200 Ageratum conyzoides L. subsp. houstonianum (Mill.) Sahu = Ageratum houstonianum Mill. ■

11201 Ageratum conyzoides L. var. mexicanum (Sims) DC. = Ageratum houstonianum Mill. ■

11202 Ageratum corymbosum Benth. et Hook. f. = Ageratum corymbosum Zuccagni ■☆

11203 Ageratum corymbosum Zuccagni;蝴蝶蓟;Butterfly Mist,Flattop Ageratum,Flat-top Whiteweed ■☆

11204 Ageratum corymbosum Zuccagni var. jaliscense B. L. Rob. = Ageratum corymbosum Zuccagni ■☆

11205 Ageratum guatemalense M. F. Johnson = Ageratum corymbosum Benth. et Hook. f. ■☆

11206 Ageratum houstonianum Mill.;紫花藿香蓟(大花藿香蓟,墨西哥藿香蓟,熊耳草);Bluemink,Floss Flower,Mexican Ageratum,Mexico Ageratum ■

11207 Ageratum houstonianum Mill. 'Blue Danube';蓝色多瑙河熊耳草■☆

11208 Ageratum houstonianum Mill. 'Blue Mink';蓝貂皮熊耳草■☆

11209 Ageratum houstonianum Mill. 'Pacific';太平洋熊耳草■☆

11210 Ageratum houstonianum Mill. 'Swing Pink';变粉熊耳草■☆

11211 Ageratum latifolium Cav. = Ageratum conyzoides L. ■

11212 Ageratum littorale A. Gray = Ageratum maritimum Kunth ■☆

11213 Ageratum littorale A. Gray var. hondurense B. L. Rob.;洪都藿香蓟■☆

11214 Ageratum littorale A. Gray var. hondurense B. L. Rob. = Ageratum maritimum Kunth ■☆

11215 Ageratum maritimum Kunth;湿地藿香蓟;Cape Sable Whiteweed ■☆

11216 Ageratum melegueta K. Schum.;美列藿香蓟■☆

11217 Ageratum mexicanum Sims = Ageratum houstonianum Mill. ■

11218 Ageratum polyphyllum Baker = Ageratinastrum polyphyllum (Baker) Mattf. ■☆

11219 Ageratum salicifolium Hemsl. = Ageratum corymbosum Zuccagni ■☆

11220 Ageratum strictum Hemsl. = Ageratum corymbosum Zuccagni ■☆

11221 Ageratum wrightii Torr. et A. Gray = Trichocoronis wrightii (Torr. et A. Gray) A. Gray ■☆

11222 Agerella Fourr. = Veronica L. ■

11223 Ageria Adans. = Ilex L. + Myrsine L. ●

11224 Ageria Raf. = Macoucoua Aubl. ●

11225 Aggeianthus Wight = Porpax Lindl. ■

11226 Aggeranthus Wight = Aggeianthus Wight ■

11227 Aggregatae Sch. Bip. = Dipsacaceae Juss. (保留科名)●■

11228 Agiabampoa Rose ex O. Hoffm. = Alvordia Brandegee ■☆

11229 Agialid Adans. (废弃属名) = Balanites Delile(保留属名)●☆

11230 Agialid Adans. = Bartonia Muhl. ex Willd. (保留属名)■☆

11231 Agialida Adans. (废弃属名) = Balanites Delile(保留属名)●☆

11232 Agialida Kuntze = Agialid Adans. (废弃属名)●☆

11233 Agialida abyssinica Tiegh. = Balanites aegyptiaca (L.) Delile ●☆

11234 Agialida arabica Tiegh. = Balanites aegyptiaca (L.) Delile ●☆

11235 Agialida chevalieri Tiegh. = Balanites aegyptiaca (L.) Delile ●☆

11236 Agialida cuneifolia Tiegh. = Balanites aegyptiaca (L.) Delile ●☆

11237 Agialida glomerata Tiegh. = Balanites aegyptiaca (L.) Delile ●☆

11238 Agialida latifolia Tiegh. = Balanites aegyptiaca (L.) Delile ●☆

11239 Agialida nigra Tiegh. = Balanites aegyptiaca (L.) Delile ●☆

11240 Agialida palestinaca Tiegh. = Balanites aegyptiaca (L.) Delile ●☆

11241 Agialida rotundifolia Tiegh. = Balanites rotundifolia (Tiegh.) Blatt. ●☆

11242 Agialida schimperi Tiegh. = Balanites aegyptiaca (L.) Delile ●☆

11243 Agialida senegalensis Tiegh. = Balanites aegyptiaca (L.) Delile var. ferox (Poir.) DC. ●☆

11244 Agialida tombouctensis Tiegh. = Balanites aegyptiaca (L.) Delile ●☆

11245 Agialidaceae Tiegh. = Balanitaceae M. Roem. (保留科名)●☆

11246 Agialidaceae Wettst. = Balanitaceae M. Roem. (保留科名)●☆

11247 Agianthus Greene = Streptanthus Nutt. ■☆

11248 Agianthus Greene(1906);加州芥属■☆

11249 Agianthus bernardinus Greene;加州芥■☆

11250 Agianthus jacobaeus Greene = Streptanthus campestris var. jacobaeus (Greene) Jeps. ■☆

11251 Agiella Tiegh. = Balanites Delile(保留属名)●☆

11252 Agiella angolensis (Welw.) Tiegh. = Balanites angolensis (Welw.) Welw. ex Mildbr. et Schltr. ●☆

11253 Agiella angolensis (Welw.) Tiegh. var. superreticulata Tiegh. = Balanites angolensis (Welw.) Welw. ex Mildbr. et Schltr. ●☆

11254 Agiella welwitschii Tiegh. = Balanites angolensis (Welw.) Welw. ex Mildbr. et Schltr. subsp. welwitschii (Tiegh.) Sands ●☆

11255 Agihalid Juss. = Agialid Adans. ■☆

11256 Agihalid Juss. = Agialida Adans. (废弃属名)●☆

11257 Agihalid Juss. = Balanites Delile(保留属名)●☆

11258 Agina Neck. = Bartonia Muhl. ex Willd. (保留属名)■☆

11259 Agiortia Quinn(2005);澳洲芒石南属●☆

11260 Agirta Baill. = Tragia L. ●☆

11261 Agistron Raf. = Uncinia Pers. ■☆

11262 Aglae Dulac = Caulinia Willd. ■

11263 Aglae Dulac = Posidonia K. D. König(保留属名)■

11264 Aglaea (Pers.) Eckl. = Melasphaerula Ker Gawl. ■☆

11265 Aglaea Post et Kuntze = Aglaia F. Allam. (废弃属名)●

11266 Aglaea Post et Kuntze = Aglaia Lour. (保留属名)●

11267 Aglaea Post et Kuntze = Aglaia Noronha ex Thouars ●☆

11268 Aglaea Steud. = Melasphaerula Ker Gawl. ■☆

11269 Aglaeopsis Post et Kuntze = Aglaia Lour. (保留属名)●

11270 Aglaeopsis Post et Kuntze = Aglaiopsis Miq. ●☆

11271 Aglaia Dumort. = Aegle Corrêa(保留属名)●

11272 Aglaia F. Allam. (废弃属名) = Aglaia Lour. (保留属名)●

11273 Aglaia Lour. (1790)(保留属名);米仔兰属(树兰属);Aglaia, Maizailan ●

11274 Aglaia Noronha ex Thouars = Hemistemma DC. ●☆

11275 Aglaia Noronha ex Thouars = Hibbertia Andréws ●☆

11276 Aglaia abbreviata C. Y. Wu;缩序米仔兰;Short Aglaia, Shortened Aglaia, Shortenpanicle Maizailan ●

11277 Aglaia aphanamixis Pellegr. = Aphanamixis grandifolia Blume ●

11278 Aglaia argentea Blume;银白米仔兰(银米兰,银米仔兰)●☆

11279 Aglaia attenuata H. L. Li = Amoora yunnanensis (H. L. Li) C. Y. Wu ●

11280 Aglaia bicolor Merr.;二色米仔兰(二色米兰)●☆

11281 Aglaia chittagonga Miq. = Amoora tetrapetala (Pierre) Pellegr. ●

11282 Aglaia cordata Hiern;心叶米仔兰(心叶米兰)●☆

11283 Aglaia cucullata (Roxb.) Pellegr.;僧帽状米仔兰●☆

11284 Aglaia dasyclada F. C. How et T. C. Chen = Amoora dasyclada (F. C. How et T. C. Chen) C. Y. Wu ●◇

11285 Aglaia diffusa Merr.;松散米仔兰(散米兰,散米仔兰,土坎米兰,土坎米仔兰)●☆

11286 Aglaia duperreana Pierre;杜珀米仔兰●☆

11287 Aglaia edulis (Roxb.) Wall.;可食米仔兰●☆

11288 Aglaia elaeagnoidea Benth.;澳大利亚米仔兰(红柴)●

11289 Aglaia elaeagnoidea Benth. var. formosana Hayata = Aglaia formosana (Hayata) Hayata ●

11290 Aglaia elaeagnoidea Benth. var. pallens Merr. = Aglaia formosana (Hayata) Hayata ●

11291 Aglaia elliptifolia Merr.;椭圆叶米仔兰(大叶树兰,椭圆米仔兰,椭圆叶树兰);Ellipticleaf Aglaia, Ellipticleaf Maizailan, Elliptic-leaved Aglaia, Large Aglaia ●

11292 Aglaia eusideroxylon Koord. et Valeton;铁米仔兰(铁米兰)●☆

11293 Aglaia formosana (Hayata) Hayata;台湾米仔兰(红柴,台湾树兰);Formosa Aglaia, Taiwan Aglaia, Taiwan Maizailan ●

11294 Aglaia gigantea Pellegr.;大米仔兰(大花米兰,大花米仔兰)●☆

11295 Aglaia laevigata Merr.;平滑米仔兰(平滑米兰)●☆

11296 Aglaia lawii (Wight) C. J. Saldanha;兰屿树兰●

11297 Aglaia llanosiana C. DC.;拉罗米仔兰(拉罗米兰)●☆

11298 Aglaia luzoniensis Merr. et Rolfe;琉球米仔兰(琉球米兰)●☆

11299 Aglaia merostela Pellegr.;莫肉米仔兰(莫肉米兰)●☆

11300 Aglaia odorata Lour.;米仔兰(兰花米,米碎兰,木珠兰,千里香,秋菊,秋兰,三叶兰,山胡椒,树兰,碎米兰,蚊惊树,午里香,暹罗花,夜兰,鱼子兰,珠兰);Aglaia Tree, Chinese Perfume Plant, Chulan Aglaia, Chulan Tree, Maizailan, Mock Lemon, Orchid Tree ●

11301 Aglaia odorata Lour. var. chaudocensis Pellegr.;红果米仔兰●☆

11302 Aglaia odorata Lour. var. microphyllina DC.;小叶米仔兰(小叶米兰);Little-leaf Chu-lan Aglaia ●

11303 Aglaia odoratissima Lour.;香米仔兰●☆

11304 Aglaia pallens Merr. = Aglaia formosana (Hayata) Hayata ●
11305 Aglaia perviridis Hiern;碧绿米仔兰;Darkgreen Maizailan,Green Aglaia ●
11306 Aglaia pirifera Hance;泰国米仔兰●☆
11307 Aglaia polystachya Wall. = Aphanamixis polystachya (Wall.) R. Parker ●
11308 Aglaia rimosa (Blanco) Merr.;大叶树兰●
11309 Aglaia roxburghiana Miq.;山椤(洛罗,洛氏米仔兰);Roxburgh Maizailan,Roxburgh Aglaia ●
11310 Aglaia sexipetala Griff.;六瓣米仔兰●☆
11311 Aglaia somalensis Chiov. = Sorindeia somalensis (Chiov.) Chiov. ●☆
11312 Aglaia tenuifolia H. L. Li = Amoora yunnanensis (H. L. Li) C. Y. Wu ●
11313 Aglaia testicularis C. Y. Wu;马肾果(马腰子果);Horsekidneyfruit,Testicular Aglaia,Testiculate Aglaia ●◇
11314 Aglaia tetrapetala Pierre = Amoora tetrapetala (Pierre) Pellegr. ●
11315 Aglaia tsangii Merr. = Amoora tsangii (Merr.) X. M. Chen ●
11316 Aglaia wangii H. L. Li = Amoora ouangliensis (H. Lév.) C. Y. Wu ●
11317 Aglaia wangii H. L. Li var. macrophylla H. L. Li = Amoora tetrapetala (Pierre) Pellegr. var. macrophylla (H. L. Li) C. Y. Wu ●
11318 Aglaia yunnanensis H. L. Li = Amoora yunnanensis (H. L. Li) C. Y. Wu ●
11319 Aglaiopsis Miq. (1868);类米仔兰属●☆
11320 Aglaiopsis Miq. = Aglaia Lour. (保留属名)●
11321 Aglaiopsis glaucescens Miq.;类米仔兰●☆
11322 Aglaiopsis lancifolia Miq.;细叶类米仔兰●☆
11323 Aglaja Endl. = Aglaia Noronha ex Thouars ●☆
11324 Aglaja Endl. = Hibbertia Andrews ●☆
11325 Aglaodendron J. Rémy = Plazia Ruiz et Pav. ●☆
11326 Aglaodendrum Post et Kuntze = Aglaodendron J. Rémy ●☆
11327 Aglaodorum Schott(1858);亮袋南星属■☆
11328 Aglaodorum griffithii Schott;亮袋南星■☆
11329 Aglaonema Schott(1829);亮丝草属(粗筋草属,粗肋草属,广东万年青属,粤万年青属);Aglaonema,China Evergreen,Chinagreen,Chinese Evergreen,Poisondart,Silver Queen ■
11330 Aglaonema 'Silver King';银王亮丝草■☆
11331 Aglaonema 'Silver Queen';银后亮丝草■☆
11332 Aglaonema commutatum Schott;细斑亮丝草(亮丝草,细斑粗肋,细斑粗肋草,银斑万年青);Chinese Evergreen,Philippine Evergreen,Poison Dart,Spotted Aglaonema ■☆
11333 Aglaonema commutatum Schott 'Albo-variegatum';白斑亮丝草■
11334 Aglaonema commutatum Schott 'Pseudobracteatum';白柄亮丝草(白斑亮丝草)■
11335 Aglaonema commutatum Schott 'San Remo';斜纹粗肋草■☆
11336 Aglaonema commutatum Schott 'Treubii';狭叶亮丝草■
11337 Aglaonema commutatum Schott 'Tricolor';三色亮丝草■
11338 Aglaonema commutatum Schott var. elegans (Engl.) Nicholson;美丽粗肋草(美丽亮丝草)■☆
11339 Aglaonema commutatum Schott var. maculatum (Engl.) Nicholson;白斑粗肋草;Aglaonema,Philippine Evergreen ■☆
11340 Aglaonema costatum N. E. Br.;爪哇亮丝草(宽肋亮丝草,心叶粗肋草,爪哇万年青);Chinese Evergreen,Spotted Evergreen ■☆
11341 Aglaonema costatum N. E. Br. 'Immaculatum';黑点爪哇亮丝草■☆
11342 Aglaonema costatum N. E. Br. 'Maculatum';斑点爪哇亮丝草■☆
11343 Aglaonema costatum N. E. Br. f. foxii (Engl.) R. N. Jervis;白肋亮丝草(白宽肋万年青,白肋万年青)■☆
11344 Aglaonema crispum (Pitcher et Manda) Nicolson;白雪粗肋草(白雪亮丝草);Painted Drop Tongue,Painted Droptongue,Painted Drop-tongue ■☆
11345 Aglaonema crispum (Pitcher et Manda) Nicolson 'Silver King';银皇■☆
11346 Aglaonema hospitum Williams;厚遇亮丝草■☆
11347 Aglaonema marantifolium Blume = Aglaonema commutatum Schott ■☆
11348 Aglaonema modestum Schott ex Engl.;广东万年青(粗肋草,大叶万年青,井干草,亮丝草,万年青,粤万年青);China Evergreen,Chinagreen,Chinese Evergreen ■
11349 Aglaonema modestum Schott ex Engl. 'Mediopictum';中道黄纹粗肋草■☆
11350 Aglaonema nitidum (Jacq.) Kunth;长叶粗肋草■
11351 Aglaonema nitidum (Jacq.) Kunth 'Curtisii';箭羽粗肋草■
11352 Aglaonema nitidum (Jacq.) Kunth = Aglaonema oblongifolium Kunth ■☆
11353 Aglaonema nitidum (Jacq.) Kunth f. curtisii (N. E. Br.) Nicolson = Aglaonema nitidum (Jacq.) Kunth 'Curtisii' ■
11354 Aglaonema oblongifolium Kunth;长椭圆叶亮丝草■☆
11355 Aglaonema pictum (Roxb.) Kunth;斑叶亮丝草(斑叶万年青,彩绘亮丝草,彩叶粗肋,绒叶粗肋草);Ornamental Aglaonema ■
11356 Aglaonema pictum (Roxb.) Kunth var. tricolor N. E. Br.;亮叶粗肋草(三色彩绘亮丝草)■
11357 Aglaonema pierreanum Engl.;皮氏万年青(越南万年青)■☆
11358 Aglaonema pseudobracteatum ?;伪苞亮丝草■☆
11359 Aglaonema robelinii Engl.;罗氏亮丝草■☆
11360 Aglaonema rotundum N. E. Br.;圆叶粗肋草■
11361 Aglaonema siamense Engl.;明脉亮丝草(白肋粗肋草);Siam Aglaonema ■
11362 Aglaonema simplex Blume;常青粗肋草(单茎万年青,绿竹)■
11363 Aglaonema tenuipes Engl.;越南万年青(观音莲,万年青);Vietnam Chinagreen ■
11364 Aglaonema treubii Engl.;特莱亮丝草■☆
11365 Aglaonema versicolor Hort. = Aglaonema pictum (Roxb.) Kunth ■
11366 Agllthela Raf. = Allium L. ■
11367 Aglossorhyncha Schltr. (1905);无舌喙兰属■☆
11368 Aglossorrhyncha Schltr. = Aglossorhyncha Schltr. ■☆
11369 Aglutoma Raf. = Aster L. ●■
11370 Aglycia Steud. = Eriochloa Kunth ■
11371 Aglyela Willd. ex Steud. = Eriochloa Kunth ■
11372 Agnesia Zuloaga et Judz. (1993);亚马逊禾属■☆
11373 Agnesia lancifolia (Mez) Zuloaga et Judz.;亚马逊禾■☆
11374 Agnirictus Schwantes = Stomatium Schwantes ■☆
11375 Agnirictus aginus (Haw.) Schwantes = Stomatium agninum (Haw.) Schwantes ■☆
11376 Agnirictus lesliei Schwantes = Stomatium lesliei (Schwantes) Volk ■☆
11377 Agnistus G. Don = Acnistus Schott ●☆
11378 Agnorhiza (Jeps.) W. A. Weber = Wyethia Nutt. ■☆
11379 Agnorhiza (Jeps.) W. A. Weber(1999);洁根菊属■☆
11380 Agnorhiza W. A. Weber = Wyethia Nutt. ■☆
11381 Agnorhiza bolanderi (A. Gray) W. A. Weber;洁根菊■☆
11382 Agnorhiza elata (H. M. Hall) W. A. Weber;大洁根菊■☆
11383 Agnorhiza invenusta (Greene) W. A. Weber;反转洁根菊■☆

11384 Agnorhiza ovata (Torr. et Gray) W. A. Weber;卵形洁根菊■☆
11385 Agnorhiza reticulata (Greene) W. A. Weber;网脉洁根菊■☆
11386 Agnostus A. Cunn. = Stenocarpus R. Br. (保留属名)●☆
11387 Agnostus G. Don ex Loudon = Stenocarpus R. Br. (保留属名)●☆
11388 Agnus-castus Carrière = Vitex L. ●
11389 Agonandra Miers = Agonandra Miers ex Benth. et Hook. f. ●☆
11390 Agonandra Miers ex Benth. = Agonandra Miers ex Benth. et Hook. f. ●☆
11391 Agonandra Miers ex Benth. et Hook. f. (1862);聚雄柚属(聚雄花属,西柚属)●☆
11392 Agonandra Miers ex Hook. f. = Agonandra Miers ex Benth. et Hook. f. ●☆
11393 Agonandra brasiliensis Benth. et Hook. f.;巴西聚雄柚(巴西聚雄花)●☆
11394 Agonandra macrocarpa L. O. Williams;大果聚雄柚(大果聚雄花)●☆
11395 Agonandra obtusifolia Standl.;钝叶聚雄柚(钝叶聚雄花)●☆
11396 Agoneissos Zoll. ex Nied. = Tristellateia Thouars ●
11397 Agonis (DC.) Sweet(1830)(保留属名);圆冠木属(柳香桃属);Willow Myrtle ●☆
11398 Agonis Lindl. = Agonis (DC.) Sweet(保留属名)●☆
11399 Agonis flexuosa (Willd.) Sweet;垂枝圆冠木(柳香桃);Australian Willow Myrtle, Juniper Myrtle, Pepper Mint Tree, Peppermint Tree, Willow Myrtle, Willowmyrtle ●☆
11400 Agonis juniperina Schauer;刺柏圆冠木;Juniper Myrtle ●☆
11401 Agonis linearifolia (DC.) Sweet;线叶圆冠木●☆
11402 Agonizanthos F. Muell. = Anigozanthos Labill. ■☆
11403 Agonolobus Rchb. = Cheiranthus L. ●■
11404 Agonomyrtus Schauer ex Rchb. = ? Leptospermum J. R. Forst. et G. Forst. (保留属名)●☆
11405 Agonon Raf. = ? Ilex L. ●
11406 Agophyllum Neck. = Zygophyllum L. ●■
11407 Agorrhinum Fourr. = Antirrhinum L. ●■
11408 Agoseris Raf. (1817);高莛苣属(高葶苣属,山羊菊属);Agoseris, False Dandelion, Mountain Dandelion ■☆
11409 Agoseris alpestris (A. Gray) Greene = Nothocalais alpestris (A. Gray) K. L. Chambers ■☆
11410 Agoseris altissima Rydb. = Agoseris glauca (Pursh) Raf. var. dasycephala (Torr. et A. Gray) Jeps. ■☆
11411 Agoseris angustissima Greene = Agoseris aurantiaca Greene ■☆
11412 Agoseris apargioides (Less.) Greene;海边高葶苣;Seaside Agoseris ■☆
11413 Agoseris apargioides (Less.) Greene subsp. maritima (E. Sheld.) Q. Jones = Agoseris apargioides (Less.) Greene var. maritima (E. Sheld.) G. I. Baird ■☆
11414 Agoseris apargioides (Less.) Greene var. eastwoodiae (Fedde) Munz;伊斯顿高葶苣■☆
11415 Agoseris apargioides (Less.) Greene var. maritima (E. Sheld.) G. I. Baird;沼泽高葶苣;Oregon Agoseris ■☆
11416 Agoseris arachnoidea Rydb. = Agoseris aurantiaca Greene ■☆
11417 Agoseris arizonica (Greene) Greene = Agoseris aurantiaca (Hook.) Greene var. purpurea (A. Gray) Cronquist ■☆
11418 Agoseris aspera (Rydb.) Rydb. = Agoseris glauca (Pursh) Raf. var. dasycephala (Torr. et A. Gray) Jeps. ■☆
11419 Agoseris attenuata Rydb. = Agoseris aurantiaca (Hook.) Greene var. purpurea (A. Gray) Cronquist ■☆
11420 Agoseris aurantiaca (Hook.) Greene;黄高葶苣(黄山羊菊);Mountain Agoseris, Orange Agoseris ■☆
11421 Agoseris aurantiaca (Hook.) Greene subsp. purpurea (A. Gray) G. W. Douglas = Agoseris aurantiaca (Hook.) Greene var. purpurea (A. Gray) Cronquist ■☆
11422 Agoseris aurantiaca (Hook.) Greene var. purpurea (A. Gray) Cronquist;紫高葶苣■☆
11423 Agoseris aurantiaca Greene = Agoseris aurantiaca (Hook.) Greene ■☆
11424 Agoseris californica (Nutt.) Hoover = Agoseris heterophylla (Nutt.) Greene var. cryptopleura Greene ■☆
11425 Agoseris carnea Rydb. = Agoseris aurantiaca (Hook.) Greene ■☆
11426 Agoseris caudata Greene = Agoseris parviflora (Nutt.) D. Dietr. ■☆
11427 Agoseris cinerea Greene;灰高葶苣■☆
11428 Agoseris confinis Greene = Agoseris aurantiaca (Hook.) Greene var. purpurea (A. Gray) Cronquist ■☆
11429 Agoseris covillei Greene = Agoseris monticola Greene ■☆
11430 Agoseris cuspidata (Pursh) D. Dietr. = Microseris cuspidata (Pursh) Sch. Bip. ■☆
11431 Agoseris cuspidata (Pursh) Raf. = Microseris cuspidata (Pursh) Sch. Bip. ■☆
11432 Agoseris cuspidata (Pursh) Steud. = Nothocalais cuspidata (Pursh) Greene ■☆
11433 Agoseris decumbens Greene = Agoseris monticola Greene ■☆
11434 Agoseris dens-leonis Greene = Agoseris parviflora (Nutt.) D. Dietr. ■☆
11435 Agoseris eastwoodiae Fedde = Agoseris apargioides (Less.) Greene var. eastwoodiae (Fedde) Munz ■☆
11436 Agoseris eisenhoweri B. Boivin = Agoseris glauca (Pursh) Raf. var. dasycephala (Torr. et A. Gray) Jeps. ■☆
11437 Agoseris elata (Nutt.) Greene;高大高葶苣■☆
11438 Agoseris frondifera Osterh. = Agoseris aurantiaca (Hook.) Greene var. purpurea (A. Gray) Cronquist ■☆
11439 Agoseris gaspensis Fernald = Agoseris aurantiaca Greene ■☆
11440 Agoseris glauca (Pursh) Raf. = Agoseris glauca Raf. ■☆
11441 Agoseris glauca (Pursh) Raf. var. cronquistii S. L. Welsh = Agoseris aurantiaca (Hook.) Greene var. purpurea (A. Gray) Cronquist ■☆
11442 Agoseris glauca (Pursh) Raf. var. dasycephala (Torr. et A. Gray) Jeps.;毛头苍白高葶苣■☆
11443 Agoseris glauca (Pursh) Raf. var. laciniata (D. C. Eaton) Kuntze = Agoseris parviflora (Nutt.) D. Dietr. ■☆
11444 Agoseris glauca (Pursh) Raf. var. monticola (Greene) Q. Jones = Agoseris monticola Greene ■☆
11445 Agoseris glauca Raf.;苍白高葶苣(苍白山羊菊);False Dandelion, Mountain Dandelion, Pale Agoseris ■☆
11446 Agoseris glauca Raf. subsp. aspera (Rydb.) Piper = Agoseris glauca (Pursh) Raf. var. dasycephala (Torr. et A. Gray) Jeps. ■☆
11447 Agoseris glauca Raf. subsp. scorzonerifolia (Schrad.) Piper = Agoseris glauca (Pursh) Raf. var. dasycephala (Torr. et A. Gray) Jeps. ■☆
11448 Agoseris glauca Raf. var. aspera (Rydb.) Cronquist = Agoseris glauca (Pursh) Raf. var. dasycephala (Torr. et A. Gray) Jeps. ■☆
11449 Agoseris glauca Raf. var. pumila (Nutt.) Garrett = Agoseris glauca (Pursh) Raf. var. dasycephala (Torr. et A. Gray) Jeps. ■☆
11450 Agoseris glauca Raf. var. villosa (Rydb.) G. L. Wittr. = Agoseris glauca (Pursh) Raf. var. dasycephala (Torr. et A. Gray) Jeps. ■☆
11451 Agoseris gracilens (A. Gray) Greene = Agoseris aurantiaca

Greene ■☆

11452 Agoseris gracilens (A. Gray) Greene var. greenei (A. Gray) S. F. Blake = Agoseris aurantiaca Greene ■☆

11453 Agoseris graminifolia Greene = Agoseris aurantiaca (Hook.) Greene var. purpurea (A. Gray) Cronquist ■☆

11454 Agoseris grandiflora (Nutt.) Greene; 大花高葶苣; Grassland Agoseris ■☆

11455 Agoseris grandiflora (Nutt.) Greene var. intermedia (Greene) Jeps. = Agoseris cinerea Greene ■☆

11456 Agoseris grandiflora (Nutt.) Greene var. laciniata (Nutt.) Jeps. = Agoseris elata (Nutt.) Greene ■☆

11457 Agoseris grandiflora (Nutt.) Greene var. leptophylla; 细叶大花高葶苣; Puget Sound Agoseris ■☆

11458 Agoseris grandiflora (Nutt.) Greene var. plebeia (Greene) G. L. Wittrock = Agoseris cinerea Greene ■☆

11459 Agoseris greenei (A. Gray) Rydb.; 格林高葶苣■☆

11460 Agoseris heterophylla (Nutt.) Greene; 一年生高葶苣; Annual Agoseris ■☆

11461 Agoseris heterophylla (Nutt.) Greene subsp. californica (Nutt.) Piper = Agoseris heterophylla (Nutt.) Greene var. cryptopleura Greene ■☆

11462 Agoseris heterophylla (Nutt.) Greene subsp. normalis Piper = Agoseris heterophylla (Nutt.) Greene ■☆

11463 Agoseris heterophylla (Nutt.) Greene var. californica (Nutt.) Davidson et Moxley = Agoseris heterophylla (Nutt.) Greene var. cryptopleura Greene ■☆

11464 Agoseris heterophylla (Nutt.) Greene var. crenulata (H. M. Hall) Jeps. = Agoseris heterophylla (Nutt.) Greene var. cryptopleura Greene ■☆

11465 Agoseris heterophylla (Nutt.) Greene var. cryptopleura Greene; 加州高葶苣; California Agoseris ■☆

11466 Agoseris heterophylla (Nutt.) Greene var. glabra (Nutt.) Howell = Agoseris heterophylla (Nutt.) Greene ■☆

11467 Agoseris heterophylla (Nutt.) Greene var. quentinii G. I. Baird; 亚利桑那高葶苣; Arizona Agoseris ■☆

11468 Agoseris heterophylla (Nutt.) Greene var. turgida (H. M. Hall) Jeps. = Agoseris heterophylla (Nutt.) Greene var. cryptopleura Greene ■☆

11469 Agoseris hirsuta (Hook.) Greene; 海岸高葶苣; Coast Range Agoseris ■☆

11470 Agoseris howellii Greene = Agoseris aurantiaca Greene ■☆

11471 Agoseris humilis (Benth.) Kuntze = Agoseris apargioides (Less.) Greene ■☆

11472 Agoseris intermedia Greene = Agoseris cinerea Greene ■☆

11473 Agoseris isomeris Greene = Agoseris glauca (Pursh) Raf. var. dasycephala (Torr. et A. Gray) Jeps. ■☆

11474 Agoseris lacera Greene = Agoseris glauca Raf. ■☆

11475 Agoseris laciniata (Nutt.) Greene = Agoseris elata (Nutt.) Greene ■☆

11476 Agoseris lackschewitzii D. M. Hend. et R. K. Moseley = Agoseris aurantiaca Greene ■☆

11477 Agoseris lanulosa Greene = Agoseris glauca (Pursh) Raf. var. dasycephala (Torr. et A. Gray) Jeps. ■☆

11478 Agoseris lapathifolia Greene = Agoseris glauca Raf. ■☆

11479 Agoseris leontodon Rydb. var. aspera Rydb. = Agoseris glauca (Pursh) Raf. var. dasycephala (Torr. et A. Gray) Jeps. ■☆

11480 Agoseris leontodon Rydb. var. pygmaea Rydb. = Agoseris glauca (Pursh) Raf. var. dasycephala (Torr. et A. Gray) Jeps. ■☆

11481 Agoseris leptocarpa Osterh. = Agoseris parviflora (Nutt.) D. Dietr. ■☆

11482 Agoseris longirostris Greene = Agoseris aurantiaca (Hook.) Greene var. purpurea (A. Gray) Cronquist ■☆

11483 Agoseris longissima Greene = Agoseris glauca Raf. ■☆

11484 Agoseris longula Greene = Agoseris glauca Raf. ■☆

11485 Agoseris maculata Rydb. = Agoseris glauca (Pursh) Raf. var. dasycephala (Torr. et A. Gray) Jeps. ■☆

11486 Agoseris major Jeps. ex Greene = Agoseris heterophylla (Nutt.) Greene var. cryptopleura Greene ■☆

11487 Agoseris maritima E. Sheld. = Agoseris apargioides (Less.) Greene var. maritima (E. Sheld.) G. I. Baird ■☆

11488 Agoseris maritima Eastw. = Agoseris apargioides (Less.) Greene var. eastwoodiae (Fedde) Munz ■☆

11489 Agoseris marshallii (Greene) Greene = Agoseris cinerea Greene ■☆

11490 Agoseris microdonta Greene = Agoseris glauca Raf. ■☆

11491 Agoseris monticola Greene; 山地高葶苣■☆

11492 Agoseris nana Rydb. = Agoseris aurantiaca Greene ■☆

11493 Agoseris naskapensis J. Rousseau et Raymond = Agoseris aurantiaca Greene ■☆

11494 Agoseris obtusifolia (Suksd.) Rydb. = Agoseris cinerea Greene ■☆

11495 Agoseris parviflora (Nutt.) D. Dietr.; 小花高葶苣; Steppe Agoseris ■☆

11496 Agoseris plebeia (Greene) Greene = Agoseris cinerea Greene ■☆

11497 Agoseris prionophylla Greene = Agoseris aurantiaca Greene ■☆

11498 Agoseris procera Greene = Agoseris glauca Raf. ■☆

11499 Agoseris pubescens Rydb. = Agoseris glauca (Pursh) Raf. var. dasycephala (Torr. et A. Gray) Jeps. ■☆

11500 Agoseris pumila (Nutt.) Rydb. = Agoseris glauca (Pursh) Raf. var. dasycephala (Torr. et A. Gray) Jeps. ■☆

11501 Agoseris purpurea (A. Gray) Greene = Agoseris aurantiaca (Hook.) Greene var. purpurea (A. Gray) Cronquist ■☆

11502 Agoseris purpurea (A. Gray) Greene var. arizonica (Greene) G. L. Wittrock = Agoseris aurantiaca (Hook.) Greene var. purpurea (A. Gray) Cronquist ■☆

11503 Agoseris retrorsa (Benth.) Greene; 矛叶高葶苣; Spearleaf Agoseris ■☆

11504 Agoseris rosea (Nutt.) D. Dietr. = Agoseris parviflora (Nutt.) D. Dietr. ■☆

11505 Agoseris roseata Rydb. = Agoseris aurantiaca (Hook.) Greene var. purpurea (A. Gray) Cronquist ■☆

11506 Agoseris scorzonerifolia (Schrad.) Greene = Agoseris glauca (Pursh) Raf. var. dasycephala (Torr. et A. Gray) Jeps. ■☆

11507 Agoseris subalpina G. N. Jones = Agoseris aurantiaca Greene ■☆

11508 Agoseris taraxacoides Greene = Agoseris parviflora (Nutt.) D. Dietr. ■☆

11509 Agoseris tenuifolia Rydb. = Agoseris elata (Nutt.) Greene ■☆

11510 Agoseris tomentosa Howell = Agoseris parviflora (Nutt.) D. Dietr. ■☆

11511 Agoseris turbinata Rydb. = Agoseris glauca (Pursh) Raf. var. dasycephala (Torr. et A. Gray) Jeps. ■☆

11512 Agoseris vestita Greene = Agoseris glauca (Pursh) Raf. var. dasycephala (Torr. et A. Gray) Jeps. ■☆

11513 Agoseris vicinalis Greene = Agoseris glauca Raf. ■☆

11514 Agoseris villosa Rydb. = Agoseris glauca (Pursh) Raf. var. dasycephala (Torr. et A. Gray) Jeps. ■☆

11515 Agoseris vulcanica Greene = Agoseris aurantiaca Greene ■☆
11516 Agostana Bute ex Gray = Agrostana Hill ●■
11517 Agostana Bute ex Gray = Bupleurum L. ●■
11518 Agouticarpa C. H. Perss. (2003);刺鼠茜属●☆
11519 Agouticarpa C. H. Perss. = Genipa L. ●☆
11520 Agraphis Link = Endymion Dumort. ■☆
11521 Agraulus P. Beauv. = Agrostis L. (保留属名)■
11522 Agrestis Bubani = Agrostis L. (保留属名)■
11523 Agrestis Raf. = Agrostis L. (保留属名)■
11524 Agretta Eckl. = Tritonia Ker Gawl. ■
11525 Agretta crispa (L. f.) Eckl. = Ixia erubescens Goldblatt ■☆
11526 Agretta grandiflora Eckl. = Ixia micrandra Baker ■☆
11527 Agretta pentandra Eckl. = Ixia scillaris L. ■☆
11528 Agretta retusa Steud. = Ixia polystachya L. ■☆
11529 Agretta stricta Eckl. = Ixia stricta (Eckl. ex Klatt) G. J. Lewis ■☆
11530 Agrianthus Mart. = Agrianthus Mart. ex DC. ●☆
11531 Agrianthus Mart. ex DC. (1836);田花菊属●☆
11532 Agrianthus almasensis D. J. N. Hind;阿尔马斯田花菊●☆
11533 Agrianthus campestris Mart.;野田花菊●☆
11534 Agrianthus corymbosus DC.;伞序田花菊●☆
11535 Agrianthus empetrifolius Mart.;田花菊●☆
11536 Agrianthus microlicioides Mattf.;野牡丹田花菊●☆
11537 Agrianthus myrtoides Mattf.;香桃木田花菊●☆
11538 Agrianthus pungens Mattf.;锐尖田花菊●☆
11539 Agrieolaea Schrank = Clerodendrum L. ●■
11540 Agrifolium Hill = Aquifolium Mill. ●
11541 Agrifolium Hill = Ilex L. ●
11542 Agrimonia L. (1753);龙牙草属(包大宁属,龙芽草属,三瓣蔷薇属);Agrimonia,Agrimony,Cocklebur ■
11543 Agrimonia × nippono-pilosa Murata;日本毛龙牙草■☆
11544 Agrimonia asiatica Juz. = Agrimonia eupatoria L. subsp. asiatica (Juz.) Skalicky ■
11545 Agrimonia bracteata E. Mey. ex C. A. Mey. = Agrimonia eupatoria L. ■
11546 Agrimonia coreana Nakai;托叶龙牙草(朝鲜龙牙草,大托叶龙牙草,仙鹤草);Korea Cocklebur,Korean Agrimonia ■
11547 Agrimonia coreana Nakai = Agrimonia pilosa Ledeb. ■
11548 Agrimonia decumbens L. f. = Acaena latebrosa Aiton ●☆
11549 Agrimonia eupatoria L.;欧洲龙牙草(草龙牙,地椒,黄花草,黄龙尾,黄龙牙,杰里花,癞茨草,龙牙,龙牙草,蛇疙瘩,线麻子花,爪香草);Aaron's Rod,Aggermony,Agrimony,Church Steeple,Church Steeples,Churchsteeples,Church-steeples,Clot Bur,Clot-burr,Cocklebur,Cockle-bur,Cockly Bur,Common Agrimony,Common Cocklebur,Eggremunny,Egremoine,Egremounde,Egri Mony,Eupator's Agrimony,European Agrimonia,European Agrimony,Fairy's Wand,Garclive,Golden Rod,Goosechite,Harvest Lice,Lemon Flower,Lemonade,Liverwort,Medicinal Agrimony,Money-in-both-pockets,Philanthropos,Rat's Tail,Rat's Tails,Salt-and-pepper,Steeple,Steeplewort,Sticklewort,Stickwort,Sweethearts,Tea Plant,While Tansy,Wild Tansy,Yellow Agrimony ■
11550 Agrimonia eupatoria L. = Agrimonia bracteata E. Mey. ex C. A. Mey. ■
11551 Agrimonia eupatoria L. = Agrimonia pilosa Ledeb. ■
11552 Agrimonia eupatoria L. subsp. asiatica (Juz.) Skalicky;亚洲龙牙草(大花龙牙草,仙鹤草,新疆龙牙草);Asia Cocklebur,Asiatic Agrimonia ■
11553 Agrimonia eupatoria L. subsp. grandis (Asch. et Graebn.) Bornm.;大欧洲龙牙草■☆
11554 Agrimonia eupatoria L. var. intermedia Batt. = Agrimonia eupatoria L. subsp. grandis (Asch. et Graebn.) Bornm. ■☆
11555 Agrimonia eupatoria L. var. japonica (Miq.) Masam. = Agrimonia pilosa Ledeb. ■
11556 Agrimonia eupatoria L. var. nepalensis (D. Don) Koidz. = Agrimonia pilosa Ledeb. var. nepalensis (D. Don) Nakai ■
11557 Agrimonia granulosa Juz.;颗粒龙牙草■☆
11558 Agrimonia gryposepala Wallr.;普通龙牙草(硬毛龙牙草);Agrimony,Common Agrimony,Hooked Agrimony,Tall Agrimony,Tall Hairy Agrimony ■☆
11559 Agrimonia japonica (Miq.) Koidz. = Agrimonia pilosa Ledeb. var. japonica (Miq.) Nakai ■☆
11560 Agrimonia japonica (Miq.) Koidz. = Agrimonia pilosa Ledeb. ■
11561 Agrimonia lanata Wall. = Agrimonia pilosa Ledeb. var. nepalensis (D. Don) Nakai ■
11562 Agrimonia mollis (Torr. et A. Gray) Britton = Agrimonia pubescens Wallr. ■☆
11563 Agrimonia nepalensis D. Don = Agrimonia bracteata E. Mey. ex C. A. Mey. ■
11564 Agrimonia nepalensis D. Don = Agrimonia pilosa Ledeb. var. nepalensis (D. Don) Nakai ■
11565 Agrimonia nepalensis D. Don var. obovata Skalicky = Agrimonia pilosa Ledeb. var. nepalensis (D. Don) Nakai ■
11566 Agrimonia nipponica Koidz.;本州龙牙草(日本龙牙草);Japanese Cocklebur ■
11567 Agrimonia nipponica Koidz. = Agrimonia pilosa Ledeb. var. nipponica (Koidz.) Kitam. ■
11568 Agrimonia nipponica Koidz. var. occidentalis Skalicky;小花龙牙草;Littleflower Agrimonia ■
11569 Agrimonia obtusifolia A. I. Baranov et Skvortsov = Agrimonia pilosa Ledeb. ■
11570 Agrimonia odorata Mill. = Agrimonia procera Wallr. ■☆
11571 Agrimonia parviflora Aiton;多花龙牙草;Harvest-lice,Many-flowered Agrimony,Small Flowered Agrimony,Southern Agrimony,Swamp Agrimony ■☆
11572 Agrimonia pilosa Ledeb.;龙牙草(草龙牙,朝鲜龙芽草,寸八节,大毛药,刀口药,地冻风,地椒,地蜈蚣,地仙草,钝齿龙芽草,多齿龙芽草,父子草,瓜香草,挂香草,过路黄,猴头草,黄花草,黄花仔,黄龙尾,黄龙牙,鸡爪沙,杰里花,金顶龙牙,金顶龙芽,狼牙草,老鹳嘴,龙头草,龙芽草,路边黄,路边鸡,马尾丝,毛鸡草,毛将军,毛脚鸡,毛脚苗,毛脚茵,牛头草,群兰败毒草,绒毛龙芽草,蛇倒退,蛇疙瘩,施州龙牙草,石打穿,疏毛龙芽草,铁胡蜂,脱力草,五蹄风,仙鹤草,线麻子花,泻痢草,新疆龙芽草,亚洲龙芽草,异风颈草,圆叶龙牙草,止血草,子不离母,子母草);Cocklebur,Hair-vein Agrimony,Hairyvein Agrimonia ■
11573 Agrimonia pilosa Ledeb. f. dahurica (Willd. ex Stev.) Nakai = Agrimonia pilosa Ledeb. ■
11574 Agrimonia pilosa Ledeb. f. nipponica (Koidz.) Ohwi = Agrimonia nipponica Koidz. ■
11575 Agrimonia pilosa Ledeb. f. typica Nakai = Agrimonia pilosa Ledeb. ■
11576 Agrimonia pilosa Ledeb. subsp. dahurica (Willd. ex Stev.) Kamelin = Agrimonia pilosa Ledeb. ■
11577 Agrimonia pilosa Ledeb. subsp. japonica (Miq.) H. Hara = Agrimonia pilosa Ledeb. var. japonica (Miq.) Nakai ■☆
11578 Agrimonia pilosa Ledeb. subsp. japonica (Miq.) H. Hara =

Agrimonia pilosa Ledeb. ■
11579 Agrimonia pilosa Ledeb. var. coreana (Nakai) Liou et W. C. Cheng = Agrimonia coreana Nakai ■
11580 Agrimonia pilosa Ledeb. var. japonica (Miq.) Nakai;日本龙牙草(龙牙草)■☆
11581 Agrimonia pilosa Ledeb. var. japonica (Miq.) Nakai = Agrimonia pilosa Ledeb. ■
11582 Agrimonia pilosa Ledeb. var. japonica (Miq.) Nakai f. nepalensis (D. Don) Murata = Agrimonia pilosa Ledeb. ■
11583 Agrimonia pilosa Ledeb. var. japonica (Miq.) Nakai f. nepalensis (D. Don) Murata = Agrimonia pilosa Ledeb. var. nepalensis (D. Don) Nakai ■
11584 Agrimonia pilosa Ledeb. var. japonica (Miq.) Nakai f. subglabra Nakai;无毛日本龙牙草■☆
11585 Agrimonia pilosa Ledeb. var. japonica (Miq.) Nakai f. subglabra Nakai = Agrimonia pilosa Ledeb. ■
11586 Agrimonia pilosa Ledeb. var. longitiolata Liou et C. Y. Li;长叶龙牙草;Longleaf Cocklebur ■
11587 Agrimonia pilosa Ledeb. var. nepalensis (D. Don) Nakai;黄龙尾(白芽蒿,产后草,地草,地洞风,地罗盘,黄牛尾,金鸡嘴壳,金仙公,九龙牙,咀草,龙牙肾,尼泊尔龙牙草,山昆菜,瘦狗还阳,仙鹤草);Nepal Agrimonia ■
11588 Agrimonia pilosa Ledeb. var. nipponica (Koidz.) Kitam. = Agrimonia nipponica Koidz. ■
11589 Agrimonia pilosa Ledeb. var. occidentalis (Skalicky) Z. Wei et Y. B. Chang;西方龙牙草■☆
11590 Agrimonia pilosa Ledeb. var. simplex T. Shimizu = Agrimonia coreana Nakai ■
11591 Agrimonia pilosa Ledeb. var. succapitata Naruh.;多汁龙牙草■☆
11592 Agrimonia pilosa Ledeb. var. zeylanica (Moon ex Hook. f.) K. M. Purohit et Panigrahi = Agrimonia pilosa Ledeb. var. nepalensis (D. Don) Nakai ■
11593 Agrimonia pilosa Maxim. = Agrimonia pilosa Ledeb. ■
11594 Agrimonia procera Wallr.;香龙牙草;Fragrant Agrimonia, Fragrant Agrimony ■☆
11595 Agrimonia pubescens Wallr.;柔毛龙牙草;Downy Agrimony, Soft Agrimony ■☆
11596 Agrimonia repens L. = Agrimonia procera Wallr. ■☆
11597 Agrimonia rostellata Wallr.;林地龙牙草;Beaked Agrimony, Woodland Agrimony ■☆
11598 Agrimonia striata Michx.;路旁龙牙草(条纹龙牙草);Agrimony, Grooved Agrimony, Roadside Agrimony, Striate Agrimony, Woodland Agrimony ■☆
11599 Agrimonia striata Michx. subsp. viscidula (Bunge) Rumjantsev = Agrimonia pilosa Ledeb. ■
11600 Agrimonia velutina Juz. = Agrimonia coreana Nakai ■
11601 Agrimonia viscidula Bunge = Agrimonia pilosa Ledeb. ■
11602 Agrimonia zeylanica Moon ex Hook. f. = Agrimonia pilosa Ledeb. var. nepalensis (D. Don) Nakai ■
11603 Agrimoniaceae Gray = Rosaceae Juss.(保留科名)●■
11604 Agrimoniaceae Gray;龙牙草科■
11605 Agrimonloides Wolf = Aremonia Neck. ex Nestl.(保留属名)■☆
11606 Agrimonoides Mill.(废弃属名) = Aremonia Neck. ex Nestl.(保留属名)■☆
11607 Agriodaphne Nees ex Meisn. = Ocotea Aubl. ●☆
11608 Agriodendron Endl. = Aloe L. ●■
11609 Agriophyllum M. Bieb. (1819);沙蓬属;Agriophyllum ■
11610 Agriophyllum Post et Kuntze = Agriphyllum Juss. ●■☆
11611 Agriophyllum arenarium M. Bieb. = Agriophyllum squarrosum (L.) Moq. ■
11612 Agriophyllum gobicum Bunge = Agriophyllum squarrosum (L.) Moq. ■
11613 Agriophyllum lateriflorum (Lam.) Moq.;侧花沙蓬;Lateralflower Agriophyllum ■
11614 Agriophyllum latifolium Fisch. et C. A. Mey.;宽叶沙蓬■
11615 Agriophyllum latifolium Fisch. et C. A. Mey. ex Fenzl. = Agropyron desertorum (Fisch. ex Link) Schult. ■
11616 Agriophyllum minus Fisch. et C. A. Mey.;小沙蓬;Small Agriophyllum ■
11617 Agriophyllum paletzkianum Litv.;帕莱沙蓬■☆
11618 Agriophyllum pungens (Vahl) Link ex A. Dietr. = Agriophyllum squarrosum (L.) Moq. ■
11619 Agriophyllum squarrosum (L.) Moq.;沙蓬(灯索,登相子,东廧子,蒺藜梗,沙米,沙蓬米);Squarrose Agriophyllum ■
11620 Agriphyllum Juss. = Berkheya Ehrh.(保留属名)●■☆
11621 Agriphyllum Juss. = Crocodilodes Adans.(废弃属名)●■☆
11622 Agrocharis Hochst. (1844);雅芹属■☆
11623 Agrocharis Hochst. = Caucalis L. ■☆
11624 Agrocharis incognita (C. Norman) Heywood et Jury;雅芹■☆
11625 Agrocharis melanantha Hochst.;黑花雅芹■☆
11626 Agrocharis pedunculata (Baker f.) Heywood et Jury;梗花雅芹■☆
11627 Agrophyllum Neck. = Zygophyllum L. ●■
11628 Agropyron Gaertn. (1770);冰草属(鹅观草属,剪棒草属);Couch-grass, Wheatgrass, Wheat-grass ■
11629 Agropyron Gaertn. = Elymus L. ■
11630 Agropyron × nakashimae Ohwi;中岛氏冰草■☆
11631 Agropyron abolinii Drobow = Elymus abolinii (Drobow) Tzvelev ■
11632 Agropyron abolinii Drobow = Roegneria abolinii (Drobow) Nevski ■
11633 Agropyron aegilopoides Drobow = Elytrigia aegilopoides (Drobow) N. R. Cui ■
11634 Agropyron aegilopoides Drobow = Elytrigia gmelinii (Trin.) Nevski ■
11635 Agropyron alaicum Drobow = Kengyilia alaica (Drobow) J. L. Yang, C. Yen et B. R. Baum ■☆
11636 Agropyron alatavicum Drobow = Elymus alatavicus (Drobow) Á. Löve ■
11637 Agropyron alatavicum Drobow = Kengyilia alatavica (Drobow) J. L. Yang, C. Yen et B. R. Baum ■
11638 Agropyron amgunense Nevski;阿姆贡■☆
11639 Agropyron amurense Drobow = Elymus ciliaris (Trin.) Tzvelev var. amurensis (Drobow) S. L. Chen ■
11640 Agropyron amurense Drobow = Roegneria amurensis (Drobow) Nevski ■
11641 Agropyron angulare Nevski;棱角冰草■☆
11642 Agropyron angustiglume Nevski = Elymus mutabilis (Drobow) Tzvelev ■
11643 Agropyron angustiglume Nevski = Roegneria angustiglumis (Nevski) Nevski ■
11644 Agropyron antiquum Nevski = Elymus antiquus (Nevski) Tzvelev ■
11645 Agropyron argenteum (Nevski) Pavlov = Kengyilia batalinii (Krasn.) J. L. Yang, C. Yen et B. R. Baum ■
11646 Agropyron arinarium W. Wang = Agropyron cristatum (L.) Gaertn. var. pectiniforme (Roem. et Schult.) H. L. Yang ■
11647 Agropyron arinarium W. Wang et Skvortsov;山冰草;Sandy

Wheatgrass ■

11648 Agropyron arinarium W. Wang et Skvortsov = Agropyron cristatum (L.) Gaertn. var. pectiniforme (Roem. et Schult.) H. L. Yang ■

11649 Agropyron armenum Nevski;亚美尼亚冰草■☆

11650 Agropyron attenuatiglume Nevski;渐狭冰草■☆

11651 Agropyron barbicallum Ohwi = Elymus barbicallus (Ohwi) S. L. Chen ■

11652 Agropyron batalinii (Krasn.) Roshev. = Elymus batalinii (Krasn.) Á. Löve ■

11653 Agropyron batalinii (Krasn.) Roshev. = Kengyilia batalinii (Krasn.) J. L. Yang, C. Yen et B. R. Baum ■

11654 Agropyron batalinii (Krasn.) Roshev. = Triticum batalinii Krasn. ■

11655 Agropyron bonaepartis (Spreng.) T. Durand et Schinz = Eremopyrum bonaepartis (Spreng.) Nevski ■

11656 Agropyron boreale (Turcz.) Drobow = Elymus kronokensis (Kom.) Tzvelev ■

11657 Agropyron brevifolium Scribn. = Elymus trachycaulus (Link) Gould ex Shinners ■☆

11658 Agropyron burchan-buddae Nevski = Elymus burchan-buddae (Nevski) Tzvelev ■

11659 Agropyron caespitosum K. Koch;丛生冰草■☆

11660 Agropyron canaliculatum Nevski = Elymus canaliculatus (Nevski) Tzvelev ■

11661 Agropyron caninum (L.) P. Beauv. = Elymus caninus (L.) L. ■

11662 Agropyron caninum (L.) P. Beauv. subsp. majus (Vasey) C. L. Hitchc. = Elymus trachycaulus (Link) Gould ex Shinners ■☆

11663 Agropyron caninum (L.) P. Beauv. var. amurense (Korsh.) Korsh. = Elymus pendulinus (Nevski) Tzvelev ■

11664 Agropyron caninum (L.) P. Beauv. var. andinum (Scribn. et J. G. Sm.) Pease et A. H. Moore = Elymus trachycaulus (Link) Gould ex Shinners ■☆

11665 Agropyron caninum (L.) P. Beauv. var. gmelinii (Ledeb.) Pease et A. H. Moore = Elymus gmelinii (Ledeb.) Tzvelev ■

11666 Agropyron caninum (L.) P. Beauv. var. hornemannii (W. D. J. Koch) Pease et A. H. Moore = Elymus trachycaulus (Link) Gould ex Shinners ■☆

11667 Agropyron caninum (L.) P. Beauv. var. mitchellii Welsh = Elymus trachycaulus (Link) Gould ex Shinners ■☆

11668 Agropyron caninum (L.) P. Beauv. var. unilaterale (Cassidy) C. L. Hitchc. = Elymus trachycaulus (Link) Gould ex Shinners subsp. subsecundus (Link) Á. Löve et D. Löve ■☆

11669 Agropyron chinense (Trin. ex Bunge) Ohwi = Leymus chinensis (Trin. ex Bunge) Tzvelev ■

11670 Agropyron chinense (Trin.) Ohwi = Leymus chinensis (Trin. ex Bunge) Tzvelev ■

11671 Agropyron chinorossicum Ohwi = Leymus secalinus (Georgi) Tzvelev ■

11672 Agropyron ciliare (Trin. ex Bunge) Franch. = Elymus ciliaris (Trin. ex Bunge) Tzvelev ■

11673 Agropyron ciliare (Trin. ex Bunge) Franch. f. japonense (Honda) Ohwi = Elymus ciliaris (Trin.) Tzvelev var. hackelianus (Honda) G. Zhu et S. L. Chen ■

11674 Agropyron ciliare (Trin. ex Bunge) Franch. f. submu-ticum (Honda) Ohwi = Elymus ciliaris (Trin.) Tzvelev var. submuticus (Honda) S. L. Chen ■

11675 Agropyron ciliare (Trin. ex Bunge) Franch. subsp. amurense (Drobow) T. Koyama = Elymus ciliaris (Trin.) Tzvelev var. amurensis (Drobow) S. L. Chen ■

11676 Agropyron ciliare (Trin. ex Bunge) Franch. subsp. amurense (Drobow) T. Koyama = Roegneria amurensis (Drobow) Nevski ■

11677 Agropyron ciliare (Trin. ex Bunge) Franch. var. hackelianum (Honda) Ohwi = Elymus ciliaris (Trin.) Tzvelev var. hackelianus (Honda) G. Zhu et S. L. Chen ■

11678 Agropyron ciliare (Trin. ex Bunge) Franch. var. hondae Keng = Elymus ciliaris (Trin.) Tzvelev var. hackelianus (Honda) G. Zhu et S. L. Chen ■

11679 Agropyron ciliare (Trin. ex Bunge) Franch. var. integrum Keng = Elymus ciliaris (Trin. ex Bunge) Tzvelev ■

11680 Agropyron ciliare (Trin. ex Bunge) Franch. var. lasiophyllum Kitag. = Elymus ciliaris (Trin.) Tzvelev var. lasiophyllus (Kitag.) S. L. Chen ■

11681 Agropyron ciliare (Trin. ex Bunge) Franch. var. pauperum Keng = Elymus ciliaris (Trin.) Tzvelev var. hackelianus (Honda) G. Zhu et S. L. Chen ■

11682 Agropyron ciliare (Trin. ex Bunge) Franch. var. pilosum (Korsh.) Honda = Elymus ciliaris (Trin.) Tzvelev var. amurensis (Drobow) S. L. Chen ■

11683 Agropyron ciliare (Trin. ex Bunge) Franch. var. submuticum Honda = Elymus ciliaris (Trin.) Tzvelev var. submuticus (Honda) S. L. Chen ■

11684 Agropyron ciliare (Trin.) Franch. = Elymus ciliaris (Trin.) Tzvelev ■

11685 Agropyron ciliare (Trin.) Franch. f. japonense (Honda) Ohwi = Agropyron racemiferum (Steud.) Koidz. var. japonense (Honda) Sugim. ■

11686 Agropyron ciliare (Trin.) Franch. f. japonense (Honda) Ohwi = Elymus ciliaris (Trin.) Tzvelev var. japonensis (Honda) S. L. Chen ■

11687 Agropyron ciliare (Trin.) Franch. f. japonense (Honda) Ohwi = Roegneria japonensis (Honda) Keng ■

11688 Agropyron ciliare (Trin.) Franch. subsp. minus (Miq.) T. Koyama = Elymus racemifer (Steud.) Tzvelev ■☆

11689 Agropyron ciliare (Trin.) Franch. subsp. submutica Honda = Roegneria ciliaris (Trin. ex Bunge) Nevski var. submutica (Honda) Keng ■

11690 Agropyron ciliare (Trin.) Franch. var. lasiophylla Kitag. = Roegneria ciliaris (Trin.) Nevski var. lasiophylla (Kitag.) Kitag. ■

11691 Agropyron ciliare (Trin.) Franch. var. minus (Miq.) Ohwi = Elymus racemifer (Steud.) Tzvelev ■☆

11692 Agropyron ciliare (Trin.) Franch. var. pilosum (Korsh.) Honda = Elymus ciliaris (Trin.) Tzvelev subsp. amurensis (Drobow) Tzvelev ■☆

11693 Agropyron cimmericum Nevski;刻赤冰草■☆

11694 Agropyron cognatum Hack. = Pseudoroegneria cognata (Hack.) Á. Löve ■

11695 Agropyron confusum Roshev. = Roegneria confusa (Roshev.) Nevski ■

11696 Agropyron cristatum (L.) Gaertn.;冰草(大麦草,山麦草);Crested Wheat Grass, Crested Wheatgrass, Crested Wheat-grass, Fairway, Fairway Wheatgrass, Standard Wheatgrass, Wheatgrass ■

11697 Agropyron cristatum (L.) Gaertn. var. brachyatherum (Maire) Maire = Agropyron cristatum (L.) Gaertn. ■

11698 Agropyron cristatum (L.) Gaertn. var. desertorum (Fisch. ex Link) Dorn = Agropyron desertorum (Fisch. ex Link) Schult. ■

11699 Agropyron cristatum (L.) Gaertn. var. pectiniforme (Roem. et Schult.) H. L. Yang;光穗冰草;Combform Wheatgrass ■

11700 Agropyron cristatum (L.) Gaertn. var. pluriflorum X. L. Yang;多花冰草;Manyflowered Wheatgrass ■

11701 Agropyron curvatum Nevski = Elymus fedtschenkoi Tzvelev ■

11702 Agropyron czilikense Drobow = Roegneria tianschanica (Drobow) Nevski ■

11703 Agropyron czimganicum Drobow = Roegneria tschimganica (Drobow) Nevski ■

11704 Agropyron dasyanthum Ledeb.;绒花冰草■☆

11705 Agropyron dasystachyum (Hook.) Scribn. = Elytrigia dasystachya (Hook.) Á. Löve et D. Löve ■☆

11706 Agropyron dasystachyum (Hook.) Scribn. subsp. psammophilum (J. M. Gillett et Senn) Dewey = Elytrigia dasystachya (Hook.) Á. Löve et D. Löve subsp. psammophila (J. M. Gillett et Senn) Dewey ■☆

11707 Agropyron dasystachyum (Hook.) Scribn. var. psammophilum (J. M. Gillett et Senn) E. G. Voss = Elytrigia dasystachya (Hook.) Á. Löve et D. Löve subsp. psammophila (J. M. Gillett et Senn) Dewey ■☆

11708 Agropyron dasystachyum (Hook.) Scribn. var. riparum (Scribn. et J. G. Sm.) Bowden = Elytrigia dasystachya (Hook.) Á. Löve et D. Löve ■☆

11709 Agropyron densiflorum P. Beauv.;密穗冰草■☆

11710 Agropyron desertorum (Fisch. ex Link) Schult.;沙生冰草(荒漠冰草);Desert Wheat Grass, Desert Wheatgrass, Standard Crested Wheatgrass, Standard Crested Wheat-grass ■

11711 Agropyron desertorum (Fisch. ex Link) Schult. = Agropyron cristatum (L.) Gaertn. ■

11712 Agropyron desertorum (Fisch. ex Link) Schult. var. pilosiusculum Melderis;毛沙生冰草(毛稃沙生冰草);Hairy Desert Wheatgrass ■

11713 Agropyron distachyon (L.) Chevall. = Brachypodium distachyon (L.) P. Beauv. ■

11714 Agropyron distans K. Koch = Eremopyrum distans (K. Koch) Nevski ■

11715 Agropyron distichum (Thunb.) P. Beauv. = Thinopyrum distichum (Thunb.) Á. Löve ■☆

11716 Agropyron donianum F. B. White = Elymus caninus (L.) L. ■

11717 Agropyron dschungaricum (Nevski) Nevski = Pseudoroegneria cognata (Hack.) Á. Löve ■

11718 Agropyron elmeri Scribn. = Elytrigia dasystachya (Hook.) Á. Löve et D. Löve ■☆

11719 Agropyron elongatiforme Drobow = Elytrigia elongatiformis (Drobow) Nevski ■

11720 Agropyron elongatiforme Drobow = Elytrigia repens (L.) Desv. ex B. D. Jacks. subsp. elongatiformis (Drobow) Tzvelev ■☆

11721 Agropyron elongatum (Host) P. Beauv. = Elymus elongatus (Host) Runemark ■☆

11722 Agropyron elongatum (Host) P. Beauv. = Elytrigia elongata (Host ex P. Beauv.) Nevski ■

11723 Agropyron elongatum (Host) P. Beauv. var. scirpeum (C. Presl) Fiori = Elytrigia scirpea (C. Presl) Holub ■☆

11724 Agropyron elongatum Host ex P. Beauv. = Elymus elongatus (Host) Runemark ■☆

11725 Agropyron elymoides (Hochst. ex A. Rich.) P. Candargy = Elymus africanus Á. Löve ■☆

11726 Agropyron embergeri Maire = Elytrigia embergeri (Maire) Dobignard ■☆

11727 Agropyron ferganense Drobow = Elytrigia ferganensis (Drobow) Nevski ■

11728 Agropyron ferganense Drobow = Pseudoroegneria cognata (Hack.) Á. Löve ■

11729 Agropyron festucoides Maire = Elymus festucoides (Maire) Ibn Tattou ■☆

11730 Agropyron festucoides Maire var. leiorrhacis ? = Elymus festucoides (Maire) Ibn Tattou ■☆

11731 Agropyron festucoides Maire var. pseudofestucoides (Emb.) Maire et Weiller = Elymus festucoides (Maire) Ibn Tattou ■☆

11732 Agropyron firmiculme Nevski;坚硬冰草■☆

11733 Agropyron formosanum Honda = Elymus formosanus (Honda) Á. Löve ■

11734 Agropyron formosanum Honda = Roegneria formosana (Honda) Ohwi ■

11735 Agropyron fragile (Roth) P. Candargy;西伯利亚冰草;Siberia Wheatgrass, Siberian Wheatgrass ■

11736 Agropyron geniculatum (Trin. ex Ledeb.) K. Koch;膝曲冰草■☆

11737 Agropyron glaucum Roem. et Schult.;蓝冰草■☆

11738 Agropyron gmelinii (Ledeb.) Scribn. et J. G. Sm. = Elymus gmelinii (Ledeb.) Tzvelev ■

11739 Agropyron gmelinii (Ledeb.) Scribn. et Sm. = Elymus gmelinii (Ledeb.) Tzvelev ■

11740 Agropyron gmelinii (Ledeb.) Scribn. et Sm. subsp. tenuisetum (Ohwi) T. Koyama = Elymus gmelinii (Ledeb.) Tzvelev var. tenuisetus (Ohwi) Osada ■☆

11741 Agropyron gmelinii (Ledeb.) Scribn. et Sm. var. tenuisetum (Ohwi) Ohwi = Elymus gmelinii (Ledeb.) Tzvelev var. tenuisetus (Ohwi) Osada ■☆

11742 Agropyron gmelinii (Trin.) P. Candargy = Elytrigia gmelinii (Trin.) Nevski ■

11743 Agropyron gracillimum Nevski;细长冰草■☆

11744 Agropyron grandiglumis (Keng et S. L. Chen) Tzvelev = Kengyilia grandiglumis (Keng et S. L. Chen) J. L. Yang, C. Yen et B. R. Baum ■

11745 Agropyron hackelianum (Honda) Beetle = Elymus ciliaris (Trin.) Tzvelev var. hackelianus (Honda) G. Zhu et S. L. Chen ■

11746 Agropyron hackelianum (Honda) Beetle var. japonicum Beetle = Elymus ciliaris (Trin.) Tzvelev var. hackelianus (Honda) G. Zhu et S. L. Chen ■

11747 Agropyron hatusimae Ohwi = Elymus mayebaranus (Honda) S. L. Chen ■

11748 Agropyron humidum Ohwi et Sakam. = Elymus humidus (Ohwi et Sakam.) Á. Löve ■☆

11749 Agropyron imbricatum Roem. et Schult.;覆瓦冰草■☆

11750 Agropyron integrum (Keng) Keng = Elymus ciliaris (Trin. ex Bunge) Tzvelev ■

11751 Agropyron intermedium (Host) P. Beauv. = Elytrigia intermedia (Host) Nevski ■

11752 Agropyron jacquemontii Hook. f. = Elymus jacquemontii (Hook. f.) Tzvelev ■

11753 Agropyron jacquemontii Hook. f. = Roegneria jacquemontii (Hook. f.) Ovcz. et Sidorenko ■

11754 Agropyron jacutorum Nevski;雅库特冰草■☆

11755 Agropyron japonense Honda = Elymus ciliaris (Trin.) Tzvelev var. hackelianus (Honda) G. Zhu et S. L. Chen ■

11756 Agropyron japonense Honda var. hackelianum (Honda) Honda =

Elymus ciliaris (Trin.) Tzvelev var. hackelianus (Honda) G. Zhu et S. L. Chen ■

11757 Agropyron japonensis Honda = Roegneria japonensis (Honda) Keng ex Keng et S. L. Chen ■

11758 Agropyron japonicum Honda = Elymus ciliaris (Trin.) Tzvelev var. hackelianus (Honda) G. Zhu et S. L. Chen ■

11759 Agropyron japonicum Honda = Roegneria japonensis (Honda) Keng ex Keng et S. L. Chen ■

11760 Agropyron japonicum Honda var. hackelianum Honda = Elymus ciliaris (Trin.) Tzvelev var. hackelianus (Honda) G. Zhu et S. L. Chen ■

11761 Agropyron japonicum Honda var. hackelianum Honda = Roegneria japonensis (Honda) Keng ex Keng et S. L. Chen var. hackeliana (Honda) Keng ■

11762 Agropyron junceum (L.) P. Beauv. = Elytrigia juncea (L.) Nevski ■

11763 Agropyron junceum (L.) P. Beauv. subsp. boreoatlanticum Simonet et Guin. = Elytrigia juncea (L.) Nevski subsp. boreoatlantica (Simonet et Guin.) Hyl. ■☆

11764 Agropyron junceum (L.) P. Beauv. subsp. mediterraneum Simonet et Guin. = Elytrigia juncea (L.) Nevski ■

11765 Agropyron junceum (L.) P. Beauv. var. glabrum (Simonet et Guin.) Maire = Elytrigia juncea (L.) Nevski ■

11766 Agropyron junceum (L.) P. Beauv. var. repens (L.) M. Marsson = Elytrigia repens (L.) Desv. ex B. D. Jacks. ■

11767 Agropyron junceum (L.) P. Beauv. var. rifense Sennen = Elytrigia juncea (L.) Nevski ■

11768 Agropyron junceum (L.) P. Beauv. var. sartorii Boiss. et Heldr. = Elytrigia juncea (L.) Nevski ■

11769 Agropyron junceum P. Beauv. = Elytrigia juncea (L.) Nevski ■

11770 Agropyron kamoji Ohwi = Elymus kamoji (Ohwi) S. L. Chen ■

11771 Agropyron kamoji Ohwi = Elymus tsukushiensis Honda var. transiens (Hack.) Osada ■

11772 Agropyron kamoji Ohwi = Roegneria kamoji (Ohwi) Ohwi ex Keng ■

11773 Agropyron kengii Tzvelev = Elymus kengii (Tzvelev) D. F. Cui ■

11774 Agropyron kengii Tzvelev = Kengyilia hirsuta (Keng) J. L. Yang, C. Yen et B. R. Baum ■

11775 Agropyron kokonoricum (Keng et S. L. Chen) Tzvelev = Kengyilia kokonorica (Keng et S. L. Chen) J. L. Yang, C. Yen et B. R. Baum ■

11776 Agropyron komarovii Nevski = Elymus komarovii (Nevski) Tzvelev ■

11777 Agropyron komarovii Nevski = Roegneria komarovii (Nevski) Nevski ■

11778 Agropyron kronokense Kom. = Elymus kronokensis (Kom.) Tzvelev ■

11779 Agropyron krylovianum Schischk.;克雷罗夫冰草■☆

11780 Agropyron lanceolatum Scribn. et J. G. Sm. = Elytrigia dasystachya (Hook.) Á. Löve et D. Löve ■☆

11781 Agropyron lasianthum Boiss. = Eremopyrum distans (K. Koch) Nevski ■

11782 Agropyron littorale (Host) Dumort. = Elytrigia atherica (Link) M. A. Carreras ex Kerguélen ■☆

11783 Agropyron lolioides Roshev.;黑麦草冰草■☆

11784 Agropyron marginatum H. Lindb. = Roegneria marginata (H. Lindb.) Dobignard ■☆

11785 Agropyron marginatum H. Lindb. subsp. kabylicum Maire et Weiller = Elymus marginatus (H. Lindb.) Á. Löve subsp. kabylicus (Maire et Weiller) Valdés et H. Scholz ■☆

11786 Agropyron marginatum H. Lindb. var. maroccanum (Font Quer et Pau) Maire et Weiller = Roegneria marginata (H. Lindb.) Dobignard ■☆

11787 Agropyron mayebaranum Honda = Elymus mayebaranus (Honda) S. L. Chen ■

11788 Agropyron mayebaranum Honda = Roegneria mayebarana (Honda) Ohwi ■

11789 Agropyron mayebaranum Honda var. intermedium Hatus. = Elymus mayebaranus (Honda) S. L. Chen ■

11790 Agropyron mayebaranum Honda var. nakashimae (Ohwi) Ohwi = Agropyron × nakashimae Ohwi ■☆

11791 Agropyron melantherum Keng = Kengyilia melanthera (Keng) J. L. Yang, C. Yen et B. R. Baum ■

11792 Agropyron melantherum Keng = Roegneria melanthera (Keng) Keng ex Keng et S. L. Chen ■

11793 Agropyron michnoi Roshev.;根茎冰草;Rhizome Wheatgrass ■

11794 Agropyron microlepis Melderis = Elymus antiquus (Nevski) Tzvelev ■

11795 Agropyron molle (Scribn. et J. G. Sm.) Rydb. = Elytrigia smithii (Rydb.) Nevski ■

11796 Agropyron mongolicum Keng;沙芦草;Mongol Wheatgrass, Mongolian Wheatgrass ■

11797 Agropyron mongolicum Keng var. villosum X. L. Yang;毛沙芦草;Hairy Mongolian Wheatgrass ■

11798 Agropyron mutabile Drobow = Elymus mutabilis (Drobow) Tzvelev ■

11799 Agropyron mutabile Drobow = Roegneria mutabilis (Drobow) Hyl. ■

11800 Agropyron muticum (Keng et S. L. Chen) Tzvelev = Kengyilia mutica (Keng et S. L. Chen) J. L. Yang, C. Yen et B. R. Baum ■

11801 Agropyron nodosum (Steven ex M. Bieb.) Nevski = Elymus nodosus (Nevski) Melderis ■☆

11802 Agropyron nutans Keng = Elymus burchan-buddae (Nevski) Tzvelev ■

11803 Agropyron nutans Keng = Roegneria nutans (Keng) Keng ex Keng et S. L. Chen ■

11804 Agropyron orientale (L.) Roem. et Schult. = Eremopyrum orientale (L.) Jaub. et Spach ■

11805 Agropyron orientale (L.) Roem. et Schult. subsp. distans (K. Koch) Maire = Eremopyrum distans (K. Koch) Nevski ■

11806 Agropyron orientale (L.) Roem. et Schult. var. lasianthum (Boiss.) Boiss. = Eremopyrum distans (K. Koch) Nevski ■

11807 Agropyron orientale (L.) Roem. et Schult. var. lasianthum Boiss. = Eremopyrum orientale (L.) Jaub. et Spach ■

11808 Agropyron orientale (L.) Roem. et Schult. var. medians Maire = Eremopyrum orientale (L.) Jaub. et Spach ■

11809 Agropyron orientale (L.) Roem. et Schult. var. sublanuginosum Drobow = Eremopyrum bonaepartis (Spreng.) Nevski ■

11810 Agropyron panormitanum Parl. var. hispanicum Boiss. = Elymus hispanicus (Boiss.) Talavera ■☆

11811 Agropyron panormitanum Parl. var. maroccanum Font Quer et Pau = Elymus hispanicus (Boiss.) Talavera ■☆

11812 Agropyron panormitanum Parl. var. pharaonis Maire = Elymus hispanicus (Boiss.) Talavera ■☆

11813 Agropyron patulum Trin. = Eremopyrum bonaepartis (Spreng.) Nevski ■

11814 Agropyron pauciflorum (Schwein. ex Keating) Hitchc. = Roegneria pauciflora (Schwein.) Hyl. ■

11815 Agropyron pauciflorum (Schwein.) Hitchc. ex Silveus = Elymus trachycaulus (Link) Gould ex Shinners ■☆

11816 Agropyron pauciflorum (Schwein.) Hitchc. ex Silveus subsp. majus (Vasey) Melderis = Elymus trachycaulus (Link) Gould ex Shinners ■☆

11817 Agropyron pauciflorum (Schwein.) Hitchc. ex Silveus subsp. novae-angliae (Scribn.) Melderis = Elymus trachycaulus (Link) Gould ex Shinners ■☆

11818 Agropyron pauciflorum (Schwein.) Hitchc. ex Silveus subsp. teslinense (Porsild et Senn) Melderis = Elymus trachycaulus (Link) Gould ex Shinners ■☆

11819 Agropyron pauciflorum (Schwein.) Hitchc. ex Silveus var. glaucum (Pease et A. H. Moore) Taylor = Elymus trachycaulus (Link) Gould ex Shinners subsp. subsecundus (Link) Á. Löve et D. Löve ■☆

11820 Agropyron pauciflorum (Schwein.) Hitchc. ex Silveus var. novae-angliae (Scribn.) R. L. Taylor et MacBryde = Elymus trachycaulus (Link) Gould ex Shinners ■☆

11821 Agropyron pectinatum (Labill.) P. Beauv.;篦穗冰草■

11822 Agropyron pectiniforme Roem. et Schult. = Agropyron cristatum (L.) Gaertn. var. pectiniforme (Roem. et Schult.) H. L. Yang ■

11823 Agropyron pinnatum (L.) Chevall. = Brachypodium pinnatum (L.) P. Beauv. ■

11824 Agropyron ponticum Nevski;黑海冰草■☆

11825 Agropyron praecaespitosum (Nevski) Nevski = Elymus praecaespitosus (Nevski) Tzvelev ■

11826 Agropyron praecaespitosum Nevski = Elymus mutabilis (Drobow) Tzvelev var. praecaespitosus (Nevski) S. L. Chen ■

11827 Agropyron propinquum Nevski = Elytrigia aegilopoides (Drobow) N. R. Cui ■

11828 Agropyron propinquum Nevski = Elytrigia gmelinii (Trin.) Nevski ■

11829 Agropyron prostratum (Pall.) P. Beauv. = Eremopyrum triticeum (Gaertn.) Nevski ■

11830 Agropyron prostratum P. Beauv.;伏生冰草■☆

11831 Agropyron pruniferum Nevski;腊霜冰草■☆

11832 Agropyron psammophilium J. M. Gillett et Senn = Elytrigia dasystachya (Hook.) Á. Löve et D. Löve subsp. psammophila (J. M. Gillett et Senn) Dewey ■☆

11833 Agropyron pseudofestucoides Emb. = Roegneria festucoides (Maire) Dobignard ■

11834 Agropyron pseudofestucoides Emb. var. acutiflorum ? = Roegneria festucoides (Maire) Dobignard ■

11835 Agropyron pseudofestucoides Emb. var. muticum ? = Roegneria festucoides (Maire) Dobignard ■

11836 Agropyron pseudonutans Keng = Elymus pseudonutans Á. Löve ■

11837 Agropyron pulcherrimum Grossh.;美丽冰草■☆

11838 Agropyron pumilum (Steud.) Nevski;矮小冰草■☆

11839 Agropyron racemiferum (Steud.) Koidz. = Elymus racemifer (Steud.) Tzvelev ■☆

11840 Agropyron racemiferum (Steud.) Koidz. var. japonense (Honda) Sugim. = Agropyron ciliare (Trin.) Franch. f. japonense (Honda) Ohwi ■

11841 Agropyron ramosum (Trin.) K. Richt.;分枝冰草■☆

11842 Agropyron ramosum (Trin.) K. Richt. = Leymus ramosus (Trin.) Tzvelev ■

11843 Agropyron ramosum K. Richt. = Leymus ramosus Tzvelev ■

11844 Agropyron reflexiaristatum Nevski;折穗冰草■☆

11845 Agropyron repens (L.) P. Beauv. = Elytrigia repens (L.) Desv. ex B. D. Jacks. ■

11846 Agropyron repens (L.) P. Beauv. f. aristatum (K. Schum.) Holmb. = Elytrigia repens (L.) Desv. ex B. D. Jacks. ■

11847 Agropyron repens (L.) P. Beauv. f. pilosum (Scribn.) Fernald = Elytrigia repens (L.) Desv. ex B. D. Jacks. ■

11848 Agropyron repens (L.) P. Beauv. f. trichorrhachis Rohlena = Elytrigia repens (L.) Desv. ex B. D. Jacks. ■

11849 Agropyron repens (L.) P. Beauv. f. vaillantianum (Wulfen) Fernald = Elymus repens (L.) Gould ■

11850 Agropyron repens (L.) P. Beauv. subsp. elongatiforme (Drobow) D. R. Dewey = Elytrigia repens (L.) Desv. ex Nevski subsp. elongatiformis (Drobow) Tzvelev ■

11851 Agropyron repens (L.) P. Beauv. var. aristatum Döll = Elytrigia repens (L.) Desv. ex B. D. Jacks. ■

11852 Agropyron repens (L.) P. Beauv. var. arvense (Schreb.) Rchb. = Elytrigia repens (L.) Desv. ex B. D. Jacks. ■

11853 Agropyron repens (L.) P. Beauv. var. atlantis Maire = Elytrigia repens (L.) Desv. ex B. D. Jacks. ■

11854 Agropyron repens (L.) P. Beauv. var. glaucum Döll = Elytrigia repens (L.) Desv. ex B. D. Jacks. ■

11855 Agropyron repens (L.) P. Beauv. var. subulatum (Schreb.) Rchb. = Elytrigia repens (L.) Desv. ex B. D. Jacks. ■

11856 Agropyron repens (L.) P. Beauv. var. subulatum (Schreb.) Roem. et Schult. = Elytrigia repens (L.) Desv. ex B. D. Jacks. ■

11857 Agropyron repens (L.) P. Beauv. var. subulatum (Schreb.) Roem. et Schult. f. setiferum Fernald = Elytrigia repens (L.) Desv. ex B. D. Jacks. ■

11858 Agropyron repens (L.) P. Beauv. var. subulatum (Schreb.) Roem. et Schult. f. heberhachis Fernald = Elytrigia repens (L.) Desv. ex B. D. Jacks. ■

11859 Agropyron repens (L.) P. Beauv. var. subulatum (Schreb.) Roem. et Schult. f. vaillantianum (Wulfen et Schreb.) Fernald = Elytrigia repens (L.) Desv. ex B. D. Jacks. ■

11860 Agropyron riparium Scribn. et J. G. Sm. = Elytrigia dasystachya (Hook.) Á. Löve et D. Löve ■☆

11861 Agropyron roshevitzii Nevski;罗舍冰草■☆

11862 Agropyron scabridulum Ohwi = Elymus scabridulus (Ohwi) Tzvelev ■

11863 Agropyron schrenkianum (Fisch. et C. A. Mey.) Drobow = Roegneria schrenkiana (Fisch. et C. A. Mey.) Nevski ■

11864 Agropyron schrenkianum (Fisch. et C. A. Mey.) P. Candargy = Elymus schrenkianus (Fisch. et C. A. Mey.) Tzvelev ■

11865 Agropyron scirpeum C. Presl = Elytrigia scirpea (C. Presl) Holub ■☆

11866 Agropyron secalinum (Georgi) Kitag. = Leymus secalinus (Georgi) Tzvelev ■

11867 Agropyron semicostatum (Nees ex Steud.) Nees ex Boiss. var. ciliare (Trin. ex Bunge) Hack. = Elymus ciliaris (Trin. ex Bunge) Tzvelev ■

11868 Agropyron semicostatum (Nees ex Steud.) Nees ex Boiss. var. transiens Hack. = Elymus kamoji (Ohwi) S. L. Chen ■

11869 Agropyron semicostatum Nees ex Steud. = Elymus semicostatus (Nees ex Steud.) Melderis ■☆

11870 Agropyron semicostatum Nees ex Steud. var. thomsonii ? = Elymus semicostatus (Nees ex Steud.) Melderis ■☆

11871 Agropyron setuliferum Nevski;刚毛冰草■☆

11872 Agropyron sibiricum (Willd.) P. Beauv. = Agropyron fragile (Roth) P. Candargy ■

11873 Agropyron sibiricum (Willd.) P. Beauv. f. pubiflorum Roshev.;毛稃冰草■

11874 Agropyron sibiricum (Willd.) P. Beauv. var. desertorum (Fisch. ex Link) Boiss. = Agropyron desertorum (Fisch. ex Link) Schult. ■

11875 Agropyron sibiricum P. Beauv. = Agropyron fragile (Roth) P. Candargy ■

11876 Agropyron sinkiangense D. F. Cui;新疆冰草;Xinjiang Wheatgrass ■

11877 Agropyron sinuatum Nevski;深波冰草■☆

11878 Agropyron smithii Rydb. = Elymus smithii (Rydb.) Gould ■

11879 Agropyron smithii Rydb. = Elytrigia smithii (Rydb.) Nevski ■

11880 Agropyron smithii Rydb. var. molle (Scribn. et J. G. Sm.) M. E. Jones = Elytrigia smithii (Rydb.) Nevski ■

11881 Agropyron smithii Rydb. var. palmeri (Scribn. et J. G. Sm.) A. Heller = Elytrigia smithii (Rydb.) Nevski ■

11882 Agropyron sosnowskyi Hack.;索斯冰草■☆

11883 Agropyron spicatum Scribn. et J. G. Sm.;穗状冰草;Bluebunch Wheat-grass ■☆

11884 Agropyron squarrosum (Roth) Link = Eremopyrum bonaepartis (Spreng.) Nevski ■

11885 Agropyron stenachyrum (Keng et S. L. Chen) Tzvelev = Kengyilia stenachyra (Keng et S. L. Chen) J. L. Yang, C. Yen et B. R. Baum ■

11886 Agropyron stipifolium Czern.;燕麦叶冰草■☆

11887 Agropyron strigosum (M. Bieb.) Boiss. subsp. aegilopoides (Drobow) Tzvelev = Elytrigia gmelinii (Trin.) Nevski ■

11888 Agropyron strigosum Boiss.;糙伏毛冰草■☆

11889 Agropyron subsecundum (Link) Hitchc. = Elymus trachycaulus (Link) Gould ex Shinners subsp. subsecundus (Link) Á. Löve et D. Löve ■☆

11890 Agropyron subsecundum (Link) Hitchc. var. andinum (Scribn. et J. G. Sm.) Hitchc. = Elymus trachycaulus (Link) Gould ex Shinners ■☆

11891 Agropyron sylvaticum (Huds.) Chevall. = Brachypodium sylvaticum (Huds.) P. Beauv. ■

11892 Agropyron tanaiticum Nevski;顿河冰草■☆

11893 Agropyron tashiroi Ohwi = Elymus koryoensis (Honda) ■☆

11894 Agropyron tenerum Vasey = Elymus trachycaulus (Link) Gould ex Shinners ■☆

11895 Agropyron teslinense Porsild et Senn = Elymus trachycaulus (Link) Gould ex Shinners ■☆

11896 Agropyron thoroldianum Oliv. = Kengyilia thoroldiana (Oliv.) J. L. Yang, C. Yen et B. R. Baum ■

11897 Agropyron thoroldianum Oliv. var. laxiuscula Melderis = Kengyilia laxiuscula (Melderis) Tzvelev ■

11898 Agropyron tianschanicum Drobow = Elymus czilikensis (Drobow) Tzvelev ■

11899 Agropyron tianschanicum Drobow = Elymus tianschanigenus Czerep. ■

11900 Agropyron tianschanicum Drobow = Roegneria tianschanica (Drobow) Nevski ■

11901 Agropyron tibeticum Melderis = Elymus tibeticus (Melderis) G. Singh ■

11902 Agropyron tibeticum Melderis = Roegneria tibetica (Melderis) H. L. Yang ■

11903 Agropyron trachycaulum (Link) Malte = Elymus trachycaulus (Link) Gould ex Shinners ■☆

11904 Agropyron trachycaulum (Link) Malte ex H. F. Lewis = Elymus trachycaulus (Link) Gould ex Shinners ■☆

11905 Agropyron trachycaulum (Link) Malte ex H. F. Lewis var. ciliatum (Scribn. et J. G. Sm.) Malte = Elymus trachycaulus (Link) Gould ex Shinners subsp. subsecundus (Link) Á. Löve et D. Löve ■☆

11906 Agropyron trachycaulum (Link) Malte ex H. F. Lewis var. glaucum (Pease et A. H. Moore) Malte = Elymus trachycaulus (Link) Gould ex Shinners subsp. subsecundus (Link) Á. Löve et D. Löve ■☆

11907 Agropyron trachycaulum (Link) Malte ex H. F. Lewis var. majus (Vasey) Fernald = Elymus trachycaulus (Link) Gould ex Shinners ■☆

11908 Agropyron trachycaulum (Link) Malte ex H. F. Lewis var. novae-angliae (Scribn.) Fernald = Elymus trachycaulus (Link) Gould ex Shinners ■☆

11909 Agropyron trachycaulum (Link) Malte ex H. F. Lewis var. unilaterale (Cassidy) Malte = Elymus trachycaulus (Link) Gould ex Shinners subsp. subsecundus (Link) Á. Löve et D. Löve ■☆

11910 Agropyron trachycaulum Steud. = Elymus trachycaulus (Link) Gould ex Shinners ■☆

11911 Agropyron trachycaulum Steud. var. glaucum (Pease et A. H. Moore) Malte = Elymus trachycaulus (Link) Gould ex Shinners ■☆

11912 Agropyron trichophorum (Link) K. Richt. = Elytrigia trichophora (Link) Nevski ■

11913 Agropyron triticeum Gaertn. = Eremopyrum triticeum (Gaertn.) Nevski ■

11914 Agropyron tschimganica Drobow = Roegneria tschimganica (Drobow) Nevski ■

11915 Agropyron tschimganicum Drobow = Elymus tschimuganicus (Drobow) Tzvelev ■

11916 Agropyron tsukushiense (Honda) Ohwi var. transiens (Hack.) Ohwi = Elymus tsukushiensis Honda var. transiens (Hack.) Osada ■

11917 Agropyron tsukushiense (Honda) Ohwi var. transiens (Hack.) Ohwi = Elymus kamoji (Ohwi) S. L. Chen ■

11918 Agropyron turczaninovii Drobow = Elymus gmelinii (Ledeb.) Tzvelev ■

11919 Agropyron turczaninovii Drobow = Roegneria turczaninovii (Drobow) Nevski ■

11920 Agropyron turczaninovii Drobow var. tenuisetum Ohwi = Elymus gmelinii (Ledeb.) Tzvelev var. tenuisetus (Ohwi) Osada ■☆

11921 Agropyron turczaninowii Drobow = Elymus gmelinii (Ledeb.) Tzvelev ■

11922 Agropyron turczaninowii Drobow var. macratherum Ohwi = Elymus gmelinii (Ledeb.) Tzvelev var. macratherus (Ohwi) S. L. Chen et G. Zhu ■

11923 Agropyron turczaninowii Drobow var. tenuisetum Ohwi = Elymus gmelinii (Ledeb.) Tzvelev ■

11924 Agropyron turkestanicum Gand. = Eremopyrum bonaepartis (Spreng.) Nevski ■

11925 Agropyron ugamica Drobow = Roegneria ugamica (Drobow) Nevski ■

11926 Agropyron ugamicum Drobow = Elymus nevskii Tzvelev ■

11927 Agropyron uninerve P. Candargy = Leymus chinensis (Trin. ex Bunge) Tzvelev ■

11928 Agropyron violaceum (Hornem.) Lange subsp. andinum (Scribn. et J. G. Sm.) Melderis = Elymus trachycaulus (Link) Gould ex Shinners ■☆

11929 Agropyron violaceum (Hornem.) Lange var. andinum Scribn. et J. G. Sm. = Elymus trachycaulus (Link) Gould ex Shinners ■☆

11930 Agropyron yezoense Honda = Elymus nipponicus Jaaska ■☆

11931 Agropyropsis (Batt. et Trab.) A. Camus(1935);类冰草属(拟冰草属)■☆

11932 Agropyropsis A. Camus = Agropyropsis (Batt. et Trab.) A. Camus ■☆

11933 Agropyropsis lolium (Balansa ex Coss. et Durieu) A. Camus;类冰草■☆

11934 Agropyropsis lolium (Balansa) A. Camus = Agropyropsis lolium (Balansa ex Coss. et Durieu) A. Camus ■☆

11935 Agropyrum Roem. et Schult. = Agropyron Gaertn. ■

11936 Agrosinapis Fourr. = Brassica L. ●■

11937 Agrostana Hill = Bupleurum L. ●■

11938 Agrostemma L. (1753);麦仙翁属(麦毒草属);Agrostemma, Cockle, Corn Cockle, Corncockle ■

11939 Agrostemma banksia Meerb. = Lychnis coronata Thunb. ■

11940 Agrostemma banksia Meerb. = Silene banksia (Meerb.) Mabberly ■

11941 Agrostemma brachylobum (Fenzl) K. Hammer;短裂麦仙翁;Narrow Corncockle ■☆

11942 Agrostemma bungeana D. Don = Lychnis senno Siebold et Zucc. ■

11943 Agrostemma bungeana D. Don = Silene bungeana (D. Don) H. Ohashi et H. Nakai ■

11944 Agrostemma chalcedonica (L.) Doellinger = Silene chalcedonica (L.) E. H. L. Krause ■

11945 Agrostemma coronaria L. = Lychnis coronaria (L.) Desr. ■

11946 Agrostemma coronaria L. = Silene coronaria (L.) Clairv. ■

11947 Agrostemma githago L.;麦仙翁(麦毒草);Bachelor's Buttons, Bastard Nigella, Billy Buttons, Cat's Ear, Cockerel, Cockle, Cockleford, Cockweed, Cokeweed, Common Cockle, Common Corn Cockle, Common Corncockle, Corn Campion, Corn Champion, Corn Cockle, Corn Pink, Corn Rose, Corncockle, Cornflower, Crown Of the Field, Curn-flower, Drawk, Gith, Githago Agrostemma, Gye, Hardhead, Little-and-pretty, Mullein Pink, Nancy Pretty, Nancy-pretty, Nele, Old Maid's Pink, Papple, Pawple, Pink, Pook-needle, Popille, Popple, Poppy, Prick Needle, Prick-needle, Puck Needle, Punk Needle, Purple Cockle, Ray, Red Cornflower, Robin Hood, Rose Campion, Rose of Heaven, Wild Nigella, Wild Savager ■

11948 Agrostemma githago L. = Githago segetum Desf. ■

11949 Agrostemma githago L. = Lychnis githago (L.) Scop. ■

11950 Agrostemma githago L. = Silene githago (L.) Clairv. ■

11951 Agrostemma githago L. var. agrostemma linicola (Terechov) K. Hammer;亚麻麦仙翁■☆

11952 Agrostemma githago L. var. microcalyx Döll = Agrostemma githago L. ■

11953 Agrostemma githago L. var. nicaeensis Willd. = Agrostemma githago L. ■

11954 Agrostemma gracilis Boiss.;小麦仙翁■☆

11955 Agrostemma linicola Terechov = Agrostemma githago L. var. agrostemma linicola (Terechov) K. Hammer ■☆

11956 Agrostemma triflorum (R. Br. ex Sommerf.) G. Don = Silene sorensenis (B. Boivin) Bocquet ■☆

11957 Agrosticula Raddi = Sporobolus R. Br. ■

11958 Agrostidaceae (Kunth) Herter = Gramineae Juss.(保留科名)●■

11959 Agrostidaceae (Kunth) Herter = Poaceae Barnhart(保留科名)●■

11960 Agrostidaceae Bercht. et J. Presl = Gramineae Juss.(保留科名)●■

11961 Agrostidaceae Bercht. et J. Presl = Poaceae Barnhart(保留科名)●■

11962 Agrostidaceae Burnett = Gramineae Juss.(保留科名)●■

11963 Agrostidaceae Burnett = Poaceae Barnhart(保留科名)●■

11964 Agrostidaceae Herter = Gramineae Juss.(保留科名)●■

11965 Agrostidaceae Herter = Poaceae Barnhart(保留科名)●■

11966 Agrostidaceae Herter;剪股颖科■

11967 Agrostis Adans. = Imperata Cyrillo ■

11968 Agrostis L. (1753)(保留属名);剪股颖属(小糠草属);Bent Grass, Bentgrass, Bent-grass ■

11969 Agrostis × dimorpholemma Ohwi = Agrostis × fouilladei P. Fourn. ■☆

11970 Agrostis × fouilladei P. Fourn.;富亚德剪股颖■☆

11971 Agrostis africana Poir. = Sporobolus africanus (Poir.) Robyns et Tournay ■☆

11972 Agrostis alba L.;小糠草(红顶草);Bentgrass, Berd's Grass, Florin, Herds-grass, Red Top, Redtop, Red-top, Stolonbearing Bentgrass, Therds-grassop ■

11973 Agrostis alba L. = Agrostis gigantea Roth ■

11974 Agrostis alba L. = Agrostis stolonifera L. ■

11975 Agrostis alba L. subsp. castellana (Boiss. et Reut.) Maire = Agrostis castellana Boiss. et Reut. ■☆

11976 Agrostis alba L. subsp. scabriglumis (Boiss. et Reut.) Jahand. et Maire = Agrostis stolonifera L. subsp. scabriglumis (Boiss. et Reut.) Maire ■☆

11977 Agrostis alba L. var. cedretorum Maire et Trab. = Agrostis castellana Boiss. et Reut. ■☆

11978 Agrostis alba L. var. coarctata (Schrad.) Coss. et Durieu = Agrostis castellana Boiss. et Reut. ■☆

11979 Agrostis alba L. var. dorsimucronata Maire et Trab. = Agrostis castellana Boiss. et Reut. ■☆

11980 Agrostis alba L. var. fontanesii Coss. et Durieu = Agrostis reuteri Boiss. ■☆

11981 Agrostis alba L. var. hostiana Rouy = Agrostis gigantea Roth ■

11982 Agrostis alba L. var. longipaleata Maire et Trab. = Agrostis stolonifera L. var. longipaleata (Maire et Trab.) Maire ■☆

11983 Agrostis alba L. var. maritima (Lam.) G. Mey. = Agrostis stolonifera L. ■

11984 Agrostis alba L. var. olivetorum (Godr.) Coss. et Durieu = Agrostis castellana Boiss. et Reut. ■☆

11985 Agrostis alba L. var. palustris (Huds.) Pers. = Agrostis stolonifera L. ■

11986 Agrostis alba L. var. palustris (Huds.) Pers. = Agrostis stolonifera L. var. palustris (Huds.) Farw. ■☆

11987 Agrostis alba L. var. scabriglumis (Boiss. et Reut.) Batt. et Trab. = Agrostis stolonifera L. subsp. scabriglumis (Boiss. et Reut.) Maire ■☆

11988 Agrostis alba L. var. schimperiana (Hochst. ex Steud.) Engl. = Polypogon schimperianus (Hochst. ex Steud.) Cope ■☆

11989 Agrostis alba L. var. simensis (Hochst. ex Steud.) Engl. = Polypogon schimperianus (Hochst. ex Steud.) Cope ■☆

11990 Agrostis alba L. var. stenantha Maire et Trab. = Agrostis castellana Boiss. et Reut. ■☆

11991 Agrostis alba L. var. stolonifera (L.) Sm. = Agrostis stolonifera L. ■

11992 Agrostis alba L. var. stolonifera Scribn. = Agrostis stolonifera L. ■

11993 Agrostis alba L. var. vulgaris (With.) Coss. et Durieu = Agrostis tenuis Sibth. ■

11994 Agrostis alba L. var. vulgaris Thurb. = Agrostis tenuis Sibth. ■

11995 Agrostis albida Trin.;白剪股颖■☆

11996 Agrostis albimontana Mez;山地白剪股颖■☆

11997 Agrostis alpicola Hochst. = Agrostis quinqueseta (Steud.) Hochst. ■☆

11998 Agrostis anadyrensis Soczava;阿纳代尔剪股颖■☆

11999 Agrostis antecedens E. P. Bicknell = Agrostis hyemalis (Walter) Britton, Sterns et Poggenb. ■☆

12000 Agrostis arisan-montana Ohwi;阿里山剪股颖(大药剪股颖);Alishan Bentgrass, Arisan Bentgrass ■

12001 Agrostis arisan-montana Ohwi = Agrostis infirma Büse var. arisan-montana (Ohwi) Veldkamp ■

12002 Agrostis arisan-montana Ohwi var. megalandra Y. C. Yang;大药剪股颖■

12003 Agrostis arisan-montana Ohwi var. megalandra Y. C. Yang = Agrostis arisan-montana Ohwi ■

12004 Agrostis arundinacea L. = Calamagrostis arundinacea (L.) Roth ■☆

12005 Agrostis arundinacea L. = Deyeuxia arundinacea (L.) P. Beauv. ■

12006 Agrostis atlantica Maire et Trab.;大西洋剪股颖■☆

12007 Agrostis atlantica Maire et Trab. var. subalpina Litard. et Maire = Agrostis atlantica Maire et Trab. ■☆

12008 Agrostis avenacea J. F. Gmel.;太平洋剪股颖;Blown-grass, Pacific Bentgrass ■☆

12009 Agrostis barbuligera Stapf;髯毛剪股颖■☆

12010 Agrostis barbuligera Stapf var. longipilosa Gooss. et Papendorf;长疏毛剪股颖■☆

12011 Agrostis beimushanica J. L. Yang;贝母山剪股颖;Beimushan Bentgrass ■

12012 Agrostis beimushanica J. L. Yang = Agrostis pilosula Trin. ■

12013 Agrostis bergiana Trin.;毕尔剪股颖■☆

12014 Agrostis bergiana Trin. var. laeviuscula Stapf;稍光毕尔剪股颖■☆

12015 Agrostis biebersteiniana Claus;毕伯氏剪股颖;Bieberstein Bentgrass ■☆

12016 Agrostis bogdanii Bogdan = Agrostis kilimandscharica Mez ■☆

12017 Agrostis borealis Hartm. = Agrostis mertensii Trin. ■☆

12018 Agrostis borealis Hartm. var. flaccida (Hack.) T. Koyama = Agrostis flaccida Hack. ■

12019 Agrostis brachiata Munro ex Hook. f.;大穗剪股颖■

12020 Agrostis brevipes Keng ex J. L. Yang;短柄剪股颖;Shortstalk Bentgrass ■

12021 Agrostis bromoides L. = Achnatherum bromoides (L.) P. Beauv. ■☆

12022 Agrostis bryophila C. E. Hubb. var. elgonensis ? = Agrostis gracilifolia C. E. Hubb. ■☆

12023 Agrostis burttii C. E. Hubb. = Agrostis kilimandscharica Mez ■☆

12024 Agrostis calamagrostis L. = Achnatherum calamagrostis (L.) P. Beauv. ■

12025 Agrostis canina L.;普通剪股颖(欧剪股颖);Brown Bent, Brown Bent Grass, Common Bentgrass, Island Rhode Bent, Rhode Island Bent, Velvet Bent, Velvet Bent Grass, Velvet Bentgrass, Velvet Bent-grass ■

12026 Agrostis canina L. = Agrostis vinealis Schreb. ■

12027 Agrostis canina L. subsp. montana (Hartm.) Hartm. = Agrostis vinealis Schreb. ■

12028 Agrostis canina L. subsp. trinii (Turcz.) Hartm. = Agrostis vinealis Schreb. ■

12029 Agrostis canina L. var. formosana Hack.;台湾剪股颖;Taiwan Bentgrass ■

12030 Agrostis canina L. var. formosana Hack. = Agrostis sozanensis Hayata ■

12031 Agrostis canina L. var. montana Hartm. = Agrostis vinealis Schreb. ■

12032 Agrostis canina L. var. tenuifolia (M. Bieb.) Boiss. = Agrostis vinealis Schreb. ■

12033 Agrostis canina L. var. tenuifolia Torr. = Agrostis vinealis Schreb. ■

12034 Agrostis capensis (L.) Lam.;细弱剪股颖(毛状剪股颖,丝状剪股颖);Brown Top, Colonial Bent, Colonial Bent Grass, Colonial Bentgrass, Colonial Bent-grass, Common Bent, Common Bent Grass, Fine Bent, Rhode Island Bent, Rhode Island Bent Grass, Weak Bentgrass ■

12035 Agrostis capillaris L. = Agrostis capensis (L.) Lam. ■

12036 Agrostis capillaris L. = Agrostis vulgaris With. ■

12037 Agrostis capillaris Schischk. = Agrostis tenuis Sibth. ■

12038 Agrostis castellana Boiss. et Reut.;丘陵剪股颖;Highland Bent ■☆

12039 Agrostis castellana Boiss. et Reut. var. cedretorum (Maire et Trab.) Jahand. et Maire = Agrostis castellana Boiss. et Reut. ■☆

12040 Agrostis castellana Boiss. et Reut. var. heterophylla Batt. = Agrostis castellana Boiss. et Reut. ■☆

12041 Agrostis castellana Boiss. et Reut. var. hispanica (Boiss. et Reut.) Hack. = Agrostis castellana Boiss. et Reut. ■☆

12042 Agrostis castellana Boiss. et Reut. var. mixta Hack. = Agrostis castellana Boiss. et Reut. ■☆

12043 Agrostis castellana Boiss. et Reut. var. mutica (Boiss. et Reut.) Romero García et al. = Agrostis castellana Boiss. et Reut. ■☆

12044 Agrostis castellana Boiss. et Reut. var. mutica Hack. = Agrostis castellana Boiss. et Reut. ■☆

12045 Agrostis castellana Boiss. et Reut. var. olivetorum (Godr.) Kerguélen = Agrostis castellana Boiss. et Reut. ■☆

12046 Agrostis castellana Boiss. et Reut. var. stenantha (Maire et Trab.) Jahand. et Maire = Agrostis castellana Boiss. et Reut. ■☆

12047 Agrostis castellana Boiss. et Reut. var. tricuspidata (Hack.) Asch. et Graebn. = Agrostis castellana Boiss. et Reut. ■☆

12048 Agrostis chaetophylla Peter = Agrostis volkensii Stapf ■☆

12049 Agrostis chinensis Kön. ex Steud. = Leptochloa chinensis (L.) Nees ■

12050 Agrostis chionogeiton Pilg. = Colpodium chionogeiton (Pilg.) Tzvelev ■☆

12051 Agrostis ciliata Trin. = Agrostis pilosula Trin. ■

12052 Agrostis clarkei Hook. f. = Agrostis nervosa Nees ex Trin. ■

12053 Agrostis clavata Trin.;华北剪股颖(剪股颖);N. China Bentgrass, North China Bentgrass ■

12054 Agrostis clavata Trin. = Agrostis perennans (Walter) Tuck. ■☆

12055 Agrostis clavata Trin. subsp. matsumurae (Hack. ex Honda) Takeoka = Agrostis matsumurae Hack. ex Honda ■

12056 Agrostis clavata Trin. subsp. matsumurae (Hack. ex Honda) Tateoka = Agrostis clavata Trin. ■

12057 Agrostis clavata Trin. subsp. micrantha (Steud.) Y. C. Tong =

Agrostis micrantha Steud. ■
12058 Agrostis clavata Trin. var. macilenta (Keng) Y. C. Yang;广东剪股颖;Guangdong Bentgrass,Kwangtung Bentgrass ■
12059 Agrostis clavata Trin. var. nukabo Ohwi = Agrostis clavata Trin. ■
12060 Agrostis clavata Trin. var. nukabo Ohwi = Agrostis matsumurae Hack. ex Honda ■
12061 Agrostis clavata Trin. var. szechuanica Y. C. Tong ex Y. C. Yang;四川剪股颖;Sichuan Bentgrass ■
12062 Agrostis composita Poir. = Sporobolus compositus (Michx.) Kunth ■☆
12063 Agrostis composita Poir. = Sporobolus compositus(Poir.) Merr. ■☆
12064 Agrostis congener Schumach. = Sporobolus virginicus (L.) Kunth ■
12065 Agrostis congesta C. E. Hubb. = Agrostis quinqueseta (Steud.) Hochst. ■☆
12066 Agrostis continentalis Hand. -Mazz. = Deyeuxia petelotii (Hitchc.) S. M. Phillips et Wen L. Chen ■
12067 Agrostis continuata Stapf;连续剪股颖■☆
12068 Agrostis contracta Y. C. Tong ex Y. C. Yang = Agrostis sinocontracta S. M. Phillips et S. L. Lu ■
12069 Agrostis contracta Y. C. Tong ex Y. C. Yang subsp. trinii (Turcz.) H. Scholz = Agrostis vinealis Schreb. ■
12070 Agrostis contracta Y. C. Tong ex Y. C. Yang subsp. trinii (Turcz.) H. Scholz = Agrostis trinii Turcz. ex Litv. ■
12071 Agrostis coromandeliana Retz. = Sporobolus coromandelianus (Retz.) Kunth ■
12072 Agrostis cryptandra Torr. = Sporobolus cryptandrus (Torr.) A. Gray ■☆
12073 Agrostis curtisii Kerguélen;刚毛剪股颖;Bristle Bent,Brittle Bent ■☆
12074 Agrostis curvifolia Hack. = Pentaschistis ecklonii (Nees) McClean ■☆
12075 Agrostis diandra Retz. = Sporobolus diander (Retz.) P. Beauv. ■
12076 Agrostis diffusa S. M. Phillips;松散剪股颖■☆
12077 Agrostis dirandra Retz. = Sporobolus indicus (L.) R. Br. var. flaccidus (Roem. et Schult.) Veldkamp ■
12078 Agrostis dissitiflora C. E. Hubb. = Agrostis gracilifolia C. E. Hubb. ■☆
12079 Agrostis distans Kuntze var. coreensis Hack. = Puccinellia coreensis (Hack.) Honda ■
12080 Agrostis divaricatissima Mez;歧序剪股颖;Divaricate Bentgrass,Most Divaricate Bentgrass ■
12081 Agrostis dregeana Steud.;德雷剪股颖■☆
12082 Agrostis dshungarica (Tzvelev) Tzvelev;线序剪股颖■
12083 Agrostis ecklonis Trin.;埃克剪股颖■☆
12084 Agrostis elatior Ehrenb. et Hemprich ex Trin. = Urochondra setulosa (Trin.) C. E. Hubb. ■☆
12085 Agrostis elegans Loisel. = Agrostis tenerrima Trin. ■☆
12086 Agrostis elegans Nees = Aira elegans Willd. ex Kunth ■
12087 Agrostis elliotii Hack.;埃利剪股颖■☆
12088 Agrostis elliottiana Schult.;埃氏剪股颖;Awned Bent Grass ■☆
12089 Agrostis elongata (R. Br.) Roem. et Schult. var. flaccida Roem. et Schult. = Sporobolus indicus (L.) R. Br. var. flaccidus (Roem. et Schult.) Veldkamp ■
12090 Agrostis elongata Keng ex Y. C. Yang var. flaccida Roth ex Roem. et Schult. = Sporobolus diander (Retz.) P. Beauv. ■
12091 Agrostis emirnensis (Baker) Bosser;埃米剪股颖■☆
12092 Agrostis eriantha Hack.;毛花剪股颖■
12093 Agrostis eriantha Hack. var. planifolia Gooss. et Papendorf;平叶毛花剪股颖■
12094 Agrostis eriolepis Keng ex Y. C. Yang = Agrostis pilosula Trin. ■
12095 Agrostis eriolepis Keng ex Y. C. Yang = Agrostis wallichiana (Steud.) Bor ■
12096 Agrostis exarata Trin.;长穗剪股颖;Spike Bent ■☆
12097 Agrostis exarata Trin. subsp. clavata (Trin.) T. Koyama = Agrostis clavata Trin. ■
12098 Agrostis exarata Trin. subsp. nukabo (Ohwi) T. Koyama = Agrostis matsumurae Hack. ex Honda ■
12099 Agrostis extensa Schumach.;伸展剪股颖■
12100 Agrostis fertilis Steud. = Sporobolus fertilis (Steud.) Clayton ■
12101 Agrostis fertilis Steud. = Sporobolus indicus (L.) R. Br. var. major (Büse) Baaijens ■
12102 Agrostis filipes Hook. f.;丝梗剪股颖■
12103 Agrostis fissa Stapf = Polypogon schimperianus (Hochst. ex Steud.) Cope ■☆
12104 Agrostis flaccida Hack.;柔软剪股颖;Flaccid Bentgrass,Soft Bentgrass ■
12105 Agrostis flaccida Hack. f. vivipara Honda;胎生剪股颖■☆
12106 Agrostis flaccida Hack. subsp. trinii (Turcz. ex Litv.) T. Koyama = Agrostis trinii Turcz. ex Litv. ■
12107 Agrostis flaccida Hack. subsp. trinii (Turcz.) T. Koyama = Agrostis vinealis Schreb. ■
12108 Agrostis flaccida Hack. var. morrisonensis (Hayata) Honda;玉山剪股颖■☆
12109 Agrostis flaccida Hack. var. trinii (Turcz. ex Litv.) Ohwi = Agrostis trinii Turcz. ex Litv. ■
12110 Agrostis flaccida Hack. var. trinii (Turcz.) Ohwi = Agrostis vinealis Schreb. ■
12111 Agrostis fontqueri Maire = Agrostis truncatula Parl. ■☆
12112 Agrostis formosana Ohwi = Agrostis clavata Trin. ■
12113 Agrostis friesiorum C. E. Hubb. = Agrostis kilimandscharica Mez var. densior Pilg. ■☆
12114 Agrostis frondosa Poir. = Muhlenbergia frondosa (Poir.) Fernald ■☆
12115 Agrostis fukuyamae Ohwi;舟颖剪股颖(伯明剪股颖);Fukuyama Bentgrass ■
12116 Agrostis fukuyamae Ohwi = Agrostis infirma Büse var. fukuyamae (Ohwi) Veldkamp ■
12117 Agrostis gaditana (Boiss. et Reut.) Nyman = Agrostis stolonifera L. ■
12118 Agrostis gigantea Roth;巨序剪股颖(大糠草);Black Bent,Black Bent Grass,Giant Bentgrass,Red Top,Redtop ■
12119 Agrostis gigantea Roth var. dispar (Michx.) Philipson = Agrostis gigantea Roth ■
12120 Agrostis gigantea Roth var. ramosa (Gray) Philipson = Agrostis gigantea Roth ■
12121 Agrostis gracilifolia C. E. Hubb.;细叶剪股颖■☆
12122 Agrostis gracilifolia C. E. Hubb. subsp. parviflora S. M. Phillips;小花细叶剪股颖■☆
12123 Agrostis gracilis Keng var. multinodis Y. Y. Qian = Deyeuxia petelotii (Hitchc.) S. M. Phillips et Wen L. Chen ■
12124 Agrostis greenwayi C. E. Hubb. = Agrostis producta Pilg. ■☆
12125 Agrostis griquensis Stapf = Polypogon griquensis (Stapf) Gibbs Russ. et L. Fish ■☆

12126 Agrostis gymnostyla Steud.;裸柱剪股颖■☆
12127 Agrostis hideoi Ohwi;秀雄剪股颖■☆
12128 Agrostis hiemalis (Walter) Britton, Stern. et Poggenb. = Agrostis hyemalis (Walter) Britton, Sterns et Poggenb. ■☆
12129 Agrostis hiemalis Britton, Stern. et Poggenb. = Agrostis hyemalis (Walter) Britton, Sterns et Poggenb. ■☆
12130 Agrostis hirtella Hochst. ex Steud. = Polypogon schimperianus (Hochst. ex Steud.) Cope ■☆
12131 Agrostis hissarica Roshev. = Polypogon hissaricus (Roshev.) Bor ■
12132 Agrostis hissarica Roshev. subsp. pamirica (Ovcz.) Tzvelev = Polypogon hissaricus (Roshev.) Bor ■
12133 Agrostis hookeriana C. B. Clarke ex Hook. f.;疏花剪股颖(贡山剪股颖,广序剪股颖);Hooker Bentgrass, Looseflower Bentgrass, Looseflowered Bentgrass ■
12134 Agrostis hookeriana C. B. Clarke ex Hook. f. var. longiflora Y. C. Tong ex Y. C. Yang;长花剪股颖;Longflower Bentgrass ■
12135 Agrostis hookeriana C. B. Clarke ex Hook. f. var. longiflora Y. C. Tong ex Y. C. Yang = Agrostis kunmingensis B. S. Sun et Y. Cai Wang ■
12136 Agrostis hugoniana Rendle;甘青剪股颖(穗序剪股颖,休氏剪股颖);Hugon Bentgrass ■
12137 Agrostis hugoniana Rendle var. aristata Keng ex Y. C. Yang;川西剪股颖;W. Sichuan Bentgrass ■
12138 Agrostis huttoniae (Hack.) C. E. Hubb. ex Gooss. et Papendorf = Agrostis lachnantha Nees ■☆
12139 Agrostis hyemalis (Walter) Britton, Sterns et Poggenb.;冬剪股颖;Hair Grass, Small Bent, Southern Hair Grass, Tickle Grass, Winter Bent Grass ■☆
12140 Agrostis hyemalis (Walter) Britton, Sterns et Poggenb. var. scabra (Willd.) Blomq.;糙冬剪股颖;Rough Bent Grass, Southern Hair Grass, Tickle Grass ■☆
12141 Agrostis hyemalis (Walter) Britton, Sterns et Poggenb. var. tenuis (Tuck.) Gleason = Agrostis hyemalis (Walter) Britton, Sterns et Poggenb. var. scabra (Willd.) Blomq. ■☆
12142 Agrostis inaequiglumis Griseb.;窄穗剪股颖;Unequalglume Bentgrass ■
12143 Agrostis inaequiglumis Griseb. var. nana Y. C. Yang = Agrostis mackliniae Bor ■
12144 Agrostis incurvata (L.) Scop. = Parapholis incurva (L.) C. E. Hubb. ■
12145 Agrostis infirma Büse;新高山剪股颖(玉山剪股颖);Morrison Bentgrass, Yushan Bentgrass ■
12146 Agrostis infirma Büse var. arisan-montana (Ohwi) Veldkamp = Agrostis arisanmontana Ohwi ■
12147 Agrostis infirma Büse var. formosana (Hack.) Veldkamp;草山剪股颖(台湾剪股颖,外玉山剪股颖);Transmorrison Bentgrass ■
12148 Agrostis infirma Büse var. fukuyamae (Ohwi) Veldkamp;伯明剪股颖■
12149 Agrostis infirma Büse var. fukuyamae (Ohwi) Veldkamp = Agrostis fukuyamae Ohwi ■
12150 Agrostis interrupta L. = Apera interrupta (L.) P. Beauv. ■☆
12151 Agrostis jacutica Schischk.;雅库特剪股颖■☆
12152 Agrostis japonica Steud. = Sporobolus pilifer (Trin.) Kunth ■
12153 Agrostis juncea Lam. = Sporobolus virginicus (L.) Kunth ■
12154 Agrostis karsensis Litv. = Agrostis stolonizans Besser ex Schult. et Schult. ■
12155 Agrostis keniensis Pilg.;肯尼亚剪股颖■☆
12156 Agrostis kentrophylla K. Schum. = Sporobolus ioclados (Trin.) Nees ■☆
12157 Agrostis kentrophylla K. Schum. = Sporobolus kentrophyllus (K. Schum.) Clayton ■☆
12158 Agrostis kilimandscharica Mez;基利剪股颖■☆
12159 Agrostis kilimandscharica Mez var. densior Pilg.;密剪股颖■☆
12160 Agrostis kilimandscharica Mez var. sororia (C. E. Hubb.) Hedberg;堆积剪股颖■☆
12161 Agrostis koelerioides Romo = Agrostis stolonifera L. ■
12162 Agrostis koreana Ohwi = Agrostis divaricatissima Mez ■
12163 Agrostis kunmingensis B. S. Sun et Y. Cai Wang;昆明剪股颖;Kunming Bentgrass ■
12164 Agrostis lachnantha Nees;非洲剪股颖;African Bent ■☆
12165 Agrostis lachnantha Nees var. glabra Gooss. et Papendorf = Agrostis lachnantha Nees ■☆
12166 Agrostis latifolia Trevir. = Cinna latifolia (Trevis.) Griseb. ■
12167 Agrostis latifolia Trevir. ex Göpp. = Cinna latifolia (Trevis.) Griseb. ■
12168 Agrostis lazica Balansa;拉扎剪股颖■☆
12169 Agrostis leioclada C. E. Hubb. = Agrostis kilimandscharica Mez ■☆
12170 Agrostis leptophylla C. E. Hubb. = Agrostis gracilifolia C. E. Hubb. ■☆
12171 Agrostis limprichtii Pilg.;侏儒剪股颖(川滇剪股颖);Limpricht Bentgrass ■
12172 Agrostis limprichtii Pilg. = Agrostis nervosa Nees ex Trin. ■
12173 Agrostis lushuiensis B. S. Sun et Y. Cai Wang = Agrostis nervosa Nees ex Trin. ■
12174 Agrostis macilenta Keng = Agrostis clavata Trin. var. macilenta (Keng) Y. C. Yang ■
12175 Agrostis macilenta Keng = Agrostis infirma Büse ■
12176 Agrostis mackliniae Bor;歧颖剪股颖;Dwarf Bentgrass ■
12177 Agrostis macrantha Schischk. = Agrostis stolonizans Besser ex Schult. et Schult. ■
12178 Agrostis macranthera C. C. Chang et Skvortsov;巨药剪股颖;Large-anther Bentgrass ■
12179 Agrostis macrothyrsa Hack.;大序剪股颖■☆
12180 Agrostis makoniensis Stent et J. M. Rattray = Agrostis continuata Stapf ■☆
12181 Agrostis mannii (Hook. f.) Stapf;曼氏剪股颖■☆
12182 Agrostis mannii (Hook. f.) Stapf subsp. aethiopica S. M. Phillips;埃塞俄比亚剪股颖■☆
12183 Agrostis maritima Lam. = Agrostis alba L. var. maritima (Lam.) G. Mey. ■
12184 Agrostis maritima Lam. = Agrostis stolonifera L. ■
12185 Agrostis matrella L. = Zoysia matrella (L.) Merr. ■
12186 Agrostis matsumurae Hack. ex Honda;剪股颖;Matsumura Bentgrass ■
12187 Agrostis matsumurae Hack. ex Honda = Agrostis clavata Trin. subsp. matsumurae (Hack. ex Honda) Tateoka ■
12188 Agrostis matsumurae Hack. ex Honda = Agrostis clavata Trin. ■
12189 Agrostis mauritii Sennen;毛里特剪股颖■☆
12190 Agrostis maxima Roxb. = Thysanolaena latifolia (Roxb. ex Hornem.) Honda ■
12191 Agrostis maxima Roxb. = Thysanolaena maxima (Roxb.) Kuntze ■
12192 Agrostis megathyrsa Keng ex P. C. Keng;大锥剪股颖;Largepanicle Bentgrass ■
12193 Agrostis megathyrsa Keng ex P. C. Keng = Agrostis brachiata

Munro ex Hook. f. ■
12194 Agrostis megathyrsa Keng ex P. C. Keng var. angustispicata D. Z. Ma et J. N. Li = Agrostis arisan-montana Ohwi ■
12195 Agrostis megathyrsa Keng var. angustispicata D. Z. Ma et J. N. Li;狭穗大锥剪股颖■
12196 Agrostis mertensii Trin.;北方剪股颖;Northern Agrostis ■☆
12197 Agrostis mertensii Trin. subsp. borealis (Hartm.) Tzvelev = Agrostis mertensii Trin. ■☆
12198 Agrostis mexicana L. = Muhlenbergia mexicana (L.) Trin. ■☆
12199 Agrostis micrandra Keng ex J. L. Yang;微药剪股颖■
12200 Agrostis micrandra Keng ex J. L. Yang = Agrostis micrantha Steud. ■
12201 Agrostis micrantha Steud.;小花剪股颖(多花剪股颖,微药剪股颖);Samllflower Bentgrass ■
12202 Agrostis mildbraedii Pilg. = Agrostis quinqueseta (Steud.) Hochst. ■☆
12203 Agrostis milioides Mez = Agrostis micrantha Steud. ■
12204 Agrostis milioides Mez = Agrostis myriantha Hook. f. ■
12205 Agrostis mongolica Roshev. = Agrostis divaricatissima Mez ■
12206 Agrostis montevidensis Spreng. ex Nees;蒙得维的亚剪股颖■☆
12207 Agrostis morrisonensis Hayata = Agrostis infirma Büse ■
12208 Agrostis muliensis J. L. Yang;木里剪股颖;Muli Bentgrass ■
12209 Agrostis muliensis J. L. Yang = Agrostis pilosula Trin. ■
12210 Agrostis muliensis J. L. Yang = Agrostis wallichiana (Steud.) Bor ■
12211 Agrostis munroana Aitch. et Hemsl.;长稃剪股颖■
12212 Agrostis myriantha Hook. f.;多花剪股颖;Manyflower Bentgrass ■
12213 Agrostis myriantha Hook. f. = Agrostis micrantha Steud. ■
12214 Agrostis myriantha Hook. f. var. yangbiensis B. S. Sun et Y. Cai Wang;漾濞剪股颖;Yangbi Bentgrass ■
12215 Agrostis myriantha Hook. f. var. yangbiensis B. S. Sun et Y. Cai Wang = Agrostis micrantha Steud. ■
12216 Agrostis natalensis Stapf = Agrostis continuata Stapf ■☆
12217 Agrostis nebulosa Boiss. et Reut.;云草;Cloudgrass,Cloud-grass ■☆
12218 Agrostis neesii Trin.;尼斯剪股颖■☆
12219 Agrostis nervosa Nees ex Trin.;泸水剪股颖(短柄剪股颖,灰岩剪股颖);Lushui Bentgrass,Shortstalk Bentgrass ■
12220 Agrostis nevadensis Boiss.;内华达剪股颖;Nevada Bentgrass ■☆
12221 Agrostis nigra With.;黑剪股颖■☆
12222 Agrostis nigra With. = Agrostis gigantea Roth ■
12223 Agrostis nipponensis Honda = Agrostis valvata Steud. ■☆
12224 Agrostis olivetorum Godr. = Agrostis castellana Boiss. et Reut. ■☆
12225 Agrostis oreades Peter = Colpodium chionogeiton (Pilg.) Tzvelev ■☆
12226 Agrostis oreophila Trin. = Agrostis perennans (Walter) Tuck. ■☆
12227 Agrostis osakae Honda;木山剪股颖■☆
12228 Agrostis pallida DC. = Agrostis pourretii Willd. ■
12229 Agrostis palustris Huds. = Agrostis alba L. ■
12230 Agrostis palustris Huds. = Agrostis stolonifera L. var. palustris (Huds.) Farw. ■☆
12231 Agrostis palustris Huds. = Agrostis stolonifera L. ■
12232 Agrostis pamirica Ovcz.;帕米尔剪股颖■☆
12233 Agrostis pamirica Ovcz. = Polypogon hissaricus (Roshev.) Bor ■
12234 Agrostis papposa Mez = Agrostis lachnantha Nees ■☆
12235 Agrostis paulsenii Hack. ex Paulsen;帕氏剪股颖■☆
12236 Agrostis perarta Keng = Agrostis arisan-montana Ohwi ■
12237 Agrostis perarta Keng ex P. C. Kuo;紧穗剪股颖;Contracted Bentgrass ■
12238 Agrostis perarta Keng ex P. C. Kuo = Agrostis arisan-montana Ohwi ■
12239 Agrostis perarta Keng ex P. C. Kuo = Agrostis matsumurae Hack. ex Honda ■
12240 Agrostis perennans (Walter) Tuck.;秋剪股颖;Autumn Bent Grass,Perennial Agrostis,Thin Grass,Upland Bent,Upland Bent Grass ■☆
12241 Agrostis perennans (Walter) Tuck. var. aestivalis Vasey = Agrostis perennans (Walter) Tuck. ■☆
12242 Agrostis perezii Sennen = Agrostis pourretii Willd. ■
12243 Agrostis perlaxa Pilg. = Agrostis hookeriana C. B. Clarke ex Hook. f. ■
12244 Agrostis petelotii (Hitchc.) Noltie = Deyeuxia petelotii (Hitchc.) S. M. Phillips et Wen L. Chen ■
12245 Agrostis petelotii (Hitchcock) Soják = Deyeuxia petelotii (Hitchc.) S. M. Phillips et Wen L. Chen ■
12246 Agrostis phalaroides Hack.;藕草剪股颖■☆
12247 Agrostis pilgeriana C. E. Hubb.;皮尔格剪股颖■☆
12248 Agrostis pilosa Retz. = Digitaria stricta Roth ex Roem. et Schult. ■
12249 Agrostis pilosula (Trin.) Hook. f. var. ciliata (Trin.) Bor = Agrostis pilosula Trin. ■
12250 Agrostis pilosula (Trin.) Hook. f. var. filifolia Bor = Agrostis pilosula Trin. ■
12251 Agrostis pilosula (Trin.) Hook. f. var. wallichiana (Steud.) Bor = Agrostis pilosula Trin. ■
12252 Agrostis pilosula Trin.;毛稃剪股颖(柔毛剪股颖);Pilose Bentgrass ■
12253 Agrostis pilosula Trin. var. ciliata (Trin.) Bor = Agrostis pilosula Trin. ■
12254 Agrostis pilosula Trin. var. filifolia Bor = Agrostis pilosula Trin. ■
12255 Agrostis pilosula Trin. var. wallichiana (Steud.) Bor = Agrostis pilosula Trin. ■
12256 Agrostis planifolia K. Koch;阔叶剪股颖■☆
12257 Agrostis platyphylla Mez = Agrostis myriantha Hook. f. ■
12258 Agrostis plumosa Ten. = Tricholaena teneriffae (L. f.) Link ■☆
12259 Agrostis poluninii Bor = Agrostis hookeriana C. B. Clarke ex Hook. f. ■
12260 Agrostis poluninii Bor var. longiflora (Y. C. Tong ex Y. C. Yang) B. S. Sun. = Agrostis kunmingensis B. S. Sun et Y. Cai Wang ■
12261 Agrostis polypogonoides Stapf;多髯毛剪股颖■☆
12262 Agrostis pourretii Willd.;普瑞剪股颖;Pourret Bentgrass ■
12263 Agrostis procera Retz. = Eriochloa procera (Retz.) C. E. Hubb. ■
12264 Agrostis producta Pilg.;铺展剪股颖■☆
12265 Agrostis pubicallis Keng ex Y. C. Yang;湖岸剪股颖;Lakeshore Bentgrass ■
12266 Agrostis pubicallis Keng ex Y. C. Yang = Agrostis hookeriana C. B. Clarke ex Hook. f. ■
12267 Agrostis pulchella Kunth;美丽剪股颖■☆
12268 Agrostis pungens Schreb. = Sporobolus pungens (Schreb.) Kunth ■☆
12269 Agrostis quinqueseta (Steud.) Hochst.;五刚毛剪股颖■☆
12270 Agrostis racemosa Michx. = Muhlenbergia racemosa (Michx.) Britton,Sterns et Poggenb. ■☆
12271 Agrostis radula Mez = Agrostis continuata Stapf ■☆
12272 Agrostis ramosa (Retz.) Poir. = Eriochloa procera (Retz.) C. E. Hubb. ■
12273 Agrostis reuteri Boiss.;路透剪股颖■☆
12274 Agrostis rigidula Steud. = Agrostis infirma Büse ■

12275 Agrostis rigidula Steud. subsp. fukuyamae (Ohwi) T. Koyama = Agrostis fukuyamae Ohwi ■

12276 Agrostis rigidula Steud. var. arisan-montana (Ohwi) Veldkamp = Agrostis infirma Büse var. arisan-montana (Ohwi) Veldkamp ■

12277 Agrostis rigidula Steud. var. formosana (Hack.) Veldkamp = Agrostis infirma Büse var. formosana (Hack.) Veldkamp ■

12278 Agrostis rigidula Steud. var. fukuyamae (Ohwi) Veldkamp = Agrostis fukuyamae Ohwi ■

12279 Agrostis royleana Trin. = Agrostis pilosula Trin. ■

12280 Agrostis rubra L.;红剪股颖;Red Bentgrass ■☆

12281 Agrostis rupestris All.;岩地剪股颖;Rocky Bentgrass ■

12282 Agrostis rupestris All. var. atlantis Maire = Agrostis rupestris All. ■

12283 Agrostis salina Dum.;盐地剪股颖■☆

12284 Agrostis salmantica (Lag.) Kunth = Agrostis pourretii Willd. ■

12285 Agrostis scabra Willd.;虾夷剪股颖; Rough Bent, Scabrous Agrostis ■☆

12286 Agrostis scabra Willd. = Agrostis hyemalis (Walter) Britton, Sterns et Poggenb. var. scabra (Willd.) Blomq. ■☆

12287 Agrostis scabrescens Sennen = Agrostis pourretii Willd. ■

12288 Agrostis scabrida Maire et Trab. = Agrostis stolonifera L. ■

12289 Agrostis scabriglumis Boiss. et Reut. = Agrostis stolonifera L. subsp. scabriglumis (Boiss. et Reut.) Maire ■☆

12290 Agrostis schimperiana Hochst. ex Steud. = Polypogon schimperianus (Hochst. ex Steud.) Cope ■☆

12291 Agrostis schlechteri Rendle;施莱剪股颖■☆

12292 Agrostis schneideri Pilg.;丽江剪股颖;Schneider Bentgrass ■

12293 Agrostis schneideri Pilg. = Agrostis nervosa Nees ex Trin. ■

12294 Agrostis schneideri Pilg. var. brevipes Keng ex Y. C. Yang = Agrostis nervosa Nees ex Trin. ■

12295 Agrostis schweinitzii Trin. = Agrostis perennans (Walter) Tuck. ■☆

12296 Agrostis sclerophylla C. E. Hubb.;硬叶剪股颖■☆

12297 Agrostis semiverticillata (Forssk.) C. Chr.;半轮生剪股颖; Water Bent ■☆

12298 Agrostis semiverticillata (Forssk.) C. Chr. = Agrostis viridis Gouan ■

12299 Agrostis semiverticillata (Forssk.) C. Chr. = Polypogon viridis (Gouan) Breistr. ■

12300 Agrostis setacea Curtis = Agrostis curtisii Kerguélen ■☆

12301 Agrostis shandongensis F. Z. Li;山东剪股颖;Shandong Bentgrass ■

12302 Agrostis shandongensis F. Z. Li = Agrostis infirma Büse ■

12303 Agrostis shensiana Mez = Agrostis hugoniana Rendle ■

12304 Agrostis sibirica Petr. = Agrostis stolonifera L. ■

12305 Agrostis sibirica Petr. = Agrostis stolonizans Besser ex Schult. et Schult. ■

12306 Agrostis sikkimensis Bor;锡金剪股颖;Sikkim Bentgrass ■

12307 Agrostis sikkimensis Bor = Agrostis nervosa Nees ex Trin. ■

12308 Agrostis simensis Hochst. ex Steud. = Polypogon schimperianus (Hochst. ex Steud.) Cope ■☆

12309 Agrostis sinkiangensis Y. C. Yang = Agrostis dshungarica (Tzvelev) Tzvelev ■

12310 Agrostis sinocontracta S. M. Phillips et S. L. Lu;紧序剪股颖; Controcted Bentgrass ■

12311 Agrostis sinorupestris L. Liou ex S. M. Phillips et S. L. Lu;岩生剪股颖■

12312 Agrostis sobolifera Muhl. ex Willd. = Muhlenbergia sobolifera (Muhl. ex Willd.) Trin. ■☆

12313 Agrostis soroira C. E. Hubb. = Agrostis kilimandscharica Mez var. sororia (C. E. Hubb.) Hedberg ■☆

12314 Agrostis sozanensis Hayata = Agrostis canina L. var. formosana Hack. ■

12315 Agrostis sozanensis Hayata = Agrostis infirma Büse var. formosana (Hack.) Veldkamp ■

12316 Agrostis sozanensis Hayata var. exaristata Hand. -Mazz. = Agrostis canina L. var. formosana Hack. ■

12317 Agrostis sozanensis Hayata var. exaristata Hand. -Mazz. = Agrostis infirma Büse ■

12318 Agrostis spicata Vahl = Sporobolus spicatus (Vahl) Kunth ■☆

12319 Agrostis spica-venti L. = Apera spica-ventii (L.) P. Beauv. ■☆

12320 Agrostis spicigera H. Lindb. = Agrostis stolonifera L. ■

12321 Agrostis stewartii Bor = Agrostis hissarica Roshev. ■

12322 Agrostis stewartii Bor = Polypogon hissaricus (Roshev.) Bor ■

12323 Agrostis stolonifera L.;西伯利亚剪股颖(滨海剪股颖,毛状剪股颖,欧洲剪股颖); Bonnet Strings, Bonnet-strings, Capillary Bentgrass, Carpet Bentgrass, Carpet Bent-grass, Common Bent Grass, Couchy Bent, Creeping Bent, Creeping Bent Grass, Creeping Bentgrass, Creeping Bent-grass, Creeping Tickle Grass, Florin, Redtop, Seaside Agrostis, Siberia Bentgrass, White Bentgrass ■

12324 Agrostis stolonifera L. subsp. alba Litard. = Agrostis stolonifera L. ■

12325 Agrostis stolonifera L. subsp. castellana (Boiss. et Reut.) Maire = Agrostis castellana Boiss. et Reut. ■☆

12326 Agrostis stolonifera L. subsp. gigantea (Roth) Maire et Weiller = Agrostis gigantea Roth ■

12327 Agrostis stolonifera L. subsp. scabrida (Maire et Trab.) Maire = Agrostis stolonifera L. subsp. scabriglumis (Boiss. et Reut.) Maire ■☆

12328 Agrostis stolonifera L. subsp. scabriglumis (Boiss. et Reut.) Maire;粗皮西伯利亚剪股颖■☆

12329 Agrostis stolonifera L. var. aristata Neilr. = Agrostis stolonifera L. ■

12330 Agrostis stolonifera L. var. cedretorum Maire et Trab. = Agrostis castellana Boiss. et Reut. ■☆

12331 Agrostis stolonifera L. var. coartata (Ehrenb.) Blytt = Agrostis stolonifera L. ■

12332 Agrostis stolonifera L. var. compacta Hartm. = Agrostis stolonifera L. var. palustris (Huds.) Farw. ■☆

12333 Agrostis stolonifera L. var. gigantea (Roth) Koch = Agrostis gigantea Roth ■

12334 Agrostis stolonifera L. var. hackelii Maire et Weiller = Agrostis castellana Boiss. et Reut. ■☆

12335 Agrostis stolonifera L. var. hispanica (Boiss. et Reut.) Maire et Weiller = Agrostis castellana Boiss. et Reut. ■☆

12336 Agrostis stolonifera L. var. jahandieziana Litard. et Maire = Agrostis stolonifera L. ■

12337 Agrostis stolonifera L. var. longipaleata (Maire et Trab.) Maire;长稃西伯利亚剪股颖■☆

12338 Agrostis stolonifera L. var. longipaleata Maire et Trab. = Agrostis stolonifera L. subsp. scabriglumis (Boiss. et Reut.) Maire ■☆

12339 Agrostis stolonifera L. var. major (Gaudin) Farw. = Agrostis gigantea Roth ■

12340 Agrostis stolonifera L. var. major Farw. = Agrostis gigantea Roth ■

12341 Agrostis stolonifera L. var. mixta (Hack.) Maire et Weiller = Agrostis castellana Boiss. et Reut. ■☆

12342 Agrostis stolonifera L. var. mixta Trab. = Agrostis stolonifera L. ■

12343 Agrostis stolonifera L. var. mutica (Hack.) Maire et Weiller = Agrostis castellana Boiss. et Reut. ■☆

12344 Agrostis stolonifera L. var. palustris (Huds.) Farw.;湿生西伯利

亚剪股颖;Creeping Bent Grass,Creeping Tickle Grass ■☆
12345 Agrostis stolonifera L. var. palustris (Huds.) Farw. = Agrostis stolonifera L. ■
12346 Agrostis stolonifera L. var. pseudopungens (Lange) Kerguélen = Agrostis castellana Boiss. et Reut. ■☆
12347 Agrostis stolonifera L. var. quinquenervosa Sauvage = Agrostis castellana Boiss. et Reut. ■☆
12348 Agrostis stolonifera L. var. ramosa (Gray) Veldkamp = Agrostis gigantea Roth ■
12349 Agrostis stolonifera L. var. scabriglumis (Boiss. et Reut.) C. E. Hubb. = Agrostis stolonifera L. subsp. scabriglumis (Boiss. et Reut.) Maire ■☆
12350 Agrostis stolonifera L. var. scabriglumis (Boiss. et Reut.) Maire = Agrostis stolonifera L. subsp. scabriglumis (Boiss. et Reut.) Maire ■☆
12351 Agrostis stolonifera L. var. stenantha Maire et Trab. = Agrostis castellana Boiss. et Reut. ■☆
12352 Agrostis stolonifera L. var. trinervata Maire et Trab. = Agrostis stolonifera L. ■
12353 Agrostis stolonizans Besser = Agrostis stolonizans Besser ex Schult. et Schult. ■
12354 Agrostis stolonizans Besser ex Schult. et Schult. = Agrostis alba L. ■
12355 Agrostis suavis Stapf;芳香剪股颖■☆
12356 Agrostis subaristata Aitch. et Hemsl.;糙颖剪股颖;Subaristate Bentgrass ■
12357 Agrostis sublaevipes Nevski;近光梗剪股颖■☆
12358 Agrostis subulifolia Stapf;钻叶剪股颖■☆
12359 Agrostis suizanensis Hayata = Deyeuxia suizanensis (Hayata) Ohwi ■
12360 Agrostis sylvatica Host;澳洲剪股颖■☆
12361 Agrostis sylvatica Huds. = Agrostis alba L. ■
12362 Agrostis sylvatica Huds. = Agrostis capillaris L. ■
12363 Agrostis sylvatica Koeler = Agrostis vulgaris With. ■
12364 Agrostis sylvatica Torr. = Muhlenbergia sylvestris (Torr.) Torr. ex A. Gray ■☆
12365 Agrostis taliensis Pilg.;大理剪股颖;Dali Bentgrass ■
12366 Agrostis taliensis Pilg. = Agrostis limprichtii Pilg. ■
12367 Agrostis taliensis Pilg. = Agrostis nervosa Nees ex Trin. ■
12368 Agrostis tandilensis (Kuntze) Parodi;肯氏剪股颖;Kennedy's Bentgrass ■☆
12369 Agrostis tateyamensis Tateoka;立山剪股颖■☆
12370 Agrostis taylori C. E. Hubb.;泰勒剪股颖■☆
12371 Agrostis teberdensis Litv.;捷别尔达剪股颖■☆
12372 Agrostis tenacissima Roxb. = Sporobolus tremulus (Willd.) Kunth ■☆
12373 Agrostis tenerrima Trin.;软剪股颖■☆
12374 Agrostis tenuifolia M. Bieb. = Agrostis vinealis Schreb. ■
12375 Agrostis tenuis Bastard ex Roem. et Schult.;细小剪股颖;Fine Bent-grass ■
12376 Agrostis tenuis Sibth. = Agrostis capillaris L. ■
12377 Agrostis tenuis Sibth. = Agrostis tenuis Bastard ex Roem. et Schult. ■
12378 Agrostis tenuis Sibth. f. aristata (Tausch.) Beldie = Agrostis capillaris L. ■
12379 Agrostis tenuis Sibth. var. aristata Druce = Agrostis capillaris L. ■
12380 Agrostis tenuis Sibth. var. hispida ? = Agrostis capillaris L. ■
12381 Agrostis tenuis Sibth. var. pumila ? = Agrostis capillaris L. ■
12382 Agrostis tianschanica Pavlov = Polypogon hissaricus (Roshev.) Bor ■
12383 Agrostis tibestica Miré et Quézel;提贝斯提剪股颖■☆
12384 Agrostis trachyphylla Pilg.;厚叶剪股颖■☆
12385 Agrostis trachyphylla Pilg. var. majuscula ? = Agrostis trachyphylla Pilg. ■☆
12386 Agrostis transcaspica Litv.;里海剪股颖■☆
12387 Agrostis transmorrisonensis Hayata;外玉山剪股颖■
12388 Agrostis transmorrisonensis Hayata = Agrostis infirma Büse var. formosana (Hack.) Veldkamp ■
12389 Agrostis transmorrisonensis Hayata = Agrostis sozanensis Hayata ■
12390 Agrostis transmorrisonensis Hayata var. kunyushanensis F. Z. Li;昆仑山剪股颖;Kunlunshan Bentgrass ■
12391 Agrostis transmorrisonensis Hayata var. opienensis Keng ex Y. C. Yang;川中剪股颖;Opien Bentgrass ■
12392 Agrostis transmorrisonensis Hayata var. opienensis Keng ex Y. C. Yang = Agrostis sozanensis Hayata ■
12393 Agrostis tremula Willd. = Sporobolus tremulus (Willd.) Kunth ■☆
12394 Agrostis tricholemma C. E. Hubb. = Agrostis keniensis Pilg. ■☆
12395 Agrostis trinii Turcz. = Agrostis vinealis Schreb. ■
12396 Agrostis trinii Turcz. ex Litv. = Agrostis flaccida Hack. var. trinii (Turcz. ex Litv.) Ohwi ■
12397 Agrostis trinii Turcz. ex Litv. = Agrostis vinealis Schreb. ■
12398 Agrostis tropica P. Beauv.;热带剪股颖■☆
12399 Agrostis truncatula Parl.;平截剪股颖■☆
12400 Agrostis truncatula Parl. var. fontqueri Maire = Agrostis truncatula Parl. ■☆
12401 Agrostis tsaratananensis A. Camus;察拉塔纳纳剪股颖■☆
12402 Agrostis tsiafajavonensis A. Camus;齐亚剪股颖■☆
12403 Agrostis turkestanica Drobow;北疆剪股颖;Turkestan Bentgrass ■
12404 Agrostis uhligii C. E. Hubb. = Agrostis volkensii Stapf ■☆
12405 Agrostis uhligii C. E. Hubb. var. contracta ? = Agrostis volkensii Stapf ■☆
12406 Agrostis umbellulata Trin.;小伞剪股颖■☆
12407 Agrostis valvata Steud.;镊合剪股颖■☆
12408 Agrostis ventricosa Gouan = Gastridium ventricosum (Gouan) Schinz et Thell. ■☆
12409 Agrostis verticillata Lam. = Vetiveria zizanioides (L.) Nash ■
12410 Agrostis verticillata Vill.;轮生剪股颖;Verticillate Bentgrass ■☆
12411 Agrostis verticillata Vill. = Polypogon viridis (Gouan) Breistr. ■
12412 Agrostis vestita A. Rich. = Agrostis lachnantha Nees ■☆
12413 Agrostis vinealis Schreb.;芒剪股颖(葡萄园剪股颖,细叶剪股颖);Brown Bent, Thinleaf Bentgrass, Trin Bentgrass, Vine-yaed Bentgrass ■
12414 Agrostis vinealis Schreb. Schreb. subsp. turkestanica (Drobow) Tzvelev = Agrostis turkestanica Drobow ■
12415 Agrostis vinealis Schreb. subsp. trinii (Turcz.) Tzvelev = Agrostis trinii Turcz. ex Litv. ■
12416 Agrostis vinealis Schreb. subsp. turkestanica (Drobow) Tzvelev. = Agrostis turkestanica Drobow ■
12417 Agrostis vinealis subsp. trinii (Turcz.) Tzvelev. = Agrostis vinealis Schreb. ■
12418 Agrostis virginica L. = Sporobolus virginicus (L.) Kunth ■
12419 Agrostis viridis Gouan = Polypogon viridis (Gouan) Breistr. ■
12420 Agrostis volkensii Stapf;福尔剪股颖■☆
12421 Agrostis volkensii Stapf var. deminuta Pilg. = Agrostis gracilifolia

C. E. Hubb. ■☆
12422 Agrostis vulgaris With. = Agrostis capillaris L. ■
12423 Agrostis vulgaris With. = Agrostis stolonifera L. ■
12424 Agrostis vulgaris With. = Agrostis tenuis Sibth. ■
12425 Agrostis wallichiana (Steud.) Bor;柔毛剪股颖;Woolly Bentgrass ■
12426 Agrostis wallichiana Steud. = Agrostis pilosula Trin. ■
12427 Agrostis whytei C. E. Hubb. = Agrostis continuata Stapf ■☆
12428 Agrostis wulingensis Honda = Agrostis clavata Trin. ■
12429 Agrostis wulingensis Honda = Agrostis infirma Büse ■
12430 Agrostistachys Dalzell(1850);剪股颖戟属(田穗戟属)■☆
12431 Agrostistachys africana Müll. Arg.;非洲田穗戟■☆
12432 Agrostistachys africana Müll. Arg. = Pseudagrostistachys africana (Müll. Arg.) Pax et K. Hoffm. ■☆
12433 Agrostistachys hookeri Benth. et Hook. f.;剪股颖戟(田穗戟)■☆
12434 Agrostistachys indica Dalzell;印度剪股颖戟(印度田穗戟)■☆
12435 Agrostistachys leptostachya Pax et K. Hoffm.;细花剪股颖戟(细花田穗戟)■☆
12436 Agrostistachys longifolia Benth. ex Hook. f.;长叶剪股颖戟(长叶田穗戟)■☆
12437 Agrostistachys ugandensis Hutch. = Pseudagrostistachys ugandensis (Hutch.) Pax et K. Hoffm. ■☆
12438 Agrostocrinum F. Muell. (1860);澳洲山菅兰属■☆
12439 Agrostocrinum stypandroides F. Muell.;澳洲山菅兰■☆
12440 Agrostomia Cerv. = Chloris Sw. ●■
12441 Agrostomia Cerv. = Panicum L. ■
12442 Agrostophyllum Blume (1825);禾叶兰属;Agrostophyllum, Grassleaf Orchis ■
12443 Agrostophyllum acutum Schltr.;尖禾叶兰■☆
12444 Agrostophyllum atrovirens J. J. Sm.;墨绿禾叶兰■☆
12445 Agrostophyllum callosum Rchb. f.;禾叶兰(胼胝唇禾叶兰,硬皮禾叶兰);Callose Agrostophyllum, Common Grassleaf orchis ■
12446 Agrostophyllum formosanum Rolfe = Agrostophyllum inocephalum (Schauer) Ames ■
12447 Agrostophyllum fragrans Schltr.;香禾叶兰■☆
12448 Agrostophyllum grandiflorum Schltr.;大花禾叶兰■☆
12449 Agrostophyllum inocephalum (Schauer) Ames;台湾禾叶兰(无头千叶兰);Taiwan Agrostophyllum, Taiwan Grassleaf orchis ■
12450 Agrostophyllum laxum J. J. Sm.;松散禾叶兰■☆
12451 Agrostophyllum leucocephalum Schltr.;白头禾叶兰■☆
12452 Agrostophyllum longifolium Rchb. f.;长叶禾叶兰■☆
12453 Agrostophyllum macrocephalum Schltr.;大头禾叶兰■☆
12454 Agrostophyllum montanum Schltr.;山地禾叶兰■☆
12455 Agrostophyllum niveum Schltr.;雪白禾叶兰■☆
12456 Agrostophyllum occidentale Schltr.;西方禾叶兰■☆
12457 Agrostophyllum parviflorum J. J. Sm.;小花禾叶兰■☆
12458 Agrostophyllum philippinense Ames;菲律宾禾叶兰■☆
12459 Agrostophyllum rigidifolium Ridl.;硬叶禾叶兰■☆
12460 Agrostophyllum seychellarum Rolfe = Agrostophyllum occidentale Schltr. ■☆
12461 Agrostophyllum tenue J. J. Sm.;细禾叶兰■☆
12462 Agrostophyllum uniflorum Schltr.;单花禾叶兰■☆
12463 Agrostophyllum zeylanicum Hook. f.;锡兰禾叶兰■☆
12464 Agroulus P. Beauv. = Agrostis L. (保留属名)■
12465 Aguava Raf. = Myrcia DC. ex Guill. ●☆
12466 Aguiaria Ducke(1935);巴西木棉属●☆
12467 Aguiaria excelsa Ducke;巴西木棉●☆
12468 Agylla Phil. = Cladium P. Browne ■
12469 Agylophora Neck. = Uncaria Schreb. (保留属名)●
12470 Agylophora Neck. ex Raf. = Ourouparia Aubl. (废弃属名)●
12471 Agylophora Neck. ex Raf. = Uncaria Schreb. (保留属名)●
12472 Agynaia Hassk. = Agyneia L. (废弃属名)●
12473 Agynaia Hassk. = Glochidion J. R. Forst. et G. Forst. (保留属名)●
12474 Agyneia L. (废弃属名) = Glochidion J. R. Forst. et G. Forst. (保留属名)●
12475 Agyneia bacciformis (L.) A. Juss. = Sauropus bacciformis (L.) Airy Shaw ■
12476 Agyneia coccinea Buch. -Ham. = Glochidion coccineum (Buch. -Ham.) Müll. Arg. ●
12477 Agyneia goniocladа (Merr. et Chun) H. Keng = Sauropus bacciformis (L.) Airy Shaw ■
12478 Agyneia pubera L. = Glochidion puberum (L.) Hutch. ●
12479 Agyneia taiwaniana H. Keng = Sauropus bacciformis (L.) Airy Shaw ■
12480 Agyneja Vent. = Synostemon F. Muell. ●■
12481 Agyneja bacciformis (L.) A. Juss. = Sauropus bacciformis (L.) Airy Shaw ■
12482 Agyneja bacciformis (L.) A. Juss. = Synostemon bacciforme (L.) Webster ■
12483 Ahernia Merr. (1909);菲柞属;Ahernia ●
12484 Ahernia glandulosa Merr.;菲柞;Ahernia, Common Ahernia ●
12485 Ahouai Mill. (废弃属名) = Thevetia L. (保留属名)●
12486 Ahouai Tourn. ex Adans. = Thevetia L. (保留属名)●
12487 Ahouai thevetia (L.) M. Gómez = Cascabela thevetia (L.) Lippold ●
12488 Ahovai Boehm. = Ahouai Mill. (废弃属名)●
12489 Ahovai Boehm. = Thevetia L. (保留属名)●
12490 Ahzolia Standl. et Steyerm. = Sechium P. Browne(保留属名)■
12491 Aichryson Webb et Berthel. (1840);爱染草属■☆
12492 Aichryson bollei Bolle;博勒爱染草■☆
12493 Aichryson brevipetalum Praeger;短瓣爱染草■☆
12494 Aichryson dichotomum Webb et Berthel.;爱染草■☆
12495 Aichryson divaricatum (Aiton) Praeger;叉开爱染草■☆
12496 Aichryson domesticum Webb et Berthel. 'Variegatum';彩叶爱染草;Cloud-grass ■☆
12497 Aichryson dumosum (Lowe) Praeger;棘丛爱染草■☆
12498 Aichryson immaculatum Christ = Aichryson pachycaulon Bolle subsp. immaculatum (Christ) Bramwell ■☆
12499 Aichryson laxum (Haw.) Bramwell;疏松爱染草■☆
12500 Aichryson pachycaulon Bolle;粗茎爱染草■☆
12501 Aichryson pachycaulon Bolle subsp. immaculatum (Christ) Bramwell;无斑粗茎爱染草■☆
12502 Aichryson pachycaulon Bolle subsp. parviflorum (Bolle) Bramwell;小花粗茎爱染草■☆
12503 Aichryson pachycaulon Bolle subsp. praetermissum Bramwell;疏忽爱染草■☆
12504 Aichryson palmense Bolle;帕尔马爱染草■☆
12505 Aichryson parviflorum Bolle = Aichryson pachycaulon Bolle subsp. parviflorum (Bolle) Bramwell ■☆
12506 Aichryson porphyrogennetos Bolle;紫色爱染草■☆
12507 Aichryson punctatum (Buch) Webb et Berthel.;斑点爱染草■☆
12508 Aichryson punctatum (Buch) Webb et Berthel. var. subvillosum

(Lowe) Bornm. = Aichryson punctatum (Buch) Webb et Berthel. ■☆
12509 Aichryson pygmaeum Webb et Berthel.;矮小爱染草■☆
12510 Aichryson radiscescens Webb et Berthel. = Aichryson tortuosum (Aiton) Webb et Berthel. ■☆
12511 Aichryson tortuosum (Aiton) Webb et Berthel.;扭曲小爱染草;Gouty Houseleek ■☆
12512 Aichryson villosum (Aiton) Webb et Berthel.;长柔毛爱染草■
12513 Aidelus Spreng. = Veronica L. ■
12514 Aidema Ravenna(2003);南美石蒜属■☆
12515 Aidia Lour. (1790);茜树属(鸡爪簕属,茜木属,山黄皮属);Maddertree,Fragrant Nanmu ●
12516 Aidia acuminatissima (Merr.) Masam. = Aidia pycnantha (Drake) Tirveng. ●
12517 Aidia canthioides (Champ. ex Benth.) Masam.;香楠(山黄皮,水棉木,台北茜草树);Fragrant nanmu, Taibei Randia, Taipei Randia ●
12518 Aidia canthioides (Champ. ex Benth.) Masam. = Randia canthioides Champ. ex Benth. ●
12519 Aidia cochinchinensis Lour.;茜树(龙虾,茜草树,山黄皮);Cochinchina Randia,Cochin-China Randia,Maddertree,Randia ●
12520 Aidia cochinchinensis Lour. = Randia cochinchinensis (Lour.) Merr. ●
12521 Aidia cochinchinensis Lour. f. angustifolia (Makino) Honda;狭叶茜树●☆
12522 Aidia densiflora (Benth.) Masam. = Aidia cochinchinensis Lour. ●
12523 Aidia henryi (E. Pritz.) T. Yamaz. = Aidia cochinchinensis Lour. ●
12524 Aidia leucocarpa (Champ. ex Benth.) T. Yamaz. = Alleizettella leucocarpa (Champ. ex Benth.) Tirveng. ●
12525 Aidia merrillii (Chun) Tirveng. = Aidia cochinchinensis Lour. ●
12526 Aidia micrantha (K. Schum.) F. White;小花茜树●☆
12527 Aidia micrantha (K. Schum.) F. White var. congolana (De Wild.) E. M. Petit;刚果茜树●☆
12528 Aidia micrantha (K. Schum.) F. White var. zenkeri (S. Moore) E. M. Petit;岑克尔茜树●☆
12529 Aidia ochroleuca (K. Schum.) E. M. Petit;绿白茜树●☆
12530 Aidia oxydonta (Drake) T. Yamaz.;尖萼茜树(尖萼山黄皮,尖叶山黄皮);Acute-leaved Randia, Sharpcalyx Randia, Tinesepal Maddertree ●
12531 Aidia oxydonta (Drake) T. Yamaz. = Randia oxydonta Drake ●
12532 Aidia pallens (Hiern) G. Taylor = Aulacocalyx pallens (Hiern) Bridson et Figueiredo ●☆
12533 Aidia pycnantha (Drake) Tirveng.;多毛茜草树(黄棉木,极尖茜草树,毛山黄皮);Hairlike Randia, Hairy Maddertree, Hairy Randia ●
12534 Aidia pycnantha (Drake) Tirveng. = Randia pycnantha Drake ●
12535 Aidia quintasii (K. Schum.) G. Taylor;昆塔斯茜树●☆
12536 Aidia racemosa (Cav.) Tirveng. = Aidia cochinchinensis Lour. ●
12537 Aidia rubens (Hiern) G. Taylor;淡红茜树●☆
12538 Aidia salicifolia (H. L. Li) T. Yamaz.;柳叶香楠(八大木,柳叶茜树);Wikkowleaf Randia ●
12539 Aidia shweliensis (J. Anthony) W. C. Chen;瑞丽茜树;Ruili Randia ●
12540 Aidia shweliensis (J. Anthony) W. C. Chen = Fosbergia shweliensis (J. Anthony) Tirveng. et Sastre ●
12541 Aidia shweliensis (J. Anthony) W. C. Chen = Randia shweliensis J. Anthony ●
12542 Aidia wallichii (Hook. f.) T. Yamaz. = Tarennoidea wallichii (Hook. f.) Tirveng. ●
12543 Aidia wattii G. Taylor;瓦特茜树●☆
12544 Aidia yunnanensis (Hutch.) T. Yamaz.;滇茜树;Yunnan Maddertree,Yunnan Randia ●
12545 Aidiopsis Tirveng. (1987);肖茜树属●☆
12546 Aidiopsis forbesii (King et Gamble) Tirveng.;肖茜树●☆
12547 Aidiopsis orophila (Miq.) Ridsdale;喜山肖茜树●☆
12548 Aidomene Stopp(1967);安哥拉萝藦属☆
12549 Aidomene parvula Stopp;安哥拉萝藦☆
12550 Aigeiros Lunell = Populus L. ●
12551 Aigiros Raf. = Populus L. ●
12552 Aigosplen Raf. = Callirhoe Nutt. ●■☆
12553 Aikinia R. Br. = Epithema Blume ■
12554 Aikinia Salisb. ex A. DC. = Wahlenbergia Schrad. ex Roth(保留属名)●■
12555 Aikinia Wall. = Ratzeburgia Kunth ■☆
12556 Ailanthaceae J. Agardh = Simaroubaceae DC.(保留科名)●
12557 Ailanthaceae J. Agardh;臭椿科●
12558 Ailanthus Desf. (1788)(保留属名);臭椿属(樗树属,樗属);Ailanthus,Tree of Heaven,Tree-of-heaven ●
12559 Ailanthus altissima (Mill.) Swingle;臭椿(臭椿皮,樗,樗树,春铃子,椿树,大果臭椿,大眼桐,凤眼草,鬼目,虎眼树);Ailanthus,Ailanto,Chinese Sumac,Chinese Sumach,Chinese tree-of-heaven,Copal Tree,Copal-tree,Downy Tree-of-heaven,Japan Varnish Tree,Slinking Cedar,Stink Tree,Stinking Shumac,Stinkweed,Tree of Heaven,Tree of Heaven Ailanthus,Tree of the Gods,Tree-of-Heaven,Tree-of-Heaven Ailanthus,Varnish Tree,Varnish-tree ●
12560 Ailanthus altissima (Mill.) Swingle var. leucoxyla B. C. Ding et T. B. Chao;白材臭椿;White-wood Ailanthus ●
12561 Ailanthus altissima (Mill.) Swingle var. microphylla B. C. Ding et T. B. Chao;小叶臭椿;Little-leaf Ailanthus ●
12562 Ailanthus altissima (Mill.) Swingle var. ramosissma B. C. Ding et T. B. Chao;千头臭椿;Manybranched Ailanthus ●
12563 Ailanthus altissima (Mill.) Swingle var. sutchuenensis (Dode) Rehder et E. H. Wilson;大果臭椿(白家香,大果樗树);Bigfruit Ailanthus,Sichuan Ailanthus,Sutchuen Ailanthus ●
12564 Ailanthus altissima (Mill.) Swingle var. tanakai (Hayata) Kaneh. et Sasaki;台湾臭椿(臭椿,凤眼草,台湾樗树);Taiwan Chinese Sumac,Tanaka Ailanthus ●
12565 Ailanthus cacodendron (Ehrh.) Schinz et Thell. = Ailanthus altissima (Mill.) Swingle ●
12566 Ailanthus cacodendron (Ehrh.) Schinz et Thell. var. sutchuenensis (Rehder) Rehder et E. H. Wilson = Ailanthus altissima (Mill.) Swingle var. sutchuenensis (Dode) Rehder et E. H. Wilson ●
12567 Ailanthus excelsa Roxb.;高樗●☆
12568 Ailanthus fauveliana Pierre = Ailanthus triphysa (Dennst.) Alston ●
12569 Ailanthus flavescens Carrière = Toona sinensis (A. Juss.) M. Roem. ●◇
12570 Ailanthus fordii Noot.;常绿臭椿(福氏臭椿,岭南樗树);Evergreen Ailanthus,Ford Ailanthus,Green Ailanthus ●
12571 Ailanthus giraldii Dode;毛臭椿;Girald Ailanthus, Hair Ailanthus,Hairy Ailanthus ●
12572 Ailanthus giraldii Dode var. duclouxii Dode = Ailanthus giraldii Dode ●

12573 Ailanthus giraldii Dode var. triphysa (Dennst.) Alston;滇毛臭椿;Yunnan Hairy Ailanthus ●

12574 Ailanthus glandulosa Desf. = Ailanthus altissima (Mill.) Swingle ●

12575 Ailanthus glandulosa Desf. var. spinosa T. Yamaz. et Bois = Ailanthus vilmoriniana Dode ●

12576 Ailanthus glandulosa Desf. var. sutchuenensis Rehder = Ailanthus altissima (Mill.) Swingle var. sutchuenensis (Dode) Rehder et E. H. Wilson ●

12577 Ailanthus glandulosa Desf. var. tanakai Hayata = Ailanthus altissima (Mill.) Swingle var. tanakai (Hayata) Kaneh. et Sasaki ●

12578 Ailanthus grandis Prain;大臭椿●☆

12579 Ailanthus guangxiensis S. L. Mo;广西樗树;Guangxi Ailanthus ●

12580 Ailanthus integrifolia Lam. = Ailanthus triphysa (Dennst.) Alston ●

12581 Ailanthus integrifolia Lam. subsp. calycina (Pierre) Noot.;全缘臭椿(全缘樗)●☆

12582 Ailanthus japonica Hort. ex K. Koch;日本臭椿●☆

12583 Ailanthus mairei Gagnep. = Ailanthus altissima (Mill.) Swingle var. sutchuenensis (Dode) Rehder et E. H. Wilson ●

12584 Ailanthus malabarica DC. = Ailanthus triphysa (Dennst.) Alston ●

12585 Ailanthus moluccana DC.;马六甲臭椿;Sky Tree ●☆

12586 Ailanthus sutchuenensis Dode = Ailanthus altissima (Mill.) Swingle var. sutchuenensis (Dode) Rehder et E. H. Wilson ●

12587 Ailanthus triphysa (Dennst.) Alston;毛叶南臭椿(岭南臭椿,岭南樗树,马拉巴臭椿,马拉巴樗,孟仑樗);Hairyleaf South Ailanthus, Hairy-leaved South Ailanthus, Lingnan Ailanthus, Mattipaul, White Bean, White Siris ●

12588 Ailanthus vilmoriniana Dode;刺臭椿(刺樗);Spine Ailanthus, Vilmorin Ailanthus ●

12589 Ailanthus vilmoriniana Dode var. henanensis J. Y. Chen et L. Y. Jin;赤叶刺臭椿;Henan Spiny Ailanthus ●

12590 Ailantopsis Gagnep. = Trichilia P. Browne(保留属名)●

12591 Ailantus DC. = Ailanthus Desf.(保留属名)●

12592 Aillya de Vriese = Goodenia Sm. ●■☆

12593 Ailuroschia Steven = Astragalus L. ●■

12594 Aimenia Comm. ex Planch. = Cissus L. ●

12595 Aimorra Raf.(1838);佛菊属☆

12596 Ainea Ravenna = Sphenostigma Baker ■☆

12597 Ainea Ravenna(1979);青铜鸢尾属■☆

12598 Ainea conzattii (R. C. Foster) Ravenna;青铜鸢尾■☆

12599 Ainsliaea DC.(1838);兔儿风属(鬼督邮属,兔耳风属);Ainsliaea, Monkeytail Plant, Rabbiten-wind ■

12600 Ainsliaea acerifolia Sch. Bip.;槭叶兔儿风;Mapleleaf Ainsliaea, Mapleleaf Rabbiten-wind ■

12601 Ainsliaea acerifolia Sch. Bip. var. affinis (Miq.) Kitam. = Ainsliaea acerifolia Sch. Bip. ■

12602 Ainsliaea acerifolia Sch. Bip. var. subapoda Nakai;无梗槭叶兔儿风■☆

12603 Ainsliaea acerifolia Sch. Bip. var. subapoda Nakai = Ainsliaea acerifolia Sch. Bip. ■

12604 Ainsliaea affinis Miq. = Ainsliaea acerifolia Sch. Bip. ■

12605 Ainsliaea angustata C. C. Chang;马边兔儿风;Angustate Ainsliaea, Angustate Rabbiten-wind ■

12606 Ainsliaea angustifolia Hook. f. et Thomson ex C. B. Clarke;狭叶兔儿风;Narrowleaf Ainsliaea, Narrowleaf Rabbiten-wind ■

12607 Ainsliaea angustifolia Hook. f. et Thomson ex C. B. Clarke var. luchunensis H. Chuang = Ainsliaea latifolia (D. Don) Sch. Bip. ■

12608 Ainsliaea apiculata Sch. Bip.;尖叶兔儿风■☆

12609 Ainsliaea apiculata Sch. Bip. f. acerifolia (Masam.) Sugim. = Ainsliaea apiculata Sch. Bip. var. acerifolia Masam. ■☆

12610 Ainsliaea apiculata Sch. Bip. f. rotundifolia (Masam.) Sugim.;圆叶兔儿风■☆

12611 Ainsliaea apiculata Sch. Bip. var. acerifolia Masam.;琉球兔儿风■☆

12612 Ainsliaea apiculata Sch. Bip. var. linearis (Makino) Masam. = Ainsliaea faurieana Beauverd ■☆

12613 Ainsliaea aptera DC.;无翅兔儿风;Wingless Ainsliaea, Wingless Rabbiten-wind ■

12614 Ainsliaea apteroides (C. C. Chang) Y. C. Tseng;狭翅兔儿风;Narrow-winged Ainsliaea, Narrow-winged Rabbiten-wind ■

12615 Ainsliaea asarifolia Hayata = Ainsliaea fragrans Champ. ■

12616 Ainsliaea bonatii Beauverd;心叶兔儿风(大俄火把,大一枝箭,双股箭,兔儿风,小接骨丹,心叶兔耳风);Cordateleaf Ainsliaea, Heartleaf Rabbiten-wind ■

12617 Ainsliaea bonatii Beauverd var. arachnoides Beauverd = Ainsliaea bonatii Beauverd ■

12618 Ainsliaea bonatii Beauverd var. glabra Beauverd = Ainsliaea bonatii Beauverd ■

12619 Ainsliaea caesia Hand.-Mazz.;蓝兔儿风;Blue Ainsliaea, Blue Rabbiten-wind ■

12620 Ainsliaea cavaleriei H. Lév.;定番兔儿风■

12621 Ainsliaea chapaensis Merr.;边地兔儿风;Chapa Ainsliaea, Chapa Rabbiten-wind ■

12622 Ainsliaea cleistogama C. C. Chang;闭花兔儿风;Closeflower Ainsliaea, Closeflower Rabbiten-wind ■

12623 Ainsliaea cordifolia Franch. et Sav.;日本心叶兔儿风■☆

12624 Ainsliaea cordifolia Franch. et Sav. var. integrifolia Maxim. = Ainsliaea fragrans Champ. var. integrifolia (Maxim.) Kitam. ■

12625 Ainsliaea cordifolia Franch. et Sav. var. maruoi (Makino) Makino ex Kitam.;马氏心叶兔儿风■☆

12626 Ainsliaea crassifolia C. C. Chang;厚叶兔儿风;Thickleaf Ainsliaea, Thickleaf Rabbiten-wind ■

12627 Ainsliaea dissecta Franch. et Sav.;裂叶兔儿风■☆

12628 Ainsliaea dissecta Franch. et Sav. f. acerifolia Sugim.;尖裂叶兔儿风■☆

12629 Ainsliaea dissecta Franch. et Sav. f. purpurascens Sugim.;紫裂叶兔儿风■☆

12630 Ainsliaea elegans Hemsl.;秀丽兔儿风(背毛红草);Elegant Ainsliaea, Elegant Rabbiten-wind ■

12631 Ainsliaea elegans Hemsl. var. strigosa Mattf.;红毛兔儿风(反背马蹄红,红毛叶,红毛叶马蹄香,红毛叶兔儿风,马蹄香,毛叶马蹄香);Redhair Ainsliaea ■

12632 Ainsliaea elegans Hemsl. var. tomentosa Mattf. = Ainsliaea elegans Hemsl. ■

12633 Ainsliaea faurieana Beauverd;法氏兔儿风■☆

12634 Ainsliaea foliosa Hand.-Mazz.;异叶兔儿风;Diverseleaf Ainsliaea, Diverseleaf Rabbiten-wind ■

12635 Ainsliaea fragrans Champ.;杏香兔儿风(巴地虎,白走马胎,朝天一炷香,大种巴地香,倒拔千金,飞针,肺形草,红金交杯,红太极图,急儿风,金边兔儿草,金边兔耳,金边兔耳风,金茶匙,毛鹿含草,毛马香,牛皮菜,牛眼珠草,扑地金钟,朴地金钟,忍冬草,山蝴蝶,天青地白,铁交杯,通天草,铜调羹,吐血草,兔耳草,兔耳箭,兔耳金边草,兔耳一枝箭,香鬼督邮,橡皮草,小鹿衔,一枝香,银茶匙,月下红,猪心草);Apricotsmell Rabbiten-wind, Fragrant

Ainsliaea ■

12636 Ainsliaea fragrans Champ. var. integrifolia (Maxim.) Kitam.;全叶杏香兔儿风■

12637 Ainsliaea fragrans Champ. var. integrifolia (Maxim.) Kitam. = Ainsliaea fragrans Champ. ■

12638 Ainsliaea fulvioides H. Chuang;拟黄毛兔儿风■

12639 Ainsliaea fulvips Jeffrey et W. W. Sm.;黄毛兔儿风;Yellowhair Ainsliaea, Yellowhair Rabbiten-wind ■

12640 Ainsliaea glabra Hemsl.;光叶兔儿风(散血草,石凤丹,兔儿风,兔耳风,心肺草,血筋草);Glabrous Ainsliaea, Glabrous Rabbiten-wind ■

12641 Ainsliaea glabra Hemsl. var. sutchuenensis (Franch.) S. E. Freire;四川兔儿风;Sichuan Ainsliaea, Sichuan Rabbiten-wind ■

12642 Ainsliaea glabra Hemsl. var. tenuicaulis (Mattf.) C. C. Chang = Ainsliaea glabra var. sutchuenensis (Franch.) S. E. Freire ■

12643 Ainsliaea glabra Hemsl. var. tenuicaulis (Mattf.) C. C. Chang = Ainsliaea tenuicaulis Mattf. ■

12644 Ainsliaea gongshanensis H. Chuang;贡山兔儿风■

12645 Ainsliaea gracilis Franch.;纤枝兔儿风(纤细兔儿风,相思草);Thin Ainsliaea, Thin Rabbiten-wind ■

12646 Ainsliaea gracilis Franch. var. robusta Diels = Ainsliaea gracilis Franch. ■

12647 Ainsliaea grossedentata Franch.;粗齿兔儿风(灯台草,灯盏七,光棍草,青菜果,一炷香);Largedentate Ainsliaea, Largedentate Rabbiten-wind ■

12648 Ainsliaea henryi Diels;长穗兔儿风(灯台草,滇桂兔儿风,二郎箭,光棍草,台湾鬼督邮,玉山鬼督邮);Henry's Ainsliaea, Longspike Rabbiten-wind ■

12649 Ainsliaea henryi Diels = Ainsliaea latifolia (D. Don) Sch. Bip. var. henryi (Diels) H. Koyama ■

12650 Ainsliaea henryi Diels var. daguanensis H. Chuang;大关长穗兔儿风■

12651 Ainsliaea henryi Diels var. daguanensis H. Chuang = Ainsliaea henryi Diels ■

12652 Ainsliaea henryi Diels var. ovatifolia C. C. Chang = Ainsliaea henryi Diels ■

12653 Ainsliaea heterantha Hand.-Mazz.;异花兔儿风;Differflower Ainsliaea, Differflower Rabbiten-wind ■

12654 Ainsliaea heterantha Hand.-Mazz. = Ainsliaea reflexa Merr. ■

12655 Ainsliaea hui Diels ex Mattf. = Ainsliaea macroclinidioides Hayata ■

12656 Ainsliaea hybrida Sugim.;杂种兔儿风■☆

12657 Ainsliaea hypoleuca Diels ex Limpr. = Ainsliaea latifolia (D. Don) Sch. Bip. ■

12658 Ainsliaea integrifolia (Maxim.) Makino = Ainsliaea fragrans Champ. ■

12659 Ainsliaea kawakamii Hayata = Ainsliaea macroclinidioides Hayata ■

12660 Ainsliaea lancangensis Y. Y. Qian;澜沧兔儿风■

12661 Ainsliaea lancifolia Franch.;穆坪兔儿风(长叶兔儿风,肺经草,披针叶兔儿风,披针叶兔耳风,小金血丹);Muping Ainsliaea, Muping Rabbiten-wind ■

12662 Ainsliaea lancifolia Franch. = Ainsliaea glabra Hemsl. ■

12663 Ainsliaea latifolia (D. Don) Sch. Bip.;宽叶兔儿风(白胡子狼毒,大叶一枝箭,刀口药,倒赤伞,宽叶兔耳风,毛叶香,牛尾一枝箭,三花兔儿风,天星地白子,小一枝箭);Broadleaf Ainsliaea, Broadleaf Rabbiten-wind ■

12664 Ainsliaea latifolia (D. Don) Sch. Bip. f. yunnanensis (Franch.) Kitam. = Ainsliaea yunnanensis Franch. ■

12665 Ainsliaea latifolia (D. Don) Sch. Bip. var. henryi (Diels) H. Koyama = Ainsliaea grossedentata Franch. ■

12666 Ainsliaea latifolia (D. Don) Sch. Bip. var. nimborum (Hand.-Mazz.) Kitam. = Ainsliaea latifolia (D. Don) Sch. Bip. var. henryi (Diels) H. Koyama ■

12667 Ainsliaea latifolia (D. Don) Sch. Bip. var. obovata (Franch.) Grierson et Lauener = Ainsliaea spicata Vaniot var. obovata (Franch.) C. Y. Wu ■

12668 Ainsliaea latifolia (D. Don) Sch. Bip. var. obovata (Franch.) Grierson et Lauener = Ainsliaea spicata Vaniot ■

12669 Ainsliaea latifolia (D. Don) Sch. Bip. var. platyphylla (Franch.) C. Y. Wu;宽穗兔儿风(白胡子狼毒,翅柄兔儿风,大叶一枝箭,刀口药,开口箭,三花兔儿风);Broadspike Ainsliaea, Triflorous Ainsliaea ■

12670 Ainsliaea latifolia (D. Don) Sch. Bip. var. ramifera H. Chuang = Ainsliaea latifolia (D. Don) Sch. Bip. ■

12671 Ainsliaea latifolia (D. Don) Sch. Bip. var. taiwanensis S. E. Freire;台湾兔儿风■

12672 Ainsliaea lijiangensis H. Chuang;漓江兔儿风■

12673 Ainsliaea linearis Makino = Ainsliaea faurieana Beauverd ■☆

12674 Ainsliaea liukiuensis Beauverd = Ainsliaea apiculata Sch. Bip. var. acerifolia Masam. ■☆

12675 Ainsliaea macrocephala (Mattf.) Y. C. Tseng;大头兔儿风;Bighead Ainsliaea, Bighead Rabbiten-wind ■

12676 Ainsliaea macroclinidioides Hayata;灯台兔儿风(阿里山鬼督邮,灯台草,铁灯兔儿风,永嘉兔儿风,中原氏鬼督邮);Integrifolius Ainsliaea, Lampstand Rabbiten-wind ■

12677 Ainsliaea macroclinidioides Hayata var. oblonga (Koidz.) Hatus. = Ainsliaea oblonga Koidz. ■

12678 Ainsliaea macroclinidioides Hayata var. okinawensis (Hayata) Kitam.;冲绳兔儿风■☆

12679 Ainsliaea macroclinidioides Hayata var. secundiflora (Hayata) Kitam.;五裂兔儿风(中原氏鬼督邮);Fivelobed Ainsliaea ■

12680 Ainsliaea macroclinidioides Hayata var. secundiflora (Hayata) Kitam. = Ainsliaea secundiflora Hayata ■

12681 Ainsliaea mairei H. Lév.;药山兔儿风;Maire Ainsliaea, Maire Rabbiten-wind ■

12682 Ainsliaea mattfeldiana Hand.-Mazz.;薄叶兔儿风;Filmleaf Ainsliaea, Filmleaf Rabbiten-wind ■

12683 Ainsliaea mattfeldiana Hand.-Mazz. = Ainsliaea bonatii Beauverd ■

12684 Ainsliaea mollis Diels ex H. Limpr.;泸定兔儿风;Softhair Ainsliaea, Softhair Rabbiten-wind ■

12685 Ainsliaea multibracteata Mattf.;多苞兔儿风;Bracteous Ainsliaea, Bracteous Rabbiten-wind ■

12686 Ainsliaea nana Y. C. Tseng;小兔儿风;Dwarf Ainsliaea, Dwarf Rabbiten-wind ■

12687 Ainsliaea nervosa Franch.;直脉兔儿风;Venose Ainsliaea, Venose Rabbiten-wind ■

12688 Ainsliaea ningpoensis Matsuda = Ainsliaea fragrans Champ. ■

12689 Ainsliaea oblonga Koidz.;阿里山鬼督邮(矩圆兔儿风)■

12690 Ainsliaea oblonga Koidz. var. latifolia Kitam. = Ainsliaea oblonga Koidz. ■

12691 Ainsliaea okinawensis Hayata = Ainsliaea macroclinidioides Hayata var. okinawensis (Hayata) Kitam. ■☆

12692 Ainsliaea ovalifolia Vaniot = Ainsliaea pertyoides Franch. var. albotomentosa Beauverd ■

12693 Ainsliaea parvifolia Merr.;小叶兔儿风;Smallleaf Ainsliaea ■

12694 Ainsliaea paucicapitata Hayata;花莲兔儿风(能高鬼督邮);Hualian Ainsliaea,Hualian Rabbiten-wind ■
12695 Ainsliaea pertyoides Franch.;腋花兔儿风(白背兔儿风,地黄连,牛毛细辛,叶下花,追风箭);Axillaryflower Rabbiten-wind,Pertybush-like Ainsliaea ■
12696 Ainsliaea pertyoides Franch. f. ovalifolia (Vaniot) Beauverd = Ainsliaea pertyoides Franch. var. albotomentosa Beauverd ●
12697 Ainsliaea pertyoides Franch. f. sparsiflora (Vaniot) Beauverd = Ainsliaea pertyoides Franch. ■
12698 Ainsliaea pertyoides Franch. var. albotomentosa Beauverd;白背兔儿风(白背叶下花,牛毛细辛,叶下花,追风箭);Whitetomentose Ainsliaea ■
12699 Ainsliaea pertyoides Franch. var. albotomentosa Beauverd f. ovalifolia (Vaniot) Beauverd = Ainsliaea pertyoides Franch. var. albotomentosa Beauverd ■
12700 Ainsliaea pertyoides Franch. var. intermedia Beauverd = Ainsliaea pertyoides Franch. var. albotomentosa Beauverd ■
12701 Ainsliaea pertyoides Franch. var. sparsiflora (Vaniot) H. Lév. = Ainsliaea pertyoides Franch. ■
12702 Ainsliaea pertyoides Vaniot var. sparsiflora (Vaniot) H. Lév. = Ainsliaea pertyoides Franch. ■
12703 Ainsliaea pingbianensis Y. C. Tseng;屏边兔儿风;Pingbian Ainsliaea,Pingbian Rabbiten-wind ■
12704 Ainsliaea pingbianensis Y. C. Tseng var. malipoensis H. Chuang = Ainsliaea pingbianensis Y. C. Tseng ■
12705 Ainsliaea plantaginifolia Mattf.;车前兔儿风;Plantainleaf Ainsliaea,Plantainleaf Rabbiten-wind ■
12706 Ainsliaea pteropoda DC. = Ainsliaea latifolia (D. Don) Sch. Bip. ■
12707 Ainsliaea pteropoda DC. f. valifolia (Vaniot) H. Lév. = Ainsliaea pertyoides Franch. var. albotomentosa Beauverd ■
12708 Ainsliaea pteropoda DC. var. apteroides C. C. Chang = Ainsliaea apteroides (C. C. Chang) Y. C. Tseng ■
12709 Ainsliaea pteropoda DC. var. incana Vaniot ex H. Lév. = Ainsliaea latifolia (D. Don) Sch. Bip. ■
12710 Ainsliaea pteropoda DC. var. leiophylla Franch. = Ainsliaea yunnanesis Franch. ■
12711 Ainsliaea pteropoda DC. var. macrocephala Mattf. = Ainsliaea macrocephala (Mattf.) Y. C. Tseng ■
12712 Ainsliaea pteropoda DC. var. obovata Franch. = Ainsliaea spicata Vaniot ■
12713 Ainsliaea pteropoda DC. var. ovalifolia (Vaniot) H. Lév. = Ainsliaea pertyoides Franch. var. albotomentosa Beauverd ■
12714 Ainsliaea pteropoda DC. var. platyphylla Franch. = Ainsliaea latifolia (D. Don) Sch. Bip. var. platyphylla (Franch.) C. Y. Wu ■
12715 Ainsliaea pteropoda DC. var. silhetensis DC. = Ainsliaea latifolia (D. Don) Sch. Bip. ■
12716 Ainsliaea qianiana S. E. Freire;钱氏兔儿风■
12717 Ainsliaea ramosa Hemsl.;莲沱兔儿风(多枝兔儿风);Branchy Ainsliaea,Branchy Rabbiten-wind ■
12718 Ainsliaea reflexa Merr.;长柄兔儿风(台湾鬼督邮);Longstalk Ainsliaea,Longstalk Rabbiten-wind ■
12719 Ainsliaea reflexa Merr. = Ainsliaea latifolia (D. Don) Sch. Bip. var. henryi (Diels) H. Koyama ■
12720 Ainsliaea reflexa Merr. var. nimborum Hand. -Mazz.;玉山鬼督邮(多极兔儿风)■
12721 Ainsliaea reflexa Merr. var. nimborum Hand. -Mazz. = Ainsliaea latifolia (D. Don) Sch. Bip. var. henryi (Diels) H. Koyama ■
12722 Ainsliaea reflexa Merr. var. subalpina Hand. -Mazz.;拟高山兔儿风■
12723 Ainsliaea rubrifolia Franch.;红背兔儿风(红走马胎,牛皮菜);Redleaf Ainsliaea,Redleaf Rabbiten-wind ■
12724 Ainsliaea rubrinervis C. C. Chang;红脉兔儿风(肺心草,红脉兔耳风,罗汉草,青毛走马胎,青兔儿风,青兔耳风,土兔儿风,土兔耳风,血筋草,紫背草,紫背金牛,走马丹,走马胎);Rednerve Ainsliaea,Redvein Rabbiten-wind ■
12725 Ainsliaea scabrida Dunn = Ainsliaea yunnanesis Franch. ■
12726 Ainsliaea secundiflora Hayata = Ainsliaea macroclinidioides Hayata var. secundiflora (Hayata) Kitam. ■
12727 Ainsliaea silhetensis (DC.) C. B. Clarke = Ainsliaea latifolia (D. Don) Sch. Bip. ■
12728 Ainsliaea smithii Mattf.;紫枝兔儿风;Purpletwig Ainsliaea,Purpletwig Rabbiten-wind ■
12729 Ainsliaea sparsiflora Vaniot = Ainsliaea pertyoides Franch. ■
12730 Ainsliaea spicata Vaniot;细穗兔儿风(倒卵叶兔儿风,肾炎草,小一枝箭,杏叶兔儿风);Spike Ainsliaea,Spike Rabbiten-wind ■
12731 Ainsliaea spicata Vaniot var. obovata (Franch.) C. Y. Wu;倒卵叶兔儿风(卵叶宽穗兔儿风,卵叶兔儿风,牛尾一枝箭,肾炎草,小一枝箭,杏叶兔耳风,一枝箭,皱叶子);Ovateleaf Ainsliaea ■
12732 Ainsliaea spicata Vaniot var. obovata (Franch.) C. Y. Wu = Ainsliaea spicata Vaniot ■
12733 Ainsliaea sutchuenensis Franch. = Ainsliaea glabra Hemsl. var. sutchuenensis (Franch.) S. E. Freire ■
12734 Ainsliaea tenuicaulis Mattf.;细茎兔儿风;Thinstem Ainsliaea,Thinstem Rabbiten-wind ■
12735 Ainsliaea tenuicaulis Mattf. = Ainsliaea glabra var. sutchuenensis (Franch.) S. E. Freire ■
12736 Ainsliaea tonkinensis Merr. = Ainsliaea angustifolia Hook. f. et Thomson ex C. B. Clarke ■
12737 Ainsliaea triflora (Buch. -Ham. ex D. Don) Druce = Ainsliaea latifolia (D. Don) Sch. Bip. ■
12738 Ainsliaea triflora (Buch. -Ham. ex D. Don) Druce var. obovata (Fr.) C. Y. Wu;倒卵叶三花兔儿风■
12739 Ainsliaea trinervis Y. C. Tseng;三脉兔儿风;Ainsliaea,Threevein Rabbiten-wind ■
12740 Ainsliaea undulata Diels = Ainsliaea henryi Diels ■
12741 Ainsliaea uniflora Sch. Bip.;单花兔儿风■☆
12742 Ainsliaea uniflora Sch. Bip. = Diaspananthus uniflorus (Sch. Bip.) Kitam. ■☆
12743 Ainsliaea walkeri Hook. f.;华南兔儿风(狭叶兔儿风);S. China Rabbiten-wind,Walker's Ainsliaea ■
12744 Ainsliaea yunnanensis Franch.;云南兔儿风(倒吊花,接骨一枝箭,双股箭,铜脚威灵,燕麦灵,羊耳草,云南兔耳风);Yunnan Ainsliaea,Yunnan Rabbiten-wind ■
12745 Ainsliaea yunnanensis Franch. var. macilenta Vaniot ex H. Lév. = Ainsliaea yunnanensis Franch. ■
12746 Ainsliea Kuntze = Ainsliaea DC. ■
12747 Ainsworthia Boiss. (1844);伊独活属■☆
12748 Ainsworthia carmeli Boiss.;伊独活■☆
12749 Ainsworthia cordata Boiss.;心形伊独活■☆
12750 Ainsworthia elegans Boiss. et Balansa;雅致伊独活■☆
12751 Ainsworthia trachycarpa Boiss.;糙果伊独活■☆
12752 Aiolon Lunell(1916);加拿大毛莨属■☆
12753 Aiolon canadense (L.) Nieuwl. et Lunell;加拿大毛莨■☆
12754 Aiolon canadense (L.) Nieuwl. et Lunell f. flavum Lunell;黄加

拿大毛茛■☆
12755 Aiolon canadense Nieuwl. et Lunell = Aiolon canadense (L.) Nieuwl. et Lunell ■☆
12756 Aiolotheca DC. = Zaluzania Pers. ■☆
12757 Aiouea Aubl. (1775);球心樟属●☆
12758 Aiouea angulata Kosterm.;窄球心樟●☆
12759 Aiouea benthamiana Mez;本瑟姆球心樟●☆
12760 Aiouea brasiliensis Meisn.;巴西球心樟●☆
12761 Aiouea densiflora Nees;密花球心樟●☆
12762 Aiouea grandiflora van der Werff;大花球心樟●☆
12763 Aiouea laevis (Mart.) Kosterm.;平滑球心樟●☆
12764 Aiouea longipetiolata van der Werff;长梗球心樟●☆
12765 Aiouea minutiflora Coe-Teix.;微花球心樟●☆
12766 Aiouea multiflora Coe-Teix.;多花球心樟●☆
12767 Aiouea parvissima (Lundell) S. S. Renner;小球心樟●☆
12768 Aiouea rubra A. C. Sm.;红球心樟●☆
12769 Aiouea tomentella (Mez) S. S. Renner;毛球心樟●☆
12770 Aiouea trinervis Meisn.;三脉球心樟●☆
12771 Aiphanes Willd. (1807);急怒棕榈属(刺孔雀椰子属,刺叶榈属,刺叶椰子属,刺叶棕属,刺鱼尾椰属,急怒棕属,马丁棕属);Coyure Palms, Ruffle Palm ●☆
12772 Aiphanes caryotifolia H. Wendl.;急怒棕榈(急怒棕);Caryoleaf Martinezia, Caryota Ruffle Palm, Chonta Palm, Ruffle Palm, Spine Palm ●☆
12773 Aiphanes erosa (Linden) Burret;齿叶急怒棕榈;Erose Martinezia, Macaw Palm ●☆
12774 Aiphanes truncata H. Wendl.;截叶急怒棕榈●☆
12775 Aipyanthus Steven = Arnebia Forssk. ●■
12776 Aira L. (1753)(保留属名);银须草属(埃若禾属,丝草属);Hair Grass, Hairgrass, Hair-grass ■
12777 Aira altaica Trin. = Eremopoa altaica (Trin.) Roshev. ■
12778 Aira altaica Trin. = Poa diaphora Trin. ■
12779 Aira aquatica L. = Catabrosa aquatica (L.) P. Beauv. ■
12780 Aira armoricana F. Albers = Aira caryophyllea L. ■
12781 Aira articulata Desf. = Corynephorus articulatus (Desf.) P. Beauv. ■☆
12782 Aira arundinacea L. = Eragrostis collina Trin. ■
12783 Aira aurea Steud. = Pentaschistis aurea (Steud.) McClean ■☆
12784 Aira bengalensis (Retz.) J. F. Gmel. = Arundo donax L. ●
12785 Aira bicolor Schumach. = Tricholaena monachne (Trin.) Stapf et C. E. Hubb. ■☆
12786 Aira caerulea L. = Molinia caerulea (L.) Moench ■☆
12787 Aira caespitosa L. = Deschampsia caespitosa (L.) P. Beauv. ■
12788 Aira capensis Steud. = Koeleria capensis (Steud.) Nees ■☆
12789 Aira capillaris (Masson) Schltr.;毛细银须草(毛细埃若禾);Annual Hair Grass, Annual Hair-grass, Capillar Hairgrass ■
12790 Aira capillaris (Masson) Schltr. = Aira elegantissima Schur ■☆
12791 Aira capillaris (Masson) Schltr. var. lensaei Coss. et Durieu = Aira elegantissima Schur ■☆
12792 Aira capillaris (Masson) Schltr. var. tenorei Coss. et Durieu = Aira elegantissima Schur ■☆
12793 Aira capillaris host = Aira capillaris (Masson) Schltr. ■
12794 Aira caryophyllea L.;银须草(石竹埃若禾);Silver Hair Grass, Silver Hairgrass, Silver Hair-grass ■
12795 Aira caryophyllea L. subsp. multicaulis (Dumort.) Bonnier et Layens;多茎银须草■☆
12796 Aira caryophyllea L. subsp. uniaristata (Lag. et Rodr.) Maire = Aira uniaristata Lag. et Rodr. ■☆
12797 Aira caryophyllea L. var. biaristata (Emb. et Maire) Maire = Aira uniaristata Lag. et Rodr. ■☆
12798 Aira caryophyllea L. var. cupaniana (Guss.) Coss. et Durieu = Aira cupaniana Guss. ■☆
12799 Aira caryophyllea L. var. edouardii Husn. = Aira caryophyllea L. ■
12800 Aira caryophyllea L. var. latigluma (Steud.) C. E. Hubb. = Aira caryophyllea L. ■
12801 Aira caryophyllea L. var. microstachya Coss. et Durieu = Aira cupaniana Guss. var. divaricata (Salis) Maire et Weiller ■☆
12802 Aira caryophyllea L. var. reverchonii (Murb.) Maire = Aira uniaristata Lag. et Rodr. ■☆
12803 Aira caryophyllea L. var. vulgaris Coss. et Durieu = Aira caryophyllea L. ■
12804 Aira cespitosa L. = Deschampsia cespitosa (L.) P. Beauv. ■
12805 Aira chinensis Retz. = Eriachne pallescens R. Br. ■
12806 Aira cristata L. = Koeleria macrantha (Ledeb.) Schult. ■
12807 Aira cupaniana Guss.;库潘银须草■☆
12808 Aira cupaniana Guss. var. biaristata Parl. = Aira cupaniana Guss. ■☆
12809 Aira cupaniana Guss. var. divaricata (Salis) Maire et Weiller = Aira cupaniana Guss. ■☆
12810 Aira divaricata Pourr. = Corynephorus divaricatus (Pourr.) Breistr. ■☆
12811 Aira elegans Gaudin = Aira elegantissima Schur ■☆
12812 Aira elegans Willd. = Aira capillaris (Masson) Schltr. ■
12813 Aira elegans Willd. ex Gaudin = Aira elegans Willd. ex Kunth ■
12814 Aira elegans Willd. ex Gaudin = Aira elegantissima Schur ■☆
12815 Aira elegans Willd. ex Gaudin subsp. ambigua (Arcang.) Holub = Aira elegantissima Schur subsp. ambigua (Arcang.) Dogan ■☆
12816 Aira elegans Willd. ex Kunth;雅致银须草(美丽银须草,银须草);Annual Hairgrass, Annual Silver Hairgrass, Hair Grass, Hairgrass ■
12817 Aira elegantissima Schur = Aira elegans Willd. ex Kunth ■
12818 Aira elegantissima Schur subsp. ambigua (Arcang.) Dogan;可疑雅致银须草■☆
12819 Aira fioriana Sennen = Aira uniaristata Lag. et Rodr. ■☆
12820 Aira flexuosa L. = Deschampsia flexuosa (L.) Trin. ■
12821 Aira fontqueriana Sennen = Aira cupaniana Guss. ■☆
12822 Aira humilis M. Bieb. = Catabrosella humilis (M. Bieb.) Tzvelev ■
12823 Aira humilis M. Bieb. = Colpodium humile (M. Bieb.) Griseb. ■
12824 Aira indica L.;印度银须草■☆
12825 Aira indica L. = Sacciolepis indica (L.) Chase ■
12826 Aira intermedia Guss. = Aira tenorei Guss. ■☆
12827 Aira kawakamii Hayata = Deschampsia flexuosa (L.) Trin. ■
12828 Aira latigluma Steud. = Aira caryophyllea L. ■
12829 Aira littoralis (Gaudin) Godet = Deschampsia littoralis (Gaudin) Reut. ■
12830 Aira macrantha Ledeb. = Koeleria macrantha (Ledeb.) Schult. ■
12831 Aira melicoides Michx. = Trisetum melicoides (Michx.) Vasey ex Scribn. ■☆
12832 Aira minuta (L.) Loefl. = Molineriella minuta (L.) Rouy ■☆
12833 Aira minuta L. = Periballia minuta (L.) Asch. et Graebn. ■☆
12834 Aira mucronulata Sennen;微凸银须草■☆
12835 Aira multiculmis Dumort. = Aira caryophyllea L. ■
12836 Aira obtusata Michx. = Sphenopholis obtusata (Michx.) Scribn. ■☆

12837 Aira paradoxa Steud.;奇异致银须草■☆
12838 Aira pictigluma Steud. = Pentaschistis pictigluma (Steud.) Pilg. ■☆
12839 Aira praecox L.;丝草(早熟埃若禾);Early Hair Grass, Early Hairgrass, Small Hair-grass, Yellow Hairgrass ■
12840 Aira pulchella (P. Beauv.) Link = Aira tenorei Guss. ■☆
12841 Aira pulchella (P. Beauv.) Link subsp. intermedia (Guss.) Jahand. et Maire = Aira tenorei Guss. ■☆
12842 Aira pulchella (P. Beauv.) Link subsp. tenorei (Guss.) Asch. et Graebn. = Aira tenorei Guss. ■☆
12843 Aira pulchella (P. Beauv.) Link var. semiaristata Godr. = Aira tenorei Guss. ■☆
12844 Aira purpurea Walter = Triplasis purpurea (Walter) Chapm. ■☆
12845 Aira reverchonii Murb. = Aira uniaristata Lag. et Rodr. ■☆
12846 Aira spectabilis Steud.;壮观银须草■☆
12847 Aira spicata L. = Sacciolepis indica (L.) Chase ■
12848 Aira spicata L. = Trisetum spicatum (L.) K. Richt. ■
12849 Aira subspicata L. = Trisetum spicatum (L.) K. Richt. ■
12850 Aira sukatschewii Popl. = Deschampsia cespitosa (L.) P. Beauv. subsp. orientalis Hultén ■
12851 Aira tenorei Guss.;美丽银须草■☆
12852 Aira tenorei Guss. subsp. intermedia (Guss.) Trab. = Aira tenorei Guss. ■☆
12853 Aira tenorei Guss. var. mutica Gren. et Godr. = Aira tenorei Guss. ■☆
12854 Aira tenorei Guss. var. semiaristata Gren. et Godr. = Aira tenorei Guss. ■☆
12855 Aira uniaristata Lag. et Rodr.;单芒银须草■☆
12856 Aira uniaristata Lag. et Rodr. var. biaristata (Emb. et Maire) Lambinon = Aira uniaristata Lag. et Rodr. ■☆
12857 Aira uniaristata Lag. et Rodr. var. incerta (Ces. et al.) Maire = Aira uniaristata Lag. et Rodr. ■☆
12858 Airampoa Frič = Opuntia Mill. ●
12859 Airella (Dumort.) Dumort. (1868);小银须草属■☆
12860 Airella (Dumort.) Dumort. = Aira L. (保留属名)■
12861 Airella Dumort. = Aira L. (保留属名)■
12862 Airella caryophyllea (L.) Dumort.;小银须草■☆
12863 Airidium Steud. = Deschampsia P. Beauv. ■
12864 Airochloa Link = Koeleria Pers. ■
12865 Airochloa alopecurus Nees var. brevifolia ? = Koeleria capensis (Steud.) Nees ■☆
12866 Airochloa cristata (L.) Link = Koeleria macrantha (Ledeb.) Schult. ■
12867 Airochloa uniflora Hochst. = Koeleria capensis (Steud.) Nees ■☆
12868 Airopsis Desv. (1809);类银须草属(拟银须草属)■☆
12869 Airopsis agrostidea DC. = Antinoria agrostidea (DC.) Parl. ■☆
12870 Airopsis aurea (Steud.) Nees = Pentaschistis aurea (Steud.) McClean ■☆
12871 Airopsis globosa (Thore) Desv. = Airopsis tenella (Cav.) Asch. et Graebn. ■☆
12872 Airopsis minuta (Loefl.) Desv. = Molineriella minuta (L.) Rouy ■☆
12873 Airopsis steudelii Nees = Pentaschistis malouinensis (Steud.) Clayton ■☆
12874 Airopsis tenella (Cav.) Asch. et Graebn.;细类银须草■☆
12875 Airosperma K. Schum. et Lauterb. = Airosperma Lauterb. et K. Schum. ■☆
12876 Airosperma Lauterb. et K. Schum. (1900);锤籽草属■☆
12877 Airosperma fuscum S. Moore;褐锤籽草■☆
12878 Airosperma grandifolium Valeton;大叶锤籽草■☆
12879 Airosperma psychotrioides K. Schum. et Lauterb.;九节锤籽草■☆
12880 Airosperma ramuense K. Schum. et Lauterb.;拉姆锤籽草■☆
12881 Airosperma trichotomum (Gillespie) A. C. Sm.;毛锤籽草■☆
12882 Airosperma vanuense S. P. Darwin;瓦努锤籽草■☆
12883 Airyantha Brummitt(1968);锤花豆属(爱丽花豆属)■☆
12884 Airyantha schweinfurthii (Taub.) Brummitt;锤花豆(爱丽花豆)■☆
12885 Airyantha schweinfurthii (Taub.) Brummitt subsp. confusa (Hutch. et Dalziel) Brummitt;铺散锤花豆(爱丽花豆)■☆
12886 Aisandra Airy Shaw = Diploknema Pierre ●
12887 Aisandra Pierre = Diploknema Pierre ●
12888 Aistocaulon Poelln. (1935);无茎番杏属■☆
12889 Aistocaulon Poelln. = Aloinopsis Schwantes ■☆
12890 Aistocaulon Poelln. ex H. Jacobsen = Aistocaulon Poelln. ■☆
12891 Aistocaulon Poelln. ex H. Jacobsen = Nananthus N. E. Br. ■☆
12892 Aistocaulon rosulatum (Kensit) Poelln.;无茎番杏■☆
12893 Aistocaulon rosulatum (Kensit) Poelln. = Aloinopsis rosulata (Kensit) Schwantes ■☆
12894 Aistopetalum Schltr. (1914);隐瓣火把树属(新几内亚火把树属)●☆
12895 Aistopetalum multiflorum Schltr.;多花隐瓣火把树●☆
12896 Aistopetalum tetramerum Kaneh. et Hatus.;隐瓣火把树●☆
12897 Aistopetalum viticoides Schltr.;藤隐瓣火把树●☆
12898 Aitchisonia Hemsl. ex Aitch. (1882);艾茜属☆
12899 Aitchlaonia roses Henri. ex Aitch.;艾茜☆
12900 Aithales Webb et Berthel. = Sedum L. ●■
12901 Aithonium Zipp. ex C. B. Clarke = Rhynchoglossum Blume(保留属名)■
12902 Aititara Endl. = Atitara Juss. ●
12903 Aititara Endl. = Evodia J. R. Forst. et G. Forst. ●
12904 Aitonia Thunb. = Nymania Lindb. ●☆
12905 Aitonia capensis Thunb. = Nymania capensis (Thunb.) Lindb. ●☆
12906 Aitoniaceae (Harvey) Harvey = Meliaceae Juss. (保留科名)●
12907 Aitoniaceae Harvey = Meliaceae Juss. (保留科名)●
12908 Aitoniaceae Harvey et Sond. = Meliaceae Juss. (保留科名)●
12909 Aitoniaceae R. A. Dyer = Meliaceae Juss. (保留科名)●
12910 Aitopsis Raf. = Salvia L. ●■
12911 Aizoaceae Martinov (1820)(保留科名);番杏科;Carpetweed Family, Dew-plant Family, Fig-marigold Family ●■
12912 Aizoaceae Rudolphi = Aizoaceae Martinov(保留科名)●■
12913 Aizoanthemum Dinter = Aizoanthemum Dinter ex Friedrich ■☆
12914 Aizoanthemum Dinter ex Friedrich(1957);隆果番杏属■☆
12915 Aizoanthemum dinteri (Schinz) Friedrich;隆果番杏■☆
12916 Aizoanthemum galenioides (Fenzl ex Sond.) Friedrich;拟小叶番杏■☆
12917 Aizoanthemum hispanicum (L.) H. E. K. Hartmann;西班牙隆果番杏■☆
12918 Aizoanthemum membrumconnectens Dinter ex Friedrich = Aizoanthemum rehmannii (Schinz) H. E. K. Hartmann ■☆
12919 Aizoanthemum mossamedense (Welw. ex Oliv.) Friedrich;莫萨梅迪隆果番杏■☆
12920 Aizoanthemum rehmannii (Schinz) H. E. K. Hartmann;拉赫曼隆果番杏■☆
12921 Aizoanthemum sphingis Dinter ex Friedrich = Aizoanthemum

dinteri (Schinz) Friedrich ■☆
12922 Aizoanthemum stellatum Dinter ex Friedrich = Aizoanthemum dinteri (Schinz) Friedrich ■☆
12923 Aizodraba Fourr. = Draba L. ■
12924 Aizoon Andrews = Sesuvium L. ■
12925 Aizoon Hill = Sempervivum L. ■☆
12926 Aizoon L. (1753);番杏属;Aizoon ■☆
12927 Aizoon angustifolium (Eckl. et Zeyh.) D. Dietr. = Acrosanthes angustifolia Eckl. et Zeyh. ■☆
12928 Aizoon argenteum Eckl. et Zeyh. = Aizoon rigidum L. f. ■☆
12929 Aizoon asbestinum Schltr.;阿斯别斯特番杏■☆
12930 Aizoon canariense L.;加那利长生番杏■☆
12931 Aizoon canariense L. var. denuatum Sond. = Aizoon canariense L. ■☆
12932 Aizoon contaminatum Eckl. et Zeyh. = Galenia secunda (L. f.) Sond. ●☆
12933 Aizoon crystallinum Eckl. et Zeyh. = Galenia crystallina (Eckl. et Zeyh.) Fenzl ●☆
12934 Aizoon dinteri Schinz = Aizoanthemum dinteri (Schinz) Friedrich ■☆
12935 Aizoon elongatum Eckl. et Zeyh. = Galenia secunda (L. f.) Sond. ●☆
12936 Aizoon fistulosum (Eckl. et Zeyh.) D. Dietr. = Acrosanthes anceps (Thunb.) Sond. ■☆
12937 Aizoon fruticosum L. f. = Galenia fruticosa (L. f.) Sond. ●☆
12938 Aizoon galenioides Fenzl ex Sond. = Aizoanthemum galenioides (Fenzl ex Sond.) Friedrich ■☆
12939 Aizoon giessii Friedrich;吉斯番杏■☆
12940 Aizoon glinoides L. f.;星粟草番杏■☆
12941 Aizoon herniarium Rchb. ex Steud. = Galenia herniariifolia (C. Presl) Fenzl ●☆
12942 Aizoon heterophylla Fenzl ex Sond. = Galenia pubescens (Eckl. et Zeyh.) Druce ●☆
12943 Aizoon hirsutum Eckl. et Zeyh. = Aizoon glinoides L. f. ■☆
12944 Aizoon hispanicum L. = Aizoanthemum hispanicum (L.) H. E. K. Hartmann ■☆
12945 Aizoon hispanicum L. var. genuinum Maire = Aizoanthemum hispanicum (L.) H. E. K. Hartmann ■☆
12946 Aizoon hispanicum L. var. minus Maire = Aizoanthemum hispanicum (L.) H. E. K. Hartmann ■☆
12947 Aizoon karooicum Compton;卡鲁番杏■☆
12948 Aizoon membrumconnectens (Dinter ex Friedrich) Adamson = Aizoanthemum rehmannii (Schinz) H. E. K. Hartmann ■☆
12949 Aizoon microphyllum Bartl. = Galenia herniariifolia (C. Presl) Fenzl ●☆
12950 Aizoon mossamedense Welw. ex Oliv. = Aizoanthemum mossamedense (Welw. ex Oliv.) Friedrich ■☆
12951 Aizoon paniculatum L.;圆锥长生番杏■☆
12952 Aizoon papulosum Eckl. et Zeyh. = Galenia papulosa (Eckl. et Zeyh.) Sond. ●☆
12953 Aizoon perfoliatum L. f. = Tetragonia decumbens Mill. ■☆
12954 Aizoon procumbens Crantz = Aizoon canariense L. ■☆
12955 Aizoon propinquum Eckl. et Zeyh. = Galenia secunda (L. f.) Sond. ●☆
12956 Aizoon pubescens Eckl. et Zeyh. = Galenia pubescens (Eckl. et Zeyh.) Druce ●■☆
12957 Aizoon quadrifidum (F. Muell.) F. Muell. = Gunniopsis quadrifida (F. Muell.) Pax ■☆
12958 Aizoon rarum N. E. Br. = Aizoon paniculatum L. ■☆
12959 Aizoon rigidum L. f.;硬长生番杏■☆
12960 Aizoon rigidum L. f. var. angustifolium Sond. = Aizoon rigidum L. f. ■☆
12961 Aizoon rigidum L. f. var. villosum Adamson = Aizoon rigidum L. f. ■☆
12962 Aizoon sarmentosum L. f.;蔓茎番杏■☆
12963 Aizoon sarmentosum L. f. var. hirsutum Eckl. et Zeyh. = Aizoon sarmentosum L. f. ■☆
12964 Aizoon sarmentosum L. f. var. strigosum Eckl. et Zeyh. = Aizoon sarmentosum L. f. ■☆
12965 Aizoon schellenbergii Adamson;谢伦番杏■☆
12966 Aizoon secundum L. f. = Galenia secunda (L. f.) Sond. ●☆
12967 Aizoon sericeum Fenzl ex Sond. = Aizoon rigidum L. f. ■☆
12968 Aizoon sessiliflorum Moench = Aizoanthemum hispanicum (L.) H. E. K. Hartmann ■☆
12969 Aizoon spathulatum Eckl. et Zeyh. = Aizoon canariense L. ■☆
12970 Aizoon squamulosum Eckl. et Zeyh. = Galenia squamulosa (Eckl. et Zeyh.) Fenzl ●☆
12971 Aizoon stellatum Lam. = Aizoon paniculatum L. ■☆
12972 Aizoon teretifolium (Eckl. et Zeyh.) D. Dietr. = Acrosanthes teretifolia Eckl. et Zeyh. ■☆
12973 Aizoon theurkauffii Maire = Mesembryanthemum cryptanthum Hook. f. ■☆
12974 Aizoon tomentosum Lam. = Aizoon paniculatum L. ■☆
12975 Aizoon virgatum Welw. ex Oliv.;细枝番杏■☆
12976 Aizoon zeyheri Sond.;泽赫番杏■☆
12977 Aizopsis Grulich = Phedimus Raf. ■
12978 Aizopsis Grulich = Sedum L. ●■
12979 Aizopsis aizoon (L.) Grulich = Phedimus aizoon (L.) 't Hart ■
12980 Aizopsis ellacombeana (Praeger) P. V. Heath = Sedum ellacombianum Praeger ■
12981 Aizopsis hybrida (L.) Grulich = Phedimus hybridum (L.) 't Hart ■
12982 Aizopsis kamtschatica (Fisch.) Grulich = Phedimus kamtschaticum (Fisch. et C. A. Mey.) 't Hart ■
12983 Aizopsis middendorffiana (Maxim.) Grulich = Phedimus middendorffianum (Maxim.) 't Hart ■
12984 Aizopsis odontophylla (Fröd.) Grulich = Phedimus odontophyllum (Fröd.) 't Hart ■
12985 Aizopsis selskiana (Regel et Maack) Grulich = Phedimus selskianum (Regel et Maack) 't Hart ■
12986 Aizopsis sikokiana (Maxim. ex Makino) Grulich = Phedimus sikokianus (Maxim.) 't Hart ■☆
12987 Aizoum L. = Aizoon L. ■☆
12988 Ajania Poljakov(1955);亚菊属(亚蒿属);Ajania ●■
12989 Ajania achilleoides (Turcz.) Poljakov ex Grubov;蓍状亚菊(蓍状艾菊,蓍状亚菊);Yarrow-like Ajania ■
12990 Ajania adenantha (Diels) Y. Ling et C. Shih;丽江亚菊;Glandflower Ajania,Lijiang Onion,Likiang Onion ■
12991 Ajania alabasica H. C. Fu;内蒙古亚菊;Inner-mongolian Ajania ■
12992 Ajania amphiseriacea (Hand.-Mazz.) C. Shih;灰叶亚菊;Greyleaf Onion ■
12993 Ajania brachyantha C. Shih;短冠亚菊;Shortcorolla Ajania ■
12994 Ajania breviloba (Franch. ex Hand.-Mazz.) Y. Ling et C. Shih;短裂亚菊;Shortlobe Ajania ■

12995 Ajania dentata X. D. Cui = Ajania potaninii (Krasch.) Poljakov ■
12996 Ajania elegantyla (W. W. Sm.) C. Shih;云南亚菊■
12997 Ajania fastigiata (C. Winkl.) Poljakov;新疆亚菊;Sinkiang Onion, Xinjiang Ajania ■
12998 Ajania fruticulosa (Ledeb.) Poljakov;灌木亚菊(灌木艾菊);Shrubby Ajania ●
12999 Ajania globularia (Besser) Poljakov = Artemisia globularia Cham. ex Besser ■☆
13000 Ajania glomerata (Ledeb.) Poljakov = Artemisia glomerata Ledeb. ■☆
13001 Ajania gracilis (Hook. f. et Thomson) Poljakov;纤细亚菊(象泉亚菊)■
13002 Ajania hypoleuca Y. Ling ex C. Shih;下白亚菊;Hypoleuca Ajania ■
13003 Ajania junnanica Poljakov;滇北亚菊■
13004 Ajania khartensis (Dunn) C. Shih;铺散亚菊;Khart Ajania ■
13005 Ajania kokanica (Krasch.) Tzvelev;考卡亚菊■☆
13006 Ajania latifolia C. Shih;宽叶亚菊;Brosdleaf Ajania ■
13007 Ajania manshurica Poljakov;东北亚菊■
13008 Ajania manshurica Poljakov = Ajania variifolia (C. C. Chang) Tzvelev ■
13009 Ajania microphylla Y. Ling = Ajania khartensis (Dunn) C. Shih ■
13010 Ajania myriantha (Bureau et Franch.) Y. Ling ex C. Shih = Ajania myriantha (Franch.) Y. Ling ex C. Shih ■
13011 Ajania myriantha (Franch.) Y. Ling ex C. Shih;多花亚菊(蜂窝菊,千花亚菊);Manyflower Ajania ■
13012 Ajania nematoloba (Hand.-Mazz.) Muldashev = Ajania nematoloba (Hand.-Mazz.) Y. Ling et C. Shih ■
13013 Ajania nematoloba (Hand.-Mazz.) Y. Ling et C. Shih;丝裂亚菊;Threadylobe Ajania ■
13014 Ajania nematoloba (Hand.-Mazz.) Y. Ling et C. Shih var. longiloba Y. Ling = Ajania nematoloba (Hand.-Mazz.) Y. Ling et C. Shih ■
13015 Ajania nitida C. Shih;光苞亚菊;Brightbract Ajania, Glabrousbract Ajania, Shining Ajania ■
13016 Ajania nubigena (Wall.) C. Shih;黄花亚菊;Yellowflower Ajania ■
13017 Ajania pacifica (Nakai) K. Bremer et Humphries = Chrysanthemum pacificum Nakai ■☆
13018 Ajania pallasiana (Fisch. ex Besser) Poljakov;亚菊■
13019 Ajania pallasiana (Fisch. ex Besser) Poljakov = Chrysanthemum pallasianum (Fisch. ex Besser) Kom. ■
13020 Ajania pallasiana (Fisch. ex Besser) Poljakov = Dendranthema pacificum (Nakai) Kitam. ■
13021 Ajania parviflora (Grüning) Y. Ling;束伞亚菊(束状亚菊,小花亚菊);Smallflower Ajania ■
13022 Ajania potaninii (Krasch.) Poljakov;川甘亚菊;Potanin Ajania ■
13023 Ajania przewalskii Poljakov;细裂亚菊(阿拉善亚菊);Przewalsk Ajania ■
13024 Ajania purpurea C. Shih;紫花亚菊;Purple Ajania, Violet Ajania ■
13025 Ajania qiraica C. H. An et Dilixiat;策勒亚菊■
13026 Ajania quercifolia (W. W. Sm.) Y. Ling et C. Shih;栎叶亚菊(山艾叶);Oakleaf Ajania ■
13027 Ajania ramosa (C. C. Chang) C. Shih;分枝亚菊(羽叶亚菊);Branchy Ajania ■
13028 Ajania remotipinna (Hand.-Mazz.) Y. Ling et C. Shih;疏齿亚菊(疏羽亚菊);Sparsepinna Ajania ■
13029 Ajania rupestris (Matsum. et Koidz.) Muldashev = Chrysanthemum rupestre Matsum. et Koidz. ■☆
13030 Ajania salicifolia (Mattf. ex Rehder et Kobuski) Poljakov;柳叶亚菊(柳叶亚菊蒿);Willowleaf Ajania, Willow-leaved Ajania ●
13031 Ajania salicifolia (Mattf.) Poljakov = Ajania salicifolia (Mattf. ex Rehder et Kobuski) Poljakov ●
13032 Ajania scharnhorstii (Regel et Schmalh.) Tzvelev;单头亚菊(单荚亚菊);Scharnhorst Ajania ■
13033 Ajania senjavinensis (Besser) Poljakov = Artemisia senjavinensis Besser ■☆
13034 Ajania sericea C. Shih;密绒亚菊;Densetomentose Ajania ■
13035 Ajania shiwogiku (Kitam.) K. Bremer et Humphries = Chrysanthemum shiwogiku Kitam. ■☆
13036 Ajania sikangensis Y. Ling = Ajania tibetica (Hook. f. et Thomson ex C. B. Clarke) Tzvelev ■
13037 Ajania tenuifolia (Jacq.) Tzvelev;细叶亚菊(细叶菊艾);Thinleaf Ajania ■
13038 Ajania tibetica (Hook. f. et Thomson ex C. B. Clarke) Tzvelev;西藏亚菊(藏艾菊);Tibet Ajania ■
13039 Ajania trifida (Turcz.) Tzvelev = Hippolytia trifida (Turcz.) Poljakov ●
13040 Ajania trilobata Poljakov ex Tzvelev;矮亚菊;Threelobed Ajania ■
13041 Ajania tripinnatisecta Y. Ling et C. Shih;多裂亚菊;Tripinnatisect Ajania ■
13042 Ajania truncata (Hand.-Mazz.) Y. Ling ex C. Shih;深裂亚菊;Truncate Ajania ■
13043 Ajania truncata (Hand.-Mazz.) Y. Ling ex C. Shih = Ajania potaninii (Krasch.) Poljakov ■
13044 Ajania variifolia (C. C. Chang) Tzvelev;异叶亚菊(变叶亚菊,太白艾);Variegateleaf Ajania ■
13045 Ajaniopsis C. Shih(1978);画笔菊属;Ajaniopsis ■★
13046 Ajaniopsis penicilliformis C. Shih;画笔菊;Brushlike Ajaniopsis ■
13047 Ajax Salisb. = Narcissus L. ■
13048 Ajaxia Raf. = Delphinium L. ■
13049 Ajovea Juss. = Aiouea Aubl. ●☆
13050 Ajuea Post et Kuntze = Ajovea Juss. ●☆
13051 Ajuga L. (1753);筋骨草属;Bugle, Bugleweed, Bugle-weed, Carpet Bugle ●■
13052 Ajuga africana (Thunb.) Pers. = Teucrium africanum Thunb. ■☆
13053 Ajuga alba (Gürke) Robyns;白筋骨草■☆
13054 Ajuga albiflora Diels ex Johnst = Ajuga ovalifolia Bureau et Franch. f. albiflora Y. Z. Sun ex C. Y. Wu ■
13055 Ajuga amurica Freyn = Ajuga multiflora Bunge ■
13056 Ajuga argyrea Stapf = Ajuga nipponensis Makino ■
13057 Ajuga boninsimae Maxim.;小笠原筋骨草■☆
13058 Ajuga brachystemon Maxim.;短丝筋骨草;Shortfilament Bugle ■
13059 Ajuga bracteosa Benth. f. alba Gürke = Ajuga alba (Gürke) Robyns ■☆
13060 Ajuga bracteosa Benth. var. alba (Gürke) Engl. = Ajuga alba (Gürke) Robyns ■☆
13061 Ajuga bracteosa Benth. var. canescens (Benth.) Engl. = Ajuga remota Benth. ■
13062 Ajuga bracteosa Benth. var. crenata (Hochst. ex Vatke) Baker = Ajuga remota Benth. ■
13063 Ajuga bracteosa Wall. ex Benth.;九味一枝蒿(地胆草,痢止蒿,赛西林,散血草,有苞筋骨草,止痢蒿);Manybracteole Bugle ■
13064 Ajuga bracteosa Wall. ex Benth. = Ajuga taiwanensis Nakai ex

Murata ■

13065 Ajuga calantha Diels = Ajuga ovalifolia Bureau et Franch. var. calantha (Diels) C. Y. Wu et C. Chen ■

13066 Ajuga calantha Diels ex Limpr. = Ajuga ovalifolia Bureau et Franch. var. calantha (Diels) C. Y. Wu et C. Chen ■

13067 Ajuga calantha Diels f. angustifolia Diels = Ajuga ovalifolia Bureau et Franch. f. angustifolia (Diels) C. Y. Wu et C. Chen ■

13068 Ajuga campylantha Diels;弯花筋骨草(止痢蒿);Bentflower Bugle ■

13069 Ajuga campylanthoides C. Y. Wu et C. Chen;康定筋骨草;Kangding Bugle ■

13070 Ajuga campylanthoides C. Y. Wu et C. Chen var. subacaulis C. Y. Wu et C. Chen;短茎康定筋骨草;Shortstem Bugle ■

13071 Ajuga capensis (Thunb.) Pers. = Teucrium trifidum Retz. ■☆

13072 Ajuga chamaecistus Ging. ex Benth.;矮筋骨草■

13073 Ajuga chamaepitys (L.) Schreb.;松筋骨草;Field Cypress, Forget-me-not, Gout Ivy, Ground Ivy, Ground Pine, Ground-pine, Hemp, Henep, Herb Eve, Herb Ive, Herb Ivy, Sicklewort, Wood Betony, Yellow Bugle ■☆

13074 Ajuga chamaepitys (L.) Schreb. subsp. grandiflora Vis. = Ajuga chamaepitys (L.) Schreb. ■☆

13075 Ajuga chamaepitys (L.) Schreb. subsp. suffrutescens (Willk.) Greuter et Burdet;灌木筋骨草■☆

13076 Ajuga chamaepitys Schreb. = Ajuga chamaepitys (L.) Schreb. ■☆

13077 Ajuga chanetii H. Lév. et Vaniot = Ajuga ciliata Bunge var. chanetii (H. Lév. et Vaniot) C. Y. Wu et C. Chen ■

13078 Ajuga chia Schreb.;奇筋骨草(贾筋骨草,筋骨草茶);L. C. Chia Bugle ■

13079 Ajuga ciliata Bunge;筋骨草(缘毛筋骨草);Chinese Bugle Weed, Ciliate Bugle ■

13080 Ajuga ciliata Bunge f. chanetii (H. Lév. et Vaniot) Kudo = Ajuga ciliata Bunge var. chanetii (H. Lév. et Vaniot) C. Y. Wu et C. Chen ■

13081 Ajuga ciliata Bunge f. pauciflora C. Y. Wu et C. Chen;少花筋骨草■

13082 Ajuga ciliata Bunge f. typica Kudo = Ajuga ciliata Bunge ■

13083 Ajuga ciliata Bunge var. chanetii (H. Lév. et Vaniot) C. Y. Wu et C. Chen;陕甘筋骨草(卵齿筋骨草)■

13084 Ajuga ciliata Bunge var. glabresces Hemsl.;微毛筋骨草■

13085 Ajuga ciliata Bunge var. glabresces Kudo = Ajuga ciliata Bunge var. glabresces Hemsl. ■

13086 Ajuga ciliata Bunge var. hirta C. Y. Wu et C. Chen;长毛筋骨草■

13087 Ajuga ciliata Bunge var. ovatisepala C. Y. Wu et C. Chen;卵齿筋骨草■

13088 Ajuga ciliata Bunge var. villosior A. Gray ex Nakai;长柔毛筋骨草■☆

13089 Ajuga crenata (Hochst. ex Vatke) Chiov. = Ajuga remota Benth. ■

13090 Ajuga decaryana Danguy ex R. A. Clement;德卡里筋骨草■☆

13091 Ajuga decumbens Thunb. ex A. Murray;金疮小草(白调羹,白喉草,白毛串,白毛夏枯草,白头翁,白夏枯草,大叶刀掀草,大叶刀焮草,地龙胆,伏地筋骨草,和胶毒草,活血草,见血青,见血清,筋骨草,苦草,苦地胆,爬爬草,朋花,破血丹,匍伏筋骨草,匍匐筋骨草,青石藤,青鱼胆草,散血草,散血丹,石灰菜,四季春草,四时春,天青地红,透骨消,土犀角,退血草,夏枯草,雪里开花,雪里青,野鹿衔花,叶下红,一盏灯,紫背金盘);Decumbent Bugle ■

13092 Ajuga decumbens Thunb. ex A. Murray = Ajuga nipponensis Makino ■

13093 Ajuga decumbens Thunb. f. albiflora Honda;白花金疮小草■☆

13094 Ajuga decumbens Thunb. f. purpurea Honda;紫金疮小草■☆

13095 Ajuga decumbens Thunb. var. obloncifolia Y. Z. Sun ex C. H. Hu;狭叶金疮小草(倒披针叶筋骨草)■

13096 Ajuga decumbens Thunb. var. pallescens (Maxim.) Hand.-Mazz. = Ajuga nipponensis Makino var. pallescens (Maxim.) C. Y. Wu et C. Chen ■

13097 Ajuga dictyocarpa Hayata;网果筋骨草(秃筋骨草);Netfruit Bugle ■

13098 Ajuga elatior Ohwi = Ajuga dictyocarpa Hayata ■

13099 Ajuga flaccida Baker;柔软筋骨草■☆

13100 Ajuga formosana Hayata = Paraphlomis formosana (Hayata) T. H. Hsieh et T. C. Huang ■

13101 Ajuga formosana Hayata = Paraphlomis gracilis (Hemsl.) Kudo ■

13102 Ajuga forrestii Diels;痢止蒿(白龙须,痢疾草,痢止草,散血草,无名草,止痢蒿);Forrest Bugle ■

13103 Ajuga furcata Link = Craniotome furcata (Link) Kuntze ■

13104 Ajuga genevensis L.;欧洲筋骨草(筋骨草,日内瓦筋骨草,散血草,下草);Blue Bugle, Blue Bugleweed, Bugle Weed, Erect Bugle, Geneva Bugle, Geneva Bugleweed, Jagged Bugle, Standing Bugle, Tufted Bugle ●■☆

13105 Ajuga genevensis L. 'Alba';白花欧洲筋骨草■☆

13106 Ajuga genevensis L. 'Rosea';粉花欧洲筋骨草■☆

13107 Ajuga genevensis L. 'Variegata';花叶欧洲筋骨草■☆

13108 Ajuga genevensis L. = Ajuga multiflora Bunge ■

13109 Ajuga genevensis L. var. pallescens Maxim. = Ajuga nipponensis Makino var. pallescens (Maxim.) C. Y. Wu et C. Chen ■

13110 Ajuga grosse-serrata Franch. et Sav. = Ajuga japonica Miq. ■☆

13111 Ajuga hildebrandtii Briq. = Ajuga robusta Baker ■☆

13112 Ajuga humilis Miq. = Lamium humile (Miq.) Maxim. ■☆

13113 Ajuga humilis Porta = Ajuga iva (L.) Schreb. ■☆

13114 Ajuga incisa Maxim.;裂叶筋骨草■☆

13115 Ajuga incisa Maxim. f. albiflora T. Yamaz.;白花裂叶筋骨草■☆

13116 Ajuga incisa Maxim. f. rosea Honda;粉花裂叶筋骨草■☆

13117 Ajuga integrifolia Buch.-Ham. = Ajuga bracteosa Wall. ex Benth. ■

13118 Ajuga integrifolia Buch.-Ham. var. canescens (Benth.) Cufod. = Ajuga remota Benth. ■

13119 Ajuga iva (L.) Schreb.;伊瓦筋骨草■☆

13120 Ajuga iva (L.) Schreb. subsp. humilis (Porta et Rigo) Sennen et Mauricio = Ajuga iva (L.) Schreb. ■☆

13121 Ajuga iva (L.) Schreb. subsp. pseudoiva (DC.) Briq. = Ajuga iva (L.) Schreb. ■☆

13122 Ajuga iva (L.) Schreb. var. cleistogama (Heldr.) Pamp. = Ajuga iva (L.) Schreb. ■☆

13123 Ajuga iva (L.) Schreb. var. grandiflora Faure et Maire = Ajuga iva (L.) Schreb. ■☆

13124 Ajuga iva (L.) Schreb. var. pseudoiva (DC.) Benth. = Ajuga iva (L.) Schreb. ■☆

13125 Ajuga iva (L.) Schreb. var. robertiana Maire = Ajuga iva (L.) Schreb. ■☆

13126 Ajuga japonica Miq.;日本筋骨草■☆

13127 Ajuga japonica Miq. f. albiflora Honda;白花日本筋骨草■☆

13128 Ajuga labordei Vaniot = Ajuga nipponensis Makino ■

13129 Ajuga lanosa Y. Z. Sun = Ajuga multiflora Bunge ■

13130 Ajuga laxmannii Benth.;拉氏筋骨草(拉曼筋骨草);Laxmann Bugle ■☆

13131 Ajuga linearifolia Pamp.;线叶筋骨草;Linearleaf Bugle ■

13132 Ajuga lobata D. Don;匍枝筋骨草(裂叶筋骨草,毛盖缘);

Lobed Bugle ■

13133 Ajuga lupulina Maxim.;白苞筋骨草(白毛夏枯草,大苞筋骨草,忽布筋骨草,基独,轮花筋骨草,甜格缩缩草,西藏筋骨草);Whitebracteole Bugle ■

13134 Ajuga lupulina Maxim. f. breviflora Y. Z. Sun ex G. H. Hu;短花白苞筋骨草;Dwarf Whitebracteole Bugle ■

13135 Ajuga lupulina Maxim. f. humilis Y. Z. Sun ex G. H. Hu;矮小白苞筋骨草;Shortflower Whitebracteole Bugle ■

13136 Ajuga lupulina Maxim. var. major Diels;齿苞筋骨草(森林氽模);Toothbract Whitebracteole Bugle ■

13137 Ajuga macrosperma Wall. ex Benth.;大籽筋骨草(散血草);Bigseed Bugle ■

13138 Ajuga macrosperma Wall. ex Benth. var. thomsonii (Maxim.) Hook. f.;无毛大籽筋骨草;Hairless Bigseed Bugle ■

13139 Ajuga mairei H. Lév. = Ajuga forrestii Diels ■

13140 Ajuga makinoi Nakai;牧野氏筋骨草■☆

13141 Ajuga matsumurana Kudo = Ajuga nipponensis Makino ■

13142 Ajuga mixta Makino;混乱筋骨草(大筋骨草)■☆

13143 Ajuga multiflora Bunge;多花筋骨草;Manyflower Bugle ■

13144 Ajuga multiflora Bunge var. brevispicata C. Y. Wu et C. Chen;短穗多花筋骨草■

13145 Ajuga multiflora Bunge var. serotina Kitag.;莲座多花筋骨草■

13146 Ajuga nipponensis Makino;紫背金盘(白毛筋骨草,白毛夏枯草,白头翁,地龙胆,见血青,筋骨草,苦草,苦地胆,破血丹,日本筋骨草,散血草,散血丹,散瘀草,石灰菜,退血草);Japan Bugle, Japanese Bugle ■

13147 Ajuga nipponensis Makino f. nivea Hiyama;白花紫背金盘■☆

13148 Ajuga nipponensis Makino var. pallescens (Maxim.) C. Y. Wu et C. Chen;矮紫背金盘(矮生紫背金盘)■

13149 Ajuga nubigena Diels;高山筋骨草;Alpine Bugle ■

13150 Ajuga oblongata M. Bieb.;长圆筋骨草;Oblong Bugle ■☆

13151 Ajuga oocephala Baker;卵头筋骨草■☆

13152 Ajuga ophrydis Burch. ex Benth.;眉兰筋骨草■☆

13153 Ajuga orientalis L.;东方筋骨草;Orientale Bugle ■☆

13154 Ajuga ovalifolia Bureau et Franch.;圆叶筋骨草;Ovalleaf Bugle, Roundleaf Bugle ■

13155 Ajuga ovalifolia Bureau et Franch. f. albiflora Y. Z. Sun ex C. Y. Wu;白花圆叶筋骨草;Whiteflower Roundleaf Bugle ■

13156 Ajuga ovalifolia Bureau et Franch. f. angustifolia (Diels) C. Y. Wu et C. Chen;狭叶圆叶筋骨草■

13157 Ajuga ovalifolia Bureau et Franch. var. angustifolia (Diels) Hand.-Mazz. = Ajuga ovalifolia Bureau et Franch. f. angustifolia (Diels) C. Y. Wu et C. Chen ■

13158 Ajuga ovalifolia Bureau et Franch. var. calantha (Diels) C. Y. Wu et C. Chen;美花圆叶筋骨草■

13159 Ajuga ovalifolia Bureau et Franch. var. calantha (Diels) C. Y. Wu et C. Chen f. albiflora Y. Z. Sun ex C. Y. Wu et C. Chen = Ajuga ovalifolia Bureau et Franch. var. calantha (Diels) C. Y. Wu et C. Chen ■

13160 Ajuga pachyrrhiza Kitag. = Ajuga linearifolia Pamp. ■

13161 Ajuga pallescens (Maxim.) Nakai = Ajuga shikotanensis Miyabe et Tatew. ■☆

13162 Ajuga pallescens (Maxim.) Price ? et Metcalf = Ajuga nipponensis Makino ■

13163 Ajuga pantantha Hand.-Mazz.;散瘀草(胆草,苦草,散血草,山苦草);Mountain Bugle ■

13164 Ajuga parviflora Benth.;小花筋骨草;Smallflower Bugle ■

13165 Ajuga pseudochia Des.-Shost.;拟奇筋骨草(拟贾筋骨草);False Chia Bugle ■

13166 Ajuga pygmaea A. Gray;矮金疮草(矮筋骨草,台湾筋骨草);Taiwan Bugle ■

13167 Ajuga pyramidalis L.;金字塔筋骨草(塔形筋骨草);Limestone Bugle, Pyramidal Ajuga, Pyramidal Bugle ■☆

13168 Ajuga pyramidalis L. 'Metallica Crispa';金泽皱叶塔形筋骨草■☆

13169 Ajuga remota Benth. = Ajuga brachystemon Maxim. ■

13170 Ajuga remota Benth. = Ajuga bracteosa Wall. ex Benth. ■

13171 Ajuga remota Benth. var. canescens ? = Ajuga remota Benth. ■

13172 Ajuga remota Benth. var. crenata Hochst. ex Vatke = Ajuga remota Benth. ■

13173 Ajuga repens L. = Ajuga macrosperma Wall. ex Benth. ■

13174 Ajuga reptans L.;匍匐筋骨草(匍匐根筋骨草);Ajuga, Ajuga 'Bronze Beauty', Ajuga 'Burgundy Glow', Baby's Rattle, Baby's Shoes, Blind Man's Hand, Brown Bugle, Bugle, Bugleweed, Carpenter's Herb, Carpet Bugle, Carpet Bugleweed, Common Bugle, Consound, Creeping Bugle, Creeping Bugleweed, Cuckoo, Dead Man's Bellows, European Bugle, Herb Carpenter, Herb-flower, Horse Peppermint, Horse-and-hounds, Middle Comfrey, Middle Consound, Nelson's Bugle, Prostrate Bugle, Sanicle, Self-heal, Sicklewort, Thunder Clover, Thunder-and-lightning, Wild Mint, Wood Betony ■☆

13175 Ajuga reptans L. 'Atropurpurea';暗紫匍匐筋骨草■☆

13176 Ajuga reptans L. 'Jungle Beauty';丛林丽匍匐筋骨错误！超链接引用无效。草■☆

13177 Ajuga reptans L. 'Multicolor';多色匍匐筋骨草■☆

13178 Ajuga reptans L. 'Rainbow' = Ajuga reptans L. 'Multicolor' ■☆

13179 Ajuga reptans L. var. japonica ? = Ajuga shikotanensis Miyabe et Tatew. ■☆

13180 Ajuga robusta Baker;粗壮筋骨草■☆

13181 Ajuga salicifolia (L.) Schreb.;柳叶筋骨草;Willowleaf Bugle ■

13182 Ajuga salzmannii (P. W. Ball) G. López et R. Morales;萨尔筋骨草■

13183 Ajuga sciaphila W. W. Sm.;喜荫筋骨草;Shadeloving Bugle ■

13184 Ajuga shikotanensis Miyabe et Tatew.;齿舞岛筋骨草■☆

13185 Ajuga shikotanensis Miyabe et Tatew. f. hirsuta (Honda) Murata;毛齿舞岛筋骨草■☆

13186 Ajuga shikotanensis Miyabe et Tatew. var. hirsuta (Honda) H. Hara = Ajuga shikotanensis Miyabe et Tatew. f. hirsuta (Honda) Murata ■☆

13187 Ajuga taiwanensis Nakai ex Murata;台湾筋骨草■

13188 Ajuga thomsoni Maxim. = Ajuga macrosperma Wall. ex Benth. var. thomsonii (Maxim.) Hook. f. ■

13189 Ajuga tsukubana (Nakai) Okuyama = Ajuga yesoensis Maxim. ex Franch. et Sav. var. tsukubana Nakai ■☆

13190 Ajuga turkestanica (Regel) Briq.;新疆筋骨草(土耳其斯坦筋骨草);Sinkiang Bugle, Xinjiang Bugle ■

13191 Ajuga yesoensis Maxim. ex Franch. et Sav.;虾夷筋骨草■☆

13192 Ajuga yesoensis Maxim. ex Franch. et Sav. f. albiflora Honda;白花虾夷筋骨草■☆

13193 Ajuga yesoensis Maxim. ex Franch. et Sav. var. tsukubana Nakai;筑波筋骨草■☆

13194 Ajuga yesoensis Maxim. ex Franch. et Sav. var. tsukubana Nakai f. alba Sugim.;白花筑波筋骨草■☆

13195 Ajugaceae Döll = Labiatae Juss.(保留科名)●■

13196 Ajugaceae Döll = Lamiaceae Martinov(保留科名)●■

13197 Ajugoides Makino(1915);拟筋骨草属■☆
13198 Ajugoides humilis (Miq.) Makino;拟筋骨草■☆
13199 Ajuvea Steud. = Aiouea Aubl. ●☆
13200 Akakia Adans. = Acacia Mill.(保留属名)●■
13201 Akania Hook. f. (1862);叠珠树属●☆
13202 Akania bidwillii (Hogg) Mabb.; 叠珠树; Turnip Wood, Turnipwood ●☆
13203 Akania hillii Hook. f.;希尔叠珠树●☆
13204 Akaniaceae Stapf(1912)(保留科名);叠珠树科●☆
13205 Akea Stokes = Blighia K. König ●☆
13206 Akeassia J. -P. Lebrun et Stork(1993);锥托田基黄属■☆
13207 Akeassia grangeoides J. -P. Lebrun et Stork;锥托田基黄☆
13208 Akebia Decne. (1837);木通属;Akebia ●
13209 Akebia cavaleriei H. Lév. = Stauntonia obovata Hemsl. ●
13210 Akebia chaffanjonii H. Lév. = Akebia trifoliata (Thunb.) Koidz. var. australis (Diels) T. Shimizu ●
13211 Akebia chingshuiensis Shimizu;清水山木通;Qingshui Akebia ●
13212 Akebia chingshuiensis Shimizu = Akebia trifoliata (Thunb.) Koidz. var. australis (Diels) T. Shimizu ●
13213 Akebia clematifolia Siebold et Zucc. = Akebia trifoliata (Thunb.) Koidz. ●
13214 Akebia lobata Decne. = Akebia trifoliata (Thunb.) Koidz. ●
13215 Akebia lobata Decne. var. australis Diels = Akebia trifoliata (Thunb.) Koidz. var. australis (Diels) T. Shimizu ●
13216 Akebia lobata Decne. var. chaffanjonii (H. Lév.) H. Lév. = Akebia trifoliata (Thunb.) Koidz. var. australis (Diels) T. Shimizu ●
13217 Akebia lobata Decne. var. clematifolia (Siebold et Zucc.) Ito = Akebia trifoliata (Thunb.) Koidz. ●
13218 Akebia lobata Decne. var. quercifolia (Siebold et Zucc.) Ito = Akebia trifoliata (Thunb.) Koidz. ●
13219 Akebia longeracemosa Matsum.;长序木通(台湾木通,台湾野木瓜,五叶长穗木通);Long-racemed Akebia, Taiwan Five-leaf Akebia ●
13220 Akebia micrantha Nakai = Akebia quinata (Thunb.) Decne. ●
13221 Akebia pentaphylla Makino;五叶木通●
13222 Akebia pentaphylla Makino var. integrifolia Y. Kimura;全缘五叶木通●☆
13223 Akebia quercifolia Siebold et Zucc. = Akebia trifoliata (Thunb.) Koidz. ●
13224 Akebia quinata (Houtt.) Decne. = Akebia quinata (Thunb.) Decne. ●
13225 Akebia quinata (Thunb.) Decne.;木通(八月瓜,八月栌,八月扎,八月炸,八月炸藤,丁年藤,丁翁,多叶木通,桴棪子,桴棪子,葍藤茎,附通子,附支,葍子,狗腰藤,海风藤,盍合子,活血藤,腊瓜,冷饭包,木通子,木蓪,拿子,山黄瓜,山通草,圣先子,圣知子,通草,蓪木,万年藤,望子,乌覆子,五风藤,五拿绳,五叶木通,狭叶八月瓜,仙沼子,畜葍子,压惊子,燕葍子,燕覆子,羊开口,野毛蛋,野木瓜,野香蕉,玉支子,预知子);Akebia, Chocolate Vine, Chocolate-vine, Five-leaf, Fiveleaf Akebia, Five-leaf Akebia, Polyleaf Akebia, Quante Akebia ●
13226 Akebia quinata (Thunb.) Decne. f. diplochlamys (Makino) T. Shimizu;双被木通●
13227 Akebia quinata (Thunb.) Decne. f. diplochlamys (Makino) T. Shimizu = Akebia quinata (Thunb.) Decne. ●
13228 Akebia quinata (Thunb.) Decne. f. polyphylla (Nakai) Hiyama = Akebia quinata (Thunb.) Decne. ●
13229 Akebia quinata (Thunb.) Decne. f. viridiflora Makino = Akebia quinata (Thunb.) Decne. ●
13230 Akebia quinata (Thunb.) Decne. var. diplochlamys Makino = Akebia quinata (Thunb.) Decne. ●
13231 Akebia quinata (Thunb.) Decne. var. leucantha Nakai;白花木通●☆
13232 Akebia quinata (Thunb.) Decne. var. longeracemosa Rehder et E. H. Wilson = Akebia longeracemosa Matsum. ●
13233 Akebia quinata (Thunb.) Decne. var. polyphylla Nakai = Akebia quinata (Thunb.) Decne. ●
13234 Akebia quinata (Thunb.) Decne. var. retusa Chun;钝叶木通;Obtuseleaf Fiveleaf Akebia ●
13235 Akebia quinata (Thunb.) Decne. var. yiehii W. C. Cheng;绿花木通;Greenflower Akebia ●
13236 Akebia quinata (Thunb.) Decne. var. yiehii W. C. Cheng = Akebia quinata (Thunb.) Decne. ●
13237 Akebia sempervirens Nakai = Akebia trifoliata (Thunb.) Koidz. ●
13238 Akebia trifoliata (Thunb.) Koidz.;三叶木通(八月瓜藤,八月栌,八月楂,八月札,八月炸,爆肚拿,葍藤,丁父,丁翁,附支,活血藤,木通,拿藤,三叶拿藤,三叶通草,甜果木通,通草,万年,万年藤,王翁,乌覆,燕覆,预知子);Threeleaf Akebia, Trifoliate Akebia ●
13239 Akebia trifoliata (Thunb.) Koidz. subsp. australis (Diels) T. Shimizu var. honanensis T. Shimizu = Akebia trifoliata (Thunb.) Koidz. subsp. australis (Diels) T. Shimizu ●
13240 Akebia trifoliata (Thunb.) Koidz. subsp. australis (Diels) T. Shimizu;白木通(八月瓜藤,八月札,八月炸,六角楂,青防己,青木香,清水木通,三叶木通,羊腰子果);Austral Akebia, Qingshui Akebia, Three-foliate Akebia ●
13241 Akebia trifoliata (Thunb.) Koidz. subsp. longisepala H. N. Qin;长萼三叶木通;Long-sepaled Threeleaf Akebia ●
13242 Akebia trifoliata (Thunb.) Koidz. var. australis (Diels) Rehder = Akebia trifoliata (Thunb.) Koidz. subsp. australis (Diels) T. Shimizu ●
13243 Akebia trifoliata (Thunb.) Koidz. var. australis (Diels) T. Shimizu = Akebia trifoliata (Thunb.) Koidz. subsp. australis (Diels) T. Shimizu ●
13244 Akebia trifoliata (Thunb.) Koidz. var. clematifolia (Siebold et Zucc.) Nakai = Akebia trifoliata (Thunb.) Koidz. ●
13245 Akebia trifoliata (Thunb.) Koidz. var. honanensis T. Shimizu = Akebia trifoliata (Thunb.) Koidz. ●
13246 Akebia trifoliata (Thunb.) Koidz. var. integrifolia T. Shimizu = Akebia trifoliata (Thunb.) Koidz. var. australis (Diels) T. Shimizu ●
13247 Akeesia Tussac = Blighia K. König ●☆
13248 Akeesia Tussac = Borzicactus Riccob. ■☆
13249 Akentra Benj. = Utricularia L. ■
13250 Aker Raf. = Acer L. ●
13251 Akersia Buining = Cleistocactus Lem. ●☆
13252 Akersia Buining(1961);秘鲁仙人掌属■☆
13253 Akersia roseiflora Buining;秘鲁仙人掌■☆
13254 Akesia Tussac = Blighia K. König ●☆
13255 Aklema Raf. = Euphorbia L. ●■
13256 Akrosida P. A. Fuertes et Fuertes(1992);巴西大叶锦葵属●☆
13257 Akrosida macrophylla (Ulbr.) P. A. Fuertes et Fuertes;巴西大叶锦葵●☆
13258 Akschindlium H. Ohashi = Desmodium Desv.(保留属名)●■
13259 Akylopsis Lehm. = Lepidotheca Nutt. ■
13260 Akylopsis Lehm. = Matricaria L. ■

13261 Ala Szlach. (1995);阿拉兰属■☆
13262 Alabella Comm. ex Baill. = Turraea L. ●
13263 Alacosperma Neck. ex Raf. = Cryptotaenia DC. (保留属名)■
13264 Alacospermum Neck. = Cryptotaenia DC. (保留属名)■
13265 Aladenia Pichon = Farquharia Stapf ●☆
13266 Aladenia jasminiflora (Hutch. et Dalziel) Pichon = Farquharia elliptica Stapf ●☆
13267 Alafia Thouars(1806);热非夹竹桃属●☆
13268 Alafia alba Pichon;白花热非夹竹桃●☆
13269 Alafia barteri Oliv.;巴尔热非夹竹桃●☆
13270 Alafia benthamii (Baill.) Stapf;本瑟姆热非夹竹桃●☆
13271 Alafia benthamii (Baill.) Stapf var. mediafra Pichon;中非热非夹竹桃●☆
13272 Alafia bequaertii De Wild. = Alafia schumannii Stapf ●☆
13273 Alafia butayei Stapf = Alafia caudata Stapf ●☆
13274 Alafia calophylla Pichon;美叶热非夹竹桃●☆
13275 Alafia caudata Stapf;尾状热非夹竹桃●☆
13276 Alafia caudata Stapf subsp. latiloba Kupicha;宽裂尾状热非夹竹桃●☆
13277 Alafia clusioides S. Moore = Alafia microstylis K. Schum. ●☆
13278 Alafia congolana Pichon;刚果热非夹竹桃●☆
13279 Alafia conica Pichon;圆锥热非夹竹桃●☆
13280 Alafia cuneata Stapf = Alafia lucida Stapf ●☆
13281 Alafia erythrophthalma (K. Schum.) Leeuwenb.;淡红热非夹竹桃●☆
13282 Alafia falcata Leeuwenb.;镰形热非夹竹桃●☆
13283 Alafia fuscata Pichon;暗棕色热非夹竹桃●☆
13284 Alafia giraudii Dubard = Alafia barteri Oliv. ●☆
13285 Alafia giraultii Dubard ex Pichon = Alafia barteri Oliv. ●☆
13286 Alafia glabriflora Pichon;光花热非夹竹桃●☆
13287 Alafia gracilis Stapf = Alafia erythrophthalma (K. Schum.) Leeuwenb. ●☆
13288 Alafia grandis Stapf = Alafia erythrophthalma (K. Schum.) Leeuwenb. ●☆
13289 Alafia insularis Pichon = Alafia perrieri Jum. ●☆
13290 Alafia intermedia Pichon = Alafia perrieri Jum. ●☆
13291 Alafia jasminiflora A. Chev. = Farquharia elliptica Stapf ●☆
13292 Alafia landolphioides (A. DC.) K. Schum. = Alafia scandens (Thonn.) De Wild. ●☆
13293 Alafia laxiflora Pierre ex Pichon = Alafia lucida Stapf ●☆
13294 Alafia lucida Stapf;光亮热非夹竹桃●☆
13295 Alafia major Stapf = Alafia lucida Stapf ●☆
13296 Alafia malouetioides K. Schum. = Alafia multiflora (Stapf) Stapf ●☆
13297 Alafia microstylis K. Schum.;小柱热非夹竹桃●☆
13298 Alafia mildbraedii Gilg et Stapf = Malouetia mildbraedii (Gilg et Stapf) J. Ploeg ●☆
13299 Alafia mirabilis A. Chev. = Farquharia elliptica Stapf ●☆
13300 Alafia multiflora (Stapf) Stapf;多花热非夹竹桃●☆
13301 Alafia nigrescens Pichon;黑热非夹竹桃●☆
13302 Alafia orientalis K. Schum. ex De Wild.;东方热非夹竹桃●☆
13303 Alafia parciflora Stapf;小花热非夹竹桃●☆
13304 Alafia parvifolia Pichon = Alafia perrieri Jum. ●☆
13305 Alafia pauciflora Radlk.;少花热非夹竹桃●☆
13306 Alafia perrieri Jum.;小叶热非夹竹桃●☆
13307 Alafia perrieri var. parvifolia (Pichon) Markgr. = Alafia perrieri Jum. ●☆
13308 Alafia reticulata K. Schum. = Alafia lucida Stapf ●☆
13309 Alafia sarmentosa Stapf = Alafia caudata Stapf ●☆
13310 Alafia scandens (Thonn.) De Wild.;攀缘热非夹竹桃●☆
13311 Alafia schumannii Stapf;舒曼热非夹竹桃●☆
13312 Alafia swynnertonii S. Moore = Alafia microstylis K. Schum. ●☆
13313 Alafia thouarsii Roem. et Schult.;图氏热非夹竹桃●☆
13314 Alafia ugandensis Pichon;乌干达热非夹竹桃●☆
13315 Alafia velutina Leeuwenb.;短绒毛热非夹竹桃●☆
13316 Alafia vermeulenii De Wild. = Cyclocotyla congolensis Stapf ●☆
13317 Alafia verschuerenii De Wild.;费许伦热非夹竹桃●☆
13318 Alafia whytei Stapf;怀特热非夹竹桃●☆
13319 Alafia zambesiaca Kupicha;赞比西热非夹竹桃●☆
13320 Alagophyla Raf. (废弃属名) = Gesneria L. ●☆
13321 Alagophyla Raf. (废弃属名) = Rechsteineria Regel(保留属名)■☆
13322 Alagophyla Raf. (废弃属名) = Sinningia Nees ●■☆
13323 Alagophylla B. D. Jacks. = Alagophyla Raf. (废弃属名)■☆
13324 Alagophylla Raf. = Gesneria L. ●☆
13325 Alagoptera Mart. = Allagoptera Nees ●☆
13326 Alaida Dvorák = Dimorphostemon Kitag. ■
13327 Alaida Dvorák = Dontostemon Andrz. ex C. A. Mey. (保留属名)■
13328 Alaida glandulosa (Kar. et Kir.) Dvorák = Dimorphostemon glandulosus (Kar. et Kir.) Golubk. ■
13329 Alaida glandulosa (Kar. et Kir.) Dvorák = Dontostemon glandulosus (Kar. et Kir.) O. E. Schulz ■
13330 Alaida pectinata (DC.) Dvorák = Dimorphostemon pinnatus (Pers.) Kitag. ■
13331 Alaida pectinata (DC.) Dvorák = Dontostemon pinnatifidus (Willd.) Al-Shehbaz et H. Ohba ■
13332 Alainanthe (Fenzl) Rchb. = Minuartia L. ■
13333 Alairia Kuntze = Mairia Nees ■☆
13334 Alajja Ikonn. (1971);菱叶元宝草属;Alajja ■
13335 Alajja anomala (Juz.) Ikonn.;异叶元宝草;Diverseleaf Alajja ■
13336 Alajja rhomboidea (Benth.) Ikonn.;菱叶元宝草;Rhomboidleaf Alajja ■
13337 Alalantia Corr. = Atalantia Corrêa(保留属名)●
13338 Alamania La Llave et Lex. = Alamania Lex. ■☆
13339 Alamania Lex. (1824);阿拉曼兰属(亚兰属)■☆
13340 Alamania punicea La Llave et Lex.;阿拉曼兰■☆
13341 Alamannia Lindl. = Alamania Lex. ■☆
13342 Alandina Neck. = Moringa Rheede ex Adans. ●
13343 Alangiaceae DC. (1828)(保留科名);八角枫科;Alangium Family ●
13344 Alangiaceae DC. (保留科名) = Cornaceae Bercht. et J. Presl(保留科名)●■
13345 Alangium Lam. (1783)(保留属名);八角枫属;Alangium ●
13346 Alangium alpinum (C. B. Clarke) W. W. Sm. et Cave;高山八角枫;Alps Alangium, Mountain Alangium ●
13347 Alangium barbatum (R. Br.) Baill. ex Kuntze;髯毛八角枫(伪八角枫);Barbate Alangium, Barbed Alangium ●
13348 Alangium barbatum (R. Br.) Baill. ex Kuntze subsp. faberi (Oliv.) Bloemb. = Alangium faberi Oliv. ●
13349 Alangium begoniifolium (Roxb.) Baill. = Alangium chinense (Lour.) Harms ●
13350 Alangium begoniifolium (Roxb.) Baill. = Alangium premnifolium Ohwi ●
13351 Alangium begoniifolium (Roxb.) Baill. subsp. tomentosum var.

typicum Wangerin = Alangium kurzii Craib ●
13352 Alangium begoniifolium (Roxb.) Baill. var. eubegoniifolium Wangerin = Alangium chinense (Lour.) Harms ●
13353 Alangium chinense (Lour.) Harms;八角枫(八角柴,八角将军,八角金盘,八角王,八角梧桐,八筋条,白尖子,白金条,白筋条,白腊金,白龙须,白绵条,包子树,鹅脚板,二珠葫芦,割舌罗,割云罗,勾儿茶,花冠木,华八角枫,华瓜木,老龙须,六角金盘,麻桐树,木八角,牛尾巴花,七角枫,山霸王,山药萸,水芒树,万字路,野罗桐,猪耳桐药);China Alangium,Chinese Alangium ●
13354 Alangium chinense (Lour.) Harms subsp. pauciflorum W. P. Fang;稀花八角枫;Fewflower Alangium ●
13355 Alangium chinense (Lour.) Harms subsp. strigosum W. P. Fang;伏毛八角枫;Strigose Alangium,Triangular Alangium ●
13356 Alangium chinense (Lour.) Harms subsp. triangulare (Wangerin) W. P. Fang;深裂八角枫;Triangular Alangium ●
13357 Alangium chinense (Lour.) Rehder = Alangium chinense (Lour.) Harms ●
13358 Alangium chinense (Lour.) Rehder var. taiwanianum (Masam.) Koidz. = Alangium chinense (Lour.) Harms ●
13359 Alangium chinense (Lour.) Rehder var. tomentosum (Blume) Merr. = Alangium kurzii Craib ●
13360 Alangium chungii H. L. Li = Alangium kurzii Craib var. laxifolium (Y. C. Wu) W. P. Fang ●
13361 Alangium decapetalum Lam. = Alangium salviifolium (L. f.) Wangerin ●
13362 Alangium faberi Oliv.;小花八角枫(西南八角枫);Faber Alangium,Little Flower Alangium,Littleflower Alangium ●
13363 Alangium faberi Oliv. var. dolicarpum Z. Y. Li;长果八角枫;Longcarp Little Flower Alangium ●
13364 Alangium faberi Oliv. var. heterophyllum Yen C. Yang;异叶八角枫;Diverseleaf Alangium,Diversisifolious Alangium ●
13365 Alangium faberi Oliv. var. perforatum (H. Lév.) Rehder;小叶八角枫;Littleleaf Alangium ●
13366 Alangium faberi Oliv. var. platyphyllum Chun et F. C. How;宽叶八角枫(阔叶八角枫);Broadleaf Alangium ●
13367 Alangium handelii Schnarf = Alangium kurzii Craib var. handelii (Schnarf) W. P. Fang ●
13368 Alangium kenyense Chiov.;肯尼亚八角枫●☆
13369 Alangium kurzii Craib;毛八角枫(长毛八角枫,毛瓜木);Kurz Alangium ●
13370 Alangium kurzii Craib var. handelii (Schnarf) W. P. Fang;云山八角枫(大花八角枫);Handel Alangium ●
13371 Alangium kurzii Craib var. laxifolium (Y. C. Wu) W. P. Fang;疏叶八角枫;Laxleaf Alangium ●
13372 Alangium kurzii Craib var. pachyphyllum W. P. Fang et H. Y. Su;厚叶八角枫;Thickleaf Alangium ●
13373 Alangium kurzii Craib var. umbellatum (Yen C. Yang) W. P. Fang;伞形八角枫;Umbellate Alangium ●
13374 Alangium kwangsiense Melch.;广西八角枫;Guangxi Alangium,Kwangsi Alangium ●
13375 Alangium lamarckii Thwaites = Alangium salviifolium (L. f.) Wangerin ●
13376 Alangium platanifolium (Siebold et Zucc.) Harms;瓜木(八角枫,八筋条,白锦条,华瓜木,麻桐树,山茱萸,水桃,筱悬叶瓜木,岩桐,猪耳桐);Planeleaf Alangium,Plane-leaved Alangium ●
13377 Alangium platanifolium (Siebold et Zucc.) Harms var. laxifolium Y. C. Wu = Alangium kurzii Craib var. laxifolium (Y. C. Wu) W. P. Fang ●
13378 Alangium platanifolium (Siebold et Zucc.) Harms var. macrophyllum Wang.;大叶瓜木●
13379 Alangium platanifolium (Siebold et Zucc.) Harms var. trilobatum (Miq.) Ohwi f. velutinum (Nakai) T. B. Lee;短毛三裂八角枫●☆
13380 Alangium platanifolium (Siebold et Zucc.) Harms var. trilobum (Miq.) Ohwi;三裂瓜木(假瓢子,灵角,三裂八角枫)●
13381 Alangium premnifolium Ohwi;日本八角枫(豆腐柴叶八角枫)●
13382 Alangium qingchuanense M. Y. He;青川八角枫●
13383 Alangium rotundifolium (Hassk.) Bloemb. var. laxifolium Y. C. Wu = Alangium kurzii Craib var. laxifolium (Y. C. Wu) W. P. Fang ●
13384 Alangium salviifolium (L. f.) Wangerin;土坛树(割舌罗,印度八角枫);Angola Alangium,Salvialeaf Alangium ●
13385 Alangium salviifolium (L. f.) Wangerin subsp. decapetalum (Lam.) Wangerin = Alangium salviifolium (L. f.) Wangerin ●
13386 Alangium shweliense W. W. Sm. = Nyssa shweliensie (W. W. Sm.) Airy Shaw ●
13387 Alangium taiwanianum Masam. = Alangium chinense (Lour.) Harms ●
13388 Alangium tetrandrum R. H. Miao;四蕊八角枫;Fourstamen Alangium ●
13389 Alangium tomentosum Lam. = Alangium salviifolium (L. f.) Wangerin ●
13390 Alangium umbellatum Yen C. Yang = Alangium kurzii Craib var. umbellatum (Yen C. Yang) W. P. Fang ●
13391 Alangium villosum (Blume) Wangerin;麝香树;Muskwood ●☆
13392 Alangium vitiense (A. Gray) Baill. ex Harms;斐济八角枫●☆
13393 Alangium yunnanense C. Y. Wu ex W. P. Fang et al.;云南八角枫;Yunnan Alangium ●
13394 Alania Colenso = Dacrydium Sol. ex J. Forst. ●
13395 Alania Endl.(1836);澳西南吊兰属■☆
13396 Alania endlicheri Kunth;澳西南吊兰■☆
13397 Alantsilodendron Villiers(1994);阿拉豆属●☆
13398 Alantsilodendron brevipes (R. Vig.) Villiers;短梗阿拉豆●☆
13399 Alantsilodendron decaryanum (R. Vig.) Villiers;德卡里阿拉豆●☆
13400 Alantsilodendron glomeratum Villiers;团集阿拉豆●☆
13401 Alantsilodendron humbertii (R. Vig.) Villiers;亨伯特阿拉豆●☆
13402 Alantsilodendron mahafalense (R. Vig.) Villiers;马哈法尔阿拉豆●☆
13403 Alantsilodendron pilosum Villiers;疏毛阿拉豆●☆
13404 Alantsilodendron ramosum Villiers;分枝阿拉豆●☆
13405 Alantsilodendron villosum (R. Vig.) Villiers;长柔毛阿拉豆●☆
13406 Alarconia DC. = Wyethia Nutt. ■☆
13407 Alarconia angustifolia DC. = Wyethia angustifolia (DC.) Nutt. ■☆
13408 Alarconia helenioides DC. = Wyethia helenioides (DC.) Nutt. ■☆
13409 Alatavia Rodion.(1999);突厥鸢尾属■☆
13410 Alatavia Rodion. = Iris L. ■
13411 Alaternoides Adans. = Phylica L. ●☆
13412 Alaternoides Fabr. = Phylica L. ●☆
13413 Alaternus Mill. = Rhamnus L. ●
13414 Alathraea Steud. = Alatraea Neck. ■☆
13415 Alathraea Steud. = Phelypaea L. ■☆
13416 Alaticaulia Luer(2006);翼茎兰属■☆
13417 Alatiglossum Baptista = Oncidium Sw.(保留属名)■☆
13418 Alatiglossum Baptista(2006);翅舌兰属■☆
13419 Alatiliparis Marg. et Szlach.(2001);翼耳蒜属■☆

13420 Alatoseta Compton(1931);南非刺菊属(细弱紫绒草属)■☆
13421 Alatoseta tenuis Compton;南非刺菊(细弱紫绒草)■☆
13422 Alatraea Neck. = Phelypaea L. ■☆
13423 Albersia Kunth = Amaranthus L. ■
13424 Albersia caudata (Jacq.) Boiss. = Amaranthus viridis L. ■
13425 Albersia caudata (Jacq.) Boiss. = Chenopodium viridis L. ■☆
13426 Alberta E. Mey. (1838);艾伯特木属(阿尔伯特木属)●☆
13427 Alberta E. Mey. = Ernestimeyera Kuntze ●☆
13428 Alberta humblotii Drake var. acuminata Cavaco = Alberta minor Baill. ●☆
13429 Alberta humblotii Drake var. obovata Cavaco = Alberta humblotii Drake ●☆
13430 Alberta isosepala Baker = Alberta minor Baill. ●☆
13431 Alberta laurifolia Baker = Alberta minor Baill. ●☆
13432 Alberta magna E. Mey.;艾伯特木(阿尔伯特木);Breekhout, Natal Flame Bush ●☆
13433 Alberta minor Baill.;桂叶艾伯特木(桂叶阿尔伯特木)●☆
13434 Alberta minor Baill. var. isaloensis Cavaco = Alberta minor Baill. ●☆
13435 Alberta orientalis Homolle ex Cavaco;东方艾伯特木(东方阿尔伯特木)●☆
13436 Alberta regalis Puff et Robbr.;大王艾伯特木●☆
13437 Alberta sambiranensis Homolle ex Cavaco;桑比朗艾伯特木●☆
13438 Albertia Regel et Schmalh. (1877);艾伯特草属■☆
13439 Albertia Regel et Schmalh. = Aulacospermum Ledeb. + Trachydinm Lindl. + Kozlovia Lipsky ■☆
13440 Albertia Regel et Schmalh. = Kozlovia Lipsky ■☆
13441 Albertia Regel ex B. Fedch. et O. Fedch. = Exochorda Lindl. ●
13442 Albertia commutata Regel et Schmalh. = Pleurospermum simplex (Rupr.) Benth. et Hook. f. ex Drude ■
13443 Albertia paleacea Regel et Schmalh.;艾伯特草■☆
13444 Albertinia DC. = Vanillosmopsis Sch. Bip. ■☆
13445 Albertinia Spreng. (1820);陷托斑鸠菊属●☆
13446 Albertinia brasiliensis Spreng.;陷托斑鸠菊■☆
13447 Albertisia Becc. (1877);崖藤属(崖爬藤属);Albertisia, Cliffvine ●
13448 Albertisia apiculata (Troupin) Forman;尖叶崖藤●☆
13449 Albertisia capituliflora (Diels) Forman;头花崖藤●☆
13450 Albertisia cordifolia (Mangenot et J. Miège) Forman;心叶崖藤●☆
13451 Albertisia cuneata (Keay) Forman;楔形崖藤●☆
13452 Albertisia delagoensis (N. E. Br.) Forman;迪拉果崖藤●☆
13453 Albertisia exelliana (Troupin) Forman;埃克塞尔崖藤●☆
13454 Albertisia ferruginea (Diels) Forman;锈色崖藤●☆
13455 Albertisia glabra (Diels ex Troupin) Forman;光崖藤●☆
13456 Albertisia laurifolia Yamam.;崖藤(崖爬藤);Laurelleaf Albertisia, Laurelleaf Cliffvine, Laurel-leaved Albertisia ●◇
13457 Albertisia mangenotii (Guillaumet et Debray) Forman;大果崖藤●☆
13458 Albertisia megacarpa Diels ex Forman = Albertisia mangenotii (Guillaumet et Debray) Forman ●☆
13459 Albertisia perryana H. L. Li = Albertisia laurifolia Yamam. ●◇
13460 Albertisia scandens (Mangenot et J. Miège) Forman;攀缘崖藤●☆
13461 Albertisia triplinervis L. L. Forman;三脉崖藤●☆
13462 Albertisia undulata (Hiern) Forman;尾叶崖藤●☆
13463 Albertisia villosa (Exell) Forman;长柔毛崖藤●☆
13464 Albertisiella Pierre ex Aubrév. = Pouteria Aubl. ●
13465 Albertokuntzea Kuntze = Seguieria Loefl. ●☆
13466 Albidella Pichon = Echinodorus Rich. ex Engelm. ■☆
13467 Albidella Pichon(1946);古巴泽泻属■☆
13468 Albidella nymphaeifolia (Griseb.) Pichon;古巴泽泻■☆
13469 Albikia J. Presl et C. Presl = Hypolytrum Rich. ex Pers. ■
13470 Albildgaardia eragrostis Nees = Fimbristylis eragrostis (Nees et Meyen) Hance ■
13471 Albina Giseke(废弃属名) = Alpinia Roxb.(保留属名)■
13472 Albinea Hombr. et Jacquinot = Pleurophyllum Hook. f. ■☆
13473 Albinea Hombr. et Jacquinot ex Decne = Pleurophyllum Hook. f. ■☆
13474 Albizia Durazz. (1772);合欢属;Albizia, Albizzia, Silk Tree, Siris ●
13475 Albizia acle (Blanco) Merr.;阿克列合欢(阿古合欢,菲律宾合欢);Akle ●☆
13476 Albizia adianthifolia (Schumach.) W. Wight;非洲合欢(南非合欢,平头合欢);African Albizzia, Flat Crown ●☆
13477 Albizia adianthifolia (Schumach.) W. Wight var. intermedia (De Wild. et T. Durand) Villiers;间型非洲合欢●☆
13478 Albizia adianthifolia W. Wight = Albizia adianthifolia (Schumach.) W. Wight ●☆
13479 Albizia adinocephala Britton et Rose = Albizia adinocephala Britton et Rose ex Record ●☆
13480 Albizia adinocephala Britton et Rose ex Record;奶酪合欢;Cream Albizia ●☆
13481 Albizia alternifoliolata (T. L. Wu) Y. Huei Huang = Cylindrokelupha alternifoliolata T. L. Wu ●
13482 Albizia altissima Hook. f. = Cathormion altissimum (Hook. f.) Hutch. et Dandy ●☆
13483 Albizia amaniensis Baker f. = Albizia schimperiana Oliv. var. amaniensis (Baker f.) Brenan ●☆
13484 Albizia amara (Roxb.) Boivin;苦合欢(阿马拉合欢)●☆
13485 Albizia amara (Roxb.) Boivin subsp. sericocephala (Benth.) Brenan;绢毛头合欢●☆
13486 Albizia androyensis Capuron;安德罗合欢●☆
13487 Albizia angolensis Welw. ex Oliv. = Albizia ferruginea (Guillaumin et Perr.) Benth. ●
13488 Albizia anthelmintica Brongn.;驱虫合欢;Musenna Albizzia ●
13489 Albizia anthelmintica Brongn. var. australis Baker f. = Albizia anthelmintica Brongn. ●
13490 Albizia anthelmintica Brongn. var. pubescens Burtt Davy = Albizia anthelmintica Brongn. ●
13491 Albizia antunesiana Harms;安图内思合欢●☆
13492 Albizia arenicola R. Vig.;沙地合欢●☆
13493 Albizia attopeuensis (Pierre) I. C. Nielsen;海南合欢(刘氏合欢);Hainan Albizia, Hainan Albizzia ●
13494 Albizia attopeuensis (Pierre) I. C. Nielsen var. laui (Merr.) I. C. Nielsen = Albizia attopeuensis (Pierre) I. C. Nielsen ●
13495 Albizia attopeuensis (Pierre) T. L. Wu = Albizia attopeuense (Pierre) I. C. Nielsen ●
13496 Albizia aurisparsa (Drake) R. Vig.;马岛合欢●☆
13497 Albizia aylmeri Hutch.;艾梅合欢●☆
13498 Albizia balansae (Oliv.) Y. Huei Huang = Archidendron balansae (Oliv.) I. C. Nielsen ●
13499 Albizia balansae (Oliv.) Y. Huei Huang = Cylindrokelupha balansae (Oliv.) Kosterm. ●
13500 Albizia bequaertii De Wild. = Albizia grandibracteata Taub. ●☆
13501 Albizia boinensis R. Vig.;博伊纳合欢●☆
13502 Albizia boivinii E. Fourn.;鲍氏合欢●☆

13503 Albizia boromoensis Aubrév. et Pellegr. = Albizia malacophylla (A. Rich.) Walp. var. ugandensis Baker f. ●☆

13504 Albizia brachycalyx Oliv. = Albizia petersiana (Bolle) Oliv. ●☆

13505 Albizia bracteata Dunn;蒙自合欢;Mengzi Albertisia, Mengzi Albizia ●

13506 Albizia bracteata Dunn = Albizia lucidior (Steud.) I. C. Nielsen ex H. Hara ●

13507 Albizia brevifolia Schinz;短叶合欢●☆

13508 Albizia brownii Oliv. = Albizia brownii Walp. ex Oliv. ●☆

13509 Albizia brownii Walp. ex Oliv. = Albizia zygia (DC.) J. F. Macbr. ●☆

13510 Albizia calcarea Y. Huei Huang;光腺合欢;Chaalky Albizzia, Limy Albizia ●

13511 Albizia carbonaria Britton;煤色合欢;Naked Albizia ●☆

13512 Albizia caribaea (Urb.) Britton et Rose;加勒比合欢●☆

13513 Albizia chevalieri (Kosterm.) Y. Huei Huang = Archidendron chevalieri (Kosterm.) I. C. Nielsen ●

13514 Albizia chevalieri (Kosterm.) Y. Huei Huang = Cylindrokelupha chevalieri Kosterm. ●

13515 Albizia chinensis (Osbeck) Merr.;楹树(华楹,母引,牛尾木,水相思,托叶合欢,盈树);China Siris, Chinese Albizia, Chinese Albizzia, Stipulete Albizzia ●

13516 Albizia commiphoroides Capuron;没药合欢●☆

13517 Albizia concinna DC.;优雅合欢●☆

13518 Albizia conjugato-pinnata Vatke = Albizia anthelmintica Brongn. ●

13519 Albizia corbisieri De Wild. = Albizia ferruginea (Guillaumin et Perr.) Benth. ●

13520 Albizia cordifolia (T. L. Wu) Y. Huei Huang = Archidendron cordifolium (T. L. Wu) I. C. Nielsen ●

13521 Albizia cordifolia (T. L. Wu) Y. Huei Huang = Zygia cordifolia T. L. Wu ●

13522 Albizia coreana Nakai = Albizia kalkora (Roxb.) Prain ●

13523 Albizia coreana Nakai = Albizia macrophylla (Bunge) P. C. Huang ●

13524 Albizia coriaria Welw. ex Oliv.;鞣料合欢●☆

13525 Albizia corniculata (Lour.) Druce;天香藤(白格,刺藤,黄豆树,藤山丝);Corniculate Albizia, Corniculate Albizzia, Corniculate Siris ●

13526 Albizia crassiramea Lace;白花合欢(滇桂合欢,滇合欢);White-flower Albizzia, White-flowered Albizia, Yunnan Albizia, Yunnan Albizzia ●

13527 Albizia croizatiana F. P. Metcalf = Cylindrokelupha turgida (Merr.) T. L. Wu ●

13528 Albizia dalatensis (Kosterm.) Y. Huei Huang = Cylindrokelupha dalatensis (Kosterm.) T. L. Wu ●

13529 Albizia dinklagei (Harms) Harms;丁克合欢●☆

13530 Albizia distachya (Vent.) J. F. Macbr. = Albizia lophantha Willd. ●☆

13531 Albizia distachya (Vent.) J. F. Macbr. = Paraserianthes lophantha (Willd.) I. C. Nielsen ●☆

13532 Albizia divaricata Capuron;叉开合欢●☆

13533 Albizia duclouxii Gagnep.;巧家合欢;Ducloux Albizia, Ducloux Albizzia ●

13534 Albizia duclouxii Gagnep. = Albizia kalkora (Roxb.) Prain ●

13535 Albizia duclouxii Gagnep. = Albizia macrophylla (Bunge) P. C. Huang var. duclouxii (Gagnep.) P. C. Huang ●

13536 Albizia ealaensis De Wild.;乌干达合欢●☆

13537 Albizia ealaensis De Wild. = Albizia adianthifolia (Schumach.) W. Wight var. intermedia (De Wild. et T. Durand) Villiers ●☆

13538 Albizia eberhardtii (I. C. Nielsen) Y. Huei Huang = Archidendron eberhardtii I. C. Nielsen ●

13539 Albizia eberhardtii (I. C. Nielsen) Y. Huei Huang = Cylindrokelupha eberhardtii (Nielsen) T. L. Wu ●

13540 Albizia eggelingii Baker f. = Albizia glaberrima (Schumach. et Thonn.) Benth. ●☆

13541 Albizia elliptica E. Fourn. = Albizia malacophylla (A. Rich.) Walp. var. ugandensis Baker f. ●☆

13542 Albizia eriorhachis Harms = Cathormion eriorhachis (Harms) Dandy ●☆

13543 Albizia esquirolii H. Lév. = Albizia kalkora (Roxb.) Prain ●

13544 Albizia euryphylla Harms;宽叶合欢●☆

13545 Albizia evansii Burtt Davy = Albizia petersiana (Bolle) Oliv. subsp. evansii (Burtt Davy) Brenan ●☆

13546 Albizia falcata (L.) Backer = Falcataria moluccana (Miq.) Barneby et J. W. Grimes ●

13547 Albizia falcata (L.) Backer = Paraserianthes falcataria (L.) I. C. Nielsen ●

13548 Albizia falcata Backer ex Merr. = Albizia falcata (L.) Backer ●

13549 Albizia falcataria (L.) Fosberg = Falcataria moluccana (Miq.) Barneby et J. W. Grimes ●

13550 Albizia falcataria (L.) Fosberg = Paraserianthes falcataria (L.) I. C. Nielsen ●

13551 Albizia fastigiata (E. Mey.) Oliv.;帚状合欢●☆

13552 Albizia fastigiata (E. Mey.) Oliv. = Albizia adianthifolia (Schumach.) W. Wight ●☆

13553 Albizia fastigiata (E. Mey.) Oliv. var. chirindensis Swynn. ex Baker f. = Albizia adianthifolia (Schumach.) W. Wight ●☆

13554 Albizia ferruginea (Guillaumin et Perr.) Benth.;锈色合欢(铁锈合欢);Rusty Albizzia, Rusty Siris, Tanga-tanga, West African Albizzia ●

13555 Albizia ferruginea Benth. = Albizia ferruginea (Guillaumin et Perr.) Benth. ●

13556 Albizia flamignii De Wild. = Samanea leptophylla (Harms) Brenan et Brummitt ●☆

13557 Albizia flavovirens Hoyle = Samanea saman (Jacq.) Merr. ●

13558 Albizia forbesii Benth.;福布斯合欢●☆

13559 Albizia garrettii I. C. Nielsen;黄毛合欢●

13560 Albizia gigantea A. Chev.;巨大合欢●☆

13561 Albizia gillardinii G. C. C. Gilbert et Boutique = Albizia grandibracteata Taub. ●☆

13562 Albizia glaberrima (Schumach. et Thonn.) Benth.;光合欢●☆

13563 Albizia glaberrima (Schumach. et Thonn.) Benth. var. mpwapwensis Brenan;姆普瓦普瓦合欢●☆

13564 Albizia glabrescens Oliv. = Albizia glaberrima (Schumach. et Thonn.) Benth. ●☆

13565 Albizia glabrior (Koidz.) Ohwi;无毛合欢●☆

13566 Albizia gracilifolia Harms = Albizia amara (Roxb.) Boivin ●☆

13567 Albizia grandibracteata Taub.;大苞合欢;Nongo ●☆

13568 Albizia greveana (Baill.) Baron;格雷弗合欢●☆

13569 Albizia gummifera (J. F. Gmel.) C. A. Sm.;平冠合欢;Flat Crown Tree, Flat-crown Tree ●☆

13570 Albizia gummifera (J. F. Gmel.) C. A. Sm. var. ealaensis (De Wild.) Brenan = Albizia adianthifolia (Schumach.) W. Wight var. intermedia (De Wild. et T. Durand) Villiers ●☆

13571 Albizia harveryi E. Fourn.;哈维合欢(哈贝合欢)●☆

13572 Albizia henryi Ricker = Albizia kalkora (Roxb.) Prain ●

13573 Albizia hypoleuca Oliv. = Albizia harveyi E. Fourn. ●☆

13574 Albizia intermedia De Wild. et T. Durand = Albizia adianthifolia (Schumach.) W. Wight ●☆

13575 Albizia intermedia De Wild. et T. Durand = Albizia adianthifolia (Schumach.) W. Wight var. intermedia (De Wild. et T. Durand) Villiers ●☆

13576 Albizia isenbergiana (A. Rich.) E. Fourn.;伊森合欢●☆

13577 Albizia jaubertiana Fourn.;若贝尔合欢●☆

13578 Albizia julibrissin (Willd.) Durazz.;合欢(芙蓉花,芙蓉花树,合欢木,合昏,黄昏,马缨,马缨花,马缨树,马樱花,萌葛,荫葛,青堂,青棠,绒花树,绒树,茸花枝,蓉花树,乌赖树,乌绒,乌绒树,乌树,乌云树,野花木,夜关门,夜合,夜合草,夜合花,夜合槐,夜合树,夜欢花,宜男);Albizia, Albizzia, Mimosa, Mimosa Tree, Mimosa-tree, Persian Acacia, Persian Silk Tree, Pink Mimosa, Pink Siris, Pink Siris Persian Acacia, Plume Albizia, Powderpuff-tree, Silk Tree, Silktree, Silk-tree, Silk-tree Albizia, Silktree Albizzia, Silk-tree Albizzia, Silktree Siris ●

13579 Albizia julibrissin Durazz. = Albizia julibrissin (Willd.) Durazz. ●

13580 Albizia julibrissin Durazz. f. albiflora J. Ohara;日本白花合欢●☆

13581 Albizia julibrissin Durazz. f. tianshuiensis T. S. Yao;天水合欢;Tianshui Albizzia ●

13582 Albizia julibrissin Durazz. var. glabrior (Koidz.) H. Ohashi = Albizia glabrior (Koidz.) Ohwi ●☆

13583 Albizia julibrissin Durazz. var. mollis (Wall.) Benth. = Albizia mollis (Wall.) Boivin ●

13584 Albizia julibrissin Durazz. var. rosea Mouill.;桃色合欢●☆

13585 Albizia julibrissin Durazz. var. speciosa Koidz.;五龙鳞(合欢)●☆

13586 Albizia kalkora (Roxb.) Prain;山合欢(白合欢,白花合欢,白夜合,刀头黄,钩龙树,马缨花,山槐,台湾合欢);Kalkora Mimosa, Lebbek Albizia, Lebbek Albizzia, Wild Siris ●

13587 Albizia katangensis De Wild. = Albizia coriaria Welw. ex Oliv. ●☆

13588 Albizia laevicorticata Zimm. = Albizia gummifera (J. F. Gmel.) C. A. Sm. ●☆

13589 Albizia lancangensis Y. Y. Qian;澜沧合欢;Lancang Albizzia ●

13590 Albizia lancangensis Y. Y. Qian = Albizia crassiramea Lace ●

13591 Albizia laotica Gagnep. = Albizia crassiramea Lace ●

13592 Albizia laui Merr. = Albizia attopeuensis (Pierre) I. C. Nielsen var. lauii (Merr.) I. C. Nielsen ●

13593 Albizia laurentii De Wild.;洛朗合欢●☆

13594 Albizia lauri Merr. = Albizia attopeuense (Pierre) I. C. Nielsen ●

13595 Albizia lebbeck (L.) Benth.;阔荚合欢(白夜合树,大叶合欢,火七树,缅甸合欢,尸利洒树,夜合,夜合树,印度合欢);Airis, E. Indian Walnut, East Indian Walnut, Indian Albizia, Kokko, Lebbek Albizzia, Lebbek Siris, Lebbek Tree, Silk Tree, Siris, Siris Tree, Siris-acacia, White Siris, Woman's Tongue, Woman's Tongue Tree ●

13596 Albizia lebbeck (L.) Benth. var. australis Burtt Davy = Albizia tanganyicensis Baker f. ●☆

13597 Albizia lebbekoides (DC.) Benth.;拟阔荚合欢●☆

13598 Albizia leptophylla Harms = Samanea leptophylla (Harms) Brenan et Brummitt ●☆

13599 Albizia letestui Pellegr. = Albizia zygia (DC.) J. F. Macbr. ●☆

13600 Albizia littoralis Teijsm. et Binn. = Albizia retusa Benth. ●

13601 Albizia longepedunculata Hayata = Albizia kalkora (Roxb.) Prain ●

13602 Albizia longepedunculata Hayata = Albizia lebbeck (L.) Benth. ●

13603 Albizia longipedata Britton et Rose ex Record;长合欢●

13604 Albizia lophantha (Willd.) Benth. = Paraserianthes lophantha (Willd.) I. C. Nielsen ●☆

13605 Albizia lophantha Willd. = Paraserianthes lophantha (Willd.) I. C. Nielsen ●☆

13606 Albizia lucida (Roxb.) Benth. = Albizia lucidior (Steud.) I. C. Nielsen ex H. Hara ●

13607 Albizia lucida Benth. = Albizia lucidior (Steud.) I. C. Nielsen ex H. Hara ●

13608 Albizia lucidior (Steud.) I. C. Nielsen = Albizia lucidior (Steud.) I. C. Nielsen ex H. Hara ●

13609 Albizia lucidior (Steud.) I. C. Nielsen ex H. Hara;亮叶合欢(光叶合欢);Glabrousleaf Siris, Lucid Albizia, Shiny Albizzia, Tapria Siris ●

13610 Albizia lugardii N. E. Br. = Acacia nigrescens Oliv. ●☆

13611 Albizia macrophylla (Bunge) P. C. Huang;大叶山合欢(白夜合,白缨,刀头黄,山合欢,山槐);Bigleaf Albizzia ●

13612 Albizia macrophylla (Bunge) P. C. Huang var. duclouxii (Gagnep.) P. C. Huang;多毛山合欢;Ducloux Bigleaf Albizzia ●

13613 Albizia malacophylla (A. Rich.) Walp.;软叶合欢●☆

13614 Albizia malacophylla (A. Rich.) Walp. var. ugandensis Baker f.;乌干达软叶合欢●☆

13615 Albizia maraguensis Taub. ex Engl. = Albizia schimperiana Oliv. ●☆

13616 Albizia mearnsii De Wild. = Albizia gummifera (J. F. Gmel.) C. A. Sm. ●☆

13617 Albizia meyeri Ricker = Albizia lucidior (Steud.) I. C. Nielsen ex H. Hara ●

13618 Albizia micrantha Boivin = Albizia odoratissima (L. f.) Benth. ●

13619 Albizia milletii Benth. = Albizia corniculata (Lour.) Druce ●

13620 Albizia minyi De Wild. = Albizia chinensis (Osbeck) Merr. ●

13621 Albizia mollis (Wall.) Boivin;毛叶合欢(大毛毛花,滇合欢,羊毛花,夜合);Ceylon Rosewood, Hairleaf Siris, Hairy-leaved Albizia, Soft Albizzia ●

13622 Albizia mollis (Wall.) Boivin = Albizia julibrissin Durazz. var. mollis (Wall.) Benth. ●

13623 Albizia mollis (Wall.) Boivin var. glabrior Koidz. = Albizia julibrissin Durazz. var. glabrior (Koidz.) H. Ohashi ●☆

13624 Albizia moluccana Miq. = Falcataria moluccana (Miq.) Barneby et J. W. Grimes ●

13625 Albizia moluccana Miq. = Paraserianthes falcataria (L.) I. C. Nielsen ●

13626 Albizia morombensis Capuron;穆龙贝合欢●☆

13627 Albizia mossambicensis Sim = Albizia versicolor Welw. ex Oliv. ●☆

13628 Albizia mossamedensis Torre;莫萨梅迪合欢●☆

13629 Albizia nanu Wild. = Albizia julibrissin Durazz. ●

13630 Albizia numidarum Capuron;努米底亚合欢●☆

13631 Albizia nyasica Dunkley = Albizia zimmermannii Harms ●☆

13632 Albizia obbadiensis (Chiov.) Brenan;奥巴迪合欢●☆

13633 Albizia obliquifoliolata De Wild. = Cathormion obliquifoliolatum (De Wild.) G. C. C. Gilbert et Boutique ●☆

13634 Albizia odorata R. Vig.;芳香合欢●☆

13635 Albizia odoratissima (L. f.) Benth.;香合欢(白丝绒,黑格,黑心木,黄豆树,乌格,细黑心,香茜藤,香须树,夜合欢);Black Siris, Ceylon Rosewood, Fragrant Albizia, Fragrant Albizzia, Fragrant Siris ●

13636 Albizia ogadensis (Chiov.) Baker f. ex Chiov. = Acacia

ogadensis Chiov. ●☆
13637 Albizia oliveri Pellegr. = Albizia adianthifolia (Schumach.) W. Wight var. intermedia (De Wild. et T. Durand) Villiers ●☆
13638 Albizia pallida Harv. = Albizia harveyi E. Fourn. ●☆
13639 Albizia parvifolia Burtt Davy = Albizia brevifolia Schinz ●☆
13640 Albizia passargei Harms = Cathormion altissimum (Hook. f.) Hutch. et Dandy ●☆
13641 Albizia perrieri (Drake) R. Vig. = Albizia perrieri (Drake) R. Vig. ex Capuron ●☆
13642 Albizia perrieri (Drake) R. Vig. ex Capuron;佩里耶香合欢●☆
13643 Albizia perrieri (Drake) R. Vig. ex Capuron var. monticola Capuron = Albizia perrieri (Drake) R. Vig. ex Capuron ●☆
13644 Albizia petersiana (Bolle) Oliv.;彼得合欢●☆
13645 Albizia petersiana (Bolle) Oliv. subsp. evansii (Burtt Davy) Brenan;埃文斯合欢●☆
13646 Albizia poissonii A. Chev. = Albizia coriaria Welw. ex Oliv. ●☆
13647 Albizia polyphylla E. Fourn.;多叶合欢●☆
13648 Albizia pospischilii Harms = Albizia harveyi E. Fourn. ●☆
13649 Albizia procera (Roxb.) Benth.;红荚合欢(白格,白其春,白相思,蕃婆树,菲律宾合欢,黄豆树);Kokko, Soybeantree, Tall Albizia, Tall Albizzia, White Siris ●
13650 Albizia procera H. L. Li = Albizia macrophylla (Bunge) P. C. Huang ●
13651 Albizia purpurea Boivin ex Fourn. = Albizia glaberrima (Schumach. et Thonn.) Benth. ●☆
13652 Albizia quartiniana (A. Rich.) Walp.;夸尔廷合欢●☆
13653 Albizia retusa Benth.;兰屿合欢;Lanyu Albizia ●
13654 Albizia rhodesica Burtt Davy;红纸树;Red Paper Tree, Red-paper Tree ●☆
13655 Albizia rhodesica Burtt Davy = Albizia tanganyicensis Baker f. ●☆
13656 Albizia rhombifolia Benth. = Cathormion rhombifolium (Benth.) Hutch. et Dandy ●☆
13657 Albizia robinsonii (Gagnep.) Y. Huei Huang = Archidendron robinsonii (Gagnep.) I. C. Nielsen ●
13658 Albizia robinsonii (Gagnep.) Y. Huei Huang = Cylindrokelupha robinsonii (Gagnep.) Kosterm. ●
13659 Albizia rogersii Burtt Davy = Albizia brevifolia Schinz ●☆
13660 Albizia sahafariensis Capuron;萨哈法利合欢●☆
13661 Albizia saman (Jacq.) F. Muell. = Samanea saman (Jacq.) Merr. ●
13662 Albizia saman F. Muell. = Samanea saman (Jacq.) Merr. ●
13663 Albizia saponaria (Lour.) Blume ex Miq.;非洲白花合欢;Whiteflower Albizia ●☆
13664 Albizia saponaria Miq. = Albizia saponaria (Lour.) Blume ex Miq. ●☆
13665 Albizia sassa (Willd.) Chiov. = Albizia gummifera (J. F. Gmel.) C. A. Sm. ●☆
13666 Albizia sassa (Willd.) Chiov. var. chirindensis (Swynn. ex Baker f.) Baker f. = Albizia adianthifolia (Schumach.) W. Wight ●☆
13667 Albizia schimperiana Oliv.;欣珀合欢●☆
13668 Albizia schimperiana Oliv. var. amaniensis (Baker f.) Brenan;阿马尼欣珀合欢●☆
13669 Albizia schimperiana Oliv. var. tephrocalyx Brenan;灰萼欣珀合欢●☆
13670 Albizia sericocephala Benth. = Albizia amara (Roxb.) Boivin subsp. sericocephala (Benth.) Brenan ●☆
13671 Albizia sherriffii Baker;藏合欢;Sherriff Albizia, Xizang Albizzia ●
13672 Albizia simeonis Harms = Albizia kalkora (Roxb.) Prain ●
13673 Albizia sinuata (Lour.) Merr. = Acacia concinna (Willd.) DC. ●■
13674 Albizia speciosa Willd. = Albizia lebbeck (L.) Benth. ●
13675 Albizia stipulata (DC.) Boivin = Albizia chinensis (Osbeck) Merr. ●
13676 Albizia stipulata Boivin = Albizia chinensis (Osbeck) Merr. ●
13677 Albizia struthiophylla Milne-Redh. = Albizia amara (Roxb.) Boivin subsp. sericocephala (Benth.) Brenan ●☆
13678 Albizia subrhombea Milne-Redh. = Albizia viridis E. Fourn. var. zygioides (Baill.) Villiers ●☆
13679 Albizia suluensis Gerstner;苏卢合欢●☆
13680 Albizia tanganyicensis Baker f.;纸皮合欢;Paperbark Albizia, Paperbark False Thorn ●☆
13681 Albizia tonkinensis (I. C. Nielsen) Y. Huei Huang = Archidendron tonkinense I. C. Nielsen ●
13682 Albizia tonkinensis (I. C. Nielsen) Y. Huei Huang = Cylindrokelupha tonkinensis (I. C. Nielsen) T. L. Wu ●
13683 Albizia trichopetala Baker = Albizia viridis E. Fourn. var. zygioides (Baill.) Villiers ●☆
13684 Albizia tulearensis R. Vig.;图莱亚尔合欢●☆
13685 Albizia turgida (Merr.) Merr. = Archidendron turgidum (Merr.) I. C. Nielsen ●
13686 Albizia turgida (Merr.) Merr. = Cylindrokelupha turgida (Merr.) T. L. Wu ●
13687 Albizia turgida (Merr.) Merr. ex Chung = Archidendron turgidum (Merr.) I. C. Nielsen ●
13688 Albizia turgida (Merr.) Merr. ex Chung = Cylindrokelupha turgida (Merr.) T. L. Wu ●
13689 Albizia umbalusiana Sim = Albizia anthelmintica Brongn. ●
13690 Albizia vaughanii Brenan;弗氏合欢●☆
13691 Albizia verrucosa Capuron;多疣合欢●☆
13692 Albizia versicolor Welw. ex Oliv.;莫桑比克合欢●☆
13693 Albizia viridis E. Fourn.;马岛绿合欢●☆
13694 Albizia viridis E. Fourn. var. zygioides (Baill.) Villiers;肖西非合欢●☆
13695 Albizia warneckei Harms = Albizia glaberrima (Schumach. et Thonn.) Benth. ●☆
13696 Albizia welwitschii Oliv.;韦尔合欢●☆
13697 Albizia welwitschii Oliv. var. pedicellata G. C. C. Gilbert et Boutique;梗花韦尔合欢●☆
13698 Albizia welwitschioides Schweinf. ex Baker f. = Albizia zygia (DC.) J. F. Macbr. ●☆
13699 Albizia yunnanensis (Kosterm.) Y. Huei Huang = Archidendron kerrii (Gagnep.) I. C. Nielsen ●
13700 Albizia yunnanensis (Kosterm.) Y. Huei Huang = Cylindrokelupha yunnanensis (Kosterm.) T. L. Wu ●
13701 Albizia yunnanensis T. L. Wu = Albizia crassiramea Lace ●
13702 Albizia zimmermannii Harms;齐默尔曼合欢●☆
13703 Albizia zygia (DC.) J. F. Macbr.;西非合欢(西非合欢木);African Walnut Albizzia, Nongo, Okuro, W. Africa Siris, West African Albizzia ●☆
13704 Albizia zygia J. F. Macbr. = Albizia zygia (DC.) J. F. Macbr. ●☆
13705 Albizia zygioides = Albizia viridis E. Fourn. var. zygioides (Baill.) Villiers ●☆
13706 Albizzia Benth. = Albizia Durazz. ●
13707 Albizzia Durazz. = Albizia Durazz. ●

13708 Albizzia aclel Merr. = Albizia acle (Blanco) Merr. ●☆
13709 Albizzia adianthifolia W. Wight = Albizia adianthifolia W. Wight ●☆
13710 Albizzia chinensis (Osbeck) Merr. = Albizia chinensis (Osbeck) Merr. ●
13711 Albizzia corniculata (Lour.) Druce = Albizia corniculata (Lour.) Druce ●
13712 Albizzia falcata (L.) Baker ex Merr. = Albizia falcata Backer ex Merr. ●☆
13713 Albizzia julibrissin Durazz. = Albizia julibrissin Durazz. ●
13714 Albizzia kalkora (Roxb.) Prain = Albizia kalkora (Roxb.) Prain ●
13715 Albizzia meyeri Kicher = Albizia lucidior (Steud.) I. C. Nielsen ex H. Hara ●
13716 Albizzia odoratissima (L. f.) Benth. = Albizia odoratissima (L. f.) Benth. ●
13717 Albizzia turgida (Merr.) Merr. = Albizia turgida (Merr.) Merr. ●
13718 Albolboa Hieron. = Abolboda Humb. ■☆
13719 Albonia Buc'hoz = Ailanthus Desf. (保留属名) ●
13720 Albonia peregrina Buc'hoz = Ailanthus altissima (Mill.) Swingle ●
13721 Albovia Schischk. (1950);肖茴芹属■☆
13722 Albovia Schischk. = Pimpinella L. ■
13723 Albovia tripartita (Kalen.) Schischk.;肖茴芹■☆
13724 Albowiodoxa Woron. ex Kolak. = Amphoricarpus Vis. ●☆
13725 Albradia D. Dietr. = Albrandia Gaudich. ●
13726 Albrandia Gaudich. = Streblus Lour. ●
13727 Albraunia Speta(1982);阿尔婆婆纳属■☆
13728 Albraunia fugax (Boiss. et Noë) Speta;阿尔婆婆纳■☆
13729 Albuca L. (1762);肋瓣花属;Sentry-boxes,Soldier-in-the-box ■☆
13730 Albuca abyssinica Jacq.;阿比西尼亚肋瓣花■☆
13731 Albuca acuminata Baker;渐尖肋瓣花■☆
13732 Albuca adlamii Baker;阿德拉姆肋瓣花■☆
13733 Albuca affinis Baker = Ornithogalum simile J. C. Manning et Goldblatt ■☆
13734 Albuca affinis J. M. Wood et M. S. Evans;近缘肋瓣花■☆
13735 Albuca allenae Baker = Albuca abyssinica Jacq. ■☆
13736 Albuca altissima Dryand. = Albuca maxima Burm. f. ■☆
13737 Albuca amboensis (Schinz) Oberm.;安博肋瓣花■☆
13738 Albuca angolensis Welw. ex Baker = Albuca abyssinica Jacq. ■☆
13739 Albuca angustibracteata De Wild.;窄苞肋瓣花■☆
13740 Albuca aperta I. Verd. = Ornithogalum concordianum (Baker) U. Müll. -Doblies et D. Müll. -Doblies ■☆
13741 Albuca asclepiadea Chiov. = Albuca abyssinica Jacq. ■☆
13742 Albuca aspera U. Müll. -Doblies = Albuca viscosa L. f. ■☆
13743 Albuca aurea Jacq.;金黄肋瓣花■☆
13744 Albuca bainesii Baker = Albuca abyssinica Jacq. ■☆
13745 Albuca batteniana Hilliard et B. L. Burtt;巴滕肋瓣花■☆
13746 Albuca baurii Baker;巴利肋瓣花■☆
13747 Albuca beguinotii Cufod. = Albuca abyssinica Jacq. ■☆
13748 Albuca bequaertii De Wild.;贝卡尔肋瓣花■☆
13749 Albuca bifolia Baker;双叶肋瓣花■☆
13750 Albuca bifoliata R. A. Dyer;双小叶肋瓣花■☆
13751 Albuca blepharophylla Cufod. = Albuca abyssinica Jacq. ■☆
13752 Albuca bontebokensis U. Müll. -Doblies = Albuca viscosa L. f. ■☆
13753 Albuca brevipes Baker = Ornithogalum suaveolens Jacq. ■☆
13754 Albuca brucebayeri U. Müll. -Doblies = Albuca hallii U. Müll. -Doblies ■☆
13755 Albuca buchananii Baker;布坎南肋瓣花■☆
13756 Albuca canadensis (L.) F. M. Leight.;加拿大肋瓣花(黄肋瓣花);Sentry in the Box ■☆
13757 Albuca canadensis (L.) F. M. Leight. = Albuca maxima Burm. f. ■☆
13758 Albuca capitata Gilli = Albuca abyssinica Jacq. ■☆
13759 Albuca caudata Jacq.;尾状肋瓣花■☆
13760 Albuca chaetopoda Chiov. = Albuca abyssinica Jacq. ■☆
13761 Albuca chlorantha Welw. ex Baker;绿花肋瓣花■☆
13762 Albuca ciliaris U. Müll. -Doblies;缘毛肋瓣花■☆
13763 Albuca circinata Baker = Albuca namaquensis Baker ■☆
13764 Albuca coarctata Dryand. = Albuca maxima Burm. f. ■☆
13765 Albuca collina Baker;山丘肋瓣花■☆
13766 Albuca concordiana Baker = Ornithogalum concordianum (Baker) U. Müll. -Doblies et D. Müll. -Doblies ■☆
13767 Albuca convoluta E. Phillips = Albuca acuminata Baker ■☆
13768 Albuca cooperi Baker;库珀肋瓣花■☆
13769 Albuca corymbosa Baker;伞序肋瓣花■☆
13770 Albuca crinifolia Baker;丝叶肋瓣花■☆
13771 Albuca crudenii Archibald;克鲁登肋瓣花■☆
13772 Albuca decipiens U. Müll. -Doblies;迷惑肋瓣花■☆
13773 Albuca dinteri U. Müll. -Doblies;丁特肋瓣花■☆
13774 Albuca donaldsonii Rendle = Ornithogalum donaldsonii (Rendle) Greenway ■☆
13775 Albuca echinosperma U. Müll. -Doblies;刺子肋瓣花■☆
13776 Albuca elliotii Baker = Albuca shawii Baker ■☆
13777 Albuca englerana K. Krause et Dinter;恩格勒肋瓣花■☆
13778 Albuca erlangeriana Engl. = Albuca abyssinica Jacq. ■☆
13779 Albuca exuviata (Jacq.) Ker Gawl. = Drimia exuviata (Jacq.) Jessop ■☆
13780 Albuca fastigiata Dryand.;帚状肋瓣花■☆
13781 Albuca fibrillosa De Wild. = Albuca abyssinica Jacq. ■☆
13782 Albuca fibrotunicata Gledhill et Oyewole;丝衣肋瓣花■☆
13783 Albuca filifolia (Jacq.) Ker Gawl. = Drimia filifolia (Jacq.) J. C. Manning et Goldblatt ■☆
13784 Albuca fischeri Engl. = Albuca abyssinica Jacq. ■☆
13785 Albuca flaccida Jacq.;柔软肋瓣花■☆
13786 Albuca fleckii Schinz;弗莱克肋瓣花■☆
13787 Albuca foetida U. Müll. -Doblies;臭肋瓣花■☆
13788 Albuca fragrans Jacq.;香肋瓣花■☆
13789 Albuca galeata Welw. ex Baker;盔形肋瓣花■☆
13790 Albuca gardenii Hook. = Speirantha gardenii (Hook.) Baill. ■
13791 Albuca garuensis Engl. et K. Krause;加鲁肋瓣花■☆
13792 Albuca gentilii De Wild.;让蒂肋瓣花■☆
13793 Albuca glandulosa Baker;具腺肋瓣花■☆
13794 Albuca glauca Baker;灰绿肋瓣花■☆
13795 Albuca granulata Baker = Albuca shawii Baker ■☆
13796 Albuca hallii U. Müll. -Doblies;霍尔肋瓣花■☆
13797 Albuca hereroensis Schinz;赫雷罗肋瓣花■☆
13798 Albuca homblei De Wild.;洪布勒肋瓣花■☆
13799 Albuca humilis Baker;矮生肋瓣花■☆
13800 Albuca hyacinthoides Chiov. = Ornithogalum viride (L.) J. C. Manning et Goldblatt ■☆
13801 Albuca hysterantha Chiov. = Albuca abyssinica Jacq. ■☆
13802 Albuca imbricata F. M. Leight. = Albuca juncifolia Baker ■☆
13803 Albuca jacquinii U. Müll. -Doblies = Albuca viscosa L. f. ■☆
13804 Albuca juncifolia Baker;灯心草叶肋瓣花■☆
13805 Albuca karasbergensis P. E. Glover;卡拉斯堡肋瓣花■☆
13806 Albuca karooica U. Müll. -Doblies = Albuca cooperi Baker ■☆

13807 Albuca kassneri Engl. et K. Krause ex De Wild. ;卡斯纳肋瓣花■☆
13808 Albuca katangensis De Wild. ;加丹加肋瓣花■☆
13809 Albuca kirkii (Baker) Brenan;柯克肋瓣花■☆
13810 Albuca kundelungensis De Wild. ;昆德龙肋瓣花■☆
13811 Albuca laxiflora Dinter;疏花肋瓣花■☆
13812 Albuca ledermannii Engl. et K. Krause;莱德肋瓣花■☆
13813 Albuca leucantha U. Müll. -Doblies;白花肋瓣花■☆
13814 Albuca longebracteata Engl. = Ornithogalum tenuifolium F. Delaroche ■☆
13815 Albuca longifolia Baker;长叶肋瓣花■☆
13816 Albuca longipes Baker;长梗肋瓣花■☆
13817 Albuca lugardii Baker;卢格德肋瓣花■☆
13818 Albuca macowanii Baker;麦克欧文肋瓣花■☆
13819 Albuca major L. = Albuca canadensis (L.) F. M. Leight. ■☆
13820 Albuca major L. = Albuca maxima Burm. f. ■☆
13821 Albuca malangensis Baker;马兰加肋瓣花■☆
13822 Albuca mankonensis A. Chev. = Albuca sudanica A. Chev. ■☆
13823 Albuca massonii Baker;马森肋瓣花■☆
13824 Albuca materfamilias U. Müll. -Doblies = Albuca flaccida Jacq ■☆
13825 Albuca maxima Burm. f. = Albuca canadensis (L.) F. M. Leight. ■☆
13826 Albuca melleri (Baker) Baker = Albuca abyssinica Jacq. ■☆
13827 Albuca melleri Baker = Albuca abyssinica Jacq. ■☆
13828 Albuca micrantha Baker = Ornithogalum secundum Jacq. ■☆
13829 Albuca minima Baker = Albuca shawii Baker ■☆
13830 Albuca minor L. = Albuca canadensis (L.) F. M. Leight. ■☆
13831 Albuca minor L. = Albuca maxima Burm. f. ■☆
13832 Albuca monophylla Baker;单叶肋瓣花■☆
13833 Albuca myogaloides Welw. ex Baker;安哥拉肋瓣花■☆
13834 Albuca namaquensis Baker;纳马夸肋瓣花■☆
13835 Albuca nana Schönland;矮小肋瓣花■☆
13836 Albuca narcissifolia A. Chev. = Albuca sudanica A. Chev. ■☆
13837 Albuca navicula U. Müll. -Doblies;舟形肋瓣花■☆
13838 Albuca nelsonii N. E. Br. ;内尔森肋瓣花■☆
13839 Albuca nemorosa Chiov. = Albuca abyssinica Jacq. ■☆
13840 Albuca nigritana (Baker) Troupin;尼格里塔肋瓣花■☆
13841 Albuca nyikensis Baker = Albuca kirkii (Baker) Brenan ■☆
13842 Albuca odoratissima Dinter = Albuca viscosa L. f. ■☆
13843 Albuca oligophylla Schltr. ;寡叶肋瓣花■☆
13844 Albuca pachychlamys Baker = Albuca setosa Jacq. ■☆
13845 Albuca papyracea J. C. Manning et Goldblatt;纸质肋瓣花■☆
13846 Albuca paradoxa Dinter;奇异肋瓣花■☆
13847 Albuca patersoniae Schönland;帕特森肋瓣花■☆
13848 Albuca polyphylla Baker;多叶肋瓣花■☆
13849 Albuca praecox Engl. et K. Krause;早肋瓣花■☆
13850 Albuca prolifera Wilson;多育肋瓣花■☆
13851 Albuca purpurascens Engl. = Albuca abyssinica Jacq. ■☆
13852 Albuca reflexa Dinter et K. Krause;反折肋瓣花■☆
13853 Albuca robertsoniana U. Müll. -Doblies;罗伯逊肋瓣花■☆
13854 Albuca rogersii Schönland;罗杰斯肋瓣花■☆
13855 Albuca rupestris Hilliard et B. L. Burtt;岩生肋瓣花■☆
13856 Albuca sassandrensis A. Chev. = Albuca sudanica A. Chev. ■☆
13857 Albuca scabromarginata De Wild. ;糙边肋瓣花■☆
13858 Albuca schlechteri Baker;施莱肋瓣花■☆
13859 Albuca schweinfurthii Engl. ;施韦肋瓣花■☆
13860 Albuca semipedalis Baker = Ornithogalum semipedale (Baker) U. Müll. -Doblies et D. Müll. -Doblies ■☆
13861 Albuca septentrionalis Quézel = Albuca sudanica A. Chev. ■☆
13862 Albuca seretii De Wild. ;赛雷肋瓣花■☆
13863 Albuca setosa Jacq. ;刚毛肋瓣花■☆
13864 Albuca shawii Baker;肖氏肋瓣花■☆
13865 Albuca sordida Bakerr;污肋瓣花■☆
13866 Albuca spiralis L. f. ;螺旋肋瓣花■☆
13867 Albuca steudneri Schweinf. et Engl. ;斯托德肋瓣花■☆
13868 Albuca stolzii K. Krause = Chlorophytum stolzii (K. Krause) Kativu ■☆
13869 Albuca stricta Engl. et K. Krause;刚直肋瓣花■☆
13870 Albuca subspicata Baker;穗状肋瓣花■☆
13871 Albuca sudanica A. Chev. ;苏丹肋瓣花■☆
13872 Albuca tayloriana Rendle = Albuca abyssinica Jacq. ■☆
13873 Albuca tenuifolia Baker;细叶肋瓣花■☆
13874 Albuca tenuis Knudtzon;小肋瓣花■☆
13875 Albuca thermarum Van Jaarsv. = Ornithogalum thermarum (Van Jaarsv.) J. C. Manning et Goldblatt ■☆
13876 Albuca tortuosa Baker;扭曲肋瓣花■☆
13877 Albuca trachyphylla U. Müll. -Doblies;糙叶肋瓣花■☆
13878 Albuca transvaalensis Mogg;德兰士瓦肋瓣花■☆
13879 Albuca trichophylla Baker = Albuca shawii Baker ■☆
13880 Albuca unifoliata G. D. Rowley = Ornithogalum unifoliatum (G. D. Rowley) Oberm. ■☆
13881 Albuca variegata De Wild. ;杂色肋瓣花■☆
13882 Albuca villosa U. Müll. ;长柔毛肋瓣花■☆
13883 Albuca viridiflora Jacq. = Albuca spiralis L. f. ■☆
13884 Albuca viscosa L. f. ;黏肋瓣花■☆
13885 Albuca viscosella U. Müll. -Doblies = Albuca viscosa L. f. ■☆
13886 Albuca vittata Ker Gawl. = Ornithogalum suaveolens Jacq. ■☆
13887 Albuca wakefieldii Baker = Albuca abyssinica Jacq. ■☆
13888 Albuca xanthocodon Hilliard et B. L. Burtt;黄铃肋瓣花■☆
13889 Albuca zebrina Baker = Ornithogalum zebrinum (Baker) Oberm. ■☆
13890 Albuca zenkeri Engl. ;岑克尔肋瓣花■☆
13891 Albucea (Rchb.) Rchb. = Honorius Gray ■☆
13892 Albucea (Rchb.) Rchb. = Ornithogalum L. ■
13893 Albucea Rchb. = Honorius Gray ■☆
13894 Albucea Rchb. = Ornithogalum L. ■
13895 Albuga Schreb. = Albuca L. ■☆
13896 Albugoides Medik. = Albuca L. ■☆
13897 Albusa gardenii Hook. = Speirantha gardenii (Hook.) Baill. ■
13898 Alcaea Burm. f. = Althaea L. ■
13899 Alcaea Hill = Althaea L. ■
13900 Alcanna Gaertn. = Lawsonia L. ●
13901 Alcanna Orph. = Alkanna Tausch(保留属名)●☆
13902 Alcantara Glaz. = Alcantara Glaz. ex G. M. Barroso ■☆
13903 Alcantara Glaz. ex G. M. Barroso = Xerxes J. R. Grant ■☆
13904 Alcantara Glaz. ex G. M. Barroso(1969);阿尔菊属■☆
13905 Alcantara petroana Glaz. ex G. M. Barroso;阿尔菊■☆
13906 Alcantarea (E. Morren ex Mez) Harms = Vriesea Lindl. (保留属名)■☆
13907 Alcantarea (E. Morren ex Mez) Harms(1929);缨凤梨属■☆
13908 Alcantarea (E. Morren) Harms = Vriesea Lindl. (保留属名)■☆
13909 Alcantarea Harms = Alcantarea (E. Morren ex Mez) Harms ■☆
13910 Alcantarea imperialis Harms;缨凤梨■☆
13911 Alcea L. (1753);蜀葵属;Hollyhock ■
13912 Alcea Mill. = Alcea L. ■

13913 Alcea Mill. = Malva L. ■
13914 Alcea acaulis (Cav.) Alef.;无茎蜀葵■☆
13915 Alcea angulata Freyn;窄叶蜀葵■☆
13916 Alcea antoninae Iljin;安氏蜀葵■☆
13917 Alcea baldshuanica (Bornm.) Iljin;巴尔德蜀葵■☆
13918 Alcea chinensis ? = Alcea rosea L. ■
13919 Alcea ficifolia L. = Alcea rosea L. ■
13920 Alcea freyniana Iljin;福氏蜀葵■☆
13921 Alcea grossheimii Iljin;格氏蜀葵■☆
13922 Alcea heldreichii Boiss.;海尔蜀葵■☆
13923 Alcea hyrcana Grossh.;希尔康蜀葵■☆
13924 Alcea indica Burm. f. = Hibiscus indicus (Burm. f.) Hochr. ●
13925 Alcea karakalensis Freyn;卡拉卡利蜀葵■☆
13926 Alcea karsiana (Bordz.) Litv.;卡尔斯蜀葵■☆
13927 Alcea kopetdaghensis Iljin;科佩特蜀葵■☆
13928 Alcea kusariensis (Iljin ex Grossh.) Iljin;库萨雷蜀葵■☆
13929 Alcea lavateraeflora (DC.) Boiss.;花葵蜀葵■☆
13930 Alcea lenkoranica Iljin;连科兰蜀葵■☆
13931 Alcea novopokrovskyi Iljin;诺沃蜀葵■☆
13932 Alcea nudiflora (Lindl.) Boiss.;裸花蜀葵;Nakedflower Althaea ■
13933 Alcea pallida (Waldst. et Kit.) Besser = Althaea pallida Waldst. et Kit. ■☆
13934 Alcea rosea L.;蜀葵(安省蜀葵,白淑气花,白蜀葵,饽饽花,侧金盏,镲钹花,大收旧花,大叔敬花,大熟钱,大蜀季,单片花,荻葵,斗莲,斗篷花,端午花,粉蜀葵,擀杖花,光光花,果木花,红牡丹,胡葵,鸡冠花,肻,荆葵,葵花,麻杆花,满长红,绵葵,木槿花,芘芣,棋盘花,苺,戎葵,舌其花,淑气花,熟季花,暑气花,蜀季,蜀季花,蜀芪花,蜀其花,蜀再,树茄花,水芙蓉,卫足,卫足葵,吴葵,小蜀芪,一丈红,栽秧花,杖红,杖葵,杖石);Antwerp Hollyhock, Billy Buttons, Hock, Hock-holler, Holliock, Holloak, Holly Oak, Hollyanders, Hollyhock, Hollyhocks, Holy Mallow, Holyoke, Jacob's Ladder, Jagged Mallow, Rose Mallow ■
13935 Alcea rosea L. 'Chater's Double';查特重瓣蜀葵■☆
13936 Alcea rosea L. 'Majorette';女指挥蜀葵■☆
13937 Alcea rosea L. 'Summer Carnival';夏日狂欢蜀葵■☆
13938 Alcea rosea L. = Althaea rosea (L.) Cav. ■
13939 Alcea rosea L. var. sibthorpii Boiss. = Alcea rosea L. ■
13940 Alcea rugosa Alef.;多皱蜀葵■☆
13941 Alcea sachsachanica Iljin;萨克萨哈蜀葵■☆
13942 Alcea sophiae Iljin;索氏蜀葵■☆
13943 Alcea sosnovskyi Iljin;锁斯诺夫斯基蜀葵■☆
13944 Alcea striata (DC.) Alef.;条纹蜀葵■☆
13945 Alcea sycophylla Iljin et Nikitin;无花果叶蜀葵■☆
13946 Alcea talassica Iljin;塔拉斯蜀葵■☆
13947 Alcea taurica Iljin;克里木蜀葵■☆
13948 Alcea turcomanica Iljin;土库曼蜀葵■☆
13949 Alcea turkeviczii Iljin;图尔蜀葵■☆
13950 Alcea woronowii (Iljin ex Grossh.) Iljin;沃氏蜀葵■☆
13951 Alchemilla L. (1753);羽衣草属(斗篷草属);Ladymantle, Lady's Mantle, Lady's-mantle, Parsley-piert ■
13952 Alchemilla abchasica Buser;阿伯哈斯羽衣草■☆
13953 Alchemilla abyssinica Fresen.;阿比西尼亚羽衣草■☆
13954 Alchemilla abyssinica Fresen. f. muscoidea Hauman et Balle = Alchemilla microbetula T. C. E. Fr. ■☆
13955 Alchemilla acutangulata Buser;锐角斗篷草■☆
13956 Alchemilla aemula Juz.;类斗篷草■☆
13957 Alchemilla alexandri Juz.;阿氏斗篷草■☆
13958 Alchemilla alpestris F. W. Schmidt;阿尔卑斯斗篷草■☆
13959 Alchemilla alpina L.;高山羽衣草(高山斗篷草);Alpine Ladymantle, Alpine Lady's Mantle, Alpine Lady's-mantle, Dew Mantle, Five-leaved Lady's Mantle, Mountain Lady's-mantle ■☆
13960 Alchemilla altaica Juz.;阿尔泰斗篷草■☆
13961 Alchemilla anisopoda Juz.;异梗斗篷草■☆
13962 Alchemilla arcuatiloba Juz.;弓裂斗篷草■☆
13963 Alchemilla argutiserrata Lindb.;具齿斗篷草■☆
13964 Alchemilla argyrophylla Oliv.;银叶羽衣草■☆
13965 Alchemilla argyrophylloides Baker f. = Alchemilla argyrophylla Oliv. ■☆
13966 Alchemilla arvensis (L.) Scop.;田斗篷草(阿韦羽衣草,野斗篷草);Argentill, Bowel-hive, Bowel-hive Grass, Breakstone Parsley, Colickwort, Field Lady's Mantle, Field Lady's-mantle, Field Parsley Piert, Fire-grass, Honewort, Parsley Breakstone, Parsley Peart, Parsley Peat, Parsley Perk, Parsley Piercestone, Parsley Pierce-stone, Parsley Piert, Parsley Vlix, Parsley-piert Parsley Piert, Percepier ■☆
13967 Alchemilla arvensis (L.) Scop. = Aphanes arvensis L. ■☆
13968 Alchemilla arvensis (L.) Scop. = Aphanes microcarpa (Boiss. et Reut.) Rothm. ■☆
13969 Alchemilla arvensis (L.) Scop. subsp. cornucopioides (Lag.) Maire = Aphanes cornucopioides Lag. ■☆
13970 Alchemilla arvensis (L.) Scop. subsp. floribunda (Murb.) Maire = Aphanes floribunda (Murb.) Rothm. ■☆
13971 Alchemilla arvensis (L.) Scop. subsp. maroccana (Hyl. et Rothm.) Maire = Aphanes maroccana Hyl. et Rothm. ■☆
13972 Alchemilla arvensis (L.) Scop. subsp. microcarpa (Boiss. et Reut.) Maire = Aphanes microcarpa (Boiss. et Reut.) Rothm. ■☆
13973 Alchemilla arvensis (L.) Scop. var. algeriensis Batt. = Aphanes floribunda (Murb.) Rothm. ■☆
13974 Alchemilla arvensis (L.) Scop. var. pusilla (Pomel) Maire = Aphanes pusilla (Pomel) Batt. ■☆
13975 Alchemilla atlantica H. Lindb.;大西洋斗篷草■☆
13976 Alchemilla bachiti Hauman et Balle = Aphanes bachiti (Hauman et Balle) Rothm. ■☆
13977 Alchemilla bakeri De Wild.;贝克斗篷草■☆
13978 Alchemilla baltica Sam. ex Juz. = Alchemilla nebulosa Sam. ■☆
13979 Alchemilla barbatiflora Juz.;毛花斗篷草■☆
13980 Alchemilla barbulata Juz.;髯毛斗篷草■☆
13981 Alchemilla bicarpellata Rothm.;双小果斗篷草■☆
13982 Alchemilla biquadrata Juz.;双四斗篷草■☆
13983 Alchemilla bolusii De Wild.;博卢斯斗篷草■☆
13984 Alchemilla brevidens Juz.;短齿斗篷草■☆
13985 Alchemilla breviloba L.;短裂斗篷草■☆
13986 Alchemilla bungei Juz.;邦奇斗篷草■☆
13987 Alchemilla buschii Juz.;布氏斗篷草;Busch Ladymantle ■☆
13988 Alchemilla camptopoda Juz.;曲柄斗篷草■☆
13989 Alchemilla capensis Thunb.;好望角斗篷草■☆
13990 Alchemilla caucasica Buser;高加索斗篷草■☆
13991 Alchemilla chlorosericea Buser;绿绢毛斗篷草■☆
13992 Alchemilla circassica Juz.;切尔卡西亚斗篷草■☆
13993 Alchemilla commutata Rothm. = Alchemilla microbetula T. C. E. Fr. ■☆
13994 Alchemilla commutata Rothm. f. muscoidea (Hauman et Balle) Rothm. = Alchemilla microbetula T. C. E. Fr. ■☆
13995 Alchemilla commutata Rothm. f. robusta ? = Alchemilla microbetula T. C. E. Fr. ■☆

13996 Alchemilla compactilis Juz.;紧密斗篷草■☆

13997 Alchemilla conglobata H. Lindb.;球形斗篷草■☆

13998 Alchemilla conjuncta Bab.;接合斗篷草;Silver Lady's-mantle ■☆

13999 Alchemilla cornucopioides (Lag.) Maire = Aphanes cornucopioides Lag. ■☆

14000 Alchemilla crebridens Juz.;密齿斗篷草■☆

14001 Alchemilla cryptantha Steud. ex A. Rich.;隐花斗篷草■☆

14002 Alchemilla cymatophylla Juz.;波叶斗篷草■☆

14003 Alchemilla cyrtopleura Juz.;弯脉斗篷草■☆

14004 Alchemilla daghestanica Juz.;达赫斯坦斗篷草■☆

14005 Alchemilla debilis Juz.;柔软斗篷草■☆

14006 Alchemilla decalvans Juz.;秃斗篷草■☆

14007 Alchemilla denticulata Juz.;细齿斗篷草■☆

14008 Alchemilla dewildemanii T. C. E. Fr.;德怀尔斗篷草■☆

14009 Alchemilla diglossa Juz.;二舌斗篷草■☆

14010 Alchemilla divaricans Buser;开展斗篷草■☆

14011 Alchemilla diversipes Juz.;异足斗篷草■☆

14012 Alchemilla dura Buser;硬斗篷草■☆

14013 Alchemilla egens Juz.;贫瘠斗篷草■☆

14014 Alchemilla elata Buser;高斗篷草■☆

14015 Alchemilla elgonensis Mildbr.;埃尔贡斗篷草■☆

14016 Alchemilla elisabethae Juz.;艾丽萨斗篷草■☆

14017 Alchemilla ellenbeckii Engl.;埃伦斗篷草■☆

14018 Alchemilla ellenbeckii Engl. subsp. nyikensis (De Wild.) R. A. Graham;尼卡斗篷草■☆

14019 Alchemilla elongata Eckl. et Zeyh.;伸长斗篷草■☆

14020 Alchemilla elongata Eckl. et Zeyh. var. platyloba Rothm.;宽裂伸长斗篷草■

14021 Alchemilla erlangeriana Engl. = Alchemilla fischeri Engl. ■☆

14022 Alchemilla erythropoda Juz.;红茎斗篷草;Redstemmed Lady's Mantle ■☆

14023 Alchemilla exsanguis Juz.;苍白斗篷草■☆

14024 Alchemilla exuens Juz.;倒斗篷草■☆

14025 Alchemilla filicaulis Buser;线茎斗篷草;Alchemilla filicaulis ■☆

14026 Alchemilla fischeri Engl.;菲舍尔斗篷草■☆

14027 Alchemilla fischeri Engl. subsp. camerunensis Letouzey;喀麦隆斗篷草■☆

14028 Alchemilla flavescens;浅黄斗篷草■☆

14029 Alchemilla floribunda Murb. = Aphanes floribunda (Murb.) Rothm. ■☆

14030 Alchemilla fontinalis Juz.;春斗篷草■☆

14031 Alchemilla frondosa Juz.;阔叶斗篷草■☆

14032 Alchemilla galpinii Hauman et Balle;盖尔斗篷草■☆

14033 Alchemilla georgica Juz.;乔治斗篷草■☆

14034 Alchemilla gerrardii De Wild. = Alchemilla incurvata Gand. ■☆

14035 Alchemilla gerrardii De Wild. var. hirsuto-petiolata ? = Alchemilla hirsuto-petiolata (De Wild.) Rothm. ■☆

14036 Alchemilla gibberulosa H. Lindb.;瘤斗篷草■☆

14037 Alchemilla glabra Neygenf.;无毛羽衣草;Alchemilla glabra, Glabrous Ladymantle, Hairless Ladymantle ■

14038 Alchemilla glabricaulis H. Lindb.;光茎斗篷草■☆

14039 Alchemilla glomerulans Buser;团集斗篷草;Alchemilla glomerulans ■☆

14040 Alchemilla gourzae Ibn Tattou;古尔扎斗篷草■☆

14041 Alchemilla gracilipes (Engl.) Engl. = Alchemilla pedata Hochst. ex A. Rich. ■☆

14042 Alchemilla gracilis Opiz;纤细羽衣草;Alchemilla gracilis, Slender Ladymantle ■

14043 Alchemilla grandidens Juz.;大齿斗篷草■☆

14044 Alchemilla grossheimii Juz.;格罗斗篷草■☆

14045 Alchemilla gunae Schweinf. = Alchemilla abyssinica Fresen. ■☆

14046 Alchemilla haraldii Juz.;哈拉尔德斗篷草■☆

14047 Alchemilla haumanii Rothm.;豪曼斗篷草■☆

14048 Alchemilla hebescens Juz.;钝化斗篷草■☆

14049 Alchemilla hendrickxii Hauman et Balle;亨德里克斯斗篷草■☆

14050 Alchemilla heptagona Juz.;七边斗篷草■☆

14051 Alchemilla hians Juz.;开裂斗篷草■☆

14052 Alchemilla hirsuticaulis H. Lindb.;毛茎斗篷草■☆

14053 Alchemilla hirsutissima Juz.;厚毛斗篷草■☆

14054 Alchemilla hirsuto-petiolata (De Wild.) Rothm.;毛柄斗篷草■☆

14055 Alchemilla hirtipedicellata Juz.;毛梗斗篷草■☆

14056 Alchemilla holotricha Juz.;全毛斗篷草■☆

14057 Alchemilla humilicaulis Juz.;矮茎斗篷草■☆

14058 Alchemilla hypochlora Juz.;下绿斗篷草■☆

14059 Alchemilla hyrcana Buser;希尔康斗篷草■☆

14060 Alchemilla imberbis Juz.;无毛斗篷草■☆

14061 Alchemilla incurvata Gand.;弯曲斗篷草■☆

14062 Alchemilla insignis Juz.;显著斗篷草■☆

14063 Alchemilla inyangensis Weim. = Alchemilla cryptantha Steud. ex A. Rich. ■☆

14064 Alchemilla jailae Juz.;艾氏斗篷草■☆

14065 Alchemilla japonica Nakai et H. Hara;羽衣草(斗篷草,日本羽衣草);Bear's Foot, Common Ladymantle, Duck's Foot, Elf-shot, Great Saniicle, Japanese Ladymantle, Ladies Mantle, Lady's Mantle, Lamb's Foot, Lion's Foot, Lion's Paw, Padelion, Parsley Breakstone, Syndow ■

14066 Alchemilla johnstonii Oliv.;约翰斯顿斗篷草■☆

14067 Alchemilla juzepszukii Alechin;朱氏斗篷草■☆

14068 Alchemilla kiwuensis Engl.;基武斗篷草■☆

14069 Alchemilla kozlowskii Juz.;考氏斗篷草■☆

14070 Alchemilla krylovii Juz.;光柄羽衣草■

14071 Alchemilla laeta Juz.;愉悦斗篷草■☆

14072 Alchemilla laeticolor Juz.;鲜色斗篷草■☆

14073 Alchemilla languescens Juz.;弱斗篷草■☆

14074 Alchemilla leiophylla Juz.;光叶斗篷草■☆

14075 Alchemilla lindbergiana Emb. = Alchemilla gourzae Ibn Tattou ■☆

14076 Alchemilla lindbergiana Juz.;林德伯氏斗篷草■☆

14077 Alchemilla lipschitzii Juz.;里普斗篷草■☆

14078 Alchemilla litardierei H. Lindb.;利塔斗篷草■☆

14079 Alchemilla lithophila Juz.;石地斗篷草■☆

14080 Alchemilla litwinowii Juz.;利特氏斗篷草■☆

14081 Alchemilla lydiae Zamells;里氏斗篷草■☆

14082 Alchemilla mairei H. Lindb. = Alchemilla atlantica H. Lindb. ■☆

14083 Alchemilla micans Buser;闪光斗篷草■☆

14084 Alchemilla microbetula T. C. E. Fr.;小白桦斗篷草■☆

14085 Alchemilla microcarpa Boiss. et Reut. = Aphanes inexspectata W. Lippert ■☆

14086 Alchemilla microdonta Juz.;小齿斗篷草■☆

14087 Alchemilla minor Huds.;小斗篷草■☆

14088 Alchemilla minusculiflora Buser;稍小花斗篷草■☆

14089 Alchemilla mollis (Buser) Rothm.;柔毛斗篷草(柔软羽衣草);Lady's-mantle, Lady's Mantle ■☆

14090 Alchemilla monticola Opiz;山地斗篷草;Hairy Lady's-mantle ■☆

14091 Alchemilla murbeckiana Buser;穆氏斗篷草■☆

14092 Alchemilla natalensis Engl. ;纳塔尔斗篷草■☆
14093 Alchemilla nebulosa Sam. ;波罗的海斗篷草■☆
14094 Alchemilla nemoralis Alechin;栎林斗篷草■☆
14095 Alchemilla neostevenii Juz. ;新司氏斗篷草■☆
14096 Alchemilla nyikensis De Wild. = Alchemilla ellenbeckii Engl. subsp. nyikensis (De Wild.) R. A. Graham ■☆
14097 Alchemilla obtegens Juz. ;覆盖斗篷草■☆
14098 Alchemilla obtusa Buser;钝斗篷草■☆
14099 Alchemilla obtusiformis Alechin;钝形斗篷草■☆
14100 Alchemilla oligotricha Juz. ;寡毛斗篷草■☆
14101 Alchemilla omalophylla Juz. ;平叶斗篷草■☆
14102 Alchemilla orbicans Juz. ;圆斗篷草■☆
14103 Alchemilla oxysepala Juz. ;尖萼斗篷草■☆
14104 Alchemilla pachyphylla Juz. ;厚叶斗篷草■☆
14105 Alchemilla palmata E. Mey. = Alchemilla elongata Eckl. et Zeyh. ■☆
14106 Alchemilla pastoralis Buser;牧场斗篷草■☆
14107 Alchemilla pedata Hochst. ex A. Rich. ;鸟足状斗篷草■☆
14108 Alchemilla pedata Hochst. ex A. Rich. var. argentea Chiov. ex Fiori = Alchemilla pedata Hochst. ex A. Rich. ■☆
14109 Alchemilla pedata Hochst. ex A. Rich. var. gracilipes Engl. = Alchemilla pedata Hochst. ex A. Rich. ■☆
14110 Alchemilla pilosiplica Juz. ;疏毛斗篷草■☆
14111 Alchemilla pinguis Juz. ;阿尔泰羽衣草■
14112 Alchemilla plicata Buser;折扇斗篷草■☆
14113 Alchemilla pogonophora Juz. ;具须斗篷草■☆
14114 Alchemilla porrectidens Juz. ;外伸齿斗篷草■☆
14115 Alchemilla propinqua H. Lindb. ex Juz. ;亲缘斗篷草■☆
14116 Alchemilla pseudocartalinica Juz. ;假木豆斗篷草■☆
14117 Alchemilla pseudomollis Juz. ;假柔毛斗篷草■☆
14118 Alchemilla psilocaulis Juz. ;裸茎斗篷草■☆
14119 Alchemilla psiloneura Juz. ;露脉斗篷草■☆
14120 Alchemilla pubescens Lam. subsp. litardierei (H. Lindb.) Maire = Alchemilla litardierei H. Lindb. ■☆
14121 Alchemilla purpurasceus Juz. ;紫斗篷草■☆
14122 Alchemilla pusilla Pomel = Aphanes pusilla (Pomel) Batt. ■☆
14123 Alchemilla pycnantha Juz. ;密花斗篷草■☆
14124 Alchemilla pycnotricha Juz. ;密毛斗篷草■☆
14125 Alchemilla quinqueloba Rothm. ;五裂斗篷草■☆
14126 Alchemilla raddeana Buser;拉德斗篷草■☆
14127 Alchemilla rehmannii Engl. ;拉赫曼斗篷草■☆
14128 Alchemilla retinervis Buser;网脉斗篷草■☆
14129 Alchemilla rigescens Juz. ;变硬斗篷草■☆
14130 Alchemilla rigida Buser;挺斗篷草■☆
14131 Alchemilla roccatii Cortesi;罗卡特斗篷草■☆
14132 Alchemilla rothii Oliv. = Alchemilla abyssinica Fresen. ■☆
14133 Alchemilla rubens Juz. ;红斗篷草■☆
14134 Alchemilla sanguinolenta Juz. ;血红斗篷草■☆
14135 Alchemilla sarmatica Juz. ;匍枝斗篷草■☆
14136 Alchemilla schischkinii Juz. ;希施斗篷草■☆
14137 Alchemilla schistophylla Juz. ;隙叶斗篷草■☆
14138 Alchemilla schlechteriana Rothm. ;施莱斗篷草■☆
14139 Alchemilla semilunaris Alechin;半月斗篷草■☆
14140 Alchemilla sericata Rchb. ;细绢毛斗篷草■☆
14141 Alchemilla sericea Willd. ;绢毛斗篷草■☆
14142 Alchemilla sessiliflora Hochst. ex Rothm. = Alchemilla microbetula T. C. E. Fr. ■☆
14143 Alchemilla sibirica Zamelis;西伯利亚羽衣草■
14144 Alchemilla speciosa Buser;美丽斗篷草■☆
14145 Alchemilla stellaris Juz. ;星状斗篷草■☆
14146 Alchemilla stellulata Juz. ;小星斗篷草■☆
14147 Alchemilla stevenii Buser;司梯文氏斗篷草■☆
14148 Alchemilla stuhlmannii Engl. ;斯图尔曼斗篷草■☆
14149 Alchemilla subcrenata Buser;凹叶斗篷草;Broadtooth Lady's Mantle ■☆
14150 Alchemilla subcrenata Buser subsp. atlantis Emb. = Alchemilla gourzae Ibn Tattou ■☆
14151 Alchemilla subcrenatiformis Juz. ;细圆齿斗篷草■☆
14152 Alchemilla suberectipila Juz. ;直立斗篷草■☆
14153 Alchemilla subnivalis Baker f. ;近雪白斗篷草■☆
14154 Alchemilla subsplendens Buser;光亮斗篷草■☆
14155 Alchemilla substrigosa Juz. ;微刺斗篷草■☆
14156 Alchemilla supina Juz. ;平卧斗篷草■☆
14157 Alchemilla taurica Juz. ;克里木斗篷草■☆
14158 Alchemilla tephroserica Buser;灰毛斗篷草■☆
14159 Alchemilla tianschanica Juz. ;天山羽衣草■
14160 Alchemilla triphylla Rothm. ;三叶斗篷草■☆
14161 Alchemilla turuchanica Juz. ;图鲁罕斗篷草■☆
14162 Alchemilla tytthantha Juz. ;小花斗篷草■☆
14163 Alchemilla urceolata Juz. ;坛状斗篷草■☆
14164 Alchemilla venosa Juz. ;多脉斗篷草■☆
14165 Alchemilla volkensii Engl. ;福尔斗篷草■☆
14166 Alchemilla vulgaris L. ;欧洲羽衣草(欧亚羽衣草,普通羽衣草,羽衣草);Common Lady's-mantle, Dew Cup, Dewcup, Lady's-mande, Lion's Foot, Parsley Breakstone, Piecestone ■☆
14167 Alchemilla vulgaris L. subsp. japonica (Nakai et H. Hara) Sugim. = Alchemilla japonica Nakai et H. Hara ■
14168 Alchemilla vulgaris L. var. pastoralis (Buser) B. Boivin = Alchemilla monticola Opiz ■☆
14169 Alchemilla vulgaris Makino = Alchemilla japonica Nakai et H. Hara ■
14170 Alchemilla wilmsii Engl. = Alchemilla woodii Kuntze ■☆
14171 Alchemilla woodii Kuntze;伍德斗篷草■☆
14172 Alchemilla woronowii Juz. ;沃氏斗篷草■☆
14173 Alchemillaceae J. Agardh = Rosaceae Juss. (保留科名)●■
14174 Alchemillaceae J. Agardh;羽衣草科(斗篷草科)●■
14175 Alchemillaceae Martinov = Rosaceae Juss. (保留科名)●■
14176 Alchimilla Mill. = Alchemilla L. ■
14177 Alchornea Sw. (1788);山麻杆属;Christmas Bush, Christmasbush, Christmas-bush, Dovewood, Xmas Bush ●
14178 Alchornea bangweolensis R. E. Fr. = Alchornea yambuyaensis De Wild. ●☆
14179 Alchornea caloneura Pax = Discoglypremna caloneura (Pax) Prain ●☆
14180 Alchornea comoensis Beille = Alchornea hirtella Benth. f. comoensis (Beille) Pax et K. Hoffm. ●☆
14181 Alchornea cordata Benth. = Alchornea cordifolia (Schumach. et Thonn.) Müll. Arg. ●☆
14182 Alchornea cordifolia (Schumach. et Thonn.) Müll. Arg. ;心叶山麻杆;Christmas Bush ●☆
14183 Alchornea cordifolia Müll. Arg. = Alchornea cordifolia (Schumach. et Thonn.) Müll. Arg. ●☆
14184 Alchornea davidii Franch. ;山麻杆(狗尾巴树,桂圆树,荷包麻,红荷叶,桐花杆,野火麻,麥包叶);David Christmas Bush,

David Christmasbush, David Christmas-bush, David Xmas Bush ●
14185 Alchornea duparquetiana Baill.;迪帕山麻杆●☆
14186 Alchornea engleri Pax = Alchornea laxiflora (Benth.) Pax et K. Hoffm. ●☆
14187 Alchornea floribunda Müll. Arg.;多花山麻杆;Many-flowers Christmasbush ●☆
14188 Alchornea floribunda Müll. Arg. var. glabrata ? = Alchornea hirtella Benth. f. glabrata (Müll. Arg.) Pax et K. Hoffm. ●☆
14189 Alchornea formosae Müll. Arg. ex Pax et K. Hoffm. = Alchornea kelungensis Hayata ●
14190 Alchornea glabrata (Müll. Arg.) Prain = Alchornea hirtella Benth. f. glabrata (Müll. Arg.) Pax et K. Hoffm. ●☆
14191 Alchornea hainanensis Pax et K. Hoffm. = Alchornea rugosa (Lour.) Müll. Arg. ●
14192 Alchornea hainanensis Pax et K. Hoffm. var. pubescens Pax et K. Hoffm. = Alchornea rugosa (Lour.) Müll. Arg. var. pubescens (Pax et K. Hoffm.) H. S. Kiu ●
14193 Alchornea hirtella Benth.;硬毛山麻杆(硬毛三稔蒟)●☆
14194 Alchornea hirtella Benth. f. comoensis (Beille) Pax et K. Hoffm.;科莫山麻杆●☆
14195 Alchornea hirtella Benth. f. cuneata Pax et K. Hoffm.;楔形硬毛山麻杆●☆
14196 Alchornea hirtella Benth. f. glabrata (Müll. Arg.) Pax et K. Hoffm.;无毛山麻杆●☆
14197 Alchornea hunanensis H. S. Kiu;湖南山麻杆;Hunan Christmasbush, Hunan Christmas-bush, Hunan Xmas Bush ●
14198 Alchornea hunanensis H. S. Kiu var. pubescens Pax et K. Hoffm. = Alchornea rugosa (Lour.) Müll. Arg. var. pubescens (Pax et K. Hoffm.) H. S. Kiu ●
14199 Alchornea kelungensis Hayata;厚柱山麻杆(台湾山麻杆);Kelung Christmas-bush, Thickstyle Xmas Bush ●
14200 Alchornea kelungensis Hayata = Alchornea liukiuensis Hayata ●
14201 Alchornea kelungensis Hayata var. formosa (Müll. Arg.) Hurus. = Alchornea kelungensis Hayata ●
14202 Alchornea latifolia Sw.;美洲山麻杆●☆
14203 Alchornea laxiflora (Benth.) Pax et K. Hoffm.;疏花山麻杆●☆
14204 Alchornea liukiuensis Hayata;琉球山麻杆●
14205 Alchornea liukiuensis Hayata = Alchornea trewioides (Benth.) Müll. Arg. ●
14206 Alchornea liukiuensis Hayata var. formosae (Müll. Arg. ex Pax et K. Hoffm.) Hurus. = Alchornea liukiuensis Hayata ●
14207 Alchornea loochooensis Hayata = Alchornea liukiuensis Hayata ●
14208 Alchornea mairei H. Lév. = Cnesmone mairei (H. Lév.) Croizat ●
14209 Alchornea mildbraedii Pax et K. Hoffm.;米氏山麻杆●☆
14210 Alchornea mollis (Benth.) Müll. Arg.;毛果山麻杆(毛叶山麻杆);Hairfruit Christmasbush, Hairfruit Xmas Bush, Hairy-fruited Christmas-bush ●
14211 Alchornea occidentalis (Müll. Arg.) Pax et K. Hoffm.;西方山麻杆●☆
14212 Alchornea rufescens Franch. = Discocleidion rufescens (Franch.) Pax et K. Hoffm. ●
14213 Alchornea rugosa (Lour.) Müll. Arg.;羽脉山麻杆(二两八树,苦茶,三稔蒟,山麻杆);Pinnatenerve Christmasbush, Pinnatenerve Xmas Bush, Wrinkled Christmas-bush ●
14214 Alchornea rugosa (Lour.) Müll. Arg. var. pubescens (Pax et K. Hoffm.) H. S. Kiu;海南山麻杆;Pubescent Pinnatenerve Christmasbush ●
14215 Alchornea schlechteri Pax = Alchornea laxiflora (Benth.) Pax et K. Hoffm. ●☆
14216 Alchornea tiliifolia (Benth.) Müll. Arg.;椴叶山麻杆(野生麻);Linden-leaf Christmasbush, Lindenleaf Xmas Bush, Linder-leaved Christmas-bush ●
14217 Alchornea trewioides (Benth.) Müll. Arg.;红背山麻杆(红背白欟树,红背娘,红背叶,红灵丹,红罗裙,红帽顶,红帽顶树,台湾山麻杆);Redback Christmasbush, Redback Christmas-bush, Redback Xmas Bush ●
14218 Alchornea trewioides (Benth.) Müll. Arg. = Alchornea liukiuensis Hayata ●
14219 Alchornea trewioides (Benth.) Müll. Arg. var. formosae (Müll. Arg.) Pax et K. Hoffm.;台湾山麻杆;Taiwan Christmasbush ●
14220 Alchornea trewioides (Benth.) Müll. Arg. var. formosae (Müll. Arg.) Pax et K. Hoffm. = Alchornea kelungensis Hayata ●
14221 Alchornea trewioides (Benth.) Müll. Arg. var. loochooensis (Hayata) H. Keng = Alchornea liukiuensis Hayata ●
14222 Alchornea trewioides (Benth.) Müll. Arg. var. sinica H. S. Kiu;绿背山麻杆;Green-back Redback Christmasbush ●
14223 Alchornea ulmifolia (Müll. Arg.) Hurus. = Discocleidion ulmifolium (Müll. Arg.) Pax et K. Hoffm. ●☆
14224 Alchornea verrucosa Pax = Alchornea yambuyaensis De Wild. ●☆
14225 Alchornea yambuyaensis De Wild.;扬布亚山麻杆●☆
14226 Alchorneopsis Müll. Arg. (1865);类山麻杆属●☆
14227 Alchorneopsis floribunda Müll. Arg.;类山麻杆●☆
14228 Alchymilla Rupp. = Alchemilla L. ■
14229 Alcimandra Dandy = Magnolia L. ●
14230 Alcimandra Dandy (1927);长蕊木兰属;Alcimandra ●
14231 Alcimandra cathcarti (Hook. f. et Thomson) Dandy;长蕊木兰;Alcimandra, Common Alcimandra ●◇
14232 Alcina Cav. = Melampodium L. ●■
14233 Alcina perfoliata Cav. = Melampodium perfoliatum (Cav.) Kunth ■☆
14234 Alcinaeanthus Merr. = Neoscortechinia Pax☆
14235 Alcinia Kunth = Alcina Cav. ●■
14236 Alcinia Kunth = Melampodium L. ●■
14237 Alciope DC. = Capelio B. Nord. ■☆
14238 Alciope DC. = Celmisia Cass. (废弃属名)■☆
14239 Alciope DC. ex Lindl. = Capelio B. Nord. ■☆
14240 Alciope DC. ex Lindl. = Celmisia Cass. (废弃属名)■☆
14241 Alciope lanata (Thunb.) DC. = Capelio tomentosa (Burm. f.) B. Nord. ■☆
14242 Alciope tabularis (Thunb.) DC. = Capelio tabularis (Thunb.) B. Nord. ■☆
14243 Alcmene Urb. = Duguetia A. St.-Hil. (保留属名)●☆
14244 Alcoceratothrix Nied. = Byrsonima Rich. ex Juss. ●☆
14245 Alcoceria Fernald = Dalembertia Baill. ☆
14246 Alcytophyllum T. Durand = Arcytophyllum Willd. ex Schult. et Schult. f. ●☆
14247 Aldaea Schltdl. = Aldea Ruiz et Pav. ■☆
14248 Aldaea Schltdl. = Phacelia Juss. ■☆
14249 Aldama La Llave (1824);齿黑药菊属■☆
14250 Aldama dentata La Llave;齿黑药菊■☆
14251 Aldasorea Hort. ex Haage et Schmidt = Aeonium Webb et Berthel. ●■☆
14252 Aldea Ruiz et Pav. = Phacelia Juss. ■☆
14253 Aldeaea Rchb. = Aldea Ruiz et Pav. ■☆

14254 Aldelaster C. Koch = Pseuderanthemum Radlk. ●■
14255 Aldelaster C. Koch = Adelaster Lindl. ex Veitch(废弃属名)■☆
14256 Aldelaster C. Koch = Adelaster Veitch ■☆
14257 Aldelaster K. Koch = Adelaster Lindl. ex Veitch(废弃属名)■☆
14258 Aldenella Greene = Cleome L. ●■
14259 Aldina Adans.(废弃属名) = Aldina Endl.(保留属名)●☆
14260 Aldina Adans.(废弃属名) = Brya P. Browne ●☆
14261 Aldina E. Mey. = Acacia Mill.(保留属名)●■
14262 Aldina Endl.(1840)(保留属名);阿尔丁豆属●☆
14263 Aldina aurea R. S. Cowan;黄阿尔丁豆■☆
14264 Aldina discolor Spruce ex Benth.;杂色阿尔丁豆■☆
14265 Aldina latifolia Benth.;宽叶阿尔丁豆■☆
14266 Aldina macrophylla Spruce ex Benth.;大叶阿尔丁豆■☆
14267 Aldina occidentalis Ducke;西方阿尔丁豆■☆
14268 Aldina polyphylla Ducke;多叶阿尔丁豆■☆
14269 Aldinia Raf. = Croton L. ●
14270 Aldinia Scop. = Justicia L. ●■
14271 Aldrovanda L.(1753);貉藻属;Aldrovanda ■
14272 Aldrovanda Monti = Aldrovanda L. ■
14273 Aldrovanda vesiculosa L.;貉藻;Aldrovanda, Water Bug Trap, Water Dionaea, Water-bug-trap, Waterwheel Plant ■
14274 Aldrovandaceae Nakai = Droseraceae Salisb.(保留科名)■
14275 Aldrovandaceae Nakai;貉藻科■
14276 Aldunatea J. Rémy = Chaetanthera Ruiz et Pav. ■☆
14277 Alectoridia A. Rich. = Arthraxon P. Beauv. ■
14278 Alectoridia quartiniana A. Rich. = Arthraxon hispidus (Thunb.) Makino ■
14279 Alectoroctonum Schltdl. = Euphorbia L. ●■
14280 Alectorolophus Mill. = Rhinanthus L. ■
14281 Alectorolophus Zinn = Rhinanthus L. ■
14282 Alectorolophus major (Ehrh.) Rchb. = Rhinanthus glaber Lam. ■
14283 Alectorolophus songaricus Sterneck = Rhinanthus glaber Lam. ■
14284 Alectorolophus vernalis N. W. Zinger = Rhinanthus glaber Lam. ■
14285 Alectorurus Makino = Comospermum Rauschert ■☆
14286 Alectorurus Makino(1908);鸡尾莲属■☆
14287 Alectorurus platypetalus Masam.;宽瓣鸡尾莲■☆
14288 Alectorurus yedoensis Makino;鸡尾莲■☆
14289 Alectra Thunb.(1784);黑蒴属■
14290 Alectra aberdarica Chiov.;阿伯德尔黑蒴■☆
14291 Alectra abyssinica (Hochst. ex Benth.) A. Rich. = Hedbergia abyssinica (Hochst. ex Benth.) Molau ■☆
14292 Alectra alba (Hepper) B. L. Burtt;白花黑蒴■☆
14293 Alectra alectroides (S. Moore) Melch.;普通黑蒴■☆
14294 Alectra angustifolia Engl. = Alectra vogelii Benth. ■☆
14295 Alectra arachidis A. Chev. = Alectra vogelii Benth. ■☆
14296 Alectra arvensis (Benth.) Merr. = Melasma arvense (Benth.) Hand. -Mazz. ■
14297 Alectra arvensis Merr.;黑蒴(红根草,化血胆,小化血草);Field Blackcapsule, Field Melasma ■
14298 Alectra arvensis Merr. = Melasma arvense (Benth.) Hand. -Mazz. ■
14299 Alectra asperrima Benth.;粗糙黑蒴■☆
14300 Alectra atrosanguinea (Hiern) Hemsl.;暗血红黑蒴■☆
14301 Alectra aurantiaca Hemsl.;橙黑蒴■☆
14302 Alectra avensis (Benth.) Merr. = Alectra sessiliflora (Vahl) Kuntze var. monticola (Engl.) Melch. ■☆
14303 Alectra bainesii Hemsl.;贝恩斯黑蒴■☆
14304 Alectra barbata (Hiern) Melch. = Alectra sessiliflora (Vahl) Kuntze ■☆
14305 Alectra capensis Thunb.;好望角黑蒴■☆
14306 Alectra communis Hemsl. = Alectra sessiliflora (Vahl) Kuntze var. monticola (Engl.) Melch. ■☆
14307 Alectra congolensis Troupin = Alectra sessiliflora (Vahl) Kuntze var. monticola (Engl.) Melch. ■☆
14308 Alectra cordata Benth. = Alectra sessiliflora (Vahl) Kuntze var. senegalensis (Benth.) Hepper ■☆
14309 Alectra dentata K. Schum. = Alectra arvensis Merr. ■
14310 Alectra dolichocalyx Philcox;长萼黑蒴■☆
14311 Alectra dunensis Hilliard et B. L. Burtt;砂丘黑蒴■☆
14312 Alectra fluminensis (Vell.) Stearn;河岸黑蒴■☆
14313 Alectra glandulosa Philcox;腺点黑蒴■☆
14314 Alectra gracilis S. Moore;纤细黑蒴■☆
14315 Alectra heyniae (Dinter) Dinter = Alectra orobanchoides Benth. ■☆
14316 Alectra hippocrepandra (Hiern) Hemsl. = Alectra vogelii Benth. ■☆
14317 Alectra hirsuta Klotzsch;粗毛黑蒴■☆
14318 Alectra hundtii Melch.;洪特黑蒴■☆
14319 Alectra indica Benth. = Alectra sessiliflora (Vahl) Kuntze var. monticola (Engl.) Melch. ■☆
14320 Alectra indica Wettst. = Alectra arvensis Merr. ■
14321 Alectra kilimandjarica Hemsl. = Alectra orobanchoides Benth. ■☆
14322 Alectra kirkii Hemsl. = Alectra orobanchoides Benth. ■☆
14323 Alectra lancifolia Hemsl.;剑叶黑蒴■☆
14324 Alectra ledermannii Engl.;莱德黑蒴■☆
14325 Alectra linearis Hepper;线状黑蒴■☆
14326 Alectra lurida Harv.;灰黄黑蒴■☆
14327 Alectra melampyroides Benth. = Alectra sessiliflora (Vahl) Kuntze ■☆
14328 Alectra merkeri Engl.;默克黑蒴■☆
14329 Alectra moeroensis Engl. = Alectra sessiliflora (Vahl) Kuntze var. monticola (Engl.) Melch. ■☆
14330 Alectra natalensis (Hiern) Melch.;纳塔尔黑蒴■☆
14331 Alectra omurambensis Dinter ex Melch. = Alectra aurantiaca Hemsl. ■☆
14332 Alectra orobanchoides Benth.;柯克黑蒴■☆
14333 Alectra parasitica A. Rich.;寄生黑蒴■☆
14334 Alectra parvifolia (Engl.) Schinz = Alectra orobanchoides Benth. ■☆
14335 Alectra petitiana A. Rich. = Hedbergia abyssinica (Hochst. ex Benth.) Molau ■☆
14336 Alectra picta (Hiern) Hemsl.;着色黑蒴■☆
14337 Alectra pseudobarleriae (Dinter) Dinter;假杜鹃黑蒴■☆
14338 Alectra pubescens Philcox;短柔毛黑蒴■☆
14339 Alectra pumila Benth.;矮小黑蒴■☆
14340 Alectra pusilla E. Phillips = Alectra schoenfelderi Dinter et Melch. ■☆
14341 Alectra rigida (Hiern) Hemsl.;坚挺黑蒴■☆
14342 Alectra schliebenii Melch.;施利本黑蒴■☆
14343 Alectra schoenfelderi Dinter et Melch.;舍恩黑蒴■☆
14344 Alectra senegalensis Benth. = Alectra sessiliflora (Vahl) Kuntze var. senegalensis (Benth.) Hepper ■☆
14345 Alectra senegalensis Benth. var. minima A. Chev. = Alectra sessiliflora (Vahl) Kuntze ■☆
14346 Alectra senegalensis Benth. var. pallescens Bonati = Alectra

sessiliflora (Vahl) Kuntze ■☆
14347 Alectra sessiliflora (Vahl) Kuntze;无梗花黑蒴■☆
14348 Alectra sessiliflora (Vahl) Kuntze f. barbata (Hiern) Hilliard et B. L. Burtt = Alectra sessiliflora (Vahl) Kuntze ■☆
14349 Alectra sessiliflora (Vahl) Kuntze var. barbata (Hiern) Hilliard et B. L. Burtt = Alectra sessiliflora (Vahl) Kuntze ■☆
14350 Alectra sessiliflora (Vahl) Kuntze var. monticola (Engl.) Melch.;山地无梗花黑蒴■☆
14351 Alectra sessiliflora (Vahl) Kuntze var. senegalensis (Benth.) Hepper;塞内加尔黑蒴■☆
14352 Alectra stolzii Engl.;斯托尔兹黑蒴■☆
14353 Alectra thyrsoidea Melch.;聚伞黑蒴■☆
14354 Alectra trinervis Hemsl.;三脉黑蒴■☆
14355 Alectra virgata Hemsl.;条纹黑蒴■☆
14356 Alectra vogelii Benth.;沃格尔黑蒴■☆
14357 Alectra welwitschii (Hiern) Hemsl.;韦尔黑蒴■☆
14358 Alectra welwitschii (Hiern) Hemsl. = Alectra orobanchoides Benth. ■☆
14359 Alectryon Gaertn. (1788);冉布檀属;Alectryon ●☆
14360 Alectryon affinis Radlk.;近缘冉布檀●☆
14361 Alectryon canescens DC.;灰冉布檀●☆
14362 Alectryon excelsus Gaertn.;冉布檀;Tikoki Alectryon,Titoki ●☆
14363 Alectryon ferrugineum Radlk.;锈色冉布檀●☆
14364 Alectryon fuscus Radlk.;褐冉布檀●☆
14365 Alectryon glabrum Radlk.;光冉布檀●☆
14366 Alectryon grandifolius A. C. Sm.;大叶冉布檀●☆
14367 Alectryon grandis Cheeseman;大冉布檀●☆
14368 Alectryon laevis Radlk.;平滑冉布檀●☆
14369 Alectryon oleifolius (Desf.) S. T. Reynolds;榄叶冉布檀;Australian Rosewood, Boonaree, Boonery, Boonery Boonaree, Western Rosewood ●☆
14370 Alectryon oleifolius (Desf.) S. T. Reynolds subsp. canescens S. T. Reynolds;灰榄叶冉布檀●☆
14371 Alectryon oleifolius (Desf.) S. T. Reynolds subsp. elongatus S. T. Reynolds;长榄叶冉布檀●☆
14372 Alectryon pubescens (S. Reyn.) S. T. Reynolds;毛冉布檀●☆
14373 Alectryon reticulatus Radlk.;网脉冉布檀●☆
14374 Alectryon subcinereus (A. Gray) Radlk.;光滑冉布檀;Australian Native Quince, Native Quince, Smooth Rambutan ●☆
14375 Alectryon tomentosus (F. Muell.) Radlk.;毛叶冉布檀;Hairy Bird's Eye, Woolly Rambutan ●☆
14376 Alegria Moc. et Sessé = Luehea Willd. (保留属名) ●☆
14377 Alegria Moc. et Sessé ex DC. = Luehea Willd. (保留属名) ●☆
14378 Aleisanthia Ridl. (1920);阿蕾茜属●☆
14379 Aleisanthia rupestris Ridl.;阿蕾茜●☆
14380 Aleisanthia sylvatica Ridl.;林地阿蕾茜●☆
14381 Aleisanthiopsis Tange(1997);加岛茜属●☆
14382 Aleome Neck. = Cleome L. ●■
14383 Alepida Kuntze = Alepidea F. Delaroche ■☆
14384 Alepidea F. Delaroche(1808);无鳞草属■☆
14385 Alepidea La Roche = Alepidea F. Delaroche ■☆
14386 Alepidea acutidens Weim.;尖无鳞草■☆
14387 Alepidea acutidens Weim. var. dispar ?;异型无鳞草■☆
14388 Alepidea amatymbica Eckl. et Zeyh.;热非无鳞草■☆
14389 Alepidea amatymbica Eckl. et Zeyh. var. aquatica (Kuntze) Weim.;水生无鳞草■☆
14390 Alepidea amatymbica Eckl. et Zeyh. var. microbracteata Weim.;小苞无鳞草■☆
14391 Alepidea angustifolia Schltr. et H. Wolff;窄叶无鳞草■☆
14392 Alepidea aquatica Kuntze = Alepidea amatymbica Eckl. et Zeyh. var. aquatica (Kuntze) Weim. ■☆
14393 Alepidea attenuata Weim.;渐狭无鳞草■☆
14394 Alepidea basinuda Pott;裸基无鳞草■☆
14395 Alepidea basinuda Pott var. subnuda Weim.;近裸基无鳞草■☆
14396 Alepidea baurii (Kuntze) Dümmer = Alepidea natalensis J. M. Wood et M. S. Evans ■☆
14397 Alepidea capensis (P. J. Bergius) R. A. Dyer;好望角无鳞草■☆
14398 Alepidea capensis (P. J. Bergius) R. A. Dyer var. tenella (Schltr. et H. Wolff) Weim.;柔软好望角无鳞草■☆
14399 Alepidea cathcartensis Kuntze = Alepidea serrata Eckl. et Zeyh. var. cathcartensis (Kuntze) Weim. ■☆
14400 Alepidea ciliaris F. Delaroche = Alepidea pilifera Weim. ■☆
14401 Alepidea ciliaris F. Delaroche var. baurii Kuntze = Alepidea natalensis J. M. Wood et M. S. Evans ■☆
14402 Alepidea ciliaris F. Delaroche var. lanceolata Kuntze = Alepidea natalensis J. M. Wood et M. S. Evans ■☆
14403 Alepidea coarctata Dümmer = Alepidea peduncularis A. Rich. ■☆
14404 Alepidea comosa Dümmer;簇毛无鳞草■☆
14405 Alepidea concinna Dümmer = Alepidea natalensis J. M. Wood et M. S. Evans ■☆
14406 Alepidea congesta Schltr. et H. Wolff = Alepidea peduncularis A. Rich. ■☆
14407 Alepidea delicatula Weim.;姣美无鳞草■☆
14408 Alepidea fischeri (Engl.) Schltr. et H. Wolff = Alepidea peduncularis A. Rich. ■☆
14409 Alepidea galpinii Dümmer;盖尔无鳞草■☆
14410 Alepidea gracilis Dümmer;纤细无鳞草■☆
14411 Alepidea gracilis Dümmer var. major Weim. = Alepidea longifolia E. Mey. ex Dümmer ■☆
14412 Alepidea insculpta Hilliard et B. L. Burtt;雕刻无鳞草■☆
14413 Alepidea longeciliata Schinz ex Dümmer;长缘毛无鳞草■☆
14414 Alepidea longifolia E. Mey. ex Dümmer = Alepidea peduncularis A. Rich. ■☆
14415 Alepidea longifolia E. Mey. ex Dümmer subsp. angusta (Dümmer) Weim. = Alepidea longifolia E. Mey. ex Dümmer var. angusta Dümmer ■☆
14416 Alepidea longifolia E. Mey. ex Dümmer subsp. coarctata (Dümmer) Weim. = Alepidea peduncularis A. Rich. ■☆
14417 Alepidea longifolia E. Mey. ex Dümmer subsp. comosa (Dümmer) Weim. = Alepidea comosa Dümmer ■☆
14418 Alepidea longifolia E. Mey. ex Dümmer subsp. lancifolia Weim. = Alepidea longifolia E. Mey. ex Dümmer ■☆
14419 Alepidea longifolia E. Mey. ex Dümmer subsp. propinqua (Dümmer) Weim. = Alepidea peduncularis A. Rich. ■☆
14420 Alepidea longifolia E. Mey. ex Dümmer subsp. swynnertonii (Dümmer) Weim. = Alepidea peduncularis A. Rich. ■☆
14421 Alepidea longifolia E. Mey. ex Dümmer var. angusta Dümmer;狭无鳞草■☆
14422 Alepidea longifolia E. Mey. ex Dümmer var. wyliei (Dümmer) Weim. = Alepidea wyliei Dümmer ■☆
14423 Alepidea longipetiolata Schltr. et H. Wolff = Alepidea capensis (P. J. Bergius) R. A. Dyer ■☆
14424 Alepidea macowani Dümmer;麦克欧文无鳞草■☆
14425 Alepidea massaica Schltr. et H. Wolff = Alepidea peduncularis A.

Rich. ■☆
14426 Alepidea multisecta B. L. Burtt;多裂无鳞草■☆
14427 Alepidea natalensis J. M. Wood et M. S. Evans;纳塔尔无鳞草■☆
14428 Alepidea peduncularis A. Rich.;长叶无鳞草■☆
14429 Alepidea peduncularis A. Rich. var. fischeri Engl. = Alepidea peduncularis A. Rich. ■☆
14430 Alepidea pilifera Weim.;纤毛无鳞草■☆
14431 Alepidea propinqua Dümmer = Alepidea peduncularis A. Rich. ■☆
14432 Alepidea pusilla Weim.;微小无鳞草■☆
14433 Alepidea reticulata Weim.;网状无鳞草■☆
14434 Alepidea serrata Eckl. et Zeyh.;具齿无鳞草■☆
14435 Alepidea serrata Eckl. et Zeyh. var. cathcartensis (Kuntze) Weim.;卡斯卡特无鳞草■☆
14436 Alepidea setifera N. E. Br.;刚毛无鳞草■☆
14437 Alepidea stellata Weim.;星状无鳞草■☆
14438 Alepidea swynnertonii Dümmer = Alepidea peduncularis A. Rich. ■☆
14439 Alepidea tenella Schltr. et H. Wolff = Alepidea capensis (P. J. Bergius) R. A. Dyer var. tenella (Schltr. et H. Wolff) Weim. ■☆
14440 Alepidea thodei Dümmer;索德无鳞草■☆
14441 Alepidea tysonii Dümmer = Alepidea woodii Oliv. ■☆
14442 Alepidea woodii Oliv.;伍氏无鳞草■☆
14443 Alepidea wyliei Dümmer;怀利无鳞草■☆
14444 Alepidixia Tiegh. ex Lecomte = Viscum L. ●
14445 Alepidocalyx Piper = Phaseolus L. ■
14446 Alepidocalyx Piper(1926);墨西哥豆属■☆
14447 Alepidocalyx amblyosepalus Piper;钝瓣墨西哥豆■☆
14448 Alepidocalyx anisophyllus Piper;异叶墨西哥豆■☆
14449 Alepidocalyx parvulus (Greene) Piper;墨西哥豆■☆
14450 Alepidocline S. F. Blake(1934);草落冠菊属■☆
14451 Alepidocline breedlovei (B. L. Turner) B. L. Turner;布利草落冠菊■☆
14452 Alepidocline macdonaldana B. L. Turner;草落冠菊■☆
14453 Alepidocline trifida (J. J. Fay) B. L. Turner;三裂草落冠菊■☆
14454 Alepis Tiegh. (1894);无苞寄生属●☆
14455 Alepis flavida Tiegh.;无苞寄生●☆
14456 Alepis polychroa Tiegh.;多色无苞寄生●☆
14457 Alepyrum Hieron. = Centrolepis Labill. ■
14458 Alepyrum Hieron. = Pseudalepyrum Dandy ■
14459 Alepyrum R. Br. = Centrolepis Labill. ■
14460 Aletes J. M. Coult. et Rose(1888);磨石草属■☆
14461 Aletes acaulis J. M. Coult. et Rose;无茎磨石草■☆
14462 Aletes bipinnata (J. M. Coult. et Rose) W. A. Weber;双羽磨石草■☆
14463 Aletes calcicola Mathias et Constance;岩地磨石草■☆
14464 Aletes filifolius Mathias, Constance et W. L. Theob.;线叶磨石草■☆
14465 Aletes humilis J. M. Coult. et Rose;矮磨石草■☆
14466 Aletes latiloba (Rydb.) W. A. Weber;宽裂磨石草■☆
14467 Aletes megarrhiza (A. Nelson) W. A. Weber;大根磨石草■☆
14468 Aletes minima (Mathias) W. A. Weber;小磨石草■☆
14469 Aletes sessiliflorus W. L. Theob. et C. C. Tseng;无梗磨石草■☆
14470 Aletes tenuifolia J. M. Coult. et Rose;细叶磨石草■☆
14471 Aletris L. (1753);粉条儿菜属(肺筋草属,粉条菜属,束心兰属);Aletris, Colic-root, Star Grass, Stargrass, Star-grass ■
14472 Aletris alpestris Diels;高山粉条儿菜(高山肺筋草,一枝箭);Alpine Aletris ■
14473 Aletris alpestris Diels var. occidentalis Hara = Aletris nana S. C. Chen ■
14474 Aletris arborea Willd. = Dracaena arborea (Willd.) Link ●☆
14475 Aletris aurea Walter;黄粉条儿菜■☆
14476 Aletris biondiana Diels = Aletris glandulifera Bureau et Franch. ■
14477 Aletris capensis L. = Veltheimia capensis (L.) DC. ■☆
14478 Aletris capitata F. T. Wang et Ts. Tang;头花粉条儿菜;Capitate Aletris ■
14479 Aletris chinensis Lam. = Cordyline fruticosa (L.) A. Chev. ●
14480 Aletris cinerascens F. T. Wang et Ts. Tang;灰鞘粉条儿菜;Graysheath Aletris ■
14481 Aletris cochinchinensis Lour. = Dracaena cochinchinensis (Lour.) S. C. Chen ●◇
14482 Aletris delavayi Franch. = Aletris pauciflora (G. Klotz) Franch. ■
14483 Aletris dickinsii Franch. = Aletris glabra Bureau et Franch. ■
14484 Aletris dielsii F. T. Wang et Ts. Tang = Aletris alpestris Diels ■
14485 Aletris elata F. T. Wang et Ts. Tang = Aletris laxiflora Bureau et Franch. ■
14486 Aletris farinosa L.;星草(北美粉条儿菜,污粉条儿菜);Ague Root, Ague-root, Blazing Star, Colic Root, Colic-root, Crow Corn, Star-grass, Unicom Root, Unicorn Root, Unicorn-root, White Colic-root, White Stargrass ■☆
14487 Aletris foliata (Maxim.) Bureau et Franch.;多叶粉条儿菜■☆
14488 Aletris foliata (Maxim.) Bureau et Franch. var. glabra (Bureau et Franch.) Yamam. = Aletris glabra Bureau et Franch. ■
14489 Aletris foliosa Bureau et Franch. var. sikkimensis (Hook. f.) Franch. = Aletris glabra Bureau et Franch. ■
14490 Aletris formosana (Hayata) Makino et Nemoto = Aletris glabra Bureau et Franch. ■
14491 Aletris formosana (Hayata) Sasaki;台湾粉条儿菜;Taiwan Aletris ■
14492 Aletris formosana (Hayata) Sasaki = Aletris glabra Bureau et Franch. ■
14493 Aletris fragrans L. = Dracaena fragrans (L.) Ker Gawl. ●
14494 Aletris glabra Bureau et Franch.;无毛粉条儿菜(光肺筋草,光叶肺筋草,蛆芽菜,小肺筋草);Glabrous Aletris ■
14495 Aletris glandulifera Bureau et Franch.;腺毛粉条儿菜;Galndularhairy Aletris ■
14496 Aletris glauca Aiton = Veltheimia capensis (L.) DC. ■☆
14497 Aletris gracilipes F. T. Wang et Ts. Tang = Aletris laxiflora Bureau et Franch. ■
14498 Aletris gracilis Rehder;星花粉条儿菜;Starryflower Aletris ■
14499 Aletris japonica Houtt. = Funkia subcordata Spreng. ■
14500 Aletris japonica Lamb. = Aletris spicata (Thunb.) Franch. ■
14501 Aletris japonica Thunb. = Funkia ovata Spreng. ■
14502 Aletris japonica Thunb. = Hosta lancifolia (Thunb.) Engl. ■
14503 Aletris lactiflora Franch. = Aletris glandulifera Bureau et Franch. ■
14504 Aletris lanuginosa Bureau et Franch. = Aletris pauciflora (G. Klotz) Franch. var. khasiana (Hook. f.) F. T. Wang et Ts. Tang ■
14505 Aletris lanuginosa Bureau et Franch. var. khasiana Hook. f. = Aletris pauciflora (G. Klotz) Franch. var. khasiana (Hook. f.) F. T. Wang et Ts. Tang ■
14506 Aletris laxiflora Bureau et Franch.;疏花粉条儿菜;Laxflower Aletris ■
14507 Aletris longibracteata T. L. Xu;长苞粉条儿菜;Longbract Aletris ■
14508 Aletris longibracteata T. L. Xu = Aletris stenoloba Franch. ■
14509 Aletris lutea Small;深黄粉条儿菜;Yellow Colic-root ■☆

14510 Aletris luteoviridis (Maxim.) Franch.;黄绿粉条儿菜■☆
14511 Aletris mairei H. Lév. = Aletris pauciflora (G. Klotz) Franch. ■
14512 Aletris makiyataroi Naruh. = Aletris scopulorum Dunn ■
14513 Aletris megalantha F. T. Wang et Ts. Tang;大花粉条儿菜;Largeflower Aletris ■
14514 Aletris nana S. C. Chen;矮粉条儿菜;Dwarf Aletris ■
14515 Aletris nepalensis Bureau et Franch. = Aletris pauciflora (G. Klotz) Franch. ■
14516 Aletris nepalensis Hook. f. = Aletris pauciflora (G. Klotz) Franch. ■
14517 Aletris nepalensis Hook. f. var. delavayi Franch. = Aletris pauciflora (G. Klotz) Franch. ■
14518 Aletris obovata Nash ex Small;矩圆粉条儿菜;White Colic-root ■☆
14519 Aletris pauciflora (G. Klotz) Franch.;少花粉条儿菜(百味参,小棕皮);Fewflower Aletris ■
14520 Aletris pauciflora (G. Klotz) Franch. var. khasiana (Hook. f.) F. T. Wang et Ts. Tang;穗花粉条儿菜(百味参,虎须草,小棕皮);Khas Aletris ■
14521 Aletris pauciflora (G. Klotz) Franch. var. lanuginosa (Bureau et Franch.) F. T. Wang et Ts. Tang;廃子草■
14522 Aletris pauciflora (G. Klotz) Franch. var. minusculata Hand.-Mazz. = Aletris pauciflora (G. Klotz) Franch. ■
14523 Aletris pedicellata F. T. Wang et Ts. Tang;长柄粉条儿菜;Longpedicel Aletris ■
14524 Aletris pumila Aiton = Kniphofia pumila (Aiton) Kunth ■☆
14525 Aletris revoluta Franch. = Aletris laxiflora Bureau et Franch. ■
14526 Aletris sarmentosa Andréws = Kniphofia sarmentosa (Andréws) Kunth ■☆
14527 Aletris scopulorum Dunn;短柄粉条儿菜(铁卵子);Cliff Aletris ■
14528 Aletris sikkimensis Hook. f. = Aletris glabra Bureau et Franch. ■
14529 Aletris spicata (Thunb.) Franch.;粉条儿菜(百味参,肺风草,肺筋草,肺痨草,肺痈草,谷穗草,金线吊白米,金线吊米,麻里草,曲折草,蛆儿草,蛆婆草,蛆芽菜,蛆芽草,束心兰,四季花,土瞿麦,细米芥,小肺金草,小肺筋草,牙虫草,一窝蛆,银针草);Spike Aletris ■
14530 Aletris spicata (Thunb.) Franch. var. fargesii Franch. = Aletris stenoloba Franch. ■
14531 Aletris spicata (Thunb.) Franch. var. micrantha Satake = Aletris spicata (Thunb.) Franch. ■
14532 Aletris stelliflora Hand.-Mazz. = Aletris gracilis Rehder ■
14533 Aletris stenoloba Franch.;狭瓣粉条儿菜(肺筋草,驱蛆草,狭叶粉条儿菜,一窝蛆);Narrowlobed Aletris ■
14534 Aletris tavelii H. Lév. = Aletris glabra Bureau et Franch. ■
14535 Aletris uvaria (L.) L. = Kniphofia uvaria (L.) Oken ■☆
14536 Aletris yaanica G. H. Yang;雅安粉条儿菜;Yaan Aletris ■
14537 Aleurites Forst. = Aleurites J. R. Forst. et G. Forst. ●
14538 Aleurites J. R. Forst. et G. Forst. (1775);石栗属(油桐属);Aleurites, Stonechestnut, Tung-oil-tree, Tung-tree ●
14539 Aleurites cordata (Thunb.) R. Br. ex Steud. = Vernicia cordata (Thunb.) Airy Shaw ●☆
14540 Aleurites cordata Steud. = Vernicia cordata (Thunb.) Airy Shaw ●☆
14541 Aleurites fordii Hemsl. = Vernicia fordii (Hemsl.) Airy Shaw ●
14542 Aleurites japonica Blume = Aleurites cordata (Thunb.) R. Br. ex Steud. ●☆
14543 Aleurites laccifera Willd.;虫胶油桐●☆
14544 Aleurites moluccana (L.) Willd.;石栗(海胡桃,黑桐油,水火树,铁果,铁桐,油桃,烛果树);Balucanat, Bancouloilplant, Belgaum Walnut, Buah Keras, Candle Berry, Candle Nut, Candle Nut Tree, Candleberry, Candleberry Tree, Candlenu Tree, Candlenut, Candle-nut Oil Tree, Candlenut Tree, Candlenut-oil Tree, Candlenutree, Indian Walnut, Indian Walnut Tree, Kekui Oil Plant, Kukui Nut, Lumbang Oil, Lumbangoil Plant, Otaheite Walnut, Stonechestnut, Tung Oil Tree, Tung Tree, Varnish Tree ●
14545 Aleurites moluccana (L.) Willd. var. floccosa Airy Shaw;丛毛石栗●☆
14546 Aleurites montana (Lour.) E. H. Wilson = Vernicia montana Lour. ●
14547 Aleurites triloba J. R. Forst. et G. Forst. = Aleurites moluccana (L.) Willd. var. floccosa Airy Shaw ●☆
14548 Aleurites triloba J. R. Forst. et G. Forst. = Aleurites moluccana (L.) Willd. ●
14549 Aleurites trisperma Blanco;菲律宾油桐(亚利伯斯油桐);Philippine Alcimandra, Soft Lumbang ●☆
14550 Aleurites verniciflua H. Lév. = Aleurites cordata Steud. ●☆
14551 Aleurites verniciflua H. Lév. = Vernicia fordii (Hemsl.) Airy Shaw ●
14552 Aleuritia (Duby) Opiz = Primula L. ■
14553 Aleuritia Spach = Primula L. ■
14554 Aleuritia miyabeana (Ito et Kawak.) Soják = Primula miyabeana Ito et Kawak. ■
14555 Aleurodendron Reinw. = Melochia L. (保留属名)●■
14556 Alevia Baill. = Bernardia L. ●
14557 Alexa Moq. (1849);护卫豆属●☆
14558 Alexa Moq. = Castanospermum A. Cunn. ex Hook. ●☆
14559 Alexa grandiflora Ducke;大花护卫豆●☆
14560 Alexa imperatricis (Schomb.) Baill.;南美护卫豆木●☆
14561 Alexa leiopetala Sandwith;平瓣护卫豆●☆
14562 Alexandra Bunge(1843);翼萼蓬属(密苞蓬属)■☆
14563 Alexandra R. H. Schomb. = Alexa Moq. ●☆
14564 Alexandra Schomb. = Alexa Moq. ●☆
14565 Alexandra lehmannii Bunge;翼萼蓬■☆
14566 Alexandrina Lindl. = Alexandra R. H. Schomb. ●☆
14567 Alexeya Pachom. = Paraquilegia J. R. Drumm. et Hutch. ■
14568 Alexfloydia B. K. Simon(1992);阿氏黍属■☆
14569 Alexfloydia repens B. K. Simon;阿氏黍■☆
14570 Alexgeorgea Carlquist(1976);亮鞘帚灯草属■☆
14571 Alexgeorgea arenicola Carlquist;沙地亮鞘帚灯草■☆
14572 Alexgeorgea subterranea Carlquist;亮鞘帚灯草■☆
14573 Alexia Wight = Alyxia Banks ex R. Br. (保留属名)●
14574 Alexis Salisb. = Amomum Roxb. (保留属名)■
14575 Alexitoxicon St.-Lag. = Cynanchum L. ●■
14576 Alexitoxicon St.-Lag. = Vincetoxicum Wolf ●■
14577 Alexitoxicon acuminatum (Decne.) Pobed. = Cynanchum acuminatifolium Hemsl. ■
14578 Alexitoxicon amplexicaule (Siebold et Zucc.) Pobed. = Cynanchum amplexicaule (Siebold et Zucc.) Hemsl. ■
14579 Alexitoxicon atratum (Bunge) Pobed. = Cynanchum atratum Bunge ■
14580 Alexitoxicon inamoenum (Maxim.) Pobed. = Cynanchum inamoenum (Maxim.) Loes. ex Gilg et Loes. ■
14581 Alexitoxicon sibiricum (L.) Pobed. = Cynanchum thesioides (Freyn) K. Schum. ■
14582 Alexitoxicon volubile (Maxim.) Pobed. = Cynanchum volubile

(Maxim.) Hemsl. ■
14583 Alfaroa Standl. (1927);哥斯达黎加胡桃属(无翅黄杞属)●☆
14584 Alfaroa colombiana Lozano et Espinal;哥伦比亚胡桃●☆
14585 Alfaroa costaricensis Standl.;哥斯达黎加胡桃●☆
14586 Alfaroa manningii J. Leon;曼宁哥斯达黎加胡桃●☆
14587 Alfaroa mexicana D. E. Stone;墨西哥胡桃●☆
14588 Alfaropsis Iljinsk. (1993);安黄杞属(拟哥斯达黎加胡桃属)●☆
14589 Alfaropsis roxburghiana (Wall.) Iljinsk.;安黄杞●☆
14590 Alfonsia Kunth = Corozo Jacq. ex Giseke ●
14591 Alfonsia Kunth = Elaeis Jacq. ●
14592 Alfonsia oleifera Kunth = Elaeis oleifera Cortes ●☆
14593 Alfredia Cass. (1816);翅膜菊属(黄飞廉属,亚飞廉属);Alfredia ■
14594 Alfredia acantholepis Kar. et Kir.;薄叶翅膜菊(土升麻,亚飞廉);Pricklyscale Alfredia ■
14595 Alfredia aspera C. Shih;糙毛翅膜菊;Hispid Alfredia, Scabrous Alfredia ■
14596 Alfredia cernua (L.) Cass.;翅膜菊;Common Alfredia ■
14597 Alfredia fetissowii Iljin;长叶翅膜菊;Longleaf Alfredia ■
14598 Alfredia karelini Ledeb. = Alfredia acantholepis Kar. et Kir. ■
14599 Alfredia nivea Kar. et Kir.;厚叶翅膜菊(白背亚飞廉);Whiteback Alfredia ■
14600 Alfredia squarosa Tausch = Alfredia cernua (L.) Cass. ■
14601 Alfredia squarrosa Kostel. = Alfredia cernua (L.) Cass. ■
14602 Alfredia stenolepis Kar. et Kir. = Alfredia cernua (L.) Cass. ■
14603 Alfredia suaveolens Rupr. = Alfredia nivea Kar. et Kir. ■
14604 Alfredia tianschanica Rupr. = Alfredia acantholepis Kar. et Kir. ■
14605 Alfredia tsianschanica Rupr. = Alfredia acantholepis Kar. et Kir. ■
14606 Alga Adans. = Zostera L. ■
14607 Alga Boehm(废弃属名) = Posidonia K. D. König(保留属名)■
14608 Alga Lam. = Zostera L. ■
14609 Alga Ludw. = Posidonia K. D. König(保留属名)■
14610 Algarobia Benth. = Prosopis L. ●
14611 Algernonia Baill. (1858);巴西大戟属☆
14612 Algernonia brasiliensis Baill.;巴西大戟☆
14613 Algernonia brasiliensis Baill. var. obovata Müll. Arg.;倒卵巴西大戟☆
14614 Algernonia gibbosa (Pax et K. Hoffm.) Emmerich;驼曲巴西大戟☆
14615 Algrizea Proença et NicLugh. (2006);巴西柽柳桃金娘属●☆
14616 Algrizea Proença et NicLugh. = Myrcia DC. ex Guill. ●☆
14617 Alguelaguen Adans. (废弃属名) = Lepechinia Willd. ●■☆
14618 Alguelaguen Adans. (废弃属名) = Sphacele Benth. (保留属名)●■☆
14619 Alguelaguen Feuill. ex Adans. = Sphacele Benth. (保留属名)●■☆
14620 Alguelagum Kuntze = Alguelaguen Adans. (废弃属名)●■☆
14621 Alhagi Adans. = Alhagi Tourn. ex Adans. ●
14622 Alhagi Gagnebin = Alhagi Tourn. ex Adans. ●
14623 Alhagi Tourn. ex Adans. (1763);骆驼刺属;Alhagi, Camelthorn ●
14624 Alhagi camelorum Fisch. = Alhagi maurorum Medik. ●
14625 Alhagi camelorum Fisch. = Alhagi pseudoalhagi Desv. ●
14626 Alhagi camelorum Fisch. ex DC. = Alhagi maurorum Medik. ●
14627 Alhagi canescens (Regel) Shap. ex Keller et Shap.;灰色骆驼刺●☆
14628 Alhagi graecorum Boiss. = Alhagi maurorum Medik. subsp. graecorum (Boiss.) Awmack et Lock ●☆
14629 Alhagi kirgnisorum Schrenk;吉尔吉斯骆驼刺●☆
14630 Alhagi mannifera Desv. = Alhagi maurorum Medik. subsp. graecorum (Boiss.) Awmack et Lock ●☆
14631 Alhagi maurorum Medik.;亚丁骆驼刺(波斯甘露树,波斯骆驼刺,西亚骆驼刺);Camel Thorn, Camelthorn ●
14632 Alhagi maurorum Medik. = Alhagi nepalensis (D. Don) Shap. ●☆
14633 Alhagi maurorum Medik. subsp. graecorum (Boiss.) Awmack et Lock;希腊骆驼刺●☆
14634 Alhagi maurorum Medik. var. sparsifolium (Shap.) Yakovlev = Alhagi pseudoalhagi Desv. ●
14635 Alhagi napaulensium DC. = Alhagi nepalensis (D. Don) Shap. ●☆
14636 Alhagi nepalensis (D. Don) Shap.;尼泊尔骆驼刺●☆
14637 Alhagi persarum Boiss. et Buhse = Alhagi maurorum Medik. ●
14638 Alhagi pseudoalhagi (M. Bieb.) Desv.;骆驼刺(假骆驼刺,疏叶骆驼刺);Camel's-thorn, Camel-thorn, Manaplant Alhagi, Monoplant Alhagi ●
14639 Alhagi pseudoalhagi (M. Bieb.) Desv. = Alhagi maurorum Medik. ●
14640 Alhagi pseudoalhagi Desv. = Alhagi maurorum Medik. ●
14641 Alhagi sparsifolium Shap. ex Keller et Shap. = Alhagi pseudoalhagi Desv. ●
14642 Alhagia Rchb. = Alhagi Tourn. ex Adans. ●
14643 Alibertia A. Rich. = Alibertia A. Rich. ex DC. ●☆
14644 Alibertia A. Rich. ex DC. (1830);阿利茜属●☆
14645 Alibertia acuminata (Benth.) Sandwith;尖阿利茜●☆
14646 Alibertia claviflora (K. Schum.) Kuntze;棒花阿利茜●☆
14647 Alibertia claviflora K. Schum = Alibertia claviflora (K. Schum.) Kuntze ●☆
14648 Alibertia curviflora K. Schum.;弯花阿利茜●☆
14649 Alibertia edulis (L. Rich.) A. Rich. ex DC.;可食阿利茜●☆
14650 Alibertia edulis A. Rich. = Alibertia edulis (L. Rich.) A. Rich. ex DC. ●☆
14651 Alibertia itayensis Standl.;意大利阿利茜●☆
14652 Alibertia latifolia K. Schum.;宽花阿利茜●☆
14653 Alibertia longiflora K. Schum.;长花阿利茜●☆
14654 Alibertia macrantha Standl.;大花阿利茜●☆
14655 Alibertia sessilis K. Schum.;无梗阿利茜●☆
14656 Alibertia stenantha Standl.;窄花阿利茜●☆
14657 Alibertia triflora K. Schum.;三花阿利茜●☆
14658 Alibertia triloba Steyerm.;三裂阿利茜●☆
14659 Alibertia uniflora Standl.;单花阿利茜●☆
14660 Alibrexia Miers = Nolana L. ex L. f. ■☆
14661 Alibum Less. = Liabum Adans. ●■☆
14662 Alicabon Raf. = Physalis L. ■
14663 Alicabon Raf. = Withania Pauquy(保留属名)●■
14664 Alicastrum P. Browne(废弃属名) = Brosimum Sw. (保留属名)●☆
14665 Alicia W. R. Anderson = Hiraea Jacq. ●☆
14666 Alicia W. R. Anderson(2006);南美藤翅果属●☆
14667 Aliciella Brand = Gilia Ruiz et Pav. ●■☆
14668 Aliciella Brand(1905);异吉莉花属■☆
14669 Aliciella formosa (Greene ex Brand) J. M. Porter;美丽异吉莉花■☆
14670 Aliciella heterostyla (Cochrane et A. G. Day) J. M. Porter;异柱吉莉花■☆
14671 Aliciella latifolia (S. Watson) J. M. Porter;宽叶异吉莉花■☆
14672 Aliciella pinnatifida (Nutt. ex A. Gray) J. M. Porter;羽裂异吉莉花■☆

14673 Aliciella triodon (Eastw.) Brand;三齿异吉莉花■☆

14674 Aliciella triodon Brand = Aliciella triodon (Eastw.) Brand ■☆

14675 Aliconia Herrera = Mikania Willd. (保留属名) ■

14676 Alicosta Dulac = Bartsia L. (保留属名) ●■☆

14677 Alicosta alpina (L.) Dulac = Bartsia alpina L. ■☆

14678 Alicteres Neck. = Helicteres L. ●

14679 Alicteres Neck. ex Schott et Endl. = Helicteres L. ●

14680 Aliella Qaiser et Lack(1986);黄鼠麴属●■☆

14681 Aliella ballii (Klatt) Greuter;鲍尔黄鼠麴■☆

14682 Aliella ballii (Klatt) Greuter subsp. nitida (Emb.) Greuter;亮鲍尔黄鼠麴■☆

14683 Aliella bracteata Anderb. = Aliella ballii (Klatt) Greuter ■☆

14684 Aliella embergeri (Humbert et Maire) Qaiser et Lack;恩贝格尔黄鼠麴■☆

14685 Aliella helichrysoides (Ball) Qaiser et Lack = Aliella ballii (Klatt) Greuter ■☆

14686 Aliella helichrysoides (Ball) Qaiser et Lack subsp. helichrysoides = Aliella ballii (Klatt) Greuter ■☆

14687 Aliella helichrysoides (Ball) Qaiser et Lack subsp. nitidum (Emb.) Qaiser et Lack = Aliella ballii (Klatt) Greuter subsp. nitida (Emb.) Greuter ■☆

14688 Aliella iminouakensis (Emb.) Dobignard et Jeanm.;莫诺黄鼠麴■☆

14689 Aliella platyphylla (Maire) Qaiser et Lack;宽叶黄鼠麴■☆

14690 Alifana Raf. = Brachyotum (DC.) Triana ex Benth. ●☆

14691 Alifanus Adans. = Rhexia L. ●■☆

14692 Alifanus Pluk. ex Adans. = Rhexia L. ●■☆

14693 Alifiola Raf. = Silene L. (保留属名) ■

14694 Aligera Suksd. (1897);翼缬草属■☆

14695 Aligera barbata Suksd.;髯毛翼缬草■☆

14696 Aligera californica Suksd.;加州翼缬草■☆

14697 Aligera ciliosa Suksd.;睫毛翼缬草■☆

14698 Aligera glabrior Suksd.;光翼缬草■☆

14699 Aligera macroptera Suksd.;大翅翼缬草■☆

14700 Aligera macroptera Suksd. var. obtusa Suksd.;钝大翅翼缬草■☆

14701 Aligera minor (Hook.) A. Heller;小翼缬草■☆

14702 Aligera minor A. Heller = Aligera minor (Hook.) A. Heller ■☆

14703 Alina Adans(废弃属名) = Hyperbaena Miers ex Benth. (保留属名) ●☆

14704 Aliniella J. Raynal = Alinula J. Raynal ■☆

14705 Aliniella lipocarphioides (Kük.) J. Raynal = Alinula lipocarphioides (Kük.) J. Raynal ■☆

14706 Alinorchis Szlach. (2002);非洲玉凤花属■☆

14707 Alinula J. Raynal(1977);阿林莎草属■☆

14708 Alinula lipocarphioides (Kük.) J. Raynal;湖瓜草状莎草■☆

14709 Alinula malawica (J. Raynal) Goetgh. et Vorster;马拉维阿林莎草■☆

14710 Alinula paradoxa (Cherm.) Goetgh. et Vorster;奇异阿林莎草■☆

14711 Alinula peteri (Kük.) Goetgh. et Vorster;彼得阿林莎草■☆

14712 Alionia Raf. = Allionia L. (保留属名) ■☆

14713 Aliopsis Omer et Qaiser = Gentianella Moench(保留属名) ■

14714 Aliopsis moorcroftiana (Wall. ex G. Don) Omer, Qaiser et Ali = Gentianella moorcroftiana (Wall. ex G. Don) Airy Shaw ■

14715 Aliopsis pygmaea (Regel et Schmalh.) Omer et Qaiser = Gentianella pygmaea (Regel et Schmalh.) Harry Sm. ■

14716 Alipendula Neck. = Filipendula Mill. ■

14717 Alipsa Hoffmanns. = Liparis Rich. (保留属名) ■

14718 Aliseta Raf. = Arnica L. ●■☆

14719 Alisma L. (1753);泽泻属;Water Plantain, Waterplantain, Water-plantain ■

14720 Alisma acanthocarpum F. Muell. = Caldesia oligococca (F. Muell.) Buchenau ■☆

14721 Alisma angustifolium Hoppe;琵琶泽泻;Angustifoliate Water-plantain ■

14722 Alisma arcuatum Michalet = Alisma plantago-aquatica L. subsp. arcuatum (Michalet) Asch. et Graebn. ■

14723 Alisma berteroi Spreng. = Echinodorus berteroi (Spreng.) Fassett ■☆

14724 Alisma brevipes Greene = Alisma triviale Pursh ■☆

14725 Alisma canaliculatum A. Braun et C. D. Bouché;窄叶泽泻(穿叶泽泻,大箭,水泽泻,狭叶泽泻,泽泻,真武剑);Channelled Waterplantain, Narrowleaf Waterplantain ■

14726 Alisma canaliculatum A. Braun et C. D. Bouché var. azuminoense Kadono et Hamashima;热见泽泻■☆

14727 Alisma canaliculatum A. Braun et C. D. Bouché var. harimense Makino;细泽泻■☆

14728 Alisma cordifolia L. = Echinodorus cordifolius (L.) Griseb. ■☆

14729 Alisma flavum L. = Limnocharis flava (L.) Buchenau ■

14730 Alisma geyeri Buchenau = Alisma geyeri Torr. ■☆

14731 Alisma geyeri Torr.;盖耶氏泽泻;Geyer Waterplantain ■☆

14732 Alisma geyeri Torr. = Alisma gramineum C. C. Gmel. ■

14733 Alisma glandulosum Thwaites = Caldesia oligococca (F. Muell.) Buchenau ■☆

14734 Alisma gramineum C. C. Gmel.;草泽泻;Alisma Graminoide, Grass Water Plantain, Grass Waterplantain, Grass Water-plantain, Grass-leaved Water Plantain, Grass-leaved Water-plantain, Narrow-leaved Water-plantain, Ribbon-leaved Water-plantain ■

14735 Alisma gramineum C. C. Gmel. var. angustissimum Hendricks = Alisma gramineum C. C. Gmel. ■

14736 Alisma gramineum C. C. Gmel. var. wahlenbergii Raymond et Kucyniak = Alisma gramineum C. C. Gmel. ■

14737 Alisma gramineum Gmel. = Alisma gramineum C. C. Gmel. ■

14738 Alisma gramineum Lej. = Alisma gramineum C. C. Gmel. ■

14739 Alisma gramineum Lej. var. angustissimum (DC.) Hendricks = Alisma gramineum C. C. Gmel. ■

14740 Alisma gramineum Lej. var. geyeri (Torr.) Lam. = Alisma gramineum C. C. Gmel. ■

14741 Alisma gramineum Lej. var. graminifolium (Wahlenb.) Hendricks = Alisma gramineum C. C. Gmel. ■

14742 Alisma humile Rich. = Ranalisma humile (Rich.) Hutch. ■☆

14743 Alisma jianshiensis J. K. Chen, X. Z. Sun et H. Q. Wang = Alisma orientale (Sam.) Juz. ■

14744 Alisma kotschyi Hochst. = Limnophyton obtusifolium (L.) Miq. ■☆

14745 Alisma lanceolatum With.;膜果泽泻(光叶泽泻,披针叶泽泻);Lanceleaf Water Plantain, Lanceolate Waterplantain, Narrow-leaved Water-plantain ■

14746 Alisma lanceolatum With. = Alisma gramineum C. C. Gmel. ■

14747 Alisma latifolium Gilib. = Alisma plantago-aquatica L. var. latifolium (Gilib.) Kunth ■

14748 Alisma loeselii Gorski = Alisma gramineum C. C. Gmel. ■

14749 Alisma loeselii Gorski ex Juz.;廖氏泽泻;Loesel Waterplantain ■

14750 Alisma loeselii Gorski ex Juz. = Alisma gramineum C. C. Gmel. ■

14751 Alisma nanum D. F. Chi;小泽泻;Small Waterplantain ■

14752 Alisma natans L. = Luronium natans Raf. ■☆
14753 Alisma oligococcum F. Muell. = Caldesia oligococca (F. Muell.) Buchenau ■☆
14754 Alisma orientale (Sam.) Juz.; 东方泽泻(泽泻); Oriental Waterplantain, Waterplantain ■
14755 Alisma orientale (Sam.) Juz. = Alisma plantago-aquatica L. var. orientale Sam. ■
14756 Alisma parnassifolium Bassi ex L. = Caldesia parnassifolia a (Bassi. ex L.) Parl. ■
14757 Alisma parnassifolium Bassi ex L. var. majus Micheli = Caldesia reniformis (D. Don) Makino ■
14758 Alisma parviflorum Pursh = Alisma subcordatum Raf. ■☆
14759 Alisma plantago L. = Alisma plantago-aquatica L. ■
14760 Alisma plantago-aquatica L.; 泽泻(车苦菜,川下,川泽泻,杳菜,耳泽,鹄泻,鹄泽,及泻,建下,建泻,建泽泻,兰江,芒芋,蔏牛唇,牛耳菜,欧泽泻,如意菜,如意花,水白菜,水车前,水杳菜,水蛤蟆叶,水泻,水泽,酸恶俞,天鹅蛋,天秃,文且,小圆泻,盐泽泻,一枝花,蕍,禹孙,禹泻,泽夕,泽下,泽泄,泽芝,宅夕,宅下); Alisma, American Water Plantain, American Water-plantain, Common Water-plantain, Devil's Spoons, Eurasian Water-plantain, European Water Plantain, European Waterplantain, Great Thunderbolts, Greater Thrumwort, Mad Dog Weed, Umbrella, Water Plantain, Water-plantain ■
14761 Alisma plantago-aquatica L. = Alisma gramineum C. C. Gmel. ■
14762 Alisma plantago-aquatica L. = Alisma lanceolatum With. ■
14763 Alisma plantago-aquatica L. sensu Boiss. = Alisma lanceolatum With. ■
14764 Alisma plantago-aquatica L. subsp. arcuatum (Michalet) Asch. et Graebn. = Alisma gramineum Lej. ■
14765 Alisma plantago-aquatica L. subsp. brevipes (Greene) Sam. = Alisma triviale Pursh ■☆
14766 Alisma plantago-aquatica L. subsp. michaletii Asch. et Graebn. = Alisma lanceolatum With. ■
14767 Alisma plantago-aquatica L. subsp. orientale (Sam.) Sam. = Alisma orientale (Sam.) Juz. ■
14768 Alisma plantago-aquatica L. subsp. subcordatum (Raf.) Hultén = Alisma subcordatum Raf. ■☆
14769 Alisma plantago-aquatica L. var. americanum Schult. = Alisma triviale Pursh ■☆
14770 Alisma plantago-aquatica L. var. brevipes (Greene) Vict. = Alisma triviale Pursh ■☆
14771 Alisma plantago-aquatica L. var. decumbens Boiss. = Alisma gramineum C. C. Gmel. ■
14772 Alisma plantago-aquatica L. var. lanceolatum (With.) Koch = Alisma lanceolatum With. ■
14773 Alisma plantago-aquatica L. var. lanceolatum (With.) Schultz = Alisma lanceolatum With. ■
14774 Alisma plantago-aquatica L. var. latifolium (Gilib.) Kunth = Alisma plantago-aquatica L. ■
14775 Alisma plantago-aquatica L. var. orientale Sam. = Alisma orientale (Sam.) Juz. ■
14776 Alisma plantago-aquatica L. var. parviflorum (Pursh) Torr. = Alisma subcordatum Raf. ■☆
14777 Alisma ranunculoides L. = Baldellia ranunculoides (L.) Parl. ■☆
14778 Alisma ranunculoides L. var. repens (Lam.) Batt. et Trab. = Baldellia repens (Lam.) Lawalrée ■☆
14779 Alisma rariflorum Sam.; 少花泽泻■☆
14780 Alisma reniforme D. Don = Caldesia parnassifolia a (Bassi. ex L.) Parl. ■
14781 Alisma reniforme D. Don = Caldesia reniformis (D. Don) Makino ■
14782 Alisma repens Lam. = Baldellia repens (Lam.) Lawalrée ■☆
14783 Alisma sagittifolium Willd. = Limnophyton obtusifolium (L.) Miq. ■☆
14784 Alisma stenophyllum Sam. = Alisma lanceolatum With. ■
14785 Alisma subcordatum Raf.; 亚心形泽泻; Alisma Subcorde, American Water-plantain, Common Water-plantain, Small Water-plantain, Southern Water Plantain, Southern Water-plantain, Subcordate Water-plantain, Water Plantain, Water-plantain ■☆
14786 Alisma subulatum L. = Sagittaria subulata (L.) Buchenau ■☆
14787 Alisma tenellum Mart. = Echinodorus tenellus (Mart.) Buchenau ■☆
14788 Alisma triviale Pursh; 北方泽泻; Alisma Commun, Northern Water Plantain, Northern Water-plantain ■☆
14789 Alisma wahlenbergii Holmb.; 瓦氏泽泻■☆
14790 Alismataceae Vent. (1799)(保留科名); 泽泻科; Alisma Family, Arrowhead Family, Water-plantain Family ■
14791 Alismographis Thouars = Eulophia R. Br. (保留属名) ■
14792 Alismographis Thouars = Limodorum Boehm. (保留属名) ■☆
14793 Alismorchis Thouars = Calanthe R. Br. (保留属名) ■
14794 Alismorkis Thouars(废弃属名) = Calanthe R. Br. (保留属名) ■
14795 Alismorkis alismatifolia (Lindl.) Kuntze = Calanthe alismifolia Lindl. ■
14796 Alismorkis alpina (Hook. f. ex Lindl.) Kuntze = Calanthe alpina Hook. f. ex Lindl. ■
14797 Alismorkis angusta (Lindl.) Kuntze = Calanthe odora Griff. ■
14798 Alismorkis angustifolia (Blume) Kuntze = Calanthe angustifolia (Blume) Lindl. ■
14799 Alismorkis aristulifera (Rchb. f.) Kuntze = Calanthe aristulifera Rchb. f. ■
14800 Alismorkis biloba (Lindl.) Kuntze = Calanthe biloba Lindl. ■
14801 Alismorkis brevicornu (Lindl.) Kuntze = Calanthe brevicornu Lindl. ■☆
14802 Alismorkis clavata (Lindl.) Kuntze = Calanthe clavata Lindl. ■
14803 Alismorkis densiflora (Lindl.) Kuntze = Calanthe densiflora Lindl. ■
14804 Alismorkis discolor (Lindl.) Kuntze = Calanthe discolor Lindl. ■
14805 Alismorkis dolichopoda Fukuy. = Cephalantheropsis longipes (Hook. f.) Ormerod ■
14806 Alismorkis foerstermannii (Rchb. f.) Kuntze = Calanthe lyroglossa Rchb. f. ■
14807 Alismorkis furcata (Bateman ex Lindl.) Kuntze = Dactylorhiza umbrosa (Kar. et Kir.) Nevski ■
14808 Alismorkis herbacea (Lindl.) Kuntze = Calanthe herbacea Lindl. ■
14809 Alismorkis japonica (Blume ex Miq.) Kuntze = Calanthe alismifolia Lindl. ■
14810 Alismorkis labrosa (Rchb. f.) Kuntze = Calanthe labrosa (Rchb. f.) Hook. f. ■
14811 Alismorkis longipes (Hook. f.) Kuntze = Cephalantheropsis longipes (Hook. f.) Ormerod ■
14812 Alismorkis lyroglossa (Rchb. f.) Kuntze = Calanthe lyroglossa Rchb. f. ■
14813 Alismorkis mannii (Hook. f.) Kuntze = Calanthe mannii Hook. f. ■
14814 Alismorkis masuca (D. Don) Kuntze = Calanthe sylvatica (Thouars) Lindl. ■
14815 Alismorkis odora (Griff.) Kuntze = Calanthe odora Griff. ■

14816 Alismorkis pachystalix (Rchb. f. ex Hook. f.) Kuntze = Calanthe davidii Franch. ■
14817 Alismorkis plantaginea (Lindl.) Kuntze = Calanthe plantaginea Lindl. ■
14818 Alismorkis puberula (Lindl.) Kuntze = Calanthe puberulla Lindl. ■
14819 Alismorkis reflexa (Maxim.) Kuntze = Calanthe reflexa (Kuntze) Maxim. ■
14820 Alismorkis speciosa (Blume) Kuntze = Calanthe speciosa (Blume) Lindl. ■
14821 Alismorkis textorii (Miq.) Kuntze = Calanthe sylvatica (Thouars) Lindl. ■
14822 Alismorkis tricarinata (Lindl.) Kuntze = Calanthe tricarinata Lindl. ■
14823 Alismorkis veratrifolia Kuntze = Dactylorhiza umbrosa (Kar. et Kir.) Nevski ■
14824 Alisson Vill. = Alyssum L. ●■
14825 Alissum Neck. = Alisson Vill. ●■
14826 Alissum Neck. = Alyssum L. ●■
14827 Alistilus N. E. Br. (1921);海柱豆属●☆
14828 Alistilus bechuanicus N. E. Br.;贝专海柱豆●☆
14829 Alistilus jumellei (R. Vig.) Verdc.;朱迈尔海柱豆●☆
14830 Alistilus magnificus Verdc.;华丽海柱豆●☆
14831 Aliteria Benoist = Clarisia Ruiz et Pav.(保留属名)●☆
14832 Alitubus Dulac = Achillea L. ■
14833 Alix Comm. ex DC. = Psiadia Jacq. ●☆
14834 Alkanna Adans.(废弃属名) = Alkanna Tausch(保留属名)●☆
14835 Alkanna Adans.(废弃属名) = Lawsonia L. ●
14836 Alkanna Tausch(1824)(保留属名);紫朱草属(红根草属,牛舌草属);Alkanet, Anchusa ●☆
14837 Alkanna cordifolia K. Koch;心叶紫朱草●☆
14838 Alkanna lehmanii (Tineo) A. DC. = Alkanna tinctoria Tausch ●☆
14839 Alkanna megacarpa DC.;大果紫朱草●☆
14840 Alkanna orientalis (L.) Boiss.;东方紫朱草●☆
14841 Alkanna tinctoria Tausch;紫朱牛舌草;Alkanet, Dyer's Alkanet, Dyer's Bugloss, Spanish Bugloss ●☆
14842 Alkanna tuberculata (Forssk.) Meikle = Alkanna tinctoria Tausch ●☆
14843 Alkekengi Mill. = Physalis L. ■
14844 Alkekengi Tourn. ex Haller = Physalis L. ■
14845 Alkibias Raf. = Aster L. ●■
14846 Alkibias Raf. = Chrysocoma L. ●☆
14847 Allaeanthus Thwaites = Broussonetia L'Hér. ex Vent.(保留属名)●
14848 Allaeanthus Thwaites(1854);落叶花桑属(阿里桑属,附尾桑属)●☆
14849 Allaeanthus glaber Warb.;光落叶花桑●☆
14850 Allaeanthus greveanus (Baill.) Capuron;格雷落叶花桑●☆
14851 Allaeanthus kurzii Hook. f. = Broussonetia kurzii (Hook. f.) Corner ●
14852 Allaeanthus luzonicus Fern.-Vill.;吕宋落叶花桑●☆
14853 Allaeanthus zeylanicus Thwaites;斯里兰卡落叶花桑●☆
14854 Allaeophania Thwaites = Hedyotis L.(保留属名)●■
14855 Allaganthera Mart. = Alternanthera Forssk. ■
14856 Allagas Raf. = Alpinia Roxb.(保留属名)■
14857 Allagopappus Cass. (1828);叉枝菊属●☆
14858 Allagopappus canariensis (Willd.) Greuter;加那利叉枝菊■☆
14859 Allagopappus dichotomus (L. f.) Cass. = Allagopappus canariensis (Willd.) Greuter ■☆
14860 Allagopappus viscosissimus Bolle;叉枝菊●☆
14861 Allagophyla Raf. = Corytholoma (Benth.) Decne. ■☆
14862 Allagoptera Nees(1821);香花棕属(互生翼棕榈属,轮羽椰属,香花椰子属)●☆
14863 Allagoptera arenaria (Gomes) Kuntze;香花棕●☆
14864 Allagoptera campestris Kuntze;群生香花棕●☆
14865 Allagosperma M. Roem. = Alternasemina Silva Manso ■
14866 Allagosperma M. Roem. = Melothria L. ■
14867 Allagostachyum Nees = Poa L. ■
14868 Allagostachyum Nees ex Steud. = Poa L. ■
14869 Allagostachyum Nees ex Steud. = Tribolium Desv. ■☆
14870 Allagostachyum Steud. = Tribolium Desv. ■☆
14871 Allamanda L. (1771);黄蝉木属(黄蝉属);Allamanda, Allemande ●
14872 Allamanda blanchetii A. DC.;紫蝉;Purple Allamanda ●☆
14873 Allamanda cathartica L.;软枝黄蝉;Allamanda, Campanilla, Common Allamanda, Golden Trumpet, Golden Trumpet Flower, Yellow Allemande ●
14874 Allamanda cathartica L. var. hendersonii (Bull. ex Dombr.) Bailey et Raf.;大花软枝黄蝉(大花黄蝉);Bigflower Allamanda, Henderson Allamanda ●
14875 Allamanda cathartica L. var. hendersonii (Bull. ex Dombr.) Bailey et Raf. = Allamanda cathartica L. ●
14876 Allamanda cathartica L. var. williamsii Gorl.;小花软枝黄蝉;William's Campanilla ●☆
14877 Allamanda hendersonii W. Bull ex Dombr. = Allamanda cathartica L. var. hendersonii (Bull ex Dombr.) Bailey et Raf. ●
14878 Allamanda hendersonii W. Bull ex Dombr. = Allamanda cathartica L. ●
14879 Allamanda neriifolia Hook. = Allamanda schottii Pohl ●
14880 Allamanda oenotherifolia Pohl;金色黄蝉;Golden Trumpet Bush ●
14881 Allamanda schottii Pohl;黄蝉(黄兰蝉,萧特黄蝉,小花黄蝉,硬枝黄蝉);Bush Allamanda, Bush Trumpet, Oleanderleaf Allamanda ●
14882 Allamanda violacea Gardner et Fielding = Allamanda blanchetii A. DC. ●☆
14883 Allamanda williamsii Hort. = Allamanda cathartica L. var. williamsii Gorl. ●☆
14884 Allanblackia Oliv. = Allanblackia Oliv. ex Benth. ●☆
14885 Allanblackia Oliv. ex Benth. (1867);阿兰藤黄属●☆
14886 Allanblackia floribunda Oliv.;多花阿兰藤黄;Kisidwe ●☆
14887 Allanblackia floribunda Oliv. var. gabonensis Pellegr. = Allanblackia gabonensis (Pellegr.) Bamps ●☆
14888 Allanblackia floribunda Oliv. var. kisonghi (Vermoesen) Pieraerts = Allanblackia kisonghi Vermoesen ●☆
14889 Allanblackia gabonensis (Pellegr.) Bamps;加蓬阿兰藤黄●☆
14890 Allanblackia kimbiliensis Spirlet;吉姆阿兰藤黄●☆
14891 Allanblackia kisonghii Vermoesen;吉氏阿兰藤黄●☆
14892 Allanblackia klainei Pierre ex A. Chev. = Allanblackia floribunda Oliv. ●☆
14893 Allanblackia marienii Staner;马里安阿兰藤黄●☆
14894 Allanblackia monticola Mildbr. ex Engl.;山生阿兰藤黄●☆
14895 Allanblackia oleifera ?;油阿兰藤黄;Kagné Butter ●☆
14896 Allanblackia parviflora A. Chev.;小花阿兰藤黄●☆
14897 Allanblackia sacleuxii Hua = Allanblackia stuhlmannii (Engl.) Engl. ●☆

14898 Allanblackia staneriana Exell et Mendonça;斯坦阿兰藤黄●☆
14899 Allanblackia stuhlmannii (Engl.) Engl.;斯图阿兰藤黄;Mkani Fat ●☆
14900 Allanblackia stuhlmannii Engl. = Allanblackia stuhlmannii (Engl.) Engl. ●☆
14901 Allanblackia ulugurensis Engl.;乌卢古尔阿兰藤黄●☆
14902 Allania Benth. = Aldina Endl.(保留属名)●☆
14903 Allania Meisn. = Alania Endl. ■☆
14904 Allantoma Miers(1874);腊肠玉蕊属●☆
14905 Allantoma aulacocarpa Miers;腊肠玉蕊●☆
14906 Allantospermum Forman(1965);腊肠木属●☆
14907 Allantospermum multicaule (Capuron) Noot.;腊肠木●☆
14908 Allardia Decne.(1841);小扁芒菊属(扁毛菊属,芒菊属)■
14909 Allardia Decne. = Waldheimia Kar. et Kir. ■
14910 Allardia glabra Decne.;光小扁芒菊■☆
14911 Allardia glabra Decne. = Allardia tridactylites (Kar. et Kir.) Sch. Bip. ■
14912 Allardia huegelii Sch. Bip. = Waldheimia huegelii (Sch. Bip.) Tzvelev ■
14913 Allardia lasiocarpa (G. X. Fu) Bremer et Humphries;毛果扁芒菊;Hairfruit Flatawndaisy,Hairfruit Waldheimia ■
14914 Allardia lasiocarpa (G. X. Fu) Bremer et Humphries = Waldheimia lasiocarpa G. X. Fu ■
14915 Allardia nivea Hook. et Roem. ex C. B. Clarke = Allardia nivea Hook. f. et Thomson ex C. B. Clarke ■
14916 Allardia nivea Hook. f. et Thomson ex C. B. Clarke;小扁毛菊(雪白小扁芒菊);Small Flatawndaisy,Small Waldheimia ■
14917 Allardia stoliczkae C. B. Clarke;光叶扁毛菊;Glabrousleaf Flatawndaisy,Glabrousleaf Waldheimia ■
14918 Allardia tomentosa Decne.;羽裂扁毛菊■
14919 Allardia tridactylites (Kar. et Kir.) Sch. Bip.;扁毛菊■
14920 Allardia vestita Hook. f. et Thomson ex C. B. Clarke;厚毛扁毛菊■
14921 Allardtia A. Dietr. = Tillandsia L. ■☆
14922 Allasia Lour. = Vitex L. ●
14923 Allasia payos Lour. = Vitex payos (Lour.) Merr. ●☆
14924 Allazia Silva Manso = Allasia Lour. ●
14925 Allazia Silva Manso = Vitex L. ●
14926 Alleizettea Dubard et Dop = Danais Comm. ex Vent. ●☆
14927 Alleizettella Pit.(1923);白香楠属(白果香楠属);Alleizettella ●
14928 Alleizettella leucocarpa (Champ. ex Benth.) Tirveng.;白果香楠(白香楠);Whitefruit Alleizettella,White-fruited Randia ●
14929 Alleizettella leucocarpa (Champ. ex Benth.) Tirveng. = Randia leucocarpa Champ. ex Benth. ●
14930 Allelotheca Steud. = Lophatherum Brongn. ■
14931 Allemanda L. = Allamanda L. ●
14932 Allemania Endl. = Allmania R. Br. ex Wight ■
14933 Allenanthus Standl.(1940);阿伦花属■☆
14934 Allenanthus erythrocarpus Standl.;阿伦花■☆
14935 Allendea La Llave = Allendea La Llave et Lex. ●■☆
14936 Allendea La Llave et Lex. = Liabum Adans. ●■☆
14937 Allenia E. Phillips = Radyera Bullock ●☆
14938 Allenia Ewart = Micrantheum Desf. ●☆
14939 Allenia urens (L. f.) Phillips = Radyera urens (L. f.) Bullock ●☆
14940 Allenrolfea Kuntze(1891);互苞盐节木属;Iodine Bush ●☆
14941 Allenrolfea occidentalis (S. Watson) Kuntze;西方互苞盐节木(西方白香楠);Burroweed,Iodine Bush ●☆
14942 Alletotheca Benth. et Hook. f. = Allelotheca Steud. ■
14943 Alletotheca Benth. et Hook. f. = Lophatherum Brongn. ■
14944 Allexis Pierre;卷瓣堇属●☆
14945 Allexis batangae (Engl.) Melch.;巴坦加卷瓣堇●☆
14946 Allexis cauliflora (Oliv.) Pierre;茎花卷瓣堇●☆
14947 Allexis cauliflora (Oliv.) Pierre = Rinorea cauliflora (Oliv.) Kuntze ●☆
14948 Allexis obanensis (Baker f.) Melch.;奥班卷瓣堇●☆
14949 Allexis zygomorpha Achoundong et Onana;对称卷瓣堇●☆
14950 Alliaceae Batsch ex Borkh. = Alliaceae Borkh.(保留科名)■
14951 Alliaceae Borkh.(1797)(保留科名);葱科■
14952 Alliaceae J. Agardh = Alliaceae Borkh.(保留科名)■
14953 Alliaria Heist. ex Fabr.(1759);葱芥属(葱臭芥属);Alliaria,Garlic Mustard,Garlicmustard,Hedge Garlic,Sauce-alone ■
14954 Alliaria Kuntze = Dysoxylum Blume ●
14955 Alliaria Scop. = Alliaria Heist. ex Fabr. ■
14956 Alliaria alliaria (L.) Britton = Alliaria petiolata (M. Bieb.) Cavara et Grande ■
14957 Alliaria auriculata Kom. = Cardamine komarovii Nakai ■
14958 Alliaria brachycarpa M. Bieb.;短果葱芥■
14959 Alliaria grandifolia C. H. An = Orychophragmus limprichtianus (Pax) Al-Shehbaz et G. Yang ■
14960 Alliaria officinalis Andrz. = Alliaria petiolata (M. Bieb.) Cavara et Grande ■
14961 Alliaria officinalis Andrz. ex DC. = Alliaria petiolata (M. Bieb.) Cavara et Grande ■
14962 Alliaria officinalis Andrz. ex M. Bieb. = Alliaria petiolata (M. Bieb.) Cavara et Grande ■
14963 Alliaria officinalis M. Bieb. = Alliaria petiolata (M. Bieb.) Cavara et Grande ■
14964 Alliaria petiolata (M. Bieb.) Cavara et Grande;葱芥(梗葱芥,西伯利亚大蒜芥,药用葱芥);Garlic Mustard,Hedge Garlic,Jack-by-the-hedge,Jack-of-the-hedge,Jack-run-along-by-the-hedge,Lady's Needlework,Lamb's Pummy,Leek Cress,Leek-cress,Medicinal Garlicmustard,Milkmaids,Penny Hedge,Penny in the Hedge,Petioled Garlicmustard,Pickpocket,Poor Man's Mustard,Poor Man's Treacle,Sauce-alone,Scabridge,Scabril,Scaybril Scabs,Siberian Wallflower,Stinking Hedge Mustard,Swarms ■
14965 Alliaria wasabi (Siebold) Prantl = Eutrema wasabii (Siebold) Maxim. ■
14966 Alliaria wasabi Prantl = Eutrema wasabii (Siebold) Maxim. ■
14967 Allibertia Marion = Agave L. ■
14968 Allibertia Marion ex Baker = Agave L. ■
14969 Allinum Neck. = Selinum L.(保留属名)■
14970 Allionia L.(1759)(保留属名);粉风车属(阿里昂花属)■☆
14971 Allionia Loefl.(废弃属名)= Allionia L.(保留属名)■☆
14972 Allionia Loefl.(废弃属名)= Mirabilis L. ■
14973 Allionia albida Walter = Mirabilis albida (Walter) Heimerl ■☆
14974 Allionia bodinii (Holz.) Morong = Mirabilis linearis (Pursh) Heimerl ■☆
14975 Allionia carletonii Standl. = Mirabilis glabra (S. Watson) Standl. ■☆
14976 Allionia choisyi Standl.;粉风车;Garapatilla,Trailing Four-o'clock,Umbrella-wort ■☆
14977 Allionia coahuilensis Standl. = Mirabilis albida (Walter) Heimerl ■☆
14978 Allionia coccinea (Torr.) Standl. = Mirabilis coccinea (Torr.) Benth. et Hook. f. ■☆

14979 Allionia comata Small = Mirabilis albida (Walter) Heimerl ■☆

14980 Allionia corymbosa (Cav.) Kuntze var. texensis J. M. Coult. = Mirabilis texensis (J. M. Coult.) B. L. Turner ■☆

14981 Allionia cristata (Standl.) Standl. = Allionia incarnata L. ■☆

14982 Allionia decipiens Standl. = Mirabilis linearis (Pursh) Heimerl var. decipiens (Standl.) S. L. Welsh ■☆

14983 Allionia decumbens (Nutt.) Spreng. = Mirabilis linearis (Pursh) Heimerl ■☆

14984 Allionia diffusa A. Heller = Mirabilis linearis (Pursh) Heimerl ■☆

14985 Allionia divaricata Rydb. = Mirabilis albida (Walter) Heimerl ■☆

14986 Allionia exaltata Standl. = Mirabilis glabra (S. Watson) Standl. ■☆

14987 Allionia gausapoides Standl. = Mirabilis linearis (Pursh) Heimerl var. subhispida (Heimerl) Spellenb. ■☆

14988 Allionia gigantea Standl. = Mirabilis gigantea (Standl.) Shinners ■☆

14989 Allionia glabra (S. Watson) Kuntze = Mirabilis glabra (S. Watson) Standl. ■☆

14990 Allionia gracillima Standl. = Mirabilis coccinea (Torr.) Benth. et Hook. f. ■☆

14991 Allionia hirsuta Pursh = Mirabilis albida (Walter) Heimerl ■☆

14992 Allionia incarnata L.;北美粉风车;Allionia, Hierba De La Hormiga, Pink Three-flower, Pink Windmills, Trailing Allionia, Trailing Four-o'clock, Trailing Windmills, Umbrella-wort ■☆

14993 Allionia incarnata L. var. glabra Choisy = Allionia choisyi Standl. ■☆

14994 Allionia incarnata L. var. glabra Choisy = Allionia incarnata L. ■☆

14995 Allionia incarnata L. var. nudata (Standl.) Munz = Allionia incarnata L. ■☆

14996 Allionia incarnata L. var. villosa (Standl.) Munz;毛北美粉风车 ■☆

14997 Allionia latifolia (A. Gray) Standl. = Mirabilis latifolia (A. Gray) Diggs, Lipscomb et O'Kennon ■☆

14998 Allionia linearis Pursh = Mirabilis linearis (Pursh) Heimerl ■☆

14999 Allionia linearis var. bodinii (Holz.) A. Nelson = Mirabilis linearis (Pursh) Heimerl ■☆

15000 Allionia melanotricha Standl. = Mirabilis melanotricha (Standl.) Spellenb. ■☆

15001 Allionia nyctaginea Michx. = Mirabilis nyctaginea (Michx.) MacMill. ■☆

15002 Allionia oblongifolia (A. Gray) Small = Mirabilis albida (Walter) Heimerl ■☆

15003 Allionia oxybaphoides (A. Gray) Kuntze = Mirabilis oxybaphoides (A. Gray) A. Gray ■☆

15004 Allionia pauciflora (Buckley) Standl. = Mirabilis albida (Walter) Heimerl ■☆

15005 Allionia pinetorum Standl. = Mirabilis linearis(Pursh) Heimerl ■☆

15006 Allionia pratensis Standl. = Mirabilis albida (Walter) Heimerl ■☆

15007 Allionia pseudaggregata (Heimerl) Standl. = Mirabilis albida (Walter) Heimerl ■☆

15008 Allionia pumila Standl. = Mirabilis albida (Walter) Heimerl ■☆

15009 Allionia rotundifolia Greene = Mirabilis rotundifolia (Greene) Standl. ■☆

15010 Allionia texensis (J. M. Coult.) Small = Mirabilis texensis (J. M. Coult.) B. L. Turner ■☆

15011 Allioniaceae Horan. = Nyctaginaceae Juss. (保留科名)●■

15012 Allioniella Rydb. (1902);小粉风车属■☆

15013 Allioniella Rydb. = Mirabilis L. ■

15014 Allioniella oxybaphoides (A. Gray) Rydb. = Mirabilis oxybaphoides (A. Gray) A. Gray ■☆

15015 Allittia P. S. Short(2004);湿地鹅河菊属■☆

15016 Allittia cardiocarpa (Benth.) P. S. Short;心果湿地鹅河菊■☆

15017 Allittia uliginosa (G. L. R. Davis) P. S. Short;湿地鹅河菊■☆

15018 Allium L. (1753);葱属;Allium, Chive, Garlic, Leek, Onion ■

15019 Allium aciphyllum J. M. Xu;针叶韭;Needleleaf Leek ■

15020 Allium acuminatum Hook.;尖瓣葱;Allium, Hooker's Onion, Wild Onion ■☆

15021 Allium acuminatum Hook. var. cuspidatum Fernald = Allium acuminatum Hook. ■☆

15022 Allium affine Ledeb.;近缘葱■☆

15023 Allium aflatunense B. Fedtsch.;阿地细茎葱;Ornamental Onion, Thinstem Onion ■☆

15024 Allium aitchisonii Baker = Allium carolinianum F. Delaroche ■

15025 Allium akaka Gmel. ex Roem. et Schult.;莲座韭(阿卡卡葱)■☆

15026 Allium alabasicum (D. S. Wen et Sh. Chen) Y. Z. Zhao;鄂尔多斯韭(阿尔巴斯韭)■

15027 Allium alaicum Vved.;阿赖葱■☆

15028 Allium alaschanicum Y. Z. Zhao = Allium flavovirens Regel ■

15029 Allium alataviense Regel = Allium platyspathum Schrenk subsp. amblyophyllum (Kar. et Kir.) Frizen ■

15030 Allium albanum Grossh.;阿尔班山葱;Alban Onion ■☆

15031 Allium alberti Regel = Allium pallasii Murray ■

15032 Allium albidum Fisch. ex Besser;白野葱(白葱,野葱);Whitish Onion ■☆

15033 Allium albopilosum C. H. Wright = Allium neapolitanum Cirillo ■☆

15034 Allium albostellerianum F. T. Wang et Ts. Tang = Allium paepalanthoides Airy Shaw ■

15035 Allium albostellerianum F. T. Wang et Ts. Tang = Allium platyspathum Schrenk subsp. amblyophyllum (Kar. et Kir.) Frizen ■

15036 Allium albovianum Vved.;阿尔勃葱■☆

15037 Allium album Santi = Allium neapolitanum Cirillo ■☆

15038 Allium album Santi var. purpurascens Maire et Weiller et Wilczek = Allium neapolitanum Cirillo ■☆

15039 Allium alexandrae Vved.;阿莱葱■☆

15040 Allium alexejanum Regel;阿莱克赛葱■☆

15041 Allium allegheniense Small = Allium cernuum Roth ■☆

15042 Allium altaicum Pall.;阿尔泰葱;Altai Onion ■

15043 Allium altissimum Regel;巨葱■☆

15044 Allium amabile Stapf = Allium mairei H. Lév. ■

15045 Allium amamianum Tawada = Allium pseudojaponicum Makino ■

15046 Allium amblyophyllum Kar. et Kir. = Allium platyspathum Schrenk subsp. amblyophyllum (Kar. et Kir.) Frizen ■

15047 Allium amblyophyllum Kar. et Kir. = Allium platyspathum Schrenk ■

15048 Allium ampeloprasum L.;南欧葱(大头葱,大头蒜,南欧蒜);Bluleek, Broadleaf Wild Leek, Bunching Pearl Onion, Elephant Garlic, Giant Garlic, Great Round-headed Garlic, Great-head Garlic, Kurrat, Levant Garlic, Pearl Onion, Round-headed Garlic, St. David's Plant, Wild Leek ■☆

15049 Allium ampeloprasum L. subsp. porrum (L.) Hayek = Allium porrum L. ■

15050 Allium ampeloprasum L. var. caudatum Pamp. = Allium ampeloprasum L. ■☆

15051 Allium ampeloprasum L. var. duriaeanum (J. Gay) Batt. = Allium baeticum Boiss. ■☆

15052 Allium ampeloprasum L. var. getulum (Batt. et Trab.) Jahand. et Maire = Allium ampeloprasum L. ■☆
15053 Allium ampeloprasum L. var. gracilis Cavara = Allium ampeloprasum L. ■☆
15054 Allium ampeloprasum L. var. porrum Regel = Allium porrum L. ■
15055 Allium ampeloprasum L. var. tortifolium (Batt.) Maire = Allium ampeloprasum L. ■☆
15056 Allium amphibolum Ledeb.;直立韭(韭葱);Amphibolous Onion ■
15057 Allium amplectens Torr.;抱茎葱;Allium ■☆
15058 Allium andersonii G. Don = Allium senescens L. ■
15059 Allium angolense Baker = Allium cepa L. ■
15060 Allium angulosum L.;角葱(棱葱);Angle Onion, Angulated Onion ■☆
15061 Allium angulosum L. var. minum Ledeb. = Allium senescens L. ■
15062 Allium angustitepalum Wendelbo = Allium rosenbachianum Regel ■☆
15063 Allium anisopetalum Vved. = Allium griffithianum Boiss. ■☆
15064 Allium anisopodium Ledeb.;矮韭(矮葱);Dwarf Leek ■
15065 Allium anisopodium Ledeb. var. zimmermannianum (Gilg) F. T. Wang et Ts. Tang;糙葶韭;Scabrousscape Leek, Scabrousscape Onion ■
15066 Allium anisopodium Ledeb. var. zimmermannianum (Gilg) Kitag. = Allium anisopodium Ledeb. var. zimmermannianum (Gilg) F. T. Wang et Ts. Tang ■
15067 Allium antiatlanticum Emb. et Maire;安蒂葱■☆
15068 Allium aonospermum Jeps. = Allium amplectens Torr. ■☆
15069 Allium arenicola Osterh. = Allium geyeri S. Watson var. tenerum M. E. Jones ■☆
15070 Allium arenicola Small = Allium canadense L. var. mobilense (Regel) Ownbey ■☆
15071 Allium argyi H. Lév. = Allium tuberosum Rootler ex Spreng. ■
15072 Allium aridum Rydb. = Allium textile A. Nelson et J. F. Macbr. ■☆
15073 Allium aroides Popov et Vved.;中亚葱■☆
15074 Allium artemisietorum Eig et Feinbrun = Allium ascalonicum L. ■
15075 Allium arvense Guss. = Allium sphaerocephalum L. var. arvense (Guss.) Parl. ■☆
15076 Allium ascalonicum L. = Allium cepa L. var. aggregatum G. Don ■
15077 Allium ascalonicum L. = Allium cepiforme G. Don ■
15078 Allium ascalonicum L. var. chinense Kunth = Allium cepiforme G. Don ■
15079 Allium aschersonianum Barbey = Allium orientale Boiss. ■☆
15080 Allium aschersonianum Barbey subsp. ambiguum Bég. et Vacc. = Allium orientale Boiss. ■☆
15081 Allium aschersonianum Barbey var. ambiguum (Bég. et Vacc.) Pamp. = Allium orientale Boiss. ■☆
15082 Allium atropurpureum Waldst. et Kit.;深紫葱;Rosepurpreum Onion ■☆
15083 Allium atrorubens S. Watson;深红葱■☆
15084 Allium atrorubens S. Watson subsp. inyonis (M. E. Jones) Traub = Allium atrorubens S. Watson var. cristatum (S. Watson) McNeal ■☆
15085 Allium atrorubens S. Watson var. cristatum (S. Watson) McNeal;冠深紫葱■☆
15086 Allium atrorubens S. Watson var. inyonis (M. E. Jones) Ownbey = Allium atrorubens S. Watson var. cristatum (S. Watson) McNeal ■☆
15087 Allium atrosanguineum Kar. et Kir. = Allium atrosanguineum Schrenk ■
15088 Allium atrosanguineum Schrenk;蓝苞葱(蓝色韭);Bluebract Onion, Bluespath Onion ■
15089 Allium atrosanguineum Schrenk var. fedschenkoanum (Regel) G. H. Zhu et N. J. Turland;费氏葱(费葱)■
15090 Allium atrosanguineum Schrenk var. tibeticum (Regel) G. H. Zhu et N. J. Turland;藏葱■
15091 Allium atroviolaceum Boiss.;黑紫葱;Broadleaf Wild Leek ■☆
15092 Allium attenuifolium Kellogg = Allium amplectens Torr. ■☆
15093 Allium attenuifolium Kellogg var. monospermum (Jeps.) Jeps. = Allium amplectens Torr. ■☆
15094 Allium aucheri Boiss.;奥氏葱■☆
15095 Allium austinae M. E. Jones = Allium campanulatum S. Watson ■☆
15096 Allium austrosibiricum Frizen = Allium spirale Willd. ■
15097 Allium azureum Ledeb. = Allium caeruleum Pall. ■
15098 Allium babingtonii Borrer = Allium ampeloprasum L. ■☆
15099 Allium baeticum Boiss.;伯蒂卡葱■☆
15100 Allium baeticum Boiss. var. laeve Maire et Weiller = Allium baeticum Boiss. ■☆
15101 Allium baeticum Boiss. var. papillosum (H. Lindb.) Maire et Weiller = Allium baeticum Boiss. ■☆
15102 Allium baicalense Willd. = Allium chinense G. Don ■
15103 Allium baicalense Willd. = Allium senescens L. ■
15104 Allium bakeri Regel = Allium chinense G. Don ■
15105 Allium bakeri Regel var. morrisonense (Hayata) Tang S. Liu et S. S. Ying;玉山蒜(野韭)■
15106 Allium bakeri Regel var. morrisonense (Hayata) Tang S. Liu et S. S. Ying = Allium thunbergii G. Don ■
15107 Allium barszczewskii Lipsky;巴尔葱■☆
15108 Allium barszczewskii Lipsky f. niveum Krassovsk.;雪白巴尔葱■☆
15109 Allium barszczewskii Lipsky f. violaceum Krassovsk.;堇色巴尔葱■☆
15110 Allium barthianum Asch. et Schweinf. = Allium ascalonicum L. ■
15111 Allium beesianum W. W. Sm.;蓝花韭;Bees Leek, Bees Onion ■
15112 Allium bellutum Prokh.;小丽葱;Pretty Onion ■☆
15113 Allium bidentatum Fisch. ex Prokh.;砂韭(双齿葱);Bidentate Onion, Sand Leek ■
15114 Allium bidentatum Fisch. ex Prokh. var. andaense Q. S. Sun;丝韭■
15115 Allium bidentatum Fisch. ex Prokh. var. andaense Q. S. Sun = Allium bidentatum Fisch. ex Prokh. ■
15116 Allium bidwelliae S. Watson = Allium campanulatum S. Watson ■☆
15117 Allium bigelovii S. Watson;毕氏葱■☆
15118 Allium bineale L.;田蒜■☆
15119 Allium bivalve (L.) Kuntze = Nothoscordum bivalve (L.) Britton ■☆
15120 Allium blandum Wall.;白韭■
15121 Allium blandum Wall. = Allium carolinianum F. Delaroche ■
15122 Allium bodeanum Regel;鲍代葱■☆
15123 Allium bodinieri H. Lév. et Vaniot = Allium chinense G. Don ■
15124 Allium bogdoicola Regel = Allium strictum Schrad. ■
15125 Allium bolanderi S. Watson;博兰德葱;Wild Onion ■☆
15126 Allium bolanderi S. Watson var. mirabile (L. F. Hend.) McNeal;奇异葱■☆
15127 Allium bolanderi S. Watson var. stenanthum (Drew) Jeps. = Allium bolanderi S. Watson ■☆
15128 Allium borszczowii Regel;鲍尔葱■☆
15129 Allium bouddhae Debeaux = Allium fistulosum L. ■
15130 Allium brachyodon Boiss. = Allium strictum Schrad. ■
15131 Allium brachyscapum Vved.;短果韭■☆
15132 Allium brahuicum Boiss. = Allium caspium M. Bieb. ■☆

15133 Allium brevidens Vved.;短齿葱■☆
15134 Allium brevidentatum F. Z. Li;矮齿韭;Shorttooth Garlic ■
15135 Allium brevistylum S. Watson;短柱葱■☆
15136 Allium breweri S. Watson = Allium falcifolium Hook. et Arn. ■☆
15137 Allium buhseanum Diels = Allium schoenoprasum L. ■
15138 Allium bulleyanum Diels = Allium wallichii Kunth ■
15139 Allium bulleyanum Diels var. tchongshanense (H. Lév.) Airy Shaw = Allium wallichii Kunth ■
15140 Allium burdickii (Hanes) A. G. Jones = Allium tricoccum Sol. var. burdickii Hanes ■☆
15141 Allium burjaticum Frizen = Allium spirale Willd. ■
15142 Allium caeruleum Pall.;棱叶韭(蓝花山蒜)■
15143 Allium caesioides Wendelbo;淡蓝韭■☆
15144 Allium caesium Schrenk;知母韭■
15145 Allium caespitosum Siev. ex Bong. et Mey.;疏生韭;Caespitose Leek,Caespitose Onion ■
15146 Allium callidictyon C. A. Mey. ex Kunth;美脉韭■☆
15147 Allium campanulatum S. Watson;北美钟花韭■☆
15148 Allium campanulatum S. Watson var. bidwelliae (S. Watson) Jeps. = Allium campanulatum S. Watson ■☆
15149 Allium canadense L.;加拿大葱(加拿大蒜);Canada Garlic, Canada Onion,Canadian Garlic,Meadow Garlic,Meadow Leek,Wild Garlic,Wild Onion,Rose Leek ■☆
15150 Allium canadense L. var. ecristatum Ownbey;无冠加拿大葱■☆
15151 Allium canadense L. var. fraseri Ownbey;弗氏加拿大葱■☆
15152 Allium canadense L. var. hyacinthoides (Bush) Ownbey;风信子葱■☆
15153 Allium canadense L. var. mobilense (Regel) Ownbey;沙丘加拿大葱■☆
15154 Allium canadense L. var. ovoideum Farw. = Allium canadense L. ■☆
15155 Allium candidissimum Cav. = Allium neapolitanum Cirillo ■☆
15156 Allium candolleanum Albov;康多勒葱■☆
15157 Allium cannifolium H. Lév. = Allium prattii C. H. Wright ■
15158 Allium caricifolium Kar. et Kir. = Allium pallasii Murray ■
15159 Allium caricoides Regel;石生韭;Stone Leek,Stone Onion ■
15160 Allium carinatum L.;龙骨葱;Keeled Garlic,Keeled Onion,Red Onion ■☆
15161 Allium carolinianum F. Delaroche;加州镰叶韭(镰叶韭); Broadstyle Onion,Carolina Leek,Carolina Onion ■
15162 Allium cascadense M. Peck = Allium crenulatum Wiegand ■☆
15163 Allium caspium M. Bieb.;里海葱■☆
15164 Allium caucasicum M. Bieb. = Allium saxatile M. Bieb. ■
15165 Allium cepa L.;洋葱(葱头,大头葱,胡葱,浑提葱,洋葱头,玉葱); Common Onion, Cultivated Onion, Egyptian Walking Onion, Garder Onion, Ine, Ineyun, Ingan, Inin, Ining, Inion, Innion, Inon, Inun,Inyun,Onion,Onions,Potato Onion,Scullion,Sour,Sour Sower, Sower,Spring Onion,St. Thomas' Onion,Winter Onion ■
15166 Allium cepa L. 'Proliferum';楼子葱(分生洋葱,红葱,楼子葱);Egyptian Onion,Garden Onion,Three Onion,Tree Onion ■
15167 Allium cepa L. var. aggregatum D. Don;火葱(分蘖葱头); Egyptian Onion,Multiplier Onion,Potato Onion,Shallot,St. Thomas' Onion ■☆
15168 Allium cepa L. var. proliferum Regel = Allium cepa L. 'Proliferum' ■
15169 Allium cepa L. var. viviparum ?;胎生葱;Egyptian Multiplier Onion,Potato Onion,Top Onion,Tree Onion ■☆
15170 Allium cepiforme G. Don;香葱(慈葱,大宫葱,冬葱,冻葱,分葱,胡葱,葫葱,回回葱,火葱,韭,科葱,蒜葱,细香葱);Chibbal, Chibble,Chiboul,Cibbols,Ciboule,Eschallot,Scallion,Shallot ■
15171 Allium ceratophyllum Besser ex Schult. et Schult. f. = Allium altaicum Pall. ■
15172 Allium ceratophyllum Besser ex Schult. f. = Allium altaicum Pall. ■
15173 Allium cernuum Roth;垂花葱;Droopingflowered Onion,Lady's Leek,Nodding Onion,Nodding Wild Onion,Wild Nodding Onion, Wild Onion ■☆
15174 Allium chalcophengos Airy Shaw = Allium atrosanguineum Schrenk var. tibeticum (Regel) G. H. Zhu et N. J. Turland ■
15175 Allium chalcophengos Airy Shaw = Allium atrosanguineum Schrenk ■
15176 Allium chamaemoly L.;矮葱■☆
15177 Allium chamaemoly L. subsp. longicaulis Rapin et Valdés;长茎葱■☆
15178 Allium chamaemoly L. var. battandieri Maire et Weiller = Allium chamaemoly L. ■☆
15179 Allium chamaemoly L. var. coloratum Batt. = Allium tourneuxii Chabert ■☆
15180 Allium chamaemoly L. var. littoralis (Jord. et Fourr.) Maire et Weiller = Allium chamaemoly L. ■☆
15181 Allium chamaemoly L. var. viridulum (Jord. et Fourr.) Maire et Weiller = Allium chamaemoly L. ■☆
15182 Allium chanetii H. Lév. = Allium macrostemon Bunge ■
15183 Allium changduense J. M. Xu;昌都韭;Changdu Leek,Changdu Onion ■
15184 Allium chienchuanense J. M. Xu;剑川韭;Jianchuan Leek, Jianchuan Onion ■
15185 Allium chinense G. Don;藠头(韭,韭白,韭白头,荞头,野韭,莜);Baker's Garlic,China Onion,Chinese Onion,Rakkyo ■
15186 Allium chinense Maxim. = Allium tuberosum Rootler ex Spreng. ■
15187 Allium chiwui F. T. Wang et Ts. Tang;冀韭;Chiwu Leek,Chiwu Onion ■
15188 Allium christophii Trautv.;波斯葱;Ornamental Onion,Star of Persia ■☆
15189 Allium chrysanthum Regel;野葱(黄花葱,黄花韭,山葱子); Wild Chive,Yellowflower Onion ■
15190 Allium chrysocephalum Regel;折被韭(黄花葱); Chrysocephalous Leek,Chrysocephalous Onion ■
15191 Allium ciliare Delaroche = Allium subhirsutum L. subsp. ciliare (Delaroche) Maire et Weiller ■☆
15192 Allium clarkei Hook. f. = Allium tuberosum Rootler ex Spreng. ■
15193 Allium clathratum Ledeb.;细叶北韭■
15194 Allium clusianum Retz.;蓝花山葱(韭白,棱叶韭,新疆韭); Ribleaf Leek,Skeyblue Onion ■
15195 Allium coerulescens G. Don = Allium caeruleum Pall. ■
15196 Allium coeruleum Pall. = Allium clusianum Retz. ■
15197 Allium columbianum (Ownbey et Mingrone) P. M. Peterson;哥伦比亚葱■☆
15198 Allium condensatum Turcz.;黄花韭(黄花葱);Crowded Chive, Crowded Onion ■
15199 Allium confragosum Vved.;粗糙葱■☆
15200 Allium congestum G. Don = Allium prostratum Trevir. ■
15201 Allium consanguineum Kunth;亲缘葱■☆
15202 Allium constrictum (Ownbey et Mingrone) P. M. Peterson;缢缩葱■☆

15203 Allium continuum Small = Allium canadense L. ■☆
15204 Allium convallarioides Grossh.;环绕葱■☆
15205 Allium cordifolium J. M. Xu = Allium ovalifolium Hand.-Mazz. var. cordifolium (J. M. Xu) J. M. Xu ■
15206 Allium coryi M. E. Jones;柯里葱;Yellow-flowered Onion ■☆
15207 Allium cowanii Lindl. = Allium neapolitanum Cirillo ■☆
15208 Allium cratericola Eastw.;火山葱■☆
15209 Allium crenulatum Wiegand;细圆齿■☆
15210 Allium cristatum Greene;冠葱;Crested Onion ■☆
15211 Allium cristatum S. Watson = Allium atrorubens S. Watson var. cristatum (S. Watson) McNeal ■☆
15212 Allium cristophii Trautv.;纸花葱■☆
15213 Allium croceum Torr. = Bloomeria crocea (Torr.) Coville ■☆
15214 Allium crystallinum Vved.;水晶葱■☆
15215 Allium cupanii Raf.;库潘葱■☆
15216 Allium cupanii Raf. subsp. hirtovaginatum (Kunth) Stearn = Allium hirtovaginatum Kunth ■☆
15217 Allium cupuliferum Regel;杯状葱■☆
15218 Allium curtum Boiss. et Gaill. = Allium sphaerocephalum L. subsp. curtum (Boiss. et Gaill.) Duyfjes ■☆
15219 Allium curtum Boiss. et Gaill. subsp. aegyptiacum Täckh. et Drar = Allium sphaerocephalum L. subsp. curtum (Boiss. et Gaill.) Duyfjes ■☆
15220 Allium cyaneum Regel;天蓝韭(白狼葱,蓝花葱,天扁韭,岩韭,野葱);Blue Onion,Skyb Leek ■
15221 Allium cyaneum Regel var. brachystemon Regel = Allium sikkimense Baker ■
15222 Allium cyathophorum Bureau et Franch.;杯花韭;Cupflower Leek,Cupflower Onion ■
15223 Allium cyathophorum Bureau et Franch. var. farreri Stearn;川甘韭(法勒氏葱);Chuangan Onion,Farrer Onion ■
15224 Allium darvasicum Regel;达尔瓦斯葱■☆
15225 Allium dauricum Frizen = Allium spurium G. Don ■
15226 Allium decipiens Fisch. ex Roem. et Schult.;星花蒜(假葱,韭白,脱苞韭);Starflower Garlic,Starflower Onion ■
15227 Allium decipiens M. E. Jones = Allium atrorubens S. Watson var. cristatum (S. Watson) McNeal ■☆
15228 Allium declinatum Rchb. = Allium prostratum Trevir. ■
15229 Allium declinatum Willd. = Allium prostratum Trevir. ■
15230 Allium deflexum Fisch. ex Kunth = Allium prostratum Trevir. ■
15231 Allium deflexum Fisch. ex Schult. et Schult. f. = Allium prostratum Trevir. ■
15232 Allium delicatulum Siev. ex Roem. et Schult. = Allium delicatulum Siev. ex Schult. et Schult. f. ■
15233 Allium delicatulum Siev. ex Schult. et Schult. f.;迷人韭(美味葱)■
15234 Allium dentifolium Webb et Berthel. = Allium longispathum Delaroche ■☆
15235 Allium dentigerum Prokh.;短齿韭;Shorttooth Leek,Shorttooth Onion ■
15236 Allium derderianum Regel;戴尔葱■☆
15237 Allium deserticola (M. E. Jones) Wooton et Standl. = Allium macropetalum Rydb. ■☆
15238 Allium deserticola Popov = Allium tekesicola Regel ■
15239 Allium desertorum Forssk.;沙漠韭■☆
15240 Allium diabolense (Ownbey et Aase ex Traub) McNeal;流苏韭■☆
15241 Allium dictyoprasum C. A. Mey. ex Kunth;网韭■☆
15242 Allium dictyoscordum Vved.;指蒜■☆
15243 Allium dictyotum Greene = Allium geyeri S. Watson ■☆
15244 Allium dolichomischum Vved.;长葱■☆
15245 Allium dolichostylum Vved.;长柱葱■☆
15246 Allium doloncarense Regel = Allium delicatulum Siev. ex Schult. et Schult. f. ■
15247 Allium douglasii Hook.;道格拉斯葱■☆
15248 Allium douglasii Hook. var. columbianum Ownbey et Mingrone = Allium columbianum (Ownbey et Mingrone) P. M. Peterson ■☆
15249 Allium douglasii Hook. var. constrictum Ownbey et Mingrone = Allium constrictum (Ownbey et Mingrone) P. M. Peterson ■☆
15250 Allium douglasii Hook. var. nevii (S. Watson) Ownbey et Mingrone = Allium nevii S. Watson ■☆
15251 Allium dregeanum Kunth;德雷葱■☆
15252 Allium drepanophyllum Vved.;镰叶韭■☆
15253 Allium drobovii Vved.;德罗葱■☆
15254 Allium drummondii Regel;德拉蒙德葱;Wild Garlic ■☆
15255 Allium dshungaricum Vved. = Allium saxatile M. Bieb. ■
15256 Allium duriaeanum Gay = Allium ampeloprasum L. var. duriaeanum (J. Gay) Batt. ■☆
15257 Allium duriaeanum J. Gay = Allium ampeloprasum L. ■☆
15258 Allium edentatum Y. P. Hsu = Allium bidentatum Fisch. ex Prokh. ■
15259 Allium eduardii Stearn;贺兰韭;Eduard Leek,Eduard Onion ■
15260 Allium elatum Regel;高葱;Tall Onion ■☆
15261 Allium elatum Regel = Allium macleanii Baker ■☆
15262 Allium elegans Drobov;西亚韭■☆
15263 Allium elegantulum Kitag.;雅韭■
15264 Allium elegantulum Kitag. = Allium tenuissimum L. ■
15265 Allium equicaeleste H. St. John = Allium macrum S. Watson ■☆
15266 Allium erubescens K. Koch;变红葱■☆
15267 Allium eugenii Vved.;欧根葱■☆
15268 Allium eusperma Airy Shaw;真籽韭;Realseed Leek,Realseed Onion ■
15269 Allium falcifolium Hook. et Arn. var. breweri (S. Watson) M. E. Jones = Allium falcifolium Hook. et Arn. ■☆
15270 Allium fallax Roem. et Schult. = Allium senescens L. ■
15271 Allium fallax Schult. f.;疑葱;Dwarf Alpine Onion ■☆
15272 Allium farreri Stearn = Allium cyathophorum Bureau et Franch. var. farreri Stearn ■
15273 Allium fasciculatum Rendle;粗根韭;Stoutroot Leek,Stoutroot Onion ■
15274 Allium feddei H. Lév. = Allium wallichii Kunth ■
15275 Allium fedschenkoanum Regel = Allium atrosanguineum Schrenk var. fedschenkoanum (Regel) G. H. Zhu et N. J. Turland ■
15276 Allium fedschenkoanum Regel var. elatum Regel = Allium atrosanguineum Schrenk var. fedschenkoanum (Regel) G. H. Zhu et N. J. Turland ■
15277 Allium ferganicum Vved.;费尔干葱■☆
15278 Allium fetisowii Regel;多籽蒜;Fetisow Garlic,Fetisow Onion ■
15279 Allium fibrosum Regel;纤维葱■☆
15280 Allium fibrosum Rydb. = Allium geyeri S. Watson var. tenerum M. E. Jones ■☆
15281 Allium filidens Regel;线齿葱■☆
15282 Allium filifolium Regel;丝叶葱;Silkleaf Onion ■☆
15283 Allium fimbriatum S. Watson var. diabolense Ownbey et Aase ex Traub = Allium diabolense (Ownbey et Aase ex Traub) McNeal ■☆

15284 Allium fimbriatum S. Watson var. parryi (S. Watson) Ownbey ex Traub = Allium parryi S. Watson ■☆
15285 Allium firmotunicatum Fomin;坚硬葱■☆
15286 Allium fischeri Besser ex Schult. et Schult. f. = Allium prostratum Trevir. ■
15287 Allium fischeri Regel = Allium eduardii Stearn ■
15288 Allium fischeri Regel = Allium prostratum Trevir. ■
15289 Allium fistulosum L.;葱(北葱,菜伯,大葱,冬葱,汉葱,火葱,龙角葱,楼葱,楼子葱,木葱,青葱,四季葱,太官,香葱);Chibbal, Chibble, Chiboul, Chinese Onion, Chipple, Cibol, Ciboul Onion, Ciboule, Fistular Onion, Fringed Onion, Gibbles, Giblets, Japanese Bunching Onion, Japanese Leek, Jibbles, Spanish Onion, Spring Onion, St. Thomas' Onion, Stone Leek, Sybie, Sybow, Welsh Onion ■
15290 Allium fistulosum L. f. viviparum (Makino) Hiroe = Allium fistulosum L. var. viviparum Makino ■
15291 Allium fistulosum L. var. caespitosum Makino;分葱■☆
15292 Allium fistulosum L. var. giganteum Makino;日本葱■☆
15293 Allium fistulosum L. var. viviparum Makino;楼葱■
15294 Allium fistulosum subsp. viviparum (Makino) Kazakova = Allium fistulosum L. var. viviparum Makino ■
15295 Allium flavescens Besser;浅黄葱■☆
15296 Allium flavidum Ledeb.;新疆韭;Xinjiang Leek, Yellowish Onion ■
15297 Allium flavovirens Regel;阿拉善韭(阿拉善葱);Alashan Onion ■
15298 Allium flavum L. subsp. ionochlorum Maire = Allium fontanesii J. Gay ■☆
15299 Allium flavum Salisb. = Allium moly L. ■☆
15300 Allium fominianum Miscz. ex Grossh. et Schischk.;佛敏葱■☆
15301 Allium fontanesii J. Gay;丰塔纳葱■☆
15302 Allium forrestii Diels;梭沙韭;Forrest Leek, Forrest Onion ■
15303 Allium fragrans Vent. = Nothoscordum borbonicum Kunth ■☆
15304 Allium fragrans Vent. = Nothoscordum gracile (Aiton) Stearn ■☆
15305 Allium fraseri (Ownbey) Shinners = Allium canadense L. var. fraseri Ownbey ■☆
15306 Allium funckiifolium Hand.-Mazz.;玉簪叶山葱(鹿耳韭,天韭,天蒜,岩蒜,玉簪叶韭);Plantainly-leaved Leek, Plantainly-leaved Onion ■
15307 Allium funiculosum A. Nelson = Allium geyeri S. Watson ■☆
15308 Allium fuscoviolaceum Fomin;褐堇葱■☆
15309 Allium gageanum W. W. Sm. = Allium fasciculatum Rendle ■
15310 Allium galanthum Kar. et Kir.;实葶葱(乳葱);Milk Chive, Milk Onion ■
15311 Allium getulum Batt. et Trab. = Allium ampeloprasum L. ■☆
15312 Allium geyeri S. Watson;盖氏葱;Geyer Onion ■☆
15313 Allium geyeri S. Watson var. graniferum Hend. = Allium geyeri S. Watson var. tenerum M. E. Jones ■☆
15314 Allium geyeri S. Watson var. tenerum M. E. Jones;极细葱■☆
15315 Allium giganteum Regel = Allium macleanii Baker ■☆
15316 Allium gilgiticum F. T. Wang et Ts. Tang;吉尔吉特葱■☆
15317 Allium gillii Wendelbo;吉尔葱■☆
15318 Allium giraudiasii H. Lév. = Allium mairei H. Lév. ■
15319 Allium glaciale Vved.;冰雪葱■☆
15320 Allium glandulosum Link et Otto;腺葱■☆
15321 Allium glaucum Schrad. ex Poir. = Allium spirale Willd. ■
15322 Allium globosum M. Bieb. ex Redoute = Allium saxatile M. Bieb. ■
15323 Allium globosum M. Bieb. ex Redoute var. albidum Regel = Allium tianschanicum Rupr. ■
15324 Allium glomeratum Prokh.;头花韭;Glonerate Leek, Glonerate Onion ■
15325 Allium gmelinianum Miscz. ex Grossh. = Allium saxatile M. Bieb. ■
15326 Allium govanianum Wall. ex Baker = Allium humile Kunth ■
15327 Allium gracile Dryand. = Nothoscordum gracile (Aiton) Stearn ■☆
15328 Allium gracilescens Sommier et H. Lév.;细葱■☆
15329 Allium graecum d'Urv. = Allium subhirsutum L. var. graecum (d'Urv.) Regel ■☆
15330 Allium grande Lipsky;利普斯基葱■☆
15331 Allium grandisceptrum Davidson = Allium unifolium Kellogg ■☆
15332 Allium grayi Regel;山蒜■
15333 Allium grayi Regel = Allium macrostemon Bunge ■
15334 Allium grayi Regel var. chanetii (H. Lév.) H. Lév. = Allium macrostemon Bunge ■
15335 Allium greuteri Brullo et Pav.;格罗特葱■☆
15336 Allium griffithianum Boiss.;格里葱■☆
15337 Allium grimmii Regel = Allium teretifolium Regel ■
15338 Allium grisellum J. M. Xu;灰皮葱(灰皮韭);Grayish Onion ■
15339 Allium guanxianense J. M. Xu;灌县韭;Guanxian Leek, Guanxian Onion ■
15340 Allium guttatum Steven;斑葱■☆
15341 Allium helicophyllum Vved.;旋叶葱■☆
15342 Allium helleri Small = Allium drummondii Regel ■☆
15343 Allium hendersonii B. L. Rob. et Seaton = Allium douglasii Hook. ■☆
15344 Allium henryi C. H. Wright;疏花韭;Henry Leek, Henry Onion ■
15345 Allium herderianum Regel;金头韭;Herder Leek, Herder Onion ■
15346 Allium herdreichii Boiss.;赫德赖克氏葱;Herdreich Onion ■☆
15347 Allium heteronema F. T. Wang et Ts. Tang;异梗韭;Heterostalk Leek, Heterostalk Onion ■
15348 Allium himalayense Regel = Allium stocksianum Boiss. ■☆
15349 Allium hirtovaginatum Kunth;毛鞘葱■☆
15350 Allium hoeltzeri Regel = Allium caricoides Regel ■
15351 Allium hookeri Thwaites;宽叶韭(大叶韭,大叶韭菜,丽江野葱);Hooker Leek, Hooker Onion ■
15352 Allium hookeri Thwaites var. muliense Airy Shaw;木里韭;Muli Leek, Muli Onion ■
15353 Allium hookeri Thwaites var. trifurcatum F. T. Wang et Ts. Tang = Allium trifurcatum (F. T. Wang et Ts. Tang) J. M. Xu ■
15354 Allium hopeiense Nakai = Allium longistylum Baker ■
15355 Allium hugonianum Rendle = Allium cyaneum Regel ■
15356 Allium humile Kunth;雪韭■
15357 Allium humile Kunth var. trifurcatum F. T. Wang et Ts. Tang = Allium trifurcatum (F. T. Wang et Ts. Tang) J. M. Xu ■
15358 Allium hyacinthoides Bush = Allium canadense L. var. hyacinthoides (Bush) Ownbey ■☆
15359 Allium hyalinum Curran var. praecox (Brandegee) Jeps. = Allium praecox Brandegee ■☆
15360 Allium hymenorhizum Ledeb. var. tianschanicum (Rupr.) Regel = Allium tianschanicum Rupr. ■
15361 Allium hymenorrhizum Ledeb.;北疆韭;Membraneroot Leek, Membraneroot Onion ■
15362 Allium hymenorrhizum Ledeb. var. dentatum J. M. Xu;旱生韭;Dentate Membraneroot Leek ■
15363 Allium iatasen H. Lév. = Allium macrostemon Bunge ■
15364 Allium iliense Regel;伊犁葱■☆
15365 Allium inaequale Janka;异型葱■☆
15366 Allium incisum A. Nelson et J. F. Macbr. = Allium lemmonii S.

Watson ■☆

15367 Allium inconspicuum Vved.;显著葱■☆

15368 Allium inderiense Fisch. ex Roem.;印得葱■☆

15369 Allium inodorum Aiton = Nothoscordum inodorum (Aiton) G. Nicholson ■☆

15370 Allium inodorum Willd. = Allium neapolitanum Cirillo ■☆

15371 Allium inops Vved.;贫弱葱■☆

15372 Allium intermedium DC. = Allium paniculatum L. subsp. intermedium (DC.) Asch. et Graebn. ■☆

15373 Allium inutile Makino;齿棱合被韭(齿棱茎合被韭,齿棱茎小蒜)■

15374 Allium inyonis M. E. Jones = Allium atrorubens S. Watson var. cristatum (S. Watson) McNeal ■☆

15375 Allium jacquemontii Kunth;高原韭(阿萨姆葱,雅奎葱);Assam Onion ■

15376 Allium jacquemontii Kunth var. grandiflorum (Boiss.) Aswal;大花高原韭■

15377 Allium jacquemontii Kunth var. parviflorum (Ledeb.) Aswal;小花高原韭■

15378 Allium jacquemontii Regel = Allium przewalskianum Regel ■

15379 Allium jacquemontii Regel = Allium stoliczkii Regel ■

15380 Allium jajlae Vved.;雅利葱■☆

15381 Allium japonicum Regel;日本韭(山韭,紫韭);Japan Leek, Japan Onion ■☆

15382 Allium japonicum Regel = Allium thunbergii G. Don ■

15383 Allium jeholense Franch. = Allium longistylum Baker ■

15384 Allium jucundum Vved.;愉悦葱■☆

15385 Allium juldusicola Regel;尤尔都斯韭;Juldus Onion ■

15386 Allium junceum Jacq. ex Bake = Allium przewalskianum Regel ■

15387 Allium junceum Sibth. et Sm. = Allium przewalskianum Regel ■

15388 Allium kansuense Regel = Allium sikkimense Baker ■

15389 Allium karataviense Regel;宽叶葱(卡拉韭);Broadleaf Onion, Ornamental Onion, Turkestan Allium ■

15390 Allium karataviense Regel var. granitovii Priszter;格氏宽叶葱■

15391 Allium karelinii P. Poljak. = Allium schoenoprasum L. var. scaberrimum Regel ■

15392 Allium karsianum Fomin;卡尔斯葱■☆

15393 Allium kaschianum Regel;草地韭;Prairie Leek, Prairie Onion ■

15394 Allium kaufmannii Regel = Allium atrosanguineum Schrenk var. fedschenkoanum (Regel) G. H. Zhu et N. J. Turland ■

15395 Allium kesselringii Regel = Allium schoenoprasoides Regel ■

15396 Allium kessleri Davidson = Allium parryi S. Watson ■☆

15397 Allium kingdonii Stearn;钟花韭;Bellflower Leek, Bellflowered Onion ■

15398 Allium kokanicum Regel;浩罕葱■☆

15399 Allium komarovianum Vved. = Allium sacculiferum Maxim. ■

15400 Allium komarovii Lipsky;科马罗夫葱■☆

15401 Allium kopetdagense Vved.;科佩特葱■☆

15402 Allium korolkowii Regel;褐皮韭;Korolkow Leek, Korolkow Onion ■

15403 Allium kujukense Vved.;库尤克葱■☆

15404 Allium kungii Nakai = Allium senescens L. ■

15405 Allium kunthianum Vved. = Allium lepidum Kunth ■☆

15406 Allium kurrat Schweinf. ex K. Krause = Allium ampeloprasum L. ■☆

15407 Allium kurssanovii Popov;条叶长喙韭;Kurssanov Onion ■

15408 Allium lacerum Freyn;撕裂韭■☆

15409 Allium lacteum (Lindl.) Benth. = Triteleia hyacinthina (Lindl.) Greene ■☆

15410 Allium lacunosum S. Watson;腔韭;Fitted Onion, Pitted Onion ■☆

15411 Allium lancifolium Stearn = Allium wallichii Kunth var. platyphyllum (Diels) J. M. Xu ■

15412 Allium lancipetalum Y. P. Hsu = Allium ramosum L. ■

15413 Allium laquetii H. Lév. = Allium thunbergii G. Don ■

15414 Allium lasiophyllum Vved.;毛叶葱■☆

15415 Allium latissimum Prokh. = Allium victorialis L. subsp. platyphyllum Hultén ■

15416 Allium latissimum Prokh. = Allium victorialis L. ■

15417 Allium laucum Schrad. = Allium senescens L. ■

15418 Allium lavendulare Bates var. fraseri (Ownbey) Shinners = Allium canadense L. var. fraseri Ownbey ■☆

15419 Allium ledebourianum Roem. et Schult. = Allium ledebourianum Schult. et Schult. f. ■

15420 Allium ledebourianum Schult. et Schult. f.;硬皮葱(麦葱,丝葱);Ledebour Chive, Ledebour Onion ■

15421 Allium lehmannianum Merck;赖氏葱■☆

15422 Allium lemmonii S. Watson;莱蒙葱■☆

15423 Allium lencoranicum Miscz. ex Grossh.;连科兰葱■☆

15424 Allium lepidum Kunth;库恩葱(鳞葱);Kunth Onion ■☆

15425 Allium lepidum Ledeb. = Allium pallasii Murray ■

15426 Allium leucanthum K. Koch;白花葱■☆

15427 Allium leucanthum K. Koch = Allium ampeloprasum L. ■☆

15428 Allium leucocephalum Turcz. ex Ledeb.;白头韭(白头葱);Whitecephalous Leek, Whitecephalous Onion ■

15429 Allium liangshanense Z. Y. Zhu;凉山韭;Liangshan Leek, Onion ■

15430 Allium liangshanense Z. Y. Zhu = Allium wallichii Kunth ■

15431 Allium libani Boiss.;黎巴嫩葱;Lebanon Onion ■☆

15432 Allium lineare L.;北韭;North Onion ■

15433 Allium lineare L. var. maackii Maxim. = Allium maackii (Maxim.) Prokh. ex Kom. et Aliss. ■

15434 Allium lineare L. var. strictum Krylov = Allium strictum Schrad. ■

15435 Allium listera Stearn;对叶山葱■

15436 Allium longanum Pamp.;朗葱■☆

15437 Allium longicuspis Regel;长尖葱■☆

15438 Allium longiradiatum Vved.;长射线葱■☆

15439 Allium longispathum Delaroche;长苞葱■☆

15440 Allium longistylum Baker;长柱韭;Longstyle Leek, Longstyle Onion ■

15441 Allium lusitanicum Delarbre;葡萄牙葱;Perennial Welsh Onion ■☆

15442 Allium lutescens Vved.;变黄葱■☆

15443 Allium luteum Dietr.;黄葱■☆

15444 Allium maackii (Maxim.) Prokh. ex Kom. et Aliss.;马克韭;Maack Onion ■

15445 Allium macleanii Baker;麦克林葱(大花葱,高葱,巨葱,硕葱,绣球葱);Giant Allium, Giant Onion, Ornamental Onion ■☆

15446 Allium macranthum Baker;大花韭;Largeflower Leek, Largeflower Onion ■

15447 Allium macropetalum Rydb.;大瓣韭;Largepetal Onion ■☆

15448 Allium macrorrhizon Regel = Allium tianschanicum Rupr. ■

15449 Allium macrorrhizum Boiss. = Allium hymenorrhizum Ledeb. ■

15450 Allium macrostemon Bunge;韭白(干韭,苦蒜果,密花小根蒜,山蒜,团葱,小独蒜,小根菜,小根蒜,小蒜,野白头,野葱,野葱果,野韭,野蒜,菞,泽蒜,宅蒜,子根蒜);Ainu Onion, Longstamen Garlic, Longstamen Onion ■

15451 Allium macrostemon Bunge var. uralense (Franch.) Airy Shaw;密花小根蒜;Denseflower Longstamen Onion ■

15452 Allium macrum S. Watson;大葱■☆
15453 Allium mairei H. Lév.;滇韭(山韭菜);Maire Onion,Yunnan Leek ■
15454 Allium maowenense J. M. Xu;茂汶韭;Maowen Leek,Maowen Onion ■
15455 Allium mareoticum Bornm. et Gauba;马雷奥特葱■☆
15456 Allium margaritaceum Sibth. et Sm. var. bulbiferum Batt. = Allium vineale L. ■☆
15457 Allium margaritaceum Sibth. et Sm. var. compactum Batt. = Allium vineale L. ■☆
15458 Allium margaritaceum Sibth. et Sm. var. faurei Maire = Allium guttatum Steven ■☆
15459 Allium margaritaceum Sibth. et Sm. var. papillosum H. Lindb. = Allium baeticum Boiss. ■☆
15460 Allium margaritaceum Sibth. et Sm. var. robustum Maire = Allium sphaerocephalum L. ■☆
15461 Allium margaritae B. Fedtsch.;马尔葱■☆
15462 Allium mariae Bordz.;玛利亚葱■☆
15463 Allium maritimum Baker = Allium maritimum Raf. ■☆
15464 Allium maritimum Raf.;滨海葱■☆
15465 Allium marschallianum Vved.;马氏葱;Marschall Onion ■☆
15466 Allium martinii H. Lév. et Vaniot = Allium chinense G. Don ■
15467 Allium maximowiczii Regel;马葱(马格葱);Maximowicz Onion ■
15468 Allium maximowiczii Regel f. leucanthum (Hara) T. Shimizu;白花马葱■☆
15469 Allium maximowiczii Regel f. shibutuense (Kitam.) Toyok. = Allium schoenoprasum L. var. orientale Regel ■
15470 Allium maximowiczii Regel var. yezomonticola (H. Hara) T. Shimizu f. leucanthum (H. Hara) T. Shimizu = Allium schoenoprasum L. var. yezomonticola H. Hara f. leucanthum H. Hara ■☆
15471 Allium maximowiczii Regel var. yezomonticola (H. Hara) T. Shimizu = Allium schoenoprasum L. var. yezomonticola H. Hara ■☆
15472 Allium megalobulbon Regel;大鳞韭■
15473 Allium microbulbum Prokh.;小鳞韭■☆
15474 Allium microdictyon Prokh. = Allium victorialis L. ■
15475 Allium microdictyum Prokh. = Allium victorialis L. ■
15476 Allium microscordion Small = Allium canadense L. var. mobilense (Regel) Ownbey ■☆
15477 Allium minutum Vved.;小葱■☆
15478 Allium mirabile L. F. Hend. = Allium bolanderi S. Watson var. mirabile (L. F. Hend.) McNeal ■☆
15479 Allium mobilense Regel = Allium canadense L. var. mobilense (Regel) Ownbey ■☆
15480 Allium moly L.;药葱(黄花葱,黄花茖葱);Garlic Onion, Golden Garlic,Lily Leek,Lily-leek,Medicinal Onion,Moly,Sorcerer's Garlic,Yellow Garlic ■☆
15481 Allium moly L. = Allium luteum Dietr. ■☆
15482 Allium monadelphum Turcz. ex Kar. et Kir. = Allium atrosanguineum Schrenk ■
15483 Allium monadelphum Turcz. ex Kar. et Kir. var. fedschenkoanum (Regel) Regel = Allium atrosanguineum Schrenk var. fedschenkoanum (Regel) G. H. Zhu et N. J. Turland ■
15484 Allium monadelphum Turcz. ex Kar. et Kir. var. kaufmannii (Regel) Regel = Allium atrosanguineum Schrenk var. fedschenkoanum (Regel) G. H. Zhu et N. J. Turland ■
15485 Allium monadelphum Turcz. ex Kar. et Kir. var. tibeticum Regel = Allium atrosanguineum Schrenk var. tibeticum (Regel) G. H. Zhu et N. J. Turland ■
15486 Allium monanthum Maxim.;单花韭(矮韭,小韭);Uniflower Leek,Uniflower Onion ■
15487 Allium monanthum Maxim. var. floribundum Z. J. Zhong et X. T. Huang;多花韭;Manyflower Leek,Manyflower Onion ■
15488 Allium monanthum Maxim. var. floribundum Z. J. Zhong et X. T. Huang = Allium monanthum Maxim. ■
15489 Allium mongolicum Regel;蒙古韭(蒙古葱);Mongol Leek, Mongolian Onion ■
15490 Allium monophyllum Vved. ex Czerniak.;单叶葱■☆
15491 Allium monospermum Jeps.;单籽葱■☆
15492 Allium monspessulanum Gouan = Allium nigrum L. ■☆
15493 Allium montanum F. W. Schmidt ex Schult. f.;阿尔卑斯山葱;Alps Onion ■☆
15494 Allium montanum F. W. Schmidt ex Schult. f. = Allium senescens L. ■
15495 Allium monticola Davidson;山生葱■☆
15496 Allium monticola Davidson var. keckii (Munz) Ownbey et Aase = Allium monticola Davidson ■☆
15497 Allium morrisonense Hayata = Allium bakeri Regel var. morrisonense (Hayata) Tang S. Liu et S. S. Ying ■
15498 Allium morrisonense Hayata = Allium thunbergii G. Don ■
15499 Allium moschatum L.;麝香葱;Musk Onion ■☆
15500 Allium moschatum L. var. brevipedunculatum Regel = Allium korolkowii Regel ■
15501 Allium moschatum L. var. dubium Regel = Allium korolkowii Regel ■
15502 Allium multiflorum Desf. = Allium ampeloprasum L. ■☆
15503 Allium mumile Kunth;矮生三柱韭■
15504 Allium mutabile Michx. = Allium canadense L. var. mobilense (Regel) Ownbey ■☆
15505 Allium mutabile Michx. = Allium canadense L. ■☆
15506 Allium myrianthum Boiss. = Allium paniculatum L. ■☆
15507 Allium nanodes Airy Shaw;短葶山葱(短葶韭);Shortscape Leek,Shortscape Onion ■
15508 Allium narcissiflorum Vill.;水仙花葱(水仙状葱,水仙状韭);Narcissus Onion ■☆
15509 Allium neapolitanum Cirillo;那波利葱(波斯葱,纸花葱);Daffodil Garlic, Guernsey Star of Bethlehem, Guernsey Star-of-Bethlehem, Naples Garlic, Naples Onion, Neapolitan Garlic, Ornamental Onion,Paperyflower Onion,White Allium,White Garlic, Wood Garlic ■☆
15510 Allium neapolitanum Cirillo var. cowanii ? = Allium neapolitanum Cirillo ■☆
15511 Allium negrianum Maire et Weiller = Allium guttatum Steven ■☆
15512 Allium nereidum Hance = Allium macrostemon Bunge ■
15513 Allium neriniflorum (Herb.) Baker;长梗合被韭(长梗葱,长梗韭,花美韭);Longpedicel Leek,Longpedicel Onion ■
15514 Allium nevadense S. Watson;内华达葱■☆
15515 Allium nevadense S. Watson subsp. cristatum (S. Watson) Traub = Allium atrorubens S. Watson var. cristatum (S. Watson) McNeal ■☆
15516 Allium nevadense S. Watson var. cristatum (S. Watson) Ownbey = Allium atrorubens S. Watson var. cristatum (S. Watson) McNeal ■☆
15517 Allium nevadense S. Watson var. macropetalum M. Peck = Allium nevadense S. Watson ■☆
15518 Allium nevii S. Watson;内瓦葱■☆

15519 Allium nigritanum A. Chev. = Allium cepa L. ■
15520 Allium nigrum L.；黑葱；Black Garlic, Black Onion, Homer's Garlic ■☆
15521 Allium nipponicum Franch. et Sav. = Allium macrostemon Bunge ■
15522 Allium nivale Jacq. ex Hook. f. et Thomson = Allium humile Kunth ■
15523 Allium nutans L.；齿丝山韭（红花葱）；Steppers Leek, Steppers Onion ■
15524 Allium nuttallii S. Watson = Allium drummondii Regel ■☆
15525 Allium obliquum L.；高葶韭；Twistedleaf Leek, Twistedleaf Onion ■
15526 Allium obtusiflorum DC. = Allium paniculatum L. subsp. obtusiflorum (DC.) Brand ■☆
15527 Allium obtusifolium Klotzsch = Allium carolinianum F. Delaroche ■
15528 Allium occidentale A. Gray = Allium amplectens Torr. ■☆
15529 Allium ochotense Prokh. = Allium victorialis L. ■
15530 Allium ochroleucum Rchb. = Allium globosum M. Bieb. ex Redoute ■
15531 Allium ochroleucum Waldst. et Kit.；淡黄韭■☆
15532 Allium odoratissimum Desf. = Allium roseum L. subsp. odoratissimum (Desf.) Murb. ■☆
15533 Allium odorum Kar. et Kir. = Allium angulosum L. ■☆
15534 Allium odorum L.；韭菜（欧洲韭菜）；Chinese Chives ■
15535 Allium odorum L. = Allium ramosum L. ■
15536 Allium odorum Lapeyr. = Allium ochroleucum Waldst. et Kit. ■☆
15537 Allium odorum Lour. = Allium thunbergii G. Don ■
15538 Allium odorum Ten. = Allium nigrum L. ■☆
15539 Allium odorum Thunb. = Allium thunbergii G. Don ■
15540 Allium oleraceum L.；菜园葱；Field Garlic, Potherb Onion ■☆
15541 Allium oliganthum Kar. et Kir.；少花葱■
15542 Allium oliganthum Kar. et Kir. var. elongatum Kar. et Kir. = Allium korolkowii Regel ■
15543 Allium omeiense Z. Y. Zhu；峨眉韭；Emei Leek, Emei Onion ■
15544 Allium omiostema Airy Shaw = Allium bidentatum Fisch. ex Prokh. ■
15545 Allium ophiophyllum Vved.；蛇叶葱■☆
15546 Allium ophiopogon H. Lév. = Allium thunbergii G. Don ■
15547 Allium ophiopogon H. Lév. ex Nakai = Allium thunbergii G. Don ■
15548 Allium ophioscoron Link = Allium sativum L. var. ophioscorodon (Link) Döll ■
15549 Allium oreophiloides Regel；假高地蒜■☆
15550 Allium oreophilum C. A. Mey.；高地蒜（山地韭）；Ornamental Onion ■
15551 Allium oreoprasoides Vved.；拟滩地韭■☆
15552 Allium oreoprasum Schrenk；滩地韭；Beach Leek, Beach Onion ■
15553 Allium oreoscordum Vved.；沙蒜■☆
15554 Allium orientale Boiss.；东方葱■☆
15555 Allium oschaninii O. Fedtsch.；奥沙葱■☆
15556 Allium ostrowskianum Regel = Allium oreophilum C. A. Mey. ■
15557 Allium ousensanense Nakai = Allium macrostemon Bunge ■
15558 Allium ovalifolium Hand.-Mazz.；卵叶山葱（鹿儿韭，鹿耳韭，卵叶茖葱，卵叶韭，天韭，天蒜）；Ovateleaf Leek, Ovateleaf Onion ■
15559 Allium ovalifolium Hand.-Mazz. var. cordifolium (J. M. Xu) J. M. Xu；心叶山葱（心叶韭）；Cordateleaf Leek ■
15560 Allium ovalifolium Hand.-Mazz. var. leuconeurum J. M. Xu；白脉山葱（白脉韭）；Whitevein Onion ■
15561 Allium oviflorum Regel = Allium macranthum Baker ■
15562 Allium oxyphilum Wherry = Allium cernuum Roth ■☆
15563 Allium paepalanthoides Airy Shaw；天蒜；Longbeak Leek, Longbeak Onion ■
15564 Allium pallasii Murray；小山韭（小山蒜）；Pallas Garlic, Pallas Onion ■
15565 Allium pallens L.；苍白葱■☆
15566 Allium pallens L. var. grandiflorum (Maire et Weiller) Pastor et Valdés = Allium pallens L. ■☆
15567 Allium paniculatum L.；散穗葱（地中海葱）；Mediterranean Onion ■☆
15568 Allium paniculatum L. subsp. antiatlanticum (Emb. et Maire) Maire et Weiller = Allium antiatlanticum Emb. et Maire ■☆
15569 Allium paniculatum L. subsp. breviscapum Litard. et Maire = Allium pallens L. ■☆
15570 Allium paniculatum L. subsp. intermedium (DC.) Asch. et Graebn. = Allium paniculatum L. ■☆
15571 Allium paniculatum L. subsp. obtusiflorum (DC.) Brand = Allium pallens L. ■☆
15572 Allium paniculatum L. subsp. tenuiflorum (Ten.) D. Löve = Allium tenuiflorum Ten. ■☆
15573 Allium paniculatum L. var. brachyspathum Faure et Maire = Allium pallens L. ■☆
15574 Allium paniculatum L. var. breviscapum (Litard. et Maire) Maire = Allium pallens L. ■☆
15575 Allium paniculatum L. var. dentiferum (Webb) Maire et Weiller = Allium longispathum Delaroche ■☆
15576 Allium paniculatum L. var. fuscum Boiss. = Allium pallens L. ■☆
15577 Allium paniculatum L. var. grandiflorum Maire et Weiller = Allium pallens L. ■☆
15578 Allium paniculatum L. var. longispathum (Delaroche) Regel = Allium longispathum Delaroche ■☆
15579 Allium paniculatum L. var. mauritii Maire et Sennen = Allium pallens L. ■☆
15580 Allium paniculatum L. var. pallens (L.) Gren. et Godr. = Allium pallens L. ■☆
15581 Allium paniculatum L. var. pseudotenuiflorum Pamp. = Allium tenuiflorum Ten. ■☆
15582 Allium paniculatum L. var. rifanum Maire = Allium pallens L. ■☆
15583 Allium paniculatum L. var. stenanthum Maire = Allium pallens L. ■☆
15584 Allium paniculatum L. var. tenoreanum Maire et Weiller = Allium tenuiflorum Ten. ■☆
15585 Allium paradoxum (M. Bieb.) G. Don；奇蒜；Few-flowered Leek, Strange Onion ■☆
15586 Allium pardoi Loscos；帕尔葱■☆
15587 Allium parishii S. Watson var. keckii Munz = Allium monticola Davidson ■☆
15588 Allium parryi S. Watson；帕里葱■☆
15589 Allium parviflorum Desf. = Allium pallens L. ■☆
15590 Allium parvulum Vved.；较小葱■☆
15591 Allium parvum Kellogg var. brucae M. E. Jones = Allium cratericola Eastw. ■☆
15592 Allium parvum Kellogg var. jacintense Munz = Allium cratericola Eastw. ■☆
15593 Allium peirsonii Jeps. = Allium monticola Davidson ■☆
15594 Allium pekinense Prokh. = Allium sativum L. ■
15595 Allium pendulinum Ten.；下垂葱；Italian Garlic ■☆
15596 Allium peninsulare Lemmon ex Greene var. crispum (Greene) Jeps. = Allium cristatum Greene ■☆

15597 Allium petraeum Kar. et Kir. ;石坡韭■

15598 Allium pevtzovii Prokh. ;昆仑韭■

15599 Allium phariense Rendle;帕里韭;Phari Leek,Phari Onion ■

15600 Allium pictum Moldenke = Allium tricoccum Sol. ■☆

15601 Allium pikeanum Rydb. = Allium geyeri S. Watson ■☆

15602 Allium platyspathum Schrenk; 宽苞韭; Broadspathe Leek, Broadspathe Onion,Obtuseleaf Onion ■

15603 Allium platyspathum Schrenk subsp. amblyophyllum (Kar. et Kir.) Frizen;钝叶宽苞韭(钝叶韭)■

15604 Allium platyspathum Schrenk var. falcatum Regel = Allium carolinianum F. Delaroche ■

15605 Allium platystemon Kar. et Kir. = Allium oreophilum C. A. Mey. ■

15606 Allium platystylum Regel = Allium carolinianum F. Delaroche ■

15607 Allium plurifoliatum Rendle; 多叶韭; Manyleaf Leek, Manyleaf Onion ■

15608 Allium plurifoliatum Rendle var. stenodon (Nakai et Kitag.) J. M. Xu = Allium stenodon Nakai et Kitag. ■

15609 Allium plurifoliatum Rendle var. zhegushanense J. M. Xu;鹧鸪韭■

15610 Allium polyanthum Schult. et Schult. f. = Allium porrum L. subsp. polyanthum (Schult. et Schult. f.) Jauzein et J. -M. Tison ■☆

15611 Allium polyastrum Diels = Allium wallichii Kunth ■

15612 Allium polyastrum Diels var. pallens F. T. Wang et Ts. Tang = Allium wallichii Kunth var. platyphyllum (Diels) J. M. Xu ■

15613 Allium polyastrum Diels var. platyphyllum Diels = Allium wallichii Kunth var. platyphyllum (Diels) J. M. Xu ■

15614 Allium polyphyllum Kar. et Kir. = Allium carolinianum F. Delaroche ■

15615 Allium polyrhizum Turcz. ex Regel;碱韭(多根葱,多根蒜,碱葱,紫花韭);Manyroot Leek,Manyroot Onion ■

15616 Allium polyrhizum Turcz. ex Regel var. alabasicum D. S. Wen et Sh. Chen = Allium alabasicum (D. S. Wen et Sh. Chen) Y. Z. Zhao ■

15617 Allium polyrhizum Turcz. ex Regel var. potaninii Regel = Allium bidentatum Fisch. ex Prokh. ■

15618 Allium polyrhizum Turcz. ex Regel var. przewalskii Regel = Allium polyrhizum Turcz. ex Regel ■

15619 Allium ponticum Miscz. ex Grossh. ;蓬特葱■☆

15620 Allium popovii Vved. ;波氏葱■☆

15621 Allium porrum L. ; 韭葱(扁葱,洋葱); Chibbal, Chibble, Chiboul, French Leek, Garden Leek, Leek, Lick, Ollick, Purret, Scallion ■

15622 Allium porrum L. subsp. polyanthum (Schult. et Schult. f.) Jauzein et J. -M. Tison;多花韭葱■☆

15623 Allium potaninii Regel = Allium ramosum L. ■

15624 Allium praecox Brandegee;早葱■☆

15625 Allium praelatitium H. Lév. = Allium wallichii Kunth ■

15626 Allium prattii C. H. Wright;太白山葱(日葱,太白韭,野葱); Prantt Leek,Prantt Onion ■

15627 Allium prattii C. H. Wright var. ellipticum F. T. Wang et Ts. Tang = Allium prattii C. H. Wright ■

15628 Allium prattii C. H. Wright var. latifoliatum F. T. Wang et Ts. Tang = Allium ovalifolium Hand. -Mazz. ■

15629 Allium prattii C. H. Wright var. vinicolor F. T. Wang et Ts. Tang = Allium prattii C. H. Wright ■

15630 Allium prokhanovii (Vorosch.) Barkalov = Allium maackii (Maxim.) Prokh. ex Kom. et Aliss. ■

15631 Allium proliferum (Moench.) Schrad. ex Willd. = Allium cepa L. 'Proliferum' ■

15632 Allium prostratum Trevir. ;蒙古野韭;Prostrate Leek,Prostrate Onion ■

15633 Allium przewalskianum Regel;青甘韭(青甘野韭);Przewalsk Leek,Przewalsk Onion ■

15634 Allium pseudoampeloprasum Miscz. ex Grossh. ;假南欧葱■☆

15635 Allium pseudocepa Schrenk = Allium galanthum Kar. et Kir. ■

15636 Allium pseudocyaneum Gruning = Allium thunbergii G. Don ■

15637 Allium pseudoflavum Vved. ;假黄葱■☆

15638 Allium pseudoglobosum Popov ex Gamajun. = Allium kurssanovii Popov ■

15639 Allium pseudojaponicum Makino = Allium thunbergii G. Don ■

15640 Allium pseudoseravschanicum Popov et Vved. ;假塞拉夫葱■☆

15641 Allium pseudoxiphopetalum Wendelbo = Allium dolichostylum Vved. ■☆

15642 Allium pskemense B. Fedtsch. ;普斯克姆葱■☆

15643 Allium pulchellum G. Don;丽葱■☆

15644 Allium pumilum Vved. ;矮小葱■☆

15645 Allium pyrrhorrhizum Airy Shaw = Allium mairei H. Lév. ■

15646 Allium pyrrhorrhizum Airy Shaw var. leucorrhizum F. T. Wang et Ts. Tang = Allium mairei H. Lév. ■

15647 Allium raddeanum Regel = Allium schoenoprasoides Regel ■

15648 Allium raddeanum Regel = Allium schoenoprasum L. ■

15649 Allium ramosum L. ;野韭(韭菜,欧洲韭菜,山韭);Branchy Leek,Branchy Onion,Chinese Allium ■

15650 Allium recurvatum Rydb. = Allium cernuum Roth ■☆

15651 Allium regelianum Becker ex Iljin;雷氏葱■☆

15652 Allium regelii Trautv. ;雷格葱■☆

15653 Allium regnieri Maire = Allium sphaerocephalum L. ■☆

15654 Allium renardii Regel = Allium caeruleum Pall. ■

15655 Allium renardii Regel = Allium caesium Schrenk ■

15656 Allium reticulatum Fraser ex G. Don = Allium textile A. Nelson et J. F. Macbr. ■☆

15657 Allium reticulatum Fraser ex G. Don var. deserticola M. E. Jones = Allium macropetalum Rydb. ■☆

15658 Allium reticulatum Fraser ex G. Don var. nuttallii (S. Watson) M. E. Jones = Allium drummondii Regel ■☆

15659 Allium reticulatum Fraser ex G. Don var. playanum M. E. Jones = Allium textile A. Nelson et J. F. Macbr. ■☆

15660 Allium rhizomatum Wooton et Standl. ;根茎葱;Wild Onion ■☆

15661 Allium rhynchogynum Diels;宽叶滇韭■

15662 Allium roborowskianum Regel;新疆蒜;Sinkiang Onion,Xinjiang Chive,Xinjiang Onion ■

15663 Allium robustum Kar. et Kir. ;健蒜■

15664 Allium robustum Kar. et Kir. var. alpestre Kar. et Kir. = Allium robustum Kar. et Kir. ■

15665 Allium rosenbachianum Regel; 罗森巴氏葱(绣球葱); Rosenbach Onion,Showy Onion ■☆

15666 Allium roseum L. ;玫瑰红葱;Pink Onion, Rose Garlic, Rosy Garlic,Rosy Onion ■☆

15667 Allium roseum L. subsp. odoratissimum (Desf.) Murb. ;芳香玫瑰红葱■☆

15668 Allium roseum L. subsp. tourneauxii Bartolo et al. ;图尔诺葱■☆

15669 Allium roseum L. var. bulbiferum Ker Gawl. = Allium roseum L. ■☆

15670 Allium roseum L. var. carneum Bertol. = Allium roseum L. ■☆

15671 Allium roseum L. var. grandiflorum Briq. = Allium roseum L. ■☆

15672 Allium roseum L. var. odoratissimum (Desf.) Coss. = Allium roseum L. subsp. odoratissimum (Desf.) Murb. ■☆

15673 Allium roseum L. var. perotii Maire = Allium roseum L. ■☆
15674 Allium roseum L. var. tourneauxii Boiss. = Allium roseum L. subsp. tourneauxii Bartolo et al. ■☆
15675 Allium rotundum L.;圆葱■☆
15676 Allium rotundum L. = Allium scorodoprasum L. subsp. rotundum (L.) Stearn ■☆
15677 Allium rotundum L. subsp. multiflorum (Desf.) Rouy = Allium ampeloprasum L. ■☆
15678 Allium rotundum L. var. polyanthum (Schult. et Schult. f.) Asch. et Graebn. = Allium polyanthum Schult. et Schult. f. ■☆
15679 Allium roxburghii Kunth = Allium tuberosum Rootler ex Spreng. ■
15680 Allium rubellum M. Bieb.;阿萨姆葱(微红葱);Assam Onion ■☆
15681 Allium rubellum M. Bieb. var. grandiflorum ? = Allium griffithianum Boiss. ■☆
15682 Allium rubens Schrad. ex Willd.;红花韭(红葱);Red Onion ■
15683 Allium rubrum Osterh. = Allium geyeri S. Watson var. tenerum M. E. Jones ■☆
15684 Allium rude J. M. Xu;野黄韭;Wild Leek,Wild Onion ■
15685 Allium ruhmerianum Asch. ex Durand et Barratte;鲁曼葱■☆
15686 Allium rupestre Steven;岩葱■☆
15687 Allium rydbergii J. F. Macbr. = Allium geyeri S. Watson var. tenerum M. E. Jones ■☆
15688 Allium sabulicola Osterh. = Allium geyeri S. Watson var. tenerum M. E. Jones ■☆
15689 Allium sabulosum Stev. ex Bunge;沙地韭(沙葱);Onion ■☆
15690 Allium sacculiferum Maxim.;朝鲜韭(球序韭);Bag Onion ■
15691 Allium sacculiferum Maxim. var. glaucum P. P. Gritz.;灰朝鲜韭■☆
15692 Allium sacculiferum Maxim. var. robustum P. P. Gritz.;粗朝鲜韭■☆
15693 Allium sairamense Regel;赛里木韭■
15694 Allium salsum Skvortsov et A. I. Baranov = Allium bidentatum Fisch. ex Prokh. ■
15695 Allium sanbornii A. W. Wood var. tuolumnense Ownbey et Aase ex Traub = Allium tuolumnense (Ownbey et Aase ex Traub) S. S. Denison et McNeal ■☆
15696 Allium sapidissimum Hedw. = Allium fistulosum L. ■
15697 Allium sapidissimum Pall. ex Schult. et Schult. f. = Allium altaicum Pall. ■
15698 Allium sativum L.;蒜(白皮蒜,大蒜,大头蒜,独头蒜,胡蒜,葫,葫蒜,青蒜,蒜头,小蒜,紫皮蒜,独蒜);Churl's Treacle, Clown's Treacle,Countryman's Treacle,Crowns' Treacle,Cultivated Garlic,Garlete,Garlic,Poor Man's Treacle,Poor-man's Treacle, Rocambole,Taper-leaved Garlic ■
15699 Allium sativum L. var. japonicum Kitam. = Allium sativum L. ■
15700 Allium sativum L. var. ophioscorodon (Link) Döll = Allium sativum L. ■
15701 Allium sativum L. var. vulgare Döll = Allium sativum L. ■
15702 Allium satoanum Kitag. = Allium prostratum Trevir. ■
15703 Allium savarinii Sennen = Allium baeticum Boiss. ■☆
15704 Allium saxatile M. Bieb.;长喙韭(博格多葱,长喙葱,圆葱);Rosestripe Leek,Rosestripe Onion ■
15705 Allium saxicola Kitag. = Allium spurium G. Don ■
15706 Allium scabrellum Boiss. et Buhse = Allium schoenoprasum L. var. scaberrimum Regel ■
15707 Allium scabriscapum Boiss. et Kotschy;糙葱■☆
15708 Allium schoenoprasoides Regel;类北韭(类北葱,细香葱);Chive-like Onion,Close to Chive ■
15709 Allium schoenoprasum L.;北葱(火葱,绵葱,细香葱,虾夷葱,香葱,小蒜);Chive,Chive Garlic,Chive Onion,Chives,Cives,Civet, Infant Onion,Rush Garlic,Rushleek,Shive,Siberian Chives,Sithes, Sives,Sweth,Syves,Wild Chives,Wild Leek ■
15710 Allium schoenoprasum L. f. albiflorum J. Rousseau;白花类北葱■
15711 Allium schoenoprasum L. subsp. sibiricum (L.) Celak. = Allium schoenoprasum L. ■
15712 Allium schoenoprasum L. var. bellum Kitam. = Allium schoenoprasum L. ■
15713 Allium schoenoprasum L. var. laurentianum Farw. = Allium schoenoprasum L. ■
15714 Allium schoenoprasum L. var. laurentianum Fernald = Allium schoenoprasum L. ■
15715 Allium schoenoprasum L. var. orientale Regel;东葱■
15716 Allium schoenoprasum L. var. orientale Regel = Allium maximowiczii Regel ■
15717 Allium schoenoprasum L. var. scaberrimum Regel;糙葶北葱(糙叶北葱);Rough Chive,Wild Chives ■
15718 Allium schoenoprasum L. var. shibutuense Kitam. = Allium schoenoprasum L. var. orientale Regel ■
15719 Allium schoenoprasum L. var. sibiricum (L.) Hartm. = Allium schoenoprasum L. ■
15720 Allium schoenoprasum L. var. yezomonticola H. Hara;北海道葱■☆
15721 Allium schoenoprasum L. var. yezomonticola H. Hara f. leucanthum H. Hara;白花北海道葱■☆
15722 Allium schrenkii Regel;单丝辉韭(单丝辉葱)■
15723 Allium schrenkii Regel = Allium strictum Schrad. ■
15724 Allium schubertii Zucc.;舒伯特葱;Schubert Onion ■☆
15725 Allium scissum A. Nelson et J. F. Macbr. = Allium lemmonii S. Watson ■☆
15726 Allium scorodoprasum L.;胡蒜(大蒜,葫,葫葱,家蒜,卵蒜,茆蒜,蒜,蒜仔,蒚子,夏蒜,小葱,小蒜);French Garlic,Giant Garlic, Great Turkey Garlic, Great Turkey-garlic, Rocambole, Roccombole, Sand Leek,Spanish Garlic,Wild Rocambole ■☆
15727 Allium scorodoprasum L. subsp. jajlae (Vved.) Stearn;贾氏胡蒜■☆
15728 Allium scorodoprasum L. subsp. rotundum (L.) Stearn;粗壮胡蒜■☆
15729 Allium scorodoprasum L. subsp. waldsteinii (G. Don) Stearn;瓦氏胡蒜■☆
15730 Allium scrobiculatum Vved.;蜂窝葱■☆
15731 Allium seirotrichum Ducell. et Maire = Allium trichocnemis J. Gay ■☆
15732 Allium semenovii Regel;管丝葱(管丝韭);Semenov Leek, Semenov Onion ■
15733 Allium semiretschenskianum Regel = Allium pallasii Murray ■
15734 Allium senescens L.;山韭(灰葱,山葱,岩葱);Aging Leek, Aging Onion,German Garlic,Japanese Onion ■
15735 Allium senescens L. f. albiflorum Q. S. Sun;白花山韭; Whiteflower Aging Onion ■
15736 Allium senescens L. f. albiflorum Q. S. Sun = Allium senescens L. ■
15737 Allium seravschanicum Regel;塞拉夫葱■☆
15738 Allium sergii Vved.;赛氏葱■☆
15739 Allium serratum S. Watson = Allium amplectens Torr. ■☆
15740 Allium setifolium Schrenk;丝叶韭;Setoseleaf Leek,Setoseleaf Onion ■
15741 Allium severtzovii Regel;赛维尔葱■☆

15742 Allium sibiricum L. = Allium schoenoprasum L. ■
15743 Allium siculum Bernardino = Nectaroscordum siculum Lindl. ■☆
15744 Allium sikkimense Baker;高山韭;Sikkim Leek,Sikkim Onion ■
15745 Allium simethis H. Lév. = Allium macranthum Baker ■
15746 Allium simethis H. Lév. et Giraudias = Allium macranthum Baker ■
15747 Allium simile Regel. = Allium fetisowii Regel ■
15748 Allium sinkiangense F. T. Wang et Y. C. Tang = Allium roborowskianum Regel ■
15749 Allium siphonanthum J. M. Xu;管花韭(管花葱);Tubeflower Chive,Tubeflower Onion ■
15750 Allium songpanicum J. M. Xu;松潘韭;Songpan Leek,Songpan Onion ■
15751 Allium spathaceum Steud. ex A. Rich. = Allium subhirsutum L. subsp. spathaceum (Steud. ex A. Rich.) Duyfjes ■☆
15752 Allium sphaerocephalum L.;圆头葱(丹顶鹤蒜,圆头大花葱);Ballhead Onion, Ball-headed Onion, Drumsticks, Globular Head Largeflower Onion,Ornamental Onion,Round-headed Garlic,Round-headed Leek ■☆
15753 Allium sphaerocephalum L. subsp. arvense (Guss.) Arcang. = Allium sphaerocephalum L. ■☆
15754 Allium sphaerocephalum L. subsp. curtum (Boiss. et Gaill.) Duyfjes;短圆头葱■☆
15755 Allium sphaerocephalum L. subsp. durandoi Batt. et Trab.;杜朗多葱■☆
15756 Allium sphaerocephalum L. var. arvense (Guss.) Parl. = Allium sphaerocephalum L. ■☆
15757 Allium sphaerocephalum L. var. durandoi Batt. et Trab. = Allium sphaerocephalum L. ■☆
15758 Allium spirale Willd.;扭叶韭■
15759 Allium splendens Willd. ex Roem. et Schult. f. var. kurilense Kitam = Allium maackii (Maxim.) Prokh. ex Kom. et Aliss. ■
15760 Allium splendens Willd. ex Schult. et Schult. f.;丽韭■
15761 Allium splendens Willd. ex Schult. et Schult. f. = Allium strictum Schrad. ■
15762 Allium splendens Willd. ex Schult. et Schult. f. subsp. prokhanovii Vorosch. = Allium maackii (Maxim.) Prokh. ex Kom. et Aliss. ■
15763 Allium spurium G. Don;岩韭■
15764 Allium stamineum Boiss.;雄蕊葱■☆
15765 Allium stellatum Fraser ex Ker Gawl.;草原葱;Autumn Onion, Pink Wild Onion,Prairie Onion,Prairie Wild Onion,Wild Onion ■☆
15766 Allium stellatum Ker Gawl. = Allium stellatum Fraser ex Ker Gawl. ■☆
15767 Allium stellerianum Willd.;司太氏葱■☆
15768 Allium stenanthum Drew = Allium bolanderi S. Watson ■☆
15769 Allium stenodon Nakai et Kitag.;雾灵韭(雾灵葱);Wuling Onion ■
15770 Allium stenophyllum Schrenk = Allium oliganthum Kar. et Kir. ■
15771 Allium stephanophorum Vved.;美冠葱■☆
15772 Allium stevenii Ledeb. = Allium saxatile M. Bieb. ■
15773 Allium stevenii Willd. ex Ledeb. = Allium saxatile M. Bieb. ■
15774 Allium stipitatum Regel;粗茎蒜■☆
15775 Allium stocksianum Boiss.;斯氏葱■☆
15776 Allium stocksianum Boiss. var. persicum ? = Allium borszczowii Regel ■☆
15777 Allium stoliczkii Regel = Allium przewalskianum Regel ■
15778 Allium striatum Jacq. = Nothoscordum bivalve (L.) Britton ■☆
15779 Allium strictum Schrad.;辉韭(短齿韭,辉葱,条纹葱,直立葱);Streakleaf Leek,Streakleaf Onion,Streaky-leaved Garlic ■
15780 Allium subangulatum Regel;紫花韭;Purple Flower Onion ■☆
15781 Allium subhirsutum L.;亚毛葱;Hairy Garlic ■☆
15782 Allium subhirsutum L. subsp. ciliare (Delaroche) Maire et Weiller = Allium subhirsutum L. ■☆
15783 Allium subhirsutum L. subsp. obtusitepalum (Svent.) G. Kunkel;钝被片葱■☆
15784 Allium subhirsutum L. subsp. spathaceum (Steud. ex A. Rich.) Duyfjes;佛焰苞葱■☆
15785 Allium subhirsutum L. var. barcense Maire et Weiller = Allium longanum Pamp. ■☆
15786 Allium subhirsutum L. var. graecum (d'Urv.) Regel = Allium subhirsutum L. subsp. obtusitepalum (Svent.) G. Kunkel ■☆
15787 Allium subhirsutum L. var. hirsutum Regel = Allium trifoliatum Cirillo ■☆
15788 Allium subhirsutum L. var. purpurascens Maire et Weiller = Allium subhirsutum L. ■☆
15789 Allium subhirsutum L. var. spathaceum Regel = Allium subhirsutum L. subsp. spathaceum (Steud. ex A. Rich.) Duyfjes ■☆
15790 Allium subhirsutum L. var. vernale (Tineo) Bonnet et Barratte = Allium subhirsutum L. ■☆
15791 Allium subtilissimum Ledeb.;蜜囊韭;Honyybag Leek,Honyybag Onion ■
15792 Allium sulvia Buch. -Ham. ex D. Don = Allium tuberosum Rootler ex Spreng. ■
15793 Allium suvorovii Regel;苏沃葱■☆
15794 Allium szchuanense C. M. Shu;四川韭;Sichuan Leek,Sichuan Onion,Szechwan Onion ■
15795 Allium szchuanicum F. T. Wang et Ts. Tang = Allium cyaneum Regel ■
15796 Allium szovitsii Regel;绍氏葱■☆
15797 Allium taeniopetalum Popov et Vved.;带瓣葱■☆
15798 Allium taishanense J. M. Xu;泰山韭;Taishan Leek,Taishan Onion ■
15799 Allium talassicum Regel;塔拉斯葱■☆
15800 Allium talyschense Miscz. ex Grossh.;塔雷葱■☆
15801 Allium tanguticum Regel;唐古特韭(唐古韭);Tangut Leek, Tangut Onion ■
15802 Allium taquetii H. Lév. et Vaniot = Allium thunbergii G. Don ■
15803 Allium tartaricum L. f. = Allium ramosum L. ■
15804 Allium tataricum H. Lév.;鞑靼葱;Black Flower Onion ■☆
15805 Allium tchefouense Debeaux = Allium anisopodium Ledeb. ■
15806 Allium tchongchanense H. Lév. = Allium wallichii Kunth ■
15807 Allium tekesicola Regel;荒漠韭(天山韭);Desert Onion, Tianshan Leek,Tianshan Mountain Onion ■
15808 Allium tel-avivense Eig = Allium orientale Boiss. ■☆
15809 Allium tenue G. Don = Allium pallasii Murray ■
15810 Allium tenuicaule Regel;细茎葱■☆
15811 Allium tenuiflorum Ten.;细花葱■☆
15812 Allium tenuiflorum Ten. var. pseudotenuiflorum Pamp. = Allium tenuiflorum Ten. ■☆
15813 Allium tenuissimum L.;细叶韭(丝葱,细丝韭,细叶葱,扎麻);Thinleaf Leek,Thinleaf Onion ■
15814 Allium tenuissimum L. f. zimmermannianum (Gilg) Q. S. Sun = Allium anisopodium Ledeb. var. zimmermannianum (Gilg) F. T. Wang et Ts. Tang ■
15815 Allium tenuissimum L. var. anisopodium Regel;矮细叶韭■

15816 Allium tenuissimum L. var. nalinicum Sh. Chen;纳林韭;Nalin Onion ■

15817 Allium tenuissimum L. var. nalinicum Sh. Chen = Allium tenuissimum L. ■

15818 Allium tenuissimum L. var. purpureum Regel = Allium anisopodium Ledeb. ■

15819 Allium teretifolium Regel;西疆韭(西藏韭);Cylindricleaf Leek, Cylindricleaf Onion ■

15820 Allium textile A. Nelson et J. F. Macbr.;编织葱■☆

15821 Allium thomsonii Baker = Allium carolinianum F. Delaroche ■

15822 Allium thunbergii G. Don;球序韭(日本韭,山韭,野韭菜,野蒜头,藿,藿菜,紫韭);Japan Leek, Japan Onion, Thunberg Leek, Thunberg Onion ■

15823 Allium tianschanicum Rupr.;天山韭;Tianshan Onion ■

15824 Allium tibeticum Rendle = Allium sikkimense Baker ■

15825 Allium tilingii Regel = Triteleia hyacinthina (Lindl.) Greene ■☆

15826 Allium tortifolium Batt. = Allium ampeloprasum L. ■☆

15827 Allium tourneuxii Chabert;图尔葱■☆

15828 Allium trachyscordum Vved.;糙蒜■☆

15829 Allium trautvetterianum Regel;特劳葱■☆

15830 Allium trichocnemis J. Gay;毛葱■☆

15831 Allium tricoccum Sol.;三果片葱(北美野韭);Ramp, Ramps, Trivalved Onion, Wild Leek ■☆

15832 Allium tricoccum Sol. var. burdickii Hanes;布氏三果片葱;Narrow-leaved Wild Leek, Ramp, Wild Leek, Wood Leek ■☆

15833 Allium triflorum Raf. = Allium tricoccum Sol. ■☆

15834 Allium trifoliatum Cirillo = Allium subhirsutum L. ■☆

15835 Allium trifoliatum Cirillo subsp. hirsutum (Regel) Kollmann = Allium subhirsutum L. ■☆

15836 Allium trifoliatum Cirillo subsp. obtusitepalum Svent. = Allium subhirsutum L. subsp. obtusitepalum (Svent.) G. Kunkel ■☆

15837 Allium trifurcatum (F. T. Wang et Ts. Tang) J. M. Xu;三柱韭;Trifurcate Leek, Trifurcate Onion ■

15838 Allium tripterum Nasir;三翅葱■☆

15839 Allium triquetrum L.;三棱茎葱;Onion Weed, Snowbells, Stinking Onion, Threecorner Leek, Three-cornered Garlic, Three-cornered Leek, Triangle Onion, Triquetrous Garlic, White Bluebell ■☆

15840 Allium tristylum Regel. = Allium semenovii Regel ■

15841 Allium tschimganicum B. Fedtsch. = Allium fetisowii Regel ■

15842 Allium tsoongii F. T. Wang et Ts. Tang = Allium hookeri Thwaites ■

15843 Allium tuberosum Rootler ex Spreng.;韭(北海道葱,扁菜,草钟乳,长生韭,大韭菜,丰本,韭菜,韭菜子,韭黄,韭子,懒人菜,欧洲韭菜,起阳草,壮阳菜);Chinese Chives, Chinese Garlic, Garlic Chives, Leek, Oriental Garlic, Tubur Onion ■

15844 Allium tuberosum Rottler ex Spreng. = Allium tuberosum Rootler ex Spreng. ■

15845 Allium tuberosum Roxb. = Allium tuberosum Rootler ex Spreng. ■

15846 Allium tubiflorum Rendle;合被韭;Tubularflower Leek, Tubularflower Onion ■

15847 Allium tui F. T. Wang et Ts. Tang = Allium cyaneum Regel ■

15848 Allium tulipifolium Ledeb.;郁金叶蒜■

15849 Allium tuolumnense (Ownbey et Aase ex Traub) S. S. Denison et McNeal;图奥勒米葱■☆

15850 Allium turcomanicum Regel;土库曼葱■☆

15851 Allium turkestanicum Regel;土耳其斯坦葱■☆

15852 Allium tytthanthum Vved.;小花葱■☆

15853 Allium tytthocephalum Roem. et Schult.;小头葱■☆

15854 Allium uliginosum G. Don = Allium tuberosum Rootler ex Spreng. ■

15855 Allium uliginosum Ledeb. = Allium ledebourianum Schult. et Schult. f. ■

15856 Allium umbilicatum Boiss.;盾葱■☆

15857 Allium unifolium Kellogg;独叶葱;American Onion ■☆

15858 Allium unifolium Kellogg var. lacteum Greene = Allium unifolium Kellogg ■☆

15859 Allium uratense Franch. = Allium macrostemon Bunge ■

15860 Allium urceolatum Regel = Allium caesium Schrenk ■

15861 Allium ursinum L.;熊葱(雄葱);Badger's Flower, Bear's Garlic, Bear's-garlic, Brandy Bottle, Brandy Bottles, Broad-leaved Garlic, Buckrams, Devil-may-care, Devil's Posy, Gipsoy Onion, Great Crow Leek, Gypsy Onion, Gypsy's Gibbles, Hog's Garlic, Iron-flower, Moly, Onion Stinker, Onion-flower, Ramps, Rams, Ram's Horn, Ram's Horns, Ramsden, Ramsey, Ramsies, Ramsins, Ramsons, Rommy, Roms, Rosamund, Rosems, Serpentine Garlic, Snake Flower, Snake Plant, Snakeflower, Snake's Flower, Snake's Food, Snake's Plant, Star-flower, Stink Plant, Stinking Jenny, Stinking Lily, Stinking Onion, Ursine Garlic, Water Leek, Wild Garlic, Wild Leek, Wild Onion, Wood Garlic ■☆

15862 Allium vancouverense Macoun = Allium crenulatum Wiegand ■☆

15863 Allium vavilovii Popov et Vved.;瓦维葱■☆

15864 Allium venustum C. H. Wright = Allium cyathophorum Bureau et Franch. ■

15865 Allium vernale Tineo = Allium subhirsutum L. var. vernale (Tineo) Bonnet et Barratte ■☆

15866 Allium verticillatum Regel;轮状葱■☆

15867 Allium veschnjakovii Regel;韦什葱■☆

15868 Allium victorialis L.;茖葱(葱,茖,茖韭,格葱,隔葱,各山葱,鹿耳葱,鹿耳韭,山葱,天韭,天蒜,崖蒜);Longroot Chive, Longroot Onion, Long-rooted Onion ■

15869 Allium victorialis L. subsp. platyphyllum Hultén = Allium victorialis L. ■

15870 Allium victorialis L. subsp. platyphyllum Hultén = Allium victorialis L. var. platyphyllum (Hultén) Makino ■

15871 Allium victorialis L. var. angustifolium Hook. f. = Allium prattii C. H. Wright ■

15872 Allium victorialis L. var. histera (Stearn) J. M. Xu;对叶韭;Twinleaf Leek, Twinleaf Onion ■

15873 Allium victorialis L. var. listera (Stearn) J. M. Xu = Allium listera Stearn ■

15874 Allium victorialis L. var. platyphyllum (Hultén) Makino;东北茖葱(茖葱)■

15875 Allium victorialis L. var. platyphyllum (Hultén) Makino = Allium victorialis L. ■

15876 Allium vineale L.;葡萄葱(葡圆雅葱);Aaron's Beard, Calf's Foot, Compact Onion, Cow Garlic, Crow Garlic, Crow Onion, Crowgarlic, Devil's Posy, False Garlic, Field Garlic, Poor Man's Treacle, Rush Leek, Rushleek, Scallions, Wild Garlic, Wild Leek, Wild Onion ■☆

15877 Allium vineale L. f. compactum (Thuill.) Asch. = Allium vineale L. ■☆

15878 Allium vineale L. subsp. compactum ? = Allium vineale L. ■☆

15879 Allium vineale L. var. compactum (Thuill.) Asch. = Allium vineale L. ■☆

15880 Allium violaceum Wall. ex Regel = Allium wallichii Kunth ■

15881 Allium viride Grossh.;绿葱■☆

15882 Allium viviparum Kar. et Kir. = Allium caeruleum Pall. ■
15883 Allium volhynicum Besser = Allium strictum Schrad. ■
15884 Allium wakegi Araki = Allium fistulosum L. ■
15885 Allium waldsteinii G. Don;瓦氏葱;Waldstein Onion ■☆
15886 Allium wallichii Kunth;多星韭(不死草,苍山韭菜,苍山野韭,长生草,黑花韭,黑花野韭,山韭菜,书带草,野韭白,野韭菜,野麦冬);Wallich Leek,Wallich Onion ■
15887 Allium wallichii Kunth var. albidum F. T. Wang et Ts. Tang = Allium wallichii Kunth ■
15888 Allium wallichii Kunth var. platyphyllum (Diels) J. M. Xu;柳叶韭;Willowleaf Onion ■
15889 Allium watsonii Howell = Allium crenulatum Wiegand ■☆
15890 Allium weichanicum Palib. = Allium ramosum L. ■
15891 Allium wenchuanense Z. Y. Zhu;汶川韭;Wenchuan Onion ■
15892 Allium wenchuanense Z. Y. Zhu = Allium victorialis L. ■
15893 Allium weschniakowii Regel;坛丝韭;Weschniakow Leek,Weschniakow Onion ■
15894 Allium willdenovii Kunth = Allium delicatulum Siev. ex Schult. et Schult. f. ■
15895 Allium winklerianum Regel;伊犁蒜(温克勒氏葱);Winkler Onion ■☆
15896 Allium xiangchengense J. M. Xu;乡城韭;Xiangcheng Leek,Xiangcheng Onion ■
15897 Allium xichuanense J. M. Xu;西川韭;Xichuan Leek,Xichuan Onion ■
15898 Allium xiphopetalum Aitch. et Baker;刀瓣■☆
15899 Allium yanchiense J. M. Xu;白花韭(白花葱);Whiteflower Onion ■
15900 Allium yesoense Nakai = Allium tuberosum Rootler ex Spreng. ■
15901 Allium yongdengense J. M. Xu;永登韭;Yongdeng Leek,Yongdeng Onion ■
15902 Allium yuanum F. T. Wang et Ts. Tang;齿被韭;Tooth-petaled Onion,Yuan Leek ■
15903 Allium yuchuanii Y. Z. Zhao et J. Y. Chao;毓泉葱;Yuchuan Onion ■
15904 Allium yuchuanii Y. Z. Zhao et J. Y. Chao = Allium anisopodium Ledeb. var. zimmermannianum (Gilg) F. T. Wang et Ts. Tang ■
15905 Allium yuchuanii Y. Z. Zhao et J. Y. Chao = Allium sacculiferum Maxim. ■
15906 Allium yunnanense Diels = Allium mairei H. Lév. ■
15907 Allium zenobiae Cory = Allium canadense L. var. mobilense (Regel) Ownbey ■☆
15908 Allium zimmermannianum Gilg = Allium anisopodium Ledeb. var. zimmermannianum (Gilg) F. T. Wang et Ts. Tang ■
15909 Allmania R. Br. = Allmania R. Br. ex Wight ■
15910 Allmania R. Br. ex Wight(1832);砂苋属(阿蔓苋属,阿蔓属,沙苋属);Allmania ■
15911 Allmania albida R. Br. = Allmania nodiflora (L.) R. Br. ■
15912 Allmania nodiflora (L.) R. Br.;砂苋(阿蔓苋,虾公草);Common Allmania,Nodalflower Allmania ■
15913 Allmania nodiflora (L.) R. Br. var. angustifolia Hook. f. = Allmania nodiflora (L.) R. Br. ■
15914 Allmania nodiflora (L.) R. Br. var. aspera Hook. f. = Allmania nodiflora (L.) R. Br. ■
15915 Allmania nodiflora (L.) R. Br. var. dicholtoma Hook. f. = Allmania nodiflora (L.) R. Br. ■
15916 Allmania nodiflora (L.) R. Br. var. esculenta Hook. f. = Allmania nodiflora (L.) R. Br. ■
15917 Allmania nodiflora (L.) R. Br. var. procumbens Hook. f. = Allmania nodiflora (L.) R. Br. ■
15918 Allmania nodiflora (L.) R. Br. var. roxberghii Hook. f. = Allmania nodiflora (L.) R. Br. ■
15919 Allmaniopsis Suess. (1950);类砂苋属●■☆
15920 Allmaniopsis fruticulosa Suess.;类砂苋■☆
15921 Allobia Raf. = Euphorbia L. ●■
15922 Allobium Miers = Phoradendron Nutt. ●☆
15923 Alloborgia Steud. = Allobrogia Tratt. ■☆
15924 Alloborgia Steud. = Paradisea Mazzuc.(保留属名)■☆
15925 Allobrogia Tratt. = Paradisea Mazzuc.(保留属名)■☆
15926 Alloburkillia Whitmore = Burkilliodendron Sastry ●☆
15927 Allocalyx Cordem. = Bacopa Aubl.(保留属名)■
15928 Allocalyx Cordem. = Monocardia Pennell ■
15929 Allocarpus Kunth = Alloispermum Willd. ●■☆
15930 Allocarpus Kunth = Calea L. ●■☆
15931 Allocarpus Kunth(1818);奇果菊属■☆
15932 Allocarpus integrifolius DC.;全叶奇果菊■☆
15933 Allocarpus lindenii Sch. Bip. = Allocarpus lindenii Sch. Bip. ex Wedd. ■☆
15934 Allocarpus lindenii Sch. Bip. ex Wedd.;林登奇果菊■☆
15935 Allocarpus lindenii Wedd. = Allocarpus lindenii Sch. Bip. ex Wedd. ■☆
15936 Allocarpus scabrifolius Hook. et Arn.;糙叶奇果菊■☆
15937 Allocarpus sillaensis Knuth;奇果菊■☆
15938 Allocarya Greene = Plagiobothrys Fisch. et C. A. Mey. ■☆
15939 Allocarya orientalis (L.) Brand = Plagiobothrys orientalis (L.) I. M. Johnst. ■☆
15940 Allocarya penicillata Greene = Plagiobothrys scouleri (Hook. et Arn.) I. M. Johnst. var. penicillatus (Greene) Cronquist ■☆
15941 Allocaryastrum Brand = Plagiobothrys Fisch. et C. A. Mey. ■☆
15942 Allocassine N. Robson(1965);异藏红卫矛属●☆
15943 Allocassine laurifolia (Harv.) N. Robson;异藏红卫矛●☆
15944 Allocassine tetragona (L. f.) N. Robson = Lauridia tetragona (L. f.) R. H. Archer ●☆
15945 Allocasuarina L. A. S. Johnson(1982);异木麻黄属●☆
15946 Allocasuarina decussata (Benth.) L. A. S. Johnson;科芮异木麻黄●☆
15947 Allocasuarina distyla (Vent.) L. A. S. Johnson;二柱异木麻黄;Scrub She-oak,She-oak ●☆
15948 Allocasuarina inophloia (F. Muell. et F. M. Bailey) L. A. S. Johnson;线皮异木麻黄●☆
15949 Allocasuarina lehmanniana (Miq.) L. A. S. Johnson;澳西南异木麻黄●☆
15950 Allocasuarina littoralis (Salisb.) L. A. S. Johnson;黑异木麻黄(栓皮木麻黄);Black She-oak,Black-she Oak,Forest Oak ●☆
15951 Allocasuarina luehmannii (R. T. Baker) L. A. S. Johnson;公牛异木麻黄;Bull Oak ●☆
15952 Allocasuarina torulosa (Aiton) L. A. S. Johnson;林生异木麻黄;Forest Oak,Forest She-oak,Forest-she Oak ●☆
15953 Allocasuarina verticillata (Lam.) L. A. S. Johnson;劲直木麻黄(垂枝异木麻黄,细直枝木麻黄,小木麻黄);Australian Pine,Beefwood,Bleeding Tree,Coast Beefwood,Coast Casuarina,Coast She Oak,Coast She-oak,Drooping She-oak,Fountain Tree,She Oak,She-oak,Strict Beefwood,Swamp Oak ●
15954 Alloceratium Hook. f. et Thomson = Diptychocarpus Trautv. ■

15955 Alloceratium strictum (Fisch. ex M. Bieb.) Hook. f. et Thomson = Diptychocarpus strictus (Fisch. ex M. Bieb.) Trautv. ■
15956 Allocheilos W. T. Wang(1983);异唇苣苔属;Allocheilos ●★
15957 Allocheilos cortusiflorum W. T. Wang;异唇苣苔;Allocheilos ●
15958 Allocheilos guangxiense H. Q. Wen, Y. G. Wei et S. H. Zhong;广西异唇苣苔(异片苣苔);Guangxi Allocheilos ●
15959 Allochilus Gagnep. = Goodyera R. Br. ■
15960 Allochlamys Moq. = Pleuropetalum Hook. f. ●☆
15961 Allochrusa Bunge = Allochrusa Bunge ex Boiss. ●■☆
15962 Allochrusa Bunge ex Boiss. (1867);凹瓣石竹属●■☆
15963 Allochrusa bungei Boiss.;邦奇凹瓣石竹■☆
15964 Allochrusa paniculata (Regel et Herder) Ovcz. et I. G. Czukavin;圆锥凹瓣石竹■☆
15965 Allochrusa pulchella Bunge ex Boiss.;凹瓣石竹●■☆
15966 Allodape Endl. (废弃属名) = Lebetanthus Endl. (保留属名)●☆
15967 Allodaphne Steud. = Allodape Endl. (废弃属名)●☆
15968 Allodaphne Steud. = Lebetanthus Endl. (保留属名)●☆
15969 Alloeochaete C. E. Hubb. (1940);非洲奇草属■☆
15970 Alloeochaete andongensis (Rendle) C. E. Hubb.;非洲奇草■☆
15971 Alloeochaete geniculata Kabuye;膝曲非洲奇草■☆
15972 Alloeochaete gracillima Kabuye;细长非洲奇草■☆
15973 Alloeochaete namuliensis Chippind.;纳木里非洲奇草■☆
15974 Alloeochaete ulugurensis Kabuye;乌卢古尔非洲奇草■☆
15975 Alloeospermum Spreng. = Alloispermum Willd. ●■☆
15976 Alloeospermum Spreng. = Calea L. ●■☆
15977 Allogyne Lewton = Alyogyne Alef. ■☆
15978 Allogyne Lewton = Fugosia Juss. ●■☆
15979 Allohemia Raf. = Oryctanthus (Griseb.) Eichler ●☆
15980 Allohemia Raf. = Phthirusa Mart. ●☆
15981 Alloiantheros Steud. = Alloiatheros Elliott ■☆
15982 Alloiatheros Elliott = Gymnopogon P. Beauv. ■☆
15983 Alloiatheros Raf. = Andropogon L. (保留属名)■
15984 Alloiatheros Raf. = Gymnopogon P. Beauv. ■☆
15985 Alloiosepalum Gilg = Purdiaea Planch. ●☆
15986 Alloiozonium Kuntze = Arctotheca J. C. Wendl. ■☆
15987 Alloiozonium Kuntze = Cryptostemma R. Br. ex W. T. Aiton ■☆
15988 Alloispermum Willd. (1807);异冠菊属●■☆
15989 Alloispermum Willd. = Calea L. ●■☆
15990 Alloispermum integrifolium (DC.) H. Rob.;全缘异冠菊■☆
15991 Alloispermum lehmannii H. Rob.;莱曼异冠菊■☆
15992 Alloispermum lindenii (Sch. Bip. ex Wedd.) H. Rob.;林登异冠菊■☆
15993 Alloispermum lindenii (Wedd.) H. Rob. = Alloispermum lindenii (Sch. Bip. ex Wedd.) H. Rob. ■☆
15994 Alloispermum longiradiatum (Urbatsch et B. L. Turner) B. L. Turner;长线异冠菊■☆
15995 Allolepis Soderstr. et H. F. Decker(1965);奇鳞草属■☆
15996 Allolepis texana (Vasey) Soderstr. et H. F. Decker;奇鳞草■☆
15997 Allolrurkillia Whitmore = Burkilliodendron Sastry ●☆
15998 Allomaieta Gleason(1929);异五月花属☆
15999 Allomaieta hirsuta (Gleason) Lozano;异五月花☆
16000 Allomarkgrafia Woodson(1932);马尔夹竹桃属●☆
16001 Allomarkgrafia brenesiana Woodson;马尔夹竹桃●☆
16002 Allomarkgrafia laxiflora A. H. Gentry;疏花马尔夹竹桃●☆
16003 Allomarkgrafia ovalis Woodson;卵形马尔夹竹桃●☆
16004 Allomarkgrafia tubiflora Woodson ex Dwyer;管花马尔夹竹桃●☆
16005 Allomia DC. = Alomia Kunth ■☆
16006 Allomorphia Blume(1831);异形木属;Allomorphia ●
16007 Allomorphia balansaei Cogn.;异形木(山暗赤,肖风木);Balansae Allomorphia ●
16008 Allomorphia baviensis Guillaumin;越南异形木(刺毛异形木);Bavi Allomorphia, Vietnam Allocheilos ●
16009 Allomorphia blinii (H. Lév.) Guillaumin = Plagiopetalum esquirolii (H. Lév.) Rehder ●
16010 Allomorphia blinii Guillaumin = Barthea blinii H. Lév. ●
16011 Allomorphia blinii Guillaumin = Plagiopetalum esquirolii (H. Lév.) Rehder ●
16012 Allomorphia bodinieri H. Lév. = Blastus cavaleriei H. Lév. et Vaniot ●
16013 Allomorphia bodinieri H. Lév. = Blastus pauciflorus (Benth.) Guillaumin ●
16014 Allomorphia caudata (Diels) H. L. Li = Styrophyton caudatum (Diels) S. Y. Hu ●◇
16015 Allomorphia cavaleriei H. Lév. et Vaniot = Phyllagathis cavaleriei (H. Lév. et Vaniot) Guillaumin ■
16016 Allomorphia curtisii (King) Ridl.;翅茎异形木(柯氏异形木)●
16017 Allomorphia eupteroton Guillaumin;东南亚异形木●☆
16018 Allomorphia eupteroton Guillaumin = Allomorphia curtisii (King) Ridl. ●
16019 Allomorphia eupteroton Guillaumin var. teretipetiolata C. Y. Wu et C. Chen = Oxyspora teretipetiolata (C. Y. Wu et C. Chen) W. H. Chen et Y. M. Shui ●
16020 Allomorphia flexuosa Hand.-Mazz. = Plagiopetalum esquirolii (H. Lév.) Rehder ●
16021 Allomorphia griffithii Hook. f. = Allomorphia griffithii Hook. f. ex Triana ●☆
16022 Allomorphia griffithii Hook. f. ex Triana;格氏异形木●☆
16023 Allomorphia howellii (Jeffrey et W. W. Sm.) Diels;腾冲异形木;Howell Allomorphia, Tengchong Allomorphia ●
16024 Allomorphia laotica Guillaumin = Allomorphia curtisii (King) Ridl. ●
16025 Allomorphia parviflora Mansf.;小花异形木●☆
16026 Allomorphia pauciflora Benth.;少花异形木●☆
16027 Allomorphia pauciflora Benth. = Blastus pauciflorus (Benth.) Guillaumin ●
16028 Allomorphia procursa Craib. = Allomorphia curtisii (King) Ridl. ●
16029 Allomorphia rosea Ridl.;粉花异形木●☆
16030 Allomorphia setosa Craib;刺毛异形木;Setose Allomorphia ●
16031 Allomorphia subsessilis Craib = Medinilla assamica (C. B. Clarke) C. Chen ●
16032 Allomorphia urophylla Diels;尾叶异形木;Caudate-leaved Allomorphia, Tailleaf Allomorphia ●
16033 Alloneuron Pilg. (1905);异脉野牡丹属☆
16034 Alloneuron maior Markgr. ex J. F. Macbr.;大异脉野牡丹☆
16035 Alloneuron ulei Pilg.;异脉野牡丹☆
16036 Allopetalum Reinw. = Labisia Lindl. (保留属名)●■☆
16037 Allophylaceae Martinov = Sapindaceae Juss. (保留科名)●■
16038 Allophyllum (Nutt.) A. D. Grant et V. E. Grant(1955);异叶花荵属■☆
16039 Allophyllum divaricatum (Nutt.) Arn. et V. E. Grant;异叶花荵(两歧吉莉花)■☆
16040 Allophyllum integrifolium (Brand) A. D. Grant et V. E. Grant;全缘异叶花荵■☆
16041 Allophyllum violaceum (A. Heller) A. D. Grant et V. E. Grant;

堇色异叶花葱■☆
16042 Allophyllus Gled. = Allophylus L. ●
16043 Allophylus L.(1753);异木患属(止宫树属);Allophylus ●
16044 Allophylus abyssinicus(Hochst.)Radlk.;阿比西尼亚异木患●☆
16045 Allophylus africanus P. Beauv.;非洲异木患●☆
16046 Allophylus africanus P. Beauv. var. griseotomentosus(Gilg)Verdc.;灰毛非洲异木患●☆
16047 Allophylus alnifolius(Baker)Radlk. = Allophylus rubifolius(Hochst. ex A. Rich.)Engl. var. alnifolius(Baker)Friis et Vollesen ●☆
16048 Allophylus alte-scandens Hauman;攀缘异木患●☆
16049 Allophylus amplissimus Hauman;膨大异木患●☆
16050 Allophylus antunesii Gilg;安图内思异木患●☆
16051 Allophylus appendiculatoserratus Gilg = Allophylus congolanus Gilg ●☆
16052 Allophylus brachycalyx Baker f. = Allophylus ferrugineus Taub. ●☆
16053 Allophylus buchananii Gilg ex Radlk. = Allophylus chaunostachys Gilg ●☆
16054 Allophylus bullatus Radlk.;泡状异木患●☆
16055 Allophylus bussei Gilg ex Engl. = Allophylus chirindensis Baker f. ●☆
16056 Allophylus calophyllus Gilg = Allophylus rubifolius(Hochst. ex A. Rich.)Engl. var. dasystachys(Gilg)Verdc. ●☆
16057 Allophylus camptoneurus Radlk.;弯脉异木患●☆
16058 Allophylus cataractarum Baker f. = Allophylus africanus P. Beauv. ●☆
16059 Allophylus caudatus Radlk.;波叶异木患(三叶茶);Caudate Allophylus,Waveleaf Allophylus ●
16060 Allophylus chartaceus(Kurz)Radlk.;大叶异木患;Bigleaf Allophylus,Largeleaf Allophylus,Papery Allophylus ●
16061 Allophylus chaunostachys Gilg;布坎南异木患●☆
16062 Allophylus chirindensis Baker f.;布瑟异木患●☆
16063 Allophylus cobbe(L.)Raeusch.;科贝异木患(止宫树)●
16064 Allophylus cobbe(L.)Raeusch. var. marinus Corner = Allophylus timorensis(DC.)Blume ●
16065 Allophylus cobbe(L.)Raeusch. var. velutinus Corner;滇南异木患;South Yunnan Allophylus ●
16066 Allophylus cominia Sw.;近异木患●☆
16067 Allophylus congolanus Gilg;刚果异木患●☆
16068 Allophylus congolanus Gilg var. monophyllus Baker f. = Allophylus congolanus Gilg ●☆
16069 Allophylus conraui Gilg ex Radlk.;康氏异木患●☆
16070 Allophylus crebriflorus Baker f. = Allophylus pseudopaniculatus Baker f. ●☆
16071 Allophylus cuneatus Baker f. = Allophylus ferrugineus Taub. ●☆
16072 Allophylus dasystachys Gilg = Allophylus rubifolius(Hochst. ex A. Rich.)Engl. var. dasystachys(Gilg)Verdc. ●☆
16073 Allophylus decipiens(Sond.)Radlk.;迷惑异木患●☆
16074 Allophylus delicatulus Verdc.;姣美异木患●☆
16075 Allophylus didymadenius Radlk. = Allophylus chaunostachys Gilg ●☆
16076 Allophylus dimorphus Radlk.;五叶异木患;Dimorphous Allomorphia,Fiveleaf Allomorphia ●
16077 Allophylus dregeanus(Sond.)De Winter;德雷异木患●☆
16078 Allophylus dummeri Baker f.;达默异木患●☆
16079 Allophylus edulis Radlk.;食用异木患●☆
16080 Allophylus elongatus Radlk.;伸长异木患●☆
16081 Allophylus erlangeri Gilg ex Chiov. = Allophylus rubifolius(Hochst. ex A. Rich.)Engl. ●☆
16082 Allophylus erlangeri Gilg ex Engl. = Allophylus rubifolius(Hochst. ex A. Rich.)Engl. ●☆
16083 Allophylus erosus Radlk. = Allophylus natalensis(Sond.)De Winter ●☆
16084 Allophylus ferrugineus Taub.;锈色异木患●☆
16085 Allophylus ferrugineus Taub. var. stipitatus Verdc.;具柄异木患●☆
16086 Allophylus fischeri Gilg = Allophylus rubifolius(Hochst. ex A. Rich.)Engl. ●☆
16087 Allophylus fulvotomentosus Gilg;褐绒毛异木患●☆
16088 Allophylus gazensis Baker f. = Allophylus chaunostachys Gilg ●☆
16089 Allophylus goetzeanus Gilg = Allophylus africanus P. Beauv. ●☆
16090 Allophylus gossweileri Baker f.;戈斯异木患●☆
16091 Allophylus griseotomentosus Gilg = Allophylus africanus P. Beauv. var. griseotomentosus(Gilg)Verdc. ●☆
16092 Allophylus grotei F. G. Davies et Verdc.;格罗特异木患●☆
16093 Allophylus hallaei Fouilloy;霍尔异木患●☆
16094 Allophylus hamatus Vermoesen ex Hauman;顶钩异木患●☆
16095 Allophylus hirsutus Radlk.;云南异木患;Hirsute Allomorphia,Yunnan Allomorphia ●
16096 Allophylus hirtellus(Hook. f.)Radlk.;多毛异木患●☆
16097 Allophylus holstii Gilg ex Engl.;霍尔斯特异木患●☆
16098 Allophylus holubii Baker f. = Allophylus africanus P. Beauv. ●☆
16099 Allophylus hylophilus Gilg;喜盐异木患●☆
16100 Allophylus imenoensis Pellegr.;伊梅诺异木患●☆
16101 Allophylus kassneri Baker f. = Allophylus fulvotomentosus Gilg ●☆
16102 Allophylus katangensis Hauman;加丹加异木患●☆
16103 Allophylus kilimandscharicus Taub. = Allophylus rubifolius(Hochst. ex A. Rich.)Engl. var. alnifolius(Baker)Friis et Vollesen ●☆
16104 Allophylus kiwuensis Gilg = Allophylus pseudopaniculatus Baker f. ●☆
16105 Allophylus laeteviridis Gilg ex Engl.;鲜绿异木患●☆
16106 Allophylus lasiopus Baker f.;毛足异木患●☆
16107 Allophylus lastoursvillensis Pellegr.;拉斯图维尔异木患●☆
16108 Allophylus latefoliolatus Baker f. = Allophylus ferrugineus Taub. ●☆
16109 Allophylus le-testui Pellegr.;勒泰斯蒂异木患●☆
16110 Allophylus longicuneatus Vermoesen ex Hauman;长楔形异木患●☆
16111 Allophylus longipes Radlk.;长柄异木患;Longstaik Allomorphia,Long-staiked Allomorphia ●
16112 Allophylus longipetiolatus Gilg;长叶柄异木患●☆
16113 Allophylus macrobotrys Gilg = Allophylus ferrugineus Taub. ●☆
16114 Allophylus macrurus Gilg = Allophylus ferrugineus Taub. ●☆
16115 Allophylus mawambensis Gilg = Allophylus africanus P. Beauv. ●☆
16116 Allophylus melanocarpus(Sond.)Radlk. = Allophylus africanus P. Beauv. ●☆
16117 Allophylus melliodorus Gilg ex Radlk.;蜜味异木患●☆
16118 Allophylus monophylla Radlk. = Allophylus dregeanus(Sond.)De Winter ●☆
16119 Allophylus mossambicensis Exell;莫桑比克异木患●☆
16120 Allophylus natalensis(Sond.)De Winter;纳塔尔异木患●☆
16121 Allophylus ngounyensis Pellegr.;恩戈尼亚异木患●☆
16122 Allophylus nigericus Baker f.;尼日利亚异木患●☆
16123 Allophylus oreophilus Gilg = Allophylus ferrugineus Taub. ●☆
16124 Allophylus oyemensis Pellegr.;奥也姆异木患●☆
16125 Allophylus paralleloneurus Gilg ex Engl.;平行脉异木患●☆

16126 Allophylus persicifolius Hauman;桃叶异木患●☆
16127 Allophylus pervillei Blume;佩尔异木患●☆
16128 Allophylus pervillei Blume. f. trifoliolatus Radlk.;三小叶佩尔异木患●☆
16129 Allophylus petelotii Merr.;广西异木患;Guangxi Allomorphia, Petelot Allomorphia ●
16130 Allophylus pierrei Pellegr. = Klaineanthus gabonii Pierre ●☆
16131 Allophylus pseudopaniculatus Baker f.;假圆锥花序异木患●☆
16132 Allophylus racemosus Radlk.;肖异木患●☆
16133 Allophylus repandifolius Merr. et Chun;单叶异木患;Repand-leaved Allomorphia, Simpleleaf Allomorphia ●
16134 Allophylus repandus (Baker) Engl. = Allophylus rubifolius (Hochst. ex A. Rich.) Engl. var. alnifolius (Baker) Friis et Vollesen ●☆
16135 Allophylus rhodesicus Exell = Allophylus africanus P. Beauv. ●☆
16136 Allophylus richardsiae Exell;理查兹异木患●☆
16137 Allophylus rubifolius (Hochst. ex A. Rich.) Engl.;悬叶异木患●☆
16138 Allophylus rubifolius (Hochst. ex A. Rich.) Engl. var. alnifolius (Baker) Friis et Vollesen;桤叶异木患●☆
16139 Allophylus rubifolius (Hochst. ex A. Rich.) Engl. var. dasystachys (Gilg) Verdc.;毛穗悬叶异木患●☆
16140 Allophylus rubifolius Engl. = Allophylus rubifolius (Hochst. ex A. Rich.) Engl. ●☆
16141 Allophylus sapinii Vermoesen ex Hauman;萨潘异木患●☆
16142 Allophylus schirensis Gilg = Allophylus ferrugineus Taub. ●☆
16143 Allophylus schweinfurthii Gilg = Allophylus africanus P. Beauv. ●☆
16144 Allophylus spectabilis Gilg;壮观异木患●☆
16145 Allophylus spicatus (Poir.) Radlk.;穗状异木患●☆
16146 Allophylus spicatus (Thunb.) Fourc. = Allophylus decipiens (Sond.) Radlk. ●☆
16147 Allophylus spragueanus Burtt Davy = Allophylus africanus P. Beauv. ●☆
16148 Allophylus stachyanthus Gilg = Allophylus rubifolius (Hochst. ex A. Rich.) Engl. var. dasystachys (Gilg) Verdc. ●☆
16149 Allophylus subcoriaceus Baker f. = Allophylus africanus P. Beauv. ●☆
16150 Allophylus talbotii Baker f.;塔尔博特异木患●☆
16151 Allophylus tanzaniensis F. G. Davies;坦桑尼亚异木患●☆
16152 Allophylus tenuifolius Radlk. = Allophylus chaunostachys Gilg ●☆
16153 Allophylus tenuis Radlk. = Allophylus rubifolius (Hochst. ex A. Rich.) Engl. var. alnifolius (Baker) Friis et Vollesen ●☆
16154 Allophylus tessmannii Gilg ex Engl.;泰斯曼异木患●☆
16155 Allophylus timorensis (DC.) Blume;滨海异木患(帝汶异木患,止宫树);Maritime Allomorphia, Sea-shore Allophylus, Timor Allophylus ●
16156 Allophylus timorensis Blume = Allophylus timorensis (DC.) Blume ●
16157 Allophylus toroensis Baker f. = Allophylus ferrugineus Taub. ●☆
16158 Allophylus torrei Exell et Mendonça;托雷异木患●☆
16159 Allophylus transvaalensis Burtt Davy = Allophylus africanus P. Beauv. ●☆
16160 Allophylus trichophyllus Merr. et Chun;毛叶异木患;Hairleaf Allomorphia, Hairy-leaved Allomorphia ●
16161 Allophylus tristis Radlk. = Allophylus rubifolius (Hochst. ex A. Rich.) Engl. ●☆
16162 Allophylus ussheri Baker f. = Allophylus dummeri Baker f. ●☆
16163 Allophylus uwembae Gilli = Allophylus chaunostachys Gilg ●☆
16164 Allophylus viridis Radlk.;异木患(大果,大果小叶枫);Green Allomorphia ●
16165 Allophylus volkensii Gilg = Allophylus ferrugineus Taub. ●☆
16166 Allophylus welwitschii Gilg = Allophylus ferrugineus Taub. ●☆
16167 Allophylus whitei Exell;怀特异木患●☆
16168 Allophylus yeru Gilg = Allophylus chaunostachys Gilg ●☆
16169 Allophylus zenkeri Gilg ex Radlk.;岑克尔异木患●☆
16170 Allophylus zeylanicus L.;锡兰异木患●☆
16171 Allophylus zeylanicus L. var. grandifolius Hiern = Allophylus chartaceus (Kurz) Radlk. ●
16172 Allophylus zimmermannianus Gilg ex Engl. = Allophylus chirindensis Baker f. ●☆
16173 Allophyton Brandegee = Tetranema Benth.(保留属名)■☆
16174 Alloplectus Mart.(1829)(保留属名);缠绕草属(金红花属);Alloplectus ●■☆
16175 Alloplectus ambiguus Urb.;可疑缠绕草■☆
16176 Alloplectus capitatus Hook.;头状缠绕草■☆
16177 Alloplectus lynchii Hook. f.;林奇缠绕草;Lynch's Alloplectus ●■☆
16178 Alloplectus martinianus J. F. Sm.;马丁缠绕草(金红花);Alloplectus ●☆
16179 Alloplectus nummularia (Hanst.) Wiehler;铜钱缠绕草;Clog Plant, Goldfish Plant ●☆
16180 Alloplectus ruacophilus Donn. Sm.;匍匐缠绕草;Repent Alloplectus ●☆
16181 Alloplectus sanguineus Mart. = Columnea sanguinea Hanst. ●☆
16182 Alloplectus schlimii Planch. et Linden;施利摩缠绕草;Schlim Alloplectus ●☆
16183 Alloplectus sparsiflorus Mart.;疏花缠绕草;Sparseflower Alloplectus ●☆
16184 Alloplectus vitatus Linden et André;墨西哥缠绕草■☆
16185 Allopleia Raf. = Sibthorpia L. ■☆
16186 Allopterigeron Dunlop = Pluchea Cass. ●■
16187 Allopterigeron Dunlop(1981);白蓬菊属■☆
16188 Allopterigeron filifolius (F. Muell.) Dunlop;白蓬菊■☆
16189 Allopythion Schott = Amorphophallus Blume ex Decne.(保留属名)●■
16190 Allopythion Schott = Thomsonia Wall.(废弃属名)●■
16191 Allosampela Raf. = Ampelopsis Michx. ●
16192 Allosampela heterophylla (Thunb.) Raf. = Ampelopsis heterophylla (Thunb.) Siebold et Zucc. ●
16193 Allosampela heterophylla Raf. = Ampelopsis heterophylla (Thunb.) Siebold et Zucc. ●
16194 Allosandra Raf. = Tragia L. ●☆
16195 Allosanthus Radlk.(1933);异花无患子属●☆
16196 Allosanthus trifoliolatus Radlk.;异花无患子●☆
16197 Alloschemone Schott(1858);异形南星属■☆
16198 Alloschemone occidentalis Baill.;西方异形南星■☆
16199 Alloschemone occidentalis Engl. et Krause = Alloschemone occidentalis Baill. ■☆
16200 Alloschemone poeppigiana Schott;异形南星■☆
16201 Alloschmidia H. E. Moore(1978);侧胚椰属(皮孔椰属)●☆
16202 Alloschmidia glabrata (Becc.) H. E. Moore;侧胚椰●☆
16203 Allosidastrum (Hochr.) Krapov., Fryxell et D. M. Bates(1988);异黄花稔属●■☆
16204 Allosidastrum dolichophyllum Krapov., Fryxell et D. M. Bates;长

叶异黄花稔●☆
16205 Allosidastrum hilarianum (C. Presl) Krapov., Fryxell et D. M. Bates;希尔异黄花稔●☆
16206 Allosidastrum interruptum (Balb. ex DC.) Krapov., Fryxell et D. M. Bates;异黄花稔●☆
16207 Allosidastrum pyramidatum (Cav.) Krapov., Fryxell et D. M. Bates;塔形异黄花稔●☆
16208 Allosperma Raf. = Commelina L. ■
16209 Allosperma Raf. = Erxlebia Medik. ■
16210 Allospondias (Pierre) Stapf = Spondias L. ●
16211 Allospondias Stapf = Spondias L. ●
16212 Allostigma W. T. Wang(1984);异片苣苔属;Allostigma ■★
16213 Allostigma guangxiensis W. T. Wang; 异片苣苔; Guangxi Allostigma ■
16214 Allostis Raf. = Baeckea L. ●
16215 Allosyncarpia S. T. Blake(1977);异合生果树属●☆
16216 Allosyncarpia ternata S. T. Blake;异合生果树●☆
16217 Alloteropsis C. Presl = Alloteropsis J. Presl ex C. Presl ■
16218 Alloteropsis J. Presl = Alloteropsis J. Presl ex C. Presl ■
16219 Alloteropsis J. Presl ex C. Presl(1830);毛颖草属;Alloteropsis ■
16220 Alloteropsis angusta Stapf;窄毛颖草■☆
16221 Alloteropsis cimicina (L.) Liou = Alloteropsis cimicina (L.) Stapf ■
16222 Alloteropsis cimicina (L.) Stapf;臭虫草;Bedbug Alloteropsis, Bug Hairgrass, Summergrass ■
16223 Alloteropsis eckloniana (Nees) Hitchc. = Alloteropsis semialata (R. Br.) Hitchc. var. eckloniana (Nees) Pilg. ■
16224 Alloteropsis gwebiense Stent et J. M. Rattray var. eckloniana (Nees) C. E. Hubb. = Alloteropsis semialata (R. Br.) Hitchc. ■
16225 Alloteropsis gwebiensis Stent et J. M. Rattray = Alloteropsis semialata (R. Br.) Hitchc. ■
16226 Alloteropsis homblei Robyns = Alloteropsis semialata (R. Br.) Hitchc. ■
16227 Alloteropsis latifolia (Peter) Pilg. = Alloteropsis cimicina (L.) Stapf ■
16228 Alloteropsis paniculata (Benth.) Stapf;圆锥毛颖草■☆
16229 Alloteropsis papillosa Clayton;乳头毛颖草■☆
16230 Alloteropsis quintasii (Mez) Pilg. = Alloteropsis cimicina (L.) Stapf ■
16231 Alloteropsis semialata (R. Br.) Hitchc.; 毛颖草; Halfwinged Alloteropsis ■
16232 Alloteropsis semialata (R. Br.) Hitchc. subsp. eckloniana (Nees) Gibbs-Russ. = Alloteropsis semialata (R. Br.) Hitchc. var. eckloniana (Nees) Pilg. ■
16233 Alloteropsis semialata (R. Br.) Hitchc. var. eckloniana (Nees) C. E. Hubb. = Alloteropsis semialata (R. Br.) Hitchc. var. eckloniana (Nees) Pilg. ■
16234 Alloteropsis semialata (R. Br.) Hitchc. var. eckloniana (Nees) Pilg.;紫纹毛颖草■
16235 Alloteropsis semialata (R. Br.) Hitchc. var. ecklonii (Stapf) Stapf = Alloteropsis semialata (R. Br.) Hitchc. subsp. eckloniana (Nees) Gibbs-Russ. ■
16236 Allotoonia J. F. Morales et J. K. Williams = Echites P. Browne ●☆
16237 Allotoonia J. F. Morales et J. K. Williams(2004);海地蛇木属●☆
16238 Allotria Raf. = Commelina L. ■
16239 Allotropa A. Gray = Allotropa Torr. et A. Gray ●☆
16240 Allotropa Torr. et A. Gray(1858);糖晶兰属●☆
16241 Allotropa virgata Torr. et Gray;糖晶兰;Candlestick, Candystick ●☆
16242 Allouya Aubl. = Calathea G. Mey. ■
16243 Allouya Plum. ex Aubl. = Calathea G. Mey. ■
16244 Allowissadula D. M. Bates(1978);异隔蒴苘属●☆
16245 Allowissadula floribunda (Schltdl.) Fryxell;多花异隔蒴苘●☆
16246 Allowissadula glandulosa (Rose) D. M. Bates;多腺异隔蒴苘●☆
16247 Allowissadula microcalyx (R. E. Fr.) D. M. Bates;小萼异隔蒴苘●☆
16248 Allowissadula rosei (R. E. Fr.) D. M. Bates;罗斯异隔蒴苘●☆
16249 Allowoodsonia Markgr. (1967);肖椰夹竹桃属●☆
16250 Allowoodsonia whitmorei Markgr.;肖椰夹竹桃●☆
16251 Alloxylon P. H. Weston et Crisp(1991);绸缎木属(瓦拉她属); Satin Oak ●☆
16252 Alloxylon flammeum P. H. Weston et Crisp; 绸缎木; Waratah Tree ●☆
16253 Alloxylon pinnatum (Maiden et Betche) P. H. Weston et Crisp; 羽状绸缎木;Dorrigo Waratah ●☆
16254 Allozygia Naudin = Oxyspora DC. ●
16255 Alluaudia (Drake) Drake(1903);亚龙木属●☆
16256 Alluaudia Drake = Alluaudia (Drake) Drake ●☆
16257 Alluaudia ascendens (Drake) Drake;直立亚龙木●☆
16258 Alluaudia comosa (Drake) Drake;簇毛亚龙木●☆
16259 Alluaudia dumosa (Drake) Drake;灌丛亚龙木●☆
16260 Alluaudia humbertii Choux;亨伯特亚龙木●☆
16261 Alluaudia montagnacii Rauh;蒙氏亚龙木●☆
16262 Alluaudia procera (Drake) Drake; 马达加斯加亚龙木; Madagascan Ocotillo ●☆
16263 Alluaudiopsis Humbert et Choux(1934);拟亚龙木属(类亚龙木属)●☆
16264 Alluaudiopsis fiherenensis Humbert et Choux;拟亚龙木●☆
16265 Alluaudiopsis marnieriana Rauh;马岛拟亚龙木●☆
16266 Allucia Klotzsch ex Petersen = Renealmia L. f. (保留属名)■☆
16267 Allughas Steud. = Alpinia Roxb. (保留属名)■
16268 Allughas Steud. = Alughas L. ■
16269 Almaleea Crisp et P. H. Weston(1991);阿尔玛豆属■☆
16270 Almaleea capitata (J. H. Willis) Crisp et P. H. Weston;头状阿尔玛豆■☆
16271 Almaleea incurvata (A. Cunn.) Crisp et P. H. Weston;内弯阿尔玛豆■☆
16272 Almana Raf. = Sinningia Nees ●■☆
16273 Almeida Cham. = Almeidea A. St. -Hil. ●☆
16274 Almeidea A. St. -Hil. (1823);阿尔芸香属●☆
16275 Almeidea acuminata A. St. -Hil. ex G. Don;尖阿尔芸香●☆
16276 Almeidea alba A. St. -Hil. ex G. Don;白阿尔芸香●☆
16277 Almeidea caerulea A. St. -Hil. ex G. Don;蓝阿尔芸香●☆
16278 Almeidea guyanensis Pulle;圭亚那阿尔芸香●☆
16279 Almeidea lanceolata A. St. -Hil.;剑叶阿尔芸香●☆
16280 Almeidea longifolia A. St. -Hil.;长叶阿尔芸香●☆
16281 Almeidea longipes C. Presl;长梗阿尔芸香●☆
16282 Almeidea macropetala Fisch. et C. A. Mey.;大瓣阿尔芸香●☆
16283 Almeidea rubra A. St. -Hil.;红阿尔芸香●☆
16284 Almeloveenia Dennst. = Moullava Adans. ■☆
16285 Almideia Rchb. = Almeidea A. St. -Hil. ●☆
16286 Almutaster Á. Löve et D. Löve = Aster L. ●■
16287 Almutaster Á. Löve et D. Löve (1982); 泽菀属; Aster, Marsh Alkali Aster ■☆
16288 Almutaster pauciflorus (Nutt.) Á. Löve et D. Löve;泽菀;Marsh

Alkali Aster ■☆

16289 Almyra Salisb. = Pancratium L. ■

16290 Alnaster Spach = Duschekia Opiz ●

16291 Alniphyllum Matsum.(1901);赤杨叶属(假赤杨属,拟赤杨属);Chinabell,Chinabells ●

16292 Alniphyllum buddleiifolium Hu et W. C. Cheng = Alniphyllum fortunei(Hemsl.) Makino ●

16293 Alniphyllum eberhardtii Guillaumin;滇赤杨叶(豆渣树,牛角树);Eberhardt Chinabells,Oxborn Tree ●

16294 Alniphyllum fauriei Perkins = Alniphyllum fortunei(Hemsl.) Makino ●

16295 Alniphyllum fortunei(Hemsl.) Makino;赤杨叶(白苍术,白花展,冬瓜木,豆渣树,福氏赤杨叶,高山望,红榄木,红皮岭麻,鹿食,鹿食皮,萝卜树,拟赤杨,水冬瓜,依白果);Fortune Chinabell,Fortune's Chinabells ●

16296 Alniphyllum fortunei(Hemsl.) Perkins = Alniphyllum fortunei(Hemsl.) Makino ●

16297 Alniphyllum fortunei(Hemsl.) Perkins f. hypoglaucum C. Y. Wu = Alniphyllum fortunei(Hemsl.) Perkins ●

16298 Alniphyllum fortunei(Hemsl.) Perkins var. hainanense(Hayata) C. Y. Wu = Alniphyllum fortunei(Hemsl.) Perkins ●

16299 Alniphyllum fortunei(Hemsl.) Perkins var. megaphyllum(Hemsl. et E. H. Wilson) C. Y. Wu = Alniphyllum fortunei(Hemsl.) Perkins ●

16300 Alniphyllum fortunei(Hemsl.) Perkins var. microcarpum C. Y. Wu = Alniphyllum fortunei(Hemsl.) Perkins ●

16301 Alniphyllum hainanensis Hayata = Alniphyllum fortunei(Hemsl.) Perkins ●

16302 Alniphyllum macranthum Perkins = Alniphyllum fortunei(Hemsl.) Makino ●

16303 Alniphyllum megaphyllum Hemsl. et E. H. Wilson = Alniphyllum fortunei(Hemsl.) Makino ●

16304 Alniphyllum pterospermum Matsum.;台湾赤杨叶(长叶赤杨叶,长叶拟赤杨,丹招树,假赤杨,冇打,冇丹树,冇圆树,翼子赤杨叶,有丹树,圆招树);Pterospermous Chinabell,Taiwan Chinabells,Winged-seed China-bell ●

16305 Alnobetula(W. D. J. Koch) Schur = Alnaster Spach ●

16306 Alnobetula(W. D. J. Koch) Schur = Duschekia Opiz ●

16307 Alnobetula Schur = Alnaster Spach ●

16308 Alnobetula Schur = Duschekia Opiz ●

16309 Alnus Mill.(1754);桤木属(赤杨属,榿属);Alder ●

16310 Alnus × hanedae Sugim.;土田桤木●☆

16311 Alnus × hosoii M. Mizush.;细井桤木●☆

16312 Alnus × moriokaensis Murai;盛冈桤木●☆

16313 Alnus × peculiaris Hiyama;特殊桤木●☆

16314 Alnus × suginoi Sugim.;杉野桤木●☆

16315 Alnus acuminata Kunth;常绿桤木(尖桤木);Evergreen Alder,Mexican Alder ●☆

16316 Alnus acuminata Kunth subsp. glabrata(Fernald) Furlow;光叶常绿桤木●☆

16317 Alnus alnobetula(Ehrh.) K. Koch = Alnus viridis(Chaix) DC. ●☆

16318 Alnus alnobetula(Ehrh.) K. Koch var. crispa(Aiton) H. J. P. Winkler = Alnus viridis(Vill.) DC. subsp. crispa(Aiton) Turrill ●☆

16319 Alnus alnus(L.) Britton = Alnus glutinosa(L.) Gaertn. ●

16320 Alnus barbata C. A. Mey.;髯毛桤木●☆

16321 Alnus borealis Koidz. = Alnus mayrii Callier nothovar. glabrescens Nakai ●

16322 Alnus boshia Buch.-Ham. ex D. Don = Alnus nepalensis D. Don ●

16323 Alnus cordata(Loisel.) Desf.;意大利桤木;Italian Alder ●☆

16324 Alnus cordifolia Ten. = Alnus cordata(Loisel.) Desf. ●☆

16325 Alnus corylifolia Kern. ex Dalla Torre = Alnus cordata(Loisel.) Desf. ●☆

16326 Alnus cremastogyne Burkill;桤木(菜壳蒜,桦茶,罗拐木,牛屎树,水冬瓜,水冬瓜树,水青冈);Longpeduncled Alder,Long-peduncled Alder ●

16327 Alnus crispa(Aiton) Pursh = Alnus viridis(Vill.) DC. subsp. crispa(Aiton) Turrill ●☆

16328 Alnus crispa(Aiton) Pursh subsp. maximowiczii(Callier) Hultén;马氏桤木;Maximowicz Alder ●☆

16329 Alnus crispa(Aiton) Pursh subsp. maximowiczii(Callier) Hultén f. grandifolia(Miyabe et Tatew.) Honda;大花马氏桤木●☆

16330 Alnus crispa(Aiton) Pursh subsp. maximowiczii(Callier) Hultén var. sachalinensis(Koidz.) H. Hara;库页马氏桤木●☆

16331 Alnus crispa(Aiton) Pursh subsp. sinuata(Regel) Hultén = Alnus sinuata(Regel ex DC.) Rydb. ●

16332 Alnus crispa(Aiton) Pursh var. elongata Raup = Alnus viridis(Vill.) DC. subsp. crispa(Aiton) Turrill ●☆

16333 Alnus crispa(Aiton) Pursh var. harricanensis Lepage = Alnus viridis(Vill.) DC. subsp. crispa(Aiton) Turrill ●☆

16334 Alnus crispa(Aiton) Pursh var. mollis(Fernald) Fernald = Alnus viridis(Vill.) DC. subsp. crispa(Aiton) Turrill ●☆

16335 Alnus crispa(Aiton) Pursh var. stragula Fernald = Alnus viridis(Vill.) DC. subsp. crispa(Aiton) Turrill ●☆

16336 Alnus cylindrostachya(H. J. P. Winkl.) Makino = Alnus fauriei H. Lév. et Vaniot ●☆

16337 Alnus dioica Roxb. = Aporusa dioica(Roxb.) Müll. Arg. ●

16338 Alnus elliptica Req.;椭圆叶桤木●☆

16339 Alnus fauriei H. Lév. et Vaniot;法氏桤木●☆

16340 Alnus ferdinandi-coburgii C. K. Schneid.;川滇桤木(滇赤杨,滇桤木);Sichuan-Yunnan Alder,Szechwan-Yunnan Alder ●

16341 Alnus firma Siebold et Zucc.;日本桤木(硬桤木);Japanese Alder ●☆

16342 Alnus firma Siebold et Zucc. var. hirtella Franch. et Sav.;毛日本桤木●☆

16343 Alnus firma Siebold et Zucc. var. sieboldiana(Matsum.) H. J. P. Winkl. = Alnus sieboldiana Matsum. ●

16344 Alnus firmifolia Fernald;硬叶桤木●☆

16345 Alnus formosana(Burkill ex Forbes et Hemsl.) Makino;台湾桤木(赤柯,赤杨,水柯,水柯仔,水柯子,水流柯,水柳柯,台湾赤杨);Formosan Alder,Taiwan Alder ●

16346 Alnus formosana(Burkill ex Forbes et Hemsl.) Makino = Alnus japonica(Thunb.) Steud. var. formosana(Burkill ex Forbes et Hemsl.) Callier ●

16347 Alnus formosana(Burkill) Makino = Alnus formosana(Burkill ex Forbes et Hemsl.) Makino ●

16348 Alnus fruticosa Rupr.;矮桤木(矮赤杨);Dwarf Alder ●

16349 Alnus fruticosa Rupr. = Alnus viridis(Chaix) DC. subsp. fruticosa(Rupr.) Nyman ●☆

16350 Alnus fruticosa Rupr. var. mandshurica Callier ex C. K. Schneid. = Alnus mandshurica(Callier ex C. K. Schneid.) Hand.-Mazz. ●

16351 Alnus glauca Michx. = Alnus incana(L.) Moench subsp. rugosa(Du Roi) R. T. Clausen ●☆

16352 Alnus glutinosa(L.) Gaertn.;欧洲桤木(欧洲赤杨);Alder,

Allar, Allen-tree, Aller, Aller-tree, Alls-bush, Arl, Aul, Aulne, Black Alder, Black Aller, Common Alder, Dog Tree, Elder, Ellen, Ellen-tree, Eller, Ell-shinders, English Alder, European Alder, European Black Alder, Halse-bush, Harrul, Howler, Irish Mahogany, Lamb's Tails, Olern Oler, Ollar, Orl, Oryelle, Owlder, Owler, Owlorn, Scotch Mahogany, Sticky Alder, Uler Otaheite Potato, Whistlewood, Wullow, Wullow Wallow, Yellow Shinders Ell ●

16353 Alnus glutinosa (L.) Gaertn. var. subrotunda Spach = Alnus glutinosa (L.) Gaertn. ●

16354 Alnus glutinosa Gaertn. 'Aurea';黄叶欧洲桤木●☆

16355 Alnus glutinosa Gaertn. 'Imperialis';壮丽欧洲桤木(帝王欧洲桤木)●☆

16356 Alnus glutinosa Gaertn. 'Laciniata';裂叶欧洲桤木●☆

16357 Alnus hakkodensis Hayashi = Alnus crispa (Aiton) Pursh subsp. maximowiczii (Callier) Hultén ●☆

16358 Alnus henryi C. K. Schneid.;台北桤木;Henry Alder ●

16359 Alnus hirsuta (Spach) Turcz. ex Rupr.;山地桤木(辽东桤木);Arrow-bearing Tree, Manchurian Alder ●

16360 Alnus hirsuta (Spach) Turcz. ex Rupr. var. microphylla (Nakai) Tatew. = Alnus inokumae Murai et Kusaka ●☆

16361 Alnus hirsuta (Spach) Turcz. ex Rupr. var. microphylla (Nakai) Tatew. f. glabrescens Tatew. = Alnus inokumae Murai et Kusaka ●☆

16362 Alnus hirsuta (Spach) Turcz. ex Rupr. var. sibirica (Spach) C. K. Schneid.;西伯利亚山地桤木(赤杨树,东北桤木,辽东桤木,色赤杨,水冬瓜,水冬瓜赤杨,水冬瓜树);Manchurian Alder, Siberia Alder, Siberian Alder ●

16363 Alnus hirsuta (Spach) Turcz. ex Rupr. var. sibirica (Spach) C. K. Schneid. = Alnus hirsuta (Spach) Turcz. ex Rupr. ●

16364 Alnus hirsuta (Spach) Turcz. ex Rupr. var. tinctoria (Sarg.) Kudo ex Murai = Alnus hirsuta (Spach) Turcz. ex Rupr. var. sibirica (Spach) C. K. Schneid. ●

16365 Alnus hirsuta Turcz. ex Rupr. = Alnus hirsuta (Spach) Turcz. ex Rupr. ●

16366 Alnus hirsuta Turcz. ex Rupr. var. sibirica (Fisch. ex Turcz.) C. K. Schneid. = Alnus hirsuta (Spach) Turcz. ex Rupr. ●

16367 Alnus hultenii Murai = Alnus viridis (Vill.) DC. subsp. crispa (Aiton) Turrill ●☆

16368 Alnus incana (L.) Moench;毛赤杨(灰赤杨,灰毛桤木,美国桤木);American Alder, American Black Alder, Black Alder, European Alder, Gray Alder, Grey Alder, Hairy Alder, Hoary Alder, Mountain Alder, Norwegian Grey Alder, Speckled Alder, White Alder ●☆

16369 Alnus incana (L.) Moench 'Aurea';金叶毛赤杨(黄叶毛赤杨)●☆

16370 Alnus incana (L.) Moench 'Laciniata';裂叶毛赤杨●☆

16371 Alnus incana (L.) Moench 'Pendula';垂枝毛赤杨●☆

16372 Alnus incana (L.) Moench 'Ramulis Coccineis';红枝毛赤杨●☆

16373 Alnus incana (L.) Moench f. tomophylla Fernald = Alnus incana (L.) Moench subsp. rugosa (Du Roi) R. T. Clausen ●☆

16374 Alnus incana (L.) Moench subsp. hirsuta (Spach) Á. Löve et D. Löve = Alnus hirsuta (Spach) Turcz. ex Rupr. ●

16375 Alnus incana (L.) Moench subsp. rugosa (Du Roi) R. T. Clausen;粗糙毛赤杨;Mountain Alder, Speckled Alder, Swamp Alder, Tag Alder ●☆

16376 Alnus incana (L.) Moench subsp. tenuifolia (Nutt.) Breitung = Alnus tenuifolia Nutt. ●

16377 Alnus incana (L.) Moench var. americana Regel = Alnus incana (L.) Moench subsp. rugosa (Du Roi) R. T. Clausen ●☆

16378 Alnus incana (L.) Moench var. glauca (F. Michx.) Loudon = Alnus hirsuta (Spach) Turcz. ex Rupr. ●

16379 Alnus incana (L.) Moench var. hirsuta Spach = Alnus hirsuta (Spach) Turcz. ex Rupr. ●

16380 Alnus incana (L.) Moench var. occidentalis (Dippel) C. L. Hitchc. = Alnus incana (L.) Moench subsp. tenuifolia (Nutt.) Breitung ●☆

16381 Alnus incana (L.) Moench var. sibirica Spach = Alnus hirsuta (Spach) Turcz. ex Rupr. ●

16382 Alnus incana (L.) Moench var. virescens S. Watson = Alnus incana (L.) Moench subsp. tenuifolia (Nutt.) Breitung ●☆

16383 Alnus incana (L.) Moench var. virescens S. Watson = Alnus tenuifolia Nutt. ●

16384 Alnus inokumae Murai et Kusaka;谷川桤木●☆

16385 Alnus jackii Hu = Alnus trabeculosa Hand. -Mazz. ●

16386 Alnus japonica (Thunb.) Steud.;赤杨(木拔树,木瓜树,日本桤木,水冬瓜,水冬果,水柯子);Japan Alder, Japanese Alder ●

16387 Alnus japonica (Thunb.) Steud. f. arguta (Regel) H. Ohba = Alnus japonica (Thunb.) Steud. var. arguta (Regel) Callier ●☆

16388 Alnus japonica (Thunb.) Steud. f. koreana (Callier) H. Ohba = Alnus japonica (Thunb.) Steud. var. koreana Callier ●☆

16389 Alnus japonica (Thunb.) Steud. var. arguta (Regel) Callier;锐齿赤杨●☆

16390 Alnus japonica (Thunb.) Steud. var. formosana (Burkill ex Forbes et Hemsl.) Callier = Alnus formosana (Burkill ex Forbes et Hemsl.) Makino ●

16391 Alnus japonica (Thunb.) Steud. var. formosana Callier = Alnus formosana (Burkill ex Forbes et Hemsl.) Makino ●

16392 Alnus japonica (Thunb.) Steud. var. genuina Callier = Alnus japonica (Thunb.) Steud. ●

16393 Alnus japonica (Thunb.) Steud. var. koreana Callier;朝鲜赤杨(毛叶赤杨)●☆

16394 Alnus japonica (Thunb.) Steud. var. latifolia Callier = Alnus japonica (Thunb.) Steud. ●

16395 Alnus japonica (Thunb.) Steud. var. rufa Nakai = Alnus japonica (Thunb.) Steud. var. koreana Callier ●☆

16396 Alnus jorullensis Benth.;常绿赤杨(乔鲁桤木);Evergreen Alder, Mexican Alder ●☆

16397 Alnus jorullensis Benth. var. ferruginea Kuntze;锈色桤木●☆

16398 Alnus kamtschatica (Callier) Kom.;勘察加桤木●☆

16399 Alnus kolaensis Orlova;科拉赤杨●☆

16400 Alnus lanata Duthie ex Bean;毛桤木(菜壳蒜,罗拐木,毛桤,牛屎树);Hairy Alder, Lanate Alder, Woolly-leaved Alder ●

16401 Alnus mandshurica (Callier ex C. K. Schneid.) Hand. -Mazz.;东北桤木(东北赤杨,水冬瓜,水冬果);Manchuri Alder, Manchurian Alder ●

16402 Alnus mandshurica (Callier ex C. K. Schneid.) Hand. -Mazz. = Alnus fruticosa Rupr. ●

16403 Alnus mandshurica (Callier ex C. K. Schneid.) Hand. -Mazz. var. pubescens Baranov;柔毛东北桤木(柔毛东北赤杨);Bubescent Manchurian Alder ●

16404 Alnus maritima (Marshall) Muhl. ex Nutt.;滨海桤木;Brook Alder, Seaside Alder ●☆

16405 Alnus maritima (Marshall) Nutt. var. arguta Regel = Alnus japonica (Thunb.) Steud. ●

16406 Alnus maritima (Marshall) Nutt. var. formosana Burkill = Alnus formosana (Burkill ex Forbes et Hemsl.) Makino ●

16407 Alnus maritima (Marshall) Nutt. var. formosana Burkill ex Forbes et Hemsl. = Alnus formosana (Burkill ex Forbes et Hemsl.) Makino ●

16408 Alnus maritima Muhl. ex Nutt. = Alnus maritima (Marshall) Muhl. ex Nutt. ●☆

16409 Alnus maritima Muhl. ex Nutt. var. arguta Regel = Alnus japonica (Thunb.) Steud. ●

16410 Alnus maritima Muhl. ex Nutt. var. formosana Burkill ex Forbes et Hemsl. = Alnus formosana (Burkill ex Forbes et Hemsl.) Makino ●

16411 Alnus maritima Muhl. ex Nutt. var. japonica Regel = Alnus japonica (Thunb.) Steud. ●

16412 Alnus matsumurae Callier;翼叶桤木;Matsumura Alder ●☆

16413 Alnus maximowiczii Callier = Alnus crispa (Aiton) Pursh subsp. maximowiczii (Callier) Hultén ●☆

16414 Alnus maximowiczii Callier var. sachalinensis (Koidz.) Nemoto = Alnus crispa (Aiton) Pursh subsp. maximowiczii (Callier) Hultén var. sachalinensis (Koidz.) H. Hara ●☆

16415 Alnus mayrii Callier;迈里桤木●☆

16416 Alnus mayrii Callier nothovar. glabrescens Nakai;北方赤杨●

16417 Alnus mitchelliana M. A. Curtis ex A. Gray = Alnus viridis (Vill.) DC. subsp. crispa (Aiton) Turrill ●☆

16418 Alnus mollis Fernald = Alnus viridis (Vill.) DC. subsp. crispa (Aiton) Turrill ●☆

16419 Alnus multinervis (Regel) Callier = Alnus pendula Matsum. ●☆

16420 Alnus nagurae Inokuma = Alnus trabeculosa Hand. -Mazz. ●

16421 Alnus nepalensis D. Don;尼泊尔桤木(冬瓜树,旱冬瓜,旱冬瓜树,蒙自赤杨,蒙自桤木,桤木,水冬瓜树);Grey Alder, Himalayan Alder, Nepal Alder ●

16422 Alnus nepalensis Wall. = Alnus nepalensis D. Don ●

16423 Alnus nitida (Spach) Endl.;亮桤木●☆

16424 Alnus noveboracensis Britton = Alnus serrulata (Aiton) Willd. ●

16425 Alnus oblongifolia Torr.;亚利桑那桤木;Aliso, Arizona Alder, Mexican Alder, New Mexican Alder ●☆

16426 Alnus obtusata (Franch. et Sav.) Makino = Alnus serrulatoides Callier ●☆

16427 Alnus occidentalis Dippel = Alnus tenuifolia Nutt. ●

16428 Alnus oregona Decne. var. pinnatisecta ? = Alnus rubra Bong. ●

16429 Alnus oregona Nutt.;俄勒冈桤木(奥雷同桤木,红桤木);Oregon Alder, Red Alder ●☆

16430 Alnus oregona Nutt. = Alnus rubra Bong. ●

16431 Alnus orientalis Decne.;东方桤木;Oriental Alder, Syrian Alder ●☆

16432 Alnus ovata (Schrank) Lodd. = Alnus viridis (Chaix) DC. ●☆

16433 Alnus pendula Matsum.;垂枝桤木(垂桤木);Pendulous Alder ●☆

16434 Alnus reginosa Nakai = Alnus japonica (Thunb.) Steud. ●

16435 Alnus rhombifolia Nutt.;菱叶桤木(美国赤杨);California Alder, Californian Alder, Sierra Alder, White Alder ●

16436 Alnus rhombifolia Nutt. var. bernardina Munz et I. M. Johnst. = Alnus rhombifolia Nutt. ●

16437 Alnus rubra Bong.;红枝桤木(红桤木);Oregon Alder, Red Alder, Smooth Alder, Tag Alder, Western Alder ●

16438 Alnus rubra Bong. var. pinnatisecta Starker = Alnus rubra Bong. ●

16439 Alnus rubra Desf. ex Spach = Alnus serrulata (Aiton) Willd. ●

16440 Alnus rugosa (Du Roi) Spreng.;齿叶桤木(斑叶桤木);Gray Alder, Hazel Alder, Smooth Alder, Speckled Alder, Tag Alder ●

16441 Alnus rugosa (Du Roi) Spreng. = Alnus incana (L.) Moench subsp. rugosa (Du Roi) R. T. Clausen ●☆

16442 Alnus rugosa (Du Roi) Spreng. var. americana (Regel) Fernald = Alnus incana (L.) Moench subsp. rugosa (Du Roi) R. T. Clausen ●☆

16443 Alnus rugosa (Du Roi) Spreng. var. serrulata (Aiton) Winkler = Alnus serrulata (Aiton) Willd. ●

16444 Alnus rugosa (Du Roi) Spreng. var. tomophylla (Fernald) Fernald = Alnus incana (L.) Moench subsp. rugosa (Du Roi) R. T. Clausen ●☆

16445 Alnus serrulata (Aiton) Willd.;锯叶桤木(齿叶桤木);Common Alder, Hazel Alder, Smooth Alder, Speckled Alder, Tag Alder ●

16446 Alnus serrulata (Aiton) Willd. f. noveboracensis (Britton) Fernald = Alnus serrulata (Aiton) Willd. ●

16447 Alnus serrulatoides Callier;拟锯叶桤木;Hazel Alder ●☆

16448 Alnus serrulatoides Callier f. katoana (Yanagita) H. Ohba;加藤桤木●☆

16449 Alnus serrulatoides Callier var. katoana (Yanagita) Sugim. = Alnus serrulatoides Callier f. katoana (Yanagita) H. Ohba ●☆

16450 Alnus sibirica (Spach) Fisch. ex Kom. = Alnus hirsuta (Spach) Turcz. ex Rupr. var. sibirica (Spach) C. K. Schneid. ●

16451 Alnus sibirica (Spach) Fisch. ex Kom. = Alnus hirsuta (Spach) Turcz. ex Rupr. ●

16452 Alnus sibirica (Spach) Fisch. ex Kom. var. hirsuta (Spach) Koidz. = Alnus hirsuta (Spach) Turcz. ex Rupr. ●

16453 Alnus sibirica (Spach) Fisch. ex Kom. var. paucinervis C. K. Schneid. = Alnus hirsuta (Spach) Turcz. ex Rupr. ●

16454 Alnus sibirica Fisch. ex Turcz. = Alnus hirsuta (Spach) Turcz. ex Rupr. ●

16455 Alnus sibirica Fisch. ex Turcz. var. hirsuta (Turcz. ex Rupr.) Koidz. = Alnus hirsuta (Spach) Turcz. ex Rupr. ●

16456 Alnus sibirica Fisch. ex Turcz. var. oxyloba C. K. Schneid. = Alnus hirsuta (Spach) Turcz. ex Rupr. ●

16457 Alnus sibirica Fisch. ex Turcz. var. paucinervis C. K. Schneid. = Alnus hirsuta (Spach) Turcz. ex Rupr. ●

16458 Alnus sieboldiana Matsum.;旅顺桤木(赤杨);Lushun Alder, Siebold Alder ●

16459 Alnus sinuata (Regel ex DC.) Rydb.;裂叶桤木;Mountain Alder, Sitka Alder, Wavyleaf Alder ●

16460 Alnus sinuata (Regel) Rydb. = Alnus sinuata (Regel ex DC.) Rydb. ●

16461 Alnus sitchensis (Regel) Sarg. = Alnus sinuata (Regel ex DC.) Rydb. ●

16462 Alnus spaethii Callier;斯佩思桤木●☆

16463 Alnus subcordata C. A. Mey.;亚心形桤木;Caucasian Alder ●☆

16464 Alnus tenuifolia Nutt.;薄叶桤木(细叶毛赤杨);Mountain Alder, River Alder, Speckled Alder, Thinleaf Alder ●

16465 Alnus tenuifolia Nutt. = Alnus incana (L.) Moench subsp. tenuifolia (Nutt.) Breitung ●☆

16466 Alnus tinctoria Sarg.;色赤杨(赤杨树,水冬瓜树)●

16467 Alnus tinctoria Sarg. = Alnus hirsuta (Spach) Turcz. ex Rupr. var. sibirica (Spach) C. K. Schneid. ●

16468 Alnus tinctoria Sarg. = Alnus hirsuta (Spach) Turcz. ex Rupr. ●

16469 Alnus tinctoria Sarg. var. glabra Callier;无毛色赤杨●☆

16470 Alnus trabeculosa Hand. -Mazz.;江南桤木(木瓜树,水冬瓜,水桶木);Trabeculate Alder ●

16471 Alnus trabeculosa Hand. -Mazz. var. hunanensis S. B. Wan;湖南桤木;Hunan Trabeculate Alder ●

16472 Alnus viridis (Chaix) DC.;绿桤木;European Green Alder, Green Alder, Mountain Alder, Sitka Alder ●☆

16473 Alnus viridis (Chaix) DC. subsp. fruticosa (Rupr.) Nyman;西伯利亚桤木;Siberian Alder ●☆

16474 Alnus viridis (Chaix) DC. subsp. sinuata (Regel) Á. Löve et D. Löve = Alnus sinuata (Regel ex DC.) Rydb. ●

16475 Alnus viridis (Chaix) DC. var. crispa (Michx.) House = Alnus viridis (Vill.) DC. subsp. crispa (Aiton) Turrill ●☆

16476 Alnus viridis (Chaix) DC. var. fruticosa (Rupr.) Regel = Alnus viridis (Chaix) DC. subsp. fruticosa (Rupr.) Nyman ●☆

16477 Alnus viridis (Chaix) DC. var. sinuata Regel = Alnus sinuata (Regel ex DC.) Rydb. ●

16478 Alnus viridis (Vill.) DC. = Alnus viridis (Chaix) DC. ●☆

16479 Alnus viridis (Vill.) DC. subsp. crispa (Aiton) Turrill;美洲绿桤木(皱绿桤木);America Alder, American Alder, American Green Alder, Aulne Vert, Green Alder, Mountain Alder ●☆

16480 Alnus viridis (Vill.) DC. var. crispa (Aiton) House = Alnus viridis (Vill.) DC. subsp. crispa (Aiton) Turrill ●☆

16481 Alnus viridis (Vill.) DC. var. fernaldii House = Alnus viridis (Vill.) DC. subsp. crispa (Aiton) Turrill ●☆

16482 Alnus vulgaris Hill = Alnus glutinosa (L.) Gaertn. ●

16483 Aloaceae Batsch = Asphodelaceae Juss. ●■

16484 Alocasia (Schott) G. Don(1839)(保留属名);海芋属(姑婆芋属,观音莲属);Alocasia, Elephant's Ear, Elephant's-ear ■

16485 Alocasia Neck. = Dracunculus L. + Arisaema Mart. ●■

16486 Alocasia Neck. ex Raf.(废弃属名) = Alocasia (Schott) G. Don (保留属名)■

16487 Alocasia Raf. = Arisaema Mart. ●■

16488 Alocasia × amazonica Hort.;非洲海芋;African Mask ■☆

16489 Alocasia × okinawensis Tawada;冲绳海芋■☆

16490 Alocasia alba Schott = Alocasia macrorhiza (L.) G. Don ■

16491 Alocasia argyroneura K. Koch;银脉海芋■☆

16492 Alocasia atropurpurea Engl.;紫色海芋■☆

16493 Alocasia brisbanensis (Bailey) Domin;昆士兰海芋;Cunjevoi ■☆

16494 Alocasia chantrieri J. Rev.;黑叶芋(大齿观音莲);Blackleaf Alocasia ■☆

16495 Alocasia cucullata (Lour.) Schott et Endl.;尖尾芋(卜芥,大虫芋,大附子,大麻芋,大叶姑婆芋,独脚莲,独足莲,番芋,狗神芋,姑婆芋,观音莲,虎耳芋,化骨丹,家海芋,假海芋,尖尾草,尖尾风,尖尾姑婆芋,狼毒,老虎耳,老虎芋,老虎掌芋,山芋,蛇芋,台湾姑婆芋,小虫芋,野山芋,猪不拱,猪管豆);Hoodshaped Alocasia ●■

16496 Alocasia cuprea K. Koch;龟甲芋(龟甲观音莲,铜红观音芋);Giant Caladium, Giant Taro ■☆

16497 Alocasia hainanensis K. Krause = Alocasia hainanica N. E. Br. ■

16498 Alocasia hainanica N. E. Br.;海南芋;Hainan Alocasia ■

16499 Alocasia hiro-beauty Hort.;丽斑观音莲■☆

16500 Alocasia indica (Lour.) Spach = Alocasia macrorhiza (L.) G. Don ■

16501 Alocasia indica (Roxb.) Schott;印度海芋(土芝,印度马来海芋,印度马来芋,芋,芋头,紫叶观音莲);Indomalayan Caladium ■

16502 Alocasia indica (Roxb.) Schott = Alocasia macrorhiza (L.) G. Don ■

16503 Alocasia indica (Roxb.) Schott var. metallica Schott;光泽印度海芋■

16504 Alocasia indica Schott = Alocasia indica (Roxb.) Schott ■

16505 Alocasia johstonii Hort. = Cyrtosperma johstonii N. E. Br. ■☆

16506 Alocasia korthalsii Schott;银脉观音莲;Silvery-nerved Caladium ■☆

16507 Alocasia lindenii Rodigas;林登观音莲■☆

16508 Alocasia longiloba Miq.;箭叶海芋(长苞芋,大叶观音莲,箭叶芋);Arrowleaf Alocasia ■

16509 Alocasia lowii Hook.;娄氏海芋(罗伟芋,罗苇芋);Low's Caladium ■☆

16510 Alocasia lowii Hook. var. grandis Hort.;大娄氏海芋■☆

16511 Alocasia lowii Hook. var. variegata Hort.;斑叶娄氏海芋■☆

16512 Alocasia macrorhiza (L.) G. Don;海芋(大虫芋,大黑附子,大麻芋,大叶野芋头,大重楼,滴水芋,毒芋头,独脚莲,隔河仙,姑婆芋,观音莲,广东狼毒,广东万年青,广狼毒,黑附子,痕芋头,尖尾野芋头,狼毒,狼毒头,老虎蒙,老虎芋,麻芋头,木芋头,朴,朴薯头,朴芋头,埔芋,青芋,茹根,山芋,山芋头,天合芋,天河芋,天荷,天蒙,天芋,土塘,奚芋头,细叶姑婆芋,羞天草,野山芋,野芋,野芋头);Alocasia, Cunjevoi, Elephant's Ear, Giant Alocasia, Giant Elephant's Ear, Giant Taro, Hawaiian Giant Taro, Taro ■

16513 Alocasia macrorhiza (L.) G. Don 'White-splashed';白斑叶大海芋■☆

16514 Alocasia macrorrhiza (L.) Schott et Endl. = Alocasia macrorhiza (L.) G. Don ■

16515 Alocasia macrorrhiza Schott = Alocasia macrorhiza (L.) G. Don ■

16516 Alocasia metallica Schott = Alocasia indica (Roxb.) Schott var. metallica Schott ■

16517 Alocasia micholitziana Sander;狭矢叶海芋;Micholitz Alocasia ■☆

16518 Alocasia odora (Lodd.) Spach;姑婆芋(海芋);Asian Taro ■

16519 Alocasia odora (Lodd.) Spach = Alocasia macrorhiza (L.) G. Don ■

16520 Alocasia odora (Roxb.) K. Koch = Alocasia macrorhiza (L.) G. Don ■

16521 Alocasia odora K. Koch = Alocasia macrorhiza (L.) G. Don ■

16522 Alocasia plumbea Van Houtte;普拉姆海芋;Metallic Taro ■☆

16523 Alocasia putzeysii N. E. Br.;波缘观音莲(波缘芋)■☆

16524 Alocasia rugosa Schott = Alocasia cucullata (Lour.) Schott et Endl. ●■

16525 Alocasia sanderiana Bull.;美叶芋(美叶观音莲);Hris Plant ■☆

16526 Alocasia sanderiana Hort. ex Bull. var. gandaviensis Hort.;朱条美叶芋■☆

16527 Alocasia sanderiana Hort. exBull. var. nobilis Hort.;高贵美叶芋■☆

16528 Alocasia sedenii Veitch;亮叶观音莲■

16529 Alocasia veitchii (Lindl.) Schott;紫背海芋■☆

16530 Alocasia veitchii Schott = Alocasia veitchii (Lindl.) Schott ■☆

16531 Alocasia villeneuvei Lindl. et Ridl.;维氏海芋■☆

16532 Alocasia watsoniana Hort.;大王海芋(大王观音莲);Watson Caladium ■☆

16533 Alocasia wentii Engl. et K. Krause;盾叶观音莲■

16534 Alocasia wenzelii Merr.;温策尔海芋■☆

16535 Alocasia zebrina K. Koch et Veitch;斑马海芋(虎斑观音莲);Zebra Caladium ■☆

16536 Alocasiophyllum Engl. = Cercestis Schott ■☆

16537 Alocasiophyllum kamerunianum Engl. = Cercestis kamerunianus (Engl.) N. E. Br. ■☆

16538 Alococarpum Riedl et Kuber(1964);沟果芹属■☆

16539 Alococarpum erianthum (DC.) Riedl et Kuber;沟果芹■☆

16540 Aloe L. (1753);芦荟属;Aloe, Bitter Aloes ●■

16541 Aloe × spinosissima Jahand.;金刺芦荟;Gold Tooth Aloe ●☆
16542 Aloe abyssinica Hook. f. = Aloe adigratana Reynolds ●☆
16543 Aloe abyssinica Lam.;阿比西尼亚芦荟●☆
16544 Aloe abyssinica Lam. var. peacockii Baker = Aloe elegans Tod. ●☆
16545 Aloe abyssinica Lam. var. percrassa Baker = Aloe percrassa Tod. ●☆
16546 Aloe acinacifolia J. Jacq. = Gasteria acinacifolia (J. Jacq.) Haw. ■☆
16547 Aloe acinacifolia J. Jacq. var. venusta (Haw.) Salm-Dyck = Gasteria acinacifolia (J. Jacq.) Haw. ■☆
16548 Aloe aculeata Pole-Evans;皮刺芦荟●☆
16549 Aloe acuminata Haw. = Aloe humilis (L.) Mill. ●☆
16550 Aloe acutissima H. Perrier;尖芦荟●☆
16551 Aloe adigratana Reynolds;阿地芦荟●☆
16552 Aloe aethiopica (Schweinf.) A. Berger = Aloe elegans Tod. ●☆
16553 Aloe affinis A. Berger;近缘芦荟●☆
16554 Aloe affinis Pole-Evans = Aloe parvibracteata Schönland ●☆
16555 Aloe africana Mill.;非洲芦荟(南非芦荟,树芦荟);Africa Aloe ●☆
16556 Aloe africana Mill. var. angustior Haw. = Aloe africana Mill. ●☆
16557 Aloe africana Mill. var. latifolia Haw. = Aloe africana Mill. ●☆
16558 Aloe agrophila Reynolds = Aloe ecklonis Salm-Dyck ●☆
16559 Aloe albicans Haw. = Haworthia marginata (Lam.) Stearn ■☆
16560 Aloe albida (Stapf) Reynolds;黄白花芦荟●☆
16561 Aloe albiflora Guillaumin;白花芦荟●☆
16562 Aloe albispina Haw. = Aloe perfoliata L. ■☆
16563 Aloe albocincta Haw. = Aloe striata Haw. ●☆
16564 Aloe albovestita S. Carter et Brandham;白被芦荟●☆
16565 Aloe altilinea (Haw.) Schult. et Schult. f. = Haworthia mucronata Haw. ■☆
16566 Aloe amaniensis A. Berger = Aloe lateritia Engl. ●☆
16567 Aloe ambrensis J. -B. Castillon;昂布尔芦荟●☆
16568 Aloe americana Crantz;美洲芦荟●☆
16569 Aloe amicorum L. E. Newton;可爱芦荟●☆
16570 Aloe ammophila Reynolds = Aloe zebrina Baker ●☆
16571 Aloe amoena Pillans = Aloe microstigma Salm-Dyck subsp. framesii (L. Bolus) Glen et D. S. Hardy ●☆
16572 Aloe andohahelensis J. -B. Castillon;安杜哈赫尔芦荟●☆
16573 Aloe andongensis Baker;安东芦荟●☆
16574 Aloe andongensis Baker var. repens L. C. Leach;匍匐芦荟■☆
16575 Aloe andringitrensis H. Perrier;安德林吉特拉山芦荟●☆
16576 Aloe angelica Pole-Evans;安杰莉卡芦荟;Wylliespoort Aloe ●☆
16577 Aloe angiensis De Wild. = Aloe wollastonii Rendle ●☆
16578 Aloe angiensis De Wild. var. kitaliensis Reynolds = Aloe wollastonii Rendle ●☆
16579 Aloe angolensis Baker;安哥拉芦荟■☆
16580 Aloe angulata Willd. = Gasteria carinata (Mill.) Duval ■☆
16581 Aloe angusta Schult. f.;狭芦荟■☆
16582 Aloe angustifolia (Aiton) Salm-Dyck = Gasteria disticha (L.) Haw. ■☆
16583 Aloe angustifolia Groenew. = Aloe africana Mill. ●☆
16584 Aloe angustifolia Groenew. = Aloe zebrina Baker ●☆
16585 Aloe angustifolia Haw. = Aloe africana Mill. ●☆
16586 Aloe antandroi (R. Decary) H. Perrier;安族芦荟●☆
16587 Aloe antonii J. -B. Castillon;安氏芦荟●☆
16588 Aloe arabica Salm-Dyck = Aloe pendens Forssk. ●☆
16589 Aloe arachnoides Thunb. = Haworthia arachnoidea (L.) Duval ■☆
16590 Aloe arachnoides Thunb. var. pellucens Salm-Dyck = Haworthia herbacea (Mill.) Stearn ■☆
16591 Aloe arborea Medik. = Aloe arborescens Mill. ●☆
16592 Aloe arborescens Mill.;单杆芦荟(鹿角芦荟,木本芦荟,木芦荟,乔木状芦荟,树芦荟,小木芦荟);Arborescent Aloe,Candalabra Aloe,Candelabra Plant,Krantz Aloe,Simple-trnk Aloe,Tree Aloe,Tree-like Aloe,Woody Aloe ●☆
16593 Aloe arborescens Mill. 'Variegata';花叶小木芦荟●☆
16594 Aloe arborescens Mill. subsp. mzimnyati Van Jaarsv. et A. E. van Wyk;姆津尼亚特芦荟●☆
16595 Aloe arborescens Mill. var. frutescens (Salm-Dyck) Link = Aloe arborescens Mill. ●☆
16596 Aloe arborescens Mill. var. milleri A. Berger = Aloe arborescens Mill. ●☆
16597 Aloe arborescens Mill. var. natalensis (J. M. Wood et M. S. Evans) A. Berger = Aloe arborescens Mill. ●☆
16598 Aloe arborescens Mill. var. natalensis Berger;南非芦荟(大芦荟);South African Aloe ●☆
16599 Aloe arborescens Mill. var. pachythyrsa A. Berger = Aloe arborescens Mill. ●☆
16600 Aloe arborescens Mill. var. viridifolia Berg;翠绿芦荟●☆
16601 Aloe archeri Lavranos;阿谢尔芦荟●☆
16602 Aloe arenicola Reynolds;沙生芦荟■☆
16603 Aloe argenticauda Merxm. et Giess;银尾芦荟●☆
16604 Aloe argyrostachys Lavranos,Rakouth et T. A. McCoy;银穗芦荟●☆
16605 Aloe aristata Haw.;点纹芦荟(长须芦荟,绫锦,木锉掌,芝麻掌);Aloe,Aristate Aloe,Lace Aloe,Torch Plant ●☆
16606 Aloe aristata Haw. var. leiophylla Baker = Aloe aristata Haw. ●☆
16607 Aloe aristata Haw. var. parviflora Baker = Aloe aristata Haw. ●☆
16608 Aloe aspera Haw. = Astroloba corrugata N. L. Mey. et G. F. Sm. ■☆
16609 Aloe asperifolia A. Berger;糙叶芦荟●☆
16610 Aloe asperiuscula (Haw.) Schult. et Schult. f. = Haworthia viscosa (L.) Haw. ■☆
16611 Aloe atherstonei Baker = Aloe pluridens Haw. ●☆
16612 Aloe atrovirens DC. = Haworthia herbacea (Mill.) Stearn ■☆
16613 Aloe attenuata Haw. = Haworthia attenuata (Haw.) Haw. ■☆
16614 Aloe aurantiaca Baker = Aloe striatula Haw. ●☆
16615 Aloe ausana Dinter = Aloe variegata L. ●
16616 Aloe babatiensis Christian et I. Verd.;巴巴蒂芦荟●☆
16617 Aloe bainesii Dyer;大树芦荟;Tree Aloe ●☆
16618 Aloe bainesii Dyer = Aloe arborescens Mill. ●☆
16619 Aloe bainesii Dyer = Aloe barberae Dyer ●☆
16620 Aloe bainesii Dyer var. barberae (Dyer) Baker = Aloe barberae Dyer ●☆
16621 Aloe bakeri Scott-Elliot;巴开尔芦荟■☆
16622 Aloe ballii Reynolds;鲍尔芦荟●☆
16623 Aloe bamangwatensis Schönland = Aloe zebrina Baker ●☆
16624 Aloe barbadensis Mill.;翠叶芦荟(库拉索芦荟)●☆
16625 Aloe barbadensis Mill. = Aloe vera (L.) Burm. f. ■
16626 Aloe barbadensis Mill. var. chinensis Haw. = Aloe vera (L.) Burm. f. ■
16627 Aloe barberae Dyer;树芦荟;Tree Aloe ●☆
16628 Aloe barbertoniae Pole-Evans = Aloe greatheadii Schönland var. davyana (Schönland) Glen et D. S. Hardy ●☆
16629 Aloe barteri Baker;巴特芦荟●☆
16630 Aloe barteri Baker = Aloe buettneri A. Berger ●☆
16631 Aloe barteri Baker var. dahomensis A. Chev. = Aloe buettneri A. Berger ●☆

16632 Aloe barteri Baker var. lutea A. Chev. = Aloe schweinfurthii Baker ●☆
16633 Aloe barteri Baker var. sudanica A. Chev. = Aloe buettneri A. Berger ●☆
16634 Aloe baumii Engl. et Gilg = Aloe zebrina Baker ●☆
16635 Aloe bella Rowley;雅致芦荟●☆
16636 Aloe bellatula Reynolds;美芦荟■☆
16637 Aloe beniensis De Wild. = Aloe dawei A. Berger ●☆
16638 Aloe bequaertii De Wild. = Aloe wollastonii Rendle ●☆
16639 Aloe betsileensis H. Perrier;贝齐尔芦荟●☆
16640 Aloe bicolor (Haw.) Schult. et Schult. f. = Gasteria bicolor Haw. ■☆
16641 Aloe boastii Letty = Aloe chortolirioides A. Berger ●☆
16642 Aloe boehmii Engl. = Aloe lateritia Engl. ●☆
16643 Aloe boiteaui Guillaumin;博特芦荟●☆
16644 Aloe bolusii Baker = Aloe africana Mill. ●☆
16645 Aloe boranensis Cufod. = Aloe otallensis Baker ●☆
16646 Aloe borziana A. Terracc. = Aloe macrocarpa Tod. ●☆
16647 Aloe bosseri J. -B. Castillon;博瑟芦荟■☆
16648 Aloe boureana Schult. et Schult. f. = Gasteria bicolor Haw. ■☆
16649 Aloe bowiea Schult. et Schult. f. ;博维芦荟■☆
16650 Aloe bowieana Salm-Dyck = Gasteria bicolor Haw. ■☆
16651 Aloe boylei Baker = Aloe ecklonis Salm-Dyck ●☆
16652 Aloe boylei Baker subsp. major Hilliard et B. L. Burtt = Aloe ecklonis Salm-Dyck ●☆
16653 Aloe brachyphylla Salm-Dyck = Gasteria brachyphylla (Salm-Dyck) Van Jaarsv. ■☆
16654 Aloe brachystachys Baker;短穗芦荟●☆
16655 Aloe bradlyana Jacq. = Haworthia herbacea (Mill.) Stearn ■☆
16656 Aloe brevifolia Mill. ;短叶芦荟;Kleinaalwyn, Short Leaf Aloe, Shortleaf Aloe ●☆
16657 Aloe brevifolia Mill. var. depressa (Haw.) Baker;不死鸟■☆
16658 Aloe brevifolia Mill. var. postgenita (Schult. et Schult. f.) Baker = Aloe brevifolia Mill. ●☆
16659 Aloe brevifolia Mill. var. serra (DC.) A. Berger = Aloe brevifolia Mill. var. depressa (Haw.) Baker ■☆
16660 Aloe breviscapa Reynolds et P. R. O. Bally;短花茎芦荟●☆
16661 Aloe broomii Schönland;布鲁姆芦荟●☆
16662 Aloe brunneo-punctata Engl. et Gilg = Aloe nuttii Baker ●☆
16663 Aloe brunneostriata Lavranos et S. Carter;褐纹芦荟●☆
16664 Aloe brunnthaleri A. Berger ex Cammerloher = Aloe microstigma Salm-Dyck ●☆
16665 Aloe buchananii Baker;布坎南芦荟●☆
16666 Aloe buchlohii Rauh;布赫芦荟●☆
16667 Aloe buettneri A. Berger;比特纳芦荟●☆
16668 Aloe buhrii Lavranos;布尔芦荟●☆
16669 Aloe bukobana Reynolds;布科巴芦荟●☆
16670 Aloe bulbicaulis Christian;球茎芦荟●☆
16671 Aloe bulbillifera H. Perrier;球根芦荟●☆
16672 Aloe bullockii Reynolds;布洛克芦荟●☆
16673 Aloe bullulata Jacq. = Astroloba bullulata (Jacq.) Uitewaal ■☆
16674 Aloe burgersfortensis Reynolds = Aloe parvibracteata Schönland ●☆
16675 Aloe bussei A. Berger;布瑟芦荟●☆
16676 Aloe calcairophila Reynolds;喜钙芦荟●☆
16677 Aloe calcairophylla Reynolds = Aloe calcairophila Reynolds ●☆
16678 Aloe calidophila Reynolds;喜岩芦荟●☆
16679 Aloe cameronii Hemsl. ;卡梅伦芦荟●☆
16680 Aloe cameronii Hemsl. var. bondana Reynolds;邦德芦荟●☆
16681 Aloe camperi Schweinf. ;羊角掌;Sheephorn Aloe ●☆
16682 Aloe campylosiphon A. Berger = Aloe lateritia Engl. ●☆
16683 Aloe canarina S. Carter;加那利芦荟●☆
16684 Aloe candelabrum A. Berger = Aloe ferox Mill. ●☆
16685 Aloe candidans (Haw.) Schult. et Schult. f. = Gasteria acinacifolia (J. Jacq.) Haw. ■☆
16686 Aloe cannellii L. C. Leach;坎内尔芦荟●☆
16687 Aloe capitata Baker;头状芦荟;Aloe ●☆
16688 Aloe caricina A. Berger = Aloe myriacantha (Haw.) Schult. et Schult. f. ●☆
16689 Aloe carinata DC. = Gasteria carinata (Mill.) Duval var. verrucosa (Mill.) Van Jaarsv. ■☆
16690 Aloe carinata Mill. = Gasteria carinata (Mill.) Duval ■☆
16691 Aloe carinata Mill. var. laevior Salm-Dyck = Gasteria carinata (Mill.) Duval ■☆
16692 Aloe carinata Mill. var. subglabra Haw. = Gasteria carinata (Mill.) Duval ■☆
16693 Aloe carnea S. Carter;肉色芦荟●☆
16694 Aloe carowii Reynolds = Aloe sladeniana Pole-Evans ●☆
16695 Aloe cascadensis Kuntze = Aloe striatula Haw. ●☆
16696 Aloe castanea Schönland;栗色芦荟;Cat's Tail Aloe ●☆
16697 Aloe cernua Tod. = Aloe capitata Baker ●☆
16698 Aloe chabaudii Schönland;沙博芦荟●☆
16699 Aloe chabaudii Schönland var. mlanjeana Christian;姆兰杰芦荟●☆
16700 Aloe chabaudii Schönland var. verekeri Christian = Aloe chabaudii Schönland ●☆
16701 Aloe cheranganiensis S. Carter et Brandham;切兰加尼芦荟●☆
16702 Aloe chimanimaniensis Christian = Aloe swynnertonii Rendle ●☆
16703 Aloe chinensis (How.) Baker = Aloe vera (L.) Burm. f. ■
16704 Aloe chloracantha (Haw.) Schult. et Schult. f. = Haworthia chloracantha Haw. ■☆
16705 Aloe chloroleuca Baker;绿白芦荟●☆
16706 Aloe chortolirioides A. Berger;肖十二卷芦荟●☆
16707 Aloe chortolirioides A. Berger var. boastii (Letty) Reynolds = Aloe chortolirioides A. Berger ●☆
16708 Aloe chrysostachys Lavranos et L. E. Newton;金穗芦荟●☆
16709 Aloe ciliaris Haw. ;缘毛芦荟(纤毛叶芦荟);Ciliate Aloe, Climbing Aloe ■
16710 Aloe ciliaris Haw. var. flanaganii Schönland = Aloe ciliaris Haw. ■
16711 Aloe cinnabarina Diels ex A. Berger;朱红芦荟●☆
16712 Aloe citrina S. Carter et Brandham;柠檬芦荟●☆
16713 Aloe clarkei L. E. Newton;克拉克芦荟●☆
16714 Aloe classenii Reynolds;克拉森芦荟●☆
16715 Aloe claviflora Burch. ;棒花芦荟;Kraalaalwyn ■☆
16716 Aloe coarctata (Haw.) Schult. et Schult. f. = Haworthia coarctata Haw. ■☆
16717 Aloe collina S. Carter;山丘芦荟●☆
16718 Aloe commelinii Willd. = Aloe perfoliata L. ■☆
16719 Aloe commixta A. Berger;混合芦荟●☆
16720 Aloe comosa Marloth et A. Berger;发芦荟●☆
16721 Aloe comosibracteata Reynolds = Aloe greatheadii Schönland var. davyana (Schönland) Glen et D. S. Hardy ●☆
16722 Aloe compacta Reynolds = Aloe macrosiphon Baker ●■☆
16723 Aloe compressa H. Perrier;扁芦荟●☆
16724 Aloe comptonii Reynolds = Aloe mitriformis Mill. subsp. comptonii (Reynolds) Zonn. ●☆

16725 Aloe concinna (Haw.) Schult. et Schult. f. = Haworthia viscosa (L.) Haw. ■☆
16726 Aloe concinna Baker = Aloe squarrosa Baker ●☆
16727 Aloe confusa Engl.;混乱芦荟●☆
16728 Aloe congesta Salm-Dyck = Astroloba congesta (Salm-Dyck) Uitewaal ■☆
16729 Aloe congolensis De Wild. et T. Durand;刚果芦荟●☆
16730 Aloe conifera H. Perrier;球果芦荟●☆
16731 Aloe conspurcata Salm-Dyck = Gasteria disticha (L.) Haw. ■☆
16732 Aloe constricta Baker = Aloe zebrina Baker ●☆
16733 Aloe contigua (H. Perrier) Reynolds = Aloe imalotensis Reynolds ●☆
16734 Aloe cooperi Baker;库珀芦荟●☆
16735 Aloe cooperi Baker subsp. pulchra Glen et D. S. Hardy;美丽库珀芦荟●☆
16736 Aloe corallina I. Verd.;珊瑚状芦荟●☆
16737 Aloe corbisieri De Wild. = Aloe nuttii Baker ●☆
16738 Aloe cordifolia (Haw.) Schult. et Schult. f. = Haworthia viscosa (L.) Haw. ■☆
16739 Aloe crassifolia (Aiton) Schult. et Schult. f. = Gasteria disticha (L.) Haw. ■☆
16740 Aloe crassipes Baker;粗梗芦荟●☆
16741 Aloe cremnophila Reynolds;悬崖芦荟●☆
16742 Aloe croucheri Hook. f. = Gasteria croucheri (Hook. f.) Baker ■☆
16743 Aloe cryptoflora Reynolds;隐花芦荟●☆
16744 Aloe cryptopoda Baker;隐足芦荟●☆
16745 Aloe cuspidata (Haw.) Schult. et Schult. f. = Haworthia cuspidata Haw. ■☆
16746 Aloe cymbifolia Schrad. = Haworthia cymbiformis (Haw.) Duval ■☆
16747 Aloe cymbiformis Haw. = Haworthia cymbiformis (Haw.) Duval ■☆
16748 Aloe cyrillei J. -B. Castillon;西里尔芦荟●☆
16749 Aloe davyana Schönland = Aloe greatheadii Schönland var. davyana (Schönland) Glen et D. S. Hardy ●☆
16750 Aloe davyana Schönland var. subolifera Groenew. = Aloe greatheadii Schönland var. davyana (Schönland) Glen et D. S. Hardy ●☆
16751 Aloe dawei A. Berger;道氏芦荟●☆
16752 Aloe decipiens (Haw.) Schult. et Schult. f. = Gasteria nitida (Salm-Dyck) Haw. ■☆
16753 Aloe decora Schönland = Aloe claviflora Burch. ■☆
16754 Aloe decurva Reynolds;下延芦荟●☆
16755 Aloe decurvidens Groenew. = Aloe komatiensis Reynolds ●☆
16756 Aloe delaetii Radl;海虎兰(杂种芦荟);Delaet Aloe, Hybrid Aloe ●☆
16757 Aloe deltoidea Hook. f. = Astroloba congesta (Salm-Dyck) Uitewaal ■☆
16758 Aloe deltoideodonta Baker;角齿芦荟■☆
16759 Aloe deltoideodonta Baker var. contigua H. Perrier = Aloe imalotensis Reynolds ●☆
16760 Aloe deltoideodonta Baker var. intermedia H. Perrier = Aloe intermedia (H. Perrier) Reynolds ●☆
16761 Aloe denticulata (Haw.) Schult. et Schult. f. = Haworthia aristata Haw. ■☆
16762 Aloe dependens Steud. = Aloe pendens Forssk. ●☆
16763 Aloe depressa Haw. = Aloe brevifolia Mill. var. depressa (Haw.) Baker ■☆
16764 Aloe descoingsii Reynolds;德斯芦荟■☆
16765 Aloe deserti A. Berger;荒漠芦荟■☆
16766 Aloe dewinteri Giess;温特芦荟●☆
16767 Aloe dichotoma L. f. = Aloe dichotoma Masson ●☆
16768 Aloe dichotoma Masson;二歧芦荟(分枝芦荟);Kokerbom, Quiver Tree, Quiver-tree, Quiver-tree Aloe ●☆
16769 Aloe dichotoma Masson subsp. pillansii (L. Guthrie) Zonn.;皮朗斯芦荟●☆
16770 Aloe dichotoma Masson subsp. ramosissima (Pillans) Zonn.;多分枝芦荟●☆
16771 Aloe dichotoma Masson var. montana (Schinz) A. Berger = Aloe dichotoma Masson ●☆
16772 Aloe dichotoma Masson var. ramosissima (Pillans) Glen et D. S. Hardy = Aloe dichotoma Masson subsp. ramosissima (Pillans) Zonn. ●☆
16773 Aloe dictyodes Schult. et Schult. f. = Gasteria bicolor Haw. ■☆
16774 Aloe dinteri A. Berger;丁特芦荟;Namibian Partridge Breast Aloe ■☆
16775 Aloe dispar A. Berger;异型芦荟●☆
16776 Aloe distans Haw. = Aloe mitriformis Mill. subsp. distans (Haw.) Zonn. ●☆
16777 Aloe disticha L. = Gasteria carinata (Mill.) Duval ■☆
16778 Aloe disticha L. = Gasteria disticha (L.) Haw. ■☆
16779 Aloe disticha L. var. plicatilis L. = Aloe plicatilis (L.) Mill. ●☆
16780 Aloe disticha Mill. = Aloe maculata All. ●☆
16781 Aloe disticha Thunb. = Gasteria carinata (Mill.) Duval var. thunbergii (N. E. Br.) Van Jaarsv. ■☆
16782 Aloe divaricata A. Berger;叉开芦荟●☆
16783 Aloe dorotheae A. Berger;多罗特娅芦荟●☆
16784 Aloe drepanophylla Baker;镰叶芦荟●☆
16785 Aloe dumetorum B. Mathew et Brandham = Aloe ellenbeckii A. Berger ●☆
16786 Aloe dyeri Schönland;戴尔芦荟●☆
16787 Aloe echinata Willd. = Aloe humilis (L.) Mill. ●☆
16788 Aloe echinata Willd. var. minor Salm-Dyck = Aloe humilis (L.) Mill. ●☆
16789 Aloe ecklonis Salm-Dyck;埃氏芦荟●☆
16790 Aloe edulis A. Chev. = Aloe macrocarpa Tod. ●☆
16791 Aloe elata S. Carter et L. E. Newton;高芦荟●☆
16792 Aloe elegans Tod.;雅丽芦荟;Barbados Aloe, Medicinal Aloe ●☆
16793 Aloe elgonica Bullock;埃尔贡芦荟●☆
16794 Aloe elizae A. Berger;伊莱扎芦荟●☆
16795 Aloe ellenbeckii A. Berger;灌丛芦荟●☆
16796 Aloe ellenbergii Guillaumin = Aloe aristata Haw. ●☆
16797 Aloe elongata Salm-Dyck = Aloe vera (L.) Burm. f. ■
16798 Aloe elongata Salm-Dyck = Gasteria trigona Haw. ■☆
16799 Aloe engleri A. Berger = Aloe secundiflora Engl. ●☆
16800 Aloe enotata L. C. Leach = Aloe veseyi Reynolds ●☆
16801 Aloe ensifolia (Haw.) Schult. et Schult. f. = Gasteria acinacifolia (J. Jacq.) Haw. ■☆
16802 Aloe erecta (Haw.) Salm-Dyck var. laetevirens Salm-Dyck = Aloe minima Baker ●☆
16803 Aloe erecta (Haw.) Schult. f. = Aloe minima Baker ●☆
16804 Aloe erinacea Hardy;刺芦荟●☆
16805 Aloe eru A. Berger = Aloe camperi Schweinf. ●☆
16806 Aloe eru A. Berger var. hookeri ? = Aloe adigratana Reynolds ●☆

16807 Aloe esculenta L. C. Leach;食用芦荟●☆
16808 Aloe esculenta L. C. Leach = Aloe angolensis Baker ■☆
16809 Aloe excavata Willd. = Gasteria carinata (Mill.) Duval ■☆
16810 Aloe excelsa A. Berger;高耸芦荟(津巴布韦芦荟);Noble Aloe, Zimbabwe Aloe ●☆
16811 Aloe excelsa A. Berger var. breviflora L. C. Leach;短叶高耸芦荟●☆
16812 Aloe eylesii Christian = Aloe rhodesiana Rendle ●☆
16813 Aloe falcata Baker;镰芦荟●☆
16814 Aloe fasciata (Willd.) Haw. = Haworthia fasciata (Willd.) Haw. ■☆
16815 Aloe fasciata (Willd.) Haw. var. major Salm-Dyck = Haworthia fasciata (Willd.) Haw. ■☆
16816 Aloe fasciata (Willd.) Haw. var. minor Salm-Dyck = Haworthia fasciata (Willd.) Haw. ■☆
16817 Aloe ferox Mill.;好望角芦荟(多产芦荟,红刺芦荟,罗帏草,青鳄,青鳄芦荟,油葱叶);Bitter Aloe, Cape Aloe, Robust Aloe, Tap Aloe ●☆
16818 Aloe ferox Mill. var. erythrocarpa ? = Aloe ferox Mill. ●☆
16819 Aloe ferox Mill. var. galpinii (Baker) Reynolds = Aloe ferox Mill. ●☆
16820 Aloe ferox Mill. var. hanburyi ? = Aloe ferox Mill. ●☆
16821 Aloe ferox Mill. var. incurva Baker = Aloe ferox Mill. ●☆
16822 Aloe ferox Mill. var. subferox ? = Aloe ferox Mill. ●☆
16823 Aloe ferox Mill. var. xanthostachys A. Berger = Aloe marlothii A. Berger ●☆
16824 Aloe fibrosa Lavranos et L. E. Newton;纤维质芦荟●☆
16825 Aloe fimbrialis S. Carter;流苏芦荟●☆
16826 Aloe flabelliformis Salisb. = Aloe plicatilis (L.) Mill. ●☆
16827 Aloe flava Pers. = Aloe vera (L.) Burm. f. ■
16828 Aloe flavispina Haw. = Aloe perfoliata L. ■☆
16829 Aloe flexilifolia Christian;弯叶芦荟●☆
16830 Aloe floramaculata Christian = Aloe secundiflora Engl. ●☆
16831 Aloe foliolosa Haw. = Astroloba foliolosa (Haw.) Uitewaal ■☆
16832 Aloe formosa Schult. et Schult. f. = Gasteria brachyphylla (Salm-Dyck) Van Jaarsv. ■☆
16833 Aloe fosteri Pillans;福斯特芦荟●☆
16834 Aloe fouriei D. S. Hardy et Glen;富里耶芦荟●☆
16835 Aloe framesii L. Bolus = Aloe microstigma Salm-Dyck subsp. framesii (L. Bolus) Glen et D. S. Hardy ●☆
16836 Aloe friisii Sebsebe et M. G. Gilbert;弗里斯芦荟●☆
16837 Aloe frutescens Salm-Dyck = Aloe arborescens Mill. ●☆
16838 Aloe fruticosa Lam. = Aloe arborescens Mill. ●☆
16839 Aloe galpinii Baker = Aloe ferox Mill. ●☆
16840 Aloe gerstneri Reynolds;格斯芦荟●☆
16841 Aloe gilbertii Reynolds ex Sebsebe et Brandham;吉尔伯特芦荟●☆
16842 Aloe gilbertii Reynolds ex Sebsebe et Brandham subsp. megalacanthoides M. G. Gilbert et Sebsebe;大刺吉尔伯特芦荟●☆
16843 Aloe gillettii S. Carter;吉莱特芦荟●☆
16844 Aloe glabra Salm-Dyck = Gasteria carinata (Mill.) Duval var. glabra (Salm-Dyck) Van Jaarsv. ■☆
16845 Aloe glabrata Salm-Dyck = Haworthia glabrata (Salm-Dyck) Baker ■☆
16846 Aloe glabrescens (Reynolds et P. R. O. Bally) S. Carter et Brandham;渐光芦荟●☆
16847 Aloe glauca Mill.;灰绿芦荟;Aloe, Blue Aloe ●☆
16848 Aloe glauca Mill. var. elatior Salm-Dyck = Aloe glauca Mill. ●☆
16849 Aloe glauca Mill. var. humilior Salm-Dyck = Aloe glauca Mill. ●☆
16850 Aloe glauca Mill. var. major Haw. = Aloe glauca Mill. ●☆
16851 Aloe glauca Mill. var. minor Haw. = Aloe glauca Mill. ●☆
16852 Aloe glauca Mill. var. muricata (Schult.) Baker = Aloe glauca Mill. ●☆
16853 Aloe glauca Mill. var. spinosior Haw. = Aloe glauca Mill. ●☆
16854 Aloe globuligemma Pole-Evans;小球芦荟●☆
16855 Aloe gloveri Reynolds et P. R. O. Bally = Aloe hildebrandtii Baker ●☆
16856 Aloe gossweileri Reynolds;戈斯芦荟●☆
16857 Aloe gracilicaulis Reynolds et P. R. O. Bally;细茎芦荟■☆
16858 Aloe graciliflora Groenew. = Aloe greatheadii Schönland var. davyana (Schönland) Glen et D. S. Hardy ●☆
16859 Aloe gracilis Haw.;纤细芦荟■☆
16860 Aloe gracilis Haw. var. decumbens Reynolds = Aloe gracilis Haw. ■☆
16861 Aloe graminicola Reynolds = Aloe lateritia Engl. var. graminicola (Reynolds) S. Carter ●☆
16862 Aloe graminifolia A. Berger = Aloe myriacantha (Haw.) Schult. et Schult. f. ●☆
16863 Aloe granata (Willd.) Haw. = Aloe minima Baker ●☆
16864 Aloe granata (Willd.) Salm-Dyck var. major Salm-Dyck = Aloe minima Baker ●☆
16865 Aloe granata (Willd.) Salm-Dyck var. minor Salm-Dyck = Aloe minima Baker ●☆
16866 Aloe grandidentata Salm-Dyck;大齿芦荟;Bontaalwyn ●☆
16867 Aloe grata Reynolds;悦人芦荟●☆
16868 Aloe greatheadii Schönland;格里芦荟●☆
16869 Aloe greatheadii Schönland var. davyana (Schönland) Glen et D. S. Hardy;粉花格里芦荟●☆
16870 Aloe greenii Baker;格林芦荟●☆
16871 Aloe greenwayi Reynolds = Aloe leptosiphon A. Berger ●☆
16872 Aloe grisea S. Carter et Brandham;灰芦荟●☆
16873 Aloe guineensis (L.) Jacq. = Sansevieria hyacinthoides (L.) Druce ■☆
16874 Aloe haemanthifolia A. Berger et Marloth;血叶芦荟●☆
16875 Aloe hanburiana Naudin = Aloe striata Haw. ●☆
16876 Aloe hardyi Glen;哈迪芦荟●☆
16877 Aloe harlana Reynolds;哈尔芦荟●☆
16878 Aloe haworthioides Baker;琉璃姬孔雀;Kaworthialike Aloe ●☆
16879 Aloe hazeliana Reynolds var. howmanii (Reynolds) S. Carter;豪曼芦荟●☆
16880 Aloe hebes Schult. et Schult. f. = Haworthia cymbiformis (Haw.) Duval ■☆
16881 Aloe helenae Danguy;海伦娜芦荟●☆
16882 Aloe hendrickxii Reynolds;亨德里克斯芦荟;Enkokorutanga ●☆
16883 Aloe herbacea DC. = Haworthia reticulata (Haw.) Haw. ■☆
16884 Aloe herbacea Mill. = Haworthia herbacea (Mill.) Stearn ■☆
16885 Aloe hereroensis Engl.;赫雷罗芦荟●☆
16886 Aloe hereroensis Engl. var. lutea A. Berger;黄赫雷罗芦荟●☆
16887 Aloe hereroensis Engl. var. orpeniae (Schönland) A. Berger = Aloe hereroensis Engl. ●☆
16888 Aloe hildebrandtii Baker;希尔德芦荟●☆
16889 Aloe hlangapies Groenew. = Aloe ecklonis Salm-Dyck ●☆
16890 Aloe horrida Haw. = Aloe ferox Mill. ●☆
16891 Aloe howmanii Reynolds = Aloe hazeliana Reynolds var. howmanii (Reynolds) S. Carter ●☆

16892 Aloe humbertii H. Perrier;亨伯特芦荟●☆
16893 Aloe humilis (L.) Haw. var. echinata (Willd.) Baker;芒刺芦荟●☆
16894 Aloe humilis (L.) Mill.;矮小芦荟(矮芦荟,帝王锦,章鱼芦荟);Dwarf Aloe,Spider Aloe ●☆
16895 Aloe humilis (L.) Mill. var. acuminata (Haw.) Baker = Aloe humilis (L.) Mill. ●☆
16896 Aloe humilis (L.) Mill. var. candollei Baker = Aloe humilis (L.) Mill. ●☆
16897 Aloe humilis (L.) Mill. var. echinata (Willd.) Baker = Aloe humilis (L.) Mill. ●☆
16898 Aloe humilis (L.) Mill. var. incurva Haw. = Aloe humilis (L.) Mill. ●☆
16899 Aloe humilis (L.) Mill. var. suberecta (Aiton) Baker = Aloe humilis (L.) Mill. ●☆
16900 Aloe humilis (L.) Mill. var. subtuberculata (Haw.) Baker = Aloe humilis (L.) Mill. ●☆
16901 Aloe hyacinthoides L. = Sansevieria hyacinthoides (L.) Druce ■☆
16902 Aloe hyacinthoides L. var. guineensis L. = Sansevieria hyacinthoides (L.) Druce ■☆
16903 Aloe hybrida Salm-Dyck = Haworthia hybrida (Salm-Dyck) Haw. ■☆
16904 Aloe ibityensis H. Perrier;伊比提芦荟●☆
16905 Aloe imalotensis Reynolds;伊马芦荟●☆
16906 Aloe imbricata Haw. = Astroloba spiralis (L.) Uitewaal ■☆
16907 Aloe immaculata Pillans = Aloe affinis A. Berger ●☆
16908 Aloe inconspicua Plowes;显著芦荟●☆
16909 Aloe incurva (Haw.) Haw. = Aloe humilis (L.) Mill. ●☆
16910 Aloe indica Royle;印度芦荟●☆
16911 Aloe indica Royle = Aloe vera (L.) Burm. f. ■
16912 Aloe indurata Schult. et Schult. f. = Haworthia viscosa (L.) Haw. ■☆
16913 Aloe inermis Forssk.;无刺芦荟●☆
16914 Aloe integra Reynolds;全缘芦荟●☆
16915 Aloe intermedia (H. Perrier) Reynolds;间型芦荟●☆
16916 Aloe intermedia Haw. = Gasteria carinata (Mill.) Duval var. verrucosa (Mill.) Van Jaarsv. ■☆
16917 Aloe intermedia Haw. var. asperrima Salm-Dyck = Gasteria carinata (Mill.) Duval var. verrucosa (Mill.) Van Jaarsv. ■☆
16918 Aloe inyangensis Christian;伊尼扬加芦荟●☆
16919 Aloe inyangensis Christian var. kimberleyana S. Carter;金伯利芦荟●☆
16920 Aloe isaloensis H. Perrier;伊萨卢芦荟●☆
16921 Aloe itremensis Reynolds;伊特雷穆芦荟●☆
16922 Aloe jacksonii Reynolds;杰克逊芦荟●☆
16923 Aloe jex-blakeae Christian = Aloe ruspoliana Baker ●☆
16924 Aloe johnstonii Baker = Aloe myriacantha (Haw.) Schult. et Schult. f. ●☆
16925 Aloe jucunda Reynolds;悦芦荟■☆
16926 Aloe juttae Dinter = Aloe microstigma Salm-Dyck ●☆
16927 Aloe kaokoensis Van Jaarsv. et al.;卡奥科芦荟●☆
16928 Aloe karasbergensis Pillans = Aloe striata Haw. subsp. karasbergensis (Pillans) Glen et D. S. Hardy ●☆
16929 Aloe keayi Reynolds;凯伊芦荟●☆
16930 Aloe kedongensis Reynolds;克东芦荟●☆
16931 Aloe kefaensis M. G. Gilbert et Sebsebe;咖法芦荟●☆
16932 Aloe keithii Reynolds = Aloe parvibracteata Schönland ●☆
16933 Aloe khamiesensis Pillans = Aloe microstigma Salm-Dyck ●☆
16934 Aloe kilifiensis Christian;基利菲芦荟●☆
16935 Aloe kirkii Baker = Aloe leptosiphon A. Berger ●☆
16936 Aloe kniphofioides Baker;火炬花芦荟●☆
16937 Aloe komaggasensis Kritzinger et Van Jaarsv. = Aloe striata Haw. subsp. komaggasensis (Kritzinger et Van Jaarsv.) Glen et D. S. Hardy ●☆
16938 Aloe komatiensis Reynolds;科马蒂芦荟●☆
16939 Aloe kraussii Baker = Aloe ecklonis Salm-Dyck ●☆
16940 Aloe kraussii Baker var. minor = Aloe albida (Stapf) Reynolds ●☆
16941 Aloe kulalensis L. E. Newton et Beentje;库拉尔芦荟●☆
16942 Aloe labiaflava Groenew. = Aloe greatheadii Schönland var. davyana (Schönland) Glen et D. S. Hardy ●☆
16943 Aloe laeta A. Berger;愉悦芦荟●☆
16944 Aloe laetepuncta (Haw.) Schult. et Schult. f. = Gasteria carinata (Mill.) Duval ■☆
16945 Aloe laetevirens (Haw.) Link = Haworthia turgida Haw. ■☆
16946 Aloe laevigata Schult. et Schult. f. = Haworthia marginata (Lam.) Stearn ■☆
16947 Aloe laevis Salm-Dyck;平滑芦荟●☆
16948 Aloe lanuriensis De Wild. = Aloe wollastonii Rendle ●☆
16949 Aloe lanzae Tod. = Aloe vera (L.) Burm. f. ■
16950 Aloe lastii Baker = Aloe brachystachys Baker ●☆
16951 Aloe lateritia Engl.;砖红芦荟●☆
16952 Aloe lateritia Engl. var. graminicola (Reynolds) S. Carter;草莺芦荟●☆
16953 Aloe lateritia Engl. var. kitaliensis (Reynolds) Reynolds = Aloe wollastonii Rendle ●☆
16954 Aloe latifolia (Haw.) Haw. = Aloe maculata All. ●☆
16955 Aloe laxiflora N. E. Br. = Aloe gracilis Haw. ■☆
16956 Aloe laxissima Reynolds = Aloe zebrina Baker ●☆
16957 Aloe leandrii Bosser;利安芦荟●☆
16958 Aloe leedalii S. Carter;利达尔芦荟●☆
16959 Aloe lepida L. C. Leach;鳞片芦荟●☆
16960 Aloe leptocaulon Bojer = Aloe antandroi (R. Decary) H. Perrier ●☆
16961 Aloe leptophylla N. E. Br. ex Baker = Aloe maculata All. ●☆
16962 Aloe leptophylla N. E. Br. ex Baker var. stenophylla Baker = Aloe maculata All. ●☆
16963 Aloe leptosiphon A. Berger;细管芦荟●☆
16964 Aloe lettyae Reynolds = Aloe zebrina Baker ●☆
16965 Aloe leucantha A. Berger;西白花芦荟●☆
16966 Aloe limpida (Haw.) Schult. f. = Haworthia mucronata Haw. ■☆
16967 Aloe lindenii Lavranos;林登芦荟●☆
16968 Aloe linearifolia A. Berger;线叶芦荟●☆
16969 Aloe lineata (Aiton) Haw.;条纹芦荟●☆
16970 Aloe lineata (Aiton) Haw. var. glaucescens Haw. = Aloe lineata (Aiton) Haw. ●☆
16971 Aloe lineata (Aiton) Haw. var. muirii (Marloth) Reynolds;缪里芦荟●☆
16972 Aloe lineata (Aiton) Haw. var. viridis Haw. = Aloe lineata (Aiton) Haw. ●☆
16973 Aloe lingua Ker Gawl. = Gasteria bicolor Haw. ■☆
16974 Aloe lingua Ker Gawl. var. angulata Haw. = Gasteria carinata (Mill.) Duval ■☆
16975 Aloe lingua Ker Gawl. var. angustifolia Aiton = Gasteria disticha (L.) Haw. ■☆
16976 Aloe lingua Ker Gawl. var. crassifolia Aiton = Gasteria disticha

(L.) Haw. ■☆

16977 Aloe lingua Ker Gawl. var. latifolia Haw. = Gasteria disticha (L.) Haw. ■☆

16978 Aloe lingua Ker Gawl. var. longifolia Haw. = Gasteria disticha (L.) Haw. ■☆

16979 Aloe lingua Ker Gawl. var. multifaria Haw. = Gasteria carinata (Mill.) Duval ■☆

16980 Aloe lingua Thunb. = Aloe plicatilis (L.) Mill. ●☆

16981 Aloe linguaeformis L. f. = Aloe plicatilis (L.) Mill. ●☆

16982 Aloe linguiformis Mill. = Gasteria disticha (L.) Haw. ■☆

16983 Aloe linita (Haw.) Schult. f. = Gasteria acinacifolia (J. Jacq.) Haw. ■☆

16984 Aloe littoralis Baker;滨海芦荟●☆

16985 Aloe lolwensis L. E. Newton;劳尔温芦荟●☆

16986 Aloe longebracteata Pole Evans.;长苞芦荟●☆

16987 Aloe longiaristata Schult. et Schult. f. = Aloe aristata Haw. ●☆

16988 Aloe longibracteata Pole-Evans = Aloe greatheadii Schönland var. davyana (Schönland) Glen et D. S. Hardy ●☆

16989 Aloe longifolia Haw. = Gasteria disticha (L.) Haw. ■☆

16990 Aloe longifolia Lam. = Kniphofia uvaria (L.) Oken ■☆

16991 Aloe longistyla Baker;长花柱芦荟;Karoo Aloe, Ramenas ●☆

16992 Aloe lugardiana Baker = Aloe zebrina Baker ●☆

16993 Aloe luntii Baker;伦特芦荟●☆

16994 Aloe lusitanica Groenew. = Aloe komatiensis Reynolds ●☆

16995 Aloe lutescens Groenew.;淡黄芦荟●☆

16996 Aloe macowanii Baker = Aloe striatula Haw. ●☆

16997 Aloe macracantha Baker = Aloe maculata All. ●☆

16998 Aloe macrocarpa Tod.;大果柱芦荟●☆

16999 Aloe macrocarpa Tod. var. major A. Berger = Aloe macrocarpa Tod. ●☆

17000 Aloe macroclada Baker;大枝芦荟●☆

17001 Aloe macrosiphon Baker;大管芦荟●■☆

17002 Aloe maculata All.;斑点芦荟;Soap Aloe, Zebra Aloe ●☆

17003 Aloe maculata Desf. = Aloe obscura Mill. ■☆

17004 Aloe maculata Ker Gawl. = Gasteria pulchra (Aiton) Haw. ■☆

17005 Aloe maculata Thunb. = Gasteria bicolor Haw. ■☆

17006 Aloe maculata Thunb. = Gasteria maculata Haw. ■☆

17007 Aloe maculata Thunb. var. obliqua Aiton = Gasteria bicolor Haw. ■☆

17008 Aloe maculata Thunb. var. pulchra Aiton = Gasteria pulchra (Aiton) Haw. ■☆

17009 Aloe maculosa Lam. = Aloe maculata All. ●☆

17010 Aloe madecassa H. Perrier;马德卡萨芦荟●☆

17011 Aloe magnidentata I. Verd. et Christian = Aloe megalacantha Baker ■☆

17012 Aloe margaritifera (L.) Burm. f. = Haworthia maxima (Haw.) Duval ■☆

17013 Aloe margaritifera (L.) Burm. f. var. maxima Haw. = Haworthia maxima (Haw.) Duval ■☆

17014 Aloe margaritifera (L.) Burm. f. var. minima Aiton = Aloe minima Baker ●☆

17015 Aloe margaritifera (L.) Burm. f. var. minor Aiton = Aloe minima Baker ●☆

17016 Aloe marginata Lam. = Haworthia marginata (Lam.) Stearn ■☆

17017 Aloe marlothii A. Berger;麦芦荟;Aloe, Bergaalwyn ●☆

17018 Aloe marlothii A. Berger subsp. orientalis Glen et D. S. Hardy;东方麦芦荟●☆

17019 Aloe marlothii A. Berger var. bicolor Reynolds = Aloe marlothii A. Berger ●☆

17020 Aloe marsabitensis I. Verd. et Christian = Aloe secundiflora Engl. ●☆

17021 Aloe marshalli J. M. Wood et M. S. Evans = Aloe kniphofioides Baker ●☆

17022 Aloe melanacantha Baker;黑刺芦荟●☆

17023 Aloe melsetterensis Christian = Aloe swynnertonii Rendle ●☆

17024 Aloe mendesii Reynolds;门代斯芦荟●☆

17025 Aloe menyharthii Baker;迈尼哈尔特芦荟●☆

17026 Aloe menyharthii Baker subsp. ensifolia S. Carter;剑叶迈尼哈尔特芦荟●☆

17027 Aloe meruana Lavranos = Aloe chrysostachys Lavranos et L. E. Newton ●☆

17028 Aloe metallica Engl. et Gilg;光泽芦荟●☆

17029 Aloe meyeri Van Jaarsv.;迈尔芦荟●☆

17030 Aloe micracantha Haw.;小刺芦荟●☆

17031 Aloe microstigma Salm-Dyck;小柱头芦荟●☆

17032 Aloe microstigma Salm-Dyck subsp. framesii (L. Bolus) Glen et D. S. Hardy;弗氏芦荟●☆

17033 Aloe millotii Reyn.;曲叶芦荟●☆

17034 Aloe minima Baker;侏儒芦荟●☆

17035 Aloe minima Baker var. blyderivierensis (Groenew.) Reynolds = Aloe minima Baker ●☆

17036 Aloe minor (Aiton) Schult. et Schult. f. = Aloe minima Baker ●☆

17037 Aloe mirabilis Haw. = Haworthia mirabilis (Haw.) Haw. ■☆

17038 Aloe mitriformis Mill.;僧帽花芦荟;Gold Tooth Aloe, Mitre-flower Aloe ●☆

17039 Aloe mitriformis Mill. = Aloe perfoliata L. ■☆

17040 Aloe mitriformis Mill. subsp. comptonii (Reynolds) Zonn.;康普顿芦荟●☆

17041 Aloe mitriformis Mill. subsp. distans (Haw.) Zonn.;远离僧帽花芦荟;Jeweled Aloe ●☆

17042 Aloe mitriformis Mill. var. albispina (Haw.) A. Berger = Aloe perfoliata L. ■☆

17043 Aloe mitriformis Mill. var. angustior Lam. = Aloe perfoliata L. ■☆

17044 Aloe mitriformis Mill. var. brevifolia (Aiton) W. T. Aiton = Aloe perfoliata L. ■☆

17045 Aloe mitriformis Mill. var. commelinii (Willd.) Baker = Aloe perfoliata L. ■☆

17046 Aloe mitriformis Mill. var. elatior Haw. = Aloe perfoliata L. ■☆

17047 Aloe mitriformis Mill. var. flavispina (Haw.) Baker = Aloe perfoliata L. ■☆

17048 Aloe mitriformis Mill. var. humilior Haw. = Aloe perfoliata L. ■☆

17049 Aloe mitriformis Mill. var. xanthacantha (Willd.) Baker = Aloe perfoliata L. ■☆

17050 Aloe mketiensis Christian = Aloe nuttii Baker ●☆

17051 Aloe modesta Reynolds;适度芦荟●☆

17052 Aloe mollis (Haw.) Schult. et Schult. f. = Gasteria disticha (L.) Haw. ■☆

17053 Aloe monotropa I. Verd.;单棱芦荟●☆

17054 Aloe montana Schinz = Aloe dichotoma L. f. ●☆

17055 Aloe monticola Reynolds;山地芦荟●☆

17056 Aloe morogoroensis Christian = Aloe bussei A. Berger ●☆

17057 Aloe mucronata (Haw.) Schult. f. = Haworthia mucronata Haw. ■☆

17058 Aloe mudenensis Reynolds;默登芦荟●☆

17059 Aloe muirii Marloth = Aloe lineata (Aiton) Haw. var. muirii

(Marloth) Reynolds ●☆
17060 Aloe multicolor L. E. Newton;多色芦荟●☆
17061 Aloe multifaria (Haw.) Schult. f. = Haworthia mirabilis (Haw.) Haw. ■☆
17062 Aloe muricata Schult. = Aloe ferox Mill. ●☆
17063 Aloe muricata Schult. = Aloe glauca Mill. ●☆
17064 Aloe murina L. E. Newton;鼠色芦荟●☆
17065 Aloe mutabilis Pillans = Aloe arborescens Mill. ●☆
17066 Aloe mutans Reynolds = Aloe greatheadii Schönland var. davyana (Schönland) Glen et D. S. Hardy ●☆
17067 Aloe mutica (Haw.) Schult. f. = Haworthia mutica Haw. ■☆
17068 Aloe mwanzana Christian = Aloe macrosiphon Baker ●■☆
17069 Aloe myriacantha (Haw.) Schult. et Schult. f.;多刺芦荟●☆
17070 Aloe myriacantha (Haw.) Schult. et Schult. f. var. minor (Baker) A. Berger = Aloe albida (Stapf) Reynolds ●☆
17071 Aloe mzimbana Christian;姆津巴芦荟●☆
17072 Aloe namibensis Giess;纳米布芦荟●☆
17073 Aloe natalensis J. M. Wood et M. S. Evans = Aloe arborescens Mill. ●☆
17074 Aloe ngobitensis Reynolds = Aloe nyeriensis Christian ●☆
17075 Aloe ngongensis Christian;恩贡芦荟●☆
17076 Aloe nigra (Haw.) Schult. et Schult. f. = Haworthia nigra (Haw.) Baker ■☆
17077 Aloe nigricans (Haw.) Duval var. fasciata Salm-Dyck = Gasteria disticha (L.) Haw. ■☆
17078 Aloe nigricans Haw. = Gasteria disticha (L.) Haw. ■☆
17079 Aloe nigricans Haw. var. denticulata Salm-Dyck = Gasteria disticha (L.) Haw. ■☆
17080 Aloe nigricans Haw. var. marmorata Salm-Dyck = Gasteria brachyphylla (Salm-Dyck) Van Jaarsv. ■☆
17081 Aloe nitens (Haw.) Schult. et Schult. f. = Gasteria acinacifolia (J. Jacq.) Haw. ■☆
17082 Aloe nitens Baker = Aloe rupestris Baker ●☆
17083 Aloe nitida Salm-Dyck = Gasteria nitida (Salm-Dyck) Haw. ■☆
17084 Aloe nitida Salm-Dyck var. grandipunctata ? = Gasteria nitida (Salm-Dyck) Haw. ■☆
17085 Aloe nitida Salm-Dyck var. major ? = Gasteria nitida (Salm-Dyck) Haw. ■☆
17086 Aloe nitida Salm-Dyck var. minor ? = Gasteria nitida (Salm-Dyck) Haw. ■☆
17087 Aloe nitida Salm-Dyck var. obtusa ? = Gasteria nitida (Salm-Dyck) Haw. ■☆
17088 Aloe nitida Salm-Dyck var. parvipunctata ? = Gasteria nitida (Salm-Dyck) Haw. ■☆
17089 Aloe nobilis Haw.;刚健芦荟;Gold Tooth Aloe, Noble Aloe ●☆
17090 Aloe nubigena Groenew.;云雾芦荟●☆
17091 Aloe nuttii Baker;纳特芦荟●☆
17092 Aloe nyeriensis Christian;涅里芦荟●☆
17093 Aloe nyeriensis Christian subsp. kedongensis (Reynolds) S. Carter = Aloe kedongensis Reynolds ●☆
17094 Aloe obliqua DC. = Gasteria pulchra (Aiton) Haw. ■☆
17095 Aloe obliqua Haw. = Gasteria bicolor Haw. ■☆
17096 Aloe obliqua Haw. var. fallax (Haw.) Schult. et Schult. f. = Gasteria bicolor Haw. ■☆
17097 Aloe obliqua Jacq. = Gasteria disticha (L.) Haw. ■☆
17098 Aloe obliqua Jacq. = Gasteria pulchra (Aiton) Haw. ■☆
17099 Aloe obscura Mill.;奥非芦荟■☆
17100 Aloe obscura Willd. = Gasteria disticha (L.) Haw. ■☆
17101 Aloe obscura Willd. = Gasteria excavata Haw. ■☆
17102 Aloe obscura Willd. var. truncata Salm-Dyck = Gasteria carinata (Mill.) Duval ■☆
17103 Aloe obtusa (Salm-Dyck) Schult. et Schult. f. = Gasteria nitida (Salm-Dyck) Haw. ■☆
17104 Aloe obtusifolia Salm-Dyck = Gasteria disticha (L.) Haw. ■☆
17105 Aloe officinalis Forssk. = Aloe vera (L.) Burm. f. ■
17106 Aloe officinalis Forssk. var. angustifolia ? = Aloe officinalis Forssk. ■
17107 Aloe oligospila Baker = Aloe percrassa Tod. ●☆
17108 Aloe orpeniae Schönland = Aloe hereroensis Engl. ●☆
17109 Aloe ortholopha Christian et Milne-Redh.;直冠芦荟●☆
17110 Aloe otallensis Baker;奥塔尔芦荟●☆
17111 Aloe otallensis Baker var. elongata A. Berger = Aloe rugosifolia M. G. Gilbert et Sebsebe ●☆
17112 Aloe pachydactylos T. A. McCoy et Lavranos;粗指芦荟●☆
17113 Aloe pachygaster Dinter;厚腹芦荟●☆
17114 Aloe paedogona A. Berger = Aloe buettneri A. Berger ●☆
17115 Aloe pallida (Haw.) Schult. et Schult. f. = Haworthia herbacea (Mill.) Stearn ■☆
17116 Aloe palmiformis Baker;掌状芦荟●☆
17117 Aloe paludicola A. Chev. = Aloe buettneri A. Berger ●☆
17118 Aloe paniculata Jacq. = Aloe striata Haw. ●☆
17119 Aloe papillosa Salm-Dyck = Haworthia maxima (Haw.) Duval ■☆
17120 Aloe parallelifolia H. Perrier;平行叶芦荟●☆
17121 Aloe parva (Haw.) Schult. et Schult. f. = Haworthia venosa (Lam.) Haw. subsp. tessellata (Haw.) M. B. Bayer ■☆
17122 Aloe parvibracteata Schönland;小苞芦荟●☆
17123 Aloe parvibracteata Schönland var. zuluensis (Reynolds) Reynolds = Aloe parvibracteata Schönland ●☆
17124 Aloe parvidens M. G. Gilbert et Sebsebe;小齿芦荟●☆
17125 Aloe parviflora Baker = Aloe minima Baker ●☆
17126 Aloe parvispina Schönland = Aloe perfoliata L. ■☆
17127 Aloe parvula A. Berger;琉璃芦荟(琉璃孔雀);Small Aloe ●☆
17128 Aloe parvula H. Perrier = Aloe perrieri Reynolds ●☆
17129 Aloe patersonii B. Mathew;帕特森芦荟●☆
17130 Aloe peacockii (Baker) A. Berger = Aloe elegans Tod. ●☆
17131 Aloe pearsonii Schönland;皮尔逊芦荟■☆
17132 Aloe peckii P. R. O. Bally et I. Verd.;佩克芦荟●☆
17133 Aloe peglerae Schönland;佩格拉芦荟●☆
17134 Aloe pellucens Haw. = Haworthia herbacea (Mill.) Stearn ■☆
17135 Aloe pembana L. E. Newton;彭贝芦荟●☆
17136 Aloe pendens Forssk.;什锦芦荟(斑叶芦荟,千代芦荟,十锦芦荟); Kanniedood Aloe, Mackerel Plant, Partridge Breast Aloe, Partridge-breast, Partridge-breasted Aloe, Pheasant's Wings, Tiger Aloe, Variegated Aloe ●☆
17137 Aloe penduliflora Baker;垂花芦荟●☆
17138 Aloe pentagona (Aiton) Haw. = Astroloba spiralis (L.) Uitewaal ■☆
17139 Aloe percrassa Schweinf. = Aloe trichosantha A. Berger ■☆
17140 Aloe percrassa Schweinf. var. albo-picta ? = Aloe trichosantha A. Berger ■☆
17141 Aloe percrassa Tod. = Aloe abyssinica Hook. f. ●☆
17142 Aloe perfoliata L.;抱茎芦荟■☆
17143 Aloe perfoliata L. var. africana ? = Aloe africana Mill. ●☆
17144 Aloe perfoliata L. var. arborescens ? = Aloe arborescens Mill. ●☆

17145 Aloe perfoliata L. var. barbadensis ? = Aloe vera (L.) Burm. f. ■
17146 Aloe perfoliata L. var. beta ? = Aloe africana Mill. ●☆
17147 Aloe perfoliata L. var. brevifolia Aiton = Aloe perfoliata L. ■☆
17148 Aloe perfoliata L. var. epsilon ? = Aloe ferox Mill. ●☆
17149 Aloe perfoliata L. var. ferox ? = Aloe ferox Mill. ●☆
17150 Aloe perfoliata L. var. gamma ? = Aloe ferox Mill. ●☆
17151 Aloe perfoliata L. var. glauca ? = Aloe glauca Mill. ●☆
17152 Aloe perfoliata L. var. humilis ? = Aloe humilis (L.) Mill. ●☆
17153 Aloe perfoliata L. var. kappa ? = Aloe glauca Mill. ●☆
17154 Aloe perfoliata L. var. lineata Aiton = Aloe lineata (Aiton) Haw. ●☆
17155 Aloe perfoliata L. var. mitriformis (Mill.) Aiton = Aloe perfoliata L. ■☆
17156 Aloe perfoliata L. var. purpurascens Aiton = Aloe succotrina Lam. ●☆
17157 Aloe perfoliata L. var. suberecta Aiton = Aloe humilis (L.) Mill. ●☆
17158 Aloe perfoliata L. var. succotrina (Lam.) Aiton = Aloe succotrina Lam. ●☆
17159 Aloe perfoliata L. var. vera L. = Aloe vera (L.) Burm. f. ■
17160 Aloe perfoliata L. var. zeta ? = Aloe ferox Mill. ●☆
17161 Aloe perfoliata Thunb. = Aloe ferox Mill. ●☆
17162 Aloe perrieri Reynolds;佩里耶芦荟●☆
17163 Aloe perryi Baker;东非芦荟;Perry's Aloe ●☆
17164 Aloe pertusa Haw. = Crassula capitella Thunb. subsp. thyrsiflora (Thunb.) Toelken ■☆
17165 Aloe petricola P. Evans;巴里锦;Aloe,Soft Aloe ●☆
17166 Aloe petrophila Pillans = Aloe swynnertonii Rendle ●☆
17167 Aloe pictifolia D. S. Hardy;叶色芦荟;Kouga Aloe ●☆
17168 Aloe pienaarii Pole-Evans = Aloe cryptopoda Baker ●☆
17169 Aloe pillansii L. Guthrie = Aloe dichotoma Masson subsp. pilllansii (L. Guthrie) Zonn. ●☆
17170 Aloe pirottae A. Berger;皮罗特芦荟●☆
17171 Aloe planifolia (Haw.) Schult. et Schult. f. = Haworthia cymbiformis (Haw.) Duval ■☆
17172 Aloe planifolia Baker = Gasteria bicolor Haw. ■☆
17173 Aloe platylepsis Baker;阔鳞芦荟●☆
17174 Aloe platyphylla Baker = Aloe zebrina Baker ●☆
17175 Aloe plicatilis (L.) Mill.;重塔芦荟(扇芦荟);Fan Aloe ●☆
17176 Aloe plicatilis (L.) Mill. var. major Salm-Dyck = Aloe plicatilis (L.) Mill. ●☆
17177 Aloe plowesii Reynolds;普洛芦荟●☆
17178 Aloe pluridens Haw.;垂叶树芦荟●☆
17179 Aloe pluridens Haw. var. beckeri Schönland = Aloe pluridens Haw. ●☆
17180 Aloe pluripuncta (Haw.) Schult. et Schult. f. = Gasteria acinacifolia (J. Jacq.) Haw. ■☆
17181 Aloe pole-evansii Christian = Aloe dawei A. Berger ●☆
17182 Aloe polyphylla Schönl. ex Pillans;碧玉冠;Manyleaf Aloe,Spiral Aloe ●☆
17183 Aloe pongolensis Reynolds = Aloe parvibracteata Schönland ●☆
17184 Aloe pongolensis Reynolds var. zuluensis ? = Aloe parvibracteata Schönland ●☆
17185 Aloe postgenita Schult. et Schult. f. = Aloe brevifolia Mill. ●☆
17186 Aloe pratensis Baker;玫瑰芦荟;Rosette Aloe ●☆
17187 Aloe pretoriensis Pole-Evans;比勒陀利亚芦荟●☆
17188 Aloe procera L. C. Leach;高大芦荟●☆
17189 Aloe prolifera Haw. = Aloe brevifolia Mill. ●☆
17190 Aloe prolifera Haw. var. major Salm-Dyck = Aloe brevifolia Mill. ●☆
17191 Aloe pruinosa Reynolds;白粉芦荟●☆
17192 Aloe pseudangulata Salm-Dyck = Gasteria carinata (Mill.) Duval ■☆
17193 Aloe pseudoafricana Salm-Dyck = Aloe africana Mill. ●☆
17194 Aloe pseudoferox Salm-Dyck = Aloe ferox Mill. ●☆
17195 Aloe pseudonigricans Salm-Dyck = Gasteria pseudonigricans (Salm-Dyck) Haw. ■☆
17196 Aloe pseudoparvula J. -B. Castillon;假小芦荟●☆
17197 Aloe pseudotortuosa Salm-Dyck = Haworthia viscosa (L.) Haw. ■☆
17198 Aloe pubescens Reynolds;短柔毛芦荟●☆
17199 Aloe pulcherrima M. G. Gilbert et Sebsebe;艳丽芦荟●☆
17200 Aloe pulchra (Aiton) Jacq. = Gasteria pulchra (Aiton) Haw. ■☆
17201 Aloe pulchra Lavranos = Aloe bella Rowley ●☆
17202 Aloe pumila L. = Haworthia herbacea (Mill.) Stearn ■☆
17203 Aloe pumila L. var. arachnoidea ? = Haworthia arachnoidea (L.) Duval ■☆
17204 Aloe pumila L. var. margaritifera ? = Haworthia maxima (Haw.) Duval ■☆
17205 Aloe pumilio Jacq. = Haworthia reticulata (Haw.) Haw. ■☆
17206 Aloe punctata Haw. = Aloe variegata L. ●
17207 Aloe purpurascens (Aiton) Haw. = Aloe succotrina Lam. ●☆
17208 Aloe quinquangularis Schult. f. = Astroloba spiralis (L.) Uitewaal ■☆
17209 Aloe rabaiensis Rendle;拉巴伊芦荟●☆
17210 Aloe racemosa Lam. = Gasteria carinata (Mill.) Duval var. verrucosa (Mill.) Van Jaarsv. ■☆
17211 Aloe radula Jacq. = Haworthia attenuata (Haw.) Haw. var. radula (Jacq.) M. B. Bayer ■☆
17212 Aloe radula Ker Gawl. = Haworthia attenuata (Haw.) Haw. ■☆
17213 Aloe ramifera (Haw.) Schult. f. = Haworthia marginata (Lam.) Stearn ■☆
17214 Aloe ramosa Haw. = Aloe dichotoma Masson ●☆
17215 Aloe ramosissima Pillans;多枝芦荟;Maiden's Quiver Tree ●☆
17216 Aloe ramosissima Pillans = Aloe dichotoma Masson subsp. ramosissima (Pillans) Zonn. ●☆
17217 Aloe rauhii Reynolds;劳氏芦荟■☆
17218 Aloe recurva Haw. = Haworthia venosa (Lam.) Haw. ■☆
17219 Aloe reinwardtii Salm-Dyck = Haworthia reinwardtii (Salm-Dyck) Haw. ■☆
17220 Aloe reitzii Reyn.;莱次芦荟;Reitz Aloe ●☆
17221 Aloe reitzii Reynolds var. vernalis D. S. Hardy;春芦荟■☆
17222 Aloe repens Schult. et Schult. f. = Gasteria carinata (Mill.) Duval var. verrucosa (Mill.) Van Jaarsv. ■☆
17223 Aloe reticulata Haw. = Haworthia reticulata (Haw.) Haw. ■☆
17224 Aloe retusa L. = Haworthia retusa (L.) Duval ■☆
17225 Aloe reynoldsii Letty;雷诺兹芦荟●☆
17226 Aloe rhodacantha DC. = Aloe glauca Mill. ●☆
17227 Aloe rhodesiana Rendle;罗得西亚芦荟●☆
17228 Aloe richardsiae Reynolds;理查兹芦荟●☆
17229 Aloe richtersveldensis Venter et Beukes = Aloe meyeri Van Jaarsv. ●☆
17230 Aloe rigens Reynolds et P. R. O. Bally;硬芦荟●☆
17231 Aloe rigens Reynolds et P. R. O. Bally var. glabrescens = Aloe glabrescens (Reynolds et P. R. O. Bally) S. Carter et Brandham ●☆
17232 Aloe rigida Salisb. = Kniphofia uvaria (L.) Oken ■☆

17233 Aloe rivae Baker;沟芦荟●☆
17234 Aloe rossii Tod. = Aloe deltoideodonta Baker ■☆
17235 Aloe rubriflora (L. Bolus) G. D. Rowley = Astroloba rubriflora (L. Bolus) G. F. Sm. et J. C. Manning ■☆
17236 Aloe rubrolutea Schinz = Aloe littoralis Baker ●☆
17237 Aloe rugosa Salm-Dyck = Haworthia rugosa (Salm-Dyck) Baker ■☆
17238 Aloe rugosifolia M. G. Gilbert et Sebsebe;褶皱芦荟●☆
17239 Aloe rupestris Baker;岩地芦荟●☆
17240 Aloe rupicola Reynolds;岩生芦荟●☆
17241 Aloe ruspoliana Baker;鲁斯波利芦荟●☆
17242 Aloe sahundra Bojer = Aloe divaricata A. Berger ●☆
17243 Aloe saponaria (Aiton) Haw.;皂草芦荟(花叶芦荟,皂芦荟);Soap Aloe, Variegated Aloe ●☆
17244 Aloe saponaria (Aiton) Haw. = Aloe maculata All. ●☆
17245 Aloe saponaria (Aiton) Haw. var. brachyphylla Baker = Aloe maculata All. ●☆
17246 Aloe saponaria (Aiton) Haw. var. ficksburgensis Reynolds = Aloe maculata All. ●☆
17247 Aloe saponaria (Aiton) Haw. var. latifolia (Haw.) Trel.;广叶芦荟●☆
17248 Aloe saponaria (Aiton) Haw. var. latifolia (Haw.) Trel. = Aloe maculata All. ●☆
17249 Aloe saundersiae (Reynolds) Reynolds;桑氏芦荟●☆
17250 Aloe scaberrima Salm-Dyck = Gasteria carinata (Mill.) Duval var. verrucosa (Mill.) Van Jaarsv. ■☆
17251 Aloe scabra (Haw.) Schult. et Schult. f. = Haworthia scabra Haw. ■☆
17252 Aloe scabrifolia L. E. Newton et Lavranos;皱叶芦荟●☆
17253 Aloe schimperi G. Karst. et Schenck = Aloe percrassa Tod. ●☆
17254 Aloe schimperi Schweinf. = Aloe percrassa Tod. ●☆
17255 Aloe schinzii Baker = Aloe littoralis Baker ●☆
17256 Aloe schlechteri Schönland = Aloe claviflora Burch. ■☆
17257 Aloe schliebenii Lavranos = Aloe brachystachys Baker ●☆
17258 Aloe schmidtiana Regel = Aloe cooperi Baker ●☆
17259 Aloe schoelleri Schweinf.;舍勒芦荟●☆
17260 Aloe schweinfurthii Baker;施韦芦荟●☆
17261 Aloe scorpioides L. C. Leach;蝎尾芦荟●☆
17262 Aloe secundiflora Engl.;侧花芦荟●☆
17263 Aloe semiglabrata (Haw.) Schult. et Schult. f. = Haworthia maxima (Haw.) Duval ■☆
17264 Aloe semimargaritifera Salm-Dyck = Haworthia maxima (Haw.) Duval ■☆
17265 Aloe semimargaritifera Salm-Dyck var. maxima Haw. = Haworthia maxima (Haw.) Duval ■☆
17266 Aloe seretii De Wild.;赛雷芦荟●☆
17267 Aloe serra DC. = Aloe brevifolia Mill. var. depressa (Haw.) Baker ■☆
17268 Aloe serrulata (Aiton) Haw.;细齿芦荟●☆
17269 Aloe sessiliflora Pole-Evans = Aloe spicata L. f. ●☆
17270 Aloe setosa Schult. et Schult. f. = Haworthia arachnoidea (L.) Duval var. setata (Haw.) M. B. Bayer ■☆
17271 Aloe simii Pole-Evans;西姆芦荟●☆
17272 Aloe sinuata Thunb. = Aloe succotrina Lam. ●☆
17273 Aloe sladeniana Pole-Evans;斯莱登芦荟●☆
17274 Aloe soccotorina Schult. et Schult. f. = Aloe succotrina Lam. ●☆
17275 Aloe soccotorina Schult. f. = Aloe ferox Mill. ●☆
17276 Aloe soccotorina Schult. f. = Aloe succotrina Lam. ●☆
17277 Aloe soccotorina Schult. f. var. purpurascens ? = Aloe succotrina Lam. ●☆
17278 Aloe socotrina DC. = Aloe succotrina Lam. ●☆
17279 Aloe socotrina Lam. = Aloe succotrina Lam. ●☆
17280 Aloe solaiana Christian = Aloe lateritia Engl. var. graminicola (Reynolds) S. Carter ●☆
17281 Aloe somaliensis W. Watson;索马里芦荟●☆
17282 Aloe somaliensis W. Watson var. marmorata Reynolds et P. R. O. Bally = Aloe somaliensis W. Watson ●☆
17283 Aloe sordida (Haw.) Schult. et Schult. f. = Haworthia sordida Haw. ■☆
17284 Aloe soutpansbergensis I. Verd.;索特潘芦荟●☆
17285 Aloe speciosa Baker;美丽芦荟;Tilt-head Aloe ●☆
17286 Aloe spectabilis Reynolds = Aloe marlothii A. Berger ●☆
17287 Aloe spicata Baker = Aloe spicata L. f. ●☆
17288 Aloe spicata L. f.;穗花芦荟●☆
17289 Aloe spinosissima Hort. ex Jahand.;金齿芦荟(七宝锦);Gold Tooth Aloe ●☆
17290 Aloe spiralis Haw. = Astroloba spiralis (L.) Uitewaal ■☆
17291 Aloe spiralis Haw. var. pentagona Aiton = Astroloba spiralis (L.) Uitewaal ■☆
17292 Aloe spiralis L. = Astroloba spiralis (L.) Uitewaal ■☆
17293 Aloe spuria A. Berger;翡翠掌;False Aloe ●☆
17294 Aloe squarrosa Baker = Aloe squarrosa Baker ex Balf. f. ●☆
17295 Aloe squarrosa Baker ex Balf. f;观赏芦荟●☆
17296 Aloe stefaninii Chiov. = Aloe ruspoliana Baker ●☆
17297 Aloe stenophylla Schult. et Schult. f. = Haworthia angustifolia Haw. ■☆
17298 Aloe steudneri Schweinf.;斯托德芦荟●☆
17299 Aloe striata Haw.;珊瑚芦荟(肋叶芦荟);Blouaalwyn, Coral Aloe ●☆
17300 Aloe striata Haw. subsp. karasbergensis (Pillans) Glen et D. S. Hardy;卡拉斯堡芦荟(乌山锦,细纹芦荟);Karasberg Aloe ●☆
17301 Aloe striata Haw. subsp. komaggasensis (Kritzinger et Van Jaarsv.) Glen et D. S. Hardy;科马芦荟●☆
17302 Aloe striata Haw. var. oligospila Baker = Aloe striata Haw. ●☆
17303 Aloe striatula Haw.;巴克拉芦荟;Basuto Kraal Aloe ●☆
17304 Aloe striatula Haw. var. caesia Reynolds;淡蓝巴克拉芦荟●☆
17305 Aloe strigata (Haw.) Schult. f. = Gasteria carinata (Mill.) Duval ■☆
17306 Aloe stuhlmannii Baker = Aloe volkensii Engl. ●☆
17307 Aloe suarezensis H. Perrier;苏亚雷斯芦荟●☆
17308 Aloe subattenuata Salm-Dyck = Haworthia subattenuata (Salm-Dyck) Baker ■☆
17309 Aloe subcarinata Salm-Dyck = Gasteria carinata (Mill.) Duval ■☆
17310 Aloe suberecta (Aiton) Haw. = Aloe humilis (L.) Mill. ●☆
17311 Aloe suberecta (Aiton) Haw. var. semiguttata Haw. = Aloe humilis (L.) Mill. ●☆
17312 Aloe subfasciata Salm-Dyck = Haworthia fasciata (Willd.) Haw. ■☆
17313 Aloe subferox Spreng. = Aloe ferox Mill. ●☆
17314 Aloe subtortuosa Schult. et Schult. f. = Haworthia viscosa (L.) Haw. ■☆
17315 Aloe subtuberculata Haw. = Aloe humilis (L.) Mill. ●☆
17316 Aloe subulata Salm-Dyck = Haworthia attenuata (Haw.) Haw. var. radula (Jacq.) M. B. Bayer ■☆
17317 Aloe subverrucosa Salm-Dyck = Gasteria carinata (Mill.) Duval

var. verrucosa (Mill.) Van Jaarsv. ■☆

17318 Aloe subverrucosa Salm-Dyck var. grandipunctata ? = Gasteria carinata (Mill.) Duval var. verrucosa (Mill.) Van Jaarsv. ■☆

17319 Aloe subverrucosa Salm-Dyck var. parvipunctata ? = Gasteria carinata (Mill.) Duval var. verrucosa (Mill.) Van Jaarsv. ■☆

17320 Aloe succotrina Lam.;索科特芦荟;Fynbos Aloe,Socotrine Aloe ●☆

17321 Aloe succotrina Lam. var. saxigena A. Berger = Aloe succotrina Lam. ●☆

17322 Aloe suffulta Reynolds;支柱芦荟●☆

17323 Aloe sulcata Salm-Dyck = Gasteria carinata (Mill.) Duval ■☆

17324 Aloe suprafoliata Pole-Evans;顶叶芦荟●☆

17325 Aloe supralaevis Haw. = Aloe ferox Mill. ●☆

17326 Aloe supralaevis Haw. var. erythrocarpa A. Berger = Aloe ferox Mill. ●☆

17327 Aloe supralaevis Haw. var. hanburyi Baker = Aloe marlothii A. Berger ●☆

17328 Aloe suzannae Decary;德卡里芦荟●☆

17329 Aloe swynnertonii Rendle;斯温纳顿芦荟●☆

17330 Aloe tauri L. C. Leach = Aloe spicata L. f. ●☆

17331 Aloe tenuior Haw.;微小芦荟●☆

17332 Aloe tenuior Haw. var. decidua Reynolds = Aloe tenuior Haw. ●☆

17333 Aloe tenuior Haw. var. densiflora Reynolds = Aloe tenuior Haw. ●☆

17334 Aloe tenuior Haw. var. glaucescens Zahlbr. = Aloe tenuior Haw. ●☆

17335 Aloe tenuior Haw. var. rubriflora Reynolds = Aloe tenuior Haw. ●☆

17336 Aloe tessellata (Haw.) Schult. et Schult. f. = Haworthia venosa (Lam.) Haw. subsp. tessellata (Haw.) M. B. Bayer ■☆

17337 Aloe tewoldei M. G. Gilbert et Sebsebe;特沃芦荟●☆

17338 Aloe thorncroftii Pole-Evans;托恩芦荟●☆

17339 Aloe tomentosa Defleurs;微毛芦荟●☆

17340 Aloe tororoana Reynolds;托罗芦荟●☆

17341 Aloe torquata (Haw.) Salm-Dyck = Haworthia viscosa (L.) Haw. ■☆

17342 Aloe torrei I. Verd. et Christian;托雷芦荟●☆

17343 Aloe torrei I. Verd. et Christian var. wildii Reynolds = Aloe wildii (Reynolds) Reynolds ●☆

17344 Aloe tortuosa Haw. = Haworthia viscosa (L.) Haw. ■☆

17345 Aloe translucens Haw. = Haworthia herbacea (Mill.) Stearn ■☆

17346 Aloe translucens W. T. Aiton = Haworthia herbacea (Mill.) Stearn ■☆

17347 Aloe transvalensis Kuntze = Aloe zebrina Baker ●☆

17348 Aloe transvalensis Kuntze et Reynolds;权士王芦荟●☆

17349 Aloe triangularis Medik. = Haworthia viscosa (L.) Haw. ■☆

17350 Aloe trichosantha A. Berger;毛花芦荟■☆

17351 Aloe trichosantha A. Berger subsp. longiflora M. G. Gilbert et Sebsebe;长毛花芦荟■☆

17352 Aloe tricolor Haw. = Haworthia venosa (Lam.) Haw. ■☆

17353 Aloe trigona (Haw.) Schult. et Schult. f. = Gasteria trigona Haw. ■☆

17354 Aloe trigona Salm-Dyck = Gasteria nitida (Salm-Dyck) Haw. ■☆

17355 Aloe trigona Salm-Dyck var. obtusa ? = Gasteria nitida (Salm-Dyck) Haw. ■☆

17356 Aloe trigonantha L. C. Leach;三角花芦荟●☆

17357 Aloe tripetala Medik. = Aloe plicatilis (L.) Mill. ●☆

17358 Aloe tristicha Medik. = Gasteria carinata (Mill.) Duval ■☆

17359 Aloe trivialis A. Chev. = Aloe schweinfurthii Baker ●☆

17360 Aloe trothae A. Berger = Aloe bulbicaulis Christian ●☆

17361 Aloe tuberculata Haw. = Aloe humilis (L.) Mill. ●☆

17362 Aloe tulearensis T. A. McCoy et Lavranos;图莱亚尔芦荟●☆

17363 Aloe turgida (Haw.) Schult. et Schult. f. = Haworthia turgida Haw. ■☆

17364 Aloe turkanensis Christian;图尔卡纳芦荟●☆

17365 Aloe tweediae Christian;特威迪芦荟●☆

17366 Aloe ukambensis Reynolds;乌卡芦荟●☆

17367 Aloe umbellata DC. = Aloe maculata All. ●☆

17368 Aloe umfoloziensis Reynolds = Aloe maculata All. ●☆

17369 Aloe uvaria L. = Kniphofia uvaria (L.) Oken ■☆

17370 Aloe vahontsohy Decorse = Aloe divaricata A. Berger ●☆

17371 Aloe vallaris L. C. Leach;河谷芦荟●☆

17372 Aloe vandermerwei Reynolds = Aloe zebrina Baker ●☆

17373 Aloe vanrooyenii G. F. Sm. et N. R. Crouch;范鲁因芦荟●☆

17374 Aloe vaotsohy Decorse et Poiss. = Aloe divaricata A. Berger ●☆

17375 Aloe variegata L.;十锦芦荟;Partridge-breasted Aloe ●

17376 Aloe variegata L. var. ausana (Dietr.) Beger;大十锦芦荟●☆

17377 Aloe variegata L. var. haworthii A. Berger = Aloe variegata L. ●

17378 Aloe venenosa Engl.;毒芦荟●☆

17379 Aloe venosa Lam. = Haworthia venosa (Lam.) Haw. ■☆

17380 Aloe venusta (Haw.) Schult. et Schult. f. = Gasteria acinacifolia (J. Jacq.) Haw. ■☆

17381 Aloe vera (L.) Burm. f.;库拉索芦荟(大芦荟,短叶芦荟,劳伟,老芦荟,龙山,卢会,芦荟,芦荟胶,罗帏草,萝蕙,纳会,讷会,奴会,象胆,羊角藤,油葱叶);Aloe,Barbados Aloe,Burn Plant,Cape Aloe,Curacao Aloe,Medicinal Aloe,Mediterranean Aloe,True Aloe,Unguentine Cactus,West Indian Aloe,Zanzibar Aloe ■

17382 Aloe vera (L.) Burm. f. var. aethiopica Schweinf. = Aloe elegans Tod. ●☆

17383 Aloe vera (L.) Burm. f. var. chinensis (Haw.) Berger;芦荟(斑纹芦荟,罗帏草,象鼻莲,油葱,中国芦荟);Burn Plant,China Aloe,Chinese Aloe,Curacao Aloe ■

17384 Aloe vera (L.) Burm. f. var. chinensis (Haw.) Berger = Aloe vera (L.) Burm. f. ■

17385 Aloe vera (L.) Burm. f. var. lanzae ? = Aloe vera (L.) Burm. f. ■

17386 Aloe vera (L.) Burm. f. var. littoralis ? = Aloe vera (L.) Burm. f. ■

17387 Aloe vera (L.) Burm. f. var. officinalis (Forssk.) Baker;药用库拉索芦荟■☆

17388 Aloe vera (L.) Burm. f. var. wratislaviensis ? = Aloe vera (L.) Burm. f. ■

17389 Aloe verdoorniae Reynolds = Aloe greatheadii Schönland var. davyana (Schönland) Glen et D. S. Hardy ●☆

17390 Aloe verecunda Pole-Evans;羞涩芦荟●☆

17391 Aloe verrucosa Mill. = Gasteria carinata (Mill.) Duval var. verrucosa (Mill.) Van Jaarsv. ■☆

17392 Aloe verrucosa Mill. var. latifolia Salm-Dyck = Gasteria carinata (Mill.) Duval var. verrucosa (Mill.) Van Jaarsv. ■☆

17393 Aloe verrucosa Mill. var. striata Salm-Dyck = Gasteria carinata (Mill.) Duval var. verrucosa (Mill.) Van Jaarsv. ■☆

17394 Aloe verrucosospinosa All. = Aloe humilis (L.) Mill. ●☆

17395 Aloe verrucula Medik. = Gasteria carinata (Mill.) Duval var. verrucosa (Mill.) Van Jaarsv. ■☆

17396 Aloe versicolor Guillaumin;异色芦荟●☆

17397 Aloe veseyi Reynolds;维西芦荟●☆

17398 Aloe viguieri H. Perrier;维基耶芦荟●☆

17399 Aloe virescens (Haw.) Schult. et Schult. f. = Haworthia marginata (Lam.) Stearn ■☆
17400 Aloe viridiflora Reynolds;绿花芦荟●☆
17401 Aloe viscosa L. = Haworthia viscosa (L.) Haw. ■☆
17402 Aloe vittata Schult. f. = Gasteria disticha (L.) Haw. ■☆
17403 Aloe vogtsii Reynolds = Aloe swynnertonii Rendle ●☆
17404 Aloe volkensii Engl.;福尔芦荟●☆
17405 Aloe volkensii Engl. subsp. multicaulis S. Carter et L. E. Newton;多茎芦荟●☆
17406 Aloe vossii Reynolds;沃斯芦荟●☆
17407 Aloe vryheidensis Groenew.;弗雷黑德芦荟●☆
17408 Aloe vulgaris Lam. = Aloe vera (L.) Burm. f. ■
17409 Aloe wickensii Pole-Evans;威肯斯芦荟●☆
17410 Aloe wickensii Pole-Evans = Aloe cryptopoda Baker ●☆
17411 Aloe wickensii Pole-Evans var. lutea Reynolds = Aloe cryptopoda Baker ●☆
17412 Aloe wildii (Reynolds) Reynolds;威尔德芦荟●☆
17413 Aloe wilsonii Reynolds;威尔逊芦荟●☆
17414 Aloe wollastonii Rendle;沃拉芦荟●☆
17415 Aloe wrefordii Reynolds;伍赖芦荟●☆
17416 Aloe xanthacantha Willd. = Aloe perfoliata L. ■☆
17417 Aloe yuccifolia A. Gray = Hesperaloe parviflora (Torr.) J. M. Coult. ■☆
17418 Aloe zanzibarica Milne-Redh. = Aloe squarrosa Baker ●☆
17419 Aloe zebrina Baker;缟芦荟●☆
17420 Aloe zebrina Baker var. coerulescens F. H. Chen;斑马芦荟;Zebra Aloe ●☆
17421 Aloe zeyheri Salm-Dyck = Gasteria bicolor Haw. ■☆
17422 Aloe zeylanica L. = Sansevieria zeylanica (L.) Willd. ■☆
17423 Aloeaceae Batsch = Asphodelaceae Juss. ●■
17424 Aloeaceae Batsch;芦荟科;Aloe Family ●■
17425 Aloeatheros Endl. = Alloiatheros Elliott ■☆
17426 Aloeatheros Endl. = Gymnopogon P. Beauv. ■☆
17427 Aloes Raf. = Aloe L. ●■
17428 Aloexylum Lour. = Aquilaria Lam.(保留属名)●
17429 Aloides (Schultz) Bruch = Aloina aloides (Schultz) Kindb. ■☆
17430 Aloides Fabr. = Stratiotes L. ■☆
17431 Aloinella (A. Berger) A. Berger ex Lemee = Aloe L. ●■
17432 Aloinella (A. Berger) Lemée = Aloe L. ●■
17433 Aloinopsis Schwantes(1926);芦荟番杏属(唐扇属)■☆
17434 Aloinopsis acuta L. Bolus;尖芦荟番杏■☆
17435 Aloinopsis albinota Schwantes;白背芦荟番杏■☆
17436 Aloinopsis aloides (Haw.) Schwantes var. striatus L. Bolus = Nananthus vittatus (N. E. Br.) Schwantes ■☆
17437 Aloinopsis broomii L. Bolus = Nananthus vittatus (N. E. Br.) Schwantes ■☆
17438 Aloinopsis crassipes (Marloth) L. Bolus = Aloinopsis spathulata (Thunb.) L. Bolus ■☆
17439 Aloinopsis dyeri (L. Bolus) L. Bolus = Aloinopsis rubrolineata (N. E. Br.) Schwantes ■☆
17440 Aloinopsis gerstneri L. Bolus = Nananthus gerstneri (L. Bolus) L. Bolus ■☆
17441 Aloinopsis hilmarii (L. Bolus) L. Bolus;希尔芦荟番杏■☆
17442 Aloinopsis jamesii L. Bolus = Aloinopsis rubrolineata (N. E. Br.) Schwantes ■☆
17443 Aloinopsis lodewykii L. Bolus = Aloinopsis luckhoffii (L. Bolus) L. Bolus ■☆
17444 Aloinopsis loganii L. Bolus;洛根芦荟番杏■☆
17445 Aloinopsis luckhoffii (L. Bolus) L. Bolus;天女裳■☆
17446 Aloinopsis luckhoffii L. Bolus = Aloinopsis luckhoffii (L. Bolus) L. Bolus ■☆
17447 Aloinopsis malherbei (L. Bolus) L. Bolus;天女云;Giant Jewel Plant ■☆
17448 Aloinopsis orpenii (N. E. Br.) L. Bolus = Prepodesma orpenii (N. E. Br.) N. E. Br. ■☆
17449 Aloinopsis pallens L. Bolus = Nananthus pallens (L. Bolus) L. Bolus ■☆
17450 Aloinopsis peersii (L. Bolus) L. Bolus;皮尔斯芦荟番杏■☆
17451 Aloinopsis rosulata (Kensit) Schwantes = Aistocaulon rosulatum (Kensit) Poelln. ■☆
17452 Aloinopsis rubrolineata (N. E. Br.) Schwantes;花锦■☆
17453 Aloinopsis schooneesii L. Bolus;唐扇■☆
17454 Aloinopsis schooneesii L. Bolus var. acutipetala ? = Aloinopsis schooneesii L. Bolus ■☆
17455 Aloinopsis schooneesii L. Bolus var. willowmorensis ? = Aloinopsis schooneesii L. Bolus ■☆
17456 Aloinopsis setifera (L. Bolus) L. Bolus = Aloinopsis luckhoffii (L. Bolus) L. Bolus ■☆
17457 Aloinopsis spathulata (Thunb.) L. Bolus;匙叶芦荟番杏■☆
17458 Aloinopsis thudichumii L. Bolus;萨迪芦荟番杏■☆
17459 Aloinopsis villetii (L. Bolus) L. Bolus;天女之舞■☆
17460 Aloinopsis villetii (L. Bolus) L. Bolus = Aloinopsis luckhoffii (L. Bolus) L. Bolus ■☆
17461 Aloinopsis wilmaniae L. Bolus = Nananthus aloides (Haw.) Schwantes ■☆
17462 Aloiozonium Lindl. = Alloiozonium Kuntze ■☆
17463 Aloiozonium Lindl. = Arctotheca J. C. Wendl. ■☆
17464 Aloiozonium Lindl. = Cryptostemma R. Br. ex W. T. Aiton ■☆
17465 Aloitis Raf. = Gentianella Moench(保留属名)■
17466 Alomia Kunth(1818);修泽兰属■☆
17467 Alomia ageratoides Kunth;修泽兰■☆
17468 Alomia alata Hemsl.;翅修泽兰■☆
17469 Alomia angustata Benth. ex Baker;窄修泽兰■☆
17470 Alomia microcarpa B. L. Rob.;小果修泽兰■☆
17471 Alomiella R. M. King et H. Rob.(1972);毛瓣尖泽兰属■☆
17472 Alomiella hatschbachii R. M. King et H. Rob.;哈氏毛瓣尖泽兰■☆
17473 Alomiella regnellii (Malme) R. M. King et H. Rob.;毛瓣尖泽兰■☆
17474 Alona Lindl. = Nolana L. ex L. f. ■☆
17475 Alonsoa Ruiz et Pav.(1798);假面花属;Alonsoa,Mask Flower ■☆
17476 Alonsoa acutifolia Ruiz et Pav.;尖叶假面花;Lanceleaf Mask Flower ■☆
17477 Alonsoa acutifolia Ruiz et Pav. var. candida Voss;白花假面花■☆
17478 Alonsoa albiflora G. Nicholson = Alonsoa acutifolia Ruiz et Pav. var. candida Voss ■☆
17479 Alonsoa incisifolia Ruiz et Pav.;刻叶假面花●☆
17480 Alonsoa linearifolia Steud. = Alonsoa linearis (Jacq.) Ruiz et Pav. ●☆
17481 Alonsoa linearis (Jacq.) Ruiz et Pav.;线叶假面花;Flaxleaf Mask Flower,Mask Flower ●☆
17482 Alonsoa linifolia Roezl = Alonsoa linearis (Jacq.) Ruiz et Pav. ●☆
17483 Alonsoa myrtifolia Roezl = Alonsoa acutifolia Ruiz et Pav. ■☆
17484 Alonsoa peduncularis (Kunze) Wettst.;梗花假面花●☆

17485 Alonsoa pumila Vilm.;矮线叶假面花●☆
17486 Alonsoa unilabiata (L. f.) Steud.;单唇假面花●☆
17487 Alonsoa unilabiata (L. f.) Steud. = Diascia dissecta Hiern ■☆
17488 Alonsoa unilabiata (L. f.) Steud. = Diascia heterandra Benth. ■☆
17489 Alonsoa urticifolia Steud. = Alonsoa incisifolia Ruiz et Pav. ●☆
17490 Alonsoa warscewiczii Regel;心叶假面花;Mask Flower ■☆
17491 Alonzoa Brongn. = Alonsoa Ruiz et Pav. ■☆
17492 Alopecias Steven = Astragalus L. ●■
17493 Alopecuraceae Martinov = Gramineae Juss.(保留科名)●■
17494 Alopecuraceae Martinov = Poaceae Barnhart(保留科名)●■
17495 Alopecuropsis Opiz = Alopecurus L. ■
17496 Alopecuro-veronica L. = Pogostemon Desf. ●■
17497 Alopecurus L.(1753);看麦娘属;Alopecurus,Fox Tail,Foxtail, Foxtail Grass,Golden Meadow Foxtail ■
17498 Alopecurus aequalis Sobol.;看麦娘;Equal Alopecurus,Floating Foxtail,Orange Foxtail,Short-awn Foxtail,Short-awned Foxtail ■
17499 Alopecurus aequalis Sobol. subsp. amurensis (Kom.) Hultén = Alopecurus aequalis Sobol. ■
17500 Alopecurus aequalis Sobol. subsp. aristulatus (Michx.) Tzvelev = Alopecurus aequalis Sobol. ■
17501 Alopecurus aequalis Sobol. var. amurensis (Kom.) Ohwi;阿穆尔看麦娘(看麦娘)■
17502 Alopecurus aequalis Sobol. var. amurensis (Kom.) Ohwi = Alopecurus aequalis Sobol. ■
17503 Alopecurus aequalis Sobol. var. aristulatus (Michx.) Tzvelev = Alopecurus aequalis Sobol. ■
17504 Alopecurus aequalis Sobol. var. natans (Wahlenb.) Fernald = Alopecurus aequalis Sobol. ■
17505 Alopecurus agrestis L. = Alopecurus myosuroides Huds. ■
17506 Alopecurus alpinus Sm.;高山看麦娘;Alpine Alopecurus,Alpine Foxtail ■☆
17507 Alopecurus alpinus Sm. var. songaricus Schrenk ex Fisch. et Meyen = Alopecurus pratensis L. ■
17508 Alopecurus alpinus Vill. = Alopecurus alpinus Sm. ■☆
17509 Alopecurus amurensis (Kom.) Kom. = Alopecurus aequalis Sobol. var. amurensis (Kom.) Ohwi ■
17510 Alopecurus amurensis Kom. = Alopecurus aequalis Sobol. ■
17511 Alopecurus aristulatus Michx. = Alopecurus aequalis Sobol. ■
17512 Alopecurus arundinaceus Poir.;苇状看麦娘(大看麦娘,东北看麦娘);Creeping Foxtail,Creeping Meadow Foxtail,Reed Foxtail, Reedlike Alopecurus ■
17513 Alopecurus aucheri Boiss.;奥氏看麦娘■☆
17514 Alopecurus borealis Trin. = Alopecurus alpinus Sm. ■☆
17515 Alopecurus brachystachyus M. Bieb.;短穗看麦娘;Shortspike Alopecurus,Shortspike Fox Tail ■
17516 Alopecurus bulbosus Gouan;球根看麦娘;Bulbous Foxtail ■☆
17517 Alopecurus bulbosus Gouan subsp. macrostachyus (Poir.) Trab.;大穗球根看麦娘■☆
17518 Alopecurus bulbosus Gouan var. macrostachyus (Poir.) Coss. et Durieu = Alopecurus bulbosus Gouan subsp. macrostachyus (Poir.) Trab. ■☆
17519 Alopecurus bulbosus Gouan var. salditanus Batt. = Alopecurus bulbosus Gouan subsp. macrostachyus (Poir.) Trab. ■☆
17520 Alopecurus capensis Thunb.;好望角看麦娘■☆
17521 Alopecurus carolinianus Walter;卡罗来纳看麦娘;Annual Foxtail,Carolina Foxtail,Common Foxtail ■☆
17522 Alopecurus creticus Trin.;克里特看麦娘;Cretan Meadow Foxtail ■☆
17523 Alopecurus dasyanthus Trautv.;毛花看麦娘■☆
17524 Alopecurus echinatus Thunb. = Tribolium echinatum (Thunb.) Renvoize ■☆
17525 Alopecurus fulvus Sm.;黄看麦娘;Yellow Alopecurus ■☆
17526 Alopecurus fulvus Sm. var. amurensis Kom. = Alopecurus aequalis Sobol. var. amurensis (Kom.) Ohwi ■
17527 Alopecurus geniculatus L.;膝曲看麦娘;Floating Foxtail,Foxtail Grass,Marsh Fox Tail,Marsh Foxtail,Water Fox Tail,Water Foxtail ■☆
17528 Alopecurus geniculatus L. subsp. fulvus (Sm.) Trab. = Alopecurus aequalis Sobol. ■
17529 Alopecurus geniculatus L. var. aequalis (Sobol.) Paunero. = Alopecurus aequalis Sobol. ■
17530 Alopecurus geniculatus L. var. amurensis (Kom.) Roshev. = Alopecurus aequalis Sobol. ■
17531 Alopecurus geniculatus L. var. aristulatus (Michx.) Torr. = Alopecurus aequalis Sobol. ■
17532 Alopecurus gerardii (All.) Vill. = Alopecurus alpinus Vill. ■☆
17533 Alopecurus gerardii Vill.;盖氏看麦娘■☆
17534 Alopecurus glacialis K. Koch;冰雪看麦娘■☆
17535 Alopecurus glaucus Less.;灰蓝看麦娘;Blue-grey Fox Tail ■☆
17536 Alopecurus himalaicus Hook. f.;喜马拉雅看麦娘;Himalayas Alopecurus,Himalayas Fox Tail ■
17537 Alopecurus japonicus Steud.;日本看麦娘;Japan Alopecurus, Japanese Alopecurus ■
17538 Alopecurus laxiflorus Ovcz.;疏花看麦娘■☆
17539 Alopecurus liouvilleanus Braun-Blanq.;利乌维尔看麦娘■☆
17540 Alopecurus litoreus All. = Rostraria litorea (All.) Holub ■☆
17541 Alopecurus longiaristatus Maxim.;长芒看麦娘;Longawn Alopecurus,Longawn Fox Tail ■
17542 Alopecurus macounii Vasey = Alopecurus carolinianus Walter ■☆
17543 Alopecurus mandshuricus Litv.;东北看麦娘;NE. China Alopecurus,Northeastern Fox Tail ■
17544 Alopecurus mandshuricus Litv. = Alopecurus longearistatus Maxim. ■
17545 Alopecurus mandshuricus Litv. var. glabratus Litv. = Alopecurus mandshuricus Litv. ■
17546 Alopecurus monspeliensis L. = Polypogon monspeliensis (L.) Desf. ■
17547 Alopecurus mucronatus Hack. ex Paulsen;短尖看麦娘■☆
17548 Alopecurus myosuroides Huds.;大穗看麦娘;Bigspike Alopecurus,Black Bent,Black Quitch,Black-grass,Field Foxtail, Hunger-weed,Hungry Grass,Mouse Fox Tail,Mouse Foxtail,Slender Foxtail,Slender Meadow Foxtail ■
17549 Alopecurus nepalensis Trin. ex Steud.;尼泊尔看麦娘■☆
17550 Alopecurus nigra L.;黑看麦娘■☆
17551 Alopecurus pallescens Piper = Alopecurus geniculatus L. ■☆
17552 Alopecurus pratensis Desf. = Alopecurus bulbosus Gouan subsp. macrostachyus (Poir.) Trab. ■☆
17553 Alopecurus pratensis L.;大看麦娘(草原看麦娘);Black Grass, Blackgrass,Meadow Alopecurus,Meadow Fox Tail,Meadow Foxtail ■
17554 Alopecurus pratensis L.'Aureomarginatus';黄斑草原看麦娘; Golden Foxtail-grass,Golden Meadow Foxtail ■☆
17555 Alopecurus pratensis L.'Aureovariegatus' = Alopecurus pratensis L.'Aureomarginatus' ■☆
17556 Alopecurus pratensis L.'Aureus';金色大看麦娘;Golden Meadow Foxtail ■☆

17557 Alopecurus pratensis L. ' Aureus ' = Alopecurus pratensis L. ' Aureomarginatus ' ■☆

17558 Alopecurus pratensis L. subsp. nigricans (Hornem.) Hartm. = Alopecurus arundinaceus Poir. ■

17559 Alopecurus pratensis L. subsp. ventricosus (Pers.) Maire = Alopecurus arundinaceus Poir. ■

17560 Alopecurus pratensis L. var. castellanus (Boiss. et Reut.) Maire et Weiller = Alopecurus arundinaceus Poir. ■

17561 Alopecurus pratensis L. var. exserens (T. Marsson) Maire = Alopecurus arundinaceus Poir. ■

17562 Alopecurus pratensis L. var. ventricosus (Pers.) Coss. et Durieu = Alopecurus arundinaceus Poir. ■

17563 Alopecurus ramosus Poir. = Alopecurus carolinianus Walter ■☆

17564 Alopecurus rendlei Eig;伦德尔看麦娘;Rendle's Meadow Foxtail ■☆

17565 Alopecurus roshevitzianus Ovcz.;罗氏看麦娘■☆

17566 Alopecurus ruthenicus Weinm. = Alopecurus arundinaceus Poir. ■

17567 Alopecurus seravschanicus Ovcz.;塞拉夫看麦娘■☆

17568 Alopecurus sericeus Albov;绢毛看麦娘■☆

17569 Alopecurus songaricus (Schrenk ex Fisch. et Meyen) Petr. = Alopecurus pratensis L. ■

17570 Alopecurus soongaricus (Schrenk) Petr. = Alopecurus pratensis L. ■

17571 Alopecurus stejnegeri Vasey;斯泰看麦娘■☆

17572 Alopecurus tenuis Kom.;细看麦娘■☆

17573 Alopecurus textilis Boiss.;编织看麦娘■☆

17574 Alopecurus tiflisiensis Grossh.;梯弗里斯看麦娘■☆

17575 Alopecurus typhoides Burm. f. = Pennisetum americanum (L.) Leeke ■

17576 Alopecurus typhoides Burm. f. = Pennisetum glaucum (L.) R. Br. ■

17577 Alopecurus utriculatus (L.) Pers. = Alopecurus rendlei Eig ■☆

17578 Alopecurus vaginatus Pall.;鞘叶看麦娘■☆

17579 Alopecurus ventricosus Pers. = Alopecurus arundinaceus Poir. ■

17580 Alophia Herb. (1840);裸柱花属;Pinewoods-lily, Purple Pleat-leaf ■☆

17581 Alophia brasiliensis (Baker) Kuntze;巴西裸柱花■☆

17582 Alophia brasiliensis Kuntze = Alophia brasiliensis (Baker) Kuntze ■☆

17583 Alophia drummondii (Graham) R. C. Foster;裸柱花;Propeller Flower ■☆

17584 Alophia lahue (Molina) Espinosa = Herbertia lahue (Molina) Goldblatt ■☆

17585 Alophia lahue (Molina) Espinosa subsp. caerulea (Herb.) Ravenna = Herbertia lahue (Molina) Goldblatt ■☆

17586 Alophia silvestris (Loes.) Goldblatt;林地裸柱花■☆

17587 Alophium Cass. = Centaurea L.(保留属名)●■

17588 Alophochloa Endl. = Koeleria Pers. ■

17589 Alophochloa Endl. = Lophochloa Rchb. ■☆

17590 Alophyllus L. = Allophylus L. ●

17591 Alopicarpus Neck. = Paris L. ■

17592 Aloranthus F. S. Voigt = Chloranthus Sw. ●■

17593 Alosemis Raf. = Rhynchanthera DC.(保留属名)●☆

17594 Aloseris Raf. = Alosemis Raf. ●☆

17595 Aloysia Juss. = Aloysia Ortega et Paláu ex Pers. ●☆

17596 Aloysia Ortega et Paláu ex Pers. = Aloysia Paláu ●☆

17597 Aloysia Paláu(1784);橙香木属(防臭木属,柠檬马鞭木属)●☆

17598 Aloysia citriodora Ortega ex Pers. = Lippia citriodora (Paláu) Kunth ●☆

17599 Aloysia citriodora Paláu = Lippia citriodora (Paláu) Kunth ●☆

17600 Aloysia gratissima (Gillies et Hook.) Tronc.;枸杞状过江藤(枸杞防臭木);Beebrush, Whitebrush ●☆

17601 Aloysia lycioides Cham. = Aloysia gratissima (Gillies et Hook.) Tronc. ●☆

17602 Aloysia macrostachya (Torr.) Moldenke;大穗过江藤(大穗防臭木);Sweet Stem ●☆

17603 Aloysia polystachya (Griseb.) Moldenke;多穗过江藤(多穗防臭木)●☆

17604 Aloysia triphylla (L'Hér.) Britton = Aloysia citriodora Paláu ●☆

17605 Aloysia triphylla Britton;三叶防臭木;Lemon Verbena ●☆

17606 Aloysia triphylla Britton = Lippia citrodora (Paláu) Kunth ●☆

17607 Aloysia wrightii A. Heller;里根过江藤;Oreganillo ●☆

17608 Alpaminia O. E. Schulz = Weberbauera Gilg et Muschl. ■☆

17609 Alpan Bose ex Raf. = Apama Lam. ●

17610 Alphandia Baill. (1873);阿尔法大戟属☆

17611 Alphandia furfuracea Baill.;阿尔法大戟☆

17612 Alphitonia Endl. = Alphitonia Reissek ex Endl. ●

17613 Alphitonia Reissek ex Endl. (1840);麦珠子属;Alphitonia, Red Almond, Red Ash, Soap Tree, Tree Buckthorn ●

17614 Alphitonia exelsa Reissek ex Endl.;高麦珠子(美丽麦珠子);Buckthorn, Ked, Red Almond, Silver-leaf, Tall Alphitonia, Tree Buekthorn ●☆

17615 Alphitonia ferruginea Merr. et L. M. Perry;锈色麦珠子●☆

17616 Alphitonia lucida Vieill. ex Guillaumin;光亮麦珠子●☆

17617 Alphitonia macrocarpa Mansf.;大果麦珠子●☆

17618 Alphitonia neocaledonica Gillies;新卡利登麦珠子●☆

17619 Alphitonia obtusifolia Braid;钝叶麦珠子☆

17620 Alphitonia petriei Braid;石生麦珠子●☆

17621 Alphitonia philippinensis Braid;麦珠子(白石松,朦朦木,蒙蒙木,山木棉,山油麻,银树);Philippine Alphitonia ●

17622 Alphitonia zizyphoides A. Gray;枣状麦珠子●☆

17623 Alphonsea Hook. f. et Thomson (1855);藤春属(阿芳属);Alphonsea ●

17624 Alphonsea arborea Merr.;乔木藤春木●☆

17625 Alphonsea boniana Finet et Gagnep.;金平藤春;Chinping Alphonsea, Jinping Alphonsea ●

17626 Alphonsea hainanensis Merr. et Chun;海南藤春(海南阿芳);Hainan Alphonsea ●

17627 Alphonsea mollis Dunn;毛叶藤春(毛阿芳,毛叶阿芳,石密);Soft Alphonsea ●

17628 Alphonsea monogyna Merr. et Chun;藤春(阿芳,单果阿芳,金榕,山坝);Monogynous Alphonsea, Onepistil Alphonsea ●

17629 Alphonsea prolifica Chun et F. C. How = Miliusa prolifica (Chun et F. C. How) P. T. Li ●◇

17630 Alphonsea prolifica Chun et F. C. How = Saccopetalum prolificum (Chun et F. C. How) Tsiang ●◇

17631 Alphonsea squamosa Finet et Gagnep.;多苞藤春;Scaled Alphonsea ●

17632 Alphonsea tsangyuanensis P. T. Li;多脉藤春;Cangyuan Alphonsea, Multivein Alphonsea, Tsangyuan Alphonsea ●

17633 Alphonseopsis Baker f. (1913);拟藤春属●☆

17634 Alphonseopsis Baker f. = Polyceratocarpus Engl. et Diels ●☆

17635 Alphonseopsis parviflora Baker f. = Polyceratocarpus parviflorus (Baker f.) Ghesq. ●☆

17636 Alpinia K. Schum. = Alpinia Roxb.(保留属名)■
17637 Alpinia L.(废弃属名)= Alpinia Roxb.(保留属名)■
17638 Alpinia L.(废弃属名)= Renealmia L. f.(保留属名)■☆
17639 Alpinia Roxb.(1753)(保留属名);山姜属(月桃属);Alpinia, Galanga, Galangal, Ginger-lily, Indian Shell-flower, Shell Ginger ■
17640 Alpinia × kiushiana Kitam.;北村山姜■☆
17641 Alpinia × okinawensis Tawada;冲绳山姜■☆
17642 Alpinia africana Ridl. = Renealmia sancti-thomae I. M. Turner ■☆
17643 Alpinia agiokuensis Hayata = Alpinia japonica (Thunb.) Miq. ■
17644 Alpinia allughas (Retz.) Roscoe = Alpinia nigra (Gaertn.) B. L. Burtt ■
17645 Alpinia allughas Roscoe = Alpinia japonica (Thunb.) Miq. ■
17646 Alpinia aquatica (König) Roscoe;水山姜;Aquatic Galangal ■
17647 Alpinia bambusifolia C. F. Liang et D. Fang;竹叶山姜(南川山姜);Bambooleaf Galangal ■
17648 Alpinia bilamellata Makino = Alpinia nigra (Gaertn.) B. L. Burtt ■
17649 Alpinia blepharocalyx K. Schum.;云南山姜(滇草蔻,小白蔻,小草蔻,云南草寇,云南野砂仁);Yunnan Galangal ■
17650 Alpinia blepharocalyx K. Schum. var. glabrior (Hand.-Mazz.) T. L. Wu;光叶云南山姜(光叶云南草寇);Shinyleaf Galangal, Smoothleaf Galangal ■
17651 Alpinia bracteata Roxb.;绿苞山姜;Greenbract Galangal ■
17652 Alpinia bracteata Roxb. = Alpinia blepharocalyx K. Schum. ■
17653 Alpinia brevis T. L. Wu et S. J. Chen;小花山姜;Smallflower Galangal ■
17654 Alpinia caerulea Benth.;澳洲山姜;Australian Ginger ■☆
17655 Alpinia calcarata Roscoe;距花山姜(小良姜);India Galangal, Indian Galangal, Indian Ginger ■
17656 Alpinia chinensis (Retz.) Roscoe;华山姜(大杆,高良姜,华良姜,见秆风,剑杆,箭杆风,姜汇,姜活,姜叶淫羊藿,九姜连,九连姜,廉姜,山姜,葰,土砂仁,座杆);China Galangal, Chinese Galangal ■
17657 Alpinia conchigera Griff.;节鞭山姜;Nodosewhip Galangal ■
17658 Alpinia conghuaensis J. P. Liao et T. L. Wu;从化山姜;Conghua Galangal ■
17659 Alpinia coplandii Ridl.;密毛山姜(川上氏月桃);Densehair Galangal ■
17660 Alpinia coriacea T. L. Wu et S. J. Chen;革叶山姜;Coriaceous Galangal, Hideleaf Galangal ■
17661 Alpinia coriandriodora D. Fang;香姜(香草);Fragrant Galangal ■
17662 Alpinia densespicata Hayata;七星月桃;Densespicate Galangal ■
17663 Alpinia densespicata Hayata = Alpinia shimadae Hayata ■
17664 Alpinia densibracteata T. L. Wu et S. J. Chen;密苞山姜;Densebract Galangal ■
17665 Alpinia densibracteata T. L. Wu et S. J. Chen = Alpinia stachyoides Hance ■
17666 Alpinia dolichocephala Hayata;紫纹山姜(长穗月桃);Purplestriate Galangal ■
17667 Alpinia elatior Jack = Etlingera elatior (Jack) R. M. Sm. ■
17668 Alpinia elwesii Turrill = Alpinia kawakamii Hayata ■
17669 Alpinia elwesii Turrill = Alpinia shimadae Hayata var. kawakamii (Hayata) J. Jung Yang et J. C. Wang ■
17670 Alpinia emaculata S. Q. Tong;无斑山姜;Spotless Galangal ■
17671 Alpinia flabellata (Ridl.) Merr.;扇唇山姜(吕宋月桃)■
17672 Alpinia flabellata Ridl. = Alpinia flabellata (Ridl.) Merr. ■
17673 Alpinia fluviatilis Hayata = Alpinia zerumbet (Pers.) B. L. Burtt et R. M. Sm. ■
17674 Alpinia formosana K. Schum.;美山姜(高良姜,台湾山姜,台湾月桃,土蔻);Taiwan Galangal ■
17675 Alpinia formosana K. Schum. f. variegata (Makino) Kitam.;斑点美山姜■☆
17676 Alpinia galanga (L.) Sw. = Alpinia galanga (L.) Willd. ■
17677 Alpinia galanga (L.) Willd.;红豆蔻(大高良姜,大良姜,高良姜,红豆,红蔻,姜,廉姜,良姜,良姜子,南姜,山姜,山姜子);Galanga Galangal, Galangal, Galingale, Great Galangal, Greater Galangal, Java Galangal, Siamese Ginger ■
17678 Alpinia galanga (L.) Willd. var. appendiculata Y. H. Chen;附尖红豆蔻;Appendiculate Galanga Galangal ■
17679 Alpinia galanga (L.) Willd. var. pyramidata (Blume) K. Schum.;有毛红豆蔻(毛红豆蔻);Pyramidate Galanga Galangal ■
17680 Alpinia galanga Lour. = Alpinia galanga (L.) Willd. ■
17681 Alpinia globosa Horan.;脆果山姜(白豆蔻,豆蔻);Globose Galangal ■
17682 Alpinia graminifolia D. Fang et G. Y. Lo;狭叶山姜;Narrowleaf Galangal ■
17683 Alpinia guangdongensis S. J. Chen et Z. Y. Chen = Alpinia maclurei Merr. var. guangdongensis (S. J. Chen et Z. Y. Chen) Z. L. Zhao et L. S. Xu ■
17684 Alpinia guinanensis D. Fang et X. X. Shen;桂南山姜;Guinan Galangal ■
17685 Alpinia hainanensis K. Schum.;海南山姜(草豆蔻);Hainan Galangal ■
17686 Alpinia henryi K. Schum.;小草寇(滇南山姜,亨氏山姜,直穗山姜);Henry Galangal, Small Galangal ■
17687 Alpinia henryi K. Schum. = Alpinia hainanensis K. Schum. ■
17688 Alpinia henryi K. Schum. var. densihispida H. Dong et G. J. Xu = Alpinia hainanensis K. Schum. ■
17689 Alpinia henryi K. Schum. var. densihispida H. Dong et G. J. Xu = Alpinia henryi K. Schum. ■
17690 Alpinia hibinoi Masam.;芽姜(日比野山姜)■
17691 Alpinia hokutensis Hayata = Alpinia formosana K. Schum. ■
17692 Alpinia hokutensis Hayata = Alpinia intermedia Gagnep. ■
17693 Alpinia intermedia Gagnep.;光叶山姜(山月桃,山月桃仔,椭圆叶月桃);Lucidleaf Galangal ■
17694 Alpinia iriomotensis Masam. = Alpinia flabellata Ridl. ■
17695 Alpinia japonica (Thunb.) Miq.;山姜(白寒果,杜若,和山姜,鸡爪莲,建砂仁,箭杆风,箭秆风,姜叶淫羊藿,九姜连,九姜莲,九节莲,九龙盘,美草,土砂仁,湘砂仁);Japan Galangal, Japanese Galangal ■
17696 Alpinia japonica (Thunb.) Miq. f. xanthocarpa Yamasiro et Maeda;黄果山姜■☆
17697 Alpinia japonica (Thunb.) Sasaki = Alpinia japonica (Thunb.) Miq. ■
17698 Alpinia jianganfeng T. L. Wu;箭杆风■
17699 Alpinia jingxiensis D. Fang;靖西山姜;Chingsi Galangal, Jingxi Galangal ■
17700 Alpinia kainantensis Masam.;开南山姜■
17701 Alpinia katsumadae Hayata = Alpinia hainanensis K. Schum. ■
17702 Alpinia katsumadaei Hayata;草豆蔻(宝蔻,草蔻,草果,草叩,草扣,草蔻,大草蔻,豆蔻,豆叩,豆蔻,飞雷子,家孽,假麻树,漏蔻,偶子);Katsumada Galangal ■
17703 Alpinia kawakamii Hayata;川上山姜(川上氏月桃,密毛山姜)■
17704 Alpinia kawakamii Hayata = Alpinia shimadae Hayata var. kawakamii (Hayata) J. Jung Yang et J. C. Wang ■

17705 Alpinia kelungensis Hayata = Alpinia intermedia Gagnep. ■
17706 Alpinia koidzumiana Kitam. = Alpinia flabellata Ridl. ■
17707 Alpinia koshunensis Hayata；高雄山姜；Gaoxiong Galangal，Koshun Galangal ■
17708 Alpinia koshunensis Hayata = Alpinia formosana K. Schum. ■
17709 Alpinia kumatake Makino = Alpinia formosana K. Schum. ■
17710 Alpinia kumatake Makino = Alpinia intermedia Gagnep. ■
17711 Alpinia kumatake Makino var. variegata Makino；斑叶高良姜■☆
17712 Alpinia kusshakuensis Hayata；菱唇山姜（屈尺月桃）；Rhombic Lip Galangal，Rhombiclip Galangal ■
17713 Alpinia kwangsiensis T. L. Wu et S. J. Chen；长柄山姜（大豆蔻）；Guangxi Galangal，Kwangsi Galangal ■
17714 Alpinia maclurei Merr.；假益智（红扣，假草果）；Maclure Galangal ■
17715 Alpinia maclurei Merr. var. guangdongensis（S. J. Chen et Z. Y. Chen）Z. L. Zhao et L. S. Xu；广东山姜（光叶假益智）；Guangdong Galangal ■
17716 Alpinia macrocephala Hayata = Alpinia sessiliflora Kitam. ■
17717 Alpinia macrocephala K. Schum.；阿里山月桃■
17718 Alpinia magnifica Blume = Amomum magnificum Benth. et Hook. f. ■☆
17719 Alpinia malaccensis（Burm.）Roscoe；毛瓣山姜（玉蔻）；Malacca Galangal ■
17720 Alpinia mediomaculata Hayata = Alpinia shimadae Hayata ■
17721 Alpinia menghaiensis S. Q. Tong et Y. M. Xia；勐海山姜；Menghai Galangal ■
17722 Alpinia mesanthera Hayata；疏花山姜（角板山高良姜，角板山月桃）；Laxflower Galangal ■
17723 Alpinia mutica Roxb.；南山姜；Small Shell Ginger ■
17724 Alpinia nakaiana Tuyama；中井氏山姜■☆
17725 Alpinia nanchuanensis Z. Y. Zhu；南川山姜■
17726 Alpinia nanchuanensis Z. Y. Zhu = Alpinia bambusifolia C. F. Liang et D. Fang ■
17727 Alpinia napoensis H. Dong et G. J. Xu；那坡山姜（土砂仁）；Napo Galangal ■
17728 Alpinia nigra（Gaertn.）B. L. Burtt；黑果山姜（益智）；Blackfruit Galangal ■
17729 Alpinia nutans Roscoe = Alpinia speciosa（Blume）D. Dietr. ■
17730 Alpinia nutans Roscoe = Alpinia zerumbet（Pers.）B. L. Burtt et R. M. Sm. ■
17731 Alpinia oblongifolia Hayata；椭圆叶月桃（华山姜）■
17732 Alpinia oblongifolia Hayata = Alpinia intermedia Gagnep. ■
17733 Alpinia officinaria Hance；高良姜（比目连理花，风姜，佛手根，高凉姜，膏凉姜，海良姜，良姜，埋光乌药，蛮姜，山羌，小良姜）；Chinese Galan，Chinese Galangal，Galangal，Lasser Galangal，Small Galangal ■
17734 Alpinia ovata Z. L. Zhao et L. S. Xu；卵唇山姜；Ovate Galangal ■
17735 Alpinia ovoidocarpa H. Dong et G. J. Xu；卵果山姜；Ovatefruit Galangal ■
17736 Alpinia oxyphylla Miq.；益智（益智子，英华库，摘艼子）；Sharpleaf Galangal ■
17737 Alpinia pesionsa K. Schum.；佩兴山姜■☆
17738 Alpinia pinnanensis T. L. Wu et S. J. Chen；柱穗山姜；Pinnan Galangal ■
17739 Alpinia platychilus K. Schum.；宽唇山姜；Broadlip Galangal ■
17740 Alpinia polyantha D. Fang；多花山姜（土白蔻）；Manyflower Galangal ■
17741 Alpinia pricei Hayata；短穗山姜（普来氏月桃）；Shortspike Galangal ■
17742 Alpinia pricei Hayata var. sessiliflora（Kitam.）J. J. Yang et F. C. Wang = Alpinia sessiliflora Kitam. ■
17743 Alpinia psilogyna D. Fang；矮山姜（产后姜）；Pygmy Galangal ■
17744 Alpinia pumila Hook. f.；花叶山姜（假砂仁，箭杆风，箭秆风，山姜，野姜黄，竹节风）；Dwarf Galangal ■
17745 Alpinia purpurea（Vieill.）K. Schum.；红花月桃（紫花山姜，紫山姜）；Purple Galangal，Red Ginger ■
17746 Alpinia pyramidata Blume = Alpinia galanga（L.）Willd. var. pyramidata（Blume）K. Schum. ■
17747 Alpinia rafflesiana Wall.；拉氏山姜■☆
17748 Alpinia rubromaculata S. Q. Tong；红斑山姜；Red-spotted Galangal ■
17749 Alpinia sanderae Sand.；桑氏山姜（斑点月桃，花叶姜，桑德山姜）■☆
17750 Alpinia sasakii Hayata = Alpinia pricei Hayata ■
17751 Alpinia schumanniana Valeton；垂花山姜（濑水月桃）；Pendulous Flower Galangal，Pendulousflower Galangal ■
17752 Alpinia schumanniana Valeton = Alpinia zerumbet（Pers.）B. L. Burtt et R. M. Sm. ■
17753 Alpinia sessiliflora Kitam.；大头山姜（阿里山月桃）；Bighead Galangal，Sessile Galangal ■
17754 Alpinia shimadae Hayata；密穗山姜（岛田氏月桃，七星山姜，新竹山姜）；Densespike Galangal ■
17755 Alpinia shimadae Hayata var. kawakamii（Hayata）J. Jung Yang et J. C. Wang = Alpinia kawakamii Hayata ■
17756 Alpinia sichuanensis Z. Y. Zhu；四川山姜；Sichuan Galangal ■
17757 Alpinia speciosa（Blume）D. Dietr. = Etlingera elatior（Jack）R. M. Sm. ■
17758 Alpinia speciosa（J. C. Wendl.）K. Schum. = Alpinia zerumbet（Pers.）B. L. Burtt et R. M. Sm. ■
17759 Alpinia stachyoides Hance；箭秆风（假砂仁，密苞山姜，一支箭，一枝箭）；Betonylike Galangal ■
17760 Alpinia stachyoides Hance var. yangchunensis Z. L. Zhao et L. S. Xu；阳春山姜；Yangchun Galangal ■
17761 Alpinia strobiliformis T. L. Wu et S. J. Chen；球穗山姜（野姜，珠穗山姜）；Coneshaped Galangal ■
17762 Alpinia strobiliformis T. L. Wu et S. J. Chen var. glabra T. L. Wu；光叶球穗山姜（光叶珠穗山姜）；Glabrous Coneshaped Galangal ■
17763 Alpinia subulatum Roxb.；印度砂仁■☆
17764 Alpinia suishaensis Hayata = Alpinia intermedia Gagnep. ■
17765 Alpinia suishaensis Hayata = Alpinia oblongifolia Hayata ■
17766 Alpinia takaminei Masam.；高峰山姜■☆
17767 Alpinia tarokoensis（Sasaki）Hayata = Alpinia pricei Hayata ■
17768 Alpinia tarokoensis Hayata = Alpinia pricei Hayata ■
17769 Alpinia tonkinensis Gagnep.；滑叶山姜（白蔻）；Tonkin Galangal ■
17770 Alpinia tonrokuensis Hayata；屯鹿月桃（敦六山姜，台北山姜，通落月桃）；Tonroku Galangal ■
17771 Alpinia tricolor Sand.；三色山姜■☆
17772 Alpinia uraiensis Hayata；大花山姜（大轮月桃，乌来月桃）；Bigflower Galangal ■
17773 Alpinia vittata Hook. = Alpinia sanderae Sand. ■☆
17774 Alpinia zerumbet（Pers.）B. L. Burtt et R. M. Sm.；艳山姜（草豆蔻，草蔻，大草蔻，大豆蔻，虎子花，良姜，裸叶，美丽山姜，土砂仁，玉桃，月桃）；Beautiful Galangal，Shell Flower，Shell Ginger，Shellflower，Shell-flower，Shellplant ■

17775 Alpinia zerumbet (Pers.) B. L. Burtt et R. M. Sm. 'Springle'; 雨花山姜■☆

17776 Alpinia zerumbet (Pers.) B. L. Burtt et R. M. Sm. 'Variegata'; 花叶艳山姜■☆

17777 Alpiniaceae F. Rudolphi = Zingiberaceae Martinov(保留科名)■

17778 Alpiniaceae F. Rudolphi;山姜科■

17779 Alpiniaceae Link = Zingiberaceae Martinov(保留科名)■

17780 Alposelinum Pimenov = Lomatocarpa Pimenov ■

17781 Alrawia (Wendelbo) K. M. Perss. et Wendelbo(1979);波斯风信子属■☆

17782 Alrawia bellii (Baker) K. M. Perss. et Wendelbo;波斯风信子■☆

17783 Alrawia nutans (Wendelbo) K. Perss. et Wendelbo;点头波斯风信子■☆

17784 Alsaton Raf. = Siler Mill. ●☆

17785 Alschingera Vis. = Physospermum Cusson ex Juss. ■☆

17786 Alseis Schott(1827);丛林茜属●☆

17787 Alseis peruviana Standl.;丛林茜●☆

17788 Alsenosmia Eudl. = Alseuosmia A. Cunn. ●☆

17789 Alseodaphne Nees(1831);油丹属(蜀楠属);Alseodaphne ●

17790 Alseodaphne andersonii (King ex Hook. f.) Kosterm.;毛叶油丹;Anderson Alseodaphne ●

17791 Alseodaphne breviflora Benth. = Machilus breviflora (Benth.) Hemsl. ●

17792 Alseodaphne camphorata (H. Lév.) C. K. Allen = Cinnamomum foveolatum (Merr.) H. W. Li et J. Li ●

17793 Alseodaphne caudata Lecomte = Cinnamomum caudiferum Kosterm. ●

17794 Alseodaphne caudata Lecomte = Cinnamomum foveolatum (Merr.) H. W. Li et J. Li ●

17795 Alseodaphne cavaleriei (H. Lév.) Kosterm. = Machilus cavaleriei (H. Lév.) H. Lév. ●

17796 Alseodaphne chinensis Benth. = Machilus chinensis (Champ. ex Benth.) Hemsl. ●

17797 Alseodaphne chinensis Champ. ex Benth. = Machilus chinensis (Champ. ex Benth.) Hemsl. ●

17798 Alseodaphne dumicola W. W. Sm. = Machilus dumicola (W. W. Sm.) H. W. Li ●

17799 Alseodaphne gracilis Kosterm.; 细梗油丹; Slenderstalk Alseodaphne, Slender-stalked Alseodaphne ●

17800 Alseodaphne hainanensis Merr.;油丹(海南峨眉楠,黄丹,黄丹公,三次香,硬壳果);Hainan Alseodaphne ●◇

17801 Alseodaphne hokouensis H. W. Li;河口油丹;Hekou Alseodaphne ●

17802 Alseodaphne huanglianshanensis H. W. Li et Y. M. Shui;黄连山油丹;Huanglianshan Alseodaphne ●

17803 Alseodaphne keenanii Gamble = Alseodaphne andersonii (King ex Hook. f.) Kosterm. ●

17804 Alseodaphne marlipoensis (H. W. Li) H. W. Li;麻栗坡油丹;Malipo Alseodaphne ●

17805 Alseodaphne medogensis H. P. Tsui; 墨脱油丹; Motuo Alseodaphne ●

17806 Alseodaphne medogensis H. P. Tsui = Alseodaphne andersonii (King ex Hook. f.) Kosterm. ●

17807 Alseodaphne mollis W. W. Sm. = Cinnamomum tenuipile Kosterm. ●

17808 Alseodaphne omeiensis Gamble = Nothaphoebe cavaleriei (H. Lév.) Yen C. Yang ●◇

17809 Alseodaphne petiolaris (Meisn.) Hook. f.;长柄油丹(石山掼槽树);Stalked Alseodaphne ●

17810 Alseodaphne rugosa Merr. et Chun;皱皮油丹(黄丹);Wrinkled Alseodaphne ●◇

17811 Alseodaphne sichourensis H. W. Li;西畴油丹(黄丹);Xichou Alseodaphne ●◇

17812 Alseodaphne yunnanensis Kosterm.; 云南油丹; Yunnan Alseodaphne ●

17813 Alseuosmia A. Cunn. (1838);岛海桐属●☆

17814 Alseuosmia atriplicifolia A. Cunn.;岛海桐●☆

17815 Alseuosmia banksii A. Cunn.;邦克岛海桐●☆

17816 Alseuosmia hookeria Colenso;胡克岛海桐●☆

17817 Alseuosmia macrophylla A. Cunn.;大叶岛海桐●☆

17818 Alseuosmia quercifolia A. Cunn.;栎叶岛海桐●☆

17819 Alseuosmiaceae Airy Shaw(1965);岛海桐科(假海桐科)●☆

17820 Alsinaceae (DC.) Bartl. = Alsinaceae Bartl. (保留科名)●■

17821 Alsinaceae (DC.) Bartl. = Caryophyllaceae Juss. (保留科名)●■

17822 Alsinaceae Adans. = Caryophyllaceae Juss. (保留科名)●■

17823 Alsinaceae Bartl. (1825)(保留科名) = Caryophyllaceae Juss. (保留科名)●■

17824 Alsinaceae Bartl. (1825)(保留科名);繁缕科■

17825 Alsinanthe (Fenzl ex Endl.) Rchb. = Minuartia L. ■

17826 Alsinanthe Rchb. = Arenaria L. ■

17827 Alsinanthe elegans (Cham. et Schltdl.) Á. Löve et D. Löve = Minuartia elegans (Cham. et Schltdl.) Schischk. ■☆

17828 Alsinanthe macrantha (Rydb.) W. A. Weber = Minuartia macrantha (Rydb.) House ■☆

17829 Alsinanthe rossii (R. Br. ex Richardson) Á. Löve et D. Löve = Minuartia rossii (R. Br. ex Richardson) Graebn. ■☆

17830 Alsinanthe stricta (Sw.) Rchb. = Minuartia stricta (Sw.) Hiern ■☆

17831 Alsinanthemos J. G. Gmel. = Alsinanthemum Fabr. ■

17832 Alsinanthemum Fabr. = Trientalis L. ■

17833 Alsinanthemum Thalius ex Greene = Trientalis L. ■

17834 Alsinanthus Desv. = Arenaria L. ■

17835 Alsinanthus Rchb. = Arenaria L. ■

17836 Alsinastraceae Rupr. = Elatinaceae Dumort. (保留科名)■

17837 Alsinastrum Quer = Elatine L. ■

17838 Alsinastrum Schur = Elatine L. ■

17839 Alsine Druce = Spergularia (Pers.) J. Presl et C. Presl(保留属名)■

17840 Alsine Gaertn. = Minuartia L. ■

17841 Alsine L. = Arenaria L. + Stellaria L. + Delia Dumort ■

17842 Alsine L. = Spergularia (Pers.) J. Presl et C. Presl(保留属名)■

17843 Alsine L. = Stellaria L. ■

17844 Alsine Scop. = Arenaria L. ■

17845 Alsine americana (Porter ex B. L. Rob.) Rydb. = Stellaria americana (Porter ex B. L. Rob.) Standl. ■☆

17846 Alsine aquatica (L.) Britton = Myosoton aquaticum (L.) Moench ■

17847 Alsine aquatica (L.) Britton = Stellaria aquatica (L.) Scop. ■☆

17848 Alsine arctica (Steven ex Ser.) Fenzl = Minuartia arctica (Steven ex Ser.) Graebn. ■

17849 Alsine baicalensis Coville = Stellaria umbellata Turcz. ■

17850 Alsine baldwinii Small = Stellaria cuspidata Willd. ex Schltdl. subsp. prostrata (Baldwin) J. K. Morton ■☆

17851 Alsine biflora Wahlenb. = Minuartia biflora (L.) Schinz et Thell. ■

17852 Alsine bocconi Scheele = Spergularia bocconei (Scheele) Asch. et Graebn. ■☆
17853 Alsine bongardiana (Fernald) Davidson et Moxley = Stellaria borealis Bigelow subsp. sitchana (Steud.) Piper et Beattie ■☆
17854 Alsine borealis (Bigelow) Britton = Stellaria borealis Bigelow ■☆
17855 Alsine calycantha (Ledeb.) Rydb. = Stellaria calycantha (Ledeb.) Bong. ■☆
17856 Alsine calycantha (Ledeb.) Rydb. var. simcoei (Howell) Fernald = Stellaria calycantha (Ledeb.) Bong. ■☆
17857 Alsine campestris (L.) Fenzl = Minuartia campestris L. ■☆
17858 Alsine canadensis House = Spergularia canadensis (Pers.) G. Don ■☆
17859 Alsine crispa (Cham. et Schltdl.) Holz. = Stellaria crispa Cham. et Schltdl. ■☆
17860 Alsine cuspidata (Willd. ex Schltdl.) Wooton et Standl. = Stellaria cuspidata Willd. ex Schltdl. ■☆
17861 Alsine filifolia (Forssk.) Schweinf. = Minuartia filifolia (Forssk.) Mattf. ■☆
17862 Alsine fontinalis (Short et R. Peter) Britton = Stellaria fontinalis (Short et R. Peter) B. L. Rob. ■☆
17863 Alsine funkii Jord. = Minuartia funkii (Jord.) Graebn. ■☆
17864 Alsine gayana Webb ex Christ = Rhodalsine gayana (Christ) Holub ■☆
17865 Alsine geniculata (Poir.) Hochr. = Rhodalsine geniculata (Poir.) F. N. Williams ■☆
17866 Alsine glauca (With.) Britton = Stellaria palustris Ehrh. ex Hoffm. ■
17867 Alsine glutinosa A. Heller = Pseudostellaria jamesiana (Torr.) W. A. Weber et R. L. Hartm. ■☆
17868 Alsine graminea (L.) Britton = Stellaria graminea L. ■
17869 Alsine hirta (Wormsk.) Hartm. var. rubella (Wahlenb.) Hartm. = Minuartia rubella (Wahlenb.) Hiern ■☆
17870 Alsine holostea (L.) Britton = Stellaria holostea L. ■
17871 Alsine humifusa (Rottb.) Britton = Stellaria humifusa Rottb. ■☆
17872 Alsine kabylica Pomel = Minuartia verna (L.) Hiern subsp. kabylica (Pomel) Maire et Weiller ■☆
17873 Alsine longifolia (Muhl. ex Willd.) Britton = Stellaria longifolia Muhl. ex Willd. ■
17874 Alsine longipes (Goldie) Coville = Stellaria longipes Goldie ■☆
17875 Alsine macrocarpa (Pursh) Fenzl var. koreana Nakai = Minuartia macrocarpa (Pursh) Ostenf. var. koreana (Nakai) H. Hara ■
17876 Alsine macrocarpa Fenzl var. koreana Nakai = Minuartia macrocarpa (Pursh) Ostenf. var. koreana (Nakai) H. Hara ■
17877 Alsine maroccana Batt. = Rhodalsine geniculata (Poir.) F. N. Williams ■☆
17878 Alsine media L. = Stellaria media (L.) Vill. ■
17879 Alsine michauxii Fenzl = Arenaria stricta Michx. ■☆
17880 Alsine michauxii Fenzl = Minuartia michauxii (Fenzl) Farw. ■☆
17881 Alsine mollugínea Lag. = Drymaria mollugínea (Ser.) Didr. ■☆
17882 Alsine montana (L.) Fenzl = Minuartia montana L. ■☆
17883 Alsine neglecta (Weihe) Á. Löve et D. Löve = Stellaria neglecta Weihe ex Bluff et Fingerh. ■
17884 Alsine obtusa (Engelm.) Rose = Stellaria obtusa Engelm. ■☆
17885 Alsine occulta Kar. et Kir. = Minuartia biflora (L.) Schinz et Thell. ■
17886 Alsine pallida Dumort. = Stellaria pallida (Dumort.) Pire ■
17887 Alsine palmeri Rydb. = Stellaria longipes Goldie ■☆
17888 Alsine palustris Kellogg = Arenaria paludicola B. L. Rob. ■☆
17889 Alsine platyphylla Christ = Minuartia platyphylla (Christ) McNeill ■☆
17890 Alsine polygonoides Greene ex Rydb. = Stellaria irrigua Bunge ■
17891 Alsine procumbens (Vahl) Fenzl = Rhodalsine geniculata (Poir.) F. N. Williams ■☆
17892 Alsine prostrata (Baldwin) A. Heller = Stellaria cuspidata Willd. ex Schltdl. subsp. prostrata (Baldwin) J. K. Morton ■☆
17893 Alsine prostrata Forssk. = Polycarpon prostratum (Forssk.) Asch. et Schweinf. ■
17894 Alsine pubera (Michx.) Britton = Stellaria pubera Michx. ■☆
17895 Alsine pubera (Michx.) Britton var. tennesseensis C. Mohr = Stellaria pubera Michx. ■☆
17896 Alsine rubella Wahlenb. = Minuartia rubella (Wahlenb.) Hiern ■☆
17897 Alsine schimperi A. Rich. = Minuartia filifolia (Forssk.) Mattf. ■☆
17898 Alsine schimperi A. Rich. var. ellenbeckii Engl. = Minuartia ellenbeckii (Engl.) M. G. Gilbert ■☆
17899 Alsine schimperi A. Rich. var. erlangeriana Engl. = Minuartia ellenbeckii (Engl.) M. G. Gilbert ■☆
17900 Alsine segetalis L. = Arenaria segetalis Lam. ■☆
17901 Alsine segetalis L. = Spergularia segetalis (L.) G. Don ■☆
17902 Alsine setacea Mert. et Koch var. tenuissima (Pomel) Batt. = Minuartia tenuissima (Pomel) Mattf. ■☆
17903 Alsine simcoei (Howell) C. L. Hitchc. = Stellaria calycantha (Ledeb.) Bong. ■☆
17904 Alsine simcoei Howell = Stellaria calycantha (Ledeb.) Bong. ■☆
17905 Alsine strictiflora Rydb. = Stellaria longipes Goldie ■☆
17906 Alsine tennesseensis (C. Mohr) Small = Stellaria corei Shinners ■☆
17907 Alsine tenuifolia (L.) Crantz = Minuartia hybrida (Vill.) Schischk. ■☆
17908 Alsine tenuifolia (L.) Crantz var. confertiflora Fenzl = Minuartia hybrida (Vill.) Schischk. ■☆
17909 Alsine tenuifolia (L.) Crantz var. munbyi (Boiss.) Batt. = Minuartia hybrida (Vill.) Schischk. ■☆
17910 Alsine tenuifolia (L.) Crantz var. regeliana Trautv. = Minuartia regeliana (Trautv.) Mattf. ■
17911 Alsine tenuifolia (L.) Crantz var. viscosa (Schreb.) Boiss. = Minuartia hybrida (Vill.) Schischk. ■☆
17912 Alsine tenuissima Pomel = Minuartia tenuissima (Pomel) Mattf. ■☆
17913 Alsine trivialis (L.) E. H. L. Krause = Moehringia trinervia (L.) Clairv. ■
17914 Alsine uliginosa (Murray) Britton = Stellaria alsine Grimm ■
17915 Alsine uniflora (Walter) A. Heller = Minuartia uniflora (Walter) Mattf. ■☆
17916 Alsine validus Goodd. = Stellaria longipes Goldie ■☆
17917 Alsine verna (L.) Wahlenb. = Minuartia verna (L.) Hiern ■
17918 Alsine verna (L.) Wahlenb. var. hirta (Wormsk.) Fenzl = Minuartia verna (L.) Hiern ■
17919 Alsine verna (L.) Wahlenb. var. umbrosa Chabert = Minuartia verna (L.) Hiern subsp. kabylica (Pomel) Maire et Weiller ■☆
17920 Alsine verna Wahlenb. = Minuartia verna (L.) Hiern ■
17921 Alsine viridula Piper = Stellaria obtusa Engelm. ■☆
17922 Alsine washingtoniana (B. L. Rob.) A. Heller = Stellaria obtusa Engelm. ■☆
17923 Alsinella Gray = Arenaria L. ■

17924 Alsinella Hill = Sagina L. ■
17925 Alsinella Hornem. = Spergularia (Pers.) J. Presl et C. Presl(保留属名)■
17926 Alsinella Moench = Cerastium L. ■
17927 Alsinella Moench = Moenchia Ehrh. (保留属名)■☆
17928 Alsinella Sw. = Stellaria L. + Arenaria L. ■
17929 Alsinidendron H. Mann = Schiedea Cham. et Schltdl. ●■☆
17930 Alsinopsis Small = Minuartia L. ■
17931 Alsinopsis biflora (L.) Rydb. = Minuartia biflora (L.) Schinz et Thell. ■
17932 Alsinopsis californica (A. Gray) A. Heller = Minuartia californica (A. Gray) Mattf. ■☆
17933 Alsinopsis caroliniana (Walter) Small = Minuartia californica (A. Gray) Mattf. ■☆
17934 Alsinopsis dawsonensis (Britton) Rydb. = Minuartia dawsonensis (Britton) House ■☆
17935 Alsinopsis glabra (Michx.) Small = Minuartia glabra (Michx.) Mattf. ■☆
17936 Alsinopsis gregaria (A. Heller) A. Heller = Minuartia nuttallii (Pax) Briq. var. gregaria (A. Heller) Rabeler et R. L. Hartm. ■☆
17937 Alsinopsis groenlandica (Retz.) Small = Minuartia groenlandica (Retz.) Ostenf. ■☆
17938 Alsinopsis howellii (S. Watson) A. Heller = Minuartia howellii (S. Watson) Mattf. ■☆
17939 Alsinopsis macrantha Rydb. = Minuartia macrantha (Rydb.) House ■☆
17940 Alsinopsis macrocarpa (Pursh) A. Heller = Minuartia macrocarpa (Pursh) Ostenf. ■
17941 Alsinopsis obtusiloba Rydb. = Minuartia obtusiloba (Rydb.) House ■☆
17942 Alsinopsis patula (Michx.) Small = Minuartia patula (Michx.) Mattf. ■☆
17943 Alsinopsis pitcheri Nutt. = Minuartia patula (Michx.) Mattf. ■☆
17944 Alsinopsis pusilla (S. Watson) Rydb. = Minuartia pusilla (S. Watson) Mattf. ■☆
17945 Alsinopsis rossii (R. Br. ex Richardson) Rydb. = Minuartia rossii (R. Br. ex Richardson) Graebn. ■☆
17946 Alsinopsis tenella (J. Gay) A. Heller = Minuartia tenella (J. Gay) Mattf. ■☆
17947 Alsinopsis uniflora (Walter) Small = Minuartia uniflora (Walter) Mattf. ■☆
17948 Alsinula Dostal = Stellaria L. ■
17949 Alsmithia H. E. Moore(1982);长柄椰属(脊籽椰属)●☆
17950 Alsmithia longipes H. E. Moore;长柄椰●☆
17951 Alsobia Hanst. (1853);肖毛毡苣苔属■☆
17952 Alsobia Hanst. = Episcia Mart. ■☆
17953 Alsobia dianthiflora (H. E. Moore et R. G. Wilson) Wiehler;肖毛毡苣苔;Lace Flower ■☆
17954 Alsobia punctata Hanst.;斑点肖毛毡苣苔■☆
17955 Alsocydia Mart. ex DC. = Bignonia L. (保留属名)●
17956 Alsocydia Mart. ex J. C. Gomes = Cuspidaria DC. (保留属名)●☆
17957 Alsocydia Mart. ex J. C. Gomes = Piriadacus Pichon ●☆
17958 Alsodeia Thouars = Rinorea Aubl. (保留属名)●
17959 Alsodeia arborea Thouars = Rinorea arborea (Thouars) Baill. ●☆
17960 Alsodeia ardisiiflora Oliv. = Rinorea angustifolia (Thouars) Baill. subsp. ardisiiflora (Oliv.) Grey-Wilson ●☆
17961 Alsodeia aucuparia Oliv. = Rinorea brachypetala (Turcz.) Kuntze ●☆
17962 Alsodeia bengalensis Wall. = Rinorea bengalensis (Wall.) Kuntze ●
17963 Alsodeia brachypetala Turcz. = Rinorea brachypetala (Turcz.) Kuntze ●☆
17964 Alsodeia castanioides Oliv. = Rinorea castanioides (Oliv.) Kuntze ●☆
17965 Alsodeia castanioides Oliv. var. strictiflora Oliv. = Rinorea strictiflora (Oliv.) Exell et Mendonça ●☆
17966 Alsodeia caudata Oliv. = Rinorea caudata (Oliv.) Kuntze ●☆
17967 Alsodeia cauliflora Oliv. = Rinorea cauliflora (Oliv.) Kuntze ●☆
17968 Alsodeia convallarioides Baker f. = Rinorea convallarioides (Baker f.) Eyles ●☆
17969 Alsodeia crassifolia Baker f. = Rinorea crassifolia (Baker f.) De Wild. ●☆
17970 Alsodeia cymulosa Oliv. = Rinorea oliveri T. Durand et Schinz ●☆
17971 Alsodeia elliptica Oliv. = Rinorea elliptica (Oliv.) Kuntze ●☆
17972 Alsodeia engleriana De Wild. et T. Durand = Rinorea angustifolia (Thouars) Baill. subsp. engleriana (De Wild. et T. Durand) Grey-Wilson ●☆
17973 Alsodeia gazensis Baker f. = Rinorea ferruginea Engl. ●☆
17974 Alsodeia ilicifolia Welw. ex Oliv. = Rinorea ilicifolia (Welw. ex Oliv.) Kuntze ●☆
17975 Alsodeia johnstonii Stapf = Rinorea johnstonii (Stapf) M. Brandt ●☆
17976 Alsodeia longiracemosa Kurz = Rinorea longiracemosa (Kurz) Craib ●
17977 Alsodeia obanensis Baker f. = Allexis obanensis (Baker f.) Melch. ●☆
17978 Alsodeia prasina Stapf = Rinorea prasina (Stapf) Chipp ●☆
17979 Alsodeia socotrana Balf. f. = Hybanthus enneaspermus (L.) F. Muell. ●
17980 Alsodeia talbotii Baker f. = Rinorea talbotii (Baker f.) De Wild. ●☆
17981 Alsodeia virgata (Thwaites) Hook. f. et Thomson = Rinorea virgata (Thwaites) Kuntze ●
17982 Alsodeia wallichiana Hook. f. et Thomson = Rinorea bengalensis (Wall.) Kuntze ●
17983 Alsodeia welwitschii Oliv. = Rinorea welwitschii (Oliv.) Kuntze ●☆
17984 Alsodeia whytei Stapf = Rinorea whytei (Stapf) M. Brandt ●☆
17985 Alsodeia woermanniana Büttner = Rinorea woermanniana (Büttner) Engl. ●☆
17986 Alsodeiaceae J. Agardh = Violaceae Batsch(保留科名)●■
17987 Alsodeiidium Engl. = Alsodeiopsis Oliv. ●☆
17988 Alsodeiidium schumannii Engl. = Alsodeiopsis schumannii (Engl.) Engl. ●☆
17989 Alsodeiopsis Oliv. = Alsodeiopsis Oliv. ex Benth. et . Hook. f. ●☆
17990 Alsodeiopsis Oliv. ex Benth. et . Hook. f. (1867);拟三角车属(热非茶茱萸属)●☆
17991 Alsodeiopsis bequaertii De Wild. = Alsodeiopsis rowlandii Engl. ●☆
17992 Alsodeiopsis chippii Hutch.;奇普拟三角车●☆
17993 Alsodeiopsis glaberrima Engl. ex Hutch. et Dalziel = Olax gambecola Baill. ●☆
17994 Alsodeiopsis holstii Engl. = Leptaulus holstii (Engl.) Engl. ●☆
17995 Alsodeiopsis laurentii De Wild.;洛朗拟三角车●☆
17996 Alsodeiopsis mannii Oliv.;曼氏拟三角车●☆

17997 Alsodeiopsis oblongifolia Engl. = Desmostachys oblongifolius (Engl.) Villiers ■☆

17998 Alsodeiopsis oddoni De Wild. = Leptaulus holstii (Engl.) Engl. ●☆

17999 Alsodeiopsis poggei Engl.;波格拟三角车●☆

18000 Alsodeiopsis poggei Engl. var. robynsii Boutique;罗宾斯拟三角车●☆

18001 Alsodeiopsis rowlandii Engl.;罗兰拟三角车●☆

18002 Alsodeiopsis rubra Engl.;红色拟三角车●☆

18003 Alsodeiopsis schumannii (Engl.) Engl.;舒曼拟三角车●☆

18004 Alsodeiopsis staudtii Engl.;施陶拟三角车●☆

18005 Alsodeiopsis villosa Keay;长柔毛拟三角车●☆

18006 Alsodeiopsis zenkeri Engl.;岑克尔拟三角车●☆

18007 Alsolinum Fourr. = Linum L. ●■

18008 Alsomitra (Blume) M. Roem. (1838);大盖瓜属(阿霜瓜属)■☆

18009 Alsomitra (Blume) M. Roem. et Hutch. = Alsomitra (Blume) M. Roem. ■☆

18010 Alsomitra (Blume) Spach = Alsomitra (Blume) M. Roem. ■☆

18011 Alsomitra Benth. et Hook. f. = Alsomitra (Blume) M. Roem. ■☆

18012 Alsomitra M. Roem. = Alsomitra (Blume) M. Roem. ■☆

18013 Alsomitra angulata M. Roem.;大盖瓜■☆

18014 Alsomitra angustipetala (Craib) Craib;窄瓣大盖瓜■☆

18015 Alsomitra cissoides M. Roem. = Gynostemma pentaphyllum (Thunb.) Makino ■

18016 Alsomitra clavigera (Wall.) M. Roem. = Zanonia indica L. ■

18017 Alsomitra graciliflora Harms = Hemsleya graciliflora (Harms) Cogn. ■

18018 Alsomitra heterosperma Roem. = Hemsleya heterosperma (Wall.) C. Jeffrey ■

18019 Alsomitra integrifoliola (Cogn.) Hayata = Zanonia indica L. ■

18020 Alsomitra laxa (Wall.) M. Roem. = Gynostemma laxum (Wall.) Cogn. ■

18021 Alsomitra pubigera Prain = Zanonia indica L. ■

18022 Alsomitra simplicifolia Merr. = Zanonia indica L. ■

18023 Alsomitra tonkinensis Gagnep. = Zanonia indica L. ■

18024 Alstonia Mutis = Symplocos Jacq. ●

18025 Alstonia Mutis ex L. f. = Praealstonia Miers ●

18026 Alstonia Mutis ex L. f. = Symplocos Jacq. ●

18027 Alstonia R. Br. (1810)(保留属名);鸡骨常山属(鸭脚树属,黑板树属,盆架树属,阿斯木属);Alstonia,Winchia ●

18028 Alstonia Scop.(废弃属名) = Landolphia P. Beauv.(保留属名)●☆

18029 Alstonia Scop.(废弃属名) = Pacouria Aubl.(废弃属名)●☆

18030 Alstonia actinophylla (A. Cunn.) K. Schum.;辐射叶鸡骨常山●☆

18031 Alstonia angustifolia Wall. ex A. DC.;狭叶鸡骨常山●☆

18032 Alstonia angustiloba Miq.;狭浅裂鸡骨常山;Common Pulai ●☆

18033 Alstonia boonei De Wild.;干酪鸡骨常山(布氏鸡骨常山);Alstonia,Boon Alstonia,Cheese Alstonia,Cheesewood,Emien ●

18034 Alstonia congensis Engl.;刚果鸡骨常山(刚果鸭脚树);Ahun,Alstonia,Awun,Pattern Wood,Stool Wood ●☆

18035 Alstonia congensis Engl. var. glabrata Hutch. et Dalziel = Alstonia congensis Engl. ●☆

18036 Alstonia constricta F. Muell.;澳洲鸡骨常山(缢缩鸭脚树,窄叶鸡骨常山);Australian Quinine,Bitter Bark,Fever Bark,Fever-bark,Quinine Tree ●☆

18037 Alstonia elliptica (Thunb.) Roem. et Schult.;椭圆鸡骨常山(水甘草)●☆

18038 Alstonia esquirolii H. Lév. = Alstonia yunnanensis Diels ●

18039 Alstonia gilletii De Wild. = Alstonia congensis Engl. ●☆

18040 Alstonia gilletii De Wild. var. laurentii ? = Alstonia congensis Engl. ●☆

18041 Alstonia glaucescens (K. Schum.) Monach. = Alstonia rostrata C. E. C. Fisch. ●

18042 Alstonia guangxiensis D. Fang et X. X. Chen = Alstonia neriifolia D. Don ●

18043 Alstonia henryi Tsiang;黄花羊角棉;Henry Alstonia ●

18044 Alstonia longifolia (A. DC.) Pichon = Carissa pichoniana Leeuwenb. ●☆

18045 Alstonia lucida D. Don = Trachelospermum lucidum (D. Don) Schum. ●☆

18046 Alstonia macrophylla Wall. et G. Don;大叶糖胶树;Deviltree,Devil-tree,Deviltree Alstonia,Devil-tree Alstonia ●

18047 Alstonia mairei H. Lév.;羊角棉(阿斯木,鸡舌头树,见血飞,闹狗药,小鸡骨常山);Maire Alstonia ●

18048 Alstonia neriifolia D. Don;广西羊角棉(水柳,竹叶羊角棉);Guangxi Alstonia ●

18049 Alstonia pachycarpa Merr. et Chun = Alstonia rostrata C. E. C. Fisch. ●

18050 Alstonia paupera Hand. -Mazz. = Alstonia mairei H. Lév. ●

18051 Alstonia pneumatophora Backer ex L. G. Berger;沼生鸡骨常山●☆

18052 Alstonia rostrata C. E. C. Fisch.;盆架树(白叶糖胶,灯架,粉叶鸭脚树,厚果鸭脚木,岭刀柄,马灯盆,山苦常,小叶灯台树,野灯台树,有限鸭脚树);Pretty-leaf Winchia,Pretty-leaved Alstonia,Pretty-leaved Winchia,Washstand Tree,Winchia ●

18053 Alstonia rostrata C. E. C. Fisch. = Winchia calophylla A. DC. ●

18054 Alstonia rupestris Kerr;岩生羊角棉;Cliff Alstonia ●

18055 Alstonia scholaris (L.) R. Br.;糖胶树(阿根木,吃力秀,大矮陀,大枯树,大树矮陀陀,大树将军,大树理肺散,灯架树,灯台木,灯台树,灯台叶,凳板风,肥猪菜,肥猪草,肥猪叶,黑板树,金爪南木,九度叶,理肺散,买担别,面架木,面条树,盆架子,象皮木,橡皮木,小灯台树,鸭脚木,鸭脚树,英台木,鹰爪木,鹰爪树);Common Alstonia,Devil Tree,Devil's-tree,Devil-tree,Dita,Dita Bark,Milk Tree,Pali-mara,Palimara Alstonia,White Cheesewood ●

18056 Alstonia sebusi (Van Heurck et Müll. Arg.) Monach. var. azemaoensis Monach. = Alstonia henryi Tsiang ●

18057 Alstonia spectabilis R. Br.;鸡树●☆

18058 Alstonia venenata R. Br.;尼尔格日鸡骨常山(印度鸭脚树)●☆

18059 Alstonia yunnanensis Diels;鸡骨常山(白虎木,滇鸡骨常山,红花岩拖,红辣椒,红辣树,三台高,四角枫,野辣椒,永固生,云南鸭脚树);Yunnan Alstonia ●

18060 Alstroemeria L. (1762);六出花属(百合水仙属);Alstroemeria,Herb Lily,Lily of the Incas,Lily-of-the-incas,Peruvian Lily,Peruvian-lily ■☆

18061 Alstroemeria aurantiaca D. Don = Alstroemeria aurea Graham ■☆

18062 Alstroemeria aurea Graham;黄六出花(六出花);Alstroemeria,Amancay,Herb Lily,Inca Lily,Peruvian Lily,Yellow Alstroemeria,Yellow Lily-of-incans ■☆

18063 Alstroemeria brasiliensis Spreng. = Alstroemeria pelegrina L. ■☆

18064 Alstroemeria brasiliensis Spreng. = Alstroemeria pulchella L. f. ■☆

18065 Alstroemeria chilensis Cree;智利百合水仙;Chilian Lily ■☆

18066 Alstroemeria edulis Hook.;可食六出花;White Jerusalem Artichoke ■☆

18067 Alstroemeria gayana Phil. = Alstroemeria pelegrina L. ■☆

18068 Alstroemeria graminea Phil. = Taltalia graminea (Phil.) Ehr. Bayer ■☆
18069 Alstroemeria haemantha Ruiz et Pav.;血红六出花;Purple-spot Parrot-lily ■☆
18070 Alstroemeria haemantha Ruiz et Pav. var. albida Herb.;白花血红六出花■☆
18071 Alstroemeria hookeri Sweet;粉色六出花■☆
18072 Alstroemeria hybrida Host.;杂种六出花;Peruvian Lily ■☆
18073 Alstroemeria ligtu L.;紫条六出花;Peruvian Lily, Purplestreak Alstroemeria ■☆
18074 Alstroemeria pelegrina L.;紫斑六出花(斑花六出花,紫红六出花);Alstroemeria,Lily of the Incas,Lily-of-the-incas,Peruvian Lily ■☆
18075 Alstroemeria psittacina Lehm. = Alstroemeria pelegrina L. ■☆
18076 Alstroemeria pulchella L. f.;鹦鹉六出花;Parrot Alstroemeria, Parrot-lily ■☆
18077 Alstroemeria pulchella Sims = Alstroemeria haemantha Ruiz et Pav. var. albida Herb. ■☆
18078 Alstroemeria simsii Spreng. = Alstroemeria haemantha Ruiz et Pav. var. albida Herb. ■☆
18079 Alstroemeria versicolor Ruiz et Pav.;花叶六出花;Spotted Alstroemeria ■☆
18080 Alstroemeriaceae Dumort.(1829)(保留科名);六出花科(彩花扭柄科,扭柄叶科)●■☆
18081 Altamirania Greenm. = Aspiliopsis Greenm. ●☆
18082 Altamiranoa Rose = Sedum L. ●■
18083 Altamiranoa Rose = Villadia Rose ■☆
18084 Altamiranoa Rose ex Britton et Rose = Altamiranoa Rose ■☆
18085 Altensteinia Kunth(1816);安第斯兰属(热美兰属)■☆
18086 Altensteinia argyrolepis Rchb. f.;白鳞安第斯兰■☆
18087 Altensteinia leucantha Rchb. f.;白花安第斯兰■☆
18088 Altensteinia nervosa Kraenzl.;多脉安第斯兰■☆
18089 Alternanthera Forssk.(1775);莲子草属(锦绣苋属,满天星属,虾钳菜属,虾钳属,织锦苋属);Alternanthera, Broad Path, Chaff-flower,Copperleaf,Joseph's Coat,Joyweed ■
18090 Alternanthera achyrantha (L.) R. Br. = Alternanthera pungens Humb. ■
18091 Alternanthera achyrantha (L.) R. Br. var. leiantha Seub. = Alternanthera pungens Humb. ■
18092 Alternanthera achyrantha (L.) Sweet = Alternanthera pungens Kunth ■
18093 Alternanthera achyrantha R. Br. = Alternanthera pungens Kunth ■
18094 Alternanthera achyranthoides Forssk. = Alternanthera sessilis (L.) R. Br. ex DC. ■
18095 Alternanthera bettzickiana (Regel) G. Nicholson;锦绣苋(红草,红节节草,红莲子草,红田乌草,毛莲子草,五色草,织锦苋);Calicoplant,Garden Alternanthera,Jacob's Coat,Joseph's Coat,Joy Weed,Parrotleaf ■
18096 Alternanthera bettzickiana (Regel) Voss = Alternanthera bettzickiana (Regel) G. Nicholson ■
18097 Alternanthera brasiliana Kuntze;巴西莲子草;Brazilian Joyweed ■☆
18098 Alternanthera caracasana Kunth;加拉加斯莲子草■☆
18099 Alternanthera denticulata R. Br. = Alternanthera sessilis (L.) R. Br. ex DC. ■
18100 Alternanthera echinata Sm. = Alternanthera pungens Humb. ■
18101 Alternanthera ficoidea (L.) P. Beauv.;艳苋草(小叶红);Jacob's Coat,Parrot Leaf,Rabbit Meat,Rabbit-meat,Sanguinarea ■☆
18102 Alternanthera ficoidea (L.) P. Beauv. var. bettzickiana (Regel) Backer = Alternanthera tenella Colla var. bettzickiana (Regel) Veldkamp ■
18103 Alternanthera ficoidea (L.) Sm. 'Tricolor';三色苋草;Joseph's Coat ■☆
18104 Alternanthera ficoidea (L.) Sm. = Alternanthera ficoidea (L.) P. Beauv. ■☆
18105 Alternanthera ficoidea (L.) Sm. var. amoena (Lem.) L. B. Sm. et Downs;美丽锦绣苋(青叶模样苋)■
18106 Alternanthera ficoidea (L.) Sm. var. bettzickiana (Regel) Backer = Alternanthera bettzickiana (Regel) G. Nicholson ■
18107 Alternanthera ficoidea (L.) Sm. var. versicolor Lem. = Alternanthera bettzickiana (Regel) G. Nicholson ■
18108 Alternanthera flavescens Kunth;黄莲子草;Yellow Joyweed ■☆
18109 Alternanthera lanceolata (Benth.) Schinz;披针莲子草■☆
18110 Alternanthera lanuginosa (Nutt.) Moq. = Tidestromia lanuginosa (Nutt.) Standl. ■☆
18111 Alternanthera littoralis P. Beauv. = Alternanthera littoralis P. Beauv. ex Moq. ■☆
18112 Alternanthera littoralis P. Beauv. ex Moq.;海岸莲子草■☆
18113 Alternanthera littoralis P. Beauv. var. guineensis Pedersen;几内亚海岸莲子草■☆
18114 Alternanthera littoralis P. Beauv. var. maritima (Mart.) Pedersen;滨海莲子草■☆
18115 Alternanthera littoralis P. Beauv. var. sparmannii (Moq.) Pedersen;斯帕曼莲子草■☆
18116 Alternanthera macrorhiza Hauman;大根莲子草■☆
18117 Alternanthera maritima (Mart.) A. St.-Hil.;海生莲子草;Seaside Joyweed ■☆
18118 Alternanthera maritima (Mart.) A. St.-Hil. var. africana Hauman = Alternanthera littoralis P. Beauv. ■☆
18119 Alternanthera maritima (Mart.) A. St.-Hil. var. sparmannii (Moq.) Mears = Alternanthera littoralis P. Beauv. var. sparmannii (Moq.) Pedersen ■☆
18120 Alternanthera micrantha Domin;小花莲子草■☆
18121 Alternanthera nodiflora R. Br. = Alternanthera sessilis (L.) R. Br. ex DC. ■
18122 Alternanthera nodiflora sensu Stewart = Alternanthera sessilis (L.) R. Br. ex DC. ■
18123 Alternanthera paronichyoides A. St.-Hil. var. amazonica Huber;光匙叶莲子草;Smooth Joyweed ■☆
18124 Alternanthera paronychioides A. St.-Hil.;匙叶莲子草(红线草,华莲子草,绿线草);Joyweed,Smooth Joyweed ■
18125 Alternanthera paronychioides A. St.-Hil. var. bettzickiana (Regel) Fosberg;贝特莲子草■☆
18126 Alternanthera peploides (Humb. et Bonpl. ex Schult.) Urb. = Alternanthera caracasana Kunth ■☆
18127 Alternanthera peploides (Humb. et Bonpl.) Urb. = Alternanthera caracasana Kunth ■☆
18128 Alternanthera philoxeroides (Mart.) Griseb.;喜旱莲子草(长梗满天星,革命菜,革命草,过塘蛇,假蕹菜,空心莲子草,空心蕹藤菜,空心苋,螃蜞菊,山蕹菜,水花生,水蕹菜);Alligator Alternanthera,Alligator Weed,Alligatorweed ■
18129 Alternanthera philoxeroides (Mart.) Griseb. = Alternanthera paronychioides A. St.-Hil. ■
18130 Alternanthera polygonoides sensu Standl. = Alternanthera paronychioides A. St.-Hil. ■
18131 Alternanthera porigens (Moq.) Kuntze;伸莲子草■☆

18132 Alternanthera prostrata D. Don. = Alternanthera sessilis (L.) R. Br. ex DC. ■
18133 Alternanthera pungens Humb. = Alternanthera pungens Kunth ■
18134 Alternanthera pungens Humb. f. pauciflora ? = Alternanthera pungens Humb. ■
18135 Alternanthera pungens Kunth;刺花莲子草;Spinyflower Alternanthera ■
18136 Alternanthera ramosissima (Mart.) Chodat = Alternanthera flavescens Kunth ■☆
18137 Alternanthera repens (L.) J. F. Gmel.;黄褐莲子草;Bur Khaki,Khaki Bur,Khaki Weed ■☆
18138 Alternanthera repens (L.) J. F. Gmel. = Alternanthera sessilis (L.) R. Br. ex DC. ■
18139 Alternanthera repens (L.) Link = Achyranthes repens L. ■
18140 Alternanthera repens (L.) Link = Alternanthera pungens Humb. ■
18141 Alternanthera repens Kuntze = Achyranthes repens L. ■
18142 Alternanthera repens Steud. = Alternanthera pungens Humb. ■
18143 Alternanthera sennii Mattei;森恩莲子草■☆
18144 Alternanthera sessilis (L.) DC. = Alternanthera sessilis (L.) R. Br. ex DC. ■
18145 Alternanthera sessilis (L.) DC. var. nodiflora (R. Br.) Kuntze = Alternanthera nodiflora R. Br. ■
18146 Alternanthera sessilis (L.) R. Br. = Alternanthera sessilis (L.) R. Br. ex DC. ■
18147 Alternanthera sessilis (L.) R. Br. ex DC.;莲子草(白花节节草,白花仔,齿叶莲子草,地扭子,地纽子,飞疔草,鲎脚菜,节节花,锦绣苋,满天星,耐惊菜,耐惊花,蟛蜞菊,曲节草,蛇痼,水金铃,水牛膝,虾蚶菜,虾蚶草,虾蠊菜,虾钳菜,一包针,猪屎草);Mukunawanna, Narrow-leaved Alternanthera, Sessile Alternanthera, Sessile Joyweed,Star-sky Alternanthera ■
18148 Alternanthera sessilis (L.) R. Br. ex Roem. et Schult. = Alternanthera sessilis (L.) R. Br. ex DC. ■
18149 Alternanthera suffruticosa Torr. = Tidestromia suffruticosa (Torr.) Standl. ●☆
18150 Alternanthera tenella Colla;娇嫩莲子草;Jacob's Coat,Joyweed ■☆
18151 Alternanthera tenella Colla var. bettzickiana (Regel) Veldkamp = Alternanthera bettzickiana (Regel) G. Nicholson ■
18152 Alternanthera versicolor (Lem.) Regel;红田乌草■
18153 Alternanthera versicolor (Lem.) Regel = Alternanthera bettzickiana (Regel) G. Nicholson ■
18154 Alternasemina Silva Manso = Melothria L. ■
18155 Althaea L. (1753);药葵属(蜀葵属);Althaea, Althea, Hollyhock,Mallow,Marsh Mallow,Marsh-mallow ■
18156 Althaea × cultorum ? = Althaea rosea (L.) Cav. ■
18157 Althaea armeniana Ten.;亚美尼亚蜀葵■☆
18158 Althaea burchellii DC. = Pavonia burchellii (DC.) R. A. Dyer ●☆
18159 Althaea cachemiriana (Cambess.) Kuntze = Lavatera cashemiriana Cambess. ■
18160 Althaea cannabina L.;大麻叶蜀葵(大麻药蜀葵,麻叶蜀葵);Hemp-leaved Hollyhock,Palm-leaf Marshmallow ■☆
18161 Althaea chinensis Wall. = Althaea rosea (L.) Cav. ■
18162 Althaea coromandelina Cav. = Alcea rosea L. ■
18163 Althaea cretica Kuntze;克里特蜀葵;Hollyhock ■☆
18164 Althaea ficifolia (L.) Cav. = Alcea rosea L. ■
18165 Althaea ficifolia Cav. = Alcea rosea L. ■
18166 Althaea hirsuta L.;硬毛蜀葵;Hairy Marshmallow, Hispid Althaea,Hispid Mallow,Hispid Marsh Mallow,Hispid Marsh-mallow, Rough Mallow,Rough Marshmallow ■☆
18167 Althaea hirsuta L. var. grandiflora Ball = Althaea hirsuta L. ■☆
18168 Althaea lavateraeflora DC. = Alcea lavateraeflora (DC.) Boiss. ■☆
18169 Althaea longiflora Boiss. et Reut.;长花药葵■☆
18170 Althaea ludwigii L.;路德维格药葵■☆
18171 Althaea ludwigii L. var. microcaliculata Dobignard = Althaea ludwigii L. ■☆
18172 Althaea nudiflora Lindl. = Alcea nudiflora (Lindl.) Boiss. ■
18173 Althaea officinalis L.;药葵(药蜀葵,药用蜀葵);Althaea, Beggarlnan Cakes, Bull Flower, Bull's Eyes, Bysmale, Common Marsh Mallow, Common Marshmallow, Drunkards, Guimauve, Marsh Mallow, Marshmallow, Mesh Mellice, Moorish Mallow, Mortification Root,Water Mallow,White Mallow,Wild Geranium,Wimote,Wymote ■
18174 Althaea pallida Waldst. et Kit.;苍白蜀葵;White Althaea ■☆
18175 Althaea pallida Waldst. et Kit. = Alcea pallida (Waldst. et Kit.) Besser ■☆
18176 Althaea rosea (L.) Cav. = Alcea rosea L. ■
18177 Althaea rosea (L.) Cav. var. sinensis (Cav.) S. Y. Hu = Althaea rosea (L.) Cav. ■
18178 Althaea rugosa Alef. = Alcea rugosa Alef. ■☆
18179 Althaea sinensis Cav. = Althaea rosea (L.) Cav. ■
18180 Althaeastrum Fabr. = Lavatera L. ●■
18181 Althea Crantz = Althaea L. ■
18182 Althenia F. Petit(1829);加利亚草属■☆
18183 Althenia filiformis F. Petit;线形加利亚草■☆
18184 Althenia orientalis (Tzvelev) Garcia Mur. et Talavera;加利亚草■☆
18185 Altheniaceae Lotsy = Zannichelliaceae Chevall. (保留科名) ■
18186 Altheria Thouars = Melochia L. (保留属名) ●■
18187 Althingia Steud. = Araucaria Juss. ●
18188 Althoffia K. Schum. = Trichospermum Blume ●☆
18189 Altingia G. Don = Altingia Noronha + Araucaria Juss. ●
18190 Altingia Noronha(1790);蕈树属(阿丁枫属);Altingia ●
18191 Altingia angustifolia Hung T. Chang = Altingia siamensis Craib ●
18192 Altingia chinensis (Champ.) Oliv. ex Hance;蕈树(阿丁枫,半边风,假荔枝,糠娘子,老虎斑,檀木,香梨,星霞树,猪肝木);China Altingia,Chinese Altingia ●
18193 Altingia chinensis (Champ.) Oliv. ex Hance f. pubescens X. H. Song;毛蕈树;Pubescent Chinese Altingia ●
18194 Altingia chingii F. P. Metcalf = Semiliquidambar chingii (F. P. Metcalf) Hung T. Chang ●
18195 Altingia chingii Metcalf = Semiliquidambar chingii (F. P. Metcalf) Hung T. Chang ●
18196 Altingia chingii Metcalf var. parvifolia Chun = Semiliquidambar cathayensis Hung T. Chang ●◇
18197 Altingia excelsa Noronha;细青皮(阿丁枫,木材树,青皮树);Rasamala,Rasamala Altingia ●
18198 Altingia gracilipes Hemsl.;细柄蕈树(窄叶阿丁枫);Slenderstalk Altingia,Slender-stalked Altingia ●
18199 Altingia gracilipes Hemsl. f. uniflora Hung T. Chang;独花蕈树;Oneflower Altingia ●
18200 Altingia gracilipes Hemsl. f. uniflora Hung T. Chang = Altingia gracilipes Hemsl. ●
18201 Altingia gracilipes Hemsl. var. serrulata Tutcher;细齿蕈树(齿叶阿丁枫,齿叶蕈树,细齿蕈);Serrulate Altingia ●
18202 Altingia multinervis W. C. Cheng;赤水蕈树;Chishui Altingia, Manynerved Altingia,Multinerved Altingia ●

18203 Altingia obovata Merr. et Chun;海南蕈树(山包蜜,山海棠);Hainan Altingia,Obovate Altingia ●
18204 Altingia siamensis Craib;窄叶蕈树(镰尖蕈树,窄叶阿丁枫);Narrowleaf Altingia,Narrow-leaved Altingia ●
18205 Altingia takhtajanii Thai;越南阿丁枫●☆
18206 Altingia tenuifolia Chun ex Hung T. Chang;薄叶蕈树(卵叶阿丁枫);Thinleaf Altingia,Thin-leaved Altingia ●
18207 Altingia yunnanensis Rehder et E. H. Wilson;云南蕈树(白皮树,桂阳渣,桂阳遮,苦梨树,蒙自阿丁枫,蒙自蕈树,青皮,青皮树);Yunnan Altingia ●
18208 Altingiaceae Horan. = Altingiaceae Lindl.(保留科名)●
18209 Altingiaceae Lindl.(1846)(保留科名);蕈树科(阿丁枫科)●
18210 Altisatis Thouars = Habenaria Willd. ■
18211 Altisatis Thouars = Satyrium Sw.(保留属名)■
18212 Altoparadisium Filg.,Davidse,Zuloaga et Morrone(2001);巴西禾属■☆
18213 Altora Adans. = Clutia L. ■☆
18214 Alughas L. = Alpinia Roxb.(保留属名)■
18215 Alus Bubani = Coris L. ●☆
18216 Aluta Rye et Trudgen(2000);澳洲假岗松属●☆
18217 Alvaradoa Liebm.(1854);短苦木属●☆
18218 Alvaradoa amorphoides (L.) Liebm.;无定短苦木●☆
18219 Alvardia Fenzl = Peucedanum L. ■
18220 Alvarezia Pav. ex Nees = Blechum P. Browne ■
18221 Alveolina Tiegh. = Loranthus Jacq.(保留属名)●
18222 Alveolina Tiegh. = Psittacanthus Mart. ●☆
18223 Alvesia Welw.(1858)(保留属名);胀萼花属(阿尔韦斯草属)●■☆
18224 Alvesia clerodendroides (T. C. E. Fr.) B. Mathew;胀萼花(阿尔韦斯草)●☆
18225 Alvesia cylindricalyx (B. Mathew) B. Mathew = Plectranthastrum cyclindricalyx B. Mathew ●☆
18226 Alvesia rosmarinifolia Welw. = Plectranthastrum rosmarinifolium (Welw.) B. Mathew ●☆
18227 Alvesia tomentosa (L.) Britton et Rose;毛胀萼花(毛阿尔韦斯草)●☆
18228 Alvimia Calderón ex Soderstr. et Londoño(1988);阿尔禾属(阿尔芬属,阿尔芬竹属,阿芬禾属)■☆
18229 Alvimia Soderstr. et Londoño = Alvimia Calderón ex Soderstr. et Londoño ■☆
18230 Alvimia auriculata Soderstr. et Londono;耳状阿尔禾■☆
18231 Alvimia gracilis Soderstr. et Londono;纤细阿尔禾■☆
18232 Alvimia lancifolia Soderstr. et Londono;剑叶阿尔禾■☆
18233 Alvimiantha Grey-Wilson(1978);阿尔花属●☆
18234 Alvimiantha tricamerata Grey-Wilson;阿尔花●☆
18235 Alvisia Lindl. = Bryobium Lindl. ■
18236 Alvisia Lindl. = Eria Lindl.(保留属名)■
18237 Alvordia Brandegee(1889);柱果菊属■☆
18238 Alvordia angusta S. F. Blake;窄柱果菊■☆
18239 Alvordia brandegeei A. M. Carter;布朗柱果菊■☆
18240 Alvordia congesta (Rose ex Hoffm.) B. L. Turner;柱果菊■☆
18241 Alvordia fruticosa Brandegee;灌木状柱果菊■☆
18242 Alvordia glomerata Brandegee;球状柱果菊■☆
18243 Alwisia Thwaites ex Lindl. = Taeniophyllum Blume ■
18244 Alycia Steud. = Eriochloa Kunth ■
18245 Alycia Willd. ex Steud. = Eriochloa Kunth ■
18246 Alymeria D. Dietr. = Aylmeria Mart. ●■
18247 Alymeria D. Dietr. = Polycarpaea Lam.(保留属名)●■
18248 Alymnia (DC.) Spach = Polymnia L. ●■☆
18249 Alymnia Neck. ex Spach = Polymniastrum Lam. ●■☆
18250 Alyogyne Alef.(1863);澳洲木槿属●☆
18251 Alyogyne Alef. = Fugosia Juss. ●■☆
18252 Alyogyne hakeifolia (Giord.) Alef.;红斑澳洲木槿;Red-centered Hibiscus ●☆
18253 Alyogyne hakeifolia Alef. = Alyogyne hakeifolia (Giord.) Alef. ●☆
18254 Alyogyne huegelii (Endl.) Fryxell;蓝紫澳洲木槿;Blue Hibiscus,Lilac Hibiscus ●☆
18255 Alypaceae Hoffmanns. et Link = Globulariaceae DC.(保留科名)●■☆
18256 Alypum Fisch. = Globularia L. ●☆
18257 Alysicarpus Desv.(1813)(保留属名);链荚豆属(炼荚豆属);Alyce Clover,Alysicarpus,Alysiclover,Chainpodpea ■
18258 Alysicarpus Neck. ex Desv. = Alysicarpus Desv.(保留属名)■
18259 Alysicarpus bupleurifolius (L.) DC.;柴胡叶链荚豆(柴胡链荚豆,长叶炼荚豆,长叶链荚豆);Bupleurum-leaf Alysicarpus ■
18260 Alysicarpus ferrugineus Hochst. et Steud. ex A. Rich.;锈色链荚豆■☆
18261 Alysicarpus ferrugineus Hochst. et Steud. var. quartinianus (A. Rich.) Schindl. = Alysicarpus quartinianus A. Rich. ■☆
18262 Alysicarpus glumaceus (Vahl) DC.;皮壳链荚豆■☆
18263 Alysicarpus glumaceus (Vahl) DC. var. hispidicarpus (Fiori) J. Léonard;硬果链荚豆■☆
18264 Alysicarpus glumaceus (Vahl) DC. var. intermedius Verdc.;间型链荚豆■☆
18265 Alysicarpus glumaceus (Vahl) DC. var. patulopedicellatus J. Léonard;张开链荚豆■☆
18266 Alysicarpus heterophyllus (Baker) Jafri et Ali;互叶链荚豆■☆
18267 Alysicarpus hochstetteri A. Rich. = Alysicarpus glumaceus (Vahl) DC. ■☆
18268 Alysicarpus longifolius (Rottler ex Spreng.) Wight et Arn.;长叶链荚豆■☆
18269 Alysicarpus longifolius Wight et Arn. = Alysicarpus longifolius (Rottler ex Spreng.) Wight et Arn. ■☆
18270 Alysicarpus monilifer (L.) DC.;串珠链荚豆■☆
18271 Alysicarpus nummularifolius A. DC. = Alysicarpus vaginalis (L.) DC. ■
18272 Alysicarpus ovalifolius (K. Schum. et Thonn.) J. Léonard = Alysicarpus ovalifolius (K. Schum.) J. Léonard ■
18273 Alysicarpus ovalifolius (K. Schum.) J. Léonard;卵叶链荚豆(圆叶炼荚豆);Alyce Clover ■
18274 Alysicarpus paradoxus Boivin ex Baill. = Alysicarpus ovalifolius (K. Schum.) J. Léonard ■
18275 Alysicarpus parviflorus Dalzell;微花链荚豆■☆
18276 Alysicarpus polygonoides Welw. ex Romariz;多节链荚豆■☆
18277 Alysicarpus porrectus Welw. ex Baker;外伸链荚豆■☆
18278 Alysicarpus quartinianus A. Rich.;夸尔廷链荚豆■☆
18279 Alysicarpus rugosus (Willd.) DC.;皱缩链荚豆(多皱链荚豆,皱果炼荚豆);Red Moneywort,Wrinkled Alysicarpus ■
18280 Alysicarpus rugosus (Willd.) DC. f. pilosus (Chiov.) Cufod.;毛皱缩链荚豆■☆
18281 Alysicarpus rugosus (Willd.) DC. subsp. reticulatus Verdc.;网状链荚豆■☆
18282 Alysicarpus rugosus (Willd.) DC. var. hispidicarpus Fiori = Alysicarpus glumaceus (Vahl) DC. var. hispidicarpus (Fiori) J.

Léonard ■☆

18283 Alysicarpus rugosus (Willd.) DC. var. quartianus (A. Rich.) Baker = Alysicarpus quartinianus A. Rich. ■☆

18284 Alysicarpus rugosus (Willd.) DC. var. styracifolius Baker = Alysicarpus scariosus Grab. ex Thwaites ■☆

18285 Alysicarpus scariosus Grab. ex Thwaites;干膜质链荚豆■☆

18286 Alysicarpus squamosus Gand. = Alysicarpus glumaceus (Vahl) DC. var. hispidicarpus (Fiori) J. Léonard ■☆

18287 Alysicarpus tetragonolobus Edgew.;四角片链荚豆■☆

18288 Alysicarpus trifoliatus Stocks = Alysicarpus heterophyllus (Baker) Jafri et Ali ■☆

18289 Alysicarpus vaginalis (L.) DC.;链荚豆(大叶春,狗蚁草,炼荚豆,山地豆,山花生,水碱草,小豆,小号野花生);Alyce Clover, Alyceclover, Sheath Chainpodpea, White Moneywort ■

18290 Alysicarpus vaginalis (L.) DC. var. diversifolius Chun;异叶链荚豆■

18291 Alysicarpus vaginalis (L.) DC. var. diversifolius Chun = Alysicarpus vaginalis (L.) DC. ■

18292 Alysicarpus vaginalis (L.) DC. var. heterophyllus Baker = Alysicarpus heterophyllus (Baker) Jafri et Ali ■☆

18293 Alysicarpus vaginalis (L.) DC. var. nummularifolius ? = Alysicarpus vaginalis (L.) DC. ■

18294 Alysicarpus vaginalis (L.) DC. var. paniculatus Baker f. = Alysicarpus vaginalis (L.) DC. ■

18295 Alysicarpus vaginalis (L.) DC. var. parvifolius Verdc.;小叶链荚豆■☆

18296 Alysicarpus vaginalis (L.) DC. var. taiwanensis S. S. Ying;黄花炼荚豆■

18297 Alysicarpus vaginalis (L.) DC. var. villosus Verdc.;长柔毛链荚豆■☆

18298 Alysicarpus violaceus (Forssk.) Schindl. = Alysicarpus glumaceus (Vahl) DC. ■☆

18299 Alysicarpus violaceus (Forssk.) Schindl. = Alysicarpus rugosus (Willd.) DC. ■

18300 Alysicarpus violaceus (Forssk.) Schindl. var. pilosus Schindl. = Alysicarpus rugosus (Willd.) DC. ■

18301 Alysicarpus wallichii Wight et Arn. = Alysicarpus rugosus (Willd.) DC. ■

18302 Alysicarpus yunnanensis Yen C. Yang et P. H. Huang;云南链荚豆;Yunnan Alysicarpus, Yunnan Chainpodpea ■

18303 Alysicarpus zeyheri Harv.;泽赫链荚豆■☆

18304 Alyssoides Adans. = Vesicaria Tourn. ex Adans. ■☆

18305 Alyssoides Mill. (1754);木庭荠属●■☆

18306 Alyssoides Tourn. ex Adans. = Vesicaria Tourn. ex Adans. ■☆

18307 Alyssoides utriculata Medik.;木庭荠;Bladderpod ●☆

18308 Alysson Crantz = Alyssum L. ●■

18309 Alyssopsis Boiss. (1842);拟庭荠属■☆

18310 Alyssopsis Rchb. = Vesicaria Tourn. ex Adans. ■☆

18311 Alyssopsis deflexa Boiss.;拟庭荠■☆

18312 Alyssopsis kotshyi Boiss.;考奇拟庭荠■☆

18313 Alyssum L. (1753);庭荠属(番荠属,庭芥属,香芥属,香荠属);Alison, Alyssum, Madwort ●■

18314 Alyssum afghanicum Rech. f. = Alyssum turkestanicum Regel et Schmalh. ex Regel ■☆

18315 Alyssum algeriense Pomel var. montanum ? = Alyssum granatense Boiss. et Reut. ■☆

18316 Alyssum alpestre Kom. = Alyssum sibiricum Willd. ■

18317 Alyssum alpestre L.;亚高山庭荠■☆

18318 Alyssum alpestre L. subsp. serpyllifolium (Desf.) Rouy et Foucaud = Alyssum serpyllifolium Desf. ■☆

18319 Alyssum alpestre L. var. djurdjurae (Chabert) Maire et Weiller = Alyssum djurjurae Chabert ■☆

18320 Alyssum alpestre L. var. macrocarpum Maire = Alyssum serpyllifolium Desf. ■☆

18321 Alyssum alpestre L. var. macrosepalum Ball = Alyssum serpyllifolium Desf. ■☆

18322 Alyssum alpestre L. var. serpyllifolium (Desf.) Coss. = Alyssum serpyllifolium Desf. ■☆

18323 Alyssum alpestre Wulf. ex Nyman = Alyssum montanum L. ■☆

18324 Alyssum altaicum C. A. Mey. = Alyssum lenens Adams ■

18325 Alyssum altaicum C. A. Mey. var. dasycarpum C. A. Mey. = Alyssum lenens Adams ■

18326 Alyssum altaicum C. A. Mey. var. leiocarpum C. A. Mey. = Alyssum lenens Adams ■

18327 Alyssum alyssoides (L.) L.;欧洲庭荠;Europe Madwort, European Madwort, Hoary Alyssum, Pale Alyssum, Pale Madwort, Small Allson, Yellow Alyssum ■

18328 Alyssum americanum Greene = Alyssum obovatum (C. A. Mey.) Turcz. ■

18329 Alyssum andinum Rupr.;安地庭荠■☆

18330 Alyssum antiatlanticum Emb. et Maire;安蒂庭荠■☆

18331 Alyssum argenteum Vitman;里白庭荠■☆

18332 Alyssum argyraeum (A. Gray) Kuntze;银叶庭荠;Yellow-tuft ■☆

18333 Alyssum armenum Boiss.;亚美尼亚庭荠■☆

18334 Alyssum artvinense N. Busch;阿尔特温庭荠■☆

18335 Alyssum atlanticum Desf.;大西洋叶庭荠■☆

18336 Alyssum atlanticum Desf. subsp. decoloratum (Pomel) Dobignard;褪色庭荠■☆

18337 Alyssum atlanticum Desf. var. clausonis (Pomel) Debeaux = Alyssum clausonis Pomel ■☆

18338 Alyssum atlanticum Desf. var. decoloratum (Pomel) Debeaux = Alyssum atlanticum Desf. subsp. decoloratum (Pomel) Dobignard ■☆

18339 Alyssum atlanticum Desf. var. speciosum (Pomel) Debeaux = Alyssum speciosum Pomel ■☆

18340 Alyssum biovulatum N. Busch = Alyssum sibiricum Willd. ■

18341 Alyssum bracteatum Boiss. et Buhse;具苞庭荠■☆

18342 Alyssum buschianum Grossh.;布什庭荠■☆

18343 Alyssum buschianum Grossh. = Alyssum stapfii Vierh. ■☆

18344 Alyssum calycinum L.;萼白庭荠;Pale Alyssum ■☆

18345 Alyssum calycinum L. = Alyssum alyssoides (L.) L. ■

18346 Alyssum calycocarpum Rupr.;萼果庭荠■☆

18347 Alyssum calycocarpum Rupr. var. edentanum H. L. Yang;无齿萼果庭荠■

18348 Alyssum calycocarpum Rupr. var. edentanum H. L. Yang = Alyssum lenens Adams ■

18349 Alyssum campestre L.;田野庭荠■☆

18350 Alyssum campestre L. var. collinum (Brot.) Cout. = Alyssum simplex Rudolphi ■

18351 Alyssum campestre L. var. edentulum Andr. = Alyssum simplex Rudolphi subsp. edentulum (Andr.) N. Galland ■☆

18352 Alyssum campestre L. var. emarginatum Andr. = Alyssum simplex Rudolphi ■

18353 Alyssum campestre L. var. genuinum Boiss. = Alyssum stapfii Vierh. ■☆

18354 Alyssum campestre L. var. minus Rouy = Alyssum simplex Rudolphi ■
18355 Alyssum campestre L. var. nadorense Sennen = Alyssum simplex Rudolphi ■
18356 Alyssum campestre L. var. nanum Pomel = Alyssum simplex Rudolphi ■
18357 Alyssum canescens DC.;灰毛庭荠■
18358 Alyssum canescens DC. = Ptilotrichum canescens (DC.) C. A. Mey. ●■
18359 Alyssum canescens DC. var. abbreviatum DC. = Alyssum canescens DC. ■
18360 Alyssum canescens DC. var. elongatum DC. = Alyssum canescens DC. ■
18361 Alyssum clausonis Pomel;克劳森庭荠■☆
18362 Alyssum clypeatum L. = Fibigia clypeata (L.) Medik. ■☆
18363 Alyssum cochleatum Coss. et Durieu = Hormathophylla cochleata (Coss. et Durieu) P. Küpfer ■☆
18364 Alyssum collinum Brot. = Alyssum simplex Rudolphi ■
18365 Alyssum cryptopetalum Bunge = Alyssum marginatum Steud. ex Boiss. ■☆
18366 Alyssum cupreum Freyn et Sint. = Alyssum linifolium Stephan ex Willd. ■
18367 Alyssum cupreum Freyn et Sint. ex Freyn = Alyssum linifolium Stephan ex Willd. ■
18368 Alyssum czernjakowskae Rech. f. = Alyssum lanceolatum Baumg. ■☆
18369 Alyssum daghestanicum Rupr.;达赫斯坦庭荠■☆
18370 Alyssum dasycarpum Stephan ex Willd.;粗果庭荠;Thickfruit Madwort ■
18371 Alyssum dasycarpum Stephan ex Willd. var. minus Bornm. ex T. R. Dudley = Alyssum dasycarpum Stephan ex Willd. ■
18372 Alyssum dasycarpum Stephan ex Willd. var. pterospermum Bordz. = Alyssum dasycarpum Stephan ex Willd. ■
18373 Alyssum decoloratum Pomel = Alyssum atlanticum Desf. subsp. decoloratum (Pomel) Dobignard ■☆
18374 Alyssum desertorum Stapf;庭荠(小庭荠);Alyssum, Desert Madwort ■
18375 Alyssum desertorum Stapf var. himalayensis T. R. Dudley;喜马拉雅庭荠;Desert Madwort ■
18376 Alyssum desertorum Stapf var. himalayensis T. R. Dudley = Alyssum desertorum Stapf ■
18377 Alyssum desertorum Stapf var. prostratum T. R. Dudley = Alyssum desertorum Stapf ■
18378 Alyssum djurjurae Chabert;朱朱拉庭荠■☆
18379 Alyssum embergeri Quézel;恩贝格尔庭荠■☆
18380 Alyssum fallax Nyar. = Alyssum obovatum (C. A. Mey.) Turcz. ■
18381 Alyssum fedtschenkoanum N. Busch;球果庭荠(珠果庭荠);Fedtschenko Madwort ■
18382 Alyssum fischerianum DC. = Alyssum lenens Adams ■
18383 Alyssum flahaultianum Emb.;弗拉奥庭荠■☆
18384 Alyssum glomeratum Burch. ex DC. = Alyssum minutum Schltdl. ex DC. ■☆
18385 Alyssum gmelinii Jord. et Fourr.;格氏庭荠;Bladderpod Mustard, Gordon's Bladderpod, Yellow Blanket ■☆
18386 Alyssum gordonii Jord. et Fourr. = Lesquerella gordonii (A. Gray) S. Watson ■☆
18387 Alyssum granatense Boiss. et Reut.;格拉庭荠■☆
18388 Alyssum granatense Boiss. et Reut. var. luteolum (Pomel) Coss. = Alyssum luteolum Pomel ■☆
18389 Alyssum granatense Boiss. et Reut. var. minutulum Batt. = Alyssum luteolum Pomel ■☆
18390 Alyssum granatense Boiss. et Reut. var. sepalinum Pomel = Alyssum granatense Boiss. et Reut. ■☆
18391 Alyssum granatense Boiss. et Reut. var. subminutulum Maire et Wilczek = Alyssum granatense Boiss. et Reut. ■☆
18392 Alyssum granatense Boiss. et Reut. var. weilleri Emb. et Maire = Alyssum granatense Boiss. et Reut. ■☆
18393 Alyssum halimifolium L. = Lobularia maritima (L.) Desv. ■
18394 Alyssum heterotrichum Boiss.;异毛庭荠■☆
18395 Alyssum hirsutum M. Bieb.;毛庭荠■☆
18396 Alyssum homalocarpum (Fisch. et C. A. Mey.) Boiss.;平果庭荠■☆
18397 Alyssum horebicum Boiss. = Alyssum homalocarpum (Fisch. et C. A. Mey.) Boiss. ■☆
18398 Alyssum idaeum Boiss. et Heldr.;横卧庭荠■☆
18399 Alyssum incanum L. = Berteroa incana (L.) DC. ■
18400 Alyssum iranicum L. = Alyssum lanceolatum Baumg. ■☆
18401 Alyssum lanceolatum Baumg.;剑叶庭荠■☆
18402 Alyssum leiocarpum Pomel = Alyssum minutum Schltdl. ex DC. ■☆
18403 Alyssum lenense Adams;北方庭荠(条叶庭荠,线叶庭荠);Linearleaf Alyssum, Linearleaf Madwort ■
18404 Alyssum lenense Adams var. dasycarpum C. A. Mey.;星毛庭荠;Thickfruit Linearleaf Alyssum Alyssum ■
18405 Alyssum lenense Adams var. leiovarpum N. Busch;光果北方庭荠■
18406 Alyssum libycum (Viv.) Coss. = Lobularia libyca (Viv.) Meisn. ■☆
18407 Alyssum linifolium Stephan ex Willd.;条叶庭荠(齿丝庭荠);Flaxleaf Alyssum, Lacinoseleaf Madwort ■
18408 Alyssum linifolium Stephan ex Willd. var. cupreum (Freyn et Sint.) T. R. Dudley = Alyssum linifolium Stephan ex Willd. ■
18409 Alyssum linifolium Stephan ex Willd. var. tehranicum Bornm. = Alyssum linifolium Stephan ex Willd. ■
18410 Alyssum linifolium Willd. = Alyssum linifolium Stephan ex Willd. ■
18411 Alyssum luteolum Pomel;淡黄庭荠■☆
18412 Alyssum luteolum Pomel var. pomelii Batt. = Alyssum luteolum Pomel ■☆
18413 Alyssum macrocalyx Coss. et Durieu;大萼庭荠■☆
18414 Alyssum macrostylum Boiss. et Huet;长柱庭荠■☆
18415 Alyssum magicum C. H. An;哈密庭荠;Hami Madwort ■
18416 Alyssum magicum C. H. An = Galitzkya potaninii (Maxim.) V. V. Botschantz. ■
18417 Alyssum marginatum (Webb) Durand et Schinz = Lobularia canariensis (DC.) L. Borgen subsp. marginata (Webb) L. Borgen ■☆
18418 Alyssum marginatum Steud. = Alyssum marginatum Steud. ex Boiss. ■☆
18419 Alyssum marginatum Steud. ex Boiss.;隐瓣庭荠■☆
18420 Alyssum maritimum (L.) Lam. = Lobularia maritima (L.) Desv. ■
18421 Alyssum maritimum (L.) Lam. var. canariense DC. = Lobularia canariensis (DC.) L. Borgen ■☆
18422 Alyssum maritimum (L.) Lam. var. humbertianum Maire = Lobularia maritima (L.) Desv. ■

18423 Alyssum maritimum (L.) Lam. var. lepidioides Ball = Lobularia maritima (L.) Desv. ■
18424 Alyssum micranthum C. A. Mey. = Alyssum simplex Rudolphi ■
18425 Alyssum minimum L. = Lobularia maritima (L.) Desv. ■
18426 Alyssum minimum Willd. = Alyssum desertorum Stapf ■
18427 Alyssum minus (L.) Rothm.;新疆庭荠;Sinkiang Madwort, Xinjiang Madwort ■
18428 Alyssum minus (L.) Rothm. = Alyssum simplex Rudolphi ■
18429 Alyssum minus Roth. = Alyssum simplex Rudolphi ■
18430 Alyssum minus Roth. var. micranthum (C. A. Mey.) T. R. Dudley = Alyssum simplex Rudolphi ■
18431 Alyssum minutum Schltdl. = Alyssum minutum Schltdl. ex DC. ■☆
18432 Alyssum minutum Schltdl. ex DC.;小庭荠■☆
18433 Alyssum montanum L.;高山庭荠(山庭荠);Mountain Alyssum ■☆
18434 Alyssum montanum L. var. atlanticum (Desf.) Boiss. = Alyssum atlanticum Desf. ■☆
18435 Alyssum montanum L. var. clausonis (Pomel) Debeaux = Alyssum clausonis Pomel ■☆
18436 Alyssum montanum L. var. decoloratum (Pomel) Debeaux = Alyssum atlanticum Desf. subsp. decoloratum (Pomel) Dobignard ■☆
18437 Alyssum montanum L. var. edentulum Andr. = Alyssum simplex Rudolphi ■
18438 Alyssum montanum L. var. numidicum (Pomel) Maire et Weiller = Alyssum numidicum Pomel ■☆
18439 Alyssum montanum L. var. patulum (Pomel) Maire et Weiller = Alyssum patulum Pomel ■☆
18440 Alyssum montanum L. var. pilosum Andr. = Alyssum atlanticum Desf. ■☆
18441 Alyssum montanum L. var. speciosum (Pomel) Debeaux = Alyssum speciosum Pomel ■☆
18442 Alyssum montanum L. var. subspeciosum Maire et Sam. = Alyssum speciosum Pomel ■☆
18443 Alyssum muelleri Boiss.;米勒庭荠■☆
18444 Alyssum murale Waldst. et Kir.;黄庭荠;Yellowtuft ■☆
18445 Alyssum musili Velen. = Alyssum homalocarpum (Fisch. et C. A. Mey.) Boiss. ■☆
18446 Alyssum mutabile Vent.;杂色庭荠■
18447 Alyssum nanum Pomel = Alyssum simplex Rudolphi ■
18448 Alyssum nomismocarpum Rech. f. et al. = Alyssum homalocarpum (Fisch. et C. A. Mey.) Boiss. ■☆
18449 Alyssum numidicum Pomel;努米底亚庭荠■☆
18450 Alyssum obovatum (C. A. Mey.) Turcz.;倒卵叶庭荠■
18451 Alyssum obtusifolium Stev.;钝叶庭荠■☆
18452 Alyssum parviflorum M. Bieb.;小花庭荠■☆
18453 Alyssum parviflorum M. Bieb. = Alyssum simplex Rudolphi ■
18454 Alyssum parviflorum M. Bieb. var. collinum (Brot.) Maire et Weiller = Alyssum simplex Rudolphi ■
18455 Alyssum pateri Nyar.;帕特庭荠■☆
18456 Alyssum patulum Pomel;张开庭荠■☆
18457 Alyssum peltarioides Boiss.;盾叶庭荠■☆
18458 Alyssum persicum Boiss.;波斯庭荠■☆
18459 Alyssum psilocarpum Boiss. = Alyssum minutum Schltdl. ex DC. ■☆
18460 Alyssum psilocarpum Boiss. var. leiocarpum (Pomel) Batt. = Alyssum minutum Schltdl. ex DC. ■☆
18461 Alyssum pyramidatum Bornm. = Alyssum szovitsianum Fisch. et C. A. Mey. ■☆
18462 Alyssum pyrenaicum Lapeyr.;比利牛斯庭荠■☆
18463 Alyssum rostratum Stev.;喙庭荠■☆
18464 Alyssum saxatile L. = Aurinia saxatilis (L.) Desv. ■
18465 Alyssum saxatile L. var. orientale ?;东方岩生庭荠;Golden Alyssum ■☆
18466 Alyssum scutigerum Durieu;盾状庭荠■☆
18467 Alyssum serpyllifolium Desf.;百里香叶庭荠■☆
18468 Alyssum serpyllifolium Desf. var. macrosepalum (Ball) Batt. = Alyssum serpyllifolium Desf. ■☆
18469 Alyssum sibiricum Willd.;西伯利亚庭荠(双胚庭荠,庭荠);Siberia Madwort, Siberian Madwort ■
18470 Alyssum simplex Rudolphi = Alyssum minus (L.) Rothm. ■
18471 Alyssum simplex Rudolphi subsp. edentulum (Andr.) N. Galland;无齿庭荠■☆
18472 Alyssum spathulatum Stephan ex Willd. = Galitzkya spathulata (Steph. ex Willd.) V. V. Botschantz. ■
18473 Alyssum speciosum Pomel;美丽庭荠■☆
18474 Alyssum spinosum L.;刺庭荠;Spiny Alyssum ●☆
18475 Alyssum spinosum L. = Hormathophylla spinosa (L.) P. Küpfer ■☆
18476 Alyssum stapfii Vierh.;施塔普夫庭荠■☆
18477 Alyssum stenostachyum Botsch. et Vved. = Alyssum szowitsianum Fisch. et C. A. Mey. ■☆
18478 Alyssum strictum Willd.;直庭荠■☆
18479 Alyssum strigosum Sol.;糙伏毛庭荠;Alyssum ■☆
18480 Alyssum szovitsianum Fisch. et C. A. Mey.;绍氏庭荠;Szowits' Madwort ■☆
18481 Alyssum szowitsianum Fisch. et C. A. Mey. = Alyssum szovitsianum Fisch. et C. A. Mey. ■☆
18482 Alyssum tenuifolium Stephan ex Willd.;细叶庭荠;Thinleaf Alyssum ■
18483 Alyssum tortuosum Waldst. et Kit. ex Willd.;扭庭荠;Tortuous Madwort ■
18484 Alyssum trichostachyum Rupr.;毛穗庭荠■☆
18485 Alyssum turkestanicum Regel et Schmalh. ex Regel;土耳其斯坦庭荠■☆
18486 Alyssum turkestanicum Regel et Schmalh. ex Regel var. desertorum (Stapf) Botsch. = Alyssum desertorum Stapf ■
18487 Alyssum turkestanicum Regel et Schmalh. var. desertorum (Stapf) Botsch. = Alyssum desertorum Stapf ■
18488 Alyssum umbellatum Desv.;小伞庭荠■☆
18489 Alyssum wulfenianum Willd.;金球庭荠■☆
18490 Alytostylis Mast. = Roydsia Roxb. ●
18491 Alyxia Banks ex R. Br. (1810)(保留属名);链珠藤属(阿莉藤属,念珠藤属);Alyxia ●
18492 Alyxia acutifolia Tsiang = Alyxia levinei Merr. ●
18493 Alyxia balansae Pit. = Alyxia siamensis Craib ●
18494 Alyxia bodinieri (H. Lév.) Woodson = Alyxia schlechteri H. Lév. ●
18495 Alyxia buxifolia R. Br.;黄杨叶念珠藤;Sea Box ●☆
18496 Alyxia erythrocarpa Vatke = Petchia erythrocarpa (Vatke) Leeuwenb. ●☆
18497 Alyxia euonymifolia Tsiang = Alyxia odorata Wall. et G. Don ●
18498 Alyxia fascicularis Benth. et Hook. f.;尾尖链珠藤;Fasciculate Alyxia ●
18499 Alyxia flavescens Pierre = Alyxia reinwardtii Blume ●

18500 Alyxia forbesii King et Gamble = Alyxia reinwardtii Blume ●
18501 Alyxia forbesii King et Gamble var. pubescens P. T. Li;那坡链珠藤;Napo Forbes Alyxia ●
18502 Alyxia funingensis Tsiang et P. T. Li = Alyxia marginata Pit. ●
18503 Alyxia hainanensis Merr. et Chun = Alyxia odorata Wall. et G. Don ●
18504 Alyxia insularis Kaneh. et Sasaki;兰屿链珠藤(兰屿阿莉藤,兰屿念珠藤,念珠藤);Insular Alyxia,Lanyu Alyxia ●
18505 Alyxia jasminea Tsiang et P. T. Li = Alyxia odorata Wall. et G. Don ●
18506 Alyxia kweichouensis Tsiang et P. T. Li;贵州链珠藤;Guizhou Alyxia ●
18507 Alyxia kweichouensis Tsiang et P. T. Li = Alyxia levinei Merr. ●
18508 Alyxia lehtungensis Tsiang = Alyxia odorata Wall. et G. Don ●
18509 Alyxia levinei Merr.;筋藤(尖叶链珠藤,坎香藤,三托藤,藤满山香);Acuteleaf Alyxia, Acute-leaved Aalyxia, Guizhou Aalyxia, Levine Aalyxia,Shrpleaf Alyxia ●
18510 Alyxia lucida Baker = Petchia erythrocarpa (Vatke) Leeuwenb. ●☆
18511 Alyxia lucida Wall. var. meiantha Stapf = Alyxia reinwardtii Blume ●
18512 Alyxia madagascariensis A. DC. = Petchia erythrocarpa (Vatke) Leeuwenb. ●☆
18513 Alyxia marginata Pit.;陷边链珠藤(富宁链珠藤);Funing Alyxia,Marginate Alyxia ●
18514 Alyxia menglungensis Tsiang et P. T. Li;勐龙链珠藤;Menglong Alyxia ●
18515 Alyxia nitens Kerr = Alyxia odorata Wall. et G. Don ●
18516 Alyxia odorata Wall. et G. Don;海南链珠藤(白骨藤,串珠子,乐东链珠藤,乐东念珠藤,茉莉链珠藤,卫矛叶链珠藤,卫矛叶念珠藤);Alyxia, Common Alyxia, Euonymus Leaf Alyxia, Hainan Alyxia,Jasmine Alyxia,Ledong Alyxia ●
18517 Alyxia olivaeformis Gaudich.;油橄榄念珠藤●☆
18518 Alyxia polysperma Scott-Elliot = Petchia erythrocarpa (Vatke) Leeuwenb. ●☆
18519 Alyxia reinwardtii Blume;长花链珠藤(兰氏念珠藤);Forbes Alyxia,Longflower Alyxia ●
18520 Alyxia reinwardtii Blume var. meiantha (Stapf) Markgr. = Alyxia reinwardtii Blume ●
18521 Alyxia ruscifolia A. Cunn.;假叶树链珠藤●☆
18522 Alyxia ruscifolia R. Br.;串果念珠藤;Chain Fruit,Prickly Alyxia ●☆
18523 Alyxia schlechteri H. Lév.;狭叶链珠藤;Narrowleaf Alyxia, Narrow-leaved Alyxia ●
18524 Alyxia schlechteri H. Lév. = Daphne tangutica Maxim. ●
18525 Alyxia schlechteri H. Lév. var. salicifolia P. T. Li;柳叶念珠藤;Wilowleaf Alyxia ●
18526 Alyxia schlechteri H. Lév. var. salicifolia P. T. Li = Alyxia schlechteri H. Lév. ●
18527 Alyxia siamensis Craib;长序链珠藤(大果链珠藤,橄榄果链珠藤,浆包藤);Longcyme Alyxia, Long-cymed Alyxia, Olivaceous Alyxia,Siam Alyxia ●
18528 Alyxia sibuyanensis Elmer;兰屿念珠藤●
18529 Alyxia sinensis Champ. ex Benth.;链珠藤(阿莉藤,阿利藤,春根藤,瓜子金,瓜子藤,瓜子英,过骨边,过滑边,过山香,鸡骨香,满山香,念珠藤,七里香,山红木,香藤,中国念珠藤);Chinese Alyxia ●
18530 Alyxia spicata R. Br.;海滩念珠藤(穗花念珠藤);Sea Box ●☆
18531 Alyxia taiwanensis S. Y. Lu et Yuen P. Yang;台湾念珠藤(台湾链珠藤);Taiwan Alyxia ●
18532 Alyxia villilimba C. Y. Wu ex Tsiang et P. T. Li;毛叶链珠藤(大叶念珠藤);Big Hairyleaf Alyxia,Hairyleaf Alyxia,Villus Alyxia ●
18533 Alyxia villilimba C. Y. Wu ex Tsiang et P. T. Li var. macrophylla P. T. Li = Alyxia villilimba C. Y. Wu ex Tsiang et P. T. Li ●
18534 Alyxia vulgaris Tsiang = Alyxia odorata Wall. et G. Don ●
18535 Alyxia yunkuniana Tsiang = Alyxia siamensis Craib ●
18536 Alzalia F. Dietr. = Alzatea Ruiz et Pav. ●☆
18537 Alzatea Ruiz et Pav. (1794);双隔果属(扭柄叶属)●☆
18538 Alzatea mexicana F. Dietr.;墨西哥双隔果(墨西哥扭柄叶)●☆
18539 Alzatea verticillata Ruiz et Pav.;双隔果(扭柄叶)●☆
18540 Alzateaceae S. A. Graham = Crypteroniaceae A. DC.(保留科名)●
18541 Alzateaceae S. A. Graham(1985);双隔果科(双翼果科)●☆
18542 Alziniana F. Dietr. ex Pfeiff. = Alzatea Ruiz et Pav. ●☆
18543 Amadea Adans. = Androsace L. ■
18544 Amagris Raf. = Calamagrostis Adans. ■
18545 Amaioua Aubl. (1775);阿迈茜属●☆
18546 Amaioua africana Spreng. = Coffea zanguebariae Lour. ●☆
18547 Amaioua edulis Baill.;可食阿迈茜●☆
18548 Amaioua grandifolia Miq.;大叶阿迈茜●☆
18549 Amaioua guianensis Aubl.;圭亚那阿迈茜●☆
18550 Amaioua longifolia Poepp. et Endl.;长叶阿迈茜●☆
18551 Amaioua oocarpa Spruce = Amaioua oocarpa Spruce ex Benth. et Hook. f. ●☆
18552 Amaioua oocarpa Spruce ex Benth. et Hook. f.;卵果阿迈茜●☆
18553 Amaioua urophylla Standl.;尾叶阿迈茜●☆
18554 Amalago Raf. = Piper L. ●■
18555 Amalia Hort. Hisp. = Tillandsia L. ■☆
18556 Amalia Hort. Hisp. ex Endl. = Tillandsia L. ■☆
18557 Amalia Rchb. = Laelia Lindl. ■☆
18558 Amalias Hoffmanns. = Amalia Rchb. ■☆
18559 Amalobatrya Kunth ex Meissa. = Symmeria Benth. ●☆
18560 Amalocalyx Pierre(1898);毛车藤属(酸果藤属);Amalocalyx ●
18561 Amalocalyx burmanicus Chatterjee = Amalocalyx microlobus Pierre ●◇
18562 Amalocalyx microlobus Pierre;毛车藤(酸果藤);Microlobed Amalocalyx,Yunnan Amalocalyx ●◇
18563 Amalocalyx yunnanensis Tsiang = Amalocalyx microlobus Pierre ●◇
18564 Amalophyllon Brandegee(1914);白岩花属■☆
18565 Amalophyllon rupestre Brandegee;白岩花■☆
18566 Amamelis Lem. = Hamamelis L. ●
18567 Amana Honda = Tulipa L. ■
18568 Amana Honda(1935);老鸦瓣属(山慈姑属)■☆
18569 Amana edulis (Miq.) Honda;老鸦瓣(光慈姑,光姑,光菇,尖慈姑,老鸦头,毛地梨,毛地栗,茅山菰,棉花包,山茨菇,山慈姑,山蛋,双鸭子,小慈姑);Edible Tulip ■☆
18570 Amana edulis (Miq.) Honda = Tulipa edulis (Miq.) Baker ■☆
18571 Amana graminifolia (Baker ex S. Moore) A. D. Hall = Tulipa edulis (Miq.) Baker ■☆
18572 Amana graminifolia (Baker) A. D. Hall. = Tulipa edulis (Miq.) Baker ■☆
18573 Amana latifolia (Makino) Honda = Tulipa erythronioides Baker ■
18574 Amannia Blume = Ammannia L. ■
18575 Amanoa Aubl. (1775);阿马木属(阿马大戟属,阿曼木属);Amanoa ●☆
18576 Amanoa bracteosa Planch.;阿马木(阿马大戟,多苞片阿马大

戟)●☆

18577 Amanoa laurifolia Pax = Pentabrachion reticulatum Müll. Arg. ●☆

18578 Amanoa potamophila Croizat;河生阿马木(河生阿曼木)●☆

18579 Amanoa schweinfurthii Baker et Hutch. = Erythroxylum fischeri Engl. ●☆

18580 Amanoa strobilacea Müll. Arg.;球果阿马木●☆

18581 Amapa Steud. = Carapa Aubl. ●☆

18582 Amaraboya Linden ex Mast. = Blakea P. Browne ■☆

18583 Amaraboya amabilis Linden = Amaraboya amabills Linden ex Mast. ●☆

18584 Amaraboya princeps Linden = Amaraboya princeps Linden ex Mast. ●☆

18585 Amaraboya splendida Linden = Amaraboya splendida Linden ex Mast. ●☆

18586 Amaracanthus Steud. = Amaracarpus Blume ●☆

18587 Amaracarpus Blume(1827);沟果茜属●☆

18588 Amaracarpus acuminatus S. Moore;尖沟果茜●☆

18589 Amaracarpus anomalus Wernham;异常沟果茜●☆

18590 Amaracarpus atrocarpus Merr. et L. M. Perry;黑果沟果茜●☆

18591 Amaracarpus bicolor Merr. et L. M. Perry;二色沟果茜●☆

18592 Amaracarpus caeruleus Merr. et L. M. Perry;蓝沟果茜●☆

18593 Amaracarpus caudatus Ridl.;尾状沟果茜●☆

18594 Amaracarpus cuneifolius Valeton;楔叶沟果茜●☆

18595 Amaracarpus grandicalyx Valeton;大萼沟果茜●☆

18596 Amaracarpus heteropus Valeton;异足沟果茜●☆

18597 Amaracarpus leucocarpus Valeton;白果沟果茜●☆

18598 Amaracarpus longifolius Elmer;长叶沟果茜●☆

18599 Amaracarpus macrophyllus Valeton;大叶沟果茜●☆

18600 Amaracarpus major (Valeton) A. P. Davis;大沟果茜●☆

18601 Amaracarpus microphyllus Miq.;小叶沟果茜●☆

18602 Amaracarpus montanus Valeton;山地沟果茜●☆

18603 Amaracarpus pubescens Blume;毛沟果茜●☆

18604 Amaracarpus saxicola Ridl.;岩地沟果茜●☆

18605 Amaracarpus trichanthus Merr. et L. M. Perry;毛花沟果茜●☆

18606 Amaracarpus trichocalyx Valeton;毛萼沟果茜●☆

18607 Amaracarpus trichocarpus Merr. et L. M. Perry;毛果沟果茜●☆

18608 Amaracarpus urophyllus Merr. et L. M. Perry;尾叶沟果茜●☆

18609 Amaracarpus xanthocarpus Merr. et L. M. Perry;黄果沟果茜●☆

18610 Amaraceae Dulac = Gentianaceae Juss.(保留科名)●■

18611 Amaracus Gled.(1764)(保留属名);阿玛草属●■☆

18612 Amaracus Hill(废弃属名) = Amaracus Gled.(保留属名)●■☆

18613 Amaracus Hill(废弃属名) = Majorana Mill.(保留属名)■☆

18614 Amaracus Hill(废弃属名) = Origanum L. ●■

18615 Amaracus brevidens Bornm.;短齿阿玛草☆

18616 Amaracus ciliatus Briq.;睫毛阿玛草☆

18617 Amaracus cordifolius Montbret et Aucher ex Benth.;心叶阿玛草☆

18618 Amaracus dictamnus Benth. = Origanum dictamnus L. ■☆

18619 Amaracus leptocladus Briq.;细枝阿玛草☆

18620 Amaracus libanoticus Briq.;黎巴嫩阿玛草☆

18621 Amaracus pulcher Briq.;美丽阿玛草☆

18622 Amaracus rotundifolius (Boiss.) Briq. = Origanum rotundifolium Boiss. ■☆

18623 Amaracus rotundifolius Briq.;圆叶阿玛草☆

18624 Amaracus tomentosus Moench;绒毛阿玛草☆

18625 Amaralia Welw. ex Benth. et Hook. f. = Sherbournia G. Don ●☆

18626 Amaralia Welw. ex Hook. f. = Sherbournia G. Don ●☆

18627 Amaralia batesii Wernham = Sherbournia batesii (Wernham) Hepper ●☆

18628 Amaralia bignoniiflora (Welw.) Hiern = Sherbournia bignoniiflora (Welw.) Hua ●☆

18629 Amaralia brazzaei (Hua) Wernham = Sherbournia bignoniiflora (Welw.) Hua ●☆

18630 Amaralia buntingii Wernham = Sherbournia bignoniiflora (Welw.) Hua ●☆

18631 Amaralia calycina (G. Don) K. Schum. = Sherbournia calycina (G. Don) Hua ●☆

18632 Amaralia ekotokicola Wernham = Sherbournia bignoniiflora (Welw.) Hua ●☆

18633 Amaralia hapalophylla (Wernham) Keay = Sherbournia hapalophylla (Wernham) Hepper ●☆

18634 Amaralia heinsioides Wernham = Sherbournia bignoniiflora (Welw.) Hua ●☆

18635 Amaralia huana Wernham = Sherbournia calycina (G. Don) Hua ●☆

18636 Amaralia micrantha Wernham = Sherbournia millenii (Wernham) Hepper ●☆

18637 Amaralia millenii Wernham = Sherbournia millenii (Wernham) Hepper ●☆

18638 Amaralia palustris Wernham = Sherbournia bignoniiflora (Welw.) Hua ●☆

18639 Amaralia penduliflora (K. Schum.) Wernham = Aoranthe penduliflora (K. Schum.) Somers ●☆

18640 Amaralia sherbourniae (Hook.) Wernham = Sherbournia calycina (G. Don) Hua ●☆

18641 Amaralia streptocaulon (K. Schum.) Keay = Sherbournia streptocaulon (K. Schum.) Hepper ●☆

18642 Amaralia zenkeri (Hua) Wernham = Sherbournia zenkeri Hua ●☆

18643 Amarantellus Speg. = Amaranthus L. ■

18644 Amarantesia Hort. ex Regel = Telanthera R. Br. ■

18645 Amaranthaceae Adans. = Amaranthaceae Juss.(保留科名)●■

18646 Amaranthaceae Juss.(1789)(保留科名);苋科;Amaranth Family, Pigweed Family ●■

18647 Amaranthoides Mill.(1754);拟苋属■☆

18648 Amaranthoides Mill. = Gomphrena L. ●■

18649 Amaranthoides decumbens (Jacq.) M. Gomez;拟苋■☆

18650 Amaranthus Adans. = Celosia L. ■

18651 Amaranthus Kunth = Amaranthus L. ■

18652 Amaranthus L.(1753);苋属(水麻属);Amaranth, Amaranthus, Bhaji, Northern Agrostis, Pigweed ■

18653 Amaranthus acanthochiton J. D. Sauer;赖特苋;Greenstripe, Greenstripe Amaranth ■☆

18654 Amaranthus adscendens Loisel. = Amaranthus lividus L. ■

18655 Amaranthus albus L.;白苋;Bushy Pigweed, Pig Weed, Pigweed, Prostrate Pigweed, Tumble Pigweed, Tumbleweed, Tumbleweed Amaranth, White Amaranth, White Amaranthus, White Pigweed, White Rollingweed ■

18656 Amaranthus albus L. var. pubescens (Uline et W. L. Bray) Fernald = Amaranthus albus L. ■

18657 Amaranthus ambigens Standl. = Amaranthus tuberculatus (Moq.) J. D. Sauer ■☆

18658 Amaranthus angustifolius Lam. = Amaranthus graecizans L. ■

18659 Amaranthus angustifolius Lam. subsp. graecizans (L.) Maire et Weiller = Amaranthus graecizans L. ■

18660 Amaranthus angustifolius Lam. subsp. polygonoides (Roxb.)

Maire et Weiller = Amaranthus graecizans L. ■

18661 Amaranthus arenicola I. M. Johnst.；沙丘苋；Sand Amaranth, Sand-hills Amaranth, Sandhills Pigweed ■☆

18662 Amaranthus ascendens Loisel. = Amaranthus blitum L. var. ascendens (Loisel.) DC. ■☆

18663 Amaranthus ascendens Loisel. = Amaranthus blitum L. ■

18664 Amaranthus ascendens Loisel. = Amaranthus lividus L. ■

18665 Amaranthus aschersonianus Chiov. = Amaranthus graecizans L. ■

18666 Amaranthus australis (A. Gray) J. D. Sauer；南苋；Southern Amaranth, Southern Water-hemp ■☆

18667 Amaranthus berlandieri (Moq.) Uline et W. L. Bray = Amaranthus polygonoides L. ■

18668 Amaranthus blitoides S. Watson；北美苋；Creeping Amaranth, Forked Beard-grass, Mat Amaranth, Matweed, Matweed Amaranth, Prostrate Amaranth, Prostrate Amaranthus, Prostrate Pigweed, Spreading Pigweed, Tumbleweed ■

18669 Amaranthus blitoides S. Watson var. scleropoides Thell. = Amaranthus blitoides S. Watson ■

18670 Amaranthus blitum L.；野苋(凹头苋，凹叶野苋菜，白莲子，刺苋菜，铅色苋，苋，苋菜，小苋，野苋菜)；Emarginate Amaranth, Emarginate Amaranthus, Livid Amaranth, Purple Amaranth ■

18671 Amaranthus blitum L. = Amaranthus lividus L. ■

18672 Amaranthus blitum L. subsp. emarginatus (Moq. ex Uline et Bray) Carretero et al.；微缺野苋■☆

18673 Amaranthus blitum L. subsp. polygonoides (Zoll. ex Moq.) Carretero；多节野苋■☆

18674 Amaranthus blitum L. var. ascendens (Loisel.) DC.；上升野苋■☆

18675 Amaranthus blitum L. var. graecizans (L.) Moq. = Amaranthus graecizans L. ■

18676 Amaranthus blitum L. var. oleraceus (L.) Hook. f. = Amaranthus blitum L. ■

18677 Amaranthus blitum L. var. polygonoides ? = Amaranthus blitum L. subsp. polygonoides (Zoll. ex Moq.) Carretero ■☆

18678 Amaranthus blitum L. var. polygonoides ? = Amaranthus thellungianus Nevski ■☆

18679 Amaranthus blitum L. var. silvestris ? = Amaranthus graecizans L. subsp. silvestris (Vill.) Brenan ■☆

18680 Amaranthus bouchonii Thell. = Amaranthus powellii S. Watson ■☆

18681 Amaranthus bracteosus Uline et W. L. Bray = Amaranthus powellii S. Watson ■☆

18682 Amaranthus californicus (Moq.) S. Watson；加州苋；California Amaranth, Californian Amaranth ■☆

18683 Amaranthus cannabinus (L.) J. D. Sauer；盐地苋；Sahmarsh Water Hemp, Salt-marsh Water-hemp, Tidal-marsh Water-hemp, Tide-marsh Water Hemp, Virginian Hemp, Water-hemp Pigweed ■☆

18684 Amaranthus cannabinus L. var. concatenata Moq. = Amaranthus tuberculatus (Moq.) J. D. Sauer ■☆

18685 Amaranthus capensis Thell.；好望角苋；Cape Pigweed ■☆

18686 Amaranthus capensis Thell. subsp. uncinatus (Thell.) Brenan；钩苋■☆

18687 Amaranthus caudatus L.；尾穗苋(高丽谷，红苋菜，红苋米草，九莲灯花，老枪谷，龙须苋，千岁谷，千穗谷，西洋谷，仙人谷，鸦谷)；Amaranth, Baldare, Bush Greens, Cat-mil, Cat's Tail, Cat's Tails, Cat-tail, Feathers, Floramor, Florimer, Flower Armour, Flower Velure, Foxtail Amaranth, Inca Wheat, Kiss-me-quick, Loblolly, Love-lies-bleeding, Nun's Scourge, Passevelours, Purple Amaranth, Purple Velvet Flower, Quilete, Soldier's Feathers, Tassel Flower, Tasseleflower Thrumwort, Thrumwort, Tumble Weed, Tumbleweed, Turkey's Snout, Velvet Flower ■

18688 Amaranthus caudatus L. f. oblongopetalus Suess. = Amaranthus hybridus L. subsp. cruentus (L.) Thell. ■

18689 Amaranthus caudatus L. var. pseudopaniculatus Suess. = Amaranthus hybridus L. subsp. cruentus (L.) Thell. ■

18690 Amaranthus chlorostachys Willd. = Amaranthus hybridus L. ■

18691 Amaranthus chlorostachys Willd. var. pseudoretroflexus Thell. = Amaranthus hybridus L. var. pseudoretroflexus (Thell.) Carretero ■☆

18692 Amaranthus crassipes Schltdl.；粗梗苋；Clubfoot Amaranth, Spreading Amaranth, Tropical Spreading Amaranth ■☆

18693 Amaranthus crassipes Schltdl. var. warnockii (I. M. Johnst.) Henrickson；瓦尔粗梗苋■☆

18694 Amaranthus crispus (Lesp. et Thévenau) A. Braun ex J. M. Coult. et S. Watson；脆叶苋；Crispleaf Amaranth, Crisp-leaved Amaranth ■☆

18695 Amaranthus crispus (Lesp.) A. Br. ex J. M. Coult. et S. Watson = Amaranthus crispus (Lesp. et Thévenau) A. Braun ex J. M. Coult. et S. Watson ■☆

18696 Amaranthus cruentus L.；老鸦谷；Blood Amaranth, Caterpillar Amaranth, Purple Amaranth, Red Amaranth ■

18697 Amaranthus cruentus L. = Amaranthus hybridus L. subsp. cruentus (L.) Thell. ■

18698 Amaranthus cruentus L. = Amaranthus paniculatus L. ■

18699 Amaranthus cruentus L. var. paniculatus (L.) Thell. = Amaranthus hybridus L. subsp. cruentus (L.) Thell. ■

18700 Amaranthus cruentus L. var. patulus (Bertol.) Lambinon = Amaranthus hybridus L. ■

18701 Amaranthus deflexus L.；外折苋；Largefruit Amaranth, Perennial Pigweed ■☆

18702 Amaranthus delilei Loret = Amaranthus retroflexus L. ■

18703 Amaranthus dililei Richt. et Loret = Amaranthus retroflexus L. var. delilei (Richt. et Loret) Thell. ■

18704 Amaranthus dinteri Schinz；丁特苋■☆

18705 Amaranthus dinteri Schinz = Amaranthus capensis Thell. ■☆

18706 Amaranthus dinteri Schinz subsp. brevipetiolatus Brenan；短叶柄苋■☆

18707 Amaranthus dinteri Schinz var. uncinatus Thell. = Amaranthus capensis Thell. subsp. uncinatus (Thell.) Brenan ■☆

18708 Amaranthus dubius Mart. ex Thell.；可疑苋；Spinach, Spleen Amaranth ■☆

18709 Amaranthus dubius Mart. ex Thell. var. crassespicatus Suess. = Amaranthus dubius Mart. ex Thell. ■☆

18710 Amaranthus edulis Michx. ex Moq. = Amaranthus caudatus L. ■

18711 Amaranthus edulis Speg. = Amaranthus caudatus L. ■

18712 Amaranthus emarginatus Moq. ex Uline et Bray = Amaranthus blitum L. subsp. emarginatus (Moq. ex Uline et Bray) Carretero et al. ■☆

18713 Amaranthus fasciatus Roxb. = Amaranthus viridis L. ■

18714 Amaranthus fimbriatus (Torr.) Benth. ex S. Watson；线苋；Fringed Amaranth ■☆

18715 Amaranthus fimbriatus (Torr.) Benth. ex S. Watson var. denticulatus (Torr.) Uline et W. L. Bray = Amaranthus fimbriatus (Torr.) Benth. ex S. Watson ■☆

18716 Amaranthus flavus L. = Amaranthus hypochondriacus L. ■

18717 Amaranthus floridanus (S. Watson) J. D. Sauer；佛罗里达苋；Florida Amaranth, Florida Water-hemp ■☆

18718 Amaranthus frumentaceus Buch. -Ham. = Amaranthus hybridus L. ■

18719 Amaranthus galii Sennen et Gonzalo = Amaranthus ozanonii Thell. ■☆

18720 Amaranthus galii Sennen et Gonzalo ex Pritzter = Amaranthus ozanonii Thell. ■☆

18721 Amaranthus gangeticus L. = Amaranthus tricolor L. ■

18722 Amaranthus gangeticus L. tricolor ? = Amaranthus tricolor L. ■

18723 Amaranthus gangeticus Wall. = Amaranthus blitum L. ■

18724 Amaranthus gracilentus H. W. Kung;细枝苋;Gracilebranch Amaranthus,Thinbrahch Amaranth,Thinbrahch Amaranthus ■

18725 Amaranthus gracilis Desf. = Amaranthus viridis L. ■

18726 Amaranthus gracilis Desf. = Chenopodium viridis L. ■☆

18727 Amaranthus gracilis Desf. ex Poir. = Amaranthus viridis L. ■

18728 Amaranthus graecizans L. = Amaranthus albus L. ■

18729 Amaranthus graecizans L. = Amaranthus blitoides S. Watson ■

18730 Amaranthus graecizans L. subsp. thellungianus (Nevski) Gusev;泰龙苋■☆

18731 Amaranthus graecizans L. var. polygonoides ? = Amaranthus thellungianus Nevski ■☆

18732 Amaranthus graecizans L. var. pubescens Uline et W. L. Bray = Amaranthus albus L. ■

18733 Amaranthus greggii S. Watson;格雷格苋;Gregg's Amaranth, Josiah Amaranth ■☆

18734 Amaranthus hybridus L.;绿穗苋(西风谷,圆穗苋);African Spinach,Green Amaranth,Green Pigweed,Hybrid Amaranth,Hybrid Amaranthus,Prince's Feather,Slender Amaranth,Slender Pigweed, Slim Amaranth,Smooth Amaranth,Smooth Pigweed ■

18735 Amaranthus hybridus L. f. acicularis Suess. = Amaranthus dubius Mart. ex Thell. ■☆

18736 Amaranthus hybridus L. f. hypochondriacus (L.) B. L. Rob. = Amaranthus paniculatus L. ■

18737 Amaranthus hybridus L. subsp. cruentus (L.) Thell. = Amaranthus cruentus L. ■

18738 Amaranthus hybridus L. subsp. cruentus (L.) Thell. var. patulus (Bertol.) Thell. = Amaranthus hybridus L. ■

18739 Amaranthus hybridus L. subsp. hypochondriacus (L.) Thell. = Amaranthus powellii S. Watson ■☆

18740 Amaranthus hybridus L. subsp. hypochondriacus (L.) Thell. = Amaranthus hypochondriacus L. ■

18741 Amaranthus hybridus L. subsp. patulus (Bertol.) Carretero = Amaranthus hybridus L. ■

18742 Amaranthus hybridus L. var. aciculatus Thell. = Amaranthus cruentus L. ■

18743 Amaranthus hybridus L. var. cruentus (L.) Moq.;血红苋■☆

18744 Amaranthus hybridus L. var. erythrostachys Moq. = Amaranthus hypochondriacus L. ■

18745 Amaranthus hybridus L. var. erythrostachys Moq. = Amaranthus powellii S. Watson ■☆

18746 Amaranthus hybridus L. var. hypochondriacus (L.) B. L. Rob. = Amaranthus hypochondriacus L. ■

18747 Amaranthus hybridus L. var. hypochondriacus (L.) B. L. Rob. = Amaranthus paniculatus L. ■

18748 Amaranthus hybridus L. var. incurvatus (Gren. et Godr.) Brenan = Amaranthus hybridus L. ■

18749 Amaranthus hybridus L. var. paniculatus (L.) Thell. = Amaranthus cruentus L. ■

18750 Amaranthus hybridus L. var. patulus (Bertol.) Thell. = Amaranthus hybridus L. ■

18751 Amaranthus hybridus L. var. pseudoretroflexus (Thell.) Carretero = Amaranthus powellii S. Watson ■☆

18752 Amaranthus hypochondriacus L.;千穗苋(白籽苋,老来红,千穗谷,玉谷);Baldar Herb,Cock's Comb,Globe Amaranth,Prince Albert's Feather, Prince-of-wales'-feather, Prince's Feather, Prince's-feather, Prince's-feather Amaranth, Prince-of-wales Feather,Red Cock's Comb,Red Cockscomb,Velvet Flower ■

18753 Amaranthus hypochondriacus L. = Amaranthus hybridus L. subsp. hypochondriacus (L.) Thell. ■

18754 Amaranthus hypochondriacus L. = Amaranthus hybridus L. var. hypochondriacus (L.) B. L. Rob. ■

18755 Amaranthus hypochondriacus L. = Amaranthus hybridus L. ■

18756 Amaranthus hypochondriacus L. var. amentaceus Suess. = Amaranthus hybridus L. ■

18757 Amaranthus hypochondriacus L. var. subdubius Suess. = Amaranthus hybridus L. subsp. cruentus (L.) Thell. ■

18758 Amaranthus inamoenus Willd. = Amaranthus tricolor L. subsp. mangostanus (L.) Aellen ■

18759 Amaranthus inamoenus Willd. = Amaranthus tricolor L. ■

18760 Amaranthus incurvatus Gren. et Godr. = Amaranthus hybridus L. ■

18761 Amaranthus lanceolatus Roxb. = Amaranthus tricolor L. ■

18762 Amaranthus leucocarpus S. Watson = Amaranthus hypochondriacus L. ■

18763 Amaranthus leucospermus S. Watson;白籽苋菜■☆

18764 Amaranthus lineatus R. Br.;条纹苋菜;Australian Amaranth ■☆

18765 Amaranthus lividus L.;凹头苋菜(凹头苋,凹叶野苋菜)■

18766 Amaranthus lividus L. = Amaranthus blitum L. ■

18767 Amaranthus lividus L. = Amaranthus thellungianus Nevski ■☆

18768 Amaranthus lividus L. subsp. polygonoides (Moq.) Probst = Amaranthus blitum L. subsp. polygonoides (Zoll. ex Moq.) Carretero ■☆

18769 Amaranthus lividus L. var. ascendens (Loisel.) Hayward et Druce = Amaranthus lividus L. ■

18770 Amaranthus lividus L. var. ascendens (Loisel.) Thell. = Amaranthus blitum L. subsp. emarginatus (Moq. ex Uline et Bray) Carretero et al. ■☆

18771 Amaranthus lividus L. var. ascendens (Loisel.) Thell. = Amaranthus blitum L. ■

18772 Amaranthus lividus L. var. ascendens (Loisel.) Thell. = Amaranthus lividus L. ■

18773 Amaranthus lividus L. var. oleraceus (L.) Thell.;蔬菜苋;Green Amaranth ■☆

18774 Amaranthus lividus L. var. oleraceus (L.) Thell. = Amaranthus blitum L. var. oleraceus (L.) Hook. f. ■☆

18775 Amaranthus luteus L.;黄苋;Golden Flower Velure,Goldilocks ■☆

18776 Amaranthus mangostanus L.;印度苋(苋,苋菜)■

18777 Amaranthus mangostanus L. = Amaranthus tricolor L. subsp. mangostanus (L.) Aellen ■

18778 Amaranthus mangostanus L. = Amaranthus tricolor L. ■

18779 Amaranthus melancholicus L.;雁来红(老少年)■

18780 Amaranthus melancholicus L. = Amaranthus tricolor L. ■

18781 Amaranthus muricatus (Moq.) Hieron.;非洲苋;African Amaranth,Muricate Amaranth ■☆

18782 Amaranthus obcordatus (A. Gray) Standl.;倒心形苋;Trans-pecos Amaranth ■☆

18783 Amaranthus oleraceus L. = Amaranthus blitum L. var. oleraceus (L.) Hook. f. ■☆

18784 Amaranthus oleraceus L. = Amaranthus lividus L. ■

18785 Amaranthus oleraceus L. = Amaranthus tricolor L. ■

18786 Amaranthus ozanonii Thell.;奥扎农苋■☆

18787 Amaranthus palmeri S. Watson;帕氏苋;Careless Weed,Palmer's Amaranth,Redroot ■☆

18788 Amaranthus paniculatus L.;紫穗苋(繁穗苋,红苋菜,红粘谷,老来红,老鸦谷,墨西哥红苋,天粟米,天雪米,西番谷,西方谷,亚谷,亚谷苋);Amaranth,Cock's Comb,Hell's Curse,Mexican Grain Amaranth,Paniculate Amaranth,Paniculate Amaranthus,Purple Amaranth,Red Amaranth,Tassel Amaranth ■

18789 Amaranthus paniculatus L. = Amaranthus cruentus L. ■

18790 Amaranthus paniculatus L. = Amaranthus hybridus L. subsp. cruentus (L.) Thell. ■

18791 Amaranthus paniculatus L. = Amaranthus hybridus L. ■

18792 Amaranthus paniculatus L. var. speciosus L. H. Bailey;美丽紫穗苋■☆

18793 Amaranthus paolii Chiov. = Amaranthus graecizans L. ■

18794 Amaranthus parvulus Peter = Amaranthus graecizans L. subsp. silvestris (Vill.) Brenan ■☆

18795 Amaranthus patulus Bertol.;青苋(台湾苋,细叶青鸡冠)■

18796 Amaranthus patulus Bertol. = Amaranthus cruentus L. ■

18797 Amaranthus patulus Bertol. = Amaranthus hybridus L. ■

18798 Amaranthus persicarioides Poir. = Acroglochin persicarioides (Poir.) Moq. ■

18799 Amaranthus polygamus Roxb. = Amaranthus thellungianus Nevski ■☆

18800 Amaranthus polygamus Roxb. = Amaranthus tricolor L. ■

18801 Amaranthus polygonoides L.;合被苋(泰山苋);Smartweed Amaranth,Taishan Amaranth,Taishan Amaranthus,Tropical Amaranth ■

18802 Amaranthus polygonoides L. = Amaranthus thellungianus Nevski ■☆

18803 Amaranthus polygonoides Roxb. = Amaranthus angustifolius Lam. subsp. polygonoides (Roxb.) Maire et Weiller ■

18804 Amaranthus powellii S. Watson;鲍威尔苋;Green Amaranth,Powell's Amaranth,Powell's Smooth Amaranth,Prince Albert's Feather,Tall Amaranth ■☆

18805 Amaranthus praetermissus Brenan;疏忽苋■☆

18806 Amaranthus pringlei S. Watson = Amaranthus torreyi (A. Gray) S. Watson ■☆

18807 Amaranthus pubescens (Uline et W. L. Bray) Rydb. = Amaranthus albus L. ■

18808 Amaranthus pumilus Raf.;小苋;Coast Amaranth,Seabeach Amaranth ■☆

18809 Amaranthus quitensis Kunth = Amaranthus caudatus L. ■

18810 Amaranthus retroflexus L.;反枝苋(粗穗绿苋,西番谷,西风古,西风谷,野千穗谷,野苋菜);Common Amaranth,Love-lies-bleeding,Pig Weed,Pigweed,Prince's Feathers,Redroot,Redroot Amaranth,Redroot Amaranthus,Redroot Pigweed,Rough Amaranth,Rough Green Amaranth,Rough Pigweed,Wild Amaranth,Wild Beet,Wild-beet Amaranth ■

18811 Amaranthus retroflexus L. var. delilei (Richt. et Loret) Thell.;短苞反枝苋;Shortbract Redroot Amaranth,Shortbract Redroot Amaranthus ■

18812 Amaranthus retroflexus L. var. powellii (S. Watson) B. Boivin = Amaranthus powellii S. Watson ■☆

18813 Amaranthus retroflexus L. var. salicifolius I. M. Johnst. = Amaranthus retroflexus L. ■

18814 Amaranthus roxburghianus H. W. Kung;腋花苋(罗氏苋);Roxburgh Amaranthus,Roxburgh Amaranth ■

18815 Amaranthus rudis J. D. Sauer;水苋;Water Hemp ■☆

18816 Amaranthus rudis J. D. Sauer = Amaranthus tuberculatus (Moq.) J. D. Sauer ■☆

18817 Amaranthus sanguineus L. = Amaranthus caudatus L. ■

18818 Amaranthus schinzianus Thell.;欣兹苋■☆

18819 Amaranthus silvestris Vill. var. graecizans ? = Amaranthus graecizans L. ■

18820 Amaranthus speciosus Sims = Amaranthus cruentus L. ■

18821 Amaranthus speciosus Sims = Amaranthus hybridus L. subsp. cruentus (L.) Thell. ■

18822 Amaranthus spinosus L.;刺苋(白刺苋,白骨刺苋,刺刺草,刺苋菜,假苋菜,筋苋,筋苋菜头,筋苋头,竻苋菜,苈苋菜,酸酸苋,土苋菜,野刺苋,野勒苋,野簕苋,野苋菜,猪母菜,猪母刺);Careless Weed,Prickly Amaranth,Prickly Caterpillar,Spiny Amaranth,Spiny Pigweed,Spyny Amaranth,Spyny Amaranthus,Thorny Amaranth,Thorny Amaranthus,Thorny Pigweed ■

18823 Amaranthus splendens Hort.;约瑟夫苋;Joseph's Coat ■☆

18824 Amaranthus standleyanus Parodi ex Covas;菱叶苋■

18825 Amaranthus taishanensis F. Z. Li et C. K. Ni;泰山苋■

18826 Amaranthus taishanensis F. Z. Li et C. K. Ni = Amaranthus polygonoides L. ■

18827 Amaranthus tamariscinus Nutt. = Amaranthus rudis J. D. Sauer ■☆

18828 Amaranthus tamariscinus Nutt. = Amaranthus tuberculatus (Moq.) J. D. Sauer ■☆

18829 Amaranthus tenuifolius Willd.;薄叶苋;Thinleaf Amaranth ■

18830 Amaranthus tenuifolius Willd. = Amaranthus thellungianus Nevski ■☆

18831 Amaranthus thellungianus Nevski;泰隆苋■☆

18832 Amaranthus thellungianus Nevski = Amaranthus graecizans L. subsp. thellungianus (Nevski) Gusev ■☆

18833 Amaranthus thunbergii Moq.;通贝里苋;Thunberg's Amaranthus,Thunberg's Pigweed ■☆

18834 Amaranthus torreyi (A. Gray) Benth. = Amaranthus arenicola I. M. Johnst. ■☆

18835 Amaranthus torreyi (A. Gray) S. Watson;托里苋;Bigelow's Amaranth,Torree's Amaranth ■☆

18836 Amaranthus tricolor L.;苋(红苋,红苋菜,后庭花,胡苋,家苋,锦西风,糠苋,蒉,老来变,老来红,老来少,老少年,蛮须,青香苋,人苋,三色苋,苋菜,雁来红);Chinese Amaranth,Chinese Spinach,Floramor,Floramour,Florimer,Flower Amor,Flower Gentle,Fountain Plant,Gangera Amaranth,Joseph's Coat,Malabar Spinach,Summer Poinsettia,Tampala,Three-coloured Amaranth,Three-coloured Amaranthus,Tricolor Amaranth ■

18837 Amaranthus tricolor L. 'Joseph's Coat';彩叶苋(雁来红)■☆

18838 Amaranthus tricolor L. 'Molten Fire';铸火苋■☆

18839 Amaranthus tricolor L. aurea ?;黄色苋;Joseph's Coat ■☆

18840 Amaranthus tricolor L. subsp. mangostanus (L.) Aellen = Amaranthus mangostanus L. ■

18841 Amaranthus tricolor Willd. var. tristis (L.) Thell. = Amaranthus tricolor L. ■

18842 Amaranthus tristis L. = Amaranthus tricolor L. ■

18843 Amaranthus tristis Willd. = Amaranthus tricolor L. ■

18844 Amaranthus tuberculatus (Moq.) J. D. Sauer; 瘤苋; Rough-fruited Amaranth, Rough-fruited Water-hemp, Seaside Agrostis, Tall Water Hemp, Tall Water-hemp, Water Hemp ■☆

18845 Amaranthus tuberculatus (Moq.) J. D. Sauer var. prostratus (Uline et W. L. Bray) B. L. Rob. = Amaranthus tuberculatus (Moq.) J. D. Sauer ■☆

18846 Amaranthus tuberculatus (Moq.) J. D. Sauer var. subnudus S. Watson = Amaranthus tuberculatus (Moq.) J. D. Sauer ■☆

18847 Amaranthus tuberculatus (Moq.) Sauer; 朱红苋; Rough-fruit Amaranth, Rough-fruited Water-hemp, Tall Water-hemp ■☆

18848 Amaranthus viridis L.; 皱果苋(白苋,假苋菜,康苋,糠苋,绿苋,细苋,野苋,野苋菜,猪母苋,猪苋); Green Amaranth, Pigweed, Slender Amaranth, Tropical Green Amaranth, White Caterpillar, Wrinkledfruit Amaranth, Wrinkledfruit Amaranthus ■

18849 Amaranthus warnockii I. M. Johnst. = Amaranthus crassipes Schltdl. var. warnockii (I. M. Johnst.) Henrickson ■☆

18850 Amaranthus watsonii Standl.; 瓦氏苋; Watson's Amaranth ■☆

18851 Amaranthus wrightii S. Watson; 赖氏苋; Wright's Amaranth ■☆

18852 Amarella Gilib. (废弃属名) = Gentianella Moench(保留属名) ■

18853 Amarella occidentalis (A. Gray) Greene = Gentianella quinquefolia (L.) Small subsp. occidentalis (A. Gray) J. M. Gillett ■☆

18854 Amarenus C. Presl = Trifolium L. ■

18855 Amaria S. Mutis ex Caldas = Bauhinia L. ●

18856 Amaridium Hort. ex Lubbers = Camaridium Lindl. ■☆

18857 Amaridium Hort. ex Lubbers = Maxillaria Ruiz et Pav. ■☆

18858 Amarolea Small = Cartrema Raf. ●

18859 Amarolea Small = Osmanthus Lour. ●

18860 Amarorta A. Gray = Soulamea Lam. ●☆

18861 Amaryllidaceae J. St.-Hil. (1805) (保留科名); 石蒜科; Amaryllis Family ●■

18862 Amaryllis L. (1753) (保留属名); 孤挺花属(鹤顶红属,朱顶兰属); Amaryllis, Jersey Lily, Knight's-star ■☆

18863 Amaryllis L. (保留属名) = Hippeastrum Herb. (保留属名) ■

18864 Amaryllis aurea L'Hér. = Lycoris aurea (L'Hér.) Herb. ■

18865 Amaryllis belladonna L.; 孤挺花(野挺花); August Lily, Belladonna Lily, Belladonnalily, Jersey Lily, Knight's Star Lily, March Lily, Naked Lady, Naked Lady Lily ■☆

18866 Amaryllis belladonna L. 'Hathor'; 爱神孤挺花■☆

18867 Amaryllis broussonetii A. DC. = Crinum broussonetii (A. DC.) Herb. ■☆

18868 Amaryllis bulbisperma Burm. f. = Crinum bulbispermum (Burm. f.) Milne-Redh. et Schweick. ■☆

18869 Amaryllis candida (Stapf) Traub et Uphof = Zephyranthes candida (Lindl.) Herb. ■

18870 Amaryllis candida Lindl. = Zephyranthes candida (Lindl.) Herb. ■

18871 Amaryllis capensis L. = Spiloxene capensis (L.) Garside ■☆

18872 Amaryllis caspia Willd. = Allium caspium M. Bieb. ■☆

18873 Amaryllis cernua L. f. ex Savage; 俯垂孤挺花■☆

18874 Amaryllis ciliaris L. = Crossyne guttata (L.) D. Müll.-Doblies et U. Müll.-Doblies ■☆

18875 Amaryllis disticha L. f. = Boophone disticha (L. f.) Herb. ■☆

18876 Amaryllis elata Jacq. = Cyrtanthus elatus (Jacq.) Traub ■☆

18877 Amaryllis equestris Aiton = Hippeastrum puniceum (Lam.) Kuntze ■☆

18878 Amaryllis flexuosa Jacq. = Nerine humilis (Jacq.) Herb. ■☆

18879 Amaryllis formosissima L. = Sprekelia formosissima (L.) Herb. ■

18880 Amaryllis guttata L. = Crossyne guttata (L.) D. Müll.-Doblies et U. Müll.-Doblies ■☆

18881 Amaryllis humilis Jacq. = Nerine humilis (Jacq.) Herb. ■☆

18882 Amaryllis josephinae Redouté = Brunsvigia josephinae (Redouté) Ker Gawl. ■☆

18883 Amaryllis laticoma Ker Gawl. = Nerine laticoma (Ker Gawl.) T. Durand et Schinz ■☆

18884 Amaryllis longifolia L. = Ammocharis longifolia (L.) M. Roem. ■☆

18885 Amaryllis lutea L. = Sternbergia lutea (L.) Ker Gawl. ex Roem. et Schult. ■☆

18886 Amaryllis marginata Jacq. = Brunsvigia marginata (Jacq.) Aiton ■☆

18887 Amaryllis montana Labill. = Ixiolirion tataricum (Pall.) Herb. ■

18888 Amaryllis mostertii Traub; 莫斯特朱顶兰■☆

18889 Amaryllis obliqua L. f. ex Savage; 偏斜孤挺花■☆

18890 Amaryllis orientalis L. = Brunsvigia orientalis (L.) Aiton ex Eckl. ■☆

18891 Amaryllis ornata L. f. ex Aiton = Crinum ornatum (L. f. ex Aiton) Bury ■☆

18892 Amaryllis punicea Lam. = Hippeastrum puniceum (Lam.) Voss ■☆

18893 Amaryllis radiata L'Hér. = Lycoris radiata (L'Hér.) Herb. ■

18894 Amaryllis radula Jacq. = Brunsvigia radula (Jacq.) Aiton ■☆

18895 Amaryllis reginae L. = Hippeastrum reginae (L.) Herb. ■☆

18896 Amaryllis reticulata L'Hér. = Hippeastrum reticulatum (L'Hér.) Herb. ■☆

18897 Amaryllis reticulata L'Hér. var. striatifolis Herb.; 白肋朱顶兰■☆

18898 Amaryllis rutila Ker Gawl. = Hippeastrum rutilum (Ker Gawl.) Herb. ■

18899 Amaryllis sarniensis L. = Nerine sarniensis (L.) Herb. ■☆

18900 Amaryllis stellaris Jacq. = Hessea stellaris (Jacq.) Herb. ■☆

18901 Amaryllis striata Jacq. = Brunsvigia striata (Jacq.) Aiton ■☆

18902 Amaryllis tatarica Pall. = Ixiolirion tataricum (Pall.) Herb. ■

18903 Amaryllis tubispatha L'Hér. = Zephyranthes tubispatha (L'Hér.) Herb. ex Traub ■☆

18904 Amaryllis undulata L. = Nerine undulata (L.) Herb. ■☆

18905 Amaryllis vittata Aiton = Hippeastrum vittatum (L'Hér.) Herb. ■

18906 Amaryllis vittata L'Hér. = Hippeastrum vittatum (L'Hér.) Herb. ■

18907 Amaryllis zeylanica L. = Crinum zeylanicum (L.) L. ■☆

18908 Amasonia L. f. (1782) (保留属名); 彩苞花属●■☆

18909 Amathea Raf. = Aphelandra R. Br. ●■☆

18910 Amatlania Lundell = Ardisia Sw. (保留属名) ●■

18911 Amatula Medik. = Lycopersicon Mill. ■

18912 Amauria Benth. (1844); 四棱菊属■☆

18913 Amauria brandegeana Rydb.; 布朗四棱菊■☆

18914 Amauria carterae A. M. Powell; 四棱菊■☆

18915 Amauria dissecta A. Gray = Amauriopsis dissecta (A. Gray) Rydb. ■☆

18916 Amauria rotundifolia Benth.; 圆叶四棱菊■☆

18917 Amauriella Rendle = Anubias Schott ■☆

18918 Amauriella auriculata (Engl.) Hepper = Anubias hastifolia Engl. ■☆

18919 Amauriella hastifolia (Engl.) Hepper = Anubias hastifolia Engl. ■☆

18920 Amauriella obanensis Rendle = Anubias hastifolia Engl. ■☆
18921 Amauriella talbotii Rendle = Anubias hastifolia Engl. ■☆
18922 Amauriopsis Rydb. (1914);橙羽菊属■☆
18923 Amauriopsia Rydb. = Bahia Lag. ■☆
18924 Amauriopsis dissecta (A. Gray) Rydb.;橙羽菊■☆
18925 Amaxitis Adans. = Dactylis L. ■
18926 Ambaiba Adans. = Cecropia Loefl. (保留属名)●☆
18927 Ambaiba Adans. = Coilotapalus P. Browne(废弃属名)●☆
18928 Ambaiba Barrere = Cecropia Loefl. (保留属名)●☆
18929 Ambaiba Barrere ex Kuntze = Cecropia Loefl. (保留属名)●☆
18930 Ambassa Steetz = Vernonia Schreb. (保留属名)●■
18931 Ambassa Steetz(1864);多肋瘦片菊属●☆
18932 Ambassa hochstetteri Steetz;多肋瘦片菊●☆
18933 Ambavia Le Thomas(1972);阿巴木属●☆
18934 Ambavia capuronii (Cav. et Keraudren) Le Thomas;阿巴木●☆
18935 Ambavia gerrardii (Baill.) Le Thomas;杰勒阿巴木●☆
18936 Ambelania Aubl. (1775);缘毛夹竹桃属●☆
18937 Ambelania grandiflora Huber;大花缘毛夹竹桃●☆
18938 Ambelania laxa (Benth.) Müll. Arg.;蓬松缘毛夹竹桃●☆
18939 Ambelania lucida (Kunth) Markgr.;光亮缘毛夹竹桃●☆
18940 Ambelania macrophylla Müll. Arg.;大叶缘毛夹竹桃●☆
18941 Ambelania occidentalis L. Zarucchi;西方缘毛夹竹桃●☆
18942 Ambelania parviflora Markgr.;小花缘毛夹竹桃●☆
18943 Ambelania tenuiflora Müll. Arg.;锡花缘毛夹竹桃●☆
18944 Amberboa (Pers.) Less. = Amberboa Vaill. (保留属名)■
18945 Amberboa Less. = Amberboa Vaill. (保留属名)■
18946 Amberboa Vaill. (1754)(保留属名);珀菊属(安波菊属,香芙蓉属);Amberboa,Sweet Sultan,Sweet-sultan ■
18947 Amberboa abyssinica A. Rich. = Volutaria abyssinica (A. Rich.) C. Jeffrey ■☆
18948 Amberboa amberboi (L.) Tzvelev;普通珀菊■☆
18949 Amberboa atlantica Pit. = Volutaria lippii (L.) Cass. ■☆
18950 Amberboa aylmeri (Baker) Soják = Volutaria abyssinica (A. Rich.) C. Jeffrey subsp. aylmeri (Baker) Wagenitz ■☆
18951 Amberboa bucharica Iljin;布哈尔珀菊■☆
18952 Amberboa crupinoides (Desf.) DC. = Volutaria crupinoides (Desf.) Cass. ex Maire ■☆
18953 Amberboa crupinoides (Desf.) DC. var. libyca (Viv.) Pamp. = Volutaria crupinoides (Desf.) Cass. ex Maire ■☆
18954 Amberboa glauca (Willd.) Grossh.;白花珀菊;White Amberboa,Whiteflower Amberboa ■
18955 Amberboa hochstetteri (Oliv. et Hiern) Soják = Volutaria abyssinica (A. Rich.) C. Jeffrey ■☆
18956 Amberboa leucantha Coss. ex Batt. = Volutaria sinaica (DC.) Wagenitz ■☆
18957 Amberboa leucantha L. Chevall. = Volutaria sinaica (DC.) Wagenitz ■☆
18958 Amberboa libyca (Viv.) Alavi = Volutaria crupinoides (Desf.) Cass. ex Maire ■☆
18959 Amberboa lippii (L.) DC. = Volutaria lippii (L.) Cass. ■☆
18960 Amberboa lippii (L.) DC. var. medians Maire = Volutaria lippii (L.) Cass. subsp. medians (Maire) Wagenitz ■☆
18961 Amberboa lippii DC. = Volutaria lippii (L.) Cass. ■☆
18962 Amberboa lippii DC. subsp. tubuliflora (Murb.) Murb. = Volutaria lippii (L.) Cass. subsp. tubuliflora (Murb.) Maire ■☆
18963 Amberboa lippii DC. var. medians Maire = Volutaria lippii (L.) Cass. subsp. medians (Maire) Wagenitz ■☆
18964 Amberboa lippii DC. var. microcephala Maire = Volutaria lippii (L.) Cass. ■☆
18965 Amberboa maroccana Barratte et Murb. = Volutaria maroccana (Barratte et Murb.) Maire ■☆
18966 Amberboa moschata (L.) DC.;珀菊(香芙蓉,香矢车菊);Amberboa,Common Amberboa,Sweet Sultan,Sweetsultan,Sweet-sultan ■
18967 Amberboa muricata (L.) DC. = Volutaria muricata (L.) Maire ■☆
18968 Amberboa muricata (L.) DC. subsp. micractis Boiss. = Volutaria muricata (L.) Maire ■☆
18969 Amberboa muricata (L.) DC. var. micractis (Boiss.) Batt. = Volutaria muricata (L.) Maire ■☆
18970 Amberboa muricata DC.;安倍菊■☆
18971 Amberboa nana (Boiss.) Iljin;小珀菊■☆
18972 Amberboa odorata DC. var. flava Trautv. = Amberboa turanica Iljin ■
18973 Amberboa omphalodes (Benth. et Hook. f.) Batt. = Stephanochilus omphalodes (Benth. et Hook. f.) Maire ■☆
18974 Amberboa omphalodes (Coss. et Durieu) Benth. et Hook. = Stephanochilus omphalodes (Benth. et Hook. f.) Maire ■☆
18975 Amberboa ramosa (Roxb.) Jofri;分枝珀菊■☆
18976 Amberboa ramosissima Pit. = Volutaria lippii (L.) Cass. ■☆
18977 Amberboa saltii (Philipson) Soják = Volutaria abyssinica (A. Rich.) C. Jeffrey ■☆
18978 Amberboa sinaica DC. = Volutaria sinaica (DC.) Wagenitz ■☆
18979 Amberboa somalensis (Oliv. et Hiern) Soják = Volutaria abyssinica (A. Rich.) C. Jeffrey ■☆
18980 Amberboa sosnovskyi Iljin;索斯珀菊■☆
18981 Amberboa subdiscolor (Lojac.) Pamp. = Volutaria lippii (L.) Cass. subsp. tubuliflora (Murb.) Maire ■☆
18982 Amberboa tubuliflora Murb. = Volutaria lippii (L.) Cass. subsp. tubuliflora (Murb.) Maire ■☆
18983 Amberboa turanica Iljin;黄花珀菊(珀菊,图兰安波菊,新疆珀菊);Purple Sweet Sultan,Sweet Sultan,Xinjiang Amberboa,Yellowflower Amberboa ■
18984 Amberboi Adans. (废弃属名) = Amberboa (Pers.) Less. ■
18985 Amberboia Kuntze = Amberboa (Pers.) Less. ■
18986 Ambianella Willis = Autranella A. Chev. ●☆
18987 Ambianella Willis = Mimusops L. ●☆
18988 Ambidopsis lasiocarpa var. micrantha W. T. Wang = Crucihimalaya lasiocarpa (Hook. f. et Thomson) Al-Shehbaz,O'Kane et R. A. Price ■
18989 Ambilobea Thulin,Beier et Razafim. (2008);马岛橄榄属●☆
18990 Ambinax B. D. Jacks. = Ambinux Comm. ex Juss. ●
18991 Ambinux Comm. ex Juss. = Vernicia Lour. ●
18992 Amblachaenium Turcz. ex DC. = Hypochaeris L. ■
18993 Amblatum G. Don = Anblatum Hill ■
18994 Amblatum G. Don = Lathraea L. ■
18995 Ambleia Spach = Stachys L. ●■
18996 Amblirion Raf. = Fritillaria L. ■
18997 Amblogyna Raf. = Amaranthus L. ■
18998 Amblogyna bigelovii Uline et W. L. Bray = Amaranthus torreyi (A. Gray) S. Watson ■☆
18999 Amblogyna torreyi A. Gray = Amaranthus torreyi (A. Gray) S. Watson ■☆
19000 Amblogyna urceolata (Benth.) A. Gray var. obcordata A. Gray = Amaranthus obcordatus (A. Gray) Standl. ■☆

19001 Amblophus Merr. = Amplophus Raf. ●■
19002 Amblophus Merr. = Valeriana L. ●■
19003 Amblostima Raf.（废弃属名）= Schoenolirion Torr.（保留属名）■☆
19004 Amblostima albiflora Raf. = Schoenolirion albiflorum（Raf.）R. R. Gates ■☆
19005 Amblostoma Scheidw. = Encyclia Hook. ■☆
19006 Amblyachyrum Hochst. ex Steud. = Apocopis Nees ■
19007 Amblyachyrum Steud. = Apocopis Nees ■
19008 Amblyanthe Rauschert = Dendrobium Sw.（保留属名）■
19009 Amblyanthe Rauschert(1983)；钝花兰属■☆
19010 Amblyanthe melanosticta（Schltr.）Rauschert；钝花兰■☆
19011 Amblyanthera Blume = Osbeckia L. ●■
19012 Amblyanthera Müll. Arg. = Mandevilla Lindl. ●
19013 Amblyanthopsis Mez(1902)；拟钝花紫金牛属●☆
19014 Amblyanthopsis membranacea Mez；膜质拟钝花紫金牛●☆
19015 Amblyanthopsis philippinensis Mez；菲律宾拟钝花紫金牛●☆
19016 Amblyanthus（Schltr.）Brieger = Amblyanthe Rauschert ■☆
19017 Amblyanthus A. DC.（1841）；钝花紫金牛属●☆
19018 Amblyanthus glandulosus（Roxb.）A. DC.；钝花紫金牛●☆
19019 Amblyanthus multiflorus Mez；多钝花紫金牛●☆
19020 Amblycarpum Lem. = Amblyocarpum Fisch. et C. A. Mey. ■☆
19021 Amblychloa Link = Sclerochloa P. Beauv. ■
19022 Amblyglottis Blume = Calanthe R. Br.（保留属名）■
19023 Amblyglottis angustifolia Blume = Calanthe angustifolia（Blume）Lindl. ■
19024 Amblyglottis speciosa Blume = Calanthe speciosa（Blume）Lindl. ■
19025 Amblyglottis veratrifolia Blume = Dactylorhiza umbrosa（Kar. et Kir.）Nevski ■
19026 Amblygonocarpus Harms(1897)；钝棱豆属；Banga-wanga ●☆
19027 Amblygonocarpus andongensis（Welw. ex Oliv.）Exell et Torre；安东钝棱豆●☆
19028 Amblygonocarpus obtusangulus（Welw. ex Oliv.）Harms = Amblygonocarpus andongensis（Welw. ex Oliv.）Exell et Torre ●☆
19029 Amblygonocarpus schweinfurthii Harms；钝棱豆（阿洞钝棱豆木）；Banga-wanga ●☆
19030 Amblygonocarpus schweinfurthii Harms = Amblygonocarpus andongensis（Welw. ex Oliv.）Exell et Torre ●☆
19031 Amblygonum（Meisn.）Rchb.（1837）；水荭属■☆
19032 Amblygonum（Meisn.）Rchb. = Polygonum L.（保留属名）●■
19033 Amblygonum Rchb. = Amblygonum（Meisn.）Rchb. ■☆
19034 Amblygonum orientale（L.）Nakai = Polygonum orientale L. ■
19035 Amblygonum orientale（L.）Nakai ex T. Mori = Polygonum orientale L. ■
19036 Amblygonum orientale（L.）Nakai ex T. Mori var. pilosum（Roxb. ex Meisn.）Nakai ex T. Mori = Polygonum orientale L. ■
19037 Amblygonum orientale Nakai；东方水荭■☆
19038 Amblygonum pilosum Nakai；水荭■☆
19039 Amblylepis Decne. = Amblyolepis DC. ■
19040 Amblylepis Decne. = Helenium L. ■
19041 Amblynotopsis J. F. Macbr. = Antiphytum DC. ex Meisn. ■☆
19042 Amblynotus（A. DC.）I. M. Johnst.（1924）；钝背草属；Amblynotus ■
19043 Amblynotus I. M. Johnst. = Amblynotus（A. DC.）I. M. Johnst. ■
19044 Amblynotus obovatus（Ledeb.）I. M. Johnst. = Amblynotus rupestris（Pall. ex Georgi）Popov ex Serg. ■
19045 Amblynotus rupestris（Pall. ex Georgi）Popov ex Serg.；钝背草（兴安齿缘草）；Common Amblynotus ■
19046 Amblyocalyx Benth.（1876）；钝萼木属●☆
19047 Amblyocalyx Benth. = Alstonia R. Br.（保留属名）●
19048 Amblyocalyx beccarii Benth.；钝萼木●☆
19049 Amblyocarpum Fisch. et C. A. Mey.（1837）；钝果菊属■☆
19050 Amblyocarpum inuloides Fisch. et C. A. Mey.；钝果菊■☆
19051 Amblyoglossum Turcz. = Tylophora R. Br. ●■
19052 Amblyolepis DC. = Helenium L. ■
19053 Amblyolepis setigera DC.；钝鳞菊■☆
19054 Amblyopappus Hook. = Amblyopappus Hook. et Arn. ■☆
19055 Amblyopappus Hook. et Arn.（1841）；钝冠菊属（钝毛菊属）■☆
19056 Amblyopappus pusillus Hook. et Arn.；钝冠菊（钝毛菊）■☆
19057 Amblyopelis Steud. = Amblyolepis DC. ■☆
19058 Amblyopelis Steud. = Helenium L. ■
19059 Amblyopetalum（Griseb.）Malme = Oxypetalum R. Br.（保留属名）●■☆
19060 Amblyopetalum（Griseb.）Malme(1927)；钝瓣萝藦属●☆
19061 Amblyopetalum Malme = Amblyopetalum（Griseb.）Malme ●☆
19062 Amblyopetalum coccineum（Griseb.）Malme；绯红钝瓣萝藦●☆
19063 Amblyopetalum coeruleum Malme；钝瓣萝藦●☆
19064 Amblyopogon（DC.）Jaub. et Spach = Centaurea L.（保留属名）●■
19065 Amblyopogon Fisch. et C. A. Mey. = Centaurea L.（保留属名）●■
19066 Amblyopogon Fisch. et C. A. Mey. ex DC. = Centaurea L.（保留属名）●■
19067 Amblyopyrum Eig = Aegilops L.（保留属名）■
19068 Amblyorhinum Turcz. = Phyllactis Pars. ■☆
19069 Amblyotropis Kitag. = Gueldenstaedtia Fisch. ■
19070 Amblysperma Benth.（1837）；钝子菊属■☆
19071 Amblysperma Benth. = Trichocline Cass. ■☆
19072 Amblysperma scapigerum Benth.；钝子菊■☆
19073 Amblysperma spathulatum（A. Cunn. ex DC.）D. J. N. Hind；小苞钝子菊■☆
19074 Amblystigma Benth.（1876）；钝子萝藦属■☆
19075 Amblystigma Post et Kuntze = Schoenolirion Torr.（保留属名）■☆
19076 Amblystigma hypoleucum Benth.；钝子萝藦■☆
19077 Amblystigma pilosum Malme；毛钝子萝藦■☆
19078 Amblystigma pulchellum（Schltr.）T. Mey.；美丽钝子萝藦■☆
19079 Amblystigma pulchellum（Schltr.）T. Mey. = Steleostemma pulchellum Schltr. ■☆
19080 Amblytes Dulac = Molinia Schrank ■
19081 Amblytropis Kitag. = Gueldenstaedtia Fisch. ■
19082 Amblytropis coelestis（Diels）C. Y. Wu ex H. P. Tsui = Tibetia coelestis（Diels）H. P. Tsui ■
19083 Amblytropis coelestis（Diels）C. Y. Wu ex H. P. Tsui = Tibetia yunnanensis（Franch.）H. P. Tsui var. coelestis（Diels）X. Y. Zhu ■
19084 Amblytropis delavayi（Franch.）C. Y. Wu ex H. P. Tsui = Gueldenstaedtia verna（Georgi）Borissov ■
19085 Amblytropis delavayi（Franch.）C. Y. Wu ex H. P. Tsui = Gueldenstaedtia delavayi Franch. ■
19086 Amblytropis diversifolia Maxim. = Tibetia himalaica（Baker）H. P. Tsui ■
19087 Amblytropis flava（Adamson）C. Y. Wu ex H. P. Tsui = Tibetia tongolensis（Ulbr.）H. P. Tsui ■
19088 Amblytropis flava Adamson = Tibetia tongolensis（Ulbr.）H. P. Tsui ■
19089 Amblytropis flava Adamson var. tongolensis（Ulbr.）Ali = Tibetia

tongolensis (Ulbr.) H. P. Tsui ■

19090 Amblytropis harmsii (Ulbr.) C. Y. Wu ex H. P. Tsui = Gueldenstaedtia harmsii Ulbr. ■

19091 Amblytropis henryi (Ulbr.) C. Y. Wu ex H. P. Tsui = Gueldenstaedtia henryi Ulbr. ■

19092 Amblytropis maritima (Maxim.) Kitag. = Gueldenstaedtia maritima Maxim. ■

19093 Amblytropis multiflora (Bunge) Kitag. = Gueldenstaedtia verna (Georgi) Borissov subsp. multiflora (Bunge) H. P. Tsui ■

19094 Amblytropis pauciflora (Pall.) Kitag. = Gueldenstaedtia verna (Georgi) Borissov ■

19095 Amblytropis santapaui Thoth. = Tibetia himalaica (Baker) H. P. Tsui ■

19096 Amblytropis stenophylla (Bunge) Kitag. = Gueldenstaedtia stenophylla Bunge ■

19097 Amblytropis uniflora (Strachey ex Jacot) Kuang et H. P. Tsui = Tibetia himalaica (Baker) H. P. Tsui ■

19098 Amblytropis uniflora Strachey ex Jacot = Tibetia himalaica (Baker) H. P. Tsui ■

19099 Amblytropis verna (Georgi) Kitag. = Gueldenstaedtia verna (Georgi) Borissov ■

19100 Amblytropis verna (Georgi) Kitag. var. longicarpa (T. H. Chung) Brieger = Gueldenstaedtia longicarpa T. H. Chung ■

19101 Amblytropis yunnanensis (Franch.) C. Y. Wu ex H. P. Tsui = Tibetia yunnanensis (Franch.) H. P. Tsui ■

19102 Ambongia Benoist(1939);阿姆爵床属■☆

19103 Ambongia perrieri Benoist;阿姆爵床■☆

19104 Ambora Juss. = Tambourissa Sonn. ●☆

19105 Ambora toxicaria Pers. = Antiaris toxicaria (Pers.) Lesch. ●◇

19106 Amborella Baill. (1873);无油樟属(互叶梅属,毛脚树属)●☆

19107 Amborella trichopoda Baill.;无油樟(互叶梅,毛脚树)●☆

19108 Amborellaceae Pichon(1948)(保留科名);无油樟科(互叶梅科,毛脚树科)●☆

19109 Amboroa Cabrera(1956);刺冠亮泽兰属(玻利维亚菊属)●☆

19110 Amboroa geminata Cabrera;刺冠亮泽兰(玻利维亚菊)●☆

19111 Amboroa wurdackii R. M. King et H. Rob.;南美刺冠亮泽兰●☆

19112 Ambotia Raf. = Annona L. ●

19113 Ambraria Cruse = Nenax Gaertn. ●☆

19114 Ambraria Fabr. = Anthospermum L. ●☆

19115 Ambraria Heist. = Danais Comm. ex Vent. ●☆

19116 Ambraria Heist. ex Fabr. = Anthospermum L. ●☆

19117 Ambraria Heist. ex Fabr. = Danais Comm. ex Vent. ●☆

19118 Ambraria acerosa (Gaertn.) Sond. = Nenax acerosa Gaertn. ●☆

19119 Ambraria glabra Cruse = Nenax acerosa Gaertn. ●☆

19120 Ambraria glabra Cruse var. papillata Sond. = Nenax acerosa Gaertn. ●☆

19121 Ambraria glabra Cruse var. tulbaghica Sond. = Nenax acerosa Gaertn. ●☆

19122 Ambraria hirta Cruse = Nenax hirta (Cruse) Salter ●☆

19123 Ambraria hirta Cruse var. macrocarpa Eckl. et Zeyh. = Nenax acerosa Gaertn. subsp. macrocarpa (Eckl. et Zeyh.) Puff ●☆

19124 Ambraria microphylla Sond. = Nenax microphylla (Sond.) T. M. Salter ●☆

19125 Ambrella H. Perrier(1934);食兰属■☆

19126 Ambrella longituba H. Perrier;食兰■☆

19127 Ambrina Moq. = Ambrina Spach ●■

19128 Ambrina Spach = Chenopodium L. ●■

19129 Ambrina Spach = Roubieva Moq. ●■

19130 Ambrina ambrosioides (L.) Spach = Chenopodium ambrosioides L. ■☆

19131 Ambrina ambrosioides (L.) Spach = Dysphania ambrosioides (L.) Mosyakin et Clemants ■

19132 Ambrina ambrosioides (L.) Spach var. pubescens (Makino) = Chenopodium ambrosioides L. ■☆

19133 Ambrina anthelmintica (Crantz) Spach = Chenopodium ambrosioides L. var. anthelminticum (Crantz) A. Gray ■

19134 Ambrina botrys (L.) Moq. = Dysphania botrys (L.) Mosyakin et Clemants ■

19135 Ambrina botrys Moq. = Chenopodium botrys L. ■

19136 Ambrina foetidum Moq. = Dysphania schraderiana (Roem. et Schult.) Mosyakin et Clemants ■

19137 Ambroma L. f. = Abroma Jacq. ●

19138 Ambroma augusta (L.) L. f. = Abroma augusta (L.) L. f. ●

19139 Ambrosia B. D. Jacks. = Ambrosina Bassi ■☆

19140 Ambrosia L. (1753);豚草属(猪草属);Ambrosia,Ragweed ●■

19141 Ambrosia acanthicarpa Hook.;尖果豚草;Annual Bursage ■☆

19142 Ambrosia ambrosioides (Cav.) W. W. Payne;类豚草(豚草);Canyon Ragweed ■☆

19143 Ambrosia aptera DC.;无翅豚草■☆

19144 Ambrosia aptera DC. = Ambrosia trifida L. ■

19145 Ambrosia artemisiifolia L.;豚草(瘤果菊,美丽豚草,猪草);American Wormwood, Annual Bur-sage, Bitter Weed, Bitterweed, Common Ragweed, Hog Brake, Hogbrake, Oak of Cappadocia, Ragweed, Roman Ragweed, Roman Wormwood, Short Ragweed, Small Ragweed ■

19146 Ambrosia artemisiifolia L. var. elatior (L.) Descourt. = Ambrosia artemisiifolia L. ■

19147 Ambrosia artemisiifolia L. var. elatior (L.) Descourt. f. villosa Fernald et Griscom = Ambrosia artemisiifolia L. ■

19148 Ambrosia artemisiifolia L. var. elator (L.) Decne. = Ambrosia artemisiifolia L. ■

19149 Ambrosia artemisiifolia L. var. paniculata (Michx.) Blank. = Ambrosia artemisiifolia L. ■

19150 Ambrosia bidentata Michx.;二裂豚草(二裂矮豚草);Lancelate Ragweed, Lanceleaf Ragweed, Ragweed, Southern Ragweed ■☆

19151 Ambrosia californica Rydb. = Ambrosia psilostachya DC. ■☆

19152 Ambrosia chamissonis (Less.) Greene;查米森豚草■☆

19153 Ambrosia chamissonis Greene = Ambrosia chamissonis (Less.) Greene ■☆

19154 Ambrosia chenopodiifolia (Benth.) W. W. Payne;藜叶豚草■☆

19155 Ambrosia confertiflora DC.;密花豚草■☆

19156 Ambrosia cordifolia (A. Gray) W. W. Payne;心叶豚草■☆

19157 Ambrosia coronopifolia Torr. et A. Gray = Ambrosia psilostachya DC. ■☆

19158 Ambrosia coronopifolia Torr. et A. Gray var. asperula A. Gray = Ambrosia psilostachya DC. ■☆

19159 Ambrosia coronopifolia Torr. et A. Gray var. gracilis A. Gray = Ambrosia psilostachya DC. ■☆

19160 Ambrosia cumanensis Kunth;库曼豚草■☆

19161 Ambrosia cumanensis Kunth = Ambrosia psilostachya DC. ■☆

19162 Ambrosia deltoidea (Torr.) W. W. Payne;三角叶豚草;Bur Sage, Rabbit Bush, Triangle Leaf Bursage, Triangleleaf Bursage ■☆

19163 Ambrosia dumosa (A. Gray) W. W. Payne;白豚草;Burro Weed, Bursage, White Bur Sage, White Bursage ■☆

19164 Ambrosia elatior L.;猪草■
19165 Ambrosia elatior L. = Ambrosia artemisiifolia L. ■
19166 Ambrosia eriocentra (A. Gray) W. W. Payne;毛距豚草;Woolly Bur Sage,Woolly Bur-sage ■☆
19167 Ambrosia glandulosa Scheele = Ambrosia artemisiifolia L. ■
19168 Ambrosia grayi (A. Nelson) Shinners;格雷豚草■☆
19169 Ambrosia ilicifolia (A. Gray) W. W. Payne;冬青叶豚草;Holly-leaf Bur Sage,Holly-leaf Bur-sage ■☆
19170 Ambrosia integrifolia Muhl. ex Willd. = Ambrosia trifida L. ■
19171 Ambrosia intergradiens W. H. Wagner;中间豚草;Intergrading Ragweed ■☆
19172 Ambrosia lindheimeriana Scheele = Ambrosia psilostachya DC. ■☆
19173 Ambrosia linearis (Rydb.) W. W. Payne;线形豚草■☆
19174 Ambrosia maritima L.;沿海豚草;Oak-of-cappadocia ■☆
19175 Ambrosia media Rydb. = Ambrosia artemisiifolia L. ■
19176 Ambrosia monogyra (Torr. et A. Gray) Strother et B. G. Baldwin;单环豚草■☆
19177 Ambrosia monophylla (Walter) Rydb. = Ambrosia artemisiifolia L. ■
19178 Ambrosia peruviana Willd.;秘鲁豚草;Peruvian Ragweed ■☆
19179 Ambrosia psilostachya DC.;裸穗花豚草(西部豚草);Black Sage,Perennial Ragweed,Western Ragweed ■☆
19180 Ambrosia psilostachya DC. var. californica (Rydb.) S. F. Blake = Ambrosia psilostachya DC. ■☆
19181 Ambrosia psilostachya DC. var. coronopifolia (Torr. et A. Gray) Farw. = Ambrosia psilostachya DC. ■☆
19182 Ambrosia psilostachya DC. var. coronopifolia (Torr. et A. Gray) Farw. ex Fernald = Ambrosia psilostachya DC. ■☆
19183 Ambrosia psilostachya DC. var. lindheimeriana (Scheele) Blank. = Ambrosia psilostachya DC. ■☆
19184 Ambrosia pumila A. Gray;矮豚草;Dwarf Ragweed ■☆
19185 Ambrosia rugelii Rydb. = Ambrosia psilostachya DC. ■☆
19186 Ambrosia salsola (Torr. et A. Gray) Strother et B. G. Baldwin;盐地豚草;Burrobush ■☆
19187 Ambrosia senegalensis DC. = Ambrosia maritima L. ■☆
19188 Ambrosia striata Rydb. = Ambrosia trifida L. ■
19189 Ambrosia tenuifolia Spreng.;细叶豚草■☆
19190 Ambrosia tomentosa Nutt.;毛豚草;Bur Ragweed,Perennial Bursage,Skeleton-leaf Bur Ragweed,Skeleton-leaf Bur-sage,White Ragweed ■☆
19191 Ambrosia trifida L.;三裂叶豚草(高豚草,三裂豚草,豚草);Buffalo Weed,Buffalo-weed,Giant Ragweed,Great Ragweed,Horse Weed,Horse-cane,Ragweed ■
19192 Ambrosia trifida L. f. integrifolia (Muhl. ex Willd.) Fernald = Ambrosia trifida L. ■
19193 Ambrosia trifida L. f. integrifolia (Muhl.) Fernald = Ambrosia trifida L. ■
19194 Ambrosia trifida L. var. integrifolia (Muhl. ex Willd.) Torr. et A. Gray = Ambrosia trifida L. ■
19195 Ambrosia trifida L. var. texana Scheele = Ambrosia trifida L. ■
19196 Ambrosia variabilis Rydb. = Ambrosia trifida L. ■
19197 Ambrosiaceae Bercht. et J. Presl(1820)(保留科名);豚草科●■
19198 Ambrosiaceae Dumort. et Link = Ambrosiaceae Bercht. et J. Presl(保留科名)●■
19199 Ambrosiaceae Dumort. et Link = Asteraceae Bercht. et J. Presl(保留科名)●■
19200 Ambrosiaceae Dumort. et Link = Compositae Giseke(保留科名)●■
19201 Ambrosiaceae Link = Ambrosiaceae Bercht. et J. Presl(保留科名)●■
19202 Ambrosiaceae Link = Asteraceae Bercht. et J. Presl(保留科名)●■
19203 Ambrosiaceae Link = Compositae Giseke(保留科名)●■
19204 Ambrosiaceae Martinov = Ambrosiaceae Bercht. et J. Presl(保留科名)●■
19205 Ambrosiaceae Martinov = Asteraceae Bercht. et J. Presl(保留科名)●■
19206 Ambrosiaceae Martinov = Compositae Giseke(保留科名)●■
19207 Ambrosina Bassi(1766);地中海南星属■☆
19208 Ambrosina bassii L.;地中海南星■☆
19209 Ambrosina bassii L. var. angustifolia Guss. = Ambrosina bassii L. ■☆
19210 Ambrosina bassii L. var. maculata (Ucria) Parl. = Ambrosina bassii L. ■☆
19211 Ambrosina bassii L. var. reticulata (Guss.) Parl. = Ambrosina bassii L. ■☆
19212 Ambrosina maculata Ucria = Ambrosina bassii L. ■☆
19213 Ambrosina reticulata Guss. = Ambrosina bassii L. ■☆
19214 Ambrosinia L. = Ambrosina Bassi ■☆
19215 Ambuli Adans.(废弃属名)= Limnophila R. Br.(保留属名)■
19216 Ambulia Lam. = Limnophila R. Br.(保留属名)■
19217 Ambulia aromatica Lam. = Limnophila aromatica (Lam.) Merr. ■
19218 Ambulia bangweolensis R. E. Fr. = Limnophila bangweolensis (R. E. Fr.) Verdc. ■☆
19219 Ambulia baumii Engl. et Gilg = Limnophila ceratophylloides (Hiern) V. Naray. ■☆
19220 Ambulia ceratophylloides (Hiern) Engl. et Gilg = Limnophila ceratophylloides (Hiern) V. Naray. ■☆
19221 Ambulia dasyantha Engl. et Gilg = Limnophila dasyantha (Engl. et Gilg) V. Naray. ■☆
19222 Ambulia gratioloides Baill. ex Wettst. = Limnophila indica (L.) Druce ■
19223 Ambulia hottonioides Wettst. = Limnophila indica (L.) Druce ■
19224 Ambulia sessiliflora (Vahl) Baill. ex Wettst. = Limnophila sessiliflora (Vahl) Blume ■
19225 Ambulia sessiliflora Vahl = Limnophila sessiliflora (Vahl) Blume ■
19226 Ambulia tenera (Hiern) Engl. et Gilg = Dopatrium tenerum (Hiern) Eb. Fisch. ■☆
19227 Amburana Schwacke et Taub.(1894);良木豆属(假商陆属)●☆
19228 Amburana acreana (Ducke) A. C. Sm.;阿雷良木豆;Cerejeira,Rebol,Soryokok ●☆
19229 Amburana cearensis (Allemão) A. C. Sm.;巴拉圭良木豆(巴拉圭豆,色拉伪香豆);Umburana ●☆
19230 Amburana cloudii Schwacke et Taub.;良木豆(假商陆)●☆
19231 Ambuya Raf. = Aristolochia L. ●■
19232 Ambuya Raf. = Howardia Klotzsch ●☆
19233 Amebia Repel = Arnebia Forssk. ●■
19234 Amecarpus Benth. ex Lindl. = Indigofera L. ●■
19235 Amechania DC. = Agarista D. Don ex G. Don ●☆
19236 Amechania DC. = Agauria (DC.) Hook. f. ●☆
19237 Amechania DC. = Leucothoe D. Don + Agauria (DC.) Hook. f. ●☆
19238 Ameghinoa Speg.(1897);腺叶钝柱菊属●☆
19239 Ameghinoa patagonica Speg.;腺叶钝柱菊●☆
19240 Amelanchier Medik.(1789);唐棣属(扶移属,红枸子木属);

Juneberry, June-berry, Service Berry, Serviceberry, Service-berry, Shad, Shad Bush, Shadblow, Shadbuah, Snowy Mespilus, Sugar Plum ●

19241 Amelanchier alnifolia (Nutt.) Nutt.;桤叶唐棣(赤杨叶唐棣,桤木叶唐棣);Alderleaf Serviceberry, Alder-leaved Service-berry, Dwarf Juneberry, Juneberry, Junebush, Saskatoon, Saskatoon Juneberry, Saskatoon Serviceberry, Saskatoon-berry, Service-berry, Shad Blow, Shadbush, Western Serviceberry, Western Service-berry, Western Shadbush ●

19242 Amelanchier alnifolia (Nutt.) Nutt. var. florida Schneid. = Amelanchier florida Lindl. ●☆

19243 Amelanchier arborea (F. Michx.) Fernald;树唐棣;Common Serviceberry, Downy Juneberry, Downy Serviceberry, June Berry, Juneberry, Sarvis, Sarviss Berry, Sarviss Tree, Service Berry, Shadblow, Shadbush, Snowy Mespilus, Sugar Plum, Tree Amelianchier ●

19244 Amelanchier arborea (F. Michx.) Fernald subsp. grandiflora (Rehder) P. Landry = Amelanchier grandiflora Rehder ●☆

19245 Amelanchier arborea (F. Michx.) Fernald subsp. laevis (Wiegand) S. McKay ex Landry = Amelanchier laevis Wiegand ●

19246 Amelanchier arborea (F. Michx.) Fernald var. cordifolia (Ashe) B. Boivin = Amelanchier laevis Wiegand ●

19247 Amelanchier arborea (F. Michx.) Fernald var. laevis (Wiegand) H. E. Ahles = Amelanchier laevis Wiegand ●

19248 Amelanchier asiatica (Siebold et Zucc.) Endl. ex Walp.;东亚唐棣(夫移,枎移,唐棣);Asia Juneberry, Asia Shadbush, Asian Service Berry, Asian Serviceberry, Asian Toddalia, Asiatic Serviceberry, Asiatic Service-berry ●

19249 Amelanchier asiatica (Siebold et Zucc.) Endl. ex Walp. var. sinica C. K. Schneid. = Amelanchier sinica (C. K. Schneid.) Chun ●

19250 Amelanchier bartramiana (Tausch) M. Roem.;灌木唐棣;Bartram's Juneberry, Mountain Serviceberry, Oblong-fruit Serviceberry ●☆

19251 Amelanchier canadensis (L.) Medik.;加拿大唐棣;Canada Serviceberry, Canada Shadbush, Canadian Serviceberry, Downy Serviceberry, Juneberry, Serviceberry, Shadblow Serviceberry, Shadblow Servlceberry, Shadbush Serviceberry, Swamp Shadbush, Thicket Juneberry, Thicket Serviceberry ●☆

19252 Amelanchier canadensis (L.) Medik. 'Micropetala';小花瓣加拿大唐棣●☆

19253 Amelanchier canadensis (L.) Medik. 'Rainbow Pillar';彩虹加拿大唐棣●☆

19254 Amelanchier canadensis (L.) Medik. 'Springtime';春天加拿大唐棣●☆

19255 Amelanchier canadensis (L.) Medik. var. asiatica Koidz. = Amelanchier asiatica (Siebold et Zucc.) Endl. ex Walp. ●

19256 Amelanchier canadensis (L.) Medik. var. subintegra Fernald = Amelanchier canadensis (L.) Medik. ●☆

19257 Amelanchier canadensis Gray = Amelanchier laevis Wiegand ●

19258 Amelanchier canadensis Medik. var. asiatica (Siebold et Zucc.) Koidz. = Amelanchier asiatica (Siebold et Zucc.) Endl. ex Walp. ●

19259 Amelanchier florida Lindl.;太平洋唐棣;Pacific Juneberry, Pacific Serviceberry ●☆

19260 Amelanchier grandiflora Rehder;大花唐棣;Apple Serviceberry, Roundleaf Serviceberry, Serviceberry ●☆

19261 Amelanchier grandiflora Rehder 'Ballerina';芭莱大花唐棣●☆

19262 Amelanchier grandiflora Rehder 'Robin Hill';罗宾大花唐棣;Serviceberry ●☆

19263 Amelanchier grandiflora Rehder 'Rubescens';红花大花唐棣●☆

19264 Amelanchier humilis Wiegand;矮唐棣;Low Service Berry, Low Shadbush ●☆

19265 Amelanchier humilis Wiegand = Amelanchier sanguinea (Pursh) DC. ●

19266 Amelanchier humilis Wiegand var. campestris Nielsen = Amelanchier sanguinea (Pursh) DC. ●

19267 Amelanchier humilis Wiegand var. compacta Nielsen = Amelanchier sanguinea (Pursh) DC. ●

19268 Amelanchier humilis Wiegand var. exserrata Nielsen = Amelanchier sanguinea (Pursh) DC. ●

19269 Amelanchier humilis Wiegand var. typica Nielsen = Amelanchier sanguinea (Pursh) DC. ●

19270 Amelanchier huronensis Wiegand = Amelanchier sanguinea (Pursh) DC. ●

19271 Amelanchier integrifolia Boiss. et Hohen.;全叶唐棣●☆

19272 Amelanchier interior Nielsen;内地唐棣;Inland Juneberry, Inland Serviceberry, Pacific Serviceberry ●☆

19273 Amelanchier intermedia Spach;间型唐棣;Juneberry ●☆

19274 Amelanchier japonica Hort. ex K. Koch = Amelanchier asiatica (Siebold et Zucc.) Endl. ex Walp. ●

19275 Amelanchier japonica K. Koch = Amelanchier asiatica (Siebold et Zucc.) Endl. ex Walp. ●

19276 Amelanchier laevis Wiegand;平滑唐棣;Allegheny Serviceberry, Allegheny Shadblow, Allsgheny Juneberry, Glabrous Shadbush, Juneberry, Service Tree, Service-berry, Shadbush, Smooth Juneberry, Smooth Serviceberry, Smooth Shadbush, Snowy Mespilus ●

19277 Amelanchier laevis Wiegand f. nitida Wiegand = Amelanchier laevis Wiegand ●

19278 Amelanchier laevis Wiegand var. nitida (Wiegand) Fernald = Amelanchier laevis Wiegand ●

19279 Amelanchier lamarckii F. G. Schroed.;加东唐棣(拉马克唐棣);Juneberry, Lamarck Serviceberry, Serviceberry, Shadblow Serviceberry, Snowy Mespilus ●☆

19280 Amelanchier lucida (Fernald) Fernald = Amelanchier canadensis (L.) Medik. ●☆

19281 Amelanchier lucida Fernald = Amelanchier canadensis (L.) Medik. ●☆

19282 Amelanchier mucronata Nielsen = Amelanchier spicata (Lam.) K. Koch ●

19283 Amelanchier neglecta Eggl. ex G. N. Jones;疏忽唐棣;Serviceberry ●☆

19284 Amelanchier oblongifolia (Torr. et A. Gray) M. Roem. = Amelanchier arborea (F. Michx.) Fernald ●

19285 Amelanchier ovalis Medik. = Amelanchier rotundifolia (Lam.) Dum. Cours. ●☆

19286 Amelanchier prunifolia Greene;李叶唐棣;Prune Leaved Serviceberry, Prune Leaved Shadberry, Prune-leaved Juneberry, Service-berry ●☆

19287 Amelanchier racemosa Lindl. = Exochorda racemosa (Lindl.) Rehder ●

19288 Amelanchier rotundifolia (Lam.) Dum. Cours.;欧洲唐棣(广椭圆叶唐棣,卵叶唐棣,欧洲圆叶唐棣);Garden Serviceberry, Juneberry, Medlar-bush, Serviceberry, Snowy Mespilus ●☆

19289 Amelanchier rotundifolia (Lam.) Dum. Cours. = Amelanchier ovalis Medik. ●☆

19290 Amelanchier sanguinea (Pursh) DC.;红皮唐棣(血红唐棣,圆叶唐棣);Juneberry, Low Shadblow, New England Serviceberry, Redbranched Amelianchier, Roundleaf Juneberry, Roundleaf Serviceberry, Round-leaved Juneberry, Round-leaved Serviceberry, Serviceberry, Shadbush, Shore Shadbush ●

19291 Amelanchier sanguinea (Pursh) DC. var. grandiflora Rehder = Amelanchier grandiflora Rehder ●☆

19292 Amelanchier sinica (C. K. Schneid.) Chun;唐棣(独摇,枎移,枎移木,高飞,红枸子,移,移杨);China Juneberry, Chinese Serviceberry, Chinese Service-berry ●

19293 Amelanchier spicata (Lam.) K. Koch;穗花唐棣(穗状唐棣);Dwarf Serviceberry ●

19294 Amelanchier spicata K. Koch = Amelanchier spicata (Lam.) K. Koch ●

19295 Amelanchier stolonifera Wiegand;匍匐唐棣;Quebec Berry, Running Shadbush, Stoloniferous Shadbush ●☆

19296 Amelanchier stolonifera Wiegand = Amelanchier spicata (Lam.) K. Koch ●

19297 Amelanchier turkestanica Litv.;土耳其斯坦唐棣●☆

19298 Amelanchier utahensis Koehne;犹他荷唐棣;Serviceberry ●☆

19299 Amelanchier vulgaris Moench = Amelanchier ovalis Medik. ●☆

19300 Amelanchier vulgaris Moench var. djurdjurae Chabert = Amelanchier ovalis Medik. ●☆

19301 Amelanchier wiegandii Nielsen = Amelanchier interior Nielsen ●☆

19302 Amelanchus Raf. = Amelanchier Medik. ●

19303 Amelancus F. Muller ex Vollm. = Amelanchier Medik. ●

19304 Amelancus Raf. = Amelanchier Medik. ●

19305 Ameletia DC. = Rotala L. ■

19306 Ameletia indica (Willd.) DC. = Rotala indica (Willd.) Koehne ■

19307 Amelia Alef. = Braxilia Raf. ●■

19308 Amelia Alef. = Pyrola L. ●■

19309 Amelia media (Sw.) Alef. = Pyrola media Sw. ●

19310 Amelia minor (L.) Alef. = Pyrola minor L. ●

19311 Amelichloa Arriaga et Barkworth = Stipa L. ■

19312 Amelichloa Arriaga et Barkworth(2006);阿根廷针茅属■☆

19313 Amelina C. B. Clarke = Aneilema R. Br. ■☆

19314 Amellus Adans. = Aster L. ●■

19315 Amellus L. (1759)(保留属名);非洲紫菀属(南非菊属)●■☆

19316 Amellus Ortegaex Willd. = Tridax L. ●■

19317 Amellus P. Browne(废弃属名) = Amellus L. (保留属名)●■☆

19318 Amellus P. Browne(废弃属名) = Melanthera Rohr ●■☆

19319 Amellus alternifolius Roth;互叶非洲紫菀■☆

19320 Amellus alternifolius Roth subsp. angustissimus (DC.) Rommel;细互叶非洲紫菀■☆

19321 Amellus annuus Willd. = Amellus alternifolius Roth ■☆

19322 Amellus arenarius S. Moore = Amellus tridactylus DC. subsp. arenarius (S. Moore) Rommel ■☆

19323 Amellus asteroides (L.) Druce;星形非洲紫菀■☆

19324 Amellus asteroides (L.) Druce subsp. mollis Rommel;柔软星形非洲紫菀■☆

19325 Amellus capensis (Walp.) Hutch.;好望角非洲紫菀■☆

19326 Amellus epaleaceus O. Hoffm.;无膜片非洲紫菀■☆

19327 Amellus flosculosus DC.;多小花非洲紫菀■☆

19328 Amellus hispidus DC. = Amellus alternifolius Roth ■☆

19329 Amellus hispidus DC. var. angustissimus ? = Amellus alternifolius Roth subsp. angustissimus (DC.) Rommel ■☆

19330 Amellus hispidus DC. var. flosculosus (DC.) Harv. = Amellus flosculosus DC. ■☆

19331 Amellus humilis Heering = Amellus tridactylus DC. subsp. arenarius (S. Moore) Rommel ■☆

19332 Amellus lychnitis L. = Amellus asteroides (L.) Druce ■☆

19333 Amellus lychnitis L. var. flosculosus Benth. ex Harv. = Amellus asteroides (L.) Druce subsp. mollis Rommel ■☆

19334 Amellus microglossus DC.;小舌非洲紫菀■☆

19335 Amellus nanus DC.;矮非洲紫菀■☆

19336 Amellus reductus Rommel;退缩非洲紫菀■☆

19337 Amellus scabridus DC. = Amellus strigosus (Thunb.) Less. subsp. scabridus (DC.) Rommel ■☆

19338 Amellus spinulosus Pursh = Xanthisma spinulosum (Pursh) D. R. Morgan et R. L. Hartm. ●■☆

19339 Amellus strigosus (Thunb.) Less.;糙伏毛非洲紫菀■☆

19340 Amellus strigosus (Thunb.) Less. subsp. pseudoscabridus Rommel;粗糙非洲紫菀■☆

19341 Amellus strigosus (Thunb.) Less. subsp. scabridus (DC.) Rommel;微糙非洲紫菀■☆

19342 Amellus strigosus (Thunb.) Less. var. thunbergii Harv. = Amellus strigosus (Thunb.) Less. ■☆

19343 Amellus strigosus (Thunb.) Less. var. tridactylus (DC.) Harv. = Amellus strigosus (Thunb.) Less. ■☆

19344 Amellus strigosus (Thunb.) Less. var. wildenovii Harv. = Amellus strigosus (Thunb.) Less. ■☆

19345 Amellus tenuifolius Burm.;矮细叶非洲紫菀■☆

19346 Amellus tridactylus DC.;三指非洲紫菀■☆

19347 Amellus tridactylus DC. subsp. arenarius (S. Moore) Rommel;沙地三指非洲紫菀■☆

19348 Amellus tridactylus DC. subsp. olivaceus Rommel;橄榄绿非洲紫菀■☆

19349 Amellus villosus Pursh = Heterotheca villosa (Pursh) Shinners ■☆

19350 Amenippis Thouars = Diplecthrum Pers. ■

19351 Amenippis Thouars = Satyrium Sw. (保留属名)■

19352 Amentaceae Dulac = Salicaceae Mirb. (保留科名)●

19353 Amentotaxaceae Kudo et Yamam.;穗花杉科●

19354 Amentotaxaceae Kudo et Yamam. = Cephalotaxaceae Neger(保留科名)●

19355 Amentotaxaceae Kudo et Yamam. = Taxaceae Gray(保留科名)●

19356 Amentotaxus Pilg. (1916);穗花杉属(紫杉属);Amentotaxus ●

19357 Amentotaxus argotaenia (Hance) Pilg.;穗花杉(华西穗花杉,水杉树);Catkin Yew, Common Amentotaxus ●◇

19358 Amentotaxus argotaenia (Hance) Pilg. var. brevifolia K. M. Lan et F. H. Zhang;短叶穗花杉;Shortleaf Common Amentotaxus ●

19359 Amentotaxus argotaenia (Hance) Pilg. var. cathayensis (H. L. Li) P. C. Keng = Amentotaxus argotaenia (Hance) Pilg. ●◇

19360 Amentotaxus argotaenia (Hance) Pilg. var. taiwanica (K. S. Hao) P. C. Keng = Amentotaxus formosana H. L. Li ●◇

19361 Amentotaxus argotaenia (Hance) Pilg. var. yunnanensis (H. L. Li) P. C. Keng = Amentotaxus yunnanensis H. L. Li ●◇

19362 Amentotaxus cathayensis H. L. Li = Amentotaxus argotaenia (Hance) Pilg. ●◇

19363 Amentotaxus formosana H. L. Li;台湾穗花杉;Formosan Amentotaxus, Taiwan Amentotaxus ●◇

19364 Amentotaxus taiwanica K. S. Hao = Amentotaxus formosana H. L. Li ●◇

19365 Amentotaxus yunnanensis H. L. Li;云南穗花杉;Yunnan Amentotaxus ●◇

19366 Amentotaxus yunnanensis H. L. Li var. formosana (H. L. Li) Silba = Amentotaxus formosana H. L. Li ●◇

19367 Amerimnon P. Browne(废弃属名) = Dalbergia L. f. (保留属名)●
19368 Amerimnum Post et Kuntze = Dalbergia L. f. (保留属名)●
19369 Amerimnum Scop. = Amerimnon P. Browne(废弃属名)●
19370 Amerimnum Scop. = Dalbergia L. f. (保留属名)●
19371 Amerina DC. = Aegiphila Jacq. ●■☆
19372 Amerina Noronha = Aglaia Lour. (保留属名)●
19373 Amerina Raf. = Salix L. (保留属名)●
19374 Amerina triphylla (Hochst.) A. DC. = Clerodendrum glabrum E. Mey. ●☆
19375 Amerix Raf. = Salix L. (保留属名)●
19376 Amerlingia Opiz = Sambucus L. ●■
19377 Ameroglossum Eb. Fisch., S. Vogel et A. V. Lopes(1999);巴西玄参木属●☆
19378 Ameroglossum pernambucense Eb. Fisch., S. Vogel et A. V. Lopes;巴西玄参木●☆
19379 Amerorchis Hultén(1968);北美兰属■☆
19380 Amerorchis rotundifolia (Banks ex Pursh) Hultén;北美兰;Round-leaved Orchid, Round-leaved Orchis ■☆
19381 Amerorchis rotundifolia (Pursh) Hultén = Amerorchis rotundifolia (Banks ex Pursh) Hultén ■☆
19382 Amerosedum Á. Löve et D. Löve = Sedum L. ●■
19383 Amesia A. Nelson et J. F. Macbr. (1913);埃姆斯兰属■☆
19384 Amesia A. Nelson et J. F. Macbr. = Epipactis Zinn(保留属名)■
19385 Amesia africana (Rendle) A. Nelson et J. F. Macbr. = Epipactis africana Rendle ■☆
19386 Amesia discolor (Kraenzl.) Hu = Epipactis discolor Kraenzl. ■
19387 Amesia discolor (Kraenzl.) Hu = Epipactis helleborine (L.) Crantz ■
19388 Amesia gigantea (Douglas ex Hook.) A. Nelson et J. F. Macbr. = Epipactis gigantea Douglas ex Hook. ■☆
19389 Amesia latifolia (L.) A. Nelson et J. F. Macbr. = Epipactis helleborine (L.) Crantz ■
19390 Amesia latifolia A. Nelson et J. F. Macbr. = Epipactis helleborine (L.) Crantz ■
19391 Amesia latifolia A. Nelson et J. F. Macbr. = Epipactis latifolia All. ■
19392 Amesia longibracteata Schweinf. = Epipactis helleborine (L.) Crantz ■
19393 Amesia mairei (Schltr.) Hu = Epipactis mairei Schltr. ■
19394 Amesia mierophylla A. Nelson et J. F. Macbr. = Epipactis microphylla Sw. ■☆
19395 Amesia monticola (Schltr.) Hu = Epipactis helleborine (L.) Crantz ■
19396 Amesia palustris A. Nelson et J. F. Macbr. = Epipactis palustris (L.) Crantz ■
19397 Amesia papillosa A. Nelson et J. F. Macbr. = Epipactis papillosa Franch. et Sav. ■
19398 Amesia royleana (Lindl.) Hu = Epipactis royleana Lindl. ■
19399 Amesia rubiginosa (Crantz) Mousley = Epipactis rubiginosa Crantz ■☆
19400 Amesia schensiana (Schltr.) Hu = Epipactis mairei Schltr. ■
19401 Amesia setschuanica (Ames et Schltr.) Hu = Epipactis mairei Schltr. ■
19402 Amesia squamellosa (Schltr.) Hu = Epipactis helleborine (L.) Crantz ■
19403 Amesia tangutica (Schltr.) Hu = Epipactis helleborine (L.) Crantz var. tangutica (Schltr.) S. C. Chen et G. H. Zhu ■
19404 Amesia tangutica (Schltr.) Hu = Epipactis helleborine (L.) Crantz ■
19405 Amesia tenii (Schltr.) Hu = Epipactis helleborine (L.) Crantz ■
19406 Amesia thunbergii (A. Gray) A. Nelson et J. F. Macbr. = Epipactis thunbergii A. Gray ■
19407 Amesia thunbergii A. Nelson et J. F. Macbr. = Epipactis thunbergii A. Gray ■
19408 Amesia trinervia A. Nelson et J. F. Macbr. = Epipactis trinervia Roxb. ■☆
19409 Amesia wilsonii (Schltr.) Hu = Epipactis mairei Schltr. ■
19410 Amesia xanthophaea (Schltr.) Hu = Epipactis xanthophaea Schltr. ■
19411 Amesia yunnanensis (Schltr.) Hu = Epipactis helleborine (L.) Crantz ■
19412 Amesiella Schltr. = Amesiella Schltr. ex Garay ■☆
19413 Amesiella Schltr. ex Garay(1972);小埃姆斯兰属(阿梅兰属)■☆
19414 Amesiella minor Senghas;小埃姆斯兰■☆
19415 Amesiella monticola Cootes et D. P. Banks;山地小埃姆斯兰■☆
19416 Amesiella philippinensis (Ames) Garay;菲律宾小埃姆斯兰■☆
19417 Amesiodendron Hu(1936);细子龙属;Amesiodendron ●★
19418 Amesiodendron chinense (Merr.) Hu;细子龙(坡露,莺哥木);China Amesiodendron, Chinese Amesiodendron ●
19419 Amesiodendron integrifoliolatum H. S. Lo;龙州细子龙(米费,米眼沙);Entireleaflet Amesiodendron, Entire-leafleted Amesiodendron, Longzhou Amesiodendron ●
19420 Amesiodendron integrifoliolatum H. S. Lo = Amesiodendron chinense (Merr.) Hu ●
19421 Amesiodendron tienlinense H. S. Lo;田林细子龙(黑仕);Tianlin Amesiodendron ●◇
19422 Amethystanthus Nakai = Isodon (Schrad. ex Benth.) Spach ●■
19423 Amethystanthus Nakai = Rabdosia (Blume) Hassk. ●■
19424 Amethystanthus daitonensis (Hayata) Nemoto = Isodon amethystoides (Benth.) H. Hara ■
19425 Amethystanthus daitonensis (Hayata) Nemoto = Rabdosia amethystoides (Benth.) H. Hara ■
19426 Amethystanthus excisus (Maxim.) Nakai = Isodon excisus (Maxim.) Kudo ■
19427 Amethystanthus excisus (Maxim.) Nakai = Rabdosia excisa (Maxim.) H. Hara ■
19428 Amethystanthus glaucocalyx (Maxim.) Nemoto = Isodon japonicus (Burm. f.) H. Hara var. glaucocalyx (Maxim.) H. W. Li ■
19429 Amethystanthus glaucocalyx (Maxim.) Nemoto = Rabdosia japonica (Burm. f.) H. Hara var. glaucocalyx (Maxim.) H. Hara ■
19430 Amethystanthus inflexus (Thunb.) Nakai = Isodon inflexus (Thunb.) Kudo ■
19431 Amethystanthus inflexus (Thunb.) Nakai = Rabdosia inflexa (Thunb.) H. Hara ■
19432 Amethystanthus japonicus (Burm. f.) Nakai = Isodon japonicus (Burm. f.) H. Hara ■
19433 Amethystanthus japonicus (Burm. f.) Nakai = Rabdosia japonica (Burm. f.) H. Hara ■
19434 Amethystanthus koroensis (Kudo) Nemoto = Isodon koroensis Kudo ■
19435 Amethystanthus koroensis (Kudo) Nemoto = Rabdosia amethystoides (Benth.) H. Hara ■
19436 Amethystanthus lasiocarpus (Hayata) Nemoto = Isodon lasiocarpus (Hayata) Kudo ■

19437 Amethystanthus lasiocarpus (Hayata) Nemoto = Isodon serrus (Maxim.) Kudo ■

19438 Amethystanthus lasiocarpus (Hayata) Nemoto = Rabdosia lasiocarpa (Hayata) H. Hara ■

19439 Amethystanthus longitubus (Miq.) Nakai = Isodon longitubus (Miq.) Kudo ■

19440 Amethystanthus longitubus (Miq.) Nakai = Rabdosia longituba (Miq.) H. Hara ■

19441 Amethystanthus macrophyllus Migo = Isodon macrophyllus (Migo) H. Hara ●■

19442 Amethystanthus macrophyllus Migo = Rabdosia macrophylla (Migo) C. Y. Wu et H. W. Li ●■

19443 Amethystanthus nakaii Migo = Isodon macrocalyx (Dunn) Kudo ■

19444 Amethystanthus nakaii Migo = Rabdosia macrocalyx (Dunn) H. Hara ■

19445 Amethystanthus racemosus (Hemsl.) Nakai = Isodon racemosus (Hemsl.) H. W. Li ■

19446 Amethystanthus racemosus (Hemsl.) Nakai = Rabdosia racemosa (Hemsl.) H. Hara ■

19447 Amethystanthus serrus (Maxim.) Nemoto = Isodon serrus (Maxim.) Kudo ■

19448 Amethystanthus serrus (Maxim.) Nemoto = Rabdosia serra (Maxim.) H. Hara ■

19449 Amethystanthus stenophyllus Migo = Isodon nervosus (Hemsl.) Kudo ■

19450 Amethystanthus stenophyllus Migo = Rabdosia nervosa (Hemsl.) C. Y. Wu et H. W. Li ■

19451 Amethystanthus taiwanensis Masam. = Isodon macrocalyx (Dunn) Kudo ■

19452 Amethystanthus taiwanensis Masam. = Rabdosia macrocalyx (Dunn) H. Hara ■

19453 Amethystanthus trichocarpus (Maxim.) Nakai = Isodon trichocarpus Kudo ●■

19454 Amethystanthus trichocarpus (Maxim.) Nakai = Rabdosia trichocarpa (Maxim.) Hara ●■

19455 Amethystanthus umbrosus (Makino) Nakai = Isodon umbrosus (Maxim.) H. Hara ■☆

19456 Amethystanthus websteri (Hemsl.) Kitag. = Isodon websteri (Hemsl.) Kudo ■

19457 Amethystanthus websteri (Hemsl.) Kitag. = Rabdosia websteri (Hemsl.) H. Hara ■

19458 Amethystea L. (1753);水棘针属;Amethystea ■

19459 Amethystea caerulea L.;水棘针(山油子,土荆芥,细叶山紫苏);Skyblue Amethystea ■

19460 Amethystina Zinn = Amethystea L. ■

19461 Ametron Raf. = Rubus L. ●■

19462 Amherstia Wall. (1829);缅甸凤凰木属(焰火树属,缨珞木属);Pride of Burma, Queen of Flowering Tree ●☆

19463 Amherstia nobilis Wall.;缅甸凤凰木(璎珞木);Orchid Tree, Pride of Burma, Pride-of-Burma ●☆

19464 Amiantanthus Kunth = Amianthium A. Gray(保留属名)■☆

19465 Amiantanthus Kunth = Cyanotris Raf. (废弃属名)■☆

19466 Amianthemum A. Gray = Zigadenus Michx. ■

19467 Amianthemum Steud. = Zigadenus Michx. ■

19468 Amianthium A. Gray (1837) (保留属名);毒蝇花属;Crow-Poison, Fly-Poison, St. Elmo's-Feather, Staggergrass ■☆

19469 Amianthium A. Gray(保留属名) = Zigadenus Michx. ■

19470 Amianthium angustifolium (Michx.) A. Gray = Zigadenus densus (Desr.) Fernald ■☆

19471 Amianthium muscaetoxicum (Walter) A. Gray;毒蝇花;Fly Poison ■☆

19472 Amianthium nuttallii A. Gray = Zigadenus nuttallii (A. Gray) S. Watson ■☆

19473 Amianthium texanum (Bush) R. R. Gates = Zigadenus densus (Desr.) Fernald ■☆

19474 Amianthum Raf. = Amianthium A. Gray(保留属名)■☆

19475 Amianthum Raf. = Zigadenus Michx. ■

19476 Amicia Kunth(1824);阿米豆属■☆

19477 Amicia zygomeris DC.;阿米豆■☆

19478 Amictonis Raf. = Callicarpa L. ●

19479 Amida Nutt. = Madia Molina ■☆

19480 Amidena Adans. = Orontium Pers. ■☆

19481 Amidena Raf. = Rohdea Roth ■

19482 Amiris La Llave = Amyris P. Browne ●☆

19483 Amirola Pers. = Llagunoa Ruiz et Pav. ●☆

19484 Amischophacelus R. S. Rao et Kammathy (1966);鞘苞花属;Amischophacelus ■

19485 Amischophacelus Rao Rolla et Kammathy = Cyanotis D. Don(保留属名)■

19486 Amischophacelus Rao Rolla et Kammathy = Tonningia Neck. ex A. Juss. ■

19487 Amischophacelus axillaris (L.) R. S. Rao et Kammathy;腋花鞘苞花■☆

19488 Amischophacelus axillaris (L.) R. S. Rao et Kammathy = Cyanotis axillaris (L.) Sweet ■

19489 Amischophacelus cucullata (Roth) R. S. Rao et Kammathy;鞘苞花■☆

19490 Amischotolype Hassk. (1863);穿鞘花属;Amischotolype ■

19491 Amischotolype chinensis (N. E. Br.) E. Walker ex Hatus.;中国穿鞘花(东陵草)■

19492 Amischotolype chinensis (N. E. Br.) E. Walker ex Hatus. = Amischotolype hispida (A. Rich.) D. Y. Hong ■

19493 Amischotolype hispida (A. Rich.) D. Y. Hong;穿鞘花(独竹草,纳闹红,鞘花,中国穿鞘花);Hispid Amischotolype ■

19494 Amischotolype hispida (Less. et A. Rich.) D. Y. Hong = Amischotolype hispida (A. Rich.) D. Y. Hong ■

19495 Amischotolype hookeri (Hassk.) Hara;尖果穿鞘花;Hooker Amischotolype ■

19496 Amischotolype tenuis (C. B. Clarke) R. S. Rao;小穿鞘花■☆

19497 Amischotopyle Pichon = Amischotolype Hassk. ■

19498 Amitostigma Schltr. (1919);无柱兰属(雏兰属,线柱兰属,锥兰属);Amitostigma, Astyleorchis ■

19499 Amitostigma alpestre Fukuy.;台湾无柱兰(高山雏兰,南湖雏兰,小黄斑兰);Amitostigma, Taiwan Astyleorchis ■

19500 Amitostigma amplexifolium Ts. Tang et F. T. Wang;抱茎叶无柱兰(抱茎无柱兰);Amplexicaul Amitostigma, Astyleorchis ■

19501 Amitostigma basifoliatum (Finet) Schltr.;四裂无柱兰;Astyleorchis, Baseleaf Amitostigma ■

19502 Amitostigma beesianum (W. W. Sm.) Ts. Tang et F. T. Wang = Ponerorchis chusua (D. Don) Soó ■

19503 Amitostigma bifoliatum Ts. Tang et F. T. Wang;棒距无柱兰(二叶无柱兰);Stickspur Astyleorchis, Twoleaveas Amitostigma ■

19504 Amitostigma capitatum Ts. Tang et F. T. Wang;头序无柱兰;Capitate Amitostigma, Capitate Astyleorchis ■

19505 Amitostigma chinense (Rolfe) Schltr. = Amitostigma gracile (Blume) Schltr. ■

19506 Amitostigma dolichacentrum Ts. Tang, F. T. Wang et K. Y. Lang;长距无柱兰;Longspur Amitostigma, Longspur Astyleorchis ■

19507 Amitostigma faberi (Rolfe) Schltr.;峨眉无柱兰;Emei Amitostigma, Emei Astyleorchis ■

19508 Amitostigma farreri Schltr.;长苞无柱兰(滇藏无柱兰);Longbract Amitostigma, Longbract Astyleorchis ■

19509 Amitostigma formosense (S. S. Ying) S. S. Ying = Amitostigma gracile (Blume) Schltr. ■

19510 Amitostigma forrestii Schltr. = Amitostigma monanthum (Finet) Schltr. var. forrestii (Schltr.) Ts. Tang et F. T. Wang ■

19511 Amitostigma gonggashanicum K. Y. Lang;贡嘎无柱兰;Gongga Amitostigma, Gongga Astyleorchis ■

19512 Amitostigma gracile (Blume) Schltr.;无柱兰(独叶一枝花,独叶一枝枪,合欢山兰,华无柱兰,双肾草,台湾红兰,细萼蒿草,细萼无柱兰,小雏兰);Astyleorchis, Slender Amitostigma ■

19513 Amitostigma gracile (Blume) Schltr. var. manshuricum Kitag.;白花无柱兰■

19514 Amitostigma hemipilioides (Finet) Ts. Tang et F. T. Wang;卵叶无柱兰;Ooleaf Amitostigma, Ooleaf Astyleorchis ■

19515 Amitostigma keiskei (Maxim. ex Franch. et Sav.) Schltr.;伊藤无柱兰■☆

19516 Amitostigma kinoshitae (Makino) Schltr.;木下氏无柱兰■☆

19517 Amitostigma kurokamianum (Ohwi et Hatus.) Ohwi = Ponerorchis graminifollia Rchb. f. var. kurokamiana (Ohwi et Hatus.) T. Hashim. ■☆

19518 Amitostigma lepidum (Rchb. f.) Schltr.;鳞片无柱兰■☆

19519 Amitostigma microhemipilia Schltr. = Amitostigma hemipilioides (Finet) Ts. Tang et F. T. Wang ■

19520 Amitostigma monanthum (Finet) Schltr.;单花无柱兰(一花无柱兰);Oneflower Astyleorchis, Singleflower Amitostigma ■

19521 Amitostigma monanthum (Finet) Schltr. var. forrestii (Schltr.) Ts. Tang et F. T. Wang;糙茎无柱兰;Forrest Oneflower Amitostigma, Forrest Oneflower Astyleorchis ■

19522 Amitostigma nivale Schltr. = Amitostigma monanthum (Finet) Schltr. ■

19523 Amitostigma papilionaceum Ts. Tang, F. T. Wang et K. Y. Lang;蝶花无柱兰;Butterfly Amitostigma, Butterfly Astyleorchis ■

19524 Amitostigma parceflorum (Finet) Schltr.;少花无柱兰;Poorflower Amitostigma, Poorflower Astyleorchis ■

19525 Amitostigma physoceras Schltr.;球距无柱兰;Ballspur Amitostigma, Ballspur Astyleorchis ■

19526 Amitostigma pingguiculum (Rchb. f. et S. Moore) Schltr.;大花无柱兰;Bigflower Astyleorchis, Largeflower Amitostigma ■

19527 Amitostigma potaninii K. V. Ivanova = Neottia camtschatea (L.) Rchb. f. ■

19528 Amitostigma potaninii K. V. Ivanova = Neottianthe camptoceras (Rolfe) Schltr. ■

19529 Amitostigma potaninii K. V. Ivanova var. macranthum K. V. Ivanova = Neottia camtschatea (L.) Rchb. f. ■

19530 Amitostigma simplex Ts. Tang et F. T. Wang;黄花无柱兰;Yellow Amitostigma, Yellow Astyleorchis ■

19531 Amitostigma taoloii S. S. Ying = Orchis tomingai (Hayata) H. J. Su ■

19532 Amitostigma taoloii S. S. Ying = Ponerorchis tominagai (Hayata) H. J. Su et J. J. Chen ■

19533 Amitostigma tetralobum (Finet) Schltr.;滇蜀无柱兰;SW. China Amitostigma, SW. China Astyleorchis ■

19534 Amitostigma tibeticum Schltr.;西藏无柱兰;Xizang Amitostigma, Xizang Astyleorchis ■

19535 Amitostigma tominagae (Hayata) Schltr. = Ponerorchis tominagai (Hayata) H. J. Su et J. J. Chen ■

19536 Amitostigma tominagai (Hayata) Schltr.;红花无柱兰(高山雏兰,红花兰);Red Amitostigma, Red Astyleorchis ■

19537 Amitostigma tominagai (Hayata) Schltr. = Ponerorchis tominagai (Hayata) H. J. Su et J. J. Chen ■

19538 Amitostigma trifurcatum Ts. Tang, F. T. Wang et K. Y. Lang;三叉无柱兰;Trident Amitostigma, Trident Astyleorchis ■

19539 Amitostigma wenshanense W. H. Chen;文山无柱兰■

19540 Amitostigma yuanum Ts. Tang et F. T. Wang;齿片无柱兰;Yuan Amitostigma, Yuan Astyleorchis ■

19541 Amitostigma yunnanense Schltr. = Amitostigma tetralobum (Finet) Schltr. ■

19542 Amitostigma yuukianum Fukuy. = Amitostigma gracile (Blume) Schltr. ■

19543 Amlanthemum Steud. = Amianthium A. Gray(保留属名)■☆

19544 Ammadenia Rupr. = Honkenya Ehrh. ■☆

19545 Ammandra O. F. Cook(1927);瘤蕊椰属(多蕊象牙椰属,砂蕊椰属,亚曼达象牙椰属)●☆

19546 Ammandra dasyneura (Burret) Barfod;毛脉瘤蕊椰●☆

19547 Ammandra decasperma O. F. Cook;瘤蕊椰●☆

19548 Ammandra natalia Balslev et A. J. Hend.;纳塔尔瘤蕊椰●☆

19549 Ammanella Miq. = Ammannia L. ■

19550 Ammannia L. (1753);水苋菜属;Ammania ■

19551 Ammannia aegyptiaca Willd. = Ammannia baccifera L. ■

19552 Ammannia anagalloides Sond. = Nesaea anagalloides (Sond.) Koehne ■☆

19553 Ammannia archboldiana A. Fern. = Ammannia prieuriana Guillaumin et Perr. ■☆

19554 Ammannia arenaria Kunth;耳基水苋菜(耳基水苋,耳水苋,耳叶水苋);Earleaf Ammania, Sand Ammania ■

19555 Ammannia arenaria Kunth = Ammannia auriculata Willd. ■

19556 Ammannia aspera Guillaumin et Perr. = Nesaea aspera (Guillaumin et Perr.) Koehne ■☆

19557 Ammannia attenuata Hochst. ex A. Rich. = Ammannia baccifera L. var. attenuata (Hochst. ex A. Rich.) Turki ■

19558 Ammannia auriculata Willd. = Ammannia arenaria Kunth ■

19559 Ammannia auriculata Willd. f. brasiliensis (A. St. -Hil.) Koehne = Ammannia auriculata Willd. ■

19560 Ammannia auriculata Willd. f. longistaminata A. Fern. = Nesaea aurita Koehne ■☆

19561 Ammannia auriculata Willd. var. arenaria (Kunth) Koehne = Ammannia arenaria Kunth ■

19562 Ammannia auriculata Willd. var. bojeriana Koehne;博耶尔苋菜■☆

19563 Ammannia auriculata Willd. var. elata (A. Fern.) A. Fern. = Ammannia auriculata Willd. ■

19564 Ammannia auriculata Willd. var. subsessilis ? = Ammannia baccifera L. ■

19565 Ammannia australasica F. Muell. = Ammannia multiflora Roxb. ■

19566 Ammannia baccifera L.;水苋菜(埃及水苋菜,浆果水苋,结筋草,细叶水苋,仙桃草);Common Ammania, Monarch Redstem ■

19567 Ammannia baccifera L. subf. contracta Koehne = Ammannia baccifera L. ■

19568 Ammannia baccifera L. subsp. aegyptiaca (Willd.) Koehne = Ammannia baccifera L. ■

19569 Ammannia baccifera L. subsp. intermedia Koehne = Ammannia senegalensis Lam. var. ondongana (Koehne) Verdc. ■☆

19570 Ammannia baccifera L. subsp. viridis (Willd. ex Hornem.) Koehne = Ammannia baccifera L. ■

19571 Ammannia baccifera L. var. attenuata (Hochst. ex A. Rich.) Turki;渐狭水苋菜■

19572 Ammannia baccifera L. var. micromerioides (Chiov.) Cufod. = Ammannia baccifera L. ■

19573 Ammannia caspica M. Bieb. = Ammannia verticillata Lam. ■☆

19574 Ammannia catholica Cham. et Schltdl. = Rotala ramosior (L.) Koehne ■☆

19575 Ammannia coccinea Rottb.;长叶水苋菜;Toothcup ■

19576 Ammannia crassicaulis Guillaumin et Perr. = Nesaea crassicaulis (Guillaumin et Perr.) Koehne ■☆

19577 Ammannia crassissima Koehne;粗水苋菜■☆

19578 Ammannia densiflora Roth = Rotala densiflora (Roth ex Roem. et Schult.) Koehne ■

19579 Ammannia densiflora Roxb. = Rotala densiflora (Roth) Koehne ■

19580 Ammannia dentelloides Kurz = Microcarpaea minima (Jos. König ex Retz.) Merr. ■

19581 Ammannia dentifera A. Gray = Rotala ramosior (L.) Koehne ■☆

19582 Ammannia discolor Nakai = Ammannia baccifera L. ■

19583 Ammannia elata A. Fern.;高水苋菜■☆

19584 Ammannia elatinoides DC. = Rotala elatinoides (DC.) Hiern ■☆

19585 Ammannia evansiana A. Fern. et Diniz = Ammannia senegalensis Lam. var. ondongana (Koehne) Verdc. ■☆

19586 Ammannia filiformis DC. = Ammannia senegalensis Lam. ■

19587 Ammannia floribunda Guillaumin et Perr. = Ammannia senegalensis Lam. ■

19588 Ammannia gracilis Guillaumin et Perr.;纤细苋菜■☆

19589 Ammannia hildebrandtii Koehne = Ammannia baccifera L. ■

19590 Ammannia indica Lam. = Ammannia baccifera L. ■

19591 Ammannia intermedia (Koehne) A. Fern. et Diniz = Ammannia senegalensis Lam. var. ondongana (Koehne) Verdc. ■☆

19592 Ammannia japonica Miq. = Ammannia multiflora Roxb. ■

19593 Ammannia latifolia L.;宽叶水苋菜■☆

19594 Ammannia leptopetala Blume = Rotala pentandra (Roxb.) Blatt. et Hallb. ■

19595 Ammannia leptopetala Blume = Rotala rosea (Poir.) C. D. K. Cook ex H. Hara ■

19596 Ammannia linearipetala A. Fern. et Diniz;线瓣苋■☆

19597 Ammannia littorea Miq. = Rotala pentandra (Roxb.) Blatt. et Hallb. ■

19598 Ammannia littorea Miq. = Rotala rosea (Poir.) C. D. K. Cook ex H. Hara ■

19599 Ammannia mexicana (Cham. et Schltdl.) Baill. = Rotala mexicana Cham. et Schltdl. ■

19600 Ammannia mexicana Bailey = Rotala mexicana Cham. et Schltdl. ■

19601 Ammannia monoflora Blanco = Rotala ramosior (L.) Koehne ■

19602 Ammannia multiflora Roxb.;多花水苋菜(多花水苋);Jerry-Jerry, Manyflower Ammania ■

19603 Ammannia multiflora Roxb. var. parviflora (DC.) Koehne = Ammannia multiflora Roxb. ■

19604 Ammannia myriophylloides Dunn;泽苋菜■

19605 Ammannia myriophylloides Dunn = Rotala wallichii (Hook. f.) Koehne ■

19606 Ammannia nana Roxb. = Ammannia indica Lam. ■

19607 Ammannia nana Roxb. = Rotala indica (Willd.) Koehne ■

19608 Ammannia octandra L. f.;八蕊水苋菜■

19609 Ammannia parviflora DC. = Ammannia multiflora Roxb. ■

19610 Ammannia passerinoides Welw. ex Hiern = Nesaea passerinoides (Welw. ex Hiern) Koehne ■☆

19611 Ammannia pentandra Roxb. = Rotala densiflora (Roth ex Roem. et Schult.) Koehne ■

19612 Ammannia pentandra Roxb. = Rotala pentandra (Roxb.) Blatt. et Hallb. ■

19613 Ammannia pentandra Roxb. = Rotala rosea (Poir.) C. D. K. Cook ex H. Hara ■

19614 Ammannia peploides Spreng. = Rotala indica (Willd.) Koehne ■

19615 Ammannia prieuriana Guillaumin et Perr.;普里厄水苋菜■☆

19616 Ammannia pubiflora (Koehne) Hewson;短毛花水苋菜■☆

19617 Ammannia pusilla Sond. = Ammannia prieuriana Guillaumin et Perr. ■☆

19618 Ammannia pygmaea A. Chev. = Rotala mexicana Cham. et Schltdl. ■

19619 Ammannia ramosior L. = Rotala ramosior (L.) Koehne ■

19620 Ammannia retusa Koehne;微凹水苋菜■☆

19621 Ammannia robusta Heer et Regel;无柄水苋菜;Grand Red-stem, Sessile Toothcup, Sessile Tooth-cup, Tooth-cup ■☆

19622 Ammannia rosea Poir. = Rotala rosea (Poir.) C. D. K. Cook ex H. Hara ■

19623 Ammannia rotundifolia Buch.-Ham. = Rotala rotundifolia (Buch.-Ham. ex Roxb.) Koehne ■

19624 Ammannia rotundifolia Buch.-Ham. ex Roxb. = Rotala rotundifolia (Buch.-Ham. ex Roxb.) Koehne ■

19625 Ammannia salicifolia Hiern = Ammannia verticillata Lam. ■☆

19626 Ammannia salsuginosa Guillaumin et Perr. = Ammannia senegalensis Lam. ■

19627 Ammannia sarcophylla Welw. ex Hiern = Nesaea sarcophylla (Welw. ex Hiern) Koehne ■☆

19628 Ammannia senegalensis Lam. = Ammannia arenaria Kunth ■

19629 Ammannia senegalensis Lam. = Ammannia auriculata Willd. ■

19630 Ammannia senegalensis Lam. f. filiformis (DC.) Hiern = Ammannia senegalensis Lam. ■

19631 Ammannia senegalensis Lam. f. patens Hiern = Ammannia prieuriana Guillaumin et Perr. ■☆

19632 Ammannia senegalensis Lam. var. brasiliensis A. St.-Hil. = Ammannia auriculata Willd. ■

19633 Ammannia senegalensis Lam. var. ondongana (Koehne) Verdc.;翁氏水苋菜■☆

19634 Ammannia subspicata Benth. = Rotala rotundifolia (Buch.-Ham. ex Roxb.) Koehne ■

19635 Ammannia tenella Guillaumin et Perr. = Rotala tenella (Guillaumin et Perr.) Hiern ■☆

19636 Ammannia uliginosa Miq. = Rotala indica (Willd.) Koehne ■

19637 Ammannia urceolata Hiern;坛状水苋菜■☆

19638 Ammannia versicatoria Roxb. = Ammannia baccifera L. ■

19639 Ammannia verticillata Lam.;轮生水苋菜■☆

19640 Ammannia vesicatoria Roxb. = Ammannia baccifera L. ■

19641 Ammannia viridis Willd. ex Hornem.;绿水苋菜■☆

19642 Ammannia viridis Willd. ex Hornem. = Ammannia baccifera L. ■

19643 Ammannia wallichii (Hook. f.) Kurz = Rotala wallichii (Hook.

f.) Koehne ■
19644 Ammannia wormskioldii Fisch. et C. A. Mey. = Ammannia baccifera L. ■
19645 Ammannia wormskioldii Fisch. et C. A. Mey. var. alata Koehne = Ammannia baccifera L. ■
19646 Ammanniaceae Horan. ;水苋菜科■
19647 Ammanniaceae Horan. = Lythraceae J. St. -Hil. (保留科名)●■
19648 Ammanthus Boiss. et Heldr. = Ammanthus Boiss. et Heldr. ex Boiss. ■☆
19649 Ammanthus Boiss. et Heldr. ex Boiss. (1849);地中海菊属■☆
19650 Ammanthus Boiss. et Heldr. ex Boiss. = Anthemis L. ■
19651 Ammanthus ageratifolius Boiss. et Heldr. ex Boiss. ;细茎地中海菊■☆
19652 Ammanthus filicaulis Boiss. et Heldr. ex Boiss. ;线茎地中海菊■☆
19653 Ammanthus glaberrimus Rech. f. ;光地中海菊■☆
19654 Ammanthus intermedius Coustur. et Gand. ;间型地中海菊■☆
19655 Ammanthus intermedius Gand. = Ammanthus intermedius Coustur. et Gand. ■☆
19656 Ammanthus maritimus Boiss. et Heldr. ;地中海菊■☆
19657 Ammanthus tomentellus Coustur. et Gand. ;毛地中海菊■☆
19658 Ammanthus tomentellus Gand. = Ammanthus tomentellus Coustur. et Gand. ■☆
19659 Ammi L. (1753);阿米芹属(阿米属,牙签草属);Ammi, Bullwort ■
19660 Ammi copticum L. = Trachyspermum ammi (L.) Sprague ■
19661 Ammi ehrenbergii (H. Wolff) M. Hiroe = Seseli strictum Ledeb. ■
19662 Ammi glaucifolium L. = Ammi majus L. var. glaucifolium (L.) Godr. ■☆
19663 Ammi huntii Watts. ;亨特阿米芹■☆
19664 Ammi majus L. ;大阿米芹(大阿米);Amy, Big Ammi, Bishop's Weed, Bullwort, False Queen Anne's Lace, Greater Ammi, Lace Flower, Large Bullwort ■
19665 Ammi majus L. var. glaucifolium (L.) Godr. ;粉绿叶阿米芹■☆
19666 Ammi majus L. var. glaucifolium (L.) Mérat = Ammi majus L. ■
19667 Ammi majus L. var. intermedium Gren. et Godr. = Ammi majus L. ■
19668 Ammi majus L. var. laciniatum Godr. = Ammi majus L. ■
19669 Ammi majus L. var. serratum Mutel = Ammi majus L. ■
19670 Ammi majus L. var. tenue Ball = Ammi majus L. ■
19671 Ammi procerum Lowe. ;高大阿米芹■☆
19672 Ammi visnaga (L.) Lam. ;阿米芹(阿米,阿密茴,齿阿米,凯刺,细叶阿米);Amee, Ameos, Bishop's Weed, Bishop's-weed, Bullwort, Cumin Royal, Ethiopian Cumin, Herb William, Khella, Tooth Ammi, Toothpick Ammi, Toothpick-plant, Toothpickweed, Visnaga ■
19673 Ammi visnaga (L.) Lam. var. hybernonis Maire = Ammi visnaga (L.) Lam. ■
19674 Ammi visnaga (L.) Lam. var. paui Maire = Ammi visnaga (L.) Lam. ■
19675 Ammiaceae (J. Presl et C. Presl) Barnhart = Apiaceae Lindl. (保留科名)●■
19676 Ammiaceae (J. Presl et C. Presl) Barnhart = Umbelliferae Juss. (保留科名)●■
19677 Ammiaceae Barnhart = Apiaceae Lindl. (保留科名)●■
19678 Ammiaceae Barnhart = Umbelliferae Juss. (保留科名)●■
19679 Ammiaceae Small = Apiaceae Lindl. (保留科名)●■
19680 Ammiaceae Small = Umbelliferae Juss. (保留科名)●■
19681 Ammiaceae Small;阿米芹科;Carrot Family ■☆
19682 Ammianthus Spruce ex Benth. = Retiniphyllum Humb. et Bonpl. ●☆
19683 Ammianthus Spruce ex Benth. et Hook. f. = Retiniphyllum Humb. et Bonpl. ●☆
19684 Ammiopsis Boiss. (1856);拟阿米芹属■☆
19685 Ammiopsis Boiss. = Daucus L. ■
19686 Ammiopsis daucoides Boiss. ;拟阿米芹■☆
19687 Ammios Moench(废弃属名) = Carum L. ■
19688 Ammios Moench(废弃属名) = Trachyspermum Link(保留属名)■
19689 Ammobium R. Br. = Ammobium R. Br. ex Sims ■☆
19690 Ammobium R. Br. ex Sims(1824);银苞菊属(苞瓣菊属,贝细工属);Ammobium ■☆
19691 Ammobium alatum R. Br. ;银包菊(合菊,铁菊);Everlasting Sand Flower, Winged Everlasting, Winged-everlasting ■☆
19692 Ammobium alatum R. Br. f. grandiflorum Siebert et Voss;大花银包菊■☆
19693 Ammobium alatum R. Br. var. grandiflorum Hort. = Ammobium alatum R. Br. f. grandiflorum Siebert et Voss ■☆
19694 Ammobium spathulatum Gaudich. = Ammobium alatum R. Br. ■☆
19695 Ammobroma Torr. = Pholisma Nutt. ex Hook. ■☆
19696 Ammobroma Torr. ex A. Gray = Pholisma Nutt. ex Hook. ■☆
19697 Ammocallis Small = Catharanthus G. Don ●■
19698 Ammocallis Small = Lochnera Rchb. ●■
19699 Ammocallis rosea (L.) Small = Catharanthus roseus (L.) G. Don ■
19700 Ammocallis rosea Small = Catharanthus roseus (L.) G. Don ■
19701 Ammocharis Herb. (1821);砂石蒜属■☆
19702 Ammocharis angolensis (Baker) Milne-Redh. et Schweick. ;安哥拉砂石蒜■☆
19703 Ammocharis baumii (Harms) Milne-Redh. et Schweick. ;鲍曼砂石蒜■☆
19704 Ammocharis falcata (Jacq.) Herb. = Ammocharis longifolia (L.) M. Roem. ■☆
19705 Ammocharis herrei F. M. Leight. = Ammocharis longifolia (L.) M. Roem. ■☆
19706 Ammocharis heterostyla (Bullock) Milne-Redh. et Schweick. = Ammocharis angolensis (Baker) Milne-Redh. et Schweick. ■☆
19707 Ammocharis longifolia (L.) M. Roem. = Crinum longifolium (L.) Thunb. ■☆
19708 Ammocharis nerinoides (Baker) Lehmiller;尼润兰砂石蒜■☆
19709 Ammocharis taveliana Schinz = Boophone disticha (L. f.) Herb. ■☆
19710 Ammochloa Boiss. (1854);燥地砂草属■☆
19711 Ammochloa involucrata Murb. ;总苞燥地砂草■☆
19712 Ammochloa palaestina Boiss. ;燥地砂草■☆
19713 Ammochloa palaestina Boiss. var. intermedia Maire et Weiller = Ammochloa palaestina Boiss. ■☆
19714 Ammochloa pungens (Schreb.) Boiss. ;刚毛地砂草■☆
19715 Ammochloa pungens (Schreb.) Boiss. var. mauritii Font Quer et Sennen = Ammochloa pungens (Schreb.) Boiss. ■☆
19716 Ammochloa pungens (Schreb.) Boiss. var. subacaulis (Balansa ex Coss. et Durieu) Pamp. = Ammochloa palaestina Boiss. ■☆
19717 Ammochloa subacaulis Balansa ex Coss. et Durieu = Ammochloa palaestina Boiss. ■☆
19718 Ammocodon Standl. (1916);沙钟花属■☆

19719 Ammocodon Standl. = Selinocarpus A. Gray ●☆

19720 Ammocodon chenopodioides (A. Gray) Standl.;沙钟花;Goosefoot Moonpo ■☆

19721 Ammocodon chenopodioides Standl. = Ammocodon chenopodioides (A. Gray) Standl. ■☆

19722 Ammocyanus (Boiss.) Dostál = Centaurea L.(保留属名)●■

19723 Ammodaucus Coss. et Durieu(1859);砂萝卜属■☆

19724 Ammodaucus leucotrichus Coss. et Durieu;砂萝卜■☆

19725 Ammodaucus leucotrichus Coss. et Durieu subsp. nanocarpus Beltrán;小果砂萝卜■☆

19726 Ammodaucus leucotrichus Coss. et Durieu var. brevipilus L. Chevall. = Ammodaucus leucotrichus Coss. et Durieu ■☆

19727 Ammodendron Fisch. = Ammodendron Fisch. ex DC. ●

19728 Ammodendron Fisch. ex DC.(1825);银砂槐属(沙树属,银沙槐属);Ammodendron, Silversandtree ●

19729 Ammodendron argenteum (Pall.) Kuntze;银砂槐(银沙槐);Bifoliate Ammodendron, Silversandtree ●◇

19730 Ammodendron bifolium (Pall.) Yakovlev = Ammodendron argenteum (Pall.) Kuntze ●◇

19731 Ammodendron conollyi Bunge ex Boiss.;考氏银砂槐●☆

19732 Ammodendron eichwaldii Ledeb. et Mey.;沙豆树(艾沃沙树);Eichwald Ammodendron ●☆

19733 Ammodendron karelinii Fisch. et C. A. Mey.;卡氏银砂槐●☆

19734 Ammodendron lehmannii Bunge ex Boiss.;赖氏银砂槐●☆

19735 Ammodendron longiracemosum Raikova;长穗银砂槐●☆

19736 Ammodendron persicum Bunge ex Boiss.;波斯银砂槐●☆

19737 Ammodendron sieversii DC. = Ammodendron bifolium (Pall.) Yakovlev ●◇

19738 Ammodendron sieversii Fisch. = Ammodendron bifolium (Pall.) Yakovlev ●◇

19739 Ammodenia J. G. Gmel. ex Rupr. = Honckenya Ehrh. ■☆

19740 Ammodenia J. G. Gmel. ex S. G. Gmel. = Honckenya Ehrh. ■☆

19741 Ammodenia Patrin. = Ammodenia Patrin. ex J. G. Gmel. ■

19742 Ammodenia Patrin. ex J. G. Gmel. = Arenaria L. ■

19743 Ammodenia major (Hook.) A. Heller = Honckenya peploides (L.) Ehrh. var. major Hook. ■☆

19744 Ammodenia major A. Heller = Honckenya peploides (L.) Ehrh. var. major Hook. ■☆

19745 Ammodenia oblongifolia (Torr. et A. Gray) A. Heller = Honckenya oblongifolia Torr. et A. Gray ■☆

19746 Ammodenia oblongifolia A. Heller = Ammodenia oblongifolia (Torr. et A. Gray) A. Heller ■☆

19747 Ammodenia oblongifolia A. Heller = Honckenya oblongifolia Torr. et A. Gray ■☆

19748 Ammodenia peploides (L.) Rupr. = Honckenya peploides (L.) Ehrh. ■☆

19749 Ammodia Nutt. = Chrysopsis (Nutt.) Elliott(保留属名)■☆

19750 Ammodia Nutt. = Heterotheca Cass. ■☆

19751 Ammodia oregona Nutt. = Heterotheca oregona (Nutt.) Shinners ■☆

19752 Ammodytes Steven = Astragalus L. ●■

19753 Ammogeton Schrad. = Troximon Gaertn. ■☆

19754 Ammoides Adans.(1763);安蒙草属■☆

19755 Ammoides atlantica (Coss. et Durieu) H. Wolff;大西洋安蒙草■☆

19756 Ammoides pusilla (Brot.) Breistr.;小安蒙草■☆

19757 Ammoides pusilla (Brot.) Breistr. var. trachysperma (Boiss.) Molero Mesa et Pérez Raya = Ammoides pusilla (Brot.) Breistr. ■☆

19758 Ammoides verticillata (Desf.) Briq. = Ammoides pusilla (Brot.) Breistr. ■☆

19759 Ammoides verticillata (Desf.) Briq. var. pusilla (Pamp.) Thell. = Ammoides pusilla (Brot.) Breistr. ■☆

19760 Ammoides verticillata (Desf.) Briq. var. trachysperma (Boiss.) Maire = Ammoides pusilla (Brot.) Breistr. ■☆

19761 Ammolirion Kar. et Kix. = Eremurus M. Bieb. ■

19762 Ammonalia Desv. = Honckenya Ehrh. ■☆

19763 Ammonalia Desv. ex Endl. = Honckenya Ehrh. ■☆

19764 Ammonia Noronha = Polyalthia Blume ●

19765 Ammophila Host(1809);砂禾属(滨草属,固沙草属,沙茅草属,砂滨草属);Beach Grass, Marram ■☆

19766 Ammophila arenaria (L.) Link;砂禾(滨草);Alkali-grass, Beach Grass, Bent Grass, European Beach Grass, European Beachgrass, European Beach-grass, Marram, Marram Grass, Marram-grass, Marrom, Matgrass, Mel Grass, Mel-grass, Reel Grass, Sea Bent, Sea Reed, Shaslagh ■☆

19767 Ammophila arenaria (L.) Link subsp. arundinacea Host = Ammophila arundinacea Host. ■☆

19768 Ammophila arenaria (L.) Link var. arundinacea (Host) Husn. = Ammophila arundinacea Host. ■☆

19769 Ammophila arenaria (L.) Link var. australis E. A. Durand et Barratte = Ammophila arenaria (L.) Link ■☆

19770 Ammophila arenaria Host = Ammophila arenaria (L.) Link ■☆

19771 Ammophila arundinacea Host.;苇状砂禾;Matweed ■☆

19772 Ammophila australis (Mabille) Porta et Rigo = Ammophila arenaria (L.) Link subsp. arundinacea Host ■☆

19773 Ammophila breviligulata Fernald;短舌砂禾;American Beachgrass, American Beach-grass, American Marram, Dune Grass, Dunegrass, Marram Grass, Marram-grass, Short-liguled Ammophila ■☆

19774 Ammophila champlainensis F. Seym. = Ammophila breviligulata Fernald ■☆

19775 Ammophila villosa (Trin.) Hand. -Mazz. = Psammochloa villosa (Trin.) Bor ■

19776 Ammopiptanthus S. H. Cheng (1959);沙冬青属;Ammopiptanthus, Sandholly ●

19777 Ammopiptanthus mongolicus (Maxim. ex Kom.) S. H. Cheng;沙冬青(冬青,蒙古黄花木,蒙古沙冬青,沙生黄花);Mongolian Ammopiptanthus, Mongolian Piptanthus, Sandholly ●◇

19778 Ammopiptanthus nanus (Popov) S. H. Cheng;新疆沙冬青(矮沙冬青,小沙冬青,小叶沙冬青);Dwarf Sandholly, Nanous Ammopiptanthus, Xinjiang Ammopiptanthus ●◇

19779 Ammopiptanthus nanus (Popov) S. H. Cheng = Ammopiptanthus mongolicus (Maxim. ex Kom.) S. H. Cheng ●◇

19780 Ammopursus Small = Liatris Gaertn. ex Schreb.(保留属名)■☆

19781 Ammopursus Small(1924);佛罗里达菊属■☆

19782 Ammopursus ohlingerae (S. F. Blake) Small = Liatris ohlingerae (S. F. Blake) B. L. Rob. ■☆

19783 Ammopursus ohlingerae Small;佛罗里达菊■☆

19784 Ammorrhiza Ehrh. = Carex L. ■

19785 Ammoselinum Torr. et A. Gray(1857);沙蛇床属■☆

19786 Ammoselinum butleri (S. Watson) Coult. et Rose;布氏沙蛇床;Sand Parsley ■☆

19787 Ammoselinum popei Torr. et Gray;沙蛇床■☆

19788 Ammoseris Endl. = Launaea Cass. ■

19789 Ammoseris Endl. = Microrhynchus Less. ■☆

19790 Ammosperma Hook. f.(1862);北非砂籽芥属■☆

19791 Ammosperma cinereum (Desf.) Baill.;北非砂籽芥■☆
19792 Ammosperma teretifolium (Desf.) Boiss. = Pseuderucaria teretifolia (Desf.) O. E. Schulz ■☆
19793 Ammosperma variabile Nègre et Le Houér.;易变北非砂籽芥■☆
19794 Ammothamnus Bunge = Sophora L. ●■
19795 Ammothamnus Bunge(1847);沙槐属●
19796 Ammothamnus lehmanni Bunge = Sophora lehmannii (Bunge) Yakovlev ●☆
19797 Ammothamnus songoricus (Schrenk) Lipsky ex Vassilcz. = Sophora songarica Schrenk ●
19798 Ammyrsine Pursh = Dendrium Desv. ●☆
19799 Ammyrsine Pursh = Leiophyllum (Pers.) R. Hedw. ●☆
19800 Amni Brongn. = Ammi L. ■
19801 Amoebophyllum N. E. Br. = Phyllobolus N. E. Br. ●☆
19802 Amoebophyllum angustum N. E. Br. = Phyllobolus roseus (L. Bolus) Gerbaulet ●☆
19803 Amoebophyllum guerichianum (Pax) N. E. Br. = Mesembryanthemum guerichianum Pax ■☆
19804 Amoebophyllum rangei N. E. Br. = Phyllobolus oculatus (N. E. Br.) Gerbaulet ●☆
19805 Amoebophyllum roseum L. Bolus = Phyllobolus roseus (L. Bolus) Gerbaulet ●☆
19806 Amogeton Neck. = Aponogeton L. f. (保留属名)■
19807 Amoleiachyris Sch. Bip. = Amphiachyris (A. DC.) Nutt. ■☆
19808 Amoleiachyris Sch. Bip. = Gutierrezia Lag. ●■☆
19809 Amolinia R. M. King et H. Rob. (1972);离苞毛泽兰属●☆
19810 Amolinia heydeana (B. L. Rob.) R. M. King et H. Rob.;离苞毛泽兰■☆
19811 Amomaceae A. Rich. = Zingiberaceae Martinov(保留科名)■
19812 Amomaceae J. St. -Hil. = Zingiberaceae Martinov(保留科名)■
19813 Amomis O. Berg = Pimenta Lindl. ●☆
19814 Amomophyllum Engl. = Spathiphyllum Schott ■☆
19815 Amomum L. (废弃属名) = Amomum Roxb. (保留属名)■
19816 Amomum L. (废弃属名) = Zingiber Mill. (保留属名)■
19817 Amomum Roxb. (1820)(保留属名);豆蔻属(沙仁属,砂仁属);Amomum, Cardamom ■
19818 Amomum aculeatum Roxb.;刺豆蔻■☆
19819 Amomum alboviolaceum Ridl. = Aframomum alboviolaceum (Ridl.) K. Schum. ■☆
19820 Amomum alpinum Gagnep. = Aframomum alpinum (Gagnep.) K. Schum. ■☆
19821 Amomum amarum F. P. Sm. = Alpinia oxyphylla Miq. ■
19822 Amomum amarum Lour.;益智子■
19823 Amomum angustifolium Sonn. = Aframomum angustifolium (Sonn.) K. Schum. ■☆
19824 Amomum aromaticum Roxb.;孟加拉豆蔻;Bengal Cardamom ■☆
19825 Amomum arundinaceum Oliv. et D. Hanb. = Aframomum arundinaceum (Oliv. et D. Hanb.) K. Schum. ■☆
19826 Amomum aurantiacum H. T. Tsai et S. W. Zhao = Amomum neoaurantiacum T. L. Wu et al. ■
19827 Amomum austrosinense D. Fang;三叶豆蔻(公天锥,土砂仁);S. China Amomum ■
19828 Amomum bitacoum Gagnep. = Aframomum alboviolaceum (Ridl.) K. Schum. ■☆
19829 Amomum capsiciforme S. Q. Tong; 辣椒砂仁; Capsiciform Amomum ■
19830 Amomum cardamomum L. = Amomum compactum Sol. ex Maton ■
19831 Amomum cardamomum L. = Elettaria cardamomum L. ■
19832 Amomum cereum Hook. f. = Aframomum sceptrum (Oliv. et D. Hanb.) K. Schum. ■☆
19833 Amomum chinense Chun ex T. L. Wu;海南假砂仁(海南土砂仁,土砂仁);China Amomum ■
19834 Amomum citratum J. Pereira ex Oliv. et D. Hanb. = Aframomum citratum (J. Pereira ex Oliv. et D. Hanb.) K. Schum. ■☆
19835 Amomum clusii D. Hanb. = Aframomum hanburyi K. Schum. ■☆
19836 Amomum compactum Sol. ex Maton;爪哇白豆蔻(白豆蔻,白蔻,草蔻,多骨,小白豆蔻,小豆蔻,圆豆蔻);Cardamon, Java Amomum, Round Cardamom ■
19837 Amomum coriandriodorum S. Q. Tong et Y. M. Xia;荽味砂仁■
19838 Amomum corrorima A. Braun = Aframomum corrorima (A. Braun) P. C. M. Jansen ■☆
19839 Amomum costatum Roxb.;豆蔻(草果)■☆
19840 Amomum crassilabium K. Schum. = Aframomum crassilabium (K. Schum.) K. Schum. ■☆
19841 Amomum cuspidatum Gagnep. = Aframomum exscapum (Sims) Hepper ■☆
19842 Amomum daniellii Hook. f. = Aframomum daniellii (Hook. f.) K. Schum. ■☆
19843 Amomum dealbatum Roxb.;长果砂仁;Longfruit Amomum ■
19844 Amomum dealbatum Roxb. var. sericeum (Roxb.) Baker = Amomum sericeum Roxb. ■
19845 Amomum dealbatum Roxb. var. sericeum Baker = Amomum sericeum Roxb. ■
19846 Amomum dolichanthum D. Fang;长花豆蔻(长花砂仁,哥卡);Longflower Amomum ■
19847 Amomum elliotii Baker = Aframomum elliottii (Baker) K. Schum. ■☆
19848 Amomum erythrocarpum Ridl. = Aframomum angustifolium (Sonn.) K. Schum. ■☆
19849 Amomum exscapum Sims = Aframomum exscapum (Sims) Hepper ■☆
19850 Amomum foetens Benth. et Hook. f.;臭豆蔻■☆
19851 Amomum fragile S. Q. Tong;脆舌砂仁;Fragile Amomum ■
19852 Amomum gagnepainii T. L. Wu, K. Larsen et Turland;长序砂仁(土砂仁);Longinflorescence Amomum ■
19853 Amomum galanga (L.) Lour. = Alpinia galanga (L.) Willd. ■
19854 Amomum giganteum Oliv. et D. Hanb. = Aframomum giganteum (Oliv. et D. Hanb.) K. Schum. ■☆
19855 Amomum glabrum S. Q. Tong;无毛砂仁;Glabrous Amomum ■
19856 Amomum glaucophyllum K. Schum. = Aframomum subsericeum (Oliv. et D. Hanb.) K. Schum. subsp. glaucophyllum (K. Schum.) Lock ■☆
19857 Amomum globosum Lour. = Alpinia globosa Horan. ■
19858 Amomum grana ?;埃及砂仁;Egyptian Paradise Seed, Grains of Paradise, Malagetta Pepper, Melegueta Pepper ■☆
19859 Amomum grana-paradisi L.;乐园子豆蔻■☆
19860 Amomum grana-paradisi L. = Aframomum exscapum (Sims) Hepper ■☆
19861 Amomum guangxiense D. Fang;广西草果■
19862 Amomum hochreutineri Valeton;霍克豆蔻■☆
19863 Amomum hongtsaoko C. F. Liang et D. Fang;红草果;Hongcaoguo Amomum ■
19864 Amomum hongtsaoko C. F. Liang et D. Fang = Amomum tsaoko Crevost et Lem. ■

19865 Amomum jingxiense D. Fang et D. H. Qin;狭叶豆蔻;Jiangxi Amomum ■
19866 Amomum kayserianum K. Schum. = Aframomum kayserianum (K. Schum.) K. Schum. ■☆
19867 Amomum kepulaga Sprague et Burkill = Amomum compactum Sol. ex Maton ■
19868 Amomum koenigii J. F. Gmel.;野草果;Koenig Amomum ■
19869 Amomum korarima J. Pereira = Aframomum corrorima (A. Braun) P. C. M. Jansen ■☆
19870 Amomum kravanh Pierre ex Gagnep.;白豆蔻(白蔻,豆蔻,豆蔻花,多骨,壳蔻,三角蔻,小豆蔻,圆豆蔻);Kervanh,Whitefruit Amomum ■
19871 Amomum kwangsiense D. Fang et X. X. Chen;广西豆蔻(广西砂仁,砂仁,土草果,土砂仁);Guangxi Amomum ■
19872 Amomum latifolium Afzel. = Aframomum alboviolaceum (Ridl.) K. Schum. ■☆
19873 Amomum laurentii De Wild. et T. Durand = Aframomum laurentii (De Wild. et T. Durand) K. Schum. ■☆
19874 Amomum leptolepis K. Schum. = Aframomum leptolepis (K. Schum.) K. Schum. ■☆
19875 Amomum limbatum Oliv. et D. Hanb. = Aframomum limbatum (Oliv. et D. Hanb.) K. Schum. ■☆
19876 Amomum littorale Jos. König = Etlingera littoralis (J. König) Giseke ■
19877 Amomum littorale Jos. König ex Retz. = Etlingera littoralis Gieseke ■
19878 Amomum longiligulare T. L. Wu;海南砂仁(海南壳砂仁,土砂仁);Hainan Amomum ■
19879 Amomum longipetiolatum Merr.;长柄豆蔻;Longpediole Amomum ■
19880 Amomum longiscapum Hook. f. = Aframomum longiscapum (Hook. f.) K. Schum. ■☆
19881 Amomum luteoalbum K. Schum. = Aframomum luteoalbum (K. Schum.) K. Schum. ■☆
19882 Amomum macroglossa K. Schum.;长舌砂仁■☆
19883 Amomum macrolepis K. Schum. = Aframomum citratum (J. Pereira ex Oliv. et D. Hanb.) K. Schum. ■☆
19884 Amomum magnificum Benth. et Hook. f.;大砂仁;Torch Ginger ■☆
19885 Amomum mala K. Schum. = Aframomum mala (K. Schum.) K. Schum. ■☆
19886 Amomum mannii Oliv. et D. Hanb. = Aframomum mannii (Oliv. et D. Hanb.) K. Schum. ■☆
19887 Amomum masuianum De Wild. et T. Durand = Aframomum sceptrum (Oliv. et D. Hanb.) K. Schum. ■☆
19888 Amomum maximum Roxb.;九翅豆蔻(邓嘎,九翅砂仁);Java Cardamom,Ninewing Amomum ■
19889 Amomum medium Lour. = Alpinia galanga (L.) Willd. ■
19890 Amomum megalocheilos (Griff.) Baker = Etlingera littoralis (J. König) Giseke ■
19891 Amomum melegueta Roscoe = Aframomum melegueta (Roscoe) K. Schum. ■☆
19892 Amomum melequeta Roscoe;非洲豆蔻■☆
19893 Amomum menglaense S. Q. Tong;勐腊砂仁;Mengla Amomum ■
19894 Amomum mengluense T. L. Wu et Senjen;云南砂仁■
19895 Amomum mengtzense H. T. Tsai et P. S. Chen;蒙自砂仁;Mengzi Amomum ■
19896 Amomum microcarpum C. F. Liang et D. Fang;细砂仁;Smallfruit Amomum ■
19897 Amomum mioga Thunb. = Zingiber mioga (Thunb.) Roscoe ■
19898 Amomum monophyllum Gagnep. = Elettariopsis monophylla (Gagnep.) Loes. ■
19899 Amomum muricarpum Elmer;疣果豆蔻(菠萝砂,大砂仁,牛牯缩蔻,牛牯缩砂);Wartyfruit Amomum ■
19900 Amomum neoaurantiacum T. L. Wu et al.;红壳砂仁(红壳砂,红砂仁);Orangeyellow Amomum ■
19901 Amomum odontocarpum D. Fang;波翅豆蔻(阪姜,野薄荷);Toothfruit Amomum ■
19902 Amomum palustre Afzel. = Aframomum strobilaceum (Sm.) Hepper ■☆
19903 Amomum paratsao-ko S. Q. Tong et Y. M. Xia;拟草果;Tsaoko-like Amomum ■
19904 Amomum petaloideum (S. Q. Tong) T. L. Wu;宽丝豆蔻■
19905 Amomum pilosum Oliv. et D. Hanb. = Aframomum pilosum (Oliv. et D. Hanb.) K. Schum. ■☆
19906 Amomum polyanthum K. Schum. = Aframomum polyanthum (K. Schum.) K. Schum. ■☆
19907 Amomum purpureorubrum S. Q. Tong et Y. M. Xia;紫红砂仁;Purple-red Amomum ■
19908 Amomum putrescens D. Fang;腐花豆蔻;Decayed Flower Amomum,Decayedflower Amomum ■
19909 Amomum quadratolaminare S. Q. Tong;方片砂仁■
19910 Amomum repens Sonn.;匍匐砂仁(小豆蔻)■☆
19911 Amomum repoeense Pierre ex Gagnep.;云南豆蔻(吕氏砂仁);Yunnan Amomum ■
19912 Amomum roseum Roxb. = Zingiber roseum (Roxb.) Roscoe ■
19913 Amomum sanguineum K. Schum. = Aframomum angustifolium (Sonn.) K. Schum. ■☆
19914 Amomum scarlatinum H. T. Tsai et P. S. Chen;红花砂仁(红花豆蔻);Redflower Amomum ■
19915 Amomum sceptrum Oliv. et D. Hanb. = Aframomum sceptrum (Oliv. et D. Hanb.) K. Schum. ■☆
19916 Amomum sericeum Roxb.;银叶砂仁;Silverleaf Amomum ■
19917 Amomum simiarum A. Chev. = Aframomum strobilaceum (Sm.) Hepper ■☆
19918 Amomum speciosum Schultz;美丽砂仁■☆
19919 Amomum stipulatum Gagnep. = Aframomum alboviolaceum (Ridl.) K. Schum. ■☆
19920 Amomum strobilaceum Sm. = Aframomum strobilaceum (Sm.) Hepper ■☆
19921 Amomum subcapitatum Y. M. Xia;头花砂仁;Subcapitate Amomum ■
19922 Amomum subsericeum Oliv. et D. Hanb. = Aframomum subsericeum (Oliv. et D. Hanb.) K. Schum. ■☆
19923 Amomum subulatum Roxb.;香豆蔻(尼泊尔豆蔻);Fragrant Amomum,Nepal Cardamom ■
19924 Amomum sulcatum Oliv. et D. Hanb. ex Baker = Aframomum sulcatum (Oliv. et D. Hanb. ex Baker) K. Schum. ■☆
19925 Amomum thyrsoideum Gagnep. = Amomum gagnepainii T. L. Wu,K. Larsen et Turland ■
19926 Amomum thysanochililum S. Q. Tong et Y. M. Xia;梳唇砂仁■
19927 Amomum tsao-ko Crevost et Lem.;草果(白草果,草果仁,草果子,麻吼);Tsaoko Amomum ■
19928 Amomum tuberculatum D. Fang;德宝豆蔻(瘤状豆蔻);Tuberculate Amomum ■

19929 Amomum verrucosum S. Q. Tong;疣子砂仁;Verrucose Amomum ■

19930 Amomum villosum Lour.;砂仁(长泰砂仁,春砂仁,缩砂密,缩砂蔤,缩砂仁,阳春砂,阳春砂仁);Villous Amomum,Amomum ■

19931 Amomum villosum Lour. var. nanum H. T. Tsai et S. W. Zhao;矮砂仁(短砂仁);Dwarf Villous Amomum ■

19932 Amomum villosum Lour. var. nanum H. T. Tsai et S. W. Zhao = Amomum villosum Lour. var. xanthioides (Wall. ex Baker) T. L. Wu et S. J. Chen ■

19933 Amomum villosum Lour. var. xanthioides (Wall. ex Baker) T. L. Wu et S. J. Chen;缩砂密(绿壳砂仁,勐仑砂仁,砂仁壳,缩砂蔤,缩砂蜜,缩砂仁);Cocklebur-like Villous Amomum ■

19934 Amomum vittatum Hance = Alpinia pumila Hook. f. ■

19935 Amomum walang Valeton;瓦兰豆蔻■☆

19936 Amomum xanthioides Wall. ex Baker = Amomum villosum Lour. var. xanthioides (Wall. ex Baker) T. L. Wu et S. J. Chen ■

19937 Amomum yingjiangense S. Q. Tong et Y. M. Xia;盈江砂仁;Yingjiang Amomum ■

19938 Amomum yunnanense (T. L. Wu et S. J. Chen) R. M. Sm.;茴香砂仁(云南砂仁);Yunnan Amomum ■

19939 Amomum yunnanense S. Q. Tong = Amomum yunnanense (T. L. Wu et S. J. Chen) R. M. Sm. ■

19940 Amomum zambesiacum Baker = Aframomum zambesiacum (Baker) K. Schum. ■☆

19941 Amomum zedoaria Christm. = Curcuma phaeocaulis Valeton ■

19942 Amomum zedoaria Christm. = Curcuma zedoaria (Christm.) Roscoe ●■

19943 Amomum zerumbet L. = Zingiber zerumbet (L.) Sm. ■

19944 Amomum zingiber L. = Zingiber officinale (Willd.) Roscoe ■

19945 Amomyrtella Kausel(1956);小智利桃金娘属●☆

19946 Amomyrtella guili (Speg.) Kausel;小智利桃金娘●☆

19947 Amomyrtus (Burret) D. Legrand et Kausel(1948);智利桃金娘属;Myrtle ●☆

19948 Amomyrtus luma (Molina) Legrand et Kausel;智利桃金娘;Cauchao,Palo Madrono ●☆

19949 Amomyrtus luma (Molina) Legrand et Kausel = Eugenia apiculata DC. ●☆

19950 Amomyrtus luma (Molina) Legrand et Kausel = Myrtus luma Molina ●☆

19951 Amonia Nestl. = Aremonia Neck. ex Nestl.(保留属名)■☆

19952 Amoora Roxb.(1820);崖摩楝属;Amoora ●

19953 Amoora Roxb. = Aglaia Lour.(保留属名)●

19954 Amoora aherniana Merr.;卡突崖摩楝;Kato ●☆

19955 Amoora calcicola C. Y. Wu et H. Li;石山崖摩楝(石山崖摩);Calcicolous Amoora,Rock Amoora ●

19956 Amoora chittagonga (Miq.) Hiern = Amoora tetrapetala (Pierre) Pellegr. ●

19957 Amoora cucullata Roxb.;兜状崖摩楝;Tasua ●☆

19958 Amoora dasyclada (F. C. How et T. C. Chen) C. Y. Wu;粗枝崖摩楝(粗枝米仔兰,粗枝木楝,粗枝崖摩);Thick-branched Amoora,Thicktwig Amoora ●◇

19959 Amoora duodecimantha H. Zhu et H. Wang;多蕊崖摩楝(多蕊崖摩);Manystamen Amoora ●

19960 Amoora elmeri Merr. = Aphanamixis polystachya (Wall.) R. Parker ●

19961 Amoora elmeri Merr. = Aphanamixis tripetala (Blanco) Merr. ●

19962 Amoora ouangliensis (H. Lév.) C. Y. Wu;望谟崖摩楝(望谟崖摩);Wangmo Amoora ●

19963 Amoora polystachya (Wall.) Steud.;多穗崖摩楝●☆

19964 Amoora rohituka (Roxb.) Pierre = Aphanamixis polystachya (Wall.) R. Parker ●

19965 Amoora rohituka Wight et Arn. = Aphanamixis polystachya (Wall.) R. Parker ●

19966 Amoora rubiginosa Hiern;褐崖摩楝●☆

19967 Amoora stellata C. Y. Wu;星毛崖摩楝(星毛崖摩);Starhair Amoora,Stellate Aglaia ●◇

19968 Amoora stellato-squamosa C. Y. Wu et H. Li;曲梗崖摩楝(曲梗崖摩);Bendstalk Amoora,Stellate-scaled Amoora ●

19969 Amoora tetrapetala (Pierre) C. Y. Wu = Amoora tetrapetala (Pierre) Pellegr. ●

19970 Amoora tetrapetala (Pierre) Pellegr.;四瓣崖摩楝(兰屿树兰,四瓣米仔兰,四瓣崖摩);Fourpetal Aglaia,Fourpetal Amoora,Quadripetalous Amoora ●

19971 Amoora tetrapetala (Pierre) Pellegr. var. macrophylla (H. L. Li) C. Y. Wu;大叶四瓣崖摩楝(大叶四瓣崖摩);Largeleaf Fourpetal Aglaia ●

19972 Amoora tsangii (Merr.) X. M. Chen;铁椤(曾氏米仔兰);Tsang Aglaia,Tsang Amoora ●

19973 Amoora wallichii King;沃利克崖摩楝●☆

19974 Amoora yunnanensis (H. L. Li) C. Y. Wu;云南崖摩楝(云南崖摩);Yunnan Aglaia,Yunnan Amoora ●

19975 Amooria Walp. = Amoria C. Presl ■

19976 Amooria Walp. = Trifolium L. ■

19977 Amordica Neck. = Momordica L. ■

19978 Amorea Moq. = Cycloloma Moq. ■☆

19979 Amorea Moq. ex Del. = Cycloloma Moq. ■☆

19980 Amoreuxia DC. = Amoreuxia Moc. et Sessé ex DC. ●☆

19981 Amoreuxia Moc. et Sessé = Amoreuxia Moc. et Sessé ex DC. ●☆

19982 Amoreuxia Moc. et Sessé ex DC.(1825);合花弯籽木属(阿莫弯籽木属)●■☆

19983 Amoreuxia Moq. = Cycloloma Moq. ■☆

19984 Amoreuxia palmatifida Moc. et Sessé ex DC.;合花弯籽木●☆

19985 Amoreuxia wrightii A. Gray;赖氏合花弯籽木●☆

19986 Amorgine Raf. = Achyranthes L.(保留属名)■

19987 Amoria C. Presl = Trifolium L. ■

19988 Amorimia W. R. Anderson(2006);美洲金虎尾属●☆

19989 Amorpha L.(1753);紫穗槐属(黑花槐树属);Amorpha,False Indigo,Falseindigo,False-indigo ●

19990 Amorpha angustifolia (Pursh) F. E. Boynton = Amorpha fruticosa L. ●

19991 Amorpha arizonica Rydb. = Amorpha fruticosa L. ●

19992 Amorpha brachycarpa E. J. Palmer = Amorpha canescens Nutt. ●☆

19993 Amorpha bushii Rydb. = Amorpha fruticosa L. ●

19994 Amorpha californica Nutt. ex Torr. et Gray;加州紫穗槐;California Falseindigo,False Indigo,Mock Locust,Stinking Willow ●☆

19995 Amorpha canescens f. glabrata (A. Gray) Fassett = Amorpha canescens Pursh ●☆

19996 Amorpha canescens Nutt. = Amorpha canescens Pursh ●☆

19997 Amorpha canescens Pursh;灰毛紫穗槐;Falseindigo,Lead Plant,Leadplant,Lead-plant,Leadplant Amorpha,Prairie Shoestring,Shoe-strings ●☆

19998 Amorpha canescens Pursh = Amorpha canescens Nutt. ●☆

19999 Amorpha canescens Pursh f. canescens = Amorpha canescens Pursh ●☆

20000 Amorpha canescens Pursh f. glabrata (A. Gray) Fassett =

Amorpha canescens Pursh ●☆
20001 Amorpha canescens Pursh var. glabrata A. Gray = Amorpha canescens Pursh ●☆
20002 Amorpha croceolanata P. Watson = Amorpha fruticosa L. ●
20003 Amorpha curtissii Rydb. = Amorpha fruticosa L. ●
20004 Amorpha dewinkeleri Small = Amorpha fruticosa L. ●
20005 Amorpha fruticosa L.;紫穗槐(槐树,椒条,棉槐,棉条,苕条,穗花槐,紫翠槐,紫花槐,紫槐);Amorpha,Bastard Indigo,Desert False Indigo, Desert Indigo-bush, False Indigo, False Indigo Bush, Falseindigo, False-indigo, Indigo Bush, Indigobush, Indigo-bush, Indigobush Amorpha,Indigo-bush Amorpha,Shrubby Amorpha ●
20006 Amorpha fruticosa L. var. angustifolia Pursh;狭叶矮紫穗槐●☆
20007 Amorpha fruticosa L. var. angustifolia Pursh = Amorpha fruticosa L. ●
20008 Amorpha fruticosa L. var. angustifolia Pursh f. glabrata E. J. Palmer = Amorpha fruticosa L. ●
20009 Amorpha fruticosa L. var. angustifolia Pursh f. latior Fassett = Amorpha fruticosa L. ●
20010 Amorpha fruticosa L. var. crispa Kirchn.;皱边紫穗槐●☆
20011 Amorpha fruticosa L. var. crispa Kirchn. = Amorpha fruticosa L. ●
20012 Amorpha fruticosa L. var. croceolanata (P. Watson) Mouill.;黄毛紫穗槐●☆
20013 Amorpha fruticosa L. var. croceolanata (P. Watson) Mouill. = Amorpha fruticosa L. ●
20014 Amorpha fruticosa L. var. croceolanata (P. Watson) P. Watson ex Mouill. = Amorpha fruticosa L. ●
20015 Amorpha fruticosa L. var. emarginata Pursh;大叶紫穗槐●☆
20016 Amorpha fruticosa L. var. emarginata Pursh = Amorpha fruticosa L. ●
20017 Amorpha fruticosa L. var. oblongifolia E. J. Palmer;长叶紫穗槐●☆
20018 Amorpha fruticosa L. var. oblongifolia E. J. Palmer = Amorpha fruticosa L. ●
20019 Amorpha fruticosa L. var. occidentalis (Abrams) Kearney et Peebles;西方紫穗槐●☆
20020 Amorpha fruticosa L. var. occidentalis (Abrams) Kearney et Peebles = Amorpha fruticosa L. ●
20021 Amorpha fruticosa L. var. tennesseensis (Shuttlew. ex Kunze) E. J. Palmer = Amorpha fruticosa L. ●
20022 Amorpha microphylla Pursh = Amorpha nana Nutt. ●☆
20023 Amorpha nana Nutt.;矮紫穗槐(短紫穗槐);Dwarf Amorpha, Dwarfindigo Amorpha,Fragrant False Indigo ●☆
20024 Amorpha occidentalis Abrams = Amorpha fruticosa L. ●
20025 Amorpha occidentalis Abrams var. arizonica (Rydb.) E. J. Palmer = Amorpha fruticosa L. ●
20026 Amorpha occidentalis Abrams var. arizonica E. J. Palmer = Amorpha fruticosa L. ●
20027 Amorpha occidentalis Abrams var. emarginata (Pursh) E. J. Palmer = Amorpha fruticosa L. ●
20028 Amorpha occidentalis Abrams var. emarginata E. J. Palmer = Amorpha fruticosa L. ●
20029 Amorpha roemeriana Scheele;得州紫穗槐;Texas Amorpha ●☆
20030 Amorpha tennesseensis Shuttlew. = Amorpha fruticosa L. ●
20031 Amorpha tennesseensis Shuttlew. ex Kunze = Amorpha fruticosa L. ●
20032 Amorpha virgata Small = Amorpha fruticosa L. ●
20033 Amorphocalyx Klotzsch = Sclerolobium Vogel ■☆
20034 Amorphocalyx Klotzsch(1848);畸萼豆属■☆
20035 Amorphocalyx roraimae Klotzsch;畸萼豆■☆
20036 Amorphophallus Blume = Amorphophallus Blume ex Decne.(保留属名)●■
20037 Amorphophallus Blume ex Decne.(1834)(保留属名);魔芋属(合药芋属,蒟蒻属,连蕊芋属,磨芋属);Amorphophallus,Devil's-tongue,Giant Arum,Giantarum,Snake-palm ●■
20038 Amorphophallus abyssinicus (A. Rich.) N. E. Br.;阿比西尼亚魔芋■☆
20039 Amorphophallus accrensis N. E. Br. = Amorphophallus johnsonii N. E. Br. ■☆
20040 Amorphophallus albus P. Y. Liu et J. F. Chen;白魔芋;White Giantarum ■
20041 Amorphophallus angolensis (Welw. ex Schott) N. E. Br.;安哥拉魔芋■☆
20042 Amorphophallus angolensis (Welw. ex Schott) N. E. Br. subsp. maculatus (N. E. Br.) Ittenb.;斑点安哥拉魔芋■☆
20043 Amorphophallus anguineus Peter = Amorphophallus goetzei (Engl.) N. E. Br. ■☆
20044 Amorphophallus anisolobus Peter = Amorphophallus abyssinicus (A. Rich.) N. E. Br. ■☆
20045 Amorphophallus ankaranus Hett.,Ittenb. et Bogner;安卡兰魔芋■☆
20046 Amorphophallus antsingyensis Bogner,Hett. et Ittenb.;安钦吉魔芋■☆
20047 Amorphophallus aphyllus (Hook.) Hutch.;无叶魔芋■☆
20048 Amorphophallus arnautovii Hett.;越滇魔芋■
20049 Amorphophallus bangkokensis Gagnep.; 天 心 壶; Bankok Giantarum ■
20050 Amorphophallus bangkokensis Gagnep. = Amorphophallus paeoniifolius (Dennst.) Nicholson ■
20051 Amorphophallus bannanensis H. Li; 勐 海 魔 芋; Menghai Giantarum ■
20052 Amorphophallus barteri N. E. Br. = Amorphophallus abyssinicus (A. Rich.) N. E. Br. ■☆
20053 Amorphophallus baumannii (Engl.) N. E. Br.;鲍曼魔芋■☆
20054 Amorphophallus bequaertii De Wild.;贝卡尔魔芋■☆
20055 Amorphophallus bulbifer (Roxb.) Blume;珠芽魔芋(珠芽磨芋);Bulbiferous Giantarum ■
20056 Amorphophallus bulbifer (Roxb.) Blume = Amorphophallus yuloensis H. Li ■
20057 Amorphophallus campanulatus Blume ex Decne.; 臭 魔 芋; Elephant Yam,Ftanley's Wash Tub,Whitespot Giant Arum ●■☆
20058 Amorphophallus campanulatus Decne. = Amorphophallus campanulatus Blume ex Decne. ●■☆
20059 Amorphophallus campanulatus Decne. = Amorphophallus paeoniifolius (Dennst.) Nicholson ■
20060 Amorphophallus campanulatus Hook. f. = Amorphophallus virosus N. E. Br. ■
20061 Amorphophallus canaliculatus Ittenb. et Hett. et Lobin;具沟魔芋■☆
20062 Amorphophallus chevalieri (Engl.) Engl. et Gehrm. = Amorphophallus abyssinicus (A. Rich.) N. E. Br. ■☆
20063 Amorphophallus coaetaneus S. Y. Liu et S. J. Wei;桂平魔芋; Guiping Giantarum ■
20064 Amorphophallus coffeatus Stapf = Amorphophallus abyssinicus (A. Rich.) N. E. Br. ■☆
20065 Amorphophallus congoensis A. D. Hawkes = Amorphophallus mossambicensis (Schott ex Garcke) N. E. Br. ■☆

20066 Amorphophallus consimilis Blume;相似魔芋■☆
20067 Amorphophallus corrugatus N. E. Br.;缅甸魔芋■
20068 Amorphophallus difformis Blume = Anchomanes difformis (Blume) Engl. ■☆
20069 Amorphophallus doryphorus Ridl. = Amorphophallus consimilis Blume ■☆
20070 Amorphophallus dracontioides (Engl.) N. E. Br.;拟小龙南星魔芋■☆
20071 Amorphophallus dunnii Tutcher;南蛇棒(大头芋,蛇枪头,土南星,岩芋);Dunn Giantarum ■
20072 Amorphophallus eichleri (Engl.) Hook. f.;艾克勒魔芋■☆
20073 Amorphophallus elliotii Hook. f.;埃利魔芋■☆
20074 Amorphophallus fischeri (Engl.) N. E. Br. = Amorphophallus maximus (Engl.) N. E. Br. subsp. fischeri (Engl.) Ittenb. ■☆
20075 Amorphophallus flavovirens N. E. Br. = Amorphophallus baumannii (Engl.) N. E. Br. ■☆
20076 Amorphophallus foetidus (Engl.) Engl. et Gehrm. = Amorphophallus abyssinicus (A. Rich.) N. E. Br. ■☆
20077 Amorphophallus fontanesii Kunth = Amorphophallus aphyllus (Hook.) Hutch. ■☆
20078 Amorphophallus gallaensis (Engl.) N. E. Br.;加拉魔芋■☆
20079 Amorphophallus gallaensis (Engl.) N. E. Br. var. major Chiov. = Amorphophallus gallaensis (Engl.) N. E. Br. ■☆
20080 Amorphophallus giganteus Blume;巨磨芋■☆
20081 Amorphophallus gigantiflorus Hayata = Amorphophallus paeoniifolius (Dennst.) Nicholson ■
20082 Amorphophallus goetzei (Engl.) N. E. Br.;格兹魔芋■☆
20083 Amorphophallus gomboczianus Pic. Serm.;根伯茨魔芋■☆
20084 Amorphophallus gracilior Hutch.;纤细魔芋■☆
20085 Amorphophallus gracilis A. Chev. = Amorphophallus gracilior Hutch. ■☆
20086 Amorphophallus gratus (Schott) N. E. Br. = Amorphophallus abyssinicus (A. Rich.) N. E. Br. ■☆
20087 Amorphophallus hayi Hett.;红河魔芋;Honghe Giantarum ■
20088 Amorphophallus henryi N. E. Br.;台湾魔芋(白毛磨芋,白毛魔芋,亨利蒟蒻,亨氏蒟蒻,亨氏芋,山薯,石薯,疏毛蒟蒻,台湾磨芋);Taiwan Giantarum ■
20089 Amorphophallus hildebrandtii (Engl.) Engl. et Gehrm.;希尔魔芋■☆
20090 Amorphophallus hirtus N. E. Br.;硬毛魔芋(毛蒟蒻,密毛蒟蒻,密毛魔芋,硬毛磨芋);Hardhair Giantarum ■
20091 Amorphophallus hirtus N. E. Br. var. kiusiams (Makino) H. Motta = Amorphophallus henryi N. E. Br. ■
20092 Amorphophallus hirtus Yamam. = Amorphophallus niimurai Yamam. ■
20093 Amorphophallus impressus Ittenb.;凹陷魔芋■☆
20094 Amorphophallus johnsonii N. E. Br.;约翰逊魔芋■☆
20095 Amorphophallus kachinensis Engl. et Gehrm.;孟海魔芋;Menghai Giantarum ■
20096 Amorphophallus kaessneri Engl. et Gehrm. = Amorphophallus gallaensis (Engl.) N. E. Br. ■☆
20097 Amorphophallus kerrii N. E. Br. = Amorphophallus yunnanensis Engl. ■
20098 Amorphophallus kiusianus (Makino) Makino;东亚魔芋■
20099 Amorphophallus konjac C. Koch = Amorphophallus rivieri Durieu ex Carrière ■
20100 Amorphophallus konjac K. Koch = Amorphophallus rivieri Durieu ex Carrière ■
20101 Amorphophallus krausei Engl.;克氏魔芋(西盟魔芋);Ximeng Giantarum ■
20102 Amorphophallus laxiflorus N. E. Br. = Amorphophallus gallaensis (Engl.) N. E. Br. ■☆
20103 Amorphophallus leonensis Lem. = Amorphophallus aphyllus (Hook.) Hutch. ■☆
20104 Amorphophallus leopoldianus (Mast.) N. E. Br. = Amorphophallus angolensis (Welw. ex Schott) N. E. Br. ■☆
20105 Amorphophallus lewallei Malaisse et Bamps;勒瓦莱魔芋■☆
20106 Amorphophallus linearilobus Peter = Amorphophallus mossambicensis (Schott ex Garcke) N. E. Br. ■☆
20107 Amorphophallus macrospadix Font Quer = Amorphophallus zenkeri (Engl.) N. E. Br. subsp. mannii (N. E. Br.) Ittenb. ■☆
20108 Amorphophallus maculatus N. E. Br. = Amorphophallus angolensis (Welw. ex Schott) N. E. Br. subsp. maculatus (N. E. Br.) Ittenb. ■☆
20109 Amorphophallus mairei H. Lév.;东川魔芋(东川磨芋);E. Sichuan Giantarum ■
20110 Amorphophallus mairei H. Lév. = Amorphophallus konjac C. Koch ■
20111 Amorphophallus mannii N. E. Br. = Amorphophallus zenkeri (Engl.) N. E. Br. subsp. mannii (N. E. Br.) Ittenb. ■☆
20112 Amorphophallus maximus (Engl.) N. E. Br.;大魔芋■☆
20113 Amorphophallus maximus (Engl.) N. E. Br. subsp. fischeri (Engl.) Ittenb.;菲舍尔魔芋■☆
20114 Amorphophallus mekongensis Engl. et Gehrm.;湄公魔芋(湄公磨芋);Mekong Giantarum ■☆
20115 Amorphophallus mellii Engl.;蛇枪头(起角莲,蛇春头,蛇蒜头,土南星);Mell Giantarum ■
20116 Amorphophallus micro-appendiculatus Engl.;灰斑魔芋(灰斑磨芋);Grayspatted Giantarum ■
20117 Amorphophallus micro-appendiculatus Engl. = Amorphophallus paeoniifolius (Dennst.) Nicholson ■
20118 Amorphophallus mildbraedii K. Krause;米尔德魔芋■☆
20119 Amorphophallus mossambicensis (Schott ex Garcke) N. E. Br.;莫桑比克魔芋■☆
20120 Amorphophallus nanus H. Li et C. L. Long;矮魔芋;Dwarf Giantarum ■
20121 Amorphophallus niimurai Yamam.;白毛磨芋(疏毛蒟蒻)■
20122 Amorphophallus niimurai Yamam. = Amorphophallus henryi N. E. Br. ■
20123 Amorphophallus odoratus Hett. et H. Li;香魔芋;Fragrant Giantarum ■
20124 Amorphophallus oncophyllus Prain ex Hook. f.;香港魔芋(香港磨芋);Hongkong Giantarum ■☆
20125 Amorphophallus paeoniifolius (Dennst.) Nicholson;疣柄魔芋(臭魔芋,大魔芋,鸡爪芋,巨花蒟蒻,雷公统,南天星,南星头,南芋,天南星,鞋板芋,疣柄磨芋);Bigflower Giantarum, Elephant Yam, Ftanley's Wash Tub, Poisonous Giantarum, Whitespot Giant Arum ■
20126 Amorphophallus petitianus Malaisse et Bamps = Amorphophallus goetzei (Engl.) N. E. Br. ■☆
20127 Amorphophallus pingbianensis H. Li et C. L. Long;结节魔芋;Pingbian Giantarum ■
20128 Amorphophallus preussii (Engl.) N. E. Br.;普罗伊斯魔芋■☆
20129 Amorphophallus purpureus (Engl.) Engl. et Gehrm. =

Amorphophallus johnsonii N. E. Br. ■☆
20130 Amorphophallus richardsiae Ittenb.;理查兹魔芋■☆
20131 Amorphophallus rivieri Durieu et Carrière var. konjac (K. Koch) Engl. = Amorphophallus konjac K. Koch ■
20132 Amorphophallus rivieri Durieu ex Carrière;魔芋(白蒟蒻,独叶一枝花,粉乌舅,鬼蜡烛,鬼头,鬼芋,黑芋头,虎掌,花杆莲,花杆莲蒟蒻,花杆南星,花梗莲,花梗天南星,花麻蛇,花魔芋,花伞把,蒟蒻,雷公铳,雷星,麻芋,麻芋子,磨芋,蘑芋,南星,蒻头,蛇六谷,蛇头草,蛇头子,蛇芋,天六谷,天南星,土南星,星芋,野魔芋);Devils-tongue, Devil's-tongue, Elephant Foot, Konjaku, Leopard-palm, Rivier Giantarum, Snake Palm ■
20133 Amorphophallus rivieri Durieu ex Carrière = Amorphophallus konjac K. Koch ■
20134 Amorphophallus schliebenii Mildbr. = Amorphophallus maximus (Engl.) N. E. Br. subsp. fischeri (Engl.) Ittenb. ■☆
20135 Amorphophallus schweinfurthii (Engl.) N. E. Br. = Amorphophallus abyssinicus (A. Rich.) N. E. Br. ■☆
20136 Amorphophallus sinensis Belval;疏毛魔芋(东亚魔芋,鬼蜡烛,华东蒟蒻,魔芋,蛇六谷,蛇头草,疏毛磨芋,土半夏,伍花莲);China Giantarum ■
20137 Amorphophallus sinensis Belval = Amorphophallus kiusianus (Makino) Mikino ■
20138 Amorphophallus staudtii (Engl.) N. E. Br.;施陶魔芋■☆
20139 Amorphophallus stipitatus Engl.;梗序魔芋(梗序磨芋);Stipitate Giantarum ■
20140 Amorphophallus stuhlmannii (Engl.) Engl. et Gehrm.;斯图尔曼魔芋■☆
20141 Amorphophallus swynnertonii Rendle = Amorphophallus mossambicensis (Schott ex Garcke) N. E. Br. ■☆
20142 Amorphophallus sylvaticus Kunth;林地魔芋;Whitespot Giantarum ■☆
20143 Amorphophallus teuszii (Engl.) N. E. Br.;托兹魔芋■☆
20144 Amorphophallus tianyangensis P. Y. Su et S. L. Zhang;田阳魔芋;Tianyang Giantarum ■
20145 Amorphophallus titanum (Becc.) Becc.;提坦魔芋;Giant Arum, Titan Arum, Titan Giant Arum ■☆
20146 Amorphophallus tonkinensis Engl. et Gehrm.;东京魔芋;Tonkin Giantarum ■
20147 Amorphophallus variabilis Blume;野魔芋(土南星,野磨芋);Variable Giantarum ■
20148 Amorphophallus variabilis Blume 'Red Leaf Form';亮泽红叶魔芋■☆
20149 Amorphophallus virosus N. E. Br. = Amorphophallus paeoniifolius (Dennst.) Nicholson ■
20150 Amorphophallus warneckei (Engl.) Engl. et Gehrm. = Amorphophallus abyssinicus (A. Rich.) N. E. Br. ■☆
20151 Amorphophallus ximengensis H. Li;西盟魔芋;Ximeng Giantarum ■
20152 Amorphophallus yuloensis H. Li;攸乐魔芋;Youle Giantarum ■
20153 Amorphophallus yunnanensis Engl.;滇魔芋(长柱魔芋,滇磨芋,滇南魔芋,岩芋);Yunnan Giantarum ■
20154 Amorphophallus zenkeri (Engl.) N. E. Br.;岑克尔魔芋■☆
20155 Amorphophallus zenkeri (Engl.) N. E. Br. subsp. mannii (N. E. Br.) Ittenb.;曼氏魔芋■☆
20156 Amorphospermum F. Muell. = Apostasia Blume ■
20157 Amorphospermum F. Muell. = Chrysophyllum L. ●
20158 Amorphospermum F. Muell. = Niemeyera F. Muell.(保留属名)●☆
20159 Amorphospermum cerasiferum (Welw.) Baehni = Synsepalum cerasiferum (Welw.) T. D. Penn. ●☆
20160 Amorphospermum msolo (Engl.) Baehni = Synsepalum msolo (Engl.) T. D. Penn. ●☆
20161 Amorphospermum natalense (Sond.) Baehni = Englerophytum natalense (Sond.) T. D. Penn. ●☆
20162 Amorphus Raf. = Amorpha L. ●
20163 Amosa Neck. = Inga Mill. ●■☆
20164 Amoureuxia C. Muell. = Amoreuxia Moc. et Sessé ex DC. ●☆
20165 Ampacus Kuntze = Euodia J. R. Forst. et G. Forst. ●
20166 Ampacus Kuntze = Evodia J. R. Forst. et G. Forst. ●
20167 Ampacus Rumph. = Evodia J. R. Forst. et G. Forst. ●
20168 Ampacus Rumph. ex Kuntze = Evodia J. R. Forst. et G. Forst. ●
20169 Ampacus aromatica (Blume) Kuntze = Melicope lunu-ankenda (Gaertn.) T. G. Hartley ●
20170 Ampacus incerta (Blume) Kuntze = Melicope triphylla (Lam.) Merr. ●
20171 Ampacus triphylla (Lam.) Kuntze = Melicope triphylla (Lam.) Merr. ●
20172 Ampalis Bojer = Streblus Lour. ●
20173 Ampalis Bojer ex Bureau = Streblus Lour. ●
20174 Ampalis Bojer ex Bureau = Trophis P. Browne(保留属名)●☆
20175 Amparoa Schltr. (1923);阿姆兰属■☆
20176 Amparoa beloglossa Schltr.;长舌阿姆兰■☆
20177 Amparoa costaricensis Schltr.;阿姆兰■☆
20178 Ampelamus Raf. (1819);丑藤属●☆
20179 Ampelamus Raf. = Enslenia Nutt. ●☆
20180 Ampelamus Raf. = Gonolobus Michx. ●☆
20181 Ampelamus albidus (Nutt.) Britton;丑藤●☆
20182 Ampelamus albidus (Nutt.) Britton = Cynanchum laeve (Michx.) Pers. ■☆
20183 Ampelamus laevis (Michx.) Krings = Cynanchum laeve (Michx.) Pers. ■☆
20184 Ampelanus B. D. Jacks. = Ampelamus Raf. ●☆
20185 Ampelanus Raf. = Enslenia Nutt. ●☆
20186 Ampelaster G. L. Nesom = Aster L. ●■
20187 Ampelaster G. L. Nesom(1994);藤菀属(加州紫菀属)●■☆
20188 Ampelaster carolinianus (Walter) G. L. Nesom;藤菀(卡罗来纳紫菀);Climbing Aster ■☆
20189 Ampelaster carolinianus (Walter) G. L. Nesom = Aster carolinianus Walter ■☆
20190 Ampelidaceae Kunth = Vitaceae Juss.(保留科名)●■
20191 Ampelocalamus S. L. Chen, T. H. Wen et G. Y. Sheng = Sinarundinaria Nakai ●
20192 Ampelocalamus S. L. Chen, T. H. Wen et G. Y. Sheng(1981);悬竹属;Ampelocalamus, Handbamboo, Pendulus-bamboo ●★
20193 Ampelocalamus actinotrichus (Merr. et Chun) S. L. Chen, T. H. Wen et G. Y. Sheng;射毛悬竹(悬竹);Rayhair Ampelocalamus, Ray-haired Ampelocalamus, Starhair Handbamboo ●
20194 Ampelocalamus anhispidis T. H. Wen;黄篱竹;Anhispid Ampelocalamus ●
20195 Ampelocalamus breviligulatus (T. P. Yi) Stapleton et D. Z. Li = Drepanostachyum breviligulatum T. P. Yi ●
20196 Ampelocalamus calcareus C. D. Chu et C. S. Chao;贵州悬竹;Guizhou Ampelocalamus, Guizhou Handbamboo, Lime-loving Ampelocalamus ●
20197 Ampelocalamus hirsutissimus (W. D. Li et Y. C. Zhong)

Stapleton et D. Z. Li;多毛悬竹●

20198 Ampelocalamus luodianense T. P. Yi et R. S. Wang = Drepanostachyum luodianense (T. P. Yi et R. S. Wang) P. C. Keng ●

20199 Ampelocalamus luodianensis T. P. Yi et R. S. Wang;小蓬竹(爬竹); Crepp Sicklebamboo, Luodian Ampelocalamus, Luodian Drepanostachyum, Luodian Handbamboo ●

20200 Ampelocalamus melicoideus (P. C. Keng) D. Z. Li et Stapleton;南川竹(南川镰序竹,小蓬竹);Meliclike Sicklebamboo, Nanchuan Drepanostachyum ●

20201 Ampelocalamus melicoideus (P. C. Keng) D. Z. Li et Stapleton = Drepanostachyum melicoideum P. C. Keng ●

20202 Ampelocalamus mianningensis (Q. Li et X. Jiang) D. Z. Li et Stapleton;冕宁悬竹(冕宁慈竹);Mianning Dendrocalamus, Mianning Dragonbamboo ●

20203 Ampelocalamus microphyllus (J. R. Xue et T. P. Yi) J. R. Xue et T. P. Yi = Drepanostachyum microphyllum (J. R. Xue et T. P. Yi) P. C. Keng ex T. P. Yi ●

20204 Ampelocalamus naibunense (Hayata) T. H. Wen = Drepanostachyum naibunense (Hayata) P. C. Keng ●

20205 Ampelocalamus naibunensis (Hayata) T. H. Wen;内门竹(恒春箭竹,恒春青篱竹,恒春矢竹,内份竹);Naibu Bamboo, Naibun Drepanostachyum, Neibun Sicklebamboo ●

20206 Ampelocalamus patellaris (Gamble) Stapleton;碟环竹;Disc-noded Bamboo ●

20207 Ampelocalamus saxatilis (J. R. Xue et T. P. Yi) J. R. Xue et T. P. Yi;羊竹子(南川镰序竹);S. Sichuan Sicklebamboo, Saxicolous Drepanostachyum, Stone Drepanostachyum ●

20208 Ampelocalamus saxatilis (J. R. Xue et T. P. Yi) J. R. Xue et T. P. Yi = Drepanostachyum scandens (J. R. Xue et W. D. Li) P. C. Keng ex T. P. Yi ●

20209 Ampelocalamus scandens J. R. Xue et W. D. Li;爬竹;Climbing Drepanostachyum, Climbing Sicklebamboo ●

20210 Ampelocalamus scandens J. R. Xue et W. D. Li = Drepanostachyum scandens (J. R. Xue et W. D. Li) P. C. Keng ex T. P. Yi ●

20211 Ampelocalamus yongshanensis J. R. Xue et D. Z. Li;永善悬竹;Yongshan Ampelocalamus ●

20212 Ampelocera Klotzsch(1847);藤榆属●☆

20213 Ampelocera hottlei (Standl.) Standl.;蜡藤榆●☆

20214 Ampelocissus Planch. (1884)(保留属名);酸蔹藤属(白粉藤属,九节铃属)●

20215 Ampelocissus abyssinica (Hochst. ex A. Rich.) Planch.;阿比西尼亚酸蔹藤●☆

20216 Ampelocissus acapulcensis (Kunth) Planch.;墨西哥酸蔹藤(墨酸蔹藤)●☆

20217 Ampelocissus aesculifolia Gilg et M. Brandt = Ampelocissus obtusata (Welw. ex Baker) Planch. ●☆

20218 Ampelocissus africana (Lour.) Merr.;非洲酸蔹藤●☆

20219 Ampelocissus angolensis (Baker) Planch.;安哥拉酸蔹藤●☆

20220 Ampelocissus artemisiifolia Planch.;酸蔹藤(艾叶酸蔹藤,大九节铃,牛角天麻,铜皮铁箍)●

20221 Ampelocissus bombycina (Baker) Planch.;丝质酸蔹藤●☆

20222 Ampelocissus brunneo-rubra Gilg;褐红酸蔹藤●☆

20223 Ampelocissus butoensis C. L. Li;四川酸蔹藤●

20224 Ampelocissus cavicaulis (Baker) Planch. = Ampelocissus abyssinica (Hochst. ex A. Rich.) Planch. ●☆

20225 Ampelocissus cinnamochroa Planch. = Ampelocissus bombycina (Baker) Planch. ●☆

20226 Ampelocissus concinna (Baker) Planch.;整洁酸蔹藤●☆

20227 Ampelocissus dekindtiana Gilg;德金酸蔹藤●☆

20228 Ampelocissus dissecta (Baker) Planch.;深裂酸蔹藤●☆

20229 Ampelocissus edulis (De Wild.) Gilg et M. Brandt;可食酸蔹藤●☆

20230 Ampelocissus elephantina Planch.;象酸蔹藤●☆

20231 Ampelocissus elisabethvilleana De Wild. = Ampelocissus obtusata (Welw. ex Baker) Planch. subsp. kirkiana (Planch.) Wild et R. B. Drumm. ●☆

20232 Ampelocissus gourmaensis A. Chev. = Ampelocissus africana (Lour.) Merr. ●☆

20233 Ampelocissus gracilipes Stapf;细梗酸蔹藤●☆

20234 Ampelocissus grantii (Baker) Planch. = Ampelocissus africana (Lour.) Merr. ●☆

20235 Ampelocissus hoabinhensis C. L. Li;红河酸蔹藤●

20236 Ampelocissus imperialis (Miq.) Planch.;王酸蔹藤●☆

20237 Ampelocissus kirkiana Planch. = Ampelocissus obtusata (Welw. ex Baker) Planch. subsp. kirkiana (Planch.) Wild et R. B. Drumm. ●☆

20238 Ampelocissus latifolia (Roxb.) Planch.;宽叶酸蔹藤(宽叶蛇葡萄)●

20239 Ampelocissus latifolia (Roxb.) Planch. = Ampelocissus xizangensis C. L. Li ●

20240 Ampelocissus leonensis (Hook. f.) Planch.;莱昂酸蔹藤●☆

20241 Ampelocissus longicuspis Mildbr. = Ampelocissus multistriata (Baker) Planch. ●☆

20242 Ampelocissus mossambicensis (Klotzsch) Planch. = Ampelocissus africana (Lour.) Merr. ●☆

20243 Ampelocissus multiloba Gilg et M. Brandt;多裂酸蔹藤●☆

20244 Ampelocissus multistriata (Baker) Planch.;多纹酸蔹藤●☆

20245 Ampelocissus obtusata (Welw. ex Baker) Planch.;钝酸蔹藤●☆

20246 Ampelocissus obtusata (Welw. ex Baker) Planch. subsp. kirkiana (Planch.) Wild et R. B. Drumm.;柯克钝酸蔹藤●☆

20247 Ampelocissus pentaphylla (Guillaumin et Perr.) Gilg et M. Brandt = Ampelocissus multistriata (Baker) Planch. ●☆

20248 Ampelocissus poggei Gilg et M. Brandt;波格酸蔹藤●☆

20249 Ampelocissus pulchra Gilg = Ampelocissus obtusata (Welw. ex Baker) Planch. subsp. kirkiana (Planch.) Wild et R. B. Drumm. ●☆

20250 Ampelocissus rhodesica Suess. = Ampelocissus obtusata (Welw. ex Baker) Planch. subsp. kirkiana (Planch.) Wild et R. B. Drumm. ●☆

20251 Ampelocissus rhodotricha Suess. = Cissus rhodotricha (Baker) Desc. ●☆

20252 Ampelocissus sapinii (De Wild.) Gilg et M. Brandt;萨潘酸蔹藤●☆

20253 Ampelocissus sarcantha Gilg et M. Brandt = Ampelocissus multistriata (Baker) Planch. ●☆

20254 Ampelocissus sarcocephala (Schweinf. ex Oliv.) Planch.;肉头酸蔹藤●☆

20255 Ampelocissus schimperiana (Hochst. ex A. Rich.) Planch.;欣珀酸蔹藤●☆

20256 Ampelocissus schliebenii Werderm. = Ampelocissus africana (Lour.) Merr. ●☆

20257 Ampelocissus sikkimensisi (M. A. Lawson) Planch.;锡金酸蔹藤(单叶酸蔹藤)●

20258 Ampelocissus sikkimensisi (M. A. Lawson) Planch. = Leea

macrophylla Roxb. ex Hornem. et Roxb. ●
20259 Ampelocissus venenosa De Wild. = Ampelocissus obtusata (Welw. ex Baker) Planch. subsp. kirkiana (Planch.) Wild et R. B. Drumm. ●☆
20260 Ampelocissus verschuerenii De Wild.;费许伦酸蔹藤●
20261 Ampelocissus volkensii Gilg = Ampelocissus africana (Lour.) Merr. ●☆
20262 Ampelocissus winkleri Lauterb.;温氏酸蔹藤●☆
20263 Ampelocissus xizangensis C. L. Li;西藏酸蔹藤●
20264 Ampelodaphne Meisn. = Endlicheria Nees(保留属名)●☆
20265 Ampelodesma P. Beauv. = Ampelodesmos Link ■☆
20266 Ampelodesma P. Beauv. ex Benth. = Ampelodesmos Link ■☆
20267 Ampelodesmos Link(1827);藤带禾属■☆
20268 Ampelodesmos mauritanicus (Poir.) T. Durand et Schinz;藤带禾;Dis Grass,Diss,Mauritanian Grass ■☆
20269 Ampelodesmos tenax Link = Ampelodesmos mauritanicus (Poir.) T. Durand et Schinz ■☆
20270 Ampelodesmos tenax Link var. bicolor (Poir.) Trab. = Ampelodesmos mauritanicus (Poir.) T. Durand et Schinz ■☆
20271 Ampelodesmos tenax Link var. microstachys Trab. = Ampelodesmos mauritanicus (Poir.) T. Durand et Schinz ■☆
20272 Ampelodesmos tenax Link var. squarrosus Coss. et Durieu = Ampelodesmos mauritanicus (Poir.) T. Durand et Schinz ■☆
20273 Ampelodesmus J. Woods = Ampelodesmos Link ■☆
20274 Ampelodonax Lojac. = Ampelodesmos Link ■☆
20275 Ampeloplis Raf. = Sageretia Brongn. ●
20276 Ampeloplis chinensis Raf. = Sageretia thea (Osbeck) M. C. Johnst. ●
20277 Ampelopsidaceae Kostel.;蛇葡萄科●
20278 Ampelopsidaceae Kostel. = Vitaceae Juss.(保留科名)●■
20279 Ampelopsis Hort. = Parthenocissus Planch.(保留属名)●
20280 Ampelopsis Michx.(1803);蛇葡萄属(白蔹属,山葡萄属);Ampelopsis,Porcelain Vine,Snakegrape ●
20281 Ampelopsis Rich. = Ampelopsis Michx. ●
20282 Ampelopsis acerifolia W. T. Wang;槭叶蛇葡萄;Mapleleaf Ampelopsis,Maple-leaved Ampelopsis ●
20283 Ampelopsis acetosa ?;微酸蛇葡萄●☆
20284 Ampelopsis aconitifolia Bunge;乌头叶蛇葡萄(草白蔹,草白蔹,草葡萄,草血藤,狗葡萄,过山龙,马葡萄,乌头叶白蔹,羊葡萄蔓,洋葡萄);Monkshoodvine,Monkshood-vine ●
20285 Ampelopsis aconitifolia Bunge f. glabra (Diels et Gilg) Kitag. = Ampelopsis aconitifolia Bunge var. glabra Diels et Gilg ●
20286 Ampelopsis aconitifolia Bunge f. glabra (Diels et Gilg) Kitag. = Ampelopsis delavayana (Franch.) Planch. var. glabra (Diels et Gilg) C. L. Li ●
20287 Ampelopsis aconitifolia Bunge var. cuneata Diels et Gilg = Ampelopsis aconitifolia Bunge ●
20288 Ampelopsis aconitifolia Bunge var. dissecta Koehne = Ampelopsis aconitifolia Bunge ●
20289 Ampelopsis aconitifolia Bunge var. glabra Diels et Gilg = Ampelopsis delavayana (Franch.) Planch. var. glabra (Diels et Gilg) C. L. Li ●
20290 Ampelopsis aconitifolia Bunge var. palmiloba (Carrière) Rehder;掌裂草葡萄(三裂草白蔹)●
20291 Ampelopsis aconitifolia Bunge var. palmiloba (Carrière) Rehder = Ampelopsis aconitifolia Bunge var. glabra Diels et Gilg ●
20292 Ampelopsis aconitifolia Bunge var. setulosa Diels et Gilg = Ampelopsis delavayana (Franch.) Planch. var. setulosa (Diels et Gilg) C. L. Li ●
20293 Ampelopsis aconitifolia Bunge var. tomentella Diels et Gilg = Ampelopsis delavayana (Franch.) Planch. var. tomentella (Diels et Gilg) C. L. Li ●
20294 Ampelopsis acutidentata W. T. Wang;尖齿蛇葡萄;Acute-dentate Ampelopsis,Acutidentate Ampelopsis,Acutidentate Snakegrape ●
20295 Ampelopsis aegirophylla Planch. = Ampelopsis vitifolia (Boiss.) Planch. ●☆
20296 Ampelopsis arborea (L.) Koehne;树状蛇葡萄(胡椒藤);Pepper Vine,Peppervine,Pepper-vine ●☆
20297 Ampelopsis bodinieri (H. Lév. et Vaniot) Rehder;蓝果蛇葡萄(扁担藤,大接骨丹,过山龙,蓝果野葡萄,闪光蛇葡萄,上山龙,蛇葡萄);Bluefruit Snakegrape,Bodinier Ampelopsis ●
20298 Ampelopsis bodinieri (H. Lév. et Vaniot) Rehder var. cinerea (Gagnep.) Rehder;灰毛蛇葡萄(毛叶蛇葡萄);Glaucousback Ampelopsis,Glaucousback Snakegrape ●
20299 Ampelopsis brevipedunculata (Maxim.) Trautv. 'Elegans';雅致蛇葡萄;Variegata Porcelain Vine ●☆
20300 Ampelopsis brevipedunculata (Maxim.) Trautv. = Ampelopsis glandulosa (Wall.) Momiy. var. brevipedunculata (Maxim.) Momiy. ●
20301 Ampelopsis brevipedunculata (Maxim.) Trautv. = Ampelopsis glandulosa (Wall.) Momiy. var. heterophylla (Thunb.) Momiy. ●
20302 Ampelopsis brevipedunculata (Maxim.) Trautv. = Ampelopsis heterophylla (Thunb.) Siebold et Zucc. var. brevipedunculata (Regel) C. L. Li ●
20303 Ampelopsis brevipedunculata (Maxim.) Trautv. = Ampelopsis humulifolia Bunge ●
20304 Ampelopsis brevipedunculata (Maxim.) Trautv. f. citrulloides (Lebas) Brieger = Ampelopsis heterophylla (Thunb.) Siebold et Zucc. ●
20305 Ampelopsis brevipedunculata (Maxim.) Trautv. f. glabrifolia (Honda) Kitam. = Ampelopsis brevipedunculata (Maxim.) Trautv. var. hancei (Planch.) Rehder ●
20306 Ampelopsis brevipedunculata (Maxim.) Trautv. f. puberula W. T. Wang;短柔毛蛇葡萄●☆
20307 Ampelopsis brevipedunculata (Maxim.) Trautv. var. ciliata (Nakai) F. Y. Lu;毛山葡萄●
20308 Ampelopsis brevipedunculata (Maxim.) Trautv. var. ciliata (Nakai) F. Y. Lu = Ampelopsis glandulosa (Wall.) Momiy. ●
20309 Ampelopsis brevipedunculata (Maxim.) Trautv. var. ciliata (Nakai) F. Y. Lu = Ampelopsis heterophylla (Thunb.) Siebold et Zucc. var. vestita Rehder ●
20310 Ampelopsis brevipedunculata (Maxim.) Trautv. var. citrulloides? = Ampelopsis heterophylla (Thunb.) Siebold et Zucc. ●
20311 Ampelopsis brevipedunculata (Maxim.) Trautv. var. glabrifolia ? = Ampelopsis brevipedunculata (Maxim.) Trautv. var. hancei (Planch.) Rehder ●
20312 Ampelopsis brevipedunculata (Maxim.) Trautv. var. hancei (Planch.) Rehder = Ampelopsis glandulosa (Wall.) Momiy. var. hancei (Planch.) Momiy. ●
20313 Ampelopsis brevipedunculata (Maxim.) Trautv. var. hancei (Planch.) Rehder = Ampelopsis humulifolia Bunge ●
20314 Ampelopsis brevipedunculata (Maxim.) Trautv. var. hancei Planch. = Ampelopsis glandulosa (Wall.) Momiy. var. hancei (Planch.) Momiy. ●

20315 Ampelopsis brevipedunculata (Maxim.) Trautv. var. heterophylla (Thunb.) H. Hara f. citrulloides (Lebas) Rehder = Ampelopsis heterophylla (Thunb.) Siebold et Zucc. ●

20316 Ampelopsis brevipedunculata (Maxim.) Trautv. var. heterophylla (Thunb.) H. Hara = Ampelopsis glandulosa (Wall.) Momiy. var. heterophylla (Thunb.) Momiy. ●

20317 Ampelopsis brevipedunculata (Maxim.) Trautv. var. heterophylla (Thunb.) H. Hara = Ampelopsis humulifolia Bunge ●

20318 Ampelopsis brevipedunculata (Maxim.) Trautv. var. heterophylla (Thunb.) H. Hara = Ampelopsis heterophylla (Thunb.) Siebold et Zucc. ●

20319 Ampelopsis brevipedunculata (Maxim.) Trautv. var. kulingensis Rehder = Ampelopsis heterophylla (Thunb.) Siebold et Zucc. var. kulingensis (Rehder) C. L. Li ●

20320 Ampelopsis brevipedunculata (Maxim.) Trautv. var. kulingensis Rehder = Ampelopsis glandulosa (Wall.) Momiy. var. kulingensis (Rehder) Momiy. ●

20321 Ampelopsis brevipedunculata (Maxim.) Trautv. var. maximowiczii (Regel) Rehder; 光叶蛇葡萄; Maximowicz Ampelopsis, Maximowicz Snakegrape ●

20322 Ampelopsis brevipedunculata (Maxim.) Trautv. var. maximowiczii (Regel) Rehder = Ampelopsis humulifolia Bunge ●

20323 Ampelopsis brevipedunculata (Maxim.) Trautv. var. maximowiczii (Regel) Rehder = Ampelopsis glandulosa (Wall.) Momiy. var. heterophylla (Thunb.) Momiy. ●

20324 Ampelopsis brevipedunculata (Maxim.) Trautv. var. vestita (Rehder) Rehder = Ampelopsis glandulosa (Wall.) Momiy. ●

20325 Ampelopsis brevipedunculata (Maxim.) Trautv. var. vestita (Rehder) Rehder = Ampelopsis heterophylla (Thunb.) Siebold et Zucc. var. vestita Rehder ●

20326 Ampelopsis cantoniensis (Hook. et Arn.) K. Koch var. grossedentata Hand. -Mazz. = Ampelopsis grossedentata (Hand. -Mazz.) W. T. Wang ●

20327 Ampelopsis cantoniensis (Hook. et Arn.) Planch.; 广东蛇葡萄(白菇茶,白劝须,赤枝山葡萄,广东山葡萄,红血龙无刺根,辣梨茶,有刺根,牛牵丝,山葡萄,蛇葡萄,田蒲茶,无莿根,粤蛇葡萄);Canton Ampelopsis, Canton Snakegrape, Guangdong Ampelopsis ●

20328 Ampelopsis cantoniensis (Hook. et Arn.) Planch. var. grossedentata Hand. -Mazz. = Ampelopsis grossedentata (Hand. -Mazz.) W. T. Wang ●

20329 Ampelopsis cantoniensis (Hook. et Arn.) Planch. var. leecoides (Maxim.) F. Y. Lu;大叶广东山葡萄(土当归叶蛇葡萄)●

20330 Ampelopsis cardiospermoides Planch. ex Franch. = Cayratia cardiospermoides (Planch.) Gagnep. ●

20331 Ampelopsis chaffanjonii (H. Lév. et Vaniot) Rehder; 羽叶蛇葡萄(鱼藤,羽叶牛果藤); Chaffanjon Ampelopsis, Chaffanjon Snakegrape ●

20332 Ampelopsis chinensis Raf. = Sageretia thea (Osbeck) M. C. Johnst. ●

20333 Ampelopsis citrulloides Dippel = Ampelopsis heterophylla (Thunb.) Siebold et Zucc. ●

20334 Ampelopsis cordata Michx.; 心叶蛇葡萄(小叶蛇葡萄); False Grape, Heartleaf Ampelopsis, Heartleaf Peppervine, Heartleaf Snakegrape, Raccoon Grape ●

20335 Ampelopsis delavayana (Franch.) Planch.; 三裂叶蛇葡萄(赤葛,赤木通,大接骨丹,德氏蛇葡萄,耳坠果,飞天蜈蚣,枫藤,红赤葛,红母猪藤,红内消,红十字创粉,见肿消,金刚散,绿葡萄,破石珠,七角藤,枪花药,三裂蛇葡萄,山葡萄,五爪金,五爪龙,野葡萄,玉葡萄); Delavay Ampelopsis, Delavay Snakegrape ●

20336 Ampelopsis delavayana (Franch.) Planch. var. gentiliana (H. Lév. et Vaniot) Hand. -Mazz. = Ampelopsis delavayana (Franch.) Planch. var. setulosa (Diels et Gilg) C. L. Li ●

20337 Ampelopsis delavayana (Franch.) Planch. var. gentiliana (H. Lév. et Vaniot) Hand. -Mazz. = Ampelopsis delavayana Planch. ex Franch. var. setulosa (Diels et Gilg) C. L. Li ●

20338 Ampelopsis delavayana (Franch.) Planch. var. glabra (Diels et Gilg) C. L. Li; 掌裂蛇葡萄(独脚蟾蜍,光叶草葡萄,过山龙,金线吊蛤蟆,石蟾蜍,五爪龙,掌裂草葡萄); Aconiteleaf Ampelopsis, Aconiteleaf Snakegrape, Glabrous Delavay Ampelopsis ●

20339 Ampelopsis delavayana (Franch.) Planch. var. mollissima C. Y. Wu; 柔毛蛇葡萄; Hairy Delavay Ampelopsis ●

20340 Ampelopsis delavayana (Franch.) Planch. var. setulosa (Diels et Gilg) C. L. Li; 毛三裂蛇葡萄(绿葡萄,刚毛蛇葡萄); Gentili's Delavay Ampelopsis, Gentili's Delavay Snakegrape ●

20341 Ampelopsis delavayana (Franch.) Planch. var. tomentella (Diels et Gilg) C. L. Li; 狭叶蛇葡萄; Tomentose Delavay Ampelopsis ●

20342 Ampelopsis delavayana Planch. ex Franch. var. setulosa (Diels et Gilg) C. L. Li = Ampelopsis delavayana (Franch.) Planch. var. setulosa (Diels et Gilg) C. L. Li ●

20343 Ampelopsis glandulosa (Franch.) Planch. var. hancei (Planch.) Momiy. = Ampelopsis sinica (Miq.) W. T. Wang var. hancei (Planch.) W. T. Wang ●

20344 Ampelopsis glandulosa (Franch.) Planch. var. vestita (Rehder) Momiy. = Ampelopsis heterophylla (Thunb.) Siebold et Zucc. var. vestita Rehder ●

20345 Ampelopsis glandulosa (Franch.) Planch. var. vestita (Rehder) Momiy. = Ampelopsis sinica (Miq.) W. T. Wang ●

20346 Ampelopsis glandulosa (Rehder) Momiy. var. kulingensis (Rehder) Momiy. = Ampelopsis glandulosa (Wall.) Momiy. var. kulingensis (Rehder) Momiy. ●

20347 Ampelopsis glandulosa (Wall.) Momiy. = Ampelopsis heterophylla (Thunb.) Siebold et Zucc. var. vestita Rehder ●

20348 Ampelopsis glandulosa (Wall.) Momiy. var. brevipedunculata (Maxim.) Momiy.; 东北蛇葡萄(大叶岩益,狗葡萄,过山龙,禾黄藤,禾稼子藤,假葡萄,见毒消,见肿消,绿葡萄,梦中消,母苦藤,内红消,爬山虎,山刺瓜,山胡烂,山葡萄,山天罗,山天萝,蛇白蔹,蛇葡萄,酸古藤,酸藤,外红消,烟火藤,野葡萄); Amur Ampelopsis, Amur Peppervine, Amur Pepper-vine, Amur Snakegrape, Porcelain Berry, Porcelain Vine, Porcelainberry, Porcelain-berry, Turquoise-berry ●

20349 Ampelopsis glandulosa (Wall.) Momiy. var. brevipedunculata (Maxim.) Momiy. = Ampelopsis brevipedunculata (Maxim.) Trautv. ●

20350 Ampelopsis glandulosa (Wall.) Momiy. var. ciliata (Nakai) Momiy. = Ampelopsis glandulosa (Wall.) Momiy. ●

20351 Ampelopsis glandulosa (Wall.) Momiy. var. glabrifolia (Honda) Momiy. = Ampelopsis sinica (Miq.) W. T. Wang var. hancei (Planch.) W. T. Wang ●

20352 Ampelopsis glandulosa (Wall.) Momiy. var. hancei (Planch.) Momiy.; 小叶蛇葡萄(汉氏山葡萄); Hance Ampelopsis, Hance Snakegrape ●

20353 Ampelopsis glandulosa (Wall.) Momiy. var. hancei (Planch.) Momiy. = Ampelopsis brevipedunculata (Maxim.) Trautv. var. hancei (Planch.) Rehder ●

20354 Ampelopsis glandulosa (Wall.) Momiy. var. heterophylla (Thunb.) Momiy.;异叶蛇葡萄(趴墙虎,趴山虎,爬山虎,七角藤,山葡萄,蛇葡萄,酸藤);Diversifolius Ampelopsis ●

20355 Ampelopsis glandulosa (Wall.) Momiy. var. heterophylla (Thunb.) Momiy. f. citrulloides (Lebas) Momiy. = Ampelopsis heterophylla (Thunb.) Siebold et Zucc. ●

20356 Ampelopsis glandulosa (Wall.) Momiy. var. kulingensis (Rehder) Momiy.;牯岭蛇葡萄(木杠藤,山葡萄,铁骨扇);Guling Ampelopsis,Guling Snakegrape,Kuling Diversifolius Ampelopsis ●

20357 Ampelopsis glandulosa (Wall.) Momiy. var. kulingensis (Rehder) Momiy. = Ampelopsis heterophylla (Thunb.) Siebold et Zucc. var. kulingensis (Rehder) C. L. Li ●

20358 Ampelopsis glandulosa (Wall.) Momiy. var. vestita (Rehder) Momiy. = Ampelopsis glandulosa (Wall.) Momiy. ●

20359 Ampelopsis gongshanensis C. L. Li;贡山蛇葡萄;Gongshan Ampelopsis ●

20360 Ampelopsis grossedentata (Hand. -Mazz.) W. T. Wang;显齿蛇葡萄(大齿蛇葡萄,红五爪金龙,苦茶,藤茶,田婆茶,甜茶藤,乌蔹);Bigdentate Ampelopsis,Big-dentate Ampelopsis,Bigdentate Snakegrape,Grossedentate Ampelopsis ●

20361 Ampelopsis hederacea DC.;美洲蛇葡萄;American Ivy,False Grape,Five-leaved Ivy,Red Clematis,Red Ivy,Woodbine ●☆

20362 Ampelopsis hederacea DC. = Parthenocissus quinquefolius (L.) Planch. ●

20363 Ampelopsis henryana (Hemsl.) Grignani = Parthenocissus henryana (Hemsl.) Diels et Gilg ●

20364 Ampelopsis henryana Rehder = Parthenocissus henryana (Hemsl.) Diels et Gilg ●

20365 Ampelopsis henryana Schele = Parthenocissus henryana (Hemsl.) Diels et Gilg ●

20366 Ampelopsis heterophylla (Thunb.) Siebold et Zucc. = Ampelopsis glandulosa (Wall.) Momiy. var. brevipedunculata (Maxim.) Momiy. ●

20367 Ampelopsis heterophylla (Thunb.) Siebold et Zucc. = Ampelopsis glandulosa (Wall.) Momiy. var. heterophylla (Thunb.) Momiy. ●

20368 Ampelopsis heterophylla (Thunb.) Siebold et Zucc. = Ampelopsis humulifolia Bunge ●

20369 Ampelopsis heterophylla (Thunb.) Siebold et Zucc. f. elegans ? = Ampelopsis heterophylla (Thunb.) Siebold et Zucc. ●

20370 Ampelopsis heterophylla (Thunb.) Siebold et Zucc. subvar. wallichii Planch. = Ampelopsis glandulosa (Wall.) Momiy. ●

20371 Ampelopsis heterophylla (Thunb.) Siebold et Zucc. var. amurensis Planch. = Ampelopsis brevipedunculata (Maxim.) Trautv. ●

20372 Ampelopsis heterophylla (Thunb.) Siebold et Zucc. var. amurensis Planch. = Ampelopsis glandulosa (Wall.) Momiy. var. brevipedunculata (Maxim.) Momiy. ●

20373 Ampelopsis heterophylla (Thunb.) Siebold et Zucc. var. brevipedunculata (Maxim.) C. L. Li = Ampelopsis brevipedunculata (Maxim.) Trautv. ●

20374 Ampelopsis heterophylla (Thunb.) Siebold et Zucc. var. brevipedunculata (Regel) C. L. Li = Ampelopsis brevipedunculata (Maxim.) Trautv. ●

20375 Ampelopsis heterophylla (Thunb.) Siebold et Zucc. var. brevipedunculata (Maxim.) C. L. Li = Ampelopsis glandulosa (Wall.) Momiy. var. brevipedunculata (Maxim.) Momiy. ●

20376 Ampelopsis heterophylla (Thunb.) Siebold et Zucc. var. bungei Planch. = Ampelopsis humulifolia Bunge ●

20377 Ampelopsis heterophylla (Thunb.) Siebold et Zucc. var. ciliata Nakai = Ampelopsis glandulosa (Wall.) Momiy. ●

20378 Ampelopsis heterophylla (Thunb.) Siebold et Zucc. var. ciliata Nakai = Ampelopsis heterophylla (Thunb.) Siebold et Zucc. var. vestita Rehder ●

20379 Ampelopsis heterophylla (Thunb.) Siebold et Zucc. var. cinerea Gagnep. = Ampelopsis bodinieri (H. Lév. et Vaniot) Rehder var. cinerea (Gagnep.) Rehder ●

20380 Ampelopsis heterophylla (Thunb.) Siebold et Zucc. var. citrulloides ? = Ampelopsis heterophylla (Thunb.) Siebold et Zucc. ●

20381 Ampelopsis heterophylla (Thunb.) Siebold et Zucc. var. delavayana (Planch.) Gagnep. = Ampelopsis delavayana (Franch.) Planch. ●

20382 Ampelopsis heterophylla (Thunb.) Siebold et Zucc. var. delavayana (Planch. ex Franch.) Gagnep. = Ampelopsis delavayana (Franch.) Planch. ●

20383 Ampelopsis heterophylla (Thunb.) Siebold et Zucc. var. gentiliana (H. Lév. et Vaniot) Gagnep. = Ampelopsis delavayana (Franch.) Planch. var. gentiliana (H. Lév. et Vaniot) Hand. -Mazz. ●

20384 Ampelopsis heterophylla (Thunb.) Siebold et Zucc. var. gentiliana (H. Lév. et Vaniot) Gagnep. = Ampelopsis delavayana Planch. ex Franch. var. setulosa (Diels et Gilg) C. L. Li ●

20385 Ampelopsis heterophylla (Thunb.) Siebold et Zucc. var. gentiliana (H. Lév. et Vaniot) Gagnep. = Ampelopsis delavayana (Franch.) Planch. var. setulosa (Diels et Gilg) C. L. Li ●

20386 Ampelopsis heterophylla (Thunb.) Siebold et Zucc. var. hancei Planch. = Ampelopsis sinica (Miq.) W. T. Wang var. hancei (Planch.) W. T. Wang ●

20387 Ampelopsis heterophylla (Thunb.) Siebold et Zucc. var. hancei Planch. subvar. walichii Planch. = Ampelopsis heterophylla (Thunb.) Siebold et Zucc. var. vestita Rehder ●

20388 Ampelopsis heterophylla (Thunb.) Siebold et Zucc. var. hancei Planch. = Ampelopsis glandulosa (Wall.) Momiy. ●

20389 Ampelopsis heterophylla (Thunb.) Siebold et Zucc. var. hancei Planch. = Ampelopsis glandulosa (Wall.) Momiy. var. hancei (Planch.) Momiy. ●

20390 Ampelopsis heterophylla (Thunb.) Siebold et Zucc. var. kulingensis (Rehder) C. L. Li = Ampelopsis glandulosa (Wall.) Momiy. var. kulingensis (Rehder) Momiy. ●

20391 Ampelopsis heterophylla (Thunb.) Siebold et Zucc. var. sinica (Miq.) Merr. = Ampelopsis sinica (Miq.) W. T. Wang ●

20392 Ampelopsis heterophylla (Thunb.) Siebold et Zucc. var. sinica (Miq.) Merr. = Ampelopsis glandulosa (Wall.) Momiy. ●

20393 Ampelopsis heterophylla (Thunb.) Siebold et Zucc. var. sinica (Miq.) Merr. = Ampelopsis heterophylla (Thunb.) Siebold et Zucc. var. hancei Planch. ●

20394 Ampelopsis heterophylla (Thunb.) Siebold et Zucc. var. vestita Rehder;锈毛蛇葡萄;Rustyhair Diversifolius Ampelopsis ●

20395 Ampelopsis heterophylla (Thunb.) Siebold et Zucc. var. vestita Rehder = Ampelopsis glandulosa (Wall.) Momiy. ●

20396 Ampelopsis heterophylla (Thunb.) Siebold et Zucc. var. vestita Rehder = Ampelopsis brevipedunculata (Maxim.) Trautv. ●

20397 Ampelopsis heterophylla (Thunb.) Siebold et Zucc. var. vestita Rehder = Ampelopsis sinica (Miq.) W. T. Wang ●

20398 Ampelopsis himalayana Royle = Parthenocissus semicordata (Wall.) Planch. var. roylei (King ex Parker) Nazim et Qaiser ●☆

20399 Ampelopsis himalayana Royle = Parthenocissus semicordata (Wall.) Planch. ●

20400 Ampelopsis humulifolia Bunge;葎叶蛇葡萄(活血丹,葎草叶蛇葡萄,葎叶白蔹,葎叶山葡萄,七角白蔹,小接骨丹);Hopleaf Ampelopsis,Hopleaf Snakegrape,Hop-leaved Ampelopsis ●

20401 Ampelopsis humulifolia Bunge = Ampelopsis heterophylla (Thunb.) Siebold et Zucc. ●

20402 Ampelopsis humulifolia Bunge f. trisecta (Nakai) Kitag.;三叶白蔹●

20403 Ampelopsis humulifolia Bunge var. heterophylla (Thunb.) K. Koch = Ampelopsis glandulosa (Wall.) Momiy. var. heterophylla (Thunb.) Momiy. ●

20404 Ampelopsis humulifolia Bunge var. heterophylla (Thunb.) K. Koch = Ampelopsis heterophylla (Thunb.) Siebold et Zucc. ●

20405 Ampelopsis humulifolia Bunge var. trisecta Nakai = Ampelopsis humulifolia Bunge f. trisecta (Nakai) Kitag. ●

20406 Ampelopsis hypoglauca (Hance) C. L. Li;粉叶蛇葡萄●

20407 Ampelopsis inconstans ? = Parthenocissus tricuspidatus (Siebold et Zucc.) Planch. ●

20408 Ampelopsis japonica (Thunb.) Makino;白蔹(八卦牛,白草,白根,白浆罐,白敛,白水罐,白天天秧,赤蔹,穿山老鼠,穿山鼠,鹅抱蛋,二无言,狗天天,旱黄钳,见肿消,箭猪腰,镜草,九牛力,昆仑,癞痢茶,老鼠瓜薯,蔹,猫儿卵,母鸡带仔,七角莲,山地瓜,山栗子,山葡萄,山葡萄秧,上竹龙,铁老鼠,菟核,五福篱,五爪藤,五爪叶,小老鸹眼,野红薯,野葡萄秧);Japanese Ampelopsis,Japanese Snakegrape ●

20409 Ampelopsis japonica (Thunb.) Makino = Parthenocissus tricuspidatus (Siebold et Zucc.) Planch. ●

20410 Ampelopsis jiangxiensis W. T. Wang = Ampelopsis megalophylla Diels et Gilg var. jiangxiensis (W. T. Wang) C. L. Li ●

20411 Ampelopsis latifolia Tausch = Ampelocissus latifolia (Roxb.) Planch. ●

20412 Ampelopsis leeoides (Maxim.) Planch. = Ampelopsis cantoniensis (Hook. et Arn.) Planch. var. leeoides (Maxim.) F. Y. Lu ●

20413 Ampelopsis loureiroi Mazel ex Planch.;楼氏蛇葡萄●☆

20414 Ampelopsis lowii L. Cook;娄氏蛇葡萄●☆

20415 Ampelopsis lucida Carrière;光亮蛇葡萄●☆

20416 Ampelopsis macrophylla Blume ex Planch. = Vitis thyrsiflora Miq. ●☆

20417 Ampelopsis macrophylla Dippel = Vitis hederacea Ehrh. ●

20418 Ampelopsis macrophylla Rehder = Parthenocissus vitacea (Knerr) Hitchc. ●☆

20419 Ampelopsis major Vilm.;大蛇葡萄●☆

20420 Ampelopsis megalophylla Diels et Gilg;大叶蛇葡萄(藤茶,大叶牛果藤);Big-leaved Ampelopsis, Largeleaf Snakegrape, Spikenard Ampelopsis ●

20421 Ampelopsis megalophylla Diels et Gilg var. jiangxiensis (W. T. Wang) C. L. Li;柔毛大叶蛇葡萄(江西蛇葡萄);Jiangxi Ampelopsis ●

20422 Ampelopsis megalophylla Diels et Gilg var. puberula W. T. Wang = Ampelopsis rubifolia (Wall.) Planch. ●

20423 Ampelopsis micans Rehder = Ampelopsis bodinieri (H. Lév. et Vaniot) Rehder ●

20424 Ampelopsis micans Rehder var. cinerea (Gagnep.) Rehder = Ampelopsis bodinieri (H. Lév. et Vaniot) Rehder var. cinerea (Gagnep.) Rehder ●

20425 Ampelopsis micans var. cinerea (Gagnep.) Rehder = Ampelopsis bodinieri (H. Lév. et Vaniot) Rehder var. cinerea (Gagnep.) Rehder ●

20426 Ampelopsis mirabilis Diels et Gilg;爬墙风●

20427 Ampelopsis mirabilis Diels et Gilg = Ampelopsis japonica (Thunb.) Makino ●

20428 Ampelopsis mollifolia W. T. Wang;毛叶蛇葡萄;Hairy-leaf Ampelopsis,Hairy-leaved Ampelopsis ●

20429 Ampelopsis napiformis Carrière = Ampelopsis japonica (Thunb.) Makino ●

20430 Ampelopsis orientalis (Lam.) Planch.;东方蛇葡萄;Oriental Ampelopsis ●☆

20431 Ampelopsis palmiloba Carrière = Ampelopsis aconitifolia Bunge var. glabra Diels et Gilg ●

20432 Ampelopsis palmiloba Carrière = Ampelopsis aconitifolia Bunge var. palmiloba (Carrière) Rehder ●

20433 Ampelopsis quinquefolia (L.) Michx. = Parthenocissus quinquefolius (L.) Planch. ●

20434 Ampelopsis quinquefolia (L.) Planch. var. vitacea Knerr = Parthenocissus vitacea (Knerr) Hitchc. ●☆

20435 Ampelopsis quinquefolia Michx.;五叶蛇葡萄;Bunch-of-grapes, Creeping Jenny,Hands-in-pockets ●☆

20436 Ampelopsis quinquefolia Michx. = Parthenocissus quinquefolia (L.) Planch. ●

20437 Ampelopsis quinquefolia Michx. var. angustifolia Dippel;窄五叶蛇葡萄●☆

20438 Ampelopsis quinquefolia Michx. var. hirsuta (Pursh) Torr. et A. Gray;毛五叶蛇葡萄●☆

20439 Ampelopsis quinquefolia Michx. var. latifolia Dippel;宽五叶蛇葡萄●☆

20440 Ampelopsis regeliana Carrière = Ampelopsis glandulosa (Wall.) Momiy. var. heterophylla (Thunb.) Momiy. ●

20441 Ampelopsis regeliana Carrière = Ampelopsis heterophylla (Thunb.) Siebold et Zucc. ●

20442 Ampelopsis regeliana Dippel = Ampelopsis heterophylla (Thunb.) Siebold et Zucc. ●

20443 Ampelopsis roylei Dippel = Parthenocissus tricuspidatus (Siebold et Zucc.) Planch. ●

20444 Ampelopsis rubifolia (Wall.) Planch.;毛枝蛇葡萄(茶藨叶牛果藤,红叶蛇葡萄);Puberulate Ampelopsis,Red-leaf Ampelopsis ●

20445 Ampelopsis sempervirens Hort. = Cissus striata Ruiz et Pav. ●☆

20446 Ampelopsis serianiifolia Bunge = Ampelopsis japonica (Thunb.) Makino ●

20447 Ampelopsis sinica (Miq.) W. T. Wang;蛇葡萄(见毒消,见肿消,母猪藤,山葡萄,山天罗,山天萝,蛇白蔹,酸藤,烟火藤,野葡萄);China Snakegrape,Chinese Ampelopsis ●

20448 Ampelopsis sinica (Miq.) W. T. Wang = Ampelopsis glandulosa (Wall.) Momiy. ●

20449 Ampelopsis sinica (Miq.) W. T. Wang = Ampelopsis heterophylla (Thunb.) Siebold et Zucc. var. vestita Rehder ●

20450 Ampelopsis sinica (Miq.) W. T. Wang var. hancei (Planch.) W. T. Wang = Ampelopsis brevipedunculata (Maxim.) Trautv. var. maximowiczii (Regel) Rehder ●

20451 Ampelopsis sinica (Miq.) W. T. Wang var. hancei (Planch.) W. T. Wang = Ampelopsis glandulosa (Wall.) Momiy. var. hancei

(Planch.) Momiy. ●

20452 Ampelopsis sinica (Miq.) W. T. Wang var. hancei (Planch.) W. T. Wang = Ampelopsis heterophylla (Thunb.) Siebold et Zucc. var. hancei Planch. ●

20453 Ampelopsis tomentosa Planch. ex Franch.;绒毛蛇葡萄;Tomentose Ampelosis ●

20454 Ampelopsis tomentosa Planch. ex Franch. var. glabrescens C. L. Li;脱绒蛇葡萄;Glabrescent Ampelopsis ●

20455 Ampelopsis tricuspidata Siebold et Zucc.;波士顿蛇葡萄;Boston Ivy,Japanese Ivy ●☆

20456 Ampelopsis tricuspidata Siebold et Zucc. = Parthenocissus tricuspidatus (Siebold et Zucc.) Planch. ●

20457 Ampelopsis tricuspidata Siebold et Zucc. f. veitchii ? = Parthenocissus tricuspidatus (Siebold et Zucc.) Planch. ●

20458 Ampelopsis tuberosa Carrière = Ampelopsis japonica (Thunb.) Makino ●

20459 Ampelopsis veitchii Hort. = Ampelopsis tricuspidata Siebold et Zucc. ●☆

20460 Ampelopsis virginiana Dippel;弗吉尼亚蛇葡萄●☆

20461 Ampelopsis vitifolia (Boiss.) Planch.;葡萄叶蛇葡萄(光叶蛇葡萄)●☆

20462 Ampelopsis watsoniana E. H. Wilson = Ampelopsis chaffanjonii (H. Lév. et Vaniot) Rehder ●

20463 Ampelosicios Thouars = Ampelosicyos Thouars ●■☆

20464 Ampelosicyos Thouars(1808);马岛葫芦属●■☆

20465 Ampelosicyos humblotii (Cogn.) Jum. et H. Perrier;洪布马岛葫芦●☆

20466 Ampelosicyos meridionalis Rabenant.;南方马岛葫芦●☆

20467 Ampelosicyos scandens Thouars;马岛葫芦●■☆

20468 Ampelothamnus Small = Pieris D. Don ●

20469 Ampelovitis Carrière = Vitis L. ●

20470 Ampelovitis davidii (Rom. Caill.) Carrière = Vitis davidii (Rom. Caill.) Foëx ●

20471 Ampelovitis romanetii (Rom. Caill.) Carrière = Vitis romanetii Rom. Caill. ●

20472 Ampelozizyphus Ducke(1935);巴西枣属(巴西鼠李属)●☆

20473 Ampelozizyphus amazonicus Ducke et Ducke;巴西枣●☆

20474 Ampelygonum Lindl. = Polygonum L.(保留属名)●■

20475 Ampelygonum chinense (L.) Lindl. = Polygonum chinense L. ■

20476 Ampelygonum malaicum (Danser) M. A. Hassan = Polygonum chinense L. var. ovalifolium Meisn. ■

20477 Ampelygonum molle (D. Don) Roberty et Vautier = Polygonum molle D. Don ●■

20478 Ampelygonum perfoliatum (L.) Roberty et Vautier = Polygonum perfoliatum L. ■

20479 Amperea A. Juss.(1824);澳洲大戟属■☆

20480 Amperea xiphoclada (Spreng.) Druce;澳洲大戟;Broom Spurge ■☆

20481 Amphania Banks ex DC. = Ternstroemia Mutis ex L. f.(保留属名)●

20482 Ampherephis Kunth = Centratherum Cass. ■☆

20483 Amphiachyris (A. DC.) Nutt.(1840);短冠帚黄花属■☆

20484 Amphiachyris Nutt. = Amphiachyris (A. DC.) Nutt. ■☆

20485 Amphiachyris amoena (Shinners) Solbrig;得州短冠帚黄花;Texas Broomweed ■☆

20486 Amphiachyris dracunculoides (DC.) Nutt.;草原短冠帚黄花;Broom Snakeroot,Broomweed,Prairie Broomweed ■☆

20487 Amphianthus Torr.(1837);岩地婆婆纳属■☆

20488 Amphianthus Torr. = Bacopa Aubl.(保留属名)■

20489 Amphianthus pusillus Torr.;岩地婆婆纳■☆

20490 Amphiasma Bremek.(1952);西南非茜草属■☆

20491 Amphiasma assimile Bremek. = Amphiasma luzuloides (K. Schum.) Bremek. ■☆

20492 Amphiasma benguellense (Hiern) Bremek.;本格拉西南非茜草■☆

20493 Amphiasma divaricatum (Engl.) Bremek.;叉开西南非茜草■☆

20494 Amphiasma gracilicaule Verdc. = Pentanopsis gracilicaulis (Verdc.) Thulin et B. Bremer ■☆

20495 Amphiasma luzuloides (K. Schum.) Bremek.;西南非茜草■☆

20496 Amphiasma micranthum (Chiov.) Bremek.;小花西南非茜草■☆

20497 Amphiasma redheadii Bremek.;雷德黑德西南非茜草■☆

20498 Amphibecis Schrank = Centratherum Cass. ■☆

20499 Amphiblemma Naudin(1850);热非野牡丹属■☆

20500 Amphiblemma acaule Cogn. = Cincinnobotrys acaulis (Cogn.) Gilg ■☆

20501 Amphiblemma acaule Cogn. var. brevipes Brenan = Cincinnobotrys acaulis (Cogn.) Gilg ■☆

20502 Amphiblemma amoenum Jacq. -Fél.;秀丽热非野牡丹■☆

20503 Amphiblemma ciliatum Cogn.;睫毛热非野牡丹■☆

20504 Amphiblemma cuneatum Jacq. -Fél.;楔形热非野牡丹■☆

20505 Amphiblemma cymosum (Schrad. et J. C. Wendl.) Naudin;聚伞热非野牡丹●☆

20506 Amphiblemma erythropodum Gilg et Ledermann ex Engl. = Amphiblemma molle Hook. f. ■☆

20507 Amphiblemma gossweileri Exell;戈斯热非野牡丹■☆

20508 Amphiblemma gossweileri Exell var. humifusum Jacq. -Fél.;平伏热非野牡丹■☆

20509 Amphiblemma grandifolium A. Chev. = Dicellandra barteri Hook. f. ●☆

20510 Amphiblemma hallei Jacq. -Fél.;哈勒热非野牡丹■☆

20511 Amphiblemma heterophyllum Jacq. -Fél.;互叶热非野牡丹■☆

20512 Amphiblemma lanceatum Jacq. -Fél.;剑形热非野牡丹■☆

20513 Amphiblemma lateriflorum Cogn. = Amphiblemma ciliatum Cogn. ■☆

20514 Amphiblemma letouzeyi Jacq. -Fél.;勒图热非野牡丹■☆

20515 Amphiblemma ludwigii Gilg ex Engl. = Calvoa hirsuta Hook. f. ■☆

20516 Amphiblemma mildbraedii Gilg ex Engl.;米尔德热非野牡丹■☆

20517 Amphiblemma molle Hook. f.;绢毛热非野牡丹■☆

20518 Amphiblemma monticole Jacq. -Fél.;山地热非野牡丹■☆

20519 Amphiblemma polyanthum Gilg = Amphiblemma mildbraedii Gilg ex Engl. ■☆

20520 Amphiblemma riparium Gilg = Amphiblemma molle Hook. f. ■☆

20521 Amphiblemma seretii (De Wild.) Brenan = Cincinnobotrys acaulis (Cogn.) Gilg ■☆

20522 Amphiblemma setosum Hook. f.;刚毛热非野牡丹■☆

20523 Amphiblemma soyauxii Cogn.;索亚热非野牡丹■☆

20524 Amphiblemma wildemanianum Cogn. = Amphiblemma ciliatum Cogn. ■☆

20525 Amphibolia L. Bolus = Amphibolia L. Bolus ex A. G. J. Herre ●☆

20526 Amphibolia L. Bolus = Eberlanzia Schwantes ●☆

20527 Amphibolia L. Bolus ex A. G. J. Herre(1971);双星番杏属●☆

20528 Amphibolia gydouwensis (L. Bolus) L. Bolus = Phiambolia incumbens (L. Bolus) Klak ■☆

20529 Amphibolia gydouwensis (L. Bolus) L. Bolus ex Toelken et

Jessop = Phiambolia incumbens (L. Bolus) Klak ■☆
20530 Amphibolia hallii (L. Bolus) L. Bolus = Phiambolia hallii (L. Bolus) Klak ■☆
20531 Amphibolia hallii (L. Bolus) L. Bolus ex Toelken et Jessop = Phiambolia hallii (L. Bolus) Klak ■☆
20532 Amphibolia hutchinsonii (L. Bolus) H. E. K. Hartmann = Amphibolia laevis (Aiton) H. E. K. Hartmann ●☆
20533 Amphibolia laevis (Aiton) H. E. K. Hartmann;平滑双星番杏●☆
20534 Amphibolia littlewoodii (L. Bolus) L. Bolus ex Toelken et Jessop = Lampranthus mutatus (G. D. Rowley) H. E. K. Hartmann ■☆
20535 Amphibolia maritima L. Bolus ex Toelken et Jessop = Amphibolia laevis (Aiton) H. E. K. Hartmann ●☆
20536 Amphibolia obscura H. E. K. Hartmann;隐匿双星番杏●☆
20537 Amphibolia rupis-arcuatae (Dinter) H. E. K. Hartmann;岩石双星番杏●☆
20538 Amphibolia saginata (L. Bolus) H. E. K. Hartmann;肥大双星番杏●☆
20539 Amphibolia stayneri L. Bolus ex Toelken et Jessop = Phiambolia stayneri (L. Bolus ex Toelken et Jessop) Klak ●☆
20540 Amphibolia succulenta (L. Bolus) H. E. K. Hartmann;多汁双星番杏●☆
20541 Amphibolis C. Agardh(1823);双星丝粉藻属■☆
20542 Amphibolis Schott et Kotschy = Hyacinthus L. ■☆
20543 Amphibolis bicornis C. Agardh;双星丝粉藻■☆
20544 Amphibolis ciliata Moldenke = Thalassodendron ciliatum (Forssk.) Hartog ■☆
20545 Amphibologyne Brand(1931);隐柱紫草属●☆
20546 Amphibologyne mexicana (Mart. et Galeotti) Brand;隐柱紫草●☆
20547 Amphibromus Nees = Helictotrichon Besser ex Schult. et Schult. f. ■
20548 Amphibromus Nees(1843);湿雀麦属(湿燕麦属)■☆
20549 Amphibromus neesii Steud.;澳洲湿雀麦(澳洲异雀麦);Australian Wallaby Grass,Swamp Wallaby-grass ■☆
20550 Amphibromus neesii Steud. = Helictotrichon neesii (Steud.) Stace ■☆
20551 Amphibromus scabrivalvis (Trin.) Swallen;湿雀麦(湿地异雀麦);Swamp Wallaby Grass ■☆
20552 Amphicalea (DC.) Gardner = Geissopappus Benth. ●■☆
20553 Amphicalea Gardner = Geissopappus Benth. ●■☆
20554 Amphicalyx Blume = Diplycosia Blume ●☆
20555 Amphicarpa Elliott = Amphicarpaea Elliott ex Nutt.(保留属名)■
20556 Amphicarpa Elliott ex Nutt. = Amphicarpaea Elliott ex Nutt.(保留属名)■
20557 Amphicarpaea Elliott = Amphicarpaea Elliott ex Nutt.(保留属名)■
20558 Amphicarpaea Elliott ex Nutt.(1818)(保留属名);两型豆属(野毛扁豆属);Amphicarpaea,Biformbean,Hogpeanut ■
20559 Amphicarpaea africana (Hook. f.) Harms;非洲两型豆■☆
20560 Amphicarpaea bracteata (L.) Fernald;野毛扁豆;American Hog-peanut,Ground Bean,Hog Peanut,Hog-peanut ■
20561 Amphicarpaea bracteata (L.) Fernald subsp. edgeworthii (Benth.) H. Ohashi = Amphicarpaea edgeworthii Benth. ■
20562 Amphicarpaea bracteata (L.) Fernald subsp. edgeworthii (Benth.) H. Ohashi var. japonica (Oliv.) H. Ohashi = Amphicarpaea japonica (Oliv.) B. Fedtsch. ■☆
20563 Amphicarpaea bracteata (L.) Fernald var. comosa (L.) Fernald = Amphicarpaea bracteata (L.) Fernald ■
20564 Amphicarpaea bracteata (L.) Fernald var. pitcheri (Torr. et A. Gray) Fassett = Amphicarpaea bracteata (L.) Fernald ■
20565 Amphicarpaea comosa (L.) G. Don = Amphicarpaea bracteata (L.) Fernald ■
20566 Amphicarpaea edgeworthii Benth.;两型豆(具苞两型豆,三籽两型豆,野扁豆,野毛扁豆,阴阳豆);Edgeworth Amphicarpaea,Edgeworth Biformbean ■
20567 Amphicarpaea edgeworthii Benth. var. japonica Oliv. = Amphicarpaea bracteata (L.) Fernald subsp. edgeworthii (Benth.) H. Ohashi var. japonica (Oliv.) H. Ohashi ■☆
20568 Amphicarpaea edgeworthii Benth. var. rufescens Franch. = Amphicarpaea rufescens (Franch.) Y. T. Wei et S. K. Lee ■
20569 Amphicarpaea edgeworthii Benth. var. trisperma (Miq.) Ohwi = Amphicarpaea bracteata (L.) Fernald subsp. edgeworthii (Benth.) H. Ohashi var. japonica (Oliv.) H. Ohashi ■☆
20570 Amphicarpaea japonica (Oliv.) B. Fedtsch.;日本两型豆■☆
20571 Amphicarpaea japonica (Oliv.) B. Fedtsch. = Amphicarpaea bracteata (L.) Fernald subsp. edgeworthii (Benth.) H. Ohashi var. japonica (Oliv.) H. Ohashi ■☆
20572 Amphicarpaea linearis Chun et F. H. Chen;线苞两型豆(线苞异型豆);Linear Biformbean ■
20573 Amphicarpaea monoica (L.) Elliott = Amphicarpaea bracteata (L.) Fernald ■
20574 Amphicarpaea monoica Elliott = Amphicarpaea bracteata (L.) Fernald ■
20575 Amphicarpaea pitcheri Torr. et A. Gray = Amphicarpaea bracteata (L.) Fernald ■
20576 Amphicarpaea rufescens (Franch.) Y. T. Wei et S. K. Lee;锈毛两型豆;Rusthair Biformbean ■
20577 Amphicarpaea trisperma (Miq.) Baker ex Jacks. = Amphicarpaea edgeworthii Benth. ■
20578 Amphicarpon Raf. = Amphicarpum Kunth ■☆
20579 Amphicarpum Kunth = Amphicarpon Raf. ■☆
20580 Amphicarpum Kunth(1829);双果雀稗属■☆
20581 Amphicarpum floridanum Chapm.;佛罗里达双果雀稗■☆
20582 Amphicarpum muehlenbergianum (Schult.) Hitchc.;双果雀稗■☆
20583 Amphicarpum muehlenbergianum Hitchc. = Amphicarpum muehlenbergianum (Schult.) Hitchc. ■☆
20584 Amphicarpum purshii Kunth;珀什双果雀稗■☆
20585 Amphicome (R. Br.) Royle ex G. Don = Incarvillea Juss. ■
20586 Amphicome (R. Br.) Royle ex Lindl. = Incarvillea Juss. ■
20587 Amphicome (Royle) G. Don f. = Incarvillea Juss. ■
20588 Amphicome Royle = Incarvillea Juss. ■
20589 Amphicome Royle ex Lindl. = Incarvillea Juss. ■
20590 Amphicome Royle(1835);两头毛属(毛子草属)■
20591 Amphicome arguta Royle = Incarvillea arguta (Royle) Royle ■
20592 Amphicome emodi Royle ex Lindl. = Incarvillea emodi (Royle ex Lindl.) Chatterjee ■☆
20593 Amphicome emodii Royle ex Lindl.;喜山两头毛(喜山毛子草)■☆
20594 Amphidasya Standl.(1936);周毛茜属●☆
20595 Amphidasya ambigua (Standl.) Standl.;周毛茜●☆
20596 Amphidasya brevidentata C. M. Taylor;短齿周毛茜●☆
20597 Amphidasya elegans C. M. Taylor;雅致周毛茜●☆
20598 Amphidasya intermedia Steyerm.;间型周毛茜●☆
20599 Amphidasya longicalycina (Dwyer) C. M. Taylor;长萼周毛茜●☆
20600 Amphiderris (R. Br.) Spach = Orites R. Br. ●☆

20601 Amphiderris Spach = Orites R. Br. ●☆
20602 Amphidetes E. Fourn. (1885);巴西萝藦属●☆
20603 Amphidetes laciniatus E. Fourn.;条裂巴西萝藦●☆
20604 Amphidetes quinquedentatus E. Fourn.;巴西萝藦●☆
20605 Amphidonax Nees = Arundo L. + Zenkeria Triu. ■☆
20606 Amphidonax Nees = Arundo L. ●
20607 Amphidonax Nees ex Steud. = Amphidonax Nees ●
20608 Amphidonax bengalensis (Retz.) Nees ex Steud. = Arundo donax L. ●
20609 Amphidoxa DC. = Gnaphalium L. ■
20610 Amphidoxa adscendens O. Hoffm. ex Zahlbr. = Gnaphalium griquense Hilliard et B. L. Burtt ■☆
20611 Amphidoxa engleriana O. Hoffm. = Gnaphalium englerianum (O. Hoffm.) Hilliard et B. L. Burtt ■☆
20612 Amphidoxa filaginea Ficalho et Hiern = Gnaphalium filagopsis Hilliard et B. L. Burtt ■☆
20613 Amphidoxa filaginea Ficalho et Hiern var. transiens Merxm. = Gnaphalium filagopsis Hilliard et B. L. Burtt ■☆
20614 Amphidoxa glandulosa Klatt = Denekia capensis Thunb. ●☆
20615 Amphidoxa gnaphalodes DC. = Gnaphalium gnaphalodes (DC.) Hilliard et B. L. Burtt ■☆
20616 Amphidoxa lasiocephala O. Hoffm. = Artemisiopsis villosa (O. Hoffm.) Schweick. ■☆
20617 Amphidoxa villosa O. Hoffm. = Artemisiopsis villosa (O. Hoffm.) Schweick. ■☆
20618 Amphiestes S. Moore = Hypoestes Sol. ex R. Br. ●■
20619 Amphiestes glandulosa S. Moore = Hypoestes glandulosa (S. Moore) Benoist ●☆
20620 Amphigena Rolfe = Disa P. J. Bergius ■☆
20621 Amphigena leptostachys (Sond.) Rolfe = Disa tenuis Lindl. ■☆
20622 Amphigena tenuis (Lindl.) Rolfe = Disa tenuis Lindl. ■☆
20623 Amphigenes Janka = Festuca L. ■
20624 Amphiglossa DC. (1838);叶苞帚鼠麹属■☆
20625 Amphiglossa alopecuroides Sch. Bip. = Stoebe alopecuroides (Lam.) Less. ●☆
20626 Amphiglossa callunoides DC.;叶苞帚鼠麹■☆
20627 Amphiglossa grisea Koekemoer;灰叶苞帚鼠麹■☆
20628 Amphiglossa nitidula DC. = Amphiglossa tomentosa (Thunb.) Harv. ■☆
20629 Amphiglossa perotrichoides DC.;拟帚鼠麹■☆
20630 Amphiglossa rudolphii Koekemoer;鲁道夫叶苞帚鼠麹■☆
20631 Amphiglossa susannae Koekemoer;苏珊娜叶苞帚鼠麹■☆
20632 Amphiglossa tecta (Brusse) Koekemoer;屋顶叶苞帚鼠麹■☆
20633 Amphiglossa tomentosa (Thunb.) Harv.;毛叶苞帚鼠麹■☆
20634 Amphiglossa tomentosa (Thunb.) Harv. var. breviligulata Merxm. = Amphiglossa tomentosa (Thunb.) Harv. ■☆
20635 Amphiglossa triflora DC.;三花叶苞帚鼠麹■☆
20636 Amphiglottis Salisb. = Epidendrum L.(保留属名)■☆
20637 Amphiglottis Salisb. = Nyctosma Raf. ■☆
20638 Amphiglottis conopsea (R. Br.) Small = Epidendrum magnoliae Muhl. ■☆
20639 Amphiglottis difformis (Jacq.) Britton = Epidendrum floridense Hágsater ■☆
20640 Amphiglottis nocturna (Jacq.) Britton = Epidendrum nocturnum Jacq. ■☆
20641 Amphilobium Loudon = Amphilophium Kunth ●☆
20642 Amphilochia Mart. = Qualea Aubl. ●☆
20643 Amphilophis Nash = Bothriochloa Kuntze ■
20644 Amphilophis glabra (Roxb.) Stapf = Bothriochloa bladhii (Retz.) S. T. Blake ■
20645 Amphilophis insculpta (A. Rich.) Stapf var. vegetior (Hack.) Stapf = Bothriochloa bladhii (Retz.) S. T. Blake ■
20646 Amphilophis insculpta (Hochst. ex A. Rich.) Stapf = Bothriochloa insculpta (Hochst. ex A. Rich.) A. Camus ■☆
20647 Amphilophis intermedia (R. Br.) Stapf = Bothriochloa bladhii (Retz.) S. T. Blake ■
20648 Amphilophis intermedia (R. Br.) Stapf var. acidula (Stapf) Stapf = Bothriochloa bladhii (Retz.) S. T. Blake ■
20649 Amphilophis intermedia Stapf = Bothriochloa bladhii (Retz.) S. T. Blake ■
20650 Amphilophis ischaemum (L.) Nash = Bothriochloa ischaemum (L.) Keng ■
20651 Amphilophis pertusa (L.) Nash ex Stapf = Bothriochloa pertusa (L.) A. Camus ■
20652 Amphilophis pertusa (L.) Stapf = Bothriochloa pertusa (L.) A. Camus ■
20653 Amphilophis radicans (Lehm.) Stapf = Bothriochloa radicans (Lehm.) A. Camus ■☆
20654 Amphilophium Kunth(1818);双冠紫葳属●☆
20655 Amphilophium macrophyllum Kunth;大叶双冠紫葳●☆
20656 Amphimas Pierre ex Dalla Torre et Harms = Amphimas Pierre ex Harms ●☆
20657 Amphimas Pierre ex Harms(1906);双雄苏木属●☆
20658 Amphimas ferrugineus Pierre ex Harms;锈色双雄苏木●☆
20659 Amphimas ferrugineus Pierre ex Pellegr. = Amphimas ferrugineus Pierre ex Harms ●☆
20660 Amphimas klaineanus Pierre ex Pellegr. = Amphimas ferrugineus Pierre ex Harms ●☆
20661 Amphimas pterocarpoides Harms;翅果双雄苏木●☆
20662 Amphimas tessmannii Harms;非洲双雄苏木●☆
20663 Amphineurion (A. DC.) Pichon = Aganosma (Blume) G. Don ●
20664 Amphinomia DC.(废弃属名) = Lotononis (DC.) Eckl. et Zeyh.(保留属名)■
20665 Amphinomia bainesii (Baker) A. Schreib. = Lotononis bainesii Baker ■
20666 Amphinomia brachyantha (Harms) A. Schreib. = Lotononis brachyantha Harms ■☆
20667 Amphinomia desertorum (Dümmer) Schreib. = Rothia hirsuta (Guillaumin et Perr.) Baker ■☆
20668 Amphinomia dichotoma Boiss. = Lotononis platycarpa (Viv.) Pic. Serm. ■☆
20669 Amphinomia dinteri (Schinz) A. Schreib. = Lotononis platycarpa (Viv.) Pic. Serm. ■☆
20670 Amphinomia furcata Merxm. et A. Schreib. = Lotononis furcata (Merxm. et A. Schreib.) A. Schreib. ■☆
20671 Amphinomia laxa (Eckl. et Zeyh.) Cufod. = Lotononis laxa Eckl. et Zeyh. ■☆
20672 Amphinomia lotoidea (Delile) Maire = Lotononis platycarpa (Viv.) Pic. Serm. ■☆
20673 Amphinomia lupinifolia (Boiss.) Pau = Lotononis lupinifolia (Boiss.) Benth. ■☆
20674 Amphinomia lupinifolia (Boiss.) Pau var. villosa (Pomel) Batt. = Lotononis lupinifolia (Boiss.) Benth. ■☆
20675 Amphinomia maroccana (Ball) Font Quer = Lotononis maroccana

Ball ■☆
20676 Amphinomia platycarpa (Viv.) Cufod. = Lotononis platycarpa (Viv.) Pic. Serm. ■☆
20677 Amphinomia riouxii Quézel = Lotononis riouxii (Quézel) Dobignard ■☆
20678 Amphinomia schoenfelderi Dinter ex Merxm. et A. Schreib. = Lotononis schoenfelderi (Dinter ex Merxm. et A. Schreib.) A. Schreib. ■☆
20679 Amphinomia steingroeveriana (Schinz) A. Schreib. = Lotononis platycarpa (Viv.) Pic. Serm. ■☆
20680 Amphinomia stipulosa (Baker f.) A. Schreib. = Lotononis stipulosa Baker f. ■☆
20681 Amphinomia strigillosa Merxm. et A. Schreib. = Lotononis strigillosa (Merxm. et A. Schreib.) A. Schreib. ■☆
20682 Amphiodon Huber = Poecilanthe Benth. ●☆
20683 Amphiolanthus Griseb. = Micranthemum Michx.(保留属名)■☆
20684 Amphion Salisb. = Semele Kunth ●☆
20685 Amphione Raf. = Ipomoea L.(保留属名)●■
20686 Amphipappus Torr. et A. Gray ex A. Gray = Amphipappus Torr. et A. Gray ●☆
20687 Amphipappus Torr. et A. Gray(1845);刺黄花属●☆
20688 Amphipappus fremontii Torr. et A. Gray = Amphipappus fremontii Torr. et A. Gray ex A. Gray ●☆
20689 Amphipappus fremontii Torr. et A. Gray ex A. Gray;刺黄花;Fremont's Chaffbush ●☆
20690 Amphipetalum Bacigalupo(1988);异瓣苋属■☆
20691 Amphipetalum paraguayense Bacigalupo;异瓣苋■☆
20692 Amphiphyllum Gleason(1931);异叶偏穗草属■☆
20693 Amphiphyllum rigidum Gleason;异叶偏穗草■☆
20694 Amphiphyllum schomburgkii Maguire;绍氏异叶偏穗草■☆
20695 Amphipleis Raf. = Nicotiana L. ●■
20696 Amphipogon R. Br.(1810);澳三芒草属■☆
20697 Amphipogon amphipogonoides (Steud.) Vickery;澳三芒草■☆
20698 Amphipogon amphipogonoides Vickery = Amphipogon amphipogonoides (Steud.) Vickery ■☆
20699 Amphipogon strictus R. Br.;劲直澳三芒草■☆
20700 Amphipogon turbinatus R. Br.;陀螺澳三芒草■☆
20701 Amphipterygium Schiede ex Standl.(1923);两翼木属●☆
20702 Amphipterygium adstringens (Schltdl.) Schiede ex Standl.;收敛两翼木●☆
20703 Amphipterygium adstringens (Schltdl.) Standl. = Amphipterygium adstringens (Schltdl.) Schiede ex Standl. ●☆
20704 Amphipterygium glaucum Hemsl. et Rose;灰两翼木●☆
20705 Amphirephis Nees et Matt. = Ampherephis Kunth ■☆
20706 Amphirephis Nees et Matt. = Centratherum Cass. ■☆
20707 Amphirhaphis cuspidata DC. = Duhaldea cuspidata (DC.) Anderb. ●
20708 Amphirhaphis rubricaulis Wall. ex DC. = Duhaldea rubricaulis (Wall. ex DC.) Anderb. ●
20709 Amphirhapis DC. = Inula L. ●■
20710 Amphirhapis DC. = Microglossa DC. ●
20711 Amphirhapis DC. = Solidago L. ■
20712 Amphirhapis albescens DC. = Aster albescens (DC.) Wall. ex Hand. -Mazz. ●
20713 Amphirhapis chinensis Sch. Bip. = Solidago decurrens Lour. ■
20714 Amphirhapis cuspidata DC. = Inula cuspidata C. B. Clarke ●
20715 Amphirhapis heterotricha DC. = Inula eupatorioides DC. ●
20716 Amphirhapis leiocarpa Benth. = Solidago decurrens Lour. ■
20717 Amphirhapis pubescens DC. = Solidago virgaurea L. ■
20718 Amphirhapis rubricaulis DC. = Inula rubricaulis (DC.) Benth. et Hook. f. ●
20719 Amphirrhox Spreng.(1827)(保留属名);尾隔堇属■☆
20720 Amphirrhox grandifolia Melch.;大叶尾隔堇■☆
20721 Amphirrhox juruana Ule = Amphirrhox juruana Ule ex Pilg. ■☆
20722 Amphirrhox juruana Ule ex Pilg.;尾隔堇■☆
20723 Amphirrhox latifolia Mart. = Amphirrhox latifolia Mart. ex Eichler ■☆
20724 Amphirrhox latifolia Mart. ex Eichler;宽叶尾隔堇■☆
20725 Amphirrhox longifolia Spreng.;长叶尾隔堇■☆
20726 Amphirrhox sprucei (Eichler ex Mart.) J. F. Macbr. ex Baehni et Weibel;斯普尾隔堇■☆
20727 Amphirrhox sprucei (Eichler) J. F. Macbr. ex Baehni et Weibel = Amphirrhox sprucei (Eichler ex Mart.) J. F. Macbr. ex Baehni et Weibel ■☆
20728 Amphirrhox surinamensis Eichler;苏里南尾隔堇■☆
20729 Amphiscirpus Oteng-Yeb.(1974);肖藨草属■☆
20730 Amphiscirpus nevadensis (S. Watson) Oteng-Yeb.;肖藨草■☆
20731 Amphiscopia Nees = Justicia L. ●■
20732 Amphiscopium St. -Lag. = Amphiscopia Nees ●■
20733 Amphisiphon W. F. Barker(1936);管丝风信子属■☆
20734 Amphisiphon stylosa W. F. Barker;管丝风信子■☆
20735 Amphisiphon stylosa W. F. Barker = Daubenya stylosa (W. F. Barker) A. M. Van der Merwe et J. C. Manning ■☆
20736 Amphistelma Griseb. = Metastelma R. Br. ●☆
20737 Amphitecna Miers(1868);中美紫葳属●☆
20738 Amphitecna apiculata A. H. Gentry;锐尖中美紫葳●☆
20739 Amphitecna latifolia (Mill.) A. H. Gentry;宽叶中美紫葳●☆
20740 Amphitecna macrophylla (Seem.) Miers ex Baill.;中美紫葳●☆
20741 Amphitecna macrophylla Miers ex Baill. = Amphitecna macrophylla (Seem.) Miers ex Baill. ●☆
20742 Amphitecna montana L. O. Williams;山地中美紫葳●☆
20743 Amphitecna nigripes Baill.;黑梗中美紫葳●☆
20744 Amphitecna parviflora A. H. Gentry;小花中美紫葳●☆
20745 Amphithalea Eckl. et Zeyh.(1836);双盛豆属■☆
20746 Amphithalea alba Granby;白双盛豆●☆
20747 Amphithalea axillaris Granby;腋花双盛豆■☆
20748 Amphithalea bodkinii Dümmer;鲍德双盛豆■☆
20749 Amphithalea bowiei (Benth.) A. L. Schutte;鲍伊双盛豆■☆
20750 Amphithalea bullata (Benth.) A. L. Schutte;泡状双盛豆■☆
20751 Amphithalea cedarbergensis (Granby) A. L. Schutte;锡达伯格双盛豆■☆
20752 Amphithalea ciliaris Eckl. et Zeyh.;缘毛双盛豆●☆
20753 Amphithalea concava Granby;凹双盛豆●☆
20754 Amphithalea cuneifolia Eckl. et Zeyh.;楔叶双盛豆■☆
20755 Amphithalea cymbifolia (C. A. Sm.) A. L. Schutte;舟叶双盛豆■☆
20756 Amphithalea dahlgrenii (Granby) A. L. Schutte;达尔双盛豆■☆
20757 Amphithalea densa Eckl. et Zeyh. = Amphithalea imbricata (L.) Druce ■☆
20758 Amphithalea densiflora Eckl. et Zeyh. = Amphithalea ericifolia (L.) Eckl. et Zeyh. ■☆
20759 Amphithalea ericifolia (L.) Eckl. et Zeyh.;密花双盛豆■☆
20760 Amphithalea ericifolia (L.) Eckl. et Zeyh. subsp. minuta Granby;微小密花双盛豆■☆

20761 Amphithalea ericifolia (L.) Eckl. et Zeyh. subsp. scoparia Granby;帚状密花双盛豆■☆
20762 Amphithalea esterhuyseniae (Granby) A. L. Schutte;埃斯特双盛豆■☆
20763 Amphithalea flava (Granby) A. L. Schutte;黄双盛豆■☆
20764 Amphithalea fourcadei Compton;富尔卡德双盛豆■☆
20765 Amphithalea hilaris Eckl. et Zeyh. = Amphithalea ericifolia (L.) Eckl. et Zeyh. ■☆
20766 Amphithalea humilis Eckl. et Zeyh. = Amphithalea intermedia Eckl. et Zeyh. ■☆
20767 Amphithalea imbricata (L.) Druce;覆瓦双盛豆■☆
20768 Amphithalea intermedia Eckl. et Zeyh.;间型双盛豆■☆
20769 Amphithalea micrantha (E. Mey.) Walp.;小花双盛豆■☆
20770 Amphithalea minima (Granby) A. L. Schutte;极小双盛豆■☆
20771 Amphithalea monticola A. L. Schutte;山地双盛豆■☆
20772 Amphithalea muirii (Granby) A. L. Schutte;缪里双盛豆■☆
20773 Amphithalea multiflora Eckl. et Zeyh. = Amphithalea ericifolia (L.) Eckl. et Zeyh. ■☆
20774 Amphithalea muraltioides (Benth.) A. L. Schutte;厚壁双盛豆■☆
20775 Amphithalea obtusiloba (Granby) A. L. Schutte;钝裂双盛豆■☆
20776 Amphithalea oppositifolia L. Bolus;对叶双盛豆■☆
20777 Amphithalea pageae (L. Bolus) A. L. Schutte;纸双盛豆■☆
20778 Amphithalea parvifolia (Thunb.) A. L. Schutte;小叶双盛豆■☆
20779 Amphithalea perplexa Eckl. et Zeyh.;缠结双盛豆■☆
20780 Amphithalea phylicoides Eckl. et Zeyh.;菲利木双盛豆■☆
20781 Amphithalea pocockiae Bolus = Amphithalea micrantha (E. Mey.) Walp. ■☆
20782 Amphithalea purpurea (Granby) A. L. Schutte;紫双盛豆■☆
20783 Amphithalea rostrata A. L. Schutte et B. -E. van Wyk;喙状双盛豆■☆
20784 Amphithalea sericea Schltr.;绢毛双盛豆■☆
20785 Amphithalea speciosa Schltr.;美丽双盛豆■☆
20786 Amphithalea spinosa (Harv.) A. L. Schutte;具刺双盛豆■☆
20787 Amphithalea stokoei L. Bolus;斯托克双盛豆■☆
20788 Amphithalea tomentosa (Thunb.) Granby;绒毛双盛豆■☆
20789 Amphithalea tortilis (E. Mey.) Steud.;螺旋状双盛豆■☆
20790 Amphithalea villosa Schltr.;长柔毛双盛豆■☆
20791 Amphithalea villosa Schltr. var. brevifolia ? = Amphithalea muraltioides (Benth.) A. L. Schutte ■☆
20792 Amphithalea violacea (E. Mey.) Benth.;堇色双盛豆■☆
20793 Amphithalea virgata Eckl. et Zeyh.;条纹双盛豆■☆
20794 Amphithalea vlokii (A. L. Schutte et B. -E. van Wyk) A. L. Schutte;弗劳克双盛豆■☆
20795 Amphithalea vogelii Walp. = Amphithalea tortilis (E. Mey.) Steud. ■☆
20796 Amphithalea williamsonii Harv.;威廉森双盛豆■☆
20797 Amphitoma Gleason = Miconia Ruiz et Pav.(保留属名)●☆
20798 Amphizoma Miers = Tontelea Miers(保留属名)●☆
20799 Amphochaeta Andersson = Pennisetum Rich. ■
20800 Amphodus Lindl. = Kennedia Vent. ●☆
20801 Amphoranthus S. Moore = Phaeoptilum Radlk. ●☆
20802 Amphoranthus spinosus S. Moore = Phaeoptilum spinosum Radlk. ●☆
20803 Amphorchis Thouars = Cynorkis Thouars ■☆
20804 Amphorchis atacorensis A. Chev. = Habenaria occidentalis (Lindl.) Summerh. ■☆
20805 Amphorchis lilacina Ridl. = Cynorkis ridleyi T. Durand et Schinz ■☆
20806 Amphorchis occidentalis Lindl. = Habenaria occidentalis (Lindl.) Summerh. ■☆
20807 Amphorella Brandegee = Matelea Aubl. ●☆
20808 Amphoricarpos Vis. = Amphoricarpus Vis. ●☆
20809 Amphoricarpus Spruce ex Miers = Cariniana Casar. ●☆
20810 Amphoricarpus Spruce ex Miers = Couratari Aubl. ●☆
20811 Amphoricarpus Vis. (1844);矮菊木属●☆
20812 Amphoricarpus elegans Alb.;雅致矮菊木●☆
20813 Amphorkis Thouars = Cynorkis Thouars ■☆
20814 Amphorocalyx Baker(1887);双耳萼属●☆
20815 Amphorocalyx albus Jum. et H. Perrier;白花双耳萼●☆
20816 Amphorocalyx auratifolius H. Perrier;双耳萼●☆
20817 Amphorocalyx latifolius H. Perrier;宽叶双耳萼●☆
20818 Amphorocalyx multiflorus Baker;多花双耳萼●☆
20819 Amphorogynaceae Nickrent et Der;长颈檀香科●☆
20820 Amphorogyne Stauffer et Hurl. (1957);长颈檀香属●☆
20821 Amphorogyne spicata Stauff. et Hurlim.;长颈檀香●☆
20822 Amphymenium Kunth = Pterocarpus Jacq.(保留属名)●
20823 Ampliglossum Campacci = Oncidium Sw.(保留属名)■☆
20824 Ampliglossum Campacci(2006);大舌兰属■☆
20825 Amplophus Raf. = Valeriana L. ●■
20826 Ampomele Raf. = Rubus L. ●■
20827 Amsinckia Lehm. (1831)(保留属名);阿氏紫草属;Fiddle-neck,Fiddleneck ■☆
20828 Amsinckia angustifolia Lehm. = Amsinckia calycina (Moris) Chater ■☆
20829 Amsinckia barbata Greene = Amsinckia lycopsoides Lehm. ex Fisch. et C. A. Mey. ■☆
20830 Amsinckia calycina (Moris) Chater = Amsinckia micrantha Suksd. ■☆
20831 Amsinckia hispida (Ruiz et Pav.) I. M. Johnst. = Amsinckia lycopsoides Lehm. ex Fisch. et C. A. Mey. ■☆
20832 Amsinckia hispida I. M. Johnst. = Amsinckia calycina (Moris) Chater ■☆
20833 Amsinckia idahoensis M. E. Jones = Amsinckia lycopsoides Lehm. ex Fisch. et C. A. Mey. ■☆
20834 Amsinckia intermedia Fisch. et C. A. Mey. = Amsinckia micrantha Suksd. ■☆
20835 Amsinckia lycopsoides (Lehm.) Lehm. = Amsinckia lycopsoides Lehm. ex Fisch. et C. A. Mey. ■☆
20836 Amsinckia lycopsoides Lehm. = Amsinckia lycopsoides Lehm. ex Fisch. et C. A. Mey. ■☆
20837 Amsinckia lycopsoides Lehm. ex Fisch. et C. A. Mey.;阿氏紫草;Fiddleneck,Scarce Fiddleneck,Tarweed,Tarweed Fiddle-neck ■☆
20838 Amsinckia menziesii (Lehm.) A. Nelson et J. F. Macbr. = Amsinckia micrantha Suksd. ■☆
20839 Amsinckia menziesii (Lehm.) A. Nelson et J. F. Macbr. var. intermedia ? = Amsinckia micrantha Suksd. ■☆
20840 Amsinckia micrantha Suksd.;小花阿氏紫草;Amsinckia, Common Fiddleneck, Fiddleneck, Fiddle-neck, Tarweed, Yellow Forget-me-not ■☆
20841 Amsinckia parviflora A. Heller = Amsinckia lycopsoides Lehm. ex Fisch. et C. A. Mey. ■☆
20842 Amsinckia retrorsa Suksd.;反向阿氏紫草;Fiddleneck,Tarweed ■☆
20843 Amsinckia tessellata A. Gray;格纹阿氏紫草;Amsinkia,Bristly

Fiddle-neck, Checker Fiddleneck, Devil's-lettuce, Fiddleneck, Tarweed ■☆

20844 Amsonia Walter(1788);水甘草属;Amsonia,Blue Stars ■

20845 Amsonia amsonia Britton = Amsonia tabernaemontana Walter ■☆

20846 Amsonia angustifolia Michx.;窄叶水甘草(狭叶水甘草);Feather Amsonia ■☆

20847 Amsonia arenaria Standl. = Amsonia tomentosa Torr. et Frem. var. stenophylla Kearney et Peebles ■☆

20848 Amsonia ciliata Walter;缘毛水甘草;Ciliate Blue Star, Fringed Blue Star Flower, Narrow Leaf Blue Star ■☆

20849 Amsonia eastwoodiana Rydb. = Amsonia tomentosa Torr. et Frem. var. stenophylla Kearney et Peebles ■☆

20850 Amsonia elliptica (Thunb.) Roem. et Schult.;水甘草(椭圆叶水甘草);China Amsonia ■

20851 Amsonia grandiflora Alexander;大叶水甘草;Arizona Slimpod ■☆

20852 Amsonia hirtella Standl.;短粗毛水甘草■☆

20853 Amsonia hubrichtii Woodson;线叶水甘草;Threadleaf Blue Star ■☆

20854 Amsonia illustris Woodson;优秀水甘草;Ozark Blue Star Flower, Shining Blue Star ■☆

20855 Amsonia orientalis Decne. = Rhazya orientalis A. DC. ■☆

20856 Amsonia palmeri A. Gray;帕氏水甘草;Palmers Slimpod ■☆

20857 Amsonia salicifolia Pursh;柳叶水甘草;Blue Star ■☆

20858 Amsonia salicifolia Pursh = Amsonia tabernaemontana Walter ■☆

20859 Amsonia sinensis Tsiang et P. T. Li = Amsonia elliptica (Thunb.) Roem. et Schult. ■

20860 Amsonia tabernaemontana Walter;蓝星水甘草(柳叶水甘草,柳叶窄叶水甘草);Blue Dogbane, Blue Star, Blue-star Amsonia, Eastern Bluestar, Willow Amsonia, Willow-leaved Amsonia ■☆

20861 Amsonia tomentosa Torr. et Frem.;毛水甘草■☆

20862 Amsonia tomentosa Torr. et Frem. var. stenophylla Kearney et Peebles;狭叶毛水甘草;Sand Slimpod, Sand Stars, Woolly Bluestar ■☆

20863 Amsora Bartl. = Amsonia Walter ■

20864 Amura Schult. = Amoora Roxb. ●

20865 Amura Schult. f. = Amoora Roxb. ●

20866 Amydrium Schott(1863);雷公莲属(雷公连属);Amydrium ●

20867 Amydrium hainanense (K. C. Ting et Y. C. Wu ex H. Li et al.) H. Li;穿心藤(穿孔藤,假万年青);Hainan Amydrium ●

20868 Amydrium sinense (Engl.) H. Li;雷公莲(大匹药,大软筋藤,风湿药,叫四门,九龙上吊,雷公连,青藤,软筋藤,下山虎,野红苕);China Amydrium, Chinese Amydrium ●

20869 Amyema Tiegh. (1894);阿米寄生属●☆

20870 Amyema Tiegh. = Dicymanthes Danser ●☆

20871 Amyema periclymenoides (Engl. et K. Krause) Danser = Helixanthera periclymenoides (Engl. et K. Krause) Balle ●☆

20872 Amyema subalata (De Wild.) Danser = Helixanthera subalata (De Wild.) Wiens et Polhill ●☆

20873 Amygdalaceae (Juss.) D. Don = Rosaceae Juss. (保留科名)●■

20874 Amygdalaceae Bartl. = Amygdalaceae Marquis(保留科名)●

20875 Amygdalaceae D. Don = Rosaceae Juss. (保留科名)●■

20876 Amygdalaceae Marquis(1820)(保留科名);桃科(拟李科)●

20877 Amygdalaceae Marquis(保留科名) = Rosaceae Juss. (保留科名)●■

20878 Amygdalopersica Daniel = Prunus L. ●

20879 Amygdalophora M. Roem. = Prunus L. ●

20880 Amygdalophora Neck. = Prunus L. ●

20881 Amygdalopsis Carrière = Louiseania Carrière ●

20882 Amygdalopsis M. Roem. = Amygdalus L. ●

20883 Amygdalopsis M. Roem. = Prunus L. ●

20884 Amygdalus Kuntze = Heritiera Aiton ●

20885 Amygdalus L. (1753);桃属(巴旦杏属,扁桃属,拟李属,榆叶梅属);Almond, Peach ●

20886 Amygdalus L. = Prunus L. ●

20887 Amygdalus afghanica Pachom.;阿富汗桃●☆

20888 Amygdalus amara Hayne = Amygdalus communis L. var. amara Ludw. ex DC. ●

20889 Amygdalus bucharica Korsh.;布哈扁桃●☆

20890 Amygdalus communis L.;扁桃(八担杏,巴达杏,巴旦杏,叭哒杏,匾桃,忽鹿麻,京杏,偏桃,婆淡树,甜欧洲李,甜杏);Almond, Almond Gum, Common Almond, Sweet Almond, Turkey Almond, Turkeyalmond ●

20891 Amygdalus communis L. = Prunus dulcis (Mill.) D. A. Webb ●

20892 Amygdalus communis L. var. amara Ludw. ex DC.;苦味扁桃(苦巴旦杏,苦扁桃);Bitter Almond ●

20893 Amygdalus communis L. var. amara Ludw. ex DC. = Prunus amygdalus Batsch var. amara (DC.) Focke ●

20894 Amygdalus communis L. var. ansu Ludw. ex DC. = Armeniaca vulgaris Lam. var. ansu (Maxim.) Te T. Yu et L. T. Lu ●

20895 Amygdalus communis L. var. dulcis (Mill.) Borkh. = Amygdalus communis L. ●

20896 Amygdalus communis L. var. dulcis (Mill.) Borkh. = Prunus amygdalus Batsch ●

20897 Amygdalus communis L. var. dulcis Borkh. ex DC.;甜味扁桃(甜巴旦杏,甜扁桃);Jordan Almond, Sweet Almond ●

20898 Amygdalus communis L. var. dulcis Borkh. ex DC. = Prunus amygdalus Batsch var. satina Focke ●

20899 Amygdalus communis L. var. fragilis (Borkh.) Ser.;软壳甜扁桃(软核甜扁桃);Fragile Almond, Fragile-shell Almond ●

20900 Amygdalus communis L. var. fragilis (Borkh.) Ser. = Prunus amygdalus Batsch var. fragilis (Borkh.) Focke ●

20901 Amygdalus communis L. var. fragilis (Borkh.) Ser. f. pendula Jaeger;垂枝软壳甜扁桃(垂枝扁桃)●

20902 Amygdalus communis L. var. fragilis (Borkh.) Ser. f. purpurea (C. K. Schneid.) Rehder;紫花扁桃●

20903 Amygdalus communis L. var. fragilis (Borkh.) Ser. f. roseoplena (C. K. Schneid.) Rehder;粉红扁桃●

20904 Amygdalus communis L. var. fragilis (Borkh.) Ser. f. variegata (C. K. Schneid.) Rehder;彩叶扁桃●

20905 Amygdalus communis L. var. tangutica Batalin = Amygdalus tangutica (Batalin) Korsh. ●

20906 Amygdalus cordifolia Roxb.;心叶桃●☆

20907 Amygdalus dasylepis Miq.;毛鳞桃●☆

20908 Amygdalus davidiana (Carrière) de Vos ex Henry;山桃(垂枝杏,普通桃,山毛桃,野桃);David Peach, David's Peach ●

20909 Amygdalus davidiana (Carrière) de Vos ex Henry = Prunus davidiana (Carrière) Franch. ●

20910 Amygdalus davidiana (Carrière) de Vos ex Henry f. alba (Carrière) Rehder;白花山桃●

20911 Amygdalus davidiana (Carrière) de Vos ex Henry var. potaninii (Batalin) Te T. Yu et A. M. Lu;陕甘山桃;Potanin David Peach ●

20912 Amygdalus davidiana (Carrière) de Vos ex Henry var. potaninii (Batalin) Te T. Yu et A. M. Lu = Prunus davidiana (Carrière) Franch. var. potaninii Rehder ●

20913 Amygdalus fenzliana (Fritsch) Lipsky;范氏桃●☆

20914 Amygdalus ferganensis (Kostina et Rjabov) Te T. Yu et A. M. Lu;新疆桃(大宛桃,费尔干桃);Xinjiang Peach ●

20915 Amygdalus ferganensis (Kostina et Rjabov) Te T. Yu et A. M. Lu = Prunus persica Siebold et Zucc. subsp. ferganensis Kostina et Rjabov ●

20916 Amygdalus fischeriana Spach;菲舍尔桃●☆

20917 Amygdalus fortunei H. Lév.;福氏桃●☆

20918 Amygdalus fragilis Borkh. = Amygdalus communis L. var. fragilis (Borkh.) Ser. ●

20919 Amygdalus fremontii (S. Watson) Abrams;弗氏桃(沙漠杏);Desert Apricot ●☆

20920 Amygdalus georgica Desf. = Amygdalus nana L. ●

20921 Amygdalus kansuensis (Rehder) Skeels;甘肃桃(毛桃,野桃);Gansu Peach,Kansu Peach ●

20922 Amygdalus kansuensis (Rehder) Skeels = Prunus kansuensis Rehder ●

20923 Amygdalus ledebouriana Schltdl.;野扁桃(野巴旦,野巴旦杏)●

20924 Amygdalus mira (Koehne) Kov. et Kostina = Amygdalus mira (Koehne) Te T. Yu et A. M. Lu ●

20925 Amygdalus mira (Koehne) Kov. et Kostina = Prunus mira Koehne ●

20926 Amygdalus mira (Koehne) Te T. Yu et A. M. Lu;光核桃(藏桃,光樱桃,康卜,毛桃,西藏桃);Smoothpit Peach ●

20927 Amygdalus mira (Koehne) Te T. Yu et A. M. Lu = Prunus mira Koehne ●

20928 Amygdalus mongolica (Maxim.) Ricker;蒙古扁桃(蒙古杏,山桃,山樱桃,土豆子);Mongol Almond ●◇

20929 Amygdalus mongolica (Maxim.) Ricker = Prunus mongolica Maxim. ●◇

20930 Amygdalus nairica Fed. et Takht.;纳伊尔扁桃●☆

20931 Amygdalus nana L.;矮扁桃(俄罗斯矮杏);Dwarf Russian Almond,Russia Almond,Russian Almond ●

20932 Amygdalus nana L. = Prunus nana (L.) Stokes ●

20933 Amygdalus pedunculata Pall.;长梗扁桃(柄扁桃,长柄扁桃,山豆子,山樱桃);Longstalk Almond,Longstalk Peach,Long-stalked Almond,Long-stalked Peach ●◇

20934 Amygdalus pedunculata Pall. = Prunus pedunculata (Pall.) Maxim. ●◇

20935 Amygdalus persica L.;桃(白桃,碧桃,冬桃,毛桃,毛桃子,旄,普通桃,气桃,桃仔);Common Peach,Early Pink Flowering Peach,Peach ●

20936 Amygdalus persica L. = Prunus persica (L.) Batsch ●

20937 Amygdalus persica L. f. alba (Lindl.) C. K. Schneid.;单瓣白桃(单瓣桃)●

20938 Amygdalus persica L. f. albo-plena C. K. Schneid.;千瓣白桃●

20939 Amygdalus persica L. f. atropurpurea C. K. Schneid.;紫叶桃花●

20940 Amygdalus persica L. f. cameliaeflora (Van Houtte) Dippel;绛桃●

20941 Amygdalus persica L. f. dianthiflora (Van Houtte) Dippel;千瓣桃●

20942 Amygdalus persica L. f. duplex Rehder;碧桃;Flowering Peach,Ornamental Peach ●

20943 Amygdalus persica L. f. magnifica C. K. Schneid.;绯桃●

20944 Amygdalus persica L. f. pendula Dippel;垂枝白桃●

20945 Amygdalus persica L. f. pyramidalis Dippel;塔形碧桃●

20946 Amygdalus persica L. f. rubro-plena C. K. Schneid.;红花碧桃●

20947 Amygdalus persica L. f. versicolor (Siebold) Voss;撒金碧桃●

20948 Amygdalus persica L. var. aganonucipersica (Schübl. et Martens) Te T. Yu et A. M. Lu;离核光桃(离核油桃);Aganonut Persian Peach ●

20949 Amygdalus persica L. var. aganonucipersica (Schübl. et Martens) Te T. Yu et A. M. Lu = Prunus persica Siebold et Zucc. var. nucipersica (Borkh.) C. K. Schneid. f. aganonucipersica (Schübeler et M. Martens) Rehder ●

20950 Amygdalus persica L. var. aganopersica Reichard;离核毛桃(离核桃,离核油桃,李光桃);Aganopersian Peach ●

20951 Amygdalus persica L. var. aganopersica Reichard = Prunus persica (L.) Batsch var. aganopersica (Reichard) Voss ●

20952 Amygdalus persica L. var. compressa (Loudon) Te T. Yu et A. M. Lu;蟠桃;Flat Peach,Saucer Peach ●

20953 Amygdalus persica L. var. compressa (Loudon) Te T. Yu et A. M. Lu = Prunus persica Siebold et Zucc. var. compressa (Loudon) Bean ●

20954 Amygdalus persica L. var. densa Makino;寿星桃●

20955 Amygdalus persica L. var. nectarina W. T. Aiton;油桃;Clingstone Nectarine,Freestone Nectarine,Nectarine ●

20956 Amygdalus persica L. var. nectarina W. T. Aiton = Prunus persica Siebold et Zucc. var. nectarina (W. T. Aiton) Maxim. ●

20957 Amygdalus persica L. var. nucipersica Suckow;蜜腺桃;Nectarine ●☆

20958 Amygdalus persica L. var. scleronucipersica (Schübl. et Martens) Te T. Yu et L. T. Lu;黏核光桃(粘核油桃);Scleronut Persian Peach ●

20959 Amygdalus persica L. var. scleronucipersica (Schübl. et Martens) Te T. Yu et L. T. Lu = Prunus persica Siebold et Zucc. var. nucipersica (Borkh.) C. K. Schneid. f. scleronucipersica (Schübeler et M. Martens) Rehder ●

20960 Amygdalus persica L. var. scleropersica (Rchb.) Te T. Yu et A. M. Lu;黏核毛桃●

20961 Amygdalus petunnikovii Litv.;排氏桃●☆

20962 Amygdalus pilosa Turcz. = Amygdalus pedunculata Pall. ●◇

20963 Amygdalus potaninii (Batalin) Te T. Yu = Amygdalus davidiana (Carrière) de Vos ex Henry var. potaninii (Batalin) Te T. Yu et A. M. Lu ●

20964 Amygdalus pseudopersica Tamamsch.;假波斯桃●☆

20965 Amygdalus pumila Lour. = Prunus persica (L.) Batsch ●

20966 Amygdalus scoparia Spach;帚状桃●☆

20967 Amygdalus spinosissima Bunge;刺桃●☆

20968 Amygdalus tangutica (Batalin) Korsh.;唐古特扁桃(四川扁桃,西康扁桃);Tangut Almond,Tangut Peach ●

20969 Amygdalus tangutica (Batalin) Korsh. = Prunus tangutica (Batalin) Koehne ●

20970 Amygdalus triloba (Lindl.) Ricker;榆叶梅(栏支,榆梅);Chinese Plum, Dwarf Flowering Almond, Flowering Almond, Flowering Plum,Rose Tree of China ●

20971 Amygdalus triloba (Lindl.) Ricker = Prunus triloba Lindl. ●

20972 Amygdalus triloba (Lindl.) Ricker f. multiplex (Bunge) Rehder = Prunus triloba Lindl. 'Multiplex' ●

20973 Amygdalus triloba (Lindl.) Ricker var. petzoldii (K. Koch) Bailey;鸾枝●

20974 Amygdalus triloba (Lindl.) Ricker var. truncata (Kom.) S. Q. Nie;截叶榆叶梅;Truncateleaf Flowering Plum ●

20975 Amygdalus triloba (Lindl.) Ricker var. truncata (Kom.) S. Q. Nie = Prunus triloba Lindl. var. truncata Kom. ●

20976 Amygdalus turcomanica Lincz.;土库曼扁桃●☆

20977 Amygdalus ulmifolia (Franch.) Popov = Aflatunia ulmifolia

(Franch.) Vassilcz. ●☆

20978 Amygdalus ulmifolia (Franch.) Popov = Amygdalus triloba (Lindl.) Ricker ●

20979 Amygdalus urartu Tamamsch.;乌拉吐扁桃;Uratu Plum ●

20980 Amygdalus vavilovii Popov;深沟扁桃●☆

20981 Amygdalus vulgaris Mill. var. compressa Loudon = Amygdalus persica L. var. compressa (Loudon) Te T. Yu et A. M. Lu ●

20982 Amygdalus vulgaris Mill. var. compressa Loudon = Prunus compressa P. Beauv. ●

20983 Amygdalus zansezura Fed. et Takht.;赞格祖尔扁桃●☆

20984 Amylocarpus Barb. Rodr. = Bactris Jacq. ●

20985 Amylocarpus Barb. Rodr. = Yuyba (Barb. Rodr.) L. H. Bailey ●

20986 Amylotheca Tiegh. (1894);粉囊寄生属●☆

20987 Amylotheca dictyophleba (F. Muell.) Tiegh.;粉囊寄生●☆

20988 Amylotheca ovatifolia Danser;卵叶粉囊寄生●☆

20989 Amylotheca parvifolia Danser;小叶粉囊寄生●☆

20990 Amyrea Léandri(1940);香胶大戟属●☆

20991 Amyrea eucleoides Radcl. -Sm.;卡柿香胶大戟●☆

20992 Amyrea gracillima Radcl. -Sm.;纤细香胶大戟●☆

20993 Amyrea grandifolia Radcl. -Sm.;大叶香胶大戟●☆

20994 Amyrea humbertii Léandri;胡氏大叶香胶大戟●☆

20995 Amyrea lancifolia Radcl. -Sm.;剑叶香胶大戟●☆

20996 Amyrea maprouneifolia Radcl. -Sm.;拟马龙戟●☆

20997 Amyrea myrtifolia Radcl. -Sm.;香桃木叶香胶大戟●☆

20998 Amyrea remotiflora Radcl. -Sm.;疏花香胶大戟●☆

20999 Amyrea sambiranensis Léandri;香胶大戟●☆

21000 Amyridaceae Kunth = Rutaceae Juss.(保留科名)●■

21001 Amyridaceae R. Br.;胶香木科●

21002 Amyridaceae R. Br. = Rutaceae Juss.(保留科名)●■

21003 Amyris P. Browne(1756);胶香木属(阿买瑞木属,胶香属,香树属);Balsam Shrub,Torcbwood ●☆

21004 Amyris anisata Willd. = Clausena anisata (Willd.) Hook. f. ex Benth. ●☆

21005 Amyris balsamifera L.;胶香木(胶香树);Balsam Amyris, Candle Wood,Candlewood,Torchwood,West Indian Sandalwood ●☆

21006 Amyris elemifera L.;榄香檀●☆

21007 Amyris gileadensis L. = Commiphora gileadensis (L.) C. Chr. ●☆

21008 Amyris kataf Forssk. = Commiphora kataf (Forssk.) Engl. ●☆

21009 Amyris opobalsamum L. = Commiphora gileadensis (L.) C. Chr. ●☆

21010 Amyris pinnata Kunth;羽状胶香木(阿买瑞木)●☆

21011 Amyris plumieri DC.;普吕米胶香木;Yucatan Elemi ●☆

21012 Amyris punctata Roxb. = Clausena excavata Burm. f. ●

21013 Amyris punctata Roxb. ex Colebr. = Clausena excavata Burm. f. ●

21014 Amyris sumatrana Roxb. = Clausena excavata Burm. f. ●

21015 Amyris zeylanica Retz. = Canarium zeylanicum (Retz.) Blume ●☆

21016 Amyrsia Raf.(废弃属名) = Eugenia L. ●

21017 Amyrsia Raf.(废弃属名) = Myrteola O. Berg(保留属名)●☆

21018 Amyxa Tiegh. (1893);裂果瑞香属●☆

21019 Amyxa kutcinensis Tiegh.;裂果瑞香●☆

21020 Anabaena A. Juss. = Romanoa Trevis. ●☆

21021 Anabaenella Pax et K. Hoffm. = Anabaena A. Juss. ●☆

21022 Anabaenella Pax et K. Hoffm. = Romanoa Trevis. ●☆

21023 Anabasis L. (1753);假木贼属;Anabasis ●■

21024 Anabasis abolinii Iljin = Anabasis brevifolia C. A. Mey. ●

21025 Anabasis affinis Fisch. et C. A. Mey. = Anabasis brevifolia C. A. Mey. ●

21026 Anabasis africana Murb. = Anabasis aphylla L. subsp. africana (Murb.) Maire ●☆

21027 Anabasis ammodendron C. A. Mey. = Haloxylon ammodendron (C. A. Mey.) Bunge ex Fenzl ●

21028 Anabasis aphylla L.;无叶假木贼;Leafless Anabasis ●

21029 Anabasis aphylla L. subsp. africana (Murb.) Maire = Anabasis syriaca Iljin ●☆

21030 Anabasis aretioides (Coss. et Moq. ex Bunge) Coss. et Moq. = Fredolia aretioides (Coss. et Moq. ex Bunge) Ulbr. ●☆

21031 Anabasis articulata (Forssk.) Moq.;关节假木贼●☆

21032 Anabasis balchaschensis Iljin;巴尔哈什假木贼●☆

21033 Anabasis brachiata Fisch. et C. A. Mey. ex Kar. et Kir.;短假木贼●☆

21034 Anabasis brevifolia C. A. Mey.;短叶假木贼(鸡爪柴);Shortleaf Anabasis ●

21035 Anabasis cretacea Pall.;白垩假木贼;Chalky Anabasis ●

21036 Anabasis ehrenbergii Boiss.;爱伦堡假木贼■☆

21037 Anabasis elatior (C. A. Mey.) Schischk.;高枝假木贼;Tall Anabasis ●

21038 Anabasis eriopoda (Schrenk) Benth. ex Volkens;毛足假木贼;Woollystalk Anabasis ●

21039 Anabasis fergansis Drobow;费尔干假木贼●☆

21040 Anabasis foliosa L. = Salsola foliosa (L.) Schrad. ex Schult. ■

21041 Anabasis glomerata M. Bieb. = Halogeton glomeratus (M. Bieb.) C. A. Mey. ■

21042 Anabasis gypsicola Iljin;喜钙假木贼■☆

21043 Anabasis heteroptera (Bunge) Jaub. et Spach = Girgensohnia oppositiflora (Pall.) Fenzl ●■

21044 Anabasis hispidula (Bunge) Benth.;小毛假木贼●☆

21045 Anabasis iliensis (Iljin) Korovin et Mironov = Arthrophytum iliense Iljin ●

21046 Anabasis jaxartica (Bunge) Benth.;锡尔假木贼●☆

21047 Anabasis korovinii Iljin;科罗假木贼●

21048 Anabasis korovinii Iljin = Anabasis elatior (C. A. Mey.) Schischk. ●

21049 Anabasis lachnantha Aellen et Rech. f.;毛花假木贼●☆

21050 Anabasis macroptera Moq.;大翅假木贼●☆

21051 Anabasis micradena Iljin;小腺假木贼●☆

21052 Anabasis oppositiflora (Pall.) M. Bieb. = Girgensohnia oppositiflora (Pall.) Fenzl ●■

21053 Anabasis pauciflora Popov;少花假木贼●☆

21054 Anabasis pelliotii Danguy;粗糙假木贼;Pelliot Anabasis ●

21055 Anabasis phyllophora Kar. et Kir. = Anabasis elatior (C. A. Mey.) Schischk. ●

21056 Anabasis prostrata Pomel;平卧假木贼●☆

21057 Anabasis ramosissima Minkw. = Anabasis salsa (C. A. Mey.) Benth. ex Volkens ●

21058 Anabasis salsa (C. A. Mey.) Benth. ex Volkens;盐生假木贼;Saltliving Anabasis ●

21059 Anabasis setifera Moq.;刚毛假木贼●☆

21060 Anabasis spinosissima L. f. = Noaea spinosissima (L. f.) Moq. ■☆

21061 Anabasis syriaca Iljin;叙利亚假木贼●☆

21062 Anabasis tatarica Pall. = Anabasis aphylla L. ●

21063 Anabasis tianschanica Botsch. = Anabasis cretacea Pall. ●

21064 Anabasis truncata (Schrenk) Bunge;展枝假木贼;Patentbranch Anabasis ●

21065 Anabasis turgaica Iljin et Krasch.;图尔嘎假木贼●☆

21066 Anabasis turkestanica Korovin;土耳其斯坦假木贼●☆

21067 Anabata Willd. ex Roem. et Schult. (1819);南美马钱属●☆

21068 Anacampseros L. (1758)(保留属名);回欢草属;Anacampseros ■☆

21069 Anacampseros Mill. (废弃属名) = Anacampseros L. (保留属名)■☆

21070 Anacampseros Mill. (废弃属名) = Hylotelephium H. Ohba ■

21071 Anacampseros Mill. (废弃属名) = Sedum L. ●■

21072 Anacampseros P. Browne = Talinum Adans. (保留属名)●■

21073 Anacampseros Sims = Anacampseros L. (保留属名)■☆

21074 Anacampseros affinis H. Pearson et Stephens = Anacampseros lanceolata (Haw.) Sweet ■☆

21075 Anacampseros albidiflora Poelln.;白花回欢草■☆

21076 Anacampseros albissima Marloth = Avonia albissima (Marloth) G. D. Rowley ■☆

21077 Anacampseros alstonii Schönl. = Avonia quinaria (E. Mey. ex Fenzl) G. D. Rowley ■☆

21078 Anacampseros alstonii Schönl. = Avonia quinaria (E. Mey. ex Fenzl) G. D. Rowley subsp. alstonii (Schönland) G. D. Rowley ■☆

21079 Anacampseros alta Poelln. = Anacampseros filamentosa (Haw.) Sims subsp. namaquensis (H. Pearson et Stephens) G. D. Rowley ■☆

21080 Anacampseros arachnoides (Haw.) Sims;回欢草;Spider's Web Anacampseros ■☆

21081 Anacampseros arachnoides (Haw.) Sims = Anacampseros rufescens (Haw.) Sweet ■☆

21082 Anacampseros australiana J. M. Black;澳洲回欢草■☆

21083 Anacampseros avasmontana Dinter ex Poelln. = Avonia albissima (Marloth) G. D. Rowley ■☆

21084 Anacampseros avasmontana Dinter ex Poelln. var. caespitosa Poelln. = Avonia albissima (Marloth) G. D. Rowley ■☆

21085 Anacampseros baeseckei Dinter;小花回欢草■☆

21086 Anacampseros bayeriana S. A. Hammer;巴耶尔回欢草■☆

21087 Anacampseros bremekampii Poelln. = Avonia rhodesica (N. E. Br.) G. D. Rowley ■☆

21088 Anacampseros buderiana Poelln. = Avonia recurvata (Schönland) G. D. Rowley subsp. buderiana (Poelln.) G. Will. ■☆

21089 Anacampseros comptonii Pillans;康普顿回欢草■☆

21090 Anacampseros crinita Dinter ex Poelln. = Anacampseros baeseckei Dinter ■☆

21091 Anacampseros decipiens Poelln. = Avonia rhodesica (N. E. Br.) G. D. Rowley ■☆

21092 Anacampseros densifolia Dinter ex Poelln. = Anacampseros filamentosa (Haw.) Sims subsp. tomentosa (A. Berger) Gerbaulet ■☆

21093 Anacampseros depauperata (A. Berger) Poelln. = Anacampseros arachnoides (Haw.) Sims ■☆

21094 Anacampseros dinteri Schinz = Avonia dinteri (Schinz) G. D. Rowley ■☆

21095 Anacampseros filamentosa (Haw.) Sims;蛛丝回欢草;Haasballetjies,Haassuring,Spider's Web Anacampseros ■☆

21096 Anacampseros filamentosa (Haw.) Sims subsp. namaquensis (H. Pearson et Stephens) G. D. Rowley;纳马夸蛛丝回欢草■☆

21097 Anacampseros filamentosa (Haw.) Sims subsp. tomentosa (A. Berger) Gerbaulet;花吹雪■☆

21098 Anacampseros filamentosa (Haw.) Sims var. depauperata A. Berger = Anacampseros arachnoides (Haw.) Sims ■☆

21099 Anacampseros fissa Poelln. = Avonia rhodesica (N. E. Br.) G. D. Rowley ■☆

21100 Anacampseros gracilis Poelln. = Anacampseros arachnoides (Haw.) Sims ■☆

21101 Anacampseros herreana Poelln. = Avonia herreana (Poelln.) G. D. Rowley ■☆

21102 Anacampseros intermedia Haw. ex G. Don = Anacampseros filamentosa (Haw.) Sims ■☆

21103 Anacampseros karasmontana Dinter ex Poelln.;卡拉斯山回欢草■☆

21104 Anacampseros lanceolata (Haw.) Sweet;剑叶回欢草■☆

21105 Anacampseros lanceolata (Haw.) Sweet subsp. nebrownii (Poelln.) Gerbaulet;尼氏剑叶回欢草■☆

21106 Anacampseros lanigera Burch. = Anacampseros filamentosa (Haw.) Sims ■☆

21107 Anacampseros mallei (G. Will.) G. Will. = Avonia mallei G. Will. ■☆

21108 Anacampseros margarethae Dinter = Anacampseros filamentosa (Haw.) Sims subsp. tomentosa (A. Berger) Gerbaulet ■☆

21109 Anacampseros marlothii Poelln.;马洛斯回欢草■☆

21110 Anacampseros maxima Haw. = Hylotelephium maximum (L.) Holub ■☆

21111 Anacampseros meyeri Poelln. = Avonia papyracea (E. Mey. ex Fenzl) G. D. Rowley subsp. namaensis (Gerbaulet) G. D. Rowley ■☆

21112 Anacampseros namaquensis H. Pearson et Stephens = Anacampseros filamentosa (Haw.) Sims subsp. namaquensis (H. Pearson et Stephens) G. D. Rowley ■☆

21113 Anacampseros nebrownii Poelln. = Anacampseros lanceolata (Haw.) Sweet subsp. nebrownii (Poelln.) Gerbaulet ■☆

21114 Anacampseros neglecta Poelln. = Avonia albissima (Marloth) G. D. Rowley ■☆

21115 Anacampseros nitida Poelln. = Anacampseros retusa Poelln. ■☆

21116 Anacampseros ombonensis Dinter ex Poelln. = Avonia dinteri (Schinz) G. D. Rowley ■☆

21117 Anacampseros papyracea E. Mey. ex Fenzl;纸回欢草(琴妆女)■☆

21118 Anacampseros papyracea E. Mey. ex Fenzl = Avonia papyracea (E. Mey. ex Fenzl) G. D. Rowley ■☆

21119 Anacampseros papyracea E. Mey. ex Fenzl subsp. namaensis Gerbaulet = Avonia papyracea (E. Mey. ex Fenzl) G. D. Rowley subsp. namaensis (Gerbaulet) G. D. Rowley ■☆

21120 Anacampseros paradoxa Poelln. = Anacampseros filamentosa (Haw.) Sims subsp. tomentosa (A. Berger) Gerbaulet ■☆

21121 Anacampseros parviflora Poelln. = Anacampseros baeseckei Dinter ■☆

21122 Anacampseros poellnitziana Dinter ex Poelln. = Anacampseros filamentosa (Haw.) Sims subsp. namaquensis (H. Pearson et Stephens) G. D. Rowley ■☆

21123 Anacampseros prominens G. Will. = Avonia prominens (G. Will.) G. Will. ■☆

21124 Anacampseros quinaria E. Mey. ex Fenzl;群蚕■☆

21125 Anacampseros quinaria E. Mey. ex Fenzl = Avonia quinaria (E. Mey. ex Fenzl) G. D. Rowley ■☆

21126 Anacampseros quinaria E. Mey. ex Fenzl var. schmidtii A. Berger = Avonia albissima (Marloth) G. D. Rowley ■☆

21127 Anacampseros quinaria E. Mey. ex Sond. = Avonia quinaria (E. Mey. ex Fenzl) G. D. Rowley ■☆

21128 Anacampseros recurvata Schönland;反折回欢草■☆

21129 Anacampseros recurvata Schönland subsp. buderiana (Poelln.) Gerbaulet = Avonia recurvata (Schönland) G. D. Rowley subsp.

buderiana (Poelln.) G. Will. ■☆

21130 Anacampseros recurvata Schönland subsp. minuta Gerbaulet = Avonia recurvata (Schönland) G. D. Rowley subsp. minuta (Gerbaulet) G. D. Rowley ■☆

21131 Anacampseros retusa Poelln.;微凹回欢草■☆

21132 Anacampseros rhodesica N. E. Br.;罗得西亚回欢草■☆

21133 Anacampseros rubens (Haw.) Sweet = Anacampseros arachnoides (Haw.) Sims ■☆

21134 Anacampseros rubens Sweet = Anacampseros rufescens (Haw.) Sweet ■☆

21135 Anacampseros rubroviridis Poelln. = Anacampseros arachnoides (Haw.) Sims ■☆

21136 Anacampseros rufescens (Haw.) Sweet;红叶回欢草(吹雪之松,厚鳞草,紫幽兰);Redleaf Anacampseros, Variegated Sand Rose ■☆

21137 Anacampseros ruschii Dinter et Poelln. = Avonia ruschii (Dinter et Poelln.) G. D. Rowley ■☆

21138 Anacampseros schmidtii (A. Berger) Poelln. = Avonia albissima (Marloth) G. D. Rowley ■☆

21139 Anacampseros schoenlandii Poelln. = Anacampseros arachnoides (Haw.) Sims ■☆

21140 Anacampseros scopata G. Will.;帚状回欢草■☆

21141 Anacampseros somalensis Poelln. = Portulaca wightiana Wall. ex Wight et Arn. ■☆

21142 Anacampseros starkiana Poelln. = Anacampseros subnuda Poelln. ■☆

21143 Anacampseros subnuda Poelln.;近裸回欢草■☆

21144 Anacampseros telephiastrum DC.;潘回欢草;Pan American love Plant ■☆

21145 Anacampseros telephioides Haw. = Hylotelephium telephioides (Michx.) H. Ohba ■☆

21146 Anacampseros tomentosa A. Berger = Anacampseros filamentosa (Haw.) Sims subsp. tomentosa (A. Berger) Gerbaulet ■☆

21147 Anacampseros tomentosa A. Berger var. crinita Dinter = Anacampseros filamentosa (Haw.) Sims subsp. tomentosa (A. Berger) Gerbaulet ■☆

21148 Anacampseros tomentosa A. Berger var. margaretae (Poelln.) Poelln. = Anacampseros filamentosa (Haw.) Sims subsp. tomentosa (A. Berger) Gerbaulet ■☆

21149 Anacampseros triphylla Haw. = Hylotelephium triphyllum (Haw.) Holub ■

21150 Anacampseros truncata Poelln. = Anacampseros retusa Poelln. ■☆

21151 Anacampseros ustulata E. Mey. ex Fenzl = Avonia ustulata (E. Mey. ex Fenzl) G. D. Rowley ■☆

21152 Anacampseros variabilis Poelln. = Avonia variabilis (Poelln.) G. Will. ■☆

21153 Anacampseros varians (Haw.) Sweet = Anacampseros telephiastrum DC. ■☆

21154 Anacampseros vespertina Thulin;夕回欢草■☆

21155 Anacampseros vulcanensis Anon = Xenia vulcanensis (Anon) Gerbaulet ■☆

21156 Anacampseros wischkonii Dinter ex Poelln. = Avonia dinteri (Schinz) G. D. Rowley ■☆

21157 Anacampseros wischkonii Dinter ex Poelln. var. levis Poelln. = Avonia dinteri (Schinz) G. D. Rowley ■☆

21158 Anacampserotaceae Eggli et Nyffeler(2010);回欢草科■☆

21159 Anacampta Miers = Tabernaemontana L. ●

21160 Anacampti-platanthera P. Fourn. (1928);高卢兰属■☆

21161 Anacampti-platanthera payotii P. Fourn.;高卢兰■☆

21162 Anacamptis Rich. (1918);倒距兰属(爱兰属);Anacamptis, Pyramidal Orchid ■☆

21163 Anacamptis champagneuxii (Barnéoud) R. M. Bateman, Pridgeon et M. W. Chase;尚帕倒距兰■☆

21164 Anacamptis collina (Banks et Sol.) R. M. Bateman, Pridgeon et M. W. Chase;山丘倒距兰■☆

21165 Anacamptis coriophora (L.) R. M. Bateman, Pridgeon et M. W. Chase;革梗倒距兰■☆

21166 Anacamptis coriophora (L.) R. M. Bateman, Pridgeon et M. W. Chase subsp. fragrans (Pollini) R. M. Bateman, Pridgeon et M. W. Chase;芳香山丘倒距兰■☆

21167 Anacamptis coriophora (L.) R. M. Bateman, Pridgeon et M. W. Chase subsp. martrinii (Timb. -Lagr.) Jacquet. et Scappat.;马特林倒距兰■☆

21168 Anacamptis cyrenaica (Durand et Barratte) Dobignard;昔兰尼倒距兰■☆

21169 Anacamptis laxiflora (Lam.) R. M. Bateman, Pridgeon et M. W. Chase;疏花倒距兰■

21170 Anacamptis longicornu (Poir.) R. M. Bateman, Pridgeon et M. W. Chase;长角倒距兰■☆

21171 Anacamptis morio (L.) R. M. Bateman, Pridgeon et M. W. Chase subsp. picta (Loisel.) Jacquet. et Scappat.;着色倒距兰■☆

21172 Anacamptis palustris (Jacq.) R. M. Bateman, Pridgeon et M. W. Chase;沼泽倒距兰■☆

21173 Anacamptis palustris (Jacq.) R. M. Bateman, Pridgeon et M. W. Chase subsp. robusta (T. Stephenson) R. M. Bateman, Pridgeon et M. W. Chase;粗壮倒距兰■☆

21174 Anacamptis papilionacea (L.) R. M. Bateman, Pridgeon et M. W. Chase;蝶形倒距兰■☆

21175 Anacamptis papilionacea (L.) R. M. Bateman, Pridgeon et M. W. Chase subsp. expansa (Ten.) Amard. et Dusak;扩展倒距兰■☆

21176 Anacamptis pyramidalis (L.) Rich.;倒距兰;Butcher Boy, Pyramid Orchid, Pyramidal Orchid, Sham Honey-flower ■☆

21177 Anacamptis pyramidalis (L.) Rich. subsp. condensata (Desf.) H. Lindb.;密集倒距兰■☆

21178 Anacamptis robusta (T. Stephenson) R. M. Bateman = Anacamptis palustris (Jacq.) R. M. Bateman, Pridgeon et M. W. Chase subsp. robusta (T. Stephenson) R. M. Bateman, Pridgeon et M. W. Chase ■☆

21179 Anacantha (Iljin) Soják = Jurinea Cass. ●■

21180 Anacantha (Iljin) Soják = Modestia Charadze et Tamamsch. ■☆

21181 Anacantha (Iljin) Soják(1982);倒刺菊属■☆

21182 Anacantha Soják = Anacantha (Iljin) Soják ■☆

21183 Anacantha Soják = Jurinea Cass. ●■

21184 Anacantha Soják = Modestia Charadze et Tamamsch. ■☆

21185 Anacaona Alain(1980);海地瓜属■☆

21186 Anacaona Liogier = Anacaona Alain ■☆

21187 Anacaona sphaerica Alain;海地葫芦■☆

21188 Anacardia St. -Lag. = Anacardium L. ●

21189 Anacardiaceae Lindl. = Anacardiaceae R. Br. (保留科名)●

21190 Anacardiaceae R. Br. (1818)(保留科名);漆树科;Cashew Family, Sumac Family, Sumach Family ●

21191 Anacardium L. (1753);腰果属(槚如树属);Cashew ●

21192 Anacardium Lam. = Anacardium L. ●

21193 Anacardium Lam. = Semecarpus L. f. ●

21194 Anacardium giganteum Hancock ex Engl.;巨大腰果●☆
21195 Anacardium occidentale L.;腰果(伯公果,都咸树,都咸子,鸡腰果,槚如树,介寿果,腰果树);Acajou, Brazilian Cashew, Cashew, Cashew Apple, Cashew Gum, Cashew Nut, Cashew Tree, Cashew-nut, Cashew-tree, Coffin Nail, Coffin-nail, Common Cashew, Maranon, Marking Nut, Marking-nut, Pajuil, Pear Cashew, Promotion-nut ●
21196 Anacharis Rich. = Elodea Michx. ■☆
21197 Anacharis canadensis (Michx.) Planch. = Elodea canadensis Michx. ■☆
21198 Anacharis canadensis (Michx.) Planch. var. planchonii (Casp.) Vict. = Elodea canadensis Michx. ■☆
21199 Anacharis canadensis A. Gray = Elodea canadensis Michx. ■☆
21200 Anacharis densa (Planch.) Vict. = Egeria densa Planch. ■
21201 Anacharis nuttallii Planch. = Elodea nuttallii (Planch.) H. St. John ■☆
21202 Anacharis occidentalis (Pursh) Vict. = Elodea nuttallii (Planch.) H. St. John ■☆
21203 Anacheilium Hoffmanns. = Epidendrum L.(保留属名)■☆
21204 Anacheilium Hoffmanns. = Prosthechea Knowles et Westc. ■☆
21205 Anacheilium Rchb. ex Hoffmanns. = Epidendrum L.(保留属名)■☆
21206 Anacheilium Reichb. ex Hoffmanns. = Epidendrum L.(保留属名)■☆
21207 Anacheilium cochleatum (L.) Hoffmanns. = Encyclia cochleata (L.) Dressler ■☆
21208 Anacheilium cochleatum (L.) Hoffmanns. var. triandrum (Ames) Sauleda, Wunderlin et B. F. Hansen = Prosthechea cochleata (L.) W. E. Higgins var. triandra (Ames) W. E. Higgins ■☆
21209 Anachortus V. Jirúsek et Chrtek = Corynephorus P. Beauv.(保留属名)■☆
21210 Anachortus V. Jirúsek et Chrtek = Hierochloe R. Br.(保留属名)■
21211 Anachyris Nees = Paspalum L. ■
21212 Anachyrium Steud. = Anachyris Nees ■
21213 Anachyrium Steud. = Paspalum L. ■
21214 Anacis Schrank = Coreopsis L. ●■
21215 Anaclanthe N. E. Br. = Babiana Ker Gawl. ex Sims(保留属名)■☆
21216 Anaclanthe namaquensis N. E. Br. = Babiana thunbergii Ker Gawl. ■☆
21217 Anaclanthe plicata (L. f.) N. E. Br. = Babiana thunbergii Ker Gawl. ■☆
21218 Anaclasmus Griff. = Nenga H. Wendl. et Drude ●☆
21219 Anacolosa (Blume) Blume(1851);短小铁青树属●☆
21220 Anacolosa Blume = Anacolosa (Blume) Blume ●☆
21221 Anacolosa frutescens (Blume) Blume;短小铁青树●☆
21222 Anacolosa uncifera Louis et Boutique;钩短小铁青树●☆
21223 Anacolus Griseb. = Hockinia Gardner ■☆
21224 Anactinia J. Rémy = Nardophyllum (Hook. et Arn.) Hook. et Arn. ●☆
21225 Anactis Cass. = Atractylis L. ■☆
21226 Anactis Raf. = Solidago L. ■
21227 Anactorion Raf. = Synnotia Sweet ■☆
21228 Anacyclia Hoffmanns. = Billbergia Thunb. ■
21229 Anacyclodon Jungh. = Leucopogon R. Br.(保留属名)●☆
21230 Anacyclus L.(1753);辐枝菊属(回环菊属);German Pellitory, Mount Atlas Daisy, Mt. Atlas Daisy, Ringflower ■☆
21231 Anacyclus alexandrinus Willd. = Anacyclus monanthos (L.) Thell. subsp. cyrtolepidioides (Pomel) Humphries ■☆
21232 Anacyclus alexandrinus Willd. var. cyrtolepidioides (Pomel) Durand et Barratte = Anacyclus monanthos (L.) Thell. subsp. cyrtolepidioides (Pomel) Humphries ■☆
21233 Anacyclus alexandrinus Willd. var. mauritanicus (Pomel) Batt. = Anacyclus monanthos (L.) Thell. subsp. cyrtolepidioides (Pomel) Humphries ■☆
21234 Anacyclus atlanticus Litard. et Maire = Heliocauta atlantica (Litard. et Maire) Humphries ■☆
21235 Anacyclus atlanticus Litard. et Maire subsp. vestitus (Humbert) Emb. = Heliocauta atlantica (Litard. et Maire) Humphries ■☆
21236 Anacyclus atlanticus Litard. et Maire var. vestitus Humbert = Heliocauta atlantica (Litard. et Maire) Humphries ■☆
21237 Anacyclus australis Sieber ex Spreng. = Cotula australis (Sieber ex Spreng.) Hook. f. ■☆
21238 Anacyclus australis Spreng. = Cotula australis (Spreng.) Hook. f. ■☆
21239 Anacyclus candollii Nyman = Anacyclus clavatus (Desf.) Pers. ■☆
21240 Anacyclus capillifolius Maire = Anacyclus clavatus (Desf.) Pers. ■☆
21241 Anacyclus ciliatus Trautv.;缘毛辐枝菊■☆
21242 Anacyclus clavatus (Desf.) Pers.;棍棒辐枝菊■☆
21243 Anacyclus clavatus (Desf.) Pers. subsp. linearilobus (Boiss. et Reut.) Batt. = Anacyclus linearilobus Boiss. et Reut. ■☆
21244 Anacyclus clavatus (Desf.) Pers. subsp. maroccanus Ball = Anacyclus maroccanus (Ball) Ball ■☆
21245 Anacyclus clavatus (Desf.) Pers. var. tomentosus (DC.) Fiori;毛棍棒辐枝菊■☆
21246 Anacyclus clavatus (Desf.) Pers. var. tomentosus (DC.) Fiori = Anacyclus maroccanus (Ball) Ball ■☆
21247 Anacyclus clavatus Pers. = Anacyclus tomentosus DC. ■☆
21248 Anacyclus cyrtolepidioides Pomel = Anacyclus monanthos (L.) Thell. subsp. cyrtolepidioides (Pomel) Humphries ■☆
21249 Anacyclus depressus Ball;白舌辐枝菊;Mount Atlas Daisy, Ringflower ■☆
21250 Anacyclus depressus Ball = Anacyclus pyrethrum (L.) Link ■☆
21251 Anacyclus dissimilis Pomel = Anacyclus homogamos (Maire) Humphries ■☆
21252 Anacyclus dissimilis Pomel var. australis Maire = Anacyclus homogamos (Maire) Humphries ■☆
21253 Anacyclus dissimilis Pomel var. eriolepis Maire = Anacyclus homogamos (Maire) Humphries ■☆
21254 Anacyclus exalatus Murb. = Anacyclus radiatus Loisel. subsp. coronatus (Murb.) Humphries ■☆
21255 Anacyclus homogamos (Maire) Humphries;同配辐枝菊■☆
21256 Anacyclus ifniensis Caball. = Anacyclus radiatus Loisel. subsp. coronatus (Murb.) Humphries ■☆
21257 Anacyclus linearilobus Boiss. et Reut.;线裂辐枝菊■☆
21258 Anacyclus maroccanus (Ball) Ball;摩洛哥辐枝菊■☆
21259 Anacyclus mauritanicus Pomel = Anacyclus monanthos (L.) Thell. subsp. cyrtolepidioides (Pomel) Humphries ■☆
21260 Anacyclus medians Murb. = Anacyclus radiatus Loisel. subsp. coronatus (Murb.) Humphries ■☆
21261 Anacyclus monanthos (L.) Thell.;单花辐枝菊■☆
21262 Anacyclus monanthos (L.) Thell. subsp. cyrtolepidioides (Pomel) Humphries;弓鳞辐枝菊■☆
21263 Anacyclus officinalis Hayne;药用辐枝菊(药用回环菊)■☆

21264 Anacyclus pedunculatus (Desf.) Pers. = Anacyclus clavatus (Desf.) Pers. ■☆
21265 Anacyclus pubescens (Willd.) Rchb. = Anacyclus clavatus (Desf.) Pers. ■☆
21266 Anacyclus pyrethrum (L.) Link;西班牙辐枝菊(南欧派利吞草);Alexander's Foot, Bertram, Longwort, Pellitory, Pellitory-of-Spain, Spanish Camomile ■☆
21267 Anacyclus pyrethrum (L.) Link var. depressus (Ball) Maire = Anacyclus pyrethrum (L.) Link ■☆
21268 Anacyclus pyrethrum (L.) Link var. microcephalus Maire = Anacyclus pyrethrum (L.) Link ■☆
21269 Anacyclus pyrethrum (L.) Link var. subdepressus Doum. = Anacyclus pyrethrum (L.) Link ■☆
21270 Anacyclus pyrethrum DC. = Anacyclus pyrethrum (L.) Link ■☆
21271 Anacyclus radiatus Loisel.;辐射回环菊(牛眼草)■☆
21272 Anacyclus radiatus Loisel. subsp. coronatus (Murb.) Humphries;冠辐射回环菊■☆
21273 Anacyclus radiatus Loisel. var. coronatus Murb. = Anacyclus radiatus Loisel. subsp. coronatus (Murb.) Humphries ■☆
21274 Anacyclus radiatus Loisel. var. sulphureus Braun-Blanq. et Maire = Anacyclus radiatus Loisel. ■☆
21275 Anacyclus submedians Maire = Anacyclus radiatus Loisel. subsp. coronatus (Murb.) Humphries ■☆
21276 Anacyclus tomentosus DC.;棒形辐枝菊(棒状回环菊);Whitebuttons ■☆
21277 Anacyclus tomentosus DC. = Anacyclus clavatus (Desf.) Pers. ■☆
21278 Anacyclus valentinus L. subsp. dissimilis (Pomel) Thell. = Anacyclus homogamos (Maire) Humphries ■☆
21279 Anacyclus valentinus L. var. homogamos Maire = Anacyclus homogamos (Maire) Humphries ■☆
21280 Anacylanthus Steud. = Ancylanthos Desf. ●☆
21281 Anadelphia Hack. (1885);兄弟草属■☆
21282 Anadelphia Hack. = Andropogon L. (保留属名)■
21283 Anadelphia Hack. = Clausospicula Lazarides ■☆
21284 Anadelphia afzeliana (Rendle) Stapf;阿芙泽尔兄弟草■☆
21285 Anadelphia arrecta (Stapf) Stapf = Anadelphia afzeliana (Rendle) Stapf ■☆
21286 Anadelphia bigeniculata Clayton;双膝曲兄弟草■☆
21287 Anadelphia chevalieri Reznik;舍瓦利耶兄弟草■☆
21288 Anadelphia funerea (Jacq.-Fél.) Clayton;墓地兄弟草■☆
21289 Anadelphia hamata Stapf;顶钩兄弟草■☆
21290 Anadelphia leptocoma (Trin.) Pilg.;细毛兄弟草■☆
21291 Anadelphia lomaense (A. Camus) Jacq.-Fél.;罗马兄弟草■☆
21292 Anadelphia longifolia Stapf = Anadelphia leptocoma (Trin.) Pilg. ■☆
21293 Anadelphia macrochaeta (Stapf) Clayton;大毛兄弟草■☆
21294 Anadelphia polychaeta Clayton;多毛兄弟草■☆
21295 Anadelphia pubiglumis Stapf = Anadelphia leptocoma (Trin.) Pilg. ■☆
21296 Anadelphia pumila Jacq.-Fél.;矮小兄弟草■☆
21297 Anadelphia scyphofera Clayton;杯状兄弟草■☆
21298 Anadelphia tenuifolia Stapf = Anadelphia leptocoma (Trin.) Pilg. ■☆
21299 Anadelphia trepidaria (Stapf) Stapf;颤动兄弟草■☆
21300 Anadelphia trichaeta (Reznik) Clayton;三毛兄弟草■☆
21301 Anadelphia triseta Reznik = Anadelphia leptocoma (Trin.) Pilg. ■☆
21302 Anadelphia trispiculata Stapf;三细刺兄弟草■☆
21303 Anademia C. Agardh = Anadenia R. Br. ●
21304 Anademia C. Agardh = Grevillea R. Br. ex Knight(保留属名)●
21305 Anadenanthera Speg. (1923);南美豆属(阿拉豆属,柯拉豆属)●☆
21306 Anadenanthera colubrina (Vell.) Brenan;南美豆●☆
21307 Anadenanthera macrocarpa (Benth.) Brenan;大果柯拉豆(大果阿拉豆,大果红心木,塞比落腺豆)●☆
21308 Anadenanthera peregrina (L.) Speg.;外来南美豆(酒醉木,外来榼藤子);Colioba, Niopo Snuff ●☆
21309 Anadendron Schott = Anadendrum Schott ●
21310 Anadendron Wight = Anadendrum Schott ●
21311 Anadendrum Schott(1857);上树南星属;Anadendrum ●
21312 Anadendrum latifolium Hook. f.;宽叶上树南星;Broadleaf Anadendrm ●
21313 Anadendrum montanum (Blume) Schott;上树南星(怀被藤);Montane Anadendrm, Mountainy Anadendrm ●
21314 Anadenia R. Br. = Grevillea R. Br. ex Knight(保留属名)●
21315 Anaectocalyx Triana = Anaectocalyx Triana ex Benth. et Hook. f. ●☆
21316 Anaectocalyx Triana ex Benth. et Hook. f. (1867);外倾萼属●☆
21317 Anaectocalyx Triana ex Hook. f. = Anaectocalyx Triana ex Benth. et Hook. f. ●☆
21318 Anaectocalyx bracteosa Triana;外倾萼☆
21319 Anaectocalyx latifolia Cogn.;宽叶外倾萼☆
21320 Anaectochilus Lindl. = Anoectochilus Blume(保留属名)■
21321 Anafrenium Arn. = Anaphrenium E. Mey. ●☆
21322 Anafrenium Arn. = Heeria Meisn. ●☆
21323 Anagallidaceae Batsch ex Borkh. = Myrsinaceae R. Br. (保留科名)●
21324 Anagallidaceae Baudo = Myrsinaceae R. Br. (保留科名)●
21325 Anagallidaceae Baudo = Primulaceae Batsch ex Borkh. (保留科名)●■
21326 Anagallidastrum Adans. = Anagallis L. ■
21327 Anagallidastrum Adans. = Centunculus L. ■
21328 Anagallidastrum Mich. ex Adans. = Centunculus L. ■
21329 Anagallidium Griseb. (1838);腺鳞草属(当药属,假海绿属)■
21330 Anagallidium Griseb. = Swertia L. ■
21331 Anagallidium dichotomum (L.) Griseb. = Swertia dichotoma L. ■
21332 Anagallidium dimorphum (Batalin) Ma ex T. N. Ho et W. L. Shih = Swertia dimorpha Batalin ■
21333 Anagallidium rubrostriatum Y. Z. Zhao, Zong Y. Zhu et L. Q. Zhao;红纹腺鳞草■
21334 Anagallidium tetrapetalum Griseb. = Swertia tetraptera Maxim. ■
21335 Anagallis L. (1753);琉璃繁缕属(海绿属);Chaffweed, John-Go-to-bed-at-noon, Pimpernel ■
21336 Anagallis aberdarica T. C. E. Fr. = Anagallis serpens Hochst. ex DC. subsp. meyeri-johannis (Engl.) P. Taylor ■☆
21337 Anagallis acuminata Welw. ex Schinz.;尖琉璃繁缕■☆
21338 Anagallis angustiloba (Engl.) Engl.;窄裂琉璃繁缕■☆
21339 Anagallis arvensis L.;琉璃繁缕(海绿,九龙吐珠,龙吐珠,四念癀,猩红海绿);Adder's Eyes, Biddy's Eyes, Bird's Eye, Bird's Tongue, Blue Pimpernel, Change-of-the-weather, Common Pimpernel, Corn Pimpernel, Countryman's Weatherglass, Cry-baby, Cry-baby Crab, Drops of Blood, Eyebright, Farmer's Weatherglass, Female Pimpernel, Grandfather's Weatherglass, Husbandman's Weather Warner, Husbandman's Weatherglass, Jack-go-to-bed-at-noon, John-that-goes-to-bed-at-noon, Ladybird, Laughter-bringer, Little Jane,

Little Peep, Little Peepers, Male Pimpernel, Merecrop, Morcrop, Numpinole, Old Man, Old Man's Friend, Old Man's Glass Eye, Old Man's Looking Glass, Old Man's Looking-glass, Old Man's Weatherglass, Orange Lily, Owl's Eye, Owls Eyes, Ox Eyes, Ox-eye, Pernel, Pheasant's Eye, Pheasant's Eyes, Pimpernel, Ploughman's Weatherglass, Poor Man's Weatherglass, Poor Man's Weather Glass, Poorman's Weatherglass, Poor-man's Weather-glass, Pumpernel, Red Bird's Eye, Red Bird's Eyes, Red Chickweed, Red Pimpernel, Redweed, Scarlet Pimpernel, Shepherd's Barometer, Shepherd's Calendar, Shepherd's Clock, Shepherd's Daylight, Shepherd's Delight, Shepherd's Dial, Shepherd's Dock, Shepherd's Glass, Shepherd's Joy, Shepherd's Sundial, Shepherd's Warning, Shepherd's Watch, Shepherd's Weatherglass, Shepherd's Weather-glass, Snapjack, Star of Bethlehem, Sunflower, Tom Pimpernel, Twelve-o'clock, Waywort, Weather Flower, Weather Teller, Weatherglass, Wincopipe, Windpipe, Wink-and-peep, Wink-a-peep ■

21340 Anagallis arvensis L. f. coerulea (Schreb.) Baumg.;蓝花琉璃繁缕(蓝海绿,蓝琉璃繁缕,琉璃繁缕,四念癀);Blue Pimpernel, Blue Scarlet Pimpernel ■

21341 Anagallis arvensis L. f. phoenicea (Scop.) Baumg. = Anagallis arvensis L. ■

21342 Anagallis arvensis L. subsp. caeulea (L.) Hartm. = Anagallis arvensis L. ■

21343 Anagallis arvensis L. subsp. foemina (Mill.) Schinz et Thell. = Anagallis foemina Mill. ■

21344 Anagallis arvensis L. subsp. foemina (Mill.) Schinz et Thell. ex Schinz et R. Keller = Anagallis arvensis L. f. coerulea (Schreb.) Baumg. ■

21345 Anagallis arvensis L. subsp. latifolia (L.) Arcang.;宽叶琉璃繁缕■☆

21346 Anagallis arvensis L. subsp. parviflora (Hoffmanns. et Link) Arcang.;小花琉璃繁缕■☆

21347 Anagallis arvensis L. subsp. phoenicea Vollm. = Anagallis arvensis L. ■

21348 Anagallis arvensis L. var. caerulea (L.) Gouan = Anagallis arvensis L. ■

21349 Anagallis arvensis L. var. coerulea (L.) Gouan = Anagallis arvensis L. ■

21350 Anagallis arvensis L. var. latifolia (L.) Lange = Anagallis arvensis L. subsp. latifolia (L.) Arcang. ■☆

21351 Anagallis arvensis L. var. phoenicea Gouan = Anagallis arvensis L. ■

21352 Anagallis arvensis L. var. platyphylloides Pau = Anagallis arvensis L. subsp. latifolia (L.) Arcang. ■☆

21353 Anagallis barbata (P. Taylor) Kupicha;髯毛琉璃繁缕■☆

21354 Anagallis baumii R. Knuth;鲍姆琉璃繁缕■☆

21355 Anagallis bella M. B. Scott = Anagallis serpens Hochst. ex DC. subsp. meyeri-johannis (Engl.) P. Taylor ■☆

21356 Anagallis brevipes P. Taylor;短梗琉璃繁缕■☆

21357 Anagallis caerulea L. = Anagallis arvensis L. ■

21358 Anagallis capensis L. = Diascia capensis (L.) Britten ■☆

21359 Anagallis churiensis T. C. E. Fr. = Anagallis angustiloba (Engl.) Engl. ■☆

21360 Anagallis coerulea L. = Anagallis arvensis L. var. coerulea (L.) Gouan ■

21361 Anagallis coerulea Schreb. = Anagallis arvensis L. f. coerulea (Schreb.) Baumg. ■

21362 Anagallis coerulea Schreb. = Anagallis arvensis L. ■

21363 Anagallis collina Schousb = Anagallis monelli L. subsp. linifolia (L.) Maire ■☆

21364 Anagallis crassifolia Thore;厚叶琉璃繁缕■☆

21365 Anagallis dekindtiana Gilg ex R. Knuth = Anagallis elegantula P. Taylor ■☆

21366 Anagallis djalonis A. Chev.;贾隆琉璃繁缕■☆

21367 Anagallis elegantula P. Taylor;雅致琉璃繁缕■☆

21368 Anagallis filifolia Engl. et Gilg;丝叶琉璃繁缕■☆

21369 Anagallis foemina Mill. = Anagallis arvensis L. f. coerulea (Schreb.) Baumg. ■

21370 Anagallis foemina Mill. = Anagallis arvensis L. ■

21371 Anagallis gracilipes P. Taylor;细梗琉璃繁缕■☆

21372 Anagallis grandiflora Andr. = Anagallis linifolia L. ■☆

21373 Anagallis granvikii T. C. E. Fr. = Anagallis serpens Hochst. ex DC. subsp. meyeri-johannis (Engl.) P. Taylor ■☆

21374 Anagallis hanningtonii Baker = Anagallis tenuicaulis Baker ■☆

21375 Anagallis hexamera P. Taylor;六数琉璃繁缕■☆

21376 Anagallis huttonii Harv.;赫顿琉璃繁缕■☆

21377 Anagallis iraruensis T. C. E. Fr. = Anagallis serpens Hochst. ex DC. subsp. meyeri-johannis (Engl.) P. Taylor ■☆

21378 Anagallis kigesiensis Good = Anagallis angustiloba (Engl.) Engl. ■☆

21379 Anagallis kilimandscharica R. Knuth = Anagallis serpens Hochst. ex DC. subsp. meyeri-johannis (Engl.) P. Taylor ■☆

21380 Anagallis kingaensis Engl.;金加琉璃繁缕■☆

21381 Anagallis kochii H. E. Hess;高知琉璃繁缕■☆

21382 Anagallis latifolia L. = Anagallis arvensis L. subsp. latifolia (L.) Arcang. ■☆

21383 Anagallis linifolia L.;亚麻叶琉璃繁缕;Blue Pimpernel, Flaxleaf Pimpernel, Shrubby Pimpernel ■☆

21384 Anagallis linifolia L. = Anagallis monelli L. subsp. linifolia (L.) Maire ■☆

21385 Anagallis linifolia L. var. collina (Schousb.) Ball = Anagallis monelli L. ■☆

21386 Anagallis linifolia L. var. hispanica (Willk.) H. Lindb. = Anagallis monelli L. ■☆

21387 Anagallis linifolia L. var. littoralis Pamp. = Anagallis monelli L. ■☆

21388 Anagallis linifolia L. var. microphylla Ball = Anagallis monelli L. ■☆

21389 Anagallis linum-stellatum (L.) Duby;星亚麻琉璃繁缕■☆

21390 Anagallis meyeri K. Schum. ex Engl. = Anagallis serpens Hochst. ex DC. subsp. meyeri-johannis (Engl.) P. Taylor ■☆

21391 Anagallis meyeri-johannis (Engl.) Engl. = Anagallis serpens Hochst. ex DC. subsp. meyeri-johannis (Engl.) P. Taylor ■☆

21392 Anagallis micrantha T. C. E. Fr. = Anagallis serpens Hochst. ex DC. subsp. meyeri-johannis (Engl.) P. Taylor ■☆

21393 Anagallis minima (L.) E. H. L. Krause = Centunculus minimus L. ■☆

21394 Anagallis minima E. H. L. Krause;微小琉璃繁缕;Chaffweed ■☆

21395 Anagallis minima E. H. L. Krause = Centunculus minimus L. ■☆

21396 Anagallis monelli L.;莫内尔琉璃繁缕;Blue Pimpernel ■☆

21397 Anagallis monelli L. = Anagallis linifolia L. ■☆

21398 Anagallis monelli L. subsp. collina (Schousb.) Maire = Anagallis monelli L. ■☆

21399 Anagallis monelli L. subsp. linifolia (L.) Maire = Anagallis linifolia L. ■☆

21400 Anagallis monelli L. var. hispanica (Willk.) Maire = Anagallis

monelli L. ■☆
21401 Anagallis monelli L. var. leptensis Chiov. = Anagallis monelli L. ■☆
21402 Anagallis monelli L. var. microphylla (Ball) Jahand. et Maire = Anagallis monelli L. subsp. linifolia (L.) Maire ■☆
21403 Anagallis monelli L. var. schousboei H. Lindb. = Anagallis monelli L. ■☆
21404 Anagallis monelli L. var. speciosa Pau et Font Quer = Anagallis monelli L. ■☆
21405 Anagallis multangularis Buch.-Ham. ex D. Don = Lysimachia pyramidalis Wall. ■☆
21406 Anagallis nana Schinz = Anagallis pumila Sw. ■☆
21407 Anagallis nummulariifolia Baker;铜钱琉璃繁缕■☆
21408 Anagallis oligantha P. Taylor;寡花琉璃繁缕■☆
21409 Anagallis peploides Baker;荸艾琉璃繁缕■☆
21410 Anagallis phoenicea Scop. = Anagallis arvensis L. ■
21411 Anagallis pulchella Welw. ex Schinz = Anagallis elegantula P. Taylor ■☆
21412 Anagallis pumila Sw.;小琉璃繁缕■☆
21413 Anagallis pumila Sw. var. barbata P. Taylor = Anagallis barbata (P. Taylor) Kupicha ■☆
21414 Anagallis pumila Sw. var. djalonis (A. Chev.) P. Taylor = Anagallis djalonis A. Chev. ■☆
21415 Anagallis pumila Sw. var. natalensis Pax et R. Knuth = Anagallis tenuicaulis Baker ■☆
21416 Anagallis quartiniana (A. Rich.) Engl. var. angustiloba Engl. = Anagallis angustiloba (Engl.) Engl. ■☆
21417 Anagallis quartiniana (A. Rich.) Engl. var. meyeri-johannis Engl. = Anagallis serpens Hochst. ex DC. subsp. meyeri-johannis (Engl.) P. Taylor ■☆
21418 Anagallis repens DC. = Anagallis arvensis L. ■
21419 Anagallis repens Pomel = Anagallis tenella (L.) L. ■☆
21420 Anagallis rhodesica R. E. Fr.;罗得西亚琉璃繁缕■☆
21421 Anagallis roberti T. C. E. Fr. = Anagallis serpens Hochst. ex DC. subsp. meyeri-johannis (Engl.) P. Taylor ■☆
21422 Anagallis ruandensis R. Knuth et Mildbr. = Anagallis angustiloba (Engl.) Engl. ■☆
21423 Anagallis rubricaulis Bojer ex Duby;茎红琉璃繁缕■☆
21424 Anagallis schliebenii R. Knuth et Mildbr.;施利本琉璃繁缕■☆
21425 Anagallis serpens Hochst. ex DC.;蛇形琉璃繁缕■☆
21426 Anagallis serpens Hochst. ex DC. subsp. meyeri-johannis (Engl.) P. Taylor;迈·约琉璃繁缕■☆
21427 Anagallis tenella (L.) L.;软枝琉璃繁缕(柔海绿);Bog Pimpernel, Moneywort, Sunflower ■☆
21428 Anagallis tenella L. 'Stadland';斯塔德兰软枝琉璃繁缕■☆
21429 Anagallis tenella L. = Anagallis tenella (L.) L. ■☆
21430 Anagallis tenuicaulis Baker;细茎琉璃繁缕■☆
21431 Anagallis tsaratananae M. Peltier;察拉塔纳纳琉璃繁缕■☆
21432 Anagallis ulugurensis R. Knuth = Anagallis angustiloba (Engl.) Engl. ■☆
21433 Anagallis verticillata All.;轮生海绿■☆
21434 Anagalloides Krock. = Lindernia All. ■
21435 Anagalloides procumbens Krock. = Lindernia procumbens (Krock.) Philcox ■
21436 Anaganthos Hook. f. = Australina Gaudich. ■☆
21437 Anaglypha DC. = Gibbaria Cass. ■☆
21438 Anaglypha acicularis Benth. = Oxylaena acicularis (Benth.) Anderb. ●☆
21439 Anaglypha aspera DC. = Gibbaria scabra (Thunb.) Norl. ■☆
21440 Anaglypha latifolia S. Moore = Anisopappus latifolius (S. Moore) B. L. Burtt ■☆
21441 Anagosperma Wettst. = Euphrasia L. ■
21442 Anagyris L. (1753);臭红豆属(红豆属,螺旋豆属,细豆属);Bean Trefoil ●☆
21443 Anagyris Lour. = Sophora L. ●■
21444 Anagyris barbata Graham = Thermopsis barbata Benth. ■
21445 Anagyris foetida L.;臭红豆(臭味红豆);Bean Trefoil, Mediterranean Stinkbush ●☆
21446 Anagyris latifolia Willd.;宽叶臭红豆●☆
21447 Anagzanthe Baudo = Lysimachia L. ●■
21448 Anaitis DC. = Sanvitalia Lam. ●■
21449 Anakasia Philipson(1973);新几内亚五加属●☆
21450 Anakasia simplicifolia Philipson;新几内亚五加●☆
21451 Analectis Juss. = Symphorema Roxb. ●
21452 Analiton Raf. = Rumex L. ●■
21453 Analyrium E. Mey. = Peucedanum L. ■
21454 Analyrium E. Mey. ex Presl = Peucedanum L. ■
21455 Anamaria V. C. Souza = Stemodia L. (保留属名)■☆
21456 Anamelis Garden = Fothergilla L. ●☆
21457 Anamenia Vent. = Knowltonia Salisb. ■☆
21458 Anamirta Colebr. (1821);印度防己属(印防己属,醉鱼藤属);Fishberry ●☆
21459 Anamirta cocculus (L.) Wight et Arn.;印度防己(印防己);Cocculua Indicus, Crow Killer, Fish Berries, Fish Killer, Fishberry, Hockle Elderberry, Indian Berry, Levant Berry, Levant Nut, Louseberry, Oriental Berry, Picrotoxin, Poison Berry ●☆
21460 Anamirta cocculus Wight et Arn. = Anamirta cocculus (L.) Wight et Arn. ●☆
21461 Anamirta paniculata Colebr. = Anamirta cocculus (L.) Wight et Arn. ●☆
21462 Anamomis Griseb. = Myrcianthes O. Berg ●☆
21463 Anamorpha H. Karst. et Triana = Melochia L. (保留属名)●■
21464 Anamtia Koidz. = Myrsine L. ●
21465 Anamtia acuminata (Merr.) Nakai = Myrsine acuminata Royle ●
21466 Anamtia elliptica (Walker) Nakai = Myrsine elliptica Walker ●
21467 Anamtia integrifolia Masam. = Myrsine integrifolia Merr. ●☆
21468 Anamtia marginata Masam. = Myrsine stolonifera (Koidz.) E. Walker ●
21469 Anamtia mezii Masam. = Myrsine stolonifera (Koidz.) E. Walker ●
21470 Anamtia stolonifera Koidz. = Myrsine stolonifera (Koidz.) E. Walker ●
21471 Ananas Gaertn. = Bromelia L. ■☆
21472 Ananas Mill. (1754);凤梨属(菠萝属,斑叶凤梨属);Pineapple, Ananas ■
21473 Ananas Tourn. ex L. = Ananas Mill. ■
21474 Ananas ananassoides (Baker) L. B. Sm.;南美凤梨;Cerrado Pineapple ■☆
21475 Ananas bracteatus (Lindl.) Schult.;红凤梨(斑叶红凤梨,苞叶红凤梨);Red Pineapple, Wild Pineapple ■☆
21476 Ananas bracteatus (Lindl.) Schult. 'Tricolor';三色红凤梨■☆
21477 Ananas bracteatus (Lindl.) Schult. var. striatus M. B. Foster;带纹凤梨(红凤梨)■☆
21478 Ananas bracteatus (Lindl.) Schult. var. tricolor (Bertoni) L. B. Sm. = Ananas bracteatus (Lindl.) Schult. 'Tricolor' ■☆
21479 Ananas comosus (L.) Merr.;凤梨(菠萝,草菠萝,地菠萝,露

兜子);Ananas,Crowa,Nana,Pineapple ■
21480 Ananas comosus (L.) Merr.'Smooth Cayenne';开印凤梨■☆
21481 Ananas comosus (L.) Merr.'Variegatus';花叶凤梨(黄边凤梨,艳凤梨)■
21482 Ananas comosus (L.) Merr. var. sativus (Schult. f.) Mez = Ananas comosus (L.) Merr. ■
21483 Ananas comosus (L.) Merr. var. variegatus (Lowe) Moldenke = Ananas comosus (L.) Merr.'Variegatus' ■
21484 Ananas erectifolius L. B. Sm.;直叶凤梨■☆
21485 Ananas lucidus Mill.;亮凤梨(野凤梨);Curagua,Wild Ananas ■☆
21486 Ananas nanus (L. B. Sm.) L. B. Sm.;小凤梨■☆
21487 Ananas sativus Lindl. = Ananas comosus (L.) Merr. ■
21488 Ananassa Lindl. = Ananas Mill. ■
21489 Anandria Less. = Leibnitzia Cass. ■
21490 Anandria laevipes Gand. = Leibnitzia anandria (L.) Turcz. ■
21491 Anangia W. J. de Wilde et Duyfjes(2006);大萼瓜属■☆
21492 Anantherix Nutt. = Asclepias L. ■
21493 Anantherix Nutt. = Asclepiodora A. Gray ■
21494 Ananthopus Raf. = Commelina L. ■
21495 Anapalina N. E. Br. = Tritoniopsis L. Bolus ■☆
21496 Anapalina burchellii (N. E. Br.) N. E. Br. = Tritoniopsis burchellii (N. E. Br.) Goldblatt ■☆
21497 Anapalina caffra (Ker Gawl. ex Baker) G. J. Lewis = Tritoniopsis caffra (Ker Gawl. ex Baker) Goldblatt ■☆
21498 Anapalina intermedia (Baker) G. J. Lewis = Tritoniopsis intermedia (Baker) Goldblatt ■☆
21499 Anapalina longituba Fourc. = Tritoniopsis antholyza (Poir.) Goldblatt ■☆
21500 Anapalina nervosa (Thunb.) G. J. Lewis = Tritoniopsis antholyza (Poir.) Goldblatt ■☆
21501 Anapalina pulchra (Baker) N. E. Br. = Tritoniopsis pulchra (Baker) Goldblatt ■☆
21502 Anapalina triticea (Burm. f.) N. E. Br. = Tritoniopsis triticea (Burm. f.) Goldblatt ■☆
21503 Anaphalioides (Benth.) Kirp. (1950);拟香青属(类香青属)●■☆
21504 Anaphalioides keriensis (A. Cunn.) Kirp.;拟香青■☆
21505 Anaphalioides keriensis (A. Cunn.) Kirp. = Gnaphalium keriense A. Cunn. ■☆
21506 Anaphalis DC. (1838);香青属(籁箫属,藾蒿属,藾箫属,山荻属);Everlasting,Pearl Everlasting,Pearleverlasting ●■
21507 Anaphalis acutifolia Hand.-Mazz.;尖叶香青;Sharpleaf Everlasting,Sharpleaf Pearleverlasting ■
21508 Anaphalis adnata DC. = Gnaphalium adnatum (Wall. ex DC.) Kitam. ■
21509 Anaphalis adnata Wall. ex DC. = Gnaphalium adnatum (Wall. ex DC.) Kitam. ■
21510 Anaphalis alata Maxim. = Anaphalis sinica Hance var. remota Y. Ling ■
21511 Anaphalis alata Maxim. var. viridis Hand.-Mazz. = Anaphalis latialata Y. Ling et Y. L. Chen ■
21512 Anaphalis alata Maxim. var. viridis Hand.-Mazz. = Anaphalis latialata Y. Ling et Y. L. Chen var. viridis (Hand.-Mazz.) Y. Ling et Y. L. Chen ■
21513 Anaphalis alpicola Makino;高山香青■☆
21514 Anaphalis araneosa DC. = Anaphalis busua (Buch.-Ham. ex D. Don) DC. ■
21515 Anaphalis araneosa Franch. = Anaphalis surculosa (Hand.-Mazz.) Hand.-Mazz. ■
21516 Anaphalis aureopunctata Lingelsh. et Borza;黄腺香青(香蒿);Yellowgland Everlasting,Yellowvariegate Pearleverlasting ■
21517 Anaphalis aureopunctata Lingelsh. et Borza f. calvescens (Pamp.) F. H. Chen;脱毛黄腺香青■
21518 Anaphalis aureopunctata Lingelsh. et Borza f. calvescens (Pamp.) F. H. Chen = Anaphalis aureopunctata Lingelsh. et Borza ■
21519 Anaphalis aureopunctata Lingelsh. et Borza var. atrata (Hand.-Mazz.) Hand.-Mazz.;黑鳞黄腺香青(黑鳞香青,五花草);Bracksquamose Everlasting,Bracksquamose Pearleverlasting ■
21520 Anaphalis aureopunctata Lingelsh. et Borza var. atrata Hand.-Mazz. = Anaphalis aureopunctata Lingelsh. et Borza var. atrata (Hand.-Mazz.) Hand.-Mazz. ■
21521 Anaphalis aureopunctata Lingelsh. et Borza var. plantaginifolia Y. L. Chen;车前叶黄腺香青(车前叶香青);Plantagoleaf Everlasting,Plantagoleaf Pearleverlasting ■
21522 Anaphalis aureopunctata Lingelsh. et Borza var. tomentosa Hand.-Mazz.;绒毛黄腺香青(茸毛黄腺香青);Tomentose Yellowgland Everlasting,Tomentose Yellowgland Pearleverlasting ■
21523 Anaphalis aurora Rech. f. et Edelb. = Anaphalis roseo-alba Krasch. ■☆
21524 Anaphalis batangensis Y. L. Chen;巴塘香青;Batang Everlasting,Batang Pearleverlasting ■
21525 Anaphalis bicolor (Franch.) Diels;二色香青(白头蒿,三轮蒿);Bicolored Everlasting,Bicolored Pearleverlasting ■
21526 Anaphalis bicolor (Franch.) Diels var. kokonorica Y. Ling;青海二色香青(青海香青);Qinghai Bicolored Everlasting,Qinghai Bicolored Pearleverlasting ■
21527 Anaphalis bicolor (Franch.) Diels var. longifolia C. C. Chang;长叶二色香青;Longleaf Bicolored Everlasting,Longleaf Bicolored Pearleverlasting ■
21528 Anaphalis bicolor (Franch.) Diels var. subconcolor Hand.-Mazz.;同色香青;Samecolor Pearleverlasting ■
21529 Anaphalis bicolor (Franch.) Diels var. subconcolor Hand.-Mazz. f. minuta Hand.-Mazz.;矮小二色香青■
21530 Anaphalis bicolor (Franch.) Diels var. undulata (Hand.-Mazz.) Y. Ling;波缘二色香青(波缘香青);Undulate Pearleverlasting ■
21531 Anaphalis bodinieri Franch. = Anaphalis hancockii Maxim. ■
21532 Anaphalis buisanensis Hayata = Anaphalis morrisonicola Hayata ■
21533 Anaphalis bulleyana (Jeffrey) C. C. Chang;黏毛香青(风蒿,昆明香青,五香草,午香草,香附草,香辣烟,香青,野辣烟);Stickyhair Everlasting,Stickyhair Pearleverlasting ■
21534 Anaphalis busua (Buch.-Ham. ex D. Don) DC.;蛛毛香青;Arachnoid Everlasting,Arachnoid Pearleverlasting ■
21535 Anaphalis busua (Ham.) DC. = Anaphalis busua (Buch.-Ham. ex D. Don) DC. ■
21536 Anaphalis chanetii (H. Lév.) H. Lév. = Anaphalis sinica Hance var. remota Y. Ling ■
21537 Anaphalis chanetii H. Lév. = Anaphalis sinica Hance var. remota Y. Ling ■
21538 Anaphalis chionantha DC. = Anaphalis royleana DC. ■
21539 Anaphalis chlamydophylla Diels;苞衣香青;Chlamydoleaf Pearleverlasting ■
21540 Anaphalis chungtienensis F. H. Chen;中甸香青;Zhongdian Everlasting,Zhongdian Pearleverlasting ■

21541 Anaphalis cinerascens Y. Ling et W. Wang;灰毛香青;Ashhairy Pearleverlasting,Grayhair Everlasting ■

21542 Anaphalis cinerascens Y. Ling et W. Wang = Anaphalis pannosa Hand. -Mazz. ■

21543 Anaphalis cinerascens Y. Ling et W. Wang var. congesta Y. Ling et W. Wang;密聚灰毛香青;Dense Grayhair Everlasting, Dense Grayhair Pearleverlasting ■

21544 Anaphalis cinerascens Y. Ling et W. Wang var. pellucida (Franch.) Y. Ling;薄叶旋叶香青■

21545 Anaphalis cinnamomea (DC.) C. B. Clarke = Anaphalis margaritacea (L.) Benth. et Hook. f. ■

21546 Anaphalis cinnamomea (DC.) C. B. Clarke = Anaphalis margaritacea (L.) Benth. et Hook. f. var. cinnamomea (DC.) Herder ex Maxim. ■

21547 Anaphalis cinnamomea (DC.) C. B. Clarke var. angustior Nakai = Anaphalis margaritacea (L.) Benth. et Hook. f. ■

21548 Anaphalis conferta C. C. Chang = Anaphalis aureopunctata Lingelsh. et Borza var. atrata (Hand. -Mazz.) Hand. -Mazz. ■

21549 Anaphalis contorta (D. Don) Hook. f.;旋叶香青(小火草);Coiledleaf Everlasting,Coiledleaf Pearleverlasting ■

21550 Anaphalis contorta (D. Don) Hook. f. morrisonicola (Hayata) Yamam. = Anaphalis morrisonicola Hayata ■

21551 Anaphalis contorta (D. Don) Hook. f. var. morrisonicola (Hayata) Yamam. = Anaphalis morrisonicola Hayata ■

21552 Anaphalis contorta (D. Don) Hook. f. var. pellucida (Franch.) Y. Ling;薄旋叶香青(薄叶香青);Thinleaf Coiledleaf Pearleverlasting ■

21553 Anaphalis contorta (D. Don) Hook. f. var. pellucida (Franch.) Y. Ling = Anaphalis contorta (D. Don) Hook. f. ■

21554 Anaphalis contortiformis Hand. -Mazz.;银衣香青;Silver-lavis Pearleverlasting,Twisty Everlasting ■

21555 Anaphalis corymbifera C. C. Chang;伞房香青;Corymb Everlasting,Corymb Pearleverlasting ■

21556 Anaphalis corymbosa (Bureau et Franch.) Diels = Anaphalis nepalensis var. corymbosa (Bureau et Franch.) Hand. -Mazz. ■

21557 Anaphalis corymbosa (Franch.) Diels = Anaphalis nepalensis (Spreng.) Hand. -Mazz. var. corymbosa (Franch.) Hand. -Mazz. ■

21558 Anaphalis cuneifolia (Wall.) Hook. f. = Anaphalis nepalensis (Spreng.) Hand. -Mazz. ■

21559 Anaphalis cuneifolia (Wall.) Hook. f. = Anaphalis triplinervis (Sims) Sims ex C. B. Clarke var. intermedia Airy Shaw ■

21560 Anaphalis darvasica Boriss.;达尔瓦斯香青■☆

21561 Anaphalis delavayi (Franch.) Diels;苍山香青;Delavay Everlasting,Delavay Pearleverlasting ■

21562 Anaphalis depauperata Boriss.;萎缩香青■☆

21563 Anaphalis desertii J. R. Drumm.;江孜香青;Jiangzi Everlasting, Jiangzi Pearleverlasting ■

21564 Anaphalis elegans Y. Ling;雅致香青;Elegant Everlasting, Elegant Pearleverlasting ■

21565 Anaphalis esquirolii H. Lév. = Gnaphalium adnatum (Wall. ex DC.) Kitam. ■

21566 Anaphalis falconeri C. B. Clarke = Anaphalis contorta (D. Don) Hook. f. ■

21567 Anaphalis flaccida Y. Ling;萎软香青;Flaccid Everlasting, Flaccid Pearleverlasting ■

21568 Anaphalis flavescens Hand. -Mazz.;淡黄香青(清明菜,铜钱花);Yellowish Everlasting,Yellowish Pearleverlasting ■

21569 Anaphalis flavescens Hand. -Mazz. var. lanata Y. Ling;绵毛淡黄香青;Woolly Yellowish Everlasting,Woolly Yellowish Pearleverlasting ■

21570 Anaphalis flavescens Hand. -Mazz. var. rosea Y. Ling;淡红香青■

21571 Anaphalis flavescens Hand. -Mazz. var. sulphurea Y. Ling;硫黄香青■

21572 Anaphalis flavescens Hand. -Mazz. var. taipeiensis J. Q. Fu;太白淡黄香青;Taibai Yellowish Everlasting, Taibai Yellowish Pearleverlasting ■

21573 Anaphalis flavescens Y. Ling f. rosea Y. Ling = Anaphalis flavescens Hand. -Mazz. ■

21574 Anaphalis flavescens Y. Ling f. sulphurea Y. Ling = Anaphalis flavescens Hand. -Mazz. ■

21575 Anaphalis franchetiana Diels = Anaphalis contorta (D. Don) Hook. f. ■

21576 Anaphalis franchetiana Diels = Anaphalis contorta (D. Don) Hook. f. var. pellucida (Franch.) Y. Ling ■

21577 Anaphalis gracilis Hand. -Mazz.;纤枝香青;Slenderbranch Pearleverlasting ■

21578 Anaphalis gracilis Hand. -Mazz. var. aspera Hand. -Mazz.;糙叶纤枝香青(糙叶香青);Roughleaf Pearleverlasting ■

21579 Anaphalis gracilis Hand. -Mazz. var. ulophylla Hand. -Mazz.;皱缘纤枝香青(皱叶香青,皱缘香青);Wrinkleleaf Everlasting, Wrinkleleaf Pearleverlasting ■

21580 Anaphalis hancockii Maxim.;铃铃香青(铃铃香,零陵香草,零陵香青,零零香,零零香青,铜钱花,五月霜,稀毛香青);Hancock Everlasting,Hancock's Pearleverlasting ■

21581 Anaphalis hondae Kitam.;多茎香青;Manystem Everlasting, Manystem Pearleverlasting ■

21582 Anaphalis hondae Kitam. = Anaphalis contorta (D. Don) Hook. f. ■

21583 Anaphalis horaimontana Masam.;大山香青(蓬莱籁箫)■

21584 Anaphalis hymenolepis Y. Ling;膜苞香青;Flimbract Everlasting,Membranaceous-phyllaris Pearleverlasting ■

21585 Anaphalis intermedia (Wall.) Duthie = Anaphalis nepalensis (Spreng.) Hand. -Mazz. ■

21586 Anaphalis lactea Maxim.;乳白香青(大白茅香,大茅香艾,高山香青);Milkywhite Everlasting,Milkywhite Pearleverlasting ■

21587 Anaphalis lactea Maxim. f. rosea Y. Ling;粉苞乳白香青;Rose Milkywhite Pearleverlasting ■

21588 Anaphalis larium Hand. -Mazz.;德钦香青;Deqin Everlasting, Deqin Pearleverlasting ■

21589 Anaphalis latialata Y. Ling et Y. L. Chen;宽翅香青;Broadwing Everlasting,Broadwing Pearleverlasting ■

21590 Anaphalis latialata Y. Ling et Y. L. Chen var. viridis (Hand. - Mazz.) Y. Ling et Y. L. Chen;青绿宽翅香青(绿宽翅香青);Green Broadwing Everlasting,Green Broadwing Pearleverlasting ■

21591 Anaphalis latialata Y. Ling et Y. L. Chen var. viridis (Hand. -Mazz.) Y. Ling et Y. L. Chen = Anaphalis latialata Y. Ling et Y. L. Chen ■

21592 Anaphalis likiangensis (Franch.) Y. Ling;丽江香青;Lijiang Everlasting,Lijiang Pearleverlasting ■

21593 Anaphalis mairei H. Lév. = Anaphalis nepalensis (Spreng.) Hand. -Mazz. ■

21594 Anaphalis mairei H. Lév. = Anaphalis triplinervis (Sims) Sims ex C. B. Clarke var. intermedia Airy Shaw ■

21595 Anaphalis margaritacea (L.) Benth. = Anaphalis margaritacea (L.) Benth. et Hook. f. ■

21596 Anaphalis margaritacea (L.) Benth. et Hook. f.;珠光香青(白头翁,抱茎藾箫,抱茎藾萧,避风草,大火草,大叶白头翁,九头艾,藾蒿,藾萧,毛女儿草,牛舌草,苹,山荻,香青,一面青,珠光香青菊); American Everlasting, Anaphale Marguerite, Common Pearleverlasting, Common Pearly Everlasting, Immortelle Blanche, Life Everlasting, Life-everlasting, Live-long, Moonshine, Old Sow, Pearl Cudweed, Pearl Everlasting, Pearl-flowered Everlasting, Pearl-flowered Life Everlasting, Pearly Everlasting, Pearly Immortelle, Shining Everlasting, Western Pearly Everlasting ■

21597 Anaphalis margaritacea (L.) Benth. et Hook. f. f. morrisonicola Hayata = Anaphalis morrisonicola Hayata ■

21598 Anaphalis margaritacea (L.) Benth. et Hook. f. f. nana Hayata = Anaphalis morrisonicola Hayata ■

21599 Anaphalis margaritacea (L.) Benth. et Hook. f. subsp. angustior (Miq.) Kitam. et H. Hara = Anaphalis margaritacea (L.) Benth. et Hook. f. ■

21600 Anaphalis margaritacea (L.) Benth. et Hook. f. subsp. angustior (Miq.) Kitam. = Anaphalis margaritacea (L.) Benth. et Hook. f. ■

21601 Anaphalis margaritacea (L.) Benth. et Hook. f. subsp. angustior (Miq.) Kitam. = Anaphalis margaritacea (L.) Benth. et Hook. f. var. japonica (Sch. Bip.) Makino ■

21602 Anaphalis margaritacea (L.) Benth. et Hook. f. subsp. angustior Kitam. = Anaphalis margaritacea (L.) Benth. et Hook. f. ■

21603 Anaphalis margaritacea (L.) Benth. et Hook. f. subsp. japonica (Sch. Bip.) Kitam. = Anaphalis margaritacea (L.) Benth. et Hook. f. var. angustifolia (Franch. et Sav.) Hayata ■☆

21604 Anaphalis margaritacea (L.) Benth. et Hook. f. subsp. japonica (Sch. Bip.) Kitam. = Anaphalis margaritacea (L.) Benth. et Hook. f. var. japonica (Sch. Bip.) Makino ■

21605 Anaphalis margaritacea (L.) Benth. et Hook. f. subsp. morrisonicola (Hayata) Kitam. = Anaphalis morrisonicola Hayata ■

21606 Anaphalis margaritacea (L.) Benth. et Hook. f. subsp. yedoensis (Franch. et Sav.) Kitam. = Anaphalis yedoensis (Franch. et Sav.) Maxim. ■☆

21607 Anaphalis margaritacea (L.) Benth. et Hook. f. var. angustifolia (Franch. et Sav.) Hayata;狭叶珠光香青■☆

21608 Anaphalis margaritacea (L.) Benth. et Hook. f. var. angustifolia (Franch. et Sav.) Hayata = Anaphalis margaritacea (L.) Benth. et Hook. f. var. japonica (Sch. Bip.) Makino ■

21609 Anaphalis margaritacea (L.) Benth. et Hook. f. var. angustifolia (Franch. et Sav.) Hayata f. morrisonicola Hayata = Anaphalis morrisonicola Hayata ■

21610 Anaphalis margaritacea (L.) Benth. et Hook. f. var. angustifolia (Franch. et Sav.) Hayata f. nana Hayata = Anaphalis morrisonicola Hayata ■

21611 Anaphalis margaritacea (L.) Benth. et Hook. f. var. angustior (Miq.) Nakai = Anaphalis margaritacea (L.) Benth. et Hook. f. ■

21612 Anaphalis margaritacea (L.) Benth. et Hook. f. var. cinnamomea (DC.) Herder ex Maxim.;黄褐珠光香青(黄褐香青,细白火草); Yellow-brown Pearleverlasting ■

21613 Anaphalis margaritacea (L.) Benth. et Hook. f. var. cinnamomea (DC.) Herder ex Maxim. = Anaphalis margaritacea (L.) Benth. et Hook. f. ■

21614 Anaphalis margaritacea (L.) Benth. et Hook. f. var. interangustifolia Hand. -Mazz. = Anaphalis margaritacea (L.) Benth. et Hook. f. ■

21615 Anaphalis margaritacea (L.) Benth. et Hook. f. var. intercedens Hara = Anaphalis margaritacea (L.) Benth. et Hook. f. ■

21616 Anaphalis margaritacea (L.) Benth. et Hook. f. var. japonica (Sch. Bip.) Makino;青线叶珠光香青(火花草,天青地白,线叶香青,线叶珠光香青)■

21617 Anaphalis margaritacea (L.) Benth. et Hook. f. var. japonica (Sch. Bip.) Makino = Anaphalis margaritacea (L.) Benth. et Hook. f. var. angustifolia (Franch. et Sav.) Hayata ■☆

21618 Anaphalis margaritacea (L.) Benth. et Hook. f. var. morrisonicola (Hayata) Kitam. = Anaphalis morrisonicola Hayata ■

21619 Anaphalis margaritacea (L.) Benth. et Hook. f. var. occidentalis Greene = Anaphalis margaritacea (L.) Benth. et Hook. f. ■

21620 Anaphalis margaritacea (L.) Benth. et Hook. f. var. revoluta Suksd. = Anaphalis margaritacea (L.) Benth. et Hook. f. ■

21621 Anaphalis margaritacea (L.) Benth. et Hook. f. var. subalpina (A. Gray) A. Gray = Anaphalis margaritacea (L.) Benth. et Hook. f. ■

21622 Anaphalis margaritacea (L.) Benth. et Hook. f. var. subalpina A. Gray = Anaphalis margaritacea (L.) Benth. et Hook. f. ■

21623 Anaphalis margaritacea (L.) Benth. et Hook. f. var. tsonngiana Y. Ling = Anaphalis margaritacea (L.) Benth. et Hook. f. var. japonica (Sch. Bip.) Makino ■

21624 Anaphalis margaritacea (L.) Benth. et Hook. f. var. yedoensis (Franch. et Sav.) Ohwi = Anaphalis margaritacea (L.) Benth. et Hook. f. subsp. yedoensis (Franch. et Sav.) Kitam. ■☆

21625 Anaphalis margaritacea (L.) Benth. et Hook. f. var. yedoensis (Franch. et Sav.) Ohwi = Anaphalis yedoensis (Franch. et Sav.) Maxim. ■☆

21626 Anaphalis margaritacea (L.) Benth. et Hook. f. var. yedoensis Franch. et Sav. = Anaphalis margaritacea (L.) Benth. et Hook. f. subsp. yedoensis (Franch. et Sav.) Kitam. ■☆

21627 Anaphalis margaritacea (L.) Benth. ex C. B. Clarke = Anaphalis margaritacea (L.) Benth. et Hook. f. ■

21628 Anaphalis monocephala DC. = Anaphalis nepalensis (Spreng.) Hand. -Mazz. var. monocephala (DC.) Hand. -Mazz. ■

21629 Anaphalis morii Nakai = Anaphalis sinica Hance var. morii (Nakai) Ohwi ■☆

21630 Anaphalis morrisonicola Hayata;玉山香青(玉山抱茎籁箫); Yushan Everlasting, Yushan Mout Pearleverlasting ■

21631 Anaphalis morrisonicola Hayata = Anaphalis margaritacea (L.) Benth. et Hook. f. var. morrisonicola (Hayata) Kitam. ■

21632 Anaphalis morrisonicola Hayata = Anaphalis margaritacea (L.) Benth. et Hook. f. subsp. morrisonicola (Hayata) Kitam. ■

21633 Anaphalis mucronata C. B. Clarke ex Hemsl. = Anaphalis nepalensis (Spreng.) Hand. -Mazz. var. monocephala (DC.) Hand. -Mazz. ■

21634 Anaphalis mucronata C. B. Clarke ex Hemsl. var. monocephala DC. = Anaphalis nepalensis (Spreng.) Hand. -Mazz. var. monocephala (DC.) Hand. -Mazz. ■

21635 Anaphalis mucronata C. B. Clarke ex Hemsl. var. polycephala DC. = Anaphalis nepalensis (Spreng.) Hand. -Mazz. ■

21636 Anaphalis mucronata C. B. Clarke var. polycephala DC. = Anaphalis nepalensis (Spreng.) Hand. -Mazz. ■

21637 Anaphalis muliensis (Hand. -Mazz.) Hand. -Mazz.;木里香青; Muli Everlasting, Muli Pearleverlasting ■

21638 Anaphalis muliensis (Hand. -Mazz.) Hand. -Mazz. = Anaphalis yunnanensis (Franch.) Diels var. muliensis Hand. -Mazz. ■

21639 Anaphalis nagasawae Hayata;永健香青■
21640 Anaphalis nagasawae Hayata = Anaphalis nepalensis (Spreng.) Hand. -Mazz. var. monocephala (DC.) Hand. -Mazz. ■
21641 Anaphalis nagasawae Hayata = Anaphalis nepalensis (Spreng.) Hand. -Mazz. ■
21642 Anaphalis nagasawae Hayata = Anaphalis triplinervis (Sims) Sims ex C. B. Clarke var. monocephala (DC.) Airy Shaw ■
21643 Anaphalis nepalensis (Spreng.) Hand. -Mazz. ;尼泊尔香青(打火草,尼泊尔籁箫,清明草,香草,香青,永健香青);Nagasawa Everlasting, Nagasawa's Pearleverlasting, Nepal Everlasting, Nepal Pearleverlasting ■
21644 Anaphalis nepalensis (Spreng.) Hand. -Mazz. var. corymbosa (Franch.) Hand. -Mazz. ;伞房尼泊尔香青(伞房清明草,伞房香青);Corymb Nepal Pearleverlasting ■
21645 Anaphalis nepalensis (Spreng.) Hand. -Mazz. var. monocephala (DC.) Hand. -Mazz. ;单头尼泊尔香青(单头三脉香青,单头香青);Singlehead Nepal Pearleverlasting ■
21646 Anaphalis nervosa Y. Ling = Anaphalis likiangensis (Franch.) Y. Ling ■
21647 Anaphalis nubigena DC. = Anaphalis nepalensis (Spreng.) Hand. -Mazz. ■
21648 Anaphalis nubigena DC. = Anaphalis nepalensis (Spreng.) Hand. -Mazz. var. monocephala (DC.) Hand. -Mazz. ■
21649 Anaphalis nubigena DC. = Anaphalis triplinervis (Sims) Sims ex C. B. Clarke var. intermedia Airy Shaw ■
21650 Anaphalis nubigena DC. f. reductanana Diels ex Limpr. = Anaphalis nepalensis (Spreng.) Hand. -Mazz. var. monocephala (DC.) Hand. -Mazz. ■
21651 Anaphalis nubigena DC. var. intermedia (Wall.) Duthie = Anaphalis nepalensis (Spreng.) Hand. -Mazz. ■
21652 Anaphalis nubigena DC. var. intermedia Hook. f. = Anaphalis nepalensis (Spreng.) Hand. -Mazz. ■
21653 Anaphalis nubigena DC. var. monocephala C. B. Clarke = Anaphalis nepalensis (Spreng.) Hand. -Mazz. var. monocephala (DC.) Hand. -Mazz. ■
21654 Anaphalis nubigena DC. var. polycephala C. B. Clarke = Anaphalis nepalensis (Spreng.) Hand. -Mazz. ■
21655 Anaphalis occidentalis (Greene) Heller = Anaphalis margaritacea (L.) Benth. et Hook. f. ■
21656 Anaphalis occidentalis A. Heller = Anaphalis margaritacea (L.) Benth. et Hook. f. ■
21657 Anaphalis oligandra DC. = Pseudognaphalium oligandrum (DC.) Hilliard et B. L. Burtt ■☆
21658 Anaphalis oxyphylla Y. Ling et C. Shih;锐叶香青;Sharpleaf Everlasting, Sharp-leaf Pearleverlasting ■
21659 Anaphalis pachylaena Y. L. Chen et Y. Ling;厚衣香青;Thickconered Everlasting, Thickening-hairs Pearleverlasting ■
21660 Anaphalis pannosa Hand. -Mazz. ;污毛香青;Dirtyhair Everlasting, Pannose Pearleverlasting ■
21661 Anaphalis patentifolia Rech. f. ;展叶香青■☆
21662 Anaphalis plicata Kitam. ;褶苞香青;Plicatebract Everlasting, Plicate-phyllaries Pearleverlasting ■
21663 Anaphalis polylepis DC. = Anaphalis royleana DC. ■
21664 Anaphalis porphyrolepis Y. Ling et Y. L. Chen;紫苞香青;Purplebract Everlasting, Purple-phyllaries Pearleverlasting ■
21665 Anaphalis possietica Kom. = Anaphalis sinica Hance ■
21666 Anaphalis pterocaulon (Franch. et Sav.) Maxim. = Anaphalis sinica Hance ■
21667 Anaphalis pterocaulon (Franch. et Sav.) Maxim. subsp. morii (Nakai) Kitam. = Anaphalis sinica Hance var. morii (Nakai) Ohwi ■☆
21668 Anaphalis pterocaulon (Franch. et Sav.) Maxim. var. atrata Hand. -Mazz. = Anaphalis aureopunctata Lingelsh. et Borza var. atrata Hand. -Mazz. ■
21669 Anaphalis pterocaulon (Franch. et Sav.) Maxim. var. calvescens Pamp. = Anaphalis aureopunctata Lingelsh. et Borza f. calvescens (Pamp.) F. H. Chen ■
21670 Anaphalis pterocaulon (Franch. et Sav.) Maxim. var. intermedia Pamp. = Anaphalis aureopunctata Lingelsh. et Borza ■
21671 Anaphalis pterocaulon (Franch. et Sav.) Maxim. var. sinica (Hance) Hand. -Mazz. = Anaphalis sinica Hance ■
21672 Anaphalis pterocaulon (Franch. et Sav.) Maxim. var. surculosa Hand. -Mazz. = Anaphalis surculosa (Hand. -Mazz.) Hand. -Mazz. ■
21673 Anaphalis racemifera Franch. ;总状香青■☆
21674 Anaphalis rhoddodactyla W. W. Sm. ;红指香青;Pinkdactil Everlasting, Pinkdactil Pearleverlasting ■
21675 Anaphalis roseo-alba Krasch. ;粉白香青■☆
21676 Anaphalis royleana DC. ;须弥香青(能高籁箫);Royle Everlasting, Royle's Pearleverlasting ■
21677 Anaphalis royleana DC. var. cana ?;灰色须弥香青■☆
21678 Anaphalis royleana DC. var. concolor ?;同色须弥香青■☆
21679 Anaphalis scopulosa Boriss. ;岩栖香青■☆
21680 Anaphalis semidecurrens (Wall. ex DC.) DC. = Anaphalis busua (Buch. -Ham. ex D. Don) DC. ■
21681 Anaphalis sericeo-albida H. Lév. = Gnaphalium adnatum (Wall. ex DC.) Kitam. ■
21682 Anaphalis sinica Hance;香青(白四棱锋,翅茎香青,荻,枫茄香,籁箫,连肠香,牛香草,通肠香);Chinese Everlasting, Chinese Pearleverlasting ■
21683 Anaphalis sinica Hance subsp. intermedia (Pamp.) Kitam. ;间型香青■☆
21684 Anaphalis sinica Hance subsp. intermedia (Pamp.) Kitam. = Anaphalis aureopunctata Lingelsh. et Borza ■
21685 Anaphalis sinica Hance subsp. intermedia (Pamp.) Kitam. var. atrata (Hand. -Mazz.) Kitam. = Anaphalis aureopunctata Lingelsh. et Borza var. atrata (Hand. -Mazz.) Hand. -Mazz. ■
21686 Anaphalis sinica Hance subsp. intermedia (Pamp.) Kitam. var. tomentosa (Hand. -Mazz.) Kitam. = Anaphalis aureopunctata Lingelsh. et Borza var. tomentosa Hand. -Mazz. ■
21687 Anaphalis sinica Hance subsp. morii (Nakai) Kitam. = Anaphalis sinica Hance var. morii (Nakai) Ohwi ■☆
21688 Anaphalis sinica Hance var. atrata (Hand. -Mazz.) Kitam. = Anaphalis aureopunctata Lingelsh. et Borza var. atrata (Hand. -Mazz.) Hand. -Mazz. ■
21689 Anaphalis sinica Hance var. calvescens (Pamp.) S. Y. Hu = Anaphalis aureopunctata Lingelsh. et Borza f. calvescens (Pamp.) F. H. Chen ■
21690 Anaphalis sinica Hance var. densata Y. Ling;密生香青;Dense Chinese Everlasting, Dense Chinese Pearleverlasting ■
21691 Anaphalis sinica Hance var. lanata Y. Ling;绵毛香青(棉毛香青);Woolly Chinese Everlasting, Woolly Chinese Pearleverlasting ■
21692 Anaphalis sinica Hance var. morii (Nakai) Ohwi;森氏香青■☆
21693 Anaphalis sinica Hance var. pernivea T. Shimizu;极白香青■☆
21694 Anaphalis sinica Hance var. remota Y. Ling;疏生香青;Lax

Chinese Everlasting, Lax Chinese Pearleverlasting ■
21695 Anaphalis sinica Hance var. remota Y. Ling f. rubra (Hand.-Mazz.) Y. Ling;红花香青■
21696 Anaphalis sinica Hance var. tomentosa (Hand.-Mazz.) Kitam. = Anaphalis aureopunctata Lingelsh. et Borza var. tomentosa Hand.-Mazz. ■
21697 Anaphalis sinica Hance var. viscosissima (Honda) Kitam.;黏绵毛香青■☆
21698 Anaphalis sinica Hance var. yakusimensis (Masam.) Yahara;屋久岛香青■☆
21699 Anaphalis souliei Diels;蜀西香青;Soulie Everlasting, Western-Sichuan Pearleverlasting ■
21700 Anaphalis spodiophylla Y. Ling et Y. L. Chen;灰叶香青;Grey Leaf Pearleverlasting, Greyleaf Everlasting ■
21701 Anaphalis staintonii Georgiadou;斯坦顿香青■☆
21702 Anaphalis stenocephala Y. Ling et C. Shih;狭苞香青;Narrowhead Everlasting, Narrowhead Pearleverlasting ●■
21703 Anaphalis stoliczkai C. B. Clarke = Anaphalis virgata Thomson ■☆
21704 Anaphalis suffruticosa Hand.-Mazz.;亚灌木香青;Suffrutescent Everlasting, Suffrutescent Pearleverlasting ●
21705 Anaphalis surculosa (Hand.-Mazz.) Hand.-Mazz.;萌条香青(五香花);Shoot Everlasting, Shoot Pearleverlasting ■
21706 Anaphalis surculosa (Hand.-Mazz.) Hand.-Mazz. = Anaphalis busua (Ham.) DC. ■
21707 Anaphalis szechuanensis Y. Ling et Y. L. Chen;四川香青;Sichuan Everlasting, Sichuan Pearleverlasting ■
21708 Anaphalis szechuanensis Y. Ling et Y. L. Chen f. humilis Y. Ling;矮小四川香青■
21709 Anaphalis tenella DC. = Anaphalis contorta (D. Don) Hook. f. ■
21710 Anaphalis tenuisissima C. C. Chang;细弱香青;Muchslender Everlasting, Muchslender Pearleverlasting ■
21711 Anaphalis tibetica Kitam.;西藏香青;Xizang Everlasting, Xizang Pearleverlasting ■
21712 Anaphalis timmua (D. Don) Hand.-Mazz. = Anaphalis margaritacea (L.) Benth. et Hook. f. ■
21713 Anaphalis todaiensis Honda = Anaphalis sinica Hance ■
21714 Anaphalis transnokoensis Sasaki;能高香青(能高籁箫);Transnoko Everlasting, Transnoko Pearleverlasting ■
21715 Anaphalis triplinervis (Sims) C. B. Clarke var. intermedia Airy Shaw = Anaphalis nepalensis (Spreng.) Hand.-Mazz. ■
21716 Anaphalis triplinervis (Sims) Sims ex C. B. Clarke;三脉香青(三脉香青菊);Pearly Everlasting, Trinerved Everlasting, Triplenerved Pearleverlasting ■
21717 Anaphalis triplinervis (Sims) Sims ex C. B. Clarke var. intermedia Airy Shaw = Anaphalis nepalensis (Spreng.) Hand.-Mazz. ■
21718 Anaphalis triplinervis (Sims) Sims ex C. B. Clarke var. monocephala (DC.) Airy Shaw = Anaphalis nepalensis (Spreng.) Hand.-Mazz. var. monocephala (DC.) Hand.-Mazz. ■
21719 Anaphalis triplinervis (Sims.) C. B. Clarke var. intermedia ? = Anaphalis nepalensis (Spreng.) Hand.-Mazz. ■
21720 Anaphalis triplinervis (Sims.) C. B. Clarke var. monocephala Airy Shaw = Anaphalis nepalensis (Spreng.) Hand.-Mazz. var. monocephala (DC.) Hand.-Mazz. ■
21721 Anaphalis undulata Hand.-Mazz. = Anaphalis bicolor (Franch.) Diels var. undulata (Hand.-Mazz.) Y. Ling ■
21722 Anaphalis velutina Krasch.;短绒毛香青■☆
21723 Anaphalis virens C. C. Chang;黄绿香青;Yellowgreen Everlasting, Yellowgreen Pearleverlasting ■
21724 Anaphalis virgata Roem.;帚枝香青(香青);Virgate Everlasting, Virgate Pearleverlasting ●■
21725 Anaphalis virgata Thomson;多枝香青■☆
21726 Anaphalis viridis Cummins;绿香青(绿黄香青);Green Everlasting, Green Pearleverlasting ■
21727 Anaphalis viridis Cummins var. acaulis Hand.-Mazz.;无茎绿香青;Stemless Green Everlasting, Stemless Green Pearleverlasting ■
21728 Anaphalis viscosissima Honda = Anaphalis sinica Hance var. viscosissima (Honda) Kitam. ■☆
21729 Anaphalis xylorhiza Sch. Bip.;木根香青;Woodyroot Everlasting, Woodyroot Pearleverlasting ■
21730 Anaphalis yangii Y. L. Chen et Y. L. Lin;竞争香青;Yang Pearleverlasting ■
21731 Anaphalis yedoensis (Franch. et Sav.) Maxim.;江户香青(荻) ■☆
21732 Anaphalis yedoensis (Franch. et Sav.) Maxim. = Anaphalis margaritacea (L.) Benth. et Hook. f. ■
21733 Anaphalis yunnanensis (Franch.) Diels;云南香青;Yunnan Everlasting, Yunnan Pearleverlasting ■
21734 Anaphalis yunnanensis (Franch.) Diels var. muliensis Hand.-Mazz. = Anaphalis muliensis (Hand.-Mazz.) Hand.-Mazz. ■
21735 Anaphalis yunnanensis Diels var. muliensis Hand.-Mazz. = Anaphalis muliensis (Hand.-Mazz.) Hand.-Mazz. ■
21736 Anaphora Gagnep. = Dienia Lindl. ■
21737 Anaphora Gagnep. = Malaxis Sol. ex Sw. ■
21738 Anaphora liparioides Gagnep. = Dienia ophrydis (J. König) Ormerod et Seidenf. ■
21739 Anaphragma Steven = Astragalus L. ●■
21740 Anaphrenium E. Mey. = Heeria Meisn. ●☆
21741 Anaphrenium E. Mey. ex Endl. = Heeria Meisn. ●☆
21742 Anaphrenium abyssinicum Hochst. = Ozoroa insignis Delile ●☆
21743 Anaphrenium abyssinicum Hochst. var. latifolium Engl. = Ozoroa insignis Delile var. latifolia (Engl.) R. Fern. ●☆
21744 Anaphrenium argenteum E. Mey. = Heeria argentea (Thunb.) Meisn. ●☆
21745 Anaphrenium concolor E. Mey. = Ozoroa concolor (C. Presl) De Winter ●☆
21746 Anaphrenium crassinervium Engl. = Ozoroa crassinervia (Engl.) R. Fern. et A. Fern. ●☆
21747 Anaphrenium dispar E. Mey. = Ozoroa dispar (C. Presl) R. Fern. et A. Fern. ●☆
21748 Anaphrenium kienerae Sacleux = Ozoroa stenophylla (Engl. et Gilg) R. Fern. et A. Fern. ●☆
21749 Anaphrenium longifolium Bernh. = Protorhus longifolia (Bernh.) Engl. ●☆
21750 Anaphrenium mucronatum Bernh. = Ozoroa mucronata (Bernh.) R. Fern. et A. Fern. ●☆
21751 Anaphrenium paniculosum (Sond.) Engl. = Ozoroa paniculosa (Sond.) R. Fern. et A. Fern. ●☆
21752 Anaphrenium pulcherrimum Schweinf. = Ozoroa pulcherrima (Schweinf.) R. Fern. et A. Fern. ●☆
21753 Anaphrenium verticillatum Engl. = Ozoroa verticillata (Engl.) R. Fern. et A. Fern. ●☆
21754 Anaphyllopsis A. Hay (1989);拟印度南星属■☆
21755 Anaphyllopsis americana (Engl.) A. Hay;拟印度南星■☆

21756 Anaphyllum Schott(1857);印度南星属■☆
21757 Anaphyllum wightii Schott;印度南星■☆
21758 Anapodophyllon Mill. = Podophyllum L. ■☆
21759 Anapodophyllum Moench = Podophyllum L. ■☆
21760 Anapodophyllum Tourn. ex Moench = Podophyllum L. ■☆
21761 Anarmodium Schott = Dracunculus Mill. ■☆
21762 Anarmosa Miers ex Hook. = Tetilla DC. ■☆
21763 Anarrhinum Desf. (1798)(保留属名);锉玄参属●■☆
21764 Anarrhinum abyssinicum Jaub. et Spach = Anarrhinum forskaohlii (J. F. Gmel.) Cufod. subsp. abyssinicum (Jaub. et Spach) Sutton ■☆
21765 Anarrhinum bellidifolium Desf.; 美 叶 锉 玄 参; Daisy-leaved Toadflax ■☆
21766 Anarrhinum brevifolium Coss. et Kralik = Anarrhinum fruticosum Desf. subsp. brevifolium (Coss. et Kralik) Sutton ■☆
21767 Anarrhinum forskaohlii (J. F. Gmel.) Cufod.;阿拉伯锉玄参■☆
21768 Anarrhinum forskaohlii (J. F. Gmel.) Cufod. subsp. abyssinicum (Jaub. et Spach) Sutton;阿比西尼亚锉玄参■☆
21769 Anarrhinum fruticosum Desf.;灌丛锉玄参■☆
21770 Anarrhinum fruticosum Desf. subsp. brevifolium (Coss. et Kralik) Sutton;短叶灌丛锉玄参■☆
21771 Anarrhinum fruticosum Desf. subsp. demnatense (Coss.) Maire; 代姆纳特锉玄参■☆
21772 Anarrhinum fruticosum Desf. var. delonii Maire = Anarrhinum fruticosum Desf. ■☆
21773 Anarrhinum laxiflorum Boiss.;疏花锉玄参■☆
21774 Anarrhinum pechuelii Kuntze = Diclis petiolaris Benth. ■☆
21775 Anarrhinum pedatum Desf.;足锉玄参■☆
21776 Anarrhinum pedatum Desf. var. longiracemosum Sennen = Anarrhinum pedatum Desf. ■☆
21777 Anarrhinum pedatum Desf. var. villicaule Maire = Anarrhinum pedatum Desf. ■☆
21778 Anarrhinum pedicellatum T. Anderson = Schweinfurthia pedicellata (T. Anderson) Balf. f. ■☆
21779 Anarrhinum veronicoides (A. Rich.) Kuntze = Diclis ovata Benth. ■☆
21780 Anarthria R. Br. (1810);刷柱草属(无柄草属)●☆
21781 Anarthria gracilis R. Br.;细刷柱草●☆
21782 Anarthria grandiflora Nees;大花刷柱草●☆
21783 Anarthria humilis Nees;矮刷柱草●☆
21784 Anarthria laevis R. Br.;平滑刷柱草●☆
21785 Anarthria pauciflora R. Br.;少花刷柱草●☆
21786 Anarthria polyphylla Nees;多叶刷柱草●☆
21787 Anarthriaceae D. F. Cutler et Airy Shaw = Restionaceae R. Br. (保留科名)■
21788 Anarthriaceae D. F. Cutler et Airy Shaw(1965);刷柱草科(无柄草科,苞穗草科)●☆
21789 Anarthrophyllum Benth. (1865);小叶金雀豆属■☆
21790 Anarthrophyllum brevistipula Phil.;短小叶金雀豆■☆
21791 Anarthrophyllum capitatum Soraru;头状小叶金雀豆■☆
21792 Anarthrophyllum macrophyllum Soraru;大小叶金雀豆■☆
21793 Anarthrosyne E. Mey. = Pseudarthria Wight et Arn. ■☆
21794 Anarthrosyne cordata Klotzsch = Desmodium velutinum (Willd.) DC. ●
21795 Anarthrosyne densiflora Klotzsch = Pseudarthria hookeri Wight et Arn. ●☆
21796 Anarthrosyne gracilis Klotzsch = Pseudarthria confertiflora (A. Rich.) Baker ●☆
21797 Anarthrosyne robusta E. Mey. = Pseudarthria hookeri Wight et Arn. ●☆
21798 Anartia Miers = Tabernaemontana L. ●
21799 Anaschovadi Adans. = Elephantopus L. ■
21800 Anaspis Rech. f. = Scutellaria L. ●■
21801 Anasser Juss. = Geniostoma J. R. Forst. et G. Forst. ●
21802 Anasser laniti Blanco = Wrightia pubescens R. Br. ●
21803 Anassera Pers. = Anasser Juss. ●
21804 Anastatica L. (1753);复活草属(安产树属,翅果荠属,含生草属);Rose of Jericho ■☆
21805 Anastatica hierochuntica L.;复活草(安产树,含生草);Jericho Resurrection Mustard,Jericho Resurrection-mustard,Mary's Flower, Palestinian Tumbleweed, Resurrection Flower, Resurrection Plant, Rose of Jericho,Rose of the Virgin,Rose-of-Jericho ■☆
21806 Anastatica syriaca L. = Euclidium syriacum (L.) R. Br. ■
21807 Anastrabe E. Mey. ex Benth. (1836);斜玄参属●☆
21808 Anastrabe integerrima E. Mey. ex Benth.;全缘斜玄参●☆
21809 Anastrabe integerrima E. Mey. ex Benth. var. serrulata (E. Mey. ex Benth.) Hiern = Anastrabe integerrima E. Mey. ex Benth. ●☆
21810 Anastrabe serrulata E. Mey. ex Benth. = Anastrabe integerrima E. Mey. ex Benth. ●☆
21811 Anastraphia D. Don(1830);南美菊属■☆
21812 Anastrephea Decne. = Anastraphia D. Don ■☆
21813 Anastrophea Wedd. = Sphaerothylax Bisch. ex Krauss ■☆
21814 Anastrophea abyssinica Wedd. = Sphaerothylax abyssinica (Wedd.) Warm. ■☆
21815 Anastrophus Schltdl. = Axonopus P. Beauv. ■
21816 Anastrophus Schltdl. = Paspalum L. ■
21817 Anasyllls E. Mey. = Loxostylis A. Spreng. ex Rchb. ●☆
21818 Anathallis Barb. Rodr. = Pleurothallis R. Br. ■☆
21819 Anatherix Steud. = Anantherix Nutt. ■
21820 Anatherix Steud. = Asclepias L. ■
21821 Anatherix Steud. = Asclepiodora A. Gray ■
21822 Anatherostipa (Hack. ex Kuntze) P. Peñailillo B. = Stipa L. ■
21823 Anatherum Nábělek = Festuca L. ■
21824 Anatherum Nábělek = Leucopoa Griseb. ■
21825 Anatherum Nábělek = Nabelekia Roshev. ■
21826 Anatherum P. Beauv. = Andropogon L. (保留属名)■
21827 Anatherum africanum (Franch.) Roberty = Andropogon africanus Franch. ■☆
21828 Anatherum cyrtocladum (Stapf) Roberty = Andropogon kelleri Hack. ■☆
21829 Anatherum muricatum (Retz.) P. Beauv. = Vetiveria zizanioides (L.) Nash ■
21830 Anatherum muricatum P. Beauv. = Vetiveria zizanioides (L.) Nash ■
21831 Anatis Sessé et Moc. ex Brongn. = Roulinia Brongn. ●■☆
21832 Anatropa Ehrenb. = Tetradiclis Steven ex M. Bieb. ■☆
21833 Anatropa tenella Ehrenb. = Tetradiclis tenella (Ehrenb.) Litv. ■☆
21834 Anatropanthus Schltr. (1908);加岛萝藦属●■☆
21835 Anatropanthus borneensis Schltr.;加岛萝藦●■☆
21836 Anatropestylia (Plitmann) Kupicha = Vicia L. ■
21837 Anatropostylia (Plitmann) Kupicha = Ormosia Jacks. (保留属名)●
21838 Anaua Miq. = Drypetes Vahl ●
21839 Anaua Miq. = Hemicyclia Wight et Arn. ●

21840 Anaueria Kosterm. (1938);巴西樟属●☆
21841 Anaueria Kosterm. = Beilschmiedia Nees ●
21842 Anaueria brasiliensis Kosterm.;巴西樟●☆
21843 Anauxanopetalum Teijsm. et Binn. = Swintonia Griff. ●☆
21844 Anavinga Adans. = Casearia Jacq. ●
21845 Anax Ravenna = Stenomesson Herb. ■☆
21846 Anaxagoraea Mart. = Anaxagorea A. St. -Hil. ●
21847 Anaxagorea A. St. -Hil. (1825);蒙蒿子属;Anaxagorea ●
21848 Anaxagorea acuminata St. Hil.;渐尖蒙蒿子●☆
21849 Anaxagorea allenii R. E. Fr.;阿伦蒙蒿子●☆
21850 Anaxagorea angustifolia Timmerman;窄叶蒙蒿子●☆
21851 Anaxagorea borneensis (Becc.) J. Sinclair;婆罗洲蒙蒿子●☆
21852 Anaxagorea brachycarpa R. E. Fr.;短果蒙蒿子●☆
21853 Anaxagorea brevipes Benth.;短梗蒙蒿子●☆
21854 Anaxagorea clavata R. E. Fr.;棒状蒙蒿子●☆
21855 Anaxagorea crassipetala Hemsl.;厚托叶蒙蒿子●☆
21856 Anaxagorea floribunda Timmerman;多花蒙蒿子●☆
21857 Anaxagorea luzonensis A. Gray;蒙蒿子(长柄灯台树);Luzon Anaxagorea ●
21858 Anaxagorea minor Diels ex R. E. Fr.;小蒙蒿子●☆
21859 Anaxagorea pallida Diels;苍白蒙蒿子●☆
21860 Anaxeton Gaertn. (1791);紫花鼠麴木属●☆
21861 Anaxeton Schrank = Helipterum DC. ex Lindl. ■☆
21862 Anaxeton Schrank = Syncarpha DC. ■☆
21863 Anaxeton angustifolium Lundgren;窄叶紫花鼠麴木●☆
21864 Anaxeton arborescens (L.) Less.;乔木紫花鼠麴木●☆
21865 Anaxeton arboreum (L.) Gaertn. = Anaxeton arborescens (L.) Less. ●☆
21866 Anaxeton asperum (Thunb.) DC.;粗糙紫花鼠麴木●☆
21867 Anaxeton asperum (Thunb.) DC. subsp. pauciflorum Lundgren;少花粗糙紫花鼠麴木●☆
21868 Anaxeton asperum (Thunb.) DC. var. laeve Harv. = Anaxeton laeve (Harv.) Lundgren ●☆
21869 Anaxeton brevipes Lundgren;短梗紫花鼠麴木●☆
21870 Anaxeton canescens (DC.) Benth. et Hook. ex B. D. Jacks. = Langebergia canescens (DC.) Anderb. ●☆
21871 Anaxeton ellipticum Lundgren;椭圆紫花鼠麴木●☆
21872 Anaxeton floridum Poir. = Anaxeton arborescens (L.) Less. ●☆
21873 Anaxeton hirsutum (Thunb.) Less.;毛紫花鼠麴木●☆
21874 Anaxeton laeve (Harv.) Lundgren;平滑紫花鼠麴木●☆
21875 Anaxeton recurvum (Lam.) DC. = Anaxeton arborescens (L.) Less. ●☆
21876 Anaxeton septentrionale Vatke = Helichrysopsis septentrionalis (Vatke) Hilliard ■☆
21877 Anaxeton virgatum DC.;条纹紫花鼠麴木●☆
21878 Anblatum Hill = Lathraea L. ■
21879 Ancalanthus Balf. f. = Angkalanthus Balf. f. ■☆
21880 Ancana F. Muell. (1865);安卡纳树属●☆
21881 Ancana F. Muell. = Fissistigma Griff. ●
21882 Ancana F. Muell. = Meiogyne Miq. ●
21883 Ancana hirsuta Jessup;毛安卡纳树●☆
21884 Ancana stenopetala F. Muell.;安卡纳树●☆
21885 Ancanthia Steud. = Ancathia DC. ■
21886 Ancathia DC. (1833);肋果蓟属;Ancathia ■
21887 Ancathia DC. = Cirsium Mill. ■
21888 Ancathia igniaria (Spreng.) DC.;肋果蓟;Ancathia, Common Ancathia ■
21889 Anchietea A. St. -Hil. (1824);囊果堇属(美堇属);Anchietea, Pirageia ■☆
21890 Anchietea parvifolia Hallier f.;小叶囊果堇■☆
21891 Anchietea pyrifolia A. St. -Hil.;梨叶囊果堇■☆
21892 Anchietea salutaris A. St. -Hil.;巴西囊果堇(健身美堇)■☆
21893 Anchionium Rchb. = Anchonium DC. ■☆
21894 Anchomanes Schott(1853);热非南星属■☆
21895 Anchomanes abbreviatus Engl.;缩短热非南星■☆
21896 Anchomanes boehmii Engl.;贝姆热非南星■☆
21897 Anchomanes dalzielii N. E. Br. = Anchomanes difformis (Blume) Engl. ■☆
21898 Anchomanes difformis (Blume) Engl.;不齐热非南星■☆
21899 Anchomanes difformis (Blume) Engl. var. obtusus (A. Chev.) Knecht = Anchomanes difformis (Blume) Engl. ■☆
21900 Anchomanes difformis (Blume) Engl. var. pallidus (Hook.) Hepper = Anchomanes difformis (Blume) Engl. ■☆
21901 Anchomanes dubius Schott = Anchomanes difformis (Blume) Engl. ■☆
21902 Anchomanes giganteus Engl.;巨大热非南星■☆
21903 Anchomanes hookeri (Kunth) Schott = Anchomanes difformis (Blume) Engl. ■☆
21904 Anchomanes hookeri (Kunth) Schott var. pallidus Hook. = Anchomanes difformis (Blume) Engl. ■☆
21905 Anchomanes hookeri Schott = Anchomanes difformis (Blume) Engl. ■☆
21906 Anchomanes nigritanus Rendle;尼格里塔热非南星■☆
21907 Anchomanes obtusus A. Chev. = Anchomanes difformis (Blume) Engl. ■☆
21908 Anchomanes welwitschii Rendle = Anchomanes difformis (Blume) Engl. ■☆
21909 Anchonium DC. (1821);木果芥属■☆
21910 Anchonium brachycarpum (Trautv.) Vassilcz.;短果木果芥■☆
21911 Anchonium elichrysifolium (DC.) Boiss.;木果芥■☆
21912 Anchonium sterigmoides Lipsky;利普斯基木果芥■☆
21913 Anchusa Hill = Alkanna Tausch(保留属名)●☆
21914 Anchusa L. (1753);牛舌草属;Alkanet, Anchusa, Blue Bugloss, Bugloss ■
21915 Anchusa aegyptiaca (L.) A. DC.;埃及牛舌草■☆
21916 Anchusa affinis R. Br.;近缘牛舌草■☆
21917 Anchusa affinis R. Br. var. magdalenae Gusul. = Anchusa affinis R. Br. ■☆
21918 Anchusa aggregata Lehm.;团集牛舌草■☆
21919 Anchusa alborosea Benoist = Anchusa atlantica Ball ■☆
21920 Anchusa arvensis (L.) M. Bieb.;小牛舌草(狼牙草,狼紫草);Annual Bugloss, Bugle, Bugloss, Corn Bugloss, Field Bugloss, Lesser Bugloss, Orchanet, Ox Tongue, Ox-tongue, Small Bugloss, Wall Bugloss, Wild Bugloss, Wolf's Eye, Wolf's Eyes ■☆
21921 Anchusa arvensis (L.) M. Bieb. subsp. orientalis (L.) Nordh. = Anchusa ovata Lehm. ■
21922 Anchusa asperrima Delile = Arnebia hispidissima (Sieber ex Lehm.) DC. ■☆
21923 Anchusa atlantica Ball;大西洋牛舌草■☆
21924 Anchusa aucheri A. DC.;奥氏牛舌草■☆
21925 Anchusa azurea Mill.;光亮牛舌草;Italian Bugloss, Showy Bugloss ■☆
21926 Anchusa azurea Mill. = Anchusa italica Retz. ■
21927 Anchusa azurea Mill. var. macrophylla (Lam.) Sauvage et Vindt

= Anchusa italica Retz. ■
21928 Anchusa barrelieri Vitman；巴氏牛舌草；Barrelier Bugloss，Barrelier's Bugloss，Early Bugloss，False Alkanet ■☆
21929 Anchusa barrelieri Vitman = Cynoglottis barrelieri（All.）Vural et Kit Tan ■☆
21930 Anchusa calcarea Batt. = Anchusa pseudogranatensis（Braun-Blanq. et Maire）Sennen et Mauricio ■☆
21931 Anchusa calcarea Boiss.；钙生牛舌草■☆
21932 Anchusa calcarea Boiss. var. scaberrima ? = Anchusa calcarea Boiss. ■☆
21933 Anchusa capensis Thunb.；南非牛舌草（非洲勿忘草，好望角牛舌草）；Alkanet，Bugloss，Cape Alkanet，Cape Bugloss，Cape Forget-me-not，Summer Forget-me-not ■☆
21934 Anchusa capensis Thunb. 'Alba'；白花非牛舌草■☆
21935 Anchusa capensis Thunb. 'Blue Angel'；蓝天使南非牛舌草（蓝天使好望角牛舌草）■☆
21936 Anchusa capensis Thunb. 'Blue Bird'；蓝鸟南非牛舌草（蓝鸟好望角牛舌草）■☆
21937 Anchusa cespitosa Lam.；簇生牛舌草■☆
21938 Anchusa claryi Batt.；克莱里牛舌草■☆
21939 Anchusa dregei A. DC. = Anchusa riparia DC. ■☆
21940 Anchusa gmelinii Ledeb. ex Spreng.；格氏牛舌草■☆
21941 Anchusa granatensis Boiss. var. albiflora Batt. = Anchusa pseudogranatensis（Braun-Blanq. et Maire）Sennen et Mauricio ■☆
21942 Anchusa hispida Forssk. = Gastrocotyle hispida（Forssk.）Bunge ■
21943 Anchusa hispida Forssk. var. songarica Trautv. = Gastrocotyle hispida（Forssk.）Bunge ■
21944 Anchusa humilis（Desf.）I. M. Johnst. = Echium humile Desf. ■☆
21945 Anchusa hybrida Ten.；杂种牛舌草■☆
21946 Anchusa hybrida Ten. var. claryi（Batt.）Gusul. = Anchusa claryi Batt. ■☆
21947 Anchusa hybrida Ten. var. proceroides Gusul. = Anchusa hybrida Ten. ■☆
21948 Anchusa hybrida Ten. var. pubescens Gusul. = Anchusa hybrida Ten. ■☆
21949 Anchusa italica Retz.；牛舌草（天蓝牛舌草，意大利牛舌草）；Alkanet，Garden Anchusa，Italian Alkanet，Italian Bugloss，Large Blue Alkanet ■
21950 Anchusa italica Retz. = Anchusa azurea Mill. ■☆
21951 Anchusa italica Retz. var. strigosa Maire = Anchusa italica Retz. ■
21952 Anchusa lanata L. = Cynoglossum mathezii Greuter et Burdet ■☆
21953 Anchusa leptophylla Roem. et Schult.；窄叶牛舌草■☆
21954 Anchusa macrophylla Lam. = Anchusa italica Retz. ■
21955 Anchusa mairei Gusul.；迈雷牛舌草■☆
21956 Anchusa milleri Spreng.；米勒牛舌草■☆
21957 Anchusa ochroleuca M. Bieb.；淡黄牛舌草；Yellow Alkanet ■☆
21958 Anchusa officinalis Desf. = Anchusa pseudogranatensis（Braun-Blanq. et Maire）Sennen et Mauricio ■☆
21959 Anchusa officinalis L.；药用牛舌草（小花牛舌草，紫草，紫朱草）；Alkanet，Common Alkanet，Common Bugloss，Eyebright，Medicinal Alkanet，Official Bugloss，Orcanette，Orchanet，Ox Tongue，Ox-tongue，Spanish Bugloss，Wild Bugloss ■
21960 Anchusa officinalis L. var. incarnata Hort.；肉花药用牛舌草■☆
21961 Anchusa orientalis（L.）Rchb. f. = Anchusa arvensis（L.）M. Bieb. subsp. orientalis（L.）Nordh. ■
21962 Anchusa orientalis（L.）Rchb. f. = Anchusa ovata Lehm. ■
21963 Anchusa orientalis L. = Alkanna orientalis（L.）Boiss. ●☆
21964 Anchusa orientalis Rchb. f. = Anchusa ovata Lehm. ■
21965 Anchusa ovata Lehm.；东方狼紫草（狼紫草，水私利，野旱烟）；Oriental Ablfgromwell，Oriental Lycopsis ■
21966 Anchusa ovata Lehm. = Lycopsis orientalis L. ■
21967 Anchusa ovate Lehm. = Anchusa arvensis（L.）M. Bieb. subsp. orientalis（L.）Nordh. ■
21968 Anchusa paniculata Aiton = Anchusa azurea Mill. ■☆
21969 Anchusa picta M. Bieb. = Nonea caspica（Willd.）G. Don ■
21970 Anchusa procera Besser = Anchusa officinalis L. ■
21971 Anchusa procera Besser ex Link = Anchusa officinalis L. ■
21972 Anchusa pseudogranatensis（Braun-Blanq. et Maire）Sennen et Mauricio = Anchusa ochroleuca M. Bieb. ■☆
21973 Anchusa pseudoochroleuca Des. -Shost.；假淡黄牛舌草■☆
21974 Anchusa pusilla Gusul.；微小牛舌草■☆
21975 Anchusa riparia DC.；河岸牛舌草■☆
21976 Anchusa saxatilis Pall. = Stenosolenium saxatile（Pall.）Turcz. ■
21977 Anchusa sikkimensis C. B. Clarke = Microula sikkimensis（C. B. Clarke）Hemsl. ■
21978 Anchusa spinocarpos Forssk. = Lappula spinocarpa（Forssk.）Asch. ex Kuntze ■
21979 Anchusa strigosa Banks et Sol.；粗毛牛舌草■☆
21980 Anchusa stylosa M. Bieb.；长花柱牛舌草■☆
21981 Anchusa tenella Hornem. = Bothriospermum tenellum（Hornem.）Fisch. et C. A. Mey. ■
21982 Anchusa tenella Hornem. = Bothriospermum zeylanicum（J. Jacq.）Druce ■
21983 Anchusa thessala Boiss. et Spreng.；费沙立牛舌草■☆
21984 Anchusa tinctoria L.；染用牛舌草；Dyer's Bugloss ■☆
21985 Anchusa tinctoria L. = Alkanna tinctoria Tausch ●☆
21986 Anchusa undulata L.；波状牛舌草■☆
21987 Anchusa undulata L. subsp. atlantica（Ball）Braun-Blanq. et Maire = Anchusa atlantica Ball ■☆
21988 Anchusa undulata L. subsp. hybrida（Ten.）Bég. = Anchusa hybrida Ten. ■☆
21989 Anchusa undulata L. subsp. lamprocarpa Braun-Blanq. et Maire；亮果波状牛舌草■☆
21990 Anchusa undulata L. subsp. pseudogranatensis（Braun-Blanq. et Maire）Ouyahya = Anchusa pseudogranatensis（Braun-Blanq. et Maire）Sennen et Mauricio ■☆
21991 Anchusa undulata L. var. alborosea（Benoist）Braun-Blanq. et Maire = Anchusa atlantica Ball ■☆
21992 Anchusa undulata L. var. atlantica（Ball）Gusul. = Anchusa atlantica Ball ■☆
21993 Anchusa undulata L. var. macrotricha Faure et Maire = Anchusa undulata L. ■☆
21994 Anchusa undulata L. var. mairei（Gusul.）Maire = Anchusa mairei Gusul. ■☆
21995 Anchusa undulata L. var. pseudogranatensis Braun-Blanq. et Maire = Anchusa pseudogranatensis（Braun-Blanq. et Maire）Sennen et Mauricio ■☆
21996 Anchusa zeylanica J. Jacq. = Bothriospermum zeylanicum（J. Jacq.）Druce ■
21997 Anchusa zeylanica Vahl ex Hornem. = Bothriospermum tenellum（Hornem.）Fisch. et C. A. Mey. ■
21998 Anchusa zeylanica Vahl ex Hornem. = Cynoglossum zeylanicum（Vahl ex Hornem.）Thunb. ex Lehm. ■☆
21999 Anchusaceae Vest = Boraginaceae Juss.（保留科名）●■

22000 Anchusella Bigazzi, E. Nardi et Selvi(1997);小牛舌草属■☆
22001 Anchusella cretica (Mill.) Bigazzi, E. Nardi et Selvi;克里特小牛舌草■☆
22002 Anchusella variegata (L.) Bigazzi, E. Nardi et Selvi;花叶小牛舌草(小牛舌草)■☆
22003 Anchusopsis Bisch. = Lindelofia Lehm. ■
22004 Anchusopsis longiflora Bisch. = Lindelofia longiflora Gurke ■☆
22005 Anchusopsis longiflora Bisch. = Lindelofia spectabilis Lehm. ■☆
22006 Ancipitia (Luer) Luer = Pleurothallis R. Br. ■☆
22007 Ancistrachne S. T. Blake(1941);钩草属■☆
22008 Ancistrachne numaeensis (Balansa) S. T. Blake;钩草■☆
22009 Ancistragrostis S. T. Blake = Deyeuxia Clarion ■
22010 Ancistragrostis S. T. Blake(1946);钩股颖属■☆
22011 Ancistranthus Lindau(1900);古巴爵床属■☆
22012 Ancistranthus harpochiloides Lindau;古巴爵床■☆
22013 Ancistrella Tiegh. = Ancistrocladus Wall. (保留属名)●
22014 Ancistrella barteri Tiegh. = Ancistrocladus abbreviatus Airy Shaw subsp. lateralis Gereau ●☆
22015 Ancistrocactus (K. Schum.) Britton et Rose = Sclerocactus Britton et Rose ●☆
22016 Ancistrocactus Britton et Rose = Sclerocactus Britton et Rose ●☆
22017 Ancistrocactus Britton et Rose(1923);罗纱锦属(短钩玉属)■☆
22018 Ancistrocactus brevihamatus (Engelm.) Britton et Rose = Sclerocactus brevihamatus (Engelm.) D. R. Hunt ●☆
22019 Ancistrocactus mathssonii (Berge ex K. Schum.) Doweld = Sclerocactus uncinatus (Galeotti) N. P. Taylor ●☆
22020 Ancistrocactus megarhizus (Rose) Britton et Rose = Sclerocactus scheeri (Salm-Dyck) N. P. Taylor ●☆
22021 Ancistrocactus scheeri (Salm-Dyck) Britton et Rose = Sclerocactus scheeri (Salm-Dyck) N. P. Taylor ●☆
22022 Ancistrocactus scheeri Britton et Rose = Sclerocactus scheeri (Salm-Dyck) N. P. Taylor ●☆
22023 Ancistrocactus tobuschii (T. Marshall) T. Marshall ex Backeb. = Sclerocactus brevihamatus (Engelm.) D. R. Hunt ●☆
22024 Ancistrocactus uncinatus (Galeotti) L. D. Benson = Sclerocactus uncinatus (Galeotti) N. P. Taylor ●☆
22025 Ancistrocactus uncinatus (Galeotti) L. D. Benson var. wrightii (Engelm.) L. D. Benson = Glandulicactus uncinatus (Galeotti) Backeb. var. wrightii (Engelm.) Backeb. ■☆
22026 Ancistrocarphus A. Gray = Stylocline Nutt. ■☆
22027 Ancistrocarphus A. Gray(1868);地星菊属;Groundstar ■
22028 Ancistrocarphus filagineus A. Gray;地星菊;Hooked Groundstar, Woolly Fishhooks ■☆
22029 Ancistrocarphus keilii Morefield;基尔地星菊;Santa Ynez Groundstar ■☆
22030 Ancistrocarpus Kunth(废弃属名) = Ancistrocarpus Oliv.(保留属名)●☆
22031 Ancistrocarpus Kunth(废弃属名) = Microtea Sw. ■☆
22032 Ancistrocarpus Oliv. (1865)(保留属名);沟果椴属●☆
22033 Ancistrocarpus bequaertii De Wild.;贝卡尔沟果椴●☆
22034 Ancistrocarpus comperei R. Wilczek;孔佩尔沟果椴●☆
22035 Ancistrocarpus densispinosus Oliv.;沟果椴●☆
22036 Ancistrocarpus wellensii De Wild. = Ancistrocarpus densispinosus Oliv. ●☆
22037 Ancistrocarya Maxim. (1872);日本紫草属■☆
22038 Ancistrocarya japonica Maxim.;日本紫草■☆
22039 Ancistrocarya japonica Maxim. f. albiflora (Honda) H. Hara;白花日本紫草■☆
22040 Ancistrochilus Rolfe(1897);沟唇兰属;Ancistrochilus ■☆
22041 Ancistrochilus rothschildianus O'Brien;洛氏沟唇兰■☆
22042 Ancistrochilus thomsonianus (Rchb. f.) Rolfe;沟唇兰■☆
22043 Ancistrochilus thomsonianus (Rchb. f.) Rolfe var. gentilii De Wild. = Ancistrochilus rothschildianus O'Brien ■☆
22044 Ancistrochloa Honda = Calamagrostis Adans. ■
22045 Ancistrochloa fauriei (Hack.) Honda = Calamagrostis fauriei Hack. ■☆
22046 Ancistrocladaceae Planch. = Ancistrocladaceae Planch. ex Walp. (保留科名)●
22047 Ancistrocladaceae Planch. ex Walp. (1851)(保留科名);钩枝藤科;Ancistrocladus Family ●
22048 Ancistrocladus Wall. (1829)(保留属名);钩枝藤属(钩枝属,钩子藤属);Ancistrocladus ●
22049 Ancistrocladus Wall. ex Wight et Arn. = Ancistrocladus Wall. (保留属名)●
22050 Ancistrocladus abbreviatus Airy Shaw;缩短钩枝藤●☆
22051 Ancistrocladus abbreviatus Airy Shaw subsp. lateralis Gereau;侧生钩枝藤●☆
22052 Ancistrocladus barteri Scott-Elliot;巴特钩枝藤●☆
22053 Ancistrocladus congolensis J. Léonard;刚果钩枝藤●☆
22054 Ancistrocladus ealaensis J. Léonard;埃阿拉钩枝藤●☆
22055 Ancistrocladus extensus Wall. ex Planch. = Ancistrocladus tectorius (Lour.) Merr. ●
22056 Ancistrocladus grandiflorus Cheek;大花钩枝藤●☆
22057 Ancistrocladus guineensis Oliv.;几内亚钩枝藤●☆
22058 Ancistrocladus hainanensis Hayata = Ancistrocladus tectorius (Lour.) Merr. ●
22059 Ancistrocladus heyneanus Wall. ex J. Graham;海尼钩枝藤●☆
22060 Ancistrocladus korupensis D. W. Thomas et Gereau;科鲁普钩枝藤●☆
22061 Ancistrocladus letestui Pellegr.;莱泰斯图钩枝藤●☆
22062 Ancistrocladus pachyrrhachis Airy Shaw = Ancistrocladus barteri Scott-Elliot ●☆
22063 Ancistrocladus pinangianus (Wall.) Planch. = Ancistrocladus tectorius (Lour.) Merr. ●
22064 Ancistrocladus pinangianus Wall. ex Planch. = Ancistrocladus tectorius (Lour.) Merr. ●
22065 Ancistrocladus robertsoniorum J. Léonard;罗伯逊钩枝藤●☆
22066 Ancistrocladus tanzaniensis Cheek et Frim.-Möll.;坦桑尼亚钩枝藤●☆
22067 Ancistrocladus tectorius (Lour.) Merr.;钩枝藤(本叶藤,车蓬藤,车叶藤);Covering Ancistrocladus ●
22068 Ancistrocladus uncinatus Hutch. et Dalziel = Ancistrocladus guineensis Oliv. ●☆
22069 Ancistrodesmus Naudin = Microlepis (DC.) Miq. (保留属名)●☆
22070 Ancistrolobus Spach = Cratoxylum Blume ●
22071 Ancistrolobus ligustrinus Spach = Cratoxylum cochinchinense (Lour.) Blume ●
22072 Ancistrophora A. Gray = Verbesina L. (保留属名)●■☆
22073 Ancistrophyllum (G. Mann et H. Wendl.) G. Mann et H. Wendl. = Ancistrophyllum (G. Mann et H. Wendl.) H. Wendl. ●☆
22074 Ancistrophyllum (G. Mann et H. Wendl.) G. Mann et H. Wendl. ex Kerch. = Ancistrophyllum (G. Mann et H. Wendl.) H. Wendl. ●☆
22075 Ancistrophyllum (G. Mann et H. Wendl.) H. Wendl. (1878);钩叶棕属●☆

22076 Ancistrophyllum (G. Mann et H. Wendl.) H. Wendl. = Laccosperma (G. Mann et H. Wendl.) Drude ●☆

22077 Ancistrophyllum G. Mann et H. Wendl. = Ancistrophyllum (G. Mann et H. Wendl.) H. Wendl. ●☆

22078 Ancistrophyllum acutiflorum Becc. = Laccosperma acutiflorum (Becc.) J. Dransf. ●☆

22079 Ancistrophyllum laeve (G. Mann et H. Wendl.) Drude = Laccosperma laeve (G. Mann et H. Wendl.) H. Wendl. ●☆

22080 Ancistrophyllum laurentii De Wild. = Laccosperma laurentii (De Wild.) J. Dransf. ●☆

22081 Ancistrophyllum opacum (G. Mann et H. Wendl.) Drude = Laccosperma opacum (G. Mann et H. Wendl.) Drude ●☆

22082 Ancistrophyllum robustum Burret = Laccosperma robustum (Burret) J. Dransf. ●☆

22083 Ancistrophyllum secundiflorum (P. Beauv.) H. Wendl. = Laccosperma secundiflorum (P. Beauv.) Kuntze ●☆

22084 Ancistrorhynchus Finet et Summerh. = Ancistrorhynchus Finet ■☆

22085 Ancistrorhynchus Finet(1907);钩喙兰属■☆

22086 Ancistrorhynchus brevifolius Finet;短叶钩喙兰■☆

22087 Ancistrorhynchus capitatus (Lindl.) Summerh.;头状钩喙兰■☆

22088 Ancistrorhynchus cephalotes (Rchb. f.) Summerh.;头序钩喙兰■☆

22089 Ancistrorhynchus clandestinus (Lindl.) Schltr.;隐匿钩喙兰■☆

22090 Ancistrorhynchus constrictus Szlach. et Olszewski;缢缩钩喙兰■☆

22091 Ancistrorhynchus glomeratus (Ridl.) Summerh. = Ancistrorhynchus cephalotes (Rchb. f.) Summerh. ■☆

22092 Ancistrorhynchus laxiflorus Mansf.;疏花钩喙兰■☆

22093 Ancistrorhynchus metteniae (Kraenzl.) Summerh.;布劳恩钩喙兰■☆

22094 Ancistrorhynchus obovata Stévart;倒卵钩喙兰■☆

22095 Ancistrorhynchus ovatus Summerh.;卵钩喙兰■☆

22096 Ancistrorhynchus parviflorus Summerh.;小花钩喙兰■☆

22097 Ancistrorhynchus recurvus Finet;反折钩喙兰■☆

22098 Ancistrorhynchus refractus (Kraenzl.) Summerh.;反曲钩喙兰■☆

22099 Ancistrorhynchus schumannii (Kraenzl.) Summerh.;舒曼钩喙兰■☆

22100 Ancistrorhynchus serratus Summerh.;具齿钩喙兰■☆

22101 Ancistrorhynchus stenophyllus (Schltr.) Schltr. = Ancistrorhynchus clandestinus (Lindl.) Schltr. ■☆

22102 Ancistrorhynchus straussii (Schltr.) Schltr.;斯特劳斯钩喙兰■☆

22103 Ancistrorhynchus tenuicaulis Summerh.;细茎钩喙兰■☆

22104 Ancistrostigma Fenzl = Trianthema L. ■

22105 Ancistrostylis T. Yamaz. (1980);钩柱玄参属■☆

22106 Ancistrostylis T. Yamaz. = Bacopa Aubl. (保留属名)■

22107 Ancistrostylis T. Yamaz. = Herpestis C. F. Gaertn. ■

22108 Ancistrostylis harmandii (Bonati) T. Yamaz.;钩柱爵床■☆

22109 Ancistrothyrsus Harms(1931);钩序西番莲属●☆

22110 Ancistrothyrsus hirtellus A. H. Gentry;毛钩序西番莲●☆

22111 Ancistrothyrsus tessmannii Harms;钩序西番莲●☆

22112 Ancistrum J. R. Forst. et G. Forst. = Acaena L. ●■☆

22113 Ancistrum decumbens Thunb. = Acaena latebrosa Aiton ●☆

22114 Ancistrum repens Vent. = Acaena ovalifolia Ruiz et Pav. ●☆

22115 Anclyanthus A. Juss. = Ancylanthos Desf. ●☆

22116 Ancouratea Tiegh. = Ouratea Aubl. (保留属名)●

22117 Ancrumia Harv. ex Baker(1877);安克葱属■☆

22118 Ancrumia cuspidata Harv. ex Baker;安克葱■☆

22119 Ancylacanthus Lindau = Ptyssiglottis T. Anderson ■☆

22120 Ancylanthos Desf. (1818);弯花茜属●☆

22121 Ancylanthos Desf. = Danais Comm. ex Vent. ●☆

22122 Ancylanthos bainesii Hiern = Ancylanthos rubiginosus Desf. ●☆

22123 Ancylanthos cinerascens Hiern = Tapiphyllum cinerascens (Hiern) Robyns ■☆

22124 Ancylanthos cistifolius Welw. ex Hiern = Tapiphyllum cistifolium (Welw. ex Hiern) Robyns ■☆

22125 Ancylanthos ferrugineus Welw. = Ancylanthos rubiginosus Desf. ●☆

22126 Ancylanthos fulgidus Welw. ex Hiern = Ancylanthos rubiginosus Desf. ●☆

22127 Ancylanthos glabrescens (Robyns) Roberty = Hutchinsonia glabrescens Robyns ●☆

22128 Ancylanthos lactiflorus (Welw. ex Hiern) Robyns = Fadogia lactiflora Welw. ex Hiern ●☆

22129 Ancylanthos monteiroi Oliv. = Lagynias monteiroi (Oliv.) Bridson ■☆

22130 Ancylanthos rhodesiacus Tennant = Tapiphyllum rhodesiacum (Tennant) Bridson ■☆

22131 Ancylanthos rogersii (Wernham) Robyns = Fadogiella rogersii (Wernham) Bridson ●☆

22132 Ancylanthos rubiginosus Desf.;锈红弯花茜●☆

22133 Ancylanthos rufescens E. A. Bruce = Lagynias rufescens (E. A. Bruce) Verdc. ■☆

22134 Ancylanthus Steud. = Ancylanthos Desf. ●☆

22135 Ancylobothrys Pierre(1898);弯穗夹竹桃属●☆

22136 Ancylobotrys Pierre = Ancylobothrys Pierre ●☆

22137 Ancylobotrys amoena Hua;秀丽弯穗夹竹桃●☆

22138 Ancylobotrys brevituba Pichon = Ancylobotrys scandens (Schumach. et Thonn.) Pichon ●☆

22139 Ancylobotrys capensis (Oliv.) Pichon;好望角弯穗夹竹桃●☆

22140 Ancylobotrys gabonensis Pichon = Ancylobotrys pyriformis Pierre ●☆

22141 Ancylobotrys mammosa Pierre = Ancylobotrys scandens (Schumach. et Thonn.) Pichon ●☆

22142 Ancylobotrys mammosa Pierre var. crassifolia (K. Schum.) Pierre = Ancylobotrys scandens (Schumach. et Thonn.) Pichon ●☆

22143 Ancylobotrys mammosa Pierre var. mucronata (Dewèvre) Pierre = Ancylobotrys scandens (Schumach. et Thonn.) Pichon ●☆

22144 Ancylobotrys petersiana (Klotzsch) Pierre;彼得斯弯穗夹竹桃●☆

22145 Ancylobotrys petersiana (Klotzsch) Pierre var. forbesiana Pierre = Ancylobotrys petersiana (Klotzsch) Pierre ●☆

22146 Ancylobotrys pyriformis Pierre;梨形弯穗夹竹桃●☆

22147 Ancylobotrys reticulata (Hallier f.) Pichon = Landolphia reticulata Hallier f. ●☆

22148 Ancylobotrys robusta Pierre = Ancylobotrys pyriformis Pierre ●☆

22149 Ancylobotrys rotundifolia (Dewèvre) Pierre = Ancylobotrys petersiana (Klotzsch) Pierre ●☆

22150 Ancylobotrys scandens (Schumach. et Thonn.) Pichon;攀缘弯穗夹竹桃●☆

22151 Ancylobotrys tayloris (Stapf) Pichon;泰勒弯穗夹竹桃●☆

22152 Ancylobotrys trichantha Pichon = Ancylobotrys pyriformis Pierre ●☆

22153 Ancylobotrys welwitschii (Stapf ex De Wild.) Pierre = Ancylobotrys scandens (Schumach. et Thonn.) Pichon ●☆

22154 Ancylocalyx Tul. = Pterocarpus Jacq. (保留属名)●

22155 Ancylocladus Wall. = Willughbeia Roxb. (保留属名)●☆

22156 Ancylocladus Wall. ex Kuntze = Willughbeia Roxb. (保留属名)●☆

22157 Ancylogyne Nees = Sanchezia Ruiz et Pav. ●■

22158 Ancylostemon Craib(1919);直瓣苣苔属;Ancylostemon ●■★
22159 Ancylostemon aureus (Franch.) B. L. Burtt;凹瓣苣苔(黄花长蒴苣苔);Concavepetal Ancylostemon, Goldenyellow Ancylostemon, Yellow-flower Didymocarpus ●
22160 Ancylostemon aureus (Franch.) Burtt var. angustifolius K. Y. Pan;窄叶直瓣苣苔;Narrowleaf Concavepetal Ancylostemon ●
22161 Ancylostemon bullatus W. T. Wang et K. Y. Pan;泡叶直瓣苣苔;Bullate Ancylostemon ●
22162 Ancylostemon concavus Craib = Ancylostemon aureus (Franch.) B. L. Burtt ●
22163 Ancylostemon convexus Craib; 凸瓣苣苔; Convexpetal Ancylostemon ●
22164 Ancylostemon flabellatus C. Y. Wu ex H. W. Li;扇叶直瓣苣苔;Fanleaf Ancylostemon ●
22165 Ancylostemon gamosepalus K. Y. Pan; 黄花直瓣苣苔; Yellowflower Ancylostemon ■
22166 Ancylostemon humilis W. T. Wang;矮直瓣苣苔(小直瓣苣苔);Dwarf Ancylostemon ■
22167 Ancylostemon lancifolius (Franch.) B. L. Burtt = Isometrum lancifolium (Franch.) K. Y. Pan ■
22168 Ancylostemon mairei (H. Lév.) Craib;滇北直瓣苣苔;Maire Ancylostemon ■
22169 Ancylostemon mairei (H. Lév.) Craib var. emeiensis K. Y. Pan;峨眉直瓣苣苔;Emei Ancylostemon ■
22170 Ancylostemon notochlaenus (H. Lév. et Vaniot) Craib;贵州直瓣苣苔;Guizhou Ancylostemon ■
22171 Ancylostemon purpureus B. L. Burtt et R. A. Davidson = Isometrum lancifolium (Franch.) K. Y. Pan ■
22172 Ancylostemon rhombifolius K. Y. Pan;菱叶直瓣苣苔(棱叶直瓣苣苔);Rhombicleaf Ancylostemon ■
22173 Ancylostemon ronganensis K. Y. Pan;融安直瓣苣苔;Rongan Ancylostemon ■
22174 Ancylostemon saxatilis (Hemsl.) Craib;直瓣苣苔(铁板还阳);Cliff Ancylostemon ■
22175 Ancylostemon saxatilis (Hemsl.) Craib var. microcalyx Hemsl. ex Craib = Ancylostemon humilis W. T. Wang ■
22176 Ancylostemon trichanthus B. L. Burtt et R. A. Davidson;毛花直瓣苣苔;Hairyflower Ancylostemon ■
22177 Ancylostemon vulpinus B. L. Burtt et R. A. Davidson;狐毛直瓣苣苔;Foxcolor Ancylostemon ■
22178 Ancylotropis B. Eriksen(1993);曲棱远志属■☆
22179 Ancylotropis malmeana (Chodat) B. Eriksen;曲棱远志■☆
22180 Ancyrossemon Poepp. et Endl. = Ancyrostemma Poepp. et Endl. ●☆
22181 Ancyrostemma Poepp. et Endl. = Sclerothrix C. Presl ■☆
22182 Ancystrochlora Ohwi = Ancistrochloa Honda ■
22183 Anda A. Juss. = Joannesia Vell. ●☆
22184 Andaca Raf. = Lotus L. ■
22185 Andenea Frič = Echinopsis Zucc. ●
22186 Andenea Kreuz. = Lobivia Britton et Rose ■
22187 Anderbergia B. Nord. (1996);外卷鼠麴木属●☆
22188 Anderbergia elsiae B. Nord.;埃尔西亚外卷鼠麴木●☆
22189 Anderbergia fallax B. Nord.;迷惑外卷鼠麴木●☆
22190 Anderbergia ustulata B. Nord.;凋萎外卷鼠麴木●☆
22191 Anderbergia vlokii (Hilliard) B. Nord.;弗劳克外卷鼠麴木●☆
22192 Andersonia Buch. -Ham. = Anogeissus (DC.) Wall. ●
22193 Andersonia Buch. -Ham. ex Wall. (1810);獐牙石南属●☆
22194 Andersonia J. König = Stylidium Sw. ex Willd. (保留属名)■
22195 Andersonia J. König ex R. Br. = Stylidium Sw. ex Willd. (保留属名)■
22196 Andersonia R. Br. = Andersonia Buch. -Ham. ex Wall. ●☆
22197 Andersonia Roxb. = Amoora Roxb. ●
22198 Andersonia Roxb. = Aphanamixis Blume ●
22199 Andersonia Willd. = Gaertnera Lam. ●
22200 Andersonia Willd. ex Roem. et Schult. = Gaertnera Lam. ●
22201 Andersonia rohituka Roxb. = Aphanamixis polystachya (Wall.) R. Parker ●
22202 Anderssoniopiper Trel. = Macropiper Miq. ●☆
22203 Anderssoniopiper Trel. = Piper L. ●■
22204 Andesia Hauman = Oxychloe Phil. ■☆
22205 Andicus Vell. = Anda A. Juss. ●☆
22206 Andicus Vell. = Joannesia Vell. ●☆
22207 Andinia (Luer) Luer = Salpistele Dressier ■☆
22208 Andinorchis Szlach., Mytnik et Górniak = Zygopetalum Hook. ■☆
22209 Andinorchis Szlach., Mytnik et Górniak(2006);秘鲁轭瓣兰属■☆
22210 Andira Juss. = Andira Lam. (保留属名)●☆
22211 Andira Lam. (1783)(保留属名);安迪尔豆属(安迪拉豆属,甘蓝豆属,甘蓝皮豆属,柯桠豆属,柯桠木属);Angelin Tree, Angelin-tree, Cabbage Tree, Cabbage-tree ●☆
22212 Andira araroba Aguiar;巴西安迪尔豆(安迪拉豆,巴西安迪拉豆,巴西柯桠豆,柯亚木,拟瓦泰豆);Araroba ●☆
22213 Andira coriacea Pulle;革质安迪尔豆(革质甘蓝皮豆)●☆
22214 Andira fraxinifolia Benth.;白蜡安迪尔豆(白蜡甘蓝皮豆)●☆
22215 Andira gabonica Baill. = Aganope gabonica (Baill.) Polhill ■☆
22216 Andira grandiflora Guillaumin et Perr. = Andira inermis (W. Wright) DC. subsp. grandiflora (Guillaumin et Perr.) J. B. Gillett ex Polhill ●☆
22217 Andira horsfieldii Lesch. = Euchresta horsfieldii (Lesch.) Benn. ●
22218 Andira inermis (W. Wright) DC.;无刺安迪尔豆(甘蓝皮豆,无刺安迪拉豆,无刺甘蓝皮豆,无刺柯桠豆);Angelin, Angelin-tree, Cabbage Angelin Tree, Cabbage Angelin-tree, Cabbage Tree, Cabbage-tree, Dog Almond, Kurara, Partridge Wood ●☆
22219 Andira inermis (W. Wright) DC. subsp. grandiflora (Guillaumin et Perr.) J. B. Gillett ex Polhill;大花无刺安迪拉豆●☆
22220 Andira inermis (W. Wright) DC. subsp. rooseveltii (De Wild.) J. B. Gillett ex Polhill;罗斯福安迪尔豆●☆
22221 Andira micrantha Ducke;小刺安迪尔豆(小刺甘蓝豆)●☆
22222 Andira parviflora Ducke;小花安迪尔豆(小花甘蓝豆,小花甘蓝皮豆)●☆
22223 Andira retusa (Poir.) DC.;微凹安迪尔豆(微凹甘蓝豆,微凹甘蓝皮豆)●☆
22224 Andiscus Vell. = Andicus Vell. ●☆
22225 Andouinia Rchb. = Audouinia Brongn. ●☆
22226 Andrachne L. (1753);黑钩叶属;Andrachne ●☆
22227 Andrachne L. = Leptopus L. ●
22228 Andrachne aspera Spreng.;粗糙黑钩叶●☆
22229 Andrachne aspera Spreng. var. glandulosa Hochst. ex A. Rich.;腺点粗糙黑钩叶●☆
22230 Andrachne attenuata Hand. -Mazz. = Leptopus esquirolii (Bunge) Pojark. ●
22231 Andrachne attenuata Hand. -Mazz. var. microcalyx Hand. -Mazz. = Leptopus esquirolii (Bunge) Pojark. ●
22232 Andrachne australis Zoll. et E. Morren = Leptopus australis (Zoll. et E. Morren) Pojark. ●
22233 Andrachne bodinieri H. Lév. = Leptopus chinensis (Bunge)

Pojark. ●
22234 Andrachne buschiana Pojark. ;布什黑钩叶●☆
22235 Andrachne capensis Baill. = Andrachne ovalis (E. Mey. ex Sond.) Müll. Arg. ●☆
22236 Andrachne capillipes Hutch. = Leptopus chinensis (Bunge) Pojark. ●
22237 Andrachne capillipes Hutch. var. pubescens Hutch. = Leptopus chinensis (Bunge) Pojark. ●
22238 Andrachne cavaleriei H. Lév. = Lysimachia capillipes Hemsl. var. cavaleriei (H. Lév.) Hand. -Mazz. ■
22239 Andrachne cerebroides Petra Hoffm. ;脑状黑钩叶●☆
22240 Andrachne chinensis Bunge = Leptopus chinensis (Bunge) Pojark. ●
22241 Andrachne chinensis Bunge var. pubescens (Hutch.) Hand. -Mazz. = Leptopus chinensis (Bunge) Pojark. ●
22242 Andrachne clarkei Hook. f. = Leptopus clarkei (Hook. f.) Pojark. ●
22243 Andrachne colchica Fisch. et C. A. Mey. ex Boiss. ;高加索黑钩叶;Caucasian Spurge ●☆
22244 Andrachne cordifolia Hemsl. = Leptopus chinensis (Bunge) Pojark. ●
22245 Andrachne dregeana (Scheele) Baill. = Andrachne ovalis (E. Mey. ex Sond.) Müll. Arg. ●☆
22246 Andrachne ephemera M. G. Gilbert;短命黑钩叶●☆
22247 Andrachne esquirolii H. Lév. = Leptopus esquirolii (Bunge) Pojark. ●
22248 Andrachne esquirolii H. Lév. var. microcalyx (Hand. -Mazz.) Rehder = Leptopus esquirolii (Bunge) Pojark. ●
22249 Andrachne fedtschenkoi Kossinsky;范氏黑钩叶●☆
22250 Andrachne filiformis Pojark. ;线叶黑钩叶●☆
22251 Andrachne fragilis M. G. Gilbert et Thulin;脆黑钩叶●☆
22252 Andrachne fruticosa L. = Breynia fruticosa (L.) Hook. f. ●
22253 Andrachne gracilipes Petra Hoffm. ;细梗黑钩叶●☆
22254 Andrachne hainanensis Merr. et Chun = Leptopus hainanensis (Merr. et Chun) Pojark. ●
22255 Andrachne hainanensis Merr. et Chun var. nummularifolia Merr. et Chun = Leptopus hainanensis (Merr. et Chun) Pojark. ●
22256 Andrachne hirsuta Spreng. ;毛黑钩叶●☆
22257 Andrachne hypoglauca H. Lév. = Leptopus esquirolii (Bunge) Pojark. ●
22258 Andrachne lolonum Hand. -Mazz. = Leptopus lolonum (Hand. -Mazz.) Pojark. ●
22259 Andrachne maroccana Ball;摩洛哥黑钩叶●☆
22260 Andrachne millietii H. Lév. = Lysimachia millietii (H. Lév.) Hand. -Mazz. ■
22261 Andrachne montana Hutch. = Leptopus chinensis (Bunge) Pojark. ●
22262 Andrachne montana Hutch. = Leptopus montanus (Hutch.) Pojark. ●
22263 Andrachne ovalis (E. Mey. ex Sond.) Müll. Arg. ;卵形黑钩叶●☆
22264 Andrachne persicariifolia H. Lév. = Leptopus esquirolii (H. Lév.) P. T. Li ●
22265 Andrachne phyllanthoides J. M. Coult. ; 叶下珠黑钩叶; Buck Brush, Buckbrush Maidenbush ●☆
22266 Andrachne pusilla Pojark. ;微小黑钩叶●☆
22267 Andrachne pygmaea Kossinsky;密集黑钩叶●☆
22268 Andrachne rotundifolia C. A. Mey. ;圆叶黑钩叶●☆
22269 Andrachne schweinfurthii (Balf. f.) Radcl. -Sm. ;索马里黑钩叶●☆
22270 Andrachne schweinfurthii (Balf. f.) Radcl. -Sm. var. papillosa Radcl. -Sm. ;乳头黑钩叶●☆
22271 Andrachne somalensis Pax = Andrachne schweinfurthii (Balf. f.) Radcl. -Sm. ●☆
22272 Andrachne stenophylla Kossinsky;狭叶黑钩叶●☆
22273 Andrachne telephioides L. ;疣黑钩叶●☆
22274 Andrachne telephioides L. var. rotundifolia C. A. Mey. = Andrachne telephioides L. ●☆
22275 Andrachne trifoliata Roxb. = Bischofia javanica Blume ●
22276 Andrachne virga-tenuis Nevski;纤细黑钩叶●☆
22277 Andracna Marnac et Reyn. = Andrachne L. ●☆
22278 Andradea Allemão(1845);巴西紫茉莉属(繁花茉莉属)●☆
22279 Andradea floribunda Allemão;巴西紫茉莉■☆
22280 Andradia Sim = Dialium L. ●☆
22281 Andradia arborea Sim = Dialium schlechteri Harms ●☆
22282 Andrastis Raf. ex Benth. = Cladrastis Raf. ●
22283 Andrea Mez = Eduandrea Leme, W. Till, G. K. Br. , J. R. Grant et Govaerts ■☆
22284 Andreadoxa Kallunki(1998);隐雄芸香属●☆
22285 Andrederaceae J. Agardh = Basellaceae Raf. (保留科名)■
22286 Andreettaea Luer = Pleurothallis R. Br. ■☆
22287 Andreoskia Boiss. = Andrzeiowskya Rchb. ■☆
22288 Andreoskia DC. = Dontostemon Andrz. ex C. A. Mey. (保留属名)■
22289 Andreoskia Spach = Andrzeiowskya Rchb. ■☆
22290 Andreoskia crassifolia Bunge ex Turcz. = Dontostemon crassifolius (Bunge) Maxim. ■
22291 Andreoskia dentata Bunge = Dontostemon dentatus (Bunge) Ledeb. ■
22292 Andreoskia eglandulosa (DC.) DC. = Dontostemon integrifolius (L.) Ledeb. ■
22293 Andreoskia eglandulosa DC. = Dontostemon integrifolius (L.) Ledeb. var. eglandulosus (DC.) Turcz. ■
22294 Andreoskia integrifolia (L.) DC. = Dontostemon integrifolius (L.) Ledeb. ■
22295 Andreoskia integrifolia DC. = Dontostemon integrifolius (L.) Ledeb. ■
22296 Andreoskia pectinata (DC.) DC. = Dontostemon pinnatifidus (Willd.) Al-Shehbaz et H. Ohba ■
22297 Andreoskia pectinata DC. = Dontostemon pinnatifidus (Willd.) Al-Shehbaz et H. Ohba ■
22298 Andresia Sleumer = Cheilotheca Hook. f. ■
22299 Andreusia Dunal = Symphysia C. Presl ●☆
22300 Andreusia Vent. = Myoporum Banks et Sol. ex G. Forst. ●
22301 Andrewaia Spreng. = Bartonia Muhl. ex Willd. (保留属名)■☆
22302 Andrewaia Spreng. = Centaurella Michx. ■☆
22303 Andriala Decne. = Andryala L. ■☆
22304 Andriana B. -E. van Wyk(1999);马岛雄蕊草属●☆
22305 Andriana coursii (Humbert) B. -E. van Wyk;库尔斯马岛雄蕊草●☆
22306 Andriana marojejyensis (Humbert) B. -E. van Wyk;马罗马岛雄蕊草●☆
22307 Andriana tsaratananensis (Humbert) B. -E. van Wyk;察拉塔纳纳马岛雄蕊草●☆
22308 Andriapetalum Pohl = Panopsis Salisb. ex Knight ●☆

22309 Andrieuxia DC. = Heliopsis Pers. (保留属名)■☆
22310 Androcalymma Dwyer(1957);小花光叶豆属■☆
22311 Androcalymma glabrifolium Dwyer;小花光叶豆■☆
22312 Androcentrum Lem. = Bravaisia DC. ●☆
22313 Androcephallum Warb. = Lunanea DC. (废弃属名)●☆
22314 Androcera Nutt. = Solanum L. ●■
22315 Androcera rostrata (Dunal) Rydb. = Solanum rostratum Dunal ■
22316 Androchilus Liebm. = Androchilus Liebm. ex Hartm. ■☆
22317 Androchilus Liebm. ex Hartm. (1844);蕊唇兰属■☆
22318 Androchilus campestris Liebm.;蕊唇兰■☆
22319 Androcoma Nees = Scirpus L. (保留属名)■
22320 Androcorys Schltr. (1919);兜蕊兰属(兜兰属);Androcorys ■
22321 Androcorys gracilis (King et Pantl.) Schltr.;细梗兜蕊兰■
22322 Androcorys japonensis F. Maek. = Androcorys pusillus (Ohwi et Fukuy.) Masam. ■
22323 Androcorys ophioglossoides Schltr.; 兜蕊兰; Common Androcorys, Ophiglossumlike Androcorys ■
22324 Androcorys orysepalus K. Y. Lang; 尖萼兜蕊兰; Tinesepal Androcorys ■
22325 Androcorys pugioniformis (Lindl. ex Hook. f.) K. Y. Lang;剑唇兜蕊兰(剑唇角盘兰);Swordlip Androcorys ■
22326 Androcorys pusillus (Ohwi et Fukuy.) Masam.;小兜蕊兰(小零余子草);Small Androcorys ■
22327 Androcorys pusillus (Ohwi et Fukuy.) S. S. Ying = Androcorys pusillus (Ohwi et Fukuy.) Masam. ■
22328 Androcorys pusillus (Ohwi et Fukuy.) Tang S. Liu et H. J. Su = Androcorys pusillus (Ohwi et Fukuy.) Masam. ■
22329 Androcorys spiralis Ts. Tang et F. T. Wang;蜀藏兜蕊兰;Helix Androcorys ■
22330 Androcymbium Willd. (1808);舟蕊秋水仙属■☆
22331 Androcymbium abyssinicum (A. Rich.) Stef. = Merendera schimperiana Hochst. ■☆
22332 Androcymbium albanense Schönland;阿尔邦舟蕊秋水仙■☆
22333 Androcymbium albomarginatum Schinz;白边舟蕊秋水仙■☆
22334 Androcymbium asteroides J. C. Manning et Goldblatt;紫菀舟蕊秋水仙■☆
22335 Androcymbium bellum Schltr. et K. Krause;雅致舟蕊秋水仙■☆
22336 Androcymbium burchellii Baker;伯切尔舟蕊秋水仙■☆
22337 Androcymbium burkei Baker;伯克舟蕊秋水仙■☆
22338 Androcymbium capense (L.) K. Krause;好望角舟蕊秋水仙■☆
22339 Androcymbium ciliolatum Schltr. et K. Krause;澳非舟蕊秋水仙■☆
22340 Androcymbium circinatum Baker;卷须舟蕊秋水仙■☆
22341 Androcymbium crispum Schinz;皱波舟蕊秋水仙■☆
22342 Androcymbium cruciatum U. Müll. -Doblies et D. Müll. -Doblies;十字形舟蕊秋水仙■☆
22343 Androcymbium cuspidatum Baker;骤尖舟蕊秋水仙■☆
22344 Androcymbium decipiens N. E. Br.;迷惑舟蕊秋水仙■☆
22345 Androcymbium dregei C. Presl;德雷舟蕊秋水仙■☆
22346 Androcymbium eucomoides (Jacq.) Willd.;美顶花舟蕊秋水仙■☆
22347 Androcymbium eucomoides Sweet = Androcymbium capense (L.) K. Krause ■☆
22348 Androcymbium exiguum Rössler;弱小舟蕊秋水仙■☆
22349 Androcymbium fenestratum Schltr. et K. Krause = Androcymbium ciliolatum Schltr. et K. Krause ■☆
22350 Androcymbium gramineum (Cav.) J. F. Macbr.;禾状舟蕊秋水仙■☆
22351 Androcymbium gramineum (Cav.) J. F. Macbr. subsp. psammophilum (Svent.) G. Kunkel = Androcymbium psammophilum Svent. ■☆
22352 Androcymbium gramineum (Cav.) J. F. Macbr. var. genuinum Maire = Androcymbium gramineum (Cav.) J. F. Macbr. ■☆
22353 Androcymbium gramineum (Cav.) J. F. Macbr. var. intermedium Gattef. et Maire = Androcymbium gramineum (Cav.) J. F. Macbr. ■☆
22354 Androcymbium gramineum (Cav.) J. F. Macbr. var. saharae (Maire) Maire = Androcymbium wyssianum Beauverd et Turrett. ■☆
22355 Androcymbium gramineum (Cav.) J. F. Macbr. var. saharae Maire = Androcymbium wyssianum Beauverd et Turrett. ■☆
22356 Androcymbium guttatum Schltr. et K. Krause = Androcymbium circinatum Baker ■☆
22357 Androcymbium hantamense Engl.;汉塔姆舟蕊秋水仙■☆
22358 Androcymbium hierrensis Santos;耶罗舟蕊秋水仙■☆
22359 Androcymbium hierrensis Santos subsp. macrospermum U. Reifenb.;大籽舟蕊秋水仙■☆
22360 Androcymbium huntleyi Pedrola, Membrives, J. M. Monts. et Caujapé;洪特兰舟蕊秋水仙■☆
22361 Androcymbium irroratum Schltr. et K. Krause;露珠舟蕊秋水仙■☆
22362 Androcymbium latifolium Schinz;宽叶舟蕊秋水仙■☆
22363 Androcymbium leistneri U. Müll. -Doblies et D. Müll. -Doblies;莱斯特纳舟蕊秋水仙■☆
22364 Androcymbium leucanthum Willd. = Androcymbium capense (L.) K. Krause ■☆
22365 Androcymbium littorale Eckl.;滨海舟蕊秋水仙■☆
22366 Androcymbium longipes Baker;长梗舟蕊秋水仙■☆
22367 Androcymbium melanthioides Willd.;黑舟蕊秋水仙■☆
22368 Androcymbium melanthioides Willd. subsp. australe U. Müll. -Doblies et D. Müll. -Doblies;南方舟蕊秋水仙■☆
22369 Androcymbium melanthioides Willd. subsp. transvaalense U. Müll. -Doblies et D. Müll. -Doblies;德兰士瓦舟蕊秋水仙■☆
22370 Androcymbium melanthioides Willd. var. acaule Baker = Androcymbium melanthioides Willd. ■☆
22371 Androcymbium melanthioides Willd. var. striatum (Hochst. ex A. Rich.) Baker = Androcymbium striatum Hochst. ex A. Rich. ■☆
22372 Androcymbium melanthioides Willd. var. subulatum (Baker) Baker = Androcymbium melanthioides Willd. subsp. transvaalense U. Müll. -Doblies et D. Müll. -Doblies ■☆
22373 Androcymbium natalense Baker;纳塔尔舟蕊秋水仙■☆
22374 Androcymbium orienticapense U. Müll. -Doblies et D. Müll. -Doblies;东好望角舟蕊秋水仙■☆
22375 Androcymbium palaestinum Baker;燥地舟蕊秋水仙■☆
22376 Androcymbium pritzelianum Diels = Androcymbium crispum Schinz ■☆
22377 Androcymbium psammophilum Svent.;喜沙舟蕊秋水仙■☆
22378 Androcymbium pulchrum Schltr. et K. Krause = Androcymbium latifolium Schinz ■☆
22379 Androcymbium punctatum (Cav.) Baker = Androcymbium gramineum (Cav.) J. F. Macbr. ■☆
22380 Androcymbium punctatum (Cav.) Baker var. genuinum Maire = Androcymbium gramineum (Cav.) J. F. Macbr. ■☆
22381 Androcymbium punctatum (Cav.) Baker var. saharae Maire = Androcymbium wyssianum Beauverd et Turrett. ■☆
22382 Androcymbium punctatum Baker = Androcymbium capense (L.) K. Krause ■☆

22383 Androcymbium rechingeri Greuter = Colchicum pusillum Sieber ■☆
22384 Androcymbium roseum Engl.;粉红舟蕊秋水仙■☆
22385 Androcymbium roseum Engl. subsp. albiflorum U. Müll. -Doblies et al.;白花粉红舟蕊秋水仙■☆
22386 Androcymbium scabromarginatum Schltr. et K. Krause;糙边舟蕊秋水仙■☆
22387 Androcymbium schlechteri K. Krause = Androcymbium albomarginatum Schinz ■☆
22388 Androcymbium striatum Hochst. ex A. Rich.;条纹舟蕊秋水仙■☆
22389 Androcymbium subulatum Baker = Androcymbium melanthioides Willd. subsp. transvaalense U. Müll. -Doblies et D. Müll. -Doblies ■☆
22390 Androcymbium swazicum U. Müll. -Doblies et D. Müll. -Doblies;斯威士舟蕊秋水仙■☆
22391 Androcymbium undulatum U. Müll. -Doblies et D. Müll. -Doblies;波状舟蕊秋水仙■☆
22392 Androcymbium villosum U. Müll. -Doblies et D. Müll. -Doblies;长柔毛舟蕊秋水仙■☆
22393 Androcymbium volutare Burch.;旋卷舟蕊秋水仙■☆
22394 Androcymbium walteri Pedrola, Membrives et J. M. Monts.;瓦尔特舟蕊秋水仙■☆
22395 Androcymbium wyssianum Beauverd et Turrett.;维斯舟蕊秋水仙■☆
22396 Androglossa Benth. = Sabia Colebr. ●
22397 Androglossum Champ. ex Benth. = Sabia Colebr. ●
22398 Androglossum reticulatum Champ. ex Benth. = Sabia limoniacea Wall. ex Hook. f. et Thomson ●
22399 Andrographis Wall. = Andrographis Wall. ex Nees ■
22400 Andrographis Wall. ex Nees(1832);穿心莲属(须药草属);Andrographis ■
22401 Andrographis echioides Nees;蓝蓟穿心莲(刺穿心莲);Echiumlike Andrographis, False Waterwillow ■☆
22402 Andrographis glomeruliflora Bremek. = Andrographis laxiflora (Blume) Lindau var. glomeruliflora (Bremek.) H. Chu ■
22403 Andrographis laxiflora (Blume) Lindau;疏花穿心莲(白花穿心莲,白花刺穿心莲,勐仑须药草,须药草,野榄核莲);Laxflower Andrographis ■
22404 Andrographis laxiflora (Blume) Lindau var. glomeruliflora (Bremek.) H. Chu;腺毛疏花穿心莲(宽丝爵床,腺毛须药草)■
22405 Andrographis monglunensis Hung T. Chang et H. Chu = Andrographis laxiflora (Blume) Lindau var. glomeruliflora (Bremek.) H. Chu ■
22406 Andrographis ovata Benth. et Hook. f.;卵叶须药草;Ovateleaf Andrographis ■
22407 Andrographis paniculata (Burm. f.) Wall. ex Nees;穿心莲(春莲秋柳,春莲夏柳,金耳钩,金香草,苦草,苦胆草,榄核莲,日行千里,四方草,四方莲,万病仙草,一见喜,印度草,印度苦草,圆锥须药草,斩蛇剑);Common Andrographis, Kariyat ■
22408 Andrographis serpyllifolia Wight;西穿心莲(西刺穿心莲)■☆
22409 Andrographis sinensis H. S. Lo = Gymnostachyum sinense (H. S. Lo) H. Chu ■
22410 Andrographis tenera (Nees) Imlay = Andrographis laxiflora (Blume) Lindau ■
22411 Andrographis tenera (Nees) Kuntze = Andrographis laxiflora (Blume) Lindau ■
22412 Andrographis tenuiflora T. Anderson = Andrographis laxiflora (Blume) Lindau ■
22413 Andrographis wightiana Arn. ex Nees;魏氏刺穿心莲;Wight Andrographis ■☆
22414 Androgyne Griff.(废弃属名)= Panisea (Lindl.) Lindl.(保留属名)■
22415 Androlepis Brongn. ex Houllet(1870);鳞蕊凤梨属(药鳞凤梨属,药鳞属)■☆
22416 Androlepis donnell-smithii Mez;鳞蕊凤梨■☆
22417 Andromachia Humb. et Bonpl. = Liabum Adans. ●■☆
22418 Andromeda L.(1753);沼迷迭香属(倒壶花属,绋木属,梫木属,青姬木属,小石楠属);Andromeda, Bog Rosemary, Bog-rosemary, Pieris ●☆
22419 Andromeda adenothrix Miq. = Gaultheria adenothrix (Miq.) Maxim. ●☆
22420 Andromeda arborea L. = Oxydendrum arboreum (L.) DC. ●☆
22421 Andromeda baccata Wangenh. = Gaylussacia baccata (Wangenh.) K. Koch ●☆
22422 Andromeda caerulea L. = Enkianthus campanulatus (Miq.) G. Nicholson ●☆
22423 Andromeda caerulea L. = Phyllodoce caerulea (L.) Bab. ●◇
22424 Andromeda calyculata L. = Cassandra calyculata (L.) Moench ●
22425 Andromeda calyculata L. = Chamaedaphne calyculata (L.) Moench ●
22426 Andromeda calyculata L. = Lyonia ovalifolia (Wall.) Drude var. formosana (Komatsu) T. Yamaz. ●
22427 Andromeda calyculata L. var. angustifolia Aiton = Chamaedaphne calyculata (L.) Moench var. angustifolia (Aiton) Rehder ●☆
22428 Andromeda cernua Miq. = Enkianthus cernuus (Siebold et Zucc.) Makino f. rubens (Maxim.) Ohwi ●☆
22429 Andromeda chinensis Lodd. = Vaccinium bracteatum Thunb. var. chinense (Lodd.) Chun ex Sleumer ●
22430 Andromeda ciliicalyx Miq. = Menziesia ciliicalyx (Miq.) Maxim. ●☆
22431 Andromeda cupressiformis Wall. ex D. Don = Cassiope fastigiata (Wall.) D. Don ●
22432 Andromeda elliptica Siebold et Zucc. = Lyonia ovalifolia (Wall.) Drude var. elliptica (Siebold et Zucc.) Hand. -Mazz. ●
22433 Andromeda fastigiata Wall. = Cassiope fastigiata (Wall.) D. Don ●
22434 Andromeda floribunda Pursh = Pieris floribunda (Pursh ex Sims) Benth. et Hook. ●
22435 Andromeda formosa Wall. = Pieris formosa (Wall.) D. Don ●
22436 Andromeda glaucophylla Link;灰叶沼迷迭香;Bog Rosemary, Bog-rosemary, Moorwort ●☆
22437 Andromeda glaucophylla Link var. iodandra Fernald = Andromeda glaucophylla Link ●☆
22438 Andromeda japonica Thunb. = Pieris japonica (Thunb.) D. Don ex G. Don ●
22439 Andromeda lanceolata Wall. = Lyonia ovalifolia (Wall.) Drude var. lanceolata (Wall.) Hand. -Mazz. ●
22440 Andromeda lycopodioides Pall. = Cassiope lycopodioides (Pall.) D. Don ●
22441 Andromeda myrtifolia Willd. ex DC.;香桃木叶迷迭香●☆
22442 Andromeda nana Maxim. = Arcterica nana (Maxim.) Makino ●☆
22443 Andromeda ovalifolia Wall. = Lyonia ovalifolia (Wall.) Drude ●
22444 Andromeda perulata Miq. = Enkianthus perulatus (Miq.) C. K. Schneid. ●
22445 Andromeda polifolia L.;沼迷迭香(倒壶花,多叶梫木,灰叶梫木,灰叶青姬木);Bog Bell, Bog Rosemary, Bog-rosemary, Bog-rosemary Andromeda, Chinese Pieris, Marsh Andromeda, Marsh

Cistus, Marsh Holy Rose, Moonwort, Moon-wort, Rosemary Bog, Wild Rosemary ●☆

22446 Andromeda polifolia L. 'Alba';洁白沼迷迭香(白色倒壶花)●☆

22447 Andromeda polifolia L. 'Compacta';致密倒壶花●☆

22448 Andromeda polifolia L. 'Kiri Kaming';日本沼迷迭香;Japanese Bog Rosemary ●☆

22449 Andromeda polifolia L. 'Macrophylla';大叶沼迷迭香●☆

22450 Andromeda polifolia L. f. acerosa Hartm.;多针沼迷迭香●☆

22451 Andromeda polifolia L. f. leucantha (Takeda) Takeda ex H. Hara;白花沼迷迭香●☆

22452 Andromeda polifolia L. subsp. glaucophylla (Link) Hultén = Andromeda glaucophylla Link ●☆

22453 Andromeda polifolia L. var. glaucophylla (Link) DC. = Andromeda glaucophylla Link ●☆

22454 Andromeda pyrifolia Pers. = Agauria salicifolia (Comm. ex Lam.) Hook. f. ex Oliv. ●☆

22455 Andromeda revoluta Steud.;外卷迷迭香●☆

22456 Andromeda salicifolia Comm. ex Lam. = Agauria salicifolia (Comm. ex Lam.) Hook. f. ex Oliv. ●☆

22457 Andromeda speciosa Michx.;大花沼迷迭香;Large-flowered Andromeda ●☆

22458 Andromeda squamulosa D. Don = Lyonia ovalifolia (Wall.) Drude var. lanceolata (Wall.) Hand. -Mazz. ●

22459 Andromeda stelleriana Pall. = Harrimanella stelleriana (Pall.) Coville ●

22460 Andromeda subsessilis Miq. = Enkianthus subsessilis (Miq.) Makino ●☆

22461 Andromeda taxifolia Pall. = Phyllodoce caerulea (L.) Bab. ●◇

22462 Andromeda villosa Wall. = Lyonia villosa (Hook. f. ex C. B. Clarke) Hand. -Mazz. ●

22463 Andromedaceae (Endl.) Schnizl. = Ericaceae Juss.(保留科名)●

22464 Andromedaceae DC. ex Schnizl. = Ericaceae Juss.(保留科名)●

22465 Andromedaceae Döll = Ericaceae Juss.(保留科名)●

22466 Andromedaceae Schnizl.;沼迷迭香科●

22467 Andromedaceae Schnizl. = Ericaceae Juss.(保留科名)●

22468 Andromycia A. Rich.(1850);古巴南星属■☆

22469 Andromycia A. Rich. = Asterostigma Fisch. et C. A. Mey. ■☆

22470 Andromycia cubensis A. Rich.;古巴南星■☆

22471 Androphilax Steud. = Cocculus DC.(保留属名)●

22472 Androphoranthus H. Karst. = Caperonia A. St. -Hil. ■☆

22473 Androphthoe Scheft. = Dendrophthoe Mart. ●

22474 Androphylax J. C. Wendl.(废弃属名) = Cocculus DC.(保留属名)●

22475 Androphysa Moq. = Halocharis Moq. ■☆

22476 Andropogon L.(1753)(保留属名);须芒草属;Beard-grass ■

22477 Andropogon abyssinicus Fresen.;阿比西尼亚须芒草■☆

22478 Andropogon achtenii Robyns = Anadelphia hamata Stapf ■☆

22479 Andropogon acicularis Retz. ex Roem. et Schult. = Chrysopogon aciculatus (Retz. ex Roem. et Schult.) Trin. ■

22480 Andropogon aciculatus Retz. = Chrysopogon aciculatus (Retz. ex Roem. et Schult.) Trin. ■

22481 Andropogon afer J. F. Gmel. = Ischaemum afrum (J. F. Gmel.) Dandy ■☆

22482 Andropogon affinis J. Presl = Sorghum propinquum (Kunth) Hitchc. ■

22483 Andropogon africanus Franch.;非洲须芒草■☆

22484 Andropogon afzelianus Rendle = Anadelphia afzeliana (Rendle) Stapf ■☆

22485 Andropogon allionii DC. = Heteropogon contortus (L.) P. Beauv. ex Roem. et Schult. ■

22486 Andropogon alopecurus (Desv.) Hack.;看麦娘须芒草■☆

22487 Andropogon altissimus Hochst. ex A. Braun = Hyparrhenia rufa (Nees) Stapf ■☆

22488 Andropogon amaurus Büse = Polytrias amaura (Büse) Kuntze ■

22489 Andropogon amaurus Büse = Polytrias indica (Houtt.) Veldkamp ■

22490 Andropogon amethystinus Steud.;紫晶须芒草■☆

22491 Andropogon amethystinus Steud. var. lima Hack. = Andropogon lima (Hack.) Stapf ■☆

22492 Andropogon amplectens Nees = Diheteropogon amplectens (Nees) Clayton ■☆

22493 Andropogon amplectens Nees var. catangensis Chiov. = Diheteropogon amplectens (Nees) Clayton var. catangensis (Chiov.) Clayton ■☆

22494 Andropogon andongensis (Rendle) K. Schum. = Hyparrhenia andongensis (Rendle) Stapf ■☆

22495 Andropogon androphilus Stapf = Elymandra androphila (Stapf) Stapf ■☆

22496 Andropogon annulatus Forssk. = Dichanthium annulatum (Forssk.) Stapf ■

22497 Andropogon annuus Hack. = Parahyparrhenia annua (Hack.) Clayton ■☆

22498 Andropogon antephoroides (Steud.) Steud. = Ischaemum antephoroides (Steud.) Miq. ■

22499 Andropogon anthephoroides (Steud.) Steud. = Ischaemum antephoroides (Steud.) Miq. ■

22500 Andropogon anthephoroides (Steud.) Steud. var. eriostachyus (Hack.) Honda = Ischaemum antephoroides (Steud.) Miq. ■

22501 Andropogon anthistirioides Hochst. ex A. Rich. = Hyparrhenia anthistirioides (Hochst. ex A. Rich.) Andersson ex Stapf ■☆

22502 Andropogon appendiculatus Nees var. polycladus Hack. = Andropogon gayanus Kunth var. polycladus (Hack.) Clayton ■☆

22503 Andropogon apricus Trin. var. chinensis (Nees) Hack. = Andropogon chinensis (Nees) Merr. ■

22504 Andropogon argenteus Vanderyst;银白须芒草■☆

22505 Andropogon ariani Edgew. = Cymbopogon jwarancusa (Jones) Schult. subsp. olivieri (Boiss.) Soenarko ■

22506 Andropogon aridus Clayton;旱生须芒草■☆

22507 Andropogon aristatus Poir. = Dichanthium aristatum (Poir.) C. E. Hubb. ■

22508 Andropogon arnottianus (Nees) Steud. = Ischaemum rugosum Salisb. ■

22509 Andropogon arrectus Stapf = Anadelphia afzeliana (Rendle) Stapf ■☆

22510 Andropogon arriani Edgew. = Cymbopogon jwarancusa (Jones) Schult. subsp. olivieri (Boiss.) Soenarko ■

22511 Andropogon arriani Edgew. = Cymbopogon jwarancusus (Jones) Schult. subsp. olivieri (Boiss.) Soenarko ■☆

22512 Andropogon arrianii Edgew. = Cymbopogon olivieri (Boiss.) Bor ■

22513 Andropogon arundinaceus P. J. Bergius = Merxmuellera arundinacea (P. J. Bergius) Conert ■☆

22514 Andropogon arundinaceus Scop. = Sorghum halepense (L.) Pers. ■

22515 Andropogon arundinaceus Willd. = Sorghum arundinaceum

(Desv.) Stapf ■☆

22516 Andropogon ascinodis C. B. Clarke; 囊节须芒草; Ascinode Bluestem ■

22517 Andropogon ascinodis C. B. Clarke = Andropogon chinensis (Nees) Merr. ■

22518 Andropogon assimilis Steud. = Capillipedium assimile (Steud.) A. Camus ■

22519 Andropogon asthenostachys Steud. = Pseudopogonatherum contortum (Brongn.) A. Camus ■

22520 Andropogon asthenostachys Steud. = Pseudopogonatherum koretrostachys (Trin.) Henrard ■

22521 Andropogon aucheri Boiss. = Chrysopogon aucheri (Boiss.) Stapf ■☆

22522 Andropogon aucheri Boiss. var. chrysopus (Coss.) Hack. = Chrysopogon aucheri (Boiss.) Stapf ■☆

22523 Andropogon aucheri Boiss. var. quinqueplumis Hack. = Chrysopogon aucheri (Boiss.) Stapf ■☆

22524 Andropogon aucheri Boiss. var. subpungens Hack. = Chrysopogon aucheri (Boiss.) Stapf ■☆

22525 Andropogon auctus Stapf = Hyparrhenia dregeana (Nees) Stapf ex Stent ■☆

22526 Andropogon aureofulvus Steud. = Eulalia leschenaultiana (Decne.) Ohwi ■

22527 Andropogon aureus Bory = Eulalia aurea (Bory) Kunth ■

22528 Andropogon aureus Vanderyst = Andropogon chinensis (Nees) Merr. ■

22529 Andropogon auriculatus Stapf; 耳形须芒草■☆

22530 Andropogon barbatus L. = Chloris barbata Sw. ■

22531 Andropogon barteri Hack. = Hyparrhenia barteri (Hack.) Stapf ■☆

22532 Andropogon bavicchii De Wild. = Schizachyrium thollonii (Franch.) Stapf ■☆

22533 Andropogon bergii Roem. et Schult. = Merxmuellera arundinacea (P. J. Bergius) Conert ■☆

22534 Andropogon besseri Kunth = Sorghum nervosum Besser ex Schult. ■

22535 Andropogon biaristatus Steud. = Microstegium biaristatum (Steud.) Keng ■

22536 Andropogon biaristatus Steud. = Microstegium ciliatum (Trin.) A. Camus ■

22537 Andropogon bicolor (L.) Roxb. = Sorghum bicolor (L.) Moench ■

22538 Andropogon bicolor Roxb. = Sorghum bicolor (L.) Moench ■

22539 Andropogon binatus Retz. = Eulaliopsis binata (Retz.) C. E. Hubb. ■

22540 Andropogon bipennatus Hack. = Sorghastrum bipennatum (Hack.) Pilg. ■☆

22541 Andropogon bladhii Retz. = Bothriochloa bladhii (Retz.) S. T. Blake ■

22542 Andropogon bootanensis Hook. f. = Eremopogon delavayi (Hack.) A. Camus ■

22543 Andropogon bootanensis Hook. f. = Schizachyrium delavayi (Hack.) Bor ■

22544 Andropogon bouangensis Franch. = Hyparrhenia rufa (Nees) Stapf ■☆

22545 Andropogon bovonei Chiov. = Elymandra lithophila (Trin.) Clayton ■☆

22546 Andropogon brachyatherus Hochst. = Ischaemum afrum (J. F. Gmel.) Dandy ■☆

22547 Andropogon bracteatus Humb. et Bonpl. ex Willd. = Hyparrhenia bracteata (Humb. et Bonpl. ex Willd.) Stapf ■

22548 Andropogon brevifolius Sw. = Schizachyrium brevifolium (Sw.) Nees ex Büse ■

22549 Andropogon brevifolius Sw. var. fragilis (R. Br.) Hack. = Schizachyrium fragile (R. Br.) A. Camus ■

22550 Andropogon brevifolius Sw. var. platyphyllus Franch. = Schizachyrium platyphyllum (Franch.) Stapf ■☆

22551 Andropogon brieyi De Wild. = Andropogon gabonensis Stapf ■☆

22552 Andropogon buchananii Stapf = Hyparrhenia poecilotricha (Hack.) Stapf ■☆

22553 Andropogon buchneri Hack. = Diheteropogon grandiflorus (Hack.) Stapf ■☆

22554 Andropogon caesius Nees ex Hook. et Arn. = Cymbopogon caesius (Nees ex Hook. et Arn.) Stapf ■

22555 Andropogon calvescens Stapf = Andropogon tenuiberbis Hack. ■☆

22556 Andropogon canaliculatus Schumach.; 具沟须芒草■☆

22557 Andropogon canaliculatus Schumach. var. fyffei Stapf = Andropogon canaliculatus Schumach. ■☆

22558 Andropogon canalicutalus Schumach. var. fastigians Stapf = Andropogon canaliculatus Schumach. ■☆

22559 Andropogon capensis Houtt. = Eustachys paspaloides (Vahl) Lanza et Mattei ■☆

22560 Andropogon capillaris Kunth; 发状须芒草■☆

22561 Andropogon caricosus L. = Dichanthium caricosum (L.) A. Camus ■

22562 Andropogon castratus Griff. = Arthraxon castratus (Griff.) Nayaran. ex Bor ■

22563 Andropogon caucasicus Trin. = Bothriochloa bladhii (Retz.) S. T. Blake ■

22564 Andropogon ceresiiformis Nees = Monocymbium ceresiiforme (Nees) Stapf ■☆

22565 Andropogon ceresiiformis Nees var. breviaristatus Hack. = Monocymbium ceresiiforme (Nees) Stapf ■☆

22566 Andropogon ceresiiformis Nees var. hirtellus Franch. = Monocymbium ceresiiforme (Nees) Stapf ■☆

22567 Andropogon ceresiiformis Nees var. submuticus Hack. = Monocymbium ceresiiforme (Nees) Stapf ■☆

22568 Andropogon chevalieri Reznik; 舍瓦利耶须芒草■☆

22569 Andropogon chinensis (Nees) Merr.; 华须芒草; China Bluestem, Chinese Bluestem ■

22570 Andropogon chrysopus Coss. = Chrysopogon aucheri (Boiss.) Stapf ■☆

22571 Andropogon chrysostachyus Steud.; 金穗须芒草■☆

22572 Andropogon ciliolatus Nees ex Steud. = Chrysopogon serrulatus Trin. ■☆

22573 Andropogon cinctus Steud. = Capillipedium parviflorum (R. Br.) Stapf ■

22574 Andropogon citratus DC. = Cymbopogon citratus (DC.) Stapf ■

22575 Andropogon claudopus Chiov. = Schizachyrium claudopus (Chiov.) Chiov. ■☆

22576 Andropogon coeruleus Nees ex Steud. = Chrysopogon serrulatus Trin. ■☆

22577 Andropogon coleotrichus Steud. = Hyparrhenia coleotricha (Steud.) Andersson ex Clayton ■☆

22578 Andropogon collinus Pilg. = Hyparrhenia collina (Pilg.) Stapf ■☆

22579 Andropogon commutatus Steud. = Cymbopogon commutatus (Steud.) Stapf ■☆

22580 Andropogon compressus Stapf = Schizachyrium rupestre (K. Schum.) Stapf ■☆

22581 Andropogon condylotrichus Hochst. ex Steud. = Euclasta condylotricha (Hochst. ex Steud.) Stapf ■☆

22582 Andropogon confertiflorus Steud. = Cymbopogon nardus (L.) Rendle ■

22583 Andropogon confinis Hochst. ex A. Rich. var. macrarrhenus Hack. = Hyparrhenia niariensis (Franch.) Clayton var. macrarrhena (Hack.) Clayton ■☆

22584 Andropogon confinis Hochst. ex A. Rich. var. nudiglumis Hack. = Hyparrhenia confinis (Hochst. ex A. Rich.) Andersson ex Stapf var. nudiglumis (Hack.) Clayton ■☆

22585 Andropogon confinis Hochst. ex A. Rich. var. pellitus Hack. = Hyparrhenia confinis (Hochst. ex A. Rich.) Andersson ex Stapf var. pellita (Hack.) Stapf ■☆

22586 Andropogon congoensis Franch. = Andropogon schirensis Hochst. ex A. Rich. ■☆

22587 Andropogon connatus Hochst. ex A. Rich. = Cymbopogon caesius (Nees ex Hook. et Arn.) Stapf ■

22588 Andropogon contortus L. = Heteropogon contortus (L.) P. Beauv. ex Roem. et Schult. ■

22589 Andropogon cornucopiae Hack. = Hyperthelia cornucopiae (Hack.) Clayton ■☆

22590 Andropogon corollatus Nees ex Steud. = Phacelurus speciosus (Steud.) C. E. Hubb. ■☆

22591 Andropogon cotulifer Thunb. = Eccoilopus cotulifer (Thunb.) A. Camus ■

22592 Andropogon cotulifer Thunb. = Spodiopogon cotulifer (Thunb.) Hack. ■

22593 Andropogon crassipes Steud. = Ischaemum aristatum L. var. glaucum (Honda) T. Koyama ■

22594 Andropogon crinitus Thunb. = Pogonatherum crinitum (Thunb.) Kunth ■

22595 Andropogon curvifolius Clayton;弯叶须芒草■☆

22596 Andropogon cuspidatus Hochst. ex A. Rich. = Arthraxon cuspidatus (Hochst. ex A. Rich.) Hochst. ■

22597 Andropogon cymbarius L. = Hyparrhenia cymbaria (L.) Stapf ■☆

22598 Andropogon cyrtocladus Stapf = Andropogon kelleri Hack. ■☆

22599 Andropogon delavayi Hack. = Eremopogon delavayi (Hack.) A. Camus ■

22600 Andropogon delavayi Hack. = Schizachyrium delavayi (Hack.) Bor ■

22601 Andropogon densiflorus Steud. = Cymbopogon densiflorus (Steud.) Stapf ■☆

22602 Andropogon dewevrei De Wild. = Andropogon gayanus Kunth var. polycladus (Hack.) Clayton ■☆

22603 Andropogon dichroos Steud. = Hyparrhenia dichroa (Steud.) Stapf ■☆

22604 Andropogon dieterlenii Stapf ex E. Phillips = Cymbopogon dieterlenii Stapf ex E. Phillips ■☆

22605 Andropogon diplandrus Hack. = Hyparrhenia diplandra (Hack.) Stapf ■

22606 Andropogon distachyos L.;二穗须芒草■☆

22607 Andropogon distachyos L. var. hirtus Chiov. = Andropogon distachyos L. ■☆

22608 Andropogon distachyos L. var. pubescens Parl. = Andropogon distachyos L. ■☆

22609 Andropogon distachyus L.;双穗须芒草■☆

22610 Andropogon distans Nees = Cymbopogon distans (Nees ex Steud.) W. Watson ■

22611 Andropogon distans Nees ex Steud. = Cymbopogon distans (Nees ex Steud.) W. Watson ■

22612 Andropogon divergens (Hack.) Andersson ex Hitchc. = Schizachyrium scoparium (Michx.) Nash var. divergens (Hack.) Gould ■☆

22613 Andropogon diversiflorus Steud. = Polytrias amaura (Büse) Kuntze ■

22614 Andropogon diversiflorus Steud. = Polytrias indica (Houtt.) Veldkamp ■

22615 Andropogon diversifolius Rendle = Diheteropogon amplectens (Nees) Clayton var. catangensis (Chiov.) Clayton ■☆

22616 Andropogon doloensis Vanderyst;多罗须芒草■☆

22617 Andropogon dregeanus Nees = Hyparrhenia dregeana (Nees) Stapf ex Stent ■☆

22618 Andropogon drummondii Nees ex Steud. = Sorghum bicolor (L.) Moench subsp. drummondii (Nees ex Steud.) de Wet ■☆

22619 Andropogon dulcis Burm. f. = Eleocharis tuberosa (Roxb.) Roem. et Schult. ■

22620 Andropogon dulcis Burm. f. = Heleocharis dulcis (Burm. f.) Trin. ex Hensch. ■

22621 Andropogon dummeri Stapf = Andropogon schirensis Hochst. ex A. Rich. ■☆

22622 Andropogon dummeri Stapf var. calvus ? = Andropogon schirensis Hochst. ex A. Rich. ■☆

22623 Andropogon dybowskii Franch. = Hyparrhenia dybowskii (Franch.) Roberty ■☆

22624 Andropogon eberhardtii (A. Camus) Merr. = Hyparrhenia diplandra (Hack.) Stapf ■

22625 Andropogon echinatus (Nees) Heyne = Arthraxon echinatus Hochst. ■

22626 Andropogon echinatus Heyne ex Steud = Arthraxon lanceolatus (Roxb.) Hochst. var. echinatus (Nees) Hack. ■

22627 Andropogon echinulatus (Nees) Steud. = Chrysopogon echinulatus (Steud.) W. Watson ■

22628 Andropogon echinulatus (Nees) Steud. = Chrysopogon gryllus (L.) Trin. subsp. echinulatus (Nees) Cope ■☆

22629 Andropogon echinulatus Nees ex Steud. = Chrysopogon echinulatus (Steud.) W. Watson ■

22630 Andropogon elegantissimus Steud. = Elionurus royleanus Nees ex A. Rich. ■☆

22631 Andropogon elegantissimus Steud. var. arabicus Steud. = Elionurus royleanus Nees ex A. Rich. ■☆

22632 Andropogon elionuroides Vanderyst = Andropogon eucomus Nees subsp. huillensis (Rendle) Sales ■☆

22633 Andropogon elliottii Chapm. = Andropogon ternarius Michx. ■☆

22634 Andropogon elliottii Chapm. var. projecta Fernald et Griscom = Andropogon gyrans Ashe ■☆

22635 Andropogon emarginatus De Wild. = Diheteropogon grandiflorus (Hack.) Stapf ■☆

22636 Andropogon eucnemis Trin. = Andropogon canaliculatus Schumach. ■☆

22637 Andropogon eucomus Nees;美顶花须芒草■☆

22638 Andropogon eucomus Nees subsp. huillensis (Rendle) Sales;威拉须芒草■☆

22639 Andropogon excavatus Hochst. = Cymbopogon caesius (Nees ex Hook. et Arn.) Stapf ■

22640 Andropogon exilis Hochst. = Schizachyrium exile (Hochst.) Pilg. ■☆

22641 Andropogon exilis Hochst. var. glabrescens Rendle = Schizachyrium exile (Hochst.) Pilg. ■☆

22642 Andropogon exothecus Hack. = Exotheca abyssinica (Hochst. ex A. Rich.) Andersson ■☆

22643 Andropogon fascicularis Roxb. = Pseudosorghum fasciculare (Roxb. A. Camus ■

22644 Andropogon fasciculatus L. = Microstegium fasciculatum (L.) Henrard ■

22645 Andropogon fastigiatus Sw.;帚状须芒草■☆

22646 Andropogon felicis Reznik = Andropogon chevalieri Reznik ■☆

22647 Andropogon festuciformis Rendle;羊茅状须芒草■☆

22648 Andropogon filifolius (Nees) Steud. = Diheteropogon filifolius (Nees) Clayton ■☆

22649 Andropogon filiformis Roxb. = Dimeria ornithopoda Trin. ■

22650 Andropogon filipendulus Hochst. = Hyparrhenia filipendula (Hochst.) Stapf ■

22651 Andropogon filipendulus Hochst. ex Steud. = Hyparrhenia filipendula (Hochst.) Stapf ■

22652 Andropogon filipendulus Hochst. var. calvescens Hack. = Hyparrhenia figariana (Chiov.) Clayton ■☆

22653 Andropogon filipendulus Hochst. var. pilosus Hochst. = Hyparrhenia filipendula (Hochst.) Stapf var. pilosa (Hochst.) Stapf ■

22654 Andropogon finitimus Hochst. ex A. Rich. = Hyparrhenia finitima (Hochst. ex A. Rich.) Stapf ■

22655 Andropogon firmandus Steud. = Polytrias indica (Houtt.) Veldkamp ■

22656 Andropogon flabellifer Pilg. = Andropogon mannii Hook. f. ■☆

22657 Andropogon flaccidus A. Rich. = Schizachyrium brevifolium (Sw.) Nees ex Büse ■

22658 Andropogon flexuosus Nees ex Steud. = Cymbopogon flexuosus (Nees ex Steud.) W. Watson ■

22659 Andropogon floccusus Schweinf. = Cymbopogon commutatus (Steud.) Stapf ■☆

22660 Andropogon formosanus Rendle var. minor Rendle = Microstegium ciliatum (Trin.) A. Camus ■

22661 Andropogon formosanus Rendle var. minor Rendle = Microstegium fasciculatum (L.) Henrard ■

22662 Andropogon formosus Hort. = Hyparrhenia formosa Stapf ■☆

22663 Andropogon foveolatus Delile = Dichanthium foveolatum (Delile) Roberty ■☆

22664 Andropogon foveolatus Delile var. plumosus A. Terracc. = Dichanthium foveolatum (Delile) Roberty ■☆

22665 Andropogon fragilis R. Br. = Schizachyrium fragile (R. Br.) A. Camus ■

22666 Andropogon fragilis R. Br. var. sinensis Rendle = Schizachyrium fragile (R. Br.) A. Camus ■

22667 Andropogon friesii Pilg.;弗里斯须芒草■☆

22668 Andropogon fulvibarbis Trin. = Vetiveria fulvibarbis (Trin.) Stapf ■☆

22669 Andropogon fulvicomus Hochst. ex A. Rich. = Hyparrhenia rufa (Nees) Stapf ■☆

22670 Andropogon fundaensis Vanderyst = Trachypogon spicatus (L. f.) Kuntze ■☆

22671 Andropogon furcatus Muhl. = Andropogon furcatus Muhl. ex Willd. ■☆

22672 Andropogon furcatus Muhl. ex Willd.;叉分须芒草■☆

22673 Andropogon furcatus Muhl. ex Willd. = Andropogon gerardii Vitman ■☆

22674 Andropogon gabonensis Stapf;加蓬须芒草■☆

22675 Andropogon gayanus Kunth;盖伊须芒草;Gamba Grass ■☆

22676 Andropogon gayanus Kunth 'Planathina';大须芒草■☆

22677 Andropogon gayanus Kunth var. bisquamulatus (Hochst.) Hack.;双鳞盖伊须芒草■☆

22678 Andropogon gayanus Kunth var. polycladus (Hack.) Clayton;多枝盖伊须芒草■☆

22679 Andropogon gayanus Kunth var. squamulatus (Hochst.) Stapf = Andropogon gayanus Kunth var. polycladus (Hack.) Clayton ■☆

22680 Andropogon gayanus Kunth var. tridentatus Hack.;三齿盖伊须芒草■☆

22681 Andropogon gerardii Vitman;吉氏须芒草;Big Bluestem, Big Blue-stem, Bluejoint, Turkey Foot, Turkey-claw, Turkeyfoot, Turkey-foot ■☆

22682 Andropogon giganteus Hochst. = Cymbopogon caesius (Nees ex Hook. et Arn.) Stapf subsp. giganteus (Chiov.) Sales ■☆

22683 Andropogon glaber Roxb. = Bothriochloa bladhii (Retz.) S. T. Blake ■

22684 Andropogon glaber Roxb. = Bothriochloa glabra (Roxb.) A. Camus ■

22685 Andropogon glabriusculus Hochst. ex A. Rich. = Hyparrhenia glabriuscula (Hochst. ex A. Rich.) Stapf ■☆

22686 Andropogon glaucopsis Steud. = Capillipedium assimile (Steud.) A. Camus ■

22687 Andropogon glauco-purpureus Stapf;灰紫须芒草■☆

22688 Andropogon glomeratus (Walter) Britton, Sterns et Poggenb.;团花须芒草;Bush Beard Grass, Bushy Broom Grass ■☆

22689 Andropogon glomeratus Britton, Stern. et Poggenb. = Andropogon glomeratus (Walter) Britton, Sterns et Poggenb. ■☆

22690 Andropogon goeringii Steud. = Cymbopogon goeringii (Steud.) A. Camus ■

22691 Andropogon golae Chiov. = Andropogon schirensis Hochst. ex A. Rich. ■☆

22692 Andropogon grandiflorus Hack. = Diheteropogon grandiflorus (Hack.) Stapf ■☆

22693 Andropogon greenwayi Napper;格林韦须芒草■☆

22694 Andropogon griffithsiae (Nees) Steud. = Ischaemum rugosum Salisb. ■

22695 Andropogon griffithsii (Nees) Steud. = Ischaemum rugosum Salisb. ■

22696 Andropogon gryllus L. = Chrysopogon echinulatus (Steud.) W. Watson ■

22697 Andropogon gryllus L. = Chrysopogon gryllus Trin. ■☆

22698 Andropogon gryllus L. subsp. echinulatus (Nees ex Steud.) Hack. = Chrysopogon echinulatus (Steud.) W. Watson ■

22699 Andropogon gryllus L. subsp. echinulatus (Nees) Hack. = Chrysopogon echinulatus (Steud.) W. Watson ■

22700 Andropogon gryllus L. subsp. echinulatus (Nees) Hack. = Chrysopogon gryllus (L.) Trin. subsp. echinulatus (Nees) Cope ■☆

22701 Andropogon gryllus L. subsp. echinulatus Hack. = Chrysopogon echinulatus (Steud.) W. Watson ■

22702 Andropogon guineensis Schumach. = Andropogon gayanus Kunth ■☆

22703 Andropogon gyirongensis L. Liou = Andropogon munroi C. B. Clarke ■

22704 Andropogon gyrans Ashe;埃利须芒草;Elliott's Broomsedge ■☆

22705 Andropogon haenkei J. Presl = Bothriochloa bladhii (Retz.) S. T. Blake ■

22706 Andropogon haenkei J. Presl ex C. Presl = Bothriochloa glabra (Roxb.) A. Camus ■

22707 Andropogon halepensis (L.) Brot. = Sorghum halepense (L.) Pers. ■

22708 Andropogon halepensis (L.) Brot. var. propinquus (Kunth) Hack. = Sorghum propinquum (Kunth) Hitchc. ■

22709 Andropogon halepensis L. = Sorghum halepense (L.) Pers. ■

22710 Andropogon halepensis L. var. genuinus (Hack.) Stapf = Sorghum halepense (L.) Pers. ■

22711 Andropogon halepensis L. var. propinquum (Kunth) Hack. = Sorghum propinquum (Kunth) Hitchc. ■

22712 Andropogon hamatulus Hook. et Arn. = Cymbopogon tortilis (J. Presl) A. Camus ■

22713 Andropogon hamatulus Nees ex Hook. et Arn. = Cymbopogon hamatulus (Nees ex Hook. et Arn.) A. Camus ■

22714 Andropogon helophilus K. Schum. = Andropogon gayanus Kunth var. polycladus (Hack.) Clayton ■☆

22715 Andropogon heterantherus Stapf;异药须芒草■☆

22716 Andropogon heteroclitus (Roxb.) L. Chen = Pseudanthistria heteroclita (Roxb.) Hook. f. ■

22717 Andropogon heteroclitus (Roxb.) Nees = Pseudanthistria heteroclita (Roxb.) Hook. f. ■

22718 Andropogon himalayensis Gand. = Cymbopogon jwarancusus (Jones) Schult. ■

22719 Andropogon himalayensis Steud. = Apocopis paleacea (Trin.) Hochr. ■

22720 Andropogon hirtiflorus (Nees) Kunth = Schizachyrium sanguineum (Retz.) Alston ■

22721 Andropogon hirtus L. = Hyparrhenia hirta (L.) Stapf ■☆

22722 Andropogon hirtus L. var. longearistatus Willk. = Hyparrhenia hirta (L.) Stapf ■☆

22723 Andropogon hirtus L. var. podotrichus (Hochst.) Hack. = Hyparrhenia hirta (L.) Stapf ■☆

22724 Andropogon hirtus L. var. pubescens (Vis.) Hack. = Hyparrhenia hirta (L.) Stapf ■☆

22725 Andropogon hispidissimus Steud. = Heteropogon contortus (L.) Roem. et Schult. ■

22726 Andropogon homblei De Wild. = Diheteropogon grandiflorus (Hack.) Stapf ■☆

22727 Andropogon homogamus Stapf = Andropogon amethystinus Steud. ■☆

22728 Andropogon hookeri Munro ex Hack. = Andropogon munroi C. B. Clarke ■

22729 Andropogon huillensis Rendle = Andropogon eucomus Nees subsp. huillensis (Rendle) Sales ■☆

22730 Andropogon humbertii A. Camus;亨伯特须芒草■☆

22731 Andropogon humilis Hochst. ex A. Rich. = Andropogon abyssinicus Fresen. ■☆

22732 Andropogon ibityensis A. Camus;伊比提须芒草■☆

22733 Andropogon imerinensis Bosser;伊梅里纳须芒草■☆

22734 Andropogon impressus Hack. = Schizachyrium impressum (Hack.) A. Camus ■☆

22735 Andropogon incanellus Clayton = Andropogon africanus Franch. ■☆

22736 Andropogon incomptus Clayton;装饰须芒草■☆

22737 Andropogon infrasulcatus Reznik = Andropogon gayanus Kunth ■☆

22738 Andropogon insculptus Hochst. ex A. Rich. = Bothriochloa insculpta (Hochst. ex A. Rich.) A. Camus ■☆

22739 Andropogon intermedia (R. Br.) Stapf = Bothriochloa bladhii (Retz.) S. T. Blake ■

22740 Andropogon intermedius (R. Br.) Stapf = Bothriochloa bladhii (Retz.) S. T. Blake ■

22741 Andropogon intermedius (R. Br.) Stapf var. caucasica (Trin.) Hack. = Bothriochloa bladhii (Retz.) S. T. Blake ■

22742 Andropogon intermedius R. Br. = Bothriochloa bladhii (Retz.) S. T. Blake ■

22743 Andropogon intermedius R. Br. var. caucasica (Trin.) Hack. = Bothriochloa bladhii (Retz.) S. T. Blake ■

22744 Andropogon intermedius R. Br. var. genuinus Hack. = Bothriochloa bladhii (Retz.) S. T. Blake ■

22745 Andropogon intermedius R. Br. var. haenkii (J. Presl) Hack. = Bothriochloa glabra (Roxb.) A. Camus ■

22746 Andropogon intermedius R. Br. var. punctatus subvar. glaber Hack. = Bothriochloa glabra (Roxb.) A. Camus ■

22747 Andropogon intumescens Pilg. = Ischaemum afrum (J. F. Gmel.) Dandy ■☆

22748 Andropogon involutus Steud. = Eulaliopsis binata (Retz.) C. E. Hubb. ■

22749 Andropogon ischaemum L. = Bothriochloa ischaemum (L.) Keng ■

22750 Andropogon ischaemum L. = Dichanthium ischaemum (L.) Roberty ■

22751 Andropogon ischaemum L. var. longearistatus Willk. = Dichanthium ischaemum (L.) Roberty ■

22752 Andropogon ischaemum L. var. somalensis Stapf = Bothriochloa radicans (Lehm.) A. Camus ■☆

22753 Andropogon ischaemum L. var. songaricus Rupr. ex Fisch. et C. A. Mey. = Bothriochloa ischaemum (L.) Keng ■

22754 Andropogon ischaemus L. = Bothriochloa ischaemum (L.) Keng ■

22755 Andropogon ischaemus L. var. redicans Hack. = Bothriochloa ischaemum (L.) Keng ■

22756 Andropogon ischyranthus Steud. = Heteropogon triticeus (R. Br.) Stapf et Craib ■

22757 Andropogon isostachyus Peter = Andropogon tenuiberbis Hack. ■☆

22758 Andropogon ivohibensis A. Camus;伊武希贝须芒草■☆

22759 Andropogon ivorensis Adjan. et Clayton;伊沃里须芒草■☆

22760 Andropogon iwarancusa Roxb. = Cymbopogon jwarancusus (Jones) Schult. ■

22761 Andropogon jeffreysii Hack. = Schizachyrium jeffreysii (Hack.) Stapf ■☆

22762 Andropogon junguensis Vanderyst = Andropogon festuciformis Rendle ■☆

22763 Andropogon jwarancusus Jones = Cymbopogon jwarancusus (Jones) Schult. ■

22764 Andropogon jwarancusus Jones subsp. laniger (Desf.) Hook. f. = Cymbopogon olivieri (Boiss.) Bor ■

22765 Andropogon jwarancusus Roxb. = Cymbopogon jwarancusus (Jones) Schult. ■

22766 Andropogon kasaiensis Vanderyst;开赛须芒草■☆
22767 Andropogon kelleri Hack.;凯勒须芒草■☆
22768 Andropogon kilimandscharicus Pilg. = Andropogon amethystinus Steud. ■☆
22769 Andropogon kindunduensis Vanderyst = Diheteropogon grandiflorus (Hack.) Stapf ■☆
22770 Andropogon kinsukaensis Vanderyst = Andropogon pseudapricus Stapf ■☆
22771 Andropogon kiwuensis Pilg. = Hyparrhenia familiaris (Steud.) Stapf ■☆
22772 Andropogon koretrostachys Trin. = Pseudopogonatherum contortum (Brongn.) A. Camus ■
22773 Andropogon koretrostachys Trin. = Pseudopogonatherum koretrostachys (Trin.) Henrard ■
22774 Andropogon kwashotensis Hayata = Capillipedium kwashotense (Hayata) C. C. Hsu ■
22775 Andropogon lacunosus J. G. Anderson;具腔须芒草■☆
22776 Andropogon lanceolatus Roxb. = Arthraxon lanceolatus (Roxb.) Hochst. ■
22777 Andropogon lancifolius Trin. = Arthraxon lanceolatus (Roxb.) Hochst. ■
22778 Andropogon lancifolius Trin. = Arthraxon lancifolius (Trin.) Hochst. ■
22779 Andropogon lancifolius Trin. var. microphyllus (Trin.) Kuntze = Arthraxon microphyllus (Trin.) Hochst. ■
22780 Andropogon laniger Desf.;绵毛须芒草■☆
22781 Andropogon laniger Desf. = Cymbopogon schoenanthus (L.) Spreng. ■
22782 Andropogon lasiobasis Pilg. = Hyparrhenia nyassae (Rendle) Stapf ■☆
22783 Andropogon laxatus Stapf = Andropogon eucomus Nees subsp. huillensis (Rendle) Sales ■☆
22784 Andropogon laxatus Stapf var. ligulatus ? = Andropogon ligulatus (Stapf) Clayton ■☆
22785 Andropogon lecomtei Franch. = Hyparrhenia newtonii (Hack.) Stapf ■
22786 Andropogon lepidus Nees = Hyparrhenia cymbaria (L.) Stapf ■☆
22787 Andropogon leprodes Cope;皮屑须芒草■☆
22788 Andropogon leptocomus Trin. = Anadelphia leptocoma (Trin.) Pilg. ■☆
22789 Andropogon leschenaultianus Decne. = Eulalia leschenaultiana (Decne.) Ohwi ■
22790 Andropogon leucostachyus Kunth;白穗须芒草■☆
22791 Andropogon liananthus Steud = Heteropogon triticeus (R. Br.) Stapf et Craib ■
22792 Andropogon ligulatus (Stapf) Clayton;舌状须芒草■☆
22793 Andropogon lima (Hack.) Stapf;侧生须芒草■☆
22794 Andropogon lindiensis Pilg. = Andropogon chinensis (Nees) Merr. ■
22795 Andropogon lindiensis Pilg. var. hirsitissimus ? = Andropogon chinensis (Nees) Merr. ■
22796 Andropogon linearis Stapf = Andropogon africanus Franch. ■☆
22797 Andropogon lithophilus Trin. = Elymandra lithophila (Trin.) Clayton ■☆
22798 Andropogon lodicularis (Nees) Steud. = Ischaemum barbatum Retz. ■
22799 Andropogon longipes Hack. = Andropogon amethystinus Steud. ■☆
22800 Andropogon lopollensis Rendle = Schizachyrium lopollense (Rendle) Sales ■☆
22801 Andropogon lugugaensis Vanderyst = Hyparrhenia nyassae (Rendle) Stapf ■☆
22802 Andropogon lumeneensis Vanderyst = Schizachyrium sanguineum (Retz.) Alston ■
22803 Andropogon luteolus Vanderyst = Hyperthelia dissoluta (Nees ex Steud.) Clayton ■☆
22804 Andropogon macleodiae Stapf = Andropogon canaliculatus Schumach. ■☆
22805 Andropogon macrolepis Hack. = Hyperthelia macrolepis (Hack.) Clayton ■☆
22806 Andropogon macrophyllus Stapf;大叶须芒草■☆
22807 Andropogon mannii Hook. f.;曼氏须芒草■☆
22808 Andropogon marginatus Steud. = Cymbopogon marginatus (Steud.) Stapf ex Burtt Davy ■☆
22809 Andropogon martinii Roxb. = Cymbopogon martinii (Roxb.) W. Watson ■
22810 Andropogon matteodanum Chiov. = Ischaemum afrum (J. F. Gmel.) Dandy ■☆
22811 Andropogon melanocarpus Elliott = Heteropogon melanocarpus (Elliott) Benth. ■
22812 Andropogon meyenianus (Nees) Steud. = Ischaemum barbatum Retz. ■
22813 Andropogon micans (Nees) Steud. = Arthraxon hispidus (Thunb.) Makino ■
22814 Andropogon micans (Nees) Steud. = Arthraxon micans (Nees) Hochst. ■
22815 Andropogon micranthus Kunth = Capillipedium parviflorum (R. Br.) Stapf ■
22816 Andropogon micranthus Kunth var. spicigerus (Benth.) Hack. = Capillipedium spicigerum S. T. Blake ■
22817 Andropogon microphyllus Trin. = Arthraxon lancifolius (Trin.) Hochst. ■
22818 Andropogon microphyllus Trin. = Arthraxon microphyllus (Trin.) Hochst. ■
22819 Andropogon miliaceus Roxb. = Sorghum halepense (L.) Pers. ■
22820 Andropogon miliformis Schult. = Sorghum halepense (L.) Pers. ■
22821 Andropogon minimus C. B. Clarke et Rendle = Schizachyrium brevifolium (Sw.) Nees ex Büse ■
22822 Andropogon mobukensis Chiov. = Hyparrhenia mobukensis (Chiov.) Chiov. ■☆
22823 Andropogon mollicomus Kunth = Dichanthium aristatum (Poir.) C. E. Hubb. ■
22824 Andropogon monandrus Roxb. = Pogonatherum crinitum (Thunb.) Kunth ■
22825 Andropogon monatherus A. Rich. = Exotheca abyssinica (Hochst. ex A. Rich.) Andersson ■☆
22826 Andropogon montanus Hack. = Capillipedium assimile (Steud.) A. Camus ■
22827 Andropogon montanus Koen. ex Trin.;山地须芒草■☆
22828 Andropogon monticola Schult. = Chrysopogon fulvus Chiov. ■☆
22829 Andropogon monticola Schult. var. trinii (Steud.) Hook. f. = Chrysopogon serrulatus Trin. ■☆
22830 Andropogon mukuluensis Vanderyst = Schizachyrium mukuluense (Vanderyst) Vanderyst ■☆
22831 Andropogon multicaulis Steud. = Arthraxon lancifolius (Trin.)

Hochst. ■

22832 Andropogon multiplex (Hochst. ex A. Rich.) Hack. = Hyparrhenia multiplex (Hochst. ex A. Rich.) Andersson ex Stapf ■☆

22833 Andropogon munroi C. B. Clarke;西藏须芒草(吉隆须芒草);Xizang Bluestem ■

22834 Andropogon muricatus Retz. = Vetiveria zizanioides (L.) Nash ■

22835 Andropogon muticus L. = Eustachys mutica (L.) Cufod. ■☆

22836 Andropogon nardus L.;香橘草;Citronella ■☆

22837 Andropogon nardus L. = Cymbopogon nardus (L.) Rendle ■

22838 Andropogon nardus L. subsp. hamatulus (Hook. et Arn.) Hack. = Cymbopogon tortilis (J. Presl) A. Camus ■

22839 Andropogon nardus L. subsp. hamatulus (Nees ex Hook. et Arn.) Hack. = Cymbopogon hamatulus (Nees ex Hook. et Arn.) A. Camus ■

22840 Andropogon nardus L. subsp. marginatus var. distans (Steud.) Hack. = Cymbopogon distans (Nees ex Steud.) W. Watson ■

22841 Andropogon nardus L. var. distans (Nees ex Steud.) Hack. = Cymbopogon distans (Nees ex Steud.) W. Watson ■

22842 Andropogon nardus L. var. fiexuosus (Nees ex Steud.) Hack. = Cymbopogon flexuosus (Nees ex Steud.) W. Watson ■

22843 Andropogon nardus L. var. goeringii (Steud.) Hack. = Cymbopogon goeringii (Steud.) A. Camus ■

22844 Andropogon nardus L. var. khasianus Munro ex Hack. = Cymbopogon khasianus (Munro ex Hack.) Stapf ex Bor ■

22845 Andropogon nardus L. var. microstachys Hook. f. = Cymbopogon microstachys (Hook. f.) Soenarko ■

22846 Andropogon nardus L. var. prolixus Stapf = Cymbopogon prolixus (Stapf) E. Phillips ■☆

22847 Andropogon nardus L. var. stracheyi Hook. f. = Cymbopogon pospischilii (K. Schum.) C. E. Hubb. ■

22848 Andropogon nardus L. var. stracheyi Hook. f. = Cymbopogon stracheyi (Hook. f.) Raizada et Jain ■

22849 Andropogon nardus L. var. validus Stapf = Cymbopogon nardus (L.) Rendle ■

22850 Andropogon nardus var. stracheyi Hook. f. = Cymbopogon pospischilii (K. Schum.) C. E. Hubb. ■

22851 Andropogon neomexicanus Nash = Schizachyrium scoparium (Michx.) Nash ■☆

22852 Andropogon nervosus Rottler = Sehima nervosa (Rottler) Stapf ■

22853 Andropogon newtonii Hack. = Hyparrhenia newtonii (Hack.) Stapf ■

22854 Andropogon niariensis Franch. = Hyparrhenia niariensis (Franch.) Clayton ■☆

22855 Andropogon nigritanus Benth. = Chrysopogon nigritanus (Benth.) Veldkamp ■☆

22856 Andropogon nitidus (Vahl) Kunth = Sorghum nitidum (Vahl) Pers. ■

22857 Andropogon nlemfuensis Vanderyst = Hyparrhenia bracteata (Humb. et Bonpl. ex Willd.) Stapf ■

22858 Andropogon nodosus (Willemet) Nash = Dichanthium annulatum (Forssk.) Stapf ■

22859 Andropogon nodosus (Willemet) Nash = Dichanthium aristatum (Poir.) C. E. Hubb. ■

22860 Andropogon nodulosus Hack. = Schizachyrium nodulosum (Hack.) Stapf ■☆

22861 Andropogon notopogon Nees ex Steud. = Eulaliopsis binata (Retz.) C. E. Hubb. ■

22862 Andropogon notopogon Steud. = Eulaliopsis binata (Retz.) C. E. Hubb. ■

22863 Andropogon nudus Nees ex Steud. = Arthraxon nudus (Nees ex Steud.) Hochst. ■

22864 Andropogon nutans L. = Sorghastrum nutans (L.) Nash ■☆

22865 Andropogon nyassae Rendle = Hyparrhenia nyassae (Rendle) Stapf ■☆

22866 Andropogon obliquiberbe Hack. = Schizachyrium obliquiberbe (Hack.) A. Camus ■

22867 Andropogon obliquiberbis Hack. = Schizachyrium fragile (R. Br.) A. Camus ■

22868 Andropogon obscurus K. Schum. = Hyparrhenia diplandra (Hack.) Stapf ■

22869 Andropogon obvallatus Steud. = Eulaliopsis binata (Retz.) C. E. Hubb. ■

22870 Andropogon odoratus (Lisboa) A. Camus = Bothriochloa bladhii (Retz.) S. T. Blake ■

22871 Andropogon odoratus Lisboa = Bothriochloa bladhii (Retz.) S. T. Blake ■

22872 Andropogon olivieri Boiss. = Cymbopogon jwarancusa (Jones) Schult. subsp. olivieri (Boiss.) Soenarko ■

22873 Andropogon olivieri Boiss. = Cymbopogon olivieri (Boiss.) Bor ■

22874 Andropogon osikensis Franch. = Hyparrhenia diplandra (Hack.) Stapf ■

22875 Andropogon pachnodes Trin. = Cymbopogon martinii (Roxb.) W. Watson ■

22876 Andropogon pachyneuros Franch. = Hyparrhenia diplandra (Hack.) Stapf ■

22877 Andropogon paleaceus (Trin.) Steud. = Apocopis paleacea (Trin.) Hochr. ■

22878 Andropogon papillipes Hochst. ex A. Rich. = Hyparrhenia papillipes (Hochst. ex A. Rich.) Andersson ex Stapf ■☆

22879 Andropogon papillosus Hochst. ex A. Rich. = Dichanthium annulatum (Forssk.) Stapf var. papillosum (Hochst. ex A. Rich.) de Wet et Harlan ■☆

22880 Andropogon pappii Gand. = Sorghum purpureo-sericeum (Hochst. ex A. Rich.) Asch. et Schweinf. ■☆

22881 Andropogon parviflorus Roxb. var. spicigerus (Benth.) Domin = Capillipedium spicigerum S. T. Blake ■

22882 Andropogon parvispicus Steud. = Capillipedium parviflorum (R. Br.) Stapf ■

22883 Andropogon patentivillosus Steud. = Ischaemum ciliare Retz. ■

22884 Andropogon patris Robyns = Andropogon chinensis (Nees) Merr. ■

22885 Andropogon pendulus Nees ex Steud. = Cymbopogon pendulus (Nees ex Steud.) W. Watson ■

22886 Andropogon pendulus Peter = Hyparrhenia pendula Peter ■☆

22887 Andropogon pertusus (L.) Willd. = Bothriochloa pertusa (L.) A. Camus ■

22888 Andropogon pertusus (L.) Willd. subvar. hirtus Chiov. = Bothriochloa insculpta (Hochst. ex A. Rich.) A. Camus ■☆

22889 Andropogon pertusus (L.) Willd. var. vegetior Hack. = Bothriochloa bladhii (Retz.) S. T. Blake ■

22890 Andropogon pertusus Willd. = Dichanthium insculptum (Hochst. ex A. Rich.) Clayton ■☆

22891 Andropogon pertusus Willd. var. maroccanus Maire = Dichanthium insculptum (Hochst. ex A. Rich.) Clayton ■☆

22892 Andropogon petiolaris (Trin.) Steud. = Microstegium petiolare (Trin.) Bor ■

22893 Andropogon phoenix (Rendle) K. Schum. = Hyparrhenia diplandra (Hack.) Stapf ■
22894 Andropogon pilipes Backer = Arthraxon castratus (Griff.) Nayaran. ex Bor ■
22895 Andropogon pilosellus Stapf = Andropogon amethystinus Steud. ■☆
22896 Andropogon pilosissimus Hack. = Hyparrhenia dregeana (Nees) Stapf ex Stent ■☆
22897 Andropogon pinguipes Stapf;肥梗须芒草■☆
22898 Andropogon platybasis J. G. Anderson = Andropogon mannii Hook. f. ■☆
22899 Andropogon platyphyllus (Franch.) Pilg. = Schizachyrium platyphyllum (Franch.) Stapf ■☆
22900 Andropogon platypus Trin. = Elionurus platypus (Trin.) Hack. ■☆
22901 Andropogon pleiarthron Stapf = Hyparrhenia poecilotricha (Hack.) Stapf ■☆
22902 Andropogon plumosus Humb. et Bonpl. ex Willd. = Trachypogon spicatus (L. f.) Kuntze ■☆
22903 Andropogon plurinodis Stapf = Cymbopogon pospischilii (K. Schum.) C. E. Hubb. ■
22904 Andropogon podotrichus Hochst. = Hyparrhenia hirta (L.) Stapf ■☆
22905 Andropogon poecilotrichus Hack. = Hyparrhenia poecilotricha (Hack.) Stapf ■☆
22906 Andropogon polyatherus A. Rich. var. plagiopus Hack. = Andropogon abyssinicus Fresen. ■☆
22907 Andropogon polyatherus Hochst. ex A. Rich. = Andropogon abyssinicus Fresen. ■☆
22908 Andropogon polystictus Steud. = Heteropogon melanocarpus (Elliott) Benth. ■
22909 Andropogon pospischilii K. Schum. = Cymbopogon pospischilii (K. Schum.) C. E. Hubb. ■
22910 Andropogon praematurus Fernald = Schizachyrium scoparium (Michx.) Nash ■☆
22911 Andropogon pratensis Hack. var. pseudoabyssinicus Chiov. = Andropogon amethystinus Steud. ■☆
22912 Andropogon pratensis Hochst. ex A. Braun = Andropogon amethystinus Steud. ■☆
22913 Andropogon princeps A. Rich. = Thelepogon elegans Roth ex Roem. et Schult. ■☆
22914 Andropogon prionodes Steud. = Arthraxon lanceolatus (Roxb.) Hochst. ■
22915 Andropogon prionodes Steud. = Arthraxon prionodes (Steud.) Dandy ■
22916 Andropogon prolixus Stapf = Andropogon africanus Franch. ■☆
22917 Andropogon propinquus Kunth = Sorghum propinquum (Kunth) Hitchc. ■
22918 Andropogon provincialis Lam. = Andropogon gerardii Vitman ■☆
22919 Andropogon proximus Hochst. ex A. Rich. = Cymbopogon schoenanthus (L.) Spreng. subsp. proximus (Hochst. ex A. Rich.) Maire et Weiller ■☆
22920 Andropogon pseudapricus Stapf;假向阳须芒草■☆
22921 Andropogon pseudauriculatus Mimeur = Andropogon chevalieri Reznik ■☆
22922 Andropogon pseudoschinzii Stapf = Andropogon chinensis (Nees) Merr. ■
22923 Andropogon pterophalis Clayton;翅鳞须芒草■☆
22924 Andropogon pterophalis Clayton var. togoensis Scholz;多哥翅鳞须芒草■☆
22925 Andropogon pubescens Vis. = Hyparrhenia hirta (L.) Stapf ■☆
22926 Andropogon pulchellus D. Don ex Benth. = Schizachyrium pulchellum (D. Don ex Benth.) Stapf ■☆
22927 Andropogon punctatus Roxb. = Bothriochloa bladhii (Retz.) S. T. Blake var. punctata (Roxb.) R. R. Stewart ■
22928 Andropogon punctatus Roxb. = Bothriochloa bladhii (Retz.) S. T. Blake ■
22929 Andropogon pungens Cope;刺须芒草■☆
22930 Andropogon purpureo-sericeus Hochst. ex A. Rich. = Sorghum purpureo-sericeum (Hochst. ex A. Rich.) Asch. et Schweinf. ■☆
22931 Andropogon purpureus Stapf = Andropogon mannii Hook. f. ■☆
22932 Andropogon pusillus Hook. f.;微小须芒草■☆
22933 Andropogon quadrivalvis L. = Themeda quadrivalvis (L.) Kuntze ■
22934 Andropogon quartinianus A. Rich. = Capillipedium parviflorum (R. Br.) Stapf ■
22935 Andropogon radicans Lehm. = Bothriochloa radicans (Lehm.) A. Camus ■☆
22936 Andropogon ravennae L. = Erianthus ravennae (L.) P. Beauv. ■
22937 Andropogon ravennae L. = Saccharum ravennae (L.) Murray ■
22938 Andropogon ravennae L. = Tripidium ravennae (L.) H. Scholz ■
22939 Andropogon rhynchophorus Stapf = Sehima ischaemoides Forssk. ■☆
22940 Andropogon ringoeti De Wild. = Andropogon gayanus Kunth var. polycladus (Hack.) Clayton ■☆
22941 Andropogon roseus Napper = Andropogon textilis Welw. ex Rendle ■☆
22942 Andropogon roxburghianus Roem. et Schult. = Dimeria ornithopoda Trin. ■
22943 Andropogon rudis Nees ex Steud. = Arthraxon castratus (Griff.) Nayaran. ex Bor ■
22944 Andropogon rufus (Nees) Kunth = Hyparrhenia rufa (Nees) Stapf ■☆
22945 Andropogon rugosus (Salisb.) Steud. = Ischaemum rugosum Salisb. ■
22946 Andropogon rupestris K. Schum. = Schizachyrium rupestre (K. Schum.) Stapf ■☆
22947 Andropogon ruprechtii Hack. = Hyperthelia dissoluta (Nees ex Steud.) Clayton ■☆
22948 Andropogon saccharoides Sw.;甘蔗须芒草;Silver Bluestem ■☆
22949 Andropogon saccharoides Sw. var. torreyanus (Steud.) Hack. = Bothriochloa laguroides (DC.) Herter ■☆
22950 Andropogon sanguineus (Retz.) Merr. = Schizachyrium sanguineum (Retz.) Alston ■
22951 Andropogon scaettai Robyns = Hyparrhenia mobukensis (Chiov.) Chiov. ■☆
22952 Andropogon scandens Roxb. = Themeda quadrivalvis (L.) Kuntze ■
22953 Andropogon schimperi Hochst. ex A. Rich. = Hyparrhenia schimperi (Hochst. ex A. Rich.) Andersson ex Stapf ■☆
22954 Andropogon schinzii Hack. = Andropogon chinensis (Nees) Merr. ■
22955 Andropogon schirensis A. Rich. var. angustifolius Stapf = Andropogon schirensis Hochst. ex A. Rich. ■☆
22956 Andropogon schirensis A. Rich. var. natalensis Hack. = Andropogon schirensis Hochst. ex A. Rich. ■☆
22957 Andropogon schirensis Hochst. ex A. Rich.;希尔须芒草■☆

22958 Andropogon schizachyrioides P. A. Duvign.;裂稃草状须芒草■☆
22959 Andropogon schlechteri Hack. = Andropogon festuciformis Rendle ■☆
22960 Andropogon schliebenii Pilg. = Andropogon tenuiberbis Hack. ■☆
22961 Andropogon schoenanthus L.;舒安须芒草(青香茅)■☆
22962 Andropogon schoenanthus L. = Cymbopogon schoenanthus (L.) Spreng. ■
22963 Andropogon schoenanthus L. var. caesius (Nees ex Hook. et Arn.) Hack. = Cymbopogon caesius (Nees ex Hook. et Arn.) Stapf ■
22964 Andropogon schoenanthus L. var. genuinus Hack. = Cymbopogon martinii (Roxb.) W. Watson ■
22965 Andropogon schoenanthus L. var. gracillimus Hook. f. = Cymbopogon caesius (Nees ex Hook. et Arn.) Stapf ■
22966 Andropogon schoenanthus L. var. martinii (Roxb.) Hook. f. = Cymbopogon martinii (Roxb.) W. Watson ■
22967 Andropogon schoenanthus L. var. proximus (Hochst. ex A. Rich.) A. Chev. = Cymbopogon schoenanthus (L.) Spreng. subsp. proximus (Hochst. ex A. Rich.) Maire et Weiller ■☆
22968 Andropogon schweinfurthii Hack. = Schizachyrium schweinfurthii (Hack.) Stapf ■☆
22969 Andropogon scoparius Michx. = Schizachyrium scoparium (Michx.) Nash ■☆
22970 Andropogon scoparius Michx. calvescens Fernald = Schizachyrium scoparium (Michx.) Nash ■☆
22971 Andropogon scoparius Michx. var. divergens Hack. = Schizachyrium scoparium (Michx.) Nash var. divergens (Hack.) Gould ■☆
22972 Andropogon scoparius Michx. var. frequens F. T. Hubb. = Schizachyrium scoparium (Michx.) Nash ■☆
22973 Andropogon scoparius Michx. var. neomexicanus (Nash) Hitchc. = Schizachyrium scoparium (Michx.) Nash ■☆
22974 Andropogon scoparius Michx. var. polycladus Scribn. et Ball = Schizachyrium scoparium (Michx.) Nash ■☆
22975 Andropogon scoparius Michx. var. septentrionalis Fernald et Griscom = Schizachyrium scoparium (Michx.) Nash ■☆
22976 Andropogon scoparius Michx. var. villosissimus Kearney = Schizachyrium scoparium (Michx.) Nash ■☆
22977 Andropogon scoparius Michx. var. virilis Shinners = Schizachyrium scoparium (Michx.) Nash var. divergens (Hack.) Gould ■☆
22978 Andropogon seemenianus Pilg. = Andropogon amethystinus Steud. ■☆
22979 Andropogon segenensis Steud. = Heteropogon triticeus (R. Br.) Stapf et Craib ■
22980 Andropogon segetum (Trin.) Steud. = Ischaemum rugosum Salisb. ■
22981 Andropogon sehima Steud. = Sehima ischaemoides Forssk. ■☆
22982 Andropogon selloanus (Hack.) Hack.;鞍须芒草;Saddle Bluestem ■☆
22983 Andropogon semiberbis (Nees) Kunth = Schizachyrium sanguineum (Retz.) Alston ■
22984 Andropogon sennarensis Hochst. = Cymbopogon schoenanthus (L.) Spreng. subsp. proximus (Hochst. ex A. Rich.) Maire et Weiller ■☆
22985 Andropogon seretii De Wild. = Hyparrhenia dybowskii (Franch.) Roberty ■☆
22986 Andropogon serratus Thunb. = Sorghum nitidum (Vahl) Pers. ■
22987 Andropogon serratus Thunb. var. nitidus (Vahl) Hack. = Sorghum nitidum (Vahl) Pers. ■
22988 Andropogon serrulatus A. Rich. = Arthraxon lanceolatus (Roxb.) Hochst. ■
22989 Andropogon serrulatus A. Rich. = Arthraxon prionodes (Steud.) Dandy ■
22990 Andropogon setifer Pilg. = Hyparrhenia bracteata (Humb. et Bonpl. ex Willd.) Stapf ■
22991 Andropogon setifolius (Thunb.) Kunth;毛叶须芒草■☆
22992 Andropogon shimadae Ohwi = Schizachyrium fragile (R. Br.) A. Camus var. shimadae (Ohwi) C. Hsu ■
22993 Andropogon sibiricus (Trin.) Steud. = Spodiopogon sibiricus Trin. ■
22994 Andropogon sibiricus Steud. = Spodiopogon sibiricus Trin. ■
22995 Andropogon sikkimensis Bor = Arthraxon microphyllus (Trin.) Hochst. ■
22996 Andropogon simplex Schumach.;简单须芒草■☆
22997 Andropogon smithianus Hook. f. = Hyparrhenia smithiana (Hook. f.) Stapf ■☆
22998 Andropogon sorghum (L.) Brot. = Holcus sorghum L. ■
22999 Andropogon sorghum (L.) Brot. = Sorghum bicolor (L.) Moench ■
23000 Andropogon sorghum (L.) Brot. = Sorghum cernuum (Ard.) Host ■
23001 Andropogon sorghum (L.) Brot. = Sorghum vulgare (L.) Pers. ■
23002 Andropogon sorghum (L.) Brot. subsp. halepensis (L.) Hack. = Sorghum halepense (L.) Pers. ■
23003 Andropogon sorghum (L.) Brot. subsp. halepensis Hack. = Sorghum halepense (L.) Pers. ■
23004 Andropogon sorghum (L.) Brot. subsp. sativus Hack. var. cafer Hack. = Sorghum caffrorum (Retz.) P. Beauv. ■
23005 Andropogon sorghum (L.) Brot. subsp. sativus Hack. var. cernuus Hack. = Sorghum cernuum (Ard.) Host ■
23006 Andropogon sorghum (L.) Brot. subsp. sativus Hack. var. durra Hack. = Sorghum durra (Forssk.) Stapf ■
23007 Andropogon sorghum (L.) Brot. subsp. sativus Hack. var. nervosus Hack. = Sorghum nervosum Besser ex Schult. ■
23008 Andropogon sorghum (L.) Brot. subsp. sativus Hack. var. saccharatus Hack. = Sorghum dochna (Forssk.) Snowden ■
23009 Andropogon sorghum (L.) Brot. subsp. sativus Hack. var. subglobosus Hack. = Sorghum bicolor (L.) Moench var. subglobosus (Hack.) Snowden ■
23010 Andropogon sorghum (L.) Brot. subsp. sativus Hack. var. technicus Hack. = Sorghum dochna (Forssk.) Snowden var. tecnicum (Körn.) Snowden ■
23011 Andropogon sorghum (L.) Brot. subsp. sativus Hack. var. vulgaris Hack. = Sorghum bicolor (L.) Moench ■
23012 Andropogon sorghum (L.) Brot. subsp. sudanensis Piper = Sorghum bicolor (L.) Moench subsp. drummondii (Nees ex Steud.) de Wet ■☆
23013 Andropogon sorghum (L.) Brot. subsp. sudanensis Piper = Sorghum sudanense (Piper) Stapf ■
23014 Andropogon sorghum (L.) Brot. subsp. verticilliflorus (Steud.) Piper = Sorghum verticilliflorum (Steud.) Stapf ■☆
23015 Andropogon sorghum (L.) Brot. var. aethiopicus Hack. = Sorghum aethiopicum (Hack.) Stapf ■☆
23016 Andropogon sorghum (L.) Brot. var. ankolib Hack. = Sorghum bicolor (L.) Moench ■

23017 Andropogon sorghum (L.) Brot. var. cafer Körn. = Sorghum caffrorum (Retz.) P. Beauv. ■
23018 Andropogon sorghum (L.) Brot. var. caudatus Hack. = Sorghum bicolor (L.) Moench ■
23019 Andropogon sorghum (L.) Brot. var. propinquus (Kunth) Hack. = Sorghum propinquum (Kunth) Hitchc. ■
23020 Andropogon sorghum (L.) Brot. var. technicus Körn. = Sorghum bicolor (L.) Moench ■
23021 Andropogon sorghum (L.) Brot. var. technicus Körn. = Sorghum dochna (Forssk.) Snowden var. tecnicum (Körn.) Snowden ■
23022 Andropogon sorghum (L.) Brot. var. virgatus Hack. = Sorghum virgatum (Hack.) Stapf ■☆
23023 Andropogon spanianthus Pilg. = Andropogon chinensis (Nees) Merr. ■
23024 Andropogon speciosus Steud. = Phacelurus speciosus (Steud.) C. E. Hubb. ■☆
23025 Andropogon spectabilis K. Schum. = Andropogon tectorum Schumach. et Thonn. ■☆
23026 Andropogon spicigerus (S. T. Blake) Reeder = Capillipedium spicigerum S. T. Blake ■
23027 Andropogon squamulatus Hochst. = Andropogon gayanus Kunth var. polycladus (Hack.) Clayton ■☆
23028 Andropogon squarrosus Hook. f. = Vetiveria zizanioides (L.) Nash ■
23029 Andropogon squarrosus L. = Vetiveria zizanioides (L.) Nash ■
23030 Andropogon squarrosus L. F. Celak. var. nigritanus (Benth.) Hack. = Chrysopogon nigritanus (Benth.) Veldkamp ■☆
23031 Andropogon stagninus Vanderyst = Schizachyrium kwiluense Vanderyst ■☆
23032 Andropogon stapfii Hook. f. = Sorghum arundinaceum (Desv.) Stapf ■☆
23033 Andropogon stipaeformis Steud. = Eulalia trispicata (Schult.) Henrard ■
23034 Andropogon stipoides Kunth = Sorghastrum stipoides (Kunth) Nash ■☆
23035 Andropogon stolzii Stapf = Andropogon mannii Hook. f. ■☆
23036 Andropogon striatus Klein ex Willd. = Sehima nervosa (Rottler) Stapf ■
23037 Andropogon strictus Roxb. = Dichanthium foveolatum (Delile) Roberty ■☆
23038 Andropogon stypticus Welw. = Cymbopogon stypticus (Welw.) Fritsch ■☆
23039 Andropogon subamplectens Berhaut = Diheteropogon amplectens (Nees) Clayton var. catangensis (Chiov.) Clayton ■☆
23040 Andropogon subcordatifolius De Wild.;亚心叶须芒草■☆
23041 Andropogon submuticus Nees ex Steud. = Arthraxon submuticus (Nees ex Steud.) Hochst. ■
23042 Andropogon subrepens Steud. = Capillipedium assimile (Steud.) A. Camus ■
23043 Andropogon sudanensis (Piper) Leppan et Bosman = Sorghum sudanense (Piper) Stapf ■
23044 Andropogon sylvaticus C. E. Hubb. = Andropogon chinensis (Nees) Merr. ■
23045 Andropogon taiwanensis Ohwi = Bothriochloa ischaemum (L.) Keng ■
23046 Andropogon tectorum Schumach. et Thonn.;屋顶须芒草■☆
23047 Andropogon tectorum Schumach. et Thonn. var. acutatus Reznik = Andropogon tectorum Schumach. et Thonn. ■☆
23048 Andropogon tectorum Schumach. et Thonn. var. falsopetiolatus Reznik = Andropogon tectorum Schumach. et Thonn. ■☆
23049 Andropogon tenuiberbis Hack.;热非须芒草■☆
23050 Andropogon tenuiculmus Reznik = Andropogon tectorum Schumach. et Thonn. ■☆
23051 Andropogon tenuiflorus Stapf = Anadelphia leptocoma (Trin.) Pilg. ■☆
23052 Andropogon ternarius Michx.;三出须芒草;Splitbeard Bluestem ■☆
23053 Andropogon textilis Welw. ex Rendle;编织须芒草■☆
23054 Andropogon thollonii Franch. = Schizachyrium thollonii (Franch.) Stapf ■☆
23055 Andropogon thomasii C. E. Hubb. = Andropogon mannii Hook. f. ■☆
23056 Andropogon timorensis (Kunth) Steud. = Ischaemum timorense Kunth ■
23057 Andropogon tong-dong Steud. = Ischaemum rugosum Salisb. ■
23058 Andropogon tonkinensis Balansa = Pseudosorghum fasciculare (Roxb. A. Camus ■
23059 Andropogon transvaalensis Stapf = Hyparrhenia hirta (L.) Stapf ■☆
23060 Andropogon tremulus Hack. = Chrysopogon serrulatus Trin. ■☆
23061 Andropogon trepidarius Stapf = Anadelphia trepidaria (Stapf) Stapf ■☆
23062 Andropogon trichopus Stapf = Sorghastrum stipoides (Kunth) Nash ■☆
23063 Andropogon trichozygus Baker;毛须芒草■☆
23064 Andropogon trinii Steud. = Chrysopogon serrulatus Trin. ■☆
23065 Andropogon trispicatus Schult. = Eulalia trispicata (Schult.) Henrard ■
23066 Andropogon tristachyus Roxb. = Eulalia trispicata (Schult.) Henrard ■
23067 Andropogon tristis Nees ex Hack. = Andropogon munroi C. B. Clarke ■
23068 Andropogon triticeus R. Br. = Heteropogon triticeus (R. Br.) Stapf et Craib ■
23069 Andropogon umbrosus Hochst. = Hyparrhenia umbrosa (Hochst.) Andersson ex Clayton ■☆
23070 Andropogon urceolatus Hack. = Schizachyrium urceolatum (Hack.) Stapf ■☆
23071 Andropogon vachellii Nees = Bothriochloa bladhii (Retz.) S. T. Blake ■
23072 Andropogon vanderystii De Wild.;范德须芒草■☆
23073 Andropogon verticilliflorus Steud. = Sorghum arundinaceum (Desv.) Stapf ■☆
23074 Andropogon verticilliflorus Steud. = Sorghum verticilliflorum (Steud.) Stapf ■☆
23075 Andropogon villosulus Nees ex Steud. = Capillipedium parviflorum (R. Br.) Stapf ■
23076 Andropogon vimineus Trin. = Microstegium vimineum (Trin.) A. Camus ■
23077 Andropogon virginicus L.;扫帚草;Beard Grass, Broomsedge, Broom-sedge, Broomsedge Bluestem, Broomstraw, Chalky Bluestem ■☆
23078 Andropogon vuilletii Mimeur = Ischaemum fasciculatum Brongn. ■☆
23079 Andropogon vulgaris Vanderyst = Hyparrhenia diplandra (Hack.) Stapf ■
23080 Andropogon welwitschii (Rendle) K. Schum. = Hyparrhenia welwitschii (Rendle) Stapf ■☆

23081 Andropogon wightianus Nees ex Steud. = Chrysopogon orientalis (Desv.) A. Camus ■
23082 Andropogon wombaliensis Vanderyst ex Robyns = Hyparrhenia wombaliensis (Vanderyst ex Robyns) Clayton ■☆
23083 Andropogon wombaliensis Vanderyst ex Robyns var. ciliatus Robyns = Hyparrhenia wombaliensis (Vanderyst ex Robyns) Clayton ■☆
23084 Andropogon xanthoblepharis Trin. = Hyparrhenia rufa (Nees) Stapf ■☆
23085 Andropogon yinduensis Vanderyst = Hyparrhenia rufa (Nees) Stapf ■☆
23086 Andropogon yunnanensis Hack.;云南须芒草(须芒草);Yunnan Bluestem ■
23087 Andropogon yunnanensis Hack. = Andropogon munroi C. B. Clarke ■
23088 Andropogon zollingeri Steud. = Pseudosorghum fasciculare (Roxb. A. Camus ■
23089 Andropogonaceae (J. Presl) Herter = Gramineae Juss.(保留科名)●■
23090 Andropogonaceae (J. Presl) Herter = Poaceae Barnhart(保留科名)●■
23091 Andropogonaceae Herter = Gramineae Juss.(保留科名)●■
23092 Andropogonaceae Herter = Poaceae Barnhart(保留科名)●■
23093 Andropogonaceae Herter;须芒草科■☆
23094 Andropogonaceae Martinov = Gramineae Juss.(保留科名)●■
23095 Andropogonaceae Martinov = Poaceae Barnhart(保留科名)●■
23096 Andropterum Stapf(1917);翼颖草属■☆
23097 Andropterum stolzii (Pilg.) C. E. Hubb.;翼颖草■☆
23098 Andropterum variegatum Stapf = Andropterum stolzii (Pilg.) C. E. Hubb. ■☆
23099 Andropus Brand = Nama L.(保留属名)■
23100 Andropus Brand(1912);新墨西哥田基麻属■☆
23101 Andropus carnosus (Wooton) Brand;新墨西哥田基麻■☆
23102 Andropus carnosus Brand = Andropus carnosus (Wooton) Brand ■☆
23103 Androrchis D. Tyteca et E. Klein = Orchis L. ■
23104 Androrchis D. Tyteca et E. Klein(2008);雄兰属■☆
23105 Androsacaceae Rchb. ex Barnhart = Primulaceae Batsch ex Borkh.(保留科名)●■
23106 Androsacaceae Rchb. ex Barnhart;点地梅科■
23107 Androsace L. (1753);点地梅属;Rock Jasmine, Rockjasmine ■
23108 Androsace acutifolia Wendelbo = Androsace himalaica (Knuth) Hand. -Mazz. subsp. kurramensis Y. J. Nasir ■☆
23109 Androsace adenocephala Hand. -Mazz.;腺序点地梅;Dotted Rockjasmine ■
23110 Androsace aizoon Duby;莲座点地梅(番杏点地梅);Aizoon Rockjasmine ■
23111 Androsace aizoon Duby var. coccinea Franch.;红花点地梅(大红花点地梅)■
23112 Androsace aizoon Duby var. coccinea Franch. = Androsace bulleyana Forrest ■
23113 Androsace aizoon Duby var. himalaica R. Knuth = Androsace aizoon Duby ■
23114 Androsace aizoon Duby var. integra Maxim. = Androsace integra (Maxim.) Hand. -Mazz. ■
23115 Androsace aizoon Duby var. purpurea Pax et K. Hoffm. = Androsace integra (Maxim.) Hand. -Mazz. ■
23116 Androsace aizoon Duby var. rosea Pax et K. Hoffm. = Androsace integra (Maxim.) Hand. -Mazz. ■
23117 Androsace alaschanica Maxim.;阿拉善点地梅;Alashan Rockjasmine ■
23118 Androsace alaschanica Maxim. var. zaduoensis Y. C. Yang et R. F. Huang;扎多点地梅(杂多点地梅);Zaduo Rockjasmine ■
23119 Androsace albana Stev.;阿尔邦点地梅■☆
23120 Androsace alchemilloides Franch.;花叶点地梅(假羽衣草);Ladymantle-like Rockjasmine ■
23121 Androsace alpina Lam.;高山点地梅;Alpine Rock Jasmine, Alpine Rock-jasmine ■☆
23122 Androsace arctica Cham. et Schltdl.;北极点地梅■☆
23123 Androsace arizonica A. Gray = Androsace occidentalis Pursh ■☆
23124 Androsace armeniaca Duby;亚美尼亚点地梅■☆
23125 Androsace aurata Petitm. = Androsace bisulca Bureau et Franch. var. aurata (Petitm.) Yen C. Yang et P. H. Huang ■
23126 Androsace axillaris (Franch.) Franch.;腋花点地梅;Axillaryflower Rockjasmine ■
23127 Androsace barbulata Ovcz.;髯毛点地梅■☆
23128 Androsace bidentata K. Koch;二齿点地梅■☆
23129 Androsace bisulca Bureau et Franch.;昌都点地梅(二槽点地梅);Doublefurrow Rockjasmine, Szechwan-tibet Rockjasmine ■
23130 Androsace bisulca Bureau et Franch. var. aurata (Petitm.) Yen C. Yang et P. H. Huang;黄花昌都点地梅;Yellowflower Doublefurrow ■
23131 Androsace brachystegia Hand. -Mazz.;玉门点地梅;Yumen Rockjasmine ■
23132 Androsace brahmaputrae Hand. -Mazz. = Androsace bisulca Bureau et Franch. ■
23133 Androsace bulleyana Forrest;景天点地梅(大红花点地梅,点地梅,红花点地梅);Bulley Rockjasmine ■
23134 Androsace bulleyana Forrest var. purpurea Hand. -Mazz. = Androsace integra (Maxim.) Hand. -Mazz. ■
23135 Androsace bungeana Schischk. et Bobrov = Androsace lehmannii Wall. ex Duby ■
23136 Androsace caduca Ovcz.;早花点地梅■☆
23137 Androsace capitata Willd. ex Roem. et Schult. = Androsace chamaejasme Host subsp. lehmanniana (Spreng.) Hultén ■
23138 Androsace capitata Willd. ex Roem. et Schult. = Androsace lehmanniana Spreng. ■
23139 Androsace carnea L.;肉色花点地梅(肉色点地梅);Rock Jasmine ■☆
23140 Androsace carnulosa Duby = Androsace umbellata (Lour.) Merr. ■
23141 Androsace cernuiflora Y. C. Yang et R. F. Huang;弯花点地梅;Curvedflower Rockjasmine ■
23142 Androsace chamaejasme Host subsp. lehmanniana (Spreng.) Hultén = Androsace lehmanniana Spreng. ■
23143 Androsace chamaejasme Host var. coronata Watt = Androsace coronata (Watt) Hand. -Mazz. ■
23144 Androsace chamaejasme Willd.;纤毛点地梅(矮点地梅,矮茉莉点地梅);Dwarfjasmine Rockjasmine, Sweetflower Rockjasmine ■☆
23145 Androsace chamaejasme Willd. var. uniflora Knuth = Androsace muscoidea Duby ■☆
23146 Androsace ciliifolia Ludlow;睫毛点地梅;Eyelash Rockjasmine ■
23147 Androsace coccinea Franch. = Androsace bulleyana Forrest ■

23148 Androsace cordifolia Wall. = Primula filipes Watt ■☆

23149 Androsace coronata (Watt) Hand. -Mazz.;环冠点地梅■

23150 Androsace crassifolia Wendelbo = Androsace harrissii Duthie subsp. crassifolia (Wendelbo) Y. J. Nasir ■☆

23151 Androsace croftii Watt;红毛点地梅■

23152 Androsace cuscutiformis Franch.;细蔓点地梅;Dodderformed Rockjasmine ■

23153 Androsace cuttingii C. E. C. Fisch.;江孜点地梅;Cutting Rockjasmine ■

23154 Androsace cylindrica DC.;柱叶点地梅■☆

23155 Androsace dasyphylla Bunge;密叶点地梅■☆

23156 Androsace delavayi Franch.;滇西北点地梅(西南点地梅);Delavay Rockjasmine ■

23157 Androsace densa Pax et K. Hoffm. = Androsace tapete Maxim. ■

23158 Androsace dielsiana R. Knuth = Androsace gmelinii (Gaertn.) Roem. et Schult. var. geophila Hand. -Mazz. ■

23159 Androsace dissecta (Franch.) Franch.;裂叶点地梅;Dissected Rockjasmine ■

23160 Androsace diversifolia C. Y. Wu = Androsace runcinata Hand. -Mazz. ■

23161 Androsace elatior Pax et K. Hoffm.;高葶点地梅;Longstock Rockjasmine ■

23162 Androsace elegans Duby = Androsace rotundifolia Hardw. ■

23163 Androsace elongata L.;长点地梅■☆

23164 Androsace engleri R. Knuth;陕西点地梅;Engler Rockjasmine ■

23165 Androsace erecta Maxim.;直立点地梅(直茎点地梅);Erect Rockjasmine ■

23166 Androsace euryantha Hand. -Mazz.;良花点地梅(大花点地梅);Broadstamen Rockjasmine ■

23167 Androsace fedtschenkoi Ovcz.;费氏点地梅■☆

23168 Androsace fedtschenkoi Ovcz. = Androsace septentrionalis L. var. breviscapa Krylov ■

23169 Androsace ferruginea Watt ex Knuth = Androsace robusta (R. Knuth) Hand. -Mazz. subsp. jacquemontii (Duby) Y. J. Nasir ■☆

23170 Androsace filiformis Retz.;东北点地梅(浑河小樱,沙河小樱,丝点地梅);Filiformis Rockjasmine ■

23171 Androsace filiformis Retz. var. glandulosa Krylov = Androsace filiformis Retz. ■

23172 Androsace flavescens Maxim.;南疆点地梅;Lightyellow Rockjasmine ■

23173 Androsace foliosa Decne. ex Duby;多叶点地梅■☆

23174 Androsace forrestiana Hand. -Mazz.;滇藏点地梅;Forrest Rockjasmine ■

23175 Androsace gagnepainiana Hand. -Mazz.;披散点地梅;Gagnepain Rockjasmine ■

23176 Androsace geraniifolia Watt;掌叶点地梅(具蔓点地梅);Geranium Rockjasmine, Palm Rockjasmine ■

23177 Androsace geraniifolia Watt var. escaposa Hand. -Mazz. = Androsace axillaris (Franch.) Franch. ■

23178 Androsace geraniifolia Watt var. setosa R. Knuth = Androsace geraniifolia Watt ■

23179 Androsace globifera Duby;球形点地梅;Globate Rockjasmine ■

23180 Androsace globifera Duby = Androsace mucronifolia Watt ■☆

23181 Androsace globifera Klatt = Androsace mucronifolia Watt ■☆

23182 Androsace gmelinii (Gaertn.) Roem. et Schult.;小点地梅(高山点地梅,兴安点地梅);Alpine Rockjasmine, Gmelin Rockjasmine ■

23183 Androsace gmelinii (Gaertn.) Roem. et Schult. var. geophila Hand. -Mazz.;短葶小点地梅■

23184 Androsace gracilis Hand. -Mazz.;细弱点地梅;Slender Rockjasmine ■

23185 Androsace graminofolia C. E. C. Fisch.;禾叶点地梅;Poaleaf Rockjasmine ■

23186 Androsace gustavii R. Knuth = Androsace tapete Maxim. ■

23187 Androsace harrissii Duthie;哈里斯点地梅■☆

23188 Androsace harrissii Duthie subsp. crassifolia (Wendelbo) Y. J. Nasir;厚叶点地梅■☆

23189 Androsace hedreantha Griseb.;巴尔干点地梅■☆

23190 Androsace helvetica All.;瑞士点地梅;Swiss Rock Jasmine, Swiss Rock-jasmine ■☆

23191 Androsace henryi Oliv.;莲叶点地梅(破头风,云雾草);Henry Rockjasmine ■

23192 Androsace henryi Oliv. var. crassifolia Hand. -Mazz. = Androsace henryi Oliv. ■

23193 Androsace henryi Oliv. var. omeiensis R. Knuth. = Androsace paxiana R. Knuth ■

23194 Androsace henryi Oliv. var. simulans C. M. Hu et Y. C. Yang;阔苞莲叶点地梅;Broadbract Henry Rockjasmine ■

23195 Androsace henryi Oliv. var. typica R. Knuth = Androsace henryi Oliv. ■

23196 Androsace himalaica (Knuth) Hand. -Mazz.;喜马拉雅点地梅■☆

23197 Androsace himalaica (Knuth) Hand. -Mazz. subsp. kurramensis Y. J. Nasir;库拉姆点地梅■☆

23198 Androsace hirtella L. Dufour;硬毛点地梅■☆

23199 Androsace hookeriana Klatt;亚东点地梅(虎克点地梅,异叶点地梅);Hooker Rockjasmine ■

23200 Androsace hookeriana Klatt var. mairei (H. Lév.) Yang et Huang = Androsace mairei H. Lév. ■

23201 Androsace hopeiensis Nakai = Androsace incana Lam. ■

23202 Androsace incana Lam.;白花点地梅;Whiteflower Rockjasmine ■

23203 Androsace incisa Wall. = Androsace rotundifolia Hardw. ■

23204 Androsace integra (Maxim.) Hand. -Mazz.;石莲叶点地梅(匙叶点地梅,高原点地梅,西藏点地梅);Spoonleaf Rockjasmine ■

23205 Androsace intermedia Ledeb.;全叶点地梅■☆

23206 Androsace jacquemontii Duby = Androsace robusta (R. Knuth) Hand. -Mazz. subsp. jacquemontii (Duby) Y. J. Nasir ■☆

23207 Androsace koso-poljanskii Ovcz.;科索点地梅■☆

23208 Androsace kouytchensis Bonati;贵州点地梅;Guizhou Rockjasmine ■

23209 Androsace lactea L.;乳白点地梅■☆

23210 Androsace lactiflora Pall.;宽花点地梅■☆

23211 Androsace lanuginosa Wall.;长绵毛点地梅(绵毛点地梅);Lanuginose Rockjasmine, Rock Jasmine ■☆

23212 Androsace laxa C. M. Hu et Y. C. Yang;秦巴点地梅;Looseflower Rockjasmine ■

23213 Androsace lehmanniana Spreng.;旱生点地梅(长毛点地梅,莱曼点地梅);Lehmann Rockjasmine, Xeric Rockjasmine ■

23214 Androsace lehmanniana Spreng. = Androsace chamaejasme Host subsp. lehmanniana (Spreng.) Hultén ■

23215 Androsace lehmannii Wall. ex Duby;钻叶点地梅(旱生点地梅);Awlshaped Rockjasmine, Bunge Rockjasmine ■

23216 Androsace limprichtii Pax et K. Hoffm.;康定点地梅(川藏点地梅);Kangding Rockjasmine, Limprecht Rockjasmine ■

23217 Androsace limprichtii Pax et K. Hoffm. var. laxiflora (Petitm.) Hand. -Mazz.;疏花康定点地梅;Looseflower Kangding Rockjasmine ■

23218 Androsace limprichtii Pax et K. Hoffm. var. laxiflora (Petitm.) Hand. -Mazz. = Androsace limprichtii Pax et K. Hoffm. ■

23219 Androsace longifolia Turcz.;长叶点地梅(矮葶点地梅);Longleaf Rockjasmine ■

23220 Androsace longifolia Turcz. var. decipiens Hand. -Mazz.;疏花长叶点地梅(疏丛长叶点地梅);Looseflower Longleaf Rockjasmine ■

23221 Androsace longifolia Turcz. var. decipiens Hand. -Mazz. = Androsace mariae Kanitz ■

23222 Androsace macrantha Boiss. et Huet;大花点地梅■☆

23223 Androsace mairei H. Lév.;绿棱点地梅;Greenridgy Rockjasmine ■

23224 Androsace mariae Kanitz;西藏点地梅;Tibet Rockjasmine, Xizang Rockjasmine ■

23225 Androsace mariae Kanitz var. tibetica (Maxim.) Hand. -Mazz. = Androsace mariae Kanitz ■

23226 Androsace mariae Kanitz var. trachyloma Hand. -Mazz. = Androsace mariae Kanitz ■

23227 Androsace maxima L.;大苞点地梅(大点地梅,大果点地梅);Bigbract Rockjasmine, Greater Rockjasmine, Largefruit Rockjasmine ■

23228 Androsace medifissa F. H. Chen et Y. C. Yang;梵净山点地梅;Fanjingshan Rockjasmine ■

23229 Androsace microphylla Hook. f. = Androsace mucronifolia Watt ■☆

23230 Androsace minor (Hand. -Mazz.) C. M. Hu et Y. C. Yang;小丛点地梅;Smallcluster Rockjasmine ■

23231 Androsace mirabilis Franch.;大叶点地梅;Largeleaf Rockjasmine ■

23232 Androsace mollis Hand. -Mazz.;柔软点地梅;Soft Rockjasmine ■

23233 Androsace mucronifolia Forrest var. stenophylla Hand. -Mazz. = Androsace yargongensis Petitm. ■

23234 Androsace mucronifolia Watt;小叶点地梅;Microphylloua Rockjasmine ■☆

23235 Androsace mucronifolia Watt var. typica R. Knuth = Androsace yargongensis Petitm. ■

23236 Androsace mucronifolia Watt var. uniflora Knuth = Androsace mucronifolia Watt ■☆

23237 Androsace muscoidea Duby;苔藓状点地梅;Mosslike Rockjasmine ■☆

23238 Androsace muscoidea Duby f. longiscapa (R. Knuth) Hand. -Mazz. = Androsace robusta (R. Knuth) Hand. -Mazz. ■

23239 Androsace muscoidea Duby f. longiscapa Knuth sensu Hand. -Mazz. = Androsace rotundifolia Hardw. ■

23240 Androsace nepalensis Derganc. = Androsace lehmannii Wall. ex Duby ■

23241 Androsace nortonii Ludlow;绢毛点地梅;Norton Rockjasmine ■

23242 Androsace obovata Wall. ex Duby = Primula floribunda Wall. ■☆

23243 Androsace occidentalis Pursh;西方点地梅;Rock-jasmine, Western Androsace, Western Rock Jasmine ■☆

23244 Androsace occidentalis Pursh var. arizonica (A. Gray) H. St. John = Androsace occidentalis Pursh ■☆

23245 Androsace occidentalis Pursh var. simplex (Rydb.) H. St. John = Androsace occidentalis Pursh ■☆

23246 Androsace ochotensis Willd. ex Roem. et Schult.;鄂霍次克点地梅■☆

23247 Androsace olgae Ovcz.;奥氏点地梅■☆

23248 Androsace orbicularis Roem. et Schult. = Androsace umbellata (Lour.) Merr. ■

23249 Androsace ovalifolia Y. C. Yang;卵叶点地梅;Ovateleaf Rockjasmine ■

23250 Androsace ovczinnikovii Schischk. et Bobrov;天山点地梅;Tianshan Rockjasmine ■

23251 Androsace patens Wright ex A. Gray = Androsace umbellata (Lour.) Merr. ■

23252 Androsace paxiana R. Knuth;峨眉点地梅;Emei Rockjasmine ■

23253 Androsace poissonii Knuth = Androsace delavayi Franch. ■

23254 Androsace pomeiensis C. M. Hu et Y. C. Yang;波密点地梅;Bomi Rockjasmine ■

23255 Androsace prattiana R. Knuth. = Androsace spinulifera (Franch.) R. Knuth ■

23256 Androsace primulina Spreng. = Primula primulina (Spreng.) Hara ■

23257 Androsace primuloides D. Don = Primula primulina (Spreng.) Hara ■

23258 Androsace pyrenaica Lam.;比利牛斯点地梅■☆

23259 Androsace raddeana Sommier et H. Lév.;拉德点地梅■☆

23260 Androsace refracta Hand. -Mazz.;折梗点地梅;Reflexedtwig Rockjasmine ■

23261 Androsace refracta Hand. -Mazz. = Androsace kouytchensis Bonati ■

23262 Androsace reverchonii Jord. = Androsace carnea L. ■☆

23263 Androsace reverchonii Jord. et Fourr. = Androsace carnea L. ■☆

23264 Androsace rigida Hand. -Mazz.;硬枝点地梅;Rigidbranch Rockjasmine ■

23265 Androsace rigida Hand. -Mazz. var. minor Hand. -Mazz. = Androsace minor (Hand. -Mazz.) C. M. Hu et Y. C. Yang ■

23266 Androsace robusta (R. Knuth) Hand. -Mazz.;雪球点地梅(粗壮点地梅);Hard Rockjasmine, Robust Rockjasmine ■

23267 Androsace robusta (R. Knuth) Hand. -Mazz. subsp. jacquemontii (Duby) Y. J. Nasir;雅克蒙点地梅■☆

23268 Androsace rockii W. E. Evans;密毛点地梅;Densehaired Rockjasmine ■

23269 Androsace rosea Jord. et Fourr. = Androsace carnea L. ■☆

23270 Androsace rotundifolia Hardw.;叶苞点地梅(圆叶点地梅);Roundleaved Rockjasmine ■

23271 Androsace rotundifolia Hardw. subsp. glandulosa (Hook. f.) Y. J. Nasir;具腺叶苞点地梅■☆

23272 Androsace rotundifolia Hardw. var. axillaris Franch. = Androsace axillaris (Franch.) Franch. ■

23273 Androsace rotundifolia Hardw. var. dissecta Franch. = Androsace dissecta (Franch.) Franch. ■

23274 Androsace rotundifolia Hardw. var. glandulosa Hook. f. = Androsace rotundifolia Hardw. subsp. glandulosa (Hook. f.) Y. J. Nasir ■☆

23275 Androsace rotundifolia Hardw. var. incisa ? = Androsace rotundifolia Hardw. ■

23276 Androsace rotundifolia Hardw. var. macrocalyx ? = Androsace rotundifolia Hardw. ■

23277 Androsace rotundifolia Hardw. var. nepalensis ? = Androsace geraniifolia Watt ■

23278 Androsace rotundifolia Hardw. var. thomsonii Watt;腺毛叶苞点地梅■

23279 Androsace rotundifolia Hardw. var. thomsonii Watt = Androsace thomsonii (Watt) Y. J. Nasir. ■☆

23280 Androsace runcinata Hand. -Mazz.;异叶点地梅(琴叶点地梅);Violinleaf Rockjasmine ■

23281 Androsace sarmentosa Wall.;匍茎点地梅(蔓茎点地梅,匍匐点

地梅,喜马拉雅点地梅);Himalayan Rockjasmine, Himalayas Rockjasmine ■☆

23282 Androsace sarmentosa Wall. var. foliosa (Decne. ex Duby) Hook. f. = Androsace foliosa Decne. ex Duby ■☆

23283 Androsace sarmentosa Wall. var. grandifolia Hook. f. = Androsace strigillosa Franch. ■

23284 Androsace sarmentosa Wall. var. laxiflora Petitm. = Androsace limprichtii Pax et K. Hoffm. ■

23285 Androsace sarmentosa Wall. var. stenophylla Petitm. = Androsace stenophylla (Petitm.) Hand. -Mazz. ■

23286 Androsace sarmentosa Wall. var. thibetensis Petitm. = Androsace wardii W. W. Sm. ■

23287 Androsace sarmentosa Wall. var. watkinsi ? = Androsace sarmentosa Wall. ■☆

23288 Androsace sarmentosa Wall. var. yunnanensis R. Knuth = Androsace mollis Hand. -Mazz. ■

23289 Androsace saxifragifolia Bunge = Androsace umbellata (Lour.) Merr. ■

23290 Androsace selago Hook. f. et Thomson ex Klatt;紫花点地梅;Purpleflower Rockjasmine ■

23291 Androsace semperviroides Jacq. var. tibetica Maxim. = Androsace mariae Kanitz ■

23292 Androsace sempervivoides Jacq. ex Duby;长生点地梅;Longliving Rockjasmine ■☆

23293 Androsace sempervivoides Jacq. ex Duby var. bracteate ? = Androsace himalaica (Knuth) Hand. -Mazz. subsp. kurramensis Y. J. Nasir ■☆

23294 Androsace septentrionalis L.;北点地梅(北方点地梅,雪山点地梅);Northern Fairy Candelabra, Northern Rockjasmine ■

23295 Androsace septentrionalis L. f. latifolia Y. Huei Huang;宽叶北点地梅(宽叶雪山点地梅);Broadleaf Northern Rockjasmine ■

23296 Androsace septentrionalis L. var. breviscapa Krylov;短葶北点地梅;Shortscape Northern Rockjasmine ■

23297 Androsace sericea Ovcz.;中亚绢毛点地梅■

23298 Androsace sessiliflora Turrill = Androsace tapete Maxim. ■

23299 Androsace simplex Rydb. = Androsace occidentalis Pursh ■☆

23300 Androsace spinulifera (Franch.) R. Knuth;刺叶点地梅;Spinyleaf Rockjasmine ■

23301 Androsace squarrosula Maxim.;鳞叶点地梅;Smallsquarose Rockjasmine ■

23302 Androsace staintonii Y. J. Nasir;斯坦顿点地梅■☆

23303 Androsace stenophylla (Petitm.) Hand. -Mazz.;狭叶点地梅;Narrowleaf Rockjasmine ■

23304 Androsace strigillosa (Franch.) R. Knuth var. spinulifera Franch. = Androsace spinulifera (Franch.) R. Knuth ■

23305 Androsace strigillosa Franch.;糙伏毛点地梅(糙伏点地梅,糙毛点地梅);Strigose Rockjasmine ■

23306 Androsace strigillosa Franch. var. canescens Marquand = Androsace strigillosa Franch. ■

23307 Androsace strigillosa Franch. var. spinulifera Franch. = Androsace spinulifera (Franch.) R. Knuth ■

23308 Androsace sublanata Hand. -Mazz.;绵毛点地梅(柔毛点地梅);Woollyhair Rockjasmine ■

23309 Androsace sutchuenensis Franch.;四川点地梅;Sichuan Rockjasmine ■

23310 Androsace sutchuenensis Franch. = Androsace axillaris (Franch.) Franch. ■

23311 Androsace sutchuenensis Franch. = Androsace cuscutiformis Franch. ■

23312 Androsace tanggulashanensis Y. C. Yang et R. F. Huang;唐古拉点地梅;Tangula Rockjasmine ■

23313 Androsace tapete Maxim.;垫状点地梅;Cushion Rockjasmine ■

23314 Androsace taurica Ovcz.;达乌尔点地梅■☆

23315 Androsace thomsonii (Watt) Y. J. Nasir. = Androsace rotundifolia Hardw. var. thomsonii Watt ■

23316 Androsace tibetica (Maxim.) R. Knuth = Androsace mariae Kanitz ■

23317 Androsace tibetica (Maxim.) R. Knuth var. himalaica ? = Androsace himalaica (Knuth) Hand. -Mazz. subsp. kurramensis Y. J. Nasir ■☆

23318 Androsace tibetica (Maxim.) R. Knuth var. mariae R. Knuth = Androsace mariae Kanitz ■

23319 Androsace tibetica R. Knuth var. mariae (Kanitz) R. Knuth. = Androsace mariae Kanitz ■

23320 Androsace tonkinensis Bonati = Mitrasacme pygmaea R. Br. ■

23321 Androsace triflora Adams;三花点地梅;Triflower Rockjasmine ■☆

23322 Androsace turczaninowii Freyn = Androsace maxima L. ■

23323 Androsace umbellata (Lour.) Merr.;点地梅(白花草,白花珍珠草,地胡椒,地钱草,点地梅草,顶珠草,佛顶珠,汉仙桃草,汉先桃草,旱仙桃草,喉痹草,喉蛾草,喉咙草,喉癣草,金牛草,清明花,索河草,索河花,天吊冬,天星花,铜钱草,五朵云,五角星草,五星草,五岳朝天,仙牛桃,小虎耳草,小一口血);Umbellate Rockjasmine ■

23324 Androsace vandellii Chiov.;迭叶点地梅■☆

23325 Androsace vegae Knuth;维加点地梅■☆

23326 Androsace villosa L.;柔毛点地梅(长毛点地梅,长柔毛点地梅);Villose Rockjasmine ■☆

23327 Androsace villosa L. = Androsace robusta (R. Knuth) Hand. -Mazz. ■

23328 Androsace villosa L. f. breviscapa ? = Androsace muscoidea Duby ■☆

23329 Androsace villosa L. f. longiscapa R. Knuth. = Androsace robusta (R. Knuth) Hand. -Mazz. ■

23330 Androsace villosa L. var. aurata Petitm. = Androsace bisulca Bureau et Franch. var. aurata (Petitm.) Yen C. Yang et P. H. Huang ■

23331 Androsace villosa L. var. bisulca (Bureau et Franch.) R. Knuth = Androsace bisulca Bureau et Franch. ■

23332 Androsace villosa L. var. incana (Lam.) Duby = Androsace incana Lam. ■

23333 Androsace villosa L. var. jacquemontii (Duby) Knuth = Androsace robusta (R. Knuth) Hand. -Mazz. subsp. jacquemontii (Duby) Y. J. Nasir ■☆

23334 Androsace villosa L. var. latifolia Bunge = Androsace lehmanniana Spreng. ■

23335 Androsace villosa L. var. robusta R. Knuth = Androsace robusta (R. Knuth) Hand. -Mazz. ■

23336 Androsace villosa L. var. robusta R. Knuth f. breviscapa R. Knuth = Androsace rotundifolia Hardw. ■

23337 Androsace villosa L. var. robusta R. Knuth f. longiscapa R. Knuth = Androsace rotundifolia Hardw. ■

23338 Androsace villosa L. var. robusta R. Knuth f. longiscapa R. Knuth = Androsace robusta (R. Knuth) Hand. -Mazz. ■

23339 Androsace villosa L. var. subexscapa Emb. et Maire = Androsace villosa L. ■☆

23340 Androsace villosa L. var. zambalensis Petitm. = Androsace zambalensis (Petitm.) Hand. -Mazz. ■

23341 Androsace villosa L. var. zambalensis Petitm. = Androsace zayulensis Hand. -Mazz. ■

23342 Androsace wardii W. W. Sm.;粗毛点地梅;Ward Rockjasmine ■

23343 Androsace wilsoniana Hand. -Mazz.;岩居点地梅;E. H. Wilson Rockjasmine ■

23344 Androsace yargongensis Petitm.;雅江点地梅;Yachiang Rockjasmine,Yajiang Rockjasmine ■

23345 Androsace yargongensis Petitm. var. stenophylla Hand. -Mazz. = Androsace yargongensis Petitm. ■

23346 Androsace zambalensis (Petitm.) Hand. -Mazz.;高原点地梅;Highland Rockjasmine ■

23347 Androsace zayulensis Hand. -Mazz.;察隅点地梅;Chayu Rockjasmine,Tsayul Rockjasmine ■

23348 Androsaces Asch. = Androsace L. ■

23349 Androsaemum Duhamel = Hypericum L. ●■

23350 Androsaemum Mill. = Hypericum L. ●■

23351 Androsaemum officinale All. = Hypericum androsaemum L. ●☆

23352 Androscepia Brongn. = Anthistiria L. f. ■

23353 Androscepia Brongn. = Calamina P. Beauv. ■

23354 Androscepia Brongn. = Themeda Forssk. ■

23355 Androscepia anathera (Nees ex Steud.) Andersson = Themeda anathera (Nees ex Steud.) Hack. ■

23356 Androscepia anathera (Nees ex Steud.) Andersson var. glabrescens Andersson = Themeda anathera (Nees ex Steud.) Hack. ■

23357 Androscepia anathera (Nees ex Steud.) Andersson var. hirsuta Andersson = Themeda anathera (Nees ex Steud.) Hack. ■

23358 Androscepia mutica Andersson = Themeda villosa (Poir.) A. Camus ■

23359 Androsemum Link = Hypericum L. ●■

23360 Androsemum Neck. = Androsaemum Duhamel ●■

23361 Androsemum Neck. = Hypericum L. ●■

23362 Androsiphon Schltr. (1924);管蕊风信子属■☆

23363 Androsiphon capense Schltr.;管蕊风信子■☆

23364 Androsiphon capense Schltr. = Daubenya capensis (Schltr.) A. M. Van der Merwe et J. C. Manning ■☆

23365 Androsiphonia Stapf = Paropsia Noronha ex Thouars ●☆

23366 Androsiphonia Stapf(1905);管蕊西番莲属●☆

23367 Androsiphonia adenostegia Stapf;腺盖管蕊西番莲●☆

23368 Androstachyaceae Airy Shaw = Euphorbiaceae Juss.(保留科名)●■

23369 Androstachyaceae Airy Shaw = Picrodendraceae Small(保留科名)●☆

23370 Androstachyaceae Airy Shaw;钉蕊科●

23371 Androstachydaceae Airy Shaw = Euphorbiaceae Juss.(保留科名)●■

23372 Androstachys Grand'Eury(废弃属名) = Androstachys Prain(保留属名)●☆

23373 Androstachys Prain(1908)(保留属名);钉蕊属●☆

23374 Androstachys johnsonii Prain;钉蕊●☆

23375 Androstachys subpeltatis (Sim) Phillips = Androstachys johnsonii Prain ●☆

23376 Androstemma Lindl. = Conostylis R. Br. ■☆

23377 Androstephanos Fern. Casas(1983);丝冠石蒜属■☆

23378 Androstephium Torr. (1859);丝冠葱属;Funnel-lily ■☆

23379 Androstephium Torr. = Bessera Schult. f.(保留属名)■☆

23380 Androstephium breviflorum S. Watson;短花丝冠葱;Pink Funnel-lily,Small-flowered Androstephium ■☆

23381 Androstephium coeruleum (Scheele) Greene;蓝丝冠葱;Blue Funnel-lily ■☆

23382 Androstoma Hook. f. (1844);岩高石南属●☆

23383 Androstoma Hook. f. = Cyathodes Labill. ●☆

23384 Androstoma empetrifolia Hook. f. = Androstoma verticillata (Hook. f.) Quinn ●☆

23385 Androstoma verticillata (Hook. f.) Quinn;岩高石南●☆

23386 Androstylanthus Ducke = Helianthostylis Baill. ●☆

23387 Androstylium Miq. = Clusia L. ●☆

23388 Androsyce Wed ex Hook f. = Elatostema J. R. Forst. et G. Forst.(保留属名)●■

23389 Androsynaceae Salisb. = Tecophilaeaceae Leyb.(保留科名)■☆

23390 Androsyne Salisb. = Walleria J. Kirk ■☆

23391 Androsyne gracilis Salisb. = Walleria gracilis (Salisb.) S. Carter ■☆

23392 Androtium Stapf(1903);雄漆属●☆

23393 Androtium astylum Stapf;雄漆●☆

23394 Androtrichum (Brongn.) Brongn. (1834);毛蕊莎草属■☆

23395 Androtrichum Brongn. = Androtrichum (Brongn.) Brongn. ■☆

23396 Androtrichum polycephalum Brongn.;毛蕊莎草■☆

23397 Androtropis R. Br. = Acranthera Arn. ex Meisn.(保留属名)●

23398 Androtropis R. Br. ex Wall. = Acranthera Arn. ex Meisn.(保留属名)●

23399 Androya H. Perrier(1952);马岛苦槛蓝属●☆

23400 Androya decaryi H. Perrier;马岛苦槛蓝●☆

23401 Andruris Schltr. (1912);东南亚霉草属■☆

23402 Andruris andajensis Schltr.;东南亚霉草■☆

23403 Andruris japonica (Makino) Giesen = Sciaphila nana Blume ■☆

23404 Andryala L. (1753);毛托菊属(毛托山柳菊属)■☆

23405 Andryala aestivalis Pomel = Andryala integrifolia L. ■☆

23406 Andryala agardhii Haens.;阿氏毛托菊■☆

23407 Andryala antonii Maire = Andryala pinnatifida Aiton subsp. antonii (Maire) Dobignard ■☆

23408 Andryala arenaria (DC.) Boiss. et Reut.;沙地毛托菊■☆

23409 Andryala arenaria (DC.) Boiss. et Reut. var. pinnatifida Lange = Andryala arenaria (DC.) Boiss. et Reut. ■☆

23410 Andryala atlanticola H. Lindb.;大西洋毛托菊■☆

23411 Andryala canariensis Lowe = Andryala pinnatifida Aiton ■☆

23412 Andryala canariensis Lowe subsp. antonii Maire = Andryala pinnatifida Aiton subsp. antonii (Maire) Dobignard ■☆

23413 Andryala canariensis Lowe subsp. ducellieri (Batt.) Maire = Andryala pinnatifida Aiton subsp. ducellieri (Batt.) Greuter ■☆

23414 Andryala canariensis Lowe subsp. jahandiezii Maire = Andryala pinnatifida Aiton subsp. jahandiezii (Maire) Greuter ■☆

23415 Andryala canariensis Lowe subsp. maroccana Maire = Andryala pinnatifida Aiton subsp. maroccana (Maire) Greuter ■☆

23416 Andryala canariensis Lowe subsp. mogadorensis (Coss. et Balansa) Maire = Andryala pinnatifida Aiton subsp. mogadorensis (Hook. f.) Greuter ■☆

23417 Andryala canariensis Lowe var. microcarpa Maire = Andryala chevallieri Barratte ■☆

23418 Andryala cheiranthifolia L'Hér. = Andryala glandulosa Lam. subsp. cheiranthifolia (L'Hér.) Greuter ■☆

23419 Andryala chevallieri Barratte;舍瓦利耶毛托菊■☆

23420 Andryala cosyrensis Guss.;考司林毛托菊■☆

23421 Andryala cosyrensis Guss. var. oligadena Maire et Weiller = Andryala cosyrensis Guss. ■☆
23422 Andryala crithmifolia Aiton;海茴香叶毛托菊■☆
23423 Andryala dentata Sibth. et Sm. var. arenaria (Boiss. et Reut.) Batt. = Andryala arenaria (DC.) Boiss. et Reut. ■☆
23424 Andryala ducellieri Batt. = Andryala pinnatifida Aiton subsp. ducellieri (Batt.) Greuter ■☆
23425 Andryala floccosa Pomel = Andryala integrifolia L. subsp. perennans Maire et Weiller ■☆
23426 Andryala glandulosa Lam.;腺点毛托菊■☆
23427 Andryala glandulosa Lam. subsp. cheiranthifolia (L'Hér.) Greuter;桂竹香叶毛托菊■☆
23428 Andryala glandulosa Lam. subsp. varia (DC.) R. Fern.;变异腺点毛托菊■☆
23429 Andryala gracilis Caball. = Andryala cosyrensis Guss. ■☆
23430 Andryala gracilis Pau = Andryala integrifolia L. ■☆
23431 Andryala humilis Pau;低矮毛托菊■☆
23432 Andryala integrifolia L.;全叶毛托菊■☆
23433 Andryala integrifolia L. subsp. perennans Maire et Weiller;多年毛托菊■☆
23434 Andryala integrifolia L. var. angustifolia DC. = Andryala integrifolia L. ■☆
23435 Andryala integrifolia L. var. arenaria (Boiss. et Reut.) Ball = Andryala arenaria (DC.) Boiss. et Reut. ■☆
23436 Andryala integrifolia L. var. corymbosa (Lam.) Willd. = Andryala integrifolia L. ■☆
23437 Andryala integrifolia L. var. floccosa Svent. = Andryala integrifolia L. ■☆
23438 Andryala integrifolia L. var. gracilis (Pau) Maire = Andryala integrifolia L. ■☆
23439 Andryala integrifolia L. var. nigricans (Poir.) Barratte = Andryala integrifolia L. ■☆
23440 Andryala integrifolia L. var. sinuata (L.) Willk. = Andryala integrifolia L. ■☆
23441 Andryala integrifolia L. var. tenuifolia (DC.) Barratte = Andryala cosyrensis Guss. ■☆
23442 Andryala jahandiezii Maire = Andryala pinnatifida Aiton subsp. jahandiezii (Maire) Greuter ■☆
23443 Andryala laxiflora DC.;疏花毛托菊■☆
23444 Andryala laxiflora DC. var. candicans Maire = Andryala laxiflora DC. ■☆
23445 Andryala maroccana Pau;摩洛哥毛托菊■☆
23446 Andryala maroccana Pau var. calendula (Doum.) Maire = Andryala maroccana Pau ■☆
23447 Andryala mogadorensis Hook. f. = Andryala pinnatifida Aiton subsp. mogadorensis (Hook. f.) Greuter ■☆
23448 Andryala nigricans Poir.;黑毛托菊■☆
23449 Andryala nigricans Poir. var. boitardii Litard. et Maire = Andryala nigricans Poir. ■☆
23450 Andryala pinnatifida Aiton;羽裂毛托菊■☆
23451 Andryala pinnatifida Aiton subsp. antonii (Maire) Dobignard;安东毛托菊■☆
23452 Andryala pinnatifida Aiton subsp. ducellieri (Batt.) Greuter;迪塞利耶毛托菊■☆
23453 Andryala pinnatifida Aiton subsp. jahandiezii (Maire) Greuter;贾汉毛托菊■☆
23454 Andryala pinnatifida Aiton subsp. latifolia (Bornm.) G. Kunkel;宽叶羽裂毛托菊■☆
23455 Andryala pinnatifida Aiton subsp. maroccana (Maire) Greuter = Andryala pinnatifida Aiton subsp. maroccana (Maire) Maire ■☆
23456 Andryala pinnatifida Aiton subsp. maroccana (Maire) Maire;摩洛哥羽裂毛托菊■☆
23457 Andryala pinnatifida Aiton subsp. mogadorensis (Hook. f.) Greuter;摩加多尔毛托菊■☆
23458 Andryala pinnatifida Aiton subsp. preauxiana (Sch. Bip.) G. Kunkel;普雷毛托菊■☆
23459 Andryala pinnatifida Aiton subsp. teydensis (Sch. Bip.) Rivas Mart. et al.;泰德羽裂毛托菊■☆
23460 Andryala pinnatifida Aiton subsp. webbii (Sch. Bip.) G. Kunkel;韦布羽裂毛托菊■☆
23461 Andryala pinnatifida Aiton var. glabrescens Sch. Bip. = Andryala pinnatifida Aiton ■☆
23462 Andryala pinnatifida Aiton var. sprengeliana G. Kunkel = Andryala pinnatifida Aiton ■☆
23463 Andryala pinnatifida Aiton var. stricta Christ = Andryala pinnatifida Aiton ■☆
23464 Andryala pinnatifida Aiton var. teydensis Sch. Bip. = Andryala pinnatifida Aiton ■☆
23465 Andryala pinnatifida Aiton var. tolpicifolia G. Kunkel = Andryala pinnatifida Aiton ■☆
23466 Andryala ragusina L. subsp. spartioides Pomel = Andryala spartioides Batt. et Trab. ■☆
23467 Andryala reboudiana Pomel;雷博毛托菊■☆
23468 Andryala rothia Pers. subsp. arenaria (DC.) Maire = Andryala arenaria (DC.) Boiss. et Reut. ■☆
23469 Andryala rothia Pers. subsp. cosyrensis (Guss.) Maire = Andryala cosyrensis Guss. ■☆
23470 Andryala rothia Pers. var. pinnatifida (Lange) Maire = Andryala arenaria (DC.) Boiss. et Reut. ■☆
23471 Andryala spartioides Batt. et Trab.;绳索毛托菊■☆
23472 Andryala webbii H. Christ;韦布毛托菊■☆
23473 Andrzeiowskia Rchb. (1824);荠状芥属■☆
23474 Andrzeiowskia cardaminifolia (DC.) Prantl;荠状芥■☆
23475 Andrzeiowskia pectinata (DC.) Turcz. = Dontostemon pinnatifidus (Willd.) Al-Shehbaz et H. Ohba ■
23476 Andrzeiowskya Rchb. = Andrzeiowskia Rchb. ■☆
23477 Anechites Griseb. (1861);异蛇木属●☆
23478 Anechites asperuginis Griseb.;异蛇木●☆
23479 Anecio Neck. = Senecio L. ●■
23480 Anectochilus Blume = Anoectochilus Blume(保留属名)■
23481 Anectron H. Winkler = Nectaropetalum Engl. ●☆
23482 Anectron H. Winkler = Peglera Bolus ●☆
23483 Aneilema R. Br. (1810);肖水竹叶属■☆
23484 Aneilema R. Br. = Murdannia Royle(保留属名)■
23485 Aneilema aequinoctiale (P. Beauv.) G. Don;昼夜肖水竹叶■☆
23486 Aneilema aequinoctiale (P. Beauv.) G. Don var. kirkii C. B. Clarke = Aneilema hockii De Wild. ■☆
23487 Aneilema aequinoctiale (P. Beauv.) Kunth = Aneilema aequinoctiale (P. Beauv.) G. Don ■☆
23488 Aneilema affine De Wild. = Aneilema silvaticum Brenan ■☆
23489 Aneilema africanum P. Beauv. = Floscopa africana (P. Beauv.) C. B. Clarke ■☆
23490 Aneilema angolense C. B. Clarke;安哥拉肖水竹叶■☆
23491 Aneilema angustifolium De Wild. = Aneilema macrorrhizum Th.

Fr. ■☆

23492 Aneilema angustifolium N. E. Br. = Murdannia loriformis (Hassk.) Rao Rolla et Kammathy ■

23493 Aneilema arenicola Faden;沙生肖水竹叶■☆

23494 Aneilema benadirense Chiov.;贝纳迪尔肖水竹叶■☆

23495 Aneilema beniniense (P. Beauv.) Kunth;贝尼尼肖水竹叶■☆

23496 Aneilema beniniense (P. Beauv.) Kunth subsp. leonense Morton = Aneilema beniniense (P. Beauv.) Kunth ■☆

23497 Aneilema beniniense (P. Beauv.) Kunth subsp. sessilifolium (Benth.) Morton = Aneilema beniniense (P. Beauv.) Kunth ■☆

23498 Aneilema beniniense (P. Beauv.) Kunth var. oxycarpum Hua = Aneilema beniniense (P. Beauv.) Kunth ■☆

23499 Aneilema beniniensis (P. Beauv.) Kunth var. sessilifolium Benth. = Aneilema beniniense (P. Beauv.) Kunth ■☆

23500 Aneilema blumei (Hassk.) Bakh. f. = Murdannia blumei (Hassk.) Brenan ■☆

23501 Aneilema bodinieri H. Lév. = Murdannia hookeri (C. B. Clarke) Brückn. ■

23502 Aneilema bodinieri H. Lév. et Vaniot = Murdannia hookeri (C. B. Clarke) Brückn. ■

23503 Aneilema bracteatum (C. B. Clarke) Kuntze;大苞水竹草■

23504 Aneilema bracteatum (C. B. Clarke) Kuntze = Murdannia bracteata (C. B. Clarke) J. K. Morton ex D. Y. Hong ■

23505 Aneilema bracteatum Kuntze = Murdannia bracteata (C. B. Clarke) J. K. Morton ex D. Y. Hong ■

23506 Aneilema bracteolatum Faden;小苞片肖水竹叶■☆

23507 Aneilema brenanianum Faden;布雷南肖水竹叶■☆

23508 Aneilema brunneospermum Faden;褐籽肖水竹叶■☆

23509 Aneilema canaliculatum Dalzell = Murdannia spirata (L.) Brückn. ■

23510 Aneilema cavaleriei H. Lév. et Vaniot = Murdannia simplex (Vahl) Brenan ■

23511 Aneilema chrysanthum K. Schum. = Aneilema petersii (Hassk.) C. B. Clarke ■☆

23512 Aneilema chrysopogon Brenan;金须肖水竹叶■☆

23513 Aneilema clarkei Rendle;克拉克肖水竹叶■☆

23514 Aneilema conspicuum (Blume) Kunth = Dictyospermum conspicuum (Blume) Hassk. ■

23515 Aneilema coreanum H. Lév. et Vaniot = Murdannia keisak (Hassk.) Hand. -Mazz. ■

23516 Aneilema densum Th. Fr. = Aneilema welwitschii C. B. Clarke ■☆

23517 Aneilema divergens (C. B. Clarke) C. B. Clarke = Murdannia divergens (C. B. Clarke) Brückn. ■

23518 Aneilema divergens C. B. Clarke = Murdannia divergens (C. B. Clarke) Brückn. ■

23519 Aneilema dregeanum Kunth;德雷肖水竹叶■☆

23520 Aneilema dregeanum Kunth subsp. mossambicense Faden;莫桑比克肖水竹叶■☆

23521 Aneilema ehrenbergii (Hassk.) C. B. Clarke = Aneilema forskalii Kunth ■☆

23522 Aneilema elatum Dalzell = Murdannia japonica (Thunb.) Faden ■

23523 Aneilema ephemerum Faden;短命肖水竹叶■☆

23524 Aneilema erectum De Wild. = Aneilema welwitschii C. B. Clarke ■☆

23525 Aneilema florentii De Wild. = Aneilema welwitschii C. B. Clarke ■☆

23526 Aneilema formosanum N. E. Br. = Murdannia edulis (Stokes) Faden ■

23527 Aneilema forskalii Kunth;福斯科尔肖水竹叶■☆

23528 Aneilema gillettii Brenan;吉莱特肖水竹叶■☆

23529 Aneilema gracile (Kotschy et Peyr.) C. B. Clarke = Aneilema lanceolatum Benth. subsp. subnudum (A. Chev.) J. K. Morton ■☆

23530 Aneilema grandibracteolatum Faden;大苞片肖水竹叶■☆

23531 Aneilema hamiltonianum Wall. = Murdannia blumei (Hassk.) Brenan ■☆

23532 Aneilema herbaceum (Roxb.) Wall. ex C. B. Clarke = Murdannia japonica (Thunb.) Faden ■

23533 Aneilema herbaceum (Roxb.) Wall. ex C. B. Clarke var. divergens C. B. Clarke = Murdannia divergens (C. B. Clarke) Brückn. ■

23534 Aneilema herbaceum Wall. = Murdannia japonica (Thunb.) Faden ■

23535 Aneilema hirtum A. Rich.;多毛肖水竹叶■☆

23536 Aneilema hispidum D. Don = Floscopa scandens Lour. ■

23537 Aneilema hockii De Wild.;霍克肖水竹叶■☆

23538 Aneilema homblei De Wild.;洪布勒肖水竹叶■☆

23539 Aneilema hookeri C. B. Clarke = Murdannia hookeri (C. B. Clarke) Brückn. ■

23540 Aneilema indehiscens Faden;开裂肖水竹叶■☆

23541 Aneilema indehiscens Faden subsp. lilacinum Faden;紫丁香色肖水竹叶■☆

23542 Aneilema japonicum (Thunb.) Kunth = Murdannia japonica (Thunb.) Faden ■

23543 Aneilema johnstonii K. Schum.;约翰斯顿肖水竹叶■☆

23544 Aneilema kainantense Masam. = Murdannia kainantensis (Masam.) D. Y. Hong ■

23545 Aneilema kainantense Masam. = Murdannia loriformis (Hassk.) R. S. Rao et Kammathy ■

23546 Aneilema katangense De Wild. = Aneilema welwitschii C. B. Clarke ■☆

23547 Aneilema keisak Hassk. = Murdannia keisak (Hassk.) Hand. - Mazz. ■

23548 Aneilema kuntzei C. B. Clarke = Murdannia bracteata (C. B. Clarke) J. K. Morton ex D. Y. Hong ■

23549 Aneilema kuntzei C. B. Clarke ex Kuntze = Murdannia bracteata (C. B. Clarke) J. K. Morton ex D. Y. Hong ■

23550 Aneilema lanceolatum Benth.;披针形肖水竹叶■☆

23551 Aneilema lanceolatum Benth. subsp. subnudum (A. Chev.) J. K. Morton;亚裸披针形肖水竹叶■☆

23552 Aneilema latifolium Wight = Murdannia japonica (Thunb.) Faden ■

23553 Aneilema leiocaule K. Schum.;光茎肖水竹叶■☆

23554 Aneilema lineolatum Kunth = Murdannia japonica (Thunb.) Faden ■

23555 Aneilema longicapsum Faden;长果肖水竹叶■☆

23556 Aneilema longifolium Hook. = Murdannia simplex (Vahl) Brenan ■

23557 Aneilema longirrhizum Faden;长根肖水竹叶■☆

23558 Aneilema loriforme Hassk. = Murdannia loriformis (Hassk.) R. S. Rao et Kammathy ■

23559 Aneilema loureiroi Hance = Murdannia spectabilis (Kurz) Faden ■

23560 Aneilema lujae De Wild. et T. Durand = Aneilema lanceolatum Benth. ■☆

23561 Aneilema macrorrhizum Th. Fr.;大根肖水竹叶■☆

23562 Aneilema malabarica (L.) Merr. = Murdannia nudiflora (L.)

Brenan ■
23563 Aneilema medicum (Lour.) R. Br. = Murdannia medica (Lour.) D. Y. Hong ■
23564 Aneilema melanostictum Hance = Murdannia spirata (L.) Brückn. ■
23565 Aneilema minutiflorum Faden;微花肖水竹叶■☆
23566 Aneilema mortehanii De Wild. = Aneilema beniniense (P. Beauv.) Kunth ■☆
23567 Aneilema mortonii Brenan;莫顿肖水竹叶■☆
23568 Aneilema nanum Kunth = Murdannia spirata (L.) Brückn. ■
23569 Aneilema nicholsonii C. B. Clarke;尼克尔森肖水竹叶■☆
23570 Aneilema nigritanum (C. B. Clarke) Hutch. = Aneilema umbrosum (Vahl) Kunth ■☆
23571 Aneilema nudiflorum (L.) R. Br. = Murdannia nudiflora (L.) Brenan ■
23572 Aneilema nudiflorum (L.) R. Br. var. bracteatum C. B. Clarke = Murdannia bracteata (C. B. Clarke) J. K. Morton ex D. Y. Hong ■
23573 Aneilema nudiflorum (L.) R. Br. var. rigidior Benth. = Murdannia loriformis (Hassk.) R. S. Rao et Kammathy ■
23574 Aneilema nudiflorum (L.) Sweet = Murdannia nudiflora (L.) Brenan ■
23575 Aneilema nudiflorum (L.) Wall. = Murdannia nudiflora (L.) Brenan ■
23576 Aneilema nudiflorum (L.) Wall. ex C. B. Clarke = Murdannia nudiflora (L.) Brenan ■
23577 Aneilema nutans H. Lév. = Murdannia triquetra (Wall.) Brückn. ■
23578 Aneilema nyasense C. B. Clarke;尼亚斯肖水竹叶■☆
23579 Aneilema obbiadense Chiov.;奥比亚德肖水竹叶■☆
23580 Aneilema obbiadensis Chiov. var. angustifolia ? = Aneilema longicapsum Faden ■☆
23581 Aneilema octospermum C. B. Clarke = Aneilema rendlei C. B. Clarke ■☆
23582 Aneilema oliganthum Franch. = Murdannia keisak (Hassk.) Hand. -Mazz. ■
23583 Aneilema ovato-oblongum P. Beauv. = Aneilema umbrosum (Vahl) Kunth subsp. ovato-oblongum (P. Beauv.) J. K. Morton ■☆
23584 Aneilema ovato-oblongum P. Beauv. var. nigritanum C. B. Clarke = Aneilema umbrosum (Vahl) Kunth ■☆
23585 Aneilema paludosum A. Chev.;沼泽肖水竹叶■☆
23586 Aneilema paludosum A. Chev. subsp. pauciflorum J. K. Morton;少花沼泽肖水竹叶■☆
23587 Aneilema paludosum A. Chev. subsp. pseudolanceolatum J. K. Morton;披针肖水竹叶■☆
23588 Aneilema paucifolium N. E. Br. = Murdannia medica (Lour.) D. Y. Hong ■
23589 Aneilema pedunculosum C. B. Clarke;梗花肖水竹叶■☆
23590 Aneilema petersii (Hassk.) C. B. Clarke;彼得肖水竹叶■☆
23591 Aneilema petersii (Hassk.) C. B. Clarke subsp. pallidiflorum Faden;苍白花肖水竹叶■☆
23592 Aneilema pomeridianum Stanf. et Brenan;午后花肖水竹叶■☆
23593 Aneilema protensum Thwaites = Dictyospermum scaberrimum (Blume) J. K. Morton ex H. Hara ■
23594 Aneilema protensum Thwaites = Rhopalephora scaberrima (Blume) Faden ■
23595 Aneilema protensum Wall. ex C. B. Clarke = Rhopalephora scaberrima (Blume) Faden ■
23596 Aneilema pusillum Chiov.;微小肖水竹叶■☆
23597 Aneilema pusillum Chiov. subsp. gypsophilum Faden;喜钙肖水竹叶■☆
23598 Aneilema pusillum Chiov. subsp. thulinii Faden;图林肖水竹叶■☆
23599 Aneilema pusillum Chiov. subsp. variabile Faden;易变微小肖水竹叶■☆
23600 Aneilema radicans D. Don = Murdannia nudiflora (L.) Brenan ■
23601 Aneilema recurvatum Faden;反折肖水竹叶■☆
23602 Aneilema rendlei Brenan = Aneilema taylori C. B. Clarke ■☆
23603 Aneilema rendlei C. B. Clarke;伦德尔肖水竹叶■☆
23604 Aneilema rhodospermum K. Schum. = Aneilema silvaticum Brenan var. pilosum ? ■☆
23605 Aneilema richardsiae Brenan;理查兹肖水竹叶■☆
23606 Aneilema ringoeti De Wild. = Aneilema hirtum A. Rich. ■☆
23607 Aneilema rivularis A. Rich. = Floscopa glomerata (Willd. ex Schult. et Schult. f.) Hassk. ■☆
23608 Aneilema rugosum H. Perrier = Rhopalephora rugosa (H. Perrier) Faden ■☆
23609 Aneilema russeggeri (Fenzl) C. B. Clarke = Aneilema lanceolatum Benth. ■☆
23610 Aneilema sacleuxii Hua = Aneilema petersii (Hassk.) C. B. Clarke ■☆
23611 Aneilema sacleuxii Hua;热非肖水竹叶■
23612 Aneilema scaberrimum (Blume) Kunth = Dictyospermum scaberrimum (Blume) J. K. Morton ex H. Hara ■
23613 Aneilema scaberrimum (Blume) Kunth = Rhopalephora scaberrima (Blume) Faden ■
23614 Aneilema scapiflorum (Roxb.) Wight = Murdannia edulis (Stokes) Faden ■
23615 Aneilema scapiflorum (Roxb.) Wight = Murdannia scapiflora (Roxb.) Royle ■
23616 Aneilema scapiflorum (Roxb.) Wight var. latifolium N. E. Br. = Murdannia edulis (Stokes) Faden ■
23617 Aneilema schlechteri K. Schum.;施莱肖水竹叶■☆
23618 Aneilema schweinfurthii C. B. Clarke = Aneilema lanceolatum Benth. ■☆
23619 Aneilema sebitense Faden;蜡质肖水竹叶■☆
23620 Aneilema semiteres Dalziel = Murdannia semiteres (Dalzell) Santapau ■☆
23621 Aneilema sepalosum C. B. Clarke = Anthericopsis sepalosa (C. B. Clarke) Engl. ■☆
23622 Aneilema serotinum Don ex C. B. Clarke = Murdannia scapiflora (Roxb.) Royle ■
23623 Aneilema setiferum A. Chev.;刚毛肖水竹叶■☆
23624 Aneilema setiferum A. Chev. var. pallidiciliatum J. K. Morton;白缘毛肖水竹叶■☆
23625 Aneilema siamense Craib = Pollia siamensis (Craib) Faden ex D. Y. Hong ■
23626 Aneilema silvaticum Brenan;森林肖水竹叶■☆
23627 Aneilema silvaticum Brenan var. pilosum ?;疏毛肖水竹叶■☆
23628 Aneilema sinicum Ker Gawl. = Murdannia simplex (Vahl) Brenan ■
23629 Aneilema smithii C. B. Clarke = Aneilema somaliense C. B. Clarke ■☆
23630 Aneilema somaliense C. B. Clarke;索马里肖水竹叶■☆
23631 Aneilema soudanicum C. B. Clarke = Aneilema lanceolatum Benth. ■☆

23632 Aneilema spectabile Kurz = Murdannia spectabilis (Kurz) Faden ■
23633 Aneilema spekei C. B. Clarke;斯皮克肖水竹叶■☆
23634 Aneilema spiratum (L.) Sweet = Murdannia spirata (L.) Brückn. ■
23635 Aneilema stenothyrsum Diels = Murdannia stenothyrsa (Diels) Hand. -Mazz. ■
23636 Aneilema subnudum A. Chev. = Aneilema lanceolatum Benth. subsp. subnudum (A. Chev.) J. K. Morton ■☆
23637 Aneilema succulentum Faden;多汁肖水竹叶■☆
23638 Aneilema tacazzeanum Hochst. ex A. Rich. = Aneilema forskalii Kunth ■☆
23639 Aneilema tacazzeanum Hochst. ex C. B. Clarke = Aneilema forskalii Kunth ■☆
23640 Aneilema tanaense Faden;塔纳肖水竹叶■☆
23641 Aneilema taquetii H. Lév. = Murdannia keisak (Hassk.) Hand. -Mazz. ■
23642 Aneilema taylori C. B. Clarke;泰勒肖水竹叶■☆
23643 Aneilema tenera Baker;极细肖水竹叶■☆
23644 Aneilema tenuissimum (A. Chev.) A. Chev. ex Hutch. et Dalziel = Murdannia tenuissima (A. Chev.) Brenan ■☆
23645 Aneilema terminale Wight = Murdannia loriformis (Hassk.) R. S. Rao et Kammathy ■
23646 Aneilema tetraspermum K. Schum. = Aneilema petersii (Hassk.) C. B. Clarke ■☆
23647 Aneilema triquetrum Wall. = Murdannia keisak (Hassk.) Hand. -Mazz. ■
23648 Aneilema triquetrum Wall. ex C. B. Clarke = Murdannia triquetra (Wall.) Brückn. ■
23649 Aneilema trispermum Faden;三籽肖水竹叶■☆
23650 Aneilema umbrosum (Vahl) Kunth;耐荫肖水竹叶■☆
23651 Aneilema umbrosum (Vahl) Kunth subsp. ovato-oblongum (P. Beauv.) J. K. Morton;卵矩圆肖水竹叶■☆
23652 Aneilema usambarense Faden;乌桑巴拉肖水竹叶■☆
23653 Aneilema vankerkhovenii De Wild. = Aneilema nyasense C. B. Clarke ■☆
23654 Aneilema welwitschii C. B. Clarke;韦尔肖水竹叶■☆
23655 Aneilema whytei C. B. Clarke = Aneilema hirtum A. Rich. ■☆
23656 Aneilema wildii Merxm. = Aneilema hockii De Wild. ■☆
23657 Aneilema woodii Faden;伍得肖水竹叶■☆
23658 Aneilema zebrinum Chiov.;条斑肖水竹叶■☆
23659 Anelasma Miers = Abuta Aubl. ●☆
23660 Anelsonia J. F. Macbr. et Payson(1917);良果芥属■☆
23661 Anelsonia eurycarpa (A. Gray) J. F. Macbr. et Payson;良果芥■☆
23662 Anelsonia eurycarpa J. F. Macbr. et Payson = Anelsonia eurycarpa (A. Gray) J. F. Macbr. et Payson ■☆
23663 Anelytrum Hack. = Avena L. ■
23664 Anemagrostis Trin. = Apera Adans. ■☆
23665 Anemanthele Veldkamp = Stipa L. ■
23666 Anemanthele Veldkamp(1985);风羽针茅属■☆
23667 Anemanthele lessoniana (Steud.) Veldkamp;风羽针茅■☆
23668 Anemanthus Fourr. = Anemone L.(保留属名)■
23669 Anemarrhena Bunge(1833);知母属;Anemarrhena ■★
23670 Anemarrhena asphodeloides Bunge;知母(昌文,昌支,莸藩,穿地龙,茨藩,大芦水,地参,东根,儿草,儿踵草,肥知母,光知母,货母,京知母,韭逢,苦心,老梗,连母,鹿列,毛知母,女雷,女理,芪母,蚳母,荨,水参,水浚,水渗,水须,蒜瓣子草,蒜辫子草,提母,蝭母,兔子油草,西陵知母,孝梗,羊胡子草,羊胡子根,野蓼,知母肉,纸母,竹连花);Anemarrhena ■
23671 Anemarrhena cavaleriei H. Lév. = Ophiopogon japonicus (L. f.) Ker Gawl. ■
23672 Anemarrhena mairei (H. Lév.) H. Lév. = Ophiopogon mairei H. Lév. ■
23673 Anemarrhenaceae Conran, M. W. Chase et Rudall = Agavaceae Dumort.(保留科名)●■
23674 Anemarrhenaceae Conran, M. W. Chase et Rudall(1997);知母科■
23675 Anemia Nutt. = Anemopsis Hook. et Arn. ■☆
23676 Anemiopsis Endl. = Anemopsis Hook. et Arn. ■☆
23677 Anemitis Raf. = Phlomis L. ●■
23678 Anemocarpa Paul G. Wilson(1992);风果彩鼠麹属■☆
23679 Anemocarpa calcicola Paul G. Wilson;风果彩鼠麹■☆
23680 Anemoclema (Franch.) W. T. Wang(1964);罂粟莲花属;Anemoclema ■★
23681 Anemoclema anhuiensis Y. K. Yang et al.;安徽罂粟莲花(安徽银莲花)■
23682 Anemoclema glaucifolium (Franch.) W. T. Wang;罂粟莲花;Hornpoppyleaf Anemoclema ■
23683 Anemoclema hofengensis W. T. Wang;鹤峰罂粟莲花(鹤峰银莲花)■
23684 Anemoclema multilobulata W. T. Wang et L. Q. Li;多裂罂粟莲花(多裂银莲花)■
23685 Anemoclema nutantiflora W. T. Wang et L. Q. Li;垂花罂粟莲花(垂花银莲花)■
23686 Anemonaceae Vest = Ranunculaceae Juss.(保留科名)●■
23687 Anemonaceae Vest;银莲花科(罂粟莲花科)■
23688 Anemonanthea (DC.) Gray = Anemone L.(保留属名)■
23689 Anemonanthea Gray = Anemone L.(保留属名)■
23690 Anemonanthera Willis = Anemonanthea (DC.) Gray ■
23691 Anemonastrum Holub = Anemone L.(保留属名)■
23692 Anemonastrum demissum (Hook. f. et Thomson) Holub = Anemone demissa Hook. f. et Thomson ■
23693 Anemonastrum elongatum (D. Don) Holub = Anemone elongata D. Don ■
23694 Anemonastrum polyanthes (D. Don) Holub = Anemone demissa Hook. f. et Thomson ■
23695 Anemonastrum polyanthes (D. Don) Holub = Anemone polyanthes D. Don ■
23696 Anemone L. (1753)(保留属名);银莲花属;Anemone, Anemony, Pasqueflower, Thimbleweed, Wind Flower, Windflower ■
23697 Anemone acutiloba (DC.) G. Lawson = Hepatica nobilis Garsault var. acuta (Pursh) Steyerm. ■☆
23698 Anemone acutiloba (DC.) G. Lawson, Procop. et Transeau;尖裂银莲花;Sharp-lobed Hepatica ■☆
23699 Anemone adamsiana Eastw. = Anemone oregana A. Gray ■☆
23700 Anemone alpina L. = Pulsatilla alpina (L.) Delarbre ■☆
23701 Anemone altaica Fisch. ex Ledeb.;阿尔泰银莲花(穿骨七,鸡爪莲,节菖蒲,京菖蒲,京玄参,九节菖蒲,九节离,菊形双瓶梅,陕西菖,太原菖,菟葵,外菖蒲,小菖蒲,玄参);Altai Windflower, Irkutsk Anemone ■
23702 Anemone ambigua Turcz. ex Hayek = Pulsatilla ambigua (Turcz. ex Pritz.) Juz. ■
23703 Anemone americana (DC.) H. Hara;美洲银莲花;Bluntleaf Hepatica, Bluntleaf Liverleaf, Round-lobed Hepatica ■☆
23704 Anemone americana (DC.) H. Hara = Hepatica nobilis Garsault

f. obtusa (Pursh) Beck ■☆
23705 Anemone amurensis (Korsh.) Kom.;黑水银莲花;Amur Anemone,Amur Windflower,European Wood Anemone ■
23706 Anemone angulosa Lam.;棱角银莲花;Hepatica ■☆
23707 Anemone apennina L.;意大利银莲花(亚平宁银莲花);Apennine Anemone,Blue Anemone,Blue Wood Anemone,Mountain Anemone,Storksbill Wind Flower,Storksbill Windflower ■☆
23708 Anemone apennina L. var. purpurea Hort.;紫意大利银莲花■☆
23709 Anemone baicalensis Turcz.;毛果银莲花;Baical Anemone,Baical Windflower ■
23710 Anemone baicalensis Turcz. var. glabrata Maxim.;光果银莲花;Glabratefruit Anemone ■
23711 Anemone baicalensis Turcz. var. kansuensis (W. T. Wang) W. T. Wang;甘肃银莲花;Gansu Baical Windflower,Kansu Anemone ■
23712 Anemone baicalensis Turcz. var. laevigata A. Garoy = Anemone flaccida F. Schmidt ■
23713 Anemone baicalensis Turcz. var. rossii (S. Moore) Kitag.;细茎银莲花;Ross Anemone,Ross Windflower ■
23714 Anemone baicalensis Turcz. var. saniculiformis (C. Y. Wu ex W. T. Wang) Ziman et B. E. Dutton;芹叶毛果银莲花(芹叶银莲花);Baoxing Windflower,Sanideleaf Anemone ■
23715 Anemone baldensis G. Don;巴尔多银莲花;Monte Baldo Anemone ■☆
23716 Anemone barbulata Turcz. = Anemone rivularis Buch. -Ham. var. flore-minore Maxim. ■
23717 Anemone batangensis Finet = Anemone rupicola Cambess. ■
23718 Anemone begoniifolia H. Lév. et Vaniot;卵叶银莲花(卵叶双瓶梅);Ovateleaf Anemone,Ovateleaf Windflower ■
23719 Anemone begoniifolioides W. T. Wang = Anemone howellii Jeffrey et W. W. Sm. ■
23720 Anemone berlandieri Pritz.;十蕊银莲花;Tenpetal Anemone ■☆
23721 Anemone bhutanica Tamura = Anemone rupestris Hook. f. et Thomson ■
23722 Anemone biarmiensis Juz.;二叠纪银莲花■☆
23723 Anemone bicolor H. Lév. = Anemone demissa Hook. f. et Thomson var. yunnanensis Franch. ■
23724 Anemone blanda Schott et Kotschy;淡色银莲花(光滑银莲花,希腊银莲花);Anemone,Flattering Anemone,Greek Anemone,Greek Thimbleweed,Windflower ■☆
23725 Anemone blanda Schott et Kotschy 'Atrocaerulea';深蓝淡色银莲花(深蓝光滑银莲花)■☆
23726 Anemone blanda Schott et Kotschy 'Ingramii';英格拉姆淡色银莲花(英格拉姆光滑银莲花)■☆
23727 Anemone blanda Schott et Kotschy 'Radar';雷达淡色银莲花(雷达光滑银莲花)■☆
23728 Anemone blanda Schott et Kotschy 'White Splendour';白辉淡色银莲花(白辉光滑银莲花)■☆
23729 Anemone bodinieri H. Lév. = Anemone begoniifolia H. Lév. et Vaniot ■
23730 Anemone boissiaei H. Lév. et Vaniot = Urophysa henryi (Oliv.) Ulbr. ■
23731 Anemone bonatiana H. Lév. = Anemone coelestina Franch. ■
23732 Anemone bonatiana H. Lév. var. geum H. Lév. = Anemone geum H. Lév. ■
23733 Anemone borealis Richardson = Anemone parviflora Michx. ■☆
23734 Anemone brevipedunculata Juz.;短梗银莲花■☆
23735 Anemone brevistyla C. C. Chang ex W. T. Wang;短柱银莲花;Brevistylar Windflower,Shortstyle Anemone ■
23736 Anemone bucharica Regel ex Finet et Gagnep.;布哈尔银莲花■☆
23737 Anemone caerulea DC. var. griffithii (Hook. f. et Thomson) Ulbr. = Anemone griffithii Hook. f. et Thomson ■
23738 Anemone caffra (Eckl. et Zeyh.) Harv.;开菲尔银莲花■☆
23739 Anemone calva Juz.;光秃银莲花■☆
23740 Anemone canadensis L.;加拿大银莲花;Canada Anemone,Canadian Anemone,Meadow Anemone,Round-leaved Anemone ■☆
23741 Anemone canadensis L. = Anemone dichotoma L. ■
23742 Anemone capensis (L.) Harv. var. tenuifolia Harv. = Anemone tenuifolia (L. f.) DC. ■☆
23743 Anemone capensis L. = Anemone tenuifolia (L. f.) DC. ■☆
23744 Anemone capensis Lam. = Anemone tenuifolia (L. f.) DC. ■☆
23745 Anemone caroliniana Walter;美国银莲花;Carolina Anemone,Prairie Anemone ■☆
23746 Anemone cathayensis Kitag.;银莲花;Cathayan Anemone,Cathayan Windflower ■
23747 Anemone cathayensis Kitag. f. hispida (Tamura) Kitag. = Anemone cathayensis Kitag. ■
23748 Anemone cathayensis Kitag. f. hispida (Tamura) Kitag. = Anemone cathayensis Kitag. var. hispida Tamura ■
23749 Anemone cathayensis Kitag. var. hispida Tamura;毛蕊银莲花;Hispid Cathayan Anemone ■
23750 Anemone caucasica Willd. ex Rupr.;高加索银莲花;Caucasia Anemone ■☆
23751 Anemone cernua Thunb. = Pulsatilla cernua (Thunb.) Bercht. et C. Presl ■
23752 Anemone cernua Thunb. ex A. Murray = Pulsatilla cernua (Thunb.) Bercht. et C. Presl ■
23753 Anemone cernua Thunb. ex A. Murray var. koreana Y. Yabe ex Nakai = Pulsatilla cernua (Thunb.) Bercht. et C. Presl ■
23754 Anemone cernua Thunb. var. koreana Y. Yabe ex Nakai = Pulsatilla cernua (Thunb.) Bercht. et C. Presl ■
23755 Anemone chinensis Bunge = Pulsatilla chinensis (Bunge) Regel ■
23756 Anemone chosenicola Ohwi;朝鲜银莲花■☆
23757 Anemone chosenicola Ohwi var. schantungensis (Hand. -Mazz.) Tamura = Anemone shikokiana (Makino) Makino ■
23758 Anemone chumulangmaensis W. T. Wang = Anemone trullifolia Hook. f. et Thomson ■
23759 Anemone coelestina Franch.;蓝匙叶银莲花;Blue Spoonleaf Anemone ■
23760 Anemone coelestina Franch. f. holophylla (Diels) Comber = Anemone coelestina Franch. ■
23761 Anemone coelestina Franch. f. holophylla (Diels) Comber = Anemone coelestina Franch. var. holophylla (Diels) Ziman et B. E. Dutton ■
23762 Anemone coelestina Franch. var. holophylla (Diels) Ziman et B. E. Dutton;拟条叶银莲花■
23763 Anemone coelestina Franch. var. linearis (Brühl) Ziman et B. E. Dutton;条叶银莲花;Linearleaf Spoonleaf Anemone ■
23764 Anemone coelestina Franch. var. polygyna Comber = Anemone coelestina Franch. ■
23765 Anemone coelestina Franch. var. truncata Comber = Anemone yulongshanica W. T. Wang var. truncata (Comber) W. T. Wang ■
23766 Anemone coerulea DC.;青色银莲花;Blue Anemone ■☆
23767 Anemone coronaria L.;冠状银莲花(罂粟草玉梅,罂粟牡丹,罂粟秋牡丹);Anemones,Crown Anemone,Garden Anemone,Lilies-of-

the-field, Nig, Nigger, Palestine Anemone, Plush Anemone, Poppy Anemone, Poppy-flowered Anemone, Robin Hood, Windflower, Windflowers ■☆

23768 Anemone coronaria L. var. alba Burnat = Anemone coronaria L. ■☆

23769 Anemone coronaria L. var. cyanea (Risso) Ardoino = Anemone coronaria L. ■☆

23770 Anemone coronaria L. var. nobilis (Jord.) Batt. = Anemone coronaria L. ■☆

23771 Anemone coronaria L. var. phoenicea Ardoino = Anemone coronaria L. ■☆

23772 Anemone coronaria L. var. rosea (Hanry) Batt. = Anemone coronaria L. ■☆

23773 Anemone crinita Juz.;长毛银莲花;Crinite Anemone, Crinite Windflower ■

23774 Anemone crinita Juz. = Anemone narcissiflora L. var. crinita (Juz.) Tamura ■

23775 Anemone crinita Juz. f. plena Y. Z. Zhao;重瓣长毛银莲花■

23776 Anemone cylindrica A. Gray;长头银莲花;Candle Anemone, Long-fruited Anemone, Long-headed Anemone, Slender-fruited Anemone, Thimbleweed ■☆

23777 Anemone cylindrica A. Gray var. alba Oakes = Anemone virginiana L. var. alba (Oakes) A. W. Wood ■☆

23778 Anemone dahurica Fisch. ex DC. = Pulsatilla dahurica (Fisch.) Spreng. ■

23779 Anemone davidii Franch.;西南银莲花(白接骨连,疗药,饿老虎,嘎嘎羊,海螺七,红接骨连,棉絮头,铜灯台,铜骨七,铜钱草,血乌,钻骨风);David Anemone, David Windflower ■

23780 Anemone debilis Fisch. ex Turcz.;柔软银莲花;Weak Anemone ■☆

23781 Anemone debilis Fisch. ex Turcz. var. yezoensis Makino = Anemone yezoensis Koidz. ■☆

23782 Anemone delavayi Franch.;滇川银莲花;Delavay Anemone, Delavay Windflower ■

23783 Anemone delavayi Franch. var. oligocarpa (C. P'ei) Ziman et B. E. Dutton;少果银莲花■

23784 Anemone deltoidea Hook.;哥伦比亚银莲花;Columbia Windflower, Western White Anemone ■☆

23785 Anemone demissa Hook. f. et Thomson;展毛银莲花(垂枝莲,高山雪莲花,高山银莲花,毛茛小将军);Patenthairy Anemone, Patenthairy Windflower ■

23786 Anemone demissa Hook. f. et Thomson subsp. villosissima (Brühl) R. P. Chaudhary = Anemone demissa Hook. f. et Thomson var. villosissima Brühl ■

23787 Anemone demissa Hook. f. et Thomson var. connectens Brühl = Anemone demissa Hook. f. et Thomson ■

23788 Anemone demissa Hook. f. et Thomson var. glabrescens Ulbr. = Anemone cathayensis Kitag. ■

23789 Anemone demissa Hook. f. et Thomson var. grandiflora C. Marquand et Airy Shaw = Anemone demissa Hook. f. et Thomson ■

23790 Anemone demissa Hook. f. et Thomson var. macrantha Brühl = Anemone demissa Hook. f. et Thomson var. major W. T. Wang ■

23791 Anemone demissa Hook. f. et Thomson var. major W. T. Wang;宽叶展毛银莲花(宽叶银莲花);Broadleaf Patenthairy Anemone ■

23792 Anemone demissa Hook. f. et Thomson var. monantha Brühl = Anemone demissa Hook. f. et Thomson ■

23793 Anemone demissa Hook. f. et Thomson var. villosa Ulbr. = Anemone demissa Hook. f. et Thomson var. villosissima Brühl ■

23794 Anemone demissa Hook. f. et Thomson var. villosissima Brühl;密毛银莲花;Densehairy Anemone ■

23795 Anemone demissa Hook. f. et Thomson var. yunnanensis Franch.;云南银莲花;Yunnan Anemone, Yunnan Windflower ■

23796 Anemone dichotoma L.;二歧银莲花(草玉梅,土黄芩);Dichotomous Anemone, Dichotomous Windflower ■

23797 Anemone dichotoma L. var. canadensis (L.) MacMill. = Anemone canadensis L. ■☆

23798 Anemone dichotoma Michx. = Anemone dichotoma L. ■

23799 Anemone discolor Royle = Anemone obtusiloba D. Don ■

23800 Anemone drummondii S. Watson;德拉蒙德银莲花;Drummond's Anemone ■☆

23801 Anemone dubia Bellardi = Anemone rivularis Buch.-Ham. ex DC. ■

23802 Anemone edwardsiana Tharp var. petraea Correll;岩地银莲花;Achenes Varnished, Edge Falls Anemone ■☆

23803 Anemone elegans Decne. = Anemone hybrida Vilm. ■☆

23804 Anemone elegans Decne. = Anemone vitifolia Buch.-Ham. ex DC. ■

23805 Anemone elegans Decne. var. tomentosa (Maxim.) Hand.-Mazz. = Anemone tomentosa (Maxim.) C. Pêi ■

23806 Anemone elongata D. Don;加长银莲花■

23807 Anemone eranthioides Regel;菟葵状银莲花(菟葵银莲花);Like Winteraconite Anemone ■☆

23808 Anemone erythrophylla Finet et Gagnep.;红叶银莲花;Redleaf Anemone, Rubrifolius Windflower ■

23809 Anemone esquirolii H. Lév. = Anemone begoniifolia H. Lév. et Vaniot ■

23810 Anemone esquirolii H. Lév. et Vaniot = Anemone rivularis Buch.-Ham. ex DC. ■

23811 Anemone exigua Maxim.;小银莲花(小双瓶梅);Small Anemone, Small Windflower ■

23812 Anemone exigua Maxim. var. shanxiensis B. L. Li et X. Y. Yu;山西银莲花;Shanxi Anemone ■

23813 Anemone extremiorientalis (Starod.) Starod. = Anemone umbrosa C. A. Mey. ■

23814 Anemone fanninii Harv. ex Mast.;范尼银莲花■☆

23815 Anemone fasciculata L.;簇生银莲花;Fascicled Anemone ■☆

23816 Anemone filisecta C. Y. Wu et W. T. Wang;细萼银莲花(细裂草玉梅,细裂银莲花);Threadsected Anemone ■

23817 Anemone flaccida F. Schmidt;鹅掌草(菜乌头,地雷,地乌,二轮七,二轻草,黑地雷,金串珠,林荫银莲花,蜈蚣三七);Flaccid Anemone, Goosepalm Windflower ■

23818 Anemone flaccida F. Schmidt var. anhuiensis (Y. K. Yang et al.) Ziman et B. E. Dutton;安徽鹅掌草(安徽银莲花)■

23819 Anemone flaccida F. Schmidt var. hirtella W. T. Wang;展毛鹅掌草;Hirtellous Anemone, Hirtellous Windflower ■

23820 Anemone flaccida F. Schmidt var. hofengensis Wuzhi;裂苞鹅掌草(鹤峰银莲花);Hefeng Anemone, Hefeng Windflower, Hofeng Anemone ■

23821 Anemone flaccida F. Schmidt var. semiplena Makino;半重瓣鹅掌草■☆

23822 Anemone flavescens Zucc. = Pulsatilla patens (L.) Mill. subsp. flavescens (Zucc.) Zämelis ■

23823 Anemone fulgens J. Gay;光亮鹅掌草;Flame Anemone, Scarlet Windflower ■☆

23824 Anemone gelida Maxim. = Anemone rupestris Hook. f. et Thomson subsp. gelida (Maxim.) Lauener ■

23825 Anemone geum H. Lév.;路边青银莲花■

23826 Anemone geum H. Lév. subsp. ovalifolia (Brühl) R. P. Chaudhary;疏齿路边青银莲花(疏齿银莲花)■

23827 Anemone glabrata (Maxim.) Juz.;光银莲花■☆

23828 Anemone glabrata (Maxim.) Juz. = Anemone baicalensis Turcz. var. glabrata Maxim. ■

23829 Anemone glaucifolia Franch. = Anemoclema glaucifolium (Franch.) W. T. Wang ■

23830 Anemone globosa Nutt. ex A. Nelson = Anemone multifida Poir. ■☆

23831 Anemone globosa Nutt. ex Pritz. = Anemone multifida Poir. ■☆

23832 Anemone gortschakovii Kar. et Kir.;块茎银莲花;Gortschakov Anemone, Gortschakov Windflower ■

23833 Anemone grayi Behr et Kellogg;格雷银莲花;Western Windflower ■☆

23834 Anemone griffithii Hook. f. et Thomson;三出银莲花;Griffith Anemone, Griffith Windflower ■

23835 Anemone groenlandica Oeder = Coptis trifolia (L.) Salisb. ■☆

23836 Anemone halleri All.;哈氏银莲花■☆

23837 Anemone henryi Oliv. = Hepatica henryi (Oliv.) Steward ■

23838 Anemone hepatica L. = Anemone americana (DC.) H. Hara ■☆

23839 Anemone hepatica L. = Hepatica nobilis Garsault f. obtusa (Pursh) Beck ■☆

23840 Anemone hepatica L. = Hepatica nobilis Garsault var. asiatica (Nakai) H. Hara ■

23841 Anemone hepatica L. = Hepatica nobilis Garsault var. nipponica Nakai ■☆

23842 Anemone hepatica L. = Hepatica nobilis Garsault ■☆

23843 Anemone heterophylla Nutt. ex A. W. Wood = Anemone berlandieri Pritz. ■☆

23844 Anemone hofengensis W. T. Wang = Anemone flaccida F. Schmidt var. hofengensis Wuzhi ■

23845 Anemone hokouensis C. Y. Wu ex W. T. Wang;河口银莲花(河口双瓶梅);Hekou Anemone, Hekou Windflower, Hokou Anemone ■

23846 Anemone hortensis L.;宽叶秋牡丹;Anemone, Broad-leaved Anemone, Garden Anemone, Rose Parsley ■☆

23847 Anemone hortensis Thore = Anemone fulgens J. Gay ■☆

23848 Anemone howellii Jeffrey et W. W. Sm.;拟卵叶银莲花(黄升麻,腾冲秋牡丹);Howell Anemone, Howell Windflower ■

23849 Anemone hudsoniana (DC.) Richardson = Anemone multifida Poir. ■☆

23850 Anemone hupehensis (Lemoine) Lemoine;打破碗花花(霸王草,遍地爬,大草花,大头翁,盖头花,拐角七,湖北秋牡丹,花升麻,火草花,老鼠棉花衣,绿升麻,满天飞,满天星,青水胆,秋牡丹,山棉花,水棉花,铁线海棠,五雷火,五匹草,野棉花,一把抓);Hubei Anemone, Hubei Windflower, Hupeh Anemone, Japanese Anemone, Japanese Tea Party, Japanese Thimbleweed ■

23851 Anemone hupehensis (Lemoine) Lemoine 'Hadspen Abundance';丰花秋牡丹■☆

23852 Anemone hupehensis (Lemoine) Lemoine 'September Charm';九月媚秋牡丹■☆

23853 Anemone hupehensis (Lemoine) Lemoine f. alba W. T. Wang;水棉花(白背湖北银莲花,白花秋牡丹,花升麻,绿升麻,满天星,野棉花,一扫光);White Hubei Anemone, White Hubei Windflower ■

23854 Anemone hupehensis (Lemoine) Lemoine var. japonica (Thunb. ex A. Murray) Bowler et Stern;秋牡丹(吹牡丹,秋芍药,日本银莲花,土牡丹,压竹花,野棉花);Autumn Hubei Anemone, Autumn Hupeh Anemone, Japan Anemone, Japan Windflower, Japanese Anemone ■

23855 Anemone hupehensis (Lemoine) Lemoine var. simplicifolia W. T. Wang = Anemone hupehensis (Lemoine) Lemoine f. alba W. T. Wang ■

23856 Anemone hybrida Vilm.;杂种银莲花;Hybrid Anemone, Japanese Anemone ■☆

23857 Anemone hybrida Vilm. 'Margarete';日本杂种银莲花;Japanese Garden Anemone ■☆

23858 Anemone imbricata Maxim.;叠裂银莲花;Imbricate Anemone, Imbricate Windflower ■

23859 Anemone integrifolia Humb. et Bonpl;全缘叶银莲花■☆

23860 Anemone japonica (Thunb.) Siebold et Zucc. = Anemone hupehensis (Lemoine) Lemoine var. japonica (Thunb. ex A. Murray) Bowler et Stern ■

23861 Anemone japonica (Thunb.) Siebold et Zucc. = Anemone hybrida Vilm. ■☆

23862 Anemone japonica Houtt. = Clematis florida Thunb. ●

23863 Anemone japonica Thunb. ex A. Murray = Anemone hupehensis (Lemoine) Lemoine var. japonica (Thunb. ex A. Murray) Bowler et Stern ■

23864 Anemone japonica Thunb. ex A. Murray var. hupehensis Lemoine = Anemone hupehensis (Lemoine) Lemoine ■

23865 Anemone japonica Thunb. ex A. Murray var. tomentosa Maxim. = Anemone tomentosa (Maxim.) C. Pĕi ■

23866 Anemone jenisseensis (Korsh.) Krylov;热尼斯银莲花■☆

23867 Anemone kansuensis W. T. Wang = Anemone baicalensis Turcz. var. kansuensis (W. T. Wang) W. T. Wang ■

23868 Anemone keiskeana Ito;伊藤银莲花■☆

23869 Anemone kostyczewii Korsh. = Pulsatilla kostyczewii (Korsh.) Juz. ■

23870 Anemone kuznetzowii Woronow ex Grossh.;库兹银莲花■☆

23871 Anemone laceratoincisa W. T. Wang;锐裂银莲花;Incised Windflower, Lacerate Anemone ■

23872 Anemone laevigata (Gray) Koidz. = Anemone flaccida F. Schmidt ■

23873 Anemone lancifolia Pursh;剑叶银莲花;Lance-leaved Anemone, Mountain Anemone ■☆

23874 Anemone laxa (Ulbr.) Juz.;疏枝银莲花■☆

23875 Anemone leveillei Ulbr. = Anemone rivularis Buch.-Ham. ex DC. ■

23876 Anemone liangshanica W. T. Wang = Anemone trullifolia Hook. f. et Thomson var. liangshanica (W. T. Wang) Ziman et B. E. Dutton ■

23877 Anemone litoralis (Litv.) Juz.;滨海银莲花■☆

23878 Anemone longipes Tamura = Anemone rivularis Buch.-Ham. ex DC. ■

23879 Anemone ludoviciana Nutt. = Anemone patens L. var. multifida Pritz. ■

23880 Anemone ludoviciana Nutt. = Pulsatilla patens (L.) Mill. subsp. multifida (Pritz.) Zämelis ■

23881 Anemone lutienensis W. T. Wang = Anemone trullifolia Hook. f. et Thomson var. lutienensis (W. T. Wang) Ziman et B. E. Dutton ■

23882 Anemone lyallii Britton;北美野银莲花;Little Mountain Anemone, Western Wood Anemone ■☆

23883 Anemone mairei H. Lév. = Pulsatilla millefolium (Hemsl. et E. H. Wilson) Ulbr. ■

23884 Anemone matsudae (Yamam.) Tamura = Anemone hupehensis

(Lemoine) Lemoine f. alba W. T. Wang ■

23885 Anemone matsudae (Yamam.) Tamura = Anemone vitifolia Buch. -Ham. ex DC. ■

23886 Anemone micrantha Klotzsch = Anemone obtusiloba D. Don ■

23887 Anemone millefolium Hemsl. et E. H. Wilson = Pulsatilla millefolium (Hemsl. et E. H. Wilson) Ulbr. ■

23888 Anemone minima DC. = Anemone quinquefolia L. var. minima (DC.) Frodin ■☆

23889 Anemone montana Hope;山地银莲花■☆

23890 Anemone multiceps (Greene) Standl.;紫色银莲花; Purple Anemone ■☆

23891 Anemone multifida Poir.;多裂银莲花; Cut-leaved Anemone, Early Anemone, Pacific Anemone, Red Anemone, Windflower ■☆

23892 Anemone multifida Poir. var. hirsuta C. L. Hitchc. = Anemone multifida Poir. var. saxicola B. Boivin ■☆

23893 Anemone multifida Poir. var. hudsoniana DC. = Anemone multifida Poir. ■☆

23894 Anemone multifida Poir. var. nowasadii B. Boivin = Anemone multifida Poir. ■☆

23895 Anemone multifida Poir. var. richardsiana Fernald = Anemone multifida Poir. ■☆

23896 Anemone multifida Poir. var. sansonii B. Boivin = Anemone multifida Poir. ■☆

23897 Anemone multifida Poir. var. saxicola B. Boivin;岩生多裂银莲花■☆

23898 Anemone multifida Poir. var. stylosa (A. Nelson) B. E. Dutton et Keener;多柱多裂银莲花■☆

23899 Anemone multilobulata W. T. Wang et L. Q. Li = Anemone narcissiflora L. var. protracta Ulbr. ■

23900 Anemone nanchuanensis W. T. Wang = Anemone griffithii Hook. f. et Thomson ■

23901 Anemone narcissiflora L.;水仙银莲花(银莲花); Narcissus Anemone, Narcissus-flowered Anemone ■☆

23902 Anemone narcissiflora L. subsp. alaskana Hultén = Anemone narcissiflora L. var. monantha DC. ■☆

23903 Anemone narcissiflora L. subsp. chinensis (Kitag.) Kitag. = Anemone cathayensis Kitag. ■

23904 Anemone narcissiflora L. subsp. crinita (Juz.) Kitag. = Anemone narcissiflora L. var. crinita (Juz.) Tamura ■

23905 Anemone narcissiflora L. subsp. interior Hultén = Anemone narcissiflora L. var. monantha DC. ■☆

23906 Anemone narcissiflora L. subsp. protracta (Ulbr.) Ziman et Fedor.;伏毛银莲花(多裂银莲花); Multifid Anemone, Multifid Windflower, Prostratehair Windflower, Protracted Anemone ■

23907 Anemone narcissiflora L. subsp. sibirica (L.) Hultén = Anemone narcissiflora L. var. monantha DC. ■☆

23908 Anemone narcissiflora L. subsp. sibirica (L.) Hultén = Anemone narcissiflora L. ■☆

23909 Anemone narcissiflora L. var. chinensis Kitag. = Anemone cathayensis Kitag. ■

23910 Anemone narcissiflora L. var. contracta (Ulbr.) Schipcz. = Anemone narcissiflora L. subsp. protracta (Ulbr.) Ziman et Fedor. ■

23911 Anemone narcissiflora L. var. crinita (Juz.) Tamura = Anemone crinita Juz. ■

23912 Anemone narcissiflora L. var. demissa (Hook. f. et Thomson) Finet et Gagnep. = Anemone demissa Hook. f. et Thomson ■

23913 Anemone narcissiflora L. var. monantha DC.;山生银莲花■☆

23914 Anemone narcissiflora L. var. pekinensis Schipcz. = Anemone cathayensis Kitag. ■

23915 Anemone narcissiflora L. var. polyanthes ? = Anemone polyanthes D. Don ■

23916 Anemone narcissiflora L. var. protracta Ulbr. = Anemone narcissiflora L. subsp. protracta (Ulbr.) Ziman et Fedor. ■

23917 Anemone narcissiflora L. var. sachalinensis Miyabe et Miyake;库页银莲花; Anemone, Sachalin Windflower ■☆

23918 Anemone narcissiflora L. var. shikokiana Makano = Anemone shikokiana (Makino) Makino ■

23919 Anemone narcissiflora L. var. sibirica (L.) Tamura;卵裂银莲花; Siberia Windflower, Siberian Anemone ■

23920 Anemone narcissiflora L. var. sibirica (L.) Tamura = Anemone narcissiflora L. var. crinita (Juz.) Tamura ■

23921 Anemone narcissiflora L. var. turkestanica Schipcz.;天山银莲花; Tianshan Windflower, Turkestan Anemone ■

23922 Anemone narcissiflora L. var. turkestanica Schipcz. = Anemone narcissiflora L. subsp. protracta (Ulbr.) Ziman et Fedor. ■

23923 Anemone narcissiflora L. var. uniflora Eastw. = Anemone narcissiflora L. var. monantha DC. ■☆

23924 Anemone narcissiflora L. var. yuldussica Schipcz. = Anemone narcissiflora L. subsp. protracta (Ulbr.) Ziman et Fedor. ■

23925 Anemone nemorosa L.;丛林银莲花(林荫银莲花,荫蔽银莲花,隐蔽银莲花); Bow Bells, Bread-and-cheese-and-cider, Candlemas Bells, Candlemas Cap, Candlemas Caps, Chimney Smock, Cowslip, Cuckoo, Cuckoo Spit, Cuckoo-flower, Cuckoo-spit, Drops of Snow, Drops-of-snow, Easter Flower, Enem, Enemy Anenemy, European Thimbleweed, European Wood Anemone, European Wood-anemone, Evening Twilight, Fairy's Wind Flower, Fairy's Windflower, Flaw-flower, Granny's Nightcap, Granny-thread-the-needle, Jack-o'-lantern, Jack-o'-lanterns, Lady's Chemise, Lady's Milkcans, Lady's Nightcap, Lady's Petticoats, Lady's Purse, Lady's Shimmey, Lady's Shimmy, Lady's Smock, Milkmaids, Moll-o'-the-woods, Moon Flower, Nancy, Nedcushion, Neminy, Shame-faced Maiden, Shoes-and-slippers, Silver Bells, Small Smock, Smeli-foxes, Smell Foxes, Snake's Flower, Snakes-and-adders, Soldier Buttons, Soldiers, Soldier's Buttons, Star of Bethlehem, White Soldiers, Wild Jasmine, Wild Jessamine, Wind Flower, Wind Plant, Wood Anemone, Wood Crowfoot, Woods Anemone, Woolly Head, Woolly Heads, Zephyr-flower ■☆

23926 Anemone nemorosa L. 'Allenii';阿伦林荫银莲花■☆

23927 Anemone nemorosa L. 'Robinsoniana';罗宾逊林荫银莲花■☆

23928 Anemone nemorosa L. 'Vestal';贞洁林荫银莲花■☆

23929 Anemone nemorosa L. 'Wilk's Giant';威尔克大花林荫银莲花■☆

23930 Anemone nemorosa L. = Anemone amurensis (Korsh.) Kom. ■

23931 Anemone nemorosa L. subsp. altaica (Fisch. ex C. A. Mey.) Korsh. = Anemone altaica Fisch. ex Ledeb. ■

23932 Anemone nemorosa L. subsp. amurensis Korsh. = Anemone amurensis (Korsh.) Kom. ■

23933 Anemone nemorosa L. var. bifolia (Farw.) B. Boivin = Anemone quinquefolia L. ■☆

23934 Anemone nemorosa L. var. fissa Ulbr. = Anemone amurensis (Korsh.) Kom. ■

23935 Anemone nemorosa L. var. quinquefolia (L.) Pursh = Anemone quinquefolia L. ■☆

23936 Anemone nikoensis Maxim.;日光银莲花■☆

23937 Anemone nutantiglora W. T. Wang et L. Q. Li;垂花银莲花;Nutantflower Anemone,Nutantflower Windflower ■

23938 Anemone nuttalliana DC. = Pulsatilla patens (L.) Mill. subsp. multifida (Pritz.) Zämelis ■

23939 Anemone obtusiloba D. Don;钝裂银莲花;Obtuselobed Anemone,Obtuselobed Windflower ■

23940 Anemone obtusiloba D. Don subsp. coelestina (Franch.) Brühl = Anemone coelestina Franch. ■

23941 Anemone obtusiloba D. Don subsp. geum (H. Lév.) Ulbr. = Anemone geum H. Lév. ■

23942 Anemone obtusiloba D. Don subsp. geum (H. Lév.) Ulbr. var. violacea Ulbr. = Anemone imbricata Maxim. ■

23943 Anemone obtusiloba D. Don subsp. imbricata (Maxim.) Brühl = Anemone imbricata Maxim. ■

23944 Anemone obtusiloba D. Don subsp. leiophylla W. T. Wang;光叶银莲花;Brightleaf Anemone ■

23945 Anemone obtusiloba D. Don subsp. megaphylla W. T. Wang;镇康银莲花;Largeleaf Anemone,Zhenkang Windflower ■

23946 Anemone obtusiloba D. Don subsp. micrantha (Klotzsch) Ulbr. = Anemone obtusiloba D. Don ■

23947 Anemone obtusiloba D. Don subsp. omalocarpella Brühl = Anemone polycarpa W. E. Evans ■

23948 Anemone obtusiloba D. Don subsp. ovalifolia Brühl;疏齿银莲花(卵叶银莲花,毛叶千紫花);Ovalleaf Anemone ■

23949 Anemone obtusiloba D. Don subsp. ovalifolia Brühl = Anemone geum H. Lév. subsp. ovalifolia (Brühl) R. P. Chaudhary ■

23950 Anemone obtusiloba D. Don subsp. ovalifolia Brühl var. angustilimba W. T. Wang = Anemone geum H. Lév. subsp. ovalifolia (Brühl) R. P. Chaudhary ■

23951 Anemone obtusiloba D. Don subsp. ovalifolia Brühl var. geochares Brühl = Anemone geum H. Lév. ■

23952 Anemone obtusiloba D. Don subsp. ovalifolia Brühl var. orthocaulon Brühl = Anemone geum H. Lév. subsp. ovalifolia (Brühl) R. P. Chaudhary ■

23953 Anemone obtusiloba D. Don subsp. ovalifolia Brühl var. polysepala W. T. Wang = Anemone geum H. Lév. subsp. ovalifolia (Brühl) R. P. Chaudhary ■

23954 Anemone obtusiloba D. Don subsp. ovalifolia Brühl var. rothocaulon Brühl = Anemone geum H. Lév. subsp. ovalifolia (Brühl) R. P. Chaudhary ■

23955 Anemone obtusiloba D. Don subsp. ovalifolia Brühl var. truncata (H. F. Comber) W. T. Wang = Anemone yulongshanica W. T. Wang var. truncata (H. F. Comber) W. T. Wang ■

23956 Anemone obtusiloba D. Don subsp. saxicola Brühl = Anemone polycarpa W. E. Evans ■

23957 Anemone obtusiloba D. Don subsp. saxicola Brühl = Anemone rupestris Hook. f. et Thomson subsp. polycarpa (W. E. Evans) W. T. Wang ■

23958 Anemone obtusiloba D. Don subsp. trullifolia (Hook. f. et Thomson) Brühl = Anemone trullifolia Hook. f. et Thomson ■

23959 Anemone obtusiloba D. Don subsp. trullifolia (Hook. f. et Thomson) Brühl var. linearis Brühl = Anemone coelestina Franch. var. linearis (Brühl) Ziman et B. E. Dutton ■

23960 Anemone obtusiloba D. Don var. angustilimba W. T. Wang;狭叶银莲花■

23961 Anemone obtusiloba D. Don var. chrysantha Ulbr. = Anemone obtusiloba D. Don ■

23962 Anemone obtusiloba D. Don var. coerulea Ulbr.;天蓝钝裂银莲花■

23963 Anemone obtusiloba D. Don var. coerulea Ulbr. = Anemone rupestris Hook. f. et Thomson ■

23964 Anemone obtusiloba D. Don var. polysepala W. T. Wang;维西银莲花■

23965 Anemone obtusiloba D. Don var. pusilla Brühl = Anemone rupestris Hook. f. et Thomson ■

23966 Anemone obtusiloba D. Don var. spatulata Brühl = Anemone trullifolia Hook. f. et Thomson ■

23967 Anemone obtusiloba D. Don var. truncata (Comber) W. T. Wang;截基钝裂银莲花■

23968 Anemone obtusiloba D. Don var. wallichii Brühl = Anemone rupestris Hook. f. et Thomson ■

23969 Anemone occidentalis S. Watson;西方山银莲花;Mountain Pasqueflower,Pulsatille,Western Pasqueflower ■☆

23970 Anemone occidentalis S. Watson var. subpilosa Hardin = Anemone occidentalis S. Watson ■☆

23971 Anemone ochotensis Fisch.;西方银莲花;Western Pasque Flower ■☆

23972 Anemone okennonii Keener et B. E. Dutton;奥氏银莲花;Okennon's Anemone ■☆

23973 Anemone oligantha Eastw. = Anemone lyallii Britton ■☆

23974 Anemone oligocarpa H. P'ei = Anemone delavayi Franch. var. oligocarpa (C. P'ei) Ziman et B. E. Dutton ■

23975 Anemone oligotoma Juz.;俄勒冈蓝银莲花;Blue Anemone ■☆

23976 Anemone oregana A. Gray;俄勒冈银莲花;Oregon Anemone,Western Wood Anemone ■☆

23977 Anemone orthocarpa Hand.-Mazz.;直果银莲花;Guizhou Windflower,Straightfruit Anemone ■

23978 Anemone ovalifolia (Brühl) Hand.-Mazz. = Anemone geum H. Lév. subsp. ovalifolia (Brühl) R. P. Chaudhary ■

23979 Anemone palmata L.;掌状银莲花;American Pasqueflower,Prairie-smoke ■☆

23980 Anemone parviflora Michx.;小花银莲花;Northern Anemone,Small-flowered Anemone ■☆

23981 Anemone parviflora Michx. var. grandiflora Ulbr. = Anemone parviflora Michx. ■☆

23982 Anemone patens L. = Pulsatilla patens (L.) Mill. ■

23983 Anemone patens L. = Pulsatilla vulgaris Mill. ■☆

23984 Anemone patens L. var. multifida Pritz. = Pulsatilla patens (L.) Mill. var. multifida (Pritz.) S. H. Li et Y. Huei Huang ■

23985 Anemone patens L. var. multifida Pritz. = Pulsatilla patens (L.) Mill. subsp. multifida (Pritz.) Zämelis ■

23986 Anemone patens L. var. nuttalliana (DC.) A. Gray = Anemone patens L. var. multifida Pritz. ■

23987 Anemone patens L. var. nuttalliana (DC.) A. Gray = Pulsatilla patens (L.) Mill. subsp. multifida (Pritz.) Zämelis ■

23988 Anemone patens L. var. wolfgangiana (Besser) Koch = Anemone patens L. var. multifida Pritz. ■

23989 Anemone patens L. var. wolfgangiana (Besser) Koch = Pulsatilla patens (L.) Mill. subsp. multifida (Pritz.) Zämelis ■

23990 Anemone patula C. C. Chang ex W. T. Wang;天全银莲花;Patulous Anemone,Tianquan Windflower ■

23991 Anemone patula C. C. Chang ex W. T. Wang var. minor W. T. Wang;鸡足叶银莲花;Small Patulous Anemone ■

23992 Anemone pavonina Lam.;孔雀银莲花;Peacock Anemone ■☆

23993 Anemone pavonina Lam. var. fulgens DC. = Anemone fulgens J. Gay ■☆

23994 Anemone petiolulata H. P'ei = Anemone davidii Franch. ■

23995 Anemone piperi Britton ex Rydb.;皮氏银莲花;Piper's Anemone ■☆

23996 Anemone polyanthes D. Don = Anemone demissa Hook. f. et Thomson ■

23997 Anemone polycarpa W. E. Evans = Anemone rupestris Hook. f. et Thomson subsp. polycarpa (W. E. Evans) W. T. Wang ■

23998 Anemone pratensis L.;草原银莲花(草甸白头翁);Meadow Pasqueflower, Meadow Pulsatilla, Small Pasqueflower, Small Pasque-flower ■☆

23999 Anemone prattii Huth ex Ulbr.;川西银莲花;Pratt Anemone, Pratt Windflower ■

24000 Anemone protracta (Ulbr.) Juz. = Anemone narcissiflora L. subsp. protracta (Ulbr.) Ziman et Fedor. ■

24001 Anemone pulsatilla L. = Anemone patens L. ■

24002 Anemone pulsatilla L. = Pulsatilla vulgaris Mill. ■☆

24003 Anemone pulsatilla L. var. chinensis (Bunge) Finet et Gagnep. = Pulsatilla chinensis (Bunge) Regel ■

24004 Anemone quinquefolia L.;木银莲花;Nightcaps, Wind Flower, Windflower, Wood Anemone ■☆

24005 Anemone quinquefolia L. var. bifolia Farw. = Anemone quinquefolia L. ■☆

24006 Anemone quinquefolia L. var. grayi (Behr et Kellogg) Jeps. = Anemone grayi Behr et Kellogg ■☆

24007 Anemone quinquefolia L. var. interior Fernald = Anemone quinquefolia L. ■☆

24008 Anemone quinquefolia L. var. lancifolia (Pursh) Fosberg = Anemone lancifolia Pursh ■☆

24009 Anemone quinquefolia L. var. lyallii (Britton) B. L. Rob. = Anemone lyallii Britton ■☆

24010 Anemone quinquefolia L. var. minima (DC.) Frodin;小木银莲花■☆

24011 Anemone quinquefolia L. var. minor (Eastw.) Munz = Anemone grayi Behr et Kellogg ■☆

24012 Anemone quinquefolia L. var. oregana (A. Gray) B. L. Rob. = Anemone oregana A. Gray ■☆

24013 Anemone raddeana Regel;多被银莲花(草乌喙,关东银莲花,红背银莲花,红被银莲花,老鼠屎,两头尖,竹节香附);Radde Anemone, Radde Windflower ■

24014 Anemone raddeana Regel subsp. glabra Ulbr. = Anemone raddeana Regel ■

24015 Anemone raddeana Regel subsp. villosa Ulbr. = Anemone raddeana Regel ■

24016 Anemone raddeana Regel var. integra Huth = Anemone raddeana Regel ■

24017 Anemone raddeana Regel var. lacerata Y. L. Xu;龙王山银莲花;Lacerate Radde Anemone ■

24018 Anemone ranunculoides L.;类莨银莲花(毛莨状银莲花);Yellow Pmemone, Yellow Wood Anemone, Yellow Wood-anemone ■☆

24019 Anemone reflexa Stephan;反萼银莲花;Refractedcalyx Anemone, Refractedcalyx Windflower ■

24020 Anemone richardsonii Hook.;理查银莲花;Yellow Anemone ■☆

24021 Anemone riparia Fernald;大白花银莲花;Large White-flowered Anemone, Riverbank Anemone, Thimbleweed ■☆

24022 Anemone riparia Fernald = Anemone virginiana L. var. alba (Oakes) A. W. Wood ■☆

24023 Anemone riparia Fernald f. inconspicua Fernald = Anemone virginiana L. var. alba (Oakes) A. W. Wood ■☆

24024 Anemone rivularis Buch. -Ham. ex DC.;草玉梅(白花虎掌草,白花舌头草,大狗脚迹,淡虎掌,狗脚迹,鬼打青,汉虎掌,虎掌草,见风黄,见风蓝,见风青,见风清,密马常,破旧草,水乌头,四大天王,土黄芪,土黄芩,乌骨鸡,五倍叶,五朵云,溪畔银莲花,小绿升麻,羊九,野鸡菜,野棉花);Brooklet Anemone, Brooklet Windflower ■

24025 Anemone rivularis Buch. -Ham. ex DC. subsp. barbulata Ulbr. = Anemone rivularis Buch. -Ham. var. flore-minore Maxim. ■

24026 Anemone rivularis Buch. -Ham. ex DC. var. barbulata Turcz. ex B. Fedtsch. = Anemone rivularis Buch. -Ham. var. flore-minore Maxim. ■

24027 Anemone rivularis Buch. -Ham. subsp. eurivularis Ulbr. = Anemone rivularis Buch. -Ham. ex DC. ■

24028 Anemone rivularis Buch. -Ham. var. flore-minore Maxim.;小花草玉梅(白头翁,河岸银莲花,破牛膝,山辣椒);Smallflower Brooklet Anemone, Smallflower Brooklet Windflower ■

24029 Anemone rivularis Ulbr. = Anemone rivularis Buch. -Ham. ex DC. ■

24030 Anemone robusta W. T. Wang;粗壮银莲花;Robust Windflower, Thick Anemone ■

24031 Anemone robustostylosa R. H. Miao;粗柱银莲花;Robust-styled Anemone ■

24032 Anemone rockii Ulbr.;岷山银莲花;Rock Anemone, Rock Windflower ■

24033 Anemone rockii Ulbr. var. multicaulis W. T. Wang;多茎银莲花;Manystem Anemone, Manystem Windflower ■

24034 Anemone rockii Ulbr. var. pilocarpa W. T. Wang;巫溪银莲花;Pilosefruit Anemone, Wuxi Windflower ■

24035 Anemone rossii S. Moore = Anemone baicalensis Turcz. var. rossii (S. Moore) Kitag. ■

24036 Anemone rupestris Hook. f. et Thomson;湿地银莲花;Wetland Anemone, Wetland Windflower ■

24037 Anemone rupestris Hook. f. et Thomson subsp. gelida (Maxim.) Lauener var. wallichii (Brühl) Lauener = Anemone rupestris Hook. f. et Thomson var. wallichii (Brühl) Lauener ■

24038 Anemone rupestris Hook. f. et Thomson subsp. gelida (Maxim.) Lauener;冻地银莲花;Icecold Anemone ■

24039 Anemone rupestris Hook. f. et Thomson subsp. polycarpa (W. E. Evans) W. T. Wang;多果银莲花;Manyfruit Anemone ■

24040 Anemone rupestris Hook. f. et Thomson subsp. polycarpa (W. E. Evans) W. T. Wang = Anemone polycarpa W. E. Evans ■

24041 Anemone rupestris Hook. f. et Thomson var. lobata Brühl = Anemone geum H. Lév. ■

24042 Anemone rupestris Hook. f. et Thomson var. pilosa Marquart et Shaw = Anemone geum H. Lév. ■

24043 Anemone rupestris Hook. f. et Thomson var. pusilla ? = Anemone rupestris Hook. f. et Thomson ■

24044 Anemone rupestris Hook. f. et Thomson var. wallichii (Brühl) Lauener;低矮银莲花■

24045 Anemone rupestris Hook. f. et Thomson var. wallichii (Brühl) Lauener = Anemone rupestris Hook. f. et Thomson ■

24046 Anemone rupicola Cambess.;岩生银莲花(石龙芮,岩秋牡丹);Cliff Anemone, Cliff Windflower ■

24047 Anemone rupicola Cambess. subsp. laceratoincisa (W. T. Wang)

R. P. Chaudhary = Anemone laceratoincisa W. T. Wang ■
24048 Anemone sachalinensis (Miyabe et Miyake) Juz. = Anemone narcissiflora L. var. sachalinensis Miyabe et Miyake ■☆
24049 Anemone saniculifolia H. Lév. = Anemone rivularis Buch. -Ham. ex DC. ■
24050 Anemone saniculiformis C. Y. Wu ex W. T. Wang = Anemone baicalensis Turcz. var. saniculiformis (C. Y. Wu ex W. T. Wang) Ziman et B. E. Dutton ■
24051 Anemone saniculiformis C. Y. Wu ex W. T. Wang = Anemone rivularis Buch. -Ham. ex DC. ■
24052 Anemone saxicola Tamura et Kitam. = Anemone rupestris Hook. f. et Thomson subsp. polycarpa (W. E. Evans) W. T. Wang ■
24053 Anemone scabiosa H. Lév. et Vaniot = Anemone hupehensis (Lemoine) Lemoine var. japonica (Thunb. ex A. Murray) Bowler et Stern ■
24054 Anemone scabriuscula W. T. Wang;糙叶银莲花;Scabridulous Windflower,Scabrousleaf Anemone ■
24055 Anemone scaposa Edgew. = Anemone polyanthes D. Don ■
24056 Anemone schantungensis Hand. -Mazz. = Anemone shikokiana (Makino) Makino ■
24057 Anemone schrenkiana Juz. = Anemone narcissiflora L. subsp. protracta (Ulbr.) Ziman et Fedor. ■
24058 Anemone shikokiana (Makino) Makino;山东银莲花;Shandong Anemone,Shandong Windflower ■
24059 Anemone sibirica L. = Anemone narcissiflora L. var. crinita (Juz.) Tamura ■
24060 Anemone sibirica L. = Anemone narcissiflora L. var. sibirica (L.) Tamura ■
24061 Anemone silvestris L.;大花银莲花(林生银莲花);Snowdrop Anemone,Snowdrop Windflower ■
24062 Anemone siuzevi Kom. = Anemone stolonifera Maxim. ■
24063 Anemone smithiana Lauener et Panigrahi;红萼银莲花;Smith Anemone,Smith Windflower ■
24064 Anemone somaliensis Hepper;索马里银莲花■☆
24065 Anemone speciosa Adams;美丽银莲花■☆
24066 Anemone stolonifera Maxim.;匍枝银莲花(三花白头翁,三花银莲花);Stolonbearing Anemone,Stolonbearing Windflower ■
24067 Anemone stolonifera Maxim. var. davidii (Franch.) Finet et Gagnep. = Anemone davidii Franch. ■
24068 Anemone stylosa A. Nelson = Anemone multifida Poir. var. stylosa (A. Nelson) B. E. Dutton et Keener ■☆
24069 Anemone subindivisa W. T. Wang;微裂银莲花;Slitlydivided Anemone,Slitlydivided Windflower ■
24070 Anemone subpinnata W. T. Wang;近羽裂银莲花;Pinnatipartite Windflower,Subpinnate Anemone ■
24071 Anemone sylvestris L.;森林银莲花(大花银莲花,雪花银莲花);Snowdrop Anemone, Snowdrop Wind Flower, Snowdrop Windflower,Snowdrop-anemone,Snowdrop-windflower ■☆
24072 Anemone sylvestris L. 'Macrantha';大花森林银莲花■☆
24073 Anemone taipaiensis W. T. Wang;太白银莲花;Taibai Anemone, Taibai Windflower,Taipai Anemone ■
24074 Anemone takasagomontana Masam.;台湾银莲花(台湾山附子,台湾双瓶梅);Taiwan Anemone,Taiwan Windflower ■
24075 Anemone takasagomontana Masam. = Anemone exigua Maxim. ■
24076 Anemone taraoi Takeda var. morii Yamam. = Ranunculus morii (Yamam.) Ohwi ■
24077 Anemone tengchongensis W. T. Wang;腾冲银莲花;Tengchong Anemone,Tengchong Windflower ■
24078 Anemone tengchongensis W. T. Wang = Anemone crinita Juz. ■
24079 Anemone tengchongensis W. T. Wang = Anemone narcissiflora L. var. crinita (Juz.) Tamura ■
24080 Anemone tenuifolia (L. f.) DC.;细叶银莲花■☆
24081 Anemone tenuiloba Hayek = Pulsatilla tenuiloba (Hayek) Juz. ■
24082 Anemone tetrasepala Royle;复伞银莲花;Foursepal Anemone, Foursepal Windflower ■
24083 Anemone thalictroides L. = Anemonella thalictroides (L.) Spach ■☆
24084 Anemone thalictroides L. = Thalictrum thalictroides (L.) A. J. Eames et B. Boivin ■☆
24085 Anemone thomsonii Oliv.;托马森银莲花■☆
24086 Anemone thomsonii Oliv. var. angustisecta Milne-Redh. et Turrill; 细毛汤姆逊银莲花■☆
24087 Anemone thomsonii Oliv. var. friesiorum Ulbr.;弗里斯银莲花■
24088 Anemone tibetica W. T. Wang;西藏银莲花;Tibet Anemone, Xizang Anemone,Xizang Windflower ■
24089 Anemone tilieecta C. Y. Wu et W. T. Wang;细裂银莲花■
24090 Anemone tomentosa (Maxim.) C. P'ei;大火草(白头翁,大头翁,火草叶,毛秋牡丹,绒毛秋牡丹,土白头翁,野棉花); Tomentose Anemone,Tomentose Windflower ■
24091 Anemone trullifolia Hook. f. et Thomson;匙叶银莲花;Spoonleaf Anemone,Spoonleaf Windflower ■
24092 Anemone trullifolia Hook. f. et Thomson var. campestris Diels = Anemone yulongshanica W. T. Wang var. truncata (H. F. Comber) W. T. Wang ■
24093 Anemone trullifolia Hook. f. et Thomson var. coelestina (Franch.) Finet et Gagnep. = Anemone coelestina Franch. ■
24094 Anemone trullifolia Hook. f. et Thomson var. holophylla Diels = Anemone coelestina Franch. var. holophylla (Diels) Ziman et B. E. Dutton ■
24095 Anemone trullifolia Hook. f. et Thomson var. liangshanica (W. T. Wang) Ziman et B. E. Dutton;凉山银莲花;Liangshan Anemone, Liangshan Windflower ■
24096 Anemone trullifolia Hook. f. et Thomson var. linearis (Brühl) Hand. -Mazz. = Anemone coelestina Franch. var. linearis (Brühl) Ziman et B. E. Dutton ■
24097 Anemone trullifolia Hook. f. et Thomson var. lutienensis (W. T. Wang) Ziman et B. E. Dutton;鲁甸银莲花;Ludian Anemone, Ludian Windflower,Lutien Anemone ■
24098 Anemone trullifolia Hook. f. et Thomson var. luxurians ? = Anemone trullifolia Hook. f. et Thomson ■
24099 Anemone trullifolia Hook. f. et Thomson var. souliei Finet et Gagnep. = Anemone coelestina Franch. var. linearis (Brühl) Ziman et B. E. Dutton ■
24100 Anemone tschernjaewii Regel;东氏银莲花(红蕊银莲花); Tschernjaew Anemone ■☆
24101 Anemone tuberosa Rydb.;沙地银莲花;Desert Anemone,Desert Wind Flower,Desert Windflower,Tuber Anemone ■☆
24102 Anemone tuberosa Rydb. var. texana Enquist et Crozier = Anemone okennonii Keener et B. E. Dutton ■☆
24103 Anemone udensis Trautv. et C. A. Mey.;乌德银莲花;Ute Anemone,Ute Windflower ■
24104 Anemone ulbrischiana Diels ex Ulbr. = Anemone baicalensis Turcz. ■
24105 Anemone umbrosa C. A. Mey.;阴地银莲花;Shady Anemone,

Shady Windflower ■

24106 Anemone umbrosa C. A. Mey. subsp. extemiorientalis Starod. = Anemone umbrosa C. A. Mey. ■

24107 Anemone uralensis Fisch. ex DC. ;乌拉尔银莲花;Ural Anemone ■☆

24108 Anemone vahlii Hornem. = Anemone richardsonii Hook. ■☆

24109 Anemone vernalis L. ; 早春银莲花(春白头翁); Pale Pasqueflower, Pale Pasque-flower, Pink Pasqueflower, Spring Anemone, Spring Pulsatilla, Thimbleweed, Tull Anemone, Vernal Anemone ■☆

24110 Anemone villosa Royle = Anemone polyanthes D. Don ■

24111 Anemone villosissima (DC.) Juz. ;长柔毛银莲花■☆

24112 Anemone virginiana L. ;弗州银莲花(弗吉尼亚银莲花);Tall Anemone, Tall Thimbleweed, Thimbleweed ■☆

24113 Anemone virginiana L. f. leucosepala Fernald = Anemone virginiana L. ■☆

24114 Anemone virginiana L. var. alba (Oakes) A. W. Wood;白弗州银莲花;Tall Anemone, Tall Thimbleweed ■☆

24115 Anemone virginiana L. var. riparia (Fernald) B. Boivin = Anemone virginiana L. var. alba (Oakes) A. W. Wood ■☆

24116 Anemone vitifolia Buch. -Ham. ex DC. ;野棉花(打破碗碗花,大星宿草,接骨莲,满天星,木棉花,清水胆,水棉花,铁蒿,土白头翁,土羌活,小白头翁,野牡丹);Grapeleaf Anemone, Grape-leaf Anemone, Grapeleaf Windflower, Scarborough Anemone ■

24117 Anemone vitifolia Buch. -Ham. ex DC. var. genuina Ulbr. ;多心皮银莲花■

24118 Anemone vitifolia Buch. -Ham. ex DC. var. japonica Finet et Gagnep. = Anemone hupehensis (Lemoine) Lemoine var. japonica (Thunb. ex A. Murray) Bowler et Stern ■

24119 Anemone vitifolia Buch. -Ham. ex DC. var. matsudai Yamam. = Anemone vitifolia Buch. -Ham. ex DC. ■

24120 Anemone vitifolia Buch. -Ham. ex DC. var. takasagomontana (Masam.) S. S. Ying = Anemone exigua Maxim. ■

24121 Anemone vitifolia Buch. -Ham. ex DC. var. tomentosa (Maxim.) Finer et Gagnep. = Anemone tomentosa (Maxim.) C. P' ei ■

24122 Anemone wardii Marquart et Shaw = Anemone geum H. Lév. ■

24123 Anemone wightiana Hook. = Anemone rivularis Buch. -Ham. ex DC. ■

24124 Anemone wilsonii Hemsl. = Anemone baicalensis Turcz. ■

24125 Anemone wolfgangiana Besser ex Koch = Anemone patens L. var. multifida Pritz. ■

24126 Anemone wolfgangiana Besser ex Koch = Pulsatilla patens (L.) Mill. subsp. multifida (Pritz.) Zämelis ■

24127 Anemone yamatutae (Nakai) Hara = Hepatica henryi (Oliv.) Steward ■

24128 Anemone yezoensis Koidz. ;北海道银莲花■☆

24129 Anemone yulongshanica W. T. Wang;玉龙山银莲花;Yulongshan Anemone, Yulongshan Windflower ■

24130 Anemone yulongshanica W. T. Wang var. truncata (Comber) W. T. Wang;截基玉龙山银莲花(截基银莲花);Truncate Yulongshan Anemone, Truncate Yulongshan Windflower ■

24131 Anemonella Spach = Thalictrum L. ■

24132 Anemonella Spach(1839);小银莲花属■☆

24133 Anemonella thalictroides (L.) Spach; 北美小银莲花; Anemonella, Rue Anemone, Rue-anemone, Windflower, Wind-flower, Wood Anemone ■☆

24134 Anemonella thalictroides (L.) Spach 'Oscar Schoaf';奥斯卡·斯科夫小银莲花■☆

24135 Anemonella thalictroides (L.) Spach = Thalictrum thalictroides (L.) A. J. Eames et B. Boivin ■☆

24136 Anemonella thalictroides (L.) Spach f. chlorantha Fassett = Anemonella thalictroides (L.) Spach ■☆

24137 Anemonella thalictroides (L.) Spach f. chlorantha Fassett = Thalictrum thalictroides (L.) A. J. Eames et B. Boivin ■☆

24138 Anemonella thalictroides (L.) Spach f. favilliana Bergseng ex Fassett = Thalictrum thalictroides (L.) A. J. Eames et B. Boivin ■☆

24139 Anemonella thalictroides (L.) Spach f. favilliana Bergseng ex Fassett = Anemonella thalictroides (L.) Spach ■☆

24140 Anemonidium (Spach) Á. Löve et D. Löve = Anemone L. (保留属名) ■

24141 Anemonidium (Spach.) Holub. = Anemone L. (保留属名) ■

24142 Anemonidium canadense (L.) Á. Löve et D. Löve = Anemone canadensis L. ■☆

24143 Anemonidium dichotomum (L.) Holub. = Anemone dichotoma L. ■

24144 Anemonidium filisectum (C. Y. Wu et W. T. Wang) Starod. = Anemone filisecta C. Y. Wu et W. T. Wang ■

24145 Anemonidium rivulare (Buch. -Ham. ex DC.) Starod. = Anemone rivularis Buch. -Ham. ex DC. ■

24146 Anemonoides Mill. = Anemone L. (保留属名) ■

24147 Anemonoides altaica (Fisch. ex C. A. Mey.) Holub = Anemone altaica Fisch. ex Ledeb. ■

24148 Anemonoides amurensis (Korsh.) Holub = Anemone amurensis (Korsh.) Kom. ■

24149 Anemonoides baicalensis (Turcz.) Holub = Anemone baicalensis Turcz. ■

24150 Anemonoides davidii (Franch.) Starod. = Anemone davidii Franch. ■

24151 Anemonoides delvayi (Franch.) Holub = Anemone delavayi Franch. ■

24152 Anemonoides exigua (Maxim.) Starod. = Anemone exigua Maxim. ■

24153 Anemonoides extremiorientalis (Starod.) Starod. = Anemone umbrosa C. A. Mey. ■

24154 Anemonoides flaccida (F. Schmidt) Holub = Anemone flaccida F. Schmidt ■

24155 Anemonoides glabrata (Maxim.) Holub = Anemone baicalensis Turcz. var. glabrata Maxim. ■

24156 Anemonoides griffithii (Hook. f. et Thomson) Holub. = Anemone griffithii Hook. f. et Thomson ■

24157 Anemonoides prattii (Huth ex Ulbr.) Holub = Anemone prattii Huth ex Ulbr. ■

24158 Anemonoides raddeana (Regel) Holub = Anemone raddeana Regel ■

24159 Anemonoides reflexa (Stephan ex Willd.) Holub = Anemone reflexa Stephan ■

24160 Anemonoides rossii (S. Moore) Holub = Anemone baicalensis Turcz. var. rossii (S. Moore) Kitag. ■

24161 Anemonoides stolonifera (Maxim.) Holub = Anemone stolonifera Maxim. ■

24162 Anemonoides udensis Trautv. et C. A. Mey. = Anemone udensis Trautv. et C. A. Mey. ■

24163 Anemonoides ulbrichiana (Diels ex Ulbr.) Holub = Anemone baicalensis Turcz. ■

24164 Anemonoides umbrosa (C. A. Mey.) Holub = Anemone umbrosa C. A. Mey. ■
24165 Anemonopsis Pritz. = Anemopsis Hook. et Arn. ■☆
24166 Anemonopsis Siebold et Zucc. (1845);拟银莲花属(假银莲花属,类银莲属,莲花升麻属);False Anemone ■☆
24167 Anemonopsis macrophylla Siebold et Zucc.;拟银莲花(莲花升麻);False Anemone,Kirengeshoma,Yellow Lanterns ■
24168 Anemonospermos Boehm. = Arctotis L. ●■☆
24169 Anemonospermos Möhring = Arctotis L. ●■☆
24170 Anemonospermum Comm. ex Steud. = Arctotheca J. C. Wendl. ■☆
24171 Anemopaegma Mart. ex DC. = Anemopaegma Mart. ex Meisn. (保留属名)●☆
24172 Anemopaegma Mart. ex Meisn. (1840)(保留属名);黄葳属(风葳木属)●☆
24173 Anemopaegma chamberlaynei (Sims) Bureau et K. Schum.;黄葳;Yellow Trumpet Vine ●☆
24174 Anemopaegma chamberlaynii (Sims) Bureau et K. Schum. = Anemopaegma chamberlaynei (Sims) Bureau et K. Schum. ●☆
24175 Anemopaegma mirandum Mart. ex DC.;奇异葳木●☆
24176 Anemopaegmia Mart. ex Meisn. = Anemopaegma Mart. ex Meisn. (保留属名)●☆
24177 Anemopsis Hook. et Arn. (1840);假蕺菜属(塔银莲属)■☆
24178 Anemopsis californica (Nutt.) Hook. et Arn.;加州假蕺菜;Apache Beads,Dill,Lizard Tail ■☆
24179 Anemopsis californica (Nutt.) Hook. et Arn. var. subglabra Kelso = Anemopsis californica (Nutt.) Hook. et Arn. ■☆
24180 Anemopsis californica Hook. et Arn. = Anemopsis californica (Nutt.) Hook. et Arn. ■☆
24181 Anemopsis macrophylla Siebold et Zucc.;大叶假蕺菜■☆
24182 Anepsa Raf. (废弃属名) = Stenanthium (A. Gray) Kunth(保留属名)■☆
24183 Anepsias Schott = Rhodospatha Poepp. ■☆
24184 Anepsias Schott(1858);威尼斯南星属●☆
24185 Anerincleistus Korth. (1844);东南亚野牡丹属●☆
24186 Anerincleistus acuminatissimus (Ridl.) M. P. Nayar;尖东南亚野牡丹●☆
24187 Anerincleistus angustifolius (Stapf) J. F. Maxwell;窄叶东南亚野牡丹●☆
24188 Anerincleistus barbatus M. P. Nayar;髯毛东南亚野牡丹●☆
24189 Anerincleistus caudatus Diels = Styrophyton caudatum (Diels) S. Y. Hu ●◇
24190 Anerincleistus floribundus King;多花东南亚野牡丹●☆
24191 Anerincleistus hirsutus Korth.;粗毛东南亚野牡丹●☆
24192 Anerincleistus macranthus King;大花东南亚野牡丹●☆
24193 Anerincleistus monticola W. W. Sm.;山地东南亚野牡丹●☆
24194 Anerincleistus pallidifolius Nayar;苍白东南亚野牡丹●☆
24195 Anerincleistus pauciflorus Ridl.;少花东南亚野牡丹●☆
24196 Anerincleistus setulosus O. Schwartz;刚毛东南亚野牡丹●☆
24197 Anerma Schrad. ex Nees = Scleria P. J. Bergius ■
24198 Aneslea Rchb. = Anneslea Wall. (保留属名)●
24199 Aneslea Rchb. = Eurycles Salisb. ■☆
24200 Anesorhiza Endl. = Annesorhiza Cham. et Schltdl. ■☆
24201 Anetanthus Benth. = Anetanthus Hiern ex Benth. ■☆
24202 Anetanthus Hiern = Anetanthus Hiern ex Benth. et Hook. f. ■☆
24203 Anetanthus Hiern ex Benth. = Anetanthus Hiern ex Benth. et Hook. f. ■☆
24204 Anetanthus Hiern ex Benth. et Hook. f. (1876);由花苣苔属■☆
24205 Anetanthus alatus (Cham. et Schltdl.) Benth. et Hook.;翅由花苣苔■☆
24206 Anetanthus gracilis Hiern;细由花苣苔■☆
24207 Anetholea Peter G. Wilson = Syzygium R. Br. ex Gaertn. (保留属名)●
24208 Anethum L. (1753);莳萝属;Dill ●■
24209 Anethum capense Thunb. = Chamarea capensis (Thunb.) Eckl. et Zeyh. ■☆
24210 Anethum foeniculoides Maire et Wilczek;假莳萝■☆
24211 Anethum foeniculoides Maire et Wilczek var. erythropotamicum Maire = Anethum foeniculoides Maire et Wilczek ■☆
24212 Anethum foeniculum L. = Foeniculum vulgare (L.) Mill. ■
24213 Anethum graveolens L.;莳萝(瘪谷茴香,黄花前胡,苗栗前胡,皮香,时美中,莳萝椒,莳萝苗,土茴香,小茴香,洋茴香,野茴香,野小茴,野药茴);Anet,Anise,Common Dill,Dill,East Indian Dill,False Fennel,Indian Dill,Meeting House Seeds,Sweet Fennel ■
24214 Anethum graveolens L. = Peucedanum graveolens Benth. et Hook. f. ■
24215 Anethum graveolens L. subsp. sowa (Roxb. ex Fleming) N. F. Koren = Anethum graveolens L. ■
24216 Anethum graveolens L. subsp. sowa (Roxb.) N. F. Koren = Anethum graveolens L. ■
24217 Anethum graveolens L. var. chevallieri Maire = Anethum graveolens L. ■
24218 Anethum japonicum (Thunb.) Koso-Pol. = Peucedanum japonicum Thunb. ■
24219 Anethum japonicum Koso-Pol. = Peucedanum japonicum Thunb. ■
24220 Anethum pannorium Roxb. = Foeniculum vulgare Mill. ■
24221 Anethum pastinaca (L.) Wibel = Pastinaca sativa L. ■
24222 Anethum piperitum Ucria = Foeniculum vulgare Mill. ■
24223 Anethum segetum L. = Ridolfia segetum (L.) Moris ■☆
24224 Anethum sowa DC. = Anethum graveolens L. ■
24225 Anethum sowa Roxb. = Anethum graveolens L. ■
24226 Anethum sowa Roxb. ex Fleming = Anethum graveolens L. ■
24227 Anetia Endl. = Byrsanthus Guill. (保留属名)●☆
24228 Anetilla Galushko = Anemone L. (保留属名)■
24229 Anettea Szlach. et Mytnik = Oncidium Sw. (保留属名)■☆
24230 Anettea Szlach. et Mytnik(2006);巴西瘤瓣兰属■☆
24231 Aneulophus Benth. (1862);无脊柯属●☆
24232 Aneulophus africanus Benth.;非洲无脊柯●☆
24233 Aneuriscus C. Presl = Moronobea Aubl. ●☆
24234 Aneuriscus C. Presl = Symphonia L. f. ●☆
24235 Aneurolepidium Nevski = Leymus Hochst. ■
24236 Aneurolepidium Nevski(1934);碱草属■
24237 Aneurolepidium aemulans Nevski = Leymus aemulans (Nevski) Tzvelev ■
24238 Aneurolepidium alaicum (Korsh.) Nevski;阿赖碱草■☆
24239 Aneurolepidium angustum (Trin.) Nevski = Leymus angustus (Trin.) Pilg. ■
24240 Aneurolepidium asiaticum Roshev. = Helictotrichon hookeri (Scribn.) Henrard ■
24241 Aneurolepidium baldschuanicum (Roshev.) Nevski;巴尔德碱草■☆
24242 Aneurolepidium chinense (Trin. ex Bunge) Kitag. = Leymus chinensis (Trin. ex Bunge) Tzvelev ■
24243 Aneurolepidium chinense (Trin.) Kitag. = Leymus chinensis (Trin. ex Bunge) Tzvelev ■

24244 Aneurolepidium chinense Trin. = Leymus chinensis (Trin. ex Bunge) Tzvelev ■

24245 Aneurolepidium condensatum (J. Presl) Nevski = Leymus condensatus (J. Presl) Á. Löve ■☆

24246 Aneurolepidium dasystachys (Trin.) Nevski = Leymus secalinus (Georgi) Tzvelev ■

24247 Aneurolepidium divaricatum (Drobow) Nevski;叉开碱草■☆

24248 Aneurolepidium fasciculatum (Roshev.) Nevski;簇生碱草■☆

24249 Aneurolepidium flexile Nevski;弯碱草■☆

24250 Aneurolepidium hissaricum Roshev. = Helictotrichon hissaricum (Roshev.) Henrard ■

24251 Aneurolepidium karataviense (Roshev.) Nevski;卡拉塔夫碱草■☆

24252 Aneurolepidium mongolicum Roshev. = Helictotrichon mongolicum (Roshev.) Henrard ■

24253 Aneurolepidium multicaule (Kar. et Kir.) Nevski = Leymus multicaulis (Kar. et Kir.) Tzvelev ■

24254 Aneurolepidium ovatum (Trin.) Nevski = Leymus ovatus (Trin.) Tzvelev ■

24255 Aneurolepidium paboanum (Claus) Nevski = Leymus paboanos (Claus) Pilg. ■

24256 Aneurolepidium petraeum Nevski;岩生碱草■☆

24257 Aneurolepidium pubescens (Huds.) Opiz = Helictotrichon pubescens (Huds.) Pilg. ■

24258 Aneurolepidium ramosum (Trin.) Nevski = Leymus ramosus (Trin.) Tzvelev ■

24259 Aneurolepidium ramosum Nevski = Leymus ramosus Tzvelev ■

24260 Aneurolepidium regelii (Roshev.) Nevski = Leymus chinensis (Trin. ex Bunge) Tzvelev ■

24261 Aneurolepidium schellianum (Hack.) Roshev. = Helictotrichon schellianum (Hack.) Kitag. ■

24262 Aneurolepidium secalinum (Georgi) Kitag. = Leymus secalinus (Georgi) Tzvelev ■

24263 Aneurolepidium secalinum (Georgi) Nevski = Leymus secalinus (Georgi) Tzvelev ■

24264 Aneurolepidium tianschanicum (Drobow) Nevski = Leymus tianschanicus (Drobow) Tzvelev ■

24265 Aneurolepidium tianschanicum Roshev. = Helictotrichon tianschanicum (Roshev.) Henrard ■

24266 Aneurolepidium tianschanicus (Drobow) Nevski = Leymus tianschanicus (Drobow) Tzvelev ■

24267 Aneurolepidium ugamicum (Drobow) Nevski;乌噶姆碱草■☆

24268 Angadenia Miers(1878);瓮腺夹竹桃属●☆

24269 Angadenia amazonica (Stadelm.) Miers;瓮腺夹竹桃●☆

24270 Angadenia amazonica Miers = Angadenia amazonica (Stadelm.) Miers ●☆

24271 Angasomyrtus Trudgen et Keighery(1983);马桃木属●☆

24272 Angasomyrtus salina Trudgen et Keighery;马桃木●☆

24273 Angeia Tidestrom = Myrica L. ●

24274 Angelandra Endl. (1843) = Engelmannia Torr. et A. Gray ex Nutt. ■☆

24275 Angelandra Endl. (1850) = Croton L. ●

24276 Angelandra Endl. (1850) = Gynamblosis Torr. ●

24277 Angelesia Korth. = Licania Aubl. ●☆

24278 Angelianthus H. Rob. et Brettell = Liabellum Rydb. ■☆

24279 Angelianthus H. Rob. et Brettell = Microliabum Cabrera ●■☆

24280 Angelianthus H. Rob. et Brettell(1974);阿根廷矮菊属■☆

24281 Angelica L. (1753);当归属;Angelica,Archangel ■

24282 Angelica acutiloba (Siebold et Zucc.) Kitag.;东当归(大和当归,日本当归,延边当归);Acutilobate Angelica ■

24283 Angelica acutiloba (Siebold et Zucc.) Kitag. f. tsukubana Hikino;筑波当归■☆

24284 Angelica acutiloba (Siebold et Zucc.) Kitag. subsp. iwatensis (Kitag.) Kitag.;仙台当归■☆

24285 Angelica acutiloba (Siebold et Zucc.) Kitag. subsp. lineariloba (Koidz.) Kitag. f. lanceolata (Tatew.) S. Watan. et Kawano = Angelica stenoloba Kitag. f. lanceolata (Tatew.) H. Hara ■☆

24286 Angelica acutiloba (Siebold et Zucc.) Kitag. subsp. lineariloba (Koidz.) Kitag. = Angelica stenoloba Kitag. ■☆

24287 Angelica acutiloba (Siebold et Zucc.) Kitag. var. iwatensis (Kitag.) Hikino = Angelica acutiloba (Siebold et Zucc.) Kitag. subsp. iwatensis (Kitag.) Kitag. ■☆

24288 Angelica acutiloba (Siebold et Zucc.) Kitag. var. lanceolata (Tatew.) Ohwi = Angelica stenoloba Kitag. f. lanceolata (Tatew.) H. Hara ■☆

24289 Angelica acutiloba (Siebold et Zucc.) Kitag. var. lineariloba (Koidz.) Hikino = Angelica stenoloba Kitag. ■☆

24290 Angelica acutiloba (Siebold et Zucc.) Kitag. var. sugiyamae Hikino;北海当归■☆

24291 Angelica amurensis Schischk.;黑水当归(阿穆尔独活,叉子芹,朝鲜白芷,朝鲜当归,黑龙江当归,碗儿芹);Amur Angelica ■

24292 Angelica angelicifolia (Franch.) Kljuykov = Ligusticum angelicifolium Franch. ■

24293 Angelica anomala Avé-Lall.;狭叶当归(白山独活,额水独活,库页白芷,库页当归,水大活,香大活,异形当归);Narrowleaf Angelica,Sachalin Angelica ■

24294 Angelica anomala Avé-Lall. subsp. sachalinensis (Maxim.) H. Ohba var. glabra (Koidz.) H. Ohba = Angelica sachalinensis Maxim. var. glabra (Koidz.) T. Yamaz. ■☆

24295 Angelica anomala Avé-Lall. subsp. sachalinensis (Maxim.) H. Ohba = Angelica sachalinensis Maxim. ■

24296 Angelica anomala Avé-Lall. subsp. sachalinensis (Maxim.) H. Ohba = Angelica sachalinensis Maxim. f. pubescens (T. Yamaz.) T. Yamaz. ■☆

24297 Angelica anomala Avé-Lall. var. kawakamii (Koidz.) Kitag. = Angelica sachalinensis Maxim. var. kawakamii (Koidz.) T. Yamaz. ■☆

24298 Angelica apaensis R. H. Shan et C. C. Yuan;阿坝当归(骚独活);Aba Angelica ■

24299 Angelica apaensis R. H. Shan et C. C. Yuan = Heracleum apaense (R. H. Shan et C. C. Yuan) R. H. Shan et T. S. Wang ■

24300 Angelica archangelica L. = Archangelica officinalis (Moench) Hoffm. ■☆

24301 Angelica archangelica L. var. decurrens (Ledeb.) Weinert = Archangelica decurrens Ledeb. ■

24302 Angelica archangelica L. var. himalaica (C. B. Clarke) Nasir;喜马拉雅当归■☆

24303 Angelica atropurpurea L.;美国当归;American Angelica, Angelica, Common Great Angelica, Great Angelica, Masterwort, Purple-stemmed Angelica ■☆

24304 Angelica atropurpurea L. var. occidentalis Fassett;西部美国当归;Purple-stemmed Angelica,Western Great Angelica ■☆

24305 Angelica balangshanensis R. H. Shan et F. T. Pu;巴狼山当归(巴郎山当归);Balangshan Angelica ■

24306 Angelica biserrata (R. H. Shan et C. Q. Yuan) C. Q. Yuan et R.

H. Shan;重齿当归(巴东独活,长生草,川独活,大活,独滑,独活,独摇草,恩施独活,绩独活,肉独活,山大活,香独活,玉活,重齿毛当归,资邱独活);Biserrate Angelica ■

24307 Angelica boissieuana Nakai;布瓦西厄当归■☆

24308 Angelica boninensis Tuyama = Angelica japonica A. Gray var. boninensis (Tuyama) T. Yamaz. ■☆

24309 Angelica brevicaulis (Rupr.) B. Fedtsch. = Archangelica brevicaulis (Rupr.) Rchb. f. ■

24310 Angelica brevicaulis Rupr. = Archangelica brevicaulis (Rupr.) Rchb. f. ■

24311 Angelica candollii Wall. = Selinum candollei DC. ■

24312 Angelica candollii Wall. = Selinum wallichianum (DC.) Raizada et H. O. Saxena ■

24313 Angelica cartilaginomarginata (Makino ex Y. Yabe) Nakai;长鞘当归(长鞘独活,骨缘当归);Longsheath Angelica ■

24314 Angelica cartilaginomarginata (Makino ex Y. Yabe) Nakai var. foliosa C. Q. Yuan et R. H. Shan;骨缘当归(对芹,山藁本,野芹菜)■

24315 Angelica cartilaginomarginata (Makino) ex Y. Yabe Nakai var. matsumurae (H. Boissieu) Kitag.;东北长鞘当归■

24316 Angelica cartilaginomarginata (Makino) Y. Yabe ex Nakai var. matsumurae (H. Boissieu) Kitag. = Angelica cartilaginomarginata (Makino ex Y. Yabe) Nakai ■

24317 Angelica chinghaiensis R. H. Shan ex K. T. Fu = Angelica nitida H. Wolff ■

24318 Angelica cincta H. Boissieu;湖北当归(围绕白芷);Hubei Angelica ■

24319 Angelica citriodora Hance = Ostericum citriodorum (Hance) C. Q. Yuan et R. H. Shan ■

24320 Angelica crucifolia Kom. = Angelica cartilaginomarginata (Makino) ex Y. Yabe Nakai var. matsumurae (H. Boissieu) Kitag. ■

24321 Angelica crucifolia Kom. = Angelica cartilaginomarginata (Makino) Y. Yabe ex Nakai ■

24322 Angelica cryptotaeniifolia Kitag.;鸭儿芹叶当归■☆

24323 Angelica cryptotaeniifolia Kitag. var. kyushiana T. Yamaz.;久七当归■☆

24324 Angelica czernaevia (Fisch. et C. A. Mey.) Kitag. = Czernaevia laevigata Turcz. ■

24325 Angelica dahurica (Fisch. ex Hoffm.) Benth. et Hook. f. ex Franch. et Sav.;白芷(白茝,大活,独活,芳香,苻蘺,河北独活,狼山芹,香白芷,香大活,蘺,兴安白芷,泽芳,芷,走马芹,走马芹筒子);Baizhi Angelica ■

24326 Angelica dahurica (Fisch. ex Hoffm.) Benth. et Hook. f. ex Franch. et Sav. 'Hangbaizhi';杭白芷(白茝,白芷,川白芷,大本山芹菜,芳香,苻蘺,台湾当归,台湾独活,香白芷,蘺,野当归,泽芳,浙白芷,芷);Taiwan Angelica ■

24327 Angelica dahurica (Fisch. ex Hoffm.) Benth. et Hook. f. ex Franch. et Sav. 'Qibaizhi';祁白芷(白芷,禹白芷);Fragrant Angelica ■

24328 Angelica dahurica (Fisch. ex Hoffm.) Benth. et Hook. f. ex Franch. et Sav. var. formosana (H. Boissieu) R. H. Shan et C. C. Yuan = Angelica dahurica (Fisch. ex Hoffm.) Benth. et Hook. f. ex Franch. et Sav. 'Hangbaizhi' ■

24329 Angelica dahurica (Fisch. ex Hoffm.) Benth. et Hook. f. ex Franch. var. formosana (H. Boissieu) D. F. Yen = Angelica dahurica (Fisch. ex Hoffm.) Benth. et Hook. f. ex Franch. et Sav. 'Hangbaizhi' ■

24330 Angelica dahurica Maxim. = Angelica dahurica (Fisch. ex Hoffm.) Benth. et Hook. f. ex Franch. et Sav. ■

24331 Angelica dahurica Maxim. var. formosana (H. Boissieu) D. F. Yen;台湾当归;Taiwan Angelica ■

24332 Angelica dahurica Maxim. var. pai-chi Kimura = Angelica dahurica (Fisch. ex Hoffm.) Benth. et Hook. f. ex Franch. et Sav. 'Hangbaizhi' ■

24333 Angelica dahurica Maxim. var. taiwaniana H. Boissieu = Angelica dahurica (Fisch. ex Hoffm.) Benth. et Hook. f. ex Franch. et Sav. 'Hangbaizhi' ■

24334 Angelica dailingensis Z. H. Pan et T. D. Zhuang;带岭当归;Dailing Angelica ■

24335 Angelica daucoides (Franch.) M. Hiroe = Ligusticum daucoides (Franch.) Franch. ■

24336 Angelica decursiva (Miq.) Franch. et Sav.;紫花前胡(艾鹰爪根,长前胡,独活,光前胡,老虎爪,前胡,羌活,麝香菜,土当归,鸭脚当归,鸭脚前胡,野当归);Purpleflower Angelica ■

24337 Angelica decursiva (Miq.) Franch. et Sav. f. albiflora (Maxim.) Nakai;白花前胡(鸭巴前胡)■

24338 Angelica decursiva (Miq.) Franch. et Sav. f. angustiloba (Makino) Hatus. et Nakash. = Angelica decursiva (Miq.) Franch. et Sav. var. angustiloba (Makino) Honda ■☆

24339 Angelica decursiva (Miq.) Franch. et Sav. f. discolor (Makino) Hatus. et Nakash.;异色紫花前胡■☆

24340 Angelica decursiva (Miq.) Franch. et Sav. var. albiflora (Maxim.) Nakai;鸭巴前胡(白花日本前胡,白花下延当归,叉子芹,大鸭巴芹,独梗芹,黑瞎子芹,日本前胡,碗儿芹,鸭巴芹)■

24341 Angelica decursiva (Miq.) Franch. et Sav. var. angustiloba (Makino) Honda;狭裂紫花前胡■☆

24342 Angelica dielsii H. Boissieu;城口当归;Diels Angelica ■

24343 Angelica dissoluta Diels = Peucedanum dissolutum (Diels) H. Wolff ■

24344 Angelica duclouxii Fedde ex H. Wolff;东川当归;Ducloux Angelica ■

24345 Angelica edulis Miyabe ex Y. Yabe;食当归■☆

24346 Angelica erythrocarpa H. Wolff = Angelica laxifoliata Diels ■

24347 Angelica fargesii H. Boissieu;曲柄当归;Farges Angelica ■

24348 Angelica fischeri Spreng. = Cenolophium denudatum (Hornem.) Tutin ■

24349 Angelica flaccida Kom. = Czernaevia laevigata Turcz. ■

24350 Angelica florentii Franch. et Sav. ex Maxim. = Ostericum florentii (Franch. et Sav. ex Maxim.) Kitag. ■☆

24351 Angelica formosana H. Boissieu = Angelica dahurica (Fisch. ex Hoffm.) Benth. et Hook. f. ex Franch. var. formosana (H. Boissieu) D. F. Yen ■

24352 Angelica forrestii Diels;雪山当归;Forrest Angelica ■

24353 Angelica forrestii Diels = Pleurospermum angelicoides (Wall. ex DC.) Benth. ex C. B. Clarke ■

24354 Angelica furcijuga Kitag. = Angelica tenuisecta (Makino) Makino var. furcijuga (Kitag.) H. Ohba ■☆

24355 Angelica genuflexa Nutt. ex Torr. et A. Gray;毛珠当归(大叶芎藭,膝曲白芷)■

24356 Angelica genuflexa Nutt. ex Torr. et A. Gray subsp. refracta (F. Schmidt) M. Hiroe = Angelica genuflexa Nutt. ex Torr. et A. Gray ■

24357 Angelica genuflexa Nutt. ex Torr. et A. Gray var. multinervis (Koidz.) M. Hiroe;多脉膝曲白芷■☆

24358 Angelica gigas Nakai;大当归(朝鲜当归,大独活,大野芹,土当

归,野当归,紫花芹);Korea Angelica,Korean Angelica ■
24359 Angelica gigas Nakai var. minor Momiy. = Angelica cryptotaeniifolia Kitag. ■☆
24360 Angelica glauca Edgew.;灰叶当归(灰绿当归)■
24361 Angelica gmelinii (DC.) A. Gray = Coelopleurum gmelinii (DC.) Ledeb. ■☆
24362 Angelica gmelinii (DC.) Ledeb. subsp. saxatilis (Turcz. ex Ledeb.) Vorosch. = Coelopleurum saxatile (Turcz. ex Ledeb.) Drude ■
24363 Angelica gmelinii subsp. saxatilis (Turcz. ex Ledeb.) Vorosch. = Coelopleurum saxatile (Turcz. ex Ledeb.) Drude ■
24364 Angelica gracilis Franch. = Czernaevia laevigata Turcz. ■
24365 Angelica grosseserrata Maxim. = Ostericum grosseserratum (Maxim.) Kitag. ■
24366 Angelica hakonensis Maxim.;箱根当归■☆
24367 Angelica hakonensis Maxim. f. nikoensis (Y. Yabe) Kitam. = Angelica hakonensis Maxim. ■☆
24368 Angelica hakonensis Maxim. var. nikoensis (Y. Yabe) H. Hara = Angelica hakonensis Maxim. ■☆
24369 Angelica henryi H. Wolff;宜昌当归;Henry Angelica ■
24370 Angelica henryi H. Wolff = Angelica setchuenensis Diels ■
24371 Angelica heterocarpa J. Lloyd;异果当归■☆
24372 Angelica hirsutiflora S. L. Liu, C. Y. Chao et T. I. Chuang;滨当归(毛独活,毛叶日本当归);Hirsuteflower Angelica ■
24373 Angelica hirsutiflora S. L. Liu, C. Y. Chao et T. I. Chuang = Angelica japonica A. Gray var. hirsutiflora (S. L. Liu, C. Y. Chao et T. I. Chuang) T. Yamaz. ■
24374 Angelica hultenii (Fernald) M. Hiroe = Ligusticum scoticum L. subsp. hultenii (Fernald) Hultén ■☆
24375 Angelica ibukiensis (Y. Yabe) Makino ex H. Hara = Dystaenia ibukiensis (Y. Yabe) Kitag. ■☆
24376 Angelica inaequalis Maxim.;不对称当归■☆
24377 Angelica involucellata Diels = Melanosciadium pimpinelloideum H. Boissieu ■
24378 Angelica iwatensis Kitag. = Angelica acutiloba (Siebold et Zucc.) Kitag. subsp. iwatensis (Kitag.) Kitag. ■☆
24379 Angelica jaluana Nakai = Angelica anomala Avé-Lall. ■
24380 Angelica japonica A. Gray;日本当归;Japan Angelica ■☆
24381 Angelica japonica A. Gray var. boninensis (Tuyama) T. Yamaz.;小笠原当归■☆
24382 Angelica japonica A. Gray var. hirsutiflora (S. L. Liu, C. Y. Chao et T. I. Chuang) T. Yamaz. = Angelica hirsutiflora S. L. Liu, C. Y. Chao et T. I. Chuang ■
24383 Angelica kangdingensis R. H. Shan et F. T. Pu;康定当归;Kangding Angelica ■
24384 Angelica keiskei Koidz.;东滨海当归■☆
24385 Angelica kiusiana Maxim.;基隆当归(咸草)■
24386 Angelica koreana Maxim.;朝鲜当归;Korean Angelica ■☆
24387 Angelica koreana Maxim. = Angelica polymorpha Maxim. ■
24388 Angelica koreana Maxim. = Ostericum grosseserratum (Maxim.) Kitag. ■
24389 Angelica laxifoliata Diels;疏叶当归(臭羌,红果当归,骚羌活,臊羌,扫羌活,山芹菜,疏叶独活,猪独活);Laxleaf Angelica ■
24390 Angelica levisticum All. = Levisticum officinale K. Koch ■
24391 Angelica likiangensis H. Wolff;丽江当归;Lijiang Angelica ■
24392 Angelica litoralis Fr.;滨海当归■☆
24393 Angelica longicaudata C. Q. Yuan et R. H. Shan;长尾叶当归(长尾当归,曲前,土羌活,尾独活,沄山当归);Longtailleaf Angelica ■
24394 Angelica longipedicellata (H. Wolff) M. Hiroe;长叶柄当归(长柄当归);Longpedicellate Angelica ■
24395 Angelica longipes H. Wolff;长序当归(长柄当归);Longumbell Angelica ■
24396 Angelica longiradiata (Maxim.) Kitag.;长射当归■☆
24397 Angelica longiradiata (Maxim.) Kitag. var. yakushimensis (Masam. et Ohwi) Kitag.;屋久岛长射当归■☆
24398 Angelica macrocarpa H. Wolff = Angelica dahurica (Fisch. ex Hoffm.) Benth. et Hook. f. ex Franch. et Sav. ■
24399 Angelica maowenensis C. Q. Yuan et R. H. Shan;茂汶当归(骚独活);Maowen Angelica ■
24400 Angelica matsumurae Y. Yabe = Angelica pubescens Maxim. var. matsumurae (Y. Yabe) Ohwi ■☆
24401 Angelica maximowiczii (F. Schmidt ex Maxim.) Benth. ex Maxim. = Ostericum maximowiczii (F. Schmidt ex Maxim.) Kitag. ■
24402 Angelica maximowiczii (F. Schmidt) Benth. ex Maxim. = Ostericum maximowiczii (F. Schmidt ex Maxim.) Kitag. ■
24403 Angelica maximowiczii (F. Schmidt) Benth. ex Maxim. f. australis Kom. = Ostericum maximowiczii (F. Schmidt ex Maxim.) Kitag. var. australe (Kom.) Kitag. ■
24404 Angelica maximowiczii (F. Schmidt) Benth. ex Maxim. var. australis (Kom.) Gorovoj = Ostericum maximowiczii (F. Schmidt ex Maxim.) Kitag. var. australe (Kom.) Kitag. ■
24405 Angelica mayebarana (Koidz.) Kitag. = Angelica shikokiana Makino ex Y. Yabe var. mayebarana (Koidz.) H. Hara ■☆
24406 Angelica megaphylla Diels;大叶当归(大叶独活);Bigleaf Angelica ■
24407 Angelica minamitanii T. Yamaz.;南谷当归■☆
24408 Angelica miqueliana Maxim. = Ostericum sieboldii (Miq.) Nakai ■
24409 Angelica mongolica Franch. = Ostericum grosseserratum (Maxim.) Kitag. ■
24410 Angelica montana Brot.;山当归■☆
24411 Angelica morii Hayata;福参(福参当归,建参,建人参,森氏当归,山芹菜,天池参,土参,土当归,土芹菜,土人参);Mor Angelica ■
24412 Angelica morii Hayata var. nanhutashanensis S. L. Liu, C. Y. Chao et T. I. Chuang = Angelica morrisonicola Hayata var. nanhutashanensis S. L. Liu, C. Y. Chao et T. I. Chuang ■
24413 Angelica morrisonicola Hayata;玉山当归(玉山芹菜);Yushan Angelica ■
24414 Angelica morrisonicola Hayata var. nanhutashanensis S. L. Liu, C. Y. Chao et T. I. Chuang;南湖当归■
24415 Angelica mukabakiensis T. Yamaz. = Angelica ubatakensis (Makino) Kitag. var. valida Kitag. ■☆
24416 Angelica multicaulis Pimenov;多茎当归;Manystem Angelica ■
24417 Angelica multisecta Maxim. = Coelopleurum multisectum (Maxim.) Kitag. ■☆
24418 Angelica nemorosa Ten.;多叶当归■☆
24419 Angelica nitida H. Wolff;青海当归(白芷,独活,麻母);Qinghai Angelica ■
24420 Angelica officinalis L. var. decurrens (Ledeb.) Avé-Lall. = Archangelica decurrens Ledeb. ■
24421 Angelica officinalis Moench = Angelica archangelica L. var. himalaica (C. B. Clarke) Nasir ■☆
24422 Angelica officinalis Moench = Archangelica officinalis (Moench) Hoffm. ■☆

24423 Angelica officinalis Moench var. decurrens (Ledeb.) Avé-Lall. = Archangelica decurrens Ledeb. ■

24424 Angelica omeiensis C. Q. Yuan et R. H. Shan;峨眉当归(当归,羌活,骚羌活,山芹菜,香白芷,岩白芷,野当归);Emei Angelica ■

24425 Angelica oncosepala Hand. -Mazz.;隆萼当归(松香痄药,土当归);Tumorsepal Angelica ■

24426 Angelica pachycarpa Lange;厚果当归■☆

24427 Angelica pachyptera Avé-Lall.;厚翼当归■☆

24428 Angelica paeoniifolia R. H. Shan et C. Q. Yuan;牡丹叶当归;Peonyleaf Angelica ■

24429 Angelica palustris Hoffm.;草原当归;Bog Angelica ■☆

24430 Angelica peucedanoides H. Wolff = Ostericum grosseserratum (Maxim.) Kitag. ■

24431 Angelica pinnatiloba R. H. Shan et F. T. Pu;羽苞当归;Pinnatebract Angelica ■

24432 Angelica polyclada Franch.;白根独活■

24433 Angelica polymorpha Maxim.;拐芹(白根独活,倒钩芹,独活,拐芹当归,拐子芹,山芹菜,紫杆芹);Polymorphic Angelica ■

24434 Angelica polymorpha Maxim. f. lineariloba T. Yamaz.;线裂拐芹■☆

24435 Angelica polymorpha Maxim. var. sinensis Oliv. = Angelica sinensis (Oliv.) Diels ■

24436 Angelica porphyrocaulis Nakai et Kitag.;紫茎独活(雾灵当归,雾灵独活)■

24437 Angelica porphyrocaulis Nakai et Kitag. = Angelica dahurica (Fisch. ex Hoffm.) Benth. et Hook. f. ex Franch. et Sav. ■

24438 Angelica porphyrocaulis Nakai et Kitag. var. albiflora (Maxim.) Makino = Angelica dahurica (Fisch. ex Hoffm.) Benth. et Hook. f. ex Franch. et Sav. ■

24439 Angelica pseudoselinum H. Boissieu;管鞘当归;Tubesheath Angelica ■

24440 Angelica pseudoshikokiana Kitag.;假四国当归■☆

24441 Angelica pubescens Maxim.;毛当归(毛独活,香独活,野独活);Pubescent Angelica ■

24442 Angelica pubescens Maxim. = Angelica biserrata (R. H. Shan et C. Q. Yuan) C. Q. Yuan et R. H. Shan ■

24443 Angelica pubescens Maxim. f. biserrata R. H. Shan et C. Q. Yuan = Angelica biserrata (R. H. Shan et C. Q. Yuan) C. Q. Yuan et R. H. Shan ■

24444 Angelica pubescens Maxim. f. glabra (Koidz.) Murata = Angelica sachalinensis Maxim. var. glabra (Koidz.) T. Yamaz. ■☆

24445 Angelica pubescens Maxim. var. matsumurae (Y. Yabe) Ohwi;松村氏毛当归■☆

24446 Angelica pubescens Maxim. var. matsumurae (Y. Yabe) Ohwi f. muratae Ohwi = Angelica sachalinensis Maxim. var. glabra (Koidz.) T. Yamaz. ■☆

24447 Angelica purpurascens (Avé-Lall.) Gilli;紫茎白芷■☆

24448 Angelica radix L.;直根当归■☆

24449 Angelica refracta F. Schmidt. = Angelica genuflexa Nutt. ex Torr. et A. Gray ■

24450 Angelica rivulorum Diels = Pleurospermum rivulorum (Diels) K. T. Fu et Y. C. Ho ■

24451 Angelica rubrivaginata H. Wolff = Notopterygium forbesii H. Boissieu ■

24452 Angelica rubrivaginata H. Wolff = Notopterygium franchetii H. Boissieu ■

24453 Angelica sachalinensis Maxim.;库页白芷■

24454 Angelica sachalinensis Maxim. = Angelica anomala Avé-Lall. ■

24455 Angelica sachalinensis Maxim. f. pubescens (T. Yamaz.) T. Yamaz.;毛库页白芷■☆

24456 Angelica sachalinensis Maxim. var. glabra (Koidz.) T. Yamaz.;光库页白芷■☆

24457 Angelica sachalinensis Maxim. var. kawakamii (Koidz.) T. Yamaz.;川上当归■☆

24458 Angelica sachalinensis Maxim. var. pubescens T. Yamaz. = Angelica sachalinensis Maxim. f. pubescens (T. Yamaz.) T. Yamaz. ■☆

24459 Angelica saxatilis Turcz. = Coelopleurum saxatile (Turcz.) Drude ■

24460 Angelica saxatilis Turcz. ex Ledeb. = Coelopleurum saxatile (Turcz. ex Ledeb.) Drude ■

24461 Angelica saxicola Makino ex Y. Yabe;岩生当归■☆

24462 Angelica saxicola Makino ex Y. Yabe var. yoshinagae (Makino) Murata et T. Yamanaka = Angelica yoshinagae Makino ■☆

24463 Angelica scaberula Franch.;微糙叶当归(粗糙当归)■

24464 Angelica scaberula Franch. = Ostericum scaberulum (Franch.) C. Q. Yuan et R. H. Shan ■

24465 Angelica setchuenensis Diels;四川当归;Sichuan Angelica ■

24466 Angelica shikokiana Makino ex Y. Yabe;四国当归■☆

24467 Angelica shikokiana Makino ex Y. Yabe subsp. yakusimensis (H. Hara) Kitam. = Angelica yakusimensis H. Hara ■☆

24468 Angelica shikokiana Makino ex Y. Yabe var. mayebarana (Koidz.) H. Hara;熊当归■☆

24469 Angelica shikokiana Makino ex Y. Yabe var. tenuisecta Makino = Angelica tenuisecta (Makino) Makino ■☆

24470 Angelica shikokiana Makino ex Y. Yabe var. yakusimensis (H. Hara) Ohwi = Angelica yakusimensis H. Hara ■☆

24471 Angelica shikokiana Makino ex Y. Yabe var. yoshinagae (Makino) H. Hara = Angelica yoshinagae Makino ■☆

24472 Angelica sibirica (Fisch. ex Spreng.) M. Hiroe. = Phlojodicarpus sibiricus (Fisch. ex Spreng.) Koso-Pol. ■

24473 Angelica sikkimensis (C. B. Clarke) P. K. Mukh. = Arcuatopterus sikkimensis (C. B. Clarke) Pimenov et Ostroumova ■

24474 Angelica silvestris L. = Angelica sylvestris L. ■

24475 Angelica sinensis (Oliv.) Diels;当归(白蕲,薜,蚕头当归,川归,大芹,地仙圆,干白,干经,甘白,胡首,马尾归,岷当归,岷归,名薜,女二天,秦归,全当归,僧庵草,山蕲,文无,西当归,西归,绤绨,象马,夷灵芝,原来头,云归);Angelica,Chinese Angelica,Dong-quai ■

24476 Angelica sinensis (Oliv.) Diels var. wilsonii (H. Wolff) Z. H. Pan et M. F. Watson;川西当归;E. H. Wilson Angelica ■

24477 Angelica sinuata H. Wolff = Angelica polymorpha Maxim. ■

24478 Angelica smithii H. Wolff = Ostericum grosseserratum (Maxim.) Kitag. ■

24479 Angelica songpanensis R. H. Shan et F. T. Pu;松潘当归;Songpan Angelica ■

24480 Angelica stenoloba Kitag.;狭裂当归■☆

24481 Angelica stenoloba Kitag. f. lanceolata (Tatew.) H. Hara;剑叶狭裂当归■☆

24482 Angelica stratoniana Aiton et Hemsl. = Angelica ternata Regel et Schmalh. ■

24483 Angelica sylvestris L.;林当归(鬼当归,林白芷,林独活,欧白芷,羌活,新疆羌活);Ait Skeiters,Bear Skeiters,Forest Angelica,Ghost Kex,Ghost-kex,Ground Ash,Ground Elder,Holy Ghost,Holy Plant, Horse Pepper, Jack-jump-about, Jeelico, Kedlock, Keglus,

Kelk-kecksy, Kerk, Kesk, Kewsies, Skytes, Smooth Kesh, Spoots, Switiks, Trumpet Keck, Water Kesh, Water Squirt, Wild Angelica, Woodland Angelica, Woody Angelica ■
24484 Angelica tarokoensis Hayata;太鲁阁当归(太鲁阁独活);Taroko Angelica ■
24485 Angelica tenuisecta (Makino) Makino;细毛当归■☆
24486 Angelica tenuisecta (Makino) Makino var. furcijuga (Kitag.) H. Ohba;分叉当归■☆
24487 Angelica tenuisecta (Makino) Makino var. mayebarana (Koidz.) H. Ohba = Angelica shikokiana Makino ex Y. Yabe var. mayebarana (Koidz.) H. Hara ■☆
24488 Angelica tenuissima Nakai = Ligusticum tenuissimum (Nakai) Kitag. ■
24489 Angelica ternata Regel et Schmalh.;三小叶当归;Trifoliolate Angelica ■
24490 Angelica tianmuensis Z. H. Pan et T. D. Zhuang;天目当归;Tianmushan Angelica ■
24491 Angelica tichomirovii V. M. Vinogr. = Angelica multicaulis Pimenov ■
24492 Angelica trichocarpa H. Hara = Coelopleurum rupestre (Koidz.) T. Yamaz. ■☆
24493 Angelica triquinata Turcz. ex Trautv. et C. A. Mey.;三五当归;Filmy Angelica ■☆
24494 Angelica tschiliensis Wolff = Angelica dahurica (Fisch. ex Hoffm.) Benth. et Hook. f. ex Franch. et Sav. ■
24495 Angelica tsinlingensis K. T. Fu;秦岭当归;Qinling Angelica ■
24496 Angelica tsinlingensis K. T. Fu = Angelica dahurica (Fisch. ex Hoffm.) Benth. et Hook. f. ex Franch. et Sav. ■
24497 Angelica ubatakensis (Makino) Kitag.;姬岳当归■☆
24498 Angelica ubatakensis (Makino) Kitag. var. valida Kitag.;刚直姬岳当归■☆
24499 Angelica uchiyamae Y. Yabe = Ostericum grosseserratum (Maxim.) Kitag. ■
24500 Angelica ursina (Rupr.) Maxim.;北白芷■☆
24501 Angelica urticifoliata H. Wolff = Ostericum sieboldii (Miq.) Nakai ■
24502 Angelica valida Diels;金山当归(叉风,差风,防风草,尖头叉风,乌当归,乌独活,岩当归);Strong Angelica ■
24503 Angelica venenosa (Greenway) Fernald;毒当归;Angelica, Hairy Angelica, Wood Angelica ■☆
24504 Angelica viridiflora (Turcz.) Benth. ex Maxim. = Ostericum viridiflorum (Turcz.) Kitag. ■
24505 Angelica wilsonii H. Wolff = Angelica sinensis (Oliv.) Diels var. wilsonii (H. Wolff) Z. H. Pan et M. F. Watson ■
24506 Angelica wulsiniana H. Wolff;洮州当归;Taozhou Angelica ■
24507 Angelica wulsiniana H. Wolff = Angelica nitida H. Wolff ■
24508 Angelica yakusimensis H. Hara;屋久岛当归■☆
24509 Angelica yoshinagae Makino;吉永当归■☆
24510 Angelicaceae Martinov = Apiaceae Lindl.(保留科名)●■
24511 Angelicaceae Martinov = Umbelliferae Juss.(保留科名)●■
24512 Angelicaceae Martinov;当归科■
24513 Angelina Pohl ex Tul. = Siparuna Aubl. ●☆
24514 Angelium (Rchb.) Opiz = Tommasinia Bertol. ■☆
24515 Angelium Opiz = Tommasinia Bertol. ■☆
24516 Angelocarpa Rupr. = Archangelica Hoffm. ■
24517 Angelocarpa brevicaulis Rupr. = Archangelica brevicaulis (Rupr.) Rchb. f. ■
24518 Angeloearpa Rupr. = Angelica L. ■
24519 Angelonia Bonpl.(1812);香彩雀属;Angelonia ●■☆
24520 Angelonia Bonpl. ex Humb. et Bonpl. = Angelonia Bonpl. ●■☆
24521 Angelonia Humb. et Bonpl. = Angelonia Bonpl. ●■☆
24522 Angelonia angustifolia Benth.;狭叶香彩雀;Narrowleaf Angelon, Narrowleaf Angelonia, Summer Snapdragon ■☆
24523 Angelonia gardneri Hook.;香彩雀;Gardner Angelonia ●☆
24524 Angelonia grandiflora C. Morren = Angelonia gardneri Hook. ●☆
24525 Angelonia salicariifolia Humb. et Bonpl.;柳叶香彩雀;Grandmother's Bonnets ■☆
24526 Angelophyllum Rupr. = Angelica L. ■
24527 Angelopogon Poepp. ex Poepp. et Endl. = Myzodendron Sol. ex DC. ●☆
24528 Angelopogon Tiegh. = Myzodendron Sol. ex DC. ●☆
24529 Angelphytum G. M. Barroso(1980);天使菊属●☆
24530 Angelphytum matogrossense G. M. Barroso;天使菊●☆
24531 Angelphytum oppositifolium (Saenz) H. Rob.;对叶天使菊●☆
24532 Angelphytum tenuifolium (Hassl.) H. Rob.;细叶天使菊●☆
24533 Angervilla Neck. = Stemodia L.(保留属名)■☆
24534 Angianthus J. C. Wendl.(1808)(保留属名);盐鼠麹属●■☆
24535 Angianthus Post et Kuntze = Aggeianthus Wight ■
24536 Angianthus Post et Kuntze = Porpax Lindl. ■
24537 Angianthus axilliflorus Ewart et Jean White;腋生盐鼠麹●☆
24538 Angianthus filifolius (Benth.) C. A. Gardner;线叶盐鼠麹●☆
24539 Angianthus flavescens (Benth.) Steetz;浅黄盐鼠麹●☆
24540 Angianthus microcephalus (F. Muell.) Benth.;小头盐鼠麹●☆
24541 Angianthus phyllocephalus Benth.;叶头盐鼠麹●☆
24542 Angianthus platycephalus Benth.;平头盐鼠麹●☆
24543 Angianthus pygmaeus (A. Gray) Benth.;密集盐鼠麹●☆
24544 Angianthus tomentosus J. C. Wendl.;毛盐鼠麹●☆
24545 Angianthus uniflorus P. S. Short;单花盐鼠麹●☆
24546 Angianthus whitei J. M. Black;瓦特盐鼠麹●☆
24547 Anginon Raf.(1840);安吉草属■☆
24548 Anginon difforme (L.) B. L. Burtt;不齐安吉草■☆
24549 Anginon fruticosum I. Allison et B. -E. van Wyk;灌丛安吉草■☆
24550 Anginon intermedium I. Allison et B. -E. van Wyk;间型安吉草■☆
24551 Anginon paniculatum (Thunb.) B. L. Burtt;圆锥安吉草■☆
24552 Anginon pumilum I. Allison et B. -E. van Wyk;矮小安吉草■☆
24553 Anginon rugosum (Thunb.) Raf.;皱纹安吉草■☆
24554 Anginon streyi (Merxm.) I. Allison et B. -E. van Wyk;施特赖安吉草■☆
24555 Anginon swellendamense (Eckl. et Zeyh.) B. L. Burtt;斯韦伦丹安吉草■☆
24556 Anginon tenuior I. Allison et B. -E. van Wyk;瘦安吉草■☆
24557 Anginon ternatum I. Allison et B. -E. van Wyk;三出安吉草■☆
24558 Anginon uitenhagense (Eckl. et Zeyh.) B. L. Burtt = Anginon rugosum (Thunb.) Raf. ■☆
24559 Anginon verticillatum (Sond.) B. L. Burtt;轮生安吉草■☆
24560 Angkalanthus Balf. f.(1883);索岛爵床属■☆
24561 Angkalanthus oligophylla Balf. f.;索岛爵床■☆
24562 Angkalanthus transvaalensis A. Meeuse = Chorisochora transvaalensis (A. Meeuse) Vollesen ■☆
24563 Angolaea Wedd.(1873);安哥拉川苔草属■☆
24564 Angolaea fluitans Wedd.;安哥拉川苔草■☆
24565 Angolam Adans.(废弃属名) = Alangium Lam.(保留属名)●
24566 Angolamia Scop. = Alangium Lam.(保留属名)●
24567 Angolamia Scop. = Angolam Adans.(废弃属名)●

24568 Angolluma R. Munster = Pachycymbium L. C. Leach ■☆
24569 Angolluma abayensis (M. G. Gilbert) Plowes = Pachycymbium abayense (M. G. Gilbert) M. G. Gilbert ■☆
24570 Angolluma baldratii (A. C. White et B. Sloane) Plowes = Orbea baldratii (A. C. White et B. Sloane) Bruyns ■☆
24571 Angolluma circes (M. G. Gilbert) Plowes = Orbea circes (M. G. Gilbert) Bruyns ■☆
24572 Angolluma commutata (A. Berger) Plowes = Orbea sprengeri (Schweinf.) Bruyns subsp. commutata (A. Berger) Bruyns ■☆
24573 Angolluma decaisneana (Lem.) L. E. Newton = Orbea decaisneana (Lehm.) Bruyns ■☆
24574 Angolluma decaisneana (Lem.) R. Munster = Orbea decaisneana (Lehm.) Bruyns ■☆
24575 Angolluma denboefii (Lavranos) Plowes = Orbea denboefii (Lavranos) Bruyns ■☆
24576 Angolluma distincta (E. A. Bruce) Plowes = Orbea distincta (E. A. Bruce) Bruyns ■☆
24577 Angolluma dummeri (N. E. Br.) Plowes = Orbea dummeri (N. E. Br.) Bruyns ■☆
24578 Angolluma foetida (M. G. Gilbert) Plowes = Orbea sprengeri (Schweinf.) Bruyns subsp. foetida (M. G. Gilbert) Bruyns ■☆
24579 Angolluma gemugofana (M. G. Gilbert) Plowes = Orbea gemugofana (M. G. Gilbert) Bruyns ■☆
24580 Angolluma gilbertii Plowes = Orbea gilbertii (Plowes) Bruyns ■☆
24581 Angolluma hesperidum (Maire) Plowes = Orbea decaisneana (Lehm.) Bruyns ■☆
24582 Angolluma huernioides (P. R. O. Bally) Plowes = Orbea huernioides (P. R. O. Bally) Bruyns ■☆
24583 Angolluma kochii (Lavranos) Plowes = Orbea sacculata (N. E. Br.) Bruyns ■☆
24584 Angolluma laikipiensis (M. G. Gilbert) Plowes = Orbea laikipiensis (M. G. Gilbert) Bruyns ■☆
24585 Angolluma laticorona (M. G. Gilbert) Plowes = Orbea laticorona (M. G. Gilbert) Bruyns ■☆
24586 Angolluma lenewtonii Lavranos = Orbea subterranea (E. A. Bruce et P. R. O. Bally) Bruyns ■☆
24587 Angolluma lugardii (N. E. Br.) Plowes = Orbea lugardii (N. E. Br.) Bruyns ■☆
24588 Angolluma miscella (N. E. Br.) Plowes = Orbea miscella (N. E. Br.) Meve ■☆
24589 Angolluma nubica L. E. Newton = Orbea laticorona (M. G. Gilbert) Bruyns ■☆
24590 Angolluma ogadensis (M. G. Gilbert) Plowes = Orbea sprengeri (Schweinf.) Bruyns subsp. ogadensis (M. G. Gilbert) Bruyns ■☆
24591 Angolluma rogersii (L. Bolus) Plowes = Orbea rogersii (L. Bolus) Bruyns ■☆
24592 Angolluma sacculata (N. E. Br.) Plowes = Orbea sacculata (N. E. Br.) Bruyns ■☆
24593 Angolluma schweinfurthii (A. Berger) Plowes = Orbea schweinfurthii (A. Berger) Bruyns ■☆
24594 Angolluma semitubiflora L. E. Newton = Orbea semitubiflora (L. E. Newton) Bruyns ■☆
24595 Angolluma sprengeri (Schweinf.) Plowes = Orbea sprengeri (Schweinf.) Bruyns ■☆
24596 Angolluma subterranea (E. A. Bruce et P. R. O. Bally) Plowes = Orbea subterranea (E. A. Bruce et P. R. O. Bally) Bruyns ■☆
24597 Angolluma sudanensis Plowes = Orbea decaisneana (Lehm.) Bruyns ■☆
24598 Angolluma tubiformis (E. A. Bruce et P. R. O. Bally) Plowes = Orbea tubiformis (E. A. Bruce et P. R. O. Bally) Bruyns ■☆
24599 Angolluma ubomboensis (I. Verd.) Plowes = Australluma ubomboensis (I. Verd.) Bruyns ■☆
24600 Angolluma venenosa (Maire) Plowes = Orbea decaisneana (Lehm.) Bruyns ■☆
24601 Angolluma vibratilis (E. A. Bruce et P. R. O. Bally) Plowes = Orbea vibratilis (E. A. Bruce et P. R. O. Bally) Bruyns ■☆
24602 Angolluma wilsonii (P. R. O. Bally) Plowes = Orbea wilsonii (P. R. O. Bally) Bruyns ■☆
24603 Angophora Cav. (1797);瓮梗桉属(安勾桉属);Eucalypt,Gum ●☆
24604 Angophora bakeri E. C. Hall;小叶瓮梗桉;Narroe-leaved Apple, Smallleved Apple ●☆
24605 Angophora cordifolia Cav.;刺毛瓮梗桉;Dwarf Apple ●☆
24606 Angophora costata Britten;红皮瓮梗桉;Angophora, Red-barked Apple Tree, Rusy Gum, Smooth-barked Apple, Sydney Red Gum ●☆
24607 Angophora floribunda (Sm.) Sweet;糙皮瓮梗桉;Rough Barked Apple, Rough-barked Apple ●☆
24608 Angophora hispida (Sm.) Blaxell = Angophora cordifolia Cav. ●☆
24609 Angophora intermedia DC.;微毛瓮梗桉●☆
24610 Angophora lanceolata Cav. = Angophora costata Britten ●☆
24611 Angophora melanoxylon R. T. Baker;黑木瓮梗桉;Coolabah Apple ●☆
24612 Angorchis Nees = Angraecum Bory ■
24613 Angorchis Thouars = Angraecum Bory ■
24614 Angorchis Thouars ex Kuntze = Angraecum Bory ■
24615 Angorchis articulata (Rchb. f.) Kuntze = Aerangis articulata (Rchb. f.) Schltr. ■☆
24616 Angorchis citrata (Thouars) Kuntze = Aerangis citrata (Thouars) Schltr. ■☆
24617 Angorchis clavigera (Ridl.) Kuntze = Angraecum clavigerum Ridl. ■☆
24618 Angorchis cowanii (Ridl.) Kuntze = Jumellea cowanii (Ridl.) Garay ■☆
24619 Angorchis crassa (Thouars) Kuntze = Angraecum crassum Thouars ■☆
24620 Angorchis cryptodon (Rchb. f.) Kuntze = Aerangis cryptodon (Rchb. f.) Schltr. ■☆
24621 Angorchis curnowiana (Rchb. f.) Kuntze = Angraecum curnowianum (Rchb. f.) T. Durand et Schinz ■☆
24622 Angorchis eburnea (Bory) Kuntze = Angraecum eburneum Bory ■
24623 Angorchis ellisii (B. S. Williams) Kuntze = Aerangis ellisii (B. S. Williams) Schltr. ■☆
24624 Angorchis falcata (Thunb.) Kuntze = Neofinetia falcata (Thunb.) Hu ■
24625 Angorchis fastuosa (Rchb. f.) Kuntze = Aerangis fastuosa (Rchb. f.) Schltr. ■☆
24626 Angorchis gladiifolia (Thouars) Kuntze = Angraecum mauritianum (Poir.) Frapp. ■☆
24627 Angorchis gracilis (Thouars) Kuntze = Chamaeangis gracilis Schltr. ■☆
24628 Angorchis hyaloides (Rchb. f.) Kuntze = Aerangis hyaloides (Rchb. f.) Schltr. ■☆
24629 Angorchis implicata (Thouars) Kuntze = Angraecum implicatum Thouars ■☆

24630 Angorchis modesta (Hook. f.) Kuntze = Aerangis modesta (Hook. f.) Schltr. ■☆
24631 Angorchis pectangis (Thouars) Kuntze = Angraecum pectinatum Thouars ■☆
24632 Angorchis pectinata (Thouars) Kuntze = Angraecum pectinatum Thouars ■☆
24633 Angorchis polystachya (Lindl.) Kuntze = Oeoniella polystachys (Thouars) Schltr. ■☆
24634 Angorchis rostrata (Ridl.) Kuntze = Angraecum rostratum Ridl. ■☆
24635 Angorchis sesquipedale (Thouars) Kuntze = Angraecum sesquipedale Thouars ■☆
24636 Angorchis spathulata (Ridl.) Kuntze = Jumellea spathulata (Ridl.) Schltr. ■☆
24637 Angorchis teretifolia (Ridl.) Kuntze = Angraecum teretifolium Ridl. ■☆
24638 Angorkis Thouars = Angraecum Bory ■
24639 Angorkis arachnopus (Rchb. f.) Kuntze = Aerangis arachnopus (Rchb. f.) Schltr. ■☆
24640 Angoseseli Chiov. (1924);瓮芹属■☆
24641 Angoseseli mossamedensis (Welw. ex Hiern) C. Norman;瓮芹■☆
24642 Angostura Roem. et Schult. (1819);安歌木属;Cusparia Bark ●☆
24643 Angostura acuminata (Pilg.) Albuq.;尖安歌木●☆
24644 Angostura adenanthera Rizzini;腺花安歌木●☆
24645 Angostura bracteata (Nees et Mart.) Kallunki;苞片安歌木●☆
24646 Angostura elegans (A. St. -Hil.) Albuq.;雅致安歌木●☆
24647 Angostura grandiflora (Engl.) Albuq.;大花安歌木●☆
24648 Angostura longiflora (K. Krause) Kallunki;长花安歌木●☆
24649 Angostura macrocarpa (Engl.) Albuq.;大果安歌木●☆
24650 Angostura macrophylla (J. C. Mikan) Albuq.;大叶安歌木●☆
24651 Angostura obovata (Nees et Mart.) Albuq.;倒卵安歌木●☆
24652 Angostura ovata (A. St. -Hil. et Tulasne) Albuq.;卵形安歌木●☆
24653 Angostura paniculata (Engl.) T. Elias;圆锥安歌木●☆
24654 Angostura trifoliata (Bild.) T. S. Elias;三叶安歌木●☆
24655 Angostura undulata (Hemsl.) Albuq.;波缘安歌木●☆
24656 Angostyles Benth. = Angostylis Benth. (保留属名)☆
24657 Angostylidium (Müll. Arg.) Pax et K. Hoffm. = Tetracarpidium Pax ●☆
24658 Angostylidium Pax et K. Hoffm. = Tetracarpidium Pax ●☆
24659 Angostylis Benth. (1854)(保留属名)('Angostyles');瓮柱大戟属☆
24660 Angostylis tabulamontana Croizat;瓮柱大戟☆
24661 Angraecopsis Kraenzl. (1900);拟风兰属(拟武夷兰属)■☆
24662 Angraecopsis amaniensis Summerh.;阿马尼拟风兰■☆
24663 Angraecopsis breviloba Summerh.;短裂拟风兰■☆
24664 Angraecopsis comorensis Summerh. = Angraecopsis trifurca (Rchb. f.) Schltr. ■☆
24665 Angraecopsis cryptantha P. J. Cribb;隐花拟风兰■☆
24666 Angraecopsis dolabriformis (Rolfe) Schltr.;斧形拟风兰■☆
24667 Angraecopsis elliptica Summerh.;椭圆拟风兰■☆
24668 Angraecopsis falcata (Thunb.) Schltr. = Neofinetia falcata (Thunb.) Hu ■
24669 Angraecopsis gassneri G. Will.;加内拟风兰■☆
24670 Angraecopsis gracillima (Rolfe) Summerh.;细长拟风兰■☆
24671 Angraecopsis hallei Szlach. et Olszewski;哈勒拟风兰■☆
24672 Angraecopsis holochila Summerh.;全缘拟风兰■☆
24673 Angraecopsis ischnopus (Schltr.) Schltr.;细长梗拟风兰■☆
24674 Angraecopsis lisowskii Szlach. et Olszewski;利索拟风兰■☆
24675 Angraecopsis lovettii P. J. Cribb;洛维特拟风兰■☆
24676 Angraecopsis macrophylla Summerh.;大叶拟风兰■☆
24677 Angraecopsis malawiensis P. J. Cribb;马拉维拟风兰■☆
24678 Angraecopsis parva (P. J. Cribb) P. J. Cribb;小拟风兰■☆
24679 Angraecopsis parviflora (Thouars) Schltr.;小花拟风兰■☆
24680 Angraecopsis pusilla Summerh.;微小拟风兰■☆
24681 Angraecopsis tenerrima Kraenzl.;极细拟风兰■☆
24682 Angraecopsis tenuicalcar Summerh. = Angraecopsis amaniensis Summerh. ■☆
24683 Angraecopsis thomensis P. J. Cribb et Stévart;爱岛拟风兰■☆
24684 Angraecopsis thouarsii (Finet) H. Perrier = Angraecopsis trifurca (Rchb. f.) Schltr. ■☆
24685 Angraecopsis tridens (Lindl.) Schltr.;三齿拟风兰■☆
24686 Angraecopsis trifurca (Rchb. f.) Schltr.;三叉拟风兰■☆
24687 Angraecum Bory (1804);风兰属(安顾兰属,茶兰属,大慧星兰属,武夷兰属);Angraecum, Bourbon Tea Orchid, Bourbon Tea-orchid ■
24688 Angraecum abietinum Schltr. = Angraecum humblotianum Schltr. ■☆
24689 Angraecum acutipetalum Schltr.;尖瓣风兰■☆
24690 Angraecum acutoemarginatum De Wild. = Tridactyle scottellii (Rendle) Schltr. var. stipulata (De Wild.) Geerinck ■☆
24691 Angraecum acutum Ridl. = Diaphananthe acuta (Ridl.) Schltr. ■☆
24692 Angraecum affine Schltr.;近缘风兰■☆
24693 Angraecum albidorubrum De Wild. = Aerangis luteoalba (Kraenzl.) Schltr. var. rhodosticta (Kraenzl.) J. Stewart ■☆
24694 Angraecum alcicorne Rchb. f. = Aerangis alcicornis (Rchb. f.) Garay ■☆
24695 Angraecum alleizettei Schltr.;阿雷风兰■☆
24696 Angraecum aloifolium Hermans et P. J. Cribb;芦荟叶风兰■☆
24697 Angraecum althoffii Kraenzl. = Diaphananthe pellucida (Lindl.) Schltr. ■☆
24698 Angraecum amaniense Kraenzl. = Angraecopsis tenerrima Kraenzl. ■☆
24699 Angraecum ambongoense Schltr. = Lemurella culicifera (Rchb. f.) H. Perrier ■☆
24700 Angraecum ambrense H. Perrier;昂布尔风兰■☆
24701 Angraecum amplexicaule Toill. -Gen. et Bosser;抱茎风兰■☆
24702 Angraecum ampullaceum Bosser;瓶形风兰■☆
24703 Angraecum andersonii Rolfe = Microcoelia caespitosa (Rolfe) Summerh. ■☆
24704 Angraecum andringitranum Schltr.;安德林吉特拉山风兰■☆
24705 Angraecum angustifolium De Wild. = Cyrtorchis aschersonii (Kraenzl.) Schltr. ■☆
24706 Angraecum angustipetalum Rendle;窄瓣风兰■☆
24707 Angraecum angustum (Rolfe) Summerh.;细风兰■☆
24708 Angraecum ankeranense H. Perrier;阿卡祖贝风兰■☆
24709 Angraecum anocentrum Schltr. = Angraecum calceolus Thouars ■☆
24710 Angraecum antennatum Kraenzl. = Homocolleticon monteriroae (Rchb. f.) Szlach. et Olszewski ■☆
24711 Angraecum aphyllum Thouars = Microcoelia aphylla (Thouars) Summerh. ■☆
24712 Angraecum apiculatum Hook. = Aerangis biloba (Lindl.) Schltr. ■☆
24713 Angraecum apiculatum Hook. var. kirkii (Rchb. f.) Rchb. f. = Aerangis kirkii (Rchb. f.) Schltr. ■☆

24714 Angraecum aporoides Summerh.;贫乏风兰■☆
24715 Angraecum appendiculoides Schltr.;附属物风兰■☆
24716 Angraecum arachnites Schltr.;蛛网风兰■☆
24717 Angraecum arachnopus Rchb. f. = Aerangis arachnopus (Rchb. f.) Schltr. ■☆
24718 Angraecum arcuatum Lindl. = Cyrtorchis arcuata (Lindl.) Schltr. ■☆
24719 Angraecum armeniacum Lindl. = Tridactyle armeniaca (Lindl.) Schltr. ■☆
24720 Angraecum arnoldianum De Wild. = Angraecum eichlerianum Kraenzl. ■☆
24721 Angraecum articulatum Rchb. f. = Aerangis articulata (Rchb. f.) Schltr. ■☆
24722 Angraecum aschersonii Kraenzl. = Cyrtorchis aschersonii (Kraenzl.) Schltr. ■☆
24723 Angraecum ashantense Lindl. = Diaphananthe bidens (Sw. ex Pers.) Schltr. ■☆
24724 Angraecum atlantica Stévart;大西洋风兰■☆
24725 Angraecum augustum Rolfe = Aerangis verdickii (De Wild.) Schltr. ■☆
24726 Angraecum avicularium Rchb. f. = Aerangis rostellaris (Rchb. f.) H. Perrier ■☆
24727 Angraecum backeri Kraenzl. = Diaphananthe bidens (Sw. ex Pers.) Schltr. ■☆
24728 Angraecum bancoense Burg;邦克风兰■☆
24729 Angraecum baronii (Finet) Schltr.;巴龙风兰■☆
24730 Angraecum batesii (Rolfe) Schltr. = Rangaeris muscicola (Rchb. f.) Summerh. ■☆
24731 Angraecum batesii Rolfe = Aerangis arachnopus (Rchb. f.) Schltr. ■☆
24732 Angraecum bathiei Schltr. = Angraecum conchoglossum Schltr. ■☆
24733 Angraecum bemarivoense Schltr.;贝马里武风兰■☆
24734 Angraecum bicallosum H. Perrier;二硬皮风兰■☆
24735 Angraecum bicaudatum Lindl. = Tridactyle bicaudata (Lindl.) Schltr. ■☆
24736 Angraecum bieleri De Wild. = Microcoelia caespitosa (Rolfe) Summerh. ■☆
24737 Angraecum biloboides De Wild. = Aerangis arachnopus (Rchb. f.) Schltr. ■☆
24738 Angraecum bilobum Lindl.;二裂风兰■☆
24739 Angraecum bilobum Lindl. = Aerangis biloba (Lindl.) Schltr. ■☆
24740 Angraecum bilobum Lindl. var. kirkii Rchb. f. = Aerangis kirkii (Rchb. f.) Schltr. ■☆
24741 Angraecum bistortum Rolfe = Homocolleticon ringens (Rchb. f.) Szlach. et Olszewski ■☆
24742 Angraecum bokoyense De Wild. = Calyptrochilum christyanum (Rchb. f.) Summerh. ■☆
24743 Angraecum bolusii Rolfe = Tridactyle tridentata (Harv.) Schltr. ■☆
24744 Angraecum boonei De Wild. = Angraecum angustipetalum Rendle ■☆
24745 Angraecum brachycarpum (A. Rich.) Rchb. f. = Aerangis brachycarpa (A. Rich.) T. Durand et Schinz ■☆
24746 Angraecum brachyrhopalon Schltr.;短风兰■☆
24747 Angraecum braunii Schltr. = Angraecum viride Kraenzl. ■☆
24748 Angraecum brevicornus Summerh.;短角风兰■☆
24749 Angraecum brunneomaculatum Rendle = Ancistrorhynchus clandestinus (Lindl.) Schltr. ■☆
24750 Angraecum bucholzianum Kraenzl. = Homocolleticon ringens (Rchb. f.) Szlach. et Olszewski ■☆
24751 Angraecum burchellii Rchb. f. = Angraecum pusillum Lindl. ■☆
24752 Angraecum buyssonii God.-Leb. = Aerangis ellisii (B. S. Williams) Schltr. ■☆
24753 Angraecum caespitosum Rolfe = Microcoelia caespitosa (Rolfe) Summerh. ■☆
24754 Angraecum caffrum Bolus = Diaphananthe caffra (Bolus) H. P. Linder ■☆
24755 Angraecum calanthum Schltr. = Aerangis calantha (Schltr.) Schltr. ■☆
24756 Angraecum calceolus Thouars;拖鞋风兰■☆
24757 Angraecum calligerum Rchb. f. = Aerangis articulata (Rchb. f.) Schltr. ■☆
24758 Angraecum campyloplectron Rchb. f. = Aerangis biloba (Lindl.) Schltr. ■☆
24759 Angraecum canaliculatum De Wild. = Angraecum subulatum Lindl. ■☆
24760 Angraecum capense (L. f.) Lindl. = Mystacidium capense (L. f.) Schltr. ■☆
24761 Angraecum capitatum Lindl. = Ancistrorhynchus capitatus (Lindl.) Summerh. ■☆
24762 Angraecum caricifolium H. Perrier;番木瓜叶风兰■☆
24763 Angraecum carpophorum Thouars = Angraecum calceolus Thouars ■☆
24764 Angraecum carusianum Severino = Aerangis brachycarpa (A. Rich.) T. Durand et Schinz ■☆
24765 Angraecum catati Baill. = Angraecum rutenbergianum Kraenzl. ■☆
24766 Angraecum caudatum Lindl. = Plectrelminthus caudatus (Lindl.) Summerh. ■☆
24767 Angraecum caulescens Thouars;具茎风兰■☆
24768 Angraecum caulescens Thouars var. multiflorum (Thouars) S. Moore = Angraecum multiflorum Thouars ■☆
24769 Angraecum cephalotes Kraenzl. = Ancistrorhynchus metteniae (Kraenzl.) Summerh. ■☆
24770 Angraecum chamaeanthus Schltr.;矮花风兰■☆
24771 Angraecum chermezonii H. Perrier;谢尔默宗风兰■☆
24772 Angraecum chevalieri Summerh. = Angraecum moandense De Wild. ■☆
24773 Angraecum chiloschistae Rchb. f. = Microcoelia exilis Lindl. ■☆
24774 Angraecum chloranthum Schltr. = Angraecum huntleyoides Schltr. ■☆
24775 Angraecum christyanum Rchb. f. = Calyptrochilum christyanum (Rchb. f.) Summerh. ■☆
24776 Angraecum citratum Thouars = Aerangis citrata (Thouars) Schltr. ■☆
24777 Angraecum claessensii De Wild.;克莱森斯风兰■☆
24778 Angraecum clandestinum Lindl. = Ancistrorhynchus clandestinus (Lindl.) Schltr. ■☆
24779 Angraecum clandestinum Lindl. var. stenophyllum Schltr. = Ancistrorhynchus clandestinus (Lindl.) Schltr. ■☆
24780 Angraecum clareae Hermans,la Croix et P. J. Cribb;克莱尔风兰■☆
24781 Angraecum clavatum Rolfe = Solenangis clavata (Rolfe) Schltr. ■☆
24782 Angraecum clavigerum Ridl.;棍棒风兰■☆
24783 Angraecum compactum Schltr.;紧密风兰■☆
24784 Angraecum compressicaule H. Perrier;扁茎风兰■☆

24785 Angraecum conchiferum Lindl.;壳风兰■☆

24786 Angraecum conchoglossum Schltr.;壳舌风兰■☆

24787 Angraecum confusum Schltr. = Jumellea confusa (Schltr.) Schltr. ■☆

24788 Angraecum conicum Schltr. = Solenangis conica (Schltr.) L. Jonss. ■☆

24789 Angraecum cordatiglandulum De Wild. = Rangaeris rhipsalisocia (Rchb. f.) Summerh. ■☆

24790 Angraecum cornucopiae H. Perrier;角状风兰■☆

24791 Angraecum corynoceras Schltr.;棒角风兰■☆

24792 Angraecum cowani Ridl. = Jumellea cowanii (Ridl.) Garay ■☆

24793 Angraecum crassiflorum H. Perrier = Angraecum crassum Thouars ■☆

24794 Angraecum crassifolium G. Will. = Angraecum sacciferum Lindl. ■☆

24795 Angraecum crassum Thouars;粗风兰■☆

24796 Angraecum crenatum Rchb. f. ex Williams;圆齿风兰■☆

24797 Angraecum crinale De Wild. = Microcoelia caespitosa (Rolfe) Summerh. ■☆

24798 Angraecum cryptodon Rchb. f. = Aerangis cryptodon (Rchb. f.) Schltr. ■☆

24799 Angraecum culiciferum Rchb. f. = Lemurella culicifera (Rchb. f.) H. Perrier ■☆

24800 Angraecum cultriforme Summerh.;刀形风兰■☆

24801 Angraecum curnowianum (Rchb. f.) T. Durand et Schinz;库氏风兰■☆

24802 Angraecum curvatum (Rolfe) Schltr. = Rhipidoglossum curvatum (Rolfe) Garay ■☆

24803 Angraecum curvicalcar Schltr.;弯距风兰■☆

24804 Angraecum curvicaule Schltr.;弯茎风兰■☆

24805 Angraecum curvipes Schltr.;弯梗风兰■☆

24806 Angraecum danguyanum H. Perrier;当吉风兰■☆

24807 Angraecum dasycarpum Schltr.;毛果风兰■☆

24808 Angraecum dauphinense (Rolfe) Schltr.;多芬风兰■☆

24809 Angraecum decaryanum H. Perrier;德卡里风兰■☆

24810 Angraecum decipiens Summerh.;迷惑风兰■☆

24811 Angraecum deflexicalcaratum De Wild. = Chauliodon deflexicalcaratum (De Wild.) L. Jonss. ■☆

24812 Angraecum descendens Rchb. f. = Aerangis articulata (Rchb. f.) Schltr. ■☆

24813 Angraecum dichaeoides Schltr. = Angraecum baronii (Finet) Schltr. ■☆

24814 Angraecum didieri (Baill. ex Finet) Schltr.;迪迪耶风兰■☆

24815 Angraecum distichum Lindl.;叉枝风兰(叉枝风兰,杈枝风兰,杈枝风兰);Distichous Angraecum ■☆

24816 Angraecum distichum Lindl. var. grandifolium (De Wild.) Summerh. = Angraecum aporoides Summerh. ■☆

24817 Angraecum dives Rolfe;富有风兰■☆

24818 Angraecum dorotheae Rendle = Diaphananthe dorotheae (Rendle) Summerh. ■☆

24819 Angraecum drouhardii H. Perrier;德鲁阿尔风兰■☆

24820 Angraecum dryadum Schltr.;森林风兰■☆

24821 Angraecum dubuyssonii God. -Leb. = Aerangis ellisii (B. S. Williams) Schltr. ■☆

24822 Angraecum ealaense De Wild. = Homocolleticon injoloensis (De Wild.) Szlach. et Olszewski ■☆

24823 Angraecum eburneum Bory;白花风兰(白花风兰,大慧星兰);Whiteflower Angraecum ■

24824 Angraecum eburneum Bory var. longicalcar Bosser = Angraecum longicalcar (Bosser) Senghas ■☆

24825 Angraecum eburneum Bory var. virens (Lindl.) Hook. = Angraecum eburneum Bory ■

24826 Angraecum egertonii Rendle;埃杰顿风兰■☆

24827 Angraecum eichlerianum Kraenzl.;艾克勒风兰■☆

24828 Angraecum eichlerianum Kraenzl. var. curvicalcaratum Szlach. et Olszewski;弯距艾克勒风兰■☆

24829 Angraecum elegans Rolfe = Aerangis flexuosa (Ridl.) Schltr. ■☆

24830 Angraecum elephantinum Schltr.;象风兰■☆

24831 Angraecum elliotii Rolfe;埃利风兰■☆

24832 Angraecum ellisii (Rchb. f.) Schltr.;艾氏风兰■☆

24833 Angraecum ellisii B. S. Williams = Aerangis ellisii (B. S. Williams) Schltr. ■☆

24834 Angraecum ellisii Rchb. f. = Angraecum ellisii (Rchb. f.) Schltr. ■☆

24835 Angraecum englerianum (Kraenzl.) Schltr. = Angraecum kranzlinianum H. Perrier ■☆

24836 Angraecum englerianum Kraenzl. = Rangaeris muscicola (Rchb. f.) Summerh. ■☆

24837 Angraecum equitans Schltr.;套折风兰■☆

24838 Angraecum erecto-calcaratum De Wild. = Diaphananthe erecto-calcarata (De Wild.) Summerh. ■☆

24839 Angraecum erectum Summerh.;直立风兰(直立风兰);Erect Angraecum ■☆

24840 Angraecum erythrurum Kraenzl. = Aerangis verdickii (De Wild.) Schltr. ■☆

24841 Angraecum evrardianum Geerinck;埃夫拉尔风兰■☆

24842 Angraecum falcatum (Thunb.) Lindl. = Neofinetia falcata (Thunb.) Hu ■

24843 Angraecum fastuosum Rchb. f. = Aerangis fastuosa (Rchb. f.) Schltr. ■☆

24844 Angraecum filicornoides De Wild. = Jumellea walleri (Rolfe) la Croix ■☆

24845 Angraecum filicornu Thouars;线角风兰■☆

24846 Angraecum filifolium Schltr. = Tridactyle filifolia (Schltr.) Schltr. ■☆

24847 Angraecum filiforme (Kraenzl.) Schltr. = Nephrangis filiformis (Kraenzl.) Summerh. ■☆

24848 Angraecum filipes Schltr. = Diaphananthe cuneata Summerh. ■☆

24849 Angraecum fimbriatipetalum De Wild. = Tridactyle fimbriatipetala (De Wild.) Schltr. ■☆

24850 Angraecum fimbriatum Rendle = Tridactyle bicaudata (Lindl.) Schltr. ■☆

24851 Angraecum finetianum Schltr. = Angraecum humblotianum Schltr. ■☆

24852 Angraecum firthii Summerh.;弗思风兰■☆

24853 Angraecum flabellifolium (Rchb. f.) Rolfe = Aerangis brachycarpa (A. Rich.) T. Durand et Schinz ■☆

24854 Angraecum flanaganii Bolus = Mystacidium flanaganii (Bolus) Bolus ■☆

24855 Angraecum flavidum Bosser;浅黄风兰■☆

24856 Angraecum floribundum Bosser;繁花风兰■☆

24857 Angraecum fournierae André = Aerangis stylosa (Rolfe) Schltr. ■☆

24858 Angraecum fournierianum Kraenzl. = Sobennikoffia fournieriana (André) Schltr. ■☆

24859 Angraecum foxii Summerh. = Angraecum rhynchoglossum Schltr. ■☆
24860 Angraecum fragrans Thouars;芳香风兰(芳香茶兰);Bourbon Tea,Faham,Faham Tea ■☆
24861 Angraecum frommianum Kraenzl. = Tridactyle tricuspis (Bolus) Schltr. ■☆
24862 Angraecum fuscatum Rchb. f. = Aerangis fuscata (Rchb. f.) Schltr. ■☆
24863 Angraecum gabonense Summerh.;加蓬风兰■☆
24864 Angraecum galeandrae Rchb. f. = Eurychone galeandrae (Rchb. f.) Schltr. ■☆
24865 Angraecum geniculatum G. Will.;膝曲风兰■☆
24866 Angraecum gentilii De Wild. = Tridactyle gentilii (De Wild.) Schltr. ■☆
24867 Angraecum gerrardii (Rchb. f.) Bolus = Diaphananthe xanthopollinia (Rchb. f.) Summerh. ■☆
24868 Angraecum gladiifolium Thouars = Angraecum mauritianum (Poir.) Frapp. ■☆
24869 Angraecum globulosocalcaratum De Wild. = Rhipidoglossum globulosocalcaratum (De Wild.) Summerh. ■☆
24870 Angraecum globulosum Hochst. = Microcoelia globulosa (Ridl.) L. Jonss. ■☆
24871 Angraecum glomeratum Ridl. = Ancistrorhynchus cephalotes (Rchb. f.) Summerh. ■☆
24872 Angraecum goetzeanum Kraenzl. = Tridactyle tridentata (Harv.) Schltr. ■☆
24873 Angraecum gracile Thouars = Chamaeangis gracilis Schltr. ■☆
24874 Angraecum gracilipes Rolfe;纤梗风兰■☆
24875 Angraecum gracilipes Rolfe = Jumellea sagittata H. Perrier ■☆
24876 Angraecum gracillimum Kraenzl. = Aerangis gracillima (Kraenzl.) Arends et J. Stewart ■☆
24877 Angraecum graminifolium (Ridl.) Schltr. = Angraecum pauciramosum Schltr. ■☆
24878 Angraecum grandidierianum Carrière = Neobathiea grandidierana (Rchb. f.) Garay ■☆
24879 Angraecum grantii Baker = Aerangis kotschyana (Rchb. f.) Schltr. ■☆
24880 Angraecum gravenreuthii (Kraenzl.) Durieu et Jacks. = Aerangis gravenreuthii (Kraenzl.) Schltr. ■☆
24881 Angraecum guillauminii H. Perrier = Angraecum calceolus Thouars ■☆
24882 Angraecum guyonianum Rchb. f. = Microcoelia globulosa (Ridl.) L. Jonss. ■☆
24883 Angraecum henriquesianum Ridl. = Homocolleticon henriquesiana (Ridl.) Szlach. et Olszewski ■☆
24884 Angraecum hislopii Rolfe = Tridactyle tridentata (Harv.) Schltr. ■☆
24885 Angraecum hologlottis Schltr. = Aerangis hologlottis (Schltr.) Schltr. ■☆
24886 Angraecum humbertii H. Perrier;亨伯特风兰■☆
24887 Angraecum humblotianum Schltr.;洪布风兰■☆
24888 Angraecum humblotii Rchb. f. ex Rolfe = Angraecum leonis (Rchb. f.) André ■☆
24889 Angraecum humile Summerh.;矮小风兰■☆
24890 Angraecum huntleyoides Schltr.;洪特兰风兰■☆
24891 Angraecum hyaloides Rchb. f. = Aerangis hyaloides (Rchb. f.) Schltr. ■☆
24892 Angraecum ichneumoneum Lindl. = Chamaeangis ichneumonea (Lindl.) Schltr. ■☆
24893 Angraecum imbricatum Lindl. = Calyptrochilum emarginatum (Sw.) Schltr. ■☆
24894 Angraecum implicatum Thouars;纠结风兰■☆
24895 Angraecum inaequilongum De Wild. = Tridactyle scottellii (Rendle) Schltr. var. stipulata (De Wild.) Geerinck ■☆
24896 Angraecum infundibulare Lindl.;漏斗风兰(漏斗风兰);Infundibular Angraecum ■☆
24897 Angraecum injoloense De Wild. = Homocolleticon injoloensis (De Wild.) Szlach. et Olszewski ■☆
24898 Angraecum ischnopus Schltr. = Angraecopsis ischnopus (Schltr.) Schltr. ■☆
24899 Angraecum ischnopus Schltr. = Angraecum tenuipes Summerh. ■☆
24900 Angraecum ivorense A. Chev. = Calyptrochilum christyanum (Rchb. f.) Summerh. ■☆
24901 Angraecum jumelleanum Schltr. = Jumellea jumelleana (Schltr.) Summerh. ■☆
24902 Angraecum kamerunense Schltr. = Rhipidoglossum kamerunense (Schltr.) Garay ■☆
24903 Angraecum keniae Kraenzl.;肯尼亚风兰■☆
24904 Angraecum kimballianum R. H. Torr. = Oeoniella polystachys (Thouars) Schltr. ■☆
24905 Angraecum kirkii (Rchb. f.) Rolfe = Aerangis kirkii (Rchb. f.) Schltr. ■☆
24906 Angraecum koehleri Schltr. = Microcoelia koehleri (Schltr.) Summerh. ■☆
24907 Angraecum konduensis De Wild. = Microcoelia konduensis (De Wild.) Summerh. ■☆
24908 Angraecum kotschyanum Rchb. f. = Aerangis kotschyana (Rchb. f.) Schltr. ■☆
24909 Angraecum kotschyi Rchb. f. = Aerangis kotschyana (Rchb. f.) Schltr. ■☆
24910 Angraecum kranzlinianum H. Perrier;粗壮风兰■☆
24911 Angraecum laciniatum Kraenzl. = Tridactyle bicaudata (Lindl.) Schltr. ■☆
24912 Angraecum laggiarae Schltr. = Angraecum calceolus Thouars ■☆
24913 Angraecum lagosense Rolfe = Tridactyle lagosensis (Rolfe) Schltr. ■☆
24914 Angraecum latibracteatum De Wild. = Homocolleticon brownii (Rolfe) Szlach. et Olszewski ■☆
24915 Angraecum laurentii De Wild. = Summerhayesia laurentii (De Wild.) P. J. Cribb ■☆
24916 Angraecum lecomtei H. Perrier;勒孔特风兰■☆
24917 Angraecum leonis (Rchb. f.) André;利昂风兰■☆
24918 Angraecum lepidotum Rolfe = Tridactyle anthomaniaca (Rchb. f.) Summerh. ■☆
24919 Angraecum letouzeyi Bosser;勒图风兰■☆
24920 Angraecum ligulatum Summerh. = Angraecum affine Schltr. ■☆
24921 Angraecum lindenii Lindl. = Dendrophylax lindenii (Lindl.) Benth. ex Rolfe ■☆
24922 Angraecum linearifolium Garay;线叶风兰■☆
24923 Angraecum linearifolium P. J. Cribb = Angraecum umbrosum P. J. Cribb ■☆
24924 Angraecum lisowskianum Szlach. et Olszewski;利索风兰■☆
24925 Angraecum litorale Schltr.;滨海风兰■☆
24926 Angraecum longicalcar (Bosser) Senghas;西方长距风兰■☆

24927 Angraecum longicaule Humbert;长茎风兰■☆

24928 Angraecum lujae De Wild. = Eurychone galeandrae (Rchb. f.) Schltr. ■☆

24929 Angraecum luridum (Sw.) Lindl. = Graphorchis lurida (Sw.) Kuntze ■☆

24930 Angraecum luteoalbum Kraenzl. = Aerangis luteoalba (Kraenzl.) Schltr. ■☆

24931 Angraecum macrocentrum Schltr. = Aerangis macrocentra (Schltr.) Schltr. ■☆

24932 Angraecum macrorrhynchium Schltr. = Microcoelia macrorrhynchia (Schltr.) Summerh. ■☆

24933 Angraecum maculatum Lindl. = Oeceoclades maculata (Lindl.) Lindl. ■☆

24934 Angraecum madagascariense (Finet) Schltr.;马岛风兰■☆

24935 Angraecum magdalenae Schltr. et H. Perrier;马格达莱纳风兰■☆

24936 Angraecum malangeanum Kraenzl. = Calyptrochilum christyanum (Rchb. f.) Summerh. ■☆

24937 Angraecum marii Geerinck = Eggelingia ligulifolia Summerh. ■☆

24938 Angraecum marsupio-calcaratum Kraenzl. = Calyptrochilum christyanum (Rchb. f.) Summerh. ■☆

24939 Angraecum maudiae Bolus = Bolusiella maudiae (Bolus) Schltr. ■☆

24940 Angraecum mauritianum (Poir.) Frapp.;毛里求斯风兰■☆

24941 Angraecum megalorrhizum Rchb. f. = Microcoelia megalorrhiza (Rchb. f.) Summerh. ■☆

24942 Angraecum melanostictum Schltr.;黑点风兰■☆

24943 Angraecum metallicum Sander = Aerangis stylosa (Rolfe) Schltr. ■☆

24944 Angraecum microcharis Schltr.;小丽风兰■☆

24945 Angraecum micropetalum Schltr. = Microcoelia caespitosa (Rolfe) Summerh. ■☆

24946 Angraecum microphyton Schltr. = Angraecum tenellum (Ridl.) Schltr. ■☆

24947 Angraecum minus Summerh.;小风兰■☆

24948 Angraecum minutissimum A. Chev.;最小风兰■☆

24949 Angraecum minutum A. Chev. = Rhipidoglossum curvatum (Rolfe) Garay ■☆

24950 Angraecum mirabile Hort. = Aerangis luteoalba (Kraenzl.) Schltr. var. rhodosticta (Kraenzl.) J. Stewart ■☆

24951 Angraecum mirabile Schltr.;奇异风兰■☆

24952 Angraecum moandense De Wild.;莫安达风兰■☆

24953 Angraecum modestum Hook. f. = Aerangis modesta (Hook. f.) Schltr. ■☆

24954 Angraecum moloneyi Rolfe = Calyptrochilum christyanum (Rchb. f.) Summerh. ■☆

24955 Angraecum mombasense Rolfe = Calyptrochilum christyanum (Rchb. f.) Summerh. ■☆

24956 Angraecum monodon Lindl. = Diaphananthe bidens (Sw. ex Pers.) Schltr. ■☆

24957 Angraecum monophyllum A. Rich. = Oeceoclades maculata (Lindl.) Lindl. ■☆

24958 Angraecum montanum Piers = Diaphananthe montana (Piers) P. J. Cribb et J. Stewart ■☆

24959 Angraecum mooreanum Rolfe ex Sander = Aerangis mooreana (Rolfe ex Sander) P. J. Cribb et J. L. Stewart ■☆

24960 Angraecum moratii Bosser;莫拉特风兰■☆

24961 Angraecum muansae Kraenzl. = Diaphananthe fragrantissima (Rchb. f.) Schltr. ■☆

24962 Angraecum multiflorum Thouars;多花风兰■☆

24963 Angraecum multinominatum Rendle;多风兰■☆

24964 Angraecum muriculatum Rendle = Tridactyle muriculata (Rendle) Schltr. ■☆

24965 Angraecum muscicola H. Perrier;圆柱风兰■☆

24966 Angraecum musculiferum H. Perrier;苔地风兰■☆

24967 Angraecum myrianthum Schltr.;丰花风兰■☆

24968 Angraecum mystacidii Rchb. f. = Aerangis mystacidii (Rchb. f.) Schltr. ■☆

24969 Angraecum nalaense De Wild. = Tridactyle nalaensis (De Wild.) Schltr. ■☆

24970 Angraecum nasutum Schltr. = Angraecum pingue Frapp. ■☆

24971 Angraecum obanense Rendle = Rhipidoglossum obanense (Rendle) Summerh. ■☆

24972 Angraecum obesum H. Perrier;肥胖风兰■☆

24973 Angraecum oblongifolium Toill.-Gen. et Bosser;矩圆叶风兰■☆

24974 Angraecum occidentale (Kraenzl.) Rolfe = Angraecopsis tridens (Lindl.) Schltr. ■☆

24975 Angraecum ochraceum (Ridl.) Schltr.;淡黄褐风兰■☆

24976 Angraecum odoratissimum Rchb. f. = Chamaeangis odoratissima (Rchb. f.) Schltr. ■☆

24977 Angraecum oeonioides Bosser;鸟花风兰■☆

24978 Angraecum onivense H. Perrier;乌尼韦风兰■☆

24979 Angraecum ovalifolium De Wild. = Calyptrochilum christyanum (Rchb. f.) Summerh. ■☆

24980 Angraecum pachyrum Kraenzl. = Jumellea pachyra (Kraenzl.) H. Perrier ■☆

24981 Angraecum pachyurum Rolfe = Aerangis mystacidii (Rchb. f.) Schltr. ■☆

24982 Angraecum pallidum W. Watson = Aerangis pallida (W. Watson) Garay ■☆

24983 Angraecum palmicolum Bosser;掌柱风兰■☆

24984 Angraecum panicifolium H. Perrier;禾叶风兰■☆

24985 Angraecum paniculatum Frapp. ex Cordem. = Angraecum calceolus Thouars ■☆

24986 Angraecum parcum Schltr. = Angraecum sacciferum Lindl. ■☆

24987 Angraecum parviflorum Thouars = Angraecopsis parviflora (Thouars) Schltr. ■☆

24988 Angraecum patens Frapp. = Angraecum calceolus Thouars ■☆

24989 Angraecum pauciramosum Schltr.;少分枝风兰■☆

24990 Angraecum pectinatum Thouars;篦状风兰■☆

24991 Angraecum pellucidum Lindl. = Diaphananthe pellucida (Lindl.) Schltr. ■☆

24992 Angraecum penzigianum Schltr.;彭西格风兰■☆

24993 Angraecum perhumile H. Perrier;低矮风兰■☆

24994 Angraecum pertusum Lindl. = Listrostachys pertusa (Lindl.) Rchb. f. ■☆

24995 Angraecum peyrotii Bosser;佩罗风兰■☆

24996 Angraecum philippinense Ames;菲律宾风兰■☆

24997 Angraecum physophora Rchb. f. = Microcoelia physophora (Rchb. f.) Summerh. ■☆

24998 Angraecum pingue Frapp.;肥厚风兰■☆

24999 Angraecum pinifolium Bosser;松叶风兰■☆

25000 Angraecum platycornu Hermans, P. J. Cribb et Bosser;扁角风兰■☆

25001 Angraecum podochiloides Schltr.;柄唇风兰■☆

25002 Angraecum polystachyum A. Rich. = Oeoniella polystachys

(Thouars) Schltr. ■☆

25003 Angraecum poophyllum Summerh. = Angraecum pauciramosum Schltr. ■☆

25004 Angraecum popowii Braem;波波夫风兰■☆

25005 Angraecum poppendickianum Szlach. et Olszewski = Angraecum distichum Lindl. ■☆

25006 Angraecum potamophilum Schltr.;河生风兰■☆

25007 Angraecum praestans Schltr.;优秀风兰■☆

25008 Angraecum primulinum Rolfe = Aerangis primulina (Rolfe) H. Perrier ■☆

25009 Angraecum protensum Schltr.;伸展风兰■☆

25010 Angraecum pseudodidieri H. Perrier;假迪迪耶风兰■☆

25011 Angraecum pseudofilicornu H. Perrier;假线角风兰■☆

25012 Angraecum pterophyllum H. Perrier;翅叶风兰■☆

25013 Angraecum pulchellum Schltr. = Aerangis pulchella (Schltr.) Schltr. ■☆

25014 Angraecum pumilio Schltr.;侏儒风兰■☆

25015 Angraecum pungens Schltr.;刚毛风兰■☆

25016 Angraecum pusillum Lindl.;微小风兰■☆

25017 Angraecum pynaertii De Wild. = Calyptrochilum christyanum (Rchb. f.) Summerh. ■☆

25018 Angraecum pyriforme Summerh.;梨形风兰■☆

25019 Angraecum quintasii Rolfe = Diaphananthe rohrii (Rchb. f.) Summerh. ■☆

25020 Angraecum ramosum Thouars;多枝风兰(多枝风兰);Manybranches Angraecum ■☆

25021 Angraecum ramosum Thouars var. arachnites (Schltr.) Schltr. = Angraecum arachnites Schltr. ■☆

25022 Angraecum ramosum Thouars var. bathiei (Schltr.) H. Perrier = Angraecum conchoglossum Schltr. ■☆

25023 Angraecum ramosum Thouars var. conchoglossum (Schltr.) H. Perrier = Angraecum conchoglossum Schltr. ■☆

25024 Angraecum ramosum var. peracuminatum H. Perrier = Angraecum conchoglossum Schltr. ■☆

25025 Angraecum ramulicolum H. Perrier = Aerangis pallidiflora H. Perrier ■☆

25026 Angraecum reygaertii De Wild.;赖氏风兰■☆

25027 Angraecum rhipsalisocium Rchb. f. = Rangaeris rhipsalisocia (Rchb. f.) Summerh. ■☆

25028 Angraecum rhizomaniacum Schltr.;根茎风兰■☆

25029 Angraecum rhodesianum Rendle = Tridactyle tricuspis (Bolus) Schltr. ■☆

25030 Angraecum rhodostictum Kraenzl. = Aerangis luteoalba (Kraenzl.) Schltr. var. rhodosticta (Kraenzl.) J. Stewart ■☆

25031 Angraecum rhopaloceras Schltr. = Angraecum calceolus Thouars ■☆

25032 Angraecum rhynchoglossum Schltr.;喙舌风兰■☆

25033 Angraecum rigidifolium H. Perrier;硬叶风兰■☆

25034 Angraecum robustum Kraenzl. = Angraecum kranzlinianum H. Perrier ■☆

25035 Angraecum rohlfsianum Kraenzl. = Aerangis brachycarpa (A. Rich.) T. Durand et Schinz ■☆

25036 Angraecum rohrii Rchb. f. = Diaphananthe rohrii (Rchb. f.) Summerh. ■☆

25037 Angraecum roseocalcaratum De Wild. = Aerangis calantha (Schltr.) Schltr. ■☆

25038 Angraecum rostellare Rchb. f. = Aerangis rostellaris (Rchb. f.) H. Perrier ■☆

25039 Angraecum rostratum Ridl.;喙状风兰■☆

25040 Angraecum rothschildianum O' Brien = Eurychone rothschildiana (O'Brien) Schltr. ■☆

25041 Angraecum rubellum Bosser;微红风兰■☆

25042 Angraecum rutenbergianum Kraenzl.;鲁滕贝格风兰■☆

25043 Angraecum sacciferum Lindl.;囊状风兰■☆

25044 Angraecum sacculatum Schltr.;小囊风兰■☆

25045 Angraecum sambiranoense Schltr.;桑比朗风兰■☆

25046 Angraecum sanderianum Rchb. f. = Aerangis modesta (Hook. f.) Schltr. ■☆

25047 Angraecum sanfordii P. J. Cribb et B. J. Pollard;桑福德风兰■☆

25048 Angraecum sankurmense De Wild. = Aerangis calantha (Schltr.) Schltr. ■☆

25049 Angraecum sarcodanthum Schltr. = Angraecum crassum Thouars ■☆

25050 Angraecum saundersiae Bolus = Aerangis mystacidii (Rchb. f.) Schltr. ■☆

25051 Angraecum scabripes Kraenzl. = Angraecum conchiferum Lindl. ■☆

25052 Angraecum scalariforme H. Perrier;梯形风兰■☆

25053 Angraecum scandens Schltr. = Solenangis scandens (Schltr.) Schltr. ■☆

25054 Angraecum schimperianum (A. Rich.) Rchb. f. = Diaphananthe schimperiana (A. Rich.) Summerh. ■☆

25055 Angraecum schoellerianum Kraenzl. = Calyptrochilum christyanum (Rchb. f.) Summerh. ■☆

25056 Angraecum schumannii Kraenzl. = Ancistrorhynchus schumannii (Kraenzl.) Summerh. ■☆

25057 Angraecum scottellii Rendle = Tridactyle scottellii (Rendle) Schltr. ■☆

25058 Angraecum scottianum Rchb. f.;斯氏风兰■☆

25059 Angraecum sedenii (Rchb. f.) G. Nicholson = Cyrtorchis arcuata (Lindl.) Schltr. subsp. variabilis Summerh. ■☆

25060 Angraecum semipedale Rendle = Aerangis kotschyana (Rchb. f.) Schltr. ■☆

25061 Angraecum seretii De Wild. = Cyrtorchis seretii (De Wild.) Schltr. ■☆

25062 Angraecum serpens (H. Perrier) Bosser;蛇形风兰■☆

25063 Angraecum sesquipedale Thouars;长距风兰(长距风兰,武夷兰,星花兰);Longspur Angraecum, Star of Bethlehem Orchid, Star-of-bethlehem Orchid ■☆

25064 Angraecum setipes Schltr.;毛梗风兰■☆

25065 Angraecum sinuatiflorum H. Perrier;深波花风兰■☆

25066 Angraecum smithii Rolfe = Microcoelia smithii (Rolfe) Summerh. ■☆

25067 Angraecum solheidi De Wild. = Rangaeris muscicola (Rchb. f.) Summerh. ■☆

25068 Angraecum somalense Schltr. = Aerangis somalensis (Schltr.) Schltr. ■☆

25069 Angraecum sororium Schltr.;堆积风兰■☆

25070 Angraecum spathulatum Ridl. = Jumellea spathulata (Ridl.) Schltr. ■☆

25071 Angraecum spectabile Summerh.;壮观风兰■☆

25072 Angraecum stella Schltr. = Aerangis gravenreuthii (Kraenzl.) Schltr. ■☆

25073 Angraecum stella-africae P. J. Cribb;非洲星风兰■☆

25074 Angraecum stipulatum De Wild. = Tridactyle scottellii (Rendle) Schltr. var. stipulata (De Wild.) Geerinck ■☆

25075 Angraecum stolzii Schltr.;斯托尔兹风兰■☆
25076 Angraecum straussii Schltr. = Ancistrorhynchus straussii (Schltr.) Schltr. ■☆
25077 Angraecum stylosum Rolfe = Aerangis stylosa (Rolfe) Schltr. ■☆
25078 Angraecum suarezense Toill.-Gen. et Bosser = Angraecum curnowianum (Rchb. f.) T. Durand et Schinz ■☆
25079 Angraecum subclavatum Rolfe = Diaphananthe acuta (Ridl.) Schltr. ■☆
25080 Angraecum subcordatum (H. Perrier) Bosser = Angraecum curnowianum (Rchb. f.) T. Durand et Schinz ■☆
25081 Angraecum subcylindricum De Wild. = Cyrtorchis aschersonii (Kraenzl.) Schltr. ■☆
25082 Angraecum subcylindrifolium De Wild. = Cyrtorchis aschersonii (Kraenzl.) Schltr. ■☆
25083 Angraecum subfalcifolium De Wild. = Diaphananthe bidens (Sw. ex Pers.) Schltr. ■☆
25084 Angraecum subulatum Lindl.;钻形风兰■☆
25085 Angraecum talbotii Rendle = Bolusiella talbotii (Rendle) Summerh. ■☆
25086 Angraecum tenellum (Ridl.) Schltr.;柔软风兰■☆
25087 Angraecum tenuipes Summerh.;细梗风兰■☆
25088 Angraecum tenuispica Schltr.;细穗风兰■☆
25089 Angraecum teretifolium Ridl.;柱叶风兰■☆
25090 Angraecum thomense Rolfe = Chamaeangis thomensis (Rolfe) Schltr. ■☆
25091 Angraecum thomsoni Rolfe = Aerangis thomsonii (Rolfe) Schltr. ■☆
25092 Angraecum trachyrrhizum Schltr. = Tridactyle anthomaniaca (Rchb. f.) Summerh. ■☆
25093 Angraecum triangulifolium Senghas;三角叶风兰■☆
25094 Angraecum trichoplectron (Rchb. f.) Schltr.;毛风兰■☆
25095 Angraecum tricuspe Bolus = Tridactyle tricuspis (Bolus) Schltr. ■☆
25096 Angraecum tridactylites Rolfe = Tridactyle tridactylites (Rolfe) Schltr. ■☆
25097 Angraecum tridens Lindl. = Angraecopsis tridens (Lindl.) Schltr. ■☆
25098 Angraecum tridentatum Harv. = Tridactyle tridentata (Harv.) Schltr. ■☆
25099 Angraecum umbrosum P. J. Cribb;耐荫风兰■☆
25100 Angraecum urschianum Toill.-Gen. et Bosser;乌尔施风兰■☆
25101 Angraecum vagans Lindl. = Chamaeangis vagans (Lindl.) Schltr. ■☆
25102 Angraecum verdickii De Wild. = Aerangis verdickii (De Wild.) Schltr. ■☆
25103 Angraecum verecundum Schltr.;羞涩风兰■☆
25104 Angraecum verrucosum Rendle = Angraecum conchiferum Lindl. ■☆
25105 Angraecum verruculosum Frapp. ex Cordem. = Angraecum implicatum Thouars ■☆
25106 Angraecum vesicatum Lindl. = Chamaeangis vesicata (Lindl.) Schltr. ■☆
25107 Angraecum vesiculatum Schltr.;多疱风兰■☆
25108 Angraecum viguieri Schltr.;维基耶风兰■☆
25109 Angraecum virens Lindl. = Angraecum eburneum Bory ■
25110 Angraecum virgulum Kraenzl. = Tridactyle virgula (Kraenzl.) Schltr. ■☆
25111 Angraecum viride Kraenzl.;绿风兰■☆
25112 Angraecum viridescens De Wild. = Tridactyle laurentii (De Wild.) Schltr. ■☆
25113 Angraecum wakefieldii Rolfe = Solenangis wakefieldii (Rolfe) P. J. Cribb et J. Stewart ■☆
25114 Angraecum waterlotii H. Perrier = Angraecum tenellum (Ridl.) Schltr. ■☆
25115 Angraecum whitfieldii Rendle = Tridactyle armeniaca (Lindl.) Schltr. ■☆
25116 Angraecum wittmackii Kraenzl. = Tridactyle anthomaniaca (Rchb. f.) Summerh. ■☆
25117 Angraecum woodianum Schltr. = Rhipidoglossum rutilum (Rchb. f.) Schltr. ■☆
25118 Angraecum zaratananae Schltr.;萨拉坦风兰■☆
25119 Angraecum zenkeri (Kraenzl.) Schltr. = Bolusiella zenkeri (Kraenzl.) Schltr. ■☆
25120 Angraecum zigzag De Wild. = Calyptrochilum christyanum (Rchb. f.) Summerh. ■☆
25121 Anguillaria Gaertn.(废弃属名) = Anguillaria R. Br.(保留属名)■☆
25122 Anguillaria Gaertn.(废弃属名) = Ardisia Sw.(保留属名)●■
25123 Anguillaria Gaertn.(废弃属名) = Heberdenia Banks ex A. DC. ●☆
25124 Anguillaria R. Br.(保留属名) = Wurmbea Thunb. ■☆
25125 Anguillaria indica (L.) R. Br. = Iphigenia indica Kunth ■
25126 Anguillaria indica R. Br. = Iphigenia indica Kunth ■
25127 Anguillicarpus Burkill = Spirorhynchus Kar. et Kir. ■
25128 Anguillicarpus bulleri Burkill = Spirorrhynchus sabulosus Kar. et Kir. ■
25129 Anguina Mill. = Trichosanthes L. ●■
25130 Anguinum (G. Don) Fourr. = Allium L. ■
25131 Anguinum (G. Don) Fourr. = Loncostemon Raf. ■
25132 Anguinum Fourr. = Allium L. ■
25133 Anguinum Fourr. = Loncostemon Raf. ■
25134 Anguloa Ruiz et Pav. (1794);安顾兰属(安古兰属);Anguloa, Babyin Cradle, Tulip Orchid ●☆
25135 Anguloa brevilabris Rolfe;短唇安顾兰;Shortlip Tulip Orchid ●☆
25136 Anguloa cliftonii Rolfe;大花安顾兰;Largeflower Tulip Orchid ●☆
25137 Anguloa clowesii Lindl.;安顾兰(安古兰);Candle Orchid, Clowes Tulip Orchid, Cradle Orchid ●☆
25138 Anguloa ruckeri Lindl.;汝氏安顾兰;Ruecker Tulip Orchid ●☆
25139 Anguloa uniflora Ruiz et Pav.;单花安顾兰;Singleflower Tulip Orchid ●☆
25140 Anguria Jacq. = Psiguria Neck. ex Arn. ■☆
25141 Anguria Mill.(废弃属名) = Citrullus Schrad. ex Eckl. et Zeyh.(保留属名)■
25142 Anguria affinis Schltdl.;近缘安瓜■☆
25143 Anguria capitata Poepp. et Endl.;头状安瓜■☆
25144 Anguria cookiana Britton;库克安瓜■☆
25145 Anguria elliptica Britton;椭圆安瓜■☆
25146 Anguria indica Noronha;印度安瓜■☆
25147 Anguria longipedicellata (Cogn.) J. F. Macbr. = Guraniopsis longipedicellata Cogn. ■☆
25148 Anguria multiflora Miq.;多花安瓜■☆
25149 Anguria oblongifolia Cogn.;矩圆叶安瓜■☆
25150 Anguria ovata Donn. Sm.;卵形安瓜■☆
25151 Anguria pachyphylla Donn. Sm.;厚叶安瓜■☆
25152 Anguria pallida Cogn.;苍白安瓜■☆
25153 Anguria parviflora Cogn.;小花安瓜■☆

25154 Anguria polyphyllos Schltdl.;多叶安瓜■☆
25155 Anguria pycnocephala (Harms) J. F. Macbr.;密头安瓜■☆
25156 Anguria rhizantha Poepp. et Endl.;根花安瓜■☆
25157 Anguria trifoliata L.;三小叶安瓜■☆
25158 Anguria triphylla Miq.;三叶安瓜■☆
25159 Anguriopsis J. R. Johnst. = Corallocarpus Welw. ex Benth. et Hook. f. ■☆
25160 Anguriopsis J. R. Johnst. = Doyerea Grosourdy ex Bello ■☆
25161 Angusta Ellis = Gardenia Ellis(保留属名)●
25162 Angustinea A. Gray = Augustinea A. St. -Hil. et Naudin ●☆
25163 Angustinea A. Gray = Miconia Ruiz et Pav.(保留属名)●☆
25164 Angutnum Fourr. = Allium L. ■
25165 Angylocalyx Taub.(1896);非洲萼豆属■☆
25166 Angylocalyx braunii Harms;布劳恩非洲萼豆■☆
25167 Angylocalyx claessensi De Wild. = Angylocalyx oligophyllus (Baker) Baker f. ■☆
25168 Angylocalyx gossweileri Baker f. = Angylocalyx pynaertii De Wild. ■☆
25169 Angylocalyx oligophyllus (Baker) Baker f.;寡叶非洲萼豆■☆
25170 Angylocalyx oligophyllus (Baker) Baker f. var. congolensis Yakovlev;刚果萼豆■☆
25171 Angylocalyx pynaertii De Wild.;皮那非洲萼豆■☆
25172 Angylocalyx ramiflorus Taub. = Angylocalyx oligophyllus (Baker) Baker f. ■☆
25173 Angylocalyx schumannianus Taub.;舒曼非洲萼豆■☆
25174 Angylocalyx talbotii Baker f. ex Hutch. et Dalziel;塔尔博特非洲萼豆■☆
25175 Angylocalyx trifoliolatus Baker f. = Angylocalyx oligophyllus (Baker) Baker f. ■☆
25176 Angylocalyx wellensii De Wild. = Angylocalyx oligophyllus (Baker) Baker f. ■☆
25177 Angylocalyx zenkeri Harms = Angylocalyx pynaertii De Wild. ■☆
25178 Angylocalyx zenkeri Harms var. gossweileri (Baker f.) Pellegr. = Angylocalyx pynaertii De Wild. ■☆
25179 Anhalonium Lem.(1839);岩掌属;Living Rock, Living-rock Cactus ●☆
25180 Anhalonium Lem. = Ariocarpus Scheidw. ●
25181 Anhalonium Lem. = Mammillaria Haw.(保留属名)●
25182 Anhalonium williamsii (Lem. ex Salm-Dyck) Lem. = Lophophora williamisii (Lem.) J. M. Coult. ■
25183 Ania Lindl.(1831);安兰属■
25184 Ania Lindl. = Tainia Blume ■
25185 Ania angustifolia Lindl. = Tainia angustifolia (Lindl.) Benth. et Hook. f. ■
25186 Ania elata (Schltr.) S. Y. Hu = Tainia ruybarrettoi (S. Y. Hu et Barretto) Z. H. Tsi ■
25187 Ania hongkongensis (Rolfe) Ts. Tang et F. T. Wang = Tainia hongkongensis Rolfe ■
25188 Ania hookeriana (King et Pantl.) Ts. Tang et F. T. Wang ex Summerh.;绿花安兰(绿花带唇兰);Greenflower Tainia ■
25189 Ania hookeriana (King et Pantl.) Ts. Tang et F. T. Wang ex Summerh. = Tainia hookeriana King et Pantl. ■
25190 Ania latifolia Lindl. = Tainia latifolia (Lindl.) Rchb. f. ■
25191 Ania maculata Thwaites = Chrysoglossum ornatum Blume ■
25192 Ania penangiana (Hook. f.) Summerh. = Tainia penangiana Hook. f. ■
25193 Ania ruybarrettoi S. Y. Hu et Barretto = Tainia ruybarrettoi (S. Y. Hu et Barretto) Z. H. Tsi ■
25194 Ania viridifusca (Hook.) Ts. Tang et F. T. Wang ex Summerh. = Tainia ruybarrettoi (S. Y. Hu et Barretto) Z. H. Tsi ■
25195 Ania viridifusca (Hook.) Ts. Tang et F. T. Wang ex Summerh. = Tainia viridifusca (Hook.) Benth. et Hook. f. ■
25196 Aniba Aubl.(1775);安尼樟属(安尼巴木属,管花楠属,蔷薇木属);Brazilian Sassafras ●☆
25197 Aniba canellila Mez;亚马逊安尼樟●☆
25198 Aniba coto (Rusby) J. F. Macbr.;科托安尼樟(科托蔷薇木);Coto Bark ●☆
25199 Aniba duckei Kosterm.;雌玫瑰安尼樟●☆
25200 Aniba gardneri Mez;加氏安尼樟(加氏蔷薇木)●☆
25201 Aniba hypoglauca Sandwith;白背安尼樟●☆
25202 Aniba panurensis Mez;圭亚那安尼樟●☆
25203 Aniba perutilis Hemsl.;考姆安尼樟●☆
25204 Aniba rosaeodora Ducke;玫瑰安尼樟(玫瑰蔷薇木,玫瑰香安尼樟)●☆
25205 Anictoclea Nimmo = Tetrameles R. Br. ●
25206 Anictoclea grahamiana Nimmo = Tetrameles nudiflora R. Br. ●◇
25207 Anidrum Neck. = Bifora Hoffm.(保留属名)■☆
25208 Anidrum Neck. ex Raf. = Bifora Hoffm.(保留属名)■☆
25209 Anigoata Salisb. = Anigozanthos Labill. ■☆
25210 Anigoazanthes Steud. = Anigozanthos Labill. ■☆
25211 Anigosanthos DC. = Anigozanthos Labill. ■☆
25212 Anigosanthos Lemée = Anigozanthos Labill. ■☆
25213 Anigosanthus Steud. = Anigozanthos Labill. ■☆
25214 Anigosia Salisb. = Anigozanthos Labill. ■☆
25215 Anigozanthes Kuntze = Anigozanthos Labill. ■☆
25216 Anigozanthos Labill.(1800);袋鼠爪属;Australian Sword Lily, Kangaroo Paw, Kangaroo-paw ■☆
25217 Anigozanthos coccineus Paxton;红花袋鼠爪■☆
25218 Anigozanthos flavidus DC.;袋鼠爪;Kangaroo Paw, Yellow Kangaroo-paw ■☆
25219 Anigozanthos humilis Lindl.;小袋鼠爪;Cat's Paw, Cat's-paw ■☆
25220 Anigozanthos manglesii D. Don;长药袋鼠爪;Red-and-green Kangaroo Paw, Red-and-green Kangaroo-paw ■☆
25221 Anigozanthos pulcherrima Hook.;黄袋鼠爪;Golden Kangaroo Paw, Golden Kangaroo-paw ■☆
25222 Anigozanthos rufus Labill.;淡红袋鼠爪■☆
25223 Anigozanthos viridis Engl.;绿袋鼠爪;Golden Kangaroo-paw ■☆
25224 Anigozanthus Labill. = Anigozanthos Labill. ■☆
25225 Anigozanthus Salisb. = Anigozanthos Labill. ■☆
25226 Anigozantos Stapf ex R. L. Massey = Anigozanthos Labill. ■☆
25227 Anigozia Endl. = Anigosia Salisb. ■☆
25228 Aniherolophus Gagnep. = Aspidistra Ker Gawl. ●■
25229 Aniketon Raf. = Smilax L. ●
25230 Anil Mill. = Indigofera L. ●■
25231 Anila Ludw. ex Kuntze = Indigofera L. ●■
25232 Anila mauritanica (L.) Kuntze var. oligantha Kuntze = Indigofera candolleana Meisn. ●☆
25233 Anila tenuifolia (Lam.) Kuntze var. filifolia Kuntze = Indigofera tenuissima E. Mey. ■☆
25234 Anila welwitschii (Baker) Kuntze = Microcharis welwitschii (Baker) Schrire ●☆
25235 Anila zeyheri (Spreng.) Kuntze var. macrophylla Kuntze = Indigofera rostrata Bolus ●☆
25236 Anila zeyheri (Spreng.) Kuntze var. normalis Kuntze = Indigofera

zeyheri Spreng. ex Eckl. et Zeyh. ●☆
25237 Anilema Kunth = Aneilema R. Br. ■☆
25238 Anilma Kunth = Aneilema R. Br. ■☆
25239 Aningeria Aubrév. et Pellegr. = Pouteria Aubl. ●
25240 Aningeria adolfi-friedericii (Engl.) Robyns et G. C. C. Gilbert = Pouteria adolfi-friedericii (Engl.) A. Meeuse ●☆
25241 Aningeria adolfi-friedericii (Engl.) Robyns et G. C. C. Gilbert subsp. australis J. H. Hemsl. = Pouteria adolfi-friedericii (Engl.) A. Meeuse subsp. australis (J. H. Hemsl.) L. Gaut. ●☆
25242 Aningeria adolfi-friedericii (Engl.) Robyns et G. C. C. Gilbert subsp. floccosa J. H. Hemsl. = Pouteria adolfi-friedericii (Engl.) A. Meeuse subsp. floccosa (J. H. Hemsl.) L. Gaut. ●☆
25243 Aningeria adolfi-friedericii (Engl.) Robyns et G. C. C. Gilbert subsp. keniensis (R. E. Fr.) J. H. Hemsl. = Pouteria adolfi-friedericii (Engl.) A. Meeuse subsp. keniensis (R. E. Fr.) L. Gaut. ●☆
25244 Aningeria adolfi-friedericii (Engl.) Robyns et G. C. C. Gilbert subsp. usambarensis J. H. Hemsl. = Pouteria adolfi-friedericii (Engl.) A. Meeuse subsp. usambarensis (J. H. Hemsl.) L. Gaut. ●☆
25245 Aningeria altissima (A. Chev.) Aubrév. et Pellegr. = Pouteria altissima (A. Chev.) Baehni ●☆
25246 Aningeria altissima (A. Chev.) Aubrév. et Pellegr. var. pierrei ? = Pouteria pierrei (A. Chev.) Baehni ●☆
25247 Aningeria pierrei (A. Chev.) Aubrév. et Pellegr. = Pouteria pierrei (A. Chev.) Baehni ●☆
25248 Aningeria pseudoracemosa J. H. Hemsl. = Pouteria pseudoracemosa (J. H. Hemsl.) L. Gaut. ●☆
25249 Aningeria superba (Vermoesen) A. Chev. = Pouteria superba (Vermoesen) L. Gaut. ●☆
25250 Aningueria Aubrév. et Pellegr. = Aningeria Aubrév. et Pellegr. ●
25251 Aningueria Aubrév. et Pellegr. = Pouteria Aubl. ●
25252 Aniotum Parkinson(废弃属名) = Idesia Maxim. (保留属名)●
25253 Aniotum Sol. ex Endl. = Inocarpus J. Forst. et G. Forst. (保留属名)●☆
25254 Aniotum Sol. ex Parkinson = Inocarpus J. Forst. et G. Forst. (保留属名)●☆
25255 Anisacantha R. Br. = Sclerolaena R. Br. ●☆
25256 Anisacanthus Nees = Idanthisa Raf. ■☆
25257 Anisacanthus Nees(1842);异刺爵床属■☆
25258 Anisacanthus andersonii T. F. Daniel; 大异刺爵床; Big Honeysuckle ■☆
25259 Anisacanthus linearis (S. H. Hagen) Henrickson et E. J. Lott;线形异刺爵床;Dwarf Anisacanthus ■☆
25260 Anisacanthus puberulus (Torr.) Henrickson et E. J. Lott;短毛异刺爵床;Pinky Anisacanthus ■☆
25261 Anisacanthus quadrifidus var. wrightii (Torr.) Henrickson;赖特异刺爵床; Flame Anisacanthus, Hummingbird Bush, Wright Anisacanthus ■☆
25262 Anisacanthus thurberi A. Gray;瑟伯异刺爵床;Chuparosa, Desert Honeysuckle, Thurber's Desert Honeysuckle ■☆
25263 Anisacanthus wrightii (Torr.) A. Gray = Anisacanthus quadrifidus var. wrightii (Torr.) Henrickson ■☆
25264 Anisachne Keng = Calamagrostis Adans. ■
25265 Anisachne Keng(1958);异颖草属;Anisachne ■★
25266 Anisachne gracilis Keng;异颖草;Slender Anisachne ■
25267 Anisachne gracilis Keng = Deyeuxia petelotii (Hitchc.) S. M. Phillips et Wen L. Chen ■
25268 Anisachne gracilis Keng var. multinodis Y. Y. Qian;多节异颖草;Manynode Anisachne ■
25269 Anisactis Dulac = Carum L. ■
25270 Anisactis Dulac = Petroselinum A. W. Hill ■
25271 Anisadenia Wall. = Anisadenia Wall. ex Meisn. ■
25272 Anisadenia Wall. ex Meisn. (1838);异腺草属;Anisadenia ■
25273 Anisadenia pubescens Griff.; 异腺草; Anisadenia, Pubescent Anisadenia ■
25274 Anisadenia saxatilis Wall. ex Meisn.; 石异腺草; Saxicolous Anisadenia ■
25275 Anisandra Bartl. = Microcorys R. Br. ●☆
25276 Anisandra Planch. ex Oilv. = Ptychopetalum Benth. ●☆
25277 Anisantha C. Koch = Anisantha K. Koch ■
25278 Anisantha C. Koch = Bromus L. (保留属名)■
25279 Anisantha K. Koch = Bromus L. (保留属名)■
25280 Anisantha K. Koch(1848);旱雀麦属;Brome, Brome Grass ■
25281 Anisantha diandra (Roth) Tutin = Bromus diandrus Roth ■
25282 Anisantha diandra (Roth) Tutin ex Tzvelev = Bromus diandrus Roth ■
25283 Anisantha diandra (Roth) Tutin ex Tzvelev subsp. rigida (Roth) Tzvelev = Bromus rigidus Roth ■
25284 Anisantha diandra (Roth) Tzvelev = Bromus diandrus Roth ■
25285 Anisantha fasciculata (C. Presl) Nevski = Bromus fasciculatus C. Presl ■
25286 Anisantha fasciculata (C. Presl) Nevski subsp. delilei (Boiss.) H. Scholz et Valdés;德利勒旱雀麦■☆
25287 Anisantha macranthera (Hack. ex Henriq.) P. Silva;大药旱雀麦■☆
25288 Anisantha madritensis (L.) Nevski = Bromus madritensis L. ■
25289 Anisantha pontica K. Koch = Bromus tectorum L. ■
25290 Anisantha rigida (Roth) Hyl. = Bromus rigidus Roth ■
25291 Anisantha rubens (L.) Nevski = Bromus rubens L. ■
25292 Anisantha rubens (L.) Nevski subsp. kunkelii (H. Scholz) H. Scholz;孔克尔旱雀麦■☆
25293 Anisantha sericea (Drobow) Nevski = Bromus sericeus Drobow ■
25294 Anisantha sericea Nevski = Bromus tectorum L. subsp. lucidus Sales ■
25295 Anisantha sterilis (L.) Nevski = Bromus sterilis L. ■
25296 Anisantha tectorum (L.) Nevski = Bromus tectorum L. ■
25297 Anisanthera Griff. = Adenosma R. Br. ■
25298 Anisanthera Raf. (1) = Crotalaria L. ●■
25299 Anisanthera Raf. (2) = Caccinia Savi ■☆
25300 Anisantherina Pennell = Agalinis Raf. (保留属名)■☆
25301 Anisantherina Pennell(1920);异药列当属■☆
25302 Anisantherina hispidula (Mart.) Pennell;异药列当■☆
25303 Anisanthus Schult. = Symphoricarpos Dill. ex Juss. ●
25304 Anisanthus Sweet = Antholyza L. ■☆
25305 Anisanthus Sweet ex Klatt = Antholyza L. ■☆
25306 Anisanthus Willd. = Symphoricarpos Dill. ex Juss. ●
25307 Anisanthus Willd. ex Roem. et Schult. = Symphoricarpos Dill. ex Juss. ●
25308 Anisanthus abyssinicus (Brongn. ex Lem.) Klatt = Gladiolus abyssinicus (Brongn. ex Lem.) Goldblatt et M. P. de Vos ■☆
25309 Anisanthus caryophyllaceus (Burm. f.) Klatt = Gladiolus caryophyllaceus (Burm. f.) Poir. ■☆
25310 Anisanthus cunonius (L.) Sweet = Gladiolus cunonius (L.) Gaertn. ■☆

25311 Anisanthus huillensis (Welw. ex Baker) Klatt = Gladiolus huillensis (Welw. ex Baker) Goldblatt ■☆
25312 Anisanthus quadrangularis (Burm. f.) Sweet = Gladiolus quadrangularis (Burm. f.) Ker Gawl. ■☆
25313 Anisanthus saccatus Klatt = Gladiolus saccatus (Klatt) Goldblatt et M. P. de Vos ■☆
25314 Anisanthus splendens Sweet = Gladiolus splendens (Sweet) Herb. ■☆
25315 Aniseia Choisy(1834);心萼薯属(叶萼薯属,异萼属);Aniseia ■
25316 Aniseia biflora (L.) Choisy;心萼薯(白花牵牛,黑面藤,华佗花,老虎豆,簕番薯,满山香,毛牵牛,亚灯堂,中华牵牛);Biflorous Morningglory,Common Aniseia,Two-flowers Aniseia ■
25317 Aniseia biflora (L.) Choisy = Ipomoea biflora (L.) Pers. ■
25318 Aniseia biflora (L.) Choisy = Ipomoea sinensis (Desr.) Choisy ■
25319 Aniseia calycina (Roxb.) Choisy = Ipomoea biflora (L.) Pers. ■
25320 Aniseia calystegioides Choisy = Ipomoea crassipes Hook. ■☆
25321 Aniseia fulvicaulis Hochst. ex Choisy = Ipomoea fulvicaulis (Hochst. ex Choisy) Boiss. ex Hallier f. ■☆
25322 Aniseia hastata Meisn. = Ipomoea fimbriosepala Choisy ■
25323 Aniseia martinicensis (Jacq.) Choisy;日本心萼薯■☆
25324 Aniseia media (L.) Choisy = Merremia medium (L.) Hallier f. ■☆
25325 Aniseia stenantha (Dunn) Y. Ling ex R. C. Fang et S. H. Huang;狭花心萼薯;Narrowflower Aniseia ■
25326 Aniseia stenantha (Dunn) Y. Ling ex R. C. Fang et S. H. Huang = Ipomoea fimbriosepala Choisy ■
25327 Aniseia stenantha (Dunn) Y. Ling ex R. C. Fang et S. H. Huang = Ipomoea stenantha Dunn ■
25328 Aniseia stenantha (Dunn) Y. Ling ex R. C. Fang et S. H. Huang var. macrostephana Y. H. Zhang;大花心萼薯;Big Narrowflower Aniseia ■
25329 Aniseia stenantha (Dunn) Y. Ling ex R. C. Fang et S. H. Huang var. macrostephana Y. H. Zhang = Ipomoea fimbriosepala Choisy ■
25330 Aniseion St. -Lag. = Aniseia Choisy ■
25331 Aniselytron Merr. (1910);沟稃草属;Aulacolepis,Furrowlemma ■
25332 Aniselytron Merr. = Calamagrostis Adans. ■
25333 Aniselytron agrostoides Merr.;小颖沟稃草(小沟稃草);Small Aulacolepis,Small Furrowlemma,Taiwan Aulacolepis ■
25334 Aniselytron agrostoides Merr. = Aulacolepis agrostoides (Merr.) Ohwi ■
25335 Aniselytron agrostoides Merr. var. formosana (Ohwi) N. X. Zhao = Aniselytron agrostoides Merr. ■
25336 Aniselytron clemensiae (Hitchc.) Soják = Aniselytron treutleri (Kuntze) Soják ■
25337 Aniselytron formosana (Ohwi) L. Liou = Aniselytron agrostoides Merr. ■
25338 Aniselytron gracilis (Keng) N. X. Zhao = Deyeuxia petelotii (Hitchc.) S. M. Phillips et Wen L. Chen ■
25339 Aniselytron japonica (Hack.) Bennet et Raizada = Aniselytron treutleri (Kuntze) Soják ■
25340 Aniselytron japonica (Hack.) Bennet et Raizada = Aniselytron treutleri (Kuntze) Soják var. japonicum (Hack.) N. X. Zhao ■
25341 Aniselytron milioides (Honda) Bennet et Raizada = Aniselytron treutleri (Kuntze) Soják ■
25342 Aniselytron pseudopoa (Jansen) Soják = Aniselytron treutleri (Kuntze) Soják ■
25343 Aniselytron treutleri (Kuntze) Soják;沟稃草;Common Aulacolepis,Common Furrowlemma ■
25344 Aniselytron treutleri (Kuntze) Soják = Aulacolepis treutleri (Kuntze) Hack. ■
25345 Aniselytron treutleri (Kuntze) Soják var. japonica (Hack.) N. X. Zhao = Aniselytron treutleri (Kuntze) Soják ■
25346 Aniselytron treutleri (Kuntze) Soják var. japonicum (Hack.) N. X. Zhao;日本沟稃草(小颖沟稃草);Japan Furrowlemma,Japanese Aulacolepis ■
25347 Anisepta Raf. = Croton L. ●
25348 Aniserica N. E. Br. (1906);异石南属●☆
25349 Aniserica N. E. Br. = Eremia D. Don ●☆
25350 Aniserica N. E. Br. = Erica L. ●☆
25351 Aniserica gracilis (Bartl.) N. E. Br. = Erica benthamiana E. G. H. Oliv. ●☆
25352 Aniserica gracilis (Bartl.) N. E. Br. var. hispida N. E. Br. = Erica benthamiana E. G. H. Oliv. ●☆
25353 Aniserica macrocalyx Salter = Erica benthamiana E. G. H. Oliv. ●☆
25354 Anisifolium Kuntze = Limonia L. ●☆
25355 Anisifolium Rumph. = Limonia L. ●☆
25356 Anisliaea bonatii Beauverd = Ainsliaea mattfeldiana Hand. -Mazz. ■
25357 Anisliaea bonatii Beauverd var. arachnoides Beauverd;蛛毛心叶兔儿风■
25358 Anisocalyx Hance = Bacopa Aubl. (保留属名) ■
25359 Anisocalyx Hance = Brami Adans. (废弃属名) ■
25360 Anisocalyx Hance = Herpestis C. F. Gaertn. ■
25361 Anisocalyx L. Bolus = Drosanthemopsis Rauschert ●☆
25362 Anisocalyx L. Bolus = Drosanthemum Schwantes ●☆
25363 Anisocalyx L. Bolus(1958);歧萼番杏属(歧萼属)●☆
25364 Anisocalyx salarius L. Bolus = Jacobsenia vaginata (L. Bolus) Ihlenf. ●☆
25365 Anisocalyx vaginatus (L. Bolus) L. Bolus = Jacobsenia vaginata (L. Bolus) Ihlenf. ●☆
25366 Anisocapparis Cornejo et Iltis = Capparis L. ●
25367 Anisocarpus Nutt. (1841);歪果菊属■☆
25368 Anisocarpus Nutt. = Madia Molina ■☆
25369 Anisocarpus bolanderi A. Gray = Kyhosia bolanderi (A. Gray) B. G. Baldwin ■☆
25370 Anisocarpus madioides Nutt.;歪果菊■☆
25371 Anisocarpus scabridus (Eastw.) B. G. Baldwin;微糙歪果菊;Raillardella Scabrida Eastwood ■☆
25372 Anisocentra Turcz. = Tropaeolum L. ■
25373 Anisocentrum Turcz. = Acisanthera P. Browne ●■☆
25374 Anisocereus Backeb. (1938);鳞花柱属■☆
25375 Anisocereus Backeb. = Escontria Rose ●☆
25376 Anisochaeta DC. (1836);芒冠鼠麹木属●☆
25377 Anisochaeta mikanioides DC.;芒冠鼠麹木●☆
25378 Anisochilus Wall. = Anisochilus Wall. ex Benth. ●■
25379 Anisochilus Wall. ex Benth. (1830);排草香属(异唇花属);Anisochilus ●■
25380 Anisochilus africanus Baker ex Scott-Elliot = Leocus africanus (Baker ex Scott-Elliot) J. K. Morton ●■☆
25381 Anisochilus carnosus (L.) Wall.;排香草(耙草,排草,排香,肉质异唇花);Common Anisochilus ●
25382 Anisochilus crassus Benth. = Anisochilus carnosus (L.) Wall. ●
25383 Anisochilus dysophylloides Benth.;水蜡烛状排草香;Dysophyllalike Anisochilus ●☆
25384 Anisochilus engleri Briq. = Leocus africanus (Baker ex Scott-

Elliot) J. K. Morton ●■☆
25385 Anisochilus eriocephalus Benth.;绵毛头排草香;Cottony Anisochilus ●☆
25386 Anisochilus pallidus Wall.;异唇花;Pallid Anisochilus ●
25387 Anisochilus paniculatus Benth.;圆锥花序排草香;Panicle Anisochilus ●☆
25388 Anisochilus plantagineus Hook. f.;车前状排草香;Plantainlike Anisochilus ●☆
25389 Anisochilus polystachys Benth.;多穗排草香;Manyspike Anisochilus ●☆
25390 Anisochilus robustus Hook. f.;粗壮排草香;Robust Anisochilus ●☆
25391 Anisochilus rupestris Wight ex Hook. f. = Anisochilus carnosus (L.) Wall. ●
25392 Anisochilus scaber Benth.;粗糙排草香;Scabrous Anisochilus ●☆
25393 Anisochilus sericeus Benth.;绢毛排草香;Silky Anisochilus ●☆
25394 Anisochilus sinensis Hance = Nosema cochinchinensis (Lour.) Merr. ■
25395 Anisochilus suffruticosus Wight;灌木排草香;Shrubby Anisochilus ●☆
25396 Anisochilus verticillatus Hook. f.;轮生排草香;Verticillate Anisochilus ●☆
25397 Anisochilus wightii Hook. f.;威特排草香;Wight Anisochilus ●☆
25398 Anisocoma Torr. = Anisocoma Torr. et A. Gray ■☆
25399 Anisocoma Torr. et A. Gray(1845);异冠苣属■☆
25400 Anisocoma acaulis Torr. et A. Gray;异冠苣;Scalebud ■☆
25401 Anisocycla Baill. (1887);异环藤属(歪环防已属)●☆
25402 Anisocycla blepharosepala Diels;缘毛萼异环藤●☆
25403 Anisocycla blepharosepala Diels subsp. tanzaniensis Vollesen;坦桑尼亚异环藤●☆
25404 Anisocycla capituliflora Diels = Albertisia capituliflora (Diels) Forman ●☆
25405 Anisocycla cymosa Troupin;聚伞异环藤●☆
25406 Anisocycla ferruginea Diels = Albertisia ferruginea (Diels) Forman ●☆
25407 Anisocycla grandidieri Baill.;异环藤(歪环防已)●☆
25408 Anisocycla jollyana (Pierre) Diels;若利异环藤●☆
25409 Anisocycla linearis Pierre ex Diels;线状异环藤●☆
25410 Anisocycla triplinervia (Pax) Diels;三脉异环藤●☆
25411 Anisodens Dulac = Scabiosa L. ●■
25412 Anisoderis Cass. = Crepis L. ■
25413 Anisoderis Cass. = Wibelia P. Gaertn., B. Mey. et Scherb. ■
25414 Anisodontea C. Presl(1845);南非葵属●■☆
25415 Anisodontea × hypomadara (Sprague) D. M. Bates;下马南非葵;Cape Mallow ●☆
25416 Anisodontea alexandri (Baker f.) Bates;亚历山大南非葵●☆
25417 Anisodontea anomala (Link et Otto) Bates;异常南非葵●☆
25418 Anisodontea biflora (Desr.) Bates;双花南非葵●☆
25419 Anisodontea bryoniifolia (L.) D. M. Bates;泻根叶南非葵●☆
25420 Anisodontea capensis (L.) D. M. Bates;好望角南非葵(南非葵);Cape African-Queen, Cape Mallow ●☆
25421 Anisodontea dissecta (Harv.) Bates;深裂南非葵●☆
25422 Anisodontea dregeana C. Presl = Anisodontea anomala (Link et Otto) Bates ●☆
25423 Anisodontea elegans (Cav.) Bates;雅致南非葵●☆
25424 Anisodontea fruticosa (P. J. Bergius) Bates;灌丛南非葵●☆
25425 Anisodontea gracilis Bates;纤细南非葵●☆
25426 Anisodontea julii (Burch. ex DC.) Bates;茸毛南非葵●☆
25427 Anisodontea julii (Burch. ex DC.) Bates subsp. pannosa (Bolus) Bates;毡状南非葵●☆
25428 Anisodontea julii (Burch. ex DC.) Bates subsp. prostrata (E. Mey. ex Turcz.) Bates;平卧南非葵●☆
25429 Anisodontea malvastroides (Baker f.) Bates;赛葵南非葵●☆
25430 Anisodontea procumbens (Harv.) Bates;平铺南非葵●☆
25431 Anisodontea pseudocapensis Bates;假好望角南非葵●☆
25432 Anisodontea racemosa (Harv.) Bates;总花南非葵●☆
25433 Anisodontea reflexa (J. C. Wendl.) Bates;反折南非葵●☆
25434 Anisodontea scabrosa (L.) D. M. Bates;南非葵;Cape Mallow ●☆
25435 Anisodontea setosa (Harv.) Bates;刚毛南非葵●☆
25436 Anisodontea triloba (Thunb.) Bates;三裂南非葵●☆
25437 Anisodus Link et Otto = Anisodus Link ex Spreng. ■
25438 Anisodus Link ex Spreng. (1824);山莨菪属(东莨菪属,赛莨菪属,三分三属);Anisodus ■
25439 Anisodus acutangulus C. Y. Wu et C. Chen;三分三(大搜山虎,锐齿东莨菪,锐萼东莨菪,山茄子,山野烟,野旱烟,野烟);Acuteangular Anisodus, Common Scopolia, Sharpangle Anisodus ■
25440 Anisodus acutangulus C. Y. Wu et C. Chen var. breviflorus C. Y. Wu et C. Chen;三分七(三分三,野莰);Shortflower Acuteangular Anisodus ■
25441 Anisodus breviflorus C. Y. Wu et C. Chen = Anisodus acutangulus C. Y. Wu et C. Chen var. breviflorus C. Y. Wu et C. Chen ■
25442 Anisodus carniolicoides (C. Y. Wu et C. Chen) D'Arcy et Zhi Y. Zhang;赛莨菪(疯药,七厘散,搜山虎,无慈);Common Scopolia, Scopolia ■
25443 Anisodus carniolicoides (C. Y. Wu et C. Chen) D'Arcy et Zhi Y. Zhang var. dentata C. Y. Wu et C. Chen;小莨菪(齿叶赛莨菪,粗齿赛莨菪,赛莨菪,三分三,小赛莨菪);Dentate Scopolia ■
25444 Anisodus carniolicoides (C. Y. Wu et C. Chen) D'Arcy et Zhi Y. Zhang var. dentata C. Y. Wu et C. Chen = Anisodus carniolicoides (C. Y. Wu et C. Chen) D'Arcy et Zhi Y. Zhang ■
25445 Anisodus caulescens (C. B. Clarke) Diels = Mandragora caulescens C. B. Clarke ■
25446 Anisodus fischerianus Pascher = Anisodus luridus Link et Otto var. fischerianus (Pascher) C. Y. Wu et C. Chen ex C. Chen et Chun L. Chen ■
25447 Anisodus luridus Link et Otto;铃铛子(藏茄,赛莨菪,三分三,山莨菪,山烟,山鰶,唐古特山莨菪,喜马拉雅东莨菪);Common Anisodus ■
25448 Anisodus luridus Link et Otto = Scopolia stramonifolia Sem. ■
25449 Anisodus luridus Link et Otto var. fischeHanus (Pascher) C. Y. Wu et C. Chen = Anisodus luridus Link et Otto ■
25450 Anisodus luridus Link et Otto var. fischerianus (Pascher) C. Y. Wu et C. Chen ex C. Chen et Chun L. Chen;丽江山莨菪(丽江莨菪,三分三);Lijiang Anisodus, Nisan Anisodus ■
25451 Anisodus mairei (H. Lév.) C. Y. Wu et C. Chen;搜山虎;Maire Anisodus ■
25452 Anisodus mariae Pascher = Mandragora caulescens C. B. Clarke ■
25453 Anisodus sinensis (Hemsl.) Pascher = Atropanthe sinensis (Hemsl.) Pascher ●■
25454 Anisodus sinensis Pascher = Atropanthe sinensis (Hemsl.) Pascher ●■
25455 Anisodus stemonifolius G. Don = Anisodus luridus Link et Otto ■
25456 Anisodus stramonifolius (Wall.) G. Don = Anisodus luridus Link et Otto ■

25457 Anisodus tanguticus (Maxim.) Pascher;山莨菪(藏茄,甘青赛莨菪,甘青山莨菪,唐古特东莨菪,唐古特莨菪,樟柳,樟柳参,樟柳柽);Tangut Anisodus ■

25458 Anisodus tanguticus (Maxim.) Pascher = Scopolia tangutica Maxim. ■

25459 Anisodus tanguticus (Maxim.) Pascher var. viridulus C. Y. Wu et C. Chen;黄花山莨菪;Yellow Tangut Anisodus ■

25460 Anisolepis Steetz = Helipterum DC. ex Lindl. ■☆

25461 Anisolepis Steetz(1845);歧鳞菊属■☆

25462 Anisolepis pyrethrum Steetz;歧鳞菊■☆

25463 Anisolobus A. DC. = Odontadenia Benth. ●☆

25464 Anisolotus Bernh. = Hosackia Douglas ex Benth. ■☆

25465 Anisomallon Baill. = Apodytes E. Mey. ex Arn. ●

25466 Anisomeles R. Br. (1810);金剑草属(防风属,广防风属);Anisomeles ●■

25467 Anisomeles R. Br. = Epimeredi Adans. ■

25468 Anisomeles candicans Benth.;白亮金剑草(白亮广防风);White Anisomeles ■☆

25469 Anisomeles disticha B. Heyne ex Roth = Anisomeles indica (L.) Kuntze ■

25470 Anisomeles furcata (Link) Sweet = Craniotome furcata (Link) Kuntze ■

25471 Anisomeles furcata Sweet = Craniotome furcata (Link) Kuntze ■

25472 Anisomeles glabrata Wall. = Anisomeles indica (L.) Kuntze ■

25473 Anisomeles heyneana Benth.;海奈金剑草(海奈广防风);Heyne Anisomeles ■☆

25474 Anisomeles indica (L.) Kuntze = Epimeredi indica (L.) Rothm. ■

25475 Anisomeles malabarica R. Br.;马拉巴金剑草(马拉巴广防风)■☆

25476 Anisomeles mollissima Wall. = Anisomeles indica (L.) Kuntze ■

25477 Anisomeles nepalensis Spreng. = Craniotome furcata (Link) Kuntze ■

25478 Anisomeles ovata R. Br. = Epimeredi indica (L.) Rothm. ■

25479 Anisomeles tonkinensis Gand. = Anisomeles indica (L.) Kuntze ■

25480 Anisomeria D. Don(1832);异商陆属●■☆

25481 Anisomeria chilensis H. Walter;智利异商陆●☆

25482 Anisomeria coriacea D. Don;异商陆●☆

25483 Anisomeria densiflora H. Walter;密花异商陆●☆

25484 Anisomeris C. Presl = Chomelia Jacq.(保留属名)●☆

25485 Anisometros Hassk. = Pimpinella L. ■

25486 Anisonema A. Juss. = Phyllanthus L. ●■

25487 Anisonema hypoleucum Miq. = Glochidion lutescens Blume ●

25488 Anisopappus Hook. et Arn. (1837);山黄菊属;Anisopapus ■

25489 Anisopappus abercornensis G. Taylor;阿伯康山黄菊■☆

25490 Anisopappus abercornensis G. Taylor subsp. anemonifolius (DC.) S. Ortiz, Paiva et Rodr. Oubina = Anisopappus anemonifolius (DC.) G. Taylor ■☆

25491 Anisopappus africanus (Hook. f.) Oliv. et Hiern = Anisopappus chinensis Hook. et Arn. subsp. africanus (Hook. f.) S. Ortiz et Paiva ■☆

25492 Anisopappus anemonifolius (DC.) G. Taylor;风花叶山黄菊■☆

25493 Anisopappus angolensis O. Hoffm. = Anisopappus anemonifolius (DC.) G. Taylor ■☆

25494 Anisopappus annuus Lawalrée = Anisopappus abercornensis G. Taylor ■☆

25495 Anisopappus athanasioides Paiva et S. Ortiz;永菊状山黄菊■☆

25496 Anisopappus aureus Hutch. et B. L. Burtt = Anisopappus chinensis (L.) Hook. et Arn. ■

25497 Anisopappus bampsianus Lisowski;邦氏山黄菊■☆

25498 Anisopappus boinensis (Humbert) Wild;博伊纳山黄菊■☆

25499 Anisopappus buchwaldii (O. Hoffm.) Wild = Anisopappus chinensis Hook. et Arn. var. buchwaldii (O. Hoffm.) S. Ortiz, Paiva et Rodr. Oubina ■☆

25500 Anisopappus buchwaldii (O. Hoffm.) Wild subsp. iodotrichus (Brenan) Wild = Anisopappus chinensis Hook. et Arn. var. buchwaldii (O. Hoffm.) S. Ortiz, Paiva et Rodr. Oubina ■☆

25501 Anisopappus burundiensis Lisowski;布隆迪山黄菊■☆

25502 Anisopappus candelabrum H. Lév. = Adenostemma lavenia (L.) Kuntze ■

25503 Anisopappus canescens Hutch. = Anisopappus chinensis (L.) Hook. et Arn. ■

25504 Anisopappus chinensis (L.) Hook. et Arn.;山黄菊(旱山菊,黄花莲,金菊花,萄涧菊,旋覆花);China Anisopapus, Chinese Anisopapus ■

25505 Anisopappus chinensis Hook. et Arn. = Anisopappus chinensis (L.) Hook. et Arn. ■

25506 Anisopappus chinensis Hook. et Arn. subsp. africanus (Hook. f.) S. Ortiz et Paiva;非洲山黄菊■☆

25507 Anisopappus chinensis Hook. et Arn. subsp. lobatus (Wild) S. Ortiz et Paiva;浅裂山黄菊■☆

25508 Anisopappus chinensis Hook. et Arn. subsp. oliveranus (Wild) S. Ortiz, Paiva et Rodr. Oubina;奥里弗山黄菊■☆

25509 Anisopappus chinensis Hook. et Arn. var. buchwaldii (O. Hoffm.) S. Ortiz, Paiva et Rodr. Oubina;布克山黄菊■☆

25510 Anisopappus chinensis Hook. et Arn. var. dentatus (DC.) S. Ortiz, Paiva et Rodr. Oubina;齿叶山黄菊■☆

25511 Anisopappus chinensis Hook. et Arn. var. macrocephala (Humbert) S. Ortiz, Paiva et Rodr. Oubina = Anisopappus chinensis Hook. et Arn. var. buchwaldii (O. Hoffm.) S. Ortiz, Paiva et Rodr. Oubina ■☆

25512 Anisopappus chinensis Hook. et Arn. var. rupestris (DC.) S. Ortiz, Paiva et Rodr. Oubina = Anisopappus chinensis Hook. et Arn. var. dentatus (DC.) S. Ortiz, Paiva et Rodr. Oubina ■☆

25513 Anisopappus corymbosus Wild;伞序山黄菊■☆

25514 Anisopappus dalzielii Hutch. = Anisopappus chinensis (L.) Hook. et Arn. ■

25515 Anisopappus davyi S. Moore;戴维山黄菊■☆

25516 Anisopappus davyi S. Moore var. bampsianus (Lisowski) S. Ortiz, Paiva et Rodr. Oubina = Anisopappus bampsianus Lisowski ■☆

25517 Anisopappus davyi S. Moore var. pumilus (Hiern) Wild = Anisopappus pumilus (Hiern) Wild ■☆

25518 Anisopappus dentatus (DC.) Wild;尖齿山黄菊■☆

25519 Anisopappus dentatus (DC.) Wild = Anisopappus chinensis Hook. et Arn. var. dentatus (DC.) S. Ortiz, Paiva et Rodr. Oubina ■☆

25520 Anisopappus dentatus (DC.) Wild subsp. lobatus Wild = Anisopappus chinensis Hook. et Arn. subsp. lobatus (Wild) S. Ortiz et Paiva ■☆

25521 Anisopappus discolor Wild;异色山黄菊■☆

25522 Anisopappus exellii Wild;埃克塞尔山黄菊■☆

25523 Anisopappus fruticosus S. Ortiz et Paiva;灌丛山黄菊■☆

25524 Anisopappus gracilis O. Hoffm. = Anisopappus chinensis (L.) Hook. et Arn. ■

25525 Anisopappus grangeoides (Vatke et Höpfner ex Klatt) Merxm.;

田基黄山黄菊■☆
25526 Anisopappus hoffmannianus Hutch. = Anisopappus chinensis (L.) Hook. et Arn. ■
25527 Anisopappus holstii (O. Hoffm.) Wild;霍尔山黄菊■☆
25528 Anisopappus inuloides Hutch. et B. L. Burtt = Anisopappus chinensis Hook. et Arn. var. buchwaldii (O. Hoffm.) S. Ortiz, Paiva et Rodr. Oubina ■☆
25529 Anisopappus iodotrichus Brenan = Anisopappus chinensis Hook. et Arn. var. buchwaldii (O. Hoffm.) S. Ortiz, Paiva et Rodr. Oubina ■☆
25530 Anisopappus junodii Hutch.;朱诺德山黄菊■☆
25531 Anisopappus kirkii (Oliv.) Brenan;柯克山黄菊■☆
25532 Anisopappus lastii (O. Hoffm.) Wild;拉斯特山黄菊■☆
25533 Anisopappus lastii (O. Hoffm.) Wild = Anisopappus chinensis Hook. et Arn. var. dentatus (DC.) S. Ortiz, Paiva et Rodr. Oubina ■☆
25534 Anisopappus lastii (O. Hoffm.) Wild subsp. welwitschii ? = Anisopappus chinensis Hook. et Arn. var. buchwaldii (O. Hoffm.) S. Ortiz, Paiva et Rodr. Oubina ■☆
25535 Anisopappus latifolius (S. Moore) B. L. Burtt;宽叶山黄菊■☆
25536 Anisopappus lawalreeanus Lisowski;拉瓦尔山黄菊■☆
25537 Anisopappus lejolyanus Lisowski;勒若利山黄菊■☆
25538 Anisopappus longipes (Comm. ex Cass.) Wild;长梗山黄菊■☆
25539 Anisopappus marianus Lawalrée;玛利亚山黄菊■☆
25540 Anisopappus oliveranus Wild = Anisopappus chinensis Hook. et Arn. subsp. oliveranus (Wild) S. Ortiz, Paiva et Rodr. Oubina ■☆
25541 Anisopappus orbicularis (Humbert) Wild;圆形山黄菊■☆
25542 Anisopappus paucidentatus Wild;少齿山黄菊■☆
25543 Anisopappus petitianus Lisowski;佩蒂蒂山黄菊■☆
25544 Anisopappus pinnatifidus (Klatt) O. Hoffm. ex Hutch.;羽裂山黄菊■☆
25545 Anisopappus pseudopinnatifidus S. Ortiz et Paiva;假羽裂山黄菊■☆
25546 Anisopappus pumilus (Hiern) Wild;矮小山黄菊■☆
25547 Anisopappus rhombifolius Wild;菱叶山黄菊■☆
25548 Anisopappus robynsianus Lisowski;罗宾斯山黄菊■☆
25549 Anisopappus rogersii G. Taylor = Anisopappus chinensis Hook. et Arn. subsp. africanus (Hook. f.) S. Ortiz et Paiva ■☆
25550 Anisopappus salviifolius (DC.) Wild;马岛山黄菊■☆
25551 Anisopappus subdiscoideus O. Hoffm. = Anisopappus chinensis (L.) Hook. et Arn. ■
25552 Anisopappus suborbicularis Hutch. et B. L. Burtt = Anisopappus chinensis Hook. et Arn. var. buchwaldii (O. Hoffm.) S. Ortiz, Paiva et Rodr. Oubina ■☆
25553 Anisopappus sylvatica (Humbert) Wild;林地山黄菊■☆
25554 Anisopappus tenerus (S. Moore) Brenan = Anisopappus chinensis Hook. et Arn. var. dentatus (DC.) S. Ortiz, Paiva et Rodr. Oubina ■☆
25555 Anisopappus triloba (Klatt) O. Hoffm. ex Merxm. = Anisopappus pinnatifidus (Klatt) O. Hoffm. ex Hutch. ■☆
25556 Anisopappus upembensis Lisowski;乌彭贝山黄菊■☆
25557 Anisopetala (Kraenzl.) M. A. Clem. = Dendrobium Sw. (保留属名)■
25558 Anisopetala Walp. = Pelargonium L'Hér. ex Aiton ●■
25559 Anisopetalon Hook. = Bulbophyllum Thouars(保留属名)■
25560 Anisopetalum Hook. = Bulbophyllum Thouars(保留属名)■
25561 Anisophyllea R. Br. = Anisophyllea R. Br. ex Sabine ●☆
25562 Anisophyllea R. Br. ex Sabine(1824);异叶木属(四柱木属,异叶红树属,异叶树属)●☆
25563 Anisophyllea boehmii Engl.;贝姆异叶木●☆
25564 Anisophyllea brachystila Engl. et Brehmer = Anisophyllea quangensis Engl. ex Henriq. ●☆
25565 Anisophyllea buchneri Engl. et Brehmer;布赫纳异叶木●☆
25566 Anisophyllea buettneri Engl. = Anisophyllea quangensis Engl. ex Henriq. ●☆
25567 Anisophyllea cavaleriei H. Lév. = Vaccinium foetidissimum H. Lév. et Vaniot ●
25568 Anisophyllea disticha (Jack.) Baill.;二裂异叶木●☆
25569 Anisophyllea exellii P. A. Duvign. et Dewit = Anisophyllea boehmii Engl. ●☆
25570 Anisophyllea fruticulosa Engl. et Gilg = Anisophyllea quangensis Engl. ex Henriq. ●☆
25571 Anisophyllea gossweileri Engl. et Brehmer = Anisophyllea boehmii Engl. ●☆
25572 Anisophyllea laurina R. Br. ex Sabine;月桂异叶木;Monkey Apple ●☆
25573 Anisophyllea mayumbensis Exell;马永巴异叶木●☆
25574 Anisophyllea myriosticta Floret;多点异叶木●☆
25575 Anisophyllea obtusifolia Engl. et Brehmer;钝叶异叶木●☆
25576 Anisophyllea pochetii P. A. Duvign. et Dewit = Anisophyllea quangensis Engl. ex Henriq. ●☆
25577 Anisophyllea poggei Engl. ex De Wild. et T. Durand = Anisophyllea quangensis Engl. ex Henriq. ●☆
25578 Anisophyllea polyneura Floret;多脉异叶木●☆
25579 Anisophyllea pomifera Engl. et Brehmer = Anisophyllea boehmii Engl. ●☆
25580 Anisophyllea purpurascens Hutch. et Dalziel;紫异叶木●☆
25581 Anisophyllea quangensis Engl. ex Henriq.;广异叶木●☆
25582 Anisophyllea setosa Mildbr.;刚毛异叶木●☆
25583 Anisophyllea sororia Pierre;团积异叶木●☆
25584 Anisophylleaceae Ridl. (1922);异叶木科(红树科,四柱木科,异形叶科,异叶红树科)●☆
25585 Anisophyllum Boivin ex Baill. = Croton L. ●
25586 Anisophyllum G. Don = Anisophyllea R. Br. ex Sabine ●☆
25587 Anisophyllum Haw. = Euphorbia L. ●■
25588 Anisophyllum aegyptiacum (Boiss.) Schweinf. = Euphorbia forsskalii J. Gay ■☆
25589 Anisophyllum arabicum (Hochst. et Steud. ex T. Anderson) Schweinf. = Euphorbia arabica Hochst. et Steud. ex T. Anderson ●■☆
25590 Anisophyllum convolvuloides (Hochst. ex Benth.) Klotzsch et Garcke = Euphorbia convolvuloides Hochst. ex Benth. ■☆
25591 Anisophyllum forsskalii (J. Gay) Klotzsch et Garcke = Euphorbia forsskalii J. Gay ■☆
25592 Anisophyllum glaucophyllum (Poir.) Klotzsch et Garcke = Euphorbia trinervia Schumach. et Thonn. ●■☆
25593 Anisophyllum indicum Schweinf. = Euphorbia indica Lam. ■☆
25594 Anisophyllum mossambicense Klotzsch et Garcke = Euphorbia mossambicensis (Klotzsch et Garcke) Boiss. ●■☆
25595 Anisophyllum polycnemoides (Hochst. ex Boiss.) Klotzsch et Garcke = Euphorbia polycnemoides Hochst. ex Boiss. ☆
25596 Anisophyllum scordifolium (Jacq.) Klotzsch et Garcke = Euphorbia scordifolia Jacq. ■☆
25597 Anisophyllum tettense Klotzsch et Garcke = Euphorbia tettensis Klotzsch ☆
25598 Anisoplectus Oerst. = Drymonia Mart. ●☆
25599 Anisopleetus Oerst. = Alloplectus Mart. (保留属名)●■☆

25600 Anisopleura Fenzl = Heptaptera Margot et Reut. ■☆
25601 Anisopoda Baker(1890);异足芹属■☆
25602 Anisopoda bupleuroides Baker;异足芹■☆
25603 Anisopogon R. Br. (1810);澳异芒草属■☆
25604 Anisopogon capensis Nees;澳异芒草■☆
25605 Anisoptera Korth. (1841);异翅香属;Krabak,Mersawa,Palosapis ●☆
25606 Anisoptera aurea Foxw.;金黄异翅香●☆
25607 Anisoptera costata Pierre;中脉异翅香●☆
25608 Anisoptera curtisii Dyer ex King;柯氏异翅香;Krabak ●☆
25609 Anisoptera glabra Kurz = Scaphula glabra (Kurz) R. Parker ●☆
25610 Anisoptera laevis Ridl.;平滑异翅香●☆
25611 Anisoptera mangachapoi (Blanco) DC. = Vatica mangachapoi Blanco ●◇
25612 Anisoptera scaphula (Roxb.) Kurz;斯卡异翅香;Kannghmu ●☆
25613 Anisoptera thurifera (Bianco) Blume;菲律宾异翅香;Palosapis ●☆
25614 Anisopus N. E. Br. (1895);异足萝藦属●☆
25615 Anisopus batesii S. Moore = Anisopus mannii N. E. Br. ●☆
25616 Anisopus bicoronatus (K. Schum.) N. E. Br. = Anisopus mannii N. E. Br. ●☆
25617 Anisopus mannii N. E. Br.;异足萝藦●☆
25618 Anisopus rostriferus (N. E. Br.) Bullock = Anisopus mannii N. E. Br. ●☆
25619 Anisopyrum (Griseb.) Gren. et Duval = Leymus Hochst. ■
25620 Anisopyrum Gren. et Duval = Agropyron Gaertn. ■
25621 Anisora Raf. = Helicteres L. ●
25622 Anisoramphus DC. = Crepis L. ■
25623 Anisoramphus hypochaeroides Hook. f. = Crepis newii Oliv. et Hiern subsp. oliveriana (Kuntze) C. Jeffrey et Beentje ■☆
25624 Anisosciadium DC. (1829);肖伞芹属■☆
25625 Anisosciadium DC. = Echinophora L. ■☆
25626 Anisosciadium orientale DC.;肖伞芹■☆
25627 Anisosepalum E. Hossain(1972);异萼爵床属■☆
25628 Anisosepalum alboviolaceum (Benoist) E. Hossain;浅堇色异萼爵床■☆
25629 Anisosepalum alboviolaceum (Benoist) E. Hossain subsp. gracilis (Heine) Champl.;大株浅堇色异萼爵床■☆
25630 Anisosepalum alboviolaceum (Benoist) E. Hossain subsp. grandiflorum (Napper) E. Hossain = Anisosepalum alboviolaceum (Benoist) E. Hossain ■☆
25631 Anisosepalum alboviolaceum (Benoist) E. Hossain var. gracilior (Heine) E. Hossain = Anisosepalum alboviolaceum (Benoist) E. Hossain subsp. gracilis (Heine) Champl. ■☆
25632 Anisosepalum humbertii (Mildbr.) E. Hossain;亨伯特异萼爵床■☆
25633 Anisosepalum humbertii (Mildbr.) E. Hossain subsp. zambiense Champl.;赞比亚异萼爵床■☆
25634 Anisosepalum lewallei Bamps;勒瓦莱异萼爵床■☆
25635 Anisosorus Trevis. = Lonchitis L. ■☆
25636 Anisosperma Silva Manso = Fevillea L. ■☆
25637 Anisosperma Silva Manso(1836);异籽葫芦属■☆
25638 Anisostachya Nees = Justicia L. ●■
25639 Anisostachya aequiloba Benoist;等裂异籽葫芦■☆
25640 Anisostachya ambositrensis Benoist;安布西特拉异籽葫芦■☆
25641 Anisostachya amoena Benoist;秀丽异籽葫芦■☆
25642 Anisostachya andringitrensis Benoist;安德林吉特拉山异籽葫芦■☆
25643 Anisostachya arida (Scott-Elliot) Benoist;旱生异籽葫芦■☆
25644 Anisostachya armandii Benoist;阿尔芒异籽葫芦■☆
25645 Anisostachya atrorubra Benoist;暗红异籽葫芦■☆
25646 Anisostachya betsiliensis Benoist;贝齐里异籽葫芦■☆
25647 Anisostachya bivalvis Benoist;双分果片异籽葫芦■☆
25648 Anisostachya bojeri Nees;博耶尔异籽葫芦■☆
25649 Anisostachya bosseri Benoist;博瑟异籽葫芦■☆
25650 Anisostachya brevibracteata Benoist;短苞异籽葫芦■☆
25651 Anisostachya breviloba Benoist;短裂异籽葫芦■☆
25652 Anisostachya capuronii Benoist;凯普伦异籽葫芦■☆
25653 Anisostachya castellana Benoist;卡地异籽葫芦■☆
25654 Anisostachya coccinea Benoist;绯红异籽葫芦■☆
25655 Anisostachya cognata Benoist;近缘异籽葫芦■☆
25656 Anisostachya colorata Benoist;着色异籽葫芦■☆
25657 Anisostachya commersonii T. Anderson;科梅逊异籽葫芦■☆
25658 Anisostachya debilis Benoist;弱小异籽葫芦■☆
25659 Anisostachya delphinensis Benoist;德尔芬异籽葫芦■☆
25660 Anisostachya denticulata Benoist;细齿异籽葫芦■☆
25661 Anisostachya elata Benoist;高异籽葫芦■☆
25662 Anisostachya elliptica Benoist;椭圆异籽葫芦■☆
25663 Anisostachya humbertii Benoist;亨伯特异籽葫芦■☆
25664 Anisostachya humblotii Benoist;洪布异籽葫芦■☆
25665 Anisostachya incisa Benoist;锐裂异籽葫芦■☆
25666 Anisostachya isalensis Benoist;伊萨卢异籽葫芦■☆
25667 Anisostachya latebracteata Benoist;宽苞异籽葫芦■☆
25668 Anisostachya littoralis Benoist;滨海异籽葫芦■☆
25669 Anisostachya maculata Benoist;斑点异籽葫芦■☆
25670 Anisostachya oblonga Benoist;矩圆异籽葫芦■☆
25671 Anisostachya parvifolia Benoist;小叶异籽葫芦■☆
25672 Anisostachya perrieri Benoist;佩里耶异籽葫芦■☆
25673 Anisostachya puberula Benoist;微毛异籽葫芦■☆
25674 Anisostachya pubescens Benoist;短毛异籽葫芦■☆
25675 Anisostachya purpurea Benoist;紫异籽葫芦■☆
25676 Anisostachya ramosa Benoist;分枝异籽葫芦■☆
25677 Anisostachya ripicola Benoist;岩地异籽葫芦■☆
25678 Anisostachya rivalis Benoist;溪边异籽葫芦■☆
25679 Anisostachya rosea Benoist;粉红异籽葫芦■☆
25680 Anisostachya sambiranensis Benoist;异籽葫芦■☆
25681 Anisostachya seyrigii Benoist;塞里格异籽葫芦■☆
25682 Anisostachya spatulata Benoist;匙形异籽葫芦■☆
25683 Anisostachya straminea Benoist;草黄异籽葫芦■☆
25684 Anisostachya tenella (Nees) Lindau = Justicia tenella (Nees) T. Anderson ■☆
25685 Anisostachya tenella Lindau = Justicia tenella (Nees) T. Anderson ■☆
25686 Anisostachya triticea (Baker) Benoist;小麦异籽葫芦■☆
25687 Anisostachya velutina Nees;短绒毛异籽葫芦■☆
25688 Anisostachya vestita Benoist;包被异籽葫芦■☆
25689 Anisostachya vohemarensis Benoist;武海马尔异籽葫芦■☆
25690 Anisostemon Turcz. = Connarus L. ●
25691 Anisostichus Bureau = Bignonia L. (保留属名)●
25692 Anisosticte Bartl. = Capparis L. ●
25693 Anisosticte Bartl. = Monoporina J. Presl ●☆
25694 Anisostictus Benth. et Hook. f. = Anisostichus Bureau ●
25695 Anisostictus Benth. et Hook. f. = Bignonia L. (保留属名)●
25696 Anisostigma Schinz = Tetragonia L. ●■

25697 Anisostigma schenckii (Schinz) Schinz = Tetragonia schenckii Schinz ■☆
25698 Anisotes Lindl. = Lythrum L. ●■
25699 Anisotes Lindl. ex Meisn. (废弃属名) = Anisotes Nees(保留属名)●☆
25700 Anisotes Nees(1847)(保留属名);异耳爵床属●☆
25701 Anisotes bracteatus Milne-Redh.;具苞异耳爵床●☆
25702 Anisotes divaricatus J. T. Daniel et al.;叉开异耳爵床●☆
25703 Anisotes dumosus Milne-Redh.;棘丛异耳爵床●☆
25704 Anisotes formosissimus (Klotzsch) Milne-Redh.;美丽异耳爵床●☆
25705 Anisotes guineensis Lindau;几内亚异耳爵床●☆
25706 Anisotes involucratus Fiori;总苞异耳爵床●☆
25707 Anisotes macrophyllus (Lindau) Heine;大叶异耳爵床●☆
25708 Anisotes madagascariensis Benoist;马岛异耳爵床●☆
25709 Anisotes nyassae Baden;尼亚萨异耳爵床●☆
25710 Anisotes parvifolius Oliv.;小叶异耳爵床●☆
25711 Anisotes pubinervis (T. Anderson) Heine = Metarungia pubinervia (T. Anderson) Baden ●☆
25712 Anisotes rogersii S. Moore;罗杰斯异耳爵床●☆
25713 Anisotes sereti De Wild. = Anisotes macrophyllus (Lindau) Heine ●☆
25714 Anisotes sessiliflorus (T. Anderson) C. B. Clarke;无花梗异耳爵床●☆
25715 Anisotes tanensis Baden;泰南异耳爵床●☆
25716 Anisotes trisulcus (Forssk.) Nees;三列异耳爵床●☆
25717 Anisotes trisulcus (Forssk.) Nees subsp. webi-schebeliensis Baden = Anisotes trisulcus (Forssk.) Nees ●☆
25718 Anisotes ukambanensis Lindau = Anisotes ukambensis Lindau ●☆
25719 Anisotes ukambensis Lindau;乌卡异耳爵床●☆
25720 Anisotes umbrosus Milne-Redh.;耐荫异耳爵床●☆
25721 Anisotes velutinus Lindau = Anisotes trisulcus (Forssk.) Nees ●☆
25722 Anisotes zenkeri (Lindau) C. B. Clarke;岑克尔异耳爵床■☆
25723 Anisothrix O. Hoffm. = Anisothrix O. Hoffm. ex Kuntze ●☆
25724 Anisothrix O. Hoffm. ex Kuntze(1898);异毛鼠麹木属(短果鼠麹木属)●☆
25725 Anisothrix integra (Compton) Anderb.;异毛鼠麹木●☆
25726 Anisothrix kuntzei O. Hoffm.;库氏异毛鼠麹木●☆
25727 Anisotoma Fenzl(1844);异片萝藦属■☆
25728 Anisotoma arnotii (Baker) Benth. et Hook. f. = Brachystelma arnotii Baker ■☆
25729 Anisotoma cordifolia Fenzl;心叶异片萝藦■☆
25730 Anisotoma mollis Schltr. = Anisotoma cordifolia Fenzl ■☆
25731 Anisotoma pedunculata N. E. Br.;梗花异片萝藦■☆
25732 Anisotomaria C. Presl = Anisotoma Fenzl ■☆
25733 Anisotome Hook. f. (1844);异片芹属■☆
25734 Anisotome acutifolia Cockayne;尖叶异片芹■☆
25735 Anisotome brevistylis Cockayne;短柱异片芹■☆
25736 Anisotome filifolia Cockayne et Laing;线叶异片芹■☆
25737 Anisotome latifolia Hook. f.;宽叶异片芹■☆
25738 Anistelma Raf. = Hedyotis L. (保留属名)●■
25739 Anistfolium Kuntze = Limonia L. ●☆
25740 Anistylis Raf. = Liparis Rich. (保留属名)■
25741 Anisum Gaertn. = Pimpinella L. ■
25742 Anisum Hill = Pimpinella L. ■
25743 Anisum caffrum Eckl. et Zeyh. = Pimpinella caffra (Eckl. et Zeyh.) D. Dietr. ■☆
25744 Anisum stadense Eckl. et Zeyh. = Pimpinella stadensis (Eckl. et Zeyh.) D. Dietr. ■☆
25745 Anisum vulgare Gaertn. = Pimpinella anisum L. ■
25746 Anithista Raf. = Carex L. ■
25747 Ankylobus Steven = Astragalus L. ●■
25748 Ankylocheilos Summerh. = Taeniophyllum Blume ■
25749 Ankyropetalum Fenzl(1843);裂瓣石头花属■☆
25750 Ankyropetalum gypsophiloides Fenzl;裂瓣石头花■☆
25751 Anna Pellegr. (1930);大苞苣苔属;Anna ■
25752 Anna mollifolia (W. T. Wang) W. T. Wang et K. Y. Pan;软叶大苞苣苔;Softhair Anna ■
25753 Anna ophiorrhizoides (Hemsl.) B. L. Burtt et R. A. Davidson;白花大苞苣苔(漏斗苣苔,青竹标);Funnel Didissandra, Whiteflower Anna ■
25754 Anna submontana Pellegr.;大苞苣苔;Anna ■
25755 Annaea Kolak. (1979);越南桔梗属■☆
25756 Annaea Kolak. = Campanula L. ●■
25757 Annaea hieracioides (Kolak.) Kolak.;越南桔梗■☆
25758 Annamia Bernh. = Smyrnium L. ■☆
25759 Annamocarya A. Chev. (1941);喙核桃属(喙嘴核桃属);Annamocarya, Billwalnut ●
25760 Annamocarya A. Chev. = Carya Nutt. (保留属名)●
25761 Annamocarya indochinensis A. Chev. = Annamocarya sinensis (Dode) Leroy ●◇
25762 Annamocarya sinensis (Dode) Leroy;喙核桃(嘴胡桃);China Billwalnut, Chinese Annamocarya, Chinese Hickory ●◇
25763 Anneliesia Brieger et Lückel = Miltonia Lindl. (保留属名)■☆
25764 Annesijoa Pax et K. Hoffm. (1919);新几内亚大戟属☆
25765 Annesijoa novoguineensis Pax et K. Hoffm.;新几内亚大戟☆
25766 Anneslea Hook. = Anneslia Salisb. (废弃属名)●
25767 Anneslea Hook. = Calliandra Benth. (保留属名)●
25768 Anneslea Roxb. ex Andrews = Anneslea Wall. (保留属名)●
25769 Anneslea Roxb. ex Andréws = Euryeles Salisb. ■☆
25770 Anneslea W. Hook. = Anneslia Salisb. (废弃属名)●
25771 Anneslea Wall. (1829)(保留属名);茶梨属(安纳士树属,红楣属);Anneslea, Anneslia ●
25772 Anneslea alpina H. L. Li = Anneslea fragrans Wall. var. alpina (H. L. Li) Kobuski ●
25773 Anneslea fragrans Wall.;茶梨(安纳士树,红楣,红香树,胖婆茶,胖婆娘,香叶树,猪头果);Common Anneslea ●
25774 Anneslea fragrans Wall. var. alpina (H. L. Li) Kobuski;高山茶梨(高山茶树,细叶茶梨);Alpine Anneslea ●
25775 Anneslea fragrans Wall. var. hainanensis Kobuski;海南茶梨(海南红楣,华柃);Hainan Anneslea ●
25776 Anneslea fragrans Wall. var. lanceolata Hayata = Anneslea lanceolata (Hayata) Kaneh. ●
25777 Anneslea fragrans Wall. var. rubriflora (Hu et Hung T. Chang) L. K. Ling;厚叶茶梨(红花安纳士树,红花红楣);Red-flower Anneslea, Red-flowered Anneslea ●
25778 Anneslea hainanensis (Kobuski) Hu = Anneslea fragrans Wall. var. hainanensis Kobuski ●
25779 Anneslea lanceolata (Hayata) Kaneh.;披针叶茶梨(披针叶红楣,台湾安纳士树,细叶安纳士树,细叶茶梨);Lanceleaf Anneslea, Slender-leaf Anneslea ●
25780 Anneslea rubriflora Hu et Hung T. Chang = Anneslea fragrans Wall. var. rubriflora (Hu et Hung T. Chang) L. K. Ling ●
25781 Annesleia Hook. = Anneslia Salisb. (废弃属名)●

25782 Annesleia Hook. = Calliandra Benth.(保留属名)●

25783 Annesleya Post et Kuntze = Anneslea Wall.(保留属名)●

25784 Anneslia Salisb.(废弃属名)= Anneslea Wall.(保留属名)●

25785 Anneslia Salisb.(废弃属名)= Calliandra Benth.(保留属名)●

25786 Anneslia californica (Benth.) Britton et Rose = Calliandra californica Benth. ●☆

25787 Annesorhiza Cham. et Schltdl.(1826);安斯草属■☆

25788 Annesorhiza abyssinica A. Braun = Heteromorpha arborescens (Spreng.) Cham. et Schltdl. var. abyssinica (Hochst. ex A. Rich.) H. Wolff ●☆

25789 Annesorhiza altiscapa Schltr.;高花茎南非草■☆

25790 Annesorhiza burttii B. -E. van Wyk;伯特南非草■☆

25791 Annesorhiza caffra (Eckl. et Zeyh.) Schönland = Stenosemis caffra (Eckl. et Zeyh.) Sond. ■☆

25792 Annesorhiza capensis Cham. et Schltdl. = Annesorhiza nuda (Aiton) B. L. Burtt ■☆

25793 Annesorhiza elata Eckl. et Zeyh. = Annesorhiza grandiflora (Thunb.) M. Hiroe ■☆

25794 Annesorhiza fibrosa B. -E. van Wyk;纤维质南非草■☆

25795 Annesorhiza filicaulis Eckl. et Zeyh.;线茎南非草■☆

25796 Annesorhiza flagellifolia Burtt Davy;鞭叶南非草■☆

25797 Annesorhiza gossweileri C. Norman = Heteromorpha gossweileri (C. Norman) C. Norman ●☆

25798 Annesorhiza grandiflora (Thunb.) M. Hiroe;大花南非草■☆

25799 Annesorhiza grossulariifolia (Eckl. et Zeyh.) M. Hiroe = Polemannia grossulariifolia Eckl. et Zeyh. ■☆

25800 Annesorhiza gummifera (L.) Jacks. = Peucedanum gummiferum (L.) Wijnands ■☆

25801 Annesorhiza hirsuta Eckl. et Zeyh. = Annesorhiza grandiflora (Thunb.) M. Hiroe ■☆

25802 Annesorhiza inebrians (Thunb.) Wijnands = Glia prolifera (Burm. f.) B. L. Burtt ■☆

25803 Annesorhiza interrupta (Thunb.) Sweet = Lichtensteinia interrupta (Thunb.) Sond. ■☆

25804 Annesorhiza lateriflora (Eckl. et Zeyh.) B. -E. van Wyk;侧花南非草■☆

25805 Annesorhiza latifolia Adamson;宽叶南非草■☆

25806 Annesorhiza macrocarpa Eckl. et Zeyh.;大果南非草■☆

25807 Annesorhiza marlothii (H. Wolff) M. Hiroe = Polemanniopsis marlothii (H. Wolff) B. L. Burtt ■☆

25808 Annesorhiza marlothii H. Wolff = Annesorhiza lateriflora (Eckl. et Zeyh.) B. -E. van Wyk ■☆

25809 Annesorhiza montana (Schltr. et H. Wolff) M. Hiroe = Polemannia montana Schltr. et H. Wolff ■☆

25810 Annesorhiza montana Eckl. et Zeyh. = Annesorhiza nuda (Aiton) B. L. Burtt ■☆

25811 Annesorhiza nuda (Aiton) B. L. Burtt;裸南非草■☆

25812 Annesorhiza schlechteri H. Wolff;施莱南非草■☆

25813 Annesorhiza thunbergii B. L. Burtt;通贝里南非草■☆

25814 Annesorhiza villosa (Thunb.) Sond. = Annesorhiza grandiflora (Thunb.) M. Hiroe ■☆

25815 Annesorhiza wilmsii H. Wolff;维尔姆斯南非草■☆

25816 Annickia Setten et Maas(1990);安尼木属;African Whitewood ●☆

25817 Annickia ambigua (Robyns et Ghesq.) Setten et Maas;可疑安尼木●☆

25818 Annickia atrocyanescens (Robyns et Ghesq.) Setten et Maas;深蓝安尼木●☆

25819 Annickia chlorantha (Oliv.) Setten et Maas;非洲安尼木;African White Wood, African Whitewood ●☆

25820 Annickia chlorantha (Oliv.) Setten et Maas = Enantia chlorantha Oliv. ●☆

25821 Annickia kummeriae (Engl. et Diels) Setten et Maas;库默里亚南非草■☆

25822 Annickia kwiluensis (Robyns et Ghesq.) Setten et Maas;奎卢安尼木●☆

25823 Annickia lebrunii (Robyns et Ghesq.) Setten et Maas;勒布伦安尼木●☆

25824 Annickia letestui (Le Thomas) Setten et Maas;莱泰斯图安尼木●☆

25825 Annickia olivacea (Robyns et Ghesq.) Setten et Maas;橄榄绿安尼木●☆

25826 Annickia pilosa (Exell) Setten et Maas;毛安尼木●☆

25827 Annickia polycarpa (DC.) Setten et Maas;多果安尼木●☆

25828 Annona L.(1753);番荔枝属;Alligator Apple, Alligator-apple, Annona, Cherimoya, Custard Apple, Custardapple, Custard-apple, Monkey Apple, Soursop, Sugar Apple, Sweet Sop, Sweet Sops ●

25829 Annona arenaria Thonn. = Annona senegalensis Pers. ●☆

25830 Annona arenaria Thonn. var. obtusa Robyns et Ghesq. = Annona senegalensis Pers. ●☆

25831 Annona asiatica L. = Annona squamosa L. ●

25832 Annona atemoya Mabb.;阿蒂番荔枝●

25833 Annona barteri Benth. = Duguetia barteri (Benth.) Chatrou ●☆

25834 Annona bullata Rich.;泡番荔枝●☆

25835 Annona chericata Mill.;毛叶番荔枝(冷子番荔枝);Cherimoya, Hairy-leaved Custardapple ●

25836 Annona cherimola Mill.;耐冷番荔枝(毛叶番荔枝,秘鲁番荔枝);Cherimoya, Cherimoya-tree, Cherimoyer, Custard Apple, Custard-apple ●

25837 Annona chrysocarpa Lepr. ex Guillaumin et Perr. = Annona glabra L. ●

25838 Annona chrysophylla Bojer;金叶番荔枝;Custard Apple, Wild Custard, Wild Custard-apple, Wild Sour Sop ●☆

25839 Annona chrysophylla Bojer = Annona senegalensis Pers. ●☆

25840 Annona chrysophylla Bojer var. porpetac (Baill.) Robyns et Ghesq. = Annona senegalensis Pers. ●☆

25841 Annona coriacea C. Mart.;巴西番荔枝●☆

25842 Annona cornifolia A. St. -Hil.;角叶番荔枝●☆

25843 Annona crassiflora C. Mart.;粗花番荔枝●☆

25844 Annona crassifolia C. Mart.;厚叶番荔枝●☆

25845 Annona cuneata (Oliv.) R. E. Fr. var. glabrescens (Oliv.) Robyns et Ghesq. = Annona stenophylla Engl. et Diels subsp. cuneata (Oliv.) N. Robson ●☆

25846 Annona cuneata (Oliv.) R. E. Fr. var. longepetiolata R. E. Fr. = Annona stenophylla Engl. et Diels subsp. longepetiolata (R. E. Fr.) N. Robson ●☆

25847 Annona cuneata (Oliv.) R. E. Fr. var. rhodesiaca (Engl. et Diels) R. E. Fr. = Annona stenophylla Engl. et Diels subsp. nana (Exell) N. Robson ●☆

25848 Annona cuneata (Oliv.) R. E. Fr. var. subsessilifolia (Engl.) R. E. Fr. = Annona stenophylla Engl. et Diels subsp. nana (Exell) N. Robson ●☆

25849 Annona densicoma Mart.;密毛番荔枝●☆

25850 Annona diversifolia Saff.;异叶番荔枝(花叶番荔枝,异形叶番荔枝);Anona BlancaWhite Anona, Cherimoya-of-the-lowlands,

Ilama, White Anona ●
25851 Annona friesii Robyns et Ghesq. = Annona stenophylla Engl. et Diels ●☆
25852 Annona glabra L.;圆滑番荔枝(牛心果,牛心梨,野番荔枝);Aligator Apple, Aligator-apple, Cork-wood, Glabrous Custardapple, Glabrous Custard-apple, Monkey Apple, Pond Apple, Pond Apple-tree, Pond-apple ●
25853 Annona glauca Schumach. et Thonn.;柔毛番荔枝●☆
25854 Annona glauca Schumach. et Thonn. var. minor Robyns et Ghesq. = Annona glauca Schumach. et Thonn. ●☆
25855 Annona grandiflora W. Bartram = Asimina obovata (Willd.) Nash ●☆
25856 Annona hexapetala L. f. = Artabotrys hexapetalus (L. f.) Bhandari ●
25857 Annona incana W. Bartram = Asimina incana (W. Bartram) Exell ●☆
25858 Annona klainei Pierre ex Engl. et Diels = Annona glabra L. ●
25859 Annona klainei Pierre ex Engl. et Diels var. moadensis De Wild. = Annona glabra L. ●
25860 Annona latifolia Scott-Elliot = Uvaria anonoides Baker f. ●☆
25861 Annona latifolia Scott-Elliot var. luluensis Engl. et Diels = Uvaria anonoides Baker f. ●☆
25862 Annona laurentii Engl. et Diels;洛朗番荔枝●☆
25863 Annona longepetiolata (R. E. Fr.) Robyns et Ghesq. = Annona stenophylla Engl. et Diels subsp. longepetiolata (R. E. Fr.) N. Robson ●☆
25864 Annona longepetiolata (R. E. Fr.) Robyns et Ghesq. var. precaria Robyns et Ghesq. = Annona stenophylla Engl. et Diels subsp. longepetiolata (R. E. Fr.) N. Robson ●☆
25865 Annona montana Macfad. = Annona montana Macfad. et R. E. Fr. ●
25866 Annona montana Macfad. et R. E. Fr.;山地番荔枝(巴西番荔枝,山地牛心果,山番荔枝);Graviola, Guanabana, Mountain Custardapple, Mountain Soursop, Mountain Soursop-tree ●
25867 Annona mucosa Jacq. = Rollinia mucosa (Jacq.) Baill. ●
25868 Annona muricata L.;刺果番荔枝(刺番荔枝,红毛榴莲,红毛石榴);Guanabana, Guanabana Custard-apple, Guanabana Soursop, Sour Sop, Sour Sops, Soursop, Soursop Custard-apple, Soursop-tree ●
25869 Annona myristica Gaertn. = Monodora myristica (Gaertn.) Dunal ●☆
25870 Annona nana Exell = Annona stenophylla Engl. et Diels subsp. nana (Exell) N. Robson ●☆
25871 Annona nana Exell var. katangensis Robyns et Ghesq. = Annona stenophylla Engl. et Diels subsp. nana (Exell) N. Robson ●☆
25872 Annona nana Exell var. oblonga Robyns et Ghesq. = Annona stenophylla Engl. et Diels subsp. nana (Exell) N. Robson ●☆
25873 Annona nana Exell var. sessilifolia ? = Annona stenophylla Engl. et Diels subsp. nana (Exell) N. Robson ●☆
25874 Annona nana Exell var. subsessilifolia (Engl.) Exell et Mendonça = Annona stenophylla Engl. et Diels subsp. nana (Exell) N. Robson ●☆
25875 Annona obovata Willd. = Asimina obovata (Willd.) Nash ●☆
25876 Annona palustris L.;沼泽番荔枝;Alligator Apple, Marsh Corkwood, Monkey Apple, Serpent Apple ●
25877 Annona palustris L. = Annona glabra L. ●
25878 Annona pendula Salisb. = Asimina triloba (L.) Dunal ●☆
25879 Annona porpetac Baill. = Annona senegalensis Pers. ●☆
25880 Annona purpurea L.;紫番荔枝;Purple Custardapple, Son Coya, Soncoya ●
25881 Annona pygmaea W. Bartram = Asimina pygmaea (W. Bartram) Dunal ●☆
25882 Annona reticulata L.;牛心番荔枝(牛心果,牛心梨,网脉番荔枝);Asiatic Tree Cotton, Bullocks Heart, Bullock's Heart, Bullocksheart, Bullock's-heart, Bullock's-heart Custard Apple, Bullocks-heart Custard-apple, Bullock's-heart Tree, Common Custard Apple, Custard Apple, Custardapple, Custard-apple, Netted Apple, Netted Custard Apple, Netted Custard-apple, Sugar Aapple, Sugar-apple ●
25883 Annona salzmanii DC.;沙氏番荔枝●☆
25884 Annona scleroderma Saff.;硬籽番荔枝;Poshte, Posh-te ●☆
25885 Annona senegalensis Pers.;塞内加尔番荔枝;Senegal Custardapple, Wild Custard-apple ●☆
25886 Annona senegalensis Pers. var. arenaria (Thonn.) Sillans = Annona senegalensis Pers. ●☆
25887 Annona senegalensis Pers. var. areolata Le Thomas;网状塞内加尔番荔枝●☆
25888 Annona senegalensis Pers. var. chrysophylla (Bojer) Sillans = Annona senegalensis Pers. ●☆
25889 Annona senegalensis Pers. var. cuneata Oliv. = Annona stenophylla Engl. et Diels subsp. cuneata (Oliv.) N. Robson ●☆
25890 Annona senegalensis Pers. var. glabrescens Oliv. = Annona stenophylla Engl. et Diels subsp. cuneata (Oliv.) N. Robson ●☆
25891 Annona senegalensis Pers. var. latifolia Oliv. = Annona senegalensis Pers. ●☆
25892 Annona senegalensis Pers. var. oncotricha Pers.;瘤毛塞内加尔番荔枝●☆
25893 Annona senegalensis Pers. var. porpetac (Baill.) Diels = Annona senegalensis Pers. ●☆
25894 Annona senegalensis Pers. var. rhodesiaca Engl. et Diels = Annona stenophylla Engl. et Diels subsp. nana (Exell) N. Robson ●☆
25895 Annona senegalensis Pers. var. subsessilifolia Engl. = Annona stenophylla Engl. et Diels subsp. nana (Exell) N. Robson ●☆
25896 Annona senegalensis Pers. var. ulotricha Le Thomas;卷毛番荔枝●☆
25897 Annona speciosa Nash = Asimina incana (W. Bartram) Exell ●☆
25898 Annona squamosa L.;番荔枝(番梨,佛顶果,佛头果,林檎,蚂蚁果,唛螺陀,麦螺陀,释加果,释迦果,释迦头,洋菠萝);Custard Apple, Custard-apple, Pinha, Scaly Apple, Sealy Custard-apple, Sugar Apple, Sugarapple, Sugar-apple, Sweet Sop, Sweetsop ●
25899 Annona stenophylla Engl. et Diels;窄叶番荔枝●☆
25900 Annona stenophylla Engl. et Diels subsp. cuneata (Oliv.) N. Robson;楔窄叶番荔枝●☆
25901 Annona stenophylla Engl. et Diels subsp. longepetiolata (R. E. Fr.) N. Robson;长梗窄叶番荔枝●☆
25902 Annona stenophylla Engl. et Diels subsp. nana (Exell) N. Robson;矮窄叶番荔枝●☆
25903 Annona stenophylla Engl. et Diels var. nana R. E. Fr. = Annona stenophylla Engl. et Diels subsp. nana (Exell) N. Robson ●☆
25904 Annona tenuiflora Mart.;细花番荔枝●☆
25905 Annona triloba L. = Asimina triloba (L.) Dunal ●☆
25906 Annona uncinata Lam. = Artabotrys hexapetalus (L. f.) Bhandari ●
25907 Annona zenkeri Engl. et Diels;岑克尔番荔枝●☆
25908 Annonaceae Adans. = Annonaceae Juss.(保留科名)●
25909 Annonaceae Juss.(1789)(保留科名);番荔枝科;Annona

Family, Custardapple Family, Custard-apple Family ●
25910 Annulaceae Dulac = Rosaceae Juss.(保留科名)●■
25911 Annularia Hochst. = Cyclostigma Hochst. ex Endl. ●
25912 Annularia Hochst. = Voacanga Thouars ●
25913 Annularia natalensis Hochst. = Voacanga thouarsii Roem. et Schult. ●☆
25914 Annulodiscus Tardieu = Salacia L.(保留属名)●
25915 Anocheile Hoffmanns. ex Rchb. = Epidendrum L.(保留属名)■☆
25916 Anochilus (Schltr.) Rolfe = Pterygodium Sw. ■☆
25917 Anochilus Rolfe = Pterygodium Sw. ■☆
25918 Anochilus flanaganii (Bolus) Rolfe = Corycium flanaganii (Bolus) Kurzweil et H. P. Linder ■☆
25919 Anochilus hallii Schelpe = Pterygodium hallii (Schelpe) Kurzweil et H. P. Linder ■☆
25920 Anochilus inversus (Thunb.) Rolfe = Pterygodium inversum (Thunb.) Sw. ■☆
25921 Anoctochilus yakushimense Yamam. = Vexillabium yakushimense (Yamam.) F. Maek. ■
25922 Anoda Cav.(1785);蔓锦葵属(无节草属);Anoda ●■☆
25923 Anoda cristata (L.) Schltdl.;鸡冠蔓锦葵(无节草);Anoda ■☆
25924 Anoda hastata Cav.;戟形蔓锦葵■☆
25925 Anoda hastata Cav. = Anoda cristata (L.) Schltdl. ■☆
25926 Anoda wrightii A. Gray;赖特蔓锦葵(赖特无节草)■☆
25927 Anodendron A. DC.(1844);鳝藤属(锦兰属,木神葛属);Anodendron, Eelvine ●
25928 Anodendron affine (Hook. et Arn.) Druce;鳝藤(安诺树,大锦兰,锦兰,榊葛,铁骨藤,小锦兰);Asian Cable Creper, Common Anodendron, Common Eelvine ●
25929 Anodendron affine (Hook. et Arn.) Druce var. effusum Tsiang;广花鳝藤(广东鳝藤);Broadflower Anodendron, Looselyflower Anodendron, Looselyflower Eelvine ●
25930 Anodendron affine (Hook. et Arn.) Druce var. effusum Tsiang = Anodendron affine (Hook. et Arn.) Druce ●
25931 Anodendron affine (Hook. et Arn.) Druce var. pingpienense Tsiang et P. T. Li = Anodendron affine (Hook. et Arn.) Druce ●
25932 Anodendron affine (Hook. et Arn.) Druce var. pingpiense Tsiang et P. T. Li;屏边鳝藤;Pingbian Anodendron, Pingbian Eelvine ●
25933 Anodendron benthamianum Hemsl.;台湾鳝藤(大锦兰);Bentham Anodendron, Bentham Cable Creper, Bentham Eelvine ●
25934 Anodendron carandus L.;纳达尔梅子(刺黄果)●☆
25935 Anodendron fangchengense Tsiang et P. T. Li;防城鳝藤;Fangcheng Anodendron, Fangcheng Eelvine ●
25936 Anodendron fangchengense Tsiang et P. T. Li = Anodendron affine (Hook. et Arn.) Druce ●
25937 Anodendron formicinum (Tsiang et P. T. Li) D. J. Middleton;平脉藤;Antshape Micrechites, Parallelvein Micrechites, Paralle-veined Anodendron ●
25938 Anodendron formicinum (Tsiang et P. T. Li) D. J. Middleton = Micrechites formicina Tsiang et P. T. Li ●
25939 Anodendron grandiflorum A. DC.;巨花假虎刺●☆
25940 Anodendron howii Tsiang;保亭鳝藤;How Anodendron, How Eelvine ●
25941 Anodendron laeve (Champ. ex Benth.) Maxim. ex Franch. et Sav. = Anodendron affine (Hook. et Arn.) Druce ●
25942 Anodendron laeve Maxim. ex Franch. et Sav. = Anodendron affine (Hook. et Arn.) Druce ●
25943 Anodendron macrocarpum (Eckl.) A. DC.;丹吾罗(大果假虎刺)●☆
25944 Anodendron paniculatum (Roxb.) A. DC.;锥花鳝藤;Paniculate Anodendron ●☆
25945 Anodendron punctatum Tsiang;腺叶鳝藤;Glandular Anodendron, Glandular Eelvine ●
25946 Anodendron salicifolium Tsiang et P. T. Li;柳叶鳝藤;Willowleaf Anodendron, Willowleaf Eelvine ●
25947 Anodendron salicifolium Tsiang et P. T. Li = Anodendron affine (Hook. et Arn.) Druce ●
25948 Anodendron suishaense Hayata = Anodendron affine (Hook. et Arn.) Druce ●
25949 Anodia Hassk. = Anoda Cav. ●■☆
25950 Anodiscus Benth.(1876);上盘苣苔属■☆
25951 Anodiscus peruvianus Benth. = Anodiscus xanthophyllus (Poepp.) Mansf. ■☆
25952 Anodiscus xanthophyllus (Poepp.) Mansf.;上盘苣苔■☆
25953 Anodontea Sweet = Alyssum L. ●■
25954 Anodopetalum A. Cunn. ex Endl.(1839);塔地火把树属●☆
25955 Anodopetalum biglandulosum (Hook.) Hook. f.;塔地火把树●☆
25956 Anoectocalyx Benth. = Anaectocalyx Triana ex Benth. et Hook. f. ●☆
25957 Anoectocalyx Hook. f. = Anaectocalyx Triana ex Benth. et Hook. f. ●☆
25958 Anoectocalyx Triana. = Anaectocalyx Triana ex Benth. et Hook. f. ●☆
25959 Anoectochilus Blume(1825)(保留属名);开唇兰属(金线兰属,金线莲属);Anoectochilus, Forkliporchis, Jewel Orchid ■
25960 Anoectochilus abbreviatus (Lindl.) Seidenf. = Rhomboda abbreviata (Lindl.) Ormerod ■
25961 Anoectochilus angulosa (Lindl.) Blume = Tropidia angulosa (Lindl.) Blume ■
25962 Anoectochilus bisaccatus Hayata = Anoectochilus lanceolatus Lindl. ■
25963 Anoectochilus bisaccatus Hayata = Odontochilus lanceolatus (Lindl.) Blume ■
25964 Anoectochilus boylei Will.;鲍伊开唇兰■☆
25965 Anoectochilus brevistylis (Hook. f.) Ridl. = Odontochilus brevistylis Hook. f. ■
25966 Anoectochilus burmannicus Rolfe;滇南开唇兰(血兰);S. Yunnan Forkliporchis ■
25967 Anoectochilus candidus (T. P. Lin et C. C. Hsu) K. Y. Lang;白齿唇兰(绿花金线莲);White Forkliporchis ■
25968 Anoectochilus candidus (T. P. Lin et C. C. Hsu) K. Y. Lang = Odontochilus brevistylis Hook. f. ■
25969 Anoectochilus chapaensis Gagnep.;滇越金线兰;Sino-Vietnam Goldlineorchis ■
25970 Anoectochilus clarkei (Hook. f.) Seidenf. = Odontochilus clarkei Hook. f. ■
25971 Anoectochilus clarkei (Hook. f.) Seidenf. et Smitinand = Odontochilus clarkei Hook. f. ■
25972 Anoectochilus crispus Lindl. = Odontochilus crispus (Lindl.) Hook. f. ■
25973 Anoectochilus dawsonianus H. Low ex Rchb. f. = Ludisia discolor (Ker Gawl.) A. Rich. ■
25974 Anoectochilus densiflorus Mansf. = Anoectochilus lanceolatus Lindl. ■
25975 Anoectochilus densiflorus Mansf. = Anoectochilus tortus (King et

Pantl.) King et Pantl. ■
25976 Anoectochilus discolor ?;宝石开唇兰;Jewel Orchid ■☆
25977 Anoectochilus elwesii (C. B. Clarke ex Hook. f.) King et Pantl. = Odontochilus elwesii C. B. Clarke ex Hook. f. ■
25978 Anoectochilus emeiensis K. Y. Lang;峨眉金线兰(峨眉开唇兰,蛇皮兰);Emei Goldlineorchis ■
25979 Anoectochilus flavus Benth. et Hook. f. = Odontochilus lanceolatus (Lindl.) Blume ■
25980 Anoectochilus formosanus Hayata;台湾银线兰(虎头蕉,金不换,金蚕,金石松,金线虎头椒,金线蕨龙,金线连,金线莲,金线屈腰,金线入骨消,金线石松,什鸡单,树草莲,台湾金线兰,台湾金线莲,台湾开唇兰,乌人参);Taiwan Silverlineorchis ■
25981 Anoectochilus formosanus Hayata = Anoectochilus roxburghii (Wall.) Lindl. ●■
25982 Anoectochilus gengmanensis K. Y. Lang = Zeuxine gengmanensis (K. Y. Lang) Ormerod ■
25983 Anoectochilus inabae Hayata = Odontochilus inabae (Hayata) Hayata ex T. P. Lin ■
25984 Anoectochilus inabae Hayata var. candidus (T. P. Lin et C. C. Hsu) S. S. Ying = Odontochilus brevistylis Hook. f. ■
25985 Anoectochilus inabae Hayata var. candidus (T. P. Lin et C. C. Hsu) S. S. Ying = Anoectochilus candidus (T. P. Lin et C. C. Hsu) K. Y. Lang ■
25986 Anoectochilus koshunensis Hayata;恒春银线兰(高雄金线莲,高雄开唇兰,恒春齿唇兰,恒春金线莲);Hengchun Forkliporchis ■
25987 Anoectochilus lanceolatus Lindl. = Odontochilus lanceolatus (Lindl.) Blume ■
25988 Anoectochilus luteus Lindl. = Odontochilus lanceolatus (Lindl.) Blume ■
25989 Anoectochilus moulmeinensis (E. C. Parish et Rchb. f.) Seidenf. et Smitinand = Rhomboda moulmeinensis (E. C. Parish et Rchb. f.) Ormerod ■
25990 Anoectochilus moulmeinensis (Parish, Rchb. f. et Sineref.) Seidenf. et Smitinand = Rhomboda moulmeinensis (E. C. Parish et Rchb. f.) Ormerod ■
25991 Anoectochilus multiflorus Rolfe ex Downie = Anoectochilus moulmeinensis (Parish, Rchb. f. et Sineref.) Seidenf. et Smitinand ■
25992 Anoectochilus multiflorus Rolfe ex Downie = Rhomboda moulmeinensis (E. C. Parish et Rchb. f.) Ormerod ■
25993 Anoectochilus nanlingensis L. P. Siu et K. Y. Lang = Odontochilus nanlingensis (L. P. Siu et K. Y. Lang) Ormerod ■
25994 Anoectochilus petolus Gentil;美丽开唇兰■☆
25995 Anoectochilus pingbianensis K. Y. Lang;屏边金线兰;Pingbian Goldlineorchis ■
25996 Anoectochilus pumilus (Hook. f.) Seidenf et Smitinand = Myrmechis pumila (Hook. f.) Ts. Tang et F. T. Wang ■
25997 Anoectochilus purpureus (C. S. Leou) S. S. Ying = Anoectochilus elwesii (C. B. Clarke ex Hook. f.) King et Pantl. ■
25998 Anoectochilus purpureus (C. S. Leou) S. S. Ying = Odontochilus elwesii C. B. Clarke ex Hook. f. ■
25999 Anoectochilus regalis Blume;王开唇兰;Royal Jewel Orchid ■☆
26000 Anoectochilus regalis Blume = Anoectochilus setaceus Blume ■☆
26001 Anoectochilus reiwardtii Blume;黄氏开唇兰■☆
26002 Anoectochilus repens (Downie) Seidenf. et Smitinand = Anoectochilus tortus (King et Pantl.) King et Pantl. ■
26003 Anoectochilus roxburghii (Wall.) Lindl.;金线兰(花叶开唇兰,金蚕,金石松,金丝线,金线枫,金线虎头蕉,金线莲,金线入骨消,鸟人参,蛇皮兰,石上藕,树草莲,小叶金耳环);Roxburgh Goldlineorchis, Roxburgh Anoectochilus ●■
26004 Anoectochilus roxburghii (Wall.) Lindl. var. baotingensis K. Y. Lang;保亭金线兰;Baoting Goldlineorchis ■
26005 Anoectochilus setaceus Blume;刺状开唇兰■☆
26006 Anoectochilus tashiroi Maxim. = Odontochilus tashiroi (Maxim.) Makino ex Kuroiwa ■☆
26007 Anoectochilus tonkinensis Gagnep. = Anoectochilus brevistylis (Hook. f.) Ridl. ■
26008 Anoectochilus tonkinensis Gagnep. = Odontochilus brevistylis Hook. f. ■
26009 Anoectochilus tortus (King et Pantl.) King et Pantl. = Odontochilus tortus King et Pantl. ■
26010 Anoectochilus vaginata Hook. f. = Chamaegastrodia vaginata (Hook. f.) Seidenf. ■
26011 Anoectochilus xingrenensis Z. H. Tsi et X. H. Jin;兴仁金线兰;Xingren Forkliporchis ■
26012 Anoectochilus yakushimensis Yamam. = Kuhlhasseltia yakushimensis (Yamam.) Ormerod ■
26013 Anoectochilus yakushinensis Yamam. = Vexillabium yakushimense (Yamam.) F. Maek. ■
26014 Anoectochilus yungianus S. Y. Hu;香港金线兰(金线枫);Hongkong Goldlineorchis ■
26015 Anoectochilus zhejiangensis Z. Wei et Y. B. Chang;浙江金线兰(浙江开唇兰);Zhejiang Goldlineorchis ■
26016 Anoegosanthos N. T. Burb. = Anigozanthos Labill. ■☆
26017 Anoegosanthus Rchb. = Anigozanthos Labill. ■☆
26018 Anogeissus (DC.) Wall. (1831);榆绿木属;Anogeissus, Indian Gum ●
26019 Anogeissus (DC.) Wall. ex Guillem. et Perr. = Anogeissus (DC.) Wall. ●
26020 Anogeissus Wall. = Anogeissus (DC.) Wall. ●
26021 Anogeissus Wall. ex Guill. et Perr. = Anogeissus (DC.) Wall. ●
26022 Anogeissus acuminata (Roxb. ex DC.) Guillaumin et al.;榆绿木;Lanceolate Anogeissus, Yon ●◇
26023 Anogeissus acuminata (Roxb. ex DC.) Guillaumin et al. var. lanceolata Wall. ex C. B. Clarke = Anogeissus acuminata (Roxb. ex DC.) Guillaumin et al. ●◇
26024 Anogeissus harmandii Pierre = Anogeissus acuminata (Roxb. ex DC.) Guillaumin et al. ●◇
26025 Anogeissus lanceolata (Wall. ex C. B. Clarke) Wall. ex Prain = Anogeissus acuminata (Roxb. ex DC.) Guillaumin et al. ●◇
26026 Anogeissus latifolius (DC.) Bedd. = Anogeissus latifolius (Roxb.) Bedd. ●☆
26027 Anogeissus latifolius (Roxb.) Bedd.;阔叶榆绿木;Batty Gum, Dhawa ●☆
26028 Anogeissus leiocarpa (DC.) Guillaumin et Perr.;光果榆绿木(马拉胶)●☆
26029 Anogeissus leiocarpa (DC.) Guillaumin et Perr. f. grandiflora Engl. et Diels = Anogeissus leiocarpa (DC.) Guillaumin et Perr. ●☆
26030 Anogeissus leiocarpa (DC.) Guillaumin et Perr. f. parviflora Hochst. ex Engl. et Diels = Anogeissus leiocarpa (DC.) Guillaumin et Perr. ●☆
26031 Anogeissus leiocarpa (DC.) Guillaumin et Perr. var. schimperi (Hochst. ex Hutch. et Dalziel) Aubrév. = Anogeissus leiocarpa (DC.) Guillaumin et Perr. ●☆
26032 Anogeissus leiocarpus Guillaumin et Perr. = Anogeissus leiocarpa

(DC.) Guillaumin et Perr. ●☆
26033 Anogeissus pierrei Gagnep. = Anogeissus acuminata (Roxb. ex DC.) Guillaumin et al. ●◇
26034 Anogeissus schimperi Hochst. ex Hutch. et Dalziel = Anogeissus leiocarpa (DC.) Guillaumin et Perr. ●☆
26035 Anogeissus sericea Brand;绢质榆绿木●☆
26036 Anogeissus tonkinensis Gagnep. = Anogeissus acuminata (Roxb. ex DC.) Guillaumin et al. ●◇
26037 Anogra Spach = Oenothera L. ●■
26038 Anogra nuttallii (Sweet) A. Nelson = Oenothera nuttallii Sweet ■☆
26039 Anogyna Nees = Lagenocarpus Nees ■☆
26040 Anoiganthus Baker = Cyrtanthus Aiton(保留属名)■☆
26041 Anoiganthus breviflorus (Harv.) Baker = Cyrtanthus breviflorus Harv. ■☆
26042 Anoiganthus gracilis Harms = Cyrtanthus breviflorus Harv. ■☆
26043 Anoiganthus luteus (Baker) Baker = Cyrtanthus breviflorus Harv. ■☆
26044 Anoma Lour. = Moringa Rheede ex Adans. ●
26045 Anomacanthus R. D. Good = Gilletiella De Wild. et T. Durand ●☆
26046 Anomacanthus R. D. Good(1923);异花爵床属●☆
26047 Anomacanthus congolanus (De Wild. et T. Durand) Brummitt;异花爵床●☆
26048 Anomacanthus drupaceus R. D. Good = Anomacanthus congolanus (De Wild. et T. Durand) Brummitt ●☆
26049 Anomalanthus Klotzsch = Erica L. ●☆
26050 Anomalanthus Klotzsch = Scyphogyne Brongn. ●☆
26051 Anomalanthus Klotzsch(1838);畸花杜鹃属●☆
26052 Anomalanthus anguliger N. E. Br. = Erica anguliger (N. E. Br.) E. G. H. Oliv. ●☆
26053 Anomalanthus collinus N. E. Br. = Erica anguliger (N. E. Br.) E. G. H. Oliv. ●☆
26054 Anomalanthus curviflorus N. E. Br. = Erica anguliger (N. E. Br.) E. G. H. Oliv. ●☆
26055 Anomalanthus discolor Klotzsch = Erica anguliger (N. E. Br.) E. G. H. Oliv. ●☆
26056 Anomalanthus galpinii N. E. Br. = Erica anguliger (N. E. Br.) E. G. H. Oliv. ●☆
26057 Anomalanthus lesliei Compton = Erica anguliger (N. E. Br.) E. G. H. Oliv. ●☆
26058 Anomalanthus marlothii N. E. Br. = Erica anguliger (N. E. Br.) E. G. H. Oliv. ●☆
26059 Anomalanthus parviflorus (Klotzsch) N. E. Br. = Erica anguliger (N. E. Br.) E. G. H. Oliv. ●☆
26060 Anomalanthus puberulus (Klotzsch) N. E. Br. = Erica anguliger (N. E. Br.) E. G. H. Oliv. ●☆
26061 Anomalanthus salteri Compton = Erica anguliger (N. E. Br.) E. G. H. Oliv. ●☆
26062 Anomalanthus scoparius Klotzsch = Erica anguliger (N. E. Br.) E. G. H. Oliv. ●☆
26063 Anomalanthus turbinatus N. E. Br. = Erica anguliger (N. E. Br.) E. G. H. Oliv. ●☆
26064 Anomalesia N. E. Br. = Cunonia Mill. (废弃属名)■
26065 Anomalesia N. E. Br. = Gladiolus L. ■
26066 Anomalesia cunonia (L.) N. E. Br. = Gladiolus cunonius (L.) Gaertn. ■☆
26067 Anomalesia saccata (Klatt) Goldblatt = Gladiolus saccatus (Klatt) Goldblatt et M. P. de Vos ■☆
26068 Anomalesia splendens (Sweet) N. E. Br. = Gladiolus splendens (Sweet) Herb. ■☆
26069 Anomalluma Plowes = Pseudolithos P. R. O. Bally ■☆
26070 Anomalluma dodsoniana (Lavranos) Plowes = Pseudolithos dodsonianus (Lavranos) Bruyns et Meve ■☆
26071 Anomalocalyx Ducke(1932);畸萼大戟属●☆
26072 Anomalocalyx uleana (Pax) Ducke;畸萼大戟●☆
26073 Anomalopteris (DC.) G. Don = Acridocarpus Guill. et Perr. (保留属名)●☆
26074 Anomalopteris G. Don = Acridocarpus Guill. et Perr. (保留属名)●☆
26075 Anomalopteris longifolia G. Don = Acridocarpus longifolius (G. Don) Hook. f. ●☆
26076 Anomalopterys (DC.) G. Don = Acridocarpus Guill. et Perr. (保留属名)●☆
26077 Anomalosicyos Gentry = Sicyos L. ■
26078 Anomalostemon Klotzsch = Cleome L. ●■
26079 Anomalostemon bororensis Klotzsch = Cleome bororensis (Klotzsch) Oliv. ■☆
26080 Anomalostylus R. C. Foster = Trimezia Salisb. ex Herb. ■☆
26081 Anomalotis Steud. = Agrostis L. (保留属名)■
26082 Anomalotis Steud. = Trisetaria Forssk. ■☆
26083 Anomalotis quinqueseta Steud. = Agrostis quinqueseta (Steud.) Hochst. ■☆
26084 Anomantha Raf. = Verbesina L. (保留属名)●■☆
26085 Anomanthodia Hook. f. (1873);乱花茜属●☆
26086 Anomanthodia Hook. f. = Randia L. ●
26087 Anomanthodia auriculata Hook. f.;乱花茜●☆
26088 Anomanthodia lancifolia (Wong) Tirveng.;披针叶乱花茜●☆
26089 Anomantia DC. = Anomantha Raf. ●■☆
26090 Anomantia DC. = Verbesina L. (保留属名)●■☆
26091 Anomantia Raf. ex DC. = Anomantha Raf. ●■☆
26092 Anomantia Raf. ex DC. = Verbesina L. (保留属名)●■☆
26093 Anomatheca Ker Gawl. (废弃属名) = Freesia Exklon ex Klatt (保留属名)■
26094 Anomatheca Ker Gawl. (废弃属名) = Lapeirousia Pourr. ■☆
26095 Anomatheca Klatt(1805);红射干属■☆
26096 Anomatheca angolensis Baker = Lapeirousia schimperi (Asch. et Klatt) Milne-Redh. ■☆
26097 Anomatheca calamifolia Klatt = Thereianthus juncifolius (Baker) G. J. Lewis ■☆
26098 Anomatheca cruenta Lindl. = Anomatheca laxa (Thunb.) Goldblatt ■☆
26099 Anomatheca cruenta Lindl. = Freesia laxa (Thunb.) Goldblatt et J. C. Manning ■☆
26100 Anomatheca fistulosa (Spreng. ex Klatt) Goldblatt = Xenoscapa fistulosa (Spreng. ex Klatt) Goldblatt et J. C. Manning ■☆
26101 Anomatheca grandiflora Baker = Freesia grandiflora (Baker) Klatt ■☆
26102 Anomatheca laxa (Thunb.) Goldblatt;红射干■☆
26103 Anomatheca laxa (Thunb.) Goldblatt = Freesia laxa (Thunb.) Goldblatt et J. C. Manning ■☆
26104 Anomatheca laxa (Thunb.) Goldblatt subsp. azurea Goldblatt et Hutchings = Freesia laxa (Thunb.) Goldblatt et J. C. Manning subsp. azurea (Goldblatt et Hutchings) Goldblatt et J. C. Manning ■☆
26105 Anomatheca verrucosa (Vogel) Goldblatt = Freesia verrucosa (Vogel) Goldblatt et J. C. Manning ■☆

26106 Anomatheca viridis (Aiton) Goldblatt;绿花红射干■☆
26107 Anomatheca viridis (Aiton) Goldblatt = Freesia viridis (Aiton) Goldblatt et J. C. Manning ■☆
26108 Anomatheca viridis (Aiton) Goldblatt subsp. crispifolia Goldblatt = Freesia viridis (Aiton) Goldblatt et J. C. Manning ■☆
26109 Anomatheca xanthospila (DC.) Ker Gawl. ex Spreng. = Freesia xanthospila (DC.) Klatt ■☆
26110 Anomaza Lawson = Lapeirousia Pourr. ■☆
26111 Anomaza Lawson ex Salisb. = Lapeirousia Pourr. ■☆
26112 Anomaza excisa Lawson ex Salisb. = Freesia verrucosa (Vogel) Goldblatt et J. C. Manning ■☆
26113 Anomeris Raf. = Actinomeris Nutt. (保留属名)■☆
26114 Anomianthus Zoll. (1858);异形花属●☆
26115 Anomianthus dulcis (Dunn) J. Sinclair;异形花■☆
26116 Anomocarpus Miers = Calycera Cav. (保留属名)■☆
26117 Anomocarpus Miers = Leucocera Turcz. ■☆
26118 Anomochloa Brongn. (1851);畸形禾属(畸苞草属)■☆
26119 Anomochloa marantoidea Brongn.;畸形禾■☆
26120 Anomochloaceae Nakai = Gramineae Juss. (保留科名)●■
26121 Anomochloaceae Nakai = Poaceae Barnhart(保留科名)●■
26122 Anomochloaceae Nakai;畸形禾科■☆
26123 Anomoctenium Pichon = Pithecoctenium Mart. ex Meisn. ●☆
26124 Anomopanax Harms = Mackinlaya F. Muell. ●☆
26125 Anomopanax Harms ex Dalla Torre et Harms = Mackinlaya F. Muell. ●☆
26126 Anomorhegmia Meisn. = Miquelia Blume(废弃属名)■
26127 Anomorhegmia Meisn. = Stauranthera Benth. ■
26128 Anomosanthes Blume = Hemigyrosa Blume ●☆
26129 Anomosanthes Blume = Lepisanthes Blume ●
26130 Anomospermum Dalzell = Actephila Blume ●
26131 Anomospermum Miers(1851);异籽藤属●☆
26132 Anomospermum axilliflorum Griseb.;腋花异籽藤●☆
26133 Anomospermum bolivianum Krukoff et Moldenke;玻利维亚异籽藤●☆
26134 Anomospermum chloranthum Diels;绿花异籽藤●☆
26135 Anomospermum excelsum Dalzell = Actephila excelsa (Dalzell) Müll. Arg. ●
26136 Anomospermum grandifolium Eichl.;大叶异籽藤●☆
26137 Anomospermum lucidum Miers;亮异籽藤●☆
26138 Anomospermum minutiflorum Diels;小花异籽藤●☆
26139 Anomostachys (Baill.) Hurus. = Excoecaria L. ●
26140 Anomostephium DC. = Aspilia Thouars ■☆
26141 Anomostephium DC. = Wedelia Jacq. (保留属名)●■
26142 Anomotassa K. Schum. (1898);厄瓜多尔萝藦属☆
26143 Anomotassa macranthus K. Schum.;厄瓜多尔萝藦☆
26144 Anona L. = Annona L. ●
26145 Anona Mill. = Annona L. ●
26146 Anona uncinata Lam. = Artabotrys hexapetalus (L. f.) Bhandari ●
26147 Anonidium Engl. et Diels = Annona L. ●
26148 Anonidium Engl. et Diels(1900);阿诺木属(阿诺属,类番荔枝属)●☆
26149 Anonidium brieyi De Wild. = Anonidium mannii (Oliv.) Engl. et Diels var. brieyi (De Wild.) R. E. Fr. ●☆
26150 Anonidium floribundum Pellegr.;多花阿诺木●☆
26151 Anonidium friesianum Exell = Anonidium mannii (Oliv.) Engl. et Diels var. brieyi (De Wild.) R. E. Fr. ●☆
26152 Anonidium letestui Pellegr.;莱泰斯图阿诺木●☆
26153 Anonidium mannii (Oliv.) Engl. et Diels;曼氏阿诺木●☆
26154 Anonidium mannii (Oliv.) Engl. et Diels var. brieyi (De Wild.) R. E. Fr.;布利阿诺木●☆
26155 Anonidium mannii Engl. et Diels = Anonidium mannii (Oliv.) Engl. et Diels ●☆
26156 Anonidium usambarense R. E. Fr.;乌桑巴拉阿诺木●☆
26157 Anoniodes Schltr. = Sloanea L. ●
26158 Anonis Mill. = Ononis L. ●■
26159 Anonis Tourn. ex Scop. = Ononis L. ●■
26160 Anonocarpus Ducke = Batocarpus H. Karst. ●☆
26161 Anonychium Schweinf. = Prosopis L. ●
26162 Anonymos Kuntze = Galax Sims(保留属名)■☆
26163 Anonymos petiolata Wall. = Mitreola petiolata (J. F. Gmel.) Torr. et A. Gray ■
26164 Anoosperma Kuntze = Oncosperma Blume ●☆
26165 Anoplanthus Endl. = Anoplon Rchb. ■☆
26166 Anoplanthus Endl. = Aphyllon Mitch. ■
26167 Anoplanthus Endl. = Phelypaea L. ■☆
26168 Anoplanthus fasciculatus (Nutt.) Walp. = Orobanche fasciculata Nutt. ■☆
26169 Anoplanthus uniflorus (L.) Endl. = Orobanche uniflora L. ■☆
26170 Anoplia Nees ex Steud. = Leptochloa P. Beauv. ■
26171 Anoplia Steud. = Leptochloa P. Beauv. ■
26172 Anoplocaryum Ledeb. (1847);平核草属■☆
26173 Anoplocaryum compessum (Turcz.) Ledeb.;平核草■☆
26174 Anoplocaryum limprichtii Brand = Microula sikkimensis (C. B. Clarke) Hemsl. ■
26175 Anoplocaryum myosotideum (Franch.) Brand = Microula myosotidea (Franch.) I. M. Johnst. ■
26176 Anoplocaryum rockii (I. M. Johnst.) Brand = Microula rockii I. M. Johnst. ■
26177 Anoplon Rchb. = Phelypaea L. + Aphyllon Mitch. ■
26178 Anoplon Rchb. = Phelypaea L. ■☆
26179 Anoplon Wallr. ex Rchb. = Aphyllon Mitch. ■
26180 Anoplophytum Beer = Tillandsia L. ■☆
26181 Anopteraceae Doweld = Iteaceae J. Agardh(保留科名)●
26182 Anopterus Labill. (1805);澳山月桂属(阿诺草属,欧洲鼠刺属)●☆
26183 Anopterus glandulosus Labill.;腺澳山月桂(欧洲鼠刺,腺阿诺,腺阿诺草);Tasmanian Laurel ●☆
26184 Anopterus glandulosus Labill. 'Woodbank Pink';沃班红腺澳山月桂;Tasmanian Laurel ●☆
26185 Anopterus macleayanus F. Muell.;大叶澳山月桂(博落回阿诺草,博落回腺阿诺);Macleay Laurel ●☆
26186 Anopyxis (Pierre) Engl. (1900);小红树属(阿诺匹斯属);Anopyxis ●☆
26187 Anopyxis Pierre ex Engl. = Anopyxis (Pierre) Engl. ●☆
26188 Anopyxis ealaensis (De Wild.) Sprague = Anopyxis klaineana (Pierre) Engl. ●☆
26189 Anopyxis klaineana (Pierre) Engl.;克莱小红树(克莱阿诺匹斯,克莱小红木);Klain Anopyxis ●☆
26190 Anopyxis klaineana Pierre = Anopyxis klaineana (Pierre) Engl. ●☆
26191 Anopyxis klaineana Pierre ex Engl. = Anopyxis klaineana (Pierre) Engl. ●☆
26192 Anopyxis occidentalis (A. Chev.) A. Chev. = Anopyxis klaineana (Pierre) Engl. ●☆
26193 Anosmia Bernh. = Smyrnium L. ■☆

26194 Anosporum Nees = Cyperus L. ■

26195 Anosporum cubense (Poepp. et Kunth) Boeck. = Oxycaryum cubense (Poepp. et Kunth) Lye ■☆

26196 Anosporum nudicaule (Poir.) Boeck. = Anosporum pectinatus (Vahl) Lye ■☆

26197 Anosporum pectinatum (Vahl) Lye = Cyperus pectinatus Vahl ■☆

26198 Anosporum schinzii Boeck. = Oxycaryum schinzii (Boeck.) Palla ■☆

26199 Anota (Lindl.) Schltr. (1914);无耳兰属■

26200 Anota (Lindl.) Schltr. = Rhynchostylis Blume ■

26201 Anota Schltr. = Anota (Lindl.) Schltr. ■

26202 Anota Schltr. = Rhynchostylis Blume ■

26203 Anota densiflora (Lindl.) Schltr. = Rhynchostylis gigantea (Lindl.) Ridl. ■

26204 Anota gigantea (Lindl.) Fukuy. = Rhynchostylis gigantea (Lindl.) Ridl. ■

26205 Anota hainanensis (Rolfe) Schltr.;无耳兰■

26206 Anota hainanensis (Rolfe) Schltr. = Rhynchostylis gigantea (Lindl.) Ridl. ■

26207 Anotea (DC.) Kunth(1846);墨西哥无耳葵属●☆

26208 Anotea Kunth = Anotea (DC.) Kunth ●☆

26209 Anotea chloranthus Kunth;墨西哥无耳葵●☆

26210 Anotea flavida (DC.) Ulbr.;淡黄墨西哥无耳葵●☆

26211 Anotea flavida Ulbr. = Anotea flavida (DC.) Ulbr. ●☆

26212 Anothea O. F. Cook = Chamaedorea Willd. (保留属名)●☆

26213 Anothea O. F. Cook(1943);墨西哥棕属●☆

26214 Anothea scandens (Liebm.) O. F. Cook;墨西哥棕●☆

26215 Anotis DC. (1830);假耳草属●

26216 Anotis DC. = Arcytophyllum Willd. ex Schult. et Schult. f. + Hedyotis L. + Oldenlandia L. ●■

26217 Anotis DC. = Arcytophyllum Willd. ex Schult. et Schult. f. ●☆

26218 Anotis DC. = Neanotis W. H. Lewis ■

26219 Anotis DC. = Panetos Raf. ●

26220 Anotis boerhaavioides (Hance) Maxim. = Neanotis boerhaavioides (Hance) W. H. Lewis ■

26221 Anotis boerhaavioides Hance = Neanotis boerhaavioides (Hance) W. H. Lewis ■

26222 Anotis calycina Wall. ex Hook. f. = Neanotis calycina (Wall. ex Hook. f.) W. H. Lewis ■

26223 Anotis chrysotricha Palib. = Hedyotis chrysotricha (Palib.) Merr. ■

26224 Anotis formosana Hayata = Neanotis formosana (Hayata) W. H. Lewis ■

26225 Anotis hirsuta (L. f.) Boerl. = Neanotis hirsuta (L. f.) W. H. Lewis ■

26226 Anotis ingrata Wall. ex Hook. f. = Neanotis ingrata (Wall. ex Hook. f.) W. H. Lewis ■

26227 Anotis kwangtungensis Merr. et F. P. Metcalf = Neanotis kwangtungensis (Merr. et F. P. Metcalf) W. H. Lewis ■

26228 Anotis thwaitesiana Hance = Neanotis thwaitesiana (Hance) W. H. Lewis ■

26229 Anotis urophylla Wall. ex Benth. et Hook. f. = Neanotis urophylla (Wall. ex Wight et Arn.) W. H. Lewis ■

26230 Anotis wightiana (Wall. ex Arn.) Hook. f. = Neanotis wightiana (Wall. ex Wight et Arn.) W. H. Lewis ■

26231 Anotis wightiana (Wall. ex Wight et Arn.) Hook. f. = Neanotis wightiana (Wall. ex Wight et Arn.) W. H. Lewis ■

26232 Anotites Greene = Silene L. (保留属名)■

26233 Anotites alsinoides Greene = Silene menziesii Hook. ■☆

26234 Anotites bakeri Greene = Silene menziesii Hook. ■☆

26235 Anotites costata Greene = Silene menziesii Hook. ■☆

26236 Anotites debilis Greene = Silene menziesii Hook. ■☆

26237 Anotites diffusa Greene = Silene menziesii Hook. ■☆

26238 Anotites discurrens Greene = Silene menziesii Hook. ■☆

26239 Anotites dorrii (Kellogg) Greene = Silene menziesii Hook. ■☆

26240 Anotites elliptica Greene = Silene menziesii Hook. ■☆

26241 Anotites halophila Greene = Silene menziesii Hook. ■☆

26242 Anotites jonesii Greene = Silene menziesii Hook. ■☆

26243 Anotites latifolia Greene = Silene menziesii Hook. ■☆

26244 Anotites macilenta Greene = Silene menziesii Hook. ■☆

26245 Anotites menziesii (Hook.) Greene = Silene menziesii Hook. ■☆

26246 Anotites nodosa Greene = Silene menziesii Hook. ■☆

26247 Anotites picta Greene = Silene menziesii Hook. ■☆

26248 Anotites seelyi (C. V. Morton et J. W. Thomps.) W. A. Weber = Silene seelyi C. V. Morton et J. W. Thomps. ■☆

26249 Anotites tenerrima Greene = Silene menziesii Hook. ■☆

26250 Anotites tereticaulis Greene = Silene menziesii Hook. ■☆

26251 Anotites villosula Greene = Silene menziesii Hook. ■☆

26252 Anotites viscosa Greene = Silene menziesii Hook. ■☆

26253 Anoumabia A. Chev. = Harpullia Roxb. ●

26254 Anoumabia cyanosperma A. Chev. = Majidea fosteri (Sprague) Radlk. ●☆

26255 Anplectrella Furtado = Creochiton Blume ●☆

26256 Anplectrella Furtado = Enchosanthera King et Stapf ●☆

26257 Anplectrum A. Gray = Diplectria (Blume) Rchb. ●■

26258 Anplectrum assamicum C. B. Clarke = Medinilla assamica (C. B. Clarke) C. Chen ●

26259 Anplectrum barbatum Triana = Diplectria barbata (Wall. ex C. B. Clarke) Franken et M. C. Roos ●■

26260 Anplectrum barbatum Wall. ex C. B. Clarke = Diplectria barbata (Wall. ex C. B. Clarke) Franken et M. C. Roos ●■

26261 Anplectrum parviflorum Benth. = Blastus cochinchinensis Lour. ●

26262 Anplectrum yunnanense Kraenzl. = Medinilla septentrionalis (W. W. Sm.) H. L. Li ●

26263 Anquetilia Decne. = Skimmia Thunb. (保留属名)●

26264 Anredera Juss. (1789);落葵薯属(藤三七属);Vineyam, Mignonettevine, Madeiravine, Mignonette Vine, Madeira Vine, Madeira-vine ●■

26265 Anredera cordifolia (Ten.) Steenis;落葵薯(马德拉藤,藤七,藤三七,藤子三七,土三七,心叶落葵薯,洋落葵);Heartleaf Madeiravine, Madeira Vine, Madeiravine, Madeira-vine, Mignonette Vine, Mignonettevine, Mignonette-vine, Vineyam ●

26266 Anredera leptostachys (Moq.) Steenis;细穗花落葵薯●☆

26267 Anredera leptostachys (Moq.) Steenis = Anredera vesicaria Gaertn. f. ●☆

26268 Anredera scandens (L.) Moq.;短序落葵薯;Shortraceme Vineyam ●

26269 Anredera vesicaria Gaertn. f.;泡状落葵薯;Sacasile ●☆

26270 Anrederaceae J. Agardh = Basellaceae Moq. ■

26271 Anrederaceae J. Agardh = Basellaceae Raf. (保留科名)■

26272 Anrederaceae J. Agardh;落葵薯科●■

26273 Ansellia Lindl. (1844);豹斑兰属(安塞丽亚兰属);Ansellia, Leopard Orchid ■☆

26274 Ansellia africana Lindl.;豹斑兰(安塞丽亚兰);Ansellia

Orchid, Leopard Orchid ■☆

26275 Ansellia africana Lindl. var. australis Summerh. = Ansellia africana Lindl. ■☆

26276 Ansellia africana Lindl. var. nilotica Baker = Ansellia africana Lindl. ■☆

26277 Ansellia confusa N. E. Br. = Ansellia africana Lindl. ■☆

26278 Ansellia congoensis Lindl.; 刚果豹斑兰■☆

26279 Ansellia congoensis Rodigas = Ansellia africana Lindl. ■☆

26280 Ansellia gigantea Rchb. f.; 大豹斑兰■☆

26281 Ansellia gigantea Rchb. f. var. nilotica (Baker) Summerh. = Ansellia africana Lindl. ■☆

26282 Ansellia humilis Bull. = Ansellia africana Lindl. ■☆

26283 Ansellia nilotica (Baker) N. E. Br. = Ansellia africana Lindl. ■☆

26284 Anselonia O. E. Schulz = Anelsonia J. F. Macbr. et Payson ■☆

26285 Anserina Dumort. = Chenopodium L. ●■

26286 Anserina candicans Montandon = Chenopodium album L. ■

26287 Ansonia Bert. ex Hemsl. = Lactoris Phil. ●☆

26288 Ansonia Raf. = Amsonia Walter ■

26289 Anstrutheria Gardner = Cassipourea Aubl. ●☆

26290 Antacanthus A. Rich. ex DC. = Scolosanthus Vahl☆

26291 Antagonia Griseb. = Cayaponia Silva Manso(保留属名)■☆

26292 Antaurea Neck. = Centaurea L. (保留属名)●■

26293 Antegibbaeum Schwantes ex C. Weber(1968); 碧玉属■☆

26294 Antegibbaeum Schwantes ex H. Wulff = Antegibbaeum Schwantes ex C. Weber ■☆

26295 Antegibbaeum fissoides (Haw.) Schwantes ex C. Weber = Gibbaeum fissoides (Haw.) Nel ■☆

26296 Antegibbaeum fissoides (Haw.) Schwantes ex H. Wulff = Gibbaeum fissoides (Haw.) Nel ■☆

26297 Antelaea Gaertn. = Melia L. ●

26298 Antelaea azadirachta (L.) Adelb. = Azadirachta indica A. Juss. ●☆

26299 Antennaria Gaertn. (1791)(保留属名); 蝶须属(蝶须菊属); Cat's Ear, Cat's Ears, Cat's-foot, Early Everlasting, Ladies Tobacco, Ladies' Tobacco, Mountain Everlasting, Pussy Toes, Pussy's Toes, Pussy's-toes, Pussytoes ●■

26300 Antennaria Link(废弃属名) = Antennaria Gaertn. (保留属名) ●■

26301 Antennaria R. Br. = Antennaria Gaertn. (保留属名)●■

26302 Antennaria acuminata Greene = Antennaria rosea Greene ■☆

26303 Antennaria acuta Rydb. = Antennaria corymbosa E. E. Nelson ■☆

26304 Antennaria affinis Fernald = Antennaria rosea Greene subsp. confinis (Greene) R. J. Bayer ■☆

26305 Antennaria aizoides Greene = Antennaria umbrinella Rydb. ■☆

26306 Antennaria alaskana Malte = Antennaria friesiana (Trautv.) E. Ekman subsp. alaskana (Malte) Hultén ■☆

26307 Antennaria albescens (E. E. Nelson) Rydb. = Antennaria rosea Greene subsp. pulvinata (Greene) R. J. Bayer ■☆

26308 Antennaria albicans Fernald = Antennaria rosea Greene subsp. confinis (Greene) R. J. Bayer ■☆

26309 Antennaria alborosea A. E. Porsild = Antennaria rosea Greene ■☆

26310 Antennaria alpina (L.) Gaertn.; 高山蝶须; Alpine Pussytoes ■☆

26311 Antennaria alpina (L.) Gaertn. var. friesiana Trautv. = Antennaria friesiana (Trautv.) E. Ekman ■☆

26312 Antennaria alpina (L.) Gaertn. var. media (Greene) Jeps. = Antennaria media Greene ■☆

26313 Antennaria alpina (L.) Gaertn. var. megacephala (Fernald ex Raup) S. L. Welsh = Antennaria monocephala DC. subsp. angustata (Greene) Hultén ■☆

26314 Antennaria alpina (L.) Gaertn. var. monocephala (DC.) Torr. et A. Gray = Antennaria monocephala DC. ■☆

26315 Antennaria alpina (L.) Gaertn. var. scabra (Greene) Jeps. = Antennaria pulchella Greene ■☆

26316 Antennaria alpina (L.) Gaertn. var. stenophylla A. Gray = Antennaria stenophylla (A. Gray) A. Gray ■☆

26317 Antennaria alpina Gaertn. = Antennaria alpina (L.) Gaertn. ■☆

26318 Antennaria alsinoides Greene = Antennaria howellii Greene subsp. neodioica (Greene) R. J. Bayer ■☆

26319 Antennaria ambigens (Greene) Fernald = Antennaria parlinii Fernald subsp. fallax (Greene) R. J. Bayer et Stebbins ■☆

26320 Antennaria ampla Bush = Antennaria parlinii Fernald subsp. fallax (Greene) R. J. Bayer et Stebbins ■☆

26321 Antennaria anaphaloides Rydb.; 贫乏蝶须; Handsome Pussytoes, Pearly Pussytoes, Tall Pussytoes ■☆

26322 Antennaria anaphaloides Rydb. var. straminea B. Boivin = Antennaria anaphaloides Rydb. ■☆

26323 Antennaria angustata Greene = Antennaria monocephala DC. subsp. angustata (Greene) Hultén ■☆

26324 Antennaria angustiarum Lunell = Antennaria neglecta Greene ■☆

26325 Antennaria angustifolia E. Ekman = Antennaria friesiana (Trautv.) E. Ekman ■☆

26326 Antennaria angustifolia Rydb. = Antennaria rosea Greene subsp. confinis (Greene) R. J. Bayer ■☆

26327 Antennaria appendiculata Fernald = Antennaria howellii Greene subsp. petaloidea (Fernald) R. J. Bayer ■☆

26328 Antennaria aprica Greene = Antennaria parvifolia Nutt. ■☆

26329 Antennaria aprica Greene var. aureola (Lunell) J. W. Moore = Antennaria parvifolia Nutt. ■☆

26330 Antennaria aprica Greene var. minuscula (B. Boivin) B. Boivin = Antennaria parvifolia Nutt. ■☆

26331 Antennaria arcuata Cronquist; 草地蝶须; Box Pussytoes, Meadow Pussytoes ■☆

26332 Antennaria argentea Benth.; 银蝶须; Silver Pussytoes ■☆

26333 Antennaria argentea Benth. subsp. aberrans E. E. Nelson = Antennaria luzuloides Torr. et A. Gray subsp. aberrans (E. E. Nelson) R. J. Bayer et Stebbins ■☆

26334 Antennaria arida E. E. Nelson = Antennaria rosea Greene subsp. arida (E. E. Nelson) R. J. Bayer ■☆

26335 Antennaria arida E. E. Nelson var. humilis (Rydb.) E. E. Nelson = Antennaria rosea Greene subsp. confinis (Greene) R. J. Bayer ■☆

26336 Antennaria arkansana Greene = Antennaria parlinii Fernald subsp. fallax (Greene) R. J. Bayer et Stebbins ■☆

26337 Antennaria arnoglossa Greene = Antennaria parlinii Fernald ■☆

26338 Antennaria arnoglossa Greene var. ambigens Greene = Antennaria parlinii Fernald subsp. fallax (Greene) R. J. Bayer et Stebbins ■☆

26339 Antennaria aromatica Evert; 芬芳蝶须; Scented Or Aromatic Pussytoes ■☆

26340 Antennaria athabascensis Greene = Antennaria neglecta Greene ■☆

26341 Antennaria aureola Lunell = Antennaria parvifolia Nutt. ■☆

26342 Antennaria austromontana E. E. Nelson = Antennaria media Greene ■☆

26343 Antennaria bifrons Greene = Antennaria parlinii Fernald subsp. fallax (Greene) R. J. Bayer et Stebbins ■☆

26344 Antennaria bracteosa Rydb. = Antennaria microphylla Rydb. ■☆

26345 Antennaria brainerdii Fernald = Antennaria parlinii Fernald subsp. fallax (Greene) R. J. Bayer et Stebbins ■☆

26346 Antennaria breitungii A. E. Porsild = Antennaria rosea Greene subsp. confinis (Greene) R. J. Bayer ■☆

26347 Antennaria brevistyla Fernald = Antennaria rosea Greene subsp. confinis (Greene) R. J. Bayer ■☆

26348 Antennaria burwellensis Malte = Antennaria monocephala DC. subsp. angustata (Greene) Hultén ■☆

26349 Antennaria callilepis Greene = Antennaria howellii Greene ■☆

26350 Antennaria calophylla Greene = Antennaria parlinii Fernald subsp. fallax (Greene) R. J. Bayer et Stebbins ■☆

26351 Antennaria campestris Rydb. = Antennaria neglecta Greene ■☆

26352 Antennaria campestris Rydb. var. athabascensis (Greene) B. Boivin = Antennaria neglecta Greene ■☆

26353 Antennaria canadensis Greene = Antennaria howellii Greene subsp. canadensis (Greene) R. J. Bayer ■☆

26354 Antennaria canadensis Greene var. randii Fernald = Antennaria howellii Greene subsp. canadensis (Greene) R. J. Bayer ■☆

26355 Antennaria canadensis Greene var. spathulata Fernald = Antennaria howellii Greene subsp. canadensis (Greene) R. J. Bayer ■☆

26356 Antennaria candida Greene = Antennaria media Greene ■☆

26357 Antennaria caroliniana Rydb. = Antennaria plantaginifolia (L.) Hook. ■☆

26358 Antennaria carpathica Greene = Antennaria howellii Greene subsp. canadensis (Greene) R. J. Bayer ■☆

26359 Antennaria carpatica (Hook.) R. Br. var. pulcherrima Hook. = Antennaria pulcherrima (Hook.) Greene ■☆

26360 Antennaria carpatica (Wahlenb.) Hook. var. humilis Hook. = Antennaria pulcherrima (Hook.) Greene subsp. eucosma (Fernald et Wiegand) R. J. Bayer ■☆

26361 Antennaria carpatica (Wahlenb.) Hook. var. lanata Hook. = Antennaria lanata (Hook.) Greene ■☆

26362 Antennaria carpatica (Wall.) Bluff et Fingerh.; 非洲蝶须; Carpathian Pussytoes, Carpatian Catisfoot ■☆

26363 Antennaria caucasica Boriss.; 高加索蝶须■☆

26364 Antennaria chelonica Lunell = Antennaria neglecta Greene ■☆

26365 Antennaria chlorantha Greene = Antennaria rosea Greene ■☆

26366 Antennaria cinnamomea C. B. Clarke var. angustior Miq. = Anaphalis margaritacea (L.) Benth. et Hook. f. ■

26367 Antennaria cinnamomea DC. = Anaphalis margaritacea (L.) Benth. et Hook. f. ■

26368 Antennaria cinnamomea DC. var. angustior Miq. = Anaphalis margaritacea (L.) Benth. et Hook. f. ■

26369 Antennaria cinnamomea DC. var. angustior Miq. = Anaphalis margaritacea (L.) Benth. et Hook. f. var. cinnamomea (DC.) Herder ex Maxim. ■

26370 Antennaria concinna E. E. Nelson = Antennaria rosea Greene subsp. confinis (Greene) R. J. Bayer ■☆

26371 Antennaria concolor Piper = Antennaria howellii Greene subsp. petaloidea (Fernald) R. J. Bayer ■☆

26372 Antennaria confinis Greene = Antennaria rosea Greene subsp. confinis (Greene) R. J. Bayer ■☆

26373 Antennaria congesta Malte = Antennaria monocephala DC. subsp. angustata (Greene) Hultén ■☆

26374 Antennaria contorta D. Don = Anaphalis contorta (D. Don) Hook. f. ■

26375 Antennaria corymbosa E. E. Nelson; 平顶蝶须; Flat-top Pussytoes, Meadow Pussytoes ■☆

26376 Antennaria decipiens Greene = Antennaria plantaginifolia (L.) Hook. ■☆

26377 Antennaria denikeana B. Boivin = Antennaria plantaginifolia (L.) Hook. ■☆

26378 Antennaria densa Greene = Antennaria media Greene ■☆

26379 Antennaria densifolia A. E. Porsild; 密花蝶须; Denseleaf Pussytoes ■☆

26380 Antennaria dimorpha (Nutt.) Torr. et A. Gray; 二型蝶须; Low Or Two-form Or Cushion Pussytoes ■☆

26381 Antennaria dimorpha (Nutt.) Torr. et A. Gray var. flagellaris A. Gray = Antennaria flagellaris (A. Gray) A. Gray ■☆

26382 Antennaria dimorpha (Nutt.) Torr. et A. Gray var. integra L. F. Hend. = Antennaria dimorpha (Nutt.) Torr. et A. Gray ■☆

26383 Antennaria dimorpha (Nutt.) Torr. et A. Gray var. macrocephala D. C. Eaton = Antennaria dimorpha (Nutt.) Torr. et A. Gray ■☆

26384 Antennaria dimorpha (Nutt.) Torr. et A. Gray var. nuttallii D. C. Eaton = Antennaria dimorpha (Nutt.) Torr. et A. Gray ■☆

26385 Antennaria dioica (L.) Gaertn.; 蝶须(兴安蝶须); Cat's Ear, Cat's Foot, Cat's Paw, Cat's-ears, Cat's-foot, Common Pussytoes, Cudweed, Gat's-toes, Lady's Tobacco, Life Everlasting, Little Mouse Ear, Live-forever, Moor Everlasting, Mountain Everlasting, Pussy Toe, Pussy's Toes, Pussytoes, Stoloniferous Pussytoes ■

26386 Antennaria dioica (L.) Gaertn. 'Rubra'; 红调蝶须; Red-tinted Pussy Toes ■☆

26387 Antennaria dioica (L.) Gaertn. var. corymbosa (E. E. Nelson) Jeps. = Antennaria corymbosa E. E. Nelson ■☆

26388 Antennaria dioica (L.) Gaertn. var. hyperborea Lange = Antennaria dioica (L.) Gaertn. ■

26389 Antennaria dioica (L.) Gaertn. var. kernensis Jeps. = Antennaria rosea Greene subsp. confinis (Greene) R. J. Bayer ■☆

26390 Antennaria dioica (L.) Gaertn. var. marginata (Greene) Jeps. = Antennaria marginata Greene ■☆

26391 Antennaria dioica (L.) Gaertn. var. parvifolia (Nutt.) Torr. et A. Gray = Antennaria parvifolia Nutt. ■☆

26392 Antennaria dioica (L.) Gaertn. var. rosea Eaton = Antennaria rosea Greene ■☆

26393 Antennaria dioiciformis Kom.; 日本蝶须■☆

26394 Antennaria ekmaniana A. E. Porsild = Antennaria friesiana (Trautv.) E. Ekman ■☆

26395 Antennaria elegans A. E. Porsild = Antennaria rosea Greene subsp. confinis (Greene) R. J. Bayer ■☆

26396 Antennaria elliptica Greene = Antennaria parlinii Fernald subsp. fallax (Greene) R. J. Bayer et Stebbins ■☆

26397 Antennaria ellyae A. E. Porsild = Antennaria densifolia A. E. Porsild ■☆

26398 Antennaria erosa Greene = Antennaria neglecta Greene ■☆

26399 Antennaria eucosma Fernald et Wiegand = Antennaria pulcherrima (Hook.) Greene subsp. eucosma (Fernald et Wiegand) R. J. Bayer ■☆

26400 Antennaria exilis Greene = Antennaria monocephala DC. ■☆

26401 Antennaria exima Greene = Antennaria howellii Greene ■☆

26402 Antennaria fallax Greene; 哥伦比亚蝶须; Large-leaved Antennaria ■☆

26403 Antennaria fallax Greene = Antennaria parlinii Fernald subsp. fallax (Greene) R. J. Bayer et Stebbins ■☆

26404 Antennaria fallax Greene var. calophylla (Greene) Fernald = Antennaria parlinii Fernald subsp. fallax (Greene) R. J. Bayer et Stebbins ■☆

26405 Antennaria farwellii Greene = Antennaria parlinii Fernald subsp. fallax (Greene) R. J. Bayer et Stebbins ■☆

26406 Antennaria fendleri Greene = Antennaria marginata Greene ■☆

26407 Antennaria fernaldiana Polunin = Antennaria monocephala DC. subsp. angustata (Greene) Hultén ■☆

26408 Antennaria flagellaris (A. Gray) A. Gray;鞭蝶须;Stoloniferous Pussytoes, Whip Pussytoes ■☆

26409 Antennaria flavescens Rydb. = Antennaria umbrinella Rydb. ■☆

26410 Antennaria foliacea Greene var. humilis Rydb. = Antennaria rosea Greene subsp. confinis (Greene) R. J. Bayer ■☆

26411 Antennaria formosa Greene = Antennaria rosea Greene ■☆

26412 Antennaria frieseana E. Ekman = Antennaria friesiana (Trautv.) E. Ekman ■☆

26413 Antennaria friesiana (Trautv.) E. Ekman;弗氏蝶须;Fries' Pussytoes ■☆

26414 Antennaria friesiana (Trautv.) E. Ekman subsp. alaskana (Malte) Hultén;阿拉斯加蝶须;Alaskan Pussytoes ■☆

26415 Antennaria friesiana (Trautv.) E. Ekman subsp. neoalaskana (A. E. Porsild) R. J. Bayer et Stebbins;新阿拉斯加蝶须;Frost Boil Pussytoes, Outcrop Pussytoes ■☆

26416 Antennaria friesiana (Trautv.) E. Ekman var. beringensis Hultén = Antennaria friesiana (Trautv.) E. Ekman subsp. alaskana (Malte) Hultén ■☆

26417 Antennaria fusca E. E. Nelson = Antennaria rosea Greene subsp. pulvinata (Greene) R. J. Bayer ■☆

26418 Antennaria gaspensis (Fernald) Fernald = Antennaria rosea Greene subsp. pulvinata (Greene) R. J. Bayer ■☆

26419 Antennaria geyeri A. Gray;松林蝶须;Pinewoods Pussytoes ■☆

26420 Antennaria grandis (Fernald) House = Antennaria howellii Greene subsp. neodioica (Greene) R. J. Bayer ■☆

26421 Antennaria greenei Bush = Antennaria parlinii Fernald subsp. fallax (Greene) R. J. Bayer et Stebbins ■☆

26422 Antennaria hendersonii Piper = Antennaria rosea Greene ■☆

26423 Antennaria holmii Greene = Antennaria parvifolia Nutt. ■☆

26424 Antennaria howellii Greene;豪氏蝶须;Howell's Pussytoes ■☆

26425 Antennaria howellii Greene subsp. canadensis (Greene) R. J. Bayer;加拿大蝶须;Canadian Antennaria, Canadian Pussytoes, Canadian Pussy-toes, Howell's Pussy-toes ■☆

26426 Antennaria howellii Greene subsp. gaspensis (Fernald) Chmiel. = Antennaria rosea Greene subsp. pulvinata (Greene) R. J. Bayer ■☆

26427 Antennaria howellii Greene subsp. neodioica (Greene) R. J. Bayer;新蝶须;Field Pussy-toes, Howell's Pussy-toes, Lesser Cat's Foot, Neodiocious Antennaria, Smaller Pussytoes ■☆

26428 Antennaria howellii Greene subsp. petaloidea (Fernald) R. J. Bayer;瓣状蝶须;Howell's Pussy-toes, Petaloid Pussytoes, Showy Antennaria, Small Pussy-toes ■☆

26429 Antennaria howellii Greene var. athabascensis (Greene) B. Boivin = Antennaria neglecta Greene ■☆

26430 Antennaria howellii Greene var. campestris (Rydb.) B. Boivin = Antennaria neglecta Greene ■☆

26431 Antennaria hudsonica Malte = Antennaria monocephala DC. subsp. angustata (Greene) Hultén ■☆

26432 Antennaria hygrophila Greene = Antennaria corymbosa E. E. Nelson ■☆

26433 Antennaria hyperborea D. Don = Antennaria dioica (L.) Gaertn. ■

26434 Antennaria imbricata E. E. Nelson = Antennaria rosea Greene ■☆

26435 Antennaria incarnata A. E. Porsild = Antennaria rosea Greene subsp. confinis (Greene) R. J. Bayer ■☆

26436 Antennaria insularis Greene = Antennaria dioica (L.) Gaertn. ■

26437 Antennaria isolepis Greene = Antennaria rosea Greene subsp. pulvinata (Greene) R. J. Bayer ■☆

26438 Antennaria japonica Sch. Bip. = Anaphalis margaritacea (L.) Benth. et Hook. f. var. japonica (Sch. Bip.) Makino ■

26439 Antennaria komarovi Juz. ex Kom.;科马罗夫蝶须■☆

26440 Antennaria laingii A. E. Porsild = Antennaria rosea Greene subsp. confinis (Greene) R. J. Bayer ■☆

26441 Antennaria lanata (Hook.) Greene;绵毛蝶须;Woolly Pussytoes ■☆

26442 Antennaria lanulosa Greene = Antennaria rosea Greene ■☆

26443 Antennaria latisquama Piper = Antennaria dimorpha (Nutt.) Torr. et A. Gray ■☆

26444 Antennaria latisquamea Greene = Antennaria parvifolia Nutt. ■☆

26445 Antennaria leontopodina DC. = Leontopodium leontopodinum (DC.) Hand.-Mazz. ■☆

26446 Antennaria leontopodioides Cody = Antennaria rosea Greene subsp. confinis (Greene) R. J. Bayer ■☆

26447 Antennaria leuchippii Porsild = Antennaria rosea Greene subsp. confinis (Greene) R. J. Bayer ■☆

26448 Antennaria leucophaea Piper = Antennaria stenophylla (A. Gray) A. Gray ■☆

26449 Antennaria longifolia Greene = Antennaria neglecta Greene ■☆

26450 Antennaria lunellii Greene = Antennaria neglecta Greene ■☆

26451 Antennaria luzuloides Torr. et A. Gray;银棕蝶须;Rush Pussytoes, Silvery Brown Pussytoes ■☆

26452 Antennaria luzuloides Torr. et A. Gray subsp. aberrans (E. E. Nelson) R. J. Bayer et Stebbins;小头蝶须;Small-headed Rush Pussytoes ■☆

26453 Antennaria luzuloides Torr. et A. Gray var. microcephala (A. Gray) Cronquist = Antennaria luzuloides Torr. et A. Gray subsp. aberrans (E. E. Nelson) R. J. Bayer et Stebbins ■☆

26454 Antennaria luzuloides Torr. et A. Gray var. oblanceolata (Rydb.) M. Peck = Antennaria luzuloides Torr. et A. Gray ■☆

26455 Antennaria macrocephala (D. C. Eaton) Rydb. = Antennaria dimorpha (Nutt.) Torr. et A. Gray ■☆

26456 Antennaria maculata Greene = Antennaria rosea Greene subsp. pulvinata (Greene) R. J. Bayer ■☆

26457 Antennaria manicouagana P. Landry = Antennaria rosea Greene subsp. pulvinata (Greene) R. J. Bayer ■☆

26458 Antennaria margaritacea (L.) R. Br. = Anaphalis margaritacea (L.) Benth. et Hook. f. ■

26459 Antennaria margaritacea R. Br. = Anaphalis margaritacea (L.) Benth. et Hook. f. ■

26460 Antennaria marginata Greene;白边蝶须;Whitemargin Pussytoes ■☆

26461 Antennaria marginata Greene var. glandulifera A. Nelson = Antennaria marginata Greene ■☆

26462 Antennaria media Greene;落基山蝶须;Rocky Mountain Pussytoes ■☆

26463 Antennaria media Greene subsp. ciliata E. E. Nelson = Antennaria pulchella Greene ■☆

26464 Antennaria media Greene subsp. fusca (E. E. Nelson) Chmiel. =

Antennaria rosea Greene subsp. pulvinata (Greene) R. J. Bayer ■☆

26465 Antennaria media Greene subsp. pulchella (Greene) Chmiel. = Antennaria pulchella Greene ■☆

26466 Antennaria megacephala Fernald ex Raup = Antennaria monocephala DC. subsp. angustata (Greene) Hultén ■☆

26467 Antennaria mesochora Greene = Antennaria parlinii Fernald subsp. fallax (Greene) R. J. Bayer et Stebbins ■☆

26468 Antennaria microcephala A. Gray = Antennaria luzuloides Torr. et A. Gray subsp. aberrans (E. E. Nelson) R. J. Bayer et Stebbins ■☆

26469 Antennaria microphylla Lunell var. solstitialis Lunell = Antennaria microphylla Rydb. ■☆

26470 Antennaria microphylla Rydb.;小叶蝶须;Littleleaf Pussytoes ■☆

26471 Antennaria minuscula B. Boivin = Antennaria parvifolia Nutt. ■☆

26472 Antennaria modesta Greene = Antennaria media Greene ■☆

26473 Antennaria monocephala (Torr. et A. Gray) Greene = Antennaria solitaria Rydb. ■☆

26474 Antennaria monocephala DC.;单头蝶须;Pygmy Pussytoes ■☆

26475 Antennaria monocephala DC. subsp. angustata (Greene) Hultén;窄叶单头蝶须;Narrow-leaved Pygmy Pussytoes ■☆

26476 Antennaria monocephala DC. subsp. philonipha (A. E. Porsild) Hultén = Antennaria monocephala DC. ■☆

26477 Antennaria monocephala DC. var. exilis (Greene) Hultén = Antennaria monocephala DC. ■☆

26478 Antennaria monocephala DC. var. latisquamea Hultén = Antennaria monocephala DC. ■☆

26479 Antennaria montana Gray = Antennaria dioica (L.) Gaertn. ■

26480 Antennaria mucronata E. E. Nelson = Antennaria media Greene ■☆

26481 Antennaria munda Fernald = Antennaria parlinii Fernald subsp. fallax (Greene) R. J. Bayer et Stebbins ■☆

26482 Antennaria muscoides Hook. f. = Leontopodium haastioides (Hand.-Mazz.) Hand.-Mazz. ■

26483 Antennaria muscoides Hook. f. et Thomson = Leontopodium nanum (Hook. f. et Thomson) Hand.-Mazz. ■

26484 Antennaria muscoides Hook. f. et Thomson = Leontopodium pusillum (Beauverd) Hand.-Mazz. ■

26485 Antennaria muscoides sensu Hook. f. = Leontopodium nanum (Hook. f. et Thomson ex C. B. Clarke) Hand.-Mazz. ■☆

26486 Antennaria nana Hook. f. et Thomson = Leontopodium nanum (Hook. f. et Thomson) Hand.-Mazz. ■

26487 Antennaria nana Hook. f. et Thomson ex C. B. Clarke = Leontopodium nanum (Hook. f. et Thomson ex C. B. Clarke) Hand.-Mazz. ■☆

26488 Antennaria nardina Greene = Antennaria corymbosa E. E. Nelson ■☆

26489 Antennaria nebrascensis Greene = Antennaria neglecta Greene ■☆

26490 Antennaria neglecta Greene;野蝶须;Cat's Foot, Cat's-foot, Field Pussytoes, Field Pussy-toes, Neglected Antennaria, Pussy Toe, Pussy Toes, Pussytoes ■☆

26491 Antennaria neglecta Greene subsp. howellii (Greene) Hultén = Antennaria howellii Greene ■☆

26492 Antennaria neglecta Greene var. argillicola (Stebbins) Cronquist = Antennaria virginica Stebbins ■☆

26493 Antennaria neglecta Greene var. athabascensis (Greene) R. L. Taylor et MacBryde = Antennaria neglecta Greene ■☆

26494 Antennaria neglecta Greene var. attenuata (Fernald) Cronquist = Antennaria howellii Greene subsp. neodioica (Greene) R. J. Bayer ■☆

26495 Antennaria neglecta Greene var. campestris (Greene) Steyerm. = Antennaria neglecta Greene ■☆

26496 Antennaria neglecta Greene var. campestris (Rydb.) Steyerm. = Antennaria neglecta Greene ■☆

26497 Antennaria neglecta Greene var. canadensis (Greene) Cronquist = Antennaria howellii Greene subsp. canadensis (Greene) R. J. Bayer ■☆

26498 Antennaria neglecta Greene var. gaspensis (Fernald) Cronquist = Antennaria rosea Greene subsp. pulvinata (Greene) R. J. Bayer ■☆

26499 Antennaria neglecta Greene var. howellii (Greene) Cronquist = Antennaria howellii Greene ■☆

26500 Antennaria neglecta Greene var. neodioica (Greene) Cronquist = Antennaria howellii Greene subsp. neodioica (Greene) R. J. Bayer ■☆

26501 Antennaria neglecta Greene var. petaloidea (Fernald) Cronquist = Antennaria howellii Greene subsp. petaloidea (Fernald) R. J. Bayer ■☆

26502 Antennaria neglecta Greene var. randii (Fernald) Cronquist = Antennaria howellii Greene subsp. canadensis (Greene) R. J. Bayer ■☆

26503 Antennaria neglecta Greene var. simplex Peck = Antennaria neglecta Greene ■☆

26504 Antennaria neglecta Greene var. subcorymbosa Fernald = Antennaria howellii Greene subsp. petaloidea (Fernald) R. J. Bayer ■☆

26505 Antennaria neglecta var. subcorymbosa Fernald = Antennaria howellii Greene subsp. petaloidea (Fernald) R. J. Bayer ■☆

26506 Antennaria nemoralis Greene = Antennaria plantaginifolia (L.) Hook. ■☆

26507 Antennaria neoalaskana A. E. Porsild = Antennaria friesiana (Trautv.) E. Ekman subsp. neoalaskana (A. E. Porsild) R. J. Bayer et Stebbins ■☆

26508 Antennaria neodioica Greene = Antennaria howellii Greene subsp. neodioica (Greene) R. J. Bayer ■☆

26509 Antennaria neodioica Greene subsp. canadensis (Greene) R. J. Bayer et Stebbins = Antennaria howellii Greene subsp. canadensis (Greene) R. J. Bayer ■☆

26510 Antennaria neodioica Greene subsp. howellii (Greene) R. J. Bayer = Antennaria howellii Greene ■☆

26511 Antennaria neodioica Greene subsp. petaloidea (Fernald) R. J. Bayer et Stebbins = Antennaria howellii Greene subsp. petaloidea (Fernald) R. J. Bayer ■☆

26512 Antennaria neodioica Greene var. argillicola (Stebbins) Fernald = Antennaria virginica Stebbins ■☆

26513 Antennaria neodioica Greene var. attenuata Fernald = Antennaria howellii Greene subsp. neodioica (Greene) R. J. Bayer ■☆

26514 Antennaria neodioica Greene var. chlorantha (Greene) B. Boivin = Antennaria rosea Greene ■☆

26515 Antennaria neodioica Greene var. chlorophylla Fernald = Antennaria howellii Greene subsp. neodioica (Greene) R. J. Bayer ■☆

26516 Antennaria neodioica Greene var. gaspensis Fernald = Antennaria rosea Greene subsp. pulvinata (Greene) R. J. Bayer ■☆

26517 Antennaria neodioica Greene var. grandis Fernald = Antennaria howellii Greene subsp. neodioica (Greene) R. J. Bayer ■☆

26518 Antennaria neodioica Greene var. interjecta Fernald = Antennaria howellii Greene subsp. neodioica (Greene) R. J. Bayer ■☆

26519 Antennaria neodioica Greene var. petaloidea Fernald = Antennaria howellii Greene subsp. petaloidea (Fernald) R. J. Bayer ■☆

26520 Antennaria neodioica Greene var. randii (Fernald) B. Boivin = Antennaria howellii Greene subsp. canadensis (Greene) R. J.

Bayer ■☆

26521 Antennaria neodioica Greene var. rupicola (Fernald) Fernald = Antennaria howellii Greene subsp. neodioica (Greene) R. J. Bayer ■☆

26522 Antennaria neodioica Greene var. typica Fernald = Antennaria howellii Greene subsp. neodioica (Greene) R. J. Bayer ■☆

26523 Antennaria nitens Greene = Antennaria monocephala DC. ■☆

26524 Antennaria nitida Greene = Antennaria microphylla Rydb. ■☆

26525 Antennaria oblanceolata Rydb. = Antennaria luzuloides Torr. et A. Gray ■☆

26526 Antennaria obovata E. E. Nelson = Antennaria howellii Greene subsp. neodioica (Greene) R. J. Bayer ■☆

26527 Antennaria obovata E. E. Nelson = Antennaria parlinii Fernald subsp. fallax (Greene) R. J. Bayer et Stebbins ■☆

26528 Antennaria occidentalis Greene = Antennaria parlinii Fernald subsp. fallax (Greene) R. J. Bayer et Stebbins ■☆

26529 Antennaria oxyphylla Greene = Antennaria rosea Greene ■☆

26530 Antennaria parlinii Fernald;帕林蝶须;Indian Tobacco, Ladies' Tobacco, Parlin's Pussytoes, Parlin's Pussy-toes, Plainleaf Pussytoes, Plantain Pussy-toes, Pussy Toes, Pussytoes, Smooth Pussytoes ■☆

26531 Antennaria parlinii Fernald subsp. fallax (Greene) R. J. Bayer et Stebbins;迷惑蝶须;Deceitful Pussytoes, Parlin's Pussy-toes, Plantain Pussy-toes ■☆

26532 Antennaria parlinii Fernald var. ambigens (Greene) Fernald = Antennaria parlinii Fernald subsp. fallax (Greene) R. J. Bayer et Stebbins ■☆

26533 Antennaria parlinii Fernald var. arnoglossa (Greene) Fernald = Antennaria parlinii Fernald ■☆

26534 Antennaria parlinii Fernald var. farwellii (Greene) B. Boivin = Antennaria parlinii Fernald subsp. fallax (Greene) R. J. Bayer et Stebbins ■☆

26535 Antennaria parviflora Nutt. = Antennaria dioica (L.) Gaertn. ■

26536 Antennaria parvifolia Nutt.;北美小叶蝶须;Nuttall's Pussytoes, Pussy's Toes, Small-leaf Pussytoes ■☆

26537 Antennaria parvula Greene = Antennaria neglecta Greene ■☆

26538 Antennaria peasei Fernald = Antennaria rosea Greene subsp. pulvinata (Greene) R. J. Bayer ■☆

26539 Antennaria pedicellata Greene = Antennaria howellii Greene subsp. petaloidea (Fernald) R. J. Bayer ■☆

26540 Antennaria peramoena Greene = Antennaria marginata Greene ■☆

26541 Antennaria petaloidea (Fernald) Fernald = Antennaria howellii Greene subsp. petaloidea (Fernald) R. J. Bayer ■☆

26542 Antennaria petaloidea (Fernald) Fernald var. novaboracensis Fernald = Antennaria howellii Greene subsp. petaloidea (Fernald) R. J. Bayer ■☆

26543 Antennaria petaloidea (Fernald) Fernald var. scariosa Fernald = Antennaria howellii Greene subsp. petaloidea (Fernald) R. J. Bayer ■☆

26544 Antennaria petaloidea (Fernald) Fernald var. subcorymbosa (Fernald) Fernald = Antennaria howellii Greene subsp. petaloidea (Fernald) R. J. Bayer ■☆

26545 Antennaria petaloidea Fernald = Antennaria howellii Greene subsp. petaloidea (Fernald) R. J. Bayer ■☆

26546 Antennaria petasites Greene = Antennaria racemosa Hook. ■☆

26547 Antennaria petiolata Fernald = Antennaria plantaginifolia (L.) Hook. ■☆

26548 Antennaria philonipha A. E. Porsild = Antennaria monocephala DC. ■☆

26549 Antennaria pinetorum Greene = Antennaria plantaginifolia (L.) Hook. ■☆

26550 Antennaria piperi Rydb. = Antennaria racemosa Hook. ■☆

26551 Antennaria plantaginifolia (L.) Hook.;车前叶蝶须;Indian Tobacco, Ladies' Tobacco, Lady's Tobacco, Life Everlasting, Love's Test, Plainleaf Pussytoes, Plantain Pussy-toes, Plantainleaf Pussytoes, Plantain-leaved Everlasting, Plantain-leaved Pussy Toes, Plantain-leaved Pussytoes, Plantain-leaved Pussy-toes, Plantaln-leaved Pussy Toe, Pussy Toes, Pussytoes, Spring Cudweed, White Plantain, Woman's Tobacco, Woman's-tobacco ■☆

26552 Antennaria plantaginifolia (L.) Hook. = Antennaria parlinii Fernald ■☆

26553 Antennaria plantaginifolia (L.) Hook. var. ambigens (Greene) Cronquist = Antennaria parlinii Fernald ■☆

26554 Antennaria plantaginifolia (L.) Hook. var. ambigens (Greene) Cronquist = Antennaria parlinii Fernald subsp. fallax (Greene) R. J. Bayer et Stebbins ■☆

26555 Antennaria plantaginifolia (L.) Hook. var. arnoglossa (Greene) Cronquist = Antennaria parlinii Fernald ■☆

26556 Antennaria plantaginifolia (L.) Hook. var. monocephala Torr. et A. Gray;单头车前叶蝶须■☆

26557 Antennaria plantaginifolia (L.) Hook. var. parlinii (Fernald) Cronquist = Antennaria parlinii Fernald ■☆

26558 Antennaria plantaginifolia (L.) Hook. var. petiolata (Fernald) A. Heller = Antennaria plantaginifolia (L.) Hook. ■☆

26559 Antennaria polyphylla Greene ex C. F. Baker = Antennaria rosea Greene subsp. confinis (Greene) R. J. Bayer ■☆

26560 Antennaria propinqua Greene = Antennaria parlinii Fernald ■☆

26561 Antennaria pulchella Greene;齿蝶须;Sierra Pussytoes ■☆

26562 Antennaria pulcherrima (Hook.) Greene;光亮蝶须;Showy Or Handsome Pussytoes ■☆

26563 Antennaria pulcherrima (Hook.) Greene subsp. anaphaloides (Rydb.) W. A. Weber = Antennaria anaphaloides Rydb. ■☆

26564 Antennaria pulcherrima (Hook.) Greene subsp. eucosma (Fernald et Wiegand) R. J. Bayer;雅致蝶须;Elegant Pussytoes ■☆

26565 Antennaria pulcherrima (Hook.) Greene var. anaphaloides (Rydb.) G. W. Douglas = Antennaria anaphaloides Rydb. ■☆

26566 Antennaria pulcherrima (Hook.) Greene var. angustisquama A. E. Porsild = Antennaria pulcherrima (Hook.) Greene ■☆

26567 Antennaria pulcherrima (Hook.) Greene var. sordida B. Boivin = Antennaria pulcherrima (Hook.) Greene ■☆

26568 Antennaria pulvinata Greene = Antennaria rosea Greene subsp. pulvinata (Greene) R. J. Bayer ■☆

26569 Antennaria pulvinata Greene subsp. albescens E. E. Nelson = Antennaria rosea Greene subsp. pulvinata (Greene) R. J. Bayer ■☆

26570 Antennaria pygmaea Fernald = Antennaria monocephala DC. subsp. angustata (Greene) Hultén ■☆

26571 Antennaria pyramidata Greene = Antennaria luzuloides Torr. et A. Gray subsp. aberrans (E. E. Nelson) R. J. Bayer et Stebbins ■☆

26572 Antennaria racemosa Hook.;总状蝶须;Raceme Pussytoes ■☆

26573 Antennaria randii Fernald = Antennaria howellii Greene subsp. canadensis (Greene) R. J. Bayer ■☆

26574 Antennaria recurva Greene = Antennaria parvifolia Nutt. ■☆

26575 Antennaria reflexa E. E. Nelson = Antennaria umbrinella Rydb. ■☆

26576 Antennaria rhodantha Fernald = Antennaria howellii Greene subsp. neodioica (Greene) R. J. Bayer ■☆

26577 Antennaria rhodantha Suksd.;粉花蝶须■☆

26578 Antennaria rhodantha Suksd. = Antennaria parvifolia Nutt. ■☆
26579 Antennaria rosea Greene;爱达荷蝶须;Pink Everlasting,Pink Pussytoes,Rosy Pussytoes ■☆
26580 Antennaria rosea Greene subsp. arida (E. E. Nelson) R. J. Bayer;荒漠蝶须;Desert Pussytoes ■☆
26581 Antennaria rosea Greene subsp. confinis (Greene) R. J. Bayer;近邻蝶须■☆
26582 Antennaria rosea Greene subsp. divaricata E. E. Nelson = Antennaria rosea Greene ■☆
26583 Antennaria rosea Greene subsp. pulvinata (Greene) R. J. Bayer;垫状蝶须;Pulvinate Pussytoes ■☆
26584 Antennaria rosea Greene var. nitida (Greene) Breitung = Antennaria microphylla Rydb. ■☆
26585 Antennaria rosea var. angustifolia (Rydb.) E. E. Nelson = Antennaria rosea Greene subsp. confinis (Greene) R. J. Bayer ■☆
26586 Antennaria rosulata Rydb.;凯巴波蝶须;Kaibab Pussytoes,Woolly Pussytoes ■☆
26587 Antennaria russellii B. Boivin = Antennaria howellii Greene subsp. neodioica (Greene) R. J. Bayer ■☆
26588 Antennaria sansonii Greene = Antennaria rosea Greene subsp. pulvinata (Greene) R. J. Bayer ■☆
26589 Antennaria scabra Greene = Antennaria pulchella Greene ■☆
26590 Antennaria scariosa E. E. Nelson;干膜质蝶须■☆
26591 Antennaria sedoides Greene = Antennaria rosea Greene subsp. confinis (Greene) R. J. Bayer ■☆
26592 Antennaria serawschanica Dunn = Anaphalis tenuisissima C. C. Chang ■
26593 Antennaria shumaginensis A. E. Porsild = Antennaria monocephala DC. ■☆
26594 Antennaria sierrae-blancae Rydb. = Antennaria rosulata Rydb. ■☆
26595 Antennaria soliceps S. F. Blake;扎尔斯顿蝶须;Charleston Mountain Pussytoes,Charleston Pussytoes ■☆
26596 Antennaria solitaria Rydb.;单头车前蝶须;Pussy-toes,Singlehead Pussytoes,Southern Single-headed Pussy-toes ■☆
26597 Antennaria solstitialis Lunell = Antennaria microphylla Rydb. ■☆
26598 Antennaria sordida Greene = Antennaria rosea Greene subsp. confinis (Greene) R. J. Bayer ■☆
26599 Antennaria spathulata (Fernald) Fernald = Antennaria howellii Greene subsp. canadensis (Greene) R. J. Bayer ■☆
26600 Antennaria speciosa E. E. Nelson = Antennaria rosea Greene ■☆
26601 Antennaria steetzeana Turcz. = Leontopodium leontopodioides (Willd.) Beauverd ■
26602 Antennaria stenolepis Greene = Antennaria howellii Greene subsp. petaloidea (Fernald) R. J. Bayer ■☆
26603 Antennaria stenophylla (A. Gray) A. Gray;狭叶蝶须;Narrowleaf Pussytoes ■☆
26604 Antennaria straminea Fernald = Antennaria rosea Greene subsp. pulvinata (Greene) R. J. Bayer ■☆
26605 Antennaria suffrutescens Greene;常绿蝶须;Evergreen Pussytoes,Everlasting Pussytoes ■☆
26606 Antennaria tansleyi Polunin = Antennaria monocephala DC. subsp. angustata (Greene) Hultén ■☆
26607 Antennaria tenella DC. = Anaphalis contorta (D. Don) Hook. f. ■
26608 Antennaria tomentella E. E. Nelson = Antennaria rosea Greene subsp. confinis (Greene) R. J. Bayer ■☆
26609 Antennaria triplinervis Sims = Anaphalis triplinervis (Sims) C. B. Clarke ■
26610 Antennaria triplinervis Sims = Anaphalis triplinervis (Sims) Sims ex C. B. Clarke ■
26611 Antennaria triplinervis Sims var. cuneifolia DC. = Anaphalis nepalensis (Spreng.) Hand. -Mazz. ■
26612 Antennaria triplinervis Sims var. intermedia DC. = Anaphalis nepalensis (Spreng.) Hand. -Mazz. ■
26613 Antennaria tweedsmuirii Polunin = Antennaria monocephala DC. subsp. angustata (Greene) Hultén ■☆
26614 Antennaria umbellata Greene = Antennaria parlinii Fernald subsp. fallax (Greene) R. J. Bayer et Stebbins ■☆
26615 Antennaria umbrinella Rydb.;褐苞蝶须;Brown Pussytoes,Brown-bracted Pussytoes,Umber Pussytoes ■☆
26616 Antennaria villifera Boriss.;长柔毛蝶须■☆
26617 Antennaria villosissima D. Don = Anaphalis busua (Ham.) DC. ■
26618 Antennaria virginica Stebbins;弗吉尼亚蝶须;Shalebarren Pussytoes ■☆
26619 Antennaria virginica Stebbins var. argillicola Stebbins = Antennaria virginica Stebbins ■☆
26620 Antennaria wiegandii Fernald = Antennaria howellii Greene subsp. canadensis (Greene) R. J. Bayer ■☆
26621 Antennaria wilsonii Greene = Antennaria neglecta Greene ■☆
26622 Antenoron Raf. (1817);金线草属;Goldthreadweed,Antenoron ■
26623 Antenoron Raf. = Persicaria (L.) Mill. ■
26624 Antenoron filiforme (Thunb.) Rob. et Vautier;金线草(白马鞭,短毛金线草,海根,红花铁菱角,化血归,鸡心七,金线蓼,九龙盘,九盘龙,蓼子七,毛蓼,人字草,山蓼,铁箍散,铁菱角三七,铁拳头,土三七,线茎蓼,蟹壳草,血经草,野蓼,一串红,重阳柳);Goldthreadweed,Longhairy Antenoron ■
26625 Antenoron filiforme (Thunb.) Rob. et Vautier var. kachinum (Nieuwl.) H. Hara;毛叶金线草(毛叶红珠七,小红袍);Hairyleaf Antenoron ■
26626 Antenoron filiforme (Thunb.) Rob. et Vautier var. neofiliforme (Nakai) A. J. Li;短毛金线草(蓼子七);Shorthairy Antenoron ■
26627 Antenoron filiforme (Thunb.) Roberty et Vautier = Persicaria filiformis (Thunb.) Nakai ex W. T. Lee ■
26628 Antenoron filiforme (Thunb.) Roberty et Vautier f. albiflorum (Hiyama) H. Hara = Persicaria filiformis (Thunb.) Nakai ex W. T. Lee f. albiflora (Hiyama) Yonek. ■☆
26629 Antenoron filiforme (Thunb.) Roberty et Vautier f. smaragdinum (Nakai ex F. Maek.) H. Hara = Persicaria filiformis (Thunb.) Nakai ex W. T. Lee ■☆
26630 Antenoron filiforme (Thunb.) Roberty et Vautier var. neofiliforme (Nakai) A. J. Li = Persicaria neofiliformis (Nakai) Ohki ■
26631 Antenoron neofiliforme (Nakai) H. Hara = Antenoron filiforme (Thunb.) Roberty et Vautier var. neofiliforme (Nakai) A. J. Li ■
26632 Antenoron neofiliforme (Nakai) H. Hara = Persicaria neofiliformis (Nakai) Ohki ■
26633 Antenoron virginianum (L.) Roberty et Vautier = Persicaria virginiana (L.) Gaertn. ■☆
26634 Antenoron virginianum (L.) Roberty et Vautier = Polygonum virginianum L. ■☆
26635 Antephora Steud. = Anthephora Schreb. ■☆
26636 Anteremanthus H. Rob. (1992);单头巴西菊属●☆
26637 Anteremanthus hatschbachii H. Rob.;单头巴西菊■☆
26638 Anteriorchis E. Klein et Strack = Orchis L. ■
26639 Anteriorchis coriophora (L.) Klein et Strack = Anacamptis coriophora (L.) R. M. Bateman,Pridgeon et Chase ■☆

26640 Anteriscium Meyen = Asteriscium Cham. et Schltdl. ■☆
26641 Anthacantha Lem. = Euphorbia L. ●■
26642 Anthacanthus Nees = Oplonia Raf. ●☆
26643 Anthacanthus vincoides (Lam.) Nees = Oplonia vincoides (Lam.) Stearn ●☆
26644 Anthactinia Bory = Passiflora L. ●■
26645 Anthactinia Bory ex M. Roem. = Passiflora L. ●■
26646 Anthactinia horsfieldii M. Roem. = Passiflora cochinchinensis Spreng. ■
26647 Anthadenia Lem. = Sesamum L. ●■
26648 Anthadenia sesamoides Lem. = Sesamum indicum L. ■
26649 Anthaea Noronha ex Thouars = Didymeles Thouars ●☆
26650 Anthaenantia P. Beauv. = Anthenantia P. Beauv. ■☆
26651 Anthaenantia glauca Hack. = Tricholaena capensis (Licht. ex Roem. et Schult.) Nees subsp. arenaria (Nees) Zizka ■☆
26652 Anthaenantiopsis Mez = Anthaenantiopsis Mez ex Pilg. ■☆
26653 Anthaenantiopsis Mez et Pilg. = Anthaenantiopsis Mez ex Pilg. ■☆
26654 Anthaenantiopsis Mez ex Pilg. (1931);拟银鳞草属■☆
26655 Anthaenantiopsis fiebrigii Mez;拟银鳞草■☆
26656 Anthaerium Schott = Anthurium Schott ■
26657 Anthagathis Harms = Jollydora Pierre ex Gilg ●☆
26658 Anthagathis monadelphia Harms = Jollydora duparquetiana (Baill.) Pierre ●☆
26659 Anthallogea Raf. = Polygala L. ●■
26660 Anthanema Raf. = Cuscuta L. ■
26661 Anthanotis Raf. = Asclepias L. ■
26662 Anthchlorophytum acrothyrsum Peter ex Poelln. = Chlorophytum polystachys Baker ■☆
26663 Anthchlorophytum afzelii Baker = Chlorophytum inornatum Ker Gawl. ■☆
26664 Anthchlorophytum asparagiflorum Engl. = Chlorophytum polystachys Baker ■☆
26665 Anthchlorophytum asphodeloides C. H. Wright = Chlorophytum nidulans (Baker) Brenan ■☆
26666 Anthchlorophytum bakeri Poelln. = Chlorophytum silvaticum Dammer ■☆
26667 Anthchlorophytum baoulense A. Chev. = Chlorophytum inornatum Ker Gawl. ■☆
26668 Anthchlorophytum carsonii Baker = Chlorophytum lancifolium Welw. ex Baker ■☆
26669 Anthchlorophytum cavalliense A. Chev. = Chlorophytum inornatum Ker Gawl. ■☆
26670 Anthchlorophytum comatum Poelln. = Chlorophytum silvaticum Dammer ■☆
26671 Anthchlorophytum cordatum Engl. = Chlorophytum lancifolium Welw. ex Baker subsp. cordatum (Engl.) A. D. Poulsen et Nordal ■☆
26672 Anthchlorophytum durbanense Kuntze = Chlorophytum modestum Baker ■☆
26673 Anthchlorophytum ealaense De Wild. = Chlorophytum lancifolium Welw. ex Baker subsp. togoense (Engl.) A. D. Poulsen et Nordal ■☆
26674 Anthchlorophytum ellenbeckii Poelln. = Chlorophytum neghellense Cufod. ■☆
26675 Anthchlorophytum fosteri A. Chev. = Chlorophytum lancifolium Welw. ex Baker subsp. togoense (Engl.) A. D. Poulsen et Nordal ■☆
26676 Anthchlorophytum gallarum Poelln. = Chlorophytum neghellense Cufod. ■☆
26677 Anthchlorophytum glabriflorum C. H. Wright = Chlorophytum nyassae (Rendle) Kativu ■☆
26678 Anthchlorophytum hispidulum Rendle = Chlorophytum scabrum Baker ■☆
26679 Anthchlorophytum holotrichum Peter ex Poelln. = Chlorophytum polystachys Baker ■☆
26680 Anthchlorophytum inexpectatum Poelln. = Chlorophytum silvaticum Dammer ■☆
26681 Anthchlorophytum longipedunculatum H. M. L. Forbes = Chlorophytum krookianum Zahlbr. ■☆
26682 Anthchlorophytum macrocladum Peter ex Poelln. = Chlorophytum viridescens Engl. ■☆
26683 Anthchlorophytum micans Engl. et K. Krause = Chlorophytum pusillum Schweinf. ex Baker ■☆
26684 Anthchlorophytum minutiflorum Poelln. = Chlorophytum lancifolium Welw. ex Baker ■☆
26685 Anthchlorophytum mossicum A. Chev. = Chlorophytum pusillum Schweinf. ex Baker ■☆
26686 Anthchlorophytum nebulosum Poelln. = Chlorophytum lancifolium Welw. ex Baker ■☆
26687 Anthchlorophytum nigericum (Hepper) Nordal = Chlorophytum angustissimum (Poelln.) Nordal ■☆
26688 Anthchlorophytum nigericum A. Chev. = Chlorophytum limosum (Baker) Nordal ■☆
26689 Anthchlorophytum norlindhii Weim. = Chlorophytum galpinii (Baker) Kativu var. norlindhii (Weim.) Kativu ■☆
26690 Anthchlorophytum palustre Engl. et K. Krause = Chlorophytum polystachys Baker ■☆
26691 Anthchlorophytum paniculosum Peter ex Poelln. = Chlorophytum pubiflorum Baker ■☆
26692 Anthchlorophytum papilliferum Poelln. = Chlorophytum silvaticum Dammer ■☆
26693 Anthchlorophytum peteri Poelln. = Chlorophytum lancifolium Welw. ex Baker ■☆
26694 Anthchlorophytum pilosissimum Engl. et K. Krause = Chlorophytum vestitum Baker ■☆
26695 Anthchlorophytum pleurostachyum Chiov. = Chlorophytum nubicum (Baker) Kativu ■☆
26696 Anthchlorophytum polyphyllum (Baker) Kativu = Chlorophytum recurvifolium (Baker) C. Archer et Kativu ■☆
26697 Anthchlorophytum pulchellum Kunth = Chlorophytum rigidum Kunth ■☆
26698 Anthchlorophytum robustum Poelln. = Chlorophytum silvaticum Dammer ■☆
26699 Anthchlorophytum schlechterianum Poelln. = Chlorophytum krookianum Zahlbr. ■☆
26700 Anthchlorophytum setosum Poelln. = Chlorophytum silvaticum Dammer ■☆
26701 Anthchlorophytum sphacelatum (Baker) Kativu subsp. hockii (De Wild.) Kativu = Chlorophytum sphacelatum (Baker) Kativu var. hockii (De Wild.) Nordal ■☆
26702 Anthchlorophytum sphacelatum (Baker) Kativu subsp. milanjianum (Rendle) Kativu = Chlorophytum sphacelatum (Baker) Kativu var. milanjianum (Rendle) Nordal ■☆
26703 Anthchlorophytum subrugosum Poelln. = Chlorophytum polystachys Baker ■☆
26704 Anthchlorophytum talbotii Rendle = Chlorophytum lancifolium Welw. ex Baker subsp. togoense (Engl.) A. D. Poulsen et Nordal ■☆

26705 Anthchlorophytum tenellum Peter ex Poelln. = Chlorophytum leptoneurum (C. H. Wright) Poelln. ■☆
26706 Anthchlorophytum tinneae Baker = Chlorophytum nubicum (Baker) Kativu ■☆
26707 Anthchlorophytum togoense Engl. = Chlorophytum lancifolium Welw. ex Baker subsp. togoense (Engl.) A. D. Poulsen et Nordal ■☆
26708 Anthchlorophytum toumodiense A. Chev. = Chlorophytum lancifolium Welw. ex Baker subsp. togoense (Engl.) A. D. Poulsen et Nordal ■☆
26709 Anthchlorophytum tuberigenum Peter ex Poelln. = Chlorophytum polystachys Baker ■☆
26710 Anthchlorophytum uhamense Poelln. = Chlorophytum lancifolium Welw. ex Baker subsp. togoense (Engl.) A. D. Poulsen et Nordal ■☆
26711 Antheeischima Korth. = Gordonia J. Ellis(保留属名)●
26712 Antheeischima Korth. = Schima Reinw. ex Blume ●
26713 Antheidosorus A. Gray = Myriocephalus Benth. ■☆
26714 Antheidosurus C. Muell. = Antheidosorus A. Gray ■☆
26715 Antheilema Raf. = Ruellia L. ●■
26716 Antheischima Korth. = Gordonia J. Ellis(保留属名)●
26717 Anthelia Schott = Epipremnum Schott ●■
26718 Antheliacanthus Ridl. = Pseuderanthemum Radlk. ●■
26719 Antheliacanthus micranthus Ridl. = Pseuderanthemum latifolium (Vahl) B. Hansen ■
26720 Anthelis Raf. = Fumana (Dunal) Spach ●☆
26721 Anthelis Raf. = Helianthemum Mill. ●■
26722 Anthelmenthia P. Browne = Spigelia L. ■☆
26723 Anthelmenthica Pfeiff. = Anthelmenthia P. Browne ■☆
26724 Anthelminthica B. D. Jacks. = Anthelmenthia P. Browne ■☆
26725 Anthelminthica P. Browne = Spigelia L. ■☆
26726 Anthema Medik. = Lavatera L. ●■
26727 Anthemidaceae Link = Asteraceae Bercht. et J. Presl(保留科名) ●■
26728 Anthemidaceae Link = Compositae Giseke(保留科名)●■
26729 Anthemidaceae Link;春黄菊科●■
26730 Anthemidaceae Martinov = Asteraceae Bercht. et J. Presl(保留科名)●■
26731 Anthemidaceae Martinov = Compositae Giseke(保留科名)●■
26732 Anthemiopsis Bojer = Wedelia Jacq. (保留属名)●■
26733 Anthemiopsis Bojer ex DC. (1836);拟春黄菊属■☆
26734 Anthemiopsis Bojer ex DC. = Wedelia Jacq. (保留属名)●■
26735 Anthemiopsis elongata Bojer;拟春黄菊■☆
26736 Anthemiopsis macrophylla Bojer = Melanthera biflora (L.) Wild ■☆
26737 Anthemiopsis macrophylla Bojer ex DC. = Melanthera biflora (L.) Wild ■☆
26738 Anthemis L. (1753);春黄菊属;Camomile, Chamomile, Dog Fennel, Dog's Fennel, Marguerite ■
26739 Anthemis Mich. ex L. = Anthemis L. ■
26740 Anthemis abylaea (Font Quer et Maire) Oberpr.;阿比拉春黄菊 ■☆
26741 Anthemis abyssinica J. Gay ex A. Rich. var. tigreensis (J. Gay ex A. Rich.) Chiov. = Anthemis tigreensis J. Gay ex A. Rich. ■☆
26742 Anthemis afra L. = Osmitopsis afra (L.) K. Bremer ●☆
26743 Anthemis altissima Boiss.;弯春黄菊;Tall Camomile ■☆
26744 Anthemis anatolica Boiss.;阿纳托里春黄菊■☆
26745 Anthemis arabicus L. = Cladanthus arabicus (L.) Cass. ●☆
26746 Anthemis arvensis L.;田春黄菊(刺干菊);Corn Chamomile, Fausse Camomille, Field Camomile, Martha, Marthus, Whitewort ■
26747 Anthemis arvensis L. subsp. incrassata (Loisel.) Nyman;粗春黄菊■☆
26748 Anthemis arvensis L. var. agrestis (Wallr.) DC. = Anthemis arvensis L. ■
26749 Anthemis arvensis L. var. ubensis (Pomel) Quézel et Santa = Anthemis ubensis Pomel ■☆
26750 Anthemis atlantica Pomel = Anthemis pedunculata Desf. subsp. atlantica (Pomel) Oberpr. ■☆
26751 Anthemis austriaca Jacq.;奥地利春黄菊;Austrian Chamomile ■☆
26752 Anthemis biebersteiniana C. Koch;高加索春黄菊■☆
26753 Anthemis boveana J. Gay;包氏春黄菊■☆
26754 Anthemis boveana J. Gay var. aguilarii Maire et Sennen = Anthemis maroccana Batt. et Pit. subsp. aguilarii (Maire et Sennen) Oberpr. ■☆
26755 Anthemis boveana J. Gay var. jahandiezii Maire = Anthemis boveana J. Gay ■☆
26756 Anthemis boveana J. Gay var. maroccana (Batt. et Pit.) Maire = Anthemis maroccana Batt. et Pit. ■☆
26757 Anthemis boveana J. Gay var. tenuisecta (Ball) Maire = Anthemis tenuisecta Ball ■☆
26758 Anthemis calcarea Sosn.;钙生春黄菊■☆
26759 Anthemis candidissima Willd. ex Spreng.;白花春黄菊■☆
26760 Anthemis carpathica Willd.;卡尔帕索斯春黄菊■☆
26761 Anthemis chrysantha J. Gay;金色春黄菊■☆
26762 Anthemis chrysantha J. Gay var. intermedia Faure et Maire = Anthemis chrysantha J. Gay ■☆
26763 Anthemis chrysantha J. Gay var. tenuisecta (Ball) Pau = Anthemis tenuisecta Ball ■☆
26764 Anthemis cinerea Pancic;灰春黄菊;Gray Camomile ■☆
26765 Anthemis clausonis Pomel = Anthemis pedunculata Desf. subsp. clausonis (Pomel) Oberpr. ■☆
26766 Anthemis clavata Desf. = Anacyclus clavatus (Desf.) Pers. ■☆
26767 Anthemis confusa Pomel;混乱春黄菊■☆
26768 Anthemis cotula L.;臭春黄菊(臭甘菊);Baldeyebrow, Barnyard Aster, Camomille Maroute, Camovyne, Chigger-weed, Common Dog Fennel, Devil Daisy, Dog Banner, Dog Binder, Dog Daisy, Dog Fennel, Dog Finkle, Dog Stinker, Dog Stinkers, Dog Stinks, Dog-fennel, Dog's Camomile, Dog's Fennel, Foefid Camomile, Horse Daisy, Hound's Fennel, Indiana Bergen, Jayweed, Madder, Maddern, Madders, Maidweed, Maithen, Maitheweed, Malden, Marg, Margan, Mathern, Mauther, Mauthern, Mavin, Mawth, Mawthem, May, May Weed, Maythem, Maythig, Mayweed, Mayweed Camomile, Mazes, Meaden, Moithern, Morgan, Murg, Pig Daisy, Pig's Cress, Pig's Daisy, Pig's Flower, Poison Daisy, Slinking Camomile, Stink Mayweed, Stinking Chamomile, Stinking Mathes, Stinking Mayweed, Stinking Nanny, Stinking-cotula, Wild Camomile ■
26769 Anthemis cotula L. var. atromarginata Vatke = Anthemis tigreensis J. Gay ex A. Rich. ■☆
26770 Anthemis cretacea Zefir.;白垩春黄菊■☆
26771 Anthemis cretica L.;克里特春黄菊■☆
26772 Anthemis cretica L. subsp. columnae (Ten.) Franzén;圆柱春黄菊■☆
26773 Anthemis cupaniana Nyman = Anthemis punctata Vahl subsp. cupania Nyman ■☆
26774 Anthemis cupaniana Nyman var. kabylica (Batt.) Pau = Anthemis punctata Vahl subsp. kabylica (Batt.) Oberpr. ■☆

26775 Anthemis cyrenaica Coss.;昔兰尼春黄菊■☆

26776 Anthemis cyrenaica Coss. var. radiata Pamp. = Anthemis cyrenaica Coss. ■☆

26777 Anthemis debilis Fed.;柔软春黄菊■☆

26778 Anthemis deserti Boiss. = Anthemis melampodina Delile subsp. deserti (Boiss.) Eig ■☆

26779 Anthemis deserticola Krasch. et Popov;荒漠春黄菊■☆

26780 Anthemis dubia Stev.;可疑春黄菊■☆

26781 Anthemis dumetorum Sosn.;灌丛春黄菊■☆

26782 Anthemis euxina Boiss.;黑春黄菊■☆

26783 Anthemis foetida Lam. = Anthemis cotula L. ■

26784 Anthemis frutescens Hort. = Argyranthemum frutescens (L.) Sch. Bip. ●

26785 Anthemis frutescens Hort. = Chrysanthemum frutescens L. ●

26786 Anthemis fruticosa L. = Relhania fruticosa (L.) K. Bremer ●☆

26787 Anthemis fruticulosa M. Bieb.;灌木状春黄菊■☆

26788 Anthemis fuscata Brot. = Chamaemelum fuscatum (Brot.) Vasc. ■☆

26789 Anthemis gharbensis Oberpr.;盖尔比春黄菊■☆

26790 Anthemis glareosa E. A. Durand et Barratte;石砾春黄菊■☆

26791 Anthemis grandiflora Ramat. = Dendranthema grandiflorum (Ramat.) Kitam. ■

26792 Anthemis grangeoides Vatke et Höpfner ex Klatt = Anisopappus grangeoides (Vatke et Höpfner ex Klatt) Merxm. ■☆

26793 Anthemis granulata Pomel = Anthemis pedunculata Desf. ■☆

26794 Anthemis grossheimii Sosn.;格罗春黄菊■☆

26795 Anthemis halimifolia Munby = Mecomischus halimifolius (Munby) Hochr. ■☆

26796 Anthemis heterophylla (Coss.) Ball = Cladanthus scariosus (Ball) Oberpr. et Vogt ●☆

26797 Anthemis hirtella C. Winkl.;硬毛春黄菊■☆

26798 Anthemis homogamos (Maire) Humphries;同配春黄菊■☆

26799 Anthemis iberica M. Bieb.;伊比利亚春黄菊■☆

26800 Anthemis incrassata (Hoffmanns. et Link) Link = Anacyclus clavatus (Desf.) Pers. ■☆

26801 Anthemis indurata Delile;坚硬春黄菊■☆

26802 Anthemis indurata Delile var. angulata Pamp. = Anthemis indurata Delile ■☆

26803 Anthemis kabylica Batt. et Trab. = Anthemis punctata Vahl subsp. kabylica (Batt.) Oberpr. ■☆

26804 Anthemis kelwayi Hort. ex L. H. Bailey et N. Taylor;凯氏春黄菊;Kelway Chamomile ■☆

26805 Anthemis kitaibelii Spreng. = Anthemis montana L. ■

26806 Anthemis kruegeriana Pamp.;克吕格尔春黄菊■☆

26807 Anthemis kruegeriana Pamp. var. discoidea Maire et Weiller = Anthemis kruegeriana Pamp. ■☆

26808 Anthemis kruegeriana Pamp. var. radiata Maire et Weiller = Anthemis kruegeriana Pamp. ■☆

26809 Anthemis laeviuscula Humbert et Maire = Anthemis pedunculata Desf. ■☆

26810 Anthemis leucantha L. = Osmitopsis afra (L.) K. Bremer ●☆

26811 Anthemis lithuanica Besser ex DC.;立陶宛春黄菊■☆

26812 Anthemis lonadioides (Coss.) Hochr. = Rhetinolepis lonadioides Coss. ■☆

26813 Anthemis macedonica Boiss. et Orph.;马其顿春黄菊;Macedonian Camomile ■☆

26814 Anthemis macranta Heuff.;大头春黄菊;Bighead Camomile ■☆

26815 Anthemis macroglossa Sommier et H. Lév.;大舌春黄菊■☆

26816 Anthemis maritima L.;滨海春黄菊■☆

26817 Anthemis maritima L. var. incana Guss. = Anthemis maritima L. ■☆

26818 Anthemis markhotensis Fed.;马尔浩特春黄菊■☆

26819 Anthemis maroccana Batt. et Pit.;摩洛哥春黄菊■☆

26820 Anthemis maroccana Batt. et Pit. subsp. aguilarii (Maire et Sennen) Oberpr.;阿圭拉里春黄菊■☆

26821 Anthemis marschalliana Willd. = Anthemis biebersteiniana C. Koch ■☆

26822 Anthemis mauritiana Maire et Sennen;毛里求斯春黄菊■☆

26823 Anthemis mauritiana Maire et Sennen subsp. faurei (Maire) Oberpr.;福雷春黄菊■☆

26824 Anthemis mauritii Maire et Sennen = Anthemis mauritiana Maire et Sennen ■☆

26825 Anthemis melampodina Delile;黑齿春黄菊■☆

26826 Anthemis melampodina Delile subsp. deserti (Boiss.) Eig;荒漠黑齿春黄菊■☆

26827 Anthemis melanoloma Trautv.;黑边春黄菊■☆

26828 Anthemis microsperma Boiss. et Kotschy;小籽春黄菊■☆

26829 Anthemis millefolia L.;千叶春黄菊(千叶菊蒿)■☆

26830 Anthemis mixta L.;混杂春黄菊■☆

26831 Anthemis mixta L. = Cladanthus mixtus (L.) Oberpr. et Vogt ●☆

26832 Anthemis monilicostata Pomel;串珠中脉春黄菊■☆

26833 Anthemis monilicostata Pomel subsp. stiparum (Pomel) Maire = Anthemis stiparum Pomel ■☆

26834 Anthemis monilicostata Pomel var. decumbens (Bonnet et Barratte) Maire et Weiller = Anthemis monilicostata Pomel ■☆

26835 Anthemis monilicostata Pomel var. sublaevis Maire = Anthemis monilicostata Pomel ■☆

26836 Anthemis montana L.;山春黄菊(白花春黄菊);Riviera Camomile ■

26837 Anthemis neglecta L. = Anthemis vulgaris L. ex Steud. ■☆

26838 Anthemis nobilis L. = Chamaenerion nobile (L.) All. ■

26839 Anthemis numidica Batt. = Anthemis cretica L. subsp. columnae (Ten.) Franzén ■☆

26840 Anthemis odontostephana Boiss.;齿冠春黄菊■☆

26841 Anthemis pedunculata Desf.;梗花春黄菊■☆

26842 Anthemis pedunculata Desf. subsp. atlantica (Pomel) Oberpr.;大西洋春黄菊■☆

26843 Anthemis pedunculata Desf. subsp. glareosa (Durand et Barratte) Le Houér. = Anthemis glareosa E. A. Durand et Barratte ■☆

26844 Anthemis pedunculata Desf. subsp. granulata (Pomel) Maire = Anthemis pedunculata Desf. ■☆

26845 Anthemis pedunculata Desf. subsp. tuberculata (Boiss.) Maire = Anthemis pedunculata Desf. ■☆

26846 Anthemis pedunculata Desf. var. clausonis (Pomel) Batt. = Anthemis pedunculata Desf. subsp. clausonis (Pomel) Oberpr. ■☆

26847 Anthemis pedunculata Desf. var. decumbens Coss. = Anthemis secundiramea Biv. ■☆

26848 Anthemis pedunculata Desf. var. discoidea (Boiss.) Oberpr. = Anthemis pedunculata Desf. ■☆

26849 Anthemis pedunculata Desf. var. faurei Maire = Anthemis mauritiana Maire et Sennen subsp. faurei (Maire) Oberpr. ■☆

26850 Anthemis pedunculata Desf. var. laevis (Emb. et Maire) Maire = Anthemis arvensis L. subsp. incrassata (Loisel.) Nyman ■☆

26851 Anthemis pedunculata Desf. var. laeviuscula (Humbert et Maire) Maire = Anthemis pedunculata Desf. ■☆

26852 Anthemis pedunculata Desf. var. microcephala (Boiss.) Jahand. et Maire = Anthemis pedunculata Desf. ■☆
26853 Anthemis pedunculata Desf. var. mucronulata Le Houér. = Anthemis pedunculata Desf. ■☆
26854 Anthemis pedunculata Desf. var. tetuanensis (Pau) Maire = Anthemis abylaea (Font Quer et Maire) Oberpr. ■☆
26855 Anthemis pedunculata Desf. var. trachycarpa Maire = Anthemis pedunculata Desf. ■☆
26856 Anthemis pontica Willd. = Anthemis montana L. ■
26857 Anthemis praecox L.;早生春黄菊■☆
26858 Anthemis pseudocotula Boiss.;假杯春黄菊■☆
26859 Anthemis pubescens Willd. = Anacyclus clavatus (Desf.) Pers. ■☆
26860 Anthemis punctata Vahl;斑点春黄菊;Sicilian Chamomile ■☆
26861 Anthemis punctata Vahl subsp. cupania Nyman;库潘春黄菊■☆
26862 Anthemis punctata Vahl subsp. cupaniana (Tod. ex Nyman) R. Fern.;白舌春黄菊■☆
26863 Anthemis punctata Vahl subsp. kabylica (Batt.) Oberpr.;卡比利亚春黄菊■☆
26864 Anthemis punctata Vahl var. abylea Font Quer et Maire = Anthemis abylaea (Font Quer et Maire) Oberpr. ■☆
26865 Anthemis punctata Vahl var. maroccana Maire = Anthemis pedunculata Desf. ■☆
26866 Anthemis punctata Vahl var. microcephala Faure et Maire = Anthemis pedunculata Desf. ■☆
26867 Anthemis pygmaea Sch. Bip. ex Oliv. et Hiern = Anthemis tigreensis J. Gay ex A. Rich. ■☆
26868 Anthemis pyrethrum L. = Anacyclus pyrethrum (L.) Link ■☆
26869 Anthemis retusa Delile;微凹春黄菊■☆
26870 Anthemis rhodocentra Iranshahr;红心春黄菊■☆
26871 Anthemis rigescens Willd.;硬直春黄菊■☆
26872 Anthemis rotata Boiss. = Anthemis pseudocotula Boiss. ■☆
26873 Anthemis rudolphiana Willd. = Anthemis biebersteiniana C. Koch ■☆
26874 Anthemis ruthenica M. Bieb.;俄罗斯春黄菊■☆
26875 Anthemis sabulicola Pomel = Anthemis stiparum Pomel subsp. sabulicola (Pomel) Oberpr. ■☆
26876 Anthemis saguramica Sosn.;萨古拉穆春黄菊■☆
26877 Anthemis sancti-johannis Stoj.,Stef. et Turrill;柔毛春黄菊■☆
26878 Anthemis santolinoides Munby = Chamaemelum nobile (L.) All. ■
26879 Anthemis scaettae Pamp. = Anthemis taubertii E. A. Durand et Barratte ■☆
26880 Anthemis schischkiniana Fed.;希施春黄菊■☆
26881 Anthemis secundiramea Biv.;单侧春黄菊;Prostrate Chamomile ■☆
26882 Anthemis secundiramea Biv. subsp. urvilleana (DC.) R. Fern. = Anthemis secundiramea Biv. ■☆
26883 Anthemis secundiramea Biv. var. cossyrensis Guss. = Anthemis secundiramea Biv. ■☆
26884 Anthemis semiensis Pic. Serm. = Anthemis tigreensis J. Gay ex A. Rich. ■☆
26885 Anthemis sosnovskyana Fed.;锁斯诺夫斯基春黄菊■☆
26886 Anthemis sterilis Stev.;不育春黄菊■☆
26887 Anthemis stiparum Pomel;托叶春黄菊■☆
26888 Anthemis stiparum Pomel subsp. intermedia Oberpr.;间型春黄菊■☆
26889 Anthemis stiparum Pomel subsp. sabulicola (Pomel) Oberpr.;砂地春黄菊■☆
26890 Anthemis stiparum Pomel var. decumbens (Bonnet et Barratte) Murb. = Anthemis confusa Pomel ■☆
26891 Anthemis stiparum Pomel var. sabulicola (Pomel) Batt. = Anthemis stiparum Pomel subsp. sabulicola (Pomel) Oberpr. ■☆
26892 Anthemis subtinctoria Dobrocz.;亚春黄菊■☆
26893 Anthemis talyschensis Fed.;塔里什春黄菊■☆
26894 Anthemis taubertii E. A. Durand et Barratte;陶贝特春黄菊■☆
26895 Anthemis taubertii E. A. Durand et Barratte subsp. arenicola (Pamp.) Brullo et Furnari;沙生陶贝特春黄菊■☆
26896 Anthemis taubertii E. A. Durand et Barratte var. subcoronata Maire et Weiller = Anthemis taubertii E. A. Durand et Barratte ■☆
26897 Anthemis tenuisecta Ball;细裂春黄菊■☆
26898 Anthemis tenuisecta Pomel = Anthemis pedunculata Desf. ■☆
26899 Anthemis tigreensis J. Gay ex A. Rich.;提格雷春黄菊■☆
26900 Anthemis tinctoria L.;春黄菊(西洋菊);Dyer's Camomile, Dyer's Chamomile, Golden Camomile, Golden Chamomile, Golden Marguerite, Ox-eye Camomile, Ox-eye Chamomile, Yellow Camomile, Yellow Chamomile, Yellow Cotula ■
26901 Anthemis tinctoria L. 'E. C. Buxton';伯克斯顿春黄菊■☆
26902 Anthemis tinctoria L. = Cota tinctoria (L.) J. Gay ex Guss. ■☆
26903 Anthemis ubensis Pomel;乌布春黄菊■☆
26904 Anthemis valentinus L. var. homogamos Maire = Anthemis homogamos (Maire) Humphries ■☆
26905 Anthemis vulgaris L. ex Steud.;普通春黄菊;Bur Chervil ■☆
26906 Anthemis wiedemanniana Fisch. et C. A. Mey.;维德曼春黄菊■☆
26907 Anthemis woronowii Sosn.;沃氏春黄菊■☆
26908 Anthemis zaianica Oberpr.;宰哈奈春黄菊■☆
26909 Anthemis zephyrovii Dobrocz.;泽氏春黄菊■☆
26910 Anthemis zyghia Woronow;齐格春黄菊■☆
26911 Anthenantia P. Beauv. (1812);银鳞草属■☆
26912 Anthenantia villosa (Michx.) P. Beauv.;银鳞草■☆
26913 Anthephora Schreb. (1810);柄花草属;phoros,具有,梗,负载,发现者■☆
26914 Anthephora acuminata (Rendle) Robyns ex Stapf et C. E. Hubb. = Anthephora ampullacea Stapf et C. E. Hubb. ■☆
26915 Anthephora aequiglumis Gooss. = Tarigidia aequiglumis (Gooss.) Stent ■☆
26916 Anthephora ampullacea Stapf et C. E. Hubb.;细颈瓶柄花草■☆
26917 Anthephora angustifolia Gooss. = Anthephora argentea Gooss. ■☆
26918 Anthephora argentea Gooss.;银白柄花草■☆
26919 Anthephora burttii Stapf et C. E. Hubb. = Anthephora elongata De Wild. ■☆
26920 Anthephora cenchroides K. Schum. var. glabra (Pilg.) Peter = Anthephora elongata De Wild. ■☆
26921 Anthephora cristata (Döll) Hack. ex De Wild. et T. Durand;冠柄花草■☆
26922 Anthephora elegans Schreb. var. acuminata Rendle = Anthephora ampullacea Stapf et C. E. Hubb. ■☆
26923 Anthephora elegans Schreb. var. cristata Döll = Anthephora cristata (Döll) Hack. ex De Wild. et T. Durand ■☆
26924 Anthephora elongata De Wild.;长柄花草■☆
26925 Anthephora elongata De Wild. var. undulata Chiov. = Anthephora elongata De Wild. ■☆
26926 Anthephora gracilis Stapf et C. E. Hubb. = Anthephora truncata Robyns ■☆
26927 Anthephora hochstetteri Hochst. var. glabra Pilg. = Anthephora

elongata De Wild. ■☆
26928 Anthephora hochstetteri Nees = Anthephora pubescens Nees ■☆
26929 Anthephora hochstetteri Nees var. serresii Dubuis et Faurel = Anthephora pubescens Nees ■☆
26930 Anthephora kotschyi Hochst. = Anthephora pubescens Nees ■☆
26931 Anthephora laevis Stapf et C. E. Hubb.;平滑长柄花草■☆
26932 Anthephora lynesii Stapf et C. E. Hubb. = Anthephora nigritana Stapf et C. E. Hubb. ■☆
26933 Anthephora nigritana Stapf et C. E. Hubb.;尼格里塔长柄花草■☆
26934 Anthephora pubescens Nees;毛长柄花草■☆
26935 Anthephora pungens Clayton;刺长柄花草■☆
26936 Anthephora ramosa Gooss. = Anthephora pubescens Nees ■☆
26937 Anthephora schinzii Hack.;施氏长柄花草■☆
26938 Anthephora truncata Robyns;平截长柄花草■☆
26939 Anthephora undulatifolia Hack. = Anthephora schinzii Hack. ■☆
26940 Anthereon Pridgeon et M. W. Chase = Pabstiella Brieger et Senghas ■☆
26941 Anthereon Pridgeon et M. W. Chase = Pleurothallis R. Br. ■☆
26942 Anthericaceae J. Agardh = Agavaceae Dumort.(保留科名)●■
26943 Anthericaceae J. Agardh = Asphodelaceae Juss. ●■
26944 Anthericaceae J. Agardh(1858);吊兰科(猴面包科,猴面包树科)●■☆
26945 Anthericlis Raf. = Tipularia Nutt. ■
26946 Anthericopsis Engl.(1895);旱竹叶属(拟花篱属)■☆
26947 Anthericopsis fischeri Engl. = Anthericopsis sepalosa (C. B. Clarke) Engl. ■☆
26948 Anthericopsis sepalosa (C. B. Clarke) Engl.;旱竹叶(拟花篱)■☆
26949 Anthericopsis tradescantioides Chiov. = Anthericopsis sepalosa (C. B. Clarke) Engl. ■☆
26950 Anthericum L.(1753);花篱属(猴面包属,鸡尾兰属,圆果吊兰属);Anthericum,Flower Hedge,Spider Plant,St. Bernard's Lily ■☆
26951 Anthericum acuminatum Rendle = Chlorophytum suffruticosum Baker ●☆
26952 Anthericum acutum C. H. Wright = Chlorophytum acutum (C. H. Wright) Nordal ■☆
26953 Anthericum adscendens Poelln. = Chlorophytum cooperi (Baker) Nordal ■☆
26954 Anthericum affine (Kunth) Baker = Chlorophytum affine Baker ■☆
26955 Anthericum aggericolum Poelln. = Trachyandra asperata Kunth var. nataglencoensis (Kuntze) Oberm. ■☆
26956 Anthericum aitonii Baker = Trachyandra filiformis (Aiton) Oberm. ■☆
26957 Anthericum albovaginatum Peter ex Poelln.;白鞘花篱■☆
26958 Anthericum albucoides Aiton = Ornithogalum suaveolens Jacq. ■☆
26959 Anthericum aloifolium Salisb. = Bulbine alooides (L.) Willd. ■☆
26960 Anthericum alooides L. = Bulbine alooides (L.) Willd. ■☆
26961 Anthericum alooides Thunb. = Bulbine praemorsa (Jacq.) Spreng. ■☆
26962 Anthericum altiscapum Poelln.;长花茎花篱■☆
26963 Anthericum altissimum Mill. = Bulbine asphodeloides (L.) Spreng. ■☆
26964 Anthericum amboense Poelln. = Trachyandra saltii (Baker) Oberm. ■☆
26965 Anthericum amplexifolium Engl. ex Poelln.;褶叶花篱■☆
26966 Anthericum anceps Baker = Chlorophytum anceps (Baker) Kativu ■☆
26967 Anthericum andongense Baker;安东花篱■☆
26968 Anthericum andongense Baker var. glabrum Poelln. = Chlorophytum fasciculatum (Baker) Kativu ■☆
26969 Anthericum andongense Baker var. pauciflorum Poelln. = Chlorophytum cameronii (Baker) Kativu var. pterocaulon (Welw. ex Baker) Nordal ■☆
26970 Anthericum angulicaule Baker = Chlorophytum angulicaule (Baker) Kativu ■☆
26971 Anthericum angustifolium Hochst. ex A. Rich.;窄叶花篱■☆
26972 Anthericum angustifolium Rendle = Chlorophytum subpetiolatum (Baker) Kativu ■☆
26973 Anthericum angustissimum Poelln. = Chlorophytum angustissimum (Poelln.) Nordal ■☆
26974 Anthericum angustovittatum Poelln.;细线花篱■☆
26975 Anthericum annuum L. = Bulbine annua (L.) Willd. ■☆
26976 Anthericum anthericoideum Hochst. ex A. Rich. = Chlorophytum tuberosum (Roxb.) Baker ■☆
26977 Anthericum apicicolum Krause = Trachyandra muricata (L. f.) Kunth ■☆
26978 Anthericum arenarium Baker = Chlorophytum sphacelatum (Baker) Kativu ■☆
26979 Anthericum aristatum Poelln. = Trachyandra saltii (Baker) Oberm. ■☆
26980 Anthericum articulatum Hutch. = Drimia indica (Roxb.) Jessop ■☆
26981 Anthericum arvense Schinz = Trachyandra arvensis (Schinz) Oberm. ■☆
26982 Anthericum arvense Schinz var. rigidum Suess. = Trachyandra laxa (N. E. Br.) Oberm. var. rigida (Suess.) Rössler ■☆
26983 Anthericum asperatum (Kunth) Baker = Trachyandra asperata Kunth ■☆
26984 Anthericum asphodeloides L. = Bulbine asphodeloides (L.) Spreng. ■☆
26985 Anthericum atacorense A. Chev. = Chlorophytum cameronii (Baker) Kativu var. pterocaulon (Welw. ex Baker) Nordal ■☆
26986 Anthericum bachmannii Kuntze = Trachyandra muricata (L. f.) Kunth ■☆
26987 Anthericum baeticum (Boiss.) Boiss.;伯蒂卡花篱■☆
26988 Anthericum bakerianum Poelln. = Chlorophytum galpinii (Baker) Kativu ■☆
26989 Anthericum basilanatum Poelln.;基毛花篱■☆
26990 Anthericum basutoense Poelln. = Trachyandra asperata Kunth var. basutoensis (Poelln.) Oberm. ■☆
26991 Anthericum benguellense Baker;本格拉花篱■☆
26992 Anthericum betschuanicum Poelln. = Trachyandra saltii (Baker) Oberm. ■☆
26993 Anthericum bichetii Hort. = Chlorophytum bichetii Baker ■☆
26994 Anthericum bicolor Desf. = Simethis mattiazzii (Vand.) G. López et Jarvis ■☆
26995 Anthericum bifoliatum Poelln. = Chlorophytum subpetiolatum (Baker) Kativu ■☆
26996 Anthericum bipedunculatum Jacq. = Chlorophytum triflorum (Aiton) Kunth ■☆
26997 Anthericum bisulcata Haw. = Bulbine cepacea (Burm. f.) Wijnands ■☆
26998 Anthericum blepharophoron Roem. et Schult. = Trachyandra ciliata (L. f.) Kunth ■☆
26999 Anthericum blepharophyllum Peter ex Poelln. = Trachyandra saltii

(Baker) Oberm. ■☆
27000 Anthericum bolusii Baker = Chlorophytum undulatum (Jacq.) Oberm. ■☆
27001 Anthericum bornmuellerianum Poelln.;博恩花篱■☆
27002 Anthericum brachyphyllum Suess. = Chlorophytum krauseanum (Dinter) Kativu ■☆
27003 Anthericum brachypodum Baker = Trachyandra brachypoda (Baker) Oberm. ■☆
27004 Anthericum brachypodum Baker var. caespitosum Adamson = Trachyandra brachypoda (Baker) Oberm. ■☆
27005 Anthericum bracteatum Thode ex Poelln. = Chlorophytum haygarthii J. M. Wood et M. S. Evans ■☆
27006 Anthericum bragae Engl. = Chlorophytum subpetiolatum (Baker) Kativu ■☆
27007 Anthericum brehmeanum (Schult. et Schult. f.) Baker = Chlorophytum triflorum (Aiton) Kunth ■☆
27008 Anthericum brehmerianum Poelln. = Chlorophytum subpetiolatum (Baker) Kativu ■☆
27009 Anthericum brehmerianum Poelln. var. longibracteatum ? = Chlorophytum subpetiolatum (Baker) Kativu ■☆
27010 Anthericum brehmerianum Poelln. var. longipedicellatum ? = Chlorophytum subpetiolatum (Baker) Kativu ■☆
27011 Anthericum breviantheratum Poelln.;短药花篱■☆
27012 Anthericum brevicaule Baker = Caesia contorta (L. f.) T. Durand et Schinz ■☆
27013 Anthericum brevifilamentatum Poelln.;短丝花篱■☆
27014 Anthericum brevifolium Thunb. = Caesia contorta (L. f.) T. Durand et Schinz ■☆
27015 Anthericum breviscapum De Wild. = Chlorophytum hysteranthum Kativu ■☆
27016 Anthericum brevitepalum Poelln. = Trachyandra saltii (Baker) Oberm. ■☆
27017 Anthericum brunneomarginatum Poelln.;褐边花篱■☆
27018 Anthericum brunneoviride Dinter ex Poelln. = Trachyandra laxa (N. E. Br.) Oberm. ■☆
27019 Anthericum brunneum Poelln.;褐色花篱■☆
27020 Anthericum brunneum Poelln. var. angustatum ? = Chlorophytum cameronii (Baker) Kativu ■☆
27021 Anthericum brunneum Poelln. var. brunneum ? = Chlorophytum cameronii (Baker) Kativu ■☆
27022 Anthericum buchananii Baker = Chlorophytum cameronii (Baker) Kativu var. pterocaulon (Welw. ex Baker) Nordal ■☆
27023 Anthericum buchubergense Poelln. = Trachyandra laxa (N. E. Br.) Oberm. ■☆
27024 Anthericum bulbine Houtt. = Onixotis stricta (Burm. f.) Wijnands ■☆
27025 Anthericum bulbinifolium Dinter = Trachyandra bulbinifolia (Dinter) Oberm. ■☆
27026 Anthericum burkei Baker = Trachyandra burkei (Baker) Oberm. ■☆
27027 Anthericum bussei Poelln. = Chlorophytum subpetiolatum (Baker) Kativu ■☆
27028 Anthericum caespitosum Dinter = Chlorophytum calyptrocarpum (Baker) Kativu ■☆
27029 Anthericum calyptrocarpum Baker = Chlorophytum calyptrocarpum (Baker) Kativu ■☆
27030 Anthericum cameronii Baker = Chlorophytum cameronii (Baker) Kativu ■☆
27031 Anthericum campestre Engl. = Chlorophytum suffruticosum Baker ●☆
27032 Anthericum canaliculatum Aiton = Trachyandra ciliata (L. f.) Kunth ■☆
27033 Anthericum capillare (Poir.) Schult. et Schult. f. = Bulbinella triquetra (L. f.) Kunth ■☆
27034 Anthericum capillatum Poelln. = Trachyandra capillata (Poelln.) Oberm. ■☆
27035 Anthericum capitatum Baker = Chlorophytum cooperi (Baker) Nordal ■☆
27036 Anthericum carnosum Baker = Ornithogalum paludosum Baker ■☆
27037 Anthericum cauda-felis L. f. = Bulbinella cauda-felis (L. f.) T. Durand et Schinz ■☆
27038 Anthericum caudatum Thunb. = Bulbinella cauda-felis (L. f.) T. Durand et Schinz ■☆
27039 Anthericum caulescens Baker = Chlorophytum caulescens (Baker) Marais et Reilly ■☆
27040 Anthericum cepifolium Dinter ex Poelln. = Trachyandra saltii (Baker) Oberm. ■☆
27041 Anthericum chamaemoly Hochst. ex A. Rich. = Chlorophytum tetraphyllum (L. f.) Baker ■☆
27042 Anthericum chandleri Greenm. et C. H. Thomps. = Echeandia chandleri (Greenm. et C. H. Thomps.) M. C. Johnst. ■☆
27043 Anthericum chlamydophylla Baker = Trachyandra chlamydophylla (Baker) Oberm. ■☆
27044 Anthericum ciliare Peter ex Poelln. = Chlorophytum affine Baker var. curviscapum (Poelln.) Hanid ■☆
27045 Anthericum ciliatum Baker = Chlorophytum blepharophyllum Schweinf. ex Baker ■☆
27046 Anthericum ciliatum L. f. = Trachyandra ciliata (L. f.) Kunth ■☆
27047 Anthericum ciliolatum (Kunth) Baker = Bulbinella ciliolata Kunth ■☆
27048 Anthericum cirrifolium Schinz = Trachyandra flexifolia (L. f.) Kunth ■☆
27049 Anthericum claessensii De Wild. = Chlorophytum subpetiolatum (Baker) Kativu ■☆
27050 Anthericum collinum Poelln. = Chlorophytum collinum (Poelln.) Nordal ■☆
27051 Anthericum colubrinum Welw. ex Baker = Chlorophytum colubrinum (Welw. ex Baker) Engl. ■☆
27052 Anthericum comosum Thunb. = Chlorophytum comosum (Thunb.) Baker ●■
27053 Anthericum congestum Adamson = Trachyandra hispida (L.) Kunth ■☆
27054 Anthericum congolense De Wild. et T. Durand = Chlorophytum cameronii (Baker) Kativu ■☆
27055 Anthericum congolense De Wild. et T. Durand var. elongatum De Wild. = Chlorophytum cameronii (Baker) Kativu ■☆
27056 Anthericum conrathii Baker = Chlorophytum fasciculatum (Baker) Kativu ■☆
27057 Anthericum contortum L. f. = Caesia contorta (L. f.) T. Durand et Schinz ■☆
27058 Anthericum cooperi Baker = Chlorophytum cooperi (Baker) Nordal ■☆
27059 Anthericum corymbosum Baker;伞序花篱■☆
27060 Anthericum corymbosum Baker var. floribundum Chiov. =

Anthericum corymbosum Baker ■☆

27061 Anthericum costatum Andr.;中脉花篱■☆

27062 Anthericum crassinerve Baker = Chlorophytum crassinerve (Baker) Oberm. ■☆

27063 Anthericum crassiusculum Dinter = Trachyandra saltii (Baker) Oberm. ■☆

27064 Anthericum crispum Thunb. = Chlorophytum crispum (Thunb.) Baker ■☆

27065 Anthericum curvifolium K. Krause = Chlorophytum calyptrocarpum (Baker) Kativu ■☆

27066 Anthericum curviscapum Poelln. = Chlorophytum affine Baker var. curviscapum (Poelln.) Hanid ■☆

27067 Anthericum dalzielii Hutch. ex Hepper = Chlorophytum dalzielii (Hutch. ex Hepper) Nordal ■☆

27068 Anthericum deflexum A. Chev. = Chlorophytum affine Baker var. curviscapum (Poelln.) Hanid ■☆

27069 Anthericum deightonii Hutch. ex Berhaut = Chlorophytum warneckei (Engl.) Marais et Reilly ■☆

27070 Anthericum delagoense Poelln. = Chlorophytum galpinii (Baker) Kativu ■☆

27071 Anthericum depauperatum Poelln.;萎缩花篱■☆

27072 Anthericum dielsii Poelln. = Chlorophytum undulatum (Jacq.) Oberm. ■☆

27073 Anthericum dilatatum Poelln.;膨大花篱■☆

27074 Anthericum dimorphum Poelln. = Chlorophytum subpetiolatum (Baker) Kativu ■☆

27075 Anthericum dinteri Poelln. = Trachyandra laxa (N. E. Br.) Oberm. ■☆

27076 Anthericum diphyllum Dinter = Trachyandra muricata (L. f.) Kunth ■☆

27077 Anthericum dissitiflorum Baker;疏花花篱■☆

27078 Anthericum divaricatum Baker ex Schinz = Chlorophytum galpinii (Baker) Kativu ■☆

27079 Anthericum divaricatum Jacq. = Trachyandra divaricata (Jacq.) Kunth ■☆

27080 Anthericum diversifolium Poelln. = Chlorophytum subpetiolatum (Baker) Kativu ■☆

27081 Anthericum djalonis A. Chev. = Aristea angolensis Baker ■☆

27082 Anthericum dregeanum (Kunth) Baker = Caesia dregeana Kunth ■☆

27083 Anthericum drepanophyllum (Baker) Schltr. = Trachyandra falcata (L. f.) Kunth ■☆

27084 Anthericum drimiopsis Baker = Chlorophytum longifolium Schweinf. ex Baker ■☆

27085 Anthericum dubium (Schult. et Schult. f.) Poelln. = Chlorophytum triflorum (Aiton) Kunth ■☆

27086 Anthericum durum Poelln. = Chlorophytum galpinii (Baker) Kativu ■☆

27087 Anthericum durum Suess. = Chlorophytum krauseanum (Dinter) Kativu ■☆

27088 Anthericum elatum Aiton = Chlorophytum capense (L.) Voss ■

27089 Anthericum elongatum Willd. var. holostachyum Baker = Trachyandra saltii (Baker) Oberm. ■☆

27090 Anthericum englerianum Poelln. = Chlorophytum affine Baker var. curviscapum (Poelln.) Hanid ■☆

27091 Anthericum ensifolium Sölch = Trachyandra ensifolia (Sölch) Rössler ■☆

27092 Anthericum erraticum Oberm. = Trachyandra laxa (N. E. Br.) Oberm. var. rigida (Suess.) Rössler ■☆

27093 Anthericum erythrorrhizum Conrath = Trachyandra erythrorrhiza (Conrath) Oberm. ■☆

27094 Anthericum excellens Poelln.;优秀花篱■☆

27095 Anthericum exellii Poelln.;埃克塞尔花篱■☆

27096 Anthericum exuviatum Jacq. = Drimia exuviata (Jacq.) Jessop ■☆

27097 Anthericum falcatum L. f. = Trachyandra falcata (L. f.) Kunth ■☆

27098 Anthericum falcatum Welw. ex Baker = Chlorophytum longifolium Schweinf. ex Baker ■☆

27099 Anthericum fallax Poelln.;迷惑花篱■☆

27100 Anthericum fasciculatum Baker = Chlorophytum fasciculatum (Baker) Kativu ■☆

27101 Anthericum favosum Thunb. = Bulbine favosa (Thunb.) Schult. et Schult. f. ■☆

27102 Anthericum fernandesii Poelln.;费尔南德花篱■☆

27103 Anthericum fibrosum Hutch. = Chlorophytum nubicum (Baker) Kativu ■☆

27104 Anthericum filifolium Jacq. = Drimia filifolia (Jacq.) J. C. Manning et Goldblatt ■☆

27105 Anthericum filiforme Aiton = Trachyandra filiformis (Aiton) Oberm. ■☆

27106 Anthericum filiforme Thunb. var. reflexipilosum Kuntze = Trachyandra reflexipilosa (Kuntze) Oberm. ■☆

27107 Anthericum fimbriatum Thunb. = Trachyandra muricata (L. f.) Kunth ■☆

27108 Anthericum fischeri Baker = Chlorophytum fischeri (Baker) Baker ■☆

27109 Anthericum flagelliforme Baker = Eriospermum flagelliforme (Baker) J. C. Manning ■☆

27110 Anthericum flavescens Schult. et Schult. f. = Echeandia flavescens (Schult. et Schult. f.) Cruden ■☆

27111 Anthericum flavoviride Baker = Trachyandra arvensis (Schinz) Oberm. ■☆

27112 Anthericum flexifolium L. f. = Trachyandra flexifolia (L. f.) Kunth ■☆

27113 Anthericum floribundum Aiton = Bulbinella floribunda (Aiton) T. Durand et Schinz ■☆

27114 Anthericum foliatum Poelln. = Chlorophytum stolzii (K. Krause) Kativu ■☆

27115 Anthericum fragrans Jacq. = Drimia fragrans (Jacq.) J. C. Manning et Goldblatt ■☆

27116 Anthericum friesii Weim. = Chlorophytum sphacelatum (Baker) Kativu var. milanjianum (Rendle) Nordal ■☆

27117 Anthericum frutescens L. = Bulbine frutescens (L.) Willd. ■☆

27118 Anthericum fruticosum Salisb. = Bulbine frutescens (L.) Willd. ■☆

27119 Anthericum galpinii Baker = Chlorophytum galpinii (Baker) Kativu ■☆

27120 Anthericum galpinii Baker var. matabelense (Baker) Oberm. = Chlorophytum galpinii (Baker) Kativu var. matabalense (Baker) Kativu ■☆

27121 Anthericum galpinii Baker var. norlindhii (Weim.) Oberm. = Chlorophytum galpinii (Baker) Kativu var. norlindhii (Weim.) Kativu ■☆

27122 Anthericum gerrardii Baker = Trachyandra gerrardii (Baker) Oberm. ■☆

27123 Anthericum giffienii F. M. Leight. = Trachyandra giffenii (F. M. Leight.) Oberm. ■☆

27124 Anthericum giganteum Poelln. = Chlorophytum polystachys Baker ■☆

27125 Anthericum gilvum K. Krause = Trachyandra arvensis (Schinz) Oberm. ■☆

27126 Anthericum gilvum K. Krause var. brunneolum Poelln. = Trachyandra arvensis (Schinz) Oberm. ■☆

27127 Anthericum giryamae Rendle = Chlorophytum cameronii (Baker) Kativu var. pterocaulon (Welw. ex Baker) Nordal ■☆

27128 Anthericum glabrum Adamson = Trachyandra tabularis (Baker) Oberm. ■☆

27129 Anthericum glandulosum Dinter = Trachyandra glandulosa (Dinter) Oberm. ■☆

27130 Anthericum glandulosum Dinter var. montis-ruschii Poelln. = Trachyandra glandulosa (Dinter) Oberm. ■☆

27131 Anthericum glutinosum Dinter = Trachyandra laxa (N. E. Br.) Oberm. ■☆

27132 Anthericum goetzei (Engl.) Poelln. = Chlorophytum goetzei Engl. ■☆

27133 Anthericum gossweileri Poelln. = Chlorophytum andongense Baker ■☆

27134 Anthericum gracile (Kunth) Baker = Bulbinella gracilis Kunth ■☆

27135 Anthericum gracilitepalum Poelln. = Trachyandra saltii (Baker) Oberm. ■☆

27136 Anthericum gramineum Poelln. = Chlorophytum goetzei Engl. ■☆

27137 Anthericum graminifolium Willd. = Anthericum ramosum L. ■☆

27138 Anthericum graminifolium Willd. = Chlorophytum undulatum (Jacq.) Oberm. ■☆

27139 Anthericum graminoides Poelln. = Chlorophytum goetzei Engl. ■☆

27140 Anthericum grantii Baker = Chlorophytum cameronii (Baker) Kativu var. grantii (Baker) Nordal ■☆

27141 Anthericum grantii Baker var. muenzneri Engl. et K. Krause = Chlorophytum cameronii (Baker) Kativu var. grantii (Baker) Nordal ■☆

27142 Anthericum gregorianum Rendle = Anthericum corymbosum Baker ■☆

27143 Anthericum hamatum Poelln. = Trachyandra ciliata (L. f.) Kunth ■☆

27144 Anthericum harrarense Poelln. = Trachyandra saltii (Baker) Oberm. ■☆

27145 Anthericum haygarthii (J. M. Wood et M. S. Evans) Oberm. = Chlorophytum haygarthii J. M. Wood et M. S. Evans ■☆

27146 Anthericum hecqii De Wild. = Chlorophytum subpetiolatum (Baker) Kativu ■☆

27147 Anthericum hereroense Schinz = Chlorophytum fasciculatum (Baker) Kativu ■☆

27148 Anthericum hereroense Schinz var. longibracteatum Poelln. = Chlorophytum fasciculatum (Baker) Kativu ■☆

27149 Anthericum hirsutiflorum Adamson = Trachyandra hirsutiflora (Adamson) Oberm. ■☆

27150 Anthericum hirsutum Thunb. = Trachyandra hirsuta (Thunb.) Kunth ■☆

27151 Anthericum hispidum L. = Trachyandra hispida (L.) Kunth ■☆

27152 Anthericum hockii De Wild. = Chlorophytum sphacelatum (Baker) Kativu var. hockii (De Wild.) Nordal ■☆

27153 Anthericum homblei De Wild. = Chlorophytum sphacelatum (Baker) Kativu ■☆

27154 Anthericum humile Hochst. ex A. Rich. = Anthericum angustifolium Hochst. ex A. Rich. ■☆

27155 Anthericum immaculatum Hepper = Chlorophytum immaculatum (Hepper) Nordal ■☆

27156 Anthericum inconspicuum Baker = Chlorophytum inconspicuum (Baker) Nordal ■☆

27157 Anthericum incurvum Thunb. = Bulbine frutescens (L.) Willd. ■☆

27158 Anthericum indutum Poelln. = Chlorophytum transvaalense (Baker) Kativu ■☆

27159 Anthericum inexpectatum Poelln. = Chlorophytum suffruticosum Baker ●☆

27160 Anthericum intricatum Baker = Drimia intricata (Baker) J. C. Manning et Goldblatt ■☆

27161 Anthericum involucratum Baker = Trachyandra involucrata (Baker) Oberm. ■☆

27162 Anthericum jacquinianum Roem. et Schult. = Trachyandra jacquiniana (Roem. et Schult.) Oberm. ■☆

27163 Anthericum jaegeri Engl. et K. Krause;耶格花篱■☆

27164 Anthericum jamesii Baker;詹姆斯花篱■☆

27165 Anthericum junciforme Poelln. ;灯心草花篱■☆

27166 Anthericum junodii Baker = Chlorophytum galpinii (Baker) Kativu ■☆

27167 Anthericum kapiriense De Wild. = Chlorophytum galpinii (Baker) Kativu var. matabalense (Baker) Kativu ■☆

27168 Anthericum kässneri Poelln. = Trachyandra saltii (Baker) Oberm. ■☆

27169 Anthericum kemoense Hua;凯莫花篱■☆

27170 Anthericum kilimandscharicum Poelln. = Chlorophytum tuberosum (Roxb.) Baker ■☆

27171 Anthericum korrowalense Engl. et K. Krause = Chlorophytum cameronii (Baker) Kativu var. pterocaulon (Welw. ex Baker) Nordal ■☆

27172 Anthericum koutiense A. Chev. = Chlorophytum cameronii (Baker) Kativu ■☆

27173 Anthericum kovismontanum Dinter = Chlorophytum viscosum Kunth ■☆

27174 Anthericum krauseanum Dinter = Chlorophytum krauseanum (Dinter) Kativu ■☆

27175 Anthericum kunthii Baker = Trachyandra asperata Kunth ■☆

27176 Anthericum kyimbilense Poelln. = Chlorophytum cameronii (Baker) Kativu var. pterocaulon (Welw. ex Baker) Nordal ■☆

27177 Anthericum kyllingioides K. Krause = Chlorophytum krauseanum (Dinter) Kativu ■☆

27178 Anthericum lacustre Poelln. = Chlorophytum subpetiolatum (Baker) Kativu ■☆

27179 Anthericum lagopus Thunb. = Bulbine lagopus (Thunb.) N. E. Br. ■☆

27180 Anthericum lanatum Dinter = Trachyandra lanata (Dinter) Oberm. ■☆

27181 Anthericum lanzae Cufod. = Trachyandra saltii (Baker) Oberm. ■☆

27182 Anthericum latifolium Jacq. = Bulbine cepacea (Burm. f.) Wijnands ■☆

27183 Anthericum latifolium L. f. = Bulbine latifolia (L. f.) Spreng. ■☆

27184 Anthericum laurentii De Wild. ;洛朗花篱■☆

27185 Anthericum laurentii De Wild. var. minor ? = Chlorophytum subpetiolatum (Baker) Kativu ■☆
27186 Anthericum laurentii De Wild. var. minus ?;小洛朗花篱■☆
27187 Anthericum laxum N. E. Br. ;疏松花篱■☆
27188 Anthericum ledermannii Engl. et K. Krause = Chlorophytum affine Baker var. curviscapum (Poelln.) Hanid ■☆
27189 Anthericum liliagastrum Engl. et Gilg;小伯纳德百合■☆
27190 Anthericum liliago L. ;伯纳德百合(高大圆果吊兰,蜘蛛百合);Saint Bernard's Lily, St. Bernard Lily, St. Bernard-lily, St. Bernard's Lily, St. -Bernard's-lily ■☆
27191 Anthericum liliago L. subsp. algeriense (Boiss. et Reut.) Maire et Weiller = Anthericum maurum Rothm. ■☆
27192 Anthericum liliago L. subsp. baeticum (Boiss.) Maire = Anthericum baeticum (Boiss.) Boiss. ■☆
27193 Anthericum liliago L. var. fontqueri (Sennen et Mauricio) Maire = Anthericum baeticum (Boiss.) Boiss. ■☆
27194 Anthericum liliago L. var. rhiphaeum (Pau et Font Quer) Maire = Anthericum baeticum (Boiss.) Boiss. ■☆
27195 Anthericum liliastrum L. ;百合花篱;St. Bruno's Lily ■☆
27196 Anthericum limbamenense Engl. et K. Krause = Chlorophytum subpetiolatum (Baker) Kativu ■☆
27197 Anthericum limosum Baker = Chlorophytum limosum (Baker) Nordal ■☆
27198 Anthericum longepedunculatum Steud. ex Roem. et Schult. = Trachyandra filiformis (Aiton) Oberm. ■☆
27199 Anthericum longibracteatum Dinter = Chlorophytum viscosum Kunth ■☆
27200 Anthericum longibracteatum Dinter var. brevibracteatum Poelln. = Chlorophytum viscosum Kunth ■☆
27201 Anthericum longiciliatum Poelln. = Chlorophytum affine Baker ■☆
27202 Anthericum longifolium A. Rich. = Chlorophytum longifolium Schweinf. ex Baker ■☆
27203 Anthericum longifolium Jacq. = Trachyandra ciliata (L. f.) Kunth ■☆
27204 Anthericum longifolium Jacq. var. burchelli Baker = Chlorophytum affine Baker ■☆
27205 Anthericum longipedicellatum Poelln. = Ornithogalum flexuosum (Thunb.) U. Müll. -Doblies et D. Müll. -Doblies ■☆
27206 Anthericum longiscapum Jacq. = Bulbine asphodeloides (L.) Spreng. ■☆
27207 Anthericum longisetosum Poelln. = Chlorophytum suffruticosum Baker ●☆
27208 Anthericum longistylum Baker = Chlorophytum recurvifolium (Baker) C. Archer et Kativu ■☆
27209 Anthericum longituberosum Poelln. = Chlorophytum comosum (Thunb.) Baker ●■
27210 Anthericum lowryense Baker = Trachyandra brachypoda (Baker) Oberm. ■☆
27211 Anthericum lukiense De Wild. = Chlorophytum sphacelatum (Baker) Kativu ■☆
27212 Anthericum lukiense De Wild. var. intermedium ? = Chlorophytum sphacelatum (Baker) Kativu ■☆
27213 Anthericum lukiense De Wild. var. kionzoense ? = Chlorophytum sphacelatum (Baker) Kativu ■☆
27214 Anthericum lunatum Poelln. = Chlorophytum longifolium Schweinf. ex Baker ■☆
27215 Anthericum lydenburgense Poelln. = Chlorophytum fasciculatum (Baker) Kativu ■☆
27216 Anthericum macowanii Baker = Trachyandra asperata Kunth var. macowanii (Baker) Oberm. ■☆
27217 Anthericum macranthum Baker;大花花篱■☆
27218 Anthericum macrophyllum A. Rich. = Chlorophytum macrophyllum (A. Rich.) Asch. ■☆
27219 Anthericum maculatum Poelln. = Trachyandra ciliata (L. f.) Kunth ■☆
27220 Anthericum magnificum Poelln. = Chlorophytum krookianum Zahlbr. ■☆
27221 Anthericum malchairii De Wild. = Chlorophytum subpetiolatum (Baker) Kativu ■☆
27222 Anthericum marginatum Thunb. = Drimia marginata (Thunb.) Jessop ■☆
27223 Anthericum matabelense Baker = Chlorophytum galpinii (Baker) Kativu var. matabalense (Baker) Kativu ■☆
27224 Anthericum mattiazzii Vand. = Simethis mattiazzii (Vand.) G. López et Jarvis ■☆
27225 Anthericum maurum Rothm. ;暗花篱■☆
27226 Anthericum mendoncai Poelln. ;门东萨花篱■☆
27227 Anthericum micranthum Baker = Trachyandra saltii (Baker) Oberm. ■☆
27228 Anthericum milanjianum Rendle = Chlorophytum sphacelatum (Baker) Kativu var. milanjianum (Rendle) Nordal ■☆
27229 Anthericum mildbraedii Poelln. ;米尔德花篱■☆
27230 Anthericum molle Baker;柔软花篱■☆
27231 Anthericum monophyllum Baker = Chlorophytum subpetiolatum (Baker) Kativu ■☆
27232 Anthericum montanum Poelln. = Chlorophytum subpetiolatum (Baker) Kativu ■☆
27233 Anthericum monticola Poelln. = Trachyandra asperata Kunth var. nataglencoensis (Kuntze) Oberm. ■☆
27234 Anthericum montium-draconis Poelln. = Trachyandra gerrardii (Baker) Oberm. ■☆
27235 Anthericum muenzneri (Engl. et K. Krause) Poelln. = Chlorophytum cameronii (Baker) Kativu var. grantii (Baker) Nordal ■☆
27236 Anthericum multiceps Poelln. = Bulbine frutescens (L.) Willd. ■☆
27237 Anthericum multisetosum Baker = Chlorophytum angulicaule (Baker) Kativu ■☆
27238 Anthericum muricatum L. f. = Trachyandra muricata (L. f.) Kunth ■☆
27239 Anthericum nataglencoense Kuntze = Trachyandra asperata Kunth var. nataglencoensis (Kuntze) Oberm. ■☆
27240 Anthericum natalense Poelln. = Trachyandra asperata Kunth var. nataglencoensis (Kuntze) Oberm. ■☆
27241 Anthericum nepalense Poelln. = Chlorophytum nepalense (Lindl.) Baker ■
27242 Anthericum nervosum Hua;多脉花篱■☆
27243 Anthericum nguluense Poelln. = Chlorophytum cameronii (Baker) Kativu ■☆
27244 Anthericum nidulans Baker = Chlorophytum nidulans (Baker) Brenan ■☆
27245 Anthericum nigericum Hepper = Chlorophytum angustissimum (Poelln.) Nordal ■☆
27246 Anthericum nigrobracteatum Dinter = Trachyandra laxa (N. E.

Br.) Oberm. ■☆

27247 Anthericum nodulosum Poelln.;多节花篱■☆

27248 Anthericum nubicum Baker = Chlorophytum nubicum (Baker) Kativu ■☆

27249 Anthericum nudicaule Baker = Chlorophytum cooperi (Baker) Nordal ■☆

27250 Anthericum nutans Jacq. = Bulbine nutans (Jacq.) Spreng. ■☆

27251 Anthericum nutans Thunb. = Bulbinella nutans (Thunb.) T. Durand et Schinz ■☆

27252 Anthericum nyassae Rendle = Chlorophytum nyassae (Rendle) Kativu ■☆

27253 Anthericum oatesii Baker = Trachyandra saltii (Baker) Oberm. ■☆

27254 Anthericum obtusifolium Poelln. = Trachyandra brachypoda (Baker) Oberm. ■☆

27255 Anthericum odoratissimum Dinter = Trachyandra bulbinifolia (Dinter) Oberm. ■☆

27256 Anthericum oehleri Engl. et K. Krause;奥勒花篱■☆

27257 Anthericum oligotrichum Baker = Trachyandra oligotricha (Baker) Oberm. ■☆

27258 Anthericum omissum Poelln. = Trachyandra saltii (Baker) Oberm. ■☆

27259 Anthericum oocarpum Schltr. ex Poelln. = Trachyandra tortilis (Baker) Oberm. ■☆

27260 Anthericum orchideum Welw. ex Baker;兰状花篱■☆

27261 Anthericum ornithogaloides (Kunth) Baker = Ornithogalum flexuosum (Thunb.) U. Müll. -Doblies et D. Müll. -Doblies ■☆

27262 Anthericum ornithogaloides Hochst. ex A. Rich. = Chlorophytum tuberosum (Roxb.) Baker ■☆

27263 Anthericum otavense Engl. et K. Krause = Chlorophytum anceps (Baker) Kativu ■☆

27264 Anthericum pachyphyllum Baker = Chlorophytum cooperi (Baker) Nordal ■☆

27265 Anthericum pachyrrhizum Dinter = Trachyandra laxa (N. E. Br.) Oberm. ■☆

27266 Anthericum pallidiflavum Engl. et Gilg = Trachyandra arvensis (Schinz) Oberm. ■☆

27267 Anthericum paludosum Engl. et K. Krause;沼泽花篱■☆

27268 Anthericum palustre Adamson = Trachyandra tabularis (Baker) Oberm. ■☆

27269 Anthericum papillosum Engl.;乳头花篱■☆

27270 Anthericum pappei Baker = Trachyandra flexifolia (L. f.) Kunth ■☆

27271 Anthericum paradoxum Schult. et Schult. f. = Trachyandra hispida (L.) Kunth ■☆

27272 Anthericum parviflorum (Wight) Benth. = Chlorophytum laxum R. Br. ■

27273 Anthericum parviflorum Benth. = Chlorophytum laxum R. Br. ■

27274 Anthericum pascuorum Poelln. = Chlorophytum cooperi (Baker) Nordal ■☆

27275 Anthericum patulum Baker = Chlorophytum galpinii (Baker) Kativu var. matabalense (Baker) Kativu ■☆

27276 Anthericum pauciflorum Thunb. = Chlorophytum triflorum (Aiton) Kunth ■☆

27277 Anthericum pauciflorum Thunb. var. minor Baker = Chlorophytum undulatum (Jacq.) Oberm. ■☆

27278 Anthericum paucinervatum Poelln. = Chlorophytum paucinervatum (Poelln.) Nordal ■☆

27279 Anthericum pauper Poelln. = Ornithogalum flexuosum (Thunb.) U. Müll. -Doblies et D. Müll. -Doblies ■☆

27280 Anthericum peculiare Dinter = Trachyandra peculiaris (Dinter) Oberm. ■☆

27281 Anthericum pendulum Engl. et K. Krause = Chlorophytum affine Baker var. curviscapum (Poelln.) Hanid ■☆

27282 Anthericum peteri Poelln. = Chlorophytum ruahense Engl. ■☆

27283 Anthericum physodes Jacq. = Drimia physodes (Jacq.) Jessop ■☆

27284 Anthericum pilosicarinatum Poelln.;毛棱花篱■☆

27285 Anthericum pilosiflorum Poelln. = Trachyandra ciliata (L. f.) Kunth ■☆

27286 Anthericum pilosiflorum Poelln. var. subpapillosum ? = Trachyandra ciliata (L. f.) Kunth ■☆

27287 Anthericum pilossisimum Poelln. = Trachyandra hirsutiflora (Adamson) Oberm. ■☆

27288 Anthericum pilosum Baker = Trachyandra capillata (Poelln.) Oberm. ■☆

27289 Anthericum planifolium Thunb. = Chlorophytum comosum (Thunb.) Baker ●■

27290 Anthericum planifolium Vand. ex L. = Simethis planifolia (L.) Gren. et Godr. ■☆

27291 Anthericum pleiophyllum Poelln. = Chlorophytum undulatum (Jacq.) Oberm. ■☆

27292 Anthericum pleiostachyum Welw. ex Baker = Chlorophytum colubrinum (Welw. ex Baker) Engl. ■☆

27293 Anthericum polyphyllum Baker = Chlorophytum recurvifolium (Baker) C. Archer et Kativu ■☆

27294 Anthericum pomeridianum (DC.) Ker Gawl. = Chlorogalum pomeridianum Kunth ■☆

27295 Anthericum porense Poelln. = Chlorophytum subpetiolatum (Baker) Kativu ■☆

27296 Anthericum praemorsum Jacq. = Bulbine praemorsa (Jacq.) Spreng. ■☆

27297 Anthericum pretoriense Baker = Chlorophytum trichophlebium (Baker) Nordal ■☆

27298 Anthericum princeae Engl. et K. Krause var. brevibracteatum Poelln.;短苞花篱■☆

27299 Anthericum princeae Engl. et K. Krause var. latifolium Poelln. = Chlorophytum sphacelatum (Baker) Kativu var. milanjianum (Rendle) Nordal ■☆

27300 Anthericum pseudofalcatum Poelln. = Chlorophytum affine Baker ■☆

27301 Anthericum pseudopapillosum Poelln. = Chlorophytum collinum (Poelln.) Nordal ■☆

27302 Anthericum pterocaulon Welw. ex Baker = Chlorophytum cameronii (Baker) Kativu var. pterocaulon (Welw. ex Baker) Nordal ■☆

27303 Anthericum puberulum Weim. = Chlorophytum pygmaeum (Weim.) Kativu ■☆

27304 Anthericum pubescens Baker = Chlorophytum affine Baker ■☆

27305 Anthericum pubiflorum Peter ex Poelln. = Chlorophytum affine Baker var. curviscapum (Poelln.) Hanid ■☆

27306 Anthericum pubirhachis Baker = Chlorophytum affine Baker ■☆

27307 Anthericum pudicum Baker = Chlorophytum affine Baker ■☆

27308 Anthericum pugioniforme Jacq. = Bulbine cepacea (Burm. f.) Wijnands ■☆

27309 Anthericum pulchellum Baker = Chlorophytum saundersiae

(Baker) Nordal ■☆

27310 Anthericum pumilum Hua;矮小花篱■☆

27311 Anthericum pungens Poelln. = Chlorophytum krauseanum (Dinter) Kativu ■☆

27312 Anthericum purpuratum Rendle = Chlorophytum cameronii (Baker) Kativu var. pterocaulon (Welw. ex Baker) Nordal ■☆

27313 Anthericum purpuratum Rendle var. longipedicellatum Poelln. = Chlorophytum cameronii (Baker) Kativu var. pterocaulon (Welw. ex Baker) Nordal ■☆

27314 Anthericum pusillum Jacq. = Drimia physodes (Jacq.) Jessop ■☆

27315 Anthericum pygmaeum Weim. = Chlorophytum pygmaeum (Weim.) Kativu ■☆

27316 Anthericum pyrenicarpum Welw. ex Baker = Trachyandra pyrenicarpa (Welw. ex Baker) Oberm. ■☆

27317 Anthericum quadrifidum Poelln. = Ornithogalum juncifolium Jacq. ■☆

27318 Anthericum radula Baker = Chlorophytum radula (Baker) Nordal ■☆

27319 Anthericum ramiferum Peter ex Poelln.;枝生花篱■☆

27320 Anthericum ramosum L.;多枝花篱(圆果吊兰);Branched St. Bernard's Lily,Manyshoot Flower Hedge ■☆

27321 Anthericum rangei Engl. et K. Krause = Chlorophytum rangei (Engl. et K. Krause) Nordal ■☆

27322 Anthericum rautanenii Schinz = Chlorophytum anceps (Baker) Kativu ■☆

27323 Anthericum recurvatum Dinter = Trachyandra ciliata (L. f.) Kunth ■☆

27324 Anthericum recurvifolium Baker = Chlorophytum recurvifolium (Baker) C. Archer et Kativu ■☆

27325 Anthericum rehmannii Baker = Chlorophytum galpinii (Baker) Kativu ■☆

27326 Anthericum remotiflorum Poelln. = Chlorophytum goetzei Engl. ■☆

27327 Anthericum revolutum L. = Trachyandra revoluta (L.) Kunth ■☆

27328 Anthericum rhodesianum Rendle = Chlorophytum pygmaeum (Weim.) Kativu subsp. rhodesianum (Rendle) Kativu ■☆

27329 Anthericum rigens Poelln. = Chlorophytum krauseanum (Dinter) Kativu ■☆

27330 Anthericum rigidifolium Poelln. = Chlorophytum rangei (Engl. et K. Krause) Nordal ■☆

27331 Anthericum rigidum (Kunth) Baker = Chlorophytum rigidum Kunth ■☆

27332 Anthericum rigidum De Wild. = Chlorophytum galpinii (Baker) Kativu ■☆

27333 Anthericum rigidum De Wild. var. breviscapum ? = Chlorophytum galpinii (Baker) Kativu ■☆

27334 Anthericum rigidum K. Krause = Chlorophytum krauseanum (Dinter) Kativu ■☆

27335 Anthericum robustulum Poelln. = Chlorophytum sphacelatum (Baker) Kativu ■☆

27336 Anthericum robustulum Poelln. var. angustum ? = Chlorophytum sphacelatum (Baker) Kativu ■☆

27337 Anthericum robustum Baker = Chlorophytum angulicaule (Baker) Kativu ■☆

27338 Anthericum rosenbrockii Poelln. = Chlorophytum crispum (Thunb.) Baker ■☆

27339 Anthericum roseum Poelln. = Chlorophytum rubribracteatum (De Wild.) Kativu ■☆

27340 Anthericum rostratum Jacq. = Bulbine frutescens (L.) Willd. ■☆

27341 Anthericum rouwenortii Gorter = Chlorophytum capense (L.) Voss ■

27342 Anthericum rubellum Baker = Chlorophytum cameronii (Baker) Kativu var. pterocaulon (Welw. ex Baker) Nordal ■☆

27343 Anthericum rubribracteatum De Wild. = Chlorophytum rubribracteatum (De Wild.) Kativu ■☆

27344 Anthericum rubrovittatum Poelln. = Chlorophytum angulicaule (Baker) Kativu ■☆

27345 Anthericum rudatisii Poelln. = Chlorophytum saundersiae (Baker) Nordal ■☆

27346 Anthericum rudatisii Poelln. var. angustum ? = Chlorophytum saundersiae (Baker) Nordal ■☆

27347 Anthericum russissiense Poelln. = Chlorophytum stolzii (K. Krause) Kativu ■☆

27348 Anthericum rustii Poelln. = Chlorophytum crispum (Thunb.) Baker ■☆

27349 Anthericum sabulosum Adamson = Trachyandra sabulosa (Adamson) Oberm. ■☆

27350 Anthericum salteri F. M. Leight. = Trachyandra tortilis (Baker) Oberm. ■☆

27351 Anthericum saltii Baker = Trachyandra saltii (Baker) Oberm. ■☆

27352 Anthericum saundersiae Baker = Chlorophytum saundersiae (Baker) Nordal ■☆

27353 Anthericum saxicola Engl. ex Poelln.;岩生花篱■☆

27354 Anthericum scabromarginatum Schltr. = Trachyandra muricata (L. f.) Kunth ■☆

27355 Anthericum scabrum L. f. = Trachyandra scabra (L. f.) Kunth ■☆

27356 Anthericum scariosum Duthie = Chlorophytum rangei (Engl. et K. Krause) Nordal ■☆

27357 Anthericum schlechteri Poelln. = Trachyandra bulbinifolia (Dinter) Oberm. ■☆

27358 Anthericum schultesii Baker = Chlorophytum rigidum Kunth ■☆

27359 Anthericum scilliflorum Eckl. ex Baker = Caesia eckloniana Schult. et Schult. f. ■☆

27360 Anthericum secundum K. Krause et Dinter = Trachyandra saltii (Baker) Oberm. ■☆

27361 Anthericum senussiorum Poelln. = Chlorophytum nubicum (Baker) Kativu ■☆

27362 Anthericum serotinum (L.) L. = Lloydia serotina (L.) Salisb. ex Rchb. ■

27363 Anthericum serpentinum Baker = Trachyandra flexifolia (L. f.) Kunth ■☆

27364 Anthericum setiferum Poelln. = Chlorophytum subpetiolatum (Baker) Kativu ■☆

27365 Anthericum setosum Schult. et Schult. f. = Bulbinella nutans (Thunb.) T. Durand et Schinz ■☆

27366 Anthericum speciosum Rendle = Chlorophytum cameronii (Baker) Kativu var. pterocaulon (Welw. ex Baker) Nordal ■☆

27367 Anthericum sphacelatum Baker = Chlorophytum sphacelatum (Baker) Kativu ■☆

27368 Anthericum spongiosum Poelln. = Trachyandra ciliata (L. f.) Kunth ■☆

27369 Anthericum squameum L. f. = Trachyandra hispida (L.) Kunth ■☆

27370 Anthericum stenocarpum Baker = Echeandia flavescens (Schult. et Schult. f.) Cruden ■☆

27371 Anthericum stenophyllum Adamson = Trachyandra brachypoda (Baker) Oberm. ■☆
27372 Anthericum stenophyllum Baker = Trachyandra asperata Kunth var. stenophylla (Baker) Oberm. ■☆
27373 Anthericum sternbergianum Roem. et Schult. = Chlorophytum comosum (Thunb.) Baker ●■
27374 Anthericum stolzii Engl. et K. Krause = Chlorophytum subpetiolatum (Baker) Kativu ■☆
27375 Anthericum stuhlmannii Engl.;斯图尔曼花篱■☆
27376 Anthericum subcontortum Baker = Trachyandra asperata Kunth ■☆
27377 Anthericum sublanatum Dinter = Trachyandra lanata (Dinter) Oberm. ■☆
27378 Anthericum submaculatum Poelln. = Trachyandra brachypoda (Baker) Oberm. ■☆
27379 Anthericum subpapillosum Poelln. = Chlorophytum cameronii (Baker) Kativu var. pterocaulon (Welw. ex Baker) Nordal ■☆
27380 Anthericum subpetiolatum Baker = Chlorophytum subpetiolatum (Baker) Kativu ■☆
27381 Anthericum subpilosum Poelln. = Trachyandra hispida (L.) Kunth ■☆
27382 Anthericum subulatum Baker = Chlorophytum fasciculatum (Baker) Kativu ■☆
27383 Anthericum subulatum Baker var. longibracteatum Poelln. = Chlorophytum fasciculatum (Baker) Kativu ■☆
27384 Anthericum succulentum Salisb. = Bulbine asphodeloides (L.) Spreng. ■☆
27385 Anthericum suessenguthii Sölch = Chlorophytum krauseanum (Dinter) Kativu ■☆
27386 Anthericum suffruticosum (Baker) Milne-Redh. = Chlorophytum suffruticosum Baker ●☆
27387 Anthericum superpositum Baker = Chlorophytum superpositum (Baker) Marais et Reilly ■☆
27388 Anthericum tabulare Baker = Trachyandra tabularis (Baker) Oberm. ■☆
27389 Anthericum taylorianum Rendle = Chlorophytum affine Baker ■☆
27390 Anthericum tenellum Welw. ex Baker = Chlorophytum calyptrocarpum (Baker) Kativu ■☆
27391 Anthericum tenuifolium Adamson = Trachyandra brachypoda (Baker) Oberm. ■☆
27392 Anthericum tenuissimum A. Chev.;极细花篱■☆
27393 Anthericum thyrsoideum Baker = Trachyandra thyrsoidea (Baker) Oberm. ■☆
27394 Anthericum torreyi Baker;托里花篱;Amber Lily ■☆
27395 Anthericum torreyi Baker = Echeandia flavescens (Schult. et Schult. f.) Cruden ■☆
27396 Anthericum torreyi Baker var. arizonicum Poelln. = Echeandia flavescens (Schult. et Schult. f.) Cruden ■☆
27397 Anthericum torreyi Baker var. lanceolatum Poelln. = Echeandia flavescens (Schult. et Schult. f.) Cruden ■☆
27398 Anthericum torreyi Baker var. neomexicanum Poelln. = Echeandia flavescens (Schult. et Schult. f.) Cruden ■☆
27399 Anthericum tortifolium Kuntze = Trachyandra gerrardii (Baker) Oberm. ■☆
27400 Anthericum tortile Baker = Trachyandra tortilis (Baker) Oberm. ■☆
27401 Anthericum transvaalense Baker = Chlorophytum transvaalense (Baker) Kativu ■☆
27402 Anthericum trichophlebium Baker = Chlorophytum trichophlebium (Baker) Nordal ■☆
27403 Anthericum triflorum Aiton = Chlorophytum triflorum (Aiton) Kunth ■☆
27404 Anthericum triflorum Aiton var. minor Baker = Chlorophytum undulatum (Jacq.) Oberm. ■☆
27405 Anthericum triflorum Baker = Chlorophytum cameronii (Baker) Kativu ■☆
27406 Anthericum triphyllum Baker = Chlorophytum subpetiolatum (Baker) Kativu ■☆
27407 Anthericum triquetrum L. f. = Bulbinella triquetra (L. f.) Kunth ■☆
27408 Anthericum triquetrum L. f. var. trinervis Baker = Bulbinella trinervis (Baker) P. L. Perry ■☆
27409 Anthericum tropicum Poelln. = Chlorophytum subpetiolatum (Baker) Kativu ■☆
27410 Anthericum tuberiferum Hutch.;块茎花篱■☆
27411 Anthericum tuberosum De Wild. = Chlorophytum subpetiolatum (Baker) Kativu ■☆
27412 Anthericum tuberosum Roxb. = Chlorophytum tuberosum (Roxb.) Baker ■☆
27413 Anthericum tubiferum Dinter = Chlorophytum rangei (Engl. et K. Krause) Nordal ■☆
27414 Anthericum tumidum Poelln. = Chlorophytum undulatum (Jacq.) Oberm. ■☆
27415 Anthericum turuense Poelln. = Chlorophytum affine Baker ■☆
27416 Anthericum ugogoense Poelln.;热非花篱■☆
27417 Anthericum ulugurense Engl.;乌卢古尔花篱■☆
27418 Anthericum undulatifolium Engl. ex Poelln.;波叶花篱■☆
27419 Anthericum undulatum Jacq. = Chlorophytum undulatum (Jacq.) Oberm. ■☆
27420 Anthericum undulatum Thunb. = Trachyandra hispida (L.) Kunth ■☆
27421 Anthericum unifolium Poelln. = Chlorophytum subpetiolatum (Baker) Kativu ■☆
27422 Anthericum urumburense Poelln. = Chlorophytum subpetiolatum (Baker) Kativu ■☆
27423 Anthericum usseramense Baker = Chlorophytum sphacelatum (Baker) Kativu var. milanjianum (Rendle) Nordal ■☆
27424 Anthericum usseramense Baker var. occidentalis A. Chev. = Chlorophytum limosum (Baker) Nordal ■☆
27425 Anthericum ustulatum Welw. ex Baker;凋萎花篱■☆
27426 Anthericum uyuiense Rendle;乌尤伊花篱■☆
27427 Anthericum uyuiense Rendle var. latifolium Poelln. = Chlorophytum cameronii (Baker) Kativu ■☆
27428 Anthericum vaginatum Baker = Chlorophytum trichophlebium (Baker) Nordal ■☆
27429 Anthericum validum Peter ex Poelln.;粗壮花篱■☆
27430 Anthericum vallistrappii Poelln. = Chlorophytum comosum (Thunb.) Baker ●■
27431 Anthericum velutinum De Wild. = Chlorophytum sphacelatum (Baker) Kativu ■☆
27432 Anthericum venulosum Baker;细脉花篱■☆
27433 Anthericum verruciferum Chiov. = Anthericum jamesii Baker ■☆
27434 Anthericum vespertinum Jacq. = Trachyandra ciliata (L. f.) Kunth ■☆
27435 Anthericum vestitum Baker ex Schinz = Chlorophytum transvaalense (Baker) Kativu ■☆

27436 Anthericum viridulobrunneum Poelln. = Chlorophytum blepharophyllum Schweinf. ex Baker ■☆
27437 Anthericum viscosum (Kunth) Baker = Chlorophytum viscosum Kunth ■☆
27438 Anthericum volkii Sölch = Chlorophytum galpinii (Baker) Kativu var. matabalense (Baker) Kativu ■☆
27439 Anthericum warneckei Engl. = Chlorophytum warneckei (Engl.) Marais et Reilly ■☆
27440 Anthericum weissianum Dinter = Trachyandra falcata (L. f.) Kunth ■☆
27441 Anthericum welwitschii Marais et Reilly = Chlorophytum stolzii (K. Krause) Kativu ■☆
27442 Anthericum whytei Baker = Chlorophytum sphacelatum (Baker) Kativu var. milanjianum (Rendle) Nordal ■☆
27443 Anthericum wilmsii Diels ex Burtt Davy et Pott-Leendertz = Chlorophytum fasciculatum (Baker) Kativu ■☆
27444 Anthericum wilmsii Diels ex Poelln. = Chlorophytum fasciculatum (Baker) Kativu ■☆
27445 Anthericum xylorrhizum Engl. et Gilg;木根花篱■☆
27446 Anthericum zanguebaricum Baker = Chlorophytum cameronii (Baker) Kativu var. pterocaulon (Welw. ex Baker) Nordal ■☆
27447 Anthericum zavattari Cufod. = Chlorophytum zavattari (Cufod.) Nordal ■☆
27448 Anthericum zebrinum Schltr. ex Poelln. = Trachyandra zebrina (Schltr. ex Poelln.) Oberm. ■☆
27449 Anthericum zenkeri Engl. = Chlorophytum staudtii Nordal ■☆
27450 Anthericum zeyheri Baker = Caesia eckloniana Schult. et Schult. f. ■☆
27451 Anthericus Asch. = Anthericum L. ■☆
27452 Antherlcus Asch. et Graetn. = Anthericum L. ■☆
27453 Antherocephala B. D. Jacks. = Antherocephala DC. ●☆
27454 Antherocephala DC. = Andersonia Buch. -Ham. ex Wall. ●☆
27455 Antheroceras Bertero = Leucocoryne Lindl. ■☆
27456 Antherolophus Gagnep. (1934);印支铃兰属■☆
27457 Antherolophus Gagnep. = Aspidistra Ker Gawl. ●■
27458 Antherolophus glandulosus Gagnep.;印支铃兰■☆
27459 Antherolophus glandulosus Gagnep. = Aspidistra glandulosa (Gagnep.) Tillich ■☆
27460 Antheropeas Rydb. (1915);北美菊属●■☆
27461 Antheropeas Rydb. = Eriophyllum Lag. ●■☆
27462 Antheropeas lanosum (A. Gray) Rydb. = Eriophyllum lanosum (A. Gray) A. Gray ■☆
27463 Antheropeas wallacei (A. Gray) Rydb. = Eriophyllum wallacei (A. Gray) A. Gray ■☆
27464 Antheroporum Gagnep. (1915);肿荚豆属;Antheroporum, Swellpod ●
27465 Antheroporum glaucum Z. Wei;粉叶肿荚豆;Glaucous Antheroporum, Glaucous Swellpod ●
27466 Antheroporum harmandii Gagnep.;肿荚豆;Harmand Antheroporum, Harmand Swellpod ●
27467 Antherosperma Poir. = Atherosperma Labill. ●☆
27468 Antherosperma Poir. ex Steud. = Atherosperma Labill. ●☆
27469 Antherostele Bremek. (1940);多花柱茜属●☆
27470 Antherostele banahaensis (Elmer) Bremek.;多花柱茜●☆
27471 Antherostylis C. A. Gardner = Velleia Sm. ■☆
27472 Antherothamnus N. E. Br. (1915);多花木玄参属●☆
27473 Antherothamnus pearsonii N. E. Br.;多花木玄参●☆
27474 Antherothamnus rigidus (L. Bolus) E. Phillips = Antherothamnus pearsonii N. E. Br. ●☆
27475 Antherotoma (Naudin) Hook. f. (1867);割花野牡丹属●■☆
27476 Antherotoma Hook. f. = Antherotoma (Naudin) Hook. f. ●■☆
27477 Antherotoma afzelii Hook. f. = Osbeckia decandra (Sm.) DC. ●☆
27478 Antherotoma angustifolia (A. Fern. et R. Fern.) Jacq. -Fél.;窄叶割花野牡丹■☆
27479 Antherotoma antherotoma (Naudin) Krasser = Antherotoma naudinii Hook. f. ■☆
27480 Antherotoma clandestina Jacq. -Fél.;隐匿割花野牡丹■☆
27481 Antherotoma debilis (Sond.) Jacq. -Fél.;弱小割花野牡丹■☆
27482 Antherotoma decandra (Sm.) A. Fern. et R. Fern. = Osbeckia decandra (Sm.) DC. ●☆
27483 Antherotoma densiflora (Gilg) Jacq. -Fél.;密割花野牡丹■☆
27484 Antherotoma gracilis (Cogn.) Jacq. -Fél.;纤细割花野牡丹■☆
27485 Antherotoma irvingiana (Hook. f.) Jacq. -Fél.;欧文割花野牡丹■☆
27486 Antherotoma irvingiana (Hook. f.) Jacq. -Fél. var. alpestris (Taub.) A. Fern. et R. Fern. = Antherotoma irvingiana (Hook. f.) Jacq. -Fél. ■☆
27487 Antherotoma naudinii Hook. f.;诺丹割花野牡丹■☆
27488 Antherotoma phaeotricha (Hochst.) Jacq. -Fél.;褐毛割花野牡丹■☆
27489 Antherotoma senegambiensis (Guillaumin et Perr.) Jacq. -Fél.;西非割花野牡丹●☆
27490 Antherotoma senegambiensis (Guillaumin et Perr.) Jacq. -Fél. var. alpestris Taub.;阿尔卑斯割花野牡丹●☆
27491 Antherotoma tenuis (A. Fern. et R. Fern.) Jacq. -Fél.;细割花野牡丹●☆
27492 Antherotoma tisserantii (Jacq. -Fél.) Jacq. -Fél.;蒂氏割花野牡丹●☆
27493 Antherotrlche Turcz. = Anisoptera Korth. ●☆
27494 Antherura Lour. = Psychotria L. (保留属名)●
27495 Antherura rubra Lour. = Psychotria asiatica L. ●
27496 Antherura rubra Lour. = Psychotria rubra (Lour.) Poir. ●
27497 Antherylium Rohr = Ginoria Jacq. ●☆
27498 Antherylium Rohr et Vahl = Ginoria Jacq. ●☆
27499 Antheryta Raf. = Tibouchina Aubl. ●■☆
27500 Anthesteria Spreng. = Anthistiria L. f. ■
27501 Anthesteria Spreng. = Themeda Forssk. ■
27502 Anthillis Neck. = Anthyllis L. ■☆
27503 Anthipsimus Raf. = Muhlenbergia Schreb. ■
27504 Anthirrinum Moench = Antirrhinum L. ●■
27505 Anthistiria L. f. = Themeda Forssk. ■
27506 Anthistiria abyssinica Hochst. ex A. Rich. = Exotheca abyssinica (Hochst. ex A. Rich.) Andersson ■☆
27507 Anthistiria anathera Nees ex Steud. = Themeda anathera (Nees ex Steud.) Hack. ■
27508 Anthistiria arundinacea Roxb. = Themeda arundinacea (Roxb.) Ridl. ■
27509 Anthistiria australis R. Br. = Themeda triandra Forssk. ■
27510 Anthistiria barteri Hack. = Hyparrhenia involucrata Stapf ■☆
27511 Anthistiria caudata Nees = Themeda caudata (Nees) A. Camus ■
27512 Anthistiria ciliata Nees var. burchellii Hack. = Themeda triandra Forssk. ■
27513 Anthistiria dissoluta Nees ex Steud. = Hyperthelia dissoluta (Nees ex Steud.) Clayton ■☆

27514 Anthistiria foliosus Kunth = Hyparrhenia bracteata (Humb. et Bonpl. ex Willd.) Stapf ■
27515 Anthistiria gigantea Cav. subsp. caudata (Nees) Hook. f. = Themeda caudata (Nees) A. Camus ■
27516 Anthistiria gigantea Cav. subsp. villosa (Poir.) Hook. f. = Themeda villosa (Poir.) A. Camus ■
27517 Anthistiria glauca Desf. = Themeda glauca (Desf.) Hack. ■
27518 Anthistiria glauca Desf. = Themeda triandra Forssk. var. glauca (Desf.) Hack. ■
27519 Anthistiria heteroclita Roxb. = Pseudanthistria heteroclita (Roxb.) Hook. f. ■
27520 Anthistiria hookeri Griseb. = Themeda hookeri (Griseb.) A. Camus ■
27521 Anthistiria imberbis Retz. = Themeda triandra Forssk. ■
27522 Anthistiria japonica Willd. = Themeda japonica (Willd.) C. Tanaka ■
27523 Anthistiria japonica Willd. = Themeda triandra Forssk. ■
27524 Anthistiria multiplex Hochst. ex A. Rich. = Hyparrhenia multiplex (Hochst. ex A. Rich.) Andersson ex Stapf ■☆
27525 Anthistiria mutica Steud. = Themeda villosa (Poir.) A. Camus ■
27526 Anthistiria paleacea (Poir.) Vahl = Themeda triandra Forssk. ■
27527 Anthistiria paleacea Ball. = Themeda triandra Forssk. ■
27528 Anthistiria pseudocymbaria Steud. = Hyparrhenia anthistirioides (Hochst. ex A. Rich.) Andersson ex Stapf ■☆
27529 Anthistiria punctata Hochst. ex A. Rich. = Themeda triandra Forssk. ■
27530 Anthistiria subsericans Nees ex Steud. = Themeda arundinacea (Roxb.) Ridl. ■
27531 Anthistiria tortilis J. Presl = Cymbopogon tortilis (J. Presl) A. Camus ■
27532 Anthistiria villosa Poir. = Themeda villosa (Poir.) A. Camus ■
27533 Anthobembix Perkins = Steganthera Perkins ●☆
27534 Anthobembix Perkins(1898);新几内亚香材树属●☆
27535 Anthobembix oligantha Perkins;新几内亚香材树●☆
27536 Anthobolaceae Dumort. = Santalaceae R. Br.(保留科名)●■
27537 Anthobolus R. Br.(1810);落花檀香属●☆
27538 Anthobolus filifolius;落花檀香●☆
27539 Anthobryum Phil.(1891);苔花属■☆
27540 Anthobryum Phil. = Frankenia L. ●■
27541 Anthobryum Phil. et Reiche = Frankenia L. ●■
27542 Anthobryum aretioides;苔花■☆
27543 Anthobryum tetragonum Phil.;四角苔花■☆
27544 Anthocarapa Pierre(1897);新喀里多尼亚楝属●☆
27545 Anthocarapa nitidula (Benth.) Mabb.;新喀里多尼亚楝●☆
27546 Anthocephalus A. Rich. = Breonia A. Rich. ex DC. ●☆
27547 Anthocephalus cadamba (Roxb.) Miq. = Anthocephalus chinensis (Lam.) Rich. ex Walp. ●
27548 Anthocephalus cadamba (Roxb.) Miq. = Neolamarckia cadamba (Roxb.) Bosser ●
27549 Anthocephalus chinensis (Lam.) Rich. ex Walp. = Breonia chinensis (Lam.) Capuron ●☆
27550 Anthocephalus chinensis (Lam.) Rich. ex Walp. = Neolamarckia cadamba (Roxb.) Bosser ●
27551 Anthocephalus indicus A. Rich. = Anthocephalus chinensis (Lam.) Rich. ex Walp. ●
27552 Anthocephalus indicus A. Rich. var. glabrescens H. L. Li = Neolamarckia cadamba (Roxb.) Bosser ●
27553 Anthoceras Baker = Antheroceras Bertero ■☆
27554 Anthoceras Baker = Leucocoryne Lindl. ■☆
27555 Anthocerastes A. Gray = Toxanthes Turcz. ■☆
27556 Anthocercis Labill.(1806);梭花茄属●■☆
27557 Anthocercis littorea Labill.;梭花茄■☆
27558 Anthoceris Steud. = Anthocercis Labill. ●■☆
27559 Anthochlamys Fenzl ex Endl. = Anthochlamys Fenzl ■☆
27560 Anthochlamys Fenzl(1837);合被虫实属■☆
27561 Anthochlamys polygaloides (Fisch. et C. A. Mey.) Fenzl.;合被虫实■☆
27562 Anthochlamys turcomanica Iljin;土库曼合被虫实■☆
27563 Anthochloa Nees et Meyen(1834);花禾属■☆
27564 Anthochloa lepidula Nees et Meyen;花禾■☆
27565 Anthochortus Endl. = Anthochortus Nees ■☆
27566 Anthochortus Endl. = Willdenowia Thunb. ■☆
27567 Anthochortus Nees = Willdenowia Thunb. ■☆
27568 Anthochortus Nees(1836);园花属■☆
27569 Anthochortus capensis Esterh.;好望角园花■☆
27570 Anthochortus crinalis (Mast.) H. P. Linder;发园花■☆
27571 Anthochortus ecklonii Nees;埃氏园花■☆
27572 Anthochortus graminifolius (Kunth) H. P. Linder;禾叶园花●■☆
27573 Anthochortus insignis (Mast.) H. P. Linder;显著园花■☆
27574 Anthochortus laxiflorus (Nees) H. P. Linder;疏花园花■☆
27575 Anthochortus singularis Esterh.;单一园花■☆
27576 Anthochytrum Rchb. = Barkhausia Moench ■
27577 Anthochytrum Rchb. = Crepis L. ■
27578 Anthocleista Afzel. ex R. Br.(1818);非洲马钱树属(闭花马钱属,花闭木属);Cabbage Tree ●☆
27579 Anthocleista amplexicaulis Baker;抱茎非洲马钱树(抱茎花闭木)●☆
27580 Anthocleista auriculata De Wild. = Anthocleista vogelii Planch. ●☆
27581 Anthocleista baertsiana De Wild. et T. Durand = Anthocleista liebrechtsiana De Wild. et T. Durand ●☆
27582 Anthocleista bequaertii De Wild. = Anthocleista vogelii Planch. ●☆
27583 Anthocleista brieyi De Wild. = Brenania brieyi (De Wild.) E. M. Petit ●☆
27584 Anthocleista buchneri Gilg = Anthocleista vogelii Planch. ●☆
27585 Anthocleista djalonensis A. Chev.;贾隆马钱树●☆
27586 Anthocleista exelliana Monod = Anthocleista scandens Hook. f. ●☆
27587 Anthocleista frezoulsii A. Chev. = Anthocleista procera Lepr. ex Bureau ●☆
27588 Anthocleista gabonensis Gentil;加蓬非洲马钱树●☆
27589 Anthocleista gigantea Gilg = Anthocleista schweinfurthii Gilg ●☆
27590 Anthocleista gossweileri Exell = Anthocleista liebrechtsiana De Wild. et T. Durand ●☆
27591 Anthocleista grandiflora Gilg;非洲马钱树(大花花闭木)●☆
27592 Anthocleista inermis Engl.;无刺非洲马钱树●☆
27593 Anthocleista insignis Galpin = Anthocleista grandiflora Gilg ●☆
27594 Anthocleista insulana S. Moore = Anthocleista schweinfurthii Gilg ●☆
27595 Anthocleista kalbreyeri Baker = Anthocleista vogelii Planch. ●☆
27596 Anthocleista kamerunensis Gilg = Anthocleista schweinfurthii Gilg ●☆
27597 Anthocleista keniensis Summerh. = Anthocleista grandiflora Gilg ●☆
27598 Anthocleista kerstingii Gilg ex Volkens;克斯廷马钱树●☆
27599 Anthocleista lanceolata Gilg = Anthocleista vogelii Planch. ●☆

27600 Anthocleista laurentii De Wild. = Anthocleista schweinfurthii Gilg ●☆
27601 Anthocleista laxiflora Baker;疏花非洲马钱树●☆
27602 Anthocleista liebrechtsiana De Wild. et T. Durand;利布非洲马钱树●☆
27603 Anthocleista macrantha Gilg = Anthocleista vogelii Planch. ●☆
27604 Anthocleista macrocalyx Philipson = Anthocleista microphylla Wernham ●☆
27605 Anthocleista macrophylla G. Don = Anthocleista nobilis G. Don ●☆
27606 Anthocleista magnifica Gilg = Anthocleista schweinfurthii Gilg ●☆
27607 Anthocleista micrantha Gilg et Mildbr. ex Hutch. et Dalziel = Anthocleista microphylla Wernham ●☆
27608 Anthocleista microphylla Wernham;小叶非洲马钱树●☆
27609 Anthocleista niamniamensis Gilg = Anthocleista schweinfurthii Gilg ●☆
27610 Anthocleista nigrescens Afzel. ex Gilg = Anthocleista nobilis G. Don ●☆
27611 Anthocleista nobilis G. Don;小花非洲马钱树●☆
27612 Anthocleista obanensis Wernham;奥班马钱树●☆
27613 Anthocleista orientalis Gilg = Anthocleista grandiflora Gilg ●☆
27614 Anthocleista oubanguiensis Aubrév. et Pellegr. = Anthocleista schweinfurthii Gilg ●☆
27615 Anthocleista parviflora Baker = Anthocleista nobilis G. Don ●☆
27616 Anthocleista procera Lepr. ex A. Chev. var. parviflora (Baker) A. Chev. = Anthocleista nobilis G. Don ●☆
27617 Anthocleista procera Lepr. ex A. Chev. var. umbellata A. Chev. = Anthocleista nobilis G. Don ●☆
27618 Anthocleista procera Lepr. ex Bureau;高大非洲马钱树●☆
27619 Anthocleista pulcherrima Gilg = Anthocleista grandiflora Gilg ●☆
27620 Anthocleista pynaertii De Wild. = Anthocleista schweinfurthii Gilg ●☆
27621 Anthocleista scandens Hook. f.;攀缘马钱树●☆
27622 Anthocleista scheffleri Gilg ex Scheffler = Anthocleista grandiflora Gilg ●☆
27623 Anthocleista schweinfurthii Gilg;施韦非洲马钱树●☆
27624 Anthocleista squamata De Wild. et T. Durand = Anthocleista schweinfurthii Gilg ●☆
27625 Anthocleista stenantha Philipson = Anthocleista microphylla Wernham ●☆
27626 Anthocleista stuhlmannii Gilg = Anthocleista schweinfurthii Gilg ●☆
27627 Anthocleista talbotii Wernham = Anthocleista vogelii Planch. ●☆
27628 Anthocleista vogelii Planch.;沃格尔马钱树●☆
27629 Anthocleista zambesiaca Baker = Anthocleista grandiflora Gilg ●☆
27630 Anthocleista zenkeri Gilg = Anthocleista vogelii Planch. ●☆
27631 Anthoclitandra (Pierre) Pichon = Landolphia P. Beauv. (保留属名)●☆
27632 Anthoclitandra nitida (Stapf) Pichon = Landolphia nitidula Pers. ●☆
27633 Anthoclitandra robustior (K. Schum.) Pichon = Landolphia robustior (K. Schum.) Pers. ●☆
27634 Anthocoma K. Koch = Rhododendron L. ●
27635 Anthocoma Zoll. et Moritzi = Cymaria Benth. ●
27636 Anthocometes Nees = Monothecium Hochst. ■☆
27637 Anthocometes aristatus (Nees) Nees = Monothecium aristatum (Nees) T. Anderson ■☆
27638 Anthocophalus indicus A. Rich. var. glabrescens Li = Neolamarckia cadamba (Roxb.) Bosser ●
27639 Anthodendron Rchb. = Rhododendron L. ●
27640 Anthodiscus Endl. = Salacia L. (保留属名)●
27641 Anthodiscus G. Mey. (1818);盘花南星属■☆
27642 Anthodiscus amazonicus Gleason et A. C. Sm.;亚马逊盘花南星■☆
27643 Anthodiscus fragrans Sleumer;香盘花南星■☆
27644 Anthodiscus glaucescens J. F. Macbr.;灰盘花南星■☆
27645 Anthodiscus guianensis G. Mey.;圭亚那盘花南星■☆
27646 Anthodiscus montanus Gleason;山地盘花南星■☆
27647 Anthodiscus obovatus Benth. ex Wittm.;倒卵盘花南星■☆
27648 Anthodiscus pilosus Ducke;毛盘花南星■☆
27649 Anthodiscus trifoliatus G. Mey.;三小叶盘花南星■☆
27650 Anthodon Ruiz et Pav. (1798);齿花卫矛属●☆
27651 Anthodon decussatum Ruiz et Pav.;齿花卫矛●☆
27652 Anthodon ellipticum Mart.;椭圆齿花卫矛●☆
27653 Anthodus Mast. ex Roem. et Schult. = Anthodon Ruiz et Pav. ●☆
27654 Anthogonium Lindl. = Anthogonium Wall. ex Lindl. ■
27655 Anthogonium Wall. ex Lindl. (1840);筒瓣兰属(红花小独蒜属,筒瓣花属);Tubepetalorchis ■
27656 Anthogonium corydaloides Schltr. = Anthogonium gracile Lindl. ■
27657 Anthogonium gracile Lindl.;筒瓣兰(红花小独蒜,小白芨);Tubepetalorchis ■
27658 Anthogonium griffithii Rchb. f. = Anthogonium gracile Lindl. ■
27659 Anthogyas Raf. = Bletia Ruiz et Pav. ■☆
27660 Anthogyas Raf. = Gyas Salisb. ■☆
27661 Antholobus Anon = Anthobolus R. Br. ●☆
27662 Antholobus Rchb. = Anthobolus R. Br. ●☆
27663 Antholoma Labill. = Sloanea L. ●
27664 Antholyza L. (1753);非洲鸢尾属(口花属);Madflower ■☆
27665 Antholyza L. = Babiana Ker Gawl. ex Sims(保留属名)■☆
27666 Antholyza L. = Gladiolus L. ■
27667 Antholyza abbreviata (Andréws) Pers. = Gladiolus abbreviatus Andréws ■☆
27668 Antholyza abyssinica Brongn. ex Lem. = Gladiolus abyssinicus (Brongn. ex Lem.) Goldblatt et M. P. de Vos ■☆
27669 Antholyza acuminata N. E. Br. = Gladiolus watsonius Thunb. ■☆
27670 Antholyza aethiopica L. = Chasmanthe aethiopica (L.) N. E. Br. ■☆
27671 Antholyza aethiopica L. var. bicolor (Gasp. ex Ten.) Baker = Chasmanthe bicolor (Gasp. ex Ten.) N. E. Br. ■☆
27672 Antholyza aethiopica L. var. immarginata Thunb. ex Baker = Chasmanthe aethiopica (L.) N. E. Br. ■☆
27673 Antholyza aethiopica L. var. minor Lindl. = Chasmanthe bicolor (Gasp. ex Ten.) N. E. Br. ■☆
27674 Antholyza aethiopica L. var. ringens (Andréws) Baker = Chasmanthe aethiopica (L.) N. E. Br. ■☆
27675 Antholyza aethiopica L. var. vittigera (Salisb.) Baker = Chasmanthe aethiopica (L.) N. E. Br. ■☆
27676 Antholyza aletroides Burm. f. = Watsonia aletroides (Burm. f.) Ker Gawl. ■☆
27677 Antholyza baguirmiensis A. Chev.;巴基米非洲鸢尾■☆
27678 Antholyza bicolor Gasp. ex Ten. = Chasmanthe bicolor (Gasp. ex Ten.) N. E. Br. ■☆
27679 Antholyza buckerveldii L. Bolus = Gladiolus buckerveldii (L. Bolus) Goldblatt ■☆
27680 Antholyza burchellii N. E. Br. = Tritoniopsis burchellii (N. E. Br.) Goldblatt ■☆

27681 Antholyza cabrae De Wild. = Gladiolus unguiculatus Baker ■☆
27682 Antholyza caffra Ker Gawl. ex Baker = Tritoniopsis caffra (Ker Gawl. ex Baker) Goldblatt ■☆
27683 Antholyza caryophyllacea Burm. f. = Gladiolus caryophyllaceus (Burm. f.) Poir. ■☆
27684 Antholyza cunonia L. = Gladiolus cunonius (L.) Gaertn. ■☆
27685 Antholyza degasparisiana Buscal. et Muschl. = Gladiolus huillensis (Welw. ex Baker) Goldblatt ■☆
27686 Antholyza descampsii De Wild. = Gladiolus gregarius Welw. ex Baker ■☆
27687 Antholyza djalonensis A. Chev. = Gladiolus unguiculatus Baker ■☆
27688 Antholyza duftii Schinz = Gladiolus saccatus (Klatt) Goldblatt et M. P. de Vos ■☆
27689 Antholyza fimbriata Klatt = Babiana fimbriata (Klatt) Baker ■☆
27690 Antholyza fleuryi A. Chev. = Gladiolus unguiculatus Baker ■☆
27691 Antholyza floribunda Salisb. = Chasmanthe aethiopica (L.) N. E. Br. ■☆
27692 Antholyza floribunda Salisb. = Chasmanthe floribunda (Salisb.) N. E. Br. ■☆
27693 Antholyza fourcadei L. Bolus = Gladiolus fourcadei (L. Bolus) Goldblatt et M. P. de Vos ■☆
27694 Antholyza fucata (Herb.) Baker = Crocosmia fucata (Herb.) M. P. de Vos ■☆
27695 Antholyza gilletii De Wild. = Gladiolus gregarius Welw. ex Baker ■☆
27696 Antholyza gracilis Pax = Gladiolus watsonioides Baker ■☆
27697 Antholyza guthriei L. Bolus = Gladiolus overbergensis Goldblatt et M. P. de Vos ■☆
27698 Antholyza hantamensis Klatt = Hesperantha cucullata Klatt ■☆
27699 Antholyza huillensis Welw. ex Baker = Gladiolus huillensis (Welw. ex Baker) Goldblatt ■☆
27700 Antholyza hypogaea (Burch.) Klatt = Babiana hypogaea Burch. ■☆
27701 Antholyza immarginata Thunb. = Chasmanthe aethiopica (L.) N. E. Br. ■☆
27702 Antholyza intermedia Baker = Tritoniopsis intermedia (Baker) Goldblatt ■☆
27703 Antholyza labiata Pax = Gladiolus unguiculatus Baker ■☆
27704 Antholyza laxiflora Baker = Gladiolus antholyzoides Baker ■☆
27705 Antholyza lucidor L. f. = Tritoniopsis triticea (Burm. f.) Goldblatt ■☆
27706 Antholyza magnifica Harms = Gladiolus magnificus (Harms) Goldblatt ■☆
27707 Antholyza meriana L. = Watsonia meriana (L.) Mill. ■☆
27708 Antholyza merianella L. = Gladiolus bonaspei Goldblatt et M. P. de Vos ■☆
27709 Antholyza muirii L. Bolus = Gladiolus teretifolius Goldblatt et M. P. de Vos ■☆
27710 Antholyza nemorosa Klatt = Tritoniopsis nemorosa (Klatt) G. J. Lewis ■☆
27711 Antholyza nervosa Thunb. = Tritoniopsis antholyza (Poir.) Goldblatt ■☆
27712 Antholyza paludosa A. Chev.;沼泽非洲鸢尾■☆
27713 Antholyza paniculata Klatt = Crocosmia paniculata (Klatt) Goldblatt ■☆
27714 Antholyza plicata L. f. = Babiana thunbergii Ker Gawl. ■☆
27715 Antholyza praealta Redouté = Chasmanthe floribunda (Salisb.) N. E. Br. ■☆
27716 Antholyza priorii N. E. Br. = Gladiolus priorii (N. E. Br.) Goldblatt et M. P. de Vos ■☆
27717 Antholyza pubescens Vaupel = Gladiolus huillensis (Welw. ex Baker) Goldblatt ■☆
27718 Antholyza pulchrum Baker = Tritoniopsis pulchra (Baker) Goldblatt ■☆
27719 Antholyza quadrangularis Burm. f. = Gladiolus quadrangularis (Burm. f.) Ker Gawl. ■☆
27720 Antholyza quinquenervia Schrank;五脉非洲鸢尾■☆
27721 Antholyza ramosa Eckl. ex Klatt = Tritoniopsis ramosa (Eckl. ex Klatt) G. J. Lewis ■☆
27722 Antholyza revoluta Burm. f. = Tritoniopsis revoluta (Burm. f.) Goldblatt ■☆
27723 Antholyza ringens L. = Babiana ringens (L.) Ker Gawl. ■☆
27724 Antholyza saccata (Klatt) Baker = Gladiolus saccatus (Klatt) Goldblatt et M. P. de Vos ■☆
27725 Antholyza schlechteri Baker = Gladiolus antholyzoides Baker ■☆
27726 Antholyza schweinfurthii Baker = Gladiolus schweinfurthii (Baker) Goldblatt et M. P. de Vos ■☆
27727 Antholyza sladeniana Pole-Evans;斯莱登非洲鸢尾■☆
27728 Antholyza speciosa Wright = Gladiolus watsonioides Baker ■☆
27729 Antholyza spectabilis Schinz = Gladiolus magnificus (Harms) Goldblatt ■☆
27730 Antholyza spicata Andréws = Watsonia laccata (Jacq.) Ker Gawl. ■☆
27731 Antholyza spicata Brehmer ex Klatt;穗状非洲鸢尾■☆
27732 Antholyza spicata Mill. = Gladiolus floribundus Jacq. ■☆
27733 Antholyza splendens (Sweet) Steud. = Gladiolus splendens (Sweet) Herb. ■☆
27734 Antholyza steingroeveri Pax = Gladiolus saccatus (Klatt) Goldblatt et M. P. de Vos ■☆
27735 Antholyza striata (Jacq.) Klatt = Babiana striata (Jacq.) G. J. Lewis ■☆
27736 Antholyza sudanica A. Chev. = Gladiolus unguiculatus Baker ■☆
27737 Antholyza thonneri De Wild. = Gladiolus unguiculatus Baker ■☆
27738 Antholyza tubulosa Andréws = Watsonia aletroides (Burm. f.) Ker Gawl. ■☆
27739 Antholyza vandermerwei L. Bolus = Gladiolus vandermerwei (L. Bolus) Goldblatt et M. P. de Vos ■☆
27740 Antholyza vittigera Salisb. = Chasmanthe aethiopica (L.) N. E. Br. ■☆
27741 Antholyza watsonia (Thunb.) Pax = Gladiolus watsonius Thunb. ■☆
27742 Antholyza watsonioides (Baker) Baker = Gladiolus watsonioides Baker ■☆
27743 Antholyza zambesiaca Baker = Gladiolus magnificus (Harms) Goldblatt ■☆
27744 Anthomeles M. Roem. = Crataegus L. ●
27745 Anthonota P. Beauv. = Macrolobium Schreb. (保留属名) ●☆
27746 Anthonotha P. Beauv. (1806);仿花苏木属●☆
27747 Anthonotha P. Beauv. = Macrolobium Schreb. (保留属名) ●☆
27748 Anthonotha acuminata (De Wild.) J. Léonard;渐尖仿花苏木●☆
27749 Anthonotha brieyi (De Wild.) J. Léonard;布里仿花苏木●☆
27750 Anthonotha cladantha (Harms) J. Léonard;枝花仿花苏木●☆
27751 Anthonotha conchyliophora (Pellegr.) J. Léonard = Englerodendron conchyliophorum (Pellegr.) Breteler ●☆
27752 Anthonotha crassifolia (Baill.) J. Léonard;厚叶仿花苏木●☆

27753 Anthonotha elongata (Hutch.) J. Léonard;伸长仿花苏木●☆

27754 Anthonotha ernae (Dinkl.) J. Léonard = Triplisomeris ernae (Dinkl.) Aubrév. et Pellegr. ●☆

27755 Anthonotha explicans (Baill.) J. Léonard = Triplisomeris explicans (Baill.) Aubrév. et Pellegr. ●☆

27756 Anthonotha ferruginea (Harms) J. Léonard;锈色仿花苏木●☆

27757 Anthonotha fragrans (Baker f.) Exell et Hillc.;香仿花苏木●☆

27758 Anthonotha gabunensis J. Léonard = Englerodendron gabunense (J. Léonard) Breteler ●☆

27759 Anthonotha gilletii (De Wild.) J. Léonard;吉勒特仿花苏木●☆

27760 Anthonotha graciliflora (Harms) J. Léonard;细花仿花苏木●☆

27761 Anthonotha hallei (Aubrév.) J. Léonard;哈勒仿花苏木●☆

27762 Anthonotha isopetala (Harms) J. Léonard;异瓣仿花苏木●☆

27763 Anthonotha lamprophylla (Harms) J. Léonard;亮叶仿花苏木●☆

27764 Anthonotha lebrunii (J. Léonard) J. Léonard;勒布伦仿花苏木●☆

27765 Anthonotha leptorrhachis (Harms) J. Léonard;细轴仿花苏木●☆

27766 Anthonotha macrophylla P. Beauv.;大叶仿花苏木●☆

27767 Anthonotha nigerica (Baker f.) J. Léonard;尼日利亚仿花苏木●☆

27768 Anthonotha noldeae (Rossberg) Exell et Hillc.;诺尔德仿花苏木●☆

27769 Anthonotha obanensis (Baker f.) J. Léonard;奥班仿花苏木●☆

27770 Anthonotha pellegrinii Aubrév.;佩尔格兰仿花苏木●☆

27771 Anthonotha pynaertii (De Wild.) Exell et Hillc.;皮那仿花苏木●☆

27772 Anthonotha sargosii (Pellegr.) J. Léonard;萨尔仿花苏木●☆

27773 Anthonotha sassandraensis Aubrév. et Pellegr.;萨桑德拉仿花苏木●☆

27774 Anthonotha stipulacea (Benth.) J. Léonard;托叶仿花苏木●☆

27775 Anthonotha trunciflora (Harms) J. Léonard;截花仿花苏木●☆

27776 Anthonotha vignei (Hoyle) J. Léonard;维涅仿花苏木●☆

27777 Anthophyllum Steud. = Scirpus L. (保留属名) ■

27778 Anthopogon Neck. = Crossopetalum Roth ■

27779 Anthopogon Neck. = Gentiana L. ■

27780 Anthopogon Neck. ex Raf. = Crossopetalum Roth ■

27781 Anthopogon Neck. ex Raf. = Gentiana L. ■

27782 Anthopogon Nutt. = Gymnopogon P. Beauv. ■☆

27783 Anthopogon crinitum (Froel.) Raf. = Gentianopsis crinita (Froel.) Ma ■☆

27784 Anthopogon virgatum Raf. = Gentianopsis procera (Holm) Ma ■☆

27785 Anthopteropsis A. C. Sm. (1941);距药莓属●☆

27786 Anthopteropsis insignis A. C. Sm.;距药莓●☆

27787 Anthopterus Hook. (1839);翼冠莓属●☆

27788 Anthopterus racemosus Hook.;翼冠莓●☆

27789 Anthora DC. = Aconitum L. ■

27790 Anthora Haller = Aconitum L. ■

27791 Anthorrhiza C. R. Huxley et Jebb (1990);根花茜属☆

27792 Anthorrhiza areolata C. R. Huxley et Jebb;根花茜●☆

27793 Anthosachne Steud. (1854);沫花禾属■☆

27794 Anthosachne Steud. = Agropyron Gaertn. ■

27795 Anthosachne Steud. = Elymus L. ■

27796 Anthosachne elymoides (Hochst. ex A. Rich.) Nevski = Elymus africanus Á. Löve ■☆

27797 Anthosachne jacquemontii (Hook. f.) Nevski = Elymus jacquemontii (Hook. f.) Tzvelev ■

27798 Anthosachne longiaristata (Boiss.) Nevski.;长芒沫花禾■☆

27799 Anthosciadium Fenzl = Selinum L. (保留属名) ■

27800 Anthoshorea Pierre = Shorea Roxb. ex C. F. Gaertn. ●

27801 Anthosiphon Schltr. (1920);哥伦比亚管花兰属■☆

27802 Anthosiphon roseans Schltr.;哥伦比亚管花兰■☆

27803 Anthospermopsis (K. Schum.) J. H. Kirkbr. (1997);拟琥珀树属●☆

27804 Anthospermum L. (1753);琥珀树属(非洲花子属);Amber Tree ●☆

27805 Anthospermum aberdaricum K. Krause = Anthospermum usambarense K. Schum. ●☆

27806 Anthospermum aethiopicum L.;埃塞俄比亚琥珀树●☆

27807 Anthospermum aethiopicum L. var. ciliare (L.) Kuntze = Anthospermum bergianum Cruse ●☆

27808 Anthospermum aethiopicum L. var. ecklonianum Cruse = Anthospermum spathulatum Spreng. subsp. ecklonianum (Cruse) Puff ●☆

27809 Anthospermum aethiopicum L. var. montanum Sond. = Anthospermum spathulatum Spreng. ●☆

27810 Anthospermum aethiopicum L. var. oppositifolium Cruse = Anthospermum spathulatum Spreng. ●☆

27811 Anthospermum aethiopicum L. var. papillatum (Sond.) Kuntze = Anthospermum galioides Rchb. ex Spreng. ●☆

27812 Anthospermum aethiopicum L. var. reflexifolium Kuntze = Anthospermum galioides Rchb. ex Spreng. subsp. reflexifolium (Kuntze) Puff ●☆

27813 Anthospermum aethiopicum L. var. ternifolium Cruse = Anthospermum aethiopicum L. ●☆

27814 Anthospermum aethiopicum L. var. tulbaghense Eckl. et Zeyh. = Anthospermum spathulatum Spreng. subsp. tulbaghense Puff ●☆

27815 Anthospermum aethiopicum L. var. uitenhagense Eckl. et Zeyh. = Anthospermum spathulatum Spreng. subsp. uitenhagense Puff ●☆

27816 Anthospermum albohirtum Mildbr. = Anthospermum whyteanum Britten ●☆

27817 Anthospermum ambiguum Greves = Anthospermum littoreum L. Bolus ●☆

27818 Anthospermum ambrosiacum Moench = Anthospermum aethiopicum L. ●☆

27819 Anthospermum ammanioides S. Moore;水苋菜琥珀树●☆

27820 Anthospermum arenicola Greves = Anthospermum hispidulum E. Mey. ex Sond. ●☆

27821 Anthospermum aromaticum Salisb. = Anthospermum aethiopicum L. ●☆

27822 Anthospermum asperuloides Hook. f.;车叶草琥珀树●☆

27823 Anthospermum bergianum Cruse;贝格琥珀树●☆

27824 Anthospermum bicorne Puff;双角琥珀树●☆

27825 Anthospermum burkei Sond. = Anthospermum hispidulum E. Mey. ex Sond. ●☆

27826 Anthospermum calycophyllum Sond. = Otiophora calycophylla (Sond.) Schltr. et K. Schum. ■☆

27827 Anthospermum cameroonense Hutch. et Dalziel = Anthospermum asperuloides Hook. f. ●☆

27828 Anthospermum ciliare L. = Anthospermum bergianum Cruse ●☆

27829 Anthospermum ciliare L. var. angustifolium Eckl. et Zeyh. = Anthospermum galioides Rchb. ex Spreng. ●☆

27830 Anthospermum ciliare L. var. glabrifolium Sond. = Anthospermum galioides Rchb. ex Spreng. ●☆

27831 Anthospermum ciliare L. var. latifolium Eckl. et Zeyh. = Anthospermum galioides Rchb. ex Spreng. ●☆

27832 Anthospermum ciliare L. var. papillatum Sond. = Anthospermum galioides Rchb. ex Spreng. ●☆

27833 Anthospermum ciliare L. var. scabrum Eckl. et Zeyh. = Anthospermum galioides Rchb. ex Spreng. ●☆

27834 Anthospermum ciliare Thunb. ex Sond. = Anthospermum hirtum Cruse ●☆

27835 Anthospermum cliffortioides K. Schum. = Anthospermum welwitschii Hiern ●☆

27836 Anthospermum comptonii Puff;康普顿琥珀树●☆

27837 Anthospermum confertum (Eckl. et Zeyh.) Cruse ex Walp. = Anthospermum paniculatum Cruse ●☆

27838 Anthospermum crocyllis Sond. = Gaillonia crocyllis (Sond.) Thulin ■☆

27839 Anthospermum dregei Sond.;德雷琥珀树●☆

27840 Anthospermum dregei Sond. subsp. ecklonis (Sond.) Puff;埃克琥珀树●☆

27841 Anthospermum ecklonis Sond. = Anthospermum dregei Sond. subsp. ecklonis (Sond.) Puff ●☆

27842 Anthospermum emirnense Baker;埃米琥珀树●☆

27843 Anthospermum erectum Suess. = Anthospermum ternatum Hiern subsp. randii (S. Moore) Puff ■☆

27844 Anthospermum ericifolium (Licht. ex Roem. et Schult.) Kuntze;毛叶琥珀树●☆

27845 Anthospermum ericoideum K. Krause = Anthospermum rigidum Eckl. et Zeyh. subsp. pumilum (Sond.) Puff ●☆

27846 Anthospermum esterhuysenianum Puff;埃斯特琥珀树●☆

27847 Anthospermum esterhuysenianum Puff var. hirsutum Puff;粗毛琥珀树●☆

27848 Anthospermum ferrugineum Eckl. et Zeyh. = Anthospermum herbaceum L. f. ●☆

27849 Anthospermum galioides Rchb. ex Spreng.;南非琥珀树●☆

27850 Anthospermum galioides Rchb. ex Spreng. subsp. reflexifolium (Kuntze) Puff;折叶琥珀树●☆

27851 Anthospermum galopina Thunb. = Galopina circaeoides Thunb. ●☆

27852 Anthospermum galpinii Schltr.;盖尔琥珀树●☆

27853 Anthospermum hedyotideum Sond. = Anthospermum herbaceum L. f. ●☆

27854 Anthospermum herbaceum L. f.;草色琥珀树●☆

27855 Anthospermum herbaceum L. f. var. villosicarpum Verdc. = Anthospermum villosicarpum (Verdc.) Puff ●☆

27856 Anthospermum hirsutum DC. = Anthospermum hirtum Cruse ●☆

27857 Anthospermum hirtum Cruse;多毛琥珀树●☆

27858 Anthospermum hispidulum E. Mey. ex Sond.;硬毛琥珀树●☆

27859 Anthospermum holtzii K. Schum. = Psychotria holtzii (K. Schum.) E. M. Petit ●☆

27860 Anthospermum humile N. E. Br. = Anthospermum rigidum Eckl. et Zeyh. subsp. pumilum (Sond.) Puff ●☆

27861 Anthospermum ibityense Puff;伊比提琥珀树●☆

27862 Anthospermum isaloense Homolle ex Puff;伊萨卢琥珀树●☆

27863 Anthospermum keilii K. Krause = Anthospermum usambarense K. Schum. ●☆

27864 Anthospermum lanceolatum Sieber ex Harv. et Sond. = Anthospermum hirtum Cruse ●☆

27865 Anthospermum lanceolatum Thunb. = Anthospermum herbaceum L. f. ●☆

27866 Anthospermum lanceolatum Thunb. var. hedyotideum (Sond.) Kuntze = Anthospermum herbaceum L. f. ●☆

27867 Anthospermum lanceolatum Thunb. var. latifolium Sond. = Anthospermum herbaceum L. f. ●☆

27868 Anthospermum latifolium E. Mey. = Anthospermum lanceolatum Thunb. var. latifolium Sond. ●☆

27869 Anthospermum leuconeuron K. Schum. = Anthospermum usambarense K. Schum. ●☆

27870 Anthospermum lichtensteinii Cruse = Anthospermum ericifolium (Licht. ex Roem. et Schult.) Kuntze ●☆

27871 Anthospermum littoreum L. Bolus;滨海琥珀树●☆

27872 Anthospermum longisepalum Homolle ex Puff;长萼琥珀树●☆

27873 Anthospermum madagascariense Homolle ex Puff;马岛琥珀树●☆

27874 Anthospermum mazzocchi-alemanii Chiov. = Spermacoce subvulgata (K. Schum.) J. G. Garcia ●☆

27875 Anthospermum mildbraedii K. Krause = Anthospermum herbaceum L. f. ●☆

27876 Anthospermum monticola Puff;山地琥珀树●☆

27877 Anthospermum muriculatum A. Rich. = Anthospermum herbaceum L. f. ●☆

27878 Anthospermum nodosum E. Mey. ex Sond. = Anthospermum herbaceum L. f. ●☆

27879 Anthospermum pachyrrhizum Hiern;粗根琥珀树●☆

27880 Anthospermum palustre Homolle ex Puff;沼泽琥珀树●☆

27881 Anthospermum paniculatum Cruse;圆锥琥珀树●☆

27882 Anthospermum paniculatum Cruse var. confertum Eckl. et Zeyh. = Anthospermum paniculatum Cruse ●☆

27883 Anthospermum paniculatum Cruse var. elongatum Eckl. et Zeyh. = Anthospermum paniculatum Cruse ●☆

27884 Anthospermum polyacanthum Baker = Galium polyacanthum (Baker) Puff ■☆

27885 Anthospermum prittwitzii K. Schum. et K. Krause = Anthospermum usambarense K. Schum. ●☆

27886 Anthospermum prostratum Sond.;平卧琥珀树●☆

27887 Anthospermum prostratum Sond. var. glabrum ? = Anthospermum prostratum Sond. ●☆

27888 Anthospermum prostratum Sond. var. velutinum ? = Anthospermum prostratum Sond. ●☆

27889 Anthospermum pumilum Sond. = Anthospermum rigidum Eckl. et Zeyh. subsp. pumilum (Sond.) Puff ●☆

27890 Anthospermum pumilum Sond. subsp. rigidum (Eckl. et Zeyh.) Puff = Anthospermum rigidum Eckl. et Zeyh. ●☆

27891 Anthospermum pumilum Sond. var. pilosum E. Phillips = Anthospermum rigidum Eckl. et Zeyh. subsp. pumilum (Sond.) Puff ●☆

27892 Anthospermum randii S. Moore = Anthospermum ternatum Hiern subsp. randii (S. Moore) Puff ■☆

27893 Anthospermum rigidum Eckl. et Zeyh.;硬琥珀树●☆

27894 Anthospermum rigidum Eckl. et Zeyh. subsp. pumilum (Sond.) Puff;小硬琥珀树●☆

27895 Anthospermum rosmarinus K. Schum.;迷迭香叶琥珀树●☆

27896 Anthospermum rubiaceum Rchb. ex Spreng. = Anthospermum hirtum Cruse ●☆

27897 Anthospermum rubricaule K. Schum. = Anthospermum hispidulum E. Mey. ex Sond. ●☆

27898 Anthospermum scabrum Thunb. = Carpacoce scabra (Thunb.) Sond. ●☆

27899 Anthospermum spathulatum Spreng.;匙形琥珀树●☆

27900 Anthospermum spathulatum Spreng. subsp. ecklonianum (Cruse)

Puff;埃氏琥珀树●☆

27901 Anthospermum spathulatum Spreng. subsp. saxatile Puff;岩栖琥珀树●☆

27902 Anthospermum spathulatum Spreng. subsp. tulbaghense Puff;塔尔巴赫琥珀树●☆

27903 Anthospermum spathulatum Spreng. subsp. uitenhagense Puff;埃滕哈赫琥珀树●☆

27904 Anthospermum spathulatum Spreng. var. ecklonianum (Cruse) Cruse = Anthospermum spathulatum Spreng. subsp. ecklonianum (Cruse) Puff ●☆

27905 Anthospermum spermacoceum Rchb. f. = Carpacoce spermacocea (Rchb. f.) Sond. ●☆

27906 Anthospermum spicatum Suess. = Anthospermum rigidum Eckl. et Zeyh. subsp. pumilum (Sond.) Puff ●☆

27907 Anthospermum streyi Puff;施特赖琥珀树●☆

27908 Anthospermum ternatum Hiern;三出琥珀树●☆

27909 Anthospermum ternatum Hiern subsp. randii (S. Moore) Puff;朗德琥珀树■☆

27910 Anthospermum thymoides Baker;百里香琥珀树●☆

27911 Anthospermum tricostatum Sond. = Anthospermum spathulatum Spreng. ●☆

27912 Anthospermum usambarense K. Schum.;乌桑巴拉琥珀树●☆

27913 Anthospermum uwembae Gilli = Anthospermum welwitschii Hiern ●☆

27914 Anthospermum vallicola S. Moore;河谷琥珀树●☆

27915 Anthospermum villosicarpum (Verdc.) Puff;长毛果琥珀树●☆

27916 Anthospermum welwitschii Hiern;韦尔琥珀树●☆

27917 Anthospermum whyteanum Britten;怀特琥珀树●☆

27918 Anthospermum zimbabwense Puff;津巴布韦琥珀树●☆

27919 Anthostema A. Juss. (1824);雄花大戟属☆

27920 Anthostema aubryanum Baill.;雄花大戟☆

27921 Anthostema senegalense A. Juss.;塞内加尔雄花大戟☆

27922 Anthostema senegalense A. Juss. f. aubrevillei Robow;奥布雄花大戟☆

27923 Anthostyrax Pierre = Styrax L. ●

27924 Anthostyrax tonkinensis Pierre = Styrax tonkinensis (Pierre) Craib ex Hartwich ●

27925 Anthotium R. Br. (1810);澳洲草海桐属■☆

27926 Anthotium humile R. Br.;澳洲草海桐■☆

27927 Anthotroche Endl. (1839);轮花茄属☆

27928 Anthoxanthaceae Link = Gramineae Juss. (保留科名)●■

27929 Anthoxanthaceae Link = Poaceae Barnhart(保留科名)●■

27930 Anthoxanthum L. (1753);黄花茅属(春茅属,黄花草属);Spring Grass, Sweet Vernalgrass, Vernal Grass, Vernalgrass, Vernal-grass ■

27931 Anthoxanthum aethiopicum I. Hedberg;埃塞俄比亚黄花茅■☆

27932 Anthoxanthum alpinum Á. Löve et D. Löve = Anthoxanthum odoratum L. ■

27933 Anthoxanthum alpinum Á. Löve et D. Löve = Anthoxanthum odoratum L. subsp. alpinum (Á. Löve et D. Löve) Tzvelev ■

27934 Anthoxanthum aristatum Boiss.;麦穗黄花草(南欧黄花草);Annual Vernal Grass, Annual Vernalgrass, Annual Vernal-grass ■☆

27935 Anthoxanthum aristatum Boiss. subsp. macranthum Valdés = Anthoxanthum ovatum Lag. subsp. macranthum (Valdés) Rivas Mart. ■☆

27936 Anthoxanthum brevifolium Stapf;短叶黄花茅■☆

27937 Anthoxanthum dregeanum (Nees ex Trin.) Stapf;德雷黄花茅■☆

27938 Anthoxanthum ecklonii (Nees ex Trin.) Stapf;埃氏黄花茅■☆

27939 Anthoxanthum elongatum (Hand.-Mazz.) Veldkamp = Anthoxanthum hookeri (Griseb.) Rendle ■

27940 Anthoxanthum formosanum Honda = Anthoxanthum horsfieldii (Kunth ex Benn.) Mez ex Reeder ■

27941 Anthoxanthum formosanum Honda = Anthoxanthum horsfieldii (Kunth ex Benn.) Mez ex Reeder var. formosanum (Honda) Veldkamp ■

27942 Anthoxanthum glabrum (Trin.) Veldkamp;光稃茅香(光稃香草);Glabrous Sweetgrass ■

27943 Anthoxanthum gracile Biv.;纤细黄花茅■☆

27944 Anthoxanthum gracillimum (Hook. f.) Mez = Anthoxanthum sikkimense (Maxim.) Ohwi ■

27945 Anthoxanthum hookeri (Griseb.) Rendle;藏黄花茅(虎克黄花茅,锡金黄花茅);Hooker Sweet Vernalgrass, Hooker Vernalgrass ■

27946 Anthoxanthum horsfieldii (Kunth ex Benn.) Mez = Anthoxanthum horsfieldii (Kunth ex Benn.) Mez ex Reeder ■

27947 Anthoxanthum horsfieldii (Kunth ex Benn.) Mez ex Reeder;台湾黄花茅(黄花茅);Taiwan Vernalgrass ■

27948 Anthoxanthum horsfieldii (Kunth ex Benn.) Mez ex Reeder var. formosanum (Honda) Veldkamp = Anthoxanthum horsfieldii (Kunth ex Benn.) Mez ex Reeder ■

27949 Anthoxanthum horsfieldii (Kunth ex Benn.) Mez ex Reeder var. japonicum (Maxim.) Veldkamp;日本野黄花茅■☆

27950 Anthoxanthum horsfieldii (Kunth ex Benn.) Mez ex Reeder var. viridescens (Honda) Veldkamp = Anthoxanthum horsfieldii (Kunth ex Benn.) Mez ex Reeder ■

27951 Anthoxanthum horsfieldii (Kunth ex Benn.) Mez var. formosanum (Honda) Veldkamp = Anthoxanthum horsfieldii (Kunth ex Benn.) Mez ex Reeder ■

27952 Anthoxanthum indicum L. = Perotis indica (L.) Kuntze ■

27953 Anthoxanthum japonicum (Maxim.) Hack. ex Matsum. = Anthoxanthum horsfieldii (Kunth ex Benn.) Mez var. japonicum (Maxim.) Veldkamp ■☆

27954 Anthoxanthum japonicum (Maxim.) Hack. ex Matsum. subsp. luzoniense (Merr.) T. Koyama = Anthoxanthum horsfieldii (Kunth ex Benn.) Mez ex Reeder ■

27955 Anthoxanthum japonicum (Maxim.) Hack. ex Matsum. var. sikokianum (Ohwi) Ohwi = Anthoxanthum horsfieldii (Kunth ex Benn.) Mez var. formosanum (Honda) Veldkamp ■

27956 Anthoxanthum japonicum (Maxim.) Hack. ex Matsumura subsp. luzoniense (Merr.) T. Koyama = Anthoxanthum horsfieldii (Kunth ex Benn.) Mez ex Reeder ■

27957 Anthoxanthum latifolium B. S. Sun et S. Wang;宽叶黄花茅;Broadleaf Vernalgrass ■

27958 Anthoxanthum latifolium B. S. Sun et S. Wang = Anthoxanthum hookeri (Griseb.) Rendle ■

27959 Anthoxanthum latifolium B. S. Sun et S. Wang var. purpurascens B. S. Sun et S. Wang;紫穗黄花茅;Purplespike Vernalgrass ■

27960 Anthoxanthum latifolium B. S. Sun et S. Wang var. purpurascens B. S. Sun et S. Wang = Anthoxanthum hookeri (Griseb.) Rendle ■

27961 Anthoxanthum luzoniense Merr. = Anthoxanthum horsfieldii (Kunth ex Benn.) Mez ex Reeder ■

27962 Anthoxanthum madagascariense Stapf;马岛茅香■☆

27963 Anthoxanthum monticola (Bigelow) Veldkamp;高山茅香(高山香草);Alpine Sweet Grass, Alpine Sweetgrass, Alpine Sweet-grass ■

27964 Anthoxanthum nipponicum Honda = Anthoxanthum odoratum L.

subsp. alpinum (Á. Löve et D. Löve) Tzvelev ■

27965 Anthoxanthum nipponicum Honda = Anthoxanthum odoratum L. var. nipponicum (Honda) Tzvelev ■

27966 Anthoxanthum nipponicum Honda var. furumii Honda = Anthoxanthum odoratum L. subsp. alpinum (Á. Löve et D. Löve) Tzvelev ■

27967 Anthoxanthum nitens (Weber) Y. Schouten et Veldkamp;茅香(布氏茅香,毛香,香草,香麻,香茅);Holy Grass,Holy-grass,Seneca Grass,Sweet Grass,Sweetgrass,Sweet-grass,Sweet-scented Grass,Sweet-scented-grass,Vanilla Grass,Vanillagrass,Vanilla-grass ■

27968 Anthoxanthum nivale K. Schum.;雪白茅香■☆

27969 Anthoxanthum odoratum L.;黄花茅(春茅,黄花草,香黄花茅);Spikenard,Spring Grass,Spring-grass,Sweet Vernal Grass,Sweet Vernalgrass,Sweet Vernal-grass ■

27970 Anthoxanthum odoratum L. subsp. alpinum (Á. Löve et D. Löve) Hultén = Anthoxanthum odoratum L. subsp. alpinum (Á. Löve et D. Löve) Tzvelev ■

27971 Anthoxanthum odoratum L. subsp. alpinum (Á. Löve et D. Löve) Tzvelev;高山黄花茅(日本黄花茅);Alpine Vernalgrass,Japan Vernalgrass,Japaneses Vernalgrass ■

27972 Anthoxanthum odoratum L. subsp. furumii (Honda) T. Koyama = Anthoxanthum odoratum L. subsp. alpinum (Á. Löve et D. Löve) Tzvelev ■

27973 Anthoxanthum odoratum L. subsp. furumii (Honda) T. Koyama = Anthoxanthum nipponicum Honda ■

27974 Anthoxanthum odoratum L. subsp. nipponicum (Honda) Tzvelev = Anthoxanthum odoratum L. subsp. alpinum (Á. Löve et D. Löve) Tzvelev ■

27975 Anthoxanthum odoratum L. subsp. nipponicum (Honda) Tzvelev = Anthoxanthum nipponicum Honda ■

27976 Anthoxanthum odoratum L. subsp. ovatum (Lag.) Trab. = Anthoxanthum ovatum Lag. ■☆

27977 Anthoxanthum odoratum L. var. alpinum (Á. Löve et D. Löve) Uechtritz = Anthoxanthum odoratum L. subsp. alpinum (Á. Löve et D. Löve) Tzvelev ■

27978 Anthoxanthum odoratum L. var. alpinum Maxim. et Uechtr. = Anthoxanthum odoratum L. subsp. alpinum (Á. Löve et D. Löve) Tzvelev ■

27979 Anthoxanthum odoratum L. var. ciliatum Emb. = Anthoxanthum ovatum Lag. ■☆

27980 Anthoxanthum odoratum L. var. exsertum (H. Lindb.) Emb. et Maire = Anthoxanthum ovatum Lag. ■☆

27981 Anthoxanthum odoratum L. var. furumii (Honda) Ohwi = Anthoxanthum nipponicum Honda ■

27982 Anthoxanthum odoratum L. var. furumii (Honda) Ohwi = Anthoxanthum odoratum L. subsp. alpinum (Á. Löve et D. Löve) Tzvelev ■

27983 Anthoxanthum odoratum L. var. glabrescens Celak. = Anthoxanthum odoratum L. ■

27984 Anthoxanthum odoratum L. var. nipponicum (Honda) Tzvelev = Anthoxanthum nipponicum Honda ■

27985 Anthoxanthum odoratum L. var. nipponicum (Honda) Tzvelev ex Y. H. Sun et P. C. Kuo = Anthoxanthum nipponicum Honda ■

27986 Anthoxanthum odoratum L. var. nipponicum (Honda) Tzvelev ex Y. H. Sun et P. C. Kuo = Anthoxanthum odoratum L. subsp. alpinum (Á. Löve et D. Löve) Tzvelev ■

27987 Anthoxanthum odoratum L. var. nipponicum (Honda) Tzvelev. = Anthoxanthum odoratum L. subsp. alpinum (Á. Löve et D. Löve) Tzvelev ■

27988 Anthoxanthum odoratum L. var. ovatum (Lag.) Coss. et Durieu = Anthoxanthum ovatum Lag. ■☆

27989 Anthoxanthum odoratum L. var. puelii Coss. et Durieu = Anthoxanthum odoratum L. ■

27990 Anthoxanthum odoratum L. var. scabrum Emb. = Anthoxanthum ovatum Lag. ■☆

27991 Anthoxanthum odoratum L. var. senenii A. Camus = Anthoxanthum odoratum L. ■

27992 Anthoxanthum odoratum L. var. strictum Asch. et Graebn. = Anthoxanthum odoratum L. ■

27993 Anthoxanthum odoratum L. var. villosum Loisel. = Anthoxanthum odoratum L. ■

27994 Anthoxanthum odoratum L. var. vulgare Coss. et Durieu = Anthoxanthum odoratum L. ■

27995 Anthoxanthum ovatum Lag.;卵形黄花茅■☆

27996 Anthoxanthum ovatum Lag. subsp. macranthum (Valdés) Rivas Mart.;大花卵形黄花茅■☆

27997 Anthoxanthum ovatum Lag. var. ciliatum Emb. = Anthoxanthum ovatum Lag. ■☆

27998 Anthoxanthum ovatum Lag. var. exsertum H. Lindb. = Anthoxanthum ovatum Lag. ■☆

27999 Anthoxanthum ovatum Lag. var. scabrum Emb. = Anthoxanthum ovatum Lag. ■☆

28000 Anthoxanthum ovatum Lag. var. sennenii A. Camus = Anthoxanthum ovatum Lag. ■☆

28001 Anthoxanthum pallidum (Hand.-Mazz.) Keng;小黄花茅(淡色黄花茅);Pallid Vernalgrass,Small Vernalgrass ■

28002 Anthoxanthum potaninii (Tzvelev) S. M. Phillips et Z. L. Wu;松序茅香草■

28003 Anthoxanthum puelii Lecoq et Lamotte = Anthoxanthum aristatum Boiss. ■☆

28004 Anthoxanthum scaposum Peter = Anthoxanthum nivale K. Schum. ■☆

28005 Anthoxanthum sikkimense (Maxim.) Ohwi;锡金黄花茅;Sikkim Vernalgrass ■

28006 Anthoxanthum tibeticum (Bor) Veldkamp;藏茅香■

28007 Anthoxanthum tongo (Nees ex Trin.) Stapf;细黄花茅■☆

28008 Anthoxanthum viridescens Honda = Anthoxanthum horsfieldii (Kunth ex Benn.) Mez ex Reeder ■

28009 Anthriscus (Pers.) Hoffm. = Anthriscus Pers.(保留属名)■

28010 Anthriscus Bernh.(废弃属名)= Anthriscus Pers.(保留属名)■

28011 Anthriscus Bernh.(废弃属名)= Torilis Adans. ■

28012 Anthriscus Hoffm. = Anthriscus Pers.(保留属名)■

28013 Anthriscus Pers.(1805)(保留属名);峨参属;Beakchervil,Beaked Chervil,Chervil,Cow Parsley ■

28014 Anthriscus Raf. = Anthriscus Pers.(保留属名)■

28015 Anthriscus aemula (Woronow) Schischk. = Anthriscus nemorosa (M. Bieb.) Spreng. ■

28016 Anthriscus aemula (Woronow) Schischk. f. hirtifructus (Ohwi) Kitag. = Anthriscus nemorosa (M. Bieb.) Spreng. ■

28017 Anthriscus aemula (Woronow) Schischk. var. hirtifructus (Ohwi) Kitag. = Anthriscus nemorosa (M. Bieb.) Spreng. ■

28018 Anthriscus aemura (Woronow) Schischk. = Anthriscus sylvestris (L.) Hoffm. ■

28019 Anthriscus aemura (Woronow) Schischk. var. hirtifructa (Ohwi)

Kitag.;东北峨参■

28020 Anthriscus africana Hook. f. = Cryptotaenia africana (Hook. f.) Drude ■☆

28021 Anthriscus boissieui H. Lév. = Chaerophyllum villosum Wall. ex DC. ■

28022 Anthriscus capensis Spreng. = Sonderina hispida (Thunb.) H. Wolff ■☆

28023 Anthriscus caucalis M. Bieb.;喙峨参;Beaked Parsley, Bur Beaked Chervil, Bur Parsley, Bur-beaked Chervil, Burr Chervil, Chervil, Rough Chervil ■☆

28024 Anthriscus cerefolius (L.) Hoffm.;蜡叶峨参;Beaked Parsley, Chervell, Chervil, Chevorell, Garden Chervil, Leaf Chervil, Salad Chervil, Salad-chervil ■☆

28025 Anthriscus cerefolius Hoffm. = Anthriscus cerefolius (L.) Hoffm. ■☆

28026 Anthriscus dissectus C. H. Wright = Peucedanum kerstenii Engl. ■☆

28027 Anthriscus glacialis Lipsky;冰雪峨参■☆

28028 Anthriscus keniensis H. Wolff = Anthriscus sylvestris (L.) Hoffm. ■

28029 Anthriscus longirostris Bertol.;长嘴峨参;Longbeak Chervil ■☆

28030 Anthriscus nemorosa (M. Bieb.) Spreng.;刺果峨参(东北峨参,峨参,胡萝卜缨子,林地峨参);Spinefruit Beakchervil, Spinefruit Chervil ■

28031 Anthriscus nemorosa (M. Bieb.) Spreng. var. glabriuscula Nasir = Anthriscus nemorosa (M. Bieb.) Spreng. ■

28032 Anthriscus nemorosa (M. Bieb.) Spreng. var. hirtifructa Ohwi = Anthriscus aemura (Woronow) Schischk. var. hirtifructa (Ohwi) Kitag. ■

28033 Anthriscus nemorosa (M. Bieb.) Spreng. var. hirtifructus Ohwi = Anthriscus nemorosa (M. Bieb.) Spreng. ■

28034 Anthriscus nemorosa (M. Bieb.) Spreng. var. hirtifructus Ohwi = Anthriscus sylvestris (L.) Hoffm. f. hirtifructus (Ohwi) H. Ohba ■

28035 Anthriscus nitida Garcke;光亮峨参■☆

28036 Anthriscus prescottii DC. = Chaerophyllum prescottii DC. ■

28037 Anthriscus ruprechtii Boiss.;卢氏峨参■☆

28038 Anthriscus scabra Koso-Pol. = Torilis scabra (Thunb.) DC. ■

28039 Anthriscus scandicina (Weber) Mansf.;欧亚峨参■☆

28040 Anthriscus scandicina (Weber) Mansf. = Anthriscus caucalis M. Bieb. ■☆

28041 Anthriscus schmalhausenii (Albov) Schischk.;施马峨参■☆

28042 Anthriscus sosnovskyi Schischk.;索思峨参■☆

28043 Anthriscus stocksiana Koso-Pol. = Torilis nodosa (L.) Gaertn. ■☆

28044 Anthriscus sylvestris (L.) Hoffm.;峨参(东北峨参,见肿消,金山田七,萝卜七,山胡萝卜缨子,田七,土当归,土田七,小叶山水芹);Adder's Meat, Ass-parsley, Bad Man's Oatmeal, Beaked Parsley, Bishop's Beard, Break-your-mother's-heart, Bun, Cashes, Caxes, Chervil, Cicely, Coney Parsley, Coney's Parsley, Cow Chervil, Cow Mumble, Cow Parsley, Cow-parsley, Cow-weed, Da-Ho, Dead Man's Flesh, Devil's Meal, Devil's Meat, Devil's Parsley, Dog Parsley, Dog's Carvi, Dog's Parsley, Eldrot, Eltrot, Gypsy Curtains, Gypsy Flower, Gypsy Laces, Gypsy's Parsley, Gypsy's Umbrella, Ha-Ho, Hare's Parsley, Hemlock, Hi-How, Honiton Lace, Humlock, June-flower, Kadle Dock, Keck, Kedlock, Keeshion, Kelk, Kellock, Kesk, Kettle Dock, Kettle-dock, Kewsies, Kex, Kill-your-mother-quick, Lady's Lace, Lady's Needlework, Mayweed, Mock Chervil, Moonlight, My Lady's Lace, Naughty Man's Oatmeal, Naughty Man's Parsley, Old Lady's Lace, Oldrot, Orchard Weed, Queen Anne's Lace, Queen Anne's Lace Handkerchief, Rabbit-meat, Rat's Bane, Satan's Bread, Scab Flower, Scabby Hands, Scabby Heads, Sheep's Parsley, Stepmother's Blessing, Sweet Ash, White Meat, Whiteweed, Wild Carraway, Wild Chervil, Wild Cicely, Wild Parsley, Woodland Beakchervil, Woodland Beaked Chervil ■

28045 Anthriscus sylvestris (L.) Hoffm. 'Meal';戴维尔峨参;Devil's Meat ■☆

28046 Anthriscus sylvestris (L.) Hoffm. f. hirtifructus (Ohwi) H. Ohba;毛果峨参■

28047 Anthriscus sylvestris (L.) Hoffm. f. hirtifructus (Ohwi) H. Ohba = Anthriscus nemorosa (M. Bieb.) Spreng. ■

28048 Anthriscus sylvestris (L.) Hoffm. subsp. aemula (Woronow) Kitam = Anthriscus sylvestris (L.) Hoffm. ■

28049 Anthriscus sylvestris (L.) Hoffm. subsp. aemula (Woronow) Kitam. var. hirtifructus (Ohwi) H. Hara = Anthriscus nemorosa (M. Bieb.) Spreng. ■

28050 Anthriscus sylvestris (L.) Hoffm. subsp. aemula (Woronow) Kitam. var. hirtifructus (Ohwi) H. Hara = Anthriscus sylvestris (L.) Hoffm. f. hirtifructus (Ohwi) H. Ohba ■

28051 Anthriscus sylvestris (L.) Hoffm. subsp. leiocarpus ? = Anthriscus sylvestris (L.) Hoffm. ■

28052 Anthriscus sylvestris (L.) Hoffm. subsp. nemorosa (M. Bieb.) C. Y. Wu et F. T. Pu = Anthriscus sylvestris (L.) Hoffm. ■

28053 Anthriscus sylvestris (L.) Hoffm. var. aemula Woronow = Anthriscus nemorosa (M. Bieb.) Spreng. ■

28054 Anthriscus sylvestris (L.) Hoffm. var. aemula Woronow = Anthriscus sylvestris (L.) Hoffm. ■

28055 Anthriscus sylvestris (L.) Hoffm. var. glabricaulis (Maire) Maire = Anthriscus sylvestris (L.) Hoffm. ■

28056 Anthriscus sylvestris (L.) Hoffm. var. hirtifructus ? = Anthriscus sylvestris (L.) Hoffm. f. hirtifructus (Ohwi) H. Ohba ■

28057 Anthriscus sylvestris (L.) Hoffm. var. mollis (Boiss. et Reut.) Batt. = Anthriscus sylvestris (L.) Hoffm. ■

28058 Anthriscus sylvestris (L.) Hoffm. var. nemonrosa Trautv. = Anthriscus nemorosa (M. Bieb.) Spreng. ■

28059 Anthriscus sylvestris (L.) Hoffm. var. nemorosa (M. Bieb.) Trautv. = Anthriscus nemorosa (M. Bieb.) Spreng. ■

28060 Anthriscus sylvestris (L.) Hoffm. var. villicaulis Maire = Anthriscus sylvestris (L.) Hoffm. ■

28061 Anthriscus velutina Sommier et H. Lév.;短绒毛峨参■☆

28062 Anthriscus vulgaris Bernh. = Torilis japonica (Houtt.) DC. ■

28063 Anthriscus vulgaris Pers. = Anthriscus caucalis M. Bieb. ■☆

28064 Anthriscus vulgaris Pers. var. baeticus Pau = Anthriscus caucalis M. Bieb. ■☆

28065 Anthriscus vulgaris Pers. var. crassus Maire et Weiller = Anthriscus caucalis M. Bieb. ■☆

28066 Anthriscus yunnanensis W. W. Sm.;滇峨参;Smith Beakchervil ■

28067 Anthriscus yunnanensis W. W. Sm. = Anthriscus sylvestris (L.) Hoffm. ■

28068 Anthrocephalus Schltdl. = Anthocephalus A. Rich. ●☆

28069 Anthropodium Sims = Arthropodium R. Br. ■☆

28070 Anthrostylis D. Dietr. = Arthrostylis R. Br. ■☆

28071 Anthurium Schott(1829);花烛属(安祖花属,红掌属,火鹤花属);Anthurium, Flamingo Flower, Flamingo Plant, Garishcandle, Tail Flower, Tailflower ■

28072 Anthurium 'Lady Jane';珍女士花烛■☆

28073 Anthurium 'Laura';劳拉花烛■☆

28074 Anthurium 'Lucifer';露氏花烛■☆
28075 Anthurium 'Renaissance';观叶花烛■☆
28076 Anthurium 'Sweetheart Red';甜心花烛■☆
28077 Anthurium 'Tropical';热带花烛■☆
28078 Anthurium 'Valentino';华伦天奴花烛■☆
28079 Anthurium amnicola Dressier;巴拿马花烛■☆
28080 Anthurium amplum Kunth = Anthurium hookeri Kunth ■☆
28081 Anthurium andraeanum André = Anthurium andraeanum Linden ex André ■☆
28082 Anthurium andraeanum Linden 'Wataru Kimura';木村亘花烛■☆
28083 Anthurium andraeanum Linden = Anthurium andraeanum Linden ex André ■☆
28084 Anthurium andraeanum Linden ex André;花烛(哥伦比亚安祖花,哥伦比亚花烛,红掌,火鹤花,烛台花);Flamingo Flower, Flamingo Lily, Painter's Palette, Palette Flower, Tail Flower, Tailflower ■☆
28085 Anthurium andraeanum Linden var. grandiflorum L. Linden et Rodigas;大花花烛■☆
28086 Anthurium armeniense Croat;危地马拉花烛■☆
28087 Anthurium bakeri Hook.;巴氏花烛■☆
28088 Anthurium clarinervium Matuda;圆叶花烛■☆
28089 Anthurium clavigerum Poepp. et Endl.;盔花烛■☆
28090 Anthurium crassinervium (Jacq.) Schott = Anthurium crassivenium Engl. ■☆
28091 Anthurium crassivenium Engl.;粗脉花烛;Pheasant's Tail, Thick-nerve Anthurium ■☆
28092 Anthurium crystallinum Linden = Anthurium crystallinum Linden et André ■☆
28093 Anthurium crystallinum Linden et André;水晶花烛(晶状安祖花,水晶安祖花);Crystal Anthurium, Strap Flower ■☆
28094 Anthurium cultorum Birdsey;大花烛;Large Flamingo Flower ■☆
28095 Anthurium digitatum (Jacq.) G. Don;指状花烛;Fingered Anthurium ■☆
28096 Anthurium dussii Engl.;杜斯花烛■☆
28097 Anthurium egregium Schott = Anthurium crassinervium (Jacq.) Schott ■☆
28098 Anthurium ferrierense Bergman;直穗花烛■☆
28099 Anthurium floribundum Linden et André = Spathiphyllum floribundum (Lindl. et André) N. E. Br. ■☆
28100 Anthurium fontanesii Schott = Anthurium crassivenium Engl. ■☆
28101 Anthurium fortunatum Bunting = Anthurium pedatum Endl. ex Kunth ■☆
28102 Anthurium harrisii (Graham) G. Don;巴西花烛■☆
28103 Anthurium hookeri Kunth;胡克花烛■☆
28104 Anthurium huegelii Schott = Anthurium hookeri Kunth ■☆
28105 Anthurium insigne Mast.;美丽花烛;Pretty Anthurium ■☆
28106 Anthurium jimenezii Matuda;希门花烛■☆
28107 Anthurium longipetiolatum Engl.;长叶柄花烛■☆
28108 Anthurium magnificum Linden;绒叶花烛(大水晶花烛);Woolly-leaf Anthurium ■☆
28109 Anthurium mariae Croat et Lingán;美苞花烛■☆
28110 Anthurium mexicanum Engl. = Anthurium schlechtendalii Kunth ■☆
28111 Anthurium miquelianum C. Koch et August;蔓花烛■☆
28112 Anthurium nymphifolium C. Koch et C. D. Bouché;睡莲叶花烛■☆
28113 Anthurium pedatoradiatum Schott;掌叶花烛(趾叶花烛);Footleaf Anthurium, Palmateleaf Garishcandle ■
28114 Anthurium pedatum Endl. ex Kunth;鸟足花烛■☆
28115 Anthurium pendulifolium N. E. Br.;热美花烛■☆
28116 Anthurium preussii Engl. = Anthurium crassivenium Engl. ■☆
28117 Anthurium rugosum Schott = Anthurium crassivenium Engl. ■☆
28118 Anthurium scandens (Aubl.) Engl.;攀缘花烛(蔓花烛);Pearl Anthurium, Pearl Laceleaf ■☆
28119 Anthurium scherzerianum Schott;火鹤花(安祖花,红苞芋,红鹤芋,花烛);Common Anthurium, Flame Plant, Flamingo Flower, Flamingo Plant, Painter's Palette, Pigtail Anthurium, Pigtail Plant, Tail Plant, Tailflower ■☆
28120 Anthurium scherzerianum Schott 'Rothschildianum';罗氏火鹤花■☆
28121 Anthurium scherzerianum Schott var. albistriatum Engl.;白条火鹤花■☆
28122 Anthurium scherzerianum Schott var. album hort;白苞火鹤花■☆
28123 Anthurium scherzerianum Schott var. atrosanguineum Engl.;紫苞火鹤花■☆
28124 Anthurium scherzerianum Schott var. flvescens Hort.;黄苞火鹤花■☆
28125 Anthurium scherzerianum Schott var. nebulosum Devansaye;星云火鹤花■☆
28126 Anthurium scherzerianum Schott var. pygmeum Hort.;矮小火鹤花■☆
28127 Anthurium scherzerianum Schott var. rotundispathaceum Engl. = Anthurium scherzerianum Schott 'Rothschildianum' ■☆
28128 Anthurium scherzerianum Schott var. viridescens Engl.;浅绿火鹤花■☆
28129 Anthurium scherzerianum Schott var. wardianum Veitch;瓦尔火鹤花■☆
28130 Anthurium schlechtendalii Kunth;谢尔花烛■☆
28131 Anthurium schlechtendalii Kunth subsp. jimenezii (Matuda) Croat = Anthurium jimenezii Matuda ■☆
28132 Anthurium signatum K. Koch et L. Mathieu;刻文花烛■☆
28133 Anthurium spathiphyllum N. E. Br.;匙叶花烛■☆
28134 Anthurium subsignatum Schott;飞鸢花烛■☆
28135 Anthurium tetragonum Hook. ex Schott = Anthurium schlechtendalii Kunth ■☆
28136 Anthurium undulatum K. Koch et C. D. Bouché;波纹花烛■☆
28137 Anthurium variabile Kunth;深裂花烛(裂叶花烛);Dissected Garishcandle ■
28138 Anthurium veitchii Mast.;维氏花烛(美居花烛,维奇安祖花,皱叶花烛);King Anthurium, Veitch's Anthurium ■☆
28139 Anthurium wallisii Mast.;瓦利斯花烛■☆
28140 Anthurium warocqueanum Moore;长叶花烛;Queen Anthurium ■☆
28141 Anthyllis Adans. = Polycarpon L. + Polycarpaea Lam.(保留属名)●■
28142 Anthyllis Adans. = Polycarpon L. ■
28143 Anthyllis L. (1753);绒毛花属(妇指豆属,岩豆属);Anthyllis, Kidney Vetch ■☆
28144 Anthyllis abyssinica (Sagorski) Becker = Anthyllis vulneraria L. subsp. abyssinica (Sagorski) Cullen ■☆
28145 Anthyllis abyssinica Sagorski = Anthyllis vulneraria L. subsp. abyssinica (Sagorski) Cullen ■☆
28146 Anthyllis affinis Britting. ex Koch;近缘绒毛花■☆
28147 Anthyllis ajmasiana (Pau) Rothm. = Anthyllis vulneraria L. subsp. ajmasiana (Pau) Raynaud et Sauvage ■☆
28148 Anthyllis alpestris Rchb.;高山绒毛花■☆

28149 Anthyllis arenaria (Rupr.) Juz.;沙地绒毛花■☆

28150 Anthyllis aspalathoides L. = Aspalathus aspalathoides (L.) R. Dahlgren ●☆

28151 Anthyllis barba-jovis L.;巴尔绒毛花;Jupiter's-beard ■☆

28152 Anthyllis bicolor Bertol. ex Colla;二色绒毛花■☆

28153 Anthyllis bidentata Munby = Genista cephalantha Spach ●☆

28154 Anthyllis caucasica (Grossh.) Juz.;高加索绒毛花■☆

28155 Anthyllis chilensis Ser.;智利绒毛花■☆

28156 Anthyllis coccinea (L.) Beck.;绯红绒毛花■☆

28157 Anthyllis colorata Juz.;着色绒毛花■☆

28158 Anthyllis cornicina L. = Hymenocarpos cornicinus (L.) Vis. ■☆

28159 Anthyllis cuneata Dum. Cours. = Lespedeza cuneata (Dum. Cours.) G. Don ●■

28160 Anthyllis cuneata Dum. Cours. = Lespedeza juncea (L. f.) Pers. var. sericea (Thunb.) Lace et Hemsl. ●

28161 Anthyllis cytisoides L.;金雀绒毛花■☆

28162 Anthyllis cytisoides L. var. laxiflora Sennen et Mauricio = Anthyllis cytisoides L. ■☆

28163 Anthyllis dillenii Schult. ex Loudon = Anthyllis vulneraria L. subsp. abyssinica (Sagorski) Cullen ■☆

28164 Anthyllis ensifolia Houtt. = Psoralea ensifolia (Houtt.) Merr. ■☆

28165 Anthyllis erinacea L. = Erinacea anthyllis Link ●☆

28166 Anthyllis gerardii L. = Dorycnopsis gerardii (L.) Boiss. ■☆

28167 Anthyllis hamosa Desf. = Hymenocarpos hamosus (Desf.) Vis. ■☆

28168 Anthyllis henoniana Coss.;埃农绒毛花■☆

28169 Anthyllis henoniana Coss. var. nalutensis Andr. = Anthyllis henoniana Coss. ■☆

28170 Anthyllis hermanniae L.;黄绒毛花(黄花岩豆)■☆

28171 Anthyllis lachnophora Juz.;绵绒毛花■☆

28172 Anthyllis lagascana Benedi;拉加绒毛花■☆

28173 Anthyllis lemanniana Lowe;赖氏绒毛花■☆

28174 Anthyllis lotoides L. = Hymenocarpos lotoides (L.) Vis. ■☆

28175 Anthyllis maritima Schweiger;沼泽绒毛花■☆

28176 Anthyllis maura Beck = Anthyllis vulneraria L. subsp. maura (Beck) Maire ■☆

28177 Anthyllis maura Beck var. ajmasiana Pau = Anthyllis vulneraria L. subsp. ajmasiana (Pau) Raynaud et Sauvage ■☆

28178 Anthyllis maura Beck var. albicans Sagorski = Anthyllis vulneraria L. subsp. maura (Beck) Maire ■☆

28179 Anthyllis montana L.;山地绒毛花(阿尔卑斯岩豆);Alp Anthyllis, Alps Anthyllis ■☆

28180 Anthyllis montana L. 'Rubra';红花阿尔卑斯岩豆■☆

28181 Anthyllis montana L. var. algerica Maire = Anthyllis montana L. ■☆

28182 Anthyllis multicaulis Pau;多茎绒毛花■☆

28183 Anthyllis polycephala Desf.;多头绒毛花■☆

28184 Anthyllis polycephala Desf. var. fontanesii Maire = Anthyllis polycephala Desf. ■☆

28185 Anthyllis polycephala Desf. var. gomarica Emb. et Maire = Anthyllis polycephala Desf. ■☆

28186 Anthyllis polycephala Desf. var. megalatlantica Emb. = Anthyllis polycephala Desf. ■☆

28187 Anthyllis polycephala Desf. var. mesatlantica Maire = Anthyllis polycephala Desf. ■☆

28188 Anthyllis polycephala Desf. var. podocephala (Boiss.) Pau et Font Quer = Anthyllis polycephala Desf. ■☆

28189 Anthyllis polyphylla Kit. ex DC.;多叶绒毛花■☆

28190 Anthyllis sericea Lag. = Anthyllis lagascana Benedi ■☆

28191 Anthyllis sericea Lag. subsp. henoniana (Coss.) Maire = Anthyllis henoniana Coss. ■☆

28192 Anthyllis sericea Lag. var. massaesyla (Riencourt) Maire = Anthyllis lagascana Benedi ■☆

28193 Anthyllis subsimplex Pomel = Anthyllis henoniana Coss. ■☆

28194 Anthyllis taurica Juz.;克里木绒毛花■☆

28195 Anthyllis terniflora (Lag.) Pau;顶花绒毛花■☆

28196 Anthyllis tetraphylla L.;四叶绒毛花(绒毛花);Fourleaf Kidney Vetch ■☆

28197 Anthyllis tetraphylla L. = Tripodion tetraphyllum (L.) Fourr. ■☆

28198 Anthyllis tetraphylla L. var. purpurea Maire et Wilczek = Tripodion tetraphyllum (L.) Fourr. ■☆

28199 Anthyllis tragacanthoides Desf. = Astragalus armatus Willd. ■☆

28200 Anthyllis vulneraria L.;疗伤绒毛花(野蚕豆);Butter Fingers, Buttered Fingers, Cat's Claws, Cheese-cake, Common Kidneyvetch, Common Kidney-vetch, Crae-nels, Dog's Paise, Double Fingers-and-thumbs, Double Lady's Fingers-and-thumbs, Double Pincushion, Fingers-and-thumbs, Fingers-and-toes, God Aimighty's Fingers-and-thumbs, Granfer-grizzle, Hen-and-chickens, Jupiter's Beard, Kidney Vetch, Kidney Vetch Anthyllis, Lady's Cushion, Lady's Double, Lady's Fingers, Lady's Pincushion, Lady's Slipper, Lamb's Foot, Lamb's Tails, Lamb's Toe, Luck, Pincushion, Silver Bush, Silver Cushion, Staunch, Twins, Woundwort, Yellow Crow Foot, Yellow Crow's Foot, Yellow Fingers-and-thumbs ■☆

28201 Anthyllis vulneraria L. subsp. abyssinica (Sagorski) Cullen;阿比西尼亚疗伤绒毛花■☆

28202 Anthyllis vulneraria L. subsp. ajmasiana (Pau) Raynaud et Sauvage;艾马斯疗伤绒毛花■☆

28203 Anthyllis vulneraria L. subsp. atlantis Emb. et Maire = Anthyllis vulneraria L. subsp. pseudoarundana H. Lindb. ■☆

28204 Anthyllis vulneraria L. subsp. fatmae Font Quer;法蒂玛疗伤绒毛花■☆

28205 Anthyllis vulneraria L. subsp. fruticans Emb.;灌木状疗伤绒毛花■☆

28206 Anthyllis vulneraria L. subsp. gandogeri (Sagorski) Maire;冈多疗伤绒毛花■☆

28207 Anthyllis vulneraria L. subsp. maura (Beck) Maire;暗淡疗伤绒毛花■☆

28208 Anthyllis vulneraria L. subsp. occidentalis (Rothm.) Sauvage;西方疗伤绒毛花■☆

28209 Anthyllis vulneraria L. subsp. pseudoarundana H. Lindb.;假芦苇疗伤绒毛花■☆

28210 Anthyllis vulneraria L. subsp. reuteri Cullen;路透疗伤绒毛花■☆

28211 Anthyllis vulneraria L. subsp. rifana (Emb. et Maire) Cullen;里夫绒毛花■☆

28212 Anthyllis vulneraria L. subsp. saharae (Sagorski) Jahand. et Maire;左原疗伤绒毛花■☆

28213 Anthyllis vulneraria L. subsp. stenophylloides Cullen;窄叶疗伤绒毛花■☆

28214 Anthyllis vulneraria L. var. ajmasiana (Pau) Maire = Anthyllis vulneraria L. subsp. ajmasiana (Pau) Raynaud et Sauvage ■☆

28215 Anthyllis vulneraria L. var. albiflora Maire = Anthyllis vulneraria L. subsp. maura (Beck) Maire ■☆

28216 Anthyllis vulneraria L. var. angustifolia Maire et Wilczek = Anthyllis vulneraria L. subsp. gandogeri (Sagorski) Maire ■☆

28217 Anthyllis vulneraria L. var. antiatlantica Emb. et Maire = Anthyllis vulneraria L. subsp. saharae (Sagorski) Jahand. et Maire ■☆

28218 Anthyllis vulneraria L. var. arenicola Pau = Anthyllis vulneraria L. subsp. maura (Beck) Maire ■☆
28219 Anthyllis vulneraria L. var. dyris Maire = Anthyllis vulneraria L. subsp. saharae (Sagorski) Jahand. et Maire ■☆
28220 Anthyllis vulneraria L. var. flaviflora Guss. = Anthyllis vulneraria L. subsp. maura (Beck) Maire ■☆
28221 Anthyllis vulneraria L. var. font-queri (Rothm.) Cullen = Anthyllis vulneraria L. subsp. gandogeri (Sagorski) Maire ■☆
28222 Anthyllis vulneraria L. var. fruticans (Emb.) Maire = Anthyllis vulneraria L. subsp. fruticans Emb. ■☆
28223 Anthyllis vulneraria L. var. hirsutissima Guss. = Anthyllis vulneraria L. subsp. maura (Beck) Maire ■☆
28224 Anthyllis vulneraria L. var. hosmarensis (Pau) Maire = Anthyllis vulneraria L. subsp. maura (Beck) Maire ■☆
28225 Anthyllis vulneraria L. var. litoralis Maire et Weiller = Anthyllis vulneraria L. subsp. maura (Beck) Maire ■☆
28226 Anthyllis vulneraria L. var. maura (Beck) Maire = Anthyllis vulneraria L. subsp. maura (Beck) Maire ■☆
28227 Anthyllis vulneraria L. var. megaphylla Pau = Anthyllis vulneraria L. subsp. maura (Beck) Maire ■☆
28228 Anthyllis vulneraria L. var. mesatlantica Maire = Anthyllis vulneraria L. subsp. maura (Beck) Maire ■☆
28229 Anthyllis vulneraria L. var. mixta Maire = Anthyllis vulneraria L. subsp. maura (Beck) Maire ■☆
28230 Anthyllis vulneraria L. var. mogadorensis Maire = Anthyllis vulneraria L. subsp. maura (Beck) Maire ■☆
28231 Anthyllis vulneraria L. var. ochroleuca Maire = Anthyllis vulneraria L. subsp. saharae (Sagorski) Jahand. et Maire ■☆
28232 Anthyllis vulneraria L. var. patulivilla Faure et Maire = Anthyllis vulneraria L. subsp. saharae (Sagorski) Jahand. et Maire ■☆
28233 Anthyllis vulneraria L. var. rifana Emb. et Maire = Anthyllis vulneraria L. subsp. rifana (Emb. et Maire) Cullen ■☆
28234 Anthyllis vulneraria L. var. rubriflora Ser. = Anthyllis vulneraria L. subsp. maura (Beck) Maire ■☆
28235 Anthyllis vulneraria L. var. sericea Maire = Anthyllis vulneraria L. subsp. saharae (Sagorski) Jahand. et Maire ■☆
28236 Anthyllis vulneraria L. var. soloitama Maire et al. = Anthyllis vulneraria L. subsp. maura (Beck) Maire ■☆
28237 Anthyllis vulneraria L. var. tangerina (Pau) Maire = Anthyllis vulneraria L. subsp. maura (Beck) Maire ■☆
28238 Anthyllis vulneraria L. var. variegata Maire = Anthyllis vulneraria L. subsp. saharae (Sagorski) Jahand. et Maire ■☆
28239 Antia O. F. Cook = Coccothrinax Sarg. ●☆
28240 Antia O. F. Cook(1941);古巴棕属●☆
28241 Antia crinita (Becc.) O. F. Cook;古巴棕●☆
28242 Antiaris Lesch. (1810)(保留属名);见血封喉属(箭毒木属);Antiaris ●
28243 Antiaris africana Engl.;非洲毒箭木(非洲见血封喉,非洲箭毒木);Africa Antiaris, African Antiaris, Bark Cloth Tree, Bark-cloth Tree, Cloth-tree, False Iroko ●
28244 Antiaris africana Engl. = Antiaris toxicaria Lesch. var. africana Scott-Elliot ex A. Chev. ●◇
28245 Antiaris africana Engl. = Antiaris toxicaria Lesch. ●◇
28246 Antiaris innoxia Blume = Antiaris toxicaria (Pers.) Lesch. ●◇
28247 Antiaris kerstingii Engl. = Antiaris toxicaria Lesch. var. africana Scott-Elliot ex A. Chev. ●◇
28248 Antiaris toxicaria (Pers.) Lesch.;见血封喉(大药树,毒箭木,加布,加毒,箭毒木);Ako, Common Antiaris, East Africa Antiaris, False Iroko, False Oroko, Ipoh, Malay Arrow-poison, Sack Tree, Upas, Upas Tree, Upas-tree ●◇
28249 Antiaris toxicaria Lesch. = Antiaris toxicaria (Pers.) Lesch. ●◇
28250 Antiaris toxicaria Lesch. subsp. africana (Engl.) C. C. Berg = Antiaris toxicaria Lesch. var. africana Scott-Elliot ex A. Chev. ●◇
28251 Antiaris toxicaria Lesch. subsp. welwitschii (Engl.) C. C. Berg;韦氏见血封喉●☆
28252 Antiaris toxicaria Lesch. var. africana Scott-Elliot ex A. Chev.;非洲见血封喉●◇
28253 Antiaris toxicaria Lesch. var. usambarensis (Engl.) C. C. Berg;乌桑巴拉见血封喉●☆
28254 Antiaris toxicaria Lesch. var. welwitschii (Engl.) C. C. Berg = Antiaris toxicaria Lesch. subsp. welwitschii (Engl.) C. C. Berg ●☆
28255 Antiaris toxicaria Lesch. var. welwitschii (Engl.) Corner = Antiaris toxicaria Lesch. subsp. welwitschii (Engl.) C. C. Berg ●☆
28256 Antiaris usambarensis Engl. = Antiaris toxicaria Lesch. var. usambarensis (Engl.) C. C. Berg ●☆
28257 Antiaris welwitschii Engl.;刚果见血封喉(魏氏箭毒木);Welwitsch Antiaris ●☆
28258 Antiaris welwitschii Engl. = Antiaris toxicaria Lesch. var. welwitschii (Engl.) C. C. Berg ●☆
28259 Antiaropsis K. Schum. (1889);类见血封喉属●☆
28260 Antiaropsis decipiens K. Schum.;类见血封喉●☆
28261 Antiaropsis uniflora C. C. Berg;单花类见血封喉●☆
28262 Anticharis Endl. (1839);劣玄参属●■☆
28263 Anticharis arabica Endl.;阿拉伯劣玄参■☆
28264 Anticharis arabica Hochst. ex Benth. = Anticharis senegalensis (Walp.) Bhandari ■☆
28265 Anticharis aschersoniana Schinz = Anticharis senegalensis (Walp.) Bhandari ■☆
28266 Anticharis dielsiana Pilg. = Anticharis scoparia (E. Mey. ex Benth.) Hiern ex Benth. et Hook. f. ■☆
28267 Anticharis ebracteata Schinz;无苞劣玄参■☆
28268 Anticharis genistoides Schinz = Anticharis juncea L. Bolus ■☆
28269 Anticharis glandulosa Asch.;具腺劣玄参■☆
28270 Anticharis glandulosa Asch. var. intermedia A. Terracc. = Anticharis arabica Endl. ■☆
28271 Anticharis imbricata Schinz;覆瓦劣玄参■☆
28272 Anticharis inflata Marloth et Engl.;膨胀劣玄参■☆
28273 Anticharis juncea L. Bolus;灯心草劣玄参■☆
28274 Anticharis linearis (Benth.) Hochst. ex Asch. = Anticharis senegalensis (Walp.) Bhandari ■☆
28275 Anticharis linearis (Benth.) Hochst. ex Asch. var. azurea Dinter ex Schinz = Anticharis senegalensis (Walp.) Bhandari ■☆
28276 Anticharis longifolia Marloth et Engl. = Anticharis senegalensis (Walp.) Bhandari ■☆
28277 Anticharis schimperi Endl. = Anticharis arabica Endl. ■☆
28278 Anticharis scoparia (E. Mey. ex Benth.) Hiern ex Benth. et Hook. f.;帚状劣玄参■☆
28279 Anticharis senegalensis (Walp.) Bhandari;塞内加尔劣玄参■☆
28280 Anticharis somalensis Vierh. = Anticharis glandulosa Asch. ■☆
28281 Anticheirostylis Fitzg. (1891);东澳兰属■☆
28282 Anticheirostylis Fitzg. = Corunastylis Fitzg. ■☆
28283 Anticheirostylis Fitzg. = Genoplesium R. Br. ■☆
28284 Anticheirostylis apostasioides Fitzg.;东澳兰■☆
28285 Antichloa Steud. = Actinochloa Willd. ■

28286 Antichloa Steud. = Botelua Lag. ■

28287 Antichloa Steud. = Chondrosum Desv. ■☆

28288 Antichorus L. = Corchorus L. ●■

28289 Antichorus depressus L. = Corchorus depressus (L.) Stocks ■☆

28290 Anticlea Kunth = Zigadenus Michx. ■

28291 Anticlea alpina A. Heller = Zigadenus elegans Pursh ■☆

28292 Anticlea chlorantha (Richardson) Rydb. = Zigadenus elegans Pursh subsp. glaucus (Nutt.) Hultén ■☆

28293 Anticlea coloradensis (Rydb.) Rydb. = Zigadenus elegans Pursh ■☆

28294 Anticlea elegans (Pursh) Rydb. = Zigadenus elegans Pursh ■☆

28295 Anticlea fremontii Torr. = Zigadenus fremontii (Torr.) Torr. ex S. Watson ■☆

28296 Anticlea glauca (Nutt.) Kunth = Zigadenus elegans Pursh ■☆

28297 Anticlea gracilenta (Greene) R. R. Gates = Zigadenus elegans Pursh ■☆

28298 Anticlea longa (Greene) A. Heller = Zigadenus elegans Pursh ■☆

28299 Anticlea mexicana Kunth = Zigadenus virescens (Kunth) J. F. Macbr. ■☆

28300 Anticlea mohinorensis (Greenm.) R. R. Gates = Zigadenus elegans Pursh ■☆

28301 Anticlea sibirica (L.) Kunth = Zigadenus sibiricus (L.) A. Gray ■

28302 Anticlea sibirica Kunth = Zigadenus sibiricus (L.) A. Gray ■

28303 Anticlea vaginata Rydb. = Zigadenus vaginatus (Rydb.) J. F. Macbr. ■☆

28304 Anticlea virescens (Kunth) Rydb. = Zigadenus virescens (Kunth) J. F. Macbr. ■☆

28305 Anticoryne Turcz. = Baeckea L. ●

28306 Antidaphne Poepp. et Endl. (1838);异瑞香属●☆

28307 Antidaphne wrightii (Griseb.) Kuijt;赖特异瑞香;Wright's Catkin Mistletoe ●☆

28308 Antidesma Burm. ex L. = Antidesma L. ●

28309 Antidesma L. (1753);五月茶属(华月桂属,橘里珍属);China Laurel,Chinalaurel,China-laurel,Chinese Laurel,Meytea ●

28310 Antidesma acidum Retz.;西南五月茶(二蕊五月茶,二药五月茶,假南五月茶,酸汤叶,酸叶树);Acid China-laurel ●

28311 Antidesma acutisepalum Hayata = Antidesma japonicum Siebold et Zucc. ●

28312 Antidesma alnifolium Hook. = Trimeria grandifolia (Hochst.) Warb. ●☆

28313 Antidesma ambiguum Pax et K. Hoffm.;蔓五月茶(拟五月茶,屏边五月茶);Pingbian China-laurel ●

28314 Antidesma barbatum C. Presl = Antidesma pentandrum (Blanco) Merr. var. barbatum (C. Presl) Merr. ●

28315 Antidesma brachyscyphum Baker = Antidesma madagascariense Lam. ●☆

28316 Antidesma bunius (L.) Spreng.;五月茶(酸味树,污槽树,五味菜,五味叶,五味子,月单);Bignay,Bignay China Laurel,Bignay Chinalaurel, Bignay China-laurel, Bignay Meytea, Chinese Laurel, Salamander Tree,Salamander-tree ●

28317 Antidesma calvescens Pax et K. Hoffm. = Antidesma montanum Blume ●

28318 Antidesma chevalieri Beille = Antidesma laciniatum Müll. Arg. ●☆

28319 Antidesma chonmon Gagnep.;滇越五月茶(越南五月茶);Viet Nam China-laurel ●

28320 Antidesma colletri Craib = Antidesma bunius (L.) Spreng. ●

28321 Antidesma comoense Beille;科莫五月茶●☆

28322 Antidesma comorense Vatke et Pax ex Pax = Antidesma madagascariense Lam. ●☆

28323 Antidesma costulatum Pax et K. Hoffm.;小肋五月茶;Fine-ribbed China-laurel ●

28324 Antidesma dallachyanum Baill. = Antidesma bunius (L.) Spreng. ●

28325 Antidesma delicatulum Hutch. = Antidesma japonicum Siebold et Zucc. ●

28326 Antidesma diandrum (Roxb.) K. Heyne ex Roth = Antidesma acidum Retz. ●

28327 Antidesma erythroxyloides Tul. = Antidesma madagascariense Lam. ●☆

28328 Antidesma filipes Hand. -Mazz. = Antidesma japonicum Siebold et Zucc. ●

28329 Antidesma fordii Hemsl.;黄毛五月茶(牛尾果,旱禾子树);Ford Chinalaurel,Ford China-laurel,Yellow-hairy China-laurel ●

28330 Antidesma fuscocinerea Beille = Antidesma venosum E. Mey. ex Tul. ●

28331 Antidesma ghaesembilla Gaertn.;方叶五月茶(田边木,圆叶早禾子);Square-leaved China-laurel ●

28332 Antidesma gracile Hemsl. = Antidesma japonicum Siebold et Zucc. ●

28333 Antidesma gracillimum Gagnep. = Antidesma japonicum Siebold et Zucc. ●

28334 Antidesma hainanense Merr.;海南五月茶(兰屿枯里珍);Hainan Chinalaurel,Hainan China-laurel ●

28335 Antidesma henryi Hemsl. = Antidesma montanum Blume ●

28336 Antidesma henryi Pax et K. Hoffm. = Antidesma acidum Retz. ●

28337 Antidesma hiiranense Hayata;南仁五月茶●

28338 Antidesma hiiranense Hayata = Antidesma japonicum Siebold et Zucc. ●

28339 Antidesma hildebrandtii Pax et K. Hoffm. var. comorense ? = Antidesma madagascariense Lam. ●☆

28340 Antidesma hontaushanensis C. E. Chang = Antidesma hainanense Merr. ●

28341 Antidesma japonicum Siebold et Zucc.;酸五月茶(禾串果,恒春五月茶,密花五月茶,南仁五月茶,南投五月茶,日本五月茶,酸味子,酸叶子,细五月茶);Hengchun China Laurel,Japan Meytea, Japanese Chinalaurel,Japanese China-laurel ●

28342 Antidesma japonicum Siebold et Zucc. f. angustissimum ? = Antidesma japonicum Siebold et Zucc. ●

28343 Antidesma japonicum Siebold et Zucc. var. acutisepalum (Hayata) Hurus.;锐萼五月茶(南投五月茶);Acule-sepal China Laurel ●

28344 Antidesma japonicum Siebold et Zucc. var. acutisepalum (Hayata) Hurus. = Antidesma japonicum Siebold et Zucc. ●

28345 Antidesma japonicum Siebold et Zucc. var. densiflorum Hurus.;密花山巴豆(枯里珍,密花五月茶);Dense-flowered China Laurel ●

28346 Antidesma japonicum Siebold et Zucc. var. densiflorum Hurus. = Antidesma japonicum Siebold et Zucc. ●

28347 Antidesma japonicum var. liukiuense ? = Antidesma japonicum Siebold et Zucc. ●

28348 Antidesma japonicum var. uncinulatum ? = Antidesma japonicum Siebold et Zucc. ●

28349 Antidesma kotoense Kaneh.;台湾五月茶;Taiwan Meytea ●

28350 Antidesma kotoense Kaneh. = Antidesma pentandrum (Blanco) Merr. var. barbatum (C. Presl) Merr. ●

28351 Antidesma kuroiwai Makino = Antidesma pentandrum (Blanco) Merr. ●
28352 Antidesma laciniatum Müll. Arg.;撕裂五月茶●☆
28353 Antidesma laciniatum Müll. Arg. subsp. membranaceum (Müll. Arg.) J. Léonard;膜质撕裂五月茶●☆
28354 Antidesma laciniatum Müll. Arg. var. membranaceum Müll. Arg. = Antidesma laciniatum Müll. Arg. subsp. membranaceum (Müll. Arg.) J. Léonard ●☆
28355 Antidesma lanceolarium (Roxb.) Wight = Antidesma acidum Retz. ●
28356 Antidesma leptobotryum Müll. Arg. = Thecacoris leptobotrya (Müll. Arg.) Brenan ●☆
28357 Antidesma longipes Pax = Maesobotrya longipes (Pax) Hutch. ●☆
28358 Antidesma maclurei Merr.;多花五月茶;Maclure China-laurel, Multiflowered China-laurel ●
28359 Antidesma madagascariense Lam.;马岛五月茶●☆
28360 Antidesma mannianum Müll. Arg. = Thecacoris manniana (Müll. Arg.) Müll. Arg. ●☆
28361 Antidesma meiocarpum J. Léonard = Antidesma membranaceum Müll. Arg. ●☆
28362 Antidesma membranaceum Müll. Arg.;膜质五月茶●☆
28363 Antidesma membranaceum Müll. Arg. var. crassifolium Pax et K. Hoffm. = Antidesma vogelianum Müll. Arg. ●☆
28364 Antidesma membranaceum Müll. Arg. var. glabrescens ? = Antidesma rufescens Tul. ●☆
28365 Antidesma membranaceum Müll. Arg. var. molle ? = Antidesma membranaceum Müll. Arg. ●☆
28366 Antidesma menasu Miq. ex Tul.;门五月茶●☆
28367 Antidesma microphyllum Hemsl. = Antidesma venosum E. Mey. ex Tul. ●
28368 Antidesma montanum Blume;山地五月茶(南五月茶,山五月茶);Mountanous China-laurel ●
28369 Antidesma moritzii F. Muell. = Antidesma montanum Blume ●
28370 Antidesma moritzii Müll. Arg. = Antidesma montanum Blume ●
28371 Antidesma neriifolium Pax et K. Hoffm. = Antidesma venosum E. Mey. ex Tul. ●
28372 Antidesma nienkui Merr. et Chun;大果五月茶(海南五月茶);Big-fruited China-laurel ●
28373 Antidesma oblongum (Hutch.) Keay;矩圆五月茶●☆
28374 Antidesma pachybotryum Pax et K. Hoffm.;粗穗五月茶●☆
28375 Antidesma paniculatum Roxb. ex Willd. = Antidesma ghaesembilla Gaertn. ●
28376 Antidesma paniculatum Willd. = Antidesma ghaesembilla Gaertn. ●
28377 Antidesma paxii F. P. Metcalf = Antidesma acidum Retz. ●
28378 Antidesma pentandrum (Blanco) Merr.;五蕊五月茶;Five-stamens Laurel ●
28379 Antidesma pentandrum (Blanco) Merr. f. kuroiwae (Makino) Hurus. = Antidesma pentandrum (Blanco) Merr. ●
28380 Antidesma pentandrum (Blanco) Merr. f. kuroiwai (Makino) Hurus. = Antidesma pentandrum (Blanco) Merr. var. barbatum (C. Presl) Merr. ●
28381 Antidesma pentandrum (Blanco) Merr. var. barbatum (C. Presl) Merr.;枯里珍五月茶(枯里珍,五蕊山巴豆);Five-stamens China Laurel ●
28382 Antidesma pentandrum (Blanco) Merr. var. hiiranense (Hayata) Hurus. = Antidesma japonicum Siebold et Zucc. ●
28383 Antidesma pentandrum (Blanco) Merr. var. pseudopentandrum Hurus. = Antidesma pentandrum (Blanco) Merr. var. barbatum (C. Presl) Merr. ●
28384 Antidesma pentandrum (Blanco) Merr. var. pseudopentandrum Hurus. = Antidesma pentandrum (Blanco) Merr. ●
28385 Antidesma pentandrum (Blanco) Merr. var. rotundisepalum (Hayata) Humsawa = Antidesma pentandrum (Blanco) Merr. var. barbatum (C. Presl) Merr. ●
28386 Antidesma pentatum Tul. var. hiieanense (Hayata) Hurus. = Antidesma japonicum Siebold et Zucc. ●
28387 Antidesma pentatum Tul. var. rotundisepalum (Hayata) Hurus. = Antidesma pentandrum (Blanco) Merr. var. barbatum (C. Presl) Merr. ●
28388 Antidesma pleuricum Tul.;河头山五月茶●
28389 Antidesma pseudolaciniatum Beille = Antidesma laciniatum Müll. Arg. subsp. membranaceum (Müll. Arg.) J. Léonard ●☆
28390 Antidesma pseudomicrophyllum Croizat;柳叶五月茶(狭叶五月茶);Narrow-leaved China-laurel ●
28391 Antidesma pseudopentandrum Hurus. = Antidesma pentandrum (Blanco) Merr. var. barbatum (C. Presl) Merr. ●
28392 Antidesma pubescens Roxb. = Antidesma ghaesembilla Gaertn. ●
28393 Antidesma rostratum Tul.;菲律宾五月茶●
28394 Antidesma rostratum Tul. var. barbatum (C. Presl) Müll. Arg. = Antidesma pentandrum (Blanco) Merr. var. barbatum (C. Presl) Merr. ●
28395 Antidesma rotundisepatum Hayata = Antidesma pentandrum (Blanco) Merr. var. barbatum (C. Presl) Merr. ●
28396 Antidesma rufescens Tul.;浅红五月茶●☆
28397 Antidesma sassandrae Beille = Antidesma rufescens Tul. ●☆
28398 Antidesma scandens Lour. = Humulus scandens (Lour.) Merr. ■
28399 Antidesma schweinfurthii Pax = Maesobotrya floribunda Benth. ●☆
28400 Antidesma sequini H. Lév. = Antidesma venosum E. Mey. ex Tul. ●
28401 Antidesma sootepense Craib;泰北五月茶;North-tailand China-laurel ●
28402 Antidesma spicatum Blanco;兰屿五月茶(穗状五月茶);Lanyu China-laurel ●
28403 Antidesma staudtii Pax = Antidesma vogelianum Müll. Arg. ●☆
28404 Antidesma stenopetalum Müll. Arg. = Thecacoris stenopetala (Müll. Arg.) Müll. Arg. ●☆
28405 Antidesma stipulare Blume;托叶五月茶●☆
28406 Antidesma thorelianum Gagnep. = Antidesma bunius (L.) Spreng. ●
28407 Antidesma venosum E. Mey. ex Tul.;小叶五月茶(沙潦木,水杨梅,蒜瓣果,小杨柳);Veined China-laurel ●
28408 Antidesma venosum Tul. f. glabrescens De Wild. = Antidesma vogelianum Müll. Arg. ●☆
28409 Antidesma vogelianum Müll. Arg.;厚叶五月茶●☆
28410 Antidesma wallichianum C. Presl = Antidesma acidum Retz. ●
28411 Antidesma yunnanense Pax et K. Hoffm. = Antidesma fordii Hemsl. ●
28412 Antidesmataceae Loudon = Euphorbiaceae Juss.(保留科名)●■
28413 Antidesmataceae Sweet ex Endl. = Phyllanthaceae J. Agardh ●■
28414 Antidesmataceae Sweet ex Endl. = Stilaginaceae C. Agardh ●
28415 Antidris Thouars = Disperis Sw. ■
28416 Antidris Thouars = Dryopeia Thouars ■
28417 Antigona Vell. = Casearia Jacq. ●
28418 Antigonon Endl. (1837);珊瑚藤属(珊瑚蓼属);Antigonon,

Confederate-vine, Coral Vine, Coralvine, Coral-vine, Mexican Creeper ●■

28419 Antigonon flavescens S. Watson;变黄珊瑚藤●☆

28420 Antigonon grandiflorum B. L. Rob.;大花珊瑚藤●☆

28421 Antigonon guatemalense Meisn.;危地马拉珊瑚藤;Bellisima Grande ●☆

28422 Antigonon leptopus Hook. et Arn.;珊瑚藤(珊瑚蓼);Confederate Vine, Coral Creeper, Coral Vine, Corallita, Coralvine, Honolulu Creeper, Love Vine, Mexican Creeper, Mountain Rose, Mountainrose Coralvine, Mountain-rose Coralvine, Mountain-rose Coral-vine, Pink Vine, Pinkvine, Queen's Wreath, Queen's-Jewels ●

28423 Antigonon macrocarpum Britton et Small;大果珊瑚藤●☆

28424 Antigonon viride S. Watson;绿花珊瑚藤●☆

28425 Antilla (Luer) Luer = Pleurothallis R. Br. ■☆

28426 Antillanorchis Garay(1974);安蒂兰属■☆

28427 Antillanorchis gundlachii (Griseb.) Garay = Antillanorchis gundlachii (Wright ex Griseb.) Garay ■☆

28428 Antillanorchis gundlachii (Wright ex Griseb.) Garay;安蒂兰■☆

28429 Antillanthus B. Nord. (2006);连柱菊属●☆

28430 Antillanthus ekmanii (Alain) B. Nord.;连柱菊●☆

28431 Antillanthus leucolepis (Greenm.) B. Nord.;白鳞连柱菊●☆

28432 Antillanthus pachylepis (Greenm.) B. Nord.;厚鳞连柱菊●☆

28433 Antillia R. M. King et H. Rob. (1971);多花亮泽兰属■☆

28434 Antillia brachychaeta (B. L. Rob.) R. M. King et H. Rob.;多花亮泽兰■☆

28435 Antimima N. E. Br. (1930);紫波属■☆

28436 Antimima alborubra (L. Bolus) Dehn;红白紫波■☆

28437 Antimima amoena (Schwantes) H. E. K. Hartmann;秀丽紫波■☆

28438 Antimima androsacea (Marloth et Schwantes) H. E. K. Hartmann;点地梅紫波■☆

28439 Antimima argentea (L. Bolus) H. E. K. Hartmann;银白紫波■☆

28440 Antimima aristulata (Sond.) Chess. et G. F. Sm.;芒紫波■☆

28441 Antimima aurasensis H. E. K. Hartmann;奥拉斯紫波■☆

28442 Antimima biformis (N. E. Br.) H. E. K. Hartmann;二形紫波■☆

28443 Antimima bina (L. Bolus) H. E. K. Hartmann = Antimima viatorum (L. Bolus) Klak ■☆

28444 Antimima bracteata (L. Bolus) H. E. K. Hartmann;具苞紫波■☆

28445 Antimima brevicarpa (L. Bolus) H. E. K. Hartmann;短果紫波■☆

28446 Antimima brevicollis (N. E. Br.) H. E. K. Hartmann;短颈紫波■☆

28447 Antimima buchubergensis (Dinter) H. E. K. Hartmann;布赫紫波■☆

28448 Antimima compacta (L. Bolus) H. E. K. Hartmann;紧密紫波■☆

28449 Antimima compressa (L. Bolus) H. E. K. Hartmann;扁紫波■☆

28450 Antimima concinna (L. Bolus) H. E. K. Hartmann;整洁紫波■☆

28451 Antimima condensa (N. E. Br.) H. E. K. Hartmann;密集紫波■☆

28452 Antimima crassifolia (L. Bolus) H. E. K. Hartmann;厚叶紫波■☆

28453 Antimima dasyphylla (Schltr.) H. E. K. Hartmann;毛叶紫波■☆

28454 Antimima dekenahi (N. E. Br.) H. E. K. Hartmann;德凯纳紫波■☆

28455 Antimima distans (L. Bolus) H. E. K. Hartmann;分离紫波■☆

28456 Antimima dolomitica (Dinter) H. E. K. Hartmann;多罗米蒂紫波■☆

28457 Antimima dualis (N. E. Br.) N. E. Br. = Ruschia dualis L. Bolus ●☆

28458 Antimima elevata (L. Bolus) H. E. K. Hartmann;隆起紫波■☆

28459 Antimima emarcescens (L. Bolus) H. E. K. Hartmann;常青紫波■☆

28460 Antimima erosa (L. Bolus) H. E. K. Hartmann;啮蚀状紫波■☆

28461 Antimima evoluta (N. E. Br.) H. E. K. Hartmann;展开紫波■☆

28462 Antimima exsurgens (L. Bolus) H. E. K. Hartmann;直立紫波■☆

28463 Antimima fenestrata (L. Bolus) H. E. K. Hartmann;窗孔紫波■☆

28464 Antimima fergusoniae (L. Bolus) H. E. K. Hartmann;费格森紫波■☆

28465 Antimima gracillima (L. Bolus) H. E. K. Hartmann;细长紫波■☆

28466 Antimima granitica (L. Bolus) H. E. K. Hartmann;花岗岩紫波■☆

28467 Antimima hallii (L. Bolus) H. E. K. Hartmann;霍尔紫波■☆

28468 Antimima hamatilis (L. Bolus) H. E. K. Hartmann;顶钩紫波■☆

28469 Antimima hantamensis (Engl.) H. E. K. Hartmann et Stüber;汉滩紫波■☆

28470 Antimima herrei (Schwantes) H. E. K. Hartmann;赫勒紫波(弥生)■☆

28471 Antimima intervallaris (L. Bolus) H. E. K. Hartmann;谷地紫波■☆

28472 Antimima ivori (N. E. Br.) H. E. K. Hartmann;伊沃里紫波■☆

28473 Antimima karroidea (L. Bolus) H. E. K. Hartmann;卡罗紫波■☆

28474 Antimima klaverensis (L. Bolus) H. E. K. Hartmann;克拉弗紫波■☆

28475 Antimima koekenaapensis (L. Bolus) H. E. K. Hartmann;库克纳普紫波■☆

28476 Antimima komkansica (L. Bolus) H. E. K. Hartmann;科姆康斯紫波■☆

28477 Antimima lawsonii (L. Bolus) H. E. K. Hartmann;劳森紫波■☆

28478 Antimima leipoldtii (L. Bolus) H. E. K. Hartmann;莱波尔德紫波■☆

28479 Antimima leucanthera (L. Bolus) H. E. K. Hartmann;白药紫波■☆

28480 Antimima limbata (N. E. Br.) H. E. K. Hartmann;具边紫波■☆

28481 Antimima lodewykii (L. Bolus) H. E. K. Hartmann;洛德紫波■☆

28482 Antimima loganii (L. Bolus) H. E. K. Hartmann;洛根紫波■☆

28483 Antimima lokenbergensis (L. Bolus) H. E. K. Hartmann;洛肯紫波■☆

28484 Antimima longipes (L. Bolus) Dehn;长梗紫波■☆

28485 Antimima luckhoffii (L. Bolus) H. E. K. Hartmann;吕克霍夫紫波■☆

28486 Antimima maleolens (L. Bolus) H. E. K. Hartmann;小锤紫波■☆

28487 Antimima mesklipensis (L. Bolus) H. E. K. Hartmann;迈斯科里普紫波■☆

28488 Antimima meyerae (Schwantes) H. E. K. Hartmann;迈尔紫波■☆

28489 Antimima microphylla (Haw.) Dehn;小叶紫波■☆

28490 Antimima minima (Tischer) H. E. K. Hartmann;小紫波■☆

28491 Antimima minutifolia (L. Bolus) H. E. K. Hartmann;微叶紫波●☆

28492 Antimima modesta (L. Bolus) H. E. K. Hartmann;适度紫波■☆

28493 Antimima mucronata (Haw.) H. E. K. Hartmann;短尖紫波■☆

28494 Antimima mutica (L. Bolus) H. E. K. Hartmann;无尖紫波■☆

28495 Antimima nobilis (Schwantes) H. E. K. Hartmann;名贵紫波■☆

28496 Antimima nordenstamii (L. Bolus) H. E. K. Hartmann;努登斯坦紫波■☆

28497 Antimima oviformis (L. Bolus) H. E. K. Hartmann;卵形紫波■☆

28498 Antimima papillata (L. Bolus) H. E. K. Hartmann;乳突紫波■☆

28499 Antimima paucifolia (L. Bolus) H. E. K. Hartmann;少叶紫波■☆

28500 Antimima pauper (L. Bolus) H. E. K. Hartmann;贫乏紫波■☆

28501 Antimima peersii (L. Bolus) H. E. K. Hartmann;皮尔斯紫波■☆

28502 Antimima perforata (L. Bolus) H. E. K. Hartmann;穿孔紫波■☆

28503 Antimima persistens H. E. K. Hartmann;宿存紫波■☆

28504 Antimima pilosula (L. Bolus) H. E. K. Hartmann;疏毛紫波■☆

28505 Antimima prolongata (L. Bolus) H. E. K. Hartmann;延长紫波■☆

28506 Antimima propinqua (N. E. Br.) H. E. K. Hartmann;邻近紫波■☆
28507 Antimima prostrata (L. Bolus) H. E. K. Hartmann;平卧紫波■☆
28508 Antimima pumila (L. Bolus ex Fedde et C. Schust.) H. E. K. Hartmann;偃伏紫波■☆
28509 Antimima pusilla (Schwantes) H. E. K. Hartmann;微小紫波■☆
28510 Antimima pygmaea (Haw.) H. E. K. Hartmann;矮小紫波■☆
28511 Antimima quarzitica (Dinter) H. E. K. Hartmann;阔茨紫波■☆
28512 Antimima roseola (N. E. Br.) H. E. K. Hartmann;粉红紫波■☆
28513 Antimima saturata (L. Bolus) H. E. K. Hartmann;富色紫波■☆
28514 Antimima saxicola (L. Bolus) H. E. K. Hartmann;岩栖紫波■☆
28515 Antimima simulans (L. Bolus) H. E. K. Hartmann;相似紫波■☆
28516 Antimima solida (L. Bolus) H. E. K. Hartmann;坚实紫波■☆
28517 Antimima stayneri (L. Bolus) H. E. K. Hartmann;斯泰纳紫波■☆
28518 Antimima stokoei (L. Bolus) H. E. K. Hartmann;斯托克紫波■☆
28519 Antimima subtruncata (L. Bolus) H. E. K. Hartmann;平截紫波■☆
28520 Antimima triquetra (L. Bolus) H. E. K. Hartmann;三棱紫波■☆
28521 Antimima tuberculosa (L. Bolus) H. E. K. Hartmann;瘤状紫波■☆
28522 Antimima turneriana (L. Bolus) H. E. K. Hartmann;特纳紫波■☆
28523 Antimima vanzylii (L. Bolus) H. E. K. Hartmann;万齐紫波■☆
28524 Antimima varians (L. Bolus) H. E. K. Hartmann;变异紫波■☆
28525 Antimima ventricosa (L. Bolus) H. E. K. Hartmann;偏肿紫波■☆
28526 Antimima verruculosa (L. Bolus) H. E. K. Hartmann;小疣紫波■☆
28527 Antimima viatorum (L. Bolus) Klak;扩散紫波■☆
28528 Antimima virgata (Haw.) Dehn = Ruschia virgata (Haw.) L. Bolus ●☆
28529 Antimima watermeyeri (L. Bolus) H. E. K. Hartmann;沃特迈耶紫波■☆
28530 Antimima wittebergensis (L. Bolus) H. E. K. Hartmann;沃特紫波■☆
28531 Antimion Raf. = Lycopersicon Mill. ■
28532 Antinisa (Tul.) Hutch. = Homalium Jacq. ●
28533 Antinoria Parl. (1845);浮燕麦属■☆
28534 Antinoria Parl. = Aira L. (保留属名)■
28535 Antinoria agrostidea (DC.) Parl.;浮燕麦■☆
28536 Antinoria agrostidea (DC.) Parl. var. algeriensis Maire = Antinoria agrostidea (DC.) Parl. ■☆
28537 Antinoria agrostidea (DC.) Parl. var. annua Lange = Antinoria agrostidea (DC.) Parl. ■☆
28538 Antinoria agrostidea (DC.) Parl. var. insularis (Parl.) Maire = Antinoria insularis Parl. ■☆
28539 Antinoria insularis Parl.;海岛燕麦■☆
28540 Antioanrus Roem. = Heptaptera Margot et Reut. ■☆
28541 Antiostelma (Tsiang et P. T. Li) P. T. Li = Micholitzia N. E. Br. ■
28542 Antiostelma lantsangense (Tsiang et P. T. Li) P. T. Li = Hoya lantsangensis Tsiang et P. T. Li ●
28543 Antiostelma lantsangense (Tsiang et P. T. Li) P. T. Li = Micholitzia obcrodata N. E. Br. ●
28544 Antiostelma manipurense (Deb) P. T. Li = Micholitzia obcrodata N. E. Br. ●
28545 Antiostelma manipurense (Debeaux) P. T. Li = Hoya lantsangensis Tsiang et P. T. Li ●
28546 Antiostelma manipurense (Debeaux) P. T. Li = Micholitzia obcrodata N. E. Br. ●
28547 Antiotrema Hand. -Mazz. (1920);长蕊斑种草属(滇牛舌草属,滇紫草属,黑阳参属);Antiotrema ■★
28548 Antiotrema dunnianum (Diels) Hand. -Mazz.;长蕊斑种草(白紫草,滇牛舌草,滇紫草,狗舌草,黑阳参,黑元参,牛舌头菜,铁打苗,土玄参);Dunn Antiotrema ■
28549 Antiphiona Merxm. (1954);修尾菊属■☆
28550 Antiphiona fragrans (Merxm.) Merxm.;香修尾菊■☆
28551 Antiphiona pinnatisecta (S. Moore) Merxm.;修尾菊■☆
28552 Antiphyla Raf. = Melochia L. (保留属名)●■
28553 Antiphyla Raf. = Riddelia Raf. ●■
28554 Antiphylla Haw. = Saxifraga L. ■
28555 Antiphylla asiatica (Hayek) Losinsk. = Saxifraga oppositifolia L. ■
28556 Antiphylla nana (Engl.) Losinsk. = Saxifraga nana Engl. ■
28557 Antiphylla octandra (Harry Sm.) Losinsk. = Saxifraga nana Engl. ■
28558 Antiphylla oppositifolia (L.) Fourr. = Saxifraga oppositifolia L. ■
28559 Antiphytum DC. = Antiphytum DC. ex Meisn. ■☆
28560 Antiphytum DC. ex Meisn. (1840);墨西哥紫草属■☆
28561 Antiphytum cruciatum DC.;墨西哥紫草■☆
28562 Antiphytum floribundum A. Gray;繁花墨西哥紫草■☆
28563 Antirhea Comm. ex Juss. (1789);毛茶属;Antirhea, Hairtea ●
28564 Antirhea Juss. = Antirhea Comm. ex Juss. ●
28565 Antirhea borbonica J. F. Gmel.;博尔翁毛茶■☆
28566 Antirhea chinensis (Champ. ex Benth.) F. B. Forbes et Hemsl.;毛茶;China Hairtea, Chinese Antirhea ●◇
28567 Antirhea madagascariensis Chaw;马岛毛茶■☆
28568 Antirhea martinii H. Lév. = Sindechites henryi Oliv. ●
28569 Antirhoea Comm. ex Juss. = Antirhea Comm. ex Juss. ●
28570 Antirhoea DC. = Antirhea Comm. ex Juss. ●
28571 Antirrhaea esquirolii H. Lév. = Urceola rosea (Hook. et Arn.) D. J. Middleton ●
28572 Antirrhaea martinii H. Lév. = Sindechites henryi Oliv. ●
28573 Antirrhinaceae DC. et Duby = Plantaginaceae Juss. (保留科名)■
28574 Antirrhinaceae DC. et Duby;金鱼草科●■
28575 Antirrhinaceae Pers. = Scrophulariaceae Juss. (保留科名)●■
28576 Antirrhinum L. (1753);金鱼草属(龙头花属);Snapdragon ●■
28577 Antirrhinum aegyptiacum L. = Kickxia aegyptiaca (L.) Nábelek ■☆
28578 Antirrhinum aphyllum L. f. = Utricularia bisquamata Schrank ■☆
28579 Antirrhinum asarinum L. = Asarina procumbens Mill. ■☆
28580 Antirrhinum australe Rothm.;南方金鱼草■☆
28581 Antirrhinum barbatum Thunb. = Nemesia barbata (Thunb.) Benth. ■☆
28582 Antirrhinum barrelieri Boreau;巴雷金鱼草■☆
28583 Antirrhinum barrelieri Boreau var. reeseanum Maire = Antirrhinum barrelieri Boreau ■☆
28584 Antirrhinum bellidifolium L.;淡紫金鱼草;Lilac Snapdragon ■☆
28585 Antirrhinum bicorne L. = Nemesia bicornis (L.) Pers. ■☆
28586 Antirrhinum calycinum Lam. = Misopates calycinum (Vent.) Rothm. ■☆
28587 Antirrhinum calycinum Lam. var. leiocarpum Pau = Misopates calycinum (Vent.) Rothm. ■☆
28588 Antirrhinum canadense L. = Nuttallanthus canadensis (L.) D. A. Sutton ■☆
28589 Antirrhinum capense Burm. f. = Nemesia bicornis (L.) Pers. ■☆
28590 Antirrhinum capense Thunb. = Nemesia fruticans (Thunb.) Benth. ■☆
28591 Antirrhinum chrysothales Font Quer = Misopates chrysothales (Font Quer) Rothm. ■☆
28592 Antirrhinum cymbalaria L. = Cymbalaria muralis P. Gaertn., B. Mey. et Scherb. ■☆

28593 Antirrhinum dalmaticum L. = Linaria dalmatica (L.) Mill. ■☆
28594 Antirrhinum diminutum Pomel;缩小金鱼草■☆
28595 Antirrhinum elatine L. = Kickxia elatine (L.) Dumort. ■☆
28596 Antirrhinum elegans G. Forst. = Kickxia elegans (G. Forst.) D. A. Sutton ■☆
28597 Antirrhinum fernandezcasasii Romo, Stübing et Peris = Antirrhinum siculum Mill. ■☆
28598 Antirrhinum filipes A. Gray;缠绕金鱼草;Twining Snapdragon, Yellow Twining Snapdragon ■☆
28599 Antirrhinum flavum Poir. = Linaria flava (Poir.) Desf. ■☆
28600 Antirrhinum flexuosum Pomel = Acanthorrhinum ramosissimum (Coss. et Durieu) Rothm. ●☆
28601 Antirrhinum fontqueri Emb. = Misopates chrysothales (Font Quer) Rothm. ■☆
28602 Antirrhinum fruticans Thunb. = Nemesia fruticans (Thunb.) Benth. ■☆
28603 Antirrhinum gebelicum Brullo et Furnari = Antirrhinum siculum Mill. ■☆
28604 Antirrhinum genistifolium L. = Linaria genistifolia (L.) Mill. ■
28605 Antirrhinum gibbosum Wall. = Misopates orontium (L.) Raf. subsp. gibbosum (Wall.) D. A. Sutton ■☆
28606 Antirrhinum glutinosum Boiss. et Reut. = Antirrhinum hispanicum Chav. ■☆
28607 Antirrhinum glutinosum Boiss. et Reut. var. africanum Pau et Font Quer = Antirrhinum hispanicum Chav. ■☆
28608 Antirrhinum glutinosum Boiss. et Reut. var. oppositifolium Maire = Antirrhinum ternatum Fern. Casas ■☆
28609 Antirrhinum heterophyllum Schousb. = Kickxia heterophylla (Schousb.) Dandy ■☆
28610 Antirrhinum hispanicum Chav.;西班牙金鱼草■☆
28611 Antirrhinum hispanicum Chav. var. faurei Maire = Antirrhinum ternatum Fern. Casas ■☆
28612 Antirrhinum hispanicum Chav. var. ochroleucum Coss. = Antirrhinum australe Rothm. ■☆
28613 Antirrhinum intricatum Ball = Acanthorrhinum ramosissimum (Coss. et Durieu) Rothm. ●☆
28614 Antirrhinum linaria L. = Linaria vulgaris Mill. ■
28615 Antirrhinum longicorne Thunb. = Diascia longicornis (Thunb.) Druce ■☆
28616 Antirrhinum macrocarpum Aiton = Nemesia macrocarpa (Aiton) Druce ■☆
28617 Antirrhinum majus L.;金鱼草(虎嘴花,龙头花,龙头木樨,洋彩雀);Bonny Rabbits, Boots-and-shoes, Breed Calf's Snout, Broad Calf's Snout, Bulldogs, Bunny Mouth, Bunny Rabbits, Bunny Rabbit's Mouth, Calf's Snout, Catchfly, Chatterbox, Chooky-pigs, Cock's Head, Common Snapdragon, Creeping Snapdragon, Devil's Ribbon, Dog Mouth, Dog Snout, Dogmouth, Dog's Mouth, Dog's Nose, Dragon's Head, Dragon's Mouth, Dragon's-mouth, Frog's Mouth, Gap-mouth, Garden Snapdragon, Granny's Bonnet, Granny's Bonnets, Granny's Nightcap, Horse's Mouth, Jacob's Ladder, Lady's Slipper, Lion's Mouth, Lion's Snap, Monkey Chops, Monkey Face, Monkey Faces, Monkey Flower, Monkey Mouth, Monkey Musk, Monkey Nose, Monkey Noses, Old Man's Face, Open Jaws, Open Mouth, Open-jaws, Open-mouth, Piggy-wiggy, Pig-o'-the-wall, Pig's Chops, Pig's Mouth, Pig's Snout, Rabbit, Rabbit's Mouth, Snapdragon, Snapjack, Snap-lion, Tiger's Mouth, Toad's Mouth, Yap Mouth, Yap-mouth ■
28618 Antirrhinum majus L. subsp. diminutum (Pomel) Batt. = Antirrhinum diminutum Pomel ■☆
28619 Antirrhinum majus L. subsp. hispanicum (Chav.) Maire = Antirrhinum hispanicum Chav. ■☆
28620 Antirrhinum majus L. subsp. tortuosum (Vent.) Rouy = Antirrhinum tortuosum Vent. ■☆
28621 Antirrhinum majus L. var. faurei (Maire) Maire = Antirrhinum ternatum Fern. Casas ■☆
28622 Antirrhinum majus L. var. lazari Sennen = Antirrhinum tortuosum Vent. ■☆
28623 Antirrhinum majus L. var. ochroleucum (Coss.) Maire;淡黄白金鱼草■☆
28624 Antirrhinum majus L. var. sagarrae (Sennen) Maire = Antirrhinum tortuosum Vent. ■☆
28625 Antirrhinum majus L. var. tortuosum (Vent.) Batt. = Antirrhinum tortuosum Vent. ■☆
28626 Antirrhinum martenii (Font Quer) Rothm.;马尔顿金鱼草■☆
28627 Antirrhinum microcarpum Pomel = Misopates microcarpum (Pomel) D. A. Sutton ■☆
28628 Antirrhinum molle L.;柔软金鱼草;Soft Snapdragon ■☆
28629 Antirrhinum nuttallanum Benth. = Antirrhinum nuttallianum Benth. ex A. DC. ■☆
28630 Antirrhinum nuttallianum Benth. ex A. DC.;纳氏金鱼草;Nuttall's Snapdragon ■☆
28631 Antirrhinum orontium L.;野金鱼草(奥龙金鱼草);Bunny Mouth, Corn Snapdragon, Hound's Head, Lesser Snapdragon, Linearleaf Snapdragon, Lion's Mouth, Oront Snapdragon, Rabbit, Rabbit Flower, Small Snapdragon, Snapdragon, Toad's Mouth, Weasel's Snout, Weasel's-snout ■☆
28632 Antirrhinum orontium L. = Misopates orontium (L.) Raf. ■☆
28633 Antirrhinum orontium L. subsp. calycinum (Lam.) Batt. = Misopates calycinum (Vent.) Rothm. ■☆
28634 Antirrhinum orontium L. subsp. microcarpum (Pomel) Batt. = Misopates orontium (L.) Raf. ■☆
28635 Antirrhinum orontium L. var. burnatii Maire et Sennen = Misopates orontium (L.) Raf. ■☆
28636 Antirrhinum orontium L. var. flavum Batt. et Pit. = Misopates orontium (L.) Raf. ■☆
28637 Antirrhinum orontium L. var. foliosum J. A. Schmidt = Misopates orontium (L.) Raf. ■☆
28638 Antirrhinum orontium L. var. grandiflorum Chav. = Misopates calycinum (Vent.) Rothm. ■☆
28639 Antirrhinum orontium L. var. indicum (Royle) Chav. = Misopates orontium (L.) Raf. ■☆
28640 Antirrhinum orontium L. var. microcarpum Pomel = Misopates microcarpum (Pomel) D. A. Sutton ■☆
28641 Antirrhinum orontium L. var. oranense Faure = Misopates oranense (Faure) D. A. Sutton ■☆
28642 Antirrhinum patens Thunb. = Diascia patens (Thunb.) Grant ex Fourc. ■☆
28643 Antirrhinum pinifolium Poir. = Linaria pinifolia (Poir.) Thell. ■☆
28644 Antirrhinum pinnatum L. f. = Nemesia pinnata (L. f.) E. Mey. ex Benth. ■☆
28645 Antirrhinum pterospermum A. Rich. = Schweinfurthia pterosperma (A. Rich.) A. Braun ■☆
28646 Antirrhinum ramosissimum Coss. et Durieu;多枝金鱼草■☆
28647 Antirrhinum ramosissimum Coss. et Durieu = Acanthorrhinum

ramosissimum (Coss. et Durieu) Rothm. ●☆
28648 Antirrhinum ramosissimum Coss. et Durieu subsp. intricatum Ball = Acanthorrhinum ramosissimum (Coss. et Durieu) Rothm. ●☆
28649 Antirrhinum ramosissimum Coss. et Durieu var. flavum Maire = Acanthorrhinum ramosissimum (Coss. et Durieu) Rothm. ●☆
28650 Antirrhinum ramosissimum Coss. et Durieu var. flexuosum (Pomel) Coss. et Durieu = Acanthorrhinum ramosissimum (Coss. et Durieu) Rothm. ●☆
28651 Antirrhinum sagittatum Poir. = Kickxia heterophylla (Schousb.) Dandy ■☆
28652 Antirrhinum scabridum Herb. Banks ex Benth. = Nemesia acuminata Benth. ■☆
28653 Antirrhinum scabrum (Spreng.) Thunb. = Nemesia macrocarpa (Aiton) Druce ■☆
28654 Antirrhinum sempervirens Lapeyr.; 岩地金鱼草; Rock Snapdragon ■☆
28655 Antirrhinum siculum Mill.; 西西里金鱼草■☆
28656 Antirrhinum spurium L. = Kickxia spuria (L.) Dumort. ■☆
28657 Antirrhinum strumosum Benth. = Nemesia strumosa (Benth.) Benth. ■☆
28658 Antirrhinum ternatum Fern. Casas = Antirrhinum siculum Mill. ■☆
28659 Antirrhinum theodori Sennen et Mauricio = Misopates calycinum (Vent.) Rothm. ■☆
28660 Antirrhinum tortuosum Bosc ex Lam.; 扭旋金鱼草■☆
28661 Antirrhinum tortuosum Vent. = Antirrhinum tortuosum Bosc ex Lam. ■☆
28662 Antirrhinum tortuosum Vent. var. hosmariense Pau = Antirrhinum tortuosum Vent. ■☆
28663 Antirrhinum unilabiatum L. f. = Alonsoa unilabiata (L. f.) Steud. ●☆
28664 Antirrhinum valentinum Font Quer = Antirrhinum martenii (Font Quer) Rothm. ■☆
28665 Antirrhinum valentinum Font Quer subsp. martenii ? = Antirrhinum martenii (Font Quer) Rothm. ■☆
28666 Antirrhinum vidalianum Pau et Font Quer = Antirrhinum tortuosum Vent. ■☆
28667 Antirrhinum virgatum Poir. = Linaria virgata (Poir.) Desf. ■☆
28668 Antirrhinum vulgare Bubani; 污金鱼草(龙头花)■☆
28669 Antirrhoa Gruel ex C. DC. = Turraea L. ●
28670 Antirrhoea Endl. = Antirhea Comm. ex Juss. ●
28671 Antisola Raf. = Miconia Ruiz et Pav. (保留属名)●☆
28672 Antistrophe A. DC. (1841); 扭带紫金牛属●☆
28673 Antitaxis Miers = Pycnarrhena Miers ex Hook. f. et Thomson ●
28674 Antitaxis calocarpa Kurz = Pycnarrhena lucida (Teijsm. et Binn.) Miq. ●
28675 Antitaxis fasciculata Miers = Pycnarrhena lucida (Teijsm. et Binn.) Miq. ●
28676 Antitaxis nodiflora (Pierre) Gagnep. = Pycnarrhena lucida (Teijsm. et Binn.) Miq. ●
28677 Antithrixia DC. (1838); 黄冠鼠麹木属●☆
28678 Antithrixia abyssinica (Sch. Bip. ex Walp.) Vatke = Macowania abyssinica (Sch. Bip. ex Walp.) B. L. Burtt ●☆
28679 Antithrixia angustifolia Oliv. et Hiern = Macowania ericifolia (Forssk.) B. L. Burtt et Grau ●☆
28680 Antithrixia flavicoma DC.; 黄冠鼠麹木●☆
28681 Antitoxicon Pobed. = Cynanchum L. ●■
28682 Antitoxicon acuminatum Pobed. = Cynanchum acuminatifolium Hemsl. ■
28683 Antitoxicon acuminatum Pobed. = Cynanchum ascyrifolium (Franch. et Sav.) Matsum. ■
28684 Antitoxicon amplexicaule (Siebold et Zucc.) Pobed. = Cynanchum amplexicaule (Siebold et Zucc.) Hemsl. ■
28685 Antitoxicon inamoenum Pobed. = Cynanchum inamoenum (Maxim.) Loes. ex Gilg et Loes. ■
28686 Antitoxicon officinale Pobed. = Cynanchum vincetoxicum (L.) Pers. ■
28687 Antitoxicum Pobed. = Alexitoxicon St. -Lag. ●■
28688 Antitoxicum Pobed. = Vincetoxicum Wolf ●■
28689 Antitoxicum acuminatum (Decne.) Pobed. = Cynanchum acuminatifolium Hemsl. ■
28690 Antitoxicum albovianum (Kusn.) Pobed. = Cynanchum albowianum Kusn. ■☆
28691 Antitoxicum amplexicaule (Siebold et Zucc.) Pobed. = Cynanchum amplexicaule (Siebold et Zucc.) Hemsl. ■
28692 Antitoxicum atratum (Bunge) Pobed. = Cynanchum atratum Bunge ■
28693 Antitoxicum boissieri (Kusn.) Pobed. = Cynanchum boissieri Kusn. ■
28694 Antitoxicum darvasicum (B. Fedtsch.) Pobed. = Vincetoxicum darvasicum B. Fedtsch. ■☆
28695 Antitoxicum huteri (Vis. et Asch.) Pobed. = Vincetoxicum huteri Vis. et Asch. ■☆
28696 Antitoxicum inamoenum (Maxim.) Pobed. = Cynanchum inamoenum (Maxim.) Loes. ex Gilg et Loes. ■
28697 Antitoxicum lanceolatum Grubov = Cynanchum laxum Bartl. ●☆
28698 Antitoxicum laxum (Bartl.) Pobed. = Cynanchum laxum Bartl. ●☆
28699 Antitoxicum rossicum (Kleopow) Pobed. = Cynanchum rossicum (Kleopow) Borhidi ■☆
28700 Antitoxicum schmalhausenii (Kusn.) Pobed. = Cynanchum schmalhausenii Kusn. ■☆
28701 Antitoxicum sibiricum (L.) Pobed. = Asclepias sibirica L. ●■
28702 Antitoxicum sibiricum (L.) Pobed. = Cynanchum thesioides (Freyn) K. Schum. ■
28703 Antitoxicum volubile (Maxim.) Pobed. = Cynanchum volubile (Maxim.) Hemsl. ■
28704 Antitragus Gaertn. = Crypsis Aiton(保留属名)■
28705 Antitypaceae Dulac = Oxalidaceae R. Br. (保留科名)●■
28706 Antizoma Miers(1851); 南非锡生藤属●☆
28707 Antizoma angolensis Exell et Mendonça; 安哥拉南非锡生藤●☆
28708 Antizoma angustifolia (Burch.) Miers ex Harv.; 窄叶南非锡生藤●☆
28709 Antizoma burchelliana Miers ex Harv. = Antizoma angustifolia (Burch.) Miers ex Harv. ●☆
28710 Antizoma capensis (L. f.) Diels = Cissampelos capensis L. f. ●☆
28711 Antizoma capensis (L. f.) Diels var. pulverulenta Harv. = Cissampelos capensis L. f. ●☆
28712 Antizoma harveyana Miers ex Harv. = Antizoma angustifolia (Burch.) Miers ex Harv. ●☆
28713 Antizoma lycioides Miers = Cissampelos lycioides (Miers) T. Durand et Schinz ●☆
28714 Antochloa Nees et Meyen ex Nees = Anthochloa Nees et Meyen ■☆
28715 Antochortus Nees = Anthochortus Nees ■☆
28716 Antodon Neck. = Leontodon L. (保留属名)■☆
28717 Antogoeringia Kuntze = Stenosiphon Spach ■☆
28718 Antoiria Raddi = Cavendishia Lindl. (保留属名)●☆
28719 Antomarchia Colla = Antommarchia Colla ex Meisn. ●☆

28720 Antommarchia Colla = Correa Sm. ●☆
28721 Antommarchia Colla ex Meisn. = Correa Sm. ●☆
28722 Antonella Caro = Tridens Roem. et Schult. ■☆
28723 Antongilia Jum. = Dypsis Noronha ex Mart. ●☆
28724 Antongilia Jum. = Neodypsis Baill. ●☆
28725 Antongilia perrieri Jum. = Dypsis perrieri (Jum.) Beentje et J. Dransf. ●☆
28726 Antonia Pohl(1829);薯蓣子属(巴圭马钱木属)●☆
28727 Antonia R. Br. = Rhynchoglossum Blume(保留属名)■
28728 Antonia ovata Pohl;薯蓣子(巴圭马钱木)●☆
28729 Antoniaceae (Endl.) J. Agardh = Loganiaceae R. Br. ex Mart. (保留科名)●■
28730 Antoniaceae Hutch. = Loganiaceae R. Br. ex Mart. (保留科名)●■
28731 Antoniaceae J. Agardh = Loganiaceae R. Br. ex Mart. (保留科名)●■
28732 Antoniaceae J. Agardh;薯蓣子科(阔柄叶科,鞘柄科)●☆
28733 Antoniana Bubani = Hesperis L. ■
28734 Antoniana Tussac = Faramea Aubl. ●☆
28735 Antoniana Tussac ex Griseb. = Faramea Aubl. ●☆
28736 Antonina Vved. = Calamintha Mill. ■
28737 Antonina Vved. = Clinopodium L. ●■
28738 Antopetitia A. Rich. (1840);热非鸟卵豆属■☆
28739 Antopetitia abyssinica A. Rich.;阿比西尼亚鸟卵豆■☆
28740 Antophylax Poir. = Androphylax J. C. Wendl. (废弃属名)●
28741 Antophylax Poir. = Cocculus DC. (保留属名)●
28742 Antoschmidtia Steud. = Schmidtia Steud. ex J. A. Schmidt(保留属名)■☆
28743 Antoschrnidtia Boiss. = Schmidtia Steud. ex J. A. Schmidt(保留属名)■☆
28744 Antriba Raf. = Loranthus Jacq. (保留属名)●
28745 Antriba Raf. = Scurrula L. (废弃属名)●
28746 Antriscus Raf. = Anthriscus Pers. (保留属名)■
28747 Antrizon Raf. = ? Antirrhinum L. ●■
28748 Antrocaryon Pierre(1898);洞果漆属●☆
28749 Antrocaryon brieyi De Wild. = Antrocaryon nannanii De Wild. ●☆
28750 Antrocaryon klaineanum Pierre;克莱恩洞果漆●☆
28751 Antrocaryon micraster A. Chev. et Guillaumin;小星洞果漆●☆
28752 Antrocaryon nannanii De Wild.;南南洞果漆●☆
28753 Antrocaryon polyneurum Mildbr. = Antrocaryon micraster A. Chev. et Guillaumin ●☆
28754 Antrocaryon schorkopfii Engl.;肖尔洞果漆●☆
28755 Antrocaryon soyauxii (Engl.) Engl. = Antrocaryon klaineanum Pierre ●☆
28756 Antrolepidaceae Welw. = Cyperaceae Juss. (保留科名)■
28757 Antrolepis Welw. = Ascolepis Nees ex Steud. (保留属名)■☆
28758 Antrophora I. M. Johnst. = Lepidocordia Ducke☆
28759 Antrospermum Sch. Bip. = Venidium Less. ■☆
28760 Antschar Horsf. = Antiaris Lesch. (保留属名)●
28761 Antunesia O. Hoffm. (1892);安哥拉菊属●☆
28762 Antunesia O. Hoffm. = Vernonia Schreb. (保留属名)●■
28763 Antunesia angolensis (O. Hoffm.) O. Hoffm. = Distephanus angolensis (O. Hoffm.) H. Rob. et B. Kahn ●☆
28764 Antuniaceae Hutch. = Strychnaceae Link ●■
28765 Antuniaceae J. Agardh = Strychnaceae Link ●■
28766 Antura Forssk. = Carissa L. (保留属名)●
28767 Antura edulis Forssk. = Carissa edulis Vahl ●
28768 Antura hadiensis J. F. Gmel. = Carissa spinarum L. ●
28769 Antura paucinervia A. DC. = Carissa spinarum L. ●
28770 Anubias Schott(1857);西非南星属■☆
28771 Anubias affinis De Wild. = Anubias heterophylla Engl. ■☆
28772 Anubias afzelii Schott;西非南星■☆
28773 Anubias auriculata Engl. = Anubias hastifolia Engl. ■☆
28774 Anubias barteri Schott;巴特西非南星■☆
28775 Anubias barteri Schott var. angustifolia (Engl.) Crusio;窄叶巴特西非南星■☆
28776 Anubias barteri Schott var. glabra N. E. Br.;光巴特西非南星■☆
28777 Anubias barteri Schott var. nana (Engl.) Crusio;矮巴特西非南星■☆
28778 Anubias bequaertii De Wild. = Anubias heterophylla Engl. ■☆
28779 Anubias congensis N. E. Br. = Anubias heterophylla Engl. ■☆
28780 Anubias congensis N. E. Br. var. crassispadix Engl. = Anubias heterophylla Engl. ■☆
28781 Anubias engleri De Wild. = Anubias heterophylla Engl. ■☆
28782 Anubias gigantea A. Chev. ex Hutch.;大西非南星■☆
28783 Anubias gigantea A. Chev. ex Hutch. var. robusta Engl. = Anubias gigantea A. Chev. ex Hutch. ■☆
28784 Anubias gigantea A. Chev. ex Hutch. var. tripartita A. Chev. = Anubias gigantea A. Chev. ex Hutch. ■☆
28785 Anubias gilletii De Wild. et T. Durand;吉莱西非南星■☆
28786 Anubias gracilis A. Chev. ex Hutch.;细西非南星■☆
28787 Anubias hastifolia Engl.;戟叶西非南星■☆
28788 Anubias haullevilleana De Wild. = Anubias hastifolia Engl. ■☆
28789 Anubias heterophylla Engl.;互叶西非南星■☆
28790 Anubias lanceolata N. E. Br. = Anubias barteri Schott var. glabra N. E. Br. ■☆
28791 Anubias lanceolata N. E. Br. f. angustifolia Engl. = Anubias barteri Schott var. angustifolia (Engl.) Crusio ■☆
28792 Anubias laurentii De Wild. = Anubias hastifolia Engl. ■☆
28793 Anubias minima A. Chev. = Anubias barteri Schott var. glabra N. E. Br. ■☆
28794 Anubias nana Engl. = Anubias barteri Schott var. nana (Engl.) Crusio ■☆
28795 Anubias pynaertii De Wild.;皮氏西非南星■☆
28796 Anubias undulata Hort. = Anubias heterophylla Engl. ■☆
28797 Anulocaulis Standl. (1909);环带草属;Ringstem ■☆
28798 Anulocaulis Standl. = Boerhavia L. ■
28799 Anulocaulis annulatus (Coville) Standl.;环带草■☆
28800 Anulocaulis eriosolenus (A. Gray) Standl.;毛环带草■☆
28801 Anulocaulis gypsogenus Waterf. = Anulocaulis leiosolenus (Torr.) Standl. var. gypsogenus (Waterf.) Spellenb. et Wootten ■☆
28802 Anulocaulis leiosolenus (Torr.) Standl.;西南环带草;Southwestern Ringstem ■☆
28803 Anulocaulis leiosolenus (Torr.) Standl. var. gypsogenus (Waterf.) Spellenb. et Wootten;喜钙环带草■☆
28804 Anulocaulis leiosolenus Standl. var. gypsogenus (Waterf.) Spellenberg et Wootten = Anulocaulis leiosolenus (Torr.) Standl. var. gypsogenus (Waterf.) Spellenb. et Wootten ■☆
28805 Anumophila Link = Ammophila Host ■☆
28806 Anura (Juz.) Tschern. = Cousinia Cass. ●■
28807 Anura Tschern. = Cousinia Cass. ●■
28808 Anura pallidivirens (Kult.) Tschern.;白绿环带草■☆
28809 Anuragia Raizada = Pogostemon Desf. ●■
28810 Anurosperma (Hook. f.) Hallier f. = Nepenthes L. ●■
28811 Anurosperma Hallier f. = Nepenthes L. ●■

28812 Anurosperma pervillei Hallier f. ;塞岛猪笼草■☆
28813 Anurus C. Presl = Lathyrus L. ■
28814 Anurusperma (Hook. f.) Hallier f. = Nepenthes L. ●■
28815 Anvillea DC. (1836);安维尔菊属(安维菊属,合杯菊属)●☆
28816 Anvillea australis L. Chevall. = Anvillea garcinii (Burm. f.) DC. subsp. radiata (Coss. et Durieu) Anderb. ●☆
28817 Anvillea faurei Gand. = Anvillea garcinii (Burm. f.) DC. subsp. radiata (Coss. et Durieu) Anderb. ●☆
28818 Anvillea garcinii (Burm. f.) DC. ;加氏安维尔菊●☆
28819 Anvillea garcinii (Burm. f.) DC. subsp. radiata (Coss. et Durieu) Anderb. ;辐射安维尔菊●☆
28820 Anvillea garcinii Burm. ;安维尔菊●☆
28821 Anvillea platycarpa (Maire) Anderb. ;宽果安维尔菊●☆
28822 Anvillea radiata Coss. et Durieu = Anvillea garcinii (Burm. f.) DC. subsp. radiata (Coss. et Durieu) Anderb. ●☆
28823 Anvillea radiata Coss. et Durieu var. australis (L. Chevall.) Diels = Anvillea garcinii (Burm. f.) DC. subsp. radiata (Coss. et Durieu) Anderb. ●☆
28824 Anvillea radiata Coss. et Durieu var. genuina Maire = Anvillea garcinii (Burm. f.) DC. subsp. radiata (Coss. et Durieu) Anderb. ●☆
28825 Anvilleina Maire = Anvillea DC. ●☆
28826 Anvilleina Maire(1939);摩洛哥菊属●☆
28827 Anvilleina platycarpa Maire = Anvillea platycarpa (Maire) Anderb. ●☆
28828 Anychia Michx. = Paronychia Mill. ■
28829 Anychia argyrocoma Michx. = Paronychia argentea Lam. ■☆
28830 Anychia baldwinii Torr. et A. Gray = Paronychia baldwinii (Torr. et A. Gray) Fenzl ex Walp. ■☆
28831 Anychia canadensis (L.) Britton, Sterns et Poggenb. = Paronychia canadensis (L.) A. W. Wood ■☆
28832 Anychia canadensis (L.) Elliott = Paronychia canadensis (L.) A. W. Wood ■☆
28833 Anychia dichotoma Michx. = Paronychia canadensis (L.) A. W. Wood ■☆
28834 Anychia fastigiata Raf. = Paronychia fastigiata (Raf.) Fernald ■☆
28835 Anychia herniarioides Michx. = Paronychia herniarioides (Michx.) Nutt. ■☆
28836 Anychia nuttallii Small = Paronychia fastigiata (Raf.) Fernald var. nuttallii (Small) Fernald ■☆
28837 Anychia polygonoides Raf. = Paronychia fastigiata (Raf.) Fernald ■☆
28838 Anychiasatrum baldwinii (Torr. et A. Gray) Small = Paronychia baldwinii (Torr. et A. Gray) Fenzl ex Walp. ■☆
28839 Anychiasatrum riparium (Chapm.) Small = Paronychia baldwinii (Torr. et A. Gray) Fenzl ex Walp. ■☆
28840 Anychiastrum Small = Paronychia Mill. ■
28841 Anychiastrum herniarioides (Michx.) Small = Paronychia herniarioides (Michx.) Nutt. ■☆
28842 Anychiastrum montanum Small = Paronychia fastigiata (Raf.) Fernald var. pumila (A. W. Wood) Fernald ■☆
28843 Anychlastrum Small = Anychia Michx. ■
28844 Anygosanthos Dum. Cours. = Anigozanthos Labill. ■☆
28845 Anygozanthes Schltdl. = Anigozanthos Labill. ■☆
28846 Anygozanthos N. T. Burb. = Anigozanthos Labill. ■☆
28847 Anzybas D. L. Jones et M. A. Clem. (2002);安尼兰属■☆
28848 Aonikena Speg. = Chiropetalum A. Juss. ●☆
28849 Aopla Lindl. = Herminium L. ■
28850 Aopla reniformis Lindl. = Habenaria reniformis (D. Don) Hook. f. ■
28851 Aoranthe Somers(1988);畸花茜属●☆
28852 Aoranthe annulata (K. Schum.) Somers;环状畸花茜●☆
28853 Aoranthe castaneofulva (S. Moore) Somers;栗褐畸花茜●☆
28854 Aoranthe cladantha (K. Schum.) Somers;枝花畸花茜●☆
28855 Aoranthe nalaensis (De Wild.) Somers;纳拉畸花茜●☆
28856 Aoranthe penduliflora (K. Schum.) Somers;垂花畸花茜●☆
28857 Aorchis Verm. (1972);异红门兰属■
28858 Aorchis Verm. = Orchis L. ■
28859 Aorchis cyclochila (Franch. et Sav.) T. Hashim. = Galearis cyclochila (Franch. et Sav.) Soó ■
28860 Aorchis cyclochila (Franch. et Sav.) T. Hashim. = Orchis cyclochila (Franch. et Sav.) Maxim. ■
28861 Aorchis roborovskii (Maxim.) Seidenf. = Orchis roborowskii Maxim. ■
28862 Aorchis roborowskyi (Maxim.) Seidenf. = Galearis roborowskyi (Maxim.) S. C. Chen, P. J. Cribb et S. W. Gale ■
28863 Aorchis spathulata (Lindl.) Verm. = Galearis spathulata (Lindl.) P. F. Hunt ■
28864 Aorchis spathulata (Lindl.) Verm. = Orchis diantha Schltr. ■
28865 Aorchis spathulata (Lindl.) Verm. var. foliosa (Finet) Soó = Galearis spathulata (Lindl.) P. F. Hunt ■
28866 Aorchis spathulata (Lindl.) Verm. var. wilsonii (Schltr.) Soó = Galearis spathulata (Lindl.) P. F. Hunt ■
28867 Aorchis spathulata (Lindl.) Verm. var. wilsonii (Schltr.) Soó = Orchis diantha Schltr. ■
28868 Aosa Weigend(1997);无苞刺莲花属●■☆
28869 Aosa gilgiana (Urb.) Weigend;大无苞刺莲花■☆
28870 Aosa parviflora (Schrad. ex DC.) Weigend;小花无苞刺莲花■☆
28871 Aosa plumieri (Urb.) Weigend;无苞刺莲花■☆
28872 Aostea Buscal. et Muschl. = Vernonia Schreb. (保留属名)●■
28873 Aotus Sm. (1805);枭豆属■☆
28874 Aotus ericoides (Vent.) G. Don;枭豆;Common Aotus ■☆
28875 Aotus gracilis W. T. Aiton ex Loudon;细枭豆■☆
28876 Aotus mollis Benth. ;毛枭豆■☆
28877 Apabuta (Griseb.) Griseb. = Hyperbaena Miers ex Benth. (保留属名)●☆
28878 Apabuta Griseb. = Hyperbaena Miers ex Benth. (保留属名)●☆
28879 Apacheria C. T. Mason(1975);亚利桑那木属●☆
28880 Apacheria chiricahuensis C. T. Mason;亚利桑那木●☆
28881 Apactis Thunb. = Xylosma G. Forst. (保留属名)●
28882 Apactis japonica Thunb. = Xylosma congesta (Lour.) Merr. ●
28883 Apalanthe Planch. (1848);柔花藻属■☆
28884 Apalanthe Planch. = Elodea Michx. ■☆
28885 Apalanthe guyanensis Planch. ;柔花藻■☆
28886 Apalantus Adans. = Callisia Loefl. ■☆
28887 Apalantus Adans. = Hapalanthus Jacq. ■☆
28888 Apalatoa Aubl. (废弃属名) = Crudia Schreb. (保留属名)●☆
28889 Apalochlamys (Cass.) Cass. (1828);锥序棕鼠麹属■☆
28890 Apalochlamys Cass. = Apalochlamys (Cass.) Cass. ■☆
28891 Apalochlamys Cass. = Cassinia R. Br. (保留属名)●☆
28892 Apalochlamys spectabilis (Labill.) J. H. Willis. ;锥序棕鼠麹●☆
28893 Apaloptera Nutt. = Abronia Juss. ■☆
28894 Apaloptera Nutt. ex A. Gray = Abronia Juss. ■☆
28895 Apaloxylon Drake = Neoapaloxylon Rauschert ●■☆
28896 Apaloxylon Drake(1903);马岛豆属■☆

28897 Apaloxylon madagascariense Drake = Neoapaloxylon madagascariense (Drake) Rauschert ●☆
28898 Apaloxylon tuberosum R. Vig. = Neoapaloxylon tuberosum (R. Vig.) Rauschert ■☆
28899 Apalus DC. = Blennosperma Less. ■☆
28900 Apama Lam. (1783);阿柏麻属;Apama ●
28901 Apama Lam. = Thottea Rottb. ●
28902 Apama hainanensis Merr. et Chun;阿柏麻(海南线果兜铃);Hainan Apama ●◇
28903 Apama hainanensis Merr. et Chun = Thottea hainanensis (Merr. et Chun) Ding Hou ●
28904 Apamaceae A. Kern.;阿柏麻科●
28905 Apamaceae A. Kern. = Aristolochiaceae Juss.(保留科名)●■
28906 Apargia Scop. = Leontodon L.(保留属名)■☆
28907 Apargia borealis Bong. = Microseris borealis (Bong.) Sch. Bip. ■☆
28908 Apargia chillensis Kunth = Hypochaeris chillensis (Kunth) Britton ■☆
28909 Apargia hieracioides (L.) Willd. = Picris hieracioides L. ■
28910 Apargia taraxaciflora Viv. = Picris sinuata (Lam.) Lack ■☆
28911 Apargia umbellata (Schrank) Schrank = Picris hieracioides L. ■
28912 Apargidium Torr. et A. Gray = Microseris D. Don ■☆
28913 Apargidium boreale (Bong.) Torr. et A. Gray = Microseris borealis (Bong.) Sch. Bip. ■☆
28914 Aparinaceae Hoffmanns. et Link = Rubiaceae Juss.(保留科名)●■
28915 Aparinanthus Fourr. = Galium L. ●■
28916 Aparine Guett. = Galium L. ●■
28917 Aparine Hill = Galium L. ●■
28918 Aparine Tourn. ex Mill. = Galium L. ●■
28919 Aparinella Fourr. = Galium L. ●■
28920 Aparisthmium Endl. (1840);拟康斯大戟属●☆
28921 Aparisthmium Endl. = Alchornea Sw. ●
28922 Aparisthmium Endl. = Conceveibum A. Rich. ex A. Juss. ●☆
28923 Apartea Pellegr. = Mapania Aubl. ■
28924 Apartea letestui Pellegr. = Mapania amplivaginata K. Schum. ■☆
28925 Apassalus Kobuski(1928);阿帕爵床属■☆
28926 Apassalus cubensis Kobuski;古巴阿帕爵床■☆
28927 Apassalus diffusus Kobuski;阿帕爵床■☆
28928 Apatales Blume ex Ridl. = Liparis Rich.(保留属名)■
28929 Apatanthus Viv. = Hieracium L. ■
28930 Apatelia DC. = Saurauia Willd.(保留属名)●
28931 Apatemone Schott = Schismatoglottis Zoll. et Moritzi ■
28932 Apatesia N. E. Br. (1927);黄苏玉属■☆
28933 Apatesia helianthoides (Aiton) N. E. Br.;黄苏玉■☆
28934 Apatesia maughanii N. E. Br. = Apatesia sabulosa (Thunb.) L. Bolus ■☆
28935 Apatesia pillansii N. E. Br.;皮朗斯黄苏玉■☆
28936 Apatesia sabulosa (Thunb.) L. Bolus;砂地黄苏玉■☆
28937 Apation Blume = Liparis Rich.(保留属名)■
28938 Apation T. Durand et Jacks. = Apatales Blume ex Ridl. ■
28939 Apation T. Durand et Jacks. = Liparis Rich.(保留属名)■
28940 Apatitia Desv. = Bellucia Neck. ex Raf.(保留属名)●☆
28941 Apatitia Desv. ex Ham. = Bellucia Neck. ex Raf.(保留属名)●☆
28942 Apatophyllum McGill. (1971);幻叶卫矛属●☆
28943 Apatophyllum constablei McGill.;幻叶卫矛●☆
28944 Apatophyllum flavovirens A. R. Bean et Jessup;黄绿幻叶卫矛●☆
28945 Apatostelis Garay = Stelis Sw.(保留属名)■☆
28946 Apaturia Lindl. = Pachystoma Blume ■
28947 Apaturia chinensis Lindl. = Pachystoma pubescens Blume ■
28948 Apaturia senilis Lindl. = Pachystoma pubescens Blume ■
28949 Apatzingania Dieterle(1974);阿帕葫芦属■☆
28950 Apatzingania arachoidea Dieterle;阿帕葫芦■☆
28951 Apegla Neck. = Ceropegia L. ■
28952 Apeiba A. Rich. = Entelea R. Br. ●☆
28953 Apeiba Aubl. (1775);热美椴属(阿帕椴属)●☆
28954 Apeiba Aubl. = Sloanea L. ●
28955 Apeiba albiflora Ducke;白花热美椴●☆
28956 Apeiba aspera Aubl.;粗糙阿帕椴●☆
28957 Apeiba discolor G. Don;杂色热美椴●☆
28958 Apeiba echinata Gaertn.;刺热美椴●☆
28959 Apeiba hirsuta Lam.;粗毛热美椴●☆
28960 Apeiba intermedia Uittien;间型热美椴●☆
28961 Apeiba macropetala Ducke;大瓣热美椴●☆
28962 Apeiba membranacea Spruce ex Benth.;膜质热美椴●☆
28963 Apeiba tibourbou Aubl.;热美椴(提布椴);Tibourbou ●☆
28964 Apella Scop. = Laurus L. ●
28965 Apella Scop. = Premna L.(保留属名)●■
28966 Apemon Raf. = Datura L. ●■
28967 Apentostera Raf. = Penstemon Schmidel ●■
28968 Apenula Neck. = Specularia Heist. ex A. DC. ●■☆
28969 Apera Adans. (1763);阿披拉草属;Apera, Silky Bent Grass, Silky-bent ■☆
28970 Apera intermedia Hack.;中间阿披拉草■☆
28971 Apera interrupta (L.) P. Beauv.;间断阿披拉草;Dense Silkybent, Dense Silky-bent, Interrupted Apera, Italian Windgrass, Silky Bent Grass ■☆
28972 Apera spica-ventii (L.) P. Beauv.;阿披拉草;Corn Grass, Corn-grass, Loose Silky Bent, Loose Silkybent, Loose Silky-bent, Silky Apera, Silky Bent-grass, Spica-venti Apera, Wind Bent Grass, Wind Bent-grass, Wind Grass, Wind-grass ■☆
28973 Aperiphracta Nees = Ocotea Aubl. ●☆
28974 Aperiphracta Nees ex Meisn. = Ocotea Aubl. ●☆
28975 Aperula Blume = Lindera Thunb.(保留属名)●
28976 Aperula Gled. = Asperula L.(保留属名)■
28977 Aperula formosana Nakai = Litsea cubeba (Lour.) Pers. var. formosana (Nakai) Yen C. Yang et P. H. Huang ●
28978 Aperula neesiana (Wall. ex Nees) Blume = Lindera neesiana (Nees) Kurz ●
28979 Apetahia Baill. (1882);背裂桔梗属●☆
28980 Apetahia raiateensis Baill.;背裂桔梗●☆
28981 Apetalon Wight = Didymoplexis Griff. ■
28982 Apetalon minutum Wight = Chamaegastrodia vaginata (Hook. f.) Seidenf. ■
28983 Apetiorhamnus Nieuwl. = Rhamnus L. ●
28984 Apetlothamnus Nieuwl. ex Lunell = Rhamnus L. ●
28985 Aphaea Mill. = Lathyrus L. ■
28986 Aphaenandra Miq. (1857);隐蕊茜属●☆
28987 Aphaenandra sumatrana Miq.;隐蕊茜●☆
28988 Aphaenandra uniflora (Wall. ex G. Don) Bremek.;单花隐蕊茜●☆
28989 Aphaerema Miers(1863);巴西大风子属●☆
28990 Aphaerema spicata Miers;巴西大风子●☆
28991 Aphanactis Wedd. (1856);隐舌菊属■☆
28992 Aphanactis jamesoniana Wedd.;隐舌菊■☆
28993 Aphanamixis Blume(1825);山楝属(红罗属,马六甲楝属,裴

赛山楝属);Aphanamixis,Wildmedia ●
28994 Aphanamixis Pierre = Aphanamixis Blume ●
28995 Aphanamixis elmeri Merr. = Aphanamixis tripetala (Blanco) Merr. ●
28996 Aphanamixis grandifolia Blume;大叶山楝(大叶沙罗,红罗木,胡桐,苦柏木,苦油木,罗浪果,山椤,叶好娇);Big-leaved Aphanamixis,Largeleaf Aphanamixis,Largeleaf Wildmedia ●
28997 Aphanamixis polystachya (Wall.) R. Parker;山楝(阿麻拉树,多穗山楝,红果树,红罗,假油桐,沙椤,山罗,穗花树兰,台湾山楝,铁罗,桐油树,小红果,油桐,云连树);Common Wildmedia,Lasua,Polystachyous Aphanamixis,Taiwan Aphanamixis,Taiwan Wildmedia ●
28998 Aphanamixis rohituka Pierre = Aphanamixis polystachya (Wall.) R. Parker ●
28999 Aphanamixis sinensis F. C. How et T. C. Chen;华山楝(中华山楝);China Wildmedia,Chinense Aphanamixis ●
29000 Aphanamixis tripetala (Blanco) Merr. = Aphanamixis polystachya (Wall.) R. Parker ●
29001 Aphanandrium Lindau = Neriacanthus Benth. ■☆
29002 Aphananthe Link(废弃属名) = Aphananthe Planch.(保留属名)●
29003 Aphananthe Link(废弃属名) = Microtea Sw. ■☆
29004 Aphananthe Planch.(1848)(保留属名);糙叶树属;Aphananthe,Roughleaftree ●
29005 Aphananthe aspera (Thunb. ex A. Murray) Planch.;糙叶树(白鸡油,糙皮树,粗叶树,加条,牛筋树,朴树,柔毛糙叶树,沙朴);Muku Tree,Pubescent Aphananthe,Pubescent Roughleaftree,Roughleaftree,Rough-leaved Aphananthe,Scabrous Aphananthe ●
29006 Aphananthe aspera (Thunb. ex A. Murray) Planch. var. pubescens C. J. Chen;柔毛糙叶树;Pubescent Scabrous Aphananthe ●
29007 Aphananthe aspera (Thunb. ex A. Murray) Planch. var. pubescens C. J. Chen = Aphananthe aspera (Thunb. ex A. Murray) Planch. ●
29008 Aphananthe aspera (Thunb.) Planch. = Aphananthe aspera (Thunb. ex A. Murray) Planch. ●
29009 Aphananthe cuspidata (Blume) Planch.;滇糙叶树(光叶白颜树,小白颜树,小叶白颜树,云南白颜树);Sharp-tipped Aphananthe,Small-leaf Gironniera,Small-leaf Villaintree,Yunnan Aphananthe,Yunnan Gironniera,Yunnan Roughleaftree,Yunnan Villaintree ●
29010 Aphananthe lissophylla Gagnep. = Aphananthe cuspidata (Blume) Planch. ●
29011 Aphananthe yunnanensis (Hu) Grudz. = Aphananthe cuspidata (Blume) Planch. ●
29012 Aphananthemum Steud. = Helianthemum Mill. ●■
29013 Aphandra Barfod(1991);隐雄棕属●☆
29014 Aphandra natalia (Balslev et Hend.) Barfod;隐雄棕●☆
29015 Aphanelytrum Hack.(1902);隐鞘草属(隐血草属)■☆
29016 Aphanelytrum Hack. ex Sodiro = Aphanelytrum Hack. ■☆
29017 Aphanelytrum procumbens Hack.;隐鞘草(隐血草)■☆
29018 Aphanes L.(1753);微花蔷薇属(隐花蔷薇属);Parsley Piert,Parsley-piert ●■☆
29019 Aphanes L. = Alchemilla L. ■
29020 Aphanes arvensis L. = Alchemilla arvensis (L.) Scop. ■☆
29021 Aphanes australis Rydb.;澳洲微花蔷薇■☆
29022 Aphanes bachiti (Hauman et Balle) Rothm.;巴氏微花蔷薇■☆
29023 Aphanes cornucopioides Lag.;非洲微花蔷薇■☆
29024 Aphanes floribunda (Murb.) Rothm.;多花微花蔷薇■☆
29025 Aphanes inexspectata W. Lippert;小果微花蔷薇;Parsley Piert,Slender Parsley Piert,Slender Parsley-piert ■☆
29026 Aphanes maroccana Hyl. et Rothm.;摩洛哥微花蔷薇■☆
29027 Aphanes microcarpa (Boiss. et Reut.) Rothm. = Aphanes inexspectata W. Lippert ■☆
29028 Aphanes minutiflora (Azn.) Holub;极微花蔷薇■☆
29029 Aphanes pusilla (Pomel) Batt.;微小微花蔷薇■☆
29030 Aphania Blume = Lepisanthes Blume ●
29031 Aphania Blume(1825);滇赤才属(滇赤材属);Aphania ●
29032 Aphania golungensis Hiern = Pancovia golungensis (Hiern) Exell et Mendonça ●☆
29033 Aphania oligophylla (Merr. et Chun) H. S. Lo = Lepisanthes oligophylla (Merr. et Chun) N. H. Xia et Gadek ●
29034 Aphania rubra (Roxb.) Radlk. = Lepisanthes senegalensis (Juss. ex Poir.) Leenh. ●
29035 Aphania senegalensis (Juss. ex Poir.) Radlk. = Lepisanthes senegalensis (Juss. ex Poir.) Leenh. ●
29036 Aphania senegalensis (Juss. ex Poir.) Radlk. var. sylvatica (A. Chev. ex Hutch. et Dalziel) Aubrév. = Lepisanthes senegalensis (Juss. ex Poir.) Leenh. ●
29037 Aphania silvatica A. Chev. ex Hutch. et Dalziel = Lepisanthes senegalensis (Juss. ex Poir.) Leenh. ●
29038 Aphanisma Nutt. = Aphanisma Nutt. ex Moq. ■☆
29039 Aphanisma Nutt. ex Moq.(1849);无针苋属(卡州藜属)■☆
29040 Aphanisma blitoides Nutt. ex Moq.;无针苋■☆
29041 Aphanocalyx Oliv.(1870);隐萼异花豆属■☆
29042 Aphanocalyx cynometroides Oliv.;茎花隐萼异花豆■☆
29043 Aphanocalyx djumaensis (De Wild.) J. Léonard;朱马隐萼异花豆■☆
29044 Aphanocalyx hedinii (A. Chev.) Wieringa;海丁隐萼异花豆●☆
29045 Aphanocalyx heitzii (Pellegr.) Wieringa;海茨隐萼异花豆●☆
29046 Aphanocalyx jenseniae (Gram) Wieringa;詹森隐萼异花豆■☆
29047 Aphanocalyx ledermannii (Harms) Wieringa;莱德隐萼异花豆■☆
29048 Aphanocalyx margininervatus J. Léonard;边脉隐萼异花豆■☆
29049 Aphanocalyx microphyllus (Harms) Wieringa;小叶隐萼异花豆■☆
29050 Aphanocalyx microphyllus (Harms) Wieringa subsp. compactus (Hutch. ex Lane-Poole) Wieringa;紧密隐萼异花豆■☆
29051 Aphanocalyx obscurus Wieringa;隐萼异花豆■☆
29052 Aphanocalyx pectinatus (A. Chev.) Wieringa;篦状隐萼异花豆■☆
29053 Aphanocalyx pteridophyllus (Harms) Wieringa;蕨叶隐萼异花豆■☆
29054 Aphanocalyx richardsiae (J. Léonard) Wieringa;理查兹隐萼异花豆●☆
29055 Aphanocalyx trapnellii (J. Léonard) Wieringa;特拉普内尔隐萼异花豆■☆
29056 Aphanocarpus Steyerm.(1965);南美隐果茜属●☆
29057 Aphanocarpus steyermarkii (Standl.) Steyerm.;南美隐果茜●☆
29058 Aphanochaeta A. Gray = Pentachaeta Nutt. ■☆
29059 Aphanochilus Benth. = Elsholtzia Willd. ●■
29060 Aphanochilus blandus Benth. = Elsholtzia blanda (Benth.) Benth. ■
29061 Aphanochilus communis Kudo = Elsholtzia cypriani (Pavol.) S. Chow ex Y. C. Hsu ■
29062 Aphanochilus eriostachyus Benth. = Elsholtzia eriostachya Benth. ■

29063 Aphanochilus flavus Benth. = Elsholtzia flava (Benth.) Benth. ●■
29064 Aphanochilus foetens Benth. = Elsholtzia stachyodes (Link) C. Y. Wu ■
29065 Aphanochilus fruticosus (D. Don) Kudo = Elsholtzia fruticosa (D. Don) Rehder ●
29066 Aphanochilus fruticosus (D. Don) Kudo var. ochroleuca (Dunn) Kudo = Elsholtzia ochroleuca Dunn ●
29067 Aphanochilus fruticosus (D. Don) Kudo var. tomentella (Rehder) Kudo = Elsholtzia myosurus Dunn ●■
29068 Aphanochilus incisus Benth. = Elsholtzia stachyodes (Link) C. Y. Wu ■
29069 Aphanochilus myosurus (Dunn) Kudo = Elsholtzia myosurus Dunn ●■
29070 Aphanochilus paniculatus Benth. = Elsholtzia stachyodes (Link) C. Y. Wu ■
29071 Aphanochilus penduliflorus (W. W. Sm.) Kudo = Elsholtzia penduliflora W. W. Sm. ●■
29072 Aphanochilus pilosus Benth. = Elsholtzia pilosa (Benth.) Benth. ■
29073 Aphanochilus polystachys Benth. = Elsholtzia fruticosa (D. Don) Rehder ●
29074 Aphanochilus rugulosus (Hemsl.) Kudo = Elsholtzia rugulosa Hemsl. ●■
29075 Aphanochilus stauntonii (Benth.) Kudo = Elsholtzia stauntonii Benth. ●
29076 Aphanococcus Radlk. = Lepisanthes Blume ●
29077 Aphanodon Naudin = Henriettella Naudin ●☆
29078 Aphanomyrtus Miq. = Syzygium R. Br. ex Gaertn. (保留属名) ●
29079 Aphanomyxis DC. = Aphanamixis Blume ●
29080 Aphanopappus Endl. = Lipochaeta DC. ■☆
29081 Aphanopetalaceae Doweld = Aphanopetalaceae Endl. ●☆
29082 Aphanopetalaceae Doweld(2001);隐瓣藤科(胶藤科)●☆
29083 Aphanopetalum Endl. (1839);隐瓣藤属(胶藤属)●☆
29084 Aphanopetalum resinosum Endl.;隐瓣藤●☆
29085 Aphanopleura Boiss. (1872);隐棱芹属;Aphanopleura ■
29086 Aphanopleura capillifolia (Regel et Schmalh.) Lipsky;细叶隐棱芹;Hairyleaf Aphanopleura ■
29087 Aphanopleura leptoclada (Aitch. et Hemsl.) Lipsky;细枝隐棱芹;Thinbranch Aphanopleura ■
29088 Aphanopleura trachysperma Boiss.;糙籽隐棱芹■☆
29089 Aphanosperma T. F. Daniel(1988);隐籽爵床属■☆
29090 Aphanosperma sinaloensis (Léonard et Gentry) T. F. Daniel;隐籽爵床■☆
29091 Aphanostelma Malme = Melinia Decne. ■☆
29092 Aphanostelma Schltr. = Metaplexis R. Br. ●■
29093 Aphanostelma chinensis Schltr. = Metaplexis hemsleyana Oliv. ●■
29094 Aphanostelma chinensis Schltr. ex H. Lév. = Metaplexis hemsleyana Oliv. ●■
29095 Aphanostemma A. St. -Hil. (1824);长萼毛茛属■☆
29096 Aphanostemma A. St. -Hil. = Ranunculus L. ■
29097 Aphanostemma Willis = Aphanostelma Schltr. ●■
29098 Aphanostemma apiifolia A. St. -Hil.;长萼毛茛■☆
29099 Aphanostephus DC. (1836);惰雏菊属;Lazydaisy ■☆
29100 Aphanostephus arizonicus A. Gray = Aphanostephus ramosissimus DC. var. humilis (Benth.) B. L. Turner et Birdsong ■☆
29101 Aphanostephus humilis (Benth.) A. Gray = Aphanostephus ramosissimus DC. var. humilis (Benth.) B. L. Turner et Birdsong ■☆
29102 Aphanostephus pilosus Buckley;毛惰雏菊;Hairy Lazydaisy ■☆
29103 Aphanostephus ramosissimus DC.;惰雏菊;Plains Lazydaisy ■☆
29104 Aphanostephus ramosissimus DC. var. humilis (Benth.) B. L. Turner et Birdsong;小惰雏菊■☆
29105 Aphanostephus riddellii Torr. et A. Gray;理德尔惰雏菊;Riddell's Lazydaisy ■☆
29106 Aphanostephus skirrhobasis (DC.) Trel.;堪萨斯惰雏菊(阿肯色惰雏菊);Arkansas Doze-daisy, Arkansas Lazydaisy ■☆
29107 Aphanostylis Pierre = Landolphia P. Beauv. (保留属名) ●☆
29108 Aphanostylis exserens (K. Schum.) Pierre = Landolphia incerta (K. Schum.) Pers. ●☆
29109 Aphanostylis flavidiflora (K. Schum.) Pierre = Landolphia flavidiflora (K. Schum.) Pers. ●☆
29110 Aphanostylis laxiflora (K. Schum.) Pierre = Landolphia incerta (K. Schum.) Pers. ●☆
29111 Aphanostylis leptantha (K. Schum.) Pierre = Landolphia leptantha (K. Schum.) Pers. ●☆
29112 Aphanostylis mammosa (Pierre) Pierre ex T. Durand et H. Durand = Ancylobotrys scandens (Schumach. et Thonn.) Pichon ●☆
29113 Aphanostylis mammosa (Pierre) Pierre ex T. Durand et H. Durand var. mucronata (Dewèvre) Pierre ex T. Durand et H. Durand = Ancylobotrys scandens (Schumach. et Thonn.) Pichon ●☆
29114 Aphanostylis mannii (Stapf) Pierre = Landolphia incerta (K. Schum.) Pers. ●☆
29115 Aphanostylis pyramidata Pierre = Landolphia pyramidata (Pierre) Pers. ●☆
29116 Aphanostylis robusta (Pierre) Pierre ex T. Durand et H. Durand = Ancylobotrys pyriformis Pierre ●☆
29117 Aphantochacta A. Gray = Pentachaeta Nutt. ■☆
29118 Aphantochaeta A. Gray = Chaetopappa DC. ■☆
29119 Aphantochaeta exilis A. Gray = Pentachaeta exilis (A. Gray) A. Gray ■☆
29120 Apharica Schltdl. = Aphania Blume ●
29121 Aphelandra R. Br. (1810);单药爵床属(单药花属,金叶木属);Aphelandra ●■☆
29122 Aphelandra aurantiaca (Scheidw.) Lindl. var. roezlii (Carrière) Nichols = Aphelandra aurantiaca Lindl. 'Roezlii' ●
29123 Aphelandra aurantiaca Lindl.;橙黄单药爵床(橙黄热美爵床,红单药花);Orange Aphelandra ●☆
29124 Aphelandra aurantiaca Lindl. 'Roezlii';红单药花(罗兹尔橙黄热美爵床)●
29125 Aphelandra leopoldii Van Houtte;金叶木单药爵床●☆
29126 Aphelandra sinclairiana Nees;红花热美爵床(辛莱单药爵床);Coral Aphelandra ●☆
29127 Aphelandra squarrosa Nees;单药爵床(斑马热美爵床,粗单药爵床,单药花,金脉单药花,银脉爵床);Saffron Spike, Saffron's Spike, Tiger Plant, Zebra Plant ■☆
29128 Aphelandra squarrosa Nees 'Claire';克莱尔单药爵床(克莱尔斑马热美爵床)●☆
29129 Aphelandra squarrosa Nees 'Dania';达尼单药爵床(达尼斑马热美爵床,戴尼亚单药爵床)●☆
29130 Aphelandra squarrosa Nees 'Lousiae';银脉单药爵床(斑马爵床,鲁依塞斑马热美爵床,银脉单药花)●☆
29131 Aphelandra squarrosa Nees 'Snow Queen';雪皇后单药爵床(雪皇后斑马热美爵床)●☆
29132 Aphelandra squarrosa Nees var. leopoldii Van Houtte = Aphelandra leopoldii Van Houtte ●☆

29133 Aphelandra squarrosa Nees var. lousiae Van Houtte = Aphelandra squarrosa Nees 'Lousiae' ●☆
29134 Aphelandrella Mildbr. (1926);小单药爵床属●■☆
29135 Aphelandrella modesta Mildbr.;小单药爵床●■☆
29136 Aphelandros St.-Lag. = Aphelandra R. Br. ●■☆
29137 Aphelexis D. Don = Edmondia Cass. ●■☆
29138 Aphelexis candollei Bojer ex DC. = Helichrysum candollei (Bojer ex DC.) R. Vig. et Humbert ■☆
29139 Aphelexis flexuosa Baker = Helichrysum hypnoides (DC.) R. Vig. et Humbert ●☆
29140 Aphelexis hypnoides DC. = Helichrysum hypnoides (DC.) R. Vig. et Humbert ●☆
29141 Aphelexis lycopodioides Bojer ex DC. = Helichrysum benthamii R. Vig. et Humbert ●☆
29142 Aphelexis selaginifolia DC. = Helichrysum selaginifolium (DC.) R. Vig. et Humbert ●☆
29143 Aphelexis stenoclada Baker = Helichrysum benthamii R. Vig. et Humbert ●☆
29144 Aphelia R. Br. (1810);独鳞草属■☆
29145 Aphelia cyperoides R. Br.;独鳞草■☆
29146 Aphelia gracilis Sond.;细独鳞草■☆
29147 Aphelia monogyna Hieron.;单蕊独鳞草■☆
29148 Aphillanthes Neck. = Aphyllanthes Tourn. ex L. ●☆
29149 Aphloia (DC.) Benn. (1840);球花柞属●☆
29150 Aphloia Benn. = Aphloia (DC.) Benn. ●☆
29151 Aphloia myrtiflora Galpin = Aphloia theiformis (Vahl) Benn. ●☆
29152 Aphloia theiformis (Vahl) Benn.;球花柞●☆
29153 Aphloiaceae Takht. (1985);球花柞科(单果树科)●☆
29154 Aphloiaceae Takht. = Flacourtiaceae Rich. ex DC. (保留科名)●
29155 Aphoma Raf. (废弃属名) = Iphigenia Kunth(保留属名)■
29156 Aphomonlx Raf. = Saxifraga L. ■
29157 Aphonina Neck. = Pariana Aubl. ■☆
29158 Aphora Neck. = Virgilia Poir. (保留属名)●☆
29159 Aphora Neck. ex Kuntze = Virgilia Poir. (保留属名)●☆
29160 Aphora Nutt. = Argythamnia P. Browne ●☆
29161 Aphora Nutt. = Ditaxis Vahl ex A. Juss. ●☆
29162 Aphragmia Nees = Ruellia L. ●■
29163 Aphragmus Andrz. = Aphragmus Andrz. ex DC. ■
29164 Aphragmus Andrz. ex DC. (1824);寒原荠属(失隔芥属,失膈荠属);Aphragmus ■
29165 Aphragmus bouffordii Al-Shehbaz;布氏寒原荠■
29166 Aphragmus himalaicus O. E. Schulz = Lignariella obscura (Dunn) Jafri ■☆
29167 Aphragmus involacratus (Bunge) O. E. Schulz;寒原荠■☆
29168 Aphragmus obscurus (Dunn) O. E. Schulz = Lignariella obscura (Dunn) Jafri ■☆
29169 Aphragmus oxycarpus (Hook. f. et Thomson) Jafri;尖果寒原荠(高山肉叶荠,尖果肉叶荠);Sharpfruit Aphragmus ■
29170 Aphragmus oxycarpus (Hook. f. et Thomson) Jafri var. glaber (Vassilcz.) C. H. An = Aphragmus oxycarpus (Hook. f. et Thomson) Jafri ■
29171 Aphragmus oxycarpus (Hook. f. et Thomson) Jafri var. glaber C. H. An;无毛寒原荠;Glabrous Sharpfruit Aphragmus, Smooth Aphragmus ■
29172 Aphragmus oxycarpus (Hook. f. et Thomson) Jafri var. microcarpus C. H. An;小果寒原荠;Smallfruit Aphragmus, Smallfruit Sharpfruit Aphragmus ■
29173 Aphragmus oxycarpus (Hook. f. et Thomson) Jafri var. microcarpus C. H. An = Aphragmus oxycarpus (Hook. f. et Thomson) Jafri ■
29174 Aphragmus oxycarpus (Hook. f. et Thomson) Jafri var. stenocarpus (O. E. O. E. Schulz) G. C. Das = Aphragmus oxycarpus (Hook. f. et Thomson) Jafri ■
29175 Aphragmus przewalskii (Maxim.) A. L. Ebel = Aphragmus oxycarpus (Hook. f. et Thomson) Jafri ■
29176 Aphragmus stewartii O. E. Schulz = Aphragmus oxycarpus (Hook. f. et Thomson) Jafri ■
29177 Aphragmus tibeticus O. E. Schulz;西藏寒原荠(寒原荠);Xizang Aphragmus ■
29178 Aphragmus tibeticus O. E. Schulz = Aphragmus oxycarpus (Hook. f. et Thomson) Jafri ■
29179 Aphragrnia Nees = Ruellia L. ●■
29180 Aphylax Salisb. = Aneilema R. Br. ■☆
29181 Aphyllangis Thouars = Angraecum Bory ■
29182 Aphyllangis Thouars = Solenangis Schltr. ■☆
29183 Aphyllanthaceae Burnett = Liliaceae Juss. (保留科名)●■
29184 Aphyllanthaceae G. T. Burnett(1835);无叶花科(星捧月科)■☆
29185 Aphyllanthaceae J. Agardh = Anthericaceae J. Agardh ●■☆
29186 Aphyllanthaceae J. Agardh = Aphyllanthaceae G. T. Burnett ■☆
29187 Aphyllanthaceae J. Agardh = Liliaceae Juss. (保留科名)●■
29188 Aphyllanthes L. (1753);无叶花属●☆
29189 Aphyllanthes Tourn. ex L. = Aphyllanthes L. ●☆
29190 Aphyllanthes monspeliensis L.;无叶花●☆
29191 Aphyllarum S. Moore = Caladium Vent. ■
29192 Aphylleia Champ. = Sciaphila Blume ■
29193 Aphyllocaulon Lag. = Gerbera L. (保留属名)■
29194 Aphyllocladus Wedd. (1855);凋叶菊属●
29195 Aphyllocladus Wedd. = Hyalis D. Don ex Hook. et Arn. ●☆
29196 Aphyllocladus spartioides Wedd.;凋叶菊●☆
29197 Aphyllodium (DC.) Gagnep. (1916);两节豆属●■
29198 Aphyllodium (DC.) Gagnep. = Dicerma DC. ●
29199 Aphyllodium (DC.) Gagnep. = Hedysarum L. (保留属名)●■
29200 Aphyllodium Gagnep. = Aphyllodium (DC.) Gagnep. ●■
29201 Aphyllodium Gagnep. = Hedysarum L. (保留属名)●■
29202 Aphyllodium australiense (Schindler) H. Ohashi = Aphyllodium biarticulatum (L.) Gagnep. ●
29203 Aphyllodium biarticulatum (L.) Gagnep.;两节豆●
29204 Aphyllodium biarticulatum Gagnep. = Aphyllodium biarticulatum (L.) Gagnep. ●
29205 Aphyllodium biarticulatum Gagnep. = Dicerma biarticulatum (L.) DC. ●
29206 Aphyllon Mitch. = Orobanche L. ■
29207 Aphyllon arenosum Suksd. = Orobanche ludoviciana Nutt. ■☆
29208 Aphyllon minutum Suksd. = Orobanche uniflora L. ■☆
29209 Aphyllon sedii Suksd. = Orobanche uniflora L. ■☆
29210 Aphyllon uniflorum (L.) A. Gray = Orobanche uniflora L. ■☆
29211 Aphyllon uniflorum (L.) A. Gray var. occidentale Greene = Orobanche uniflora L. ■☆
29212 Aphyllorchis Blume (1825);无叶兰属;Aphyllorchis, Leaflessorchis ■
29213 Aphyllorchis alpina King et Pantl.;高山无叶兰;Alp Leaflessorchis ■
29214 Aphyllorchis caudata Rolfe ex Downie;尾萼无叶兰;Tailcalyx Leaflessorchis ■

29215 Aphyllorchis gollanii Duthie; 大花无叶兰; Bigflower Leaflessorchis ■
29216 Aphyllorchis montana Rchb. f.;无叶兰(山林无叶兰);Common Leaflessorchis, Montane Aphyllorchis ■
29217 Aphyllorchis parviflora King et Pantl. = Neottia acuminata Schltr. ■
29218 Aphyllorchis prainii Hook. f. = Aphyllorchis montana Rchb. f. ■
29219 Aphyllorchis purpurea Fukuy. = Aphyllorchis montana Rchb. f. ■
29220 Aphyllorchis simplex Ts. Tang et F. T. Wang;单唇无叶兰(梅兰);Singlelip Leaflessorchis ■
29221 Aphyllorchis simplex Ts. Tang et F. T. Wang = Sinorchis simplex (Ts. Tang et F. T. Wang) S. C. Chen ●
29222 Aphyllorchis tanegashimensis Hayata = Aphyllorchis montana Rchb. f. ■
29223 Aphyllorchis unguiculata Rolfe ex Downie = Aphyllorchis montana Rchb. f. ■
29224 Aphyllorchis vaginata Hook. f. = Chamaegastrodia vaginata (Hook. f.) Seidenf. ■
29225 Aphyteia L. = Hydnora Thunb. ■☆
29226 Aphyteia acharii Steud. = Hydnora africana Thunb. ■☆
29227 Aphyteia africana (Thunb.) Oken = Hydnora africana Thunb. ■☆
29228 Aphyteia hydnora L. = Hydnora africana Thunb. ■☆
29229 Aphyteia multiceps Burch. = Cytinus sanguineus (Thunb.) Fourc. ■☆
29230 Apiaceae Lindl. (1836)(保留科名);伞形花科(伞形科);Carrot Family ●■
29231 Apiaceae Lindl. (保留科名) = Umbelliferae Juss. (保留科名)●■
29232 Apiastrum Nutt. = Apiastrum Nutt. ex Torr. et A. Gray ■☆
29233 Apiastrum Nutt. ex Torr. et A. Gray(1840);拟芹属■☆
29234 Apiastrum angustifolium Nutt.;窄叶拟芹■☆
29235 Apiastrum latifolium Nutt.;宽叶拟芹■☆
29236 Apiastrum patens (Nutt. ex DC.) J. M. Coult. et Rose;拟芹■☆
29237 Apicra Willd. = Haworthia Duval(保留属名)■☆
29238 Apicra albicans Willd. = Haworthia marginata (Lem.) Stearn ■☆
29239 Apicra arachnoides Willd. = Haworthia arachnoidea (L.) Duval ■☆
29240 Apicra aspera (Haw.) Willd. = Astroloba corrugata N. L. Mey. et G. F. Sm. ■☆
29241 Apicra aspera (Haw.) Willd. var. major Haw. = Haworthia aspera Haw. var. major (Haw.) Parr ■☆
29242 Apicra atrovirens Willd. = Haworthia herbacea (Mill.) Stearn ■☆
29243 Apicra bullulata (Jacq.) Willd. = Astroloba bullulata (Jacq.) Uitewaal ■☆
29244 Apicra congesta (Salm-Dyck) Baker = Astroloba congesta (Salm-Dyck) Uitewaal ■☆
29245 Apicra cymbifolia Willd. = Haworthia cymbiformis (Haw.) Duval ■☆
29246 Apicra deltoidea (Hook. f.) Baker = Astroloba congesta (Salm-Dyck) Uitewaal ■☆
29247 Apicra deltoidea (Hook. f.) Baker var. intermedia A. Berger = Astroloba congesta (Salm-Dyck) Uitewaal ■☆
29248 Apicra egregia Poelln. = Astroloba bullulata (Jacq.) Uitewaal ■☆
29249 Apicra fasciata Willd. = Haworthia fasciata (Willd.) Haw. ■☆
29250 Apicra foliolosa (Haw.) Willd. = Astroloba foliolosa (Haw.) Uitewaal ■☆
29251 Apicra granata Willd. = Haworthia minima (Aiton) Haw. ■☆
29252 Apicra jacobseniana Poelln. = Astroloba rubriflora (L. Bolus) G. F. Sm. et J. C. Manning ■☆
29253 Apicra margaritifera Willd. = Haworthia maxima (Haw.) Duval ■☆
29254 Apicra maxima Steud. = Haworthia maxima (Haw.) Duval ■☆
29255 Apicra minor Steud. = Haworthia minima (Aiton) Haw. ■☆
29256 Apicra mirabilis Willd. = Haworthia mirabilis Haw. ■☆
29257 Apicra nigra Haw. = Haworthia nigra Baker ■☆
29258 Apicra pentagona (Aiton) Willd. var. bullulata (Jacq.) Baker = Astroloba bullulata (Jacq.) Uitewaal ■☆
29259 Apicra pentagona (Aiton) Willd. var. torulosa Haw. = Astroloba spiralis (L.) Uitewaal ■☆
29260 Apicra pumilio Willd. = Haworthia pumila (L.) M. B. Bayer ■☆
29261 Apicra radula Willd. = Haworthia attenuata Haw. var. radula (Jacq.) M. B. Bayer ■☆
29262 Apicra recurva Willd. = Haworthia venosa (Lam.) Haw. ■☆
29263 Apicra reticulata Willd. = Haworthia reticulata (Haw.) Haw. ■☆
29264 Apicra retusa Willd. = Haworthia retusa Duval ■☆
29265 Apicra rubriflora L. Bolus = Astroloba rubriflora (L. Bolus) G. F. Sm. et J. C. Manning ■☆
29266 Apicra spiralis (L.) Baker = Astroloba spiralis (L.) Uitewaal ■☆
29267 Apicra spiralis (L.) Willd. = Astroloba spiralis (L.) Uitewaal ■☆
29268 Apicra tortuosa Willd. = Haworthia viscosa Haw. ■☆
29269 Apicra translucens Willd. = Haworthia herbacea (Mill.) Stearn ■☆
29270 Apicra tricolor Willd. = Haworthia venosa (Lam.) Haw. ■☆
29271 Apicra turgida Baker = Astroloba congesta (Salm-Dyck) Uitewaal ■☆
29272 Apicra viscosa Willd. = Haworthia viscosa Haw. ■☆
29273 Apilia Raf. = Fraxinus L. ●
29274 Apinagia Tul. emend. P. Royen(1849);南美川苔草属■☆
29275 Apinagia latifolia (K. I. Goebel) P. Royen;宽叶南美川苔草■☆
29276 Apinagia longifolia (Tul.) P. Royen;长叶南美川苔草■☆
29277 Apinagia minor P. Royen;小南美川苔草■☆
29278 Apinagia parvifolia P. Royen;小叶南美川苔草■☆
29279 Apinella Kuntze = Trinia Hoffm. (保留属名)■☆
29280 Apinella Neck. = Trinia Hoffm. (保留属名)■☆
29281 Apinella Neck. ex Raf. = Trinia Hoffm. (保留属名)■☆
29282 Apinus Neck. = Pinus L. ●
29283 Apinus Neck. ex Rydb. = Pinus L. ●
29284 Apinus albicaulis (Engelm.) Rydb. = Pinus albicaulis Engelm. ●☆
29285 Apinus flexilis (E. James) Rydb. = Pinus flexilis E. James ●☆
29286 Apinus koraiensis (Siebold et Zucc.) Moldenke = Pinus koraiensis Siebold et Zucc. ●◇
29287 Apiocarpus Montrouz. = ? Harpullia Roxb. ●
29288 Apiocarpus Montrouz. = Akania Hook. f. ●☆
29289 Apiopetalum Baill. (1878);梨瓣五加属●☆
29290 Apiopetalum arboreum Baker f.;北方梨瓣五加●☆
29291 Apiopetalum glabratum Baill.;光梨瓣五加●☆
29292 Apiopetalum penneli R. Vig.;梨瓣五加●☆
29293 Apiopetalum velutinum Baill.;黏梨瓣五加●☆
29294 Apios Boehm. = Glycine Willd. (保留属名)■
29295 Apios Fabr. (1759)(保留属名);土圞儿属(九子羊属,九子洋属,块茎豆属);Apios, Groudnut, Potato Bean, Potatobean ●
29296 Apios Medik. = Apios Fabr. (保留属名)●
29297 Apios Moench = Apios Fabr. (保留属名)●
29298 Apios americana Medik.;北美土圞儿(美洲土圞儿);American Groundnut, American Potato Bean, American Potatobean, Common Groundnut, Dakota Potato, Ground Nut, Groundnut, Indian Potato,

Indian-potato, Potato Bean, Potatobean, Potato-bean, Rosary Root, Wild Bean ●☆
29299 Apios americana Medik. f. pilosa Steyerm. = Apios americana Medik. ●☆
29300 Apios americana Medik. var. turrigera Fernald = Apios americana Medik. ●☆
29301 Apios carnea (Wall.) Benth.;肉色土圞儿(满塘红,肉土圞儿,鸭嘴花);Fleshcolor Apios,Fleshcolour Apios ●
29302 Apios cavaleriei H. Lév. = Apios fortune Maxim. ●
29303 Apios delavayi Franch.;云南土圞儿(德氏土圞儿,菊架豆);Delavay Apios,Yunnan Apios,Yunnan Potatobean ●
29304 Apios fortunei Maxim.;土圞儿(地栗子,黄皮狗圞,金线吊葫芦,九连珠,九莲珠,九牛子,九子草,九子羊,疬子薯,罗汉参,三叶青,食用土圞儿,土蛋,土鸡蛋,土凉薯,土子,野凉薯,野绿豆,子鸡生蛋);Fortune Apios ●
29305 Apios gracillima Dunn;纤细土圞儿;Apios, Gracile, Gracile Potatobean,Slender Apios ●
29306 Apios macrantha Oliv.;大花土圞儿;Bigflower Apios, Long Flowered Potatobean ●
29307 Apios taiwaniana Hosok.;台湾土圞儿;Taiwan Apios ●
29308 Apios tuberosa Moench = Apios americana Medik. ●☆
29309 Apiospermum Klotzsch = Pistia L. ■
29310 Apirophorum Neck. = Pyrus L. ●
29311 Apista Blume = Podochilus Blume ■
29312 Apium L. (1753);芹属(旱芹属);Celery,Marshwort ■
29313 Apium ammi (Jacq.) Urb. = Apium leptophyllum (Pers.) F. Muell. ex Benth. ■
29314 Apium ammi Crantz = Ammi majus L. ■
29315 Apium ammi Urb. = Apium leptophyllum (Pers.) F. Muell. ex Benth. ■
29316 Apium ammi Urb. var. genuinum H. Wolff = Apium leptophyllum (Pers.) F. Muell. ex Benth. ■
29317 Apium ammi Urb. var. leptophyllum Chodat = Apium leptophyllum (Pers.) F. Muell. ex Benth. ■
29318 Apium anisum (L.) Crantz = Pimpinella anisum L. ■
29319 Apium australe Thouars;南芹;Green Celery, Maori Celery, Prostrate Celery ■☆
29320 Apium cicutifolium (Schrenk) Benth. et Hook. ex Forbes et Hemsl. = Sium suave Walter ■
29321 Apium cicutifolium Benth. et Hook. ex Forbes et Hemsl. = Sium suave Walter ■
29322 Apium crassipes (Rchb.) Rchb. f.;粗梗旱芹■☆
29323 Apium crispum Mill. = Petroselinum crispum (Mill.) Nyman ex A. W. Hill ■
29324 Apium decumbens Eckl. et Zeyh. = Apium graveolens L. ■
29325 Apium distachyus L. f.;双穗芹;Water Hawthorn ■☆
29326 Apium fenestralis Hook. f.;窗格芹;Lace-leaf,Lattice-leaf ■☆
29327 Apium graveolens L.;旱芹(旱菜,荷兰鸭儿芹,堇,蒲芹,芹,芹菜,细叶芹菜,香芹,药芹,药芹菜,野芹,野园荽);Celery, Dry Celery, Italian Upright Celery, Least Marshwort, March, Marsh Parsley, Marshwort, Merch, Mile, Mudweed, Salry, Smalach, Smallache, Smalladge, Smallage, Smalledge, Sollery, Upright Italian Celery, Water Parsley, Wild Celery ■
29328 Apium graveolens L. var. dulce (Mill.) Pers.;野芹;Celery, Garden Celery ■☆
29329 Apium graveolens L. var. rapaceum (Mill.) DC.;球根塘芹;Celeriac,German Celery,Turniprooted Celery,Turnip-rooted Celery ■☆
29330 Apium humile (Meisn.) Benth. et Hook. f. = Sonderina humilis (Meisn.) H. Wolff ■☆
29331 Apium integrilobum Hayata = Apium graveolens L. ■
29332 Apium inundatum (L.) Rchb. f.;小欧芹;Lesser Marshwort ■☆
29333 Apium inundatum Rchb. f. = Apium inundatum (L.) Rchb. f. ■☆
29334 Apium involucratum Roxb. = Trachyspermum roxburghianum (DC.) H. Wolff ■
29335 Apium leptophyllum (Pers.) F. Muell. = Apium leptophyllum (Pers.) F. Muell. ex Benth. ■
29336 Apium leptophyllum (Pers.) F. Muell. = Cyclospermum leptophyllum (Pers.) Sprague ■
29337 Apium leptophyllum (Pers.) F. Muell. ex Benth. = Cyclospermum leptophyllum (Pers.) Sprague ex Britton et P. Wilson ■
29338 Apium leptophyllum (Pers.) F. Muell. ex Benth. = Cyclospermum leptophyllum (Pers.) Sprague ■
29339 Apium nodiflorum (L.) Lag. = Apium nodiflorum (L.) Rchb. f. ■☆
29340 Apium nodiflorum (L.) Lag. subsp. mairei Molina Abril et Sardinero = Apium nodiflorum (L.) Lag. ■☆
29341 Apium nodiflorum (L.) Rchb. f.;匍匐芹;Bilders, Billers, Brooklime, Cow Cress, European Marshwort, Fool's Watercress, Fool's Water-cress, Pie Cress, Pie-cress, Procumbent Marshwort, Sion,Swine's Cress,Water Case ■☆
29342 Apium petroselinum L. = Petroselinum crispum (Mill.) Nyman ex A. W. Hill ■
29343 Apium prostratum Labill. = Apium prostratum Vent. ■☆
29344 Apium prostratum Vent.;匍匐旱芹;Prostrate Marshwort ■☆
29345 Apium repens (Jacq.) Lag.;扩散旱芹;Creeping Marshwort ■☆
29346 Apium repens Rchb. f. = Apium repens (Jacq.) Lag. ■☆
29347 Apium tenuifolium (Moench) Tehll. = Apium leptophyllum (Pers.) F. Muell. ex Benth. ■
29348 Apium tenuifolium (Moench) Tehll. = Cyclospermum leptophyllum (Pers.) Sprague ■
29349 Apium visnaga (L.) Crantz = Ammi visnaga (L.) Lam. ■
29350 Apium visnaga L. = Ammi visnaga (L.) Lam. ■
29351 Apivea Steud. = Aiouea Aubl. ●☆
29352 Aplactia Raf. = Solidago L. ■
29353 Aplanodes Marais(1966);土蓍荠属■☆
29354 Aplanodes doidgeana Marais;土蓍荠■☆
29355 Aplanodes sisymbrioides (Schltr.) Marais;大蒜芥土蓍荠■☆
29356 Aplarina Raf. = Euphorbia L. ●■
29357 Aplectra Raf. = Aplectrum (Nutt.) Torr. ■☆
29358 Aplectrocapnos Boiss. et Reut. = Sarcocapnos DC. ■☆
29359 Aplectrum (Nutt.) Torr. (1826);北美无距兰属(拟杜鹃兰属);Adam-and-eve,Putty-root ■☆
29360 Aplectrum Blume = Anplectrum A. Gray ●■
29361 Aplectrum Blume = Diplectria (Blume) Rchb. ●■
29362 Aplectrum Nutt. = Aplectrum (Nutt.) Torr. ■☆
29363 Aplectrum Torr. = Aplectrum (Nutt.) Torr. ■☆
29364 Aplectrum appendiculatum (Blume) F. Maek. = Cremastra appendiculata (D. Don) Makino ■
29365 Aplectrum hyemale (Muhl. ex Willd.) Torr.;北美无距兰;Adam and Eve, Adam And Eve Orchid, Adam-and-eve, Putty Root, Puttyroot, Putty-root ■☆
29366 Aplectrum spicatum Britton, Sterns et Poggenb. = Aplectrum hyemale (Muhl. ex Willd.) Torr. ■☆
29367 Aplectrum unguiculatum (Finet) F. Maek. = Cremastra

unguiculata (Finet) Finet ■
29368 Apleura Phil. = Azorella Lam. ■☆
29369 Aplexia Raf. = Leersia Sw. (保留属名) ■
29370 Aplilia Raf. = Fraxinus L. ●
29371 Aplina Raf. = Staehelina L. ●☆
29372 Aploca Neck. = Periploca L. ●
29373 Aploca Neck. ex Kuntze = Oxystelma R. Br. ●■
29374 Aploca Neck. ex Kuntze = Periploca L. ●
29375 Aplocarya Lindl. = Nolana L. ex L. f. ■☆
29376 Aplocera Raf. = Ctenium Panz. (保留属名) ■☆
29377 Aplochlamis Steud. = Apalochlamys Cass. ●☆
29378 Aplochlamis Steud. = Cassinia R. Br. (保留属名) ●☆
29379 Aploleia Raf. = Callisia Loefl. ■☆
29380 Aplolophium Cham. = Haplolophium Cham. (保留属名) ●☆
29381 Aplopappus Cass. = Haplopappus Cass. (保留属名) ●■☆
29382 Aplophyllum Cass. (废弃属名) = Haplophyllum A. Juss. (保留属名) ●■
29383 Aplophyllum Cass. (废弃属名) = Mutisia L. f. ●☆
29384 Aplostellis A. Rich. = Nervilia Comm. ex Gaudich. (保留属名) ■
29385 Aplostellis A. Rich. = Stellorkis Thouars(废弃属名) ■
29386 Aplostellis Thouars = Nervilia Comm. ex Gaudich. (保留属名) ■
29387 Aplostellis ambigua A. Rich. = Nervilia simplex (Thouars) Schltr. ■☆
29388 Aplostellis flabelliformis (Lindl.) Ridl. = Nervilia aragoana Gaudich. ■
29389 Aplostellis velutina (E. C. Parish et Rchb. f.) Ridl. = Nervilia plicata (Andréws) Schltr. ■
29390 Aplostemon Raf. = Fimbristylis Vahl(保留属名) ■
29391 Aplostemon Raf. = Scirpus L. (保留属名) ■
29392 Aplostylis Raf. = Cuscuta L. ■
29393 Aplotaxia circioides DC. = Cirsium lanatum (Roxb. ex Willd.) Spreng. ■
29394 Aplotaxis DC. = Saussurea DC. (保留属名) ●■
29395 Aplotaxis andryaloides DC. = Saussurea andryaloides (DC.) Sch. Bip. ■
29396 Aplotaxis auriculata DC. = Saussurea auriculata (DC.) Sch. Bip. ■
29397 Aplotaxis bungei DC. = Hemistepta lyrata (Bunge) Bunge ■
29398 Aplotaxis carthamoides (Buch.-Ham. ex DC.) DC. = Hemistepta lyrata (Bunge) Bunge ■
29399 Aplotaxis circioides DC. = Cirsium lanatum (Roxb. ex Willd.) Spreng. ■
29400 Aplotaxis deltoidea DC. = Saussurea deltoidea (DC.) Sch. Bip. ■
29401 Aplotaxis denticulata Wall. ex DC. = Saussurea fastuosa (Decne.) Sch. Bip. ■
29402 Aplotaxis denticulata Wall. ex DC. var. glabrata DC. = Saussurea glabrata (DC.) C. Shih ■
29403 Aplotaxis denticulata Wall. ex DC. var. hypoleuca DC. = Saussurea fastuosa (Decne.) Sch. Bip. ■
29404 Aplotaxis fastuosa Decne. = Saussurea fastuosa (Decne.) Sch. Bip. ■
29405 Aplotaxis gnaphaloides Royle = Saussurea gnaphaloides (Royle) Sch. Bip. ■
29406 Aplotaxis gossypina DC. = Saussurea gossypiphora D. Don ■
29407 Aplotaxis gossypina DC. var. minor DC. = Saussurea simpsoniana (Fielding et Gardner) Lipsch. ■
29408 Aplotaxis involucrata Kar. et Kir. = Saussurea involucrata (Kar. et Kir.) Sch. Bip. ■
29409 Aplotaxis lappa Decne. = Saussurea costus (Falc.) Lipsch. ■
29410 Aplotaxis lenontodontoides DC. = Saussurea leontodontoides (DC.) Sch. Bip. ■
29411 Aplotaxis leontodontoides DC. = Saussurea leontodontoides (DC.) Sch. Bip. ■
29412 Aplotaxis nepalensis (Spreng.) DC. = Saussurea nepalensis Spreng. ■
29413 Aplotaxis nivea DC. = Saussurea crispa Vaniot ■
29414 Aplotaxis obvallata DC. = Saussurea obvallata (DC.) Edgew. ■
29415 Aplotaxis simpsoniana Fielding et Gardner = Saussurea simpsoniana (Fielding et Gardner) Lipsch. ■
29416 Aplotaxis sorocephala (Schrenk) Schrenk = Saussurea gnaphaloides (Royle) Sch. Bip. ■
29417 Aplotaxis sorocephala Schrenk = Saussurea gnaphalodes (Royle) Sch. Bip. ■
29418 Aplotaxis uniflora DC. = Saussurea uniflora (DC.) Wall. ex Sch. Bip. ■
29419 Aplotaxis uniflora Wall. ex DC. = Saussurea uniflora (Wall. ex DC.) Sch. Bip. ■
29420 Aplotaxis uniflora Wall. ex DC. var. sinensis J. Anthony = Saussurea uniflora (Wall. ex DC.) Sch. Bip. ■
29421 Aplotheca Mart. ex Cham. = Froelichia Moench ■☆
29422 Apluda L. (1753);水蔗草属;Apluda ■
29423 Apluda P. Beauv. = Anadelphia Hack. ■☆
29424 Apluda aristata L. = Apluda mutica L. ■
29425 Apluda communis Nees = Apluda mutica L. ■
29426 Apluda digitata L. f. = Polytoca digitata (L. f.) Druce ■
29427 Apluda geniculata Roxb. = Apluda mutica L. ■
29428 Apluda inermis Regel = Apluda mutica L. ■
29429 Apluda microstachya Nees = Apluda mutica L. ■
29430 Apluda mutica L.;水蔗草(崩疮草,假雀麦,米草,糯米草,秋米草,水蔗,丝线草,牙尖草,竹子草);Apluda, Common Apluda, Mauritian Grass ■
29431 Apluda mutica L. var. aristata (L.) Hack. = Apluda mutica L. ■
29432 Apluda mutica L. var. aristata (L.) Hack. ex Barker = Apluda mutica L. ■
29433 Apluda varia Hack. = Apluda mutica L. ■
29434 Apoballis Sehott = Schismatoglottis Zoll. et Moritzi ■
29435 Apocaulon R. S. Cowan(1953);离茎芸香属●☆
29436 Apocaulon carnosum R. S. Cowan;离茎芸香●☆
29437 Apochaete (C. E. Hubb.) J. B. Phipps = Tristachya Nees ■☆
29438 Apochaete auronitens (P. A. Duvign.) J. B. Phipps = Tristachya auronitens P. A. Duvign. ■☆
29439 Apochaete hispida (L. f.) J. B. Phipps = Tristachya leucothrix Trin. ex Nees ■☆
29440 Apochaete thollonii (Franch.) J. B. Phipps = Tristachya thollonii Franch. ■☆
29441 Apochiton C. E. Hubb. (1936);离颖草属■☆
29442 Apochiton burttii C. E. Hubb.;离颖草■☆
29443 Apochloa Zuloaga et Morrone(2008);离禾属■☆
29444 Apochoris Duby = Lysimachia L. ●■
29445 Apochoris pentapetala (Bunge) Duby = Lysimachia pentapetala Bunge ■
29446 Apochoris pentapetala Duby = Lysimachia pentapetala Bunge ■
29447 Apoclada McClure(1967);离枝竹属●☆
29448 Apoclada arenicola McClure;离枝竹●☆

29449 Apocopis Nees(1841);楔颖草属;Cunealglume,Apocopis ■

29450 Apocopis breviglumis Keng et S. L. Chen;短颖楔颖草;Shortglume Cunealglume ■

29451 Apocopis heterogamus Keng et S. L. Chen = Apocopis intermedia (A. Camus) Chai-Anan ■

29452 Apocopis himalayensis (Steud.) W. Watson = Apocopis paleacea (Trin.) Hochr. ■

29453 Apocopis intermedia (A. Camus) Chai-Anan;异穗楔颖草;Twosexflower Cunealglume ■

29454 Apocopis paleacea (Trin.) Hochr.;楔颖草;Cunealglume,Pallet Apocopis ■

29455 Apocopis royleana Nees = Apocopis paleacea (Trin.) Hochr. ■

29456 Apocopis tridentata Benth. var. intermedia (A. Camus) Roberty = Apocopis intermedia (A. Camus) Chai-Anan ■

29457 Apocopis wrightii Munro;瑞氏楔颖草(曲芒楔颖草);Wright Apocopis,Wright Cunealglume ■

29458 Apocopis wrightii Munro var. macrantha S. L. Chen;大花楔颖草;Bigflower Wright Cunealglume ■

29459 Apocopis wrightii Munro var. macrantha S. L. Chen = Apocopis intermedia (A. Camus) Chai-Anan ■

29460 Apocopsis Meisn. = Apocopis Nees ■

29461 Apocynaceae Adans. = Apocynaceae Juss.(保留科名)●■

29462 Apocynaceae Juss.(1789)(保留科名);夹竹桃科;Dogbane Family,Periwinkle Family ●■

29463 Apocynastrum Fabr. = Apocynum L. ●■

29464 Apocynastrum Heist. ex Fabr. = Apocynum L. ●■

29465 Apocynum L.(1753);罗布麻属(草夹竹桃属,茶叶花属,红麻属);Dog Bane,Dogbane,Indian Hemp ●■

29466 Apocynum album Greene var. hypericifolium A. Gray = Apocynum cannabinum L. ●☆

29467 Apocynum alterniflorum Lour. = Gymnema sylvestre (Retz.) Schult. ●

29468 Apocynum androsaemifolium L.;美国罗布麻(美国茶叶花);Angel's Turnip,Bitter Root,Dogsbane,Flytrap,Milkweed,Pink-flowered Dogbane,Spreading Dogbane ●■

29469 Apocynum androsaemifolium L. subsp. androsaemifolium var. incanum A. DC. = Apocynum androsaemifolium L. ●■

29470 Apocynum armenum Pobed.;亚美尼亚罗布麻●☆

29471 Apocynum basikurumon H. Hara = Apocynum venetum L. var. basikurumon (H. Hara) H. Hara ●☆

29472 Apocynum basikurumon H. Hara = Apocynum venetum L. ●

29473 Apocynum cannabinum L.;加拿大麻(夹竹桃麻,大麻叶罗布麻);American Hemp,Amyroot,Bitter Root,Black Indian Hemp,Bowman's Root,Canadian Hemp,Choctaw Root,Clasping-leaved Dogbane,Dogbane,Hemp Dogbane,Hemp-dogbane,Indian Hemp,Indian Physic,Milkweed,Ogbane,Prairie Dogbane,Rheumatism Weed,Wild Cotton ●☆

29474 Apocynum cannabinum L. var. angustifolium (Wooton) N. H. Holmgren = Apocynum cannabinum L. ●☆

29475 Apocynum cannabinum L. var. glaberrimum A. DC. = Apocynum cannabinum L. ●☆

29476 Apocynum cannabinum L. var. greeneanum (Bég. et Bél.) Woodson = Apocynum cannabinum L. ●☆

29477 Apocynum cannabinum L. var. hypericifolium (Aiton) A. Gray = Apocynum sibiricum Jacq. ●☆

29478 Apocynum cannabinum L. var. nemorale (G. S. Mill.) Fernald = Apocynum cannabinum L. ●☆

29479 Apocynum cannabinum L. var. pubescens (Mitch. ex R. Br.) A. DC. = Apocynum cannabinum L. ●☆

29480 Apocynum cannabinum L. var. suksdorfii (Greene) Bég. et Bél. = Apocynum cannabinum L. ●☆

29481 Apocynum cordatum Thunb. = Tylophora cordata (Thunb.) Druce ●☆

29482 Apocynum filiforme L. f. = Eustegia filiformis (L. f.) Schult. ■☆

29483 Apocynum floribundum Greene;杂种繁花罗布麻;Hybrid Dogbane,Intermediate Dogbane ●☆

29484 Apocynum frutescens L. = Apocynum pictum Schrenk ●

29485 Apocynum frutescens L. = Ichnocarpus frutescens (L.) W. T. Aiton ●

29486 Apocynum hastatum Thunb. = Eustegia minuta (L. f.) R. Br. ■☆

29487 Apocynum hendersonii Hook. f. = Apocynum pictum Schrenk ●

29488 Apocynum hypericifolium Aiton = Apocynum sibiricum Jacq. ●☆

29489 Apocynum jonesii Woodson = Apocynum floribundum Greene ●☆

29490 Apocynum juventas Lour. = Streptocaulon juventas (Lour.) Merr. ■

29491 Apocynum lanceolatum Thunb. = Oncinema lineare (L. f.) Bullock ●☆

29492 Apocynum lancifolium Russanov;红麻(披针叶罗布麻)●☆

29493 Apocynum lancifolium Russanov = Apocynum venetum L. ●

29494 Apocynum lineare L. f. = Oncinema lineare (L. f.) Bullock ●☆

29495 Apocynum medium Greene = Apocynum floribundum Greene ●☆

29496 Apocynum medium Greene var. floribundum (Greene) Woodson = Apocynum floribundum Greene ●☆

29497 Apocynum medium Greene var. leuconeuron (Greene) Woodson = Apocynum floribundum Greene ●☆

29498 Apocynum medium Greene var. lividum (Greene) Woodson = Apocynum floribundum Greene ●☆

29499 Apocynum medium Greene var. sarniense (Greene) Woodson = Apocynum floribundum Greene ●☆

29500 Apocynum medium Greene var. vestitum (Greene) Woodson = Apocynum floribundum Greene ●☆

29501 Apocynum milleri Britton = Apocynum floribundum Greene ●☆

29502 Apocynum minutum L. f. = Eustegia minuta (L. f.) R. Br. ■☆

29503 Apocynum mucronatum Blanco = Jasminanthes mucronata (Blanco) W. D. Stevens et P. T. Li ●

29504 Apocynum mucronatum Blanco = Jasminanthes pilosa (Kerr) W. D. Stevens et P. T. Li ●

29505 Apocynum mucronatum Blanco = Stephanotis mucronata (Blanco) Merr. ●

29506 Apocynum pictum Schrenk;白麻(大花罗布麻,大叶白麻,大叶罗布麻,野麻,着色罗布麻,紫斑罗布麻);Common Poacynum,Largeleaf Poacynum ●

29507 Apocynum pubescens Mitch. ex R. Br. = Apocynum cannabinum L. ●☆

29508 Apocynum pumilum (A. Gray) Greene var. rhomboideum (Greene) Bég. et Bél. = Apocynum androsaemifolium L. ●■

29509 Apocynum rusaanovii Pobed.;鲁氏罗布麻●☆

29510 Apocynum sarmatiense (Woodson) O. D. Wissjul.;萨马提罗布麻(萨尔马罗布麻)●☆

29511 Apocynum scabrum Russanov;糙麻●☆

29512 Apocynum scabrum Russanov = Trachomitum venetum (L.) Woodson subsp. scabrum (Russanov) Rech. f. ●☆

29513 Apocynum scopulorum Greene ex Rydb. = Apocynum androsaemifolium L. ●■

29514 Apocynum sibiricum Jacq.；西伯利亚罗布麻；Clasping Dogbane，Dogbane，Indian Hemp ●☆
29515 Apocynum sibiricum Jacq. = Apocynum cannabinum L. ●☆
29516 Apocynum sibiricum Jacq. var. cordigerum (Greene) Fernald = Apocynum sibiricum Jacq. ●☆
29517 Apocynum sibiricum Jacq. var. farwellii (Greene) Fernald = Apocynum sibiricum Jacq. ●☆
29518 Apocynum sibiricum Jacq. var. salignum (Greene) Fernald = Apocynum sibiricum Jacq. ●☆
29519 Apocynum sibiricum Pall. ex Roem. et Schult. = Apocynum venetum L. ●
29520 Apocynum suksdorfii Greene = Apocynum cannabinum L. ●☆
29521 Apocynum suksdorfii Greene var. angustifolium (Wooton) Woodson = Apocynum cannabinum L. ●☆
29522 Apocynum tauricum Pobed.；克里木罗布麻●☆
29523 Apocynum triflorum L. f. = Astephanus triflorus (L. f.) Schult. ■☆
29524 Apocynum venetum L. = Trachomitum venetum (L.) Woodson ●
29525 Apocynum venetum L. var. basikurumon (H. Hara) H. Hara = Apocynum venetum L. ●
29526 Apocynum venetum L. var. ellipticifolium Bég. et Bél. = Apocynum venetum L. ●
29527 Apocynum venetum L. var. microphyllum Bég. et Bél. = Apocynum venetum L. ●
29528 Apocynum venetum L. var. scabrum Bég. et Belosersky = Trachomitum venetum (L.) Woodson subsp. scabrum (Russanov) Rech. f. ●☆
29529 Apodandra Pax et K. Hoffm. (1919)；梗蕊大戟属●☆
29530 Apodandra buchtienii Pax；梗蕊大戟●☆
29531 Apodanthaceae (R. Br.) Takht. = Rafflesiaceae Dumort.(保留科名)■
29532 Apodanthaceae Takht. = Rafflesiaceae Dumort.(保留科名)■
29533 Apodanthaceae Tiegh. = Apodanthaceae Tiegh. ex Takht. ■☆
29534 Apodanthaceae Tiegh. = Rafflesiaceae Dumort.(保留科名)■
29535 Apodanthaceae Tiegh. ex Takht. (1987)；无柄花科(离花科)■☆
29536 Apodanthera Arn. (1841)；温美葫芦属■☆
29537 Apodanthera undulata A. Gray；温美葫芦；Melon Loco，Melon-Loco ■☆
29538 Apodanthes Poit. (1824)；无柄花属■☆
29539 Apodanthes caseariae Poit.；无柄花■☆
29540 Apoda-prorepentia (Luer) Luer = Pleurothallis R. Br. ■☆
29541 Apodasmia B. G. Briggs et L. A. S. Johnson(1998)；短被帚灯草属■☆
29542 Apodasmia brownii (Hook. f.) B. G. Briggs et L. A. S. Johnson；布朗短被帚灯草■☆
29543 Apodasmia chilensis (Gay) B. G. Briggs et L. A. S. Johnson；智利短被帚灯草■☆
29544 Apodasmia similis (Edgar) B. G. Briggs et L. A. S. Johnson；短被帚灯草■☆
29545 Apodicarpum Makino(1891)；无梗果芹属■☆
29546 Apodicarpum ikenoi Makino；无梗果芹■☆
29547 Apodina Tiegh. = Loranthus Jacq.(保留属名)●
29548 Apodina Tiegh. = Psittacanthus Mart. ●☆
29549 Apodiscus Hutch. (1912)；盘柄大戟属☆
29550 Apodiscus chevalieri Hutch.；盘柄大戟☆
29551 Apodocephala Baker(1885)；马达加斯加菊属●☆
29552 Apodocephala angustifolia Humbert；窄叶马达加斯加菊●☆
29553 Apodocephala begueana Humbert；布氏马达加斯加菊●☆
29554 Apodocephala minor Scott-Elliot；小马达加斯加菊●☆
29555 Apodocephala multiflora Humbert；多花马达加斯加菊●☆
29556 Apodocephala oliganthoides Humbert；拟少花马达加斯加菊●☆
29557 Apodocephala pauciflora Baker；少花马达加斯加菊●☆
29558 Apodocephala radula Humbert；刮刀马达加斯加菊●☆
29559 Apodocephala urschiana Humbert；马达加斯加菊●☆
29560 Apodolirion Baker(1878)；无梗石蒜属■☆
29561 Apodolirion bolusii Baker；博卢斯无梗石蒜■☆
29562 Apodolirion buchananii Baker；布坎南无梗石蒜■☆
29563 Apodolirion cedarbergense D. Müll. -Doblies；锡达伯格无梗石蒜■☆
29564 Apodolirion ettae Baker = Apodolirion buchananii Baker ■☆
29565 Apodolirion lanceolatum (L. f.) Benth. et Hook. f.；披针形无梗石蒜■☆
29566 Apodolirion mackenii Baker = Apodolirion buchananii Baker ■☆
29567 Apodolirion macowanii Baker；麦克欧文无梗石蒜■☆
29568 Apodostachys Turcz. = Ercilla A. Juss. ●☆
29569 Apodostigma R. Wilczek(1956)；无梗柱卫矛属■☆
29570 Apodostigma pallens (Oliv.) R. Wilczek = Apodostigma pallens (Planch. ex Oliv.) R. Wilczek ■☆
29571 Apodostigma pallens (Planch. ex Oliv.) R. Wilczek；变苍白柱卫矛■☆
29572 Apodostigma pallens (Planch. ex Oliv.) R. Wilczek var. buchholzii (Loes.) N. Hallé；布赫无梗柱卫矛■☆
29573 Apodostigma pallens (Planch. ex Oliv.) R. Wilczek var. dummeri Kovács-Lang；杜默变苍白柱卫矛■☆
29574 Apodynomene E. Mey. = Tephrosia Pers.(保留属名)●■
29575 Apodynomene aemula E. Mey. = Tephrosia macropoda (E. Mey.) Harv. var. diffusa (E. Mey.) Schrire ■☆
29576 Apodynomene diffusa E. Mey. = Tephrosia macropoda (E. Mey.) Harv. var. diffusa (E. Mey.) Schrire ■☆
29577 Apodynomene grandiflora E. Mey. = Tephrosia grandiflora (Aiton) Pers. ●☆
29578 Apodynomene macropoda E. Mey. = Tephrosia macropoda (E. Mey.) Harv. ■☆
29579 Apodytes Arn. = Apodytes E. Mey. ex Arn. ●
29580 Apodytes E. Mey. = Apodytes E. Mey. ex Arn. ●
29581 Apodytes E. Mey. ex Arn. (1840)；柴龙树属(柴龙属)；Apodytes ●
29582 Apodytes abbottii Potg. et A. E. van Wyk；阿巴特柴龙树●☆
29583 Apodytes acutifolia Hochst. ex A. Rich. = Apodytes dimidiata E. Mey. ex Arn. ●
29584 Apodytes beninensis Hook. f. ex Planch. = Rhaphiostylis beninensis (Hook. f. ex Planch.) Planch. ex Benth. ●☆
29585 Apodytes bequaertii De Wild. = Apodytes dimidiata E. Mey. ex Arn. ●
29586 Apodytes cambodiana Pierre = Apodytes dimidiata E. Mey. ex Arn. ●
29587 Apodytes dimidiata E. Mey. ex Arn.；柴龙树(白梨柴龙树，柴龙木)；Cambodia Apodytes，Halved Apodytes，Mugonione，White Pear ●
29588 Apodytes dimidiata E. Mey. ex Arn. f. farinosa H. Perrier = Apodytes dimidiata E. Mey. ex Arn. ●
29589 Apodytes dimidiata E. Mey. ex Arn. f. microphylla H. Perrier = Apodytes dimidiata E. Mey. ex Arn. ●
29590 Apodytes dimidiata E. Mey. ex Arn. subsp. acutifolia (Hochst. ex A. Rich.) Cufod. = Apodytes dimidiata E. Mey. ex Arn. ●
29591 Apodytes dimidiata E. Mey. ex Arn. var. acutifolia (Hochst. ex A.

Rich.) Boutique = Apodytes dimidiata E. Mey. ex Arn. ●
29592 Apodytes dimidiata E. Mey. ex Arn. var. hazomaitso (Danguy) H. Perrier = Apodytes dimidiata E. Mey. ex Arn. ●
29593 Apodytes dimidiata E. Mey. ex Arn. var. ikongoensis H. Perrier = Apodytes dimidiata E. Mey. ex Arn. ●
29594 Apodytes dimidiata E. Mey. ex Arn. var. inversa (Baill. ex Grandid.) H. Perrier = Apodytes dimidiata E. Mey. ex Arn. ●
29595 Apodytes dimidiata E. Mey. ex Arn. var. thouvenotii (Danguy) H. Perrier = Apodytes thouvenotii Danguy ●☆
29596 Apodytes emirnensis Baker = Apodytes dimidiata E. Mey. ex Arn. ●
29597 Apodytes frappieri Cordem. = Apodytes dimidiata E. Mey. ex Arn. ●
29598 Apodytes grandifolia Benth. et Hook. f. ex B. D. Jacks.;大叶柴龙树●☆
29599 Apodytes hazomaitso Danguy = Apodytes dimidiata E. Mey. ex Arn. ●
29600 Apodytes inversa Baill. = Apodytes dimidiata E. Mey. ex Arn. ●
29601 Apodytes inversa Baill. ex Grandid. = Apodytes dimidiata E. Mey. ex Arn. ●
29602 Apodytes macrocarpa Capuron = Apodytes dimidiata E. Mey. ex Arn. ●
29603 Apodytes macrocarpa Capuron = Apodytes grandifolia Benth. et Hook. f. ex B. D. Jacks. ●☆
29604 Apodytes mauritiana Benth. et Hook. f. = Apodytes dimidiata E. Mey. ex Arn. ●
29605 Apodytes stuhlmannii Engl. = Apodytes dimidiata E. Mey. ex Arn. ●
29606 Apodytes thouarsiana Baill. = Apodytes dimidiata E. Mey. ex Arn. ●
29607 Apodytes thouarsiana Baill. = Potameia thouarsiana (Baill.) Capuron ●☆
29608 Apodytes thouvenotii Danguy;图弗诺柴龙树●☆
29609 Apodytes yunnanensis Hu = Apodytes dimidiata E. Mey. ex Arn. ●
29610 Apogandrum Neck. = Erica L. ●☆
29611 Apogeton Schrad. ex Steud. = Aponogeton L. f. (保留属名)■
29612 Apogon Elliott = Krigia Schreb. (保留属名)■☆
29613 Apogon Elliott = Serinia Raf. ■☆
29614 Apogon Steud. = Chloris Sw. ●■
29615 Apogon gracilis DC. = Krigia cespitosa (Raf.) K. L. Chambers var. gracilis (DC.) K. L. Chambers ■☆
29616 Apogon wrightii A. Gray = Krigia wrightii (A. Gray) K. L. Chambers ex K. J. Kim ■☆
29617 Apogonia (Nutt.) E. Fourn. = Coelorachis Brongn. ■
29618 Apogonia E. Fourn. = Coelorachis Brongn. ■
29619 Apogonia E. Fourn. = Rottboellia L. f. (保留属名)■
29620 Apoia Merr. = Sarcosperma Hook. f. ●
29621 Apolanesia Rchb. = Apoplanesia C. Presl ■☆
29622 Apolepsis (Blume) Hassk. = Lepidagathis Willd. ●■
29623 Apolepsis Hassk. = Lepidagathis Willd. ●■
29624 Apoleya Gleason = Apuleia Mart. (保留属名)●☆
29625 Apolgusa Raf. = Lecokia DC. ■☆
29626 Apollonias Nees(1833);太阳楠属(印度樟属)●☆
29627 Apollonias barbujana (Cav.) Bornm.;太阳楠●☆
29628 Apollonias grandiflora Kosterm. = Beilschmiedia velutina (Kosterm.) Kosterm. ●☆
29629 Apollonias madagascariensis (Baill.) Kosterm. = Beilschmiedia madagascariensis (Baill.) Kosterm. ●☆
29630 Apollonias microphylla Kosterm. = Beilschmiedia microphylla (Kosterm.) Kosterm. ●☆
29631 Apollonias oppositifolia Kosterm. = Beilschmiedia opposita Kosterm. ●☆
29632 Apollonias sericea Kosterm. = Beilschmiedia sericans Kosterm. ●☆
29633 Apollonias velutina Kosterm. = Beilschmiedia velutina (Kosterm.) Kosterm. ●☆
29634 Apomaea Neck. = Ipomoea L. (保留属名)●■
29635 Apomoea Steud. = Apomaea Neck. ●■
29636 Apomoea Steud. = Ipomoea L. (保留属名)●■
29637 Apomuria Bremek. = Psychotria L. (保留属名)●
29638 Apomuria punctata (Vatke) Bremek. = Psychotria punctata Vatke ●☆
29639 Aponoa Raf. = Columnea L. ●■☆
29640 Aponoa Raf. = Limnophila R. Br. (保留属名)■
29641 Aponogeton Hill(废弃属名) = Aponogeton L. f. (保留属名)■
29642 Aponogeton Hill(废弃属名) = Zannichellia L. ■
29643 Aponogeton L. f. (1782) (保留属名);水蕹属(田干草属);Cape Pondweed, Lacewort, Water Hawthorn, Waterhawthorn, Water-hawthorn ■
29644 Aponogeton abyssinicum Hochst. ex A. Rich.;阿比西尼亚水蕹■☆
29645 Aponogeton abyssinicum Hochst. ex A. Rich. var. albiflorum Lye;白花阿比西尼亚水蕹■☆
29646 Aponogeton abyssinicum Hochst. ex A. Rich. var. cordatum Lye;心叶阿比西尼亚水蕹■☆
29647 Aponogeton abyssinicum Hochst. ex A. Rich. var. glanduliferum Lye;腺点阿比西尼亚水蕹■☆
29648 Aponogeton abyssinicum Hochst. ex A. Rich. var. graminifolium Lye;禾叶阿比西尼亚水蕹■☆
29649 Aponogeton afroviolaceum Lye;暗堇色水蕹■☆
29650 Aponogeton afroviolaceum Lye var. angustifolium Lye;窄叶暗堇色水蕹■☆
29651 Aponogeton angustifolium Aiton;窄叶水蕹■☆
29652 Aponogeton appendiculatum H. Bruggen;具附属体水蕹;Appendiculate Waterhawthorn ■☆
29653 Aponogeton azureum H. Bruggen;天蓝水蕹■☆
29654 Aponogeton boehmii Engl. = Aponogeton abyssinicum Hochst. ex A. Rich. ■☆
29655 Aponogeton bogneri H. Bruggen;博格纳水蕹■☆
29656 Aponogeton braunii K. Krause = Aponogeton abyssinicum Hochst. ex A. Rich. ■☆
29657 Aponogeton crinifolium Lehm. ex Schltdl. = Aponogeton angustifolium Aiton ■☆
29658 Aponogeton desertorum Zeyh. ex A. Spreng.;荒漠水蕹■☆
29659 Aponogeton dinteri Engl. et K. Krause = Aponogeton desertorum Zeyh. ex A. Spreng. ■☆
29660 Aponogeton distachyos L. f.;二穗水蕹(长柄浪草,长柄水蕹);Cape Asparagus, Cape Pondweed, Cape Waterhawthorn, Cape-pondweed, Water Hawthorn, Water Uintje, Water-hawthorn ■
29661 Aponogeton distachyos L. f. var. lagrangei André = Aponogeton distachyos L. f. ■
29662 Aponogeton eylesii Rendle = Aponogeton desertorum Zeyh. ex A. Spreng. ■☆
29663 Aponogeton fenestralis (Pers.) Hook. f.;膜孔水蕹;Lace-leaf, Lattice-leaf ■☆
29664 Aponogeton fenestralis (Pers.) Hook. f. = Aponogeton madagascariense (Mirb.) Bruggen ■☆

29665 Aponogeton gracilis Schinz ex A. Benn. = Aponogeton stuhlmannii Engl. ■☆
29666 Aponogeton gramineum Lye = Aponogeton stuhlmannii Engl. ■☆
29667 Aponogeton hereroense Schinz = Aponogeton rehmannii Oliv. ■☆
29668 Aponogeton heudelotii (Kunth) Engl. = Aponogeton subconjugatum Schumach. et Thonn. ■
29669 Aponogeton holubii Oliv. = Aponogeton desertorum Zeyh. ex A. Spreng. ■☆
29670 Aponogeton junceum Lehm. ;灯心草水蕹■☆
29671 Aponogeton junceum Lehm. subsp. natalense (Oliv.) Oberm. = Aponogeton natalensis Oliv. ■☆
29672 Aponogeton kraussianum Hochst. ex Krauss = Aponogeton desertorum Zeyh. ex A. Spreng. ■☆
29673 Aponogeton lakhonensm A. Camus; 水蕹(田干菜); Common Waterhawthorn, Waterhawthorn ■
29674 Aponogeton leptostachyum E. Mey. ex Engl. = Aponogeton desertorum Zeyh. ex A. Spreng. ■☆
29675 Aponogeton leptostachyum E. Mey. ex Engl. var. minor Baker = Aponogeton abyssinicus Hochst. ex A. Rich. ■☆
29676 Aponogeton madagascariense (Mirb.) Bruggen; 马达加斯加水蕹; Lace-leaf ■☆
29677 Aponogeton natalensis Oliv. ;纳塔尔水蕹■☆
29678 Aponogeton natans (L.) Engl. et Krause; 田干草(水蕹); Water Hawthorn ■
29679 Aponogeton nudiflorus Peter; 裸花水蕹■☆
29680 Aponogeton nudiflorus Peter var. angustifolius ? = Aponogeton nudiflorus Peter ■☆
29681 Aponogeton oblongus Peter = Aponogeton abyssinicus Hochst. ex A. Rich. ■☆
29682 Aponogeton oblongus Troupin = Aponogeton rehmannii Oliv. ■☆
29683 Aponogeton pygmaeum Krause = Aponogeton lakhonensis A. Camus ■
29684 Aponogeton ranunculiflorum Jacot Guillaumin et Marais; 毛茛花水蕹■☆
29685 Aponogeton rehmannii Oliv. ;拉赫曼水蕹■☆
29686 Aponogeton rehmannii Oliv. var. hereroensis (Schinz) Engl. et Krause = Aponogeton rehmannii Oliv. ■☆
29687 Aponogeton spathaceum E. Mey. ex Hook. = Aponogeton junceum Lehm. ■☆
29688 Aponogeton stuhlmannii Engl. ;斯图尔曼水蕹■☆
29689 Aponogeton subconjugatum Schumach. et Thonn. ;成对水蕹■
29690 Aponogeton taiwanense Masam. ;台湾水蕹■
29691 Aponogeton taiwanense Masam. = Aponogeton lakhonensis A. Camus ■
29692 Aponogeton troupinii J. Raynal; 特鲁皮尼水蕹■☆
29693 Aponogeton vallisnerioides Baker; 苦草水蕹■☆
29694 Aponogeton violaceum Lye = Aponogeton afroviolaceum Lye ■☆
29695 Aponogetonaceae J. Agardh = Aponogetonaceae Planch. (保留科名)■
29696 Aponogetonaceae Planch. (1856)(保留科名); 水蕹科; Aponogeton Family, Cape-pondweed Family, Waterhawthorn Family ■
29697 Aponogiton Kuntze = Aponogeton L. f. (保留属名)■
29698 Apopetalum Pax = Brunellia Ruiz et Pav. ●☆
29699 Apopetalum pinnatum Pax = Brunellia boliviana Britton ●☆
29700 Apophragma Griseb. = Curtia Cham. et Schltdl. ■☆
29701 Apophyllum F. Muell. (1857); 澳洲白花菜属; Apophyllum ●☆
29702 Apophyllum anomalum F. Muell. ;澳洲白花菜●☆
29703 Apoplanesia C. Presl(1832); 微红血豆属■☆
29704 Apoplanesia paniculata C. Presl; 微红血豆■☆
29705 Apopleumon Raf. = Ipomoea L. (保留属名)●■
29706 Apopyros G. L. Nesom(1994); 柱果白酒草属■☆
29707 Apopyros warmingii (Baker) G. L. Nesom; 柱果白酒草■☆
29708 Apopyros warmingii Baker = Apopyros warmingii (Baker) G. L. Nesom ■☆
29709 Aporanthus Bromf. = Trigonella L. ■
29710 Aporetia Walp. = Aporetica J. R. Forst. et G. Forst. ●
29711 Aporetica J. R. Forst. et G. Forst. = Allophylus L. ●
29712 Aporocactus Lem. (1860); 鼠尾掌属(鼠尾鞭属); Rattail Cactus ●■
29713 Aporocactus Lem. = Disocactus Lindl. ●☆
29714 Aporocactus conzattii Britton et Rose; 康氏鼠尾掌; Conzatt Rattail Cactus ■☆
29715 Aporocactus flagelliformis (L.) Lem. ;鼠尾掌(倒吊仙人鞭,倒挂仙人鞭,金纽,鼠尾鞭); Rat's Tail Cactus, Rat's-tail Cactus, Rattail Cactus, Rat-tail Cactus ●
29716 Aporocactus flagriformis Britton et Rose; 鞭形鼠尾掌; Whipforme Rattail Cactus ■☆
29717 Aporocactus leptophis (DC.) Britton et Rose; 细蛇鼠尾掌; Thin-snake Rattail Cactus ■☆
29718 Aporocereus Fričet Kreuz. = Aporocactus Lem. ●■
29719 Aporocereus Fričet Kreuz. = Disocactus Lindl. ●☆
29720 Aporodes (Schltr.) W. Suarez et Cootes = Eria Lindl. (保留属名)■
29721 Aporopsis (Schltr.) M. A. Clem. et D. L. Jones = Dendrobium Sw. (保留属名)■
29722 Aporosa Blume = Aporusa Blume ●
29723 Aporosaceae Lindl. ex Miq. = Euphorbiaceae Juss. (保留科名)●■
29724 Aporosaceae Lindl. ex Planch. = Euphorbiaceae Juss. (保留科名)●■
29725 Aporosaceae Planch. = Euphorbiaceae Juss. (保留科名)●■
29726 Aporosella Chodat = Phyllanthus L. ●■
29727 Aporosella Chodat et Hassl. (1905); 小银柴属; Aporoseila ●☆
29728 Aporosella Chodat et Hassl. = Phyllanthus L. ●■
29729 Aporosella hassleriana Chodat; 小银柴●☆
29730 Aporostylis Rupp et Hatch(1946); 弱柱兰属■☆
29731 Aporostylis bifolia (Hook. f.) Rupp et Hatch; 弱柱兰■☆
29732 Aporrhiza Radlk. (1878); 离根无患子属●■☆
29733 Aporrhiza lastoursvillensis Pellegr. ;拉斯图维尔无患子●☆
29734 Aporrhiza letestui Pellegr. ;莱泰斯图离根无患子●☆
29735 Aporrhiza multijuga Gilg; 多对离根无患子●☆
29736 Aporrhiza nitida Gilg = Aporrhiza paniculata Radlk. ●☆
29737 Aporrhiza paniculata Radlk. ;锥序离根无患子●☆
29738 Aporrhiza talbotii Baker f. ;塔尔博特离根无患子●☆
29739 Aporrhiza tessmannii Gilg ex Radlk. ;泰斯曼离根无患子●☆
29740 Aporrhiza urophylla Gilg; 尾叶离根无患子●☆
29741 Aporuellia C. B. Clarke = Pararuellia Bremek. et Nann. -Bremek. ■
29742 Aporuellia C. B. Clarke = Ruellia L. ●■
29743 Aporuellia flagelliformis (Roxb.) C. B. Clarke = Pararuellia alata H. P. Tsui ■
29744 Aporum Blume = Dendrobium Sw. (保留属名)■
29745 Aporum acinaciforme (Roxb.) Griff. = Dendrobium acinaciforme Roxb. ■
29746 Aporum banaense (Gagnep.) Rauschert = Dendrobium spatella Rchb. f. ■

29747 Aporum crumenatum (Sw.) Brieger = Dendrobium crumenatum Sw. ■

29748 Aporum equitans (Kraenzl.) Brieger = Dendrobium equitans Kraenzl. ■

29749 Aporum hainanense (Rolfe) Rauschert = Dendrobium hainanense Rolfe ■

29750 Aporum jenkinsii Griff. = Dendrobium parciflorum Rchb. f. ex Lindl. ■

29751 Aporum kwashotense (Hayata) Rauschert = Dendrobium crumenatum Sw. ■

29752 Aporum pendulicaule (Hayata) Rauschert = Thrixspermum pendulicaule (Hayata) Schltr. ■

29753 Aporum pendulicaule Hayata = Thrixspermum pendulicaule (Hayata) Schltr. ■

29754 Aporum rivesii (Gagnep.) Rauschert = Dendrobium chryseum Rolfe ■

29755 Aporum spatella (Rchb. f.) M. A. Clem. = Dendrobium spatella Rchb. f. ■

29756 Aporum terminale (E. C. Parish et Rchb. f.) M. A. Clem. = Dendrobium terminale Parl. et Rchb. f. ■

29757 Aporum verlaquii (Costantin) Rauschert = Dendrobium terminale Parl. et Rchb. f. ■

29758 Aporusa Blume = Aporosa Blume ●

29759 Aporusa Blume(1828);银柴属(阿孛属);Aporosa ●

29760 Aporusa chinensis (Champ. ex Benth.) Merr. = Aporusa dioica (Roxb.) Müll. Arg. ●

29761 Aporusa dioica (Roxb.) Airy Shaw;银柴(大沙叶,厚皮稔,山咖啡,甜糖木,香港银柴,异叶银柴,占米赤树);China Aporosa, Chinese Aporosa ●

29762 Aporusa dioica (Roxb.) Airy Shaw var. yunnanensis (Pax et K. Hoffm.) H. S. Kiu;滇银柴●

29763 Aporusa dioica (Roxb.) Müll. Arg. = Aporusa dioica (Roxb.) Airy Shaw ●

29764 Aporusa dioica (Roxb.) Müll. Arg. var. yunnanensis (Pax et K. Hoffm.) H. S. Kiu = Aporusa dioica (Roxb.) Airy Shaw var. yunnanensis (Pax et K. Hoffm.) H. S. Kiu ●

29765 Aporusa frutescens Blume;香银柴●☆

29766 Aporusa frutescens Blume = Aporusa dioica (Roxb.) Airy Shaw ●

29767 Aporusa glabrifolia Kurz = Aporusa dioica (Roxb.) Airy Shaw ●

29768 Aporusa glabrifolia Kurz = Aporusa villosa (Lindl.) Baill. ●

29769 Aporusa lanceolata var. murtonii F. N. Williams = Aporusa planchoniana Baill. ex Müll. Arg. ●

29770 Aporusa leptostachya Benth. = Aporusa dioica (Roxb.) Airy Shaw ●

29771 Aporusa microcalyx (Hassk.) Hassk. = Aporusa dioica (Roxb.) Airy Shaw ●

29772 Aporusa microcalyx (Hassk.) Hassk. var. chinensis (Champ. ex Benth.) Müll. Arg. = Aporusa dioica (Roxb.) Müll. Arg. ●

29773 Aporusa microcalyx (Hassk.) Hassk. var. intermedia Pax. et Hoffm. = Aporusa dioica (Roxb.) Airy Shaw ●

29774 Aporusa microcalyx (Hassk.) Hassk. var. yunnanensis Pax et K. Hoffm. = Aporusa dioica (Roxb.) Müll. Arg. var. yunnanensis (Pax et K. Hoffm.) H. S. Kiu ●

29775 Aporusa microcalyx (Hassk.) Hassk. var. yunnanensis Pax et K. Hoffm. = Aporusa villosa (Lindl.) Baill. ●

29776 Aporusa planchoniana Baill. ex Müll. Arg.;全缘叶银柴(披针叶银柴);Entire-leaved Aporosa ●

29777 Aporusa roxburghii Baill. ex Müll. Arg. = Aporusa dioica (Roxb.) Airy Shaw ●

29778 Aporusa villosa (Lindl.) Baill.;毛银柴(毛大沙叶);Villose Aporosa ●

29779 Aporusa wallichii var. yunnanensis Pax et K. Hoffm. = Aporusa dioica (Roxb.) Airy Shaw var. yunnanensis (Pax et K. Hoffm.) H. S. Kiu ●

29780 Aporusa yunnanensis (Pax et K. Hoffm.) F. P. Metcalf;云南银柴(滇银柴,橄树);Yunnan Aporosa ●

29781 Aporusaceae Lindl. ex Miq. = Phyllanthaceae J. Agardh ●■

29782 Aposeridaceae Raf. = Asteraceae Bercht. et J. Presl(保留科名)●■

29783 Aposeridaceae Raf. = Compositae Giseke(保留科名)●■

29784 Aposeris Neck. = Hyoseris L. ■☆

29785 Aposeris Neck. ex Cass. (1827);齿叶羊苣属;Aposeris ■☆

29786 Aposeris foetida Less.;齿叶羊苣■☆

29787 Apostasia Blume(1825);拟兰属(假兰属);Apostasia ■

29788 Apostasia angustielliptica Thorel;假兰树■☆

29789 Apostasia nipponica Masam. = Apostasia wallichii R. Br. var. nipponica (Masam.) Masam. ■☆

29790 Apostasia odorata Blume;拟兰(假兰);Apostasia, Fragrant Apostasia ■

29791 Apostasia ramifera S. C. Chen et K. Y. Lang;多枝拟兰;Branchy Apostasia ■

29792 Apostasia thorelii Gagnep. = Apostasia odorata Blume ■

29793 Apostasia wallichii R. Br.;剑叶拟兰(假兰);Swordleaf Apostasia, Wallich Apostasia ■

29794 Apostasia wallichii R. Br. var. nipponica (Masam.) Masam.;日本拟兰■☆

29795 Apostasiaceae Blume = Apostasiaceae Lindl. (保留科名)■

29796 Apostasiaceae Lindl. (1833)(保留科名);拟兰科(假兰科);Apostasia Family ■

29797 Apostasiaceae Lindl. (保留科名) = Orchidaceae Juss. (保留科名)■

29798 Apostates Lander(1989);腺药菊属■☆

29799 Apostates rapae (F. Br.) Lander;腺药菊■☆

29800 Apotaenium Koso-Pol. = Chaerophyllum L. ■

29801 Apoterium Blume = Calophyllum L. ●

29802 Apoxyanthera Hochst. = Raphionacme Harv. ■☆

29803 Apoxyanthera pubescens Hochst. = Raphionacme hirsuta (E. Mey.) R. A. Dyer ■☆

29804 Apozia Willd. ex Benth. = Micromeria Benth. (保留属名)●■

29805 Apozia Willd. ex Steud. = Micromeria Benth. (保留属名)●■

29806 Appella Adans. (废弃属名) = Premna L. (保留属名)●■

29807 Appendicula Blume (1825);牛齿兰属(竹叶兰属);Appendicula ■

29808 Appendicula bifaria Lindl. ex Benth. = Appendicula cornuta (Blume) Schltr. ■

29809 Appendicula cornuta (Blume) Schltr.;牛齿兰(石壁兰);Common Appendicula, Spurred Appendicula ■

29810 Appendicula cornuta Blume var. formosana (Hayata) S. S. Ying = Appendicula formosana Hayata ■

29811 Appendicula cornuta Blume var. kotoensis (Hayata) S. S. Ying = Appendicula formosana Hayata ■

29812 Appendicula cristata Blume;厚叶牛齿兰;Thickleaf Appendicula ■☆

29813 Appendicula fenixii (Ames) Schltr.;长叶竹节兰■

29814 Appendicula formosana Hayata;台湾牛齿兰(台湾竹节兰,台湾

竹叶兰）;Taiwan Appendicula ■
29815 Appendicula formosana Hayata var. kotoensis (Hayata) T. P. Lin = Appendicula formosana Hayata ■
29816 Appendicula kotoensis Hayata = Appendicula formosana Hayata ■
29817 Appendicula lucida Ridl.;亮叶牛齿兰;Shiningleaf Appendicula ■☆
29818 Appendicula micrantha Lindl.;小花牛齿兰;Littleflower Appendicula ■
29819 Appendicula reflexa Blume;卷唇牛齿兰(竹节兰,竹叶兰);Reflexed Appendicula ■☆
29820 Appendicula teres Griff. = Ceratostylis subulata Blume ■
29821 Appendicula terrestris Fukuy.;长叶牛齿兰(长叶竹节兰);Longleaf Appendicula ■
29822 Appendiculana Kuntze = Appendicularia DC. ■☆
29823 Appendicularia DC. (1828);肖牛齿兰属■☆
29824 Appendicularia thymifolia DC.;肖牛齿兰■☆
29825 Appendiculopsis (Schltr.) Szlach. (1995);拟牛齿兰属■☆
29826 Appendiculopsis bicuspidata (J. J. Sm.) Szlach.;拟牛齿兰■☆
29827 Appendiculopsis trifida (Schltr.) Szlach.;三裂拟牛齿兰■☆
29828 Appertiella C. D. K. Cook et Triest(1982);六蕊藻属■☆
29829 Appertiella hexandra C. D. K. Cook et Triest;六蕊藻■☆
29830 Appunettia R. D. Good = Morinda L. ●■
29831 Appunettia angolensis R. D. Good = Morinda angolensis (R. D. Good) F. White ●☆
29832 Appunia Hook. f. = Morinda L. ●■
29833 Apradus Adans. = Arctopus L. ☆
29834 Aprella Steud. = Asprella Schreb. ■
29835 Aprella Steud. = Leersia Sw. (保留属名) ■
29836 Aprevalia Baill. = Delonix Raf. ●
29837 Aprevalia floribunda Baill. = Delonix floribunda (Baill.) Capuron ●☆
29838 Aprevalia perrieri R. Vig. = Delonix floribunda (Baill.) Capuron ●☆
29839 Apsanthea Jord. = Scilla L. ■
29840 Apseudes Raf. = Palimbia Besser ex DC. ■
29841 Apseudes Raf. = Peucedanum L. ■
29842 Aptandra Miers(1851);丝管花属;Aptandra ●☆
29843 Aptandra gora Hua = Ongokea gore (Hua) Pierre ●☆
29844 Aptandra spruceana Miers;丝管花●☆
29845 Aptandra zenkeri Engl.;岑克尔丝管花●☆
29846 Aptandraceae Miers = Olacaceae R. Br. (保留科名) ●
29847 Aptandraceae Tiegh.;丝管花科(油籽树科) ●☆
29848 Aptandraceae Tiegh. = Olacaceae R. Br. (保留科名) ●
29849 Aptandropsis Ducke = Heisteria Jacq. (保留属名) ●☆
29850 Aptandropsis Ducke(1945);丝管木属●☆
29851 Aptandropsis amphoricarpa Ducke;丝管木(巴西铁青树) ●☆
29852 Aptandropsis discophora Ducke;盘梗丝管木●☆
29853 Aptenia N. E. Br. (1925);露草属(露花属);Aptenia, Heartleaf, Ice-plant ●☆
29854 Aptenia cordifolia (L. f.) N. E. Br. = Mesembryanthemum cordifolium L. f. ■
29855 Aptenia cordifolia (L. f.) Schwantes;露草(花蔓草,花藤草,露花,心叶日中花);Baby Sun-rose, Bady Sun Rose, Dew-plant, Heartleaf Aptenia, Heartleaf Figmarigold, Heartleaf Iceplant, Heartleaf Ice-plant, Ice Plant, Red Apple ■
29856 Aptenia cordifolia (L. f.) Schwantes 'Variegata';白边露草(白边露花) ■☆
29857 Aptenia cordifolia (L. f.) Schwantes = Mesembryanthemum cordifolium L. f. ■
29858 Aptenia geniculiflora (L.) Bittrich ex Gerbaulet;膝花露草(膝花露花) ■☆
29859 Aptenia haeckeliana (A. Berger) Bittrich ex Gerbaulet;海克露草(海克露花) ■☆
29860 Aptenia lancifolia L. Bolus;剑叶露草■☆
29861 Apterantha C. H. Wright = Lagrezia Moq. ●■☆
29862 Apterantha C. H. Wright(1918);无翼苋属■☆
29863 Apterantha oligomeroides C. H. Wright;无翼苋■☆
29864 Apteranthe F. Muell. = Kochia Roth ●■
29865 Apteranthes Mik. = Boucerosia Wight et Arn. ■☆
29866 Apteranthes burchardii (N. E. Br.) Plowes = Caralluma burchardii N. E. Br. var. purpurascens Gatt. et Maire ■☆
29867 Apteranthes burchardii (N. E. Br.) Plowes = Caralluma burchardii N. E. Br. var. sventenii E. Lamb et B. M. Lamb ■☆
29868 Apteranthes burchardii (N. E. Br.) Plowes = Caralluma burchardii N. E. Br. ■☆
29869 Apteranthes burchardii (N. E. Br.) Plowes subsp. maura (Maire) Meve et F. Albers = Caralluma burchardii N. E. Br. subsp. maura (Maire) Meve et F. Albers ■☆
29870 Apteranthes burchardii (N. E. Br.) Plowes subsp. maura (Maire) Meve et F. Albers = Caralluma burchardii N. E. Br. var. maura Maire ■☆
29871 Apteranthes europaea (Guss.) Plowes = Caralluma europaea (Guss.) N. E. Br. ■☆
29872 Apteranthes europaea (Guss.) Plowes subsp. gussoneana (J. C. Mikan) Plowes = Apteranthes europaea (Guss.) Plowes ■☆
29873 Apteranthes europaea (Guss.) Plowes subsp. maroccana (Hook. f.) Plowes = Apteranthes europaea (Guss.) Plowes ■☆
29874 Apteranthes europaea (Guss.) Plowes var. affinis (De Wild.) Plowes = Apteranthes europaea (Guss.) Plowes ■☆
29875 Apteranthes europaea (Guss.) Plowes var. albotigrina (Maire) Plowes = Apteranthes europaea (Guss.) Plowes ■☆
29876 Apteranthes europaea (Guss.) Plowes var. barrueliana (Maire) Plowes = Apteranthes europaea (Guss.) Plowes ■☆
29877 Apteranthes europaea (Guss.) Plowes var. decipiens (Maire) Plowes = Apteranthes europaea (Guss.) Plowes ■☆
29878 Apteranthes europaea (Guss.) Plowes var. gattefossei (Maire) Plowes = Apteranthes europaea (Guss.) Plowes ■☆
29879 Apteranthes europaea (Guss.) Plowes var. marmaricensis (A. Berger) Plowes = Apteranthes europaea (Guss.) Plowes ■☆
29880 Apteranthes europaea (Guss.) Plowes var. micrantha (Maire) Plowes = Apteranthes europaea (Guss.) Plowes ■☆
29881 Apteranthes europaea (Guss.) Plowes var. schmuckiana (Gatt. et Maire) Plowes = Apteranthes europaea (Guss.) Plowes ■☆
29882 Apteranthes europaea (Guss.) Plowes var. simonis (A. Berger) Plowes = Apteranthes europaea (Guss.) Plowes ■☆
29883 Apteranthes europaea (Guss.) Plowes var. tristis (Maire) Plowes = Apteranthes europaea (Guss.) Plowes ■☆
29884 Apteranthes gussoneana J. C. Mikan = Apteranthes europaea (Guss.) Plowes ■☆
29885 Apteranthes joannis (Maire) Plowes = Caralluma joannis Maire ■☆
29886 Apteranthes munbyana (Decne.) Meve et Liede;芒比肖单脉青葙■☆
29887 Apteranthes tessellata Decne. = Echidnopsis cereiformis Hook. f. ■☆

29888 Apteria Nutt.(1834);无翼簪属;Nodding-nixie ■☆
29889 Apteria aphylla (Nutt.) Barnhart ex Small;无翼簪;Nodding-nixie ■☆
29890 Apteria setacea Nutt. = Apteria aphylla (Nutt.) Barnhart ex Small ■☆
29891 Apterigia (Ledeb.) Galushko = Thlaspi L. ■
29892 Apterigia Galushko = Thlaspi L. ■
29893 Apterocaryon Opiz = Betula L. ●
29894 Apteroearyon (Spach) Opiz = Betula L. ●
29895 Apterokarpos Rizzini = Loxopterygium Hook. f. ●☆
29896 Apteron Kurz = Ventilago Gaertn. ●
29897 Apterosperma Hung T. Chang(1976);圆籽荷属(圆子荷属);Apterosperma ●★
29898 Apterosperma oblata Hung T. Chang;圆籽荷;Apterosperma, Common Apterosperma ●◇
29899 Apterygia Baehni = Sideroxylon L. ●☆
29900 Apteuxis Griff. = Pternandra Jack ●
29901 Aptllon Raf. = Serinia Raf. ■☆
29902 Aptosimum Burch. = Aptosimum Burch. ex Benth.(保留属名)● ■☆
29903 Aptosimum Burch. ex Benth.(1836)(保留属名);直玄参属●■☆
29904 Aptosimum abietinum Burch. ex Benth. = Aptosimum spinescens (Thunb.) F. E. Weber ■☆
29905 Aptosimum abietinum Burch. ex Benth. varelongatum Benth. = Aptosimum neglectum F. E. Weber ■☆
29906 Aptosimum albomarginatum Marloth et Engl.;白边直玄参■☆
29907 Aptosimum angustifolium F. E. Weber et Schinz;窄叶直玄参■☆
29908 Aptosimum arenarium Engl.;沙地直玄参■☆
29909 Aptosimum decumbens Schinz;外倾直玄参■☆
29910 Aptosimum depressum Burch. ex Benth. = Aptosimum procumbens (Lehm.) Steud. ■☆
29911 Aptosimum depressum Burch. ex Benth. var. elongatum Hiern = Aptosimum elongatum Engl. ■☆
29912 Aptosimum dinteri F. E. Weber = Aptosimum glandulosum F. E. Weber et Schinz ■☆
29913 Aptosimum elongatum Engl.;短柔毛直玄参■☆
29914 Aptosimum eriocephalum E. Mey. ex Benth.;红头直玄参■☆
29915 Aptosimum eriocephalum E. Mey. ex Benth. var. pubescens Diels ex Weber = Aptosimum elongatum Engl. ■☆
29916 Aptosimum feddeanum Pilg. = Aptosimum glandulosum F. E. Weber et Schinz ■☆
29917 Aptosimum glandulosum F. E. Weber et Schinz;具腺直玄参■☆
29918 Aptosimum gossweileri V. Naray.;戈斯直玄参■☆
29919 Aptosimum indivisum Burch. ex Benth.;全裂直玄参;Karroo Violet ■☆
29920 Aptosimum junceum (Hiern) Philcox;灯心草直玄参■☆
29921 Aptosimum laricinum Dinter = Aptosimum spinescens (Thunb.) F. E. Weber ■☆
29922 Aptosimum laricinum Dinter var. spinosior ? = Aptosimum spinescens (Thunb.) F. E. Weber ■☆
29923 Aptosimum leucorrhizum (E. Mey. ex Benth.) E. Phillips = Peliostomum leucorrhizum E. Mey. ex Benth. ■☆
29924 Aptosimum lineare Marloth et Engl.;线形直玄参■☆
29925 Aptosimum lineare Marloth et Engl. var. acaule F. E. Weber;无茎线形直玄参■☆
29926 Aptosimum lineare Marloth et Engl. var. ciliatum Schinz ex F. Weber;缘毛直玄参■☆
29927 Aptosimum lineare Marloth et Engl. var. randii (S. Moore) F. E. Weber = Aptosimum lineare Marloth et Engl. ■☆
29928 Aptosimum lugardiae (N. E. Br. ex Hemsl. et V. Naray.) E. Phillips;卢格德直玄参■☆
29929 Aptosimum marlothii (Engl.) Hiern;马洛斯直玄参■☆
29930 Aptosimum molle V. Naray.;柔软直玄参■☆
29931 Aptosimum nanum Engl. = Aptosimum indivisum Burch. ex Benth. ■☆
29932 Aptosimum neglectum F. E. Weber;忽视直玄参■☆
29933 Aptosimum nelsii F. E. Weber = Aptosimum lineare Marloth et Engl. ■☆
29934 Aptosimum oppositifolium (Engl.) E. Phillips = Jamesbrittenia fruticosa (Benth.) Hilliard ■☆
29935 Aptosimum patulum Bremek.;张开直玄参■☆
29936 Aptosimum procumbens (Lehm.) Steud.;平铺直玄参■☆
29937 Aptosimum procumbens (Lehm.) Steud. var. elongatum (Hiern) Codd = Aptosimum elongatum Engl. ■☆
29938 Aptosimum pubescens F. E. Weber = Aptosimum elongatum Engl. ■☆
29939 Aptosimum pumilum (Hochst.) Benth.;矮小直玄参■☆
29940 Aptosimum randii S. Moore = Aptosimum lineare Marloth et Engl. ■☆
29941 Aptosimum scaberrimum Schinz = Aptosimum spinescens (Thunb.) F. E. Weber ■☆
29942 Aptosimum scaberrimum Schinz var. tenuifolium F. E. Weber = Aptosimum spinescens (Thunb.) F. E. Weber ■☆
29943 Aptosimum schinzii F. E. Weber = Aptosimum angustifolium F. E. Weber et Schinz ■☆
29944 Aptosimum spinescens (Thunb.) F. E. Weber;小刺直玄参■☆
29945 Aptosimum steingroeveri Engl. = Aptosimum spinescens (Thunb.) F. E. Weber ■☆
29946 Aptosimum steingroeveri Engl. var. glabrum F. E. Weber et Schinz = Aptosimum spinescens (Thunb.) F. E. Weber ■☆
29947 Aptosimum suberosum F. E. Weber;木栓质直玄参■☆
29948 Aptosimum tragacanthoides E. Mey. ex Benth.;羊角刺直玄参■☆
29949 Aptosimum transvaalense F. E. Weber;德兰士瓦直玄参■☆
29950 Aptosimum viscosum (E. Mey. ex Benth.) E. Phillips = Peliostomum viscosum E. Mey. ex Benth. ■☆
29951 Aptosimum viscosum Benth.;黏直玄参■☆
29952 Aptosimum weberianum Pilg. = Aptosimum elongatum Engl. ■☆
29953 Aptosimum welwitschii Hiern;韦尔直玄参■☆
29954 Aptotheca Miers = Forsteronia G. Mey. ●☆
29955 Apuleia Gaertn. = Berkheya Ehrh.(保留属名)●■☆
29956 Apuleia Mart.(1837)(保留属名);铁苏木属;Apuleia ●☆
29957 Apuleia leiocarpa (J. Vogel) J. F. Macbr.;平滑果铁苏木■☆
29958 Apuleia leiocarpa J. F. Macbr. = Apuleia leiocarpa (J. Vogel) J. F. Macbr. ■☆
29959 Apuleia molaris Spruce ex Benth.;臼齿苏木■☆
29960 Apuleja Gaertn.(废弃属名) = Apuleia Mart.(保留属名)●☆
29961 Apuleja Gaertn.(废弃属名) = Berkheya Ehrh.(保留属名)●■☆
29962 Apurimacia Harms(1923);安第斯山豆属(阿普里豆属)■☆
29963 Apurimacia boliviana (Britton) Lavin;玻利维亚安第斯山豆■☆
29964 Apurimacia incarum Harms;安第斯山豆■☆
29965 Aquartia Jacq. = Solanum L. ●■
29966 Aquifoliaceae A. Rich. = Aquifoliaceae Bercht. et J. Presl(保留科名)●
29967 Aquifoliaceae Bartl. = Aquifoliaceae Bercht. et J. Presl(保留科

名)●

29968 Aquifoliaceae Bercht. et J. Presl(1825)(保留科名);冬青科;Holly Family ●

29969 Aquifoliaceae DC. ex A. Rich. = Aquifoliaceae Bercht. et J. Presl(保留科名)●

29970 Aquifolium Mill. = Ilex L. ●

29971 Aquifolium Tourn. ex Mill. = Ilex L. ●

29972 Aquilaria Lam.(1783)(保留属名);沉香属;Eagle Wood, Eaglewood ●

29973 Aquilaria agallocha (Lour.) Roxb.;沉香(伽南香,没香,蜜香,泰国沉香,印度沉香);Agalawood, Agalloch, Agalwood, Agilawood, Aloe Wood, Aloes-wood, Aloeswood Eaglewood, Chinese Eaglewood, Eaglewood ●

29974 Aquilaria borneensis Tiegh. ex Gilg;婆罗州沉香●☆

29975 Aquilaria crassa Pierre ex Lecomte;粗厚沉香●☆

29976 Aquilaria grandiflora Benth. = Aquilaria sinensis (Lour.) Spreng. ●◇

29977 Aquilaria malaccensis Benth.;马六甲沉香(马来沉香);Agarwood, Aloe Wood, Calambac, Lign-aloes ●☆

29978 Aquilaria microcarpa Baill.;小果沉香●☆

29979 Aquilaria ophispermum Poir. = Aquilaria sinensis (Lour.) Spreng. ●◇

29980 Aquilaria secundaria DC.;印度沉香●☆

29981 Aquilaria sinensis (Lour.) Gilg = Aquilaria grandiflora Benth. ●◇

29982 Aquilaria sinensis (Lour.) Merr. = Ophiospermum sinense Lour. ●◇

29983 Aquilaria sinensis (Lour.) Spreng.;土沉香(白木香,沉水香,沉香,伽楠香,六麻树,蜜香,女儿香,奇南香,奇楠,琪楠,清桂香,外贡顺,香材,牙香树,崖香,芫香,栈香);China Eaglewood, Chinese Eagle Wood, Chinese Eaglewood ●◇

29984 Aquilaria sinensis Merr. = Ophiospermum sinense Lour. ●◇

29985 Aquilaria yunnanensis S. C. Huang;云南沉香;Yunnan Eaglewood ●

29986 Aquilariaceae R. Br. = Aquilariaceae R. Br. ex DC. ●

29987 Aquilariaceae R. Br. = Thymelaea Mill.(保留属名)●■

29988 Aquilariaceae R. Br. ex DC.;沉香科●

29989 Aquilariaceae R. Br. ex DC. = Thymelaea Mill.(保留属名)●■

29990 Aquilariella Tiegh. = Aquilaria Lam.(保留属名)●

29991 Aquilariella borneensis Tiegh. = Aquilaria borneensis Tiegh. ex Gilg ●☆

29992 Aquilariella malaccensis Tiegh. = Aquilaria malaccensis Benth. ●☆

29993 Aquilariella microcarpa Tiegh. = Aquilaria microcarpa Baill. ●☆

29994 Aquilegia L.(1753);耧斗菜属;Aquilegia, Columbine, Granny-bonnets ■

29995 Aquilegia adoxoides (DC.) Ohwi = Semiaquilegia adoxoides (DC.) Makino ■

29996 Aquilegia akitensis Huth = Aquilegia flabellata Siebold et Zucc. ■☆

29997 Aquilegia alpina L.;高山耧斗菜;Alpine Columbine ■☆

29998 Aquilegia amurensis Kom.;阿穆尔耧斗菜;Amur Columbine ■

29999 Aquilegia anemonoides Willd. = Paraquilegia anemonoides (Willd.) Engl. ex Ulbr. ■

30000 Aquilegia atropurpurea Willd. = Aquilegia viridiflora Pall. f. atropurpurea (Willd.) Kitag. ■

30001 Aquilegia atrovinosa Popov ex Gamajun.;暗紫耧斗菜;Darkpurple Columbine ■

30002 Aquilegia australis Small = Aquilegia canadensis L. ■☆

30003 Aquilegia borodinii Schischk.;波罗氏耧斗菜;Borod Columbine ■☆

30004 Aquilegia brevicalcarata Kolok. ex Serg.;短距耧斗菜;Short-spur Columbine ■

30005 Aquilegia burgeriana Siebold et Zucc.;布氏耧斗菜■☆

30006 Aquilegia caerulea James = Aquilegia leptoceras Fisch. et C. A. Mey. ■

30007 Aquilegia canadensis L.;加拿大耧斗菜(美洲耧斗菜,山羊七);American Columbine, Canadian Columbine, Columbine, Common American Columbine, Eastern Columbine, Honeysuckle, Meeting Houses, Red Columbine, Rock Bells, Wild Columbine ■☆

30008 Aquilegia canadensis L. f. flaviflora Britton = Aquilegia canadensis L. ■☆

30009 Aquilegia canadensis L. var. australis (Small) Munz = Aquilegia canadensis L. ■☆

30010 Aquilegia canadensis L. var. coccinea (Small) Munz = Aquilegia canadensis L. ■☆

30011 Aquilegia canadensis L. var. eminens (Greene) B. Boivin = Aquilegia canadensis L. ■☆

30012 Aquilegia canadensis L. var. flaviflora Britton = Aquilegia canadensis L. ■☆

30013 Aquilegia canadensis L. var. formosa (Fisch. ex DC.) J. G. Cooper = Aquilegia formosa Fisch. ex DC. ■☆

30014 Aquilegia canadensis L. var. hybrida Hook. = Aquilegia canadensis L. ■☆

30015 Aquilegia canadensis L. var. latiuscula (Greene) Munz = Aquilegia canadensis L. ■☆

30016 Aquilegia chaplinei Standl. ex Payson;查氏耧斗菜■☆

30017 Aquilegia chrysantha A. Gray;黄花耧斗菜;Golden Columbine, Yellow Columbine, Yellow-flowered Columbine ■☆

30018 Aquilegia chrysantha A. Gray var. chaplinei (Payson) E. J. Lott = Aquilegia chaplinei Standl. ex Payson ■☆

30019 Aquilegia chrysantha A. Gray var. hinckleyana (Munz) E. J. Lott;欣克利耧斗菜;Hinckley's Golden Columbine ■☆

30020 Aquilegia chrysantha A. Gray var. hinckleyana (Munz) E. J. Lott = Aquilegia hinckleyana Munz ■☆

30021 Aquilegia chrysantha A. Gray var. longissima ?;哥伦比亚耧斗菜;Longspur Golden Columbine ■☆

30022 Aquilegia chrysantha A. Gray var. rydbergii Munz = Aquilegia chrysantha A. Gray ■☆

30023 Aquilegia coccinea Small = Aquilegia canadensis L. ■☆

30024 Aquilegia ecalcarata Maxim.;无距耧斗菜(官柴胡,千年耗子屎,无距天葵,野柴胡,野前胡);Spurless Columbine ■

30025 Aquilegia ecalcarata Maxim. f. semicalcarata (Schipcz.) Hand.-Mazz. = Aquilegia ecalcarata Maxim. ■

30026 Aquilegia ecalcarata Maxim. f. semicalcerata (Schipcz.) Hand.-Mazz.;细距耧斗菜■

30027 Aquilegia eminens Greene = Aquilegia canadensis L. ■☆

30028 Aquilegia eximia Van Houtte = Aquilegia eximia Van Houtte ex Planch. ■☆

30029 Aquilegia eximia Van Houtte ex Planch.;加州耧斗菜■☆

30030 Aquilegia fauriei H. Lév. = Dictamnus dasycarpus Turcz. ■

30031 Aquilegia flabellata Siebold et Zucc.;扇形耧斗菜(秋田氏耧斗菜,洋牡丹);Fan Columbine, Fanshaped Columbine ■☆

30032 Aquilegia flabellata Siebold et Zucc. var. humiliata Makino;曲距扇形耧斗菜■☆

30033 Aquilegia flaviflora Tenney = Aquilegia canadensis L. ■☆

30034 Aquilegia formosa Fisch. ex DC.;红花耧斗菜;Columbine, Red

Columbine, Sikta Columbine, Western Columbine ■☆

30035 Aquilegia formosa Fisch. ex DC. var. communis B. Boivin = Aquilegia formosa Fisch. ex DC. ■☆

30036 Aquilegia formosa Fisch. ex DC. var. megalantha B. Boivin = Aquilegia formosa Fisch. ex DC. ■☆

30037 Aquilegia formosa Fisch. ex DC. var. wawawensis (Payson) H. St. John = Aquilegia formosa Fisch. ex DC. ■☆

30038 Aquilegia glandulosa Fisch. ex Link; 大花耧斗菜; Altai Columbine, Siberian Columbine ■

30039 Aquilegia henryi (Oliv.) Finet et Gagnep. = Urophysa henryi (Oliv.) Ulbr. ■

30040 Aquilegia hinckleyana Munz; 欣可利耧斗菜; Hinckley's Columbine ■☆

30041 Aquilegia hybrida Sims; 杂种耧斗菜; Columbine, Hybrid Columbine ■☆

30042 Aquilegia incurvata P. K. Hsiao; 秦岭耧斗菜(灯笼草, 银扁担); Chinling Mountain Columbine, Qinling Columbine ■

30043 Aquilegia japonica Nakai et H. Hara; 白山耧斗菜; Japan Columbine, Japanese Columbine ■

30044 Aquilegia jonesii Parry; 琼斯耧斗菜■☆

30045 Aquilegia jonesii Parry var. elatior Boothman = Aquilegia jonesii Parry ■☆

30046 Aquilegia karatavica Mikeschin; 卡拉塔夫耧斗菜■☆

30047 Aquilegia karelinii (Baker) O. Fedtsch. et B. Fedtsch.; 长距耧斗菜; Long-spur Columbine ■

30048 Aquilegia kozakii Masam.; 台湾耧斗菜; Taiwan Columbine ■

30049 Aquilegia lactiflora Kar. et Kir.; 白花耧斗菜; Milkywhiteflower Columbine ■

30050 Aquilegia latiuscula Greene = Aquilegia canadensis L. ■☆

30051 Aquilegia leptoceras Fisch. et C. A. Mey.; 细角耧斗菜(细距耧斗菜); Colorado Columbine, Delicatespur Columbine, Rocky Mountain Columbine, Thinspur Columbine ■

30052 Aquilegia leptoceras var. chrysantha (A. Gray) Hook. f. = Aquilegia chrysantha A. Gray ■☆

30053 Aquilegia longissima A. Gray = Aquilegia longissima A. Gray ex S. Watson ■☆

30054 Aquilegia longissima A. Gray ex S. Watson; 黄长距耧斗菜; Longspur Columbine, Long-spur Columbine, Yellow Bonnet ■☆

30055 Aquilegia macrantha Hook. et Arn. = Aquilegia leptoceras Fisch. et C. A. Mey. ■

30056 Aquilegia micrantha Eastw.; 美洲小花耧斗菜■☆

30057 Aquilegia micrantha Eastw. var. mancosana Eastw. = Aquilegia micrantha Eastw. ■☆

30058 Aquilegia mohavensis Munz = Aquilegia formosa Fisch. ex DC. ■☆

30059 Aquilegia moorcroftiana Wall.; 腺毛耧斗菜; Glandularhair Columbine ■

30060 Aquilegia olympica Boiss.; 高加索耧斗菜(奥林帕斯耧斗菜); Caucasia Columbine, Columbine Olympicus ■☆

30061 Aquilegia oxysepala Trautv. et C. A. Mey.; 尖萼耧斗菜(血见愁); Early Columbine ■

30062 Aquilegia oxysepala Trautv. et C. A. Mey. f. pallidiflora (Nakai) Kitag.; 黄花尖萼耧斗菜; Pallidflower Columbine ■

30063 Aquilegia oxysepala Trautv. et C. A. Mey. f. pallidiflora (Nakai) Kitag. = Aquilegia oxysepala Trautv. et C. A. Mey. ■

30064 Aquilegia oxysepala Trautv. et C. A. Mey. f. pallidiflora Kitag. = Aquilegia oxysepala Trautv. et C. A. Mey. ■

30065 Aquilegia oxysepala Trautv. et C. A. Mey. var. kansuensis Brühl; 甘肃耧斗菜(石蛋七, 乌全胡); Gansu Columbine, Kansu Columbine ■

30066 Aquilegia oxysepala Trautv. et C. A. Mey. var. pallidiflora Nakai ex Mori = Aquilegia oxysepala Trautv. et C. A. Mey. ■

30067 Aquilegia oxysepala Trautv. et C. A. Mey. var. yabeana (Kitag.) Munz = Aquilegia yabeana Kitag. ■

30068 Aquilegia parviflora Ledeb.; 小花耧斗菜(耧斗菜, 漏斗菜, 猫爪花, 血见愁); Smallflower Columbine ■

30069 Aquilegia phoenicantha Cory = Aquilegia canadensis L. ■☆

30070 Aquilegia pyrenaica DC.; 比利牛斯耧斗菜; Pyrenean Columbine ■☆

30071 Aquilegia rockii Munz; 直距耧斗菜; Rock Columbine ■

30072 Aquilegia scopulorum Tidestr.; 岩壁耧斗菜■☆

30073 Aquilegia scopulorum Tidestr. var. calcarea (M. E. Jones) Munz = Aquilegia scopulorum Tidestr. ■☆

30074 Aquilegia sibirica Lam.; 西伯利亚耧斗菜; Siberia Columbine, Siberian Columbine ■

30075 Aquilegia skinneri Hook.; 墨西哥耧斗菜; Mexican Columbine, Skinner Columbine ■☆

30076 Aquilegia tianshanica Butkov; 天山耧斗菜; Tianshan Columbine ■☆

30077 Aquilegia viridiflora Pall.; 耧斗菜(绿花耧斗菜, 血见愁); Greenflower Columbine ■

30078 Aquilegia viridiflora Pall. f. atropurpurea (Willd.) Kitag.; 紫花耧斗菜(石头花, 紫花菜); Purpleflower Columbine ■

30079 Aquilegia viridiflora Pall. var. atropurpurea (Willd.) Finet et Gagnep. = Aquilegia viridiflora Pall. f. atropurpurea (Willd.) Kitag. ■

30080 Aquilegia vitalii Gamojun.; 维氏耧斗菜■☆

30081 Aquilegia vulgaris L.; 普通耧斗菜(耧斗菜, 欧洲耧斗菜, 山羊七); Baby's Shoes, Bachelor's Buttons, Bluebell, Bonnets, Boots-and-shoes, Cain-and-abel, Capon's Feather, Capon's Feathers, Capon's Tail, Capon's Tails, Cock's Foot, Collarbind, Colourbine, Columbine, Cork's Foot, Cullavine, Cullenbeam, Culverkeys, Culverwort, Curranbine, Dolly's Bonnets, Dolly's Shoes, Dove-plant, Dove's Foot, Dove's Plant, Doves-at-the-fountain, Doves-in-the-ark, Dovesround-aodish, European Columbine, Folly's Flower, Fool's Cap, Garden Columbine, Grandmother's Bonnets, Granny Bonnets, Granny Hood, Granny-bonnets, Granny-jump-out-of-bed, Granny's Bonnet, Granny's Bonnets, Granny's Cap, Granny's Nightcap, Granny's Thimble, Granny's Thimbles, Hawk's Foot, Hen-and-chickens, Lady's Bonnet, Lady's Bonnets, Lady's Petticoats, Lady's Purse, Lady's Shoes, Lady's Slipper, Lion's Herb, Nightcap, Noah's Ark, Old Lady's Bonnets, Old Maid's Basket, Old Woman's Bonnets, Primrose Soldiers, Rags-and-tatters, Shoes-and-socks, Shoes-and-stocking, Shoes-and-stockings, Skullcap, Snapdragon, Soldier Buttons, Soldier's Buttons, Stockings-and-shoes, Straw Bonnets, Thimbles, Two-faces-under-a-hat, Widow's Weeds ■

30082 Aquilegia vulgaris L. 'Nora Barlow'; 诺娜·巴洛普通耧斗菜■☆

30083 Aquilegia vulgaris L. subsp. ballii (Litard. et Maire) Dobignard et D. Jord.; 鲍尔耧斗菜■☆

30084 Aquilegia vulgaris L. subsp. cossoniana (Maire et Sennen) Dobignard et D. Jord.; 科森耧斗菜■☆

30085 Aquilegia vulgaris L. var. ballii Litard. et Maire = Aquilegia vulgaris L. subsp. ballii (Litard. et Maire) Dobignard et D. Jord. ■☆

30086 Aquilegia vulgaris L. var. cossoniana Maire et Sennen = Aquilegia vulgaris L. subsp. cossoniana (Maire et Sennen) Dobignard et D. Jord. ■☆

30087 Aquilegia vulgaris L. var. olympica (Boiss.) Parsa;奥林匹克耧斗菜■☆

30088 Aquilegia vulgaris L. var. oxysepala (Trautv. et C. A. Mey.) Regel = Aquilegia oxysepala Trautv. et C. A. Mey. ■

30089 Aquilegia vulgaris L. var. oxysepala Regel = Aquilegia oxysepala Trautv. et C. A. Mey. ■

30090 Aquilegia yabeana Kitag.;华北耧斗菜(亮亮草,五铃花,紫霞耧斗菜);N. China Columbine,Yabe Columbine ■

30091 Aquilegia yabeana Kitag. f. luteola S. H. Li et Y. Huei Huang;黄花华北耧斗菜;Yellowflower Yabe Columbine ■

30092 Aquilegia yabeana Kitag. f. luteola S. H. Li et Y. Huei Huang = Aquilegia yabeana Kitag. ■

30093 Aquilegiaceae Lilja = Ranunculaceae Juss. (保留科名)●■

30094 Aquilicia L. = Leea D. Royen ex L. (保留属名)●■

30095 Aquilicia sambucina L. = Leea indica (Burm. f.) Merr. ●

30096 Aquilina Bubani = Aquilegia L. ■

30097 Arabidaceae Döll = Brassicaceae Burnett(保留科名)●■

30098 Arabidaceae Döll = Cruciferae Juss. (保留科名)●■

30099 Arabidella (F. Muell.) O. E. Schulz(1924);小鼠耳芥属(澳小南芥属)■☆

30100 Arabidella O. E. Schulz = Arabidella (F. Muell.) O. E. Schulz ■☆

30101 Arabidella trisecta (F. Muell.) O. E. Schulz;小鼠耳芥■☆

30102 Arabidium Spach = Arabis L. ●■

30103 Arabidopsis (DC.) Heynh. = Arabidopsis Heynh. (保留属名)■

30104 Arabidopsis (DC.) Heynh. = Arabis L. ●■

30105 Arabidopsis Heynh. (1842)(保留属名);鼠耳芥属(拟筷子芥属,拟南芥菜属,拟南芥属);Mouseear Cress,Thale Cress ■

30106 Arabidopsis Heynh. (保留属名) = Arabis L. ●■

30107 Arabidopsis Schur = Sisymbrium L. ■

30108 Arabidopsis brevicaulis (Jafri) Jafri = Crucihimalaya himalaica (Edgew.) Al-Shehbaz,O'Kane et R. A. Price ■

30109 Arabidopsis campestris O. E. Schulz = Arabidopsis wallichii (Hook. f. et Thomson) N. Busch ■

30110 Arabidopsis griffithiana (Boiss.) N. Busch = Olimarabidopsis pumila (Stephan) Al-Shehbaz,O'Kane et R. A. Price ■

30111 Arabidopsis halleri (L.) O'Kane et Al-Shehbaz subsp. gemmifera (Matsum.) O'Kane et Al-Shehbaz;叶芽鼠耳芥■

30112 Arabidopsis himalaica (Edgew.) O. E. Schulz = Crucihimalaya himalaica (Edgew.) Al-Shehbaz,O'Kane et R. A. Price ■

30113 Arabidopsis himalaica (Edgew.) O. E. Schulz var. harrissii O. E. Schulz = Crucihimalaya himalaica (Edgew.) Al-Shehbaz,O'Kane et R. A. Price ■

30114 Arabidopsis himalaica (Edgew.) O. E. Schulz var. rupestris (Edgew.) O. E. Schulz = Crucihimalaya himalaica (Edgew.) Al-Shehbaz,O'Kane et R. A. Price ■

30115 Arabidopsis kamchatica (DC.) K. Shimizu et Kudoh = Arabidopsis lyrata (L.) O'Kane et Al-Shehbaz subsp. kamchatica (Fisch. ex DC.) O'Kane et Al-Shehbaz ■

30116 Arabidopsis kamchatica (Fisch. ex DC.) Shimizu et Kudoh = Arabidopsis lyrata (L.) O'Kane et Al-Shehbaz subsp. kamchatica (Fisch. ex DC.) O'Kane et Al-Shehbaz ■

30117 Arabidopsis kneuckeri O. E. Schulz = Arabidopsis taraxacifolia (T. Anderson) Jafri ■

30118 Arabidopsis korshinskyi Botsch. = Olimarabidopsis cabulica (Hook. f. et Thomson) Al-Shehbaz,O'Kane et R. A. Price ■

30119 Arabidopsis lasiocarpa (Hook. f. et Thomson) O. E. Schulz = Crucihimalaya lasiocarpa (Hook. f. et Thomson) Al-Shehbaz, O'Kane et R. A. Price ■

30120 Arabidopsis lasiocarpa (Hook. f. et Thomson) O. E. Schulz var. micrantha W. T. Wang;小花毛果拟南芥■

30121 Arabidopsis lyrata (L.) O'Kane et Al-Shehbaz subsp. kamchatica (Fisch. ex DC.) O'Kane et Al-Shehbaz;琴叶鼠耳芥■

30122 Arabidopsis minutiflora (Hook. f. et Thomson) Busch = Microsisymbrium minutiflorum (Hook. f. et Thomson) O. E. Schulz ■

30123 Arabidopsis mollissima (C. A. Mey.) N. Busch = Crucihimalaya mollissima (C. A. Mey.) Al-Shehbaz,O'Kane et R. A. Price ■

30124 Arabidopsis mollissima (C. A. Mey.) N. Busch var. afghanica O. E. Schulz = Crucihimalaya wallichii (Hook. f. et Thomson) Al-Shehbaz,O'Kane et R. A. Price ■

30125 Arabidopsis mollissima (C. A. Mey.) N. Busch var. dentata O. E. Schulz = Crucihimalaya mollissima (C. A. Mey.) Al-Shehbaz,O'Kane et R. A. Price ■

30126 Arabidopsis mollissima (C. A. Mey.) N. Busch var. glaberrima (Hook. f. et Thomson) O. E. Schulz = Crucihimalaya mollissima (C. A. Mey.) Al-Shehbaz,O'Kane et R. A. Price ■

30127 Arabidopsis mollissima (C. A. Mey.) N. Busch var. pamirica (Korsh.) O. E. Schulz = Crucihimalaya mollissima (C. A. Mey.) Al-Shehbaz,O'Kane et R. A. Price ■

30128 Arabidopsis mollissima (C. A. Mey.) N. Busch var. thomsonii (Hook. f.) O. E. Schulz = Crucihimalaya mollissima (C. A. Mey.) Al-Shehbaz,O'Kane et R. A. Price ■

30129 Arabidopsis mollissima (C. A. Mey.) N. Busch var. yunnanensis O. E. Schulz = Arabis paniculata Franch. ■

30130 Arabidopsis monachorum (W. W. Sm.) O. E. Schulz = Crucihimalaya lasiocarpa (Hook. f. et Thomson) Al-Shehbaz, O'Kane et R. A. Price ■

30131 Arabidopsis nuda (Bél.) Bornm. = Drabopsis nuda (Bél.) Stapf ■

30132 Arabidopsis nuda (Bél.) Bornm. = Drabopsis verna K. Koch ■

30133 Arabidopsis parvula (Schrenk) O. E. Schulz = Thellungiella parvula (Schrenk) Al-Shehbaz et O'Kane ■

30134 Arabidopsis pumila (Steph.) N. Busch var. alpina (Korsh.) O. E. Schulz = Olimarabidopsis cabulica (Hook. f. et Thomson) Al-Shehbaz,O'Kane et R. A. Price ■

30135 Arabidopsis pumila (Steph.) N. Busch var. griffithiana (Boiss.) Jafri = Olimarabidopsis pumila (Stephan) Al-Shehbaz, O'Kane et R. A. Price ■

30136 Arabidopsis pumila (Stephan) N. Busch = Olimarabidopsis pumila (Stephan) Al-Shehbaz,O'Kane et R. A. Price ■

30137 Arabidopsis qiranica C. H. An = Sisymbriopsis mollipila (Maxim.) Botsch. ■

30138 Arabidopsis quqiranica C. H. An;策勒鼠耳芥;Cele Mouseear Cress ■

30139 Arabidopsis salsuginea (Pall.) N. Busch = Thellungiella salsuginea (Pall.) O. E. Schulz ■

30140 Arabidopsis schimperi (Boiss.) N. Busch = Robeschia schimperi (Boiss.) O. E. Schulz ■☆

30141 Arabidopsis stricta (Cambess.) N. Busch;直鼠耳芥;Strict Mouseear Cress ■

30142 Arabidopsis stricta (Cambess.) N. Busch = Crucihimalaya stricta (Cambess.) Al-Shehbaz,O'Kane et R. A. Price ■

30143 Arabidopsis stricta (Cambess.) N. Busch var. bracteata O. E. Schulz = Crucihimalaya stricta (Cambess.) Al-Shehbaz,O'Kane et R. A. Price ■

30144 Arabidopsis suecica (Fr.) Norrl.;瑞典鼠耳芥■☆

30145 Arabidopsis taraxacifolia (T. Anderson) Jafri = Crucihimalaya wallichii (Hook. f. et Thomson) Al-Shehbaz, O'Kane et R. A. Price ■

30146 Arabidopsis thaliana (L.) Heynh.;鼠耳芥(拟南芥菜,扎氏大蒜芥);Common Mouseear Cress, Common Wall Cress, Mouse Ear Cress, Mouseear Cress, Mouse-ear Cress, Mouse-ear-cress, Podded Mouse Ear, Thale Cress, Thale-cress, Touch-me-not ■

30147 Arabidopsis thaliana (L.) Heynh. var. apetala O. E. Schulz = Arabidopsis thaliana (L.) Heynh. ■

30148 Arabidopsis thaliana (L.) Heynh. var. pusilla (Hochst. ex A. Rich.) O. E. Schulz = Arabidopsis thaliana (L.) Heynh. ■

30149 Arabidopsis tibetica (Hook. f. et Thomson) Y. C. Lan et C. H. An ex K. C. Kuan = Arabis tibetica Hook. f. et Thomson ■

30150 Arabidopsis toxophylla (M. Bieb.) N. Busch = Pseudoarabidopsis toxophylla (M. Bieb.) Al-Shehbaz, O'Kane et R. A. Price ■

30151 Arabidopsis trichocarpa R. F. Huang = Neotorularia humilis (C. A. Mey.) Hedge et J. Léonard ■

30152 Arabidopsis tuemurica K. C. Kuan et C. H. An = Neotorularia humilis (C. A. Mey.) Hedge et J. Léonard ■

30153 Arabidopsis verna (K. Koch) N. Busch = Drabopsis nuda (Bél.) Stapf ■

30154 Arabidopsis verna (K. Koch) N. Busch = Drabopsis verna K. Koch ■

30155 Arabidopsis wallichii (Hook. f. et Thomson) N. Busch;卵叶鼠耳芥;Wallich Mouseear Cress ■

30156 Arabidopsis wallichii (Hook. f. et Thomson) N. Busch = Crucihimalaya wallichii (Hook. f. et Thomson) Al-Shehbaz, O'Kane et R. A. Price ■

30157 Arabidopsis yadongensis K. C. Kuan et C. H. An;亚东鼠耳芥(亚东拟南芥);Yadong Mouseear Cress ■

30158 Arabidopsis yadungensis K. C. Kuan et C. H. An = Arabis pterosperma Edgew. ■

30159 Arabis Adans. = Iberis L. ●■

30160 Arabis L. (1753);南芥属(筷子芥属);Mountain Rock-cress, Rock Cress, Rockcress, Rock-cress, Wall Cress ●■

30161 Arabis acutina Greene = Arabis divaricarpa A. Nelson ■☆

30162 Arabis alaschanica Maxim.;贺兰山南芥(阿拉善南芥,疙瘩七,钮子七,雪三七,珠参,珠儿参);Alashan Rockcress, Helanshan Rockcress ■

30163 Arabis albida Steven = Arabis caucasica Willd. ■☆

30164 Arabis alpina L.;高山南芥(筷子芥,筷子芥草);Alpine Rock Cress, Alpine Rockcress, Alpine Rock-cress, Aunt Hannah, Bishop's Wig, Dusty Husband, March And May, March-and-may, Mountain Rock Cress, Mountain Rockcress, Rockcress, Rocky Wall-cress, Snow-on-the-mountain, Sweet Alice, White Alisson, White Lacey, White May, White Rock ■☆

30165 Arabis alpina L. = Arabis pterosperma Edgew. ■

30166 Arabis alpina L. = Arabis serrata Franch. et Sav. ■

30167 Arabis alpina L. subsp. caucasica (Willd.) Briq. = Arabis caucasica Willd. ■☆

30168 Arabis alpina L. var. albida ? = Arabis caucasica Willd. ■☆

30169 Arabis alpina L. var. commutata Pau et Font Quer = Arabis alpina L. ■☆

30170 Arabis alpina L. var. formosana (Masam.) Tang S. Liu et H. Ying = Arabis formosana (Masam. ex S. F. Huang) Tang S. Liu et S. S. Ying ■

30171 Arabis alpina L. var. formosana Masam. ex S. F. Huang = Arabis formosana (Masam. ex S. F. Huang) Tang S. Liu et S. S. Ying ■

30172 Arabis alpina L. var. formosana Masam. ex S. F. Huang = Arabis serrata Franch. et Sav. ■

30173 Arabis alpina L. var. japonica A. Gray = Arabis stelleri DC. ■

30174 Arabis alpina L. var. parviflora Franch.;小花南芥(小花山南芥);Smallflower Alpine Rockcress ■

30175 Arabis alpina L. var. parviflora Franch. = Arabis paniculata Franch. ■

30176 Arabis alpina L. var. purpurea W. Sm. = Arabis paniculata Franch. ■

30177 Arabis alpina L. var. rigida Franch. = Arabis paniculata Franch. ■

30178 Arabis alpina L. var. rubrocalyx Franch. = Arabis paniculata Franch. ■

30179 Arabis alticola O. E. Schulz = Arabis amplexicaulis Edgew. ■

30180 Arabis ambigua DC. = Arabis paniculata Franch. ■

30181 Arabis ambigua DC. var. glabra DC. = Arabis lyrata L. ■

30182 Arabis amplexicaulis Edgew.;抱茎南芥;Amplexicaul Rockcress ■

30183 Arabis amplexicaulis Edgew. var. japonica H. Boissieu = Arabis serrata Franch. et Sav. ■

30184 Arabis amplexicaulis Edgew. var. serrata (Franch. et Sav.) Makino = Arabis serrata Franch. et Sav. ■

30185 Arabis amurensis N. Busch;阿穆尔南芥■☆

30186 Arabis arenosa (L.) Scop.;沙筷子芥;Rock Cress, Rock-cress, Sand Rock-cress, Tall Rock-cress ■☆

30187 Arabis arvensis Edgew. = Malcolmia africana (L.) R. Br. ■

30188 Arabis attenuata Royle ex Hook. f.;尖果南芥;Attenuate Rockcress ■

30189 Arabis auriculata Lam.;耳叶南芥(耳筷子芥);Annual Rock-cress, Auriculate Rockcress ■

30190 Arabis auriculata Lam. var. dasycarpa Andrz. = Arabis auriculata Lam. ■

30191 Arabis auriculata Lam. var. malinvaldiana (Rouy et Coincy) Batt. = Arabis auriculata Lam. ■

30192 Arabis auriculata Lam. var. puberula Koch = Arabis auriculata Lam. ■

30193 Arabis axillaris Kom. = Neotorularia humilis (C. A. Mey.) Hedge et J. Léonard ■

30194 Arabis axilliflora (Jafri) H. Hara;腋花南芥■

30195 Arabis axilliflora (Jafri) H. Hara = Parryodes axilliflora Jafri ■

30196 Arabis axilliflora (Jafri) H. Hara var. brevistyla H. Hara = Arabis axilliflora (Jafri) H. Hara ■

30197 Arabis balansae Boiss. et Reut. = Arabis hirsuta (L.) Scop. ■

30198 Arabis bijuga Watt;大花南芥;Largeflower Rockcress ■

30199 Arabis billardieri DC.;毕拉尔南芥■☆

30200 Arabis blepharophylla Hook. et Arn.;加州南芥(毛叶南芥);California Rockcress, Rose Rockcress ■☆

30201 Arabis boissieuana Nakai = Arabis serrata Franch. et Sav. var. japonica (H. Boissieu) Ohwi ■

30202 Arabis boissieuana Nakai = Arabis serrata Franch. et Sav. ■

30203 Arabis boissieuana Nakai var. glauca (H. Boissieu) Koidz. = Arabis serrata Franch. et Sav. ■

30204 Arabis boissieuana Nakai var. sikokiana Nakai = Arabis serrata Franch. et Sav. ■

30205 Arabis borealis Andrz. ex Ledeb.;新疆南芥(北方筷子芥);Sinkiang Rockcress, Xinjiang Rockcress ■

30206 Arabis brachycarpa (Torr. et A. Gray) Britton = Arabis divaricarpa A. Nelson ■☆

30207 Arabis brachycarpa Rupr.;短果南芥;Short-fruited Arabis ■☆

30208 Arabis brevicaulis Jafri = Arabidopsis brevicaulis (Jafri) Jafri ■

30209 Arabis brevicaulis Jafri = Crucihimalaya himalaica (Edgew.) Al-Shehbaz, O'Kane et R. A. Price ■

30210 Arabis brownii Jord. = Arabis hirsuta (L.) Scop. ■

30211 Arabis bucharica (Lipsky) N. Busch;布哈尔南芥■☆

30212 Arabis bulbosa Schreb. ex Muhl. = Cardamine bulbosa (Schreb. ex Muhl.) Britton, Sterns et Poggenb. ■☆

30213 Arabis cadmea Boiss. = Arabis auriculata Lam. ■

30214 Arabis canadensis L.;加拿大南芥;Sickle-pod, Sicktepod ■☆

30215 Arabis caucasica Willd. ex Schltdl. 'Plena';重瓣高加索南芥■☆

30216 Arabis caucasica Willd. ex Schltdl. 'Variegata';黄边高加索南芥■☆

30217 Arabis caucasica Willd. ex Schltdl. = Arabis alpina L. subsp. caucasica (Willd.) Briq. ■☆

30218 Arabis caucasica Willd. ex Schltdl. = Arabis caucasica Wind. ex Schltdl. ■☆

30219 Arabis caucasica Willd. ex Schltdl. var. leiopoda Pau = Arabis alpina L. subsp. caucasica (Willd.) Briq. ■☆

30220 Arabis caucasica Wind. ex Schltdl.;高加索南芥(淡白南芥,高加索筷子芥,高加索南芥菜);Caucasia Rockcress, Garden Arabis, Gray Rockcress, Gray Rock-cress, Lady's Cushion, Mountain Snow, Shepherd's Flock, Snow-in-summer, Wall Cress, Wall Rock Cress, Wall Rockcress, Wall Rock-cress ■☆

30221 Arabis cebennensis DC. var. coreana H. Lév. = Cardamine komarovii Nakai ■

30222 Arabis chanetii H. Lév.;短梗南芥;Chanet Rockcress, Shortstalk Rockcress ■

30223 Arabis chanetii H. Lév. = Orychophragmus violaceus (L.) O. E. Schulz ■

30224 Arabis charbonnelii H. Lév. = Sisymbrium irio L. ■

30225 Arabis christianii N. Busch;克里南芥■☆

30226 Arabis ciliata R. Br. = Arabis hirsuta (L.) Scop. ■

30227 Arabis ciliata R. Br. var. balansae (Boiss. et Reut.) Coss. = Arabis hirsuta (L.) Scop. ■

30228 Arabis clarkei O. E. Schulz = Arabis tibetica Hook. f. et Thomson ■

30229 Arabis collina Ten.;山丘南芥;Italian Rock-cress, Rosy Cress ■☆

30230 Arabis collinsii Fernald;柯林斯南芥;Colins' Arabis ■☆

30231 Arabis confinis S. Watson = Arabis divaricarpa A. Nelson ■☆

30232 Arabis confinis S. Watson var. interposita (Greene) Welsh et Reveal = Arabis divaricarpa A. Nelson ■☆

30233 Arabis connexa Greene = Arabis drummondii A. Gray ■☆

30234 Arabis conringioides Ball;拟线果芥■☆

30235 Arabis conringioides Ball subsp. humbertii (Quézel) Dobignard et D. Jord.;亨伯特南芥■☆

30236 Arabis conringioides Ball var. aphanostyla Maire = Arabis conringioides Ball subsp. humbertii (Quézel) Dobignard et D. Jord. ■☆

30237 Arabis conringioides Ball var. stylosa Maire = Arabis conringioides Ball ■☆

30238 Arabis conringioides Ball var. werneri (Emb. et Maire) N. Galland et Favarger = Arabis conringioides Ball ■☆

30239 Arabis coronata Nakai = Arabidopsis halleri (L.) O'Kane et Al-Shehbaz subsp. gemmifera (Matsum.) O'Kane et Al-Shehbaz ■

30240 Arabis coronata Nakai = Arabis halleri L. ■

30241 Arabis cuneifolia Hochst. ex A. Rich. = Arabis alpina L. ■☆

30242 Arabis decumbens Ball = Arabis pubescens (Desf.) Poir. subsp. decumbens Ball ■☆

30243 Arabis decumbens Ball var. brachypoda Font Quer et Pau = Arabis pubescens (Desf.) Poir. subsp. decumbens Ball ■☆

30244 Arabis dentata (Torr.) Torr. et A. Gray = Arabis shortii (Fernald) Gleason ■☆

30245 Arabis dentata (Torr.) Torr. et A. Gray var. phalacrocarpa M. Hopkins = Arabis shortii (Fernald) Gleason ■☆

30246 Arabis divaricarpa A. Nelson;阔足南芥;Spreading-pod Rock-cress ■☆

30247 Arabis divaricarpa A. Nelson var. dechamplainii B. Boivin = Arabis divaricarpa A. Nelson ■☆

30248 Arabis divaricarpa A. Nelson var. interposita (Greene) Rollins = Arabis divaricarpa A. Nelson ■☆

30249 Arabis divaricarpa A. Nelson var. stenocarpa M. Hopkins = Arabis divaricarpa A. Nelson ■☆

30250 Arabis divaricarpa A. Nelson var. typica Rollins = Arabis divaricarpa A. Nelson ■☆

30251 Arabis douglassii Torr. = Cardamine douglassii (Torr.) Britton ■☆

30252 Arabis drummondii A. Gray;德拉蒙德南芥;Drummond's Arabis, Drummond's Rock-cress ■☆

30253 Arabis drummondii A. Gray var. connexa (Greene) Fernald = Arabis drummondii A. Gray ■☆

30254 Arabis drummondii A. Gray var. oxyphylla (Greene) M. Hopkins = Arabis drummondii A. Gray ■☆

30255 Arabis erubescens Ball;变红南芥■☆

30256 Arabis erysimoides Kar. et Kir.;糖芥状南芥■☆

30257 Arabis erysimoides Kar. et Kir. = Prionotrichon erysimoides (Kar. et Kir.) Botsch. et Vved. ■☆

30258 Arabis falcata Hochst. ex A. Rich. = Oreophyton falcatum (Hochst. ex A. Rich.) O. E. Schulz ■☆

30259 Arabis farinacea Rupr.;粉质南芥■☆

30260 Arabis fauriei H. Boissieu = Arabis serrata Franch. et Sav. ■

30261 Arabis fauriei H. Lév. = Arabis stelleri DC. ■

30262 Arabis fauriei H. Lév. var. grandiflora Nakai = Arabis serrata Franch. et Sav. ■

30263 Arabis flagellosa Miq.;匍匐南芥;Creeping Rockcress ■

30264 Arabis flagellosa Miq. f. lasiocarpa (Matsum.) Ohwi;光果匍匐南芥■☆

30265 Arabis flagellosa Miq. var. kawachiensis S. Fujii;河内南芥■☆

30266 Arabis flagellosa Miq. var. lasiocarpa Matsum. = Arabis flagellosa Miq. ■

30267 Arabis flaviflora Bunge;黄花南芥■☆

30268 Arabis formosana (Masam. ex S. F. Huang) Tang S. Liu et S. S. Ying;台湾南芥(台湾筷子芥);Taiwan Rockcress ■

30269 Arabis formosana (Masam. ex S. F. Huang) Tang S. Liu et S. S. Ying = Arabis serrata Franch. et Sav. ■

30270 Arabis fruticulosa C. A. Mey.;小灌木南芥(半灌木南芥);Smallshrub Rockcress ●

30271 Arabis fruticulosa C. A. Mey. var. albescens N. Busch = Arabis fruticulosa C. A. Mey. ●

30272 Arabis gemmifera (Matsum.) Makino;叶芽南芥(蔓田芥,水芹菜,叶芽筷子芥);Leafbud Rockcress ■

30273 Arabis gemmifera (Matsum.) Makino = Arabidopsis halleri (L.) O'Kane et Al-Shehbaz subsp. gemmifera (Matsum.) O'Kane et Al-Shehbaz ■

30274 Arabis gemmifera (Matsum.) Makino f. alpicola (H. Hara) Ohwi = Arabis gemmifera (Matsum.) Makino var. alpicola H. Hara ■☆

30275 Arabis gemmifera (Matsum.) Makino var. alpicola H. Hara;山地

叶芽南芥■☆

30276 Arabis gemmifera (Matsum.) Makino var. alpicola H. Hara = Arabidopsis halleri (L.) O' Kane et Al-Shehbaz subsp. gemmifera (Matsum.) O' Kane et Al-Shehbaz ■

30277 Arabis gerardii Besser;格氏筷子芥;Gerald Rockcress,Gerard's Rock Cress,Gerrard's Rock-cress ■☆

30278 Arabis glabra (L.) Bernh.;光筷子芥;Tower Mustard,Tower Rock-cress ■☆

30279 Arabis glabra (L.) Bernh. = Turritis glabra L. ■

30280 Arabis glandulosa Kar. et Kir. = Dontostemon glandulosus (Kar. et Kir.) O. E. Schulz ■

30281 Arabis glauca H. Boissieu = Arabis serrata Franch. et Sav. var. glauca (H. Boissieu) Ohwi ■☆

30282 Arabis glauca H. Boissieu = Arabis serrata Franch. et Sav. ■

30283 Arabis glauca H. Boissieu subsp. pseudocauriculata (H. Boissieu) Vorosch. = Arabis serrata Franch. et Sav. ■

30284 Arabis greatrexii (Miyabe et Kudo) Miiyabe et Tatewaki = Arabidopsis halleri (L.) O' Kane et Al-Shehbaz subsp. gemmifera (Matsum.) O' Kane et Al-Shehbaz ■

30285 Arabis hallaisanensis Nakai = Arabis serrata Franch. et Sav. ■

30286 Arabis halleri L.;圆叶南芥;Roundleaf Rockcress ■

30287 Arabis halleri L. var. senanensis Franch. et Sav. = Arabidopsis halleri (L.) O' Kane et Al-Shehbaz subsp. gemmifera (Matsum.) O' Kane et Al-Shehbaz ■

30288 Arabis halleri L. var. senanensis Franch. et Sav. = Arabis gemmifera (Matsum.) Makino ■

30289 Arabis heliophila DC. = Farsetia jacquemontii Hook. f. et Thomson ■☆

30290 Arabis heterophylla Nutt. = Arabis laevigata (Muhl. ex Willd.) Poir. ■☆

30291 Arabis himalaica Edgew. = Crucihimalaya himalaica (Edgew.) Al-Shehbaz,O' Kane et R. A. Price ■

30292 Arabis hirsuta (L.) Scop.;硬毛南芥(毛筷子芥,毛南芥,山地南芥,野南芥,野南芥菜);Fringed Rock-cress,Hairy Arabis,Hairy Rock Cress,Hairy Rock-cress,Hirsute Rockcress,Rock Fringed Cress ■

30293 Arabis hirsuta (L.) Scop. subsp. balansae (Boiss. et Reut.) Maire = Arabis hirsuta (L.) Scop. ■

30294 Arabis hirsuta (L.) Scop. subsp. pycnocarpa (M. Hopkins) Hultén = Arabis hirsuta (L.) Scop. var. pycnocarpa (M. Hopkins) Rollins ■☆

30295 Arabis hirsuta (L.) Scop. subsp. sagittata (Bertol.) Nyman = Arabis sagittata (Bertol.) DC. ■

30296 Arabis hirsuta (L.) Scop. subsp. sessilifolia Gaudin = Arabis hirsuta (L.) Scop. ■

30297 Arabis hirsuta (L.) Scop. subsp. tunetana (Murb.) Maire = Arabis tunetana Murb. ■☆

30298 Arabis hirsuta (L.) Scop. var. adpressipilis (M. Hopkins) Rollins;伏毛南芥;Hairy Rock-cress ■☆

30299 Arabis hirsuta (L.) Scop. var. glabrata Torr. et A. Gray;无毛南芥;Hairy Rock-cress,Mountain Rock-cress,Western Rock-cress ■☆

30300 Arabis hirsuta (L.) Scop. var. mesatlantica Maire = Arabis hirsuta (L.) Scop. ■

30301 Arabis hirsuta (L.) Scop. var. nipponica (Franch. et Sav.) C. C. Yuan et T. Y. Cheo;卵叶硬毛南芥;Japan Hirsute Rockcress, Ovate Hirsute Rockcress ■

30302 Arabis hirsuta (L.) Scop. var. nipponica (Franch. et Sav.) C. C. Yuan et T. Y. Cheo = Arabis hirsuta (L.) Scop. ■

30303 Arabis hirsuta (L.) Scop. var. ovata (Poir.) Wallr. = Arabis hirsuta (L.) Scop. ■

30304 Arabis hirsuta (L.) Scop. var. purpurea Y. C. Lan et T. Y. Cheo;紫花硬毛南芥;Purpleflower Hirsute Rockcress ■

30305 Arabis hirsuta (L.) Scop. var. purpurea Y. C. Lan et T. Y. Cheo = Arabis hirsuta (L.) Scop. ■

30306 Arabis hirsuta (L.) Scop. var. pycnocarpa (M. Hopkins) Rollins;密果硬毛南芥(乳花硬毛南芥);Cream-flower Rock-cress, Hairy Rock-cress ■☆

30307 Arabis holanshanica Y. Z. Lan et T. Y. Cheo = Arabis alaschanica Maxim. ■

30308 Arabis holboellii Hornem.;霍尔南芥;Holboell's Arabis ■☆

30309 Arabis humbertii Quézel = Arabis conringioides Ball subsp. humbertii (Quézel) Dobignard et D. Jord. ■☆

30310 Arabis incarnata Pall. = Stevenia cheiranthoides DC. ■

30311 Arabis intermedia Hoppe ex Steud.;中型南芥■☆

30312 Arabis interposita Greene = Arabis divaricarpa A. Nelson ■☆

30313 Arabis ionocalyx Boiss.;堇萼南芥■☆

30314 Arabis iwatensis Makino = Arabis serrata Franch. et Sav. ■

30315 Arabis japonica (A. Gray.) A. Gray = Arabis stelleri DC. ■

30316 Arabis japonica A. Gray = Arabis stelleri DC. var. japonica (A. Gray) F. Schmidt ■

30317 Arabis josiae Jahand. et Maire;约西亚南芥■☆

30318 Arabis josiae Jahand. et Maire var. leptopoda Pau et Font Quer = Arabis josiae Jahand. et Maire ■☆

30319 Arabis kamchatica (Fisch. ex DC.) Ledeb. = Arabidopsis lyrata (L.) O' Kane et Al-Shehbaz subsp. kamchatica (Fisch. ex DC.) O' Kane et Al-Shehbaz ■

30320 Arabis kamtchatica (Fisch. ex DC.) Ledeb. = Arabis lyrata L. subsp. kamtschatica (Fisch. ex DC.) Hultén ■

30321 Arabis kamtschatica Fisch. = Arabis lyrata L. subsp. kamtschatica (Fisch. ex DC.) Hultén ■

30322 Arabis kamtschatica Fisch. ex DC. = Arabis lyrata L. subsp. kamtschatica (Fisch. ex DC.) Hultén ■

30323 Arabis kangdingensis Y. H. Zhang = Sisymbrium yunnanense W. W. Sm. ■

30324 Arabis karatavica (Lipsch.) Botsch. et Vved.;卡拉塔夫南芥■☆

30325 Arabis karatesina Lipsky;卡拉特金南芥■☆

30326 Arabis kawasakiana Makino = Arabidopsis kamchatica (DC.) K. Shimizu et Kudoh ■☆

30327 Arabis kawasakiana Makino = Arabidopsis lyrata (L.) O' Kane et Al-Shehbaz subsp. kamchatica (Fisch. ex DC.) O' Kane et Al-Shehbaz ■

30328 Arabis kelung-insulari Hayata = Arabis stelleri DC. ■

30329 Arabis kelung-insulari Hayata = Arabis stelleri DC. var. japonica (A. Gray) F. Schmidt ■

30330 Arabis kishidae Nakai = Arabis serrata Franch. et Sav. ■

30331 Arabis kokanica Regel et Schmalh.;浩罕南芥■☆

30332 Arabis laevigata (Muhl. ex Willd.) Poir.;平滑南芥;Smooth Arabis,Smooth Bank-cress,Smooth Rock Cress,Smooth Rock-cress ■☆

30333 Arabis laevigata (Muhl. ex Willd.) Poir. var. missouriensis (Greene) Ahles = Arabis missouriensis Greene ■☆

30334 Arabis latialata Y. Z. Lan et T. Y. Cheo = Arabis pterosperma Edgew. ■

30335 Arabis latifolia Durieu = Arabis parvula DC. ■☆

30336 Arabis laxa Sibth. et Sm.;疏松南芥■☆

30337 Arabis ligulifolia Nakai;舌叶南芥■☆

30338 Arabis lithophila Hayata = Arabis kelung-insularis Hayata ■
30339 Arabis lithophila Hayata = Arabis stelleri DC. ■
30340 Arabis lithophila Hayata = Arabis stelleri DC. var. japonica (A. Gray) F. Schmidt ■
30341 Arabis lyrata Kom. = Arabis lyrata L. subsp. kamtschatica (Fisch. ex DC.) Hultén ■
30342 Arabis lyrata L.;深山南芥;Lyrate Rockcress, Lyrate Rock-cress, Lyre-leaved Rock Cress, Lyre-leaved Rock-cress, Lyre-shaped Rockcress, Sand Cress ■
30343 Arabis lyrata L. subsp. kamchatica (Fisch. ex DC.) Hultén = Arabidopsis lyrata (L.) O'Kane et Al-Shehbaz subsp. kamchatica (Fisch. ex DC.) O'Kane et Al-Shehbaz ■
30344 Arabis lyrata L. subsp. kamchatica (Fisch. ex DC.) Hultén = Arabidopsis lyrata (L.) O'Kane et Al-Shehbaz subsp. kamchatica (Fisch. ex DC.) O'Kane et Al-Shehbaz ■
30345 Arabis lyrata L. subsp. kamtschatica (Fisch. ex DC.) Hultén;琴叶南芥(勘察加南芥,玉山筷子芥,玉山南芥);Kamtschatic Rockcress, Morrison Rockcress, Yushan Rockcress ■
30346 Arabis lyrata L. var. glabra (DC.) M. Hopkins = Arabis lyrata L. ■
30347 Arabis lyrata L. var. kamchatica DC. = Arabis lyrata L. subsp. kamtschatica (Fisch. ex DC.) Hultén ■
30348 Arabis lyrata L. var. kamchatica Fisch. ex DC. = Arabidopsis lyrata (L.) O'Kane et Al-Shehbaz subsp. kamchatica (Fisch. ex DC.) O'Kane et Al-Shehbaz ■
30349 Arabis lyrata L. var. kamchatica Fisch. ex DC. = Arabidopsis lyrata (L.) O'Kane et Al-Shehbaz subsp. kamchatica (Fisch. ex DC.) O'Kane et Al-Shehbaz ■
30350 Arabis lyrata L. var. occidentalis S. Watson = Arabis lyrata L. ■
30351 Arabis lyrata L. var. typica M. Hopkins = Arabis lyrata L. ■
30352 Arabis lyrata L. var. typica M. Hopkins f. parvisiliqua M. Hopkins = Arabis lyrata L. ■
30353 Arabis lyrifolia DC. = Arabis laevigata (Muhl. ex Willd.) Poir. ■☆
30354 Arabis macrantha C. C. Yuan et T. Y. Cheo = Arabis bijuga Watt ■
30355 Arabis malinvaldiana Rouy et Coincy = Arabis auriculata Lam. ■
30356 Arabis maximowiczii N. Busch = Arabidopsis halleri (L.) O'Kane et Al-Shehbaz subsp. gemmifera (Matsum.) O'Kane et Al-Shehbaz ■
30357 Arabis media N. Busch;中间南芥■☆
30358 Arabis missouriensis Greene;密苏里南芥;Green Rock-cress, Missouri Rock Cress, Missouri Rock-cress, Rock Cress ■☆
30359 Arabis missouriensis Greene var. deamii (M. Hopkins) M. Hopkins = Arabis missouriensis Greene ■☆
30360 Arabis montbretiana Boiss.;蒙氏南芥■☆
30361 Arabis morrisonensis Hayata = Arabidopsis lyrata (L.) O'Kane et Al-Shehbaz subsp. kamchatica (Fisch. ex DC.) O'Kane et Al-Shehbaz ■
30362 Arabis multicaulis Pamp. = Arabis tibetica Hook. f. et Thomson ■
30363 Arabis muralis Bertol.;意大利南芥;Italian Rock-cress ■☆
30364 Arabis muralis Bertol. = Arabis collina Ten. ■☆
30365 Arabis nepetifolia Boiss.;荆芥叶南芥■☆
30366 Arabis nipponica (Franch.) H. Boissieu = Arabis hirsuta (L.) Scop. ■
30367 Arabis nordmanaiana Rupr.;诺尔德南芥■☆
30368 Arabis nova Vill.;诺瓦南芥■☆
30369 Arabis nova Vill. subsp. iberica Rivas Mart.;伊比利亚南芥■☆
30370 Arabis nuda Bél. = Drabopsis nuda (Bél.) Stapf ■
30371 Arabis nuda Bél. ex Boiss. = Drabopsis verna C. Koch ■
30372 Arabis nudicaulis (L.) DC. = Parrya nudicaulis (L.) Regel ■
30373 Arabis nudicaulis = Parrya nudicaulis (L.) Regel ■
30374 Arabis nudiuscula E. Mey. ex Sond. = Rorippa nudiuscula (E. Mey. ex Sond.) Thell. ■☆
30375 Arabis nuristanica Kitam. = Arabis amplexicaulis Edgew. ■
30376 Arabis officinalis Andrz. ex M. Bieb. = Alliaria petiolata (M. Bieb.) Cavara et Grande ■
30377 Arabis ovata Poir. = Arabis hirsuta (L.) Scop. var. ovata (Poir.) Wallr. ■
30378 Arabis ovata sensu Small = Arabis hirsuta (L.) Scop. var. pycnocarpa (M. Hopkins) Rollins ■☆
30379 Arabis oxyphylla Greene = Arabis drummondii A. Gray ■☆
30380 Arabis pachyrhiza Kar. et Kir.;粗根南芥■☆
30381 Arabis pamirica Y. C. Lan et C. H. An = Sisymbriopsis pamirica (Y. C. Lan et C. H. An) Al-Shehbaz ■
30382 Arabis pangiensis Watt = Arabis bijuga Watt ■
30383 Arabis paniculata Franch.;圆锥南芥;Paniculate Rockcress ■
30384 Arabis paniculata Franch. var. parviflora (Franch.) W. T. Wang = Arabis paniculata Franch. ■
30385 Arabis parvula DC.;较小南芥■☆
30386 Arabis parvula DC. var. paui Sennen et Mauricio = Arabis parvula DC. ■☆
30387 Arabis pendula L.;垂果南芥(扁担蒿,垂果南芥菜,大蒜芥,唐芥,野白菜);Pendentfruit Rockcress ■
30388 Arabis pendula L. = Catolobus pendula (L.) Al-Shehbaz ■☆
30389 Arabis pendula L. var. glabrescens Franch.;疏毛垂果南芥;Glabrous Pendentfruit Rockcress, Slightlyglabrous Pendentfruit Rockcress ■
30390 Arabis pendula L. var. glabrescens Franch. = Arabis pendula L. ■
30391 Arabis pendula L. var. hebecarpa Y. Z. Lan et T. Y. Cheo;毛垂果南芥(毛果垂果南芥,毛果南芥);Hairy Pendentfruit Rockcress, Hairyfruit Pendentfruit Rockcress ■
30392 Arabis pendula L. var. hebecarpa Y. Z. Lan et T. Y. Cheo = Arabis pendula L. ■
30393 Arabis pendula L. var. hypoglauca Franch.;粉绿垂果南芥;Glaucous Pendentfruit Rockcress ■
30394 Arabis pendula L. var. hypoglauca Franch. = Arabis pendula L. ■
30395 Arabis perfoliata Lam. = Turritis glabra L. ■
30396 Arabis perstellata E. L. Braun var. phalacrocarpa (M. Hopkins) Fernald = Arabis shortii (Fernald) Gleason ■☆
30397 Arabis perstellata E. L. Braun var. shortii Fernald = Arabis shortii (Fernald) Gleason ■☆
30398 Arabis petiolata M. Bieb. = Alliaria petiolata (M. Bieb.) Cavara et Grande ■
30399 Arabis petraea Hook.;岩生南芥;Northern Rock-cress ■☆
30400 Arabis pinnatifida Lam. = Murbeckiella boryi (Boiss.) Rothm. ■☆
30401 Arabis popovii Botsch. et Vved.;波氏南芥■☆
30402 Arabis procurrens Waldst. et Kit.;铺展南芥;Running Rockcress ■☆
30403 Arabis pseudoauriculata H. Boissieu = Arabis serrata Franch. et Sav. ■
30404 Arabis pseudodecumbens Emb. et Maire = Arabis pubescens (Desf.) Poir. ■☆
30405 Arabis pseudoturritis H. Boissieu et Heldr. = Turritis glabra L. ■
30406 Arabis pterosperma Edgew.;窄翅南芥(宽翅南芥,亚东拟南芥,亚东鼠耳芥);Broadwing Rockcress, Wingedseed Rockcress, Yadong Mouseear Cress ■

30407 Arabis pubescens (Desf.) Poir.;短绒毛南芥■☆

30408 Arabis pubescens (Desf.) Poir. subsp. decumbens Ball;外倾短绒毛南芥■☆

30409 Arabis pubescens (Desf.) Poir. var. brachycarpa Batt. = Arabis pubescens (Desf.) Poir. ■☆

30410 Arabis pubescens (Desf.) Poir. var. brachypoda Font Quer et Pau = Arabis pubescens (Desf.) Poir. ■☆

30411 Arabis pubescens (Desf.) Poir. var. bracteosa Maire = Arabis pubescens (Desf.) Poir. ■☆

30412 Arabis pubescens (Desf.) Poir. var. gracilis Emb. et Maire = Arabis pubescens (Desf.) Poir. ■☆

30413 Arabis pubescens (Desf.) Poir. var. longisiliqua Coss. = Arabis tunetana Murb. ■☆

30414 Arabis pubescens (Desf.) Poir. var. pseudodecumbens (Emb. et Maire) Maire = Arabis pubescens (Desf.) Poir. ■☆

30415 Arabis pulchra M. E. Jones ex S. Watson;美丽南芥;Prince's Rock Cress,Prince's Rock-cress ■☆

30416 Arabis pycnocarpa M. Hopkins = Arabis hirsuta (L.) Scop. var. pycnocarpa (M. Hopkins) Rollins ■☆

30417 Arabis pycnocarpa M. Hopkins var. adpressipilis M. Hopkins = Arabis hirsuta (L.) Scop. var. adpressipilis (M. Hopkins) Rollins ■☆

30418 Arabis quinqueloba O. E. Schulz = Arabis tibetica Hook. f. et Thomson ■

30419 Arabis recta Vill. = Arabis auriculata Lam. ■

30420 Arabis rhomboidea Pers. = Cardamine bulbosa (Schreb. ex Muhl.) Britton,Sterns et Poggenb. ■☆

30421 Arabis rosea DC. = Arabis muralis Bertol. ■☆

30422 Arabis rupestris Edgew. = Crucihimalaya himalaica (Edgew.) Al-Shehbaz,O'Kane et R. A. Price ■

30423 Arabis rupicola Krylov = Arabidopsis mollissima (C. A. Mey.) N. Busch ■☆

30424 Arabis sagittata (Bertol.) DC.;箭叶南芥;Sagittate Rockcress ■

30425 Arabis sagittata (Bertol.) DC. var. exauriculata Lange = Arabis sagittata (Bertol.) DC. ■

30426 Arabis sagittata (Bertol.) DC. var. nipponica Franch. et Sav. = Arabis hirsuta (L.) Scop. var. nipponica (Franch. et Sav.) C. C. Yuan et T. Y. Cheo ■

30427 Arabis sagittata (Bertol.) DC. var. nipponica Franch. et Sav. = Arabis hirsuta (L.) Scop. ■

30428 Arabis saxicola Edgew.;岩栖南芥■☆

30429 Arabis scabra All.;布港南芥;Bristol Rock Cress,Bristol Rock-cress ■☆

30430 Arabis scapigera Boiss. = Drabopsis nuda (Bél.) Stapf ■

30431 Arabis scapigera Boiss. = Drabopsis verna C. Koch ■

30432 Arabis secunda N. Busch;单侧南芥■☆

30433 Arabis senanensis (Franch. et Sav.) Makino = Arabidopsis halleri (L.) O'Kane et Al-Shehbaz subsp. gemmifera (Matsum.) O'Kane et Al-Shehbaz ■

30434 Arabis septentrionalis N. Busch;北极筷子芥;Arctic Rockcress ■☆

30435 Arabis serrata Franch. et Sav.;齿叶南芥(齿叶筷子芥);Serrate Rockcress ■

30436 Arabis serrata Franch. et Sav. var. glabrescens Ohwi = Arabis serrata Franch. et Sav. ■

30437 Arabis serrata Franch. et Sav. var. glauca (H. Boissieu) Ohwi;灰齿叶南芥(粉叶南芥);Glaucous Rockcress ■☆

30438 Arabis serrata Franch. et Sav. var. glauca (H. Boissieu) Ohwi = Arabis serrata Franch. et Sav. ■

30439 Arabis serrata Franch. et Sav. var. japonica (H. Boissieu) Ohwi;日本齿叶南芥■

30440 Arabis serrata Franch. et Sav. var. japonica (H. Boissieu) Ohwi = Arabis serrata Franch. et Sav. ■

30441 Arabis serrata Franch. et Sav. var. japonica (H. Boissieu) Ohwi f. fauriei (H. Boissieu) Ohwi;法氏日本齿叶南芥■☆

30442 Arabis serrata Franch. et Sav. var. japonica (H. Boissieu) Ohwi f. glabrescens (Ohwi) Ohwi;光日本齿叶南芥■☆

30443 Arabis serrata Franch. et Sav. var. japonica (H. Boissieu) Ohwi f. glabrescens (Ohwi) Ohwi = Arabis serrata Franch. et Sav. ■

30444 Arabis serrata Franch. et Sav. var. japonica (H. Boissieu) Ohwi f. grandiflora (Nakai) Ohwi;大花日本齿叶南芥■☆

30445 Arabis serrata Franch. et Sav. var. japonica (H. Boissieu) Ohwi f. grandiflora (Nakai) Ohwi = Arabis serrata Franch. et Sav. ■

30446 Arabis serrata Franch. et Sav. var. platycarpa Ohwi = Arabis serrata Franch. et Sav. ■

30447 Arabis serrata Franch. et Sav. var. shikokiana (Nakai) Ohwi;四国南芥■☆

30448 Arabis serrata Franch. et Sav. var. sikokiana (Nakai) Ohwi = Arabis serrata Franch. et Sav. ■

30449 Arabis setosifolia Al-Shehbaz;毛叶南芥■

30450 Arabis shikokiana (Nakai) Honda = Arabis serrata Franch. et Sav. var. shikokiana (Nakai) Ohwi ■☆

30451 Arabis shortii (Fernald) Gleason;绍氏南芥;Rock Cress,Short's Rock-cress,Toothed Cress ■☆

30452 Arabis shortii (Fernald) Gleason var. phalacrocarpa (M. Hopkins) Steyerm. = Arabis shortii (Fernald) Gleason ■☆

30453 Arabis sikokiana (Nakai) Honda = Arabis serrata Franch. et Sav. ■

30454 Arabis sinuata Turcz.;深波南芥■☆

30455 Arabis sogdiana Kom. = Arabis auriculata Lam. ■

30456 Arabis stelleri DC.;基隆南芥(斯氏南芥)■

30457 Arabis stelleri DC. subsp. japonica (A. Gray) Vorosch. = Arabis stelleri DC. ■

30458 Arabis stelleri DC. var. japonica (A. Gray) F. Schmidt;日本南芥(基隆筷子芥,基隆南芥);Jilong Rockcress,Kelung Rockcress ■

30459 Arabis stelleri DC. var. japonica (A. Gray) F. Schmidt = Arabis serrata Franch. et Sav. ■

30460 Arabis stelleri DC. var. japonica (A. Gray) F. Schmidt = Arabis stelleri DC. ■

30461 Arabis stelleri DC. var. japonica (A. Gray) F. Schmidt f. calvescens Hiyama;渐光基隆南芥■☆

30462 Arabis stelleri DC. var. japonica (A. Gray) F. Schmidt f. purpurascens H. Nakai et H. Ohashi;紫基隆南芥■☆

30463 Arabis stricta Huds. = Arabis scabra All. ■☆

30464 Arabis subpendula Ohwi = Arabis ligulifolia Nakai ■☆

30465 Arabis subpendula Ohwi = Arabis pendula L. ■

30466 Arabis subpendula Ohwi = Catolobus pendula (L.) Al-Shehbaz ■☆

30467 Arabis tanakae Makino;田中氏南芥■☆

30468 Arabis taraxacifolia T. Anderson = Arabidopsis wallichii (Hook. f. et Thomson) N. Busch ■

30469 Arabis tenuirostris O. E. Schulz = Arabis tibetica Hook. f. et Thomson ■

30470 Arabis thaliana L. = Arabidopsis thaliana (L.) Heynh. ■

30471 Arabis thomsonii Hook. f. et Thomson = Arabis tibetica Hook. f. et Thomson ■

30472 Arabis tibetica Hook. f. et Thomson;西藏南芥(西藏鼠耳芥);Tibet Mouseear Cress,Xizang Mouseear Cress ■

30473 Arabis toxophylla M. Bieb. = Pseudoarabidopsis toxophylla (M. Bieb.) Al-Shehbaz, O'Kane et R. A. Price ■

30474 Arabis trichopoda Turcz.;毛足南芥■☆

30475 Arabis tunetana Murb.;长荚南芥■☆

30476 Arabis turczaninovii Ledeb.;图尔南芥■☆

30477 Arabis turrita L.;塔形筷子芥;Tower Cress, Tower Mustard, Tower Rock Cress, Tower Rock-cress ■☆

30478 Arabis verdieri Quézel;贝迭尔南芥■☆

30479 Arabis verna (K. Koch) N. Busch = Drabopsis verna K. Koch ■

30480 Arabis verna (L.) R. Br.;春南芥■☆

30481 Arabis verna (L.) R. Br. var. dasycarpa Godr. ex Rouy et Foucaud = Arabis verna (L.) R. Br. ■☆

30482 Arabis viridis Harger = Arabis missouriensis Greene ■☆

30483 Arabis viridis Harger var. deamii M. Hopkins = Arabis missouriensis Greene ■☆

30484 Arabis viridis Harger var. heterophylla (Nutt.) Farw. = Arabis laevigata (Muhl. ex Willd.) Poir. ■☆

30485 Arabis werneri Emb. et Maire = Arabis conringioides Ball var. werneri (Emb. et Maire) N. Galland et Favarger ■☆

30486 Arabis yokoscensis Franch. et Sav. = Arabis stelleri DC. ■

30487 Arabisa Rchb. = Arabis L. ●■

30488 Aracamunia Carnevali et I. Ramirez(1989);阿拉卡兰属■☆

30489 Aracamunia liesneri Carnevali et I. Ramirez;阿拉卡兰■☆

30490 Araceae Adans. = Araceae Juss.(保留科名)●■

30491 Araceae Juss.(1789)(保留科名);天南星科;Arum Family, Calla Family, Lords-and-ladies Family ■

30492 Arachidna Boehm. = Arachis L. ■

30493 Arachidna Plum. ex Moench = Arachis L. ■

30494 Arachis L.(1753);落花生属(花生属);Goober, Peanut ■

30495 Arachis fruticosa Retz. = Stylosanthes fruticosa (Retz.) Alston ●☆

30496 Arachis glabrata Benth.;多年落花生;Perennial Peanut ■☆

30497 Arachis hagenbeckii Harms ex Kuntze;哈根落花生;Rhizoma Peanut ■☆

30498 Arachis hypogaea L.;落花生(长生果,地豆,番豆,番果,花生,及地果,落地生,落地松,落花参,南京豆,土豆,土露子,香芋);Earth Nut, Earth Pea, Earthnut, Earth-nut, Goober, Goober Pea, Ground Nut, Ground Pea, Groundnut, Monkey Nut, Monkey-nut, Peanut, Pindar, Pinder ■

30499 Arachis prostrata Benth.;平卧落花生;Grassnut ■☆

30500 Arachna Noronha = Hedychium J. König ■

30501 Arachnabenis Thouars = Habenaria Willd. ■

30502 Arachnanthe Blume = Arachnis Blume ■

30503 Arachnanthe annamensis Rolfe = Arachnis annamensis (Rolfe) J. J. Sm. ■☆

30504 Arachnanthe cathcarthii Benth. et Hook. f. = Arachnis cathcarthii J. J. Sm. ■☆

30505 Arachnanthe clarkei (Rchb. f.) Rolfe = Esmeralda clarkei Rchb. f. ■

30506 Arachnanthe flos-aeris J. J. Sm. = Arachnis moschifera Blume ■☆

30507 Arachnanthe moschifera Blume = Arachnis moschifera Blume ■☆

30508 Arachnaria Szlach.(2003);类蜘蛛兰属■☆

30509 Arachnaria Szlach. = Habenaria Willd. ■

30510 Arachne (Endl.) Pojark. = Leptopus Decne. ●

30511 Arachne Endl. = Leptopus Decne. ●

30512 Arachne Neck. = Breynia J. R. Forst. et G. Forst.(保留属名)●

30513 Arachne Neck. = Leptopus Decne. ●

30514 Arachne australis (Zoll. et Moritz) Humsawa = Leptopus australis (Zoll. et E. Morren) Pojark. ●

30515 Arachne australis (Zoll. et Moritz) Pojark. = Leptopus australis (Zoll. et E. Morren) Pojark. ●

30516 Arachne capillipes (Pax) Pojark. = Leptopus chinensis (Bunge) Pojark. ●

30517 Arachne chinensis (Bunge) Humsawa = Leptopus chinensis (Bunge) Pojark. ●

30518 Arachne chinensis (Bunge) Pojark. = Leptopus chinensis (Bunge) Pojark. ●

30519 Arachne clarkei (Hook. f.) Pojark. = Leptopus clarkei (Hook. f.) Pojark. ●

30520 Arachne hirsuta (Hutch.) Hurusawa = Leptopus chinensis (Bunge) Pojark. var. hirsutus (Hutch.) P. T. Li ●

30521 Arachne hirsuta (Hutch.) Pojark. = Leptopus chinensis (Bunge) Pojark. var. hirsutus (Hutch.) P. T. Li ●

30522 Arachne hirsuta Hutch. = Leptopus chinensis (Bunge) Pojark. var. hirsutus (Hutch.) P. T. Li ●

30523 Arachne montana (Hutch.) Humsawa = Leptopus chinensis (Bunge) Pojark. ●

30524 Arachne montana (Hutch.) Pojark. = Leptopus chinensis (Bunge) Pojark. ●

30525 Arachnimorpha Desv. = Rondeletia L. ●

30526 Arachnis Blume(1825);蜘蛛兰属(龙爪兰属);Arachnis, Scorpion Orchid, Spiderorchis ■

30527 Arachnis annamensis (Rolfe) J. J. Sm.;越南蜘蛛兰;Vietnam Arachnis ■☆

30528 Arachnis bella (Rchb. f.) J. J. Sm. = Esmeralda bella Rchb. f. ■

30529 Arachnis breviscapa (J. J. Sm.) J. J. Sm.;短序蜘蛛兰;Shortscape Arachnis ■☆

30530 Arachnis cathcarthii J. J. Sm.;卡氏蜘蛛兰■☆

30531 Arachnis clarker (Rchb. f.) J. J. Sm. = Esmeralda clarkei Rchb. f. ■

30532 Arachnis flos-aeris (L.) Rchb. f.;指甲兰花蜘蛛兰;Flower Arachnis, Fox-tail Orchid, Scorpion Orchid ■☆

30533 Arachnis hookeriana (Rchb. f.) Rchb. f.;香花蜘蛛兰;Fragrantflower Arachnis ■☆

30534 Arachnis labrosa (Lindl. et Paxton) Rchb. f.;窄唇蜘蛛兰(假肾药兰,龙爪兰,蔓生阿芒多兰);Labrose Spiderorchis, Narrowlip Arachnis, Prostrate Armodorum ■

30535 Arachnis labrosa (Lindl. ex Paxton) Rchb. f. = Renanthera labrosa (Lindl.) Rchb. f. ■

30536 Arachnis moschifera Blume;麝香蜘蛛兰■☆

30537 Arachnites F. W. Schmidt = Ophrys L. + Chamorchis Rich. + Aceras R. Br. ■☆

30538 Arachnites F. W. Schmidt = Ophrys L. ■☆

30539 Arachnites monorchis (L.) Hoffm. = Herminium monorchis (L.) R. Br. ■

30540 Arachnitidaceae Munoz = Corsiaceae Becc.(保留科名)■

30541 Arachnitidaceae Munoz;智利腐蛛草科■

30542 Arachnitis Phil.(1864)(保留属名);智利腐蛛草属■☆

30543 Arachnitis uniflora Phil.;智利腐蛛草■☆

30544 Arachnocalyx Compton = Erica L. ●☆

30545 Arachnocalyx Compton(1935);蛛萼杜鹃属(南非杜鹃属)●☆

30546 Arachnocalyx cereris Compton;蛛萼杜鹃●☆

30547 Arachnocalyx viscidus (N. E. Br.) E. G. H. Oliv. = Erica arachnocalyx E. G. H. Oliv. ●☆

30548 Arachnodendris Thouars = Aeranthes Lindl. ■☆

30549 Arachnodendris Thouars = Dendrobium Sw.(保留属名)■
30550 Arachnodes Gagnep. = Phyllanthodendron Hemsl. ●
30551 Arachnodes Gagnep. = Phyllanthus L. ●■
30552 Arachnopogon Berg ex Haberl = Hypochaeris L. ■
30553 Arachnopogon Berg ex Steud. = Hypochaeris L. ■
30554 Arachnorchis D. L. Jones et M. A. Clem. = Caladenia R. Br. ■☆
30555 Arachnospermum Berg ex Haberl = Hypochaeris L. ■
30556 Arachnospermum Berg. = Hypochaeris L. ■
30557 Arachnospermum F. W. Schmidt.(废弃属名) = Podospermum DC.(保留属名)■
30558 Arachnothrix Walp. = Arachnothryx Planch. ●☆
30559 Arachnothryx Planch.(1849);蜘蛛茜属●☆
30560 Arachnothryx Planch. = Rondeletia L. ●
30561 Arachnothryx leucophylla (Kunth) Planch.;蜘蛛茜●☆
30562 Arachus Medlk. = Vicia L. ■
30563 Aracium (Neck.) Monnier = Crepis L. ■
30564 Aracium Monnier = Crepis L. ■
30565 Aracium Neck. = Crepis L. ■
30566 Aracium multicaule (Ledeb.) D. Dietr. = Crepis multicaulis Ledeb. ■
30567 Aracium sibiricum (L.) Sch. Bip. = Crepis sibirica L. ■
30568 Araeoandra Lefor = Viviania Cav. ■☆
30569 Araeococcus Brongn.(1841);多穗凤梨属(阿来果属,鞭叶凤梨属)■☆
30570 Araeococcus flagellifolius Harms;鞭叶多穗凤梨■☆
30571 Araeococcus pectinatus L. B. Sm.;栉齿多穗凤梨■☆
30572 Arafoe Pimenov et Lavrova(1989);高加索香草属☆
30573 Arafoe aromatica Pimenov et Lavrova;高加索香草☆
30574 Aragallus Neck. = Astragalus L. ●■
30575 Aragallus Neck. ex Greene = Astragalus L. ●■
30576 Aragallus Neck. ex Greene = Oxytropis DC.(保留属名)●■
30577 Arago Endl. = Aragoa Kunth ●☆
30578 Aragoa Kunth(1819);阿拉戈婆婆纳属●☆
30579 Aragoa abietina Kunth;阿拉戈婆婆纳●☆
30580 Aragoa lucidula S. F. Blake;亮阿拉戈婆婆纳●☆
30581 Aragoa occidentalis Pennell;西方阿拉戈婆婆纳●☆
30582 Aragoa parviflora Fern. Alonso et Castrov.;小花阿拉戈婆婆纳●☆
30583 Aragoaceae D. Don = Orobanchaceae Vent.(保留科名)●■
30584 Aragoaceae D. Don = Scrophulariaceae Juss.(保留科名)●■
30585 Aragus Steud. = Aragallns Neck. ●■
30586 Aragus Steud. = Astragalus L. ●■
30587 Aralia L.(1753);楤木属(刺楤属,独活属,土当归属);Angelica, Angelica Tree, Angelica-tree, Aralia, Mountain-angelica ●■
30588 Aralia abyssinica Hochst. ex A. Rich. = Schefflera abyssinica (Hochst. ex A. Rich.) Harms ●☆
30589 Aralia apioides Hand.-Mazz.;芹叶龙眼独活(黑羌活,丽江土当归,龙头羌活,牛角七,牛尾当归,肉五加,血秦归);Celeryleaf Aralia ■
30590 Aralia armata (Wall. ex G. Don) Seem.;野楤头(百鸟不落,刺老包,楤木,广东楤木,虎刺楤木,酒合木,雷公木,鸟不宿,小郎伞,小鸟不企,野楤木,鹰不扑);Spine Aralia ●
30591 Aralia atropurpurea Franch.;浓紫龙眼独活;Darkpurple Aralia ■
30592 Aralia balfouriana André = Polyscias scutellaria (Burm. f.) Fosberg ●
30593 Aralia bipinnata Blanco;台湾楤木(里白楤木,三七);Taiwan Angelica-tree, Taiwan Aralia ●
30594 Aralia bipinnata Blanco = Aralia ryukyuensis (J. Wen) T. Yamaz. ●
30595 Aralia bipinnata Blanco var. inermis (Yanagita) T. Yamaz. = Aralia ryukyuensis (J. Wen) T. Yamaz. var. inermis (Yanagita) T. Yamaz. ●☆
30596 Aralia bipinnatifida (Seem.) C. B. Clarke = Panax bipinnatifidus Seem. ■
30597 Aralia bipinnatifida (Seem.) C. B. Clarke = Panax pseudoginseng Wall. var. bipinnatifidus (Seem.) H. L. Li ■
30598 Aralia bodinieri H. Lév. = Nothopanax delavayi (Franch.) Harms ex Diels ●
30599 Aralia cachemirica Decne.;大叶楤木■☆
30600 Aralia caesia Hand.-Mazz.;圆叶楤木;Rotundleaf Aralia ●
30601 Aralia caesia Hand.-Mazz. = Pentapanax caesius (Hand.-Mazz.) C. B. Shang ●
30602 Aralia canescens Siebold et Zucc. = Aralia elata (Miq.) Seem. ●
30603 Aralia castanopsidicola (Hayata) J. Wen = Pentapanax castanopseicola Hayata ●
30604 Aralia chinensis L.;楤木(百鸟不落,刺包头,刺椿头,刺老包,刺老苞,刺龙柏,刺龙包,刺龙袍,刺树椿,飞天蜈蚣,海桐皮,虎阳刺,黄龙苞,黄毛楤木,箭当树,鸟不企,鸟不宿,鹊不踏,山通花,树头菜,通刺,吻头);Angelica Tree, China Aralia, Chinese Angelica Tree, Chinese Aralia, Dimorphantis ●
30605 Aralia chinensis L. var. aureovariegata Rehder = Aralia elata (Miq.) Seem. f. aureovariegata (Rehder) Nakai ●☆
30606 Aralia chinensis L. var. dasyphylloides Hand.-Mazz.;毛叶楤木;Tomentoseleaf Aralia ●
30607 Aralia chinensis L. var. dasyphylloides Hand.-Mazz. = Aralia dasyphylla Miq. ●
30608 Aralia chinensis L. var. elata (Miq.) H. Lév. = Aralia elata (Miq.) Seem. ●
30609 Aralia chinensis L. var. glabrescens (Franch. et Sav.) C. K. Schneid. = Aralia elata (Miq.) Seem. var. glabrescens (Franch. et Sav.) Pojark. ●
30610 Aralia chinensis L. var. glabrescens C. K. Schneid. = Aralia elata (Miq.) Seem. ●
30611 Aralia chinensis L. var. mandshurica (Rupr. et Maxim.) Rehder = Aralia elata (Miq.) Seem. var. glabrescens (Franch. et Sav.) Pojark. ●
30612 Aralia chinensis L. var. mandshurica Rehder = Aralia elata (Miq.) Seem. ●
30613 Aralia chinensis L. var. nuda Nakai;白背叶楤木(苍蓝楤木);Whiteback Aralia ●
30614 Aralia chinensis L. var. nuda Nakai = Aralia elata (Miq.) Seem. ●
30615 Aralia cissifolia Griff. = Acanthopanax cissifolius (Griff.) Harms ●
30616 Aralia cissifolia Griff. ex C. B. Clarke = Eleutherococcus cissifolius (Griff. ex C. B. Clarke) Nakai ●
30617 Aralia cochleata Lam. = Nothopanax cochleatus (Lam.) Miq. ●☆
30618 Aralia continentalis Kitag.;东北土当归(长白楤木,牛尾大活,香秸颗);Continent Aralia, Continental Aralia ■
30619 Aralia cordata Thunb.;土当归(当归,独活,杜当归,鬼眼独活,九眼独活,食用楤木,食用土当归,台湾楤木,心叶九眼独活);Japanese Udo Salad, Udo, Udo Salad Plant ■
30620 Aralia cordata Thunb. f. biternata Nakai;三出楤木●☆
30621 Aralia cordata Thunb. var. continentalis (Kitag.) Y. C. Zhu = Aralia continentalis Kitag. ■
30622 Aralia cordata Thunb. var. sachalinensis (Regel) Nakai;库页土当归■☆

30623 Aralia dasyphylla Miq.;头序楤木(鸡姆盼,雷公种,毛叶楤木,牛尾木,伞坝菜);Hairyleaf Aralia,Hairy-leaved Aralia ●

30624 Aralia dasyphylloides (Hand.-Mazz.) J. Wen = Aralia chinensis L. var. dasyphylloides Hand.-Mazz. ●

30625 Aralia dasyphylloides (Hand.-Mazz.) J. Wen. = Aralia dasyphylla Miq. ●

30626 Aralia debilis J. Wen;秀丽楤木;Beautiful Aralia,Elegant Aralia ●

30627 Aralia debilis J. Wen = Aralia elegans C. N. Ho ●

30628 Aralia decaisneana Hance;台湾毛楤木(刺楤,大鹰不扑,黄毛楤木,鸟不企,鹊不踏,台湾楤木);Decaisne Angelica Tree,Decaisne Angelica-tree,Yellowhair Aralia,Yellow-haired Aralia ●

30629 Aralia delavayi J. Wen = Pentapanax yunnanensis Franch. ●

30630 Aralia disperma Blume = Macropanax dispermus (Blume) Kuntze ●

30631 Aralia dumetorum Hand.-Mazz. = Aralia melanocarpa (H. Lév.) Lauener ■

30632 Aralia echinocaulis Hand.-Mazz.;棘茎楤木(刺茎楤木,红刺党,红刺桐,红刺筒,红楤木,红老虎刺,红毛刺桐,红鸟不宿,红鸟不踏刺,红叶大猫刺,红叶雨伞刺,虎椒刺,鸟不踏,千枚针);Spinystem Aralia,Spiny-stemmed Aralia ●

30633 Aralia edulis Siebold et Zucc. = Aralia cordata Thunb. ■

30634 Aralia elata (Miq.) Seem.;龙牙楤木(白背叶楤木,苍蓝楤木,刺老鸦,刺龙牙,楤木,楤木子,矬木枝,虎阳刺,辽东楤木,辽宁楤木,鹊不踏,日本楤木);Hercules-club,Japanese Angelica,Japanese Angelica Tree,Japanese Angelica-tree,Japanese Aralia,Whiteback Aralia ●

30635 Aralia elata (Miq.) Seem. 'Albomarginata' = Aralia elata (Miq.) Seem. 'Variegata' ●☆

30636 Aralia elata (Miq.) Seem. 'Aureo-marginata';金边龙牙楤木(金边楤木);Japanese Angelica ●☆

30637 Aralia elata (Miq.) Seem. 'Silver Umbrella';银伞龙牙楤木●☆

30638 Aralia elata (Miq.) Seem. 'Variegata';花叶龙牙楤木(银边楤木);Variegated Japanese Aralia ●☆

30639 Aralia elata (Miq.) Seem. = Aralia ryukyuensis (J. Wen) T. Yamaz. ●

30640 Aralia elata (Miq.) Seem. f. aureovariegata (Rehder) Nakai;黄斑龙牙楤木●☆

30641 Aralia elata (Miq.) Seem. f. canescens (Siebold et Zucc.) T. Yamaz. = Aralia elata (Miq.) Seem. f. subinermis (Ohwi) Jotani ●☆

30642 Aralia elata (Miq.) Seem. f. canescens (Siebold et Zucc.) T. Yamaz. = Aralia elata (Miq.) Seem. ●

30643 Aralia elata (Miq.) Seem. f. subinermis (Ohwi) Jotani;近无刺龙牙楤木●☆

30644 Aralia elata (Miq.) Seem. f. variegata (Rehder) Nakai = Aralia elata (Miq.) Seem. 'Variegata' ●☆

30645 Aralia elata (Miq.) Seem. f. variegata Nakai = Aralia elata (Miq.) Seem. 'Variegata' ●☆

30646 Aralia elata (Miq.) Seem. var. canescens (Franch. et Sav.) Nakai = Aralia elata (Miq.) Seem. f. subinermis (Ohwi) Jotani ●☆

30647 Aralia elata (Miq.) Seem. var. canescens (Franch. et Sav.) Nakai = Aralia elata (Miq.) Seem. ●

30648 Aralia elata (Miq.) Seem. var. glabrescens (Franch. et Sav.) C. K. Schneid. = Aralia elata (Miq.) Seem. ●

30649 Aralia elata (Miq.) Seem. var. glabrescens (Franch. et Sav.) Pojark.;辽东楤木;Liaodongapanese Aralia ●

30650 Aralia elata (Miq.) Seem. var. glabrescens (Franch. et Sav.) Pojark. = Aralia elata (Miq.) Seem. var. mandshurica (Rupr. et Maxim.) J. Wen ●

30651 Aralia elata (Miq.) Seem. var. inermis (Yanagita) J. Wen = Aralia ryukyuensis (J. Wen) T. Yamaz. var. inermis (Yanagita) T. Yamaz. ●☆

30652 Aralia elata (Miq.) Seem. var. mandshurica (Rupr. et Maxim.) J. Wen = Aralia elata (Miq.) Seem. var. glabrescens (Franch. et Sav.) Pojark. ●

30653 Aralia elata (Miq.) Seem. var. ryukyuensis J. Wen = Aralia ryukyuensis (J. Wen) T. Yamaz. ●

30654 Aralia elata (Miq.) Seem. var. subinermis Ohwi = Aralia elata (Miq.) Seem. ●

30655 Aralia elata (Miq.) Seem. var. subinermis Y. C. Chu;少刺辽宁楤木;Poorspine Japanese Aralia ●

30656 Aralia elegans C. N. Ho = Aralia debilis J. Wen ●

30657 Aralia elegantissima Veitch = Dizygotheca elegantissima (Veitch) R. Vig. et Guillaumin ●☆

30658 Aralia emeiensis Z. Y. Zhu = Aralia elata (Miq.) Seem. ●

30659 Aralia fargesii Franch.;龙眼独活(川独活,独活,九眼独活);Farges Aralia ■

30660 Aralia fargesii Franch. var. yunnanensis H. L. Li = Aralia yunnanensis Franch. ■

30661 Aralia farinosa Delile = Polyscias farinosa (Delile) Harms ●☆

30662 Aralia filicifolia Moore = Polyscias cumingiana Fern.-Vill. ●

30663 Aralia filicifolia Moore ex E. Fourn. = Polyscias cumingiana Fern.-Vill. ●

30664 Aralia finlaysoniana (Wall. ex G. Don) Seem.;虎刺楤木(费氏楤木);Finlayson Aralia ●

30665 Aralia foliolosa (Wall.) Seem. = Aralia foliolosa (Wall.) Seem. ex C. B. Clarke ●

30666 Aralia foliolosa (Wall.) Seem. ex C. B. Clarke;小叶楤木;Littleleaf Aralia,Little-leaved Aralia ●

30667 Aralia foliolosa Seem. ex C. B. Clarke = Aralia foliolosa (Wall.) Seem. ex C. B. Clarke ●

30668 Aralia fragrans (D. Don) Jebb et J. Wen = Pentapanax fragrans (D. Don) T. D. Ha ●

30669 Aralia franchetii J. Wen = Pentapanax henryi Harms ●

30670 Aralia gaoshania Z. Y. Zhu = Aralia elata (Miq.) Seem. ●

30671 Aralia gigantea J. Wen = Pentapanax racemosus Seem. ●

30672 Aralia ginseng (C. A. Mey.) Baill. = Panax ginseng C. A. Mey. ■

30673 Aralia ginseng Baill. = Panax ginseng C. A. Mey. ■

30674 Aralia gintungensis C. Y. Wu;景东楤木;Jingdong Aralia ●

30675 Aralia glabra Matsum.;光楤木●☆

30676 Aralia glabrifoliolata (C. B. Shang) J. Wen = Pentapanax glabrifoliolatus C. B. Shang ●

30677 Aralia glauca Merr. = Aralia bipinnata Blanco ●

30678 Aralia glomerulata Blume = Brassaiopsis glomerulata (Blume) Regel ●

30679 Aralia guilfoylei W. Bull = Polyscias guilfoylei (Bull. ex Cogn. et J. F. Macbr.) L. H. Bailey ●

30680 Aralia henryi Harms;柔毛龙眼独活(短序楤木,短序龙眼独活,鬼眼独活,龙眼独活,小叶龙眼独活,心叶龙眼独活);Henry Aralia ■

30681 Aralia hispida Vent.;硬毛楤木;Bristly Aralia,Bristly Sarsaparilla,Bristly Spikenard ●

30682 Aralia houheensis W. X. Wang,W. Y. Guo et Y. S. Fu;后河龙眼独活;Houhe Sarsaparilla ■

30683 Aralia houheensis W. X. Wang,W. Y. Guo et Y. S. Fu = Aralia henryi Harms ■

30684 Aralia humilis Cav. ;矮楤木●☆

30685 Aralia hupehensis G. Hoo;湖北楤木(刺包头,飞天蜈蚣);Hubei Aralia,Hupeh Aralia ●

30686 Aralia hupehensis G. Hoo = Aralia elata (Miq.) Seem. ●

30687 Aralia hypoglauca (C. J. Qi et T. R. Cao) J. Wen et Y. F. Deng = Pentapanax hypoglaucus (C. J. Qi et T. R. Cao) C. B. Shang et X. P. Li ●

30688 Aralia hypoleuca C. Presl = Aralia bipinnata Blanco ●

30689 Aralia hypoleuca C. Presl var. inermis ? = Aralia ryukyuensis (J. Wen) T. Yamaz. var. inermis (Yanagita) T. Yamaz. ●☆

30690 Aralia japonica Thunb. = Fatsia japonica (Thunb.) Decne. et Planch. ●

30691 Aralia kansuensis G. Hoo;甘肃土当归;Gansu Aralia,Kansu Aralia ■

30692 Aralia kingdon-wardii J. Wen, Lowry et Esser = Pentapanax longipes (Merr.) C. B. Shang et C. F. Ji ●

30693 Aralia labordei H. Lév. = Toddalia asiatica (L.) Lam. ●

30694 Aralia lantsangensis G. Hoo;澜沧楤木;Lancang Aralia ●

30695 Aralia lantsangensis G. Hoo. = Aralia foliolosa (Wall.) Seem. ex C. B. Clarke ●

30696 Aralia leschenaultii (DC.) J. Wen = Pentapanax fragrans (D. Don) T. D. Ha ●

30697 Aralia lihengiana J. Wen et al. = Pentapanax racemosue Seem. ●

30698 Aralia macrophylla Lindl. = Aralia cachemirica Decne. ■☆

30699 Aralia mairei H. Lév. = Tetrapanax papyrifer (Hook.) K. Koch ●

30700 Aralia mandshurica Kom. = Aralia elata (Miq.) Seem. ●

30701 Aralia mandshurica Maxim. = Aralia elata (Miq.) Seem. ●

30702 Aralia mandshurica Rupr. et Maxim. = Aralia elata (Miq.) Seem. var. glabrescens (Franch. et Sav.) Pojark. ●

30703 Aralia mandshurica Rupr. et Maxim. = Aralia elata (Miq.) Seem. ●

30704 Aralia maralia Roem. et Schult. = Polyscias maralia (Roem. et Schult.) Bernardi ●☆

30705 Aralia maximowiczi Van Houtte = Kalopanax septemlobus (Thunb.) Koidz. f. maximowiczii (Van Houtte) H. Ohashi ●

30706 Aralia maximowiczii Van Houtte = Kalopanax septemlobus (Thunb.) Koidz. ●

30707 Aralia melanocarpa (H. Lév.) Lauener;黑果土当归(白九股牛,丛林楤木,丛枝楤木,丛枝土当归,九股牛,蜜油参,牛角七,牛尾独活,松皮九股牛);Blackfruit Aralia ■

30708 Aralia mitsde Siebold = Dendropanax trifidus (Thunb.) Makino ex H. Hara ●

30709 Aralia nantouensis S. S. Ying = Aralia armata (Wall. ex G. Don) Seem. ●

30710 Aralia nantouensis S. S. Ying. = Aralia officinalis Z. Z. Wang et H. C. Zheng ●

30711 Aralia nodosa Blume = Polyscias nodosa Blanco ●

30712 Aralia nudicaulis L. ;裸茎楤木;American Sarsaparilla, False Sarsaparilla, Rabbit-root, Shotbush, Small Spikenard, Wild Liquorice, Wild Sarsaparilla, Wild Sarsaparilla Aralia ●

30713 Aralia nutans Franch. et Sav. = Aralia cordata Thunb. ■

30714 Aralia octophylla Lour. = Schefflera bodinieri (H. Lév.) Rehder ●

30715 Aralia octophylla Lour. = Schefflera heptaphylla (L.) Frodin ●

30716 Aralia octophylla Lour. = Schefflera octophylla (Lour.) Harms ●

30717 Aralia officinalis Z. Z. Wang et H. C. Zheng;陕鄂楤木(陕甘楤木)●

30718 Aralia palmata Lour. = Eleutherococcus nodiflorus (Dunn) S. Y. Hu ●

30719 Aralia palmata Lour. = Kalopanax septemlobus (Thunb.) Nakai ●

30720 Aralia papyrifera Hook. = Tetrapanax papyrifer (Hook.) K. Koch ●

30721 Aralia parasitica (D. Don) J. Wen = Pentapanax parasiticus (D. Don) Seem. ●

30722 Aralia pentaphylla Siebold et Zucc. = Acanthopanax sieboldianus Makino ●

30723 Aralia pentaphylla Thunb. = Acanthopanax gracilistylus W. W. Sm. ●

30724 Aralia pentaphylla Thunb. = Eleutherococcus sieboldianus (Makino) Koidz. ●

30725 Aralia pilosa Franch. = Aralia henryi Harms ■

30726 Aralia pinnata Hochst. = Polyscias farinosa (Delile) Harms ●☆

30727 Aralia planchoniana Hance = Aralia decaisneana Hance ●

30728 Aralia planchoniana Hance = Aralia elata (Miq.) Seem. ●

30729 Aralia plumosa H. L. Li;羽叶楤木;Pinnateleaf Aralia, Plumose Aralia ●

30730 Aralia plumosa H. L. Li = Pentapanax plumosus (H. L. Li) C. B. Shang ●

30731 Aralia pseudoginseng (Wall.) Benth. ex C. B. Clarke = Panax pseudoginseng Wall. ■

30732 Aralia quinquefolia (L.) Decne. et Planch. = Panax quinquefolius L. ■☆

30733 Aralia quinquefolia (L.) Decne. et Planch. var. ginseng (C. A. Mey.) ? = Panax ginseng C. A. Mey. ■

30734 Aralia quinquefolia (L.) Decne. et Planch. var. notoginseng Burkill = Panax notoginseng (Burkill) F. H. Chen ex C. Chow et W. G. Huang ■

30735 Aralia quinquefolia (L.) Decne. et Planch. var. pseudoginseng (Wall.) Burkill = Panax pseudoginseng Wall. ■

30736 Aralia quinquefolia (L.) Decne. et Planch. var. repens (Maxim.) Burkill = Panax japonicus (T. Nees) C. A. Mey. ■

30737 Aralia quinquefolia Decne. et Planch. var. angustifolia Burkill = Panax pseudoginseng Wall. var. angustifolius (Burkill) H. L. Li ■

30738 Aralia quinquefolia Decne. et Planch. var. angustifolia Burkill = Panax japonicus (T. Nees) C. A. Mey. var. angustifolius (Burkill) C. Y. Cheng et Y. C. Chu ■

30739 Aralia quinquefolia Decne. et Planch. var. elegantior Burkill = Panax pseudoginseng Wall. var. elegantior (Burkill) G. Hoo et C. J. Tseng ■

30740 Aralia quinquefolia Decne. et Planch. var. elegantior Burkill = Panax bipinnatifidus Seem. ■

30741 Aralia quinquefolia Decne. et Planch. var. ginseng (C. A. Mey.) Regel et Maack = Panax ginseng C. A. Mey. ■

30742 Aralia quinquefolia Decne. et Planch. var. major Burkill = Panax japonicus (T. Nees) C. A. Mey. var. major (Burkill) C. Y. Wu et K. M. Feng ■

30743 Aralia quinquefolia Decne. et Planch. var. major Burkill = Panax pseudoginseng Wall. var. japonicus (C. A. Mey.) G. Hoo et C. J. Tseng ■

30744 Aralia quinquefolia Decne. et Planch. var. notoginseng Burkill = Panax pseudoginseng Wall. var. notoginseng (Burkill) G. Hoo et C. J. Tseng ■

30745 Aralia quinquefolia Decne. et Planch. var. pseudoginseng (Wall.) Burkill = Panax pseudoginseng Wall. ■

30746 Aralia quinquefolia Decne. et Planch. var. repens (Maxim.) Burkill = Panax pseudoginseng Wall. var. japonicus (C. A. Mey.) G.

Hoo et C. J. Tseng ■

30747 Aralia quinquefolia Decne. et Planch. var. repens (Maxim.) Burkill = Panax japonicus C. A. Mey. ■

30748 Aralia racemosa L.;美楤木;American Spikenard, Beautiful Aralia, Indian-root, King of the Woods, Life-of-man, Petty Morrel, Spikenard ●☆

30749 Aralia ryukyuensis (J. Wen) T. Yamaz.;琉球楤木●

30750 Aralia ryukyuensis (J. Wen) T. Yamaz. var. inermis (Yanagita) T. Yamaz.;无刺琉球楤木●☆

30751 Aralia scaberula G. Hoo;糙叶楤木;Ruggedleaf Aralia, Scabrousleaf Aralia, Scabrous-leaved Aralia ●◇

30752 Aralia scaberula G. Hoo = Aralia elata (Miq.) Seem. ●

30753 Aralia schmidtii Pojark.;远东楤木;Schmidt Aralia ●☆

30754 Aralia schmidtii Pojark. = Aralia cordata Thunb. var. sachalinensis (Regel) Nakai ■☆

30755 Aralia searelliana Dunn;粗毛楤木;Hisute Acanthopanax, Searell Aralia, Thickhair Aralia ●

30756 Aralia shangiana J. Wen = Pentapanax yunnanensis Franch. ●

30757 Aralia sieboldii Hort. ex K. Koch = Fatsia japonica (Thunb.) Decne. et Planch. ●

30758 Aralia sieboldii K. Koch = Fatsia japonica (Thunb.) Decne. et Planch. ●

30759 Aralia spinifolia Merr.;长刺楤木(刺叶楤木,鸡云木,雷公木,乌不企,鹰不扒);Spinyleaf Aralia, Spiny-leaved Aralia ●

30760 Aralia spinosa L.;多刺楤木(刺楤木,楤木,美国楤木);American Angelica Tree, Angelica Tree, Devil Walking Stick, Devil's Walking Stick, Devil's Walkingstick, Devil's Walking-stick, Devil's-walking-stick, Hercules Club, Hercules' Club, Hercules'-club, Prickly Ash, Prickly Elder, Prickly-ash, Tear-blanket, Toothache Tree, Tree Aralia, Virginian Angelica Tree ●

30761 Aralia spinosa L. var. elata (Miq.) Sarg. = Aralia elata (Miq.) Seem. ●

30762 Aralia spinosa L. var. glabrescens Franch. et Sav. = Aralia elata (Miq.) Seem. var. glabrescens (Franch. et Sav.) Pojark. ●

30763 Aralia staphyleina Hand. -Mazz. = Aralia caesia Hand. -Mazz. ●

30764 Aralia staphyleina Hand. -Mazz. = Pentapanax caesius (Hand. -Mazz.) C. B. Shang ●

30765 Aralia stipulata Franch.;披针叶楤木●

30766 Aralia stipulata Franch. = Aralia chinensis L. var. nuda Nakai ●

30767 Aralia strigosa C. Y. Wu ex C. B. Shang. = Aralia vietnamensis T. D. Ha ●

30768 Aralia subcapitata G. Hoo;安徽楤木;Subcapitate Aralia ●

30769 Aralia subcapitata G. Hoo = Aralia elata (Miq.) Seem. ●

30770 Aralia subcordata (Wall. ex G. Don) J. Wen = Pentapanax subcordatus (Wall.) Seem. ●

30771 Aralia taibaiensis Z. Z. Wang et H. C. Zheng;太白楤木●

30772 Aralia taibaiensis Z. Z. Wang et H. C. Zheng. = Aralia elata (Miq.) Seem. ●

30773 Aralia taiwaniana Y. C. Liu et F. Y. Lu;台湾土当归■

30774 Aralia taiwaniana Y. C. Liu et F. Y. Lu = Aralia cordata Thunb. ■

30775 Aralia tengyuehensis C. Y. Wu;腾冲楤木;Tengchong Aralia ●

30776 Aralia tengyuehensis C. Y. Wu = Aralia armata (Wall. ex G. Don) Seem. ●

30777 Aralia thomsonii Seem. ex C. B. Clarke;云南楤木;Thomson Aralia ●

30778 Aralia thomsonii Seem. ex C. B. Clarke var. brevipedicellata K. M. Feng;短柄云南楤木;Shortstalk Aralia ●

30779 Aralia thomsonii Seem. ex C. B. Clarke var. brevipedicellata K. M. Feng = Aralia thomsonii Seem. ex C. B. Clarke ●

30780 Aralia thomsonii Seem. ex C. B. Clarke var. glabrescens C. Y. Wu;少毛云南楤木;Glabrous Thomson Aralia ●

30781 Aralia thomsonii Seem. ex C. B. Clarke var. glabrescens C. Y. Wu. = Aralia armata (Wall. ex G. Don) Seem. ●

30782 Aralia thomsonii Seem. ex C. B. Clarke var. integerrina Ha = Aralia thomsonii Seem. ex C. B. Clarke ●

30783 Aralia tibetana G. Hoo;西藏土当归;Tibetan Aralia, Xizang Aralia ■

30784 Aralia tomentella Franch. = Pentapanax henryi Harms ●

30785 Aralia tomentella Franch. = Pentapanax tomentellus (Franch.) C. B. Shang ●

30786 Aralia toranensis T. D. Ha = Aralia finlaysoniana (Wall. ex G. Don) Seem. ●

30787 Aralia toranensis T. D. Ha var. pubescens T. D. Ha = Aralia finlaysoniana (Wall. ex G. Don) Seem. ●

30788 Aralia undulata Hand. -Mazz.;波缘楤木(顶天刺,董睡,红刺老包,三百棒,紫红伞);Undulate Aralia, Undulateleaf Aralia, Undulate-leaved Aralia ●

30789 Aralia undulata Hand. -Mazz. var. cirrhifolia Z. Z. Wang = Aralia undulata Hand. -Mazz. ●

30790 Aralia undulata Hand. -Mazz. var. nudifolia Z. Z. Wang. = Aralia undulata Hand. -Mazz. ●

30791 Aralia veitchii Hort. ex T. Moore = Dizygotheca veitchii N. Taylor ●☆

30792 Aralia verticillata (Dunn) J. Wen = Pentapanax verticillatus Dunn ●

30793 Aralia vietnamensis T. D. Ha;偃毛楤木(越南楤木);Bristly Aralia, Strigose Aralia, Vietnam Aralia ●

30794 Aralia wilsonii Harms;西南楤木(川西楤木);E. H. Wilson Aralia ●

30795 Aralia wilsonii Harms = Pentapanax wilsonii (Harms) C. B. Shang ●

30796 Aralia wilsonii Harms var. plumosa (H. L. Li) K. M. Feng = Pentapanax plumosus (H. L. Li) C. B. Shang ●

30797 Aralia wilsonii Harms var. plumosa H. L. Li;羽毛状楤木●

30798 Aralia yunnanensis Franch.;云南龙眼独活(草独活,大九股牛,龙眼独活,牛角七,松香痱药,珠钱草);Yunnan Aralia ■

30799 Aralia yunnanensis Franch. = Aralia wilsonii Harms ●

30800 Araliaceae Juss. (1789)(保留科名);五加科;Ginseng Family ●■

30801 Aralidiaceae Philipson et B. C. Stone = Toricelliaceae Hu ●

30802 Aralidiaceae Philipson et B. C. Stone(1980);沟子树科(假茱萸科)●☆

30803 Aralidium Miq. (1856);沟子树属●☆

30804 Aralidium pinnatifidum (Jungh. et de Vriese) Miq.;沟子树●☆

30805 Araliopsis Engl. (1896)(保留属名);类五加芸香属●☆

30806 Araliopsis Kurz = Euaraliopsis Hutch. ●

30807 Araliopsis soyauxii Engl. = Vepris soyauxii (Engl.) Mziray ●☆

30808 Araliopsis tabouensis Aubrév. et Pellegr.;类五加芸香(五加芸香)●☆

30809 Araliopsis tabouensis Aubrév. et Pellegr. = Vepris tabouensis (Aubrév. et Pellegr.) Mziray ●☆

30810 Araliopsis trifoliolata Engl. = Vepris trifoliolata (Engl.) Mziray ●☆

30811 Araliorhamnus H. Perrier = Berchemia Neck. ex DC. (保留属名)●

30812 Araliorhamnus H. Perrier(1943);楤木鼠李属●☆

30813 Araliorhamnus punctulata H. Perrier;斑点楤木鼠李●☆

30814 Araliorhamnus vaginata H. Perrier;楤木鼠李●☆
30815 Aralodendron Oerst. ex Marchal = Oreopanax Decne. et Planch. ●☆
30816 Aranella Barnhart = Utricularia L. ■
30817 Aranella Barnhart ex Small = Utricularia L. ■
30818 Arapabaca Adans. = Spigelia L. ■☆
30819 Arapatiella Rizzini et A. Mattos(1972);小阿拉苏木属●☆
30820 Arapatiella trepocarpa Rizzini et A. Mattos;小阿拉苏木●☆
30821 Araracuara Fern. Alonso(2008);哥伦比亚鼠李属●☆
30822 Arariba Mart. = Sickingia Willd. ■☆
30823 Arariba Mart. = Simira Aubl. ■☆
30824 Ararocarpus Scheff. = Meiogyne Miq. ●
30825 Araschcoolia Sch. Bip. = Geigeria Griess. ●■☆
30826 Araschcoolia Sch. Bip. ex Benth. et Hook. f. = Geigeria Griess. ●■☆
30827 Aratitiyopea Steyerm. = Aratitiyopea Steyerm. et P. E. Berry ■☆
30828 Aratitiyopea Steyerm. et P. E. Berry(1984);立花黄眼草属■☆
30829 Aratitiyopea lopezii (L. B. Sm.) Steyerm. et P. Berry;立花黄眼草■☆
30830 Araucaria Juss. (1789);南洋杉属;Araucaria, Monkey Puzzle, Monkey Puzzle Tree, Monkey-puzzle ●
30831 Araucaria angustifolia (Bertol.) Kuntze;窄叶南洋杉(巴西南洋杉);Brasilian Araucaria, Brazilian Araucaria, Brazilian Pine, Candelabar Tree, Candelabra Tree, Parana Araucaria, Parana Pine ●☆
30832 Araucaria araucana (Molina) K. Koch;智利南洋杉(覆瓦南洋杉,南美杉);Chile Nut, Chile Nut Tree, Chile Pine, Chilean Pine, Chili Pine, Monkey Puzzle, Monkey Puzzle Tree, Monkey-puzzle, Monkey-puzzle Araucaria, Monkey-puzzle Tree, Puzzle-Monkey ●☆
30833 Araucaria araucana K. Koch = Araucaria araucana (Molina) K. Koch ●☆
30834 Araucaria balansae Brongn. et Gris;新喀里多尼亚南洋杉●☆
30835 Araucaria bidwillii Hook.;大叶南洋杉(澳洲南洋杉,广叶南洋杉,宽叶南洋杉,阔叶南洋杉,披针叶南洋杉,塔杉,洋刺杉);Bidwill's Araucaria, Bunya Bunya, Bunya Pine, Bunya-bunya, Bunyabunya Araucaria, Bunya-bunya Araucaria, Bunya-bunya Pine, Bunya-pine, Monkey Puzzle, Mr. Bidwill's Araucaria ●
30836 Araucaria brasiliana A. Rich. = Araucaria angustifolia (Bertol.) Kuntze ●☆
30837 Araucaria columnaris (G. Forst.) Hook. = Araucaria columnaris (J. Forst.) Hook. ●☆
30838 Araucaria columnaris (J. Forst.) Hook.;柱状南洋杉(新喀里多尼亚杉);Captain Cook's Araucaria, Captain Cook's Pine, Columnar Araucaria, Cook Pine, New Caledonia Pine, Pine Colonnaire ●☆
30839 Araucaria cookii R. Br. ex D. Don.;古氏南洋杉(库氏南洋杉);Arauria-da-caledomia, Columnar Araucaria ●
30840 Araucaria cookii R. Br. ex D. Don. = Araucaria columnaris (J. Forst.) Hook. ●☆
30841 Araucaria cunninghamii Aiton ex D. Don;南洋杉(花旗杉,肯氏南洋杉,鳞叶南洋杉,狭叶南洋杉);Cunningham Araucaria, Cunningham's Araucaria, Hoop Pine, Moreton Bay Pine, Richmond River Pine ●
30842 Araucaria cunninghamii Aiton ex Sweet = Araucaria cunninghamii Aiton ex D. Don ●
30843 Araucaria cunninghamii D. Don = Araucaria cunninghamii Aiton ex D. Don ●
30844 Araucaria cunninghamii Sweet = Araucaria cunninghamii Aiton ex D. Don ●
30845 Araucaria cunninghamii Sweet ex Courtois = Araucaria cunninghamii Aiton ex D. Don ●
30846 Araucaria excelsa (Lamb.) R. Br.;高大南洋杉(大南洋杉,小叶南洋杉);Norfolk Island Pine ●☆
30847 Araucaria excelsa (Lamb.) R. Br. = Araucaria heterophylla (Salisb.) Franco ●
30848 Araucaria heterophylla (Salisb.) Franco;异叶南洋杉(猴子杉,南洋杉,诺福克南洋杉,诺和克南洋杉,细叶南洋杉,小叶南洋杉);Australian Pine, Differentleaf Araucaria Pine, House Pine, Norfolk Island Pine, Norfolk Pine, Norfolk-Island Pine ●
30849 Araucaria heterophylla (Salisb.) Franco = Araucaria excelsa (Lamb.) R. Br. ●☆
30850 Araucaria hunsteinii K. Schum.;亮叶南洋杉(高大南洋杉);Klinki Pine ●☆
30851 Araucaria imbricata Pav. = Araucaria araucana (Molina) K. Koch ●☆
30852 Araucaria klinkii Lauterb.;克林克南洋杉(克氏南洋杉);Klinki Pine ●☆
30853 Araucaria klinkii Lauterb. = Araucaria hunsteinii K. Schum. ●☆
30854 Araucaria luxurians (Brongn. et Gris) de Laub.;繁茂南洋杉●☆
30855 Araucaria muelleri Brongn. et Gris;三角叶南洋杉●☆
30856 Araucaria rulei F. Muell. ex Lindl.;鲁莱南洋杉●☆
30857 Araucariaceae Henkel et W. Hochst. (1865)(保留科名);南洋杉科;Araucaria Family, Chile Pine Family, Monkey-puzzle Family ●
30858 Araucasia Benth. et Hook. f. = Arausiaca Blume ●☆
30859 Araucasia Benth. et Hook. f. = Orania Zipp. ●☆
30860 Araujia Brot. (1817);阿鲁藤属(白蛾藤属);Araujia, Bladder Flower ●☆
30861 Araujia hortorum E. Fourn. = Araujia sericofera Brot. ●☆
30862 Araujia sericifera Brot.;阿鲁藤(白蛾藤);Araujia, Bladder Vine, Cruel Plant, Cruel Vine, White Bladder Flower, White Bladderflower, White Bladder-flower ●☆
30863 Arausiaca Blume = Orania Zipp. ●☆
30864 Arbelaezaster Cuatrec. (1986);革茼蒿属■☆
30865 Arbelaezaster ellsworthii (Cuatrec.) Cuatrec.;革茼蒿■☆
30866 Arberella Soderstr. et C. E. Calderón(1979);阿波禾属■☆
30867 Arberella costaricensis (Hitchc.) Soderstr. et C. E. Calderón;阿波禾■☆
30868 Arberella flaccida (Döll) Soderstr. et C. E. Calderón;柔软阿波禾■☆
30869 Arberella lancifolia Soderstr. et Zuloaga;披针叶阿波禾■☆
30870 Arbulocarpus Tennant = Spermacoce L. ●■
30871 Arbulocarpus somalensis (Chiov.) Cufod. = Diodia aulacosperma K. Schum. ■☆
30872 Arbulocarpus sphaerostigma (A. Rich.) Tennant = Spermacoce sphaerostigma (A. Rich.) Vatke ■☆
30873 Arbutaceae Bromhead = Ericaceae Juss. (保留科名)●
30874 Arbutaceae J. Agardh = Ericaceae Juss. (保留科名)●
30875 Arbutaceae J. Agardh;草莓树科●
30876 Arbutaceae Miers = Ericaceae Juss. (保留科名)●
30877 Arbutus L. (1753);草莓树属(荔莓属,乔杜鹃属,乔鹃属,洋杨梅属);Arbutus, Madrona, Madrone, Manzanita, Strawberry Tree, Strawberry-tree ●☆
30878 Arbutus × andrachnoides Link;杂种草莓树(杂交荔莓);Hybrid Strawberry Tree ●☆
30879 Arbutus alpina L. = Arctous alpinus (L.) Nied. ●
30880 Arbutus andrachne L.;南欧草莓树(希腊荔莓);Grecian

Strawberry Tree, Greek Strawberry Tree, Southern Europe Strawberry Tree ●☆

30881 Arbutus arizonica Sarg.;亚墨草莓树(亚墨乔杜鹃,亚墨乔鹃);Arizona Madrone ●☆

30882 Arbutus canariensis Veill.;加那利草莓树(康纳利岛草莓树);Canary Island Strawberry Tree ●☆

30883 Arbutus glandulosa M. Martens et Galeotti;腺草莓树(格兰乔杜鹃,格兰乔鹃)●☆

30884 Arbutus glandulosa M. Martens et Galeotti 'Marina';玛丽娜腺草莓树●☆

30885 Arbutus menziesii Pursh;优材草莓树(浆果鹃,美国荔莓,太平洋乔杜鹃,太平洋乔鹃,直花树莓);Madrona, Madrona Laurel, Madrone, Madrono, Madrono Laurel, Pacific Madrone ●☆

30886 Arbutus pavarii Pamp.;帕氏草莓树●☆

30887 Arbutus texana Buckley = Arbutus xalapensis Kunth var. texana (Buckley) A. Gray ●☆

30888 Arbutus unedo L.;草莓树(垂花树莓,荔莓,莓实树);Arbeset, Arbute-tree, Cane Apple, Common Strawberry Tree, Irish Strawberry Tree, Killarney Strawberry-tree, One-I-eat, Strawberry Bush, Strawberry Madrone, Strawberry Tree, Winter Strawberry ●☆

30889 Arbutus unedo L. 'Compacta';紧凑草莓树;Compacte Strawberry Bush ●☆

30890 Arbutus unedo L. var. integrrima Sims;全缘叶草莓树●☆

30891 Arbutus unedo L. var. rubra Aiton;红花草莓树●☆

30892 Arbutus ursina L. = Arctostaphylos uva-ursi (L.) Spreng. ●☆

30893 Arbutus xalapensis Kunth;墨西哥草莓树(萨拉草莓树);Madrone, Xalapen Strawberry Tree ●☆

30894 Arbutus xalapensis Kunth var. texana (Buckley) A. Gray;得州草莓树(得州乔杜鹃,得州乔鹃);Texas Madrone ●☆

30895 Arcangelina Kuntze = Kralikia Coss. et Durieu ■

30896 Arcangelina Kuntze = Tripogon Roem. et Schult. ■

30897 Arcangelisia Becc. (1877);古山龙属;Arcang. sia, Gushanlong, Garden Angelica ●

30898 Arcangelisia flava Merr.;黄古山龙●☆

30899 Arcangelisia gusanlung H. S. Lo;古山龙(黄胆榄,黄连藤,黄藤);Garden Angelica, Gushanlung ●◇

30900 Arcangelisia loureiri (Pierre) Diels = Arcangelisia gusanlung H. S. Lo ●◇

30901 Arcaula Raf. = Lithocarpus Blume ●

30902 Arceuthidaceae A. V. Bobrov et Melikyan = Cupressaceae Gray (保留科名)●

30903 Arceuthobiaceae Tiegh. = Santalaceae R. Br. (保留科名)●■

30904 Arceuthobiaceae Tiegh. ex Nakai = Viscaceae Miq. ●

30905 Arceuthobium Griseb. = Dendrophthora Eichler ●☆

30906 Arceuthobium M. Bieb. (1819)(保留属名);油杉寄生属(油松寄生属);Dwarf Mistletoe, Dwarfmistletoe, Dwarf-mistletoe, Parasite ●

30907 Arceuthobium americanum Nutt. ex A. Gray = Arceuthobium oxycedri (DC.) M. Bieb. ●

30908 Arceuthobium chinense Lecomte;油杉寄生(小莲枝);Chinese Dwarfmistletoe, Chinese Dwarf-mistletoe, Keteleeria Parasite ●

30909 Arceuthobium gillii Hawksw. et Wiens;吉尔油杉寄生●☆

30910 Arceuthobium juniperi-procerae Chiov.;高大油杉寄生●☆

30911 Arceuthobium minutissimum Hook. f.;微小油杉寄生●☆

30912 Arceuthobium oxycedri (DC.) M. Bieb.;圆柏寄生;Chinese Juniper Dwarfmistletoe, Chinese Juniper Dwarf-mistletoe, Juniper Parasite ●

30913 Arceuthobium oxycedri M. Bieb. = Arceuthobium oxycedri (DC.) M. Bieb. ●

30914 Arceuthobium petadendra (L.) Miq. = Dendrophthoe pentandra (L.) Miq. ●

30915 Arceuthobium pini Hawksw. et Wiens;高山松寄生;Alpine Pine Dwarfmistletoe, Alpine Pine Dwarf-mistletoe, Alpine Pine Parasite ●

30916 Arceuthobium pini Hawksw. et Wiens var. sichuanense H. S. Kiu = Arceuthobium sichuanense (H. S. Kiu) Hawksw. et Wiens ●

30917 Arceuthobium pusillum Peck;东部油杉寄生;Eastern Dwarf Mistletoe, Small Mistletoe ●☆

30918 Arceuthobium sichuanense (H. S. Kiu) Hawksw. et Wiens;云杉寄生;Sichuan Alpine Pine Dwarf-mistletoe, Sichuan Alpinepine Dwarfmistletoe, Sichuan Dwarfmistletoe ●

30919 Arceuthobium tibetense H. S. Kiu;冷杉寄生(冷杉矮槲寄生);Tibet Dwarfmistletoe, Tibet Dwarf-mistletoe, Xizang Parasite ●

30920 Arceutholobium Steud. = Arceuthobium M. Bieb. (保留属名)●

30921 Arceuthos Antoine et Kotschy = Juniperus L. ●

30922 Arceuthos drupacea Antoine et Kotschy = Juniperus drupacea Labill. ●☆

30923 Archaeocarex Börner = Schoenoxiphium Neas ■☆

30924 Archaetogeron Greenm. = Achaetogeron A. Gray ■

30925 Archakebia C. Y. Wu, F. H. Chen et H. N. Qin (1995);长萼木通属(长蕊木通属,古木通属)●

30926 Archakebia apelata (Q. Xia, J. Z. Suen et Z. X. Peng) C. Y. Wu, T. C. Chen et H. N. Qin;长萼木通(长蕊木通,古木通,缺瓣牛姆瓜)●

30927 Archangelica Hoffm. = Angelica L. ■

30928 Archangelica Wolf = Angelica L. ■

30929 Archangelica Wolf (1781);古当归属;Archangelica ■

30930 Archangelica brevicaulis (Rupr.) Rchb. f.;短茎古当归(短茎独活,水防风);Shortstem Archangelica ■

30931 Archangelica decurrens Ledeb.;下延叶古当归(下延古当归,走马芹);Decurrent Archangelica ■

30932 Archangelica officinalis (Moench) Hoffm.;药用古当归(古当归,欧白芷,园当归);Angelica, Archangel, Garden Angelica, Holy Ghost, Jack-jump-about, Lingwort, Masterwort, Medicinal Archangelica ■☆

30933 Archangelica officinalis (Moench) Hoffm. = Angelica archangelica L. ■☆

30934 Archangelica officinalis (Moench) Hoffm. = Angelica archangelica L. var. himalaica (C. B. Clarke) Nasir ■☆

30935 Archangelica officinalis (Moench) Hoffm. var. himalaica C. B. Clarke = Angelica archangelica L. var. himalaica (C. B. Clarke) Nasir ■☆

30936 Archangelica roylei Lindl. = Angelica archangelica L. var. himalaica (C. B. Clarke) Nasir ■☆

30937 Archangelica tschimganica (Korovin) Schischk.;契穆干古当归 ■☆

30938 Archboldia E. Beer et H. J. Lam = Clerodendrum L. ●■

30939 Archboldiodendron Kobuski (1940);阿奇山茶属●☆

30940 Archboldiodendron calosericeum Kobuski;阿奇山茶●☆

30941 Archemera Raf. = Archemora DC. ■☆

30942 Archemora DC. = Tiedemannia DC. ■☆

30943 Archeria Hook. f. (1857);狼毒石南属●☆

30944 Archeria eriocarpa Hook. f.;狼毒石南●☆

30945 Archeria minor Hook. f.;小狼毒石南●☆

30946 Archiatriplex G. L. Chu (1987);单性滨藜属(古滨藜属,始滨藜属);Archiatriplex ●■★

30947 Archiatriplex nanpinensis G. L. Chu;单性滨藜;Archiatriplex ●
30948 Archibaccharis Heering(1904);近单性紫菀属●■☆
30949 Archibaccharis hieraciifolia Heering;近单性紫菀●☆
30950 Archiboehmeria C. J. Chen(1980);舌柱麻属;Archiboehmeria, Linguaramie ●★
30951 Archiboehmeria atrata (Gagnep.) C. J. Chen;舌柱麻;Blacken Archiboehmeria, Linguaramie ●◇
30952 Archiclematis (Tamura) Tamura = Clematis L. ●■
30953 Archiclematis (Tamura) Tamura(1968);互叶铁线莲属(五叶铁线莲属);Archiclematis ●■★
30954 Archiclematis Tamura = Clematis L. ●■
30955 Archiclematis alternata (Kitam. et Tamura) Tamura;互叶铁线莲(五叶铁线莲);Alternate Archiclematis, Alternateleaf Archiclematis ●
30956 Archiclematis alternata (Kitam. et Tamura) Tamura = Clematis alternata Kitam. et Tamura ●
30957 Archidendron F. Muell. (1865);颔垂豆属(古木属,猴耳环属)●
30958 Archidendron F. Muell. = Cylindrokelupha Kosterm. ●
30959 Archidendron alternifoliolatum (T. L. Wu) I. C. Nielsen = Cylindrokelupha alternifoliolata T. L. Wu ●
30960 Archidendron balansae (Oliv.) I. C. Nielsen = Cylindrokelupha balansae (Oliv.) Kosterm. ●
30961 Archidendron chevalieri (Kosterm.) I. C. Nielsen = Cylindrokelupha chevalieri Kosterm. ●
30962 Archidendron clypearia (Jack) I. C. Nielsen = Pithecellobium clypearia (Jack) Benth. ●
30963 Archidendron cordifolia (T. L. Wu) I. C. Nielsen;心叶大合欢(心叶合欢,心叶猴耳环);Cordifoliate Zygia, Heartleaf Zygia, Heart-leaved Zygia ●
30964 Archidendron cordifolia (T. L. Wu) I. C. Nielsen = Zygia cordifolia T. L. Wu ●
30965 Archidendron dalatensis (Kosterm.) I. C. Nielsen = Cylindrokelupha dalatensis (Kosterm.) T. L. Wu ●
30966 Archidendron eberhardtii I. C. Nielsen = Cylindrokelupha eberhardtii (Nielsen) T. L. Wu ●
30967 Archidendron ellipticum (Blume) I. C. Nielsen;椭圆叶猴耳环●
30968 Archidendron glabrifolium (T. L. Wu) I. C. Nielsen = Archidendron alternifoliolatum (T. L. Wu) I. C. Nielsen ●
30969 Archidendron glabrifolium (T. L. Wu) I. C. Nielsen = Cylindrokelupha glabrifolia T. L. Wu ●
30970 Archidendron kerrii (Gagnep.) I. C. Nielsen = Cylindrokelupha kerrii (Gagnep.) T. L. Wu ●
30971 Archidendron laoticum (Gagnep.) I. C. Nielsen = Cylindrokelupha laoticum (Gagnep.) C. Chen et H. Sun ●
30972 Archidendron lucidum (Benth.) I. C. Nielsen = Pithecellobium lucidum Benth. ●
30973 Archidendron robinsonii (Gagnep.) I. C. Nielsen = Cylindrokelupha robinsonii (Gagnep.) Kosterm. ●
30974 Archidendron robinsonii I. C. Nielsen = Cylindrokelupha tonkinensis (I. C. Nielsen) T. L. Wu ●
30975 Archidendron tonkinense I. C. Nielsen = Cylindrokelupha tonkinensis (I. C. Nielsen) T. L. Wu ●
30976 Archidendron turgida (Merr.) I. C. Nielsen = Cylindrokelupha tonkinensis (I. C. Nielsen) T. L. Wu ●
30977 Archidendron utile (Chun et F. C. How) I. C. Nielsen = Pithecellobium utile Chun et F. C. How ●
30978 Archidendron xichouensis (C. Chen et H. Sun) T. L. Wu;巨腺棋子豆;Xichou Cylindrokelupha ●
30979 Archidendron yunnanense (Kosterm.) I. C. Nielsen = Archidendron kerrii (Gagnep.) I. C. Nielsen ●
30980 Archidendron yunnanensis (Kosterm.) I. C. Nielsen = Cylindrokelupha yunnanensis (Kosterm.) T. L. Wu ●
30981 Archidendropsis I. C. Nielsen(1983);拟颔垂豆属(拟古木属)●☆
30982 Archidendropsis fulgens (Labill.) I. C. Nielsen;拟颔垂豆●☆
30983 Archidernatis (Tamura) Tamura = Clematis L. ●■
30984 Archihyoscyamus A. M. Lu = Hyoscyamus L. ■
30985 Archihyoscyamus A. M. Lu(1997);细萼天仙子属■☆
30986 Archileptopus P. T. Li = Leptopus Decne. ●
30987 Archileptopus P. T. Li(1991);方鼎木属;Archileptopus ●★
30988 Archileptopus fangdingianus P. T. Li;方鼎木;Archileptopus, Fangding Archileptopus ●
30989 Archimedea Leandro = Lophophytum Schott et Endl. ■☆
30990 Archimedea Leandro ex A. St. -Hil. = Lophophytum Schott et Endl. ■☆
30991 Archimedia Raf. = Iberis L. ●■
30992 Archineottia S. C. Chen = Holopogon Kom. et Nevski ■
30993 Archineottia S. C. Chen = Neottia Guett. (保留属名)■
30994 Archineottia gaudissartii (Hand. -Mazz.) S. C. Chen = Holopogon gaudissartii (Hand. -Mazz.) S. C. Chen ■
30995 Archineottia smithiana (Schltr.) S. C. Chen = Holopogon smithianus (Schltr.) S. C. Chen ■
30996 Archiphyllum Tiegh. = Myzodendron Sol. ex DC. ●☆
30997 Archiphysalis Kuang = Physaliastrum Makino ■
30998 Archiphysalis Kuang(1966);地海椒属;Archiphysalis ●■
30999 Archiphysalis chamaesarachoides (Makino) Kuang = Physaliastrum chamaesarachoides (Makino) Makino ●■
31000 Archiphysalis kwangsiensis Kuang = Physaliastrum chamaesarachoides (Makino) Makino ●■
31001 Archiphysalis linii Y. C. Liu et C. H. Ou = Physaliastrum chamaesarachoides (Makino) Makino ●■
31002 Archiphysalis sinensis (Hemsl.) Kuang;地海椒;Chinese Archiphysalis ●
31003 Archiphysalis sinensis (Hemsl.) Kuang = Physaliastrum sinense (Hemsl.) D'Arcy et Zhi Y. Zhang ●
31004 Archirhodomyrtus (Nied.) Burret = Rhodomyrtus (DC.) Rchb. ●
31005 Archirhodomyrtus (Nied.) Burret(1941);原始桃金娘属●☆
31006 Archirhodomyrtus baladensis (Brongn. et Gris) Burret;原始桃金娘●☆
31007 Archirnedea Leandro = Lophophytum Schott et Endl. ■☆
31008 Archiserratula L. Martins = Serratula L. ■
31009 Archiserratula L. Martins(2006);滇麻花头属■
31010 Architaea Mart. = Archytaea Mart. ●☆
31011 Archivea Christenson et Jenny(1996);巴西爱尔兰属■☆
31012 Archontophoenix H. Wendl. et Drude(1875);假槟榔属(亚历山大椰子属);Bangalow Palm, Butterfly Palm, Falseareca, King Palm, Kingpalm, King-palm ●
31013 Archontophoenix alexandrae (F. Muell.) H. Wendl. et Drude;假槟榔(槟榔葵,亚历山大椰子);Alexandra, Alexandra King Palm, Alexandra Palm, Alexandran King-palm, Falseareca, King Palm, Kingpalm, Northern Bungalow Palm, Piccabeen ●
31014 Archontophoenix alexandrae (F. Muell.) H. Wendl. et Drude var. beatricae Bailey;台阶假槟榔;Step Palm ●☆
31015 Archontophoenix cunninghamiana H. Wendl. et Drude;肯宁安氏假槟榔(垂序假槟榔,肯氏假槟榔,肯氏椰子,阔叶假槟榔,紫花假槟榔);Bangalow, Bangalow Palm, Cunningham Seaforthia,

Illawarra Palm, King Palm, Piccabeen Bangalow Palm, Piccabeen Palm ●

31016 Archontophoenix purpurea Hodel et Dowe;紫假槟榔;Mount Lewis Palm ●☆

31017 Archytaea Mart.(1826);阿奇藤属●☆

31018 Archytaea angustifolia Maguire;窄叶阿奇藤●☆

31019 Archytaea triflora Mart.;阿奇藤●☆

31020 Arcion Bubani = Arctium L. ■

31021 Arcoa Urb.(1923);海地豆属☆

31022 Arcoa gonavensis Urb.;海地豆☆

31023 Arctagrostis Griseb.(1852);寒剪股颖属■☆

31024 Arctagrostis arundinacea (Trin.) Beal;苇状寒剪股颖■☆

31025 Arctagrostis festucacea Petr.;羊茅寒剪股颖■☆

31026 Arctagrostis latifolia Griseb.;宽叶寒剪股颖■☆

31027 Arctanthemum (Tzvelev) Tzvelev (1985);极地菊属;Arctic Daisy ■☆

31028 Arctanthemum arcticum (L.) Tzvelev;极地菊;Arctic Daisy ■☆

31029 Arctanthemum arcticum (L.) Tzvelev = Chrysanthemum arcticum L. ■☆

31030 Arctanthemum arcticum (L.) Tzvelev subsp. kurilense (Tzvelev) Tzvelev = Chrysanthemum arcticum L. subsp. yezoense (Maek.) H. Ohashi et Yonek. ■☆

31031 Arctanthemum integrifolium (Richardson) Tzvelev = Hulteniella integrifolia (Richardson) Tzvelev ■☆

31032 Arctanthemum kurilense (Tzvelev) Tzvelev = Chrysanthemum arcticum L. subsp. yezoense (Maek.) H. Ohashi et Yonek. ■☆

31033 Arcteranthis Greene = Ranunculus L. ■

31034 Arcteranthis Greene(1897);极地毛茛属■☆

31035 Arcteranthis cooleyae (Vasey et Rose ex Rose) Greene = Ranunculus cooleyae Vasey et Rose ex Rose ■☆

31036 Arcterica Coville = Pieris D. Don ●

31037 Arcterica Coville(1901);北石南属●☆

31038 Arcterica nana (Maxim.) Makino;小北石南(矮生马醉木);Dwarf Pieris ●☆

31039 Arcterica nana (Maxim.) Makino = Pieris nana (Maxim.) Makino ●☆

31040 Arctio Lam. = Berardia Vill. ■☆

31041 Arctiodracon A. Gray = Lysichitum Schott ■☆

31042 Arction Cass. = Berardia Vill. ■☆

31043 Arction Lam. = Berardia Vill. ■☆

31044 Arctium L. (1753);牛蒡属;Bur Dock, Burdock, Clotbur, Flapper Bags ■

31045 Arctium L. emend. ex Kuntze = Arctium L. ■

31046 Arctium Lam. = Berardia Vill. ■☆

31047 Arctium affine Kuntze = Cousinia affinis Schrenk ■

31048 Arctium atlanticum (Pomel) H. Lindb.;大西洋牛蒡■☆

31049 Arctium dissectum (Kar. et Kir.) Kuntze = Cousinia dissecta Kar. et Kir. ■

31050 Arctium eriophorum Kuntze = Schmalhausenia nidulans (Regel) Petr. ■

31051 Arctium lappa L.;牛蒡(蒡翁菜,蝙蝠刺,便牵牛,大九子,大力子,大牛子,恶实,疙瘩菜,黑风子,老母猪耳朵,老鼠愁,毛然然子,毛锥子,牛榜,牛蒡子,牛菜,牛旁,牛子,然娃娃,黍粘子,鼠尖子,鼠见愁,鼠粘草,鼠粘根,鼠粘子,天龙子,土大桐子,弯巴钩子,弯把钩子,万把钩,象耳朵,夜叉头,粘苍子);Bachelor's Buttons, Bardana, Bardane, Bazzies, Beggar's Bur, Beggar's Burr, Beggar's Buttons, Billy Buttons, Bobby's Buttons, Bores, Buddy-bud, Buddy-buss, Bulldock, Bur Thistle, Bur Tree, Burdock, Burtons, Butter Dock, Cackle Buttons, Cackle Dock, Ciote Bur, Cleavers, Clip-me-dick, Clitch Buttons, Clite, Clod Bur, Clod-burr, Clogweed, Close Sciences, Clotbur, Clot-burr, Clote-burr, Clout, Cockle, Cockle Bells, Cockle Bur, Cockle Buttons, Cockle-bell, Cocklebur, Cockle-bur, Cockle-buttons, Cockly Bur, Cockly-bur, Cocoa-buttons, Coses Sciences, Crnekelty Bur, Crockelty-bur, Cuckle, Cuckle Buttons, Cuckle Dock, Cucklemoors, Cuckold, Cuckold Buttons, Cuckold Dock, Cuckoldy Buttons, Cuckoo Buttons, Cuckoo-buttons, Ditch Bur, Donkey, Eddick, Edible Burdock, Errick, Flapper Bags, Fox's Clote, Gobo, Great Bur, Great Burdock, Great Ciote Bur, Greater Burdock, Gypsy Comb, Gypsy's Comb, Gypsy's Rhubarb, Hardock, Hardoke, Hareburr, Harlock, Hedgehogs, Hurr Bur, Hurr-burr, Kisses, Loppy-major, Love-leaves, Old Man's Buttons, Pig's Rhubarb, Reaf-a-robber, Snake's Rhubarb, Soldier Buttons, Soldier's Buttons, Stick Buttons, Stick-button, Stickers, Sticky Ball, Sticky Buttons, Sticky Jack, Sticky Willow, Sticky-back, Sticky-balls, Sticky-buttons, Sweethearts, Thistle, Thorny Bur, Thorny Burr, Thor's Mantle, Touch-me-not, Turkey Rhubarb, Tuzzy-muzzy, Water Dock, Wild Rhubarb ■

31052 Arctium lappa L. subsp. majus Arènes = Arctium lappa L. ■

31053 Arctium leiospermum Juz. = Arctium lappa L. ■

31054 Arctium leiospermum Juz. et Ye. V. Serg. = Arctium lappa L. ■

31055 Arctium majus Bernh. = Arctium lappa L. ■

31056 Arctium minus (Hill) Bernh.;小牛蒡(美牛蒡);Bourrier, Burdock, Chou Bourache, Cibourroche, Common Burdock, Cuckoo Buttons, Cuckoo-buttons, Kiss-me-quick, Lesser Bur, Lesser Burdock, Louse Bur, Petite Bardane, Small Bur, Small Burdock ■☆

31057 Arctium minus (Hill) Bernh. subsp. atlanticum (Pomel) Maire = Arctium atlanticum (Pomel) H. Lindb. ■☆

31058 Arctium minus (Hill) Bernh. var. grandiceps Maire = Arctium atlanticum (Pomel) H. Lindb. ■☆

31059 Arctium minus Bernh. = Arctium minus (Hill) Bernh. ■☆

31060 Arctium minus Bernh. f. laciniatum Clute = Arctium minus Bernh. ■☆

31061 Arctium nemorosum Lej.;木牛蒡;Wood Burdock ■☆

31062 Arctium niveum Kuntze = Alfredia nivea Kar. et Kir. ■

31063 Arctium palladinii (Marcow.) Grossh.;帕拉丁牛蒡■☆

31064 Arctium platylepis (Boiss. et Balansa) Sosn. ex Grossh. = Cousinia platylepis Schrenk ex Fisch., C. A. Mey. et Avé-Lall. ■

31065 Arctium platylepis (Fisch., C. A. Mey. et Avé-Lall.) Kuntze = Cousinia platylepis Schrenk ex Fisch., C. A. Mey. et Avé-Lall. ■

31066 Arctium polycephalum (Rupr.) Kuntze = Cousinia polycephala Rupr. ■

31067 Arctium radula Juz. et Ye. V. Serg.;刮刀牛蒡■☆

31068 Arctium tomentosum Mill.;毛头牛蒡(绒毛牛蒡);Cotton Burdock, Cottony Burdock, Hiry Brdock, Tomentose Burdock, Woolly Burdock ■

31069 Arctium vulgare Druce;林地牛蒡;Woodland Burdock ■☆

31070 Arctocalyx Fenzl = Solenophora Benth. ●☆

31071 Arctocarpus Blanco = Artocarpus J. R. Forst. et G. Forst.(保留属名)●

31072 Arctocrania (Endl.) Nakai = Chamaepericlymenum Asch. et Graebn. ■

31073 Arctocrania (Endl.) Nakai = Cornus L. ●

31074 Arctocrania Nakai = Chamaepericlymenum Asch. et Graebn. ■

31075 Arctocrania Nakai = Cornus L. ●

31076 Arctocrania canadensis (L.) Nakai = Chamaepericlymenum canadense (L.) Asch. et Graebn. ■
31077 Arctocrania suecica (L.) Nakai = Chamaepericlymenum suecicum (L.) Asch. et Graebn. ●☆
31078 Arctogentia Á. Löve = Gentianella Moench(保留属名)■
31079 Arctogeron DC. (1836);莎菀属;Arctogeron ■
31080 Arctogeron gramineum (L.) DC.;莎菀(禾矮翁);Arctogeron, Common Arctogeron ■
31081 Arctomecon Torr. et Frém. (1845);北美罂粟属(沙漠罂粟属);Desert Bearclaw-poppy ■☆
31082 Arctomecon californica Torr. et Frém.;加州北美罂粟;Golden Bearclaw-poppy ■☆
31083 Arctomecon humilis Coville;小北美罂粟;Dwarf Bearclaw-poppy ■☆
31084 Arctomecon merriami Coville;北美罂粟;Great Bearclaw-poppy, Great Desert Poppy ■☆
31085 Arctophila (Rupr.) Andersson(1852);耐寒禾属(喜极禾属)■☆
31086 Arctophila Rupr. = Arctophila (Rupr.) Andersson ■☆
31087 Arctophila Rupr. = Poa L. + Colpodium Trin. ■
31088 Arctophila fulva Nyman;耐寒禾■☆
31089 Arctopoa (Griseb.) Prob. (1974);寒地禾属(寒早熟禾属)■☆
31090 Arctopoa (Griseb.) Prob. = Poa L. ■
31091 Arctopoa eminens (C. Presl) Prob. = Poa eminens J. Presl et C. Presl ■
31092 Arctopoa eminens (J. Presl) Prob.;寒地禾■☆
31093 Arctopoa schischkinii (Tzvelev) Prob. = Poa schischkinii Tzvelev ■
31094 Arctopoa subfastigiata (Trin. ex Ledeb.) Prob. = Poa subfastigiata Trin. ex Ledeb. ■
31095 Arctopoa subfastigiata (Trin.) Prob. = Poa subfastigiata Trin. ex Ledeb. ■
31096 Arctopus L. (1753);熊足芹属☆
31097 Arctopus dregei Sond.;熊足芹☆
31098 Arctopus echinatus L.;多刺熊足芹☆
31099 Arctopus monacanthus Carmich. ex Sond.;单刺熊足芹☆
31100 Arctostaphylaceae J. Agardh = Ericaceae Juss. (保留科名)●
31101 Arctostaphylaceae J. Agardh;熊果科●
31102 Arctostaphylos Adans. (1763)(保留属名);熊果属(熊葡萄属);Bearberry, Manzanita ●☆
31103 Arctostaphylos adenotricha (Fern et MacBryde) Á. Löve, D. Löve et B. M. Kapoor = Arctostaphylos uva-ursi (L.) Spreng. ●☆
31104 Arctostaphylos alpina (L.) Spreng. = Arctous alpinus (L.) Nied. ●
31105 Arctostaphylos alpina (L.) Spreng. subsp. ruber (Rehder et E. H. Wilson) Hultén = Arctous alpinus (L.) Nied. var. ruber Rehder et E. H. Wilson ●
31106 Arctostaphylos alpina (L.) Spreng. var. japonicus (Nakai) Hultén = Arctous alpinus (L.) Nied. var. japonicus (Nakai) Ohwi ●
31107 Arctostaphylos arguta Zucc.;尖熊果●☆
31108 Arctostaphylos bakeri Eastw.;贝克熊果;Louis Edmunds ●☆
31109 Arctostaphylos canescens Eastw.;灰白熊果;Hoary Manzanita, Silver Manzanita ●☆
31110 Arctostaphylos columbiana Piper;毛熊果;Hairy Manzanita ●☆
31111 Arctostaphylos columbiana Piper var. tracyi (Eastw.) J. E. Adams ex McMinn = Arctostaphylos columbiana Piper ●☆
31112 Arctostaphylos densiflora M. S. Baker;密花熊果;Vine Hill Manzanita ●☆
31113 Arctostaphylos diversifolia Parry;夏熊果(浅裂萼熊果);Summer Holly ●☆
31114 Arctostaphylos edmundsii J. T. Howell;匍匐熊果;Little Sur Manzanita ●☆
31115 Arctostaphylos edmundsii J. T. Howell f. parvifolia (Roof) P. V. Wells = Arctostaphylos edmundsii J. T. Howell ●☆
31116 Arctostaphylos glauca Lindl.;大果熊果(粉绿,灰叶熊果);Bigberry Manzanita ●☆
31117 Arctostaphylos glauca Lindl. f. puberula (J. T. Howell) P. V. Wells = Arctostaphylos glauca Lindl. ●☆
31118 Arctostaphylos hookeri G. Don;垫状熊果(虎克熊果);Hooker's Manzanita, Monterey Manzanita ●☆
31119 Arctostaphylos hookeri G. Don 'Monterey Carpet';蒙特雷毯胡克熊果●☆
31120 Arctostaphylos insularis Greene et Parry;岛生熊果;Island Manzanita ●☆
31121 Arctostaphylos insularis Greene et Parry var. pubescens Eastw. = Arctostaphylos insularis Greene ex Parry ●☆
31122 Arctostaphylos insularis Greene ex Parry f. pubescens (Eastw.) P. V. Wells = Arctostaphylos insularis Greene ex Parry ●☆
31123 Arctostaphylos manzanita Parry;北加州熊果(加州熊果,帕立熊果);Common Manzanita, Manzanita, Whiteleaf Manzanita ●☆
31124 Arctostaphylos myrtifolia Parry;香桃木熊果;Lone Manzanita ●☆
31125 Arctostaphylos nevadensis A. Gray;内华达熊果;Pinemat, Pinemat Manzanita, Pine-mat Manzanita ●☆
31126 Arctostaphylos nummularia A. Gray;圆叶熊果;Fort Bragg Manzanita ●☆
31127 Arctostaphylos nummularia A. Gray subsp. sensitiva (Jeps.) P. V. Wells = Arctostaphylos nummularia A. Gray ●☆
31128 Arctostaphylos nummularia A. Gray var. sensitiva (Jeps.) McMinn = Arctostaphylos nummularia A. Gray ●☆
31129 Arctostaphylos obispoensis Eastw.;蛇纹岩土熊果;Serpentina Manzanita ●☆
31130 Arctostaphylos officinalis Wimm. et Grab. = Arctostaphylos uva-ursi (L.) Spreng. ●☆
31131 Arctostaphylos pajaroensis J. E. Adams;帕哈罗熊果;Pajaro Manzanita ●☆
31132 Arctostaphylos patula Greene;绿叶熊果(稍展熊果,展枝熊果);Green Manzanita, Greenleaf Manzanita, Green-leaf Manzanita ●☆
31133 Arctostaphylos pumila Nutt.;沙丘熊果;Dune Manzanita ●☆
31134 Arctostaphylos pungens A. Gray;具刺熊果(点叶熊果);Manzanita, Pointleaf Manzanita ●☆
31135 Arctostaphylos purissima P. V. Wells;珑泊熊果;Lompoc Manzanita ●☆
31136 Arctostaphylos ruber (Rehder et E. H. Wilson) Fernald = Arctous alpinus (L.) Nied. var. ruber Rehder et E. H. Wilson ●
31137 Arctostaphylos rubra (Rehder et E. H. Wilson) Fernald = Arctous ruber (Rehder et E. H. Wilson) Nakai ●
31138 Arctostaphylos sensitiva Jeps. = Arctostaphylos nummularia A. Gray ●☆
31139 Arctostaphylos stanfordiana Parry;斯坦福熊果;Stanford Manzanita ●☆
31140 Arctostaphylos tomentosa Lindl.;长毛熊果;Downy Manzanita, Shaggy-bark Manzanita ●☆
31141 Arctostaphylos tracyi Eastw. = Arctostaphylos columbiana Piper ●☆
31142 Arctostaphylos uva-ursi (L.) Spreng.;熊果;Alpine Bearberry, Barren Myrtle, Bear Bilberry, Bear Whortleberry, Bearberry, Bear's Grape, Bilberry, Brawlins, Burren Myrtle, Cranberry, Craneberry

Wire, Creashak, Crowbars, Dogberry, Fox Plum, Guashacks, Guashicks, Hog Cranberry, Indian Turnip, Kinnikinnick, Kinnikinnik, Manzanita, Meal Plum, Mealberry, Mealy Plum, Mealyplum, Mountain Box, Nashag, Rapper Dandy, Rapper-dandy, Red Bearberry, Rockberry, Sandberry, Trailing Arbutus, Trailing Strawberry Tree, Universe Plant, Upland Cranberry, Uva-ursi ●☆

31143 Arctostaphylos uva-ursi (L.) Spreng. 'Point Reyes';长枝熊果●☆

31144 Arctostaphylos uva-ursi (L.) Spreng. 'Vancouver Jade';白玉熊果●☆

31145 Arctostaphylos uva-ursi (L.) Spreng. subsp. adenotricha (Fernald et J. F. Macbr.) Calder et R. L. Taylor = Arctostaphylos uva-ursi (L.) Spreng. ●☆

31146 Arctostaphylos uva-ursi (L.) Spreng. subsp. edmundsii (J. T. Howell) Roof = Arctostaphylos edmundsii J. T. Howell ●☆

31147 Arctostaphylos uva-ursi (L.) Spreng. subsp. longipilosa Packer et Denford = Arctostaphylos uva-ursi (L.) Spreng. ●☆

31148 Arctostaphylos uva-ursi (L.) Spreng. subsp. monoensis Roof = Arctostaphylos uva-ursi (L.) Spreng. ●☆

31149 Arctostaphylos uva-ursi (L.) Spreng. subsp. sensitiva (Jeps.) Roof = Arctostaphylos nummularia A. Gray ●☆

31150 Arctostaphylos uva-ursi (L.) Spreng. subsp. stipitata Packer et Denford = Arctostaphylos uva-ursi (L.) Spreng. ●☆

31151 Arctostaphylos uva-ursi (L.) Spreng. var. adenotricha Fernald et J. F. Macbr. = Arctostaphylos uva-ursi (L.) Spreng. ●☆

31152 Arctostaphylos uva-ursi (L.) Spreng. var. coactilis Fernald et J. F. Macbr. = Arctostaphylos uva-ursi (L.) Spreng. ●☆

31153 Arctostaphylos uva-ursi (L.) Spreng. var. leobreweri Roof = Arctostaphylos uva-ursi (L.) Spreng. ●☆

31154 Arctostaphylos uva-ursi (L.) Spreng. var. marinensis Roof = Arctostaphylos uva-ursi (L.) Spreng. ●☆

31155 Arctostaphylos uva-ursi (L.) Spreng. var. pacifica Hultén = Arctostaphylos uva-ursi (L.) Spreng. ●☆

31156 Arctostaphylos uva-ursi (L.) Spreng. var. parvifolia (Roof.) Roof. = Arctostaphylos edmundsii J. T. Howell ●☆

31157 Arctostaphylos uva-ursi (L.) Spreng. var. stipitata (Packer et Denford) Dorn = Arctostaphylos uva-ursi (L.) Spreng. ●☆

31158 Arctostaphylos uva-ursi (L.) Spreng. var. suborbiculata W. Knight = Arctostaphylos uva-ursi (L.) Spreng. ●☆

31159 Arctostaphylos viscida Parry;黏熊果;Whiteleaf Manzanita ●☆

31160 Arctotheca J. C. Wendl. (1798);赛金盏属;Capeweed, Plain Treasureflower ■☆

31161 Arctotheca Vaill. = Arctotheca J. C. Wendl. ■☆

31162 Arctotheca calendula (L.) Levyns;赛金盏;Cape Dandelion, Cape Marigold, Cape Weed, Capeweed, Plain Treasureflower ■☆

31163 Arctotheca calendulacea (L.) K. Lewin = Arctotheca calendula (L.) Levyns ■☆

31164 Arctotheca forbesiana (DC.) K. Lewin;福布斯赛金盏■☆

31165 Arctotheca grandiflora Schrad. = Arctotheca prostrata (Salisb.) Britten ■☆

31166 Arctotheca marginata Beyers;具边赛金盏■☆

31167 Arctotheca nivea (L. f.) K. Lewin = Arctotheca populifolia (P. J. Bergius) Norl. ■☆

31168 Arctotheca populifolia (P. J. Bergius) Norl.;杨叶赛金盏■☆

31169 Arctotheca prostrata (Salisb.) Britten;平卧赛金盏■☆

31170 Arctotheca repens J. C. Wendl. = Arctotheca prostrata (Salisb.) Britten ■☆

31171 Arctotidaceae Bercht. et J. Presl = Asteraceae Bercht. et J. Presl (保留科名)●■

31172 Arctotidaceae Bercht. et J. Presl = Compositae Giseke(保留科名)●■

31173 Arctotidaceae Bessey = Asteraceae Bercht. et J. Presl(保留科名)●■

31174 Arctotidaceae Bessey = Compositae Giseke(保留科名)●■

31175 Arctotidaceae Bessey;灰毛菊科■

31176 Arctotis L. (1753);灰毛菊属(非洲菊属,蓝目菊属,熊耳菊属);African-Daisy, Arctotis ●■☆

31177 Arctotis acaulis Jacq. = Arctotis adpressa DC. ■☆

31178 Arctotis acaulis L.;无茎灰毛菊;African Daisy ■☆

31179 Arctotis acuminata K. Lewin;渐尖灰毛菊■☆

31180 Arctotis adpressa DC.;匍匐灰毛菊■☆

31181 Arctotis angustifolia Jacq. = Arctotis aspera L. ■☆

31182 Arctotis angustifolia L.;窄叶灰毛菊■☆

31183 Arctotis anthemoides L. = Ursinia anthemoides (L.) Poir. ■☆

31184 Arctotis arborescens Willd. = Arctotis aspera L. ■☆

31185 Arctotis arctotoides (L. f.) O. Hoffm.;南非灰毛菊■☆

31186 Arctotis argentea Thunb.;银色灰毛菊■☆

31187 Arctotis aspera L.;粗糙灰毛菊■☆

31188 Arctotis aspera L. var. angustifolia (Jacq.) Less. = Arctotis aspera L. ■☆

31189 Arctotis aspera L. var. scabra P. J. Bergius;糙灰毛菊■☆

31190 Arctotis aurea (DC.) Beauverd = Arctotis fastuosa Jacq. ■☆

31191 Arctotis aureola Edwards = Arctotis aspera L. ■☆

31192 Arctotis auriculata Jacq.;耳形灰毛菊■☆

31193 Arctotis bellidiastrum (S. Moore) Lewin;雅致灰毛菊■☆

31194 Arctotis bellidifolia P. J. Bergius;雅叶灰毛菊■☆

31195 Arctotis bolusii (S. Moore) Lewin;博卢斯灰毛菊■☆

31196 Arctotis breviscapa Thunb.;近无茎灰毛菊;Stemless Arctotis ●☆

31197 Arctotis breviscapa Thunb. = Arctotis leptorhiza DC. ■☆

31198 Arctotis calendula L. = Arctotheca calendula (L.) Levyns ■☆

31199 Arctotis campanulata DC.;风铃草状灰毛菊■☆

31200 Arctotis candida Thunb.;纯白灰毛菊■☆

31201 Arctotis canescens DC. = Arctotis diffusa Thunb. ■☆

31202 Arctotis caudata K. Lewin;尾状灰毛菊■☆

31203 Arctotis caulescens Thunb. = Arctotis incisa Thunb. ■☆

31204 Arctotis cineraria Jacq. = Arctotis cuprea Jacq. ■☆

31205 Arctotis crispata Hutch.;皱波灰毛菊■☆

31206 Arctotis crithmoides P. J. Bergius = Ursinia paleacea (L.) Moench ●☆

31207 Arctotis cuneata DC.;楔形灰毛菊■☆

31208 Arctotis cuprea Jacq.;铜色灰毛菊■☆

31209 Arctotis decumbens Jacq. = Arctotis angustifolia L. ■☆

31210 Arctotis decumbens Thunb. = Arctotis stoechadifolia P. J. Bergius ●☆

31211 Arctotis decurrens Jacq.;下延灰毛菊■☆

31212 Arctotis dentata L. = Ursinia dentata (L.) Poir. ■☆

31213 Arctotis denudata Thunb. = Arctotis laevis Thunb. ■☆

31214 Arctotis diffusa Thunb.;铺散灰毛菊■☆

31215 Arctotis discolor (Less.) Beauverd;异色灰毛菊■☆

31216 Arctotis dregei Turcz.;德雷灰毛菊■☆

31217 Arctotis echinata DC. = Haplocarpha nervosa (Thunb.) Beauverd ■☆

31218 Arctotis elatior Jacq. = Arctotis laevis Thunb. ■☆

31219 Arctotis elongata Thunb.;伸长灰毛菊■☆

31220 Arctotis erosa (Harv.) Beauverd;啮蚀状灰毛菊■☆

31221 Arctotis fastuosa Jacq.;骄傲灰毛菊;Monarch-of-the-veld ■☆
31222 Arctotis flaccida Jacq.;柔弱灰毛菊■☆
31223 Arctotis foeniculacea Jacq. = Ursinia anthemoides (L.) Poir. ■☆
31224 Arctotis formosa Thunb. = Arctotis incisa Thunb. ■☆
31225 Arctotis fosteri N. E. Br.;福斯特灰毛菊●☆
31226 Arctotis frutescens Norl.;灌木灰毛菊●☆
31227 Arctotis gigantea A. Rich.;巨大灰毛菊■☆
31228 Arctotis glabrata Jacq. = Arctotis laevis Thunb. ■☆
31229 Arctotis glandulosa Thunb. = Arctotis aspera L. var. scabra P. J. Bergius ■☆
31230 Arctotis glutinosa Sims = Dimorphotheca cuneata (Thunb.) Less. ■☆
31231 Arctotis graminea K. Lewin;禾状灰毛菊■☆
31232 Arctotis grandiflora Jacq. = Arctotis laevis Thunb. ■☆
31233 Arctotis grandis Thunb. = Arctotis stoechadifolia P. J. Bergius ●☆
31234 Arctotis hirsuta (Harv.) Beauverd;粗毛灰毛菊■☆
31235 Arctotis hispidula (Less.) Beauverd;细毛灰毛菊■☆
31236 Arctotis hybrida Hort.;杂种灰毛菊;African Daisy ●☆
31237 Arctotis incisa Thunb.;锐裂灰毛菊■☆
31238 Arctotis karasmontana Dinter = Arctotis leiocarpa Harv. ■☆
31239 Arctotis laevis Thunb.;平滑灰毛菊■☆
31240 Arctotis lanata Thunb. = Haplocarpha lanata Less. ■☆
31241 Arctotis lanceolata Harv.;剑叶灰毛菊■☆
31242 Arctotis leiocarpa Harv.;光果灰毛菊■☆
31243 Arctotis leptorhiza DC.;细根灰毛菊■☆
31244 Arctotis linearis Thunb.;线形灰毛菊■☆
31245 Arctotis macrosperma (DC.) Beauverd;大籽灰毛菊■☆
31246 Arctotis macrostylis K. Lewin = Arctotis leiocarpa Harv. ■☆
31247 Arctotis maculata Jacq. = Arctotis aspera L. ■☆
31248 Arctotis maximiliani Schltr. ex Dinter = Arctotis fastuosa Jacq. ■☆
31249 Arctotis melanocycla Willd. ex Harv. = Arctotis auriculata Jacq. ■☆
31250 Arctotis merxmuelleri Friedrich = Arctotis decurrens Jacq. ■☆
31251 Arctotis microcephala (DC.) Beauverd;小头灰毛菊■☆
31252 Arctotis microcephala S. Moore = Arctotis leiocarpa Harv. ■☆
31253 Arctotis muricata Thunb. = Arctotis bellidifolia P. J. Bergius ■☆
31254 Arctotis namaquana Schltr. ex Hutch. = Arctotis auriculata Jacq. ■☆
31255 Arctotis nodosa Thunb. = Leucoptera nodosa (Thunb.) B. Nord. ●☆
31256 Arctotis nudicaulis Thunb. = Ursinia nudicaulis (Thunb.) N. E. Br. ●☆
31257 Arctotis oocephala DC. = Haplocarpha oocephala (DC.) Beyers ■☆
31258 Arctotis paleacea L. = Ursinia paleacea (L.) Moench ●☆
31259 Arctotis paleacea Thunb. = Ursinia subflosculosa (DC.) Prassler ●☆
31260 Arctotis paniculata Jacq. = Arctotis bellidifolia P. J. Bergius ■☆
31261 Arctotis parvifolia Schltr. = Haplocarpha parvifolia (Schltr.) Beauverd ■☆
31262 Arctotis pectinata Thunb. = Ursinia dentata (L.) Poir. ■☆
31263 Arctotis perfoliata (Less.) Beauverd;穿叶灰毛菊■☆
31264 Arctotis petiolata Thunb.;柄叶灰毛菊■☆
31265 Arctotis pilifera P. J. Bergius = Ursinia pilifera (P. J. Bergius) Poir. ■☆
31266 Arctotis pinnata Thunb. = Ursinia pinnata (Thunb.) Prassler ■☆
31267 Arctotis pinnatifida Thunb.;羽裂灰毛菊■☆
31268 Arctotis populifolia P. J. Bergius = Arctotheca populifolia (P. J. Bergius) Norl. ■☆
31269 Arctotis prostrata Salisb. = Arctotheca prostrata (Salisb.) Britten ■☆
31270 Arctotis punctata Thunb. = Ursinia punctata (Thunb.) N. E. Br. ●☆
31271 Arctotis pusilla DC.;微小灰毛菊■☆
31272 Arctotis pygmaea A. Rich. = Haplocarpha rueppellii (Sch. Bip.) Beauverd ■☆
31273 Arctotis revoluta DC. = Arctotis candida Thunb. ■☆
31274 Arctotis revoluta Jacq.;外卷灰毛菊■☆
31275 Arctotis rigida Burm. f. = Gazania rigida (Burm. f.) Rössler ■☆
31276 Arctotis rogersii (Benson) M. C. Johnst.;罗杰斯灰毛菊■☆
31277 Arctotis rosea Less. = Arctotis stoechadifolia P. J. Bergius ●☆
31278 Arctotis rotundifolia K. Lewin;圆叶灰毛菊■☆
31279 Arctotis rueppellii (Sch. Bip.) O. Hoffm. = Haplocarpha rueppellii (Sch. Bip.) Beauverd ■☆
31280 Arctotis scapigera Thunb. = Arctotis acaulis L. ■☆
31281 Arctotis scaposa (Harv.) O. Hoffm. = Haplocarpha thunbergii Less. ■☆
31282 Arctotis scariosa Aiton = Ursinia scariosa (Aiton) Poir. ●☆
31283 Arctotis schlechteri K. Lewin;施莱灰毛菊■☆
31284 Arctotis scullyi Dümmer = Arctotis decurrens Jacq. ■☆
31285 Arctotis semipapposa (DC.) Beauverd;半冠毛灰毛菊■☆
31286 Arctotis sericea Thunb. = Ursinia sericea (Thunb.) N. E. Br. ●☆
31287 Arctotis serpens (S. Moore) Lewin;蛇形灰毛菊■☆
31288 Arctotis serrata L. f. = Ursinia serrata (L. f.) Poir. ●☆
31289 Arctotis sessilifolia K. Lewin;无梗灰毛菊■☆
31290 Arctotis setosa K. Lewin;刚毛灰毛菊■☆
31291 Arctotis speciosa Jacq. = Arctotis acaulis L. ■☆
31292 Arctotis splendens Muschl. = Arctotis diffusa Thunb. ■☆
31293 Arctotis squarrosa Jacq. = Arctotis laevis Thunb. ■☆
31294 Arctotis stoechadifolia P. J. Bergius;非洲灰毛菊(大花熊耳菊,非洲雏菊,灰毛菊,蓝目菊);African Arctotis, African Daisy, Blue-eyed African-daisy ●☆
31295 Arctotis stoechadifolia P. J. Bergius var. grandis (Thunb.) Less.;灰白菊●☆
31296 Arctotis stoechadifolia P. J. Bergius var. grandis (Thunb.) Less. = Arctotis stoechadifolia P. J. Bergius ●☆
31297 Arctotis suffruticosa K. Lewin;亚灌木灰毛菊●☆
31298 Arctotis sulcocarpa K. Lewin;沟果灰毛菊●☆
31299 Arctotis tenuifolia L. = Ursinia tenuifolia (L.) Poir. ●☆
31300 Arctotis tenuifolia Poir. = Dimorphotheca nudicaulis (L.) DC. var. graminifolia (L.) Harv. ■☆
31301 Arctotis tricolor Jacq.;三色灰毛菊■☆
31302 Arctotis trifida Thunb. = Ursinia trifida (Thunb.) N. E. Br. ■☆
31303 Arctotis undulata Jacq.;波叶灰毛菊■☆
31304 Arctotis undulata Thunb. = Arctotis cuprea Jacq. ■☆
31305 Arctotis venidioides DC.;凉菊状灰毛菊;African Daisy ■☆
31306 Arctotis venusta Norl. = Arctotis stoechadifolia P. J. Bergius ●☆
31307 Arctotis verbascifolia Harv. = Arctotheca populifolia (P. J. Bergius) Norl. ■☆
31308 Arctotis virgata Jacq.;条纹灰毛菊■☆
31309 Arctottonia Trel. (1930);柄花胡椒属●☆
31310 Arctottonia pittieri Trel.;柄花胡椒●☆
31311 Arctous (A. Gray) Nied. (1889);北极果属(当年枯属,天栌属);Arctous, North Pole Fruit, Ptarmiganberry, Ptarmigan-berry ●
31312 Arctous Nied. = Arctostaphylos Adans. (保留属名)●☆

31313 Arctous Nied. = Arctous (A. Gray) Nied. ●
31314 Arctous alpinus (L.) Nied.;北极果(阿尔卑斯熊果,阿尔卑斯熊葡萄,高山当年枯,高山天栌,高山熊果);Alpine Bearberry, Alpine Ptarmigan Berry, Alpine Ptarmiganberry, Alpine Ptarmiganberry, Arctic Bearberry, Black Arctous, Black Bearberry, Mountain Bearberry ●
31315 Arctous alpinus (L.) Nied. subsp. japonicus (Nakai) Sugim. = Arctous alpinus (L.) Nied. var. japonicus (Nakai) Ohwi ●
31316 Arctous alpinus (L.) Nied. var. japonicus (Nakai) Ohwi;黑北极果(黑果天栌,日本北极果);Black Arctous, Japanese Ptarmiganberry ●
31317 Arctous alpinus (L.) Nied. var. japonicus (Nakai) Ohwi = Arctous alpinus (L.) Nied. ●
31318 Arctous alpinus (L.) Nied. var. ruber Rehder et E. H. Wilson = Arctous ruber (Rehder et E. H. Wilson) Nakai ●
31319 Arctous erythrocarpus Small;红果当年枯●☆
31320 Arctous japonicus Nakai = Arctous alpinus (L.) Nied. var. japonicus (Nakai) Ohwi ●
31321 Arctous microphyllus C. Y. Wu;小叶当年枯;Smallleaf North Pole Fruit, Small-leaf Ptarmiganberry, Small-leaved Ptarmigan-berry ●
31322 Arctous ruber (Rehder et E. H. Wilson) Nakai;红北极果(当年枯,天栌);Red Arctous, Redfruit North Pole Fruit, Redfruit Ptarmiganberry, Red-fruited Ptarmigan-berry ●
31323 Arctous ruber (Rehder et E. H. Wilson) Nakai = Arctous alpinus (L.) Nied. var. ruber Rehder et E. H. Wilson ●
31324 Arctuus (A. Gray) Nied. = Arctostaphylos Adans.(保留属名)●☆
31325 Arcuatopterus M. L. Sheh et R. H. Shan(1986);弓翅芹属;Bowwingpaesley ■★
31326 Arcuatopterus filipedicellus M. L. Sheh et R. H. Shan = Arcuatopterus sikkimensis (C. B. Clarke) Pimenov et Ostroumova ■
31327 Arcuatopterus linearifolis M. L. Sheh et R. H. Shan;条叶弓翅芹;Linearleaf Bowwingpaesley ■
31328 Arcuatopterus sikkimensis (C. B. Clarke) Pimenov et Ostroumova;弓翅芹;Bowwingpaesley ■
31329 Arcuatopterus thalictrioideus M. L. Sheh et R. H. Shan;唐松叶弓翅芹;Meadowruelike Bowwingpaesley ■
31330 Arculus Tiegh. = Amylotheca Tiegh. ●☆
31331 Arcyna Wiklund(2003);西班牙网菊属■☆
31332 Arcyna tournefortii (Boiss. et Reut.) Wiklund;西班牙网菊■☆
31333 Arcynospermum Turcz.(1858);网籽锦葵属■☆
31334 Arcynospermum nodiflorum Turcz.;网籽锦葵☆
31335 Arcyosperma O. E. Schulz(1924);网籽芥属■☆
31336 Arcyosperma primulifolium (Thomson) O. E. Schulz;网籽芥■☆
31337 Arcyphyllum Elliott = Rhynchosia Lour.(保留属名)●■
31338 Arcythophyllum Schltdl. = Arcytophyllum Willd. ex Schult. et Schult. f. ●☆
31339 Arcythophyllum Willd. ex Schltdl. = Arcytophyllum Willd. ex Schult. et Schult. f. ●☆
31340 Arcytophyllum Roem. et Schult. = Arcytophyllum Willd. ex Schult. et Schult. f. ●☆
31341 Arcytophyllum Willd. ex Schult. et Schult. f.(1827);网叶茜属●☆
31342 Ardernia Salisb. = Ornithogalum L. ■
31343 Ardinghalia Comm. ex A. Juss. = Phyllanthus L. ●■
31344 Ardinghella Thouars = Mammea L. ●
31345 Ardinghella Thouars = Ochrocarpos Thouars ●
31346 Ardisia Gaertn. = Ardisia Sw.(保留属名)●■
31347 Ardisia Sw.(1788)(保留属名);紫金牛属;Ardisia, Spearflower, Spiceberry ●■
31348 Ardisia aberrans (E. Walker) C. Y. Wu et C. Chen;狗骨头;Aberrant Ardisia, Abnormal Ardisia, Dogbone Ardisia ●◇
31349 Ardisia adenopes R. H. Miao;腺梗紫金牛●
31350 Ardisia adenopes R. H. Miao = Ardisia lindleyana D. Dietr. ●
31351 Ardisia affinis Hemsl.;西罗伞(波叶紫金牛,细罗伞);Undulate-leaved Ardisia ●
31352 Ardisia affinis Hemsl. = Ardisia sinoaustralis C. Chen ●
31353 Ardisia alutacea C. Y. Wu ex C. Chen;显脉紫金牛;Clearvein Ardisia, Distinctvein Ardisia, Leather-colored Ardisia ●
31354 Ardisia alyxiifolia Tsiang ex C. Chen;少年红(念珠藤叶紫金牛);Alyxialeaf Ardisia, Alyxia-leaved Ardisia ●
31355 Ardisia aquifolioides W. Z. Fang et K. Yao = Ardisia crassinervosa E. Walker ●
31356 Ardisia arborescens Wall. et A. DC. = Ardisia garrettii H. R. Fletcher ●
31357 Ardisia argenticaulis Yuen P. Yang et Dwyer;五花紫金牛●
31358 Ardisia atrobullata Taton;暗泡紫金牛●☆
31359 Ardisia austroasiatica E. Walker = Ardisia thyrsiflora D. Don ●
31360 Ardisia balansana Yuen P. Yang;束花紫金牛(滇东紫金牛)●
31361 Ardisia bampsiana Taton;邦氏紫金牛●☆
31362 Ardisia baotingensis C. M. Hu;保亭紫金牛;Baoting Ardisia ●
31363 Ardisia batangaensis Taton;巴坦加紫金牛●☆
31364 Ardisia beibeinensis Z. Y. Zhu;北碚紫金牛(藤八爪);Beibei Ardisia ●
31365 Ardisia beibeinensis Z. Y. Zhu = Ardisia alyxiifolia Tsiang ex C. Chen ●
31366 Ardisia bicolor E. Walker = Ardisia crenata Sims var. bicolor (E. Walker) C. Y. Wu et C. Chen ●
31367 Ardisia bicolor E. Walker = Ardisia crenata Sims ●
31368 Ardisia bodinieri H. Lév. = Ardisia brevicaulis Diels ●
31369 Ardisia botryosa E. Walker;簇花紫金牛(束花紫金牛);Bunch-like Ardisia, Racemose Ardisia ●
31370 Ardisia bracteata Baker;具苞紫金牛●☆
31371 Ardisia brevicaulis Diels;九管血(矮八爪金龙,矮凉伞子,矮陀陀,八爪金龙,八爪龙,大郎伞,地柑子,短茎朱砂根,短茎紫金牛,猴爪,活血胎,开喉箭,散血丹,山豆根,团叶八爪金龙,屯鹿紫金牛,乌肉鸡,小罗伞,血党,血猴爪,猪总管);Brevicaulinary Ardisia, Shortstem Ardisia, Short-stemmed Ardisia ●
31372 Ardisia brevicaulis Diels var. violacea (T. Suzuki) E. Walker = Ardisia violacea (T. Suzuki) W. Z. Fang et K. Yao ●
31373 Ardisia brunneo-purpurea Gilg = Ardisia staudtii Gilg ●☆
31374 Ardisia brunnescens E. Walker;凹脉紫金牛(山脑根,石狮子,棕紫金牛);Sunkenvein Ardisia, Sunken-veined Ardisia ●
31375 Ardisia buesgenii (Gilg et G. Schellenb.) Taton;比斯根紫金牛●☆
31376 Ardisia carnosi-caulis C. Chen et D. Fang;肉茎紫金牛●
31377 Ardisia castaneifolia H. Lév. = Ardisia faberi Hemsl. ●
31378 Ardisia caudata Hemsl.;尾叶紫金牛(峨眉紫金牛);Caudate Ardisia ●
31379 Ardisia cavaleriei H. Lév. = Ardisia faberi Hemsl. ●
31380 Ardisia chinensis Benth.;小紫金牛(产后草,黑果凉伞,华紫金牛,三花紫金牛,衫纽根,石狮子,五花紫金牛,小凉伞,小狮子);China Ardisia, Chinese Ardisia, Fiveflower Ardisia, Five-flowered Ardisia ●
31381 Ardisia citrifolia Hayata = Ardisia brevicaulis Diels ●
31382 Ardisia cochinchinensis (Pit.) C. M. Hu;东南亚紫金牛(越南

紫金牛)●
31383 Ardisia colorata Roxb.;具色紫金牛(开喉箭);Coloured Ardisia ●
31384 Ardisia comosa (de Wit) Taton;簇毛紫金牛●☆
31385 Ardisia conraui Gilg;康氏紫金牛●☆
31386 Ardisia conspersa E. Walker;散花紫金牛;Consperseflower Ardisia,Consperse-flowered Ardisia ●
31387 Ardisia cornudentata Mez;腺齿紫金牛(万两金,雨伞仔,雨伞子,玉山紫金牛);Cornudentate Ardisia,Glandular-toothed Ardisia,Honytoothed Ardisia ●
31388 Ardisia cornudentata Mez var. morrisonensis (Hayata) Yuen P. Yang;玉山紫金牛●
31389 Ardisia corymbifera Mez;伞形紫金牛(不待劳,毛高,西南紫金牛,紫背绿,紫绿西南紫金牛);Corymb Ardisia,Corymbous Ardisia ●
31390 Ardisia corymbifera Mez var. tuberifera C. Chen;块根紫金牛;Tuber-bearing Ardisia ●
31391 Ardisia crassinervosa E. Walker;粗脉紫金牛(多脉紫金牛,小罗伞树);Many-veins Ardisia, Thick-nerved Ardisia, Thick-nerves Ardisia,Thickvein Ardisia ●
31392 Ardisia crassipes C. Y. Wu et C. Chen;粗梗紫金牛;Thick-pedicelled Ardisia,Thickstalk Ardisia ●◇
31393 Ardisia crassipes C. Y. Wu et C. Chen = Ardisia hokouensis Yuen P. Yang ●
31394 Ardisia crassirhiza Z. X. Li et F. W. Xing ex C. M. Hu;粗根紫金牛;Thick-rooted Ardisia ●
31395 Ardisia crassirhiza Z. X. Li et F. W. Xing ex C. M. Hu = Ardisia crassinervosa E. Walker ●
31396 Ardisia crenata Sims;朱砂根(八爪金龙,豹子眼睛果,大凉伞,大罗伞,地杨梅,凤凰肠,凤凰翔,高茶风,高脚金鸡,高脚罗伞,高脚铜盘,桂笃油,红铜盘,火龙珠,金鸡凉伞,金鸡爪,金锁匙,开喉箭,朗伞树,浪伞根,老鼠尾,凉伞遮金珠,龙山子,苗栗紫金牛,平地木,青红草,三两金,三条根,散血丹,散血胆,山豆根,石青子,水龙珠,铁凉伞,铁雨伞,土丹皮,万龙,万雨金,小罗伞,雪里开花,硬脚金鸡,珍珠伞,真珠凉伞,珠砂根);Cinnabarroot,Coral Ardisia,Coralberry,Coral-berry,Crenate-leaved Ardisia,Hen's Eyes,Hilo Holly,Spiceberry ●
31397 Ardisia crenata Sims 'Takarabune';宝船朱砂根●☆
31398 Ardisia crenata Sims = Ardisia crispa (Thunb.) A. DC. ●
31399 Ardisia crenata Sims f. hortensis (Migo) W. Z. Fang et K. Yao;红凉伞(园圃朱砂根)●
31400 Ardisia crenata Sims f. leucocarpa (Nakai) T. Yamanaka;白果朱砂根●☆
31401 Ardisia crenata Sims f. taquetii (H. Lév.) Ohwi = Ardisia crenata Sims ●
31402 Ardisia crenata Sims f. xanthocarpa (Nakai) H. Ohashi;黄果朱砂根●☆
31403 Ardisia crenata Sims var. bicolor (E. Walker) C. Y. Wu et C. Chen;二色朱砂根(红凉伞,两色紫金牛,绿天红地,天青地红,铁凉伞,铁伞,叶下红);Bicoloured Ardisia,Twocolored Ardisia ●
31404 Ardisia crenata Sims var. bicolor (E. Walker) C. Y. Wu et C. Chen = Ardisia crenata Sims ●
31405 Ardisia crenata Sims var. lanceolata Masam. = Ardisia crenata Sims ●
31406 Ardisia crenulata Lodd. = Ardisia crenata Sims ●
31407 Ardisia crenulata Lodd. = Ardisia punctata Lindl. ●
31408 Ardisia crispa (Thunb.) A. DC.;百两金(矮茶,八爪根,八爪金,八爪金龙,八爪龙,白八爪,地杨梅,高八爪,高脚凉茶,开喉剑,开喉箭,山豆根,铁雨伞,野猴枣,叶下藏珠,珍珠凉伞,珍珠伞,真珠凉伞,朱砂根,状元红);Coral Berry, Coralberry, Crispateleaf Ardisia, Crispate-leaved Ardisia, Crisped Ardisia, Spiceberry ●
31409 Ardisia crispa (Thunb.) A. DC. = Ardisia crenata Sims ●
31410 Ardisia crispa (Thunb.) A. DC. f. leucocarpa (Nakai) H. Ohashi;白果百两金●☆
31411 Ardisia crispa (Thunb.) A. DC. f. xanthocarpa (Nakai) H. Ohashi;黄果百两金●☆
31412 Ardisia crispa (Thunb.) A. DC. var. amplifolia E. Walker;大叶百两金(八爪金,高八爪);Bigleaf Ardisia,Largeleaf Ardisia ●
31413 Ardisia crispa (Thunb.) A. DC. var. amplifolia E. Walker = Ardisia crispa (Thunb.) A. DC. ●
31414 Ardisia crispa (Thunb.) A. DC. var. brevifolia Sugim.;小叶百两金●☆
31415 Ardisia crispa (Thunb.) A. DC. var. dielsii (H. Lév.) E. Walker;细柄百两金(山豆根,台湾百两金,玉兰草);Diels Ardisia,Diels Coral Ardisia ●
31416 Ardisia crispa (Thunb.) A. DC. var. dielsii (H. Lév.) E. Walker = Ardisia crispa (Thunb.) A. DC. ●
31417 Ardisia crispa (Thunb.) A. DC. var. elegans A. DC. = Ardisia elegans Andréws ●
31418 Ardisia crispa (Thunb.) A. DC. var. taquetii H. Lév. = Ardisia crenata Sims f. taquetii (H. Lév.) Ohwi ●
31419 Ardisia crispa (Thunb.) A. DC. var. taquetii H. Lév. = Ardisia crispa (Thunb.) A. DC. ●
31420 Ardisia curvula C. Y. Wu et C. Chen;折梗紫金牛(弯梗紫金牛);Bentstalk Ardisia,Curved Ardisia ●
31421 Ardisia cymosa Baker = Ardisia staudtii Gilg ●☆
31422 Ardisia dasyrhizomatica C. Y. Wu et C. Chen;粗茎紫金牛;Curved Ardisia,Thickstem Ardisia,Thick-stemmed Ardisia ●
31423 Ardisia densilepidotula Merr.;密鳞紫金牛(黑度,罗芒树,山马皮,仙人血树);Densescale Ardisia,Dense-squamaceous Ardisia ●
31424 Ardisia depressa C. B. Clarke;圆果罗伞(开展紫金牛,痨病木,拟罗伞树);Depressed Ardisia ●
31425 Ardisia depressa C. B. Clarke = Ardisia crispa (Thunb.) A. DC. ●
31426 Ardisia depressa C. B. Clarke = Ardisia thyrsiflora D. Don ●
31427 Ardisia devredii Taton;德夫雷紫金牛●☆
31428 Ardisia didymopora (H. Perrier) Capuron;双孔紫金牛●☆
31429 Ardisia dielsii H. Lév. = Ardisia crispa (Thunb.) A. DC. ●
31430 Ardisia discolor H. Lév. = Cornus oblonga Wall. ex Roxb. ●☆
31431 Ardisia discolor H. Lév. = Swida controversa (Hemsl. ex Prain) Soják ●
31432 Ardisia dolichocalyx Taton;长萼紫金牛●☆
31433 Ardisia dumetosa Tutcher = Ardisia villosa Roxb. ●
31434 Ardisia ebolowensis Taton;埃博洛瓦紫金牛●☆
31435 Ardisia elegans Andréws;郎伞木(大罗伞,高脚鸡眼,花针木,美丽紫金牛,雀儿肾,小罗伞,胭脂木);Elegant Ardisia ●
31436 Ardisia elegantissima H. Lév. = Ardisia hanceana Mez ●
31437 Ardisia elliptica Thunb.;东方紫金牛(春不老,兰屿树杞,兰屿紫金牛);Ceylon Ardisia, Oriental Ardisia, Shoebutton, Shoebutton Ardisia ●
31438 Ardisia elliptisepala E. Walker = Ardisia quinquegona Blume ●
31439 Ardisia ensifolia E. Walker;剑叶紫金牛(开喉箭);Swordleaf Ardisia,Sword-leaved Ardisia ●
31440 Ardisia erythroxyloides Thouars ex Roem. et Schult. = Rapanea erythroxyloides (Thouars ex Roem. et Schult.) Mez ●☆
31441 Ardisia escallonioides Schltdl. et Cham.;鼠刺紫金牛;Marlberry ●☆

31442 Ardisia esquirolii H. Lév. = Lysimachia fooningensis C. Y. Wu ■
31443 Ardisia faberi Hemsl.;月月红(峨眉紫金牛,红毛走马胎,江南紫金牛,毛虫草,毛青岗,毛青杠,木布马胎);Faber Ardisia ●
31444 Ardisia faberi Hemsl. var. oblanceifolia C. Chen;短柄月月红;Oblance-leaf Ardisia ●
31445 Ardisia faberi Hemsl. var. oblanceifolia C. Chen = Ardisia faberi Hemsl. ●
31446 Ardisia filiformis E. Walker;狭叶紫金牛(喘咳木,石龙胴,竹叶凉伞);Filiform Ardisia,Narrowleaf Ardisia ●
31447 Ardisia flaviflora C. Chen et D. Fang;黄花紫金牛;Yellowflower Ardisia ●
31448 Ardisia flaviflora C. Chen et D. Fang = Ardisia virens Kurz ●
31449 Ardisia floribunda Wall. = Ardisia thyrsiflora D. Don ●
31450 Ardisia fordii Hemsl.;灰色紫金牛(两广紫金牛,全缘紫金牛,细罗伞);Ford Ardisia ●
31451 Ardisia formosana Rolfe = Ardisia sieboldii Miq. ●
31452 Ardisia fuliginosa Blume;黑褐紫金牛●☆
31453 Ardisia fuscopilosa Baker = Oncostemum leprosum Mez ●☆
31454 Ardisia garrettii H. R. Fletcher;小乔木紫金牛(大叶紫金牛,地打果树,石狮子);Arborescent Ardisia ●
31455 Ardisia gigantifolia Stapf;走马胎(白马胎,大发药,大叶紫金牛,马胎,山鼠,山猪药,束花紫金牛,血枫,走马风,走马藤);Giantleaf Ardisia,Giant-leaved Ardisia ●
31456 Ardisia glabra (Thunb.) A. DC. = Sarcandra glabra (Thunb.) Nakai ●
31457 Ardisia glauca Pit. = Ardisia brunnescens E. Walker ●
31458 Ardisia graciliflora Pit.;小花紫金牛;Thinflower Ardisia,Thin-flowered Ardisia ●
31459 Ardisia haemantha Gilg = Ardisia staudtii Gilg ●☆
31460 Ardisia hainanensis Mez = Ardisia humilis Vahl ●
31461 Ardisia hallei Taton;哈勒紫金牛●☆
31462 Ardisia hanceana Mez;大罗伞树(大叶紫金牛);Hance Ardisia ●
31463 Ardisia henryi Hemsl. = Ardisia crispa (Thunb.) A. DC. ●
31464 Ardisia henryi Hemsl. var. dielsii E. Walker = Ardisia crispa (Thunb.) A. DC. ●
31465 Ardisia hokouensis Yuen P. Yang;河口紫金牛(粗梗紫金牛)●
31466 Ardisia hortorum Maxim. = Ardisia crispa (Thunb.) A. DC. ●
31467 Ardisia hortorum Maxim. = Ardisia punctata Lindl. ●
31468 Ardisia hortorum Maxim. ex Regel = Ardisia crispa (Thunb.) A. DC. ●
31469 Ardisia hortorum Maxim. ex Regel = Ardisia punctata Lindl. ●
31470 Ardisia hortorum Maxim. ex Regel var. brachysepala Hand.-Mazz. = Ardisia crispa (Thunb.) A. DC. ●
31471 Ardisia humilis Vahl;矮紫金牛(春不老);Dwarf Ardisia,Low Ardisia,Low Shoebutton ●
31472 Ardisia humilis Vahl var. arborescens C. B. Clarke = Ardisia arborescens Wall. et A. DC. ●
31473 Ardisia humilis Vahl var. arborescens C. B. Clarke = Ardisia garrettii H. R. Fletcher ●
31474 Ardisia hypargyrea C. Y. Wu et C. Chen;柳叶紫金牛;Willowleaf Ardisia,Willow-leaved Ardisia ●
31475 Ardisia impressa H. R. Fletcher = Ardisia hanceana Mez ●
31476 Ardisia iwahigensis Elmer;岩髯紫金牛●☆
31477 Ardisia japonica (Thunb.) Blume;紫金牛(矮茶,矮茶风,矮茶荷,矮茶子,矮地茶,矮脚草,矮脚茶,矮脚三郎,矮脚樟,矮脚樟菜,矮郎伞,矮山茶,不出林,茶果,赤玉树,大地风消,地茶,地红消,地青杠,火炭酸,金牛草,老不大,老勿大,凉伞盖珍珠,马台剪,平地木,破血珠,铺地凉伞,千年矮,千年不大,千年茶,山橘,四叶茶,薮柑子,薮橘,五托香,小接骨茶,小青,雪里珠,野枇杷叶,叶底红,叶下红,叶下珍珠,阴山红,映山红);Japan Ardisia,Japanese Ardisia,Marlberry ●
31478 Ardisia japonica (Thunb.) Blume = Ardisia montana (Miq.) Siebold ex Franch. et Sav. ●
31479 Ardisia japonica (Thunb.) Blume f. leucocarpa Sugim. ex T. Yamaz.;白果紫金牛●☆
31480 Ardisia japonica (Thunb.) Blume var. angusta Makino et Nemoto;窄叶紫金牛●☆
31481 Ardisia jiajiangensis Z. Y. Zhu;夹江紫金牛;Jiajiang Ardisia ●
31482 Ardisia jiajiangensis Z. Y. Zhu = Ardisia chinensis Benth. ●
31483 Ardisia jinyunensis Z. Y. Zhu;缙云紫金牛;Jinyun Ardisia ●
31484 Ardisia jinyunensis Z. Y. Zhu = Ardisia quinquegona Blume ●
31485 Ardisia kivuensis Taton;基伍紫金牛●☆
31486 Ardisia konishii Hayata = Ardisia crenata Sims ●
31487 Ardisia konishii Hayata = Ardisia elegans Andréws ●
31488 Ardisia kotoensis Hayata = Ardisia elliptica Thunb. ●
31489 Ardisia kusukusensis Hayata;高士佛紫金牛●
31490 Ardisia kusukusensis Hayata = Ardisia crenata Sims ●
31491 Ardisia kwangtungensis E. Walker;防城紫金牛;Guangdong Ardisia ●
31492 Ardisia kwangtungensis E. Walker = Ardisia lindleyana D. Dietr. ●
31493 Ardisia labordei H. Lév. = Ardisia crenata Sims ●
31494 Ardisia laurifolia (Bojer ex A. DC.) Baker = Oncostemum laurifolium (Bojer ex A. DC.) Mez ●☆
31495 Ardisia lentiginosa Ker Gawl. = Ardisia crenata Sims ●
31496 Ardisia lentiginosa Ker Gawl. var. rectangularis Hatus. = Ardisia crenata Sims ●
31497 Ardisia leptoclada Baker = Oncostemum leptocladum (Baker) Mez ●☆
31498 Ardisia letestui Taton;莱泰斯图紫金牛●☆
31499 Ardisia lethomasiae Taton;托马斯紫金牛●☆
31500 Ardisia letouzeyi Taton;勒图紫金牛●☆
31501 Ardisia linangensis C. M. Hu = Ardisia crenata Sims ●
31502 Ardisia lindleyana D. Dietr.;山血丹(百两金,斑叶朱砂根,斑叶紫金牛,活血胎,郎伞,马胎,山马胎,铁雨伞,细罗伞树,腺点紫金牛,小凉伞,小罗伞,血党,沿海紫金牛,珍珠盖凉伞);Coral Ardisia,Punctate Ardisia ●
31503 Ardisia lindleyana D. Dietr. var. angustifolia C. M. Hu et X. J. Ma;狭叶山血丹●
31504 Ardisia linearifolia X. W. Wei et M. H. Xiao;条叶紫金牛;Line-leaf Ardisia ●
31505 Ardisia linearifolia X. W. Wei et M. H. Xiao = Ardisia ensifolia E. Walker ●
31506 Ardisia lisowskii Taton;利索紫金牛●☆
31507 Ardisia longipedunculata C. Y. Wu et C. Chen = Ardisia pingbienensis Yuen P. Yang ●
31508 Ardisia longipes Baker = Oncostemum longipes (Baker) Mez ●☆
31509 Ardisia maclurei Merr.;心叶紫金牛(红云草,麦氏紫金牛);Heartleaf Ardisia,Maclure Ardisia ●
31510 Ardisia macroscyphon Baker = Oncostemum macroscyphon (Baker) Mez ●☆
31511 Ardisia maculosa Mez;珍珠伞(多斑紫金牛,山豆根,天青地红,小罗伞,紫背绿,紫绿果,紫青绿);Pear Umbrella,Spotted Ardisia ●

31512 Ardisia maculosa Mez = Ardisia botryosa E. Walker ●

31513 Ardisia maculosa Mez = Ardisia virens Kurz ●

31514 Ardisia maculosa Mez var. symplocifolia C. Chen;黄叶珍珠伞;Yellow Spotted Ardisia ●

31515 Ardisia maculosa Mez var. symplocifolia C. Chen = Ardisia virens Kurz ●

31516 Ardisia malipoensis C. M. Hu;麻栗坡紫金牛(麻栗坡罗伞);Malipo Ardisia ●

31517 Ardisia malouiana (Lindau et Rodigas) Markgr.;马氏紫金牛;Malou Ardisia ●☆

31518 Ardisia mamillata Hance;虎舌红(矮朵朵,白毛毡,豺狗舌,红八枣,红八爪,红胆,红地毯,红地毡,红毛过江,红毛山豆根,红毛毡,红毛针,红毛紫金牛,红毛走马胎,红毡,红毡草,红毡毯,虎舌草,老虎利,老虎舌,毛虫药,毛地红,毛凉伞,毛罗伞,毛青杠,铺地毡,蛐蜍皮,肉八枣,肉八爪,乳毛紫金牛,山猪耳,山猪怕);Mamillate Ardisia,Teatshaped Ardisia ●

31519 Ardisia marginata Blume;兰屿紫金牛;Lanyu Ardisia ●

31520 Ardisia marojejyensis J. S. Mill. et Pipoly;马罗紫金牛●☆

31521 Ardisia mayumbensis (R. D. Good) Taton;马永巴紫金牛●☆

31522 Ardisia merrillii E. Walker;白花紫金牛;Merrill Ardisia,White Ardisia ●

31523 Ardisia meziana H. Lév. = Ardisia thyrsiflora D. Don ●

31524 Ardisia miaoliensis S. Y. Lu = Ardisia crenata Sims ●

31525 Ardisia microphylla Roem. et Schult. = Oncostemum microphyllum (Roem. et Schult.) Mez ●☆

31526 Ardisia mildbraedii (Gilg et G. Schellenb.) Taton;米尔德紫金牛●☆

31527 Ardisia montana (Miq.) Siebold ex Franch. et Sav.;山地紫金牛●

31528 Ardisia montana (Miq.) Siebold ex Franch. et Sav. = Ardisia walkeri Y. P. Yang ●

31529 Ardisia morrisonensis Hayata = Ardisia cornudentata Mez var. morrisonensis (Hayata) Yuen P. Yang ●

31530 Ardisia morrisonensis Hayata = Ardisia cornudentata Mez ●

31531 Ardisia mouretii Pit. = Ardisia crenata Sims ●

31532 Ardisia multicaulis Z. Y. Zhu;多茎紫金牛;Many-stem Ardisia,Multicaulinary Ardisia ●

31533 Ardisia multicaulis Z. Y. Zhu = Ardisia crispa (Thunb.) A. DC. ●

31534 Ardisia neriifolia Wall. = Ardisia thyrsiflora D. Don ●

31535 Ardisia nervosa E. Walker;多脉紫金牛;Manyveins Ardisia,Muchvein Ardisia,Nervate Ardisia ●

31536 Ardisia nervosa E. Walker = Ardisia aquifolioides W. Z. Fang et K. Yao ●

31537 Ardisia nervosa E. Walker = Ardisia crassinervosa E. Walker ●

31538 Ardisia nigropilosa Pit.;星毛紫金牛;Starhair Ardisia,Stellar Ardisia ●

31539 Ardisia nitidula Baker = Oncostemum nitidulum (Baker) Mez ●☆

31540 Ardisia obtusa Mez;铜盆花(钝叶紫金牛);Bronzybasin Ardisia,Nlunt-leaved Ardisia,Obtuse Ardisia ●

31541 Ardisia obtusa Mez subsp. pachyphylla (Dunn) Pipoly et C. Chen;厚叶铜盆花●

31542 Ardisia oldhamii Mez = Ardisia virens Kurz ●

31543 Ardisia oligantha (Gilg et G. Schellenb.) Taton;寡花紫金牛●☆

31544 Ardisia oligantha Baker = Oncostemum oliganthum (Baker) Mez ●☆

31545 Ardisia olivacea E. Walker;榄色紫金牛;Olivaceous Ardisia,Olivecolor Ardisia,Olive-green Ardisia ●

31546 Ardisia omissa C. M. Hu;光萼紫金牛;Smooth-calyxed Ardisia ●

31547 Ardisia ordinata E. Walker;轮叶紫金牛;Whorlleaf Ardisia,Whorl-leaved Ardisia ●◇

31548 Ardisia oxyphylla Wall. et A. DC. var. cochinchinensis Pit. = Ardisia waitakii C. M. Hu ●

31549 Ardisia pachyphylla Dunn = Ardisia obtusa Mez subsp. pachyphylla (Dunn) Pipoly et C. Chen ●

31550 Ardisia patens Mez = Ardisia virens Kurz ●

31551 Ardisia pauciflora Heyne = Ardisia quinquegona Blume ●

31552 Ardisia pedalis E. Walker;矮短紫金牛●

31553 Ardisia pedunculata Bojer ex Baker = Oncostemum pauciflorum A. DC. ●☆

31554 Ardisia penduliflora Mez = Ardisia crispa (Thunb.) A. DC. ●

31555 Ardisia pentagona A. DC. = Ardisia quinquegona Blume ●

31556 Ardisia perforata H. Lév. = Alangium faberi Oliv. var. perforatum (H. Lév.) Rehder ●

31557 Ardisia perpendicularis E. Walker = Ardisia gigantifolia Stapf ●

31558 Ardisia perreticulata C. Chen;花脉紫金牛(假血党);Thin-reticulate-veined Ardisia,Thinreticulateveins Ardisia ●

31559 Ardisia pierreana Taton;皮埃尔紫金牛●☆

31560 Ardisia pingbienensis Yuen P. Yang;长穗紫金牛;Longspike Ardisia,Long-spiked Ardisia ●

31561 Ardisia platyphylla (Gilg et G. Schellenb.) Taton;宽叶紫金牛●☆

31562 Ardisia polyadenia Gilg;多腺紫金牛●☆

31563 Ardisia polycephala Wight;多头紫金牛(多头花紫金牛)●☆

31564 Ardisia porifera E. Walker;细孔紫金牛;Smallhole Ardisia,Thin-pored Ardisia ●

31565 Ardisia primulifolia Gardner et Champ.;莲座紫金牛(赫地涩,脚皮,咳嗽草,老虎毛虫药,老虎舌,莲座叶紫金牛,落地紫金牛,毛虫药,毛虫药公,毛脚皮,铺地罗伞);Primrose-leaf Ardisia,Primulaleaf Ardisia,Primula-leaved Ardisia,Rosula Ardisia ●

31566 Ardisia pseudoverticillata Merr. = Ardisia gigantifolia Stapf ●

31567 Ardisia pubivenula E. Walker;毛脉紫金牛;Hairyvein Ardisia,Hairy-veined Ardisia ●

31568 Ardisia punctata Lindl. = Ardisia lindleyana D. Dietr. ●

31569 Ardisia punctata Lindl. var. latifolia E. Walker = Ardisia perreticulata C. Chen ●

31570 Ardisia purpureovillosa C. Y. Wu et C. Chen ex C. M. Hu;紫肋紫金牛(紫脉紫金牛);Purple-veined Ardisia ●

31571 Ardisia pusilla A. DC.;九节龙(矮茶子,刺毛藤,地茶,红刺毛藤,猴接骨,轮叶紫金牛,毛茎紫金牛,毛青杠,蛇药,狮子头,细叶紫金牛,小紫金牛,斩龙剑);Pretty Ardisia,Tiny Ardisia,Whorl-leaved Ardisia ●

31572 Ardisia pusilla A. DC. f. liukiuensis (Nakai) Ohwi = Ardisia pusilla A. DC. ●

31573 Ardisia pusilla A. DC. f. liukiuensis (Nakai) Ohwi = Ardisia pusilla A. DC. var. liukiuensis (Nakai) Okuyama ●☆

31574 Ardisia pusilla A. DC. var. liukiuensis (Nakai) Okuyama;琉球九节龙●☆

31575 Ardisia pyrgina St. -Lag. = Ardisia humilis Vahl ●

31576 Ardisia pyrgus Roem. et Schult. = Ardisia humilis Vahl ●

31577 Ardisia pyrifolia Willd. ex Roem. et Schult. = Embelia pyrifolia (Willd. ex Roem. et Schult.) Mez ●☆

31578 Ardisia quinquegona Blume;罗伞树(大罗伞,高脚凉伞,高脚罗伞,高脚罗伞树,火泡树,火屎炭树,火炭树,鸡眼树,筷子根,棱果紫金牛,提枯肠,小叶树杞);Asiatic Ardisia,Five-angular Ardisia,Pentagonous Ardisia,Pretty Ardisia ●

31579 Ardisia quinquegona Blume var. hainanensis E. Walker;海南罗伞树(火灰树);Hainan Ardisia ●

31580 Ardisia quinquegona Blume var. hainanensis E. Walker = Ardisia quinquegona Blume ●

31581 Ardisia quinquegona Blume var. linearifolia Pit. = Ardisia hypargyrea C. Y. Wu et C. Chen ●

31582 Ardisia quinquegona Blume var. oblonga E. Walker;长萼罗伞树;Longsepal Ardisia ●

31583 Ardisia quinquegona Blume var. oblonga E. Walker = Ardisia quinquegona Blume ●

31584 Ardisia radians Hemsl. et Mez ex Mez = Ardisia virens Kurz ●

31585 Ardisia ramondiiformis Pit.;海南梯脉紫金牛(梯脉紫金牛)●

31586 Ardisia rectangularis Hayata = Ardisia virens Kurz ●

31587 Ardisia remotiserrata Hayata = Ardisia cornudentata Mez ●

31588 Ardisia replicata E. Walker;卷边紫金牛;Replicate Ardisia ●

31589 Ardisia retroflexa E. Walker;弯梗紫金牛;Flexstalk Ardisia, Reflexed Ardisia ●

31590 Ardisia rigida Kurz = Ardisia humilis Vahl ●

31591 Ardisia roseiflora Pit. = Ardisia elegans Andréws ●

31592 Ardisia sadebeckiana Gilg;萨德拜克紫金牛●☆

31593 Ardisia salicifolia E. Walker = Ardisia hypargyrea C. Y. Wu et C. Chen ●

31594 Ardisia scalaxinervis E. Walker;云南梯脉紫金牛(梯脉紫金牛);Ladder-veins Ardisia, Scalariform-nerved Ardisia, Scalarvein Ardisia ●

31595 Ardisia schlechteri Gilg;施莱紫金牛●☆

31596 Ardisia sciophila T. Suzuld = Ardisia maclurei Merr. ●

31597 Ardisia scuamulosa C. Presl = Ardisia squamulosa C. Presl ●

31598 Ardisia shweliensis W. W. Sm.;瑞丽紫金牛(麂子扣甘树);Ruili Ardisia ●

31599 Ardisia sieboldii Miq.;多枝紫金牛(东南紫金牛,树杞);Siebold Ardisia ●

31600 Ardisia sieboldii Miq. f. nigrocarpa (Tuyama ex Nakai) H. Ohashi;黑果多枝紫金牛●

31601 Ardisia sieboldii Miq. f. rubricarpa (Tuyama ex Nakai) H. Ohashi;红果多枝紫金牛●

31602 Ardisia silvestris Pit.;短柄紫金牛;Shortstalk Ardisia, Short-stalked Ardisia ●

31603 Ardisia simplicicaulis Hayata = Ardisia crispa (Thunb.) A. DC. ●

31604 Ardisia sinoaustralis C. Chen;细罗伞(矮脚罗伞,波叶紫金牛,小郎伞);Undulate-leaf Ardisia, Waveleaf Ardisia ●

31605 Ardisia sinoaustralis C. Chen et D. Fang var. longicalyx C. Chen et D. Fang;长萼细罗伞;Longcalyx Ardisia ●

31606 Ardisia sino-australis C. Chen et D. Fang var. longicalyx C. Chen et D. Fang = Ardisia sinoaustralis C. Chen et D. Fang ●

31607 Ardisia solanacea Roxb.;酸苔菜(茄花紫金牛);China-shrub, Shoebutton Ardisia, Shoe-button Ardisia ●

31608 Ardisia solanacea Roxb. = Ardisia humilis Vahl ●

31609 Ardisia squamulosa C. Presl = Ardisia elliptica Thunb. ●

31610 Ardisia staudtii Gilg;施陶紫金牛●☆

31611 Ardisia stellata E. Walker = Ardisia nigropilosa Pit. ●

31612 Ardisia stellifera Pit. = Ardisia virens Kurz ●

31613 Ardisia stenosepala Hayata;狭萼紫金牛(阿里山雨伞仔,阿里山雨伞子,阿里山紫金牛);Narrowsepaled Ardisia, Stenosepalous Ardisia ●

31614 Ardisia stenosepala Hayata = Ardisia cornudentata Mez ●

31615 Ardisia suishaensis Hayata = Ardisia cornudentata Mez var. morrisonensis (Hayata) Yuen P. Yang ●

31616 Ardisia suishaensis Hayata = Ardisia cornudentata Mez ●

31617 Ardisia tenera Mez;细柄罗伞;Slender-stalked Ardisia, Smallstipe Ardisia, Thin-stalk Ardisia ●

31618 Ardisia tenera Mez = Ardisia thyrsiflora D. Don ●

31619 Ardisia thorelii Pit. = Ardisia hanceana Mez ●

31620 Ardisia thyrsiflora D. Don;南方紫金牛;Oleanderleaf Ardisia, Oleander-leaved Ardisia ●

31621 Ardisia tonkinensis A. DC. = Ardisia virens Kurz ●

31622 Ardisia trichocarpa Merr. = Ardisia villosa Roxb. ●

31623 Ardisia triflora Hemsl. = Ardisia chinensis Benth. ●

31624 Ardisia tsangii E. Walker = Ardisia lindleyana D. Dietr. ●

31625 Ardisia umbellata Baker = Oncostemum umbellatum (Baker) Mez ●☆

31626 Ardisia undulata Mez = Ardisia conspersa E. Walker ●

31627 Ardisia uregaensis Taton;乌雷加紫金牛●☆

31628 Ardisia velutina Pit.;紫脉紫金牛;Purplevein Ardisia, Purple-veined Ardisia ●

31629 Ardisia verbascifolia Mez;长毛紫金牛;Long-haired Ardisia, Longhairy Ardisia ●

31630 Ardisia villosa Roxb.;雪下红(矮茶风,矮脚罗伞,矮脚三郎,地茶,短脚三郎,猴接骨,脚龙子,九节龙,卷毛雪下红,卷毛紫金牛,毛茶,毛茎紫金牛,山毛茶,珊瑚树,珊瑚珠,小罗伞,医药师);Villose Ardisia ●

31631 Ardisia villosa Roxb. = Ardisia pusilla A. DC. ●

31632 Ardisia villosa Roxb. var. ambovestita E. Walker;毛叶雪下红;Hainryleaf Ardisia ●

31633 Ardisia villosa Roxb. var. ambovestita E. Walker = Ardisia villosa Roxb. ●

31634 Ardisia villosa Roxb. var. latifolia E. Walker = Ardisia villosa Roxb. ●

31635 Ardisia villosa Roxb. var. oblanceolata E. Walker;狭叶雪下红;Narrowleaf Ardisia ●

31636 Ardisia villosa Roxb. var. oblanceolata E. Walker = Ardisia villosa Roxb. ●

31637 Ardisia villosoides E. Walker. = Ardisia verbascifolia Mez ●

31638 Ardisia violacea (T. Suzuki) W. Z. Fang et K. Yao;锦花紫金牛(锦花九管血,里堇紫金牛);Violet Short-stemmed Ardisia ●

31639 Ardisia virens Kurz;纽子果(大罗伞,黑星紫金牛,厚皮树,扣子果,绿叶紫金牛,米汤果,圆齿紫金牛);Black-spotted Ardisia, Green Ardisia, Green-leaf Ardisia, Green-leaved Ardisia ●

31640 Ardisia virens Kurz var. annamensis Pit.;长叶纽子果;Longleaf Ardisia ●

31641 Ardisia virens Kurz var. annamensis Pit. = Ardisia virens Kurz ●

31642 Ardisia waitakii C. M. Hu;越南紫金牛;Cochinchina Ardisia, Vietnam Ardisia ●

31643 Ardisia walkeri Y. P. Yang = Ardisia montana (Miq.) Siebold ex Franch. et Sav. ●

31644 Ardisia yunnanensis Mez;滇紫金牛(云南紫金牛);Yunnan Ardisia ●

31645 Ardisia yunnanensis Mez = Ardisia thyrsiflora D. Don ●

31646 Ardisia zenkeri Gilg;岑克尔紫金牛●☆

31647 Ardisiaceae Bartl. = Myrsinaceae R. Br.(保留科名)●

31648 Ardisiaceae Juss. = Myrsinaceae R. Br.(保留科名)●

31649 Ardisiandra Hook. f.(1864);紫金花属■☆

31650 Ardisiandra engleri Weim. = Ardisiandra sibthorpioides Hook. f. ■☆

31651 Ardisiandra engleri Weim. var. microphylla ? = Ardisiandra

sibthorpioides Hook. f. ■☆
31652 Ardisiandra orientalis Weim. = Ardisiandra wettsteinii R. Wagner ■☆
31653 Ardisiandra orientalis Weim. var. hirsuta ? = Ardisiandra wettsteinii R. Wagner ■☆
31654 Ardisiandra primuloides R. Knuth;非洲紫金花■☆
31655 Ardisiandra sibthorpioides Hook. f. ;紫金花■☆
31656 Ardisiandra stolzii Weim. = Ardisiandra wettsteinii R. Wagner ■☆
31657 Ardisiandra wettsteinii R. Wagner;东方紫金花■☆
31658 Ardlsia Gaertn. = Cyathodes Labill. ●☆
31659 Arduina Adans. (废弃属名) = Kundmannia Scop. ■☆
31660 Arduina Mill. = Carissa L. (保留属名)●
31661 Arduina Mill. ex L. = Carissa L. (保留属名)●
31662 Arduina acuminata E. Mey. = Carissa bispinosa (L.) Desf. ex Brenan ●☆
31663 Arduina bispinosa L. = Carissa bispinosa (L.) Desf. ex Brenan ●☆
31664 Arduina carandas (L.) K. Schum. = Carissa carandas L. ●
31665 Arduina edulis (Forssk.) Spreng. = Carissa edulis Vahl ●
31666 Arduina erythrocarpa Eckl. = Carissa bispinosa (L.) Desf. ex Brenan ●☆
31667 Arduina ferox E. Mey. = Carissa bispinosa (L.) Desf. ex Brenan ●☆
31668 Arduina grandiflora E. Mey. = Carissa macrocarpa (Eckl.) A. DC. ●
31669 Arduina haematocarpa Eckl. = Carissa bispinosa (L.) Desf. ex Brenan ●☆
31670 Arduina macrocarpa Eckl. = Carissa macrocarpa (Eckl.) A. DC. ●
31671 Arduina megaphylla Gand. = Carissa bispinosa (L.) Desf. ex Brenan ●☆
31672 Arduina ouabaiocornu ex Holmes = Acokanthera schimperi (A. DC.) Schweinf. ●☆
31673 Arduina schimperi (A. DC.) Baill. = Acokanthera schimperi (A. DC.) Schweinf. ●☆
31674 Arduina tetramera Sacleux = Carissa tetramera (Sacleux) Stapf ●☆
31675 Arduina venenata (Thunb.) Baill. = Acokanthera oppositifolia (Lam.) Codd ●
31676 Area gracilis Roxb. = Pinanga gracilis (Roxb.) Blume ●
31677 Areca L. (1753);槟榔属;Areca, Areca Palm, Arecapalm, Betel Palms ●
31678 Areca alicae F. Muell. ;澳洲槟榔;Alice Areca Palm, Alice Palm ●☆
31679 Areca borbonica Hort. = Chrysalidocarpus lutescens H. Wendl. ●
31680 Areca catechu L. ;槟榔(白槟,白槟榔,宾门,宾门药饯,槟药,槟榔衣,槟榔玉,槟榔子,槟楠,槟玉,梹榔,大白槟,大腹,大腹槟榔,大腹毛,大腹皮,大腹绒,大腹子,儿茶,苻毛,茯毛,橄榄子,国马,花槟榔,花大白,鸡心槟榔,尖槟,榔玉,吕宋槟,马金南,蒳子,青仔,青子,箐仔,仁榔,仁频,洗瘴丹,枣槟榔,枣儿槟榔);Areca, Areca Nut, Areca Palm, Areca-nut Palm, Betel, Betel Nut, Betel Palm, Betelnut Palm, Betel-nut Palm, Betle-nut Palm, Catechu, Catechu Palm, Fansel, Indian Filbert, Pinang, Ragonet ●
31681 Areca catechu Willd. = Areca catechu L. ●
31682 Areca crinita Bory = Acanthophoenix crinita H. Hendl. ●☆
31683 Areca dicksonii Roxb. ;大腹槟榔(大腹子)●
31684 Areca erythropoda Miq. = Cyrtostachys renda Blume ●☆
31685 Areca hortensis Lour. = Areca catechu L. ●
31686 Areca ipot Becc. ;菲律宾槟榔●☆
31687 Areca latiloba Ridl. ;阔裂片槟榔;Broadlobe Arecapalm ●☆
31688 Areca lutescens Bory = Chrysalidocarpus lutescens H. Wendl. ●
31689 Areca madagascariensis Mart. = Chrysalidocarpus lutescens H. Wendl. ●
31690 Areca montana Ridl. ;山地槟榔;Montane Arecapalm ●☆
31691 Areca nobilis Hort. = Deckenia nobilis H. Wendl. ●☆
31692 Areca oleracea Jacq. = Roystonea oleracea (Jacq.) O. F. Cook ●
31693 Areca passalacquae Kunth = Medemia argun (Mart.) Württemb. ex H. Wendl. ●☆
31694 Areca ridleyana Becc. ex Furtado;瑞得莱槟榔;Ridley Arecapalm ●☆
31695 Areca rubra Bory = Acanthophoenix rubra H. Hendl. ●☆
31696 Areca triandra Roxb. ex Buch. -Ham. ;三药槟榔(丛立槟榔,丛立槟榔子,三雄蕊槟榔);Bungua, Bungua Areca Palm, Bungua Arecapalm, Trianther Areca ●
31697 Areca vestiaria Giseke;橙冠轴槟榔●☆
31698 Arecaceae Bercht. et J. Presl(1820)(保留科名);棕榈科(槟榔科)●
31699 Arecaceae Bercht. et J. Presl(保留科名) = Palmae Juss. (保留科名)●
31700 Arecaceae Schultz Sch. = Arecaceae Bercht. et J. Presl(保留科名)●
31701 Arecaceae Schultz Sch. = Palmae Juss. (保留科名)●
31702 Arecastrum (Drude) Becc. (1916);山葵属(槟榔星属,皇后葵属,克利巴椰子属,女王椰属,女王椰子属);Arecastrum, Queen Palm ●
31703 Arecastrum (Drude) Becc. = Syagrus Mart. ●
31704 Arecastrum Becc. = Arecastrum (Drude) Becc. ●
31705 Arecastrum romanzoffianum (Cham.) Becc. ;金山葵(后棕,皇后葵,皇后椰,克利巴椰子,女王椰子);Cocos Palm, Geriba Palm, Jelly Palm, Jinshanpalm, Pindo Palm, Queen Palm, Romanzoff Syagrus, Romanzov's Syagrus ●
31706 Arecastrum romanzoffianum (Cham.) Becc. = Syagrus romanzoffiana (Cham.) Glassman ●
31707 Arecastrum romanzoffianum (Cham.) Becc. var. australe (Mart.) Becc. ;山葵(南方山葵,南皇后葵);Austral Arecastrum, Pindo Palm, Southern Queen Palm ●
31708 Arecastrum romanzoffianum Becc. = Arecastrum romanzoffianum (Cham.) Becc. ●
31709 Arecastrum romanzoffianum Becc. = Syagrus romanzoffiana (Cham.) Glassman ●
31710 Arechavaletaia Speg. (1899);乌拉圭大风子属●☆
31711 Arechavaletaia Speg. = Azara Ruiz et Pav. ●☆
31712 Arechavaletaia uruguayensis Speg. ;乌拉圭大风子●☆
31713 Aregelia Kuntze = Neoregelia L. B. Sm. ■☆
31714 Aregelia Kuntze = Nidularium Lem. ■☆
31715 Aregelia Kuntze(1891);热美凤梨属(阿瑞盖利属)■☆
31716 Aregelia Mez = Neoregelia L. B. Sm. ■☆
31717 Aregelia spectabilis Mez;巴西凤梨■☆
31718 Areldia Luer = Pleurothallis R. Br. ■☆
31719 Arelina Neck. = Berkheya Ehrh. (保留属名)●■☆
31720 Aremonia Neck. = Aremonia Neck. ex Nestl. (保留属名)■☆
31721 Aremonia Neck. ex Nestl. (1816)(保留属名);龙牙蔷薇属;Bastard Agrimony ■☆
31722 Aremonia Nestl. = Aremonia Neck. ex Nestl. (保留属名)■☆
31723 Aremonia agrimonoides (L.) DC. ;龙牙蔷薇■☆
31724 Aremonia agrimonoides DC. = Aremonia agrimonoides (L.) DC. ■☆

31725 Arenaria Adans. = Phaloe Dumort. ■
31726 Arenaria Adans. = Sagina L. ■
31727 Arenaria L.(1753);无心菜属(蚤缀属);Sandwort ■
31728 Arenaria aberrans M. E. Jones;异常无心菜;Mount Dellenbaugh Sandwort ■☆
31729 Arenaria aberrans M. E. Jones = Eremogone aberrans (M. E. Jones) Ikonn. ■☆
31730 Arenaria acicularis F. N. Williams ex Keissl.;针叶老牛筋(针叶雪灵芝,针叶蚤缀);Needle Sandwort ■
31731 Arenaria aculeata S. Watson;多刺无心菜;Prickly Sandwort ■☆
31732 Arenaria aculeata S. Watson = Eremogone aculeata (S. Watson) Ikonn. ■☆
31733 Arenaria aculeata S. Watson var. uintahensis (A. Nelson) M. Peck = Arenaria fendleri A. Gray var. glabrescens S. Watson ■☆
31734 Arenaria aculeata S. Watson var. uintahensis (A. Nelson) M. Peck = Eremogone kingii (S. Watson) Ikonn. var. glabrescens (S. Watson) Dorn ■☆
31735 Arenaria africana Hook. f. = Cerastium indicum Wight et Arn. ■☆
31736 Arenaria aggregata (L.) Loisel.;团集无心菜■☆
31737 Arenaria aggregata (L.) Loisel. subsp. mauritanica (Batt.) Maire;毛里塔尼亚团集无心菜■☆
31738 Arenaria aksayqingensis L. H. Zhou;阿克赛钦雪灵芝;Aksaichin Sandwort ■
31739 Arenaria alabamensis (J. F. McCormick, Bozeman et Spongberg) R. E. Wyatt = Minuartia uniflora (Walter) Mattf. ■☆
31740 Arenaria amdoensis L. H. Zhou;安多无心菜;Amdo Sandwort ■
31741 Arenaria androsacea Grubov;点地梅状老牛筋(点地梅蚤缀);Rockjasmine Sandwort ■
31742 Arenaria arctica Steven ex Ser.;北极无心菜■☆
31743 Arenaria arctica Steven ex Ser. = Minuartia arctica (Steven ex Ser.) Graebn. ■
31744 Arenaria arctica Steven ex Ser. var. hondoensis (Ohwi) H. Hara;本州无心菜■☆
31745 Arenaria arctica Steven ex Ser. var. rebunensis T. Shimizu;礼文无心菜■☆
31746 Arenaria arenarioides (Crantz) Maire = Arenaria cerastioides Poir. ■☆
31747 Arenaria arenarioides (Crantz) Maire subsp. cerastioides (Poir.) Maire = Arenaria cerastioides Poir. ■☆
31748 Arenaria arenarioides (Crantz) Maire var. fallax (Batt.) Maire = Arenaria hispanica Spreng. ■☆
31749 Arenaria arenarioides (Crantz) Maire var. macrosperma (Batt.) Maire = Arenaria cerastioides Poir. ■☆
31750 Arenaria arenarioides (Crantz) Maire var. microsperma Maire = Arenaria cerastioides Poir. ■☆
31751 Arenaria arenarioides (Crantz) Maire var. oranensis (Batt.) Maire;奥兰无心菜■☆
31752 Arenaria arenarioides (Crantz) Maire var. parviflora Maire = Arenaria cerastioides Poir. ■☆
31753 Arenaria arenarioides (Crantz) Maire var. spathulata (Desf.) Maire = Arenaria cerastioides Poir. ■☆
31754 Arenaria armeriastrum Boiss. = Arenaria armerina Bory ■☆
31755 Arenaria armerina Bory;耐寒无心菜■☆
31756 Arenaria armerina Bory var. elongata (Boiss.) Pau = Arenaria armerina Bory ■☆
31757 Arenaria armerina Bory var. frigida (Boiss.) Pau = Arenaria armerina Bory ■☆
31758 Arenaria asiatica Schischk.;亚洲无心菜;Asia Sandwort ■
31759 Arenaria atuntziensis C. Y. Wu ex L. H. Zhou;德钦无心菜;Deqin Sandwort ■
31760 Arenaria atuntziensis C. Y. Wu ex L. H. Zhou = Arenaria roseiflora Sprague ■
31761 Arenaria atuntziensis C. Y. Wu ex L. H. Zhou var. stenopetala Y. W. Tsui ex L. H. Zhou = Arenaria roseiflora Sprague ■
31762 Arenaria atuntziensis C. Y. Wu var. stenopetala Y. W. Tsui;狭瓣无心菜;Narrowpetal Deqin Sandwort ■
31763 Arenaria atuntziensis C. Y. Wu var. stenopetala Y. W. Tsui = Arenaria roseiflora Sprague ■
31764 Arenaria aureocaulis C. Y. Wu ex L. H. Zhou;黄茎无心菜;Yellowstem Sandwort ■
31765 Arenaria aureocaulis C. Y. Wu ex L. H. Zhou = Arenaria debilis Hook. f. ■
31766 Arenaria auricoma Y. W. Tsui ex L. H. Zhou;黄毛无心菜;Yellowhair Sandwort ■
31767 Arenaria balearica L.;科西嘉蚤缀(匐雪草);Balearic Pearlwort, Balearic Sandwort, Corsican Sandwort, Mossy Sandwort ■☆
31768 Arenaria barbata Franch.;髯毛无心菜(鸡肠子,髯毛蚤缀);Barbate Sandwort ■
31769 Arenaria barbata Franch. var. hirsutissima W. W. Sm.;硬毛无心菜(须花参,硬髯毛蚤缀)■
31770 Arenaria baxoiensis L. H. Zhou;八宿雪灵芝;Basu Sandwort ■
31771 Arenaria benthamii Edgew. = Arenaria debilis Hook. f. ■
31772 Arenaria benthamii Fenzl ex Torr. et A. Gray;山丘无心菜;Hilly Sandwort ■☆
31773 Arenaria blinkworthii Schltdl. = Arenaria debilis Hook. f. ■
31774 Arenaria bomiensis L. H. Zhou;波密无心菜;Bomi Sandwort ■
31775 Arenaria brevifolia Nutt. = Minuartia uniflora (Walter) Mattf. ■☆
31776 Arenaria brevifolia Nutt. var. californica A. Gray = Minuartia californica (A. Gray) Mattf. ■☆
31777 Arenaria brevipetala Y. W. Tsui et L. H. Zhou;雪灵芝(短瓣雪灵芝);Shortpetal Sandwort ■
31778 Arenaria bryophylla Fernald;藓状雪灵芝(苔藓状蚤缀);Bryoleaf Sandwort, Mooseform Sandwort ■
31779 Arenaria burkei Howell = Arenaria fendleri A. Gray var. subcongesta S. Watson ■☆
31780 Arenaria burkei Howell = Eremogone congesta (Nutt.) Ikonn. var. subcongesta (S. Watson) R. L. Hartm. et Rabeler ■☆
31781 Arenaria caespitosa (Cambess.) Kozhevn. = Thylacospermum caespitosum (Cambess.) Schischk. ■
31782 Arenaria caespitosa J. Vahl = Sagina caespitosa (J. Vahl) Lange ■☆
31783 Arenaria californica (A. Gray) Brewer = Minuartia californica (A. Gray) Mattf. ■☆
31784 Arenaria calycantha Ledeb. = Stellaria calycantha (Ledeb.) Bong. ■☆
31785 Arenaria calycina Poir. = Moenchia erecta (L.) P. Gaertn., B. Mey. et Scherb. ■☆
31786 Arenaria canadensis Pers. = Spergularia canadensis (Pers.) G. Don ■☆
31787 Arenaria capillaris Poir.;毛叶老牛筋(毛梗蚤缀,细毛蚤缀,线叶蚤缀,兴安鹅不食,蚤缀);Beautiful Sandwort, Hairleaf Sandwort, Slender Mountain Sandwort ■
31788 Arenaria capillaris Poir. = Arenaria formosa Fisch. ex Ser. ■
31789 Arenaria capillaris Poir. = Arenaria haitzeshanensis Y. W. Tsui ex L. H. Zhou ■

31790 Arenaria capillaris Poir. = Eremogone capillaris (Poir.) Fenzl ■☆

31791 Arenaria capillaris Poir. subsp. americana Maguire;美洲毛叶无心菜;Fescue Sandwort ■☆

31792 Arenaria capillaris Poir. subsp. americana Maguire = Eremogone capillaris (Poir.) Fenzl var. americana (Maguire) R. L. Hartm. et Rabeler ■☆

31793 Arenaria capillaris Poir. var. glabra Fenzl = Arenaria capillaris Poir. ■

31794 Arenaria capillaris Poir. var. glabrata (Ser.) Schischk. = Arenaria capillaris Poir. ■

31795 Arenaria capillaris Poir. var. glandulifera (Ser.) I. Schischk. et Knorring;腺毛叶老牛筋(腺毛鹅不食,腺毛蚤缀)■

31796 Arenaria capillaris Poir. var. nardifolia (Ledeb.) Regel = Arenaria capillaris Poir. ■

31797 Arenaria capillaris Poir. var. nardifolia (Ledeb.) Regel = Eremogone capillaris (Poir.) Fenzl ■☆

31798 Arenaria capitata Lam. = Arenaria aggregata (L.) Loisel. ■☆

31799 Arenaria capitata Lam. subsp. mauritanica (Batt.) Jahand. et Maire = Arenaria aggregata (L.) Loisel. subsp. mauritanica (Batt.) Maire ■☆

31800 Arenaria capitata Lam. var. austro-oranensis Maire = Arenaria aggregata (L.) Loisel. ■☆

31801 Arenaria caroliniana Walter; 卡罗来纳蚤缀; Pine-barren Sandwort ■☆

31802 Arenaria caroliniana Walter = Minuartia californica (A. Gray) Mattf. ■☆

31803 Arenaria cephalotes M. Bieb.;头状蚤缀■☆

31804 Arenaria cerastioides Poir.;角无心菜■☆

31805 Arenaria cerastioides Poir. subsp. saxigena (Humbert et Maire) Maire = Arenaria saxigena (Humbert et Maire) Dobignard ■☆

31806 Arenaria cerastioides Poir. var. macrosperma (Batt.) Maire = Arenaria cerastioides Poir. ■☆

31807 Arenaria cerastioides Poir. var. oranensis (Batt.) Maire = Arenaria cerastioides Poir. ■☆

31808 Arenaria cerastioides Poir. var. parviflora (Maire) Maire = Arenaria cerastioides Poir. ■☆

31809 Arenaria cerastioides Poir. var. prostrata Pau = Arenaria cerastioides Poir. ■☆

31810 Arenaria cerastioides Poir. var. saxigena Humbert et Maire = Arenaria saxigena (Humbert et Maire) Dobignard ■☆

31811 Arenaria cerastioides Poir. var. spathulata (Desf.) Maire = Arenaria cerastioides Poir. ■☆

31812 Arenaria cerastioides Poir. var. uliginosa Maire = Arenaria cerastioides Poir. ■☆

31813 Arenaria chamdoensis C. Y. Wu ex L. H. Zhou;昌都无心菜;Changdu Sandwort ■

31814 Arenaria chamissonis Maguire = Stellaria dicranoides (Cham. et Schltdl.) Fenzl ■☆

31815 Arenaria cherleriae Fisch. ex Ser. = Stellaria cherleriae (Fisch. ex Ser.) F. N. Williams ■

31816 Arenaria ciliata L.;爱尔兰无心菜;Fringed Sandwort, Hairy Sandwort, Irish Sandwort ■☆

31817 Arenaria ciliata L. subsp. pseudofrigida Ostenf. et Dahl = Arenaria pseudofrigida (Ostenf. et Dahl) Schischk. et Knorring ■☆

31818 Arenaria ciliata L. var. pseudofrigida (Ostenf. et Dahl) B. Boivin = Arenaria pseudofrigida (Ostenf. et Dahl) Schischk. et Knorring ■☆

31819 Arenaria ciliolata Edgew. et Hook. f.;缘毛无心菜;Ciliolate Sandwort ■

31820 Arenaria compacta Coville = Arenaria fendleri A. Gray var. glabrescens S. Watson ■☆

31821 Arenaria compacta Coville = Eremogone kingii (S. Watson) Ikonn. var. glabrescens (S. Watson) Dorn ■☆

31822 Arenaria compressa McNeill;扁翅无心菜;Flatweb Sandwort ■

31823 Arenaria confusa Rydb. = Arenaria lanuginosa (Michx.) Rohrb. var. saxosa (A. Gray) Zarucchi ■☆

31824 Arenaria congesta Nutt.;球头无心菜;Ballhead Sandwort ■☆

31825 Arenaria congesta Nutt. = Eremogone congesta (Nutt.) Ikonn. ■☆

31826 Arenaria congesta Nutt. var. charlestonensis Maguire;查尔斯顿无心菜;Charleston Sandwort ■☆

31827 Arenaria congesta Nutt. var. charlestonensis Maguire = Eremogone congesta (Nutt.) Ikonn. var. charlestonensis (Maguire) R. L. Hartm. et Rabeler ■☆

31828 Arenaria congesta Nutt. var. glandulifera Maguire;腺点无心菜■☆

31829 Arenaria congesta Nutt. var. glandulifera Maguire = Arenaria congesta Nutt. var. prolifera Maguire ■☆

31830 Arenaria congesta Nutt. var. parishiorum (B. L. Rob.) B. L. Rob. = Arenaria macradenia S. Watson ■☆

31831 Arenaria congesta Nutt. var. parishiorum (B. L. Rob.) B. L. Rob. = Eremogone macradenia (S. Watson) Ikonn. ■☆

31832 Arenaria congesta Nutt. var. prolifera Maguire;多育无心菜■☆

31833 Arenaria congesta Nutt. var. prolifera Maguire = Eremogone congesta (Nutt.) Ikonn. var. prolifera (Maguire) R. L. Hartm. et Rabeler ■☆

31834 Arenaria congesta Nutt. var. simulans Maguire;相似无心菜■☆

31835 Arenaria congesta Nutt. var. simulans Maguire = Eremogone congesta (Nutt.) Ikonn. var. simulans (Maguire) R. L. Hartm. et Rabeler ■☆

31836 Arenaria congesta Nutt. var. subcongesta (S. Watson) S. Watson = Arenaria fendleri A. Gray var. subcongesta S. Watson ■☆

31837 Arenaria congesta Nutt. var. subcongesta (S. Watson) S. Watson = Eremogone congesta (Nutt.) Ikonn. var. subcongesta (S. Watson) R. L. Hartm. et Rabeler ■☆

31838 Arenaria congesta Nutt. var. suffrutescens (A. Gray) B. L. Rob. = Arenaria suffrutescens (A. Gray) A. Heller ■☆

31839 Arenaria congesta Nutt. var. suffrutescens (A. Gray) B. L. Rob. = Eremogone congesta (Nutt.) Ikonn. var. suffrutescens (A. Gray) R. L. Hartm. et Rabeler ■☆

31840 Arenaria congesta Nutt. var. wheelerensis Maguire = Arenaria congesta Nutt. var. simulans Maguire ■☆

31841 Arenaria congesta Nutt. var. wheelerensis Maguire = Eremogone congesta (Nutt.) Ikonn. var. simulans (Maguire) R. L. Hartm. et Rabeler ■☆

31842 Arenaria cucubaloides Sm.;狗筋蔓无心菜■☆

31843 Arenaria cumberlandensis Wofford et Kràl = Minuartia cumberlandensis (Wofford et Kràl) McNeill ■☆

31844 Arenaria dahurica Fisch. ex Ser. = Arenaria juncea M. Bieb. ■

31845 Arenaria dawsonensis Britton = Arenaria stricta Michx. subsp. dawsonensis (Britton) Maguire ■☆

31846 Arenaria dawsonensis Britton = Minuartia dawsonensis (Britton) House ■☆

31847 Arenaria dawuensis A. J. Li;道孚无心菜;Dawu Sandwort ■

31848 Arenaria debilis Hook. f.;柔软无心菜(比伯史坦氏蚤缀,毕伯蚤缀);Bieberstein Sandwort, Bieberstein's Mouse-ear Chickweed, Soft Sandwort ■

31849 Arenaria delavayi Franch.;大理无心菜(川西无心菜,戴氏蚤缀);Delavay Sandwort ■

31850 Arenaria densissima Wall. ex Edgew. et Hook. f.;密生福禄草(密生雪灵芝);Dense Sandwort ■

31851 Arenaria diandra Guss. = Spergularia diandra (Guss.) Heldr. ■

31852 Arenaria dianthoides Sm.;石竹无心菜■☆

31853 Arenaria dicranoides (Cham. et Schltdl.) Hultén = Stellaria dicranoides (Cham. et Schltdl.) Fenzl ■☆

31854 Arenaria diffusa (Hornem.) Wormskjöld = Honckenya peploides (L.) Ehrh. subsp. diffusa (Hornem.) Hultén ex V. V. Petrovsky ■☆

31855 Arenaria diffusa Elliott;松散无心菜■☆

31856 Arenaria dimorphitricha C. Y. Wu ex L. H. Zhou;滇蜀无心菜;Dian-Shu Sandwort ■

31857 Arenaria douglasii Fenzl ex Torr. et A. Gray = Minuartia douglasii (Fenzl ex Torr. et A. Gray) Mattf. ■☆

31858 Arenaria drummondii Shinners = Minuartia drummondii (Shinners) McNeill ■☆

31859 Arenaria dsharaensis Pax et K. Hoffm.;察龙无心菜;Chalong Sandwort ■

31860 Arenaria dyris Humbert;荒地无心菜■☆

31861 Arenaria eastwoodiae Rydb.;伊斯伍德无心菜;Eastwood's Sandwort ■☆

31862 Arenaria eastwoodiae Rydb. = Eremogone eastwoodiae (Rydb.) Ikonn. ■☆

31863 Arenaria eastwoodiae Rydb. var. adenophora Kearney et Peebles;腺梗伊斯伍德无心菜■☆

31864 Arenaria eastwoodiae Rydb. var. adenophora Kearney et Peebles = Eremogone eastwoodiae (Rydb.) Ikonn. var. adenophora (Kearney et Peebles) R. L. Hartm. et Rabeler ■☆

31865 Arenaria edgeworthiana Majumdar;山居雪灵芝;Edgeworth Sandwort ■

31866 Arenaria elegans Cham. et Schltdl. = Minuartia elegans (Cham. et Schltdl.) Schischk. ■☆

31867 Arenaria emarginata Brot.;微缺无心菜■☆

31868 Arenaria emarginata Brot. subsp. salzmannii (Willk.) Maire;萨尔无心菜■☆

31869 Arenaria emarginata Brot. var. macrosperma Maire = Arenaria emarginata Brot. ■☆

31870 Arenaria emarginata Brot. var. salzmannii Willk. = Arenaria emarginata Brot. subsp. salzmannii (Willk.) Maire ■☆

31871 Arenaria euodonta W. W. Sm.;真齿无心菜(真齿蚤缀);Realtooth Sandwort ■

31872 Arenaria euodonta W. W. Sm. = Arenaria roseiflora Sprague ■

31873 Arenaria fallax Batt. = Arenaria hispanica Spreng. ■☆

31874 Arenaria farganica Schischk.;费尔干无心菜■☆

31875 Arenaria fendleri A. Gray;芬德勒无心菜;Fendler's Sandwort ■☆

31876 Arenaria fendleri A. Gray = Eremogone fendleri (A. Gray) Ikonn. ■☆

31877 Arenaria fendleri A. Gray subsp. brevifolia Maguire = Arenaria fendleri A. Gray ■☆

31878 Arenaria fendleri A. Gray subsp. brevifolia Maguire = Eremogone fendleri (A. Gray) Ikonn. ■☆

31879 Arenaria fendleri A. Gray var. aculeata (S. Watson) S. L. Welsh = Arenaria aculeata S. Watson ■☆

31880 Arenaria fendleri A. Gray var. aculeata (S. Watson) S. L. Welsh = Eremogone aculeata (S. Watson) Ikonn. ■☆

31881 Arenaria fendleri A. Gray var. brevifolia (Maguire) Maguire = Arenaria fendleri A. Gray ■☆

31882 Arenaria fendleri A. Gray var. brevifolia (Maguire) Maguire = Eremogone fendleri (A. Gray) Ikonn. ■☆

31883 Arenaria fendleri A. Gray var. diffusa Porter et Coult. = Arenaria fendleri A. Gray ■☆

31884 Arenaria fendleri A. Gray var. diffusa Porter et Coult. = Eremogone fendleri (A. Gray) Ikonn. ■☆

31885 Arenaria fendleri A. Gray var. eastwoodiae (Rydb.) S. L. Welsh = Arenaria eastwoodiae Rydb. ■☆

31886 Arenaria fendleri A. Gray var. eastwoodiae (Rydb.) S. L. Welsh = Eremogone eastwoodiae (Rydb.) Ikonn. ■☆

31887 Arenaria fendleri A. Gray var. glabrescens S. Watson;变光金氏无心菜■☆

31888 Arenaria fendleri A. Gray var. glabrescens S. Watson = Eremogone kingii (S. Watson) Ikonn. var. glabrescens (S. Watson) Dorn ■☆

31889 Arenaria fendleri A. Gray var. porteri Rydb. = Arenaria fendleri A. Gray ■☆

31890 Arenaria fendleri A. Gray var. porteri Rydb. = Eremogone fendleri (A. Gray) Ikonn. ■☆

31891 Arenaria fendleri A. Gray var. subcongesta S. Watson;亚球无心菜■☆

31892 Arenaria fendleri A. Gray var. subcongesta S. Watson = Eremogone congesta (Nutt.) Ikonn. var. subcongesta (S. Watson) R. L. Hartm. et Rabeler ■☆

31893 Arenaria fendleri A. Gray var. tweedyi (Rydb.) Maguire = Arenaria fendleri A. Gray ■☆

31894 Arenaria fendleri A. Gray var. tweedyi (Rydb.) Maguire = Eremogone fendleri (A. Gray) Ikonn. ■☆

31895 Arenaria festucoides Benth.;狐茅状雪灵芝(狐茅蚤缀);Festuelike Sandwort ■

31896 Arenaria festucoides Benth. var. imbricata Edgew. et Hook. f.;小狐茅状雪灵芝;Imbricate Sandwort ■

31897 Arenaria filifolia Forssk. = Minuartia filifolia (Forssk.) Mattf. ■☆

31898 Arenaria filiorum Maguire = Minuartia macrantha (Rydb.) House ■☆

31899 Arenaria filipes C. Y. Wu ex L. H. Zhou;细柄无心菜;Thinstalk Sandwort ■

31900 Arenaria fimbriata (E. Pritz.) Mattf.;缝瓣无心菜(缝瓣蚤缀);Fimbriate Sandwort ■

31901 Arenaria foliacea Turrill;多叶无心菜■☆

31902 Arenaria fontinalis (Short et R. Peter) Shinners = Stellaria fontinalis (Short et R. Peter) B. L. Rob. ■☆

31903 Arenaria formosa Fisch. ex Ser.;美丽老牛筋(美丽蚤缀);Beautiful Sandwort ■

31904 Arenaria formosa Fisch. ex Ser. = Arenaria capillaris Poir. ■

31905 Arenaria formosa Fisch. ex Ser. = Eremogone capillaris (Poir.) Fenzl ■☆

31906 Arenaria formosa Fisch. ex Ser. var. angustipetala Maxim. = Arenaria grueningiana Pax et K. Hoffm. ■

31907 Arenaria formosa Fisch. ex Ser. var. latipetala Maxim. = Arenaria formosa Fisch. ex Ser. ■

31908 Arenaria forrestii Diels;西南无心菜(福氏蚤缀,硬尖叶蚤缀);Forrest Sandwort ■

31909 Arenaria forrestii Diels f. cernua (Williams) C. Y. Wu;垂花无心菜;Nutantflower Forrest Sandwort ■

31910 Arenaria forrestii Diels f. cernua (Williams) C. Y. Wu = Arenaria forrestii Diels ■

31911 Arenaria forrestii Diels f. micrantha (Williams) C. Y. Wu;小花无心菜;Smallflower Forrest Sandwort ■

31912 Arenaria forrestii Diels f. micrantha (Williams) C. Y. Wu = Arenaria forrestii Diels ■

31913 Arenaria forrestii Diels f. roseotincta (W. W. Sm.) C. Y. Wu;粉晕无心菜(粉晕蚤缀,玫瑰无心菜);Rose Forrest Sandwort ■

31914 Arenaria forrestii Diels f. roseotincta (W. W. Sm.) C. Y. Wu = Arenaria forrestii Diels ■

31915 Arenaria franklinii Douglas ex Hook.;富兰克林无心菜;Franklin's Sandwort ■☆

31916 Arenaria franklinii Douglas ex Hook. = Eremogone franklinii (Douglas ex Hook.) R. L. Hartm. et Rabeler ■☆

31917 Arenaria franklinii Douglas ex Hook. var. thompsonii M. Peck;汤普森无心菜;Thompson's Sandwort ■☆

31918 Arenaria franklinii Douglas ex Hook. var. thompsonii M. Peck = Eremogone franklinii (Douglas ex Hook.) R. L. Hartm. et Rabeler var. thompsonii (M. Peck) R. L. Hartm. et Rabeler ■☆

31919 Arenaria fridericae Hand.-Mazz.;玉龙山无心菜(玉龙山蚤缀);Yulongshan Sandwort ■

31920 Arenaria galliformis C. Y. Wu;轮叶无心菜;Whorlleaf Sandwort ■

31921 Arenaria geniculata Poir. = Rhodalsine geniculata (Poir.) F. N. Williams ■☆

31922 Arenaria gerzensis L. H. Zhou;改则雪灵芝;Gaize Sandwort ■

31923 Arenaria giraldii (Diels) Mattf.;秦岭无心菜(秦岭蚤缀);Girald Sandwort ■

31924 Arenaria glabra Michx. = Minuartia glabra (Michx.) Mattf. ■☆

31925 Arenaria glanduligera Edgew. ex Edgew. et Hook.;小腺无心菜(腺毛蚤缀);Glandular Sandwort ■

31926 Arenaria glanduligera F. N. Williams var. cernua F. N. Williams = Arenaria forrestii Diels ■

31927 Arenaria glanduligera F. N. Williams var. micrantha F. N. Williams = Arenaria forrestii Diels ■

31928 Arenaria glandulosa (Benth. ex G. Don) F. N. Williams = Arenaria debilis Hook. f. ■

31929 Arenaria godfreyi Shinners = Minuartia godfreyi (Shinners) McNeill ■☆

31930 Arenaria gorgonea J. A. Schmidt = Arenaria leptoclados (Rchb.) Guss. ■

31931 Arenaria gothica Fr.;约克无心菜;Fries' Sandwort, Yorkshire Sandwort ■☆

31932 Arenaria graminea C. A. Mey.;禾叶蚤缀■☆

31933 Arenaria graminifolia Schrad.;禾叶鹅不食;Grass-leaf Sandwort, Grass-leaved Sandwort ■☆

31934 Arenaria grandiflora L.;大花蚤缀;Largeflowered Sandwort, Showy Sandwort ■☆

31935 Arenaria gregaria A. Heller = Minuartia nuttallii (Pax) Briq. var. gregaria (A. Heller) Rabeler et R. L. Hartm. ■☆

31936 Arenaria griffithii Boiss.;裸茎老牛筋(裸茎雪灵芝);Griffith Sandwort ■

31937 Arenaria groenlandica (Retz.) Spreng. = Minuartia groenlandica (Retz.) Ostenf. ■☆

31938 Arenaria groenlandica (Retz.) Spreng. var. glabra (Michx.) Fernald = Minuartia glabra (Michx.) Mattf. ■☆

31939 Arenaria grueningiana Pax et K. Hoffm.;华北老牛筋(高原福禄草);N. China Sandwort ■

31940 Arenaria gypsophiloides L.;喜钙无心菜■☆

31941 Arenaria haitzeshanensis Y. W. Tsui ex L. H. Zhou;海子山老牛筋(狐茅状雪灵芝);Haizishan Sandwort ■

31942 Arenaria herniariifolia Desf. = Rhodalsine geniculata (Poir.) F. N. Williams ■☆

31943 Arenaria heterosperma Guss. = Spergularia salina J. Presl et C. Presl ■☆

31944 Arenaria hispanica Spreng.;西班牙无心菜■☆

31945 Arenaria holosteoides (C. A. Mey.) Edgew. = Lepyrodiclis holosteoides (C. A. Mey.) Fenzl ex Fisch. et C. A. Mey. ■

31946 Arenaria holosteoides (C. A. Mey.) Edgew. var. stellarioides (Schrenk ex Fisch. et C. A. Mey.) F. N. Williams = Lepyrodiclis stellarioides Schrenk ex Fisch. et C. A. Mey. ■

31947 Arenaria holosteoides (C. A. Mey.) Edgew. var. stellarioides Williams = Lepyrodiclis stellarioides Fisch. et C. A. Mey. ■

31948 Arenaria hookeri Nutt.;胡克无心菜;Hooker's Sandwort ■☆

31949 Arenaria hookeri Nutt. = Eremogone hookeri (Nutt.) W. A. Weber ■☆

31950 Arenaria hookeri Nutt. subsp. desertorum (Maguire) W. A. Weber = Arenaria hookeri Nutt. ■☆

31951 Arenaria hookeri Nutt. subsp. desertorum (Maguire) W. A. Weber = Eremogone hookeri (Nutt.) W. A. Weber ■☆

31952 Arenaria hookeri Nutt. subsp. pinetorum (A. Nelson) Maguire = Arenaria pinetorum A. Nelson ■☆

31953 Arenaria hookeri Nutt. subsp. pinetorum (A. Nelson) Maguire = Eremogone hookeri (Nutt.) W. A. Weber var. pinetorum (A. Nelson) Dorn ■☆

31954 Arenaria hookeri Nutt. var. desertorum Maguire = Arenaria hookeri Nutt. ■☆

31955 Arenaria hookeri Nutt. var. desertorum Maguire = Eremogone hookeri (Nutt.) W. A. Weber ■☆

31956 Arenaria howellii S. Watson = Minuartia howellii (S. Watson) Mattf. ■☆

31957 Arenaria humifusa (Sw.) Wahlenb. et Nordhagen;匍匐蚤缀;Creeping Sandwort ■☆

31958 Arenaria hybrida Vill. = Minuartia hybrida (Vill.) Schischk. ■☆

31959 Arenaria inconspicua Hand.-Mazz.;不显无心菜(不显蚤缀);Inconspicuous Sandwort ■

31960 Arenaria inornata W. W. Sm.;无饰无心菜(无饰蚤缀);Inornate Sandwort ■

31961 Arenaria insignis Litv.;显著无心菜■☆

31962 Arenaria iochanensis C. Y. Wu;药山无心菜;Yaoshan Sandwort ■

31963 Arenaria ionaodra Diels;紫蕊无心菜(紫蕊蚤缀);Purpleanther Sandwort ■

31964 Arenaria ionaodra Diels var. melanotricha Comber;黑毛无心菜(具毛紫蕊蚤缀);Hairy Purpleanther Sandwort ■

31965 Arenaria ischnophylla F. N. Williams;瘦叶雪灵芝(瘦叶蚤缀);Thinleaf Sandwort ■

31966 Arenaria jamesiana (Torr.) Shinners = Pseudostellaria jamesiana (Torr.) W. A. Weber et R. L. Hartm. ■☆

31967 Arenaria juncea M. Bieb.;灯心草蚤缀(老牛筋,毛轴鹅不食,毛轴蚤缀,山银柴胡,追风箭);Junc-like Sandwort, Rush Sandwort ■

31968 Arenaria juncea M. Bieb. var. abbreviata Kitag.;短灯心草蚤缀(矮茎鹅不食草,小无心菜);Small Rush Sandwort ■

31969 Arenaria juncea M. Bieb. var. glabra Regel;光轴蚤缀(光轴鹅不食,无毛老牛筋);Glabrous Rush Sandwort ■

31970 Arenaria juniperina Vill. = Arenaria grandiflora L. ■☆

31971 Arenaria kansuensis Maxim.;甘肃雪灵芝(甘草蚤缀,甘肃蚤缀,雪灵芝);Gansu Sandwort, Kansu Sandwort ■

31972 Arenaria kansuensis Maxim. var. acropetala Y. W. Tsui et L. H. Zhou = Arenaria kansuensis Maxim. ■

31973 Arenaria kansuensis Maxim. var. ovatipetala Y. W. Tsui et L. H. Zhou;卵瓣雪灵芝(卵瓣蚤缀);Ovatipetal Sandwort ■

31974 Arenaria kansuensis Maxim. var. oxypetala Y. W. Tsui et L. H. Zhou;尖瓣雪灵芝;Sharppetal Gansu Sandwort ■

31975 Arenaria karakorensis Em. Schmid; 克拉克无心菜; Karakor Sandwort ■

31976 Arenaria kashmirica Edgew. = Minuartia kashmirica (Edgew.) Mattf. ■

31977 Arenaria katoana Makino;加藤蚤缀■☆

31978 Arenaria katoana Makino var. lanceolata Tatew.;细叶加藤蚤缀■☆

31979 Arenaria kingii (S. Watson) M. E. Jones;金氏无心菜;King's Sandwort ■☆

31980 Arenaria kingii (S. Watson) M. E. Jones = Eremogone kingii (S. Watson) Ikonn. ■☆

31981 Arenaria kingii S. Watson subsp. compacta (Coville) Maguire = Arenaria fendleri A. Gray var. glabrescens S. Watson ■☆

31982 Arenaria kingii S. Watson subsp. compacta (Coville) Maguire = Eremogone kingii (S. Watson) Ikonn. var. glabrescens (S. Watson) Dorn ■☆

31983 Arenaria kingii S. Watson subsp. plateauensis Maguire = Arenaria fendleri A. Gray var. glabrescens S. Watson ■☆

31984 Arenaria kingii S. Watson subsp. plateauensis Maguire = Eremogone kingii (S. Watson) Ikonn. var. glabrescens (S. Watson) Dorn ■☆

31985 Arenaria kingii S. Watson var. glabrescens (S. Watson) Maguire = Arenaria fendleri A. Gray var. glabrescens S. Watson ■☆

31986 Arenaria kingii S. Watson var. glabrescens (S. Watson) Maguire = Eremogone kingii (S. Watson) Ikonn. var. glabrescens (S. Watson) Dorn ■☆

31987 Arenaria kingii S. Watson var. plateauensis (Maguire) Reveal = Arenaria fendleri A. Gray var. glabrescens S. Watson ■☆

31988 Arenaria kingii S. Watson var. plateauensis (Maguire) Reveal = Eremogone kingii (S. Watson) Ikonn. var. glabrescens (S. Watson) Dorn ■☆

31989 Arenaria kingii S. Watson var. uintahensis (A. Nelson) C. L. Hitchc. = Arenaria fendleri A. Gray var. glabrescens S. Watson ■☆

31990 Arenaria kingii S. Watson var. uintahensis (A. Nelson) C. L. Hitchc. = Eremogone kingii (S. Watson) Ikonn. var. glabrescens (S. Watson) Dorn ■☆

31991 Arenaria koriniana Fisch. ex Ledeb.;朝鲜蚤缀;Korean Sandwort ■☆

31992 Arenaria kumaonensis Maxim.;库莽雪灵芝;Kumang Sandwort ■

31993 Arenaria kumaonensis Maxim. = Arenaria kansuensis Maxim. ■

31994 Arenaria kuschei Eastw. = Arenaria macradenia S. Watson var. arcuifolia Maguire ■☆

31995 Arenaria kuschei Eastw. = Eremogone macradenia (S. Watson) Ikonn. var. arcuifolia (Maguire) R. L. Hartm. et Rabeler ■☆

31996 Arenaria lancangensis L. H. Zhou; 澜沧雪灵芝; Lancang Sandwort ■

31997 Arenaria lanceolatifolia L. H. Zhou;披针叶无心菜;Lanceoleaf Sandwort ■

31998 Arenaria lanceolatifolia L. H. Zhou = Arenaria compressa McNeill ■

31999 Arenaria lanuginosa (Michx.) Rohrb.; 绵毛蚤缀; Seabeach Sandwort, Spreading Sandwort ■☆

32000 Arenaria lanuginosa (Michx.) Rohrb. subsp. saxosa (A. Gray) Maguire = Arenaria lanuginosa (Michx.) Rohrb. var. saxosa (A. Gray) Zarucchi ■☆

32001 Arenaria lanuginosa (Michx.) Rohrb. var. longipedunculata W. H. Duncan = Arenaria lanuginosa (Michx.) Rohrb. ■☆

32002 Arenaria lanuginosa (Michx.) Rohrb. var. saxosa (A. Gray) Zarucchi;岩石绵毛蚤缀■☆

32003 Arenaria lapanshanensis Y. W. Tsui; 万年松; Lapanshan Sandwort ■

32004 Arenaria laricifolia L.;落叶松叶蚤缀;Grove Sandwort, Larchleaf Sandwort, Wood Sandwort ■☆

32005 Arenaria laricifolia L. var. marcescens (Fernald) B. Boivin = Minuartia marcescens (Fernald) House ■☆

32006 Arenaria lateriflora L. = Moehringia lateriflora (L.) Fenzl ■

32007 Arenaria lateriflora L. var. angustifolia (Regel) H. St. John = Arenaria lateriflora L. ■

32008 Arenaria lateriflora L. var. angustifolia (Regel) H. St. John = Moehringia lateriflora (L.) Fenzl ■

32009 Arenaria lateriflora L. var. angustifolia H. St. John = Moehringia lateriflora (L.) Fenzl ■

32010 Arenaria lateriflora L. var. taylorae H. St. John = Moehringia lateriflora (L.) Fenzl ■

32011 Arenaria lateriflora L. var. tayloriae H. St. John = Arenaria lateriflora L. ■

32012 Arenaria lateriflora L. var. tayloriae H. St. John = Moehringia lateriflora (L.) Fenzl ■

32013 Arenaria lateriflora L. var. tenuicaulis Blank. = Moehringia lateriflora (L.) Fenzl ■

32014 Arenaria ledebouriana Fenzl;赖氏无心菜■☆

32015 Arenaria leptoclados (Rchb.) Guss.;细枝无心菜(细茎鹅不食);Thinstem Sandwort ■

32016 Arenaria leptoclados (Rchb.) Guss. = Arenaria serpyllifolia L. var. tenuior Mert. et W. D. J. Koch ■

32017 Arenaria leptoclados (Rchb.) Guss. = Arenaria serpyllifolia L. ■

32018 Arenaria leptoclados (Rchb.) Guss. var. lindbergii Sennen et Mauricio = Arenaria serpyllifolia L. subsp. leptoclados (Rchb.) Nyman ■

32019 Arenaria leptophylla C. Y. Wu ex L. H. Zhou = Arenaria pseudostellaria C. Y. Wu ■

32020 Arenaria leptophylla Cham. et Schltdl. = Drymaria leptophylla (Cham. et Schltdl.) Fenzl ex Rohrb. ■☆

32021 Arenaria leucasteria Mattf.;毛萼无心菜;Haircalyx Sandwort ■

32022 Arenaria lichiangensis W. W. Sm. = Arenaria oreophila Hook. f. ex Edgew. et Hook. f. ■

32023 Arenaria linearifolia Franch. = Arenaria pseudostellaria C. Y. Wu ■

32024 Arenaria lithophila (Rydb.) Rydb. = Arenaria subcongesta Nutt. var. lithophila Rydb. ■☆

32025 Arenaria lithophila (Rydb.) Rydb. = Eremogone congesta (Nutt.) Ikonn. var. lithophila (Rydb.) Dorn ■☆

32026 Arenaria litorea Fernald = Arenaria stricta Michx. subsp. dawsonensis (Britton) Maguire ■☆

32027 Arenaria litorea Fernald = Minuartia dawsonensis (Britton) House ■☆

32028 Arenaria littledalei Hemsl.;古临无心菜;Gulin Sandwort ■

32029 Arenaria livermorensis Correll; 利夫莫尔无心菜; Livermore Sandwort ■☆

32030 Arenaria longicaulis C. Y. Wu ex L. H. Zhou;长茎无心菜(长茎蚤缀);Longstem Sandwort ■

32031 Arenaria longifolia M. Bieb. ;长叶蚤缀(长花蚤缀);Longflower Sandwort ■☆
32032 Arenaria longipedunculata Hultén;长花梗无心菜;Longstem Sandwort ■☆
32033 Arenaria longipes C. Y. Wu ex L. H. Zhou;长梗无心菜;Longstalk Sandwort ■
32034 Arenaria longipetiolata C. Y. Wu ex L. H. Zhou;长叶柄无心菜;Longpetiole Sandwort ■
32035 Arenaria longiseta C. Y. Wu;长刚毛无心菜;Longsetose Sandwort ■
32036 Arenaria longistyla Franch. ;长柱无心菜(长柱蚤缀);Longstyle Sandwort ■
32037 Arenaria longistyla Franch. var. eugonophylla Fernald;棱长柱无心菜■
32038 Arenaria longistyla Franch. var. pleurogynoides Diels;侧长柱无心菜■
32039 Arenaria lychnidea M. Bieb. ;剪秋罗无心菜■☆
32040 Arenaria macra A. Nelson et J. F. Macbr. = Arenaria stricta Michx. ■☆
32041 Arenaria macra A. Nelson et J. F. Macbr. = Minuartia tenella (J. Gay) Mattf. ■☆
32042 Arenaria macradenia S. Watson;莫哈维无心菜;Mohave Sandwort ■☆
32043 Arenaria macradenia S. Watson = Eremogone macradenia (S. Watson) Ikonn. ■☆
32044 Arenaria macradenia S. Watson subsp. ferrisiae Abrams;费氏莫哈维无心菜;Ferris' Sandwort ■☆
32045 Arenaria macradenia S. Watson subsp. ferrisiae Abrams = Eremogone ferrisiae (Abrams) R. L. Hartm. et Rabeler ■☆
32046 Arenaria macradenia S. Watson var. arcuifolia Maguire;弯叶无心菜■☆
32047 Arenaria macradenia S. Watson var. arcuifolia Maguire = Eremogone macradenia (S. Watson) Ikonn. var. arcuifolia (Maguire) R. L. Hartm. et Rabeler ■☆
32048 Arenaria macradenia S. Watson var. kuschei (Eastw.) Maguire = Arenaria macradenia S. Watson var. arcuifolia Maguire ■☆
32049 Arenaria macradenia S. Watson var. kuschei (Eastw.) Maguire = Eremogone macradenia (S. Watson) Ikonn. var. arcuifolia (Maguire) R. L. Hartm. et Rabeler ■☆
32050 Arenaria macradenia S. Watson var. parishiorum B. L. Rob. = Arenaria macradenia S. Watson ■☆
32051 Arenaria macradenia S. Watson var. parishiorum B. L. Rob. = Eremogone macradenia (S. Watson) Ikonn. ■☆
32052 Arenaria macrantha (Rydb.) A. Nelson = Minuartia macrantha (Rydb.) House ■☆
32053 Arenaria macrantha Schischk. ;大花无心菜■☆
32054 Arenaria macrocarpa Pursh;大果无心菜■
32055 Arenaria macrocarpa Pursh = Minuartia macrocarpa (Pursh) Ostenf. ■
32056 Arenaria macrocarpa Pursh var. jooi (Makino) H. Hara;约氏无心菜■☆
32057 Arenaria macrocarpa Pursh var. yezoalpina (H. Hara) H. Hara;北海道大果无心菜■☆
32058 Arenaria macrophylla Hook. ;大叶蚤缀;Large-leaved Sandwort ■☆
32059 Arenaria macrophylla Hook. = Moehringia macrophylla (Hook.) Fenzl ■☆
32060 Arenaria mairei Emb. = Arenaria iochanensis C. Y. Wu ■
32061 Arenaria mairei Emb. ex Jahand. et Maire = Arenaria iochanensis C. Y. Wu ■
32062 Arenaria marcescens Fernald = Minuartia marcescens (Fernald) House ■☆
32063 Arenaria marginata DC. = Spergularia media (L.) C. Presl ex Griseb. ■
32064 Arenaria mearnsii Wooton et Standl. = Arenaria lanuginosa (Michx.) Rohrb. var. saxosa (A. Gray) Zarucchi ■☆
32065 Arenaria media L. = Spergularia media (L.) C. Presl ex Griseb. ■
32066 Arenaria melanandra (Maxim.) Mattf. ex Hand. -Mazz. ;黑蕊无心菜(大板山蚤缀,黑蕊蚤缀);Blackanther Sandwort ■
32067 Arenaria melandryiformis F. N. Williams;女娄无心菜;Melandrium Sandwort ■
32068 Arenaria melandryoides Edgew. ex Edgew. et Hook. f. ;桃色无心菜;Pinkcolor Sandwort ■
32069 Arenaria membranisepala C. Y. Wu;膜萼无心菜;Filmcalyx Sandwort ■
32070 Arenaria merckioides Maxim. ;蚤缀无心菜■☆
32071 Arenaria meyeri Fenzl ex Ledeb. ;高山蚤缀(高山老牛筋,麦氏蚤缀)■
32072 Arenaria michauxii (Fenzl) Hook. f. = Arenaria stricta Michx. ■☆
32073 Arenaria micradenia C. C. Davis;草甸蚤缀■☆
32074 Arenaria microstella C. Y. Wu ex L. H. Zhou;小星无心菜;Stellulate Sandwort ■
32075 Arenaria minima C. Y. Wu ex L. H. Zhou;微无心菜(小无心菜);Mini Sandwort ■
32076 Arenaria minutiflora Loscos = Arenaria serpyllifolia L. subsp. leptoclados (Rchb.) Nyman ■
32077 Arenaria modesta L. Dufour;适度无心菜■☆
32078 Arenaria modesta L. Dufour subsp. africana (Pau) Dobignard;非洲适度无心菜■☆
32079 Arenaria modesta L. Dufour var. africana Pau = Arenaria modesta L. Dufour subsp. africana (Pau) Dobignard ■☆
32080 Arenaria mollugínea Ser. = Drymaria mollugínea (Ser.) Didr. ■☆
32081 Arenaria monantha F. N. Williams;山地无心菜(山地雪灵芝,山蚤缀);Montane Sandwort,Mountain Sandwort ■
32082 Arenaria mongolica Schischk. ;蒙古无心菜■☆
32083 Arenaria monilifera Mattf. ;念珠无心菜;Beads Sandwort ■
32084 Arenaria monosperma F. N. Williams;单子无心菜(单子蚤缀);Monoseed Sandwort ■
32085 Arenaria montana L. ;山蚤缀;Alpine Sandwort, Mountain Sandwort ■☆
32086 Arenaria montana L. var. eglandulosa Maire = Arenaria montana L. ■☆
32087 Arenaria montana L. var. glandulosa Maire = Arenaria montana L. ■☆
32088 Arenaria monticola Edgew. = Arenaria edgeworthiana Majumdar ■
32089 Arenaria muliensis C. Y. Wu ex L. H. Zhou;木里无心菜;Muli Sandwort ■
32090 Arenaria muliensis C. Y. Wu ex L. H. Zhou = Arenaria spathulifolia C. Y. Wu ex L. H. Zhou ■
32091 Arenaria muriculata Maguire = Arenaria muscorum (Fassett) Shinners ■☆
32092 Arenaria muriculata Maguire = Minuartia muscorum (Fassett) Rabeler ■☆
32093 Arenaria musciformis Wall. = Arenaria bryophylla Fernald ■
32094 Arenaria musciformis Wall. ex Edgew. et Hook. f. = Arenaria

bryophylla Fernald ■
32095 Arenaria muscorum (Fassett) Shinners;藓地无心菜;Sandwort ■☆
32096 Arenaria napuligera Franch.;滇藏无心菜(小块根无心菜);Dian-Zang Sandwort ■
32097 Arenaria napuligera Franch. var. monocephela W. W. Sm.;单头无心菜;Monocephal Sandwort ■
32098 Arenaria neelgherrensis Wight et Arn.;尼盖无心菜;Neelgerr Sandwort ■
32099 Arenaria nepalensis Spreng. = Brachystemma calycinum D. Don ■
32100 Arenaria nigricans Hand. -Mazz.;变黑无心菜;Nigrescent Sandwort ■
32101 Arenaria nigricans Hand. -Mazz. var. zhengkangensis (C. Y. Wu ex L. H. Zhou) C. Y. Wu;镇康无心菜;Zhenkang Sandwort ■
32102 Arenaria nivalomontana C. Y. Wu ex L. H. Zhou;大雪山无心菜(里瓦弄无心菜);Big Jokul Sandwort ■
32103 Arenaria norvegica Gunn.;挪威无心菜(北极无心菜);Arctic Sandwort, Norwegian Sandwort, Scottish Sandwort ■☆
32104 Arenaria norvegica Gunn. var. anglica ?;英国无心菜;English Sandwort ■☆
32105 Arenaria nuttallii Pax = Minuartia nuttallii (Pax) Briq. ■☆
32106 Arenaria nuttallii Pax subsp. fragilis Maguire et A. H. Holmgren = Minuartia nuttallii (Pax) Briq. var. fragilis (Maguire et A. H. Holmgren) Rabeler et R. L. Hartm. ■☆
32107 Arenaria nuttallii Pax subsp. gracilis (B. L. Rob.) Maguire = Minuartia nuttallii (Pax) Briq. var. gracilis (B. L. Rob.) Rabeler et R. L. Hartm. ■☆
32108 Arenaria nuttallii Pax subsp. gregaria (A. Heller) Maguire = Minuartia nuttallii (Pax) Briq. var. gregaria (A. Heller) Rabeler et R. L. Hartm. ■☆
32109 Arenaria nuttallii Pax var. fragilis (Maguire et A. H. Holmgren) C. L. Hitchc. = Minuartia nuttallii (Pax) Briq. var. fragilis (Maguire et A. H. Holmgren) Rabeler et R. L. Hartm. ■☆
32110 Arenaria nuttallii Pax var. gracilis B. L. Rob. = Minuartia nuttallii (Pax) Briq. var. gracilis (B. L. Rob.) Rabeler et R. L. Hartm. ■☆
32111 Arenaria nuttallii Pax var. gregaria (A. Heller) Jeps. = Minuartia nuttallii (Pax) Briq. var. gregaria (A. Heller) Rabeler et R. L. Hartm. ■☆
32112 Arenaria obtusa Torr. = Minuartia obtusiloba (Rydb.) House ■☆
32113 Arenaria obtusiloba (Rydb.) Fernald = Minuartia obtusiloba (Rydb.) House ■☆
32114 Arenaria omeiensis C. Y. Wu ex L. H. Zhou;峨眉无心菜;Emei Sandwort ■
32115 Arenaria orbiculata Royle ex Edgew. et Hook. f.;圆叶无心菜(圆叶蚤缀);Roundleaf Sandwort ■
32116 Arenaria oreophila Hook. f. ex Edgew. et Hook. f.;山生福禄草(丽江雪灵芝,丽江蚤缀);Mountane Sandwort ■
32117 Arenaria oresbia W. W. Sm. = Arenaria smithiana Mattf. ■
32118 Arenaria paludicola B. L. Rob.;湿地无心菜;Marsh Sandwort ■☆
32119 Arenaria patula F. Michx. media Steyerm. = Arenaria muscorum (Fassett) Shinners ■☆
32120 Arenaria patula Michx.;纤细蚤缀;Sandwort, Slender Sandwort ■☆
32121 Arenaria patula Michx. = Minuartia patula (Michx.) Mattf. ■☆
32122 Arenaria patula Michx. var. robusta (Steyerm.) Maguire = Arenaria muscorum (Fassett) Shinners ■☆
32123 Arenaria patula Michx. var. robusta (Steyerm.) Maguire = Minuartia muscorum (Fassett) Rabeler ■☆
32124 Arenaria paulsenii H. Winkl.;保尔森无心菜■☆
32125 Arenaria pentandra Maxim. = Arenaria potaninii Schischk. ■
32126 Arenaria peploides L.;荸艾蚤缀;Seabeach Sandwort ■☆
32127 Arenaria peploides L. = Honckenya peploides (L.) Ehrh. ■☆
32128 Arenaria peploides L. subsp. major (Hook.) Calder et R. L. Taylor = Honckenya peploides (L.) Ehrh. subsp. major (Hook.) Hultén ■☆
32129 Arenaria peploides L. var. diffusa Hornem. = Honckenya peploides (L.) Ehrh. subsp. diffusa (Hornem.) Hultén ex V. V. Petrovsky ■☆
32130 Arenaria peploides L. var. major (Hook.) Abrams = Honckenya peploides (L.) Ehrh. subsp. major (Hook.) Hultén ■☆
32131 Arenaria peploides L. var. major Hook. = Honckenya peploides (L.) Ehrh. subsp. major (Hook.) Hultén ■☆
32132 Arenaria peploides L. var. maxima Fernald = Honckenya peploides (L.) Ehrh. subsp. major (Hook.) Hultén ■☆
32133 Arenaria peploides L. var. oblongifolia (Torr. et A. Gray) S. Watson = Honckenya peploides (L.) Ehrh. subsp. major (Hook.) Hultén ■☆
32134 Arenaria peploides L. var. robusta Fernald = Honckenya peploides (L.) Ehrh. subsp. robusta (Fernald) Hultén ■☆
32135 Arenaria perlevis (F. N. Williams) Hand. -Mazz. = Arenaria pulvinata Edgew. ■
32136 Arenaria petiolata Hayata;台湾蚤缀(具柄无心菜);Taiwan Sandwort ■
32137 Arenaria petiolata Hayata = Arenaria serpyllifolia L. ■
32138 Arenaria pharensis McNeill et Majumdar;帕里无心菜;Phar Sandwort ■
32139 Arenaria physodes Fisch. ex Ser. = Wilhelmsia physodes (Fisch. ex Ser.) McNeill ■☆
32140 Arenaria pinetorum A. Nelson;松林无心菜■☆
32141 Arenaria pinetorum A. Nelson = Eremogone hookeri (Nutt.) W. A. Weber var. pinetorum (A. Nelson) Dorn ■☆
32142 Arenaria pogonantha W. W. Sm.;须花无心菜;Pogonflower Sandwort ■
32143 Arenaria polaris Schischk.;极地无心菜(极地蚤缀)■☆
32144 Arenaria polysperma C. Y. Wu ex L. H. Zhou;多子无心菜;Seedy Sandwort ■
32145 Arenaria polytrichoides Edgew. ex Edgew. et Hook.;团状福禄草(金法藓蚤缀,马紫无心菜,团状雪灵芝);Group Sandwort ■
32146 Arenaria polytrichoides Edgew. ex Edgew. et Hook. var. perlevis F. N. Williams = Arenaria pulvinata Edgew. ■
32147 Arenaria pomelii Munby;波梅尔无心菜■☆
32148 Arenaria potaninii Schischk.;五蕊老牛筋;Fivepistil Sandwort ■
32149 Arenaria procumbens Vahl = Rhodalsine geniculata (Poir.) F. N. Williams ■☆
32150 Arenaria propinqua Richardson = Minuartia rubella (Wahlenb.) Hiern ■☆
32151 Arenaria przewalskii Maxim.;福禄草(高原蚤缀,西北蚤缀);Przewalsk Sandwort ■
32152 Arenaria pseudofrigida (Ostenf. et Dahl) Juz. = Arenaria pseudofrigida (Ostenf. et Dahl) Schischk. et Knorring ■☆
32153 Arenaria pseudofrigida (Ostenf. et Dahl) Schischk. et Knorring;拟霜蚤缀■☆
32154 Arenaria pseudostellaria C. Y. Wu;线叶无心菜(薄叶无心菜);Linearleaf Sandwort ■
32155 Arenaria pulvinata Edgew.;垫状雪灵芝;Pulvinate Sandwort ■
32156 Arenaria pumicola Coville et Leiberg;湖无心菜;Crater Lake

Sandwort ■☆
32157 Arenaria pumicola Coville et Leiberg = Eremogone pumicola (Coville et Leiberg) Ikonn. ■☆
32158 Arenaria pumicola Coville et Leiberg var. californica Maguire = Arenaria aculeata S. Watson ■☆
32159 Arenaria pumicola Coville et Leiberg var. californica Maguire = Eremogone aculeata (S. Watson) Ikonn. ■☆
32160 Arenaria pungens Lag.;刺无心菜■☆
32161 Arenaria pungens Lag. subsp. boissieri Emb.;布瓦西耶刺无心菜■☆
32162 Arenaria pungens Lag. subsp. parviflora Quézel = Arenaria pungens Lag. subsp. boissieri Emb. ■☆
32163 Arenaria pungens Lag. var. eriosepala Maire = Arenaria pungens Lag. ■☆
32164 Arenaria pungens Lag. var. glabrescens Ball = Arenaria pungens Lag. ■☆
32165 Arenaria pungens Lag. var. leiosepala Maire = Arenaria pungens Lag. ■☆
32166 Arenaria pungens Lag. var. microsperma Maire = Arenaria pungens Lag. ■☆
32167 Arenaria pungens Nutt. = Minuartia nuttallii (Pax) Briq. ■☆
32168 Arenaria puranensis L. H. Zhou;普兰无心菜;Pulan Sandwort, Puran Sandwort ■
32169 Arenaria purpurascens Ramond;紫花蚤缀;Purple-flowered Sandwort ■☆
32170 Arenaria pusilla S. Watson = Minuartia pusilla (S. Watson) Mattf. ■☆
32171 Arenaria pusilla S. Watson var. diffusa Maguire = Minuartia californica (A. Gray) Mattf. ■☆
32172 Arenaria qinghaiensis Y. W. Tsui et L. H. Zhou;青海雪灵芝;Qinhai Sandwort ■
32173 Arenaria quadridentata (Maxim.) F. N. Williams;四齿无心菜(四齿蚤缀);Four-lobe Sandwort, Fourtooth Sandwort ■
32174 Arenaria ramellata F. N. Williams;嫩枝无心菜(嫩枝蚤缀);Tendertwig Sandwort ■
32175 Arenaria redowskii Cham. et Schltdl.;雷德无心菜■☆
32176 Arenaria reducta Hand.-Mazz.;缩减无心菜(退化蚤缀);Reduced Sandwort ■
32177 Arenaria rhodantha Pax et K. Hoffm.;红花无心菜(红花蚤缀);Redflower Sandwort ■
32178 Arenaria rigida M. Bieb.;硬蚤缀■☆
32179 Arenaria roborowskii Maxim.;青藏雪灵芝(洛氏蚤缀);Roborowsk ■
32180 Arenaria rockii Diels;紫红无心菜(洛克氏蚤缀);Rock Sandwort ■
32181 Arenaria roseiflora Sprague;粉花无心菜(粉花蚤缀);Rose Sandwort ■
32182 Arenaria roseiflora Sprague f. albiflora C. Y. Wu;白粉花无心菜;Whiteflower Rose Sandwort ■
32183 Arenaria roseiflora Sprague f. labiflora C. Y. Wu = Arenaria roseiflora Sprague ■
32184 Arenaria roseii Maguire et Barneby = Minuartia rossii (R. Br. ex Richardson) Graebn. ■☆
32185 Arenaria roseoticta W. W. Sm. = Arenaria forrestii Diels ■
32186 Arenaria rossii R. Br. ex Richardson = Minuartia rossii (R. Br. ex Richardson) Graebn. ■☆
32187 Arenaria rossii R. Br. ex Richardson subsp. columbiana (Raup) Maguire = Minuartia elegans (Cham. et Schltdl.) Schischk. ■☆
32188 Arenaria rossii R. Br. ex Richardson subsp. elegans (Cham. et Schltdl.) Maguire = Minuartia elegans (Cham. et Schltdl.) Schischk. ■☆
32189 Arenaria rossii R. Br. ex Richardson var. apetala Maguire = Minuartia rossii (R. Br. ex Richardson) Graebn. ■☆
32190 Arenaria rossii R. Br. ex Richardson var. columbiana Raup = Minuartia elegans (Cham. et Schltdl.) Schischk. ■☆
32191 Arenaria rossii R. Br. ex Richardson var. elegans (Cham. et Schltdl.) S. L. Welsh = Minuartia elegans (Cham. et Schltdl.) Schischk. ■☆
32192 Arenaria rotundifolia M. Bieb.;欧洲圆叶无心菜■☆
32193 Arenaria rubella (Wahlenb.) Sm.;红蚤缀;Vernal Sandwort ■☆
32194 Arenaria rubella (Wahlenb.) Sm. = Minuartia rubella (Wahlenb.) Hiern ■☆
32195 Arenaria rubella (Wahlenb.) Sm. var. filiorum (Maguire) S. L. Welsh = Minuartia macrantha (Rydb.) House ■☆
32196 Arenaria rubra L. = Spergularia rubra (L.) J. Presl et C. Presl ■
32197 Arenaria rubra L. var. campestris L. = Spergularia rubra (L.) J. Presl et C. Presl ■
32198 Arenaria rubra L. var. marina L. = Spergularia marina (L.) Griseb. ■
32199 Arenaria rupifraga (Kar. et Kir.) Fenzl = Thylacospermum caespitosum (Cambess.) Schischk. ■
32200 Arenaria rupifraga (Kar. et Kir.) Fenzl ex Ledeb. = Thylacospermum caespitosum (Cambess.) Schischk. ■
32201 Arenaria saginoides Maxim.;漆姑无心菜(漆姑草蚤缀);Pearlweedlike Sandwort ■
32202 Arenaria sajanensis Willd. = Minuartia biflora (L.) Schinz et Thell. ■
32203 Arenaria salweenensis W. W. Sm.;怒江无心菜(怒江蚤缀);Nujiang Sandwort ■
32204 Arenaria saxigena (Humbert et Maire) Dobignard;岩生无心菜■☆
32205 Arenaria saxosa A. Gray = Arenaria lanuginosa (Michx.) Rohrb. var. saxosa (A. Gray) Zarucchi ■☆
32206 Arenaria saxosa A. Gray var. cinerascens B. L. Rob. = Arenaria lanuginosa (Michx.) Rohrb. var. saxosa (A. Gray) Zarucchi ■☆
32207 Arenaria saxosa A. Gray var. mearnsii (Wooton et Standl.) Kearney et Peebles = Arenaria lanuginosa (Michx.) Rohrb. var. saxosa (A. Gray) Zarucchi ■☆
32208 Arenaria schimperi (A. Rich.) Oliv. = Minuartia filifolia (Forssk.) Mattf. ■☆
32209 Arenaria schneideriana Hand.-Mazz.;雪山无心菜(雪山蚤缀);Jokul Sandwort ■
32210 Arenaria segetalis Lam.;谷地无心菜■☆
32211 Arenaria sericea Ser. = Gypsophila sericea (Ser.) Krylov ■
32212 Arenaria serpyllifolia L.;无心菜(百里香叶蚤缀,大叶米栖草,地胡椒,鹅不食,鹅不食草,鸡肠子草,铃铃草,卵叶蚤缀,雀儿蛋,小无心菜,蚤缀);Sandwort, Thymeleaf Sandwort, Thyme-leaved Chickweed, Thyme-leaved Sandwort ■
32213 Arenaria serpyllifolia L. subsp. leptoclados (Rchb.) Nyman = Arenaria leptoclados (Rchb.) Guss. ■
32214 Arenaria serpyllifolia L. subsp. leptoclados (Rchb.) Nyman = Arenaria serpyllifolia L. ■
32215 Arenaria serpyllifolia L. subsp. leptoclados (Rchb.) Nyman = Arenaria serpyllifolia L. var. tenuior Mert. et W. D. J. Koch ■
32216 Arenaria serpyllifolia L. subsp. minutiflora H. Lindb. = Arenaria serpyllifolia L. subsp. leptoclados (Rchb.) Nyman ■

32217 Arenaria serpyllifolia L. var. leptoclados Rchb. = Arenaria leptoclados (Rchb.) Guss. ■
32218 Arenaria serpyllifolia L. var. minutiflora Loscos = Arenaria serpyllifolia L. subsp. leptoclados (Rchb.) Nyman ■
32219 Arenaria serpyllifolia L. var. scabra Fenzl = Arenaria serpyllifolia L. ■
32220 Arenaria serpyllifolia L. var. tenuior Mert. et W. D. J. Koch;小无心菜■
32221 Arenaria serpyllifolia L. var. tenuior Mert. et W. D. J. Koch = Arenaria serpyllifolia L. ■
32222 Arenaria serpyllifolia L. var. viscida (R. J. Loisel) DC.;黏无心菜■☆
32223 Arenaria serpyllifolia L. var. viscidula Rouy et Foucaud = Arenaria serpyllifolia L. subsp. leptoclados (Rchb.) Nyman ■
32224 Arenaria setacea Thuill. var. atlantica Ball = Minuartia tenuissima (Pomel) Mattf. ■☆
32225 Arenaria setifera C. Y. Wu ex L. H. Zhou;刚毛无心菜;Setose Sandwort ■
32226 Arenaria shannanensis L. H. Zhou;粉花雪灵芝;Pinkflower Sandwort ■
32227 Arenaria shennongjiaensis Z. E. Zhao et Z. H. Shen;神农架无心菜■
32228 Arenaria sikkimensis Majumdar = Arenaria debilis Hook. f. ■
32229 Arenaria smithiana Mattf.;大花福禄草(云南雪灵芝);Bigflower Sandwort ■
32230 Arenaria spathulata Desf. = Arenaria cerastioides Poir. ■☆
32231 Arenaria spathulata Desf. var. macrosperma Batt. = Arenaria cerastioides Poir. ■☆
32232 Arenaria spathulata Desf. var. oranensis Batt. = Arenaria cerastioides Poir. ■☆
32233 Arenaria spathulata Desf. var. parviflora Pomel = Arenaria cerastioides Poir. ■☆
32234 Arenaria spathulifolia C. Y. Wu ex L. H. Zhou;匙叶无心菜;Spponleaf Sandwort ■
32235 Arenaria stellarioides C. Y. Wu ex L. H. Zhou;繁缕状无心菜;Stellaria-leaf Sandwort ■
32236 Arenaria stellarioides C. Y. Wu ex L. H. Zhou = Arenaria debilis Hook. f. ■
32237 Arenaria stenomeres Eastw.;牧场无心菜;Meadow Valley Sandwort ■☆
32238 Arenaria stenomeres Eastw. = Eremogone stenomeres (Eastw.) Ikonn. ■☆
32239 Arenaria stephaniana (Willd. ex Schltdl.) Shinners var. americana (Porter ex B. L. Rob.) Shinners = Stellaria americana (Porter ex B. L. Rob.) Standl. ■☆
32240 Arenaria steveniana Boiss.;斯梯文无心菜■☆
32241 Arenaria stracheyi Edgew.;藏西无心菜;W. Xizang Sandwort ■
32242 Arenaria stricta Michx.;岩生蚤缀;Rock Sandwort, Stiff Sandwort ■☆
32243 Arenaria stricta Michx. = Minuartia michauxii (Fenzl) Farw. ■☆
32244 Arenaria stricta Michx. = Minuartia tenella (J. Gay) Mattf. ■☆
32245 Arenaria stricta Michx. subsp. dawsonensis (Britton) Maguire;北方岩生蚤缀;Northern Rock Sandwort, Rock Stitchwort ■☆
32246 Arenaria stricta Michx. subsp. macra (A. Nelson et J. F. Macbr.) Maguire = Arenaria stricta Michx. ■☆
32247 Arenaria stricta Michx. subsp. macra (A. Nelson et J. F. Macbr.) Maguire = Minuartia tenella (J. Gay) Mattf. ■☆
32248 Arenaria stricta Michx. subsp. texana (B. L. Rob.) Maguire = Minuartia michauxii (Fenzl) Farw. ■☆
32249 Arenaria stricta Michx. var. dawsonensis (Britton) Scoggan = Arenaria stricta Michx. subsp. dawsonensis (Britton) Maguire ■☆
32250 Arenaria stricta Michx. var. dawsonensis (Britton) Scoggan = Minuartia dawsonensis (Britton) House ■☆
32251 Arenaria stricta Michx. var. litorea (Fernald) B. Boivin = Arenaria stricta Michx. subsp. dawsonensis (Britton) Maguire ■☆
32252 Arenaria stricta Michx. var. litorea (Fernald) B. Boivin = Minuartia dawsonensis (Britton) House ■☆
32253 Arenaria stricta Michx. var. puberulenta (M. Peck) C. L. Hitchc. = Arenaria stricta Michx. ■☆
32254 Arenaria stricta Michx. var. puberulenta (M. Peck) C. L. Hitchc. = Minuartia tenella (J. Gay) Mattf. ■☆
32255 Arenaria stricta Michx. var. texana (Britton) B. L. Rob. = Minuartia michauxii (Fenzl) Farw. ■☆
32256 Arenaria stricta Michx. var. uliginosa (Schleich. ex Lam. et DC.) B. Boivin = Minuartia stricta (Sw.) Hiern ■☆
32257 Arenaria stricta Michx. var. uliginosa (Schleich. ex Lam. et DC.) B. Boivin = Arenaria stricta Michx. ■☆
32258 Arenaria subcongesta (S. Watson) Rydb. = Arenaria fendleri A. Gray var. subcongesta S. Watson ■☆
32259 Arenaria subcongesta (S. Watson) Rydb. = Eremogone congesta (Nutt.) Ikonn. var. subcongesta (S. Watson) R. L. Hartm. et Rabeler ■☆
32260 Arenaria subcongesta Nutt. var. lithophila Rydb.;疏松无心菜;Loosehead Sandwort ■☆
32261 Arenaria subcongesta Nutt. var. lithophila Rydb. = Eremogone congesta (Nutt.) Ikonn. var. lithophila (Rydb.) Dorn ■☆
32262 Arenaria subpilosa (Hayata) Ohwi = Cerastium subpilosum Hayata ■
32263 Arenaria subulata (Hayata) Ohwi var. glabrata Ser. = Arenaria capillaris Poir. ■
32264 Arenaria suffrutescens (A. Gray) A. Heller;半灌木状无心菜;Suffrutescent Sandwort ■☆
32265 Arenaria suffrutescens (A. Gray) A. Heller = Eremogone congesta (Nutt.) Ikonn. var. suffrutescens (A. Gray) R. L. Hartm. et Rabeler ■☆
32266 Arenaria syriistschikowii C. C. Davis;赛氏蚤缀■☆
32267 Arenaria szechuanensis Williams;四川无心菜;Sichuan Sandwort ■
32268 Arenaria szowitsii Boiss.;绍氏无心菜■☆
32269 Arenaria taibaishanensis L. H. Zhou;太白雪灵芝(万年松);Taibaishan Sandwort ■
32270 Arenaria takasagomontana (Masam.) S. S. Ying;高山无心菜■
32271 Arenaria takasagomontana (Masam.) S. S. Ying = Cerastium takasagomontanum Masam ■
32272 Arenaria tapanshanensis Y. W. Tsui;大板山蚤缀■
32273 Arenaria tatewakii Nakai = Arenaria katoana Makino var. lanceolata Tatew. ■☆
32274 Arenaria tenella Nutt. = Minuartia tenella (J. Gay) Mattf. ■☆
32275 Arenaria tenuifolia (L.) DC. = Minuartia hybrida (Vill.) Schischk. ■☆
32276 Arenaria tenuifolia (L.) DC. var. glandulosa Ball = Minuartia hybrida (Vill.) Schischk. ■☆
32277 Arenaria tetraquatra L.;方茎蚤缀;Squarestemmed Sandwort ■☆
32278 Arenaria tetraquetra L. subsp. mauritanica Batt. = Arenaria

aggregata (L.) Loisel. subsp. mauritanica (Batt.) Maire ■☆
32279 Arenaria texana Britton = Minuartia michauxii (Fenzl) Farw. ■☆
32280 Arenaria tonsa Kitag. = Arenaria juncea M. Bieb. var. glabra Regel ■
32281 Arenaria trichophora Franch.;具毛无心菜(具毛蚤缀);Hairy Sandwort ■
32282 Arenaria trichophora Franch. var. angustifolia Franch. = Arenaria yulongshanensis L. H. Zhou ■
32283 Arenaria trichophylla C. Y. Wu ex L. H. Zhou;毛叶无心菜;Hairleaf Sandwort ■
32284 Arenaria trichotoma Royle ex Edgew. et Hook. f. = Arenaria compressa McNeill ■
32285 Arenaria trigyna (Vill.) Shinners = Cerastium cerastoides (L.) Britton ■
32286 Arenaria trinervia (L.) DC. = Moehringia trinervia (L.) Clairv. ■
32287 Arenaria trinervia L. = Moehringia trinervia (L.) Clairv. ■
32288 Arenaria tschuktschorum Regel;邱氏无心菜■☆
32289 Arenaria tumengelaensis L. H. Zhou;土门无心菜;Tumengela Sandwort ■
32290 Arenaria turkestanica Schischk.;土耳其斯坦无心菜■☆
32291 Arenaria tweedyi Rydb. = Arenaria fendleri A. Gray ■☆
32292 Arenaria tweedyi Rydb. = Eremogone fendleri (A. Gray) Ikonn. ■☆
32293 Arenaria uintahensis A. Nelson = Arenaria fendleri A. Gray var. glabrescens S. Watson ■☆
32294 Arenaria uintahensis A. Nelson = Eremogone kingii (S. Watson) Ikonn. var. glabrescens (S. Watson) Dorn ■☆
32295 Arenaria uliginosa Schleich. ex Lam. et DC. = Arenaria stricta Michx. ■☆
32296 Arenaria uliginosa Schleich. ex Lam. et DC. = Minuartia stricta (Sw.) Hiern ■☆
32297 Arenaria umbrosa Bunge = Moehringia umbrosa (Bunge) Fenzl ■
32298 Arenaria ursina B. L. Rob.;熊无心菜;Bear Valley Sandwort ■☆
32299 Arenaria ursina B. L. Rob. = Eremogone ursina (B. L. Rob.) Ikonn. ■☆
32300 Arenaria velutina Pax et K. Hoffm. = Stellaria infracta Maxim. ■
32301 Arenaria verna L.;钻形漆姑草(钻叶漆姑草);Awl-leaf Pearlwort, Awl-leaved Pearlwort, Awloleaved Pearlwort, Corsican Pearlwort, Heath Pearlwort, Irish Moss, Scotch Moss, Tufted Sandwort ■
32302 Arenaria verna L. = Minuartia verna (L.) Hiern ■
32303 Arenaria verna L. = Sagina subulata C. Presl ■
32304 Arenaria verna L. var. brachypetala Ball = Minuartia verna (L.) Hiern ■
32305 Arenaria verna L. var. caespitosa (Ehrh.) Pers. = Arenaria verna L. ■
32306 Arenaria verna L. var. japonica (H. Hara) H. Hara;日本钻叶漆姑草■☆
32307 Arenaria verna L. var. propinqua (Richardson) Fernald = Minuartia rubella (Wahlenb.) Hiern ■☆
32308 Arenaria verna L. var. pubescens (Cham. et Schltdl.) Fernald = Minuartia rubella (Wahlenb.) Hiern ■☆
32309 Arenaria verna L. var. rubella (Wahlenb.) S. Watson = Minuartia rubella (Wahlenb.) Hiern ■☆
32310 Arenaria vestita Baker = Minuartia vestita (Baker) McNeill ■☆
32311 Arenaria villosa Ledeb. = Minuartia verna (L.) Hiern ■
32312 Arenaria weissiana Hand. -Mazz.;多柱无心菜(维西无心菜,中甸蚤缀);Weiss Sandwort ■
32313 Arenaria weissiana Hand. -Mazz. var. bifida C. Y. Wu et H. Chuang;裂瓣多柱无心菜(裂瓣无心菜)■
32314 Arenaria weissiana Hand. -Mazz. var. puberula C. Y. Wu ex L. H. Zhou;微毛无心菜(毛维西无心菜)■
32315 Arenaria xerophila W. W. Sm.;旱生无心菜(旱生蚤缀);Dryland Sandwort ■
32316 Arenaria xerophila W. W. Sm. var. xiangchengensis (L. H. Zhou) C. Y. Wu;乡城无心菜■
32317 Arenaria xiangchengensis L. H. Zhou = Arenaria xerophila W. W. Sm. var. xiangchengensis (L. H. Zhou) C. Y. Wu ■
32318 Arenaria yulongshanensis L. H. Zhou;狭叶无心菜;Narrowleaf Sandwort ■
32319 Arenaria yunnanensis Franch.;云南无心菜(云南蚤缀);Yunnan Sandwort ■
32320 Arenaria yunnanensis Franch. f. angustifolia F. N. Williams = Arenaria debilis Hook. f. ■
32321 Arenaria yunnanensis Franch. f. robusta C. Y. Wu ex L. H. Zhou = Arenaria yunnanensis Franch. ■
32322 Arenaria yunnanensis Franch. var. caespitosa C. Y. Wu;簇生无心菜;Fascicullate Yunnan Sandwort ■
32323 Arenaria yunnanensis Franch. var. linearifolia C. Y. Wu ex L. H. Zhou;细叶云南无心菜;Linearleaf Yunnan Sandwort ■
32324 Arenaria yunnanensis Franch. var. linearifolia C. Y. Wu ex L. H. Zhou = Arenaria iochanensis C. Y. Wu ■
32325 Arenaria yunnanensis Franch. var. trichophora (Franch.) Williams = Arenaria trichophora Franch. ■
32326 Arenaria zadoiensis L. H. Zhou;杂多雪灵芝;Zaduo Sandwort ■
32327 Arenaria zhengkangensis C. Y. Wu ex L. H. Zhou = Arenaria nigricans Hand. -Mazz. var. zhengkangensis (C. Y. Wu ex L. H. Zhou) C. Y. Wu ■
32328 Arenaria zhongdianensis C. Y. Wu;中甸无心菜;Zhongdian Sandwort ■
32329 Arenbergia Mart. et Galeotti = Eustoma Salisb. ■☆
32330 Arenga Labill. (1800)(保留属名);桄榔属(桄榔子属,南椰属,砂糖椰子属,莎木属,山棕属,糖椰子属,羽棕属);Arenga, Sugar Palm, Sugarpalm ●
32331 Arenga ambong Becc.;安汶羽棕●☆
32332 Arenga australasica (H. Wendl. et Drude) S. T. Blake;澳洲桄榔●☆
32333 Arenga caudata (Lour.) H. E. Moore;双籽桄榔(大幅棕,山棕,双籽藤,双籽棕,双子棕,尾状羽棕,野棕);Caudate Sugarpalm, Caudate Two-seeded Palm, Doubleseed Sugarpalm ●
32334 Arenga caudata (Lour.) H. E. Moore = Didymosperma caudatum (Lour.) H. Wendl. et Drude ●
32335 Arenga caudata (Lour.) H. E. Moore var. stenodkylla ?;尖尾状羽棕●☆
32336 Arenga engleri Becc.;山棕(矮桄榔,桄榔子,散尾棕,椶);Engler Sugarpalm, Formosan Sugarpalm, Wild Sugarpalm ●
32337 Arenga hastata (Becc.) Whitmore;戟形桄榔;Hastate Sugarpalm ●☆
32338 Arenga hookeriana (Becc.) Whitmore;胡克桄榔(虎克桄榔,虎克棕);Hooker Sugarpalm ●☆
32339 Arenga longicarpa C. F. Wei;长果桄榔;Longfruit Sugarpalm ●
32340 Arenga micrantha C. F. Wei;小花桄榔;Littleleaf Sugarpalm, Smallflower Sugarpalm, Small-flowered Sugarpalm ●
32341 Arenga microsperma Becc.;细籽棕●☆
32342 Arenga obtusifolia Mart.;苏门答腊桄榔;Sumatra Sugarpalm ●☆

32343 Arenga pinnata (Wurmb) Merr.;桄榔(董棕,姑榔木,面木,南椰子,砂糖椰子,莎木,莎木面,山椰子,糖树,糖棕,铁木,樣木);Aren, Areng Palm, Blackfibre Palm, Ejow, Ejow Palm, Gomuti, Gomuti Palm, Gomuti Sugar Palm, Gomuti Sugarpalm, Gomuti Sugarpalm, Kabong, Sugar Palm, Sugarpalm ●

32344 Arenga porphyrocarpa (Blume) H. E. Moore;紫果桄榔●☆

32345 Arenga saccharifera Labill. = Arenga pinnata (Wurmb) Merr. ●

32346 Arenga tremula (Blanco) Becc.;鱼骨葵(山棕)●

32347 Arenga tremula (Blanco) Becc. var. engleri (Becc.) Hatus. ex Shimabuku = Arenga engleri Becc. ●

32348 Arenga undulatifolia Becc.;波叶桄榔●☆

32349 Arenga westerhoutii Griff.;维氏桄榔;Curry Sugarpalm ●☆

32350 Arenifera A. G. J. Herre(1948);紫沙玉属●☆

32351 Arenifera pillansii (L. Bolus) Herre;紫沙玉■☆

32352 Arenifera pungens H. E. K. Hartmann;刺紫沙玉■☆

32353 Arenifera spinescens (L. Bolus) H. E. K. Hartmann;小刺紫沙玉■☆

32354 Arenifera stylosa (L. Bolus) H. E. K. Hartmann;多柱紫沙玉■☆

32355 Arequipa Bntton et Rose = Borzicactus Riccob. ■☆

32356 Arequipa Britton et Rose = Oreocereus (A. Berger) Riccob. ●

32357 Arequipa Britton et Rose(1922);醉翁玉属(阿雷魁帕属)■☆

32358 Arequipa australis F. Ritter = Oreocereus hempelianus (Gürke) D. R. Hunt ●☆

32359 Arequipa erectocylindrica Rauh et Backeb. = Oreocereus hempelianus (Gürke) D. R. Hunt ●☆

32360 Arequipa hempeliana (Gürke) Oehme = Oreocereus hempelianus (Gürke) D. R. Hunt ●☆

32361 Arequipa leucotricha (Phil.) Britton et Rose;醉翁玉;Borzi Cactus ■☆

32362 Arequipa rettigii (Quehl) Oehme;醉眉玉■☆

32363 Arequipa rettigii (Quehl) Oehme = Oreocereus hempelianus (Gürke) D. R. Hunt ●☆

32364 Arequipa soehrensii Backeb. = Oreocereus hempelianus (Gürke) D. R. Hunt ●☆

32365 Arequipa spinosissima F. Ritter = Oreocereus hempelianus (Gürke) D. R. Hunt ●☆

32366 Arequipa weingartiana Backeb.;醉熊玉■☆

32367 Arequipa weingartiana Backeb. = Oreocereus hempelianus (Gürke) D. R. Hunt ●☆

32368 Arequipiopsis Kreuz. et Buining = Borzicactus Riccob. ■☆

32369 Arequipiopsis Kreuz. et Buining = Matucana Britton et Rose ●☆

32370 Arequipiopsis Kreuz. et Buining = Oreocereus (A. Berger) Riccob. ●

32371 Arethusa L. (1753);龙嘴兰属(北美湿地兰属,泽兰属);Arethusa, Dragon's-mouth ■☆

32372 Arethusa alaris (L. f.) Thunb. = Pterygodium catholicum (L.) Sw. ■☆

32373 Arethusa bulbosa L.;龙嘴兰(北美球茎湿地兰);Arethusa, Bogrose Orchid, Calopogon Grass Pink, Dragons Month, Dragon's Mouth, Dragon's-mouth, Moss Nymph, Swamp Pink, Swamp-pink ■☆

32374 Arethusa capensis L. f. = Disperis capensis (L. f.) Sw. ■☆

32375 Arethusa ciliaris L. f. = Bartholina burmanniana (L.) Ker Gawl. ■☆

32376 Arethusa crispa Thunb. = Corycium crispum (Thunb.) Sw. ■☆

32377 Arethusa divaricata L. = Cleistes divaricata (L.) Ames ■☆

32378 Arethusa ecristata Griff. = Chamaegastrodia vaginata (Hook. f.) Seidenf. ■

32379 Arethusa medeoloides Pursh = Isotria medeoloides (Pursh) Raf. ■☆

32380 Arethusa ophioglossoides L. = Pogonia ophioglossoides (L.) Ker Gawl. ■☆

32381 Arethusa petraea Afzel. ex Sw. = Nervilia petraea (Afzel. ex Schrad.) Summerh. ■☆

32382 Arethusa petraea Sw. ex Pers. = Nervilia petraea (Sw. ex Pers.) Summerh. ■☆

32383 Arethusa plicata Andréws = Nervilia plicata (Andréws) Schltr. ■

32384 Arethusa racemosa Walter = Ponthieva racemosa (Walter) C. Mohr ■☆

32385 Arethusa simplex Thouars = Nervilia simplex (Thouars) Schltr. ■☆

32386 Arethusa sinensis Rolfe = Bletilla sinensis (Rolfe) Schltr. ■

32387 Arethusa spicata Walter = Hexalectris spicata (Walter) Barnhart ■☆

32388 Arethusa trianthophoros Sw. = Triphora trianthophora (Sw.) Rydb. ■☆

32389 Arethusa verticillata Muhl. ex Willd. = Isotria verticillata (Muhl. ex Willd.) Raf. ■☆

32390 Arethusa villosa L. f. = Disperis villosa (L. f.) Sw. ■☆

32391 Arethusantha Finet = Cymbidium Sw. ■

32392 Aretia Haller = Androsace L. ■

32393 Aretia L. = Androsace L. ■

32394 Aretia Link = Auricula-ursi Ség. ■

32395 Aretia Link = Primula L. ■

32396 Aretiastrum (DC.) Spach = Valeriana L. ●■

32397 Aretiastrum Spach = Valeriana L. ●■

32398 Arfeuillea Pierre ex Radlk. (1895);阿福木属●☆

32399 Arfeuillea arborescens Pierre;阿福木●☆

32400 Argan Dryand. = Argania Roem. et Schult. (保留属名)●☆

32401 Argania Roem. et Schult. (1819)(保留属名);摩洛哥山榄属●☆

32402 Argania sideroxylon Roem. et Schult. = Argania spinosa (L.) Skeels ●☆

32403 Argania spinosa (L.) Skeels;具刺摩洛哥山榄;Argan Tree, Argan-tree ●☆

32404 Argania spinosa (L.) Skeels var. apiculata Maire = Argania spinosa (L.) Skeels ●☆

32405 Argania spinosa (L.) Skeels var. mutica Maire = Argania spinosa (L.) Skeels ●☆

32406 Argantoniella G. López et R. Morales = Satureja L. ●■

32407 Argantoniella salzmanii (P. W. Ball) G. López et R. Morales;萨尔摩洛哥山榄●☆

32408 Argelasia Fourr. = Genista L. ●

32409 Argelia Decne. = Solenostemma Hayne ●☆

32410 Argemisia eriocephala Pamp. = Artemisia roxburghiana Besser ■

32411 Argemone L. (1753);蓟罂粟属(刺罂粟属,蓟叶罂粟属);Argemony, Mexican Poppy, Pricklepoppy, Prickly Poppy, Prickly-poppy, Yellow Thistle ■

32412 Argemone Tourn. ex L. = Argemone L. ■

32413 Argemone aenea Ownbey;青铜蓟罂粟■☆

32414 Argemone alba T. Lestib. = Argemone albiflora Hornem. ■☆

32415 Argemone albiflora Hornem.;白花蓟罂粟;Blue-stem Prickly Poppy, Prickly Poppy, White Prickly Poppy, Whiteflower Pricklepoppy ■☆

32416 Argemone albiflora Hornem. subsp. texana G. B. Ownbey;得州白花蓟罂粟■☆

32417 Argemone albiflora Hornem. var. texana (G. B. Ownbey) Shinners = Argemone albiflora Hornem. subsp. texana G. B. Ownbey ■☆

32418 Argemone bipinnatifida Greene = Argemone hispida A. Gray ■☆
32419 Argemone corymbosa Greene;多刺蓟罂粟;Prickly Poppy ■☆
32420 Argemone corymbosa Greene subsp. arenicola G. B. Ownbey;沙丘多刺蓟罂粟■☆
32421 Argemone corymbosa Greene var. arenicola (G. B. Ownbey) Shinners = Argemone corymbosa Greene subsp. arenicola G. B. Ownbey ■☆
32422 Argemone gracilenta Greene;细黏蓟罂粟;Prickly Poppy ■☆
32423 Argemone grandiflora Sweet;大花蓟罂粟(大蓟罂粟);Argemony, Bigflower Pricklepoppy, Mexican Poppy, Mexican Prickly Poppy, Prickly Poppy ■☆
32424 Argemone hispida A. Gray;毛蓟罂粟;Hedgehog Prickly Poppy ■☆
32425 Argemone intermedia Sweet var. corymbosa (Greene) Eastw. = Argemone corymbosa Greene ■☆
32426 Argemone intermedia Sweet var. polyanthemos Fedde = Argemone polyanthemos (Fedde) G. B. Ownbey ■☆
32427 Argemone leiocarpa Greene = Argemone mexicana L. f. leiocarpa (Greene) G. B. Ownbey ■
32428 Argemone leiocarpa Greene = Argemone mexicana L. ■
32429 Argemone mexicana L.;蓟罂粟(刺罂粟,老鼠艻);Devil's Fig, Mexican Poppy, Mexican Prickly Poppy, Mexican Thistle, Prickly Poppy, Thornapple, Yellow Hollyhock, Yellow Poppy, Yellow Thistle ■
32430 Argemone mexicana L. f. lanata N. Robson;绵毛蓟罂粟■☆
32431 Argemone mexicana L. f. leiocarpa (Greene) G. B. Ownbey;光果蓟罂粟■
32432 Argemone mexicana L. var. ochroleuca (Sweet) Lindl. = Argemone mexicana L. ■
32433 Argemone mexicana L. var. ochroleuca (Sweet) Lindl. = Argemone ochroleuca Sweet ■☆
32434 Argemone mucronata Dum. Cours. ex Steud. = Argemone mexicana L. ■
32435 Argemone munita Durand et Hilg.;刺蓟罂粟■☆
32436 Argemone munita Durand et Hilg. subsp. robusta G. B. Ownbey;粗壮刺蓟罂粟■☆
32437 Argemone munita Durand et Hilg. subsp. rotundata (Rydb.) G. B. Ownbey;圆刺蓟罂粟■☆
32438 Argemone munita Durand et Hilg. var. robusta (G. B. Ownbey) Shinners = Argemone munita Durand et Hilg. subsp. robusta G. B. Ownbey ■☆
32439 Argemone munita Durand et Hilg. var. rotundata (Rydb.) Shinners = Argemone munita Durand et Hilg. subsp. rotundata (Rydb.) G. B. Ownbey ■☆
32440 Argemone ochroleuca Sweet;淡黄蓟罂粟■☆
32441 Argemone platyceras Link et Otto;阔果蓟罂粟(白花蓟罂粟);Rough Prickly Poppy ■☆
32442 Argemone platyceras Link et Otto var. hispida (A. Gray) Prain = Argemone hispida A. Gray ■☆
32443 Argemone platyceras Link et Otto var. rosea J. M. Coult. = Argemone sanguinea Greene ■☆
32444 Argemone pleiacantha Greene;西南蓟罂粟(多刺蓟罂粟);Southwestern Pricklypoppy ■☆
32445 Argemone pleiacantha Greene subsp. ambigua G. B. Ownbey;可疑蓟罂粟■☆
32446 Argemone pleiacantha Greene subsp. pinnatisecta G. B. Ownbey;羽状可疑蓟罂粟■☆
32447 Argemone pleiacantha Greene var. ambigua (G. B. Ownbey) Shinners = Argemone pleiacantha Greene subsp. ambigua G. B. Ownbey ■☆
32448 Argemone pleiacantha Greene var. pinnatisecta (G. B. Ownbey) Shinners = Argemone pleiacantha Greene subsp. pinnatisecta G. B. Ownbey ■☆
32449 Argemone polyanthemos (Fedde) G. B. Ownbey;多花蓟罂粟;Crested Prickly Poppy ■☆
32450 Argemone rotundata Rydb. = Argemone munita Durand et Hilg. subsp. rotundata (Rydb.) G. B. Ownbey ■☆
32451 Argemone sanguinea Greene;血红蓟罂粟■☆
32452 Argemone sexvalvis Stokes = Argemone mexicana L. ■
32453 Argemone squarrosa Greene;粗鳞蓟罂粟;Hedgehog Pricklypoppy ■☆
32454 Argemone subfusiformis G. B. Ownbey;近纺锤状蓟罂粟■☆
32455 Argemone subfusiformis G. B. Ownbey = Argemone ochroleuca Sweet ■☆
32456 Argenope Salisb. = Narcissus L. ■
32457 Argentacer Small = Acer L. ●
32458 Argentacer saccharinum (L.) Small = Acer saccharinum L. ●
32459 Argentia Lam. = Potentilla L. ●■
32460 Argentina Hill = Potentilla L. ●■
32461 Argentina Lam. = Potentilla L. ●■
32462 Argentina anserina (L.) Rydb. = Potentilla anserina L. ■
32463 Argentina anserina (L.) Rydb. var. concolor Rydb. = Argentina anserina (L.) Rydb. ■
32464 Argentina anserina (L.) Rydb. var. concolor Rydb. = Potentilla anserina L. ■
32465 Argentina argentea (L.) Rydb. = Argentina anserina (L.) Rydb. ■
32466 Argentina argentea (L.) Rydb. = Potentilla anserina L. ■
32467 Argentina pacifica (Howell) Rydb. = Potentilla anserina L. subsp. pacifica (Howell) Rousi ■☆
32468 Argentipallium Paul G. Wilson(1992);彩鼠麹属■☆
32469 Argentipallium dealbatum (Labill.) Paul G. Wilson;彩鼠麹■☆
32470 Argentipallium obtusifolium (F. Muell. et Sond. ex Sond.) Paul G. Wilson;钝叶彩鼠麹■☆
32471 Argeta N. E. Br. = Gibbaeum Haw. ex N. E. Br. ●☆
32472 Argeta petrense N. E. Br. = Gibbaeum petrense (N. E. Br.) Tischer ■☆
32473 Argillochloa W. A. Weber = Festuca L. ■
32474 Argithamnia Sw. = Argythamnia P. Browne ●☆
32475 Argocoffea (Pierre ex De Wild.) Lebrun = Coffea L. ●
32476 Argocoffea (Pierre ex De Wild.) Lebrun(1941);阿尔加咖啡属●☆
32477 Argocoffea Lebrun = Argocoffeopsis (Pierre ex De Wild.) Lebrun ●☆
32478 Argocoffea afzelii (Hiern) J. -F. Leroy = Argocoffeopsis afzelii (Hiern) Robbr. ●☆
32479 Argocoffea jasminoides (Welw. ex Hiern) Lebrun = Argocoffeopsis eketensis (Wernham) Robbr. ●☆
32480 Argocoffea lemblinii (A. Chev.) J. -F. Leroy = Argocoffeopsis lemblinii (A. Chev.) Robbr. ●☆
32481 Argocoffea nudiflora (Stapf) J. -F. Leroy = Argocoffeopsis rupestris (Hiern) Robbr. ●☆
32482 Argocoffea pulchella (K. Schum.) J. -F. Leroy = Argocoffeopsis pulchella (K. Schum.) Robbr. ●☆
32483 Argocoffea rupestris (Hiern) Lebrun = Argocoffeopsis rupestris (Hiern) Robbr. ●☆

32484 Argocoffea scandens (K. Schum.) J. -F. Leroy = Argocoffeopsis scandens (K. Schum.) Lebrun ●☆
32485 Argocoffea subcordata (Hiern) J. -F. Leroy = Argocoffeopsis subcordata (Hiern) Lebrun ●☆
32486 Argocoffeopsis (Pierre ex De Wild.) Lebrun = Coffea L. ●
32487 Argocoffeopsis (Pierre ex De Wild.) Lebrun(1941);拟阿尔加咖啡属●☆
32488 Argocoffeopsis Lebrun = Argocoffeopsis (Pierre ex De Wild.) Lebrun ●☆
32489 Argocoffeopsis Lebrun = Coffea L. ●
32490 Argocoffeopsis afzelii (Hiern) Robbr.;阿芙泽尔拟阿尔加咖啡●☆
32491 Argocoffeopsis eketensis (Wernham) Robbr.;茉莉拟阿尔加咖啡●☆
32492 Argocoffeopsis jasminoides (Welw. ex Hiern) Robbr. = Argocoffeopsis eketensis (Wernham) Robbr. ●☆
32493 Argocoffeopsis kivuensis Robbr.;基伍拟阿尔加咖啡●☆
32494 Argocoffeopsis lemblinii (A. Chev.) Robbr.;勒梅林拟阿尔加咖啡●☆
32495 Argocoffeopsis pulchella (K. Schum.) Robbr.;美丽拟阿尔加咖啡●☆
32496 Argocoffeopsis rupestris (Hiern) Robbr.;岩生拟阿尔加咖啡●☆
32497 Argocoffeopsis rupestris (Hiern) Robbr. subsp. thonneri (Lebrun) Robbr.;托内拟阿尔加咖啡●☆
32498 Argocoffeopsis scandens (K. Schum.) Lebrun;攀缘拟阿尔加咖啡●☆
32499 Argocoffeopsis subcordata (Hiern) Lebrun;亚心形拟阿尔加咖啡●☆
32500 Argocoffeopsis subcordata (Hiern) Lebrun var. claessensii (Lebrun) Lebrun = Argocoffeopsis subcordata (Hiern) Lebrun ●☆
32501 Argolasia Juss.(废弃属名) = Lanaria Aiton ■☆
32502 Argomuellera Pax(1894);白雪叶属(雪叶属)●☆
32503 Argomuellera basicordata Peter ex Radcl. -Sm.;基心白雪叶●☆
32504 Argomuellera lancifolia (Pax) Pax;剑叶白雪叶●☆
32505 Argomuellera macrophylla Pax;白雪叶●☆
32506 Argomuellera pierlotiana J. Léonard;皮氏白雪叶●☆
32507 Argomuellera sessilifolia Prain;无梗白雪叶●☆
32508 Argophilum Blanco = Aglaia Lour.(保留属名)●
32509 Argophyllaceae (Engl.) Takht. = Grossulariaceae DC.(保留科名)●
32510 Argophyllaceae Takht.(1987);雪叶木科(雪叶科)●☆
32511 Argophyllaceae Takht. = Grossulariaceae DC.(保留科名)●
32512 Argophyllum Blanco = Argophilum Blanco ●
32513 Argophyllum J. R. Forst. et G. Forst.(1775);雪叶木属●☆
32514 Argophyllum nitidum J. R. Forst. et G. Forst.;雪叶木●☆
32515 Argopogon Mimeur = Ischaemum L. ■
32516 Argopogon vuilletii Mimeur = Ischaemum amethystinum J. -P. Lebrun ■☆
32517 Argorips Raf. = Salix L.(保留属名)●
32518 Argostemma Wall.(1824);水冠草属(雪花属);Argostemma, Snowflake ■
32519 Argostemma africanum K. Schum.;非洲水冠草■☆
32520 Argostemma discolor Merr.;异色雪花;Discolor Argostemma, Twocolor Snowflake ■
32521 Argostemma hainanicum H. S. Lo;海南雪花;Hainan Argostemma, Hainan Snowflake ■
32522 Argostemma iriomotense Masam. = Argostemma solanifolium Elmer ■
32523 Argostemma pumilum Benn.;偃伏水冠草■☆
32524 Argostemma saxatile Chun et F. C. How;岩雪花;Cliff Argostemma, Cliff Snowflake ■
32525 Argostemma solaniflorum Elmer;水冠草;Nightshadeflower Snowflake ■
32526 Argostemma taiwanense S. S. Ying = Argostemma solaniflorum Elmer ■
32527 Argostemma verticillatum Wall.;小雪花(雪花);Snow Argostemma, Verticillate Snowflake ■
32528 Argostemma yunnanense F. C. How ex H. S. Lo;滇雪花;Yunnan Argostemma, Yunnan Snowflake ■
32529 Argostemmella Ridl. = Argostemma Wall. ■
32530 Argothamnia Spreng. = Argythamnia P. Browne ●☆
32531 Argusia Boehm. = Tournefortia L. ●■
32532 Argusia Boehm. ex Ludw. = Tournefortia L. ●■
32533 Argusia argentea (L. f.) Heine = Heliotropium foertherianum Diane et Hilger ●
32534 Argusia argentea (L. f.) Heine = Tournefortia argentea L. f. ●
32535 Argusia sibirica (L.) Boehmer = Heliotropium japonicum A. Gray ■☆
32536 Argusia sibirica (L.) Dandy = Tournefortia sibirica L. ●■
32537 Argussiera Bubani = Hippophae L. ●
32538 Arguzia Amm. ex Steud. = Messerschmidia L. ex Hebenstr. ●■
32539 Arguzia Amm. ex Steud. = Tournefortia L. ●■
32540 Arguzia Raf. = Argusia Boehm. ●■
32541 Argylia D. Don(1823);阿盖紫葳属●☆
32542 Argylia australis Phil.;南方阿盖紫葳●☆
32543 Argylia bifrons Phil.;双花阿盖紫葳●☆
32544 Argylia chrysantha Phil.;金花阿盖紫葳●☆
32545 Argylia tenuifolia C. Presl;细叶阿盖紫葳●☆
32546 Argyra Noronha ex Baill. = Croton L. ●
32547 Argyranthemum Webb = Argyranthemum Webb ex Sch. Bip. ●
32548 Argyranthemum Webb ex Sch. Bip.(1839);木茼蒿属(木筒蒿属);Argyranthemum, Marguerite ●
32549 Argyranthemum adauctum (Link) Humphries;膨胀木茼蒿●☆
32550 Argyranthemum adauctum (Link) Humphries subsp. canariense (Sch. Bip.) Humphries;加那利膨胀木茼蒿●☆
32551 Argyranthemum adauctum (Link) Humphries subsp. dugourii (Bolle) Humphries;迪古尔木茼蒿●☆
32552 Argyranthemum adauctum (Link) Humphries subsp. gracile (Sch. Bip.) Humphries;细膨胀木茼蒿●☆
32553 Argyranthemum adauctum (Link) Humphries subsp. jacobiifolium (Sch. Bip.) Humphries;雅各菊木茼蒿●☆
32554 Argyranthemum adauctum (Link) Humphries subsp. palmensis Santos;帕尔马木茼蒿●☆
32555 Argyranthemum anethifolium Webb = Argyranthemum foeniculaceum (Willd.) Webb ex Sch. Bip. ●☆
32556 Argyranthemum broussonetii (Pers.) Humphries;布鲁索内木茼蒿●☆
32557 Argyranthemum broussonetii (Pers.) Humphries subsp. gomerensis Humphries;戈梅拉木茼蒿●☆
32558 Argyranthemum coronopifolium (Willd.) Humphries;乌足叶木茼蒿●☆
32559 Argyranthemum dissectum (Lowe) Lowe;深裂木茼蒿●☆
32560 Argyranthemum filifolium (Sch. Bip.) Humphries;丝叶木茼蒿●☆

32561 Argyranthemum foeniculaceum (Willd.) Sch. Bip. = Argyranthemum foeniculaceum (Willd.) Webb ex Sch. Bip. ●☆

32562 Argyranthemum foeniculaceum (Willd.) Webb ex Sch. Bip.;剑叶木茼蒿;Lanceleaf Marguerite ●☆

32563 Argyranthemum frutescens (L.) Sch. Bip.;木茼蒿(木春菊,蓬蒿菊,茼蒿菊);Bash Daisy, Butterfly, Dill Daisy, Marguerite, Marguerite Daisy, Paris Daisy, Shrubby Argyranthemum, Shrubby Marguerite, White Marguerite ●

32564 Argyranthemum frutescens (L.) Sch. Bip. 'Blizzard';大风雪木茼蒿●☆

32565 Argyranthemum frutescens (L.) Sch. Bip. 'Butterfly';蝴蝶木茼蒿●☆

32566 Argyranthemum frutescens (L.) Sch. Bip. 'California Gold';加州金黄木茼蒿●☆

32567 Argyranthemum frutescens (L.) Sch. Bip. 'Gill Pink';吉儿粉木茼蒿●☆

32568 Argyranthemum frutescens (L.) Sch. Bip. 'Jamaica Primrose';牙买加报春木茼蒿●☆

32569 Argyranthemum frutescens (L.) Sch. Bip. 'Mary Wootton';沃顿玛丽木茼蒿●☆

32570 Argyranthemum frutescens (L.) Sch. Bip. 'Mrs F. Sander';桑德夫人木茼蒿●☆

32571 Argyranthemum frutescens (L.) Sch. Bip. 'Tauranga Star';陶兰加之星木茼蒿●☆

32572 Argyranthemum frutescens (L.) Sch. Bip. 'Vancouver';温哥华木茼蒿●☆

32573 Argyranthemum frutescens (L.) Sch. Bip. subsp. canariae (Christ) Humphries;加那利木茼蒿●☆

32574 Argyranthemum frutescens (L.) Sch. Bip. subsp. foeniculaceum (Pit. et Proust) Humphries = Argyranthemum frutescens (L.) Sch. Bip. ●

32575 Argyranthemum frutescens (L.) Sch. Bip. subsp. gracilescens (Christ) Humphries;纤细木茼蒿●☆

32576 Argyranthemum frutescens (L.) Sch. Bip. subsp. parviflorum (Pit. et Proust) Humphries;小花木茼蒿●☆

32577 Argyranthemum frutescens (L.) Sch. Bip. subsp. pumilum Humphries;小木茼蒿●☆

32578 Argyranthemum frutescens (L.) Sch. Bip. var. canariae (Christ) Pit. et Proust = Argyranthemum frutescens (L.) Sch. Bip. ●

32579 Argyranthemum frutescens (L.) Sch. Bip. var. crithmifolium (Link) Pit. et Proust = Argyranthemum frutescens (L.) Sch. Bip. ●

32580 Argyranthemum frutescens (L.) Sch. Bip. var. foeniculaceum Pit. et Proust = Argyranthemum frutescens (L.) Sch. Bip. ●

32581 Argyranthemum frutescens (L.) Sch. Bip. var. parviflorum Pit. et Proust = Argyranthemum frutescens (L.) Sch. Bip. subsp. parviflorum (Pit. et Proust) Humphries ●☆

32582 Argyranthemum gracile Sch. Bip.;细瘦木茼蒿●☆

32583 Argyranthemum haematomma (Lowe) Lowe;血红木茼蒿●☆

32584 Argyranthemum hierrense Humphries;耶罗木茼蒿●☆

32585 Argyranthemum jacobiifolium G. Kunkel = Argyranthemum adauctum (Link) Humphries subsp. jacobiifolium (Sch. Bip.) Humphries ●☆

32586 Argyranthemum lidii Humphries;利德木茼蒿●☆

32587 Argyranthemum maderense (D. Don) Humphries;马德拉木茼蒿●☆

32588 Argyranthemum ochroleucum Sch. Bip.;绿白木茼蒿●☆

32589 Argyranthemum pinnatifidum (L. f.) Lowe;羽裂木茼蒿●☆

32590 Argyranthemum pinnatifidum (L. f.) Lowe subsp. montanum Rustan;山地羽裂木茼蒿●☆

32591 Argyranthemum pinnatifidum (L. f.) Lowe subsp. succulentum (Lowe) Humphries;多汁羽裂木茼蒿●☆

32592 Argyranthemum pumilum G. Kunkel = Argyranthemum frutescens (L.) Sch. Bip. subsp. pumilum Humphries ●☆

32593 Argyranthemum sundingii L. Borgen;松德林木茼蒿●☆

32594 Argyranthemum sventenii Humphries et Aldridge;斯文顿木茼蒿●☆

32595 Argyranthemum teneriffae Humphries;特纳木茼蒿●☆

32596 Argyranthemum thalassophilum (Svent.) Humphries;海洋木茼蒿●☆

32597 Argyranthemum webbii Sch. Bip.;韦布木茼蒿●☆

32598 Argyranthemum winteri (Svent.) Humphries;温特木茼蒿●☆

32599 Argyranthus Neck. = Helipterum DC. ex Lindl. ■☆

32600 Argyreia Lour. (1790);银背藤属(白鹤藤属,朝颜属,木旋花属,银叶花属);Argyreia, Asia Glory, Asiaglory, Asia-glory, Silver Weed, Silverweed ●

32601 Argyreia abyssinica Choisy = Ipomoea abyssinica (Choisy) Hochst. ■☆

32602 Argyreia acuta Lour.;白鹤藤(白背绸,白背丝绸,白背藤,白背叶,白底丝绸,白面割鸡藤,白面水鸡,白牡丹,绸缎木叶,绸缎藤,绸缎叶,女菀,一匹绸,银背藤,银背叶);Acute Asiaglory, Common Argyreia ●

32603 Argyreia aggregata (Roxb.) Choisy = Argyreia osyrensis (Roth) Choisy ●

32604 Argyreia aggregata (Roxb.) Choisy var. osyrensis (Roth) Gagnep. et Courchet = Argyreia osyrensis (Roth) Choisy ●

32605 Argyreia alulata Miq. = Convolvulus turpethum L. ■

32606 Argyreia alulata Miq. = Operculina turpetha (L.) Silva Manso ■

32607 Argyreia ampla (Wall.) Choisy = Argyreia mastersii (Prain) Raizada ●

32608 Argyreia ampla (Wall.) Choisy = Argyreia roxburghii (Wall.) Arn. ex Choisy var. ampla (Wall.) C. B. Clarke ●

32609 Argyreia androyensis Deroin;安德罗银背藤●☆

32610 Argyreia bagshavei Rendle = Stictocardia beraviensis (Vatke) Hallier f. ●☆

32611 Argyreia baoshanensis S. H. Huang;保山银背藤;Baoshan Argyreia ■

32612 Argyreia barbigera Choisy;髯毛银背藤(髯毛白叶藤)●☆

32613 Argyreia baronii Deroin;巴龙银背藤●☆

32614 Argyreia beraviensis (Vatke) Baker = Stictocardia beraviensis (Vatke) Hallier f. ●☆

32615 Argyreia campanulata (L.) Alston = Stictocardia tiliifolia (Desr.) Hallier f. ●■

32616 Argyreia capitata (Vahl) Choisy = Argyreia capitiformis (Poir.) Ooststr. ●

32617 Argyreia capitata (Wahl) Arn. ex Choisy = Argyreia capitiformis (Poir.) Ooststr. ●

32618 Argyreia capitiformis (Poir.) Ooststr.;头花银背藤(硬毛白鹤藤);Capitate Argyreia, Capitate-flower Asiaglory ●

32619 Argyreia championii Benth. = Argyreia mollis (Burm. f.) Choisy ●

32620 Argyreia cheliensis C. Y. Wu;车里银背藤;Cheli Argyreia, Jinghong Argyreia ●

32621 Argyreia eriocephala C. Y. Wu;毛头银背藤;Cottonhead Argyreia, Eriocephalous Argyreia, Hairy-head Argyreia ●

32622 Argyreia festiva Wall. = Argyreia acuta Lour. ●

32623 Argyreia formosana Ishig. ex T. Yamaz.;台湾银背藤(钝叶朝颜);Formosan Argyreia ●
32624 Argyreia fulvo-cymosa C. Y. Wu;黄伞白鹤藤;Yellowcyme Argyreia,Yellow-cymose Argyreia ●
32625 Argyreia fulvo-cymosa C. Y. Wu var. pauciflora C. Y. Wu;少花黄伞白鹤藤;Fewflower Argyreia ●
32626 Argyreia fulvo-villosa C. Y. Wu et S. H. Huang;黄背藤;Yellow Argyreia,Yellow-back Argyreia,Yellow-villosed Argyreia ●
32627 Argyreia grantii Baker = Ipomoea hildebrandtii Vatke subsp. grantii (Baker) Verdc. ■☆
32628 Argyreia hanningtonii Baker = Ipomoea macrosepala Brenan ■☆
32629 Argyreia henryi (Craib) Craib;长叶银背藤;Henry Argyreia ●
32630 Argyreia henryi (Craib) Craib var. hypochrysa C. Y. Wu;金背长叶藤(金银背藤);Golden-back Argyreia ●
32631 Argyreia laxiflora Baker = Stictocardia laxiflora (Baker) Hallier f. ●☆
32632 Argyreia liliiflora C. Y. Wu = Argyreia pierreana Bois ●
32633 Argyreia lineariloba C. Y. Wu;线叶银背藤;Linearleaf Argyreia,Linear-leaved Argyreia ●
32634 Argyreia macrocalyx Baker = Ipomoea macrosepala Brenan ■☆
32635 Argyreia marlipoensis C. Y. Wu et S. H. Huang;麻栗坡银背藤;Malipo Argyreia ●
32636 Argyreia mastersii (Prain) Raizada;叶苞银背藤●
32637 Argyreia maymyo (W. W. Sm.) Raizada;思茅银背藤●
32638 Argyreia mollis (Burm. f.) Choisy;银背藤(白背绸缎,白底丝绸,白鹤藤,白面水鸡,绸缎根,钝叶超颜,钝叶朝颜,一匹绸);Bluntleaf Asiaglory,Obtuseleaf Argyreia,Obtuse-leaved Argyreia ●
32639 Argyreia monglaensis C. Y. Wu et S. H. Huang;勐腊银背藤;Mengla Argyreia ●
32640 Argyreia monosperma C. Y. Wu;单籽银背藤;Monospermous Argyreia,Singleseed Argyreia ●
32641 Argyreia multiflora Baker;多花银背藤●☆
32642 Argyreia nervosa (Burm. f.) Bojer;美丽银背藤(脉叶超颜,脉叶朝颜,木旋花);Elephant Climber,Elephant Creeper,Neined Argyreia,Veined Argyreia,Wood Rose,Wooden Rose,Woolly Asiaglory,Woolly Morning Glory ●
32643 Argyreia obtecta (Choisy) C. B. Clarke = Argyreia mollis (Burm. f.) Choisy ●
32644 Argyreia obtecta (Wall.) C. B. Clarke = Argyreia mollis (Burm. f.) Choisy ●
32645 Argyreia obtusifolia Lour.;钝叶朝颜(银背藤)●
32646 Argyreia obtusifolia Lour. = Argyreia mollis (Burm. f.) Choisy ●
32647 Argyreia obtusifolia Lour. = Argyreia obtecta (Choisy) C. B. Clarke ●
32648 Argyreia onilahiensis Deroin;乌尼拉希银背藤●☆
32649 Argyreia osyrensis (Roth) Choisy;聚花银背藤(聚花白鹤藤);Clusteredflower Argyreia,Cluster-flowered Argyreia,Thyrsiferous Argyreia ●
32650 Argyreia osyrensis (Roth) Choisy var. cinerea Hand.-Mazz.;灰毛银背藤(合苞叶,红心果,灰毛白鹤藤,藤本夜关门,猪叶菜);Greyhair Clusteredflower Argyreia ●
32651 Argyreia pierreana Bois;东京银背藤(滇一匹绸,个吉芸,牛白藤,一匹绸,紫苞银背藤);Pierre Argyreia,Purplebract Argyreia ●
32652 Argyreia populifolia Choisy;杨叶银背藤;Poplar-leaf Argyreia ●☆
32653 Argyreia roxburghii (Wall.) Arn. ex Choisy;细苞银背藤;Roxburgh Argyreia,Thinbract Argyreia ●
32654 Argyreia roxburghii (Wall.) Arn. ex Choisy var. ampla (Wall.) C. B. Clarke = Argyreia mastersii (Prain) Raizada ●
32655 Argyreia roxburghii (Wall.) Arn. ex Choisy var. ampla (Wall.) C. B. Clarke = Argyreia roxburghii (Wall.) Arn. ex Choisy ●
32656 Argyreia rufohirsuta H. Lév. = Argyreia capitiformis (Poir.) Ooststr. ●
32657 Argyreia seguinii (H. Lév.) Vaniot ex H. Lév.;白花银背藤(白背藤,白面水鸡,白牛藤,跌打王,葛藤,黄藤,山牡丹,藤续断,旋花藤);Seguin Argyreia,White Argyreia,White-flower Asiaglory ●
32658 Argyreia seguinii (H. Lév.) Vaniot ex H. Lév. = Argyreia pierreana Bois ●
32659 Argyreia speciosa (L. f.) Sweet = Argyreia nervosa (Burm. f.) Bojer ●
32660 Argyreia splendens (Roxb.) Sweet;亮叶银背藤;Shinyleaf Argyreia,Shiny-leaved Argyreia,Silver Morning-glory ●
32661 Argyreia strigillosa C. Y. Wu;细毛银背藤;Strigillose Argyreia,Strigose Argyreia ●
32662 Argyreia tiliifolia (Desr.) Wight = Stictocardia tiliifolia (Desr.) Hallier f. ●■
32663 Argyreia tomentosa Choisy = Argyreia formosana Ishig. ex T. Yamaz. ●
32664 Argyreia ubanghensis A. Chev.;乌班吉银背藤●☆
32665 Argyreia velutina C. Y. Wu;黄毛银背藤;Velvet-like Argyreia,Yellowhairs Argyreia ●
32666 Argyreia verrucosochispida Y. Y. Qian;瘤毛银背藤;Verrucosohispid Argyreia ●
32667 Argyreia verrucosochispida Y. Y. Qian = Argyreia capitiformis (Poir.) Ooststr. ●
32668 Argyreia wallichii Choisy;大叶银背藤(猴子烟袋花,小团叶,羊角藤);Bigleaf Argyreia,Wallich Argyreia ●
32669 Argyrella Naudin = Dissotis Benth.(保留属名)●☆
32670 Argyrella canescens (E. Mey. ex R. A. Graham) Harv. = Heterotis canescens (E. Mey. ex R. A. Graham) Jacq.-Fél. ●☆
32671 Argyrella incana (Walp.) Naudin = Heterotis canescens (E. Mey. ex R. A. Graham) Jacq.-Fél. ●☆
32672 Argyrella phaeotricha (Hochst.) Naudin = Antherotoma phaeotricha (Hochst.) Jacq.-Fél. ■☆
32673 Argyreon St.-Lag. = Argyreia Lour. ●
32674 Argyrexlas Raf. = Echium L. ●■
32675 Argyrocalymma K. Schum. et Lauterb. = Carpodetus J. R. Forst. et G. Forst. ●☆
32676 Argyrochaeta Cav. = Parthenium L. ●■
32677 Argyrochlymma K. Schum. et Lauterb. = Carpodetus J. R. Forst. et G. Forst. ●☆
32678 Argyrocoma Raf. = Paronychia Mill. ■
32679 Argyrocome Breyne = Helipterum DC. ex Lindl. ■☆
32680 Argyrocome Breyne ex Kuntze = Helipterum DC. ex Lindl. ■☆
32681 Argyrocome Gaertn. = Helipterum DC. ex Lindl. ■☆
32682 Argyrocytisus (Maire) Raynaud(1975);银雀儿属;Silver Broom ●☆
32683 Argyrocytisus battandieri (Maire) Raynaud;银雀儿;Silver Broom ●☆
32684 Argyrodendron (Endl.) Klotzsch = Croton L. ●
32685 Argyrodendron F. Muell. = Heritiera Aiton ●
32686 Argyrodendron Klotzsch = Croton L. ●
32687 Argyrodendron petersii Klotzsch = Combretum imberbe Wawra ●☆
32688 Argyrodendron trifoliolatum F. Muell. = Heritiera trifoliolata (F. Muell.) Kosterm. ●☆

32689 Argyroderma N. E. Br. (1922);银叶花属(银皮属,银石属);Argyroderma ●☆
32690 Argyroderma amoenum Schwantes = Argyroderma pearsonii (N. E. Br.) Schwantes ●☆
32691 Argyroderma angustipetalum L. Bolus;狭瓣银叶花(碧铃);Narrowpetal Argyroderma ●☆
32692 Argyroderma angustipetalum L. Bolus = Argyroderma congregatum L. Bolus ●☆
32693 Argyroderma aureum L. Bolus = Argyroderma delaetii C. A. Maass ●☆
32694 Argyroderma blandum L. Bolus = Argyroderma delaetii C. A. Maass ●☆
32695 Argyroderma boreale L. Bolus = Argyroderma delaetii C. A. Maass ●☆
32696 Argyroderma braunsii (Schwantes) Schwantes;绿管银叶花(碧管玉,佛指草);Brauns Argyroderma ●
32697 Argyroderma braunsii (Schwantes) Schwantes = Argyroderma fissum (Haw.) L. Bolus ●☆
32698 Argyroderma brevipes (Schltr.) L. Bolus = Argyroderma fissum (Haw.) L. Bolus ●☆
32699 Argyroderma brevipes L. Bolus = Argyroderma fissum L. Bolus ●☆
32700 Argyroderma brevitubum L. Bolus = Argyroderma delaetii C. A. Maass ●☆
32701 Argyroderma carinatum L. Bolus = Argyroderma delaetii C. A. Maass ●☆
32702 Argyroderma citrinum L. Bolus = Argyroderma delaetii C. A. Maass ●☆
32703 Argyroderma concinnum Schwantes = Argyroderma delaetii C. A. Maass ●☆
32704 Argyroderma congregatum L. Bolus;聚集银叶花●☆
32705 Argyroderma crateriforme (L. Bolus) N. E. Br.;杯状银叶花●☆
32706 Argyroderma cuneatipetalum L. Bolus = Argyroderma delaetii C. A. Maass ●☆
32707 Argyroderma delaetii C. A. Maass;德氏银石●☆
32708 Argyroderma delaetii C. A. Maass var. purpureum ? = Argyroderma delaetii C. A. Maass ●☆
32709 Argyroderma densipetalum L. Bolus = Argyroderma delaetii C. A. Maass ●☆
32710 Argyroderma digitifolium (N. E. Br.) Schwantes ex L. Bolus = Argyroderma fissum (Haw.) L. Bolus ●☆
32711 Argyroderma duale (N. E. Br.) N. E. Br. = Antimima dualis (N. E. Br.) N. E. Br. ●☆
32712 Argyroderma fissum (Haw.) L. Bolus;宝槌石(宝槌玉)●☆
32713 Argyroderma fissum L. Bolus = Argyroderma fissum (Haw.) L. Bolus ●☆
32714 Argyroderma formosum L. Bolus = Argyroderma delaetii C. A. Maass ●☆
32715 Argyroderma framesii L. Bolus;弗雷斯银叶花●☆
32716 Argyroderma framesii L. Bolus subsp. hallii (L. Bolus) H. E. K. Hartmann;霍尔银叶花●☆
32717 Argyroderma framesii L. Bolus var. minus ? = Argyroderma framesii L. Bolus ●☆
32718 Argyroderma gregarium L. Bolus = Argyroderma delaetii C. A. Maass ●☆
32719 Argyroderma hallii L. Bolus = Argyroderma framesii L. Bolus subsp. hallii (L. Bolus) H. E. K. Hartmann ●☆
32720 Argyroderma hutchinsonii L. Bolus = Argyroderma fissum (Haw.) L. Bolus ●☆
32721 Argyroderma jacobsenianum Schwantes = Argyroderma congregatum L. Bolus ●☆
32722 Argyroderma latifolium L. Bolus = Argyroderma delaetii C. A. Maass ●☆
32723 Argyroderma latipetalum L. Bolus;新桩玉●☆
32724 Argyroderma latipetalum L. Bolus = Argyroderma fissum (Haw.) L. Bolus ●☆
32725 Argyroderma latipetalum L. Bolus var. longitubum ? = Argyroderma fissum (Haw.) L. Bolus ●☆
32726 Argyroderma lesliei R. Br. = Argyroderma delaetii C. A. Maass ●☆
32727 Argyroderma leucanthum L. Bolus;白银叶花●☆
32728 Argyroderma leucanthum L. Bolus = Argyroderma delaetii C. A. Maass ●☆
32729 Argyroderma litorale L. Bolus = Argyroderma fissum (Haw.) L. Bolus ●☆
32730 Argyroderma longipes L. Bolus = Argyroderma delaetii C. A. Maass ●☆
32731 Argyroderma luckoffii L. Bolus = Argyroderma pearsonii (N. E. Br.) Schwantes ●☆
32732 Argyroderma nortieri L. Bolus = Argyroderma congregatum L. Bolus ●☆
32733 Argyroderma octophyllum Schwantes;银叶花(金铃,银铃);Eightleaves Argyroderma ●☆
32734 Argyroderma orientale L. Bolus;东方银叶花●☆
32735 Argyroderma orientale L. Bolus = Argyroderma fissum (Haw.) L. Bolus ●☆
32736 Argyroderma ovale L. Bolus;卵叶银叶花●☆
32737 Argyroderma ovale L. Bolus = Argyroderma pearsonii (N. E. Br.) Schwantes ●☆
32738 Argyroderma patens L. Bolus;铺展银叶花●☆
32739 Argyroderma pearsonii (N. E. Br.) Schwantes;皮尔逊银石●☆
32740 Argyroderma pearsonii Schwantes = Argyroderma pearsonii (N. E. Br.) Schwantes ●☆
32741 Argyroderma peersii L. Bolus = Argyroderma congregatum L. Bolus ●☆
32742 Argyroderma planum L. Bolus = Argyroderma delaetii C. A. Maass ●☆
32743 Argyroderma productum L. Bolus;秀眉玉●☆
32744 Argyroderma productum L. Bolus = Argyroderma delaetii C. A. Maass ●☆
32745 Argyroderma pulvinare L. Bolus = Argyroderma crateriforme (L. Bolus) N. E. Br. ●☆
32746 Argyroderma reniforme L. Bolus = Argyroderma delaetii C. A. Maass ●☆
32747 Argyroderma ringens L. Bolus;张开银叶花●☆
32748 Argyroderma rooipanense L. Bolus = Argyroderma congregatum L. Bolus ●☆
32749 Argyroderma roseatum N. E. Br. = Lapidaria margaretae (Schwantes) Dinter et Schwantes ■☆
32750 Argyroderma roseum Schwantes;红银叶花(赤花金铃);Red Argyroderma ●☆
32751 Argyroderma roseum Schwantes = Argyroderma delaetii C. A. Maass ●☆
32752 Argyroderma roseum Schwantes. f. delaetii G. D. Rowley = Argyroderma delaetii C. A. Maass ●☆
32753 Argyroderma schlechteri Schwantes;光彩银叶花;Schlechter

Argyroderma ●☆
32754 Argyroderma schlechteri Schwantes = Argyroderma pearsonii (N. E. Br.) Schwantes ●☆
32755 Argyroderma schuldtii Schwantes = Argyroderma delaetii C. A. Maass ●☆
32756 Argyroderma speciosum L. Bolus = Argyroderma delaetii C. A. Maass ●☆
32757 Argyroderma splendens L. Bolus = Argyroderma delaetii C. A. Maass ●☆
32758 Argyroderma strictum L. Bolus = Argyroderma framesii L. Bolus subsp. hallii (L. Bolus) H. E. K. Hartmann ●☆
32759 Argyroderma subalbum (N. E. Br.) N. E. Br.;浅白银叶花●☆
32760 Argyroderma subrotundum L. Bolus = Argyroderma crateriforme (L. Bolus) N. E. Br. ●☆
32761 Argyroderma testiculare (Aiton) N. E. Br.;双球银叶花(银铃)●☆
32762 Argyroderma testiculare (Aiton) N. E. Br. var. luteum N. E. Br. = Argyroderma pearsonii (N. E. Br.) Schwantes ●☆
32763 Argyroderma testiculare (Aiton) N. E. Br. var. pearsonii N. E. Br. = Argyroderma pearsonii (N. E. Br.) Schwantes ●☆
32764 Argyroderma testiculare N. E. Br. = Argyroderma testiculare (Aiton) N. E. Br. ●☆
32765 Argyroderma villetii L. Bolus = Argyroderma subalbum (N. E. Br.) N. E. Br. ●☆
32766 Argyroglottis Turcz. (1851);银舌鼠麴木属●☆
32767 Argyroglottis Turcz. = Helichrysum Mill. (保留属名)●■
32768 Argyroglottis turbinata Turcz.;银舌鼠麴木●☆
32769 Argyrolobium Eckl. et Zeyh. (1836)(保留属名);银豆属;Argyrolobium, Silverleaf ●☆
32770 Argyrolobium aberdaricum Harms = Argyrolobium rupestre (E. Mey.) Walp. subsp. aberdaricum (Harms) Polhill ●☆
32771 Argyrolobium abyssinicum (Decne.) Jaub. et Spach var. caespitosum (Lanza) Fiori = Argyrolobium arabicum (Decne.) Jaub. et Spach ●☆
32772 Argyrolobium abyssinicum (Decne.) Jaub. et Spach var. diffusum (Lanza) Fiori = Argyrolobium arabicum (Decne.) Jaub. et Spach ●☆
32773 Argyrolobium abyssinicum Jaub. et Spach = Argyrolobium arabicum (Decne.) Jaub. et Spach ●☆
32774 Argyrolobium abyssinicum Jaub. et Spach var. garamantum Quézel = Argyrolobium arabicum (Decne.) Jaub. et Spach ●☆
32775 Argyrolobium aciculare Dümmer;针形银豆●☆
32776 Argyrolobium aequinoctiale Welw. ex Baker;昼夜银豆●☆
32777 Argyrolobium amplexicaule (E. Mey.) Dümmer;抱茎银豆●☆
32778 Argyrolobium andrewsianum (E. Mey.) Steud. = Argyrolobium tomentosum (Andréws) Druce ●☆
32779 Argyrolobium angustifolium Eckl. et Zeyh. = Argyrolobium tuberosum Eckl. et Zeyh. ●☆
32780 Argyrolobium angustissimum (E. Mey.) T. J. Edwards;极窄银豆●☆
32781 Argyrolobium angustistipulatum De Wild. = Argyrolobium tomentosum (Andréws) Druce ●☆
32782 Argyrolobium arabicum (Decne.) Jaub. et Spach;阿拉伯银豆●☆
32783 Argyrolobium argenteum (Jacq.) Eckl. et Zeyh.;银豆;Silvery Argyrolobium ●☆
32784 Argyrolobium argenteum (L.) Willk. subsp. fallax (Ball) Murb. = Argyrolobium zanonii (Turra) P. W. Ball subsp. fallax (Ball) Greuter et Burdet ●☆
32785 Argyrolobium ascendens (E. Mey.) Walp.;上升银豆●☆
32786 Argyrolobium baptisioides (E. Mey.) Walp.;赛靛银豆●☆
32787 Argyrolobium barbatum (Meisn.) Walp.;髯毛银豆●☆
32788 Argyrolobium bequaertii De Wild. = Argyrolobium fischeri Taub. ●☆
32789 Argyrolobium biflorum Eckl. et Zeyh. = Argyrolobium pauciflorum Eckl. et Zeyh. ●☆
32790 Argyrolobium brevicalyx C. H. Stirt. = Polhillia brevicalyx (C. H. Stirt.) B. -E. van Wyk et A. L. Schutte ●☆
32791 Argyrolobium buaricum Harms = Argyrolobium aequinoctiale Welw. ex Baker ●☆
32792 Argyrolobium calycinum Jaub. et Spach;萼状银豆●☆
32793 Argyrolobium campicola Harms;平原银豆●☆
32794 Argyrolobium candicans Eckl. et Zeyh.;纯白银豆●☆
32795 Argyrolobium catatii (Drake) M. Peltier;卡他银豆●☆
32796 Argyrolobium collinum Eckl. et Zeyh.;山丘银豆●☆
32797 Argyrolobium collinum Eckl. et Zeyh. var. seminudum Harv. = Argyrolobium argenteum (Jacq.) Eckl. et Zeyh. ●☆
32798 Argyrolobium confertum Polhill;密集银豆●☆
32799 Argyrolobium connatum Harv. = Polhillia connata (Harv.) C. H. Stirt. ●☆
32800 Argyrolobium crassifolium (E. Mey.) Eckl. et Zeyh.;厚叶银豆●☆
32801 Argyrolobium crinitum (E. Mey.) Walp.;长软毛银豆●☆
32802 Argyrolobium deflexiflorum Baker = Lotononis angolensis Welw. ex Baker ■☆
32803 Argyrolobium dekindtii Harms = Argyrolobium fischeri Taub. ●☆
32804 Argyrolobium dimidiatum Schinz;对开银豆●☆
32805 Argyrolobium dorycnoides Baker = Argyrolobium ramosissimum Baker ●☆
32806 Argyrolobium emirnense Baker = Argyrolobium pedunculare Benth. ●☆
32807 Argyrolobium eylesii Baker f.;艾尔斯银豆●☆
32808 Argyrolobium filiforme (Thunb.) Eckl. et Zeyh.;丝状银豆●☆
32809 Argyrolobium fischeri Taub.;菲舍尔银豆●☆
32810 Argyrolobium flaccidum (Royle) Jaub. et Spach;柔软银豆●☆
32811 Argyrolobium friesianum Harms;弗里斯银豆●☆
32812 Argyrolobium frutescens Burtt Davy;灌木银豆●☆
32813 Argyrolobium fulvicaule Hochst. ex Engl. = Argyrolobium ramosissimum Baker ●☆
32814 Argyrolobium glaucum Schinz;灰绿银豆●☆
32815 Argyrolobium goodioides (Meisn.) Walp. = Argyrolobium crassifolium (E. Mey.) Eckl. et Zeyh. ●☆
32816 Argyrolobium grandiflorum Boiss. et Reut. = Argyrolobium zanonii (Turra) P. W. Ball subsp. grandiflorum (Boiss. et Reut.) Greuter ●☆
32817 Argyrolobium harmsianum Schltr. ex Harms;哈姆斯银豆●☆
32818 Argyrolobium harveyanum Oliv.;哈维银豆●☆
32819 Argyrolobium helenae Buscal. et Muschl. = Argyrolobium fischeri Taub. ●☆
32820 Argyrolobium hirsuticaule Harms = Argyrolobium lotoides Bunge ex Trautv. ●☆
32821 Argyrolobium humile E. Phillips;矮小银豆●☆
32822 Argyrolobium incanum Eckl. et Zeyh.;灰毛银豆●☆
32823 Argyrolobium involucratum (Thunb.) Harv. = Polhillia involucrata (Thunb.) B. -E. van Wyk et A. L. Schutte ●☆
32824 Argyrolobium itremoense Du Puy et Labat;伊特雷穆银豆●☆
32825 Argyrolobium keniense Harms = Argyrolobium fischeri Taub. ●☆
32826 Argyrolobium kilimandscharicum Taub. = Argyrolobium rupestre

(E. Mey.) Walp. subsp. kilimandscharicum (Taub.) Polhill ●☆
32827 Argyrolobium krebsianum C. Presl;克雷布斯银豆●☆
32828 Argyrolobium lanceolatum (E. Mey.) Eckl. et Zeyh.;披针形银豆●☆
32829 Argyrolobium lancifolium Burtt Davy = Argyrolobium transvaalense Schinz ●☆
32830 Argyrolobium lejeunei R. Wilczek = Argyrolobium vaginiferum Harms ●☆
32831 Argyrolobium leptocladum Harms = Argyrolobium lotoides Bunge ex Trautv. ●☆
32832 Argyrolobium leucophyllum Baker = Argyrolobium fischeri Taub. ●☆
32833 Argyrolobium linneanum Walp. = Argyrolobium zanonii (Turra) P. W. Ball ●☆
32834 Argyrolobium linneanum Walp. subsp. fallax Ball = Argyrolobium zanonii (Turra) P. W. Ball subsp. fallax (Ball) Greuter et Burdet ●☆
32835 Argyrolobium linneanum Walp. subsp. stipulaceum Ball = Argyrolobium zanonii (Turra) P. W. Ball subsp. stipulaceum (Ball) Greuter ●☆
32836 Argyrolobium linneanum Walp. var. fallax (Ball) Ball = Argyrolobium zanonii (Turra) P. W. Ball subsp. fallax (Ball) Greuter et Burdet ●☆
32837 Argyrolobium linneanum Walp. var. grandiflorum (Boiss. et Reut.) Batt. = Argyrolobium zanonii (Turra) P. W. Ball subsp. grandiflorum (Boiss. et Reut.) Greuter ●☆
32838 Argyrolobium linneanum Walp. var. saharae (Pomel) Batt. = Argyrolobium saharae Pomel ●☆
32839 Argyrolobium longifolium (Meisn.) Walp.;长叶银豆●☆
32840 Argyrolobium longipes N. E. Br. = Argyrolobium ascendens (E. Mey.) Walp. ●☆
32841 Argyrolobium lotoides Bunge ex Trautv.;君迁子银豆●☆
32842 Argyrolobium lotoides Harv. = Argyrolobium lotoides Bunge ex Trautv. ●☆
32843 Argyrolobium lunare (L.) Druce;新月银豆●☆
32844 Argyrolobium lunare (L.) Druce subsp. sericeum (Thunb.) T. J. Edwards;绢毛新月银豆●☆
32845 Argyrolobium lydenburgense Harms;莱登堡银豆●☆
32846 Argyrolobium macrophyllum Harms;大叶银豆●☆
32847 Argyrolobium macrophyllum Harms var. mendesii Torre = Argyrolobium macrophyllum Harms ●☆
32848 Argyrolobium marginatum Bolus;具边银豆●☆
32849 Argyrolobium megarhizum Bolus;大根银豆●☆
32850 Argyrolobium microphyllum Ball;小叶银豆●☆
32851 Argyrolobium microphyllum Ball var. racemosum Maire = Argyrolobium microphyllum Ball ●☆
32852 Argyrolobium mildbraedii Harms = Argyrolobium fischeri Taub. ●☆
32853 Argyrolobium modestum Hochst. = Argyrolobium arabicum (Decne.) Jaub. et Spach ●☆
32854 Argyrolobium molle Eckl. et Zeyh.;绵银豆●☆
32855 Argyrolobium monticola Baker f. = Argyrolobium fischeri Taub. ●☆
32856 Argyrolobium muddii Dümmer;马德银豆●☆
32857 Argyrolobium muirii L. Bolus = Argyrolobium filiforme (Thunb.) Eckl. et Zeyh. ●☆
32858 Argyrolobium nanum Burtt Davy = Argyrolobium molle Eckl. et Zeyh. ●☆
32859 Argyrolobium nanum Walp. ex Harms = Argyrolobium nigrescens Dümmer ●☆
32860 Argyrolobium natalense Dümmer = Argyrolobium longifolium (Meisn.) Walp. ●☆
32861 Argyrolobium nigrescens Dümmer;变黑银豆●☆
32862 Argyrolobium nitens Burtt Davy = Argyrolobium wilmsii Harms ●☆
32863 Argyrolobium obcordatum (E. Mey.) Steud. = Argyrolobium trifoliatum (Thunb.) Druce ●☆
32864 Argyrolobium obovatum (E. Mey.) Eckl. et Zeyh. = Argyrolobium argenteum (Jacq.) Eckl. et Zeyh. ●☆
32865 Argyrolobium obsoletum Harv. = Polhillia obsoleta (Harv.) B. -E. van Wyk ●☆
32866 Argyrolobium ornithopodioides Jaub. et Spach = Argyrolobium roseum (Cambess.) Jaub. et Spach subsp. ornithopodioides (Jaub. et Spach) Jafri et Ali ●☆
32867 Argyrolobium pachyphyllum Schltr.;澳非厚叶银豆●☆
32868 Argyrolobium parviflorum T. J. Edwards;小花银豆●☆
32869 Argyrolobium patens Eckl. et Zeyh. = Argyrolobium molle Eckl. et Zeyh. ●☆
32870 Argyrolobium pauciflorum Eckl. et Zeyh.;少花银豆●☆
32871 Argyrolobium pauciflorum Eckl. et Zeyh. var. semiglabrum Harv.;半光少花银豆●☆
32872 Argyrolobium pedunculare Benth.;梗花银豆●☆
32873 Argyrolobium petiolare (E. Mey.) Steud.;柄叶银豆●☆
32874 Argyrolobium petitianum A. Rich. = Argyrolobium schimperianum Hochst. ex A. Rich. ●☆
32875 Argyrolobium pilosum Harv. = Argyrolobium amplexicaule (E. Mey.) Dümmer ●☆
32876 Argyrolobium podalyrioides Dümmer = Argyrolobium collinum Eckl. et Zeyh. ●☆
32877 Argyrolobium polyphyllum Eckl. et Zeyh.;多叶银豆●☆
32878 Argyrolobium pseudotuberosum T. J. Edwards;假块状银豆●☆
32879 Argyrolobium pumilum Eckl. et Zeyh.;低矮银豆●☆
32880 Argyrolobium pumilum Eckl. et Zeyh. var. pilosum (E. Mey.) Harv. = Argyrolobium argenteum (Jacq.) Eckl. et Zeyh. ●☆
32881 Argyrolobium ramosissimum Baker;多枝银豆●☆
32882 Argyrolobium rarum Dümmer;稀银豆●☆
32883 Argyrolobium reflexum N. E. Br. = Dichilus reflexus (N. E. Br.) A. L. Schutte ■☆
32884 Argyrolobium remotum Hochst. ex A. Rich. = Argyrolobium rupestre (E. Mey.) Walp. subsp. remotum (Hochst. ex A. Rich.) Polhill ●☆
32885 Argyrolobium rhodesicum Baker f. = Argyrolobium rupestre (E. Mey.) Walp. ●☆
32886 Argyrolobium rivae (Harms) Cufod. = Argyrolobium fischeri Taub. ●☆
32887 Argyrolobium robustum T. J. Edwards;粗壮银豆●☆
32888 Argyrolobium rogersii N. E. Br. = Argyrolobium rupestre (E. Mey.) Walp. ●☆
32889 Argyrolobium roseum (Cambess.) Jaub. et Spach;粉红银豆●☆
32890 Argyrolobium roseum (Cambess.) Jaub. et Spach subsp. ornithopodioides (Jaub. et Spach) Jafri et Ali;鸟爪银豆●☆
32891 Argyrolobium rotundifolium T. J. Edwards;圆叶银豆●☆
32892 Argyrolobium rufopilosum De Wild. = Argyrolobium fischeri Taub. ●☆
32893 Argyrolobium rupestre (E. Mey.) Walp.;岩生银豆●☆
32894 Argyrolobium rupestre (E. Mey.) Walp. subsp. aberdaricum (Harms) Polhill;阿伯德尔岩生银豆●☆
32895 Argyrolobium rupestre (E. Mey.) Walp. subsp. kilimandscharicum (Taub.) Polhill;基利岩生银豆●☆

32896 Argyrolobium rupestre (E. Mey.) Walp. subsp. remotum (Hochst. ex A. Rich.) Polhill;散岩生银豆●☆

32897 Argyrolobium saharae Pomel;左原银豆●☆

32898 Argyrolobium sandersonii Harv. = Argyrolobium baptisioides (E. Mey.) Walp. ●☆

32899 Argyrolobium sankeyi Dümmer = Argyrolobium marginatum Bolus ●☆

32900 Argyrolobium sankeyi Harms;桑基银豆●☆

32901 Argyrolobium schimperianum Hochst. ex A. Rich.;欣珀银豆●☆

32902 Argyrolobium sericeum (Spreng.) Eckl. et Zeyh. = Argyrolobium trifoliatum (Thunb.) Druce ●☆

32903 Argyrolobium sericosemium Harms;绢毛银豆●☆

32904 Argyrolobium shirense Taub. = Argyrolobium tomentosum (Andréws) Druce ●☆

32905 Argyrolobium shirensé Taub. var. elgonense Harms = Argyrolobium fischeri Taub. ●☆

32906 Argyrolobium speciosum Eckl. et Zeyh.;美丽亮银豆●☆

32907 Argyrolobium splendens (Meisn.) Walp.;光亮银豆●☆

32908 Argyrolobium stenophyllum Boiss;窄叶银豆●☆

32909 Argyrolobium stenorrhizon Oliv. = Argyrolobium filiforme (Thunb.) Eckl. et Zeyh. ●☆

32910 Argyrolobium stipulaceum (Ball) Ball = Argyrolobium zanonii (Turra) P. W. Ball subsp. stipulaceum (Ball) Greuter ●☆

32911 Argyrolobium stipulaceum Eckl. et Zeyh.;托叶银豆●☆

32912 Argyrolobium stolzii Harms;斯托尔兹银豆●☆

32913 Argyrolobium strictum (E. Mey.) Steud. = Argyrolobium pauciflorum Eckl. et Zeyh. ●☆

32914 Argyrolobium stuhlmannii Taub. = Argyrolobium tomentosum (Andréws) Druce ●☆

32915 Argyrolobium summomontanum Hilliard et B. L. Burtt = Argyrolobium candicans Eckl. et Zeyh. ●☆

32916 Argyrolobium sutherlandii Harv. = Argyrolobium baptisioides (E. Mey.) Walp. ●☆

32917 Argyrolobium tenue (E. Mey.) Walp.;细银豆●☆

32918 Argyrolobium thodei Harms = Argyrolobium lotoides Bunge ex Trautv. ●☆

32919 Argyrolobium tomentosum (Andréws) Druce;绒毛银豆●☆

32920 Argyrolobium tortum Suess.;旋扭银豆●☆

32921 Argyrolobium transvaalense Schinz;德兰士瓦银豆●☆

32922 Argyrolobium trifoliatum (Thunb.) Druce;三小叶银豆●☆

32923 Argyrolobium trigonelloides Jaub. et Spach;三棱银豆●☆

32924 Argyrolobium tuberosum Eckl. et Zeyh.;块状银豆●☆

32925 Argyrolobium tysonii Harms = Argyrolobium rupestre (E. Mey.) Walp. ●☆

32926 Argyrolobium umbellatum (Walp.) Steud.;小伞银豆●☆

32927 Argyrolobium uniflorum (Decne.) Jaub. et Spach;单花银豆●☆

32928 Argyrolobium uniflorum Harv. = Argyrolobium harveyanum Oliv. ●☆

32929 Argyrolobium vaginiferum Harms;具鞘银豆●☆

32930 Argyrolobium variopile N. E. Br. = Argyrolobium lotoides Bunge ex Trautv. ●☆

32931 Argyrolobium velutinum Eckl. et Zeyh.;短绒毛银豆●☆

32932 Argyrolobium venustum Eckl. et Zeyh. = Argyrolobium pumilum Eckl. et Zeyh. ●☆

32933 Argyrolobium virgatum Baker = Argyrolobium rupestre (E. Mey.) Walp. subsp. remotum (Hochst. ex A. Rich.) Polhill ●☆

32934 Argyrolobium wilmsii Harms;维尔姆斯银豆●☆

32935 Argyrolobium woodii Dümmer;伍得银豆●☆

32936 Argyrolobium zanonii (Turra) P. W. Ball;扎农银豆●☆

32937 Argyrolobium zanonii (Turra) P. W. Ball subsp. fallax (Ball) Greuter et Burdet;迷惑扎农银豆●☆

32938 Argyrolobium zanonii (Turra) P. W. Ball subsp. grandiflorum (Boiss. et Reut.) Greuter;大花扎农银豆●☆

32939 Argyrolobium zanonii (Turra) P. W. Ball subsp. stipulaceum (Ball) Greuter;托叶扎农银豆●☆

32940 Argyronerium Pit. = Epigynum Wight ●

32941 Argyrophanes Schltdl. = Chrysocephalum Walp. ■☆

32942 Argyrophanes Schltdl. = Helichrysum Mill. (保留属名)●■

32943 Argyrophyllum Pohl ex Baker = Soaresia Sch. Bip. (保留属名)●☆

32944 Argyrophyllum Pohl ex Baker(1873);银叶菊属●☆

32945 Argyrophyllum ovali-ellipticum Pohl ex Baker;银叶菊●☆

32946 Argyrophyton Hook. = Argyroxiphium DC. ■☆

32947 Argyropsis M. Poem. = Plectronema Raf. ■

32948 Argyropsis M. Poem. = Zephyranthes Herb. (保留属名)■

32949 Argyropsis candida (Lindl.) M. Roem. = Zephyranthes candida (Lindl.) Herb. ■

32950 Argyrorchis Blume = Macodes (Blume) Lindl. ■☆

32951 Argyrostachys Lopriore = Achyropsis (Moq.) Benth. et Hook. f. ■☆

32952 Argyrostachys splendens Lopr. = Pandiaka carsonii (Baker) C. B. Clarke ■☆

32953 Argyrotegium J. M. Ward et Breitw. (2003);银盖鼠麴草属■☆

32954 Argyrotegium nitidulum (Hook. f.) J. M. Ward et Breitw.;银盖鼠麴草■☆

32955 Argyrothamnia Müll. Arg. = Argythamnia P. Browne ●☆

32956 Argyrothamnia canthonensis Hance = Speranskia cantonensis (Hance) Pax et K. Hoffm. ■

32957 Argyrothamnia tuberculata (Bunge) Müll. Arg. = Speranskia tuberculata (Bunge) Baill. ■

32958 Argyrovernonia MacLeish(废弃属名) = Chresta Vell. ex DC. ●■☆

32959 Argyroxiphium DC. (1836);星银菊属(银剑草属);Silversword ■☆

32960 Argyroxiphium sandwicense DC.;桑威奇星银菊(银剑草);Silver Sword, Silversword ■☆

32961 Argyroxiphium sandwicense DC. subsp. macrocephalum (Gray) Meyrat;大花星银菊(大花银剑草)■☆

32962 Argyroxyphium DC. = Argyroxiphium DC. ■☆

32963 Argytamnia Duchesne = Argythamnia P. Browne ●☆

32964 Argythamnia P. Browne(1756);银灌戟属;Silverbush ●☆

32965 Argythamnia acaulis (Herter) J. W. Ingram;无茎银灌戟●☆

32966 Argythamnia argentea Millsp.;银白银灌戟●☆

32967 Argythamnia bicolor M. E. Jones;二色银灌戟●☆

32968 Argythamnia brasiliensis Müll. Arg.;巴西银灌戟●☆

32969 Argythamnia cubensis Britton et P. Wilson;古巴银灌戟●☆

32970 Argythamnia gracilis Brandegee;细银灌戟●☆

32971 Argythamnia heteropetala Kuntze;异瓣银灌戟●☆

32972 Argythamnia humilis (Engelm. et A. Gray) Müll. Arg.;矮银灌戟●☆

32973 Argythamnia intermedia (Pax et K. Hoffm.) Allem et Irgang;间型银灌戟●☆

32974 Argythamnia laevis (A. Gray ex Torr.) Müll. Arg.;平滑银灌戟●☆

32975 Argythamnia lanceolata (Müll. Arg. ex DC.) Pax et K. Hoffm.;披针叶银灌戟●☆

32976 Argythamnia macrobotrys (Pax et K. Hoffm.) J. W. Ingram;大穗

银灌戟●☆
32977 Argythamnia mollis Müll. Arg. ;软银灌戟●☆
32978 Argythamnia oblongifolia Urb. ;矩圆叶银灌戟●☆
32979 Argythamnia rubricaulis (Pax et K. Hoffm.) Croizat;红茎银灌戟●☆
32980 Argythamnia tinctoria Millsp. ;染色银灌戟●☆
32981 Argythamnia triandra (Griseb.) Allem et Irgang;三蕊银灌戟●☆
32982 Arhynchium Lindl. = Armodorum Breda ■
32983 Arhynchium Lindl. et Paxton = Armodorum Breda ■☆
32984 Aria (Pers.) Host = Sorbus L. ●
32985 Aria (Pers.) Host(1831);赤杨叶梨属●
32986 Aria Host = Sorbus L. ●
32987 Aria J. Jacq. = Sorbus L. ●
32988 Aria alnifolia (Siebold et Zucc.) Decne. ;赤杨叶梨(枫榆,花楸,黄山榆,糯米珠,桤叶花楸,千筋树,水榆,水榆花楸,粘枣子);Dense-head Mountain Ash, Densehead Mountain-ash, Dense-headed Mountain-ash, Korean Mountain Ash, Rose-acacia ●
32989 Aria alnifolia (Siebold et Zucc.) Decne. = Sorbus alnifolia (Siebold et Zucc.) K. Koch ●
32990 Aria alnifolia (Siebold et Zucc.) Decne. f. lobulata Koidz. = Sorbus alnifolia (Siebold et Zucc.) K. Koch var. lobulata (Koidz.) Rehder ●
32991 Aria alnifolia Decne. = Sorbus alnifolia (Siebold et Zucc.) K. Koch ●
32992 Aria aronioides (Rehder) H. Ohashi et Iketani = Sorbus aronioides Rehder ●
32993 Aria astateria (Cardot) H. Ohashi et Iketani = Sorbus astateria (Cardot) Hand. -Mazz. ●
32994 Aria caloneura (Stapf) H. Ohashi et Iketani = Sorbus caloneura (Stapf) Rehder ●
32995 Aria caloneura (Stapf) H. Ohashi et Iketani var. kwangtungensis (Te T. Yu) H. Ohashi et Iketani = Sorbus caloneura (Stapf) Rehder var. kwangtungensis Te T. Yu ●
32996 Aria carpinifolia H. Ohashi et Iketani = Sorbus yunnanensis L. T. Lu ●
32997 Aria chengii (C. J. Qi) H. Ohashi et Iketani = Sorbus folgneri (C. K. Schneid.) Rehder var. duplicatodentata Te T. Yu et A. M. Lu ●
32998 Aria coronata (Cardot) H. Ohashi et Iketani = Sorbus coronata (Cardot) Te T. Yu et H. T. Tsai ●
32999 Aria corymbifera (Miq.) H. Ohashi et Iketani = Sorbus corymbifera (Miq.) T. H. Nguyên et Yakovlev ●
33000 Aria detergibilis (Merr.) H. Ohashi et Iketani = Sorbus epidendron Hand. -Mazz. ●
33001 Aria dunnii (Rehder) H. Ohashi et Iketani = Sorbus dunnii Rehder ●
33002 Aria epidendron (Hand. -Mazz.) H. Ohashi et Iketani = Sorbus epidendron Hand. -Mazz. ●
33003 Aria ferruginea (Wenz.) H. Ohashi et Iketani = Sorbus ferruginea (Wenz.) Rehder ●
33004 Aria folgneri (C. K. Schneid.) H. Ohashi et Iketani = Sorbus folgneri (C. K. Schneid.) Rehder ●
33005 Aria globosa (Te T. Yu et H. T. Tsai) H. Ohashi et Iketani = Sorbus globosa Te T. Yu et H. T. Tsai ●
33006 Aria hemsleyi (C. K. Schneid.) H. Ohashi et Iketani = Sorbus hemsleyi (C. K. Schneid.) Rehder ●
33007 Aria hunanica (C. J. Qi) H. Ohashi et Iketani = Sorbus zahlbruckneri C. K. Schneid. ●
33008 Aria japonica Decne. ;日本赤杨叶梨(日本花楸);Japan Mountainash, Japanese Mountain Ash ●☆
33009 Aria japonica Decne. = Sorbus japonica (Decne.) Hedl. ●☆
33010 Aria japonica Decne. f. calocarpa (Rehder) Yonek. ;美果日本赤杨叶梨●☆
33011 Aria japonica Decne. f. denudata (Nakai) Yonek. ;裸果日本赤杨叶梨●☆
33012 Aria keissleri (C. K. Schneid.) H. Ohashi et Iketani = Sorbus keissleri (C. K. Schneid.) Rehder ●
33013 Aria megalocarpa (Rehder) H. Ohashi et Iketani = Sorbus megalocarpa Rehder ●
33014 Aria megalocarpa (Rehder) H. Ohashi et Iketani var. cuneata (Rehder) H. Ohashi et Iketani = Sorbus megalocarpa Rehder var. cuneata Rehder ●
33015 Aria meliosmifolia (Rehder) H. Ohashi et Iketani = Sorbus meliosmifolia Rehder ●
33016 Aria ochracea (Hand. -Mazz.) H. Ohashi et Iketani = Sorbus ochracea (Hand. -Mazz.) J. E. Vidal ●
33017 Aria pallescens (Rehder) H. Ohashi et Iketani = Sorbus pallescens Rehder ●
33018 Aria rhamnoides (Decne.) H. Ohashi et Iketani = Sorbus rhamnoides (Decne.) Rehder ●
33019 Aria subochracea (Te T. Yu et A. M. Lu) H. Ohashi et Iketani = Sorbus subochracea Te T. Yu et A. M. Lu ●
33020 Aria thibetica (Cardot) H. Ohashi et Iketani = Sorbus thibetica (Cardot) Hand. -Mazz. ●
33021 Aria thomsonii (King ex Hook. f.) H. Ohashi et Iketani = Sorbus thomsonii (King) Rehder ●
33022 Aria tsinlingensis (C. L. Tang) H. Ohashi et Iketani = Sorbus tsinlingensis C. L. Tang ●
33023 Aria xanthoneura (Rehder) H. Ohashi et Iketani = Sorbus hemsleyi (C. K. Schneid.) Rehder ●
33024 Aria yuana (Spongberg) H. Ohashi et Iketani = Sorbus yuana Spongberg ●
33025 Aria yuarguta H. Ohashi et Iketani = Sorbus arguta Te T. Yu ●
33026 Aria zahlbruckneri (C. K. Schneid.) H. Ohashi et Iketani = Sorbus zahlbruckneri C. K. Schneid. ●
33027 Ariadne Urb. (1922);蛛形茜属☆
33028 Ariadne ekmanii Urb. ;蛛形茜☆
33029 Ariaria Cuervo = Bauhinia L. ●
33030 Ariaria Cuervo(1893);哥伦比亚豆属●☆
33031 Ariaria superba Cuervo;哥伦比亚豆●☆
33032 Aribis toxophylla M. Bieb. = Arabidopsis toxophylla (M. Bieb.) N. Busch ■
33033 Aribis toxophylla M. Bieb. = Pseudoarabidopsis toxophylla (M. Bieb.) Al-Shehbaz, O' Kane et R. A. Price ■
33034 Arida (R. L. Hartm.) D. R. Morgan et R. L. Hartm. (2003);沙蒿菀属;Desert Tansy-aster ■☆
33035 Arida arizonica (R. C. Jacks. et R. R. Johnson) D. R. Morgan et R. L. Hartm. ;亚利桑那沙蒿菀;Desert Tansy-aster ■☆
33036 Arida blepharophylla (A. Gray) D. R. Morgan et R. L. Hartm. ;睫毛叶沙蒿菀■☆
33037 Arida carnosa (A. Gray) D. R. Morgan et R. L. Hartm. ;肉质沙蒿菀;Shrubby Alkali Tansy-aster ■☆
33038 Arida parviflora (A. Gray) D. R. Morgan et R. L. Hartm. ;小花沙蒿菀;Small-flower Tansy-aster ■☆
33039 Arida riparia (Kunth) D. R. Morgan et R. L. Hartm. ;河岸;

Alkali Aster, Chiricahua Mountain Tansy-aster ■☆
33040 Aridaria N. E. Br. (1925);干番杏属●☆
33041 Aridaria N. E. Br. = Phyllobolus N. E. Br. ●☆
33042 Aridaria abbreviata L. Bolus = Phyllobolus abbreviatus (L. Bolus) Gerbaulet ●☆
33043 Aridaria acuminata (Haw.) Schwantes = Phyllobolus splendens (L.) Gerbaulet ●☆
33044 Aridaria albertensis L. Bolus = Phyllobolus pumilus (L. Bolus) Gerbaulet ●☆
33045 Aridaria albicaulis (Haw.) N. E. Br. = Phyllobolus splendens (L.) Gerbaulet ●☆
33046 Aridaria anguinea L. Bolus = Phyllobolus oculatus (N. E. Br.) Gerbaulet ●☆
33047 Aridaria arcuata L. Bolus = Aridaria noctiflora (L.) Schwantes ●☆
33048 Aridaria arenicola L. Bolus = Phyllobolus oculatus (N. E. Br.) Gerbaulet ●☆
33049 Aridaria aurea (Thunb.) L. Bolus = Phyllobolus nitidus (Haw.) Gerbaulet ●☆
33050 Aridaria ausana (Dinter et A. Berger) Dinter et Schwantes = Prenia tetragona (Thunb.) Gerbaulet ■☆
33051 Aridaria barkerae L. Bolus = Aridaria noctiflora (L.) Schwantes subsp. defoliata (Haw.) Gerbaulet ●☆
33052 Aridaria beaufortensis L. Bolus = Aridaria noctiflora (L.) Schwantes subsp. straminea (Haw.) Gerbaulet ●☆
33053 Aridaria bijliae N. E. Br. = Phyllobolus splendens (L.) Gerbaulet ●☆
33054 Aridaria blanda L. Bolus = Phyllobolus splendens (L.) Gerbaulet ●☆
33055 Aridaria brevicarpa L. Bolus;短果干番杏●☆
33056 Aridaria brevifolia L. Bolus = Phyllobolus splendens (L.) Gerbaulet ●☆
33057 Aridaria brevisepala L. Bolus = Phyllobolus spinuliferus (Haw.) Gerbaulet ●☆
33058 Aridaria calycina L. Bolus = Aridaria noctiflora (L.) Schwantes ●☆
33059 Aridaria canaliculata (Haw.) Friedrich = Phyllobolus canaliculatus (Haw.) Bittrich ●☆
33060 Aridaria caudata (L. Bolus) L. Bolus = Phyllobolus caudatus (L. Bolus) Gerbaulet ●☆
33061 Aridaria celans L. Bolus = Phyllobolus splendens (L.) Gerbaulet ●☆
33062 Aridaria compacta L. Bolus = Aridaria serotina L. Bolus ●☆
33063 Aridaria congesta L. Bolus = Phyllobolus congestus (L. Bolus) Gerbaulet ●☆
33064 Aridaria constricta L. Bolus = Phyllobolus splendens (L.) Gerbaulet ●☆
33065 Aridaria debilis L. Bolus = Aridaria noctiflora (L.) Schwantes ●☆
33066 Aridaria decidua L. Bolus = Phyllobolus deciduus (L. Bolus) Gerbaulet ●☆
33067 Aridaria decurvata L. Bolus = Phyllobolus decurvatus (L. Bolus) Gerbaulet ●☆
33068 Aridaria defoliata (Haw.) Schwantes = Aridaria noctiflora (L.) Schwantes subsp. defoliata (Haw.) Gerbaulet ●☆
33069 Aridaria dejagerae L. Bolus = Aridaria noctiflora (L.) Schwantes subsp. straminea (Haw.) Gerbaulet ●☆
33070 Aridaria dela (L. Bolus) L. Bolus = Phyllobolus delus (L. Bolus) Gerbaulet ■☆
33071 Aridaria dinteri L. Bolus = Phyllobolus melanospermus (Dinter et Schwantes) Gerbaulet ●☆
33072 Aridaria dyeri L. Bolus = Phyllobolus splendens (L.) Gerbaulet ●☆
33073 Aridaria ebracteata N. E. Br. = Phyllobolus trichotomus (Thunb.) Gerbaulet ●☆
33074 Aridaria ebracteata N. E. Br. var. brevipetala L. Bolus = Phyllobolus trichotomus (Thunb.) Gerbaulet ●☆
33075 Aridaria elongata L. Bolus = Phyllobolus prasinus (L. Bolus) Gerbaulet ●☆
33076 Aridaria englishiae (L. Bolus) N. E. Br. = Prenia englishiae (L. Bolus) Gerbaulet ■☆
33077 Aridaria esterhuyseniae L. Bolus = Aridaria noctiflora (L.) Schwantes subsp. straminea (Haw.) Gerbaulet ●☆
33078 Aridaria fastigiata Schwantes = Phyllobolus splendens (L.) Gerbaulet ●☆
33079 Aridaria flexuosa (Haw.) Schwantes = Phyllobolus splendens (L.) Gerbaulet ●☆
33080 Aridaria floribunda L. Bolus = Aridaria noctiflora (L.) Schwantes ●☆
33081 Aridaria fourcadei L. Bolus = Phyllobolus splendens (L.) Gerbaulet ●☆
33082 Aridaria fragilis (N. E. Br.) Friedrich = Phyllobolus oculatus (N. E. Br.) Gerbaulet ●☆
33083 Aridaria framesii L. Bolus = Phyllobolus spinuliferus (Haw.) Gerbaulet ●☆
33084 Aridaria fulva (Haw.) Schwantes = Aridaria noctiflora (L.) Schwantes subsp. straminea (Haw.) Gerbaulet ●☆
33085 Aridaria geniculiflora (L.) N. E. Br. = Aptenia geniculiflora (L.) Bittrich ex Gerbaulet ■☆
33086 Aridaria gibbosa L. Bolus = Phyllobolus spinuliferus (Haw.) Gerbaulet ●☆
33087 Aridaria glandulifera L. Bolus = Phyllobolus sinuosus (L. Bolus) Gerbaulet ●☆
33088 Aridaria globosa L. Bolus = Aridaria serotina L. Bolus ●☆
33089 Aridaria godmaniae L. Bolus = Phyllobolus sinuosus (L. Bolus) Gerbaulet ●☆
33090 Aridaria gracilis L. Bolus = Aridaria serotina L. Bolus ●☆
33091 Aridaria gratiae L. Bolus = Phyllobolus grossus (Aiton) Gerbaulet ■☆
33092 Aridaria grossa (Aiton) Friedrich = Phyllobolus grossus (Aiton) Gerbaulet ■☆
33093 Aridaria herbertii (N. E. Br.) Friedrich = Phyllobolus herbertii (N. E. Br.) Gerbaulet ●☆
33094 Aridaria hesperantha (L. Bolus) N. E. Br. = Mesembryanthemum longistylum DC. ■☆
33095 Aridaria horizontalis (Haw.) Schwantes = Aridaria noctiflora (L.) Schwantes subsp. defoliata (Haw.) Gerbaulet ●☆
33096 Aridaria inaequalis L. Bolus = Phyllobolus nitidus (Haw.) Gerbaulet ●☆
33097 Aridaria intricata L. Bolus = Aridaria brevicarpa L. Bolus ●☆
33098 Aridaria klaverensis L. Bolus = Aridaria brevicarpa L. Bolus ●☆
33099 Aridaria latipetala L. Bolus = Phyllobolus latipetalus (L. Bolus) Gerbaulet ●☆
33100 Aridaria laxa L. Bolus = Phyllobolus decurvatus (L. Bolus) Gerbaulet ●☆
33101 Aridaria laxipetala L. Bolus = Phyllobolus grossus (Aiton) Gerbaulet ■☆
33102 Aridaria leipoldtii L. Bolus = Aridaria noctiflora (L.) Schwantes

●☆

33103 Aridaria leptopetala L. Bolus = Phyllobolus splendens (L.) Gerbaulet ●☆

33104 Aridaria lignea L. Bolus = Phyllobolus melanospermus (Dinter et Schwantes) Gerbaulet ●☆

33105 Aridaria littlewoodii L. Bolus = Aridaria noctiflora (L.) Schwantes subsp. straminea (Haw.) Gerbaulet ●☆

33106 Aridaria longisepala L. Bolus = Aridaria noctiflora (L.) Schwantes subsp. defoliata (Haw.) Gerbaulet ●☆

33107 Aridaria longispinula (Haw.) L. Bolus = Phyllobolus grossus (Aiton) Gerbaulet ■☆

33108 Aridaria longistyla (DC.) Schwantes = Mesembryanthemum longistylum DC. ■☆

33109 Aridaria longituba L. Bolus = Phyllobolus tenuiflorus (Jacq.) Gerbaulet ●☆

33110 Aridaria luteoalba L. Bolus = Prenia tetragona (Thunb.) Gerbaulet ■☆

33111 Aridaria macrosiphon L. Bolus = Phyllobolus tenuiflorus (Jacq.) Gerbaulet ●☆

33112 Aridaria meridiana L. Bolus = Aridaria noctiflora (L.) Schwantes subsp. defoliata (Haw.) Gerbaulet ●☆

33113 Aridaria meyeri L. Bolus = Aridaria brevicarpa L. Bolus ●☆

33114 Aridaria muirii N. E. Br. = Aridaria noctiflora (L.) Schwantes subsp. defoliata (Haw.) Gerbaulet ●☆

33115 Aridaria multiseriata L. Bolus = Phyllobolus prasinus (L. Bolus) Gerbaulet ●☆

33116 Aridaria mutans L. Bolus = Prenia tetragona (Thunb.) Gerbaulet ■☆

33117 Aridaria nevillei L. Bolus = Aridaria noctiflora (L.) Schwantes ●☆

33118 Aridaria nitida (Haw.) N. E. Br. = Phyllobolus nitidus (Haw.) Gerbaulet ●☆

33119 Aridaria nocta (N. E. Br.) N. E. Br. = Phyllobolus splendens (L.) Gerbaulet ●☆

33120 Aridaria noctiflora (L.) Schwantes;夜花干番杏●☆

33121 Aridaria noctiflora (L.) Schwantes subsp. defoliata (Haw.) Gerbaulet;落叶夜花干番杏●☆

33122 Aridaria noctiflora (L.) Schwantes subsp. straminea (Haw.) Gerbaulet;草黄夜花干番杏●☆

33123 Aridaria noctiflora (L.) Schwantes var. fulva (Haw.) Herre et Friedrich = Aridaria noctiflora (L.) Schwantes subsp. straminea (Haw.) Gerbaulet ●☆

33124 Aridaria obtusa L. Bolus = Phyllobolus decurvatus (L. Bolus) Gerbaulet ●☆

33125 Aridaria oculata (N. E. Br.) L. Bolus = Phyllobolus oculatus (N. E. Br.) Gerbaulet ●☆

33126 Aridaria odorata (L. Bolus) Schwantes ex H. Jacobsen = Hereroa odorata (L. Bolus) L. Bolus ●☆

33127 Aridaria oubergensis L. Bolus = Phyllobolus pumilus (L. Bolus) Gerbaulet ●☆

33128 Aridaria ovalis L. Bolus = Aridaria serotina L. Bolus ●☆

33129 Aridaria parvisepala L. Bolus = Phyllobolus spinuliferus (Haw.) Gerbaulet ●☆

33130 Aridaria paucandra L. Bolus;寡蕊干番杏●☆

33131 Aridaria paucandra L. Bolus var. gracillima = Aridaria serotina L. Bolus ●☆

33132 Aridaria paucandra L. Bolus var. paucandra = Aridaria serotina L. Bolus ●☆

33133 Aridaria peersii L. Bolus = Phyllobolus tenuiflorus (Jacq.) Gerbaulet ●☆

33134 Aridaria pentagona L. Bolus = Phyllobolus splendens (L.) Gerbaulet subsp. pentagonus (L. Bolus) Gerbaulet ●☆

33135 Aridaria pentagona L. Bolus var. occidentalis = Phyllobolus splendens (L.) Gerbaulet subsp. pentagonus (L. Bolus) Gerbaulet ●☆

33136 Aridaria pillansii L. Bolus = Aridaria noctiflora (L.) Schwantes ●☆

33137 Aridaria platysepala L. Bolus = Phyllobolus grossus (Aiton) Gerbaulet ■☆

33138 Aridaria plenifolia (N. E. Br.) Stearn = Phyllobolus splendens (L.) Gerbaulet ●☆

33139 Aridaria pomonae L. Bolus = Phyllobolus oculatus (N. E. Br.) Gerbaulet ●☆

33140 Aridaria prasina L. Bolus = Phyllobolus prasinus (L. Bolus) Gerbaulet ●☆

33141 Aridaria primulina L. Bolus = Phyllobolus splendens (L.) Gerbaulet ●☆

33142 Aridaria pumila L. Bolus = Phyllobolus pumilus (L. Bolus) Gerbaulet ●☆

33143 Aridaria quartzitica L. Bolus = Phyllobolus quartziticus (L. Bolus) Gerbaulet ●☆

33144 Aridaria quaterna L. Bolus = Phyllobolus spinuliferus (Haw.) Gerbaulet ●☆

33145 Aridaria rabiei L. Bolus = Phyllobolus rabiei (L. Bolus) Gerbaulet ●☆

33146 Aridaria rabiesbergensis L. Bolus = Phyllobolus splendens (L.) Gerbaulet ●☆

33147 Aridaria radicans L. Bolus = Prenia radicans (L. Bolus) Gerbaulet ●☆

33148 Aridaria rangei (N. E. Br.) Friedrich = Phyllobolus oculatus (N. E. Br.) Gerbaulet ●☆

33149 Aridaria recurva L. Bolus = Phyllobolus sinuosus (L. Bolus) Gerbaulet ●☆

33150 Aridaria reflexa (Haw.) N. E. Br. = Phyllobolus splendens (L.) Gerbaulet ●☆

33151 Aridaria resurgens (Kensit) L. Bolus = Phyllobolus resurgens (Kensit) Schwantes ●☆

33152 Aridaria rhodandra L. Bolus = Phyllobolus nitidus (Haw.) Gerbaulet ●☆

33153 Aridaria rosea L. Bolus = Phyllobolus splendens (L.) Gerbaulet ●☆

33154 Aridaria saturata L. Bolus = Phyllobolus saturatus (L. Bolus) Gerbaulet ●☆

33155 Aridaria scintillans (Dinter) Friedrich = Phyllobolus oculatus (N. E. Br.) Gerbaulet ●☆

33156 Aridaria serotina L. Bolus;迟花干番杏●☆

33157 Aridaria spinulifera (Haw.) N. E. Br. = Phyllobolus spinuliferus (Haw.) Gerbaulet ●☆

33158 Aridaria splendens (L.) Schwantes = Phyllobolus splendens (L.) Gerbaulet ●☆

33159 Aridaria straminea (Haw.) Schwantes = Aridaria noctiflora (L.) Schwantes subsp. straminea (Haw.) Gerbaulet ●☆

33160 Aridaria straminea L. Bolus = Phyllobolus sinuosus (L. Bolus) Gerbaulet ●☆

33161 Aridaria straminicolor L. Bolus = Phyllobolus sinuosus (L. Bolus) Gerbaulet ●☆

33162 Aridaria striata L. Bolus = Phyllobolus splendens (L.) Gerbaulet

●☆
33163 Aridaria stricta L. Bolus = Phyllobolus spinuliferus (Haw.) Gerbaulet ●☆
33164 Aridaria subaequans L. Bolus = Phyllobolus splendens (L.) Gerbaulet subsp. pentagonus (L. Bolus) Gerbaulet ●☆
33165 Aridaria subpatens L. Bolus = Phyllobolus splendens (L.) Gerbaulet subsp. pentagonus (L. Bolus) Gerbaulet ●☆
33166 Aridaria subpetiolata L. Bolus = Phyllobolus grossus (Aiton) Gerbaulet ■☆
33167 Aridaria subtruncata L. Bolus = Aridaria noctiflora (L.) Schwantes subsp. straminea (Haw.) Gerbaulet ●☆
33168 Aridaria suffusa L. Bolus = Prenia tetragona (Thunb.) Gerbaulet ■☆
33169 Aridaria sulcata (Haw.) Schwantes = Phyllobolus splendens (L.) Gerbaulet ●☆
33170 Aridaria tenuifolia L. Bolus;细叶干番杏●☆
33171 Aridaria tenuifolia L. Bolus = Aridaria serotina L. Bolus ●☆
33172 Aridaria tenuifolia L. Bolus var. speciosa (L. Bolus) L. Bolus = Aridaria serotina L. Bolus ●☆
33173 Aridaria tetragona (Thunb.) L. Bolus = Prenia tetragona (Thunb.) Gerbaulet ■☆
33174 Aridaria tetramera L. Bolus;四数干番杏●☆
33175 Aridaria tetramera L. Bolus = Phyllobolus trichotomus (Thunb.) Gerbaulet ●☆
33176 Aridaria tetramera L. Bolus var. parviflora ? = Phyllobolus trichotomus (Thunb.) Gerbaulet ●☆
33177 Aridaria trichosantha (A. Berger) N. E. Br. = Phyllobolus viridiflorus (Aiton) Gerbaulet ●☆
33178 Aridaria trichotoma (Thunb.) L. Bolus = Phyllobolus trichotomus (Thunb.) Gerbaulet ●☆
33179 Aridaria umbelliflora (Jacq.) Schwantes = Phyllobolus splendens (L.) Gerbaulet ●☆
33180 Aridaria varians L. Bolus = Phyllobolus oculatus (N. E. Br.) Gerbaulet ●☆
33181 Aridaria vernalis L. Bolus = Phyllobolus splendens (L.) Gerbaulet ●☆
33182 Aridaria vespertina L. Bolus;夕干番杏●☆
33183 Aridaria viridiflora (Aiton) L. Bolus;绿花干番杏;Green Fig Marigold ●☆
33184 Aridaria viridiflora (Aiton) L. Bolus = Phyllobolus viridiflorus (Aiton) Gerbaulet ●☆
33185 Aridaria watermeyeri L. Bolus = Phyllobolus spinuliferus (Haw.) Gerbaulet ●☆
33186 Aridaria willowmorensis L. Bolus = Phyllobolus grossus (Aiton) Gerbaulet ■☆
33187 Aridarum Ridl. (1913);异疆南星属■☆
33188 Aridarum minimum H. Okada;小异疆南星■☆
33189 Aridarum montanum Ridl.;山地异疆南星■☆
33190 Arietinum Beck = Criosanthes Raf. ■
33191 Arietinum Beck = Cypripedium L. ■
33192 Arietinum L. C. Beck = Criosanthes Raf. ■
33193 Arietinum L. C. Beck = Cypripedium L. ■
33194 Arikuriroba Barb. Rodr. = Arikuryroba Barb. Rodr. ●☆
33195 Arikury Becc. = Arikuryroba Barb. Rodr. ●☆
33196 Arikury Becc. = Syagrus Mart. ●
33197 Arikuryroba Barb. Rodr. (1891);阿利棕属(巴西棕属);Arikury Palm ●☆
33198 Arikuryroba Barb. Rodr. = Syagrus Mart. ●
33199 Arikuryroba capanemae Barb. Rodr.;阿利棕●☆
33200 Arikuryroba schizophylla L. H. Bailey;裂叶阿利棕●☆
33201 Arillaria S. Kurz = Ormosia Jacks. (保留属名)●
33202 Arillastrum Pancher ex Baill. (1877);假皮桃金娘属●☆
33203 Arillastrum Pancher ex Baill. = Stereocaryum Burret ●☆
33204 Arillastrum gummiferum (Brongn. et Gris) Pancher ex Baill.;假皮桃金娘●☆
33205 Arinemia Raf. = Ilex L. ●
33206 Ariocarpus Scheidw. (1838);岩牡丹属(牡丹球属,玉牡丹属);Living-rock Cactus, Living Rock ●
33207 Ariocarpus agavoides (Castaneda) E. F. Anderson;龙舌兰岩牡丹(龙舌岩牡丹);Agavelike Living-rock Cactus ●
33208 Ariocarpus elongatus (Salm-Dyck) Wettst. = Ariocarpus retusus Scheidw. ●
33209 Ariocarpus fissuratus (Engelm.) K. Schum.;龟甲岩牡丹;Chautle Living Rock, False Peyote, Living Rock, Spliting Living-rock Cactus ●☆
33210 Ariocarpus fissuratus (Engelm.) K. Schum. var. lloydii (Rose) Marshall = Ariocarpus fissuratus (Engelm.) K. Schum. ●☆
33211 Ariocarpus fissuratus K. Schum. = Ariocarpus fissuratus (Engelm.) K. Schum. ●☆
33212 Ariocarpus furfuraceus C. H. Thomps.;微凹岩牡丹(龟甲岩牡丹,花牡丹,微凹牡丹)●☆
33213 Ariocarpus kotschoubeyanus (Lem.) K. Schum.;黑岩牡丹(黑牡丹)●☆
33214 Ariocarpus lloydii Rose;娄氏龟甲岩牡丹●☆
33215 Ariocarpus lloydii Rose = Ariocarpus fissuratus (Engelm.) K. Schum. ●☆
33216 Ariocarpus pulvilligeris K. Schum. = Ariocarpus retusus Scheidw. ●
33217 Ariocarpus retusus Scheidw.;岩牡丹;Living Rock, Seven Stars, Seven-stars Living-rock Cactus ●
33218 Ariocarpus retusus Scheidw. subsp. trigonus (F. A. C. Weber) E. F. Anderson et W. A. Fitz Maur. = Ariocarpus trigonus (A. Weber) K. Schum. ●
33219 Ariocarpus retusus Scheidw. var. furfuraceus (Watson) G. Frank = Ariocarpus furfuraceus C. H. Thomps. ●☆
33220 Ariocarpus scapharostrus Boed.;龙角岩牡丹(龙角牡丹);Boed Living-rock Cactus ●
33221 Ariocarpus trigonus (A. Weber) K. Schum.;三角岩牡丹(三角牡丹);Triangular Living-rock Cactus ●
33222 Ariodendron Meisn. = Agriodendron Endl. ●■
33223 Ariodendron Meisn. = Aloe L. ●■
33224 Ariona Pers. = Arjona Comm. ex Cav. ☆
33225 Arionaceae Tiegh. = Arjonaceae Tiegh. ●■
33226 Arionaceae Tiegh. = Santalaceae R. Br. (保留科名)●■
33227 Ariopsis J. Graham = Ariopsis Nimmo ■☆
33228 Ariopsis Nimmo (1839);假赤杨叶梨属■☆
33229 Ariopsis peltata Nimmo;假赤杨叶梨●☆
33230 Ariosorbus Koidz. = Sorbus L. ●
33231 Aripuana Struwe, Maas et V. A. Albert (1997);阿利龙胆属■☆
33232 Arirolobium Desv. = Coronilla L. (保留属名)●■
33233 Arisacontis Schott = Cyrtosperma Griff. ■
33234 Arisaema Mart. (1831);天南星属;Arisaema, Cobra Lily, Dragon Arum, Indian Turnip, Jack-in-the-pulpit, Snake Lily, Southstar ●■
33235 Arisaema abbreviatum Schott = Arisaema flavum (Forssk.) Schott ■
33236 Arisaema abei Seriz.;阿拜天南星■☆

33237 Arisaema aequinoctiale Nakai et F. Maek.;昼夜天南星■☆
33238 Arisaema affine Schott = Arisaema concinnum Schott ■
33239 Arisaema akiense Nakai = Arisaema iyoanum Makino ■☆
33240 Arisaema akiense Nakai var. nakaianum Kitag. et Ohba = Mitella stylosa H. Boissieu var. makinoi (H. Hara) Wakab. ■☆
33241 Arisaema alcareum H. Li = Arisaema calcareum H. Li ■
33242 Arisaema alienatum Schott = Arisaema concinnum Schott ■
33243 Arisaema alienatum Schott var. formosanum Hayata = Arisaema formosanum (Hayata) Hayata ■
33244 Arisaema ambiguum Engl. = Arisaema heterophyllun Blume ■
33245 Arisaema amurense Maxim.;东北南星(长虫苞米,大参,大天落星,大头参,东北天南星,虎掌,山苞米,天老星,天南星);Amur Arisaema,Amur Southstar ■
33246 Arisaema amurense Maxim. f. purpureum (Nakai) Kitag.;齿叶紫苞东北天南星(齿缘紫苞天南星);Purple Amur Arisaema ■
33247 Arisaema amurense Maxim. f. serratum Kitag. = Arisaema amurense Maxim. var. serratum Nakai ■
33248 Arisaema amurense Maxim. f. violaceum (Engl.) Kitag.;紫苞东北天南星(紫苞天南星);Violet Amur Arisaema ■
33249 Arisaema amurense Maxim. subsp. robustum (Engl.) H. Ohashi et J. Murata var. ovale (Nakai) H. Ohashi et J. Murata = Arisaema ovale Nakai ■☆
33250 Arisaema amurense Maxim. var. denticulatum Engl. = Arisaema amurense Maxim. var. serratum Nakai ■
33251 Arisaema amurense Maxim. var. inaense Seriz. = Arisaema ovale Nakai var. inaense (Seriz.) J. Murata ■☆
33252 Arisaema amurense Maxim. var. robustum Engl. = Arisaema amurense Maxim. ■
33253 Arisaema amurense Maxim. var. serratum Nakai;齿叶东北南星(锯叶天南星);Serrate Amur Arisaema,Serrate Amur Southstar ■
33254 Arisaema amurense Maxim. var. typicum Engl. = Arisaema amurense Maxim. ■
33255 Arisaema amurense Maxim. var. violaceum Engl. = Arisaema amurense Maxim. f. violaceum (Engl.) Kitag. ■
33256 Arisaema amurense Maxim. var. violaceum Engl. = Arisaema amurense Maxim. ■
33257 Arisaema angustatum Engl. = Arisaema angustatum Franch. et Sav. ■
33258 Arisaema angustatum Franch. et Sav.;狭叶南星;Narrowleaf Arisaema,Narrowleaf Southstar ■
33259 Arisaema angustatum Franch. et Sav. = Arisaema serratum (Thunb.) Schott ■
33260 Arisaema angustatum Franch. et Sav. var. peninsulae (Nakai) Nakai ex Miyabe et Kudo;朝鲜南星(长虫苞米,朝鲜天南星,大参,山苞米,天老星,天南星);Korea Arisaema,Korea Southstar ■
33261 Arisaema angustatum Franch. et Sav. var. peninsulae (Nakai) Nakai ex Miyabe et Kudo = Arisaema serratum (Thunb.) Schott ■
33262 Arisaema angustatum Franch. et Sav. var. peninsulae (Nakai) Nakai = Arisaema angustatum Franch. et Sav. var. peninsulae (Nakai) Nakai ex Miyabe et Kudo ■
33263 Arisaema aridum H. Li;旱生南星;Xeric Arisaema,Xeric Southstar ■
33264 Arisaema arisanense Hayata;阿里山南星(虎掌,天南星);Alishan Arisaema,Alishan Southstar ■
33265 Arisaema arisanense Hayata = Arisaema ringens (Thunb.) Schott ●■
33266 Arisaema asperatum N. E. Br.;刺柄南星(白南星,刺梗天南星,绿南星,南星,南星七,三步跳,三甫莲,三角莲,山苞谷,天南星);Scabrous Arisaema,Scabrous Southstar ■
33267 Arisaema atrorubens (Aiton) Blume = Arisaema triphyllum (L.) Schott ■☆
33268 Arisaema atrorubens (Aiton) Blume f. viride (Engl.) Fernald = Arisaema triphyllum (L.) Schott ■☆
33269 Arisaema atrorubens (Aiton) Blume f. zebrinum (Sims) Fernald = Arisaema triphyllum (L.) Schott ■☆
33270 Arisaema atrorubens Blume;深红天南星;Jack-in-the-pulpit ■
33271 Arisaema atrorubens Blume = Arisaema triphyllum (L.) Schott ■☆
33272 Arisaema auriculatum Buchet;长耳南星(半夏,大耳南星);Auricled Arisaema,Auriculate Southstar ■
33273 Arisaema austro-yunnanense H. Li;滇南星;S. Yunnan Arisaema,S. Yunnan Southstar ■
33274 Arisaema balansae Engl.;越南红根南星(红根)■☆
33275 Arisaema bannaense H. Li;版纳南星;Xishuangbanna Arisaema ■
33276 Arisaema bathycoleum Hand. -Mazz.;银南星(半夏,地球半夏,高鞘南星,麻芋子,银半夏);Argentate Southstar,Highsheath Arisaema ■
33277 Arisaema bequaertii De Wild. = Arisaema mildbraedii Engl. ■☆
33278 Arisaema biauriculatum W. W. Sm. ex Hand. -Mazz.;双耳南星(半夏);Doubleear Southstar,Two-auricled Arisaema ■
33279 Arisaema biliferum ?;胆天南星■☆
33280 Arisaema biradiatifoliatum Kitam.;大关山南星■
33281 Arisaema biradiatifoliatum Kitam. = Arisaema consanguineum Schott ■
33282 Arisaema bockii Engl. = Arisaema sikokianum Franch. et Sav. var. serratum (Makino) Hand. -Mazz. ■
33283 Arisaema bonatianum Engl.;沧江南星;Cangjiang Arisaema ■
33284 Arisaema brachyspathum Hayata;短檐南星;Shortlimbate Arisaema,Shortlimbate Southstar ■
33285 Arisaema brevipes Engl.;短柄南星;Shortpetiole Arisaema,Shortpetiole Southstar ■
33286 Arisaema brevispathum Buchet;短苞南星;Shortbract Arisaema ■
33287 Arisaema brevistipitatum Merr. = Arisaema cordatum N. E. Br. ■
33288 Arisaema calcareum H. Li;红根南星(长虫包谷,红根,见血飞,山磨芋,小独角莲,小独脚莲,野磨芋);Redroot Arisaema,Redroot Southstar ■
33289 Arisaema candidissimum W. W. Sm.;白苞南星(白南星,极白南星);Whitebract Arisaema,Whitespathe Southstar ■
33290 Arisaema canshanense X. D. Dong;苍山南星■
33291 Arisaema ciliatum H. Li;缘毛南星;Greenhair Arisaema,Greenhair Southstar ■
33292 Arisaema clavatum Buchet;棒头南星(虎掌,麻芋子,南星,七寸胆,蛇包谷);Clavate Arisaema,Clavate Southstar ■
33293 Arisaema concinnum Schott;皱序南星;Wrinklespadix Arisaema,Wrinklespadix Southstar ■
33294 Arisaema concinnum Schott var. alienatum (Schott) Engl. = Arisaema concinnum Schott ■
33295 Arisaema consanguineum Schott;长行天南星(白南星,斑杖,半夏精,长尾叶天南星,长须南星,刺天南星,大扁老鸦芋头,大关山南星,大头参,独角莲,独叶一枝枪,狗爪半夏,狗爪南星,鬼蒟蒻,虎膏,虎掌,虎掌半夏,虎掌南星,基隆南星,雷公统,拟天南星,三棒子,山棒子,山苞米,蛇包谷,蛇六谷,蛇头草,蛇头天南星,蛇芋,天老星,天南星,象天南星,药狗丹,野芋头,异叶天南星,禹南星);Biradiateleaf Arisaema,Biradiateleaf Southstar,Jack in the Pulpit,Jilong Arisaema,Jilong Southstar ■

33296 Arisaema consanguineum Schott = Arisaema erubescens (Wall.) Schott ■
33297 Arisaema consanguineum Schott var. divaricatum Engl. = Atisaema erubescens (Wall.) Schott ■
33298 Arisaema consanguineum Schott var. kelung-insulare (Hayata) T. C. Huang = Arisaema consanguineum Schott ■
33299 Arisaema cordatum N. E. Br.;心檐南星(七叶莲);Cordate Arisaema,Cordate Southstar ■
33300 Arisaema cornutum Schott = Arisaema jacquemontii Blume ■
33301 Arisaema costatum (Wall.) Mart.;多脉南星(长尾天南星);Manyvein Arisaema,Manyvein Southstar ■
33302 Arisaema costatum (Wall.) Mart. f. propinquum (Schott) H. Hara = Arisaema propinquum Schott ■
33303 Arisaema costatum (Wall.) Mart. var. sikkimense (Stapf) Hara = Arisaema propinquum Schott ■
33304 Arisaema cucullatum M. Hotta;僧帽南星■☆
33305 Arisaema curvatum (Roxb.) Kunth = Arisaema tortuosum (Wall.) Schott var. curvatum (Roxb.) Engl. ■☆
33306 Arisaema curvatum Kunth = Arisaema tortuosum (Wall.) Schott ■
33307 Arisaema dahaiense H. Li;会泽南星;Dahai Southstar,Huize Arisaema ■
33308 Arisaema daochengense P. C. Kao;稻城南星;Daocheng Arisaema ■
33309 Arisaema decipiens Schott;奇异南星(青脚莲,蛇饭,铁灯台);Peculiar Arisaema,Peculiar Southstar ■
33310 Arisaema delavayi Buchet;大理南星;Delavay Arisaema,Delavay Southstar ■
33311 Arisaema dilatatum Buchet;粗序南星(南星);Dilatate Arisaema,Dilatate Southstar ■
33312 Arisaema divaricatum Engl. = Arisaema erubescens (Wall.) Schott ■
33313 Arisaema dolosum Schott = Arisaema intermedium Blume ■
33314 Arisaema dracontium (L.) Schott;龙根天南星(龙南星);Dragon Arum,Dragon Root,Dragon-root,Green Dragon,Green-dragon ■☆
33315 Arisaema dracontium Schott = Arisaema dracontium (L.) Schott ■☆
33316 Arisaema du-bois-reymondiae Engl.;云台南星(江南南星,江苏南星,江苏天南星,细齿天南星);Jiangsu Arisaema,Yuntai Southstar ■
33317 Arisaema dulongense H. Li;独龙南星;Dulong Arisaema ■
33318 Arisaema echinatum (Wall.) Schott;刺棒南星;Echinate Arisaema,Echinate Southstar ■
33319 Arisaema elephas Buchet;象南星(半节烂,大半夏,大麻芋子,黑南星,虎掌,麻芋子,三步跳,水包谷,天南星,象鼻南星,象鼻子,象天南星,银半夏);Elephant Arisaema,Elephant Southstar ■
33320 Arisaema elephas Buchet var. jishaense H. Li et A. M. Li;吉沙南星;Jisha Arisaema,Jisha Southstar ■
33321 Arisaema eminens Schott = Arisaema speciosum (Wall.) Mart. ■
33322 Arisaema engleri Pamp. = Arisaema sikokianum Franch. et Sav. var. serratum (Makino) Hand. -Mazz. ■
33323 Arisaema engleri Pamp. = Arisaema sikokianum Franch. et Sav. ■
33324 Arisaema enneaphyllum Hochst. ex A. Rich.;九叶天南星■☆
33325 Arisaema erubescens (Wall.) Schott;一把伞南星(白南星,斑杖,半夏,半夏精,打蛇棒,大扁老鸦芋头,刀口药,独角莲,独脚莲,法夏,粉南星,狗爪半夏,虎膏,虎掌,虎掌南星,黄狗卵,老蛇包谷,麻蛇饭,麻芋杆,麻芋子,南星,闹狗药,三棒子,伞南星,山包谷,山苞米,山蕃芋,山魔芋,蛇包谷,蛇六谷,蛇木芋,蛇舌草,蛇蒜头,蛇头天南星,蛇芋,蛇子麦,蛇钻头,社芋头,天南星,铁骨伞,血南星,药狗丹,野魔芋,野芋头,一把伞);Blush Red Arisaema,One Umbrella Southstar ■
33326 Arisaema erubescens (Wall.) Schott f. dongyyangense ?;紫序一把伞南星■
33327 Arisaema erubescens (Wall.) Schott var. consanguineum Engl. = Arisaema erubescens (Wall.) Schott ■
33328 Arisaema exappendiculatum Hara;圈药南星;Appendageless Arisaema,Coiledanther Southstar ■
33329 Arisaema exile Schott = Arisaema jacquemontii Blume ■
33330 Arisaema fargesii Buchet;螃蟹七(白南星,城口天南星,红南星,虎掌南星,郎毒,天南星);Farges Arisaema,Farges Southstar ■
33331 Arisaema flavum (Forssk.) Schott;黄苞南星(达黄,狗角莲,黄花南星,天南星);Yellowflower Arisaema,Yellowspathe Southstar ■
33332 Arisaema formosanum (Hayata) Hayata;台南星(台湾南星,台湾天南星,溪南山南星,狭叶台南星,狭叶天南星);Narrowleaf Taiwan Arisaema,Oblanceolate Arisaema,Oblanceolate Southstar,Taiwan Arisaema,Taiwan Southstar ■
33333 Arisaema formosanum Hayata f. stenophyllum Hayata = Arisaema formosanum (Hayata) Hayata ■
33334 Arisaema formosanum Hayata var. bicolorifolium T. C. Huang = Arisaema formosanum (Hayata) Hayata ■
33335 Arisaema franchetianum Engl.;象头花(半夏,大半夏,独叶半夏,狗爪南星,黑南星,红半夏,红南星,虎掌,老母猪半夏,母猪半夏,南星,三步莲,三步跳,山半夏,天南星,象鼻花,小独角莲,小独脚莲,岩芋,野魔芋,野芋头,紫盔南星);Franchet Arisaema,Franchet Southstar ■
33336 Arisaema fraternum Schott = Arisaema erubescens (Wall.) Schott ■
33337 Arisaema grapsospadix Hayata;二色南星(毛笔天南星);Bicolored Arisaema,Bicolored Southstar ■
33338 Arisaema griffithii Schott;翼檐南星(格里氏南星,天南星);Griffith Arisaema,Griffith Southstar ■
33339 Arisaema griffithii Schott var. verrucosum (Schott) Hara;疣柄翼檐南星(多疣南星,多疣天南星,疣柄南星)■
33340 Arisaema guixiense S. Y. Liu;桂西南星;Guixi Arisaema ■
33341 Arisaema hainanense C. Y. Wu ex H. Li et al.;黎婆花;Hainan Arisaema,Hainan Southstar ■
33342 Arisaema handelii Stapf ex Hand. -Mazz.;疣序南星;Handel Arisaema,Handel Southstar ■
33343 Arisaema hatizyoense Nakai;八丈南星■☆
33344 Arisaema helleborifolium Schott = Arisaema tortuosum (Wall.) Schott ■
33345 Arisaema heterocephalum Koidz.;互头天南星■☆
33346 Arisaema heterocephalum Koidz. subsp. majus (Seriz.) J. Murata;大互头天南星■☆
33347 Arisaema heterocephalum Koidz. subsp. okinawense H. Ohashi et J. Murata;兴津川南星■☆
33348 Arisaema heterocephalum Koidz. var. okinawense (H. Ohashi et J. Murata) Seriz. = Arisaema heterocephalum Koidz. subsp. okinawense H. Ohashi et J. Murata ■☆
33349 Arisaema heterophyllum Blume;天南星(白南星,半边莲,不求人,大半夏,独角莲,独脚莲,独叶一枝枪,独足伞,逢人不见面,狗爪半夏,虎掌,虎掌半夏,麻芋子,母子半夏南星,南星,青杆独叶一枝枪,山魔芋,蛇棒头,蛇包谷,蛇六谷,蛇头草,蛇头蒜,双隆芋,锁喉莲,天老咋,天凉伞,土南星,小苞米,异叶天南星,羽叶天南星);Differentleaved Arisaema,Diverseleaf Southstar ■
33350 Arisaema heterophyllum Blume var. typicum Makino = Arisaema

heterophyllum Blume ■
33351 Arisaema hookeri Schott = Arisaema griffithii Schott ■
33352 Arisaema hookerianum Schott = Arisaema griffithii Schott ■
33353 Arisaema hunanense Hand. -Mazz. ; 湘南星(南星); Hunan Arisaema, Hunan Southstar ■
33354 Arisaema hungyaense H. Li;洪雅南星;Hongya Arisaema,Hongya Southstar ■
33355 Arisaema hypoglaucum Craib. = Arisaema erubescens (Wall.) Schott ■
33356 Arisaema ilanense J. C. Wang;宜兰天南星;Yilan Southstar ■
33357 Arisaema inkiangense H. Li;三匹箭(三叶半夏,盈江南星); Yingjiang Southstar ■
33358 Arisaema inkiangense H. Li var. maculatum H. Li;斑叶三匹箭(斑叶盈江南星);Spot-leaf Southstar ■
33359 Arisaema intermedium Blume;高原南星(土半夏,中南星); Plateau Arisaema, Plateau Southstar ■
33360 Arisaema intermedium Blume var. propinquum (Schott) Engl. = Arisaema propinquum Schott ■
33361 Arisaema ishizuchiense Murata;石锤南星■☆
33362 Arisaema ishizuchiense Murata subsp. brevicollum (H. Ohashi et J. Murata) Seriz. ;短颈石锤南星■☆
33363 Arisaema ishizuchiense Murata subsp. brevicollum (H. Ohashi et J. Murata) Seriz. var. alpicola Seriz. ;山地短颈石锤南星■☆
33364 Arisaema ishizuchiense Murata var. brevicollum H. Ohashi et J. Murata = Arisaema ishizuchiense Murata subsp. brevicollum (H. Ohashi et J. Murata) Seriz. ■☆
33365 Arisaema iyoanum Makino;伊予南星■☆
33366 Arisaema iyoanum Makino subsp. nakaianum (Kitag. et Ohba) H. Ohashi et J. Murata;中井氏南星■☆
33367 Arisaema iyoanum Makino var. nakaianum (Kitag. et Ohba) Kitag. et Ohba = Mitella stylosa H. Boissieu var. makinoi (H. Hara) Wakab. ■☆
33368 Arisaema jacquemontii Blume;藏南绿南星(杰氏天南星); Jasquemont Arisaema, Jasquemont Southstar ■
33369 Arisaema japonicum Blume;蛇头草(日本南星,天南星);Japan Southstar, Japaneses Arisaema ■
33370 Arisaema japonicum Blume = Arisaema serratum (Thunb.) Schott ■
33371 Arisaema japonicum Blume var. atropurpureum (Engl.) Kitam. = Arisaema serratum (Thunb.) Schott ■
33372 Arisaema japonicum Blume var. hatizyoense (Nakai) Sugim. = Arisaema hatizyoense Nakai ■☆
33373 Arisaema japonicum Blume var. serratum Engl. = Arisaema serratum (Thunb.) Schott ■
33374 Arisaema japonicum Kom. = Arisaema angustatum Franch. et Sav. var. peninsulae (Nakai) Nakai ex Miyabe et Kudo ■
33375 Arisaema jingdongense H. Peng et H. Li;景东南星(景东天南星);Jingdong Arisaema, Jingdong Southstar ■
33376 Arisaema jinshajiangense H. Li; 金沙江南星; Jinshajiang Southstar, Jinshajiang Arisaema ■
33377 Arisaema kawashimae Seriz. ;川岛氏南星■☆
33378 Arisaema kelung-insulare Hayata;基隆天南星(基隆南星)■
33379 Arisaema kerrii Craib. = Arisaema erubescens (Wall.) Schott ■
33380 Arisaema kerrii Gagnep. = Arisaema erubescens (Wall.) Schott ■
33381 Arisaema kishidae Makino ex Nakai;岸田南星■☆
33382 Arisaema kishidae Makino ex Nakai var. minus Seriz. = Arisaema minus (Seriz.) J. Murata ■☆
33383 Arisaema kiushianum Makino;九州天南星■☆
33384 Arisaema konjac Siebold ex K. Kcoh = Amorphophallus rivieri Durieu ex Carrière ■
33385 Arisaema kuratae Seriz. ;仓田天南星■☆
33386 Arisaema kwangtungense Merr. = Arisaema heterophyllun Blume ■
33387 Arisaema laminatum Benth. = Arisaema penicillatum N. E. Br. ■
33388 Arisaema lichiangense W. W. Sm. ;丽江南星;Lijiang Arisaema, Lijiang Southstar ■
33389 Arisaema limbatum Nakai et F. Maek. ;具边南星■☆
33390 Arisaema limbatum Nakai et F. Maek. f. viridiflavum Hayashi;绿黄南星■☆
33391 Arisaema limbatum Nakai et F. Maek. var. aequinoctiale (Nakai et F. Maek.) Seriz. = Arisaema aequinoctiale Nakai et F. Maek. ■☆
33392 Arisaema limbatum Nakai et F. Maek. var. conspicuum Seriz. = Arisaema limbatum Nakai et F. Maek. ■☆
33393 Arisaema limbatum Nakai et F. Maek. var. stenophyllum (Nakai et F. Maek.) Seriz. = Arisaema aequinoctiale Nakai et F. Maek. ■☆
33394 Arisaema limprichtii Krause = Arisaema heterophyllun Blume ■
33395 Arisaema lineare Buchet; 线叶南星; Linear Arisaema, Linear Southstar ■
33396 Arisaema lingyunense H. Li; 凌云南星(三步跳); Lingyun Arisaema, Lingyun Southstar ■
33397 Arisaema lobatum Engl. ;花南星(白南星,半边莲,大半夏,大麻芋子,大麦冬,独角莲,狗爪半夏,狗爪南星,黑南星,红包谷,虎芋,虎掌,虎爪南星,花包谷,金半夏,烂屁股,狼毒,绿南星,麻芋子,南星,南星七,浅裂南星,蛇包谷,蛇杵棒,蛇魔芋,蛇芋头,天南星,血理箭,芋儿南星);Lobed Arisaema, Lobed Southstar ■
33398 Arisaema lobatum Engl. var. latisectum Engl. = Arisaema lobatum Engl. ■
33399 Arisaema lobatum Engl. var. rosthornianum Engl. = Arisaema lobatum Engl. ■
33400 Arisaema longipedunculatum M. Hotta;长梗南星■☆
33401 Arisaema longipedunculatum M. Hotta var. yakumontanum Seriz. ; 屋久岛南星■☆
33402 Arisaema maireanum Engl. = Arisaema saxatile Buchet ■
33403 Arisaema mairei H. Lév. = Arisaema saxatile Buchet ■
33404 Arisaema manshuricum Nakai; 单叶朝鲜天南星; Manchurian Arisaema ■
33405 Arisaema masisiense De Wild. = Arisaema mildbraedii Engl. ■☆
33406 Arisaema matsudae Hayata; 线花南星(绒花南星); Matsuda Arisaema, Matsuda Southstar ■
33407 Arisaema maximowiczii Nakai;马氏南星■☆
33408 Arisaema maximowiczii Nakai subsp. tashiroi (Kitam.) Seriz. ; 田代氏南星■☆
33409 Arisaema mayebarae Nakai = Arisaema serratum (Thunb.) Schott var. mayebarae (Nakai) H. Ohashi et J. Murata ■☆
33410 Arisaema meleagris Buchet; 褐斑南星; Brownspotted Arisaema, Brownspotted Southstar ■
33411 Arisaema meleagris Buchet var. sinuatum Buchet;具齿褐斑南星(白附子,半夏);Sinuate Arisaema, Sinuate Southstar ■
33412 Arisaema mildbraedii Engl. ;米尔德南星■☆
33413 Arisaema minamitanii Seriz. ;南谷南星■☆
33414 Arisaema minus (Seriz.) J. Murata;微小南星■☆
33415 Arisaema monbeigii Gamble ex Fisch. = Arisaema franchetianum Engl. ■
33416 Arisaema monophyllum Nakai;单叶南星■
33417 Arisaema monophyllum Nakai f. akitense (Nakai) H. Ohashi;秋

田单叶南星■☆
33418 Arisaema monophyllum Nakai f. atrolinguum (F. Maek.) Kitam. ex H. Ohashi et J. Murata;暗单叶南星■☆
33419 Arisaema monophyllum Nakai f. variegatum (Honda) H. Ohashi et J. Murata = Arisaema monophyllum Nakai f. akitense (Nakai) H. Ohashi ■☆
33420 Arisaema monophyllum Nakai var. akitense (Nakai) H. Ohashi = Arisaema monophyllum Nakai f. akitense (Nakai) H. Ohashi ■☆
33421 Arisaema monophyllum Nakai var. atrolinguum (F. Maek.) Sa. Kurata = Arisaema monophyllum Nakai f. atrolinguum (F. Maek.) Kitam. ex H. Ohashi et J. Murata ■☆
33422 Arisaema mooneyanum M. G. Gilbert et Mayo;穆尼南星■☆
33423 Arisaema multisectum Engl.;多裂南星;Manylobed Arisaema, Manylobed Southstar ■
33424 Arisaema nakaianum (Kitag. et Ohba) M. Hotta = Mitella stylosa H. Boissieu var. makinoi (H. Hara) Wakab. ■☆
33425 Arisaema nambae Kitam.;难波南星■☆
33426 Arisaema nanjenense T. C. Huang et M. J. Wu;南仁山天南星;Nanrensahan Arisaema ■
33427 Arisaema nantciangense Pamp.;南漳南星;Nanzhang Arisaema, Nanzhang Southstar ■
33428 Arisaema nanum Nakai = Arisaema sazensoo (Buerger ex Blume) Makino ■☆
33429 Arisaema negishii Makino;根岸南星■☆
33430 Arisaema nepenthoides (Wall.) Mart.;猪笼南星(猪笼草状南星);Nepenthes Southstar, Pitcher-like Arisaema ■
33431 Arisaema nikoense Nakai;日光南星■☆
33432 Arisaema nikoense Nakai f. kubotae H. Ohashi et J. Murata;久保南星■☆
33433 Arisaema nikoense Nakai subsp. australe (M. Hotta) Seriz.;南日光山南星■☆
33434 Arisaema nikoense Nakai var. australe M. Hotta = Arisaema nikoense Nakai subsp. australe (M. Hotta) Seriz. ■☆
33435 Arisaema nikoense Nakai var. brevicollum (H. Ohashi et J. Murata) Seriz. = Arisaema ishizuchiense Murata subsp. brevicollum (H. Ohashi et J. Murata) Seriz. ■☆
33436 Arisaema nikoense Nakai var. ishizuchiense (Murata) M. Hotta ex Seriz. = Arisaema ishizuchiense Murata ■☆
33437 Arisaema nikoense Nakai var. kaimontanum Seriz.;贝田南星■☆
33438 Arisaema oblanceolatum Kitam. = Arisaema formosanum (Hayata) Hayata ■
33439 Arisaema ochraceum Schott = Arisaema nepenthoides (Wall.) Mart. ■
33440 Arisaema ogatae Koidz.;绪方南星■☆
33441 Arisaema omeiense P. C. Kao;峨眉南星;Emei Arisaema ■
33442 Arisaema onoticum Buchet;驴耳南星;Assear Arisaema, Assear Southstar ■
33443 Arisaema ostiolatum Hara = Arisaema propinquum Schott ■
33444 Arisaema ovale Nakai;卵形南星■☆
33445 Arisaema ovale Nakai var. inaense (Seriz.) J. Murata;伊那南星■☆
33446 Arisaema ovale Nakai var. sadoense (Nakai) J. Murata;佐渡南星■☆
33447 Arisaema paichuanense Z. Y. Zhu;北川南星;Beichuan Arisaema, Beichuan Southstar ■
33448 Arisaema pangii H. Li;潘南星;Pan Arisaema ■
33449 Arisaema parvum N. E. Br. ex Hemsl.;小南星;Small Arisaema, Small Southstar ■
33450 Arisaema penicillatum N. E. Br.;画笔南星(顶刷南星,广东土南星,花伞柄,三叶天南星,蛇姜头,蛇香头,蛇钻头);Brush Arisaema, Pencil Southstar ■
33451 Arisaema peninsulae Nakai = Arisaema angustatum Franch. et Sav. var. peninsulae (Nakai) Nakai ex Miyabe et Kudo ■
33452 Arisaema peninsulae Nakai = Arisaema serratum (Thunb.) Schott ■
33453 Arisaema peninsulae Nakai atropurpureum Y. C. Chu et D. C. Wu;紫苞朝鲜南星■
33454 Arisaema peninsulae Nakai var. manshuricum (Nakai) Y. C. Chu et T. K. Cheng = Arisaema manshuricum Nakai ■
33455 Arisaema pertusum Riedl;孔洞南星■☆
33456 Arisaema pianmaense H. Li;片马南星;Pianma Arisaema ■
33457 Arisaema picture N. E. Br. = Arisaema lobatum Engl. ■
33458 Arisaema pingbianense H. Li;屏边南星;Pingbian Arisaema ■
33459 Arisaema planilaminum J. Murata;扁南星■☆
33460 Arisaema polydactylum Riedl;多指南星■☆
33461 Arisaema pradhanii C. E. C. Fisch. = Arisaema griffithii Schott var. verrucosum (Schott) Hara ■
33462 Arisaema praecax de Vriese = Arisaema ringens (Thunb.) Schott ●■
33463 Arisaema praecox de Vriese ex K. Koch = Arisaema ringens (Thunb.) Schott ●■
33464 Arisaema prazeri Hook. f.;河谷南星(半夏);Valley Arisaema, Valley Southstar ■
33465 Arisaema prazeri Hook. f. var. variegatum Engl. = Arisaema prazeri Hook. f. ■
33466 Arisaema prazeri Hook. f. var. viride Engl. = Arisaema prazeri Hook. f. ■
33467 Arisaema propinquum Schott;藏南星;Xizang Arisaema, Xizang Southstar ■
33468 Arisaema pseudojaponicum Nakai = Arisaema japonicum Blume ■
33469 Arisaema purpureogaleatum Engl. = Arisaema fargesii Buchet ■
33470 Arisaema purpureogaleatum Engl. = Arisaema franchetianum Engl. ■
33471 Arisaema quinquefoliolum Hayata = Arisaema grapsospadix Hayata ■
33472 Arisaema rhizomamm C. E. C. Fisch. var. viride C. E. C. Fisch. = Arisaema rhizomatum C. E. C. Fisch. ■
33473 Arisaema rhizomatum C. E. C. Fisch.;雪里见(半截烂,大半夏,大麻药,独角莲,躲雷草,花脸,麻醉药,蛇包谷,铁灯台,野包谷);Rhizome Arisaema, Rootstock Southstar ■
33474 Arisaema rhizomatum C. E. C. Fisch. var. nudum C. E. C. Fisch.;绥阳雪里见;Suiyang Arisaema ■
33475 Arisaema rhombiforme Buchet;黑南星;Black Arisaema, Black Southstar ■
33476 Arisaema ringens (Thunb.) Schott;普陀南星(狗爪南星,黑南星,衢南星,申跋,台北南星,小南星,小天南星,由跋,油跋);Putuo Southstar, Rigid Arisaema, Taibei Arisaema, Taibei Southstar ●■
33477 Arisaema ringens (Thunb.) Schott var. praecox (de Vriese) Engl. = Arisaema ringens (Thunb.) Schott ●■
33478 Arisaema ringens (Thunb.) Schott var. sieboldii Engl. = Arisaema ringens (Thunb.) Schott ●■
33479 Arisaema robustum (Engl.) Nakai var. ovale (Nakai) Kitam. = Arisaema ovale Nakai ■☆
33480 Arisaema robustum (Engl.) Nakai var. shikokumontanum H.

Ohashi = Arisaema longipedunculatum M. Hotta ■☆

33481 Arisaema robustum Nakai = Arisaema ovale Nakai var. sadoense (Nakai) J. Murata ■☆

33482 Arisaema ruwenzoricum N. E. Br.;鲁文佐里南星■☆

33483 Arisaema sachalinense (Miyabe et Kudo) J. Murata;库页南星■☆

33484 Arisaema salwinense Hand. -Mazz. = Arisaema griffithii Schott var. verrucosum (Schott) Hara ■

33485 Arisaema saxatile Buchet;岩生南星;Cliff Arisaema, Cliff Southstar ■

33486 Arisaema sazensoo (Buerger ex Blume) Makino;佐善南星■☆

33487 Arisaema sazensoo (Buerger ex Blume) Makino f. viride Sugim.;绿南星■☆

33488 Arisaema sazensoo (Buerger ex Blume) Makino var. henryanum Engl. = Arisaema sikokianum Franch. et Sav. var. henryanum (Engl.) H. Li ■

33489 Arisaema sazensoo (Buerger ex Blume) Makino var. integrifolium Makino = Arisaema sikokianum Franch. et Sav. ■

33490 Arisaema sazensoo (Buerger ex Blume) Makino var. magnidens N. E. Br. = Arisaema sikokianum Franch. et Sav. var. serratum (Makino) Hand. -Mazz. ■

33491 Arisaema sazensoo (Buerger ex Blume) Makino var. serrato-dentatum Engl. = Arisaema sikokianum Franch. et Sav. var. serratum (Makino) Hand. -Mazz. ■

33492 Arisaema sazensoo (Buerger ex Blume) Makino var. serratum Makino = Arisaema sikokianum Franch. et Sav. var. serratum (Makino) Hand. -Mazz. ■

33493 Arisaema sazensoo (Buerger) Makino = Arisaema sikokianum Franch. et Sav. ■

33494 Arisaema schimperianum Schott = Arisaema enneaphyllum Hochst. ex A. Rich. ■☆

33495 Arisaema seppikoense Kitam.;雪彦南星■☆

33496 Arisaema serratum (Thunb.) Schott;细齿南星(斑杖);Serrate Arisaema, Serrate Southstar ■

33497 Arisaema serratum (Thunb.) Schott subsp. amplissimum (Blume) Kitam. = Arisaema serratum (Thunb.) Schott ■

33498 Arisaema serratum (Thunb.) Schott var. atropurpureum Engl. = Arisaema angustatum Franch. et Sav. var. peninsulae (Nakai) Nakai ex Miyabe et Kudo ■

33499 Arisaema serratum (Thunb.) Schott var. euserratum Engl. = Arisaema angustatum Franch. et Sav. var. peninsulae (Nakai) Nakai ex Miyabe et Kudo ■

33500 Arisaema serratum (Thunb.) Schott var. mayebarae (Nakai) H. Ohashi et J. Murata;马屋原南星■☆

33501 Arisaema serratum (Thunb.) Schott var. suwoense (Nakai) H. Ohashi et J. Murata;中井南星■☆

33502 Arisaema serratum (Thunb.) Schott var. viridescens Nakai;绿苞细齿南星;Greenbract Serrate Arisaema ■

33503 Arisaema shihmienense H. Li;石棉南星(麻芋子);Shimian Arisaema, Shimian Southstar ■

33504 Arisaema sikkimense Stapf ex Chatterjee = Arisaema propinquum Schott ■

33505 Arisaema sikokianum Franch. et Sav.;全缘灯台莲(半边莲,大叶天南星,灯台莲,阔叶天南星,路边黄,绿南星,全缘灯台树,日本灯台莲,蛇包谷,蛇根头,蛇壳南星,蛇磨芋,天南星,细齿灯台莲);Sikoku Arisaema, Sikoku Southstar ■

33506 Arisaema sikokianum Franch. et Sav. var. albescens ?;白苞灯台莲■

33507 Arisaema sikokianum Franch. et Sav. var. henryanum (Engl.) H. Li;七叶灯台莲(七叶灯台树);Sevenleaf Arisaema, Sevenleaf Southstar ■

33508 Arisaema sikokianum Franch. et Sav. var. magnidens (N. E. Br.) Hand. -Mazz.;粗齿南星■

33509 Arisaema sikokianum Franch. et Sav. var. serratum (Makino) Hand. -Mazz.;灯台莲(半边莲,齿叶南星,粗齿灯台莲,大叶天南星,狗爪南星,欢喜草,路边黄,绿南星,南星,山苞米,蛇包谷,蛇根头,蛇魔芋,蛇芋头,天南星,蜗壳南星);Serrate Sikoku Arisaema, Serrate Sikoku Southstar ■

33510 Arisaema silvestrii Pamp.;鄂西南星;W. Hubei Arisaema, W. Hubei Southstar ■

33511 Arisaema sinanoense Nakai = Arisaema serratum (Thunb.) Schott ■

33512 Arisaema sinii Krause;瑶山南星(独角莲,三角龙,三角条);Yaoshan Arisaema, Yaoshan Southstar ■

33513 Arisaema smithii K. Krause;相岭南星;Smith Arisaema, Smith Southstar ■

33514 Arisaema solenochlamis Nakai ex F. Maek. = Arisaema serratum (Thunb.) Schott ■

33515 Arisaema somalense M. G. Gilbert et Mayo;索马里南星■☆

33516 Arisaema souliei Buchet;东俄洛南星;Soulie Southstar, Soulie Arisaema ■

33517 Arisaema speciosum (Wall.) Mart.;美丽南星;Beautiful Arisaema, Beautiful Southstar, Cobra Lily ■

33518 Arisaema speciosum (Wall.) Mart. var. eminens (Schott) Engl. = Arisaema speciosum (Wall.) Mart. ■

33519 Arisaema sprengerianum Pamp. var. dentatum Pamp. = Arisaema sikokianum Franch. et Sav. var. serratum (Makino) Hand. -Mazz. ■

33520 Arisaema stenophyllum Nakai et F. Maek. = Arisaema aequinoctiale Nakai et F. Maek. ■☆

33521 Arisaema stenospathum Hand. -Mazz. = Arisaema heterophyllun Blume ■

33522 Arisaema stewardsonii Britton;斯图尔森南星;Northern Jack-in-the-pulpit, Stewardson's Jack-in-the-pulpit ■☆

33523 Arisaema stracheyanum Schott = Arisaema intermedium Blume ■

33524 Arisaema suwoense Nakai = Arisaema serratum (Thunb.) Schott var. suwoense (Nakai) H. Ohashi et J. Murata ■☆

33525 Arisaema taihokense Hosok. = Arisaema ringens (Thunb.) Schott ●■

33526 Arisaema taiwanense J. Murata;蓬莱天南星■

33527 Arisaema taiwanense J. Murata var. brevipedunculatum J. Murata;短梗天南星;Shortstal Arisaema ■

33528 Arisaema takedae Makino = Arisaema serratum (Thunb.) Schott ■

33529 Arisaema takeoi Hayata = Arisaema heterophyllum Blume ■

33530 Arisaema taliense Engl. = Arisaema yunnanense Buchet ■

33531 Arisaema taliense Engl. var. latisectum Engl. = Arisaema yunnanense Buchet ■

33532 Arisaema tashiroi Kitam. = Arisaema maximowiczii Nakai subsp. tashiroi (Kitam.) Seriz. ■☆

33533 Arisaema tatarinowii Schott = Arisaema erubescens (Wall.) Schott ■

33534 Arisaema tengtsungense H. Li;腾冲南星;Tengchong Arisaema, Tengchong Southstar ■

33535 Arisaema ternatipartitum Makino;顶深裂南星■☆

33536 Arisaema thunbergii Blume;通贝里南星■☆

33537 Arisaema thunbergii Blume subsp. autumnale J. C. Wang, J.

Murata et H. Ohashi;台东天南星■

33538 Arisaema thunbergii Blume subsp. urashima (H. Hara) H. Ohashi et J. Murata;浦岛南星■☆

33539 Arisaema thunbergii Blume var. heterophyllum Engl. = Arisaema heterophyllun Blume ■

33540 Arisaema tortuosum (Wall.) Schott;曲序南星(鼠尾南星,弯曲天南星);Arisaema, Curv Arisaema, Twisted Arisaema, Twistedspadix Southstar ■

33541 Arisaema tortuosum (Wall.) Schott var. curvatum (Roxb.) Engl.;内折南星■☆

33542 Arisaema tortuosum (Wall.) Schott var. helleborifolium (Schott) Engl. = Arisaema tortuosum (Wall.) Schott ■

33543 Arisaema tosaense Makino;土佐南星■☆

33544 Arisaema triphyllum (L.) Schott;三叶南星(印度天南星);American Wake Robin, American Wake-robin, Devil's Ear, Dragon-root, Indian Almond, Indian Jack-in-the-pulpit, Indian Turnip, Iroquois Breadfruit, Iroquois Breadroot, Jack in the Pulpit, Jack-in-the-pulpit, Northern Jack-in-the-pulpit, Parson-in-the-pulpit, Peace Plant, Pepper Turnip, Small Jack-in-the-pulpit, Trileaf Southstar, Wild Turnip ■☆

33545 Arisaema triphyllum (L.) Schott f. viride (Engl.) Farw. = Arisaema triphyllum (L.) Schott ■☆

33546 Arisaema ulugurense M. G. Gilbert et Mayo;乌卢古尔南星■☆

33547 Arisaema undulatifolium Nakai;波叶南星■☆

33548 Arisaema undulatifolium Nakai subsp. nambae (Kitam.) H. Ohashi et J. Murata = Arisaema nambae Kitam. ■☆

33549 Arisaema undulatifolium Nakai subsp. uwajimense Tom. Kobay. et J. Murata;宇和岛南星■☆

33550 Arisaema undulatifolium Nakai var. ionostemma (Nakai et F. Maek.) H. Ohashi et J. Murata = Arisaema limbatum Nakai et F. Maek. ■☆

33551 Arisaema undulatifolium Nakai var. limbatum (Nakai et F. Maek.) H. Ohashi = Arisaema limbatum Nakai et F. Maek. ■☆

33552 Arisaema undulatifolium Nakai var. yoshinagae (Nakai) Seriz. = Arisaema aequinoctiale Nakai et F. Maek. ■☆

33553 Arisaema undulatum K. Krause;洱南南星;Undulate Arisaema, Undulate Southstar ■

33554 Arisaema unzenense Seriz.;运天南星■☆

33555 Arisaema urashima H. Hara = Arisaema thunbergii Blume subsp. urashima (H. Hara) H. Ohashi et J. Murata ■☆

33556 Arisaema urashima H. Hara var. giganteum Konta = Arisaema thunbergii Blume subsp. urashima (H. Hara) H. Ohashi et J. Murata ■☆

33557 Arisaema utile Hook. f. ex Engl.;网檐南星;Useful Arisaema, Useful Southstar ■

33558 Arisaema utile Hook. f. ex Schott = Arisaema utile Hook. f. ex Engl. ■

33559 Arisaema verrucosum Schott = Arisaema griffithii Schott var. verrucosum (Schott) Hara ■

33560 Arisaema vituperatum Schott = Arisaema erubescens (Wall.) Schott ■

33561 Arisaema wallichianum Hook. f. = Arisaema propinquum Schott ■

33562 Arisaema wallichianum Hook. f. f. propinquum (Schott) H. Hara = Arisaema propinquum Schott ■

33563 Arisaema wallichianum Hook. f. var. sikkimense (Stapf) H. Hara = Arisaema propinquum Schott ■

33564 Arisaema wardii C. Marquand et Airy Shaw;隐序南星(天南星);Ward Arisaema, Ward Southstar ■

33565 Arisaema wilsonii Engl.;川中南星(川南星);E. H. Wilson Arisaema, E. H. Wilson Southstar ■

33566 Arisaema wilsonii Engl. var. forrestii Engl.;短柄川中南星;Forrest E. H. Wilson Arisaema, Forrest E. H. Wilson Southstar ■

33567 Arisaema xiangchengense H. Li et A. M. Li;乡城南星;Xiangcheng Arisaema, Xiangcheng Southstar ■

33568 Arisaema yakusimense Nakai = Arisaema serratum (Thunb.) Schott ■

33569 Arisaema yamatense (Nakai) Nakai;山手南星■☆

33570 Arisaema yamatense (Nakai) Nakai subsp. sugimotoi (Nakai) H. Ohashi et J. Murata;杉本南星■☆

33571 Arisaema yamatense (Nakai) Nakai var. intermedium Sugim. = Arisaema yamatense (Nakai) Nakai subsp. sugimotoi (Nakai) H. Ohashi et J. Murata ■☆

33572 Arisaema yamatense (Nakai) Nakai var. sugimotoi (Nakai) Kitam. = Arisaema yamatense (Nakai) Nakai subsp. sugimotoi (Nakai) H. Ohashi et J. Murata ■☆

33573 Arisaema yoshinagae Nakai = Arisaema aequinoctiale Nakai et F. Maek. ■☆

33574 Arisaema yunnanense Buchet;山珠南星(半夏,长虫魔芋,刀口药,滇南星,狗闹子,山包米,山珠半夏,蛇饭果,小南星);Yunnan Arisaema, Yunnan Southstar ■

33575 Arisaema zanlanscianense Pamp.;樟瑯乡南星;Zhanglangxiang Arisaema ■

33576 Arisanorchis Hayata = Cheirostylis Blume ■

33577 Arisanorchis tairae Fukuy. = Cheirostylis takeoi (Hayata) Schltr. ■

33578 Arisanorchis takeoi Hayata = Cheirostylis takeoi (Hayata) Schltr. ■

33579 Arisaraceae Raf.;老鼠芋科■

33580 Arisaraceae Raf. = Araceae Juss.(保留科名)●■

33581 Arisaraceae Raf. = Aristolochiaceae Juss.(保留科名)●■

33582 Arisaron Adans. = Arisarum Mill. ■☆

33583 Arisarum Haller = Calla L. ■

33584 Arisarum Mill. (1754);老鼠芋属(盔苞芋属,鼠尾南星属);Mousetail-plant ■☆

33585 Arisarum Targ. Tozz. = Arisarum Mill. ■☆

33586 Arisarum clusii Schott = Arisarum simorrhinum Durieu var. clusii (Schott) Talavera ■☆

33587 Arisarum hastatum Pomel = Arisarum vulgare Targ. Tozz. subsp. hastatum (Pomel) Dobignard ■☆

33588 Arisarum proboscideum Savi;象鼻老鼠芋(盔苞芋);Mouse Plant, Mouseplant, Mousetail-plant ■☆

33589 Arisarum simorrhinum Durieu;阿尔及利亚老鼠芋■☆

33590 Arisarum simorrhinum Durieu var. clusii (Schott) Talavera;克卢斯老鼠芋■☆

33591 Arisarum simorrhinum Durieu var. subexsertum (Webb et Berthel.) Talavera = Arisarum vulgare Targ. Tozz. subsp. subexsertum (Webb et Berthel.) G. Kunkel ■☆

33592 Arisarum subexertum Webb et Berthel. = Arisarum vulgare Targ. Tozz. subsp. subexsertum (Webb et Berthel.) G. Kunkel ■☆

33593 Arisarum vulgare Targ. Tozz.;老鼠芋;Friar's Cowl ■☆

33594 Arisarum vulgare Targ. Tozz. subsp. exsertum Maire et Weiller = Arisarum vulgare Targ. Tozz. ■☆

33595 Arisarum vulgare Targ. Tozz. subsp. hastatum (Pomel) Dobignard;戟形老鼠芋■☆

33596 Arisarum vulgare Targ. Tozz. subsp. simorrhinum (Durieu) Maire et Weiller = Arisarum simorrhinum Durieu ■☆

33597 Arisarum vulgare Targ. Tozz. subsp. subexsertum (Webb et Berthel.) G. Kunkel;伸出老鼠芋■☆
33598 Arisarum vulgare Targ. Tozz. var. clusii (Schott) Engl. = Arisarum simorrhinum Durieu var. clusii (Schott) Talavera ■☆
33599 Arisarum vulgare Targ. Tozz. var. subexsertum (Webb et Berthel.) Engl. = Arisarum vulgare Targ. Tozz. subsp. subexsertum (Webb et Berthel.) G. Kunkel ■☆
33600 Arischrada Pobed.(废弃属名)= Salvia L. ●■
33601 Aristaea A. Rich. = Aristea Sol. ex Aiton ■☆
33602 Aristaria Jungh. = Anthistiria L. f. ■
33603 Aristaria Jungh. = Themeda Forssk. ■
33604 Aristavena F. Albers et Butzin = Deschampsia P. Beauv. ■
33605 Aristea Aiton = Aristea Sol. ex Aiton ■☆
33606 Aristea Aiton = Ixia L.(保留属名)■☆
33607 Aristea Sol. = Aristea Sol. ex Aiton ■☆
33608 Aristea Sol. ex Aiton(1789);蓝星花属;Aristea,Blue Corn-lily ■☆
33609 Aristea abyssinica Pax;阿比西尼亚蓝星花■☆
33610 Aristea affinis N. E. Br. = Aristea woodii N. E. Br. ■☆
33611 Aristea africana (L.) Hoffmanns.;非洲蓝星花■☆
33612 Aristea alata Baker;翅蓝星花■☆
33613 Aristea alata Baker subsp. abyssinica (Pax) Weim. = Aristea abyssinica Pax ■☆
33614 Aristea alata Baker subsp. bequaertii (De Wild.) Weim. = Aristea abyssinica Pax ■☆
33615 Aristea anceps Eckl. ex Klatt;二棱蓝星花■☆
33616 Aristea angolensis Baker;安哥拉蓝星花■☆
33617 Aristea angolensis Baker subsp. acutivalvis Weim.;尖果片安哥拉蓝星花■☆
33618 Aristea angolensis Baker subsp. pulchella Weim.;美丽安哥拉蓝星花■☆
33619 Aristea angolensis Baker var. robusta Marais;粗壮安哥拉蓝星花■☆
33620 Aristea angustifolia Baker;窄叶蓝星花■☆
33621 Aristea bakeri Klatt;贝克蓝星花■☆
33622 Aristea bequaertii De Wild. = Aristea abyssinica Pax ■☆
33623 Aristea biflora Weim.;双花蓝星花■☆
33624 Aristea bracteata Pers.;具苞蓝星花■☆
33625 Aristea capitata (L.) Ker Gawl.;头状蓝星花■☆
33626 Aristea cistiflora J. C. Manning et Goldblatt;岩蔷薇蓝星花■☆
33627 Aristea cladocarpa Baker;枝果蓝星花■☆
33628 Aristea coerulea (Thunb.) Vahl = Aristea bracteata Pers. ■☆
33629 Aristea cognata N. E. Br. ex Weim. = Aristea abyssinica Pax ■☆
33630 Aristea cognata N. E. Br. ex Weim. subsp. abyssinica (Pax) Marais = Aristea abyssinica Pax ■☆
33631 Aristea compressa Buchinger ex Baker;扁蓝星花■☆
33632 Aristea confusa Goldblatt = Aristea bakeri Klatt ■☆
33633 Aristea congesta N. E. Br. = Aristea woodii N. E. Br. ■☆
33634 Aristea corymbosa (Ker Gawl.) Benth. et Hook. f. = Nivenia corymbosa (Ker Gawl.) Baker ●☆
33635 Aristea curvata N. E. Br. = Aristea schizolaena Harv. ex Baker ■☆
33636 Aristea cuspidata Schinz;骤尖蓝星花■☆
33637 Aristea cyanea De Wild. = Aristea ecklonii Baker ■☆
33638 Aristea cyanea Sol. = Aristea africana (L.) Hoffmanns. ■☆
33639 Aristea dichotoma (Thunb.) Ker Gawl.;松散蓝星花■☆
33640 Aristea dichotoma Eckl. ex Klatt = Aristea ecklonii Baker ■☆
33641 Aristea diffusa Eckl. = Aristea dichotoma (Thunb.) Ker Gawl. ■☆
33642 Aristea djalonis Hutch. = Aristea angolensis Baker ■☆
33643 Aristea ecklonii Baker;蓝星花(蓝星);Aristea,Blue Corn-lily ■☆
33644 Aristea elliptica Goldblatt et A. P. Dold;椭圆蓝星花■☆
33645 Aristea ensifolia J. Muir bis;剑叶蓝星花■☆
33646 Aristea eriophora Pers. = Aristea africana (L.) Hoffmanns. ■☆
33647 Aristea fimbriata Goldblatt et J. C. Manning;流苏蓝星花■☆
33648 Aristea flexicaulis Baker;曲茎蓝星花■☆
33649 Aristea fruticosa (L. f.) Pers. = Nivenia fruticosa (L. f.) Baker ●☆
33650 Aristea galpinii N. E. Br. ex Weim.;盖尔蓝星花■☆
33651 Aristea gerrardii Weim.;吉氏蓝星花;Gerrard's Aristea ■☆
33652 Aristea glauca Klatt;灰蓝蓝星花■☆
33653 Aristea goetzei Harms;格兹蓝星花■☆
33654 Aristea gracilis N. E. Br. = Aristea woodii N. E. Br. ■☆
33655 Aristea grandis Weim.;大蓝星花■☆
33656 Aristea hockii De Wild. = Aristea nyikensis Baker ■☆
33657 Aristea homblei De Wild. = Aristea abyssinica Pax ■☆
33658 Aristea humbertii H. Perrier;亨伯特蓝星花■☆
33659 Aristea inaequalis Goldblatt et J. C. Manning;不对称蓝星花■☆
33660 Aristea intermedia Eckl. ex Klatt = Aristea dichotoma (Thunb.) Ker Gawl. ■☆
33661 Aristea johnstoniana Rendle = Aristea abyssinica Pax ■☆
33662 Aristea juncifolia Baker;灯心草叶蓝星花■☆
33663 Aristea kitchingii Baker;基钦蓝星花■☆
33664 Aristea lastii Baker = Aristea ecklonii Baker ■☆
33665 Aristea latifolia G. J. Lewis;宽叶蓝星花■☆
33666 Aristea longifolia Baker = Aristea alata Baker ■☆
33667 Aristea lugens (L. f.) Steud.;黑蓝星花■☆
33668 Aristea lukwangulensis Marais;卢夸古尔蓝星花■☆
33669 Aristea macrocarpa G. J. Lewis = Aristea bakeri Klatt ■☆
33670 Aristea madagascariensis Baker;马岛蓝星花■☆
33671 Aristea maitlandii Hutch. = Aristea ecklonii Baker ■☆
33672 Aristea major Andréws;聚穗蓝星花■☆
33673 Aristea major Andréws = Aristea capitata (L.) Ker Gawl. ■☆
33674 Aristea melaleuca (Thunb.) Ker Gawl. = Aristea lugens (L. f.) Steud. ■☆
33675 Aristea montana Baker;山地蓝星花■☆
33676 Aristea monticola Goldblatt = Aristea bracteata Pers. ■☆
33677 Aristea nana Goldblatt et J. C. Manning;矮蓝星花■☆
33678 Aristea nandiensis Baker = Aristea angolensis Baker ■☆
33679 Aristea nitida Weim. = Aristea goetzei Harms ■☆
33680 Aristea nyikensis Baker;尼卡蓝星花■☆
33681 Aristea oligocephala Baker;寡头蓝星花■☆
33682 Aristea palustris Schltr.;沼泽蓝星花■☆
33683 Aristea paniculata Baker = Aristea bakeri Klatt ■☆
33684 Aristea paniculata Pax = Aristea ecklonii Baker ■☆
33685 Aristea parviflora Baker = Aristea compressa Buchinger ex Baker ■☆
33686 Aristea pauciflora Wolley-Dod;少花蓝星花■☆
33687 Aristea platycaulis Baker;平茎蓝星花■☆
33688 Aristea polycephala Harms;多头蓝星花■☆
33689 Aristea pusilla (Thunb.) Ker Gawl.;微小蓝星花■☆
33690 Aristea pusilla (Thunb.) Ker Gawl. subsp. robustior Weim.;粗微小蓝星花■☆
33691 Aristea racemosa Baker;多枝蓝星花■☆
33692 Aristea racemosa Baker var. inflata Weim.;膨胀蓝星花■☆
33693 Aristea ramosa De Wild. = Aristea angolensis Baker ■☆
33694 Aristea ranomafana Goldblatt;拉努马法纳蓝星花■☆

33695 Aristea repens A. Dietr.;匍匐蓝星花■☆
33696 Aristea rigidifolia G. J. Lewis;硬叶蓝星花■☆
33697 Aristea rupicola Goldblatt et J. C. Manning;岩生蓝星花■☆
33698 Aristea schizolaena Harv. ex Baker;裂被蓝星花■☆
33699 Aristea simplex Weim.;简单蓝星花■☆
33700 Aristea singularis Weim.;单一蓝星花■☆
33701 Aristea spiralis (L. f.) Ker Gawl.;螺旋蓝星花■☆
33702 Aristea stipitata R. C. Foster = Aristea ecklonii Baker ■☆
33703 Aristea stokoei L. Guthrie = Nivenia stokoei (L. Guthrie) N. E. Br. ●☆
33704 Aristea tayloriana Rendle = Aristea abyssinica Pax ■☆
33705 Aristea teretifolia Goldblatt et J. C. Manning;柱叶蓝星花■☆
33706 Aristea thyrsiflora N. E. Br. = Aristea major Andréws ■☆
33707 Aristea torulosa Baker = Aristea woodii N. E. Br. ■☆
33708 Aristea torulosa Baker var. monostachya ? = Aristea woodii N. E. Br. ■☆
33709 Aristea torulosa Klatt;结节蓝星花■☆
33710 Aristea uhehensis Harms ex Engl. = Aristea nyikensis Baker ■☆
33711 Aristea umbellata Spreng. = Moraea fugax (D. Delaroche) Jacq. ■☆
33712 Aristea woodii N. E. Br.;伍得蓝星花■☆
33713 Aristea wredowia Steud. = Pillansia templemannii (Baker) L. Bolus ■☆
33714 Aristea zeyheri Baker;泽赫蓝星花■☆
33715 Aristea zombensis Baker = Aristea angolensis Baker ■☆
33716 Aristega Miers = Tiliacora Colebr.(保留属名)●☆
33717 Aristeguietia R. M. King et H. Rob. (1975);尖苞亮泽兰属●☆
33718 Aristeguietia amethystina (Rob.) R. M. King et H. Rob.;紫水晶亮泽兰●☆
33719 Aristeguietia arborea (Kunth) R. M. King et H. Rob.;树尖苞亮泽兰●☆
33720 Aristeguietia cacalioides (Kunth) R. M. King et H. Rob.;蟹甲草尖苞亮泽兰●☆
33721 Aristeguietia dielsii (Rob.) R. M. King et H. Rob.;迪尔斯尖苞亮泽兰●☆
33722 Aristeguietia lamiifolia (Lam.) R. M. King et H. Rob.;黏性尖苞亮泽兰●☆
33723 Aristeguietia persicifolia (Kunth) R. M. King et H. Rob.;桃叶尖苞亮泽兰●☆
33724 Aristella Bertol. (1833);类蓝星花属■☆
33725 Aristella Bertol. = Achnatherum P. Beauv. ■
33726 Aristella Bertol. = Stipa L. ■
33727 Aristella bromoides (L.) Bertol.;类蓝星花■☆
33728 Aristella longiflora Regel;长花类蓝星花■☆
33729 Aristeyera H. E. Moore = Asterogyne H. Wendl. ex Benth. et Hook. f. ●☆
33730 Aristida L. (1753);三芒草属(三枪茅属);Three-awn, Three-awned Grass, Three-awned-grass, Threeawngrass, Triawn ■
33731 Aristida abnormis Chiov.;异常三芒草■☆
33732 Aristida acutiflora Trin. et Rupr. = Stipagrostis acutiflora (Trin. et Rupr.) De Winter ■☆
33733 Aristida acutiflora Trin. et Rupr. subsp. brachyathera (Coss. et Balansa) Trab. = Stipagrostis acutiflora (Trin. et Rupr.) De Winter ■☆
33734 Aristida acutiflora Trin. et Rupr. subsp. zittelii (Asch.) Maire et Weiller = Stipagrostis zittelii (Asch.) De Winter ■☆
33735 Aristida acutiflora Trin. et Rupr. var. algeriensis (Henrard) Maire et Weiller = Stipagrostis acutiflora (Trin. et Rupr.) De Winter ■☆
33736 Aristida adoensis Hochst.;阿多三芒草■☆
33737 Aristida adscensionis L.;三芒草(三枪茅);Sixweeks Threeawn, Sixweeks Threeawngrass, Sixweeks Triawn ■
33738 Aristida adscensionis L. subsp. coerulescens (Desf.) Bourreil et Trouin = Aristida adscensionis L. ■
33739 Aristida adscensionis L. subsp. coerulescens (Desf.) Bourreil et Trouin = Aristida coerulescens Desf. ■☆
33740 Aristida adscensionis L. subsp. gfaruineensis (Trin. et Rupr.) Henrard = Aristida adscensionis L. ■
33741 Aristida adscensionis L. var. aethiopica (Trin. et Rupr.) T. Durand et Schinz;埃塞俄比亚三芒草■☆
33742 Aristida adscensionis L. var. coerulescens (Desf.) Durand et Schinz = Aristida coerulescens Desf. ■☆
33743 Aristida adscensionis L. var. ehrenbergii (Trin. et Rupr.) Henrard = Aristida adscensionis L. ■
33744 Aristida adscensionis L. var. ehrenbergii Henrard = Aristida adscensionis L. ■
33745 Aristida adscensionis L. var. festucoides (Poir.) Henrard;羊茅状三芒草■☆
33746 Aristida adscensionis L. var. glabricallis Maire et Weiller = Aristida adscensionis L. ■
33747 Aristida adscensionis L. var. pumila (Decne.) Coss. et Durieu = Aristida adscensionis L. ■
33748 Aristida adscensionis L. var. pumila Coss. et Durieu = Aristida adscensionis L. ■
33749 Aristida adscensionis L. var. vulpioides (Hance) Hack. ex Henrard = Aristida adscensionis L. ■
33750 Aristida aemulans Melderis;匹敌三芒草■☆
33751 Aristida aequiglumis Hack.;等颖三芒草■☆
33752 Aristida alopecuroides Hack. = Aristida congesta Roem. et Schult. ■☆
33753 Aristida alpina L. Liou;高原三芒草;Plateau Threeawngrass, Plateau Triawn ■
33754 Aristida amabilis Schweick. = Stipagrostis amabilis (Schweick.) De Winter ■☆
33755 Aristida ambongensis A. Camus;安邦三芒草■☆
33756 Aristida amplissima Trin. et Rupr. = Aristida stipoides Lam. ■☆
33757 Aristida angolensis C. E. Hubb. = Sartidia angolensis (C. E. Hubb.) De Winter ■☆
33758 Aristida angustata Stapf = Aristida junciformis Trin. et Rupr. ■☆
33759 Aristida anisochaeta Clayton;异毛三芒草■☆
33760 Aristida arachnoidea Litv.;蛛毛三芒草■☆
33761 Aristida argentea Schweick. = Aristida mollissima Pilg. subsp. argentea (Schweick.) Melderis ■☆
33762 Aristida aristidis Trab. = Aristida sieberiana Trin. ■☆
33763 Aristida aristidis Trab. var. chudaei Batt. et Trab. = Aristida sieberiana Trin. ■☆
33764 Aristida articulata Edgew. = Aristida mutabilis Trin. et Rupr. ■☆
33765 Aristida arundinacea L. = Neyraudia arundinacea (L.) Henrard ■
33766 Aristida astroclada Chiov. = Aristida mutabilis Trin. et Rupr. ■☆
33767 Aristida atroviolacea Hack. = Aristida recta Franch. ■☆
33768 Aristida basiramea Engelm. ex Vasey;叉三芒草;Forktip Three-awn, Fork-tip Three-awn Grass ■☆
33769 Aristida basiramea Engelm. ex Vasey var. curtissii (A. Gray) Shinners = Aristida dichotoma Michx. var. curtissii A. Gray ■☆
33770 Aristida basiramea Engelm. var. curtissii (A. Gray) Shinners =

Aristida dichotoma Michx. ■☆

33771 Aristida batangensis Z. X. Tang et H. X. Liu;巴塘三芒草;Batang Threeawngrass ■

33772 Aristida beyrichiana Trin. et Rupr.;贝氏三芒草;Wire Grass ■☆

33773 Aristida bifida Karl = Stipagrostis foexiana (Maire et Wilczek) De Winter ■☆

33774 Aristida bipartita (Nees) Trin. et Rupr.;二深裂三芒草■☆

33775 Aristida boninensis Ohwi et Tuyama;小笠原三芒草■☆

33776 Aristida borumensis Henrard = Aristida scabrivalvis Hack. subsp. borumensis (Henrard) Melderis ■☆

33777 Aristida brachypoda Tausch = Stipagrostis brachypoda (Tausch) De Winter ■☆

33778 Aristida brachypoda Tausch = Stipagrostis plumosa (L.) Munro ex T. Anderson ■

33779 Aristida brachypthera Coss. et Balansa = Stipagrostis brachyathera (Coss. et Balansa) De Winter ■☆

33780 Aristida brainii Melderis;布雷恩三芒草■☆

33781 Aristida brevifolia (Nees) Steud. = Stipagrostis brevifolia (Nees) De Winter ■☆

33782 Aristida brevifolia (Nees) Steud. var. floccosum Nees = Stipagrostis brevifolia (Nees) De Winter ■☆

33783 Aristida brevissima L. Liou;短三芒草(短芒草);Short Threeawngrass, Short Triawn ■

33784 Aristida brevisubulata Maire = Aristida sieberiana Trin. ■☆

33785 Aristida burkei Stapf = Aristida diffusa Trin. subsp. burkei (Stapf) Melderis ■☆

33786 Aristida caerulescens Desf. = Aristida adscensionis L. ■

33787 Aristida caloptila (Jaub. et Spach) Boiss. = Stipagrostis paradisea (Edgew.) De Winter ■☆

33788 Aristida canescens Henrard;灰白三芒草■☆

33789 Aristida canescens Henrard subsp. ramosa De Winter;分枝三芒草■☆

33790 Aristida capensis Thunb. = Stipagrostis zeyheri (Nees) De Winter subsp. macropus ? ■☆

33791 Aristida capensis Thunb. var. barbata Stapf = Stipagrostis zeyheri (Nees) De Winter subsp. barbata (Stapf) De Winter ■☆

33792 Aristida capensis Thunb. var. canescens Trin. et Rupr. = Stipagrostis zeyheri (Nees) De Winter ■☆

33793 Aristida capensis Thunb. var. dieterleniana Schweick. = Stipagrostis zeyheri (Nees) De Winter subsp. sericans (Hack.) De Winter ■☆

33794 Aristida capensis Thunb. var. genuina Henrard = Stipagrostis zeyheri (Nees) De Winter subsp. macropus ? ■☆

33795 Aristida capensis Thunb. var. macropus (Nees) Trin. et Rupr. = Stipagrostis zeyheri (Nees) De Winter subsp. macropus ? ■☆

33796 Aristida cassanellii A. Terracc. = Aristida mutabilis Trin. et Rupr. ■☆

33797 Aristida chinensis Munro;华三芒草;China Triawn, Chineses Threeawngrass ■

33798 Aristida chrysochlaena Henrard;金三芒草■☆

33799 Aristida ciliata Desf. = Stipagrostis ciliata (Desf.) De Winter ■☆

33800 Aristida ciliata Desf. var. capensis Trin. et Rupr. = Stipagrostis ciliata (Desf.) De Winter var. capensis (Trin. et Rupr.) De Winter ■☆

33801 Aristida ciliata Desf. var. pectinata Henrard = Stipagrostis ciliata (Desf.) De Winter var. capensis (Trin. et Rupr.) De Winter ■☆

33802 Aristida ciliata Desf. var. tricholaena Hack. = Stipagrostis ciliata (Desf.) De Winter var. capensis (Trin. et Rupr.) De Winter ■☆

33803 Aristida ciliata Desf. var. villosa Hack. = Stipagrostis ciliata (Desf.) De Winter var. capensis (Trin. et Rupr.) De Winter ■☆

33804 Aristida coerulescens Desf.;天蓝三芒草■☆

33805 Aristida coerulescens Desf. var. laevilemma Maire = Aristida coerulescens Desf. ■☆

33806 Aristida coerulescens Desf. var. scabrilemma Maire = Aristida coerulescens Desf. ■☆

33807 Aristida coma-ardeae Mez = Stipagrostis dinteri (Hack.) De Winter ■☆

33808 Aristida concinna Sond. ex J. A. Schmidt = Stipagrostis uniplumis (Licht. ex Roem. et Schult.) De Winter ■☆

33809 Aristida congesta Roem. et Schult.;密集三芒草■☆

33810 Aristida congesta Roem. et Schult. var. megatostachya Henrard = Aristida congesta Roem. et Schult. ■☆

33811 Aristida congesta Roem. et Schult. var. pilifera Chiov. = Aristida congesta Roem. et Schult. ■☆

33812 Aristida congesta Roem. et Schult. var. tunetana (Coss.) Bourreil = Aristida congesta Roem. et Schult. ■☆

33813 Aristida contractinodis Stent et J. M. Rattray = Aristida junciformis Trin. et Rupr. subsp. macilenta (Henrard) Melderis ■☆

33814 Aristida corradii Chiov. ex Chiarugi = Stipagrostis foexiana (Maire et Wilczek) De Winter ■☆

33815 Aristida corythroides Karl = Stipagrostis lutescens (Nees) De Winter ■☆

33816 Aristida cumingiana Trin. et Rupr. var. diminuta (Mez) Jacq.-Fél. = Aristida diminuta (Mez) C. E. Hubb. ■☆

33817 Aristida cumningiana Trin. et Rupr.;黄草毛;Cuming Threeawngrass, Cuming Triawn ■

33818 Aristida curtisii (A. Gray) Nash = Aristida dichotoma Michx. ■☆

33819 Aristida curtissii (A. Gray) Nash = Aristida dichotoma Michx. var. curtissii A. Gray ■☆

33820 Aristida curvata (Nees) T. Durand et Schinz = Aristida adscensionis L. ■

33821 Aristida cyanantha Nees ex Steud.;蓝花三芒草■☆

33822 Aristida damarensis Mez = Stipagrostis damarensis (Mez) De Winter ■☆

33823 Aristida dasydesmis (Pilg.) Mez;粗毛三芒草■☆

33824 Aristida denudata Pilg.;裸露三芒草■☆

33825 Aristida depressa Retz.;异颖三芒草;Depressa Threeawngrass, Depressa Triawn ■

33826 Aristida depressa Retz. = Aristida adscensionis L. ■

33827 Aristida desmantha Trin. et Rupr.;德斯曼三芒草;Sand Three-awn ■☆

33828 Aristida dewildemani Henrard;德怀尔三芒草■☆

33829 Aristida dewinteri Giess;德温三芒草■☆

33830 Aristida dichotoma Michx.;二歧三芒草;Churchmouse Three-awn, Church-mouse Three-awn, Pigbutt Three-awn, Poverty Grass, Shinners' Three-awn, Shinners' Three-awned Grass ■☆

33831 Aristida dichotoma Michx. var. curtissii A. Gray;柯氏三芒草;Church-mouse Three-awn, Curtiss' Three-awn ■☆

33832 Aristida diffusa Trin.;松散三芒草■☆

33833 Aristida diffusa Trin. subsp. burkei (Stapf) Melderis;伯克三芒草■☆

33834 Aristida diffusa Trin. var. burkei (Stapf) Schweick. = Aristida diffusa Trin. subsp. burkei (Stapf) Melderis ■☆

33835 Aristida diffusa Trin. var. genuina Henrard = Aristida diffusa

Trin. ■☆

33836 Aristida diffusa Trin. var. pseudohystrix (Trin. et Rupr.) Henrard = Aristida diffusa Trin. ■☆

33837 Aristida diminuta (Mez) C. E. Hubb.;缩小三芒草■☆

33838 Aristida dinteri Hack. = Stipagrostis dinteri (Hack.) De Winter ■☆

33839 Aristida dregeana (Nees) Trin. et Rupr. = Stipagrostis dregeana Nees ■☆

33840 Aristida effusa Henrard;铺散三芒草■☆

33841 Aristida ehrenbergii Trin. et Rupr. = Aristida adscensionis L. ■

33842 Aristida elliotii A. Chev. = Aristida recta Franch. ■☆

33843 Aristida elytrophoroides Chiov. = Aristida congesta Roem. et Schult. ■☆

33844 Aristida engleri Mez;恩格勒三芒草■☆

33845 Aristida engleri Mez var. ramosissima De Winter;密枝三芒草■☆

33846 Aristida fastigiata Hack. = Stipagrostis fastigiata (Hack.) De Winter ■☆

33847 Aristida festucoides Poir. = Aristida adscensionis L. var. festucoides (Poir.) Henrard ■☆

33848 Aristida flocciculmis Mez = Aristida vestita Thunb. ■☆

33849 Aristida foexiana Maire et Wilczek = Stipagrostis foexiana (Maire et Wilczek) De Winter ■☆

33850 Aristida fontismagni Schweick. = Aristida stipoides Lam. ■☆

33851 Aristida formosana Honda = Aristida chinensis Munro ■

33852 Aristida funicularis Trin. ex Steud. = Aristida funiculata Trin. et Rupr. ■☆

33853 Aristida funiculata Trin. et Rupr.;种柄三芒草■☆

33854 Aristida funiculata Trin. et Rupr. var. brevis Maire = Aristida funiculata Trin. et Rupr. ■☆

33855 Aristida funiculata Trin. et Rupr. var. mallica (Edgew.) Henrard = Aristida funiculata Trin. et Rupr. ■☆

33856 Aristida funiculata Trin. et Rupr. var. royleana (Trin. et Rupr.) Hook. f. = Aristida funiculata Trin. et Rupr. ■☆

33857 Aristida furfurosa Henrard = Aristida adoensis Hochst. ■☆

33858 Aristida galpinii Stapf = Aristida junciformis Trin. et Rupr. subsp. galpinii (Stapf) De Winter ■☆

33859 Aristida garubensis Pilg. = Stipagrostis garubensis (Pilg.) De Winter ■☆

33860 Aristida geminifolia (Nees) Trin. et Rupr. = Stipagrostis geminifolia Nees ■☆

33861 Aristida geniculata Raf. = Aristida longespica Poir. var. geniculata (Raf.) Fernald ■☆

33862 Aristida gigantea L. f. var. arabica (Trin. et Rupr.) Cufod. = Aristida adscensionis L. ■

33863 Aristida gonatostachys Pilg. = Stipagrostis gonatostachys (Pilg.) De Winter ■☆

33864 Aristida gossweileri Pilg. = Aristida recta Franch. ■☆

33865 Aristida graciliflora Pilg. = Aristida stipitata Hack. subsp. graciliflora (Pilg.) Melderis ■☆

33866 Aristida graciliflora Pilg. var. robusta Stent et J. M. Rattray = Aristida stipitata Hack. subsp. robusta (Stent et J. M. Rattray) Melderis ■☆

33867 Aristida gracilior Pilg.;纤细三芒草■☆

33868 Aristida gracilior Pilg. = Stipagrostis hirtigluma (Steud. ex Trin. et Rupr.) De Winter subsp. patula (Hack.) De Winter ■☆

33869 Aristida gracilior Pilg. var. intermedia Schweick. = Stipagrostis uniplumis (Licht. ex Roem. et Schult.) De Winter var. intermedia (Schweick.) De Winter ■☆

33870 Aristida gracilior Pilg. var. pearsonii Henrard = Stipagrostis hirtigluma (Steud. ex Trin. et Rupr.) De Winter subsp. pearsonii (Henrard) De Winter ■☆

33871 Aristida gracilis Elliott = Aristida longespica Poir. ■☆

33872 Aristida gracillima Oliv. = Aristida stipoides Lam. ■☆

33873 Aristida grandiglumis Roshev. = Stipagrostis grandiglumis (Roshev.) Tzvelev ■

33874 Aristida guineensis Trin. et Rupr. = Aristida adscensionis L. ■

33875 Aristida hemmingii Clayton = Aristida paoliana (Chiov.) Henrard ■☆

33876 Aristida hermannii Mez = Stipagrostis hermannii (Mez) De Winter ■☆

33877 Aristida heymannii Regel = Aristida adscensionis L. ■

33878 Aristida hirtigluma Steud. ex Trin. et Rupr. = Stipagrostis hirtigluma (Steud. ex Trin. et Rupr.) De Winter ■☆

33879 Aristida hirtigluma Steud. ex Trin. et Rupr. var. patula Hack. = Stipagrostis hirtigluma (Steud. ex Trin. et Rupr.) De Winter subsp. patula (Hack.) De Winter ■☆

33880 Aristida hirtigluma Steud. var. gymnobasis Maire = Stipagrostis hirtigluma (Steud. ex Trin. et Rupr.) De Winter ■☆

33881 Aristida hirtigluma Steud. var. uzzararum Maire = Stipagrostis hirtigluma (Steud. ex Trin. et Rupr.) De Winter ■☆

33882 Aristida hispidula Henrard;硬毛三芒草■☆

33883 Aristida hochstetteriana Beck ex Hack. = Stipagrostis hochstetteriana (Beck ex Hack.) De Winter ■☆

33884 Aristida hockii De Wild. = Aristida recta Franch. ■☆

33885 Aristida hoggariensis Batt. et Trab. = Aristida mutabilis Trin. et Rupr. ■☆

33886 Aristida hubbardiana Schweick.;哈伯德三芒草■☆

33887 Aristida huillensis Rendle = Aristida junciformis Trin. et Rupr. ■☆

33888 Aristida humbertii Bourreil;亨伯特三芒草■☆

33889 Aristida humidicola S. M. Phillips;湿地三芒草■☆

33890 Aristida hystricula Edgew.;豪猪三芒草■☆

33891 Aristida intermedia Scribn. et Ball = Aristida longespica Poir. var. geniculata (Raf.) Fernald ■☆

33892 Aristida intermedia Scribn. et C. R. Ball = Aristida longespica Poir. ■☆

33893 Aristida jucunda Schweick. = Sartidia jucunda (Schweick.) De Winter ■☆

33894 Aristida junciformis Trin. et Rupr.;灯心草三芒草■☆

33895 Aristida junciformis Trin. et Rupr. subsp. galpinii (Stapf) De Winter;盖尔三芒草■☆

33896 Aristida junciformis Trin. et Rupr. subsp. macilenta (Henrard) Melderis;贫弱三芒草■☆

33897 Aristida junciformis Trin. et Rupr. subsp. welwitschii (Rendle) Melderis;韦氏三芒草■☆

33898 Aristida karelinii (Trin. et Rupr.) Roshev.;卡氏三芒草■☆

33899 Aristida kelleri Hack.;凯勒三芒草■☆

33900 Aristida kenyensis Henrard;肯尼亚三芒草■☆

33901 Aristida kerstingii Pilg.;克斯廷三芒草■☆

33902 Aristida kotschyi Hochst. ex Steud. = Aristida funiculata Trin. et Rupr. ■☆

33903 Aristida kunthiana Trin. et Rupr.;孔斯三芒草■☆

33904 Aristida lanata Forssk. = Stipagrostis lanata (Forssk.) De Winter ■☆

33905 Aristida lanata Forssk. var. maroccana Sauvage et Vindt =

Stipagrostis lanata (Forssk.) De Winter ■☆
33906 Aristida lanipes Mez = Stipagrostis lanipes (Mez) De Winter ■☆
33907 Aristida lanosa Muhl. ex Elliott;绵毛三芒草;Woolly Three-awn ■☆
33908 Aristida lanuginosa Burch. = Aristida vestita Thunb. ■☆
33909 Aristida lauriolii Maire = Aristida mutabilis Trin. et Rupr. ■☆
33910 Aristida leiocalycina Trin. et Rupr.;光萼三芒草■☆
33911 Aristida libyca H. Scholz = Stipagrostis libyca (H. Scholz) H. Scholz ■☆
33912 Aristida lisowskii Richel;利索三芒草■☆
33913 Aristida longeradiata Steud. = Aristida mutabilis Trin. et Rupr. ■☆
33914 Aristida longespica Poir.;长穗三芒草;Slender Three-awn, Slimspike Three-awn, Slim-spike Three-awn Grass ■☆
33915 Aristida longespica Poir. var. geniculata (Raf.) Fernald;膝曲长穗三芒草;Kearney's Three-awn, Red Three-awn, Slender Three-awn, Slim-spike Three-awn Grass ■☆
33916 Aristida longicauda Hack. = Aristida congesta Roem. et Schult. ■☆
33917 Aristida longiflora Schumach. = Aristida sieberiana Trin. ■☆
33918 Aristida longiflora Schumach. var. brevisubulata Maire = Aristida sieberiana Trin. ■☆
33919 Aristida lutescens (Nees) Trin. et Rupr. = Stipagrostis lutescens (Nees) De Winter ■☆
33920 Aristida macilenta Henrard = Aristida junciformis Trin. et Rupr. subsp. macilenta (Henrard) Melderis ■☆
33921 Aristida macrathera A. Rich. = Aristida funiculata Trin. et Rupr. ■☆
33922 Aristida macrochloa Hochst. = Aristida adscensionis L. ■
33923 Aristida mallica Edgew. = Aristida funiculata Trin. et Rupr. ■☆
33924 Aristida marlothii Hack. = Stipagrostis lutescens (Nees) De Winter var. marlothii (Hack.) De Winter ■☆
33925 Aristida mauritiana Hochst. ex A. Rich. = Aristida adscensionis L. ■
33926 Aristida meccana Hochst. ex Trin. et Rupr. = Aristida mutabilis Trin. et Rupr. ■☆
33927 Aristida meccana Hochst. ex Trin. et Rupr. var. cassanellii Bourreil = Aristida mutabilis Trin. et Rupr. ■☆
33928 Aristida meccana Hochst. ex Trin. et Rupr. var. genuina Sauvage = Aristida mutabilis Trin. et Rupr. ■☆
33929 Aristida meccana Hochst. ex Trin. et Rupr. var. lauriolii (Maire) Maire = Aristida mutabilis Trin. et Rupr. ■☆
33930 Aristida meccana Hochst. ex Trin. et Rupr. var. schweinfurthii (Boiss.) Maire et Weiller = Aristida mutabilis Trin. et Rupr. ■☆
33931 Aristida meridionalis Henrard;南方三芒草■☆
33932 Aristida migiurtina Chiov.;米朱蒂三芒草■☆
33933 Aristida moandaensis Vanderyst;莫安达三芒草■☆
33934 Aristida modatica Steud. = Aristida adscensionis L. ■
33935 Aristida mollissima Pilg.;柔软三芒草■☆
33936 Aristida mollissima Pilg. subsp. argentea (Schweick.) Melderis;银白柔软三芒草■☆
33937 Aristida monticola Henrard;山地三芒草■☆
33938 Aristida mutabilis Trin. et Rupr.;易变三芒草■☆
33939 Aristida mutabilis Trin. et Rupr. subsp. nigritiana (Hack.) Bourreil = Aristida mutabilis Trin. et Rupr. ■☆
33940 Aristida mutabilis Trin. et Rupr. var. aequilonga ? = Aristida mutabilis Trin. et Rupr. ■☆
33941 Aristida mutabilis Trin. et Rupr. var. glabricollaris Bourreil = Aristida mutabilis Trin. et Rupr. ■☆
33942 Aristida mutabilis Trin. et Rupr. var. hoggariensis (Batt. et Trab.) Henrard = Aristida mutabilis Trin. et Rupr. ■☆
33943 Aristida mutabilis Trin. et Rupr. var. laeviglumis Henrard ex Henrard = Aristida mutabilis Trin. et Rupr. ■☆
33944 Aristida mutabilis Trin. et Rupr. var. longiflora ? = Aristida mutabilis Trin. et Rupr. ■☆
33945 Aristida mutabilis Trin. et Rupr. var. nigritiana (Hack.) Bourreil = Aristida mutabilis Trin. et Rupr. ■☆
33946 Aristida mutabilis Trin. et Rupr. var. senegalensis ? = Aristida mutabilis Trin. et Rupr. ■☆
33947 Aristida mutabilis Trin. et Rupr. var. tangensis Henrard = Aristida mutabilis Trin. et Rupr. ■☆
33948 Aristida namaquensis (Nees) Trin. et Rupr. = Stipagrostis namaquensis (Nees) De Winter ■☆
33949 Aristida namaquensis (Nees) Trin. et Rupr. var. vagans ? = Stipagrostis namaquensis (Nees) De Winter ■☆
33950 Aristida necopina Shinners = Aristida longespica Poir. var. geniculata (Raf.) Fernald ■☆
33951 Aristida nemorivaga Henrard;尼莫三芒草■☆
33952 Aristida nigritiana Hack. = Aristida mutabilis Trin. et Rupr. ■☆
33953 Aristida obtusa Delile = Stipagrostis obtusa (Delile) Nees ■☆
33954 Aristida obtusa Delile subsp. pubescens Andr. = Stipagrostis foexiana (Maire et Wilczek) De Winter ■☆
33955 Aristida obtusa Delile var. pubescens Andr. = Stipagrostis foexiana (Maire et Wilczek) De Winter ■☆
33956 Aristida obtusa Delile. f. araneosa Corti = Stipagrostis foexiana (Maire et Wilczek) De Winter ■☆
33957 Aristida oligantha Michx.;少花三芒草;Old Field Three-awn, Old-field Three-awn, Plains Three-awn Grass, Prairie Three-awn ■☆
33958 Aristida oranensis Henrard = Stipagrostis oranensis (Henrard) De Winter ■☆
33959 Aristida pallida Steud. = Aristida sieberiana Trin. ■☆
33960 Aristida pallida Steud. var. chudaei (Batt. et Trab.) Maire et Weiller = Aristida sieberiana Trin. ■☆
33961 Aristida pallida Steud. var. glabriglumis Maire = Aristida sieberiana Trin. ■☆
33962 Aristida paoliana (Chiov.) Henrard;保尔三芒草■☆
33963 Aristida papposa Trin. et Rupr. = Stipagrostis uniplumis (Licht. ex Roem. et Schult.) De Winter ■☆
33964 Aristida papposa Trin. et Rupr. var. senegalensis ? = Stipagrostis uniplumis (Licht. ex Roem. et Schult.) De Winter ■☆
33965 Aristida paradisea Edgew. = Stipagrostis paradisea (Edgew.) De Winter ■☆
33966 Aristida paradoxa Steud. ex J. A. Schmidt = Aristida funiculata Trin. et Rupr. ■☆
33967 Aristida pardyi Stent et J. M. Rattray = Aristida junciformis Trin. et Rupr. subsp. welwitschii (Rendle) Melderis ■☆
33968 Aristida parvula (Nees) De Winter;较小三芒草■☆
33969 Aristida pennata Trin. = Stipagrostis pennata (Trin.) De Winter ■
33970 Aristida pennei Chiov.;彭尼三芒草■☆
33971 Aristida pilgeri Henrard;皮尔格三芒草■☆
33972 Aristida plumosa Desf. = Stipagrostis ciliata (Desf.) De Winter ■☆
33973 Aristida plumosa L.;羽状三芒草■☆
33974 Aristida plumosa L. = Stipagrostis plumosa (L.) Munro ex T. Anderson ■☆
33975 Aristida plumosa L. subsp. lanuginosa (Trab.) Maire = Stipagrostis plumosa (L.) Munro ex T. Anderson subsp. seminuda

(Trin. et Rupr.) H. Scholz ■☆
33976 Aristida plumosa L. subsp. sahelica Trab. = Stipagrostis sahelica (Trab.) De Winter ■☆
33977 Aristida plumosa L. var. aethiopica Trin. et Rupr. = Stipagrostis plumosa (L.) Munro ex T. Anderson subsp. seminuda (Trin. et Rupr.) H. Scholz ■☆
33978 Aristida plumosa L. var. australis Maire = Stipagrostis plumosa (L.) Munro ex T. Anderson ■☆
33979 Aristida plumosa L. var. berberica Trin. et Rupr. = Stipagrostis plumosa (L.) Munro ex T. Anderson subsp. seminuda (Trin. et Rupr.) H. Scholz ■☆
33980 Aristida plumosa L. var. brachypoda (Tausch) Trin. et Rupr. = Stipagrostis plumosa (L.) Munro ex T. Anderson ■
33981 Aristida plumosa L. var. dubia Maire = Stipagrostis oranensis (Henrard) De Winter ■☆
33982 Aristida plumosa L. var. floccosa (Coss. et Durieu) T. Durand et Schinz = Stipagrostis plumosa (L.) Munro ex T. Anderson subsp. seminuda (Trin. et Rupr.) H. Scholz ■☆
33983 Aristida plumosa L. var. lanuginosa Trab. = Stipagrostis plumosa (L.) Munro ex T. Anderson subsp. seminuda (Trin. et Rupr.) H. Scholz ■☆
33984 Aristida plumosa L. var. oranensis (Henrard) Maire = Stipagrostis oranensis (Henrard) De Winter ■☆
33985 Aristida plumosa L. var. seminuda Trin. et Rupr. = Stipagrostis plumosa (L.) Munro ex T. Anderson subsp. seminuda (Trin. et Rupr.) H. Scholz ■☆
33986 Aristida plumosa L. var. superciliata Henrard = Stipagrostis rigidifolia H. Scholz ■☆
33987 Aristida plumosa var. brachypoda (Tausch) Trin. et Rupr. = Stipagrostis plumosa (L.) Munro ex T. Anderson ■
33988 Aristida pogonoptila (Jaub. et Spach) Boiss. = Stipagrostis pogonoptila (Jaub. et Spach) De Winter ■☆
33989 Aristida pogonoptila Jaub. et Spach subsp. tibestica Maire = Stipagrostis pogonoptila (Jaub. et Spach) De Winter subsp. tibestica (Maire) J. -P. Lebrun et Stork ■☆
33990 Aristida protensa Henrard;伸展三芒草■☆
33991 Aristida proxima Steud. = Stipagrostis proxima (Steud.) De Winter ■☆
33992 Aristida pseudobromus Chiov.;假雀麦三芒草■☆
33993 Aristida pseudohystrix (Trin. et Rupr.) Steud. = Aristida diffusa Trin. ■☆
33994 Aristida pumila Decne. = Aristida adscensionis L. ■
33995 Aristida pungens Desf. = Stipagrostis pungens (Desf.) De Winter ■☆
33996 Aristida pungens Desf. var. gaditanus (Boiss. et Reut.) Ball = Stipagrostis pungens (Desf.) De Winter ■☆
33997 Aristida pungens Desf. var. pennata (Trin.) Trautv. = Aristida pennata Trin. ■
33998 Aristida pungens Desf. var. pubescens Henrard = Stipagrostis pungens (Desf.) De Winter ■☆
33999 Aristida pungens Desf. var. transiens Maire = Stipagrostis pungens (Desf.) De Winter subsp. transiens (Maire) H. Scholz ■☆
34000 Aristida purpurascens Poir.;浅紫三芒草;Arrowfeather, Arrow-feather, Arrow-feather Three-awn ■☆
34001 Aristida purpurascens Poir. var. minor Vasey = Aristida purpurascens Poir. ■☆
34002 Aristida purpurea Nutt.;紫三芒草;Purple Threeawn ■☆
34003 Aristida pusilla Trin. et Rupr. = Aristida adscensionis L. ■
34004 Aristida pycnostachya Cope;密穗三芒草■☆
34005 Aristida ramifera Pilg. = Aristida stipitata Hack. subsp. ramifera (Pilg.) Melderis ■☆
34006 Aristida ramosissima Engelm. ex A. Gray;多枝三芒草;Scurve Three-awn, Slender Three-awn ■☆
34007 Aristida rangei Pilg. = Aristida congesta Roem. et Schult. ■☆
34008 Aristida recta Franch.;直立三芒草■☆
34009 Aristida rigidifolia Scholz = Stipagrostis rigidifolia H. Scholz ■☆
34010 Aristida royleana Trin. et Rupr. = Aristida funiculata Trin. et Rupr. ■☆
34011 Aristida rufescens Steud.;浅红三芒草■☆
34012 Aristida sabulicola Pilg. = Stipagrostis sabulicola (Pilg.) De Winter ■☆
34013 Aristida sahelica Trab. = Stipagrostis sahelica (Trab.) De Winter ■☆
34014 Aristida sahelica Trab. var. scabra Leredde = Stipagrostis sahelica (Trab.) De Winter ■☆
34015 Aristida scabrescens L. Liou;糙三芒草;Rough Threeawngrass, Rough Triawn ■
34016 Aristida scabrivalvis Hack.;糙果片三芒草■☆
34017 Aristida scabrivalvis Hack. subsp. borumensis (Henrard) Melderis;博鲁姆三芒草■☆
34018 Aristida scabrivalvis Hack. subsp. contracta (De Winter) Melderis;紧缩三芒草■☆
34019 Aristida scabrivalvis Hack. var. contracta De Winter = Aristida scabrivalvis Hack. subsp. contracta (De Winter) Melderis ■☆
34020 Aristida schaeferi Mez;谢弗三芒草■☆
34021 Aristida schaeferi Mez = Stipagrostis schaeferi (Mez) De Winter ■☆
34022 Aristida schaeferi Mez var. biseriata Henrard = Stipagrostis schaeferi (Mez) De Winter ■☆
34023 Aristida schimperi Hochst. et Steud. ex Steud. = Stipagrostis ciliata (Desf.) De Winter ■☆
34024 Aristida schlechteri Henrard;施莱三芒草■☆
34025 Aristida schliebenii Henrard = Aristida junciformis Trin. et Rupr. ■☆
34026 Aristida schweinfurthii Boiss. = Aristida mutabilis Trin. et Rupr. ■☆
34027 Aristida schweinfurthii Boiss. var. boissieri Schweinf. = Aristida mutabilis Trin. et Rupr. ■☆
34028 Aristida sciurus Stapf;松鼠三芒草■☆
34029 Aristida scoparia Trin. et Rupr. = Stipagrostis scoparia (Trin. et Rupr.) De Winter ■☆
34030 Aristida sericans Hack. = Stipagrostis zeyheri (Nees) De Winter subsp. sericans (Hack.) De Winter ■☆
34031 Aristida shawii Scholz = Stipagrostis shawii (H. Scholz) H. Scholz ■☆
34032 Aristida sieberiana Trin.;西伯尔三芒草■☆
34033 Aristida similis Steud.;相似三芒草■☆
34034 Aristida spectabilis Hack.;壮观三芒草■☆
34035 Aristida stenophylla Henrard;窄叶三芒草■☆
34036 Aristida stenostachya Clayton;窄穗三芒草■☆
34037 Aristida stipiformis Poir. var. paoliana Chiov. = Aristida paoliana (Chiov.) Henrard ■☆
34038 Aristida stipitata Hack.;具柄三芒草■☆
34039 Aristida stipitata Hack. subsp. graciliflora (Pilg.) Melderis;细花具柄三芒草■☆

34040 Aristida stipitata Hack. subsp. ramifera (Pilg.) Melderis;枝生具柄三芒草■☆

34041 Aristida stipitata Hack. subsp. robusta (Stent et J. M. Rattray) Melderis;粗壮具柄三芒草■☆

34042 Aristida stipitata Hack. subsp. spicata (De Winter) Melderis;穗状具柄三芒草■☆

34043 Aristida stipitata Hack. var. graciliflora (Pilg.) De Winter = Aristida stipitata Hack. subsp. graciliflora (Pilg.) Melderis ■☆

34044 Aristida stipitata Hack. var. robusta (Stent et J. M. Rattray) De Winter = Aristida stipitata Hack. subsp. robusta (Stent et J. M. Rattray) Melderis ■☆

34045 Aristida stipitata Hack. var. stipitata De Winter = Aristida stipitata Hack. subsp. spicata (De Winter) Melderis ■☆

34046 Aristida stipoides Lam.;托叶状三芒草■☆

34047 Aristida subacaulis (Nees) Steud. = Stipagrostis subacaulis (Nees) De Winter ■☆

34048 Aristida submucronata Schumach. = Aristida adscensionis L. ■

34049 Aristida swartziana Steud.;斯瓦茨三芒草;Swartz's Threeawn ■☆

34050 Aristida takeoi Ohwi;竹生三芒草■☆

34051 Aristida tenuiflora Steud. = Aristida mutabilis Trin. et Rupr. ■☆

34052 Aristida tenuirostris Henrard = Stipagrostis tenuirostris (Henrard) De Winter ■☆

34053 Aristida tenuis Hochst. = Aristida mutabilis Trin. et Rupr. ■☆

34054 Aristida tenuiseta Cope;细刚毛三芒草■☆

34055 Aristida tenuissima A. Camus;细三芒草■☆

34056 Aristida textilis Mez;编织三芒草■☆

34057 Aristida thonningii Trin. et Rupr. = Aristida adscensionis L. ■

34058 Aristida transvaalensis Henrard;德兰士瓦三芒草■☆

34059 Aristida triseta Keng;三刺草;Threebristle Threeawngrass, Threespine Triawn ■

34060 Aristida triticoides Henrard;小麦三芒草■☆

34061 Aristida tsangpoensis L. Liou;藏布三芒草;Zangbu Threeawngrass, Zangbu Triawn ■

34062 Aristida tuberculosa Nutt.;瘤状三芒草;Beach Three-awn, Dune Three-awn Grass, Seaside Three-awn ■☆

34063 Aristida tunetana Coss. = Aristida congesta Roem. et Schult. ■☆

34064 Aristida tunetana Coss. var. intermedia Maire = Aristida congesta Roem. et Schult. ■☆

34065 Aristida uniplumis Licht. ex Roem. et Schult.;单羽三芒草■☆

34066 Aristida uniplumis Licht. ex Roem. et Schult. var. neesii Trin. et Rupr. = Stipagrostis uniplumis (Licht. ex Roem. et Schult.) De Winter var. neesii (Trin. et Rupr.) De Winter ■☆

34067 Aristida uniplumis Licht. ex Roem. et Schult. var. pearsonii Henrard = Stipagrostis uniplumis (Licht. ex Roem. et Schult.) De Winter ■☆

34068 Aristida vanderystii De Wild. = Sartidia vanderystii (De Wild.) De Winter ■☆

34069 Aristida vestita Thunb.;包被三芒草■☆

34070 Aristida vestita Thunb. f. amplior Hack. = Aristida vestita Thunb. ■☆

34071 Aristida vestita Thunb. var. pseudohystrix Trin. et Rupr. = Aristida diffusa Trin. ■☆

34072 Aristida vinosa Henrard = Aristida stipitata Hack. subsp. graciliflora (Pilg.) Melderis ■☆

34073 Aristida vulgaris Trin. et Rupr. = Aristida adscensionis L. ■

34074 Aristida vulgaris Trin. et Rupr. var. aethiopica ? = Aristida adscensionis L. var. aethiopica (Trin. et Rupr.) T. Durand et Schinz ■☆

34075 Aristida vulgaris Trin. et Rupr. var. arabica ? = Aristida adscensionis L. ■

34076 Aristida vulgaris Trin. et Rupr. var. depressa (Retz.) Trin. et Rupr. = Aristida depressa Retz. ■

34077 Aristida vulpinoides Hance = Aristida adscensionis L. ■

34078 Aristida wachteri Henrard = Aristida stipitata Hack. subsp. robusta (Stent et J. M. Rattray) Melderis ■☆

34079 Aristida waibeliana Henrard = Aristida effusa Henrard ■☆

34080 Aristida walteri Suess. = Stipagrostis fastigiata (Hack.) De Winter ■☆

34081 Aristida welwitschii Rendle = Aristida junciformis Trin. et Rupr. subsp. welwitschii (Rendle) Melderis ■☆

34082 Aristida wildii Melderis;维尔德三芒草■☆

34083 Aristida zeyheri (Nees) Steud. = Stipagrostis zeyheri (Nees) De Winter ■☆

34084 Aristida zitteli Asch. = Stipagrostis zittelii (Asch.) De Winter ■☆

34085 Aristida zitteli Asch. var. algeriensis Henrard = Stipagrostis acutiflora (Trin. et Rupr.) De Winter ■☆

34086 Aristidium (Endl.) Lindl. = Bouteloua Lag.(保留属名)■

34087 Aristidium Lindl. = Botelua Lag. ■

34088 Aristocapsa Reveal et Hardham (1989);谷刺蓼属;Valley Spinycape ■☆

34089 Aristocapsa insignis (Curran) Reveal et Hardham;谷刺蓼■☆

34090 Aristoclesia Coville = Platonia Mart.(保留属名)●☆

34091 Aristogeitonia Prain(1908);邻刺大戟属☆

34092 Aristogeitonia gabonica Breteler;加蓬邻刺大戟☆

34093 Aristogeitonia limonifolia Prain;邻刺大戟☆

34094 Aristogeitonia magnistipula Radcl. -Sm.;大托叶邻刺大戟☆

34095 Aristogeitonia monophylla Airy Shaw;单叶邻刺大戟☆

34096 Aristolelea Lour. = Spiranthes Rich.(保留属名)■

34097 Aristolochia L. (1753);马兜铃属;Aristolochia, Birthwort, Dutchman's Pipe, Dutchmanspipe, Dutchman's-pipe, Snake-root ●■

34098 Aristolochia L. = Pararistolochia Hutch. et Dalziel ●☆

34099 Aristolochia abyssinica Klotzsch = Aristolochia bracteolata Lam. ●■☆

34100 Aristolochia acuminata Roxb. = Aristolochia tagala Champ. ●■

34101 Aristolochia acutifolia Duch.;尖叶马兜铃●☆

34102 Aristolochia aethiopica Welw. = Aristolochia albida Duch. ●☆

34103 Aristolochia albida Duch.;微白马兜铃●☆

34104 Aristolochia angulata Bojer ex Duch. = Aristolochia albida Duch. ●☆

34105 Aristolochia antihysterica Mart. ex Duch.;治癔马兜铃●☆

34106 Aristolochia arborea Lindau;木本马兜铃(木马兜铃)●☆

34107 Aristolochia argentea Ule ex O. Schmidt;阿根廷马兜铃●☆

34108 Aristolochia atlantica Pomel = Aristolochia fontanesii Boiss. et Reut. ●☆

34109 Aristolochia atropurpurea Parish ex Hook. f.;缅南马兜铃●☆

34110 Aristolochia aurita Duch. = Aristolochia albida Duch. ●☆

34111 Aristolochia austrochinensis C. Y. Cheng et J. S. Ma;华南马兜铃;S. China Dutchmanspipe ●■

34112 Aristolochia austroszechuanica C. B. Chien et C. Y. Cheng ex C. Y. Cheng et J. L. Wu;川南马兜铃(大叶青木香);S. Sichuan Dutchmanspipe ●

34113 Aristolochia austroszechuanica C. B. Chien et C. Y. Cheng ex C. Y. Cheng et J. L. Wu = Aristolochia kwangsiensis Chun et F. C. How ex C. F. Liang ●

34114 Aristolochia austroyunnanensis S. M. Hwang = Aristolochia petelotii O. C. Schmidt ●

34115 Aristolochia baetica L.;攀木马兜铃●☆

34116 Aristolochia baetica L. var. bicolor Maire = Aristolochia baetica L. ●☆

34117 Aristolochia bainesii Burtt Davy = Aristolochia albida Duch. ●☆

34118 Aristolochia bambusifolia C. F. Liang ex H. Q. Wen;竹叶马兜铃;Bamboo-leaf Birthwort ●

34119 Aristolochia benadiriana Fiori = Aristolochia bracteolata Lam. ●■☆

34120 Aristolochia benadiriana Fiori var. longilabia Chiov. = Aristolochia bracteolata Lam. ●■☆

34121 Aristolochia bernieri Duch. = Aristolochia albida Duch. ●☆

34122 Aristolochia bilobata L.;双裂马兜铃;Twolobe Dutchman's Pipe ●☆

34123 Aristolochia blinii H. Lév. = Ceropegia mairei (H. Lév.) H. Huber ■

34124 Aristolochia bonatii H. Lév. = Aristolochia moupinensis Franch. ●

34125 Aristolochia bongoensis Engl. = Aristolochia albida Duch. ●☆

34126 Aristolochia bottae Jaub. et Spach;鲍他马兜铃●☆

34127 Aristolochia bracteata Retz. = Aristolochia bracteolata Lam. ●■☆

34128 Aristolochia bracteata Retz. var. basitruncata Hauman = Aristolochia bracteolata Lam. ●■☆

34129 Aristolochia bracteolata Lam.;具苞马兜铃●☆

34130 Aristolochia brasiliensis Mart. et Zucc.;巴西马兜铃;Rooster Flower ●☆

34131 Aristolochia brevipes Benth. var. wrightii (Seem.) Duch. = Aristolochia wrightii Seem. ■☆

34132 Aristolochia calcicola C. Y. Wu;青香藤(青木香);Rock Birthwort ●

34133 Aristolochia californica Torr.;加州马兜铃;California Dutchman's Pipe, California Snakeroot, Dutchman's Pipe ●☆

34134 Aristolochia carinata Merr. et Chun = Aristolochia hainanensis Merr. ●◇

34135 Aristolochia cathcartii Hook. f.;管兰香(白背马兜铃,萝卜防己);Cathcart Birthwort, Cathcart Dutchman's-pipe ●

34136 Aristolochia cathcartii Hook. f. = Aristolochia saccata Wall. ●

34137 Aristolochia caulialata C. Y. Wu ex C. Y. Cheng et J. S. Ma;翅茎马兜铃;Caulialate Dutchmanspipe ●

34138 Aristolochia ceropegioides S. Moore = Pararistolochia ceropegioides (S. Moore) Hutch. et Dalziel ●☆

34139 Aristolochia championii Merr. et Chun;长叶马兜铃(白解藤,白金古榄,百解薯,百解藤,绊藤香,金银带,千金薯,青藤,三筒管,山总管,竹叶薯);Champion Birthwort, Champion Dutchmanspipe, Champion Dutchman's-pipe, Longleaf Dutchmanspipe ●

34140 Aristolochia chilensis Bridges ex Lindl.;智利马兜铃●☆

34141 Aristolochia chlamydophylla C. Y. Wu ex S. M. Hwang;苞叶马兜铃(十八钻);Coatedleaf Birthwort ■

34142 Aristolochia chrysops (Stapf) E. H. Wilson ex Rehder = Aristolochia kaempferi Willd. f. heterophylla (Hemsl.) S. M. Hwang ●

34143 Aristolochia chrysops (Stapf) E. H. Wilson ex Rehder = Aristolochia kaempferi Willd. ●■

34144 Aristolochia chuandianensis Z. L. Yang;川滇马兜铃;Chuan-Dian Dutchmanspipe ●

34145 Aristolochia chuii C. Y. Wu;云南土木香●

34146 Aristolochia ciliata Hook. = Aristolochia fimbriata Cham. ●☆

34147 Aristolochia cinnabarina C. Y. Zheng et J. L. Wu;四川马兜铃(四川朱砂莲,朱砂莲);Sichuan Dutchmanspipe ●

34148 Aristolochia cinnabarina C. Y. Zheng et J. L. Wu = Aristolochia tuberosa C. F. Liang et S. M. Hwang ■

34149 Aristolochia clematitis L.;铁线莲状马兜铃(草马兜铃,欧马兜铃,欧洲马兜铃);Birthwort, Birthwort Dutchman's-pipe, Climbing Birthwort, Dutchman's-pipe, European Snakeroot, Heartwort, Saracen Birthwort, Saracen's Birthwort, Small Heartwort, Smearwort, Smerewort ●☆

34150 Aristolochia clematitis L. 'Saracen';撒拉逊马兜铃;Saracen Birthwort ●☆

34151 Aristolochia compressicaulis Z. L. Yang;扁茎马兜铃;Flatstem Dutchmanspipe ●

34152 Aristolochia compressicaulis Z. L. Yang = Aristolochia championii Merr. et Chun ●

34153 Aristolochia congolana Hauman = Pararistolochia promissa (Mast.) Keay ●☆

34154 Aristolochia contorta Bunge;北马兜铃(茶叶包,臭瓜旦,臭瓜蛋,臭瓜篓,臭罐罐,臭葫芦,臭铃铛,大叶马兜铃,刁铃,吊挂篮子,兜铃,斗铃,河沟精,后老婆罐,葫芦罐,麻丢铃,马兜果,马兜苓,马兜铃,马兜零,马斗铃,马斗令,马铃果,蜜马兜铃,南马兜铃,青木香,蛇参果,水马香果,天仙藤,铁扁担,土青木香,万丈龙,王宝瓜,王黄瓜,王室瓜,野木香果,圆叶马兜铃);Northern Dutchmanspipe ■

34155 Aristolochia convolvulacea Small = Aristolochia serpentaria L. ●■☆

34156 Aristolochia coryi I. M. Johnst.;科里马兜铃;Cory's Dutchman's-pipe ●☆

34157 Aristolochia cucurbitifolia Hayata;瓜叶马兜铃(黄藤,青木香,青香木);Melonleaf Dutchmanspipe ●

34158 Aristolochia cucurbitoides C. F. Liang;葫芦叶马兜铃(齿背马兜铃,齿被马兜铃);Calabashleaf Dutchmanspipe, Melonleaf Birthwort ■

34159 Aristolochia cymbifera Mart. et Zucc.;舟叶马兜铃(舟花马兜铃)●☆

34160 Aristolochia dabieshanensis C. Y. Cheng et W. Yu;大别山马兜铃;Dabieshan Dutchmanspipe ●■

34161 Aristolochia dabieshanensis C. Y. Cheng et W. Yu = Aristolochia kaempferi Willd. ●■

34162 Aristolochia debilis Siebold et Zucc.;马兜铃(百两金,长痧藤,臭罐罐,臭拉秧子,臭铃铛,定海根,都淋藤,兜铃根,独行根,独行木香,葫芦罐,马兜苓,马兜铃根,马兜铃藤,马兜零,南马兜铃,青木香,青木香藤,青藤香,三白银药,三百两金,痧药,蛇参根,蛇参果,水马香果,天仙藤,铁扁担,土木香,土青木香,土麝,万丈龙,痒辣菜,野木香根,一点气,云南根);Slender Dutchmans Pipe, Slender Dutchmanspipe ■

34163 Aristolochia delavayi Franch.;贯叶马兜铃(山草果);Delavay Birthwort ■

34164 Aristolochia delavayi Franch. var. micrantha W. W. Sm.;山草果(山胡椒,山蔓草);Smallflower Dutchmanspipe ■

34165 Aristolochia delavayi Franch. var. micrantha W. W. Sm. = Aristolochia delavayi Franch. ■

34166 Aristolochia densivenia Engl. = Aristolochia albida Duch. ●☆

34167 Aristolochia dewevrei De Wild. et T. Durand = Aristolochia albida Duch. ●☆

34168 Aristolochia districha Mast.;并毛马兜铃●☆

34169 Aristolochia duchartrei André;迪氏马兜铃●☆

34170 Aristolochia durior Hill = Aristolochia sipho L'Hér. ●☆

34171 Aristolochia elegans Mast.;烟斗花藤■

34172 Aristolochia elegans Mast. = Aristolochia littoralis D. Parodi ●

34173 Aristolochia embergeri Nozeran et N. Hallé;恩贝格尔马兜铃●☆

34174 Aristolochia erecta L.;直立马兜铃;Swanflower ■☆

34175 Aristolochia esperanzae Kuntze;尹氏马兜铃●☆

34176 Aristolochia fangchi Y. C. Wu ex L. D. Chow et S. M. Hwang;广防己(百解头,大痧药,防己,防己马兜铃,木防己,水防己,藤防己);Fangchi,Guangdong Birthwort ●

34177 Aristolochia faucimaculata H. Zhang et C. K. Hsieh;斑喉马兜铃;Faucimaculate Dutchmanspipe ●

34178 Aristolochia feddei H. Lév. = Aristolochia kaempferi Willd. f. thibetica (Franch.) S. M. Hwang ●

34179 Aristolochia feddei H. Lév. = Aristolochia thibetica Franch. ●

34180 Aristolochia fimbriata Cham. = Howardia fimbriata (Cham.) Klotzsch ●☆

34181 Aristolochia flagellata Stapf = Pararistolochia promissa (Mast.) Keay ●☆

34182 Aristolochia flavimollissima Y. K. Yang et al.;大暗消●

34183 Aristolochia flos-avis A. Chev. = Pararistolochia macrocarpa (Duch.) Poncy ●☆

34184 Aristolochia fontanesii Boiss. et Reut.;丰塔纳马兜铃●☆

34185 Aristolochia fordiana Hemsl.;通城虎(大散血,定心草,福德铃,福德马兜铃,天然草,万丈藤,五虎通城,血蒟,血娄,血藤暗消,一点血);Ford Birthwort ■

34186 Aristolochia forrestiana J. S. Ma;大囊马兜铃;Forrest Dutchmanspipe ●

34187 Aristolochia foveolata Merr.;蜂窝马兜铃(蜂巢马兜铃,蜂窝叶马兜铃,高氏马兜铃);Faveolate Birthwort ■

34188 Aristolochia fujianensis S. M. Hwang;福建马兜铃(小号山东瓜,一条根);Fujian Birthwort ■

34189 Aristolochia fulvicoma Merr. et Chun;黄毛马兜铃;Yellowhair Birthwort, Yellowhair Dutchmanspipe, Yellow-haired Dutchman's-pipe ●

34190 Aristolochia galeata Mart. et Zucc.;锥籽马兜铃●☆

34191 Aristolochia gentilis Franch.;优贵马兜铃(高贵马兜铃);Gentile Birthwort ■

34192 Aristolochia gibbosa Duch.;浅囊马兜铃●☆

34193 Aristolochia gigantea Mart. et Zucc.;巨大马兜铃;Dutchman's Pipe,Pelican Flower ●☆

34194 Aristolochia gigas Lindl. = Aristolochia grandiflora Sw. ●☆

34195 Aristolochia glaucescens Kunth = Howardia glaucescens (Kunth) Klotzsch ●☆

34196 Aristolochia goldieana Hook. f. = Pararistolochia goldieana (Hook. f.) Hutch. et Dalziel ●☆

34197 Aristolochia gracillima Hemsl. = Aristolochia gentilis Franch. ■

34198 Aristolochia grandiflora Sw.;大花马兜铃(大花孔雀花);Pelican Dutchman's-pipe, Pelican Flower, Pelicanflower, Pelican-flower Fleurs,Poison Hog Meat,Swan Flower ●☆

34199 Aristolochia grandiflora Sw. = Howardia grandiflora (Sw.) Klotzsch ●☆

34200 Aristolochia griffithii Hook. f. et Thomson ex Duch.;西藏马兜铃(藏木通,格氏马兜铃);Griffith Dutchman's-pipe, Tibet Dutchmanspipe,Xizang Birthwort,Xizang Dutchmanspipe ●

34201 Aristolochia hainanensis Merr.;海南马兜铃(假青黄藤);Hainan Birthwort,Hainan Dutchmanspipe,Hainan Dutchman's-pipe ●◇

34202 Aristolochia hastata Nutt. = Aristolochia serpentaria L. ●■☆

34203 Aristolochia heterophylla Hemsl.;异叶马兜铃(防己,汉中防己,痢药草,木防己,青木香,山豆根,台湾马兜铃,天仙藤,小南木香);Taiwan Dutchmanspipe,Yellowmouth Dutchmanspipe ●

34204 Aristolochia heterophylla Hemsl. = Aristolochia kaempferi Willd. f. heterophylla (Hemsl.) S. M. Hwang ●

34205 Aristolochia heterophylla Hemsl. = Aristolochia kaempferi Willd. ●■

34206 Aristolochia heterophylla Hemsl. var. linearifolia S. M. Hwang;狭异叶马兜铃(狭叶马兜铃)●

34207 Aristolochia hirta Peter = Aristolochia hockii De Wild. ●☆

34208 Aristolochia hockii De Wild.;霍克马兜铃●☆

34209 Aristolochia hockii De Wild. subsp. tuberculata Verdc.;多疣马兜铃●☆

34210 Aristolochia howii Merr. et Chun;南粤马兜铃(侯氏马兜铃,汪喉和);How Birthwort,How Dutchmanspipe,How Dutchman's-pipe ●

34211 Aristolochia iberica Fisch. et C. A. Mey.;伊比利亚马兜铃●☆

34212 Aristolochia impresinervis C. F. Liang;凹脉马兜铃(穿石藤);Impressvein Dutchmanspipe ■

34213 Aristolochia incisiloba Jongkind;锐裂马兜铃●☆

34214 Aristolochia indica L.;印度马兜铃●☆

34215 Aristolochia jinjiangensis H. Zhang et C. K. Hsieh;金江马兜铃;Jinjiang Dutchmanspipe ●

34216 Aristolochia jinshanensis Z. L. Yang et S. X. Tan;金山马兜铃;Jinfoshan Dutchmanspipe ●

34217 Aristolochia jinshanensis Z. L. Yang et S. X. Tan = Aristolochia moupinensis Franch. ●

34218 Aristolochia ju-ju S. Moore = Pararistolochia mannii (Hook. f.) Keay ●☆

34219 Aristolochia kaempferi Willd.;大叶马兜铃(背蛇生,避蛇生,地黄蒲,躲蛇生,淮通藤,金狮藤,金腰带,痢药草,琉球马兜铃,马兜铃,南木香,南投马兜铃,青木香,台湾马兜铃,香里藤,薰骨藤,寻骨风,一点血,异叶马兜铃,朱砂莲);Largeleaf Birthwort ●■

34220 Aristolochia kaempferi Willd. f. heterophylla (Hemsl.) S. M. Hwang = Aristolochia kaempferi Willd. ●■

34221 Aristolochia kaempferi Willd. f. heterophylla S. M. Hwang = Aristolochia kaempferi Willd. ●■

34222 Aristolochia kaempferi Willd. f. heterophylla S. M. Hwang = Aristolochia heterophylla Hemsl. ●

34223 Aristolochia kaempferi Willd. f. lineata Makino;条纹马兜铃●☆

34224 Aristolochia kaempferi Willd. f. longifolia Makino;长大叶马兜铃●☆

34225 Aristolochia kaempferi Willd. f. mirabilis S. M. Hwang;奇异马兜铃;Curious Dutchmanspipe ●

34226 Aristolochia kaempferi Willd. f. mirabilis S. M. Hwang = Aristolochia kaempferi Willd. ●■

34227 Aristolochia kaempferi Willd. f. thibetica (Franch.) S. M. Hwang;川西马兜铃(大寒药,躲蛇生,费氏马兜铃,山豆根);Western Sichuan Dutchmanspipe ●

34228 Aristolochia kaempferi Willd. f. thibetica (Franch.) S. M. Hwang = Aristolochia thibetica Franch. ●

34229 Aristolochia kaempferi Willd. f. trilobata Makino;三裂大叶马兜铃●☆

34230 Aristolochia kaempferi Willd. var. tanzawana Kigawa;丹氏马兜铃●☆

34231 Aristolochia kankauensis (Sasaki) Nakai ex Masam. = Aristolochia zollingeriana Miq. ●■

34232 Aristolochia kankauensis Sasaki = Aristolochia zollingeriana Miq. ●■

34233 Aristolochia kaoi Tang S. Liu et M. J. Lai;高氏马兜铃●

34234 Aristolochia kaoi Tang S. Liu et M. J. Lai = Aristolochia foveolata Merr. ■

34235 Aristolochia kirkii Baker = Aristolochia albida Duch. ●☆

34236 Aristolochia kotschyi A. Rich. = Aristolochia bracteolata Lam. ●☆

34237 Aristolochia kunmingensis C. Y. Cheng et J. S. Ma;昆明马兜铃;Kunming Dutchmanspipe ●

34238 Aristolochia kwangsiensis Chun et F. C. How ex C. F. Liang;广西马兜铃(川南马兜铃,大百解薯,大青木香,大叶马兜铃,大叶山总管,大总管,滇南马兜铃,管南香,金银袋,萝卜防己,南蛇藤,青木香,圆叶马兜铃,圆叶山总管,总管);Guangxi Birthwort, Guangxi Dutchmanspipe, Guangxi Dutchman's-pipe, Kwangsi Dutchmanspipe,Kwangsi Dutchman's-pipe ●

34239 Aristolochia labiata Willd.;斑驳马兜铃;Mottled Dutchman's Pipe ●☆

34240 Aristolochia lasiops Stapf;鄂西马兜铃●

34241 Aristolochia ledermannii Engl. = Aristolochia albida Duch. ●☆

34242 Aristolochia leonensis Mast. = Pararistolochia leonensis (Mast.) Hutch. et Dalziel ●☆

34243 Aristolochia leuconeura Linden;白脉马兜铃●☆

34244 Aristolochia liangshanensis Z. L. Yang;凉山马兜铃;Liangshan Dutchmanspipe ●

34245 Aristolochia liangshanensis Z. L. Yang = Aristolochia thibetica Franch. ●

34246 Aristolochia littoralis D. Parodi;美丽马兜铃(彩花马兜铃,烟斗花藤,优美马兜铃);Calico Flower, Dutchman's Pipe, Elegant Dutchman's Pipe ●

34247 Aristolochia liukiuensis Hatus.;琉球马兜铃●

34248 Aristolochia liukiuensis Hatus. = Aristolochia kaempferi Willd. ●■

34249 Aristolochia longa Desf. = Aristolochia fontanesii Boiss. et Reut. ●☆

34250 Aristolochia longa Georgi = Aristolochia clematitis L. ●☆

34251 Aristolochia longa L.;长柄马兜铃;Long Birthwort, Long Heartwort ●☆

34252 Aristolochia longa L. subsp. atlantica (Pomel) Batt. = Aristolochia fontanesii Boiss. et Reut. ●☆

34253 Aristolochia longa L. subsp. fontanesii (Boiss. et Reut.) Batt. = Aristolochia fontanesii Boiss. et Reut. ●☆

34254 Aristolochia longa L. subsp. paucinervis (Pomel) Batt. = Aristolochia paucinervis Pomel ●☆

34255 Aristolochia longa L. var. longilabiata Maire et Weiller = Aristolochia navicularis Nardi ●☆

34256 Aristolochia longa Thunb. = Aristolochia debilis Siebold et Zucc. ■

34257 Aristolochia longa Woodv. = Aristolochia baetica L. ●☆

34258 Aristolochia longgangensis C. F. Liang;弄岗马兜铃(弄岗通城虎);Longgang Birthwort ■

34259 Aristolochia longifolia Champ. ex Benth. = Aristolochia championii Merr. et Chun ●

34260 Aristolochia longilinqua C. Y. Cheng et W. Yu;长花马兜铃;Longflower Dutchmanspipe ●

34261 Aristolochia macedonica Bornm.;马其顿马兜铃●☆

34262 Aristolochia macrocarpa C. Y. Wu et S. K. Wu ex D. D. Tao;大果马兜铃;Bigcarp Dutchmanspipe,Dutchman's Pipe ●

34263 Aristolochia macrocarpa Duch. = Pararistolochia macrocarpa (Duch.) Poncy ●☆

34264 Aristolochia macrophylla Lam.;杜氏马兜铃;Dutchman's Pipe, Dutchman's-pipe ●☆

34265 Aristolochia macrophylla Lam. = Aristolochia sipho L'Hér. ●☆

34266 Aristolochia mairei H. Lév. = Ceropegia mairei (H. Lév.) H. Huber ■

34267 Aristolochia mannii Hook. f. = Pararistolochia mannii (Hook. f.) Keay ●☆

34268 Aristolochia manshuriensis Kom.;关木通(东北木通,淮通,苦木通,马木通,木通,木通马兜铃,棺木香,万年藤);Akebi Birthwort,Manchurian Dutchmanspipe,Manchurian Dutchman's-pipe ●◇

34269 Aristolochia marshii Standl. = Aristolochia pentandra Jacq. ●☆

34270 Aristolochia maurorum L.;莫尔马兜铃●☆

34271 Aristolochia maxima Jacq.;大马兜铃;Florida Dutchman's Pipe, Florida Dutchman's-pipe ●☆

34272 Aristolochia minutissima C. Y. Cheng;避蛇生●

34273 Aristolochia mollis Dunn;柔毛马兜铃(金狮藤,金丝藤,金腰带,青香藤,香里陈,香里藤);Softhair Dutchmanspipe ●

34274 Aristolochia mollis Dunn = Aristolochia kaempferi Willd. ●■

34275 Aristolochia mollissima Hance;绵毛马兜铃(白毛藤,白面风,穿地草,穿地筋,地丁香,鹅婆娘,猴儿草,猴耳草,黄木耳,黄木香,猫耳朵,猫耳朵草,猫香,毛白藤,毛风草,毛香,清骨风,兔子耳,寻骨风,巡骨风,烟袋锅);Hairy Dutchmanspipe, Woolly Dutchmanspipe,Woolly Dutchman's-pipe ●

34276 Aristolochia moupinensis Franch.;宝兴马兜铃(大半药,大内消,大条请木香,关木通,淮木通,淮通,淮通马兜铃,老蛇藤,理防己,木通,木香,木香马兜铃,穆坪马兜铃,南木香,青木香,藤藤黄);Mouping Dutchman's-pipe,Muping Dutchmanspipe,Muping Dutchman's-pipe,Woolly Birthwort ●

34277 Aristolochia multiflora Duch.;多花马兜铃■☆

34278 Aristolochia multiflora Duch. = Aristolochia albida Duch. ●☆

34279 Aristolochia multinervis Pomel = Aristolochia fontanesii Boiss. et Reut. ●☆

34280 Aristolochia nashii Kearney = Aristolochia serpentaria L. ●■☆

34281 Aristolochia navicularis Nardi;船状马兜铃●☆

34282 Aristolochia neolongifolia J. L. Wu et Z. L. Yang;线叶马兜铃;Linearleaf Dutchmanspipe ●■

34283 Aristolochia neolongifolia J. L. Wu et Z. L. Yang = Aristolochia kaempferi Willd. ●■

34284 Aristolochia nipponica Makino = Aristolochia contorta Bunge ■

34285 Aristolochia obliqua S. M. Hwang;偏花马兜铃(汉防己);Obliqueflower Dutchmanspipe ●

34286 Aristolochia odoratissima Benth.;芳香马兜铃;Contrayerva, Fragrant Dutchman's Pipe,Junction Vine,Sweet-scented Birthwort ●☆

34287 Aristolochia onoei Franch. et Sav. ex Koidz. = Aristolochia shimadae Hayata ●

34288 Aristolochia ovatifolia S. M. Hwang;卵叶马兜铃(大寒药,木防己);Eggleaf Birthwort, Eggleaf Dutchmanspipe, Ovale-leaved Dutchman's-pipe,Ovate-leaf Dutchmanspipe ●

34289 Aristolochia pallida Salzm. ex Ball;白马兜铃●☆

34290 Aristolochia pandurata Jacq.;琴叶马兜铃●☆

34291 Aristolochia parensis Engl. ex Peter = Aristolochia bracteolata Lam. ●☆

34292 Aristolochia paucinervis Pomel;少脉马兜铃●☆

34293 Aristolochia peltata L.;钝状马兜铃;Peltate Dutchman's Pipe ●☆

34294 Aristolochia pentandra Jacq.;马什马兜铃;Marsh's Dutchman's-pipe ●☆

34295 Aristolochia petelotii O. C. Schmidt;滇南马兜铃;S. Yunnan Birthwort,South Yunnan Dutchmanspipe,South Yunnan Dutchman's-pipe ●

34296 Aristolochia petelotii O. C. Schmidt = Aristolochia austroyunnanensis S. M. Hwang ●

34297 Aristolochia petersiana Klotzsch = Aristolochia albida Duch. ●☆

34298 Aristolochia platanifolia Duch.;掌叶马兜铃●

34299 Aristolochia polymorpha S. M. Hwang;多型马兜铃(多型叶马兜铃);Manyform Dutchmanspipe,Polumorph Dutchmanspipe ■

34300 Aristolochia pontica Lam.;黑海马兜铃●☆

34301 Aristolochia porphyrophylla Pfeifer = Aristolochia watsonii Wooton et Standl. ●■☆

34302 Aristolochia preussii Engl. = Pararistolochia preussii (Engl.) Hutch. et Dalziel ●☆

34303 Aristolochia promissa Mast. = Pararistolochia promissa (Mast.) Keay ●☆

34304 Aristolochia punjabensis Lace;旁遮普马兜铃●☆

34305 Aristolochia racemosa Brandegee = Aristolochia pentandra Jacq. ●☆

34306 Aristolochia recurvilabra Hance = Aristolochia debilis Siebold et Zucc. ■

34307 Aristolochia reticulata Nutt.;网叶马兜铃;Texas Dutchman's-pipe,Texas Snakeroot ●☆

34308 Aristolochia rhodesica R. E. Fr. = Aristolochia hockii De Wild. ●☆

34309 Aristolochia rigida Duch.;坚挺马兜铃●☆

34310 Aristolochia ringens Vahl = Howardia ringens (Vahl) Klotzsch ●

34311 Aristolochia rotunda Desf. = Aristolochia fontanesii Boiss. et Reut. ●☆

34312 Aristolochia rotunda L.;圆根马兜铃(圆叶马兜铃);Smearwort ●☆

34313 Aristolochia rotunda L. var. grandiflora H. Duch. = Aristolochia fontanesii Boiss. et Reut. ●☆

34314 Aristolochia roxburghiana Klotzsch = Aristolochia tagala Champ. ●■

34315 Aristolochia roxburghiana Klotzsch subsp. kankauensis (Sasaki) Kitam. = Aristolochia zollingeriana Miq. ●■

34316 Aristolochia roxburghiana Klotzsch subsp. kankauensis (Sasaki) T. Yamaz. = Aristolochia zollingeriana Miq. ●■

34317 Aristolochia ruiziana Duch. = Aristolochia duchartrei André ●☆

34318 Aristolochia saccata Wall.;袋形马兜铃(管兰香,囊花马兜铃);Bag-shaped Dutchmanspipe ●

34319 Aristolochia saccata Wall. var. angustifolia (G. Klotz) Duch.;狭叶马兜铃;Narrowlea Bag-shaped Dutchmanspipe ●

34320 Aristolochia salweenensis C. Y. Cheng et J. S. Ma;怒江马兜铃;Nujiang Dutchmanspipe ●

34321 Aristolochia salweenensis C. Y. Cheng et J. S. Ma = Aristolochia kunmingensis C. Y. Cheng et J. S. Ma ●

34322 Aristolochia schweinfurthii Engl. = Pararistolochia triactina (Hook. f.) Hutch. et Dalziel ●☆

34323 Aristolochia scytophylla S. M. Hwang et D. Y. Chen;革叶马兜铃(银带);Coriaceous Dutchman's-pipe, Coriaceousleaf Dutchmanspipe,Leatherleaf Dutchmanspipe ●

34324 Aristolochia sempervirens L.;常绿马兜铃●☆

34325 Aristolochia serpentaria L.;蛇根马兜铃;Serpentaria, Serpentary,Virginia Snake Root,Virginian Snakeroot ●■☆

34326 Aristolochia serpentaria L. var. hastata (Nutt.) Duch. = Aristolochia serpentaria L. ●■☆

34327 Aristolochia serrata ?;锯齿马兜铃●☆

34328 Aristolochia setchuenensis Franch. = Aristolochia kaempferi Willd. f. thibetica (Franch.) S. M. Hwang ●

34329 Aristolochia setchuenensis Franch. = Aristolochia thibetica Franch. ●

34330 Aristolochia setchuenensis Franch. var. holotricha Diels = Aristolochia thibetica Franch. ●

34331 Aristolochia setchuenensis Franch. var. holotricha Diels = Aristolochia kaempferi Willd. f. thibetica (Franch.) S. M. Hwang ●

34332 Aristolochia shimadae Hayata;台湾马兜铃(岛田马兜铃)●

34333 Aristolochia shimadae Hayata = Aristolochia heterophylla Hemsl. ●

34334 Aristolochia shimadae Hayata = Aristolochia kaempferi Willd. ●■

34335 Aristolochia shukangii Chun et F. C. How = Aristolochia kwangsiensis Chun et F. C. How ex C. F. Liang ●

34336 Aristolochia sinarum Lindl. = Aristolochia debilis Siebold et Zucc. ■

34337 Aristolochia sipho L'Hér.;欧洲马兜铃(榴莲马兜铃);Common Dutchman's-pipe, Dutchman's Pipe, Dutchman's-pipe, Pipe Vine,Pipe-vine ●☆

34338 Aristolochia sipho L'Hér. = Aristolochia macrophylla Lam. ●☆

34339 Aristolochia sipho L'Hér. f. grandiflora Franch.;城口马兜铃(亚美马兜铃)●

34340 Aristolochia somalensis Oliv. = Aristolochia rigida Duch. ●☆

34341 Aristolochia soyauxiana Oliv. = Pararistolochia macrocarpa (Duch.) Poncy subsp. soyauxiana (Oliv.) Poncy ●☆

34342 Aristolochia staudtii Engl. = Pararistolochia macrocarpa (Duch.) Poncy ●☆

34343 Aristolochia steupii Woronow;斯托普马兜铃●☆

34344 Aristolochia stuhlmannii Engl. = Pararistolochia triactina (Hook. f.) Hutch. et Dalziel ●☆

34345 Aristolochia sylvicola Standl. = Aristolochia gigantea Mart. et Zucc. ●☆

34346 Aristolochia sylvicola Standl. = Aristolochia leuconeura Linden ●☆

34347 Aristolochia szemaoense C. Y. Wu;思茅马兜铃●

34348 Aristolochia tagala Champ.;耳叶马兜铃(暗消,槌果马兜铃,黑面防己,假大薯,假通城虎,卵叶雷公藤,卵叶马兜铃,麻疯龙,木防己,青木香,土木香);Ovalleaf Dutchmanspipe ●■

34349 Aristolochia tagala Champ. var. kankauensis (Sasaki) T. Yamaz. = Aristolochia zollingeriana Miq. ●■

34350 Aristolochia talbotii S. Moore = Pararistolochia promissa (Mast.) Keay ●☆

34351 Aristolochia taliscana Hook. et Arn.;它里斯马兜铃●☆

34352 Aristolochia tenuicauda S. Moore = Pararistolochia promissa (Mast.) Keay ●☆

34353 Aristolochia tessmannii Engl. = Pararistolochia macrocarpa (Duch.) Poncy ●☆

34354 Aristolochia thibetica Franch. = Aristolochia kaempferi Willd. f. thibetica (Franch.) S. M. Hwang ●

34355 Aristolochia thwaitesii Hook.;海边马兜铃(石蟾蜍,钟花马兜铃); Seacoast Dutchmanspipe, Seashore Birthwort, Thwaites Dutchmanspipe ●

34356 Aristolochia tomentosa Sims;绒毛马兜铃;Dutchman's Pipe, Pipevine,Pipe-vine,Woolly Dutchman's-pipe,Woolly Pipe-vine ●☆

34357 Aristolochia transsecta (Chatterjee) C. Y. Wu ex S. M. Hwang;粉花马兜铃(粉质花马兜铃,粉质青木香,蝴蝶暗消,黄木香,青藤香,细尖马兜铃,朱砂莲);Powderflower Birthwort,Powderflower Dutchmanspipe,Transsecte Dutchmanspipe,Transvers Dutchman's-pipe ●

34358 Aristolochia triactina Hook. f. = Pararistolochia triactina (Hook. f.) Hutch. et Dalziel ●☆

34359 Aristolochia triangulifolia W. Yu;角叶马兜铃(三角叶青木香);Triangular Dutchmanspipe ●

34360 Aristolochia tribrachiata S. Moore = Pararistolochia macrocarpa (Duch.) Poncy ●☆

34361 Aristolochia trilobata Lam.;三裂马兜铃;Bejuco De Santiago ●☆

34362 Aristolochia truncata Peter = Aristolochia albida Duch. ●☆

34363 Aristolochia tuberosa C. F. Liang et S. M. Hwang;背蛇生(避蛇药,毒蛇药,躲蛇生,广西朱砂莲,块茎马兜铃,牛血莲,万丈龙,朱砂莲);Tuberous Birthwort ■

34364 Aristolochia tubiflora Dunn;管花马兜铃(逼血雷,鼻血雷,鼻血连,鼻血莲,毕石牛,辟蛇雷,碧血雷,避蛇灵,独一味,红白药,金丝丸,一点血,钟铃藤细辛);Tubeflower Dutchmanspipe ■

34365 Aristolochia utriformis S. M. Hwang;囊花马兜铃(马兜铃);Utricularflower Dutchmanspipe,Utriculateflower Birthwort ●

34366 Aristolochia veraguensis Duch. = Aristolochia leuconeura Linden ●☆

34367 Aristolochia versicolor S. M. Hwang;变色马兜铃(白金古榄,过石珠,苦凉藤,青香藤,银袋);Versicolor Birthwort, Versicolor Dutchmanspipe,Versicolous Dutchman's-pipe ●

34368 Aristolochia viridiflora H. Lév. = Ceropegia mairei (H. Lév.) H. Huber ■

34369 Aristolochia viridiflora H. Lév. var. occlusa H. Lév. = Ceropegia mairei (H. Lév.) H. Huber ■

34370 Aristolochia watsonii Wooton et Standl.;瓦氏马兜铃;Indian Root,Indianroot ●■☆

34371 Aristolochia westlanda Hemsl. = Aristolochia fangchi Y. C. Wu ex L. D. Chow et S. M. Hwang ●

34372 Aristolochia westlandii Hemsl.;香港马兜铃(白金果榄,百解马兜铃,百解薯,苦凉藤,山总管,银袋);Hongkong Birthwort, Hongkong Dutchmanspipe,Hongkong Dutchman's-pipe ●

34373 Aristolochia westlandii Hemsl. = Aristolochia versicolor S. M. Hwang ●

34374 Aristolochia wrightii Seem.;赖特马兜铃;Wright's Dutchman's-pipe ■☆

34375 Aristolochia yunnanensis Franch.;云南马兜铃(白防己,串石藤,打鼓藤,地檀香,金不换,南木香,楠木香,青木香,藤七,藤子暗消,土木香,小南木香,追风散);Yunnan Birthwort, Yunnan Dutchmanspipe,Yunnan Dutchman's-pipe ●

34376 Aristolochia yunnanensis Franch. = Aristolochia griffithii Hook. f. et Thomson ex Duch. ●

34377 Aristolochia yunnanensis Franch. var. meionantha Hand.-Mazz.;小花马兜铃;Littleflower Dutchmanspipe, Small-flower Yunnan Dutchmanspipe ●

34378 Aristolochia yunnanensis Franch. var. meionantha Hand.-Mazz. = Aristolochia griffithii Hook. f. et Thomson ex Duch. ●

34379 Aristolochia zenkeri Engl.;赛克氏马兜铃●☆

34380 Aristolochia zenkeri Engl. = Pararistolochia zenkeri (Engl.) Hutch. et Dalziel ●☆

34381 Aristolochia zhongdianensis J. S. Ma;中甸马兜铃;Zhongdian Dutchmanspipe ■

34382 Aristolochia zollingeriana Miq.;港口马兜铃;Zollinger Birthwort,Zollinger Dutchmanspipe ●■

34383 Aristolochiaceae Adans. = Aristolochiaceae Juss.(保留科名)●■

34384 Aristolochiaceae Juss.(1789)(保留科名);马兜铃科;Birthwort Family,Dutchmanspipe Family,Dutchman's-pipe Family ●■

34385 Aristomenia Vell. = Stifftia J. C. Mikan(保留属名)●☆

34386 Aristopetalum Willis = Aistopetalum Schltr. ●☆

34387 Aristopsis Catasus = Aristida L. ■

34388 Aristotela Adans.(废弃属名) = Aristotelia L'Hér.(保留属名)●☆

34389 Aristotela Adans.(废弃属名) = Othonna L. ●■☆

34390 Aristotela J. F. Gmel. = Aristotelia L'Hér.(保留属名)●☆

34391 Aristotelea Lour. = Spiranthes Rich.(保留属名)■

34392 Aristotelea Spreng. = Aristotelia L'Hér.(保留属名)●☆

34393 Aristotelia Comm. ex Lam. = Terminalia L.(保留属名)●

34394 Aristotelia L'Hér.(1786)(保留属名);酒果属;Aristotelia, Wineberry ●☆

34395 Aristotelia australasica F. Muell.;山酒果;Mountain Wineberry ●☆

34396 Aristotelia chilensis Stuntz;智利酒果(酒果)●☆

34397 Aristotelia chilensis Stuntz 'Variegata';斑叶智利酒果●☆

34398 Aristotelia glandulosa Ruiz et Pav. = Aristotelia macqui L'Hér. ●☆

34399 Aristotelia macqui L'Hér. = Aristotelia chilensis Stuntz ●☆

34400 Aristotelia serrata Oliv.;齿叶酒果;Makomako,Wineberry ●☆

34401 Aristoteliaceae Dumort.;酒果科●☆

34402 Aristoteliaceae Dumort. = Elaeocarpaceae Juss.(保留科名)●

34403 Arivela Raf.(1838);黄花草属■

34404 Arivela Raf. = Cleome L. ●■

34405 Arivela Raf. = Polanisia Raf. ■

34406 Arivela viscosa (L.) Raf.;黄花草(臭点菜,臭矢菜,黄花菜,向天黄,羊角草);Wild Mustard,Yellow Spiderflower,Yellowflower Spiderflower ■

34407 Arivela viscosa (L.) Raf. var. deglabrata (Backer) M. L. Zhang et G. C. Tucker;无毛黄花草;Hairless Yellow Spiderflower ■

34408 Arivona Steud. = Arjona Comm. ex Cav. ☆

34409 Arjona Cav. = Arjona Comm. ex Cav. ☆

34410 Arjona Comm. ex Cav.(1798);阿霍檀香属☆

34411 Arjona tuberosa Cav.;阿霍檀香☆

34412 Arjonaceae Tiegh. = Olacaceae R. Br.(保留科名)●

34413 Arjonaceae Tiegh. = Santalaceae R. Br.(保留科名)●■

34414 Arjonaea Kuntze = Arjona Comm. ex Cav. ☆

34415 Arkezostis Raf. = Cayaponia Silva Manso(保留属名)■☆

34416 Arkopoda Raf. = Reseda L. ■

34417 Armania Bert. ex DC. = Encelia Adans. ●■☆

34418 Armarintea Bubani = Cachrys L. ■

34419 Armatocereus Backeb.(1938);花铠柱属●☆

34420 Armatocereus cartwrightianus Backeb.;铁干●☆

34421 Armatocereus laetus (Kunth) Backeb.;花铠柱●☆

34422 Armatocereus matucanensis Backeb.;摩天楼●☆

34423 Armeniaca Mill. = Armeniaca Scop. ●

34424 Armeniaca Mill. = Prunus L. ●

34425 Armeniaca Scop.(1754);杏属;Apricot,Common Apricot ●

34426 Armeniaca Tourn. ex Mill. = Armeniaca Scop. ●

34427 Armeniaca ansu (Maxim.) Kostina = Armeniaca vulgaris Lam. var. ansu (Maxim.) Te T. Yu et A. M. Lu ●

34428 Armeniaca ansu (Maxim.) Kostina = Prunus armeniaca L. var. ansu Maxim. ●

34429 Armeniaca atropurpurea Loisel. = Armeniaca dasycarpa (Ehrh.) Borkh. ●

34430 Armeniaca brigantiaca Pers.;布里康杏;Briancon Apricot ●☆

34431 Armeniaca dasycarpa (Ehrh.) Borkh.;紫杏(加拿大李);Black Apricot,Canada Plum,Canadian Plum,Horse Plum,Purple Apricot, Red Plum ●

34432 Armeniaca dasycarpa (Ehrh.) Borkh. = Prunus dasycarpa Ehrh. ●

34433 Armeniaca dasycarpa (Ehrh.) Pers. = Armeniaca dasycarpa (Ehrh.) Borkh. ●

34434 Armeniaca davidiana Carrière = Prunus davidiana (Carrière) Franch. ●

34435 Armeniaca fusca Turpin et Poit. = Armeniaca dasycarpa (Ehrh.) Borkh. ●

34436 Armeniaca holosericea (Batalin) Kostina;藏杏(毛叶杏);Hairyleaf Apricot,Silky Apricot ●◇
34437 Armeniaca holosericea (Batalin) Kostina = Prunus armeniaca L. var. holosericea Batalin ●◇
34438 Armeniaca holosericea (Batalin) Kostina var. xupuensis T. Z. Li;叙浦杏;Xupu Silky Apricot ●
34439 Armeniaca hongpingensis Te T. Yu et C. L. Li;洪平杏;Hongping Apricot ●
34440 Armeniaca hypotrichodes (Cardot) C. L. Li et S. Y. Jiang;背毛杏(背毛樱);Hairy-back Cherry ●
34441 Armeniaca hypotrichodes (Cardot) C. L. Li et S. Y. Jiang = Prunus hypotricha Rehder ●
34442 Armeniaca limeixing J. Y. Zhang et Z. M. Wang;李梅杏(酸梅,杏梅,转子红);Limeixing Apricot ●
34443 Armeniaca mandshurica (Maxim.) Skvortsov;东北杏(辽杏,满洲杏);Manchurian Apricot ●
34444 Armeniaca mandshurica (Maxim.) Skvortsov = Prunus mandshurica (Maxim.) Koehne ●
34445 Armeniaca mandshurica (Maxim.) Skvortsov f. major T. Z. Li;大果东北杏(大果辽杏);Bigfruit Manchurian Apricot ●
34446 Armeniaca mandshurica (Maxim.) Skvortsov var. glabra (Nakai) Te T. Yu et A. M. Lu;光叶东北杏;Glabrous Manchurian Apricot,Glabrous-leaf Manchurian Apricot ●
34447 Armeniaca mandshurica (Maxim.) Skvortsov var. glabra (Nakai) Te T. Yu et A. M. Lu = Prunus mandshurica (Maxim.) Koehne var. glabra Nakai ●
34448 Armeniaca mume (Siebold et Zucc.) de Vriese;梅(白梅,白梅花,白霜梅,春梅,干枝梅,杲,合汉梅,合溪梅,鹤顶梅,黑梅,红梅,黄仔,建梅,江梅,桔梅,丽枝梅,六瓣梅,绿萼梅,绿梅花,梅干,梅果,梅柟,梅实,梅树,梅诸,梅仔,梅子,枏,品字梅,青梅,青竹梅,霜梅,酸梅,乌梅,乌梅炭,杏梅,杏叶梅,熏梅,盐梅,野梅,鸳鸯梅,早梅,照水梅,重叶梅);Flowering Apricot,Japanese Apricot,Mei,Mei Flower,Mei Hua,Mume,Mume Plant,Mumeplant,Mume-plant ●
34449 Armeniaca mume (Siebold et Zucc.) de Vriese f. albo-plena (Bailey) Rehder;玉碟梅●
34450 Armeniaca mume (Siebold et Zucc.) de Vriese f. alphandii (Carrière) Rehder;宫粉梅●
34451 Armeniaca mume (Siebold et Zucc.) de Vriese f. purpurea (Makino) T. Y. Chen;朱砂梅●
34452 Armeniaca mume (Siebold et Zucc.) de Vriese f. rubriflora T. Y. Chen;大红梅●
34453 Armeniaca mume (Siebold et Zucc.) de Vriese f. simpliciflora T. Y. Chen;江梅●
34454 Armeniaca mume (Siebold et Zucc.) de Vriese f. versicolor T. Y. Chen et H. H. Lu;洒金梅●
34455 Armeniaca mume (Siebold et Zucc.) de Vriese f. viridicalyx (Makino) T. Y. Chen;绿萼梅(绿梅);Greencalyx Mumeplant ●
34456 Armeniaca mume (Siebold et Zucc.) de Vriese f. viridicalyx (Makino) T. Y. Chen = Prunus mume (Siebold) Siebold et Zucc. f. viridicalyx (Makino) T. Y. Chen ●
34457 Armeniaca mume (Siebold et Zucc.) de Vriese var. cernua (Franch.) Te T. Yu et A. M. Lu = Prunus mume (Siebold) Siebold et Zucc. var. cernua Franch. ●
34458 Armeniaca mume (Siebold et Zucc.) de Vriese var. pallescens (Franch.) Te T. Yu et A. M. Lu;厚叶梅(野梅);Thickleaf Mume ●
34459 Armeniaca mume (Siebold et Zucc.) de Vriese var. pallescens (Franch.) Te T. Yu et A. M. Lu = Prunus mume (Siebold) Siebold et Zucc. var. pellescens Franch. ●
34460 Armeniaca mume (Siebold et Zucc.) de Vriese var. pendula Siebold;照水梅●
34461 Armeniaca mume (Siebold et Zucc.) de Vriese var. pendula Siebold f. albiflora T. Y. Chen;残雪照水梅●
34462 Armeniaca mume (Siebold et Zucc.) de Vriese var. pendula Siebold f. atropurpurea T. Y. Chen;骨红照水梅●
34463 Armeniaca mume (Siebold et Zucc.) de Vriese var. pendula Siebold f. marmorata T. Y. Chen;五宝照水梅●
34464 Armeniaca mume (Siebold et Zucc.) de Vriese var. pendula Siebold f. modesta T. Y. Chen;双粉照水梅●
34465 Armeniaca mume (Siebold et Zucc.) de Vriese var. pendula Siebold f. simplex T. Y. Chen;单粉照水梅●
34466 Armeniaca mume (Siebold et Zucc.) de Vriese var. pendula Siebold f. viridiflora T. Y. Chen;白碧照水梅●
34467 Armeniaca mume (Siebold et Zucc.) de Vriese var. pubicaulina C. Z. Qiao et H. M. Shen;毛茎梅●
34468 Armeniaca mume (Siebold et Zucc.) de Vriese var. tortuosa T. Y. Chen et H. H. Lu;龙游梅●
34469 Armeniaca mume Siebold = Armeniaca mume (Siebold et Zucc.) de Vriese ●
34470 Armeniaca sibirica (L.) Lam.;西伯利亚杏(蒙古杏,山杏);Siberian Apricot ●
34471 Armeniaca sibirica (L.) Lam. = Prunus sibirica L. ●
34472 Armeniaca sibirica (L.) Lam. var. multipetala G. S. Liu et L. B. Zhang;重瓣山杏●
34473 Armeniaca sibirica (L.) Lam. var. pleniflora J. Y. Zhang,T. Z. Li et Y. He;辽梅杏;Liaoning Apricot ●
34474 Armeniaca sibirica (L.) Lam. var. pubescens Kostina;毛杏(毛山杏,毛枝西伯利亚杏);Pubescent Siberian Apricot ●
34475 Armeniaca vulgaris Lam.;杏(没落子,普通杏,甜梅,杏花,杏仁树,杏实,杏树,杏子);Abricok,Aprecock,Apricock,Apricot,Common Apricot,Hasty Peach,Precocious Tree ●
34476 Armeniaca vulgaris Lam. = Prunus armeniaca L. ●☆
34477 Armeniaca vulgaris Lam. var. ansu (Maxim.) Te T. Yu et L. T. Lu;山杏(安苏杏,野杏);Ansu Apricot,Apricot,Wild Apricot ●
34478 Armeniaca vulgaris Lam. var. ansu (Maxim.) Te T. Yu et L. T. Lu = Prunus armeniaca L. var. ansu Maxim. ●
34479 Armeniaca vulgaris Lam. var. meixianensis J. Y. Zhang,T. Z. Li,X. J. Li et Y. He;陕梅杏;Meixian Apricot ●
34480 Armeniaca vulgaris Lam. var. xiongyueensis T. Z. Li,J. Y. Zhang,X. J. Li et Y. He;熊岳大扁杏;Xiongyue Apricot ●
34481 Armeniaca vulgaris Lam. var. zhidanensis (C. Z. Qiao et Y. P. Zhu) L. T. Lu;志丹杏;Zhidan Apricot ●
34482 Armeniaca zhengheensis J. Y. Zhang et M. N. Lu;政和杏(红梅杏);Zhenghe Apricot ●
34483 Armeniaca zhidanensis C. Z. Qiao et Y. P. Zhu = Armeniaca vulgaris Lam. var. zhidanensis (C. Z. Qiao et Y. P. Zhu) L. T. Lu ●
34484 Armeniastrum Lem. = Espadaea A. Rich. ●☆
34485 Armeria (DC.) Willd. = Armeria Willd.(保留属名)■☆
34486 Armeria Kuntze = Phlox L. ■
34487 Armeria L. = Armeria Willd.(保留属名)■☆
34488 Armeria Willd.(1809)(保留属名);海石竹属;Armeria,Sea Pink,Sea Thrift,Thrift ■☆
34489 Armeria alliacea (Cav.) Hoffmanns. et Link;韭状石竹■☆
34490 Armeria alliacea (Cav.) Hoffmanns. et Link var. gracilis Sauvage

et Vindt = Armeria alliacea (Cav.) Hoffmanns. et Link ■☆

34491 Armeria alliacea (Cav.) Hoffmanns. et Link var. yebalina Pau et Font Quer = Armeria masguindalii (Pau) Nieto Fel. ■☆

34492 Armeria alliacea Roem. et Schult. = Armeria pseudarmeria (Murray) Mansf. ■☆

34493 Armeria allioides Boiss. = Armeria alliacea (Cav.) Hoffmanns. et Link ■☆

34494 Armeria alpina Willd.;高山海石竹■☆

34495 Armeria alpinifolia Pau et Font Quer;松叶海石竹■☆

34496 Armeria amplifoliata Pau = Armeria tingitana Boiss. et Reut. ■☆

34497 Armeria andina Poepp. ex Boiss. var. californica Boiss. = Armeria maritima (Mill.) Willd. subsp. californica (Boiss.) A. E. Porsild ■☆

34498 Armeria arctica (Cham.) Wallr. = Armeria sibirica Turcz. ■☆

34499 Armeria arctica (Cham.) Wallr. subsp. californica (Boiss.) Abrams = Armeria maritima (Mill.) Willd. subsp. californica (Boiss.) A. E. Porsild ■☆

34500 Armeria arctica Wallr. = Armeria arenaria (Mill.) Willd. subsp. arctica (Cham.) Hultén ■☆

34501 Armeria arenaria (Mill.) Willd.;阔叶海石竹;Broad-leaved Thrift,Jersey Thrift,Plantain-leaved Thrift ■☆

34502 Armeria arenaria (Mill.) Willd. subsp. arctica (Cham.) Hultén;北极海石竹■☆

34503 Armeria atlantica Pomel;大西洋海石竹■☆

34504 Armeria atlantica Pomel var. fibrosa (Pomel) Batt. = Armeria atlantica Pomel ■☆

34505 Armeria atlantica Pomel var. major Batt. = Armeria atlantica Pomel ■☆

34506 Armeria baetica Boiss. var. africana Batt. = Armeria simplex Pomel ■☆

34507 Armeria boissieriana Coss. = Armeria mauritanica Wallr. ■☆

34508 Armeria caespitosa Boiss. = Armeria juniperifolia Koch ■☆

34509 Armeria cephalotes Hoffmanns. et Link = Armeria latifolia Willd. ■☆

34510 Armeria cephalotes Hoffmanns. et Link = Armeria pseudarmeria (Murray) Mansf. ■☆

34511 Armeria choulettiana Pomel;舒莱海石竹■☆

34512 Armeria ebracteata Pomel;无苞海石竹■☆

34513 Armeria ebracteata Pomel var. laevis Maire = Armeria ebracteata Pomel ■☆

34514 Armeria elongata Koch;伸长海石竹■☆

34515 Armeria fasciculata Willd.;多刺海石竹;Spiny Thrift ■☆

34516 Armeria fibrosa Pomel;纤维海石竹■☆

34517 Armeria filicaulis (Boiss.) Boiss.;丝茎海石竹■☆

34518 Armeria filicaulis (Boiss.) Boiss. var. maroccana Pau et Font Quer = Armeria filicaulis (Boiss.) Boiss. ■☆

34519 Armeria formosa Vilm.;美丽海石竹■☆

34520 Armeria gaditana Boiss. var. chamaeropicola Pau = Armeria tingitana Boiss. et Reut. ■☆

34521 Armeria gaditana Boiss. var. tingitana (Boiss. et Reut.) Ball = Armeria tingitana Boiss. et Reut. ■☆

34522 Armeria juniperifolia (Vahl) Hoff. et Link;杜松叶海石竹(密生海石竹);Juniper Thrift ■☆

34523 Armeria juniperifolia Koch = Armeria juniperifolia (Vahl) Hoff. et Link ■☆

34524 Armeria labradorica Wallr. = Armeria sibirica Turcz. ■☆

34525 Armeria labradorica Wallr. var. submutica (S. F. Blake) H. F. Lewis = Armeria sibirica Turcz. ■☆

34526 Armeria lachnolepis Pomel = Armeria ebracteata Pomel ■☆

34527 Armeria lachnolepis Pomel var. bracteolata Emb. et Maire = Armeria ebracteata Pomel ■☆

34528 Armeria latifolia Willd. = Armeria pseudarmeria (Murray) Mansf. ■☆

34529 Armeria longevaginata Batt. = Armeria choulettiana Pomel ■☆

34530 Armeria maderensis Lowe;梅德海石竹■☆

34531 Armeria maghrebensis Donad. = Armeria simplex Pomel ■☆

34532 Armeria maghrebensis Donad. var. ebracteolata Donad. = Armeria simplex Pomel ■☆

34533 Armeria maghrebensis Donad. var. minor (Batt.) Donad. = Armeria simplex Pomel ■☆

34534 Armeria maghrebensis Donad. var. simplex (Pomel) Donad. = Armeria simplex Pomel ■☆

34535 Armeria maghrebensis Donad. var. soloitana (Maire) Donad. = Armeria simplex Pomel ■☆

34536 Armeria maritima (Mill.) Willd.;海石竹(普通海石竹);Arbyroot Arby, Brittons, California Thrift, Cliff Rose, Common Thrift, Cushings Curshins, Cushion, Cushion Pink, Edging, French Pink, Gilliflower, Jersev Thrift, Lady Cushion, Lady's Cushion, Lady's Pincushion, Maritime Thrift, Marsh Daisy, Midsummer Fairmaid, Our Lady's Cushion, Pincushion, Plantain Thrift, Profolium, Rock Rose, Salt Rose, Sand Flower, Sea Cushion, Sea Daisy, Sea Gilliflower, Sea Pink, Sea Rose, Sea Thrift, Sea Turf, Seagrass, Seawfell Pink, Swift, Tab-mawn, Thrift ■☆

34537 Armeria maritima (Mill.) Willd. 'Vindictive';玫红海石竹■☆

34538 Armeria maritima (Mill.) Willd. subsp. arctica (Cham.) Hultén = Armeria sibirica Turcz. ■☆

34539 Armeria maritima (Mill.) Willd. subsp. californica (Boiss.) A. E. Porsild;加州海石竹■☆

34540 Armeria maritima (Mill.) Willd. subsp. labradorica (Wallr.) Hultén = Armeria sibirica Turcz. ■☆

34541 Armeria maritima (Mill.) Willd. subsp. sibirica (Turcz.) Nyman = Armeria sibirica Turcz. ■☆

34542 Armeria maritima (Mill.) Willd. var. californica (Boiss.) G. H. M. Lawr. = Armeria maritima (Mill.) Willd. subsp. californica (Boiss.) A. E. Porsild ■☆

34543 Armeria maritima (Mill.) Willd. var. labradorica (Wallr.) G. H. M. Lawr. = Armeria sibirica Turcz. ■☆

34544 Armeria maritima (Mill.) Willd. var. sibirica (Turcz.) G. H. M. Lawr. = Armeria sibirica Turcz. ■☆

34545 Armeria masguindalii (Pau) Nieto Fel.;马斯海石竹■☆

34546 Armeria mauritanica Wallr.;毛里塔尼亚海石竹■☆

34547 Armeria mauritanica Wallr. var. boissierana (Coss.) Quézel et Santa = Armeria mauritanica Wallr. ■☆

34548 Armeria mauritanica Wallr. var. simplex (Pomel) Quézel et Santa = Armeria simplex Pomel ■☆

34549 Armeria mauritanica Wallr. var. soloitana Maire = Armeria mauritanica Wallr. ■☆

34550 Armeria plantaginea (All.) Willd.;桃花钗;Plantain Thrift ■

34551 Armeria plantaginea (All.) Willd. = Armeria pseudarmeria (Murray) Mansf. ■☆

34552 Armeria plantaginea Willd. = Armeria plantaginea (All.) Willd. ■

34553 Armeria plantaginea Willd. subsp. choulettiana (Pomel) Maire = Armeria choulettiana Pomel ■☆

34554 Armeria plantaginea Willd. subsp. leucantha (Boiss.) Maire = Armeria atlantica Pomel ■☆

34555 Armeria plantaginea Willd. subsp. medians Maire = Armeria atlantica Pomel ■☆

34556 Armeria plantaginea Willd. var. atlantica (Pomel) Maire = Armeria atlantica Pomel ■☆

34557 Armeria plantaginea Willd. var. ifranensis Sauvage et Vindt = Armeria atlantica Pomel ■☆

34558 Armeria plantaginea Willd. var. masguindalii Pau = Armeria masguindalii (Pau) Nieto Fel. ■☆

34559 Armeria plantaginea Willd. var. microcephala Maire = Armeria choulettiana Pomel ■☆

34560 Armeria pseudarmeria (Murray) Mansf.;宽叶海石竹(桃花钗);Estoril Thrift,Jersey Thrift ■☆

34561 Armeria repens ? = Trifolium repens L. ■

34562 Armeria sibirica Turcz.;西伯利亚海石竹■☆

34563 Armeria sibirica Turcz. ex Boiss. = Armeria sibirica Turcz. ■☆

34564 Armeria simplex Pomel;单枝海石竹■☆

34565 Armeria spinulosa Boiss.;细刺海石竹■☆

34566 Armeria suffocata ? = Trifolium suffocatum L. ■☆

34567 Armeria tingitana Boiss. et Reut.;丹吉尔海石竹■☆

34568 Armeria tingitana Boiss. et Reut. var. chamaeropicola (Pau) Donad. = Armeria tingitana Boiss. et Reut. ■☆

34569 Armeria vulgaris Willd. = Armeria maritima (Mill.) Willd. ■☆

34570 Armeriaceae Horan.;海石竹科■

34571 Armeriaceae Horan. = Plumbaginaceae Juss.(保留科名)●■

34572 Armeriastrum (Jaub. et Spach) Lindl.(废弃属名) = Acantholimon Boiss.(保留属名)●

34573 Armeriastrum Lindl. = Acantholimon Boiss.(保留属名)●

34574 Arminia Bronner(1857);德国葡萄属●☆

34575 Armodorum Breda (1829);阿芒多兰属(蜘蛛兰属);Armodorum ■

34576 Armodorum labrosum (Lindl. et Paxton) Schltr. = Arachnis labrosa (Lindl. et Paxton) Rchb. f. ■

34577 Armodorum labrosum (Lindl. ex Paxton) Schltr. = Renanthera labrosa (Lindl. et Paxton) Rchb. f. ■

34578 Armodorum sulingi Schltr.;苏苓氏阿芒多兰;Suling Armodorum ■☆

34579 Armola (Kirschl.) Montandon = Atriplex L. ●■

34580 Armola Friche-Joset et Montandon = Atriplex L. ●■

34581 Armoracia Fabr. = Armoracia P. Gaertn., B. Mey. et Scherb.(保留属名)■

34582 Armoracia P. Gaertn., B. Mey. et Scherb. (1800)(保留属名);辣根属(马萝卜属);Horseradish ■

34583 Armoracia aquatica (Eaton) Wiegand;水辣根;Lake Cress ■☆

34584 Armoracia aquatica (Eaton) Wiegand = Armoracia lacustris (A. Gray) Al-Shehbaz et V. M. Bates ■☆

34585 Armoracia aquatica (Eaton) Wiegand = Neobeckia aquatica (Eaton) Greene ■☆

34586 Armoracia armoracia (L.) Cockerell ex Daniels = Armoracia rusticana (Lam.) Gaertn., B. Mey. et Scherb. ■

34587 Armoracia lacustris (A. Gray) Al-Shehbaz et V. M. Bates;湖畔辣根;Lake Cress ■☆

34588 Armoracia lacustris (A. Gray) Al-Shehbaz et V. M. Bates = Neobeckia aquatica (Eaton) Greene ■☆

34589 Armoracia lapathifolia Gilib. = Armoracia rusticana (Lam.) Gaertn., B. Mey. et Scherb. ■

34590 Armoracia rusticana (Lam.) Gaertn., B. Mey. et Scherb.;辣根(马萝卜,牛蒡叶辣根,山葥菜,香辣根);Clown's Mustard, Dock, Great Raifort, Green Radish, Horse Radish, Horseradish, Horse-radish, Mountain Radish, Racadal, Red Cole, Redco, Rotcoll ■

34591 Armoracia rusticana (Lam.) Gaertn., B. Mey. et Scherb. 'Variegata';斑叶辣根;Variegated Horseradish ■☆

34592 Armoracia rusticana Gaertn., B. Mey. et Scherb. = Armoracia rusticana (Lam.) Gaertn., B. Mey. et Scherb. ■

34593 Armoracia sativa Bernh. = Armoracia rusticana (Lam.) Gaertn., B. Mey. et Scherb. ■

34594 Armoracia sisymbrioides N. Busch ex Ganesh;草地辣根■☆

34595 Armourea Lewton = Thespesia Sol. ex Corrêa(保留属名)●

34596 Arnaldoa Cabrera(1962);同花刺菊木属●☆

34597 Arnaldoa argentea C. Ulloa, P. M. Jörg. et M. O. Dillon;银白同花刺菊木●☆

34598 Arnaldoa magnifica Cabrera;同花刺菊木●☆

34599 Arnanthus Baehni = Pichonia Pierre ●☆

34600 Arnebia Forssk. (1775);软紫草属(阿纳芘属,光喉草属,假紫草属);Arabian Primrose, Arnebia, Friar's Cowl ●■

34601 Arnebia asperrima (Delile) Hutchinson et Dalziel = Arnebia hispidissima (Sieber ex Lehm.) DC. ■☆

34602 Arnebia baldschuanica (Lipsky) Schischk.;巴地软紫草■☆

34603 Arnebia benthami (Wall. ex G. Don) I. M. Johnst.;本瑟姆软紫草■☆

34604 Arnebia bungei Boiss. = Arnebia fimbriopetala Stocks ■☆

34605 Arnebia coerulea Schipcz.;灰蓝软紫草■☆

34606 Arnebia cornuta Fisch. et C. A. Mey.;阿拉伯软紫草;Arabian Primrose, Arabian-primrose, Pipe Vine ■☆

34607 Arnebia cornuta Fisch. et C. A. Mey. = Arnebia decumbens (Vent.) Coss. et Kralik ●■

34608 Arnebia cornuta Fisch. et C. A. Mey. var. grandiflora Trautv. = Arnebia grandiflora (Trautv.) Popov ■☆

34609 Arnebia decumbens (Vent.) Coss. et Kralik;硬萼软紫草(俯仰假紫草,俯仰紫草);Arabian-primrose, Hardcalyx Arnebia ●■

34610 Arnebia decumbens (Vent.) Coss. et Kralik subsp. macrocalyx (Coss. et Kralik) Riedl;大硬萼软紫草■☆

34611 Arnebia decumbens (Vent.) Coss. et Kralik var. macrocalyx Coss. et Kralik = Arnebia decumbens (Vent.) Coss. et Kralik subsp. macrocalyx (Coss. et Kralik) Riedl ■☆

34612 Arnebia decumbens (Vent.) Coss. et Kralik var. microcalyx Coss. et Kralik = Arnebia decumbens (Vent.) Coss. et Kralik ●■

34613 Arnebia densiflora (Nordm.) Ledeb.;密花软紫草■☆

34614 Arnebia echioides A. DC. = Arnebia pulchra (Willd. ex Roem. et Schult.) J. R. Edm. ■☆

34615 Arnebia euchroma (Royle) I. M. Johnst.;软紫草(茈草,地血,红石根,藐,山紫草,新藏假紫草,新疆紫草,鸦衔草,紫芙,紫草,紫草茸,紫丹);Xinjiang-Xizang Arnebia ■

34616 Arnebia euchroma (Royle) I. M. Johnst. subsp. caespitosa ? = Arnebia euchroma (Royle) I. M. Johnst. ■

34617 Arnebia euchroma (Royle) I. M. Johnst. var. grandis (Bornm.) Kazmi = Arnebia euchroma (Royle) I. M. Johnst. ■

34618 Arnebia fimbriata Maxim.;灰毛软紫草(灰毛假紫草,新疆紫草);Greyhair Arnebia, Greyhairy Arnebia ■

34619 Arnebia fimbriopetala Stocks;线瓣软紫草■☆

34620 Arnebia fimbriopetala Stocks var. bungei ? = Arnebia fimbriopetala Stocks ■☆

34621 Arnebia flavescens Boiss. = Arnebia linearifolia DC. ■☆

34622 Arnebia grandiflora (Trautv.) Popov;大花软紫草■☆

34623 Arnebia griffithii Boiss.;格氏软紫草■☆

34624 Arnebia guttata Bunge;黄花软紫草(黄花紫草,假紫草,蒙紫草,内蒙古紫草,内蒙紫草,帕米尔假紫草,帕米尔紫草,西藏软紫草,新疆紫草);Common Arnebia,Yellow Arnebia ■

34625 Arnebia hispidissima (Sieber ex Lehm.) DC.;刚毛软紫草■☆

34626 Arnebia hispidissima DC. = Arnebia hispidissima (Sieber ex Lehm.) DC. ■☆

34627 Arnebia johnstonii Riedl;约翰斯顿软紫草■☆

34628 Arnebia linearifolia DC.;线叶软紫草■☆

34629 Arnebia longiflora K. Koch;长花软紫草;Prophet Flower ■☆

34630 Arnebia lutea (A. Rich.) Armari = Arnebia hispidissima (Sieber ex Lehm.) DC. ■☆

34631 Arnebia macrocalyx (Coss. et Kralik) Boulos = Arnebia decumbens (Vent.) Coss. et Kralik subsp. macrocalyx (Coss. et Kralik) Riedl ■☆

34632 Arnebia nobilis Rech. f.;华美假紫草(假紫草)■☆

34633 Arnebia obovata Bunge;倒卵形软紫草■☆

34634 Arnebia perennis A. DC. = Arnebia euchroma (Royle) I. M. Johnst. ■

34635 Arnebia pulchra (Willd. ex Roem. et Schult.) J. R. Edm.;美丽软紫草;Arabian Primrose,Prophet Bower,Prophet Flower,Prophet-flower,Prophet's Flower ■☆

34636 Arnebia purpurascens (A. Rich.) Baker;紫色软紫草■☆

34637 Arnebia saxatile (Turcz.) Benth. et Hook. = Stenosolenium saxatile (Pall.) Turcz. ■

34638 Arnebia saxatile (Turcz.) Benth. et Hook. f. = Stenosolenium saxatile (Pall.) Turcz. ■

34639 Arnebia szechenyi Kanitz;疏花软紫草(疏花假紫草);Sparseflower Arnebia ■

34640 Arnebia tetrastigma Forssk. = Arnebia tinctoria Forssk. ■☆

34641 Arnebia thomsonii C. B. Clarke;帕米尔假紫草(帕米尔紫草);Thomson Arnebia ■☆

34642 Arnebia thomsonii C. B. Clarke = Arnebia guttata Bunge ■

34643 Arnebia tibetana Kurz = Arnebia guttata Bunge ■

34644 Arnebia tinctoria Forssk.;染色软紫草■☆

34645 Arnebia tingens A. DC. = Arnebia euchroma (Royle) I. M. Johnst. ■

34646 Arnebia transcaspica Popov;特拉软紫草■☆

34647 Arnebia tschimganica (B. Fedtsch.) G. L. Zhu;天山软紫草;Tianshan Arnebia ■

34648 Arnebiola Chiov. = Arnebia Forssk. ●■

34649 Arnebiola migiurtina Chiov. = Arnebia hispidissima (Sieber ex Lehm.) DC. ■☆

34650 Arnedina Rchb. = Arundina Blume ■

34651 Arnedina Rchb. f. = Arundina Blume ■

34652 Arnhemia Airy Shaw(1978);澳洲瑞香属●☆

34653 Arnhemia cryptantha Airy Shaw;澳洲瑞香●☆

34654 Arnica Boehm. = Doronicum L. ■

34655 Arnica L. (1753);山金车属(阿尼菊属,金车菊属,山烟菊属,兔菊属,羊菊属);Arnica ●■☆

34656 Arnica Rupp. ex L. = Arnica L. ●■☆

34657 Arnica acaulis (Walter) Britton,Sterns et Poggenb.;无茎山金车;Common Leopardbane,Leopard's Bane ●☆

34658 Arnica alata Rydb. = Arnica discoidea Benth. ■☆

34659 Arnica alpina (L.) Olin et Ladau;山地山金车●☆

34660 Arnica alpina (L.) Olin et Ladau subsp. angustifolia (Vahl) Maguire = Arnica angustifolia Vahl ■☆

34661 Arnica alpina (L.) Olin et Ladau subsp. attenuata (Greene) Maguire = Arnica angustifolia Vahl ■☆

34662 Arnica alpina (L.) Olin et Ladau subsp. iljinii Maguire = Arnica angustifolia Vahl ■☆

34663 Arnica alpina (L.) Olin et Ladau subsp. lonchophylla (Greene) G. W. Douglas et Ruyle-Douglas = Arnica lonchophylla Greene ■☆

34664 Arnica alpina (L.) Olin et Ladau subsp. sornborgeri (Fernald) Maguire = Arnica angustifolia Vahl ■☆

34665 Arnica alpina (L.) Olin et Ladau subsp. tomentosa (J. M. Macoun) Maguire = Arnica angustifolia Vahl subsp. tomentosa (Macoun) G. W. Douglas et Ruyle-Douglas ■☆

34666 Arnica alpina (L.) Olin et Ladau var. angustifolia (Vahl) Fernald = Arnica angustifolia Vahl ■☆

34667 Arnica alpina (L.) Olin et Ladau var. attenuata (Greene) Ediger et T. M. Barkley = Arnica angustifolia Vahl ■☆

34668 Arnica alpina (L.) Olin et Ladau var. linearis Hultén = Arnica angustifolia Vahl ■☆

34669 Arnica alpina (L.) Olin et Ladau var. tomentosa (J. M. Macoun) Cronquist = Arnica angustifolia Vahl subsp. tomentosa (Macoun) G. W. Douglas et Ruyle-Douglas ■☆

34670 Arnica alpina (L.) Olin et Ladau var. vestita Hultén = Arnica angustifolia Vahl ■☆

34671 Arnica alpina Olin = Arnica alpina (L.) Olin et Ladau ●☆

34672 Arnica altaica Turcz. = Doronicum altaicum Pall. ■

34673 Arnica amplexicaulis Nutt. = Arnica lanceolata Nutt. subsp. prima (Maguire) Strother et S. J. Wolf ■☆

34674 Arnica amplexicaulis Nutt. subsp. prima (Maguire) Maguire = Arnica lanceolata Nutt. subsp. prima (Maguire) Strother et S. J. Wolf ■☆

34675 Arnica amplexicaulis Nutt. var. piperi H. St. John et F. Warren = Arnica lanceolata Nutt. subsp. prima (Maguire) Strother et S. J. Wolf ■☆

34676 Arnica amplexicaulis Nutt. var. prima (Maguire) B. Boivin = Arnica lanceolata Nutt. subsp. prima (Maguire) Strother et S. J. Wolf ■☆

34677 Arnica angustifolia Vahl;窄叶山金车;Narrowleaf Arnica ■☆

34678 Arnica angustifolia Vahl subsp. eradiata A. Gray = Arnica parryi A. Gray ■☆

34679 Arnica angustifolia Vahl subsp. lonchophylla (Greene) G. W. Douglas et Ruyle-Douglas = Arnica lonchophylla Greene ■☆

34680 Arnica angustifolia Vahl subsp. tomentosa (Macoun) G. W. Douglas et Ruyle-Douglas;毛窄叶山金车■☆

34681 Arnica angustifolia Vahl var. lessingii Torr. et A. Gray = Arnica lessingii (Torr. et A. Gray) Greene ■☆

34682 Arnica arnoglossa Greene = Arnica lonchophylla Greene ■☆

34683 Arnica attenuata Greene = Arnica angustifolia Vahl ■☆

34684 Arnica bernardina Greene = Arnica chamissonis Less. ■☆

34685 Arnica cernua Howell;蜿蜒山金车;Serpentine Arnica ■☆

34686 Arnica chamissonis Greene subsp. foliosa (Nutt.) Maguire = Arnica chamissonis Less. ■☆

34687 Arnica chamissonis Greene subsp. incana (A. Gray) Maguire = Arnica chamissonis Less. ■☆

34688 Arnica chamissonis Greene var. bernardina (Greene) Jeps. ex Maguire = Arnica chamissonis Less. ■☆

34689 Arnica chamissonis Greene var. foliosa (Nutt.) Maguire = Arnica chamissonis Less. ■☆

34690 Arnica chamissonis Greene var. incana (A. Gray) Hultén = Arnica chamissonis Less. ■☆

34691 Arnica chamissonis Greene var. interior Maguire = Arnica chamissonis Less. ■☆
34692 Arnica chamissonis Greene var. jepsoniana Maguire = Arnica chamissonis Less. ■☆
34693 Arnica chamissonis Less.;卡密松山金车(密叶山金车);Chamisso Arnica,Leafy Arnica ■☆
34694 Arnica chamissonis Less. var. sachalinensis Regel = Arnica sachalinensis (Regel) A. Gray ●☆
34695 Arnica chandleri Rydb. = Arnica cernua Howell ■☆
34696 Arnica chionopappa Fernald = Arnica lonchophylla Greene ■☆
34697 Arnica ciliata Thunb. = Hypochaeris ciliata (Thunb.) Makino ■
34698 Arnica cordata Thunb. = Gerbera cordata (Thunb.) Less. ■☆
34699 Arnica cordifolia Hook.;心叶山金车;Heartleaf Arnica ■☆
34700 Arnica cordifolia Hook. var. eradiata A. Gray = Arnica discoidea Benth. ■☆
34701 Arnica cordifolia Hook. var. pumila (Rydb.) Maguire = Arnica cordifolia Hook. ■☆
34702 Arnica crenata Thunb. = Mairia crenata (Thunb.) Nees ■☆
34703 Arnica crocea L. = Gerbera crocea (L.) Kuntze ■☆
34704 Arnica dealbata (A. Gray) B. G. Baldwin;白色山金车;Mock Leopardbane ■☆
34705 Arnica discoidea Benth.;昏暗山金车;Rayless Arnica ■☆
34706 Arnica discoidea Benth. var. alata (Rydb.) Cronquist = Arnica discoidea Benth. ■☆
34707 Arnica discoidea Benth. var. eradiata (A. Gray) Cronquist = Arnica discoidea Benth. ■☆
34708 Arnica diversifolia Greene = Arnica ovata Greene ■☆
34709 Arnica eastwoodiae Rydb. = Arnica spathulata Greene ■☆
34710 Arnica foliosa Nutt.;多叶山金车●☆
34711 Arnica foliosa Nutt. = Arnica chamissonis Less. ■☆
34712 Arnica frigida C. A. Mey. ex Iljin;硬山金车●☆
34713 Arnica frigida C. A. Mey. ex Iljin = Arnica griscomii Fernald subsp. frigida (C. A. Mey. ex Iljin) S. J. Wolf ■☆
34714 Arnica fulgens Pursh;光亮山金车;Foothill Arnica ■☆
34715 Arnica fulgens Pursh var. sororia (Greene) G. W. Douglas et Ruyle-Douglas = Arnica sororia Greene ■☆
34716 Arnica gaspensis Fernald = Arnica lonchophylla Greene ■☆
34717 Arnica gerbera L. = Gerbera linnaei Cass. ■☆
34718 Arnica gracilis Rydb.;小头山金车;Smallhead Arnica ■☆
34719 Arnica grandis Thunb. = Oldenburgia grandis (Thunb.) Baill. ■☆
34720 Arnica griscomii Fernald;雪地山金车;Griscom's Arnica,Snow Arnica ■☆
34721 Arnica griscomii Fernald subsp. frigida (C. A. Mey. ex Iljin) S. J. Wolf;冷地山金车;Snow Arnica ■☆
34722 Arnica hirsuta Forssk. = Gerbera piloselloides (L.) Cass. ■
34723 Arnica hirsuta Forssk. = Piloselloides hirsuta (Forssk.) C. Jeffrey ex Cufod. ■
34724 Arnica iljinii (Maguire) Iljin;伊氏山金车●☆
34725 Arnica intermedia Turcz.;间型山金车●☆
34726 Arnica japonica Thunb. = Ligularia japonica (Thunb.) Less. ■
34727 Arnica lanata Thunb. = Capelio tomentosa (Burm. f.) B. Nord. ■☆
34728 Arnica lanceolata Nutt.;剑叶山金车;Lanceleaf Arnica ■☆
34729 Arnica lanceolata Nutt. subsp. prima (Maguire) Strother et S. J. Wolf;紧密山金车;Clasping Arnica ■☆
34730 Arnica latifolia Bong.;宽叶山金车;Arnica,Broadleaf Arnica ■☆
34731 Arnica latifolia Bong. var. gracilis (Rydb.) Cronquist = Arnica gracilis Rydb. ■☆
34732 Arnica lessingii (Torr. et A. Gray) Greene;悬垂山金车;Nodding Arnica ■☆
34733 Arnica lessingii Green;莱辛山金车■☆
34734 Arnica lessingii Green subsp. norbergii Hultén et Maguire = Arnica lessingii (Torr. et A. Gray) Greene ■☆
34735 Arnica lonchophylla Greene;北方山金车;Longleaf Arnica,Northern Arnica ■☆
34736 Arnica lonchophylla Greene subsp. arnoglossa (Greene) Maguire = Arnica lonchophylla Greene ■☆
34737 Arnica lonchophylla Greene subsp. chionopappa (Fernald) Maguire = Arnica lonchophylla Greene ■☆
34738 Arnica longifolia D. C. Eaton;长叶山金车(矛叶山金车);Spearleaf Arnica ■☆
34739 Arnica longifolia D. C. Eaton subsp. myriadenia (Piper) Maguire = Arnica longifolia D. C. Eaton ■☆
34740 Arnica louiseana Farr;路易斯山金车;Lake Louise Arnica,Snow Arnica ■☆
34741 Arnica louiseana Farr subsp. frigida (C. A. Mey. ex Iljin) Maguire = Arnica griscomii Fernald subsp. frigida (C. A. Mey. ex Iljin) S. J. Wolf ■☆
34742 Arnica louiseana Farr subsp. griscomii (Fernald) Maguire = Arnica griscomii Fernald ■☆
34743 Arnica louiseana Farr var. frigida (C. A. Mey. ex Iljin) S. L. Welsh = Arnica griscomii Fernald subsp. frigida (C. A. Mey. ex Iljin) S. J. Wolf ■☆
34744 Arnica louiseana Farr var. mendenhallii (Rydb.) Maguire = Arnica griscomii Fernald subsp. frigida (C. A. Mey. ex Iljin) S. J. Wolf ■☆
34745 Arnica louiseana Farr var. pilosa Maguire = Arnica griscomii Fernald subsp. frigida (C. A. Mey. ex Iljin) S. J. Wolf ■☆
34746 Arnica mallotopus Makino;日本山金车●☆
34747 Arnica maritima L. = Senecio pseudoarnica Less. ■
34748 Arnica mollis Hook.;北美山金车;Hairy Arnica ●☆
34749 Arnica mollis Hook. = Arnica chamissonis Less. ■☆
34750 Arnica mollis Hook. var. petiolaris Fernald = Arnica lanceolata Nutt. ■☆
34751 Arnica monocephala Rydb. = Arnica fulgens Pursh ■☆
34752 Arnica montana L.;山金车(高山阿尼卡菊,山生阿尼菊,羊菊);Arnica,Arnica Root,Leopard's Bane,Leopard's-bane,Mountain Arnica,Mountain Snuff,Mountain Tobacco ●☆
34753 Arnica myriadenia Piper = Arnica longifolia D. C. Eaton ■☆
34754 Arnica nevadensis A. Gray;内华达山金车;Nevada Arnica,Sierra Arnica ■☆
34755 Arnica ovata Greene;卵叶山金车;Sticky Leaf Arnica ■☆
34756 Arnica paniculata A. Nelson = Arnica cordifolia Hook. ■☆
34757 Arnica parryi A. Gray;帕里山金车;Nodding Arnica,Parry's Arnica ■☆
34758 Arnica parryi A. Gray subsp. sonnei (Greene) Maguire = Arnica parryi A. Gray ■☆
34759 Arnica parryi A. Gray var. sonnei (Greene) Cronquist = Arnica parryi A. Gray ■☆
34760 Arnica parviflora A. Gray = Arnica discoidea Benth. ■☆
34761 Arnica pedunculata Rydb. = Arnica fulgens Pursh ■☆
34762 Arnica piloselloides L. = Gerbera piloselloides (L.) Cass. ■
34763 Arnica piloselloides L. = Piloselloides hirsuta (Forssk.) C. Jeffrey ex Cufod. ■
34764 Arnica plantaginea Pursh = Arnica angustifolia Vahl ■☆

34765 Arnica rydbergii Greene;雷氏山金车;Rydberg's Arnica ■☆
34766 Arnica sachalinensis (Regel) A. Gray;库页山金车●☆
34767 Arnica serrata Thunb. = Gerbera serrata (Thunb.) Druce ■☆
34768 Arnica sinuata Thunb. = Gerbera crocea (L.) Kuntze ■☆
34769 Arnica sornborgeri Fernald = Arnica angustifolia Vahl ■☆
34770 Arnica sororia Greene;姊妹山金车(姐妹山金车);Twin Arnica ■☆
34771 Arnica spathulata Greene;匙叶山金车;Klamath Arnica ■☆
34772 Arnica spathulata Greene subsp. eastwoodiae (Rydb.) Maguire = Arnica spathulata Greene ■☆
34773 Arnica spathulata Greene var. eastwoodiae (Rydb.) Ediger et T. M. Barkley = Arnica spathulata Greene ■☆
34774 Arnica tabularis Thunb. = Capelio tabularis (Thunb.) B. Nord. ■☆
34775 Arnica terrae-novae Fernald = Arnica angustifolia Vahl ■☆
34776 Arnica tomentella Greene = Arnica nevadensis A. Gray ■☆
34777 Arnica tomentosa J. M. Macoun = Arnica angustifolia Vahl subsp. tomentosa (Macoun) G. W. Douglas et Ruyle-Douglas ■☆
34778 Arnica tschonoskyi Iljin = Arnica unalaschcensis Less. var. tschonoskyi (Iljin) Kitam. et H. Hara ●☆
34779 Arnica tussilaginea Burm. f. = Farfugium japonicum (L.) Kitam. ■
34780 Arnica unalaschcensis Less.;阿拉斯加山金车;Alaska Arnica ■☆
34781 Arnica unalaschcensis Less. var. tschonoskyi (Iljin) Kitam. et H. Hara;须川氏山金车●☆
34782 Arnica unalaschcensis Less. var. tschonoskyi (Iljin) Kitam. et H. Hara f. semiplena Nakai;重瓣须川氏山金车●☆
34783 Arnica venosa H. M. Hall;沙斯塔山金车;Shasta County Arnica ■☆
34784 Arnica viscosa A. Gray;沙斯塔锥山金车;Mt. Shasta Arnica ■☆
34785 Arnica whitneyi Fernald = Arnica cordifolia Hook. ■☆
34786 Arnicastrum Greenm. (1903);肖羊菊属●■☆
34787 Arnicastrum glandulosum Greenm.;多腺肖羊菊●☆
34788 Arnicastrum guerrerense Villaseñor;肖羊菊●☆
34789 Arnicratea N. Hallé(1984);羊头卫矛属●☆
34790 Arnicratea cambodiana (Pierre) N. Hallé;羊头卫矛●☆
34791 Arnicratea ferruginea (King) N. Hallé;锈色羊头卫矛●☆
34792 Arnicratea grahamii (Wight) N. Hallé;格氏羊头卫矛●☆
34793 Arnicula Kuntze = Arnica L. ●■☆
34794 Arnocrinum Endl. et Lehm. (1846);毛兰草属■☆
34795 Arnocrinum drummondii Endl. ex Lehm.;毛兰草■☆
34796 Arnocrinum glabrum Baker;光毛兰草■☆
34797 Arnocrinum gracillimum Keighery;细毛兰草■☆
34798 Arnoglissium Gray = Plantago L. ●■
34799 Arnoglossum Raf. (1817);美蟹甲属;Indian Plantain ■☆
34800 Arnoglossum Raf. = Cacalia L. ●■
34801 Arnoglossum Raf. = Plantago L. ●■
34802 Arnoglossum atriplicifolium (L.) H. Rob.;暗沟美蟹甲(滨藜叶蟹甲草);Indian Plantain, Pale Indian Plantain, Pale Indian-plantain ■☆
34803 Arnoglossum diversifolium (Torr. et A. Gray) H. Rob.;美蟹甲■☆
34804 Arnoglossum floridanum (A. Gray) H. Rob.;佛罗里达美蟹甲■☆
34805 Arnoglossum muhlenbergii (Sch. Bip.) H. Rob. = Arnoglossum reniforme (Hook.) H. Rob. ■☆
34806 Arnoglossum ovatum (Walter) H. Rob.;卵叶美蟹甲■☆
34807 Arnoglossum ovatum (Walter) H. Rob. var. lanceolatum (Nutt.) D. B. Ward = Arnoglossum ovatum (Walter) H. Rob. ■☆
34808 Arnoglossum plantagineum Raf.;车前状美蟹甲;Prairie Indian-plantain ■☆
34809 Arnoglossum reniforme (Hook.) H. Rob.;肾叶美蟹甲(巨蟹甲草);Cacalia Muhlenbergii, Great Indian Plantain, Great Indian-plantain, Muhlenberg's Cacalia ■☆
34810 Arnoglossum sulcatum (Fernald) H. Rob.;凹陷美蟹甲■☆
34811 Arnoldia Blume = Weinmannia L. (保留属名)●☆
34812 Arnoldia Cass. = Dimorphotheca Vaill. (保留属名)●■☆
34813 Arnoldoschultzea Mildbr. (1922);喀麦隆山榄属●☆
34814 Arnopogon Willd. = Urospermum Scop. ■☆
34815 Arnoseris Gaertn. (1791);羊莴苣属(阿诺菊属,羊苣属);Lamb's Succory ■☆
34816 Arnoseris minima (L.) Schweigg. et Korte;羊苣(阿诺菊);Lamb's Succory, Small Lamb's-succory, Swine Succory, Swine's Succory ■☆
34817 Arnoseris pusilla Gaertn. = Arnoseris minima (L.) Schweigg. et Korte ■☆
34818 Arnottia A. Rich. (1828);阿尔兰属■☆
34819 Arnottia mauritiana A. Rich.;阿尔兰■☆
34820 Arodendron Werth = Typhonodorum Schott ■☆
34821 Arodendron engleri Werth = Typhonodorum lindleyanum Schott ■☆
34822 Arodes Heist. = Richardia L. ■
34823 Arodes Heist. ex Fabr. = Richardia L. ■
34824 Arodes Heist. ex Kuntze = Richardia L. ■
34825 Arodes Kuntze = Aroides Fabr. ■
34826 Arodes Kuntze = Zantedeschia Spreng. (保留属名)■
34827 Arodes aethiopicum (L.) Kuntze = Zantedeschia aethiopica (L.) Spreng. ■
34828 Arodia Raf. = Rubus L. ●■
34829 Aroides Fabr. = Calla L. ■
34830 Aroides Heist. ex Fabr. = Calla L. ■
34831 Aroides Heist. ex Fabr. = Richardia L. ■
34832 Aromadendron Andréws ex Steud. = Aromadendrum W. Anderson ex R. Br. ●
34833 Aromadendron Andréws ex Steud. = Eucalyptus L'Hér. ●
34834 Aromadendron Blume = Magnolia L. ●
34835 Aromadendron Blume(1825);香木兰属(香兰属)●☆
34836 Aromadendron baillonii (Pierre) Craib = Paramichelia baillonii (Pierre) Hu ●◇
34837 Aromadendron elegans Blume;香木兰(香兰)●☆
34838 Aromadendron spongiocarpum (King) Craib = Paramichelia baillonii (Pierre) Hu ●◇
34839 Aromadendron yunnanense Hu = Paramichelia baillonii (Pierre) Hu ●◇
34840 Aromadendrum Blume = Aromadendron Blume ●☆
34841 Aromadendrum W. Anderson ex R. Br. = Eucalyptus L'Hér. ●
34842 Aromia Nutt. = Amblyopappus Hook. et Arn. ■☆
34843 Aron Adans. = Colocasla Schott + Dracunculus Mill. ■☆
34844 Arongana Cholay = Haronga Thouars ■☆
34845 Aronia Medik. (1789)(保留属名);苦味果属(涩果属,腺肋花楸属);Chokeberry ●☆
34846 Aronia Medik. = Amelanchier Medik. ●
34847 Aronia Medik. = Photinia Lindl. ●
34848 Aronia Mitch. (废弃属名) = Aronia Medik. (保留属名)●☆
34849 Aronia Mitch. (废弃属名) = Orontium L. ■☆
34850 Aronia Mitch. (废弃属名) = Orontium Pers. ■☆
34851 Aronia Pers. = Amelanchier Medik. ●
34852 Aronia arbutifolia (L.) Elliott var. atropurpurea (Britton) F.

Seym. = Aronia prunifolia (Marshall) Rehder ●☆
34853 Aronia arbutifolia (L.) Elliott var. nigra (Willd.) F. Seym. = Aronia melanocarpa (Michx.) Elliott ●☆
34854 Aronia arbutifolia (L.) Pers.;红苦味果(荔莓叶涩果);Amelachier,Red Chokeberry ●☆
34855 Aronia arbutifolia (L.) Pers. 'Brilliantisma';极美红苦味果;Red Chokeberry ●☆
34856 Aronia arbutifolia (L.) Pers. = Photinia pyrifolia (Lam.) K. R. Robertson et J. B. Phipps ●☆
34857 Aronia arbutifolia (L.) Pers. var. glabra Elliott = Photinia pyrifolia (Lam.) K. R. Robertson et J. B. Phipps ●☆
34858 Aronia arbutifolia (L.) Pers. var. nigra (Willd.) F. Seym. = Photinia melanocarpa (Michx.) K. R. Robertson et J. B. Phipps ●☆
34859 Aronia asiatica Siebold et Zucc. = Amelanchier asiatica (Siebold et Zucc.) Endl. ex Walp. ●
34860 Aronia atropurpurea Britton = Aronia prunifolia (Marshall) Rehder ●☆
34861 Aronia floribunda (Lindl.) Spach = Aronia prunifolia (Marshall) Rehder ●☆
34862 Aronia japonica Hort. ex K. Koch;日本苦味果●☆
34863 Aronia melanocarpa (Michx.) Elliott = Aronia melanocarpa (Michx.) Nutt. et Elliott ●☆
34864 Aronia melanocarpa (Michx.) Nutt. et Elliott;黑苦味果;Black Chokeberry ●☆
34865 Aronia melanocarpa (Michx.) Nutt. et Elliott = Photinia melanocarpa (Michx.) K. R. Robertson et J. B. Phipps ●☆
34866 Aronia melanocarpa (Michx.) Nutt. et Elliott var. elata Rehder;高黑苦味果;Black Chokeberry ●☆
34867 Aronia nigra (Willd.) Koehne = Aronia melanocarpa (Michx.) Elliott ●☆
34868 Aronia nigra (Willd.) Koehne = Photinia melanocarpa (Michx.) K. R. Robertson et J. B. Phipps ●☆
34869 Aronia prunifolia (Marshall) Rehder;紫苦味果(稠李叶涩果);Black Chokecherry,Hybrid Chokeberry,Purple Chokeberry ●☆
34870 Aronicum Neck. = Doronicum L. ■
34871 Aronicum Neck. ex Rchb. = Doronicum L. ■
34872 Aronicum Neck. ex Rchb. = Grammarthron Cass. ■
34873 Aronicum altaicum (Pall.) A. DC. = Doronicum altaicum Pall. ■
34874 Aronicum altaicum DC. = Doronicum altaicum Pall. ■
34875 Aronicum atlanticum Chabert = Doronicum plantagineum L. subsp. atlanticum (Rouy) Greuter ■☆
34876 Arophyton Jum. (1928);拟白星海芋属■☆
34877 Arophyton buchetii Bogner;比谢拟白星海芋■☆
34878 Arophyton crassifolium (Buchet) Bogner;厚叶拟白星海芋■☆
34879 Arophyton humbertii Bogner;亨伯特拟白星海芋■☆
34880 Arophyton pedatum Buchet;鸟足状拟白星海芋■☆
34881 Arophyton rhizomatosum (Buchet) Bogner;根茎拟白星海芋■☆
34882 Arophyton simplex Buchet;简单拟白星海芋■☆
34883 Arophyton tripartitum Jum.;拟白星海芋■☆
34884 Aropsis Rojas = Spathicarpa Hook. ■☆
34885 Arosma Raf. = Philodendron Schott(保留属名)●■
34886 Aroton Neck. = Croton L. ●
34887 Arouna Aubl. = Dialium L. ●☆
34888 Arpitium Neck. = Ligusticum L. ■
34889 Arpitium Neck. ex Sweet = Endressia J. Gay ■☆
34890 Arpitium Neck. ex Sweet = Ligusticum L. ■
34891 Arpitium Neck. ex Sweet = Pachypleurum Ledeb. ■
34892 Arpitium alpinum (Ledeb.) Koso-Pol. = Pachypleurum alpinum Ledeb. ■
34893 Arpitium alpinum Koso-Pol. = Pachypleurum alpinum Ledeb. ■
34894 Arpophyllum La Llave et Lex. = Arpophyllum Lex. ■☆
34895 Arpophyllum Lex. (1825);风信子兰属(镰叶兰属);Hyacinth Orchid ■☆
34896 Arpophyllum alpinum Lindl.;高山风信子兰;Alpine Hyacinth Orchid ■☆
34897 Arpophyllum cardinale Lindl. et Rchb. f.;绯红风信子兰■☆
34898 Arpophyllum giganteum Hartw. ex Lindl.;大风信子兰■☆
34899 Arpophyllum spicatum La Llave et Lex.;穗花风信子兰;Spiked Hyacinth Orchid ■☆
34900 Arrabidaea DC. (1838);阿拉树属●☆
34901 Arrabidaea Steud. = Cormonema Reissek ex Endl. ●
34902 Arrabidaea chica (Humb. et Bonpl.) Verl.;阿拉树;Chica ●☆
34903 Arrabidaea magnifica (W. Bull) Sprague ex Steenis = Bignonia magnifica W. Bull ●
34904 Arrabidaea magnifica Sprague ex Steenis = Bignonia magnifica W. Bull ●
34905 Arrabidaea magnifica Sprague ex Steenis = Saritaea magnifica (Sprague ex Steenis) Dugand ●
34906 Arrabidaea rotundata (DC.) Schum.;巴西阿拉树●☆
34907 Arrabidaea selloi (Spreng.) Sandwith;塞罗阿拉树●☆
34908 Arracacha DC. = Arracacia Bancr. ■☆
34909 Arracacia Bancr. (1828);秘鲁胡萝卜属;Arracacia ■☆
34910 Arracacia delavayi Franch. = Physospermopsis delavayi (Franch.) H. Wolff ■
34911 Arracacia esculenta DC. = Arracacia xanthorrhiza Bancr. ■☆
34912 Arracacia peucedanifolia Franch. = Cyclorhiza peucedanifolia (Franch.) Constance ■
34913 Arracacia xanthorhiza Bancr. = Arracacia xanthorrhiza Bancr. ■☆
34914 Arracacia xanthorrhiza Bancr.;秘鲁胡萝卜;Apio, Apio Arracacia,Arracacha,Peruvian Parsnip ■☆
34915 Arraschkoolia Hochst. = Araschcoolia Sch. Bip. ●■☆
34916 Arraschkoolia Hochst. = Geigeria Griess. ●■☆
34917 Arraschkoolia Sch. Bip. ex Hochst. = Araschcoolia Sch. Bip. ●■☆
34918 Arraschkoolia Sch. Bip. ex Hochst. = Geigeria Griess. ●■☆
34919 Arrhenaehne Cass. = Baccharis L.(保留属名)●■☆
34920 Arrhenatherum P. Beauv. (1812);燕麦草属(大蟹钓属);Bulbous Oat Grass,False Oat-grass,Oat Grass,Oatgrass ■
34921 Arrhenatherum album (Vahl) Clayton;白燕麦草■☆
34922 Arrhenatherum album (Vahl) Clayton var. erianthum (Boiss. et Reut.) Romero Zarco = Arrhenatherum album (Vahl) Clayton ■☆
34923 Arrhenatherum avenaceum P. Beauv. var. nodosum Rchb. = Arrhenatherum elatium (L.) P. Beauv. ex J. Presl et C. Presl var. bulbosum (Willd.) Spenn. ■
34924 Arrhenatherum bulbosum (Willd.) C. Presl = Arrhenatherum elatium (L.) P. Beauv. ex J. Presl et C. Presl var. bulbosum (Willd.) Spenn. ■
34925 Arrhenatherum darius (L.) J. Presl et C. Presl = Arrhenatherum elatium (L.) P. Beauv. ex J. Presl et C. Presl ■
34926 Arrhenatherum elatium (L.) J. Presl et C. Presl subsp. bulbosum (Willd.) Schübl. et G. Martens = Arrhenatherum elatium (L.) P. Beauv. ex J. Presl et C. Presl var. bulbosum (Willd.) Spenn. ■
34927 Arrhenatherum elatium (L.) J. Presl et C. Presl subsp. erianthum (Boiss. et Reut.) Trab. = Arrhenatherum album (Vahl) Clayton ■☆
34928 Arrhenatherum elatium (L.) J. Presl et C. Presl var. vulgare

(Fr.) Koch = Arrhenatherum elatium (L.) J. Presl et C. Presl ■
34929 Arrhenatherum elatium (L.) Mert. et Koch = Arrhenatherum elatium (L.) J. Presl et C. Presl ■
34930 Arrhenatherum elatium (L.) P. Beauv. ex J. Presl et C. Presl;燕麦草(大蟹钓);False Oat, False Oat Grass, False Oat-grass, French Rye Grass, French Rye-grass, Knot Oat-grass, Oat-grass, Pearl Grass, Tall Oat Grass, Tall Oatgrass, Tall Oat-grass ■
34931 Arrhenatherum elatium (L.) P. Beauv. ex J. Presl et C. Presl subsp. bulbosum (Willd.) Schüber et Martens = Arrhenatherum elatium (L.) P. Beauv. ex J. Presl et C. Presl var. bulbosum (Willd.) Spenn. ■
34932 Arrhenatherum elatium (L.) P. Beauv. ex J. Presl et C. Presl subsp. bulbosum (Willd.) Schübl. et G. Martens = Arrhenatherum elatium (L.) P. Beauv. ex J. Presl et C. Presl var. bulbosum (Willd.) Spenn. ■
34933 Arrhenatherum elatium (L.) P. Beauv. ex J. Presl et C. Presl subsp. nodosum (Parl.) Arcang. = Arrhenatherum elatium (L.) P. Beauv. ex J. Presl et C. Presl var. bulbosum (Willd.) Spenn. ■
34934 Arrhenatherum elatium (L.) P. Beauv. ex J. Presl et C. Presl var. biaristatum (Peterm.) Peterm.;二芒燕麦草;Onion Couch, Onion Twitch ■☆
34935 Arrhenatherum elatium (L.) P. Beauv. ex J. Presl et C. Presl var. bulbosum (Willd.) Spenn.;鳞茎燕麦(球茎燕麦,银边草);Bulbous Oat Grass, Bulbous Oatgrass, Onion Couch, Tall Oatgrass, Tuber Oat Grass ■
34936 Arrhenatherum elatium (L.) P. Beauv. ex J. Presl et C. Presl var. bulbosum (Willd.) Spenn. 'Variegatum';变叶燕麦草■
34937 Arrhenatherum elatium (L.) P. Beauv. ex J. Presl et C. Presl var. bulbosum (Willd.) Spenn. f. variegatum Hitchc. = Arrhenatherum elatium (L.) P. Beauv. ex J. Presl et C. Presl var. bulbosum (Willd.) Spenn. 'Variegatum' ■
34938 Arrhenatherum elatium (L.) P. Beauv. ex J. Presl et C. Presl var. nodosum Hubbard = Arrhenatherum elatium (L.) P. Beauv. ex J. Presl et C. Presl var. bulbosum (Willd.) Spenn. ■
34939 Arrhenatherum elatium (L.) P. Beauv. ex J. Presl et C. Presl var. nodosum f. striatum Hubb. = Arrhenatherum elatium (L.) P. Beauv. ex J. Presl et C. Presl var. bulbosum (Willd.) Spenn. 'Variegatum' ■
34940 Arrhenatherum elatium (L.) P. Beauv. ex J. Presl et C. Presl var. variegatum Hitchc. = Arrhenatherum elatium (L.) P. Beauv. ex J. Presl et C. Presl var. bulbosum (Willd.) Spenn. 'Variegatum' ■
34941 Arrhenatherum elatium (L.) Presl = Arrhenatherum elatium (L.) P. Beauv. ex J. Presl et C. Presl ■
34942 Arrhenatherum elongatum (Hochst. ex A. Rich.) Potztal = Helictotrichon elongatum (Hochst. ex A. Rich.) C. E. Hubb. ■☆
34943 Arrhenatherum erianthum Boiss. et Reut. = Arrhenatherum album (Vahl) Clayton ■☆
34944 Arrhenatherum friesiorum (Pilg.) Potztal = Helictotrichon umbrosum (Hochst. ex Steud.) C. E. Hubb. ■☆
34945 Arrhenatherum kotschyi Boiss.;考奇燕麦草■☆
34946 Arrhenatherum lachnanthum (Hochst. ex A. Rich.) Potztal = Helictotrichon lachnanthum (Hochst. ex A. Rich.) C. E. Hubb. ■☆
34947 Arrhenatherum longifolium (Thore) Dulac = Pseudarrhenatherum longifolium (Thore) Rouy ■☆
34948 Arrhenatherum mannii (Pilg.) Potztal = Helictotrichon mannii (Pilg.) C. E. Hubb. ■☆
34949 Arrhenatherum milanjianum (Rendle) Potztal = Helictotrichon milanjianum (Rendle) C. E. Hubb. ■☆
34950 Arrhenatherum mongolicum (Roshev.) Potztal = Helictotrichon mongolicum (Roshev.) Henrard ■
34951 Arrhenatherum phaneroneuron (C. E. Hubb.) Potztal = Helictotrichon elongatum (Hochst. ex A. Rich.) C. E. Hubb. ■☆
34952 Arrhenatherum riofrioi Sennen = Arrhenatherum album (Vahl) Clayton ■☆
34953 Arrhenatherum tuberosum (Gilib.) F. W. Schultz = Arrhenatherum elatium (L.) J. Presl et C. Presl subsp. bulbosum (Willd.) Schübl. et G. Martens ■
34954 Arrhenatherum umbrosum (Hochst. ex Steud.) Potztal = Helictotrichon umbrosum (Hochst. ex Steud.) C. E. Hubb. ■☆
34955 Arrhenechthites Mattf. (1938);紫芹菊属●■☆
34956 Arrhenechthites alba H. Kost.;白紫芹菊■☆
34957 Arrhenechthites dolichophylla Mattf.;紫芹菊■☆
34958 Arrhenechthites tomentella Mattf.;毛紫芹菊■☆
34959 Arrhostoxylon Mart. ex Nees = Ruellia L. ●■
34960 Arrhostoxylon Nees = Ruellia L. ●■
34961 Arrhostoxylum Mart. ex Nees = Ruellia L. ●■
34962 Arrhostoxylum Nees = Ruellia L. ●■
34963 Arrhynchium Lindl. = Arachnis Blume ■
34964 Arrhynchium labrosa Lindl. et Paxton = Arachnis labrosa (Lindl. et Paxton) Rchb. f. ■
34965 Armentaria Thouars ex Baill. = Uvaria L. ●
34966 Arrojadoa Britton et Rose (1920);猩猩冠柱属●☆
34967 Arrojadoa Mattf. = Arrojadocharis Mattf. ■☆
34968 Arrojadoa albiflora Buining et Brederoo;白花猩猩冠柱●☆
34969 Arrojadoa aureispina Buining et Brederoo;黄刺猩猩冠柱●☆
34970 Arrojadoa penicillata Britton et Rose;猩猩冠柱●☆
34971 Arrojadoa rhodantha (Gurke) Britton et Rose;由贵柱●☆
34972 Arrojadocharis Mattf. (1930);密叶柄泽兰属■☆
34973 Arrojadocharis praxeloides (Mattf.) Mattf.;密叶柄泽兰■☆
34974 Arrojadocharis praxeloides Mattf. = Arrojadocharis praxeloides (Mattf.) Mattf. ■☆
34975 Arrojadoopsis Guiggi = Arrojadoa Britton et Rose ●☆
34976 Arrojadoopsis Guiggi (2008);拟猩猩冠柱属●☆
34977 Arrostia Raf. = Gypsophila L. ●■
34978 Arrowsmithia DC. (1838);毛柱鼠麹木属●☆
34979 Arrowsmithia styphelioides DC.;毛柱鼠麹木●☆
34980 Arrozia Kunth = Luziola Juss. ■☆
34981 Arrozia Schrad. ex Kunth = Caryochloa Trin. ■☆
34982 Arrozia Schrad. ex Kunth = Luziola Juss. ■☆
34983 Arrudaria Macedo = Copernicia Mart. ex Endl. ●☆
34984 Arrudea A. St. -Hil. et Camb. = Clusia L. ●☆
34985 Arsaee Fourr. = Erica L. ●☆
34986 Arsenia Noronha = Uvaria L. ●
34987 Arsenjevia Starod. = Anemone L. (保留属名) ■
34988 Arsenjevia baicalensis (Turcz.) Starod. = Anemone baicalensis Turcz. ■
34989 Arsenjevia flaccida (F. Schmidt) Starod. = Anemone flaccida F. Schmidt ■
34990 Arsenjevia glabrata (Maxim.) Starod. = Anemone baicalensis Turcz. ■
34991 Arsenjevia prattii (Huth ex Ulbr.) Starod. = Anemone prattii Huth ex Ulbr. ■
34992 Arsenjevia rossii (S. Moore) Starod. = Anemone baicalensis Turcz. var. rossii (S. Moore) Kitag. ■

34993 Arsenoeoecus Small = Lyonia Nutt.(保留属名)●

34994 Arsis Lour. = Microcos Burm. ex L. ●

34995 Artabotrys R. Br.(1820);鹰爪花属(莺爪花属,鹰爪属);Eagleclaw,Tail Grape,Tailgrape,Tail-grape ●

34996 Artabotrys R. Br. ex Ker Gawl. = Artabotrys R. Br. ●

34997 Artabotrys antunesii Engl. et Diels;安图内思鹰爪花●☆

34998 Artabotrys aurantiacus Engl. et Diels;橙黄鹰爪花●☆

34999 Artabotrys aurantiacus Engl. et Diels var. multiflorus Pellegr. ex Le Thomas;多花橙黄鹰爪花●☆

35000 Artabotrys aurantiodorus(De Wild. et T. Durand) Engl. et Diels = Xylopia aurantiiodora De Wild. et T. Durand ●☆

35001 Artabotrys boonei De Wild. = Artabotrys velutinus Scott-Elliot ●☆

35002 Artabotrys brachypetalus Benth.;短瓣鹰爪花●☆

35003 Artabotrys claessensii De Wild. = Artabotrys aurantiacus Engl. et Diels ●☆

35004 Artabotrys coccineus Keay;绯红鹰爪花●☆

35005 Artabotrys collinus Hutch.;山丘鹰爪花●☆

35006 Artabotrys concolor Pellegr. = Artabotrys insignis Engl. et Diels var. concolor(Pellegr.) Le Thomas ●☆

35007 Artabotrys congolensis De Wild. et T. Durand;刚果鹰爪花●☆

35008 Artabotrys crassipetalus Pellegr.;厚瓣鹰爪花●☆

35009 Artabotrys dielsianus Le Thomas;迪尔斯鹰爪花●☆

35010 Artabotrys djalonis A. Chev. = Artabotrys velutinus Scott-Elliot ●☆

35011 Artabotrys esquirolii H. Lév. = Desmos chinensis Lour. ●

35012 Artabotrys esquirolii H. Lév. = Holboellia coriacea Diels ●

35013 Artabotrys fragrans Ast;香鹰爪花(东南亚鹰爪花);Fragrant Eagleclaw ●

35014 Artabotrys gossweileri Baker f. ex Exell;戈斯鹰爪花●☆

35015 Artabotrys hainanensis R. E. Fr.;狭瓣鹰爪花(海南鹰爪花,狭瓣鹰爪);Hainan Eagleclaw,Hainan Tailgrape ●

35016 Artabotrys hexapetalus(L. f.) Bhandari;鹰爪花(芳香鹰爪花,鸡爪兰,五爪兰,莺爪,莺爪花,鹰爪,鹰爪兰,鹰爪桃);Climbing Ilang-Ilang,Climbing Ylang-Ylang,Hexapetalous Tailgrape,Sixpetal Eagleclaw,Sixpetal Tailgrape,Tailgrape,Tail-grape ●

35017 Artabotrys hispidus Sprague et Hutch.;硬毛鹰爪花●☆

35018 Artabotrys hongkongensis Hance;香港鹰爪花(港鹰爪,钩枝藤,香港鹰爪,野鹰爪花,野鹰爪藤);Hongkong Eagleclaw,Hongkong Tailgrape ●

35019 Artabotrys insignis Engl. et Diels;显著鹰爪花●☆

35020 Artabotrys insignis Engl. et Diels var. concolor(Pellegr.) Le Thomas;同色鹰爪花●☆

35021 Artabotrys insignis Engl. et Diels var. latifolius Pellegr. = Artabotrys insignis Engl. et Diels ●☆

35022 Artabotrys jacques-felicis Pellegr.;雅凯鹰爪花●☆

35023 Artabotrys jollyanus Pierre;若利鹰爪花●☆

35024 Artabotrys lastoursvillensis Pellegr.;拉斯图维尔鹰爪花●☆

35025 Artabotrys letestui Pellegr.;莱泰斯图鹰爪花●☆

35026 Artabotrys libericus Diels;离生鹰爪花●☆

35027 Artabotrys lucidus A. Chev. = Artabotrys insignis Engl. et Diels ●☆

35028 Artabotrys mabifolius Diels = Artabotrys monteiroae Oliv. ●☆

35029 Artabotrys macrophyllus Hook. f.;大叶鹰爪花●☆

35030 Artabotrys malchairi De Wild. = Artabotrys insignis Engl. et Diels ●☆

35031 Artabotrys modestus Diels;适度鹰爪花●☆

35032 Artabotrys modestus Diels subsp. macranthus Verdc.;大花鹰爪花●☆

35033 Artabotrys monteiroae Oliv.;山鹰爪花●☆

35034 Artabotrys multiflorus C. E. C. Fisch.;多花鹰爪花;Multiflorous Tailgrape ●

35035 Artabotrys nigericus Hutch. = Artabotrys velutinus Scott-Elliot ●☆

35036 Artabotrys nitidus Engl. = Artabotrys monteiroae Oliv. ●☆

35037 Artabotrys odoratissimus R. Br. ex Ker Gawl. = Artabotrys hexapetalus(L. f.) Bhandari ●

35038 Artabotrys oliganthus Engl. et Diels;寡花鹰爪花●☆

35039 Artabotrys olivaeformis A. Chev. = Artabotrys aurantiacus Engl. et Diels ●☆

35040 Artabotrys oliveri(Engl.) Roberty = Polyalthia oliveri Engl. ●☆

35041 Artabotrys palustris Louis ex Boutique;沼泽鹰爪花●☆

35042 Artabotrys parviflorus Miq.;小花鹰爪花●☆

35043 Artabotrys petelotii Merr.;皮氏鹰爪花●☆

35044 Artabotrys pierreanus Engl. et Diels;皮埃尔鹰爪花●☆

35045 Artabotrys pilosus Merr. et Chun;毛叶鹰爪花;Piloseleaf Eagleclaw,Piloseleaf Tailgrape ●

35046 Artabotrys pleurocarpus Maingay ex Hook. f. et Thomson;侧果鹰爪花●☆

35047 Artabotrys punctulatus C. Y. Wu = Artabotrys punctulatus C. Y. Wu ex S. H. Yuan ●

35048 Artabotrys punctulatus C. Y. Wu ex S. H. Yuan;点叶鹰爪花;Punctulate Tailgrape ●

35049 Artabotrys pynaertii De Wild. = Artabotrys aurantiacus Engl. et Diels ●☆

35050 Artabotrys rhopalocarpus Le Thomas = Artabotrys congolensis De Wild. et T. Durand ●☆

35051 Artabotrys rhynchocarpus C. Y. Wu = Artabotrys rhynchocarpus C. Y. Wu ex S. H. Yuan ●

35052 Artabotrys rhynchocarpus C. Y. Wu ex S. H. Yuan;喙果鹰爪花;Beak-fruited Tailgrape,Rostratefruit Tailgrape ●

35053 Artabotrys robustus Louis ex Boutique;粗壮鹰爪花●☆

35054 Artabotrys rolfei S. Vidal;罗氏鹰爪花●☆

35055 Artabotrys rubicundus A. Chev. = Artabotrys oliganthus Engl. et Diels ●☆

35056 Artabotrys rufus De Wild.;浅红鹰爪花●☆

35057 Artabotrys rupestris Diels;岩生鹰爪花●☆

35058 Artabotrys setulosus Mildbr. et Diels = Artabotrys rufus De Wild. ●☆

35059 Artabotrys stenopetalus Engl. et Diels;热非鹰爪花●☆

35060 Artabotrys stenopetalus Engl. et Diels var. parviflorus Pellegr. = Artabotrys stenopetalus Engl. et Diels ●☆

35061 Artabotrys stenopetalus Merr. et Chun = Artabotrys hainanensis R. E. Fr. ●

35062 Artabotrys stolzii Diels;斯托尔兹鹰爪花●☆

35063 Artabotrys suaveolens Blume;芳香鹰爪花;Fragrant Tailgrape ●☆

35064 Artabotrys sumatranus Miq.;苏门答腊鹰爪花●☆

35065 Artabotrys thomsonii Oliv.;汤氏鹰爪花●☆

35066 Artabotrys trichopetalus Merr.;毛瓣鹰爪花●☆

35067 Artabotrys uncatus(Lour.) Baill. = Artabotrys hexapetalus(L. f.) Bhandari ●

35068 Artabotrys uncinatus(Lam.) Merr. = Artabotrys hexapetalus(L. f.) Bhandari ●

35069 Artabotrys velutinus Scott-Elliot;短绒毛鹰爪花●☆

35070 Artabotrys velutinus Scott-Elliot var. sphaerocarpa Sillans;球果短绒毛鹰爪花●☆

35071 Artabotrys zeylanicus Hook. f. et Thomson;锡兰鹰爪花●☆

35072 Artanacetum (Rzazade) Rzazade = Artemisia L. ●■

35073 Artanema D. Don(1834)(保留属名);悬丝参属■☆

35074 Artanema cabrae De Wild. et T. Durand = Artanema longifolium (L.) Vatke ■☆

35075 Artanema longiflorum Wettst. = Artanema longifolium (L.) Vatke ■☆

35076 Artanema longifolium (L.) Vatke;长叶悬丝参■☆

35077 Artanema longifolium (L.) Vatke var. amplexicaule Vatke = Artanema longifolium (L.) Vatke ■☆

35078 Artanema sesamoides (Vahl) Benth. = Artanema longifolium (L.) Vatke ■☆

35079 Artanthe Miq. = Oxodium Raf. ●■

35080 Artanthe Miq. = Piper L. ●■

35081 Artaphaxis Mill. = Atraphaxis L. ●

35082 Artedia L. (1753);阿特迪草属■☆

35083 Artedia muricata L.;阿特迪草■☆

35084 Artemisia L. (1753);蒿属(艾蒿属,艾属);Artemisia, Felon-herb, Mugwort, Ragweed, Sage, Sagebrush, Sailor's-tobacco, Wormwood ●■

35085 Artemisia abaensis Y. R. Ling et S. Y. Zhao;阿坝蒿;Aba Sagebrush ■

35086 Artemisia ablida Willd. ex Spreng. = Artemisia marschalliana Spreng. ■

35087 Artemisia abrotanoides Nutt. = Artemisia californica Less. ●☆

35088 Artemisia abrotanum L.;南蒿(长蒿,欧亚艾蒿,青蒿,雅艾,茵陈,茵蔯蒿);Apple Ringie, Averoyne, Boy's Love, Boy's-love, Cedar-wood, Cithernwood, Citherwood, Fine-leaved Mugwort, Garden Sagebrush, Girl's Curly Love, Girl's Delight, God's Tree, Kiss-I'-my-corner, Kiss-me-and-go, Kiss-me-quick, Kiss-me-quick-and-go, Lad Savour, Lad-love-lass, Lad's Love, Ladsavvur, London Pride, Maiden's Delight, Maiden's Ruin, Maid's Love, Maids' love, Maidwort, Malden's Delight, Motherwood, Old Man, Old Man Wormwood, Old Man's Beard, Old Man's Love, Old-man Wormwood, Overenyie, Slovenwood, Smelling-wood, Southernwood, Stalewort, Suthywood ■☆

35089 Artemisia abrotanum Thunb. = Artemisia carvifolia Buch.-Ham. ex Roxb. ■

35090 Artemisia absinthium L.;中亚苦蒿(苦艾,苦蒿,欧洲艾,啤酒蒿,洋艾);Absinth, Absinth Sagewort, Absinth Sage-wort, Absinth Wormwood, Absinthe, Absinthium, Allienne, Armoise Absinthe, Artemisia, Bitter Sagebrush, Common Sagebrush, Common Wormwood, Green Ginger, Mingwort, Mugwort, Old Woman, Oldman, Sage-wort, Ullimer, Ware-moth, Warmot, Wermout, Wermud, Wormit, Wormod, Wormwood ■

35091 Artemisia absinthium L. 'Lambrook Silver';银毛洋艾(灰叶苦艾)■☆

35092 Artemisia abyssinica Sch. Bip. ex A. Rich.;阿比西尼亚蒿■☆

35093 Artemisia abyssinica Sch. Bip. ex A. Rich. var. eriocephala Sch. Bip. ex Engl.;毛头阿比西尼亚■☆

35094 Artemisia achilleoides Turcz. = Ajania achilloides (Turcz.) Poljakov ex Grubov ■

35095 Artemisia adamsii Besser;东北丝裂蒿(阿达姆斯蒿,阿氏蒿,丝裂蒿,丝叶蒿);Adams Sagebrush ■

35096 Artemisia adamsii Kitag. = Artemisia brachyloba Franch. ■

35097 Artemisia afra Jacq. ex Willd.;非洲野蒿;Wild Wormwood ■☆

35098 Artemisia afra Jacq. ex Willd. var. arussorum Chiov.;阿鲁斯蒿■☆

35099 Artemisia afra Jacq. ex Willd. var. friesiorum Chiov. = Artemisia afra Jacq. ex Willd. ■☆

35100 Artemisia aksaiensis Y. R. Ling;阿克塞蒿;Aksai Sagebrush ●■

35101 Artemisia alaskana Rydb.;西伯利亚蒿;Siberian Wormwood ■☆

35102 Artemisia alba Turra;非洲白蒿■☆

35103 Artemisia alba Turra subsp. kabylica (Chabert) Greuter;卡比利亚蒿■☆

35104 Artemisia alba Turra var. mesatlantica Quézel = Artemisia alba Turra ■☆

35105 Artemisia albicerata Krasch.;白角蒿■☆

35106 Artemisia albida Ledeb. = Seriphidium compactum (Fisch. ex Besser) Poljakov ■

35107 Artemisia albida Maxim. = Seriphidium nitrosum (Weber ex Stechm.) Poljakov ■

35108 Artemisia albida Willd. ex Spreng. = Artemisia marschalliana Spreng. ■

35109 Artemisia albida Willd. ex Spreng. = Seriphidium nitrosum (Weber ex Stechm.) Poljakov ■

35110 Artemisia albula Wooton = Artemisia ludoviciana Nutt. subsp. albula (Wooton) D. D. Keck ■☆

35111 Artemisia aleutica Hultén;阿留申蒿;Aleutian Wormwood ■☆

35112 Artemisia algeriensis Filatova;阿尔及利亚蒿■☆

35113 Artemisia altaiensis Krasch.;阿尔泰蒿■☆

35114 Artemisia ambigua Thunb. = Seriphium plumosum L. ●☆

35115 Artemisia amoena Poljakov = Seriphidium amoenum (Poljakov) Poljakov ■

35116 Artemisia amritima L.;驱蛔蒿(蛔蒿,蒙古蛔蒿)■☆

35117 Artemisia amygdalina Decne.;膀胱蒿■☆

35118 Artemisia anethifolia Poljakov = Artemisia anethoides Mattf. ■

35119 Artemisia anethifolia Weber ex Stechm.;碱蒿(臭蒿,大莳萝蒿,糜糜蒿,伪茵陈,盐蒿);Dillleaf Sagebrush, Dillleaf Wormwood ■

35120 Artemisia anethifolia Weber ex Stechm. f. gracilis Pamp. = Artemisia anethifolia Weber ex Stechm. ■

35121 Artemisia anethifolia Weber ex Stechm. f. shansiensis Pamp. = Artemisia anethifolia Weber ex Stechm. ■

35122 Artemisia anethifolia Weber ex Stechm. var. anethoides (Mattf.) Pamp. = Artemisia anethoides Mattf. ■

35123 Artemisia anethifolia Weber ex Stechm. var. cum f. gracilis Pamp. = Artemisia anethifolia Weber ex Stechm. ■

35124 Artemisia anethifolia Weber ex Stechm. var. cum? f. shansiensis Pamp. = Artemisia anethifolia Weber ex Stechm. ■

35125 Artemisia anethifolia Weber ex Stechm. var. erectiflora DC. = Artemisia anethifolia Weber ex Stechm. ■

35126 Artemisia anethifolia Weber ex Stechm. var. multicaulis DC. = Artemisia anethifolia Weber ex Stechm. ■

35127 Artemisia anethifolia Weber ex Stechm. var. stelleriana DC. = Artemisia anethifolia Weber ex Stechm. ■

35128 Artemisia anethoides Mattf.;莳萝蒿(伪茵陈,小碱蒿,茵陈,肇东蒿);Dill-like Sagebrush, Dill-like Wormwood ■

35129 Artemisia angustifolia (A. Gray) Rydb. = Artemisia tridentata Nutt. ■

35130 Artemisia angustissima Nakai;狭叶牡蒿(狭叶牡葛);Muchnarrowleaf Sagebrush ■

35131 Artemisia angustissima Nakai = Artemisia japonica Thunb. var. angustissima (Nakai) Kitam. ■

35132 Artemisia annua L.;黄花蒿(白染艮,草蒿,草蒿子,臭蒿,臭黄蒿,臭青蒿,方溃,蒿,蒿子,黑蒿,黄蒿,黄色土因呈,黄香蒿,鸡虱草,假香菜,酒饼草,苦蒿,苦香蒿,良蒿,马尿蒿,青蒿,秋蒿,三庚草,茼蒿,细青蒿,细叶蒿,香蒿,香蒿花,香苦草,香青蒿,香丝草,邪蒿,药用青蒿,野蒿,野苦草,野兰蒿,野茼蒿);Ambrosia, Annual Mugwort, Annual Sage-wort, Annual Wormwood, Armoise Annuelle, Chinese Fragrant Fern, Huanghuahaosu, Sweet Annie, Sweet Sagebrush, Sweet Sagewort, Sweet Wormwood ■

35133 Artemisia annua L. f. genuina Pamp. = Artemisia annua L. ■

35134 Artemisia annua L. f. macrocephala Pamp.;大花黄花蒿(大头黄花蒿,青蒿);Largehead Sweet Sagebrush, Largehead Sweet Wormwood ■

35135 Artemisia annua L. f. macrocephala Pamp. = Artemisia annua L. ■

35136 Artemisia anomala S. Moore;奇蒿(白花尾,斑枣子,大叶蒿子,红陈艾,化石丹,化食丹,寄奴,金寄奴,九里光,九牛草,苦连婆,苦婆菜,刘寄奴,六月白,六月霜,六月雪,芦蒿,南刘寄奴,千粒米,炭包包,乌藤菜,细白花草,狭叶艾,鸭脚,野马兰头,一枝梅,异形蒿,珍珠蒿);Diverse Sagebrush, Diverse Wormwood ■

35137 Artemisia anomala S. Moore var. acuminatissima Y. R. Ling;尖奇蒿■

35138 Artemisia anomala S. Moore var. tomentella Hand. -Mazz.;密毛奇蒿;Tomentose Diverse Sagebrush, Tomentose Diverse Wormwood ■

35139 Artemisia anthriscifolia C. C. Chang = Artemisia emeiensis Y. R. Ling ■

35140 Artemisia apiacea Hance = Artemisia carvifolia Buch. -Ham. ex Roxb. ■

35141 Artemisia apiacea Hance var. schochii (Mattf.) Hand. -Mazz. = Artemisia carvifolia Buch. -Ham. ex Roxb. var. schochii (Mattf.) Pamp. ■

35142 Artemisia arachnoidea E. Sheld. = Artemisia ludoviciana Nutt. subsp. incompta (Nutt.) D. D. Keck ■☆

35143 Artemisia aragonensis Lam. = Artemisia herba-alba Asso ■☆

35144 Artemisia aralensis Krasch.;阿拉里蒿■☆

35145 Artemisia araneosa Kitam. = Artemisia lavandulifolia DC. ●■

35146 Artemisia arborescens L.;灌木蒿(乔木蒿,小木艾);Shrubby Wormwood, Tree Wormwood ●☆

35147 Artemisia arbuscula Nutt.;北美矮蒿;Low Sagebrush ■☆

35148 Artemisia arbuscula Nutt. subsp. longiloba (Osterh.) L. M. Shultz;蒿■☆

35149 Artemisia arbuscula Nutt. subsp. nova (A. Nelson) G. H. Ward = Artemisia nova A. Nelson ■☆

35150 Artemisia arbuscula Nutt. subsp. thermopola Beetle;温泉蒿;Hot Springs Sagebrush ■☆

35151 Artemisia arbuscula Nutt. var. nova (A. Nelson) Cronquist = Artemisia nova A. Nelson ■☆

35152 Artemisia arctica Besser = Artemisia borealis Pall. subsp. richardsoniana (Besser) Korobkov ■☆

35153 Artemisia arctica Less.;北极蒿;Arctic Sagebrush ■☆

35154 Artemisia arctica Less. = Artemisia norvegica Fr. subsp. saxatilis (Besser) H. M. Hall et Clem. ■☆

35155 Artemisia arctica Less. subsp. beringensis (Hultén) Hultén = Artemisia norvegica Fr. subsp. saxatilis (Besser) H. M. Hall et Clem. ■☆

35156 Artemisia arctica Less. subsp. comata (Rydb.) Hultén = Artemisia norvegica Fr. subsp. saxatilis (Besser) H. M. Hall et Clem. ■☆

35157 Artemisia arctica Less. subsp. sachalinensis (F. Schmidt) Hultén;库页蒿■☆

35158 Artemisia arctica Less. subsp. sachalinensis (F. Schmidt) Hultén f. villosa (Koidz.) Kitam.;毛库页蒿■☆

35159 Artemisia arctica Less. var. sachalinensis F. Schmidt = Artemisia arctica Less. subsp. sachalinensis (F. Schmidt) Hultén ■☆

35160 Artemisia arctica Less. var. saxatilis (Besser) Y. R. Ling = Artemisia norvegica Fr. subsp. saxatilis (Besser) H. M. Hall et Clem. ■☆

35161 Artemisia arctica Less. var. villosa (Koidz.) Tatew. = Artemisia arctica Less. subsp. sachalinensis (F. Schmidt) Hultén f. villosa (Koidz.) Kitam. ■☆

35162 Artemisia arenaria DC. = Artemisia oxycephala Kitag. ■

35163 Artemisia arenaria H. C. Fu = Artemisia oxycephala Kitag. ■

35164 Artemisia argentea L'Hér.;欧洲银蒿■☆

35165 Artemisia argillosa Beetle = Artemisia cana Pursh subsp. viscidula (Osterh.) Beetle ■☆

35166 Artemisia argyi H. Lév. = Artemisia argyi H. Lév. et Vaniot ■

35167 Artemisia argyi H. Lév. et Vaniot;艾(阿及艾,艾蒿,艾蓬,艾绒,艾叶,白艾,白陈艾,白蒿,半茸,北艾,冰台,病草,草艾,草蓬,陈艾,大艾,大叶艾,肚里屏风,海艾,红艾,黄草,火艾,家艾,家陈艾,狼尾蒿子,祁艾,甜艾,五月艾,香艾,野艾,野莲头,医草,炙草);Argy Sagebrush, Argy's Wormwood, Chinese Mugwort ■

35168 Artemisia argyi H. Lév. et Vaniot f. gracilis (Pamp.) Kitag. = Artemisia argyi H. Lév. et Vaniot var. gracilis Pamp. ■

35169 Artemisia argyi H. Lév. et Vaniot var. com f. genuina Pamp. = Artemisia argyi H. Lév. et Vaniot ■

35170 Artemisia argyi H. Lév. et Vaniot var. eximia (Pamp.) Kitam.;无齿艾蒿;Toothless Argy's Wormwood, Toothless Sagebrush ■

35171 Artemisia argyi H. Lév. et Vaniot var. gracilis Pamp.;朝鲜艾(朝鲜艾蒿,内蒙野艾,深裂叶艾蒿,野艾)■

35172 Artemisia argyi H. Lév. et Vaniot var. incana Pamp. = Artemisia argyi H. Lév. et Vaniot ■

35173 Artemisia argyi H. Lév. et Vaniot var. incana Pamp. f. microcephala Pamp. = Artemisia argyi H. Lév. et Vaniot ■

35174 Artemisia argyrophylla Ledeb.;银叶蒿;Silverleaf Sagebrush ■

35175 Artemisia argyrophylla Ledeb. var. brevis (Pamp.) Y. R. Ling;小银叶蒿■

35176 Artemisia armeniaca Lam.;亚美尼亚蒿(灰色白艾);Silver Sagebrush ■☆

35177 Artemisia aromatica A. Nelson = Artemisia dracunculus L. ■

35178 Artemisia aschurbajewii C. Winkl.;褐头蒿;Brownhead Sagebrush ■

35179 Artemisia asiatica (Pamp.) Nakai ex Kitam. = Artemisia indica Willd. ■

35180 Artemisia asiatica Nakai ex Pamp. = Artemisia indica Willd. ■

35181 Artemisia atlantica Coss.;大西洋蒿■☆

35182 Artemisia atlantica Coss. var. maroccana Maire = Artemisia atlantica Coss. ■☆

35183 Artemisia atomifera Piper = Artemisia ludoviciana Nutt. subsp. incompta (Nutt.) D. D. Keck ■☆

35184 Artemisia atrata Lam.;黑苞蒿■

35185 Artemisia atrovirens Hand. -Mazz.;暗绿蒿(白艾蒿,白蒿,白毛蒿,大蒿,青蒿,水蒿,铁蒿);Dimgreen Sagebrush ■

35186 Artemisia aucheri Boiss. = Seriphidium aucheri (Boiss.) Y. Ling et Y. R. Ling ■

35187 Artemisia aurata Kom.;金黄蒿(黄金蒿);Goldnyellow Sagebrush, Goldnyellow Wormwood ■

35188 Artemisia austriaca Jacq.;银蒿(奥地利蒿,银叶蒿);Austrian Sagebrush, Austrian Wormwood ■

35189 Artemisia austriaca Jacq. var. jacquiniana DC. = Artemisia austriaca Jacq. ■

35190 Artemisia austriaca Jacq. var. jacquiniana DC. f. microcephala Pamp. = Artemisia austriaca Jacq. ■

35191 Artemisia austriaca Jacq. var. orientalis DC. = Artemisia austriaca Jacq. ■

35192 Artemisia austroyunnanensis Y. Ling et Y. R. Ling;滇南艾(滇南艾蒿);S. Yunnan Sagebrush ■

35193 Artemisia badgbysi Krasch. et Lincz. ex Poljakov;巴德蒿■☆

35194 Artemisia baimaensis Y. R. Ling et Z. C. Zhuo;班玛蒿;Banma Sagebrush ■

35195 Artemisia bakeri Greene = Artemisia carruthii A. W. Wood ex Carruth. ■☆

35196 Artemisia baldshuanica Krasch. et Zopf;巴尔德蒿■☆

35197 Artemisia bargusinensis Spreng.;巴尔古津蒿(穗花蒿);Bargusinian Sagebrush ■

35198 Artemisia barrelieri Besser;巴雷艾■☆

35199 Artemisia batakensis Hayata = Artemisia somai Hayata var. batakensis (Hayata) Kitam. ■

35200 Artemisia besseriana Ledeb. = Artemisia lagocephala (Fisch. ex Besser) Fisch. ex DC. ●■

35201 Artemisia besseriana Ledeb. var. integrifolia Ledeb. = Artemisia lagocephala (Fisch. ex Besser) Fisch. ex DC. ●■

35202 Artemisia besseriana Ledeb. var. triloba Ledeb. = Artemisia lagocephala (Fisch. ex Besser) Fisch. ex DC. ●■

35203 Artemisia biennis Hook. f. = Artemisia hedinii Ostenf. et Paulsen ■

35204 Artemisia biennis Willd.;二年生蒿;Biennial Sage-wort, Biennial Wormwood ■

35205 Artemisia biennis Willd. = Artemisia hedinii Ostenf. et Paulsen ■

35206 Artemisia bigelovii A. Gray;比氏蒿;Bigelow Sagebrush ■☆

35207 Artemisia blepharolepis Bunge;白莎蒿(白里蒿,白沙蒿,糜蒿);Ciliate Scale Wormwood, Ciliatescale Wormwood ■

35208 Artemisia bolanderi A. Gray = Artemisia cana Pursh subsp. bolanderi (A. Gray) G. H. Ward ■☆

35209 Artemisia borealis Kitam. = Artemisia oligocarpa Hayata ■

35210 Artemisia borealis Liou et al. = Artemisia bargusinensis Spreng. ■

35211 Artemisia borealis Pall.;北方蒿;Boreal Sage, Northern Wormwood, Old Lady ■

35212 Artemisia borealis Pall. = Artemisia baimaensis Y. R. Ling et Z. C. Zhuo ■

35213 Artemisia borealis Pall. subsp. richardsoniana (Besser) Korobkov;理氏蒿;Richardson's Sagewort ■☆

35214 Artemisia borealis Pall. var. ledebouri S. Y. Hu = Artemisia bargusinensis Spreng. ■

35215 Artemisia borealis Pall. var. oligocarpa (Hayata) Kitam. = Artemisia oligocarpa Hayata ■

35216 Artemisia borealis Pall. var. willdenovii Besser = Artemisia bargusinensis Spreng. ■

35217 Artemisia borotalensis Poljakov = Seriphidium borotalense (Poljakov) Y. Ling et Y. R. Ling ■

35218 Artemisia botschantzevii Filatova;包兹蒿■☆

35219 Artemisia bracenathemoides C. Winkl. = Kaschgaria brachanthemoides (C. Winkl.) Poljakov ●■

35220 Artemisia brachyloba Franch.;山蒿(骆驼蒿,岩蒿);Shortlobed Wormwood ■

35221 Artemisia brachyphylla Kitam.;高岭蒿(长白山蒿,绒叶蒿);Shortleaf Sagebrush ■

35222 Artemisia brevifolia Wall. ex DC. = Seriphidium brevifolium (Wall. ex DC.) Y. Ling et Y. R. Ling ■

35223 Artemisia brevis Pamp. = Artemisia argyrophylla Ledeb. var. brevis (Pamp.) Y. R. Ling ■

35224 Artemisia brittonii Rydb. = Artemisia ludoviciana Nutt. ■☆

35225 Artemisia burmanica Pamp. = Artemisia myriantha Wall. ex Besser ■

35226 Artemisia burmanica Pamp. = Artemisia zhongdianensis Y. R. Ling ■

35227 Artemisia burmanica Pamp. f. latifolia Pamp. = Artemisia austroyunnanensis Y. Ling et Y. R. Ling ■

35228 Artemisia burmanica Y. R. Ling = Artemisia zhongdianensis Y. R. Ling ■

35229 Artemisia caerulescens L.;天蓝蒿■☆

35230 Artemisia caespitosa Ledeb.;矮丛蒿(丛蒿,灰莲蒿);Dwarf Fascicular Sagebrush ■

35231 Artemisia californica Less.;加州蒿;California Sagebrush ●☆

35232 Artemisia californica Less. var. insularis (Rydb.) Munz = Artemisia nesiotica P. H. Raven ■

35233 Artemisia calophylla Pamp.;美叶蒿;Beautifulleaf Sagebrush ■

35234 Artemisia campbellii Hook. f. et Thomson ex Clarke;绒毛蒿;Campbell Sagebrush ■

35235 Artemisia campbellii Hook. f. et Thomson ex Clarke var. limprichtii Pamp. = Artemisia tainingensis Hand. -Mazz. ■

35236 Artemisia campestris Kitam. = Artemisia oxycephala Kitag. ■

35237 Artemisia campestris L.;荒野蒿(细叶山艾,野蒿,野生蒿,玉山艾);Breckland Mugwort, Field Mugwort, Field Sagebrush, Field Sagewort, Field Sage-wort, Field Southernwood, Field Wormwood, Sagewort Wormwood, Sand Wormwood, Western Sagewort, Wild Wormwood, Wormwood ■

35238 Artemisia campestris L. = Artemisia morrisonensis Hayata ■

35239 Artemisia campestris L. = Artemisia oxycephala Kitag. ■

35240 Artemisia campestris L. = Artemisia pubescens Ledeb. ■

35241 Artemisia campestris L. subsp. borealis (Pall.) H. M. Hall et Clem. = Artemisia borealis Pall. ■

35242 Artemisia campestris L. subsp. canadensis (Michx.) Scoggan = Artemisia canadensis Michx. ■☆

35243 Artemisia campestris L. subsp. canescens Le Houér.;灰白蒿■☆

35244 Artemisia campestris L. subsp. caudata (Michx.) H. M. Hall et Clem.;尾状荒野蒿;Field Sage-wort, Field Wormwood, Green Sage ■☆

35245 Artemisia campestris L. subsp. glutinosa (DC.) Batt.;黏性蒿■☆

35246 Artemisia campestris L. subsp. pacifica (Nutt.) H. M. Hall et Clem.;西部荒野蒿;Western Sagewort ■☆

35247 Artemisia campestris L. subsp. pycnocephala (Less.) H. M. Hall et Clem. = Artemisia pycnocephala (Less.) DC. ■☆

35248 Artemisia campestris L. subsp. variabilis (Ten.) Greuter;易变蒿■☆

35249 Artemisia campestris L. var. caudata (Michx.) E. J. Palmer et Steyerm. = Artemisia campestris L. subsp. caudata (Michx.) H. M. Hall et Clem. ■☆

35250 Artemisia campestris L. var. clausonis (Pomel) Batt. = Artemisia campestris L. ■

35251 Artemisia campestris L. var. douglasiana (Besser) B. Boivin = Artemisia douglasiana Besser ■☆

35252 Artemisia campestris L. var. gmeliniana Besser = Artemisia marschalliana Spreng. ■

35253 Artemisia campestris L. var. macilenta Maxim. = Artemisia

macilenta (Maxim.) Krasch. ●■
35254 Artemisia campestris L. var. marschalliana (Spreng.) Poljakov = Artemisia marschalliana Spreng. ■
35255 Artemisia campestris L. var. odoratissima (Desf.) Batt. = Artemisia campestris L. ■
35256 Artemisia campestris L. var. petiolata S. L. Welsh = Artemisia campestris L. subsp. pacifica (Nutt.) H. M. Hall et Clem. ■☆
35257 Artemisia campestris L. var. purshii (Besser) Cronquist = Artemisia borealis Pall. ■
35258 Artemisia campestris L. var. scouleriana (Besser) Cronquist = Artemisia campestris L. subsp. pacifica (Nutt.) H. M. Hall et Clem. ■☆
35259 Artemisia campestris L. var. sericophylla (Rupr.) Poljakov = Artemisia sericophylla Rupr. ■☆
35260 Artemisia campestris L. var. sericophylla (Rupr.) Poljakov = Artemisia marschalliana Spreng. var. sericophylla (Rupr.) Y. R. Ling ■☆
35261 Artemisia campestris L. var. spithamaea (Pursh) M. Peck = Artemisia borealis Pall. ■
35262 Artemisia campestris L. var. steveniana Besser = Artemisia marschalliana Spreng. ■
35263 Artemisia campestris L. var. strutziae S. L. Welsh = Artemisia borealis Pall. ■
35264 Artemisia campestris Ledeb. var. glutinosa (Gay ex Bess.) Y. R. Ling = Artemisia campestris L. ■
35265 Artemisia campestris Ledeb. var. pubescens (Ledeb.) Trautv. et C. A. Mey. = Artemisia pubescens Ledeb. ■
35266 Artemisia camphorata Vill.;樟脑蒿;Camphor Wormwood ■☆
35267 Artemisia camphorata Vill. = Artemisia alba Turra ■☆
35268 Artemisia cana Pursh;灰色蒿;Silver Wormwood ■☆
35269 Artemisia cana Pursh subsp. bolanderi (A. Gray) G. H. Ward;博兰德蒿;Bolander Sagebrush ■☆
35270 Artemisia cana Pursh subsp. viscidula (Osterh.) Beetle;微黏蒿;Sticky Sagebrush ■☆
35271 Artemisia cana Pursh var. viscidula Osterh. = Artemisia cana Pursh subsp. viscidula (Osterh.) Beetle ■☆
35272 Artemisia canabina Jacquem. ex Besser = Artemisia dubia Wall. ex Besser var. subdigitata (Mattf.) Y. R. Ling ■
35273 Artemisia canadensis Michx.;加拿大蒿;Canadian Sagebrush ■☆
35274 Artemisia canadensis Michx. = Artemisia campestris L. subsp. canadensis (Michx.) Scoggan ■☆
35275 Artemisia canariensis (Besser) Less.;加那利艾■☆
35276 Artemisia canariensis (Besser) Less. = Artemisia thuscula Cav. ■☆
35277 Artemisia canariensis (Besser) Less. var. eleta Bolle = Artemisia thuscula Cav. ■☆
35278 Artemisia candicans Rydb. = Artemisia ludoviciana Nutt. subsp. candicans (Rydb.) D. D. Keck ■☆
35279 Artemisia canescens Willd. = Artemisia armeniaca Lam. ■☆
35280 Artemisia cannabifolia H. Lév. = Artemisia selengensis Turcz. ex Besser ■
35281 Artemisia cannabifolia H. Lév. var. nigrescens H. Lév. = Artemisia selengensis Turcz. ex Besser ■
35282 Artemisia capillaris Thunb.;茵陈蒿(安吕草,白蒿,白茵陈,臭蒿,猴子毛,家茵陈,绵茵陈,青蒿,青蒿草,日本茵陈,绒蒿,桐蒿草,土茵陈,萧,小马尿蒿,因尘,因陈,茵陈);Capillary Sagebrush, Capillary Wormwood, Mosquito Wormwood, Yin-Chen Wormwood ■
35283 Artemisia capillaris Thunb. f. villosa Korsh.;柔毛茵陈蒿■
35284 Artemisia capillaris Thunb. var. acaulis Pamp. = Artemisia capillaris Thunb. ■
35285 Artemisia capillaris Thunb. var. arbuscula Miq. = Artemisia capillaris Thunb. ■
35286 Artemisia capillaris Thunb. var. arbuscula Miq. f. genuina Pamp. = Artemisia capillaris Thunb. ■
35287 Artemisia capillaris Thunb. var. arbuscula Miq. f. glabra Pamp. = Artemisia capillaris Thunb. ■
35288 Artemisia capillaris Thunb. var. arbuscula Miq. f. sericea Pamp. = Artemisia capillaris Thunb. ■
35289 Artemisia capillaris Thunb. var. grandiflora (Pamp.) Pamp. = Artemisia capillaris Thunb. ■
35290 Artemisia capillaris Thunb. var. grandiflora (Pamp.) Pamp. = Artemisia scoparia Waldst. et Kit. ■
35291 Artemisia capillaris Thunb. var. grandiflora (Pamp.) Pamp. f. genuina Pamp. = Artemisia scoparia Waldst. et Kit. ■
35292 Artemisia capillaris Thunb. var. grandiflora (Pamp.) Pamp. f. genuina Pamp. subf. angustissecta Pamp. = Artemisia scoparia Waldst. et Kit. ■
35293 Artemisia capillaris Thunb. var. grandiflora (Pamp.) Pamp. f. latifolia Pamp. subf. tenuifolia Pamp. = Artemisia scoparia Waldst. et Kit. ■
35294 Artemisia capillaris Thunb. var. grandiflora (Pamp.) Pamp. f. latifolia Pamp. = Artemisia scoparia Waldst. et Kit. ■
35295 Artemisia capillaris Thunb. var. grandiflora Pamp. = Artemisia scoparia Waldst. et Kit. ■
35296 Artemisia capillaris Thunb. var. sachalinensis (Tiles) Pamp. = Artemisia scoparia Waldst. et Kit. ■
35297 Artemisia capillaris Thunb. var. scopana (Waldst. et Kit.) Pamp. f. grandiflora Pamp. = Artemisia scoparia Waldst. et Kit. ■
35298 Artemisia capillaris Thunb. var. scopana (Waldst. et Kit.) Pamp. f. kohatica (Klatt) Pamp. = Artemisia scoparia Waldst. et Kit. ■
35299 Artemisia capillaris Thunb. var. scoparia (Waldst. et Kit.) Pamp. = Artemisia scoparia Waldst. et Kit. ■
35300 Artemisia capillaris Thunb. var. scoparia (Waldst. et Kit.) Pamp. f. elegans (Roxb.) Pamp. = Artemisia scoparia Waldst. et Kit. ■
35301 Artemisia capillaris Thunb. var. scoparia (Waldst. et Kit.) Pamp. f. grandiflora Pamp. = Artemisia eriopoda Bunge ■
35302 Artemisia capillaris Thunb. var. scoparia (Waldst. et Kit.) Pamp. f. myriocephala Pamp. = Artemisia scoparia Waldst. et Kit. ■
35303 Artemisia capillaris Thunb. var. scoparia (Waldst. et Kit.) Pamp. f. villosa Korsh. = Artemisia scoparia Waldst. et Kit. ■
35304 Artemisia capillaris Thunb. var. scoparia (Waldst. et Kit.) Pamp. f. williamsonii Pamp. = Artemisia scoparia Waldst. et Kit. ■
35305 Artemisia capillaris Thunb. var. simplex Maxim. = Artemisia pubescens Ledeb. ■
35306 Artemisia capillifolia Lam. = Eupatorium capillifolium (Lam.) Small ■☆
35307 Artemisia carruthii A. W. Wood ex Carruth.;卡鲁斯蒿;Carruth Wormwood, Wormwood ■☆
35308 Artemisia carvifolia Buch.-Ham. ex Roxb.;青蒿(白染艮,鳖血青蒿,草蒿,方溃,蒿,黑蒿,黄花蒿,凛蒿,廪蒿,苹蒿,菣,三庚草,香蒿,香青蒿,邪蒿,野兰蒿,茵陈蒿);Celery Sagebrush, Celery Wormwood ■

35309 Artemisia carvifolia Buch. -Ham. ex Roxb. var. apiacea (Hance) Pamp. = Artemisia carvifolia Buch. -Ham. ex Roxb. ■

35310 Artemisia carvifolia Buch. -Ham. ex Roxb. var. apiacea Pamp. = Artemisia carvifolia Buch. -Ham. ex Roxb. ■

35311 Artemisia carvifolia Buch. -Ham. ex Roxb. var. schochii (Mattf.) Pamp.;大头青蒿(大花青蒿);Largehead Celery Wormwood ■

35312 Artemisia carvifolia Buch. -Ham. ex Roxb. var. typtca Pamp. = Artemisia carvifolia Buch. -Ham. ex Roxb. ■

35313 Artemisia caspia B. Keller et Kom.;里海蒿■☆

35314 Artemisia caucasica Willd.;高加索蒿(绵毛蒿);Caucasia Sagebrush,Silver Spreader ■☆

35315 Artemisia caudata Michx.;尾蒿;Tall Wormwood,Caudate Wormwood ■☆

35316 Artemisia caudata Michx. = Artemisia campestris L. subsp. caudata (Michx.) H. M. Hall et Clem. ■☆

35317 Artemisia caudata Michx. = Artemisia campestris L. ■

35318 Artemisia caudata Michx. var. calvens Lunell = Artemisia campestris L. subsp. caudata (Michx.) H. M. Hall et Clem. ■☆

35319 Artemisia caudata Michx. var. douglasiana (Besser) B. Boivin = Artemisia douglasiana Besser ■☆

35320 Artemisia caudata Michx. var. richardsoniana (Besser) B. Boivin = Artemisia borealis Pall. subsp. richardsoniana (Besser) Korobkov ■☆

35321 Artemisia centiflora Maxim. var. pilifera Y. Ling = Stilpnolepis centiflora (Maxim.) Krasch. ■

35322 Artemisia centtflora Maxim. = Stilpnolepis centiflora (Maxim.) Krasch. ■

35323 Artemisia chamaemelifolia Vill.;甘菊叶蒿;Lady's Maid ■☆

35324 Artemisia chamissoniana Besser var. saxatilis Besser = Artemisia norvegica Fr. subsp. saxatilis (Besser) H. M. Hall et Clem. ■☆

35325 Artemisia chamomilla C. Winkl. = Artemisia annua L. ■

35326 Artemisia changaica Krasch. = Artemisia dracunculus L. var. changaica (Krasch.) Y. R. Ling ■

35327 Artemisia chiarugii Pamp. = Artemisia argyi H. Lév. et Vaniot var. gracilis Pamp. ■

35328 Artemisia chienshanica Y. Ling et W. Wang;千山蒿;Qianshan Sagebrush ■

35329 Artemisia chinense L. = Crossostephium chinense (L.) Makino ●

35330 Artemisia chingii Pamp.;南毛蒿;Ching Sagebrush ■

35331 Artemisia chrysolepis Kitag. = Artemisia sieversiana Ehrh. ex Willd. ■

35332 Artemisia cina Berg. = Artemisia cina Berg. ex Poljakov ■☆

35333 Artemisia cina Berg. ex Poljakov;土耳其斯坦蒿;Levant Wormseed,Santonica ■☆

35334 Artemisia cina Berg. ex Poljakov = Seriphidium cinum (Berger ex Poljakov) Poljakov ■

35335 Artemisia clausonis Pomel = Artemisia campestris L. ■

35336 Artemisia clemensiana Pamp. = Artemisia lavandulifolia DC. ●■

35337 Artemisia codonocephala Diels = Artemisia lavandulifolia DC. ●■

35338 Artemisia codonocephala Diels var. maireana Pamp. = Artemisia lavandulifolia DC. ●■

35339 Artemisia coloradensis Osterh. = Artemisia carruthii A. W. Wood ex Carruth. ■☆

35340 Artemisia columbiensis Nutt. = Artemisia cana Pursh ■☆

35341 Artemisia comaiensis Y. Ling et Y. R. Ling;高山矮蒿;Alpine Dwarf Sagebrush ■

35342 Artemisia comata Rydb. = Artemisia norvegica Fr. subsp. saxatilis (Besser) H. M. Hall et Clem. ■☆

35343 Artemisia commutata Besser;变蒿;Changed Sagebrush,Changed Wormwood ■

35344 Artemisia commutata Besser = Artemisia pubescens Ledeb. ■

35345 Artemisia commutata Besser var. acutiloba W. Wang et C. Y. Li;尖叶变蒿;Acuminate Changed Sagebrush ■

35346 Artemisia commutata Besser var. coracina W. Wang = Artemisia pubescens Ledeb. var. coracina (W. Wang) Y. Ling et Y. R. Ling ■

35347 Artemisia commutata Besser var. douglasiana (Besser) Besser = Artemisia douglasiana Besser ■☆

35348 Artemisia commutata Besser var. gebleriana Besser = Artemisia pubescens Ledeb. var. gebleriana (Besser) Y. R. Ling ■

35349 Artemisia commutata Besser var. helmiana Besser = Artemisia pubescens Ledeb. ■

35350 Artemisia commutata Besser var. pallasiana Besser = Artemisia pubescens Ledeb. ■

35351 Artemisia commutata Besser var. pubescens (Ledeb.) Poljakov = Artemisia pubescens Ledeb. ■

35352 Artemisia commutata Besser var. pumila H. C. Fu et C. Y. Li = Artemisia pubescens Ledeb. ■

35353 Artemisia commutata Besser var. rotundifolia W. Wang et C. Y. Li;圆叶变蒿;Rotundileaf Changed Sagebrush ■

35354 Artemisia compacta Fisch. ex Besser = Seriphidium compactum (Fisch. ex DC.) Poljakov ■

35355 Artemisia conaensis Y. Ling et Y. R. Ling;错那蒿;Cuona Sagebrush ■

35356 Artemisia congesta Kitam.;密集蒿■☆

35357 Artemisia copa Phil.;科帕蒿■☆

35358 Artemisia coracina W. Wang = Artemisia pubescens Ledeb. var. coracina (W. Wang) Y. Ling et Y. R. Ling ■

35359 Artemisia cuneata Rydb. = Artemisia ludoviciana Nutt. ■☆

35360 Artemisia cuneifolia DC. = Artemisia japonica Thunb. ■

35361 Artemisia cuspidata Krasch.;骤尖蒿■☆

35362 Artemisia daghestanica Krasch. et Poretzky;达赫斯坦蒿■☆

35363 Artemisia dahurica (Turcz.) Poljakov = Seriphidium nitrosum (Weber ex Stechm.) Poljakov ■

35364 Artemisia dalai-lamae Krasch.;米蒿(达赖蒿,碱蒿,驴驴蒿);Dalai Sagebrush ■

35365 Artemisia demissa Krasch.;纤杆蒿(纤杆沙蒿);Thinstalk Sagebrush ■

35366 Artemisia demissa Krasch. = Artemisia stricta Edgew. ■

35367 Artemisia densifolia Filatova;密叶蒿■☆

35368 Artemisia dentata Willd. = Artemisia rupestris L. ■

35369 Artemisia depauperata Krasch.;中亚草原蒿;Poor Sagebrush ■

35370 Artemisia deserti Krasch.;荒漠蒿■☆

35371 Artemisia desertorum C. B. Clarke = Artemisia desertorum Spreng. var. foetida (Jacq. ex Besser) Y. Ling et Y. R. Ling ■

35372 Artemisia desertorum Hook. f. = Artemisia dubia Wall. ex Besser var. subdigitata (Mattf.) Y. R. Ling ■

35373 Artemisia desertorum Spreng.;沙蒿(薄蒿,草蒿,荒地蒿,荒漠蒿,漠蒿);Desert Sagebrush,Desert Wormwood ■

35374 Artemisia desertorum Spreng. = Artemisia littoricola Kitam. ■

35375 Artemisia desertorum Spreng. f. latifolia Pamp. = Artemisia desertorum Spreng. ■

35376 Artemisia desertorum Spreng. var. douglasiana Besser = Artemisia douglasiana Besser ■☆

35377 Artemisia desertorum Spreng. var. foetida (Jacq. ex Besser) Y.

Ling et Y. R. Ling;矮沙蒿■
35378 Artemisia desertorum Spreng. var. foetida (Jacq. ex DC.) Y. Ling et Y. R. Ling = Artemisia desertorum Spreng. var. foetida (Jacq. ex Besser) Y. Ling et Y. R. Ling ■
35379 Artemisia desertorum Spreng. var. jacquemontiana (Besser) DC. = Artemisia dubia Wall. ex Besser var. subdigitata (Mattf.) Y. R. Ling ■
35380 Artemisia desertorum Spreng. var. lineata G. Y. Chang;线叶沙蒿;Linear Desert Sagebrush ■
35381 Artemisia desertorum Spreng. var. lineata G. Y. Chang = Artemisia desertorum Spreng. ■
35382 Artemisia desertorum Spreng. var. macilenta (Maxim.) Pamp. = Artemisia macilenta (Maxim.) Krasch. ●■
35383 Artemisia desertorum Spreng. var. macrocephala Franch. = Artemisia dracunculus L. ■
35384 Artemisia desertorum Spreng. var. macrocephala Spreng. = Artemisia dracunculus L. ■
35385 Artemisia desertorum Spreng. var. richardsoniana (Besser) Besser = Artemisia borealis Pall. subsp. richardsoniana (Besser) Korobkov ■☆
35386 Artemisia desertorum Spreng. var. scouleriana Besser = Artemisia campestris L. subsp. pacifica (Nutt.) H. M. Hall et Clem. ■☆
35387 Artemisia desertorum Spreng. var. sprengeliana Besser = Artemisia desertorum Spreng. ■
35388 Artemisia desertorum Spreng. var. sprengeliana Besser f. gebleriana (Besser) Pamp. = Artemisia pubescens Ledeb. var. gebleriana (Besser) Y. R. Ling ■
35389 Artemisia desertorum Spreng. var. sprengeliana Besser f. gebleriana Pamp. = Artemisia pubescens Ledeb. var. gebleriana (Besser) Y. R. Ling ■
35390 Artemisia desertorum Spreng. var. sprengeliana Besser f. helmiana (Besser) Pamp. = Artemisia pubescens Ledeb. ■
35391 Artemisia desertorum Spreng. var. sprengeliana Besser f. helmiana Pamp. = Artemisia pubescens Ledeb. ■
35392 Artemisia desertorum Spreng. var. tongolensis Pamp.;东俄洛沙蒿(沙蒿);Tongol Desert Sagebrush, Tongol Desert Wormwood ■
35393 Artemisia desertorum Spreng. var. tongolensis Pamp. f. glabm Pamp. = Artemisia desertorum Spreng. var. tongolensis Pamp. ■
35394 Artemisia desertorum Spreng. var. tongolensis Pamp. f. latifolia Pamp. = Artemisia eriopoda Bunge ■
35395 Artemisia desertorum Spreng. var. willdenowiana Mattf. = Artemisia desertorum Spreng. ■
35396 Artemisia desertorum Takeda = Artemisia littoricola Kitam. ■
35397 Artemisia deversa Diels;侧蒿(笋花蒿);Declivious Sagebrush, Declivious Wormwood ■
35398 Artemisia discolor Douglas ex Besser = Artemisia michauxiana Besser ■☆
35399 Artemisia disjuncta Krasch.;矮丛光蒿;Dwarf Nitid Sagebrush ■
35400 Artemisia divaricata (Pamp.) Pamp.;叉枝蒿;Forkbranch Sagebrush ■
35401 Artemisia diversifolia Rydb. = Artemisia ludoviciana Nutt. ■☆
35402 Artemisia dolichocephala Pamp. = Artemisia myriantha Wall. ex Besser ■
35403 Artemisia dolichocephala Pamp. f. yunnanensis Pamp. = Artemisia myriantha Wall. ex Besser ■
35404 Artemisia douglasiana Besser;道格拉斯蒿(道氏蒿);Douglas Sagewort, Northwest Mugwort, Western Mugwort ■☆
35405 Artemisia dracunculiformis Krasch.;龙芋蒿■☆
35406 Artemisia dracunculina S. Watson = Artemisia dracunculus L. ■
35407 Artemisia dracunculoides Pursh;俄罗斯龙蒿;False Tarragon, Fuzzy-weed, Mugwort, Russian Tarragon ■☆
35408 Artemisia dracunculoides Pursh = Artemisia dracunculus L. ■
35409 Artemisia dracunculoides Pursh subsp. dracunculina (S. Watson) H. M. Hall et Clem. = Artemisia dracunculus L. ■
35410 Artemisia dracunculoides Pursh var. dracunculina (S. Watson) S. F. Blake = Artemisia dracunculus L. ■
35411 Artemisia dracunculus L.;龙蒿(草蒿,方溃,蒿莀,椒蒿,南星,青蒿,蛇蒿,无味蒿,狭叶青蒿,香蒿,茵陈蒿);Dragon, Dragon Sage-wort, Dragon Wormwood, Dragons, Dragon's Mugwort, Dragonwort, Estragon, False Tarragon, French Tarragon, Silky Wormwood, Tarragon, True Tarragon, Wild Tarragon ■
35412 Artemisia dracunculus L. subsp. dracunculina (S. Watson) H. M. Hall et Clem. = Artemisia dracunculus L. ■
35413 Artemisia dracunculus L. subsp. glauca (Pall. ex Willd.) H. M. Hall et Clem. = Artemisia dracunculus L. ■
35414 Artemisia dracunculus L. var. changaica (Krasch.) Y. R. Ling;杭爱龙蒿;Hang'ai Sagebrush ■
35415 Artemisia dracunculus L. var. glauca (Pall. ex Willd.) Besser = Artemisia dracunculus L. ■
35416 Artemisia dracunculus L. var. inodora Besser;无味龙蒿;Russian Tarragon ■
35417 Artemisia dracunculus L. var. inodora Besser = Artemisia dracunculus L. ■
35418 Artemisia dracunculus L. var. inodora? f. minor Kom. = Artemisia dracunculus L. ■
35419 Artemisia dracunculus L. var. inodora? f. pinnata Pamp. = Artemisia dubia Wall. ex Besser var. subdigitata (Mattf.) Y. R. Ling ■
35420 Artemisia dracunculus L. var. inodora Pamp. = Artemisia dubia Wall. ex Besser ■
35421 Artemisia dracunculus L. var. pamirica (C. Winkl.) Y. R. Ling et Humphries;帕米尔蒿(龙蒿);Pamir Tarragon ■
35422 Artemisia dracunculus L. var. qinghaiensis Y. R. Ling;青海龙蒿;Qinghai Sagebrush ■
35423 Artemisia dracunculus L. var. subdigitata (Mattf.) Y. R. Ling f. thomsonii Pamp. = Artemisia dubia Wall. ex Besser ■
35424 Artemisia dracunculus L. var. subdigitata Pamp. = Artemisia dubia Wall. ex Besser var. subdigitata (Mattf.) Y. R. Ling ■
35425 Artemisia dracunculus L. var. subdigitata Pamp. = Artemisia subdigitata Mattf. ■
35426 Artemisia dracunculus L. var. subdigitata Pamp. f. intermedia Pamp. = Artemisia dubia Wall. ex Besser var. subdigitata (Mattf.) Y. R. Ling ■
35427 Artemisia dracunculus L. var. subdigitata Pamp. subf. oblonga Pamp. = Artemisia dubia Wall. ex Besser var. subdigitata (Mattf.) Y. R. Ling ■
35428 Artemisia dracunculus L. var. turkestanica Krasch.;宽裂龙蒿●■
35429 Artemisia dubia Hara = Artemisia lavandulifolia DC. ●■
35430 Artemisia dubia Wall. ex Besser;牛尾蒿(艾蒿,荻蒿,米蒿,水蒿,野艾,浙野艾,指叶蒿,紫杆蒿);Oxtail Sagebrush ■
35431 Artemisia dubia Wall. ex Besser = Artemisia codonocephala Diels ■
35432 Artemisia dubia Wall. ex Besser = Artemisia lavandulifolia DC. ●■
35433 Artemisia dubia Wall. ex Besser f. asiatica Pamp. = Artemisia asiatica (Pamp.) Nakai ex Kitam. ■☆
35434 Artemisia dubia Wall. ex Besser f. asiatica Pamp. = Artemisia

indica Willd. ■
35435 Artemisia dubia Wall. ex Besser var. acuminata Pamp. = Artemisia indica Willd. ■
35436 Artemisia dubia Wall. ex Besser var. acuminata Pamp. f. congesta Pamp. = Artemisia indica Willd. ■
35437 Artemisia dubia Wall. ex Besser var. compacta Pamp. = Artemisia indica Willd. ■
35438 Artemisia dubia Wall. ex Besser var. gracilis Pamp. = Artemisia indica Willd. ■
35439 Artemisia dubia Wall. ex Besser var. grata (Wall. ex Besser) Pamp. = Artemisia indica Willd. ■
35440 Artemisia dubia Wall. ex Besser var. legitima (Besser) Pamp. = Artemisia indica Willd. ■
35441 Artemisia dubia Wall. ex Besser var. legitima Pamp. = Artemisia verlotorum Lamotte ■
35442 Artemisia dubia Wall. ex Besser var. legitima Pamp. f. communis subf. intermedia Pamp. = Artemisia myriantha Wall. ex Besser ■
35443 Artemisia dubia Wall. ex Besser var. legitima Pamp. f. genuina Pamp. = Artemisia myriantha Wall. ex Besser ■
35444 Artemisia dubia Wall. ex Besser var. legitima Pamp. subf. pauciflora Pamp. = Artemisia myriantha Wall. ex Besser var. pleiocephala (Pamp.) Y. R. Ling ■
35445 Artemisia dubia Wall. ex Besser var. longeracemulosa Pamp. = Artemisia myriantha Wall. ex Besser ■
35446 Artemisia dubia Wall. ex Besser var. longiracemulosa Pamp. f. tonkinensis Pamp. = Artemisia austroyunnanensis Y. Ling et Y. R. Ling ■
35447 Artemisia dubia Wall. ex Besser var. multtflora (Wall. ex Besser) Pamp. = Artemisia indica Willd. ■
35448 Artemisia dubia Wall. ex Besser var. myriantha (Wall. ex Besser) Pamp. f. meridionalis Pamp. = Artemisia myriantha Wall. ex Besser var. pleiocephala (Pamp.) Y. R. Ling ■
35449 Artemisia dubia Wall. ex Besser var. myriantha Pamp. = Artemisia robusta (Pamp.) Y. Ling et Y. R. Ling ■
35450 Artemisia dubia Wall. ex Besser var. orientalis Pamp. = Artemisia indica Willd. ■
35451 Artemisia dubia Wall. ex Besser var. orientalis Pamp. = Artemisia verlotorum Lamotte ■
35452 Artemisia dubia Wall. ex Besser var. septentrionalis Pamp. = Artemisia indica Willd. ■
35453 Artemisia dubia Wall. ex Besser var. subdigitata (Mattf.) Y. R. Ling = Artemisia subdigitata Mattf. ■
35454 Artemisia dubia Wall. ex Besser var. tegitama? f. communis? subf. puberula Pamp. = Artemisia verlotorum Lamotte ■
35455 Artemisia duthreuil-de-rhinsi Krasch.; 青藏蒿; Qingzang Sagebrush ■
35456 Artemisia edgeworthii N. P. Balakr.;直茎蒿(短叶蒿,劲直蒿); Edgeworth Sagebrush ■
35457 Artemisia edgeworthii N. P. Balakr. = Artemisia stricta Edgew. ■
35458 Artemisia edgeworthii N. P. Balakr. var. diffusa (Pamp.) Y. Ling et Y. R. Ling;披散直茎蒿■
35459 Artemisia elegans Roxb. = Artemisia scoparia Waldst. et Kit. ■
35460 Artemisia elegantissima (Pamp.) Y. R. Ling et Humphries = Artemisia indica Willd. var. elegantissima (Pamp.) Y. R. Ling et Humparies ■
35461 Artemisia elegantissima Pamp. = Artemisia indica Willd. var. elegantissima (Pamp.) Y. R. Ling et Humparies ■
35462 Artemisia elengans Roxb. = Artemisia scoparia Waldst. et Kit. ■
35463 Artemisia emeiensis Y. R. Ling; 峨眉蒿(峨参叶蒿); Emei Sagebrush ■
35464 Artemisia eranthema Bunge;中亚球序蒿(球序蒿)■
35465 Artemisia eriocephala Pamp. = Artemisia roxburghiana Besser ■
35466 Artemisia eriopoda Bunge;南牡蒿(拔拉蒿,黄蒿,米蒿,牡蒿,青蒿,田蒿,一枝蒿); Woollystalk Sagebrush, Woollystalk Wormwood ■
35467 Artemisia eriopoda Bunge subsp. jiagedaqiensis G. Y. Chang et X. J. Liu;北牡蒿;Jiagedaqi Wormwood ■
35468 Artemisia eriopoda Bunge var. gansuensis Y. R. Ling;甘肃南牡蒿;Gansu Sagebrush,Kansu Wormwood ■
35469 Artemisia eriopoda Bunge var. maritima Y. Ling et Y. R. Ling;渤海滨南牡蒿■
35470 Artemisia eriopoda Bunge var. rotundifolia (Debeaux) Y. R. Ling;圆叶南牡蒿;Roundleaf Woollystalk Wormwood ■
35471 Artemisia eriopoda Bunge var. shanxiensis Y. R. Ling;山西南牡蒿;Shanxi Sagebrush ■
35472 Artemisia erlangshanensis Y. Ling et Y. R. Ling; 二郎山蒿; Erlangshan Sagebrush ■
35473 Artemisia fadtschenkoana Krasch. = Seriphidium fedtschenkoanum (Krasch.) Poljakov ■
35474 Artemisia fadtschenkoana Krasch. var. issykkulensis Poljakov = Seriphidium issykkulense (Poljakov) Poljakov ■
35475 Artemisia falcata Rydb. = Artemisia longifolia Nutt. ■
35476 Artemisia falcata Rydb. = Artemisia ludoviciana Nutt. ■☆
35477 Artemisia falconeri C. B. Clarke = Artemisia rutifolia Stephan ex Spreng. var. altaica (Krylov) Krasch. ■
35478 Artemisia falconeri C. B. Clarke = Artemisia rutifolia Stephan ex Spreng. ●
35479 Artemisia fasciculata M. Bieb.;丛生蒿■☆
35480 Artemisia fastigiata C. Winkl. = Ajania fastigiata (C. Winkl.) Poljakov ■
35481 Artemisia fauriei Nakai;海州蒿(矮青蒿,苏北碱蒿,茵陈); Faurie Sagebrush,Haizhou Sagebrush ■
35482 Artemisia fauriei Nakai = Artemisia nakaii Pamp. ■
35483 Artemisia feddei H. Lév. et Vaniot;矮蒿(青蒿,细叶艾,小艾,小叶艾);Fedde Sagebrush,Fedde Wormwood ■
35484 Artemisia feddei H. Lév. et Vaniot = Artemisia lancea Vaniot ■
35485 Artemisia fedtschenkoana Krasch. = Seriphidium fedtschenkoanum (Krasch.) Poljakov ■
35486 Artemisia fedtschenkoana Krasch. var. issykkulense Poljakov = Seriphidium issykkulense (Poljakov) Poljakov ■
35487 Artemisia ferganense Krasch. ex Poljakov = Seriphidium ferganense (Krasch. ex Poljakov) Poljakov ■
35488 Artemisia filifolia Torr.; 沙苦艾; Sand Sage, Sand Sagebrush, Silver Sage, Silver Sagebrush, Silvery Wormwood ■☆
35489 Artemisia finita Kitag. = Seriphidium finitum (Kitag.) Y. Ling et Y. R. Ling ■
35490 Artemisia fischeriana Besser = Artemisia californica Less. ●☆
35491 Artemisia flaccidda Hand. -Mazz.;垂叶蒿;Nutantleaf Sagebrush ■
35492 Artemisia flaccidda Hand. -Mazz. var. meiguensis Y. R. Ling;齿裂垂叶蒿■
35493 Artemisia flahaultii Emb. et Maire;弗拉奥蒿■☆
35494 Artemisia flava Jurtzev = Artemisia globularia Cham. ex Besser subsp. lutea (Hultén) L. M. Shultz ■☆
35495 Artemisia foetida Jacq. ex Besser = Artemisia desertorum Spreng.

var. foetida (Jacq. ex Besser) Y. Ling et Y. R. Ling ■
35496 Artemisia foetida Jacq. ex DC. = Artemisia desertorum Spreng. var. foetida (Jacq. ex DC.) Y. Ling et Y. R. Ling ■
35497 Artemisia foliosa Nutt. = Artemisia californica Less. ●☆
35498 Artemisia forrestii W. W. Sm.;亮苞蒿;Forrest Sagebrush ■
35499 Artemisia forwoodii A. Gray = Artemisia campestris L. subsp. caudata (Michx.) H. M. Hall et Clem. ■☆
35500 Artemisia forwoodii S. Watson = Artemisia campestris L. subsp. caudata (Michx.) H. M. Hall et Clem. ■☆
35501 Artemisia fragrans Willd.;香蒿■☆
35502 Artemisia franserioides Greene;豚草蒿;Bursage Mugwort ■☆
35503 Artemisia freyniana (Pamp.) Krasch.;绿栉齿叶蒿(宽裂叶莲蒿);Freyn Sagebrush ■
35504 Artemisia freyniana (Pamp.) Krasch. f. discolor (Kom.) Kitag. = Artemisia gmelinii Weber ex Stechm. ■
35505 Artemisia frigida Willd.;冷蒿(白蒿,刚蒿,寒地蒿,兔毛蒿,小白蒿,茵陈蒿);Arctic Sagebrush, Armoise Douce, Colorado Sage, Fringed Sage, Fringed Sagebrush, Fringed Sage-wort, Fringed Wormwood, Little Wild Sage, Mountain Fringe, Mountain Sagebrush, Pasture Sagebrush, Prairie Sagebrush, Prairie Sagewort, Prairie Sage-wort, Sage Colorado, Silky Perennial Sagebr ■
35506 Artemisia frigida Willd. f. atropurpurea (Pamp.) W. Wang et C. Y. Li = Artemisia frigida Willd. var. atropurpurea Pamp. ■
35507 Artemisia frigida Willd. var. argyrophylla (Ledeb.) Trautv. = Artemisia argyrophylla Ledeb. ■
35508 Artemisia frigida Willd. var. argyrophylla Trautv. = Artemisia argyrophylla Ledeb. ■
35509 Artemisia frigida Willd. var. atropurpurea Pamp.;紫花冷蒿(黑紫冷蒿,紫冷蒿);Blackpurple Fringed Sagebrush ■
35510 Artemisia frigida Willd. var. fischeriana (Besser) DC. = Artemisia frigida Willd. ■
35511 Artemisia frigida Willd. var. intermedia Trautv. = Artemisia frigida Willd. ■
35512 Artemisia frigida Willd. var. mongolica Kitam. = Artemisia frigida Willd. ■
35513 Artemisia frigida Willd. var. typica Pamp. = Artemisia frigida Willd. ■
35514 Artemisia frigida Willd. var. willdenowiana (Besser) DC. = Artemisia frigida Willd. ■
35515 Artemisia frigida Willd. var. willdiana (Besser) DC. = Artemisia frigida Willd. ■
35516 Artemisia frigidioides H. C. Fu et Z. Y. Chu = Artemisia caespitosa Ledeb. ■
35517 Artemisia fukudo Makino;滨艾;Shore Sagebrush ■
35518 Artemisia fukudo Makino var. mokpensis Pamp. = Artemisia fauriei Nakai ■
35519 Artemisia fulgens Pamp.;亮蒿;Bright Wormwood ■
35520 Artemisia furcata M. Bieb.;叉蒿■☆
35521 Artemisia furcata M. Bieb. var. heterophylla (Besser) Hultén = Artemisia furcata M. Bieb. ■☆
35522 Artemisia gansuensis Y. Ling et Y. R. Lin var. oligantha Y. Ling et Y. R. Ling = Artemisia gansuensis Y. Ling et Y. R. Ling ■
35523 Artemisia gansuensis Y. Ling et Y. R. Ling;甘肃蒿;Gansu Sagebrush, Kansu Sagebrush ■
35524 Artemisia gansuensis Y. Ling et Y. R. Ling var. oligantha Y. Ling et Y. R. Ling;小甘肃蒿■
35525 Artemisia genipi Stechm.;吉蒿■☆
35526 Artemisia gigantea Kitam. = Artemisia montana (Nakai) Pamp. ■☆
35527 Artemisia gilvescens Miq.;湘赣艾(点叶艾,湘赣艾蒿,湘赣蒿);Lightyellow Sagebrush ■
35528 Artemisia giraldii Pamp.;华北米蒿(艾蒿,灰蒿,吉氏蒿,米棉蒿);Girald Sagebrush ■
35529 Artemisia giraldii Pamp. var. longipedunculata Y. R. Ling;长梗米蒿■
35530 Artemisia glabella Kar. et Kir.;亮绿蒿;Hairless Sagebrush ■
35531 Artemisia glabrata Wall. ex Besser = Artemisia japonica Thunb. ■
35532 Artemisia glabrata Wall. ex DC. = Artemisia parviflora Buch.-Ham. ex Roxb. ■
35533 Artemisia glabrata Wight = Artemisia parviflora Buch.-Ham. ex Roxb. ■
35534 Artemisia glacialis L.;冰川蒿(丝蒿);Mugwort, Wormwood ■☆
35535 Artemisia glauca Forbes et Hemsl. = Artemisia giraldii Pamp. ■
35536 Artemisia glauca Hook. f. = Artemisia dubia Wall. ex Besser var. subdigitata (Mattf.) Y. R. Ling ■
35537 Artemisia glauca Maxim. = Artemisia dracunculus L. ■
35538 Artemisia glauca Pall.;灰蒿(灰绿蒿,狭叶青蒿);Greyblue Sagebrush, Greyblue Wormwood ■☆
35539 Artemisia glauca Pall. = Artemisia dracunculus L. ■
35540 Artemisia glauca Pall. = Artemisia dubia Wall. ex Besser var. subdigitata (Mattf.) Y. R. Ling ■
35541 Artemisia glauca Pall. ex Willd. = Artemisia dracunculus L. ■
35542 Artemisia glauca Pall. ex Willd. = Artemisia giraldii Pamp. ■
35543 Artemisia glauca Pall. ex Willd. var. dracunculina (S. Watson) Fernald = Artemisia dracunculus L. ■
35544 Artemisia glauca Pall. ex Willd. var. megacephala B. Boivin = Artemisia dracunculus L. ■
35545 Artemisia glauca Pall. var. dracunculina (S. Watson) Fernald = Artemisia dracunculus L. ■
35546 Artemisia globosoides Y. Ling et Y. R. Ling;假球蒿;Ballshape Sagebrush ■
35547 Artemisia globularia Cham. ex Besser;小球蒿■☆
35548 Artemisia globularia Cham. ex Besser subsp. lutea (Hultén) L. M. Shultz;黄小球蒿■☆
35549 Artemisia globularia Cham. ex Besser var. lutea Hultén = Artemisia globularia Cham. ex Besser subsp. lutea (Hultén) L. M. Shultz ■☆
35550 Artemisia glomerata Ledeb.;聚头蒿;Congested Sagewort ■☆
35551 Artemisia glomerata Ledeb. var. leontopodioides (Fisch. ex Besser) Kitam. = Artemisia glomerata Ledeb. ■☆
35552 Artemisia glomerata Ledeb. var. subglabra Hultén = Artemisia glomerata Ledeb. ■☆
35553 Artemisia glutinosa DC. = Artemisia campestris L. subsp. glutinosa (DC.) Batt. ■☆
35554 Artemisia glutinosa Gay ex Bess. = Artemisia campestris L. ■
35555 Artemisia gmelinii Fisch. ex Besser = Artemisia sacrorum Ledeb. ■
35556 Artemisia gmelinii Fisch. ex Besser var. discolor (Kom.) Nakai = Artemisia gmelinii Weber ex Stechm. ■
35557 Artemisia gmelinii Fisch. ex Besser var. incana (Besser) H. C. Fu = Artemisia sacrorum Ledeb. var. incana (Besser) Y. R. Ling ■
35558 Artemisia gmelinii Fisch. ex Besser var. legitima Besser = Artemisia gmelinii Weber ex Stechm. ■
35559 Artemisia gmelinii Fisch. ex Besser var. messerschmidtiana (Besser) Poljakov = Artemisia sacrorum Ledeb. var. messerschmidtiana (Besser) Y. R. Ling ●

35560 Artemisia gmelinii Weber ex Stechm.;细裂叶莲蒿(白莅蒿,白莲蒿,两色万年蒿,铁杆蒿,万年蒿,小裂齿蒿,茵陈);Gmelin Sagebrush,Gmelin's Wormwood,Thinlobed Sagebrush ■

35561 Artemisia gmelinii Weber ex Stechm. = Artemisia sacrorum Ledeb. ■

35562 Artemisia gmelinii Weber ex Stechm. var. biebersteiniana Besser = Artemisia gmelinii Weber ex Stechm. ■

35563 Artemisia gmelinii Weber ex Stechm. var. discolor (Kom.) Nakai = Artemisia sacrorum Ledeb. var. incana (Besser) Y. R. Ling ■

35564 Artemisia gmelinii Weber ex Stechm. var. discolor (Kom.) Nakai = Artemisia gmelinii Weber ex Stechm. ■

35565 Artemisia gmelinii Weber ex Stechm. var. incana (Besser) H. C. Fu = Artemisia sacrorum Ledeb. var. incana (Besser) Y. R. Ling ■

35566 Artemisia gmelinii Weber ex Stechm. var. incana (Besser) H. C. Fu = Artemisia sacrorum Ledeb. var. messerschmidtiana (Besser) Y. R. Ling ■

35567 Artemisia gmelinii Weber ex Stechm. var. legitima Besser = Artemisia gmelinii Weber ex Stechm. ■

35568 Artemisia gmelinii Weber ex Stechm. var. messerschmidtiana (Besser) Poljakov = Artemisia sacrorum Ledeb. var. messerschmidtiana (Besser) Y. R. Ling ■

35569 Artemisia gmelinii Weber ex Stechm. var. vestita (Kom.) Nakai = Artemisia sacrorum Ledeb. var. incana (Besser) Y. R. Ling ■

35570 Artemisia gnaphaloides Nutt.;鼠麴草蒿;Prairie Sage, White Mugwort,Wild Sage ■☆

35571 Artemisia gnaphaloides Nutt. = Artemisia ludoviciana Nutt. ■☆

35572 Artemisia gobica (Krasch. ex Poljakov) Grubov = Seriphidium nitrosum (Weber ex Stechm.) Poljakov var. gobicum (Krasch.) Y. R. Ling ●■

35573 Artemisia gongshanensis Y. R. Ling et Humphries;贡山蒿;Gongshan Sagebrush ■

35574 Artemisia gracilenta A. Nelson = Artemisia ludoviciana Nutt. subsp. candicans (Rydb.) D. D. Keck ■☆

35575 Artemisia gracilescens Krasch. et Iljin = Seriphidium gracilescens (Krasch. et Iljin) Poljakov ■

35576 Artemisia granatensis Boiss. ex DC.;格兰蒿■☆

35577 Artemisia grata Heyne ex Steud. = Artemisia indica Willd. ■

35578 Artemisia grata Wall. ex Besser = Artemisia indica Willd. ■

35579 Artemisia grenardii Franch. = Seriphidium grenardii (Franch.) Y. R. Ling et Humphries ■

35580 Artemisia griffithiana Boiss. = Artemisia macrocephala Jacq. ex Besser ■

35581 Artemisia griffithiana Boiss. = Seriphidium aucheri (Boiss.) Y. Ling et Y. R. Ling ■

35582 Artemisia griffithiana C. B. Clarke = Seriphidium aucheri (Boiss.) Y. Ling et Y. R. Ling ■

35583 Artemisia grisea Pamp. = Artemisia lavandulifolia DC. ●■

35584 Artemisia gyangzeensis Y. Ling et Y. R. Ling;江孜蒿;Jiangzi Sagebrush ■

35585 Artemisia gyitangensis Y. Ling et Y. R. Ling;吉塘蒿;Jitang Sagebrush ■

35586 Artemisia gypsacea Krasch.,Popov et Lincz. ex Poljakov;钙蒿■☆

35587 Artemisia haichowensis C. C. Chang = Artemisia fauriei Nakai ■

35588 Artemisia halimodendron Ledeb. ex Hook. f. = Artemisia xigazeensis Y. Ling et Y. R. Ling ■

35589 Artemisia halimodendron Ledeb. ex Hook. f. var. salsoloides? f. genuina Pamp. = Artemisia xigazeensis Y. Ling et Y. R. Ling ■

35590 Artemisia hallaisanensis Nakai var. formosana Pamp. = Artemisia capillaris Thunb. ■

35591 Artemisia hallaisanensis Nakai var. hancei Pamp. = Artemisia hancei (Pamp.) Y. Ling et Y. R. Ling ■

35592 Artemisia hallaisanensis Nakai var. philippinensis Pamp. = Artemisia capillaris Thunb. ■

35593 Artemisia hallaisanensis Nakai var. philippinensis Pamp. f. parvula Pamp. = Artemisia capillaris Thunb. ■

35594 Artemisia hallaisanensis Nakai var. philippinensis Pamp. f. swatowiana Pamp. = Artemisia capillaris Thunb. ■

35595 Artemisia halodendron Turcz. ex Besser;盐蒿(差把嘎蒿,差不嘎蒿,褐沙蒿,沙蒿,沙漠嘎,沙漠蒿);Halophilous Wormwood, Saltliving Sagebrush,Saltliving Wormwood ●

35596 Artemisia hancei (Pamp.) Y. Ling et Y. R. Ling;雷琼牡蒿;Hance Sagebrush ■

35597 Artemisia handel-mazzetii Pamp. = Artemisia argyi H. Lév. et Vaniot ■

35598 Artemisia hedinii Ostenf. et Paulsen;臭蒿(海定蒿,狼尾巴蒿,牛尾蒿);Armoise Bisannuelle, Biannual Wormwood, Biennial Wormwood,Hedin Sagebrush,Hedin's Wormwood,Slender Mugwort ■

35599 Artemisia heptapotamica Poljakov = Seriphidium heptapotamicum (Poljakov) Y. Ling et Y. R. Ling ■

35600 Artemisia herba-alba Asso;白草蒿(草白蒿);Bible Wormwood ■☆

35601 Artemisia herba-alba Asso var. aurasiaca Maire = Artemisia herba-alba Asso ■☆

35602 Artemisia herba-alba Asso var. huguetii (Caball.) Maire = Artemisia huguetii Caball. ■☆

35603 Artemisia herba-alba Asso var. laxiflora Boiss. = Artemisia herba-alba Asso ■☆

35604 Artemisia herba-alba Asso var. oranensis Debeaux = Artemisia oranensis (Debeaux) Filatova ●☆

35605 Artemisia herba-alba Asso var. saharae (Pomel) Quézel et Santa = Artemisia saharae Pomel ■☆

35606 Artemisia herbacea Ehrh.;绿蒿■☆

35607 Artemisia herriotii Rydb. = Artemisia ludoviciana Nutt. ■☆

35608 Artemisia heterophylla Besser;异叶蒿■☆

35609 Artemisia heterophylla Nutt. = Artemisia douglasiana Besser ■☆

35610 Artemisia heterophylla Nutt. = Artemisia suksdorfii Piper ■☆

35611 Artemisia hirsuta Rottler = Cyathocline purpurea (Buch. -Ham. ex De Don) Kuntze ■

35612 Artemisia hololeuca M. Bieb. ex Besser;白蒿■☆

35613 Artemisia holosericea Ledeb. = Artemisia sericea Weber ex Stechm. ■

35614 Artemisia holosericea Ledeb. var. grandiflora Ledeb. = Artemisia sericea Weber ex Stechm. ■

35615 Artemisia holosericea Ledeb. var. parviflora Ledeb. = Artemisia sericea Weber ex Stechm. ■

35616 Artemisia hookeriana Besser = Artemisia tilesii Ledeb. ■

35617 Artemisia huguetii Caball.;于盖蒿■☆

35618 Artemisia hultenii M. M. Maximova = Artemisia tilesii Ledeb. ■

35619 Artemisia hyperborea Rydb. = Artemisia furcata M. Bieb. ■☆

35620 Artemisia hypoleuca Edgew. = Artemisia roxburghiana Besser ■

35621 Artemisia ifranensis Didier;伊夫尼蒿■☆

35622 Artemisia igniaria Maxim.;歧茎蒿(白艾,锯叶家蒿,蒌蒿,篓蒿,歧茎艾,野艾);Forkstem Sagebrush,Forkstem Wormwood ■

35623 Artemisia igniaria Maxim. var. typica Pamp. = Artemisia igniaria Maxim. ■

35624 Artemisia igniaria Maxim. var. typica Pamp. f. pubescens Pamp. = Artemisia igniaria Maxim. ■

35625 Artemisia igniaria Maxim. var. yunnanensis (Jeffrey) Pamp. = Artemisia yunnanensis Jeffrey ex Diels ■

35626 Artemisia imponens Pamp.;锈苞蒿;Rustbract Sagebrush ■

35627 Artemisia incana (L.) Druce;灰色蛔蒿(灰白毛蒿,蛔蒿)■☆

35628 Artemisia incana Druce = Artemisia incana (L.) Druce ■☆

35629 Artemisia incana Keller = Artemisia incana Druce ■☆

35630 Artemisia incisa Pamp.;尖裂叶蒿■

35631 Artemisia incompta Nutt. = Artemisia ludoviciana Nutt. subsp. incompta (Nutt.) D. D. Keck ■☆

35632 Artemisia inculta Delile;荒地蒿■☆

35633 Artemisia inculta Sieber ex DC. = Artemisia inculta Delile ■☆

35634 Artemisia indica Willd.;五月艾(艾,艾叶,白艾,白蒿,草蓬,黑蒿,鸡脚艾,蒌,蒌蒿,蔏蒌,生艾,狭叶艾,小艾蒿,小北艾,小野艾,野艾蒿,印度蒿,指叶艾);India Sagebrush ■

35635 Artemisia indica Willd. var. elegantissima (Pamp.) Y. R. Ling et Humparies;雅致艾;Elegant Sagebrush ■

35636 Artemisia indica Willd. var. exilis Pamp. = Artemisia roxburghiana Besser ■

35637 Artemisia indica Willd. var. heyneana Wall. ex Besser = Artemisia indica Willd. ■

35638 Artemisia indica Willd. var. maximowiczii (Nakai) H. Hara;马氏五月艾■☆

35639 Artemisia indica Willd. var. momiyamae (Kitam.) H. Hara = Artemisia momiyamae Kitam. ■☆

35640 Artemisia indica Willd. var. multiflora Wall. ex Besser = Artemisia indica Willd. ■

35641 Artemisia indica Willd. var. nepalensis Besser = Artemisia indica Willd. ■

35642 Artemisia indica Willd. var. orientalis (Pamp.) H. Hara = Artemisia asiatica (Pamp.) Nakai ex Kitam. ■☆

35643 Artemisia indica Willd. var. orientalis (Pamp.) H. Hara = Artemisia indica Willd. ■

35644 Artemisia inodora M. Bieb. = Artemisia dracunculus L. ■

35645 Artemisia inodora M. Bieb. = Artemisia marschalliana Spreng. ■

35646 Artemisia inodora Willd. = Artemisia dracunculus L. ■

35647 Artemisia insularis Kitam. = Artemisia borealis Pall. ■

35648 Artemisia integrifolia L.;柳叶蒿(九牛草,科马罗夫蒿,柳蒿);Willowleaf Sagebrush,Willowleaf Wormwood ■

35649 Artemisia integrifolia L. f. subulata (Nakai) Kitag. = Artemisia subulata Nakai ■

35650 Artemisia integrifolia L. var. stolonifera (Maxim.) Pamp. = Artemisia stolonifera (Maxim.) Kom. ■

35651 Artemisia integrifolia L. var. stolonifera Pamp. = Artemisia stolonifera (Maxim.) Kom. ■

35652 Artemisia integrifolia L. var. subulata (Nakai) Pamp. = Artemisia subulata Nakai ■

35653 Artemisia integrifolia L. var. subulata Pamp. = Artemisia subulata Nakai ■

35654 Artemisia integrifolia L. var. typica ? f. bohnhofii Pamp. = Artemisia integrifolia L. ■

35655 Artemisia integrifolia L. var. typica ? f. genuina Pamp. = Artemisia integrifolia L. ■

35656 Artemisia integrifolia L. var. typica ? f. siuzievii Pamp. = Artemisia integrifolia L. ■

35657 Artemisia integrifolia L. var. typica ? f. transiens Pamp. = Artemisia integrifolia L. ■

35658 Artemisia integrifolia Nakai = Artemisia viridissima (Kom.) Pamp. ■

35659 Artemisia intramongolica H. C. Fu = Artemisia halodendron Turcz. ex Besser ●■

35660 Artemisia intramongolica H. C. Fu var. microphylla H. C. Fu = Artemisia halodendron Turcz. ex Besser ●■

35661 Artemisia intricata Franch. = Elachanthemum intricatum (Franch.) Y. Ling et Y. R. Ling ■

35662 Artemisia intricata Franch. = Stilpnolepis intricata (Franch.) C. Shih ■

35663 Artemisia intromongolica H. C. Fu;褐沙蒿;Brun Wormwood ■

35664 Artemisia intromongolica H. C. Fu = Artemisia halodendron Turcz. ex Besser ●■

35665 Artemisia intromongolica H. C. Fu var. microphylla H. C. Fu;小叶褐沙蒿■

35666 Artemisia intromongolica H. C. Fu var. microphylla H. C. Fu = Artemisia halodendron Turcz. ex Besser ●■

35667 Artemisia issykkulensis Poljakov = Seriphidium issykkulense (Poljakov) Poljakov ■

35668 Artemisia iwayomogi Kitam. = Artemisia gmelinii Weber ex Stechm. ■

35669 Artemisia iwayomogi Kitam. = Artemisia sacrorum Ledeb. ■

35670 Artemisia iwayomogi Kitam. f. laciniiformis (Nakai) Kitam. = Artemisia sacrorum Ledeb. ■

35671 Artemisia jacutica Drobow;屋久岛蒿(雅库蒿)■☆

35672 Artemisia japonica Lauener = Artemisia parviflora Buch. -Ham. ex Roxb. ■

35673 Artemisia japonica Schmidt = Artemisia selengensis Turcz. ex Besser var. shansiensis Y. R. Ling ■

35674 Artemisia japonica Thunb.;牡蒿(白花蒿,布菜,匙叶艾,臭艾,鹅草药,猴掌草,花艾草,花等草,鸡肉菜,假柴胡,脚板蒿,菊叶柴胡,老鸦青,流尿蒿,流水蒿,六月雪,马根柴,马连蒿,牡莪,奶疳药,牛尾蒿,齐头蒿,青蒿,日本牡蒿,沙祖叶,水辣菜,铁菜子,铁蒿,茼蒿,土柴胡,碗头青,蔚,细艾,香蒿,香青蒿,熊掌草,熊掌蒿,野塘蒿,油艾,油蒿,油蓬,油蓬白花蒿);Japan Sagebrush,Japanese Wormwood ■

35675 Artemisia japonica Thunb. = Artemisia keiskeana Miq. ■

35676 Artemisia japonica Thunb. = Artemisia parviflora Buch. -Ham. ex Roxb. ■

35677 Artemisia japonica Thunb. f. eriopoda (Bunge) Pamp. = Artemisia eriopoda Bunge ■

35678 Artemisia japonica Thunb. f. eriopoda Pamp. = Artemisia angustissima Nakai ■

35679 Artemisia japonica Thunb. f. eriopoda Pamp. = Artemisia parviflora Buch. -Ham. ex Roxb. subf. tongtchuanensis (H. Lév.) Pamp. ■

35680 Artemisia japonica Thunb. f. eriopoda Pamp. = Artemisia parviflora Buch. -Ham. ex Roxb. ■

35681 Artemisia japonica Thunb. f. eriopoda Pamp. subf. angustissima (Nakai) Pamp. = Artemisia angustissima Nakai ■

35682 Artemisia japonica Thunb. f. eriopoda Pamp. subf. tongtchouanensis Pamp. = Artemisia parviflora Buch. -Ham. ex Roxb. ■

35683 Artemisia japonica Thunb. f. manshurica Kom. = Artemisia japonica Thunb. ■

35684 Artemisia japonica Thunb. f. resedifolia Takeda = Artemisia littoricola Kitam. ■

35685 Artemisia japonica Thunb. f. rotundifolia (Debeaux) Franch. = Artemisia eriopoda Bunge var. rotundifolia (Debeaux) Y. R. Ling ■
35686 Artemisia japonica Thunb. f. rotundifolia Franch. = Artemisia eriopoda Bunge var. rotundifolia (Debeaux) Y. R. Ling ■
35687 Artemisia japonica Thunb. f. typica Nakai = Artemisia japonica Thunb. ■
35688 Artemisia japonica Thunb. subf. angustissima Pamp. = Artemisia angustissima Nakai ■
35689 Artemisia japonica Thunb. subf. intermidia Pamp. = Artemisia japonica Thunb. ■
35690 Artemisia japonica Thunb. subf. laxiflora Pamp. = Artemisia japonica Thunb. ■
35691 Artemisia japonica Thunb. subf. resedifolia Pamp. = Artemisia eriopoda Bunge ■
35692 Artemisia japonica Thunb. subf. spathulata Pamp. = Artemisia japonica Thunb. ■
35693 Artemisia japonica Thunb. subsp. littoricola (Kitam.) Kitam. = Artemisia littoricola Kitam. ■
35694 Artemisia japonica Thunb. var. angustissima (Nakai) Kitam. = Artemisia angustissima Nakai ■
35695 Artemisia japonica Thunb. var. desertorum (Spreng.) Maxim. = Artemisia desertorum Spreng. ■
35696 Artemisia japonica Thunb. var. desertorum Matsum. = Artemisia littoricola Kitam. ■
35697 Artemisia japonica Thunb. var. desertorum Maxim. = Artemisia littoricola Kitam. ■
35698 Artemisia japonica Thunb. var. eriopoda (Besser) Kom. = Artemisia eriopoda Bunge ■
35699 Artemisia japonica Thunb. var. eriopoda (Bunge) Kom. = Artemisia eriopoda Bunge ■
35700 Artemisia japonica Thunb. var. grandifolia f. vestita Pamp. = Artemisia capillaris Thunb. ■
35701 Artemisia japonica Thunb. var. hainanensis Y. R. Ling;海南牡蒿;Hainan Sagebrush ■
35702 Artemisia japonica Thunb. var. lanata Pamp. = Artemisia japonica Thunb. ■
35703 Artemisia japonica Thunb. var. macrocephala Pamp. = Artemisia japonica Thunb. subsp. littoricola (Kitam.) Kitam. ■
35704 Artemisia japonica Thunb. var. macrocephala Pamp. = Artemisia japonica Thunb. ■
35705 Artemisia japonica Thunb. var. macrocephala Pamp. f. chinensis Pamp. = Artemisia eriopoda Bunge ■
35706 Artemisia japonica Thunb. var. macrocephala Pamp. f. sachalinensis Pamp. = Artemisia littoricola Kitam. ■
35707 Artemisia japonica Thunb. var. manshurica (Kom.) Kitag. = Artemisia japonica Thunb. ■
35708 Artemisia japonica Thunb. var. manshurica Kom. = Artemisia manshurica (Kom.) Kom. ■
35709 Artemisia japonica Thunb. var. microcephala Pamp. = Artemisia japonica Thunb. ■
35710 Artemisia japonica Thunb. var. myriocephala Pamp. = Artemisia japonica Thunb. ■
35711 Artemisia japonica Thunb. var. myriocephala Pamp. f. silvestris Pamp. = Artemisia japonica Thunb. ■
35712 Artemisia japonica Thunb. var. parviflora (Buch. -Ham. ex Roxb.) Pamp. = Artemisia parviflora Buch. -Ham. ex Roxb. ■
35713 Artemisia japonica Thunb. var. parviflora (Buch. -Ham. ex Roxb.) Pamp. = Artemisia japonica Thunb. ■
35714 Artemisia japonica Thunb. var. parviflora Pamp. = Artemisia parviflora Buch. -Ham. ex Roxb. ■
35715 Artemisia japonica Thunb. var. rotundifolia Debeaux = Artemisia eriopoda Bunge var. rotundifolia (Debeaux) Y. R. Ling ■
35716 Artemisia japonica Thunb. var. rotundifolia Debeaux f. genuina Pamp. = Artemisia eriopoda Bunge var. rotundifolia (Debeaux) Y. R. Ling ■
35717 Artemisia japonica Thunb. var. rotundifolia Debeaux subf. elata Pamp. = Artemisia eriopoda Bunge var. rotundifolia (Debeaux) Y. R. Ling ■
35718 Artemisia jeffreyana H. Lév. = Artemisia yunnanensis Jeffrey ex Diels ■
35719 Artemisia jilongensis Y. R. Ling et Humphries;吉隆蒿;Jilong Sagebrush ■
35720 Artemisia juncea Kar. et Kir. = Seriphidium junceum (Kar. et Kir.) Poljakov ■
35721 Artemisia juncea Kar. et Kir. var. macrosciadium Poljakov = Seriphidium junceum (Kar. et Kir.) Poljakov var. macrosciadium (Poljakov) Y. Ling et Y. R. Ling ■
35722 Artemisia kabylica Chabert = Artemisia alba Turra subsp. kabylica (Chabert) Greuter ■☆
35723 Artemisia kanashiroi Kitam.;狭裂白蒿(白蒿);Kanashiro Sagebrush ■
35724 Artemisia kangmarensis Y. Ling et Y. R. Ling;康马蒿;Kangmar Sagebrush ■
35725 Artemisia kansana Britton = Artemisia carruthii A. W. Wood ex Carruth. ■☆
35726 Artemisia karatavica Krasch. et Abolin ex Poljakov = Seriphidium karatavicum (Krasch. et Abolin ex Poljakov) Y. Ling et Y. R. Ling ■
35727 Artemisia kaschgarica (Krasch.) Poljakov var. dshungaricum Filatova = Seriphidium kaschgaricum (Krasch.) Poljakov var. dshungaricum (Filatova) Y. R. Ling ■
35728 Artemisia kaschgarica Krasch. = Seriphidium kaschgaricum (Krasch.) Poljakov ■
35729 Artemisia kaschgarica Krasch. = Seriphidium kaschgaricum (Krasch.) Poljakov var. dshungaricum (Filatova) Y. R. Ling ■
35730 Artemisia kaschgarica Krasch. var. dshungaricum Filatova = Seriphidium kaschgaricum (Krasch.) Poljakov var. dshungaricum (Filatova) Y. R. Ling ■
35731 Artemisia kaschgarica var. dshungarica Filatova = Seriphidium kaschgaricum (Krasch.) Poljakov var. dshungaricum (Filatova) Y. R. Ling ■
35732 Artemisia kawakamii Hayata;山艾(白艾,川上氏艾);Kawakami Sagebrush ■
35733 Artemisia keiskeana Miq.;庵闾(庵蒿,庵芦,庵闾草,庵闾蒿,庵闾子,臭蒿,覆闾,狗乳花,淹闾);Keiske Sagebrush, Keiske Wormwood ■
35734 Artemisia keiskeana Miq. f. hirtella Nakai = Artemisia keiskeana Miq. ■
35735 Artemisia keiskeana Miq. f. typica Nakai = Artemisia keiskeana Miq. ■
35736 Artemisia keiskeana Miq. subf. rotundifolia Pamp. = Artemisia keiskeana Miq. ■
35737 Artemisia kitadakensis H. Hara et Kitam.;北岳蒿■☆
35738 Artemisia klementzae Krasch. = Artemisia klementzae Krasch. ex Leonova ●■

35739 Artemisia klementzae Krasch. ex Leonova;蒙古沙地蒿(小叶褐沙蒿);Mongolsandy Sagebrush ●■

35740 Artemisia knorringiana Krasch.;克诺林蒿■☆

35741 Artemisia kohatica K. W. Klatt = Artemisia scoparia Waldst. et Kit. ■

35742 Artemisia koidzumii Nakai;小泉蒿■☆

35743 Artemisia koidzumii Nakai var. laciniata (Nakai) Kitam. = Artemisia stolonifera (Maxim.) Kom. ■

35744 Artemisia koidzumii Nakai var. manchurica Pamp. = Artemisia brachyphylla Kitam. ■

35745 Artemisia koidzumii Nakai var. manchurica Pamp. = Artemisia stolonifera (Maxim.) Kom. ■

35746 Artemisia koidzumii Nakai var. megaphylla Kitam.;大叶小泉氏蒿■☆

35747 Artemisia komarovii Poljakov = Artemisia integrifolia L. ■

35748 Artemisia kopetdaghensis Krasch., Popov et Lincz. ex Poljakov;科佩特蒿■☆

35749 Artemisia koreana Nakai = Artemisia sieversiana Ehrh. ex Willd. ■

35750 Artemisia korovinii Poljakov = Seriphidium korovinii (Poljakov) Poljakov ■

35751 Artemisia kruhsiana Besser = Artemisia lagocephala (Fisch. ex Besser) Fisch. ex DC. ●■

35752 Artemisia kulbadica Boiss. et Buhse;库尔巴德蒿■☆

35753 Artemisia kurramensis Qazilb.;夸朗蛔蒿(巴基斯坦蛔蒿)■☆

35754 Artemisia kuschakewiczii C. Winkl.;掌裂蒿;Kuschakewicz Sagebrush ■

35755 Artemisia laciniata C. B. Clarke = Artemisia vestita Wall. ex Besser ●■

35756 Artemisia laciniata C. B. Clarke var. glabriuscula Ledeb. = Artemisia tanacetifolia L. ■

35757 Artemisia laciniata C. B. Clarke var. latifolia (Ledeb.) Maxim. = Artemisia latifolia Ledeb. ■

35758 Artemisia laciniata C. B. Clarke var. turczaninowiana Besser = Artemisia phaeolepis Krasch. ■

35759 Artemisia laciniata Willd.;细裂叶蒿■

35760 Artemisia laciniata Willd. = Artemisia tanacetifolia L. ■

35761 Artemisia laciniata Willd. f. racemosa Krylov = Artemisia phaeolepis Krasch. ■

35762 Artemisia laciniata Willd. subsp. parryi (A. Gray) W. A. Weber;帕里蒿;Parry Sagewort ■☆

35763 Artemisia laciniata Willd. var. latifolia (Ledeb.) Maxim. = Artemisia latifolia Ledeb. ■

35764 Artemisia laciniata Willd. var. latifolia (Ledeb.) Maxim. f. maximoviczii Pamp. = Artemisia maximowicziana (F. Schumach.) Krasch. ex Poljakov ●■

35765 Artemisia laciniata Willd. var. latifolia Maxim. = Artemisia tanacetifolia L. ■

35766 Artemisia laciniata Willd. var. turtschaninoviana Besser = Artemisia phaeolepis Krasch. ■

35767 Artemisia laciniata Willd. var. turtschaninoviana DC. = Artemisia phaeolepis Krasch. ■

35768 Artemisia laciniatiformis Kom. = Artemisia laciniata Willd. ■

35769 Artemisia lactiflora Wall. = Artemisia lactiflora Wall. ex DC. ■

35770 Artemisia lactiflora Wall. ex DC.;白苞蒿(白苍蒿,白花艾,白花蒿,白米蒿,肺痨草,广东刘寄奴,红姨妈菜,鸡甜菜,鸡鸭脚艾,角菜,刘寄奴,秦州庵闾子,四季菜,四季花,甜艾,甜菜子,土鳅菜,土三七,鸭脚艾,鸭脚菜,野红芹菜,野勒菜,野芹菜,珍珠菜,珍珠菊,真珠菜,真珠花菜);Ghstplant Sagebrush,Ghstplant Wormwood,Lady's Smock,White Mugwort,White Wormwood ■

35771 Artemisia lactiflora Wall. ex DC. f. genuina Pamp. = Artemisia lactiflora Wall. ex DC. ■

35772 Artemisia lactiflora Wall. ex DC. f. incisa Pamp. = Artemisia lactiflora Wall. ex DC. var. incisa (Pamp.) Y. Ling et Y. R. Ling ■

35773 Artemisia lactiflora Wall. ex DC. f. septemlobata (H. Lév.) Pamp. = Artemisia lactiflora Wall. ex DC. ■

35774 Artemisia lactiflora Wall. ex DC. var. incisa (Pamp.) Y. Ling et Y. R. Ling;细裂叶白苞蒿(细裂白苞蒿);Thinlobed Ghstplant Sagebrush ■

35775 Artemisia lactiflora Wall. ex DC. var. taibaishanensis X. D. Cui;长叶羽裂蒿;Taibaishan Ghstplant Sagebrush ■

35776 Artemisia lactiflora Willd. var. taibaishanensis X. D. Cui = Artemisia lactiflora Wall. ex DC. var. taibaishanensis X. D. Cui ■

35777 Artemisia lagocephala (Fisch. ex Besser) DC. = Artemisia lagocephala (Fisch. ex Besser) Fisch. ex DC. ●■

35778 Artemisia lagocephala (Fisch. ex Besser) DC. var. tafelii (Mattf.) Pamp. = Artemisia tafelii Mattf. ■

35779 Artemisia lagocephala (Fisch. ex Besser) Fisch. ex DC.;白山蒿(白蒿,北亚蒿,狼毒黄蒿,石艾,狭叶蒿,银叶艾);Beach Wormwood, Bud Sagebrush, Dusty Miller, Hairyhead Sagebrush, Hairyhead Wormwood, Hoary Mugwort, Old Woman, Steller's Wormwood ●■

35780 Artemisia lagocephala (Fisch. ex Besser) Fisch. ex DC. f. triloba (Ledeb.) Kitag. = Artemisia lagocephala (Fisch. ex Besser) Fisch. ex DC. ●■

35781 Artemisia lagocephala (Fisch. ex Besser) Fisch. ex DC. var. besseriana Pamp. = Artemisia lagocephala (Fisch. ex Besser) Fisch. ex DC. ●■

35782 Artemisia lagocephala (Fisch. ex Besser) Fisch. ex DC. var. lithophila (Turcz. ex DC.) Y. R. Ling = Artemisia lagocephala (Fisch. ex Besser) DC. ●■

35783 Artemisia lagocephala (Fisch. ex Besser) Fisch. ex DC. var. tafelii (Mattf.) Pamp. = Artemisia tafelii Mattf. ■

35784 Artemisia lagocephala (Fisch. ex Besser) Fisch. ex DC. var. triloba (Ledeb.) Herder = Artemisia lagocephala (Fisch. ex Besser) Fisch. ex DC. ●■

35785 Artemisia lagopus Fisch. ex Besser;免足蒿■☆

35786 Artemisia lanata DC. = Artemisia caucasica Willd. ■☆

35787 Artemisia lanata DC. var. alpina DC. = Artemisia frigida Willd. ■

35788 Artemisia lancea Vaniot;细艾叶(矮蒿,牛尾蒿,青蒿,小艾,小蓬蒿,野艾蒿);Short Sagebrush ■

35789 Artemisia latifolia Ledeb.;宽叶蒿;Broadleaf Sagebrush ■

35790 Artemisia latifolia Ledeb. subsp. maximowiczii (F. Schmidt) Vorosch. = Artemisia maximowicziana (F. Schumach.) Krasch. ex Poljakov ●■

35791 Artemisia latifolia Ledeb. var. maximoviczii F. Schmidt = Artemisia maximowicziana (F. Schumach.) Krasch. ex Poljakov ●■

35792 Artemisia latifolia Maxim. = Artemisia medioxima Krasch. ex Poljakov ■

35793 Artemisia latiloba (Nutt.) Rydb. = Artemisia ludoviciana Nutt. subsp. candicans (Rydb.) D. D. Keck ■☆

35794 Artemisia lavandulifolia DC.;野艾蒿(艾叶,陈艾,苦艾,祈艾,蕲艾,细叶艾,狭叶艾,小叶艾,野艾,荫地蒿);Lavenderleaf Sagebrush,Lavenderleaf Wormwood ●■

35795 Artemisia lavandulifolia DC. = Artemisia codonocephala Diels ■

35796 Artemisia lavandulifolia DC. = Artemisia feddei H. Lév. et Vaniot ■

35797 Artemisia lavandulifolia DC. var. feddei (H. Lév. et Vaniot) Pamp. = Artemisia lancea Vaniot ■

35798 Artemisia lavandulifolia DC. var. feddei (H. Lév. et Vaniot) Pamp. f. effusa Pamp. = Artemisia lancea Vaniot ■

35799 Artemisia lavandulifolia DC. var. feddei (H. Lév. et Vaniot) Pamp. f. effusa Pamp. subf. angusta Pamp. = Artemisia lancea Vaniot ■

35800 Artemisia lavandulifolia DC. var. feddei (H. Lév. et Vaniot) Pamp. f. genuina Pamp. = Artemisia lancea Vaniot ■

35801 Artemisia lavandulifolia DC. var. feddei (H. Lév. et Vaniot) Pamp. f. stenocephala Pamp. = Artemisia lancea Vaniot ■

35802 Artemisia lavandulifolia DC. var. feddei (H. Lév. et Vaniot) Pamp. f. stenocephala Pamp. subsf. minutiflora Pamp. = Artemisia lancea Vaniot ■

35803 Artemisia lavandulifolia DC. var. lancea (Vaniot) Pamp. = Artemisia lancea Vaniot ■

35804 Artemisia lavandulifolia DC. var. maximowiczii Pamp. = Artemisia lavandulifolia DC. ●■

35805 Artemisia lavandulifolia DC. var. pekinensis Pamp. = Artemisia lavandulifolia DC. ●■

35806 Artemisia lavandulifolia Miq. = Artemisia lancea Vaniot ■

35807 Artemisia lavandulifolia Nakai = Artemisia subulata Nakai ■

35808 Artemisia ledebouriana Besser;红点草蒿■☆

35809 Artemisia lehmaniana Hook. f. = Seriphidium thomsonianum (C. B. Clarke) Y. Ling et Y. R. Ling ■

35810 Artemisia lehmanniana Bunge = Seriphidium lehmannianum (Bunge) Poljakov ■

35811 Artemisia leontopodioides Fisch. ex Besser = Artemisia glomerata Ledeb. ■☆

35812 Artemisia leontopodioides Fisch. ex Besser = Artemisia glomerata Ledeb. var. leontopodioides (Fisch. ex Besser) Kitam. ■☆

35813 Artemisia leptostachya Don = Artemisia indica Willd. ■

35814 Artemisia lessingiana Krylov = Seriphidium sublessingianum (B. Keller) Poljakov ■

35815 Artemisia lessingina Besser;李氏蒿■☆

35816 Artemisia leucophylla (Turcz. ex Besser) C. B. Clarke;白叶蒿(白蒿,白毛蒿,朝鲜艾,蒂氏蒿,茭蒿,苦蒿,野艾蒿);Whiteleaf Sagebrush ■

35817 Artemisia leucophylla Kitag. = Artemisia lavandulifolia DC. ●■

35818 Artemisia leucophylla Kom. = Artemisia argyi H. Lév. ■

35819 Artemisia leucophylla Turcz. ex Pavlov var. pusilla Pamp. = Artemisia leucophylla (Turcz. ex Besser) C. B. Clarke ■

35820 Artemisia leucophylla Turcz. ex Pavlov var. pusilla Pamp. f. genuina Pamp. = Artemisia leucophylla (Turcz. ex Besser) C. B. Clarke ■

35821 Artemisia leucophylla Turcz. ex Pavlov var. pusilla Pamp. f. minuta Pamp. = Artemisia leucophylla (Turcz. ex Besser) C. B. Clarke ■

35822 Artemisia leucophylla Turcz. ex Pavlov var. typica Pamp. = Artemisia leucophylla (Turcz. ex Besser) C. B. Clarke ■

35823 Artemisia leucophylla Turcz. ex Pavlov var. typica Pamp. f. genuina Pamp. = Artemisia leucophylla (Turcz. ex Besser) C. B. Clarke ■

35824 Artemisia leucophylla Turcz. ex Pavlov var. typica Pamp. f. simplicifolia Pamp. = Artemisia leucophylla (Turcz. ex Besser) C. B. Clarke ■

35825 Artemisia liaotungensis Kitag. = Artemisia verbenacea (Kom.) Kitag. ■

35826 Artemisia licentii Pamp. = Artemisia brachyloba Franch. ■

35827 Artemisia limosa Koidz.;耐湿蒿■☆

35828 Artemisia lindheimeriana Scheele = Artemisia ludoviciana Nutt. ■☆

35829 Artemisia lindleyana Besser = Artemisia ludoviciana Nutt. subsp. incompta (Nutt.) D. D. Keck ■☆

35830 Artemisia lipskyi Poljakov;利普斯基蒿■☆

35831 Artemisia littoricola Kitam.;滨海牡蒿;Seashore Sagebrush ■

35832 Artemisia liukiuensis Kitam. = Artemisia morrisonensis Hayata ■

35833 Artemisia lobulifolia Boiss.;浅裂叶蒿■☆

35834 Artemisia longiflora Pamp. = Artemisia indica Willd. ■

35835 Artemisia longifolia Nutt.;长叶蒿;Long-leaved Sage ■

35836 Artemisia longiloba (Osterh.) Beetle = Artemisia arbuscula Nutt. subsp. longiloba (Osterh.) L. M. Shultz ■☆

35837 Artemisia ludoviciana Nutt.;路得威蒿(陆得威蒿);Artemisia, Lobed Cudweed, Louisiana Sage-wort, Louisiana Wormwood, Mexican Sagebrush, Native Wormwood, Sagewort, Silver Sage, Silver Wormwood, Western Mugwort, Western Sage, White Sage, White Sagebrush, Wormwood ■☆

35838 Artemisia ludoviciana Nutt. = Artemisia gnaphaloides Nutt. ■☆

35839 Artemisia ludoviciana Nutt. subsp. albula (Wooton) D. D. Keck;白路得威蒿;Silver King Artemisia, White Wormwood ■☆

35840 Artemisia ludoviciana Nutt. subsp. candicans (Rydb.) D. D. Keck;变白路得威蒿■☆

35841 Artemisia ludoviciana Nutt. subsp. incompta (Nutt.) D. D. Keck;山地路得威蒿;Mountain Wormwood ■☆

35842 Artemisia ludoviciana Nutt. subsp. mexicana (Willd. ex Spreng.) D. D. Keck;墨西哥路得威蒿;Mexican Wormwood ■☆

35843 Artemisia ludoviciana Nutt. subsp. typica D. D. Keck = Artemisia ludoviciana Nutt. ■☆

35844 Artemisia ludoviciana Nutt. var. albula (Wooton) Shinners = Artemisia ludoviciana Nutt. subsp. albula (Wooton) D. D. Keck ■☆

35845 Artemisia ludoviciana Nutt. var. americana (Besser) Fernald = Artemisia ludoviciana Nutt. ■☆

35846 Artemisia ludoviciana Nutt. var. brittonii (Rydb.) Fernald = Artemisia ludoviciana Nutt. ■☆

35847 Artemisia ludoviciana Nutt. var. candicans (Rydb.) H. St. John = Artemisia ludoviciana Nutt. subsp. candicans (Rydb.) D. D. Keck ■☆

35848 Artemisia ludoviciana Nutt. var. douglasiana (Besser) D. C. Eaton = Artemisia douglasiana Besser ■☆

35849 Artemisia ludoviciana Nutt. var. gnaphalodes (Nutt.) Torr. et A. Gray = Artemisia ludoviciana Nutt. ■☆

35850 Artemisia ludoviciana Nutt. var. incompta (Nutt.) Cronquist = Artemisia ludoviciana Nutt. subsp. incompta (Nutt.) D. D. Keck ■☆

35851 Artemisia ludoviciana Nutt. var. integrifolia A. Nelson = Artemisia longifolia Nutt. ■

35852 Artemisia ludoviciana Nutt. var. latifolia (Besser) Torr. et A. Gray = Artemisia ludoviciana Nutt. ■☆

35853 Artemisia ludoviciana Nutt. var. latiloba Nutt. = Artemisia ludoviciana Nutt. subsp. candicans (Rydb.) D. D. Keck ■☆

35854 Artemisia ludoviciana Nutt. var. pabularis (A. Nelson) Fernald = Artemisia ludoviciana Nutt. ■☆

35855 Artemisia macarosciadia Poljakov = Seriphidium junceum (Kar. et Kir.) Poljakov var. macrosciadium (Poljakov) Y. Ling et Y. R. Ling ■

35856 Artemisia macilenta (Maxim.) Krasch.;细杆沙蒿(细叶蒿,小砂蒿);Thinstalk Sagebrush ●■

35857 Artemisia macrantha Ledeb.; 亚洲大沙蒿(大花蒿); Asiabigflower Sagebrush ●■

35858 Artemisia macrobotrys Ledeb. = Artemisia laciniata Willd. ■

35859 Artemisia macrocephala Jacq. ex Besser;大花蒿(草蒿,戈壁蒿); Bigflower Sagebrush, Largehead Wormwood ■

35860 Artemisia macrorhiza Turcz.;大根蒿■☆

35861 Artemisia macrosciadia Poljakov = Seriphidium junceum (Kar. et Kir.) Poljakov var. macrosciadium (Poljakov) Y. Ling et Y. R. Ling ■

35862 Artemisia maderaspatana L. = Grangea maderaspatana (L.) Poir. ■

35863 Artemisia mairei H. Lév.;小亮苞蒿(滇茵陈,东川蒿); Maire Sagebrush ■

35864 Artemisia mairei H. Lév. var. latifolia Pamp. = Artemisia mairei H. Lév. ■

35865 Artemisia manshurica (Kom.) Kom.; 东北牡蒿; NE. China Sagebrush ■

35866 Artemisia maritima C. B. Clarke = Seriphidium brevifolium (Wall. ex DC.) Y. Ling et Y. R. Ling ■

35867 Artemisia maritima Kitag. = Seriphidium finitum (Kitag.) Y. Ling et Y. R. Ling ■

35868 Artemisia maritima L.;海蒿(滨海绢蒿,海滨蒿,蛔蒿,蒙古蛔蒿,驱蛔蒿); French Wormwood, Garden Cypress, Levant Wormseed, Maritime Wormwood, Old Woman, Santonica, Savin, Sea Mugwort, Sea Wormwood, Southernwood, Wormseed ■

35869 Artemisia maritima L. = Cerbera manghas L. ●

35870 Artemisia maritima L. = Seriphidium brevifolium (Wall. ex DC.) Y. Ling et Y. R. Ling ■

35871 Artemisia maritima L. = Seriphidium finitum (Kitag.) Y. Ling et Y. R. Ling ■

35872 Artemisia maritima L. = Seriphidium maritimum (L.) Poljakov ■

35873 Artemisia maritima L. = Seriphidium schrenkianum (Ledeb.) Poljakov ■

35874 Artemisia maritima L. subsp. gmeliniana (Besser) Krasch. = Seriphidium compactum (Fisch. ex Besser) Poljakov ■

35875 Artemisia maritima L. var. aucheri (Boiss.) Pamp. = Seriphidium aucheri (Boiss.) Y. Ling et Y. R. Ling ■

35876 Artemisia maritima L. var. compacta (Fisch. ex Besser) Ledeb. = Seriphidium compactum (Fisch. ex DC.) Poljakov ■

35877 Artemisia maritima L. var. compacta Pamp. = Seriphidium thomsonianum (C. B. Clarke) Y. Ling et Y. R. Ling ■

35878 Artemisia maritima L. var. fischeriana Besser = Seriphidium compactum (Fisch. ex DC.) Poljakov ■

35879 Artemisia maritima L. var. gmeliniana Besser = Seriphidium compactum (Fisch. ex DC.) Poljakov ■

35880 Artemisia maritima L. var. lercheana Besser = Seriphidium nitrosum (Weber ex Stechm.) Poljakov ■

35881 Artemisia maritima L. var. lercheana f. dahurica Turcz. = Seriphidium nitrosum (Weber ex Stechm.) Poljakov ■

35882 Artemisia maritima L. var. lercheana f. gmeliniana (Besser) Ledeb. = Seriphidium nitrosum (Weber ex Stechm.) Poljakov ■

35883 Artemisia maritima L. var. lercheana f. humilis Ledeb. = Seriphidium nitrosum (Weber ex Stechm.) Poljakov ■

35884 Artemisia maritima L. var. sublessingiana B. Keller = Seriphidium sublessingianum (B. Keller) Poljakov ■

35885 Artemisia maritima L. var. thomsoniana C. B. Clarke = Seriphidium thomsonianum (C. B. Clarke) Y. Ling et Y. R. Ling ■

35886 Artemisia maritima L. var. thomsonianum C. B. Clarke = Seriphidium thomsonianum (C. B. Clarke) Y. Ling et Y. R. Ling ■

35887 Artemisia maritima subsp. gmeliniana (Besser) Krasch. = Seriphidium nitrosum (Weber ex Stechm.) Poljakov ■

35888 Artemisia marschalliana Spreng.;中亚旱蒿;Marschall Sagebrush ■

35889 Artemisia marschalliana Spreng. var. sericophylla (Rupr.) Y. R. Ling = Artemisia sericophylla Rupr. ■☆

35890 Artemisia matricarioides Less. = Matricaria discoidea DC. ■

35891 Artemisia matricarioides Less. = Matricaria matricarioides (Less.) Ced. Porter ex Britton ■

35892 Artemisia mattfeldii Pamp.; 黏毛蒿; Mattfeld Sagebrush, Mattfeld's Wormwood ■

35893 Artemisia mattfeldii Pamp. var. etomentosa Hand. -Mazz.;无绒黏毛蒿(光黏毛蒿);Glabrous Mattfeld Sagebrush, Glabrous Mattfeld's Wormwood ■

35894 Artemisia maxa DC. = Artemisia sieversiana Ehrh. ex Willd. ■

35895 Artemisia maximowicziana (F. Schumach.) Krasch. ex Poljakov; 东亚栉齿蒿;Maximowicz Sagebrush ●■

35896 Artemisia maximowicziana Krasch. ex Poljakov = Artemisia maximowicziana (F. Schumach.) Krasch. ex Poljakov ●■

35897 Artemisia medioxima Krasch. ex Poljakov;尖栉齿叶蒿(宽裂叶蒿);Broadlobed Wormwood, Central Sagebrush ■

35898 Artemisia megalobotrys Nakai = Artemisia stolonifera (Maxim.) Kom. ■

35899 Artemisia mesatlantica Maire;梅萨蒿■☆

35900 Artemisia mesatlantica Maire var. subsimplex ? = Artemisia negrei Ouyahya ■☆

35901 Artemisia messerschmidtiana Besser = Artemisia sacrorum Ledeb. var. messerschmidtiana (Besser) Y. R. Ling ■

35902 Artemisia messerschmidtiana Besser var. incana Besser = Artemisia sacrorum Ledeb. var. messerschmidtiana (Besser) Y. R. Ling ■

35903 Artemisia messerschmidtiana Besser var. incana Besser = Artemisia sacrorum Ledeb. var. incana (Besser) Y. R. Ling ■

35904 Artemisia mexicana Willd. ex Spreng.; 墨西哥蒿; Mexican Mugwort ■☆

35905 Artemisia mexicana Willd. ex Spreng. = Artemisia ludoviciana Nutt. subsp. mexicana (Willd. ex Spreng.) D. D. Keck ■☆

35906 Artemisia mexicana Willd. ex Spreng. var. angustifolia ?;狭叶墨西哥蒿■☆

35907 Artemisia mexicana Willd. ex Spreng. var. latifolia Sch. Bip.;宽叶墨西哥蒿■☆

35908 Artemisia meyeriana Besser;梅氏蒿;Mayer Wormwood ■☆

35909 Artemisia michauxiana Besser;柠檬蒿;Lemon Sagewort ■☆

35910 Artemisia migoana Kitam. = Artemisia stolonifera (Maxim.) Kom. ■

35911 Artemisia minima L. = Centipeda minima (L.) A. Braun et Asch. ■

35912 Artemisia minor Jacq. ex Besser;垫型蒿(小灰蒿);Padshaped Sagebrush ■

35913 Artemisia minutiflora Nakai = Artemisia lancea Vaniot ■

35914 Artemisia mnacrantha Ledeb.;亚洲大花蒿(大花蒿)■

35915 Artemisia momiyamae Kitam.;雪蒿■☆

35916 Artemisia mongolica (Fisch. ex Besser) Fisch. ex Nakai;蒙古蒿(狼尾蒿,蒙蒿,水红蒿,狭叶蒿);Mongolian Sagebrush, Mongolian Wormwood ■

35917 Artemisia mongolica (Fisch. ex Besser) Fisch. ex Nakai subsp. genuina Kitag. = Artemisia mongolica (Fisch. ex Besser) Fisch. ex

Nakai ■

35918 Artemisia mongolica (Fisch. ex Besser) Fisch. ex Nakai subsp. orientalis Kitag. = Artemisia mongolica (Fisch. ex Besser) Fisch. ex Nakai ■

35919 Artemisia mongolica (Fisch. ex Besser) Fisch. ex Nakai subsp. orientalis Kitag. = Artemisia verbenacea (Kom.) Kitag. ■

35920 Artemisia mongolica (Fisch. ex Besser) Fisch. ex Nakai var. interposita Kitag. = Artemisia integrifolia L. ■

35921 Artemisia mongolica (Fisch. ex Besser) Fisch. ex Nakai var. krascheninnikovii f. debilis Pamp. = Artemisia mongolica (Fisch. ex Besser) Fisch. ex Nakai ■

35922 Artemisia mongolica (Fisch. ex Besser) Fisch. ex Nakai var. leucophylla (Turcz. ex Besser) W. Wang = Artemisia leucophylla (Turcz. ex Besser) C. B. Clarke ■

35923 Artemisia mongolica (Fisch. ex Besser) Fisch. ex Nakai var. leucophylla (Turcz. ex Besser) W. Wang et H. T. Ho ex H. C. Fu = Artemisia leucophylla (Turcz. ex Besser) C. B. Clarke ■

35924 Artemisia mongolica (Fisch. ex Besser) Fisch. ex Nakai var. parviflora (Maxim.) Kitag. = Artemisia rubripes Nakai ■

35925 Artemisia mongolica (Fisch. ex Besser) Fisch. ex Nakai var. parviflora (Maxim.) Kitag. f. luxurians (Pamp.) Kitag. = Artemisia rupestris L. ■

35926 Artemisia mongolica (Fisch. ex Besser) Fisch. ex Nakai var. pseudovulgaris Pamp. = Artemisia rupestris L. ■

35927 Artemisia mongolica (Fisch. ex Besser) Fisch. ex Nakai var. tenuifolia f. genuina Pamp. = Artemisia mongolica (Fisch. ex Besser) Fisch. ex Nakai ■

35928 Artemisia mongolica (Fisch. ex Besser) Fisch. ex Nakai var. verbenacea (Kom.) Pamp. f. genuina Pamp. = Artemisia verbenacea (Kom.) Kitag. ■

35929 Artemisia mongolica (Fisch. ex Besser) Fisch. ex Nakai var. verbenacea (Kom.) Pamp. f. viscosa subf. glabrescnes Pamp. = Artemisia verbenacea (Kom.) Kitag. ■

35930 Artemisia mongolica (Fisch. ex Besser) Fisch. ex Nakai var. verbenacea (Kom.) Pamp. f. williamsonii Pamp. = Artemisia verbenacea (Kom.) Kitag. ■

35931 Artemisia mongolica (Fisch. ex Besser) Fisch. ex Nakai var. verbenacea (Kom.) Pamp. = Artemisia verbenacea (Kom.) Kitag. ■

35932 Artemisia mongolica (Fisch. ex Besser) Nakai var. krascheninnikovii Pamp. = Artemisia leucophylla (Turcz. ex Besser) C. B. Clarke ■

35933 Artemisia mongolica (Fisch. ex Besser) Nakai var. krascheninnikovii Pamp. = Artemisia mongolica (Fisch. ex Besser) Nakai ■

35934 Artemisia mongolica (Fisch. ex Besser) Nakai var. leucophylla (Turcz. ex Besser) W. Wang et H. R. Ho = Artemisia leucophylla (Turcz. ex Besser) C. B. Clarke ■

35935 Artemisia mongolica (Fisch. ex Besser) Nakai var. pseudovulgaris Pamp. = Artemisia rubripes Nakai ■

35936 Artemisia mongolica (Fisch. ex Besser) Nakai var. tenuifolia Pamp. = Artemisia mongolica (Fisch. ex Besser) Nakai ■

35937 Artemisia mongolica (Fisch. ex Besser) Nakai var. verbenacea (Kom.) Pamp. = Artemisia verbenacea (Kom.) Kitag. ■

35938 Artemisia mongolica C. C. Chang = Artemisia princeps Pamp. ■

35939 Artemisia mongolica C. C. Chang = Artemisia verlotorum Lamotte ■

35940 Artemisia mongolorum Krasch. = Seriphidium mongolorum (Krasch.) Y. Ling et Y. R. Ling ■

35941 Artemisia mongolorum Krasch. subsp. gobicum Krasch. = Seriphidium nitrosum (Weber ex Stechm.) Poljakov var. gobicum (Krasch.) Y. R. Ling ●■

35942 Artemisia mongolorum Krasch. var. salsuginosa Krasch. = Seriphidium nitrosum (Weber ex Stechm.) Poljakov var. gobicum (Krasch.) Y. R. Ling ●■

35943 Artemisia monocephala (A. Gray) A. Heller = Artemisia pattersonii A. Gray ■☆

35944 Artemisia monogyna Waldst. et Kit.;单雌蕊蛔蒿■☆

35945 Artemisia monophylla Kitam.;单叶蒿■☆

35946 Artemisia monosperma Delile;单子蒿■☆

35947 Artemisia monosperma Delile var. libyca Chiov. = Artemisia monosperma Delile ■☆

35948 Artemisia montana (Nakai) Pamp.;山地蒿(山蒿)■☆

35949 Artemisia montana (Nakai) Pamp. var. nipponica f. occidentalis Pamp. = Artemisia princeps Pamp. ■

35950 Artemisia montana (Nakai) Pamp. var. shiretokoensis Koji Ito;知床蒿■☆

35951 Artemisia montana Pamp. var. latiloba Pamp. = Artemisia montana (Nakai) Pamp. ■☆

35952 Artemisia montana Pamp. var. nipponica Pamp. f. occidentalis Pamp. = Artemisia princeps Pamp. ■

35953 Artemisia moorcroftiana Mattf. = Artemisia imponens Pamp. ■

35954 Artemisia moorcroftiana Pamp. = Artemisia viscida (Mattf.) Pamp. ■

35955 Artemisia moorcroftiana Pamp. var. nitida (Pamp.) Y. Ling et Y. R. Ling = Artemisia tainingensis Hand. -Mazz. var. nitida (Pamp.) Y. R. Ling ■

35956 Artemisia moorcroftiana Pamp. var. viscida Mattf. = Artemisia viscida (Mattf.) Pamp. ■

35957 Artemisia moorcroftiana Wall. ex DC.;小球花蒿(大叶青蒿,芳枝蒿,小白蒿);Moorcroft Sagebrush,Moorcroft' s Wormwood ■

35958 Artemisia moorcroftiana Wall. ex DC. = Artemisia imponens Pamp. ■

35959 Artemisia moorcroftiana Wall. ex DC. f. nitida Pamp. = Artemisia tainingensis Hand. -Mazz. var. nitida (Pamp.) Y. R. Ling ■

35960 Artemisia moorcroftiana Wall. ex DC. var. campanulata Pamp. = Artemisia moorcroftiana Wall. ex DC. ■

35961 Artemisia moorcroftiana Wall. ex DC. var. campanulata Pamp. f. tenuifolia Pamp. = Artemisia moorcroftiana Wall. ex DC. ■

35962 Artemisia moorcroftiana Wall. ex DC. var. typia Pamp. = Artemisia moorcroftiana Wall. ex DC. ■

35963 Artemisia moorcroftiana Wall. ex DC. var. typia Pamp. f. genuina Pamp. = Artemisia moorcroftiana Wall. ex DC. ■

35964 Artemisia morrisonensis Hayata;细叶山艾;Morrison Sagebrush ■

35965 Artemisia morrisonensis Hayata var. minima Pamp. = Artemisia japonica Thunb. ■

35966 Artemisia moxa DC. = Artemisia sieversiana Ehrh. ex Willd. ■

35967 Artemisia multicaulis Ledeb. = Artemisia anethifolia Weber ex Stechm. ■

35968 Artemisia myriantha Wall. ex Besser;多花蒿(艾蒿,蒿枝,黑蒿,苦蒿);Manyflower Sagebrush ■

35969 Artemisia myriantha Wall. ex Besser = Artemisia indica Willd. ■

35970 Artemisia myriantha Wall. ex Besser var. pleiocephala (Pamp.) Y. R. Ling;白毛多花蒿■

35971 Artemisia myriantha Y. R. Ling = Artemisia indica Willd. ■

35972 Artemisia nakaii Pamp.;矮滨蒿;Nakai Sagebrush ■

35973 Artemisia nakaii Pamp. = Artemisia fauriei Nakai ■
35974 Artemisia namanganica Poljakov;纳曼干蒿■☆
35975 Artemisia nanschanica Krasch.;昆仑蒿(南山蒿,祁连山蒿);Kunlunshan Sagebrush ■
35976 Artemisia natronensis A. Nelson = Artemisia longifolia Nutt. ■
35977 Artemisia negrei Ouyahya;内格里蒿■☆
35978 Artemisia neomexicana Greene ex Rydb.;新墨西哥蒿■☆
35979 Artemisia neomexicana Greene ex Rydb. = Artemisia ludoviciana Nutt. subsp. mexicana (Willd. ex Spreng.) D. D. Keck ■☆
35980 Artemisia nesiotica P. H. Raven;结血蒿(岛蒿);Island Sagebrush ■
35981 Artemisia niitakayamensis Hayata;玉山艾;Yushan Sagebrush ■
35982 Artemisia niitakayamensis Hayata var. tsugitakaensis Kitam. = Artemisia tsugitakaensis (Kitam.) Y. Ling et Y. R. Ling ■
35983 Artemisia nilagarica (C. B. Clarke) Pamp.;南亚蒿;Nilagirian Sagebrush ■
35984 Artemisia nipponica Pamp. var. rubripes Pamp. = Artemisia rubripes Nakai ■
35985 Artemisia nitens (Steven et Besser) Krasch. = Artemisia sericea Weber ex Stechm. ■
35986 Artemisia nitrosa Weber ex Stechm. = Seriphidium nitrosum (Weber ex Stechm.) Poljakov ■
35987 Artemisia nitrosa Weber ex Stechm. subsp. kasakorum Krasch. = Seriphidium nitrosum (Weber ex Stechm.) Poljakov var. gobicum (Krasch.) Y. R. Ling ●■
35988 Artemisia nitrosa Weber ex Stechm. var. gobica Krasch. ex Poljakov = Seriphidium nitrosum (Weber ex Stechm.) Poljakov var. gobicum (Krasch.) Y. R. Ling ●■
35989 Artemisia nitrosa Weber ex Stechm. var. gobicum Krasch. ex Poljakov = Seriphidium nitrosum (Weber ex Stechm.) Poljakov var. gobicum (Krasch.) Y. R. Ling ■
35990 Artemisia nitrosa Weber ex Stechm. var. subglabra Krasch. = Seriphidium nitrosum (Weber ex Stechm.) Poljakov var. gobicum (Krasch.) Y. R. Ling ●■
35991 Artemisia nivea Redow. ex Willd. = Artemisia austriaca Jacq. ■
35992 Artemisia nortonii Pamp.;藏旱蒿;Norton Sagebrush ■
35993 Artemisia norvegica Fr.;苏格兰蒿;Alpine Sagewort, Norwegian Mugwort, Scottish Wormwood ■☆
35994 Artemisia norvegica Fr. subsp. globularia (Besser) H. M. Hall et Clem. = Artemisia globularia Cham. ex Besser ■☆
35995 Artemisia norvegica Fr. subsp. saxatilis (Besser) H. M. Hall et Clem.;岩生苏格兰蒿■☆
35996 Artemisia norvegica Fr. var. glomerata (Ledeb.) H. M. Hall et Clem. = Artemisia glomerata Ledeb. ■☆
35997 Artemisia norvegica Fr. var. piceetorum S. L. Welsh et Goodrich = Artemisia norvegica Fr. subsp. saxatilis (Besser) H. M. Hall et Clem. ■☆
35998 Artemisia nova A. Nelson;新蒿;Black Sage, Black Sagebrush ■☆
35999 Artemisia nubigena Wall. = Ajania nubigena (Wall.) C. Shih ■
36000 Artemisia nujiangensis (Y. Ling et Y. R. Ling) Y. R. Ling;怒江蒿(云南蒿);Nujiang Sagebrush ■
36001 Artemisia nutans Willd. = Artemisia argyi H. Lév. et Vaniot ■
36002 Artemisia nutantiflora Nakai = Artemisia argyi H. Lév. et Vaniot ■
36003 Artemisia obscura Pamp. = Artemisia mongolica (Fisch. ex Besser) Fisch. ex Nakai ■
36004 Artemisia obscura Pamp. = Artemisia vulgaris L. ■
36005 Artemisia obscura Pamp. f. genuina Pamp. = Artemisia mongolica (Fisch. ex Besser) Fisch. ex Nakai ■
36006 Artemisia obscura Pamp. var. congesta Pamp. = Artemisia verbenacea (Kom.) Kitag. ■
36007 Artemisia obscura Pamp. var. regina Pamp. = Artemisia mongolica (Fisch. ex Besser) Fisch. ex Nakai ■
36008 Artemisia obscura Pamp. var. rigida Pamp. = Artemisia leucophylla (Turcz. ex Besser) C. B. Clarke ■
36009 Artemisia obscura Pamp. var. tenuifolia (Turcz.) Pamp. = Artemisia mongolica (Fisch. ex Besser) Fisch. ex Nakai ■
36010 Artemisia obscura Pamp. var. tenuifolia Pamp. = Artemisia mongolica (Fisch. ex Besser) Nakai ■
36011 Artemisia obscura Pamp. var. typica Pamp. = Artemisia mongolica (Fisch. ex Besser) Fisch. ex Nakai ■
36012 Artemisia obtusiloba Ledeb.;钝裂蒿(小裂蒿);Obtuselobate Sagebrush ■
36013 Artemisia obtusiloba Ledeb. var. glabella (Kar. et Kir.) Poljakov = Artemisia giraldii Pamp. var. longipedunculata Y. R. Ling ■
36014 Artemisia obtusiloba Ledeb. var. glabella (Kar. et Kir.) Poljakov = Artemisia glabella Kar. et Kir. ■
36015 Artemisia obtusiloba Ledeb. var. glabra Ledeb. = Artemisia glabella Kar. et Kir. ■
36016 Artemisia occidentali-sichuanensis Y. R. Ling et S. Y. Zhao;川西腺毛蒿;W. Sichuan Sagebrush ■
36017 Artemisia occidentali-sinensis Y. R. Ling;华西蒿;W. China Sagebrush ■
36018 Artemisia occidentali-sinensis Y. R. Ling var. denticulata Y. R. Ling;齿裂华西蒿■
36019 Artemisia odoratissima Desf. = Artemisia saharae Pomel ■☆
36020 Artemisia oligocarpa Hayata;高山艾;Poorfruit Sagebrush ■
36021 Artemisia opulenta Pamp. = Artemisia vulgaris L. ■
36022 Artemisia oranensis (Debeaux) Filatova;奥兰蒿●☆
36023 Artemisia ordosica Krasch.;黑沙蒿(鄂尔多斯蒿,沙蒿,油蒿,籽蒿);Ordos Sagebrush, Ordos Wormwood ●
36024 Artemisia ordosica Krasch. var. furva H. C. Fu;乌油蒿●
36025 Artemisia ordosica Krasch. var. furva H. C. Fu = Artemisia ordosica Krasch. ●
36026 Artemisia ordosica Krasch. var. montana H. C. Fu;山油蒿●
36027 Artemisia ordosica Krasch. var. montana H. C. Fu = Artemisia ordosica Krasch. ●
36028 Artemisia orientali-hengduangensis Y. Ling et Y. R. Ling;东方蒿;Oriental Sagebrush ■
36029 Artemisia orientali-xizangensis Y. R. Ling et Humphries;昌都蒿;Changdu Sagebrush ■
36030 Artemisia orientali-yunnanensis Y. R. Ling;滇东蒿;E. Yunnan Sagebrush ■
36031 Artemisia orthobotrys Kitag. = Artemisia tanacetifolia L. ■
36032 Artemisia oxycephala Kitag.;光沙蒿(红杆蒿,沙蒿,小白蒿);Sharphead Sagebrush ■
36033 Artemisia oxycephala Kitag. f. taiyangensis G. Y. Chang, L. S. Wang et H. X. Ma;太阳岛光沙蒿;Harbin Sharphead Sagebrush ●■
36034 Artemisia oxycephala Kitag. f. taiyangensis G. Y. Chang, L. S. Wang et H. X. Ma = Artemisia oxycephala Kitag. ■
36035 Artemisia oxycephala Kitag. subsp. shanhaiensis G. Y. Chang, L. S. Wang et H. X. Ma = Artemisia oxycephala Kitag. ■
36036 Artemisia oxycephala Kitag. var. aureinitens W. Wang;金沙蒿■
36037 Artemisia oxycephala Kitag. var. shanhaiensis G. Y. Chang, L. S. Wang et H. X. Ma;山海关沙蒿;Shanhaiguan Sharphead Sagebrush ■

36038 Artemisia oxycephala Kitag. var. sporadantha W. Wang;疏花光沙蒿■

36039 Artemisia oxycephala Kitag. var. xinkaiensis G. Y. Chang et L. S. Wang;兴凯光沙蒿;Xingkaihu Sharphead Sagebrush ■

36040 Artemisia oxycephala Kitag. var. xinkaiensis G. Y. Chang et L. S. Wang = Artemisia oxycephala Kitag. ■

36041 Artemisia pabularis (A. Nelson) Rydb. = Artemisia ludoviciana Nutt. ■☆

36042 Artemisia pacifica Nutt. = Artemisia campestris L. subsp. pacifica (Nutt.) H. M. Hall et Clem. ■☆

36043 Artemisia packardiae J. W. Grimes et Ertter; 帕卡蒿; Succor Creek Mugwort ■☆

36044 Artemisia palens Wall.;白艾■☆

36045 Artemisia pallasiana Fisch. ex Besser = Ajania pallasiana (Fisch. ex Besser) Poljakov ■

36046 Artemisia pallasiana Fisch. ex Besser = Dendranthema pallasianum (Fisch. ex Besser) Vorosch. ■

36047 Artemisia palmeri A. Gray;帕默蒿;Palmer Sagewort ■☆

36048 Artemisia palustris L.;黑蒿(沼泽蒿);Swampy Sagebrush, Swampy Wormwood ■

36049 Artemisia palustris L. var. aurata (Kom.) Pamp. = Artemisia aurata Kom. ■

36050 Artemisia pamirica C. Winkl. = Artemisia dracunculus L. var. pamirica (C. Winkl.) Y. R. Ling et Humphries ■

36051 Artemisia pamirica C. Winkl. f. trifida Pamp. = Artemisia dracunculus L. var. pamirica (C. Winkl.) Y. R. Ling et Humphries ■

36052 Artemisia pamirica C. Winkl. var. aschurbazewi C. Winkl. = Artemisia dracunculus L. var. pamirica (C. Winkl.) Y. R. Ling et Humphries ■

36053 Artemisia pamiricum O. Fedtsch. = Artemisia macrocephala Jacq. ex Besser ■

36054 Artemisia pannosa Krasch.;毡状蒿■☆

36055 Artemisia papposa S. F. Blake et Cronquist;冠毛蒿;Owyhee Sage ■☆

36056 Artemisia parishii A. Gray = Artemisia tridentata Nutt. subsp. parishii (A. Gray) H. M. Hall et Clem. ■☆

36057 Artemisia parryi A. Gray = Artemisia laciniata Willd. subsp. parryi (A. Gray) W. A. Weber ■☆

36058 Artemisia parviflora Aitch. = Artemisia japonica Thunb. ■

36059 Artemisia parviflora Buch. -Ham. ex Roxb.;西南牡蒿(青蒿,小花蒿,小花牡蒿);Smallflower Sagebrush ■

36060 Artemisia parviflora Buch. -Ham. ex Roxb. subf. tongtchuanensis (H. Lév.) Pamp. = Artemisia parviflora Buch. -Ham. ex Roxb. ■

36061 Artemisia parvula Pamp. = Artemisia princeps Pamp. ■

36062 Artemisia pattersonii A. Gray;岩栖蒿;Patterson Sagewort ■☆

36063 Artemisia paucicephala A. Nelson = Artemisia ludoviciana Nutt. ■☆

36064 Artemisia pauciflora Krylov = Seriphidium gracilescens (Krasch. et Iljin) Poljakov ■

36065 Artemisia pauciflora Weber ex Stechm. = Seriphidium gracilescens (Krasch. et Iljin) Poljakov ■

36066 Artemisia pectinata Pall. = Neopallasia pectinata (Pall.) Poljakov ■

36067 Artemisia pectinata Pall. var. typica Pamp. = Neopallasia pectinata (Pall.) Poljakov ■

36068 Artemisia pectinata Pall. var. yunnanensis Pamp. = Neopallasia pectinata (Pall.) Poljakov ■

36069 Artemisia pedatifida Nutt.;鸟足蒿;Matted Sagewort ■☆

36070 Artemisia pedunculosa Miq.;梗花蒿■☆

36071 Artemisia pengchuoensis Y. R. Ling et S. Y. Zhao; 彭错蒿; Pengcuo Sagebrush ■

36072 Artemisia persica Boiss.;伊朗蒿(波斯蒿);Persian Sagebrush ●■

36073 Artemisia persica Boiss. var. subspinescens (Boiss.) Boiss.;微刺伊朗蒿●■

36074 Artemisia petrophila Wooton et Standl. = Artemisia bigelovii A. Gray ■☆

36075 Artemisia pewzowii C. Winkl.;纤梗蒿;Pewzow Sagebrush ■

36076 Artemisia phaeolepis Krasch.;褐苞蒿(褐鳞蒿);Brownbract Sagebrush ■

36077 Artemisia phyllobotrys (Hand. -Mazz.) Y. Ling et Y. R. Ling;叶苞蒿;Leaflikebract Sagebrush ■

36078 Artemisia plattensis Nutt. = Artemisia filifolia Torr. ■☆

36079 Artemisia pleiocephala Pamp. = Artemisia myriantha Wall. ex Besser var. pleiocephala (Pamp.) Y. R. Ling ■

36080 Artemisia pleiocephala Pamp. f. yunnanensis Pamp. = Artemisia myriantha Wall. ex Besser ■

36081 Artemisia pleiocephala Pamp. var. grandis Pamp. = Artemisia indica Willd. ■

36082 Artemisia pleiocephala Pamp. var. insularis Pamp. = Artemisia rupestris L. ■

36083 Artemisia pleiocephala Pamp. var. typica Pamp. = Artemisia myriantha Wall. ex Besser var. pleiocephala (Pamp.) Y. R. Ling ■

36084 Artemisia pleiocephala Pamp. var. typica Pamp. f. discolor Pamp. = Artemisia myriantha Wall. ex Besser var. pleiocephala (Pamp.) Y. R. Ling ■

36085 Artemisia pleiocephala Pamp. var. typica Pamp. f. yunnanensis Pamp. = Artemisia myriantha Wall. ex Besser ■

36086 Artemisia poljakovii Filatova;波尔蒿蒿■☆

36087 Artemisia polybotryoidea Y. R. Ling; 甘新青蒿; Manyraceme Sagebrush ■

36088 Artemisia pontica L.;西北蒿(本都山蒿,黑海蒿,罗马蒿,宁新叶莲蒿); Cypress Wormwood, French Wormwood, Garden Wormwood, Green-ginger, Old Warrior, Old Woman, Pontic Wormwood, Pontus Sagebrush, Roman Rocket, Roman Wormwood ●■

36089 Artemisia porrecta Krasch. ex Poljakov;外伸蒿■☆

36090 Artemisia porteri Cronquist;波特蒿;Porter Mugwort ■☆

36091 Artemisia potentillifolia H. Lév.;委陵菜叶蒿■☆

36092 Artemisia potentillifolia H. Lév. = Artemisia vestita Wall. ex Besser ●■

36093 Artemisia potentilloides A. Gray = Sphaeromeria potentilloides (A. Gray) A. Heller ■☆

36094 Artemisia prasina Krasch. ex Poljakov;草绿蒿■☆

36095 Artemisia prattii (Pamp.) Y. Ling et Y. R. Ling;藏岩蒿■

36096 Artemisia prescottiana Besser = Artemisia ludoviciana Nutt. subsp. incompta (Nutt.) D. D. Keck ■☆

36097 Artemisia princeps Pamp.;魁蒿(艾,艾叶,端午艾,黄花艾,蓬,王侯蒿,五月艾,野艾,野艾蒿);First Sagebrush, First Wormwood ■

36098 Artemisia princeps Pamp. = Artemisia indica Willd. var. maximowiczii (Nakai) H. Hara ■☆

36099 Artemisia princeps Pamp. var. candicans Pamp. = Artemisia argyi H. Lév. ■

36100 Artemisia princeps Pamp. var. candicans Pamp. = Artemisia verbenacea (Kom.) Kitag. ■

36101 Artemisia princeps Pamp. var. orientalis (Pamp.) H. Hara = Artemisia indica Willd. ■

36102 Artemisia princeps Pamp. var. typia Pamp. = Artemisia princeps Pamp. ■
36103 Artemisia princeps Pamp. var. typica f. dissecta Pamp. = Artemisia princeps Pamp. ■
36104 Artemisia princeps Pamp. var. typica f. genuina Pamp. = Artemisia princeps Pamp. ■
36105 Artemisia procera Willd.;高蒿■☆
36106 Artemisia procera Willd. = Artemisia abrotanum L. ■☆
36107 Artemisia proceriformis Krasch. = Artemisia abrotanum L. ■☆
36108 Artemisia prolixa Krasch. ex Poljakov;伸展蒿■☆
36109 Artemisia pronutans Kitag. = Artemisia brachyphylla Kitam. ■
36110 Artemisia przewalskii Krasch.;甘青小蒿;Przewalsk Sagebrush ■
36111 Artemisia pubescens Ledeb.;柔毛蒿(变蒿,立沙蒿,立砂蒿,麻蒿,米拉蒿,转蒿);Velvent Sagebrush ■
36112 Artemisia pubescens Ledeb. subsp. eriopoda (Bunge) Kitam. = Artemisia eriopoda Bunge ■
36113 Artemisia pubescens Ledeb. var. coracina (W. Wang) Y. Ling et Y. R. Ling;黑柔毛蒿(黑砂蒿)■
36114 Artemisia pubescens Ledeb. var. gebleriana (Besser) Y. R. Ling;大头柔毛蒿(大头变蒿)■
36115 Artemisia pubescens Ledeb. var. monostachya (Bunge ex Maxim.) Y. R. Ling = Artemisia pubescens Ledeb. ■
36116 Artemisia pubescens Ledeb. var. oxycephala (Kitag.) Kitag. = Artemisia oxycephala Kitag. ■
36117 Artemisia pubescens Ledeb. var. pallasiana (Besser) Kitag. = Artemisia desertorum Spreng. ■
36118 Artemisia pubescens Ledeb. var. pumila (H. C. Fu et C. Y. Li) H. C. Fu et C. Y. Li = Artemisia pubescens Ledeb. ■
36119 Artemisia pubescens Ledeb. var. taheensis G. Y. Chang et L. S. Wang;塔河柔毛蒿■
36120 Artemisia pubescens Ledeb. var. taheensis G. Y. Chang et L. S. Wang = Artemisia pubescens Ledeb. ■
36121 Artemisia pudica Rydb. = Artemisia ludoviciana Nutt. ■☆
36122 Artemisia punctigera Krasch. ex Poljakov;斑蒿■☆
36123 Artemisia purpurascens Jacq. ex Besser = Artemisia roxburghiana Besser var. purpurascens (Jacq. ex Besser) Hook. f. ■
36124 Artemisia purshiana Besser = Artemisia ludoviciana Nutt. ■☆
36125 Artemisia purshiana Besser var. latifolia Besser = Artemisia ludoviciana Nutt. ■☆
36126 Artemisia purshii Besser = Artemisia borealis Pall. ■
36127 Artemisia pycnocephala (Less.) DC.;沙丘蒿;Coastal Sagewort,Sandhill Sage,Sandhill Wormwood ■☆
36128 Artemisia pycnocephala DC. = Artemisia pycnocephala (Less.) DC. ■☆
36129 Artemisia pycnorhiza Ledeb.;密根蒿■☆
36130 Artemisia pycnorhiza Ledeb. var. depauperata (Krasch.) Poljakov = Artemisia depauperata Krasch. ■
36131 Artemisia pygmaea A. Gray;矮小蒿;Pygmy Sage ■☆
36132 Artemisia qinlingensis Y. Ling et Y. R. Ling;秦岭蒿;Qinling Sagebrush ■
36133 Artemisia quadriauriculata Chen = Artemisia integrifolia L. ■
36134 Artemisia quinqueloba Trautv.;五裂蒿■☆
36135 Artemisia ramosa C. Sm.;分枝蒿■☆
36136 Artemisia rehan Chiov. = Artemisia absinthium L. ■
36137 Artemisia remotiloba Krasch. ex Poljakov;疏裂片蒿■☆
36138 Artemisia repens Pall. ex Willd. = Artemisia austriaca Jacq. ■
36139 Artemisia reptans C. Sm.;匍匐蒿■☆
36140 Artemisia revoluta Edgew. = Artemisia roxburghiana Besser ■
36141 Artemisia revoluta Rydb. = Artemisia ludoviciana Nutt. subsp. mexicana (Willd. ex Spreng.) D. D. Keck ■☆
36142 Artemisia rhizomata A. Nelson = Artemisia ludoviciana Nutt. ■☆
36143 Artemisia rhizomata A. Nelson var. pabularis A. Nelson = Artemisia ludoviciana Nutt. ■☆
36144 Artemisia rhodantha Rupr. = Seriphidium rhodanthum (Rupr.) Poljakov ■
36145 Artemisia richardsoniana Besser = Artemisia borealis Pall. subsp. richardsoniana (Besser) Korobkov ■☆
36146 Artemisia rigida (Nutt.) A. Gray;硬蒿;Scabland Sagebrush ■☆
36147 Artemisia robusta (Pamp.) Y. Ling et Y. R. Ling;粗茎蒿;Sturdy Sagebrush ■
36148 Artemisia rosthornii Pamp.;川南蒿;Rosthorn Sagebrush ■
36149 Artemisia rothrockii A. Gray;罗思罗克蒿;Rothrock Sagebrush, Sticky Sagebrush ■☆
36150 Artemisia rothrockii A. Gray = Artemisia tridentata Nutt. subsp. rothrockii (A. Gray) H. M. Hall et Clem. ■☆
36151 Artemisia rothrockii A. Gray = Seriphidium rothrockii (A. Gray) W. A. Weber ■☆
36152 Artemisia rotundifolia (Debeaux) Krasch. = Artemisia eriopoda Bunge var. rotundifolia (Debeaux) Y. R. Ling ■
36153 Artemisia roxburghiana Besser;灰苞蒿(白蒿子,土艾叶);Roxburgh Sagebrush,Roxburgh's Wormwood ■
36154 Artemisia roxburghiana Besser var. acutiloba Pamp. = Artemisia roxburghiana Besser ■
36155 Artemisia roxburghiana Besser var. acutiloba Pamp. f. forrestii Pamp. = Artemisia roxburghiana Besser ■
36156 Artemisia roxburghiana Besser var. divaricata Pamp. = Artemisia divaricata (Pamp.) Pamp. ■
36157 Artemisia roxburghiana Besser var. kasuensis Pamp. = Artemisia roxburghiana Besser ■
36158 Artemisia roxburghiana Besser var. orientalis Pamp. = Artemisia orientali-hengduangensis Y. Ling et Y. R. Ling ■
36159 Artemisia roxburghiana Besser var. orientalis Pamp. f. angustisecta Pamp. = Artemisia orientali-hengduangensis Y. Ling et Y. R. Ling ■
36160 Artemisia roxburghiana Besser var. purpurascens (Jacq. ex Besser) Hook. f.;紫苞蒿■
36161 Artemisia royleana DC. = Artemisia dubia Wall. ex Besser var. subdigitata (Mattf.) Y. R. Ling ■
36162 Artemisia rubripes Nakai;红足蒿(大狭叶蒿,红茎蒿,小香艾);Alpine Yarrow, Redfoot Sagebrush, Redfoot Wormwood, Rock Wormwood ■
36163 Artemisia rubripes Nakai f. grancilis Kitag. = Artemisia rupestris L. ■
36164 Artemisia rubripes Nakai f. luxurians (Pamp.) Kitag.;茂盛蒿■☆
36165 Artemisia rubripes Nakai f. tomentosa Kitag. = Artemisia mongolica (Fisch. ex Besser) Fisch. ex Nakai ■
36166 Artemisia rupestris L.;岩蒿(鹿角蒿,新疆一支蒿,岩生蒿,一支蒿,一枝蒿);Rock Wormwood,Sagebrush ■
36167 Artemisia rupestris L. subsp. woodii Nelson = Artemisia rupestris L. ■
36168 Artemisia rupestris L. var. oelandica (Besser) DC. = Artemisia rupestris L. ■
36169 Artemisia rupestris L. var. thuringiaca (Besser) DC. = Artemisia rupestris L. ■
36170 Artemisia rupestris L. var. viridifolia (Besser) DC. = Artemisia

rupestris L. ■

36171 Artemisia rupestris L. var. viridifolia DC. = Artemisia rupestris L. ■

36172 Artemisia rupestris L. var. viridis (Besser) DC. = Artemisia rupestris L. ■

36173 Artemisia rutifolia Stephan ex Spreng.;香叶蒿(芸香叶蒿);Rueleaf Sagebrush ■

36174 Artemisia rutifolia Stephan ex Spreng. var. altaica (Krylov) Krasch.;阿尔泰香叶蒿;Altai Sagebrush ■

36175 Artemisia rutifolia Stephan ex Spreng. var. ruoqiangensis Y. R. Ling;诺羌香叶蒿;Nuoqiang Rueleaf Sagebrush ■

36176 Artemisia sachalinensis Tiles ex Besser = Artemisia capillaris Thunb. ■

36177 Artemisia sacrorum Ledeb.;白莲蒿(白蒿,白莲毫,供蒿,僧蒿,铁秆蒿,万年蒿,蚊艾,香蒿);Holy Sagebrush,Russian Wormwood ■

36178 Artemisia sacrorum Ledeb. f. incana Pamp. = Artemisia sacrorum Ledeb. ■

36179 Artemisia sacrorum Ledeb. subsp. laxiflora (Nakai) Kitag.;疏花白莲蒿■

36180 Artemisia sacrorum Ledeb. subsp. laxiflora (Nakai) Kitag. = Artemisia sacrorum Ledeb. ■

36181 Artemisia sacrorum Ledeb. subsp. laxiflora (Nakai) Kitag. var. laciniiformis Nakai f. platyphylla Pamp. = Artemisia sacrorum Ledeb. ■

36182 Artemisia sacrorum Ledeb. subsp. manshurica (Kom.) Kitam. = Artemisia sacrorum Ledeb. ■

36183 Artemisia sacrorum Ledeb. subsp. manshurica Kitam. = Artemisia gmelinii Weber ex Stechm. ■

36184 Artemisia sacrorum Ledeb. subsp. manshurica Kitam. = Artemisia sacrorum Ledeb. var. incana (Besser) Y. R. Ling ■

36185 Artemisia sacrorum Ledeb. var. incana (Besser) Y. R. Ling;灰白莲蒿(供蒿,灰莲蒿,铁杆蒿,万年蒿,万年蓬)■

36186 Artemisia sacrorum Ledeb. var. intermedia Ledeb. = Artemisia sacrorum Ledeb. ■

36187 Artemisia sacrorum Ledeb. var. laciniaeformis Nakai = Artemisia sacrorum Ledeb. ■

36188 Artemisia sacrorum Ledeb. var. latiloba Ledeb. = Artemisia sacrorum Ledeb. ■

36189 Artemisia sacrorum Ledeb. var. latiloba Ledeb. f. freyniana Pamp. = Artemisia freyniana (Pamp.) Krasch. ■

36190 Artemisia sacrorum Ledeb. var. laxiflora (Nakai) Kitag. = Artemisia sacrorum Ledeb. var. messerschmidtiana (Besser) Y. R. Ling ●

36191 Artemisia sacrorum Ledeb. var. major f. japonica Pamp. = Artemisia tanacetifolia L. ■

36192 Artemisia sacrorum Ledeb. var. major Pamp. = Artemisia tanacetifolia L. ■

36193 Artemisia sacrorum Ledeb. var. messerschmidtiana (Besser) Y. R. Ling;密毛白莲蒿(白万年蒿)■

36194 Artemisia sacrorum Ledeb. var. minor Ledeb. = Artemisia gmelinii Weber ex Stechm. ■

36195 Artemisia sacrorum Ledeb. var. minor Ledeb. f. discolor Kom. = Artemisia sacrorum Ledeb. var. incana (Besser) Y. R. Ling ■

36196 Artemisia sacrorum Ledeb. var. minor Ledeb. f. vestota Kom. = Artemisia sacrorum Ledeb. var. incana (Besser) Y. R. Ling ■

36197 Artemisia sacrorum Ledeb. var. minor Ledeb. f. walliochiana Pamp. = Artemisia vestita Wall. ex Besser ●■

36198 Artemisia sacrorum Ledeb. var. santolinifolia (Turcz. ex Besser) Pamp. = Artemisia gmelinii Weber ex Stechm. ■

36199 Artemisia sacrorum Ledeb. var. vestita (Wall. ex Besser) Kitam. = Artemisia vestita Wall. ex Besser ●■

36200 Artemisia sacrorum Ledeb. var. viridis f. minor Freyn = Artemisia freyniana (Pamp.) Krasch. ■

36201 Artemisia saharae Pomel;左原蒿■☆

36202 Artemisia salina Willd.;盐土蒿■☆

36203 Artemisia salsoloides Pamp. = Artemisia ordosica Krasch. ●

36204 Artemisia salsoloides Pamp. var. mongolica Pamp. = Artemisia ordosica Krasch. ●

36205 Artemisia salsoloides Pamp. var. mongolica Pamp. = Artemisia sphaerocephala Krasch. ●

36206 Artemisia salsoloides Pamp. var. paniculata Hook. f. = Artemisia xigazeensis Y. Ling et Y. R. Ling ■

36207 Artemisia salsoloides Pamp. var. prattii Pamp. = Artemisia prattii (Pamp.) Y. Ling et Y. R. Ling ■

36208 Artemisia salsoloides Pamp. var. salsoloides Hook. f. = Artemisia xigazeensis Y. Ling et Y. R. Ling ■

36209 Artemisia salsoloides Pamp. var. salsoloides Hook. f. f. paniculata (Hook. f.) Pamp. = Artemisia xigazeensis Y. Ling et Y. R. Ling ■

36210 Artemisia salsoloides Pamp. var. wellbyi (Hemsl. et Pears.) Ostenf. et Paulsen = Artemisia wellbyi Hemsl. et Pears. ■

36211 Artemisia salsoloides Willd.;猪毛菜蒿(籽蒿);Russianthistle-like Sagebrush,Russianthistle-like Wormwood ■☆

36212 Artemisia salsoloides Willd. var. prattii Pamp. = Artemisia prattii (Pamp.) Y. Ling et Y. R. Ling ■

36213 Artemisia salsoloides Willd. var. wellbyi (Hemsl. et Pearson) Ostenf. et Paulsen = Artemisia wellbyi Hemsl. et Pears. ■

36214 Artemisia salsoloides Y. Ling = Artemisia sphaerocephala Krasch. ●

36215 Artemisia samamisica Besser = Artemisia vulgaris L. ■

36216 Artemisia samoiedorum Pamp.;萨摩蒿■☆

36217 Artemisia santolina Schrenk = Seriphidium santolinum (Schrenk) Poljakov ■

36218 Artemisia santolinifolia Turcz. = Artemisia gmelinii Weber ex Stechm. ■

36219 Artemisia santolinifolia Turcz. ex Besser = Artemisia gmelinii Weber ex Stechm. ■

36220 Artemisia santonica Lam.;山道年蛔蒿(山道年蒿)■☆

36221 Artemisia saposhnikovii Krasch. ex Poljakov;昆仑沙蒿;Saposhinikov Sagebrush ●■

36222 Artemisia saxicola Rydb. var. parryi A. Nelson = Artemisia laciniata Willd. subsp. parryi (A. Gray) W. A. Weber ■☆

36223 Artemisia schimperi Sch. Bip. ex Engl.;欣珀蒿■☆

36224 Artemisia schischkini Krasch. = Seriphidium nitrosum (Weber ex Stechm.) Poljakov ■

36225 Artemisia schmidtiana Maxim.;日本山蒿(蕨叶蒿);Angel's Hair,Artemisia,Silver Mound Artemisia ■☆

36226 Artemisia schmidtiana Maxim. 'Nana';矮生蕨叶蒿(银雾)■☆

36227 Artemisia schochii Mattf. = Artemisia carvifolia Buch.-Ham. ex Roxb. var. schochii (Mattf.) Pamp. ■

36228 Artemisia schrenkiana Ledeb. = Seriphidium schrenkianum (Ledeb.) Poljakov ■

36229 Artemisia scopaeformis Ledeb. = Seriphidium scopiforme (Ledeb.) Poljakov ■

36230 Artemisia scopaeformis Ledeb. f. longiramosa Poljakov = Seriphidium scopiforme (Ledeb.) Poljakov ■

36231 Artemisia scoparia Maxim. = Artemisia capillaris Thunb. ■

36232 Artemisia scoparia Waldst. et Kit.;猪毛蒿(安吕草,白蒿,白毛

蒿,白青蒿,白头蒿,白茵陈,北茵陈,滨蒿,臭蒿,东北茵陈蒿,猴子毛,黄蒿,黄毛蒿,灰毛蒿,马先,毛滨蒿,毛毛蒿,米蒿,绵茵陈,棉蒿,婆婆蒿,绒蒿,扫帚艾,沙蒿,山茵陈,石茵陈,土茵陈,西茵陈,细叶青蒿,香蒿,小白蒿,野兰蒿,野茼蒿,因尘,因陈蒿,茵陈,茵陈蒿,迎春蒿);Oriental Wormwood,Redstem Wormwood,Vigate Sagebrush,Vigated Wormwood ■

36233 Artemisia scoparia Waldst. et Kit. f. sericea Kom. = Artemisia scoparia Waldst. et Kit. ■

36234 Artemisia scoparia Waldst. et Kit. f. villosa Korsh.;绢毛东北茵陈蒿■

36235 Artemisia scoparia Waldst. et Kit. f. villosa Korsh. = Artemisia scoparia Waldst. et Kit. ■

36236 Artemisia scoparia Waldst. et Kit. var. heteromorpha Kitag. = Artemisia scoparia Waldst. et Kit. ■

36237 Artemisia scopariaeformis Popov = Artemisia scoparia Waldst. et Kit. ■

36238 Artemisia scopiforme Ledeb. = Seriphidium scopiforme (Ledeb.) Poljakov ■

36239 Artemisia scopulorum A. Gray var. monocephala A. Gray = Artemisia pattersonii A. Gray ■☆

36240 Artemisia selengensis Turcz. = Artemisia verlotiorum Lamotte ■

36241 Artemisia selengensis Turcz. ex Besser;蒌蒿(白蒿,蘩,高茎蒿,购,蒿蒌,红艾,红陈艾,刘寄奴,柳叶蒿,蒌,蒌蒿子,芦蒿,闾蒿,三叉叶蒿,水艾,水陈艾,水蒿,蒌,蒌蒿,狭叶艾,狭叶蒿,香艾,小蒿子,由胡);Seleng Sagebrush,Seleng Wormwood ■

36242 Artemisia selengensis Turcz. ex Besser f. simplicifolia Naiai = Artemisia selengensis Turcz. ex Besser ■

36243 Artemisia selengensis Turcz. ex Besser var. canabifolia Pamp. = Artemisia selengensis Turcz. ex Besser ■

36244 Artemisia selengensis Turcz. ex Besser var. canabifolia Pamp. f. dielsii Pamp. = Artemisia selengensis Turcz. ex Besser ■

36245 Artemisia selengensis Turcz. ex Besser var. canabifolia Pamp. f. genuina Pamp. = Artemisia selengensis Turcz. ex Besser ■

36246 Artemisia selengensis Turcz. ex Besser var. canabifolia Pamp. f. integerrima (Kom.) Kitag. = Artemisia selengensis Turcz. ex Besser ■

36247 Artemisia selengensis Turcz. ex Besser var. canabifolia Pamp. f. simplicifolia Pamp. = Artemisia selengensis Turcz. ex Besser ■

36248 Artemisia selengensis Turcz. ex Besser var. canabifolia Pamp. f. suingegra Pamp. = Artemisia selengensis Turcz. ex Besser ■

36249 Artemisia selengensis Turcz. ex Besser var. integerrima (Kom.) Y. Ling et Y. R. Ling ex X. D. Cui = Artemisia selengensis Turcz. ex Besser ■

36250 Artemisia selengensis Turcz. ex Besser var. pannosa Pamp. = Artemisia selengensis Turcz. ex Besser ■

36251 Artemisia selengensis Turcz. ex Besser var. shansiensis Y. R. Ling;无齿蒌蒿(柳叶蒿);Shaanxi Seleng Wormwood ■

36252 Artemisia selengensis Turcz. ex Besser var. typica Pamp. = Artemisia selengensis Turcz. ex Besser ■

36253 Artemisia selengensis Turcz. ex Besser var. typica Pamp. f. amurensis Pamp. = Artemisia selengensis Turcz. ex Besser ■

36254 Artemisia selengensis Turcz. ex Besser var. typica Pamp. f. genuina Pamp. = Artemisia selengensis Turcz. ex Besser ■

36255 Artemisia selengensis Turcz. ex Besser var. typica Pamp. f. serratifolia Pamp. = Artemisia selengensis Turcz. ex Besser ■

36256 Artemisia selengensis Turcz. ex Besser var. umbrosa Ledeb. = Artemisia lavandulifolia DC. ●■

36257 Artemisia semiarida (Krasch. et Lavrenko) Filatova = Seriphidium semiaridum (Krasch. et Lavrova) Y. Ling et Y. R. Ling ■

36258 Artemisia senecionis Jacquem. ex Besser = Hippolytia senecionis (Jacquem. ex Besser) Poljakov ex Tzvelev ■

36259 Artemisia senjavinensis Besser;塞尼亚蒿■☆

36260 Artemisia septemlobata H. Lév. et Vaniot = Artemisia lactiflora Wall. ex DC. ■

36261 Artemisia sericea (Besser) Weber = Artemisia sericea Weber ex Stechm. ■

36262 Artemisia sericea Weber ex Stechm.;绢毛蒿(丝蒿);Sericeous Sagebrush ■

36263 Artemisia sericea Weber ex Stechm. f. parviflora (DC.) Pamp. = Artemisia sericea Weber ex Stechm. ■

36264 Artemisia sericea Weber ex Stechm. var. gemliniana Besser = Artemisia sericea Weber ex Stechm. ■

36265 Artemisia sericea Weber ex Stechm. var. grandiflora DC. = Artemisia sericea Weber ex Stechm. ■

36266 Artemisia sericea Weber ex Stechm. var. ledebouriana Besser = Artemisia sericea Weber ex Stechm. ■

36267 Artemisia sericea Weber ex Stechm. var. nitens (Stev.) DC. = Artemisia sericea Weber ex Stechm. ■

36268 Artemisia sericea Weber ex Stechm. var. nitens DC. = Artemisia sericea Weber ex Stechm. ■

36269 Artemisia sericea Weber ex Stechm. var. pallsiana Besser = Artemisia sericea Weber ex Stechm. ■

36270 Artemisia sericea Weber ex Stechm. var. parviflora DC. = Artemisia sericea Weber ex Stechm. ■

36271 Artemisia sericea Weber ex Stechm. var. steveniana Besser = Artemisia sericea Weber ex Stechm. ■

36272 Artemisia sericea Weber ex Stechm. var. turkestanica C. Winkl. = Artemisia aschurbajewii C. Winkl. ■

36273 Artemisia sericophylla Rupr.;纤叶蒿(绢毛旱蒿)■☆

36274 Artemisia sericophylla Rupr. = Artemisia marschalliana Spreng. var. sericophylla (Rupr.) Y. R. Ling ■☆

36275 Artemisia seriphidium var. borotalense Poljakov = Seriphidium borotalense (Poljakov) Y. Ling et Y. R. Ling ■

36276 Artemisia serotina Bunge;晚熟蒿■☆

36277 Artemisia serrata Nutt.;齿叶蒿;Saw-leaf Mugwort,Saw-tooth Wormwood,Saw-toothed Sagebrush,Serrate-leaved Sage,Toothed Sage ■☆

36278 Artemisia serreana Pamp. = Artemisia tanacetifolia L. ■

36279 Artemisia shangnanensis Y. Ling et Y. R. Ling;商南蒿;Shangnan Sagebrush ■

36280 Artemisia shansiensis Pamp. = Artemisia lavandulifolia DC. ●■

36281 Artemisia shennongjiaensis Y. Ling et Y. R. Ling;神农架蒿;Shennongjia Sagebrush ■

36282 Artemisia sibirica (L.) Maxim. = Filifolium sibiricum (L.) Kitam. ■

36283 Artemisia sichuanensis Y. Ling et Y. R. Ling;四川艾;Sichuan Sagebrush ■

36284 Artemisia sichuanensis Y. Ling et Y. R. Ling var. tomentosa Y. Ling et Y. R. Ling;密毛四川艾;Tomentose Sichuan Sagebrush ■

36285 Artemisia sieberi Besser;西伯尔蒿■☆

36286 Artemisia sieversiana Ehrh. ex Willd.;大籽蒿(白艾蒿,白蒿,臭蒿子,大白蒿,大蓬蒿,大头蒿,蘩,蘩母,苦蒿,蓬蒿,皤蒿,山艾,习威氏蒿,小艾叶,洋艾,由胡);Bigseed Sagebrush,Sievers Wormwood ■

36287 Artemisia sieversiana Ehrh. ex Willd. var. blinii H. Lév. =

Artemisia sieversiana Ehrh. ex Willd. ■

36288 Artemisia sieversiana Ehrh. ex Willd. var. grandis Pamp. = Artemisia sieversiana Ehrh. ex Willd. ■

36289 Artemisia sieversiana Ehrh. ex Willd. var. koreana (Nakai) W. Wang et C. Y. Li = Artemisia sieversiana Ehrh. ex Willd. ■

36290 Artemisia sieversiana Ehrh. ex Willd. var. tibetica C. B. Clarke = Artemisia minor Jacq. ex Besser ■

36291 Artemisia sievesriana Ehrh. ex Willd. var. pygmaea Krylov = Artemisia macrocephala Jacq. ex Besser ■

36292 Artemisia silvatica Maxim.;森林蒿■☆

36293 Artemisia simplicifolia Pamp. = Artemisia dracunculus L. var. pamirica (C. Winkl.) Y. R. Ling et Humphries ■

36294 Artemisia simulans Pamp.;中南蒿(南艾);Similar Sagebrush ■

36295 Artemisia sinanensis Y. Yabe;西南蒿■☆

36296 Artemisia sinencionis Jacq. ex Besser = Hippolytia senecionis (Jacquem. ex Besser) Poljakov ex Tzvelev ■

36297 Artemisia sinensis (Pamp.) Y. Ling et Y. R. Ling;西南圆头蒿;SW. China Sagebrush ■

36298 Artemisia smithii Mattf.;球花蒿(高山蒿);Smith Sagebrush, Smith's Wormwood ■

36299 Artemisia smithii Mattf. f. paniculata Pamp. = Artemisia speciosa (Pamp.) Y. Ling et Y. R. Ling ■

36300 Artemisia smithii Mattf. var. speciosa Pamp. = Artemisia speciosa (Pamp.) Y. Ling et Y. R. Ling ■

36301 Artemisia smithii Mattf. var. speciosa Pamp. f. paniculata Pamp. = Artemisia speciosa (Pamp.) Y. Ling et Y. R. Ling ■

36302 Artemisia somai Hayata;台湾狭叶艾(相马氏艾);Taiwan Sagebrush ■

36303 Artemisia somai Hayata var. batakensis (Hayata) Kitam.;太鲁阁艾■

36304 Artemisia soongarica Schrenk;准噶尔沙蒿(黄沙蒿,沙蒿,中亚砂蒿);Dzungar sand Sagebrush, Songar Wormwood ●■

36305 Artemisia sparsa Kitag. = Artemisia sieversiana Ehrh. ex Willd. ■

36306 Artemisia speciosa (Pamp.) Y. Ling et Y. R. Ling;西南大头蒿;Brilliant Sagebrush ■

36307 Artemisia sphaerocephala Krasch.;圆头蒿(白杆子砂蒿,白沙蒿,白砂蒿,黄蒿,黄毛菜子,米蒿,香蒿,油砂蒿,圆头沙蒿,籽蒿);Round-cephaloid Wormwood, Roundhead Wormwood ●

36308 Artemisia spiciformis Osterh.;雪地蒿;Snowfield Sagebrush ■☆

36309 Artemisia spiciformis Osterh. var. longiloba Osterh. = Artemisia arbuscula Nutt. subsp. longiloba (Osterh.) L. M. Shultz ■☆

36310 Artemisia spinescens D. C. Eaton = Picrothamnus desertorum Nutt. ●☆

36311 Artemisia spithamaea Pursh = Artemisia borealis Pall. ■

36312 Artemisia splendens Willd.;光亮蒿■☆

36313 Artemisia stechmanniana Bess. var. sibirica Besser = Artemisia gmelinii Weber ex Stechm. ■

36314 Artemisia stelleriana Besser;海滩蒿;Beach Sage-wort, Beach Wormwood, Dusty-miller Sage-wort ■☆

36315 Artemisia stelleriana Besser = Artemisia lagocephala (Fisch. ex Besser) Fisch. ex DC. ●■

36316 Artemisia stelleriana Kom. = Artemisia lagocephala (Fisch. ex Besser) DC. ●■

36317 Artemisia stenocephala Krasch. ex Poljakov;窄头蒿■☆

36318 Artemisia stenophylla Kitam. = Artemisia integrifolia L. ■

36319 Artemisia stepposa B. Keller;草原蒿■☆

36320 Artemisia stewartii C. B. Clarke = Artemisia annua L. ■

36321 Artemisia stolonifera (Maxim.) Kom.;宽叶山蒿(天目蒿);Stolonbearing Sagebrush, Stolonbearing Wormwood ■

36322 Artemisia stolonifera (Maxim.) Kom. var. laciniata Nakai = Artemisia stolonifera (Maxim.) Kom. ■

36323 Artemisia stolonifera (Maxim.) Kom. var. microcephala Kitam. = Artemisia stolonifera (Maxim.) Kom. ■

36324 Artemisia strachereyi Hook. f. et Thomson ex C. B. Clarke var. grenardii (Franch.) Y. R. Ling = Seriphidium grenardii (Franch.) Y. R. Ling et Humphries ■

36325 Artemisia stracheyi Hook. f. et Thomson ex C. B. Clarke;冻原白蒿;Starchey Sagebrush, Starchey's Wormwood ■

36326 Artemisia stracheyi Hook. f. et Thomson ex C. B. Clarke var. grenardii (Franch.) Y. R. Ling = Seriphidium grenardii (Franch.) Y. R. Ling et Humphries ■

36327 Artemisia stricta Edgew.;劲直蒿(藏茵陈蒿,短叶蒿)■

36328 Artemisia stricta Edgew. = Artemisia edgeworthii N. P. Balakr. ■

36329 Artemisia stricta Edgew. f. diffusa Pamp. = Artemisia edgeworthii N. P. Balakr. var. diffusa (Pamp.) Y. Ling et Y. R. Ling ■

36330 Artemisia stricta Edgew. f. genuina Pamp. = Artemisia edgeworthii N. P. Balakr. ■

36331 Artemisia stricta Fisch. = Artemisia desertorum Spreng. ■

36332 Artemisia strongylocephala Pamp. = Artemisia roxburghiana Besser ■

36333 Artemisia strongylocephala Pamp. var. phyllobotrys Hand.-Mazz. = Artemisia phyllobotrys (Hand.-Mazz.) Y. Ling et Y. R. Ling ■

36334 Artemisia strongylocephala Pamp. var. sinensis Pamp. = Artemisia sinensis (Pamp.) Y. Ling et Y. R. Ling ■

36335 Artemisia strongylocephala Pamp. var. sinensis Pamp. f. genuina Pamp. = Artemisia sinensis (Pamp.) Y. Ling et Y. R. Ling ■

36336 Artemisia strongylocephala Pamp. var. sinensis Pamp. f. robusta Pamp. = Artemisia robusta (Pamp.) Y. Ling et Y. R. Ling ■

36337 Artemisia strongylocephala Pamp. var. sinensis Pamp. f. virgata Pamp. = Artemisia sinensis (Pamp.) Y. Ling et Y. R. Ling ■

36338 Artemisia suaveolens Poljakov = Seriphidium terrae-albae (Krasch.) Poljakov ■

36339 Artemisia suavis Jord.;法国蒿■☆

36340 Artemisia subdigitata Mattf.;无毛牛尾蒿(茶绒,牛尾蒿,指叶蒿);Hairless Oxtail Sagebrush, Subdigitate Wormwood ■

36341 Artemisia subdigitata Mattf. = Artemisia dubia Wall. ex Besser var. subdigitata (Mattf.) Y. R. Ling ■

36342 Artemisia subdigitata Mattf. var. falciloba Mattf. = Artemisia dubia Wall. ex Besser var. subdigitata (Mattf.) Y. R. Ling ■

36343 Artemisia subdigitata Mattf. var. intermedia Kitag. = Artemisia dubia Wall. ex Besser var. subdigitata (Mattf.) Y. R. Ling ■

36344 Artemisia subdigitata Mattf. var. thomsonii (C. B. Clarke ex Pamp.) S. Y. Hu = Artemisia dubia Wall. ex Besser ■

36345 Artemisia subintegra Kitam. = Artemisia japonica Thunb. ■

36346 Artemisia sublessingiana (B. Keller) Krasch. ex Poljakov = Seriphidium sublessingianum (B. Keller) Poljakov ■

36347 Artemisia subspinescens Boiss. = Artemisia persica Boiss. var. subspinescens (Boiss.) Boiss. ●■

36348 Artemisia subulata Nakai;线叶蒿;Libearleaf Sagebrush ■

36349 Artemisia subviscosa Besser;黏蒿■☆

36350 Artemisia succulenta Ledeb.;苏联肉质叶蒿;Sarcoleaf Sagebrush ■

36351 Artemisia succulentoides Y. Ling et Y. R. Ling;肉质叶蒿;Similar Sagebrush ■

36352 Artemisia suksdorfii Piper;萨克蒿;Suksdorf Sagewort ■☆

36353 Artemisia superba Pamp. = Artemisia vulgaris L. ■

36354 Artemisia sylvatica Maxim.;阴地蒿(白蒿,白脸蒿,茶绒蒿,火绒蒿,林地艾,林地蒿,林下艾,林中艾,毛蒿,山艾叶,野蒿);Woodland Sagebrush,Woodland Wormwood ■

36355 Artemisia sylvatica Maxim. var. meridionalis Pamp.;密序阴地蒿(阴地蒿);Densehead Woodland Sagebrush,Densehead Woodland Wormwood ■

36356 Artemisia sylvatica Maxim. var. typica Pamp. = Artemisia sylvatica Maxim. ■

36357 Artemisia szovitziana (Besser) Grossh.;曹氏蒿■☆

36358 Artemisia tacomensis Rydb. = Artemisia furcata M. Bieb. ■☆

36359 Artemisia tafelii Mattf.;波密蒿(寒漠蒿);Bomi Sagebrush ■

36360 Artemisia taibaishanensis Y. R. Ling et Humphries;太白山蒿;Taibaishan Sagebrush ●■

36361 Artemisia tainingensis Hand.-Mazz.;川藏蒿;Taining Sagebrush ■

36362 Artemisia tainingensis Hand.-Mazz. var. nitida (Pamp.) Y. R. Ling;无毛川藏蒿■

36363 Artemisia tanacetifolia L.;裂叶蒿(菊叶蒿,深山菊蒿,条蒿,细裂叶蒿);Tansyleaf Sagebrush,Tansyleaf Wormwood ■

36364 Artemisia tanacetifolia L. var. laxa Kitam. = Artemisia latifolia Ledeb. ■

36365 Artemisia tangutica Pamp.;甘青蒿;Tangut Sagebrush,Tangut Wormwood ■

36366 Artemisia tangutica Pamp. var. tomentosa Hand.-Mazz.;绒毛甘青蒿;Tomentose Tangut Sagebrush,Tomentose Tangut Wormwood ■

36367 Artemisia taurica Willd.;牛蒿(克里米亚蒿,克里木蒿);Tauri Wormwood ■☆

36368 Artemisia tenuisecta Nevski;细毛蒿■☆

36369 Artemisia tenuisecta Nevski var. karatavicum Krasch. et Abolin ex Poljakov = Seriphidium karatavicum (Krasch. et Abolin ex Poljakov) Y. Ling et Y. R. Ling ■

36370 Artemisia terrae-albae Krasch. = Seriphidium terrae-albae (Krasch.) Poljakov ■

36371 Artemisia terrae-albae Krasch. subsp. semiarida Krasch. = Seriphidium semiaridum (Krasch. et Lavrova) Y. Ling et Y. R. Ling ■

36372 Artemisia terrae-albae Krasch. subsp. semiarida Krasch. et Lavrova = Seriphidium semiaridum (Krasch. et Lavrova) Y. Ling et Y. R. Ling ■

36373 Artemisia terrae-albae Krasch. subsp. semiaridum Krasch. et Lavrenko = Seriphidium semiaridum (Krasch. et Lavrova) Y. Ling et Y. R. Ling ■

36374 Artemisia terrae-albae Krasch. var. heptapotamica Poljakov = Seriphidium heptapotamicum (Poljakov) Y. Ling et Y. R. Ling ■

36375 Artemisia terrae-albae Krasch. var. heptapotamicum Poljakov = Seriphidium heptapotamicum (Poljakov) Y. Ling et Y. R. Ling ■

36376 Artemisia terrae-albae Krasch. var. massagetovii Krasch. = Seriphidium terrae-albae (Krasch.) Poljakov ■

36377 Artemisia thellungiana Pamp.;藏腺毛蒿;Thellung Sagebrush ■

36378 Artemisia thomsoniana C. B. Clarke = Seriphidium thomsonianum (C. B. Clarke) Y. Ling et Y. R. Ling ■

36379 Artemisia thomsonii C. B. Clarke ex Pamp. = Artemisia dubia Wall. ex Besser ■

36380 Artemisia thunbergiana Maxim. = Artemisia carvifolia Buch.-Ham. ex Roxb. ■

36381 Artemisia thuscula Cav.;岛蒿■☆

36382 Artemisia tibetica Hook. f. et Thomson = Artemisia minor Jacq. ex Besser ■

36383 Artemisia tilesii Ledeb.;蒂氏蒿(蒂耳西艾);Tiles Sagebrush ■

36384 Artemisia tilesii Ledeb. = Artemisia leucophylla (Turcz. ex Besser) C. B. Clarke ■

36385 Artemisia tilesii Ledeb. subsp. gormanii (Rydb.) Hultén = Artemisia tilesii Ledeb. ■

36386 Artemisia tilesii Ledeb. subsp. hultenii (M. M. Maximova) V. G. Sergienko = Artemisia tilesii Ledeb. ■

36387 Artemisia tilesii Ledeb. var. aleutica (Hultén) S. L. Welsh = Artemisia tilesii Ledeb. ■

36388 Artemisia tilesii Ledeb. var. elatior Torr. et A. Gray = Artemisia tilesii Ledeb. ■

36389 Artemisia tilesii Ledeb. var. unalaschcensis Besser = Artemisia tilesii Ledeb. ■

36390 Artemisia togusbulakensis O. Fedtsch. = Artemisia persica Boiss. ●■

36391 Artemisia tomentella Trautv.;欧洲绒毛蒿■☆

36392 Artemisia tomentella Trautv. var. subglabra Krasch. = Artemisia marschalliana Spreng. ■

36393 Artemisia tongtchouanensis H. Lév. = Artemisia parviflora Buch.-Ham. ex Roxb. ■

36394 Artemisia tournefortiana Rchb.;湿地蒿;Wetland Sagebrush,Wetland Wormwood ■

36395 Artemisia transiliense Poljakov = Seriphidium transiliense (Poljakov) Poljakov ■

36396 Artemisia trautvetteriana Besser;特劳氏蒿;Trautvetter Sagebrush ■☆

36397 Artemisia trichophylla Wall. ex DC. = Artemisia scoparia Waldst. et Kit. ■

36398 Artemisia tridactyla Hand.-Mazz.;指裂蒿(三裂蒿);Fingershapesplit Sagebrush ■

36399 Artemisia tridactyla Hand.-Mazz. var. minima Y. R. Ling;小指裂蒿;Small Fingershapesplit Sagebrush ■

36400 Artemisia tridentata Nutt.;三齿蒿;Basin Sagebrush,Big Sage,Big Sagebrush,Bill Sagebrush,Black Sagebrush,Common Sagebrush,Great Basin Sagebrush,Rocky Mountain Sage,Sage Brush,Sage Plant,Sagebrush,Three-toothed Sagebrush ■

36401 Artemisia tridentata Nutt. subsp. arbuscula (Nutt.) H. M. Hall et Clem. = Artemisia arbuscula Nutt. ■☆

36402 Artemisia tridentata Nutt. subsp. bolanderi (A. Gray) H. M. Hall et Clem. = Artemisia cana Pursh subsp. bolanderi (A. Gray) G. H. Ward ■☆

36403 Artemisia tridentata Nutt. subsp. nova (A. Nelson) H. M. Hall et Clem. = Artemisia nova A. Nelson ■☆

36404 Artemisia tridentata Nutt. subsp. parishii (A. Gray) H. M. Hall et Clem.;莫哈韦三齿蒿;Mojave Sagebrush ■☆

36405 Artemisia tridentata Nutt. subsp. rothrockii (A. Gray) H. M. Hall et Clem. = Artemisia rothrockii A. Gray ■☆

36406 Artemisia tridentata Nutt. subsp. spiciformis (Osterh.) Kartesz et Gandhi = Artemisia spiciformis Osterh. ■☆

36407 Artemisia tridentata Nutt. subsp. trifida H. M. Hall et Clem. = Artemisia tripartita (Nutt.) Rydb. ■☆

36408 Artemisia tridentata Nutt. subsp. vaseyana (Rydb.) Beetle;山地三齿蒿;Mountain Sagebrush ■☆

36409 Artemisia tridentata Nutt. subsp. wyomingensis Beetle et A. M. Young;怀俄明三齿蒿;Wyoming Sagebrush ■☆

36410 Artemisia tridentata Nutt. subsp. xericensis Winward ex

Rosentreter et R. G. Kelsey = Artemisia tridentata Nutt. ■

36411 Artemisia tridentata Nutt. var. arbuscula (Nutt.) McMinn = Artemisia arbuscula Nutt. ■☆

36412 Artemisia tridentata Nutt. var. parishii (A. Gray) Jeps. = Artemisia tridentata Nutt. subsp. parishii (A. Gray) H. M. Hall et Clem. ■☆

36413 Artemisia tridentata Nutt. var. pauciflora Winward et Goodrich = Artemisia tridentata Nutt. subsp. vaseyana (Rydb.) Beetle ■☆

36414 Artemisia tridentata Nutt. var. vaseyana (Rydb.) B. Boivin = Artemisia tridentata Nutt. subsp. vaseyana (Rydb.) Beetle ■☆

36415 Artemisia tridentata Nutt. var. wyomingensis (Beetle et A. M. Young) S. L. Welsh = Artemisia tridentata Nutt. subsp. wyomingensis Beetle et A. M. Young ■☆

36416 Artemisia trifida Nutt.;三节蒿;Three-lobed Sagebrush, Three-lobed Wormwood ■☆

36417 Artemisia trifida Nutt. = Artemisia tripartita (Nutt.) Rydb. ■☆

36418 Artemisia trifida Nutt. var. rigida Nutt. = Artemisia rigida (Nutt.) A. Gray ■☆

36419 Artemisia trifida Turcz. = Hippolytia trifida (Turcz.) Poljakov ●

36420 Artemisia trifurcata Stephan ex Spreng. = Artemisia furcata M. Bieb. ■☆

36421 Artemisia trifurcata Stephan ex Spreng. var. pedunculosa (Koidz.) Kitam. = Artemisia furcata M. Bieb. ■☆

36422 Artemisia trifurcata Stephani ex Spreng. = Artemisia furcata M. Bieb. ■☆

36423 Artemisia tripartita (Nutt.) Rydb.;三尖蒿;Three-tip Sagebrush, Three-tipped Sagebrush ■☆

36424 Artemisia tripartita Rydb. = Artemisia tripartita (Nutt.) Rydb. ■☆

36425 Artemisia tristis Pamp. = Artemisia lavandulifolia DC. ●■

36426 Artemisia tschernieviana Besser;切氏蒿;Tscherniev Sagebrush ■☆

36427 Artemisia tsugitakaensis (Kitam.) Y. Ling et Y. R. Ling;雪山艾;Snowmountain Sagebrush ■

36428 Artemisia tsuneoi Tatew. et Kitam.;常尾蒿■☆

36429 Artemisia turcomanica Gand.;土库曼蒿■☆

36430 Artemisia turczaninoviana Besser var. altaica Krylov = Artemisia rutifolia Stephan ex Spreng. var. altaica (Krylov) Krasch. ■

36431 Artemisia turczaninoviana Besser var. dasyantha Schrenk = Artemisia rutifolia Stephan ex Spreng. ■

36432 Artemisia turczaninowiana Besser = Artemisia rutifolia Stephan ex Spreng. ●

36433 Artemisia turczaninowiana Besser var. altaica Krylov = Artemisia rutifolia Stephan ex Spreng. var. altaica (Krylov) Krasch. ■

36434 Artemisia turczaninowiana Besser var. falconeri (C. B. Clarke) O. Fedtsch. = Artemisia rutifolia Stephan ex Spreng. ■

36435 Artemisia turschaninowiana Krasn. = Artemisia gmelinii Weber ex Stechm. ■

36436 Artemisia tyrrellii Rydb. = Artemisia alaskana Rydb. ■☆

36437 Artemisia umbelliformis Lam.;阿尔卑斯蒿;Alps Wormwood ■

36438 Artemisia umbrosa (Besser) Turcz. ex DC. = Artemisia lavandulifolia DC. ●■

36439 Artemisia umbrosa (Besser) Turcz. ex Pamp. = Artemisia codonocephala Diels ■

36440 Artemisia unalaskensis Rydb.;乌那拉斯卡蒿■☆

36441 Artemisia unalaskensis Rydb. var. aleutica Hultén = Artemisia tilesii Ledeb. ■

36442 Artemisia ussuriensis Poljakov;乌苏里蒿■☆

36443 Artemisia vachanica Krasch. ex Poljakov;瓦哈恩蒿■☆

36444 Artemisia valida Krasch. ex Poljakov;刚直蒿■☆

36445 Artemisia variabilis Ten. = Artemisia campestris L. subsp. variabilis (Ten.) Greuter ■☆

36446 Artemisia vaseyana Rydb. = Artemisia tridentata Nutt. subsp. vaseyana (Rydb.) Beetle ■☆

36447 Artemisia velutina Pamp.;黄毛蒿;Yellowhair Sagebrush ■

36448 Artemisia velutina Pamp. f. foliosa Pamp. = Artemisia velutina Pamp. ■

36449 Artemisia velutina Pamp. f. genuina Pamp. = Artemisia velutina Pamp. ■

36450 Artemisia venusta Pamp. = Artemisia rubripes Nakai ■

36451 Artemisia venusta Pamp. var. microcephala Pamp. = Artemisia rupestris L. ■

36452 Artemisia venusta Pamp. var. typica Pamp. = Artemisia rupestris L. ■

36453 Artemisia verbenacea (Kom.) Kitag.;辽东蒿(蒿,小花蒙古蒿);Liaodong Sagebrush ■

36454 Artemisia verlotorum Lamotte;南艾蒿(白蒿,大青蒿,红陈艾,苦蒿,刘寄奴,紫蒿);Chinese Mugwort, Southern Sagebrush, Verlot's Mugwort ■

36455 Artemisia verlotorum Lamotte var. lobata Pamp. = Artemisia verlotorum Lamotte ■

36456 Artemisia verlotorum Lamotte var. rigida Pamp. = Artemisia verlotorum Lamotte ■

36457 Artemisia vermiculata L. = Seriphium plumosum L. ●☆

36458 Artemisia vestita Kitag. = Artemisia sacrorum Ledeb. ■

36459 Artemisia vestita Kitag. var. viridis (Besser) W. Wang et H. T. Ho ex H. C. Fu = Artemisia sacrorum Ledeb. ■

36460 Artemisia vestita Wall. ex Besser;毛莲蒿(白蒿,洁白蒿,结血蒿,老羊蒿,普尔那,山蒿,万年蓬,蚊子艾);Hairy Wormwood ●■

36461 Artemisia vestita Wall. ex Besser var. discolor (Kom.) Kitag. = Artemisia gmelinii Weber ex Stechm. ■

36462 Artemisia vestita Wall. ex Besser var. discolor (Kom.) Kitag. = Artemisia sacrorum Ledeb. var. incana (Besser) Y. R. Ling ■

36463 Artemisia vestita Wall. ex Besser var. viridis (Besser) W. Wang et H. T. Ho ex H. C. Fu = Artemisia sacrorum Ledeb. ■

36464 Artemisia vexans Pamp.;藏东蒿;Troublesome Sagebrush ■

36465 Artemisia viridifolia Spreng. = Artemisia rupestris L. ■

36466 Artemisia viridis Willd. = Artemisia rupestris L. ■

36467 Artemisia viridis Willd. ex DC. = Artemisia rupestris L. ■

36468 Artemisia viridisquama Kitam.;绿苞蒿;Greenbract Sagebrush ■

36469 Artemisia viridissima (Kom.) Pamp.;林艾蒿(绿蒿,一叶蒿,一枝蒿);Green Sagebrush, Green Wormwood ■

36470 Artemisia viridissima (Kom.) Pamp. var. japonica Pamp. = Artemisia monophylla Kitam. ■☆

36471 Artemisia viscida (Mattf.) Pamp.;腺毛蒿;Sticky Sagebrush ■

36472 Artemisia viscidissima Y. Ling et Y. R. Ling;密腺毛蒿;Muchsticky Sagebrush ■

36473 Artemisia viscidula (Osterh.) Rydb. = Artemisia cana Pursh subsp. viscidula (Osterh.) Beetle ■☆

36474 Artemisia vulgaris L.;北艾(艾,艾蒿,艾叶,艾子,菴闾,白蒿,白蒿子,冰台,眗,火艾,灸草,蒌,蒌蒿,蕲艾,蔄蒌,生艾,田艾蒿,五月艾,细叶艾,野艾,野艾蒿,野蒿,医草);Apple Pie, Bollane-bane, Bowlocks, Bulwand, Common Mugwort, Common Wormwood, Docko, Dog's Ear, Dog's Ears, Fat Hen, Fellon-herb, Fellon-wort, Felon Herb, Gallwood, Green Ginger, Green-ginger, Grey Bulwand, John's Feast-day Wort, Maderwort, Maidenwort, Maywort,

Midge Plant, Midgewort, Migwort, Mogvurd, Motherwort, Mugger, Muggert, Muggert Kale, Muggins, Muggons, Muggurth, Muggwith, Mugweed, Mugwood, Mugwort, Mugwort Wormwood, Naughty Man, Old Uncle Harry, Sage, Sage-wort, Sailor's Tobacco, Smotherwood, St. John's Herb, St. John's Wort, Wormwood ■

36475 Artemisia vulgaris L. = Artemisia indica Willd. ■

36476 Artemisia vulgaris L. subsp. candicans (Rydb.) H. M. Hall et Clem. = Artemisia ludoviciana Nutt. subsp. candicans (Rydb.) D. D. Keck ■☆

36477 Artemisia vulgaris L. subsp. longifolia (Nutt.) H. M. Hall et Clem. = Artemisia longifolia Nutt. ■

36478 Artemisia vulgaris L. subsp. ludoviciana (Nutt.) H. M. Hall et Clem. = Artemisia ludoviciana Nutt. ■☆

36479 Artemisia vulgaris L. subsp. michauxiana (Besser) H. St. John = Artemisia michauxiana Besser ■☆

36480 Artemisia vulgaris L. subsp. serrata (Nutt.) H. M. Hall et Clem. = Artemisia serrata Nutt. ■☆

36481 Artemisia vulgaris L. subsp. tilesii (Ledeb.) H. M. Hall et Clem. = Artemisia tilesii Ledeb. ■

36482 Artemisia vulgaris L. subsp. wrightii (A. Gray) H. M. Hall et Clem. = Artemisia carruthii A. W. Wood ex Carruth. ■☆

36483 Artemisia vulgaris L. var. candicans (Rydb.) M. Peck = Artemisia ludoviciana Nutt. subsp. candicans (Rydb.) D. D. Keck ■☆

36484 Artemisia vulgaris L. var. coarctica Forbes ex Besser = Artemisia mongolica (Fisch. ex Besser) Nakai ■

36485 Artemisia vulgaris L. var. coarctica Forbes ex Besser = Artemisia vulgaris L. ■

36486 Artemisia vulgaris L. var. douglasiana (Besser) H. St. John = Artemisia douglasiana Besser ■☆

36487 Artemisia vulgaris L. var. gilvescens (Miq.) Nakai = Artemisia gilvescens Miq. ■

36488 Artemisia vulgaris L. var. glabra Ledeb. = Artemisia vulgaris L. ■

36489 Artemisia vulgaris L. var. heterophylla (Nutt.) Jeps.;异叶北艾;Californian Mugwort ■☆

36490 Artemisia vulgaris L. var. incana Maxim. = Artemisia argyi H. Lév. ■

36491 Artemisia vulgaris L. var. incanescens Franch. = Artemisia argyi H. Lév. et Vaniot ■

36492 Artemisia vulgaris L. var. incompta (Nutt.) H. St. John = Artemisia ludoviciana Nutt. subsp. incompta (Nutt.) D. D. Keck ■☆

36493 Artemisia vulgaris L. var. indica (Willd.) Maxim. = Artemisia indica Willd. ■

36494 Artemisia vulgaris L. var. indica (Willd.) Maxim. f. montana Nakai = Artemisia montana (Nakai) Pamp. ■☆

36495 Artemisia vulgaris L. var. indica (Willd.) Maxim. f. nipponica Nakai = Artemisia princeps Pamp. ■

36496 Artemisia vulgaris L. var. indica Hayata f. nipponica Nakai = Artemisia princeps Pamp. ■

36497 Artemisia vulgaris L. var. indica Maxim. = Artemisia indica Willd. ■

36498 Artemisia vulgaris L. var. indica Maxim. = Artemisia princeps Pamp. ■

36499 Artemisia vulgaris L. var. integerrima Kom. = Artemisia selengensis Turcz. ex Besser ■

36500 Artemisia vulgaris L. var. integerrima Liou et al. = Artemisia subulata Nakai ■

36501 Artemisia vulgaris L. var. integrifolia Franch. = Artemisia anomala S. Moore ■

36502 Artemisia vulgaris L. var. integrifolia Ledeb. = Artemisia integrifolia L. ■

36503 Artemisia vulgaris L. var. integrifolia Makino et Nemoto = Artemisia viridissima (Kom.) Pamp. ■

36504 Artemisia vulgaris L. var. kamtschatica Besser = Artemisia leucophylla (Turcz. ex Besser) C. B. Clarke ■

36505 Artemisia vulgaris L. var. kamtschatica Besser = Artemisia vulgaris L. ■

36506 Artemisia vulgaris L. var. kiusiana Makino = Artemisia stolonifera (Maxim.) Kom. ■

36507 Artemisia vulgaris L. var. latifolia Fisch. ex Besser = Artemisia vulgaris L. ■

36508 Artemisia vulgaris L. var. latifolia Tanaka = Artemisia gilvescens Miq. ■

36509 Artemisia vulgaris L. var. latiloba Ledeb. = Artemisia princeps Pamp. ■

36510 Artemisia vulgaris L. var. latiloba Nakai = Artemisia princeps Pamp. ■

36511 Artemisia vulgaris L. var. latiloba Nakai. = Artemisia montana (Nakai) Pamp. ■☆

36512 Artemisia vulgaris L. var. leucophylla Turcz. ex Besser = Artemisia leucophylla (Turcz. ex Besser) C. B. Clarke ■

36513 Artemisia vulgaris L. var. littoralis Suksd. = Artemisia suksdorfii Piper ■☆

36514 Artemisia vulgaris L. var. longifolia (Nutt.) M. Peck = Artemisia longifolia Nutt. ■

36515 Artemisia vulgaris L. var. ludoviciana (Nutt.) Kuntze = Artemisia ludoviciana Nutt. ■☆

36516 Artemisia vulgaris L. var. maximoviczii Nakai = Artemisia princeps Pamp. ■

36517 Artemisia vulgaris L. var. maximowiczii Nakai = Artemisia indica Willd. var. maximowiczii (Nakai) H. Hara ■☆

36518 Artemisia vulgaris L. var. maximowiczii Nakai = Artemisia lancea Vaniot ■

36519 Artemisia vulgaris L. var. maximowiczii Nakai = Artemisia rubripes Nakai ■

36520 Artemisia vulgaris L. var. minor Ledeb. = Artemisia leucophylla (Turcz. ex Besser) C. B. Clarke ■

36521 Artemisia vulgaris L. var. mongolica Fisch. ex Besser = Artemisia mongolica (Fisch. ex Besser) Fisch. ex Nakai ■

36522 Artemisia vulgaris L. var. mongolica Fisch. ex Besser = Artemisia verbenacea (Kom.) Kitag. ■

36523 Artemisia vulgaris L. var. myriantha (Wall. ex Besser) C. B. Clarke = Artemisia myriantha Wall. ex Besser ■

36524 Artemisia vulgaris L. var. nilgarica C. B. Clarke = Artemisia nilagarica (C. B. Clarke) Pamp. ■

36525 Artemisia vulgaris L. var. parviflora Besser = Artemisia rupestris L. ■

36526 Artemisia vulgaris L. var. parviflora Matsum. = Artemisia lancea Vaniot ■

36527 Artemisia vulgaris L. var. parviflora Maxim. = Artemisia lancea Vaniot ■

36528 Artemisia vulgaris L. var. parviflora Maxim. = Artemisia rubripes Nakai ■

36529 Artemisia vulgaris L. var. racemulosa Pamp. = Artemisia lavandulifolia DC. ●■

36530 Artemisia vulgaris L. var. selegensis (Turcz. ex Besser) Maxim. = Artemisia selengensis Turcz. ex Besser ■
36531 Artemisia vulgaris L. var. stolonifera Loes. = Artemisia igniaria Maxim. ■
36532 Artemisia vulgaris L. var. stolonifera Maxim. = Artemisia igniaria Maxim. ■
36533 Artemisia vulgaris L. var. stolonifera Maxim. = Artemisia stolonifera (Maxim.) Kom. ■
36534 Artemisia vulgaris L. var. subulata Nakai = Artemisia subulata Nakai ■
36535 Artemisia vulgaris L. var. subulata Nakai ex Pamp. = Artemisia subulata Nakai ■
36536 Artemisia vulgaris L. var. tenuifolia Turcz. ex Besser = Artemisia mongolica (Fisch. ex Besser) Fisch. ex Nakai ■
36537 Artemisia vulgaris L. var. umbrosa Turcz. ex Besser = Artemisia lavandulifolia DC. ●■
36538 Artemisia vulgaris L. var. verbenacea Kom. = Artemisia verbenacea (Kom.) Kitag. ■
36539 Artemisia vulgaris L. var. viridissima Kom. = Artemisia viridissima (Kom.) Pamp. ■
36540 Artemisia vulgaris L. var. vulgatissima Besser = Artemisia igniaria Maxim. ■
36541 Artemisia vulgaris L. var. vulgatissima Besser = Artemisia indica Willd. ■
36542 Artemisia vulgaris L. var. vulgatissima Besser = Artemisia vulgaris L. ■
36543 Artemisia vulgaris L. var. vulgatissima Liou et al. = Artemisia igniaria Maxim. ■
36544 Artemisia vulgaris L. var. xizangensis Y. Ling et Y. R. Ling;藏北艾(北艾);Xizang Mugwort Wormwood ■
36545 Artemisia vulgaris Mattf. = Artemisia divaricata (Pamp.) Pamp. ■
36546 Artemisia vulgaris Mattf. = Artemisia orientali-hengduangensis Y. Ling et Y. R. Ling ■
36547 Artemisia wadei Edgew. = Artemisia annua L. ■
36548 Artemisia waltonii J. R. Drumm. ex Pamp.;藏龙蒿;Walton Wormwood ■
36549 Artemisia waltonii J. R. Drumm. ex Pamp. var. yushuensis Y. R. Ling;玉树龙蒿;Yushu Sagebrush ■
36550 Artemisia wellbyi Hemsl. et Pears.;藏沙蒿;Wellby Sagebrush ■
36551 Artemisia wellbyi Hemsl. et Pears. ex Deasy = Artemisia wellbyi Hemsl. et Pears. ■
36552 Artemisia wrightii A. Gray = Artemisia carruthii A. W. Wood ex Carruth. ■☆
36553 Artemisia wudanica Liou et W. Wang;乌丹蒿(大头蒿,圆头蒿);Wudan Wormwood ●
36554 Artemisia wulingschanensis A. I. Baranov et Skvortsov = Artemisia dubia Wall. ex Besser var. subdigitata (Mattf.) Y. R. Ling ■
36555 Artemisia wulingschanensis A. I. Baranov et Skvortsov ex Liou = Artemisia dubia Wall. ex Besser var. subdigitata (Mattf.) Y. R. Ling ■
36556 Artemisia xanthochloa Krasch.;黄绿蒿(黄沙蒿,沙蒿);Yellowgreen Sagebrush ■
36557 Artemisia xerophytica Krasch.;内蒙古旱蒿(旱蒿,内蒙古蒿,小砂蒿);Inner Mongol Sagebrush ●■
36558 Artemisia xigazeensis Y. Ling et Y. R. Ling;日喀则蒿;Rikeze Sagebrush ■
36559 Artemisia xylorrhiza Krasch. = Artemisia klementzae Krasch. ex Leonova ●■
36560 Artemisia yadongensis Y. Ling et Y. R. Ling;亚东蒿;Yadong Sagebrush ■
36561 Artemisia younghusbandii J. R. Drumm. ex Pamp.;藏白蒿;Xizang White Sagebrush ■
36562 Artemisia youngii Y. R. Ling;高原蒿;Young Sagebrush ■
36563 Artemisia yunnanensis (Pamp.) Krasch. = Neopallasia pectinata (Pall.) Poljakov ■
36564 Artemisia yunnanensis Jeffrey ex Diels;云南蒿(滇艾,戟叶蒿);Yunnan Sagebrush ■
36565 Artemisia yunnanensis Jeffrey ex Diels = Artemisia myriantha Wall. ex Besser ■
36566 Artemisia yunnanensis Jeffrey ex Diels var. nujiangensis Y. Ling et Y. R. Ling = Artemisia nujiangensis (Y. Ling et Y. R. Ling) Y. R. Ling ■
36567 Artemisia zayuensis Y. Ling et Y. R. Ling;察隅蒿;Chayu Sagebrush ■
36568 Artemisia zayuensis Y. Ling et Y. R. Ling var. pienmaensis Y. Ling et Y. R. Ling;片马蒿;Panma Sagebrush ■
36569 Artemisia zhaodongensis G. Y. Chang et M. Y. Liou = Artemisia anethoides Mattf. ■
36570 Artemisia zhongdianensis Y. R. Ling;中甸艾;Zhongdian Sagebrush ■
36571 Artemisiaceae Mertinov = Asteraceae Bercht. et J. Presl(保留科名)●■
36572 Artemisiaceae Mertinov = Compositae Giseke(保留科名)●■
36573 Artemisiaceae Mertinov;蒿科●■
36574 Artemisiastrum Rydb. = Artemisia L. ●■
36575 Artemisiastrum palmeri (A. Gray) Rydb. = Artemisia palmeri A. Gray ■☆
36576 Artemisiella Ghafoor(1992);小蒿属(冻原白蒿属)■
36577 Artemisiella stracheyii (Hook. f. et Thomson ex C. B. Clarke) Ghafoor;小蒿■
36578 Artemisiella stracheyii (Hook. f. et Thomson ex C. B. Clarke) Ghafoor = Arthraxon lancifolius (Trin.) Hochst. ■
36579 Artemisiopsis S. Moore(1902);蒿绒草属■☆
36580 Artemisiopsis linearis S. Moore = Artemisiopsis villosa (O. Hoffm.) Schweick. ■☆
36581 Artemisiopsis villosa (O. Hoffm.) Schweick.;蒿绒草■☆
36582 Artenema G. Don = Artanema D. Don(保留属名)■☆
36583 Arthostema Neck. = Gnetum L. ●
36584 Arthraerua (Kuntze) Schinz. (1893);无叶苋属■☆
36585 Arthraerua Schinz = Arthraerua (Kuntze) Schinz. ■☆
36586 Arthraerua leubnitziae (Kuntze) Schinz;无叶苋■☆
36587 Arthragrostis Lazarides(1985);北澳黍属■☆
36588 Arthragrostis deschampsioides (Domin) Lazarides;北澳黍■☆
36589 Arthratherum P. Beauv. = Aristida L. ■
36590 Arthratherum acutiflorum Trin. et Rupr. = Stipagrostis acutiflora (Trin. et Rupr.) De Winter ■☆
36591 Arthratherum brachyatherum Coss. et Balansa = Stipagrostis brachyathera (Coss. et Balansa) De Winter ■☆
36592 Arthratherum brachyatherum Coss. et Balansa var. acutiflorum (Trin. et Rupr.) Coss. et Balansa = Stipagrostis acutiflora (Trin. et Rupr.) De Winter ■☆
36593 Arthratherum brevifolium Nees = Stipagrostis brevifolia (Nees) De Winter ■☆
36594 Arthratherum caloptilum Jaub. et Spach = Stipagrostis paradisea (Edgew.) De Winter ■☆

36595 Arthratherum capense Nees var. macropus ? = Stipagrostis zeyheri (Nees) De Winter subsp. macropus ? ■☆
36596 Arthratherum ciliatum (Desf.) Nees = Stipagrostis ciliata (Desf.) De Winter ■☆
36597 Arthratherum lutescens Nees = Stipagrostis lutescens (Nees) De Winter ■☆
36598 Arthratherum namaquense Nees = Stipagrostis namaquensis (Nees) De Winter ■☆
36599 Arthratherum namaquense Nees var. fruticuliforme ? = Stipagrostis namaquensis (Nees) De Winter ■☆
36600 Arthratherum namaquense Nees var. vagans ? = Stipagrostis namaquensis (Nees) De Winter ■☆
36601 Arthratherum obtusum (Delile) Nees = Stipagrostis obtusa (Delile) Nees ■☆
36602 Arthratherum pennatum (Trin.) Tzvelev. = Stipagrostis pennata (Trin.) De Winter ■
36603 Arthratherum plumosum (L.) Nees = Stipagrostis plumosa (L.) Munro ex T. Anderson ■☆
36604 Arthratherum plumosum (L.) Nees var. floccosum Coss. et Durieu = Stipagrostis plumosa (L.) Munro ex T. Anderson subsp. seminuda (Trin. et Rupr.) H. Scholz ■☆
36605 Arthratherum pogonoptilum Jaub. et Spach = Stipagrostis pogonoptila (Jaub. et Spach) De Winter ■☆
36606 Arthratherum pungens (Desf.) P. Beauv. = Stipagrostis pungens (Desf.) De Winter ■☆
36607 Arthratherum subacaule Nees = Stipagrostis subacaulis (Nees) De Winter ■☆
36608 Arthratherum zeyheri Nees = Stipagrostis zeyheri (Nees) De Winter ■☆
36609 Arthraxella Nakai = Psittacanthus Mart. ●☆
36610 Arthraxon (Eichler) Tiegh. = Arthraxella Nakai ●☆
36611 Arthraxon (Eichler) Tiegh. = Psittacanthus Mart. ●☆
36612 Arthraxon P. Beauv. (1812);荩草属;Arthraxon, Ungeargrass ■
36613 Arthraxon Tiegh. = Arthraxella Nakai ●☆
36614 Arthraxon Tiegh. = Psittacanthus Mart. ●☆
36615 Arthraxon antsirabensis A. Camus = Arthraxon micans (Nees) Hochst. ■
36616 Arthraxon batangensis S. L. Zhong; 巴塘荩草; Batang Ungeargrass ■
36617 Arthraxon batangensis S. L. Zhong = Microstegium batangense (S. L. Zhong) S. M. Phillips et S. L. Chen ■
36618 Arthraxon breviaristatus Hack.;短芒荩草;Shortawn Ungeargrass ■
36619 Arthraxon breviaristatus Hack. = Arthraxon hispidus (Thunb.) Makino ■
36620 Arthraxon breviaristatus Hack. = Arthraxon typicus (Büse) Koord. ■
36621 Arthraxon breviaristatus Hack. var. brevisetus (Regel) Hara;短裂荩草;Shortseta Shortawn Ungeargrass ■
36622 Arthraxon breviaristatus Hack. var. cryptatheri (Hack.) Honda;匿芒荩草■
36623 Arthraxon breviaristatus Hack. var. echinatum (Nees) Hochst.;刺荩草;Spine Shortawn Ungeargrass ■
36624 Arthraxon breviaristatus Hack. var. hookeri (Hack.) Honda;毛轴荩草;Hooker Shortawn Ungeargrass ■
36625 Arthraxon castratus (Griff.) Nayaran. ex Bor;海南荩草;Castrate Carpgrass, Hainan Ungeargrass ■
36626 Arthraxon centrasiaticus (Griseb.) Gamajun. = Arthraxon hispidus (Thunb.) Makino var. centrasiaticus (Griseb.) Tzvelev ■
36627 Arthraxon ciliaris P. Beauv. = Arthraxon hispidus (Thunb.) Makino ■
36628 Arthraxon ciliaris P. Beauv. subsp. langsdorffii var. centrasiaticus Hochst. = Arthraxon hispidus (Thunb.) Makino var. centrasiaticus (Griseb.) Tzvelev ■
36629 Arthraxon ciliaris P. Beauv. subsp. langsdorfii var. cryptacherus Hack. = Arthraxon hispidus (Thunb.) Makino var. cryptatherus (Hack.) Honda ■
36630 Arthraxon ciliaris P. Beauv. subsp. nudus (Nees ex Steud.) Hack. = Arthraxon nudus (Nees ex Steud.) Hochst. ■
36631 Arthraxon ciliaris P. Beauv. subsp. submuticus (Nees ex Steud.) Hack. = Arthraxon submuticus (Nees ex Steud.) Hochst. ■
36632 Arthraxon ciliaris P. Beauv. var. centrasiaticus (Griseb.) Hack. = Arthraxon hispidus (Thunb.) Makino var. centrasiaticus (Griseb.) Tzvelev ■
36633 Arthraxon ciliaris P. Beauv. var. cryptatherus Hack. = Arthraxon hispidus (Thunb.) Makino ■
36634 Arthraxon ciliaris P. Beauv. var. hookeri Hack. = Arthraxon hispidus (Thunb.) Makino ■
36635 Arthraxon comorensis A. Camus = Arthraxon lancifolius (Trin.) Hochst. ■
36636 Arthraxon cryptatherus (Hack.) Koidz. = Arthraxon hispidus (Thunb.) Makino ■
36637 Arthraxon cryptacherus (Hack.) Koidz. = Arthraxon hispidus (Thunb.) Makino var. cryptatherus (Hack.) Honda ■
36638 Arthraxon cuspidatus (A. Rich.) Hack. var. micans (Nees) Hack. = Arthraxon micans (Nees) Hochst. ■
36639 Arthraxon cuspidatus (Hochst. ex A. Rich.) Hochst.;骤尖荩草■
36640 Arthraxon cuspidatus Hochst. ex A. Rich. var. micans (Nees) Hack. = Arthraxon hispidus (Thunb.) Makino ■
36641 Arthraxon cuspidatus Hochst. var. micans Hack. = Arthraxon micans (Nees) Hochst. ■
36642 Arthraxon echinatus Hochst. = Arthraxon lanceolatus (Roxb.) Hochst. var. echinatus (Nees) Hochst. ■
36643 Arthraxon epectinatus B. S. Sun et H. Peng;光脊荩草■
36644 Arthraxon guizhouensis S. L. Chen et Y. X. Jin; 贵州荩草; Guizhou Ungeargrass ■
36645 Arthraxon guizhouensis S. L. Chen et Y. X. Jin = Arthraxon epectinatus B. S. Sun et H. Peng ■
36646 Arthraxon hainanensis Keng et S. L. Chen = Arthraxon castratus (Griff.) Nayaran. ex Bor ■
36647 Arthraxon hispidus (Thunb.) Makino;荩草(鸱脚莎,黄草,戾草,菉草,菉蓐草,菉竹,绿竹,马耳草,马耳朵草,马牙草,毛竹,炮竹草,蓐,王刍,细叶秀竹,细叶莠竹);Hispid Arthraxon, Small Carpgrass, Ungeargrass ■
36648 Arthraxon hispidus (Thunb.) Makino f. brevisetus (Regel) Ohwi;短毛荩草■☆
36649 Arthraxon hispidus (Thunb.) Makino f. centrasiaticus Ohwi = Arthraxon hispidus (Thunb.) Makino var. centrasiaticus (Griseb.) Tzvelev ■
36650 Arthraxon hispidus (Thunb.) Makino f. japonicus (Regel) Ohwi;日本荩草■☆
36651 Arthraxon hispidus (Thunb.) Makino subsp. centrasiaticus (Griseb.) Tzvelev = Arthraxon hispidus (Thunb.) Makino var. centrasiaticus (Griseb.) Tzvelev ■
36652 Arthraxon hispidus (Thunb.) Makino subsp. centrasiaticus

Tzvelev = Arthraxon hispidus (Thunb.) Makino var. centrasiaticus (Griseb.) Tzvelev ■

36653 Arthraxon hispidus (Thunb.) Makino subsp. ciliaris (P. Beauv.) Masam. et Yamam. = Arthraxon hispidus (Thunb.) Makino ■

36654 Arthraxon hispidus (Thunb.) Makino subsp. langsdorffii (Trin.) Tzvelev = Arthraxon hispidus (Thunb.) Makino f. japonicus (Regel) Ohwi ■☆

36655 Arthraxon hispidus (Thunb.) Makino var. brevisetus (Regel) H. Hara = Arthraxon hispidus (Thunb.) Makino f. brevisetus (Regel) Ohwi ■☆

36656 Arthraxon hispidus (Thunb.) Makino var. centrasiaticus (Griseb.) Honda = Arthraxon hispidus (Thunb.) Makino var. centrasiaticus (Griseb.) Tzvelev ■

36657 Arthraxon hispidus (Thunb.) Makino var. centrasiaticus (Griseb.) Tzvelev;中亚荩草(荩草);Central-asian Arthraxon ■

36658 Arthraxon hispidus (Thunb.) Makino var. cryptatherus (Hack.) Honda = Arthraxon hispidus (Thunb.) Makino ■

36659 Arthraxon hispidus (Thunb.) Makino var. hookeri (Hack.) Honda;虎氏荩草;Small Carpgrass ■

36660 Arthraxon hispidus (Thunb.) Makino var. junnarensis (S. K. Jain et Hemadri) Welzen = Arthraxon junnarensis S. K. Jain et Hemadri ■

36661 Arthraxon hispidus (Thunb.) Makino var. muticus (Honda) Ohwi = Arthraxon hispidus (Thunb.) Makino ■

36662 Arthraxon hispidus (Thunb.) Makino var. nudus (Nees ex Steud.) Ohwi.;裸荩草■☆

36663 Arthraxon hispidus (Thunb.) Makino var. nudus (Nees ex Steud.) Ohwi = Arthraxon nudus (Nees ex Steud.) Hochst. ■

36664 Arthraxon hispidus (Thunb.) Makino var. quartinianus (A. Rich.) Backer = Arthraxon micans (Nees) Hochst. ■

36665 Arthraxon hispidus (Thunb.) Makino var. robustior Welzen = Arthraxon typicus (Büse) Koord. ■

36666 Arthraxon hispidus Humb. et Bonpl. ex Willd. subsp. langsdorffii (Thunb.) Tzvelev = Arthraxon hispidus (Thunb.) Makino ■

36667 Arthraxon hispidus Humb. et Bonpl. ex Willd. var. cryptatherus (Hack.) Honda = Arthraxon hispidus (Thunb.) Makino ■

36668 Arthraxon hispidus Humb. et Bonpl. ex Willd. var. muticus (Honda) Ohwi = Arthraxon hispidus (Thunb.) Makino ■

36669 Arthraxon hookeri (Hack.) Henrard = Arthraxon hispidus (Thunb.) Makino ■

36670 Arthraxon japonicus Miq. = Arthraxon hispidus (Thunb.) Makino ■

36671 Arthraxon junghuhnii (Steud.) Hochst. = Arthraxon typicus (Büse) Koord. ■

36672 Arthraxon junnarensis S. K. Jain et Hemadri;微穗荩草■

36673 Arthraxon lanceolatus (Roxb.) Hochst.;矛叶荩草(鸡窝乱,莠竹);Lanceleaf Arthraxon, Pikeleaf Ungeargrass ■

36674 Arthraxon lanceolatus (Roxb.) Hochst. f. glaberrimus Chiov. = Arthraxon lanceolatus (Roxb.) Hochst. ■

36675 Arthraxon lanceolatus (Roxb.) Hochst. f. puberulus Chiov. = Arthraxon lanceolatus (Roxb.) Hochst. ■

36676 Arthraxon lanceolatus (Roxb.) Hochst. subvar. serrulatus (Hochst.) Hack. = Arthraxon lanceolatus (Roxb.) Hochst. ■

36677 Arthraxon lanceolatus (Roxb.) Hochst. var. echinatus (Nees) Hack. = Arthraxon echinatus Hochst. ■

36678 Arthraxon lanceolatus (Roxb.) Hochst. var. echinatus (Nees) Hochst.;粗刺荩草;Thikspine Lanceleaf Arthraxon ■

36679 Arthraxon lanceolatus (Roxb.) Hochst. var. glabratus S. L. Chen et Y. X. Jin = Arthraxon prionodes (Steud.) Dandy ■

36680 Arthraxon lanceolatus (Roxb.) Hochst. var. raizadae (S. K. Jain, Hemadri et Deshp.) Welzen;毛颖荩草■

36681 Arthraxon lanceolatus (Roxb.) Hochst. var. serrulatus (Hochst.) T. Durand et Schinz = Arthraxon lanceolatus (Roxb.) Hochst. ■

36682 Arthraxon lancifolius (Trin.) Hochst.;小叶荩草;Littleleaf Ungeargrass ■

36683 Arthraxon lancifolius (Trin.) Hochst. var. eremophilus Bor = Arthraxon lancifolius (Trin.) Hochst. ■

36684 Arthraxon lancifolius (Trin.) Hochst. var. microphyllus Kuntze = Arthraxon microphyllus (Trin.) Hochst. ■

36685 Arthraxon langsdorffii (Trin.) Roshev. = Arthraxon hispidus (Thunb.) Makino f. japonicus (Regel) Ohwi ■☆

36686 Arthraxon langsdorffii (Trin.) Roshev. = Arthraxon hispidus (Thunb.) Makino ■

36687 Arthraxon maopingensis S. L. Chen et Y. X. Jin;茅坪荩草;Maoping Ungeargrass ■

36688 Arthraxon maopingensis S. L. Chen et Y. X. Jin = Arthraxon typicus (Büse) Koord. ■

36689 Arthraxon mauritianus Stapf ex C. E. Hubb. = Arthraxon micans (Nees) Hochst. ■

36690 Arthraxon micans (Nees) Hochst.;光亮荩草;Shine Ungeargrass ■

36691 Arthraxon micans (Nees) Hochst. = Arthraxon hispidus (Thunb.) Makino ■

36692 Arthraxon microphyllus (Trin.) Hochst.;小荩草;Small Ungeargrass, Small-leaf Arthraxon ■

36693 Arthraxon microphyllus (Trin.) Hochst. = Arthraxon lancifolius (Trin.) Hochst. ■

36694 Arthraxon microphyllus (Trin.) Hochst. var. genuinus Hook. f. = Arthraxon lancifolius (Trin.) Hochst. ■

36695 Arthraxon microphyllus (Trin.) Hochst. var. lancifolius (Trin.) Hack. = Arthraxon lancifolius (Trin.) Hochst. ■

36696 Arthraxon microphyllus (Trin.) Hochst. var. lancifolius (Trin.) Hack. = Arthraxon lanceolatus (Roxb.) Hochst. ■

36697 Arthraxon microphyllus (Trin.) Hochst. var. lancifolius Hochst. = Arthraxon lancifolius (Trin.) Hochst. ■

36698 Arthraxon microphyllus Hack. = Arthraxon lancifolius (Trin.) Hochst. ■

36699 Arthraxon microphyllus Hook. f. = Arthraxon lancifolius (Trin.) Hochst. ■

36700 Arthraxon minor Hochst. = Arthraxon lancifolius (Trin.) Hochst. ■

36701 Arthraxon molle (Nees) Duthie = Arthraxon lancifolius (Trin.) Hochst. ■

36702 Arthraxon molle Duttie = Arthraxon lancifolius (Trin.) Hochst. ■

36703 Arthraxon mollis (Nees) Duthie = Arthraxon lanceolatus (Roxb.) Hochst. ■

36704 Arthraxon multinervus S. L. Chen et Y. X. Jin;多脉荩草;Vervose Ungeargrass ■

36705 Arthraxon nodosus Kom. = Microstegium nodosum (Kom.) Tzvelev ■

36706 Arthraxon nodosus Kom. = Microstegium vimineum (Trin.) A. Camus ■

36707 Arthraxon nudus (Nees ex Steud.) Hochst.;光轴荩草;Glabrous

Lanceleaf Arthraxon ■
36708 Arthraxon okamotoi Ohwi = Arthraxon hispidus (Thunb.) Makino ■
36709 Arthraxon okamotoi Ohwi = Arthraxon quartinianus (A. Rich.) Nash ■
36710 Arthraxon pauciflorus Honda;粗梗荩草;Pauciflorous Ungeargrass ■
36711 Arthraxon pauciflorus Honda = Arthraxon hispidus (Thunb.) Makino ■
36712 Arthraxon pauciflorus Honda var. muticus Honda = Arthraxon hispidus (Thunb.) Makino ■
36713 Arthraxon pilo-phorus B. S. Sun. = Arthraxon prionodes (Steud.) Dandy ■
36714 Arthraxon prionodes (Steud.) Dandy = Arthraxon lanceolatus (Roxb.) Hochst. ■
36715 Arthraxon quartinianus (A. Rich.) Nash;暖地荩草■
36716 Arthraxon quartinianus (A. Rich.) Nash = Arthraxon hispidus (Thunb.) Makino ■
36717 Arthraxon quartinianus (A. Rich.) Nash = Arthraxon micans (Nees) Hochst. ■
36718 Arthraxon quartinianus (A. Rich.) Nash var. montanus Jacq.-Fél. = Arthraxon micans (Nees) Hochst. ■
36719 Arthraxon raizadae Jain Hemadri et Deshpande = Arthraxon lanceolatus (Roxb.) Hochst. var. raizadae (S. K. Jain, Hemadri et Deshp.) Welzen ■
36720 Arthraxon rudis (Nees ex Steud.) Hochst. = Arthraxon castratus (Griff.) Nayaran. ex Bor ■
36721 Arthraxon schimperi (Hochst. ex A. Rich.) Hochst. = Arthraxon lanceolatus (Roxb.) Hochst. ■
36722 Arthraxon schimperi Hochst. = Arthraxon lancifolius (Trin.) Hochst. ■
36723 Arthraxon serrulatus Hochst. = Arthraxon lanceolatus (Roxb.) Hochst. ■
36724 Arthraxon serrulatus Hochst. = Arthraxon prionodes (Steud.) Dandy ■
36725 Arthraxon sikkimensis Bor = Arthraxon microphyllus (Trin.) Hochst. ■
36726 Arthraxon spathacens Hook. f. = Arthraxon lanceolatus (Roxb.) Hochst. var. echinatus (Nees) Hack. ■
36727 Arthraxon spathaceus Hook. f. = Arthraxon echinatus Hochst. ■
36728 Arthraxon submuticus (Nees ex Steud.) Hochst.;无芒荩草■
36729 Arthraxon typicus (Büse) Koord.;洱源荩草■
36730 Arthraxon xinanensis S. L. Chen et Y. X. Jin;西南荩草;SW. China Ungeargrass ■
36731 Arthraxon xinanensis S. L. Chen et Y. X. Jin = Arthraxon epectinatus B. S. Sun et H. Peng ■
36732 Arthraxon xinanensis S. L. Chen et Y. X. Jin var. laxiflorus S. L. Chen et Y. X. Jin;疏序荩草;Laxflower SW. China Ungeargrass ■
36733 Arthraxon xinanensis S. L. Chen et Y. X. Jin var. laxiflorus S. L. Chen et Y. X. Jin = Arthraxon epectinatus B. S. Sun et H. Peng ■
36734 Arthrocarpum Balf. f. (1882);节果豆属■☆
36735 Arthrocarpum gracile Balf. f.;节果豆■☆
36736 Arthrocarpum somalense Hillc. et J. B. Gillett = Chapmannia somalensis (Hillc. et J. B. Gillett) Thulin ■☆
36737 Arthrocereus (A. Berger) A. Berger = Arthrocereus A. Berger(保留属名)●☆
36738 Arthrocereus A. Berger et F. M. Knuth = Arthrocereus A. Berger (保留属名)●☆
36739 Arthrocereus A. Berger(1929)(保留属名);关节柱属(关节仙人柱属)●☆
36740 Arthrochilium (Irmisch) Beck = Epipactis Zinn(保留属名)■
36741 Arthrochilium Beck = Epipactis Zinn(保留属名)■
36742 Arthrochilium Irmisch = Epipactis Zinn(保留属名)■
36743 Arthrochilium mairei (Schltr.) Szlach. = Epipactis mairei Schltr. ■
36744 Arthrochilium palustre (L.) Beck = Epipactis palustris (L.) Crantz ■
36745 Arthrochilium royleanum (Lindl.) Szlach. = Epipactis royleana Lindl. ■
36746 Arthrochilium schensianum (Schltr.) Szlach. = Epipactis mairei Schltr. ■
36747 Arthrochilium setschuanicum (Ames et Schltr.) Szlach. = Epipactis mairei Schltr. ■
36748 Arthrochilium thunbergii (A. Gray) Szlach. = Epipactis thunbergii A. Gray ■
36749 Arthrochilium veratrifolium (Boiss. et Hohenacker) Szlach. = Epipactis veratrifolia Boiss. et Hohen. ■
36750 Arthrochilium wallichii (Schltr.) Szlach. = Epipactis veratrifolia Boiss. et Hohen. ■
36751 Arthrochilium wilsonii (Schltr.) Szlach. = Epipactis mairei Schltr. ■
36752 Arthrochilium xanthophaeum (Schltr.) Szlach. = Epipactis xanthophaea Schltr. ■
36753 Arthrochilus F. Muell. (1858);节唇兰属■☆
36754 Arthrochilus F. Muell. = Spiculaea Lindl. ■☆
36755 Arthrochilus irritabilis F. Muell.;节唇兰■☆
36756 Arthrochilus latipes D. L. Jones;宽梗节唇兰■☆
36757 Arthrochilus stenophyllus D. L. Jones;窄叶节唇兰■☆
36758 Arthrochlaena Benth. = Sclerodactylon Stapf ■☆
36759 Arthrochlaena Boiv. ex Benth. = Sclerodactylon Stapf ■☆
36760 Arthrochloa Lorch = Acrachne Wight et Arn. ex Chiov. ■
36761 Arthrochloa Lorch = Dactyloctenium Willd. ■
36762 Arthrochloa Lorch = Normanboria Butzin ■
36763 Arthrochloa R. Br. = Holcus L. (保留属名)■
36764 Arthrochloa Schult. = Holcus L. (保留属名)■
36765 Arthrochortus Lowe = Lolium L. ■
36766 Arthroclianthus Baill. (1870);节花豆属(新喀豆属,新耀花豆属)■☆
36767 Arthroclianthus macrophyllus Schindl.;大叶节花豆■☆
36768 Arthroclianthus maximus Schindl;大节花豆■☆
36769 Arthroclianthus sanguineus Baill.;节花豆■☆
36770 Arthrocnemum Moq. (1840);大苞盐节木属(节藜属)●
36771 Arthrocnemum africanum Moss = Sarcocornia natalensis (Bunge ex Ung.-Sternb.) A. J. Scott ●☆
36772 Arthrocnemum ambiguum (Micheli) Moq. = Sarcocornia perennis (Mill.) A. J. Scott ■☆
36773 Arthrocnemum belangerianum Moq. = Halostachys caspica C. A. Mey. ex Schrenk ●
36774 Arthrocnemum capense Moss = Sarcocornia capensis (Moss) A. J. Scott ●☆
36775 Arthrocnemum decumbens Toelken;匍匐大苞盐节木●☆
36776 Arthrocnemum dunense Moss = Sarcocornia pillansii (Moss) A. J. Scott var. dunensis (Moss) O'Call. ●☆
36777 Arthrocnemum fruticosum (L.) Moq. = Sarcocornia fruticosa (L.) A. J. Scott ●☆
36778 Arthrocnemum glaucum Ung.-Sternb. = Arthrocnemum

macrostachyum (Moric.) K. Koch ●☆
36779 Arthrocnemum glaucum Ung. -Sternb. var. fasciculatum Sennen = Arthrocnemum macrostachyum (Moric.) K. Koch ●☆
36780 Arthrocnemum hottentoticum Moss = Sarcocornia pillansii (Moss) A. J. Scott ●☆
36781 Arthrocnemum indicum (Willd.) Moq.;印度大苞盐节木●☆
36782 Arthrocnemum indicum (Willd.) Moq. subsp. glaucum Maire et Weiller = Arthrocnemum macrostachyum (Moric.) K. Koch ●☆
36783 Arthrocnemum littoreum Moss = Sarcocornia littorea (Moss) A. J. Scott ●☆
36784 Arthrocnemum macrostachyum (Moric.) K. Koch;大穗大苞盐节木●☆
36785 Arthrocnemum macrostachyum K. Koch = Arthrocnemum macrostachyum (Moric.) K. Koch ●☆
36786 Arthrocnemum mossianum Toelken = Sarcocornia mossiana (Toelken) A. J. Scott ●☆
36787 Arthrocnemum namaquense Moss = Sarcocornia pillansii (Moss) A. J. Scott ●☆
36788 Arthrocnemum natalense (Bunge ex Ung. -Sternb.) Moss;纳塔尔大苞盐节木●☆
36789 Arthrocnemum natalense (Bunge ex Ung. -Sternb.) Moss var. affine (Moss) Toelken = Sarcocornia natalensis (Bunge ex Ung. -Sternb.) A. J. Scott var. affinis (Moss) O'Call. ●☆
36790 Arthrocnemum perenne (Mill.) Moss ex Fourc.;多年生盐节木●☆
36791 Arthrocnemum perenne (Mill.) Moss ex Fourc. var. lignosum (Woods) Moss = Sarcocornia perennis (Mill.) A. J. Scott var. lignosa (Woods) O'Call. ●☆
36792 Arthrocnemum pillansii Moss;皮朗斯盐节木●☆
36793 Arthrocnemum pillansii Moss var. dunense (Moss) Toelken = Sarcocornia pillansii (Moss) A. J. Scott var. dunensis (Moss) O'Call. ●☆
36794 Arthrocnemum subterminale (Parish) Standl.;北美盐节木(北美节藜)●☆
36795 Arthrocnemum terminale Toelken = Sarcocornia terminalis (Toelken) A. J. Scott ●☆
36796 Arthrocnemum variiflorum Moss = Sarcocornia perennis (Mill.) A. J. Scott ■☆
36797 Arthrocnemum xerophilum Toelken = Sarcocornia xerophila (Toelken) A. J. Scott ●☆
36798 Arthroiophis (Trin.) Chiov. = Andropogon L. (保留属名) ■
36799 Arthrolepis Boiss. = Achillea L. ■
36800 Arthrolobium Rchb. = Artrolobium Desv. ■☆
36801 Arthrolobium Rchb. = Ornithopus L. ■☆
36802 Arthrolobus Andrz. ex DC. = Rapistrum Crantz(保留属名)■☆
36803 Arthrolobus Steven ex DC. = Sterigma DC. ■
36804 Arthrolobus Steven ex DC. = Sterigmostemum M. Bieb. ■
36805 Arthrolophis Chiov. = Andropogon L. (保留属名) ■
36806 Arthrolophis fazoglensis Chiov. = Ischaemum afrum (J. F. Gmel.) Dandy ■☆
36807 Arthromischus Thwaites = Paramignya Wight ●
36808 Arthrophyllum Blume(1826);节叶属(节叶五加属)●☆
36809 Arthrophyllum Bojer = Phyllarthron DC. ex Meisn. ●☆
36810 Arthrophyllum Bojer ex A. DC. = Phyllarthron DC. ex Meisn. ●☆
36811 Arthrophyllum bojerianum DC.;节叶(节叶五加)●☆
36812 Arthrophyllum madagascariense Bojer = Phyllarthron madagascariense K. Schum. ●☆
36813 Arthrophytum Schrenk (1845); 节节木属; Arthrophytum, Nodosetree ●
36814 Arthrophytum acutifolium (Minkw.) Minkw. = Haloxylon persicum Bunge ex Boiss. et Buhse ●◇
36815 Arthrophytum ammodendron (C. A. Mey.) Litv. = Haloxylon ammodendron (C. A. Mey.) Bunge ex Fenzl ●
36816 Arthrophytum ammodendron (C. A. Mey.) Litv. var. acutifolium Minkw. = Haloxylon persicum Bunge ex Boiss. et Buhse ●◇
36817 Arthrophytum ammodendron (C. A. Mey.) Litv. var. aphyllum Minkw. = Haloxylon ammodendron (C. A. Mey.) Bunge ex Fenzl ●
36818 Arthrophytum ammodendron Litv. var. acutifolium Minkw. = Haloxylon persicum Bunge ex Boiss. et Buhse ●◇
36819 Arthrophytum ammodendron Litv. var. aphyllum Minkw. = Haloxylon ammodendron (C. A. Mey.) Bunge ex Fenzl ●
36820 Arthrophytum arborescens Litv.;土耳其斯坦节节木●☆
36821 Arthrophytum balchaschense (Iljin) Botsch.;鳞叶节节木●
36822 Arthrophytum iliense Iljin; 长枝节节木; Longbranch Arthrophytum, Longbranch Nodosetree ●
36823 Arthrophytum korovinii Botsch.; 棒叶节节木; Korovin Arthrophytum, Korovin Nodosetree ●
36824 Arthrophytum longibracteatum Korovin; 长叶节节木; Longleaf Arthrophytum, Longleaf Nodosetree ●
36825 Arthrophytum persicum (Bunge) Sav. -Rycz. = Haloxylon persicum Bunge ex Boiss. et Buhse ●◇
36826 Arthrophytum persicum Sav. -Rycz. = Haloxylon persicum Bunge ex Boiss. et Buhse ●◇
36827 Arthrophytum regelii (Bunge) Litv. = Iljinia regelii (Bunge) Korovin ●
36828 Arthrophytum regelii Litv. = Iljinia regelii (Bunge) Korovin ●
36829 Arthrophytum schmittianum (Pomel) Maire et Weiller = Hammada schmittiana (Pomel) Botsch. ●☆
36830 Arthrophytum schmittianum (Pomel) Maire et Weiller var. prostratum Le Houér. = Hammada schmittiana (Pomel) Botsch. ●☆
36831 Arthrophytum scoparium (Pomel) Iljin ex Jahand. et Maire = Hammada scoparia (Pomel) Iljin ●☆
36832 Arthrophytum scoparium (Pomel) Iljin var. articulatum (Moq.) Le Houér. = Hammada scoparia (Pomel) Iljin ●☆
36833 Arthropodium R. Br. (1810);龙舌百合属;Chocolate Lily ■☆
36834 Arthropodium caesioides H. Perrier;淡蓝龙舌百合■☆
36835 Arthropodium cirrhatum R. Br.; 龙舌百合; Rengarenga, Riengalily, Rock Lily ■☆
36836 Arthropodium hispidum (L.) Spreng. = Trachyandra hispida (L.) Kunth ■☆
36837 Arthropodium milleflorum (DC.) J. F. Macbr.;乳花龙舌百合; Pale Vanilla-lily ■☆
36838 Arthropodium muricatum (L. f.) Spreng. = Trachyandra muricata (L. f.) Kunth ■☆
36839 Arthropogon Nees(1829);节芒草属■☆
36840 Arthropogon bolivianus Filg.;玻利维亚节芒草■☆
36841 Arthropogon filifolius Filg.;线叶节芒草■☆
36842 Arthropogon lanceolatus Filg.;披针叶节芒草■☆
36843 Arthropogon villosus Nees;节芒草■☆
36844 Arthrosamanea Britton et Rose = Albizia Durazz. ●
36845 Arthrosamanea Britton et Rose ex Britton et Killip(1936);节雨树属●☆
36846 Arthrosamanea altissima (Oliv.) G. C. C. Gilbert et Boutique = Cathormion altissimum (Hook. f.) Hutch. et Dandy ●☆
36847 Arthrosamanea eriorhachis (Harms) Aubrév. = Cathormion

eriorhachis (Harms) Dandy ●☆
36848 Arthrosamanea leptophylla (Harms) G. C. C. Gilbert et Boutique = Samanea leptophylla (Harms) Brenan et Brummitt ●☆
36849 Arthrosamanea leptophylla (Harms) G. C. C. Gilbert et Boutique var. guineensis G. C. C. Gilbert et Boutique = Samanea leptophylla (Harms) Brenan et Brummitt ●☆
36850 Arthrosamanea obliquifoliolata (De Wild.) G. C. C. Gilbert et Boutique = Cathormion obliquifoliolatum (De Wild.) G. C. C. Gilbert et Boutique ●☆
36851 Arthrosamanea pistaciifolia (Willd.) Britton et Rose;节雨树●☆
36852 Arthrosia (Luer) Luer = Pleurothallis R. Br. ■☆
36853 Arthrosolen C. A. Mey. = Gnidia L. ●☆
36854 Arthrosolen calocephalus C. A. Mey. = Gnidia calocephala (C. A. Mey.) Gilg ●☆
36855 Arthrosolen chrysanthus Solms = Gnidia chrysantha Gilg ●☆
36856 Arthrosolen chrysanthus Solms var. ignea (Gilg) H. Pearson = Gnidia chrysantha Gilg ●☆
36857 Arthrosolen compactus C. H. Wright = Gnidia compacta (C. H. Wright) J. H. Ross ●☆
36858 Arthrosolen foliosus H. Pearson = Gnidia foliosa (H. Pearson) Engl. ●☆
36859 Arthrosolen fraternus N. E. Br. = Gnidia fraterna (N. E. Br.) E. Phillips ●☆
36860 Arthrosolen gymnostachys C. A. Mey. = Gnidia gymnostachya (C. A. Mey.) Gilg ●☆
36861 Arthrosolen inconspicuus (Meisn.) Meisn. = Gnidia inconspicua Meisn. ●☆
36862 Arthrosolen latifolius Oliv. = Gnidia latifolia (Oliv.) Gilg ●☆
36863 Arthrosolen laxa (L. f.) C. A. Mey. = Gnidia laxa (L. f.) Gilg ●☆
36864 Arthrosolen microcephalus (Meisn.) E. Phillips = Gnidia microcephala Meisn. ●☆
36865 Arthrosolen ornatus Meisn. = Gnidia ornata (Meisn.) Gilg ●☆
36866 Arthrosolen phaeotrichus C. H. Wright = Gnidia phaeotricha Gilg ●☆
36867 Arthrosolen polycephalus C. A. Mey. = Gnidia polycephala (C. A. Mey.) Gilg ex Engl. ●☆
36868 Arthrosolen sericocephalus Meisn. = Gnidia sericocephala (Meisn.) Gilg ex Engl. ●☆
36869 Arthrosolen somalensis Franch. = Gnidia somalensis (Franch.) Gilg ●☆
36870 Arthrosolen sphaerantha H. Pearson = Gnidia chrysantha Gilg ●☆
36871 Arthrosolen sphaerocephalus Baker = Gnidia somalensis (Franch.) Gilg ●☆
36872 Arthrosolen spicatus (L. f.) C. A. Mey. = Gnidia spicata (L. f.) Gilg ●☆
36873 Arthrosolen variabilis C. H. Wright = Gnidia variabilis (C. H. Wright) Engl. ●☆
36874 Arthrosprion Hassk. = Acacia Mill. (保留属名)●■
36875 Arthrostachya Link = Gaudinia P. Beauv. ■☆
36876 Arthrostachys Desv. = Andropogon L. (保留属名)■
36877 Arthrostemma DC. = Brachyotum (DC.) Triana ex Benth. ●☆
36878 Arthrostemma Naudin = Pterolepis (DC.) Miq. (保留属名)●■☆
36879 Arthrostemma Pav. ex D. Don(1823);节冠野牡丹属■☆
36880 Arthrostemma ciliatum ?;缘毛节冠野牡丹;Arthrostemma, Pinkfringe ■☆
36881 Arthrostemma paniculatum D. Don = Oxyspora paniculata (D. Don) DC. ●
36882 Arthrostygma Steud. = Petrophile R. Br. ex Knight ●☆
36883 Arthrostylidium Rupr. (1840);芦柱竹属(内门竹属)●☆
36884 Arthrostylidium naibunense (Hayata) W. C. Lin = Drepanostachyum naibunense (Hayata) P. C. Keng ●
36885 Arthrostylis Boeck. = Actinoschoenus Benth. ■
36886 Arthrostylis R. Br. (1810);节柱莎草属■☆
36887 Arthrostylis chinensis Benth. = Fimbristylis chinensis (Benth.) Ts. Tang et F. T. Wang ■
36888 Arthrotaxidaceae Lotsy = Cupressaceae Gray(保留科名)●
36889 Arthrotaxis Endl. = Athrotaxis D. Don ●☆
36890 Arthrothamnus Klotzsch et Garcke = Euphorbia L. ●■
36891 Arthrothamnus bergii Klotzsch et Garcke = Euphorbia burgeri M. G. Gilbert ●■☆
36892 Arthrothamnus densiflorus Klotzsch et Garcke = Euphorbia mundii N. E. Br. ●☆
36893 Arthrothamnus ecklonii Klotzsch et Garcke = Euphorbia burgeri M. G. Gilbert ●■☆
36894 Arthrothamnus scopiformis Klotzsch et Garcke = Euphorbia arceuthobioides Boiss. ☆
36895 Arthrotrichum F. Muell. = Trichinium R. Br. ●■☆
36896 Arthrozamia Rchb. = Encephalartos Lehm. ●☆
36897 Artia Guillaumin(1941);南亚夹竹桃属●☆
36898 Artocarpaceae Bercht. et J. Presl = Moraceae Gaudich. (保留科名)●■
36899 Artocarpaceae R. Br.;波罗蜜科●
36900 Artocarpaceae R. Br. = Moraceae Gaudich. (保留科名)●■
36901 Artocarpus Forst. = Artocarpus J. R. Forst. et G. Forst. (保留属名)●
36902 Artocarpus J. R. Forst. et G. Forst. (1775)(保留属名);波罗蜜属(菠萝蜜属,桂木属,面包树属,木波罗属,木菠萝属);Artocarpus, Bread Fruit Tree, Bread-fruit, Roman Wormwood ●
36903 Artocarpus africana Sim = Treculia africana Decne. ex Trécul ●☆
36904 Artocarpus altilis (Parkinson) Fosberg = Artocarpus communis J. R. Forst. et G. Forst. ●
36905 Artocarpus altilis (Parkinson) Fosberg = Artocarpus incisus (Thunb.) L. f. ●
36906 Artocarpus altilis (Parkinson) Fosberg var. seminiferus (Duss) Fournet = Artocarpus communis J. R. Forst. et G. Forst. ●
36907 Artocarpus altilis G. Forst. = Artocarpus incisus (Thunb.) L. f. ●
36908 Artocarpus bicolor Merr. et Chun = Artocarpus styracifolius Pierre ●
36909 Artocarpus brevisericeus C. Y. Wu et W. T. Wang = Artocarpus petelotii Gagnep. ●
36910 Artocarpus chama Buch. -Ham. ex Wall.;野树波罗(山波罗,榅桲木波罗蜜);Chaplash, Wild Artocarpus ●
36911 Artocarpus champeden (Lour.) Spreng.;尖笔拉●☆
36912 Artocarpus chaplasha Roxb. = Artocarpus chama Buch. -Ham. ex Wall. ●
36913 Artocarpus communis J. R. Forst. et G. Forst.;面包树(兰屿面包树,面磅树,面包果,面包果树,缺刻木波罗,锐裂波罗蜜);Bread Fruit, Bread Fruit Tree, Breadfruit, Bread-fruit, Bread-fruit Tree, Breadnut ●
36914 Artocarpus elasticus Reinw. ex Blume;马来波罗蜜(弹性木波罗);Antipdo Gumihan ●
36915 Artocarpus ficifolius W. T. Wang = Artocarpus lacucha Buch. -Ham. ex D. Don ●◇
36916 Artocarpus gomezianus Wall. = Artocarpus gomezianus Wall. ex Trécul ●

36917 Artocarpus gomezianus Wall. ex Trécul;长圆叶波罗蜜;Oblong-leaved Artocarpus ●

36918 Artocarpus gomezianus Wall. ex Trécul var. griffithii King = Artocarpus nitidus Trécul subsp. griffithii (King ex Hook. f.) F. M. Jarrett ●

36919 Artocarpus gongshanensis S. K. Wu ex C. Y. Wu et S. S. Chang;贡山波罗蜜;Gongshan Artocarpus ●

36920 Artocarpus heterophyllus Lam.;波罗蜜(阿萨躯,包蜜,波罗蜜树,木波罗,木波萝,曩伽结,曩伽结树,牛肚子果,婆罗,婆那娑,树波罗,树菠萝,树婆罗,天菠萝,天婆罗,优珠昙);Breadfruit, Diversileaf Artocarpus, Heterophyllous Artocarpus, Jaca, Jack, Jack Fruit, Jack Nut, Jack Tree, Jackfruit, Jack-fruit, Jak, Jakfruit, Nangka Langka ●

36921 Artocarpus hirsutus Lam.;硬毛面包果;Ainee ●☆

36922 Artocarpus hypargyreus Hance ex Benth.;白桂木(红桂木,红桂树,将军木,将军树,胭脂木);Red Cassia Tree, Silverback Artocarpus, Silver-backed Artocarpus ●

36923 Artocarpus incisus (Thunb.) L. f. = Artocarpus communis J. R. Forst. et G. Forst. ●

36924 Artocarpus integer (Thunb.) Merr.;全缘桂木;Chempedak, Integerleaf Artocarpus, Jakfrujt, Jeckfrult ●

36925 Artocarpus integrifolia L. f. = Artocarpus heterophyllus Lam. ●

36926 Artocarpus integrifolia L. f. = Artocarpus integer (Thunb.) Merr. ●

36927 Artocarpus jambolana ? = Artocarpus heterophyllus Lam. ●

36928 Artocarpus lacucha Buch. -Ham. ex D. Don;滇波罗蜜(猴面果,拉口沙面包果,拉口沙木波罗,泰国波罗蜜,野波罗蜜);Lacoocha, Lakoocha, Monkey Jack, Yunnan Artocarpus ●◇

36929 Artocarpus lakoocha Roxb. = Artocarpus lacucha Buch. -Ham. ex D. Don ●◇

36930 Artocarpus lakoocha Wall. ex Roxb. = Artocarpus lacucha Buch. -Ham. ex D. Don ●◇

36931 Artocarpus lakucha Buch. -Ham. = Artocarpus lacucha Buch. -Ham. ex D. Don ●◇

36932 Artocarpus lanceolata Trécul = Artocarpus nitidus Trécul subsp. griffithii (King ex Hook. f.) F. M. Jarrett ●

36933 Artocarpus lingnanensis Merr. = Artocarpus nitidus Trécul subsp. lingnanensis (Merr.) F. M. Jarrett ●

36934 Artocarpus melinoxylus Gagnep. = Artocarpus chama Buch. -Ham. ex Wall. ●

36935 Artocarpus nanchuanensis S. S. Chang, S. C. Tan et Z. Y. Liu;南川波罗蜜(南川木波罗);Nanchuan Artocarpus ●

36936 Artocarpus nigrifolius C. Y. Wu;牛李;Black-leaf Artocarpus ●

36937 Artocarpus nitidus Trécul;光叶桂木(滇光叶桂木,披针叶桂木);Glabrous Artocarpus, Shining Artocarpus ●

36938 Artocarpus nitidus Trécul subsp. griffithii (King ex Hook. f.) F. M. Jarrett;披针叶桂木;Griffith Glabrous Artocarpus ●

36939 Artocarpus nitidus Trécul subsp. lingnanensis (Merr.) F. M. Jarrett;桂木(白桂木,大叶胭脂,狗果,狗果树,红桂木,胭脂公);Lingnan Artocarpus ●

36940 Artocarpus odoratissimus Blanco;芳香波罗蜜;Indian Fountain-bamboo, Marang, Terap ●☆

36941 Artocarpus parva Gagnep. = Artocarpus nitidus Trécul subsp. lingnanensis (Merr.) F. M. Jarrett ●

36942 Artocarpus petelotii Gagnep.;短绢毛波罗蜜(短绢毛桂木,猴欢喜,马蛋果,糖包果,仙桃);Pepelot's Artocarpus ●

36943 Artocarpus petiolaris Miq. = Artocarpus gomezianus Wall. ex Trécul ●

36944 Artocarpus pithecogallus C. Y. Wu;猴子瘿袋●

36945 Artocarpus rigidus Blume;猴面果;Monkey Jack, Rigid Artocarpus ●

36946 Artocarpus styracifolius Pierre;二色波罗蜜(二色菠萝蜜,二色桂木,红枫荷,红山梅,将军木,木皮,奶浆果,沙蕾木,英杜);Bicolor Artocarpus ●

36947 Artocarpus tonkinensis A. Chev. ex Gagnep.;胭脂树(北部湾桂木,果,鸡脖子,鸡嗉果,胭脂,胭脂木,越南桂木);Tonkin Artocarpus ●

36948 Artocarpus xanthocarpa Merr.;黄果波罗蜜(兰屿面包树);Yellow-fruit Artocarpus ●

36949 Artocarpus yunnanensis Hu = Artocarpus lacucha Buch. -Ham. ex D. Don ●◇

36950 Artomeria Breda = Eria Lindl.(保留属名)■

36951 Artorhiza Raf. = Battata Hill ●■

36952 Artorhiza Raf. = Solanum L. ●■

36953 Artorhlza Raf. = Parmentiera Raf. ●■

36954 Artorima Dressler et G. E. Pollard(1971);裂盘兰属■☆

36955 Artorima erubescens (Lindl.) Dressler et G. E. Pollard;裂盘兰■☆

36956 Artrolobium Desv.(1813);地中海豆属■☆

36957 Artrolobium Desv. = Coronilla L.(保留属名)●■

36958 Artrolobium Desv. = Ornithopus L. ■☆

36959 Artrolobium Desv. = Scorpius Medik. ●■

36960 Artrolobium durum Desv.;地中海豆☆

36961 Artrolobium micranthum Benth.;小花地中海豆☆

36962 Aruana Burm. f. = Myristica Gronov.(保留属名)●

36963 Aruba Aubl. = Quassia L. ●☆

36964 Aruba Nees et Mart. = Almeidea A. St. -Hil. ●☆

36965 Arudinaria naibunense Hayata = Drepanostachyum naibunense (Hayata) P. C. Keng ●

36966 Aruebia pulchra (Roem. et Schult.) Edm. = Arnebia pulchra (Willd. ex Roem. et Schult.) J. R. Edm. ■☆

36967 Arum L.(1753);疆南星属(黄苞芋属,箭芋属);Arum, Arum Lily, Cuckoo Pint, Lords and Ladies, Lords-and-ladies, Wild Ginger, Yellow Calla ■☆

36968 Arum abyssinicum A. Rich. = Amorphophallus abyssinicus (A. Rich.) N. E. Br. ■☆

36969 Arum albispathum Stev.;白苞疆南星■☆

36970 Arum aphyllum Hook. = Amorphophallus aphyllus (Hook.) Hutch. ■☆

36971 Arum bicolor Aiton = Caladium bicolor (Aiton) Vaniot ■

36972 Arum bulbiferum Roxb. = Amorphophallus bulbifer (Roxb.) Blume ■

36973 Arum canariense Webb et Berthel. = Arum italicum Mill. subsp. canariense (Webb et Berthel.) P. C. Boyce ■☆

36974 Arum colocasia L. = Colocasia antiquorum Schott ■

36975 Arum colocasia L. = Colocasia esculenta (L.) Schott ■

36976 Arum cornutum ? = Sauromatum guttatum (Wall.) Schott ■

36977 Arum cornutum ? = Sauromatum venosum (Aiton) Kunth ■

36978 Arum costatum Wall. = Arisaema costatum (Wall.) Mart. ■

36979 Arum creticum Boiss. et Heldr.;绿苞疆南星(黄棒箭芋);Green Calla ■☆

36980 Arum crinitum Aiton = Helicodiceros crinitus Schott ■☆

36981 Arum cucullatum Lour. = Alocasia cucullata (Lour.) Schott et Endl. ●■

36982 Arum curvatum Roxb. = Arisaema tortuosum (Wall.) Schott var.

curvatum (Roxb.) Engl. ■☆

36983 Arum cuspitatum Blume = Typhonium flagelliforme (Lodd.) Blume ■

36984 Arum cyrenaicum Hruby;昔兰尼疆南星■☆

36985 Arum dioscoridis Sibth. et Sm.;薯蓣疆南星(薯蓣箭芋);Yam Arum ■☆

36986 Arum divaricatum L. =Typhonium divaricatum (L.) Decne. ■

36987 Arum diversifolium Blume = Typhonium divaricatum (L.) Decne. ■

36988 Arum dracontium L. =Arisaema dracontium (L.) Schott ■☆

36989 Arum dracunculus L. =Dracunculus vulgaris Schott ■☆

36990 Arum echinatum Wall. =Arisaema echinatum (Wall.) Schott ■

36991 Arum elongatum Stev.;伸长疆南星■☆

36992 Arum erubescens Wall. =Arisaema erubescens (Wall.) Schott ■

36993 Arum esculentum L. =Colocasia antiquorum Schott ■

36994 Arum esculentum L. =Colocasia esculenta (L.) Schott ■

36995 Arum flagelliforme Lodd. = Typhonium flagelliforme (Lodd.) Blume ■

36996 Arum flavum Forssk. =Arisaema flavum (Forssk.) Schott ■

36997 Arum griffithii Schott =Arum jacquemontii Blume ■☆

36998 Arum guttatum Wall. =Sauromatum guttatum Wall. ■

36999 Arum guttatum Wall. =Sauromatum venosum (Aiton) Kunth ■

37000 Arum guttatum Wall. =Typhonium venosum (Dryand. ex Aiton) Hett. et P. C. Boyce ■☆

37001 Arum hygrophilum Boiss.;喜水疆南星■☆

37002 Arum hygrophilum Boiss. subsp. maurum Braun-Blanq. et Maire = Arum hygrophilum Boiss. ■☆

37003 Arum hygrophilum Boiss. var. maurum (Braun-Blanq. et Maire) Maire et Weiller =Arum hygrophilum Boiss. ■☆

37004 Arum indicum Lour. =Alocasia macrorhiza (L.) G. Don ■

37005 Arum italicum Mill.;意大利疆南星;Italian Arum, Italian Lords and Ladies, Italian Lords-and-ladies ■☆

37006 Arum italicum Mill. 'Marmoratum';云纹美果芋■☆

37007 Arum italicum Mill. 'Pictum' = Arum italicum Mill. 'Marmoratum' ■☆

37008 Arum italicum Mill. subsp. canariense (Webb et Berthel.) P. C. Boyce;加那利疆南星■☆

37009 Arum italicum Mill. subsp. neglectum (F. Towns.) Prime;忽视疆南星;Large Cuckoo-pint ■☆

37010 Arum italicum Mill. var. intermedium Mutel = Arum italicum Mill. subsp. neglectum (F. Towns.) Prime ■☆

37011 Arum italicum Mill. var. neglectum F. Towns. = Arum italicum Mill. subsp. neglectum (F. Towns.) Prime ■☆

37012 Arum jacquemontii Blume;亚克疆南星■☆

37013 Arum korolkowii Regel;疆南星;Kolorkow Arum ■

37014 Arum macrorhizum L. =Alocasia macrorhiza (L.) G. Don ■

37015 Arum maculatum L.;点纹疆南星(欧海芋);Aaron's Rod, Adam and Eve, Adam-and-eve, Adder's Food, Adder's Meat, Adder's Tongue, Adderwort, Angels-and-devil, Angels-and-devils, Aron, Arrowroot, Arum, Arum Lily, Babe-in-the-cradle, Bloody Fingers, Bloody Man's Fingers, Bobbin And Joan, Bobbin Joan, Bobbin-and-joan, Bobbins, Bobby-and-joan, British Arrowroot, Buckrams, Bullocks, Bulls, Bull's Cocks, Bulls-and-cows, Bulls-and-wheys, Calf's Foot, Cocky Baby, Cocky-baby, Coke Pintel, Cows And Calves, Cow's Parsnip, Cows-and-bulls, Cows-and-calves, Cows-and-kies, Cuckoo Babies, Cuckoo Baby, Cuckoo Cock, Cuckoo Lily, Cuckoo Pint, Cuckoo Pintel, Cuckoo Point, Cuckoo Spit, Cuckoo-cock, Cuckoopint, Cuckoo-pint, Cuckoo-pintle, Cuckoo-point, Cuckoo-spit, Cuckow Pint, Cukoo-pint, Darmell Goddard, Dead Man's Fingers, Devil-and-angels, Devil's Ladies And Gentlemen, Devil's Ladies-and-gentlemen, Devil's Men And Women, Devil's Men-andwomen, Devils-and-angels, Dog Bobbins, Dog Cocks, Dog Spear, Dog-and-bobbin, Dog-and-bobbins, Dog's Dibble, Dog's Tassel, Dog's Tausle, Dog's Thistle, Dragons, English Passion Flower, Fairies, Fairy Lamps, Fairy's Lamp, Flycatcher, Friar's Cowl, Frog's Meat, Gentleman's Finger, Gentleman's Fingers, Gentlemen-and-ladies, Gentlemen-and-ladies' Fingers, Gentlemen's and Ladies' Fingers, Gentlemen's-and-ladies' Fingers, Gethsemane, Great Dragon, Heal-all, Hobble-gobbles, Jack-in-the-box, Jack-in-the-green, Jack-in-the-pulpit, Karnip, Kings-and-queens, Kitty-come-down-the-lane-jumpand-kiss-me, Knights-and-ladles, Ladies' Lords, Ladies-and-gentlemen, Lady-my-lord, Lady's Fingers, Lady's Keys, Lady's Slipper, Lady's Smock, Lady's Tresses, Lamb-in-a-pulpit, Lamb-lakins, Lamb's Lakens, Lily, Lily-grass, Long Purples, Lords and Ladies, Lords' and Ladies' Fingers, Lords-and-ladies, Lords-and-ladies' Fingers, Lorts-and-ladies, Mandrake, Man-in-the Pulpit, Man-in-ti-the-pulpit, Men-and-women, Moll-o'- the-woods, Nightingale, Old Man's Pulpit, Oxberry, Parson-and-clerk, Parson-in-hissmock, Parson-in-his-smock, Parson-in-the-pulpit, Parson's Billycock, Passion Flower, Pig Lily, Pintelwort, Poison Berry, Poison Fingers, Poison Root, Poker, Portland Arrowroot, Portland Sago, Portland Starch, Preacher-in-the-pulpit, Priestand-pulpit, Priest-and-pulpit, Priesties, Priest-in-the-pulpit, Priest's Hood, Priest's Pilly, Priest's Pintel, Priest's Pintle, Quakers, Rampe, Ram's Horn, Ram's Horns, Ramsons, Shiners, Shoes-and-stocking, Silly Lovers, Small Dragon, Snake Plant, Snake's Food, Snake's Meat, Snake's Victuals, Soldiers-and-angels, Soldiers-and-sailors, Stallions, Stallions-and-mares, Standing Gusses, Starchwort, Stockings-and-shoes, Sucky Calves, Sweethearts, Toad's Meat, Viper's Victuals, Wake Pintel, Wake Robin, Wake Robine, Wake-robin, White-and-red, Wild Arum, Wild Lily ■

37016 Arum margaritiferum Roxb.;珍珠疆南星■☆

37017 Arum neglectum (Towns.) Ridl. = Arum italicum Mill. subsp. neglectum (F. Towns.) Prime ■☆

37018 Arum nepenthoides Walter = Arisaema nepenthoides (Wall.) Mart. ■

37019 Arum nigrum Vell. =Xanthosoma sagittifolium (L.) Schott ■

37020 Arum nymphaeifolium Roxb. =Colocasia esculenta (L.) Schott ■

37021 Arum odorum Roxb. =Alocasia macrorhiza (L.) G. Don ■

37022 Arum orientale M. Bieb.;东方疆南星;Oriental Arum ■☆

37023 Arum orixense Roxb. =Typhonium trilobatum (L.) Schott ■

37024 Arum palaestinum Boiss.;黑疆南星;Black Calla, Loof, Solomon's Lily ■☆

37025 Arum pedatum Willd. = Sauromatum venosum (Dryand. ex Aiton) Kunth ■

37026 Arum pedatum Willd. =Typhonium venosum (Dryand. ex Aiton) Hett. et P. C. Boyce ■☆

37027 Arum pictum L. f.;花叶疆南星(白网箭芋);Ornamental Arum ■☆

37028 Arum ringens Thunb. =Arisaema ringens (Thunb.) Schott ●■

37029 Arum roxburghii Thwaites =Typhonium roxburghii Schott ■

37030 Arum sagittifolium L. =Xanthosoma sagittifolium (L.) Schott ■

37031 Arum sanctum Damm. =Arum palaestinum Boiss. ■☆

37032 Arum sarmentosum Fisch. = Gonatanthus pumilus (D. Don)

Engl. et Krause ■
37033 Arum sazensoo Buerger ex Blume = Arisaema sikokianum Franch. et Sav. ■
37034 Arum seguine Jacq. = Dieffenbachia seguine (Jacq.) Schott ●■
37035 Arum seguinum L. = Dieffenbachia seguina (L.) Schott ●■
37036 Arum serratum Thunb. = Arisaema serratum (Thunb.) Schott ■
37037 Arum sessiliflorum Roxb. = Sauromatum venosum (Dryand. ex Aiton) Kunth ■
37038 Arum speciosum Wall. = Arisaema speciosum (Wall.) Mart. ■
37039 Arum spiculatum Blume = Eminium spiculatum (Blume) Schott ■☆
37040 Arum tenuifolium L. = Biarum tenuifolium (L.) Schott ■☆
37041 Arum ternatum Thunb. = Pinellia ternata (Thunb.) Breitenb. ■
37042 Arum tortuosum Wall. = Arisaema tortuosum (Wall.) Schott ■
37043 Arum trilobatum L. = Typhonium trilobatum (L.) Schott ■
37044 Arum trilobatum Roxb. = Typhonium roxburghii Schott ■
37045 Arum trilobatum Thunb. = Typhonium divaricatum (L.) Decne. ■
37046 Arum venosum Aiton = Sauromatum venosum (Aiton) Kunth ■
37047 Arum venosum Dryand. ex Aiton = Typhonium venosum (Dryand. ex Aiton) Hett. et P. C. Boyce ■☆
37048 Arum virginicum L. = Peltandra virginica (L.) Schott ■☆
37049 Arum viviparum Roxb. = Remusatia vivipara (Lodd.) Schott ■
37050 Aruna Schreb. = Arouna Aubl. ●☆
37051 Aruna Schreb. = Dialium L. ●☆
37052 Aruncus Adans. = Aruncus L. ●■
37053 Aruncus L. (1758); 假升麻属(棣棠升麻); Buck's-beard, Goat's Beard, Goatsbeard, Goat's-beard ●■
37054 Aruncus acuminatus (Rydb.) Rydb. = Aruncus dioicus (Walter) Fernald var. acuminatus (Rydb.) Rydb. ex Hara ■☆
37055 Aruncus arthusifolius (H. Lév.) Nakai = Aruncus dioicus (Walter) Fernald var. arthusifolius (H. Lév.) H. Hara ■☆
37056 Aruncus asiaticus Pojark.; 亚洲假升麻; Asian Goatsbeard ■☆
37057 Aruncus asiaticus Pojark. = Aruncus sylvester Kostel. ex Maxim. ■
37058 Aruncus astilboides Maxim. = Aruncus dioicus (Walter) Fernald var. astilboides (Maxim.) H. Hara ■☆
37059 Aruncus astilboides Maxim. = Aruncus dioicus (Walter) Fernald var. insularis H. Hara ■☆
37060 Aruncus dioicus (Walter) Fernald; 异株假升麻; Aurunсus, Bride's Feathers, Bride's-feathers, Buck's-beard, Goat's Beard, Goatsbeard, Goat's-beard, Sylvan Goat's-beard ■☆
37061 Aruncus dioicus (Walter) Fernald subsp. vulgaris (Raf.) Tzvelev; 新娘异株假升麻; Bride's Feathers ■☆
37062 Aruncus dioicus (Walter) Fernald var. acuminatus (Rydb.) Rydb. ex Hara; 尖异株假升麻; Bride's-feathers, Goat's-beard ■☆
37063 Aruncus dioicus (Walter) Fernald var. arthusifolius (H. Lév.) H. Hara; 节叶假升麻; Dwarf Goats-beard ■☆
37064 Aruncus dioicus (Walter) Fernald var. astilboides (Maxim.) H. Hara; 落新妇假升麻■☆
37065 Aruncus dioicus (Walter) Fernald var. insularis H. Hara; 海岛假升麻■☆
37066 Aruncus dioicus (Walter) Fernald var. kamtschaticus (Maxim.) H. Hara f. tomentosus (Koidz.) H. Hara; 毛勘察加假升麻■☆
37067 Aruncus dioicus (Walter) Fernald var. kamtschaticus (Maxim.) H. Hara f. laciniatus (H. Hara) H. Ikeda; 条裂勘察加假升麻■☆
37068 Aruncus dioicus (Walter) Fernald var. kamtschaticus (Maxim.) H. Hara f. latilobus (H. Hara) H. Ikeda = Aruncus dioicus (Walter) Fernald var. subrotundus (Tatew.) H. Hara ■☆
37069 Aruncus dioicus (Walter) Fernald var. kamtschaticus (Maxim.) H. Hara; 勘察加假升麻■☆
37070 Aruncus dioicus (Walter) Fernald var. kamtschaticus (Maxim.) H. Hara = Aruncus sylvester Kostel. ex Maxim. ■
37071 Aruncus dioicus (Walter) Fernald var. laciniatus (H. Hara) H. Hara = Aruncus dioicus (Walter) Fernald var. kamtschaticus (Maxim.) H. Hara f. laciniatus (H. Hara) H. Ikeda ■☆
37072 Aruncus dioicus (Walter) Fernald var. rotundifoliolatus H. Hara = Aruncus gombalanus (Hand.-Mazz.) Hand.-Mazz. ■
37073 Aruncus dioicus (Walter) Fernald var. subrotundus (Tatew.) H. Hara; 圆叶假升麻■☆
37074 Aruncus dioicus (Walter) Fernald var. tenuifolius (Nakai ex H. Hara) H. Hara = Aruncus sylvester Kostel. ex Maxim. ■
37075 Aruncus dioicus (Walter) Fernald var. tenuifolius (Nakai ex H. Hara) H. Hara f. laciniatus (H. Hara) Sugim. = Aruncus dioicus (Walter) Fernald var. kamtschaticus (Maxim.) H. Hara f. laciniatus (H. Hara) H. Ikeda ■☆
37076 Aruncus dioicus (Walter) Fernald var. tenuifolius (Nakai ex H. Hara) H. Hara = Aruncus dioicus (Walter) Fernald ■☆
37077 Aruncus dioicus (Walter) Fernald var. triternatus (Wall. ex Maxim.) H. Hara = Aruncus sylvester Kostel. ex Maxim. ■
37078 Aruncus dioicus (Walter) Fernald var. vulgaris (Maxim.) H. Hara = Aruncus sylvester Kostel. ex Maxim. ■
37079 Aruncus gombalanus (Hand.-Mazz.) Hand.-Mazz.; 贡山假升麻; Gongshan Goatsbeard ■
37080 Aruncus gombalanus Hand.-Mazz. = Aruncus gombalanus (Hand.-Mazz.) Hand.-Mazz. ■
37081 Aruncus kamtschaticus (Maxim.) Rydb. = Aruncus sylvester Kostel. ex Maxim. ■
37082 Aruncus kamtschaticus (Maxim.) Rydb. var. tomentosus (Koidz.) Miyabe et Tatewaki = Aruncus sylvester Kostel. ex Maxim. ■
37083 Aruncus parvulus Kom.; 小假升麻■☆
37084 Aruncus silvester Kostel. = Aruncus sylvester Kostel. ex Maxim. ■
37085 Aruncus sylvester Kostel. ex Maxim.; 假升麻(棣棠升麻, 金毛三七, 升麻草, 竹土子); Goat's Beard, Goatsbeard, Goat's-beard, Sylvan Goatsbeard ■
37086 Aruncus sylvester Kostel. ex Maxim. = Aruncus dioicus (Walter) Fernald ■☆
37087 Aruncus sylvester Kostel. ex Maxim. = Aruncus dioicus (Walter) Fernald var. tenuifolius (Nakai ex H. Hara) H. Hara ■☆
37088 Aruncus sylvester Kostel. ex Maxim. var. acuminatus (Rydb.) Jeps.; 渐尖假升麻■☆
37089 Aruncus sylvester Kostel. ex Maxim. var. americanus (Pers.) Maxim.; 北美假升麻■☆
37090 Aruncus sylvester Kostel. ex Maxim. var. kamtschaticus Maxim. = Aruncus sylvester Kostel. ex Maxim. ■
37091 Aruncus sylvester Kostel. ex Maxim. var. subrotundus (Tatew.) Ohwi = Aruncus dioicus (Walter) Fernald var. subrotundus (Tatew.) H. Hara ■☆
37092 Aruncus sylvester Kostel. ex Maxim. var. tenuifolius Nakai ex H. Hara = Aruncus sylvester Kostel. ex Maxim. ■
37093 Aruncus sylvester Kostel. ex Maxim. var. tomentosus Koidz. = Aruncus sylvester Kostel. ex Maxim. ■
37094 Aruncus sylvester Kostel. ex Maxim. var. triternatus Wall. ex Maxim. = Aruncus sylvester Kostel. ex Maxim. ■
37095 Aruncus sylvester Kostel. ex Maxim. var. vulgaris Maxim. = Aruncus sylvester Kostel. ex Maxim. ■
37096 Aruncus tomentosus (Koidz.) Koidz. = Aruncus sylvester

Kostel. ex Maxim. ■

37097 Aruncus vulgaris Raf. = Aruncus sylvester Kostel. ex Maxim. ■

37098 Arundarbor Kuntze = Bambusa Schreb. (保留属名)●

37099 Arundarbor Rumph. = Bambusa Schreb. (保留属名)●

37100 Arundarbor Rumph. = Donax Lour. + Clinogyne Salisb. ex Benth. + Marantohloa Brongn. ex Gris ●

37101 Arundarbor Rumph. ex Kuntze = Bambusa Schreb. (保留属名)●

37102 Arundarbor Rumph. ex Kuntze = Donax Lour. + Clinogyne Salisb. ex Benth. + Marantohloa Brongn. ex Gris ●

37103 Arundarbor cantorii (Munro) Kuntze = Pseudosasa cantorii (Munro) P. C. Keng ex S. L. Chen et al. ●

37104 Arundarbor remotiflora Kuntze = Bambusa remotiflora (Kuntze) L. C. Chia et H. L. Fung ●

37105 Arundastrum Kuntze = Donax Lour. + Clinogyne Salisb. ex Benth. + Marantohloa Brongn. ex Gris ●

37106 Arundastrum Kuntze = Donax Lour. ■

37107 Arundastrum Rumph. ex Kuntze = Arundastrum Kuntze ●

37108 Arundastrum schweinfurthianum Kuntze = Sarcophrynium schweinfurthianum (Kuntze) Milne-Redh. ■☆

37109 Arundina Blume(1825);竹叶兰属(苇草兰属);Arundina ■

37110 Arundina bambusifolia Lindl. = Arundina graminifolia (D. Don) Hochr. ■

37111 Arundina chinensis Blume;竹叶兰■

37112 Arundina chinensis Blume = Arundina graminifolia (D. Don) Hochr. ■

37113 Arundina densa Lindl.;密生竹叶兰■☆

37114 Arundina densiflora Hook. f. = Arundina bambusifolia Lindl. ■

37115 Arundina graminifolia (D. Don) Hochr.;苇草兰(扁竹兰,草姜,长杆兰,大叶寮刁竹,地黄草,禾叶竹叶兰,胡连,鸟仔花,鸟子花,山荸荠,山姜,石玉,土白芨,文蛤海,小竹叶兰,野兰花,幽涧兰,竹兰,竹叶兰);Bamboo Orchid, Chinese Arundina, Common Arundina, Grassleaf Arundina ■

37116 Arundina graminifolia (D. Don) Hochr. var. chinensis (Blume) S. S. Ying = Arundina graminifolia (D. Don) Hochr. ■

37117 Arundina speciosa Blume = Arundina bambusifolia Lindl. ■

37118 Arundina stenopetala Gagnep. = Arundina graminifolia (D. Don) Hochr. ■

37119 Arundinaceae (Dumort.) Herter = Gramineae Juss. (保留科名) ●■

37120 Arundinaceae (Dumort.) Herter = Poaceae Barnhart(保留科名) ●■

37121 Arundinaceae (Kunth) Herter = Gramineae Juss. (保留科名)●■

37122 Arundinaceae (Kunth) Herter = Poaceae Barnhart(保留科名)●■

37123 Arundinaceae Döll = Gramineae Juss. (保留科名)●■

37124 Arundinaceae Döll = Poaceae Barnhart(保留科名)●■

37125 Arundinaceae Herter = Gramineae Juss. (保留科名)●■

37126 Arundinaceae Herter = Poaceae Barnhart(保留科名)●■

37127 Arundinaria Michx. (1803);青篱竹属(北美箭竹属);Bamboo, Cane, Canebrake, Fern-leaf Bamboo ●

37128 Arundinaria acerba W. T. Lin;苦篱竹●

37129 Arundinaria actinotricha Merr. et Chun = Ampelocalamus actinotrichus (Merr. et Chun) S. L. Chen, T. H. Wen et G. Y. Sheng ●

37130 Arundinaria acutissima Keng = Fargesia melanostachys (Hack. ex Hand. -Mazz.) T. P. Yi ●

37131 Arundinaria alpina K. Schum.;山地青篱竹(高山箭竹);Mountain Bamboo ●☆

37132 Arundinaria alpina K. Schum. = Sinarundinaria alpina (K. Schum.) C. S. Chao et Renvoize ●☆

37133 Arundinaria amabilis McClure = Pseudosasa amabilis (McClure) P. C. Keng ●

37134 Arundinaria amara Keng = Pleioblastus amarus (Keng) P. C. Keng ●

37135 Arundinaria ambositrensis A. Camus;安布西特拉青篱竹●☆

37136 Arundinaria anceps Mitford;二棱箭竹;Hidian Fountain-bamboo ●☆

37137 Arundinaria anceps Mitford = Sinarundinaria anceps (Mitford) C. S. Chao et Renvoize ●☆

37138 Arundinaria andropogonoides (Hand. -Mazz.) Hand. -Mazz. = Yushania andropogonoides (Hand. -Mazz.) T. P. Yi ●

37139 Arundinaria argenteostriata (Regel) E. G. Camus = Pleioblastus argenteostriatus (Regel) Nakai 'Argenteostriatus' ●☆

37140 Arundinaria argenteostriata (Regel) E. G. Camus 'Disticha' = Pleioblastus argenteostriatus (Regel) Nakai 'Distichus' ●

37141 Arundinaria aristata Gamble = Thamnocalamus aristatus (Gamble) E. G. Camus ●

37142 Arundinaria aristata Gamble = Thamnocalamus spathiflorus (Trin.) Munro ●

37143 Arundinaria armata Gamble = Chimonobambusa armata (Gamble) J. R. Xue et T. P. Yi ●

37144 Arundinaria basiaurita W. T. Lin = Pseudosasa cantorii (Munro) P. C. Keng ex S. L. Chen et al. ●

37145 Arundinaria basiaurita W. T. Lin et X. B. Ye;耳叶青篱竹●

37146 Arundinaria basigibbosa McClure = Pseudosasa cantorii (Munro) P. C. Keng ex S. L. Chen et al. ●

37147 Arundinaria bitchuensis (Makino) Koidz. = Sasaella bitchuensis (Makino) Makino ex Koidz. ●☆

37148 Arundinaria brevipaniculata Hand. -Mazz. = Yushania brevipaniculata (Hand. -Mazz.) T. P. Yi ●

37149 Arundinaria brevipes McClure = Fargesia brevipes (McClure) T. P. Yi ●

37150 Arundinaria brilletii A. Camus = Acidosasa brilletii (A. Camus) C. S. Chao et Renvoize ●

37151 Arundinaria cantorii (Munro) L. C. Chia = Pseudosasa cantorii (Munro) P. C. Keng ex S. L. Chen et al. ●

37152 Arundinaria cantorii (Munro) L. C. Chia ex C. S. Chao et G. Y. Yang = Pseudosasa cantorii (Munro) P. C. Keng ex S. L. Chen et al. ●

37153 Arundinaria cerata McClure = Pseudosasa hindsii (Munro) S. L. Chen et G. Y. Sheng ex T. G. Liang ●

37154 Arundinaria chinensis C. S. Chao et G. Y. Yang = Arundinaria oleosa (T. H. Wen) C. S. Chao et G. Y. Yang ●

37155 Arundinaria chinensis C. S. Chao et G. Y. Yang = Pleioblastus maculatus (McClure) C. D. Chu et C. S. Chao ●

37156 Arundinaria chinensis C. S. Chao et G. Y. Yang = Pleioblastus oleosus T. H. Wen ●

37157 Arundinaria chino (Franch. et Sav.) Makino = Pleioblastus chino (Franch. et Sav.) Makino ●

37158 Arundinaria chino (Franch. et Sav.) Makino f. angustifolia (Mitford) C. S. Chao et Renvoize;窄叶青篱竹;Thin-leaved Bamboo ●☆

37159 Arundinaria chino (Franch. et Sav.) Makino var. viridis Makino = Pleioblastus argenteostriatus (Regel) Nakai f. glaber (Makino) Murata ●☆

37160 Arundinaria chino (Franch. et Sav.) Makino var. viridistriata (Siebold ex André) Makino = Pleioblastus viridistriatus (Siebold ex André) Makino ●☆

37161 Arundinaria chungii Keng = Yushania brevipaniculata (Hand.-Mazz.) T. P. Yi ●

37162 Arundinaria concava C. D. Chu et H. Y. Zou;南平青篱竹;Nanping Cane,Nanping Canebrake ●

37163 Arundinaria cuspidata Keng = Fargesia cuspidata (Keng) Z. P. Wang et G. H. Ye ●

37164 Arundinaria densiflora Rendle = Brachystachyum densiflorum (Rendle) Keng ●

37165 Arundinaria densiflora Rendle = Semiarundinaria densiflora (Rendle) T. H. Wen ●

37166 Arundinaria dolichantha Keng = Sinobambusa tootsik (Siebold) Makino ex Nakai ●

37167 Arundinaria dumetosa Rendle = Arundinaria fargesii (E. G. Camus) P. C. Keng et T. P. Yi ●

37168 Arundinaria dushanensis C. D. Chu et J. Q. Zhang = Sinobambusa dushanensis (C. D. Chu et J. Q. Zhang) T. H. Wen ●

37169 Arundinaria elegans Kurz;雅致箭竹●

37170 Arundinaria emeiensis (C. D. Chu et C. S. Chao) Demoly = Indocalamus emeiensis C. D. Chu et C. S. Chao ●

37171 Arundinaria faberi Rendle;冷箭竹;Fang Bashanbamboo,Fang Bashania ●

37172 Arundinaria falcata Nees;镰形青篱竹;Tropical Blue Bamboo ●☆

37173 Arundinaria falconeri (Hook. f. ex Munro) Duthie = Himalayacalamus falconeri (Hook. f. ex Munro) P. C. Keng ●

37174 Arundinaria fangiana A. Camus;冷青篱竹(冷箭竹);Fang Cane,Fang Canebrake ●

37175 Arundinaria fangiana A. Camus = Arundinaria faberi Rendle ●

37176 Arundinaria fargesii (E. G. Camus) P. C. Keng et T. P. Yi;巴山青篱竹(巴山木竹,法氏箬竹);Farges Cane,Farges Canebrake,Farges Indocalamus ●

37177 Arundinaria fargesii (E. G. Camus) P. C. Keng et T. P. Yi var. grandifolia E. G. Camus = Arundinaria fargesii (E. G. Camus) P. C. Keng et T. P. Yi ●

37178 Arundinaria fargesii (E. G. Camus) P. C. Keng et T. P. Yi var. grandifolia E. G. Camus = Bashania fargesii (E. G. Camus) P. C. Keng et T. P. Yi ●

37179 Arundinaria fargesii E. G. Camus = Bashania fargesii (E. G. Camus) P. C. Keng et T. P. Yi ●

37180 Arundinaria fargesii E. G. Camus var. grandifolia E. G. Camus = Bashania fargesii (E. G. Camus) P. C. Keng et T. P. Yi ●

37181 Arundinaria fastuosa (Mitford) J. Houz. = Semiarundinaria fastuosa (Mitford) Makino ex Nakai ●

37182 Arundinaria ferax Keng = Fargesia ferax (Keng) T. P. Yi ●

37183 Arundinaria fischeri K. Schum. = Sinarundinaria alpina (K. Schum.) C. S. Chao et Renvoize ●☆

37184 Arundinaria flexuosa Hance = Pseudosasa hindsii (Munro) S. L. Chen et G. Y. Sheng ex T. G. Liang ●

37185 Arundinaria forrestii Keng = Fargesia melanostachys (Hack. ex Hand.-Mazz.) T. P. Yi ●

37186 Arundinaria fortunei (Van Houtte) Riv. = Pleioblastus variegatus (Siebold ex Miq.) Makino 'Fortunei' ●☆

37187 Arundinaria fortunei (Van Houtte) Riv. = Pleioblastus variegatus (Siebold ex Miq.) Makino ●

37188 Arundinaria fortunei (Van Houtte) Riv. = Sasa fortunei (Van Houtte) Fiori ●

37189 Arundinaria funghomii McClure;小篱竹(冯氏青篱竹);Fungohom Cane,Fungohom Canebrake ●

37190 Arundinaria funghomii McClure = Pseudosasa cantorii (Munro) P. C. Keng ex S. L. Chen et al. ●

37191 Arundinaria gigantea (Walter) Muehlenbeck;大青篱竹;Giant Cane,Switch Cane,Switchcane ●☆

37192 Arundinaria glaucescens (Willd.) P. Beauv. = Bambusa multiplex (Lour.) Raeusch. ex Schult. et Schult. f. ●

37193 Arundinaria gracilipes (McClure) C. D. Chu et C. S. Chao = Oligostachyum gracilipes (McClure) G. H. Ye et Z. P. Wang ●

37194 Arundinaria graminea (Bean) Makino = Pleioblastus gramineus (Bean) Nakai ●

37195 Arundinaria griffithiana Munro = Chimonocalamus griffithianus (Munro) J. R. Xue et T. P. Yi ●

37196 Arundinaria hakonensis Nakai;箱根青篱竹●☆

37197 Arundinaria hakonensis Nakai = Sasaella hisauchii (Makino) Makino ●☆

37198 Arundinaria hashimotoi (Makino) Koidz. = Sasaella hashimotoi (Makino) Makino ex Koidz. ●☆

37199 Arundinaria hebechlamys Nakai var. yoshinoi (Koidz.) Murata = Sasaella kogasensis (Nakai) Nakai ex Koidz. var. yoshinoi (Koidz.) Sad. Suzuki ●☆

37200 Arundinaria heterolodicula (W. T. Lin et Z. J. Feng) W. T. Lin = Oligostachyum scabriflorum (McClure) Z. P. Wang et G. H. Ye ●

37201 Arundinaria hidaensis (Makino) Nakai = Sasaella hidaensis (Makino) Makino ●☆

37202 Arundinaria hindsii Munro = Pleioblastus hindsii (Munro) Nakai ●

37203 Arundinaria hindsii Munro = Pseudosasa hindsii (Munro) S. L. Chen et G. Y. Sheng ex T. G. Liang ●

37204 Arundinaria hirtivaginata W. T. Lin;毛鞘青篱竹●

37205 Arundinaria hirtivaginata W. T. Lin = Pseudosasa hindsii (Munro) S. L. Chen et G. Y. Sheng ex T. G. Liang ●

37206 Arundinaria hisauchii (Makino) Nakai = Arundinaria hakonensis Nakai ●☆

37207 Arundinaria hisauchii (Makino) Nakai = Sasaella hisauchii (Makino) Makino ●☆

37208 Arundinaria hsienchuensis (T. H. Wen) C. S. Chao et G. Y. Yang = Pleioblastus hsienchuensis T. H. Wen ●

37209 Arundinaria hsienchuensis (T. H. Wen) C. S. Chao et G. Y. Yang var. subglabrata (S. Y. Chen) C. S. Chao et G. Y. Yang = Pleioblastus hsienchuensis T. H. Wen var. subglabratus (S. Y. Chen) C. S. Chao et G. Y. Yang ●

37210 Arundinaria humbertii A. Camus;亨伯特青篱竹●☆

37211 Arundinaria humilis Mitford = Pleioblastus humilis (Mitford) Nakai ●☆

37212 Arundinaria hupehensis (J. L. Lu) C. S. Chao et G. Y. Yang = Oligostachyum hupehense (J. L. Lu) Z. P. Wang et G. H. Ye ●

37213 Arundinaria ibityensis A. Camus;伊比提青篱竹●☆

37214 Arundinaria ikegamii Nakai = Sasaella ikegamii (Nakai) Sad. Suzuki ●☆

37215 Arundinaria japonica A. Gray = Pleioblastus simonii (Carrière) Nakai ●

37216 Arundinaria japonica Franch. et Sav. = Pleioblastus chino (Franch. et Sav.) Makino ●

37217 Arundinaria japonica Siebold et Zucc. = Pseudosasa japonica (Siebold et Zucc. ex Steud.) Makino ex Nakai ●

37218 Arundinaria japonica Siebold et Zucc. ex Steud. = Pseudosasa japonica (Siebold et Zucc. ex Steud.) Makino ex Nakai ●

37219 Arundinaria jaunsanensis Gamble = Sinarundinaria anceps

(Mitford) C. S. Chao et Renvoize ●☆

37220 Arundinaria kindsii var. graminea Bean = Pleioblastus gramineus (Bean) Nakai ●

37221 Arundinaria kongosanensis Makino = Pleioblastus kongosanensis Makino ●☆

37222 Arundinaria kunishii (Hayata) P. C. Keng et T. H. Wen = Gelidocalamus kunishii (Hayata) P. C. Keng et T. H. Wen ●

37223 Arundinaria kunishii Hayata = Gelidocalamus kunishii (Hayata) P. C. Keng et T. H. Wen ●

37224 Arundinaria kwangsiensis (W. Y. Hsiung et C. S. Chao) C. S. Chao et G. Y. Yang = Pleioblastus maculatus (McClure) C. D. Chu et C. S. Chao ●

37225 Arundinaria lanshanensis (T. H. Wen) T. H. Wen = Pseudosasa pubiflora (Keng) P. C. Keng ex D. Z. Li et L. M. Gao ●

37226 Arundinaria latifolia Keng = Indocalamus latifolius (Keng) McClure ●

37227 Arundinaria leucorhoda (Koidz.) Koidz. = Elaeagnus arakiana Koidz. ●☆

37228 Arundinaria leucorhoda (Koidz.) Koidz. var. kanayamensis (Nakai) Murata = Sasaella leucorhoda (Koidz.) Koidz. var. kanayamensis (Nakai) Sad. Suzuki ●☆

37229 Arundinaria lima (McClure) C. D. Chu et C. S. Chao;海南青篱竹;Hainan Bamboo,Hainan Cane,Hainan Canebrake ●

37230 Arundinaria lima (McClure) C. D. Chu et C. S. Chao = Oligostachyum nuspiculum (McClure) Z. P. Wang et G. H. Ye ●

37231 Arundinaria linearis Hack. = Pleioblastus linearis (Hack.) Nakai ●

37232 Arundinaria longiaurita (Hand.-Mazz.) Hand.-Mazz. = Indocalamus longiauritus Hand.-Mazz. ●

37233 Arundinaria longifimbriata (S. Y. Chen) T. H. Wen = Sinobambusa intermedia McClure ●

37234 Arundinaria longiramea Munro = Indocalamus sinicus (Hance) Nakai ●

37235 Arundinaria lubrica (T. H. Wen) C. S. Chao et G. Y. Yang = Oligostachyum lubricum (T. H. Wen) P. C. Keng ●

37236 Arundinaria macclureana Bor = Fargesia macclureana (Bor) Stapleton ●

37237 Arundinaria maculata (McClure) C. D. Chu et C. S. Chao ex K. M. Lan = Pleioblastus maculatus (McClure) C. D. Chu et C. S. Chao ●

37238 Arundinaria maculata C. D. Chu et C. S. Chao = Oligostachyum scabriflorum (McClure) Z. P. Wang et G. H. Ye ●

37239 Arundinaria maculosa C. D. Chu et C. S. Chao var. breviligulata (Z. P. Wang et G. H. Ye) C. S. Chao et G. Y. Yang = Oligostachyum scabriflorum (McClure) Z. P. Wang et G. H. Ye var. breviligulatum Z. P. Wang et G. H. Ye ●

37240 Arundinaria maculoxa C. D. Chun et C. S. Chao;白眼竹●

37241 Arundinaria madagascariensis A. Camus;马岛青篱竹●☆

37242 Arundinaria mairei Hack. ex Hand.-Mazz. = Fargesia mairei (Hack. ex Hand.-Mazz.) T. P. Yi ●

37243 Arundinaria mannii Gamble = Cephalostachyum mannii (Gamble) Stapleton et D. Z. Li ●

37244 Arundinaria marmorea (Mitford) Makino = Chimonobambusa marmorea (Mitford) Makino ex Nakai ●◇

37245 Arundinaria marojejyensis A. Camus;马罗青篱竹●☆

37246 Arundinaria masamuneana (Makino) Masam. = Sasaella masamuneana (Makino) Hatus. et Muroi ●☆

37247 Arundinaria matsumurae Hack. = Chimonobambusa marmorea (Mitford) Makino ex Nakai ●◇

37248 Arundinaria maudiae (Dunn) Keng = Pseudosasa hindsii (Munro) S. L. Chen et G. Y. Sheng ex T. G. Liang ●

37249 Arundinaria megalothyrsa (Makino) Hand.-Mazz. = Yushania megalothyrsa (Hand.-Mazz.) T. H. Wen ●

37250 Arundinaria megalothyrsa Hand.-Mazz. = Gaoligongshania megathyrsa (Hand.-Mazz.) D. Z. Li,J. R. Xue et N. H. Xia ●

37251 Arundinaria melanostachys Hand.-Mazz. = Fargesia melanostachys (Hack. ex Hand.-Mazz.) T. P. Yi ●

37252 Arundinaria multifloscula W. T. Lin;多花青篱竹;Manyflowers Canebrake ●

37253 Arundinaria multifloscula W. T. Lin = Pseudosasa hindsii (Munro) S. L. Chen et G. Y. Sheng ex T. G. Liang ●

37254 Arundinaria murielae Gamble = Fargesia murielae (Gamble) T. P. Yi ●

37255 Arundinaria murielae Gamble ex Bean = Fargesia murielae (Gamble) T. P. Yi ●

37256 Arundinaria murielae Gamble ex Bean = Thamnocalamus spathiflorus (Trin.) Munro ●

37257 Arundinaria nagashima Mitford = Pleioblastus nagashima (Mitford) Nakai ●☆

37258 Arundinaria naibunensis Hayata = Ampelocalamus naibunensis (Hayata) T. H. Wen ●

37259 Arundinaria naibunensis Hayata = Drepanostachyum naibunense (Hayata) P. C. Keng ●

37260 Arundinaria nana Makino = Chimonobambusa marmorea (Mitford) Makino ex Nakai ●◇

37261 Arundinaria nanningensis Q. H. Dai = Sinobambusa intermedia McClure ●

37262 Arundinaria nanunica (McClure) C. D. Chu et C. S. Chao = Acidosasa nanunica (McClure) C. S. Chao et G. Y. Yang ●

37263 Arundinaria nanunica (McClure) C. D. Chu et C. S. Chao = Pseudosasa nanunica (McClure) Z. P. Wang et G. H. Ye ●

37264 Arundinaria narihira (Bean) Makino = Semiarundinaria fastuosa (Lat.-Marl. ex Mitford) Makino ex Nakai ●

37265 Arundinaria narihira Makino = Semiarundinaria fastuosa (Mitford) Makino ex Nakai ●

37266 Arundinaria niitakayamensis Hayata = Yushania niitakayamensis (Hayata) P. C. Keng ●

37267 Arundinaria nitida Mitford = Fargesia nitida (Mitford ex Stapf) P. C. Keng ex T. P. Yi ●

37268 Arundinaria nuspicula (McClure) C. D. Chu et C. S. Chao = Oligostachyum nuspiculum (McClure) Z. P. Wang et G. H. Ye ●

37269 Arundinaria oedogonata (Z. P. Wang et G. H. Ye) G. Y. Yang et C. S. Chao = Oligostachyum oedogonatum (Z. P. Wang et G. H. Ye) Q. F. Zheng et K. F. Huang ●

37270 Arundinaria oiwakensis Hayata = Yushania niitakayamensis (Hayata) P. C. Keng ●

37271 Arundinaria oleosa (T. H. Wen) C. S. Chao et G. Y. Yang;油苦竹(斑苦竹,长穗苦竹,秋竹);Maculate Bitter Bamboo,Oil Bitter Bamboo,Oil Bitterbamboo,Oily Bitter-bamboo ●

37272 Arundinaria oleosa (T. H. Wen) Demoly = Pleioblastus oleosus T. H. Wen ●

37273 Arundinaria orthotropoides (W. T. Lin) W. T. Lin = Pseudosasa hindsii (Munro) S. L. Chen et G. Y. Sheng ex T. G. Liang ●

37274 Arundinaria pallidiflora (McClure) T. H. Wen = Pseudosasa pubiflora (Keng) P. C. Keng ex D. Z. Li et L. M. Gao ●

37275 Arundinaria panda Keng = Pseudosasa hindsii (Munro) S. L. Chen et G. Y. Sheng ex T. G. Liang ●

37276 Arundinaria pauciflora Keng = Fargesia pauciflora (Keng) T. P. Yi ●

37277 Arundinaria pedalis Keng = Indocalamus auriculatus (H. R. Zhao et Y. L. Yang) Y. L. Yang ●

37278 Arundinaria pedalis Keng = Indocalamus pedalis (Keng) P. C. Keng ●

37279 Arundinaria perrieri A. Camus;佩里耶竹●☆

37280 Arundinaria pleniculmis Hand. -Mazz. = Fargesia pleniculmis (Hand. -Mazz.) T. P. Yi ●

37281 Arundinaria prainii (Gamble) Gamble = Racemobambos prainii (Gamble) P. C. Keng et T. H. Wen ●

37282 Arundinaria prainii Gamble = Neomicrocalamus prainii (Gamble) P. C. Keng ●

37283 Arundinaria projecta W. T. Lin;真子竹●

37284 Arundinaria pubiannula W. T. Lin et Z. J. Feng;水仔竹●

37285 Arundinaria pubiannula W. T. Lin et Z. J. Feng = Pseudosasa cantorii (Munro) P. C. Keng ex S. L. Chen et al. ●

37286 Arundinaria pubiflora Keng;毛花青篱竹(毛花茶秆竹);Pubescent Cane,Pubescent Canebrake ●

37287 Arundinaria pubiflora Keng = Pseudosasa pubiflora (Keng) P. C. Keng ex D. Z. Li et L. M. Gao ●

37288 Arundinaria pygmaea (Miq.) Mitford = Pleioblastus pygmaeus (Miq.) Nakai ●

37289 Arundinaria pygmaea (Miq.) Mitford var. disticha (Mitford) C. S. Chao et Renvoize = Sasa pygmaea (Miq.) E. G. Camus var. disticha (Mitford) C. S. Chao et G. G. Tang ●

37290 Arundinaria pygmaea (Miq.) Mitford var. disticha (Mitford) C. S. Chao et Renvoize = Pleioblastus argenteostriatus (Regel) Nakai 'Distichus' ●

37291 Arundinaria pygmaea (Miq.) Mitford var. disticha (Mitford) C. S. Chao et Renvoize = Pleioblastus distichus (Mitford) Nakai ●

37292 Arundinaria pygmaea (Miq.) Mitford = Sasa pygmaea (Miq.) E. G. Camus ●

37293 Arundinaria qingchengshanensis (P. C. Keng et T. P. Yi) D. Z. Li;饱竹子;Qingchengshan Bashanbamboo,Qingchengshan Bashania ●

37294 Arundinaria quadrangula W. T. Lin et Z. J. Feng = Pseudosasa hindsii (Munro) S. L. Chen et G. Y. Sheng ex T. G. Liang ●

37295 Arundinaria quadrangularis (Franceschi) Makino = Chimonobambusa quadrangularis (Franceschi) Makino ex Nakai ●

37296 Arundinaria racemosa Munro;总花冷箭竹●

37297 Arundinaria racemosa Munro subsp. fangiana A. Camus = Bashania fangiana (A. Camus) P. C. Keng et T. H. Wen ●

37298 Arundinaria ramosa Makino = Pleioblastus chino (Franch. et Sav.) Makino var. hisauchii Makino ●

37299 Arundinaria ramosa Makino = Sasaella ramosa (Makino) Makino ●

37300 Arundinaria ramosa Makino f. tomikusensis (Nakai) Murata = Sasaella ramosa (Makino) Makino f. tomikusensis (Nakai) Sad. Suzuki ●☆

37301 Arundinaria ramosa Makino var. latifolia Nakai = Sasaella ramosa (Makino) Makino var. latifolia (Nakai) Sad. Suzuki ●☆

37302 Arundinaria rectirama W. T. Lin;蒲竹●

37303 Arundinaria rugata (T. H. Wen et S. Y. Chen) C. S. Chao et G. Y. Yang = Pleioblastus rugatus T. H. Wen et S. Y. Chen ●

37304 Arundinaria sadoensis (Makino ex Koidz.) Koidz. = Sasaella sadoensis (Nakai) Sad. Suzuki ●☆

37305 Arundinaria sadoensis (Makino ex Koidz.) Koidz. var. infrapilosa Koidz.;背毛竹●☆

37306 Arundinaria sasakiana (Makino et Uchida) Nakai = Sasaella sasakiana Makino et Uchida ●☆

37307 Arundinaria sawadae (Makino) Nakai = Sasaella sawadae (Makino) Makino ex Koidz. ●☆

37308 Arundinaria scabriflora (McClure) C. D. Chu et C. S. Chao = Oligostachyum scabriflorum (McClure) Z. P. Wang et G. H. Ye ●

37309 Arundinaria scopula (McClure) C. D. Chu et C. S. Chao = Oligostachyum scopulum (McClure) Z. P. Wang et G. H. Ye ●

37310 Arundinaria shiuyingiana L. C. Chia et But = Oligostachyum shiuyingianum (L. C. Chia et But) G. H. Ye et Z. P. Wang ●

37311 Arundinaria simonii (Carrière) Rivière et C. Rivière = Pleioblastus simonii (Carrière) Nakai ●

37312 Arundinaria sinica Hance = Indocalamus sinicus (Hance) Nakai ●

37313 Arundinaria solida (S. Y. Chen) C. S. Chao et G. Y. Yang = Pleioblastus solidus S. Y. Chen ●

37314 Arundinaria spanostachya (T. P. Yi) D. Z. Li;峨热竹(峨热巴山竹);Poorspike Bashanbamboo,Rare-spiked Bashania ●◇

37315 Arundinaria sparsiflora Rendle = Fargesia murielae (Gamble) T. P. Yi ●

37316 Arundinaria spathacea (Franch.) D. C. McClint. = Fargesia spathacea Franch. ●

37317 Arundinaria spathacea (Franch.) D. C. McClint. = Thamnocalamus spathaceus (Franch.) Soderstr. ●

37318 Arundinaria spathiflora Trin. = Thamnocalamus spathiflorus (Trin.) Munro ●

37319 Arundinaria spongiosa C. D. Chu et C. S. Chao = Oligostachyum spongiosum (C. D. Chu et C. S. Chao) G. H. Ye et Z. P. Wang ●

37320 Arundinaria subsolida (S. L. Chen et G. Y. Sheng) C. S. Chao et G. Y. Yang = Pseudosasa subsolida S. L. Chen et G. Y. Sheng ●

37321 Arundinaria sulcata (Z. P. Wang et G. H. Ye) C. S. Chao et G. Y. Yang = Oligostachyum sulcatum Z. P. Wang et G. H. Ye ●

37322 Arundinaria szechuanensis Rendle = Chimonobambusa szechuanensis (Rendle) P. C. Keng ●◇

37323 Arundinaria tanegashimensis (Makino et Koidz.) Masam. = Sasaella masamuneana (Makino) Hatus. et Muroi ●☆

37324 Arundinaria tashirozentaroana Koidz. = Sasaella bitchuensis (Makino) Makino ex Koidz. var. tashirozentaroana (Koidz.) Sad. Suzuki ●☆

37325 Arundinaria tenuivagina W. T. Lin = Pseudosasa pubiflora (Keng) P. C. Keng ex D. Z. Li et L. M. Gao ●

37326 Arundinaria tessellata (Nees) Munro = Thamnocalamus tessellatus (Nees) Soderstr. et R. P. Ellis ●☆

37327 Arundinaria tolange K. Schum. = Sinarundinaria alpina (K. Schum.) C. S. Chao et Renvoize ●☆

37328 Arundinaria tootsik (Siebold) Makino = Sinobambusa tootsik (Siebold) Makino ex Nakai ●

37329 Arundinaria tootsik Makino = Sinobambusa tootsik (Siebold) Makino ex Nakai ●

37330 Arundinaria triangulata (J. R. Xue et T. P. Yi) C. S. Chao et G. Y. Yang = Indosasa triangulata J. R. Xue et T. P. Yi ●

37331 Arundinaria usawae Hayata = Pseudosasa japonica (Siebold et Zucc. ex Steud.) Makino ex Nakai ●

37332 Arundinaria usawai Hayata = Pseudosasa usawai (Hayata) Makino et Nemoto ●

37333 Arundinaria vaginata Hack. = Pleioblastus chino (Franch. et

Sav.) Makino ●

37334 Arundinaria vaginata Hack. = Pleioblastus simonii (Carrière) Nakai ●

37335 Arundinaria varia Keng = Pleioblastus amarus (Keng) P. C. Keng ●

37336 Arundinaria variabilis Makino ex Vilm. var. disticha (Mitford) J. Houz. = Sasa pygmaea (Miq.) E. G. Camus var. disticha (Mitford) C. S. Chao et G. G. Tang ●

37337 Arundinaria variabilis Makino ex Vilm. var. disticha (Mitford) J. Houz. = Pleioblastus distichus (Mitford) Nakai ●

37338 Arundinaria variabilis Makino ex Vilm. var. fortunei (Van Houtte) J. Houz. = Sasa fortunei (Van Houtte) Fiori ●

37339 Arundinaria variegata (Siebold ex Miq.) Makino = Pleioblastus variegatus (Siebold ex Miq.) Makino ●

37340 Arundinaria vicina Keng = Fargesia vicina (Keng) T. P. Yi ●

37341 Arundinaria violascens Keng = Yushania violascens (Keng) T. P. Yi ●

37342 Arundinaria vulgata (W. T. Lin et X. B. Ye) W. T. Lin = Indocalamus longiauritus Hand. -Mazz. ●

37343 Arundinaria wilsonii Rendle = Indocalamus wilsonii (Rendle) C. S. Chao et C. D. Chu ●

37344 Arundinaria yangshanensis W. T. Lin = Pseudosasa hindsii (Munro) S. L. Chen et G. Y. Sheng ex T. G. Liang ●

37345 Arundinaria yangshanensis W. T. Lin = Pseudosasa yangshanensis (W. T. Lin) T. P. Yi ●

37346 Arundinaria yixingensis (S. L. Chen et S. Y. Chen) C. S. Chao et G. Y. Yang = Pleioblastus yixingensis S. L. Chen et S. Y. Chen ●

37347 Arundinariaceae Baum. -Bod.;青篱竹科●

37348 Arundinariaceae Baum. -Bod. = Gramineae Juss.(保留科名)●■

37349 Arundinariaceae Baum. -Bod. = Poaceae Barnhart(保留科名)●■

37350 Arundinella Raddi(1823);野古草属(野牯草属);Arundinella ■

37351 Arundinella acratherum Nees ex Steud. = Arundinella nepalensis Trin. ■

37352 Arundinella anomala Steud.;野古草(白牛公,乌骨草,硬骨草);Common Arundinella ■

37353 Arundinella anomala Steud. = Arundinella hirta (Thunb.) Tanaka f. ciliata (Thunb.) T. Koyama ■

37354 Arundinella anomala Steud. = Arundinella hirta (Thunb.) Tanaka ■

37355 Arundinella anomala Steud. var. depauperata Rendle = Arundinella anomala Steud. ■

37356 Arundinella barbinodis Keng ex B. S. Sun et Z. H. Hu;毛节野古草;Beardednode Arundinella, Hairrode Arundinella ■

37357 Arundinella bengalensis (Spreng.) Druce;孟加拉野古草(密序野古草);Bengal Arundinella ■

37358 Arundinella bidentata Keng = Arundinella setosa Trin. var. esetosa Bor ■

37359 Arundinella bidentata Keng = Arundinella setosa Trin. ■

37360 Arundinella brasiliensis Raddi;巴西野古草;Brazil Arundinella ■

37361 Arundinella caespitosa Janowski = Arundinella pubescens Merr. et Hack. ex Hack. ■

37362 Arundinella capillaris Hook. f. = Arundinella setosa Trin. ■

37363 Arundinella chenii Keng;陈谋野古草;Chen Arundinella ■

37364 Arundinella chenii Keng = Arundinella hookeri Munro ex Keng ■

37365 Arundinella chevalieri (A. Camus et C. E. Hubb.) Roberty = Danthoniopsis chevalieri A. Camus et C. E. Hubb. ■☆

37366 Arundinella clarkei Hook. f. = Arundinella decempedalis (Kuntze) Janowski ■

37367 Arundinella cochinchinensis Keng;大序野古草(交趾野古草,越南野古草);Bigspike Arundinella ■

37368 Arundinella decempedalis (Kuntze) Janowski;大野古草(丈野古草);Tenfeed Arundinella, Zhang Arundinella ■

37369 Arundinella ecklonii Nees = Arundinella nepalensis Trin. ■

37370 Arundinella effusa C. E. Hubb. = Arundinella pumila (Hochst. ex A. Rich.) Steud. ■☆

37371 Arundinella elegantula Hook. f. = Trichopteryx elegantula (Hook. f.) Stapf ■☆

37372 Arundinella filiformis Janowski = Arundinella pubescens Merr. et Hack. ex Hack. ■

37373 Arundinella flavida Keng;硬叶野古草;Hardleaf Arundinella ■

37374 Arundinella fluviatilis Hand. -Mazz.;溪边野古草;Stream Arundinella, Stream Shore Arundinella ■

37375 Arundinella fluviatilis Hand. -Mazz. var. pachyathera Hand. -Mazz. = Arundinella rupestris A. Camus var. pachyathera (Hand. -Mazz.) B. S. Sun et Z. H. Hu ■

37376 Arundinella funaensis Vanderyst = Trichopteryx dregeana Nees ■☆

37377 Arundinella glabra Nees ex Hook. et Arn. = Arundinella nepalensis Trin. ■

37378 Arundinella grandiflora Hack.;大花野古草;Largeflower Arundinella ■

37379 Arundinella hirta (Thunb.) Tanaka;毛秆野古草(迭茅草,红眼疤,淮草,鸡子杆,麦穗草,田草,野古草,野罐草);Hisute Arundinella ■

37380 Arundinella hirta (Thunb.) Tanaka = Arundinella anomala Steud. ■

37381 Arundinella hirta (Thunb.) Tanaka f. ciliata (Thunb.) T. Koyama = Arundinella anomala Steud. ■

37382 Arundinella hirta (Thunb.) Tanaka f. hondana (Koidz.) T. Koyama = Arundinella hirta (Thunb.) Tanaka var. hondana Koidz. ■

37383 Arundinella hirta (Thunb.) Tanaka f. koryuensis (Honda) Kitag. = Arundinella hirta (Thunb.) Tanaka ■

37384 Arundinella hirta (Thunb.) Tanaka f. shotokuensis (Honda) Kitag. = Arundinella hirta (Thunb.) Tanaka ■

37385 Arundinella hirta (Thunb.) Tanaka subsp. anomala (Steud.) Tanaka = Arundinella anomala Steud. ■

37386 Arundinella hirta (Thunb.) Tanaka subsp. anomala (Steud.) Tzvelev = Arundinella hirta (Thunb.) Tanaka ■

37387 Arundinella hirta (Thunb.) Tanaka subsp. hirta Tzvelev = Arundinella hirta (Thunb.) Tanaka ■

37388 Arundinella hirta (Thunb.) Tanaka subsp. riparia (Honda) T. Koyama = Arundinella riparia Honda ■☆

37389 Arundinella hirta (Thunb.) Tanaka var. ciliata (Thunb.) Koidz.;睫毛野古草;Ciliate Arundinella ■

37390 Arundinella hirta (Thunb.) Tanaka var. ciliata (Thunb.) Koidz. = Arundinella anomala Steud. ■

37391 Arundinella hirta (Thunb.) Tanaka var. depauperata Rendle = Arundinella anomala Steud. ■

37392 Arundinella hirta (Thunb.) Tanaka var. glauca (Koidz.) Honda;灰蓝野古草■☆

37393 Arundinella hirta (Thunb.) Tanaka var. hondana Koidz. = Arundinella hondana (Koidz.) B. S. Sun et Z. H. Hu ■

37394 Arundinella hirta (Thunb.) Tanaka var. koryuensis Honda = Arundinella hirta (Thunb.) Tanaka ■

37395 Arundinella hirta (Thunb.) Tanaka var. riparia (Honda) Ohwi = Arundinella riparia Honda ■☆

37396 Arundinella hispida (Humb. et Bonpl. ex Willd.) Kuntze f. humilior Hack. = Arundinella pubescens Merr. et Hack. ex Hack. ■
37397 Arundinella hispida (Humb. et Bonpl. ex Willd.) Kuntze subsp. humilior (Hack.) Hack. = Arundinella pubescens Merr. et Hack. ex Hack. ■
37398 Arundinella hondana (Koidz.) B. S. Sun et Z. H. Hu;庐山野古草;Lushan Arundinella ■
37399 Arundinella hondana (Koidz.) B. S. Sun et Z. H. Hu = Arundinella hirta (Thunb.) Tanaka var. hondana Koidz. ■
37400 Arundinella hookeri Munro ex Keng;西南野古草(陈谋野古草,穗序野古草,喜马拉雅野古草);Hooker Arundinella ■
37401 Arundinella hordeiformis (Stapf) Roberty = Loudetia hordeiformis (Stapf) C. E. Hubb. ■☆
37402 Arundinella hubeiensis D. M. Chen;湖北野古草;Hubei Arundinella ■
37403 Arundinella hubeiensis D. M. Chen = Arundinella setosa Trin. var. esetosa Bor ■
37404 Arundinella hubeiensis D. M. Chen = Arundinella setosa Trin. ■
37405 Arundinella humilior (Hack.) Jansen = Arundinella pubescens Merr. et Hack. ex Hack. ■
37406 Arundinella intricata Hughes;错立野古草■
37407 Arundinella kengiana N. X. Zhao;大别山野古草;Dabieshan Arundinella ■
37408 Arundinella kengiana N. X. Zhao = Arundinella hirta (Thunb.) Tanaka ■
37409 Arundinella khaseana Nees ex Steud.;滇西野古草;W. Yunnan Arundinella ■
37410 Arundinella longispicata B. S. Sun;长序野古草■
37411 Arundinella macauensis Bor = Arundinella setosa Trin. var. esetosa Bor ■
37412 Arundinella marungensis (Chiov.) Chiov. = Trichopteryx marungensis Chiov. ■☆
37413 Arundinella miliacea (Link) Nees = Arundinella nepalensis Trin. ■
37414 Arundinella nepalensis Trin.;石芒草(吹鸡秆,石清草,石珍芒,硬骨草);Nepal Arundinella ■
37415 Arundinella nervosa (Roxb.) Nees;具脉野古草■☆
37416 Arundinella nodosa B. S. Sun et Z. H. Hu;多节野古草;Manynode Arundinella ■
37417 Arundinella parviflora B. S. Sun et Z. H. Hu;小花野古草;Smallflower Arundinella ■
37418 Arundinella pilaxilis B. S. Sun et Z. H. Hu;毛轴野古草;Hairaxle Arundinella ■
37419 Arundinella pilaxilis B. S. Sun et Z. H. Hu = Arundinella nepalensis Trin. ■
37420 Arundinella pilomarginata B. S. Sun = Arundinella nepalensis Trin. ■
37421 Arundinella pubescens Merr. et Hack. ex Hack.;毛野古草;Hairy Arundinella ■
37422 Arundinella pumila (Hochst. ex A. Rich.) Steud.;矮小野古草■☆
37423 Arundinella rigida Nees = Arundinella nepalensis Trin. ■
37424 Arundinella riparia Honda;河岸野古草■☆
37425 Arundinella ritchiei Munro ex Lisboa = Arundinella nepalensis Trin. ■
37426 Arundinella rupestris A. Camus;岩生野古草;Rocky Arundinella ■
37427 Arundinella rupestris A. Camus var. pachyathera (Hand. -Mazz.) B. S. Sun et Z. H. Hu;粗芒野古草■
37428 Arundinella rupestris A. Camus var. pachyathera (Hand. -Mazz.) B. S. Sun et Z. H. Hu. = Arundinella rupestris A. Camus ■
37429 Arundinella setosa Trin.;刺芒野古草;Setose Arundinella ■
37430 Arundinella setosa Trin. var. esetosa Bor;无刺野古草;Spineless Arundinella ■
37431 Arundinella setosa Trin. var. tengchongensis B. S. Sun et Z. H. Hu ex S. L. Chen;腾冲野古草;Tengchong Arundinella ■
37432 Arundinella simplex (Nees) Roberty = Loudetia simplex (Nees) C. E. Hubb. ■☆
37433 Arundinella sinensis Rendle = Arundinella setosa Trin. ■
37434 Arundinella stipoides Hack. = Loudetia simplex (Nees) C. E. Hubb. ■☆
37435 Arundinella suniana S. M. Phillips et S. L. Chen = Arundinella khaseana Nees ex Steud. ■
37436 Arundinella togoensis (Pilg.) Roberty = Loudetia togoensis (Pilg.) C. E. Hubb. ■☆
37437 Arundinella tricholepis B. S. Sun et Z. H. Hu;毛颖野古草;Hairglume Arundinella ■
37438 Arundinella tristachyoides (Trin.) Roberty = Loudetiopsis tristachyoides (Trin.) Conert ■☆
37439 Arundinella villosa Arn. ex Steud. var. himalaica Hook. f. = Arundinella hookeri Munro ex Keng ■
37440 Arundinella virgata Janowski = Arundinella nepalensis Trin. ■
37441 Arundinella wallichii Nees ex Steud. = Arundinella bengalensis (Spreng.) Druce ■
37442 Arundinella yunnanensis Keng ex B. S. Sun et Z. H. Hu;云南野古草;Yunnan Arundinella ■
37443 Arundinellaceae Herter = Gramineae Juss. (保留科名)●■
37444 Arundinellaceae Herter = Poaceae Barnhart(保留科名)●■
37445 Arundinellaceae Herter;野古草科■
37446 Arundinellaceae Stapf = Gramineae Juss. (保留科名)●■
37447 Arundinellaceae Stapf = Poaceae Barnhart(保留科名)●■
37448 Arundlnellaceae (Stapf) Herter = Gramineae Juss. (保留科名)●■
37449 Arundlnellaceae (Stapf) Herter = Poaceae Barnhart(保留科名)●■
37450 Arundo L. (1753);芦竹属(荻芦竹属);Giant Reed, Giantreed, Great Reed ●
37451 Arundo P. Beauv. = Phragmites Adans. ■
37452 Arundo Tourn. ex L. = Arundo L. ●
37453 Arundo aegyptiaca Vilm. = Arundo donax L. ●
37454 Arundo altissima Benth. = Phragmites australis (Cav.) Steud. subsp. altissimus (Benth.) Clayton ■
37455 Arundo arenaria L. = Ammophila arenaria (L.) Link ■☆
37456 Arundo australis Cav. = Phragmites australis (Cav.) Trin. ex Steud. ■
37457 Arundo bambos L. = Bambusa arundinacea (Retz.) Willd. ●
37458 Arundo bengalensis Retz. = Arundo donax L. ●
37459 Arundo bicolor Desf. = Ampelodesmos mauritanicus (Poir.) T. Durand et Schinz ■☆
37460 Arundo bicolor Poir. = Ampelodesmos mauritanicus (Poir.) T. Durand et Schinz ■☆
37461 Arundo bifaria Retz. = Arundo donax L. ●
37462 Arundo canescens F. H. Wigg. = Calamagrostis canescens (F. H. Wigg.) Roth ■☆
37463 Arundo coleotricha (Hack.) Honda = Arundo donax L. var. coleotricha Hack. ●
37464 Arundo coleotricha (Hack.) Honda = Arundo donax L. ●
37465 Arundo colorata Aiton = Phalaris arundinacea L. ■

37466 Arundo communis ? = Phragmites australis (Cav.) Trin. ex Steud. ■

37467 Arundo donax L.;芦竹(荻芦竹,楼梯杆,芦荻,芦荻头,芦荻竹,篆竹笋,绿竹,毛鞘芦竹);Cana Brava,Carrizo,Giant Cane,Giant Reed,Giantreed,Long-leaved Reed,Provence Reed,Spanish Cane,Spanish Reed,Wild Cane ●

37468 Arundo donax L. 'Variegata' = Arundo donax L. 'Versicolor' ●

37469 Arundo donax L. 'Versicolor';彩叶芦竹(变叶芦竹,花叶芦竹);Giant Variegated Mediterranean,Rush ●

37470 Arundo donax L. var. barbigera (Honda) Ohwi;髯毛芦竹●☆

37471 Arundo donax L. var. coleotricha Hack.;毛鞘芦竹●

37472 Arundo donax L. var. coleotricha Hack. = Arundo donax L. ●

37473 Arundo donax L. var. versicolor Stokes = Arundo donax L. 'Versicolor' ●

37474 Arundo epigejos L. = Calamagrostis epigejos (L.) Roth ■

37475 Arundo festucacea Willd. = Scolochloa festucacea (Willd.) Link ■

37476 Arundo festucoides Desf. = Ampelodesmos mauritanicus (Poir.) T. Durand et Schinz ■☆

37477 Arundo formosana Hack.;台湾芦竹;Taiwan Giantreed ■

37478 Arundo formosana Hack. var. gracilis Hack. = Arundo formosana Hack. ■

37479 Arundo formosana Hack. var. robusta Conert = Arundo formosana Hack. ■

37480 Arundo glauca M. Bieb. = Calamagrostis pseudophragmites (Hallier f.) Koeler ■

37481 Arundo henslowiana Nees = Neyraudia reynaudiana (Kunth) Keng ex Hitchc. ■

37482 Arundo isiaca Delile = Phragmites australis (Cav.) Steud. subsp. altissimus (Benth.) Clayton ■

37483 Arundo karka Retz. = Phragmites karka (Retz.) Trin. ex Steud. ■

37484 Arundo langsdorffii Link = Calamagrostis purpurea (Trin.) Trin. subsp. langsdorfii (Link) Tzvelev ■

37485 Arundo langsdorffii Link = Deyeuxia langsdorffii (Link) Kunth ■

37486 Arundo lapponica Wahlenb. = Deyeuxia lapponica (Wahlenb.) Kunth ■

37487 Arundo littorea Schrad. = Calamagrostis pseudophragmites (Hallier f.) Koeler ■

37488 Arundo longifolia Salisb. ex Hook. f. = Arundo donax L. ●

37489 Arundo madagascariensis Kunth = Neyraudia arundinacea (L.) Henrard ■

37490 Arundo mauritanica Desf. = Arundo mediterranea Danin ●☆

37491 Arundo mauritanica Poir. = Ampelodesmos mauritanicus (Poir.) T. Durand et Schinz ■☆

37492 Arundo maxima Forssk. = Phragmites australis (Cav.) Steud. subsp. altissimus (Benth.) Clayton ■

37493 Arundo maxima Lour. = Gigantochloa verticillata (Willd.) Munro ●

37494 Arundo mediterranea Danin;地中海芦竹●☆

37495 Arundo multiplex Lour. = Bambusa multiplex (Lour.) Raeusch. ex Schult. et Schult. f. ●

37496 Arundo neglecta Ehrh. = Calamagrostis neglecta (Ehrh.) Gaertn., Mey. et Scherb. ■

37497 Arundo neglecta Ehrh. = Deyeuxia neglecta (Ehrh.) Kunth ■

37498 Arundo parviflora Ohwi = Arundo formosana Hack. ■

37499 Arundo phragmites L. = Phragmites australis (Cav.) Trin. ex Steud. ■

37500 Arundo plinii Turra;普林芦竹●☆

37501 Arundo pseudophragmites Hallier f. = Calamagrostis pseudophragmites (Hallier f.) Koeler ■

37502 Arundo reynaudiana Kunth = Neyraudia reynaudiana (Kunth) Keng ex Hitchc. ■

37503 Arundo riparia Salisb. = Phalaris arundinacea L. ■

37504 Arundo roxburghii Kunth = Phragmites karka (Retz.) Trin. ex Steud. ■

37505 Arundo selloana Schult. et Schult. f. = Cortaderia selloana (Schult. et Schult. f.) Asch. et Graebn. ■

37506 Arundo sylvatica Schrad. = Calamagrostis arundinacea (L.) Roth ■☆

37507 Arundo sylvatica Schrad. = Deyeuxia arundinacea (L.) P. Beauv. ■

37508 Arundo triflora Roxb. ex Hook. f. = Arundo donax L. ●

37509 Arundo villosa Trin. = Psammochloa mongolica (Hitchc.) Roshev. ■

37510 Arundo villosa Trin. = Psammochloa villosa (Trin.) Bor ■

37511 Arundo webbiana Steud.;韦布芦竹●☆

37512 Arundo zollingeri Büse = Neyraudia reynaudiana (Kunth) Keng ex Hitchc. ■

37513 Arundoclaytonia Davidse et R. P. Ellis(1987);克莱东芦竹属■☆

37514 Arundoclaytonia dissimilis Davidse et R. P. Ellis;克莱东芦竹■☆

37515 Arungana Pers. = Haronga Thouars ■☆

37516 Arunia Pers. = Brunia Lam.(保留属名)●☆

37517 Arupsis Rojas = Spathicarpa Hook. ■☆

37518 Arversia Cambess. = Polycarpon L. ■

37519 Arversia depressa (L.) Klotzsch = Polycarpon prostratum (Forssk.) Asch. et Schweinf. ■

37520 Arviela Salisb. = Zephyranthes Herb.(保留属名)■

37521 Arytera Blume(1849);滨木患属;Arytera ●

37522 Arytera littoralis Blume;滨木患;Arytera,Common Arytera ●

37523 Asaeara Raf. = Gleditsia L. ●

37524 Asaemia (Harv.) Benth. = Athanasia L. ●☆

37525 Asaemia (Harv.) Benth. et Hook. f. = Stilpnophyton Less. ●☆

37526 Asaemia Harv. = Athanasia L. ●☆

37527 Asaemia Harv. ex Benth. et Hook. f. = Athanasia L. ●☆

37528 Asaemia axillaris (Thunb.) Harv. ex Hoffm. = Athanasia minuta (L. f.) Källersjö ●☆

37529 Asaemia inermis E. Phillips = Athanasia minuta (L. f.) Källersjö subsp. inermis (E. Phillips) Källersjö ●☆

37530 Asaemia minuta (L. f.) K. Bremer = Athanasia minuta (L. f.) Källersjö ●☆

37531 Asaemia minuta (L. f.) K. Bremer subsp. inermis (E. Phillips) K. Bremer = Athanasia minuta (L. f.) Källersjö subsp. inermis (E. Phillips) Källersjö ●☆

37532 Asagraea Baill. = Dalea L.(保留属名)●■☆

37533 Asagraea Baill. = Psorothamnus Rydb. ●☆

37534 Asagraea Lindl. = Sabadilla Brandt et Ratzeb. ■☆

37535 Asagraea Lindl. = Schoenocaulon A. Gray ■☆

37536 Asamanthia (Stapf) Ridl. = Mussaenda L. ●■

37537 Asanthus R. M. King et H. Rob.(1972);鳞叶肋泽兰属●☆

37538 Asanthus R. M. King et H. Rob. = Steviopsis R. M. King et H. Rob. ■☆

37539 Asanthus squamulosus (A. Gray) R. M. King et H. Rob.;鳞叶肋泽兰;Mule Mountain False Brickellbush ■☆

37540 Asaphes DC. = Toddalia Juss.(保留属名)●

37541 Asaphes DC. = Vepris Comm. ex A. Juss. ●☆

37542 Asaphes Spreng. = Morina L. ■
37543 Asaraceae Vent.;细辛科(杜蘅科)■
37544 Asaraceae Vent. = Aristolochiaceae Juss.(保留科名)●■
37545 Asarca Lindl. = Chloraea Lindl. ■☆
37546 Asarca Lindl. = Gavilea Poepp. ■☆
37547 Asarca Poepp. ex Lindl. = Asarca Lindl. ■☆
37548 Asarina Mill.(1757);金鱼藤属(腋花金鱼草属);Asarina, Trailing Snapdragon, Wild Ginger ■☆
37549 Asarina Tourn. ex Mill. = Asarina Mill. ■☆
37550 Asarina antirrhiniflora (Humb. et Bonpl. ex Willd.) Pennell = Maurandella antirrhiniflora (Humb. et Bonpl. ex Willd.) Rothm. ■☆
37551 Asarina antirrhinifolia (Humb. et Bonpl. ex Willd.) Pennell = Asarina procumbens Mill. ■☆
37552 Asarina erubescens (D. Don) Pennell = Kickxia petiolata D. A. Sutton ■☆
37553 Asarina lobelii Quer;罗氏金鱼藤■☆
37554 Asarina procumbens Mill.;金鱼藤(匍生金鱼草,腋花金鱼草);Creeping Snapdragon, Procumbent Asarina, Trailing Snapdragon ■☆
37555 Asarina scandens (Cav.) Pennell = Kickxia scalarum D. A. Sutton ■☆
37556 Asarum L.(1753);细辛属;Asarabaeca, Wild Ginger, Wildginger, Wild-ginger ■
37557 Asarum acuminatum (Ashe) E. P. Bicknell = Asarum canadense L. ●■☆
37558 Asarum acuminatum E. P. Bicknell = Asarum canadense L. ●■☆
37559 Asarum albomaculatum Hayata;白斑细辛■
37560 Asarum albomaculatum Hayata = Asarum macranthum Hook. f. ■
37561 Asarum arifolium Michx.;北美细辛;Wild Ginger ■☆
37562 Asarum arifolium Michx. = Hexastylis arifolia (Michx.) Small ■☆
37563 Asarum arifolium Michx. var. callifolium (Small) Barringer = Hexastylis arifolia (Michx.) Small var. callifolia (Small) H. L. Blomq. ■☆
37564 Asarum arifolium Michx. var. ruthii (Ashe) Barringer = Hexastylis arifolia (Michx.) Small var. ruthii (Ashe) H. L. Blomq. ■☆
37565 Asarum arrhizoma H. Lév. et Vaniot = Asarum caudigerum Hance ■
37566 Asarum asaroides (C. Morren et Decne.) Makino;异细辛■☆
37567 Asarum asaroides (C. Morren et Decne.) Makino = Heterotropa asaroides C. Morren et Decne. ■☆
37568 Asarum asperum F. Maek.;粗糙细辛■☆
37569 Asarum balansae Franch.;巴兰细辛(花脸细辛)■☆
37570 Asarum bashanense Z. L. Yang;巴山细辛;Bashan Wildginger ■
37571 Asarum blumei Duch.;布氏细辛(钹儿草,杜衡,杜蘅,马蹄细辛,马蹄香)■☆
37572 Asarum brevistylum Franch.;短柱细辛;Shortstyle Wildginger ■
37573 Asarum callifolium Small = Hexastylis arifolia (Michx.) Small var. callifolia (Small) H. L. Blomq. ■☆
37574 Asarum canadense L.;加拿大细辛;American Wild Ginger, Canada Snake-root, Canada Wildginger, Canadian Snakeroot, Canadian Wild Ginger, Canadian Wild-ginger, Coltsfoot, Eastern Wild Ginger, Indian Ginger, Wild Ginger, Wild-ginger ●■☆
37575 Asarum canadense L. var. acuminatum Ashe = Asarum canadense L. ●■☆
37576 Asarum canadense L. var. acuminatum Ashe f. prattii Fassett = Asarum canadense L. ●■☆
37577 Asarum canadense L. var. ambiguum (E. P. Bicknell) Farw. = Asarum canadense L. ●■☆
37578 Asarum canadense L. var. ambiguum Farw. = Asarum canadense L. ●■☆
37579 Asarum canadense L. var. reflexum (E. P. Bicknell) B. L. Rob. = Asarum canadense L. ●■☆
37580 Asarum canadense L. var. reflexum B. L. Rob. = Asarum canadense L. ●■☆
37581 Asarum cardiophyllum Franch.;花叶细辛(红三百棒,花叶尾花细辛);Cordateleaf Caudate Wildginger ■
37582 Asarum caudatum Lindl.;尾萼细辛(尾状细辛);British Columbia Wild Ginger, Long-tailed Wild Ginger, Western Wild Ginger, Wild Ginger ■☆
37583 Asarum caudatum Lindl. var. viridiflorum M. Peck = Asarum wagneri K. L. Lu et Mesler ■☆
37584 Asarum caudigerellum C. Y. Cheng et C. S. Yang;短尾细辛(接气草,圆叶细辛);Short Caudate Wildginger, Shorttail Wildginger ■
37585 Asarum caudigerum Hance;尾花细辛(白三百棒,白细辛,薄叶细辛,花乌金草,顺河香,四两淋,土细辛,香菇草,圆叶细辛);Caudate Wildginger ■
37586 Asarum caudigerum Hance var. cardiophyllum (Franch.) C. Y. Cheng et C. S. Yang = Asarum cardiophyllum Franch. ■
37587 Asarum caudigerum Hance var. leptophyllum (Hayata) C. Y. Cheng et C. S. Yang;窄叶尾花细辛;Smallleaf Caudate Wildginger ■
37588 Asarum caudigerum Hance var. leptophyllum (Hayata) C. Y. Cheng et C. S. Yang = Asarum caudigerum Hance ■
37589 Asarum caudigerum Hance var. triangulare (Hayata) S. S. Ying = Asarum caudigerum Hance ■
37590 Asarum caulescens Maxim.;双叶细辛(草马蹄香,乌金草);Caulescent Wildginger ■
37591 Asarum caulescens Maxim. f. geroensis J. Ohara;下吕细辛■☆
37592 Asarum caulescens Maxim. var. setchuenense Franch. = Asarum caulescens Maxim. ■
37593 Asarum cavaleriei H. Lév. et Vaniot = Asarum geophilum Hemsl. ■
37594 Asarum cavaleriei H. Lév. et Vaniot var. esquirolii H. Lév. = Asarum geophilum Hemsl. ■
37595 Asarum celsum F. Maek. ex Hatus. et Yamahata;高细辛■☆
37596 Asarum chengkouense Z. L. Yang;城口细辛;Chengkou Wildginger ■
37597 Asarum chinense Franch.;川北细辛(花叶细辛,中国细辛);China Wildginger, North Sichuan Wildginger, North Szechwan Wildginger ■
37598 Asarum chinense Franch. f. fargesii (Franch.) D. Y. Cheng et C. S. Yang;绿背白脉细辛;Farges China Wildginger ■
37599 Asarum chingchengense C. Y. Cheng et C. S. Yang = Asarum splendens (Maek.) C. Y. Cheng et C. S. Yang ■
37600 Asarum constrictum F. Maek.;缢缩细辛■☆
37601 Asarum contractum (H. L. Blomq.) Barringer = Hexastylis contracta H. L. Blomq. ■☆
37602 Asarum costatum (F. Maek.) F. Maek.;粗细辛■☆
37603 Asarum crassisepalum S. F. Huang, T. H. Hsieh et T. C. Huang;鸳鸯湖细辛■
37604 Asarum crassum F. Maek.;粗醉魂藤■☆
37605 Asarum crispulatum C. Y. Cheng et C. S. Yang;皱花细辛(盆草细辛);Crispedflower Wildginger ■
37606 Asarum curvistigma F. Maek.;曲柱细辛■☆
37607 Asarum dabieshanense D. Q. Wang et S. H. Hwang;大别山细辛;Dabieshan Wildginger ■

37608 Asarum debile Franch.;铜钱细辛(胡椒七,毛细辛,铜钱乌金);Copper-cash Wildginger,Thin Wildginger ■
37609 Asarum delavayi C. Y. Cheng et C. S. Yang = Asarum splendens (Maek.) C. Y. Cheng et C. S. Yang ■
37610 Asarum delavayi Franch.;川滇细辛(滇细辛,牛蹄细辛);Delavay Wildginger ■
37611 Asarum delavayi Franch. var. omeiense C. Y. Cheng et C. S. Yang;峨眉牛蹄细辛■
37612 Asarum dimidiatum F. Maek.;对开细辛■☆
37613 Asarum dissitum Hatus.;疏细辛■☆
37614 Asarum elegans Duch.;雅致细辛■☆
37615 Asarum epigynum Hayata;台湾细辛(上花细辛);Taiwan Wildginger ■
37616 Asarum europaeum L.;欧细辛(常绿蛇根草,欧洲细辛);Asarabacca,Cabaret,European Asarum,European Ginger,European Wild Ginger,European Wildginger,Foalfoot,Hazelwort,Wild Nard ■☆
37617 Asarum fargesii Franch.;莲花细辛(江南细辛,小叶细辛草);Farges Wildginger ■
37618 Asarum fargesii Franch. = Asarum chinense Franch. ■
37619 Asarum fauriei Franch.;法氏细辛■☆
37620 Asarum fauriei Franch. var. nakaianum (F. Maek.) Ohwi;中井氏细辛■☆
37621 Asarum fauriei Franch. var. serpens F. Maek.;蛇细辛■☆
37622 Asarum forbesii Maxim.;杜衡(钵儿草,楚蘅,杜衡葵,杜蘅,杜葵,杜细辛,福氏细辛,蘅薇草,蘅薇香,怀,蘹香,金锁匙,马蹄细辛,马蹄香,马蹄莕,马细辛,马辛,南细辛,泥里花,双龙麻消,水马蹄,土里开花,土卤,土细辛,土杏,土荇);Forbes Wildginger ■
37623 Asarum franchetianum Diels = Asarum caulescens Maxim. ■
37624 Asarum fukienense C. Y. Cheng et C. S. Yang;福建细辛(马脚蹄,薯叶细辛,土里开花);Fujian Wildginger,Fukien Wildginger ■
37625 Asarum gelasinum Hatus.;笑细辛■☆
37626 Asarum geophilum Hemsl.;地花细辛(矮细辛,大块瓦,大瓦块,花叶细辛,铺地细辛);Earthloving Wildginger,Mattae Wildginger ■
37627 Asarum gracilipes C. S. Yang ex C. F. Liang;纤梗细辛(金耳环);Slender Wildginger ■
37628 Asarum gracilipes C. S. Yang ex C. F. Liang = Asarum insigne Diels ■
37629 Asarum grandiflorum Maek.;西方大花细辛■☆
37630 Asarum grandiflorum Maek. var. colocasiifolium Hayata;芋叶大花细辛;Colocasiifoliate Slender Wildginger ■
37631 Asarum hartwegii S. Watson;哈氏细辛■☆
37632 Asarum hatsushimae F. Maek. ex Hatus. et Yamahata;初岛细辛■☆
37633 Asarum hayatanum (Maek.) Masam.;芋叶细辛■
37634 Asarum hayatanum (Maek.) Masam. = Asarum hypogynum Hayata ■
37635 Asarum heterotropoi F. Schmidt;库页细辛;Heterotropalike Wildginger,Kuye Wildginger ■☆
37636 Asarum heterotropoi F. Schmidt var. mandshuricum (Maxim.) Kitag.;辽细辛(北细辛,细辛,烟袋锅花)■
37637 Asarum heterotropoides F. Schmidt;拟库页细辛■
37638 Asarum heterotropoides F. Schmidt var. mandshuricum (Maxim.) Kitag.;北细辛(白细辛,东北细辛,独叶草,华细辛,金盆草,辽细辛,盆草细辛,山人参,少辛,细草,细辛,小辛,烟袋锅花);Manchur Wildginger,Manchurian Wildginger ■
37639 Asarum hexalobum (F. Maek.) F. Maek.;六裂细辛■☆
37640 Asarum hexalobum F. Maek. var. controversum Hatus. et Yamahata;疑惑细辛■☆
37641 Asarum hexalobum F. Maek. var. perfectum F. Maek. = Asarum perfectum (F. Maek.) F. Maek. ■☆
37642 Asarum himalaicum Hook. f. et Thomson ex Klotzsch;单叶细辛(马蹄细辛,毛细辛,盆草细辛,水细辛,土癞蜘蛛香,乌金七,西南细辛,细辛);Himalayas Wildginger,Simpleleaf Wildginger ■
37643 Asarum hirsutisepalum Hatus. = Asarum yakusimense Masam. ■☆
37644 Asarum hisauchii F. Maek. = Asarum takaoi F. Maek. var. hisauchii (F. Maek.) F. Maek. ■☆
37645 Asarum hongkongense S. M. Hwang et T. P. Wang;香港细辛;Hongkong Wildginger ■
37646 Asarum hypogynum Hayata;下花细辛(芋叶细辛);Hayata Wildginger,Hypogynousflower Wildginger,Taroleaf Wildginger ■
37647 Asarum ibericum Steven ex Ledeb.;伊伯利亚细辛■☆
37648 Asarum ichangense C. Y. Cheng et C. S. Yang;小叶马蹄香(马蹄香,土细辛,小叶马蹄细辛,小叶细辛,宜昌细辛,窄叶细辛);Smallleaf Wildginger ■
37649 Asarum ikegamii (F. Maek. ex Y. Maek.) T. Sugaw.;池上马蹄香■☆
37650 Asarum ikegamii (F. Maek. ex Y. Maek.) T. Sugaw. var. fujimakii T. Sugaw.;藤卷马蹄香■☆
37651 Asarum inflatum C. Y. Cheng et C. S. Yang;灯笼细辛;Inflated Wildginger,Lantern Wildginger ■
37652 Asarum infrapurpureum Hayata = Asarum macranthum Hook. f. ■
37653 Asarum insigne Diels;金耳环细辛(长花轴细辛,长叶细辛,慈姑叶细辛,大叶山茨菇,大叶细辛,金耳环,马蹄细辛,盘山草,山薯,苕叶细辛,土细辛,细辛,小犁头,瑶山金耳环,一块瓦);Goldenearrring,Notable Wildginger ■
37654 Asarum kiusianum F. Maek.;九州细辛■☆
37655 Asarum kiusianum F. Maek. var. melanosiphon (F. Maek.) F. Maek.;黑管九州细辛■☆
37656 Asarum kiusianum F. Maek. var. tubulosum (F. Maek.) F. Maek.;块状九州细辛■☆
37657 Asarum kooyanum Makino;高野山细辛■☆
37658 Asarum kooyanum Makino var. brachypodion (F. Maek.) Kitam.;短梗九州细辛■☆
37659 Asarum kooyanum Makino var. nipponicum (F. Maek.) Kitam.;本州细辛(日本细辛);Japan Wildginger ■☆
37660 Asarum kooyanum Makino var. nipponicum (F. Maek.) Kitam. = Asarum nipponicum F. Maek. ■☆
37661 Asarum kooyanum Makino var. rigescens (F. Maek.) Kitam. = Asarum rigescens F. Maek. ■☆
37662 Asarum kumageanum Masam.;熊毛细辛■☆
37663 Asarum kumageanum Masam. var. satakeanum (F. Maek.) Hatus.;佐竹细辛■☆
37664 Asarum kurokawanum Makino = Asarum viridiflorum Regel ■☆
37665 Asarum kurosawae Sugim.;黑泽细辛■☆
37666 Asarum leptophyllum Hayata = Asarum caudigerum Hance ■
37667 Asarum leptophyllum Hayata var. triangulare Hayata = Asarum caudigerum Hance ■
37668 Asarum leucosepalum Hatus. ex Yamahata;白萼细辛■☆
37669 Asarum lewisii Fernald = Hexastylis lewisii (Fernald) H. L. Blomq. et Oosting ■☆
37670 Asarum longepedunculatum O. C. Schmidt = Asarum insigne Diels ■
37671 Asarum longerhizomatosum C. F. Liang et C. S. Yang;长茎金耳环(金耳环,一块瓦);Longstem Wildginger ■
37672 Asarum longiflorum C. Y. Cheng et C. S. Yang;长花细辛(黄细

辛,土细辛);Longflower Wildginger ■
37673 Asarum lutchuense T. Ito;琉球细辛■☆
37674 Asarum macranthum Hook. f.;大花细辛(白斑细辛,大屯细辛,里紫细辛,马蹄香,毛柄细辛,台东细辛,下紫细辛,小花细辛);Datun Wildginger, Infrapurple Wildginger, Largeflower Wildginger, Whitemaculate Wildginger ■
37675 Asarum maculatum Nakai = Asarum sieboldii Miq. f. maculatum (Nakai) Yamaji ■☆
37676 Asarum maekawae H. Hara = Asarum delavayi C. Y. Cheng et C. S. Yang ■
37677 Asarum maekawae H. Hara = Asarum splendens (Maek.) C. Y. Cheng et C. S. Yang ■
37678 Asarum magnificum Tsiang ex C. Y. Cheng et C. S. Yang;祁阳细辛(南细辛,山慈姑,山慈菇);Qiyang Wildginger, Robust Wildginger ■
37679 Asarum magnificum Tsiang ex C. Y. Cheng et C. S. Yang var. dinghuense C. Y. Cheng et C. S. Yang;鼎湖细辛(山慈姑,山慈菇);Dinghu Wildginger ■
37680 Asarum maximum Hemsl.;大叶马蹄香(大花细辛,大叶细辛,杜衡,翻天印,花脸猫,花脸细辛,花叶细辛,马蹄细辛,马蹄香,水马蹄,土细辛);Largeleaf Wildginger ■
37681 Asarum megacalyx (F. Maek.) T. Sugaw.;大萼细辛■☆
37682 Asarum melanosiphon F. Maek. = Asarum kiusianum F. Maek. var. melanosiphon (F. Maek.) F. Maek. ■☆
37683 Asarum memmingeri Ashe = Hexastylis virginica (L.) Small ■☆
37684 Asarum minamitanianum Hatus.;南谷细辛■☆
37685 Asarum minus Ashe = Hexastylis minor (Ashe) H. L. Blomq. ■☆
37686 Asarum mitoanum T. Sugaw.;三刀细辛■☆
37687 Asarum monodoriflorum Hatus. et Yamahata;单兜花细辛■☆
37688 Asarum muramatsui Makino;松村细辛■☆
37689 Asarum muramatsui Makino var. shimodanum (F. Maek.);下田细辛■☆
37690 Asarum nakaianum F. Maek. = Asarum fauriei Franch. var. nakaianum (F. Maek.) Ohwi ■☆
37691 Asarum nanchuanense C. S. Yang et J. L. Wu;南川细辛(山花椒);Nanchuan Wildginger ■
37692 Asarum nankaiense F. Maek.;南海道细辛■☆
37693 Asarum nipponicum F. Maek. = Asarum kooyanum Makino var. nipponicum (F. Maek.) Kitam. ■☆
37694 Asarum nipponicum F. Maek. var. brachypodion F. Maek. = Asarum kooyanum Makino var. brachypodion (F. Maek.) Kitam. ■☆
37695 Asarum nobilissimum Z. L. Yang;奉节细辛;Fengjie Wildginger ■
37696 Asarum nomadakense Hatus.;野间岳细辛■☆
37697 Asarum oblongum (F. Maek.) F. Maek.;矩圆细辛■☆
37698 Asarum okinawense Hatus.;冲绳细辛■☆
37699 Asarum parviflorum (Hook.) Regel;小花细辛■☆
37700 Asarum pellucidum Hatus. et Yamahata;透明细辛■☆
37701 Asarum perfectum (F. Maek.) F. Maek.;完美细辛■☆
37702 Asarum petelotii O. C. Schmidt;红金耳环;Petelot Wildginger, Red Goldenearrring ■
37703 Asarum porphyronotum C. Y. Cheng et C. S. Yang;紫背细辛;Purple Wildginger ■
37704 Asarum porphyronotum C. Y. Cheng et C. S. Yang var. atrovirens C. Y. Cheng et C. S. Yang;深绿细辛;Deepgreen Wildginger, Green Wildginger ■
37705 Asarum pseudosavatieri F. Maek.;假萨瓦细辛■☆
37706 Asarum pulchellum Hemsl.;长毛细辛(白毛细辛,白三百棒,毛乌金,牛毛细辛,乌金草);Longhair Wildginger ■
37707 Asarum reflexum E. P. Bicknell = Asarum canadense L. ●■☆
37708 Asarum reflexum E. P. Bicknell var. ambiguum E. P. Bicknell = Asarum canadense L. ●■☆
37709 Asarum renicordatum C. Y. Cheng et C. S. Yang;肾叶细辛(马蹄香);Kidneyleaf Wildginger, Reniformleaf Wildginger ■
37710 Asarum rigescens F. Maek.;稍硬细辛■☆
37711 Asarum rubrocinctum Peattie = Asarum canadense L. ●■☆
37712 Asarum ruthii Ashe = Hexastylis arifolia (Michx.) Small var. ruthii (Ashe) H. L. Blomq. ■☆
37713 Asarum sagittarioides C. F. Liang;岩慈姑(慈姑叶细辛,山慈姑);Arrowheadlike Wildginger, Wild Wildginger ■
37714 Asarum sakawanum Makino;佐川细辛■☆
37715 Asarum satsumense F. Maek.;萨摩细辛■☆
37716 Asarum savatieri (Franch.) F. Maek. subsp. pseudosavatieri (F. Maek.) T. Sugaw. var. iseanum T. Sugaw.;伊势细辛■☆
37717 Asarum savatieri Franch.;萨瓦细辛■☆
37718 Asarum senkakuinsulare Hatus.;尖阁细辛■☆
37719 Asarum shuttleworthii Britten et Baker;舒氏细辛;Shuttleworth's Ginger ■☆
37720 Asarum shuttleworthii Britten et Baker f. = Hexastylis shuttleworthii (Britten et Baker f.) Small ■☆
37721 Asarum shuttleworthii Britten et Baker f. var. harperi (Gaddy) Barringer = Hexastylis shuttleworthii (Britten et Baker f.) Small var. harperi Gaddy ■☆
37722 Asarum sieboldii Miq.;细辛(白细辛,薄叶细辛,独叶草,汉城细辛,华细辛,金盆草,盆草细辛,山人参,少辛,细草,小辛);Siebold Wildginger ■
37723 Asarum sieboldii Miq. f. maculatum (Nakai) Yamaji;斑点细辛■☆
37724 Asarum sieboldii Miq. f. seoulense (Nakai) C. Y. Chang et C. S. Yang = Asarum sieboldii Miq. ■
37725 Asarum sieboldii Miq. f. seoulense (Nakai) C. Y. Cheng et C. S. Yang = Asarum sieboldii Miq. var. seoulense Nakai ■
37726 Asarum sieboldii Miq. subsp. heterotropoides (F. Schmidt) Kitam. = Asarum heterotropoides F. Schmidt ■
37727 Asarum sieboldii Miq. var. dimidiatum (F. Maek.) T. Sugaw. = Asarum dimidiatum F. Maek. ■☆
37728 Asarum sieboldii Miq. var. mandshuricum Maxim. = Asarum heterotropoides F. Schmidt var. mandshuricum (Maxim.) Kitag. ■
37729 Asarum sieboldii Miq. var. seoulense Nakai;汉城细辛(朝鲜细辛);Seoul Siebold Wildginger ■
37730 Asarum simile Hatus.;相似细辛■☆
37731 Asarum speciosum (R. M. Harper) Barringer = Hexastylis speciosa R. M. Harper ■☆
37732 Asarum splendens (Maek.) C. Y. Cheng et C. S. Yang;青城细辛(翻天印,花脸王,花脸细辛);Chinese Wild Ginger, Splendid Wildginger ■
37733 Asarum sprengeri Pamp.;南漳细辛;Sprenger Wildginger ■
37734 Asarum stellatum F. Maek. ex Akasawa;星状细辛■☆
37735 Asarum stoloniferum F. Maek.;匍匐细辛■☆
37736 Asarum subglobosum F. Maek. ex Hatus. et Yamahata;亚球细辛■☆
37737 Asarum taipingshanianum S. F. Huang, T. H. Hsieh et T. C. Huang;太平山细辛■
37738 Asarum taitonense Hayata;大屯细辛■
37739 Asarum taitonense Hayata = Asarum macranthum Hook. f. ■
37740 Asarum taiwanense S. S. Ying = Asarum epigynum Hayata ■

37741 Asarum takaoi F. Maek.;高雄细辛(黑川细辛)■
37742 Asarum takaoi F. Maek. var. dilatatum (F. Maek.) F. Maek.;膨大高雄细辛■☆
37743 Asarum takaoi F. Maek. var. hisauchii (F. Maek.) F. Maek.;久内细辛■☆
37744 Asarum tamaense Makino;多摩细辛■☆
37745 Asarum thunbergii A. Br.;粗根细辛■☆
37746 Asarum tokarense Hatus.;陶卡尔细辛■☆
37747 Asarum tongjiangense Z. L. Yang;通江细辛(同江细辛);Tongjiang Wildginger ■
37748 Asarum trigynum (F. Maek.) Araki;三柱细辛■☆
37749 Asarum trinacriforme Yamahata;三珠细辛■☆
37750 Asarum tubulosum F. Maek. = Asarum kiusianum F. Maek. var. tubulosum (F. Maek.) F. Maek. ■☆
37751 Asarum unzen (F. Maek.) Kitam. et Murata;运天细辛■☆
37752 Asarum unzen (F. Maek.) Kitam. et Murata f. luteoviride Yonek. et H. Ohashi;黄绿九州细辛■☆
37753 Asarum variegatum A. Braun et C. D. Bouché;斑叶细辛■☆
37754 Asarum virginicum L.;心叶细辛;Heart Snakeroot, Heartleaf Snakeroot, Southern Wild Ginger, Virginia Wild Ginger ■☆
37755 Asarum virginicum L. = Hexastylis virginica (L.) Small ■☆
37756 Asarum viridiflorum Regel;绿花细辛■☆
37757 Asarum wagneri K. L. Lu et Mesler;俄勒冈细辛■☆
37758 Asarum wulingense C. F. Liang;五岭细辛(山慈姑,山慈菇);Wuling Wildginger ■
37759 Asarum wulongense Z. L. Yang;武隆细辛;Wulong Wildginger ■
37760 Asarum yaeyamense Hatus.;八重山细辛■☆
37761 Asarum yakusimense Masam.;屋久岛细辛■☆
37762 Asarum yamashiroi Hatus.;山代细辛■☆
37763 Asarum yoshikawae T. Sugaw.;吉川细辛■☆
37764 Asarum yunnanense T. Sugaw., Ogisu et C. Y. Chen;云南细辛■
37765 Ascalea Hill = Carduus L. + Cirsium Mill. ■
37766 Ascalea Hill = Carduus L. ■
37767 Ascalea lanceolata (L.) Hill. = Cirsium vulgare (Savi) Ten. ■
37768 Ascalonicum P. Renault = Allium L. ■
37769 Ascania Crantz = Patagonula L. ●☆
37770 Ascanica B. D. Jacks. = Ascania Crantz ●☆
37771 Ascanica Crantz = Patagonula L. ●☆
37772 Ascaricida (Cass.) Cass. = Baccharoides Moench ●■
37773 Ascaricida Cass. = Baccharoides Moench ●■
37774 Ascaricida Cass. = Vernonia Schreb.(保留属名)●■
37775 Ascaricida buchingeri Steetz = Vernonia buchingeri (Steetz) Oliv. et Hiern ■☆
37776 Ascaricida mossambiquensis Steetz = Vernonia adoensis Sch. Bip. ex Walp. var. mossambiquensis (Steetz) G. V. Pope ■☆
37777 Ascaridia Rchb. = Ascaricida Cass. ●■
37778 Ascarina J. R. Forst. et G. Forst. (1775);蛔囊花属●☆
37779 Ascarina coursii (Humbert et Capuron) J. -F. Leroy et Jérémie;蛔囊花●☆
37780 Ascarina serrata Blume = Sarcandra glabra (Thunb.) Nakai ●
37781 Ascarina serrata Blume = Sarcandra glabra Nakai subsp. brachystachys (Blume) Verdc. ●
37782 Ascarinopsis Humbert et Capuron = Ascarina J. R. Forst. et G. Forst. ●☆
37783 Ascarinopsis Humbert et Capuron(1955);类蛔囊花属●☆
37784 Ascarinopsis coursii Humbert et Capuron;类蛔囊花●☆
37785 Ascarinopsis coursii Humbert et Capuron = Ascarina coursii (Humbert et Capuron) J. -F. Leroy et Jérémie ●☆
37786 Aschamia Salisb. = Hippeastrum Herb.(保留属名)■
37787 Aschenbornia S. Schauer = Calea L. ●■☆
37788 Aschenfeldtia F. Muell. = Pimelea Banks ex Gaertn.(保留属名)●☆
37789 Aschenfeldtia F. Muell. ex Meisn. = Pimelea Banks ex Gaertn.(保留属名)●☆
37790 Aschersonia F. Muell. (1878) = Halophila Thouars ■
37791 Aschersonia F. Muell. ex Benth. = Halophila Thouars ■
37792 Aschersoniodoxa Gilg et Muschl. (1909);山白花芥属■☆
37793 Aschersoniodoxa chimborazensis Gilg et Muschl.;山白花芥■☆
37794 Aschersoniodoxa pilosa Al-Shehbaz;毛山白花芥■☆
37795 Aschistanthera C. Hansen(1987);全药野牡丹属☆
37796 Aschistanthera cristanthera C. Hansen;全药野牡丹☆
37797 Asciadium Griseb. (1866);古巴草属☆
37798 Ascidieria Seidenf. (1984);囊兰属■☆
37799 Ascidieria longifolia (Hook. f.) Seidenf.;囊兰■☆
37800 Ascidieria maculiflora J. J. Wood;斑叶囊兰■☆
37801 Ascidieria verticillaris (Kraenzl.) Garay;轮生囊兰■☆
37802 Ascidiogyne Cuatrec. (1965);瓶实菊属■☆
37803 Ascidiogyne sanchezvegae Cuatrec.;瓶实菊■☆
37804 Ascium Schreb. = Norantea Aubl. ●☆
37805 Ascleia Raf. = Hydrolea L.(保留属名)■
37806 Asclepiadaceae Borkh. (1797)(保留科名);萝藦科(萝摩科);Milkweed Family ●■
37807 Asclepiadaceae Borkh.(保留科名) = Apocynaceae Juss.(保留科名)●■
37808 Asclepiadaceae Medikus ex Borkh. = Apocynaceae Juss.(保留科名)●■
37809 Asclepiadaceae Medikus ex Borkh. = Asclepiadaceae Borkh.(保留科名)●■
37810 Asclepiadaceae R. Br. = Asclepiadaceae Borkh.(保留科名)●■
37811 Asclepias L. (1753);马利筋属(尖尾凤属,莲生桂子花属,莲生桂子属);Asclepias, Butterfly Flower, Mildweed, Milkweed, Milkweed Rubber, Silkweed ■
37812 Asclepias abyssinica (Decne.) N. E. Br. = Gomphocarpus abyssinicus Decne. ●☆
37813 Asclepias acerateoides (Schltr.) Schltr. = Xysmalobium acerateoides (Schltr.) N. E. Br. ■☆
37814 Asclepias acida Roxb. = Sarcostemma acidum (Roxb.) Voigt ■
37815 Asclepias acida Roxb. = Sarcostemma viminale (L.) R. Br. ■
37816 Asclepias adscendens (Schltr.) Schltr.;上举马利筋■☆
37817 Asclepias affinis (Schltr.) Schltr. = Asclepias albens (E. Mey.) Schltr. ■☆
37818 Asclepias affinis De Wild. = Trachycalymma buchwaldii (Schltr. et K. Schum.) Goyder ■☆
37819 Asclepias alatas Schltr. = Pachycarpus dealbatus E. Mey. ■☆
37820 Asclepias alba Mill. = Cynanchum vincetoxicum (L.) Pers. ■
37821 Asclepias albens (E. Mey.) Schltr.;变白马利筋■☆
37822 Asclepias albicans S. Watson;白茎马利筋(白马利筋,变白马利筋);White Milkweed, White-stemmed Milkweed ■☆
37823 Asclepias albida N. E. Br. = Gomphocarpus purpurascens A. Rich. ●☆
37824 Asclepias amabilis N. E. Br. = Gomphocarpus amabilis (N. E. Br.) Bullock ●☆
37825 Asclepias ameliae S. Moore = Trachycalymma pulchellum (Decne.) Bullock ■☆

37826 Asclepias amplexicaulis Michx. = Asclepias amplexicaulis Small ■☆
37827 Asclepias amplexicaulis Small;钝叶马利筋;Bluntleaf Milkweed, Blunt-leaved Milkweed, Clasping Milkweed, Sand Milkweed ■☆
37828 Asclepias angustata N. E. Br. = Stathmostelma angustatum K. Schum. ■☆
37829 Asclepias angustifolia Schweigg. = Gomphocarpus fruticosus (L.) W. T. Aiton ●
37830 Asclepias anisophylla Conrath et Schltr. = Xysmalobium brownianum S. Moore ■☆
37831 Asclepias annularia Roxb. = Holostemma annularium (Roxb.) K. Schum. ●■
37832 Asclepias annularis Roxb. = Holostemma ada-kodien Schult. ●■
37833 Asclepias aphylla Thunb. = Sarcostemma viminale (L.) R. Br. ■
37834 Asclepias appendiculata (E. Mey.) Schltr. = Pachycarpus appendiculatus E. Mey. ■☆
37835 Asclepias arborea Salisb. = Gomphocarpus cancellatus (Burm. f.) Bruyns ●☆
37836 Asclepias arborescens L. = Gomphocarpus cancellatus (Burm. f.) Bruyns ●☆
37837 Asclepias asperula (Decne.) Woodson;粗糙马利筋;Spider Milkweed ■☆
37838 Asclepias aurea (Schltr.) Schltr.;黄马利筋■☆
37839 Asclepias aurea (Schltr.) Schltr. var. brevicuspis S. Moore = Asclepias aurea (Schltr.) Schltr. ■☆
37840 Asclepias aurea (Schltr.) Schltr. var. vittata N. E. Br. = Asclepias aurea (Schltr.) Schltr. ■☆
37841 Asclepias bagshawei S. Moore = Stathmostelma welwitschii Britten et Rendle var. bagshawei (S. Moore) Goyder ■☆
37842 Asclepias baumii Schltr.;鲍姆马利筋■☆
37843 Asclepias bicknellii Vail = Asclepias exaltata L. ■☆
37844 Asclepias bicolor Moench = Asclepias curassavica L. ■
37845 Asclepias bicuspis N. E. Br.;双尖马利筋■☆
37846 Asclepias brachystephana Engelm. ex Torr.;短马利筋;Bract Milkweed ■☆
37847 Asclepias brasiliensis (E. Fourn.) Schltr. = Gomphocarpus physocarpus E. Mey. ●
37848 Asclepias brevicuspis (E. Mey.) Schltr.;短尖马利筋●☆
37849 Asclepias brevipes (Schltr.) Schltr.;短梗马利筋●☆
37850 Asclepias browniana S. Moore = Pachycarpus lineolatus (Decne.) Bullock ■☆
37851 Asclepias buchenaviana Schinz = Gomphocarpus filiformis (E. Mey.) D. Dietr. ●☆
37852 Asclepias buchwaldii (Schltr. et K. Schum.) De Wild. = Trachycalymma buchwaldii (Schltr. et K. Schum.) Goyder ■☆
37853 Asclepias burchellii Schltr. = Gomphocarpus tomentosus Burch. ●☆
37854 Asclepias cabrae De Wild. = Glossostelma cabrae (De Wild.) Goyder ■☆
37855 Asclepias calceolus S. Moore = Asclepias dregeana Schltr. var. calceola (S. Moore) N. E. Br. ■☆
37856 Asclepias cancellata Burm. f. = Gomphocarpus cancellatus (Burm. f.) Bruyns ●☆
37857 Asclepias canescens Willd. = Cynanchum canescens (Willd.) K. Schum. ■
37858 Asclepias canescens Willd. = Vincetoxicum canescens (Willd.) Decne. ■
37859 Asclepias carnosa L. f. = Hoya carnosa (L. f.) R. Br. ●
37860 Asclepias carsonii (N. E. Br.) Schltr. = Glossostelma carsonii (N. E. Br.) Bullock ■☆
37861 Asclepias chloroglossa Schltr. = Xysmalobium involucratum (E. Mey.) Decne. ■☆
37862 Asclepias coarctata S. Moore = Kanahia laniflora (Forssk.) R. Br. ■☆
37863 Asclepias coccinea N. E. Br. = Stathmostelma incarnatum K. Schum. ■☆
37864 Asclepias cognata N. E. Br. = Aspidonepsis cognata (N. E. Br.) Nicholas et Goyder ●☆
37865 Asclepias compressidens (N. E. Br.) Nicholas;扁马利筋■☆
37866 Asclepias concinna (Schltr.) Schltr.;整洁马利筋■☆
37867 Asclepias concolor (E. Mey.) Schltr. = Pachycarpus concolor E. Mey. ●☆
37868 Asclepias confusa (Scott-Elliot) Schltr. = Xysmalobium confusum Scott-Elliot ■☆
37869 Asclepias congolensis De Wild. = Glossostelma lisianthoides (Decne.) Bullock ■☆
37870 Asclepias conspicua N. E. Br. = Pachycarpus lineolatus (Decne.) Bullock ■☆
37871 Asclepias convolvulacea Willd. = Pergularia daemia (Forssk.) Chiov. ■☆
37872 Asclepias cooperi N. E. Br.;库珀马利筋■☆
37873 Asclepias cordata Burm. f. = Hoya cordata P. T. Li et S. Z. Huang ●
37874 Asclepias cordata Burm. f. = Telosma cordata (Burm. f.) Merr. ●
37875 Asclepias cordata Forssk. = Pergularia tomentosa L. ●☆
37876 Asclepias cornuta (Decne.) Cordem. = Gomphocarpus fruticosus (L.) W. T. Aiton ●
37877 Asclepias cornutii Decne. = Asclepias syriaca L. ■☆
37878 Asclepias crassifolia Decne. = Gomphocarpus fruticosus (L.) W. T. Aiton ●
37879 Asclepias crassinervis N. E. Br.;粗脉马利筋■☆
37880 Asclepias crinita (G. Bertol.) N. E. Br. = Gomphocarpus fruticosus (L.) W. T. Aiton ●
37881 Asclepias crispa P. J. Bergius;皱波马利筋■☆
37882 Asclepias crispa P. J. Bergius var. plana N. E. Br.;扁平马利筋■☆
37883 Asclepias crispa P. J. Bergius var. pseudocrispa (Schltr.) N. E. Br.;假皱波马利筋■☆
37884 Asclepias cristata S. Moore;冠状马利筋■☆
37885 Asclepias cucullata (Schltr.) Schltr.;僧帽马利筋■☆
37886 Asclepias cucullata (Schltr.) Schltr. subsp. scabrifolia (S. Moore) Goyder;糙叶马利筋■☆
37887 Asclepias cultriformis (Harv. ex Schltr.) Schltr.;刀形马利筋■☆
37888 Asclepias curassavica L.;马利筋(半天花,草木棉,刀口药,对叶莲,芳草花,红花矮陀陀,黄花仔,尖尾凤,见肿消,金凤花,金银花台,金盏银台,辣子七,老鸦咀,莲生桂子草,莲生桂子花,女金丹,七姊妹,山桃花,水羊角,唐绵,唐棉,土常山,细牛角仔树,羊角丽,野鹤嘴,野辣子,竹林标,状元红);Annual Milkweed, Bastard Ipecac, Bastard Ipecacuanha, Blood Flower, Bloodflower, Blood-flower, Bloodflower Milkweed, Bloodweed, Indian Root, Matac, Red Head, Swallow Wort, Swallow-wort, Tropical Milkweed, West Indian Ipecacuanha, Wild Ipecacuanha ■
37889 Asclepias curassavica L. 'Flaviflora';黄冠马利筋■
37890 Asclepias curassavica L. f. flaviflora Tawada = Asclepias curassavica L. 'Flaviflora' ■
37891 Asclepias daemia Forssk. = Pergularia daemia (Forssk.) Chiov. ■☆
37892 Asclepias davurica Willd. = Cynanchum purpureum (Pall.) K.

Schum. ■

37893 Asclepias dealbata (E. Mey.) Schltr. = Pachycarpus dealbatus E. Mey. ■☆

37894 Asclepias decipiens N. E. Br. = Gomphocarpus fruticosus (L.) W. T. Aiton subsp. decipiens (N. E. Br.) Goyder et Nicholas ●☆

37895 Asclepias densiflora N. E. Br.;密花马利筋■☆

37896 Asclepias denticulata Schltr. = Gomphocarpus physocarpus E. Mey. ●

37897 Asclepias dependens (K. Schum.) N. E. Br.;悬垂马利筋■☆

37898 Asclepias depressa (Schltr.) Schltr. = Asclepias multicaulis (E. Mey.) Schltr. ■☆

37899 Asclepias dewevrei De Wild. = Stathmostelma katangense (De Wild.) Goyder ■☆

37900 Asclepias diploglossa (Turcz.) Druce = Aspidonepsis diploglossa (Turcz.) Nicholas et Goyder ●☆

37901 Asclepias disparilis N. E. Br.;异型马利筋■☆

37902 Asclepias dissoluta (K. Schum.) Schltr. = Glossostelma lisianthoides (Decne.) Bullock ■☆

37903 Asclepias dregeana Schltr.;德雷马利筋■☆

37904 Asclepias dregeana Schltr. var. calceola (S. Moore) N. E. Br.;鞋状德雷马利筋■☆

37905 Asclepias dregeana Schltr. var. sordida N. E. Br. = Asclepias dregeana Schltr. var. calceola (S. Moore) N. E. Br. ■☆

37906 Asclepias echinata Roxb.;刺马利筋■☆

37907 Asclepias eminens (Harv.) Schltr.;显著马利筋■☆

37908 Asclepias endotrachys Schltr. = Trachycalymma foliosum (K. Schum.) Goyder ■☆

37909 Asclepias erecta De Wild. = Glossostelma erectum (De Wild.) Goyder ■☆

37910 Asclepias eriocarpa Torr.;毛果马利筋;Woolly-pod Milkweed ■☆

37911 Asclepias erosa Torr.;沙地马利筋;Desert Milkweed ■☆

37912 Asclepias euphorbioides A. Chev. = Gomphocarpus abyssinicus Decne. ●☆

37913 Asclepias eustegioides (E. Mey.) Schltr. = Schizoglossum eustegioides (E. Mey.) Druce ■☆

37914 Asclepias exaltata L.;高马利筋;Poke Milkweed, Tall Milkweed ■☆

37915 Asclepias exaltata L. = Asclepias syriaca Blanco ■☆

37916 Asclepias exaltata Muhl. ex Bigelow = Asclepias phytolaccoides Lyon ex Pursh ■☆

37917 Asclepias eximia Schltr. = Pachycarpus eximius (Schltr.) Bullock ■☆

37918 Asclepias expansa (E. Mey.) Schltr.;扩展马利筋■☆

37919 Asclepias extenta S. Moore = Stathmostelma katangense (De Wild.) Goyder ■☆

37920 Asclepias fallax (Schltr.) Schltr.;迷惑马利筋■☆

37921 Asclepias filiformis (E. Mey.) Benth. et Hook. f. = Gomphocarpus filiformis (E. Mey.) D. Dietr. ●☆

37922 Asclepias filiformis (E. Mey.) Benth. et Hook. f. ex Kuntze = Gomphocarpus filiformis (E. Mey.) D. Dietr. ●☆

37923 Asclepias filiformis (E. Mey.) Kuntze var. buchenaviana (Schinz) N. E. Br. = Gomphocarpus filiformis (E. Mey.) D. Dietr. ●☆

37924 Asclepias filiformis L. f. = Schizoglossum linifolium Schltr. ■☆

37925 Asclepias fimbriata Weim. = Trachycalymma fimbriatum (Weim.) Bullock ■☆

37926 Asclepias firma (N. E. Br.) Schltr. ex Hiern = Pachycarpus firmus (N. E. Br.) Goyder ■☆

37927 Asclepias flagellaris Bolus ex Schltr. = Gomphocarpus filiformis (E. Mey.) D. Dietr. ●☆

37928 Asclepias flanaganii Schltr. = Asclepias hastata (E. Mey.) Schltr. ■☆

37929 Asclepias flava N. E. Br. = Aspidonepsis flava (N. E. Br.) Nicholas et Goyder ●☆

37930 Asclepias flavida N. E. Br. = Gomphocarpus fruticosus (L.) W. T. Aiton subsp. flavidus (N. E. Br.) Goyder ●☆

37931 Asclepias flexuosa (E. Mey.) Schltr.;曲折马利筋■☆

37932 Asclepias fluviatilis A. Chev. = Kanahia laniflora (Forssk.) R. Br. ■☆

37933 Asclepias foliosa (K. Schum.) Hiern = Trachycalymma foliosum (K. Schum.) Goyder ■☆

37934 Asclepias fornicata N. E. Br. = Stathmostelma fornicatum (N. E. Br.) Bullock ■☆

37935 Asclepias forskalii Roem. et Schult. = Pentatropis nivalis (J. F. Gmel.) D. V. Field et J. R. I. Wood ■☆

37936 Asclepias frederici Hiern = Gomphocarpus tomentosus Burch. subsp. frederici (Hiern) Goyder et Nicholas ●☆

37937 Asclepias friesii Schltr. = Gomphocarpus praticola (S. Moore) Goyder et Nicholas ●☆

37938 Asclepias fruticosa L. = Gomphocarpus fruticosus (L.) W. T. Aiton ●

37939 Asclepias fruticosa L. var. angustissima (Engl.) Schltr. = Gomphocarpus phillipsiae (N. E. Br.) Goyder ●☆

37940 Asclepias fruticosa Schltr. = Gomphocarpus fruticosus (L.) W. T. Aiton subsp. rostratus (N. E. Br.) Goyder et Nicholas ●☆

37941 Asclepias fulva N. E. Br. = Asclepias dregeana Schltr. ■☆

37942 Asclepias galpinii (Schltr.) Schltr. = Pachycarpus galpinii (Schltr.) N. E. Br. ■☆

37943 Asclepias geminata Roxb. = Gymnema sylvestre (Retz.) Schult. ●

37944 Asclepias geminiflora (Schltr.) Schltr. = Pachycarpus concolor E. Mey. ●☆

37945 Asclepias gerrardii (Harv.) Schltr. = Pachycarpus campanulatus (Harv.) N. E. Br. var. sutherlandii N. E. Br. ■☆

37946 Asclepias gibba (E. Mey.) Schltr.;浅囊马利筋■☆

37947 Asclepias gibba (E. Mey.) Schltr. var. media N. E. Br.;中间马利筋■☆

37948 Asclepias gigantea Jacq. = Calotropis procera (L.) Dryand. ex W. T. Aiton ●

37949 Asclepias gigantea L. = Calotropis gigantea (L.) Dryand. ex W. T. Aiton ●

37950 Asclepias gigantiflora (K. Schum.) N. E. Br. = Stathmostelma gigantiflorum K. Schum. ■☆

37951 Asclepias glaberrima (Oliv.) Schltr. = Kanahia laniflora (Forssk.) R. Br. ■☆

37952 Asclepias glabra Forssk. = Pergularia glabra (Forssk.) Chiov. ■☆

37953 Asclepias glabra Mill. = Gomphocarpus fruticosus (L.) W. T. Aiton ●

37954 Asclepias glaucophylla (Schltr.) Schltr. = Gomphocarpus glaucophyllus Schltr. ●☆

37955 Asclepias gomphocarpoides (E. Mey.) Schltr. = Xysmalobium gomphocarpoides (E. Mey.) D. Dietr. ■☆

37956 Asclepias gossweileri S. Moore = Trachycalymma amoenum (K. Schum.) Goyder ■☆

37957 Asclepias grandiflora L. f. = Pachycarpus grandiflorus (L. f.) E.

Mey. ●☆
37958 Asclepias grandiflora L. f. var. chrysantha Schltr. = Pachycarpus grandiflorus (L. f.) E. Mey. ●☆
37959 Asclepias grantii (Oliv.) Schltr. = Pachycarpus grantii (Oliv.) Bullock ●☆
37960 Asclepias hallii A. Gray;哈利马利筋■☆
37961 Asclepias harveyana (Schltr.) Schltr. = Xysmalobium prunelloides Turcz. ■☆
37962 Asclepias hastata (E. Mey.) Schltr.;戟形马利筋■☆
37963 Asclepias hastata Bunge = Cynanchum bungei Decne. ●■
37964 Asclepias hirtella (Pennell) Woodson;草原大马利筋;Prairie Milkweed, Tall Green Milkweed ■☆
37965 Asclepias humilis (E. Mey.) Schltr.;低矮马利筋■☆
37966 Asclepias humistrata Walter;松林马利筋;Pinewoods Milkweed ■☆
37967 Asclepias incarnata L.;喜湿马利筋(肉红马利筋);Incarnate Swallowwort, Swamp Milkweed ■☆
37968 Asclepias incarnata L. var. pulchra (Ehrh. ex Willd.) Pers.;毛喜湿马利筋;Hairy Milkweed, Hairy Swamp Milkweed ■☆
37969 Asclepias inflexa S. Moore;内折马利筋■☆
37970 Asclepias insignis (Schltr.) Schltr. = Pachycarpus transvaalensis (Schltr.) N. E. Br. ■☆
37971 Asclepias integra N. E. Br. = Gomphocarpus integer (N. E. Br.) Bullock ●☆
37972 Asclepias intermedia Vail = Asclepias syriaca L. ■☆
37973 Asclepias kaessneri N. E. Br. = Gomphocarpus kaessneri (N. E. Br.) Goyder et Nicholas ●☆
37974 Asclepias kamerunensis Schltr. = Gomphocarpus kamerunensis (Schltr.) Bullock ●☆
37975 Asclepias kansana Vail = Asclepias syriaca L. ■☆
37976 Asclepias katangensis De Wild. = Stathmostelma katangense (De Wild.) Goyder ■☆
37977 Asclepias katangensis S. Moore = Gomphocarpus praticola (S. Moore) Goyder et Nicholas ●☆
37978 Asclepias lanata (E. Mey.) Druce = Gomphocarpus tomentosus Burch. ●☆
37979 Asclepias lanceolata E. Ives;剑叶马利筋;Asclepias, Fewflower Milkweed, Lanceolated Milkwort ■☆
37980 Asclepias lanceolata E. Ives = Asclepias viridiflora Raf. ■☆
37981 Asclepias laniflora Delile = Kanahia laniflora (Forssk.) R. Br. ■☆
37982 Asclepias laniflora Forssk. = Kanahia laniflora (Forssk.) R. Br. ■☆
37983 Asclepias lanuginosa Nutt.;绵毛马利筋;Side-cluster Milkweed, Woolly Milkweed ■☆
37984 Asclepias latifolia Raf.;宽叶马利筋;Broadleaf Milkweed ■☆
37985 Asclepias laurentiana (Dewèvre) N. E. Br. = Stathmostelma welwitschii Britten et Rendle ■☆
37986 Asclepias laurifolius Roxb. = Genianthus laurifolius (Roxb.) Hook. f. ●
37987 Asclepias lepida S. Moore = Trachycalymma foliosum (K. Schum.) Goyder ■☆
37988 Asclepias leucocarpa Schltr. = Gomphocarpus stenophyllus Oliv. ●☆
37989 Asclepias leucotricha Schltr. = Xysmalobium undulatum (L.) W. T. Aiton ■☆
37990 Asclepias lilacina Weim. = Gomphocarpus glaucophyllus Schltr. ●☆
37991 Asclepias linaria Cav.;松叶马利筋(柳穿鱼马利筋);Pineleaf Milkweed, Pineneedle Milkweed, Threadleaf Milkweed ■☆
37992 Asclepias linearis (E. Mey.) Schltr. = Pachycarpus linearis (E. Mey.) N. E. Br. ●☆
37993 Asclepias lineolata (Decne.) Schltr. = Pachycarpus lineolatus (Decne.) Bullock ■☆
37994 Asclepias lineolatus S. Moore = Pachycarpus lineolatus (Decne.) Bullock ■☆
37995 Asclepias lisianthoides (Decne.) N. E. Br. = Glossostelma lisianthoides (Decne.) Bullock ■☆
37996 Asclepias litocarpa Chiov. = Gomphocarpus integer (N. E. Br.) Bullock ●☆
37997 Asclepias longifolia Michx.;长叶马利筋;Florida Milkweed ■☆
37998 Asclepias longifolia Michx. subsp. hirtella (Pennell) J. Farmer et C. R. Bell = Asclepias hirtella (Pennell) Woodson ■☆
37999 Asclepias longifolia Michx. var. hirtella (Pennell) J. Farmer et C. R. Bell = Asclepias hirtella (Pennell) Woodson ■☆
38000 Asclepias longissima (K. Schum.) N. E. Br. = Gomphocarpus longissimus K. Schum. ●☆
38001 Asclepias mackenii (Harv.) Schltr. = Pachycarpus mackenii (Harv.) N. E. Br. ■☆
38002 Asclepias macra Schltr. = Pachycarpus suaveolens (Schltr.) Nicholas et Goyder ■☆
38003 Asclepias macrantha Hochst. ex Oliv. = Stathmostelma pedunculatum (Decne.) K. Schum. ●☆
38004 Asclepias macrochila Schltr. = Pachycarpus macrochilus (Schltr.) N. E. Br. ■☆
38005 Asclepias macropetala (Schltr. et K. Schum.) N. E. Br. = Stathmostelma spectabile (N. E. Br.) Schltr. ■☆
38006 Asclepias macropus (Schltr.) Schltr.;大足马利筋■☆
38007 Asclepias margaritacea Hoffmanns. = Asclepias curassavica L. ■
38008 Asclepias mashonensis Schltr. = Glossostelma carsonii (N. E. Br.) Bullock ■☆
38009 Asclepias meadii Torr. ex A. Gray;米氏马利筋;Mead's Milkweed ■☆
38010 Asclepias meliodora (Schltr.) Schltr.;蜜味马利筋■☆
38011 Asclepias meliodora (Schltr.) Schltr. var. brevicoronata N. E. Br. = Asclepias meliodora (Schltr.) Schltr. ■☆
38012 Asclepias mexicana Cav.;墨西哥马利筋;Mexican Whorled Milkweed ■☆
38013 Asclepias meyeriana (Schltr.) Schltr.;迈尔马利筋■☆
38014 Asclepias microphylla Roth = Pentatropis capensis (L. f.) Bullock ■☆
38015 Asclepias microphylla Roxb. = Pentatropis capensis (L. f.) Bullock ■☆
38016 Asclepias minuta A. Chev. = Trachycalymma foliosum (K. Schum.) Goyder ■☆
38017 Asclepias modesta N. E. Br.;适度马利筋■☆
38018 Asclepias modesta N. E. Br. = Trachycalymma foliosum (K. Schum.) Goyder ■☆
38019 Asclepias modesta N. E. Br. var. foliosa (K. Schum.) N. E. Br. = Trachycalymma foliosum (K. Schum.) Goyder ■☆
38020 Asclepias moorei De Wild. = Gomphocarpus praticola (S. Moore) Goyder et Nicholas ●☆
38021 Asclepias mucronata Thunb. = Woodia mucronata (Thunb.) N. E. Br. ■☆
38022 Asclepias muhindensis N. E. Br. = Stathmostelma gigantiflorum K. Schum. ■☆
38023 Asclepias multicaulis (E. Mey.) Schltr.;多茎马利筋■☆

38024 Asclepias multiflora (Decne.) N. E. Br.;多花马利筋■☆
38025 Asclepias muricata Schumach. et Thonn. = Pergularia daemia (Forssk.) Chiov. ■☆
38026 Asclepias nana I. Verd.;矮小马利筋■☆
38027 Asclepias navicularis (E. Mey.) Schltr.;船状马利筋■☆
38028 Asclepias navicularis (E. Mey.) Schltr. var. compressidens N. E. Br. = Asclepias compressidens (N. E. Br.) Nicholas ■☆
38029 Asclepias negrii Chiov. = Gomphocarpus integer (N. E. Br.) Bullock ●☆
38030 Asclepias nemorensis S. Moore = Glossostelma lisianthoides (Decne.) Bullock ■☆
38031 Asclepias nigra L. = Vincetoxicum nigrum (L.) Moench ■☆
38032 Asclepias nivalis J. F. Gmel. = Pentatropis nivalis (J. F. Gmel.) D. V. Field et J. R. I. Wood ■☆
38033 Asclepias nivea Forssk. = Pentatropis nivalis (J. F. Gmel.) D. V. Field et J. R. I. Wood ■☆
38034 Asclepias nivea L. var. curassavica (L.) Kuntze = Asclepias curassavica L. ■
38035 Asclepias nuda Schumach. et Thonn. = Sarcostemma viminale (L.) R. Br. ■
38036 Asclepias nutans (Klotzsch) N. E. Br.;点头马利筋■☆
38037 Asclepias nuttalliana Torr. = Asclepias lanuginosa Nutt. ■☆
38038 Asclepias nuttii N. E. Br. = Stathmostelma nuttii (N. E. Br.) Bullock ■☆
38039 Asclepias nyctaginifolia A. Gray;紫茉莉马利筋;Four-o' clock Milkweed ■☆
38040 Asclepias nyikana Schltr. = Gomphocarpus swynnertonii (S. Moore) Goyder et Nicholas ●☆
38041 Asclepias ochroleuca (Schltr.) Schltr. = Xysmalobium gerrardii Scott-Elliot ■☆
38042 Asclepias odorata (K. Schum.) N. E. Br. = Stathmostelma spectabile (N. E. Br.) Schltr. ■☆
38043 Asclepias orbicularis (E. Mey.) Schltr. = Xysmalobium orbiculare (E. Mey.) D. Dietr. ■☆
38044 Asclepias oreophila Nicholas;喜山马利筋■☆
38045 Asclepias ovalifolia Decne.;卵叶马利筋;Dwarf Milkweed, Oval Milkweed, Oval-leaved Milkweed ■☆
38046 Asclepias oxytropis (Turcz.) Schltr. = Asclepias gibba (E. Mey.) Schltr. ■☆
38047 Asclepias pachyclada (K. Schum.) N. E. Br. = Stathmostelma spectabile (N. E. Br.) Schltr. ■☆
38048 Asclepias pachyglossa (Schltr.) Schltr. = Xysmalobium parviflorum Harv. ex Scott-Elliot ●☆
38049 Asclepias pachystephana Schltr. = Schizoglossum linifolium Schltr. ■☆
38050 Asclepias pallida Roxb. = Telosma cordata (Burm. f.) Merr. ●
38051 Asclepias pallida Roxb. = Telosma pallida (Roxb.) Craib ●
38052 Asclepias palustris (K. Schum.) Schltr. = Trachycalymma cristatum (Decne.) Bullock ●☆
38053 Asclepias paniculata Bunge = Cynanchum paniculatum (Bunge) Kitag. ex H. Hara ■
38054 Asclepias patens N. E. Br.;铺展马利筋■☆
38055 Asclepias pauciflora (Klotzsch) E. A. Bruce = Stathmostelma pauciflorum (Klotzsch) K. Schum. ●☆
38056 Asclepias pedunculata (Decne.) Dandy = Stathmostelma pedunculatum (Decne.) K. Schum. ●☆
38057 Asclepias peltigera (E. Mey.) Schltr.;盾状马利筋■☆
38058 Asclepias perennis Walter;湿地白马利筋;Aquatic Milkweed, White Milkweed ■☆
38059 Asclepias pethericklana (Oliv.) Schltr. = Pachycarpus petherickianus (Oliv.) Goyder ■☆
38060 Asclepias petherickiana (Oliv.) Schltr. var. cordatum S. Moore = Pachycarpus petherickianus (Oliv.) Goyder ■☆
38061 Asclepias phillipsiae N. E. Br. = Gomphocarpus phillipsiae (N. E. Br.) Goyder ●☆
38062 Asclepias physocarpa (E. Mey.) Schltr.;囊果马利筋;Balloonplant ■☆
38063 Asclepias physocarpa (E. Mey.) Schltr. = Gomphocarpus physocarpus E. Mey. ●
38064 Asclepias phytolaccoides Lyon ex Pursh = Asclepias exaltata L. ■☆
38065 Asclepias phytolaccoides Pursh = Asclepias exaltata L. ■☆
38066 Asclepias praemorsa Schltr.;啮蚀马利筋■☆
38067 Asclepias praticola S. Moore = Gomphocarpus praticola (S. Moore) Goyder et Nicholas ●☆
38068 Asclepias procera Aiton = Calotropis procera (Aiton) R. Br. ●
38069 Asclepias procera Aiton = Calotropis procera (L.) Dryand. ex W. T. Aiton ●
38070 Asclepias propinqua N. E. Br. = Stathmostelma propinquum (N. E. Br.) Schltr. ■☆
38071 Asclepias pseudocrispa Schltr. = Asclepias crispa P. J. Bergius var. pseudocrispa (Schltr.) N. E. Br. ■☆
38072 Asclepias pubescens L. = Gomphocarpus cancellatus (Burm. f.) Bruyns ●☆
38073 Asclepias pubiseta N. E. Br. = Gomphocarpus purpurascens A. Rich. ●☆
38074 Asclepias pulchella (Decne.) N. E. Br. = Trachycalymma pulchellum (Decne.) Bullock ■☆
38075 Asclepias pulchella Roxb. = Raphistemma pulchellum (Roxb.) Wall. ●
38076 Asclepias pumila (A. Gray) Vail;小马利筋;Low Whorled Milkweed, Plains Whorled Milkweed ■☆
38077 Asclepias purpurascens L.;紫马利筋;Purple Milkweed ■☆
38078 Asclepias purpurea Pall. = Cynanchum purpureum (Pall.) K. Schum. ■
38079 Asclepias pygmaea N. E. Br.;微小马利筋■☆
38080 Asclepias quadrifolia Jacq.;四叶马利筋;Fourleaf Milkweed, Four-leaved Milkweed, Whorled Milkweed ■☆
38081 Asclepias radians Forssk. = Odontanthera radians (Forssk.) D. V. Field ■☆
38082 Asclepias radiata S. Moore;辐射马利筋■☆
38083 Asclepias randii S. Moore;兰德马利筋■☆
38084 Asclepias rara N. E. Br.;珍稀马利筋■☆
38085 Asclepias rectinervis (Schltr.) Schltr. = Xysmalobium confusum Scott-Elliot ■☆
38086 Asclepias reenensis N. E. Br. = Aspidonepsis reenensis (N. E. Br.) Nicholas et Goyder ●☆
38087 Asclepias reflectens (E. Mey.) Schltr. = Pachycarpus reflectens E. Mey. ■☆
38088 Asclepias reflexa (Britten et Rendle) N. E. Br. = Stathmostelma pauciflorum (Klotzsch) K. Schum. ●☆
38089 Asclepias rhodesica Weim. = Gomphocarpus tenuifolius (N. E. Br.) Bullock ●☆
38090 Asclepias rigidus (E. Mey.) Schltr. = Pachycarpus rigidus E. Mey. ■☆

38091 Asclepias rivalis S. Moore = Kanahia laniflora (Forssk.) R. Br. ■☆
38092 Asclepias rivularis (Schltr.) Schltr. = Gomphocarpus rivularis Schltr. ●☆
38093 Asclepias robusta (A. Rich.) N. E. Br. = Pachycarpus robustus (A. Rich.) Bullock ■☆
38094 Asclepias rosea Roxb. = Oxystelma esculentum (L. f.) Sm. ■
38095 Asclepias rostrata N. E. Br. = Gomphocarpus fruticosus (L.) W. T. Aiton subsp. rostratus (N. E. Br.) Goyder et Nicholas ●☆
38096 Asclepias rotundifolia Mill. = Gomphocarpus cancellatus (Burm. f.) Bruyns ●☆
38097 Asclepias rubella N. E. Br. = Trachycalymma amoenum (K. Schum.) Goyder ■☆
38098 Asclepias rubicunda Schltr. = Asclepias dregeana Schltr. ■☆
38099 Asclepias rubra L.;红马利筋;Red Milkweed ■☆
38100 Asclepias rusbyi (Vail) Woodson;鲁氏马利筋;Rusby's Milkweed ■☆
38101 Asclepias sabulosa Schltr. = Asclepias crispa P. J. Bergius ■☆
38102 Asclepias salicifolia Salisb. = Gomphocarpus fruticosus (L.) W. T. Aiton ●
38103 Asclepias scabra (Harv.) Schltr. = Pachycarpus scaber (Harv.) N. E. Br. ■☆
38104 Asclepias scabridifolia Schltr. = Xysmalobium acerateoides (Schltr.) N. E. Br. ■☆
38105 Asclepias scabrifolia S. Moore = Asclepias cucullata (Schltr.) Schltr. subsp. scabrifolia (S. Moore) Goyder ■☆
38106 Asclepias scandens P. Beauv. = Pergularia daemia (Forssk.) Chiov. ■☆
38107 Asclepias schinziana (Schltr.) Schltr. = Pachycarpus schinzianus (Schltr.) N. E. Br. ■☆
38108 Asclepias schizoglossoides Schltr. = Aspidonepsis diploglossa (Turcz.) Nicholas et Goyder ●☆
38109 Asclepias schlechteri (K. Schum.) N. E. Br.;施莱马利筋■☆
38110 Asclepias schumanniana Hiern = Trachycalymma amoenum (K. Schum.) Goyder ■☆
38111 Asclepias schweinfurthii N. E. Br. = Pachycarpus lineolatus (Decne.) Bullock ■☆
38112 Asclepias semiamplectens (K. Schum.) Hiern = Gomphocarpus semiamplectens K. Schum. ●☆
38113 Asclepias semilunata (A. Rich.) N. E. Br. = Gomphocarpus semilunatus A. Rich. ●☆
38114 Asclepias setosa Forssk. = Gomphocarpus fruticosus (L.) W. T. Aiton subsp. setosus (Forssk.) Goyder et Nicholas ●☆
38115 Asclepias sibirica L. = Cynanchum thesioides (Freyn) K. Schum. ■
38116 Asclepias simplex (Schltr.) Schltr. = Asclepias stellifera Schltr. ■☆
38117 Asclepias sinaica (Boiss.) Muschl.;西奈马利筋■☆
38118 Asclepias sinuosa Burm. f. = Asclepias crispa P. J. Bergius ■☆
38119 Asclepias speciosa Torr.;艳丽马利筋;Showy Milkweed ■☆
38120 Asclepias spectabile N. E. Br. = Stathmostelma spectabile (N. E. Br.) Schltr. ■☆
38121 Asclepias sphacelata (K. Schum.) N. E. Br.;毒马利筋■☆
38122 Asclepias spiralis Forssk. = Blyttia spiralis (Forssk.) D. V. Field et J. R. I. Wood ■☆
38123 Asclepias spiralis Forssk. = Pentatropis spiralis (Forssk.) Decne. ■☆
38124 Asclepias stellifera Schltr.;星状马利筋■☆
38125 Asclepias stenophylla A. Gray;窄叶马利筋;Narrow-leaved Milkweed ■☆
38126 Asclepias stockenstromense (Scott-Elliot) Schltr. = Xysmalobium stockenstromense Scott-Elliot ■☆
38127 Asclepias stolzianus (K. Schum.) N. E. Br. = Gomphocarpus stolzianus K. Schum. ●☆
38128 Asclepias suaveolens (Schltr.) Schltr. = Pachycarpus suaveolens (Schltr.) Nicholas et Goyder ■☆
38129 Asclepias subulata Larranaga;钻形马利筋;Ajamente, Desert Milkweed, Rush Milkweed ■☆
38130 Asclepias subverticillata (A. Gray) Vail;近轮生马利筋;Poison Milkweed ■☆
38131 Asclepias subviridis S. Moore;浅绿马利筋■☆
38132 Asclepias sullivantii Engelm. ex A. Gray;沙利文特马利筋;Prairie Milkweed, Smooth Milkweed, Sullivant's Milkweed ■☆
38133 Asclepias swynnertonii S. Moore = Gomphocarpus swynnertonii (S. Moore) Goyder et Nicholas ●☆
38134 Asclepias syriaca Blanco = Asclepias syriaca L. ■☆
38135 Asclepias syriaca L.;叙利亚马利筋(大马利筋,角马利筋);Budgerigar Flower, Common Milkweed, Jamaica Liquorice, Milkweed, Silk Weed, Silken Cicely, Silkweed, Swallow-wort, Virginian Silk ■☆
38136 Asclepias syriaca L. var. kansana (Vail) E. J. Palmer et Steyerm. = Asclepias syriaca L. ■☆
38137 Asclepias tanganyikensis E. A. Bruce = Gomphocarpus tanganyikensis (E. A. Bruce) Bullock ●☆
38138 Asclepias tenacissima Roxb. = Marsdenia tenacissima (Roxb.) Moon ●
38139 Asclepias tenuifolia N. E. Br. = Gomphocarpus tenuifolius (N. E. Br.) Bullock ●☆
38140 Asclepias tenuis (E. Mey.) Schltr. = Schizoglossum linifolium Schltr. ■☆
38141 Asclepias tenuissima Roxb. = Tylophora flexuosa R. Br. ■
38142 Asclepias tetrapetala Dennst. = Tylophora flexuosa R. Br. ■
38143 Asclepias tingens Roxb. = Gymnema inodorum (Lour.) Decne. ●
38144 Asclepias transvaalensis (Schltr.) Schltr. = Pachycarpus transvaalensis (Schltr.) N. E. Br. ■☆
38145 Asclepias tricorniculata (K. Schum.) Schltr. = Xysmalobium andongense Hiern ■☆
38146 Asclepias truncata Harv. = Asclepias peltigera (E. Mey.) Schltr. ■☆
38147 Asclepias tuberosa L.;块茎马利筋(块根马利筋);Butter Weed, Butterfly Milkweed, Butterfly Weed, Butterfly-weed, Chigger Flower, Chigger-flower, Chigger-plant, Indian Paintbrush, Milkweed, Orange Milkweed, Pleurisy Root, Pleurisy-root, Swallow-wort, Tuber-root, Wind Root ■☆
38148 Asclepias tuberosa L. f. lutea (Clute) Steyerm. = Asclepias tuberosa L. ■☆
38149 Asclepias tuberosa L. subsp. interior Woodson = Asclepias tuberosa L. ■☆
38150 Asclepias tuberosa L. subsp. terminalis Woodson = Asclepias tuberosa L. ■☆
38151 Asclepias tuberosa L. var. interior (Woodson) Shinners = Asclepias tuberosa L. ■☆
38152 Asclepias tysoniana Schltr. = Xysmalobium tysonianum (Schltr.) N. E. Br. ■☆
38153 Asclepias ulophylla Schltr.;尾叶马利筋■☆
38154 Asclepias undulata L. = Xysmalobium undulatum (L.) W. T.

Aiton ■☆

38155 Asclepias uvirensis S. Moore;乌维拉马利筋■☆

38156 Asclepias valida (Schltr.) Schltr. = Pachycarpus asperifolius Meisn. ■☆

38157 Asclepias variegata L.;白马利筋;Variegated Milkweed, White Milkweed, White-flowered Milkweed ■☆

38158 Asclepias velutina (Schltr.) Schltr.;短绒毛马利筋■☆

38159 Asclepias verdickii De Wild. = Stathmostelma wildemanianum Durand ■☆

38160 Asclepias verticillata L.;轮叶马利筋;Eastern Whorled Milkweed, Horsetail Milkweed, Whorled Milkweed, Whorled Silkweed ■☆

38161 Asclepias vexillare (E. Mey.) Schltr. = Pachycarpus vexillaris E. Mey. ■☆

38162 Asclepias vicaria N. E. Br.;替代马利筋■☆

38163 Asclepias villosa Mill. = Gomphocarpus tomentosus Burch. ●☆

38164 Asclepias vincetoxicum L. = Cynanchum vincetoxicum (L.) Pers. ■

38165 Asclepias vincetoxicum L. = Vincetoxicum hirundinaria Medik. ■☆

38166 Asclepias vincetoxicum L. = Vincetoxicum officinale Moench ■☆

38167 Asclepias viridiflora (E. Mey.) Goyder = Asclepias dregeana Schltr. ■☆

38168 Asclepias viridiflora Raf.;绿花马利筋;Green Milkweed, Short Green Milkweed ■☆

38169 Asclepias viridiflora Raf. var. lanceolata (E. Ives) Torr. = Asclepias viridiflora Raf. ■☆

38170 Asclepias viridiflora Raf. var. linearis (A. Gray) Fernald = Asclepias viridiflora Raf. ■☆

38171 Asclepias viridis Walter;绿马利筋;Antelope Horn, Green Milkweed, Green-flowered Milkweed, Spider Milkweed ■☆

38172 Asclepias volubilis L. f. = Dregea volubilis (L. f.) Benth. ex Hook. f. ●

38173 Asclepias volubilis L. f. = Wattakaka volubilis (L. f.) Stapf ●

38174 Asclepias vomeriformis S. Moore = Stathmostelma angustatum K. Schum. subsp. vomeriforme (S. Moore) Goyder ■☆

38175 Asclepias welwitschii (Britten et Rendle) Britten et Rendle = Stathmostelma welwitschii Britten et Rendle ■☆

38176 Asclepias woodii (Schltr.) Schltr.;伍得马利筋■☆

38177 Asclepias xysmalobioides Hilliard et B. L. Burtt;止泻萝藦马利筋■☆

38178 Asclepias xysmalobioides S. Moore = Glossostelma xysmalobioides (S. Moore) Bullock ■☆

38179 Asclepiodella Small = Asclepias L. ■

38180 Asclepiodora A. Gray = Anantherix Nutt. ■

38181 Asclepiodora A. Gray = Asclepias L. ■

38182 Asclerum Tiegh. = Gonystylus Teijsm. et Binn. ●

38183 Ascocarydion G. Taylor = Plectranthus L'Hér. (保留属名)●■

38184 Ascocarydion mirabile (Briq.) G. Taylor = Plectranthus mirabilis (Briq.) Launert ■☆

38185 Ascocentropsis Senghas et Schildh. = Ascocentrum Schltr. ex J. J. Sm. ■

38186 Ascocentrum Schltr. = Ascocentrum Schltr. ex J. J. Sm. ■

38187 Ascocentrum Schltr. ex J. J. Sm. (1914);鸟舌兰属(百代兰属,假囊距兰属,鹿角兰属);Ascocentrum ■

38188 Ascocentrum ampullaceum (Roxb.) Schltr.;鸟舌兰(囊距兰);Ampule Ascocentrum, Common Ascocentrum ■

38189 Ascocentrum curvifolium (Lindl.) Schltr.;弯叶鸟舌兰(弯叶囊距兰);Curvedleaf Ascocentrum ■☆

38190 Ascocentrum hendersonianum (Rchb. f.) Schltr.;美花鸟舌兰;Henderson Ascocentrum ■☆

38191 Ascocentrum himalaicum (Deb et Malick) Christenson;圆柱叶鸟舌兰(短茎槽舌兰,小花槽舌兰);Column Ascocentrum, Shortstem Holcoglossum ■

38192 Ascocentrum himalaicum (Deb, Sengupta et Malick) Christensen = Ascocentrum himalaicum (Deb et Malick) Christenson ■

38193 Ascocentrum miniatum (Lindl.) Schltr.;朱红鸟舌兰(小囊距兰);Red Ascocentrum ■☆

38194 Ascocentrum pumilum (Hayata) Schltr.;尖叶鸟舌兰(鹿角兰,小鹿角兰);Dwarf Ascocentrum ■

38195 Ascochilopsis Carr(1929);类囊唇兰属■☆

38196 Ascochilopsis myosurus Carr;类囊唇兰■☆

38197 Ascochilus Blume = Cistella Blume ■

38198 Ascochilus Blume = Geodorum Jacks. ■

38199 Ascochilus Ridl. (1896);肖囊唇兰属■☆

38200 Ascochilus Ridl. = Pteroceras Hasselt ex Hassk. ■

38201 Ascochilus annamensis Guillaumin = Thrixspermum annamense (Guillaumin) Garay ■

38202 Ascochilus loratus Rolfe ex Downie = Staurochilus loratus (Rolfe ex Downie) Seidenf. ■

38203 Ascochilus minutiflorus Ridl.;小花肖囊唇兰■☆

38204 Ascochilus siamensis Ridl.;肖囊唇兰■☆

38205 Ascoglossum Schltr. (1913);袋舌兰属■☆

38206 Ascoglossum calopterum Schltr.;袋舌兰■☆

38207 Ascoglossum purpureum Schltr.;紫袋舌兰■☆

38208 Ascolabiura S. S. Ying = Ascocentrum Schltr. ex J. J. Sm. ■

38209 Ascolepis Nees = Ascolepis Nees ex Steud. (保留属名)■☆

38210 Ascolepis Nees ex Steud. (1855)(保留属名);囊鳞莎草属■☆

38211 Ascolepis Nees ex Steud. (保留属名) = Lipocarpha R. Br. (保留属名)■

38212 Ascolepis ampullacea J. Raynal;瓶形囊鳞莎草■☆

38213 Ascolepis anthemiflora Welw. = Ascolepis protea Welw. subsp. anthemiflora (Welw.) Lye ■☆

38214 Ascolepis bellidiflora (Welw.) Cherm. = Ascolepis protea Welw. subsp. bellidiflora (Welw.) Lye ■☆

38215 Ascolepis brasiliensis (Kunth) Benth.;巴西囊鳞莎草■☆

38216 Ascolepis capensis (Kunth) Ridl.;好望角囊鳞莎草■☆

38217 Ascolepis densa Goetgh.;密集囊鳞莎草■☆

38218 Ascolepis dipsacoides (Schumach.) J. Raynal;蛇状囊鳞莎草■☆

38219 Ascolepis elata Welw. = Ascolepis speciosa Welw. ■☆

38220 Ascolepis eriocauloides (Steud.) Nees ex Steud.;毛茎囊鳞莎草■☆

38221 Ascolepis erythrocephala S. S. Hooper;淡红头囊鳞莎草■☆

38222 Ascolepis fibrillosa Goetgh.;须毛囊鳞莎草■☆

38223 Ascolepis gracilis Turrill = Ascolepis dipsacoides (Schumach.) J. Raynal ■☆

38224 Ascolepis hemisphaerica Peter ex Goetgh.;半球形囊鳞莎草■☆

38225 Ascolepis lineariglumis Lye;线颖囊鳞莎草■☆

38226 Ascolepis lineariglumis Lye var. pulcherrima Lye;美丽线颖囊鳞莎草■☆

38227 Ascolepis majestuosa P. A. Duvign. et G. Léonard;壮丽囊鳞莎草■☆

38228 Ascolepis metallorum P. A. Duvign. et G. Léonard;光泽囊鳞莎草■☆

38229 Ascolepis neglecta Goetgh.;忽视囊鳞莎草■☆

38230 Ascolepis peteri Kük. = Alinula peteri (Kük.) Goetgh. et Vorster ■☆

38231 Ascolepis pinguis C. B. Clarke;肥厚囊鳞莎草■☆
38232 Ascolepis protea Welw.;易变囊鳞莎草■☆
38233 Ascolepis protea Welw. subsp. anthemiflora (Welw.) Lye;春花囊鳞莎草■☆
38234 Ascolepis protea Welw. subsp. atropurpurea Lye;暗紫囊鳞莎草■☆
38235 Ascolepis protea Welw. subsp. bellidiflora (Welw.) Lye;匙叶囊鳞莎草■☆
38236 Ascolepis protea Welw. subsp. chrysocephala Lye;金囊鳞莎草■☆
38237 Ascolepis protea Welw. subsp. rhizomatosa Lye;根茎囊鳞莎草■☆
38238 Ascolepis protea Welw. var. bellidiflora Welw. = Ascolepis protea Welw. subsp. bellidiflora (Welw.) Lye ■☆
38239 Ascolepis protea Welw. var. santolinoides;银香菊囊鳞莎草■☆
38240 Ascolepis protea Welw. var. splendida K. Schum. = Ascolepis leariglumis Lye ■☆
38241 Ascolepis pseudopeteri Goetgh.;假彼得囊鳞莎草■☆
38242 Ascolepis pusilla Ridl.;微小囊鳞莎草■☆
38243 Ascolepis pusilla Ridl. var. cylindrica S. S. Hooper;柱形囊鳞莎草■☆
38244 Ascolepis pusilla Ridl. var. echinata S. S. Hooper;具刺囊鳞莎草■☆
38245 Ascolepis pusilla Ridl. var. microcuspis Lye;小囊鳞莎草■☆
38246 Ascolepis pusilla Ridl. var. ochracea (Meneses) Goetgh.;淡黄褐囊鳞莎草■☆
38247 Ascolepis setigera Hutch. = Ascolepis dipsacoides (Schumach.) J. Raynal ■☆
38248 Ascolepis speciosa Welw.;美丽囊鳞莎草■☆
38249 Ascolepis spinulosa Goetgh.;多刺囊鳞莎草■☆
38250 Ascolepis trigona Goetgh.;三棱囊鳞莎草■☆
38251 Ascopholis C. E. C. Fisch. (1931);南印度莎草属■☆
38252 Ascopholis C. E. C. Fisch. = Cyperus L. ■
38253 Ascopholis gamblei C. E. C. Fisch.;南印度莎草■☆
38254 Ascotainia Ridl. = Ania Lindl. ■☆
38255 Ascotainia Ridl. = Tainia Blume ■
38256 Ascotainia angustifolia (Lindl.) Schltr. = Tainia angustifolia (Lindl.) Benth. et Hook. f. ■
38257 Ascotainia elata Schltr. = Tainia ruybarrettoi (S. Y. Hu et Barretto) Z. H. Tsi ■
38258 Ascotainia elata Schltr. = Tainia viridifusca (Hook.) Benth. et Hook. f. ■
38259 Ascotainia hongkongensis (Rolfe) Schltr. = Tainia hongkongensis Rolfe ■
38260 Ascotainia hookeriana (King et Pantl.) Ridl. = Tainia penangiana Hook. f. ■
38261 Ascotainia hookeriana Ridl. = Tainia hookeriana King et Pantl. ■
38262 Ascotainia laxiflora (Makino) C. D. Darl. et Wylie = Tainia laxiflora Makino ■
38263 Ascotainia penangiana (Hook. f.) Ridl. = Tainia penangiana Hook. f. ■
38264 Ascotainia viridifusca (Hook.) Schltr. = Tainia ruybarrettoi (S. Y. Hu et Barretto) Z. H. Tsi ■
38265 Ascotainia viridifusca (Hook.) Schltr. = Tainia viridifusca (Hook.) Benth. et Hook. f. ■
38266 Ascotheca Heine(1966);少脉孩儿草属■☆
38267 Ascotheca paucinervia (T. Anderson ex C. B. Clarke) Heine;少脉孩儿草■☆
38268 Ascra Schott = Banara Aubl. ●☆
38269 Asculaceae Martinov = Hippocastanaceae A. Rich.(保留科名)●
38270 Ascyraceae Martinov = Ascyraceae Plenck ●
38271 Ascyraceae Martinov = Clusiaceae Lindl.(保留科名)●■
38272 Ascyraceae Martinov = Guttiferae Juss.(保留科名)●■
38273 Ascyraceae Martinov = Hypericaceae Juss.(保留科名)●■
38274 Ascyraceae Plenck = Clusiaceae Lindl.(保留科名)●■
38275 Ascyraceae Plenck = Guttiferae Juss.(保留科名)●■
38276 Ascyraceae Plenck;四数金丝桃科●
38277 Ascyroides Lippi ex Adans. = Bergia L. ●■
38278 Ascyroides Lippi ex Adans. = Bistorta (L.) Adans. ■
38279 Ascyrum L. (1753);四数金丝桃属●☆
38280 Ascyrum L. = Hypericum L. ●■
38281 Ascyrum Mill. = Hypericum L. ●■
38282 Ascyrum erux-andreae ? = Ascyrum hypericoides L. ●☆
38283 Ascyrum filicaule Dyer = Hypericum filicaule (Dyer) N. Robson ■
38284 Ascyrum filicaule Dyer = Hypericum monanthemum Hook. f. et Thomson ex Dyer subsp. filicaule (Dyer) N. Robson ■
38285 Ascyrum hypericoides L.;四数金丝桃;St. Andréw's Cross, St. Andréw's-cross ●☆
38286 Ascyrum hypericoides L. = Hypericum hypericoides (L.) Crantz ●☆
38287 Ascyrum hypericoides L. var. multicaule (Michx. ex Willd.) Fernald = Hypericum hypericoides (L.) Crantz ●☆
38288 Ascyrum involutum Labill. = Hypericum lalandii Choisy ■☆
38289 Ascyrum sibiricum Lam. ex Poir. = Hypericum ascyron L. ●■
38290 Ascyrum spathulatum Spach = Hypericum prolificum L. ●☆
38291 Ascyrum stans Michx.;北美四数金丝桃;St. Peter's-wort ■☆
38292 Ascyrum stans Michx. = Hypericum stans (Michx.) Adams et Robson ●☆
38293 Ascyum Vahl = Ascium Schreb. ●☆
38294 Ascyum Vahl = Norantea Aubl. ●☆
38295 Ascyum Vahl ex DC. = Ascyum Vahl ●☆
38296 Asemanthia (Stapf) Ridl. = Mussaenda L. ●■
38297 Asemanthia Ridl. = Mussaenda L. ●■
38298 Asemeia Raf. = Polygala L. ●■
38299 Asemnantha Hook. f. (1873);阿塞茜属☆
38300 Asemnantha pubescens Hook. f.;阿塞茜☆
38301 Asepalum Marais(1981);无萼木属●☆
38302 Asepalum eriantherum (Vatke) Marais;无萼木●☆
38303 Asephananthes Bory = Passiflora L. ●■
38304 Asephananthes Bory ex DC. = Passiflora L. ●■
38305 Ashtonia Airy Shaw(1968);阿什顿大戟属☆
38306 Ashtonia excelsa Airy Shaw;阿什顿大戟☆
38307 Asiasarum F. Maek. (1936);东亚细辛属(萍叶细辛属,亚洲细辛属)■☆
38308 Asiasarum F. Maek. = Asarum L. ■
38309 Asiasarum dimidiatum (F. Maek.) F. Maek. = Asarum dimidiatum F. Maek. ■☆
38310 Asiasarum heterotropoi F. Schmidt = Asarum heterotropoi F. Schmidt ■☆
38311 Asiasarum heterotropoides (F. Schmidt) F. Maek. = Asarum heterotropoides F. Schmidt ■
38312 Asiasarum heterotropoides (F. Schmidt) F. Maek. f. viridis Sugaya;绿花东亚细辛■☆
38313 Asiasarum heterotropoides (F. Schmidt) F. Maek. var. mandshuricum (Maxim.) F. Maek. = Asarum heterotropoides F. Schmidt var. mandshuricum (Maxim.) Kitag. ■
38314 Asiasarum heterotropoides (F. Schmidt) F. Maek. var. seoulense

(Nakai) F. Maek. = Asarum sieboldii Miq. ■
38315 Asiasarum heterotropoides (F. Schmidt) F. Maek. var. seoulense (Nakai) F. Maek. = Asarum sieboldii Miq. var. seoulense Nakai ■
38316 Asiasarum patens Yamaki;铺展东亚细辛■☆
38317 Asiasarum sieboldii (Miq.) F. Maek. = Asarum sieboldii Miq. ■
38318 Asiasarum sieboldii (Miq.) F. Maek. var. seoulense Nakai = Asarum sieboldii Miq. ■
38319 Asiasarum sieboldii (Miq.) F. Maek. var. versicolor Yamaki;变色东亚细辛■☆
38320 Asicaria Neck. = Persicaria (L.) Mill. ■
38321 Asimia Kunth = Asimina Adans. ●☆
38322 Asimina Adans. (1763);泡泡果属(巴婆果属,巴婆属,泡泡属,万寿果属);Papaw,Paw Paw,Pawpaw ●☆
38323 Asimina angustifolia A. Gray = Asimina longifolia Král ●☆
38324 Asimina cuneata Shuttlew. ex A. Gray = Asimina reticulata Shuttlew. ex Chapm. ●☆
38325 Asimina glabra Horta ex K. Koch = Asimina triloba (L.) Dunal ●☆
38326 Asimina grandiflora (W. Bartram) Dunal = Asimina obovata (Willd.) Nash ●☆
38327 Asimina incana (W. Bartram) Exell;臭泡泡果;Flag-pawpaw, Polecat-bush ●☆
38328 Asimina longifolia Král;长叶泡泡果;Polecat-bush ●☆
38329 Asimina obovata (Willd.) Nash;矩圆泡泡果;Flag-pawpaw ●☆
38330 Asimina parviflora (Michx.) Dunal;小花泡泡果(小花巴婆);Dwarf Pawpaw,Small-flowered Pawpaw,Small-fruited Pawpaw ●☆
38331 Asimina parviflora Dunal = Asimina parviflora (Michx.) Dunal ●☆
38332 Asimina pulchella (Small) Rehder et Dayton = Deeringothamnus pulchellus Small ●☆
38333 Asimina pygmaea (W. Bartram) Dunal;矮泡泡果;Dwarf Pawpaw,Gopher-berry ●☆
38334 Asimina reticulata Shuttlew. ex Chapm.;网脉泡泡果(网脉泡泡);Dog Apple,Flag-pawpaw,Seminole Tea ●☆
38335 Asimina rugelii B. L. Rob. = Deeringothamnus rugelii (B. L. Rob.) Small ●☆
38336 Asimina secundiflora Shuttlew. ex Exell = Asimina pygmaea (W. Bartram) Dunal ●☆
38337 Asimina tetramera Small;四瓣泡泡果;Four-petaled Pawpaw ●☆
38338 Asimina triloba (L.) Dunal;泡泡果(巴婆);American Pawpaw, Common Papaw,Common Pawpaw,Papaw,Papaya,Pawpaw ●☆
38339 Asiphonia Griff. (1844);胡椒兜铃属●☆
38340 Asiphonia Griff. = Apama Lam. ●
38341 Asiphonia Griff. = Thottea Rottb. ●
38342 Asiphonia piperiformis Griff.;胡椒兜铃●☆
38343 Asisadenia Hutch. = Anisadenia Wall. ex Meisn. ■
38344 Askellia W. A. Weber = Crepis L. ■
38345 Askellia W. A. Weber(1984);假苦菜属■☆
38346 Askellia alaica (Krasch.) W. A. Weber;阿赖假苦菜■☆
38347 Askellia elegans (Hook.) W. A. Weber;雅致假苦菜■☆
38348 Askellia nana (Richards.) W. A. Weber;小假苦菜■☆
38349 Askellia sogdiana (Krasch.) W. A. Weber;假苦菜■☆
38350 Asketanthera Woodson(1932);美药夹竹桃属●☆
38351 Asketanthera calycosa (A. Rich.) Woodson;美药夹竹桃●☆
38352 Asketanthera obtusifolia Alain;钝叶美药夹竹桃●☆
38353 Askidiosperma Steud. (1855);瓶子帚灯草属(南非帚灯草属)■☆
38354 Askidiosperma albo-aristatum (Pillans) H. P. Linder;白芒南非帚灯草■☆
38355 Askidiosperma alticola (Esterh.) H. P. Linder;高原南非帚灯草■☆
38356 Askidiosperma capitatum Steud.;头状南非帚灯草■☆
38357 Askidiosperma chartaceum (Pillans) H. P. Linder;纸质南非帚灯草■☆
38358 Askidiosperma chartaceum (Pillans) H. P. Linder subsp. alticola Esterh. = Askidiosperma alticola (Esterh.) H. P. Linder ■☆
38359 Askidiosperma delicatulum H. P. Linder;姣美南非帚灯草■☆
38360 Askidiosperma esterhuyseniae (Pillans) H. P. Linder;埃斯特南非帚灯草■☆
38361 Askidiosperma insigne (Pillans) H. P. Linder;显著南非帚灯草■☆
38362 Askidiosperma longiflorum (Pillans) H. P. Linder;长花南非帚灯草■☆
38363 Askidiosperma nitidum (Mast.) H. P. Linder;光亮南非帚灯草■☆
38364 Askidiosperma paniculatum (Mast.) H. P. Linder;圆锥瓶子帚灯草■☆
38365 Askidiosperma rugosum Esterh.;皱纹南非帚灯草■☆
38366 Askofake Raf. = Utricularia L. ■
38367 Askolame Raf. = Milla Cav. ■☆
38368 Askolame biflora (Cav.) Raf. = Milla biflora Cav. ■☆
38369 Asophila Neck. = Gypsophila L. ●■
38370 Aspalanthus cuneata D. Don = Lespedeza cuneata (Dum. Cours.) G. Don ●■
38371 Aspalathaceae Martinov = Fabaceae Lindl. (保留科名)●■
38372 Aspalathaceae Martinov = Leguminosae Juss. (保留科名)●■
38373 Aspalathaceae Martinov;芳香木科●
38374 Aspalathoides (DC.) K. Koch = Anthyllis L. ■☆
38375 Aspalathoides K. Koch = Anthyllis L. ■☆
38376 Aspalathus Amm. = Caragana Lam. ●
38377 Aspalathus Kuntze = Caragana Lam. ●
38378 Aspalathus L. (1753);芳香木属(骆驼刺属,南非香豆属)●☆
38379 Aspalathus L. = Scaligera Adans. (废弃属名)●☆
38380 Aspalathus abietina Thunb.;冷杉芳香木●☆
38381 Aspalathus acanthes Eckl. et Zeyh.;刺芳香木●☆
38382 Aspalathus acanthiloba R. Dahlgren;尖裂芳香木●☆
38383 Aspalathus acanthoclada R. Dahlgren;刺枝芳香木●☆
38384 Aspalathus acanthophylla Eckl. et Zeyh.;刺叶芳香木●☆
38385 Aspalathus acicularis E. Mey.;针形芳香木●☆
38386 Aspalathus acicularis E. Mey. subsp. planifolia R. Dahlgren;平叶芳香木●☆
38387 Aspalathus aciloba R. Dahlgren;锐裂芳香木●☆
38388 Aspalathus aciphylla Harv.;尖叶芳香木●☆
38389 Aspalathus acocksii (R. Dahlgren) R. Dahlgren;阿氏芳香木●☆
38390 Aspalathus aculeata Thunb.;皮刺芳香木●☆
38391 Aspalathus acuminata Lam.;渐尖芳香木●☆
38392 Aspalathus acuminata Lam. subsp. magniflora R. Dahlgren = Aspalathus tulbaghensis R. Dahlgren ●☆
38393 Aspalathus acuminata Lam. subsp. pungens (Thunb.) R. Dahlgren;锐尖芳香木●☆
38394 Aspalathus acuminata Lam. var. subinermis E. Mey. = Aspalathus acuminata Lam. ●☆
38395 Aspalathus acutiflora R. Dahlgren;尖花芳香木●☆
38396 Aspalathus aemula E. Mey.;匹敌芳香木●☆
38397 Aspalathus affinis Thunb. = Aspalathus pinguis Thunb. ●☆
38398 Aspalathus agardhiana DC. = Aspalathus albens L. ●☆

38399 Aspalathus albanensis Eckl. et Zeyh. = Aspalathus frankenioides DC. ●☆

38400 Aspalathus albens L;淅白芳香木●☆

38401 Aspalathus albiflora Eckl. et Zeyh. = Aspalathus hispida Thunb. subsp. albiflora (Eckl. et Zeyh.) R. Dahlgren ●☆

38402 Aspalathus alopecuroides E. Mey. = Aspalathus setacea Eckl. et Zeyh. ●☆

38403 Aspalathus alpestris (Benth.) R. Dahlgren;高山芳香木●☆

38404 Aspalathus alpina Eckl. et Zeyh. = Aspalathus asparagoides L. f. subsp. rubro-fusca (Eckl. et Zeyh.) R. Dahlgren ●☆

38405 Aspalathus alternifolia Harv. = Aspalathus nudiflora Harv. ●☆

38406 Aspalathus altissima R. Dahlgren;极高芳香木●☆

38407 Aspalathus amoena (R. Dahlgren) R. Dahlgren;秀丽芳香木●☆

38408 Aspalathus angustifolia (Lam.) R. Dahlgren;窄叶芳香木●☆

38409 Aspalathus angustifolia (Lam.) R. Dahlgren subsp. robusta (E. Phillips) R. Dahlgren;粗壮窄叶芳香木●☆

38410 Aspalathus angustissima E. Mey. = Aspalathus filicaulis Eckl. et Zeyh. ●☆

38411 Aspalathus anthylloides L. = Aspalathus aspalathoides (L.) R. Dahlgren ●☆

38412 Aspalathus appendiculata E. Mey. = Aspalathus ciliaris L. ●☆

38413 Aspalathus arachnoidea Otto ex Walp. = Aspalathus setacea Eckl. et Zeyh. ●☆

38414 Aspalathus araneosa L.;喜沙芳香木●☆

38415 Aspalathus arborescens Amm. = Caragana arborescens Lam. ●

38416 Aspalathus arenaria R. Dahlgren;沙地芳香木●☆

38417 Aspalathus argentea L. = Aspalathus caledonensis R. Dahlgren ●☆

38418 Aspalathus argyraea DC. = Aspalathus pedunculata Houtt. ●☆

38419 Aspalathus argyrella MacOwan;银色芳香木●☆

38420 Aspalathus arida E. Mey.;旱生芳香木●☆

38421 Aspalathus arida E. Mey. subsp. erecta (E. Mey.) R. Dahlgren;直立芳香木●☆

38422 Aspalathus arida E. Mey. subsp. procumbens (E. Mey.) R. Dahlgren;平铺芳香木●☆

38423 Aspalathus arida E. Mey. var. erecta E. Mey. = Aspalathus arida E. Mey. subsp. erecta (E. Mey.) R. Dahlgren ●☆

38424 Aspalathus arida E. Mey. var. procumbens E. Mey. = Aspalathus arida E. Mey. subsp. procumbens (E. Mey.) R. Dahlgren ●☆

38425 Aspalathus aristata Compton;具芒芳香木●☆

38426 Aspalathus aristifolia R. Dahlgren;芒叶芳香木●☆

38427 Aspalathus armata Thunb. = Aspalathus albens L. ●☆

38428 Aspalathus ascendens E. Mey. = Aspalathus quinquefolia L. subsp. virgata (Thunb.) R. Dahlgren ●☆

38429 Aspalathus aspalathoides (L.) R. Dahlgren;灌木芳香木●☆

38430 Aspalathus asparagoides L. f.;天门冬芳香木●☆

38431 Aspalathus asparagoides L. f. subsp. rubrofusca (Eckl. et Zeyh.) R. Dahlgren;红褐芳香木●☆

38432 Aspalathus astroites Thunb. = Aspalathus racemosa E. Mey. ●☆

38433 Aspalathus attenuata R. Dahlgren;渐狭芳香木●☆

38434 Aspalathus aulonogena Eckl. et Zeyh. = Aspalathus ciliaris L. ●☆

38435 Aspalathus aurantiaca R. Dahlgren;橙黄芳香木●☆

38436 Aspalathus barbata (Lam.) R. Dahlgren;髯毛芳香木●☆

38437 Aspalathus barbigera R. Dahlgren;胡须芳香木●☆

38438 Aspalathus batodes Eckl. et Zeyh.;肉穗芳香木(肉穗果芳香木)●☆

38439 Aspalathus batodes Eckl. et Zeyh. subsp. spinulifolia R. Dahlgren;刺叶肉穗芳香木●☆

38440 Aspalathus benthamii Harv. = Aspalathus spicata Thunb. ●☆

38441 Aspalathus bicolor Eckl. et Zeyh. = Aspalathus retroflexa L. subsp. bicolor (Eckl. et Zeyh.) R. Dahlgren ●☆

38442 Aspalathus biflora E. Mey.;双花芳香木●☆

38443 Aspalathus biflora E. Mey. subsp. longicarpa R. Dahlgren;长果双花芳香木●☆

38444 Aspalathus bodkinii Bolus;博德金芳香木●☆

38445 Aspalathus bowieana (Benth.) R. Dahlgren;鲍伊芳香木●☆

38446 Aspalathus bracteata Thunb.;具苞芳香木●☆

38447 Aspalathus brevicarpa (R. Dahlgren) R. Dahlgren;短果芳香木●☆

38448 Aspalathus burchelliana Benth.;伯切尔芳香木●☆

38449 Aspalathus caerulescens E. Mey. = Lotononis caerulescens (E. Mey.) B. -E. van Wyk ■☆

38450 Aspalathus caespitosa R. Dahlgren;丛生芳香木●☆

38451 Aspalathus calcarata Harv.;距芳香木●☆

38452 Aspalathus calcarea R. Dahlgren;钙生芳香木●☆

38453 Aspalathus caledonensis R. Dahlgren;卡利登芳香木●☆

38454 Aspalathus callosa L.;硬皮芳香木●☆

38455 Aspalathus callosa L. var. fusca (Thunb.) Harv. = Aspalathus fusca Thunb. ●☆

38456 Aspalathus campestris R. Dahlgren;田野芳香木●☆

38457 Aspalathus canaliculata E. Mey. = Aspalathus stenophylla Eckl. et Zeyh. ●☆

38458 Aspalathus candicans W. T. Aiton;纯白芳香木●☆

38459 Aspalathus candidula R. Dahlgren;浅白芳香木●☆

38460 Aspalathus canescens L. = Aspalathus laricifolia P. J. Bergius subsp. canescens (L.) R. Dahlgren ●☆

38461 Aspalathus canescens L. f. elongata E. Mey. = Aspalathus laricifolia P. J. Bergius subsp. canescens (L.) R. Dahlgren ●☆

38462 Aspalathus capensis (Walp.) R. Dahlgren;好望角芳香木●☆

38463 Aspalathus capillaris (Thunb.) Benth. = Aspalathus bracteata Thunb. ●☆

38464 Aspalathus capitata L.;头状芳香木●☆

38465 Aspalathus capitella Burch. ex Benth. = Aspalathus inops Eckl. et Zeyh. ●☆

38466 Aspalathus carnosa P. J. Bergius;肉质芳香木●☆

38467 Aspalathus cephalotes Thunb.;大头芳香木●☆

38468 Aspalathus cephalotes Thunb. subsp. obscurifolia R. Dahlgren;隐叶芳香木●☆

38469 Aspalathus chamissonis Vogel = Aspalathus acanthophylla Eckl. et Zeyh. ●☆

38470 Aspalathus chamlagu (Lam.) Kuntze = Caragana sinica (Buc'hoz) Rehder ●

38471 Aspalathus chenopoda L.;芳香木●☆

38472 Aspalathus chenopoda L. subsp. gracilis (Eckl. et Zeyh.) R. Dahlgren;纤细芳香木●☆

38473 Aspalathus chenopus Spreng. = Aspalathus chenopoda L. ●☆

38474 Aspalathus chortophila Eckl. et Zeyh.;庭院芳香木●☆

38475 Aspalathus chortophila Eckl. et Zeyh. subsp. congesta R. Dahlgren = Aspalathus congesta (R. Dahlgren) R. Dahlgren ●☆

38476 Aspalathus chortophila Eckl. et Zeyh. subsp. kougaensis R. Dahlgren = Aspalathus kougaensis (Garab. ex R. Dahlgren) R. Dahlgren ●☆

38477 Aspalathus chrysantha R. Dahlgren;金花芳香木●☆

38478 Aspalathus ciliaris L.;缘毛芳香木●☆

38479 Aspalathus ciliatistyla L. Bolus = Aspalathus mundiana Eckl. et Zeyh. ●☆

38480 Aspalathus cinerascens E. Mey. ;灰色芳香木●☆
38481 Aspalathus cinerea Thunb. = Aspalathus cytisoides Lam. ●☆
38482 Aspalathus citrina R. Dahlgren;柠檬芳香木●☆
38483 Aspalathus cliffortiifolia R. Dahlgren;可利果叶芳香木●☆
38484 Aspalathus cliffortioides Bolus;可利果状芳香木●☆
38485 Aspalathus cognata C. Presl = Aspalathus linearis (Burm. f.) R. Dahlgren ●☆
38486 Aspalathus collina Eckl. et Zeyh. ;山丘芳香木●☆
38487 Aspalathus commutata (Vogel) R. Dahlgren;变异芳香木●☆
38488 Aspalathus comosa Thunb. = Aspalathus parviflora P. J. Bergius ●☆
38489 Aspalathus compacta R. Dahlgren;紧密芳香木●☆
38490 Aspalathus complicata (Benth.) R. Dahlgren;折叠芳香木●☆
38491 Aspalathus comptonii R. Dahlgren;康普顿芳香木●☆
38492 Aspalathus concava Bolus;凹芳香木●☆
38493 Aspalathus concavifolia (Eckl. et Zeyh.) R. Dahlgren = Aspalathus cytisoides Lam. ●☆
38494 Aspalathus condensata R. Dahlgren;密集芳香木●☆
38495 Aspalathus confusa R. Dahlgren;混乱芳香木●☆
38496 Aspalathus congesta (R. Dahlgren) R. Dahlgren;堆积芳香木●☆
38497 Aspalathus cordata (L.) R. Dahlgren;心形芳香木●☆
38498 Aspalathus corniculata R. Dahlgren;小角芳香木●☆
38499 Aspalathus corymbosa E. Mey. = Aspalathus linearis (Burm. f.) R. Dahlgren ●☆
38500 Aspalathus crassifolia Andréws = Aspalathus fusca Thunb. ●☆
38501 Aspalathus crassisepala R. Dahlgren;粗萼芳香木●☆
38502 Aspalathus crenata (L.) R. Dahlgren;圆齿芳香木●☆
38503 Aspalathus cuneata D. Don = Lespedeza juncea (L. f.) Pers. var. sericea (Thunb.) Lace et Hemsl. ●
38504 Aspalathus cuneata E. Mey. = Lotononis alpina (Eckl. et Zeyh.) B. -E. van Wyk subsp. multiflora (Eckl. et Zeyh.) B. -E. van Wyk ■☆
38505 Aspalathus cuspidata R. Dahlgren;骤尖芳香木●☆
38506 Aspalathus cuspidata R. Dahlgren subsp. stricticlada ? = Aspalathus stricticlada (R. Dahlgren) R. Dahlgren ●☆
38507 Aspalathus cymbiformis DC. ;船状芳香木●☆
38508 Aspalathus cymbiformis DC. var. hirta ? = Aspalathus cymbiformis DC. ●☆
38509 Aspalathus cytisoides Lam. ;金雀花芳香木●☆
38510 Aspalathus dasyantha Eckl. et Zeyh. ;毛花芳香木●☆
38511 Aspalathus deciduifolia Eckl. et Zeyh. = Aspalathus nigra L. ●☆
38512 Aspalathus densifolia Benth. ;密花芳香木●☆
38513 Aspalathus desertorum Bolus;荒漠芳香木●☆
38514 Aspalathus dianthopora E. Phillips;石竹芳香木●☆
38515 Aspalathus diffusa Eckl. et Zeyh. ;松散芳香木●☆
38516 Aspalathus digitifolia R. Dahlgren;指叶芳香木●☆
38517 Aspalathus divaricata Thunb. ;叉开芳香木●☆
38518 Aspalathus divaricata Thunb. subsp. brevicarpa R. Dahlgren = Aspalathus brevicarpa (R. Dahlgren) R. Dahlgren ●☆
38519 Aspalathus divaricata Thunb. subsp. gracilior R. Dahlgren;纤细叉开芳香木●☆
38520 Aspalathus divaricata Thunb. subsp. horizontalis R. Dahlgren = Aspalathus horizontalis (R. Dahlgren) R. Dahlgren ●☆
38521 Aspalathus divaricata Thunb. subsp. leptocoma (Eckl. et Zeyh.) R. Dahlgren = Aspalathus leptocoma Eckl. et Zeyh. ●☆
38522 Aspalathus divergens Willd. ex E. Mey. = Aspalathus microphylla DC. ●☆
38523 Aspalathus divergens Willd. ex E. Mey. f. microphylla (DC.) E. Mey. = Aspalathus microphylla DC. ●☆
38524 Aspalathus dregeana Walp. = Aspalathus incurva Thunb. ●☆
38525 Aspalathus dubia E. Mey. = Aspalathus ciliaris L. ●☆
38526 Aspalathus echinata E. Mey. = Aspalathus setacea Eckl. et Zeyh. ●☆
38527 Aspalathus elliptica (E. Phillips) R. Dahlgren;椭圆芳香木●☆
38528 Aspalathus elongata E. Mey. f. virgata Benth. = Aspalathus linguiloba R. Dahlgren ●☆
38529 Aspalathus elongata Eckl. et Zeyh. = Aspalathus quinquefolia L. subsp. virgata (Thunb.) R. Dahlgren ●☆
38530 Aspalathus empetrifolia (R. Dahlgren) R. Dahlgren;岩高兰叶芳香木●☆
38531 Aspalathus ericifolia L. ;毛叶芳香木●☆
38532 Aspalathus ericifolia L. subsp. minuta R. Dahlgren;微小毛叶芳香木●☆
38533 Aspalathus ericifolia L. subsp. puberula (Eckl. et Zeyh.) R. Dahlgren = Aspalathus puberula (Eckl. et Zeyh.) R. Dahlgren ●☆
38534 Aspalathus ericifolia L. subsp. pusilla R. Dahlgren;微小芳香木●☆
38535 Aspalathus eriophylla Walp. = Aspalathus setacea Eckl. et Zeyh. ●☆
38536 Aspalathus erythrodes Eckl. et Zeyh. ;淡红芳香木●☆
38537 Aspalathus esterhuyseniae R. Dahlgren;埃斯特芳香木●☆
38538 Aspalathus excelsa R. Dahlgren;高大芳香木●☆
38539 Aspalathus exigua Eckl. et Zeyh. = Aspalathus securifolia Eckl. et Zeyh. ●☆
38540 Aspalathus exilis Harv. = Aspalathus albens L. ●☆
38541 Aspalathus falcata Benth. = Aspalathus lanata E. Mey. ●☆
38542 Aspalathus fascicularis Burm. f. ;带状芳香木●☆
38543 Aspalathus fasciculata (Thunb.) Druce;簇生芳香木●☆
38544 Aspalathus ferox Harv. ;多刺芳香木●☆
38545 Aspalathus ferruginea H. P. Banks ex Benth. = Aspalathus ternata (Thunb.) Druce ●☆
38546 Aspalathus filicaulis Eckl. et Zeyh. ;线茎芳香木●☆
38547 Aspalathus filifolia E. Mey. = Aspalathus abietina Thunb. ●☆
38548 Aspalathus flavispina C. Presl = Aspalathus spinosa L. subsp. flavispina (C. Presl ex Benth.) R. Dahlgren ●☆
38549 Aspalathus flexuosa Thunb. ;曲折芳香木●☆
38550 Aspalathus floribunda Benth. = Aspalathus subulata Thunb. ●☆
38551 Aspalathus florulenta R. Dahlgren;多花芳香木●☆
38552 Aspalathus forbesii Harv. ;福布斯芳香木●☆
38553 Aspalathus fornicata Benth. = Aspalathus abietina Thunb. ●☆
38554 Aspalathus fourcadei L. Bolus;富尔卡德芳香木●☆
38555 Aspalathus frankenioides DC. ;瓣鳞花芳香木●☆
38556 Aspalathus frankenioides DC. var. albanensis (Eckl. et Zeyh.) Harv. = Aspalathus frankenioides DC. ●☆
38557 Aspalathus frankenioides DC. var. alpina Harv. = Aspalathus intermedia Eckl. et Zeyh. ●☆
38558 Aspalathus frankenioides DC. var. chortophila (Eckl. et Zeyh.) Harv. = Aspalathus chortophila Eckl. et Zeyh. ●☆
38559 Aspalathus frankenioides DC. var. poliotes (Eckl. et Zeyh.) Harv. = Aspalathus intermedia Eckl. et Zeyh. ●☆
38560 Aspalathus fusca Thunb. ;棕色芳香木●☆
38561 Aspalathus galeata E. Mey. ;盔形芳香木●☆
38562 Aspalathus galioides P. J. Bergius = Aspalathus retroflexa L. ●☆
38563 Aspalathus garipensis E. Mey. ;加里普芳香木●☆
38564 Aspalathus gerrardii Bolus;杰勒德芳香木●☆
38565 Aspalathus gillii Benth. = Aspalathus setacea Eckl. et Zeyh. ●☆
38566 Aspalathus glabrata R. Dahlgren;光滑芳香木●☆

38567 Aspalathus glabrescens R. Dahlgren;渐光芳香木●☆
38568 Aspalathus glauca Eckl. et Zeyh. = Aspalathus spinosa L. subsp. glauca (Eckl. et Zeyh.) R. Dahlgren ●☆
38569 Aspalathus globosa Andréws;球形芳香木●☆
38570 Aspalathus globulosa E. Mey.;小球芳香木●☆
38571 Aspalathus glomerata L. f. = Aspalathus capitata L. ●☆
38572 Aspalathus glossoides R. Dahlgren;舌状芳香木●☆
38573 Aspalathus gracilifolia R. Dahlgren = Aspalathus juniperina Thunb. subsp. gracilifolia (R. Dahlgren) R. Dahlgren ●☆
38574 Aspalathus gracilis Garab. ex Fourc. = Aspalathus rubens Thunb. ●☆
38575 Aspalathus grandiflora Benth.;大花芳香木●☆
38576 Aspalathus granulata R. Dahlgren;颗粒芳香木●☆
38577 Aspalathus heterophylla L. f.;互叶芳香木●☆
38578 Aspalathus heterophylla L. f. subsp. lagopus (Thunb.) R. Dahlgren = Aspalathus lotoides Thunb. subsp. lagopus (Thunb.) R. Dahlgren ●☆
38579 Aspalathus heterophylla L. f. subsp. lotoides (Thunb.) R. Dahlgren = Aspalathus lotoides Thunb. ●☆
38580 Aspalathus hiatuum Eckl. et Zeyh. = Aspalathus intermedia Eckl. et Zeyh. ●☆
38581 Aspalathus hirta E. Mey.;多毛芳香木●☆
38582 Aspalathus hirta E. Mey. subsp. stellaris R. Dahlgren;星状芳香木●☆
38583 Aspalathus hispida Thunb.;粗毛芳香木●☆
38584 Aspalathus hispida Thunb. subsp. albiflora (Eckl. et Zeyh.) R. Dahlgren;白花粗毛芳香木●☆
38585 Aspalathus holosericea E. Mey. = Lotononis holosericea (E. Mey.) B. -E. van Wyk ■☆
38586 Aspalathus horizontalis (R. Dahlgren) R. Dahlgren;平展芳香木●☆
38587 Aspalathus horrida Eckl. et Zeyh. = Aspalathus spinosa L. ●☆
38588 Aspalathus humilis Bolus;低矮芳香木●☆
38589 Aspalathus hypnoides R. Dahlgren;藓芳香木●☆
38590 Aspalathus hystrix L. f.;豪猪芳香木●☆
38591 Aspalathus incana R. Dahlgren;灰毛芳香木●☆
38592 Aspalathus incompta Thunb.;装饰芳香木●☆
38593 Aspalathus incurva Thunb.;内折芳香木●☆
38594 Aspalathus incurvifolia Vogel ex Walp.;折叶芳香木●☆
38595 Aspalathus inops Eckl. et Zeyh.;贫弱芳香木●☆
38596 Aspalathus intermedia Eckl. et Zeyh.;间型芳香木●☆
38597 Aspalathus intervallaris Bolus;谷地芳香木●☆
38598 Aspalathus intricata Compton;缠结芳香木●☆
38599 Aspalathus intricata Compton subsp. anthospermoides (R. Dahlgren) R. Dahlgren;琥珀芳香木●☆
38600 Aspalathus intricata Compton subsp. oxyclada (Compton) R. Dahlgren;锐枝缠结芳香木●☆
38601 Aspalathus involucrata E. Mey. = Aspalathus fasciculata (Thunb.) Druce ●☆
38602 Aspalathus joubertiana Eckl. et Zeyh.;朱伯特芳香木●☆
38603 Aspalathus joubertiana Eckl. et Zeyh. subsp. glabripetala R. Dahlgren = Aspalathus shawii L. Bolus subsp. glabripetala (R. Dahlgren) R. Dahlgren ●☆
38604 Aspalathus joubertiana Eckl. et Zeyh. subsp. longispica R. Dahlgren = Aspalathus shawii L. Bolus subsp. longispica (R. Dahlgren) R. Dahlgren ●☆
38605 Aspalathus joubertiana Eckl. et Zeyh. subsp. shawii (L. Bolus) R. Dahlgren = Aspalathus shawii L. Bolus ●☆
38606 Aspalathus juniperina Thunb.;刺柏状芳香木●☆
38607 Aspalathus juniperina Thunb. subsp. gracilifolia (R. Dahlgren) R. Dahlgren;细叶芳香木●☆
38608 Aspalathus juniperina Thunb. subsp. grandis R. Dahlgren;大芳香木●☆
38609 Aspalathus juniperina Thunb. subsp. monticola R. Dahlgren;山地刺柏状芳香木●☆
38610 Aspalathus kannaensis Eckl. et Zeyh. = Aspalathus flexuosa Thunb. ●☆
38611 Aspalathus karrooensis R. Dahlgren;卡卢芳香木●☆
38612 Aspalathus katbergensis (R. Dahlgren) R. Dahlgren;卡特贝赫芳香木●☆
38613 Aspalathus kougaensis (Garab. ex R. Dahlgren) R. Dahlgren;科加芳香木●☆
38614 Aspalathus kraussiana Meisn. = Aspalathus aspalathoides (L.) R. Dahlgren ●☆
38615 Aspalathus lactea Thunb.;乳白芳香木●☆
38616 Aspalathus lactea Thunb. subsp. breviloba R. Dahlgren;短裂乳白芳香木●☆
38617 Aspalathus laeta Bolus;愉悦芳香木●☆
38618 Aspalathus lamarckiana R. Dahlgren;拉马克芳香木●☆
38619 Aspalathus lanata E. Mey.;绵毛芳香木●☆
38620 Aspalathus lanceicarpa R. Dahlgren;披针果芳香木●☆
38621 Aspalathus lanceifolia R. Dahlgren;披针叶芳香木●☆
38622 Aspalathus lanceolata E. Mey. = Lotononis lanceolata (E. Mey.) Benth. ■☆
38623 Aspalathus lanifera R. Dahlgren;软毛芳香木●☆
38624 Aspalathus laricifolia Lam. = Aspalathus uniflora L. ●☆
38625 Aspalathus laricifolia P. J. Bergius;落叶松叶芳香木●☆
38626 Aspalathus laricifolia P. J. Bergius subsp. canescens (L.) R. Dahlgren;灰白芳香木●☆
38627 Aspalathus laricina DC. = Aspalathus laricifolia P. J. Bergius ●☆
38628 Aspalathus latibracteata (Kuntze) K. Schum.;宽苞芳香木●☆
38629 Aspalathus latifolia Bolus;宽叶芳香木●☆
38630 Aspalathus laxata L. = Lotononis involucrata (P. J. Bergius) Benth. ●☆
38631 Aspalathus lebeckioides R. Dahlgren;针叶豆芳香木●☆
38632 Aspalathus leiantha (E. Phillips) R. Dahlgren = Aspalathus crenata (L.) R. Dahlgren ●☆
38633 Aspalathus leipoldtii Schltr. = Aspalathus spinescens Thunb. subsp. lepida (E. Mey.) R. Dahlgren ●☆
38634 Aspalathus lenticula Bolus;凸镜芳香木●☆
38635 Aspalathus lepida E. Mey. = Aspalathus spinescens Thunb. subsp. lepida (E. Mey.) R. Dahlgren ●☆
38636 Aspalathus leptophylla Eckl. et Zeyh. = Aspalathus uniflora L. ●☆
38637 Aspalathus leptoptera Bolus;细翅芳香木●☆
38638 Aspalathus leptothria Eckl. et Zeyh. = Aspalathus intermedia Eckl. et Zeyh. ●☆
38639 Aspalathus leucocephala E. Mey. = Aspalathus quinquefolia L. subsp. virgata (Thunb.) R. Dahlgren ●☆
38640 Aspalathus leucophaea Harv. = Aspalathus ciliaris L. ●☆
38641 Aspalathus leucophylla R. Dahlgren;白叶芳香木●☆
38642 Aspalathus leucophylla R. Dahlgren subsp. septentrionalis = Aspalathus petersonii R. Dahlgren ●☆
38643 Aspalathus linearifolia DC.;线叶芳香木●☆
38644 Aspalathus linearifolia DC. f. discreta Drège = Aspalathus

linearifolia DC. ●☆
38645 Aspalathus linearis (Burm. f.) R. Dahlgren;线状芳香木;Rooibos Tea ●☆
38646 Aspalathus linearis (Burm. f.) R. Dahlgren subsp. latipetala R. Dahlgren = Aspalathus lebeckioides R. Dahlgren ●☆
38647 Aspalathus linearis (Burm. f.) R. Dahlgren subsp. pinifolia (Marloth) R. Dahlgren = Aspalathus linearis (Burm. f.) R. Dahlgren ●☆
38648 Aspalathus linguiloba R. Dahlgren;大萼芳香木●☆
38649 Aspalathus longifolia Benth.;长叶芳香木●☆
38650 Aspalathus longipes Harv.;长梗芳香木●☆
38651 Aspalathus lotoides Thunb.;君迁子芳香木●☆
38652 Aspalathus lotoides Thunb. subsp. lagopus (Thunb.) R. Dahlgren;兔足芳香木●☆
38653 Aspalathus lotoides Thunb. var. stachyera (Eckl. et Zeyh.) Harv. = Aspalathus heterophylla L. f. ●☆
38654 Aspalathus macrocarpa Eckl. et Zeyh.;大果芳香木●☆
38655 Aspalathus macrosepala Steud. = Aspalathus linguiloba R. Dahlgren ●☆
38656 Aspalathus marginalis Eckl. et Zeyh.;边生芳香木●☆
38657 Aspalathus marginata Harv.;具边芳香木●☆
38658 Aspalathus melanoides Eckl. et Zeyh. = Aspalathus nigra L. ●☆
38659 Aspalathus meyeri Harv. = Aspalathus quinquefolia L. subsp. virgata (Thunb.) R. Dahlgren ●☆
38660 Aspalathus meyeriana Eckl. et Zeyh. = Aspalathus ciliaris L. ●☆
38661 Aspalathus micrantha E. Mey. = Aspalathus hispida Thunb. ●☆
38662 Aspalathus microcarpa DC. = Aspalathus hispida Thunb. ●☆
38663 Aspalathus microdon Benth. = Aspalathus steudeliana Brongn. ●☆
38664 Aspalathus microphylla DC.;小叶芳香木●☆
38665 Aspalathus millefolia R. Dahlgren;粟草叶芳香木●☆
38666 Aspalathus monosperma (DC.) R. Dahlgren;单籽芳香木●☆
38667 Aspalathus mucronata L. f. = Wiborgia mucronata (L. f.) Druce ■☆
38668 Aspalathus multiflora Thunb. = Aspalathus vermiculata Lam. ●☆
38669 Aspalathus mundiana Eckl. et Zeyh.;蒙德芳香木●☆
38670 Aspalathus muraltioides Eckl. et Zeyh.;厚壁芳香木●☆
38671 Aspalathus myrtillifolia Benth.;黑果越橘叶芳香木●☆
38672 Aspalathus neanthes Eckl. et Zeyh. = Aspalathus laricifolia P. J. Bergius subsp. canescens (L.) R. Dahlgren ●☆
38673 Aspalathus neglecta T. M. Salter;忽视芳香木●☆
38674 Aspalathus nervosa E. Mey. = Aspalathus spicata Thunb. ●☆
38675 Aspalathus nigra L.;黑芳香木●☆
38676 Aspalathus nigrescens E. Mey. = Aspalathus nigra L. ●☆
38677 Aspalathus nivalis Schltr. ex Marloth = Aspalathus pedicellata Harv. ●☆
38678 Aspalathus nivea Thunb.;雪白芳香木●☆
38679 Aspalathus nodosa Vogel ex Walp.;结节芳香木●☆
38680 Aspalathus nudiflora Harv.;裸花芳香木●☆
38681 Aspalathus obliqua R. Dahlgren;偏斜芳香木●☆
38682 Aspalathus oblongifolia R. Dahlgren;矩圆叶芳香木●☆
38683 Aspalathus obtusata Thunb.;钝芳香木●☆
38684 Aspalathus obtusifolia R. Dahlgren;钝叶芳香木●☆
38685 Aspalathus odontoloba R. Dahlgren;齿裂片芳香木●☆
38686 Aspalathus oliveri R. Dahlgren;奥里弗芳香木●☆
38687 Aspalathus opaca Eckl. et Zeyh.;暗色芳香木●☆
38688 Aspalathus opaca Eckl. et Zeyh. subsp. pappeana (Harv.) R. Dahlgren;帕珀暗色芳香木●☆
38689 Aspalathus opaca Eckl. et Zeyh. subsp. rostriloba R. Dahlgren;喙裂暗色芳香木●☆
38690 Aspalathus orbiculata Benth.;圆芳香木●☆
38691 Aspalathus oresigena Eckl. et Zeyh. = Aspalathus ciliaris L. ●☆
38692 Aspalathus oxyclada Compton = Aspalathus intricata Compton subsp. oxyclada (Compton) R. Dahlgren ●☆
38693 Aspalathus pachyloba Benth.;深裂芳香木●☆
38694 Aspalathus pachyloba Benth. subsp. macroclada R. Dahlgren;大枝深裂芳香木●☆
38695 Aspalathus pachyloba Benth. subsp. succulentifolia R. Dahlgren = Aspalathus pachyloba Benth. subsp. villicaulis R. Dahlgren ●☆
38696 Aspalathus pachyloba Benth. subsp. villicaulis R. Dahlgren;毛茎深裂芳香木●☆
38697 Aspalathus pallens Eckl. et Zeyh. = Aspalathus nigra L. ●☆
38698 Aspalathus pallescens Eckl. et Zeyh.;变苍白芳香木●☆
38699 Aspalathus pallidiflora R. Dahlgren;苍白花芳香木●☆
38700 Aspalathus papillosa Eckl. et Zeyh. = Aspalathus ciliaris L. ●☆
38701 Aspalathus pappeana Harv. = Aspalathus opaca Eckl. et Zeyh. subsp. pappeana (Harv.) R. Dahlgren ●☆
38702 Aspalathus parviflora P. J. Bergius;小花芳香木●☆
38703 Aspalathus patens Garab. ex R. Dahlgren;铺展芳香木●☆
38704 Aspalathus pedicellata Harv.;梗花芳香木●☆
38705 Aspalathus pedunculata Houtt.;序花芳香木●☆
38706 Aspalathus pedunculata L'Hér. = Aspalathus bracteata Thunb. ●☆
38707 Aspalathus pendula R. Dahlgren;下垂芳香木●☆
38708 Aspalathus pentheri Gand. = Aspalathus pinguis Thunb. ●☆
38709 Aspalathus perfoliata (Lam.) R. Dahlgren;贯叶芳香木●☆
38710 Aspalathus perfoliata (Lam.) R. Dahlgren subsp. phillipsii R. Dahlgren;菲利芳香木●☆
38711 Aspalathus perforata (Thunb.) R. Dahlgren;穿孔芳香木●☆
38712 Aspalathus persica Burm. f. = Lotus garcinii DC. ■☆
38713 Aspalathus petersonii R. Dahlgren;彼得逊芳香木●☆
38714 Aspalathus phylicoides Compton = Aspalathus shawii L. Bolus ●☆
38715 Aspalathus pileata L. Bolus = Aspalathus globosa Andréws ●☆
38716 Aspalathus pilosa L. = Aspalathus tridentata L. ●☆
38717 Aspalathus pinea Thunb.;松林芳香木●☆
38718 Aspalathus pinea Thunb. subsp. caudata R. Dahlgren;尾状芳香木●☆
38719 Aspalathus pinguis Thunb.;肥厚芳香木●☆
38720 Aspalathus pinguis Thunb. subsp. australis R. Dahlgren;南方肥厚芳香木●☆
38721 Aspalathus pinguis Thunb. subsp. longissima R. Dahlgren;极长西方芳香木●☆
38722 Aspalathus pinguis Thunb. subsp. occidentalis R. Dahlgren;西方芳香木●☆
38723 Aspalathus pinnata L.;羽状芳香木●☆
38724 Aspalathus plukenetiana Eckl. et Zeyh. = Aspalathus rugosa Thunb. ●☆
38725 Aspalathus poliotes Eckl. et Zeyh. = Aspalathus intermedia Eckl. et Zeyh. ●☆
38726 Aspalathus polyantha Walp. = Lotononis alpina (Eckl. et Zeyh.) B. -E. van Wyk subsp. multiflora (Eckl. et Zeyh.) B. -E. van Wyk ■☆
38727 Aspalathus polycephala E. Mey.;多头芳香木●☆
38728 Aspalathus polycephala E. Mey. subsp. lanatifolia R. Dahlgren;毛叶多头芳香木●☆
38729 Aspalathus polycephala E. Mey. subsp. rigida (Schltr.) R. Dahlgren;硬多头芳香木●☆

38730 Aspalathus polycephala E. Mey. var. pauciflora ? = Aspalathus polycephala E. Mey. subsp. lanatifolia R. Dahlgren ●☆
38731 Aspalathus potbergensis R. Dahlgren;贝尔热芳香木●☆
38732 Aspalathus priorii Harv. = Aspalathus forbesii Harv. ●☆
38733 Aspalathus proboscidea R. Dahlgren;长角芳香木●☆
38734 Aspalathus procumbens E. Mey. = Aspalathus lotoides Thunb. ●☆
38735 Aspalathus propinqua E. Mey. = Aspalathus triquetra Thunb. ●☆
38736 Aspalathus prostrata Eckl. et Zeyh.;平卧芳香木●☆
38737 Aspalathus psoraleoides (C. Presl) Benth.;补骨脂芳香木●☆
38738 Aspalathus puberula (Eckl. et Zeyh.) R. Dahlgren;微毛芳香木●☆
38739 Aspalathus pulchella E. Mey. = Lotononis pulchella (E. Mey.) B. -E. van Wyk ■☆
38740 Aspalathus pungens Thunb. = Aspalathus acuminata Lam. subsp. pungens (Thunb.) R. Dahlgren ●☆
38741 Aspalathus purpurascens E. Mey. = Aspalathus ternata (Thunb.) Druce ●☆
38742 Aspalathus purpurea Eckl. et Zeyh. = Aspalathus ternata (Thunb.) Druce ●☆
38743 Aspalathus pycnantha R. Dahlgren;非洲密花芳香木●☆
38744 Aspalathus quadrata L. Bolus;四方形芳香木●☆
38745 Aspalathus quinquefolia L.;五叶芳香木●☆
38746 Aspalathus quinquefolia L. subsp. acocksii R. Dahlgren = Aspalathus acocksii (R. Dahlgren) R. Dahlgren ●☆
38747 Aspalathus quinquefolia L. subsp. compacta R. Dahlgren;紧密五叶芳香木●☆
38748 Aspalathus quinquefolia L. subsp. virgata (Thunb.) R. Dahlgren;条纹芳香木●☆
38749 Aspalathus racemosa E. Mey.;总花芳香木●☆
38750 Aspalathus radiata Garab. ex R. Dahlgren;辐射芳香木●☆
38751 Aspalathus radiata Garab. ex R. Dahlgren subsp. pseudosericea R. Dahlgren;假绢毛辐射芳香木●☆
38752 Aspalathus ramosissima R. Dahlgren;多分枝芳香木●☆
38753 Aspalathus ramulosa E. Mey.;多枝芳香木●☆
38754 Aspalathus rectistyla R. Dahlgren;直柱芳香木●☆
38755 Aspalathus recurva Benth.;弯芳香木●☆
38756 Aspalathus recurvispina R. Dahlgren;弯刺芳香木●☆
38757 Aspalathus remota Eckl. et Zeyh. = Aspalathus spicata Thunb. ●☆
38758 Aspalathus remota L. Bolus = Aspalathus calcarea R. Dahlgren ●☆
38759 Aspalathus repens R. Dahlgren;匍匐芳香木●☆
38760 Aspalathus retroflexa L.;反折芳香木●☆
38761 Aspalathus retroflexa L. subsp. amoena R. Dahlgren = Aspalathus amoena (R. Dahlgren) R. Dahlgren ●☆
38762 Aspalathus retroflexa L. subsp. angustipetala R. Dahlgren;窄瓣芳香木●☆
38763 Aspalathus retroflexa L. subsp. bicolor (Eckl. et Zeyh.) R. Dahlgren;二色芳香木●☆
38764 Aspalathus retroflexa L. subsp. empetrifolia R. Dahlgren = Aspalathus empetrifolia (R. Dahlgren) R. Dahlgren ●☆
38765 Aspalathus rigescens E. Mey. = Aspalathus setacea Eckl. et Zeyh. ●☆
38766 Aspalathus rigida Schltr. = Aspalathus polycephala E. Mey. subsp. rigida (Schltr.) R. Dahlgren ●☆
38767 Aspalathus rigidifolia R. Dahlgren;挺叶芳香木●☆
38768 Aspalathus robusta Bolus = Aspalathus ciliaris L. ●☆
38769 Aspalathus rosea Garab. ex R. Dahlgren;粉红状芳香木●☆
38770 Aspalathus rostrata Benth.;喙状芳香木●☆
38771 Aspalathus rostripetala R. Dahlgren;喙瓣芳香木●☆
38772 Aspalathus rubens Thunb.;变淡红芳香木●☆
38773 Aspalathus rubiginosa R. Dahlgren;锈红芳香木●☆
38774 Aspalathus rubrocalyx Garab. ex Compton = Aspalathus laeta Bolus ●☆
38775 Aspalathus rubrofusca Eckl. et Zeyh. = Aspalathus asparagoides L. f. subsp. rubrofusca (Eckl. et Zeyh.) R. Dahlgren ●☆
38776 Aspalathus rugosa Thunb.;皱纹芳香木●☆
38777 Aspalathus rugosa Thunb. subsp. linearifolia (DC.) Dahlgren = Aspalathus linearifolia DC. ●☆
38778 Aspalathus rupestris R. Dahlgren;岩生芳香木●☆
38779 Aspalathus salicifolia R. Dahlgren;柳叶芳香木●☆
38780 Aspalathus salteri L. Bolus;索尔特芳香木●☆
38781 Aspalathus sanguinea Thunb.;血红芳香木●☆
38782 Aspalathus sanguinea Thunb. subsp. foliosa R. Dahlgren;多叶血红芳香木●☆
38783 Aspalathus sarcantha Vogel ex Walp. = Aspalathus carnosa P. J. Bergius ●☆
38784 Aspalathus sarcodes Vogel ex Benth. = Aspalathus capensis (Walp.) R. Dahlgren ●☆
38785 Aspalathus scaphoides Eckl. et Zeyh. = Aspalathus cymbiformis DC. ●☆
38786 Aspalathus sceptrum-aureum R. Dahlgren;王杖芳香木●☆
38787 Aspalathus schlechteri Bolus = Aspalathus opaca Eckl. et Zeyh. ●☆
38788 Aspalathus scholliana C. Presl = Aspalathus lotoides Thunb. ●☆
38789 Aspalathus secunda E. Mey.;单侧芳香木●☆
38790 Aspalathus securifolia Eckl. et Zeyh.;斧形芳香木●☆
38791 Aspalathus securifolia Eckl. et Zeyh. subsp. crassa R. Dahlgren = Aspalathus securifolia Eckl. et Zeyh. ●☆
38792 Aspalathus sericea Lam. = Aspalathus candicans W. T. Aiton ●☆
38793 Aspalathus sericea P. J. Bergius;绢毛芳香木●☆
38794 Aspalathus sericea P. J. Bergius subsp. aemula (E. Mey.) R. Dahlgren = Aspalathus aemula E. Mey. ●☆
38795 Aspalathus serpens R. Dahlgren;蛇形芳香木●☆
38796 Aspalathus setacea Eckl. et Zeyh.;刚毛芳香木●☆
38797 Aspalathus setacea Eckl. et Zeyh. var. arachnoidea Walp. = Aspalathus setacea Eckl. et Zeyh. ●☆
38798 Aspalathus setacea Eckl. et Zeyh. var. canescens Walp. = Aspalathus setacea Eckl. et Zeyh. ●☆
38799 Aspalathus shawii L. Bolus;肖氏芳香木●☆
38800 Aspalathus shawii L. Bolus subsp. glabripetala (R. Dahlgren) R. Dahlgren;光瓣芳香木●☆
38801 Aspalathus shawii L. Bolus subsp. longispica (R. Dahlgren) R. Dahlgren;长穗芳香木●☆
38802 Aspalathus simii Bolus;西姆芳香木●☆
38803 Aspalathus simsiana Eckl. et Zeyh. = Aspalathus confusa R. Dahlgren ●☆
38804 Aspalathus singuliflora R. Dahlgren;单花芳香木●☆
38805 Aspalathus smithii R. Dahlgren;史密斯芳香木●☆
38806 Aspalathus spectabilis R. Dahlgren;壮观芳香木●☆
38807 Aspalathus sphaerocephala Schltr. = Aspalathus truncata Eckl. et Zeyh. ●☆
38808 Aspalathus spicata Thunb.;穗状芳香木●☆
38809 Aspalathus spicata Thunb. subsp. cliffortioides (Bolus) R. Dahlgren = Aspalathus cliffortioides Bolus ●☆
38810 Aspalathus spicata Thunb. subsp. neglecta (T. M. Salter) R. Dahlgren = Aspalathus neglecta T. M. Salter ●☆

38811 Aspalathus spiculata R. Dahlgren;小刺芳香木●☆
38812 Aspalathus spinescens Thunb.;细刺芳香木●☆
38813 Aspalathus spinescens Thunb. subsp. lepida (E. Mey.) R. Dahlgren;鳞片芳香木●☆
38814 Aspalathus spinosa L.;显刺芳香木●☆
38815 Aspalathus spinosa L. subsp. flavispina (C. Presl ex Benth.) R. Dahlgren;黄刺芳香木●☆
38816 Aspalathus spinosa L. subsp. glauca (Eckl. et Zeyh.) R. Dahlgren;光刺芳香木●☆
38817 Aspalathus spinosa L. subsp. obtusata (Thunb.) R. Dahlgren = Aspalathus obtusata Thunb. ●☆
38818 Aspalathus spinosissima R. Dahlgren;大刺芳香木●☆
38819 Aspalathus spinosissima R. Dahlgren subsp. tenuiflora ?;细叶大刺芳香木●☆
38820 Aspalathus squarrosa Thunb. = Aspalathus bracteata Thunb. ●☆
38821 Aspalathus stachyera Eckl. et Zeyh. = Aspalathus heterophylla L. f. ●☆
38822 Aspalathus staurantha Eckl. et Zeyh. = Aspalathus tridentata L. subsp. staurantha (Eckl. et Zeyh.) R. Dahlgren ●☆
38823 Aspalathus stellaris Eckl. et Zeyh. = Aspalathus aspalathoides (L.) R. Dahlgren ●☆
38824 Aspalathus stenophylla Eckl. et Zeyh.;狭叶芳香木●☆
38825 Aspalathus stenophylla Eckl. et Zeyh. subsp. colorata R. Dahlgren = Aspalathus stenophylla Eckl. et Zeyh. ●☆
38826 Aspalathus stenophylla Eckl. et Zeyh. subsp. garciana R. Dahlgren = Aspalathus inops Eckl. et Zeyh. ●☆
38827 Aspalathus steudeliana Brongn.;斯托芳香木●☆
38828 Aspalathus stokoei L. Bolus;斯托克芳香木●☆
38829 Aspalathus stricticlada (R. Dahlgren) R. Dahlgren;直枝芳香木●☆
38830 Aspalathus strictus Steud. = Aspalathus quinquefolia L. subsp. virgata (Thunb.) R. Dahlgren ●☆
38831 Aspalathus suaveolens Eckl. et Zeyh.;芳香芳香木●☆
38832 Aspalathus subinermis (E. Mey.) Benth. = Aspalathus acuminata Lam. ●☆
38833 Aspalathus subulata Thunb.;钻形芳香木●☆
38834 Aspalathus suffruticosa DC. = Aspalathus biflora E. Mey. ●☆
38835 Aspalathus sulphurea R. Dahlgren;硫色芳香木●☆
38836 Aspalathus taylori R. Dahlgren;泰勒芳香木●☆
38837 Aspalathus tenuifolia DC. = Aspalathus linearis (Burm. f.) R. Dahlgren ●☆
38838 Aspalathus tenuissima R. Dahlgren;极细芳香木●☆
38839 Aspalathus teres Eckl. et Zeyh.;圆柱芳香木●☆
38840 Aspalathus teres Eckl. et Zeyh. subsp. thodei R. Dahlgren;索德芳香木●☆
38841 Aspalathus ternata (Thunb.) Druce;三出芳香木●☆
38842 Aspalathus thymifolia L. = Aspalathus cymbiformis DC. ●☆
38843 Aspalathus thymifolia L. f. albiflora (Eckl. et Zeyh.) Benth. = Aspalathus hispida Thunb. subsp. albiflora (Eckl. et Zeyh.) R. Dahlgren ●☆
38844 Aspalathus tomentosa E. Mey. = Aspalathus frankenioides DC. ●☆
38845 Aspalathus trichodes E. Mey. = Lotononis trichodes (E. Mey.) B. -E. van Wyk ■☆
38846 Aspalathus tridentata L.;三齿芳香木●☆
38847 Aspalathus tridentata L. subsp. fragilis R. Dahlgren;纤细三齿芳香木●☆
38848 Aspalathus tridentata L. subsp. rotunda R. Dahlgren;圆形三齿芳香木(圆形芳香木)●☆
38849 Aspalathus tridentata L. subsp. staurantha (Eckl. et Zeyh.) R. Dahlgren;十字花芳香木●☆
38850 Aspalathus trigona Thunb.;三角芳香木●☆
38851 Aspalathus triquetra Thunb.;三棱芳香木●☆
38852 Aspalathus truncata Eckl. et Zeyh.;平截芳香木●☆
38853 Aspalathus truncata Eckl. et Zeyh. subsp. sphaerocephala (Schltr.) R. Dahlgren = Aspalathus truncata Eckl. et Zeyh. ●☆
38854 Aspalathus tuberculata Walp.;多疣芳香木●☆
38855 Aspalathus tulbaghensis R. Dahlgren;塔尔巴赫芳香木●☆
38856 Aspalathus ulicina Eckl. et Zeyh.;荆豆芳香木●☆
38857 Aspalathus undulata Eckl. et Zeyh. = Aspalathus fasciculata (Thunb.) Druce ●☆
38858 Aspalathus uniflora L.;独花芳香木●☆
38859 Aspalathus uniflora L. subsp. willdenowiana (Benth.) R. Dahlgren = Aspalathus willdenowiana Benth ●☆
38860 Aspalathus vacciniifolia R. Dahlgren;越橘芳香木●☆
38861 Aspalathus variegata Eckl. et Zeyh.;杂色芳香木●☆
38862 Aspalathus venosa E. Mey.;多脉芳香木●☆
38863 Aspalathus verbasciformis R. Dahlgren;毛蕊花芳香木●☆
38864 Aspalathus vermiculata Lam.;虫状芳香木●☆
38865 Aspalathus verrucosa L. = Aspalathus tuberculata Walp. ●☆
38866 Aspalathus versicolor E. Mey. = Aspalathus collina Eckl. et Zeyh. ●☆
38867 Aspalathus villosa Thunb.;长柔毛芳香木●☆
38868 Aspalathus viminea N. E. Br. = Lotononis viminea (E. Mey.) B. -E. van Wyk ■☆
38869 Aspalathus virgata Thunb. = Aspalathus quinquefolia L. subsp. virgata (Thunb.) R. Dahlgren ●☆
38870 Aspalathus vulpina Garab. ex R. Dahlgren;狐色芳香木●☆
38871 Aspalathus willdenowiana Benth.;威尔芳香木●☆
38872 Aspalathus wittebergensis Compton et P. E. Barnes;维特贝格芳香木●☆
38873 Aspalathus wittebergensis Compton et P. E. Barnes subsp. anthospermoides R. Dahlgren = Aspalathus intricata Compton subsp. anthospermoides (R. Dahlgren) R. Dahlgren ●☆
38874 Aspalathus wittebergensis Compton et P. E. Barnes subsp. intricata (Compton) R. Dahlgren = Aspalathus intricata Compton ●☆
38875 Aspalathus wittebergensis Compton et P. E. Barnes subsp. oxyclada (Compton) R. Dahlgren = Aspalathus intricata Compton subsp. oxyclada (Compton) R. Dahlgren ●☆
38876 Aspalathus wurmbeana E. Mey.;乌姆布芳香木●☆
38877 Aspalathus zeyheri (Harv.) R. Dahlgren;泽耶尔芳香木●☆
38878 Aspalatus A. St. -Hil. = Aspalathus L. ●☆
38879 Aspalthium Medik. = Asphalthium Medik. ●■
38880 Aspalthium Medik. = Bituminaria Heist. ex Fabr. ■☆
38881 Aspalthium Medik. = Psoralea L. ●■
38882 Aspalthium acaulis (Stev.) Hutch. = Bituminaria acaulis (Stev.) C. H. Stirt. ■☆
38883 Aspalthium bituminosum (L.) Fourr. = Bituminaria bituminosa (L.) C. H. Stirt. ■☆
38884 Aspalthium frutescens Medik. = Bituminaria bituminosa (L.) C. H. Stirt. ■☆
38885 Aspalthium herbaceum Medik. = Bituminaria bituminosa (L.) C. H. Stirt. ■☆
38886 Asparagaceae Juss. (1789)(保留科名);天门冬科●■
38887 Asparagopsis (Kunth) Kunth = Asparagus L. ■

38888 Asparagopsis (Kunth) Kunth = Protasparagus Oberm. ■
38889 Asparagopsis Kunth = Asparagus L. ■
38890 Asparagopsis L. = Asparagus L. ■
38891 Asparagopsis alba (L.) Kunth = Asparagus albus L. ■☆
38892 Asparagopsis alba (L.) Kunth var. pastorianus (Webb et Berthel.) Ball = Asparagus pastorianus Webb et Berthel. ■☆
38893 Asparagopsis consanguinea Kunth = Asparagus fasciculatus Thunb. ■☆
38894 Asparagopsis densiflora Kunth = Asparagus densiflorus (Kunth) Jessop ■
38895 Asparagopsis denudata Kunth = Asparagus denudatus (Kunth) Baker ■☆
38896 Asparagopsis flagellaris Kunth = Asparagus flagellaris (Kunth) Baker ■☆
38897 Asparagopsis floribunda Kunth = Asparagus racemosus Willd. ■
38898 Asparagopsis krebsiana Kunth = Asparagus krebsianus (Kunth) Jessop ■☆
38899 Asparagopsis lamarckii Kunth = Asparagus africanus Lam. ■☆
38900 Asparagopsis microraphis Kunth = Asparagus microraphis (Kunth) Baker ■☆
38901 Asparagopsis minutiflora Kunth = Asparagus minutiflorus (Kunth) Baker ■☆
38902 Asparagopsis setacea Kunth = Asparagus setaceus (Kunth) Jessop ■
38903 Asparagopsis sinica Miq. = Asparagus cochinchinensis (Lour.) Merr. ■
38904 Asparagus L. (1753);天门冬属(天冬属);Asparagus, Asparagus-fern, Asperge ■
38905 Asparagus abyssinicus Hochst. ex A. Rich. = Asparagus flagellaris (Kunth) Baker ■☆
38906 Asparagus acicularis F. T. Wang et S. C. Chen;山文竹(假天冬,千条蜈蚣赶条蛇,天冬);Needlelike Asparagus ■
38907 Asparagus acocksii Jessop;阿氏天门冬■☆
38908 Asparagus acutifolius L.;尖叶天门冬;Sharp-leaved Asparagus ■☆
38909 Asparagus acutifolius L. var. achhalii Valdés = Asparagus acutifolius L. ■☆
38910 Asparagus acutifolius L. var. gracilis Baker = Asparagus acutifolius L. ■☆
38911 Asparagus adscendens Roxb.;上举天门冬■☆
38912 Asparagus aethiopicus L. = Asparagus densiflorus (Kunth) Jessop 'Sprengeri' ■
38913 Asparagus aethiopicus L. f. 'Sprengeri' = Asparagus densiflorus (Kunth) Jessop 'Sprengeri' ■
38914 Asparagus aethiopicus L. var. natalensis Baker = Asparagus natalensis (Baker) J. -P. Lebrun et Stork ■☆
38915 Asparagus aethiopicus L. var. ternifolius Baker = Asparagus falcatus L. ■☆
38916 Asparagus africanus Lam.;非洲天门冬■☆
38917 Asparagus africanus Lam. var. abyssinicus (Hochst. ex A. Rich.) Fiori = Asparagus flagellaris (Kunth) Baker ■☆
38918 Asparagus africanus Lam. var. concinnus Baker = Asparagus concinnus (Baker) Kies ■☆
38919 Asparagus africanus Lam. var. puberulus (Baker) Sebsebe;微毛天门冬■☆
38920 Asparagus aggregatus (Oberm.) Fellingham et N. L. Mey.;聚集天门冬■☆
38921 Asparagus albus L.;白天门冬;White Asparagus ■☆
38922 Asparagus alopecurus (Oberm.) Malcomber et Sebsebe;看麦娘天门冬■☆
38923 Asparagus altiscandens Engl. et Gilg;攀缘天门冬■☆
38924 Asparagus altissimus Munby;高大天门冬■☆
38925 Asparagus altissimus Munby var. asperulus Maire;粗糙芳香木●☆
38926 Asparagus altissimus Munby var. foeniculaceus (Lowe) Maire = Asparagus altissimus Munby ■☆
38927 Asparagus angolensis Baker = Asparagus laricinus Burch. ■☆
38928 Asparagus angulofractus Iljin;折枝天门冬(荒漠贝母,准噶尔天门冬);Bentshoot Asparagus ■
38929 Asparagus angulofractus Iljin var. scabridus Kitag. = Asparagus gobicus Ivan. ex Grubov ■
38930 Asparagus angusticladus (Jessop) Fellingham et N. L. Mey. = Asparagus angusticladus (Jessop) J. -P. Lebrun et Stork ■☆
38931 Asparagus angusticladus (Jessop) J. -P. Lebrun et Stork;细枝天门冬■☆
38932 Asparagus aphyllus L.;无叶天门冬■☆
38933 Asparagus arborescens Willd.;树状天门冬■☆
38934 Asparagus aridicola Sebsebe;旱生天门冬■☆
38935 Asparagus asiaticus L. subsp. amharicus Pic. Serm. = Asparagus setaceus (Kunth) Jessop ■
38936 Asparagus asiaticus L. var. mitis (A. Rich.) Chiov. = Asparagus africanus Lam. ■☆
38937 Asparagus asiaticus L. var. scaberulus (A. Rich.) Engl. = Asparagus scaberulus A. Rich. ■☆
38938 Asparagus asparagoides (L.) Druce;拟天冬草(卵叶天门冬);African Asparagus Fern, Cape Smilax, Smilax, Smilax Asparagus, Smilax of Florists ■☆
38939 Asparagus asparagoides W. Wight = Asparagus asparagoides (L.) Druce ■☆
38940 Asparagus aspergillus Jessop;刷状天门冬■☆
38941 Asparagus baguirmiensis A. Chev.;巴地天门冬■☆
38942 Asparagus baumii Engl. et Gilg;鲍姆天门冬■☆
38943 Asparagus bayeri (Oberm.) Fellingham et N. L. Mey.;巴耶尔天门冬■☆
38944 Asparagus bechuanicus Baker;贝专天门冬■☆
38945 Asparagus benguellensis Baker;本格拉天门冬■☆
38946 Asparagus bequaertii De Wild.;贝卡尔天门冬■☆
38947 Asparagus biflorus (Oberm.) Fellingham et N. L. Mey.;双花天门冬■☆
38948 Asparagus borealis S. C. Chen;北天门冬;Northern Asparagus ■
38949 Asparagus brachiatus Thulin;短天门冬■☆
38950 Asparagus brachyphyllus Turcz.;短叶天门冬(椴叶石刁柏,海滨天冬,寄马桩,攀援天门冬);Climbing Asparagus ■
38951 Asparagus breslerianus Schult. et Schult. f.;西天门冬■
38952 Asparagus buchananii Baker;布坎南天门冬■☆
38953 Asparagus bucharicus Iljin;布哈尔天门冬■☆
38954 Asparagus burchellii Baker;伯切尔天门冬■☆
38955 Asparagus burkei Baker = Asparagus cooperi Baker ■☆
38956 Asparagus buruensis Engl.;布鲁天门冬■☆
38957 Asparagus capensis L.;好望角天门冬■☆
38958 Asparagus capensis L. var. litoralis Suess. et Karl;滨海好望角天门冬■☆
38959 Asparagus clareae (Oberm.) Fellingham et N. L. Mey.;克莱尔天门冬■☆
38960 Asparagus cochinchinensis (Lour.) Merr.;天门冬(八百崽,白罗杉,波罗树,大当门根,地门冬,颠棘,颠勒,多儿母,多仔婆,肥天冬,赶条蛇,管松浣草,假天冬,金华,九子十弟,老虎尾巴,满冬,虋冬,明天冬,倪铃,牛虱仔,蘠蘼,三百棒,商棘,十二根,石薯子,丝冬,天冬,天冬草,天棘,天门,天虋冬,天文冬,万岁藤,无不愈,小叶青,筳门冬);Chinese Asparagus,Cochichinense Asparagus,Indochina Asparagus ■

38961 Asparagus cochinchinensis (Lour.) Merr. var. dolichoclados (Merr. et Rolfe) F. T. Wang et Ts. Tang = Asparagus cochinchinensis (Lour.) Merr. ■

38962 Asparagus cochinchinensis (Lour.) Merr. var. longifoliatus F. T. Wang et Ts. Tang = Asparagus cochinchinensis (Lour.) Merr. ■

38963 Asparagus cochinchinensis (Lour.) Merr. var. longifolius F. T. Wang et Ts. Tang = Asparagus cochinchinensis (Lour.) Merr. ■

38964 Asparagus cochinchinensis (Lour.) Merr. var. lucidus (Lindl.) Hatus. = Asparagus lucidus Lindl. ■

38965 Asparagus cochinchinensis (Lour.) Merr. var. pygmaeus (Makino) Ohwi = Asparagus pygmaeus Makino ■☆

38966 Asparagus coddii (Oberm.) Fellingham et N. L. Mey.;科德天门冬■☆

38967 Asparagus compactus T. M. Salter = Asparagus lignosus Burm. f. ■☆

38968 Asparagus concinnus (Baker) Kies;整洁天门冬■☆

38969 Asparagus confertus K. Krause;密集天门冬■☆

38970 Asparagus conglomeratus Baker = Asparagus cooperi Baker ■☆

38971 Asparagus consanguineus (Kunth) Baker = Asparagus fasciculatus Thunb. ■☆

38972 Asparagus cooperi Baker;库珀天门冬;Cooper Asparagus ■☆

38973 Asparagus crassicladus Jessop;粗枝天门冬■☆

38974 Asparagus crispus Lam.;香天冬草;Drooping Asparagus, Fragrant Asparagus ■☆

38975 Asparagus crispus Lam. = Asparagus declinatus L. ■☆

38976 Asparagus cuscutoides Burch. ex Baker = Drimia cuscutoides (Burch. ex Baker) J. C. Manning et Goldblatt ■☆

38977 Asparagus dauricus Fisch. ex Link;兴安天门冬(镰叶天门冬,山天冬);Dahur Asparagus,Dahurian Asparagus ■

38978 Asparagus dauricus Fisch. ex Link var. elongatus Pamp. = Asparagus cochinchinensis (Lour.) Merr. ■

38979 Asparagus debilis A. Chev.;弱小天门冬■☆

38980 Asparagus declinatus L.;外曲天门冬■☆

38981 Asparagus decumbens Jacq. = Asparagus crispus Lam. ■☆

38982 Asparagus deflexus Baker;外折天门冬■

38983 Asparagus densiflorus (Kunth) Jessop;密花天门冬(非洲天门冬,密花天冬,天冬草,万年青);Africa Asparagus, African Asparagus,Asparagus Fern,Myers Asparagus,Sprenger's Asparagus Fern ■

38984 Asparagus densiflorus (Kunth) Jessop 'Myers';狐尾天冬(狐尾武竹)■☆

38985 Asparagus densiflorus (Kunth) Jessop 'Sprengeri';天冬草(施氏天门冬,天冬);Asparagus Fern, Emerald-fern, Garden Asparagus, Myers Asparagus, Sprenger Asparagus, Sprenger Asparagus-fern ■

38986 Asparagus densiflorus (Kunth) Jessop var. myriocladus Hort.;多枝文竹;Manyshoot Asparagus ■☆

38987 Asparagus densiflorus (Kunth) Jessop var. pyramidalis Hort.;塔形天冬;Pyramidal Asparagus ■☆

38988 Asparagus densiflorus (Kunth) Jessop var. sprrengeri Hort. = Asparagus densiflorus (Kunth) Jessop 'Sprengeri' ■

38989 Asparagus densus Sol. ex Baker = Asparagus stipulaceus Lam. ■☆

38990 Asparagus denudatus (Kunth) Baker;裸露天门冬■☆

38991 Asparagus denudatus (Kunth) Baker subsp. nudicaulis (Baker) Sebsebe;裸茎天门冬■

38992 Asparagus dependens Thunb. = Asparagus cooperi Baker ■☆

38993 Asparagus dinteri Engl. et K. Krause = Asparagus cooperi Baker ■☆

38994 Asparagus divaricatus (Oberm.) Fellingham et N. L. Mey.;叉开天门冬■

38995 Asparagus dolichorhizhomatus J. M. Ni et R. N. Zhao;长根茎天门冬■

38996 Asparagus draco L. = Dracaena draco (L.) L. ●

38997 Asparagus drepanophyllus Welw. ex Baker;镰叶天门冬;Falcateleaf Asparagus ■☆

38998 Asparagus drepanophyllus Welw. ex Baker var. warneckei Engl. = Asparagus warneckei (Engl.) Hutch. ■☆

38999 Asparagus duchesnei L. Linden;迪谢纳天门冬■☆

39000 Asparagus ecklonii Baker = Asparagus exuvialis Burch. f. ecklonii (Baker) Fellingham et N. L. Mey. ■☆

39001 Asparagus edulis (Oberm.) Fellingham et N. L. Mey. = Asparagus edulis (Oberm.) J. -P. Lebrun et Stork ■☆

39002 Asparagus edulis (Oberm.) J. -P. Lebrun et Stork;可食天门冬■☆

39003 Asparagus engleri K. Krause = Asparagus striatus (L. f.) Thunb. ■☆

39004 Asparagus equisetoides Welw. ex Baker;木贼天门冬■☆

39005 Asparagus erectus (L. f.) Thunb. = Asparagus striatus (L. f.) Thunb. ■☆

39006 Asparagus exsertus (Oberm.) Fellingham et N. L. Mey.;伸出天门冬■☆

39007 Asparagus exuvialis Burch.;剥落天门冬■☆

39008 Asparagus exuvialis Burch. f. ecklonii (Baker) Fellingham et N. L. Mey.;埃氏天门冬■☆

39009 Asparagus falcatus L.;优香天门冬(镰叶天冬);Falcated-leaved Asparagus,Sweet-scented Asparagus ■☆

39010 Asparagus falcatus L. var. ternifolius (Baker) Jessop = Asparagus falcatus L. ■☆

39011 Asparagus fallax Svent.;迷惑天门冬■☆

39012 Asparagus fasciculatus Thunb.;簇生天门冬■☆

39013 Asparagus faulkneri Sebsebe;福克纳天门冬■☆

39014 Asparagus filicinus Buch. -Ham. ex D. Don;羊齿天门冬(百部,滇百部,广麦冬,九重根,蕨叶天门冬,丽江百部,蓬莱竹,千锤打,石刁柏,天门冬,土百部,小百部,小天冬,月牙一枝蒿);Fernlike Asparagus ●■

39015 Asparagus filicinus Buch. -Ham. ex D. Don var. brevifolius Diels = Asparagus filicinus Buch. -Ham. ex D. Don ●■

39016 Asparagus filicinus Buch. -Ham. ex D. Don var. brevipes Baker = Asparagus lycopodineus Wall. ex Baker ■

39017 Asparagus filicinus Buch. -Ham. ex D. Don var. giraldii C. H. Wright = Asparagus filicinus Buch. -Ham. ex D. Don ●■

39018 Asparagus filicinus Buch. -Ham. ex D. Don var. lycopodineus Baker = Asparagus lycopodineus Wall. ex Baker ■

39019 Asparagus filicinus Buch. -Ham. ex D. Don var. megaphyllus F. T. Wang et Ts. Tang = Asparagus filicinus Buch. -Ham. ex D. Don ●■

39020 Asparagus filicladus (Oberm.) Fellingham et N. L. Mey.;丝枝天门冬■☆

39021 Asparagus flagellaris (Kunth) Baker;鞭状天门冬■☆

39022 Asparagus flavicaulis (Oberm.) Fellingham et N. L. Mey.;黄茎天门冬■☆

39023 Asparagus fleckii Schinz = Asparagus cooperi Baker ■☆

39024 Asparagus foeniculaceus Lowe = Asparagus altissimus Munby ■☆

39025 Asparagus fourei (Oberm.) Fellingham et N. L. Mey.;富里耶天

门冬■☆

39026 Asparagus fractiflexus (Oberm.) Fellingham et N. L. Mey.;弯曲天门冬■☆

39027 Asparagus francisci K. Krause = Asparagus cooperi Baker ■☆

39028 Asparagus gaudichaudianus Kunth = Asparagus cochinchinensis (Lour.) Merr. ■

39029 Asparagus gibbus Bunge = Asparagus dauricus Fisch. ex Link ■

39030 Asparagus gillettii Chiov. = Asparagus leptocladodius Chiov. ■☆

39031 Asparagus glaucus Kies;灰绿天门冬■☆

39032 Asparagus gobicus Ivan. ex Grubov;戈壁天门冬(戈壁天冬,鸡麻抓,寄马桩,寄子桩);Desertliving Asparagus ■

39033 Asparagus gourmacus A. Chev. = Asparagus africanus Lam. ■☆

39034 Asparagus graminifolius L. = Liriope graminifolia (L.) Baker ■

39035 Asparagus graniticus (Oberm.) Fellingham et N. L. Mey.;花岗岩天门冬■☆

39036 Asparagus greveanus H. Perrier;格雷弗天门冬■☆

39037 Asparagus hereroensis Schinz = Asparagus exuvialis Burch. ■☆

39038 Asparagus homblei De Wild.;洪布勒天门冬■☆

39039 Asparagus horridus L. f. = Asparagus stipularis Forssk. ■☆

39040 Asparagus humilis Engl.;低矮天门冬■☆

39041 Asparagus insularis Hance = Asparagus cochinchinensis (Lour.) Merr. ■

39042 Asparagus intangibilis Dinter = Asparagus suaveolens Burch. ■☆

39043 Asparagus intricatus (Oberm.) Fellingham et N. L. Mey.;缠结天门冬■☆

39044 Asparagus irregularis Baker = Asparagus africanus Lam. ■☆

39045 Asparagus judtii Schinz = Asparagus cooperi Baker ■☆

39046 Asparagus juniperoides Engl.;刺柏状天门冬■☆

39047 Asparagus kaessneri De Wild.;卡斯纳天门冬■☆

39048 Asparagus kansuensis F. T. Wang et Ts. Tang ex S. C. Chen;甘肃天门冬;Gansu Asparagus, Kansu Asparagus ■

39049 Asparagus kasakstanicus Iljin;哈萨克斯坦天门冬■☆

39050 Asparagus katangensis De Wild. et T. Durand;加丹加天门冬■☆

39051 Asparagus kiusianus Makino;九州天门冬■☆

39052 Asparagus klinghardtianus Dinter = Asparagus undulatus (L. f.) Thunb. ■☆

39053 Asparagus krausii Baker = Asparagus kraussianus (Kunth) J. F. Macbr. ■☆

39054 Asparagus kraussianus (Kunth) J. F. Macbr.;克劳斯天门冬■☆

39055 Asparagus krebsianus (Kunth) Jessop;克雷布斯天门冬■☆

39056 Asparagus kuisibensis Dinter = Asparagus asparagoides (L.) Druce ■☆

39057 Asparagus lanceus Thunb. = Asparagus aethiopicus L. ■

39058 Asparagus laricinus Burch.;南非天门冬;South African Asparagus ■☆

39059 Asparagus lecardii De Wild.;莱卡德天门冬■☆

39060 Asparagus leptocladodius Chiov.;非洲细枝天门冬■☆

39061 Asparagus leviniae Kink.;莱维天门冬■☆

39062 Asparagus lignosus Burm. f.;木质天门冬■☆

39063 Asparagus linearis (Burm. f.) R. Dahlgren;线形天门冬■☆

39064 Asparagus littoralis Stev.;海岸天门冬■☆

39065 Asparagus longicladus N. E. Br.;长枝天门冬■☆

39066 Asparagus longiflorus Franch.;长花天门冬;Longflower Asparagus ■

39067 Asparagus longipes Baker;长梗天门冬■☆

39068 Asparagus lowei Kunth = Asparagus umbellatus Link subsp. lowei (Kunth) Valdés ■☆

39069 Asparagus lucidus Lindl.;光亮天门冬(天门冬)■

39070 Asparagus lucidus Lindl. = Asparagus cochinchinensis (Lour.) Merr. ■

39071 Asparagus lucidus Lindl. var. pygmaeus Makino = Asparagus pygmaeus Makino ■☆

39072 Asparagus lugardii Baker = Asparagus cooperi Baker ■☆

39073 Asparagus lujae De Wild.;卢加天门冬■☆

39074 Asparagus lutzii Hort.;卢茨天门冬■☆

39075 Asparagus lycopodineus Wall. ex Baker;短梗天门冬(滇百部,山百部,山漏芦,山扫帚,石松状天门冬,铁扫把,土百部,乌小天冬,一窝鸡);Shortpedicel Asparagus ■

39076 Asparagus lycopodineus Wall. ex Baker var. sessilis F. T. Wang et Ts. Tang = Asparagus lycopodineus Wall. ex Baker ■

39077 Asparagus macowanii Baker;麦克欧文天门冬■☆

39078 Asparagus macowanii Baker var. zuluensis (N. E. Br.) Jessop = Asparagus macowanii Baker ■☆

39079 Asparagus madagascariensis Baker;马达加斯加天门冬■☆

39080 Asparagus madagascariensis Baker = Asparagus simulans Baker ■☆

39081 Asparagus madecassus H. Perrier;马德卡萨天门冬■☆

39082 Asparagus mairei H. Lév.;昆明天门冬;Kunming Asparagus, Maire Asparagus ■

39083 Asparagus mariae (Oberm.) Fellingham et N. L. Mey.;玛利亚天门冬■☆

39084 Asparagus maritimus (L.) Mill.;海天冬■☆

39085 Asparagus medeoloides (L. f.) Thunb. = Asparagus asparagoides W. Wight ■☆

39086 Asparagus medeoloides (Thunb.) Baker var. angustifolius (Mill.) Baker = Asparagus asparagoides (L.) Druce ■☆

39087 Asparagus meioclados H. Lév.;密齿天门冬(天冬,小茎叶天门冬,小天冬);Denthtooth Asparagus ■

39088 Asparagus meioclados H. Lév. var. trichoclados F. T. Wang et Ts. Tang = Asparagus trichoclados (F. T. Wang et Ts. Tang) F. T. Wang et S. C. Chen ■

39089 Asparagus merkeri K. Krause;默克天门冬■☆

39090 Asparagus meyeri Harv. = Asparagus densiflorus (Kunth) Jessop 'Myers' ■☆

39091 Asparagus microphyllus Burch. ex Baker = Asparagus retrofractus L. ■☆

39092 Asparagus microraphis (Kunth) Baker;小天门冬■☆

39093 Asparagus migeodii Sebsebe;米容德天门冬■☆

39094 Asparagus minutiflorus (Kunth) Baker;微花天门冬■☆

39095 Asparagus misczenkoi Iljin;米谢天门冬■☆

39096 Asparagus mitis A. Rich. = Asparagus africanus Lam. ■☆

39097 Asparagus mollis (Oberm.) Fellingham et N. L. Mey.;柔毛天门冬■☆

39098 Asparagus mozambicus Kunth;莫桑比克天门冬■☆

39099 Asparagus mucronatus Jessop;短尖天门冬■☆

39100 Asparagus multiflorus Baker;多花天门冬■☆

39101 Asparagus multituberosus R. A. Dyer;多瘤天门冬■☆

39102 Asparagus munitus F. T. Wang et S. C. Chen;西南天门冬;Provided Asparagus ■

39103 Asparagus myersii ? = Asparagus densiflorus (Kunth) Jessop 'Myers' ■☆

39104 Asparagus myriacanthus F. T. Wang et S. C. Chen;多棘天门冬(多刺天门冬);Manyspine Asparagus, Manythorny Asparagus ■

39105 Asparagus myriocladus Baker;松叶文竹■☆

39106 Asparagus myriocladus Baker = Asparagus densiflorus (Kunth)

Jessop ■
39107 Asparagus namaensis Schinz = Asparagus retrofractus L. ■☆
39108 Asparagus natalensis (Baker) Fellingham et N. L. Mey. = Asparagus natalensis (Baker) J. -P. Lebrun et Stork ■☆
39109 Asparagus natalensis (Baker) J. -P. Lebrun et Stork;纳塔尔天门冬■☆
39110 Asparagus ndelleensis A. Chev. ;恩代尔天门冬■☆
39111 Asparagus neglectus Kar. et Kir. ;新疆天门冬;Sinkiang Asparagus,Xinjiang Asparagus ■
39112 Asparagus nelsii Schinz;内尔斯天门冬■☆
39113 Asparagus nelsonii Baker = Asparagus capensis L. ■☆
39114 Asparagus nivenianus Schult. et Schult. f. = Asparagus rubicundus P. J. Bergius ■☆
39115 Asparagus nodosus Sol. ex Baker = Adenogramma teretifolia (Thunb.) Adamson ■☆
39116 Asparagus nodulosus (Oberm.) Fellingham et N. L. Mey. = Asparagus nodulosus (Oberm.) J. -P. Lebrun et Stork ■☆
39117 Asparagus nodulosus (Oberm.) J. -P. Lebrun et Stork;多节天门冬■☆
39118 Asparagus nudicaulis Baker = Asparagus denudatus (Kunth) Baker subsp. nudicaulis (Baker) Sebsebe ■
39119 Asparagus obermeyerae Jessop = Asparagus schroederi Engl. ■☆
39120 Asparagus officinalis L. ;石刁柏(多叶天门冬,龙须菜,露笋,芦笋,门冬薯,山文竹,索罗罗,细叶百部,小百部);Asparagus,Asparagus Fern,Common Asparagus,Culinary Asparagus,Garden Asparagus,Paddock Cheese,Sea Asparagus,Sparage,Sparagus,Sparra Grace,Sparra Grass,Sparrow-grass,Sperage ■
39121 Asparagus officinalis L. var. altilis L. = Asparagus officinalis L. ■
39122 Asparagus officinalis L. var. campestris Gren. et Godr. = Asparagus officinalis L. ■
39123 Asparagus officinalis L. var. maritimus L. = Asparagus maritimus (L.) Mill. ■☆
39124 Asparagus officinalis L. var. prostratus ?;平卧天门冬;Wild Asparagus ■☆
39125 Asparagus officinalis L. var. tenuifolius L. = Asparagus tenuifolius Lam. ■☆
39126 Asparagus oligoclonos Maxim. ;南玉带(南立带);Fewtwig Asparagus ■
39127 Asparagus oligoclonos Maxim. var. purpurascens X. J. Xue et H. Yao;紫花南玉带;Purpleflower Fewtwig Asparagus ■
39128 Asparagus oligoclonos Maxim. var. purpurascens X. J. Xue et H. Yao = Asparagus oligoclonos Maxim. ■
39129 Asparagus oliveri (Oberm.) Fellingham et N. L. Mey. ;奥里弗天门冬■☆
39130 Asparagus omahekensis K. Krause = Asparagus suaveolens Burch. ■☆
39131 Asparagus ovatus T. M. Salter;卵形天门冬■☆
39132 Asparagus oxyacanthus Baker;尖刺天门冬■☆
39133 Asparagus pallasii Printz;多刺龙须菜■
39134 Asparagus parviflorus Turcz. = Asparagus schoberioides Kunth ■
39135 Asparagus pastorianus Webb et Berthel. ;牧场天门冬■☆
39136 Asparagus patens K. Krause = Asparagus cooperi Baker ■☆
39137 Asparagus pauli-guilelmii Solms = Asparagus flagellaris (Kunth) Baker ■☆
39138 Asparagus pearsonii Kies;皮尔逊天门冬■☆
39139 Asparagus pendulus (Oberm.) Fellingham et N. L. Mey. = Asparagus pendulus (Oberm.) J. -P. Lebrun et Stork ■☆
39140 Asparagus pendulus (Oberm.) J. -P. Lebrun et Stork;下垂天门冬■☆
39141 Asparagus persicus Baker;西北天门冬;Iran Asparagus ■
39142 Asparagus petersianus Kunth;彼得斯天门冬■☆
39143 Asparagus petitianus A. Rich. = Asparagus racemosus Willd. ■
39144 Asparagus pilosus Baker = Asparagus africanus Lam. var. puberulus (Baker) Sebsebe ■☆
39145 Asparagus planiusculus Burm. f. ;稍扁天门冬■☆
39146 Asparagus plocamoides Webb ex Svent. ;卷毛茜天门冬■☆
39147 Asparagus plumosus Baker = Asparagus setaceus (Kunth) Jessop ■
39148 Asparagus plumosus Baker var. comorensis Hort. ;纤美文竹;Delicate Asparagus-fren ■☆
39149 Asparagus plumosus Baker var. compactus Hort. ;密丛文竹;Compact Asparagus-fren ■☆
39150 Asparagus plumosus Baker var. nanus Nicholson;矮文竹(鸡绒芒);Dwarf Asparagus-fren ■☆
39151 Asparagus plumosus Baker var. robustus Hort. ;粗壮文竹;Robust Asparagus-fren ■☆
39152 Asparagus poissonii H. Perrier;普瓦松天门冬■☆
39153 Asparagus polyphyllus Stev. = Asparagus officinalis L. ■
39154 Asparagus popovii Iljin;波氏天门冬■☆
39155 Asparagus przewalskyi N. A. Ivanova ex Grubov et T. V. Egorova;颇氏天门冬(北天门冬);Przewalsky Asparagus-fren ■
39156 Asparagus pseudofilicinus F. T. Wang et Ts. Tang;丽江百部(小天冬);Lijiang Asparagus ■
39157 Asparagus pseudoscaber Grecescu;粗糙天门冬■☆
39158 Asparagus puberulus Baker = Asparagus africanus Lam. var. puberulus (Baker) Sebsebe ■☆
39159 Asparagus pubescens Baker;短柔毛天门冬■
39160 Asparagus pubescens Baker = Asparagus africanus Lam. var. puberulus (Baker) Sebsebe ■☆
39161 Asparagus pygmaeus Makino;直立天门冬(立天门冬)■☆
39162 Asparagus pyramidalis ? = Asparagus setaceus (Kunth) Jessop ■
39163 Asparagus qinghaiensis Y. Wan;青海天门冬;Qinghai Asparagus ■
39164 Asparagus qinghaiensis Y. Wan = Asparagus filicinus Buch. -Ham. ex D. Don ●■
39165 Asparagus racemosus Willd. ;总序天冬(长刺天门冬,喜马拉雅天冬,总状花天冬)■
39166 Asparagus racemosus Willd. var. longicladodius Chiov. = Asparagus racemosus Willd. ■
39167 Asparagus racemosus Willd. var. ruspolii Engl. = Asparagus leptocladodius Chiov. ■☆
39168 Asparagus racemosus Willd. var. tetragonus (Bresler) Baker = Asparagus aethiopicus L. ■
39169 Asparagus ramosissimus Baker;多枝天门冬■☆
39170 Asparagus recurvispinus (Oberm.) Fellingham et N. L. Mey. ;反折天门冬■☆
39171 Asparagus retrofractus L. ;绣球松(反曲天冬,蓬莱松);Asparagus Fern,Fern Asparagus,Ming Asparagus,Refracted Asparagus ■☆
39172 Asparagus retrofractus Schousb. = Asparagus pastorianus Webb et Berthel. ■☆
39173 Asparagus rigidus Jessop;硬天门冬■☆
39174 Asparagus ritschardii De Wild. ;里恰德天门冬■☆
39175 Asparagus rivalis Burch. ex Kies = Asparagus cooperi Baker ■☆
39176 Asparagus rogersii R. E. Fr. ;罗杰斯天门冬■☆
39177 Asparagus ruber Burm. f. = Asparagus rubicundus P. J. Bergius ■☆

39178 Asparagus rubicundus P. J. Bergius;稍红天门冬■☆
39179 Asparagus sapinii De Wild.;萨潘天门冬■☆
39180 Asparagus sarmentosus L.;匍茎天冬■☆
39181 Asparagus sarmentosus L. var. densiflorus (Kunth) Baker = Asparagus densiflorus (Kunth) Jessop ■
39182 Asparagus saundersiae Baker = Asparagus racemosus Willd. ■
39183 Asparagus scaber Brign. = Asparagus maritimus (L.) Mill. ■☆
39184 Asparagus scaberulus A. Rich.;微糙天门冬■☆
39185 Asparagus scandens Thunb.;攀缘天冬(蔓天冬);Basket Asparagus,Scandent Asparagus ■☆
39186 Asparagus scandens Thunb. var. deflexus Baker;外折天冬;Deflexed Asparagus ■☆
39187 Asparagus schoberioides Kunth;龙须菜(玉带天门冬,雉隐天冬);Schoberia-like Asparagus ■
39188 Asparagus schoberioides Kunth var. subsetaceus Franch. = Asparagus schoberioides Kunth ■
39189 Asparagus schroederi Engl.;施罗德天门冬■☆
39190 Asparagus schweinfurthii Baker = Asparagus flagellaris (Kunth) Baker ■☆
39191 Asparagus scoparius Ball = Asparagus altissimus Munby ■☆
39192 Asparagus scoparius Lowe;帚状天门冬■☆
39193 Asparagus sekukuniensis (Oberm.) Fellingham et N. L. Mey.;塞库库尼天门冬■☆
39194 Asparagus sennii Chiov. = Asparagus africanus Lam. ■☆
39195 Asparagus setaceus (Kunth) Jessop;文竹(蓬莱竹,小百部,羽毛天门冬);Asparagus Fern,Asparagus-fern,Climbing Asparagus Fern,Climbing Asparagus-fern,Common Asparagus-fern,Fern Asparagus,King's Spear,Lace Fern,Lace-fern,Plume Fern,Setose Asparagus,Sprenger Asparagus ■
39196 Asparagus setiformis Krylov;刚毛天门冬■☆
39197 Asparagus shirensis Baker = Asparagus africanus Lam. var. puberulus (Baker) Sebsebe ■☆
39198 Asparagus sichuanicus S. C. Chen et D. Q. Liu;四川天门冬;Sichuan Asparagus ■
39199 Asparagus sidamensis Cufod. = Asparagus africanus Lam. ■☆
39200 Asparagus sieboldi Maxim. = Asparagus schoberioides Kunth ■
39201 Asparagus simulans Baker;相似天门冬■☆
39202 Asparagus sinicus (Miq.) C. H. Wright = Asparagus cochinchinensis (Lour.) Merr. ■
39203 Asparagus somalensis Chiov. = Asparagus flagellaris (Kunth) Baker ■☆
39204 Asparagus soongoricus Iljin = Asparagus angulofractus Iljin ■
39205 Asparagus spinescens Steud. ex Roem. et Schult.;细刺天门冬■☆
39206 Asparagus spinosissimus Dinter = Asparagus capensis L. ■☆
39207 Asparagus spinosissimus F. T. Wang et S. C. Chen = Asparagus myriacanthus F. T. Wang et S. C. Chen ■
39208 Asparagus spinosissimus Kuntze = Asparagus suaveolens Burch. ■☆
39209 Asparagus sprengeri Regel = Asparagus aethiopicus L. ■
39210 Asparagus sprengeri Regel = Asparagus densiflorus (Kunth) Jessop 'Sprengeri' ■
39211 Asparagus sprengeri Regel = Asparagus densiflorus (Kunth) Jessop ■
39212 Asparagus squarrosus J. A. Schmidt;粗鳞天门冬■☆
39213 Asparagus stachyoides Spreng. ex Baker = Asparagus aethiopicus L. ■
39214 Asparagus stellatus Baker;星状天门冬■☆
39215 Asparagus stipulaceus Lam.;托叶状天门冬■☆
39216 Asparagus stipularis Forssk.;托叶天门冬■☆
39217 Asparagus stipularis Forssk. var. brachyclados Boiss. = Asparagus stipularis Forssk. ■☆
39218 Asparagus stipularis Forssk. var. horridus (L. f.) Maire et Weiller = Asparagus stipularis Forssk. ■☆
39219 Asparagus striatus (L. f.) Thunb.;条纹天门冬■☆
39220 Asparagus striatus De Wild. = Asparagus schroederi Engl ■☆
39221 Asparagus suaveolens Burch.;多刺天门冬(多棘天门冬)■☆
39222 Asparagus subfalcatus De Wild.;亚镰形天门冬■☆
39223 Asparagus subscandens F. T. Wang et S. C. Chen;滇南天门冬(稍攀援天门冬,天门冬,土天冬,小茎叶天冬);Climbinglike Asparagus,South Yunnan Asparagus ■
39224 Asparagus subulatus Thunb.;钻形天门冬■☆
39225 Asparagus taliensis F. T. Wang et Ts. Tang ex S. C. Chen;大理天门冬;Dali Asparagus,Tali Asparagus ■
39226 Asparagus tamaboki Yatabe = Asparagus oligoclonos Maxim. ■
39227 Asparagus tamariscinus Ivan. ex Grubov = Asparagus oligoclonos Maxim. ■
39228 Asparagus tamariscinus Ivan. ex Grubov = Asparagus persicus Baker ■
39229 Asparagus tenuifolius Lam.;细叶天门冬■☆
39230 Asparagus terminalis L. = Cordyline fruticosa (L.) A. Chev. ●
39231 Asparagus ternifolius (Baker) Hook. f. = Asparagus densiflorus (Kunth) Jessop ■
39232 Asparagus tetragonus Bresler = Asparagus aethiopicus L. ■
39233 Asparagus thunbergianus Schult. = Asparagus rubicundus P. J. Bergius ■☆
39234 Asparagus tibeticus F. T. Wang et S. C. Chen;西藏天门冬;Tibet Asparagus,Xizang Asparagus ■
39235 Asparagus transvaalensis (Oberm.) Fellingham et N. L. Mey.;德兰士瓦天门冬■☆
39236 Asparagus triacanthus Burm. f. = Asparagus capensis L. ■☆
39237 Asparagus triacanthus Roem. et Schult. = Asparagus suaveolens Burch. ■☆
39238 Asparagus trichoclados (F. T. Wang et Ts. Tang) F. T. Wang et S. C. Chen;毛枝天门冬(霸天王,糙叶天冬,毛叶天冬,细枝天门冬,抓地龙);Slenderbranch Asparagus ■
39239 Asparagus trichophyllus Bunge;曲枝天门冬;Hairyleaf Asparagus ■
39240 Asparagus trichophyllus Bunge var. trachyphyllus Kunth = Asparagus brachyphyllus Turcz. ■
39241 Asparagus tuberculatus Bunge = Asparagus dauricus Fisch. ex Link ■
39242 Asparagus turkestanicus Popov;土耳其斯坦天门冬■☆
39243 Asparagus uhligii K. Krause;乌里希天门冬■☆
39244 Asparagus umbellatus Link;小伞天门冬■☆
39245 Asparagus umbellatus Link subsp. lowei (Kunth) Valdés;洛氏天门冬■☆
39246 Asparagus umbellatus Link var. flavescens Svent. = Asparagus umbellatus Link ■☆
39247 Asparagus undulatus (L. f.) Thunb.;波状天门冬■☆
39248 Asparagus usambarensis Sebsebe;乌桑巴拉天门冬■☆
39249 Asparagus vaginellatus Bojer ex Baker;具鞘天门冬■☆
39250 Asparagus vanioti H. Lév. = Asparagus meiocladus H. Lév. ■
39251 Asparagus verticillatus L.;轮生天门冬(轮生天冬)■☆
39252 Asparagus vincentinus Welw. ex Cout. = Asparagus squarrosus J. A. Schmidt ■☆
39253 Asparagus virgatus Baker;扫状天门冬(细枝天冬)■☆

39254 Asparagus virgatus Baker var. capillaris ? = Asparagus virgatus Baker ■☆
39255 Asparagus volubilis Thunb.;缠绕天门冬■☆
39256 Asparagus warneckei (Engl.) Hutch.;沃内克天门冬■☆
39257 Asparagus wildemanii Weim. = Asparagus schroederi Engl. ■☆
39258 Asparagus yanbianensis S. C. Chen;盐边天门冬;Yanbian Asparagus ■
39259 Asparagus yanyuanensis S. C. Chen;盐源天门冬;Yanyuan Asparagus ■
39260 Asparagus yuanus F. T. Wang et Ts. Tang;俞氏天门冬(于氏天门冬);Yu Asparagus ■
39261 Asparagus yunnanensis H. Lév. = Asparagus meioclados H. Lév. ■
39262 Asparagus zanzibaricus Baker = Asparagus setaceus (Kunth) Jessop ■
39263 Asparagus zeyheri Kunth = Asparagus suaveolens Burch. ■☆
39264 Asparagus zuluensis N. E. Br. = Asparagus macowanii Baker ■☆
39265 Aspasia E. Mey. = Stachys L. ●■
39266 Aspasia E. Mey. ex Pfeiff. = Stachys L. ●■
39267 Aspasia Lindl. (1832);美乐兰属■☆
39268 Aspasia Salisb. = Ornithogalum L. ■
39269 Aspasia aurea Salisb.;黄美乐兰■☆
39270 Aspasia epidendroides Lindl.;柱瓣兰状美乐兰■☆
39271 Aspasia epidendroides Lindl. var. principissa (Rchb. f.) P. H. Allen = Aspasia principissa Rchb. f. ■☆
39272 Aspasia fragrans Klotzsch = Aspasia epidendroides Lindl. ■☆
39273 Aspasia lunata Lindl.;斑舌美乐兰■☆
39274 Aspasia principissa Rchb. f.;大花美乐兰■☆
39275 Aspasia rousseauae Schltr. = Aspasia principissa Rchb. f. ■☆
39276 Aspasia variegata Lindl.;彩色美乐兰■☆
39277 Aspazoma N. E. Br. (1925);大花日中花属●☆
39278 Aspazoma amplectens (L. Bolus) N. E. Br.;大花日中花■☆
39279 Aspegrenia Poepp. et Endl. = Octomeria R. Br. ■☆
39280 Aspelina Cass. = Senecio L. ●■
39281 Aspera Columna ex Moench = Galium L. ●■
39282 Aspera Moench = Galium L. ●■
39283 Asperella Humb. = Hystrix Moench ■
39284 Asperella Schreb. = Hystrix Moench ■
39285 Asperella coreana (Honda) Nevski = Hystrix coreana (Honda) Ohwi ■
39286 Asperella duthiei Stapf ex Hook. f. = Hystrix duthiei (Stapf ex Hook. f.) Bor ■
39287 Asperella hystrix (L.) Humb. = Elymus hystrix L. ■☆
39288 Asperella hystrix (L.) Humb. var. bigeloviana Fernald = Elymus hystrix L. var. bigeloviana (Fernald) Bowden ■☆
39289 Asperella japonica Hack. = Hystrix duthiei (Stapf) Bor subsp. japonica (Hack.) Baden, Fred. et Seberg ■☆
39290 Asperella komarovii Roshev. = Hystrix komarovii (Roshev.) Ohwi ■
39291 Asperella longe-aristata (Hack.) Ohwi = Hystrix duthiei (Stapf) Bor subsp. longearistata (Hack.) Baden, Fred. et Seberg ■☆
39292 Asperellaceae Link = Gramineae Juss. (保留科名)●■
39293 Asperellaceae Link = Poaceae Barnhart(保留科名)●■
39294 Asperifoliaceae Rchb. = Boraginaceae Juss. (保留科名)●■
39295 Asperifoliae Batsch = Asperifoliaceae Rchb. ●■
39296 Asperuginoides Rauschert(1982);糙芥属■☆
39297 Asperuginoides axillaris (Boisset Hohen.) Rauschert;糙芥■☆
39298 Asperugo L. (1753);糙草属(糙芥属);German Madwort, Madwort, Roughstraw ■
39299 Asperugo procumbens L.;糙草(平卧糙草);German Madwort, German-madwort, Madderwort, Madwort, Roughstraw ■
39300 Asperula L. (1753)(保留属名);车叶草属;Asphodel, Woodruff ■
39301 Asperula L. = Galium L. ●■
39302 Asperula abchasica Krecz.;阿伯哈斯车叶草■☆
39303 Asperula aemulans Krecz. ex Klokov;匹敌车叶草■☆
39304 Asperula affinis Boiss. et Huet;近缘车叶草■☆
39305 Asperula albovii Manden.;阿氏车叶草■☆
39306 Asperula alpina M. Bieb.;高山车叶草■☆
39307 Asperula aparine Besser = Galium rivale (Sibth. et Sm.) Griseb. ■
39308 Asperula aparine M. Bieb. = Galium rivale (Sibth. et Sm.) Griseb. ■
39309 Asperula aristata L. f.;具芒车叶草■☆
39310 Asperula aristata L. f. subsp. longiflora (Waldst. et Kit.) Hayek;长花具芒车叶草■☆
39311 Asperula aristata L. f. subsp. scabra (J. Presl et C. Presl) Nyman = Asperula aristata L. f. ■☆
39312 Asperula aristata L. f. var. breviflora Batt. = Asperula aristata L. f. ■☆
39313 Asperula aristata L. f. var. longiflora (Waldst. et Kit.) Batt. = Asperula aristata L. f. subsp. longiflora (Waldst. et Kit.) Hayek ■☆
39314 Asperula arvensis L.;野车叶草;Blue Woodruff, Blue-flowered Woodrush ■☆
39315 Asperula attenuata Klokov;渐狭车叶草■☆
39316 Asperula azurea Jaub. et Spach = Asperula orientalis Boiss. et Hohen. ■
39317 Asperula azurea Jaub. et Spach var. setosa ?;刚毛蓝车叶草;Blue Brocade ■☆
39318 Asperula azurea-setosa ? = Asperula orientalis Boiss. et Hohen. ■
39319 Asperula beatica Rouy = Galium baeticum (Rouy) Ehrend. et Krendl ■☆
39320 Asperula bidentata Klokov;双齿车叶草■☆
39321 Asperula biebersteinii Krecz.;毕氏车叶草■☆
39322 Asperula calabrica L. f. = Putoria calabrica (L. f.) Pers. ●☆
39323 Asperula caucasica Pobed.;高加索车叶草■☆
39324 Asperula conferta (K. Koch) Stankov;紧密车叶草■☆
39325 Asperula conferta (K. Koch) Stankov = Asperula humifusa (M. Bieb.) Besser ■
39326 Asperula creticola Klokov;克里特车叶草■☆
39327 Asperula cristata (Sommier et H. Lév.) Krecz.;冠状车叶草■☆
39328 Asperula cynanchica L.;牛皮消车叶草(多枝阿福花,拟阿福花,小果阿福花,小果拟阿福花);Affodill, Dutch Daffodil, Herb of Vine, Herb Squinantyke, King's Spear, Quinsey-wort, Quinsywort, Royal Staff, Shepherd's Bedstraw, Silver Rod, Squinancy Woodruff, Squinancywort, Squinaney, Squinaney Woodrush, Tchirisch, Waxflower, White Affodill, White Asphodel, Woodruff ■☆
39329 Asperula cynanchica L. subsp. aristata (L. f.) Bég. = Asperula aristata L. f. ■☆
39330 Asperula cynanchica L. var. breviflora Batt. = Asperula aristata L. f. ■☆
39331 Asperula cynanchica L. var. gracilis Batt. = Asperula aristata L. f. ■☆
39332 Asperula cynanchica L. var. longiflora Gren. et Godr. = Asperula aristata L. f. ■☆
39333 Asperula cynanchica L. var. scabra (J. Presl et C. Presl) Lange = Asperula aristata L. f. ■☆

39334 Asperula cynanchica L. var. scabridula Batt. = Asperula aristata L. f. ■☆
39335 Asperula cyrenaica (E. A. Durand et Barratte) Pamp.;昔兰尼车叶草■☆
39336 Asperula danilewskiana Basiner = Asperula diminuta Klokov ■☆
39337 Asperula dasyantha Klokov;毛花车叶草■☆
39338 Asperula debilis Ledeb.;小车叶草■☆
39339 Asperula debilis Ledeb. = Asperula humifusa (M. Bieb.) Besser ■
39340 Asperula debilis Ledeb. = Galium humifusum (M. Bieb.) Besser ■
39341 Asperula diffusum Champ. = Coptosapelta diffusa (Champ. ex Benth.) Steenis ●
39342 Asperula diminuta Klokov;丹尼氏车叶草■☆
39343 Asperula dolichophylla Klokov;长叶车叶草■☆
39344 Asperula elongata Schrenk = Microphysa elongata (Schrenk) Pobed. ■
39345 Asperula exasperata V. I. Krecz.;糙刺车叶草■☆
39346 Asperula ferganica Pobed.;费尔干车叶草■☆
39347 Asperula galioides M. Bieb.;灰蓝车叶草;Bedstraw Asperula, Bedstruw Woodruff ■☆
39348 Asperula glomerata Griseb.;团集车叶草■☆
39349 Asperula gracilis C. A. Mey.;纤细车叶草■☆
39350 Asperula graniticola Klokov;花岗岩车叶草■☆
39351 Asperula graveolens M. Bieb. ex Besser;辛辣车叶草■☆
39352 Asperula hexaphylla All.;六叶车叶草■☆
39353 Asperula hirsuta Desf.;非洲毛车叶草■☆
39354 Asperula hirsuta Desf. var. breviflora Batt. = Asperula hirsuta Desf. ■☆
39355 Asperula hirsuta Desf. var. cyrenaica E. A. Durand et Barratte = Asperula cyrenaica (E. A. Durand et Barratte) Pamp. ■☆
39356 Asperula hirsuta Desf. var. glabrescens Pau = Asperula hirsuta Desf. ■☆
39357 Asperula hirsuta Desf. var. longiflora Maire = Asperula hirsuta Desf. ■☆
39358 Asperula hirsuta Desf. var. mairei Sennen = Asperula hirsuta Desf. ■☆
39359 Asperula hirsuta Desf. var. prostrata Maire et al. = Asperula hirsuta Desf. ■☆
39360 Asperula hirsuta Desf. var. squarrosa Sennen et Maire = Asperula hirsuta Desf. ■☆
39361 Asperula hirsuta Desf. var. villosissima Jahand. et al. = Asperula hirsuta Desf. ■☆
39362 Asperula hirsutiuscula Pobed.;粗毛车叶草■☆
39363 Asperula hoffmeisteri Klotzsch = Galium asperuloides Edgew. subsp. hoffmeisteri (Klotzsch) H. Hara ■
39364 Asperula humifusa (M. Bieb.) Besser = Galium humifusum (M. Bieb.) Besser ■
39365 Asperula infracta Klokov;内折车叶草■☆
39366 Asperula insuavis Pobed.;芳香车叶草■☆
39367 Asperula karataviensis Pavlov;卡拉塔夫车叶草■☆
39368 Asperula kemulariae Manden.;凯穆拉利亚车叶草■☆
39369 Asperula kryloviana Serg.;克雷罗夫车叶草■☆
39370 Asperula laevis Schischk.;平滑车叶草■☆
39371 Asperula laevissima Klokov;光滑车叶草■☆
39372 Asperula leiograveolens Popov et Chrshanovsky;烈味车叶草■☆
39373 Asperula leucanthera Becker;白花车叶草■☆
39374 Asperula lipskyana Krecz.;里普车叶草■☆
39375 Asperula litardierei Humbert;利塔车叶草■☆
39376 Asperula longiflora Waldst. et Kit. = Asperula aristata L. f. subsp. longiflora (Waldst. et Kit.) Hayek ■☆
39377 Asperula markothensis Klokov;马尔考特车叶草■☆
39378 Asperula maximowiczii Kom. = Galium maximowiczii (Kom.) Pobed. ■
39379 Asperula molluginoides Rchb.;粟米草车叶草■☆
39380 Asperula occidentalis Rouy;西方车叶草;Dune Squincywort ■☆
39381 Asperula odorata L. = Galium odoratum (L.) Scop. ■
39382 Asperula oppositifolia Regel et Schmalh. ex Regel;对叶车叶草■
39383 Asperula orientalis Boiss. et Hohen.;蓝花车叶草(车叶草);Blue Woodruff, Blue Woodrush, Oriental Asperula, Oriental Woodruff ■
39384 Asperula pamirica Pobed.;帕米尔车叶草■☆
39385 Asperula paniculata Bunge = Galium paniculatum (Bunge) Pobed. ■
39386 Asperula papillosa Lange = Asperula aristata L. f. ■☆
39387 Asperula pedicellata Klokov;梗花车叶草■☆
39388 Asperula pendula Boiss. = Galium baeticum (Rouy) Ehrend. et Krendl ■☆
39389 Asperula pendula Boiss. var. concatenata (Coss.) Pau = Galium concatenatum Coss. ■☆
39390 Asperula pendula Boiss. var. glaberrima Emb. et Maire = Galium baeticum (Rouy) Ehrend. et Krendl ■☆
39391 Asperula pendula Boiss. var. glabrescens Emb. et Maire = Galium concatenatum Coss. ■☆
39392 Asperula pendula Boiss. var. viridifolia Emb. et Maire = Galium concatenatum Coss. ■☆
39393 Asperula petraea V. I. Krecz. ex Klokov;石生车叶草■☆
39394 Asperula platygalium Maxim. = Galium platygalium (Maxim.) Pobed. ■
39395 Asperula pontica Boiss.;蓬特车叶草■☆
39396 Asperula popovii Schischk.;波氏车叶草■☆
39397 Asperula praevestita Klokov;原包被车叶草■☆
39398 Asperula propinqua Pobed.;邻近车叶草■☆
39399 Asperula prostrata (Adams) K. Koch;平卧车叶草■☆
39400 Asperula rivalis Sibth. = Galium rivale (Sibth. et Sm.) Griseb. ■
39401 Asperula rivalis Sibth. et Sm. = Galium rivale (Sibth. et Sm.) Griseb. ■
39402 Asperula rumelica Boiss.;鲁迈尔车叶草■☆
39403 Asperula scabra J. Presl et C. Presl = Asperula aristata L. f. ■☆
39404 Asperula semiamicta Klokov;半被车叶草■☆
39405 Asperula setosa Jaub. et Spach;刚毛车叶草■☆
39406 Asperula setulosa Boiss.;细刺车叶草■☆
39407 Asperula stevenii Krecz.;斯氏车叶草■☆
39408 Asperula stylosa Boiss. = Phuopsis stylosa (Trin.) Hook. f. ■
39409 Asperula suberosa Sibth. et Sm.;粉花车叶草■☆
39410 Asperula supina M. Bieb.;低车叶草■☆
39411 Asperula taurica Pacz.;克里木车叶草;Pink Woodruff, Pink Woodrush ■☆
39412 Asperula tephrocarpa Czern. ex Popov et Chrshan.;灰果车叶草■☆
39413 Asperula tinctoria L.;染料车叶草;Dyer Bedstraw, Dyer's Weed, Dyer's Woodruff, Dyer's Woodrush ■☆
39414 Asperula tinctoria L. = Galium tinctorium L. ■☆
39415 Asperula tragacanthoides Brullo;羊角刺车叶草■☆
39416 Asperula trichoides J. Gay = Leptunis trichoides (J. Gay ex DC.) Schischk. ■
39417 Asperula trifida Makino = Galium shikokianum Nakai ■☆
39418 Asperula turcomanica Pobed.;土库曼车叶草■☆

39419 Asperula vestita Krecz. ;包被车叶草■☆

39420 Asperula woronowii Krecz. ;沃氏车叶草■☆

39421 Asperula xerotica Klokov;干地车叶草■☆

39422 Asperulaceae Cham. ex Spenn. ;车叶草科■

39423 Asperulaceae Cham. ex Spenn. = Rubiaceae Juss. (保留科名)●■

39424 Asperulaceae Spenn. = Asperulaceae Cham. ex Spenn. ■

39425 Asperulaceae Spenn. = Rubiaceae Juss. (保留科名)●■

39426 Asphalathus Burm. f. = Aspalathus L. ●☆

39427 Asphalthium Medik. = Psoralea L. ●■

39428 Asphaltium Fourr. = Asphalthium Medik. ●■

39429 Asphelexis fasciculata (Andréws) D. Don = Edmondia fasciculata (Andréws) Hilliard ●☆

39430 Asphelexis filiformis D. Don = Edmondia sesamoides (L.) Hilliard ●☆

39431 Asphelexis humilis (Andréws) D. Don = Edmondia pinifolia (Lam.) Hilliard ●☆

39432 Asphelexis sesamoides (L.) D. Don = Edmondia sesamoides (L.) Hilliard ●☆

39433 Asphodelaceae Juss. (1789);阿福花科(芦荟科,日光兰科)●■

39434 Asphodelaceae Juss. = Liliaceae Juss. (保留科名)●■

39435 Asphodeline Rchb. (1830);阿福花属(金穗花属,矛百合属,日光兰属,香阿福花属);Asphodel,Jacob's Rod,Jacobs-rod ■☆

39436 Asphodeline dendroides (Hoffm.) Woronow ex Grossh. ;树状阿福花■☆

39437 Asphodeline liburnica Rchb. ;小苞金穗花■☆

39438 Asphodeline lutea (L.) Rchb. ;阿福花(黄日光兰,金穗花,日光兰,香阿福花);Asphodel,Common Jacob's-rod,Jacob's Rod,Jacob's Staff,King's Spear,Lily Potato,True Asphodel of the Ancients,Yellow Asphodel ■☆

39439 Asphodeline taurica (Pall.) Kunth;克里木阿福花■☆

39440 Asphodeline tenuiflora (K. Koch) Miscz. ;细花阿福花■☆

39441 Asphodeline tenuior (Fisch.) Ledeb. ;小阿福花■☆

39442 Asphodeline tenuior (Fisch.) Ledeb. subsp. tenuiflora (K. Koch) E. Tuzlaci = Asphodeline tenuiflora (K. Koch) Miscz. ■☆

39443 Asphodeline tenuior Ledeb. = Asphodeline tenuior (Fisch.) Ledeb. ■☆

39444 Asphodeliris Kuntze = Tofieldia Huds. ■

39445 Asphodeloides Moench = Asphodelus L. ■☆

39446 Asphodelopsis Steud. ex Baker = Chlorophytum Ker Gawl. ■

39447 Asphodelopsis Steud. ex Baker(1876);类阿福花属■☆

39448 Asphodelopsis arangadinensis Steud. ex Baker;类阿福花■☆

39449 Asphodelus L. (1753);拟阿福花属(阿福花属);Asphodel,White Asphodel ■☆

39450 Asphodelus acaulis Desf. ;无茎拟阿福花(无茎阿福花)■☆

39451 Asphodelus aestivus Brot. var. gracilis (Braun-Blanq. et Maire) Maire = Asphodelus gracilis Braun-Blanq. et Maire ■☆

39452 Asphodelus aestivus Rchb. = Asperula cynanchica L. ■☆

39453 Asphodelus africanus Jord. = Asphodelus microcarpus Viv. var. africanus (Jord.) Batt. et Trab. ■☆

39454 Asphodelus albus Willd. ;白拟阿福花(白阿福花);Asphodelus,Branching Asphodel,King's Spear,Pyrenean Asphodel,White Asphodel ■☆

39455 Asphodelus albus Willd. var. pyrenaeus ?;比利牛斯拟阿福花;Pyrenean Asphodel ■☆

39456 Asphodelus altaicus Pall. = Eremurus altaicus (Pall.) Steven ■

39457 Asphodelus ayardii Jahand. et Maire;阿亚德拟阿福花■☆

39458 Asphodelus capensis L. = Chlorophytum capense (L.) Voss ■

39459 Asphodelus cedretorum Sennen et Mauricio = Asphodelus macrocarpus Parl. subsp. rubescens Z. Díaz et Valdés ■☆

39460 Asphodelus cerasiferus J. Gay;樱桃拟阿福花■☆

39461 Asphodelus cicerae Sennen = Asphodelus ayardii Jahand. et Maire ■☆

39462 Asphodelus comosus Houtt. = Eucomis comosa (Houtt.) Wehrh. ■☆

39463 Asphodelus fistulosus L. ;管状拟阿福花(管状阿福花);Onion Weed,Onion-leaved Asphodel,Onionweed,Onion-weed ■☆

39464 Asphodelus fistulosus L. subsp. tenuifolius (Cav.) Batt. = Asphodelus tenuifolius Cav. ■☆

39465 Asphodelus fistulosus L. var. atlanticus Jahand. et al. = Asphodelus ayardii Jahand. et Maire ■☆

39466 Asphodelus fistulosus L. var. mauritii Sennen = Asphodelus ayardii Jahand. et Maire ■☆

39467 Asphodelus fistulosus L. var. roseus (Humbert et Maire) Pau = Asphodelus roseus Humbert et Maire ■☆

39468 Asphodelus fistulosus L. var. tenuifolius (Cav.) Kunth = Asphodelus tenuifolius Cav. ■☆

39469 Asphodelus gracilis Braun-Blanq. et Maire;纤细拟阿福花■☆

39470 Asphodelus inderiensis Steven = Eremurus inderiensis (M. Bieb.) Regel ■

39471 Asphodelus jacobii Sennen et Mauricio = Asphodelus ramosus L. ■☆

39472 Asphodelus luteus L. = Asphodeline lutea (L.) Rchb. ■☆

39473 Asphodelus macrocarpus Parl. ;大花拟阿福花■☆

39474 Asphodelus macrocarpus Parl. subsp. rubescens Z. Díaz et Valdés;红大花拟阿福花■☆

39475 Asphodelus maroccanus Gand. = Asphodelus tenuifolius Cav. ■☆

39476 Asphodelus mauritii Sennen = Asphodelus ayardii Jahand. et Maire ■☆

39477 Asphodelus microcarpus Rchb. = Asphodelus aestivus Rchb. ■☆

39478 Asphodelus microcarpus Viv. = Asphodelus ramosus L. ■☆

39479 Asphodelus microcarpus Viv. subsp. nervosus (Pomel) Maire = Asphodelus ramosus L. ■☆

39480 Asphodelus microcarpus Viv. var. africanus (Jord.) Batt. et Trab. = Asphodelus ramosus L. ■☆

39481 Asphodelus microcarpus Viv. var. jacobii Maire et Sennen = Asphodelus ramosus L. var. nervosus (Pomel) Z. Díaz et Valdés ■☆

39482 Asphodelus microcarpus Viv. var. nervosus (Pomel) Maire = Asphodelus ramosus L. ■☆

39483 Asphodelus nervosus Pomel = Asphodelus ramosus L. ■☆

39484 Asphodelus occidentalis Cout. ;西方拟阿福花■☆

39485 Asphodelus occidentalis Jord. = Asphodelus albus Willd. ■☆

39486 Asphodelus pendulinus Coss. et Durieu = Asphodelus refractus Boiss. ■☆

39487 Asphodelus ramosus L. = Asperula cynanchica L. ■☆

39488 Asphodelus ramosus L. var. africanus (Jord.) Z. Díaz et Valdés = Asphodelus ramosus L. ■☆

39489 Asphodelus ramosus L. var. nervosus (Pomel) Z. Díaz et Valdés;多脉分枝拟阿福花■☆

39490 Asphodelus refractus Boiss. ;反折拟阿福花■☆

39491 Asphodelus repens Pomel = Asphodelus cerasiferus J. Gay ■☆

39492 Asphodelus roseus Humbert et Maire;粉红拟阿福花■☆

39493 Asphodelus tenuifolius Cav. ;薄叶拟阿福花(薄叶阿福花)■☆

39494 Asphodelus tenuifolius Cav. var. micranthus Boiss. = Asphodelus tenuifolius Cav. ■☆

39495 Asphodelus viscidulus Boiss. ;微黏拟阿福花■☆

39496 Asphodelus viscidulus Boiss. var. gabesianus J. Gay = Asphodelus viscidulus Boiss. ■☆
39497 Aspicaria D. Dietr. = Aspicarpa Rich. ●☆
39498 Aspicarpa Rich.(1815);盾果金虎尾属●☆
39499 Aspicarpa affinis Hassl.;近缘盾果金虎尾●☆
39500 Aspicarpa argentea Nied.;银白盾果金虎尾●☆
39501 Aspicarpa axillaris Hassl.;腋生盾果金虎尾●☆
39502 Aspicarpa boliviensis Nied.;玻利维亚盾果金虎尾●☆
39503 Aspicarpa brevipes (DC.) W. R. Anderson;短梗盾果金虎尾●☆
39504 Aspicarpa diandra Hassl.;双蕊盾果金虎尾●☆
39505 Aspicarpa gracilis Hassl.;细盾果金虎尾●☆
39506 Aspicarpa humilis Nied.;矮盾果金虎尾●☆
39507 Aspicarpa linearifolia (A. St. -Hil.) Nied.;线叶盾果金虎尾●☆
39508 Aspicarpa mollis Hassl.;软盾果金虎尾●☆
39509 Aspicarpa pentandra Hassl.;五蕊盾果金虎尾●☆
39510 Aspicarpa salicifolia (Chodat) Nied.;柳叶盾果金虎尾●☆
39511 Aspicarpa triphylla Hassl.;三叶盾果金虎尾●☆
39512 Aspidalis Gaertn. = Cuspidia Gaertn. ■☆
39513 Aspidandra Hassk. = Ryparosa Blume ●☆
39514 Aspidanthera Benth. = Ferdinandusa Pohl ●☆
39515 Aspideium Zollik. ex DC. = Chondrilla L. ■
39516 Aspidistra Ker Gawl.(1822);蜘蛛抱蛋属(叶兰属);Aspidistra ●■
39517 Aspidistra acetabuliformis Y. Wan et C. C. Huang;碟柱蜘蛛抱蛋;Dishstyle Aspidistra ■
39518 Aspidistra alternativa D. Fang et L. Y. Yu;忻城蜘蛛抱蛋;Xincheng Aspidistra ■
39519 Aspidistra altostamina S. Z. Huang;高雄蕊蜘蛛抱蛋;Gaoxiong Aspidistra ■
39520 Aspidistra attenuata Hayata;薄叶蜘蛛抱蛋;Thinleaf Aspidistra ■
39521 Aspidistra attenuata Hayata = Aspidistra elatior Blume var. attenuata (Hayata) S. S. Ying ■
39522 Aspidistra austrosinensis Y. Wan et C. C. Huang;华南蜘蛛抱蛋;S. China Aspidistra ■
39523 Aspidistra caespitosa C. P'ei;丛生蜘蛛抱蛋;Clustered Aspidistra ■
39524 Aspidistra carinata Y. Wan et X. H. Lu;天峨蜘蛛抱蛋(花叶蜘蛛抱蛋);Tian'e Aspidistra ■
39525 Aspidistra cavicola D. Fang et K. C. Yen;洞生蜘蛛抱蛋;Cavicolous Aspidistra ■
39526 Aspidistra claviformis Y. Wan;棒蕊蜘蛛抱蛋;Claviform Aspidistra ■
39527 Aspidistra cruciformis Y. Wan et X. H. Lu;十字蜘蛛抱蛋;Crisscross Aspidistra ■
39528 Aspidistra cyathiflora Y. Wan et C. C. Huang;杯花蜘蛛抱蛋;Cupflower Aspidistra ■
39529 Aspidistra daibuensis Hayata;大武蜘蛛抱蛋;Dawu Aspidistra ■
39530 Aspidistra dolichanthera X. X. Chen;长药蜘蛛抱蛋;Longstyle Aspidistra ■
39531 Aspidistra ebianensis K. Y. Lang et Z. Y. Zhu;峨边蜘蛛抱蛋;Ebian Aspidistra ■
39532 Aspidistra elatior Blume;蜘蛛抱蛋(大九龙盘,大伸筋,大叶万年青,单枝白叶,地蜈蚣,飞天蜈蚣,甘心蜈蚣,赶山鞭,哈萨喇,九节龙,九龙盘,狸角叶,蓼叶伸筋,龙骨草,入地蜈蚣,蛇退草,铁马鞭,土里蜈蚣,蜈蚣草,摇边竹,一寸十八节,一帆青,一叶,一叶兰,斩龙剑,竹根七,竹节伸筋,竹叶盘,竹叶伸筋);Aspidistra,Bar-room Plant,Cannon-ball Plant,Cast lron Plant,Cast-iron Plant,Common Aspidistra ●■
39533 Aspidistra elatior Blume 'Variegata';花叶蜘蛛抱蛋(花叶一叶兰,嵌玉蜘蛛抱蛋)■☆
39534 Aspidistra elatior Blume var. attenuata (Hayata) S. S. Ying;台湾蜘蛛抱蛋(薄叶蜘蛛抱蛋,蜘蛛抱蛋);Taiwan Aspidistra ■
39535 Aspidistra elatior Blume var. minor Hort.;点叶蜘蛛抱蛋■☆
39536 Aspidistra elatior Blume var. punctata Lindl.;斑叶蜘蛛抱蛋(洒金蜘蛛抱蛋)■☆
39537 Aspidistra elatior Blume var. variegata Hort. = Aspidistra elatior Blume 'Variegata' ■☆
39538 Aspidistra fasciaria G. Z. Li;带叶蜘蛛抱蛋;Fascileaf Aspidistra ■
39539 Aspidistra fenghuangensis K. Y. Lang;凤凰蜘蛛抱蛋;Fenghuang Aspidistra ■
39540 Aspidistra fimbriata F. T. Wang et K. Y. Lang;流苏蜘蛛抱蛋;Fimbriate Aspidistra ■
39541 Aspidistra flaviflora K. Y. Lang et Z. Y. Zhu;黄花蜘蛛抱蛋;Yellowflower Aspidistra ■
39542 Aspidistra fungilliformis Y. Wan;伞柱蜘蛛抱蛋;Fungistyle Aspidistra ■
39543 Aspidistra glandulosa (Gagnep.) Tillich;多腺蜘蛛抱蛋■☆
39544 Aspidistra hainanensis Chun et F. C. How;海南蜘蛛抱蛋;Hainan Aspidistra ■
39545 Aspidistra hekouensis H. Li et al.;河口蜘蛛抱蛋;Hekou Aspidistra ■
39546 Aspidistra huanjiangensis G. Z. Li et Y. G. Wei;环江蜘蛛抱蛋;Huanjiang Aspidistra ■
39547 Aspidistra kouytchensis H. Lév. et Vaniot = Aspidistra lurida Ker Gawl. ■
39548 Aspidistra kouytchensis H. Lév. et Vaniot var. aucubaemaculata H. Lév. et Vaniot = Aspidistra lurida Ker Gawl. ■
39549 Aspidistra leshanensis K. Y. Lang et Z. Y. Zhu;乐山蜘蛛抱蛋;Leshan Aspidistra ■
39550 Aspidistra leyeensis Y. Wan et C. C. Huang;乐业蜘蛛抱蛋;Leye Aspidistra ■
39551 Aspidistra linearifolia Y. Wan et C. C. Huang;线叶蜘蛛抱蛋(线萼蜘蛛抱蛋);Linearleaf Aspidistra ■
39552 Aspidistra longanensis Y. Wan;隆安蜘蛛抱蛋;Longan Aspidistra ■
39553 Aspidistra longiloba G. Z. Li;巨型蜘蛛抱蛋;Giant Aspidistra ■
39554 Aspidistra longipedunculata D. Fang;长梗蜘蛛抱蛋;Longpediculate Aspidistra ■
39555 Aspidistra longipetala S. Z. Huang;长瓣蜘蛛抱蛋;Longpetal Aspidistra ■
39556 Aspidistra luodianensis D. D. Tao;罗甸蜘蛛抱蛋;Luodian Aspidistra ■
39557 Aspidistra lurida Ker Gawl.;九龙盘(寸八节,地蜈蚣,赶山鞭,褐黄蜘蛛抱蛋,花棕叶,爬地蜈蚣,盘龙七,千年竹,青蛇莲,蛇莲,蛇退,蜈蚣草,俞莲,竹叶根,竹叶盘,棕巴叶,走石马);Brownishpurple Aspidistra ■
39558 Aspidistra lurida Ker Gawl. = Aspidistra elatior Blume ●■
39559 Aspidistra marginella D. Fang et L. Zeng;啮边蜘蛛抱蛋■
39560 Aspidistra minutiflora Stapf;小花蜘蛛抱蛋(毛知母);Smallflower Aspidistra ■
39561 Aspidistra minutiflora Stapf 'Punetata';点斑小花一叶兰■☆
39562 Aspidistra muricata F. C. How ex K. Y. Lang;糙果蜘蛛抱蛋;Muricate Aspidistra ■
39563 Aspidistra mushaensis Hayata;雾社蜘蛛抱蛋(穆沙蜘蛛抱蛋)■

39564 Aspidistra oblanceifolia F. T. Wang et K. Y. Lang;棕叶草蜘蛛抱蛋(棕叶草);Oblanceleaf Aspidistra ■
39565 Aspidistra obliquipeltata D. Fang et L. Y. Yu;歪盾蜘蛛抱蛋■
39566 Aspidistra oblongifolia F. T. Wang et K. Y. Lang;长圆叶蜘蛛抱蛋;Oblongleaf Aspidistra ■
39567 Aspidistra omeiensis Z. Y. Zhu et J. L. Zhang;峨眉蜘蛛抱蛋(赶山鞭);Emei Aspidistra ■
39568 Aspidistra papillata G. Z. Li;乳突蜘蛛抱蛋;Papillate Aspidistra ■
39569 Aspidistra patentiloba Y. Wan et C. C. Huang;柳江蜘蛛抱蛋;Liujiang Aspidistra ■
39570 Aspidistra pileata D. Fang et L. Y. Yu;帽状蜘蛛抱蛋■
39571 Aspidistra punctata Lindl.;紫点蜘蛛抱蛋(斑点蜘蛛抱蛋);Punctate Aspidistra ■
39572 Aspidistra punctata Lindl. var. albomaculata Hook. = Aspidistra elatior Blume ●■
39573 Aspidistra retusa K. Y. Lang et S. Z. Huang;广西蜘蛛抱蛋;Guangxi Aspidistra ■
39574 Aspidistra saxicola Y. Wan;石山蜘蛛抱蛋;Shishan Aspidistra ■
39575 Aspidistra sichuanensis K. Y. Lang et Z. Y. Zhu;四川蜘蛛抱蛋;Sichuan Aspidistra ■
39576 Aspidistra spinula S. Z. He;刺果蜘蛛抱蛋;Spiny-fruited Aspidistra ■
39577 Aspidistra subrotata Y. Wan et C. C. Huang;辐花蜘蛛抱蛋;Subrotate Aspidistra ■
39578 Aspidistra tonkinensis (Gagnep.) F. T. Wang et K. Y. Lang;大花蜘蛛抱蛋;Largeflower Aspidistra ■
39579 Aspidistra triloba F. T. Wang et K. Y. Lang;湖南蜘蛛抱蛋;Hunan Aspidistra ■
39580 Aspidistra typica Baill.;卵叶蜘蛛抱蛋(蛇退,俞莲,蜘蛛抱蛋,棕巴叶,棕粑叶);Ovalleaf Aspidistra ■
39581 Aspidistra urceolata F. T. Wang et K. Y. Lang;坛花蜘蛛抱蛋;Urceolate Aspidistra ■
39582 Aspidistra xilinensis Y. Wan et X. H. Lu;西林蜘蛛抱蛋;Xilin Aspidistra ■
39583 Aspidistra yingjiangensis L. J. Peng;盈江蜘蛛抱蛋;Yingjiang Aspidistra ■
39584 Aspidistra zongbayi K. Y. Lang et Z. Y. Zhu;棕粑叶(棕子叶);Zongbaye Aspidistra ■
39585 Aspidistraceae Endl.;蜘蛛抱蛋科■
39586 Aspidistraceae Endl. = Convallariaceae L. ■
39587 Aspidistraceae Endl. = Ruscaceae M. Roem.(保留科名)●
39588 Aspidistraceae Hassk. = Aspidistraceae Endl. ■
39589 Aspidistraceae Hassk. = Convallariaceae L. ■
39590 Aspidistraceae Hassk. = Ruscaceae M. Roem.(保留科名)●
39591 Aspidistraceae J. Agardh = Aspidistraceae Endl. ■
39592 Aspidixia (Korth.) Tiegh. = Viscum L. ●
39593 Aspidixia Tiegh. = Viscum L. ●
39594 Aspidixia anceps E. Mey. ex Tiegh. = Viscum anceps E. Mey. ex Sprague ●☆
39595 Aspidixia angulata J. M. Chao = Viscum diospyrosicolum Hayata ●
39596 Aspidixia articulata (Burm. f.) Tiegh. = Viscum articulatum Burm. f. ●
39597 Aspidixia articulatum (Burm. f.) Tiegh. = Viscum articulatum Burm. f. ●
39598 Aspidixia articulatum (Burm. f.) Tiegh. = Viscum liquidambaricola Hayata ●
39599 Aspidixia bivalvis Tiegh. = Viscum obscurum Thunb. ●☆
39600 Aspidixia capensis (L. f.) Tiegh. = Viscum capense L. f. ●☆
39601 Aspidixia grandidieri Tiegh. = Viscum echinocarpum Baker ●☆
39602 Aspidixia junodii Tiegh. = Viscum shirense Sprague ●☆
39603 Aspidixia minima (Harv.) Tiegh. = Viscum minimum Harv. ●☆
39604 Aspidixia robusta (Eckl. et Zeyh.) Tiegh. = Viscum capense L. f. ●☆
39605 Aspidixia semiplana Tiegh. = Viscum schimperi Engl. ●☆
39606 Aspidocarpus Neck. = Rhamnus L. ●
39607 Aspidocarya Hook. f. et Thomson (1855);球果藤属(盾核藤属);Aspidocarya, Conevine ●
39608 Aspidocarya uvifera Hook. f. et Thomson;球果藤(盾核藤);Berry-bearing Aspidocarya, Conevine ●
39609 Aspidogenia Burret = Myrcianthes O. Berg ●☆
39610 Aspidogenia Burret = Reichea Kausel ●☆
39611 Aspidoglossum E. Mey. (1838);盾舌萝藦属■☆
39612 Aspidoglossum E. Mey. = Schizoglossum E. Mey. ■☆
39613 Aspidoglossum angustissimum (K. Schum.) Bullock;狭盾舌萝藦■☆
39614 Aspidoglossum biflorum E. Mey.;双花盾舌萝藦■☆
39615 Aspidoglossum biflorum E. Mey. var. gwelense (N. E. Br.) N. E. Br. = Aspidoglossum biflorum E. Mey. ■☆
39616 Aspidoglossum carinatum (Schltr.) Kupicha;龙骨盾舌萝藦■☆
39617 Aspidoglossum connatum (N. E. Br.) Bullock;合生盾舌萝藦■☆
39618 Aspidoglossum crebrum Kupicha;密集盾舌萝藦■☆
39619 Aspidoglossum delagoense (Schltr.) Kupicha;迪拉果盾舌萝藦■☆
39620 Aspidoglossum demissum Kupicha;下垂盾舌萝藦■☆
39621 Aspidoglossum elliotii (Schltr.) Kupicha;埃利盾舌萝藦■☆
39622 Aspidoglossum erubescens (Schltr.) Bullock;变红盾舌萝藦■☆
39623 Aspidoglossum eylesii (S. Moore) Kupicha;艾尔斯盾舌萝藦■☆
39624 Aspidoglossum fasciculare E. Mey.;带状盾舌萝藦■☆
39625 Aspidoglossum flanaganii (Schltr.) Kupicha;弗拉纳根盾舌萝藦■☆
39626 Aspidoglossum glabellum Kupicha;光滑盾舌萝藦■☆
39627 Aspidoglossum glabrescens (Schltr.) Kupicha;渐光盾舌萝藦■☆
39628 Aspidoglossum glanduliferum (Schltr.) Kupicha;腺点盾舌萝藦■☆
39629 Aspidoglossum gracile (E. Mey.) Kupicha;纤细盾舌萝藦■☆
39630 Aspidoglossum grandiflorum (Schltr.) Kupicha;大花盾舌萝藦■☆
39631 Aspidoglossum heterophyllum E. Mey.;互叶盾舌萝藦■☆
39632 Aspidoglossum interruptum (E. Mey.) Bullock;间断盾舌萝藦■☆
39633 Aspidoglossum kulsii Cufod. = Aspidoglossum masaicum (N. E. Br.) Kupicha ■☆
39634 Aspidoglossum lamellatum (Schltr.) Kupicha;片状盾舌萝藦■☆
39635 Aspidoglossum lanatum (Weim.) Kupicha;绵毛盾舌萝藦■☆
39636 Aspidoglossum masaicum (N. E. Br.) Kupicha;马萨盾舌萝藦■☆
39637 Aspidoglossum nyasae (Britten et Rendle) Kupicha;尼亚萨盾舌萝藦■☆
39638 Aspidoglossum ovalifolium (Schltr.) Kupicha;椭圆叶盾舌萝藦■☆
39639 Aspidoglossum restioides (Schltr.) Kupicha;绳盾舌萝藦■☆
39640 Aspidoglossum rhodesicum (Weim.) Kupicha;罗得西亚盾舌萝藦■☆
39641 Aspidoglossum uncinatum (N. E. Br.) Kupicha;具钩盾舌萝藦■☆
39642 Aspidoglossum validum Kupicha;刚直盾舌萝藦■☆
39643 Aspidoglossum virgatum (E. Mey.) Kupicha;条纹盾舌萝藦■☆
39644 Aspidoglossum whytei (N. E. Br.) Bullock = Aspidoglossum

angustissimum (K. Schum.) Bullock ■☆

39645 Aspidoglossum woodii (Schltr.) Kupicha;伍得盾舌萝藦■☆

39646 Aspidoglossum xanthosphaerum Hilliard;黄球形盾舌萝藦■☆

39647 Aspidogyne Garay(1977);盾柱兰属■☆

39648 Aspidogyne argentea (Vell.) Garay;银白盾柱兰■☆

39649 Aspidogyne bicolor (Barb. Rodr.) Garay;二色盾柱兰■☆

39650 Aspidogyne boliviensis (Cogn.) Garay;玻利维亚盾柱兰■☆

39651 Aspidogyne brachyrrhyncha (Rchb. f.) Garay;短喙盾柱兰■☆

39652 Aspidogyne foliosa (Poepp. et Endl.) Garay;多叶盾柱兰■☆

39653 Aspidogyne grandis (Ormerod) Ormerod;大盾柱兰■☆

39654 Aspidogyne longicornu (Cogn.) Garay;长角盾柱兰■☆

39655 Aspidogyne robusta (C. Schweinf.) Garay;粗壮盾柱兰■☆

39656 Aspidonepsis Nicholas et Goyder(1992);盾萝藦属●☆

39657 Aspidonepsis cognata (N. E. Br.) Nicholas et Goyder;近缘盾萝藦●☆

39658 Aspidonepsis diploglossa (Turcz.) Nicholas et Goyder;双舌盾萝藦●☆

39659 Aspidonepsis flava (N. E. Br.) Nicholas et Goyder;黄盾萝藦●☆

39660 Aspidonepsis reenensis (N. E. Br.) Nicholas et Goyder;里恩盾萝藦●☆

39661 Aspidophyllum Ulbr. (1922);盾叶毛茛属■☆

39662 Aspidophyllum Ulbr. = Ranunculus L. ■

39663 Aspidophyllum clypeatum Ulbr.;盾叶毛茛■☆

39664 Aspidopteris hypoglaucum H. Lév. = Tripterygium hypoglaucum (H. Lév.) Hutch. ●

39665 Aspidopterys A. Juss. = Aspidopterys A. Juss. ex Endl. ●

39666 Aspidopterys A. Juss. ex Endl. (1840);盾翅藤属(盾翅果属);Aspidopterys ●

39667 Aspidopterys cavaleriei H. Lév. = Combretum wallichii DC. ●

39668 Aspidopterys cavaleriei H. Lév. var. pubinerve C. Y. Wu = Combretum wallichii DC. var. pubinerve C. Y. Wu ex T. Z. Hsu ●

39669 Aspidopterys concava (Wall.) A. Juss.;广西盾翅藤;Guangxi Aspidopterys ●

39670 Aspidopterys concava (Wall.) A. Juss. var. dasyphylla Arènes;毛叶广西盾翅藤(毛叶盾翅藤);Concave Aspidopterys ●

39671 Aspidopterys concava (Wall.) A. Juss. var. dasyphylla Arènes = Aspidopterys nutans (Roxb. ex DC.) A. Juss. ●

39672 Aspidopterys dunniana H. Lév. = Aspidopterys cavaleriei H. Lév. ●

39673 Aspidopterys elliptica H. Lév.;椭圆叶盾翅藤●☆

39674 Aspidopterys esquirolii H. Lév. et Arènes;花江盾翅藤;Esquirol Aspidopterys ●◇

39675 Aspidopterys floribunda Hutch.;多花盾翅藤;Flowery Aspidopterys, Many-flower Aspidopterys, Multiflowered Aspidopterys ●

39676 Aspidopterys glabriuscula (Wall.) A. Juss.;盾翅藤(盾翅果);Aspidopterys, Common Aspidopterys ●

39677 Aspidopterys glabriuscula (Wall.) A. Juss. = Aspidopterys esquirolii H. Lév. et Arènes ●◇

39678 Aspidopterys glabriuscula (Wall.) A. Juss. var. subrotunda Nied. = Aspidopterys henryi Hutch. ●

39679 Aspidopterys henryi Hutch.;蒙自盾翅藤;Henry Aspidopterys, Mengzi Aspidopterys ●

39680 Aspidopterys henryi Hutch. var. tokinensis Arènes;越南盾翅藤;Tokin Aspidopterys ●

39681 Aspidopterys heterocarpa Arènes = Aspidopterys glabriuscula (Wall.) A. Juss. ●

39682 Aspidopterys hypoglauca H. Lév. = Tripterygium hypoglaucum (H. Lév.) Hutch. ●

39683 Aspidopterys lanuginosa (Wall.) A. Juss. = Aspidopterys nutans (Roxb. ex DC.) A. Juss. ●

39684 Aspidopterys microcarpa H. W. Li ex S. K. Chen;小果盾翅藤;Littlefruit Aspidopterys, Minifruit Aspidopterys ●

39685 Aspidopterys nutans (Roxb. ex DC.) A. Juss.;毛叶盾翅藤;Hairleaf Aspidopterys ●

39686 Aspidopterys nutans Hook. f. = Aspidopterys nutans (Roxb. ex DC.) A. Juss. ●

39687 Aspidopterys obcordata Hemsl.;倒心叶盾翅藤(倒心叶盾翅果,盾翅藤);Invert-hearted Aspidopterys, Obcordateleaf Aspidopterys, Obcordate-shaped Aspidopterys ●

39688 Aspidopterys obcordata Hemsl. var. hainanensis Arènes;海南盾翅藤;Hainan Aspidopterys ●

39689 Aspidopterys stipulacea Nied. = Aspidopterys esquirolii H. Lév. et Arènes ●◇

39690 Aspidopterys tomentosa Hemsl. var. obcordata (Hemsl.) Nied. = Aspidopterys obcordata Hemsl. ●

39691 Aspidopterys tomentosa Hemsl. var. obocordata Nied. = Aspidopterys obcordata Hemsl. ●

39692 Aspidopteryx Dalla Torre et Harms = Aspidopterys A. Juss. ex Endl. ●

39693 Aspidosperma Mart. et Zucc. (1824)(保留属名);白坚木属(盾籽木属,楯籽木属);Peroba Rosa, White Quebracho ●☆

39694 Aspidosperma condylocarpon Müll. Arg. = Diplorhynchus condylocarpon (Müll. Arg.) Pichon ●☆

39695 Aspidosperma ellipticum Rusby;椭圆白坚木●☆

39696 Aspidosperma excelsum Benth.;高白坚木;Paddlewood ●☆

39697 Aspidosperma nitidum Benth. ex Müll. Arg.;光泽白坚木●☆

39698 Aspidosperma olivaceum Müll. Arg.;褐绿白坚木●☆

39699 Aspidosperma polyneuron Müll. Arg.;多脉白坚木●☆

39700 Aspidosperma quebracho-blanco Schltdl.;白坚木;Common White Quebracho ●☆

39701 Aspidosperma rhombeosignatum Markgr.;菱痣白坚木(菱形白坚木)●☆

39702 Aspidosperma tomentosum Mart. et Zucc.;毛白坚木;Quebracha ●☆

39703 Aspidostemon Rohwer et H. G. Richt. (1987);盾蕊樟属(盾蕊厚壳桂属)●☆

39704 Aspidostemon andohahelense van der Werff;安杜哈赫尔盾蕊樟●☆

39705 Aspidostemon antongilense van der Werff;安通吉尔盾蕊樟●☆

39706 Aspidostemon apiculatum van der Werff;细尖盾蕊樟●☆

39707 Aspidostemon capuronii van der Werff;凯普伦盾蕊樟●☆

39708 Aspidostemon caudatum Rohwer;尾状盾蕊樟●☆

39709 Aspidostemon dolichocarpum (Kosterm.) Rohwer;长果盾蕊樟●☆

39710 Aspidostemon glandulosum Rohwer;具腺盾蕊樟●☆

39711 Aspidostemon grayi van der Werff;格雷盾蕊樟●☆

39712 Aspidostemon humbertianum (Kosterm.) Rohwer;亨伯特盾蕊樟●☆

39713 Aspidostemon inconspicuum Rohwer;大穗盾蕊樟●☆

39714 Aspidostemon insigne van der Werff;显著盾蕊樟●☆

39715 Aspidostemon litorale van der Werff;滨海盾蕊樟●☆

39716 Aspidostemon longipedicellatum van der Werff;长梗盾蕊樟●☆

39717 Aspidostemon lucens van der Werff;光亮盾蕊樟●☆

39718 Aspidostemon macrophyllum van der Werff;大叶盾蕊樟●☆

39719 Aspidostemon manongarivense van der Werff;马农加盾蕊樟●☆

39720 Aspidostemon masoalense van der Werff;马苏阿拉盾蕊樟●☆

39721 Aspidostemon microphyllum van der Werff;马岛小叶盾蕊樟●☆
39722 Aspidostemon occultum van der Werff;隐蔽盾蕊樟●☆
39723 Aspidostemon parvifolium (Scott-Elliot) van der Werff;小叶盾蕊樟●☆
39724 Aspidostemon percoriaceum (Kosterm.) Rohwer;厚革盾蕊樟●☆
39725 Aspidostemon perrieri (Danguy) Rohwer;佩里耶盾蕊樟●☆
39726 Aspidostemon reticulatum van der Werff;网状盾蕊樟●☆
39727 Aspidostemon synandra Rohwer;合蕊盾蕊樟●☆
39728 Aspidostemon trianthera (Kosterm.) Rohwer;三花盾蕊樟●☆
39729 Aspidostemon trichandra van der Werff;三蕊盾蕊樟●☆
39730 Aspidostigma Hochst.(废弃属名)= Teclea Delile(保留属名)●☆
39731 Aspilia Thouars = Wedelia Jacq.(保留属名)●■
39732 Aspilia Thouars(1806);阿斯皮菊属■☆
39733 Aspilia abyssinica (Sch. Bip.) Vatke = Aspilia helianthoides (Schumach. et Thonn.) Oliv. et Hiern subsp. ciliata (Schumach.) C. D. Adams ■☆
39734 Aspilia abyssinica Oliv. et Hiern;阿比西尼亚阿斯皮而菊■☆
39735 Aspilia africana (P. Beauv.) C. D. Adams;非洲阿斯皮菊■☆
39736 Aspilia africana (P. Beauv.) C. D. Adams subsp. australis Wild;南非阿斯皮菊■☆
39737 Aspilia africana (P. Beauv.) C. D. Adams subsp. magnifica (Chiov.) Wild;华丽阿斯皮菊■☆
39738 Aspilia africana (P. Beauv.) C. D. Adams var. ambigua C. D. Adams = Aspilia africana (P. Beauv.) C. D. Adams ■☆
39739 Aspilia africana (P. Beauv.) C. D. Adams var. guineensis (O. Hoffm. et Muschl.) C. D. Adams = Aspilia africana (P. Beauv.) C. D. Adams ■☆
39740 Aspilia africana (P. Beauv.) C. D. Adams var. minor C. D. Adams = Aspilia africana (P. Beauv.) C. D. Adams ■☆
39741 Aspilia africana (P. Beauv.) C. D. Adams var. ubanguensis O. Hoffm. et Muschl. = Aspilia africana (P. Beauv.) C. D. Adams ■☆
39742 Aspilia africana (Pers.) C. D. Adams = Aspilia africana (P. Beauv.) C. D. Adams ■☆
39743 Aspilia angolensis (Klatt) Muschl.;安哥拉阿斯皮菊■☆
39744 Aspilia angustifolia Oliv. et Hiern;窄叶阿斯皮菊■☆
39745 Aspilia asperifolia O. Hoffm. = Aspilia pluriseta Schweinf. ■☆
39746 Aspilia aspilioides (Baker) S. Moore = Aspilia mossambicensis (Oliv.) Wild ■☆
39747 Aspilia attrivittata Merxm. = Aspilia eenii S. Moore ■☆
39748 Aspilia baoulensis A. Chev. = Aspilia rudis Oliv. et Hiern ■☆
39749 Aspilia baronii Baker = Aspilia bojeri DC. ■☆
39750 Aspilia baumii O. Hoffm.;鲍姆阿斯皮菊■☆
39751 Aspilia bipartita O. Hoffm.;二深裂阿斯皮菊■☆
39752 Aspilia bojeri DC.;博耶尔阿斯皮菊■☆
39753 Aspilia brachyphylla S. Moore = Aspilia pluriseta Schweinf. ■☆
39754 Aspilia bracteosa C. D. Adams = Aspilia helianthoides (Schumach. et Thonn.) Oliv. et Hiern subsp. prieuriana (DC.) C. D. Adams ■☆
39755 Aspilia bussei O. Hoffm. et Muschl.;布瑟阿斯皮菊■☆
39756 Aspilia bussei O. Hoffm. et Muschl. var. kitsonii (S. Moore) Adams = Aspilia bussei O. Hoffm. et Muschl. ■☆
39757 Aspilia chevalieri O. Hoffm. et Muschl.;舍瓦利耶阿斯皮菊■☆
39758 Aspilia chrysops S. Moore = Aspilia mossambicensis (Oliv.) Wild ■☆
39759 Aspilia ciliata (Schumach.) Wild = Aspilia helianthoides (Schumach. et Thonn.) Oliv. et Hiern subsp. ciliata (Schumach.) C. D. Adams ■☆
39760 Aspilia congoensis S. Moore = Aspilia africana (P. Beauv.) C. D. Adams subsp. magnifica (Chiov.) Wild ■☆
39761 Aspilia courtetii O. Hoffm. et Muschl. = Aspilia kotschyi (Sch. Bip.) Oliv. ■☆
39762 Aspilia culuensis S. Moore = Aspilia angolensis (Klatt) Muschl. ■☆
39763 Aspilia dewevrei O. Hoffm. = Aspilia helianthoides (Schumach. et Thonn.) Oliv. et Hiern subsp. ciliata (Schumach.) C. D. Adams ■☆
39764 Aspilia eenii S. Moore;埃恩阿斯皮菊■☆
39765 Aspilia elegans (C. D. Adams) J.-P. Lebrun et Stork;雅致阿斯皮菊■☆
39766 Aspilia engleriana Muschl. = Aspilia angolensis (Klatt) Muschl. ■☆
39767 Aspilia eylesii S. Moore = Melanthera pungens Oliv. et Hiern var. albinervia (O. Hoffm.) Beentje ■☆
39768 Aspilia fischeri O. Hoffm.;菲舍尔阿斯皮菊■☆
39769 Aspilia fontinalis Hiern = Aspilia natalensis (Sond.) Wild ■☆
39770 Aspilia gillettii Wild = Aspilia mossambicensis (Oliv.) Wild ■☆
39771 Aspilia gondensis O. Hoffm. = Aspilia pluriseta Schweinf. ■☆
39772 Aspilia guineensis O. Hoffm. et Muschl. = Aspilia africana (P. Beauv.) C. D. Adams ■☆
39773 Aspilia helianthoides (Schumach. et Thonn.) Oliv. et Hiern;向日葵阿斯皮菊■☆
39774 Aspilia helianthoides (Schumach. et Thonn.) Oliv. et Hiern subsp. ciliata (Schumach.) C. D. Adams;睫毛阿斯皮菊■☆
39775 Aspilia helianthoides (Schumach. et Thonn.) Oliv. et Hiern subsp. papposa (O. Hoffm. et Muschl.) C. D. Adams = Aspilia helianthoides (Schumach. et Thonn.) Oliv. et Hiern subsp. prieuriana (DC.) C. D. Adams ■☆
39776 Aspilia helianthoides (Schumach. et Thonn.) Oliv. et Hiern subsp. prieuriana (DC.) C. D. Adams;普里厄阿斯皮菊■☆
39777 Aspilia helianthoides (Schumach. et Thonn.) Oliv. et Hiern var. papposa O. Hoffm. et Muschl. = Aspilia helianthoides (Schumach. et Thonn.) Oliv. et Hiern subsp. prieuriana (DC.) C. D. Adams ■☆
39778 Aspilia holstii O. Hoffm. = Aspilia mossambicensis (Oliv.) Wild ■☆
39779 Aspilia huillensis (Hiern) S. Moore = Aspilia angolensis (Klatt) Muschl. ■☆
39780 Aspilia involucrata O. Hoffm. = Aspilia pluriseta Schweinf. ■☆
39781 Aspilia kakondensis S. Moore = Aspilia natalensis (Sond.) Wild ■☆
39782 Aspilia kitsonii S. Moore = Aspilia bussei O. Hoffm. et Muschl. ■☆
39783 Aspilia kotschyi (Sch. Bip.) Oliv.;科奇阿斯皮菊■☆
39784 Aspilia kotschyi (Sch. Bip.) Oliv. var. alba Berhaut;白色科奇阿斯皮菊■☆
39785 Aspilia latifolia Oliv. et Hiern = Aspilia africana (P. Beauv.) C. D. Adams ■☆
39786 Aspilia linearifolia Oliv. et Hiern = Aspilia angustifolia Oliv. et Hiern ■☆
39787 Aspilia macrorrhiza Chiov.;大根阿斯皮菊■☆
39788 Aspilia malaissei Lisowski;马莱泽阿斯皮菊■☆
39789 Aspilia mendoncae Wild;门东萨阿斯皮菊■☆
39790 Aspilia mildbraedii Muschl.;米尔德阿斯皮菊■☆
39791 Aspilia minima Humbert;极小阿斯皮菊■☆
39792 Aspilia monocephala Baker = Aspilia mossambicensis (Oliv.) Wild ■☆
39793 Aspilia mortonii C. D. Adams = Aspilia angustifolia Oliv. et Hiern

■☆
39794 Aspilia mossambicensis (Oliv.) Wild;莫桑比克阿斯皮菊■☆
39795 Aspilia multiflora Fenzl ex Oliv. et Hiern = Aspilia helianthoides (Schumach. et Thonn.) Oliv. et Hiern subsp. prieuriana (DC.) C. D. Adams ■☆
39796 Aspilia natalensis (Sond.) Wild;纳塔尔阿斯皮菊■☆
39797 Aspilia paludosa Berhaut;沼泽阿斯皮菊■☆
39798 Aspilia pluriseta Schweinf.;短叶阿斯皮菊■☆
39799 Aspilia pluriseta Schweinf. subsp. gondensis (O. Hoffm.) Wild = Aspilia pluriseta Schweinf. ■☆
39800 Aspilia polycephala S. Moore = Aspilia kotschyi (Sch. Bip.) Oliv. var. alba Berhaut ■☆
39801 Aspilia ritellii Chiov. = Aspilia mossambicensis (Oliv.) Wild ■☆
39802 Aspilia rudis Oliv. et Hiern;粗糙阿斯皮菊■☆
39803 Aspilia rudis Oliv. et Hiern subsp. fontinaloides C. D. Adams;春阿斯皮菊■☆
39804 Aspilia rugulosa Humbert;稍皱阿斯皮菊■☆
39805 Aspilia samariensis O. Hoffm. et Muschl.;撒马利亚阿斯皮菊■☆
39806 Aspilia schimperi (Sch. Bip. ex A. Rich.) Oliv. et Hiern = Aspilia helianthoides (Schumach. et Thonn.) Oliv. et Hiern subsp. prieuriana (DC.) C. D. Adams ■☆
39807 Aspilia smithiana Oliv. et Hiern = Aspilia helianthoides (Schumach. et Thonn.) Oliv. et Hiern subsp. ciliata (Schumach.) C. D. Adams ■☆
39808 Aspilia spenceriana Muschl. = Aspilia rudis Oliv. et Hiern ■☆
39809 Aspilia subpandurata O. Hoffm.;亚琴形阿斯皮菊■☆
39810 Aspilia tanganyikensis Lawalrée = Aspilia mossambicensis (Oliv.) Wild ■☆
39811 Aspilia thouarsii DC. = Wedelia thouarsii (DC.) H. Rob. ■☆
39812 Aspilia trichodesmoides O. Hoffm.;束毛阿斯皮菊■☆
39813 Aspilia vernayi Brenan = Aspilia mossambicensis (Oliv.) Wild ■☆
39814 Aspilia vulgaris N. E. Br. = Aspilia pluriseta Schweinf. ■☆
39815 Aspilia wedeliaeformis Vatke = Aspilia mossambicensis (Oliv.) Wild ■☆
39816 Aspilia welwitschii O. Hoffm. = Aspilia natalensis (Sond.) Wild ■☆
39817 Aspilia welwitschii O. Hoffm. var. serrata Hiern = Aspilia natalensis (Sond.) Wild ■☆
39818 Aspilia zombensis Baker = Melanthera pungens Oliv. et Hiern var. albinervia (O. Hoffm.) Beentje ■☆
39819 Aspilia zombensis Baker var. longifolia S. Moore = Melanthera pungens Oliv. et Hiern var. albinervia (O. Hoffm.) Beentje ■☆
39820 Aspiliopsis Greenm. = Podachaenium Benth. ex Oerst. ●☆
39821 Aspilobium Sol. = Geniostoma J. R. Forst. et G. Forst. ●
39822 Aspilobium Sol. ex A. Cunn. = Geniostoma J. R. Forst. et G. Forst. ●
39823 Aspilotum Sol. ex Steud. = Aspilobium Sol. ●
39824 Aspitium Neck. ex Steud. = Laserpitium L. ●☆
39825 Aspla Rchb. = Aopla Lindl. ■
39826 Aspla Rchb. = Habenaria Willd. ■
39827 Aspla Rchb. = Herminium L. ■
39828 Asplundia Harling(1954)(保留属名);阿斯草属■☆
39829 Asplundia acuminata (Ruiz et Pav.) Harling;渐尖阿斯草■☆
39830 Asplundia alata Harling;翅阿斯草■☆
39831 Asplundia albicarpa Hammel;白果阿斯草■☆
39832 Asplundia allenii Hammel;阿伦阿斯草■☆
39833 Asplundia australis Harling;南方阿斯草■☆
39834 Asplundia brachyphylla Harling;短叶阿斯草■☆
39835 Asplundia brasiliensis Harling;巴西阿斯草■☆
39836 Asplundia guianensis Harling;圭亚那阿斯草■☆
39837 Asplundia latifolia (Ruiz et Pav.) Harling;宽叶阿斯草■☆
39838 Asplundia latifrons (Drude) Harling;宽花阿斯草■☆
39839 Asplundia longistyla Harling;长柱阿斯草■☆
39840 Asplundia longitepala Harling;长瓣阿斯草■☆
39841 Asplundia lutea Harling;黄阿斯草■☆
39842 Asplundia microphylla (Oerst.) Harling;小叶阿斯草■☆
39843 Asplundia parviflora Harling;小花阿斯草■☆
39844 Asplundia pycnantha Harling;密花阿斯草■☆
39845 Asplundia rigida (Aubl.) Harling;硬阿斯草■☆
39846 Asplundia stenophylla (Standl.) Harling;窄叶阿斯草■☆
39847 Asplundia truncata Harling;平截阿斯草■☆
39848 Asplundianthus R. M. King et H. Rob. (1975);平托亮泽兰属●☆
39849 Asplundianthus pseudoglomeratus (Sodiro) R. M. King et H. Rob.;平托亮泽兰●☆
39850 Asprella Host = Psilurus Trin. ■☆
39851 Asprella Schreb. = Homalocenchrus Mieg. ex Haller ■
39852 Asprella Schreb. = Leersia Sw. (保留属名)■
39853 Asprella Willd. = Asperella Humb. ■
39854 Asprella oryzoides (L.) Lam. = Leersia oryzoides (L.) Sw. ■
39855 Aspris Adans. = Aira L. (保留属名)■
39856 Asraoa J. Joseph = Wallichia Roxb. ●
39857 Assa Houtt. = Tetracera L. ●
39858 Assaracus Haw. = Narcissus L. ■
39859 Assidora A. Chev. = Schumanniophyton Harms ●☆
39860 Assidora problematica A. Chev. = Schumanniophyton problematicum (A. Chev.) Aubrév. ●☆
39861 Assoella J. M. Monts. = Arenaria L. ■
39862 Assoella J. M. Monts. = Dufourea Gren. ■
39863 Assonia Cav. (废弃属名) = Dombeya Cav. (保留属名)●☆
39864 Assonia cuanzensis Hiern = Dombeya rotundifolia (Hochst.) Planch. ●☆
39865 Assonia huillensis Hiern = Dombeya rotundifolia (Hochst.) Planch. ●☆
39866 Assonia sparmannioides Hiern = Dombeya burgessiae Gerr. -Corn. ex Harv. et Sond. ●☆
39867 Asta Klotzsch ex O. E. Schulz(1933);阿斯塔芥属■☆
39868 Asta schaffineri (S. Watson) O. E. Schulz;阿斯塔芥■☆
39869 Astartea DC. (1828);澳洲桃金娘属●☆
39870 Astartea clavifolia C. A. Gardner;棒叶澳洲桃金娘●☆
39871 Astartea fascicularis (Labill.) DC.;澳洲桃金娘;Winter Pink ●☆
39872 Astelia Banks = Astelia Banks et Sol. ex R. Br. (保留属名)■☆
39873 Astelia Banks et Sol. = Astelia Banks et Sol. ex R. Br. (保留属名)■☆
39874 Astelia Banks et Sol. ex R. Br. (1810)(保留属名);聚星草属(芳香草属,无柱花属);Astelia ■☆
39875 Astelia nervosa Banks et Sol. ex Hook. f.;聚星草;Bush Flax, Bush Lily, Kakaha ■☆
39876 Asteliaceae Dumort. (1829);聚星草科(芳香草科,无柱花科)●■☆
39877 Astelma R. Br. = Helichrysum Mill. (保留属名)●■
39878 Astelma R. Br. = Helipterum DC. ex Lindl. ■☆
39879 Astelma Schltr. = Papuastelma Bullock ●☆
39880 Astemma Less. = Monactis Kunth ●☆
39881 Astemon Regel = Lepechinia Willd. + Sphacele Benth. (保留属

名)●■☆
39882 Astemon Regel = Lepechinia Willd. ●■☆
39883 Astenolobium Nevski = Astragalus L. ●■
39884 Astephananthes Bory = Passiflora L. ●■
39885 Astephania Oliv. (1886);隐冠菊属■☆
39886 Astephania Oliv. = Anisopappus Hook. et Arn. ■
39887 Astephania africana Oliv.;隐冠菊■☆
39888 Astephania africana Oliv. = Anisopappus chinensis Hook. et Arn. subsp. oliveranus (Wild) S. Ortiz, Paiva et Rodr. Oubina ■☆
39889 Astephanocarpa Baker = Syncephalum DC. ●☆
39890 Astephanocarpa arbutifolia Baker = Syncephalum arbutifolium (Baker) Humbert ●☆
39891 Astephanus R. Br. (1810);无冠萝藦属■☆
39892 Astephanus arenarius Decne. = Pleurostelma cernuum (Decne.) Bullock ■☆
39893 Astephanus badius E. Mey. = Tylophora badia (E. Mey.) Schltr. ●☆
39894 Astephanus cernuus Decne. = Pleurostelma cernuum (Decne.) Bullock ■☆
39895 Astephanus cordatus (Thunb.) R. Br. = Tylophora cordata (Thunb.) Druce ●☆
39896 Astephanus dregei (E. Mey.) D. Dietr. = Microloma armatum (Thunb.) Schltr. ■☆
39897 Astephanus frutescens E. Mey. = Secamone filiformis (L. f.) J. H. Ross ●☆
39898 Astephanus lanceolatus (Thunb.) R. Br. = Oncinema lineare (L. f.) Bullock ●☆
39899 Astephanus linearis (Thunb.) R. Br. = Oncinema lineare (L. f.) Bullock ●☆
39900 Astephanus marginatus Decne. = Astephanus zeyheri Turcz. ■☆
39901 Astephanus massoni Schult. = Microloma armatum (Thunb.) Schltr. ■☆
39902 Astephanus neglectus Schltr. = Astephanus triflorus (L. f.) Schult. ■☆
39903 Astephanus ovatus Decne. = Pleurostelma cernuum (Decne.) Bullock ■☆
39904 Astephanus pauciflorus E. Mey. = Astephanus triflorus (L. f.) Schult. ■☆
39905 Astephanus recurvatus Klotzsch = Pleurostelma cernuum (Decne.) Bullock ■☆
39906 Astephanus schimperi Vatke = Pleurostelma schimperi (Vatke) Liede ■☆
39907 Astephanus stenolobus K. Schum. = Tylophora stenoloba (K. Schum.) N. E. Br. ●☆
39908 Astephanus triflorus (L. f.) Schult.;三花无冠萝藦■☆
39909 Astephanus zeyheri Turcz.;泽赫无冠萝藦■☆
39910 Aster L. (1753);紫菀属;Aster, Easter Daisy, Frost Flowers, Michaelmas Daisy, Michaelmas-daisy, Starwort ●■
39911 Aster × frikartii Silva Tar. et C. K. Schneid.;大头紫菀;Aster ■☆
39912 Aster × frikartii Silva Tar. et C. K. Schneid. 'Mönch';修道士大头紫菀■☆
39913 Aster × frikartii Silva Tar. et C. K. Schneid. 'Wunder von Stafa';奇异大头紫菀■☆
39914 Aster × hashimotoi Kitam. = Aster sekimotoi Makino ■☆
39915 Aster × koshikiensis Kitam.;五色紫菀■☆
39916 Aster abatus S. F. Blake;莫哈维紫菀;Desert Aster, Mohave Aster ■☆
39917 Aster abyssinicus Sch. Bip. = Felicia dentata (A. Rich.) Dandy ■☆
39918 Aster acadiensis Shinners = Aster lateriflorus (L.) Britton var. hirsuticaulis (Lindl. ex DC.) Porter ■☆
39919 Aster acuminatus Michx. = Oclemena acuminata (Michx.) Greene ■☆
39920 Aster acuminatus Nees;山木紫菀;Acuminate Aster, Mountain Aster, Whorled Aster, Whorled Wood Aster ■☆
39921 Aster acuminatus var. magdalenensis Fernald = Oclemena acuminata (Michx.) Greene ■☆
39922 Aster adfinis Less. = Felicia dubia Cass. ■☆
39923 Aster adfinis Less. var. strictus (DC.) Harv. = Felicia amoena (Sch. Bip.) Levyns subsp. stricta (DC.) Grau ■☆
39924 Aster adnatus Nutt. = Symphyotrichum adnatum (Nutt.) G. L. Nesom ■☆
39925 Aster adustus Koidz. ex Nakai = Aster ageratoides Turcz. ■
39926 Aster aethiopicus Burm. f. = Felicia aethiopica (Burm. f.) Bolus et Wolley-Dod ex Adamson et T. M. Salter ■☆
39927 Aster aethiopicus Burm. f. var. glandulosus Harv. = Felicia aethiopica (Burm. f.) Bolus et Wolley-Dod ex Adamson et T. M. Salter ■☆
39928 Aster ageratoides Turcz.;三脉紫菀(八月白,八月霜,白花千里光,白马兰,白升麻,常年青,红管药,鸡儿肠,马兰,三脉叶马兰,三褶脉紫菀,山白菊,山白兰,山马兰,山雪花,田边菊,消食花,小雪花,野白菊,野白菊花);Threevein Aster ■
39929 Aster ageratoides Turcz. f. leucanthus Kitam. = Aster ageratoides Turcz. var. leiophyllus (Franch. et Sav.) Y. Ling ■
39930 Aster ageratoides Turcz. f. ovalifolius (Kitam.) Ohwi;卵叶三脉紫菀■☆
39931 Aster ageratoides Turcz. f. purpurascens Kitam.;紫三脉紫菀■☆
39932 Aster ageratoides Turcz. subsp. amplexifolius Kitam. = Aster semiamplexicaulis (Makino) Makino ex Koidz. ■☆
39933 Aster ageratoides Turcz. subsp. angustifolius (Kitam.) Kitam. = Aster microcephalus (Miq.) Franch. et Sav. var. angustifolius (Kitam.) Nor. Tanaka ■☆
39934 Aster ageratoides Turcz. subsp. lasioeladus (Hayata) Kitam. = Aster ageratoides Turcz. var. lasiocladus (Hayata) Hand. -Mazz. ■
39935 Aster ageratoides Turcz. subsp. leiophyllus (Franch. et Sav.) Kitam. = Aster ageratoides Turcz. ■
39936 Aster ageratoides Turcz. subsp. leiophyllus (Franch. et Sav.) Kitam. = Aster ageratoides Turcz. var. leiophyllus (Franch. et Sav.) Y. Ling ■
39937 Aster ageratoides Turcz. subsp. megalocephalus Kitam. = Aster ageratoides Turcz. ■
39938 Aster ageratoides Turcz. subsp. microcephalus (Miq.) Kitam. = Aster microcephalus (Miq.) Franch. et Sav. ■☆
39939 Aster ageratoides Turcz. subsp. ovatus (Franch. et Sav.) Kitam. 'Hortensis' = Aster microcephalus (Miq.) Franch. et Sav. var. ovatus (Franch. et Sav.) Soejima et Mot. Ito 'Hortensis' ■☆
39940 Aster ageratoides Turcz. subsp. ovatus (Franch. et Sav.) Kitam. = Aster microcephalus (Miq.) Franch. et Sav. var. ovatus (Franch. et Sav.) Soejima et Mot. Ito ■☆
39941 Aster ageratoides Turcz. subsp. ovatus (Franch. et Sav.) Kitam. f. leucanthus Kitam. = Aster ageratoides Turcz. subsp. leiophyllus (Franch. et Sav.) Kitam. ■
39942 Aster ageratoides Turcz. subsp. ovatus (Franch. et Sav.) Kitam. var. hortensis (Makino) Kitam. = Aster microcephalus (Miq.)

Franch. et Sav. var. ovatus (Franch. et Sav.) Soejima et Mot. Ito 'Hortensis' ■☆

39943 Aster ageratoides Turcz. subsp. ovatus (Franch. et Sav.) Kitam. var. humilis Nakai;矮卵叶三脉紫菀■☆

39944 Aster ageratoides Turcz. subsp. ovatus (Franch. et Sav.) Kitam. var. littoricola Kitam. = Aster microcephalus (Miq.) Franch. et Sav. var. littoricola (Kitam.) Nor. Tanaka ■☆

39945 Aster ageratoides Turcz. subsp. ovatus (Franch. et Sav.) Kitam. var. yezoensis Kitam. et H. Hara = Aster microcephalus (Miq.) Franch. et Sav. var. yezoensis (Kitam. et H. Hara) Soejima et Mot. Ito ■☆

39946 Aster ageratoides Turcz. subsp. ripensis (Makino) Kitam. = Aster microcephalus (Miq.) Franch. et Sav. var. ripensis Makino ■☆

39947 Aster ageratoides Turcz. subsp. sugimotoi (Kitam.) Kitam. = Aster sugimotoi Kitam. ■☆

39948 Aster ageratoides Turcz. subsp. trinervius (Roxb. ex D. Don) Grierson = Aster trinervius Roxb. ex D. Don ■

39949 Aster ageratoides Turcz. subsp. trinervius Grierson = Aster trinervius D. Don ■

39950 Aster ageratoides Turcz. subsp. tubulosus (Makino) Kitam. = Aster microcephalus (Miq.) Franch. et Sav. var. ripensis Makino f. tubulosus (Makino) Makino ■☆

39951 Aster ageratoides Turcz. subsp. yakushimensis Kitam. = Aster yakushimensis (Kitam.) Soejima et Yahara ■☆

39952 Aster ageratoides Turcz. subsp. yoshinaganus Kitam. = Aster yoshinaganus (Kitam.) Mot. Ito et Soejima ■☆

39953 Aster ageratoides Turcz. var. adustus Maxim. = Aster ageratoides Turcz. ■

39954 Aster ageratoides Turcz. var. alpinus Koidz. = Aster viscidulus (Makino) Makino var. alpina (Koidz.) Kitam. ■☆

39955 Aster ageratoides Turcz. var. angustifolius Kitam. = Aster microcephalus (Miq.) Franch. et Sav. var. angustifolius (Kitam.) Nor. Tanaka ■☆

39956 Aster ageratoides Turcz. var. firmus (Diels) Hand. -Mazz.;坚叶三脉紫菀(坚叶紫菀);Firm Threevein Aster ■

39957 Aster ageratoides Turcz. var. gerlachii (Hance) C. C. Chang = Aster ageratoides Turcz. var. gerlachii (Hance) C. C. Chang ex Y. Ling ■

39958 Aster ageratoides Turcz. var. gerlachii (Hance) C. C. Chang ex Y. Ling;狭叶三脉紫菀;Narrowleaf Threevein Aster ■

39959 Aster ageratoides Turcz. var. harae (Makino) Kitam. = Aster ageratoides Turcz. f. purpurascens Kitam. ■☆

39960 Aster ageratoides Turcz. var. harae (Makino) Kitam. f. leucanthus Honda = Aster ageratoides Turcz. ■

39961 Aster ageratoides Turcz. var. harae (Makino) Kitam. f. sawadanus (Kitam.) Ohwi = Aster ageratoides Turcz. var. oligocephalus (Nakai ex H. Hara) Nor. Tanaka ■☆

39962 Aster ageratoides Turcz. var. harae (Makino) Kitam. f. stenophyllus (Kitam.) Ohwi = Aster ageratoides Turcz. ■

39963 Aster ageratoides Turcz. var. harae (Makino) Kitam. f. tenuifolius (Kitam.) Ohwi = Aster ageratoides Turcz. var. tenuifolius Kitam. ■☆

39964 Aster ageratoides Turcz. var. heterophyllus Maxim.;异叶三脉紫菀(异叶三褶脉紫菀,异叶紫菀,玉米托子花,紫叶紫菀);Heteroleaf Threevein Aster ■

39965 Aster ageratoides Turcz. var. intermedius (Soejima) Mot. Ito et Soejima;间型三脉紫菀■☆

39966 Aster ageratoides Turcz. var. lasiocladus (Hayata) Hand. -Mazz.;毛枝三脉紫菀(大柴胡,大鱼鳅串,马兰头,毛茎马兰,毛枝三脉紫菀毛枝,毛枝紫菀,青箭杆草,绒山白兰,细叶六月雪,银柴胡);Hairtwig Threevein Aster ■

39967 Aster ageratoides Turcz. var. laticorymbus (Vaniot) Hand. -Mazz.;宽伞三脉紫菀(红管药,宽伞紫菀,宽序三褶脉紫菀,宽序紫菀,宽叶三褶脉马兰);Broadcorymb Threevein Aster ■

39968 Aster ageratoides Turcz. var. leiophyllus (Franch. et Sav.) Y. Ling;光叶三脉紫菀;Smoothleaf Threevein Aster ■

39969 Aster ageratoides Turcz. var. leiophyllus (Franch. et Sav.) Y. Ling = Aster ageratoides Turcz. ■

39970 Aster ageratoides Turcz. var. micranthus Y. Ling;小花三脉紫菀(山白菊,小花紫菀,野鸡尾巴);Smallflower Threevein Aster ■

39971 Aster ageratoides Turcz. var. microcephalus (Miq.) Ohwi = Aster microcephalus (Miq.) Franch. et Sav. ■☆

39972 Aster ageratoides Turcz. var. oligocephalus (Nakai ex H. Hara) Nor. Tanaka;寡头三脉紫菀■☆

39973 Aster ageratoides Turcz. var. oophyllus Y. Ling = Aster ageratoides Turcz. var. oophyllus Y. Ling ex J. Q. Fu ■

39974 Aster ageratoides Turcz. var. oophyllus Y. Ling ex J. Q. Fu;卵叶山白菊(卵叶三脉紫菀,卵叶紫菀,山白菊);Ovalleaf Threevein Aster ■

39975 Aster ageratoides Turcz. var. ovalifolius Kitam. = Aster ageratoides Turcz. f. ovalifolius (Kitam.) Ohwi ■☆

39976 Aster ageratoides Turcz. var. ovatus (Franch. et Sav.) Nakai = Aster microcephalus (Miq.) Franch. et Sav. var. ovatus (Franch. et Sav.) Soejima et Mot. Ito ■☆

39977 Aster ageratoides Turcz. var. ovatus (Franch. et Sav.) Nakai f. albovariegatus Kitam. = Aster microcephalus (Miq.) Franch. et Sav. var. ovatus (Franch. et Sav.) Soejima et Mot. Ito ■☆

39978 Aster ageratoides Turcz. var. ovatus (Franch. et Sav.) Nakai f. humilis (Nakai) Ohwi = Aster ageratoides Turcz. subsp. ovatus (Franch. et Sav.) Kitam. var. humilis Nakai ■☆

39979 Aster ageratoides Turcz. var. ovatus (Franch. et Sav.) Nakai f. littoricola (Kitam.) Ohwi = Aster microcephalus (Miq.) Franch. et Sav. var. littoricola (Kitam.) Nor. Tanaka ■☆

39980 Aster ageratoides Turcz. var. ovatus (Franch. et Sav.) Nakai f. vernalis (Honda) Ohwi;春三脉紫菀■☆

39981 Aster ageratoides Turcz. var. ovatus (Franch. et Sav.) Nakai f. yezoensis (Kitam. et H. Hara) Ohwi = Aster microcephalus (Miq.) Franch. et Sav. var. yezoensis (Kitam. et H. Hara) Soejima et Mot. Ito ■☆

39982 Aster ageratoides Turcz. var. ovatus Hand. -Mazz. = Aster ageratoides Turcz. var. oophyllus Y. Ling ex J. Q. Fu ■

39983 Aster ageratoides Turcz. var. pendulus W. P. Li et G. X. Chen;垂茎三脉紫菀■

39984 Aster ageratoides Turcz. var. pilosus (Diels) Hand. -Mazz.;长毛三脉紫菀;Pilose Threevein Aster ■

39985 Aster ageratoides Turcz. var. pilosus Hand. -Mazz. = Aster ageratoides Turcz. var. pilosus (Diels) Hand. -Mazz. ■

39986 Aster ageratoides Turcz. var. ripensis (Makino) Ohwi = Aster microcephalus (Miq.) Franch. et Sav. var. ripensis Makino ■☆

39987 Aster ageratoides Turcz. var. robustus (Koidz.) Makino et Nemoto = Aster ageratoides Turcz. ■

39988 Aster ageratoides Turcz. var. sawadanus Kitam. = Aster ageratoides Turcz. var. oligocephalus (Nakai ex H. Hara) Nor. Tanaka ■☆

39989 Aster ageratoides Turcz. var. scaberulus (Miq.) Y. Ling;微糙三

脉紫菀(鸡儿肠,马兰,山白菊,微糙山白菊,微糙紫菀,野粉团儿);Scabrous Threevein Aster ■
39990 Aster ageratoides Turcz. var. scandens (Hayata) Makino et Nemoto = Aster ageratoides Turcz. ■
39991 Aster ageratoides Turcz. var. semiamplexicaulis (Makino) Makino et Nemoto = Aster semiamplexicaulis (Makino) Makino ex Koidz. ■☆
39992 Aster ageratoides Turcz. var. sugimotoi (Kitam.) Kitam. = Aster sugimotoi Kitam. ■☆
39993 Aster ageratoides Turcz. var. tenuifolius Kitam.;细叶三脉紫菀■☆
39994 Aster ageratoides Turcz. var. tubulosus (Makino) Ohwi = Aster microcephalus (Miq.) Franch. et Sav. var. ripensis Makino f. tubulosus (Makino) Makino ■☆
39995 Aster ageratoides Turcz. var. tubulosus (Makino) Ohwi f. albiflorus (Makino) Kitam. = Aster microcephalus (Miq.) Franch. et Sav. var. ripensis Makino f. albiflorus (Makino) H. Hara ■☆
39996 Aster ageratoides Turcz. var. vernalis Honda = Aster ageratoides Turcz. var. ovatus (Franch. et Sav.) Nakai f. vernalis (Honda) Ohwi ■☆
39997 Aster ageratoides Turcz. var. viscidulus (Makino) Kitam. = Aster viscidulus (Makino) Makino ■☆
39998 Aster ageratoides Turcz. var. wattii (C. B. Clarke) Grierson;瓦特三脉紫菀(毛三脉紫菀)■
39999 Aster ageratoides Turcz. var. yoshinaganus (Kitam.) Ohwi = Aster yoshinaganus (Kitam.) Mot. Ito et Soejima ■☆
40000 Aster alatipes Hemsl.;翼柄紫菀(柴胡,大柴胡,伏花,红柴胡,九灵光);Earpetiole Aster,Wingstipe Aster ■
40001 Aster albescens (DC.) Hand.-Mazz. = Aster albescens (DC.) Wall. ex Hand.-Mazz. ●
40002 Aster albescens (DC.) Wall. ex Hand.-Mazz.;小舌紫菀(白背紫菀);Smallligulate Aster,Smallligulatecorolla Aster,Small-ligulated Aster ●
40003 Aster albescens (DC.) Wall. ex Hand.-Mazz. var. discolor Y. Ling;白背小舌紫菀;Discolor Smallligulate Aster ●
40004 Aster albescens (DC.) Wall. ex Hand.-Mazz. var. glandulosus Hand.-Mazz.;腺点小舌紫菀(腺点紫菀);Glandule Smallligulate Aster ●
40005 Aster albescens (DC.) Wall. ex Hand.-Mazz. var. gracilior Hand.-Mazz.;狭叶小舌紫菀;Narrowleaf Smallligulate Aster ●
40006 Aster albescens (DC.) Wall. ex Hand.-Mazz. var. levissimus Hand.-Mazz.;无毛小舌紫菀;Smooth Smallligulate Aster ●
40007 Aster albescens (DC.) Wall. ex Hand.-Mazz. var. limprichtii (Diels) Hand.-Mazz.;椭叶小舌紫菀;Elliptic Smallligulate Aster ●
40008 Aster albescens (DC.) Wall. ex Hand.-Mazz. var. limprichtii Hand.-Mazz. = Aster albescens (DC.) Wall. ex Hand.-Mazz. var. limprichtii (Diels) Hand.-Mazz. ●
40009 Aster albescens (DC.) Wall. ex Hand.-Mazz. var. megaphyllus Y. Ling;大叶小舌紫菀;Bigleaf Smallligulate Aster ●
40010 Aster albescens (DC.) Wall. ex Hand.-Mazz. var. niveus Hand.-Mazz.;白雪小舌紫菀●
40011 Aster albescens (DC.) Wall. ex Hand.-Mazz. var. pilosus Hand.-Mazz.;长毛小舌紫菀;Longhair Smallligulate Aster ●
40012 Aster albescens (DC.) Wall. ex Hand.-Mazz. var. rugosus Y. Ling;糙叶小舌紫菀;Rugose Smallligulate Aster ●
40013 Aster albescens (DC.) Wall. ex Hand.-Mazz. var. salignus Hand.-Mazz.;柳叶小舌紫菀;Willowleaf Smallligulate Aster ●
40014 Aster alpigenus (Torr. et A. Gray) A. Gray = Oreostemma alpigenum (Torr. et A. Gray) Greene ■☆
40015 Aster alpigenus (Torr. et A. Gray) A. Gray var. haydenii (Porter) Cronquist = Oreostemma alpigenum (Torr. et A. Gray) Greene var. haydenii (Porter) G. L. Nesom ■☆
40016 Aster alpinus L.;高山紫菀(高岭紫菀);Alpine Aster,Blue Alpine Daisy,Rock Aster ■
40017 Aster alpinus L. 'Dark Beauty' = Aster alpinus L. 'Dunkle Schöne' ■☆
40018 Aster alpinus L. 'Dunkle Schöne';黑丽高山紫菀■☆
40019 Aster alpinus L. var. diversisquamus Y. Ling;异苞高山紫菀;Diversibract Alpine Aster ■
40020 Aster alpinus L. var. dolomiticus (S. S. Beck) Onno = Aster alpinus L. ■
40021 Aster alpinus L. var. fallax (Tamamsch.) Y. Ling;伪形高山紫菀;False Alpine Aster ■
40022 Aster alpinus L. var. serpentimontanus (Tamamsch.) Y. Ling;蛇岩高山紫菀■
40023 Aster altacus Willd. = Heteropappus altaicus (Willd.) Novopokr. ■
40024 Aster altaicus (Willd.) Novopokr. var. taitoensis Kitam.;台东铁杆蒿■
40025 Aster altaicus (Willd.) Novopokr. var. taitoensis Kitam. = Aster altaicus Willd. ■
40026 Aster altaicus L. var. hirsutus Hand.-Mazz. = Heteropappus altaicus (Willd.) Novopokr. var. hirsutus (Hand.-Mazz.) Y. Ling ■
40027 Aster altaicus Willd. = Aster moupinensis (Franch.) Hand.-Mazz. ■
40028 Aster altaicus Willd. = Heteropappus altaicus (Willd.) Novopokr. ■
40029 Aster altaicus Willd. var. canescens (Nees) Serg. = Heteropappus altaicus (Willd.) Novopokr. var. canescens (Nees) Serg. ■
40030 Aster altaicus Willd. var. hirsutus Hand.-Mazz. = Heteropappus altaicus (Willd.) Novopokr. var. hirsutus (Hand.-Mazz.) Y. Ling ■
40031 Aster altaicus Willd. var. millefolius (Vaniot) Hand.-Mazz. = Heteropappus altaicus (Willd.) Novopokr. var. millefolius (Vaniot) Hand.-Mazz. ■
40032 Aster altaicus Willd. var. scaber (Avé-Lall) Hand.-Mazz. = Heteropappus altaicus (Willd.) Novopokr. var. scaber (Avé-Lall.) Wang-Wei ■
40033 Aster altaicus Willd. var. taitoensis Kitam. = Aster altaicus Willd. ■
40034 Aster altaicus Willd. var. taitoensis Kitam. = Heteropappus altaicus (Willd.) Novopokr. var. taitoensis (Kitam.) Y. Ling ■
40035 Aster alyssoides Turcz. = Asterothamnus alyssoides (Turcz.) Novopokr. ●
40036 Aster alyssoides Turcz. var. achnolepis Hand.-Mazz. = Asterothamnus centrali-asiaticus Novopokr. ●
40037 Aster amelloides Besser;假意大利紫菀■☆
40038 Aster amellus L.;意大利紫菀(雅美紫菀);Italian Aster,Italian Starwort ■☆
40039 Aster amellus L. 'King George';乔治王蓝菀■☆
40040 Aster amellus L. 'Mauve Beauty';紫丽蓝菀■☆
40041 Aster amellus L. 'Nocturne';夜曲蓝菀■☆
40042 Aster amellus L. 'Rudolph Goethe';歌德蓝菀■☆
40043 Aster amellus L. 'Sonia';索尼亚蓝菀■☆
40044 Aster amellus L. 'Veilchenkonigin';紫后蓝菀■☆
40045 Aster amellus L. 'Violet Queen' = Aster amellus L. 'Veilchenkonigin' ■☆

40046 Aster amellus L. = Aster indamellus Grierson ■
40047 Aster amethystinus Nutt.;紫罗兰紫菀;Amethyst Aster ■☆
40048 Aster amethystinus Nutt. = Symphyotrichum amethystinum (Nutt.) G. L. Nesom ■☆
40049 Aster amethystinus Nutt. f. leucerythros Bemis = Aster amethystinus Nutt. ■☆
40050 Aster amethystinus Nutt. f. leucos Bemis = Aster amethystinus Nutt. ■☆
40051 Aster amoenus (Sch. Bip.) Harv. = Felicia amoena (Sch. Bip.) Levyns ■☆
40052 Aster amplexicaulis Michx. = Symphyotrichum patens (Aiton) G. L. Nesom ■☆
40053 Aster amplexifolius Rydb. = Eurybia integrifolia (Nutt.) G. L. Nesom ■☆
40054 Aster amygdalinus Lam. = Doellingeria umbellata (Mill.) Nees ■☆
40055 Aster andersonii (A. Gray) A. Gray;安氏紫菀■☆
40056 Aster andohahelensis Humbert = Madagaster andohahelensis (Humbert) G. L. Nesom ●☆
40057 Aster andringitrensis Humbert = Rochonia aspera Humbert ●☆
40058 Aster angustifolius C. C. Chang = Gymnaster angustifolius (C. C. Chang) Y. Ling ■
40059 Aster angustifolius Jacq. = Felicia hyssopifolia (P. J. Bergius) Nees ■☆
40060 Aster angustifolius Lindl. ex Royle = Heteropappus altaicus (Willd.) Novopokr. ■
40061 Aster angustissimus Tausch. = Galatella angustissima (Tausch) Novopokr. ■
40062 Aster angustus (Lindl.) Torr. et A. Gray = Brachyactis ciliata (Ledeb.) Ledeb. subsp. angusta (Lindl.) A. G. Jones ■☆
40063 Aster angustus (Lindl.) Torr. et A. Gray = Symphyotrichum ciliatum (Ledeb.) G. L. Nesom ■☆
40064 Aster angustus Tort. et Gray = Brachyactis ciliata Ledeb. ■
40065 Aster angustus Tort. et Gray = Symphyotrichum ciliatum (Ledeb.) G. L. Nesom ■☆
40066 Aster annectens Harv. = Felicia annectens (Harv.) Grau ■☆
40067 Aster annuus L. = Erigeron annuus (L.) Pers. ■
40068 Aster anomalus Engelm. ex Torr. et A. Gray = Symphyotrichum anomalum (Engelm. ex Torr. et A. Gray) G. L. Nesom ■☆
40069 Aster anticostensis Fernald = Symphyotrichum anticostense (Fernald) G. L. Nesom ■☆
40070 Aster arenarioides D. C. Eaton ex A. Gray = Erigeron arenarioides (D. C. Eaton ex A. Gray) A. Gray ex Rydb. ■☆
40071 Aster arenarius (Kitam.) Nemoto = Heteropappus arenarius Kitam. ■
40072 Aster arenosus (A. Heller) S. F. Blake = Chaetopappa ericoides (Torr.) G. L. Nesom ■☆
40073 Aster argunensis DC. = Aster flaccidus Bunge ■
40074 Aster argutus (Aiton) Kuntze = Solidago arguta Aiton ■☆
40075 Aster argyi H. Lév. = Aster panduratus Nees ex Walp. ■
40076 Aster argyropholis Hand.-Mazz.;银鳞紫菀;Silvery Aster ●
40077 Aster argyropholis Hand.-Mazz. var. niveus Y. Ling;白雪银鳞紫菀;Snow Silvery Aster ●
40078 Aster argyropholis Hand.-Mazz. var. paradoxus Y. Ling;奇形银鳞紫菀●
40079 Aster arnottii Nees ex Torr. et A. Gray = Symphyotrichum patens (Aiton) G. L. Nesom var. patentissimum (Lindl. ex DC.) G. L. Nesom ■☆
40080 Aster asagrayi Makino = Heteropappus ciliosus (Turcz.) Y. Ling ■
40081 Aster asagrayi Makino var. walkeri Kitam.;沃克紫菀■☆
40082 Aster ascendens Lindl. = Symphyotrichum ascendens (Lindl.) G. L. Nesom ■☆
40083 Aster ascendens Lindl. var. parryi D. C. Eaton = Symphyotrichum foliaceum (Lindl. ex DC.) G. L. Nesom var. parryi (D. C. Eaton) G. L. Nesom ■☆
40084 Aster ascendens Lindl. var. yosemitanus A. Gray = Symphyotrichum spathulatum (Lindl.) G. L. Nesom var. yosemitanum (A. Gray) G. L. Nesom ■☆
40085 Aster asper J. M. Wood et M. S. Evans = Aster bakerianus Burtt Davy ex C. A. Sm. ■☆
40086 Aster asperifolius E. S. Burgess = Symphyotrichum undulatum (L.) G. L. Nesom ■☆
40087 Aster asperrimus Nees = Aster trinervius D. Don ■
40088 Aster asperugineus D. C. Eaton = Erigeron asperugineus (D. C. Eaton) A. Gray ■☆
40089 Aster associatus Kitag. = Kalimeris lautureana (Debeaux) Kitam. ■
40090 Aster associatus Kitag. var. stenolobus Kitag. = Kalimeris lautureana (Debeaux) Kitam. ■
40091 Aster asteroides (DC.) Kuntze;星舌紫菀(块根紫菀);Startongue Aster ■
40092 Aster asteroides (DC.) Kuntze subsp. costei (H. Lév.) Grierson = Aster likiangensis Franch. ■
40093 Aster asteroides (L.) MacMill. = Sericocarpus asteroides (L.) Nees ■☆
40094 Aster attenuatus Lindl. = Symphyotrichum laeve (L.) G. L. Nesom var. purpuratum (Nees) G. L. Nesom ■☆
40095 Aster auriculatus Franch.;耳叶紫菀(毛叶子,散药,蓑衣莲,银钱菊);Auriculateleaf Aster ■
40096 Aster auriculatus Franch. f. crenatus Y. Ling;怒江耳叶紫菀;Nujiang Auriculateleaf Aster ■
40097 Aster auriculatus Franch. var. oligocephalus Y. Ling = Aster veitchianus Hutch. et Drumm. ■
40098 Aster azureus Lindl. = Aster oolentangiensis Riddell ■☆
40099 Aster azureus Lindl. = Symphyotrichum oolentangiense (Riddell) G. L. Nesom ■☆
40100 Aster azureus Lindl. f. incarnatus Farw. = Symphyotrichum oolentangiense (Riddell) G. L. Nesom ■☆
40101 Aster azureus Lindl. f. laevicaulis Fernald = Aster oolentangiensis Riddell ■☆
40102 Aster azureus Lindl. var. poaceus (E. S. Burgess) Fernald = Aster oolentangiensis Riddell ■☆
40103 Aster azureus Lindl. var. poaceus (E. S. Burgess) Fernald = Symphyotrichum oolentangiense (Riddell) G. L. Nesom ■☆
40104 Aster azureus Lindl. var. scabrior Engelm. ex E. S. Burgess = Symphyotrichum oolentangiense (Riddell) G. L. Nesom ■☆
40105 Aster baccharoides (Benth.) Steetz;白舌紫菀;Whiteligulate Aster, Whiteligulatecorolla Aster ■
40106 Aster baccharoides (Benth.) Steetz = Aster ovalifolius Kitam. ■
40107 Aster baccharoides (Benth.) Steetz var. kanehirai Yamam. = Aster taiwanensis Kitam. ■
40108 Aster baccharoides (Benth.) Steetz var. sinianus (Hand.-Mazz.) Y. Ling;长苞白舌紫菀■
40109 Aster baccharoides Steetz = Aster ovalifolius Kitam. ■
40110 Aster baccharoides Steetz var. kanehirae Yamam. = Aster

taiwanensis Kitam. ■

40111 Aster bahamensis Britton = Symphyotrichum subulatum (Michx.) G. L. Nesom var. elongatum (Bosser. ex A. G. Jones et Lowry) S. D. Sundb. ■☆

40112 Aster bakerianus Burtt Davy ex C. A. Sm.;贝克紫菀■☆

40113 Aster bakerianus Burtt Davy ex C. A. Sm. subsp. albiflorus W. Lippert = Aster bakerianus Burtt Davy ex C. A. Sm. ■☆

40114 Aster bakerianus Burtt Davy ex C. A. Sm. subsp. angustifolius W. Lippert = Aster bakerianus Burtt Davy ex C. A. Sm. ■☆

40115 Aster bakerianus Burtt Davy ex C. A. Sm. subsp. intermedius W. Lippert = Aster bakerianus Burtt Davy ex C. A. Sm. ■☆

40116 Aster bakerianus Burtt Davy ex C. A. Sm. subsp. ovalis W. Lippert = Aster bakerianus Burtt Davy ex C. A. Sm. ■☆

40117 Aster bakerianus Burtt Davy ex C. A. Sm. subsp. septentrionalis W. Lippert = Aster bakerianus Burtt Davy ex C. A. Sm. ■☆

40118 Aster baldwinii Torr. et A. Gray = Symphyotrichum undulatum (L.) G. L. Nesom ■☆

40119 Aster barbatus (DC.) Harv. = Felicia ovata (Thunb.) Compton ●☆

40120 Aster barbellatus Grierson;髯毛紫菀;Barbate Aster,Beard Aster ■

40121 Aster baronii Humbert = Madagaster senecionoides (Baker) G. L. Nesom ●☆

40122 Aster batakensis Hayata = Heteropappus hispidus (Thunb.) Less. ■

40123 Aster batangensis Bureau et Franch.;巴塘紫菀(万年青);Batang Aster ■

40124 Aster batangensis Bureau et Franch. var. staticefolius (Franch.) Y. Ling;匙叶巴塘紫菀(打毒根);Spoonleaf Batang Aster ■

40125 Aster bellidiastrum Nees;禾鼠麹紫菀;Michel Aster ■☆

40126 Aster bellidiastrum Nees ex Walp. = Aster bellidiastrum Nees ■☆

40127 Aster bellidiastrum Scop. = Aster bellidiastrum Nees ■☆

40128 Aster bellidiflorus Willd. = Symphyotrichum lanceolatum (Willd.) G. L. Nesom ■☆

40129 Aster bergerianus (Spreng.) Harv. = Felicia bergeriana (Spreng.) O. Hoffm. ■☆

40130 Aster beringensis Gand. = Eurybia sibirica (L.) G. L. Nesom ■☆

40131 Aster bernardinus H. M. Hall = Symphyotrichum defoliatum (Parish) G. L. Nesom ■☆

40132 Aster bicolor (L.) Nees = Solidago bicolor L. ■☆

40133 Aster bicolor (L.) Nees var. lanatus (Hook.) Kuntze = Solidago hispida Muhl. ex Willd. ■☆

40134 Aster biennis (Lindl.) Ledeb. = Heteropappus tataricus (Lindl.) Tamamsch. ■

40135 Aster bietii Franch.;线舌紫菀;Lineartongue Aster ■

40136 Aster bifoliatus (Walter) H. E. Ahles = Sericocarpus tortifolius (Michx.) Nees ■☆

40137 Aster bigelovii A. Gray = Dieteria bigelovii (A. Gray) D. R. Morgan et R. L. Hartm. ■☆

40138 Aster bipinnatisectus Ludlow ex Grierson;重羽紫菀;Bipinnatisect Aster ■

40139 Aster blakei (Porter) House = Oclemena blakei (Porter) G. L. Nesom ■☆

40140 Aster blepharophyllus A. Gray = Arida blepharophylla (A. Gray) D. R. Morgan et R. L. Hartm. ■☆

40141 Aster blinii H. Lév. = Aster ageratoides Turcz. var. oophyllus Y. Ling ex J. Q. Fu ■

40142 Aster bloomeri A. Gray = Symphyotrichum campestre (Nutt.) G. L. Nesom ■☆

40143 Aster bodinieri H. Lév. = Aster brachytrichus Franch. ■

40144 Aster bodkinii Compton = Felicia filifolia (Vent.) Burtt Davy subsp. bodkinii (Compton) Grau ■☆

40145 Aster boltoniae Greene = Psilactis asteroides A. Gray ■☆

40146 Aster bonplandii Kuntze = Solidago simplex Kunth ■☆

40147 Aster borealis (Torr. et A. Gray) Prov.;北方紫菀;Northern Bog Aster,Rush Aster ■☆

40148 Aster borealis (Torr. et A. Gray) Prov. = Symphyotrichum boreale (Torr. et A. Gray) Á. Löve et D. Löve ■☆

40149 Aster bowerii Hemsl. = Heteropappus bowerii (Hemsl.) Grierson ■

40150 Aster bowerii Hemsl. = Heteropappus crenatifolius (Hand.-Mazz.) Grierson ■

40151 Aster bowiei Harv.;鲍伊紫菀■☆

40152 Aster boykinii E. S. Burgess = Eurybia divaricata (L.) G. L. Nesom ■☆

40153 Aster bracei Britton = Symphyotrichum tenuifolium (L.) G. L. Nesom var. aphyllum (R. W. Long) S. D. Sundb. ■☆

40154 Aster brachyactis S. F. Blake;短线紫菀;Rayless Aster ■☆

40155 Aster brachyactis S. F. Blake = Brachyactis ciliata (Ledeb.) Ledeb. subsp. angusta (Lindl.) A. G. Jones ■☆

40156 Aster brachyactis S. F. Blake = Brachyactis ciliata Ledeb. ■

40157 Aster brachyactis S. F. Blake = Symphyotrichum ciliatum (Ledeb.) G. L. Nesom ■☆

40158 Aster brachypholius Small = Symphyotrichum racemosum (Elliott) G. L. Nesom ■☆

40159 Aster brachyphylius C. C. Chang = Aster falcifolius Hand.-Mazz. ■

40160 Aster brachytrichus Franch.;短毛紫菀;Hankow Aster,Shorthair Aster ■

40161 Aster brachytrichus Franch. var. angustisquamus Y. Ling;狭苞短毛紫菀;Narrowbract Shorthair Aster ■

40162 Aster brachytrichus Franch. var. denticulatus Onno = Aster brachytrichus Franch. ■

40163 Aster brachytrichus Franch. var. latifolius Y. Ling;宽叶短毛紫菀;Broadleaf Shorthair Aster ■

40164 Aster brachytrichus Franch. var. oreaster Onno = Aster brachytrichus Franch. ■

40165 Aster brachytrichus Franch. var. tenuiligulatus Y. Ling;细舌短毛紫菀;Thintongue Shorthair Aster ■

40166 Aster bracteolatus Nutt. = Symphyotrichum eatonii (A. Gray) G. L. Nesom ■☆

40167 Aster brevilingulatus (Sch. Bip. ex Hemsl.) McVaugh = Psilactis brevilingulata Sch. Bip. ex Hemsl. ■☆

40168 Aster brevipedunculatus Hutch.;短梗紫菀■☆

40169 Aster brevipes Benth. = Aster baccharoides (Benth.) Steetz ■

40170 Aster brevis Hand.-Mazz.;短茎紫菀;Shortstem Aster ■

40171 Aster breviscapus Vaniot = Erigeron breviscapus (Vaniot) Hand.-Mazz. ■

40172 Aster breweri (A. Gray) Semple = Eucephalus breweri (A. Gray) G. L. Nesom ■☆

40173 Aster brickellioides Greene = Eucephalus tomentellus (Greene) Greene ■☆

40174 Aster brickellioides Greene var. glabratus Greene = Eucephalus glabratus (Greene) Greene ■☆

40175 Aster brittonii Kuntze = Solidago wrightii A. Gray ■☆

40176 Aster buchubergensis Dinter = Felicia hirsuta DC. ●☆

40177 Aster buckleyi (Torr. et A. Gray) Kuntze = Solidago buckleyi

Torr. et A. Gray ■☆
40178 Aster bulleyanus Jeffrey;扁毛紫菀;Flathair Aster ■
40179 Aster burkei Harv. = Felicia burkei (Harv.) L. Bolus ■☆
40180 Aster butleri Rydb. = Symphyotrichum subspicatum (Nees) G. L. Nesom ■☆
40181 Aster cabulicus Lindl. = Aster albescens (DC.) Wall. ex Hand. -Mazz. ●
40182 Aster caesius (L.) Kuntze = Solidago caesia L. ■☆
40183 Aster caffrorum Less. = Microglossa caffrorum (Less.) Grau ●☆
40184 Aster calderi B. Boivin = Symphyotrichum puniceum (L.) Á. Löve et D. Löve ■☆
40185 Aster californicus (Nutt.) Kuntze = Solidago velutina DC. subsp. californica (Nutt.) Semple ■☆
40186 Aster campestris Nutt. = Symphyotrichum campestre (Nutt.) G. L. Nesom ■☆
40187 Aster campestris var. bloomeri (A. Gray) A. Gray = Symphyotrichum campestre (Nutt.) G. L. Nesom ■☆
40188 Aster camptosorus Small = Aster shortii Lindl. ■☆
40189 Aster camptosorus Small = Symphyotrichum shortii (Lindl.) G. L. Nesom ■☆
40190 Aster canadensis (L.) Kuntze = Solidago canadensis L. ■
40191 Aster candelabrum Vaniot = Aster panduratus Nees ex Walp. ■
40192 Aster candollei Harv. = Felicia cymbalarioides (DC.) Grau ■☆
40193 Aster candollei Kuntze = Solidago spathulata DC. ■☆
40194 Aster canescens (Nees) Fisjun;灰色紫菀;Hairy Aster, Hairy Starwort ■☆
40195 Aster canescens Pursh = Dieteria canescens (Pursh) Nutt. ■☆
40196 Aster canescens Pursh var. aristatus Eastw. = Dieteria canescens (Pursh) Nutt. var. aristata (Eastw.) D. R. Morgan et R. L. Hartm. ■☆
40197 Aster canescens Pursh var. tephrodes A. Gray = Dieteria asteroides Torr. ■☆
40198 Aster canescens Pursh var. viridis A. Gray = Dieteria canescens (Pursh) Nutt. var. glabra (A. Gray) D. R. Morgan et R. L. Hartm. ■☆
40199 Aster cantonensis Lour. = Kalimeris indica (L.) Sch. Bip. ■
40200 Aster capensis Less. = Felicia amelloides (L.) Voss ■☆
40201 Aster capensis Less. = Felicia amelloides Schltr. ■☆
40202 Aster capensis Less. var. rotundifolius (Thunb.) Harv. = Felicia amelloides (L.) Voss ■☆
40203 Aster capillaceus E. S. Burgess = Symphyotrichum oolentangiense (Riddell) G. L. Nesom ■☆
40204 Aster carmesinus E. S. Burgess ex Britton et A. Br. = Eurybia divaricata (L.) G. L. Nesom ■☆
40205 Aster carnosus Gilib. = Tripolium vulgare Nees ■
40206 Aster carolinianus Walter = Ampelaster carolinianus (Walter) G. L. Nesom ■☆
40207 Aster castaneus E. S. Burgess = Eurybia divaricata (L.) G. L. Nesom ■☆
40208 Aster cavaleriei Vaniot et H. Lév. = Aster albescens (DC.) Wall. ex Hand. -Mazz. ●
40209 Aster centrali-asiaticus Novopokr. = Asterothamnus centrali-asiaticus Novopokr. ●
40210 Aster chapmanii Torr. et A. Gray = Symphyotrichum chapmanii (Torr. et A. Gray) Semple et Brouillet ■☆
40211 Aster charieis Less. = Felicia ovata (Thunb.) Compton ●☆
40212 Aster chilensis Nees = Symphyotrichum chilense (Nees) G. L. Nesom ■☆
40213 Aster chilensis Nees subsp. ascendens (Lindl.) Cronquist = Symphyotrichum ascendens (Lindl.) G. L. Nesom ■☆
40214 Aster chilensis Nees subsp. hallii (A. Gray) Cronquist = Symphyotrichum hallii (A. Gray) G. L. Nesom ■☆
40215 Aster chilensis Nees var. bernardinus (H. M. Hall) Cronquist = Symphyotrichum defoliatum (Parish) G. L. Nesom ■☆
40216 Aster chilensis Nees var. invenustus (Greene) Jeps. = Symphyotrichum chilense (Nees) G. L. Nesom ■☆
40217 Aster chilensis Nees var. lentus (Greene) Jeps. = Symphyotrichum lentum (Greene) G. L. Nesom ■☆
40218 Aster chilensis Nees var. sonomensis (Greene) Jeps. = Symphyotrichum lentum (Greene) G. L. Nesom ■☆
40219 Aster chimanimaniensis W. Lippert;奇马尼马尼紫菀■☆
40220 Aster chinensis L. = Callistephus chinensis (L.) Nees ■
40221 Aster chingshuiensis Y. C. Liu et C. H. Ou;清水马兰■
40222 Aster chlorolepis E. S. Burgess = Eurybia chlorolepis (E. S. Burgess) G. L. Nesom ■☆
40223 Aster chromopappus Novopokr. = Galatella chromopappa Novopokr. ■
40224 Aster ciliatus (Turcz.) Hand. -Mazz. = Heteropappus ciliosus (Turcz.) Y. Ling ■
40225 Aster ciliatus Muhl. ex Willd. = Symphyotrichum ericoides (L.) G. L. Nesom ■☆
40226 Aster ciliolatus Lindl. = Symphyotrichum ciliolatum (Lindl.) Á. Löve et D. Löve ■☆
40227 Aster ciliolatus Lindl. var. comatus (Fernald) A. G. Jones = Symphyotrichum ciliolatum (Lindl.) Á. Löve et D. Löve ■☆
40228 Aster ciliosus (Turcz.) Hand. -Mazz. = Heteropappus ciliosus (Turcz.) Y. Ling ■
40229 Aster ciliosus Kitam. = Aster meyendorfii (Regel et Maack) Voss ■
40230 Aster ciliosus Kitam. = Heteropappus meyendorffii (Regel et Maack) Kom. ■
40231 Aster claviger E. S. Burgess = Symphyotrichum undulatum (L.) G. L. Nesom ■☆
40232 Aster claytonii E. S. Burgess = Eurybia divaricata (L.) G. L. Nesom ■☆
40233 Aster coerulescens DC. = Symphyotrichum praealtum (Poir.) G. L. Nesom var. texicola (Wiegand) G. L. Nesom ■☆
40234 Aster coerulescens DC. var. wootonii (Greene) Wiegand = Symphyotrichum lanceolatum (Willd.) G. L. Nesom var. hesperium (A. Gray) G. L. Nesom ■☆
40235 Aster cognatus H. M. Hall;麦加紫菀;Mecca Aster, Meccaaster ■☆
40236 Aster cognatus H. M. Hall = Xylorhiza cognata (H. M. Hall) T. J. Watson ●☆
40237 Aster coloradoensis A. Gray = Xanthisma coloradoense (A. Gray) D. R. Morgan et R. L. Hartm. ■☆
40238 Aster commixtus (Nees) Kuntze = Eurybia spectabilis (Aiton) G. L. Nesom ■☆
40239 Aster commutatus (Torr. et A. Gray) A. Gray;草地白紫菀;White Prairie Aster ■☆
40240 Aster commutatus (Torr. et A. Gray) A. Gray = Symphyotrichum falcatum (Lindl.) G. L. Nesom var. commutatum (Torr. et A. Gray) G. L. Nesom ■☆
40241 Aster commutatus (Torr. et A. Gray) A. Gray var. crassulus (Rydb.) S. F. Blake = Symphyotrichum falcatum (Lindl.) G. L. Nesom var. commutatum (Torr. et A. Gray) G. L. Nesom ■☆
40242 Aster commutatus (Torr. et A. Gray) A. Gray var. polycephalus (Rydb.) S. F. Blake = Symphyotrichum falcatum (Lindl.) G. L.

Nesom var. commutatum (Torr. et A. Gray) G. L. Nesom ■☆
40243 Aster commutatus (Torr. et A. Gray) A. Gray var. polycephalus S. F. Blake = Symphyotrichum falcatum (Lindl.) G. L. Nesom var. commutatum (Torr. et A. Gray) G. L. Nesom ■☆
40244 Aster comptonii W. Lippert;康普顿紫菀■☆
40245 Aster concinnus Willd. = Symphyotrichum laeve (L.) G. L. Nesom var. concinnum (Willd.) G. L. Nesom ■☆
40246 Aster concolor L. = Symphyotrichum concolor (L.) G. L. Nesom ■☆
40247 Aster concolor L. var. simulatus (Small) R. W. Long = Symphyotrichum concolor (L.) G. L. Nesom ■☆
40248 Aster conduplicatus E. S. Burgess = Aster puniceus L. ■☆
40249 Aster conduplicatus E. S. Burgess = Symphyotrichum puniceum (L.) Á. Löve et D. Löve ■☆
40250 Aster confertiflorus (DC.) Kuntze = Solidago simplex Kunth ■☆
40251 Aster confusus Harv. = Felicia hirsuta DC. ●☆
40252 Aster consanguineus Ledeb. = Erigeron komarovii Botsch. ■
40253 Aster conspicuus Lindl. = Eurybia conspicua (Lindl.) G. L. Nesom ■☆
40254 Aster continuus Small = Symphyotrichum patens (Aiton) G. L. Nesom var. patentissimum (Lindl. ex DC.) G. L. Nesom ■☆
40255 Aster cordifolium Michx. = Symphyotrichum cordifolium (L.) G. L. Nesom ■☆
40256 Aster cordifolius L.;心叶紫菀;Beewood, Blue Wood Aster, Common Blue Heart-leaved Aster, Common Blue Wood Aster, Heart-leaf Aster, Heart-leafed Aster, Heart-shaped Aster, Wood Aster ■☆
40257 Aster cordifolius L. 'Silver Spray';银雾心叶紫菀■
40258 Aster cordifolius L. subsp. sagittifolius (Wedem. ex Willd.) A. G. Jones = Symphyotrichum urophyllum (Lindl. ex DC.) G. L. Nesom ■☆
40259 Aster cordifolius L. subsp. sagittifolius (Wedem. ex Willd.) A. G. Jones = Aster sagittifolius Wedem. ex Willd. ■☆
40260 Aster cordifolius L. subsp. sagittifolius (Wedem. ex Willd.) A. G. Jones = Symphyotrichum cordifolium (L.) G. L. Nesom ■☆
40261 Aster cordifolius L. var. furbishiae Fernald = Symphyotrichum cordifolium (L.) G. L. Nesom ■☆
40262 Aster cordifolius L. var. incisus Britton = Symphyotrichum cordifolium (L.) G. L. Nesom ■☆
40263 Aster cordifolius L. var. lanceolatus Porter = Symphyotrichum cordifolium (L.) G. L. Nesom ■☆
40264 Aster cordifolius L. var. moratus (Shinners) Shinners = Aster cordifolius L. ■☆
40265 Aster cordifolius L. var. moratus Shinners = Symphyotrichum cordifolium (L.) G. L. Nesom ■☆
40266 Aster cordifolius L. var. polycephalus Porter = Symphyotrichum cordifolium (L.) G. L. Nesom ■☆
40267 Aster cordifolius L. var. racemiflorus Fernald = Symphyotrichum cordifolium (L.) G. L. Nesom ■☆
40268 Aster cordifolius L. var. sagittifolius (Wedem. ex Willd.) A. G. Jones = Symphyotrichum urophyllum (Lindl. ex DC.) G. L. Nesom ■☆
40269 Aster cordifolius L. var. sagittifolius (Wedem. ex Willd.) A. G. Jones = Aster sagittifolius Wedem. ex Willd. ■☆
40270 Aster cordifolius Lam. = Aster paniculatus Lam. ■☆
40271 Aster cordineri A. Nelson = Symphyotrichum falcatum (Lindl.) G. L. Nesom var. commutatum (Torr. et A. Gray) G. L. Nesom ■☆
40272 Aster coriaceifolius H. Lév. et Vaniot;岗斑鸠菊■
40273 Aster coriaceifolius H. Lév. et Vaniot = Vernonia clivorum Hance ■
40274 Aster coridifolius Michx. = Symphyotrichum dumosum (L.) G. L. Nesom ■☆
40275 Aster cornifolius Muhl. ex Willd. = Doellingeria infirma (Michx.) Greene ■☆
40276 Aster corrigiatus E. S. Burgess = Symphyotrichum undulatum (L.) G. L. Nesom ■☆
40277 Aster corymbosus Aiton = Eurybia divaricata (L.) G. L. Nesom ■☆
40278 Aster corymbosus Aiton var. alatus W. P. C. Barton = Eurybia divaricata (L.) G. L. Nesom ■☆
40279 Aster corymbosus Sol. ex Aiton = Doellingeria scabra (Thunb.) Nees ■
40280 Aster costei H. Lév. = Aster likiangensis Franch. ■
40281 Aster covillei (Greene) S. F. Blake ex M. Peck = Eucephalus ledophyllus (A. Gray) Greene var. covillei (Greene) G. L. Nesom ■☆
40282 Aster crassulus Rydb. = Symphyotrichum falcatum (Lindl.) G. L. Nesom var. commutatum (Torr. et A. Gray) G. L. Nesom ■☆
40283 Aster crenatifolius Hand.-Mazz. = Heteropappus crenatifolius (Hand.-Mazz.) Grierson ■
40284 Aster crenatus (Thunb.) Less. = Mairia crenata (Thunb.) Nees ■☆
40285 Aster crenifolius (Fernald) Cronquist = Aster foliaceus Lindl. var. crenifolius Fernald ■☆
40286 Aster crenifolius (Fernald) Cronquist var. arcuans (Fernald) Cronquist = Aster novi-belgii L. ■
40287 Aster crenulatus Hutch. = Felicia boehmii O. Hoffm. ■☆
40288 Aster crinitus L. = Athrixia crinita (L.) Druce ●☆
40289 Aster crinitus Thunb. = Athrixia capensis Ker Gawl. ●☆
40290 Aster crispus Forssk. = Pulicaria undulata (L.) C. A. Mey. ■☆
40291 Aster curtisii Torr. et A. Gray = Symphyotrichum retroflexum (Lindl. ex DC.) G. L. Nesom ■☆
40292 Aster curtus Cronquist = Sericocarpus rigidus Lindl. ■☆
40293 Aster curvatus Vaniot = Aster ageratoides Turcz. var. gerlachii (Hance) C. C. Chang ex Y. Ling ■
40294 Aster cusickii A. Gray = Symphyotrichum cusickii (A. Gray) G. L. Nesom ■☆
40295 Aster cymbalariae Aiton = Felicia cymbalariae (Aiton) Bolus et Wolley-Dod ex Adamson et T. M. Salter ■☆
40296 Aster cymbalariae Aiton var. ionops Harv. = Felicia cymbalariae (Aiton) Bolus et Wolley-Dod ex Adamson et T. M. Salter subsp. ionops (Harv.) Grau ■☆
40297 Aster dahuricus Benth. ex Baker = Galatella dahurica DC. ■
40298 Aster dahuricus Benth. ex Baker subsp. yamatsudanus (Kitag.) Kitag. = Galatella dahurica DC. ■
40299 Aster decemflora Kuntze = Solidago radula Nutt. ■☆
40300 Aster decumbens (Schltr.) G. L. Nesom;外倾紫菀■☆
40301 Aster defoliatus Parish = Symphyotrichum defoliatum (Parish) G. L. Nesom ■☆
40302 Aster delavayi Franch. = Aster diplostephioides Benth. et Hook. f. ■
40303 Aster demissus Harv. = Felicia diffusa (DC.) Grau ■☆
40304 Aster dentatus Thunb. = Felicia australis (Alston) E. Phillips ■☆
40305 Aster depauperatus Fernald = Symphyotrichum depauperatum (Fernald) G. L. Nesom ■☆
40306 Aster depauperatus H. Lév. et Vaniot = Heteropappus meyendorffii (Regel et Maack) Kom. ■
40307 Aster dichotomus Elliott = Oclemena reticulata (Pursh) G. L. Nesom ■☆

40308 Aster diffusus Aiton var. thyrsoides A. Gray = Symphyotrichum ontarionis (Wiegand) G. L. Nesom ■☆
40309 Aster dimorphophyllus Franch. et Sav.;二型叶紫菀■☆
40310 Aster dimorphophyllus Franch. et Sav. = Sinosenecio oldhamianus (Maxim.) B. Nord. ■
40311 Aster dimorphophyllus Franch. et Sav. f. roseus Hayashi;粉花二型叶紫菀■☆
40312 Aster diplostephioides (DC.) C. B. Clarke = Aster tsarungensis (Grierson) Y. Ling ■
40313 Aster diplostephioides (DC.) C. B. Clarke subsp. farreri (W. W. Sm. et Jeffrey) Onno = Aster farreri W. W. Sm. et Jeffrey ■
40314 Aster diplostephioides (DC.) C. B. Clarke var. yunnanensis (Franch.) Onno = Aster yunnanensis Franch. ■
40315 Aster diplostephioides Benth. et Hook. f.;重冠紫菀(寒风参,美多类,太阳花);Doublecorolla Aster ■
40316 Aster diplostephioides Benth. et Hook. f. subsp. farreri (W. W. Sm. et Jeffrey) Onno = Aster farreri W. W. Sm. et Jeffrey ■
40317 Aster diplostephioides Benth. et Hook. f. var. delavayi (Franch.) Onno = Aster diplostephioides (DC.) C. B. Clarke ■
40318 Aster diplostephioides Benth. var. yunnanensis Onno = Aster bulleyanus Jeffrey ■
40319 Aster diplostephioides C. B. Clarke = Aster tsarungensis (Grierson) Y. Ling ■
40320 Aster discoideus Elliott = Brintonia discoidea (Elliott) Greene ■☆
40321 Aster discoideus Sond. = Chrysocoma ciliata L. ■☆
40322 Aster divaricatus L.;宽叉紫菀;White Wood Aster ■☆
40323 Aster divaricatus L. = Eurybia divaricata (L.) G. L. Nesom ■☆
40324 Aster divaricatus L. var. chlorolepis (E. S. Burgess) H. E. Ahles = Eurybia chlorolepis (E. S. Burgess) G. L. Nesom ■☆
40325 Aster dolichophyllus Y. Ling;长叶紫菀;Longleaf Aster ■
40326 Aster dolichopodus Y. Ling;长梗紫菀;Long Peduncled Aster, Longstalk Aster ■
40327 Aster doronicifolius H. Lév. = Aster fuscescens Burret et Franch. ■
40328 Aster dregei (DC.) Harv. = Felicia dregei DC. ■☆
40329 Aster dregei (DC.) Harv. var. dentata ? = Felicia brevifolia (DC.) Grau ■☆
40330 Aster drummondii Lindl.;德拉蒙德紫菀;Drummond's Aster, Hairy Heart-leaved Aster ■☆
40331 Aster drummondii Lindl. = Symphyotrichum drummondii (Lindl.) G. L. Nesom ■☆
40332 Aster drummondii Lindl. var. rhodactis Benke = Aster drummondii Lindl. ■☆
40333 Aster drummondii Lindl. var. texanus (E. S. Burgess) A. G. Jones = Symphyotrichum drummondii (Lindl.) G. L. Nesom var. texanum (E. S. Burgess) G. L. Nesom ■☆
40334 Aster dubius (Thunb.) Onno;东菊(兰铁草,踏地莲花菜,细牛舌片,细药,牙陷药,野菠菜)■
40335 Aster dubius (Thunb.) Onno = Erigeron thunbergii A. Gray ■
40336 Aster dubius (Thunb.) Onno f. leucanthus (H. Hara) Kitam. = Erigeron thunbergii A. Gray f. leucanthus H. Hara ■☆
40337 Aster dubius (Thunb.) Onno subsp. glabratus (A. Gray) Kitam. et H. Hara var. angustifolius (Tatew.) H. Hara ex Kitam. = Erigeron thunbergii A. Gray subsp. glabratus (A. Gray) H. Hara var. angustifolius (Tatew.) H. Hara ■☆
40338 Aster dubius (Thunb.) Onno subsp. glabratus (A. Gray) Kitam. et H. Hara = Erigeron thunbergii A. Gray subsp. glabratus (A. Gray) H. Hara ■☆
40339 Aster dubius (Thunb.) Onno subsp. glabratus Kitam. et H. Hara ex Kitam. = Erigeron komarovii Botsch. ■
40340 Aster dubius (Thunb.) Onno var. glabratus (A. Gray) T. Shimizu = Erigeron thunbergii A. Gray subsp. glabratus (A. Gray) H. Hara ■☆
40341 Aster dubius Onno subsp. glabratus Kitam. = Erigeron komarovii Botsch. ■
40342 Aster dumosus L. = Symphyotrichum dumosum (L.) G. L. Nesom ■☆
40343 Aster dumosus L. var. dodgei Fernald = Aster fragilis Willd. ■☆
40344 Aster dumosus L. var. strictior Torr. et A. Gray = Symphyotrichum dumosum (L.) G. L. Nesom var. strictior (Torr. et A. Gray) G. L. Nesom ■☆
40345 Aster durus Lunell = Symphyotrichum lanceolatum (Willd.) G. L. Nesom var. hesperium (A. Gray) G. L. Nesom ■☆
40346 Aster dysentericus Scop. = Pulicaria dysenterica (L.) Gaertn. ■
40347 Aster eatonii (A. Gray) Howell = Symphyotrichum eatonii (A. Gray) G. L. Nesom ■☆
40348 Aster echinatus (Thunb.) Less. = Felicia echinata (Thunb.) Nees ■☆
40349 Aster echinatus (Thunb.) Less. var. paralia (DC.) Harv. = Felicia echinata (Thunb.) Nees ■☆
40350 Aster ecklonis Less. = Felicia aethiopica (Burm. f.) Bolus et Wolley-Dod ex Adamson et T. M. Salter subsp. ecklonis (Less.) Grau ■☆
40351 Aster elatus (Greene) Cronquist = Oreostemma elatum (Greene) Greene ■☆
40352 Aster elegans (Nutt.) Torr. et A. Gray var. engelmannii D. C. Eaton = Eucephalus engelmannii (D. C. Eaton) Greene ■☆
40353 Aster elegans Hook. f. et Thomson = Aster neoelegans Grierson ■
40354 Aster elegantulus Porsild = Symphyotrichum falcatum (Lindl.) G. L. Nesom ■☆
40355 Aster eliasii A. Nelson = Eurybia radulina (A. Gray) G. L. Nesom ■☆
40356 Aster elliottii Torr. et A. Gray = Symphyotrichum elliottii (Torr. et A. Gray) G. L. Nesom ■☆
40357 Aster elmeri Greene = Erigeron elmeri (Greene) Greene ■☆
40358 Aster elongatus Thunb. = Felicia elongata (Thunb.) O. Hoffm. ■☆
40359 Aster elongatus Thunb. var. barbiger (DC.) Harv. = Felicia hispida (DC.) Grau ■☆
40360 Aster elongatus Thunb. var. candollei Harv. = Felicia namaquana (Harv.) Merxm. ■☆
40361 Aster elongatus Thunb. var. crassifolius Harv. = Felicia ovata (Thunb.) Compton ●☆
40362 Aster elongatus Thunb. var. pappei Harv. = Felicia minima (Hutch.) Grau ■☆
40363 Aster elongatus Thunb. var. spathulifolius Harv. = Felicia amoena (Sch. Bip.) Levyns ■☆
40364 Aster elongatus Thunb. var. thunbergii Harv. = Felicia elongata (Thunb.) O. Hoffm. ■☆
40365 Aster eminens Willd. = Aster novi-belgii L. ■
40366 Aster eminens Willd. = Symphyotrichum lanceolatum (Willd.) G. L. Nesom ■☆
40367 Aster eminens Willd. = Symphyotrichum longifolium (Lam.) G. L. Nesom ■☆
40368 Aster engelmannii (D. C. Eaton) A. Gray;恩氏紫菀;Engelmann Aster ■☆

40369 Aster engelmannii (D. C. Eaton) A. Gray = Eucephalus engelmannii (D. C. Eaton) Greene ■☆
40370 Aster engelmannii (D. C. Eaton) A. Gray var. glaucescens A. Gray = Eucephalus glaucescens (A. Gray) Greene ■☆
40371 Aster engelmannii (D. C. Eaton) A. Gray var. ledophyllus A. Gray = Eucephalus ledophyllus (A. Gray) Greene ■☆
40372 Aster engelmannii (D. C. Eaton) A. Gray var. paucicapitatus B. L. Rob. = Eucephalus paucicapitatus (B. L. Rob.) Greene ■☆
40373 Aster eremophilus Bunge = Krylovia eremophila (Bunge) Schischk. ■
40374 Aster ericifolius Forssk. = Macowania ericifolia (Forssk.) B. L. Burtt et Grau ●☆
40375 Aster ericoides L. 'Golden Spray';金雾毛紫菀■☆
40376 Aster ericoides L. 'White Heather';白石南毛紫菀■☆
40377 Aster ericoides L. = Symphyotrichum ericoides (L.) G. L. Nesom ■☆
40378 Aster ericoides L. f. caeruleus (Benke) S. F. Blake = Symphyotrichum ericoides (L.) G. L. Nesom ■☆
40379 Aster ericoides L. f. gramsii Benke = Symphyotrichum ericoides (L.) G. L. Nesom ■☆
40380 Aster ericoides L. f. prostratus (Kuntze) Fernald = Symphyotrichum ericoides (L.) G. L. Nesom var. prostratum (Kuntze) G. L. Nesom ■☆
40381 Aster ericoides L. subsp. ericoides = Symphyotrichum ericoides (L.) G. L. Nesom ■☆
40382 Aster ericoides L. subsp. pansus (S. F. Blake) A. G. Jones = Symphyotrichum ericoides (L.) G. L. Nesom var. pansum (S. F. Blake) G. L. Nesom ■☆
40383 Aster ericoides L. var. commutatus (Torr. et A. Gray) B. Boivin = Symphyotrichum falcatum (Lindl.) G. L. Nesom var. commutatum (Torr. et A. Gray) G. L. Nesom ■☆
40384 Aster ericoides L. var. depauperatus Porter = Symphyotrichum depauperatum (Fernald) G. L. Nesom ■☆
40385 Aster ericoides L. var. pansus (S. F. Blake) B. Boivin = Symphyotrichum ericoides (L.) G. L. Nesom var. pansum (S. F. Blake) G. L. Nesom ■☆
40386 Aster ericoides L. var. parviceps E. S. Burgess = Symphyotrichum parviceps (E. S. Burgess) G. L. Nesom ■☆
40387 Aster ericoides L. var. pilosus (Willd.) Porter = Symphyotrichum pilosum (Willd.) G. L. Nesom ■☆
40388 Aster ericoides L. var. platyphyllus Torr. et A. Gray = Symphyotrichum pilosum (Willd.) G. L. Nesom ■☆
40389 Aster ericoides L. var. polycephalus (Rydb.) F. C. Gates = Symphyotrichum falcatum (Lindl.) G. L. Nesom var. commutatum (Torr. et A. Gray) G. L. Nesom ■☆
40390 Aster ericoides L. var. pringlei A. Gray = Symphyotrichum pilosum (Willd.) G. L. Nesom var. pringlei (A. Gray) G. L. Nesom ■☆
40391 Aster ericoides L. var. prostratus (Kuntze) S. F. Blake = Symphyotrichum ericoides (L.) G. L. Nesom ■☆
40392 Aster ericoides L. var. prostratus (Kuntze) S. F. Blake = Symphyotrichum ericoides (L.) G. L. Nesom var. prostratum (Kuntze) G. L. Nesom ■☆
40393 Aster ericoides L. var. pusillus A. Gray = Symphyotrichum depauperatum (Fernald) G. L. Nesom ■☆
40394 Aster ericoides L. var. stricticaulis (Torr. et A. Gray) F. C. Gates = Symphyotrichum ericoides (L.) G. L. Nesom var. pansum (S. F. Blake) G. L. Nesom ■☆
40395 Aster ericoides L. var. strictus Porter = Symphyotrichum porteri (A. Gray) G. L. Nesom ■☆
40396 Aster ericoides L. var. villosus (Michx.) Torr. et A. Gray = Symphyotrichum pilosum (Willd.) G. L. Nesom ■☆
40397 Aster erigeroides (DC.) Harv. = Felicia erigeroides DC. ■☆
40398 Aster erigeroides (DC.) Harv. var. schultesii Harv. = Felicia erigeroides DC. ■☆
40399 Aster erigeroides (DC.) Harv. var. trinervius (Turcz.) Harv. = Felicia erigeroides DC. ■☆
40400 Aster erucifolius (Thell.) Lippert;芥叶紫菀■☆
40401 Aster eryngiifolius Torr. et A. Gray = Eurybia eryngiifolia (Torr. et A. Gray) G. L. Nesom ■☆
40402 Aster esquirolii H. Lév. = Picris japonica Thunb. ■
40403 Aster eulae Shinners = Symphyotrichum eulae (Shinners) G. L. Nesom ■☆
40404 Aster excavatus E. S. Burgess = Eurybia divaricata (L.) G. L. Nesom ■☆
40405 Aster exiguus Rydb. = Symphyotrichum ericoides (L.) G. L. Nesom var. prostratum (Kuntze) G. L. Nesom ■☆
40406 Aster exiguus Rydb. = Symphyotrichum ericoides (L.) G. L. Nesom ■☆
40407 Aster exilis Elliott = Aster subulatus Michx. ■
40408 Aster exscapus Richardson = Townsendia exscapa Porter ■☆
40409 Aster eylesii (S. Moore) Milne-Redh. = Felicia welwitschii (Hiern) Grau ●☆
40410 Aster faberi Franch.;青药(水白菊);Faber Aster ■
40411 Aster fabri Hook. f. = Gymnaster angustifolius (C. C. Chang) Y. Ling ■
40412 Aster falcatus Lindl. = Symphyotrichum falcatum (Lindl.) G. L. Nesom ■☆
40413 Aster falcatus Lindl. subsp. commutatus (Torr. et A. Gray) A. G. Jones = Symphyotrichum falcatum (Lindl.) G. L. Nesom var. commutatum (Torr. et A. Gray) G. L. Nesom ■☆
40414 Aster falcatus Lindl. var. commutatus (Torr. et A. Gray) A. G. Jones = Symphyotrichum falcatum (Lindl.) G. L. Nesom var. commutatum (Torr. et A. Gray) G. L. Nesom ■☆
40415 Aster falcatus Lindl. var. crassulus (Rydb.) Cronquist = Symphyotrichum falcatum (Lindl.) G. L. Nesom var. commutatum (Torr. et A. Gray) G. L. Nesom ■☆
40416 Aster falcifolius Hand.-Mazz.;镰叶紫菀;Falcate Leaf Aster, Sickleleaf Aster ■
40417 Aster fallax Tamamsch. = Aster alpinus L. var. fallax (Tamamsch.) Y. Ling ■
40418 Aster falri Hook. f. = Miyamayomena angustifolius (C. C. Chang) Y. L. Chen ■
40419 Aster fanjingshanicus Y. L. Chen et D. J. Liu;梵净山紫菀;Fanjingshan Aster ■
40420 Aster farreri W. W. Sm. et Jeffrey;狭苞紫菀(法勒氏紫菀,线叶紫菀,羊眼花);Farrer Aster ■
40421 Aster fastigiatus Fisch. = Turczaninovia fastigiata (Fisch.) DC. ■
40422 Aster fastigiatus Ledeb. = Galatella hauptii (Ledeb.) Lindl. ex DC. ■
40423 Aster fastigiatus Lehm. ex Nees = Galatella hauptii (Ledeb.) Lindl. ex DC. ■
40424 Aster fastigiiformis Novopokr. = Galatella fastigiiformis Novopokr. ■
40425 Aster fauriei H. Lév. et Vaniot;法氏紫菀■☆
40426 Aster faxonii Porter = Symphyotrichum pilosum (Willd.) G. L.

Nesom var. pringlei (A. Gray) G. L. Nesom ■☆
40427 Aster fendleri A. Gray = Symphyotrichum fendleri (A. Gray) G. L. Nesom ■☆
40428 Aster ferrugineus Edgew. = Aster albescens (DC.) Wall. ex Hand. -Mazz. ●
40429 Aster ficoideus (DC.) Harv. = Poecilolepis ficoidea (DC.) Grau ■☆
40430 Aster filaginifolius Hook. et Arn. = Corethrogyne filaginifolia (Hook. et Arn.) Nutt. ●■☆
40431 Aster filifolius Vent. = Felicia filifolia (Vent.) Burtt Davy ●☆
40432 Aster filiformis Eckl. ex DC. = Zyrphelis taxifolia (L.) Nees ■☆
40433 Aster filipes J. Q. Fu;丝柄紫菀;Filistalk Aster ■
40434 Aster finkii Rydb. var. moratus Shinners = Aster cordifolius L. ■☆
40435 Aster finkii Rydb. var. moratus Shinners = Symphyotrichum cordifolium (L.) G. L. Nesom ■☆
40436 Aster firmus Nees = Aster puniceus L. ■☆
40437 Aster firmus Nees = Symphyotrichum firmum (Nees) G. L. Nesom ■☆
40438 Aster fistulosus (Mill.) Kuntze = Solidago fistulosa Mill. ■☆
40439 Aster flabellum Vaniot = Turczaninovia fastigiata (Fisch.) DC. ■
40440 Aster flaccidus Bunge;萎软紫菀(肺经草,千花紫菀,柔软紫菀,太白菊);Flaccid Aster ■
40441 Aster flaccidus Bunge f. gracilis J. Q. Fu;纤细柔软紫菀;Thin Flaccid Aster ■
40442 Aster flaccidus Bunge f. griseo-barbatus Grierson;灰毛萎软紫菀;Greyhair Flaccid Aster ■
40443 Aster flaccidus Bunge f. ovatifolius J. Q. Fu;卵叶柔软紫菀;Ovateleaf Flaccid Aster ■
40444 Aster flaccidus Bunge f. stolonifer Onno = Aster flaccidus Bunge ■
40445 Aster flaccidus Bunge f. ternicatus Onno = Aster brevis Hand. -Mazz. ■
40446 Aster flaccidus Bunge f. tunicatus Onno = Aster brevis Hand. -Mazz. ■
40447 Aster flaccidus Bunge subsp. fructa-glandulosus (Oestenf.) Onno = Aster flaccidus Bunge ■
40448 Aster flaccidus Bunge subsp. fructu-glandulosus (Oestenf.) Onno = Aster flaccidus Bunge subsp. glandulosus (Keissl.) Onno ■
40449 Aster flaccidus Bunge subsp. glandulosus (Keissl.) Onno;腺毛萎软紫菀■
40450 Aster flaccidus Bunge subsp. tsarungensis Grierson = Aster tsarungensis (Grierson) Y. Ling ■
40451 Aster flaccidus Bunge var. atropurpureus Onno = Aster alpinus L. var. fallax (Tamamsch.) Y. Ling ■
40452 Aster flaccidus Bunge var. frutu-glandulosus Oestenfeld = Aster flaccidus Bunge subsp. glandulosus (Keissl.) Onno ■
40453 Aster flaccidus Bunge var. glandulosus (Keissl.) Hand. -Mazz. = Aster flaccidus Bunge subsp. glandulosus (Keissl.) Onno ■
40454 Aster flaccidus Bunge var. glandulosus Keissl. = Aster flaccidus Bunge subsp. glandulosus (Keissl.) Onno ■
40455 Aster flavovirens Kuntze = Solidago stricta Aiton ■☆
40456 Aster flexilis E. S. Burgess = Eurybia divaricata (L.) G. L. Nesom ■☆
40457 Aster fluvialis Osterh. = Symphyotrichum lanceolatum (Willd.) G. L. Nesom var. hesperium (A. Gray) G. L. Nesom ■☆
40458 Aster foliaceus Lindl. = Aster foliaceus Lindl. ex DC. ■☆
40459 Aster foliaceus Lindl. ex DC. = Symphyotrichum foliaceum (Lindl. ex DC.) G. L. Nesom ■☆
40460 Aster foliaceus Lindl. ex DC. subsp. lyallii (A. Gray) Cronquist = Symphyotrichum hendersonii (Fernald) G. L. Nesom ■☆
40461 Aster foliaceus Lindl. ex DC. var. apricus A. Gray = Symphyotrichum foliaceum (Lindl. ex DC.) G. L. Nesom var. apricum (A. Gray) G. L. Nesom ■☆
40462 Aster foliaceus Lindl. ex DC. var. arcuans Fernald = Aster novi-belgii L. ■
40463 Aster foliaceus Lindl. ex DC. var. burkei A. Gray = Symphyotrichum foliaceum (Lindl. ex DC.) G. L. Nesom var. canbyi (A. Gray) G. L. Nesom ■☆
40464 Aster foliaceus Lindl. ex DC. var. crenifolius Fernald;圆齿荷兰菊■☆
40465 Aster foliaceus Lindl. ex DC. var. crenifolius Fernald = Symphyotrichum novi-belgii (L.) G. L. Nesom var. crenifolium (Fernald) Labrecque et Brouillet ■☆
40466 Aster foliaceus Lindl. ex DC. var. cusickii (A. Gray) Cronquist = Symphyotrichum cusickii (A. Gray) G. L. Nesom ■☆
40467 Aster foliaceus Lindl. ex DC. var. eatonii A. Gray = Symphyotrichum eatonii (A. Gray) G. L. Nesom ■☆
40468 Aster foliaceus Lindl. ex DC. var. parryi (D. C. Eaton) A. Gray = Symphyotrichum foliaceum (Lindl. ex DC.) G. L. Nesom var. parryi (D. C. Eaton) G. L. Nesom ■☆
40469 Aster foliaceus Lindl. ex DC. var. sublinearis Griscom et R. J. Eaton = Aster novi-belgii L. ■
40470 Aster foliaceus Lindl. ex DC. var. subpetiolatus Fernald = Aster foliaceus Lindl. var. crenifolius Fernald ■☆
40471 Aster foliaceus Lindl. var. crenifolius Fernald = Aster foliaceus Lindl. ex DC. var. crenifolius Fernald ■☆
40472 Aster fontialis Alexander = Symphyotrichum fontinale (Alexander) G. L. Nesom ■☆
40473 Aster fordii Hemsl. = Aster panduratus Nees ex Walp. ■
40474 Aster formosanus Hayata;台岩紫菀(台湾白山兰,台湾山白兰);Taiwan Aster ■
40475 Aster forrestii Stapf = Aster souliei Franch. var. limitaneus (W. W. Sm. et Farr.) Hand. -Mazz. ■
40476 Aster forrestii Stapf = Aster souliei Franch. ■
40477 Aster forwoodii S. Watson = Aster puniceus L. ■☆
40478 Aster forwoodii S. Watson = Symphyotrichum puniceum (L.) Á. Löve et D. Löve ■☆
40479 Aster fragilis Willd.;纤细紫菀;Fragile-stem Aster, Frost Flower, Small White Aster ■☆
40480 Aster fragilis Willd. var. subdumosus (Wiegand) A. G. Jones;灌丛纤细紫菀;Fragile-stemmed Aster ■☆
40481 Aster franchetianus H. Lév. = Kalimeris integrifolia Turcz. ex DC. ■
40482 Aster frankliniauus Rydb. = Aster borealis (Torr. et A. Gray) Prov. ■☆
40483 Aster frankliniauus Rydb. = Symphyotrichum boreale (Torr. et A. Gray) Á. Löve et D. Löve ■☆
40484 Aster frondosus (Nutt.) Torr. et A. Gray = Symphyotrichum frondosum (Nutt.) G. L. Nesom ■☆
40485 Aster fruticosus L. = Asterothamnus fruticosus (C. Winkl.) Novopokr. ●
40486 Aster fruticosus L. = Felicia fruticosa (L.) G. Nicholson ●
40487 Aster fruticulosus Willd. = Felicia fruticosa (L.) G. Nicholson ●
40488 Aster fulgidulus Grierson;辉叶紫菀;Brightleaf Aster, Glistening Leaf Aster, Shining-leaved Aster ●
40489 Aster furcatus E. S. Burgess;叉状紫菀;Forked Aster, Midwestern

White Heart-leaved Aster ■☆

40490 Aster furcatus E. S. Burgess = Eurybia furcata (E. S. Burgess) G. L. Nesom ■☆

40491 Aster furcatus E. S. Burgess f. elaciniatus (Benke) Shinners = Aster furcatus E. S. Burgess ■☆

40492 Aster furcatus E. S. Burgess f. erythractis Benke = Aster furcatus E. S. Burgess ■☆

40493 Aster furcatus E. S. Burgess var. elaciniatus Benke = Aster furcatus E. S. Burgess ■☆

40494 Aster fusanensis H. Lév. et Vaniot = Heteropappus hispidus (Thunb.) Less. ■

40495 Aster fuscescens Burret et Franch.;褐毛紫菀;Brownhair Aster ■

40496 Aster fuscescens Burret et Franch. var. oblongifolius Grierson;长圆叶褐毛紫菀(长圆叶紫菀);Oblongleaf Brownhair Aster ■

40497 Aster fuscescens Burret et Franch. var. scaberoides C. C. Chang;少毛褐毛紫菀(少毛紫菀)■

40498 Aster ganlun Kitam. = Aster souliei Franch. ■

40499 Aster gaspensis Vict. = Symphyotrichum anticostense (Fernald) G. L. Nesom ■☆

40500 Aster gattingeri (Chapm. ex A. Gray) Kuntze = Solidago gattingeri Chapm. ex A. Gray ■☆

40501 Aster gattingeri Alexander = Eurybia hemispherica (Alexander) G. L. Nesom ■☆

40502 Aster gentryi Standl. = Psilactis gentryi (Standl.) D. R. Morgan ■☆

40503 Aster georgianus Alexander = Symphyotrichum georgianum (Alexander) G. L. Nesom ■☆

40504 Aster geralachii Hance = Aster ageratoides Turcz. var. gerlachii (Hance) C. C. Chang ex Y. Ling ■

40505 Aster geyeri (A. Gray) Howell = Symphyotrichum laeve (L.) G. L. Nesom var. geyeri (A. Gray) G. L. Nesom ■☆

40506 Aster giganteus (Hook.) Rydb. = Eurybia sibirica (L.) G. L. Nesom ■☆

40507 Aster giraldii Diels;秦中紫菀(中紫菀);Girald Aster ■

40508 Aster glacialis Nutt. = Erigeron glacialis (Nutt.) A. Nelson ■☆

40509 Aster glandulosus (Keissl.) Hand.-Mazz. = Aster flaccidus Bunge subsp. glandulosus (Keissl.) Onno ■

40510 Aster glarearum W. W. Sm. et Farr. = Aster flaccidus Bunge ■

40511 Aster glaucescens (A. Gray) S. F. Blake = Eucephalus glaucescens (A. Gray) Greene ■☆

40512 Aster glaucodes S. F. Blake = Herrickia glauca (Nutt.) Brouillet ■☆

40513 Aster glaucodes S. F. Blake subsp. pulcher S. F. Blake;美丽灰紫菀■☆

40514 Aster glaucodes S. F. Blake var. formosus (Greene) Kittell = Herrickia glauca (Nutt.) Brouillet ■☆

40515 Aster glaucophyllus (Piper) Frye et Rigg = Eucephalus glaucescens (A. Gray) Greene ■☆

40516 Aster glaucus (Nutt.) Torr. et A. Gray = Herrickia glauca (Nutt.) Brouillet ■☆

40517 Aster glaucus (Nutt.) Torr. et A. Gray var. wasatchensis M. E. Jones = Herrickia wasatchensis (M. E. Jones) Brouillet ■☆

40518 Aster glehnii F. Schmidt;格氏紫菀■☆

40519 Aster glehnii F. Schmidt var. hondoensis Kitam.;本州紫菀■☆

40520 Aster glehnii F. Schmidt. = Aster ageratoides Turcz. var. scaberulus (Miq.) Y. Ling ■

40521 Aster gmelini Tausch. = Heteropappus altaicus (Willd.) Novopokr. ■

40522 Aster gormanii (Piper) S. F. Blake = Eucephalus gormanii Piper ■☆

40523 Aster gossypiphorus Y. Ling = Aster prainii (J. R. Drumm.) Y. L. Chen ■

40524 Aster gouldii C. E. C. Fisch. = Heteropappus gouldii (C. E. C. Fisch.) Grierson ■

40525 Aster gracilescens E. S. Burgess = Symphyotrichum undulatum (L.) G. L. Nesom ■☆

40526 Aster gracilicaulis Y. Ling;细茎紫菀;Slenderstem Aster ■

40527 Aster gracilis Nutt. = Eurybia compacta G. L. Nesom ■☆

40528 Aster gracillimus (Torr. et A. Gray) Kuntze = Solidago stricta Aiton subsp. gracillima (Torr. et A. Gray) Semple ■☆

40529 Aster gramineus (L.) Kom. = Arctogeron gramineum (L.) DC. ■

40530 Aster grandiflorus L. = Symphyotrichum grandiflorum (L.) G. L. Nesom ■☆

40531 Aster grauii W. Lippert = Aster bakerianus Burtt Davy ex C. A. Sm. ■☆

40532 Aster greatae Parish = Symphyotrichum greatae (Parish) G. L. Nesom ■☆

40533 Aster grossedentatus Dinter = Felicia brevifolia (DC.) Grau ■☆

40534 Aster guelpaertensis H. Lév. et Vaniot = Aster ageratoides Turcz. ■

40535 Aster guiradonis (A. Gray) Kuntze = Solidago guiradonis A. Gray ■☆

40536 Aster hallii A. Gray = Symphyotrichum hallii (A. Gray) G. L. Nesom ■☆

40537 Aster handelii Onno;红冠紫菀;Redpappo Aster, Red-pappus Aster ■

40538 Aster harrowianus Diels = Aster albescens (DC.) Wall. ex Hand.-Mazz. var. gracilior Hand.-Mazz. ●

40539 Aster harrowianus Diels var. glabratus Diels = Aster albescens (DC.) Wall. ex Hand.-Mazz. var. salignus Hand.-Mazz. ●

40540 Aster harrowianus Diels var. glabratus Diels = Aster albescens (DC.) Wall. ex Hand.-Mazz. var. levissimus Hand.-Mazz. ●

40541 Aster harrowianus Diels var. pycnophyllus (Franch.) H. Lév. = Aster pycnophyllus W. W. Sm. ■

40542 Aster harrowianus Diels var. slabratus Diels = Aster albescens (DC.) Wall. ex Hand.-Mazz. var. levissimus Hand.-Mazz. ●

40543 Aster harveyanus Kuntze;哈维紫菀■☆

40544 Aster harveyanus Kuntze subsp. corymbosus W. Lippert = Aster harveyanus Kuntze ■☆

40545 Aster harveyanus Kuntze subsp. gracilis W. Lippert = Aster harveyanus Kuntze ■☆

40546 Aster harveyanus Kuntze subsp. nyikensis W. Lippert = Aster harveyanus Kuntze ■☆

40547 Aster harveyanus Kuntze subsp. robustus W. Lippert = Aster harveyanus Kuntze ■☆

40548 Aster harveyanus Kuntze subsp. xylophyllus (Klatt) Lippert = Aster harveyanus Kuntze ■☆

40549 Aster hashimotoi Kitam. = Aster sekimotoi Makino ■☆

40550 Aster hauptii Ledeb. = Galatella hauptii (Ledeb.) Lindl. ex DC. ■

40551 Aster haydenii Porter = Oreostemma alpigenum (Torr. et A. Gray) Greene var. haydenii (Porter) G. L. Nesom ■☆

40552 Aster hedinii Ostenf. = Aster asteroides (DC.) Kuntze ■

40553 Aster helenius Scop. = Inula helenium L. ■

40554 Aster hemisphericus Alexander = Eurybia hemispherica (Alexander) G. L. Nesom ■☆

40555 Aster hendersonii Fernald = Symphyotrichum hendersonii (Fernald) G. L. Nesom ■☆

40556 Aster henryi Hemsl. = Aster moupinensis (Franch.) Hand.-Mazz. ■

40557 Aster hersileoides C. K. Schneid.;横斜紫菀;Across Aster, Ascending-branched Aster ●

40558 Aster herveyi A. Gray = Eurybia herveyi (A. Gray) G. L. Nesom ■☆

40559 Aster hesperius A. Gray = Aster lanceolatus Willd. ■☆

40560 Aster hesperius A. Gray = Symphyotrichum lanceolatum (Willd.) G. L. Nesom var. hesperium (A. Gray) G. L. Nesom ■☆

40561 Aster hesperius A. Gray var. gaspensis (Vict.) B. Boivin = Symphyotrichum anticostense (Fernald) G. L. Nesom ■☆

40562 Aster hesperius A. Gray var. laetevirens (Greene) Cronquist = Symphyotrichum lanceolatum (Willd.) G. L. Nesom var. hesperium (A. Gray) G. L. Nesom ■☆

40563 Aster hesperius A. Gray var. wootonii Greene = Symphyotrichum lanceolatum (Willd.) G. L. Nesom var. hesperium (A. Gray) G. L. Nesom ■☆

40564 Aster heterochaeta Benth. ex C. B. Clarke = Aster flaccidus Bunge ■

40565 Aster heterolepis Hand.-Mazz.;异苞紫菀;Differentbract Aster, Heterophyllaries Aster ■

40566 Aster heterophyllus Thunb. = Athrixia heterophylla (Thunb.) Less ■☆

40567 Aster heterotrichus (H. Hara) H. Hara ex Kitam. = Erigeron thunbergii A. Gray subsp. glabratus (A. Gray) H. Hara var. heterotrichus (H. Hara) H. Hara ■☆

40568 Aster himalaicus C. B. Clarke;须弥紫菀(喜马拉雅紫菀); Himalayan Aster ■

40569 Aster hirsuticaulis Lindl. ex DC. = Aster lateriflorus (L.) Britton var. hirsuticaulis (Lindl. ex DC.) Porter ■☆

40570 Aster hirsutus (Vent.) Harv. = Felicia cymbalariae (Aiton) Bolus et Wolley-Dod ex Adamson et T. M. Salter ■☆

40571 Aster hirtellus Lindl. ex DC. = Aster sagittifolius Wedem. ex Willd. ■☆

40572 Aster hirtellus Lindl. ex DC. = Symphyotrichum urophyllum (Lindl. ex DC.) G. L. Nesom ■☆

40573 Aster hirtus (DC.) Harv. = Felicia aculeata Grau ■☆

40574 Aster hirtus Thunb. = Felicia hirta (Thunb.) Grau ●☆

40575 Aster hispidulus C. C. Chang = Aster trichoneurus Y. Ling ■

40576 Aster hispidus (Thunb.) Baker = Aster bakerianus Burtt Davy ex C. A. Sm. ■☆

40577 Aster hispidus Maxim. var. koidzumianus (Kitam.) Okuyama;小泉氏紫菀■☆

40578 Aster hispidus Thunb. = Heteropappus hispidus (Thunb.) Less. ■

40579 Aster hispidus Thunb. var. arenarius (Kitam.) Okuyama = Aster arenarius (Kitam.) Nemoto ■

40580 Aster hispidus Thunb. var. decipiens (Maxim.) Y. Ling = Heteropappus hispidus (Thunb.) Less. ■

40581 Aster hispidus Thunb. var. heterochaeta Franch. et Sav. = Heteropappus hispidus (Thunb.) Less. ■

40582 Aster hispidus Thunb. var. insularis (Makino) Okuyama = Heteropappus hispidus (Thunb.) Less. var. insularis (Makino) Kitam. ex Ohwi ■☆

40583 Aster hispidus Thunb. var. leptocladus (Makino) Okuyama = Heteropappus hispidus (Thunb.) Less. subsp. leptocladus (Makino) Kitam. ■☆

40584 Aster hispidus Thunb. var. mesochaeta Franch. et Sav. = Heteropappus hispidus (Thunb.) Less. ■

40585 Aster hispidus Thunb. var. microphyllus Pamp. = Heteropappus hispidus (Thunb.) Less. ■

40586 Aster hispidus Thunb. var. oldhamii (Hemsl.) S. S. Ying = Aster oldhamii Hemsl. ■

40587 Aster hololachnus Y. Ling;全茸紫菀;Cottony Aster, Whole Woolly Aster ■

40588 Aster holophyllus Hemsl. = Kalimeris integrifolia Turcz. ex DC. ■

40589 Aster homochlamydeus Hand.-Mazz.;等苞紫菀;Equalbract Aster, Equalphyllaries Aster ■

40590 Aster homochlamydeus Hand.-Mazz. f. filipes Y. Ling;腺梗等苞紫菀■

40591 Aster homolepis C. C. Chang = Aster homochlamydeus Hand.-Mazz. ■

40592 Aster horizontalis Desf. = Aster lateriflorus (L.) Britton var. horizontalis (Desf.) Farw. ■☆

40593 Aster horridifolius H. Lév. et Vaniot = Aster maackii Regel ■

40594 Aster horridus (Wooton et Standl.) S. F. Blake = Herrickia horrida Wooton et Standl. ■☆

40595 Aster houghtonii (Torr. et A. Gray) Kuntze = Solidago houghtonii Torr. et A. Gray ■☆

40596 Aster hualiensis S. S. Ying = Aster chingshuiensis Y. C. Liu et C. H. Ou ■

40597 Aster humilis Willd. = Doellingeria infirma (Michx.) Greene ■☆

40598 Aster hunanensis Hand.-Mazz.;湖南紫菀;Hunan Aster ■

40599 Aster hypoleucus Hand.-Mazz.;白背紫菀;White Beneath Aster, Whiteback Aster, White-backed Aster ●

40600 Aster hyssopifolius P. J. Bergius = Felicia hyssopifolia (P. J. Bergius) Nees ■☆

40601 Aster hyssopifolius P. J. Bergius var. canescens Harv. = Chrysocoma rigidula (DC.) Ehr. Bayer ■☆

40602 Aster hyssopifolius P. J. Bergius var. canus (DC.) Harv. = Felicia cana DC. ■☆

40603 Aster hyssopifolius P. J. Bergius var. hirtus (Thunb.) Harv. = Felicia hirta (Thunb.) Grau ●☆

40604 Aster hyssopifolius P. J. Bergius var. linifolius Harv. = Felicia zeyheri (Less.) Nees subsp. linifolia (Harv.) Grau ●☆

40605 Aster hyssopifolius P. J. Bergius var. rigidulus (DC.) Harv. = Chrysocoma rigidula (DC.) Ehr. Bayer ■☆

40606 Aster ianthinus E. S. Burgess = Aster macrophyllus L. ■☆

40607 Aster ibericus Stev. ex M. Bieb.;伊比利亚紫菀■☆

40608 Aster ignoratus Kunth et C. D. Bouché = Aster albescens (DC.) Wall. ex Hand.-Mazz. ●

40609 Aster iinumae Kitam.;饭沼紫菀■☆

40610 Aster iinumae Kitam. 'Hortensis';庭院饭沼紫菀■☆

40611 Aster iinumae Kitam. f. hortensis (Makino) H. Hara = Aster iinumae Kitam. 'Hortensis' ■☆

40612 Aster imbricatus (DC.) Harv. = Polyarrhena imbricata (DC.) Grau ●☆

40613 Aster imbricatus L. = Polyarrhena reflexa (L.) Cass. ●☆

40614 Aster incanopilosus (Lindl.) E. Sheld. = Symphyotrichum falcatum (Lindl.) G. L. Nesom var. commutatum (Torr. et A. Gray) G. L. Nesom ■☆

40615 Aster incisus Fisch. = Kalimeris incisa (Fisch.) DC. ■

40616 Aster incisus Fisch. var. australis Kitag. = Kalimeris incisa (Fisch.) DC. ■

40617 Aster incisus Fisch. var. pinnatifidus (Maxim. ex Makino) Nakai

= Aster iinumae Kitam. ■☆
40618 Aster indamellus Grierson;叶苞紫菀(蓝菀);Leaf-like Bracteal Aster,Leaflikebract Aster ■
40619 Aster indica (L.) Blume = Kalimeris indica (L.) Sch. Bip. ■
40620 Aster indicus (L.) Sch. Bip. var. lautureana Yamam. = Kalimeris shimadae (Kitam.) Kitam. ■
40621 Aster indicus L.;鸡儿肠(马兰)■
40622 Aster indicus L. = Kalimeris incisa (Fisch.) DC. ■
40623 Aster indicus L. var. lautureana Yamam. = Kalimeris shimadae (Kitam.) Kitam. ■
40624 Aster indicus L. var. pinnatifidus Maxim. ex Makino = Aster iinumae Kitam. ■☆
40625 Aster infirmus Michx.;山茱萸叶紫菀;Cornel-leaved Aster ■☆
40626 Aster infirmus Michx. = Doellingeria infirma (Michx.) Greene ■☆
40627 Aster insularis Makino = Aster hispidus Thunb. var. insularis (Makino) Okuyama ■☆
40628 Aster intdamellus Grierson = Aster indamellus Grierson ■
40629 Aster integrifolius Franch. = Kalimeris integrifolia Turcz. ex DC. ■
40630 Aster integrifolius Nutt. = Eurybia integrifolia (Nutt.) G. L. Nesom ■☆
40631 Aster interior Wiegand = Symphyotrichum lanceolatum (Willd.) G. L. Nesom var. interior (Wiegand) G. L. Nesom ■☆
40632 Aster intermedius Turcz. ex DC. = Aster flaccidus Bunge ■
40633 Aster inuloides D. Don = Erigeron multiradiatus (Lindl. ex DC.) Benth. ■
40634 Aster inuloides D. Don = Erigeron multiradiatus (Lindl.) Benth. et Hook. f. ex Hook. f. ■
40635 Aster ionoglossus Y. Ling;堇舌紫菀;Violettongue Aster ■
40636 Aster ircutianus DC. = Aster sibiricus L. ■
40637 Aster itsunboshii Kitam.;大埔紫菀(大武山紫菀);Dapu Aster ■
40638 Aster jacksonii Kuntze = Solidago rigida L. subsp. glabrata (E. L. Braun) S. B. Heard et Semple ■☆
40639 Aster jeffreyanus Diels;滇西北紫菀;Jeffrey Aster ■
40640 Aster jessicae Piper = Symphyotrichum jessicae (Piper) G. L. Nesom ■☆
40641 Aster jessicae Piper = Symphyotrichum laeve (L.) Á. Löve et D. Löve ■☆
40642 Aster johannensis Fernald;约翰紫菀;Lake St. John Aster ■☆
40643 Aster johannensis Fernald = Aster novi-belgii L. ■
40644 Aster johannensis Fernald var. villicaulis (A. Gray) Fernald = Aster longifolius Lam. var. villicaulis A. Gray ■☆
40645 Aster jonesiae Lamboy = Eurybia jonesiae (Lamboy) G. L. Nesom ■☆
40646 Aster juchaihu Z. Y. Zhu et B. Q. Min;菊柴胡;Juchaihu Aster ■
40647 Aster junceus Aiton = Aster novi-belgii L. ■
40648 Aster junceus Aiton = Symphyotrichum longifolium (Lam.) G. L. Nesom ■☆
40649 Aster junciformis Rydb. = Aster borealis (Torr. et A. Gray) Prov. ■☆
40650 Aster junciformis Rydb. = Symphyotrichum boreale (Torr. et A. Gray) Á. Löve et D. Löve ■☆
40651 Aster kansuensis Farr. = Aster flaccidus Bunge ■
40652 Aster kantoensis Kitam.;河原野菊■☆
40653 Aster kawanguchi Kitam. = Aster yunnanensis Franch. var. labrangensis (Hand.-Mazz.) Y. Ling ■
40654 Aster kentuckiensis Britton = Symphyotrichum priceae (Britton) G. L. Nesom ■☆
40655 Aster kingii D. C. Eaton = Herrickia kingii (D. C. Eaton) Brouillet ■☆
40656 Aster kingii D. C. Eaton var. barnebyana (S. L. Welsh et Goodrich) S. L. Welsh = Herrickia kingii (D. C. Eaton) Brouillet, Urbatsch et R. P. Roberts var. barnebyana (S. L. Welsh et Goodrich) Brouillet ■☆
40657 Aster kiusianus Sugim.;九州紫菀■☆
40658 Aster koidzumianus (Kitam.) Nemoto = Aster hispidus Maxim. var. koidzumianus (Kitam.) Okuyama ■☆
40659 Aster koidzumianus Makino = Aster maackii Regel ■
40660 Aster komarovii H. Lév. = Doellingeria scabra (Thunb.) Nees ■
40661 Aster komarovii H. Lév. et Vaniot = Aster scaber Thunb. ■
40662 Aster komarovii H. Lév. et Vaniot = Doellingeria scabra (Thunb.) Nees ■
40663 Aster komonoensis Makino;菰野山紫菀■☆
40664 Aster koraiensis Nakai;朝鲜紫菀(朝鲜裸菀)■☆
40665 Aster korshinskyi Tamamsch.;考尔紫菀■☆
40666 Aster kraussii (Sch. Bip.) Harv. = Felicia aethiopica (Burm. f.) Bolus et Wolley-Dod ex Adamson et T. M. Salter ■☆
40667 Aster kumleinii Fr. ex A. Gray = Aster oblongifolius Nutt. ■☆
40668 Aster kumleinii Fr. ex A. Gray f. roseoligulatus Benke = Aster oblongifolius Nutt. ■☆
40669 Aster kumleinii Fr. ex A. Gray var. oliganthemos Lunell = Aster oblongifolius Nutt. ■☆
40670 Aster labrangensis Hand.-Mazz. = Aster yunnanensis Franch. var. labrangensis (Hand.-Mazz.) Y. Ling ■
40671 Aster laetevirens Greene = Symphyotrichum lanceolatum (Willd.) G. L. Nesom var. hesperium (A. Gray) G. L. Nesom ■☆
40672 Aster laevigatus (Sond.) Kuntze;光滑紫菀■☆
40673 Aster laevis L. = Symphyotrichum laeve (L.) Á. Löve et D. Löve ■☆
40674 Aster laevis L. f. amplifolius (Porter) Fernald = Symphyotrichum laeve (L.) Á. Löve et D. Löve ■☆
40675 Aster laevis L. f. latifolius (Porter) Shinners = Symphyotrichum laeve (L.) Á. Löve et D. Löve ■☆
40676 Aster laevis L. f. purpurascens (Farw.) Shinners = Symphyotrichum laeve (L.) Á. Löve et D. Löve ■☆
40677 Aster laevis L. subsp. geyeri (A. Gray) Piper = Symphyotrichum laeve (L.) G. L. Nesom var. geyeri (A. Gray) G. L. Nesom ■☆
40678 Aster laevis L. var. amplifolius Porter = Symphyotrichum laeve (L.) Á. Löve et D. Löve ■☆
40679 Aster laevis L. var. concinnus (Willd.) House = Symphyotrichum laeve (L.) G. L. Nesom var. concinnum (Willd.) G. L. Nesom ■☆
40680 Aster laevis L. var. falcatus Farw. = Symphyotrichum laeve (L.) Á. Löve et D. Löve ■☆
40681 Aster laevis L. var. falcatus Farw. f. purpurascens Farw. = Symphyotrichum laeve (L.) Á. Löve et D. Löve ■☆
40682 Aster laevis L. var. geyeri A. Gray = Symphyotrichum laeve (L.) G. L. Nesom var. geyeri (A. Gray) G. L. Nesom ■☆
40683 Aster laevis L. var. guadalupensis A. G. Jones = Symphyotrichum laeve (L.) G. L. Nesom var. geyeri (A. Gray) G. L. Nesom ■☆
40684 Aster laevis L. var. latifolius Porter = Symphyotrichum laeve (L.) Á. Löve et D. Löve ■☆
40685 Aster laevis L. var. purpuratus (Nees) A. G. Jones = Symphyotrichum laeve (L.) G. L. Nesom var. purpuratum (Nees) G. L. Nesom ■☆
40686 Aster laevis L. var. thyrsoideus Farw. = Symphyotrichum laeve

(L.) Á. Löve et D. Löve ■☆

40687 Aster lamarckianus Nees = Symphyotrichum lanceolatum (Willd.) G. L. Nesom ■☆

40688 Aster lanceolatus Willd. = Symphyotrichum lanceolatum (Willd.) G. L. Nesom ■☆

40689 Aster lanceolatus Willd. subsp. hesperius (A. Gray) Semple et Chmiel. = Symphyotrichum lanceolatum (Willd.) G. L. Nesom var. hesperium (A. Gray) G. L. Nesom ■☆

40690 Aster lanceolatus Willd. subsp. interior (Wiegand) A. G. Jones = Symphyotrichum lanceolatum (Willd.) G. L. Nesom var. interior (Wiegand) G. L. Nesom ■☆

40691 Aster lanceolatus Willd. subsp. simplex (Willd.) A. G. Jones = Symphyotrichum simplex (Willd.) Á. Löve et D. Löve ■☆

40692 Aster lanceolatus Willd. subsp. simplex (Willd.) A. G. Jones = Symphyotrichum lanceolatum (Willd.) G. L. Nesom ■☆

40693 Aster lanceolatus Willd. var. hirsuticaulis Semple et Chmiel. = Symphyotrichum lanceolatum (Willd.) G. L. Nesom var. hirsuticaule (Semple et Chmiel.) G. L. Nesom ■☆

40694 Aster lanceolatus Willd. var. interior (Wiegand) A. G. Jones = Symphyotrichum lanceolatum (Willd.) G. L. Nesom var. interior (Wiegand) G. L. Nesom ■☆

40695 Aster lanceolatus Willd. var. interior (Wiegand) Semple et Chmiel. = Symphyotrichum lanceolatum (Willd.) G. L. Nesom var. interior (Wiegand) G. L. Nesom ■☆

40696 Aster lanceolatus Willd. var. latifolius Semple et Chmiel. = Symphyotrichum lanceolatum (Willd.) G. L. Nesom var. latiflorum (Semple et Chmiel.) G. L. Nesom ■☆

40697 Aster lanceolatus Willd. var. simplex (Willd.) A. G. Jones = Symphyotrichum simplex (Willd.) Á. Löve et D. Löve ■☆

40698 Aster lancifolius (Torr. et A. Gray) Kuntze = Solidago lancifolia (Torr. et A. Gray) Chapm. ■☆

40699 Aster langaoensis J. Q. Fu;兰皋紫菀;Langao Aster ■

40700 Aster lanuginosus (J. Small) Y. Ling;绵毛紫菀(棉毛紫菀);Denselanose Aster, Woolly Aster ■

40701 Aster lasiocarpus (DC.) Harv. = Felicia lasiocarpa DC. ■☆

40702 Aster lasiocladus Hayata;绒山白兰■

40703 Aster lasiocladus Hayata = Aster ageratoides Turcz. var. lasiocladus (Hayata) Hand. -Mazz. ■

40704 Aster lateriflorus (L.) Britton;宽花紫菀;Calico Aster, Goblet Aster, Side-flowering Aster ■☆

40705 Aster lateriflorus (L.) Britton = Symphyotrichum lateriflorum (L.) Á. Löve et D. Löve ■☆

40706 Aster lateriflorus (L.) Britton var. angustifolius Wiegand;窄叶宽花紫菀;Calico Aster, Goblet Aster, Side-flowering Aster ■☆

40707 Aster lateriflorus (L.) Britton var. glomerellus (Torr. et A. Gray) E. S. Burgess;团集宽花紫菀■☆

40708 Aster lateriflorus (L.) Britton var. hirsuticaulis (Lindl. ex DC.) Porter;毛茎宽花紫菀;Calico Aster, Goblet Aster, Side-flowering Aster ■☆

40709 Aster lateriflorus (L.) Britton var. horizontalis (Desf.) Farw.;水平宽花紫菀;Calico Aster, Goblet Aster, Side-flowering Aster ■☆

40710 Aster lateriflorus (L.) Britton var. pendulus (Aiton) E. S. Burgess = Aster lateriflorus (L.) Britton var. horizontalis (Desf.) Farw. ■☆

40711 Aster lateriflorus (L.) Britton var. tenuipes Wiegand = Aster lateriflorus (L.) Britton var. hirsuticaulis (Lindl. ex DC.) Porter ■☆

40712 Aster lateriflorus (L.) Britton var. thyrsoides (A. Gray) E. Sheld. = Symphyotrichum ontarionis (Wiegand) G. L. Nesom ■☆

40713 Aster lateriflorus Britton 'Horizontalis';平顶侧花紫菀■☆

40714 Aster lateriflorus Britton = Symphyotrichum lateriflorum (L.) Á. Löve et D. Löve ■☆

40715 Aster latibracteatus Franch.;宽苞紫菀;Broadbract Aster, Broadphyllaries Aster ■

40716 Aster laticorymbus Vaniot = Aster ageratoides Turcz. var. laticorymbus (Vaniot) Hand. -Mazz. ■

40717 Aster latisquamus (Maxim.) Hand. -Mazz. = Brachyactis ciliata Ledeb. ■

40718 Aster latissimifolius (Mill.) Kuntze = Solidago latissimifolia Mill. ■☆

40719 Aster latissimifolius (Mill.) Kuntze var. serotinus Kuntze = Solidago gigantea Aiton ■☆

40720 Aster laurentianus Fernald = Symphyotrichum laurentianum (Fernald) G. L. Nesom ■☆

40721 Aster laurentianus Fernald var. contiguus Fernald = Symphyotrichum laurentianum (Fernald) G. L. Nesom ■☆

40722 Aster laurentianus Fernald var. magdalenensis Fernald = Symphyotrichum laurentianum (Fernald) G. L. Nesom ■☆

40723 Aster lautureanus (Debeaux) Franch. = Kalimeris lautureana (Debeaux) Kitam. ■

40724 Aster lautureanus (Debeaux) Franch. var. mangtaoensis (Kitag.) Kitam. = Kalimeris lautureana (Debeaux) Kitam. ■

40725 Aster lautureanus (Debeaux) Franch. var. mongolicus (Franch.) Kitag. = Kalimeris mongolica (Franch.) Kitam. ■

40726 Aster lauturensis (Debeaux) Franch. var. mangtaoensis Kitag. = Kalimeris lautureana (Debeaux) Kitam. ■

40727 Aster lautus Lunell = Symphyotrichum lanceolatum (Willd.) G. L. Nesom var. hesperium (A. Gray) G. L. Nesom ■☆

40728 Aster lavandulifolius Hand. -Mazz.;线叶紫菀(狭叶紫菀);Lavenderleaf Aster, Lavender-leaved Aster ●

40729 Aster laxifolius Lindl. var. borealis Torr. et A. Gray = Aster borealis (Torr. et A. Gray) Prov. ■☆

40730 Aster laxifolius Lindl. var. borealis Torr. et A. Gray = Symphyotrichum boreale (Torr. et A. Gray) Á. Löve et D. Löve ■☆

40731 Aster laxifolius Nees var. laetiflorus Torr. et A. Gray = Aster borealis (Torr. et A. Gray) Prov. ■☆

40732 Aster laxifolius Nees var. laetifolius Torr. et A. Gray = Symphyotrichum lanceolatum (Willd.) G. L. Nesom ■☆

40733 Aster laxus Willd. = Symphyotrichum lanceolatum (Willd.) G. L. Nesom ■☆

40734 Aster leavenworthii (Torr. et A. Gray) Kuntze = Solidago leavenworthii Torr. et A. Gray ■☆

40735 Aster ledebourianus Novopokr. = Galatella punctata (Kar. et Kir.) Nees ■

40736 Aster ledophyllus (A. Gray) A. Gray = Eucephalus ledophyllus (A. Gray) Greene ■☆

40737 Aster ledophyllus (A. Gray) A. Gray var. covillei (Greene) Cronquist = Eucephalus ledophyllus (A. Gray) Greene var. covillei (Greene) G. L. Nesom ■☆

40738 Aster leiocarpus (DC.) Harv. = Zyrphelis montana (Schltr.) G. L. Nesom ■☆

40739 Aster leiophyllus Franch. et Sav.;白花马兰(山白兰)■

40740 Aster leiophyllus Franch. et Sav. = Aster ageratoides Turcz. var. leiophyllus (Franch. et Sav.) Y. Ling ■

40741 Aster leiophyllus Franch. et Sav. = Aster ageratoides Turcz. ■

40742 Aster leiophyllus Franch. et Sav. var. harae (Makino) H. Hara = Aster ageratoides Turcz. f. purpurascens Kitam. ■☆

40743 Aster leiophyllus Franch. et Sav. var. ovalifolius (Kitam.) H. Hara = Aster ageratoides Turcz. f. ovalifolius (Kitam.) Ohwi ■☆

40744 Aster leiophyllus Franch. et Sav. var. robustus (Koidz.) H. Hara = Aster ageratoides Turcz. ■

40745 Aster leiophyllus Franch. et Sav. var. sawadanus (Kitam.) H. Hara = Aster ageratoides Turcz. var. oligocephalus (Nakai ex H. Hara) Nor. Tanaka ■☆

40746 Aster leiophyllus Franch. et Sav. var. stenophyllus (Kitam.) H. Hara = Aster ageratoides Turcz. ■

40747 Aster leiophyllus Franch. et Sav. var. tenuifolius (Kitam.) H. Hara = Aster ageratoides Turcz. var. tenuifolius Kitam. ■☆

40748 Aster lemmonii A. Gray = Symphyotrichum potosinum (A. Gray) G. L. Nesom ■☆

40749 Aster lentus Greene = Symphyotrichum lentum (Greene) G. L. Nesom ■☆

40750 Aster lepidus (DC.) Kuntze = Solidago lepida DC. ■☆

40751 Aster leptocladus Makino = Aster hispidus Thunb. var. leptocladus (Makino) Okuyama ■☆

40752 Aster leucanthemifolius Greene = Dieteria canescens (Pursh) Nutt. var. leucanthemifolia (Greene) D. R. Morgan et R. L. Hartm. ■☆

40753 Aster leucelene S. F. Blake;白紫菀;White Aster ■☆

40754 Aster likiangensis Franch.;丽江紫菀(肥儿草);Lijiang Aster, Likiang Aster ■

40755 Aster likiangensis Franch. = Aster asteroides (DC.) Kuntze ■

40756 Aster likiangensis Franch. f. polianthus Y. Ling;多花丽江紫菀■

40757 Aster likiangensis Franch. subsp. costei ? = Aster asteroides (DC.) Kuntze ■

40758 Aster likiangensis Franch. subsp. hedinii (Ostenf.) Onno = Aster asteroides (DC.) Kuntze ■

40759 Aster likiangensis Franch. subsp. redinii (Ostenf.) Onno = Aster asteroides (DC.) Kuntze ■

40760 Aster likiangensis Franch. subsp. typicus Onno = Aster likiangensis Franch. ■

40761 Aster limitaneus W. W. Sm. et Farr = Aster souliei Franch. ■

40762 Aster limitaneus W. W. Sm. et Farr. = Aster souliei Franch. var. limitaneus (W. W. Sm. et Farr.) Hand. -Mazz. ■

40763 Aster limoniifolius Fedtsch. = Krylovia limoniifolia (Less.) Schischk. ■

40764 Aster limosus Hemsl.;湿生紫菀;Moist Aster ■

40765 Aster limprichtii Diels = Aster albescens (DC.) Wall. ex Hand. - Mazz. var. limprichtii (Diels) Hand. -Mazz. ●

40766 Aster limprichtii Diels var. gracilior Hand. -Mazz. = Aster albescens (DC.) Wall. ex Hand. -Mazz. var. gracilior Hand. -Mazz. ●

40767 Aster linariifolius L.;丝叶紫菀;Flax-leaved Aster, Savory-leaved Aster, Stiff Aster, Stiff-leaved Aster ■☆

40768 Aster linariifolius L. = Ionactis linariifolia (L.) Greene ■☆

40769 Aster linariifolius L. f. lateralis House = Aster linariifolius L. ■☆

40770 Aster linariifolius L. f. leucactis Benke = Aster linariifolius L. ■☆

40771 Aster linariifolius L. f. monocephalus House = Aster linariifolius L. ■☆

40772 Aster linariifolius L. var. victorinii Fernald = Aster linariifolius L. ■☆

40773 Aster linariifolius L. var. victorinii Fernald = Ionactis linariifolia (L.) Greene ■☆

40774 Aster lindheimeranus (Scheele) Kuntze = Solidago petiolaris Aiton ■☆

40775 Aster lindleyanus Torr. et A. Gray = Aster ciliolatus Lindl. ■☆

40776 Aster lindleyanus Torr. et A. Gray = Symphyotrichum ciliolatum (Lindl.) Á. Löve et D. Löve ■☆

40777 Aster lindleyanus Torr. et Gray;林德利紫菀(林德利氏紫菀);Lindley Aster, Lindley's Aster ■☆

40778 Aster linguiformis E. S. Burgess = Symphyotrichum undulatum (L.) G. L. Nesom ■☆

40779 Aster lingulatus Franch.;舌叶紫菀;Tongueleaf Aster, Tongue-like Aster ■

40780 Aster linifolius Harv. = Felicia linifolia (Harv.) Grau ■☆

40781 Aster linosyris (L.) Bernh.;麻紫菀(麻菀,无毛紫菀);Geldilocks, Goldilocks Aster ■☆

40782 Aster lipskyi Kom.;青海紫菀;Lipsky Aster ■

40783 Aster lithospermifolius DC. = Heteropappus altaicus (Willd.) Novopokr. var. canescens (Nees) Serg. ■

40784 Aster lofouensis H. Lév. et Vaniot = Pentanema indicum (L.) Y. Ling var. hypoleucum (Hand. -Mazz.) Y. Ling ■

40785 Aster longicaulis Desf. ex DC. = Aster tripolium L. subsp. longicaulis (DC.) Nyman ■☆

40786 Aster longifolius Lam. = Aster novi-belgii L. ■

40787 Aster longifolius Lam. = Symphyotrichum longifolium (Lam.) G. L. Nesom ■☆

40788 Aster longifolius Lam. var. villicaulis A. Gray;毛茎荷兰菊■☆

40789 Aster longifolius Lam. var. villicaulis A. Gray = Symphyotrichum novi-belgii (L.) G. L. Nesom var. villicaule (A. Gray) Labrecque et Brouillet ■☆

40790 Aster longipetiolatus C. C. Chang = Kalimeris longipetiolata (C. C. Chang) Y. Ling ■

40791 Aster loriformis E. S. Burgess = Symphyotrichum undulatum (L.) G. L. Nesom ■☆

40792 Aster lowrieanus Porter;楼氏紫菀;Lowrie's Aster ■☆

40793 Aster lowrieanus Porter = Aster leiophyllus Franch. et Sav. ■

40794 Aster lowrieanus Porter var. incisus Porter = Symphyotrichum cordifolium (L.) G. L. Nesom ■☆

40795 Aster lowrieanus Porter var. lanceolatus Porter = Symphyotrichum cordifolium (L.) G. L. Nesom ■☆

40796 Aster lucens Kitam. = Aster taiwanensis Kitam. var. lucens (Kitam.) Kitam. ■☆

40797 Aster lucidulus (A. Gray) Wiegand;光亮紫菀;Shining Aster ■☆

40798 Aster lucidulus (A. Gray) Wiegand = Aster puniceus L. ■☆

40799 Aster lucidulus (A. Gray) Wiegand = Symphyotrichum firmum (Nees) G. L. Nesom ■☆

40800 Aster lucidulus (A. Gray) Wiegand f. albiflorus (R. Hoffm.) Benke = Symphyotrichum firmum (Nees) G. L. Nesom ■☆

40801 Aster lucidulus (A. Gray) Wiegand f. firmus (Nees) Deam = Symphyotrichum firmum (Nees) G. L. Nesom ■☆

40802 Aster luteus (N. E. Br.) Hutch. = Felicia mossamedensis (Hiern) Mendonça ■☆

40803 Aster luxurifolius Tamamsch.;疏花紫菀■☆

40804 Aster lydenburgensis W. Lippert;莱登堡紫菀■☆

40805 Aster maackii Regel;圆苞紫菀(麻氏紫菀);Maack Aster ■

40806 Aster maccallae Rydb. = Symphyotrichum subspicatum (Nees) G. L. Nesom ■☆

40807 Aster macilentus Vaniot = Inula nervosa Wall. ■

40808 Aster macrodon H. Lév. et Vaniot = Kalimeris incisa (Fisch.) DC. ■

40809 Aster macrolophus H. Lév. et Vaniot = Aster tripolium L. ■☆
40810 Aster macrolophus H. Lév. et Vaniot = Tripolium vulgare Nees ■
40811 Aster macrophyllus (L.) Cass. var. apricensis E. S. Burgess = Eurybia macrophylla (L.) Cass. ■☆
40812 Aster macrophyllus (L.) Cass. var. excelsior E. S. Burgess = Eurybia macrophylla (L.) Cass. ■☆
40813 Aster macrophyllus (L.) Cass. var. ianthinus (E. S. Burgess) Fernald = Eurybia macrophylla (L.) Cass. ■☆
40814 Aster macrophyllus (L.) Cass. var. pinguifolius E. S. Burgess = Eurybia macrophylla (L.) Cass. ■☆
40815 Aster macrophyllus (L.) Cass. var. sejunctus E. S. Burgess = Eurybia macrophylla (L.) Cass. ■☆
40816 Aster macrophyllus (L.) Cass. var. velutinus E. S. Burgess = Eurybia macrophylla (L.) Cass. ■☆
40817 Aster macrophyllus L.; 大叶紫菀; Bigleaf Aster, Big-leaved Aster, Large Leaf Aster, Large-leaved Aster, Starwort, Tongue ■☆
40818 Aster macrophyllus L. = Eurybia macrophylla (L.) Cass. ■☆
40819 Aster macrophyllus L. f. eglandulosus Shinners = Aster macrophyllus L. ■☆
40820 Aster macrophyllus L. var. excelsior E. S. Burgess = Aster macrophyllus L. ■☆
40821 Aster macrophyllus L. var. ianthinus (E. S. Burgess) Fernald = Aster macrophyllus L. ■☆
40822 Aster macrophyllus L. var. pinguifolius E. S. Burgess = Aster macrophyllus L. ■☆
40823 Aster macrophyllus L. var. sejunctus E. S. Burgess = Aster macrophyllus L. ■☆
40824 Aster macrophyllus L. var. velutinus E. S. Burgess = Aster macrophyllus L. ■☆
40825 Aster macrorrhizus Thunb. = Felicia macrorrhiza (Thunb.) DC. ●☆
40826 Aster madagascariensis (Humbert) Humbert = Madagaster madagascariensis (Humbert) G. L. Nesom ●☆
40827 Aster mairei H. Lév. = Aster vestitus Franch. ■
40828 Aster major (Hook.) Porter = Canadanthus modestus (Lindl.) G. L. Nesom ■☆
40829 Aster mandrarensis Humbert = Madagaster mandrarensis (Humbert) G. L. Nesom ●☆
40830 Aster mangshanensis Y. Ling;莽山紫菀;Mangshan Aster ■
40831 Aster mangtaoensis Kitag. = Kalimeris lautureana (Debeaux) Kitam. ■
40832 Aster marchandii H. Lév. = Doellingeria marchandii (H. Lév.) Y. Ling ■
40833 Aster maritimus Lam. = Tripolium vulgare Nees ■
40834 Aster marshallii (Rothr.) Kuntze = Solidago missouriensis Nutt. ■☆
40835 Aster maruyamae Kitam.;丸山紫菀■☆
40836 Aster megalanthus Y. Ling; 大花紫菀; Bigflower Aster, Large Head Aster ■
40837 Aster mekongensis Onno = Aster bietii Franch. ■
40838 Aster mekongensis Onno = Aster himalaicus C. B. Clarke ■
40839 Aster menelii H. Lév.;黔中紫菀;Menel Aster, Ouizhou Aster ■
40840 Aster meritus A. Nelson = Eurybia merita (A. Nelson) G. L. Nesom ■☆
40841 Aster mespilifolius Less. = Microglossa mespilifolia (Less.) B. L. Rob. ●☆
40842 Aster mexicanus (L.) Kuntze = Solidago sempervirens L. subsp. mexicana (L.) Semple ■☆
40843 Aster meyendorffii (Regel et Maack) Voss = Heteropappus meyendorffii (Regel et Maack) Kom. ■
40844 Aster micranthus H. Lév. et Vaniot = Turczaninovia fastigiata (Fisch.) DC. ■
40845 Aster microcephalus (Miq.) Franch. et Sav.;小头紫菀■☆
40846 Aster microcephalus (Miq.) Franch. et Sav. var. angustifolius (Kitam.) Nor. Tanaka;狭叶小头紫菀■☆
40847 Aster microcephalus (Miq.) Franch. et Sav. var. littoricola (Kitam.) Nor. Tanaka;海滨小头紫菀■☆
40848 Aster microcephalus (Miq.) Franch. et Sav. var. ovatus (Franch. et Sav.) Soejima et Mot. Ito 'Hortensis';田园卵形小头紫菀■☆
40849 Aster microcephalus (Miq.) Franch. et Sav. var. ovatus (Franch. et Sav.) Soejima et Mot. Ito;卵形小头紫菀■☆
40850 Aster microcephalus (Miq.) Franch. et Sav. var. ripensis Makino;河岸小头紫菀■☆
40851 Aster microcephalus (Miq.) Franch. et Sav. var. ripensis Makino f. albiflorus (Makino) H. Hara;白花河岸小头紫菀■☆
40852 Aster microcephalus (Miq.) Franch. et Sav. var. ripensis Makino f. tubulosus (Makino) Makino;块状河岸小头紫菀■☆
40853 Aster microcephalus (Miq.) Franch. et Sav. var. tubulosus (Makino) H. Hara = Aster microcephalus (Miq.) Franch. et Sav. var. ripensis Makino f. tubulosus (Makino) Makino ■☆
40854 Aster microcephalus (Miq.) Franch. et Sav. var. yezoensis (Kitam. et H. Hara) Soejima et Mot. Ito;北海道小头紫菀■☆
40855 Aster microcephalus (Miq.) Franch. et Sav. var. yoshinaganus (Kitam.) H. Hara = Aster yoshinaganus (Kitam.) Mot. Ito et Soejima ■☆
40856 Aster microspermus (DC.) Harv. = Felicia microsperma DC. ■☆
40857 Aster milanjiensis S. Moore;米兰吉紫菀■☆
40858 Aster millefolius Vaniot = Heteropappus altaicus (Willd.) Novopokr. var. millefolius (Vaniot) Hand. -Mazz. ■
40859 Aster minimus Hutch. = Felicia minima (Hutch.) Grau ■☆
40860 Aster miquelianus H. Hara;米克尔紫菀■☆
40861 Aster mirabilis Torr. et A. Gray = Eurybia mirabilis (Torr. et A. Gray) G. L. Nesom ■☆
40862 Aster miser L. var. glomerellus Torr. et A. Gray = Aster lateriflorus (L.) Britton ■☆
40863 Aster missouriensis (Nutt.) Kuntze = Solidago missouriensis Nutt. ■☆
40864 Aster missouriensis Britton = Symphyotrichum ontarionis (Wiegand) G. L. Nesom ■☆
40865 Aster missouriensis Britton var. thyrsoides (A. Gray) Wiegand = Symphyotrichum ontarionis (Wiegand) G. L. Nesom ■☆
40866 Aster miyabeanus Tatew. et Kitam. = Erigeron miyabeanus (Tatew. et Kitam.) Tatew. et Kitam. ex H. Hara ■☆
40867 Aster miyagii Koidz.;宫木紫菀■☆
40868 Aster modestus Lindl. = Canadanthus modestus (Lindl.) G. L. Nesom ■☆
40869 Aster modestus Lindl. var. major (Hook.) Muenscher = Canadanthus modestus (Lindl.) G. L. Nesom ■☆
40870 Aster mohrii E. S. Burgess = Symphyotrichum undulatum (L.) G. L. Nesom ■☆
40871 Aster mollis Rydb. = Symphyotrichum molle (Rydb.) G. L. Nesom ■☆
40872 Aster molliusculus (DC.) C. B. Clarke; 软毛紫菀; Softhair

Aster, Weakly-hair Aster ■
40873 Aster molliusculus Novopokr. = Asterothamnus molliusculus Novopokr. ●
40874 Aster mongolicus Franch. = Kalimeris mongolica (Franch.) Kitam. ■
40875 Aster montanus R. Br. = Eurybia sibirica (L.) G. L. Nesom ■☆
40876 Aster montanus R. Br. var. giganteus (Hook.) Torr. et A. Gray = Eurybia sibirica (L.) G. L. Nesom ■☆
40877 Aster monticola (Torr. et A. Gray) Kuntze = Solidago roanensis Porter ■☆
40878 Aster morrisonensis Hayata;玉山紫菀(玉山铁杆蒿);Jade Mountain Aster, Morrison Aster ■
40879 Aster motuoensis Y. L. Chen;墨脱紫菀;Motuo Aster ■
40880 Aster moupinensis (Franch.) Hand. -Mazz.;川鄂紫菀(穆坪紫菀);Baoxing Aster ■
40881 Aster muehlenbergianus Kuntze = Solidago squarrosa Muhl. ■☆
40882 Aster multiflorus Aiton = Aster ericoides L. ■☆
40883 Aster multiflorus Aiton = Symphyotrichum ericoides (L.) G. L. Nesom ■☆
40884 Aster multiflorus Aiton var. caeruleus Benke = Symphyotrichum ericoides (L.) G. L. Nesom ■☆
40885 Aster multiflorus Aiton var. ciliatus Barton = Symphyotrichum ericoides (L.) G. L. Nesom ■☆
40886 Aster multiflorus Aiton var. commutatus Torr. et A. Gray = Symphyotrichum falcatum (Lindl.) G. L. Nesom var. commutatum (Torr. et A. Gray) G. L. Nesom ■☆
40887 Aster multiflorus Aiton var. exiguus Fernald = Symphyotrichum ericoides (L.) G. L. Nesom ■☆
40888 Aster multiflorus Aiton var. incanopilosus (Lindl.) Rydb. = Symphyotrichum falcatum (Lindl.) G. L. Nesom var. commutatum (Torr. et A. Gray) G. L. Nesom ■☆
40889 Aster multiflorus Aiton var. pansus S. F. Blake = Symphyotrichum ericoides (L.) G. L. Nesom var. pansum (S. F. Blake) G. L. Nesom ■☆
40890 Aster multiflorus Aiton var. prostratus Kuntze = Symphyotrichum ericoides (L.) G. L. Nesom var. prostratum (Kuntze) G. L. Nesom ■☆
40891 Aster multiflorus Aiton var. stricticaulis Torr. et A. Gray = Symphyotrichum ericoides (L.) G. L. Nesom var. pansum (S. F. Blake) G. L. Nesom ■☆
40892 Aster multiformis E. S. Burgess = Eurybia macrophylla (L.) Cass. ■☆
40893 Aster multiradiatus (Aiton) Kuntze = Solidago multiradiata Aiton ■☆
40894 Aster multiradiatus Wall. = Erigeron multiradiatus (Lindl.) Benth. et Hook. f. ■
40895 Aster muricatus Thunb. = Felicia muricata (Thunb.) Nees ■☆
40896 Aster muricatus Thunb. var. chrysocomoides Sond. = Felicia muricata (Thunb.) Nees ■☆
40897 Aster muricatus Thunb. var. fascicularis (DC.) E. Mey. ex Harv. = Felicia fascicularis DC. ■☆
40898 Aster nahanniensis Cody = Symphyotrichum nahanniense (Cody) Semple ■☆
40899 Aster nakaii H. Lév. et Vaniot = Aster tataricus L. f. ■
40900 Aster namaquanus Harv. = Felicia namaquana (Harv.) Merxm. ■☆
40901 Aster nanus (Nutt.) Kuntze = Solidago nana Nutt. ■☆
40902 Aster natalensis (Sch. Bip.) Harv. = Felicia rosulata Yeo ●☆
40903 Aster nemoralis Aiton = Oclemena nemoralis (Aiton) Greene ■☆
40904 Aster nemoralis Aiton var. blakei Porter = Oclemena blakei (Porter) G. L. Nesom ■☆
40905 Aster nemoralis Aiton var. major Peck = Oclemena blakei (Porter) G. L. Nesom ■☆
40906 Aster nemoralis Sol. ex Aiton;湿地紫菀;Bog Aster, Wood Aster ■☆
40907 Aster neoelegans Grierson;新雅紫菀;Graceful Aster ■
40908 Aster nigrescens Vaniot = Aster ageratoides Turcz. var. heterophyllus Maxim. ■
40909 Aster nigromontanus Dunn;黑山紫菀;Blackmountain Aster ■
40910 Aster nigrotinctus Y. Ling = Aster farreri W. W. Sm. et Jeffrey ■
40911 Aster nitidus C. C. Chang;亮叶紫菀;Light Leaf Aster, Nitidleaf Aster, Shining Aster, Shinyleaf Aster ●
40912 Aster novae-angliae L. 'Alma Pötschke' = Aster novae-angliae L. 'Andenken an Alma Pötschke' ■☆
40913 Aster novae-angliae L. 'Andenken an Alma Pötschke';纪念美国紫菀■☆
40914 Aster novae-angliae L. 'Autumn Snow' = Aster novae-angliae L. 'Herbstschnee' ■☆
40915 Aster novae-angliae L. 'Barr's Pink';巴尔粉红美国紫菀■☆
40916 Aster novae-angliae L. 'Harrington's Pink';哈林顿粉红美国紫菀■☆
40917 Aster novae-angliae L. 'Herbstschnee';秋雪美国紫菀■☆
40918 Aster novae-angliae L. = Symphyotrichum novae-angliae (L.) G. L. Nesom ■☆
40919 Aster novae-angliae L. f. geneseensis House = Aster novi-belgii L. ■
40920 Aster novae-angliae L. f. rosarius House = Aster novi-belgii L. ■
40921 Aster novae-angliae L. f. roseus (Desf.) Britton = Aster novi-belgii L. ■
40922 Aster novae-angliae L. var. monocephalus Farw. = Aster novi-belgii L. ■
40923 Aster novi-belgii L.;荷兰菊(荷兰紫菀);Confused Michaelmas-daisy, Long-leaved Aster, Michaemas Daisy, New Belgium Aster, New England Aster, New York Aster, New-York Aster, Rush Aster ■
40924 Aster novi-belgii L. 'Apple Blossom';苹果花荷兰菊■☆
40925 Aster novi-belgii L. 'Carnival';狂欢节荷兰菊■☆
40926 Aster novi-belgii L. 'Chequers';栀子花荷兰菊■☆
40927 Aster novi-belgii L. 'Climax';顶点荷兰菊■☆
40928 Aster novi-belgii L. 'Fellowship';友谊荷兰菊■☆
40929 Aster novi-belgii L. 'Freda Ballard';弗瑞达·巴利达荷兰菊■☆
40930 Aster novi-belgii L. 'Kristina';克里斯蒂娜荷兰菊■☆
40931 Aster novi-belgii L. 'Lassie';少女荷兰菊■☆
40932 Aster novi-belgii L. 'Little Pink Beauty';小粉丽荷兰菊■☆
40933 Aster novi-belgii L. 'Marie Ballard';玛丽·巴拉德荷兰菊■☆
40934 Aster novi-belgii L. 'Orlando';奥兰多荷兰菊■☆
40935 Aster novi-belgii L. 'Patricia Ballard';高贵荷兰菊■☆
40936 Aster novi-belgii L. 'Raspberry Ripple';树莓紫荷兰菊■☆
40937 Aster novi-belgii L. 'Royal Ruby';品红宝石荷兰菊■☆
40938 Aster novi-belgii L. 'Royal Velvet';天鹅绒荷兰菊■☆
40939 Aster novi-belgii L. 'Sandford White Swan';斯坦福白天鹅荷兰菊■☆
40940 Aster novi-belgii L. = Symphyotrichum novi-belgii (L.) G. L. Nesom ■
40941 Aster novi-belgii L. subsp. johannensis (Fernald) A. G. Jones = Aster novi-belgii L. ■
40942 Aster novi-belgii L. subsp. laevigatus (Lam.) Thell. = Aster

versicolor Steud. ■☆

40943 Aster novi-belgii L. var. crenifolius (Fernald) Labrecque et Brouillet = Aster foliaceus Lindl. var. crenifolius Fernald ■☆

40944 Aster novi-belgii L. var. johannensis (Fernald) A. G. Jones = Aster novi-belgii L. ■

40945 Aster novi-belgii L. var. rosaceus J. Rousseau = Aster novi-belgii L. ■

40946 Aster novi-belgii L. var. villicaulis (A. Gray) B. Boivin = Aster longifolius Lam. var. villicaulis A. Gray ■☆

40947 Aster nubimontis W. Lippert;云雾紫菀■☆

40948 Aster oblongifolius Nutt.;芳香紫菀;Aromatic Aster, Oblong-leaved Aster, Romatic Aster ■☆

40949 Aster oblongifolius Nutt. = Symphyotrichum oblongifolium (Nutt.) G. L. Nesom ■☆

40950 Aster oblongifolius Nutt. f. roseoligulatus (Benke) Shinners = Aster oblongifolius Nutt. ■☆

40951 Aster oblongifolius Nutt. var. angustatus Shinners;狭叶芳香紫菀;Eastern aromatic Aster ■☆

40952 Aster oblongifolius Nutt. var. angustatus Shinners = Aster oblongifolius Nutt. ■☆

40953 Aster oblongifolius Nutt. var. orientis Shinners = Aster oblongifolius Nutt. ■☆

40954 Aster oblongifolius Nutt. var. rigidulus A. Gray = Aster oblongifolius Nutt. ■☆

40955 Aster obovatus Ledeb. = Krylovia limoniifolia (Less.) Schischk. ■

40956 Aster obtusatus Thunb. = Chrysocoma obtusata (Thunb.) Ehr. Bayer ■☆

40957 Aster occidentalis (Nutt.) Torr. et A. Gray = Symphyotrichum spathulatum (Lindl.) G. L. Nesom ■☆

40958 Aster occidentalis (Nutt.) Torr. et A. Gray var. intermedius A. Gray = Symphyotrichum spathulatum (Lindl.) G. L. Nesom var. intermedium (A. Gray) G. L. Nesom ■☆

40959 Aster occidentalis (Nutt.) Torr. et A. Gray var. yosemitanus (A. Gray) Cronquist = Symphyotrichum spathulatum (Lindl.) G. L. Nesom var. yosemitanum (A. Gray) G. L. Nesom ■☆

40960 Aster odorus (Aiton) Kuntze = Solidago odora Aiton ■☆

40961 Aster ohioensis (Riddell) Kuntze = Solidago ohioensis Riddell ■☆

40962 Aster oldhamii Hemsl. = Heteropappus oldhamii (Hemsl.) Kitam. ■

40963 Aster omerophyllus Hayata = Heteropappus hispidus (Thunb.) Less. ■

40964 Aster ontarionis Wiegand = Symphyotrichum ontarionis (Wiegand) G. L. Nesom ■☆

40965 Aster ontarionis Wiegand var. glabratus Semple = Symphyotrichum ontarionis (Wiegand) G. L. Nesom var. glabratum (Semple) Brouillet et Bouchard ■☆

40966 Aster oolentangiensis Riddell;奥地紫菀■☆

40967 Aster oolentangiensis Riddell = Symphyotrichum oolentangiense (Riddell) G. L. Nesom ■☆

40968 Aster oppositifolius L. = Felicia cymbalariae (Aiton) Bolus et Wolley-Dod ex Adamson et T. M. Salter ■☆

40969 Aster orcuttii Vasey et Rose = Xylorhiza orcuttii (Vasey et Rose) Greene ●☆

40970 Aster oregonensis (Nutt.) Cronquist = Sericocarpus oregonensis Nutt. ■☆

40971 Aster oregonensis (Nutt.) Cronquist subsp. californicus (Durand) D. D. Keck = Sericocarpus oregonensis Nutt. subsp. californicus (Durand) Ferris ■☆

40972 Aster oreophilus Franch.;石生紫菀(菊花暗消,肋痛草,毛脉一枝蒿,野冬菊);Rocky Aster, Stony Living Aster ■

40973 Aster oreophilus Franch. f. inaequisquamus Y. Ling;昆明石生紫菀;Kunming Rocky Aster ■

40974 Aster oreophilus Franch. f. umbrosus Y. Ling;鹤庆石生紫菀(阴地石生紫菀);Heqing Rocky Aster ■

40975 Aster osterhoutii Rydb. = Symphyotrichum lanceolatum (Willd.) G. L. Nesom var. hesperium (A. Gray) G. L. Nesom ■☆

40976 Aster ovalifolius Kitam.;卵叶紫菀(卵形紫菀,台湾绀菊);Ovate Aster, Ovateleaf Aster ■

40977 Aster ovatus (Franch. et Sav.) Mot. Ito et Soejima = Aster microcephalus (Miq.) Franch. et Sav. var. ovatus (Franch. et Sav.) Soejima et Mot. Ito ■☆

40978 Aster ovatus (Franch. et Sav.) Mot. Ito et Soejima var. microcepahlus (Miq.) Mot. Ito et Soejima = Aster microcephalus (Miq.) Franch. et Sav. ■☆

40979 Aster ovatus (Franch. et Sav.) Mot. Ito et Soejima var. ripensis (Makino) Mot. Ito et Soejima = Aster microcephalus (Miq.) Franch. et Sav. var. ripensis Makino ■☆

40980 Aster ovatus (Franch. et Sav.) Mot. Ito et Soejima var. yezoensis (Kitam. et H. Hara) Mot. Ito et Soejima = Aster microcephalus (Miq.) Franch. et Sav. var. yezoensis (Kitam. et H. Hara) Soejima et Mot. Ito ■☆

40981 Aster ovovatus Ledeb. = Krylovia limoniifolia (Less.) Schischk. ■

40982 Aster palmeri A. Gray = Neonesomia palmeri (A. Gray) Urbatsch et R. P. Roberts ●☆

40983 Aster paludicola Piper = Symphyotrichum spathulatum (Lindl.) G. L. Nesom var. yosemitanum (A. Gray) G. L. Nesom ■☆

40984 Aster paludosus Aiton = Eurybia paludosa (Aiton) G. L. Nesom ■☆

40985 Aster paludosus Aiton subsp. hemisphericus (Alexander) Cronquist = Eurybia hemispherica (Alexander) G. L. Nesom ■☆

40986 Aster paludosus Aiton var. hemisphericus (Alexander) Waterf. = Eurybia hemispherica (Alexander) G. L. Nesom ■☆

40987 Aster palustris Lam. = Tripolium vulgare Nees ■

40988 Aster panduratus Nees ex Walp.;琴叶紫菀(大风草,福氏紫菀,岗边菊,鱼鳅串);Fiddleleaf Aster ■

40989 Aster paniculatus Lam.;圆锥紫菀;Farewell Summer, Farewell-summer, Michaelmas Daisy, Panicled Aster ■☆

40990 Aster paniculatus Lam. = Symphyotrichum lanceolatum (Willd.) G. L. Nesom ■☆

40991 Aster paniculatus Lam. var. bellidiflorus (Willd.) E. S. Burgess = Symphyotrichum lanceolatum (Willd.) G. L. Nesom ■☆

40992 Aster paniculatus Lam. var. polychrous Lunell = Symphyotrichum lanceolatum (Willd.) G. L. Nesom ■☆

40993 Aster paniculatus Lam. var. simplex (Willd.) E. S. Burgess = Symphyotrichum simplex (Willd.) Á. Löve et D. Löve ■☆

40994 Aster pannonicus Jacq. = Aster tripolium L. ■☆

40995 Aster pansus (S. F. Blake) Cronquist = Symphyotrichum ericoides (L.) G. L. Nesom var. pansum (S. F. Blake) G. L. Nesom ■☆

40996 Aster pantotrichus S. F. Blake = Symphyotrichum ontarionis (Wiegand) G. L. Nesom ■☆

40997 Aster pantotrichus S. F. Blake var. thyrsoides (A. Gray) S. F. Blake = Symphyotrichum ontarionis (Wiegand) G. L. Nesom ■☆

40998 Aster pappei Harv. = Felicia amoena (Sch. Bip.) Levyns ■☆

40999 Aster papposissimus H. Lév. et Vaniot = Aster tripolium L. ■☆

41000 Aster papposissimus H. Lév. et Vaniot = Tripolium vulgare Nees ■

41001 Aster papposus Willd. ex Spreng. = Printzia polifolia (L.) Hutch. ■☆
41002 Aster parryi A. Gray = Xylorhiza glabriuscula Nutt. ●☆
41003 Aster parviceps (E. S. Burgess) Mack. et Bush = Symphyotrichum parviceps (E. S. Burgess) G. L. Nesom ■☆
41004 Aster parviceps (E. S. Burgess) Mack. et Bush var. pusillus (A. Gray) Fernald = Symphyotrichum depauperatum (Fernald) G. L. Nesom ■☆
41005 Aster patens Aiton = Symphyotrichum patens (Aiton) G. L. Nesom ■☆
41006 Aster patens Aiton var. floridanus R. W. Long = Symphyotrichum fontinale (Alexander) G. L. Nesom ■☆
41007 Aster patens Aiton var. georgianus (Alexander) Cronquist = Symphyotrichum georgianum (Alexander) G. L. Nesom ■☆
41008 Aster patens Aiton var. gracilis Hook. = Symphyotrichum patens (Aiton) G. L. Nesom var. gracile (Hook.) G. L. Nesom ■☆
41009 Aster patens Aiton var. patentissimus (Lindl. ex DC.) Torr. et A. Gray = Symphyotrichum patens (Aiton) G. L. Nesom var. patentissimum (Lindl. ex DC.) G. L. Nesom ■☆
41010 Aster patens Aiton var. phlogifolius (Muhl. ex Willd.) Nees = Symphyotrichum phlogifolium (Muhl. ex Willd.) G. L. Nesom ■☆
41011 Aster patens Aiton var. tenuicaulis C. Mohr = Symphyotrichum patens (Aiton) G. L. Nesom var. gracile (Hook.) G. L. Nesom ■☆
41012 Aster patentissimus Lindl. ex DC. = Symphyotrichum patens (Aiton) G. L. Nesom var. patentissimum (Lindl. ex DC.) G. L. Nesom ■☆
41013 Aster paternus Cronquist = Sericocarpus asteroides (L.) Nees ■☆
41014 Aster patulus Lam. = Symphyotrichum patulum (Lam.) Karlsson ■☆
41015 Aster paucicapitatus (B. L. Rob.) B. L. Rob. = Eucephalus paucicapitatus (B. L. Rob.) Greene ■☆
41016 Aster pauciflorus Nutt. = Almutaster pauciflorus (Nutt.) Á. Löve et D. Löve ■☆
41017 Aster pedionomus Alexander = Eurybia hemispherica (Alexander) G. L. Nesom ■☆
41018 Aster peglerae Bolus;佩格拉紫菀■☆
41019 Aster peglerae Bolus var. longipes Thell. = Aster comptonii W. Lippert ■☆
41020 Aster peirsonii Sharsm. = Oreostemma peirsonii (Sharsm.) G. L. Nesom ■☆
41021 Aster pekinensis (Hance) Chen = Kalimeris integrifolia Turcz. ex DC. ■
41022 Aster pekinensis (Hance) Kitag. = Kalimeris integrifolia Turcz. ex DC. ■
41023 Aster pekinensis Hance = Kalimeris integrifolia Turcz. ex DC. ■
41024 Aster pendulus Aiton = Aster lateriflorus (L.) Britton var. horizontalis (Desf.) Farw. ■☆
41025 Aster pensauensis Tamamsch.;潘萨乌紫菀■☆
41026 Aster peregrinus Banks ex Pursh = Erigeron peregrinus (Pursh) Greene ■☆
41027 Aster perelegans A. Nelson et J. F. Macbr. = Eucephalus elegans Nutt. ■☆
41028 Aster perezioides Less. = Zyrphelis perezioides (Less.) G. L. Nesom ■☆
41029 Aster perfoliatus Oliv.;贯叶紫菀■☆
41030 Aster petiolaris (Aiton) Kuntze = Solidago petiolaris Aiton ■☆
41031 Aster petiolatus Harv. = Felicia petiolata (Harv.) N. E. Br. ●☆
41032 Aster phlogifolius Muhl. ex Willd. = Symphyotrichum phlogifolium (Muhl. ex Willd.) G. L. Nesom ■☆
41033 Aster phyllolepis Torr. et A. Gray = Symphyotrichum pratense (Raf.) G. L. Nesom ■☆
41034 Aster piccolii Hook. f. = Gymnaster piccolii (Hook. f.) Kitam. ■
41035 Aster piccolii Hook. f. = Miyamayomena piccolii (Hook. f.) Kitam. ■
41036 Aster pilosus Willd. = Aster ericoides L. ■☆
41037 Aster pilosus Willd. = Symphyotrichum pilosum (Willd.) G. L. Nesom ■☆
41038 Aster pilosus Willd. f. pulchellus Benke = Symphyotrichum pilosum (Willd.) G. L. Nesom ■☆
41039 Aster pilosus Willd. subsp. parviceps (E. S. Burgess) A. G. Jones = Symphyotrichum parviceps (E. S. Burgess) G. L. Nesom ■☆
41040 Aster pilosus Willd. var. demotus S. F. Blake = Symphyotrichum pilosum (Willd.) G. L. Nesom ■☆
41041 Aster pilosus Willd. var. demotus S. F. Blake = Symphyotrichum pilosum (Willd.) G. L. Nesom var. pringlei (A. Gray) G. L. Nesom ■☆
41042 Aster pilosus Willd. var. platyphyllus (Torr. et A. Gray) S. F. Blake = Symphyotrichum pilosum (Willd.) G. L. Nesom ■☆
41043 Aster pilosus Willd. var. priceae (Britton) Cronquist = Symphyotrichum priceae (Britton) G. L. Nesom ■☆
41044 Aster pilosus Willd. var. pringlei (A. Gray) S. F. Blake = Symphyotrichum pilosum (Willd.) G. L. Nesom var. pringlei (A. Gray) G. L. Nesom ■☆
41045 Aster pilosus Willd. var. pusillus (A. Gray) A. G. Jones = Symphyotrichum depauperatum (Fernald) G. L. Nesom ■☆
41046 Aster pinnatifidus (Hook.) Kuntze;羽裂紫菀■☆
41047 Aster pinnatifidus (Hook.) Kuntze f. robustus Makino = Kalimeris incisa (Fisch.) DC. ■
41048 Aster pinnatifidus Makino f. robustus Makino = Kalimeris incisa (Fisch.) DC. ■
41049 Aster platylepis Y. L. Chen;阔苞紫菀■
41050 Aster pleiocephalus (Harv.) Hutch.;多头紫菀■☆
41051 Aster plumarius Burgess = Symphyotrichum cordifolium (L.) G. L. Nesom ■☆
41052 Aster plumosus Small = Symphyotrichum plumosum (Small) Semple ■☆
41053 Aster poaceus E. S. Burgess = Symphyotrichum oolentangiense (Riddell) G. L. Nesom ■☆
41054 Aster poliifolius L. = Printzia polifolia (L.) Hutch. ■☆
41055 Aster poliifolius Novopokr. = Asterothamnus poliifolius Novopokr. ●
41056 Aster poliothamnus Diels;灰枝紫菀(灰木紫菀);Greybranch Aster ■
41057 Aster polius C. K. Schneid.;灰毛紫菀;Grayhair Aster ■
41058 Aster polycephalus Rydb. = Aster ericoides L. ■☆
41059 Aster polycephalus Rydb. = Symphyotrichum ericoides (L.) G. L. Nesom ■☆
41060 Aster polyphyllus Willd. = Symphyotrichum pilosum (Willd.) G. L. Nesom var. pringlei (A. Gray) G. L. Nesom ■☆
41061 Aster poncinsii Franch. = Psychrogeton poncinsii (Franch.) Y. Ling et Y. L. Chen ■
41062 Aster porteri A. Gray = Symphyotrichum porteri (A. Gray) G. L. Nesom ■☆
41063 Aster potaninii Novopokr. = Asterothamnus centrali-asiaticus Novopokr. var. potaninii (Novopokr.) Y. Ling et Y. L. Chen ●

41064 Aster potosinus A. Gray = Symphyotrichum potosinum (A. Gray) G. L. Nesom ■☆

41065 Aster praealtus Poir. = Symphyotrichum praealtum (Poir.) G. L. Nesom ■☆

41066 Aster praealtus Poir. var. angustior Wiegand = Symphyotrichum praealtum (Poir.) G. L. Nesom var. angustior (Wiegand) G. L. Nesom ■☆

41067 Aster praealtus Poir. var. coerulescens (DC.) A. G. Jones = Symphyotrichum praealtum (Poir.) G. L. Nesom var. texicola (Wiegand) G. L. Nesom ■☆

41068 Aster praealtus Poir. var. subasper (Lindl.) Wiegand = Symphyotrichum praealtum (Poir.) G. L. Nesom var. angustior (Wiegand) G. L. Nesom ■☆

41069 Aster praealtus Poir. var. texicola Wiegand = Symphyotrichum praealtum (Poir.) G. L. Nesom var. texicola (Wiegand) G. L. Nesom ■☆

41070 Aster praetermissus Drumm. = Heteropappus crenatifolius (Hand. -Mazz.) Grierson ■

41071 Aster praetermissus Drumm. ex Hand. -Mazz. = Heteropappus crenatifolius (Hand. -Mazz.) Grierson ■

41072 Aster prainii (J. R. Drumm.) Y. L. Chen;厚棉紫菀■

41073 Aster prascottii Lindl. ex DC. = Aster sibiricus L. ■

41074 Aster prascottii Lindl. ex DC. = Chlamydites prainii J. R. Drumm. ●■

41075 Aster pratensis Raf. = Symphyotrichum pratense (Raf.) G. L. Nesom ■☆

41076 Aster prenanthoides Muhl. ex Willd. = Symphyotrichum prenanthoides (Muhl. ex Willd.) G. L. Nesom ■☆

41077 Aster prenanthoides Muhl. ex Willd. f. milwaukeensis Benke = Symphyotrichum pratense (Raf.) G. L. Nesom ■☆

41078 Aster priceae Britton = Symphyotrichum priceae (Britton) G. L. Nesom ■☆

41079 Aster pringlei (A. Gray) Britton = Symphyotrichum pilosum (Willd.) G. L. Nesom var. pringlei (A. Gray) G. L. Nesom ■☆

41080 Aster procerus Hemsl.;高茎紫菀;High Stem Aster, Muchhigh Aster ■

41081 Aster procumbens Houst. ex Mill. = Erigeron procumbens (Houst. ex Mill.) G. L. Nesom ■☆

41082 Aster proteus E. S. Burgess = Symphyotrichum undulatum (L.) G. L. Nesom ■☆

41083 Aster pseudamellus Hook. f. = Aster indamellus Grierson ■

41084 Aster pseudoasagrayi Makino;假华南铁杆蒿■

41085 Aster pseudobakeranus W. Lippert;假贝克紫菀■☆

41086 Aster ptarmicoides (Nees) Torr. et A. Gray;白花紫菀;Upland White Aster, White Upland Aster ■☆

41087 Aster ptarmicoides (Nees) Torr. et A. Gray = Solidago ptarmicoides (Nees) B. Boivin ■☆

41088 Aster ptarmicoides (Nees) Torr. et A. Gray var. georgianus A. Gray = Solidago ptarmicoides (Nees) B. Boivin ■☆

41089 Aster ptarmicoides Torr. et A. Gray = Solidago ptarmicoides (Torr. et A. Gray) B. Boivin ■☆

41090 Aster pubens (M. A. Curtis ex Torr. et A. Gray) Kuntze = Solidago bicolor L. ■☆

41091 Aster pubentior Cronquist = Aster umbellatus Mill. var. pubens A. Gray ■☆

41092 Aster pubentior Cronquist = Aster umbellatus Mill. ■☆

41093 Aster pubentior Cronquist = Doellingeria umbellata (Mill.) Nees var. pubens (A. Gray) Britton ■☆

41094 Aster pujosii Quézel;皮若斯紫菀■☆

41095 Aster pulchellus Willd. = Aster alpinus L. var. fallax (Tamamsch.) Y. Ling ■

41096 Aster pulchellus Willd. = Aster alpinus L. ■

41097 Aster pulicarius Scop. = Pulicaria prostrata (Gilib.) Asch. ■

41098 Aster pulverulentus (Nutt.) Kuntze = Solidago puberula Nutt. subsp. pulverulenta (Nutt.) Semple ■☆

41099 Aster punctatus Waldst. et Kit. = Galatella punctata (Kar. et Kir.) Nees ■

41100 Aster puniceus L. = Symphyotrichum puniceum (L.) Á. Löve et D. Löve ■☆

41101 Aster puniceus L. f. albiflorus (Farw.) Shinners = Aster puniceus L. ■☆

41102 Aster puniceus L. f. albiflorus (Farw.) Shinners = Symphyotrichum puniceum (L.) Á. Löve et D. Löve ■☆

41103 Aster puniceus L. f. colbyi (Benke) Shinners = Aster puniceus L. ■☆

41104 Aster puniceus L. f. colbyi (Benke) Shinners = Symphyotrichum puniceum (L.) Á. Löve et D. Löve ■☆

41105 Aster puniceus L. f. glabratus Shinners = Aster puniceus L. ■☆

41106 Aster puniceus L. f. glabratus Shinners = Symphyotrichum puniceum (L.) Á. Löve et D. Löve ■☆

41107 Aster puniceus L. f. lucidulus (A. Gray) Fernald = Aster puniceus L. ■☆

41108 Aster puniceus L. subsp. firmus (Nees) A. G. Jones = Symphyotrichum firmum (Nees) G. L. Nesom ■☆

41109 Aster puniceus L. var. albiflorus Farw. = Aster puniceus L. ■☆

41110 Aster puniceus L. var. albiflorus Farw. = Symphyotrichum puniceum (L.) Á. Löve et D. Löve ■☆

41111 Aster puniceus L. var. calderi (B. Boivin) Lepage = Symphyotrichum puniceum (L.) Á. Löve et D. Löve ■☆

41112 Aster puniceus L. var. calvus Shinners = Aster puniceus L. ■☆

41113 Aster puniceus L. var. calvus Shinners = Symphyotrichum puniceum (L.) Á. Löve et D. Löve ■☆

41114 Aster puniceus L. var. colbyi Benke = Aster puniceus L. ■☆

41115 Aster puniceus L. var. colbyi Benke = Symphyotrichum puniceum (L.) Á. Löve et D. Löve ■☆

41116 Aster puniceus L. var. compactus Fernald = Aster puniceus L. ■☆

41117 Aster puniceus L. var. compactus Fernald = Symphyotrichum puniceum (L.) Á. Löve et D. Löve ■☆

41118 Aster puniceus L. var. demissus Lindl. = Aster puniceus L. ■☆

41119 Aster puniceus L. var. demissus Lindl. = Symphyotrichum puniceum (L.) Á. Löve et D. Löve ■☆

41120 Aster puniceus L. var. firmus (Nees) Torr. et A. Gray = Symphyotrichum firmum (Nees) G. L. Nesom ■☆

41121 Aster puniceus L. var. firmus (Nees) Torr. et A. Gray f. lucidulus (A. Gray) Fernald = Symphyotrichum firmum (Nees) G. L. Nesom ■☆

41122 Aster puniceus L. var. lucidulus A. Gray = Symphyotrichum firmum (Nees) G. L. Nesom ■☆

41123 Aster puniceus L. var. monocephalus Farw. = Aster puniceus L. ■☆

41124 Aster puniceus L. var. monocephalus Farw. = Symphyotrichum puniceum (L.) Á. Löve et D. Löve ■☆

41125 Aster puniceus L. var. oligocephalus Fernald = Aster puniceus L. ■☆

41126 Aster puniceus L. var. oligocephalus Fernald = Symphyotrichum puniceum (L.) Á. Löve et D. Löve ■☆

41127 Aster puniceus L. var. perlongus Fernald = Symphyotrichum puniceum (L.) Á. Löve et D. Löve ■☆
41128 Aster puniceus L. var. scabricaulis (Shinners) A. G. Jones = Symphyotrichum puniceum (L.) Á. Löve et D. Löve var. scabricaule (Shinners) G. L. Nesom ■☆
41129 Aster purdomii Hutch. = Aster flaccidus Bunge ■
41130 Aster purpuratus Nees = Symphyotrichum laeve (L.) G. L. Nesom var. purpuratum (Nees) G. L. Nesom ■☆
41131 Aster pycnophyllus W. W. Sm.;密叶紫菀;Denseleaf Aster ■
41132 Aster pygmaeus Lindl. = Symphyotrichum pygmaeum (Lindl.) Brouillet et S. Selliah ■☆
41133 Aster pyrenaeus Desf. ex DC.;卑利牛斯紫菀;Pyrenean Aster, Pyrenees Aster ■☆
41134 Aster pyropappus Boiss. = Heteropappus altaicus (Willd.) Novopokr. var. canescens (Nees) Serg. ■
41135 Aster quinquenervius Klatt = Felicia quinquenervia (Klatt) Grau ●☆
41136 Aster racemosus Elliott = Symphyotrichum racemosum (Elliott) G. L. Nesom ■☆
41137 Aster radula Aiton;糙叶紫菀;Rough Aster, Rough-leaved Aster ■☆
41138 Aster radula Aiton = Eurybia radula (Aiton) G. L. Nesom ■☆
41139 Aster radula Aiton var. strictus A. Gray = Eurybia radula (Aiton) G. L. Nesom ■☆
41140 Aster radulinus A. Gray = Eurybia radulina (A. Gray) G. L. Nesom ■☆
41141 Aster rafinesquii Kuntze = Solidago shortii Torr. et A. Gray ■☆
41142 Aster ramsbottomi Hand. -Mazz. = Aster poliothamnus Diels ■
41143 Aster ramulosus Lindl. = Symphyotrichum falcatum (Lindl.) G. L. Nesom ■☆
41144 Aster reflexus L. = Polyarrhena reflexa (L.) Cass. ●☆
41145 Aster reflexus L. var. brachyphyllus Sond. ex Harv. = Polyarrhena reflexa (L.) Cass. subsp. brachyphylla (Sond. ex Harv.) Grau ●☆
41146 Aster reticulatus Pursh = Oclemena reticulata (Pursh) G. L. Nesom ■☆
41147 Aster retroflexus Lindl. ex DC. = Symphyotrichum retroflexum (Lindl. ex DC.) G. L. Nesom ■☆
41148 Aster retusus Ludlow;凹叶紫菀;Concave Leaf Aster, Concaveleaf Aster ■
41149 Aster rhomboideus Lindl. ex DC. = Aster tataricus L. f. ■
41150 Aster richardsonii Spreng.;理氏紫菀■☆
41151 Aster richardsonii Spreng. = Eurybia sibirica (L.) G. L. Nesom ■☆
41152 Aster richardsonii Spreng. var. giganteus Hook. = Eurybia sibirica (L.) G. L. Nesom ■☆
41153 Aster richardsonii Spreng. var. meritus (A. Nelson) Raup = Eurybia merita (A. Nelson) G. L. Nesom ■☆
41154 Aster riciniatus E. S. Burgess = Aster macrophyllus L. ■☆
41155 Aster riciniatus E. S. Burgess = Eurybia macrophylla (L.) Cass. ■☆
41156 Aster riddellii (Frank) Kuntze = Solidago riddellii Frank ■☆
41157 Aster rigidus (L.) Kuntze = Solidago rigida L. ■☆
41158 Aster riparius Kunth = Arida riparia (Kunth) D. R. Morgan et R. L. Hartm. ■☆
41159 Aster robustus (Makino) Yonek.;粗壮紫菀■☆
41160 Aster robynsianus J. Rousseau = Symphyotrichum robynsianum (J. Rousseau) Brouillet et Labrecque ■☆
41161 Aster rockianus Hand. -Mazz.;腾越紫菀;Rock Aster, Tengyue Aster ■
41162 Aster rolandii Shinners = Aster novi-belgii L. ■
41163 Aster roscidus E. S. Burgess = Aster macrophyllus L. ■☆
41164 Aster roseus Desf. = Aster novi-belgii L. ■
41165 Aster rothrockii Kuntze = Solidago spectabilis (D. C. Eaton) A. Gray ■☆
41166 Aster rotundifolius Thunb. = Felicia amelloides (L.) Voss ■☆
41167 Aster roylei Onno = Erigeron multiradiatus (Lindl.) Benth. et Hook. f. et Hook. f. ■
41168 Aster rufibarbis Harv. = Felicia burkei (Harv.) L. Bolus ■☆
41169 Aster rufopappus Hayata = Heteropappus hispidus (Thunb.) Less. ■
41170 Aster rugosus (Mill.) Kuntze = Solidago rugosa Mill. ■☆
41171 Aster rugulosus Maxim.;稍皱紫菀■☆
41172 Aster rugulosus Maxim. var. shibukawaensis Kitam. et Murata;涉川紫菀■☆
41173 Aster ruoqiangensis Y. Wei et C. H. An;若羌紫菀;Ruoqiang Aster ■
41174 Aster rupestris (Raf.) Kuntze = Solidago rupestris Raf. ■☆
41175 Aster saboureaui Humbert = Madagaster saboureaui (Humbert) G. L. Nesom ●☆
41176 Aster sachalinensis Kudo = Aster sibiricus L. ■
41177 Aster sagittifolius Wedem. ex Willd. = Aster cordifolius L. ■☆
41178 Aster sagittifolius Wedem. ex Willd. = Symphyotrichum cordifolium (L.) G. L. Nesom ■☆
41179 Aster sagittifolius Wedem. ex Willd. = Symphyotrichum urophyllum (Lindl. ex DC.) G. L. Nesom ■☆
41180 Aster sagittifolius Wedem. ex Willd. f. hirtellus (Lindl. ex DC.) Shinners = Symphyotrichum urophyllum (Lindl. ex DC.) G. L. Nesom ■☆
41181 Aster sagittifolius Wedem. ex Willd. f. hirtellus (Lindl. ex DC.) Shinners = Aster sagittifolius Wedem. ex Willd. ■☆
41182 Aster sagittifolius Wedem. ex Willd. f. hirtellus (Lindl.) Shinners = Aster urophyllus Lindl. ■☆
41183 Aster sagittifolius Wedem. ex Willd. var. dissitiflorus E. S. Burgess = Symphyotrichum urophyllum (Lindl. ex DC.) G. L. Nesom ■☆
41184 Aster sagittifolius Wedem. ex Willd. var. drummondii (Lindl.) Shinners = Aster drummondii Lindl. ■☆
41185 Aster sagittifolius Wedem. ex Willd. var. drummondii (Lindl.) Shinners = Symphyotrichum drummondii (Lindl.) G. L. Nesom ■☆
41186 Aster sagittifolius Wedem. ex Willd. var. glomerellus Farw. = Aster sagittifolius Wedem. ex Willd. ■☆
41187 Aster sagittifolius Wedem. ex Willd. var. glomerellus Farw. = Symphyotrichum urophyllum (Lindl. ex DC.) G. L. Nesom ■☆
41188 Aster sagittifolius Wedem. ex Willd. var. urophyllus (Lindl.) E. S. Burgess = Symphyotrichum urophyllum (Lindl. ex DC.) G. L. Nesom ■☆
41189 Aster sagittifolius Wedem. ex Willd. var. urophyllus (Lindl.) E. S. Burgess = Aster sagittifolius Wedem. ex Willd. ■☆
41190 Aster salicifolius Aiton;柳叶紫菀;Common Michaelmas-daisy, Willowleaf Aster, Willow-leaved Aster ■☆
41191 Aster salicifolius Aiton = Symphyotrichum praealtum (Poir.) G. L. Nesom ■☆
41192 Aster salicifolius sensu Willd. = Symphyotrichum longifolium (Lam.) G. L. Nesom ■☆
41193 Aster salicinus Scop. = Inula salicina L. ■
41194 Aster salinus Schrad. = Tripolium vulgare Nees ■

41195 Aster salwinensis Onno;怒江紫菀;Nujiang Aster,Salwin Aster ■
41196 Aster sampsonii (Hance) Hemsl.;短舌紫菀(黑根紫菀,接骨草,桑氏紫菀,小儿还魂草);Sampson Aster ■
41197 Aster sampsonii (Hance) Hemsl. var. isochaetus C. C. Chang;等毛短舌紫菀■
41198 Aster satsumensis Soejima;萨摩紫菀■☆
41199 Aster saundersii E. S. Burgess = Aster ciliolatus Lindl. ■☆
41200 Aster saundersii E. S. Burgess = Symphyotrichum ciliolatum (Lindl.) Á. Löve et D. Löve ■☆
41201 Aster savatieri Makino;萨氏紫菀(日本裸菀);Japan Gymnaster,Japan Nakeaster ■☆
41202 Aster savatieri Makino var. pygmaeus Makino;春寿菊■☆
41203 Aster saxicastelli J. J. N. Campb. et Medley = Eurybia saxicastelli (J. J. N. Campb. et Medley) G. L. Nesom ■☆
41204 Aster sayianus Nutt. = Canadanthus modestus (Lindl.) G. L. Nesom ■☆
41205 Aster scaber Thunb. = Doellingeria scabra (Thunb.) Nees ■
41206 Aster scaberrimus Hayata = Aster taiwanensis Kitam. ■
41207 Aster scaberulus Miq. = Aster ageratoides Turcz. var. scaberulus (Miq.) Y. Ling ■
41208 Aster scabricaulis Shinners = Symphyotrichum puniceum (L.) Á. Löve et D. Löve var. scabricaule (Shinners) G. L. Nesom ■☆
41209 Aster scabridus (DC.) E. Mey. ex Harv. et Sond. = Felicia scabrida (DC.) Range ●☆
41210 Aster scabridus (DC.) E. Mey. ex Harv. et Sond. var. brevifolius (DC.) Harv. = Felicia brevifolia (DC.) Grau ■☆
41211 Aster scabridus C. B. Clarke = Aster trinervius Roxb. ex D. Don ■
41212 Aster scabrimus Hayata = Aster taiwanensis Kitam. ■
41213 Aster scandens J. Jacq. ex Spreng. = Ampelaster carolinianus (Walter) G. L. Nesom ■☆
41214 Aster scaposus Klatt. = Aster molliusculus (DC.) C. B. Clarke ■
41215 Aster schaeferi Dinter = Felicia filifolia (Vent.) Burtt Davy subsp. schaeferi (Dinter) Grau ■☆
41216 Aster schlechteri Compton = Felicia filifolia (Vent.) Burtt Davy subsp. schlechteri (Compton) Grau ■☆
41217 Aster schreberi Nees;荨麻叶紫菀(施雷紫菀);Nettle-leaved Michaelmas-daisy,Schreber's Aster ■☆
41218 Aster schreberi Nees = Eurybia schreberi (Nees) Nees ■☆
41219 Aster scoparius (Kar. et Kir.) B. Fedtsch. = Galatella scoparia (Kar. et Kir.) Novopokr. ■
41220 Aster scoparius DC. = Symphyotrichum ericoides (L.) G. L. Nesom ■☆
41221 Aster scopulorum A. Gray = Ionactis alpina (Nutt.) Greene ■☆
41222 Aster sedifolius L.;多叶紫菀■☆
41223 Aster sedifolius L. 'Nana';矮生多叶紫菀■☆
41224 Aster see-burejensis Tamamsch.;布列亚山紫菀■☆
41225 Aster sekimotoi Makino;关本紫菀■☆
41226 Aster semiamplexicaulis (Makino) Makino ex Koidz.;半抱茎紫菀■☆
41227 Aster sempervirens (L.) Kuntze = Solidago sempervirens L. ■☆
41228 Aster senecioides Franch.;狗舌紫菀(狗舌草);Groundsel-like Aster ■
41229 Aster senecioides Franch. var. latisquamus Y. Ling;阔苞狗舌紫菀;Broadbract Groundsel-like Aster ■
41230 Aster sericeus Vent.;丝紫菀;Silky Aster,Western Silvery Aster ■☆
41231 Aster sericeus Vent. = Symphyotrichum sericeum (Vent.) G. L. Nesom ■☆
41232 Aster sericeus Vent. f. albiligulatus Fassett = Aster sericeus Vent. ■☆
41233 Aster sericeus Vent. var. microphyllus DC. = Symphyotrichum pratense (Raf.) G. L. Nesom ■☆
41234 Aster sericocarpoides (Small) K. Schum. = Aster umbellatus Mill. var. pubens A. Gray ■☆
41235 Aster sericocarpoides (Small) K. Schum. = Doellingeria sericocarpoides Small ■☆
41236 Aster serpentimontanus Tamamsch. = Aster alpinus L. var. serpentimontanus (Tamamsch.) Y. Ling ■
41237 Aster serratus Thunb. = Felicia serrata (Thunb.) Grau ●☆
41238 Aster serrulatus (Greene) Frye et Rigg = Eucephalus glaucescens (A. Gray) Greene ■☆
41239 Aster serrulatus Harv. var. densus ? = Felicia hyssopifolia (P. J. Bergius) Nees subsp. polyphylla (Harv.) Grau ●☆
41240 Aster serrulatus Harv. var. glaber (DC.) Harv. = Felicia hyssopifolia (P. J. Bergius) Nees subsp. glabra (DC.) Grau ■☆
41241 Aster serrulatus Harv. var. polyphyllus ? = Felicia hyssopifolia (P. J. Bergius) Nees subsp. polyphylla (Harv.) Grau ●☆
41242 Aster serrulatus Harv. var. setosus ? = Felicia zeyheri (Less.) Nees ●☆
41243 Aster serrulatus Harv. var. xylophyllus (Klatt) Schinz = Aster harveyanus Kuntze ■☆
41244 Aster setchuenensis Franch.;四川紫菀;Sichuan Aster ■
41245 Aster shastensis (A. Gray) A. Gray = Dieteria canescens (Pursh) Nutt. var. shastensis (A. Gray) D. R. Morgan et R. L. Hartm. ■☆
41246 Aster sherriffianus Hand.-Mazz. = Aster vestitus Franch. ■
41247 Aster shimadae (Kitam.) Nemoto = Kalimeris shimadae (Kitam.) Kitam. ■
41248 Aster shortii Lindl. = Symphyotrichum shortii (Lindl.) G. L. Nesom ■☆
41249 Aster shortii Lindl. f. asper Shinners = Aster shortii Lindl. ■☆
41250 Aster shortii Lindl. f. asper Shinners = Symphyotrichum shortii (Lindl.) G. L. Nesom ■☆
41251 Aster shortii Lindl. f. candidus Benke = Aster shortii Lindl. ■☆
41252 Aster shortii Lindl. f. candidus Benke = Symphyotrichum shortii (Lindl.) G. L. Nesom ■☆
41253 Aster shortii Lindl. f. gronemanni Benke = Aster shortii Lindl. ■☆
41254 Aster shortii Lindl. f. gronemanni Benke = Symphyotrichum shortii (Lindl.) G. L. Nesom ■☆
41255 Aster shortii Lindl. var. monroei Benke = Aster shortii Lindl. ■☆
41256 Aster shortii Lindl. var. monroei Benke = Symphyotrichum shortii (Lindl.) G. L. Nesom ■☆
41257 Aster sibiricus L.;西伯利亚紫菀(黑水紫菀,鲜卑紫菀);Siberian Aster ■
41258 Aster sibiricus L. = Eurybia sibirica (L.) G. L. Nesom ■☆
41259 Aster sibiricus L. subsp. pygmaeus (Lindl.) Á. Löve et D. Löve = Symphyotrichum pygmaeum (Lindl.) Brouillet et S. Selliah ■☆
41260 Aster sibiricus L. subsp. richardsonii (Spreng.) Á. Löve et D. Löve = Eurybia sibirica (L.) G. L. Nesom ■☆
41261 Aster sibiricus L. var. giganteus (Hook.) A. Gray = Eurybia sibirica (L.) G. L. Nesom ■☆
41262 Aster sibiricus L. var. meritus (A. Nelson) Raup = Eurybia merita (A. Nelson) G. L. Nesom ■☆
41263 Aster sibiricus L. var. pygmaeus (Lindl.) Cody = Symphyotrichum pygmaeum (Lindl.) Brouillet et S. Selliah ■☆

41264 Aster sikkimmensis Hook. f. et Thomson;锡金紫菀;Sikkim Aster ■
41265 Aster sikuensis W. W. Sm. et Farrer;西固紫菀;Siku Aster, Xigu Aster ■
41266 Aster simmondsii Small = Symphyotrichum simmondsii (Small) G. L. Nesom ■☆
41267 Aster simplex C. C. Chang = Gymnaster simplex (C. C. Chang) Y. Ling ex J. Q. Fu ■
41268 Aster simplex Willd. = Aster lanceolatus Willd. ■☆
41269 Aster simplex Willd. = Symphyotrichum lanceolatum (Willd.) G. L. Nesom ■☆
41270 Aster simplex Willd. = Symphyotrichum simplex (Willd.) Á. Löve et D. Löve ■☆
41271 Aster simplex Willd. var. estuarinus B. Boivin = Symphyotrichum lanceolatum (Willd.) G. L. Nesom ■☆
41272 Aster simplex Willd. var. interior (Wiegand) Cronquist = Aster lanceolatus Willd. ■☆
41273 Aster simplex Willd. var. interior (Wiegand) Cronquist = Symphyotrichum lanceolatum (Willd.) G. L. Nesom var. interior (Wiegand) G. L. Nesom ■☆
41274 Aster simplex Willd. var. ramosissimus (Torr. et A. Gray) Cronquist = Aster lanceolatus Willd. ■☆
41275 Aster simplex Willd. var. ramosissimus (Torr. et A. Gray) Cronquist = Symphyotrichum lanceolatum (Willd.) G. L. Nesom ■☆
41276 Aster simulans Harv. = Felicia hyssopifolia (P. J. Bergius) Nees subsp. glabra (DC.) Grau ■☆
41277 Aster simulatus Small = Symphyotrichum concolor (L.) G. L. Nesom ■☆
41278 Aster sinianus Hand. -Mazz.;岳麓紫菀;Chinese Aster, Yuelu Aster ■
41279 Aster siskiyouensis A. Nelson et J. F. Macbr. = Eucephalus glabratus (Greene) Greene ■☆
41280 Aster smithianus Hand. -Mazz.;甘川紫菀;Smith Aster ■
41281 Aster smithianus Hand. -Mazz. var. pilosior Hand. -Mazz. = Aster dolichopodus Y. Ling ■
41282 Aster sohayakiensis Koidz.;小泉紫菀■☆
41283 Aster solidagineus Michx. = Sericocarpus linifolius Britton, Sterns et Poggenb. ■☆
41284 Aster songoricus Novopokr. = Galatella songorica (Kar. et Kir.) Novopokr. ■
41285 Aster souliei Franch.;缘毛紫菀(西藏紫菀);Soulie Aster ■
41286 Aster souliei Franch. var. limitaneus (W. W. Sm. et Farr.) Hand. -Mazz. = Aster souliei Franch. ■
41287 Aster sparsiflorus (A. Gray) Kuntze = Solidago velutina DC. subsp. sparsiflora (A. Gray) Semple ■☆
41288 Aster spathulatus Lindl. = Symphyotrichum spathulatum (Lindl.) G. L. Nesom ■☆
41289 Aster spathulifolius Maxim.;匙叶紫菀;Spoonlaef Aster ■
41290 Aster spathulifolius Maxim. var. oharae (Nakai) Nakai ex Kitam.;小原紫菀■☆
41291 Aster spatioides C. B. Clarke = Heteropappus altaicus (Willd.) Novopokr. var. canescens (Nees) Serg. ■
41292 Aster speciosus (Nutt.) Kuntze = Solidago speciosa Nutt. ■☆
41293 Aster spectabilis Aiton;海滨紫菀;Late Aster, Seaside Aster, Showy Aster ■☆
41294 Aster spectabilis Aiton = Eurybia spectabilis (Aiton) G. L. Nesom ■☆
41295 Aster spectabilis Aiton var. suffultus Fernald = Eurybia spectabilis (Aiton) G. L. Nesom ■☆
41296 Aster sphaerotus Y. Ling;圆耳紫菀;Roundear Aster ■
41297 Aster spinosus Benth.;多刺紫菀;Mexican Devilweed, Spiny Aster ■☆
41298 Aster spinosus Benth. = Chloracantha spinosa (Benth.) G. L. Nesom ●■☆
41299 Aster spinulosus Chapm. = Eurybia spinulosa (Chapm.) G. L. Nesom ■☆
41300 Aster spithamaeus (M. A. Curtis ex A. Gray) Kuntze = Solidago spithamaea M. A. Curtis ex A. Gray ■☆
41301 Aster squamatus (Spreng.) Hieron.;鳞片紫菀■☆
41302 Aster squarrosus Walter = Symphyotrichum walteri (Alexander) G. L. Nesom ■☆
41303 Aster staticefolius Franch. = Aster batangensis Bureau et Franch. var. staticefolius (Franch.) Y. Ling ■
41304 Aster stenophyllus Lindl. ex DC. = Symphyotrichum lanceolatum (Willd.) G. L. Nesom ■☆
41305 Aster stilettiformis E. S. Burgess = Eurybia divaricata (L.) G. L. Nesom ■☆
41306 Aster stracheyi Hook. f.;匍生紫菀;Procumbent Aster ■
41307 Aster striatus Champ. ex Benth.;接骨紫菀■☆
41308 Aster strigosus (A. Spreng.) Harv. = Felicia ovata (Thunb.) Compton ●☆
41309 Aster strigosus Thunb. = Amellus strigosus (Thunb.) Less. ■☆
41310 Aster subcaerulens S. Moore = Aster tongolensis Franch. ■
41311 Aster subintegerrimus (Trautv.) Ostenf.;全缘紫菀■☆
41312 Aster sublitoralis (Torr. et A. Gray) Kuntze = Solidago latissimifolia Mill. ■☆
41313 Aster subsessilis E. S. Burgess = Symphyotrichum patens (Aiton) G. L. Nesom var. patentissimum (Lindl. ex DC.) G. L. Nesom ■☆
41314 Aster subspicatus Nees = Symphyotrichum subspicatum (Nees) G. L. Nesom ■☆
41315 Aster subspicatus Nees var. apricus (A. Gray) B. Boivin = Symphyotrichum foliaceum (Lindl. ex DC.) G. L. Nesom var. apricum (A. Gray) G. L. Nesom ■☆
41316 Aster subspicatus Nees var. grayi (Suksd.) Cronquist = Symphyotrichum subspicatum (Nees) G. L. Nesom ■☆
41317 Aster subulatus Michx.;钻叶紫菀(瑞连草,扫帚菊,土柴胡,帚马兰,钻形紫菀);Annual Salmarsh Aster, Eastern Annual Saltmarsh Aster, Saltmarsh Aster, Small Saltmarsh Aster ■
41318 Aster subulatus Michx. = Symphyotrichum subulatum (Michx.) G. L. Nesom ■☆
41319 Aster subulatus Michx. var. bahamensis (Britton) Bosser. = Symphyotrichum subulatum (Michx.) G. L. Nesom var. elongatum (Bosser. ex A. G. Jones et Lowry) S. D. Sundb. ■☆
41320 Aster subulatus Michx. var. cubensis (DC.) Shinners = Symphyotrichum subulatum (Michx.) G. L. Nesom var. parviflorum (Nees) S. D. Sundb. ■☆
41321 Aster subulatus Michx. var. elongatus Bosser. = Aster subulatus Michx. var. elongatus Bosser. ex A. G. Jones et Lowry ■☆
41322 Aster subulatus Michx. var. elongatus Bosser. ex A. G. Jones et Lowry;长钻叶紫菀(伸长紫菀)■☆
41323 Aster subulatus Michx. var. ligulatus Shinners;蒂马兰■
41324 Aster subulatus Michx. var. ligulatus Shinners = Aster subulatus Michx. var. sandwicensis (A. Gray et H. Mann) A. G. Jones ■
41325 Aster subulatus Michx. var. ligulatus Shinners = Symphyotrichum subulatum (Michx.) G. L. Nesom var. ligulatum S. D. Sundb. ■☆

41326 Aster subulatus Michx. var. obtusifolius Fernald = Aster subulatus Michx. ■
41327 Aster subulatus Michx. var. obtusifolius Fernald = Symphyotrichum subulatum (Michx.) G. L. Nesom ■☆
41328 Aster subulatus Michx. var. sandwicensis (A. Gray et H. Mann) A. G. Jones;泽扫帚菊■
41329 Aster sugimotoi Kitam.;杉本紫菀■☆
41330 Aster surculosus Michx. = Eurybia surculosa (Michx.) G. L. Nesom ■☆
41331 Aster surculosus Michx. var. gracilis (Nutt.) A. Gray = Eurybia compacta G. L. Nesom ■☆
41332 Aster sylvestris E. S. Burgess = Symphyotrichum undulatum (L.) G. L. Nesom ■☆
41333 Aster tagasagomontanus Sasaki;山紫菀;Mountain Aster ■
41334 Aster tagetinus (Greene) S. F. Blake = Machaeranthera tagetina Greene ■☆
41335 Aster taiwanensis Kitam.;台湾紫菀(台湾马兰);Taiwan Aster ■
41336 Aster taiwanensis Kitam. var. lucens (Kitam.) Kitam.;光亮台湾紫菀■☆
41337 Aster takasagomontanus Sasaki;雪山马兰■
41338 Aster taliangshanensis Y. Ling;凉山紫菀;Daliangshan Aster, Taliangshan Aster ■
41339 Aster tanacetifolius Kunth = Machaeranthera tanacetifolia (Kunth) Nees ■☆
41340 Aster tansaniensis W. Lippert;坦萨尼紫菀■☆
41341 Aster taoyuenensis S. S. Ying;桃园马兰;Taoyuan Aster ■
41342 Aster tataricus L. f.;紫菀(白菀,白羊须草,北紫菀,辫紫菀,茈菀,返魂草,关公须,还魂草,夹板菜,驴耳朵菜,驴耳朵草,驴夹板菜,青牛舌头花,青菀,软紫菀,山白菜,甜紫菀,万金茸,小辫儿,小辫子,夜牵牛,子菀,子元,紫旧,紫蒨);Purple Aster, Tatarian Aster ■
41343 Aster tataricus L. f. var. fauriei ? = Aster tataricus L. f. ■
41344 Aster tataricus L. f. var. hortensis Nakai = Aster tataricus L. f. ■
41345 Aster tataricus L. f. var. minor Makino = Aster tataricus L. f. ■
41346 Aster tataricus L. f. var. nakaii (H. Lév. et Vaniot) Kitam. = Aster tataricus L. f. ■
41347 Aster tataricus L. f. var. persianus Bailey;药用紫菀■
41348 Aster tataricus L. f. var. robustus Nakai = Aster tataricus L. f. ■
41349 Aster tataricus L. f. var. vernalis Nakai = Aster tataricus L. f. ■
41350 Aster taxifolius L. = Zyrphelis taxifolia (L.) Nees ■☆
41351 Aster techinensis Y. Ling;德钦紫菀;Deqin Aster, Techin Aster ■
41352 Aster tenebrosus E. S. Burgess = Eurybia divaricata (L.) G. L. Nesom ■☆
41353 Aster tenellus L. = Felicia tenella (L.) Nees ●☆
41354 Aster tenellus L. var. cotuloides (DC.) Harv. = Felicia tenella (L.) Nees subsp. cotuloides (DC.) Grau ■☆
41355 Aster tenellus L. var. glaber Harv. = Felicia australis (Alston) E. Phillips ■☆
41356 Aster tenellus L. var. longifolius (DC.) Harv. = Felicia tenella (L.) Nees subsp. longifolia (DC.) Grau ●☆
41357 Aster tenellus L. var. pusillus Harv. = Felicia tenella (L.) Nees subsp. pusilla (Harv.) Grau ●☆
41358 Aster tenellus L. var. robustus Harv. = Felicia tenella (L.) Nees ●☆
41359 Aster tener (DC.) Harv. = Felicia tenera (DC.) Grau ●☆
41360 Aster tenuicaulis (C. Mohr) E. S. Burgess = Symphyotrichum patens (Aiton) G. L. Nesom var. gracile (Hook.) G. L. Nesom ■☆
41361 Aster tenuifolius L. = Symphyotrichum tenuifolium (L.) G. L. Nesom ■☆
41362 Aster tenuifolius L. var. aphyllus R. W. Long = Symphyotrichum tenuifolium (L.) G. L. Nesom var. aphyllum (R. W. Long) S. D. Sundb. ■☆
41363 Aster tenuifolius L. var. bellidifolius (Willd.) Torr. et A. Gray = Symphyotrichum lanceolatum (Willd.) G. L. Nesom ■☆
41364 Aster tenuifolius L. var. ramosissimus Torr. et A. Gray = Symphyotrichum lanceolatum (Willd.) G. L. Nesom ■☆
41365 Aster tenuifolius sensu Torr. et A. Gray var. ramosissimus Torr. et A. Gray = Symphyotrichum lanceolatum (Willd.) G. L. Nesom ■☆
41366 Aster tenuipes Makino;细柄紫菀■☆
41367 Aster tephrodes (A. Gray) S. F. Blake = Dieteria asteroides Torr. ■☆
41368 Aster terrae-novae (Torr. et A. Gray) Kuntze = Solidago uliginosa Nutt. ■☆
41369 Aster texanus E. S. Burgess = Symphyotrichum drummondii (Lindl.) G. L. Nesom var. texanum (E. S. Burgess) G. L. Nesom ■☆
41370 Aster texanus E. S. Burgess var. parviceps Shinners = Symphyotrichum drummondii (Lindl.) G. L. Nesom var. texanum (E. S. Burgess) G. L. Nesom ■☆
41371 Aster thomsonii C. B. Clarke;喜马拉雅紫菀■☆
41372 Aster thomsonii C. B. Clarke 'Nana';矮生喜马拉雅紫菀■☆
41373 Aster thyrsoideus (E. Mey.) Kuntze = Solidago macrophylla Banks ex Pursh ■☆
41374 Aster tianschanicus Novopokr. = Galatella tianschanica Novopokr. ■
41375 Aster tibeticus Hook. f. = Aster flaccidus Bunge ■
41376 Aster tientschuanensis Hand.-Mazz.;天全紫菀;Tianquan Aster, Tienchuan Aster ■
41377 Aster tolmatschevii Tamamsch.;陶氏紫菀■☆
41378 Aster tolmieanus (A. Gray) Kuntze = Solidago missouriensis Nutt. ■☆
41379 Aster tomentellus (Greene) Frye et Rigg = Eucephalus tomentellus (Greene) Greene ■☆
41380 Aster tongolensis Franch.;东俄洛紫菀(川紫菀,东菊,兰铁草,踏地莲花菜,细牛舌片,细药,牙陷药,野菠菜);Tongol Aster ■
41381 Aster tongolensis Franch. f. humilis Diels;低小东俄洛紫菀■
41382 Aster tongolensis Franch. f. ramosus Y. Ling;多枝东俄洛紫菀■
41383 Aster tongolensis Franch. subsp. forrestii (Stapf) Onno = Aster souliei Franch. var. limitaneus (W. W. Sm. et Farr.) Hand.-Mazz. ■
41384 Aster torreyi Kuntze = Solidago drummondii Torr. et A. Gray ■☆
41385 Aster tortifolius Michx. = Sericocarpus tortifolius (Michx.) Nees ■☆
41386 Aster tradescanti Hoffm. = Aster patulus Lam. ■☆
41387 Aster tradescanti L.;北美紫菀;Tradescant Aster ■☆
41388 Aster tradescantii L. = Aster lateriflorus (L.) Britton var. hirsuticaulis (Lindl. ex DC.) Porter ■☆
41389 Aster tradescantii L. = Symphyotrichum tradescantii (L.) G. L. Nesom ■☆
41390 Aster tradescantii L. var. saxatilis (Fernald) House = Aster lateriflorus (L.) Britton var. hirsuticaulis (Lindl. ex DC.) Porter ■☆
41391 Aster tradescantii L. var. thyrsoides (A. Gray) B. Boivin = Symphyotrichum ontarionis (Wiegand) G. L. Nesom ■☆
41392 Aster tradescantii Michx. = Aster diffusus DC. ■☆
41393 Aster tradescantii Nees = Aster paniculatus Lam. ■☆
41394 Aster triangularis E. S. Burgess = Symphyotrichum undulatum

(L.) G. L. Nesom ■☆
41395 Aster tricapitatus Vaniot = Aster oreophilus Franch. ■
41396 Aster tricephalus C. B. Clarke;三头紫菀;Three Head Aster, Threehead Aster ■
41397 Aster trichanthus Hand.-Mazz. = Kalimeris longipetiolata (C. C. Chang) Y. Ling ■
41398 Aster trichoneurus Y. Ling;毛脉紫菀;Hairvein Aster ■
41399 Aster tricolor (Nees) Harv. = Felicia elongata (Thunb.) O. Hoffm. ■☆
41400 Aster trinervis Ruiz et Pav. ex DC. = Aster trinervius D. Don ■
41401 Aster trinervis Ruiz et Pav. ex DC. subsp. ageratoides (Turcz.) Grierson = Aster ageratoides Turcz. ■
41402 Aster trinervis Ruiz et Pav. ex DC. var. lasiocladus (Hayata) Yamam. = Aster lanuginosus (J. Small) Y. Ling ■
41403 Aster trinervis Ruiz et Pav. ex DC. var. wattii (C. B. Clarke) Grierson;密毛三基脉紫菀■
41404 Aster trinervius D. Don;三基脉紫菀(绀菊,马兰,三脉山白菊,三脉叶马兰,三脉叶紫菀,三脉紫菀);Three Basinerved Aster, Threebasevein Aster ■
41405 Aster trinervius D. Don f. pubescens Kuntze = Aster ageratoides Turcz. var. scaberulus (Miq.) Y. Ling ■
41406 Aster trinervius D. Don var. adustus Kom. = Aster ageratoides Turcz. ■
41407 Aster trinervius D. Don var. firmus Diels = Aster ageratoides Turcz. var. firmus (Diels) Hand.-Mazz. ■
41408 Aster trinervius D. Don var. grossedentatus Franch. ex Diels = Aster homochlamydeus Hand.-Mazz. ■
41409 Aster trinervius D. Don var. hayatae Yamam. = Aster taiwanensis Kitam. ■
41410 Aster trinervius D. Don var. lasiocladus (Hayata) Yamam. = Aster ageratoides Turcz. var. lasiocladus (Hayata) Hand.-Mazz. ■
41411 Aster trinervius D. Don var. longifolius Franch. et Sav. = Aster tataricus L. f. ■
41412 Aster trinervius D. Don var. ovatus f. pubeescens Kuntze = Aster ageratoides Turcz. var. scaberulus (Miq.) Y. Ling ■
41413 Aster trinervius D. Don var. pilosus Diels = Aster ageratoides Turcz. var. pilosus (Diels) Hand.-Mazz. ■
41414 Aster trinervius D. Don var. potaninii Diels = Aster ageratoides Turcz. ■
41415 Aster trinervius D. Don var. rosthornii Diels = Aster ageratoides Turcz. ■
41416 Aster trinervius D. Don var. scandens Hayata = Aster taiwanensis Kitam. ■
41417 Aster trinervius Forbes et Hemsl. = Aster ageratoides Turcz. ■
41418 Aster trinervius Gilih. = Aster amellus L. ■☆
41419 Aster trinervius Hayata = Aster taiwanensis Kitam. ■
41420 Aster trinervius Roxb. ex D. Don subsp. ageratoides (Turcz.) Grierson = Aster ageratoides Turcz. ■
41421 Aster trinervius Roxb. ex D. Don var. adustus Kom. = Aster ageratoides Turcz. ■
41422 Aster trinervius Roxb. ex D. Don var. firmus Diels = Aster ageratoides Turcz. var. firmus (Diels) Hand.-Mazz. ■
41423 Aster trinervius Roxb. ex D. Don var. grossedentatus Franch. ex Diels = Aster homochlamydeus Hand.-Mazz. ■
41424 Aster trinervius Roxb. ex D. Don var. hayata Yamam. = Aster taiwanensis Kitam. ■
41425 Aster trinervius Roxb. ex D. Don var. lasiocladus (Hayata) Yamomoto = Aster ageratoides Turcz. var. lasiocladus (Hayata) Hand.-Mazz. ■
41426 Aster trinervius Roxb. ex D. Don var. ovatus Franch. et Sav. = Aster ageratoides Turcz. var. oophyllus Y. Ling ex J. Q. Fu ■
41427 Aster trinervius Roxb. ex D. Don var. pilosus Diels = Aster ageratoides Turcz. var. pilosus (Diels) Hand.-Mazz. ■
41428 Aster trinervius Roxb. ex D. Don var. potaninii Diels = Aster ageratoides Turcz. ■
41429 Aster trinervius Roxb. ex D. Don var. rosthornii Diels = Aster ageratoides Turcz. ■
41430 Aster trinervius Roxb. ex D. Don var. scandens Hayata = Aster taiwanensis Kitam. ■
41431 Aster trinervius Roxb. ex DC. var. microcephalus (Miq.) Makino = Aster microcephalus (Miq.) Franch. et Sav. ■☆
41432 Aster tripolium L.;麝香紫菀■☆
41433 Aster tripolium L. = Tripolium vulgare Nees ■
41434 Aster tripolium L. f. albiflorus S. Toyama;白花碱菀(白果麝香紫菀)■☆
41435 Aster tripolium L. subsp. longicaulis (DC.) Nyman;长茎麝香紫菀■☆
41436 Aster tripolium L. var. integrifolius Miyabe et Kudo = Aster tripolium L. ■☆
41437 Aster tripolium L. var. integrifolius Miyabe et Kudo = Tripolium vulgare Nees ■
41438 Aster truellius E. S. Burgess = Symphyotrichum undulatum (L.) G. L. Nesom ■☆
41439 Aster tsarungensis (Grierson) Y. Ling;察瓦龙紫菀(滇藏紫菀);Tsarung Aster ■
41440 Aster tsuneoi Miyabe et Tatew. = Erigeron thunbergii A. Gray subsp. glabratus (A. Gray) H. Hara var. angustifolius (Tatew.) H. Hara ■☆
41441 Aster turbinatus S. Moore;陀螺紫菀(百条根,打风草,单头紫菀,喉风草,牛舌草,野白菊,一枝香);Turbinate Aster ■
41442 Aster turbinatus S. Moore var. chekiangensis C. Ling ex Y. Ling;仙白紫菀(白仙草,仙白草,陀螺紫菀);Zhejiang Turbinate Aster ■
41443 Aster turbinellus Lindl. = Symphyotrichum turbinellum (Lindl.) G. L. Nesom ■☆
41444 Aster ujiinsularis Kitam.;北村紫菀■☆
41445 Aster uliginosus (Nutt.) Kuntze = Solidago uliginosa Nutt. ■☆
41446 Aster uliginosus J. M. Wood et M. S. Evans = Felicia uliginosa (J. M. Wood et M. S. Evans) Grau ●☆
41447 Aster ulmifolius (Muhl. ex Willd.) Kuntze = Solidago ulmifolia Muhl. ex Willd. ■☆
41448 Aster umbellatus Mill.;平头紫菀(伞状东风菜);Flat-top Aster, Flat-topped Aster, Flat-topped White Aster, Parasol Aster, Tall Flat-topped White Aster, Umbellate Aster ■☆
41449 Aster umbellatus Mill. = Doellingeria umbellata (Mill.) Nees ■☆
41450 Aster umbellatus Mill. var. brevisquamus Fernald = Doellingeria sericocarpoides Small ■☆
41451 Aster umbellatus Mill. var. latifolius A. Gray = Aster umbellatus Mill. var. pubens A. Gray ■☆
41452 Aster umbellatus Mill. var. latifolius A. Gray = Doellingeria sericocarpoides Small ■☆
41453 Aster umbellatus Mill. var. pubens A. Gray;绒毛平头紫菀(毛伞状东风菜);Hairy Flat-top Aster, Parasol Aster ■☆
41454 Aster umbellatus Mill. var. pubens A. Gray = Doellingeria umbellata (Mill.) Nees var. pubens (A. Gray) Britton ■☆

41455 Aster unalaschensis Less. ex Bong. var. major Hook. = Canadanthus modestus (Lindl.) G. L. Nesom ■☆
41456 Aster undulatus (L.) G. L. Nesom var. diversifolius (Michx.) A. Gray = Symphyotrichum undulatum (L.) G. L. Nesom ■☆
41457 Aster undulatus (L.) G. L. Nesom var. loriformis E. S. Burgess = Symphyotrichum undulatum (L.) G. L. Nesom ■☆
41458 Aster undulatus L. = Symphyotrichum undulatum (L.) G. L. Nesom ■☆
41459 Aster uniligulata (DC.) Kuntze = Solidago uliginosa Nutt. ■☆
41460 Aster urophyllus Lindl. = Aster sagittifolius Wedem. ex Willd. ■☆
41461 Aster urophyllus Lindl. = Symphyotrichum urophyllum (Lindl. ex DC.) G. L. Nesom ■☆
41462 Aster urophyllus Lindl. ex DC. = Symphyotrichum urophyllum (Lindl. ex DC.) G. L. Nesom ■☆
41463 Aster urophyllus Lindl. var. glomerellus Farw. = Aster sagittifolius Wedem. ex Willd. ■☆
41464 Aster urophyllus Lindl. var. glomerellus Farw. = Symphyotrichum urophyllum (Lindl. ex DC.) G. L. Nesom ■☆
41465 Aster ursinus E. S. Burgess = Aster indicus L. ■
41466 Aster ursinus H. Lév. = Kalimeris indica (L.) Sch. Bip. ■
41467 Aster vallicola Greene = Symphyotrichum spathulatum (Lindl.) G. L. Nesom var. intermedium (A. Gray) G. L. Nesom ■☆
41468 Aster vaniotii H. Lév. = Aster oreophilus Franch. ■
41469 Aster vaurealis J. Rousseau = Aster novi-belgii L. ■
41470 Aster veitchianus Hutch. et Drumm.;峨眉紫菀;Emei Aster ■
41471 Aster veitchianus Hutch. et Drumm. f. yamatzutae (Matsuda) Y. Ling;单头峨眉紫菀;Onehead Emei Aster ■
41472 Aster vellereus Franch. = Inula nervosa Wall. ■
41473 Aster velutinosus Y. Ling;毡毛紫菀;Felthair Aster, Velutinous Aster ■
41474 Aster velutinus (DC.) Kuntze = Solidago velutina DC. ■☆
41475 Aster venustus Fourc. = Felicia linifolia (Harv.) Grau ■☆
41476 Aster venustus M. E. Jones = Xylorhiza venusta (M. E. Jones) A. Heller ●■☆
41477 Aster vernalis Engelm. ex E. S. Burgess = Symphyotrichum oolentangiense (Riddell) G. L. Nesom ■☆
41478 Aster vernus (M. A. Curtis ex Torr. et A. Gray) Kuntze = Solidago verna M. A. Curtis ex Torr. et A. Gray ■☆
41479 Aster vernus L. = Erigeron vernus (L.) Torr. et A. Gray ■☆
41480 Aster versicolor Steud.;彩叶紫菀;Late Michaelmas-daisy ■☆
41481 Aster verutifolius Alexander = Eurybia hemispherica (Alexander) G. L. Nesom ■☆
41482 Aster vestitus Franch.;密毛紫菀(菊花暗消,烧盏花);Densehair Aster ■
41483 Aster vialis (Bradshaw) S. F. Blake = Eucephalus vialis Bradshaw ■☆
41484 Aster villosus Michx. = Symphyotrichum pilosum (Willd.) G. L. Nesom ■☆
41485 Aster villosus Thunb. = Felicia hyssopifolia (P. J. Bergius) Nees ■☆
41486 Aster vilmorinii Franch. = Aster yunnanensis Franch. var. angustior Hand. -Mazz. ■
41487 Aster vilmorinii Franch. var. nigrotinctus (Y. Ling) Y. Ling = Aster farreri W. W. Sm. et Jeffrey ■
41488 Aster vimineus Lam. = Aster fragilis Willd. ■☆
41489 Aster vimineus Lam. = Aster lateriflorus (L.) Britton ■☆
41490 Aster vimineus Lam. var. saxatilis Fernald = Aster lateriflorus (L.) Britton var. hirsuticaulis (Lindl. ex DC.) Porter ■☆
41491 Aster vimineus Lam. var. subdumosus Wiegand = Aster fragilis Willd. ■☆
41492 Aster vimineus Lam. var. subdumosus Wiegand = Aster fragilis Willd. var. subdumosus (Wiegand) A. G. Jones ■☆
41493 Aster violaris E. S. Burgess = Aster macrophyllus L. ■☆
41494 Aster virgatus Elliott = Symphyotrichum patens (Aiton) G. L. Nesom ■☆
41495 Aster viscidulus (Makino) Makino;微黏紫菀■☆
41496 Aster viscidulus (Makino) Makino var. alpina (Koidz.) Kitam.;山地三脉紫菀■☆
41497 Aster walkeri (Kitam.) Kitam. ex Shimabuku = Aster asagrayi Makino var. walkeri Kitam. ■☆
41498 Aster walteri Alexander = Symphyotrichum walteri (Alexander) G. L. Nesom ■☆
41499 Aster wasatchensis (M. E. Jones) S. F. Blake = Herrickia wasatchensis (M. E. Jones) Brouillet ■☆
41500 Aster wasatchensis (M. E. Jones) S. F. Blake var. pulcher (S. F. Blake) S. L. Welsh;美丽紫菀■☆
41501 Aster watsonii A. Gray = Erigeron watsonii (A. Gray) Cronquist ■☆
41502 Aster welshii Cronquist = Symphyotrichum welshii (Cronquist) G. L. Nesom ■☆
41503 Aster westae Fourc. = Felicia westae (Fourc.) Grau ●☆
41504 Aster wilsonii Rydb. = Symphyotrichum ciliolatum (Lindl.) Á. Löve et D. Löve ■☆
41505 Aster winkleri Novopokr. = Asterothamnus fruticosus (C. Winkl.) Novopokr. ●
41506 Aster woodii Klatt = Felicia quinquenervia (Klatt) Grau ●☆
41507 Aster wootonii (Greene) Greene = Symphyotrichum lanceolatum (Willd.) G. L. Nesom var. hesperium (A. Gray) G. L. Nesom ■☆
41508 Aster worcesterensis Kuntze = Polyarrhena imbricata (DC.) Grau ●☆
41509 Aster wrightii A. Gray = Xylorhiza wrightii (A. Gray) Greene ●■☆
41510 Aster xylophyllus Klatt = Aster harveyanus Kuntze ■☆
41511 Aster yakushimensis (Kitam.) Soejima et Yahara;屋久岛紫菀■☆
41512 Aster yamatsudanus Kitag. = Galatella dahurica DC. ■
41513 Aster yamatzutae Matsuda = Aster veitchlanus Hutch. et Drumm. f. yamatzutae (Matsuda) Y. Ling ■
41514 Aster yangtzensis Migo = Kalimeris indica (L.) Sch. Bip. ■
41515 Aster yokusaianus Kitam. ex H. Hara = Aster savatieri Makino var. pygmaeus Makino ■☆
41516 Aster yomena (Kitam.) Honda;本田紫菀■☆
41517 Aster yomena (Kitam.) Honda var. dentatus (Kitam.) H. Hara;尖齿紫菀■☆
41518 Aster yoshinaganus (Kitam.) Mot. Ito et Soejima;吉永紫菀■☆
41519 Aster yukonensis Cronquist = Symphyotrichum yukonense (Cronquist) G. L. Nesom ■☆
41520 Aster yunnanensis Franch.;云南紫菀;Yunnan Aster ■
41521 Aster yunnanensis Franch. var. angustior Hand. -Mazz.;狭苞云南紫菀;Narrowbract Yunnan Aster ■
41522 Aster yunnanensis Franch. var. labrangensis (Hand. -Mazz.) Y. Ling;夏河云南紫菀■
41523 Aster zayuensis Y. L. Chen;察隅紫菀;Chayu Aster ■
41524 Aster zazureus Lindl. ex Hook.;天蓝花紫菀■☆
41525 Aster zeyheri Less. = Felicia zeyheri (Less.) Nees ●☆
41526 Aster zuluensis W. Lippert;祖卢紫菀■☆
41527 Asteracantha Nees = Hygrophila R. Br. ●■

41528 Asteracantha lindaviana De Wild. et T. Durand = Hygrophila lindaviana (De Wild. et T. Durand) Burkill ■☆

41529 Asteracantha longifolia (L.) Nees = Hygrophila auriculata (Schumach.) Heine ●☆

41530 Asteraceae Bercht. et J. Presl(1820)(保留科名);菊科●■

41531 Asteraceae Bercht. et J. Presl(保留科名) = Compositae Giseke(保留科名)●■

41532 Asteraceae Dumort. = Asteraceae Bercht. et J. Presl(保留科名)●■

41533 Asteraceae Dumort. = Compositae Giseke(保留科名)●■

41534 Asteraceae Martinov = Asteraceae Bercht. et J. Presl(保留科名)●■

41535 Asteraceae Martinov = Compositae Giseke(保留科名)●■

41536 Asterandra Klotzsch = Phyllanthus L. ●■

41537 Asterantha Rchb. = Asteracantha Nees ●■

41538 Asterantha Rchb. = Hygrophila R. Br. ●■

41539 Asteranthaceae R. Knuth(1939)(保留科名);星花科(合玉蕊科)●☆

41540 Asteranthaceae R. Knuth(保留科名) = Lecythidaceae A. Rich.(保留科名)●

41541 Asteranthe Engl. et Diels(1901);菀花木属●☆

41542 Asteranthe asterias (S. Moore) Engl. et Diels;菀花木●☆

41543 Asteranthe asterias (S. Moore) Engl. et Diels subsp. triangularis Verdc.;三角菀花木●☆

41544 Asteranthe lutea Vollesen;黄菀花木●☆

41545 Asteranthe trollii Diels;特洛尔菀花木●☆

41546 Asteranthemum Kunth = Smilacina Desf.(保留属名)■

41547 Asteranthemum dahuricum (Turcz. ex Fisch. et C. A. Mey.) Kunth = Maianthemum dahuricum (Turcz. ex Fisch. et C. A. Mey.) LaFrankie ■

41548 Asteranthemum dahuricum Kunth = Maianthemum dahuricum (Turcz. ex Fisch. et C. A. Mey.) LaFrankie ■

41549 Asteranthemum trifoliatum Kunth = Maianthemum trifolium (L.) Slobada ■

41550 Asteranthemum trifolium (L.) Kunth = Maianthemum trifolium (L.) Slobada ■

41551 Asteranthera Hanst.(1854);星花苣苔属(智利苣苔属,紫菀花苣苔属);Asteranthera ●☆

41552 Asteranthera Klotzsch et Hanst. = Asteranthera Hanst. ●☆

41553 Asteranthera chiloensis Hanst.;星花苣苔●☆

41554 Asteranthera ovata (Cav.) Hanst.;紫菀花苣苔(卵状紫菀花苣苔,智利苣苔);Ovate Asteranthera ●☆

41555 Asteranthera ovata Hanst. = Asteranthera ovata (Cav.) Hanst. ●☆

41556 Asteranthopsis Kuntze = Asteranthe Engl. et Diels ●☆

41557 Asteranthopsis asterias Pilg.;拟星花●☆

41558 Asteranthos Desf.(1820);星花属●☆

41559 Asteranthos brasiliensis Desf.;星花●☆

41560 Asteranthus Endl. = Astranthus Lour. ●

41561 Asteranthus Endl. = Homalium Jacq. ●

41562 Asteranthus Spreng. = Asteranthos Desf. ●☆

41563 Asterias Borkh. = Gentiana L. ■

41564 Asteriastigma Bedd. = Hydnocarpus Gaertn. ●

41565 Asteridea Lindl.(1839);星绒草属■☆

41566 Asteridea Lindl. = Athrixia Ker Gawl. ●■☆

41567 Asteridea gracilis A. Gray;细星绒草■☆

41568 Asteridea multiceps A. Gray;多头星绒草■☆

41569 Asteridea nivea (Steetz) Kroner;雪白星绒草■☆

41570 Asteridea pulverulenta Lindl.;星绒草■☆

41571 Asteridia N. T. Burb. = Asteridea Lindl. ■☆

41572 Asteridium Engelm. ex Walp. = Chaetopappa DC. ■☆

41573 Asterigeron Rydb. = Aster L. ●■

41574 Asteringa E. Mey. ex DC. = Pentzia Thunb. ●■☆

41575 Asteriscium Cham. et Schltdl.(1826);星箱草属■☆

41576 Asteriscium chilense Cham. et Schltdl.;星箱草■☆

41577 Asteriscium glaucum Hieron. et H. Wolff;灰星箱草■☆

41578 Asteriscium polycephalum Gill. et Hook.;多都星箱草■☆

41579 Asteriscodes Kuntze = Callistephus Cass.(保留属名)■

41580 Asteriscus Mill.(1754);金币花属●■☆

41581 Asteriscus Moench = Odontospermum Neck. ex Sch. Bip. ■☆

41582 Asteriscus Rchb. = Asteriscium Cham. et Schltdl. ■☆

41583 Asteriscus Sch. Bip. = Pallenis Cass.(保留属名)●■☆

41584 Asteriscus Siegesb. = Odontospermum Neck. ex Sch. Bip. ■☆

41585 Asteriscus Tourn. ex Sch. Bip. = Pallenis Cass.(保留属名)●■☆

41586 Asteriscus aquaticus (L.) Less.;水生金币花■☆

41587 Asteriscus aquaticus (L.) Less. var. pygmaeus DC. = Pallenis hierochuntica (Michon) Greuter ■☆

41588 Asteriscus cavanillesi Caball. = Asteriscus graveolens (Forssk.) Less. subsp. odorus (Schousb.) Greuter ■☆

41589 Asteriscus cuspidatus (Pomel) Aurich et Podlech = Pallenis cuspidata Pomel ■☆

41590 Asteriscus cuspidatus (Pomel) Aurich et Podlech = Pallenis cuspidata Pomel subsp. canescens (Maire) Greuter ■☆

41591 Asteriscus cuspidatus (Pomel) Aurich et Podlech subsp. canescens (Maire) Aurich et Podlech = Pallenis cuspidata Pomel subsp. canescens (Maire) Greuter ■☆

41592 Asteriscus cyrenaicus (Alavi) Dobignard = Pallenis cyrenaica Alavi ■☆

41593 Asteriscus graveolens (Forssk.) Less.;异味金币花■☆

41594 Asteriscus graveolens (Forssk.) Less. subsp. odorus (Schousb.) Greuter;尖齿异味金币花■☆

41595 Asteriscus graveolens (Forssk.) Less. subsp. stenophyllus (Link) Greuter;窄叶异味金币花■☆

41596 Asteriscus graveolens (Forssk.) Less. var. pygmaeus Maire = Asteriscus graveolens (Forssk.) Less. ■☆

41597 Asteriscus graveolens (Forssk.) Less. var. scaber (Thell.) Emb. et Maire = Asteriscus graveolens (Forssk.) Less. ■☆

41598 Asteriscus graveolens (Forssk.) Less. var. scaber Thell. = Asteriscus graveolens (Forssk.) Less. ■☆

41599 Asteriscus graveolens (Forssk.) Less. var. villosus (Thell.) Emb. et Maire = Asteriscus graveolens (Forssk.) Less. ■☆

41600 Asteriscus graveolens (Forssk.) Less. var. villosus Thell. = Asteriscus graveolens (Forssk.) Less. ■☆

41601 Asteriscus hierochunticus (Michon) Wiklund = Pallenis hierochuntica (Michon) Greuter ■☆

41602 Asteriscus imbricatus (Cav.) DC.;覆瓦金币花■☆

41603 Asteriscus intermedius (DC.) Pit. et Proust;间型金币花■☆

41604 Asteriscus maritimus (L.) Less. = Pallenis maritima (L.) Greuter ■☆

41605 Asteriscus maritimus (L.) Less. var. lanuginosus Sennen = Pallenis maritima (L.) Greuter ■☆

41606 Asteriscus maritimus (L.) Less. var. mauritanicus (Jord. et Fourr.) Maire = Pallenis maritima (L.) Greuter ■☆

41607 Asteriscus maritimus (L.) Less. var. microphyllus (Ball) Batt. = Pallenis maritima (L.) Greuter ■☆

41608 Asteriscus maritimus (L.) Less. var. parviflorus Sennen et

Mauricio = Pallenis maritima (L.) Greuter ■☆

41609 Asteriscus maritimus (L.) Less. var. perpusillus Batt. = Pallenis maritima (L.) Greuter ■☆

41610 Asteriscus mauritanicus Jord. et Fourr. = Pallenis maritima (L.) Greuter ■☆

41611 Asteriscus odorus (Schousb.) DC. = Asteriscus graveolens (Forssk.) Less. subsp. odorus (Schousb.) Greuter ■☆

41612 Asteriscus paui Caball. = Asteriscus graveolens (Forssk.) Less. subsp. odorus (Schousb.) Greuter ■☆

41613 Asteriscus pinifolius Maire et Wilczek = Ighermia pinifolia (Maire et Wilczek) Wiklund ■☆

41614 Asteriscus pygmaeus (DC.) Coss. et Durieu = Pallenis hierochuntica (Michon) Greuter ■☆

41615 Asteriscus schimperi (Boiss.) Boiss. = Asteriscus graveolens (Forssk.) Less. ■☆

41616 Asteriscus schultzii (Bolle) Pit. et Proust;舒尔茨金币花(舒尔茨北非菊)■☆

41617 Asteriscus sericeus (L. f.) DC.;绢毛金币花(绢毛北非菊);Canary Island Daisy ■☆

41618 Asteriscus smithii (Webb) Walp.;史密斯金币花(史密斯北非菊)■☆

41619 Asteriscus spinosus (L.) Sch. Bip. = Pallenis spinosa (L.) Cass. ■☆

41620 Asteriscus spinosus (L.) Sch. Bip. subsp. asteroideus (Viv.) Aurich et Podlech = Pallenis spinosa (L.) Cass. subsp. asteroidea (Viv.) Greuter ■☆

41621 Asteriscus spinosus (L.) Sch. Bip. subsp. aureus (Willk.) Aurich et Podlech = Pallenis spinosa (L.) Cass. subsp. aurea (Willk.) Nyman ■☆

41622 Asteriscus spinosus (L.) Sch. Bip. subsp. maroccanus Aurich et Podlech = Pallenis spinosa (L.) Cass. subsp. maroccana (Aurich et Podlech) Greuter ■☆

41623 Asteriscus spinosus (L.) Sch. Bip. var. intermedius Alleiz. = Pallenis spinosa (L.) Cass. ■☆

41624 Asteriscus stenophyllus (Link) Kuntze = Asteriscus graveolens (Forssk.) Less. subsp. stenophyllus (Link) Greuter ■☆

41625 Asteriscus stenophyllus (Link) Kuntze var. filifolius (G. Kunkel) A. Hansen et Sunding = Asteriscus graveolens (Forssk.) Less. subsp. stenophyllus (Link) Greuter ■☆

41626 Asteriscus stenophyllus (Link) Kuntze var. villososericeus Kuntze = Asteriscus graveolens (Forssk.) Less. subsp. odorus (Schousb.) Greuter ■☆

41627 Asteriscus teknensis Dobignard et Jacquemoud = Pallenis teknensis (Dobignard et Jacquemoud) Greuter et Jury ■☆

41628 Asterocarpus Eckl. et Zeyh. = Pterocelastrus Meisn. ●☆

41629 Asterocarpus Rchb. = Astrocarpa Dumort. ■☆

41630 Asterocarpus Rchb. = Sesamoides Ortega ■☆

41631 Asterocephalus Adans. = Scabiosa L. ●■

41632 Asterocephalus Zinn = Scabiosa L. ●■

41633 Asterocephalus altissimus Spreng. = Scabiosa africana L. ■☆

41634 Asterocephalus cochinchinensis Spreng. = Elephantopus scaber L. ■

41635 Asterocephalus induratus Spreng. = Scabiosa africana L. ■☆

41636 Asterochaete Nees = Carpha Banks et Sol. ex R. Br. ■☆

41637 Asterochaete angustifolia Nees = Carpha capitellata (Nees) Boeck. ■☆

41638 Asterochaete capitellata Nees = Carpha capitellata (Nees) Boeck. ■☆

41639 Asterochaete glomerata (Thunb.) Nees = Carpha glomerata (Thunb.) Nees ■☆

41640 Asterochaete ludwigii Hochst. = Carpha capitellata (Nees) Boeck. ■☆

41641 Asterochaete tenuis Kunth = Carpha capitellata (Nees) Boeck. ■☆

41642 Asterochiton Turcz. = Thomasia J. Gay ●☆

41643 Asterochlaena Garcke = Pavonia Cav.(保留属名)●■☆

41644 Asterocytisus (W. D. J. Koch) Schur ex Fuss = Genista L. ●

41645 Asterocytisus Schur ex Fuss = Genista L. ●

41646 Asterogeum Gray = Plantago L. ●■

41647 Asterogyne H. Wendl. ex Benth. et Hook. f. (1883);星蕊棕属(单叶棕属,星雌椰属,星蕊榈属)●☆

41648 Asterogyne H. Wendl. ex Hook. f. = Asterogyne H. Wendl. ex Benth. et Hook. f. ●☆

41649 Asterogyne martians (H. Wendl.) Hemsl.;星蕊棕(单叶棕)●☆

41650 Asterohyptis Epling(1932);星香属●☆

41651 Asterohyptis mociniana (Benth.) Epling;星香●☆

41652 Asterohyptis seemannii (A. Gray) Epling;西曼星香●☆

41653 Asteroides Mill. = Buphthalmum L. ■

41654 Asterolasia F. Muell. (1854);星毛芸香属●☆

41655 Asterolasia nivea (Paul G. Wilson) Paul G. Wilson;雪白星毛芸香●☆

41656 Asterolasia trymalioides F. Muell.;星毛芸香●☆

41657 Asterolepidion Ducke = Dendrobangia Rusby ●☆

41658 Asterolinion Brongn. = Asterolinon Hoffmanns. et Link ■☆

41659 Asterolinon Hoffmanns. et Link = Borissa Raf. ●■

41660 Asterolinon Hoffmanns. et Link = Lysimachia L. ●■

41661 Asterolinon Hoffmanns. et Link(1813-1820);星亚麻属■☆

41662 Asterolinon adoense Kunze;阿多星亚麻■☆

41663 Asterolinon linum-stellatum (L.) Duby;星亚麻■☆

41664 Asterolinum Duby = Asterolinon Hoffmanns. et Link ■☆

41665 Asteroloma Kuntze = Astroloma R. Br. ●☆

41666 Asteromaea DC. = Asteromoea Blume ●■

41667 Asteromoea Blume = Aster L. ●■

41668 Asteromoea Blume = Kalimeris (Cass.) Cass. ■

41669 Asteromoea angustifolia (C. C. Chang) Hand. -Mazz. = Gymnaster angustifolius (C. C. Chang) Y. Ling ■

41670 Asteromoea cantoniensis (Lour.) Matsum. = Kalimeris indica (L.) Sch. Bip. ■

41671 Asteromoea incisa Koidz. = Kalimeris incisa (Fisch.) DC. ■

41672 Asteromoea indica (L.) Blume = Kalimeris indica (L.) Sch. Bip. ■

41673 Asteromoea indica (L.) Blume var. lautureana Yamam. = Kalimeris shimadae (Kitam.) Kitam. ■

41674 Asteromoea indica (L.) Blume var. stenolepis Hand. -Mazz. = Kalimeris indica (L.) Sch. Bip. var. stenolepis (Hand. -Mazz.) Kitam. ■

41675 Asteromoea indica Blume var. lantureana Yamam. = Kalimeris shimadae (Kitam.) Kitam. ■

41676 Asteromoea integrifolia (Turcz.) Loes. = Kalimeris integrifolia Turcz. ex DC. ■

41677 Asteromoea integrifolia Loes. = Kalimeris indica (L.) Sch. Bip. ■

41678 Asteromoea lautureana (Debeaux) Hand. -Mazz. = Kalimeris lautureana (Debeaux) Kitam. ■

41679 Asteromoea lautureana Debeaux = Kalimeris shimadae (Kitam.) Kitam. ■

41680 Asteromoea mongolica (Franch.) Kitam. = Kalimeris mongolica (Franch.) Kitam. ■
41681 Asteromoea pekinensis Hance = Kalimeris integrifolia Turcz. ex DC. ■
41682 Asteromoea piccolii (Hook. f.) Hand.-Mazz. = Gymnaster piccolii (Hook. f.) Kitam. ■
41683 Asteromoea piccolii (Hook. f.) Mand.-Mazz. = Miyamayomena piccolii (Hook. f.) Kitam. ■
41684 Asteromoea procera (Hemsl.) Y. Ling = Aster procerus Hemsl. ■
41685 Asteromoea shimadai Kitam. = Kalimeris shimadae (Kitam.) Kitam. ■
41686 Asteromoea simplex (C. C. Chang) Hand.-Mazz. = Gymnaster simplex (C. C. Chang) Y. Ling ex J. Q. Fu ■
41687 Asteromoea simplex (C. C. Chang) Hand.-Mazz. = Miyamayomena simplex (C. C. Chang) Y. L. Chen ■
41688 Asteromyrtus Schauer = Melaleuca L. (保留属名) ●
41689 Asteromyrtus Schauer(1843);菀桃木属●☆
41690 Asteromyrtus angustifolia (Gaertner) Craven;窄叶菀桃木●☆
41691 Asteropeia Thouars(1805);翼萼茶属●☆
41692 Asteropeia amblyocarpa Tul.;钝果翼萼茶●☆
41693 Asteropeia amblyocarpa Tul. var. longifolia H. Perrier = Asteropeia amblyocarpa Tul. ●☆
41694 Asteropeia bakeri Scott-Elliot = Asteropeia multiflora Thouars ●☆
41695 Asteropeia densiflora Baker;密花翼萼茶●☆
41696 Asteropeia labatii G. E. Schatz, Lowry et A.-E. Wolf;拉巴翼萼茶●☆
41697 Asteropeia micraster Hallier f.;小星翼萼茶●☆
41698 Asteropeia micraster Hallier f. var. angustifolia H. Perrier = Asteropeia labatii G. E. Schatz, Lowry et A.-E. Wolf ●☆
41699 Asteropeia multiflora Thouars;多花翼萼茶●☆
41700 Asteropeia rhopaloides (Baker) Baill.;棒状翼萼茶●☆
41701 Asteropeia rhopaloides (Baker) Baill. var. angustata H. Perrier = Asteropeia amblyocarpa Tul. ●☆
41702 Asteropeia sphaerocarpa Baker = Asteropeia densiflora Baker ●☆
41703 Asteropeiaceae (Szyszyl.) Takht. ex Reveal et Hoogland = Theaceae Mirb. (保留科名) ●
41704 Asteropeiaceae (Szyszyl.) Takht. ex Reveal et Hoogland(1990);翼萼茶科●☆
41705 Asteropeiaceae Reveal et Hoogland = Asteropeiaceae (Szyszyl.) Takht. ex Reveal et Hoogland ●
41706 Asteropeiaceae Reveal et Hoogland = Theaceae Mirb. (保留科名) ●
41707 Asteropeiaceae Takht. = Asteropeiaceae (Szyszyl.) Takht. ex Reveal et Hoogland ●
41708 Asteropeiaceae Takht. = Theaceae Mirb. (保留科名) ●
41709 Asteropeiaceae Takht. ex Reveal et Hoogland = Asteropeiaceae (Szyszyl.) Takht. ex Reveal et Hoogland ●☆
41710 Asteropeiaceae Takht. ex Reveal et Hoogland = Theaceae Mirb. (保留科名) ●
41711 Asterophorum Sprague = Christiana DC. ●☆
41712 Asterophorum Sprague(1908);厄瓜多尔椴属●☆
41713 Asterophorum eburneum Sprague;厄瓜多尔椴●☆
41714 Asterophyllum Schimp. et Spenn. = Asperula L. (保留属名) ■
41715 Asterophyllum Schimp. et Spenn. = Galium L. ●■
41716 Asterophyllum Schimp. et Spenn. = Sherardia L. ■☆
41717 Asterophyllum Schimp. et Spenn. = Valantia L. ■☆
41718 Asteropsis Less. (1832);大头菀属■☆
41719 Asteropsis Less. = Podocoma Cass. ■☆
41720 Asteropsis macrocephala Less.;大头菀■☆
41721 Asteropterus Adans. (1763);星翅菊属●☆
41722 Asteropterus Adans. = Leysera L. ●■☆
41723 Asteropterus Vaillant(废弃属名) = Printzia Cass. (保留属名) ●■☆
41724 Asteropterus dinteri Rothm. = Leysera gnaphalodes (L.) L. ●☆
41725 Asteropterus gnaphalodes (L.) Rothm. = Leysera gnaphalodes (L.) L. ●☆
41726 Asteropterus gracilis Rothm. = Leysera gnaphalodes (L.) L. ●☆
41727 Asteropterus incanus (Thunb.) Rothm. = Leysera gnaphalodes (L.) L. ●☆
41728 Asteropterus leyseroides (Desf.) Rothm. = Leysera leyseroides (Desf.) Maire ●☆
41729 Asteropterus tenellus (DC.) Rothm. = Leysera tenella DC. ●☆
41730 Asteropus Schult. = Astropus Spreng. ●■
41731 Asteropus Schult. = Waltheria L. ●■
41732 Asteropyrum J. R. Drumm. et Hutch. (1920);星果草属;Starfruit, Starfruitestraw ■★
41733 Asteropyrum cavaleriei (H. Lév. et Vaniot) Drumm. et Hutch.;裂叶星果草(水八角,水黄莲,五角莲,鸭脚黄连);Lobedleaf Starfruit, Lobedleaf Starfruitestraw ■
41734 Asteropyrum hederifolium Schipcz. = Asteropyrum cavaleriei (H. Lév. et Vaniot) Drumm. et Hutch. ■
41735 Asteropyrum peltatum (Franch.) J. R. Drumm. et Hutch.;星果草(五角莲);Starfruit, Starfruitestraw ■
41736 Asteroschoenus Nees = Rhynchospora Vahl(保留属名) ■
41737 Asterosedum Grulich = Phedimus Raf. ■
41738 Asterosedum stellatum (L.) Grulich = Sedum stellatum L. ■☆
41739 Asterosperma Less. = Felicia Cass. (保留属名) ●■
41740 Asterostemma Decne. (1838);星冠萝藦属☆
41741 Asterostemma repandum Decne.;星冠萝藦☆
41742 Asterostigma Fisch. et C. A. Mey. (1845);星柱南星属■☆
41743 Asterostigma langsdorffianum Fisch. et C. A. Mey.;星柱南星■☆
41744 Asterostoma Blume = Osbeckia L. ●■
41745 Asterostoma repens (Desr.) Blume = Melastoma dodecandrum Lour. ●■
41746 Asterostoma repens Blume = Melastoma dodecandrum Lour. ●■
41747 Asterothamnus Novopokr. (1950);紫菀木属;Asterbush, Asterothamnus ●■
41748 Asterothamnus alyssoides (Turcz.) Novopokr.;紫菀木(庭荠紫菀木);Alyssus-like Asterothamnus, Common Asterbush ●
41749 Asterothamnus centrali-asiaticus Novopokr.;中亚紫菀木;Central Asia Asterbush, Central Asia Asterothamnus ●
41750 Asterothamnus centrali-asiaticus Novopokr. var. potaninii (Novopokr.) Y. Ling et Y. L. Chen;波氏中亚紫菀木(短叶中亚紫菀木) ●
41751 Asterothamnus centrali-asiaticus Novopokr. var. procerior Novopokr.;高大紫菀木;High Asterbush ●
41752 Asterothamnus fruticosus (C. Winkl.) Novopokr.;灌木紫菀木(灌木费利菊,紫菀木);Shrub Aster, Shrubby Asterbush, Shrubby Asterothamnus ●
41753 Asterothamnus fruticosus (C. Winkl.) Novopokr. = Felicia fruticosa G. Nicholson ●
41754 Asterothamnus fruticosus (C. Winkl.) Novopokr. f. discoideus Novopokr.;无舌灌木紫菀木(无舌花紫菀木) ●
41755 Asterothamnus heteropappoides Novopokr.;异型紫菀木●☆
41756 Asterothamnus molliusculus Novopokr.;软叶紫菀木;Softleaf

Asterbush, Softleaf Asterothamnus ●

41757 Asterothamnus poliifolius Novopokr.; 毛叶紫菀木; Hairleaf Asterbush, Hairyleaf Asterothamnus ●

41758 Asterothamnus potaninii Novopokr. = Asterothamnus centrali-asiaticus Novopokr. var. potaninii (Novopokr.) Y. Ling et Y. L. Chen ●

41759 Asterothamnus schischkinii Tamamsch.; 希施紫菀木●☆

41760 Asterothrix Cass. = Leontodon L.(保留属名)■☆

41761 Asterothryx hispanica (Poir.) Batt. = Leontodon hispanicus Poir. ■☆

41762 Asterotricha Kuntze = Astrotricha DC. ●☆

41763 Asterotricha V. V. Botschantz. = Fibigia Medik. ■☆

41764 Asterotricha V. V. Botschantz. = Pterygostemon V. V. Botsch. ■☆

41765 Asterotrichion Klotzsch = Plagianthus J. R. Forst. et G. Forst. ●☆

41766 Asterotrichion Klotzsch(1840); 星毛锦葵属●☆

41767 Asterotrichion discolor (Hook.) Melville; 异色星毛锦葵●☆

41768 Asterotrichion sidoides Klotzsch; 星毛锦葵●☆

41769 Asterotrichon N. T. Burb. = Asterotrichion Klotzsch ●☆

41770 Asthenatherum Nevski = Centropodia (R. Br.) Rchb. ■☆

41771 Asthenatherum forskalii (Vahl) Nevski = Centropodia forskalii (Vahl) Cope ■☆

41772 Asthenatherum fragile (Guinet et Sauvage) Monod = Centropodia fragilis (Guinet et Sauvage) Cope ■☆

41773 Asthenatherum glaucum (Nees) Nevski = Centropodia glauca (Nees) Cope ■☆

41774 Asthenatherum glaucum (Nees) Nevski var. lasiophyllum (Pilg.) Conert = Centropodia glauca (Nees) Cope ■☆

41775 Asthenatherum mossamedense (Rendle) Conert = Centropodia mossamedensis (Rendle) Cope ■☆

41776 Asthenochloa Büse(1854); 柔草属■☆

41777 Asthenochloa tenera Büse; 柔草■☆

41778 Asthotheca Miers ex Planch. et Triana = Clusia L. ●☆

41779 Astianthus D. Don(1823); 美花属●☆

41780 Astianthus viminalis Baill.; 柳枝美花●☆

41781 Astiella Jovet(1941); 小美茜属☆

41782 Astiella delicatula Jovet; 小美茜☆

41783 Astilbaceae Krach = Saxifragaceae Juss.(保留科名)●■

41784 Astilbaceae Krach; 落新妇科■

41785 Astilbe Buch.-Ham. = Astilbe Buch.-Ham. ex D. Don ■

41786 Astilbe Buch.-Ham. ex D. Don(1825); 落新妇属(红升麻属); Astilbe, False Buck's-beard, False Goat's Beard, False Spirea, Goat's-beard ■

41787 Astilbe × amabilis H. Hara; 秀丽落新妇■☆

41788 Astilbe × photeinophylla Koidz.; 丝脉落新妇■☆

41789 Astilbe arendsii H. Junge; 阿氏落新妇; Astilbe, False Spiraea, Red False Buck's-beard ■☆

41790 Astilbe aruncus (L.) Trev. = Aruncus sylvester Kostel. ex Maxim. ■

41791 Astilbe austro-sinensis Hand.-Mazz. = Astilbe grandis Stapf ex E. H. Wilson ■

41792 Astilbe bandaica (Honda) Koidz. = Astilbe odontophylla Miq. var. bandaica (Honda) H. Hara ■☆

41793 Astilbe biternata Britton; 二回三出落新妇; False Goat's Beard, False Goatsbeard Astilbe ■☆

41794 Astilbe chinensis (Maxim.) Franch. et Sav. = Astilbe chinensis (Maxim.) Maxim. ex Franch. et Sav. ■

41795 Astilbe chinensis (Maxim.) Maxim. ex Franch. et Sav.; 落新妇(阿根八,红花落新妇,红三七,红升麻,虎麻,金猫儿,金毛狗,金毛三七,金尾鳝,马尾参,毛三七,山花七,升麻,术活,水三七,水升麻,铁杆升麻,铁火钳,土升麻,小升麻,野升麻,阴阳虎); Chinese Astilbe, False Spiraea ■

41796 Astilbe chinensis (Maxim.) Maxim. ex Franch. et Sav. = Astilbe rubra Hook. f. et Thomson ■

41797 Astilbe chinensis (Maxim.) Maxim. ex Franch. et Sav. 'Pumila'; 矮落新妇; Dwarf Pink Astilbe ■☆

41798 Astilbe chinensis (Maxim.) Maxim. ex Franch. et Sav. var. davidii Franch. = Astilbe rubra Hook. f. et Thomson ■

41799 Astilbe chinensis (Maxim.) Maxim. ex Franch. et Sav. var. davidii Franch. = Astilbe chinensis (Maxim.) Maxim. ex Franch. et Sav. ■

41800 Astilbe chinensis (Maxim.) Maxim. ex Franch. et Sav. var. koreana Kom. = Astilbe grandis Stapf ex E. H. Wilson ■

41801 Astilbe chinensis (Maxim.) Maxim. ex Franch. et Sav. var. longicarpa Hayata = Astilbe longicarpa (Hayata) Hayata ■

41802 Astilbe chinensis (Maxim.) Maxim. ex Franch. et Sav. var. pumila Hort. = Astilbe chinensis (Maxim.) Maxim. ex Franch. et Sav. 'Pumila' ■☆

41803 Astilbe chinensis (Maxim.) Maxim. ex Franch. et Sav. var. taquetii (H. Lév.) H. Hara = Astilbe rubra Hook. f. et Thomson var. taquetii (H. Lév.) H. Hara ■☆

41804 Astilbe chinensis Maxim. var. davidii Franch. = Astilbe chinensis (Maxim.) Maxim. ex Franch. et Sav. ■

41805 Astilbe congesta (H. Boissieu) Nakai = Astilbe odontophylla Miq. ■☆

41806 Astilbe davidii (Franch.) Henry; 北红升麻(大卫氏落新妇); David Astilbe ■

41807 Astilbe davidii (Franch.) Henry = Astilbe chinensis (Maxim.) Maxim. ex Franch. et Sav. ■

41808 Astilbe formosa Nakai; 美丽落新妇(岩地光秃落新妇)■☆

41809 Astilbe fujisanensis Nakai = Astilbe thunbergii (Siebold et Zucc.) Miq. var. fujisanensis (Nakai) Ohwi ■☆

41810 Astilbe glaberrima Nakai; 光秃落新妇■☆

41811 Astilbe glaberrima Nakai var. saxatilis (Nakai) H. Ohba; 岩地光秃落新妇■☆

41812 Astilbe grandis Stapf ex E. H. Wilson; 大落新妇(大花落新妇,大濇疙瘩,红升麻,华南落新妇,马尾参,毛头寒药,山花七,水高粱,水红柳,水升麻,野红稗); Great Astilbe, South China Astilbe ■

41813 Astilbe hachijoensis Nakai; 八丈岛落新妇■☆

41814 Astilbe henricii Franch. = Rodgersia aesculifolia Batalin var. henricii (Franch.) C. Y. Wu ex J. T. Pan ■

41815 Astilbe heteropetala Mattf. = Astilbe rivularis Buch.-Ham. ex D. Don var. myriantha (Diels) J. T. Pan ■

41816 Astilbe hybrida Host.; 杂种落新妇; False Spiraea ■☆

41817 Astilbe japonica (C. Morren et Decne.) A. Gray; 泡盛落新妇(泡盛草); False Buck's-beard, Florist's Spiraea, Japanese Astilbe ■☆

41818 Astilbe japonica (C. Morren et Decne.) A. Gray f. rubeola H. Hara; 淡红泡盛落新妇■☆

41819 Astilbe japonica (C. Morren et Decne.) A. Gray subsp. glaberrima (Nakai) Kitam. = Astilbe glaberrima Nakai ■☆

41820 Astilbe japonica (C. Morren et Decne.) A. Gray var. terrestris (Nakai) Murata ex Ohwi = Astilbe glaberrima Nakai ■☆

41821 Astilbe kiusiana H. Hara = Astilbe thunbergii (Siebold et Zucc.) Miq. var. kiusiana (H. Hara) H. Hara ex H. Ohba ■☆

41822 Astilbe koreana (Kom.) Nakai; 朝鲜落新妇; Korean Astilbe ■

41823 Astilbe koreana (Kom.) Nakai = Astilbe grandis Stapf ex E. H. Wilson ■
41824 Astilbe leucantha Knoll = Astilbe grandis Stapf ex E. H. Wilson ■
41825 Astilbe longicarpa (Hayata) Hayata;长果落新妇(落新妇);Longfruit Astilbe ■
41826 Astilbe macrocarpa Knoll;大果落新妇;Bigfruit Astilbe ■
41827 Astilbe macroflora Hayata;阿里山落新妇;Alishan Astilbe ■
41828 Astilbe microphylla Knoll;小叶落新妇;Little Leaf Astilbe ■☆
41829 Astilbe microphylla Knoll var. riparia Hatus.;河岸小叶落新妇■☆
41830 Astilbe myriantha Diels = Astilbe rivularis Buch.-Ham. ex D. Don var. myriantha (Diels) J. T. Pan ■
41831 Astilbe myriantha Diels = Astilbe rivularis Buch.-Ham. ex D. Don ■
41832 Astilbe odontophylla Miq.;齿叶落新妇(密集落新妇);Crowded Astilbe ■☆
41833 Astilbe odontophylla Miq. f. rosea Honda;粉花齿叶落新妇■☆
41834 Astilbe odontophylla Miq. var. bandaica (Honda) H. Hara;班达落新妇■☆
41835 Astilbe odontophylla Miq. var. oblongifolia H. Hara;矩圆齿叶落新妇■☆
41836 Astilbe odontophylla Miq. var. okuyamae (H. Hara) H. Hara;亿山落新妇■☆
41837 Astilbe okuyamae H. Hara = Astilbe odontophylla Miq. var. okuyamae (H. Hara) H. Hara ■☆
41838 Astilbe pinnata (Franch.) Franch. = Rodgersia pinnata Franch. ■
41839 Astilbe platyphylla H. Boissieu;宽叶落新妇■☆
41840 Astilbe polyandra Hemsl.;多蕊落新妇■☆
41841 Astilbe rivularis Buch.-Ham. ex D. Don;溪畔落新妇(红升麻,假升麻,假淫羊藿,水滨落新妇,水滨升麻,野高粱,野泽兰);Brooklet Astilbe,Tall False Buck's-beard ■
41842 Astilbe rivularis Buch.-Ham. ex D. Don var. angustata C. Y. Wu et J. T. Pan = Astilbe rivularis Buch.-Ham. ex D. Don var. angustifoliolata H. Hara ■
41843 Astilbe rivularis Buch.-Ham. ex D. Don var. angustifoliolata H. Hara;狭叶落新妇■
41844 Astilbe rivularis Buch.-Ham. ex D. Don var. myriantha (Diels) J. T. Pan;多花落新妇(多花红升麻,红升麻,金毛七,铁杆升麻,小牛胃花);Manyflower Astilbe ■
41845 Astilbe rosea ? = Astilbe arendsii H. Junge ■☆
41846 Astilbe rubra Hook. f. et Thomson = Astilbe chinensis (Maxim.) Maxim. ex Franch. et Sav. ■
41847 Astilbe rubra Hook. f. et Thomson ex Hook.;腺萼落新妇(红花升麻);Chinese Astilbe,Glandcalyx Astilbe ■
41848 Astilbe rubra Hook. f. et Thomson var. taquetii (H. Lév.) H. Hara;塔克特落新妇■☆
41849 Astilbe shikokiana Nakai = Astilbe thunbergii (Siebold et Zucc.) Miq. var. shikokiana (Nakai) Ohwi ■☆
41850 Astilbe sikokumontana Koidz. = Astilbe thunbergii (Siebold et Zucc.) Miq. var. sikokumontana (Koidz.) Murata ■☆
41851 Astilbe simplicifolia Makino;单叶落新妇;Dwarf Astilbe,Star Astilbe ■☆
41852 Astilbe thunbergii (Siebold et Zucc.) Miq.;通贝里落新妇(鬼箭羽,童氏落新妇,西南落新妇);Thunberg Astilbe,False Spiraea ●■☆
41853 Astilbe thunbergii (Siebold et Zucc.) Miq. f. rosea Hiyama;粉花通贝里落新妇■☆
41854 Astilbe thunbergii (Siebold et Zucc.) Miq. nothovar. oblongifolia (H. Hara) Murata = Astilbe odontophylla Miq. var. oblongifolia H. Hara ■☆
41855 Astilbe thunbergii (Siebold et Zucc.) Miq. var. bandaica (Honda) Murata = Astilbe odontophylla Miq. var. bandaica (Honda) H. Hara ■☆
41856 Astilbe thunbergii (Siebold et Zucc.) Miq. var. congesta H. Boissieu f. rosea (Honda) Sugim. = Astilbe odontophylla Miq. f. rosea Honda ■☆
41857 Astilbe thunbergii (Siebold et Zucc.) Miq. var. congesta H. Boissieu f. bandaica (Honda) Ohwi = Astilbe odontophylla Miq. var. bandaica (Honda) H. Hara ■☆
41858 Astilbe thunbergii (Siebold et Zucc.) Miq. var. congesta H. Boissieu = Astilbe odontophylla Miq. ■☆
41859 Astilbe thunbergii (Siebold et Zucc.) Miq. var. formosa (Nakai) Ohwi f. nipponica Hiyama = Astilbe formosa Nakai ■☆
41860 Astilbe thunbergii (Siebold et Zucc.) Miq. var. formosa (Nakai) Ohwi = Astilbe formosa Nakai ■☆
41861 Astilbe thunbergii (Siebold et Zucc.) Miq. var. fujisanensis (Nakai) Ohwi;富士山落新妇■☆
41862 Astilbe thunbergii (Siebold et Zucc.) Miq. var. hachijoensis (Nakai) Ohwi = Astilbe hachijoensis Nakai ■☆
41863 Astilbe thunbergii (Siebold et Zucc.) Miq. var. kiusiana (H. Hara) H. Hara ex H. Ohba;九州落新妇■☆
41864 Astilbe thunbergii (Siebold et Zucc.) Miq. var. longipedicellata Hatus.;长梗通贝里落新妇●■☆
41865 Astilbe thunbergii (Siebold et Zucc.) Miq. var. okuyamae (H. Hara) Ohwi = Astilbe odontophylla Miq. var. okuyamae (H. Hara) H. Hara ■☆
41866 Astilbe thunbergii (Siebold et Zucc.) Miq. var. shikokiana (Nakai) Ohwi;四国落新妇■☆
41867 Astilbe thunbergii (Siebold et Zucc.) Miq. var. sikokumontana (Koidz.) Murata;四国山地落新妇■☆
41868 Astilbe thunbergii (Siebold et Zucc.) Miq. var. terrestris (Nakai) Ohwi = Astilbe glaberrima Nakai ■☆
41869 Astilbe virescens Hutch. = Astilbe rivularis Buch.-Ham. ex D. Don var. myriantha (Diels) J. T. Pan ■
41870 Astilboides (Hemsl.) Engl. (1930);大叶子属(山荷叶属);Astilboides,Rodgersia ■
41871 Astilboides Engl. = Astilboides (Hemsl.) Engl. ■
41872 Astilboides tabularis (Hemsl.) Engl.;大叶子(大脖梗子,佛爷伞,山荷叶);Astilboides,Common Astilboides ■
41873 Astiria Lindl. (1844);毛梧桐属●☆
41874 Astiria rosea Lindl.;毛梧桐●☆
41875 Astoma DC. = Astomaea Rchb. ●☆
41876 Astomaea Rchb. (1837);无口草属●☆
41877 Astomaea galiocarpa (Korovin) Govaerts;乳果无口草●☆
41878 Astomaea seselifolia (DC.) Rchb.;无口草●☆
41879 Astomatopsis Korovin = Astomaea Rchb. ●☆
41880 Astonia S. W. L. Jacobs(1997);阿氏泽泻属■☆
41881 Astonia australiensis (Aston) S. W. L. Jacobs;阿氏泽泻■☆
41882 Astorganthus Endl. = Melicope J. R. Forst. et G. Forst. ●
41883 Astracantha Podlech (1983);星刺豆属(云英花属);Dragon Gum,Gum Dragon,Gum Tragacanth ●☆
41884 Astracantha granatensis (Lam.) Podlech = Astragalus granatensis Lam. ●☆
41885 Astracantha gummifera (Labill.) Podlech;产胶星刺豆;Dragon Gum,Gum Sarcoculla,Sarcoculla ●☆
41886 Astradelphus J. Rémy = Erigeron L. ●■

41887 Astradelphus J. Rémy = Gusmania J. Rémy ●■
41888 Astradelphus J. Rémy(1849);兄弟星属■☆
41889 Astradelphus chilensis Remy;兄弟星■☆
41890 Astraea Klotzsch = Croton L. ●
41891 Astraea Schauer = Gomphotis Raf.(废弃属名)●☆
41892 Astraea Schauer = Thryptomene Endl.(保留属名)●☆
41893 Astraea lobata (L.) Klotzsch = Croton lobatus L. ●☆
41894 Astragalaceae Bercht. et J. Presl = Fabaceae Lindl.(保留科名)●■
41895 Astragalaceae Bercht. et J. Presl = Leguminosae Juss.(保留科名)●■
41896 Astragalaceae Martinov = Fabaceae Lindl.(保留科名)●■
41897 Astragalaceae Martinov = Leguminosae Juss.(保留科名)●■
41898 Astragalaceae Martinov;黄耆科●■
41899 Astragalina Bubani = Astragalus L. ●■
41900 Astragalina Bubani = Phaca L. ●■
41901 Astragaloides Adans. = Astragalus L. ●■
41902 Astragaloides Boehm.(1760);拟黄耆属●■☆
41903 Astragaloides Boehm. = Astragalus L. ●■
41904 Astragaloides Boehm. = Phaca L. ●■
41905 Astragaloides Quer. = Astragalus L. ●■
41906 Astragalus L.(1753);黄耆属(黄花属,黄芪属,黄蓍属,紫云英属);Coat's Thorn,Loco,Loeoweed,Milk Vetch,Milkvetch,Milk-vetch,Poisonvetch ●■
41907 Astragalus aberrans Förther et Podlech;异常黄耆■☆
41908 Astragalus abramovii Gontsch.;阿勃黄耆■
41909 Astragalus abramovii Gontsch. = Astragalus peterae H. T. Tsai et T. T. Yu ■
41910 Astragalus abyssinicus Steud. ex A. Rich. = Astragalus atropilosulus (Hochst.) Bunge var. abyssinicus (Hochst.) J. B. Gillett ■☆
41911 Astragalus acanthocarpus Boriss.;尖果黄耆■☆
41912 Astragalus acaulis Baker;无茎黄耆(无茎黄芪,无茎黄蓍);Stemless Milkvetch ■
41913 Astragalus acceptus Podlech et L. R. Xu;德令哈黄耆(心叶猴耳环)■
41914 Astragalus adesmiifolius Bunge = Astragalus hoffmeisteri (Klotzsch) Ali ■
41915 Astragalus adscendens Boiss. et Hausskn. ex Boiss.;上举黄耆(上举黄芪)■☆
41916 Astragalus adsurgens Pall. 'Shadawang';沙打旺■
41917 Astragalus adsurgens Pall. = Astragalus laxmannii Jacq. ■
41918 Astragalus adsurgens Pall. f. leucanthus Takeda;白花斜茎黄耆■☆
41919 Astragalus adsurgens Pall. subsp. fujisanensis (Miyabe et Tatew.) Kitag. = Astragalus adsurgens Pall. ■
41920 Astragalus aduncus Willd.;钩状黄耆■☆
41921 Astragalus aegacanthoides R. Parker = Caragana aegacanthoides (R. Parker) L. B. Chaudhary et S. K. Srivast. ●
41922 Astragalus aemulans (Nevski) Gontsch.;匹敌黄耆■☆
41923 Astragalus affghanus Boiss.;阿富汗黄耆■☆
41924 Astragalus africanus Bunge = Astragalus alopecuroides L. ■☆
41925 Astragalus agrestis Douglas ex G. Don;毛喉黄耆(毛喉黄芪);Field Milk-vetch,Hairthroat Milkvetch ■
41926 Astragalus agrestis G. Don = Astragalus agrestis Douglas ex G. Don ■
41927 Astragalus aitchisonii ? = Astragalus ophiocarpus Benth. et Bunge ■
41928 Astragalus akkensis Coss.;阿克黄耆■☆
41929 Astragalus akkensis Coss. subsp. occidentalis Maire = Astragalus akkensis Coss. ■☆
41930 Astragalus akkensis Coss. subsp. uzzararum Maire = Astragalus akkensis Coss. ■☆
41931 Astragalus akkensis Coss. var. integrifolius Maire = Astragalus akkensis Coss. ■☆
41932 Astragalus akkensis Coss. var. maurorum (Murb.) Maire = Astragalus maurorum Murb. ■☆
41933 Astragalus akkensis Coss. var. pinoyi Maire = Astragalus akkensis Coss. ■☆
41934 Astragalus akkensis Coss. var. uzzararum (Maire) Maire = Astragalus akkensis Coss. ■☆
41935 Astragalus aksuensis Bunge;阿克苏黄耆(阿克苏黄芪,阿克苏黄蓍,阿拉套黄耆);Aksu Milkvetch ■
41936 Astragalus aksuensis Bunge = Astragalus moellendorffii Bunge ex Maxim. ■
41937 Astragalus alaicus Freyn;阿赖黄耆■☆
41938 Astragalus alaschanensis H. C. Fu = Astragalus dengkouensis H. C. Fu ■
41939 Astragalus alaschanensis H. C. Fu = Astragalus grubovii Sanchir ■
41940 Astragalus alaschanus Bunge = Astragalus alaschanus Bunge ex Maxim. ■
41941 Astragalus alaschanus Bunge ex Maxim.;阿拉善黄耆(阿拉善黄芪,阿拉善黄蓍,陀螺棘豆);Alashan Milkvetch ■
41942 Astragalus alatavicus Kar. et Kir.;阿拉套黄耆(阿尔泰黄耆,阿拉套黄芪,阿勒泰黄芪,雀儿豆);Alatao Milkvetch,Altai Milkvetch ■
41943 Astragalus alatavicus Kar. et Kir. var. pamirensis (Franch.) B. Fedtsch. = Astragalus pamirensis Franch. ■
41944 Astragalus albanicus Grossh.;阿尔邦黄耆■☆
41945 Astragalus albertii Bunge;阿尔伯特黄耆■☆
41946 Astragalus albescens Boriss.;变白黄耆■☆
41947 Astragalus albicans Bong.;革果黄耆■
41948 Astragalus albicaulis DC.;白茎黄耆;Whitestem Milkvetch ■☆
41949 Astragalus albido-flavus K. T. Fu;黄白黄耆(黄白黄芪,黄白黄蓍);Paleyellow Milkvetch,Yellow-white Milkvetch ■
41950 Astragalus albovillosus Kitam.;白长柔毛黄耆■☆
41951 Astragalus albovillosus Kitam. var. nigrescens ?;变黑黄耆■☆
41952 Astragalus albus Sirj. = Astragalus laxmannii Jacq. ■
41953 Astragalus alexandri Charadze;阿莱黄耆■☆
41954 Astragalus alexandrinus Boiss. = Astragalus caprinus L. ■☆
41955 Astragalus algerianus E. Sheld.;阿尔及利亚黄耆■☆
41956 Astragalus allochrous A. Gray;彩黄耆;Many-coloured Rattlepodd ■☆
41957 Astragalus alopecias Pall.;长果黄耆(长尾黄芪,长尾黄蓍);Longtail Milkvetch ■
41958 Astragalus alopeculoides L. = Astragalus vulpinus Willd. ■
41959 Astragalus alopecuroides L.;非洲黄耆■☆
41960 Astragalus alopecurus Pall. = Astragalus alopecurus Pall. ex DC. ■
41961 Astragalus alopecurus Pall. ex DC.;狐尾黄耆(狐尾黄芪,狐尾黄蓍,狭叶猪屎豆);Foxtail Milkvetch ■
41962 Astragalus alpinus L.;高山黄耆(鲍氏黄耆,高山黄芪,高山黄蓍);Alp Milkvetch,Alpine Milk Vetch,Alpine Milkvetch,Alpine Milk-vetch ■
41963 Astragalus alpinus L. subsp. alaskanus Hultén = Astragalus alpinus L. ■
41964 Astragalus alpinus L. subsp. arcticus Hultén = Astragalus alpinus L. ■

41965 Astragalus alsugineus Kar. et Kir.;喜盐黄耆(喜盐黄芪,喜盐黄耆)■

41966 Astragalus alsugineus Kar. et Kir. var. multijugus S. B. Ho = Astragalus lang-ranii Podlech ■

41967 Astragalus altaicola Podlech;阿尔泰黄耆■

41968 Astragalus altaicus Bunge = Astragalus altaicola Podlech ■

41969 Astragalus altaicus Pall. = Oxytropis altaica (Pall.) Pers. ■

41970 Astragalus alyssoides Lam.;庭荠黄耆■☆

41971 Astragalus amarus Pall.;西方苦黄耆■☆

41972 Astragalus ambigens Popov;可疑黄耆■☆

41973 Astragalus ambiguus Pall. = Oxytropis ambigua (Pall.) DC. ■

41974 Astragalus ammodendron Bunge;沙生树黄耆(沙生树黄芪,沙树黄耆)■☆

41975 Astragalus ammodytes Pall.;喜沙黄耆(沙生黄芪,沙生黄耆,喜沙黄芪);Sand-living Milkvetch,Sandt Milkvetch ■

41976 Astragalus ammophilus Kar. et Kir.;沙生黄耆(沙生黄芪,喜沙黄芪,喜沙黄耆);Lovesand Milkvetch,Sand-loving Milkvetch ■

41977 Astragalus ammophilus Kar. et Kir. var. persepolitanus (Boiss.) Ali = Astragalus ammophilus Kar. et Kir. ■

41978 Astragalus ampullatus Pall. = Oxytropis ampullata (Pall.) Pers. ■

41979 Astragalus amygdalinus Bunge;膀胱黄耆■☆

41980 Astragalus ancistron Pomel = Astragalus hamosus L. ■☆

41981 Astragalus ancocarpus Pomel = Astragalus hamosus L. ■☆

41982 Astragalus androssovianus Gontsch.;安得罗黄耆■☆

41983 Astragalus anfractuosus Bunge;曲之黄耆■

41984 Astragalus anfractuosus Bunge = Astragalus subuliformis DC. ■

41985 Astragalus angustidens Freyn et Sint.;细齿黄耆■☆

41986 Astragalus angustiflorus C. Koch;狭花黄耆■☆

41987 Astragalus angustifoliolatus K. T. Fu;狭叶黄耆(细叶黄耆,狭叶黄芪);Narrowleaf Milkvetch,Narrowleaflet Milkvetch ■

41988 Astragalus anisacanthus Bunge;异刺黄耆■☆

41989 Astragalus ankylotus Fisch. et C. A. Mey.;锚黄耆■☆

41990 Astragalus annularis Forssk.;环黄耆■☆

41991 Astragalus anomalus Bunge = Astragalus rhizanthus Benth. ●

41992 Astragalus anrachaicus Golosk. = Astragalus lanuginosus Kar. et Kir. ■

41993 Astragalus antiatlanticus Emb. et Maire;安蒂黄耆■☆

41994 Astragalus antoninae Grig.;安托黄耆■☆

41995 Astragalus apiculatus Gontsch.;细尖黄耆■☆

41996 Astragalus arabicus Ehrenb. ex Bunge var. congestus Schweinf. = Astragalus fatmensis Hochst. ex Chiov. ■☆

41997 Astragalus arbuscula Pall.;木黄耆(木黄芪);Tree Milkvetch,Woody Milkvetch ●

41998 Astragalus arcuatus Kar. et Kir.;弓形黄耆(弯弓黄耆)■

41999 Astragalus arenarius L. = Astragalus danicus Retz. ■

42000 Astragalus arenicola Pomel = Astragalus stella Gouan ■☆

42001 Astragalus argentatus Pall. = Oxytropis argentata (Pall.) Pers. ■

42002 Astragalus argillophilus Cory;白土黄耆;Halfmoon Loco,Half-moon Loco ■☆

42003 Astragalus arguteasis Bunge;亮黄耆■☆

42004 Astragalus argyroides Beck ex Stapf;银色黄耆■☆

42005 Astragalus aridicola P. C. Li;旱生黄耆(旱生黄芪)■

42006 Astragalus aridicola P. C. Li = Astragalus maowenensis Podlech et L. R. Xu ■

42007 Astragalus aridovallicola P. C. Li;干谷黄耆(干谷黄芪,旱谷黄耆)■

42008 Astragalus aristidis Coss. = Astragalus saharae Pomel ■☆

42009 Astragalus arkalycensis Bunge;边塞黄耆(阿尔卡黄芪,边塞黄芪,草原黄芪,草原黄耆);Border Milkvetch,Frontier Milkvetch ■

42010 Astragalus armatus Willd.;食用黄耆■☆

42011 Astragalus armatus Willd. subsp. numidicus (Murb.) Tietz;努米底亚黄耆■☆

42012 Astragalus armatus Willd. subsp. tragacanthoides (Desf.) Emb. et Maire = Astragalus armatus Willd. ■☆

42013 Astragalus armatus Willd. var. libycus Pamp. = Astragalus armatus Willd. subsp. numidicus (Murb.) Tietz ■☆

42014 Astragalus arnacantha M. Bieb.;多刺黄耆■☆

42015 Astragalus arnacanthoides (Boriss.) Boriss.;假多刺黄耆■☆

42016 Astragalus arnoldianus N. D. Simpson = Astragalus simpsonii E. Peter ■

42017 Astragalus arnoldii Hemsl. et H. Pearson;团垫黄耆(团垫黄芪);Arnold Milkvetch,Culciform Milkvetch ■

42018 Astragalus arnoldii Hemsl. et H. Pearson f. albiflorus Y. H. Wu;白花团垫黄耆(白花团垫黄芪);Whiteflower Arnold Milkvetch ■

42019 Astragalus arnoldii Hemsl. et H. Pearson f. albiflorus Y. H. Wu = Astragalus arnoldii Hemsl. et H. Pearson ■

42020 Astragalus aronarius L. = Astragalus danicus Retz. ■

42021 Astragalus arpilobus Kar. et Kir.;镰荚黄耆(镰荚黄芪);Falcate-pod Milkvetch,Sicklepod Milkvetch ■

42022 Astragalus artvinensis Popov;阿尔特温黄耆■☆

42023 Astragalus asper Jacq.;糙黄耆■☆

42024 Astragalus asterias Hohen.;星黄耆■☆

42025 Astragalus asterias Hohen. subsp. aristidis (Batt.) Greuter = Astragalus saharae Pomel ■☆

42026 Astragalus asterias Hohen. subsp. astraboides (Pomel) Greuter = Astragalus longicaulis Pomel ■☆

42027 Astragalus asterias Hohen. subsp. polyactinus (Boiss.) Greuter = Astragalus stella Gouan ■☆

42028 Astragalus asterias Hohen. subsp. radiatus (Batt.) Greuter = Astragalus asterias Hohen. ■☆

42029 Astragalus astraboides Pomel = Astragalus longicaulis Pomel ■☆

42030 Astragalus athranthus Podlech et L. R. Xu;黑药黄耆■

42031 Astragalus atlanticus (Ball) Ball = Astragalus alopecuroides L. ■☆

42032 Astragalus atripurpureus Boiss.;暗紫黄耆■☆

42033 Astragalus atropilosulus (Hochst.) Bunge;暗毛紫黄耆■☆

42034 Astragalus atropilosulus (Hochst.) Bunge var. aberdaricus J. B. Gillett;阿伯德尔黄耆■☆

42035 Astragalus atropilosulus (Hochst.) Bunge var. abyssinicus (Hochst.) J. B. Gillett;阿比西尼亚暗毛黄耆■☆

42036 Astragalus atropilosulus (Hochst.) Bunge var. bequaertii (De Wild.) J. B. Gillett;贝卡尔暗毛黄耆■☆

42037 Astragalus atropilosulus (Hochst.) Bunge var. burkeanus (Harv.) J. B. Gillett;伯克暗毛黄耆■☆

42038 Astragalus atropilosulus (Hochst.) Bunge var. coerulescens (Chiov.) J. B. Gillett;天蓝暗毛黄耆■☆

42039 Astragalus atropilosulus (Hochst.) Bunge var. elgonensis (Bullock) J. B. Gillett;埃尔贡暗毛黄耆■☆

42040 Astragalus atropilosulus (Hochst.) Bunge var. longeracemosus J. B. Gillett;长总花暗毛黄耆■☆

42041 Astragalus atropilosulus (Hochst.) Bunge var. mooneyi J. B. Gillett;穆尼暗毛黄耆■☆

42042 Astragalus atropilosulus (Hochst.) Bunge var. platycarpus J. B. Gillett;宽果暗毛黄耆■☆

42043 Astragalus atropilosulus (Hochst.) Bunge var. pubescens J. B.

Gillett;软暗毛紫黄耆■☆

42044 Astragalus atropilosulus (Hochst.) Bunge var. venosus (Hochst.) J. B. Gillett;显脉暗毛黄耆■☆

42045 Astragalus atrosanguineus Murb. = Astragalus reinii Ball ■☆

42046 Astragalus aurantiacus Hand. -Mazz. = Astragalus dependens Bunge ex Maxim. var. aurantiacus (Hand. -Mazz.) Y. C. Ho ■

42047 Astragalus australis (L.) Lam.;南方黄耆(南黄芪)■☆

42048 Astragalus austriacus L.;奥地利黄耆;Austrian Milkvetch ■☆

42049 Astragalus austriacus L. = Astragalus satoi Kitag. ■

42050 Astragalus austrodshungaricus Golosk.;南准葛尔黄耆■

42051 Astragalus austro-sibiricus Schischk.;漠北黄耆(漠北黄芪);N. Desert Milkvetch, South-siberian Milkvetch ■

42052 Astragalus austrotibetanus Podlech et L. R. Xu;藏南黄耆■

42053 Astragalus badachschanicus Boriss. = Astragalus falconeri Bunge ■

42054 Astragalus badrinathensis M. P. Sharma = Astragalus rhizanthus Benth. ●

42055 Astragalus bahrakianus Grey-Wilson;巴拉克黄耆■

42056 Astragalus baicalensis Pall. = Oxytropis coerulea (Pall.) DC. ■

42057 Astragalus baischinticus N. Ulziykh. = Astragalus pseudoborodinii S. B. Ho ■

42058 Astragalus baisensis Sumnev. = Astragalus sphaerocystis Bunge ■

42059 Astragalus bakaliensis Bunge;巴卡利黄耆(巴卡利黄芪)■

42060 Astragalus bakeri Ali;贝克黄耆■☆

42061 Astragalus bakeristrobiliferus H. Ohashi = Astragalus chlorostachys Lindl. ■

42062 Astragalus bakuensis Bunge = Astragalus flemingii Ali ■☆

42063 Astragalus balfourianus N. D. Simpson;长小苞黄耆(长小苞黄芪,滇西北黄芪,雪山芪);Balfour Milkvetch ■

42064 Astragalus balfourianus Simpson = Astragalus tribulifolius Benth. ex Bunge ■

42065 Astragalus banzragczii N. Ulziykh. = Astragalus hamiensis S. B. Ho ■

42066 Astragalus baotouensis H. C. Fu;包头黄耆(包头黄芪);Baotou Milkvetch ■

42067 Astragalus basiflorus E. Peter;地花黄耆(地花黄芪);Baseflower Milkvetch ■

42068 Astragalus batangensis E. Peter;巴塘黄耆(巴塘黄芪);Batang Milkvetch ■

42069 Astragalus battiscombei Baker f. = Galega battiscombei (Baker f.) J. B. Gillett ■☆

42070 Astragalus baxoiensis Podlech et L. R. Xu;八宿黄耆■

42071 Astragalus beckerianus Trautv.;拜克黄耆■☆

42072 Astragalus beketovii (Krasn.) B. Fedtsch.;斑果黄耆(斑果黄芪);Beketov Milkvetch ■

42073 Astragalus bequaertii De Wild. = Astragalus atropilosulus (Hochst.) Bunge var. bequaertii (De Wild.) J. B. Gillett ■☆

42074 Astragalus bhotanensis Baker;地八角(八角花,不丹黄芪,不丹黄耆,地皂角,旱皂角,黄鳝草,球花紫云英,土牛膝);Bhotan Milkvetch ■

42075 Astragalus bhotanensis Baker var. minor Pamp. = Astragalus bhotanensis Baker ■

42076 Astragalus bhotanensis Baker var. montigenus Hand. -Mazz. = Astragalus bhotanensis Baker ■

42077 Astragalus bicolor Lam.;二色黄耆■☆

42078 Astragalus bicuspis Fisch.;二尖齿黄耆(二尖齿黄芪);Bicusped Milkvetch, Bisharptooth Milkvetch, Two-cusp Milkvetch ●

42079 Astragalus bicuspis Fisch. = Astragalus multiceps Benth. ■

42080 Astragalus biflorus L. = Gueldenstaedtia verna (Georgi) Borissov ■

42081 Astragalus biondianus Ulbr. = Astragalus discolor Bunge ex Maxim. ■

42082 Astragalus bisulcatus (Hook.) A. Gray;双沟黄耆;Two-grooved Milk-vetch ■☆

42083 Astragalus bisulcatus A. Gray = Astragalus bisulcatus (Hook.) A. Gray ■☆

42084 Astragalus blandulus Podlech et L. R. Xu;温和黄耆■☆

42085 Astragalus bodinieri H. Lév. = Astragalus graveolens Benth. ■

42086 Astragalus boeticus L.;贝蒂黄耆(贝蒂黄芪);Swedish Coffee, Swedish-coffee ■☆

42087 Astragalus boeticus L. var. subinflatus Rouy = Astragalus boeticus L. ■☆

42088 Astragalus boissieri Fisch. = Astragalus granatensis Lam. ●☆

42089 Astragalus bombycinus Boiss.;丝质黄耆■☆

42090 Astragalus bomiensis C. C. Ni et P. C. Li;波密黄耆;Bomi Milkvetch ■

42091 Astragalus borodinii Krasn.;东天山黄耆(东天山黄芪);Borodin Milkvetch, E. Tianshan Milkvetch ■

42092 Astragalus bourgaeanus Coss.;布尔黄耆■☆

42093 Astragalus bourgaeanus Coss. var. adpressipilis Maire = Astragalus bourgaeanus Coss. ■☆

42094 Astragalus bourgaeanus Coss. var. heterotrichus Maire = Astragalus bourgaeanus Coss. ■☆

42095 Astragalus brachybius Stev.;短距黄耆(短距黄芪)■☆

42096 Astragalus brachycalyx Fisch. ex Boiss.;短萼黄耆(短萼黄芪)■☆

42097 Astragalus brachycarpus M. Bieb.;短果黄耆■☆

42098 Astragalus brachycephalus Franch. = Astragalus bhotanensis Baker ■

42099 Astragalus brachycephalus Franch. var. minor Pamp. = Astragalus bhotanensis Baker ■

42100 Astragalus brachylobus DC.;短裂黄耆■☆

42101 Astragalus brachymorphus Nikif. = Astragalus stalinskyi Sirj. ■

42102 Astragalus brachypetalus Trautv.;短瓣黄耆■☆

42103 Astragalus brachypus Schrenk;盐木黄耆■

42104 Astragalus brachysemia Podlech et L. R. Xu;短籽黄耆■

42105 Astragalus brachytrichus Podlech et L. R. Xu;短毛黄耆■

42106 Astragalus brachytropis (DC.) C. A. Mey. f. gigantea Lipsky = Astragalus peterae H. T. Tsai et T. T. Yu ■

42107 Astragalus bracteosus Boiss. et Noë = Astragalus melanostachys Benth. ex Bunge ■

42108 Astragalus bracteosus Klotzsch = Astragalus melanostachys Benth. ex Bunge ■

42109 Astragalus brahuicus Bunge = Astragalus affghanus Boiss. ■☆

42110 Astragalus brevialatus H. T. Tsai et Te T. Yu;短翼黄耆(短翼黄芪);Shortwing Milkvetch ■

42111 Astragalus brevicarinatus DC. = Gueldenstaedtia verna (Georgi) Borissov ■

42112 Astragalus brevidens Freyn et Sint.;短齿黄耆■☆

42113 Astragalus brevifolius Ledeb.;短叶黄耆■☆

42114 Astragalus breviscapus B. Fedtsch.;短梗黄耆■

42115 Astragalus brevivexillatus Podlech et L. R. Xu;短旗瓣黄耆■

42116 Astragalus brunetianus (Fernald) Rousseau;布氏黄耆;Brunet's Milk-vetch ■☆

42117 Astragalus bubaloceras Maire = Astragalus solandri Lowe ■☆

42118 Astragalus bucharicus Regel.;布哈尔黄耆■☆

42119 Astragalus buchtormensis Pall.;布河黄耆(布河黄芪);

Buchtorm Milkvetch ■

42120 Astragalus bungeanus Boiss.;布吉氏黄耆;Bunge Milkvetch ■☆

42121 Astragalus burchan-buddaicus N. Ulziykh.;布尔卡黄耆■

42122 Astragalus burkeanus Harv. = Astragalus atropilosulus (Hochst.) Bunge var. burkeanus (Harv.) J. B. Gillett ■☆

42123 Astragalus burqinensis Podlech et L. R. Xu;布尔津黄耆■

42124 Astragalus burtschumensis Sumnev.;布尔楚黄耆■

42125 Astragalus caerulea Pall. = Oxytropis caerulea (Pall.) DC. ■

42126 Astragalus caeruleopetalinus Y. C. Ho;蓝花黄耆(蓝花黄芪);Blue Milkvetch,Blue-petal Milkvetch ■

42127 Astragalus caeruleopetalinus Y. C. Ho var. glabricarpus Y. C. Ho;光果蓝花黄耆(光果蓝花黄芪);Glabrous-fruit Milkvetch, Smoothcarp Blue Milkvetch ■

42128 Astragalus caerulescens Chiov. = Astragalus atropilosulus (Hochst.) Bunge var. coerulescens (Chiov.) J. B. Gillett ■☆

42129 Astragalus caeruleus H. T. Tsai et Te T. Yu = Astragalus caeruleopetalinus Y. C. Ho ■

42130 Astragalus calycosus Torr. ex S. Watson;灰萼黄耆;Gray Loco ■☆

42131 Astragalus campanulatus R. Br.;扁茎黄耆(沙蒺藜,外沙苑)■☆

42132 Astragalus camptodontoides N. D. Simpson;类芒齿黄耆(类芒齿黄芪);Bentdentate-like Milkvetch,Sham Bendtooth Milkvetch ■

42133 Astragalus camptodontus Franch.;弯齿黄耆(弯齿黄芪);Bendtooth Milkvetch,Bentdentate Milkvetch ■

42134 Astragalus camptodontus Franch. var. lichiangensis (Simpson) K. T. Fu;丽江黄耆(丽江黄芪);Lijiang Milkvetch ■

42135 Astragalus campylorhynchus Fisch. et C. A. Mey.;弯喙黄耆(弯喙黄芪);Bendbeak Milkvetch ■

42136 Astragalus campylotrichus Bunge;弯毛黄耆■☆

42137 Astragalus canadensis L.;加拿大黄耆;Canadian Milk Vetch, Canadian Milk-vetch,Milk-vetch,Rattle Weed ■

42138 Astragalus canadensis L. var. carolinianus (L.) M. E. Jones = Astragalus canadensis L. ■

42139 Astragalus canadensis L. var. longilobus Fassett = Astragalus canadensis L. ■

42140 Astragalus candidissimus Ledeb.;亮白黄耆(亮白黄芪);Shining-white Milkvetch,Verywhite Milkvetch ■

42141 Astragalus candidissimus Ledeb. var. pauciflorus Krylov et Sarg. = Astragalus steinhergianus Sumner ■

42142 Astragalus candolleanus Benth. = Astragalus rhizanthus Benth. subsp. candolleanus (Benth.) Podlech ■

42143 Astragalus candolleanus Benth. var. pindreensis Baker = Astragalus pindreensis (Baker) Ali ■

42144 Astragalus candolleanus Royle;康多勒黄耆■☆

42145 Astragalus capillipes Fisch. ex Bunge;草珠黄耆(草珠黄芪,毛细柄黄芪,毛细柄黄耆);Hairstalk Milkvetch ■

42146 Astragalus caprinus L.;山羊黄耆■☆

42147 Astragalus caprinus L. subsp. alexandrinus Boiss. = Astragalus caprinus L. ■☆

42148 Astragalus caprinus L. subsp. glaber (DC.) Podlech;光山羊黄耆■☆

42149 Astragalus caprinus L. subsp. lanigerus (Desf.) Maire = Astragalus caprinus L. ■☆

42150 Astragalus caprinus L. var. brevifolius Maire = Astragalus caprinus L. ■☆

42151 Astragalus caprinus L. var. dictyocarpus Pomel = Astragalus caprinus L. ■☆

42152 Astragalus caprinus L. var. glaber (DC.) Pomel = Astragalus caprinus L. subsp. glaber (DC.) Podlech ■☆

42153 Astragalus caprinus L. var. macrocarpus Maire = Astragalus caprinus L. ■☆

42154 Astragalus caprinus L. var. reboudii (Coss.) Maire = Astragalus caprinus L. ■☆

42155 Astragalus caprinus L. var. subglabratus DC. = Astragalus caprinus L. ■☆

42156 Astragalus carolinianus L.;卡罗来纳黄耆;Buffalo Bean, Canadian Milk Vetch,Ground Plum,Little Rattlepod ■☆

42157 Astragalus carolinianus L. = Astragalus canadensis L. ■

42158 Astragalus caryocarpus Ker Gawl. = Astragalus crassicarpus Nutt. ■☆

42159 Astragalus caucasicus Pall.;高加索黄耆■☆

42160 Astragalus caudiculosus Kom. = Astragalus nematodioides H. Ohba,Akiyama et S. K. Wu ■

42161 Astragalus cavaleriei H. Lév. = Gueldenstaedtia delavayi Franch. ■

42162 Astragalus cavaleriei H. Lév. = Gueldenstaedtia verna (Georgi) Borissov ■

42163 Astragalus centraligobicus Z. Y. Chu et Y. Z. Zhao = Astragalus hamiensis S. B. Ho ■

42164 Astragalus cephalotes Pall.;头花黄耆■☆

42165 Astragalus cerasocrenus Bunge;小樱黄耆(小樱黄芪)■☆

42166 Astragalus ceratoides M. Bieb.;角黄耆(角黄芪);Horn Milkvetch,Hornlike Milkvetch ■

42167 Astragalus ceratoides M. Bieb. var. montanus Ledeb. = Astragalus ceratoides M. Bieb. ■

42168 Astragalus chadjanensis Franch. = Astragalus tibetanus Bunge ■

42169 Astragalus chagyabensis P. C. Li et C. C. Ni;察雅黄耆(察雅黄芪);Chaya Milkvetch ■

42170 Astragalus chaidamuensis (S. B. Ho) Podlech et L. R. Xu;柴达木黄耆(柴达木黄芪);Chaidamu Milkvetch ■

42171 Astragalus chamaephyton Podlech et L. R. Xu;低矮黄耆■

42172 Astragalus changduensis Y. C. Ho;昌都黄耆(昌都黄芪);Changdu Milkvetch ■

42173 Astragalus changmuicus C. C. Ni et P. C. Li;樟木黄耆(樟木黄芪);Changmu Milkvetch,Zhangmu Milkvetch ■

42174 Astragalus charguschanus Freyn;卡尔古斯黄耆■

42175 Astragalus chengkangensis Podlech et L. R. Xu;成康黄耆■

42176 Astragalus chilienshanensis Y. C. Ho;祁连山黄耆(祁连山黄芪);Chilienshan Milkvetch,Qilianshan Milkvetch ■

42177 Astragalus chinensis L. f.;华黄耆(地黄芪,地黄耆,华黄芪,忙牛花,芒牛旦,沙苑子,天津沙苑子,中国黄耆);China Milkvetch, Chinese Milkvetch ■

42178 Astragalus chingianus E. Peter;鄂尔多斯黄耆(鄂尔多斯黄芪,华黄耆,秦氏黄芪,仁昌黄芪);Ching's Milkvetch ■

42179 Astragalus chingianus E. Peter = Astragalus alaschanus Bunge ■

42180 Astragalus chingianus P. C. Li = Astragalus chingianus E. Peter ■

42181 Astragalus chingianus Stibal = Astragalus alaschanus Bunge ex Maxim. ■

42182 Astragalus chionanthus Popov = Astragalus olgae Bunge ■

42183 Astragalus chiukiangensis H. T. Tsai et Te T. Yu;俅江黄耆(俅江黄芪);Chiukiang Milkvetch,Qiujiang Milkvetch ■

42184 Astragalus chlorocyaneus Boiss. et Reut. = Astragalus monspessulanus L. subsp. gypsophilus Rouy ■☆

42185 Astragalus chlorostachys Lindl.;绿穗黄耆■

42186 Astragalus chomutovii B. Fedtsch.;中天山黄耆(中山黄芪,中天山黄芪);Choumutov Milkvetch ■

42187 Astragalus chorgosicus Lipsky;霍尔果斯黄耆●

42188 Astragalus chrysopterus Bunge = Astragalus chrysopterus Bunge ex Maxim. ■

42189 Astragalus chrysopterus Bunge ex Maxim.;金翼黄耆(金黄芪,金翼黄芪,小白芪,小黄芪);Goldenwing Milkvetch ■

42190 Astragalus chrysopterus Bunge ex Maxim. var. pilosus S. C. Yueh ex Z. Q. He;毛果金翼黄耆(毛果金翼黄芪)■

42191 Astragalus chrysopterus Bunge ex Maxim. var. wutaicus Hand. -Mazz. = Astragalus chrysopterus Bunge ex Maxim. ■

42192 Astragalus chudaei Batt. et Trab. = Astragalus eremophilus Boiss. ■☆

42193 Astragalus cicer L.;山黧豆黄耆;Chickpea Milk-Vetch, Chick-pea Vetch, Cicer, Cicer Milkvetch, Mountain Chick-pea, Wild Lentil ■☆

42194 Astragalus cicerifolius Bunge = Astragalus oplites Parker ■

42195 Astragalus cicerifolius Royle ex Bunge = Astragalus oplites Benth. ex Parker ■

42196 Astragalus cinerascens H. T. Tsai et Te T. Yu = Astragalus batangensis E. Peter ■

42197 Astragalus clerceanus Iljin et Krasch.;克乐西黄耆■☆

42198 Astragalus clulanensis Y. H. Wu;都兰黄耆(都兰黄芪);Dulan Milkvetch ■

42199 Astragalus cobresiiphilus Podlech et L. R. Xu;雅鲁黄耆■

42200 Astragalus coelestis Diels = Tibetia coelestis (Diels) H. P. Tsui ■

42201 Astragalus coelestis Diels = Tibetia yunnanensis (Franch.) H. P. Tsui var. coelestis (Diels) X. Y. Zhu ■

42202 Astragalus coeruleus Pall. = Oxytropis coerulea (Pall.) DC. ■

42203 Astragalus cognatus C. A. Mey.;沙丘黄芪;Dune Milkvetch, Sand-dune Milkvetch ■

42204 Astragalus cognatus C. A. Mey. var. brachybotrys Trautv. = Astragalus cognatus Schrenk ■

42205 Astragalus cognatus Schrenk = Astragalus cognatus C. A. Mey. ■

42206 Astragalus coluteocarpus Boiss.;膀胱果黄耆■☆

42207 Astragalus commixitus Bunge;混合黄耆(混合黄芪);Mix Milkvetch, Mix-together Milkvetch ■

42208 Astragalus complanatus R. Br. ex Bunge;背扁黄耆(背扁黄芪,扁茎黄芪,大沙苑,蔓黄芪,沙蒺藜,沙苑白蒺藜,沙苑蒺藜,沙苑子,梭果黄芪,同州白蒺藜,潼蒺藜,夏黄草,夏黄芪,夏黄耆);Flat Milkvetch, Flatstem Milkvetch ■

42209 Astragalus complanatus R. Br. ex Bunge var. eutrichus Hand. -Mazz.;真毛黄耆(真毛黄芪);Hairy Milkvetch ■

42210 Astragalus compositus Pavlov;复合黄耆■☆

42211 Astragalus compressus Ledeb.;扁序黄耆(扁序黄芪);Compressed Milkvetch, Flatraceme Milkvetch ■

42212 Astragalus conaensis Podlech et L. R. Xu;错那黄耆■

42213 Astragalus concinnus Benth. ex Bunge;整洁黄耆■☆

42214 Astragalus concretus Benth.;合生黄耆■

42215 Astragalus confertus Benth. et Bunge f. albiflorus R. F. Huang et Y. H. Wu;白花丛生黄耆(白花丛生黄芪)■

42216 Astragalus confertus Benth. ex Bunge;丛生黄耆(丛生黄芪);Conferted Milkvetch, Crowded Milkvetch ■

42217 Astragalus consanguineus Bong.;亚黄耆■

42218 Astragalus contortuplicatus L.;环荚黄耆(环荚黄芪);Hungarian Milkvetch, Ringpod Milkvetch, Twisted Milkvetch ■

42219 Astragalus cooperi A. Gray = Astragalus neglectus (Torr. et A. Gray) E. Sheld. ■☆

42220 Astragalus corniculatus M. Bieb.;小角黄耆■☆

42221 Astragalus cornutus Pall.;角形黄耆;Cornute Milkvetch ■☆

42222 Astragalus coronilloides Ulbr. = Astragalus chrysopterus Bunge ex Maxim. ■

42223 Astragalus corrugatus Bertol.;皱褶黄耆■☆

42224 Astragalus corrugatus Bertol. = Astragalus crenatus Schult. ■☆

42225 Astragalus corrugatus Bertol. subsp. tenuirugis (Boiss.) Eig = Astragalus crenatus Schult. ■☆

42226 Astragalus corrugatus Bertol. var. tenuirugis (Boiss.) Coss. et Kralik = Astragalus crenatus Schult. ■☆

42227 Astragalus corydalinus Bunge = Astragalus peduncularis Benth. ■☆

42228 Astragalus craibianus N. D. Simpson;川西黄耆(川西黄芪);Craib Milkvetch ■

42229 Astragalus craibianus Simpson var. baimashanensis C. Chen et Z. G. Qian;无毛川西黄耆(无毛川西黄芪);Baimashan Milkvetch ■

42230 Astragalus craibianus Simpson var. baimashanensis C. Chen et Z. G. Qian = Astragalus craibianus N. D. Simpson ■

42231 Astragalus crassicarpus Nutt.;厚果黄耆;Bladderpod Locoweed, Buffalo Bean, Ground Plum, Ground-plum, Ground-plum Milk-Vetch, Prairie-plum ■☆

42232 Astragalus crassicaulis Graham = Chesneya nubigena (D. Don) Ali ●☆

42233 Astragalus crassiffolius Ulbr.;厚叶黄耆(厚叶黄芪);Thickleaf Milkvetch ■

42234 Astragalus crenatus Schult.;圆齿黄耆■☆

42235 Astragalus creticus Lam. var. boissieri (Fisher) Pau = Astragalus granatensis Lam. ●☆

42236 Astragalus cruciatus Link;十字形黄耆(十字形黄芪);Crisscross Milkvetch, Cross-shaped Milkvetch ■

42237 Astragalus cruciatus Link = Astragalus corrugatus Bertol. ■☆

42238 Astragalus cruciatus Link subsp. aristidis Batt. = Astragalus saharae Pomel ■☆

42239 Astragalus cruciatus Link subsp. astraboides (Pomel) Batt. = Astragalus longicaulis Pomel ■☆

42240 Astragalus cruciatus Link subsp. linkeanus Maire = Astragalus stella Gouan ■☆

42241 Astragalus cruciatus Link subsp. radians (Pomel) Batt. = Astragalus longicaulis Pomel ■☆

42242 Astragalus cruciatus Link subsp. radiatus Batt. = Astragalus asterias Hohen. ■☆

42243 Astragalus cruciatus Link subsp. trabutianus Batt. = Astragalus longicaulis Pomel ■☆

42244 Astragalus cruciatus Link var. aster Jahand. et Maire = Astragalus saharae Pomel ■☆

42245 Astragalus cruciatus Link var. garamantum Maire = Astragalus corrugatus Bertol. ■☆

42246 Astragalus cruciatus Link var. longicaulis (Pomel) Batt. = Astragalus longicaulis Pomel ■☆

42247 Astragalus cruciatus Link var. polyactinus (Boiss.) Hochr. = Astragalus stella Gouan ■☆

42248 Astragalus cruciatus Link var. pterolobus Maire = Astragalus longicaulis Pomel ■☆

42249 Astragalus cruciatus Link var. radians (Pomel) Emb. et Maire = Astragalus longicaulis Pomel ■☆

42250 Astragalus culciformis P. C. Li et C. C. Ni = Astragalus arnoldii Hemsl. et H. Pearson ■

42251 Astragalus cupulicalycinus S. B. Ho et Y. C. Ho;杯萼黄耆(杯萼黄芪);Cupcalyx Milkvetch, Cupilar-calyx Milkvetch ■

42252 Astragalus cymbaecarpos Brot.;舟果黄耆■☆

42253 Astragalus cymbaecarpos Brot. var. brevipes Willk. = Astragalus cymbaecarpos Brot. ■☆

42254 Astragalus cyrenaicus Coss. = Astragalus graecus Boiss. et Spruner ■☆

42255 Astragalus cysticalyx Ledeb. ;囊萼黄耆(囊萼黄芪);Bagcalyx Milkvetch,Saccate-calyx Milkvetch,Saccate-calyxed Milkvetch ●

42256 Astragalus cytisodes Bunge;金雀黄耆(金雀黄芪)■☆

42257 Astragalus dabanshanicus Y. H. Wu;大板山黄耆(大板山黄芪);Dabanshan Milkvetch ■

42258 Astragalus dahuricus (Pall.) DC. ;达乌里黄耆(达呼里黄芪,达乌里黄芪,驴干粮,兴安黄芪,兴安黄耆,鸭食花,野豆角花);Dahur Milkvetch,Dahurian Milkvetch ■

42259 Astragalus dahuricus (Pall.) DC. f. albiflorus H. Wei Jen et Y. J. Chang;白花达乌里黄耆;Whiteflower Dahurian Milkvetch ■

42260 Astragalus dalaiensis Kitag. ;草原黄耆(草原黄芪);Grassland Milkvetch ■

42261 Astragalus damxungensis Podlech et L. R. Xu;当雄黄耆■

42262 Astragalus danicus Retz. ;丹麦黄耆(丹黄芪,丹黄耆,丹麦黄芪,旱生黄耆,沙地黄耆,舌下黄芪,舌下黄耆);Purple Cock's Head,Purple Milk Vetch,Purple Milkvetch,Purple Milk-vetch,Sandy Milkvetch,Tongue-under-tongue ■

42263 Astragalus daqingshanicus Z. G. Jiang et Z. T. Yin;大青山黄耆(大青山黄芪);Daqingshan Milkvetch ■

42264 Astragalus dasyanthus Pall. ;毛花黄耆(毛花黄芪);Hairy-flowered Milkvetch ■☆

42265 Astragalus dasycephalus Besser ex Stev. = Astragalus roseus Ledeb. ■

42266 Astragalus dasyglottis DC. = Astragalus agrestis G. Don ■

42267 Astragalus dasyglottis Fisch. ex DC. = Astragalus agrestis G. Don ■

42268 Astragalus datunensis Y. C. Ho;大通黄耆(大通黄芪);Datong Milkvetch,Datung Milkvetch ■

42269 Astragalus davidii Franch. ;宝兴黄耆(宝兴黄芪);David Milkvetch ■

42270 Astragalus davidii Franch. var. acutidentatus P. C. Li;尖齿宝兴黄耆;Sharp-toothed David Milkvetch ■

42271 Astragalus davidii Franch. var. acutidentatus P. C. Li = Astragalus sutchuensis Franch. ■☆

42272 Astragalus davuricus (Pall.) DC. = Astragalus dahuricus (Pall.) DC. ■

42273 Astragalus decumbens Kom. = Astragalus polycladus Bureau et Franch. ■

42274 Astragalus deflexus Pall. = Oxytropis deflexa (Pall.) DC. ■

42275 Astragalus degensis Ulbr. ;窄翼黄耆(窄翼黄芪);Deg Milkvetch,Narrowwing Milkvetch ■

42276 Astragalus degensis Ulbr. var. rockianus E. Peter;大花窄翼黄耆(大花窄叶黄芪,大花窄翼黄芪);Largeflower Narrowwing Milkvetch ■

42277 Astragalus dendroides Kar. et Kir. ;树黄耆(树黄芪);Tree-like Milkvetch,Wood Milkvetch ●

42278 Astragalus dengkouensis H. C. Fu;荒漠黄耆(荒漠黄芪,宁夏黄芪)■

42279 Astragalus dengkouensis H. C. Fu = Astragalus grubovii Sanchir ■

42280 Astragalus densiflorus Kar. et Kir. ;密花黄耆(密花黄芪);Denseflower Milkvetch,Flowery Milkvetch ■

42281 Astragalus densiflorus Kar. et Kir. var. konlonicus H. Ohba, S. Akiyama et S. K. Wu = Astragalus mieheorum Podlech et L. R. Xu ■

42282 Astragalus depauperatus Ledeb. ;疆西北黄耆■

42283 Astragalus dependens Bunge ex Maxim. ;悬垂黄耆(悬垂黄芪);Hang Milkvetch,Hanging-down Milkvetch ■

42284 Astragalus dependens Bunge ex Maxim. var. aurantiacus (Hand. -Mazz.) Y. C. Ho;橙黄花黄耆(橙黄花黄芪)■

42285 Astragalus dependens Bunge ex Maxim. var. flavescens Y. C. Ho;黄白悬垂黄耆(黄白花黄芪,黄白悬垂黄芪);Yellowish Hanging-down Milkvetch ■

42286 Astragalus dependens Bunge ex Maxim. var. flavescens Y. C. Ho = Astragalus aurantiacus Hand. -Mazz. ■

42287 Astragalus dependens Bunge ex Maxim. var. sericeus K. T. Fu;绢毛黄耆;Sericeous Hanging-down Milkvetch ■

42288 Astragalus dependens Bunge ex Maxim. var. sericeus K. T. Fu = Astragalus aurantiacus Hand. -Mazz. ■

42289 Astragalus dependens Bunge var. aurantiacus (Hand. -Mazz.) Y. C. Ho = Astragalus aurantiacus Hand. -Mazz. ■

42290 Astragalus depressus L. ;平扁黄耆(平扁黄芪)■

42291 Astragalus depressus L. subsp. atlantis Maire;大西洋黄耆■☆

42292 Astragalus despectus Podlech et L. R. Xu;合托叶黄耆■

42293 Astragalus dicystis Bunge = Astragalus lithophilus Kar. et Kir. ■

42294 Astragalus dilutus Bunge;浅黄耆(淡黄芪,浅黄芪);Paleyellow Milkvetch ■

42295 Astragalus dilutus Bunge = Astragalus ellipsoideus Ledeb. ■

42296 Astragalus dingjiensis C. C. Ni et P. C. Li;定结黄耆(定结黄芪);Dingjie Milkvetch ■

42297 Astragalus discolor Bunge = Astragalus discolor Bunge ex Maxim. ■

42298 Astragalus discolor Bunge ex Maxim. ;灰叶黄耆(灰叶黄芪);Differentcolour Milkvetch,Greyleaf Milkvetch ■

42299 Astragalus distortus Torr. et A. Gray;旋扭黄耆;Bent Milk Vetch,Ozark Milk Vetch ●☆

42300 Astragalus divnogorskajae N. Ulziykh. ;疆西黄耆■

42301 Astragalus dolichocllaete Diels;芒齿黄耆(芒齿黄芪);Awn-tooth Milkvetch,Longseta Milkvetch ■

42302 Astragalus dolichophyllus Pall. ;长叶黄耆(长叶黄芪);Long-leaved Milkvetch ■☆

42303 Astragalus donianus DC. ;亚东黄耆(亚东黄芪);Yadong Milkvetch,Yatung Milkvetch ■

42304 Astragalus dschangartensis Sumner;詹加尔特黄耆(詹加尔特黄芪);Changart Milkvetch,Dschangar Milkvetch ■

42305 Astragalus dschinensis Gontsch. ;边垂黄耆■

42306 Astragalus dsharkenticus Popov;托木尔黄耆(托木尔黄芪);Dsharkent Milkvetch ■

42307 Astragalus dsharkenticus Popov var. gongliuensis S. B. Ho = Astragalus dsharkenticus Popov ■

42308 Astragalus dshimensis Gontsch. = Astragalus hoantchy Franch. subsp. dshimensis (Gontsch.) K. T. Fu ■

42309 Astragalus duclouxii N. D. Simpson = Astragalus khasianus Bunge ■

42310 Astragalus duclouxii Simpson = Astragalus englerianus Ulbr. ■

42311 Astragalus dulanensis Y. H. Wu = Astragalus variabilis Bunge ex Maxim. ■

42312 Astragalus dulungkiangensis P. C. Li;独龙黄耆(独龙黄芪,独龙江黄耆);Dulong Milkvetch ■

42313 Astragalus dumetorum Hand. -Mazz. ;灌丛黄耆(灌丛黄芪);Shrub Milkvetch,Ushy Milkvetch ■

42314 Astragalus dutreuilii (Franch.) Grubov et N. Ulziykh. ;中昆仑黄耆■

42315 Astragalus echinatus Murray;刺黄耆■☆

42316 Astragalus echinatus Murray var. pentaglottis (L.) Maire =

Astragalus echinatus Murray ■☆
42317 Astragalus echinatus Murray var. stenorrhinus (Pau) Maire = Astragalus echinatus Murray ■☆
42318 Astragalus edulis Bunge;可食黄耆■☆
42319 Astragalus efoliatus Hand. -Mazz.;单叶黄耆(单叶黄芪);Simpleleaf Milkvetch,Singleleaf Milkvetch ■
42320 Astragalus elatus Boiss. et Balansa;高黄耆(高黄芪)■☆
42321 Astragalus elgonensis Bullock = Astragalus atropilosulus (Hochst.) Bunge var. elgonensis (Bullock) J. B. Gillett ■☆
42322 Astragalus ellipsoideus Ledeb.;胀萼黄耆(胀萼黄芪);Elliptoid Milkvetch,Turgid-calyx Milkvetch ■
42323 Astragalus ellipsoideus Ledeb. = Astragalus arkalycensis Bunge ■
42324 Astragalus ellipsoideus Ledeb. var. abbreviatus Ledeb. = Astragalus arkalycensis Bunge ■
42325 Astragalus ellipsoideus Ledeb. var. kuldshensis Basil. = Astragalus ellipsoideus Ledeb. ■
42326 Astragalus embergeri Jahand. et Maire et Weiller = Astragalus hamosus L. subsp. embergeri (Jahand. et al.) Maire ■☆
42327 Astragalus englerianus Ulbr.;长果柄黄耆(长果柄黄芪,长果茎黄芪,长果颈黄芪,恩氏黄芪,恩氏黄耆,野黄芪,野黄耆);Engler Milkvetch ■
42328 Astragalus englerianus Ulbr. = Astragalus khasianus Bunge ■
42329 Astragalus englerianus Ulbr. var. gingdongensis Z. G. Qian;景东长果柄黄耆(景东长果颈黄芪)■
42330 Astragalus englerianus Ulbr. var. gingdongensis Z. G. Qian = Astragalus khasianus Bunge ■
42331 Astragalus englerianus Ulbr. var. longiflorus C. Chen et Z. G. Qian;长花长果柄黄耆(长花长果柄黄芪);Longflower Engler Milkvetch ■
42332 Astragalus englerianus Ulbr. var. longiflorus C. Chen et Z. G. Qian = Astragalus khasianus Bunge ■
42333 Astragalus enneaphyllus P. C. Li;九叶黄耆(九叶黄芪)■
42334 Astragalus ephippium Pomel = Astragalus epiglottis L. ■☆
42335 Astragalus ephippium Pomel var. ephippium (Pomel) Murb. = Astragalus epiglottis L. ■☆
42336 Astragalus epiglottis L.;马鞍黄耆■☆
42337 Astragalus epiglottis L. subsp. asperulus (Dufour) Nyman = Astragalus epiglottis L. ■☆
42338 Astragalus epiglottis L. var. ephippium (Pomel) Maire = Astragalus epiglottis L. ■☆
42339 Astragalus epiglottis L. var. intermedius Faure et Maire = Astragalus epiglottis L. ■☆
42340 Astragalus epiglottis L. var. longipes Lange = Astragalus epiglottis L. ■☆
42341 Astragalus epiglottis L. var. pedunculatus Sennen et Mauricio = Astragalus epiglottis L. ■☆
42342 Astragalus epiglottoides Willk. = Astragalus epiglottis L. ■☆
42343 Astragalus eremophilus Boiss.;沙漠黄耆■☆
42344 Astragalus eremophilus Boiss. var. astacurus Maire = Astragalus eremophilus Boiss. ■☆
42345 Astragalus eremophilus Boiss. var. falcinellus ?;镰沙漠黄耆■☆
42346 Astragalus eremothamnus Kar. et Kir. = Astragalus arbuscula Pall. ●
42347 Astragalus eriolobus Bunge = Astragalus depauperatus Ledeb. ■
42348 Astragalus ernestii Comber;梭果黄耆(棱果黄芪,梭果黄芪);Ernest Milkvetch ■
42349 Astragalus ervoides Turcz. = Astragalus miniatus Bunge ■
42350 Astragalus euchlorus K. T. Fu;深绿黄耆(深绿黄芪);Darkgreen Milkvetch ■
42351 Astragalus eugeniae B. Fedtsch. = Astragalus mongutensis Lipsky ■
42352 Astragalus exscapus L.;欧洲无茎黄耆■☆
42353 Astragalus exscapus L. subsp. maurus Humbert et Maire = Astragalus maurus (Humbert et Maire) Pau ■☆
42354 Astragalus exscapus L. subsp. mesatlanticus ? = Astragalus maurus (Humbert et Maire) Pau ■☆
42355 Astragalus falcatus Lam.;镰叶黄耆;Falcate Milkvetch,Russian Milkvetch ■☆
42356 Astragalus falciformis Desf.;镰形黄耆■☆
42357 Astragalus falconeri Bunge;侧扁黄耆(侧扁黄芪);Falconer Milkvetch ■
42358 Astragalus falconeri Bunge var. paucistrigosus K. T. Fu;伏毛黄耆(伏毛黄芪,少毛侧扁黄芪);Few-strigose Milkvetch ■
42359 Astragalus falconeri Bunge var. paucistrigosus K. T. Fu = Astragalus falconeri Bunge ■
42360 Astragalus fangensis N. D. Simpson;房县黄耆(房县黄芪);Fangxian Milkvetch ■
42361 Astragalus fatmensis Hochst. ex Chiov.;法蒂玛黄耆■☆
42362 Astragalus faurei Maire;福雷黄耆■☆
42363 Astragalus fenzelianus E. Peter;西北黄耆(西北黄芪);Fenzel Milkvetch ■
42364 Astragalus fenzelianus E. Peter = Astragalus yunnanensis Franch. ■
42365 Astragalus filicaulis Fisch. et C. A. Mey. ex Kar. = Astragalus filicaulis Kar. et Kir. ■
42366 Astragalus filicaulis Kar. et Kir.;丝茎黄耆(丝茎黄芪);Filiform-stem Milkvetch,Silkstem Milkvetch ■
42367 Astragalus filicaulis Kar. et Kir. subsp. rytilobus (Bunge) Popov = Astragalus filicaulis Kar. et Kir. ■
42368 Astragalus filidens Podlech et L. R. Xu;丝齿黄耆■
42369 Astragalus flavovirens K. T. Fu;黄绿黄耆(黄绿黄芪);Yellowgreen Milkvetch ■
42370 Astragalus flemingii Ali;福来明黄耆■☆
42371 Astragalus flexus Fisch.;弯花黄耆(弯花黄芪);Bendflower Milkvetch,Flexed-flower Milkvetch ■
42372 Astragalus floccosifolius Sumnev.;丛毛叶黄耆■
42373 Astragalus floribundus Pall. = Oxytropis floribunda (Pall.) DC. ■
42374 Astragalus floridulus Podlech;多花黄耆(多花黄芪);Flowery Milkvetch,Manyflower Milkvetch ■
42375 Astragalus floridus Benth. ex Bunge = Astragalus floridulus Podlech ■
42376 Astragalus floridus Benth. ex Bunge var. multipilus Y. H. Wu;多毛多花黄耆(多毛多花黄芪);Manyhair Flowery Milkvetch ■
42377 Astragalus fontanesii Coss. et Durieu = Coronilla valentina L. ●☆
42378 Astragalus fontanesii Coss. et Durieu subsp. numidicus (Coss. et Durieu) Maire = Astragalus armatus Willd. subsp. numidicus (Murb.) Tietz ■☆
42379 Astragalus fontanesii Coss. et Durieu subsp. tragacanthoides (Desf.) Maire = Astragalus armatus Willd. ■☆
42380 Astragalus fontianus Maire = Astragalus incanus L. subsp. nummularioides (Desf.) Maire ■☆
42381 Astragalus fontqueri Maire et Sennen;丰特黄耆■☆
42382 Astragalus forrestii N. D. Simpson;中甸黄耆(密花争黄芪,中甸黄芪);Forrest Milkvetch,Zhongdian Milkvetch ■
42383 Astragalus forrestii N. D. Simpson var. minor H. T. Tsai et Te T. Yu = Astragalus kialensis N. D. Simpson ■

42384 Astragalus frigidus (L.) A. Gray;广布黄耆(广布黄芪);Blazon Milkvetch, Cold Milkvetch, Yellow Alpine Milk Vetch ■

42385 Astragalus frigidus (L.) A. Gray subsp. parviflorus (Turcz.) Hultén;小花广布黄耆■☆

42386 Astragalus froedinii Murb.;弗勒丁黄耆■☆

42387 Astragalus fruticosus Forssk. subsp. gombo (Bunge) Jafri = Astragalus gombo Bunge ■☆

42388 Astragalus fruticosus Pall.;丛枝黄耆■☆

42389 Astragalus fujisanensis Miyabe et Tatew. = Astragalus laxmannii Jacq. ■

42390 Astragalus fukangensis Podlech et L. R. Xu;阜康黄耆■

42391 Astragalus galactites Pall.;乳白黄耆(白花黄芪,白花黄耆,乳白花黄芪,乳白花黄耆,乳白黄芪);Milk-white Milkvetch, Milky Milkvetch ■

42392 Astragalus galegiformis L.;山羊豆黄耆■☆

42393 Astragalus gautieri Batt. et Trab. = Astragalus vogelii (Webb) Bornm. ■☆

42394 Astragalus gebleri Bong. = Astragalus gebleri Fisch. ex Bong. ■☆

42395 Astragalus gebleri Fisch. ex Bong.;准噶尔黄耆(准噶尔黄芪);Dzungar Milkvetch ■☆

42396 Astragalus geerwusuensis H. C. Fu;格尔乌苏黄耆(格尔乌苏黄芪)■

42397 Astragalus geniculatus Desf.;膝曲黄耆■☆

42398 Astragalus geniorum Maire = Astragalus eremophilus Boiss. ■☆

42399 Astragalus genistoides Boiss. = Astragalus lasiopetalus Bunge ■

42400 Astragalus gerardianus Graham = Caragana gerardiana Benth. ●

42401 Astragalus gilgitensis Ali;吉尔吉特黄耆■☆

42402 Astragalus giraldianus Ulbr. = Astragalus scaberrimus Bunge ■

42403 Astragalus glaber Lam. = Oxytropis glabra (Lam.) DC. ■

42404 Astragalus glabritubus Podlech et L. R. Xu;秃萼筒黄耆■

42405 Astragalus gladiatus Boiss.;歧枝黄耆(剑叶黄芪,剑叶黄耆,歧枝黄芪);Forktwig Milkvetch ■

42406 Astragalus gladiatus Boiss. = Astragalus subuliformis DC. ■

42407 Astragalus glanduliferus Debeaux = Glycyrrhiza squamulosa Franch. ■

42408 Astragalus glauciformis Pomel = Astragalus glaux L. ■☆

42409 Astragalus glaucus M. Bieb.;灰兰黄耆;Blue-grey Milkvetch ■☆

42410 Astragalus glaux L.;银光黄耆■☆

42411 Astragalus glaux L. var. glauciformis (Pomel) Batt. = Astragalus glaux L. ■☆

42412 Astragalus glaux L. var. granatensis (Lange) Pau = Astragalus glaux L. ■☆

42413 Astragalus glaux L. var. macrocephalus Alleiz. = Astragalus glaux L. ■☆

42414 Astragalus glaux L. var. purpureus Maire = Astragalus glaux L. ■☆

42415 Astragalus glaux L. var. rostratus Ball = Astragalus glaux L. ■☆

42416 Astragalus glycyphylloides DC.;拟甜叶黄耆;Glycyphyllous-like Milkvetch ■☆

42417 Astragalus glycyphyllos L.;甜叶黄耆(草地黄芪);Black Milkwort, Fitsroot, Glycyphyllous Milkvetch, Licorice, Licorice Milkvetch, Liquorice Vetch, Milk Vetch, Milk-vetch, Sweet Milk Vetch, Wild Liquorice ■

42418 Astragalus gobi-altaicus N. Ulziykh. = Astragalus laguroides Pall. ■☆

42419 Astragalus golmuensis Y. C. Ho;格尔木黄耆(格尔木黄芪);Golmu Milkvetch ■

42420 Astragalus golmuensis Y. C. Ho var. paucipilus Y. H. Wu;少毛格尔木黄耆(少毛格尔木黄芪);Paucipilose Golmu Milkvetch ■

42421 Astragalus golmunensis Podlech et L. R. Xu var. paucipilus Y. H. Wu = Astragalus golmuensis Y. C. Ho ■

42422 Astragalus golubojensis Podlech et L. R. Xu = Astragalus pseudoversicolor Y. C. Ho ■

42423 Astragalus gombo Bunge = Astragalus gombo Coss. et Durieu ex Bunge ■☆

42424 Astragalus gombo Bunge subsp. gomboeformis (Pomel) Eug. Ott;戈姆黄耆■☆

42425 Astragalus gombo Bunge subsp. pseudogombo (Fern. Casas) Romo = Astragalus maurorum Murb. ■☆

42426 Astragalus gombo Bunge var. discedens Faure = Astragalus gombo Bunge ■☆

42427 Astragalus gombo Bunge var. gomboeformis (Pomel) Batt. = Astragalus gombo Bunge subsp. gomboeformis (Pomel) Eug. Ott ■☆

42428 Astragalus gombo Coss. et Durieu ex Bunge;戈波黄耆■☆

42429 Astragalus gomboeformis Pomel = Astragalus gombo Bunge subsp. gomboeformis (Pomel) Eug. Ott ■☆

42430 Astragalus gomboeformis Pomel var. amblyocarpus Maire = Astragalus gombo Bunge subsp. gomboeformis (Pomel) Eug. Ott ■☆

42431 Astragalus gomboeformis Pomel var. oranensis (Barratte) Maire = Astragalus akkensis Coss. ■☆

42432 Astragalus gonggamontis P. C. Li;贡嘎黄耆(贡嘎黄芪);Gongga Milkvetch ■

42433 Astragalus gonggamontis P. C. Li = Astragalus sutchuensis Franch. ■☆

42434 Astragalus gongliuensis Podlech et L. R. Xu;巩留黄耆●

42435 Astragalus gongshanensis Podlech et L. R. Xu;贡山黄耆■

42436 Astragalus gontscharovii Vassilcz.;半灌黄耆●

42437 Astragalus gracilidentatus S. B. Ho;纤齿黄耆(纤齿黄芪);Minitooth Milkvetch, Slender-tooth Milkvetch ■

42438 Astragalus gracilipes Benth. ex Bunge;细柄黄耆(细柄黄芪);Thin-stalk Milkvetch, Thinstipe Milkvetch ■

42439 Astragalus graecus Boiss. et Spruner;希腊黄耆■☆

42440 Astragalus grahamianus Fisch.;格雷厄姆黄耆■☆

42441 Astragalus granatensis Lam.;格拉黄耆●☆

42442 Astragalus granatensis Lam. subsp. maroccanus (Font Quer) Raynaud et Sauvage;摩洛哥格拉黄耆●☆

42443 Astragalus grandiflorus Pall. = Oxytropis grandiflora (Pall.) DC. ■

42444 Astragalus graveolens Benth.;烈香黄耆(烈香黄芪);Hot-scented Milkvetch, Mushfragrant Milkvetch ■

42445 Astragalus gregorii B. Fedtsch. et Basil.;格热高尔黄耆■

42446 Astragalus grubovii Sanchir;卵果黄耆(卵果黄芪);Ovate-fruited Milkvetch ■

42447 Astragalus grubovii Sanchir var. angustifolia H. C. Fu;细叶卵果黄耆(细叶卵果黄芪);Narrow-leaved Ovate-fruited Milkvetch ■

42448 Astragalus grum-grshimailoi Palibin;胶黄耆■

42449 Astragalus gryphus Bunge subsp. embergeri (Jahand. et al.) Maire = Astragalus hamosus L. ■☆

42450 Astragalus gryphus Bunge var. maroccanus Maire = Astragalus gryphus Bunge ■☆

42451 Astragalus gryphus Bunge var. oranensis Maire = Astragalus gryphus Bunge ■☆

42452 Astragalus guinanicus Y. H. Wu;贵南黄耆(贵南黄芪);Guinan Milkvetch ■

42453 Astragalus gummifer Labill.;西黄耆(胶黄芪,西黄芪,西黄芪胶树,真黄耆);Tragacanth Milk-vetch, Tragaeanth Milkvetch ●☆

42454 Astragalus gyzensis Delile = Astragalus hamosus L. ■☆
42455 Astragalus gyzensis Delile ex Boiss. = Astragalus hamosus L. ■☆
42456 Astragalus habaheensis Y. X. Liou;哈巴河黄耆●
42457 Astragalus habamontis K. T. Fu;哈巴山黄耆(哈巴山黄芪);Haba Mountain Milkvetch,Habashan Milkvetch ■
42458 Astragalus halei Rydb. = Astragalus canadensis L. ■
42459 Astragalus halodendron Bunge = Astragalus brachypus Schrenk ■
42460 Astragalus hamiensis S. B. Ho;哈密黄耆(哈密黄芪);Hami Milkvetch ■
42461 Astragalus hamosus L.;多钩黄耆;European Milkvetch,Hamose Milkvetch ■☆
42462 Astragalus hamosus L. subsp. brachyceras (Boiss.) Batt. = Astragalus hamosus L. ■☆
42463 Astragalus hamosus L. subsp. embergeri (Jahand. et al.) Maire;恩贝格尔黄耆■☆
42464 Astragalus hamosus L. subsp. verus Emb. et Maire = Astragalus hamosus L. ■☆
42465 Astragalus hamosus L. var. ancistron (Pomel) Batt. = Astragalus hamosus L. ■☆
42466 Astragalus hamosus L. var. ancocarpus (Pomel) Emb. et Maire = Astragalus hamosus L. ■☆
42467 Astragalus hamosus L. var. brevipes Faure et Maire = Astragalus hamosus L. ■☆
42468 Astragalus hamosus L. var. volubilitanus (Braun-Blanq. et Maire) Maire = Astragalus hamosus L. ■☆
42469 Astragalus hamulosus H. Lév. = Astragalus bhotanensis Baker ■
42470 Astragalus hancockii Bunge ex Maxim.;短花梗黄耆(短花梗黄芪);Hancock Milkvetch ■
42471 Astragalus handelii H. T. Tsai et Te T. Yu;头序黄耆(头序黄芪);Hondel Milkvetch ■
42472 Astragalus harmsii Ulbr. = Astragalus scaberrimus Bunge ■
42473 Astragalus hauarensis Boiss. var. brachycarpus (Širj. et Rech. f.) Ali;短果多钩黄耆■☆
42474 Astragalus havianus E. Peter;华山黄耆(华山黄芪);Huashan Milkvetch ■
42475 Astragalus havianus E. Peter var. pallidiflorus Y. C. Ho;白花华山黄耆(白花华山黄芪);Pallid-flower Huashan Milkvetch,Whiteflower Huashan Milkvetch ■
42476 Astragalus havianus E. Peter var. pallidiflorus Y. C. Ho = Astragalus havianus E. Peter ■
42477 Astragalus hebecarpus H. S. Cheng ex S. B. Ho;茸毛果黄耆(茸毛果黄芪,茸毛黄耆);Downy-fruit Milkvetch,Softhair Milkvetch ■
42478 Astragalus hedinii Ulbr.;鱼鳔黄耆(鱼鳔黄芪);Hedin Milkvetch,Sound Milkvetch ■
42479 Astragalus hedinii Ulbr. = Astragalus hoantchy Franch. ■
42480 Astragalus hegingensis Y. X. Liou = Astragalus mongutensis Lipsky ■
42481 Astragalus hejingensis Y. X. Liou;和靖黄耆(和靖黄芪);Hejing Milkvetch ■
42482 Astragalus hemiphaca Kar. et Kir.;扁豆黄耆(扁豆黄芪)■☆
42483 Astragalus hendersonii Baker;绒毛黄耆(绒毛黄芪);Floss Milkvetch,Henderson Milkvetch ■
42484 Astragalus henryi Oliv.;秦岭黄耆(黄芪,黄蓍,秦岭黄芪);Henry Milkvetch ●
42485 Astragalus heptapotamicus Sumnev.;七溪黄耆(七溪黄芪,七叶黄芪,七叶黄耆);Seven-leaves Milkvetch,Sevenrivulet Milkvetch ■
42486 Astragalus hesiensis N. Ulziykh.;河西黄耆■
42487 Astragalus heterodontus Boriss.;异齿黄耆(异齿黄芪);Diferttooth Milkvetch,Different-tooth Milkvetch ■
42488 Astragalus heydei Baker;毛柱黄耆(毛柱黄芪);Hairstyle Milkvetch,Hairy-style Milkvetch ■
42489 Astragalus himalayanus Klotzsch;喜马拉雅黄耆■☆
42490 Astragalus hispidulus DC.;细毛黄耆■☆
42491 Astragalus hispidulus DC. subsp. kralikianus Täckh. et Boulos = Astragalus kralikii Batt. ■☆
42492 Astragalus hispidulus DC. subsp. kralikii (Batt.) Boulos = Astragalus kralikii Batt. ■☆
42493 Astragalus hoantchy Franch.;乌拉特黄耆(百本,粗壮黄芪,粗壮黄耆,戴椹,贺兰山黄芪,贺兰山黄耆,黄芪,黄蓍,乌拉特黄芪);Robust Milkvetch,Wulate Milkvetch ■
42494 Astragalus hoantchy Franch. subsp. dshimensis (Gontsch.) K. T. Fu;边陲黄耆(边陲黄芪);Dshimen Milkvetch ■
42495 Astragalus hoantschy Franch. subsp. dschinensis (Gontsch.) K. T. Fu = Astragalus dschinensis Gontsch. ■
42496 Astragalus hoffmeisteri (Klotzsch) Ali;疏花黄耆(疏花黄芪,札达黄芪,札达黄耆);Looseflower Milkvetch,Tsata Milkvetch,Zhada Milkvetch ■
42497 Astragalus hoffmeisteri (Klotzsch) Ali var. pilosulus Ali = Astragalus falconeri Bunge ■
42498 Astragalus horizontalis Kar. et Kir. = Astragalus arbuscula Pall. ●
42499 Astragalus hoshanbaoensis Podlech et L. R. Xu;善宝黄耆■
42500 Astragalus hotianensis S. B. Ho;和田黄耆(和田黄芪);Hetian Milkvetch,Hotan Milkvetch ■
42501 Astragalus hsinbaticus P. Y. Fu et Y. A. Chen;新巴黄耆(新巴黄芪);Hsinba Milkvetch,Xinba Milkvetch ■
42502 Astragalus hsinbaticus P. Y. Fu et Y. A. Chen = Astragalus grubovii Sanchir ■
42503 Astragalus huiningensis Y. C. Ho;会宁黄耆(会宁黄芪);Huining Milkvetch ■
42504 Astragalus huiningensis Y. C. Ho var. psilocarpus K. T. Fu;毛果会宁黄耆(盐地黄芪,盐地黄耆)■
42505 Astragalus hulunensis P. Y. Fu et Y. A. Chen;小叶黄耆(小叶黄芪);Smallleaf Milkvetch ■
42506 Astragalus hulunensis P. Y. Fu et Y. A. Chen = Astragalus zacharensis Bunge ■
42507 Astragalus huochengensis Podlech et L. R. Xu;金沟河黄耆■
42508 Astragalus hypogaeus Ledeb.;留土黄耆■
42509 Astragalus hypogaeus Ledeb. var. borodinii Krasn. = Astragalus borodinii Krasn. ■
42510 Astragalus hypoglottis L. = Astragalus danicus Retz. ■
42511 Astragalus hypoglottis L. var. dasyglottis (DC.) Ledeb. = Astragalus agrestis G. Don ■
42512 Astragalus hysophilus Podlech et L. R. Xu;高地黄耆■
42513 Astragalus ibrahimianus Maire;易卜黄耆■☆
42514 Astragalus ibrahimianus Maire var. cossonianus Emb. et Maire = Astragalus ibrahimianus Maire ■☆
42515 Astragalus ibrahimianus Maire var. mesatlanticus Emb. et Maire = Astragalus ibrahimianus Maire ■☆
42516 Astragalus iliensis Bunge;伊犁黄耆(伊犁黄芪);Yili Milkvetch ■
42517 Astragalus iliensis Bunge var. macrostephanus S. B. Ho;大花伊犁黄耆(大花伊犁黄芪);Large-flower Yili Milkvetch ■
42518 Astragalus iliensis Bunge var. macrostephanus S. B. Ho = Astragalus macrostephanus (S. B. Ho) Podlech et L. R. Xu ■
42519 Astragalus immersus Baker ex Aitch. = Oxytropis immersa (Baker

ex Aitch.) Bunge ex B. Fedtsch. ■
42520 Astragalus incanus L. ;灰毛黄耆■☆
42521 Astragalus incanus L. subsp. incurvus (Desf.) Maire;曲黄耆■☆
42522 Astragalus incanus L. subsp. nummularioides (Desf.) Maire;铜钱灰毛黄耆■☆
42523 Astragalus incanus L. var. mesatlanticus Maire = Astragalus incanus L. ■☆
42524 Astragalus incanus L. var. occidentalis Maire = Astragalus incanus L. ■☆
42525 Astragalus incanus L. var. pinguefactus (Pau) Maire = Astragalus incanus L. ■☆
42526 Astragalus incanus L. var. stenocarpus Pau et Sennen = Astragalus incanus L. subsp. incurvus (Desf.) Maire ■☆
42527 Astragalus inconspicuus Baker = Astragalus sikkimensis Bunge ■
42528 Astragalus incurvus Desf. = Astragalus incanus L. subsp. incurvus (Desf.) Maire ■☆
42529 Astragalus incurvus Desf. var. pinguefactus Pau = Astragalus incanus L. subsp. incurvus (Desf.) Maire ■☆
42530 Astragalus inderiensis Claus = Astragalus pallasii Spreng. ■
42531 Astragalus infestus Boiss. ;有害黄耆■☆
42532 Astragalus inopinatus Boriss. ;北黄耆(北黄芪)■
42533 Astragalus inopinatus Boriss. = Astragalus adsurgens Pall. ■
42534 Astragalus inopinatus Boriss. = Astragalus laxmannii Jacq. ■
42535 Astragalus intercedens Rchb. f. ;中间黄耆■☆
42536 Astragalus irkeschtami B. Fedtsch. = Astragalus petraeus Kar. et Kir. ■
42537 Astragalus japonicus H. Boissieu;日本黄耆■☆
42538 Astragalus jiazaensis Podlech et L. R. Xu;加扎黄耆■
42539 Astragalus jiuquanensis S. B. Ho;酒泉黄耆(酒泉黄芪);Jiuquan Milkvetch ■
42540 Astragalus josephi E. Peter;沙基黄耆(沙基黄芪);Joseph Milkvetch ■
42541 Astragalus jubatus (Pall.) Kuntze = Caragana jubata (Pall.) Poir. ●
42542 Astragalus junatovii Sanchir;圆果黄耆(尤那托夫黄耆,圆果黄芪)■
42543 Astragalus karaculensis Ovcz. et Rassulova = Astragalus macropterus DC. ■
42544 Astragalus karkarensis Popov;直荚黄耆(霍城黄芪,卡尔卡尔黄耆);Karker Milkvetch ■
42545 Astragalus kasachstanicus Golosk. ;哈萨克黄耆■
42546 Astragalus kawakamii Matsum. ;川上黄耆■☆
42547 Astragalus kendyrlyki Popov;塔城黄耆(塔城黄芪)■☆
42548 Astragalus kessleri Trautv. ;凯斯列黄耆■
42549 Astragalus khasianus Bunge;长果颈黄耆■
42550 Astragalus kialensis N. D. Simpson;苦黄耆(苦黄芪,茸毛黄耆,西康黄芪,西康黄耆);Bitter Milkvetch ■
42551 Astragalus kifonsanicus Ulbr. ;鸡峰山黄耆(鸡峰黄芪,鸡峰黄耆,鸡峰山黄芪);Jifengshan Milkvetch,Kifonsan Milkvetch ■
42552 Astragalus kirghisicus Stschegl. ;柯什黄耆(柯什黄芪)■☆
42553 Astragalus kongrensis Baker;深紫萼黄耆■
42554 Astragalus kralikianus Coss. = Astragalus kralikii Batt. ■☆
42555 Astragalus kralikii Batt. ;克拉利克黄耆■☆
42556 Astragalus kronenburgii B. Fedtsch. ex Kneuck. ;古利恰黄耆(古利恰黄芪);Kroneburg Milkvetch ■
42557 Astragalus kronenburgii B. Fedtsch. ex Kneuck. var. chaidamuensis S. B. Ho = Astragalus chaidamuensis (S. B. Ho) Podlech et L. R. Xu ■
42558 Astragalus kronenburgii B. Fedtsch. var. chaidamuensis S. B. Ho = Astragalus chaidamuensis (S. B. Ho) Podlech et L. R. Xu ■
42559 Astragalus kukunoricus N. Ulziykh. ;青海黄耆■
42560 Astragalus kuldshensis Bunge;伊宁黄耆■
42561 Astragalus kunlunensis H. Ohba;昆仑黄耆■
42562 Astragalus kurdaicus Soposchn. ;库尔德黄耆(库尔德黄芪)■☆
42563 Astragalus kurtschumensis Bunge;库尔楚黄耆(库尔楚黄芪)■
42564 Astragalus kuschakewiczii B. Fedtsch. ;库萨克黄耆■
42565 Astragalus labradoricus DC. ;拉布拉多黄耆;Labrador Milk-vetch ■☆
42566 Astragalus laceratus Lipsky;裂翼黄耆(裂翼黄芪);Cleft-wing Milkvetch,Splitwing Milkvetch ■
42567 Astragalus ladakensis Balakr = Astragalus strictus Benth. ■
42568 Astragalus laetabilis Podlech et L. R. Xu;丝叶黄耆●
42569 Astragalus lagocephalus Fisch. et C. A. Mey. = Astragalus vulpinus Willd. ■
42570 Astragalus laguroides Pall. ;兔尾黄耆(兔尾状黄芪,兔尾状黄耆);Rabbit Milkvetch,Rabittail Milkvetch ■☆
42571 Astragalus laguroides Pall. var. micranthus S. B. Ho;小花兔尾状黄耆(小花兔黄芪,小花兔尾状黄芪);Smallflower Rabbit Milkvetch,Smallflower Rabbittail Milkvetch ■
42572 Astragalus laguroides Pall. var. micranthus S. B. Ho = Astragalus novissimus Podlech et L. R. Xu ■
42573 Astragalus lagurus Pall. = Astragalus laguroides Pall. ■☆
42574 Astragalus lamalaensis C. C. Ni;拉木拉黄耆(拉马拉黄耆,拉木拉黄芪);Lamula Milkvetch ■
42575 Astragalus lang-ranii Podlech;盐生黄耆■
42576 Astragalus lanigerus Desf. = Astragalus caprinus L. ■☆
42577 Astragalus lanigerus Desf. var. reboudii (Coss.) Maire = Astragalus caprinus L. ■☆
42578 Astragalus lanigerus Desf. var. salinus Pomel = Astragalus caprinus L. ■☆
42579 Astragalus lanigerus Desf. var. subglabratus DC. = Astragalus caprinus L. ■☆
42580 Astragalus lanuginosus Kar. et Kir. ;绵毛黄耆(棉毛黄芪,新疆黄芪,新疆黄耆);Cottony Milkvetch,Woolly Milkvetch,Xinjiang Milkvetch ■
42581 Astragalus lanzhouensis Podlech et L. R. Xu;兰州黄耆■
42582 Astragalus lapponicus (DC.) Schischk. ;拉普兰黄耆;Lapulan Milkvetch ■☆
42583 Astragalus lasaensis C. C. Ni et P. C. Li;拉萨黄耆(拉萨黄芪);Lasa Milkvetch ■
42584 Astragalus lasianthus C. A. Mey. = Astragalus lasiopetalus Bunge ■
42585 Astragalus lasiopetalus Bunge;毛瓣黄耆(毛瓣黄芪);Hairpetal Milkvetch,Hairy-petal Milkvetch ■
42586 Astragalus lasiophyllus Ledeb. ;毛叶黄耆(毛叶黄芪);Hairleaf Milkvetch,Hairy-leaf Milkvetch ■
42587 Astragalus lasiophyllus Ledeb. = Astragalus pallasii Spreng. ■
42588 Astragalus lasiosemius Boiss. ;毛果黄芪;Hairyfruit Milkvetch,Hairy-fruited Milkvetch ●
42589 Astragalus lasius Blatt. = Astragalus scorpiurus Bunge ■☆
42590 Astragalus laspurensis Ali;西巴黄耆(西巴黄芪);Laspur Milkvetch,Siba Milkvetch ■
42591 Astragalus latistylus Freyn = Astragalus lasiosemius Boiss. ●
42592 Astragalus latistylus Freyn subsp. aridus Freyn = Astragalus lasiosemius Boiss. ●

42593 Astragalus latiunguiculatus Y. C. Ho；宽爪黄耆（宽爪黄芪）；Brosdclaw Milkvetch ■

42594 Astragalus laxmannii Jacq.；斜茎黄耆（马拌肠，沙打旺，斜茎黄芪，新疆黄耆，直茎黄芪，直茎黄耆，直立黄芪，直立黄耆）；Erect Milkvetch ■

42595 Astragalus laxmannii Nutt. var. adsurgens (Pall.) Kitag. = Astragalus adsurgens Pall. ■

42596 Astragalus leansanicus Ulbr.；莲山黄耆（历安山黄芪，历安山黄耆，莲山黄芪）；Leanshan Milkvetch，Lianshan Milkvetch ■

42597 Astragalus lehmannianus Bunge；茧荚黄耆（茧荚黄芪）；Lehmann Milkvetch ■

42598 Astragalus leioclados Boiss.；平滑枝黄耆（平滑枝黄芪）●☆

42599 Astragalus lentiginosus Douglas ex Hook.；斑点黄耆；Freckled Milk-vetch，Speckled Loco ●☆

42600 Astragalus lepsensis Bunge；天山黄耆（天山黄芪，伊犁黄芪，伊犁黄耆）；Tianshan Milkvetch ■

42601 Astragalus lepsensis Bunge var. leduensis Y. H. Wu；乐都黄耆（乐都黄芪）；Ledu Milkvetch ■

42602 Astragalus leptocephalus Batt.；细头黄耆■☆

42603 Astragalus leptocladus Podlech et L. R. Xu；细枝黄耆■

42604 Astragalus leptodermus Bunge = Astragalus filicaulis Kar. et Kir. ■

42605 Astragalus leptophyllus Desf. = Astragalus falciformis Desf. ■☆

42606 Astragalus leptophyllus Pall. = Oxytropis leptophylla (Pall.) DC. ■

42607 Astragalus leptostachys Pall. = Astragalus sulcatus L. ■

42608 Astragalus leptus Boiss.；细黄耆■☆

42609 Astragalus lessertioides Benth. ex Bunge；秃萼黄耆（秃萼黄芪）■

42610 Astragalus lessertioides Bunge = Astragalus lessertioides Benth. ex Bunge ■

42611 Astragalus leucacanthus Boiss. = Astragalus trigonus DC. ■☆

42612 Astragalus leucocephalus Graham ex Benth.；白序黄耆（白花黄芪，白花黄耆，白序黄芪）；White-flower Milkvetch，Whitehead Milkvetch ■

42613 Astragalus leucocladus Bunge；白枝黄芪；White-branch Milkvetch，White-branched Milkvetch，Whitetwig Milkvetch ●■

42614 Astragalus levidensis Podlech et L. R. Xu；光萼齿黄耆■

42615 Astragalus levitubus H. T. Tsai et Te T. Yu；光萼筒黄耆（光萼筒黄芪）；Smooth-tube Milkvetch，Velvettube Milkvetch ■

42616 Astragalus lhorongensis P. C. Li et C. C. Ni；洛隆黄耆（洛隆黄芪）；Luolong Milkvetch ■

42617 Astragalus licentianus Hand. -Mazz.；甘肃黄耆（甘肃黄芪）；Gansu Milkvetch，Kansu Milkvetch ■

42618 Astragalus lichiangensis Simpson = Astragalus camptodontus Franch. var. lichiangensis (Simpson) K. T. Fu ■

42619 Astragalus limprichtii Ulbr.；长管萼黄耆（长管萼黄芪，长管黄芪）；Limpricht Milkvetch，Longtube Milkvetch ■

42620 Astragalus lineariaurifer P. C. Li；线耳黄耆（线耳黄芪）■

42621 Astragalus linii E. Gomez-Sosa = Astragalus arnoldii Hemsl. et H. Pearson ■

42622 Astragalus lioui H. T. Tsai et Te T. Yu = Astragalus pavlovii B. Fedtsch. et Basil. ■

42623 Astragalus litangensis Bureau et Franch. = Astragalus acaulis Baker ■

42624 Astragalus lithophilus Kar. et Kir.；岩生黄耆（岩生黄芪）；Rock Milkvetch，Rock-loving Milkvetch ■

42625 Astragalus liushaensis C. Y. Chang et L. R. Xu；流沙黄耆■

42626 Astragalus loczyi Kanitz = Astragalus variabilis Bunge ex Maxim. ■

42627 Astragalus loczyi Kanitz var. scaposa Kaintz = Oxytropis merkensis Bunge ■

42628 Astragalus loczyi Kanitz var. scaposa Kanitz = Oxytropis imbricata Kom. ■

42629 Astragalus longicalyx C. C. Ni et P. C. Li；长萼黄耆（长萼黄芪）；Longcalyx Milkvetch ■

42630 Astragalus longicalyx C. C. Ni et P. C. Li = Astragalus munroi Benth. ex Bunge ■

42631 Astragalus longicaulis Pomel；长茎黄耆■☆

42632 Astragalus longidentatus Chater；长齿黄耆■☆

42633 Astragalus longiflorus Pall.；长花黄耆；Longflower Milkvetch ■☆

42634 Astragalus longilobus E. Peter；长萼裂黄耆（长萼裂黄芪，秃萼黄耆）；Longlobe Milkvetch，Longlobed Milkvetch ■

42635 Astragalus longipes Kar. et Kir. = Astragalus macropterus DC. ■

42636 Astragalus longiracemosus N. Ulzijkh.；长序黄耆■

42637 Astragalus longiscapus C. C. Ni et P. C. Li；长梗黄耆（长梗黄芪，长序黄芪，长序黄耆）；Long-scape Milkvetch，Longscapus Milkvetch ■

42638 Astragalus longispicatus Ulbr. = Astragalus adsurgens Pall. ■

42639 Astragalus longispicatus Ulbr. = Astragalus laxmannii Jacq. ■

42640 Astragalus lotoides Pall. = Astragalus sinicus L. ●■

42641 Astragalus lucidus H. T. Tsai et Te T. Yu；光亮黄耆（光萼黄芪，光萼黄耆）；Shining-calyx Milkvetch，Velvetcalyx Milkvetch ■

42642 Astragalus luculentus Podlech et L. R. Xu；光滑黄耆■

42643 Astragalus ludlowii Wenninger = Astragalus nanfengensis C. C. Ni ■

42644 Astragalus lupulinus Pall. var. laguroides (Pall.) Basil. = Astragalus laguroides Pall. ■☆

42645 Astragalus lusitanicus Lam. = Erophaca baetica (L.) Boiss. ●☆

42646 Astragalus lustricola Podlech et L. R. Xu；荒野黄耆■

42647 Astragalus luteiflorus N. Ulzijkh.；黄花黄耆■☆

42648 Astragalus luteolus H. T. Tsai et Te T. Yu；浅黄花黄耆（黄花黄芪，秃萼黄耆）；Yellow Milkvetch，Yellow-flower Milkvetch ■

42649 Astragalus luteus Ulbr. = Astragalus monadelphus Bunge ex Maxim. ■

42650 Astragalus lychnobius Podlech et L. R. Xu；喜光黄耆■

42651 Astragalus macriculus Podlech et L. R. Xu；裕民黄耆■

42652 Astragalus macrocarpus DC.；大果黄耆■☆

42653 Astragalus macroceras Bong. = Astragalus macrolobus M. Bieb. ■

42654 Astragalus macroceras C. A. Mey. ex Bong.；大荚黄耆（长荚黄芪）；Longpod Milkvetch ■

42655 Astragalus macrolobus M. Bieb.；长荚黄耆；Big-lobed Milkvetch ■

42656 Astragalus macrolobus M. Bieb. = Astragalus macroceras C. A. Mey. ex Bong. ■

42657 Astragalus macropterus DC.；大翼黄耆（大翼黄芪）；Bigwing Milkvetch ■

42658 Astragalus macropus Bunge；大脚黄耆■☆

42659 Astragalus macrorhizus Cav. = Astragalus donianus DC. ■

42660 Astragalus macrorhizus Cav. = Astragalus incanus L. ■☆

42661 Astragalus macrostegius Rech. f. = Astragalus melanostachys Benth. ex Bunge ■

42662 Astragalus macrostephanus (S. B. Ho) Podlech et L. R. Xu；青海大花黄耆（大花黄耆）■

42663 Astragalus macrotrichus E. Peter = Astragalus monophyllus Bunge ex Maxim. ■

42664 Astragalus macrotropis Bunge；长龙骨黄耆（长龙骨黄芪，阔叶杭子梢）；Longkeel Milkvetch ■

42665 Astragalus mahoschanicus Hand. -Mazz.；马衔山黄耆（马河山黄

芪,马河山黄耆,马衔山黄芪);Mahoschan Milkvetch, Maxianshan Milkvetch ■

42666 Astragalus mahoschanicus Hand. -Mazz. var. mengdaensis Y. H. Wu;孟达黄耆(孟达黄芪);Mengda Milkvetch ■

42667 Astragalus mahoschanicus Hand. -Mazz. var. multipilosus Y. H. Wu;多毛马衔山黄耆(多毛马衔山黄芪);Manyhair Maxianshan Milkvetch ■

42668 Astragalus mahoschanicus Hand. -Mazz. var. subeicus K. T. Fu;肃北黄耆;Subei Milkvetch ■

42669 Astragalus mahoschanicus Hand. -Mazz. var. subeicus K. T. Fu = Astragalus luteiflorus N. Ulzijkh. ■☆

42670 Astragalus maireanus Greuter et Burdet;迈雷黄耆■☆

42671 Astragalus mairei (Emb.) Emb. et Maire = Astragalus maireanus Greuter et Burdet ■☆

42672 Astragalus majevskianus Krylov;富蕴黄芪(哈巴河黄芪,青河黄芪);Majevski Milkvetch ■

42673 Astragalus majusculus Podlech et L. R. Xu;买依尔黄耆■

42674 Astragalus malacophyllus Benth. ex Bunge;软叶黄耆■☆

42675 Astragalus malacophyllus Bunge = Astragalus rhizanthus Benth. ●

42676 Astragalus malcolmii Hemsl. et H. Pearson;短茎黄耆(短茎黄芪);Shortstem Milkvetch ■

42677 Astragalus maowenensis Podlech et L. R. Xu;茂汶黄耆(茂汶黄芪);Maowen Milkvetch ■

42678 Astragalus mareoticus Delile;马雷奥特黄耆■☆

42679 Astragalus mareoticus Delile var. handiensis Bolle = Astragalus mareoticus Delile ■☆

42680 Astragalus maritima (Maxim.) Kitag. = Gueldenstaedtia verna (Georgi) Borissov ■

42681 Astragalus maroccanus Braun-Blanq. et Maire;摩洛哥黄耆■☆

42682 Astragalus masenderanus Bunge;乌恰黄耆■

42683 Astragalus massiliensis (Mill.) Lam. = Astragalus tragacantha L. ■☆

42684 Astragalus matiensis P. C. Li;马蹄黄耆(马蹄黄芪);Mati Milkvetch ■

42685 Astragalus mattam H. T. Tsai et Te T. Yu;茵垫黄耆(茵垫黄芪);Mattam Milkvetch ■

42686 Astragalus mattam H. T. Tsai et Te T. Yu var. macroflorus Y. H. Wu;大花茵垫黄耆(大花茵垫黄芪);Bigflower Mattam Milkvetch ■

42687 Astragalus mauritanicus Coss. = Astragalus longidentatus Chater ■☆

42688 Astragalus maurorum Murb.;莫尔黄耆■☆

42689 Astragalus maurus (Humbert et Maire) Pau;暗黄耆■☆

42690 Astragalus maximus Willd. = Astragalus alopecurus Pall. ■

42691 Astragalus medius Schrenk;中型黄耆■☆

42692 Astragalus megalanthus DC.;大花黄耆■

42693 Astragalus melanophrurius Boiss.;黑蛛黄耆(黑蛛黄芪)■☆

42694 Astragalus melanostachys Benth. ex Bunge;黑穗黄耆(黑穗黄芪);Blackspike Milkvetch ■

42695 Astragalus melanostachys Bunge = Astragalus melanostachys Benth. ex Bunge ■

42696 Astragalus melilotoides Pall.;草木樨状黄耆(草木犀黄芪,草木樨状黄芪,草木樨状紫云英,苦豆根,秦头,扫帚苗,山胡麻,樨状黄芪,樨状黄耆,小马层子,紫云英);Sweetcloverlike Milkvetch ■

42697 Astragalus melilotoides Pall. var. tenuis (Turcz.) Ledeb.;细叶黄耆(细叶黄芪);Slender-leaf Milkvetch ■

42698 Astragalus melilotoides Pall. var. tenuis (Turcz.) Ledeb. = Astragalus tenuis Turcz. ■

42699 Astragalus membranaceus (Fisch. ex Link) Bunge;黄耆(百本,百药绵,戴糁,戴椹,东山黄芪,独根,独椹,二人抬,黄芪,芰草,箭杆花,箭芪,绵黄耆,绵芪,膜荚黄芪,膜荚黄耆,山爆仗,蜀脂,土山爆仗根,王孙);Membranous Milkvetch, Milkvetch, Milkvetch Huangchi ■

42700 Astragalus membranaceus (Fisch. ex Link) Bunge f. pallidipurpureus Y. C. Ho = Astragalus purpurinus (Y. C. Ho) Podlech et L. R. Xu ■

42701 Astragalus membranaceus (Fisch. ex Link) Bunge f. purpurinus (Y. C. Ho) Y. C. Ho = Astragalus purpurinus (Y. C. Ho) Podlech et L. R. Xu ■

42702 Astragalus membranaceus (Fisch. ex Link) Bunge f. purpurinus (Y. C. Ho) K. T. Fu = Astragalus purpurinus (Y. C. Ho) Podlech et L. R. Xu ■

42703 Astragalus membranaceus (Fisch. ex Link) Bunge subsp. pallidipurpureus (Y. C. Ho) X. Y. Zhu et C. J. Chen = Astragalus purpurinus (Y. C. Ho) Podlech et L. R. Xu ■

42704 Astragalus membranaceus (Fisch. ex Link) Bunge var. mongholicus (Bunge) P. K. Hsiao;蒙古黄耆(白皮芪,百本,百药绵,戴糁,戴椹,独根,独椹,二人抬,黄芪,芰草,箭芪,蒙古黄芪,绵黄芪,绵黄耆,绵芪,内蒙黄芪,内蒙黄耆,蜀脂,土山爆仗根,王孙);Inner Mongolian Milkvetch, Mongol Milkvetch, Mongolian Milkvetch ■

42705 Astragalus membranaceus (Fisch. ex Link) Bunge var. obtusus Makino;钝头黄芪■

42706 Astragalus membranaceus (Fisch. ex Link) Bunge var. obtusus Makino = Astragalus membranaceus (Fisch. ex Link) Bunge ■

42707 Astragalus membranaceus (Fisch. ex Link) Bunge var. purpurinus (Y. C. Ho) Y. C. Ho = Astragalus membranaceus (Fisch. ex Link) Bunge f. purpurinus (Y. C. Ho) Y. C. Ho ■

42708 Astragalus membranaceus Bunge var. purpurinus Y. C. Ho = Astragalus purpurinus (Y. C. Ho) Podlech et L. R. Xu ■

42709 Astragalus mendax Freyn = Astragalus taldicensis Franch. ■

42710 Astragalus mesatlanticus Andr. = Astragalus echinatus Murray ■☆

42711 Astragalus meuselii Romo = Astragalus granatensis Lam. subsp. maroccanus (Font Quer) Raynaud et Sauvage ●☆

42712 Astragalus mexicanus A. DC. var. trichocalyx (Nutt.) DC. = Astragalus crassicarpus Nutt. ■☆

42713 Astragalus microcephalus Willd.;小头黄耆(小头黄芪);Samll-head Milkvetch ■☆

42714 Astragalus microdontus Baker = Astragalus chlorostachys Lindl. ■

42715 Astragalus microphyllus (Pall.) Pall. = Oxytropis microphylla (Pall.) DC. ■

42716 Astragalus microphyllus Pall. = Oxytropis microphylla (Pall.) DC. ■

42717 Astragalus mieheorum Podlech et L. R. Xu;青东黄耆■

42718 Astragalus milingensis C. C. Ni et P. C. Li;米林黄耆(米林黄芪);Milin Milkvetch ■

42719 Astragalus milingensis C. C. Ni et P. C. Li var. heydeoides K. T. Fu;类毛柱黄耆(类毛柱黄芪);Hairy-style-like Milkvetch ■

42720 Astragalus milingensis C. C. Ni et P. C. Li var. paucijugus (K. T. Fu) K. T. Fu = Astragalus prodigiosus K. T. Fu var. paucijugus K. T. Fu ■

42721 Astragalus milingensis C. C. Ni et P. C. Li var. prodigiosus (K. T. Fu) K. T. Fu = Astragalus prodigiosus K. T. Fu ■

42722 Astragalus minheensis X. Y. Zhu et C. J. Chen;民和黄耆;Minhe Milkvetch ■

42723 Astragalus miniatus Bunge;细弱黄耆(红花黄芪,细茎黄芪,细

茎黄耆,细弱黄芪);Slender Milkvetch,Weak Milkvetch ■
42724 Astragalus minshanensis K. T. Fu;岷山黄耆;Minshan Milkvetch ■
42725 Astragalus minudentatus Y. C. Ho;小齿黄耆(小齿黄芪);Minitooth Milkvetch ■
42726 Astragalus minustebracteolatus Simpson = Astragalus prattii Simpson ■
42727 Astragalus minutifoliolatus Wendelbo;微小叶黄耆■
42728 Astragalus minutifoliolatus Wendelbo = Astragalus webbianus Benth. ■
42729 Astragalus mirabilis Lipsky;中亚奇异黄耆■☆
42730 Astragalus mirpoureanus Cambess. = Gueldenstaedtia verna (Georgi) Borissov ■
42731 Astragalus miser Douglas;贫弱黄耆;Timber Milk-vetch ■☆
42732 Astragalus miyalomontis P. C. Li;米亚罗黄耆(米亚罗黄芪);Missouri Milkvetch,Miyaluo Milkvetch ■
42733 Astragalus moellendorffii Bunge ex Maxim.;边向花黄耆(边向花黄芪);Moellendorff Milkvetch ■
42734 Astragalus moellendorffii Bunge ex Maxim. var. kansuensis E. Peter;莲花山黄耆(莲花山黄芪);Lianhuashan Milkvetch ■
42735 Astragalus moellendorffii Bunge var. kansuensis E. Peter = Astragalus moellendorffii Bunge ex Maxim. var. kansuensis E. Peter ■
42736 Astragalus moellendorfii Bunge = Astragalus moellendorffii Bunge ex Maxim. ■
42737 Astragalus mohavensis S. Watson;莫哈维黄耆;Mohave Loco ■☆
42738 Astragalus mokurensis S. Watson = Astragalus koschukensis ■☆
42739 Astragalus mollissimus Gontsch.;软毛黄耆(软毛黄芪);Purple Loco,Woolly Loco,Woolly Locoweed ■☆
42740 Astragalus monadelphus Bunge ex Maxim.;单蕊黄耆(单蕊黄芪,单体蕊黄芪,单体蕊黄耆);Monadelphus Milkvetch,Singlepistil Milkvetch ■
42741 Astragalus monadelphus Bunge ex Maxim. subsp. xitaibaicus K. T. Fu = Astragalus xitaibaicus (K. T. Fu) Podlech et L. R. Xu ■
42742 Astragalus monanthus K. T. Fu;单花黄耆(单花黄芪);Singleflower Milkvetch ■
42743 Astragalus monanthus K. T. Fu = Astragalus donianus DC. ■
42744 Astragalus monbeigii N. D. Simpson;异长齿黄耆(异长齿黄芪);Monbeig Milkvetch ■
42745 Astragalus mongholicus Bunge = Astragalus membranaceus (Fisch. ex Link) Bunge var. mongholicus (Bunge) P. K. Hsiao ■
42746 Astragalus mongutensis Lipsky;蒙古特黄耆(蒙古特黄芪);Mongut Milkvetch ■
42747 Astragalus monophyllus Bunge ex Maxim.;长毛荚黄耆(长毛黄芪,长毛荚黄芪,一叶黄芪);Longhair Milkvetch,Long-hairy-pod Milkvetch ■
42748 Astragalus monspessulanus L. subsp. gypsophilus Rouy;喜钙黄耆■☆
42749 Astragalus monspessulanus L. var. aurasiacus Maire = Astragalus monspessulanus L. subsp. gypsophilus Rouy ■☆
42750 Astragalus monspessulanus L. var. chlorocyaneus (Boiss. et Reut.) Costa = Astragalus monspessulanus L. subsp. gypsophilus Rouy ■☆
42751 Astragalus monspessulanus L. var. cossonii (Bunge) Batt. = Astragalus monspessulanus L. subsp. gypsophilus Rouy ■☆
42752 Astragalus monticola P. C. Li et C. C. Ni;山地黄耆(山地黄芪);Montane Milkvetch,Mountainous Milkvetch ■
42753 Astragalus monticola P. C. Li et C. C. Ni = Astragalus arnoldii Hemsl. et H. Pearson ■
42754 Astragalus montivagus Podlech et L. R. Xu;如多黄耆■
42755 Astragalus moorcroftiana Benth. = Sophora moorcroftiana (Benth.) Baker ●
42756 Astragalus moorcroftiana Wall. = Sophora moocroftiana (Graham) Benth. ex Baker ●
42757 Astragalus moupinensis Franch.;天全黄耆(天全黄芪);Moupin Milkvetch,Mupin Milkvetch ■
42758 Astragalus muliensis Hand.-Mazz.;木里黄耆(木里黄芪);Muli Milkvetch ■
42759 Astragalus multicaulis Ledeb. = Astragalus macropterus DC. ■
42760 Astragalus multiceps Benth.;多头黄耆(二尖齿黄耆)■
42761 Astragalus munroi Benth. ex Bunge;细梗黄耆(细梗黄芪);Slenderstalk Milkvetch,Thin-pedicel Milkvetch ■
42762 Astragalus munroi Bunge = Astragalus munroi Benth. ex Bunge ■
42763 Astragalus muricatus (Pall.) Pall. = Oxytropis muricata (Pall.) DC. ■
42764 Astragalus muschketowii B. Fedtsch.;木斯克黄耆■
42765 Astragalus myriophyllus (Pall.) Pall. = Oxytropis myriophylla (Pall.) DC. ■
42766 Astragalus myriophyllus Bunge = Astragalus pamirensis Franch. ■
42767 Astragalus myriophyllus Pall. = Oxytropis myriophylla (Pall.) DC. ■
42768 Astragalus nanellus H. T. Tsai et Te T. Yu;极矮黄耆(极矮黄芪);Most-dwarf Milkvetch,Pygmean Milkvetch ■
42769 Astragalus nanfengensis Z. C. Ni;南峰黄耆(南峰黄芪);Nanfeng Milkvetch ■
42770 Astragalus nangxianensis P. C. Li et C. C. Ni;朗县黄耆(朗县黄芪);Langxian Milkvetch ■
42771 Astragalus nanjiangianus K. T. Fu;南疆黄耆(南疆黄芪);Nanjiang Milkvetch ■
42772 Astragalus nankotaizanensis Sasaki;南湖大山黄耆(南湖大山黄芪,南湖大山紫云英,南口台黄耆)■
42773 Astragalus nankotaizanensis Sasaki = Astragalus sinicus L. ●■
42774 Astragalus nanshanicus Podlech et L. R. Xu;南山黄耆(南峰黄耆)■
42775 Astragalus narbonensis Gouan = Astragalus alopecuroides L. ■☆
42776 Astragalus narbonensis Gouan subsp. africanus (Bunge) Ball = Astragalus alopecuroides L. ■☆
42777 Astragalus narbonensis Gouan subsp. atlanticus Ball = Astragalus alopecuroides L. ■☆
42778 Astragalus narbonensis Gouan var. africanus (Bunge) Batt. = Astragalus alopecuroides L. ■☆
42779 Astragalus narbonensis Gouan var. claryi (Batt.) Pamp. = Astragalus alopecuroides L. ■☆
42780 Astragalus narbonensis Gouan var. claryi Batt. = Astragalus alopecuroides L. ■☆
42781 Astragalus narynensis Freyn = Astragalus tibetanus Bunge ■
42782 Astragalus nathaliae Meffert = Astragalus nivalis Kar. et Kir. ■
42783 Astragalus neglectus (Torr. et A. Gray) E. Sheld.;库氏黄耆;Cooper's Milk-vetch ■☆
42784 Astragalus nematodes Bunge ex Boiss.;线叶黄耆(线叶黄芪);Linearleaf Milkvetch,Thread-like-leaf Milkvetch ■
42785 Astragalus nematodioides H. Ohba,Akiyama et S. K. Wu;类线叶黄耆■
42786 Astragalus nemorosus Batt. = Astragalus reinii Ball subsp. nemorosus (Batt.) Maire ■☆
42787 Astragalus neochorgosicus Podlech;新霍尔果斯■

42788 Astragalus neomonadelphus H. T. Tsai et Te T. Yu;新单蕊黄耆(新单蕊黄芪);Neosinglelepistil Milkvetch, New-monadelphus Milkvetch ■

42789 Astragalus neospinosus ?;新刺黄耆■☆

42790 Astragalus nertschinskensis Freyn = Astragalus uliginosus L. ■

42791 Astragalus nicolai Boriss.;木垒黄耆(木垒黄芪);Mulei Milkvetch,Nicola Milkvetch ■

42792 Astragalus nigrescens Franch. = Astragalus polycladus Bureau et Franch. ■

42793 Astragalus nigrescens Franch. = Astragalus polycladus Bureau et Franch. var. nigrescens (Franch.) E. Peter ■

42794 Astragalus nigrodentatus Podlech et L. R. Xu;黑齿黄耆■

42795 Astragalus ningxiaensis Podlech et L. R. Xu;宁夏黄耆■

42796 Astragalus nivalis Kar. et Kir.;雪地黄耆(雪地黄芪);Lovesnow Milkvetch,Snow Milkvetch ■

42797 Astragalus nivalis Kar. et Kir. var. aureocalycatus S. B. Ho;黄萼雪地黄耆(黄萼雪地黄芪)■

42798 Astragalus nivalis Kar. et Kir. var. aureocalycatus S. B. Ho = Astragalus kukunoricus N. Ulziykh. ■

42799 Astragalus nivelleanus Braun-Blanq. = Astragalus incanus L. ■☆

42800 Astragalus nobilis B. Fedtsch. var. obtusifoliolus S. B. Ho = Astragalus obtusifoliolus (S. B. Ho) Podlech et L. R. Xu ■

42801 Astragalus nobilis Bunge et B. Fedtsch. ex B. Fedtsch.;华贵黄耆(华贵黄芪);Noble Milkvetch ■

42802 Astragalus nobilis Bunge et B. Fedtsch. ex B. Fedtsch. var. obtusifoliolatus S. B. Ho;钝叶华贵黄耆(钝叶华贵黄芪);Obtuseleaf Noble Milkvetch ■

42803 Astragalus nokoensis Sasaki;台湾黄耆(能高大山紫云英,能高紫云英)■

42804 Astragalus nokoensis Sasaki = Astragalus sinicus L. ●■

42805 Astragalus novissimus Podlech et L. R. Xu;小花兔尾黄耆■

42806 Astragalus nubigenus D. Don = Chesneya nubigena (D. Don) Ali ●☆

42807 Astragalus numidicus Murb. = Astragalus armatus Willd. subsp. numidicus (Murb.) Tietz ■☆

42808 Astragalus nummularioides Desf. = Astragalus incanus L. subsp. nummularioides (Desf.) Maire ■☆

42809 Astragalus nummularioides Desf. var. atlanticus Pomel = Astragalus incanus L. subsp. nummularioides (Desf.) Maire ■☆

42810 Astragalus nuttallianus DC.;小花紫云英;Annual Locoweed, Smallflowered Milkvetch ■☆

42811 Astragalus obtusifoliolus (S. B. Ho) Podlech et L. R. Xu;钝叶黄耆■

42812 Astragalus occultus Podlech et L. R. Xu;克郎河黄耆■

42813 Astragalus ochrias Bunge;中宁黄耆(中宁黄芪);Chungning Milkvetch,Zhongning Milkvetch ■

42814 Astragalus ochroleucus Coss. = Astragalus ibrahimianus Maire ■☆

42815 Astragalus odoratus Lam.;香黄耆(香黄芪);Lesser Milk-vetch ■☆

42816 Astragalus ohbaensis Podlech;奥巴黄耆■

42817 Astragalus olgae Bunge;奥尔格黄耆■

42818 Astragalus oligophyllus Schrenk = Astragalus unijugus Bunge ●

42819 Astragalus olufsenii Freyn = Astragalus tibetanus Bunge ■

42820 Astragalus olygophyllus Schrenk = Astragalus unijugus Bunge ●

42821 Astragalus onobrychis L.;驴喜豆黄耆(驴豆黄芪,驴豆黄耆,驴喜豆黄芪)■☆

42822 Astragalus onobrychis L. var. numidarum Maire = Astragalus onobrychis L. ■☆

42823 Astragalus oophorus Freyn = Astragalus breviscapus B. Fedtsch. ■

42824 Astragalus oostachys E. Peter = Astragalus adsurgens Pall. ■

42825 Astragalus oostachys E. Peter = Astragalus laxmannii Jacq. ■

42826 Astragalus ophiocarpus Benth. et Bunge;蛇荚黄耆(蛇荚黄芪);Snakepod Milkvetch ■

42827 Astragalus oplites Benth. ex Parker;刺叶柄黄芪(黎豆叶黄芪,黎豆叶黄耆);Spineleaf Milkvetch,Spine-stalk Milkvetch ■

42828 Astragalus oplites Parker;刺叶柄黄耆■

42829 Astragalus orbicularifolius P. C. Li et C. C. Ni;圆叶黄耆(圆叶黄芪);Roundleaf Milkvetch ■

42830 Astragalus orbiculatus Ledeb.;圆形黄耆(圆形黄芪);Rotund Milkvetch,Rotundity Milkvetch ■

42831 Astragalus ordosicus H. C. Fu;鄂托克黄耆(鄂托克黄芪);Etuoke Milkvetch ■

42832 Astragalus oreocharis Podlech et L. R. Xu;山黄耆■

42833 Astragalus ornithopodioides Lam.;鸟爪黄耆(鸟爪黄芪)■☆

42834 Astragalus ornithorrhynchus Popov;雀喙黄耆(雀喙黄芪);Bird-beak Milkvetch,Sharpbill Milkvetch ■

42835 Astragalus oroboides Hornem.;欧洲扁豆黄耆■☆

42836 Astragalus oroboides Ulbr. = Astragalus hancockii Bunge ex Maxim. ■

42837 Astragalus orthanthoides Boriss. = Astragalus nivalis Kar. et Kir. ■

42838 Astragalus orthanthus Freyn = Astragalus nivalis Kar. et Kir. ■

42839 Astragalus orthocarpus Boiss.;直果黄耆■☆

42840 Astragalus ortholobiformis Sumner;直荚草黄耆(直荚草黄芪);Straightpod Milkvetch,Straighty-lobe-form Milkvetch ■

42841 Astragalus ortholobus Bunge;直裂黄耆■☆

42842 Astragalus otosemius Kitag. = Astragalus galactites Pall. ■

42843 Astragalus oxyglottis Steven;尖舌黄耆(尖舌黄芪);Sharptongue Milkvetch ■

42844 Astragalus oxyodon Baker;尖齿黄耆(尖齿黄芪);Sharptooth Milkvetch ■

42845 Astragalus pallasii Fisch.;帕氏黄耆;Pallas Milkvetch ■☆

42846 Astragalus pallasii Spreng. = Astragalus pallasii Fisch. ■☆

42847 Astragalus pallescens M. Bieb.;白色黄耆■☆

42848 Astragalus pamirensis Franch.;帕米尔黄耆(帕米尔黄芪);Pamir Milkvetch ■

42849 Astragalus pamiricus (B. Fedtsch.) B. Fedtsch. = Astragalus charguschanus Freyn ■

42850 Astragalus pamiroalaicus Lipsky = Astragalus mendax Freyn ■

42851 Astragalus pamiroalaicus Lipsky = Astragalus taldicensis Franch. ■

42852 Astragalus parrowianus Boiss. et Haussk. ex Boiss.;帕罗黄耆(帕罗黄芪)■☆

42853 Astragalus parvicarinatus S. B. Ho;短龙骨黄耆(短龙骨黄芪);Shortkeel Milkvetch ■

42854 Astragalus parviflorus Lam. = Oxytropis deflexa (Pall.) DC. ■

42855 Astragalus pastorius H. T. Tsai et Te T. Yu;牧场黄耆(多花黄芪,多花黄耆,茂汶黄芪,茂汶黄耆,牧场黄芪);Maowen Milkvetch,Multiflower Smallbract Milkvetch,Pasture Milkvetch ■

42856 Astragalus pastorius H. T. Tsai et Te T. Yu var. linearibracteatus K. T. Fu;线苞黄耆(线苞黄芪);Linearbract Milkvetch ■

42857 Astragalus patulepilosus Sirj. et Rech. f.;展毛黄耆■☆

42858 Astragalus pauciflorus (Pall.) Kitag. = Gueldenstaedtia verna (Georgi) Boriss. ■

42859 Astragalus pauciflorus Pall. = Gueldenstaedtia verna (Georgi) Borissov ■

42860 Astragalus pavlovianus Gamajun.;萨雷古拉黄耆(萨雷古拉黄芪);Pavlov Milkvetch ■

42861 Astragalus pavlovianus Gamajun. var. longirostris S. B. Ho = Astragalus yanerwoensis Podlech et L. R. Xu ■

42862 Astragalus pavlovii B. Fedtsch. et Basil.;了墩黄耆(甘新黄芪,了墩黄芪,刘氏黄芪,刘氏黄耆);Liou's Milkvetch ■

42863 Astragalus peduncularis Benth.;梗花黄耆■☆

42864 Astragalus peduncularis Royle;青藏黄耆(青藏黄芪);Peduncular Milkvetch,Qing-Zang Milkvetch ■

42865 Astragalus pelecinus (L.) Barneby = Biserrula pelecinus L. ■☆

42866 Astragalus pelecinus (L.) Barneby subsp. leiocarpus (A. Rich.) Podlech = Biserrula pelecinus L. subsp. leiocarpa (A. Rich.) J. B. Gillett ■☆

42867 Astragalus pendulatopetalus S. B. Ho et Z. H. Wu;琴瓣黄耆■

42868 Astragalus penduliflorus Lam. = Astragalus membranaceus (Fisch. ex Link) Bunge ■

42869 Astragalus penduliflorus Lam. var. pallidipurpureus (Y. C. Ho) X. Y. Zhu = Astragalus purpurinus (Y. C. Ho) Podlech et L. R. Xu ■

42870 Astragalus pentaglottis L. = Astragalus echinatus Murray ■☆

42871 Astragalus pentapetaloides Bunge = Astragalus flexus Fisch. ■

42872 Astragalus perbrevis Podlech et L. R. Xu;紫色黄耆■

42873 Astragalus peregrinus Vahl;外来黄耆■☆

42874 Astragalus persepolitanus Boiss. = Astragalus ammophilus Kar. et Kir. var. persepolitanus (Boiss.) Ali ■

42875 Astragalus persepolitanus Boiss. = Astragalus ammophilus Kar. et Kir. ■

42876 Astragalus persimilis Podlech et L. R. Xu;类中天山黄耆■

42877 Astragalus peterae H. T. Tsai et Te T. Yu;川青黄耆(川青黄芪);Chuan-Qing Milkvetch,Peter Milkvetch ■

42878 Astragalus petraeus Kar. et Kir.;喜石黄耆(喜石黄芪);Lovestone Milkvetch,Rock-loving Milkvetch ■

42879 Astragalus petrovii N. Ulziykh.;苏南黄耆(南山黄耆)■

42880 Astragalus physocalyx Kar. et Kir. = Astragalus cysticalyx Ledeb. ●

42881 Astragalus physodes L.;泡果黄耆(泡囊黄耆)■☆

42882 Astragalus physodes L. = Astragalus skorniakovii B. Fedtsch. ■

42883 Astragalus piletocladus Freyn et Sint.;毛枝黄耆(毛枝黄芪);Hairy-branched Milkvetch ■☆

42884 Astragalus pilosus L. = Oxytropis pilosa (L.) DC. ■

42885 Astragalus pilutschensis N. Ulziykh.;皮鲁斯黄耆■

42886 Astragalus pindreensis (Baker) Ali;明铁盖黄耆■

42887 Astragalus pishanxianensis Podlech;皮山黄耆■

42888 Astragalus platyphyllus Kar. et Kir.;宽叶黄耆(宽叶黄芪);Broadleaf Milkvetch ■

42889 Astragalus poljakovii Popov = Astragalus heptapotamicus Sumnev. ■

42890 Astragalus polyactinus Boiss. = Astragalus stella Gouan ■☆

42891 Astragalus polybotrys Boiss.;多穗黄耆■☆

42892 Astragalus polychromus Freyn = Astragalus beketovii (Krasn.) B. Fedtsch. ■

42893 Astragalus polycladus Bureau et Franch.;多枝黄耆(鞑靼黄芪,多枝黄芪,小果黄芪,小果黄耆,小叶黄芪,皱黄芪);Branchy Milkvetch, Manybranch Milkvetch, Smallfruit Milkvetch, Tatarian Milkvetch ■

42894 Astragalus polycladus Bureau et Franch. var. glabricarpus Y. H. Wu;光果多枝黄耆(光果多枝黄芪);Smoothcarp Branchy Milkvetch ■

42895 Astragalus polycladus Bureau et Franch. var. glabricarpus Y. H. Wu = Astragalus polycladus Bureau et Franch. ■

42896 Astragalus polycladus Bureau et Franch. var. magniflorus Y. H. Wu;大花多枝黄耆(大花多枝黄芪);Bigflower Branchy Milkvetch ■

42897 Astragalus polycladus Bureau et Franch. var. magniflorus Y. H. Wu = Astragalus sungpanensis E. Peter ■

42898 Astragalus polycladus Bureau et Franch. var. nigrescens (Franch.) E. Peter;黑毛多枝黄耆(黑多枝黄芪,黑毛多枝黄芪);Blackhair Branchy Milkvetch ■

42899 Astragalus polycladus Bureau et Franch. var. nigrescens (Franch.) E. Peter = Astragalus polycladus Bureau et Franch. ■

42900 Astragalus polycladus Bureau et Franch. var. nigrescens E. Peter = Astragalus polycladus Bureau et Franch. var. nigrescens (Franch.) E. Peter ■

42901 Astragalus polygalus Pall.;远志黄耆(远志黄芪)■☆

42902 Astragalus ponticus Pall.;黑海黄耆■☆

42903 Astragalus porphyreus Podlech et L. R. Xu;博乐黄耆■

42904 Astragalus porphyrocalyx Y. C. Ho;紫萼黄耆(紫萼黄芪);Purple Calyx Milkvetch,Purplecalyx Milkvetch ■

42905 Astragalus potaninii Kom. = Astragalus tongolensis Ulbr. ■

42906 Astragalus praeteritus Podlech et L. R. Xu;贡觉黄耆■

42907 Astragalus praetermissus Ball = Astragalus gryphus Bunge ■☆

42908 Astragalus pratensis Ulbr. = Astragalus complanatus R. Br. ex Bunge ■

42909 Astragalus prattii Simpson;小苞黄耆(小苞黄芪);Pratt Milkvetch,Smallbract Milkvetch ■

42910 Astragalus prattii Simpson var. multiflorus K. T. Fu = Astragalus pastorius H. T. Tsai et Te T. Yu var. linearibracteatus K. T. Fu ■

42911 Astragalus prattii Simpson var. pastorioides K. T. Fu = Astragalus pastorius H. T. Tsai et Te T. Yu var. linearibracteatus K. T. Fu ■

42912 Astragalus prattii Simpson var. uniflorus E. Peter;一花黄耆(一花黄芪);One Flower Pratt Milkvetch ■

42913 Astragalus prodigiosus K. T. Fu;奇异黄耆(奇异黄芪);Unusual Milkvetch ■

42914 Astragalus prodigiosus K. T. Fu var. paucijugus K. T. Fu;减缩黄耆(减缩黄芪)■

42915 Astragalus projecturus Sumnev. = Astragalus borodinii Krasn. ■

42916 Astragalus prolixus Sieber ex Bunge = Astragalus vogelii (Webb) Bornm. ■☆

42917 Astragalus propinquus Schischk. = Astragalus membranaceus (Fisch. ex Link) Bunge ■

42918 Astragalus przevalskianus Podlech et N. Ulziykh.;波氏黄耆■

42919 Astragalus przewalskii Bunge;黑紫花黄耆(黑紫花黄芪);Black Purple Flower Milkvetch,Przewalsk Milkvetch ■

42920 Astragalus przewalskii Hand. -Mazz. = Astragalus floridus Benth. ex Bunge ■

42921 Astragalus pseudoborodinii S. B. Ho;西域黄耆(西域黄芪);Occidental Milkvetch,Sham Borodin Milkvetch ■

42922 Astragalus pseudobrachytropis Gontsch = Astragalus peterae H. T. Tsai et T. T. Yu ■

42923 Astragalus pseudobrachytropis Gontsch.;类短肋黄耆(类短肋黄芪);False-short-keel Milkvetch,Sham Shootkeel Milkvetch ■

42924 Astragalus pseudochlorostachys Ali;假绿穗黄耆■☆

42925 Astragalus pseudogombo Fern. Casas = Astragalus maurorum Murb. ■☆

42926 Astragalus pseudohypogaeus S. B. Ho;类留土黄耆(类留土黄芪);Baseflower Milkvetch,False-underground Milkvetch ■

42927 Astragalus pseudojagnobicus Podlech et L. R. Xu;喀什黄耆■

42928 Astragalus pseudomacropterus Karmysch. = Astragalus

macropterus DC. ■
42929 Astragalus pseudoroseus N. Ulziykh.;类毛冠黄耆■
42930 Astragalus pseudoscaberrimus F. T. Wang et Ts. Tang ex S. B. Ho;拟糙叶黄耆(拟糙叶黄芪);False-scabrous-leaf Milkvetch, Sham Roughleaf Milkvetch ■
42931 Astragalus pseudoscaberrimus F. T. Wang et Ts. Tang ex S. B. Ho = Astragalus grubovii Sanchir ■
42932 Astragalus pseudoscaberrimus S. B. Ho = Astragalus pseudoscaberrimus F. T. Wang et Ts. Tang ex S. B. Ho ■
42933 Astragalus pseudoscoparius Gontsch.;类帚黄耆(类帚黄芪);Broom-like Milkvetch, Sham Broom Milkvetch ■
42934 Astragalus pseudosinaicus Gazer et Podlech;假西奈黄耆■☆
42935 Astragalus pseudostella Delile = Astragalus asterias Hohen. ■☆
42936 Astragalus pseudostella Delile var. saharae (Pomel) Batt. = Astragalus saharae Pomel ■☆
42937 Astragalus pseudotrigonus Batt. et Trab. = Astragalus trigonus DC. ■☆
42938 Astragalus pseudoversicolor Y. C. Ho;类变色黄耆(类变色黄芪);False-varied-colour Milkvetch, Sham Variant Milkvetch ■
42939 Astragalus pseudoxytropis Ulbr. = Astragalus acaulis Baker ■
42940 Astragalus psilacanthus Boiss.;裸刺黄耆■☆
42941 Astragalus psilocentros Fisch.;央光黄耆■☆
42942 Astragalus psilocentros Fisch. var. pilosus ?;疏毛黄耆■☆
42943 Astragalus psilocephalus Baker ex Aitch.;毛头黄耆■☆
42944 Astragalus psiloglottis Stev. ex DC. = Astragalus oxyglottis Steven ■
42945 Astragalus psilopterus Bunge = Astragalus lasiosemius Boiss. ●
42946 Astragalus psilosepalus Podlech et L. R. Xu;光萼黄耆■
42947 Astragalus puberulus Ledeb.;茸毛黄耆■
42948 Astragalus pubiflorus DC.;密毛花黄耆■☆
42949 Astragalus pullus Simpson;黑毛黄耆(黑毛黄芪);Blackhair Milkvetch ■
42950 Astragalus pullus Simpson var. pubifolius C. C. Ni et P. C. Li;黑毛叶黄耆(毛叶黄芪,毛叶黄耆);Hairleaf Blackhair Milkvetch ■
42951 Astragalus pullus Simpson var. pubifolius C. C. Ni et P. C. Li = Astragalus monbeigii N. D. Simpson ■
42952 Astragalus pulvinalis P. C. Li et C. C. Ni = Astragalus mongutensis Lipsky ■
42953 Astragalus pulvinalis P. C. Li et H. L. Li;小垫黄耆(小垫黄芪);Small-cushion Milkvetch ■
42954 Astragalus pulvinalis P. C. Li et H. L. Li = Astragalus kuschakewiczii B. Fedtsch. ■
42955 Astragalus purdomii Simpson;紫花黄耆(紫花黄芪);Pueple Milkvetch, Pueple-flower Milkvetch ■
42956 Astragalus purpurascens Bunge;紫黄耆■☆
42957 Astragalus purpurinus (Y. C. Ho) Podlech et L. R. Xu;淡紫花黄耆(淡紫花黄芪)■
42958 Astragalus pycnorhizus Wall. ex Benth.;密根黄耆(密根黄芪);Dense-root Milkvetch, Rooty Milkvetch ■
42959 Astragalus pyrrhotrichus Boiss.;火红毛黄耆■☆
42960 Astragalus qingheensis Y. X. Liou;清河黄耆(清河黄芪);Qinghe Milkvetch ■
42961 Astragalus qitaiensis Podlech et L. R. Xu;奇台黄耆■
42962 Astragalus quasitestinculatus Bar. et Chu ? = Astragalus grubovii Sanchir ■
42963 Astragalus quasitestinculatus Bar. et Chu ? = Astragalus hsinbaticus P. Y. Fu et Y. A. Chen ■
42964 Astragalus radians Pomel = Astragalus longicaulis Pomel ■☆
42965 Astragalus radiatus Bunge = Astragalus asterias Hohen. ■☆
42966 Astragalus reboudii Bunge = Astragalus caprinus L. ■☆
42967 Astragalus reduncus Pall.;钩刺黄耆■☆
42968 Astragalus reesei Maire;里斯黄耆■☆
42969 Astragalus reflexistipulus Miq.;反折木黄耆(木黄芪,木黄耆)●☆
42970 Astragalus reinii Ball;赖因黄耆■☆
42971 Astragalus reinii Ball subsp. atrosanguineus (Murb.) Maire = Astragalus reinii Ball ■☆
42972 Astragalus reinii Ball subsp. mairei Emb. = Astragalus maireanus Greuter et Burdet ■☆
42973 Astragalus reinii Ball subsp. nemorosus (Batt.) Maire;森林黄耆■☆
42974 Astragalus reticulatus M. Bieb.;网黄耆;Reticulate Milkvetch ■☆
42975 Astragalus retroflexus Pall. = Oxytropis deflexa (Pall.) DC. ■
42976 Astragalus retusifoliatus Y. C. Ho;凹叶黄耆(凹叶黄芪);Dentedleaf Milkvetch, Notched Leaf Milkvetch ■
42977 Astragalus rhizanthus Benth.;畸形黄耆;Abnomal Milkvetch, Irregular Milkvetch ●
42978 Astragalus rhizanthus Benth. subsp. candolleanus (Benth.) Podlech;短毛畸形黄耆(短梗黄耆,短毛黄耆)■
42979 Astragalus rhizanthus Benth. var. pindreensis (Ali) Podlech = Astragalus pindreensis (Baker) Ali ■
42980 Astragalus rhizocephalus Baker ex Aitch.;根头黄耆■☆
42981 Astragalus rhododendrophilus Podlech et L. R. Xu;杜鹃黄耆■
42982 Astragalus rigidulus Benth. ex Bunge;坚硬黄耆(坚硬黄芪);Hard Milkvetch, Rigid Milkvetch ■
42983 Astragalus roborovskyi N. Ulziykh. = Astragalus arnoldii Hemsl. et H. Pearson ■
42984 Astragalus rockii Marquart et Airy Shaw = Astragalus camptodontus Franch. var. lichiangensis (Simpson) K. T. Fu ■
42985 Astragalus roseus Ledeb.;毛冠黄耆(毛冠黄芪);Roseal Milkvetch, Roseus Milkvetch ■
42986 Astragalus rotundifolius Benth. = Astragalus graveolens Benth. ■
42987 Astragalus rupifragus Pall.;石黄耆■☆
42988 Astragalus rytidocarpus Ledeb.;橙果黄耆■
42989 Astragalus rytilobus Bunge = Astragalus filicaulis Kar. et Kir. ■
42990 Astragalus sabuletorum Ledeb.;粗沙黄耆■
42991 Astragalus saccatocarpus K. T. Fu;囊果黄耆(囊果黄芪);Bagfruit Milkvetch, Saccate Fruit Milkvetch ■
42992 Astragalus saccatocarpus K. T. Fu = Astragalus craibianus N. D. Simpson ■
42993 Astragalus saccocalyx Schrenk = Astragalus saccocalyx Schrenk ex Fisch. ■
42994 Astragalus saccocalyx Schrenk ex Fisch.;袋萼黄耆(袋萼黄芪);Bagcalyx Milkvetch, Saccate Calyx Milkvetch ■
42995 Astragalus sachalinensis Bunge;库页黄耆;Sachalin Milkvetch ■☆
42996 Astragalus sadiensis Podlech, L. R. Xu et C. Y. Chang;沙地黄耆■
42997 Astragalus sagastaigolensis N. Ulziykh. ex Podlech et L. R. Xu;萨格斯台黄耆■
42998 Astragalus saharae Pomel;左原黄耆■☆
42999 Astragalus salicetorum Kom. = Astragalus alpinus L. ■
43000 Astragalus salsugineus Kar. et Kir.;盐地黄耆(喜盐黄芪,喜盐黄耆);Lovesalt Milkvetch, Salt-loving Milkvetch ■☆
43001 Astragalus salsugineus Kar. et Kir. var. hetaoensis H. C. Fu;河套盐生黄耆(河套盐生黄芪)■
43002 Astragalus salsugineus Kar. et Kir. var. multijugus S. B. Ho = Astragalus lang-ranii Podlech ■

43003 Astragalus sanbilingensis H. T. Tsai et Te T. Yu;乡城黄耆(乡城黄芪);Sanbiling Milkvetch ■
43004 Astragalus saratagius Bunge;阿赖山黄耆(阿赖山黄芪);Alaishan Milkvetch ■
43005 Astragalus saratagius Bunge var. minutiflorus S. B. Ho;小花阿赖山黄耆(小花阿赖山黄芪);Smallflower Alaishan Milkvetch ■
43006 Astragalus saratagius Bunge var. minutiflorus S. B. Ho = Astragalus saratagius Bunge ■
43007 Astragalus sarcocolla Dymock;肉质黄耆(肉质黄芪)■☆
43008 Astragalus satoi Kitag.;小米黄耆(小米黄芪);Sato Milkvetch ■
43009 Astragalus satoi Kitag. = Astragalus capillipes Bunge ■
43010 Astragalus saxicola Ulbr. = Astragalus hancockii Bunge ex Maxim. ■
43011 Astragalus saxorum Simpson;石生黄耆(石生黄芪);Rock-living Milkvetch,Saxicolous Milkvetch ■
43012 Astragalus scaberrimus Bunge;糙叶黄耆(糙叶黄芪,春黄芪,春黄耆,粗糙紫云英,掐不齐);Coslseleaf Milkvetch,Scabrous-leaf Milkvetch ■
43013 Astragalus scabrisetus Bong.;粗毛黄耆(粗毛黄芪);Scabrous-hair Milkvetch,Shag Milkvetch ■
43014 Astragalus scabrisetus Bong. var. multijugus Hand.-Mazz. = Astragalus dengkouensis H. C. Fu ■
43015 Astragalus scabrisetus Bong. var. multijugus Hand.-Mazz. = Astragalus grubovii Sanchir ■
43016 Astragalus schanginianus Pall.;卡通黄耆(卡通黄芪);Katong Milkvetch ■
43017 Astragalus schanginianus Pall. subsp. neoschanginianus Golosk. = Astragalus schanginianus Pall. ■
43018 Astragalus schelichovii Turcz.;舍氏黄耆■☆
43019 Astragalus schimperi Boiss.;欣珀黄耆■☆
43020 Astragalus schimperi Boiss. var. subsessilis Eig = Astragalus schimperi Boiss. ■☆
43021 Astragalus schneideri Ulbr. = Astragalus balfourianus N. D. Simpson ■
43022 Astragalus sciadophorus Franch.;辽西黄耆(辽西黄芪);Shady Milkvetch,Umbrella Milkvetch ■
43023 Astragalus scleropodius Ledeb.;硬柄黄耆●
43024 Astragalus scoparius Schrenk;帚黄耆(帚黄芪);Broom Milkvetch ■
43025 Astragalus scoparius Schrenk f. minutus Lipsky = Astragalus scoparius Schrenk ■
43026 Astragalus scorpioides Willd.;蝎尾状黄耆■☆
43027 Astragalus scorpiurus Bunge;蝎尾黄耆■☆
43028 Astragalus scorpiurus Bunge var. glaber Ali;光滑蝎尾黄耆■☆
43029 Astragalus secretus Podlech et L. R. Xu;黏线黄耆■
43030 Astragalus secundus DC.;侧花黄耆■☆
43031 Astragalus sedaensis Y. C. Ho;色达黄耆(色达黄芪);Seda Milkvetch ■
43032 Astragalus semibilocularis Fisch. ex Bunge = Astragalus austro-sibiricus Schischk. ■
43033 Astragalus semicircularis P. C. Li;半圆黄耆(半圆黄芪);Semi-circular Milkvetch ■
43034 Astragalus sericostachyus Stocks;绢毛穗黄耆■☆
43035 Astragalus sesameus L.;芝麻黄耆■☆
43036 Astragalus sesameus L. var. anoectotrichus Faure et Maire = Astragalus sesameus L. ■☆
43037 Astragalus sesameus L. var. faurei Maire = Astragalus sesameus L. ■☆
43038 Astragalus sesameus L. var. substellaris Maire = Astragalus sesameus L. ■☆
43039 Astragalus sesamoides Boiss.;胡麻黄耆(胡麻黄芪);Sesame Milkvetch ■
43040 Astragalus severzovii Bunge;无毛黄耆(塞氏黄芪,塞氏黄耆,无毛黄芪);Hairless Milkvetch ■
43041 Astragalus shinanensis Ohwi = Astragalus membranaceus (Fisch. ex Link) Bunge ■
43042 Astragalus shiroumensis Makino;白马岳黄耆■☆
43043 Astragalus siccaneus P. C. Li;耐寒黄耆(耐寒黄芪)■
43044 Astragalus sichuanensis L. Meng;四川黄耆(四川黄芪);Sichuan Milkvetch ■
43045 Astragalus sieberi DC.;西伯尔黄耆■☆
43046 Astragalus sieversianus Pall.;绵果黄耆(绵果黄芪,绵毛黄芪,绵毛黄耆);Cottonfruit Milkvetch ■
43047 Astragalus sikkimensis Bunge;锡金黄耆■
43048 Astragalus sikokianus Nakai;四国黄耆■☆
43049 Astragalus simpsonii E. Peter;灌县黄耆(灌县黄芪,辛氏黄芪,辛氏黄耆);Simpson Milkvetch ■
43050 Astragalus sinaicus Boiss. var. arenicola (Pomel) Maire = Astragalus stella Gouan ■☆
43051 Astragalus sinaicus Boiss. var. pedunculatus Pamp. = Astragalus stella Gouan ■☆
43052 Astragalus sinaicus Boiss. var. saharae (Pomel) Batt. = Astragalus saharae Pomel ■☆
43053 Astragalus sinicus L.;紫云英(斑鸠花,草蒺藜,灯笼花,滚龙珠,荷花苢,荷花郎,荷花紫草,红花菜,红花草,红花花,红花郎,花菜,花草,蒺藜子,莲花草,莲华草,米布袋,米伞花,米筛花草,螃蟹花,翘翘花,翘摇,翘摇车,沙蒺藜,沙苑子,苕,苕翘,苕子菜,碎米芥,碎米荠,铁马豆,摇车,野蚕豆,野鸭草,柱夫);China Milkvetch,Chinese Clover,Chinese Milk Vetch,Chinese Milkvetch ●■
43054 Astragalus sinicus L. f. albiflorus S. Okamoto;白花紫云英■☆
43055 Astragalus sinicus L. var. macrocalyx Ulbr. = Astragalus sinicus L. ●■
43056 Astragalus sinkiangensis Podlech et L. R. Xu;新疆黄耆■
43057 Astragalus sitchsonii Baker = Astragalus ophiocarpus Benth. et Bunge ■
43058 Astragalus skorniakovii B. Fedtsch.;戈尔诺黄耆(戈尔诺黄芪);Skorniakov Milkvetch ■
43059 Astragalus skorniakovii B. Fedtsch. var. wuqiaensis S. B. Ho;乌恰戈尔诺黄耆(乌恰黄芪);Wuqia Milkvetch ■
43060 Astragalus skorniakowii B. Fedtsch. = Astragalus masenderanus Bunge ■
43061 Astragalus skorniakowii B. Fedtsch. = Astragalus skorniakovii B. Fedtsch. ■
43062 Astragalus skorniakowii B. Fedtsch. var. wuqiaensis S. B. Ho = Astragalus masenderanus Bunge ■
43063 Astragalus skythropos Bunge;肾形子黄耆(肾形子黄芪);Kidneyshaped Seed Milkvetch,Sullen Milkvetch ■
43064 Astragalus skythropos Bunge var. acaulis Danguy = Astragalus skythropos Bunge ■
43065 Astragalus smithianus E. Peter;无毛叶黄耆(无毛叶黄芪);Hairless Leaf Milkvetch,Smith Milkvetch ■
43066 Astragalus sogotensis Lipsky;索戈塔黄耆(戟塔黄芪,索戈塔黄芪);Sogot Milkvetch ■
43067 Astragalus solandri Lowe;索兰德黄耆■☆

43068 Astragalus solandri Lowe var. bubaloceras (Maire) Emb. et Maire = Astragalus solandri Lowe ■☆

43069 Astragalus somalensis Taub. ex Harms = Galega somalensis (Taub. ex Harms) J. B. Gillett ■

43070 Astragalus somalensis Taub. ex Harms var. lindblomii Harms = Galega lindblomii (Harms) J. B. Gillett ■☆

43071 Astragalus songaricus Pall. = Oxytropis songarica (Pall.) DC. ■

43072 Astragalus songolicus Gontsch. = Astragalus nicolai Boriss. ■

43073 Astragalus souliei Simpson;蜀西黄耆(蜀西黄芪);Soulie Milkvetch ■

43074 Astragalus sphaeranthus Boiss.;圆花黄耆■☆

43075 Astragalus sphaerocystis Bunge;球囊黄耆(球囊黄芪);Ballbag Milkvetch,Sphaere-sac Milkvetch ■

43076 Astragalus sphaerophysa Kar. et Kir.;球稃黄耆(球稃黄芪);Ballbullate Milkvetch,Sphaere-inflation Milkvetch ■

43077 Astragalus spinosissimus Wall. = Caragana gerardiana Benth. ●

43078 Astragalus spinosus (Forssk.) Muschl.;具刺黄耆■☆

43079 Astragalus squarrosus Bunge;粗鳞黄耆■☆

43080 Astragalus staintonianus Ali;斯坦顿黄耆■☆

43081 Astragalus staintonianus Ali = Astragalus charguschanus Freyn ■

43082 Astragalus stalinskyi Sirj.;矮型黄耆(矮型黄芪);Dwarf Milkvetch,Stalinsky Milkvetch ■

43083 Astragalus steinhergianus Sumner;蒙西黄耆(蒙西黄芪);Steinberg Milkvetch ■

43084 Astragalus stella Gouan;星状黄耆■☆

43085 Astragalus stenoceras C. A. Mey.;狭荚黄耆(狭荚黄芪);Narrow Pod Milkvetch,Narrowpod Milkvetch ■

43086 Astragalus stenoceras C. A. Mey. var. longidentatus S. B. Ho;长齿狭荚黄耆(长齿狭荚黄芪);Longtooth Narrowpod Milkvetch ■

43087 Astragalus stenoceras C. A. Mey. var. longidentatus S. B. Ho = Astragalus lanzhouensis Podlech et L. R. Xu ■

43088 Astragalus stenoceras C. A. Mey. var. macranthus Bunge = Astragalus macrotropis Bunge ■

43089 Astragalus stenolobus Bunge = Astragalus ceratoides M. Bieb. ■

43090 Astragalus stenophyllus Bunge = Gueldenstaedtia verna (Georgi) Borissov ■

43091 Astragalus stenorrhinus Pau = Astragalus echinatus Murray ■☆

43092 Astragalus stevenianus DC. = Astragalus leansanicus Ulbr. ■

43093 Astragalus stewartii Baker;斯图尔特黄耆■☆

43094 Astragalus stipulatus D. Don ex Simpson;大托叶黄耆(大托叶黄芪);Bigstipule Milkvetch,Large-stipule Milkvetch ■

43095 Astragalus stocksii Benth. ex Bunge;斯托克斯黄耆■☆

43096 Astragalus striatellus M. Bieb.;条纹黄耆;Striate Milkvetch ■☆

43097 Astragalus strictus Benth. = Astragalus strictus Graham ex Benth. ■

43098 Astragalus strictus Graham ex Benth.;笔直黄耆(笔直黄芪,劲直黄芪,劲直黄耆);Straight Milkvetch,Strict Milkvetch ■

43099 Astragalus strobiliferus Lindl.;球果黄耆(黄芪球果)■☆

43100 Astragalus subarcuatus Popov;弧果黄耆(弧果黄芪);Arcfruit Milkvetch,Arcuate-fruit Milkvetch ■

43101 Astragalus subulatus Desf. = Astragalus annularis Forssk. ■☆

43102 Astragalus subulatus Desf. = Astragalus gladiatus Boiss. ■

43103 Astragalus subulatus M. Bieb.;锥黄耆■☆

43104 Astragalus subulatus M. Bieb. = Astragalus gladiatus Boiss. ■

43105 Astragalus subulatus M. Bieb. = Astragalus subuliformis DC. ■

43106 Astragalus subulatus M. Bieb. var. altaicus Pall. = Astragalus stenoceras C. A. Mey. ■

43107 Astragalus subuliformis DC. = Astragalus gladiatus Boiss. ■

43108 Astragalus subumbellatus Klotzsch;小伞黄耆■☆

43109 Astragalus succulentus Richardson = Astragalus crassicarpus Nutt. ■☆

43110 Astragalus suffalcatus Bunge = Astragalus subuliformis DC. ■

43111 Astragalus suidenensis Bunge;水定黄耆(萨伊墩黄芪,萨伊墩黄耆,水定黄芪);Shuiding Milkvetch ■

43112 Astragalus sulcatus L.;纹茎黄耆(纹茎黄芪);Furrowstem Milkvetch,Sulcatestem Milkvetch ■

43113 Astragalus sungpanensis E. Peter;松潘黄耆(松潘黄芪);Songpan Milkvetch,Sungpan Milkvetch ■

43114 Astragalus sungpanensis E. Peter f. albiflorus Y. H. Wu;白花松潘黄耆(白花松潘黄芪);Whiteflower Songpan Milkvetch ■

43115 Astragalus sungpanensis E. Peter f. albiflorus Y. H. Wu = Astragalus sungpanensis E. Peter ■

43116 Astragalus supraglaber Kitam. = Astragalus melanostachys Benth. ex Bunge ■

43117 Astragalus supralaevis Podlech et L. R. Xu;德钦黄耆■

43118 Astragalus sutchuensis Franch.;苏黄耆■☆

43119 Astragalus suwuiformis DC.;剑叶黄耆(剑叶黄芪)■☆

43120 Astragalus tachdirtensis Andr. = Astragalus depressus L. subsp. atlantis Maire ■☆

43121 Astragalus taipaishanensis Y. C. Ho et S. B. Ho ex C. W. Chang;太白山黄耆(太白山黄芪);Taibaishan Milkvetch, Taipaishan Milkvetch ■

43122 Astragalus taiyuanensis S. B. Ho;太原黄耆(太原黄芪);Taiyuan Milkvetch ■

43123 Astragalus taldicensis Franch.;假黄耆(假黄芪);False Milkvetch,Sham Milkvetch ■

43124 Astragalus talievii Sirj. = Astragalus tibetanus Bunge ■

43125 Astragalus tanguticus Batalin;甘青黄耆(甘青黄芪,青海黄芪,青海黄耆);Tangut Milkvetch ■

43126 Astragalus tanguticus Batalin f. albiflorus (S. W. Liu ex K. T. Fu) K. T. Fu;白花甘青黄耆(白花甘青黄芪);Whiteflower Tangut Milkvetch ■

43127 Astragalus tanguticus Batalin var. albiflorus S. W. Liu ex K. T. Fu = Astragalus tanguticus Batalin f. albiflorus (S. W. Liu ex K. T. Fu) K. T. Fu ■

43128 Astragalus taschkendicus Bunge;塔什干黄耆(塔什干黄芪)■☆

43129 Astragalus tataricus Franch. = Astragalus polycladus Bureau et Franch. ■

43130 Astragalus tataricus Franch. = Astragalus zacharensis Bunge ■

43131 Astragalus tataricus Franch. var. major H. T. Tsai et Te T. Yu = Astragalus havianus E. Peter ■

43132 Astragalus tatsienensis Bureau et Franch.;康定黄耆(康定黄芪);Kangding Milkvetch ■

43133 Astragalus tatsienensis Bureau et Franch. = Astragalus yunnanensis Franch. ■

43134 Astragalus tatsienensis Bureau et Franch. f. incanus E. Peter = Astragalus tatsienensis Bureau et Franch. var. incanus (E. Peter) Y. C. Ho ■

43135 Astragalus tatsienensis Bureau et Franch. var. incanus (E. Peter) Y. C. Ho;灰毛康定黄耆(灰毛康定黄芪);Greyhair Kangding Milkvetch ■

43136 Astragalus tatsienensis Bureau et Franch. var. incanus (E. Peter) Y. C. Ho = Astragalus yunnanensis Franch. subsp. incanus (E. Peter) Podlech et L. R. Xu ■☆

43137 Astragalus tatsienensis Bureau et Franch. var. kangrenbuchiensis

(C. C. Ni et P. C. Li) Y. C. Ho;岗仁布齐黄耆(岗仁布齐黄芪);Gangrenbuchi Milkvetch,Gangrenbuqi Milkvetch ■

43138 Astragalus tatsienensis Bureau et Franch. var. kangrenbuchiensis (Ni et P. C. Li) Y. C. Ho = Astragalus yunnanensis Franch. ■

43139 Astragalus tauricus Pall.;克里木黄耆;Kelimu Milkvetch ■☆

43140 Astragalus tecti-mundi Freyn;屋脊黄耆■

43141 Astragalus tecti-mundi Freyn subsp. orientalis Podlech;东方屋脊黄耆■

43142 Astragalus tekesensis S. B. Ho;特克斯黄耆(特克斯黄芪);Tekes Milkvetch ■

43143 Astragalus tenuicaulis Benth. ex Bunge;细茎黄耆(细茎黄芪);Tenuous-stem Milkvetch,Thinstem Milkvetch ■

43144 Astragalus tenuifoliolosus Maire = Astragalus algerianus E. Sheld. ■☆

43145 Astragalus tenuifolius Desf. = Astragalus algerianus E. Sheld. ■☆

43146 Astragalus tenuifolius Desf. var. austro-oranensis Hochr. = Astragalus algerianus E. Sheld. ■☆

43147 Astragalus tenuirugis Boiss. = Astragalus corrugatus Bertol. ■☆

43148 Astragalus tenuis Turcz. = Astragalus melilotoides Pall. var. tenuis (Turcz.) Ledeb. ■

43149 Astragalus tephrosioides Boiss.;灰黄耆■☆

43150 Astragalus tesquorum Podlech et L. R. Xu;干草原黄耆■

43151 Astragalus testiculatus Pall.;西方卵果黄耆■☆

43152 Astragalus thomsonii Podlech;汤姆森黄耆■

43153 Astragalus tianschanicus Bunge var. pamiricus B. Fedtsch. = Astragalus charguschanus Freyn ■

43154 Astragalus tibetanus Benth. ex Bunge;藏新黄耆(藏黄芪,藏黄耆,藏新黄芪,春黄芪,春黄耆);Tibet Milkvetch,Xizang Milkvetch ■

43155 Astragalus tibetanus Benth. ex Bunge var. patentipilus K. T. Fu;展毛藏新黄耆(展毛藏黄耆,展毛黄芪,展毛黄耆)■

43156 Astragalus tibetanus Bunge = Astragalus tibetanus Benth. ex Bunge ■

43157 Astragalus tibeticola Podlech et L. R. Xu;藏黄耆■

43158 Astragalus tibeticus var. patentipilus K. T. Fu = Astragalus tibetanus Bunge ■

43159 Astragalus tingriensis C. C. Ni et P. C. Li;定日黄耆(定日黄芪);Dingri Milkvetch,Tingri Milkvetch ■

43160 Astragalus tokachiensis T. Yamaz. et Kadota;十胜黄耆■☆

43161 Astragalus toksunensis S. B. Ho;托克逊黄耆(托克逊黄芪);Toksun Milkvetch ■

43162 Astragalus toktjenensis Ulbr. = Hedysarum tibeticum (Benth.) B. H. Choi et H. Ohashi ●■◇

43163 Astragalus toktjensis Ulbr. = Stracheya tibetica Benth. ●■◇

43164 Astragalus tongolensis Ulbr.;东俄洛黄耆(东俄洛黄芪,连七,毛黑果黄芪,毛黑果黄耆,唐谷耳黄芪,塘谷耳黄芪,塘谷耳黄耆);Tongol Milkvetch ■

43165 Astragalus tongolensis Ulbr. var. breviflorus H. T. Tsai et Te T. Yu;小花黄耆(小花东俄洛黄芪,小花黄芪);Short-flower Tongol Milkvetch ■

43166 Astragalus tongolensis Ulbr. var. breviflorus H. T. Tsai et Te T. Yu = Astragalus tongolensis Ulbr. ■

43167 Astragalus tongolensis Ulbr. var. glaber E. Peter;光东俄洛黄耆(光东俄洛黄芪);Glabrous Tongol Milkvetch ■

43168 Astragalus tongolensis Ulbr. var. glaber E. Peter = Astragalus tongolensis Ulbr. ■

43169 Astragalus tongolensis Ulbr. var. lanceolato-dentatus E. Peter;剑齿黄耆(长齿东俄洛黄芪,长齿黄芪,长齿黄耆);Longtooth Tongol Milkvetch ■

43170 Astragalus tongolensis Ulbr. var. lanceolato-dentatus E. Peter = Astragalus tongolensis Ulbr. ■

43171 Astragalus tongolensis Ulbr. var. longibracteatus Y. H. Wu;长苞东俄洛黄耆(长苞东俄洛黄芪);Longbract Tongol Milkvetch ■

43172 Astragalus trabutianus Batt. = Astragalus asterias Hohen. ■☆

43173 Astragalus tragacantha L.;西方胶黄耆;Bishop's Mitre,Tragacanth ■☆

43174 Astragalus transecticola Podlech et L. R. Xu;路边黄耆■

43175 Astragalus transiliensis Gontsch.;外伊犁黄耆(外伊犁黄芪);Transe Ili Milkvetch ■

43176 Astragalus transiliensis Gontsch. = Astragalus ellipsoideus Ledeb. ■

43177 Astragalus transiliensis Gontsch. var. microphyllus S. B. Ho = Astragalus ochrias Bunge ■

43178 Astragalus tribulifolius Benth. ex Bunge;蒺藜叶黄耆(蒺藜叶黄芪);Caltropleaf Milkvetch,Caltropus Milkvetch ■

43179 Astragalus tribulifolius Benth. ex Bunge var. pauciflorus C. Marquand et Airy Shaw;少花黄耆(少花黄芪);Fewflower Caltropus Milkvetch,Fewflower Milkvetch ■

43180 Astragalus tribuloides Delile;蒺藜黄耆(蒺藜黄芪);Caltroplike Milkvetch ■

43181 Astragalus tribuloides Delile var. arenicola (Pomel) Murb. = Astragalus stella Gouan ■☆

43182 Astragalus tribuloides Delile var. leiocarpus Boiss. = Astragalus tribuloides Delile ■

43183 Astragalus tribuloides Delile var. mareoticus Sirj. = Astragalus tribuloides Delile ■

43184 Astragalus tribuloides Delile var. minutus (Boiss.) Boiss. = Astragalus tribuloides Delile ■

43185 Astragalus tribuloides Delile var. platycarpus Pamp. = Astragalus tribuloides Delile ■

43186 Astragalus tridens Bullock et C. A. Sm. ex Jex-Blake = Galega lindblomii (Harms) J. B. Gillett ■☆

43187 Astragalus trigonus DC.;三角黄耆■☆

43188 Astragalus trigonus DC. var. franchii (Trotter) Maire et Weiller = Astragalus trigonus DC. ■☆

43189 Astragalus trijugus Podlech et L. R. Xu;三棱黄耆■

43190 Astragalus trimestris L.;三月黄耆■☆

43191 Astragalus trojanus Stev.;特洛伊黄耆(特洛伊黄芪)■☆

43192 Astragalus tsangpoensis Podlech et L. R. Xu;藏布黄耆■

43193 Astragalus tsataensis C. C. Ni et P. C. Li = Astragalus hoffmeisteri (Klotzsch) Ali ■

43194 Astragalus tulinovii O. Fedtsch.;土力黄耆■

43195 Astragalus tumbatsica C. Marquand et Airy Shaw;东坝子黄耆(东坝子黄芪);Dongbazi Milkvetch,Tumbatsi Milkvetch ■

43196 Astragalus tungensis N. D. Simpson;洞川黄耆(洞川黄芪);Dongchuan Milkvetch,Tungchuan Milkvetch ■

43197 Astragalus turgidocarpus K. T. Fu;膨果黄耆(绵耆,膨果黄芪);Inflatefruit Milkvetch,Turgid Fruit Milkvetch ■

43198 Astragalus turolensis Pau;图罗尔黄耆■☆

43199 Astragalus tyttocarpus Gontsch.;细果黄耆;Smallfruit Milkvetch,Tenuous Fruit Milkvetch,Tenuous-fruited Milkvetch ●

43200 Astragalus ui-eilakensis Gharemani-Nejad = Astragalus nematodioides H. Ohba,Akiyama et S. K. Wu ■

43201 Astragalus ulachanensis Franch. = Astragalus discolor Bunge ex Maxim. ■

43202 Astragalus ulacholensis B. Fedtsch. = Astragalus lasiopetalus

Bunge ■
43203 Astragalus ulaschanensis Franch. = Astragalus discolor Bunge ex Maxim. ■
43204 Astragalus uliginosus L.;湿地黄耆(湿地黄芪);Everglade Milkvetch,Swamp Milkvetch ■
43205 Astragalus umbellatus Bunge;伞形黄耆■☆
43206 Astragalus uncinatus Bertol. = Astragalus scorpioides Willd. ■☆
43207 Astragalus unijugus Bunge;对叶黄耆(对叶黄芪);Pairleaf Milkvetch,Unijugous Milkvetch ●
43208 Astragalus urunguensis N. Ulziykh.;乌伦古黄耆■
43209 Astragalus utriger Pall.;袋状黄耆■☆
43210 Astragalus vaginatus Pall.;鞘叶黄耆■
43211 Astragalus valerii N. Ulzijkh.;瓦来黄耆■
43212 Astragalus vallestris Kamelin;线沟黄耆■
43213 Astragalus variabilis Bunge = Astragalus variabilis Bunge ex Maxim. ■
43214 Astragalus variabilis Bunge ex Maxim.;变异黄耆(变异黄芪);Variant Milkvetch,Varied Milkvetch ■
43215 Astragalus veitchianus N. D. Simpson = Astragalus tongolensis Ulbr. ■
43216 Astragalus veitchianus Simpson = Astragalus tongolensis Ulbr. ■
43217 Astragalus venosus Hochst. ex A. Rich. = Astragalus atropilosulus (Hochst.) Bunge ■☆
43218 Astragalus verna (Georgi) Kitag. = Gueldenstaedtia verna (Georgi) Borissov ■
43219 Astragalus vernus Georgi = Gueldenstaedtia verna (Georgi) Borissov ■
43220 Astragalus versicolor Pall. = Astragalus pseudoversicolor Y. C. Ho ■
43221 Astragalus verus DC. ex Bunge;真实黄耆(真实黄芪)■☆
43222 Astragalus vescus Podlech et L. R. Xu;辛辣黄耆■
43223 Astragalus vesicarius L.;具泡黄耆■☆
43224 Astragalus vicarius Lipsky;替代黄耆■
43225 Astragalus vicioides Baker = Astragalus concretus Benth. ■
43226 Astragalus vicioides Ledeb. = Astragalus macropterus DC. ■
43227 Astragalus vicioides Ledeb. var. longipes Trautv. = Astragalus macropterus DC. ■
43228 Astragalus virgatus Pall.;柳条黄耆(多枝黄耆)■☆
43229 Astragalus visibilis Podlech et L. R. Xu;明媚黄耆■
43230 Astragalus vladimiri-komarovii B. Fedtsch.;卡乌洛夫黄耆■
43231 Astragalus vogelii (Webb) Bornm.;沃格尔黄耆■☆
43232 Astragalus vogelii (Webb) Bornm. subsp. fatmensis (Hochst. ex Chiov.) Maire = Astragalus fatmensis Hochst. ex Chiov. ■☆
43233 Astragalus vogelii (Webb) Bornm. subsp. prolixus (Sieber ex Bunge) Maire = Astragalus vogelii (Webb) Bornm. ■☆
43234 Astragalus vogelii (Webb) Bornm. var. congestus (Schweinf.) Cufod. = Astragalus fatmensis Hochst. ex Chiov. ■☆
43235 Astragalus volubilitanus Braun-Blanq. et Maire = Astragalus hamosus L. ■☆
43236 Astragalus vulpinus Willd.;拟狐尾黄耆(拟狐尾黄芪,似狐尾黄芪);Foxtail Milkvetch,Foxtail-like Milkvetch ■
43237 Astragalus webbianus Benth.;藏西黄耆■
43238 Astragalus weigoldianus Hand.-Mazz. = Astragalus skythropos Bunge ■
43239 Astragalus weilleri Emb. et al. = Astragalus caprinus L. ■☆
43240 Astragalus weixiensis Y. C. Ho;维西黄耆(维西黄芪);Weixi Milkvetch ■
43241 Astragalus wenquanensis S. B. Ho;温泉黄耆(温泉黄芪);Wenquan Milkvetch ■
43242 Astragalus wensuensis S. B. Ho;温宿黄耆(温宿黄芪);Wensu Milkvetch ■
43243 Astragalus wensuensis S. B. Ho = Astragalus heptapotamicus Sumnev. ■
43244 Astragalus wenxianensis Y. C. Ho;文县黄耆(文县黄芪);Wenxian Milkvetch ■
43245 Astragalus wenxianensis Y. C. Ho = Astragalus sutchuensis Franch. ■☆
43246 Astragalus wilsonii Simpson = Astragalus ernestii Comber ■
43247 Astragalus woldemari Juz.;中亚黄芪(线叶黄芪);Central Asia Milkvetch,Linear Leaf Milkvetch ●☆
43248 Astragalus woldemari Juz. = Astragalus tyttocarpus Gontsch. ●
43249 Astragalus woldemari Juz. var. atrotrichocladus S. B. Ho;黑枝黄芪(黑枝线叶黄芪);Black-branch Milkvetch, Black-branched Milkvetch ●
43250 Astragalus woldemari Juz. var. atrotrichocladus S. B. Ho = Astragalus tyttocarpus Gontsch. ●
43251 Astragalus wolgensis Bunge;伏尔加黄耆;Volga Milkvetch ■☆
43252 Astragalus wolungensis P. C. Li;卧龙黄耆(卧龙黄芪);Wolong Milkvetch ■
43253 Astragalus wootonii E. Sheld.;伍氏黄耆;Wooton Loco ■☆
43254 Astragalus wulumuqianus F. T. Wang et Ts. Tang ex K. T. Fu;乌鲁木齐黄耆(乌市黄芪,乌市黄耆);Urumqi Milkvetch,Wulumuqi Milkvetch ■
43255 Astragalus wulumuquianus K. T. Fu = Astragalus wulumuqianus F. T. Wang et Ts. Tang ex K. T. Fu ■
43256 Astragalus wushanicus N. D. Simpson;巫山黄耆(巫山黄芪);Wushan Snakemushroom ■
43257 Astragalus xanthotrichos Ledeb.;黄毛黄耆●
43258 Astragalus xiaojinensis Y. C. Ho;小金黄耆(小金黄芪);Xiaojin Milkvetch ■
43259 Astragalus xinjiangensis Y. C. Ho = Astragalus lanuginosus Kar. et Kir. ■
43260 Astragalus xiphocarpus Bunge = Astragalus concretus Benth. ■
43261 Astragalus xiqingshanicus Y. H. Wu;西倾山黄耆(西倾山黄芪);Xiqingshan Milkvetch ■
43262 Astragalus xiqingshanicus Y. H. Wu = Astragalus minshanensis K. T. Fu ■
43263 Astragalus xitaibaicus (K. T. Fu) Podlech et L. R. Xu;西太白黄耆(西太白黄芪);Xitaibai Milkvetch ■
43264 Astragalus xylorrhizus Bunge = Astragalus petraeus Kar. et Kir. ■
43265 Astragalus yamamotoi Miyabe et Tatew.;山本氏黄耆■☆
43266 Astragalus yanerwoensis Podlech et L. R. Xu;长喙黄耆(长喙黄芪);Long-beaked Pavlov Milkvetch ■
43267 Astragalus yang-changii Podlech et L. R. Xu;托里黄耆■
43268 Astragalus yangii C. Chen et Z. G. Qian;竟生黄耆■
43269 Astragalus yangtzeanus N. D. Simpson;扬子黄耆(扬子黄芪);Yangtze Milkvetch,Yangzi Milkvetch ■
43270 Astragalus yatungensis C. C. Ni et P. C. Li = Astragalus donianus DC. ■
43271 Astragalus yechengensis Podlech et L. R. Xu;叶城黄耆■
43272 Astragalus yezoensis Miyabe et Tatew. = Astragalus membranaceus (Fisch. ex Link) Bunge ■
43273 Astragalus yumenensis S. B. Ho = Astragalus yumenensis S. B. Ho ex Podlech et L. R. Xu ■
43274 Astragalus yumenensis S. B. Ho ex Podlech et L. R. Xu;玉门黄

耆(玉门黄芪);Yumen Milkvetch ■
43275 Astragalus yunnanensis Franch.;云南黄耆(云南黄芪);Yunnan Milkvetch ■
43276 Astragalus yunnanensis Franch. f. elongatus N. D. Simpson = Astragalus yunnanensis Franch. ■
43277 Astragalus yunnanensis Franch. subsp. incanus (E. Peter) Podlech et L. R. Xu;灰毛云南黄耆■☆
43278 Astragalus yunnanensis Franch. var. kangrenbuchiensis C. C. Ni et P. C. Li = Astragalus tatsienensis Bureau et Franch. var. kangrenbuchiensis (C. C. Ni et P. C. Li) Y. C. Ho ■
43279 Astragalus yunnanensis Franch. var. kangrenbuchiensis C. C. Ni et P. C. Li = Astragalus yunnanensis Franch. ■
43280 Astragalus yunnanensis Franch. var. tatsienensis (Bureau et Franch.) P. C. Li et C. C. Li = Astragalus yunnanensis Franch. ■
43281 Astragalus yunningensis H. T. Tsai et Te T. Yu;永宁黄耆■
43282 Astragalus yunningensis H. T. Tsai et Te T. Yu = Astragalus polycladus Bureau et Franch. ■
43283 Astragalus yutianensis Podlech et L. R. Xu;于田黄耆■
43284 Astragalus zacharensis Bunge;小果黄耆■
43285 Astragalus zadaensis Podlech et L. R. Xu;札达黄耆■
43286 Astragalus zaissanensis Sumnev.;斋桑黄耆(斋桑黄芪);Zaissan Milkvetch,Zhaisang Milkvetch ■
43287 Astragalus zanskarensis Bunge subsp. oplites (Parker) I. Deml = Astragalus oplites Parker ■
43288 Astragalus zayuensis C. C. Ni et P. C. Li;察隅黄耆(察隅黄芪);Chayu Milkvetch,Zayu Milkvetch ■
43289 Astragalus zhaolingicus K. T. Fu;昭陵黄耆(昭陵黄芪);Zhaoling Milkvetch ■
43290 Astragalus zhaolingicus K. T. Fu = Astragalus galactites Pall. ■
43291 Astragalus zhouquinus K. T. Fu;舟曲黄耆(舟曲黄芪);Zhouqu Milkvetch ■
43292 Astragalus zingeri Korsh.;金氏黄耆;Zinger Milkvetch ■☆
43293 Astragalus zinjiangensis Y. C. Ho = Astragalus nicolai Boriss. ■
43294 Astralagus Curran = Astragalus L. ●■
43295 Astranthium Nutt. (1840);西雏菊属;Western-daisy ■☆
43296 Astranthium ciliatum (Raf.) G. L. Nesom;缘毛西雏菊;Comanche Western-daisy ■☆
43297 Astranthium integrifolium (Michx.) Nutt.;东部西雏菊;Eastern Western-daisy ■☆
43298 Astranthium integrifolium (Michx.) Nutt. subsp. ciliatum (Raf.) De Jong = Astranthium ciliatum (Raf.) G. L. Nesom ■☆
43299 Astranthium integrifolium (Michx.) Nutt. var. ciliatum (Raf.) Larsen = Astranthium ciliatum (Raf.) G. L. Nesom ■☆
43300 Astranthium integrifolium (Michx.) Nutt. var. robustum Shinners = Astranthium robustum (Shinners) De Jong ■☆
43301 Astranthium integrifolium (Michx.) Nutt. var. triflorum (Raf.) Shinners = Astranthium ciliatum (Raf.) G. L. Nesom ■☆
43302 Astranthium robustum (Shinners) De Jong;得州西雏菊;Texas Western-daisy ■☆
43303 Astranthus Lour. = Homalium Jacq. ●
43304 Astranthus cochinchinensis Lour. = Homalium cochinchinense (Lour.) Druce ●
43305 Astrantia Ehth. = Astrantia L. ■☆
43306 Astrantia L. (1753);大星芹属(粉珠花属,星芹属);Astrantia, Hattie's Pincushion,Master Wort,Masterwort ■☆
43307 Astrantia capensis (P. J. Bergius) Druce = Alepidea capensis (P. J. Bergius) R. A. Dyer ■☆
43308 Astrantia colchica Albov;黑海大星芹■☆
43309 Astrantia major L.;大星芹(粉珠花);Astrantia, Black Hellebore,Black Masterwort,Black Sanicle,Great Masterwort,Master Wort, Masterwort, Melancholy Gentlemen, Mountain Sanicle, Mountain Sannicle,Pink Masterwort ■☆
43310 Astrantia maxima Pall.;大粉珠花;Masterwort ■☆
43311 Astrantia minor L.;小星芹;Dwarf Masterwort ■☆
43312 Astrantia pontica Albov;蓬特大星芹■☆
43313 Astrantia trifida Hoffm.;三裂大星芹■☆
43314 Astrapaea Lindl. = Dombeya Cav.(保留属名)●☆
43315 Astrapaea wallichii Lindl. = Dombeya wallichii (Lindl.) K. Schum. ●☆
43316 Astrebla F. Muell. (1878);阿司禾属(阿司吹禾属);Mitchell Grass ■☆
43317 Astrebla pectinata (Lindl.) Benth.;阿司禾(阿司吹禾);Mitchell Grass ■☆
43318 Astrephia Dufr. = Valeriana L. ●■
43319 Astridia Dinter et Schwantes = Astridia Dinter ●☆
43320 Astridia Dinter(1926);鹿角海棠属;Astridia ●☆
43321 Astridia alba (L. Bolus) L. Bolus;白鹿角海棠;White Astridia ●☆
43322 Astridia alba L. Bolus = Astridia alba (L. Bolus) L. Bolus ●☆
43323 Astridia blanda L. Bolus;光滑鹿角海棠●☆
43324 Astridia blanda L. Bolus = Astridia velutina Dinter ■☆
43325 Astridia blanda L. Bolus. f. angusta L. Bolus ex H. Jacobsen = Astridia velutina Dinter ■☆
43326 Astridia blanda L. Bolus. f. latipetala L. Bolus ex H. Jacobsen = Astridia velutina Dinter ■☆
43327 Astridia citrina (L. Bolus) L. Bolus;柠檬鹿角海棠●☆
43328 Astridia dulcis L. Bolus;甜鹿角海棠●☆
43329 Astridia hallii L. Bolus;明日帆●☆
43330 Astridia herrei L. Bolus;明日春●☆
43331 Astridia latisepala L. Bolus = Astridia longifolia (L. Bolus) L. Bolus ■☆
43332 Astridia longifolia (L. Bolus) L. Bolus;长花鹿角海棠;Longflower Astridia ■☆
43333 Astridia maxima (Haw.) Schwantes = Ruschia maxima (Haw.) L. Bolus ●☆
43334 Astridia rubra (L. Bolus) L. Bolus;明日火(红鹿角海棠)●☆
43335 Astridia rubra (L. Bolus) L. Bolus var. alba L. Bolus = Astridia alba (L. Bolus) L. Bolus ●☆
43336 Astridia rubra (L. Bolus) L. Bolus var. citrina L. Bolus = Astridia citrina (L. Bolus) L. Bolus ●☆
43337 Astridia ruschii L. Bolus = Astridia hallii L. Bolus ●☆
43338 Astridia speciosa L. Bolus;美丽鹿角海棠●☆
43339 Astridia swartpoortensis L. Bolus = Astridia speciosa L. Bolus ●☆
43340 Astridia vanbredai L. Bolus = Astridia longifolia (L. Bolus) L. Bolus ■☆
43341 Astridia vanheerdei L. Bolus;黑尔德鹿角海棠●☆
43342 Astridia velutina (Dinter) Dinter;鹿角海棠(明日红);Hairy Astridia,Velvety Astridia ■☆
43343 Astridia velutina Dinter = Astridia velutina (Dinter) Dinter ■☆
43344 Astripomoea A. Meeuse(1958);星牵牛属■☆
43345 Astripomoea cephalantha (Hallier f.) Verdc.;头花星牵牛■☆
43346 Astripomoea delamereana (Rendle) Verdc.;德拉米尔星牵牛■☆
43347 Astripomoea grantii (Rendle) Verdc.;格兰特星牵牛■☆
43348 Astripomoea hyoscyamoides (Vatke) Verdc.;天仙子星牵牛■☆
43349 Astripomoea hyoscyamoides (Vatke) Verdc. var. melandrioides

(Hallier f.) Verdc.;女娄菜星牵牛■☆
43350 Astripomoea lachnosperma (Choisy) A. Meeuse;绵毛籽星牵牛■☆
43351 Astripomoea longituba Verdc.;长管星牵牛■☆
43352 Astripomoea malvacea (Klotzsch) A. Meeuse;锦葵星牵牛■☆
43353 Astripomoea malvacea (Klotzsch) A. Meeuse var. epedunculata (Rendle) Verdc.;无梗花星牵牛■☆
43354 Astripomoea malvacea (Klotzsch) A. Meeuse var. floccosa (Vatke) Verdc.;丛毛星牵牛■☆
43355 Astripomoea malvacea (Klotzsch) A. Meeuse var. involuta (Rendle) Verdc.;内卷星牵牛■☆
43356 Astripomoea malvacea (Klotzsch) A. Meeuse var. volkensii (Dammer) Verdc.;福尔内卷星牵牛■☆
43357 Astripomoea nogalensis (Chiov.) Verdc.;诺加尔星牵牛■☆
43358 Astripomoea polycephala (Hallier f.) Verdc.;多头星牵牛■☆
43359 Astripomoea procera Thulin;高大星牵牛■☆
43360 Astripomoea rotundata (Pilg.) A. Meeuse;圆形星牵牛■☆
43361 Astripomoea tubiflora (Hallier f.) Verdc.;管花星牵牛■☆
43362 Astrocalyx Merr. (1910);星萼野牡丹属●☆
43363 Astrocalyx pleiosandra Merr.;星萼野牡丹●☆
43364 Astrocarpa Dumort. = Sesamoides Ortega ■☆
43365 Astrocarpa Neck. ex Dumort. = Sesamoides Ortega ■☆
43366 Astrocarpa clusii J. Gay = Sesamoides purpurascens (L.) G. López ■☆
43367 Astrocarpa sesamoides (L.) Duby = Sesamoides purpurascens (L.) G. López ■☆
43368 Astrocarpa sesamoides (L.) Duby subsp. purpurascens (L.) Rouy et Foucaud = Sesamoides purpurascens (L.) G. López ■☆
43369 Astrocarpaceae A. Kern. = Resedaceae Martinov(保留科名)●■
43370 Astrocarpus Duby = Sesamoides Ortega ■☆
43371 Astrocarpus Neck. ex DC. = Astrocaryum G. Mey. (保留属名)●☆
43372 Astrocaryum G. Mey. (1818)(保留属名);星果棕属(星果刺椰子属,星果榈属,星果椰属,星果椰子属,星坚果棕属,星棕属);Astrocaryum, Astrucarpus, Guere Palm, Star Nut Palm, Star-nut Plam, Tucum Palm, Tucuma ●☆
43373 Astrocaryum aculeatum G. Mey.;皮刺星果棕(星果椰子);Tucuma ●☆
43374 Astrocaryum ayri Mart.;艾尔星果棕;Ayr Astrocaryum ●☆
43375 Astrocaryum macrocalyx Burret;大萼星果棕●☆
43376 Astrocaryum mexicanum Liebm. ex Mart.;星果棕;Mexican Astrocaryum ●☆
43377 Astrocaryum murumuru Mart.;木鲁星果棕;Murumuru, Murumuru Astrocaryum ●☆
43378 Astrocaryum tucuma Mart.;刚毛星果棕(刚毛棕榈,南美洲图皮棕榈);Tucum Palm ●☆
43379 Astrocaryum tucumoides Drude;阿瓦拉星果棕;Awarra, Awarra Palm ●☆
43380 Astrocasia B. L. Rob. et Millsp. (1905);纤梗珠属☆
43381 Astrocasia phyllanthoides B. L. Rob. et Millsp.;纤梗珠☆
43382 Astrocephalus Raf. = Asterocephalus Zinn ●■
43383 Astrocephalus Raf. = Scabiosa L. ●■
43384 Astrochlaena Hallier f. = Astripomoea A. Meeuse ■☆
43385 Astrochlaena annua Rendle = Astripomoea hyoscyamoides (Vatke) Verdc. ■☆
43386 Astrochlaena cephalantha Hallier f. = Astripomoea cephalantha (Hallier f.) Verdc. ■☆
43387 Astrochlaena chariensis A. Chev. = Astripomoea malvacea (Klotzsch) A. Meeuse ■☆
43388 Astrochlaena delamereana Rendle = Astripomoea delamereana (Rendle) Verdc. ■☆
43389 Astrochlaena engleriana Dammer = Astripomoea malvacea (Klotzsch) A. Meeuse ■☆
43390 Astrochlaena floccosa (Vatke) Hallier f. = Astripomoea malvacea (Klotzsch) A. Meeuse var. floccosa (Vatke) Verdc. ■☆
43391 Astrochlaena grantii Rendle = Astripomoea grantii (Rendle) Verdc. ■☆
43392 Astrochlaena hyoscyamoides (Vatke) Hallier f. = Astripomoea hyoscyamoides (Vatke) Verdc. ■☆
43393 Astrochlaena involuta Rendle = Astripomoea malvacea (Klotzsch) A. Meeuse var. involuta (Rendle) Verdc. ■☆
43394 Astrochlaena kaessneri Rendle = Astripomoea malvacea (Klotzsch) A. Meeuse var. floccosa (Vatke) Verdc. ■☆
43395 Astrochlaena lachnosperma (Choisy) Hallier f. = Astripomoea lachnosperma (Choisy) A. Meeuse ■☆
43396 Astrochlaena ledermannii Pilg. = Astripomoea malvacea (Klotzsch) A. Meeuse var. volkensii (Dammer) Verdc. ■☆
43397 Astrochlaena malvacea (Klotzsch) Hallier f. = Astripomoea malvacea (Klotzsch) A. Meeuse ■☆
43398 Astrochlaena malvacea (Klotzsch) Hallier f. var. epedunculata Rendle = Astripomoea malvacea (Klotzsch) A. Meeuse var. epedunculata (Rendle) Verdc. ■☆
43399 Astrochlaena melandrioides Hallier f. = Astripomoea hyoscyamoides (Vatke) Verdc. var. melandrioides (Hallier f.) Verdc. ■☆
43400 Astrochlaena menispermoides Standl. = Astripomoea malvacea (Klotzsch) A. Meeuse var. volkensii (Dammer) Verdc. ■☆
43401 Astrochlaena mildbraedii Pilg. = Astripomoea malvacea (Klotzsch) A. Meeuse var. volkensii (Dammer) Verdc. ■☆
43402 Astrochlaena nogalensis Chiov. = Astripomoea nogalensis (Chiov.) Verdc. ■☆
43403 Astrochlaena phillipsiae (Baker) Rendle = Astripomoea malvacea (Klotzsch) A. Meeuse var. volkensii (Dammer) Verdc. ■☆
43404 Astrochlaena polycephala Hallier f. = Astripomoea polycephala (Hallier f.) Verdc. ■☆
43405 Astrochlaena rotundata Pilg. = Astripomoea rotundata (Pilg.) A. Meeuse ■☆
43406 Astrochlaena solanacea Haller f. = Astripomoea lachnosperma (Choisy) A. Meeuse ■☆
43407 Astrochlaena stuhlmannii Dammer ex Hallier f. = Astripomoea malvacea (Klotzsch) A. Meeuse var. floccosa (Vatke) Verdc. ■☆
43408 Astrochlaena tubiflora Hallier f. = Astripomoea tubiflora (Hallier f.) Verdc. ■☆
43409 Astrochlaena ugandensis Rendle = Astripomoea grantii (Rendle) Verdc. ■☆
43410 Astrochlaena volkensii Dammer = Astripomoea malvacea (Klotzsch) A. Meeuse var. volkensii (Dammer) Verdc. ■☆
43411 Astrochlaena whytei Rendle = Astripomoea hyoscyamoides (Vatke) Verdc. ■☆
43412 Astrococcus Benth. (1854);星果大戟属☆
43413 Astrococcus cornutus Benth.;星果大戟☆
43414 Astrocodon Fed. = Campanula L. ●■
43415 Astrocoma Neck. = Staavia Dahl ●☆
43416 Astrodaucus Drude = Ageomoron Raf. ■☆
43417 Astrodaucus Drude(1898);星萝卜属■☆
43418 Astrodaucus littoralis (M. Bieb.) Drude;星萝卜■☆

43419 Astrodaucus orientalis (L.) Drude;东方星萝卜■☆
43420 Astrodaucus persicus (Boiss.) Drude;波斯星萝卜■☆
43421 Astrodendrum Dennst. = Sterculia L. ●
43422 Astroglossus Rchb. ex Benth. = Stellilabium Schltr. ■☆
43423 Astroglossus Rchb. f. = Trichoceros Kunth ■☆
43424 Astrogyne Benth. = Croton L. ●
43425 Astrogyne Wall. ex M. A. Lawson = Siphonodon Griff. ●☆
43426 Astrolinon Baudo = Asterolinon Hoffmanns. et Link ■☆
43427 Astroloba Uitewaal = Haworthia Duval(保留属名)■☆
43428 Astroloba Uitewaal(1947);松塔掌属;Astroloba ■☆
43429 Astroloba aspera (Haw.) Uitewaal = Astroloba corrugata N. L. Mey. et G. F. Sm. ■☆
43430 Astroloba bullulata (Jacq.) Uitewaal;泡状松塔掌■☆
43431 Astroloba congesta (Salm-Dyck) Uitewaal;密集松塔掌■☆
43432 Astroloba corrugata N. L. Mey. et G. F. Sm.;皱折松塔掌(粗糙十二卷)■☆
43433 Astroloba deltoidea (Hook. f.) Uitewaal = Astroloba congesta (Salm-Dyck) Uitewaal ■☆
43434 Astroloba deltoidea (Hook. f.) Uitewaal var. intermedia (A. Berger) Uitewaal = Astroloba congesta (Salm-Dyck) Uitewaal ■☆
43435 Astroloba dodsoniana Uitewaal = Astroloba herrei Uitewaal ■☆
43436 Astroloba egregia (Poelln.) Uitewaal;优秀松塔掌■☆
43437 Astroloba egregia (Poelln.) Uitewaal var. fardeniana Uitewaal = Astroloba bullulata (Jacq.) Uitewaal ■☆
43438 Astroloba foliolosa (Haw.) Uitewaal;多小叶松塔掌■☆
43439 Astroloba herrei Uitewaal;赫勒松塔掌■☆
43440 Astroloba muricata L. E. Groen = Astroloba corrugata N. L. Mey. et G. F. Sm. ■☆
43441 Astroloba pentagona (Aiton) Uitewaal = Astroloba spiralis (L.) Uitewaal ■☆
43442 Astroloba pentagona (Haw.) Uitewaal;五角松塔掌(松塔掌)■☆
43443 Astroloba rubriflora (L. Bolus) G. F. Sm. et J. C. Manning = Poellnitzia rubriflora (L. Bolus) Uitewaal ■☆
43444 Astroloba rugosa Roberts = Astroloba corrugata N. L. Mey. et G. F. Sm. ■☆
43445 Astroloba spiralis (L.) Uitewaal;螺旋松塔掌■☆
43446 Astroloba spiralis (L.) Uitewaal subsp. foliolosa (Haw.) Groen = Astroloba foliolosa (Haw.) Uitewaal ■☆
43447 Astroloba turgida (Baker) H. Jacobsen = Astroloba congesta (Salm-Dyck) Uitewaal ■☆
43448 Astrolobium DC. = Artrolobium Desv. ■☆
43449 Astrolobium DC. = Coronilla L. (保留属名)●■
43450 Astrolobium DC. = Ornithopus L. ■☆
43451 Astroloma R. Br. (1810);松石南属●☆
43452 Astroloma conostephioides F. Muell. ex Benth.;松石南;Flame Heath ●☆
43453 Astroloma conostephioides F. Muell. ex Benth. = Styphelia behrii (Schltdl.) Sleumer ●☆
43454 Astroloma humifusum (Cav.) R. Br.;平伏松石南●☆
43455 Astromerremia Pilg. = Merremia Dennst. ex Endl. (保留属名)●■
43456 Astromerremia digitata Pilg. = Merremia stellata Rendle ■☆
43457 Astronia Blume(1827);褐鳞木属(大野牡丹属);Astronia ●
43458 Astronia Noronha = Murraya J. König ex L. (保留属名)●
43459 Astronia Noronha ex Blume = Astronia Noronha ●
43460 Astronia Noronha ex Blume = Murraya J. König ex L. (保留属名)●
43461 Astronia cumingiana Vidal = Astronia ferruginea Elmer ●
43462 Astronia ferruginea Elmer;台湾褐鳞木(大野牡丹,褐鳞木,锈叶野牡丹);Rusty Astronia, Rustyleaf Astronia, Rusty-leaved Astronia, Taiwan Astronia ●
43463 Astronia ferruginea Elmer = Astronia formosana Kaneh. ●
43464 Astronia formosana Kaneh. = Astronia ferruginea Elmer ●
43465 Astronia pulchra Vidal = Astronia ferruginea Elmer ●
43466 Astronidium A. Gray(1853)(保留属名);小褐鳞木属●☆
43467 Astronidium acutifolium (Mansf.) Markgr.;尖叶小褐鳞木●☆
43468 Astronidium floribundum (Gillespie) A. C. Sm.;繁花小褐鳞木●☆
43469 Astronidium montanum Merr. et L. M. Perry;山地小褐鳞木●☆
43470 Astronidium parviflorum A. Gray;小花小褐鳞木●☆
43471 Astronidium robustum (Seem.) A. C. Sm.;粗壮小褐鳞木●☆
43472 Astronium Jacq. (1760);星漆木属(斑纹漆属,星漆属);Star Tree, Star-tree ●☆
43473 Astronium balansae Engl.;巴氏星漆木(巴氏斑纹漆木)●☆
43474 Astronium concinnum Schott;优雅星漆木(优雅斑纹漆木)●☆
43475 Astronium fraxinifolium Schott;梣叶星漆木(梣叶斑纹漆木);Goncalo Alves, King Wood, Locust Wood, Tiger Wood, Zebrawood ●☆
43476 Astronium gracile Engl.;南美洲星漆木(南美洲斑纹漆木)●☆
43477 Astronium grande Engl.;巴西星漆木●☆
43478 Astronium graveeolens Jacq.;裂叶星漆木(裂叶星漆)●☆
43479 Astronium lecointei Ducke;莱蔻星漆木(莱蔻斑纹漆木)●☆
43480 Astronium urundeuva Engl.;乌隆迪星漆木(乌隆斑纹漆木);Locust, Locust Bean, Urunday ●☆
43481 Astropanax Seem. = Schefflera J. R. Forst. et G. Forst. (保留属名)●
43482 Astropanax abyssinicus (Hochst. ex A. Rich.) Seem. = Schefflera abyssinica (Hochst. ex A. Rich.) Harms ●☆
43483 Astropanax baikiei Seem. = Schefflera barteri (Seem.) Harms ●☆
43484 Astropanax mannii (Hook. f.) Seem. = Schefflera mannii (Hook. f.) Harms ●☆
43485 Astropetalum Griff. = Swintonia Griff. ●☆
43486 Astrophea DC. = Passiflora L. ●■
43487 Astrophea Rchb. = Passiflora L. ●■
43488 Astrophia Nutt. = Lathyrus L. ■
43489 Astrophyllum Torr. et A. Gray = Choisya Kunth ●☆
43490 Astrophyton Lawr. = Astrophytum Lem. ●
43491 Astrophyton Lawr. et Lem. = Astrophytum Lem. ●
43492 Astrophytum Lem. (1839);星状仙人球属(星冠属,星球属,有星属);Star Cactus, Star-cactus ●
43493 Astrophytum asterias (Siebold et Zucc.) Lem.;星球(兜丸,海胆仙人掌,甲丸,星兜,星冠,星红仙人球,星叶球);Sand Dollar Cactus, Sand-dollar, Sea Urchin Cactus, Sea-urchin Cactus, Silver-dollar, Star Cactus, Star-peyote ■
43494 Astrophytum asterias (Zucc.) Lem. = Astrophytum asterias (Siebold et Zucc.) Lem. ■
43495 Astrophytum capricorne (A. Dietr.) Britton et Rose;瑞凤玉(群凤玉,羊角仙人球);Capricorn Star Cactus, Goat's Horns Cactus, Goat's-horn cactus ■
43496 Astrophytum capricorne (A. Dietr.) Britton et Rose 'Aureum';黄凤玉;Capricorn Star Cactus ■
43497 Astrophytum capricorne (A. Dietr.) Britton et Rose 'Crassipinum';大凤玉;Capricorn Star Cactus ■
43498 Astrophytum capricorne (A. Dietr.) Britton et Rose 'Minus';凤凰玉;Capricorn Star Cactus ■
43499 Astrophytum capricorne (A. Dietr.) Britton et Rose 'Senile';群

凤玉;Capricorn Star Cactus ■
43500 Astrophytum capricorne (A. Dietr.) Britton et Rose subsp. senile (Frič) Doweld = Astrophytum capricorne (A. Dietr.) Britton et Rose 'Senile' ■
43501 Astrophytum capricorne (A. Dietr.) Britton et Rose subsp. senile (Frič) Doweld = Astrophytum senile Frič ■
43502 Astrophytum capricorne (A. Dietr.) Britton et Rose var. crassispinum (H. Moeller) Backeb. = Astrophytum capricorne (A. Dietr.) Britton et Rose 'Crassipinum' ■
43503 Astrophytum capricorne (A. Dietr.) Britton et Rose var. minus Frič = Astrophytum capricorne (A. Dietr.) Britton et Rose 'Minus' ■
43504 Astrophytum capricorne (A. Dietr.) Britton et Rose var. niveum (K. Kayser) Backeb.;雪白瑞凤玉■☆
43505 Astrophytum coahuilense (H. Moeller) K. Kayser = Astrophytum myriostigma (Salm-Dyck) Lem. ■
43506 Astrophytum columnare (Schum.) Sadovsky et Schütz = Astrophytum myriostigma (Salm-Dyck) Lem. ■
43507 Astrophytum glabrescens F. A. C. Weber = Astrophytum ornatum (DC.) Britton et Rose ■
43508 Astrophytum hybridum Hort.;杂交僧帽■☆
43509 Astrophytum myriostigma (Salm-Dyck) Lem.;鸾凤玉(多蕊仙人球);Bishop's Cap, Bishop's Cap Cactus, Bishop's Hood, Bishop's Miter, Monkshood ■
43510 Astrophytum myriostigma (Salm-Dyck) Lem. 'Columnare';鸾凤阁;Columnar ■
43511 Astrophytum myriostigma (Salm-Dyck) Lem. 'Nudum';裸鸾凤阁;Naked Bishop's Cap ■
43512 Astrophytum myriostigma (Salm-Dyck) Lem. subsp. glabrum Backeb.;碧琉璃鸾凤玉■☆
43513 Astrophytum myriostigma (Salm-Dyck) Lem. subsp. potosinum K. Kayser;三角鸾凤玉■☆
43514 Astrophytum myriostigma (Salm-Dyck) Lem. subsp. potosinum K. Kayser = Astrophytum myriostigma (Salm-Dyck) Lem. ■
43515 Astrophytum myriostigma (Salm-Dyck) Lem. subsp. quadricostatum (H. Moeller) K. Kayser = Astrophytum myriostigma (Salm-Dyck) Lem. var. quadricostatum (H. Moeller) Borg ■
43516 Astrophytum myriostigma (Salm-Dyck) Lem. subsp. tulense (K. Kayser) Backeb. = Astrophytum myriostigma (Salm-Dyck) Lem. var. tulense (K. Kayser) Y. Ito ■☆
43517 Astrophytum myriostigma (Salm-Dyck) Lem. var. columnaris (Schum.) Tsuda = Astrophytum myriostigma (Salm-Dyck) Lem. 'Columnare' ■
43518 Astrophytum myriostigma (Salm-Dyck) Lem. var. multicostatus ? = Astrophytum myriostigma (Salm-Dyck) Lem. ■
43519 Astrophytum myriostigma (Salm-Dyck) Lem. var. potosinum (K. Kayser) Borg = Astrophytum myriostigma (Salm-Dyck) Lem. ■
43520 Astrophytum myriostigma (Salm-Dyck) Lem. var. quadricostatum (H. Moeller) Borg;四角鸾凤玉;Gaudrangular Bishop's Cap, Bishop's Mitre Cactus ■
43521 Astrophytum myriostigma (Salm-Dyck) Lem. var. strongylogonum Backeb.;圆棱鸾凤玉■☆
43522 Astrophytum myriostigma (Salm-Dyck) Lem. var. tulense (K. Kayser) Y. Ito;图拉鸾凤玉■☆
43523 Astrophytum myriostigma (Salm-Dyck) Lem. var. viescaensis ? = Astrophytum myriostigma (Salm-Dyck) Lem. ■
43524 Astrophytum myriostigma (Salm-Dyck) Lem. var. wahuilense (H. Moeller) Y. Ito;白鸾凤玉■
43525 Astrophytum nudum ? = Astrophytum myriostigma (Salm-Dyck) Lem. ■
43526 Astrophytum ornatum (DC.) Britton et Rose;般若(美丽星球,装饰仙人球);Bishop's Cap, Bishop's Cap Cactus, Monk's Hood, Ornamental Monkshood, Star Cactus ■
43527 Astrophytum ornatum (DC.) Britton et Rose 'Mirbellii';金刺般若(黄刺般若);Mirbel Star Cactus ■
43528 Astrophytum ornatum (DC.) Britton et Rose var. pubescens Y. Ito;白云般若■
43529 Astrophytum ornatum F. A. C. Weber f. glabrescens (F. A. C. Weber) Krainz;裸般若(光秃般若)■
43530 Astrophytum ornatum F. A. C. Weber var. mirbelii (Lem.) Frič;黄刺般若(金刺般若)■☆
43531 Astrophytum ornatum F. A. C. Weber var. niveum Schütz et Z. Fleisch.;白瑞凤玉■☆
43532 Astrophytum ornatum F. A. C. Weber var. spirale ? = Astrophytum ornatum (DC.) Britton et Rose ■
43533 Astrophytum ornatum F. A. C. Weber var. virens Schütz et Z. Fleisch. = Astrophytum ornatum (DC.) Britton et Rose ■
43534 Astrophytum prismaticum Lem. = Astrophytum myriostigma (Salm-Dyck) Lem. ■
43535 Astrophytum senile Frič = Astrophytum capricorne (A. Dietr.) Britton et Rose ■
43536 Astrophytum senile Frič = Astrophytum capricorne (A. Dietr.) Britton et Rose 'Senile' ■
43537 Astrophytum tulense (K. Kayser) Sadovsky et Schütz = Astrophytum myriostigma (Salm-Dyck) Lem. var. tulense (K. Kayser) Y. Ito ■☆
43538 Astrophytum tulense (K. Kayser) Sadovsky et Schütz = Astrophytum myriostigma (Salm-Dyck) Lem. ■
43539 Astrophytum virens ? = Astrophytum ornatum (DC.) Britton et Rose ■
43540 Astropus Spreng. = Waltheria L. ●■
43541 Astroschoenus Lindl. = Asteroschoenus Nees ■
43542 Astrostemma Benth. = Absolmsia Kuntze ■
43543 Astrothalamus C. B. Rob. (1911);星室麻属●☆
43544 Astrothalamus reticulatus (Wedd.) C. B. Rob.;星室麻●☆
43545 Astrotheca Miers ex Planch. et Triana = Asthotheca Miers ex Planch. et Triana ●☆
43546 Astrotheca Miers ex Planch. et Triana = Clusia L. ●☆
43547 Astrotheca Vesque = Asthotheca Miers ex Planch. et Triana ●☆
43548 Astrotheca Vesque = Clusia L. ●☆
43549 Astrotricha DC. (1829);澳洲五加属●☆
43550 Astrotricha floccosa DC.;澳洲五加;Flannel Leaf ●☆
43551 Astrotriche Benth. = Astrotricha DC. ●☆
43552 Astrotrichia Rchb. = Astrotricha DC. ●☆
43553 Astrotrichilia (Harms) J. -F. Leroy = Astrotrichilia (Harms) J. -F. Leroy ex T. D. Penn. et Styles ●☆
43554 Astrotrichilia (Harms) J. -F. Leroy ex T. D. Penn. et Styles (1975);星毛楝属●☆
43555 Astrotrichilia (Harms) T. D. Penn. et Styles = Astrotrichilia (Harms) J. -F. Leroy ex T. D. Penn. et Styles ●☆
43556 Astrotrichilia asterotricha (Radlk.) Cheek;星毛楝●☆
43557 Astus Trudgen et Rye = Baeckea L. ●
43558 Astus Trudgen et Rye(2005);澳洲鳞叶树属●☆
43559 Astydamia DC. (1829);星隔芹属■☆
43560 Astydamia canariensis DC.;加那利星隔芹■☆

43561 Astydamia canariensis DC. = Astydamia latifolia (L. f.) Kuntze ■☆
43562 Astydamia ifniensis Caball. = Astydamia latifolia (L. f.) Kuntze ■☆
43563 Astydamia latifolia (L. f.) Kuntze;宽叶星隔芹■☆
43564 Astydamia latifolia L. f. = Astydamia latifolia (L. f.) Kuntze ■☆
43565 Astylis Wight = Drypetes Vahl ●
43566 Astylus Dulac = Hutchinsia R. Br. ■☆
43567 Astyposanthea Herter = Stylosanthes Sw. ●■
43568 Astyria Lindl. = Astiria Lindl. ●☆
43569 Asyneuma Griseb. et Schenk(1852);异牧根草属(喉节草属);Asyneuma ■
43570 Asyneuma amplexicaule (Willd.) Hand. -Mazz.;抱茎牧根草●☆
43571 Asyneuma anhuiense B. A. Shen;安徽牧根草;Anhui Asyneuma ●
43572 Asyneuma anhuiense B. A. Shen = Triodanis biflora (Ruiz et Pav.) Greene ■
43573 Asyneuma argutum Bornm.;亮牧根草■☆
43574 Asyneuma babadaghense Yildiz et Kit Tan;巴巴塔戈牧根草■☆
43575 Asyneuma campanuloides Bornm.;钟状牧根草■☆
43576 Asyneuma canescens Griseb. et Schenk;灰牧根草●☆
43577 Asyneuma chinense D. Y. Hong;球果牧根草(喉节草,鸡肉参,咳嗽草,牧根草,土沙参,止咳草);China Asyneuma, Chinese Asyneuma ●
43578 Asyneuma cichoriforme Bornm.;裂牧根草■☆
43579 Asyneuma fulgens (Wall.) Briq.;长果牧根草(鸡肉参,亮牧根草,西南牧根草);Longfruit Asyneuma ●
43580 Asyneuma japonicum (Miq.) Briq.;牧根草;Japanese Asyneuma ●
43581 Asyneuma lanceolatum Hand. -Mazz.;披针形牧根草■☆
43582 Asyneuma leianthum Bornm.;光花牧根草■☆
43583 Asyneuma lobelioides Hand. -Mazz.;拟洛贝尔■☆
43584 Asyneuma pulchellum Bornm.;美丽牧根草■☆
43585 Asyneuma ramosum Pavlov;分枝牧根草■☆
43586 Asyneuma rigidum (Willd.) Grossh.;硬牧根草■☆
43587 Asyneuma rigidum (Willd.) Grossh. subsp. aurasiacum (Batt. et Trab.) Damboldt;奥拉斯牧根草■☆
43588 Asyneuma salignum (Waldst. et Kit. ex Besser) Fed.;柳牧根草■☆
43589 Asyneuma strictum Wendelbo;刚直牧根草■☆
43590 Asyneuma talyschense Fed.;塔里什牧根草■☆
43591 Asyneuma thomsonii (Hook. f.) Bornm.;托马森牧根草■☆
43592 Asyneuma thomsonii (Hook. f.) Bornm. var. strictum (Wendelbo) Kitamura = Asyneuma strictum Wendelbo ■☆
43593 Asyneuma trautvetteri Bornm.;特劳特牧根草■☆
43594 Asyneuma urceolatum (Fomin) Fed.;坛状牧根草■☆
43595 Asyneumopsis Contandr., Quézel et Pamukç. (1972);土耳其牧根草属(喉节草属)■☆
43596 Asystasia Blume(1826);十万错属;Asystasia ●■
43597 Asystasia acuminata Klotzsch = Asystasia gangetica (L.) T. Anderson subsp. micrantha (Nees) Ensermu ●☆
43598 Asystasia africana (S. Moore) C. B. Clarke;非洲十万错■☆
43599 Asystasia albiflora Ensermu;白花十万错■☆
43600 Asystasia ammophila Ensermu;喜沙十万错■☆
43601 Asystasia amoena Turrill;秀丽十万错■☆
43602 Asystasia ansellioides C. B. Clarke;豹斑兰十万错■☆
43603 Asystasia ansellioides C. B. Clarke var. lanceolata Fiori = Asystasia gangetica (L.) T. Anderson subsp. micrantha (Nees) Ensermu ●☆
43604 Asystasia atriplicifolia Bremek.;暗沟十万错■☆
43605 Asystasia axillaria Lindau ex Engl. = Asystasia excellens Lindau ●☆
43606 Asystasia bella Benth. et Hook. f. = Mackaya bella Harv. ●☆
43607 Asystasia buettneri Lindau;比特纳十万错■☆
43608 Asystasia calycina Benth. = Asystasia buettneri Lindau ■☆
43609 Asystasia calycina Nees = Pteracanthus calycinus (Nees) Bremek. ●
43610 Asystasia chelonoides Nees;十万错(盗偷草,跌打草);Common Asystasia ●
43611 Asystasia chinensis S. Moore = Asystasiella neesiana (Wall.) Lindau ●
43612 Asystasia coleae Rolfe = Asystasia guttata (Forssk.) Brummitt ●☆
43613 Asystasia congensis C. B. Clarke;康格十万错■☆
43614 Asystasia coromandeliana Nees;盗偷草(十万错花)●
43615 Asystasia coromandeliana Nees = Asystasia gangetica (L.) T. Anderson subsp. micrantha (Nees) Ensermu ●☆
43616 Asystasia coromandeliana Nees var. micrantha ? = Asystasia gangetica (L.) T. Anderson subsp. micrantha (Nees) Ensermu ●☆
43617 Asystasia coromandeliana Wight ex Nees = Asystasia gangetica (L.) T. Anderson ●
43618 Asystasia decipiens Heine;迷惑十万错●☆
43619 Asystasia drake-brockmanii Turrill. f. lejogyna Chiov. = Asystasia excellens Lindau ●☆
43620 Asystasia dryadum S. Moore = Asystasia buettneri Lindau ■☆
43621 Asystasia excellens Lindau;优秀十万错●☆
43622 Asystasia floribunda Klotzsch = Asystasia gangetica (L.) T. Anderson subsp. micrantha (Nees) Ensermu ●☆
43623 Asystasia fuchsiifolia Lindau;倒挂金钟叶十万错●☆
43624 Asystasia gangetica (L.) T. Anderson;宽叶十万错(恒河十万错);Chinese Violet, Ganges Asystasia, Ganges Primrose ●
43625 Asystasia gangetica (L.) T. Anderson subsp. micrantha (Nees) Ensermu;小花宽叶十万错●☆
43626 Asystasia glandulifera Lindau;腺体十万错●☆
43627 Asystasia glandulosa Lindau;具腺十万错●☆
43628 Asystasia guttata (Forssk.) Brummitt;斑点十万错●☆
43629 Asystasia hedbergii Ensermu;赫德十万错●☆
43630 Asystasia henryi C. B. Clarke ex C. Y. Wu = Asystasia salicifolia Craib ●
43631 Asystasia kalantha Lindau = Asystasia vogeliana Benth. ●☆
43632 Asystasia kerri Craib = Chroesthes lanceolata (T. Anderson) B. Hansen ●
43633 Asystasia lanceolata T. Anderson = Chroesthes lanceolata (T. Anderson) B. Hansen ●
43634 Asystasia laticapsula C. B. Clarke ex Karlström;宽十万错●☆
43635 Asystasia leptostachya Lindau;细穗十万错●☆
43636 Asystasia lindauiana Hutch. et Dalziel;林达十万错●☆
43637 Asystasia linearis S. Moore;线形十万错●☆
43638 Asystasia longituba Lindau = Asystasia vogeliana Benth. ●☆
43639 Asystasia lorata Ensermu;舌状十万错●☆
43640 Asystasia macrophylla (T. Anderson) Lindau;大叶十万错●☆
43641 Asystasia malawiana Brummitt et Chisumpa;马拉维十万错●☆
43642 Asystasia masaiensis Lindau;吗西十万错●☆
43643 Asystasia multiflora Klotzsch = Asystasia gangetica (L.) T. Anderson subsp. micrantha (Nees) Ensermu ●☆
43644 Asystasia mysurensis (Roth) T. Anderson;印度十万错●☆
43645 Asystasia natalensis C. B. Clarke = Salpinctium natalense (C. B. Clarke) T. J. Edwards ●☆

43646 Asystasia neesiana (Wall.) Nees = Asystasiella neesiana (Wall.) Lindau ●
43647 Asystasia parvula C. B. Clarke = Asystasia gangetica (L.) T. Anderson subsp. micrantha (Nees) Ensermu ●☆
43648 Asystasia pauciflora Nees = Codonacanthus pauciflorus (Nees) Nees ■
43649 Asystasia pinguifolia T. J. Edwards;厚叶十万错●☆
43650 Asystasia podostachys Klotzsch = Asystasia gangetica (L.) T. Anderson subsp. micrantha (Nees) Ensermu ●☆
43651 Asystasia pubescens Klotzsch = Asystasia gangetica (L.) T. Anderson subsp. micrantha (Nees) Ensermu ●☆
43652 Asystasia quarterna Nees = Asystasia gangetica (L.) T. Anderson subsp. micrantha (Nees) Ensermu ●☆
43653 Asystasia querimbensis Klotzsch = Asystasia gangetica (L.) T. Anderson subsp. micrantha (Nees) Ensermu ●☆
43654 Asystasia retrocarpa T. J. Edwards;折果十万错●☆
43655 Asystasia richardsiae Ensermu;理查兹十万错●☆
43656 Asystasia riparia Lindau;溪畔十万错●☆
43657 Asystasia ritellii Chiov. = Asystasia riparia Lindau ●☆
43658 Asystasia rostrata Solms = Asystasia mysurensis (Roth) T. Anderson ●☆
43659 Asystasia salicifolia Craib;囊管花;Henry Asystasia ●
43660 Asystasia scabrida Klotzsch = Asystasia gangetica (L.) T. Anderson subsp. micrantha (Nees) Ensermu ●☆
43661 Asystasia scandens (Lindau) Hook.;攀缘十万错●☆
43662 Asystasia schimperi T. Anderson;欣珀十万错●☆
43663 Asystasia schimperi T. Anderson = Asystasia mysurensis (Roth) T. Anderson ●☆
43664 Asystasia schimperi T. Anderson var. grantii C. B. Clarke = Asystasia mysurensis (Roth) T. Anderson ●☆
43665 Asystasia schliebenii Mildbr.;施利本十万错●☆
43666 Asystasia silvicola W. W. Sm. = Chroesthes lanceolata (T. Anderson) B. Hansen ●
43667 Asystasia somalensis (Franch.) Lebrun et L. Touss. = Asystasia guttata (Forssk.) Brummitt ●☆
43668 Asystasia somalica Gand. = Asystasia riparia Lindau ●☆
43669 Asystasia stenosiphon C. B. Clarke = Salpinctium stenosiphon (C. B. Clarke) T. J. Edwards ●☆
43670 Asystasia striata S. Moore = Whitfieldia liebrechtsiana De Wild. et T. Durand ■☆
43671 Asystasia subbiflora C. B. Clarke;亚双花十万错●☆
43672 Asystasia subhastata Klotzsch = Asystasia gangetica (L.) T. Anderson subsp. micrantha (Nees) Ensermu ●☆
43673 Asystasia trichotogyne Lindau = Justicia biokoensis V. A. W. Graham ■☆
43674 Asystasia varia N. E. Br.;变异十万错●☆
43675 Asystasia violacea Dalzell ex C. B. Clarke = Asystasia chelonoides Nees ●
43676 Asystasia vogeliana Benth.;沃格尔十万错●☆
43677 Asystasia welwitschii S. Moore;韦尔十万错●☆
43678 Asystasia zambiana Brummitt et Chisumpa;赞比亚十万错●☆
43679 Asystasiella Lindau (1895);白接骨属(拟马偕花属);Asystasiella ●■☆
43680 Asystasiella africana S. Moore;非洲白接骨●☆
43681 Asystasiella chinensis (S. Moore) E. Hossain = Asystasiella neesiana (Wall.) Lindau ●
43682 Asystasiella neesiana (Wall.) Lindau;白接骨(白龙骨,猢狲节根,假牛膝,接骨草,接骨丹,金不换,六厘草,尼氏拟马偕花,拟马偕花,麒麟草,无骨苎麻,五钱草,橡皮草,小阿西达,血见愁,玉梗半枝莲,玉接骨,玉连环,玉龙半枝莲,玉龙盘,玉钱草,蛀木虫);China Asystasiella,Chinese Asystasiella ●
43683 Atacca Lem. = Ataccia J. Presl ■
43684 Ataccia J. Presl = Tacca J. R. Forst. et G. Forst.(保留属名)■
43685 Ataccia integrifolia (Ker Gawl.) C. Presl = Tacca integrifolia Ker Gawl. ■
43686 Ataccia integrifolia C. Presl = Tacca integrifolia Ker Gawl. ■
43687 Atadinus Raf. = Rhamnus L. ●
43688 Ataenia Endl. = Atenia Hook. et Arn. ■☆
43689 Ataenia Endl. = Perideridia Rchb. ■☆
43690 Ataenidia Gagnep.(1908);簇叶竹芋属■☆
43691 Ataenidia Gagnep. = Phrynium Willd.(保留属名)■
43692 Ataenidia conferta (Benth.) Milne-Redh.;簇叶竹芋■☆
43693 Ataenidia gabonensis Gagnep. = Ataenidia conferta (Benth.) Milne-Redh. ■☆
43694 Ataenidium Gagnep. = Phrynium Willd.(保留属名)■
43695 Atalanta (Nutt.) Raf. = Atalanthus D. Don ■☆
43696 Atalanta (Nutt.) Raf. = Cleome L. ●■
43697 Atalanta (Nutt.) Raf. = Peritoma DC. ●■
43698 Atalanta Nutt. = Cleome L. ●■
43699 Atalanta Raf. = Atalanthus D. Don ■☆
43700 Atalanthus D. Don = Sonchus L. ■
43701 Atalanthus D. Don(1829);多枝苦荬菜属■☆
43702 Atalanthus acanthodes (Boiss.) Kirp.;多枝苦荬菜■☆
43703 Atalanthus angustifolius (Desf.) Pomel = Launaea angustifolia (Desf.) Kuntze ■☆
43704 Atalanthus arboreus (DC.) Sw. = Sonchus arboreus DC. ■☆
43705 Atalanthus canariensis (Boulos) A. Hansen et Sunding = Sonchus filifolius N. Kilian et Greuter ■☆
43706 Atalanthus candolleanus Pomel = Launaea fragilis (Asso) Pau ■☆
43707 Atalanthus capillaris (Svent.) A. Hansen et Sunding = Sonchus capillaris Svent. ■☆
43708 Atalanthus divaricatus (Desf.) Pomel = Launaea nudicaulis (L.) Hook. f. ■☆
43709 Atalanthus longilobus (Boiss. et Reut.) Pomel = Launaea fragilis (Asso) Pau ■☆
43710 Atalanthus microcarpus (Boulos) A. Hansen et Sunding = Sonchus microcarpus (Boulos) U. Reifenb. et A. Reifenb. ■☆
43711 Atalanthus nudicaulis (L.) Pomel = Launaea nudicaulis (L.) Hook. f. ■☆
43712 Atalanthus pinnatus (L. f.) Don = Sonchus pinnatus Aiton ■☆
43713 Atalanthus quercifolius (Desf.) Pomel = Launaea quercifolia (Desf.) Pamp. ■☆
43714 Atalanthus regis-jubae (Pit.) A. Hansen et Sunding = Sonchus regis-jubae Pit. ■☆
43715 Atalanthus resedifolius (Coss.) Pomel = Launaea fragilis (Asso) Pau ■☆
43716 Atalanthus spinosus (Lam.) D. Don = Launaea lanifera Pau ■☆
43717 Atalanthus webbii Sch. Bip. = Sonchus webbii Sch. Bip. ■☆
43718 Atalantia Corrêa(1805)(保留属名);酒饼簕属(狗橘属,蠔壳刺属,绿黄柑属,乌柑属);Atalantia ●
43719 Atalantia acuminata C. C. Huang;尖叶酒饼簕;Acuminate Atalantia,Acuminate-leaved Atalantia,Sharpleaf Atalantia ●
43720 Atalantia bilocularis Wall. = Atalantia buxifolia (Poir.) Oliv. ●
43721 Atalantia buxifolia (Poir.) Oliv.;酒饼簕(半天钓,蚌壳刺,单

叶乌柑,东风桔,东风橘,儿针簕,狗骨簕,狗吉,狗橘,狗橘刺,蠔壳刺,假花椒,酒饼勒,酒饼药,雷公簕,梅橘,牛屎橘,山柑簕,山柑子,铜将军,乌柑,乌柑仔,针仔簕,猪钓簕公);Boxleaf Atalantia,Box-leaved Atalantia,Chinese Boxrange ●

43722 Atalantia buxifolia Oliv. = Atalantia buxifolia (Poir.) Oliv. ●

43723 Atalantia dasycarpa C. C. Huang;厚皮酒饼簕;Thickpericarp Atalantia,Thick-pericarped Atalantia ●

43724 Atalantia disticha Merr. = Atalantia guillauminii Swingle ●

43725 Atalantia fongkaica C. C. Huang;封开酒饼簕;Fengkai Atalantia ●

43726 Atalantia guillauminii Swingle;大果酒饼簕;Bigfruit Atalantia,Guillaumin Atalantia ●

43727 Atalantia henryi (Swingle) C. C. Huang;薄皮酒饼簕;Hanry Atalantia,Thin-pericarped Atalantia ●

43728 Atalantia hindsii (Champ. ex Benth.) Oliv. ex Benth. = Citrus japonica Thunb. ●

43729 Atalantia hindsii Oliv. ex Benth. = Fortunella hindsii (Champ. ex Benth.) Swingle ●

43730 Atalantia kwangtungensis Merr.;广东酒饼簕(亨利酒饼簕);Guangdong Atalantia,Kwangtung Atalantia ●

43731 Atalantia monophylla DC.;单叶酒饼簕;India Atalantia,Indian Atalantia ●☆

43732 Atalantia pseudoracemosa Guillaumin = Glycosmis pseudoracemosa (Guillaumin) Swingle ●

43733 Atalantia racemosa Wight et Arn. = Glycosmis pseudoracemosa (Guillaumin) Swingle ●

43734 Atalantia racemosa Wight et Arn. var. henryi Swingle = Atalantia henryi (Swingle) C. C. Huang ●

43735 Atalantia roxburghiana Hook. f. var. kwangtungensis Swingle = Atalantia kwangtungensis Merr. ●

43736 Atalantia simplicifolia (Roxb.) Engl. = Atalantia dasycarpa C. C. Huang ●

43737 Atalaya Blume(1849);阿塔木属●☆

43738 Atalaya alata (Sim) H. M. L. Forbes;高大阿塔木●☆

43739 Atalaya capensis R. A. Dyer;好望角阿塔木●☆

43740 Atalaya hemiglauca (F. Muell.) F. Muell.;灰绿阿塔木;White Wood,Whitewood ●☆

43741 Atalaya natalensis R. A. Dyer;纳塔尔阿塔木●☆

43742 Atamasco Raf. = Zephyranthes Herb.(保留属名)■

43743 Atamisquea Miers = Atamisquea Miers ex Hook. et Arn. ●☆

43744 Atamisquea Miers ex Hook. et Arn.(1833);阿根廷山柑属●☆

43745 Atamisquea emarginata Miers;阿根廷山柑●☆

43746 Atamosco Adans.(废弃属名) = Zephyranthes Herb.(保留属名)■

43747 Atamosco candida (Lindl.) Sasaki = Zephyranthes candida (Lindl.) Herb. ■

43748 Atamosco carinata P. Wilson = Zephyranthes carinata Herb. ■

43749 Atamosco carinata P. Wilson = Zephyranthes grandiflora Lindl. ■

43750 Atamosco simpsonii (Chapm.) Greene = Zephyranthes simpsonii Chapm. ■☆

43751 Atamosco texana (Herb.) Greene = Habranthus tubispathus (L'Hér.) Traub ■☆

43752 Atamosco treatiae (S. Watson) Greene = Zephyranthes treatiae S. Watson ■☆

43753 Atanara Raf. = Annona L. ●

43754 Atasites Neck. = Gerbera L.(保留属名)■

43755 Ataxia R. Br. = Hierochloe R. Br.(保留属名)■

43756 Ataxia ecklonii Nees ex Trin. = Anthoxanthum ecklonii (Nees ex Trin.) Stapf ■☆

43757 Ataxia hookeri Griseb. = Anthoxanthum hookeri (Griseb.) Rendle ■

43758 Ataxia horsfieldii Kunth ex Benn. = Anthoxanthum horsfieldii (Kunth ex Benn.) Mez ex Reeder ■

43759 Ataxia tenuis Trin. = Anthoxanthum tongo (Nees ex Trin.) Stapf ■☆

43760 Ate Lindl. = Habenaria Willd. ■

43761 Atecosa Raf. = Rumex L. ●■

43762 Ateixa Ravenna = Sarcodraba Gilg et Muschl. ■☆

43763 Atelandra Bello = Meliosma Blume ●

43764 Atelandra Lindl. = Hemigenia R. Br. ●☆

43765 Atelanthera Hook. f. et Thomson(1861);异药芥属(无尾药属,异药荠属);Diversianther ■

43766 Atelanthera contorta Gilli = Atelanthera perpusilla Hook. f. et Thomson ■

43767 Atelanthera pentandra Jafri = Atelanthera perpusilla Hook. f. et Thomson ■

43768 Atelanthera perpusilla Hook. f. et Thomson;异药芥;Diversianther ■

43769 Atelea A. Rich. = Ateleia (Moc. et Sessé ex DC.) D. Dietr. ■☆

43770 Ateleia (DC.) Benth.(1837);美瑕豆属■☆

43771 Ateleia (Moc. et Sessé ex DC.) D. Dietr. = Ateleia (DC.) Benth. ■☆

43772 Ateleia DC. = Ateleia (DC.) Benth. ■☆

43773 Ateleia Moc. et Sessé ex DC. = Ateleia (DC.) Benth. ■☆

43774 Ateleia herbert-smithii Pittier;史密斯美瑕豆■☆

43775 Ateleia peltaria D. Dietr. = Wiborgia fusca Thunb. ■☆

43776 Ateleste Sond. = Doryalis E. Mey. ex Arn. ●

43777 Atellanthus Nutt. ex Benth. = Synthyris Benth. ■☆

43778 Atelophragma Rydb. = Astragalus L. ●■

43779 Atelophragma alpinum (L.) Rydb. = Astragalus alpinus L. ■

43780 Atemnosiphon Léandri = Gnidia L. ●☆

43781 Atemnosiphon Léandri(1947);革质瑞香属●☆

43782 Atemnosiphon coriaceus (Léandri) Léandri;革质瑞香●☆

43783 Atenia Hook. et Arn. = Perideridia Rchb. ■☆

43784 Ateramnus P. Browne = Gymnanthes Sw. ●☆

43785 Ateramnus P. Browne = Sapium Jacq.(保留属名)●

43786 Atevala Raf. = Aloe L. ●■

43787 Athalmum Neck. = Pallenis Cass.(保留属名)●■☆

43788 Athalmus B. D. Jacks. = Athalmum Neck. ●■☆

43789 Athalmus Neck. = Pallenis Cass.(保留属名)●■☆

43790 Athamanta L.(1753);糖胡萝卜属■☆

43791 Athamanta achilleifolia Wall. = Pimpinella achilleifolia (Wall.) C. B. Clarke ■

43792 Athamanta capensis Burm. f. = Torilis arvensis (Huds.) Link ■☆

43793 Athamanta chinensis L. = Conioselinum chinense (L.) Britton, Sterns et Poggenb. ■

43794 Athamanta condensata L. = Libanotis condensata Crantz ■

43795 Athamanta cretensis L.;克里特糖胡萝卜;Candy Carrot ■☆

43796 Athamanta crinita (Pall.) Ledeb. = Schulzia crinita (Pall.) Spreng. ■

43797 Athamanta denudata Fisch. ex Hornem. = Cenolophium denudatum (Hornem.) Tutin ■

43798 Athamanta depressa D. Don = Cortia depressa (D. Don) C. Norman ■

43799 Athamanta incana Stephan ex Willd. = Libanotis incana (Stephan) O. Fedtsch. et B. Fedtsch. ■

43800 Athamanta lateriflora Eckl. et Zeyh. = Annesorhiza lateriflora (Eckl. et Zeyh.) B. -E. van Wyk ■☆

43801 Athamanta libanotis L. = Seseli libanotis (L.) W. D. J. Koch ■☆

43802 Athamanta macedonica (L.) Spreng.;马西登糖胡萝卜■☆

43803 Athamanta montana (Christ) Spalik et Wojew. et Downie;山生糖胡萝卜■☆

43804 Athamanta sibirica L. = Libanotis sibirica (L.) C. A. Mey. ■

43805 Athamanta sicula L.;西西里糖胡萝卜■☆

43806 Athamanta stricta (Ledeb.) Ledeb. ex Steud. = Seseli strictum Ledeb. ■

43807 Athamanta turbith Brot.;糖胡萝卜;Candy Carrot ■☆

43808 Athamantha L. (废弃属名) = Killinga Adans. (废弃属名) ■

43809 Athamantha Raf. = Athamanta L. ■☆

43810 Athamantha canescens DC. = Psammogeton canescens (DC.) Vatke ■☆

43811 Athamus Neck. = Carlina L. ●■

43812 Athanasia L. (1763);永菊属●☆

43813 Athanasia L. = Inulanthera Källersjö ●☆

43814 Athanasia acerosa (DC.) Harv. = Phymaspermum acerosum (DC.) Källersjö ■☆

43815 Athanasia adenantha (Harv.) Källersjö;腺花永菊●☆

43816 Athanasia alba Källersjö;白永菊●☆

43817 Athanasia annua L. = Lonas inodora Gaertn. ■☆

43818 Athanasia aspera Thunb. = Athanasia dentata (L.) L. ●☆

43819 Athanasia brownii Hochr. = Inulanthera brownii (Hochr.) Källersjö ●☆

43820 Athanasia calophylla Källersjö;美叶永菊●☆

43821 Athanasia canescens Thunb. = Athanasia pubescens (L.) L. ●☆

43822 Athanasia capitata (L.) L.;头状永菊●☆

43823 Athanasia capitata (L.) L. var. glabrata Harv. = Athanasia capitata (L.) L. ●☆

43824 Athanasia cinerea L. f. = Athanasia filiformis L. f. ●☆

43825 Athanasia cochlearifolia Källersjö;螺叶永菊●☆

43826 Athanasia coronopifolia Harv. = Inulanthera coronopifolia (Harv.) Källersjö ●☆

43827 Athanasia crenata (L.) L.;圆齿永菊●☆

43828 Athanasia crithmifolia (L.) L.;海茴香叶永菊●☆

43829 Athanasia crithmifolia (L.) L. subsp. palmatifida (DC.) Källersjö;掌状半裂永菊●☆

43830 Athanasia cuneifolia Lam.;楔叶永菊●☆

43831 Athanasia cuneiformis DC. = Athanasia trifurcata (L.) L. ●☆

43832 Athanasia dentata (L.) L.;具齿永菊●☆

43833 Athanasia dentata L. var. pachyphylla (Sch. Bip.) Harv. = Athanasia dentata (L.) L. ●☆

43834 Athanasia dentata L. var. subintegrifolia Harv. = Athanasia dentata (L.) L. ●☆

43835 Athanasia dimorpha DC. = Athanasia quinquedentata Thunb. ●☆

43836 Athanasia dimorpha DC. var. minor Harv. = Athanasia quinquedentata Thunb. ●☆

43837 Athanasia dimorpha DC. var. obovata Harv. = Athanasia quinquedentata Thunb. ●☆

43838 Athanasia dregeana (DC.) Harv. = Inulanthera dregeana (DC.) Källersjö ■☆

43839 Athanasia ebracteata E. Mey. ex DC. = Athanasia crenata (L.) L. ●☆

43840 Athanasia ebracteata E. Mey. ex DC. var. araneosa DC. = Athanasia flexuosa Thunb. ●☆

43841 Athanasia ebracteata E. Mey. ex DC. var. brachypoda DC. = Athanasia flexuosa Thunb. ●☆

43842 Athanasia ebracteata E. Mey. ex DC. var. hirsuta DC. = Athanasia crenata (L.) L. ●☆

43843 Athanasia elsiae Källersjö;埃尔西亚永菊●☆

43844 Athanasia eriopoda DC. = Athanasia pachycephala DC. subsp. eriopoda (DC.) Källersjö ●☆

43845 Athanasia eriopoda DC. var. depauperata ? = Athanasia pachycephala DC. ●☆

43846 Athanasia fasciculata (Less.) D. Dietr. = Athanasia vestita (Thunb.) Druce ●☆

43847 Athanasia filiformis L. f.;线状永菊●☆

43848 Athanasia filiformis L. f. var. cinerea (L. f.) Harv. = Athanasia filiformis L. f. ●☆

43849 Athanasia flexuosa E. Mey. ex DC. = Athanasia microcephala (DC.) D. Dietr. ●☆

43850 Athanasia flexuosa Thunb.;曲折永菊●☆

43851 Athanasia flexuosa Thunb. var. oligocephala DC. = Athanasia crenata (L.) L. ●☆

43852 Athanasia flexuosa Thunb. var. tomentella Hutch. = Athanasia flexuosa Thunb. ●☆

43853 Athanasia genistifolia L. = Oedera genistifolia (L.) Anderb. et K. Bremer ●☆

43854 Athanasia glabra Thunb. = Athanasia trifurcata (L.) L. ●☆

43855 Athanasia glabrescens DC. = Athanasia flexuosa Thunb. ●☆

43856 Athanasia graminifolia Walter = Marshallia graminifolia (Walter) Small ■☆

43857 Athanasia grandiceps Hilliard et B. L. Burtt;大头永菊●☆

43858 Athanasia hameri E. Phillips = Athanasia crithmifolia (L.) L. subsp. palmatifida (DC.) Källersjö ●☆

43859 Athanasia hirsuta Thunb.;粗毛永菊●☆

43860 Athanasia humilis Källersjö;低矮永菊●☆

43861 Athanasia imbricata Harv.;覆瓦永菊●☆

43862 Athanasia incisa (DC.) Harv. = Hymenolepis incisa DC. ■☆

43863 Athanasia indivisa Harv. = Hymenolepis indivisa (Harv.) Källersjö ●☆

43864 Athanasia inopinata (Hutch.) Källersjö;意外永菊●☆

43865 Athanasia juncea (DC.) D. Dietr.;灯心草永菊●☆

43866 Athanasia laevigata (L.) L. = Athanasia dentata (L.) L. ●☆

43867 Athanasia lanuginosa Cav. = Athanasia capitata (L.) L. ●☆

43868 Athanasia leptocephala Källersjö;细头永菊●☆

43869 Athanasia leucoclada (DC.) Harv. = Inulanthera leucoclada (DC.) Källersjö ●☆

43870 Athanasia linifolia Burm.;亚麻叶永菊●☆

43871 Athanasia linifolia Harv. = Athanasia microcephala (DC.) D. Dietr. ●☆

43872 Athanasia linifolia L. f. = Athanasia linifolia Burm. ●☆

43873 Athanasia longifolia Lam. = Athanasia linifolia Burm. ●☆

43874 Athanasia microcephala (DC.) D. Dietr.;小头永菊●☆

43875 Athanasia microphylla DC.;小叶永菊●☆

43876 Athanasia minuta (L. f.) Källersjö;微小永菊●☆

43877 Athanasia minuta (L. f.) Källersjö subsp. inermis (E. Phillips) Källersjö;无刺微小永菊●☆

43878 Athanasia montana J. M. Wood = Inulanthera montana (J. M. Wood) Källersjö ●☆

43879 Athanasia mundtii Harv. = Athanasia quinquedentata Thunb. ●☆

43880 Athanasia natalensis Schltr. = Phymaspermum acerosum (DC.)

Källersjö ■☆
43881 Athanasia obovata Walter = Marshallia obovata (Walter) Beadle et F. E. Boynton ■☆
43882 Athanasia obtusa Compton = Athanasia cuneifolia Lam. ●☆
43883 Athanasia oligocephala (DC.) Harv. = Athanasia crenata (L.) L. ●☆
43884 Athanasia oligocephala (DC.) Harv. var. adenantha Harv. = Athanasia adenantha (Harv.) Källersjö ●☆
43885 Athanasia oligocephala (DC.) Harv. var. araneosa ? = Athanasia flexuosa Thunb. ●☆
43886 Athanasia oligocephala (DC.) Harv. var. brachypoda ? = Athanasia flexuosa Thunb. ●☆
43887 Athanasia oligocephala (DC.) Harv. var. hirsuta ? = Athanasia crenata (L.) L. ●☆
43888 Athanasia oocephala (DC.) Källersjö;卵头永菊●☆
43889 Athanasia pachycephala DC.;粗头永菊●☆
43890 Athanasia pachycephala DC. subsp. eriopoda (DC.) Källersjö;红足永菊●☆
43891 Athanasia palmatifida DC. = Athanasia crithmifolia (L.) L. subsp. palmatifida (DC.) Källersjö ●☆
43892 Athanasia paniculata Walter = Verbesina walteri Shinners ■☆
43893 Athanasia parviflora L. = Hymenolepis parviflora (L.) DC. ■☆
43894 Athanasia pauciflora (DC.) D. Dietr. = Athanasia microcephala (DC.) D. Dietr. ●☆
43895 Athanasia pectinata L. f.;篦状永菊●☆
43896 Athanasia pinnata L. f.;羽状永菊●☆
43897 Athanasia pinnatifida (Oliv.) Hilliard = Phymaspermum pinnatifidum (Oliv.) Källersjö ●☆
43898 Athanasia pubescens (L.) L.;短柔毛永菊●☆
43899 Athanasia pumila L. f. = Rhynchopsidium pumilum (L. f.) DC. ■☆
43900 Athanasia punctata (DC.) Harv. = Inulanthera dregeana (DC.) Källersjö ■☆
43901 Athanasia punctata P. J. Bergius = Athanasia capitata (L.) L. ●☆
43902 Athanasia quinquedentata Thunb.;五齿永菊●☆
43903 Athanasia ramosa Klatt = Sphaeranthus ramosus (Klatt) Mesfin ■☆
43904 Athanasia recurvifolia Salisb. = Athanasia dentata (L.) L. ●☆
43905 Athanasia rotundifolia DC. = Athanasia quinquedentata Thunb. ●☆
43906 Athanasia rugulosa E. Mey. ex DC.;稍皱永菊●☆
43907 Athanasia scabra Thunb.;粗糙永菊●☆
43908 Athanasia scariosa DC. = Athanasia trifurcata (L.) L. ●☆
43909 Athanasia schistostephioides Hiern = Inulanthera schistostephioides (Hiern) Källersjö ●☆
43910 Athanasia schizolepis Harv. = Hymenolepis dentata (DC.) Källersjö ●☆
43911 Athanasia sessiliflora L. f. = Rhynchopsidium sessiliflorum (L. f.) DC. ■☆
43912 Athanasia spathulata (DC.) D. Dietr.;匙形永菊●☆
43913 Athanasia speciosa Hutch. = Hymenolepis speciosa (Hutch.) Källersjö ■☆
43914 Athanasia thodei Bolus = Inulanthera thodei (Bolus) Källersjö ●☆
43915 Athanasia tomentella Hutch. = Athanasia flexuosa Thunb. ●☆
43916 Athanasia tomentosa Thunb.;绒毛永菊●☆
43917 Athanasia tricuspis Poir. = Athanasia trifurcata (L.) L. ●☆
43918 Athanasia tridens Oliv. = Inulanthera tridens (Oliv.) Källersjö ●☆
43919 Athanasia tridentata Salisb. = Athanasia trifurcata (L.) L. ●☆
43920 Athanasia trifurcata (L.) L.;三叉永菊●☆
43921 Athanasia trifurcata (L.) L. var. cuneiformis (DC.) Harv. = Athanasia trifurcata (L.) L. ●☆
43922 Athanasia trifurcata (L.) L. var. glabra (Thunb.) Harv. = Athanasia trifurcata (L.) L. ●☆
43923 Athanasia trifurcata (L.) L. var. thunbergii Harv. = Athanasia trifurcata (L.) L. ●☆
43924 Athanasia trifurcata (L.) L. var. tricuspis (Poir.) DC. = Athanasia trifurcata (L.) L. ●☆
43925 Athanasia trifurcata (L.) L. var. virgata (Jacq.) Harv. = Athanasia virgata Jacq. ●☆
43926 Athanasia triloba Klatt = Anisopappus pinnatifidus (Klatt) O. Hoffm. ex Hutch. ■☆
43927 Athanasia trinervia Walter = Marshallia trinervia (Walter) Trel. ■☆
43928 Athanasia turbinata Burtt Davy = Gymnopentzia bifurcata Benth. ■☆
43929 Athanasia uniflora L. f. = Oedera uniflora (L. f.) Anderb. et K. Bremer ●☆
43930 Athanasia velutina (DC.) D. Dietr. = Athanasia tomentosa Thunb. ●☆
43931 Athanasia vestita (Thunb.) Druce;包被永菊●☆
43932 Athanasia villosa Hilliard = Phymaspermum villosum (Hilliard) Källersjö ■☆
43933 Athanasia virgata Jacq.;条纹永菊●☆
43934 Athanasia virgata Jacq. var. glabra (Thunb.) DC. = Athanasia trifurcata (L.) L. ●☆
43935 Athanasia viridis Källersjö;绿永菊●☆
43936 Athanasia woodii (Thell.) Hilliard = Phymaspermum woodii (Thell.) Källersjö ■☆
43937 Athanasiaceae Martinov = Asteraceae Bercht. et J. Presl(保留科名)●■
43938 Athanasiaceae Martinov = Compositae Giseke(保留科名)●■
43939 Athecia Gaertn. = Forstera L. f. ■☆
43940 Athenaea Adans.(废弃属名) = Athenaea Sendtn.(保留属名)●☆
43941 Athenaea Adans.(废弃属名) = Struchium P. Browne ■☆
43942 Athenaea Schreb. = Athenaea Sendtn.(保留属名)●☆
43943 Athenaea Schreb. = Casearia Jacq. ●
43944 Athenaea Schreb. = Iroucana Aubl. ●
43945 Athenaea Sendtn.(1846)(保留属名);阿西娜茄属●☆
43946 Athenaea affinis C. V. Morton = Brachistus affinis (C. V. Morton) D'Arcy, J. L. Gentry et Averett ●☆
43947 Athenaea anonacea Sendtn.;阿西娜茄●☆
43948 Athenaea brasiliana Hunz.;巴西阿西娜茄●☆
43949 Athenaea hirsuta Sendtn.;毛阿西娜茄●☆
43950 Athenanthia Kunth = Anthaenantia P. Beauv. ■☆
43951 Athenoea Schreb. = Casearia Jacq. ●
43952 Atheolaena Rchb. = Aetheolaena Cass. ●☆
43953 Atheolaena Rchb. = Senecio L. ●■
43954 Atherandra Decne.(1844);芒蕊萝藦属■☆
43955 Atherandra acutifolia Decne.;芒蕊萝藦■☆
43956 Atheranthera Mast. = Gerrardanthus Harv. ex Benth. et Hook. f. ■☆
43957 Atheranthera paniculata Mast. = Gerrardanthus paniculatus (Mast.) Cogn. ■☆
43958 Athernotus Dulac = Calamagrostis Adans. ■
43959 Atherocephala DC. = Andersonia Buch. -Ham. ex Wall. ●☆
43960 Atherolepis Hook. f.(1883);芒鳞萝藦属☆
43961 Atherolepis pierrei Costantin;芒鳞萝藦☆
43962 Atherolepsis Willis = Atherolepis Hook. f. ☆
43963 Atherophora Steud. = Aegopogon Humb. et Bonpl. ex Willd. ■☆

43964 Atherophora Willd. ex Steud. = Aegopogon Humb. et Bonpl. ex Willd. ■☆
43965 Atheropogon Muhlenb. ex Willd. = Botelua Lag. ■
43966 Atheropogon Muhlenb. ex Willd. = Bouteloua Lag.（保留属名）■
43967 Atheropogon Willd. = Botelua Lag. ■
43968 Atheropogon Willd. = Bouteloua Lag.（保留属名）■
43969 Atheropogon curtipendulus (Michx.) E. Fourn. = Bouteloua curtipendula (Michx.) Torr. ■
43970 Atheropogon gracilis (Kunth) Spreng. = Bouteloua gracilis (Kunth) Lag. ex Steud. ■
43971 Atherosperma Labill.(1806)；黑檫木属(蔻香木属，芒籽香属，芒籽属，麝香芒籽属，香皮茶属)●☆
43972 Atherosperma moschatum Labill.；黑檫木(蔻香木，麝香芒籽)；Australian Sassafras, Black Sassafras, Southern Sassafras, Tasmanian Sassafras ●☆
43973 Atherospermataceae R. Br.(1814)；黑檫木科(芒籽科，芒籽香科，芒子科，香皮茶科，异籽木科)●☆
43974 Atherospermataceae R. Br. = Monimiaceae Juss.（保留科名）●■☆
43975 Atherostemon Blume = Atherandra Decne. ■☆
43976 Atherstonea Pappe = Strychnos L. ●
43977 Atherstonea decussata Pappe = Strychnos decussata (Pappe) Gilg ●☆
43978 Athertonia L. A. S. Johnson et B. G. Briggs(1975)；昆士兰龙眼属●☆
43979 Athertonia diversifolia (C. T. White) L. A. S. Johnson et B. G. Briggs；昆士兰龙眼●☆
43980 Atherurus Blume(废弃属名) = Pinellia Ten.（保留属名）■
43981 Athesiandra Miers ex Benth. et Hook. f. = Ptychopetalum Benth. ●☆
43982 Athlianthus Endl. = Justicia L. ●■
43983 Athrixia Ker Gawl.(1823)；紫绒草属●■☆
43984 Athrixia angustissima DC.；窄紫绒草●☆
43985 Athrixia arachnoidea J. M. Wood et M. S. Evans ex J. M. Wood；蛛网紫绒草●☆
43986 Athrixia asteroides Bolus et Schltr. = Lepidostephium asteroides (Bolus et Schltr.) Kroner ■
43987 Athrixia capensis Ker Gawl.；好望角紫绒草●☆
43988 Athrixia capensis Ker Gawl. var. latifolia DC. = Athrixia crinita (L.) Druce ●☆
43989 Athrixia crinita (L.) Druce；宽叶紫绒草●☆
43990 Athrixia debilis DC.；弱小紫绒草●☆
43991 Athrixia diffusa Baker = Hirpicium gracile (O. Hoffm.) Rössler ■☆
43992 Athrixia elata Sond.；高紫绒草●☆
43993 Athrixia felicioides Hiern = Nicolasia felicioides (Hiern) S. Moore ■☆
43994 Athrixia foliosa S. Moore = Athrixia rosmarinifolia (Sch. Bip. ex Walp.) Oliv. et Hiern var. foliosa (S. Moore) Kroner ●☆
43995 Athrixia fontana MacOwan；泉紫绒草●☆
43996 Athrixia fontinalis Wild；春紫绒草●☆
43997 Athrixia gerrardii Harv.；杰勒德紫绒草●☆
43998 Athrixia glandulosa Kunth = Athrixia crinita (L.) Druce ●☆
43999 Athrixia heterophylla (Thunb.) Less.；互叶紫绒草●☆
44000 Athrixia heterophylla (Thunb.) Less. subsp. sessilifolia (DC.) Kroner；无柄互叶紫绒草●☆
44001 Athrixia kassneri Muschl. = Athrixia rosmarinifolia (Sch. Bip. ex Walp.) Oliv. et Hiern ●☆
44002 Athrixia nyassana S. Moore；尼亚萨紫绒草●☆
44003 Athrixia oblonga S. Moore；矩圆紫绒草●☆
44004 Athrixia phylicoides DC.；菲利木紫绒草●☆
44005 Athrixia pinifolia N. E. Br. = Macowania pinifolia (N. E. Br.) Kroner ●☆
44006 Athrixia rosmarinifolia (Sch. Bip. ex Walp.) Oliv. et Hiern；迷迭香叶紫绒草●☆
44007 Athrixia rosmarinifolia (Sch. Bip. ex Walp.) Oliv. et Hiern var. foliosa (S. Moore) Kroner；多迷迭香叶紫绒草●☆
44008 Athrixia sessilifolia DC. = Athrixia heterophylla (Thunb.) Less. subsp. sessilifolia (DC.) Kroner ●☆
44009 Athrixia stenophylla Baker = Dewildemania stenophylla (Baker) B. L. Burtt ■☆
44010 Athrixia subsimplex Brenan；简单紫绒草●☆
44011 Athrixia tomentosa (Thunb.) Less. = Amphiglossa tomentosa (Thunb.) Harv. ■☆
44012 Athroandra (Hook. f.) Pax et K. Hoffm. = Erythrococca Benth. ●☆
44013 Athroandra Pax et K. Hoffm. = Erythrococca Benth. ●☆
44014 Athroandra atrovirens (Pax) Pax et K. Hoffm. = Erythrococca atrovirens (Pax) Prain ●☆
44015 Athroandra dewevrei (Pax ex De Wild.) Pax et K. Hoffm. = Erythrococca dewevrei (Pax ex De Wild.) Prain ●☆
44016 Athroandra hispida Pax et K. Hoffm. = Erythrococca macrophylla (Prain) Prain ●☆
44017 Athroandra inopinata (Prain) Pax et K. Hoffm. = Erythrococca macrophylla (Prain) Prain ●☆
44018 Athroandra pallidifolia Pax et K. Hoffm. = Erythrococca pallidifolia (Pax et K. Hoffm.) Keay ●☆
44019 Athroandra rivularis (Müll. Arg.) Pax et K. Hoffm. = Erythrococca rivularis (Müll. Arg.) Prain ●☆
44020 Athroandra welwitschiana (Müll. Arg.) Pax et K. Hoffm. = Erythrococca welwitschiana (Müll. Arg.) Prain ●☆
44021 Athrodactylis J. R. Forst. et G. Forst. = Keura Forssk. ●■
44022 Athrodactylis J. R. Forst. et G. Forst. = Pandanus Parkinson ex Du Roi ●■
44023 Athroisma DC.(1833)；黑果菊属●■☆
44024 Athroisma Griff. = Trigonostemon Blume(保留属名)●
44025 Athroisma boranense Cufod.；博兰黑果菊●☆
44026 Athroisma boranense Cufod. var. hispidum ? = Athroisma boranense Cufod. ●☆
44027 Athroisma fanshawei Wild；范肖黑果菊●☆
44028 Athroisma gracile (Oliv.) Mattf.；纤细黑果菊●☆
44029 Athroisma gracile (Oliv.) Mattf. subsp. psyllioides (Oliv.) T. Erikss.；补血草黑果菊●☆
44030 Athroisma haareri (Dandy) Mattf. = Athroisma gracile (Oliv.) Mattf. subsp. psyllioides (Oliv.) T. Erikss. ●☆
44031 Athroisma hastifolium Mattf.；戟叶黑果菊●☆
44032 Athroisma lobatum (Klatt) Mattf.；浅裂黑果菊●☆
44033 Athroisma pinnatifidum T. Erikss.；羽裂黑果菊●☆
44034 Athroisma proteiforme (Humbert) Mattf.；易变黑果菊●☆
44035 Athroisma psilocarpum T. Erikss.；光果黑果菊●☆
44036 Athroisma psyllioides (Oliv.) Mattf. = Athroisma gracile (Oliv.) Mattf. subsp. psyllioides (Oliv.) T. Erikss. ●☆
44037 Athroisma pusillum T. Erikss.；微小黑果菊●☆
44038 Athroisma stuhlmannii (O. Hoffm.) Mattf.；斯图尔曼黑果菊●☆
44039 Athrolophis (Trin.) Chiov. = Andropogon L.（保留属名）■
44040 Athronia Neck. = Spilanthes Jacq. ■
44041 Athroostachys Benth. ex Benth. et Hook. f.(1883)；密穗竹属●☆

44042 Athroostachys capitata (Hook.) Benth.;密穗竹●☆
44043 Athroostachys capitata Benth. = Athroostachys capitata (Hook.) Benth. ●☆
44044 Athrotaxidaceae Doweld;澳洲杉科●☆
44045 Athrotaxidaceae Nakai = Taxodiaceae Saporta(保留科名)●
44046 Athrotaxis D. Don(1838);澳洲杉属(密叶杉属);Pencil Pine, Tasmanian Cedar ●☆
44047 Athrotaxis cupressoides D. Don;澳洲杉;Pencil-pine, Smooth Tasmanian Cedar,Tasmania Pencil Pine ●☆
44048 Athrotaxis latifolia ?;宽叶澳洲杉;Summit Cedar ●☆
44049 Athrotaxis laxifolia Hook.;疏叶澳洲杉;Summit Cedar,Tasmanian Cedar ●☆
44050 Athrotaxis selaginoides D. Don;大澳洲杉(密叶杉);King Billy, King Billy Pine, King William Billy Pine, King William Pine, Tasmanian Cedar,William Pine ●☆
44051 Athruphyllum Lour. (1790);密叶掌属●☆
44052 Athruphyllum Lour. = Rapanea Aubl. ●
44053 Athruphyllum lineare Lour. = Myrsine linearis (Lour.) Poir. ●
44054 Athruphyllum neriifolium (Siebold et Zucc.) H. Hara = Myrsine seguinii H. Lév. ●
44055 Athruphyllum neriifolium (Siebold et Zucc.) H. Hara = Rapanea neriifolia (Siebold et Zucc.) Mez ●
44056 Athruphyllum neriifolium H. Hara = Athruphyllum neriifolium (Siebold et Zucc.) H. Hara ●
44057 Athruphyllum seguinii (H. Lév.) Nakai = Myrsine seguinii H. Lév. ●
44058 Athruphyllum taiwanianum Nakai = Myrsine seguinii H. Lév. ●
44059 Athruphyllum yunnanensis (Mez) Nakai = Myrsine seguinii H. Lév. ●
44060 Athyana (Griseb.) Radlk. (1887);阿西无患子属●☆
44061 Athyana Radlk. = Athyana (Griseb.) Radlk. ●☆
44062 Athymalus Neck. = Euphorbia L. ●■
44063 Athyrocarpus Schltdl. = Commelina L. ■
44064 Athyrocarpus Schltdl. = Phaeosphaerion Hassk. ■
44065 Athyrocarpus Schltdl. ex Benth. = Commelina L. ■
44066 Athyrocarpus Schltdl. ex Benth. = Phaeosphaerion Hassk. ■
44067 Athyrus Neck. = Lathyrus L. ■
44068 Athysanus Greene(1885);小盾芥属■☆
44069 Athysanus pusillus Greene;小盾芥■☆
44070 Atimeta Schott = Rhodospatha Poepp. ■☆
44071 Atirbesia Raf. = Marrubium L. ■
44072 Atitara Juss. = Evodia J. R. Forst. et G. Forst. ●
44073 Atitara Kuntze = Desmoncus Mart. (保留属名)●☆
44074 Atitara Marcgr. ex Juss. = Evodia J. R. Forst. et G. Forst. ●
44075 Atitara O. F. Cook = Desmoncus Mart. (保留属名)●☆
44076 Atkinsia R. A. Howard = Thespesia Sol. ex Corrêa(保留属名)●
44077 Atkinsonia F. Muell. (1865);西南澳寄生属●☆
44078 Atkinsonia ligustrina (F. Muell.) F. Muell.;西南澳寄生●☆
44079 Atlanthemum Raynaud = Helianthemum Mill. ●■
44080 Atlanthemum sanguineum (Lag.) Raynaud = Helianthemum sanguineum (Lag.) Dunal ●☆
44081 Atlantia Kurz = Atalantia Corrêa(保留属名)●
44082 Atocion Adans. = Melandrium Röhl. ■
44083 Atocion Adans. = Silene L. (保留属名)■
44084 Atocion armeria (L.) Raf. = Silene armeria L. ■
44085 Atolaria Neck. = Crotalaria L. ●■
44086 Atomostigma Kuntze(1898);巴西蔷薇属■☆
44087 Atomostylis Steud. = Cyperus L. ■
44088 Atomostylis cyperiformis Steud. = Anosporum pectinatus (Vahl) Lye ■☆
44089 Atomostylis flavescens Steud. = Anosporum pectinatus (Vahl) Lye ■☆
44090 Atopocarpus Cuatrec. = Clonodia Griseb. ●☆
44091 Atopoglossum Luer(2004);异舌兰属■☆
44092 Atopostema Boutique = Monanthotaxis Baill. ●☆
44093 Atopostema Boutique(1951);肖单花番荔枝属●☆
44094 Atopostema angustifolia Boutique = Monanthotaxis klainei (Engl.) Verdc. var. angustifolia (Boutique) Verdc. ●☆
44095 Atopostema klainii (Engl.) Boutique = Monanthotaxis klainei (Engl.) Verdc. ●☆
44096 Atossn Alef. = Vicia L. ■
44097 Atractantha McClure(1973);纺锤花竹属●☆
44098 Atractantha falcata McClure;镰形纺锤花竹●☆
44099 Atractantha radiata McClure;纺锤花竹●☆
44100 Atractocarpa Franch. = Puelia Franch. ■☆
44101 Atractocarpa olyriformis Franch. = Puelia olyriformis (Franch.) Clayton ■☆
44102 Atractocarpus Schltr. et K. Krause(1908);纺锤果茜属●☆
44103 Atractocarpus hirtus (F. Muell.) Puttock;毛纺锤果茜●☆
44104 Atractocarpus rotundifolius Guillaumin;圆叶纺锤果茜●☆
44105 Atractogyne Pierre(1896);梭柱茜属●☆
44106 Atractogyne batesii Wernham = Atractogyne gabonii Pierre ●☆
44107 Atractogyne bracteata (Wernham) Hutch. et Dalziel;具苞梭柱茜●☆
44108 Atractogyne gabonii Pierre;加蓬梭柱茜●☆
44109 Atractogyne melongenifolia A. Chev. ex Hutch. et Dalziel = Atractogyne bracteata (Wernham) Hutch. et Dalziel ●☆
44110 Atractogyne stenocarpa K. Schum.;狭果梭柱茜●☆
44111 Atractylia Rchb. = Carthamus L. ■
44112 Atractylis Boelun. = Carthamus L. ■
44113 Atractylis Boelun. = Phonus Hill ●☆
44114 Atractylis L. (1753);纺锤菊属(苍术属,羽叶苍术属);Distaff Thistle ■☆
44115 Atractylis amplexicaulis Nakai ex Mori = Atractylodes coreana (Nakai) Kitam. ■
44116 Atractylis angustifolia Houtt. = Berkheya angustifolia (Houtt.) Merr. ■☆
44117 Atractylis arbuscula Svent. et Michaelis;树状纺锤菊■☆
44118 Atractylis arbuscula Svent. et Michaelis var. schizogynophylla Svent. et Kahne = Atractylis arbuscula Svent. et Michaelis ■☆
44119 Atractylis aristata Batt.;具芒纺锤菊■☆
44120 Atractylis babelii Hochr.;巴贝尔纺锤菊■☆
44121 Atractylis babelii Hochr. var. monodii (Arènes) D. P. Petit;莫诺纺锤菊■☆
44122 Atractylis caerulea Batt.;天蓝纺锤菊■☆
44123 Atractylis caespitosa Desf.;簇生纺锤菊■☆
44124 Atractylis caespitosa Desf. var. incana Maire = Atractylis caespitosa Desf. ■☆
44125 Atractylis caespitosa Desf. var. radians Batt. = Atractylis caespitosa Desf. ■☆
44126 Atractylis cancellata L.;格纹纺锤菊■☆
44127 Atractylis cancellata L. var. eremophila Braun-Blanq. et Maire = Atractylis cancellata L. ■☆
44128 Atractylis candida Cuénod = Atractylis carduus (Forssk.) H.

Christ ■☆

44129 Atractylis carduus (Forssk.) H. Christ;刺纺锤菊■☆

44130 Atractylis carduus (Forssk.) H. Christ var. angustifolia Täckh. et Boulos = Atractylis carduus (Forssk.) H. Christ ■☆

44131 Atractylis carduus (Forssk.) H. Christ var. glabrescens (Boiss.) Täckh. et Boulos = Atractylis carduus (Forssk.) H. Christ ■☆

44132 Atractylis carduus (Forssk.) H. Christ var. latifolia Täckh. et Boulos = Atractylis carduus (Forssk.) H. Christ ■☆

44133 Atractylis carduus (Forssk.) H. Christ var. marmarica Täckh. et Boulos = Atractylis carduus (Forssk.) H. Christ ■☆

44134 Atractylis carlinoides Hand. -Mazz. = Atractylodes carlinoides (Hand. -Mazz.) Kitam. ■

44135 Atractylis chinensis (Bunge) DC. = Atractylodes lancea (Thunb.) DC. ■

44136 Atractylis chinensis (Bunge) DC. = Atractylodes lancea (Thunb.) DC. var. chinensis (Bunge) Kitam. ■☆

44137 Atractylis chinensis (Bunge) DC. f. erossodentata (Koidz.) Hand. -Mazz. = Atractylodes lancea (Thunb.) DC. ■

44138 Atractylis chinensis (Bunge) DC. f. simplicifolia (Loes.) Hand. -Mazz. = Atractylodes lancea (Thunb.) DC. ■

44139 Atractylis chinensis (Bunge) DC. f. stapfii (Baroni) Hand. -Mazz. = Atractylodes lancea (Thunb.) DC. ■

44140 Atractylis chinensis (Bunge) DC. var. coreana (Nakai) Chu ? = Atractylodes coreana (Nakai) Kitam. ■

44141 Atractylis chinensis (Bunge) DC. var. liaotungensis Kitag. = Atractylodes coreana (Nakai) Kitam. ■

44142 Atractylis chinensis (Bunge) DC. var. liaotungensis Kitag. = Atractylodes lancea (Thunb.) DC. ■

44143 Atractylis chinensis (Bunge) DC. var. loeseneri Kitag. = Atractylodes lancea (Thunb.) DC. ■

44144 Atractylis chinensis (Bunge) DC. var. quiqueloba A. I. Baranov et Skvortsov = Atractylodes lancea (Thunb.) DC. ■

44145 Atractylis chinensis (Bunge) DC. var. simplicifolia (Loesen) Chu = Atractylodes lancea (Thunb.) DC. ■

44146 Atractylis coreana Nakai = Atractylodes coreana (Nakai) Kitam. ■

44147 Atractylis cryptocephalus (Baker) F. G. Davies;隐头纺锤菊■☆

44148 Atractylis delicatula Batt. ex L. Chevall.;姣美纺锤菊■☆

44149 Atractylis echinata Pomel;具刺纺锤菊■☆

44150 Atractylis flava Desf. = Atractylis carduus (Forssk.) H. Christ ■☆

44151 Atractylis flava Desf. var. candida (Cuénod) Burollet = Atractylis carduus (Forssk.) H. Christ ■☆

44152 Atractylis flava Desf. var. citrina (Coss. et Kralik) Hochr. = Atractylis carduus (Forssk.) H. Christ ■☆

44153 Atractylis flava Desf. var. glabrescens Boiss. = Atractylis carduus (Forssk.) H. Christ ■☆

44154 Atractylis fruticosa L. = Berkheya fruticosa (L.) Ehrh. ■☆

44155 Atractylis glomerata Caball. = Atractylis cancellata L. ■☆

44156 Atractylis gummifera L. = Carlina gummifera (L.) Less. ■☆

44157 Atractylis gummifera L. var. macrocephala (Desf.) Pott. -Alap. = Chamaeleon gummifer (L.) Cass. ■☆

44158 Atractylis humilis L. subsp. caespitosa (Desf.) Maire = Atractylis caespitosa Desf. ■☆

44159 Atractylis humilis L. var. incana Maire = Atractylis caespitosa Desf. ■☆

44160 Atractylis humilis L. var. radians Batt. = Atractylis caespitosa Desf. ■☆

44161 Atractylis japonica (Koidz. ex Kitam.) Kitag. = Atractylodes japonica Koidz. ex Kitam. ■

44162 Atractylis japonica (Koidz. ex Kitam.) Kitag. = Atractylodes ovata (Thunb.) DC. ■☆

44163 Atractylis japonica (Koidz.) Kitag. = Atractylodes japonica Koidz. ex Kitam. ■

44164 Atractylis lancea Thunb. = Atractylodes lancea (Thunb.) DC. ■

44165 Atractylis lyrata (Siebold et Zucc.) Hand. -Mazz. f. ternata (Kom.) Nakai = Atractylodes japonica Koidz. ex Kitam. ■

44166 Atractylis macrocephala (Koidz.) Hand. -Mazz. = Atractylodes macrocephala Koidz. ■

44167 Atractylis macrocephala (Koidz.) Hand. -Mazz. var. hunanensis Y. Ling = Atractylodes macrocephala Koidz. ■

44168 Atractylis macrocephala Desf. = Chamaeleon gummifer (L.) Cass. ■☆

44169 Atractylis macrophylla Desf.;大叶纺锤菊■☆

44170 Atractylis macrophylla Desf. var. incana Maire = Atractylis macrophylla Desf. ■☆

44171 Atractylis microcephala Coss. et Durieu = Atractylis serratuloides (Cass.) DC. ■☆

44172 Atractylis monodii Arènes = Atractylis babelii Hochr. var. monodii (Arènes) D. P. Petit ■☆

44173 Atractylis ovata Thunb. = Atractylodes lancea (Thunb.) DC. ■

44174 Atractylis ovata Thunb. = Atractylodes macrocephala Koidz. ■

44175 Atractylis ovata Thunb. = Atractylodes ovata (Thunb.) DC. ■☆

44176 Atractylis ovata Thunb. f. amurensis Freyn ex Kom. = Atractylodes lancea (Thunb.) DC. ■

44177 Atractylis ovata Thunb. f. lyratifolia Kom. = Atractylodes lancea (Thunb.) DC. ■

44178 Atractylis ovata Thunb. f. pinnatifolia Kom. = Atractylodes japonica Koidz. ex Kitam. ■

44179 Atractylis ovata Thunb. f. simplicifolia (Loes.) Kom. = Atractylodes lancea (Thunb.) DC. ■

44180 Atractylis ovata Thunb. var. simplicifolia Loes. = Atractylodes lancea (Thunb.) DC. ■

44181 Atractylis ovata Thunb. var. ternata Kom. = Atractylodes japonica Koidz. ex Kitam. ■

44182 Atractylis phaeolepis Pomel;大苞纺锤菊■☆

44183 Atractylis pinnatifolia (Kom.) S. Y. Hu = Atractylodes japonica Koidz. ex Kitam. ■

44184 Atractylis polycephala Coss.;多头纺锤菊■☆

44185 Atractylis preauxiana Sch. Bip.;普雷纺锤菊■☆

44186 Atractylis prolifera Boiss.;多育纺锤菊■☆

44187 Atractylis prolifera Boiss. var. albiflora Cavara = Atractylis prolifera Boiss. ■☆

44188 Atractylis prolifera Boiss. var. sulfurea Petit = Atractylis prolifera Boiss. ■☆

44189 Atractylis separata Bailey = Atractylodes lancea (Thunb.) DC. ■

44190 Atractylis serrata Pomel;锯齿纺锤菊■☆

44191 Atractylis serratuloides (Cass.) DC.;拟锯齿纺锤菊■☆

44192 Atractylodes DC. (1838);苍术属;Atractylodes ■

44193 Atractylodes amurensis (Freyn ex Kom.) H. S. Pak;阿穆尔苍术■☆

44194 Atractylodes carlinoides (Hand. -Mazz.) Kitam.;鄂西苍术;W. Hubei Atractylodes, Western Hupeh Atractylodes ■

44195 Atractylodes chinensis (Bunge) Koidz. = Atractylodes lancea (Thunb.) DC. ■

44196 Atractylodes chinensis (Bunge) Koidz. = Atractylodes lancea

(Thunb.) DC. var. chinensis (Bunge) Kitam. ■☆
44197 Atractylodes chinensis (DC.) Koidz. = Atractylodes lancea (Thunb.) DC. ■
44198 Atractylodes chinensis Koidz. f. quinqueloba (A. I. Baranov et Skvortsov) Y. C. Zhu;赤峰苍术■
44199 Atractylodes chinensis Koidz. f. simplicifolia (Loes.) Y. C. Zhu = Atractylodes lancea (Thunb.) DC. ■
44200 Atractylodes chinensis Koidz. var. liaotungensis (Kitag.) Y. C. Zhu;辽东苍术■
44201 Atractylodes chinensis Koidz. var. simplicifolia (Loes.) Kitag. = Atractylodes lancea (Thunb.) DC. ■
44202 Atractylodes coreana (Nakai) Kitam.;朝鲜苍术;Korea Atractylodes ■
44203 Atractylodes erosodentata Koidz. = Atractylodes lancea (Thunb.) DC. ■
44204 Atractylodes gummifera L.;欧苍术■☆
44205 Atractylodes japonica Koidz. ex Kitam.;关苍术(白术,东苍术,吴苍术);Japan Atractylodes,Japanese Atractylodes ■
44206 Atractylodes japonica Koidz. ex Kitam. = Atractylodes ovata (Thunb.) DC. ■☆
44207 Atractylodes koreana (Nakai) Kitam. = Atractylodes coreana (Nakai) Kitam. ■
44208 Atractylodes lancea (Thunb.) DC.;苍术(北苍术,赤术,京苍术,马蓟,茅苍术,茅山苍术,南苍术,枪头菜,青术,山刺菜,西北苍术,仙术);Cang-zhu Atractylodes,China Atractylodes,Common Atractylodes,Swordlike Atractylodes ■
44209 Atractylodes lancea (Thunb.) DC. subsp. luotianensis S. L. Hu et X. F. Feng;罗田苍术;Luotian Atractylodes ■
44210 Atractylodes lancea (Thunb.) DC. var. chinensis (Bunge) Kitam.;日本苍术(中国苍术)■☆
44211 Atractylodes lancea (Thunb.) DC. var. simplicifolia (Loes.) Kitam. = Atractylodes lancea (Thunb.) DC. ■
44212 Atractylodes lancea DC. subsp. luotianensis S. L. Hu et X. F. Feng;罗甸苍术■
44213 Atractylodes lancea DC. var. chiensis (Bunge) Kitam. = Atractylodes lancea (Thunb.) DC. ■
44214 Atractylodes lancea DC. var. simplicifolia (Loes.) Kitam. = Atractylodes lancea (Thunb.) DC. ■
44215 Atractylodes lyrata Siebold et Zucc. = Atractylodes lancea (Thunb.) DC. ■
44216 Atractylodes lyrata Siebold et Zucc. f. ternata (Kom.) Nakai = Atractylodes japonica Koidz. ex Kitam. ■
44217 Atractylodes lyrata Siebold et Zucc. var. ternata (Kom.) Koidz. = Atractylodes japonica Koidz. ex Kitam. ■
44218 Atractylodes macrocephala Koidz.;白术(白大寿,吃力伽,大头苍术,东境术,冬白术,冬术,枹蓟,桴蓟,贡术,狗头术,杭术,马蓟,乞力伽,乞力佳,山蓟,山姜,山芥,山精,山连,山莲,狮子术,术,台术,天蓟,天芥,天生术,土白术,歙术,羊枹蓟,杨枹,杨枹蓟,杨桴,于术,云术,云头术,浙江白术,浙术,种术);Chang-zhu Atractylodes,Largehead Atractylodes ■
44219 Atractylodes ovata (Thunb.) DC.;卵白术■☆
44220 Atractylodes ovata (Thunb.) DC. = Atractylodes lancea (Thunb.) DC. ■
44221 Atractylodes rubra Dekker;红苍术■☆
44222 Atragene L. (1753);赛铁线莲属(瓣铁线莲属)●☆
44223 Atragene L. = Clematis L. ●■
44224 Atragene alpina L. = Clematis alpina Mill. ●☆
44225 Atragene alpina L. var. ochotensis (Pall.) Regel et Tiling = Clematis sibirica (L.) Mill. var. ochotensis (Pall.) S. H. Li et Y. Huei Huang ●
44226 Atragene americana Sims;赛铁线莲(瓣铁线莲);Purple Virgin's-bower ■☆
44227 Atragene americana Sims = Clematis occidentalis (Hornem.) DC. ●☆
44228 Atragene columbiana Nutt. = Clematis columbiana (Nutt.) Torr. et A. Gray ■☆
44229 Atragene dianae Serov = Clematis macropetala Ledeb. ●
44230 Atragene florida Pers. = Clematis florida Thunb. ●
44231 Atragene grosseserrata Rydb. = Clematis occidentalis (Hornem.) DC. var. grosseserrata (Rydb.) J. S. Pringle ●☆
44232 Atragene indica Desf. = Clematis florida Thunb. ●
44233 Atragene japonica Thunb. = Anemone hupehensis (Lemoine) Lemoine var. japonica (Thunb. ex A. Murray) Bowler et Stern ■
44234 Atragene koreana (Kom.) Kom. = Clematis koreana Kom. ●
44235 Atragene koreana Kom. = Clematis koreana Kom. ●
44236 Atragene macropetala (Ledeb.) Ledeb. = Clematis macropetala Ledeb. ●
44237 Atragene moisseenkoi Serov = Clematis moisseenkoi (Serov) W. T. Wang ●
44238 Atragene occidentalis Hornem. = Clematis occidentalis (Hornem.) DC. ●☆
44239 Atragene ochotensis Pall. = Clematis sibirica (L.) Mill. var. ochotensis (Pall.) S. H. Li et Y. Huei Huang ●
44240 Atragene platysepala Trautv. et C. A. Mey. = Clematis sibirica (L.) Mill. var. ochotensis (Pall.) S. H. Li et Y. Huei Huang ●
44241 Atragene sibirica L. = Clematis sibirica (L.) Mill. ●
44242 Atragene tenuifolia L. f. = Anemone tenuifolia (L. f.) DC. ■☆
44243 Atragene tianschanica Pavlov = Clematis sibirica (L.) Mill. ●
44244 Atragene tianschanica Pavlov = Clematis tianschanica Pavlov ●
44245 Atragene zeylanica L. = Naravelia zeylanica (L.) DC. ●
44246 Atraphax Scop. = Atragene L. ●☆
44247 Atraphaxis L. (1753);木蓼属(针枝蓼属,针枝属);Goat's-Wheat,Goatwheat,Knotwood ●
44248 Atraphaxis afghanica Meisn. = Atraphaxis spinosa L. ●
44249 Atraphaxis angustifolia Jaub. et Spach;狭叶木蓼●
44250 Atraphaxis badghysi Kult.;巴德木蓼●☆
44251 Atraphaxis bracteata Losinsk.;沙木蓼;Bracteate Goatwheat,Sandy Knotwood ●
44252 Atraphaxis bracteata Losinsk. var. angustifolia Losinsk.;狭叶沙木蓼;Narrowleaf Bracteate Goatwheat ●
44253 Atraphaxis bracteata Losinsk. var. angustifolia Losinsk. = Atraphaxis bracteata Losinsk. ●
44254 Atraphaxis bracteata Losinsk. var. latifolia H. C. Fu et M. H. Zhao;宽叶沙木蓼;Broadleaf Bracteate Goatwheat ●
44255 Atraphaxis bracteata Losinsk. var. latifolia H. C. Fu et M. H. Zhao = Atraphaxis bracteata Losinsk. ●
44256 Atraphaxis canescens Bunge;糙叶木蓼;Hoary Goatwheat,Roughleaf Knotwood ●
44257 Atraphaxis caucusca (Hoffm.) Pavlov;高加索木蓼●
44258 Atraphaxis compacta Ledeb.;拳木蓼;Fist Knotwood ●
44259 Atraphaxis decipiens Jaub. et Spach;细枝木蓼(反折木蓼,美丽木蓼);Beautiful Goatwheat,Deceiving Goatwheat,Thintwig Knotwood ●
44260 Atraphaxis frutescens (L.) Eversm.;兴安木蓼(灌木蓼,木蓼,

沙木蓼）；Dahurian Goatwheat, Knotwood, Shrubby Atraphaxis, Shrubby Goatwheat ●

44261 Atraphaxis frutescens (L.) Eversm. var. papillosa Y. L. Liu；乳头叶木蓼；Papillseleaf Knotwood ●

44262 Atraphaxis frutescens (L.) K. Koch = Atraphaxis frutescens Eversm. ●

44263 Atraphaxis frutescens Eversm. = Atraphaxis frutescens (L.) Eversm. ●

44264 Atraphaxis jrtyschensis Chang Y. Yang et Y. L. Han；额河木蓼；Ehe Goatwheat, Emur Goatwheat, Xinjiang Knotwood ●

44265 Atraphaxis karataviensis Lipsch. et Pavlov；卡拉塔夫●☆

44266 Atraphaxis laetevirens (Ledeb.) Jaub. et Spach；绿叶木蓼；Greenleaf Goatwheat, Greenleaf Knotwood, Green-leaved Goatwheat ●

44267 Atraphaxis lanceolata (M. Bieb.) Meisn. = Atraphaxis frutescens (L.) Eversm. ●

44268 Atraphaxis lanceolata (M. Bieb.) Meisn. var. virgata Regel = Atraphaxis virgata (Regel) Krasn. ●

44269 Atraphaxis manshurica Kitag.；东北木蓼（东北针枝蓼，木蓼）；Manchurian Goatwheat, NE. China Knotwood ●

44270 Atraphaxis muschketowii Krasn.；穆什木蓼；Shrubby Buckwheat ●☆

44271 Atraphaxis pungens (M. Bieb.) Jaub. et Spach；锐枝木蓼（刺针木蓼，坚针木蓼，针木蓼）；Punge Knotwood, Sharp Goatwheat ●

44272 Atraphaxis pyrifolia Bunge；梨叶木蓼；Pearleaf Knotwood, Pear-leaved Goatwheat ●

44273 Atraphaxis replicata Lam.；反折木蓼（扁果木蓼）；Replicate Knotwood ●

44274 Atraphaxis replicata Lam. = Atraphaxis spinosa L. ●

44275 Atraphaxis seravschanica Pavlov；塞拉夫木蓼●☆

44276 Atraphaxis spinosa L.；刺木蓼（针枝木蓼）；Spine Knotwood, Spiny Goatwheat ●

44277 Atraphaxis spinosa L. var. angustifolia Chang Y. Yang et Y. L. Han；狭叶刺木蓼（窄叶刺木蓼）；Narrow-leaf Spiny Goatwheat ●

44278 Atraphaxis spinosa L. var. angustifolia Chang Y. Yang et Y. L. Han = Atraphaxis spinosa L. ●

44279 Atraphaxis teretifolia (Popov) Kom.；圆柱叶木蓼●☆

44280 Atraphaxis tortuosa Losinsk.；圆叶木蓼；Roundleaf Goatwheat ●

44281 Atraphaxis tortuosa Losinsk. = Polygonum intramongolicum Borodina ●

44282 Atraphaxis tournefortii Jaub. et Spach；图氏木蓼●☆

44283 Atraphaxis undulata L. = Polygonum undulatum (L.) P. J. Bergius ■☆

44284 Atraphaxis virgata (Regel) Krasn.；帚枝木蓼（长枝木蓼）●

44285 Atrategia Bedd. ex Hook. f. = Atrutegia Bedd. ●

44286 Atrategia Bedd. ex Hook. f. = Goniothalamus (Blume) Hook. f. et Thomson ●

44287 Atrategia Hook. f. = Atrutegia Bedd. ●

44288 Atrategia Hook. f. = Goniothalamus (Blume) Hook. f. et Thomson ●

44289 Atrema DC. = Bifora Hoffm.（保留属名）■☆

44290 Atrichantha Hilliard et B. L. Burtt (1981)；疏毛鼠麹木属●☆

44291 Atrichantha elsiae Hilliard = Hydroidea elsiae (Hilliard) P. O. Karis ■☆

44292 Atrichantha gemmifera (Bolus) Hilliard et B. L. Burtt；疏毛鼠麹木●☆

44293 Atrichodendron Gagnep. (1950)；无毛茄属●☆

44294 Atrichodendron Gagnep. = Lycium L. ●

44295 Atrichodendron tonkinense Gagnep.；无毛茄●☆

44296 Atrichoseris A. Gray (1884)；无冠苣属；Gravel-Ghost, Parachute Plant, Tobacco-weed ■☆

44297 Atrichoseris platyphylla (A. Gray) A. Gray；无冠苣；Tobacco Weed ■☆

44298 Atriplex L. (1753)；滨藜属（海滨藜属）；Orach, Orache, Saltbush, Salt-Bush ●■

44299 Atriplex acanthocarpa (Torr.) S. Watson；尖果滨藜；Burscale ■☆

44300 Atriplex acanthocarpa (Torr.) S. Watson subsp. coahuilensis Henrickson = Atriplex acanthocarpa (Torr.) S. Watson var. coahuilensis (Henrickson) S. L. Welsh et Crompton ■☆

44301 Atriplex acanthocarpa (Torr.) S. Watson var. coahuilensis (Henrickson) S. L. Welsh et Crompton；科阿韦拉滨藜；Coahuila Orach ■☆

44302 Atriplex acanthocarpa (Torr.) S. Watson var. cuneata (A. Nelson) M. E. Jones = Atriplex gardneri (Moq.) D. Dietr. var. cuneata (A. Nelson) S. L. Welsh ■☆

44303 Atriplex alaschanica Y. Z. Zhao；阿拉善滨藜；Alashan Saltbush ●

44304 Atriplex alaskensis S. Watson = Atriplex gmelinii C. A. Mey. ex Bong. var. alaskensis (S. Watson) S. L. Welsh ■☆

44305 Atriplex alba Crantz = Chenopodium album L. ■

44306 Atriplex alba Scop. = Atriplex rosea L. ●☆

44307 Atriplex albicans Aiton = Manochlamys albicans (Aiton) Aellen ●☆

44308 Atriplex amblyostegia Turcz. = Atriplex aucheri Moq. ●

44309 Atriplex amboensis Schinz；安博滨藜■☆

44310 Atriplex ambrosioides (L.) Crantz = Dysphania ambrosioides (L.) Mosyakin et Clemants ■

44311 Atriplex ambrosioides Crantz. = Chenopodium ambrosioides L. ■☆

44312 Atriplex amnicola Paul G. Wilson；沼泽滨藜；Swamp Saltbush ■☆

44313 Atriplex amplyostegia Turcz. = Atriplex aucheri Moq. ●

44314 Atriplex aptera A. Nelson = Atriplex gardneri (Moq.) D. Dietr. var. aptera (A. Nelson) S. L. Welsh et Crompton ■☆

44315 Atriplex arenaria Nutt. = Atriplex maximowicziana Makino ●

44316 Atriplex arenaria Nutt. = Atriplex mucronata Raf. ■☆

44317 Atriplex argentea Nutt.；银滨藜；Orache, Saltbush, Silver Orache, Silverscale, Silver-scale, Silver-scale Saltbush ●☆

44318 Atriplex argentea Nutt. var. cornuta (M. E. Jones) M. E. Jones = Atriplex saccaria S. Watson var. cornuta (M. E. Jones) S. L. Welsh ■☆

44319 Atriplex argentea Nutt. var. hillmanii M. E. Jones；希尔曼滨藜；Hillman's Orach ■☆

44320 Atriplex argentea Nutt. var. longitrichoma (Stutz, G. L. Chu et S. C. Sand.) S. L. Welsh；长毛银滨藜；Pahrump Orach ●☆

44321 Atriplex argentea Nutt. var. mohavensis (M. E. Jones) S. L. Welsh；莫哈维银滨藜；Mohave Orach ■☆

44322 Atriplex argentea Nutt. var. rydbergii (Standl.) S. L. Welsh；雷氏滨藜；Rydberg's Orach ■☆

44323 Atriplex asterocarpa Stutz, G. L. Chu et S. C. Sand. = Atriplex saccaria S. Watson var. asterocarpa (Stutz, G. L. Chu et S. C. Sand.) S. L. Welsh ■☆

44324 Atriplex aucheri Moq.；野榆钱菠菜；Aucher Saltbush ●

44325 Atriplex belangeri Boiss. = Atriplex repens Roth ●

44326 Atriplex billardierei (Moq.) Hook. f. = Theleophyton billardierei (Moq.) Moq. ●☆

44327 Atriplex brandegeei (A. Gray) Collotzi ex W. A. Weber = Zuckia brandegeei (A. Gray) S. L. Welsh et Stutz ●☆

44328 Atriplex brandegeei (A. Gray) W. A. Weber = Zuckia brandegeei (A. Gray) S. L. Welsh et Stutz ●☆

44329 Atriplex breweri S. Watson = Atriplex lentiformis (Torr.) S.

Watson var. breweri (S. Watson) McMinn ●☆
44330 Atriplex breweri S. Watson = Atriplex lentiformis (Torr.) S. Watson ●☆
44331 Atriplex californica Moq.;加州滨藜;California Orach ■☆
44332 Atriplex calotheca Fr.;美果滨藜●☆
44333 Atriplex cana C. A. Mey.;白滨藜;Greywhite Saltbush ●
44334 Atriplex canescens (Pursh) Nutt.;灰毛滨藜;Cenizo, Chamisa, Chamizo, Fourwing Saltbush, Four-wing Saltbush, Gray Sagebrush, Hoary Saltbush, Sagebrush, Saltbush, Shadscale Wingscale, Wing Scale ●
44335 Atriplex canescens (Pursh) Nutt. subsp. aptera (A. Nelson) H. M. Hall et Clem. = Atriplex gardneri (Moq.) D. Dietr. var. aptera (A. Nelson) S. L. Welsh et Crompton ■☆
44336 Atriplex canescens (Pursh) Nutt. subsp. garrettii (Rydb.) H. M. Hall et Clem. = Atriplex garrettii Rydb. ■☆
44337 Atriplex canescens (Pursh) Nutt. subsp. linearis (S. Watson) H. M. Hall et Clem. = Atriplex linearis S. Watson ■☆
44338 Atriplex canescens (Pursh) Nutt. var. angustifolia (Torr.) S. Watson = Atriplex canescens (Pursh) Nutt. ●
44339 Atriplex canescens (Pursh) Nutt. var. garrettii (Rydb.) L. D. Benson = Atriplex garrettii Rydb. ■☆
44340 Atriplex canescens (Pursh) Nutt. var. gigantea S. L. Welsh et Stutz;大灰毛滨藜;Lynndyl Fourwing ●
44341 Atriplex canescens (Pursh) Nutt. var. laciniata Parish;裂叶灰毛滨藜;Caleb Saltbush ■☆
44342 Atriplex canescens (Pursh) Nutt. var. linearis (S. Watson) Munz = Atriplex linearis S. Watson ■☆
44343 Atriplex canescens (Pursh) Nutt. var. macilenta Jeps.;贫弱灰毛滨藜;Salton Saltbush ■☆
44344 Atriplex canescens (Pursh) Nutt. var. occidentalis (Torr. et FrÉMont) S. L. Welsh et Stutz = Atriplex canescens (Pursh) Nutt. ●
44345 Atriplex capensis Moq. = Atriplex vestita (Thunb.) Aellen var. appendiculata Aellen ●☆
44346 Atriplex caput-medusae Eastw. = Atriplex saccaria S. Watson ■☆
44347 Atriplex centralasiatica Iljin;中亚滨藜(旱蒺藜,麻落粒,马灰条,中亚粉藜);Central Asia Saltbush ●
44348 Atriplex centralasiatica Iljin var. macrobracteata H. C. Fu et Z. Y. Chu;大苞中亚滨藜●
44349 Atriplex centralasiatica Iljin var. megalotheca (Popov) G. L. Chu;大苞滨藜;Bigbracted Saltbush, Largebract Saltbush ●
44350 Atriplex chenopodioides Batt.;藜状滨藜●☆
44351 Atriplex cinerea Poir. = Atriplex cinerea Poir. et Aellen ●☆
44352 Atriplex cinerea Poir. et Aellen;蓝灰滨藜●☆
44353 Atriplex cinerea Poir. var. adamsonii Aellen;亚当森滨藜●☆
44354 Atriplex collina Wooton et Standl. = Atriplex confertifolia S. Watson ●☆
44355 Atriplex confertifolia S. Watson;密叶滨藜;Shadscale, Sheep Fat, Sheep-fat, Spiny Saltbush ●☆
44356 Atriplex cordulata Jeps.;心形滨藜;Heart-leaf Orach ●☆
44357 Atriplex cordulata Jeps. var. erecticaulis (Stutz, G. L. Chu et S. C. Sand.) S. L. Welsh;直茎心形滨藜;Earlimart Orach ■☆
44358 Atriplex cordulata Jeps. var. tularensis (Coville) Jeps. = Atriplex tularensis Coville ■☆
44359 Atriplex coriacea Forssk.;革质滨藜●☆
44360 Atriplex cornuta M. E. Jones = Atriplex saccaria S. Watson var. cornuta (M. E. Jones) S. L. Welsh ■☆
44361 Atriplex coronata S. Watson;冠滨藜;Crownscale, Wedgescale ■☆
44362 Atriplex coronata S. Watson var. notatior Jeps.;斑纹滨藜;San Jacinto Valley Crownscale ●☆
44363 Atriplex coronata S. Watson var. vallicola (Hoover) S. L. Welsh;谷地滨藜;Valley Crownscale ●☆
44364 Atriplex corrugata S. Watson;皱折滨藜;Mat-atriplex, Mat-saltbush, Matscale ●☆
44365 Atriplex coulteri (Moq.) D. Dietr.;寇氏滨藜;Coulter's Orach ■☆
44366 Atriplex crassifolia C. A. Mey.;厚叶滨藜●☆
44367 Atriplex cristata Humb. et Bonpl. ex Willd. var. arenaria (Nutt.) Kuntze = Atriplex mucronata Raf. ■☆
44368 Atriplex cuneata A. Nelson = Atriplex gardneri (Moq.) D. Dietr. var. cuneata (A. Nelson) S. L. Welsh ■☆
44369 Atriplex cuneata A. Nelson var. introgressa C. A. Hanson = Atriplex gardneri (Moq.) D. Dietr. var. utahensis (M. E. Jones) Dorn ■☆
44370 Atriplex curvidens Brandegee = Atriplex polycarpa S. Watson ●☆
44371 Atriplex davidsonii Standl. = Atriplex serenana A. Nelson ex Abrams var. davidsonii (Standl.) Munz ■☆
44372 Atriplex decumbens S. Watson = Atriplex watsonii A. Nelson ex Abrams ■☆
44373 Atriplex depressa Jeps. = Atriplex parishii S. Watson var. depressa (Jeps.) S. L. Welsh ■☆
44374 Atriplex dimorphostegia Kar. et Kir.;犁苞滨藜;Ploughbracted Saltbush ●
44375 Atriplex dimorphostegia Kar. et Kir. var. sagittiformis Aellen;箭苞滨藜;Sagittate Saltbush ●
44376 Atriplex dimorphostegia Kar. et Kir. var. sagittiformis Aellen = Atriplex dimorphostegia Kar. et Kir. ●
44377 Atriplex dioica (Nutt.) J. F. Macbr. = Atriplex suckleyi (Torr.) Rydb. ■☆
44378 Atriplex dioica (Nutt.) Standl. = Atriplex suckleyi (Torr.) Rydb. ■☆
44379 Atriplex dioica Raf.;纽约厚叶滨藜;Thickleaf Orach ■☆
44380 Atriplex draconis M. E. Jones = Atriplex phyllostegia (Torr. ex S. Watson) S. Watson ■☆
44381 Atriplex drymarioides Standl. = Atriplex gmelinii C. A. Mey. ex Bong. ●
44382 Atriplex eardleyae Aellen;小滨藜;Small Saltbush ●☆
44383 Atriplex elegans (Moq.) D. Dietr.;白鳞滨藜;Wheelscale Orach, White-scale Saltbush ■☆
44384 Atriplex elegans (Moq.) D. Dietr. subsp. fasciculata (S. Watson) H. M. Hall et Clem. = Atriplex elegans (Moq.) D. Dietr. var. fasciculata (S. Watson) M. E. Jones ■☆
44385 Atriplex elegans (Moq.) D. Dietr. var. coronata (S. Watson) M. E. Jones = Atriplex coronata S. Watson ■☆
44386 Atriplex elegans (Moq.) D. Dietr. var. fasciculata (S. Watson) M. E. Jones;簇生白鳞滨藜;Mecca Orach, Wheelscale ■☆
44387 Atriplex elegans (Moq.) D. Dietr. var. thornberi M. E. Jones = Atriplex elegans (Moq.) D. Dietr. ■☆
44388 Atriplex endolepis S. Watson = Atriplex suckleyi (Torr.) Rydb. ■☆
44389 Atriplex erecticaulis Stutz = Atriplex cordulata Jeps. var. erecticaulis (Stutz, G. L. Chu et S. C. Sand.) S. L. Welsh ■☆
44390 Atriplex erosa G. Brückn. et I. Verd.;啮蚀状滨藜■☆
44391 Atriplex expansa S. Watson;扩展滨藜;Fogweed ■☆
44392 Atriplex expansa S. Watson var. mohavensis M. E. Jones = Atriplex argentea Nutt. var. mohavensis (M. E. Jones) S. L. Welsh ■☆

44393 Atriplex expansa S. Watson var. trinervata (Jeps.) J. F. Macbr. = Atriplex argentea Nutt. var. mohavensis (M. E. Jones) S. L. Welsh ■☆

44394 Atriplex falcata (M. E. Jones) Standl. = Atriplex gardneri (Moq.) D. Dietr. var. falcata (M. E. Jones) S. L. Welsh ■☆

44395 Atriplex farinosa Forssk.;被粉滨藜■☆

44396 Atriplex farinosa Forssk. subsp. keniensis (Brenan) Friis et M. G. Gilbert;肯尼亚滨藜■☆

44397 Atriplex farinosa Forssk. var. keniensis Brenan = Atriplex farinosa Forssk. subsp. keniensis (Brenan) Friis et M. G. Gilbert ■☆

44398 Atriplex farinosa Moq. = Atriplex patula L. subsp. verreauxii Aellen ●☆

44399 Atriplex fasciculata S. Watson = Atriplex elegans (Moq.) D. Dietr. var. fasciculata (S. Watson) M. E. Jones ■☆

44400 Atriplex fera (L.) Bunge;野滨藜(粉藜,三齿滨藜,三齿粉藜,咸卜子菜);Wild Saltbush ●

44401 Atriplex fera (L.) Bunge var. commixta H. C. Fu et Z. Y. Chu;角果野滨藜(角果野藜);Cornfruit Wild Saltbush ●

44402 Atriplex flabellum Bunge;扇状滨藜●☆

44403 Atriplex flagellaris Wooton et Standl. = Atriplex semibaccata R. Br. ●☆

44404 Atriplex flexuosa Moq.;之字滨藜●☆

44405 Atriplex fruticulosa Jeps.;灌木状滨藜■☆

44406 Atriplex gardneri (Moq.) D. Dietr.;加德纳滨藜;Gardner's Saltbush ■☆

44407 Atriplex gardneri (Moq.) D. Dietr. var. aptera (A. Nelson) S. L. Welsh et Crompton;灰白滨藜;Nelson's Saltbush ■☆

44408 Atriplex gardneri (Moq.) D. Dietr. var. cuneata (A. Nelson) S. L. Welsh;楔形滨藜;Castle Valley Saltbush ■☆

44409 Atriplex gardneri (Moq.) D. Dietr. var. falcata (M. E. Jones) S. L. Welsh;镰形加德纳滨藜;Jones' Saltbush ■☆

44410 Atriplex gardneri (Moq.) D. Dietr. var. tridentata (Kuntze) J. F. Macbr. = Atriplex gardneri (Moq.) D. Dietr. var. utahensis (M. E. Jones) Dorn ■☆

44411 Atriplex gardneri (Moq.) D. Dietr. var. utahensis (M. E. Jones) Dorn;犹他荷滨藜;Basin Saltbush ■☆

44412 Atriplex gardneri (Moq.) D. Dietr. var. welshii (C. A. Hanson) S. L. Welsh;沃尔什滨藜;Welsh's Saltbush ■☆

44413 Atriplex garrettii Rydb.;加勒特滨藜;Garrett's Saltbush ■☆

44414 Atriplex glabriuscula Edmondston;近光滨藜;Babington's Orache, Glabrous Atriplex, Glabrous Orach ■☆

44415 Atriplex glauca C. H. Wright = Atriplex vestita (Thunb.) Aellen var. appendiculata Aellen ●☆

44416 Atriplex glauca L.;灰绿滨藜■☆

44417 Atriplex glauca L. subsp. alexandrina (Boiss.) Lambinon et Dobignard;亚历山大滨藜■☆

44418 Atriplex glauca L. subsp. ifniensis (Caball.) Rivas Mart. et al.;伊夫尼大滨藜■☆

44419 Atriplex glauca L. subsp. loweana Dobignard;洛氏滨藜■☆

44420 Atriplex glauca L. subsp. mauritanica (Boiss. et Reut.) Dobignard;毛里塔尼亚滨藜■☆

44421 Atriplex glauca L. subsp. palaestina (Boiss.) Dobignard;叙利亚滨藜■☆

44422 Atriplex glauca L. subsp. parvifolia (Lowe) Dobignard;小叶滨藜■☆

44423 Atriplex glauca L. var. embergeri Maire = Atriplex glauca L. subsp. parvifolia (Lowe) Dobignard ■☆

44424 Atriplex glauca L. var. ifniensis (Caball.) Maire = Atriplex glauca L. subsp. ifniensis (Caball.) Rivas Mart. et al. ■☆

44425 Atriplex glauca L. var. mauritanica (Boiss. et Reut.) Maire = Atriplex glauca L. subsp. mauritanica (Boiss. et Reut.) Dobignard ■☆

44426 Atriplex glauca L. var. rotundifolia Bonnet et Barratte = Atriplex glauca L. ■☆

44427 Atriplex glauca Pall. = Atriplex verrucifera M. Bieb. ●

44428 Atriplex gmelinii C. A. Mey. = Atriplex gmelinii C. A. Mey. ex Bong. ●

44429 Atriplex gmelinii C. A. Mey. = Atriplex patens (Litv.) Iljin ●

44430 Atriplex gmelinii C. A. Mey. ex Bong.;北滨藜;Gmelin's Orach ●

44431 Atriplex gmelinii C. A. Mey. ex Bong. var. alaskensis (S. Watson) S. L. Welsh;阿拉斯加滨藜;Alaska Orach ■☆

44432 Atriplex gmelinii C. A. Mey. ex Bong. var. zosterifolia (Hook.) Moq. = Atriplex gmelinii C. A. Mey. ex Bong. ●

44433 Atriplex gmelinii C. A. Mey. subsp. dilatata (Franch. et Sav.) Kitam. = Atriplex subcordata Kitag. ●☆

44434 Atriplex gmelinii C. A. Mey. var. saccata H. C. Fu et Z. Y. Chu;囊苞北滨藜●

44435 Atriplex graciliflora M. E. Jones;细花滨藜;Blue Valley Orach ■☆

44436 Atriplex grayi Collotzi ex W. A. Weber = Grayia spinosa (Hook.) Moq. ■☆

44437 Atriplex greenei A. Nelson = Atriplex wolfii S. Watson var. tenuissima (A. Nelson) S. L. Welsh ■☆

44438 Atriplex greggii S. Watson = Atriplex obovata Moq. ■☆

44439 Atriplex griffithii Moq. = Atriplex griffithsii Standl. ●☆

44440 Atriplex griffithii Moq. subsp. stocksii (Boiss.) Boulos;斯托克斯滨藜●☆

44441 Atriplex griffithii Moq. var. stocksii (Boiss.) Boiss. = Atriplex griffithii Moq. subsp. stocksii (Boiss.) Boulos ●☆

44442 Atriplex griffithsii Standl. = Atriplex torreyi (S. Watson) S. Watson var. griffithsii (Standl.) G. D. Br. ●☆

44443 Atriplex halimoides Lindl. = Atriplex lindleyi Moq. subsp. inflata (F. Muell.) Paul G. Wilson ■☆

44444 Atriplex halimoides Lindl. = Atriplex lindleyi Moq. ●☆

44445 Atriplex halimus C. H. Wright = Atriplex vestita (Thunb.) Aellen var. appendiculata Aellen ●☆

44446 Atriplex halimus L.;树滨藜(地中海滨藜);Mediterranean Saltbush, Saltbush, Sea Purslane, Shrubby Orache, Spanish Sea Purslane, Tree Purslane ●☆

44447 Atriplex halimus L. f. granulata (L. Chevall.) Maire = Atriplex halimus L. var. granulata L. Chevall. ●☆

44448 Atriplex halimus L. var. glaucoidea Maire = Atriplex halimus L. ●☆

44449 Atriplex halimus L. var. granulata L. Chevall. = Atriplex halimus L. f. granulata (L. Chevall.) Maire ●☆

44450 Atriplex halimus L. var. hastulata Maire = Atriplex halimus L. ●☆

44451 Atriplex halimus L. var. intermedia L. Chevall. = Atriplex halimus L. ●☆

44452 Atriplex halimus L. var. ramosissima L. Chevall. = Atriplex halimus L. var. glaucoidea Maire ●☆

44453 Atriplex halimus L. var. rifea Sennen et Mauricio = Atriplex halimus L. ●☆

44454 Atriplex halimus L. var. schweinfurthii Boiss. = Atriplex halimus L. ●☆

44455 Atriplex halimus L. var. venosa L. Chevall. = Atriplex halimus L. ●☆

44456 Atriplex hastata L. = Atriplex prostrata Boucher ex DC. ●

44457 Atriplex hastata L. sensu Aellen = Atriplex prostrata Boucher ex DC. ●
44458 Atriplex hastata L. subsp. patula (L.) S. Pons = Atriplex patula L. ●
44459 Atriplex hastata L. var. heterocarpa Fenzl = Atriplex micrantha C. A. Mey. ●
44460 Atriplex hastata L. var. littoralis (L.) Farw. = Atriplex littoralis L. ●☆
44461 Atriplex hastata L. var. macrotheca Raf. = Atriplex prostrata DC. ●
44462 Atriplex hastata L. var. microtheca Schum. = Atriplex prostrata DC. ●
44463 Atriplex hastata L. var. patula (L.) Farw. = Atriplex patula L. ●
44464 Atriplex hastatum Boiss. = Atriplex micrantha C. A. Mey. ●
44465 Atriplex heteropserma Bunge = Atriplex micrantha C. A. Mey. ●
44466 Atriplex hillmanii (M. E. Jones) Standl. = Atriplex argentea Nutt. var. hillmanii M. E. Jones ■☆
44467 Atriplex holocarpa F. Muell.;全果滨藜;Pop Saltbush ●☆
44468 Atriplex hortensis L.;榆钱菠菜(法国菠菜,山菠菜,山菠薐草,洋菠菜);Arach, Arage, Areche, Asiatic Orach, Asiatic Orache, Butter Leaves, French Spinach, Garden Orach, Garden Orache, Mountain Spinach, Orach, Orache, Orage, Red mountain Spinach, Red Orache, Sea Purslane ●
44469 Atriplex hortensis L. 'Rubra';红榆钱菠菜;Red Mountain-spinach ●☆
44470 Atriplex hortensis L. subsp. desertorum (Iljin) Aellen = Atriplex aucheri Moq. ●
44471 Atriplex hymenelytra (Torr.) S. Watson;沙地滨藜;Desert Holly ●☆
44472 Atriplex hymenelytra Torr. = Atriplex hymenelytra (Torr.) S. Watson ●☆
44473 Atriplex hymenotheca Moq. = Atriplex vesicaria Heward ex Benth. ■☆
44474 Atriplex ifniensis Caball. = Atriplex glauca L. subsp. ifniensis (Caball.) Rivas Mart. et al. ■☆
44475 Atriplex inamoena Aellen = Atriplex leucoclada Boiss. ■☆
44476 Atriplex inflata F. Muell. = Atriplex lindleyi Moq. subsp. inflata (F. Muell.) Paul G. Wilson ■☆
44477 Atriplex joaquiniana A. Nelson;拉古纳滨藜;San Joaquin Orach ■☆
44478 Atriplex johnstonii C. B. Wolf = Atriplex nummularia Lindl. ●
44479 Atriplex jonesii Standl. = Atriplex obovata Moq. ■☆
44480 Atriplex kuzenevae Semenova;库氏滨藜●☆
44481 Atriplex laciniata L. = Atriplex erosa G. Brückn. et I. Verd. ■☆
44482 Atriplex laciniata L. = Atriplex tatarica L. ●
44483 Atriplex laciniata L. var. turcomanica Moq. = Atriplex leucoclada Boiss. var. turcomanica (Moq.) Zohary ■☆
44484 Atriplex laevis C. A. Mey.;光滑滨藜;Smooth Saltbush ●
44485 Atriplex laevis C. A. Mey. var. patens (Litv.) Grubov = Atriplex patens (Litv.) Iljin ●
44486 Atriplex lampa Gillies ex Moq.;南美滨藜;South American Saltbush ●☆
44487 Atriplex lasiantha ?;毛花滨藜■☆
44488 Atriplex latifolia Wahlenb. = Atriplex prostrata Boucher ex DC. ●
44489 Atriplex lehmanniana Bunge = Atriplex tatarica L. ●
44490 Atriplex lentiformis (Torr.) S. Watson;大滨藜;Big Salt Bush, Lenscale, Quail Bush, White Thistle ●☆
44491 Atriplex lentiformis (Torr.) S. Watson subsp. breweri (S. Watson) H. M. Hall et Clem. = Atriplex lentiformis (Torr.) S. Watson ●☆
44492 Atriplex lentiformis (Torr.) S. Watson subsp. griffithsii (Standl.) H. M. Hall et Clem. = Atriplex torreyi (S. Watson) S. Watson var. griffithsii (Standl.) G. D. Br. ●☆
44493 Atriplex lentiformis (Torr.) S. Watson subsp. griffithsii (Standl.) H. M. Hall et Clem. = Atriplex griffithsii Standl. ●☆
44494 Atriplex lentiformis (Torr.) S. Watson subsp. torreyi (S. Watson) H. M. Hall et Clem. = Atriplex torreyi (S. Watson) S. Watson ●☆
44495 Atriplex lentiformis (Torr.) S. Watson var. breweri (S. Watson) McMinn;布鲁尔大滨藜;Brewer's Saltbrush, Quail Brush ●☆
44496 Atriplex lentiformis (Torr.) S. Watson var. breweri (S. Watson) McMinn = Atriplex lentiformis (Torr.) S. Watson ●☆
44497 Atriplex lentiformis (Torr.) S. Watson var. griffithsii (Standl.) L. D. Benson = Atriplex griffithsii Standl. ●☆
44498 Atriplex lentiformis (Torr.) S. Watson var. griffithsii (Standl.) L. D. Benson = Atriplex torreyi (S. Watson) S. Watson var. griffithsii (Standl.) G. D. Br. ●☆
44499 Atriplex lentiformis (Torr.) S. Watson var. torreyi (S. Watson) McMinn = Atriplex torreyi (S. Watson) S. Watson ●☆
44500 Atriplex leptostachys L. Chevall. = Atriplex coriacea Forssk. ●☆
44501 Atriplex leucoclada Boiss.;白枝滨藜■☆
44502 Atriplex leucoclada Boiss. var. inamoena (Aellen) Zohary = Atriplex leucoclada Boiss. ■☆
44503 Atriplex leucoclada Boiss. var. turcomanica (Moq.) Zohary = Atriplex leucoclada Boiss. ■☆
44504 Atriplex leucophylla (Moq.) D. Dietr.;白叶滨藜;Whiteleaf Orach ■☆
44505 Atriplex lindleyi Moq.;林德利滨藜;Lindley's Saltbush ●☆
44506 Atriplex lindleyi Moq. subsp. inflata (F. Muell.) Paul G. Wilson;膨胀滨藜■☆
44507 Atriplex lindleyi Moq. subsp. quadripartita Paul G. Wilson;五深裂滨藜■☆
44508 Atriplex linearis S. Watson;线形滨藜;Slenderleaf Saltbush ■☆
44509 Atriplex littoralis C. H. Wright = Atriplex patula L. subsp. verreauxii Aellen ●☆
44510 Atriplex littoralis L.;禾叶滨藜;Grass-leaved Orache, Narrow-leaved Atriplex, Seashore Orache, Shore Orach, Shore Orache ●☆
44511 Atriplex littoralis L. subsp. stepposa Kitag. = Atriplex patens (Litv.) Iljin ●
44512 Atriplex littoralis L. var. patens Litv. = Atriplex patens (Litv.) Iljin ●
44513 Atriplex longipes Drejer et Fr.;长梗滨藜;Long-stalked Orache ●☆
44514 Atriplex longipes Drejer subsp. praecox (Hülph.) Turesson = Atriplex nudicaulis Bogusl. ●☆
44515 Atriplex longitrichoma Stutz, G. L. Chu et S. C. Sand. = Atriplex argentea Nutt. var. longitrichoma (Stutz, G. L. Chu et S. C. Sand.) S. L. Welsh ●☆
44516 Atriplex malvana Aellen et Sauvage;锦葵滨藜●☆
44517 Atriplex matamorensis A. Nelson;对叶藜;Quelite Cenizo ■☆
44518 Atriplex mauritanica Boiss. et Reut. = Atriplex glauca L. subsp. mauritanica (Boiss. et Reut.) Dobignard ■☆
44519 Atriplex mauritanica Boiss. et Reut. var. ifniensis (Caball.) Maire = Atriplex glauca L. subsp. ifniensis (Caball.) Rivas Mart. et al. ■☆
44520 Atriplex maximowicziana Makino;马氏滨藜(海滨藜);

Maximowicz Saltbush, Maximowicz's Saltbush, Seabeach Orach ●

44521 Atriplex megalotheca Popov = Atriplex centralasiatica Iljin var. megalotheca (Popov) G. L. Chu ●

44522 Atriplex megalotheca Popov ex Iljin = Atriplex centralasiatica Iljin var. megalotheca (Popov) G. L. Chu ●

44523 Atriplex micrantha C. A. Mey.; 异苞滨藜(杂果滨藜); Russian Atriplex, Smallflower Saltbush, Twoscale Saltbush, Two-seeded Orach ●

44524 Atriplex microphylla (Thunb.) Willd. = Exomis microphylla (Thunb.) Aellen ■☆

44525 Atriplex microsperma Waldst. et Kit. = Atriplex hastata L. ●

44526 Atriplex microsperma Waldst. et Kit. = Atriplex prostrata Boucher ex DC. ●

44527 Atriplex minuscula Standl. = Atriplex parishii S. Watson var. minuscula (Standl.) S. L. Welsh ■☆

44528 Atriplex minuticarpa Stutz et G. L. Chu = Atriplex powellii S. Watson var. minuticarpa (Stutz et G. L. Chu) S. L. Welsh ■☆

44529 Atriplex mollis Desf.; 柔软滨藜■☆

44530 Atriplex moneta Bunge; 天后滨藜■☆

44531 Atriplex mucronata Raf.; 短尖滨藜; Quelite ■☆

44532 Atriplex muelleri Benth.; 米勒滨藜; Mueller's Saltbush ■☆

44533 Atriplex multicolora Aellen = Atriplex prostrata Boucher ex DC. ●

44534 Atriplex multicolora Aellen = Atriplex tatarica L. ●

44535 Atriplex nelsonii M. E. Jones = Atriplex powellii S. Watson ■☆

44536 Atriplex nitens Boiss. = Atriplex aucheri Moq. ●

44537 Atriplex nitens Schkuhr; 光亮滨藜●☆

44538 Atriplex nitens Schkuhr = Atriplex hortensis L. ●

44539 Atriplex nitens Schkuhr subsp. desertorum Iljin = Atriplex aucheri Moq. ●

44540 Atriplex nogalensis Friis et M. G. Gilbert; 诺加尔滨藜■☆

44541 Atriplex nudicaulis Bogusl.; 裸茎滨藜; Nude Orach ●☆

44542 Atriplex nummularia Lindl.; 大洋洲滨藜(台湾滨藜); Bluegreen Saltbush, Coinshaped Saltbush, Giant Saltbush, Old Man Saltbush, Old-man Saltbush ●

44543 Atriplex nuttallii S. Watson; 努塌滨藜●☆

44544 Atriplex nuttallii S. Watson = Atriplex canescens (Pursh) Nutt. ●

44545 Atriplex nuttallii S. Watson subsp. cuneata (A. Nelson) H. M. Hall et Clem. = Atriplex gardneri (Moq.) D. Dietr. var. cuneata (A. Nelson) S. L. Welsh ■☆

44546 Atriplex nuttallii S. Watson subsp. falcata (M. E. Jones) H. M. Hall et Clem. = Atriplex gardneri (Moq.) D. Dietr. var. falcata (M. E. Jones) S. L. Welsh ■☆

44547 Atriplex nuttallii S. Watson subsp. gardneri (Moq.) H. M. Hall et Clem. = Atriplex gardneri (Moq.) D. Dietr. ■☆

44548 Atriplex nuttallii S. Watson var. anomala M. E. Jones = Atriplex gardneri (Moq.) D. Dietr. var. falcata (M. E. Jones) S. L. Welsh ■☆

44549 Atriplex nuttallii S. Watson var. corrugata (S. Watson) A. Nelson = Atriplex corrugata S. Watson ●☆

44550 Atriplex nuttallii S. Watson var. falcata M. E. Jones = Atriplex gardneri (Moq.) D. Dietr. var. falcata (M. E. Jones) S. L. Welsh ■☆

44551 Atriplex nuttallii S. Watson var. gardneri (Moq.) R. J. Davis = Atriplex gardneri (Moq.) D. Dietr. ■☆

44552 Atriplex nuttallii S. Watson var. utahensis M. E. Jones = Atriplex gardneri (Moq.) D. Dietr. var. utahensis (M. E. Jones) Dorn ■☆

44553 Atriplex oblanceolata Rydb. = Atriplex gardneri (Moq.) D. Dietr. var. cuneata (A. Nelson) S. L. Welsh ■☆

44554 Atriplex oblongifolia Waldst. et Kit.; 草地滨藜(椭叶滨藜); Oblong-leaf Orach, Oblongleaf Orache, Oblongleaf Saltbush ●

44555 Atriplex obovata Moq.; 新墨西哥滨藜; Broadscale, New Mexico Saltbush ■☆

44556 Atriplex obovata Moq. var. tuberata J. F. Macbr. = Atriplex obovata Moq. ■☆

44557 Atriplex oppositifolia S. Watson = Atriplex matamorensis A. Nelson ■☆

44558 Atriplex ornata Iljin; 装饰滨藜●☆

44559 Atriplex ovata (Rydb.) Clem. et E. G. Clem. = Atriplex suckleyi (Torr.) Rydb. ■☆

44560 Atriplex ovata Rydb. = Atriplex suckleyi (Torr.) Rydb. ■☆

44561 Atriplex pachypoda Stutz et G. L. Chu = Atriplex argentea Nutt. var. rydbergii (Standl.) S. L. Welsh ■☆

44562 Atriplex pacifica A. Nelson; 太平洋滨藜; Pacific Orach ■☆

44563 Atriplex palaestina Boiss. = Atriplex glauca L. subsp. palaestina (Boiss.) Dobignard ■☆

44564 Atriplex pamirica Iljin; 帕米尔滨藜(帕米尔鞑靼滨藜); Pamir Saltbush ●☆

44565 Atriplex parishii S. Watson; 帕里什滨藜■☆

44566 Atriplex parishii S. Watson var. depressa (Jeps.) S. L. Welsh; 凹陷滨藜; Depressed Orach ■☆

44567 Atriplex parishii S. Watson var. minuscula (Standl.) S. L. Welsh; 稍小滨藜; Lesser Orach ■☆

44568 Atriplex parishii S. Watson var. persistens (Stutz et G. L. Chu) S. L. Welsh; 宿存滨藜; Refuge Orach ■☆

44569 Atriplex parishii S. Watson var. subtilis (Stutz et G. L. Chu) S. L. Welsh; 纤细滨藜; Subtle Orach ■☆

44570 Atriplex parryi S. Watson; 帕里滨藜; Parishes' Orach, Parry's Saltbush ●☆

44571 Atriplex parvifolia Lowe = Atriplex glauca L. subsp. loweana Dobignard ■☆

44572 Atriplex parvifolia Lowe var. mauritanica (Boiss. et Reut.) Maire = Atriplex glauca L. subsp. mauritanica (Boiss. et Reut.) Dobignard ■☆

44573 Atriplex parvifolia Lowe var. melillensis Maire et Sennen = Atriplex glauca L. subsp. mauritanica (Boiss. et Reut.) Dobignard ■☆

44574 Atriplex parvifolia Lowe var. palaestina (Boiss.) Durand et Barratte = Atriplex glauca L. subsp. palaestina (Boiss.) Dobignard ■☆

44575 Atriplex patens (Litv.) Iljin; 滨藜; Patent Saltbush ●

44576 Atriplex patula L.; 平伏滨藜(草地滨藜); Allseed, Arage, Arrach, Common Orache, Delt-orach, Delt-orache, Fat Hen, Fat Hen Saltbush, Fat-hen, Iron-root, Lamb's Quarters, Meedles, Melde, Mercury, Motherwort, Orache, Orech, Orege, Spear Orach, Spear Saltbush, Spearscale, Spear-scale, Spreading Orache, Wild Arrach ●

44577 Atriplex patula L. = Atriplex oblongifolia Waldst. et Kit. ●

44578 Atriplex patula L. subsp. alaskensis (S. Watson) H. M. Hall et Clem. = Atriplex gmelinii C. A. Mey. ex Bong. var. alaskensis (S. Watson) S. L. Welsh ■☆

44579 Atriplex patula L. subsp. austro-africana Aellen; 南非平伏滨藜●☆

44580 Atriplex patula L. subsp. hastata sensu H. M. Hall et Clem. = Atriplex prostrata Boucher ex DC. ●

44581 Atriplex patula L. subsp. littoralis (L.) H. M. Hall et Clem. = Atriplex littoralis L. ●☆

44582 Atriplex patula L. subsp. obtusa (Cham.) H. M. Hall et Clem. = Atriplex gmelinii C. A. Mey. ex Bong. ●

44583 Atriplex patula L. subsp. spicata H. M. Hall et Clem. = Atriplex joaquiniana A. Nelson ■☆

44584 Atriplex patula L. subsp. verreauxii Aellen; 韦罗滨藜●☆

44585 Atriplex patula L. subsp. zosterifolia (Hook.) H. M. Hall et Clem. = Atriplex gmelinii C. A. Mey. ex Bong. ●

44586 Atriplex patula L. var. alaskensis (S. Watson) S. L. Welsh = Atriplex gmelinii C. A. Mey. ex Bong. var. alaskensis (S. Watson) S. L. Welsh ■☆

44587 Atriplex patula L. var. angustifolia Bolus et Wolley-Dod = Atriplex patula L. subsp. austro-africana Aellen ●☆

44588 Atriplex patula L. var. angustifolia Coss. et Germ. = Atriplex patula L. ●

44589 Atriplex patula L. var. hastata (L.) A. Gray = Atriplex patula L. ●

44590 Atriplex patula L. var. hastata (L.) A. Gray = Atriplex prostrata Boucher ex DC. ●

44591 Atriplex patula L. var. japonica H. Lév. = Atriplex patula L. ●

44592 Atriplex patula L. var. littoralis (L.) A. Gray = Atriplex littoralis L. ●☆

44593 Atriplex patula L. var. oblanceolata (Vict. et J. Rousseau) B. Boivin = Atriplex glabriuscula Edmondston ■☆

44594 Atriplex patula L. var. oblongifolia (Waldst. et Kit.) Westerl. = Atriplex oblongifolia Waldst. et Kit. ●

44595 Atriplex patula L. var. obtusa (Cham.) C. L. Hitchc. = Atriplex gmelinii C. A. Mey. ex Bong. ●

44596 Atriplex patula L. var. subspicata (Nutt.) S. Watson = Atriplex dioica Raf. ■☆

44597 Atriplex patula L. var. triangularis (Willd.) K. H. Thorne et S. L. Welsh = Atriplex prostrata Boucher ex DC. ●

44598 Atriplex patula L. var. triangularis (Willd.) Thorne et Welsh = Atriplex prostrata Boucher ex DC. ●

44599 Atriplex patula L. var. zosterifolia (Hook.) C. L. Hitchc. = Atriplex gmelinii C. A. Mey. ex Bong. ●

44600 Atriplex patutum Boiss. = Atriplex oblongifolia Waldst. et Kit. ●

44601 Atriplex pedunculata L.;梗花滨藜;Pedunculate Sea-purslane, Stalked Orach ●☆

44602 Atriplex pentandra (Jacq.) Standl.;海滨藜;Seashore Orach ■☆

44603 Atriplex pentandra (Jacq.) Standl. subsp. arenaria (Nutt.) H. M. Hall et Clem. = Atriplex mucronata Raf. ■☆

44604 Atriplex persistens Stutz et G. L. Chu = Atriplex parishii S. Watson var. persistens (Stutz et G. L. Chu) S. L. Welsh ■☆

44605 Atriplex philonitra A. Nelson = Atriplex powellii S. Watson ■☆

44606 Atriplex phyllostegia (Torr. ex S. Watson) S. Watson;叶盖滨藜;Arrowscale, Truckee Orach ■☆

44607 Atriplex phyllostegia (Torr. ex S. Watson) S. Watson var. draconis (M. E. Jones) Fosberg = Atriplex phyllostegia (Torr. ex S. Watson) S. Watson ■☆

44608 Atriplex pleiantha W. A. Weber;四角滨藜;Four-corners Orach ■☆

44609 Atriplex polycarpa S. Watson;多果滨藜;Allscale, Cattle Spinach, Cattle-spinach, Desert Saltbush, Sage, Sagebrush ●☆

44610 Atriplex portulacoides C. H. Wright = Atriplex vestita (Thunb.) Aellen var. appendiculata Aellen ●☆

44611 Atriplex portulacoides L.;马齿苋滨藜;Grey Mat, Sea Purslane, Sea-purslane ●☆

44612 Atriplex portulacoides L. = Halimione portulacoides (L.) Aellen ●☆

44613 Atriplex portulacoides L. var. complanata Maire et Weiller = Halimione portulacoides (L.) Aellen ●☆

44614 Atriplex portulacoides L. var. erecta Gaudin = Halimione portulacoides (L.) Aellen ●☆

44615 Atriplex portulacoides L. var. murex Maire = Halimione portulacoides (L.) Aellen ●☆

44616 Atriplex portulacoides L. var. subhastata Maire = Halimione portulacoides (L.) Aellen ●☆

44617 Atriplex powellii S. Watson;鲍厄尔滨藜;Powell's Orach ■☆

44618 Atriplex powellii S. Watson var. minuticarpa (Stutz et G. L. Chu) S. L. Welsh;小果鲍厄尔滨藜;Smallbract Orach ■☆

44619 Atriplex praecox Hülph.;早滨藜;Early Orache ■☆

44620 Atriplex praecox Hülph. = Atriplex nudicaulis Bogusl. ●☆

44621 Atriplex prostrata Boucher ex DC.;薄叶滨藜(戟形滨藜,戟叶滨藜);Fat-hen, Halberdleaf Saltbush, Halberd-leaved Orache, Hastate Atriplex, Hastate Orache, Spear-leaved Fat Hen Saltbush, Spear-leaved Fat-hen Saltbush, Spear-leaved Orache, Spear-scale, Thinleaf Orach, Thin-leaf Orach, Triangle Orache ●

44622 Atriplex prostrata DC. = Atriplex prostrata Boucher ex DC. ●

44623 Atriplex pumilio R. Br.;弱小滨藜■☆

44624 Atriplex pungens Trautv.;刺滨藜■☆

44625 Atriplex pusilla (Torr.) S. Watson;矮小滨藜;Dwarf Orach ■☆

44626 Atriplex ramosissima Nutt. ex Moq. = Atriplex pacifica A. Nelson ■☆

44627 Atriplex repens Roth;匍匐滨藜;Creeping Saltbush ●

44628 Atriplex rosea C. H. Wright = Atriplex erosa G. Brückn. et I. Verd. ■☆

44629 Atriplex rosea L.;玫瑰滨藜;Red Orache, Red Scale, Redscale, Tumbling Orach, Tumbling Saltweed ●☆

44630 Atriplex rosea L. var. subintegra C. A. Mey. = Atriplex prostrata Boucher ex DC. ●

44631 Atriplex rosea L. var. subintergra C. A. Mey. = Atriplex tatarica L. ●

44632 Atriplex rydbergii Standl. = Atriplex argentea Nutt. var. rydbergii (Standl.) S. L. Welsh ■☆

44633 Atriplex saccaria S. Watson;高茎滨藜;Stalked Orach ■☆

44634 Atriplex saccaria S. Watson var. asterocarpa (Stutz, G. L. Chu et S. C. Sand.) S. L. Welsh;星苞滨藜;Starbract Orach ■☆

44635 Atriplex saccaria S. Watson var. caput-medusae (Eastw.) S. L. Welsh = Atriplex saccaria S. Watson ■☆

44636 Atriplex saccaria S. Watson var. cornuta (M. E. Jones) S. L. Welsh;平角滨藜;Flat-horn Orach ■☆

44637 Atriplex saltonensis Parish = Atriplex elegans (Moq.) D. Dietr. var. fasciculata (S. Watson) M. E. Jones ■☆

44638 Atriplex salzmaniana Bunge = Atriplex chenopodioides Batt. ●☆

44639 Atriplex sarcocarpa Dinter = Manochlamys albicans (Aiton) Aellen ●☆

44640 Atriplex schugnanica Iljin;舒格南滨藜■☆

44641 Atriplex semibaccata R. Br.;半浆果滨藜(澳大利亚滨藜);Australian Saltbush, Berry Saltbush, Creeping Saltbush ●☆

44642 Atriplex semibaccata R. Br. subsp. erecta Le Houér. et Franch.;直立半浆果滨藜■☆

44643 Atriplex semibaccata R. Br. var. appendiculata Aellen;附物滨藜■☆

44644 Atriplex serenana A. Nelson ex Abrams;臭滨藜;Bracteate Orach, Bractscale, Stinking Orach ■☆

44645 Atriplex serenana A. Nelson ex Abrams var. davidsonii (Standl.) Munz;戴氏滨藜;Davidson's Orach ■☆

44646 Atriplex sibirica L.;西伯利亚滨藜(白蒺藜,刺果粉藜,大灰条,灰菜,碱灰菜,麻落粒,软蒺藜);Siberia Saltbush, Siberian Saltbush ●

44647 Atriplex sibirica L. var. centralasiatica (Iljin) Grubov = Atriplex

centralasiatica Iljin ●
44648 Atriplex sokotranum Vierh. = Atriplex griffithii Moq. subsp. stocksii (Boiss.) Boulos ●☆
44649 Atriplex sordida S. Watson Standl. = Atriplex argentea Nutt. var. mohavensis (M. E. Jones) S. L. Welsh ■☆
44650 Atriplex sphaeromorpha Iljin;球形滨藜●☆
44651 Atriplex sphaeromorpha Iljin = Atriplex tatarica L. ●
44652 Atriplex spicata S. Watson = Atriplex joaquiniana A. Nelson ■☆
44653 Atriplex spicata S. Watson var. lagunita Jeps. = Atriplex joaquiniana A. Nelson ■☆
44654 Atriplex spinifera J. F. Macbr.;莫哈维滨藜;Mohave Saltbush, Spiny Saltbush ■☆
44655 Atriplex stocksii Boiss. = Atriplex griffithii Moq. subsp. stocksii (Boiss.) Boulos ●☆
44656 Atriplex stocksii Boiss. f. sokotranum (Vierh.) Vierh. = Atriplex griffithii Moq. subsp. stocksii (Boiss.) Boulos ●☆
44657 Atriplex stylosa Viv. = Atriplex glauca L. subsp. palaestina (Boiss.) Dobignard ■☆
44658 Atriplex subconferta Rydb. = Atriplex confertifolia S. Watson ●☆
44659 Atriplex subcordata Kitag.;亚心形滨藜●☆
44660 Atriplex subdecumbens M. E. Jones = Atriplex truncata (Torr. ex S. Watson) A. Gray ●☆
44661 Atriplex suberecta I. Verd.;澳洲滨藜;Australian Orache, Peregrine Saltbush, Sprawling Saltbush ●☆
44662 Atriplex subspicata (Nutt.) Rydb. = Atriplex dioica Raf. ■☆
44663 Atriplex subspicata (Nutt.) Rydb. = Atriplex prostrata Boucher ex DC. ●
44664 Atriplex subtilis Stutz et G. L. Chu = Atriplex parishii S. Watson var. subtilis (Stutz et G. L. Chu) S. L. Welsh ■☆
44665 Atriplex suckleyana (Torr.) S. Watson = Suckleya suckleyana (Torr.) Rydb. ■☆
44666 Atriplex suckleyi (Torr.) Rydb.;萨克滨藜;Suckley's Orach ■☆
44667 Atriplex tatarica Aellen = Atriplex erosa G. Brückn. et I. Verd. ■☆
44668 Atriplex tatarica L.;鞑靼滨藜;Belgian Orach, Frosted Orache, Jagged Sea Orache, Tatar Saltbush, Tatarian Orach, Tatarian Orache, Tatarian Saltbush ●
44669 Atriplex tatarica L. var. pamirica (Iljin) G. L. Chu = Atriplex pamirica Iljin ●☆
44670 Atriplex tenuissima A. Nelson = Atriplex wolfii S. Watson var. tenuissima (A. Nelson) S. L. Welsh ■☆
44671 Atriplex tenuissima A. Nelson var. greenei (A. Nelson) Fosberg = Atriplex wolfii S. Watson var. tenuissima (A. Nelson) S. L. Welsh ■☆
44672 Atriplex texana S. Watson = Atriplex pentandra (Jacq.) Standl. ■☆
44673 Atriplex thornberi (M. E. Jones) Standl. = Atriplex elegans (Moq.) D. Dietr. ■☆
44674 Atriplex thunbergiifolia Boiss.;通贝里滨藜●☆
44675 Atriplex torreyi (S. Watson) S. Watson;托里滨藜;Torree Saltbush, Torree's Saltbush ●☆
44676 Atriplex torreyi (S. Watson) S. Watson var. griffithsii (Standl.) G. D. Br.;格氏滨藜;Griffiths' Saltbush ●☆
44677 Atriplex torreyi (S. Watson) S. Watson var. griffithsii (Standl.) G. D. Br. = Atriplex griffithsii Standl. ●☆
44678 Atriplex triangularis Willd. = Atriplex prostrata Boucher ex DC. ●
44679 Atriplex trinervata Jeps. = Atriplex argentea Nutt. var. mohavensis (M. E. Jones) S. L. Welsh ■☆
44680 Atriplex truncata (Torr. ex S. Watson) A. Gray;平截滨藜;Silver Scale, Silver Scale Saltbrush, Wedge Orach, Wedgescale ●☆
44681 Atriplex truncata (Torr. ex S. Watson) A. Gray var. saccaria (S. Watson) M. E. Jones = Atriplex saccaria S. Watson ■☆
44682 Atriplex truncata (Torr. ex S. Watson) A. Gray var. stricta A. Gray = Atriplex truncata (Torr. ex S. Watson) A. Gray ●☆
44683 Atriplex tularensis Coville;图拉尔滨藜;Tulare Orach ■☆
44684 Atriplex turcomanica Fisch. et C. A. Mey.;土库曼滨藜●☆
44685 Atriplex vallicola Hoover = Atriplex coronata S. Watson var. vallicola (Hoover) S. L. Welsh ●☆
44686 Atriplex veneta Willd. = Atriplex tatarica L. ●
44687 Atriplex verreauxii Moq. = Atriplex patula L. subsp. verreauxii Aellen ●☆
44688 Atriplex verrucifera M. Bieb.;疣苞滨藜(多疣滨藜);Verrucousbract Saltbush, Waxy Saltbush ●
44689 Atriplex vesicaria Heward ex Benth.;土著滨藜;Aboriginal Saltbush, Saltbush ■☆
44690 Atriplex vestita (Thunb.) Aellen;包被滨藜■☆
44691 Atriplex vestita (Thunb.) Aellen var. appendiculata Aellen;附属物包被滨藜●☆
44692 Atriplex vestita (Thunb.) Aellen var. inappendiculata Aellen;无附属物包被滨藜●☆
44693 Atriplex volutans A. Nelson = Atriplex argentea Nutt. ●☆
44694 Atriplex wardii Standl. = Atriplex pentandra (Jacq.) Standl. ■☆
44695 Atriplex watsonii A. Nelson ex Abrams;瓦氏滨藜;Watson's Orach ■☆
44696 Atriplex welshii C. A. Hanson = Atriplex gardneri (Moq.) D. Dietr. var. welshii (C. A. Hanson) S. L. Welsh ■☆
44697 Atriplex wolfii S. Watson;沃尔夫滨藜;Slender Orach, Wolf's Orach ■☆
44698 Atriplex wolfii S. Watson var. tenuissima (A. Nelson) S. L. Welsh;纤细沃尔夫滨藜■☆
44699 Atriplex wrightii S. Watson;赖氏滨藜;Wright's Orach ■☆
44700 Atriplex zosterifolia (Hook.) S. Watson = Atriplex gmelinii C. A. Mey. ex Bong. ●
44701 Atriplicaceae Juss.;滨藜科●■
44702 Atriplicaceae Juss. = Amaranthaceae Juss.(保留科名)●■
44703 Atriplicaceae Juss. = Chenopodiaceae Vent.(保留科名)●■
44704 Atropa L. (1753);颠茄属;Atropa, Deadly Nightshade ■
44705 Atropa acuminata Royle ex Lindl. = Atropa belladonna L. ■
44706 Atropa acuminata Royle ex Miers;渐尖颠茄■☆
44707 Atropa baetica Willk.;伯蒂卡颠茄■☆
44708 Atropa belladonna L.;颠茄(颠茄草);Apple of Sodom, Atterlothe, Belladonna, Belladonna Atropa, Black Cherry, Common Atropa, Daft Berries, Deadly Dwale, Deadly Nightshade, Death's Herb, Deathweed, Devil's Berries, Devil's Cherries, Devil's Cherry, Devil's Herb, Devil's Rhubarb, Dog Berries, Doleful Bells, Dwale, Dwayoberry, Fair Lady, Great Morel, Houndsberry, Jacob's Ladder, Jacob's Stee, Manicon, Man's Naughty Cherry, Mekilwort, Morette, Naughty Man's Cherries, Satan's Cherry, Sleeping Nightshade, Tetra Mad ■
44709 Atropa caucasica Kreyer;高加索颠茄;Caucasia Atropa ■☆
44710 Atropa frutescens L. = Withania frutescens (L.) Pauquy ●☆
44711 Atropa komarovii Blin. et Shaler;科马罗夫颠茄■☆
44712 Atropa lutescens Jacquem. ex C. B. Clarke;印度颠茄(淡黄颠茄)■☆
44713 Atropa mandragora L. = Mandragora officinalis L. ■☆
44714 Atropa physalodes L. = Nicandra physaloides (L.) Gaertn. ■

44715 Atropaceae Martinov = Solanaceae Juss. (保留科名)●■
44716 Atropaceae Martinov;颠茄科●■
44717 Atropanthe Pascher(1909);天蓬子属;Atropanthe ●■★
44718 Atropanthe mairei (H. Lév.) H. Lév. = Cyananthus flavus C. Marquand subsp. montanus (C. Y. Wu) D. Y. Hong et L. M. Ma ■
44719 Atropanthe sinensis (Hemsl.) Pascher;天蓬子(白商陆,拟颠茄,搜山虎,小独活,新莨菪);Atropanthe, China Atropanthe, Chinese Atropanthe ●■
44720 Atropatenia F. K. Mey. = Thlaspi L. ■
44721 Atropis (Trin.) Griseb. = Puccinellia Parl. (保留属名)■
44722 Atropis (Trin.) Rupr. ex Griseb. (废弃属名) = Puccinellia Parl. (保留属名)■
44723 Atropis Rupr. = Puccinellia Parl. (保留属名)■
44724 Atropis alascana (Scribn. et Merr.) V. I. Krecz. = Puccinellia kurilensis (Takeda) Honda ■
44725 Atropis angusta (Nees) Stapf = Puccinellia angusta (Nees) C. A. Sm. et C. E. Hubb. ■☆
44726 Atropis angustata (R. Br.) Griseb. = Puccinellia angustata (R. Br.) E. L. Rand et Redfield ■
44727 Atropis anisoclada V. I. Krecz. = Puccinellia gigantea (Grossh.) Grossh. ■
44728 Atropis aquatica ? = Catabrosa aquatica (L.) P. Beauv. ■
44729 Atropis beltranii Sennen = Puccinellia fasciculata (Torr.) E. P. Bicknell ■☆
44730 Atropis borreri Stapf = Puccinellia fasciculata (Torr.) E. P. Bicknell ■☆
44731 Atropis bulbosa Grossh. = Puccinellia bulbosa (Grossh.) Grossh. ■
44732 Atropis chilochloa V. I. Krecz. = Puccinellia poecilantha (K. Koch) V. I. Krecz. ■
44733 Atropis convoluta (Hornem.) Griseb. = Puccinellia convoluta (Hornem.) Fourr. ■
44734 Atropis convoluta (Hornem.) Griseb. var. expansa (Crép.) Batt. et Trab. = Puccinellia expansa (Crép.) Julià et J. M. Monts. ■☆
44735 Atropis convoluta (Hornem.) Griseb. var. gussonei (Parl.) Batt. et Trab. = Puccinellia convoluta (Hornem.) Fourr. ■
44736 Atropis diffusa V. I. Krecz. = Puccinellia diffusa V. I. Krecz. ■
44737 Atropis distans (Jacq.) Griseb. = Puccinellia distans (L.) Parl. ■
44738 Atropis distans (Jacq.) Griseb. f. pamirica Roshev. = Puccinellia pamirica (Roshev.) V. I. Krecz. ex Roshev. et Czukav. ■
44739 Atropis distans (Jacq.) Griseb. f. pauciramea Hack. = Puccinellia pauciramea (Hack.) V. I. Krecz. ex Ovcz. et Czukav. ■
44740 Atropis distans (Jacq.) Griseb. var. glauca Regel = Puccinellia glauca (Regel) V. I. Krecz. ex Drobow ■
44741 Atropis distans (L.) Parl. = Puccinellia distans (Jacq.) Parl. ■
44742 Atropis distans (L.) Parl. subsp. embergeri (H. Lindb.) Maire = Puccinellia festuciformis (Host) Parl. ■
44743 Atropis distans (L.) Parl. subsp. fontqueri Maire = Puccinellia festuciformis (Host) Parl. ■
44744 Atropis distans (L.) Parl. var. festuciformis (Host) Coss. et Durieu = Puccinellia festuciformis (Host) Parl. ■
44745 Atropis distans (L.) Parl. var. halophila Trab. = Puccinellia festuciformis (Host) Parl. ■
44746 Atropis distans (L.) Parl. var. permixta (Guss.) Trab. = Puccinellia fasciculata (Torr.) E. P. Bicknell ■☆
44747 Atropis distans (L.) Parl. var. tenuifolia (Boiss. et Reut.) Coss. et Durieu = Puccinellia tenuifolia (Boiss. et Reut.) H. Lindb. ■☆
44748 Atropis distans (L.) Parl. var. vulgaris Coss. et Durieu = Puccinellia distans (Jacq.) Parl. ■
44749 Atropis distans (Wahlb.) Griseb. f. pamirica Roshev. = Puccinellia pamirica (Roshev.) V. I. Krecz. ex Roshev. et Czukav. ■
44750 Atropis distans (Wahlb.) Griseb. var. crassifolia Roshev. = Puccinellia sclerodes V. I. Krecz. ■
44751 Atropis distans (Wahlb.) Griseb. var. glauca Regel = Puccinellia glauca (Regel) V. I. Krecz. ex Drobow ■
44752 Atropis distans (Wahlb.) Griseb. var. pauciramea Hack. = Puccinellia pauciramea (Hack.) V. I. Krecz. ex Ovcz. et Czukav. ■
44753 Atropis dolicholepis V. I. Krecz. = Puccinellia dolicholepis V. I. Krecz. ■
44754 Atropis embergeri H. Lindb. = Puccinellia festuciformis (Host) Parl. ■
44755 Atropis festuciformis (Host) Trab. = Puccinellia festuciformis (Host) Parl. ■
44756 Atropis festuciformis (Host) Trab. var. expansa Trab. = Puccinellia festuciformis (Host) Parl. ■
44757 Atropis festuciformis (Host) Trab. var. gussonei Trab. = Puccinellia festuciformis (Host) Parl. ■
44758 Atropis fontqueri Maire = Puccinellia festuciformis (Host) Parl. ■
44759 Atropis gigantea Grossh. = Puccinellia gigantea (Grossh.) Grossh. ■
44760 Atropis glauca (Regel) V. I. Krecz. = Puccinellia distans (Wahlb.) Parl. subsp. glauca (Regel) Tzvelev ■☆
44761 Atropis glauca (Regel) V. I. Krecz. = Puccinellia glauca (Regel) V. I. Krecz. ex Drobow ■
44762 Atropis grossheimiana V. I. Krecz. = Puccinellia grossheimiana (V. I. Krecz.) V. I. Krecz. ■
44763 Atropis hackeliana V. I. Krecz. = Puccinellia hackeliana (V. I. Krecz.) V. I. Krecz. ex Drobow ■
44764 Atropis hauptiana Trin. ex V. I. Krecz. = Puccinellia hauptiana (Trin.) V. I. Krecz. ■
44765 Atropis humilis Litv. ex V. I. Krecz. = Puccinellia hackeliana (Roshev.) Pers. subsp. humilis (Litv. ex V. I. Krecz.) Tzvelev ■
44766 Atropis humilis Litv. ex V. I. Krecz. = Puccinellia humilis (Litv. ex V. I. Krecz.) Bor ■
44767 Atropis iberica Wolley-Dod = Puccinellia iberica (Wolley-Dod) Tzvelev ■
44768 Atropis iliensis V. I. Krecz. = Puccinellia iliensis (V. I. Krecz.) Serg. ■
44769 Atropis intermedia Schur. = Puccinellia intermedia (Schrad.) Janch. ■
44770 Atropis kamtschatica (Holmb.) V. I. Krecz. = Puccinellia kamtschatica (Holmb.) V. I. Krecz. ■
44771 Atropis kurilensis Takeda = Puccinellia kurilensis (Takeda) Honda ■
44772 Atropis macranthera V. I. Krecz. = Puccinellia macranthera V. I. Krecz. ■
44773 Atropis maritima (Huds.) Griseb. = Puccinellia maritima (Huds.) Parl. ■
44774 Atropis palustris (Seenus) Hayek = Puccinellia festuciformis (Host) Parl. ■
44775 Atropis palustris (Seenus) Hayek subsp. festuciformis Briq. = Puccinellia festuciformis (Host) Parl. ■
44776 Atropis palustris (Seenus) Hayek subsp. tenuifolia (Boiss. et

Reut.) Maire = Puccinellia tenuifolia (Boiss. et Reut.) H. Lindb. ■☆
44777 Atropis pamirica (Roshev.) V. I. Krecz. = Puccinellia pamirica (Roshev.) V. I. Krecz. ex Roshev. et Czukav. ■
44778 Atropis pamirica V. I. Krecz. = Puccinellia pamirica (Roshev.) V. I. Krecz. ex Roshev. et Czukav. ■
44779 Atropis pauciramea (Hack.) V. I. Krecz. = Puccinellia pauciramea (Hack.) V. I. Krecz. ex Ovcz. et Czukav. ■
44780 Atropis phryganodes (Trin.) V. I. Krecz. = Puccinellia phryganodes (Trin.) Scribn. et Merr. ■
44781 Atropis poecilantha (K. Koch) V. I. Krecz. = Puccinellia poecilantha (K. Koch) V. I. Krecz. ■
44782 Atropis roshevitsiana Schischk. = Puccinellia roshevitsiana (Schischk.) V. I. Krecz. ex Tzvelev ■
44783 Atropis sclerodes V. I. Krecz. = Puccinellia gigantea (Grossh.) Grossh. ■
44784 Atropis sevangensis (Grossh.) V. I. Krecz. = Puccinellia sevengensis Grossh. ■
44785 Atropis sibirica (Holmb.) V. I. Krecz. = Puccinellia sibirica Holmb. ■
44786 Atropis subspicata V. I. Krecz. = Puccinellia subspicata V. I. Krecz. ex Ovcz. et Czukav. ■
44787 Atropis tenella (Lange) V. I. Krecz. = Puccinellia tenella (Lange) Holmb. ex A. E. Porsild ■
44788 Atropis tenuiflora Griseb. = Puccinellia tenuiflora (Griseb.) Scribn. et Merr. ■
44789 Atropis tenuiflora Griseb. ex Ledeb. = Puccinellia tenuiflora (Griseb.) Scribn. et Merr. ■
44790 Atropis tenuiflora Turcz. = Puccinellia tenuiflora (Turcz.) Scribn. et Merr. ■
44791 Atropis tenuifolia (Boiss. et Reut.) Trab. = Puccinellia tenuifolia (Boiss. et Reut.) H. Lindb. ■☆
44792 Atropis tenuissima Korsh. = Puccinellia tenuissima Litv. ex V. I. Krecz. ■
44793 Atropis tenuissima Litv. ex V. I. Krecz. = Puccinellia tenuissima (Litv. ex V. I. Krecz.) Litv. ex Pavlov ■
44794 Atropis thomsonii (Stapf) Pamp. = Puccinellia thomsonii (Stapf) R. R. Stewart ■
44795 Atroxima Stapf(1905);黑远志属●☆
44796 Atroxima afzeliana (Oliv.) Stapf;阿芙泽尔黑远志●☆
44797 Atroxima congolana E. M. Petit = Atroxima afzeliana (Oliv.) Stapf ●☆
44798 Atroxima gossweileri Exell = Carpolobia gossweileri (Exell) E. M. Petit ●☆
44799 Atroxima liberica Stapf;离生黑远志●☆
44800 Atrutegia Bedd. = Goniothalamus (Blume) Hook. f. et Thomson ●
44801 Attalea Kunth(1816);亚塔棕属(阿他利属,奥达尔椰子属,奥达椰子属,巴西榈属,巴西棕属,帝王椰属,刷棕属,亚达利亚棕属,亚达利椰属,直叶椰子属,直叶棕属);Attalea ●☆
44802 Attalea amygdalina Kunth;扁桃亚塔棕(杏叶直叶榈);Amygdaline Attalea ●☆
44803 Attalea boehmii Drude;非洲亚塔棕●☆
44804 Attalea colenda (Cook) Balslev et A. Hend.;厄瓜多尔亚塔棕●☆
44805 Attalea excelsa Mart.;高亚塔棕;Tall Attalea ●☆
44806 Attalea funifera Mart. = Attalea funifera Mart. ex Spreng. ●☆
44807 Attalea funifera Mart. ex Spreng.;亚塔棕(巴西棕);Bahia Piassava,Coquiila Palm,Coquina Nut,Piassaba Attalea ●☆
44808 Attalea gomphococca Mart.;丁状果片亚塔棕;Gomphococcous Attalea ●☆
44809 Attalea macrocarpa Drude;大果直叶榈●☆
44810 Attalea spectabilis Mart.;美丽亚塔棕(美丽直叶榈);Showy Attalea ●☆
44811 Attalea spiciosa Mart.;奥达尔椰子;Uauassu Palm ●
44812 Attilaea E. Martínez et Ramos(2007);阿特漆树属●☆
44813 Attractilis Haller ex Scop. = Atractylis L. ■☆
44814 Atulandra Raf. = Rhamnus L. ●
44815 Atuna Raf. (1838);阿顿果属●☆
44816 Atuna cordata Cockburn ex Prance;心形阿顿果●☆
44817 Atuna elata (King) Kosterm.;高阿顿果●☆
44818 Atuna elliptica (Kosterm.) Kosterm.;椭圆阿顿果●☆
44819 Atuna excelsa (Jack) Kosterm.;阿顿果●☆
44820 Atuna indica (Bedd.) Kosterm.;印度阿顿果●☆
44821 Atuna latifolia (M. R. Hend.) Kosterm.;宽叶阿顿果●☆
44822 Atuna latifrons (Kosterm.) Prance et F. White;宽花阿顿果●☆
44823 Atuna nitida (Hook. f.) Panigrahi et K. M. Purohit;光亮阿顿果●☆
44824 Atylosia Wight et Arn. (1834);虫豆属(蝙蝠豆属);Atylosia ●■
44825 Atylosia Wight et Arn. = Cajanus Adans. (保留属名)●
44826 Atylosia barbarta (Benth.) Baker = Cajanus goensis Dalzell ●
44827 Atylosia circinalis Benth. = Dunbaria circinalis (Benth.) Baker ●
44828 Atylosia crassa Prain ex King = Atylosia volubilis (Blanco) Gamble ●
44829 Atylosia crassa Prain ex King = Cajanus crassus (Prain ex King) Maesen ●
44830 Atylosia crinita Dunn = Dunbaria fusca (Wall.) Kurz ■
44831 Atylosia geminiflora Dalzell = Atylosia platycarpa Benth. ■☆
44832 Atylosia goensis (Dalzell) Dalzell = Cajanus goensis Dalzell ●
44833 Atylosia grandiflora Benth. ex Baker = Cajanus grandiflorus (Benth. ex Baker) Maesen ●
44834 Atylosia mollis (Willd.) Benth. = Cajanus mollis (Benth.) Maesen ●
44835 Atylosia nivea Benth. = Cajanus niveus (Benth.) Maesen ●
44836 Atylosia platycarpa Benth.;宽果虫豆■☆
44837 Atylosia scarabaeoides (L.) Benth. = Cajanus scarabaeoides (L.) Thouars ●
44838 Atylosia scarabaeoides (L.) Benth. var. argyrophyllus Y. T. Wei et S. K. Lee = Cajanus scarabaeoides (L.) Thouars var. argyrophyllus (Y. T. Wei et S. K. Lee) Y. T. Wei et S. K. Lee ●
44839 Atylosia volubilis (Blanco) Gamble = Atylosia mollis (Willd.) Benth. ●
44840 Atylosia volubilis (Blanco) Gamble = Cajanus crassus (Prain ex King) Maesen ●
44841 Atylus Salisb. (废弃属名) = Isopogon R. Br. ex Knight(保留属名)●☆
44842 Atylus Salisb. (废弃属名) = Petrophile R. Br. ex Knight ●☆
44843 Atyson Raf. = Aectyson Raf. ●■
44844 Atyson Raf. = Sedum L. ●■
44845 Aubentonla Dombey ex Steud. = Waltheria L. ●■
44846 Aubertia Bory = Zanthoxylum L. ●
44847 Aubertia Chapel. ex Baill. = Croton L. ●
44848 Aubertiella Briq. = Audibertiella Briq. ●■
44849 Aubertiella Briq. = Salvia L. ●■
44850 Aubion Raf. = Cleome L. ●■
44851 Aubletella Pierre = Chrysophyllum L. ●
44852 Aubletia Gaertn. = Sonneratia L. f. (保留属名)●

44853 Aubletia Le Monn. ex Rozier = Obletia Lemonn. ex Rozier ■
44854 Aubletia Le Monn. ex Rozier = Verbena L. ●■
44855 Aubletia Lour. = Paliurus Tourn. ex Mill. ●
44856 Aubletia Neck. = Ruellia L. ●■
44857 Aubletia Pers. = Moniera Loefl. ●☆
44858 Aubletia Schreb. = Apeiba Aubl. ●☆
44859 Aubletia ramosissima Lour. = Paliurus ramosissimus (Lour.) Poir. ●
44860 Aubletiana J. Murillo = Conceveiba Aubl. ●☆
44861 Aubletiana J. Murillo(2000);非洲大戟属●☆
44862 Aubregrinia Heine(1960);西非单性榄属(西非榄属)●☆
44863 Aubregrinia taiensis (Aubrév. et Pellegr.) Heine;西非榄●☆
44864 Aubrevillea Pellegr. (1933);奥布雷豆属■☆
44865 Aubrevillea kerstingii (Harms) Pellegr.;奥布雷豆●☆
44866 Aubrevillea platycarpa Pellegr.;宽果奥布雷豆●☆
44867 Aubrieta Adans. (1763);南庭荠属(紫荠属);Aubretia, Aubrieta, Aubrietia, Purple Rock-cress, Rock-cress ■☆
44868 Aubrieta deltoidea (L.) DC.;南庭荠;Aubrietia, Common Aubrieta, False Rockcress, Lilacbush, Purple Rockcress, Purple Rock-cress ■☆
44869 Aubrieta deltoidea (L.) DC. 'Argenteovariegata';银边南庭荠■☆
44870 Aubrieta purpurea DC.;紫南庭荠;Wall Cress ■☆
44871 Aubrietia DC. = Aubrieta Adans. ■☆
44872 Aubrietia deltoidea (L.) DC. = Aubrieta deltoidea (L.) DC. ■☆
44873 Aubrietia deltoidea DC. = Aubrieta deltoidea (L.) DC. ■☆
44874 Aubrya Baill. = Sacoglottis Mart. ●☆
44875 Aubrya gabonensis Baill. = Sacoglottis gabonensis (Baill.) Urb. ●☆
44876 Aubrya occidentalis A. Chev. = Sacoglottis gabonensis (Baill.) Urb. ●☆
44877 Auchera DC. = Cousinia Cass. ●■
44878 Aucklandia Falc. (1841);云木香属(云木香菊属);Aucklandia ■
44879 Aucklandia Falc. = Saussurea DC. (保留属名)●■
44880 Aucklandia costus Falc. = Saussurea costus (Falc.) Lipsch. ■
44881 Aucklandia lappa Decne.;云木香(广木香,蜜香,木香,南木香,青木香,五木香,五香,中脉风毛菊);Common Aucklandia, Indian Orris, Yun Muxiang ■
44882 Aucklandia lappa Decne. = Saussurea costus (Falc.) Lipsch. ■
44883 Aucoumea Pierre(1896);假榇木属(奥克橄榄属);Gaboon Mahogany, Okoume, Tasmanian Sassafras ●☆
44884 Aucoumea klaineana Pierre;假榇木;Gaboon Mahogany ●☆
44885 Aucuba Cham. = Aruba Nees et Mart. ●☆
44886 Aucuba Cham. = Raputia Aubl. ●☆
44887 Aucuba Thunb. (1784);桃叶珊瑚属;Aucuba, Gold Dust Shrub, Spotted-laurel ●
44888 Aucuba albo-punctifolia F. T. Wang;斑叶珊瑚;Spot-leaf Aucuba, Spot-leaved Aucuba, Variegateleaf Aucuba ●
44889 Aucuba albo-punctifolia F. T. Wang var. angustula W. P. Fang et Soong;窄斑叶珊瑚(窄叶斑叶珊瑚);Narrow Spot-leaf Aucuba ●
44890 Aucuba albo-punctifolia F. T. Wang var. obcordata Rehder = Aucuba obcordata (Rehder) S. H. Fu ex W. K. Hu et Soong ●
44891 Aucuba cavinervis C. Y. Wu ex Soong;凹脉桃叶珊瑚;Sunkenvein Aucuba ●
44892 Aucuba cavinervis C. Y. Wu ex Soong = Aucuba filicauda Chun et F. C. How ●
44893 Aucuba chinensis Benth.;桃叶珊瑚(软叶罗伞,天脚板,狭叶桃叶珊瑚,植楠树);China Aucuba, Chinese Aucuba, Concave-veined Aucuba ●
44894 Aucuba chinensis Benth. = Aucuba himalaica Hook. f. et Thomson ●
44895 Aucuba chinensis Benth. f. angustifolia Rehder = Aucuba chinensis Benth. var. angusta F. T. Wang ●
44896 Aucuba chinensis Benth. f. obcordata Rehder = Aucuba obcordata (Rehder) S. H. Fu ex W. K. Hu et Soong ●
44897 Aucuba chinensis Benth. f. subintegra H. L. Li = Aucuba eriobotryifolia F. T. Wang ●
44898 Aucuba chinensis Benth. subsp. omeiensis (W. P. Fang) W. P. Fang et Soong;峨眉桃叶珊瑚(青皮树);Emei Aucuba, Fang Aucuba, Omei Aucuba ●
44899 Aucuba chinensis Benth. subsp. omeiensis (W. P. Fang) W. P. Fang et Soong = Aucuba chinensis Benth. ●
44900 Aucuba chinensis Benth. var. angusta F. T. Wang;狭叶桃叶珊瑚●
44901 Aucuba chinensis Benth. var. fongfangshanensis J. C. Liao et al.;凤凰山珊瑚;Fenghuangshan Aucuba ●
44902 Aucuba chinensis Benth. var. fongfangshanensis J. C. Liao et al. = Aucuba obcordata (Rehder) S. H. Fu ex W. K. Hu et Soong ●
44903 Aucuba chinensis Benth. var. variegata Dombrain = Aucuba chinensis Benth. ●
44904 Aucuba chlorascens F. T. Wang;细齿桃叶珊瑚(绿花桃叶珊瑚);Serrulate Aucuba ●
44905 Aucuba confertiflora W. P. Fang et Soong;密花桃叶珊瑚;Denseflower Aucuba, Densiflowered Aucuba ●
44906 Aucuba eriobotryifolia F. T. Wang;枇杷叶珊瑚(琵琶叶珊瑚);Broadleaf Aucuba, Loquatleaf Aucuba, Loquat-leaved leaf Aucuba ●
44907 Aucuba filicauda Chun et F. C. How;纤尾桃叶珊瑚;Filicaudate Aucuba, Fine-caudated Aucuba, Thin Tailleaf Aucuba ●
44908 Aucuba filicauda Chun et F. C. How var. pauciflora W. P. Fang et Soong;少花桃叶珊瑚;Fewflower Aucuba ●
44909 Aucuba grandiflora C. Y. Wu;大花桃叶珊瑚;Large-flower Aucuba ●
44910 Aucuba grandiflora C. Y. Wu = Aucuba chinensis Benth. ●
44911 Aucuba himalaica Hook. f. et Thomson;喜马拉雅桃叶珊瑚(软叶罗伞,桃叶珊瑚,西藏桃叶珊瑚,狭叶桃叶珊瑚);Himalayan Aucuba, Himalayas Aucuba, Narrowleaf Aucuba ●
44912 Aucuba himalaica Hook. f. et Thomson var. dolichophylla W. P. Fang et Soong;长叶珊瑚(长叶岩马桑);Longleaf Himalayan Aucuba ●
44913 Aucuba himalaica Hook. f. et Thomson var. oblanceolata W. P. Fang et Soong;倒披针叶珊瑚(木珊瑚);Oblanceleaf Aucuba ●
44914 Aucuba himalaica Hook. f. et Thomson var. pilosissima W. P. Fang et Soong;密毛桃叶珊瑚;Densehairy Aucuba ●
44915 Aucuba japonica Thunb.;日本桃叶珊瑚(桎菜,东瀛珊瑚,东瀛珊瑚木,青木,珊瑚,桃叶珊瑚);Aucuba, Dog Laurel, Gold Dust Plant, Gold Dust Tree, Japan Aucuba, Japanese Aucuba, Japanese Laurel, Sparked Laurel, Spotted Laurel, Spotted-laurel ●
44916 Aucuba japonica Thunb. 'Crotonifolia';巴豆叶东瀛珊瑚●☆
44917 Aucuba japonica Thunb. 'Gold Dust';金粉东瀛珊瑚●☆
44918 Aucuba japonica Thunb. 'Picturata';金心东瀛珊瑚●☆
44919 Aucuba japonica Thunb. 'Variegata';花叶青木(斑叶青木,黄叶日本桃叶珊瑚,金沙树,洒金珊瑚,洒金叶珊瑚);Gold Dust Shrub, Yellowleaf Aucuba ●
44920 Aucuba japonica Thunb. f. amanogawa Nakai;斑叶青木;Maculate Japan Aucuba ●☆
44921 Aucuba japonica Thunb. f. brachyphylla (Honda) H. Hara;短叶

青木●☆
44922 Aucuba japonica Thunb. f. castaneopedicellata Makino;黑梗青木;Black-stalked Japan Aucuba ●☆
44923 Aucuba japonica Thunb. f. leucocarpa (Matsum. et Nakai) Ohwi;白果青木●☆
44924 Aucuba japonica Thunb. f. longifolia Schelle;长叶青木;Longleaf Japan Aucuba ●☆
44925 Aucuba japonica Thunb. f. luteocarpa (Rehder) Rehder;黄果青木●☆
44926 Aucuba japonica Thunb. f. tagayasan Nakai;皱叶青木●☆
44927 Aucuba japonica Thunb. f. viridiflora Makino; 绿花青木; Greenflower Japan Aucuba ●☆
44928 Aucuba japonica Thunb. var. angustifolia Regel = Aucuba japonica Thunb. f. longifolia Schelle ●☆
44929 Aucuba japonica Thunb. var. aureo-maculata Hibberd;黄斑青木●☆
44930 Aucuba japonica Thunb. var. australis H. Hara et S. Kuros. = Aucuba japonica Thunb. var. ovoidea Koidz. ●☆
44931 Aucuba japonica Thunb. var. bicolor Carrière;二色青木●☆
44932 Aucuba japonica Thunb. var. borealis Miyabe et Kudo;北方青木;Northern Japan Aucuba ●☆
44933 Aucuba japonica Thunb. var. borealis Miyabe et Kudo f. rugosa Sugim.;粗糙北方青木●☆
44934 Aucuba japonica Thunb. var. concolor Regel;小叶青木●☆
44935 Aucuba japonica Thunb. var. longifolia Veitch = Aucuba japonica Thunb. f. longifolia Schelle ●☆
44936 Aucuba japonica Thunb. var. ovata Siebold;卵叶青木●☆
44937 Aucuba japonica Thunb. var. ovoidea Koidz.;卵形青木;Ovate-leaved Japan Aucuba ●☆
44938 Aucuba japonica Thunb. var. sulphrea Regel;硫叶青木●☆
44939 Aucuba japonica Thunb. var. variegata Dombr. = Aucuba japonica Thunb. 'Variegata' ●
44940 Aucuba japonica Thunb. var. variegata Regel = Aucuba japonica Thunb. 'Variegata' ●
44941 Aucuba japonica Thunb. var. versicolor Regel;彩色青木●☆
44942 Aucuba mollifolia C. Y. Wu;软叶桃叶珊瑚;Soft-leaf Aucuba ●
44943 Aucuba mollifolia C. Y. Wu = Aucuba himalaica Hook. f. et Thomson var. oblanceolata W. P. Fang et Soong ●
44944 Aucuba obcordata (Rehder) S. H. Fu ex W. K. Hu et Soong;倒心叶珊瑚(青竹叶);Inversley Heart-shaped Aucuba, Obcordateleaf Aucuba, Obcordate-leaved Aucuba ●
44945 Aucuba oblanceolata (W. P. Fang et Soong) C. J. Qi. = Aucuba himalaica Hook. f. et Thomson var. oblanceolata W. P. Fang et Soong ●
44946 Aucuba omeiensis W. P. Fang = Aucuba chinensis Benth. subsp. omeiensis (W. P. Fang) W. P. Fang et Soong ●
44947 Aucuba omeiensis W. P. Fang = Aucuba chinensis Benth. ●
44948 Aucuba robusta W. P. Fang et Soong; 粗梗桃叶珊瑚; Robust Aucuba, Robuste Aucuba, Thickstalk Aucuba ●
44949 Aucuba yunnanensis C. Y. Wu;云南桃叶珊瑚;Yunnan Aucuba ●
44950 Aucuba yunnanensis C. Y. Wu = Aucuba chinensis Benth. var. angusta F. T. Wang ●
44951 Aucuba yunnanensis C. Y. Wu var. pubiger C. Y. Wu et S. Y. Pao;软毛桃叶珊瑚;Soft-hair Aucuba ●
44952 Aucuba yunnanensis C. Y. Wu var. pubigera C. Y. Wu et S. Y. Pao = Aucuba chinensis Benth. var. angusta F. T. Wang ●
44953 Aucubaceae Bercht. et J. Presl = Cornaceae Bercht. et J. Presl(保留科名)●■
44954 Aucubaceae Bercht. et J. Presl = Garryaceae Lindl. (保留科名)●☆
44955 Aucubaceae J. Agardh = Cornaceae Bercht. et J. Presl(保留科名)●■
44956 Aucubaceae J. Agardh = Garryaceae Lindl. (保留科名)●☆
44957 Aucubaceae J. Agardh(1858);桃叶珊瑚科●
44958 Aucubaephyllum Ahlburg = Grumilea Gaertn. ●☆
44959 Aucubaephyllum Ahlburg = Psychotria L. (保留属名)●
44960 Aucuparia Medik. (1789);捕鸟蔷薇属;Cuirn Sorbus ●☆
44961 Aucuparia Medik. = Sorbus L. ●
44962 Aucuparia americana Nieuwl.;美洲捕鸟蔷薇●☆
44963 Aucuparia pinnata Fourr.;羽状捕鸟蔷薇●☆
44964 Aucuparia sylvestris Medik.;捕鸟蔷薇●☆
44965 Audibertia Benth. (1829) = Mentha L. ●■
44966 Audibertia Benth. (1832) = Audibertiella Briq. ●■
44967 Audibertia Benth. (1832) = Salvia L. ●■
44968 Audibertiella Briq. = Salvia L. ●■
44969 Audouinia Brongn. (1826);奥丁鳞叶树属●☆
44970 Audouinia capitata (L.) Brongn.;奥丁鳞叶树●☆
44971 Auerodendron Urb. (1924);奥尔鼠李属●☆
44972 Auerodendron paucifiorum Alain;奥尔鼠李;Puerto Rica ●☆
44973 Auganthus Link = Primula L. ■
44974 Auganthus praenitens Link = Primula sinensis Sabine ex Lindl. ■
44975 Augea Thunb. (1794)(保留属名);日光藜属■☆
44976 Augea Thunb. ex Retz. = Augea Thunb. (保留属名)■☆
44977 Augea Thunb. ex Retz. = Lanaria Aiton ■☆
44978 Augea capensis Thunb.;日光藜■☆
44979 Augia Lour. (废弃属名) = Augea Thunb. (保留属名)■☆
44980 Augia Lour. (废弃属名) = Calophyllum L. ●
44981 Augia Lour. (废弃属名) = Rhus L. ●
44982 Augia sinensis Lour. = Rhus succedanea L. ●
44983 Augia sinensis Lour. = Toxicodendron succedaneum (L.) Kuntze ●
44984 Augouardia Pellegr. (1924);日光豆属■☆
44985 Augouardia letestui Pellegr.;日光豆■☆
44986 Augusta Ellis = Gardenia Ellis(保留属名)●
44987 Augusta Ellis = Warneria Mill. ●
44988 Augusta Leandro(废弃属名) = Augusta Pohl(保留属名)■☆
44989 Augusta Leandro(废弃属名) = Stifftia J. C. Mikan(保留属名)●☆
44990 Augusta Pohl(1828)(保留属名);巴西茜草属■☆
44991 Augusta attenuata Pohl;巴西茜草■☆
44992 Augustea DC. = Augusta Pohl(保留属名)■☆
44993 Augustia Klotzsch = Begonia L. ●■
44994 Augustinea A. St.-Hil. et Naudin = Miconia Ruiz et Pav. (保留属名)●☆
44995 Augustinea H. Karst. = Pyrenoglyphis H. Karst. ●
44996 Augustinea Mart. = Bactris Jacq. ●
44997 Aukuba Kochne = Aucuba Thunb. ●
44998 Aukuba Thunb. = Aucuba Thunb. ●
44999 Aukuba japonica Thunb. = Aucuba japonica Thunb. ●
45000 Aulacia Lour. (废弃属名) = Micromelum Blume(保留属名)●
45001 Aulacia falcata Lour. = Micromelum falcatum (Lour.) Tanaka ●
45002 Aulacidium Rich. ex DC. = Macrocentrum Hook. f. ■☆
45003 Aulacinthus E. Mey. = Lotononis (DC.) Eckl. et Zeyh. (保留属名)■
45004 Aulacinthus gracilis E. Mey. = Lotononis densa (Thunb.) Harv. subsp. gracilis (E. Mey.) B.-E. van Wyk ■☆
45005 Aulacinthus rigidus E. Mey. = Lotononis rigida (E. Mey.) Benth. ■☆

45006 Aulacocalyx Hook. f. (1873);沟萼茜属●☆
45007 Aulacocalyx auriculata K. Schum.;耳形沟萼茜●☆
45008 Aulacocalyx brevilobus Hutch. et Dalziel = Aulacocalyx jasminiflora Hook. f. ●☆
45009 Aulacocalyx camerooniana Sonké et S. E. Dawson;喀麦隆沟萼茜●☆
45010 Aulacocalyx caudata (Hiern) Keay;尾状沟萼茜●☆
45011 Aulacocalyx diervilleoides (K. Schum.) E. M. Petit = Heinsenia diervilleoides K. Schum. ●☆
45012 Aulacocalyx divergens (Hutch. et Dalziel) Keay;稍叉沟萼茜●☆
45013 Aulacocalyx infundibuliflora E. M. Petit = Heinsenia diervilleoides K. Schum. ●☆
45014 Aulacocalyx jasminiflora Hook. f.;茉莉沟萼茜●☆
45015 Aulacocalyx jasminiflora Hook. f. f. brevis N. Hallé = Aulacocalyx jasminiflora Hook. f. ●☆
45016 Aulacocalyx jasminiflora Hook. f. subsp. kivuensis Figueiredo;基伍沟萼茜●☆
45017 Aulacocalyx jasminiflora Hook. f. var. brevis N. Hallé = Aulacocalyx jasminiflora Hook. f. ●☆
45018 Aulacocalyx jasminiflora Hook. f. var. latifolia De Wild. et T. Durand = Aulacocalyx jasminiflora Hook. f. ●☆
45019 Aulacocalyx lamprophylla K. Krause;亮叶沟萼茜●☆
45020 Aulacocalyx laxiflora E. M. Petit;疏花沟萼茜●☆
45021 Aulacocalyx leptactinoides K. Schum. = Aulacocalyx caudata (Hiern) Keay ●☆
45022 Aulacocalyx letestui (Pellegr.) E. M. Petit = Aulacocalyx pallens (Hiern) Bridson et Figueiredo subsp. letestui (Pellegr.) Figueiredo ●☆
45023 Aulacocalyx lujae De Wild.;卢亚沟萼茜●☆
45024 Aulacocalyx lujae De Wild. var. subulata N. Hallé = Aulacocalyx subulata (N. Hallé) Figueiredo ●☆
45025 Aulacocalyx pallens (Hiern) Bridson et Figueiredo;苍白沟萼茜●☆
45026 Aulacocalyx pallens (Hiern) Bridson et Figueiredo subsp. letestui (Pellegr.) Figueiredo;莱泰斯图沟萼茜●☆
45027 Aulacocalyx subulata (N. Hallé) Figueiredo;钻形沟萼茜●☆
45028 Aulacocalyx subulata (N. Hallé) Figueiredo subsp. glabra Figueiredo;光滑钻形沟萼茜●☆
45029 Aulacocalyx talbotii (Wernham) Keay;塔尔博特沟萼茜●☆
45030 Aulacocalyx trilocularis Scott-Elliot = Sericanthe trilocularis (Scott-Elliot) Robbr. ●☆
45031 Aulacocarpus O. Berg = Mouriri Aubl. ●☆
45032 Aulacocarpus O. Berg(1856);沟果野牡丹属●☆
45033 Aulacocarpus crassifolius O. Berg;厚叶沟果野牡丹●☆
45034 Aulacocarpus sellowianus O. Berg;沟果野牡丹●☆
45035 Aulacodiscus Hook. f. = Pleiocarpidia K. Schum. ●☆
45036 Aulacolepis Hack. = Aniselytron Merr. ■
45037 Aulacolepis Hack. = Calamagrostis Adans. ■
45038 Aulacolepis Hack. = Neoaulacolepis Rauschert ■
45039 Aulacolepis agrostoides (Merr.) Ohwi = Aniselytron agrostoides Merr. ■
45040 Aulacolepis agrostoides (Merr.) Ohwi var. formosana Ohwi = Aulacolepis agrostoides (Merr.) Ohwi ■
45041 Aulacolepis clemensiae Hitchc. = Aniselytron treutleri (Kuntze) Soják ■
45042 Aulacolepis formosana (Ohwi) L. Liou = Aulacolepis agrostoides (Merr.) Ohwi var. formosana Ohwi ■
45043 Aulacolepis formosana (Ohwi) L. Liou = Aulacolepis agrostoides (Merr.) Ohwi ■
45044 Aulacolepis japonica Hack. = Aniselytron treutleri (Kuntze) Soják var. japonicum (Hack.) N. X. Zhao ■
45045 Aulacolepis japonica Hack. = Aniselytron treutleri (Kuntze) Soják ■
45046 Aulacolepis milioides (Honda) Ohwi = Aniselytron treutleri (Kuntze) Soják ■
45047 Aulacolepis petelotii Hitchc. = Deyeuxia petelotii (Hitchc.) S. M. Phillips et Wen L. Chen ■
45048 Aulacolepis pseudopoa (Jansen) Ohwi = Aniselytron treutleri (Kuntze) Soják ■
45049 Aulacolepis treutleri (Kuntze) Hack. = Aniselytron treutleri (Kuntze) Soják ■
45050 Aulacolepis treutleri (Kuntze) Hack. subsp. japonica (Hack.) T. Koyama = Aniselytron treutleri (Kuntze) Soják ■
45051 Aulacolepis treutleri (Kuntze) Hack. subsp. japonica (Hack.) T. Koyama = Aniselytron treutleri (Kuntze) Soják var. japonicum (Hack.) N. X. Zhao ■
45052 Aulacolepis treutleri (Kuntze) Hack. var. japonica (Hack.) Ohwi = Aniselytron treutleri (Kuntze) Soják var. japonicum (Hack.) N. X. Zhao ■
45053 Aulacolepis treutleri (Kuntze) Hack. var. japonica (Hack.) Ohwi = Aniselytron treutleri (Kuntze) Soják ■
45054 Aulacolepis treutleri (Kuntze) Hack. var. japonica (Hack.) Ohwi = Aulacolepis japonica Hack. ■
45055 Aulacolepis treutleri (Kuntze) Hack. var. milioides (Honda) Ohwi = Aniselytron treutleri (Kuntze) Soják ■
45056 Aulacophyllum Regel = Zamia L. ●☆
45057 Aulacophyllum Regel(1876);沟叶苏铁属●☆
45058 Aulacophyllum montanum Regel;山地沟叶苏铁●☆
45059 Aulacophyllum skinneri Regel;沟叶苏铁●☆
45060 Aulacorhynchus Nees = Tetraria P. Beauv. ■☆
45061 Aulacorhynchus crinifolius Nees = Tetraria crinifolia (Nees) C. B. Clarke ■☆
45062 Aulacospermum Ledeb. (1833);沟子芹属(种沟芹属)■
45063 Aulacospermum Ledeb. = Pleurospermum Hoffm. ■
45064 Aulacospermum anomalum Ledeb.;异常沟子芹■☆
45065 Aulacospermum darvasicum (Lipsky) Schischk.;达尔瓦斯沟子芹■☆
45066 Aulacospermum hookeri (C. B. Clarke) Farille et S. B. Malla = Pleurospermum hookeri C. B. Clarke ■
45067 Aulacospermum pulchrum (Aitch. et Hemsl.) Rech. f. et Riedl. = Pleurospermum stylosum C. B. Clarke ■
45068 Aulacospermum rupestre Popov = Pleurospermum rupestre (Popov) K. T. Fu et Y. C. Ho ■
45069 Aulacospermum simplex Rupr. = Pleurospermum simplex (Rupr.) Benth. et Hook. f. ex Drude ■
45070 Aulacospermum stylosum (C. B. Clarke) Rech. f. et Riedl = Pleurospermum stylosum C. B. Clarke ■
45071 Aulacospermum tianschanicum (Korovin) C. Norman;天山沟子芹■☆
45072 Aulacospermum tianschanicum (Korovin) C. Norman = Trachydium tianschanicum Korovin ■
45073 Aulacospermum turkestanicum (Franch.) Schischk.;土耳其斯坦沟子芹■☆
45074 Aulacostigma Turcz. = Rhynchotheca Ruiz et Pav. ●☆

45075 Aulacothelae Lem. = Coryphantha (Engelm.) Lem.(保留属名)●■
45076 Aulacothele Monv.(废弃属名) = Coryphantha (Engelm.) Lem.(保留属名)●■
45077 Aulacothele Monv. ex Lem. = Coryphantha (Engelm.) Lem.(保留属名)●■
45078 Aulandra H. J. Lam(1927);管蕊榄属●☆
45079 Aulandra cauliflora H. J. Lam;管蕊榄●☆
45080 Aulax P. J. Bergius(1767);南非山龙眼属●☆
45081 Aulax cancellata (L.) Druce;南非山龙眼●☆
45082 Aulax cneorifolia Salisb. ex Knight = Aulax umbellata (Thunb.) R. Br. ●☆
45083 Aulax pallasia Stapf;帕拉斯南非山龙眼●☆
45084 Aulax pinifolia P. J. Bergius = Aulax cancellata (L.) Druce ●☆
45085 Aulax umbellata (Thunb.) R. Br.;伞形南非山龙眼●☆
45086 Aulaxanthus Elliott = Anthenantia P. Beauv. ■☆
45087 Aulaxia Nutt. = Anthenantia P. Beauv. ■☆
45088 Aulaxia Nutt. = Aulaxanthus Elliott ■☆
45089 Aulaxis Haw. = Saxifraga L. ■
45090 Aulaxis Steud. = Anthenantia P. Beauv. ■☆
45091 Aulaxis Steud. = Aulaxia Nutt. ■☆
45092 Aulaya Harv. = Harveya Hook. ■☆
45093 Aulaya coccinea Harv. = Harveya pauciflora (Benth.) Hiern ■☆
45094 Aulaya grandiflora Benth. = Harveya purpurea (L. f.) Harv. ex Hook. ■☆
45095 Aulaya obtusifolia Benth. = Harveya obtusifolia (Benth.) Vatke ■☆
45096 Aulaya pauciflora Benth. = Harveya pauciflora (Benth.) Hiern ■☆
45097 Aulaya purpurea (L. f.) Benth. = Harveya purpurea (L. f.) Harv. ex Hook. ■☆
45098 Aulaya scarlatina Benth. = Harveya scarlatina (Benth.) Hiern ■☆
45099 Aulaya squamosa (Thunb.) Harv. = Harveya squamosa (Thunb.) Steud. ■☆
45100 Aulea amicorum (J. B. Hall) C. Cusset = Saxicolella amicorum J. B. Hall ■☆
45101 Aulea submersa (J. B. Hall) C. Cusset = Saxicolella submersa (J. B. Hall) C. D. K. Cook et Rutish. ■☆
45102 Auleya D. Dietr. = Aulaya Harv. ■☆
45103 Aulica Raf. = Hippeastrum Herb.(保留属名)■
45104 Auliphas Raf. = Miconia Ruiz et Pav.(保留属名)●☆
45105 Aulisconema Hua = Disporopsis Hance ■
45106 Aulisconema aspera Hua = Disporopsis aspera (Hua) Engl. ex Krause ■
45107 Aulisconema pernyi Hua = Disporopsis pernyi (Hua) Diels ■
45108 Auliza Salisb. = Epidendrum L.(保留属名)■☆
45109 Auliza Small = Epidendrum L.(保留属名)■☆
45110 Auliza nocturna (Jacq.) Small = Epidendrum nocturnum Jacq. ■☆
45111 Aulocaulis Standl. = Boerhavia L. ■
45112 Aulojusticia Lindau = Justicia L. ●■
45113 Aulojusticia linifolia Lindau = Siphonoglossa linifolia (Lindau) C. B. Clarke ●☆
45114 Aulomyrcia O. Berg = Myrcia DC. ex Guill. ●☆
45115 Aulonemia Goudot = Arthrostylidium Rupr. ●☆
45116 Aulonemia Goudot(1846);牧笛竹属●☆
45117 Aulonemia gueko Goudot;牧笛竹●☆
45118 Aulonix Raf. = Cytisus Desf.(保留属名)●
45119 Aulosema Walp. = Astragalus L. ●■
45120 Aulosepalum Garay(1982);管萼兰属■☆
45121 Aulosepalum hemichrea (Lindl.) Garay;管萼兰■☆
45122 Aulosepalum pulchrum (Schltr.) Catling;美丽管萼兰■☆
45123 Aulosepalum tenuiflorum (Greenm.) Garay;细花管萼兰■☆
45124 Aulosolena Koso-Pol. = Sanicula L. ■
45125 Aulospermum J. M. Coult. et Rose = Cymopterus Raf. ■☆
45126 Aulospermum J. M. Coult. et Rose(1900);管籽芹属(管子芹属)■☆
45127 Aulospermum purpureum (S. Watson) J. M. Coult. et Rose;紫管籽芹■☆
45128 Aulostephanus Schltr. = Brachystelma R. Br.(保留属名)■
45129 Aulostephanus natalensis Schltr. = Brachystelma natalense (Schltr.) N. E. Br. ■☆
45130 Aulostylis Schltr.(1912);管柱兰属■☆
45131 Aulostylis Schltr. = Calanthe R. Br.(保留属名)■
45132 Aulostylis papuana Schltr.;管柱兰■☆
45133 Aulotandra Gagnep.(1902);管蕊姜属■☆
45134 Aulotandra kamerunensis Loes.;管蕊姜■☆
45135 Aurantiaceae Juss. = Rutaceae Juss.(保留科名)●■
45136 Auranticarpa L. W. Cayzer, Crisp et I. Telford = Pittosporum Banks ex Gaertn.(保留属名)●
45137 Auranticarpa L. W. Cayzer, Crisp et I. Telford(2000);澳洲海桐花属●☆
45138 Aurantium Mill. = Citrus L. ●
45139 Aurantium Tourn. ex Mill. = Citrus L. ●
45140 Aurantium acre Mill. = Citrus aurantium L. ●
45141 Aurantium corniculatum Mill. = Citrus aurantium L. ●
45142 Aurantium decumanum (L.) Mill. = Citrus maxima (Burm.) Merr. ●
45143 Aurantium distortum Mill. = Citrus aurantium L. ●
45144 Aurantium humile Mill. = Citrus aurantium L. ●
45145 Aurantium maximum Burm. = Citrus maxima (Burm.) Merr. ●
45146 Aurantium myrtifolium Descourtilz = Citrus aurantium L. ●
45147 Aurantium sinense Mill. = Citrus aurantium L. ●
45148 Aurantium sinensis Mill. = Citrus sinensis (L.) Osbeck ●
45149 Aurantium vulgare (Risso) M. Gómez = Citrus aurantium L. ●
45150 Aureilobivia Frič = Echinopsis Zucc. ●
45151 Aurelia Cass. = Donia R. Br. ■☆
45152 Aurelia Cass. = Grindelia Willd. ●■☆
45153 Aurelia J. Gay = Narcissus L. ■
45154 Aurelia broussonetii (Lag.) J. Gay = Narcissus broussonetii Lag. ■☆
45155 Aureliana Boehm. = Panax L. ■
45156 Aureliana Lafit. ex Catesb. = Aralia L. ●■
45157 Aureliana Sendtn.(1846);金蛹茄属●■☆
45158 Aureliana Sendtn. = Bassovia Aubl. ●■
45159 Aureliana Sendtn. = Solanum L. ●■
45160 Aureliana angustifolia Alm. -Lafetá;窄叶金蛹茄■☆
45161 Aureliana brasiliana (Hunz.) Barboza et Hunz.;巴西金蛹茄■☆
45162 Aureliana fasciculata Sendtn.;簇生金蛹茄■☆
45163 Aureliana lucida Sendtn.;亮金蛹茄■☆
45164 Aureliana tomentosa Sendtn.;毛金蛹茄■☆
45165 Aureolaria Raf.(1837);类毛地黄属;Oakleeeh ■☆
45166 Aureolaria auriculata (Michx.) Farw. = Agalinis auriculata (Michx.) S. F. Blake ■☆
45167 Aureolaria flava (L.) Farw.;光滑类毛地黄(光滑假毛地黄);Smooth False Foxglove, Smooth Yellow False Foxglove ■☆

45168 Aureolaria gerardia ?;杰勒德类毛地黄(杰勒德假毛地黄);False Foxglove ■☆

45169 Aureolaria grandiflora (Benth.) Pennell;大花类毛地黄(大花假毛地黄);Big-flowered Gerardia, Large-flowered Yellow False Foxglove, Western False Foxglove ■☆

45170 Aureolaria grandiflora (Benth.) Pennell subsp. pulchra (Pennell) Pennell = Aureolaria grandiflora (Benth.) Pennell var. pulchra Pennell ■☆

45171 Aureolaria grandiflora (Benth.) Pennell var. pulchra Pennell;美丽大花类毛地黄(美丽大花假毛地黄);Large-flowered Yellow False Foxglove ■☆

45172 Aureolaria grandiflora Pennell = Aureolaria grandiflora (Benth.) Pennell ■☆

45173 Aureolaria levigata Raf.;平滑类毛地黄(平滑假毛地黄);Smooth False Foxglove ■☆

45174 Aureolaria pectinata (Nutt.) Pennell;篦齿类毛地黄(篦齿假毛地黄);False Foxglove ■☆

45175 Aureolaria pedicularia (L.) Raf.;一年生类毛地黄(蕨叶假毛地黄,蕨叶杰勒草,一年生假毛地黄);Annual False Foxglove, Clammy False Foxglove, Downy False Foxglove, Fern-leaf Yellow False Foxglove, Fern-leaved False Foxglove, Smooth False Foxglove ■☆

45176 Aureolaria pedicularia (L.) Raf. subsp. ambigens (Fernald) Pennell = Aureolaria pedicularia (L.) Raf. var. ambigens (Fernald) Farw. ■☆

45177 Aureolaria pedicularia (L.) Raf. subsp. caesariensis Pennell = Aureolaria pedicularia (L.) Raf. ■☆

45178 Aureolaria pedicularia (L.) Raf. subsp. carolinensis Pennell = Aureolaria pedicularia (L.) Raf. ■☆

45179 Aureolaria pedicularia (L.) Raf. subsp. intercedens (Pennell) Pennell = Aureolaria pedicularia (L.) Raf. var. intercedens Pennell ■☆

45180 Aureolaria pedicularia (L.) Raf. subsp. typica Pennell = Aureolaria pedicularia (L.) Raf. ■☆

45181 Aureolaria pedicularia (L.) Raf. var. ambigens (Fernald) Farw.;蕨叶一年生类毛地黄(蕨叶一年生假毛地黄);Annual False Foxglove, Clammy False Foxglove, Fern-leaf Yellow False Foxglove ■☆

45182 Aureolaria pedicularia (L.) Raf. var. caesariensis (Pennell) Pennell = Aureolaria pedicularia (L.) Raf. ■☆

45183 Aureolaria pedicularia (L.) Raf. var. carolinensis (Pennell) Pennell = Aureolaria pedicularia (L.) Raf. ■☆

45184 Aureolaria pedicularia (L.) Raf. var. intercedens Pennell;内向一年生类毛地黄(内向一年生假毛地黄);Annual False Foxglove, Clammy False Foxglove, Fern-leaf Yellow False Foxglove ■☆

45185 Aureolaria pedicularia (L.) Raf. var. typica (Pennell) Deam = Aureolaria pedicularia (L.) Raf. ■☆

45186 Auricula Hill = Primula L. ■

45187 Auricula Tourn. ex Spach = Primula L. ■

45188 Auriculardisia Lundell = Ardisia Sw. (保留属名)●■

45189 Auricula-ursi Ség. = Auricula Hill ■

45190 Aurinia Desv. (1815);奥林荠属(金庭荠属)■☆

45191 Aurinia petraeum Fuss;奥林荠;Goldentuft ■☆

45192 Aurinia saxatilis (L.) Desv.;金庭荠(岩生庭荠);Alison, Basket of Gold, Basket-of-gold, Egg-in-the-pan, Gold Basket, Gold Dust, Gold-dust, Golden Alison, Golden Dust, Golden Tuft, Goldentuft, Golden-tuft, Goldentuft Alyssum, Golden-tuft Alyssum, Madwort, Rock Alyssum, Rock Madwort, Yellow Alyssum ■

45193 Aurinia saxatilis (L.) Desv. 'Citrina';柠檬金庭荠■☆

45194 Aurinia saxatilis (L.) Desv. 'Dudley Nevill';达德利·内维尔金庭荠■☆

45195 Aurinia saxatilis (L.) Desv. 'Variegata';镶边金庭荠■☆

45196 Aurinia saxatilis (L.) Desv. = Alyssum saxatile L. ■

45197 Aurinocidium Romowicz et Szlach. = Oncidium Sw. (保留属名)■☆

45198 Aurora Noronha = Quisqualis L. ●

45199 Aurota Raf. = Curculigo Gaertn. ■

45200 Austerium Poir. ex DC. = Rhynchosia Lour. (保留属名)●■

45201 Australina Gaudich. (1830);澳洲单蕊麻属■☆

45202 Australina acuminata Wedd. = Didymodoxa caffra (Thunb.) Friis et Wilmot-Dear ■☆

45203 Australina caffra (Thunb.) Prain = Didymodoxa caffra (Thunb.) Friis et Wilmot-Dear ■☆

45204 Australina capensis Wedd. = Didymodoxa capensis (L. f.) Friis et Wilmot-Dear var. integrifolia (Wedd.) Friis et Wilmot-Dear ■☆

45205 Australina flaccida (A. Rich.) Wedd.;澳洲单蕊麻■☆

45206 Australina integrifolia Wedd. = Didymodoxa capensis (L. f.) Friis et Wilmot-Dear var. integrifolia (Wedd.) Friis et Wilmot-Dear ■☆

45207 Australina procumbens N. E. Br. = Didymodoxa capensis (L. f.) Friis et Wilmot-Dear ■☆

45208 Australina schimperiana Wedd. = Australina flaccida (A. Rich.) Wedd. ■☆

45209 Australina thunbergii N. E. Br. = Didymodoxa capensis (L. f.) Friis et Wilmot-Dear ■☆

45210 Australluma Plowes = Caralluma R. Br. ■

45211 Australluma Plowes(1995);澳非水牛角属■☆

45212 Australluma peschii (Nel) Plowes;佩施单蕊麻■☆

45213 Australluma ubomboensis (I. Verd.) Bruyns;乌邦博单蕊麻■☆

45214 Australopyrum (Tzvelev) Á. Löve(1984);澳麦草属(澳大利亚冰草属)■☆

45215 Australopyrum Á. Löve = Agropyron Gaertn. ■

45216 Australorchis Brieger = Dendrobium Sw. (保留属名)■

45217 Austroamericium Hendrych = Thesium L. ■

45218 Austrobaileya C. T. White(1933);木兰藤属(对叶藤属,昆士兰樟属)●☆

45219 Austrobaileya maculata C. T. White;斑点木兰藤●☆

45220 Austrobaileya scandens C. T. White;木兰藤●☆

45221 Austrobaileyaceae Croizat(1943)(保留科名);木兰藤科(对叶藤科,昆士兰樟科);匍匐灌木●☆

45222 Austrobassia Ulbr. (1934);澳洲雾冰藜属■☆

45223 Austrobassia Ulbr. = Sclerolaena R. Br. ●☆

45224 Austrobassia costata (R. H. Anderson) Ulbr.;澳洲雾冰藜■☆

45225 Austrobrickellia R. M. King et H. Rob. (1972);南美肋泽兰属●☆

45226 Austrobrickellia arnottii (Baker) R. M. King et H. Rob.;南美肋泽兰●☆

45227 Austrobryonia H. Schaef. (1855);澳泻根属■☆

45228 Austrobryonia H. Schaef. = Cucurbita L. ■

45229 Austrobuxus Miq. (1861);黄杨大戟属●☆

45230 Austrobuxus nitidus Miq.;黄杨大戟●☆

45231 Austrocactus Britton et Rose(1922);狼爪玉属(狼爪球属,狼爪属)●☆

45232 Austrocactus bertinii (Cels ex Hérincq) Britton et Rose;狼爪玉■☆

45233 Austrocactus coxii (K. Schum.) Backeb.;豹爪玉■☆

45234 Austrocactus duseni Speg. = Austrocactus bertinii (Cels ex Hérincq) Britton et Rose ■☆

45235 Austrocactus gracilis Backeb. = Austrocactus bertinii (Cels ex Hérincq) Britton et Rose ■☆
45236 Austrocactus patagonicus (F. A. C. Weber ex Speg.) Hosseus;熊爪玉■☆
45237 Austrocedrus Florin et Boutelje = Libocedrus Endl. ●☆
45238 Austrocedrus Florin et Boutelje(1954);智利柏属(南美柏属);Chilean Cedar,Chilean Incense Cedar ●☆
45239 Austrocedrus chilensis (D. Don) Florin et Boutelje;智利柏(智利香松);Alerce,Chilean Cedar,Chilean Incense Cedar,Chilean Incense-cedar ●☆
45240 Austrocedrus chilensis (D. Don) Florin et Boutelje = Libocedrus chilensis (D. Don) Endl. ●☆
45241 Austrocephalocereus (Backeb.) Backeb. (1938);白丽翁属(南美翁柱属)●☆
45242 Austrocephalocereus (Backeb.) Backeb. = Espostoopsis Buxb. ●☆
45243 Austrocephalocereus Backeb. = Micranthocereus Backeb. ●☆
45244 Austrocephalocereus Backeb., Buxb. et Buining = Micranthocereus Backeb. ●☆
45245 Austrocephalocereus dybowskii (Rol.-Goss.) Backeb.;白丽翁●☆
45246 Austrocephalocereus dybowskii (Rol.-Goss.) Backeb. = Espostoopsis dybowskii (Rol.-Goss.) Buxb. ●☆
45247 Austrocephalocereus purouruus (Gurke) Backeb.;紫红白丽翁●☆
45248 Austrochloris Lazarides(1972);澳洲禾属(澳洲虎尾草属)■☆
45249 Austrochloris dichanthioides (Everist) Lazarides;澳洲禾■☆
45250 Austrocritonia R. M. King et H. Rob. (1975);巴西亮泽兰属●☆
45251 Austrocritonia angulicaulis (Sch. Bip. ex Baker) R. M. King et H. Rob.;窄叶巴西亮泽兰●☆
45252 Austrocritonia rosea (Gardner) R. M. King et H. Rob.;粉红巴西亮泽兰●☆
45253 Austrocritonia velutina (Gardner) R. M. King et H. Rob.;巴西亮泽兰●☆
45254 Austrocylindropuntia Backeb. (1938);南方圆筒仙人掌属■☆
45255 Austrocylindropuntia Backeb. = Opuntia Mill. ●
45256 Austrocylindropuntia albiflora (K. Schum.) Backeb. = Austrocylindropuntia salmiana (Parm. ex Pfeiff.) Backeb. ●■☆
45257 Austrocylindropuntia cylindrica (Lam.) Backeb. = Opuntia cylindrica (Lam.) DC. ■☆
45258 Austrocylindropuntia exaltata (A. Berger) Backeb.;极高南方圆筒仙人掌■☆
45259 Austrocylindropuntia ipatiana (Cárdenas) Backeb. = Austrocylindropuntia salmiana (Parm. ex Pfeiff.) Backeb. ■☆
45260 Austrocylindropuntia salmiana (Parm. ex Pfeiff.) Backeb.;南方圆筒仙人掌(圆筒仙人掌)■☆
45261 Austrocylindropuntia salmiana (Parm. ex Sweet) Backeb. = Austrocylindropuntia salmiana (Parm. ex Pfeiff.) Backeb. ■☆
45262 Austrocylindropuntia salmiana (Parm. ex Sweet) Backeb. var. albiflora (K. Schum.) Backeb. = Austrocylindropuntia salmiana (Parm. ex Sweet) Backeb. ■☆
45263 Austrocylindropuntia subulata (Engelm.) Backeb. = Opuntia subulata Engelm. ■☆
45264 Austrocylindropuntia subulata (Muehlenpf.) Backeb. = Opuntia subulata Engelm. ■☆
45265 Austrocylindropuntia vestita (Salm-Dyck) Backeb.;翁团扇■☆
45266 Austrocynoglossum Popov ex R. R. Mill(1989);南琉璃草属■☆
45267 Austrocynoglossum latifolium (R. Br.) R. R. Mill;南琉璃草■☆
45268 Austrodanthonia H. P. Linder(1997);澳洲扁芒草属■☆
45269 Austrodolichos Verdc. (1970);澳扁豆属■☆
45270 Austrodolichos errabundus (M. B. Scott) Verdc.;澳扁豆■☆
45271 Austrodolichos errabundus Verdc. = Austrodolichos errabundus (M. B. Scott) Verdc. ■☆
45272 Austrodrimys Doweld(2000);澳洲辛酸木属●☆
45273 Austroeupatorium R. M. King et H. Rob. (1970);南泽兰属●■
45274 Austroeupatorium albescens (Gardner) H. Rob.;白南泽兰●☆
45275 Austroeupatorium decemflorum (DC.) R. M. King et H. Rob.;十花南泽兰●☆
45276 Austroeupatorium inulifolium (Kunth) R. M. King et H. Rob.;南泽兰●■
45277 Austrofestuca (Tzvelev) E. B. Alexeev(1976);澳羊茅属■☆
45278 Austrofestuca hookeriana (F. Muell.) S. W. L. Jacobs;虎克澳羊茅■☆
45279 Austrofestuca triticoides (Trin.) E. B. Alexeev;澳羊茅■☆
45280 Austrogambeya Aubrév. et Pellegr. = Chrysophyllum L. ●
45281 Austrogambeya bangweolensis (R. E. Fr.) Aubrév. et Pellegr. = Chrysophyllum bangweolense R. E. Fr. ●☆
45282 Austroliabum H. Rob. et Brettell = Microliabum Cabrera ●■☆
45283 Austroliabum H. Rob. et Brettell(1974);南黄安菊属■☆
45284 Austroliabum candidum (Griseb.) H. Rob. et Brettell;南黄安菊■☆
45285 Austromatthaea L. S. Sm. (1969);南黑檫木属(南圣马太属)●☆
45286 Austromimusops A. Meeuse = Vitellariopsis Baill. ex Dubard ●☆
45287 Austromimusops cuneata (Engl.) A. Meeuse = Vitellariopsis cuneata (Engl.) Aubrév. ●☆
45288 Austromimusops dispar (N. E. Br.) A. Meeuse = Vitellariopsis dispar (N. E. Br.) Aubrév. ●☆
45289 Austromimusops marginata (N. E. Br.) A. Meeuse = Vitellariopsis marginata (N. E. Br.) Aubrév. ●☆
45290 Austromimusops sylvestris (S. Moore) A. Meeuse = Vitellariopsis marginata (N. E. Br.) Aubrév. ●☆
45291 Austromuellera C. T. White(1930);矛果豆山龙眼属●☆
45292 Austromuellera trinervia C. T. White;矛果豆山龙眼●☆
45293 Austromyrtus (Nied.) Burret(1941);南香桃木属●☆
45294 Austromyrtus acutiuscula Burret;南香桃木●☆
45295 Austromyrtus floribunda (A. J. Scott) Guymer;繁花南香桃木●☆
45296 Austromyrtus glabra N. Snow et Guymer;无毛南香桃木●☆
45297 Austromyrtus lucida L. S. Sm.;光亮南香桃木●☆
45298 Austromyrtus luteo-viridis (Baker f.) Burret;黄绿南香桃木●☆
45299 Austromyrtus minutiflora Burret;小花南香桃木●☆
45300 Austromyrtus nigripes (Guillaumin) Burret;黑梗南香桃木●☆
45301 Austromyrtus tenuifolia (Sm.) Burret;细叶南香桃木;Narrow-leaf Myrtle ●☆
45302 Austropeucedanum Mathias et Constance(1952);澳前胡属■☆
45303 Austropeucedanum oreopansil (Griseb.) Mathias et Constance;澳前胡■☆
45304 Austroplenckia Lundell = Plenckia Reissek(保留属名)●☆
45305 Austrosteenisia R. Geesink(1984);澳矛果豆属■☆
45306 Austrosteenisia blackii (F. Muell.) Geesink;布拉克澳矛果豆■☆
45307 Austrosteenisia glabristyla Jessup;光柱澳矛果豆■☆
45308 Austrosteenisia stipularis (C. T. White) Jessup;澳矛果豆■☆
45309 Austrostipa S. W. L. Jacobs et J. Everett(1996);澳针茅属■☆
45310 Austrostipa mollis (R. Br.) S. W. L. Jacobs et J. Everett;澳针茅■☆
45311 Austrosynotis C. Jeffrey(1986);非洲合耳菊属■☆
45312 Austrosynotis rectirama (Baker) C. Jeffrey;非洲合耳菊■☆
45313 Austrotaxaceae Nakai = Austrotaxaceae Nakai ex Takht. et Reveal ●☆

45314 Austrotaxaceae Nakai = Taxaceae Gray(保留科名)●
45315 Austrotaxaceae Nakai ex Takht. et Reveal = Taxaceae Gray(保留科名)●
45316 Austrotaxaceae Nakai ex Takht. et Reveal(1938);澳洲红豆杉科●☆
45317 Austrotaxus Compton(1922);澳洲红豆杉属;Australianyew, Austrotaxus ●
45318 Austrotaxus spicata Compton;澳洲红豆杉;Australianyew, Spike Austrotaxus ●
45319 Auticoryne Turcz. = Baeckea L. ●
45320 Autogenes Raf. = Narcissus L. ■
45321 Autonoe (Webb et Berthel.) Speta = Scilla L. ■
45322 Autonoe (Webb et Berthel.) Speta(1998);光梗风信子属●☆
45323 Autonoe dasyantha (Webb et Berthel.) Speta;毛花光梗风信子●☆
45324 Autonoe iridifolia (Webb et Berthel.) Speta;鸢尾叶光梗风信子●☆
45325 Autonoe latifolia (Willd.) Speta;宽叶光梗风信子●☆
45326 Autonoe madeirensis (Menezes) Speta;马德拉光梗风信子●☆
45327 Autrandra Pierre ex Prain = Athroandra (Hook. f.) Pax et K. Hoffm. ●☆
45328 Autrandra Pierre ex Prain = Erythrococca Benth. ●☆
45329 Autranea C. Winkl. et Barbey;叙利亚菊属■☆
45330 Autranea C. Winkl. et Barbey = Centaurea L. (保留属名)●■
45331 Autranella A. Chev. (1917);奥特山榄属●☆
45332 Autranella A. Chev. et Aubrév. = Autranella A. Chev. ●☆
45333 Autranella boonei (De Wild.) A. Chev. = Autranella congolensis (De Wild.) A. Chev. ●☆
45334 Autranella congolensis (De Wild.) A. Chev.;奥特山榄(肉穗果)●☆
45335 Autranella congolensis A. Chev. = Autranella congolensis (De Wild.) A. Chev. ●☆
45336 Autranella letestui (Lecomte) A. Chev. = Autranella congolensis (De Wild.) A. Chev. ●☆
45337 Autrania C. Willis = Autranea C. Winkl. et Barbey ■☆
45338 Autrania C. Winkl. et Barbey = Autranea C. Winkl. et Barbey ■☆
45339 Autrania C. Winkl. et Barbey = Jurinea Cass. ●■
45340 Autrania pulchella C. Winkl. et Barbey;奥特拉菊■☆
45341 Autumnalia Pimenov(1989);秋芹属■☆
45342 Autumnalia botschantzevii Pimenov;秋芹■☆
45343 Autunesia Dyer = Antunesia O. Hoffm. ●☆
45344 Autunesia O. Hoffm. = Newtonia Baill. ●☆
45345 Auxemma Miers(1875);巴西紫草属●☆
45346 Auxemma gardneriana Miers;巴西紫草●☆
45347 Auxopus Schltr. (1900);大足兰属■☆
45348 Auxopus kamerunensis Schltr.;大足兰■☆
45349 Auxopus macranthus Summerh.;大花大足兰■☆
45350 Auxopus madagascariensis Schltr.;马岛大足兰■☆
45351 Auzuba Juss. = Sideroxylon L. ●☆
45352 Aveledoa Pittier = Metteniusa H. Karst. ●☆
45353 Avellanita Phil. (1864);榛色大戟属☆
45354 Avellanita bustillosi Phil.;榛色大戟☆
45355 Avellara Blanca et C. Diaz = Scorzonera L. ■
45356 Avellinia Parl. = Colobanthium Rchb. ■☆
45357 Avellinia Parl. = Trisetaria Forssk. ■☆
45358 Avellinia michelii (Savi) Parl. = Rostraria festucoides (Link) Romero Zarco ■☆
45359 Avellinia michelii (Savi) Parl. var. brevipila Maire = Rostraria festucoides (Link) Romero Zarco ■☆
45360 Avellinia michelii (Savi) Parl. var. warionis (Sennen et Mauricio) Maire = Rostraria festucoides (Link) Romero Zarco ■☆
45361 Avellinia warionis Sennen et Mauricio = Rostraria festucoides (Link) Romero Zarco ■☆
45362 Avena Haller ex Scop. = Agrostis L. (保留属名)■
45363 Avena L. (1753);燕麦属(乌麦属);Avena Grass, Oat, Oat Grass, Oats ■
45364 Avena Scop. = Lagurus L. ■☆
45365 Avena abietetorum Ohwi = Helictotrichon abietorum (Ohwi) Ohwi ■
45366 Avena abyssinica Hochst;阿比西尼亚燕麦(衣索匹亚燕麦);Abyssinian Oat ■☆
45367 Avena abyssinica Hochst. var. granulata Chiov. = Avena sterilis L. subsp. ludoviciana (Durieu) Nyman ■
45368 Avena aenea Hook. f. = Trisetum aeneum (Hook. f.) R. R. Stewart ■☆
45369 Avena alba Vahl = Arrhenatherum album (Vahl) Clayton ■☆
45370 Avena alba Vahl subsp. abyssinica (Hochst.) Á. Löve et D. Löve = Avena abyssinica Hochst. ■☆
45371 Avena alba Vahl var. barbata (Pott ex Link) Maire et Weiller = Avena barbata Pott ex Link ■
45372 Avena alba Vahl var. wiestii (Steud.) Maire et Weiller = Avena barbata Pott ex Link subsp. wiestii (Steud.) Mansf. ■☆
45373 Avena albinervis Boiss. = Helictotrichon albinerve (Boiss.) Henrard ■☆
45374 Avena algeriensis Trab. = Avena byzantina K. Koch ■☆
45375 Avena altaica Steph. ex Roshev. = Trisetum altaicum (Stephan) Roshev. ■
45376 Avena altior Hitchc. = Helictotrichon altius (Hitchc.) Ohwi ■
45377 Avena andropogonoides Steud. = Triraphis andropogonoides (Steud.) E. Phillips ■☆
45378 Avena antartica Thunb. = Helictotrichon leoninum (Steud.) Schweick. ■☆
45379 Avena argaea Boiss. = Helictotrichon pratense (L.) Pilg. ■☆
45380 Avena aristidoides Thunb. = Pentaschistis aristidoides (Thunb.) Stapf ■☆
45381 Avena arvensis (L.) Salisb. = Bromus arvensis L. ■
45382 Avena aspera Munro ex Thwaites = Helictotrichon virescens (Nees ex Steud.) Henrard ■
45383 Avena aspera Munro ex Thwaites var. parviflora Hook. f. = Helictotrichon parviflorum (Hook. f.) Bor ■
45384 Avena aspera Munro ex Thwaites var. parviflora Hook. f. = Helictotrichon virescens (Nees ex Steud.) Henrard ■
45385 Avena aspera Munro ex Thwaites var. roylei Hook. f. = Helictotrichon virescens (Nees ex Steud.) Henrard ■
45386 Avena aspera Munro ex Thwaites var. roylei Hook. f. = Helictotrichon junghuhnii (Büse) Henrard. ■
45387 Avena aspera Munro ex Thwaites var. schmidii Hook. f. = Helictotrichon schmidii (Hook. f.) Henrard ■
45388 Avena atlantica B. R. Baum et Fedak;大西洋燕麦■☆
45389 Avena barbata Pott ex Link;裂稃燕麦(髯毛燕麦);Bearded Oat, Slender Oat ■
45390 Avena barbata Pott ex Link subsp. castellana Romero Zarco;卡地裂稃燕麦■☆

45391 Avena barbata Pott ex Link subsp. hirtula (Lag.) Tab. Morais = Avena barbata Pott ex Link ■

45392 Avena barbata Pott ex Link subsp. lusitanica (Tab. Morais) Romero Zarco = Avena lusitanica (Tab. Morais) B. R. Baum ■☆

45393 Avena barbata Pott ex Link subsp. wiestii (Steud.) Mansf.;维斯特燕麦■☆

45394 Avena barbata Pott ex Link var. hoppeana (Scheele) Richt. = Avena barbata Pott ex Link ■

45395 Avena barbata Pott ex Link var. minor Lange = Avena barbata Pott ex Link ■

45396 Avena barbata Pott ex Link var. sallentiana Pau = Avena barbata Pott ex Link ■

45397 Avena barbata Pott ex Link var. sublongesubulata Sennen et Mauricio = Avena barbata Pott ex Link ■

45398 Avena beguinotiana Pamp. = Avena ventricosa Coss. ■☆

45399 Avena bifida (Thunb.) P. Beauv. = Trisetum bifidum (Thunb.) Ohwi ■

45400 Avena bifida P. Beauv. = Trisetum bifidum (Thunb.) Ohwi ■

45401 Avena breviaristata Barratte = Helictotrichon breviaristatum (Barratte) Henrard ■☆

45402 Avena brevis Roth;短燕麦;Little Oat,Short Oat ■☆

45403 Avena bromoides Gouan = Helictotrichon bromoides (Gouan) C. E. Hubb. ■☆

45404 Avena bromoides Gouan subsp. australis (Parl.) Trab. = Helictotrichon cincinnatum (Ten.) Röser ■☆

45405 Avena bromoides Gouan subsp. bromoides (Gouan) Trab. = Helictotrichon bromoides (Gouan) C. E. Hubb. ■☆

45406 Avena bromoides Gouan subsp. gouanii St. -Yves = Helictotrichon gervaisii (Holub) Röser ■☆

45407 Avena bromoides Gouan subsp. letourneuxii Trab. = Helictotrichon bromoides (Gouan) C. E. Hubb. ■☆

45408 Avena bromoides Gouan subsp. pruinosa (Hack.) Trab. = Helictotrichon gervaisii (Holub) Röser subsp. arundanum (Romero Zarco) Röser ■☆

45409 Avena bromoides Gouan subsp. requienii Trab. = Helictotrichon gervaisii (Holub) Röser ■☆

45410 Avena bromoides Gouan var. dubia Maire = Helictotrichon bromoides (Gouan) C. E. Hubb. ■☆

45411 Avena bromoides Gouan var. hirsuta Trab. = Helictotrichon bromoides (Gouan) C. E. Hubb. ■☆

45412 Avena bromoides Gouan var. letourneuxii (Trab.) St. -Yves = Helictotrichon cincinnatum (Ten.) Röser ■☆

45413 Avena bromoides Gouan var. oranensis Trab. = Helictotrichon gervaisii (Holub) Röser ■☆

45414 Avena bromoides Gouan var. parlatorei St. -Yves = Helictotrichon bromoides (Gouan) C. E. Hubb. ■☆

45415 Avena bromoides Gouan var. pruinosa (Hack. et Trab.) St. -Yves = Helictotrichon gervaisii (Holub) Röser subsp. arundanum (Romero Zarco) Röser ■☆

45416 Avena bromoides Gouan var. pruinosa (Hack. et Trab.) Trab. = Helictotrichon gervaisii (Holub) Röser subsp. arundanum (Romero Zarco) Röser ■☆

45417 Avena bruhnsiana Gruner;布氏燕麦■☆

45418 Avena bulbosa Willd. = Arrhenatherum elatium (L.) J. Presl et C. Presl subsp. bulbosum (Willd.) Schübl. et G. Martens ■

45419 Avena bulbosa Willd. = Arrhenatherum elatium (L.) P. Beauv. ex J. Presl et C. Presl var. bulbosum (Willd.) Spenn. 'Variegatum' ■

45420 Avena bulbosa Willd. = Arrhenatherum elatium (L.) P. Beauv. ex J. Presl et C. Presl var. bulbosum (Willd.) Spenn. ■

45421 Avena byzantina C. Koch = Avena byzantina K. Koch ■☆

45422 Avena byzantina C. Koch var. pseudovilis (Hausskn.) Maire et Weiller = Avena byzantina K. Koch ■☆

45423 Avena byzantina C. Koch var. solida (Hausskn.) Maire et Weiller = Avena byzantina K. Koch ■☆

45424 Avena byzantina K. Koch;比赞燕麦;Algerian Oat,Byzant Oat,Red Oat ■☆

45425 Avena byzantina K. Koch = Avena sativa L. ■

45426 Avena caffra Stapf = Helictotrichon longifolium (Nees) Schweick. ■☆

45427 Avena callosa Turcz. ex Griseb. = Schizachne callosa (Turcz. ex Griseb.) Ohwi ■

45428 Avena canariensis B. R. Baum,Rajhathy et Sampson;加那利燕麦■☆

45429 Avena candida ? = Helictotrichon sempervirens (Vill.) Pilg. ■☆

45430 Avena cavanillesii (Trin.) Koch = Trisetaria loeflingiana (L.) Paunero ■☆

45431 Avena chinensis (Fisch. ex Roem. et Schult.) Metzg.;莜麦(华莜麦,尤麦,油麦);China Oat,Chinese Oat ■

45432 Avena clarkei Hook. f. = Trisetum clarkei (Hook. f.) R. R. Stewart ■

45433 Avena clauda Durieu;闭燕麦■☆

45434 Avena clauda Durieu var. eriantha Coss. = Avena clauda Durieu ■☆

45435 Avena clauda Durieu var. leiantha Malzev = Avena clauda Durieu ■☆

45436 Avena colorata Steud. = Pentaschistis colorata (Steud.) Stapf ■☆

45437 Avena cultiformis (Malzev) Malzev et al.;类燕麦;Cultiform Oat ■☆

45438 Avena delavayi Hack. = Helictotrichon delavayi (Hack.) Henrard ■

45439 Avena desertorum Less.;沙燕麦■☆

45440 Avena desertorum Podp. = Avena desertorum Less. ■☆

45441 Avena donax L.;芦苇状燕麦■☆

45442 Avena dubia Leers = Ventenata dubia (Leers) Coss. et Durieu ■☆

45443 Avena elatior L. = Arrhenatherum elatium (L.) P. Beauv. ex J. Presl et C. Presl ■

45444 Avena elatior L. subsp. eriantha (Boiss. et Reut.) Litard. = Arrhenatherum album (Vahl) Clayton ■☆

45445 Avena elatius L. var. bulbosa ? = Arrhenatherum elatium (L.) P. Beauv. ex J. Presl et C. Presl var. bulbosum (Willd.) Spenn. ■

45446 Avena eriantha Durieu;异颖燕麦;Woollyflower Oat,Woolly-flowered Oat ■

45447 Avena eriantha Durieu var. acuminata Coss. = Avena eriantha Durieu ■

45448 Avena fatua L.;野燕麦(蒿,青稞麦,雀麦,乌麦,燕麦,燕麦草,野麦草,野麦子);Bearded Oat-grass,Havers,Pillus,Poor Oat,Poor Oats,Sowlers,Tartarian Oats,Wild Oat,Wild Oats,Wild-oat ■

45449 Avena fatua L. f. glaberrima Thell. = Avena sativa L. var. glaberrima (Thell.) Maire et Weiller ■

45450 Avena fatua L. subsp. meridionalis Malzev;南方野燕麦■☆

45451 Avena fatua L. subsp. meridionalis Malzev = Avena fatua L. ■

45452 Avena fatua L. subsp. meridionalis Malzev = Avena meridionalis (Malzev) Roshev. ■

45453 Avena fatua L. subsp. sativa (L.) Thell. = Avena sativa L. ■

45454 Avena fatua L. var. glabrata Peterm.;光稃野燕麦(光野燕麦);

Glabrous Wild Oat ■
45455 Avena fatua L. var. glabrata Peterm. = Avena fatua L. ■
45456 Avena fatua L. var. glabrescens Coss. = Avena byzantina K. Koch ■☆
45457 Avena fatua L. var. mollis Keng;光轴野燕麦;Soft Wild Oat ■
45458 Avena fatua L. var. mollis Keng = Avena fatua L. var. glabrata Peterm. ■
45459 Avena fatua L. var. pilosissima Gray = Avena fatua L. ■
45460 Avena fatua L. var. sativa (L.) Hausskn. = Avena sativa L. ■
45461 Avena fatua L. var. vilis (Wallr.) Hausskn. = Avena fatua L. ■
45462 Avena festuciformis Hochst. = Helictotrichon elongatum (Hochst. ex A. Rich.) C. E. Hubb. ■☆
45463 Avena filifolia Lag. = Helictotrichon filifolium (Lag.) Henrard ■☆
45464 Avena filifolia Lag. var. glabra Boiss. = Helictotrichon filifolium (Lag.) Henrard ■☆
45465 Avena filifolia Lag. var. lagascae St. -Yves = Helictotrichon filifolium (Lag.) Henrard ■☆
45466 Avena flaniculmis Schreb.;扁秆燕麦;Flat-culm Oat ■
45467 Avena flavescens Hook. f. = Trisetum scitulum Bor ■
45468 Avena flavescens Hook. f. var. virescens Regel = Trisetum spicatum (L.) K. Richt. subsp. virescens (Regel) Tzvelev ■
45469 Avena flavescens L. = Trisetum flavescens (L.) P. Beauv. ■☆
45470 Avena flavescens L. = Trisetum pratense Pers. ■☆
45471 Avena flavescens P. Beauv. = Avena flavescens L. ■☆
45472 Avena flxuosa (L.) Mert. et Koch = Deschampsia flexuosa (L.) Trin. ■
45473 Avena forskalii Vahl = Asthenatherum forskalii (Vahl) Nevski ■☆
45474 Avena forskalii Vahl = Centropodia forskalii (Vahl) Cope ■☆
45475 Avena fragilis L. = Gaudinia fragilis (L.) P. Beauv. ■☆
45476 Avena hirsuta Moench. = Avena barbata Pott ex Link ■
45477 Avena hirta Schrad. = Helictotrichon hirtulum (Steud.) Schweick. ■☆
45478 Avena hirtula Lag. = Avena barbata Pott ex Link subsp. hirtula (Lag.) Tab. Morais ■☆
45479 Avena hispida L. f. = Tristachya leucothrix Trin. ex Nees ■☆
45480 Avena hookeri Scribn. = Helictotrichon hookeri (Scribn.) Henrard ■
45481 Avena hybrida Peterm.;杂种燕麦;Hybrid Oat ■☆
45482 Avena hybrida Peterm. ex Rchb. = Avena fatua L. subsp. meridionalis Malzev ■☆
45483 Avena involucrata Schrad. = Chaetobromus involucratus (Schrad.) Nees ■☆
45484 Avena jahandiezii Litard. = Helictotrichon jahandiezii (Litard.) Potztal ■☆
45485 Avena junghuhnii Büse = Helictotrichon junghuhnii (Büse) Henrard. ■
45486 Avena lachnantha (Hochst. ex A. Rich.) Hook. f. = Helictotrichon lachnanthum (Hochst. ex A. Rich.) C. E. Hubb. ■☆
45487 Avena laevis Hack. = Helictotrichon laeve (Hack.) Potztal ■☆
45488 Avena lanata (L.) Koeler = Holcus lanatus L. ■
45489 Avena lanata Schrad. = Merxmuellera rufa (Nees) Conert ■☆
45490 Avena leiantha Keng = Helictotrichon leianthum (Keng) Ohwi ■
45491 Avena leonina Steud. = Helictotrichon leoninum (Steud.) Schweick. ■☆
45492 Avena letourneuxii Trab. = Helictotrichon cincinnatum (Ten.) Röser ■☆
45493 Avena loeflingiana L. = Trisetaria loeflingiana (L.) Paunero ■☆
45494 Avena longa Stapf = Helictotrichon longum (Stapf) Schweick. ■☆
45495 Avena longiglumis Durieu;长颖燕麦(长颖状燕麦,陆氏燕麦,绿多燕麦,乌麦);Longglume Oat,Ludovician Oat ■
45496 Avena longiglumis Durieu var. tripolitana Maire et Weiller = Avena longiglumis Durieu ■
45497 Avena ludoviciana Durieu = Avena sterilis L. subsp. ludoviciana (Durieu) Gillet et Magne ■☆
45498 Avena ludoviciana Durieu = Avena sterilis L. ■☆
45499 Avena ludoviciana Durieu = Avena trichophylla C. Koch ■☆
45500 Avena ludoviciana Durieu subsp. glabrescens Thell. = Avena sativa L. var. glaberrima (Thell.) Maire et Weiller ■
45501 Avena lupulina Thunb. = Merxmuellera lupulina (Thunb.) Conert ■☆
45502 Avena lusitanica (Tab. Morais) B. R. Baum;葡萄牙燕麦■☆
45503 Avena macrantha (Hack.) Nevski;大花燕麦■☆
45504 Avena macrocalycina Steud. = Pentameris macrocalycina (Steud.) Schweick. ■☆
45505 Avena macrocalyx Sennen;大萼燕麦■☆
45506 Avena macrocarpa Moench = Avena sterilis L. ■☆
45507 Avena macrostachya Coss. et Durieu;大穗燕麦■☆
45508 Avena magna Murphy et Terrell;大燕麦■☆
45509 Avena maroccana Gand. = Avena sterilis L. ■☆
45510 Avena matritensis B. R. Baum;马德里燕麦■☆
45511 Avena maxima C. Presl;最大燕麦■☆
45512 Avena melillensis Sennen = Avena sterilis L. subsp. ludoviciana (Durieu) Gillet et Magne ■☆
45513 Avena meridionalis (Malzev) Roshev.;南燕麦;Southern Oat ■
45514 Avena meridionalis (Malzev) Roshev. = Avena fatua L. subsp. meridionalis Malzev ■☆
45515 Avena meridionalis (Malzev) Roshev. = Avena fatua L. ■
45516 Avena mollis (L.) Salisb. = Bromus hordeaceus L. ■
45517 Avena mongolica Roshev. = Helictotrichon mongolicum (Roshev.) Henrard ■
45518 Avena montana Brot. = Pseudarrhenatherum longifolium (Thore) Rouy ■☆
45519 Avena montana Vill. = Helictotrichon sedenense (DC.) Holub ■☆
45520 Avena montana Vill. var. teretifolia Willk. = Helictotrichon sedenense (DC.) Holub ■☆
45521 Avena muriculata Stapf = Helictotrichon elongatum (Hochst. ex A. Rich.) C. E. Hubb. ■☆
45522 Avena neesii (Steud.) Hook. f. = Helictotrichon elongatum (Hochst. ex A. Rich.) C. E. Hubb. ■☆
45523 Avena nervosa R. Br.;脉燕麦■
45524 Avena newtonii Stapf = Helictotrichon newtonii (Stapf) C. E. Hubb. ■☆
45525 Avena nitida Desf. = Trisetaria nitida (Desf.) Maire ■☆
45526 Avena nodipilosa Malzev;毛节燕麦;Hairy-noded Oat ■☆
45527 Avena nuda L.;裸燕麦■
45528 Avena nuda L. = Avena chinensis (Fisch. ex Roem. et Schult.) Metzg. ■
45529 Avena nuda L. = Hordeum vulgare L. var. nudum Hook. f. ■
45530 Avena nuda L. var. chinensis Fisch. ex Roem. et Schult. = Avena chinensis (Fisch. ex Roem. et Schult.) Metzg. ■
45531 Avena occidentalis Durieu;西方狐尾草;Western Oat ■☆
45532 Avena oligostachya Munro ex Aitch. = Duthiea oligostachya (Munro ex Aitch.) Stapf ■☆
45533 Avena orientalis Schreb.;东方狐尾草(狐尾草);Oriental Oat,

Side Oat ■
45534 Avena orientalis Schreb. = Avena sativa L. ■
45535 Avena pallida Thunb. = Pentaschistis pallida (Thunb.) H. P. Linder ■☆
45536 Avena panicea Lam. = Trisetaria panicea (Lam.) Maire ■☆
45537 Avena papillosa Steud. = Pentaschistis papillosa (Steud.) H. P. Linder ■☆
45538 Avena parviflora Desf. = Trisetaria parviflora (Desf.) Maire ■☆
45539 Avena persica Steud. = Avena ludoviciana Durieu ■
45540 Avena persica Steud. = Avena sterilis L. subsp. ludoviciana (Durieu) Gillet et Magne ■☆
45541 Avena pilosa (Roem. et Schult.) M. Bieb. = Avena eriantha Durieu ■
45542 Avena pilosa M. Bieb.;柔毛燕麦■☆
45543 Avena planiculmis Schrad. subsp. dahurica Kom. = Helictotrichon dahuricum (Kom.) Kitag. ■
45544 Avena planiculmis Schreb.;西方扁秆燕麦■☆
45545 Avena polyneurum Hook. f. = Helictotrichon polyneurum (Hook. f.) Henrard ■
45546 Avena praegravis Roshev. ex Kom.;重燕麦■☆
45547 Avena pratensis Gouan = Avena pubescens L. ■☆
45548 Avena pratensis L.;草地燕麦(草地异燕麦);Meadow Oat Grass,Meadow Oat-grass ■☆
45549 Avena pratensis L. = Helictotrichon pratense (L.) Pilg. ■☆
45550 Avena pratensis L. subsp. laevis (Hack.) St.-Yves = Helictotrichon laeve (Hack.) Potztal ■☆
45551 Avena pratensis L. subsp. sulcata (Gay) St.-Yves = Helictotrichon albinerve (Boiss.) Henrard ■☆
45552 Avena pratensis L. var. albinervis (Boiss.) Husn. = Helictotrichon albinerve (Boiss.) Henrard ■☆
45553 Avena pratensis Roshev. ex Kom. = Avena pratensis L. ■☆
45554 Avena prostrata Ladiz.;平卧燕麦■☆
45555 Avena pruinosa Hack. et Trab. = Helictotrichon gervaisii (Holub) Röser subsp. arundanum (Romero Zarco) Röser ■☆
45556 Avena pubescens (Huds.) Pilg. = Helictotrichon pubescens (Huds.) Pilg. ■
45557 Avena pubescens Huds. = Helictotrichon pubescens (Huds.) Pilg. ■
45558 Avena pubescens L.;毛燕麦;Downy Oat-grass, Hairy Oat, Pubescent Oat ■☆
45559 Avena pumila Desf. = Lophochloa pumila (Desf.) Bor ■☆
45560 Avena pumila Desf. = Rostraria pumila (Desf.) Tzvelev ■☆
45561 Avena quinqueseta Steud. = Helictotrichon quinquesetum (Steud.) Schweick. ■☆
45562 Avena requienii Mutel = Helictotrichon gervaisii (Holub) Röser ■☆
45563 Avena requienii Mutel var. oranensis Trab. = Helictotrichon gervaisii (Holub) Röser ■☆
45564 Avena rigida Steud. = Pseudopentameris obtusifolia (Hochst.) N. P. Barker ■☆
45565 Avena rothii Stapf = Helictotrichon lachnanthum (Hochst. ex A. Rich.) C. E. Hubb. ■☆
45566 Avena sativa L.;燕麦(大麦,铃铛麦,香麦);Ailes, Ails, Common Oat,Cowflop,Cultivated Oats,Groats,Hav,Haw,Oat,Oats, Oils,Oils Hoyles,Plackett,Wets,Wild Oats ■
45567 Avena sativa L. = Avena fatua L. ■
45568 Avena sativa L. orientalis ? = Avena orientalis Schreb. ■
45569 Avena sativa L. subsp. byzantina (K. Koch) Romero Zarco = Avena byzantina K. Koch ■☆
45570 Avena sativa L. subsp. chinensis (Fisch. ex Roem. et Schult.) Janch. ex Holub. = Avena chinensis (Fisch. ex Roem. et Schult.) Metzg. ■
45571 Avena sativa L. var. abyssinica (Hochst.) Engl. = Avena abyssinica Hochst. ■☆
45572 Avena sativa L. var. glaberrima (Thell.) Maire et Weiller = Avena sativa L. ■
45573 Avena sativa L. var. nuda (L.) Körn. = Avena nuda L. ■
45574 Avena sativa L. var. orientalis (Schreb.) Alef. = Avena sativa L. ■
45575 Avena sativa L. var. praegravis Krause = Avena praegravis Roshev. ex Kom. ■☆
45576 Avena sativa L. var. sericea Hook. f. = Avena fatua L. ■
45577 Avena sativa L. var. subuniflora (Trab.) Maire et Weiller = Avena sativa L. ■
45578 Avena schelliana Hack. = Helictotrichon schellianum (Hack.) Kitag. ■
45579 Avena secalina (L.) Salisb. = Bromus secalinus L. ■
45580 Avena sempervirens Vill. = Helictotrichon sempervirens (Vill.) Pilg. ■☆
45581 Avena septentrionalis Malzev;北燕麦■☆
45582 Avena serratiglumis Sennen et Mauricio = Avena barbata Pott ex Link ■
45583 Avena sibirica L. = Achnatherum sibiricum (L.) Keng ■
45584 Avena sibirica L. = Stipa sibirica (L.) Lam. ■
45585 Avena smithii Porter ex A. Gray = Melica smithii (Porter ex A. Gray) Vasey ■☆
45586 Avena sterilis L.;不实燕麦;Animated Oat,Sterile Oat,Winter Wild-oat ■☆
45587 Avena sterilis L. subsp. ludoviciana (Durieu) Gillet et Magne = Avena ludoviciana Durieu ■
45588 Avena sterilis L. subsp. ludoviciana (Durieu) Gillet et Magne = Avena sterilis L. ■☆
45589 Avena sterilis L. subsp. ludoviciana (Durieu) Nyman = Avena sterilis L. subsp. ludoviciana (Durieu) Gillet et Magne ■☆
45590 Avena sterilis L. subsp. macrocarpa (Moench) Briq. = Avena sterilis L. ■☆
45591 Avena sterilis L. subsp. trichophylla (C. Koch) Malzev = Avena sterilis L. subsp. trichophylla (K. Koch) Malzev ■☆
45592 Avena sterilis L. subsp. trichophylla (K. Koch) Malzev;毛叶燕麦;Hairleaf Oat ■☆
45593 Avena sterilis L. var. calvescens Trab. et Thell. = Avena sterilis L. ■☆
45594 Avena sterilis L. var. glabrescens (Durieu) Malzev = Avena sterilis L. subsp. ludoviciana (Durieu) Gillet et Magne ■☆
45595 Avena sterilis L. var. ludoviciana (Durieu) Husn. = Avena sterilis L. subsp. ludoviciana (Durieu) Gillet et Magne ■☆
45596 Avena sterilis L. var. mauritiana Maire et Sennen = Avena sterilis L. ■☆
45597 Avena sterilis L. var. maxima Pérez Lara = Avena sterilis L. ■☆
45598 Avena sterilis L. var. psilathera Thell. = Avena sterilis L. ■☆
45599 Avena sterilis L. var. scabriuscula Pérez Lara = Avena sterilis L. ■☆
45600 Avena sterilis L. var. setigera Malzev = Avena sterilis L. ■☆
45601 Avena stipaeformis L.;托叶状燕麦■☆
45602 Avena strigosa Schreb.;糙伏燕麦(毛燕麦,德国野燕麦);Black Oat,Bristle Oat,Bristle-pointed Oat,Hispid Oat,Lopside Oat, Lopsided Oat ■☆

45603 Avena strigosa Schreb. subsp. abyssinica (Hochst.) Thell. = Avena abyssinica Hochst. ■☆
45604 Avena strigosa Schreb. var. abyssinica (Hochst.) Hausskn. = Avena abyssinica Hochst. ■☆
45605 Avena subspicata Clairv. = Trisetum spicatum (L.) K. Richt. ■
45606 Avena suffusca Hitchc. = Helictotrichon tibeticum (Roshev.) P. C. Keng ■
45607 Avena sulcata L.; 畦燕麦■☆
45608 Avena tentoensis Honda = Helictotrichon hookeri (Scribn.) Henrard ■
45609 Avena tibestica Miré et Quézel = Helictotrichon elongatum (Hochst. ex A. Rich.) C. E. Hubb. ■☆
45610 Avena tibetica Roshev. = Helictotrichon tibeticum (Roshev.) P. C. Keng ■
45611 Avena torreyi Nash = Schizachne purpurascens (Torr.) Swallen ■☆
45612 Avena trichophylla C. Koch = Avena sterilis L. subsp. trichophylla (K. Koch) Malzev ■☆
45613 Avena trichophylla K. Koch = Avena sterilis L. subsp. trichophylla (K. Koch) Malzev ■☆
45614 Avena triseta Thunb. = Pentaschistis triseta (Thunb.) Stapf ■☆
45615 Avena turgidula Stapf = Helictotrichon turgidulum (Stapf) Schweick. ■☆
45616 Avena uniflora Webb = Avena brevis Roth ■☆
45617 Avena vaviloviana (Malzev) Mordv. var. pseudoabyssinica (Thell.) C. E. Hubb. = Avena abyssinica Hochst. ■☆
45618 Avena ventricosa Coss.; 偏肿燕麦■☆
45619 Avena versicolor Vill.; 多色燕麦■☆
45620 Avena virescens (Regel) Regel = Trisetum spicatum (L.) K. Richt. subsp. virescens (Regel) Tzvelev ■
45621 Avena wiestii Steud.; 韦氏燕麦■☆
45622 Avena wiestii Steud. = Avena barbata Pott ex Link subsp. wiestii (Steud.) Mansf. ■☆
45623 Avena wiestii Steud. var. pseudoabyssinica Thell. = Avena abyssinica Hochst. ■☆
45624 Avenaceae (Kunth) Hotter = Gramineae Juss. (保留科名)●■
45625 Avenaceae (Kunth) Hotter = Poaceae Barnhart(保留科名)●■
45626 Avenaceae Martinov = Gramineae Juss. (保留科名)●■
45627 Avenaceae Martinov = Poaceae Barnhart(保留科名)●■
45628 Avenaceae Martinov; 燕麦科■
45629 Avenaria Fabr. = Bromus L. (保留属名)■
45630 Avenaria Heist. ex Fabr. = Bromus L. (保留属名)■
45631 Avenastrum (Koch) Opiz = Helictotrichon Besser ex Schult. et Schult. f. ■
45632 Avenastrum Jess. = Avenula (Dumort.) Dumort. ■
45633 Avenastrum Jess. = Helictotrichon Besser ex Schult. et Schult. f. ■
45634 Avenastrum Opiz = Helictotrichon Besser ex Schult. et Schult. f. ■
45635 Avenastrum antarticum (Thunb.) Stapf = Helictotrichon leoninum (Steud.) Schweick. ■☆
45636 Avenastrum asiaticum Roshev. = Helictotrichon hookeri (Scribn.) Henrard ■
45637 Avenastrum asperum (Munro ex Thwaites) Vierh. var. schmidii (Hook. f.) C. E. C. Fisch. = Helictotrichon schmidii (Hook. f.) Henrard ■
45638 Avenastrum caffrum (Stapf) Stapf = Helictotrichon longifolium (Nees) Schweick. ■☆
45639 Avenastrum caffrum Stapf var. natalensis ? = Helictotrichon natalense (Stapf) Schweick. ■☆
45640 Avenastrum dahuricum (Kom.) Roshev. = Helictotrichon dahuricum (Kom.) Kitag. ■
45641 Avenastrum dodii Stapf = Helictotrichon dodii (Stapf) Schweick. ■☆
45642 Avenastrum dregeanum (Steud.) Stapf = Helictotrichon namaquense Schweick. ■☆
45643 Avenastrum elongatum (A. Rich.) Pilg. var. friesiorum Pilg. = Helictotrichon umbrosum (Hochst. ex Steud.) C. E. Hubb. ■☆
45644 Avenastrum flabellatum Peter = Bewsia biflora (Hack.) Gooss. ■☆
45645 Avenastrum longum (Stapf) Stapf = Helictotrichon longum (Stapf) Schweick. ■☆
45646 Avenastrum majus Pilg. = Helictotrichon milanjianum (Rendle) C. E. Hubb. ■☆
45647 Avenastrum mannii Pilg. = Helictotrichon mannii (Pilg.) C. E. Hubb. ■☆
45648 Avenastrum mannii Pilg. var. angustior ? = Helictotrichon milanjianum (Rendle) C. E. Hubb. ■☆
45649 Avenastrum mongolicum (Roshev.) Roshev. = Helictotrichon mongolicum (Roshev.) Henrard ■
45650 Avenastrum mongolicum Roshev. = Helictotrichon mongolicum (Roshev.) Henrard ■
45651 Avenastrum nervosum (R. Br.) Vierh. = Avena nervosa R. Br. ■
45652 Avenastrum planiculme (Schrad.) Opiz = Avena planiculmis Schreb. ■☆
45653 Avenastrum planiculme Jess. = Avena planiculmis Schreb. ■☆
45654 Avenastrum praecox Jess. = Aira praecox L. ■
45655 Avenastrum pratense (L.) Opiz = Avena pratensis L. ■☆
45656 Avenastrum pratense (L.) Opiz = Helictotrichon pratense (L.) Pilg. ■☆
45657 Avenastrum pratense Jess. = Avena pratensis Roshev. ex Kom. ■☆
45658 Avenastrum pubescens (Huds.) Opiz = Helictotrichon pubescens (Huds.) Pilg. ■
45659 Avenastrum pubescens (L.) Opiz = Avena pubescens (Huds.) Pilg. ■
45660 Avenastrum pubescens Jess. = Avena pubescens (Huds.) Pilg. ■
45661 Avenastrum quinquenerve Stent et J. M. Rattray = Helictotrichon elongatum (Hochst. ex A. Rich.) C. E. Hubb. ■☆
45662 Avenastrum quinquesetum (Steud.) Stapf = Helictotrichon quinquesetum (Steud.) Schweick. ■☆
45663 Avenastrum rigidulum Pilg. = Helictotrichon elongatum (Hochst. ex A. Rich.) C. E. Hubb. ■☆
45664 Avenastrum schellianum (Hack.) Roshev. = Avena schelliana Hack. ■
45665 Avenastrum schellianum (Hack.) Roshev. = Helictotrichon schellianum (Hack.) Kitag. ■
45666 Avenastrum sulcatum (J. Gay) Vierh. = Avena sulcata L. ■☆
45667 Avenastrum sulcatum C. E. Hubb. et Sandwith = Avena sulcata L. ■☆
45668 Avenastrum tentoense (Honda) Cretz. = Avena tentoensis Honda ■☆
45669 Avenastrum tentoense (Honda) Kitag. = Helictotrichon hookeri (Scribn.) Henrard ■
45670 Avenastrum tianschanicum Roshev. = Helictotrichon tianschanicum (Roshev.) Henrard ■
45671 Avenastrum trisetoides Kitag. = Helictotrichon altius (Hitchc.) Ohwi ■
45672 Avenastrum turgidulum (Stapf) Stapf = Helictotrichon turgidulum

(Stapf) Schweick. ■☆
45673 Avenastrum umbrosum Pilg. = Trisetum umbrosum Hochst. ex Steud. ■☆
45674 Avenastrum versicolor Fritsch = Avena versicolor Vill. ■☆
45675 Avenella (Bluff et Fingerh.) Drejer = Deschampsia P. Beauv. ■
45676 Avenella Koch = Deschampsia P. Beauv. ■
45677 Avenella Koch ex Steud. = Deschampsia P. Beauv. ■
45678 Avenella Parl. = Deschampsia P. Beauv. ■
45679 Avenella flexuosa (L.) Drejer = Deschampsia flexuosa (L.) Trin. ■
45680 Avenella flexuosa (L.) Drejer subsp. iberica (Rivas Mart.) Valdés et H. Scholz = Deschampsia flexuosa (L.) Trin. subsp. iberica Rivas Mart. et al. ■☆
45681 Avenochloa Holub = Helictotrichon Besser ex Schult. et Schult. f. ■
45682 Avenochloa albinervis (Boiss.) Holub = Helictotrichon albinerve (Boiss.) Henrard ■☆
45683 Avenochloa breviaristata (Barratte) Holub = Helictotrichon breviaristatum (Barratte) Henrard ■☆
45684 Avenochloa bromoides (Gouan) Holub = Helictotrichon bromoides (Gouan) C. E. Hubb. ■☆
45685 Avenochloa cincinnata (Ten.) Holub = Helictotrichon cincinnatum (Ten.) Röser ■☆
45686 Avenochloa dahurica (Kom.) Holub = Helictotrichon dahuricum (Kom.) Kitag. ■
45687 Avenochloa jahandiiezii (Litard.) Holub = Helictotrichon jahandiezii (Litard.) Potztal ■☆
45688 Avenochloa laevis (Hack.) Holub = Helictotrichon laeve (Hack.) Potztal ■☆
45689 Avenochloa letourneuxii (Trab.) Holub = Helictotrichon cincinnatum (Ten.) Röser ■☆
45690 Avenochloa pruinosa (Batt. et Trab.) Holub = Helictotrichon pruinosum (Hack. et Trab.) Henrard ■☆
45691 Avenochloa pubescens (Huds.) Holub = Helictotrichon pubescens (Huds.) Pilg. ■
45692 Avenochloa schelliana (Hack.) Tzvelev = Helictotrichon schellianum (Hack.) Kitag. ■
45693 Avenochloa sulcata (Boiss.) Dumort. = Helictotrichon albinerve (Boiss.) Henrard ■☆
45694 Avenula (Dumort.) Dumort. = Helictotrichon Besser ex Schult. et Schult. f. ■
45695 Avenula albinervis (Boiss.) Lainz = Helictotrichon albinerve (Boiss.) Henrard ■☆
45696 Avenula breviaristata (Barratte) Holub = Helictotrichon breviaristatum (Barratte) Henrard ■☆
45697 Avenula bromoides (Gouan) Scholz = Helictotrichon bromoides (Gouan) C. E. Hubb. ■☆
45698 Avenula bromoides (Gouan) Scholz subsp. australis (Breistr.) Scholz = Helictotrichon cincinnatum (Ten.) Röser ■☆
45699 Avenula bromoides (Gouan) Scholz subsp. cincinnata (Ten.) Romo = Helictotrichon cincinnatum (Ten.) Röser ■☆
45700 Avenula cincinnata (Ten.) Holub = Helictotrichon cincinnatum (Ten.) Röser ■☆
45701 Avenula dahurica (Kom.) W. Sauer et Chmel. = Helictotrichon dahuricum (Kom.) Kitag. ■
45702 Avenula gervaisii Holub = Helictotrichon gervaisii (Holub) Röser ■☆
45703 Avenula hookeri (Scribn.) Holub subsp. schelliana (Hack.) M. N. Lomon. = Helictotrichon schellianum (Hack.) Kitag. ■
45704 Avenula jahandiezii (Litard.) Holub = Helictotrichon jahandiezii (Litard.) Potztal ■☆
45705 Avenula laevis (Hack.) Holub = Helictotrichon laeve (Hack.) Potztal ■☆
45706 Avenula letourneuxii (Trab.) Scholz = Helictotrichon cincinnatum (Ten.) Röser ■☆
45707 Avenula lodunensis (Delastre) Kerguélen = Helictotrichon marginatum (Lowe) Röser ■☆
45708 Avenula marginata (Lowe) Holub = Helictotrichon marginatum (Lowe) Röser ■☆
45709 Avenula marginata (Lowe) Holub subsp. albinervis (Boiss.) Romero Zarco = Helictotrichon albinerve (Boiss.) Henrard ■☆
45710 Avenula pratensis (L.) Dumort. = Avena pratensis L. ■☆
45711 Avenula pratensis (L.) Dumort. = Helictotrichon pratense (L.) Pilg. ■☆
45712 Avenula pruinosa (Batt. et Trab.) Holub = Helictotrichon pruinosum (Hack. et Trab.) Henrard ■☆
45713 Avenula pubescens (Huds.) Dumort. = Helictotrichon pubescens (Huds.) Pilg. ■
45714 Avenula schelliana (Hack.) W. Sauer et Chmel. = Helictotrichon schellianum (Hack.) Kitag. ■
45715 Avenula sulcata (Boiss.) Dumort. = Helictotrichon albinerve (Boiss.) Henrard ■☆
45716 Avenula sulcata (Boiss.) Dumort. subsp. albinervis (Boiss.) Rivas Mart. = Helictotrichon albinerve (Boiss.) Henrard ■☆
45717 Averia Léonard = Justicia L. ●■
45718 Averia Léonard(1940);奥弗涅爵床属●☆
45719 Averia longipes (Standl.) Léonard;奥弗涅爵床●☆
45720 Averis Léonard = Tetramerium Nees(保留属名)●☆
45721 Averrhoa L. (1753);阳桃属(羊桃属,杨桃属);Averrhoa, Carambola ●
45722 Averrhoa acida L. = Phyllanthus acidus (L.) Skeels ●☆
45723 Averrhoa acutangula Stokes = Averrhoa carambola L. ●
45724 Averrhoa bilimbi L.;三敛(长叶五敛子,长叶羊桃,黄瓜树,木胡瓜,三桧,三捻);Bilimb Carambola, Bilimbi, Biling, Bilumbing, Blimbing, Blimbing Kamia, Camias, Cucumber Tree, Cucumber-tree, Eastern Strawberry-tree, Pickle Fruit, Tree Sorrel ●
45725 Averrhoa carambola L.;阳桃(风鼓,鬼桃,马槟榔,木踏子,三棱子,三帘,三廉子,三敛,三敛子,三稔,山敛,酸三姩,酸五棱,五棱果,五棱子,五敛子,羊桃,杨桃,洋桃);Averrhoa, Blimbing, Caramba, Carambola, Carambole, Chinese Gooseberry, Common Averrhoa, Country Gooseberry, Star Fruit, Starfruit ●
45726 Averrhoa obtusangula Stokes = Averrhoa bilimbi L. ●
45727 Averrhoaceae Hutch.;阳桃科(羊桃科,捻子科)■
45728 Averrhoaceae Hutch. = Oxalidaceae R. Br. (保留科名)●■
45729 Averrhoidium Baill. (1874);阳桃无患子属●☆
45730 Aversia G. Don = Arversia Cambess. ■
45731 Aversia G. Don = Polycarpon L. ■
45732 Avesicaria (Kamienski) Barnhart = Utricularia L. ■
45733 Avesicaria Barnhart = Utricularia L. ■
45734 Avetra H. Perrier = Trichopus Gaertn. ■☆
45735 Avetra H. Perrier(1924);马达藤属(木本薯蓣属)●☆
45736 Avetra sempervirens H. Perrier;马达藤(木本薯蓣)●☆
45737 Avetra sempervirens H. Perrier = Trichopus sempervirens (H. Perrier) Caddick et Wilkin ●☆
45738 Avetraceae Takht.;马达藤科●☆

45739 Avetraceae Takht. = Dioscoreaceae R. Br.(保留科名)●■

45740 Avicennia L. (1753);海榄雌属(海茄冬属);Avicennia, Mangrove ●

45741 Avicennia africana P. Beauv. = Avicennia germinans (L.) L. ●☆

45742 Avicennia alba Blume;白海榄雌●☆

45743 Avicennia germinans (L.) L.;黑海榄雌(发芽海榄雌);Black Mangrove ●☆

45744 Avicennia intermedia Griff.;居间海榄雌(居间白骨壤)●☆

45745 Avicennia lanata Ridl.;绵毛海榄雌(绵毛白骨壤)●☆

45746 Avicennia marina (Forssk.) Vierh.;海榄雌(白骨浪,白骨壤,海榄钱,海茄冬,海茄苳,黑海榄雌,茄藤,茄藤树,咸水矮让木,药用海榄雌木);Black Mangrove, Blackmangrove, Black-mangrove, Coastal Avicennia, Grey Mangrove, Indian Mangrove, White Mangrove ●

45747 Avicennia marina (Forssk.) Vierh. var. acutissima Stapf et Moldenke = Avicennia marina (Forssk.) Vierh. ●

45748 Avicennia marina (Forssk.) Vierh. var. resinifera (G. Forst.) Bakh.;灰海榄雌;Gray Mangrove ●☆

45749 Avicennia marina Forssk. = Avicennia marina (Forssk.) Vierh. ●

45750 Avicennia nitida Jacq.;亮叶海榄雌(海茄冬);White Mangrove ●☆

45751 Avicennia nitida Jacq. = Avicennia germinans (L.) L. ●☆

45752 Avicennia officinalis L. = Avicennia marina (Forssk.) Vierh. ●

45753 Avicennia officinalis Schauer f. flaviflora Kuntze = Avicennia marina (Forssk.) Vierh. ●

45754 Avicennia officinalis Schauer f. tomentosa Kuntze = Avicennia marina (Forssk.) Vierh. ●

45755 Avicennia tomentosa Jacq. = Avicennia germinans (L.) L. ●☆

45756 Avicennia tomentosa Sieber var. arabica Walp. = Avicennia marina (Forssk.) Vierh. ●

45757 Avicenniaceae Endl. = Acanthaceae Juss.(保留科名)●■

45758 Avicenniaceae Endl. = Avicenniaceae Miq.(保留科名)●

45759 Avicenniaceae Endl. ex Schnizl. = Avicenniaceae Miq.(保留科名)●

45760 Avicenniaceae Miq. (1845)(保留科名);海榄雌科●

45761 Aviceps Lindl. = Satyrium Sw.(保留属名)■

45762 Aviceps pumila (Thunb.) Lindl. = Satyrium pumilum Thunb. ■☆

45763 Avicularia (Meisn.) Börner = Polygonum L.(保留属名)●■

45764 Avicularia Steud. = Polygonum L.(保留属名)●■

45765 Aviunculus Fourr. = Coronilla L.(保留属名)●■

45766 Avoira Giseke(废弃属名) = Astrocaryum G. Mey.(保留属名)●☆

45767 Avonia (E. Mey. ex Fenzl) G. D. Rowley = Anacampseros L.(保留属名)■☆

45768 Avonia (E. Mey. ex Fenzl) G. D. Rowley(1994);阿冯苋属■☆

45769 Avonia albissima (Marloth) G. D. Rowley;白阿冯苋■☆

45770 Avonia dinteri (Schinz) G. D. Rowley;丁特阿冯苋■☆

45771 Avonia herreana (Poelln.) G. D. Rowley;赫勒阿冯苋■☆

45772 Avonia mallei G. Will.;马勒阿冯苋■☆

45773 Avonia papyracea (E. Mey. ex Fenzl) G. D. Rowley;纸质阿冯苋■☆

45774 Avonia papyracea (E. Mey. ex Fenzl) G. D. Rowley subsp. namaensis (Gerbaulet) G. D. Rowley;纳马阿冯苋■☆

45775 Avonia prominens (G. Will.) G. Will.;突起阿冯苋■☆

45776 Avonia quinaria (E. Mey. ex Fenzl) G. D. Rowley;五出阿冯苋■☆

45777 Avonia quinaria (E. Mey. ex Fenzl) G. D. Rowley subsp. alstonii (Schönland) G. D. Rowley;奥尔斯顿阿冯苋■☆

45778 Avonia quinaria (E. Mey. ex Fenzl) G. D. Rowley var. schmidtii A. Berger = Avonia albissima (Marloth) G. D. Rowley ■☆

45779 Avonia recurvata (Schönland) G. D. Rowley;反折阿冯苋■☆

45780 Avonia recurvata (Schönland) G. D. Rowley subsp. buderiana (Poelln.) G. Will.;比德阿冯苋■☆

45781 Avonia recurvata (Schönland) G. D. Rowley subsp. minuta (Gerbaulet) G. D. Rowley;微小阿冯苋■☆

45782 Avonia rhodesica (N. E. Br.) G. D. Rowley;罗得西亚阿冯苋■☆

45783 Avonia ruschii (Dinter et Poelln.) G. D. Rowley;鲁施阿冯苋■☆

45784 Avonia ustulata (E. Mey. ex Fenzl) G. D. Rowley;凋萎阿冯苋■☆

45785 Avonia variabilis (Poelln.) G. Will.;易变阿冯苋■☆

45786 Avonsera Speta(1998);多棱被风信子属■☆

45787 Avonsera lachenalioides (Baker) Speta;多棱被风信子■☆

45788 Avornela Raf. = Chamaespartium Adans. ●

45789 Awayus Raf. = Spiraea L. ●

45790 Axanthes Blume = Urophyllum Jack ex Wall. ●

45791 Axanthopsis Korth. = Axanthes Blume ●

45792 Axanthopsis Korth. = Urophyllum Jack ex Wall. ●

45793 Axenfeldia Baill. = Mallotus Lour. ●

45794 Axia Lour. = Boerhavia L. ■

45795 Axiana Raf. = Axia Lour. ■

45796 Axillaria Raf. = Polygonatum Mill. ■

45797 Axinaea Ruiz et Pav. (1794);斧丹属●☆

45798 Axinaea affinis (Naudin) Cogn.;近缘斧丹●☆

45799 Axinaea macrophylla (Naudin) Triana;大叶斧丹●☆

45800 Axinaea oblongifolia (Cogn.) Wurdack;椭圆叶斧丹●☆

45801 Axinaea pauciflora Cogn.;少花斧丹●☆

45802 Axinaea sclerophylla Triana;硬叶斧丹●☆

45803 Axinandra Thwaites(1854);楔蕊牡丹属(斧药属)●☆

45804 Axinandra zeylanica Thwaites;楔蕊牡丹●☆

45805 Axinanthera H. Karst. = Bellucia Neck. ex Raf.(保留属名)●☆

45806 Axinea Juss. = Axinaea Ruiz et Pav. ●☆

45807 Axiniphyllum Benth. (1872);斧叶菊属(箭叶菊属)■☆

45808 Axiniphyllum corymbosum Benth.;斧叶菊■☆

45809 Axiniphyllum tomentosum Benth.;毛斧叶菊■☆

45810 Axinopus Kunth = Axonopus P. Beauv. ■

45811 Axinopus Kunth = Paspalum L. ■

45812 Axiris L. = Axyris L. ■

45813 Axiron Raf. = Cytisus Desf.(保留属名)●

45814 Axolopha Alef. = Lavatera L. ●■

45815 Axolus Raf. = Acrodryon Spreng. ●

45816 Axolus Raf. = Cephalanthus L. ●

45817 Axonopus (Steud.) Chase = Axonopus P. Beauv. ■

45818 Axonopus (Steud.) Chase = Lappagopsis Steud. ■

45819 Axonopus Hook. f. = Alloteropsis J. Presl ex C. Presl ■

45820 Axonopus P. Beauv. (1812);地毯草属;Carpet Grass, Carpetgrass ■

45821 Axonopus affinis A. Camus;近缘地毯草(类地毯草);Common Carpetgrass ■

45822 Axonopus affinis Chase = Axonopus fissifolius (Raddi) Kuhlm. ■

45823 Axonopus arenosus Gledhill = Axonopus flexuosus (Peter) C. E. Hubb. ■☆

45824 Axonopus brevipedunculatus (Gledhill) Gledhill = Axonopus compressus (Sw.) P. Beauv. ■

45825 Axonopus cimicinus (L.) P. Beauv. = Alloteropsis cimicina (L.) Stapf ■

45826 Axonopus cimicinus P. Beauv. = Alloteropsis cimicina (L.) Stapf ■

45827 Axonopus compressus (Sw.) P. Beauv.;地毯草;Broad-leaved

Carpetgrass, Carpetgrass ■
45828 Axonopus compressus (Sw.) P. Beauv. subsp. brevipedunculatus Gledhill = Axonopus compressus (Sw.) P. Beauv. ■
45829 Axonopus compressus (Sw.) P. Beauv. subsp. congoensis Henrard = Axonopus flexuosus (Peter) C. E. Hubb. ■☆
45830 Axonopus compressus (Sw.) P. Beauv. var. affinis (Chase) M. R. Hend. = Axonopus affinis Chase ■☆
45831 Axonopus compressus (Sw.) P. Beauv. var. affinis (Chase) M. R. Hend. = Axonopus fissifolius (Raddi) Kuhlm. ■
45832 Axonopus compressus P. Beauv. = Axonopus compressus (Sw.) P. Beauv. ■
45833 Axonopus fissifolius (Raddi) Kuhlm.;普通地毯草(类地毯草,西藏雀稗);Common Carpetgrass ■
45834 Axonopus flexuosus (Peter) C. E. Hubb.;之字地毯草■☆
45835 Axonopus kisantuensis Vanderyst = Axonopus compressus (Sw.) P. Beauv. ■
45836 Axonopus latifolius Peter = Alloteropsis cimicina (L.) Stapf ■
45837 Axonopus semialatus (R. Br.) Hook. f. = Alloteropsis semialata (R. Br.) Hitchc. ■
45838 Axonopus semialatus (R. Br.) Hook. f. var. ecklonianus (Nees) Peter = Alloteropsis semialata (R. Br.) Hitchc. var. eckloniana (Nees) Pilg. ■
45839 Axonopus semialatus (R. Br.) Hook. f. var. ecklonii Stapf = Alloteropsis semialata (R. Br.) Hitchc. subsp. eckloniana (Nees) Gibbs-Russ. ■
45840 Axonopus semialatus (R. Br.) Hook. f. var. ecklonii Stapf = Alloteropsis semialata (R. Br.) Hitchc. var. eckloniana (Nees) Pilg. ■
45841 Axonotechium Fenzl = Orygia Forssk. ■☆
45842 Axonotechium trianthemiodes (F. Heyne) Fenzl = Corbichonia decumbens (Forssk.) Exell ■☆
45843 Axyris L. (1753);轴藜属;Axyris, Russian Pigweed ■
45844 Axyris amaranthoides L.;轴藜(苋轴藜);Common Axyris, Russian Pigweed, Upright Axyris ■
45845 Axyris amaranthoides L. f. dentata (Baranov) Kitag. = Axyris amaranthoides L. ■
45846 Axyris amaranthoides L. f. nana (Wang-Wei et P. Y. Fu) Kitag. = Axyris hybrida L. ■
45847 Axyris amaranthoides L. var. dentata Baranov = Axyris amaranthoides L. ■
45848 Axyris amaranthoides L. var. nana Wang-Wei et P. Y. Fu;小轴藜;Dwarf Axyris ■
45849 Axyris amaranthoides L. var. nana Wang-Wei et P. Y. Fu = Axyris hybrida L. ■
45850 Axyris amaranthoides Wang-Wei et P. Y. Fu = Axyris hybrida L. ■
45851 Axyris caucasica (Sommier et H. Lév.) Lipsky;高加索轴藜■☆
45852 Axyris ceratoides L. = Ceratoides latens (J. F. Gmel.) Reveal et N. H. Holmgren ●
45853 Axyris hybrida L.;杂配轴藜;Hybrid Axyris ■
45854 Axyris pamirica B. Fedtsch. = Axyris prostrata L. ■
45855 Axyris pentandra Jacq. = Atriplex pentandra (Jacq.) Standl. ■☆
45856 Axyris prostrata L.;平卧轴藜;Prostrate Axyris ■
45857 Axyris prostrata L. f. ovatifolia Soong;卵叶平卧轴藜;Ovatileaf Prostrate Axyris ■
45858 Axyris prostrata L. f. ovatifolia Soong = Axyris prostrata L. ■
45859 Axyris sphaerosperma Fisch. et C. A. Mey.;球籽轴藜■
45860 Ayapana Spach (1841);尖泽兰属■☆
45861 Ayapana triplinervis (Vahl) R. M. King et H. Rob.;尖泽兰■☆
45862 Ayapanopsis R. M. King et H. Rob. (1972);显药尖泽兰属●■☆
45863 Ayapanopsis adenophora R. M. King et H. Rob.;腺梗显药尖泽兰●☆
45864 Ayapanopsis beckii H. Rob.;贝克显药尖泽兰●☆
45865 Ayapanopsis latipaniculata (Rusby) R. M. King et H. Rob.;宽锥显药尖泽兰●☆
45866 Aydendron Nees = Aniba Aubl. ●☆
45867 Ayenia Griseb. = Ayenia L. ●☆
45868 Ayenia L. (1756);阿延梧桐属●☆
45869 Ayenia Loefl. = Ayenia L. ●☆
45870 Ayenia acuminata Rusby;阿延梧桐●☆
45871 Ayenia cordifolia Sessé ex DC.;心叶阿延梧桐●☆
45872 Ayensua L. B. Sm. (1969);委内瑞拉凤梨属■☆
45873 Ayensua uaipanensis (Maguire) L. B. Sm.;委内瑞拉凤梨■☆
45874 Aylacophora Cabrera (1953);沙黄菀属●☆
45875 Aylacophora deserticola Cabrera;沙黄菀■☆
45876 Aylanthus Raf. = Aylantus Juss. ●
45877 Aylantus Juss. = Ailanthus Desf. (保留属名)●
45878 Aylmeria Mart. = Polycarpaea Lam. (保留属名)●■
45879 Aylostera Speg. (1923);红笠属■
45880 Aylostera Speg. = Rebutia K. Schum. ●
45881 Aylostera albiflora (F. Ritter) Backeb.;白花红笠■☆
45882 Aylostera albiflora Backeb. = Aylostera albiflora (F. Ritter) Backeb. ■☆
45883 Aylostera deminuta (F. A. C. Weber) Backeb. = Rebutia deminuta Britton et Rose ●☆
45884 Aylostera deminuta (F. A. C. Weber) Backeb. var. pseudominuscula (Speg.) Backeb. = Rebutia deminuta Britton et Rose ●☆
45885 Aylostera fiebrigii (Gurke) Backeb. = Rebutia fiebrigii (Gurke) Britton et Rose ●☆
45886 Aylostera kupperiana (Boed.) Backeb.;优宝球(优宝丸)■☆
45887 Aylostera pseudodeminuta (Backeb.) Backeb.;夕照球(夕照丸)■☆
45888 Aylostera pseudominuscula (Speg.) Speg.;艳丽球(艳丽丸)■☆
45889 Aylostera pseudominuscula (Speg.) Speg. = Rebutia deminuta Britton et Rose ●☆
45890 Aylostera rubiginosa (F. Ritter) Backeb. = Rebutia spegazziniana Backeb. ■☆
45891 Aylostera spegazziniana (Backeb.) Backeb. = Rebutia spegazziniana Backeb. ■☆
45892 Aylostera spinosissima Backeb.;红照球(红照丸)■☆
45893 Aylostera steinmannii (Solms) Backeb.;周天球(周天丸)■☆
45894 Aylostera tuberosa (F. Ritter) Backeb. = Rebutia spegazziniana Backeb. ■☆
45895 Aylostera waltheriana (Backeb.) Y. Ito;寿宝球(寿宝丸)■☆
45896 Aylthonia N. L. Menezes = Barbacenia Vand. ■☆
45897 Aynia H. Rob. (1988);叶苞斑鸠菊属■☆
45898 Aynia H. Rob. = Vernonia Schreb. (保留属名)●■
45899 Aynia pseudascaricida H. Rob.;叶苞斑鸠菊■☆
45900 Ayparia Raf. = Elaeocarpus L. ●
45901 Aytonia L. = Aitonia Thunb. ●☆
45902 Aytonia L. = Nymania Lindb. ●☆
45903 Aytonia L. f. = Nymania Lindb. ●☆
45904 Azadehdelia Braem = Cribbia Senghas ■☆
45905 Azadehdelia brachyceras (Summerh.) Braem = Cribbia

brachyceras (Summerh.) Senghas ■☆
45906 Azadirachta A. Juss. (1830);蒜楝木属(蒜楝属,印度楝属,印楝属)●☆
45907 Azadirachta excelsa (Jack) Jacobs;高大蒜楝木(大楝树)●☆
45908 Azadirachta indica A. Juss.;蒜楝木(楝树,印度楝,印度楝树);Bead-tree, Cape Lilac, Margosa, Margosa Tree, Neem, Neem Tree, Nim, Pride of Indian ●☆
45909 Azadirachta indica A. Juss. subsp. siamensis Valeton;暹罗苦楝●☆
45910 Azalea Desv. = Rhododendron L. ●
45911 Azalea Gaertn. = Loiseleuria Desv. (保留属名)●☆
45912 Azalea L. (废弃属名) = Loiseleuria Desv. (保留属名)●☆
45913 Azalea L. (废弃属名) = Loiseleuria Desv. + Rhododendron L. ●
45914 Azalea L. (废弃属名) = Rhododendron L. ●
45915 Azalea farrerae K. Koch = Rhododendron farrerae Tate ex Sweet ●
45916 Azalea ferruginosa Pall. = Rhododendron lapponicum (L.) Wahlenb. ●
45917 Azalea indica L. = Rhododendron indicum (L.) Sweet ●
45918 Azalea indica L. var. alba Lindl. = Rhododendron mucronatum (Blume) G. Don ●
45919 Azalea indica L. var. simsii (Planch.) Rehder = Rhododendron simsii Planch. ●
45920 Azalea indica L. var. simsii Rehder = Rhododendron simsii Planch. ●
45921 Azalea lapponica L. = Rhododendron lapponicum (L.) Wahlenb. ●
45922 Azalea macrantha Bunge = Rhododendron indicum (L.) Sweet ●
45923 Azalea mollis Blume = Rhododendron molle (Blume) G. Don ●
45924 Azalea mucronata Blume = Rhododendron mucronatum (Blume) G. Don ●
45925 Azalea myrtifolia Champ. = Rhododendron hongkongense Hutch. ●
45926 Azalea obtusa Lindl. = Rhododendron obtusum (Lindl.) Planch. ●
45927 Azalea oldhamii (Maxim.) Mast. = Rhododendron oldhamii Maxim. ●
45928 Azalea oldhamii Hort. = Rhododendron oldhamii Maxim. ●
45929 Azalea ovata Lindl. = Rhododendron ovatum Planch. ●
45930 Azalea parvifolia (Adams) Kuntze = Rhododendron lapponicum (L.) Wahlenb. ●
45931 Azalea pontica L. = Rhododendron luteum Sweet ●
45932 Azalea procumbens L. = Rhododendron procumbens (L.) A. W. Wood ●☆
45933 Azalea rosmarinifolia Burm. f. = Rhododendron mucronatum (Blume) G. Don ●
45934 Azalea schlippenbachii (Maxim.) Kuntze = Rhododendron schlippenbachii Maxim. ●
45935 Azalea schlippenbachii Kuntze = Rhododendron schlippenbachii Maxim. ●
45936 Azalea sinensis Lodd. = Rhododendron molle (Blume) G. Don ●
45937 Azalea squamata Lindl. = Rhododendron farrerae Tate ex Sweet ●
45938 Azalea viscosa L. = Rhododendron viscosum (L.) Torr. ●☆
45939 Azaleaceae Vest = Ericaceae Juss. (保留科名)●
45940 Azaleastrum (Maxim.) Rydb. = Rhododendron L. ●
45941 Azaleastrum Rydb. = Rhododendron L. ●
45942 Azaltea Walp. = Alzatea Ruiz et Pav. ●☆
45943 Azamara Hochst. ex Rchb. = Schmidelia L. ●
45944 Azanza Alef. = Thespesia Sol. ex Corrêa(保留属名)●
45945 Azanza Moc. et Sessé ex DC. = Hibiscus L. (保留属名)●■
45946 Azanza garckeana (F. Hoffm.) Exell et Hillc. = Thespesia garckeana F. Hoffm. ●☆
45947 Azanza lampas (Cav.) Alef. = Thespesia lampas (Cav.) Dalzell et A. Gibson ●
45948 Azanza lampas Alef. = Thespesia lampas (Cav.) Dalzell et A. Gibson ●
45949 Azaola Blanco = Madhuca Buch. -Ham. ex J. F. Gmel. ●
45950 Azaola Blanco = Payena A. DC. ●☆
45951 Azaola leerii Teijsm. et Binn. = Payena leerii (Teijsm. et Binn.) Kurz ●☆
45952 Azara Ruiz et Pav. (1794);阿氏木属(阿查拉属);Azara ●☆
45953 Azara dentata Ruiz;齿叶阿氏木●☆
45954 Azara lanceolata Hook. f.;披针叶阿氏木(披针叶阿查拉);Box-leaf Azara, Lanceleaf Azara ●☆
45955 Azara microphylla Hook. f.;小叶阿氏木(小叶阿查拉);Boxleaf Azara, Box-leaf Azara ●☆
45956 Azara salicifolia Griseb.;柳叶阿氏木●☆
45957 Azara serrata Ruiz et Pav.;锯齿阿氏木(锯齿阿查拉)●☆
45958 Azaraea Post et Kuntze = Azara Ruiz et Pav. ●☆
45959 Azarolus Borkh. = Crataegus L. ●
45960 Azarolus Borkh. = Sorbus L. ●
45961 Azedara Raf. = Azedarach Mill. ●
45962 Azedarac Adans. = Melia L. ●
45963 Azedarach Adans. = Melia L. ●
45964 Azedarach Mill. = Melia L. ●
45965 Azedarae Adans. = Azedarach Mill. ●
45966 Azedaraea Raf. = Azedarach Mill. ●
45967 Azeredia Allemão = Maximilianea Mart. (废弃属名)●
45968 Azeredia Arruda ex Allemão = Cochlospermum Kunth (保留属名)●☆
45969 Azilia Hedge et Lamond(1987);伊朗草属■☆
45970 Azilia eryngioides (Pau) Hedge et Lamond;伊朗草■☆
45971 Azima Lam. (1783);刺茉莉属;Azima ●
45972 Azima angustifolia A. DC.;狭叶刺茉莉;Narrowleaf Azima ●☆
45973 Azima pubescens Suess. = Carissa spinarum L. ●
45974 Azima sarmentosa (Blume) Benth. et Hook. f.;刺茉莉(牙刷树);Sarmentose Azima ●
45975 Azima tetracantha Lam.;四花刺茉莉●☆
45976 Azimaceae Gardner = Salvadoraceae Lindl. (保留科名)●
45977 Azimaceae Wight et Gardner = Salvadoraceae Lindl. (保留科名)●
45978 Azophora Neck. = Rhizophora L. ●
45979 Azorella Lam. (1783);小鹰芹属(南美芹属,牵环花属)■☆
45980 Azorella caespitosa Cav. = Azorella monantha Clos ex Gay ■☆
45981 Azorella compacta Phil.;紧密小鹰芹■☆
45982 Azorella monantha Clos ex Gay.;丛生小鹰芹;Balsam Bog, Balsam-bog, Common Burdock, Smaller Burdock ■☆
45983 Azorella nivalis Phil. = Azorella trifuscata Hook. ■☆
45984 Azorella trifuscata Hook.;南美芹;Baby Blue Eyes ■☆
45985 Azorella yareta Hauman = Azorella compacta Phil. ■☆
45986 Azorellopsis H. Wolff = Mulinum Pers. ■☆
45987 Azorina Feer = Campanula L. ●■
45988 Azorina Feer(1890);风铃木属●☆
45989 Azorina vidalii Feer;风铃木●☆
45990 Aztecaster G. L. Nesom(1993);异株菀属●☆
45991 Aztecaster matudae (Rzed.) G. L. Nesom;异株菀●☆
45992 Aztecaster pyramidatus (B. L. Rob. et Greenm.) G. L. Nesom;塔形异株菀●☆
45993 Aztekium Boed. (1929);皱棱球属;Aztekium ●☆

45994 Aztekium ritteri (Boed.) Boed.;皱棱球(花笠,花笼,皱棱仙人球);Aztekium,Ritter Aztekium ■☆
45995 Azukia Takah. ex Ohwi = Vigna Savi(保留属名)■
45996 Azukia angularis (Willd.) Ohwi = Vigna angularis (Willd.) Ohwi et H. Ohashi ■
45997 Azukia angularis (Willd.) Ohwi var. nippoensis (Ohwi) Ohwi = Vigna angularis (Willd.) Ohwi et Ohashi var. nippoensis (Ohwi) Ohwi et H. Ohashi ■
45998 Azukia minima (Roxb.) Ohwi = Vigna minima (Roxb.) Ohwi et H. Ohashi ■
45999 Azukia mungo (L.) Masam. = Vigna mungo (L.) Hepper ■
46000 Azukia nakashimae (Ohwi) Ohwi = Vigna minima (Roxb.) Ohwi et H. Ohashi ■
46001 Azukia radiata (L.) Ohwi = Vigna radiata (L.) R. Wilczek ■
46002 Azukia reflexo-pilosa (Hayata) Ohwi = Vigna reflexopilosa Hayata ■
46003 Azukia riukiuensis (Ohwi) Ohwi = Vigna minima (Roxb.) Ohwi et H. Ohashi var. minor (Matsum.) Tateishi ■
46004 Azukia riukiuensis (Ohwi) Ohwi = Vigna riukiuensis (Ohwi) Ohwi et H. Ohashi ■
46005 Azukia umbellata (Thunb.) Ohwi = Vigna umbellata (Thunb.) Ohwi et H. Ohashi ■
46006 Azukia umbellata (Thunb.) Ohwi et Ohashi = Vigna umbellata (Thunb.) Ohwi et H. Ohashi ■
46007 Azureocereus Akers et H. Johnson = Browningia Britton et Rose ●☆
46008 Azureocereus Akers et H. Johnson(1949);佛塔柱属●☆
46009 Azureocereus hertlingianus (Backeb.) Backeb. = Browningia hertlingiana (Backeb.) Buxb. ●☆
46010 Azureocereus nobilis Akers = Browningia hertlingiana (Backeb.) Buxb. ●☆
46011 Azureocereus viridis Rauh et Backeb.;绿佛头●☆
46012 Azurinia Fourr. = Veronica L. ■
46013 Babactes A. DC. = Chirita Buch. -Ham. ex D. Don ●■
46014 Babactes A. DC. ex Meisn. = Chirita Buch. -Ham. ex D. Don ●■
46015 Babactes oblongifolia (Roxb.) DC. ex Meisn. = Chirita oblongifolia (Roxb.) J. Sinclair ■
46016 Babbagia F. Muell. (1858);翅果澳藜属●☆
46017 Babbagia F. Muell. = Osteocarpum F. Muell. ■☆
46018 Babbagia F. Muell. = Threlkeldia R. Br. ●☆
46019 Babcockia Boulos = Sonchus L. ■
46020 Babcockia Boulos(1965);加那利菊属■☆
46021 Babcockia platylepis (Webb) Boulos = Sonchus platylepis Webb ■☆
46022 Babiana Ker Gawl. ex Sims(1801)(保留属名);狒狒花属(狒狒草属,穗花溪荪属);Babiana,Baboon Flower,Ballon-root,Blue Freesias ■☆
46023 Babiana Sims = Babiana Ker Gawl. ex Sims(保留属名)■☆
46024 Babiana adpressa G. J. Lewis = Babiana scabrifolia Brehmer ex Klatt ■☆
46025 Babiana ambigua (Roem. et Schult.) G. J. Lewis;可疑狒狒花■☆
46026 Babiana angusta N. E. Br. = Babiana nana (Andréws) Spreng. var. maculata (Klatt) B. Nord. ■☆
46027 Babiana angustifolia Eckl. = Babiana nana (Andréws) Spreng. var. maculata (Klatt) B. Nord. ■☆
46028 Babiana angustifolia Sweet;窄叶狒狒花■☆
46029 Babiana atrocyanea Eckl. = Babiana angustifolia Sweet ■☆
46030 Babiana atrodeltoidea Eckl. = Babiana angustifolia Sweet ■☆
46031 Babiana attenuata G. J. Lewis;渐狭狒狒花■☆
46032 Babiana aurea Klotzsch = Crocosmia aurea (Pappe ex Hook.) Planch. ■
46033 Babiana auriculata G. J. Lewis;耳形狒狒花■☆
46034 Babiana bainesii Baker;贝恩斯狒狒花■☆
46035 Babiana bakeri Schinz = Babiana hypogaea Burch. ■☆
46036 Babiana blanda (L. Bolus) G. J. Lewis;光滑狒狒花■☆
46037 Babiana brachystachys (Baker) G. J. Lewis;短穗狒狒花■☆
46038 Babiana buchubergensis Dinter = Babiana namaquensis Baker ■☆
46039 Babiana caesia Eckl. = Babiana stricta (Aiton) Ker Gawl. ■☆
46040 Babiana cedarbergensis G. J. Lewis;锡达伯格狒狒花■☆
46041 Babiana crispa G. J. Lewis;皱波狒狒花■☆
46042 Babiana cuneata J. C. Manning et Goldblatt;楔形狒狒花■☆
46043 Babiana cuneifolia Baker = Babiana flabellifolia Harv. ex Klatt ■☆
46044 Babiana curviscapa G. J. Lewis;弯花茎狒狒花■☆
46045 Babiana densiflora Klatt = Babiana spathacea (L. f.) Ker Gawl. ■☆
46046 Babiana disticha Ker Gawl. = Babiana fragrans (Jacq.) Goldblatt et J. C. Manning ■☆
46047 Babiana disticha Ker Gawl. = Babiana plicata (Thunb.) Ker Gawl. ■☆
46048 Babiana dregei Baker;德雷狒狒花■☆
46049 Babiana ecklonii Klatt;埃氏狒狒花■☆
46050 Babiana ecklonii Klatt var. latifolia (L. Bolus) G. J. Lewis;宽叶埃氏狒狒花■☆
46051 Babiana erectifolia G. J. Lewis = Babiana stricta (Aiton) Ker Gawl. ■☆
46052 Babiana falcata G. J. Lewis = Babiana hypogaea Burch. ■☆
46053 Babiana fastigiata L. Bolus = Babiana ecklonii Klatt ■☆
46054 Babiana fimbriata (Klatt) Baker;流苏狒狒花■☆
46055 Babiana flabellifolia Harv. ex Klatt;扇叶狒狒花■☆
46056 Babiana flavida G. J. Lewis = Babiana hypogaea Burch. ■☆
46057 Babiana flavocaesia Eckl. = Babiana stricta (Aiton) Ker Gawl. var. sulphurea (Jacq.) Baker ■☆
46058 Babiana foliosa G. J. Lewis;多叶狒狒花■☆
46059 Babiana fourcadei G. J. Lewis;富尔卡德狒狒花■☆
46060 Babiana fragrans (Jacq.) Goldblatt et J. C. Manning;芳香狒狒花;Baboon Root ■☆
46061 Babiana framesii L. Bolus;弗雷斯狒狒花■☆
46062 Babiana gawleri N. E. Br. = Babiana sambucina (Jacq.) Ker Gawl. ■☆
46063 Babiana geniculata G. J. Lewis;膝曲狒狒花■☆
46064 Babiana hiemalis L. Bolus = Babiana villosula (J. F. Gmel.) Ker Gawl. ex Steud. ■☆
46065 Babiana horizontalis G. J. Lewis;平展狒狒花■☆
46066 Babiana hypogaea Burch.;地下狒狒花■☆
46067 Babiana hypogaea Burch. var. ensifolia G. J. Lewis = Babiana bainesii Baker ■☆
46068 Babiana hypogaea Burch. var. longituba G. J. Lewis = Babiana bainesii Baker ■☆
46069 Babiana intermedia L. Bolus = Babiana angustifolia Sweet ■☆
46070 Babiana klaverensis G. J. Lewis;克拉弗狒狒花■☆
46071 Babiana latifolia L. Bolus = Babiana ecklonii Klatt var. latifolia (L. Bolus) G. J. Lewis ■☆
46072 Babiana leipoldtii G. J. Lewis;莱波尔德狒狒花■☆
46073 Babiana lewisiana B. Nord.;刘易斯狒狒花■☆
46074 Babiana lilacina Eckl. = Babiana fragrans (Jacq.) Goldblatt et J. C. Manning ■☆
46075 Babiana lineolata Klatt;小花狒狒花■☆

46076 Babiana lobata G. J. Lewis;浅裂狒狒花■☆
46077 Babiana longibracteata G. J. Lewis = Babiana sambucina (Jacq.) Ker Gawl. var. longibracteata (G. J. Lewis) G. J. Lewis ■☆
46078 Babiana longicollis Dinter;长狒狒花■☆
46079 Babiana longiflora Goldblatt et J. C. Manning;长花狒狒花■☆
46080 Babiana macrantha MacOwan = Babiana pygmaea (Burm. f.) N. E. Br. ■☆
46081 Babiana macrantha MacOwan var. blanda L. Bolus = Babiana blanda (L. Bolus) G. J. Lewis ■☆
46082 Babiana maculata Klatt = Babiana nana (Andréws) Spreng. var. maculata (Klatt) B. Nord. ■☆
46083 Babiana minuta G. J. Lewis;微小狒狒花■☆
46084 Babiana montana G. J. Lewis;山地狒狒花■☆
46085 Babiana mucronata (Jacq.) Ker Gawl.;短尖狒狒花■☆
46086 Babiana mucronata (Jacq.) Ker Gawl. var. longituba G. J. Lewis;长管短尖狒狒花■☆
46087 Babiana mucronata (Jacq.) Ker Gawl. var. minor G. J. Lewis;小短尖狒狒花■☆
46088 Babiana multiflora Klatt;多花狒狒花■☆
46089 Babiana multiflora Klatt = Wachendorfia multiflora (Klatt) J. C. Manning et Goldblatt ■☆
46090 Babiana namaquensis Baker;纳马夸狒狒花■☆
46091 Babiana nana (Andréws) Spreng.;矮小乱狒狒花■☆
46092 Babiana nana (Andréws) Spreng. var. angustifolia (Eckl.) G. J. Lewis = Babiana nana (Andréws) Spreng. var. maculata (Klatt) B. Nord. ■☆
46093 Babiana nana (Andréws) Spreng. var. confusa G. J. Lewis;混乱狒狒花■☆
46094 Babiana nana (Andréws) Spreng. var. maculata (Klatt) B. Nord.;斑点矮小乱狒狒花■☆
46095 Babiana obliqua E. Phillips;偏斜狒狒花■☆
46096 Babiana occidentalis Baker = Babiana scabrifolia Brehmer ex Klatt ■☆
46097 Babiana odorata L. Bolus;香狒狒花■☆
46098 Babiana orthosantha Baker = Babiana villosula (J. F. Gmel.) Ker Gawl. ex Steud. ■☆
46099 Babiana parviflora Brehmer ex Klatt = Babiana lineolata Klatt ■☆
46100 Babiana patersoniae L. Bolus;帕特森狒狒花■☆
46101 Babiana patula N. E. Br.;张开狒狒花■☆
46102 Babiana pauciflora G. J. Lewis;少花狒狒花■☆
46103 Babiana pilosa G. J. Lewis;疏毛狒狒花■☆
46104 Babiana planifolia (G. J. Lewis) Goldblatt et J. C. Manning;平叶狒狒花■☆
46105 Babiana plicata (Thunb.) Ker Gawl.;皱迭狒狒花(皱迭狒狒草);Baboon Root ■☆
46106 Babiana plicata Ker Gawl. = Babiana fragrans (Jacq.) Goldblatt et J. C. Manning ■☆
46107 Babiana praemorsa Goldblatt et J. C. Manning;啮蚀狒狒花■☆
46108 Babiana pubescens (Lam.) G. J. Lewis;短柔毛狒狒花■☆
46109 Babiana pulchra (Salisb.) G. J. Lewis = Babiana angustifolia Sweet ■☆
46110 Babiana punicea Eckl. = Babiana villosa (Aiton) Ker Gawl. ■☆
46111 Babiana purpurea (Jacq.) Ker Gawl.;紫狒狒花■☆
46112 Babiana pygmaea (Burm. f.) N. E. Br.;瘦小狒狒花■☆
46113 Babiana pygmaea Baker = Babiana nana (Andréws) Spreng. var. confusa G. J. Lewis ■☆
46114 Babiana pygmaea Spreng. ex Steud. = Babiana nana (Andréws) Spreng. ■☆
46115 Babiana reflexa Eckl. = Babiana secunda (Thunb.) Ker Gawl. ■☆
46116 Babiana regia (G. J. Lewis) Goldblatt et J. C. Manning;硬狒狒花■☆
46117 Babiana ringens (L.) Ker Gawl.;张口狒狒花■☆
46118 Babiana rosea Eckl.;粉红狒狒花■☆
46119 Babiana rubrocyanea (Jacq.) Ker Gawl.;红蓝狒狒花(红蓝狒狒草);Winecups ■☆
46120 Babiana rubrocyanea Ker Gawl. = Babiana rubrocyanea (Jacq.) Ker Gawl. ■☆
46121 Babiana salteri G. J. Lewis;索尔特狒狒花■☆
46122 Babiana sambucina (Jacq.) Ker Gawl.;异味狒狒花;Sambucus Baboon Flower ■☆
46123 Babiana sambucina (Jacq.) Ker Gawl. var. longibracteata (G. J. Lewis) G. J. Lewis;长苞异味狒狒花■☆
46124 Babiana sambucina (Jacq.) Ker Gawl. var. undulato-venosa (Klatt) G. J. Lewis;波状异味狒狒花■☆
46125 Babiana sambucina (Jacq.) Ker Gawl. var. unguiculata G. J. Lewis;具爪狒狒花■☆
46126 Babiana sambucina Ker Gawl. = Babiana sambucina (Jacq.) Ker Gawl. ■☆
46127 Babiana scabrifolia Brehmer ex Klatt;糙叶狒狒花■☆
46128 Babiana scabrifolia Brehmer ex Klatt var. acuminata G. J. Lewis;渐尖糙叶狒狒花■☆
46129 Babiana scabrifolia Brehmer ex Klatt var. declinata G. J. Lewis;外折糙叶狒狒花■☆
46130 Babiana scariosa G. J. Lewis;干膜质狒狒花■☆
46131 Babiana schlechteri Baker = Babiana hypogaea Burch. ■☆
46132 Babiana schlechteri Baker = Babiana nana (Andréws) Spreng. var. maculata (Klatt) B. Nord. ■☆
46133 Babiana secunda (Thunb.) Ker Gawl.;单侧狒狒花■☆
46134 Babiana sinuata G. J. Lewis;深波狒狒花■☆
46135 Babiana spathacea (L. f.) Ker Gawl.;佛焰苞狒狒花■☆
46136 Babiana spiralis Baker;螺旋狒狒花■☆
46137 Babiana sprengelii Baker = Babiana nana (Andréws) Spreng. ■☆
46138 Babiana stellata Schltr. = Babiana sambucina (Jacq.) Ker Gawl. ■☆
46139 Babiana stenomera Schltr.;狭果狒狒花■☆
46140 Babiana stenophylla Baker = Babiana lineolata Klatt ■☆
46141 Babiana striata (Jacq.) G. J. Lewis;条纹狒狒花■☆
46142 Babiana striata (Jacq.) G. J. Lewis var. planifolia G. J. Lewis = Babiana planifolia (G. J. Lewis) Goldblatt et J. C. Manning ■☆
46143 Babiana stricta (Aiton) Ker Gawl. = Babiana stricta (Sol. ex Aiton) Ker Gawl. ■☆
46144 Babiana stricta (Aiton) Ker Gawl. var. erectifolia (G. J. Lewis) G. J. Lewis = Babiana stricta (Aiton) Ker Gawl. ■☆
46145 Babiana stricta (Aiton) Ker Gawl. var. grandiflora G. J. Lewis = Babiana longiflora Goldblatt et J. C. Manning ■☆
46146 Babiana stricta (Aiton) Ker Gawl. var. regia G. J. Lewis = Babiana regia (G. J. Lewis) Goldblatt et J. C. Manning ■☆
46147 Babiana stricta (Aiton) Ker Gawl. var. sulphurea (Jacq.) Baker;硫色直立狒狒花■☆
46148 Babiana stricta (Sol. ex Aiton) Ker Gawl.;直立狒狒花(狒狒草,穗花溪苏,条叶狒狒花,直立鸢尾);Stripedleaf Baboon Flower, Upright Baboon Flower ■☆
46149 Babiana stricta (Sol. ex Aiton) Ker Gawl. var. rubrocyanea Ker Gawl.;二色狒狒花;Twocolour Baboon Flower ■☆

46150 Babiana stricta Ker Gawl. var. rubrocyanea Ker Gawl. = Babiana stricta (Sol. ex Aiton) Ker Gawl. var. rubrocyanea Ker Gawl. ■☆

46151 Babiana stricta Ker Gawl. var. sulphrea (Ker Gawl.) Baker = Babiana stricta (Aiton) Ker Gawl. var. sulphurea (Jacq.) Baker ■☆

46152 Babiana subglabra G. J. Lewis = Babiana scabrifolia Brehmer ex Klatt ■☆

46153 Babiana sulphurea (Jacq.) Ker Gawl. = Babiana stricta (Aiton) Ker Gawl. var. sulphurea (Jacq.) Baker ■☆

46154 Babiana thunbergii Ker Gawl.;通贝里狒狒花■☆

46155 Babiana torta G. J. Lewis;缠扭狒狒花■☆

46156 Babiana tritonioides G. J. Lewis;观音兰狒狒花■☆

46157 Babiana truncata G. J. Lewis = Babiana flabellifolia Harv. ex Klatt ■☆

46158 Babiana tubata (Jacq.) Sweet = Babiana tubulosa (Burm. f.) Ker Gawl. ■☆

46159 Babiana tubiflora (L. f.) Ker Gawl. = Babiana tubulosa (Burm. f.) Ker Gawl. var. tubiflora (L. f.) G. J. Lewis ■☆

46160 Babiana tubiflora Eckl. = Babiana ecklonii Klatt ■☆

46161 Babiana tubulosa (Burm. f.) Ker Gawl.;管状狒狒花■☆

46162 Babiana tubulosa (Burm. f.) Ker Gawl. var. tubiflora (L. f.) G. J. Lewis;管花狒狒花■☆

46163 Babiana undulato-venosa Klatt = Babiana sambucina (Jacq.) Ker Gawl. var. undulato-venosa (Klatt) G. J. Lewis ■☆

46164 Babiana unguiculata G. J. Lewis;爪状狒狒花■☆

46165 Babiana vanzyliae L. Bolus;万齐狒狒花■☆

46166 Babiana velutina Schltr. = Babiana ecklonii Klatt ■☆

46167 Babiana villosa (Aiton) Ker Gawl.;长柔毛狒狒花■☆

46168 Babiana villosa (Aiton) Ker Gawl. var. grandis G. J. Lewis;大狒狒花■☆

46169 Babiana villosula (J. F. Gmel.) Ker Gawl. ex Steud.;毛狒狒花■☆

46170 Babiana virginea Goldblatt;纯白狒狒花■☆

46171 Babingtonia Lindl. = Baeckea L. ●

46172 Babiron Raf. = Spermolepis Raf. ■☆

46173 Baca Raf. = Boea Comm. ex Lam. ■

46174 Bacasia Ruiz et Pav. = Barnadesia Mutis ex L. f. ●☆

46175 Baccataceae Dulac = Caprifoliaceae Juss.(保留科名) + Sambucaceae Link + Adoxaceae E. Mey.(保留科名)●■

46176 Baccaurea Lour.(1790);木奶果属(黄果树属);Baccaurea ●

46177 Baccaurea barteri (Baill.) Hutch. = Maesobotrya barteri (Baill.) Hutch. ●☆

46178 Baccaurea bipendensis Pax = Maesobotrya bipindensis (Pax) Hutch. ●☆

46179 Baccaurea bonnetii Beille = Maesobotrya barteri (Baill.) Hutch. var. sparsiflora (Scott-Elliot) Keay ●☆

46180 Baccaurea caillei Beille = Maesobotrya barteri (Baill.) Hutch. var. sparsiflora (Scott-Elliot) Keay ●☆

46181 Baccaurea cauliflora Lour. = Baccaurea ramiflora Lour. ●

46182 Baccaurea cavaleriei H. Lév. = Cleidiocarpon cavalerei (H. Lév.) Airy Shaw ●◇

46183 Baccaurea cavalliensis Beille = Maesobotrya barteri (Baill.) Hutch. var. sparsiflora (Scott-Elliot) Keay ●☆

46184 Baccaurea edulis A. Chev. = Maesobotrya barteri (Baill.) Hutch. var. sparsiflora (Scott-Elliot) Keay ●☆

46185 Baccaurea esquirolii H. Lév. = Sapium rotundifolium Hemsl. ●

46186 Baccaurea gagnepainii Beille = Maesobotrya barteri (Baill.) Hutch. var. sparsiflora (Scott-Elliot) Keay ●☆

46187 Baccaurea glaziovii Beille = Maesobotrya barteri (Baill.) Hutch. var. sparsiflora (Scott-Elliot) Keay ●☆

46188 Baccaurea gracilis Merr. = Richeriella gracilis (Merr.) Pax et K. Hoffm. ●

46189 Baccaurea griffoniana (Baill.) Müll. Arg. = Maesobotrya griffoniana (Baill.) Hutch. ●☆

46190 Baccaurea lanceolata Müll. Arg.;披针木奶果●☆

46191 Baccaurea longispicata Beille = Maesobotrya barteri (Baill.) Hutch. var. sparsiflora (Scott-Elliot) Keay ●☆

46192 Baccaurea macrophylla Pax = Protomegabaria macrophylla (Pax) Hutch. ●☆

46193 Baccaurea motleyana (Müll. Arg.) Müll. Arg.;多脉木奶果;Many-veins Baccaurea, Rambai ●

46194 Baccaurea oxycarpa Gagnep. = Baccaurea ramiflora Lour. ●

46195 Baccaurea parviflora (Müll. Arg.) Müll. Arg.;小花木奶果●☆

46196 Baccaurea poissonii Beille = Maesobotrya barteri (Baill.) Hutch. var. sparsiflora (Scott-Elliot) Keay ●☆

46197 Baccaurea pynaertii De Wild. = Maesobotrya pynaertii (De Wild.) Pax et K. Hoffm. ●☆

46198 Baccaurea ramiflora Lour.;木奶果(白皮,大连果,黄果树,火果,麦穗,美味木奶果,木来果,木荔枝,山豆,山萝卜,树葡萄,蒜瓣果,铁东木,野黄皮树,枝花木奶果);Common Baccaurea, Lutqua, Rambai, Ramiflorous Baccaurea ●

46199 Baccaurea sapida (Roxb.) Müll. Arg. = Baccaurea ramiflora Lour. ●

46200 Baccaurea sparsiflora Scott-Elliot = Maesobotrya barteri (Baill.) Hutch. var. sparsiflora (Scott-Elliot) Keay ●☆

46201 Baccaurea staudtii Pax = Maesobotrya staudtii (Pax) Hutch. ●☆

46202 Baccaurea vermeulenii De Wild. = Maesobotrya vermeulenii (De Wild.) J. Léonard ●☆

46203 Baccaureopsis Pax = Thecacoris A. Juss. ●☆

46204 Baccaureopsis lucida Pax = Thecacoris lucida (Pax) Hutch. ●☆

46205 Baccharidastrum Cabrera = Baccharis L.(保留属名)●■☆

46206 Baccharidastrum Cabrera(1937);小种棉木属●☆

46207 Baccharidastrum triplinervium (Less.) Cabrera;小种棉木●☆

46208 Baccharidiopsis G. M. Barroso = Aster L. ●■

46209 Baccharidiopsis G. M. Barroso = Baccharis L.(保留属名)●■☆

46210 Baccharis L.(1753)(保留属名);种棉木属(酒神菊属,无舌紫菀属);Baccharis, Coyote Bush, Groundsel-tree, Tree Groundsel ●■☆

46211 Baccharis alamosana S. F. Blake = Baccharis thesioides Kunth ●☆

46212 Baccharis angustifolia Michx.;窄叶种棉木;Narrowleaf baccharis, Saltwater False Willow ●☆

46213 Baccharis bigelovii A. Gray;毕氏种棉木;Bigelow's False Willow ●☆

46214 Baccharis brachyphylla A. Gray;短叶种棉木;Shortleaf Baccharis or False Willow ●☆

46215 Baccharis chinensis Lour. = Duhaldea chinensis DC. ●■

46216 Baccharis chinensis Lour. = Inula cappa (Buch. -Ham.) DC. ●■

46217 Baccharis consanguinea DC. = Baccharis pilularis DC. subsp. consanguinea (DC.) C. B. Wolf ●☆

46218 Baccharis crispa Spreng.;皱酒神菊●☆

46219 Baccharis dioica Vahl;帚状种棉木;Broombush False Willow ●☆

46220 Baccharis dioscoridis L. = Pluchea dioscoridis (L.) DC. ●☆

46221 Baccharis douglasii DC.;道氏酒神菊;Douglas' Falsewillow, Saltmarsh Baccharis ■☆

46222 Baccharis emoryi A. Gray = Baccharis salicina Torr. et Gray ●☆

46223 Baccharis foetida L. = Pluchea foetida (L.) DC. ■☆

46224 Baccharis glomeruliflora Pers.;银叶种棉木;Silverling ●☆

46225 Baccharis glutinosa Pers.;水种棉木(胶质酒神菊);Seep Willow,Water Wally,Water Willow,Watermotie ●☆

46226 Baccharis glutinosa Pers. = Baccharis salicifolia (Ruiz et Pav.) Pers. ●☆

46227 Baccharis halimifolia L.;倒卵叶种棉木(倒卵叶无舌紫菀,酒神菊);Bush Groundsel,Consumption Weed,Consumption-weed,Cotton-seed Tree,Eastern Baccharis,Giant Groundsel,Groundsel Bush,Groundsel Tree,Groundsel-tree,Salt Marsh Elder,Sea Myrtle,Sea-myrtle,Tree Groundsel ●☆

46228 Baccharis halimifolia var. angustior DC. = Baccharis halimifolia L. ●☆

46229 Baccharis havardii A. Gray;哈氏种棉木;Havard's False Willow ●☆

46230 Baccharis ilicifolia Lam. = Brachylaena ilicifolia (Lam.) E. Phillips et Schweick. ●☆

46231 Baccharis indica L. = Pluchea indica (L.) Less. ●■

46232 Baccharis ivifolia L. = Conyza scabrida DC. ■☆

46233 Baccharis magellanica (Lam.) Pers.;麦哲伦种棉木●☆

46234 Baccharis malibuensis R. M. Beauch. et Henrickson;种棉木;Coyote Brush,Malibu Baccharis ●☆

46235 Baccharis megapotamica Spreng.;旱地菊●☆

46236 Baccharis neglecta Britton;疏忽种棉木;Linear-leaved False Willow,New Deal Weed,Roosevelt Weed ●☆

46237 Baccharis neriifolia L. = Brachylaena neriifolia (L.) R. Br. ●☆

46238 Baccharis ovalis Pers. = Pluchea ovalis (Pers.) DC. ■☆

46239 Baccharis ovata Sieber ex DC. = Pluchea ovalis (Pers.) DC. ■☆

46240 Baccharis pedunculata (Mill.) Cabrera;花梗酒神菊●☆

46241 Baccharis pilularis DC.;小球种棉木(小球花酒神菊);Chaparal Broom,Coyote Bush,Dwarf Baccharis,Dwarf Chaparral False Willow,Dwarf Coyote Brush ●☆

46242 Baccharis pilularis DC. subsp. consanguinea (DC.) C. B. Wolf;亲缘小球种棉木●☆

46243 Baccharis pilularis DC. var. consanguinea (DC.) Kuntze = Baccharis pilularis DC. subsp. consanguinea (DC.) C. B. Wolf ●☆

46244 Baccharis pingraea DC.;智利种棉木●☆

46245 Baccharis plummerae A. Gray;普氏种棉木;Plummer's Baccharis ●☆

46246 Baccharis plummerae A. Gray subsp. glabrata Hoover;光滑种棉木;San Simeon,Smooth Baccharis ●☆

46247 Baccharis polyantha Kunth;多花种棉木(多花酒神菊)●☆

46248 Baccharis pteronioides DC.;巴拉圭种棉木●☆

46249 Baccharis ramulosa (DC.) A. Gray = Baccharis pteronioides DC. ●☆

46250 Baccharis resiniflua Hochst. et Steud. ex DC. = Psiadia punctulata (DC.) Vatke ●☆

46251 Baccharis salicifolia (Ruiz et Pav.) Pers.;柳叶种棉木;Mule Fat,Mule's Fat,Seep Willow,Seepwillow,Water Wally ●☆

46252 Baccharis salicifolia Nutt. = Baccharis salicina Torr. et Gray ●☆

46253 Baccharis salicina Torr. et Gray;柳状种棉木;Great Plains False Willow,Willow Baccharis,Willow-baccharis ●☆

46254 Baccharis salvia Lour. = Blumea balsamifera (L.) DC. ■

46255 Baccharis sarothroides A. Gray;沙地种棉木(普通种棉木);Broom Baccharis,Desert Broom,Desertbroom,Greasewood,Groundsel,Rosin-brush ●☆

46256 Baccharis senegalensis Pers. = Vernonia colorata (Willd.) Drake ■☆

46257 Baccharis sergiloides A. Gray;沙生种棉木;Desert Baccharis,Squaw False Willow ●☆

46258 Baccharis serrifolia DC.;锯叶酒神菊●☆

46259 Baccharis sessiliflora Michx. = Baccharis glomeruliflora Pers. ●☆

46260 Baccharis texana (Torr. et A. Gray) A. Gray;草地种棉木;Prairie Baccharis,Prairie False Willow ●■☆

46261 Baccharis thesioides Kunth;亚利桑那种棉木;Arizona Baccharis ●☆

46262 Baccharis tricuneata Pers.;三楔旱地菊●☆

46263 Baccharis trimaera (Less.) DC.;三数旱地菊(三数酒神菊)●☆

46264 Baccharis trinermis (Lam.) Pers.;三脉酒神菊●☆

46265 Baccharis tucumanensis Hook. et Arn.;土库曼酒神菊(土可曼酒神菊)●☆

46266 Baccharis ulmifolia Burm. f. = Conyza ulmifolia (Burm. f.) Kuntze ■☆

46267 Baccharis vaccinioides Gardner;越橘酒神菊●☆

46268 Baccharis vanessae R. M. Beauch.;巴涅斯种棉木;Encinitas Baccharis,Encinitas False Willow ●☆

46269 Baccharis viminea DC.;柔枝酒神菊;Mule Fat,Mule-fat ●☆

46270 Baccharis viminea DC. = Baccharis salicifolia (Ruiz et Pav.) Pers. ●☆

46271 Baccharis viminea DC. var. atwoodii S. L. Welsh = Baccharis salicifolia (Ruiz et Pav.) Pers. ●☆

46272 Baccharis wrightii A. Gray;赖氏种棉木;False Willow,Wright's Baccharis ●☆

46273 Baccharodes Kuntze = Baccharoides Moench ●■

46274 Baccharodes Kuntze = Vernonia Schreb.(保留属名)●■

46275 Baccharoides Moench = Vernonia Schreb.(保留属名)●■

46276 Baccharoides Moench(1794);驱虫菊属●■

46277 Baccharoides adoensis (Sch. Bip. ex Walp.) H. Rob. = Vernonia adoensis Sch. Bip. ex Walp. ■☆

46278 Baccharoides adoensis (Sch. Bip. ex Walp.) H. Rob. var. kotschyana (Sch. Bip. ex Walp.) Isawumi,El-Ghazaly et B. Nord. = Vernonia adoensis Sch. Bip. ex Walp. ■☆

46279 Baccharoides adoensis (Sch. Bip. ex Walp.) H. Rob. var. mossambiquensis (Steetz) Isawumi,El-Ghazaly et B. Nord.;莫桑比克驱虫菊●☆

46280 Baccharoides anthelmintica (L.) Moench;驱虫菊●

46281 Baccharoides ballyi (C. Jeffrey) Isawumi,El-Ghazaly et B. Nord. = Vernonia ballyi C. Jeffrey ■☆

46282 Baccharoides bracteosa (O. Hoffm.) Isawumi et El-Ghazaly et B. Nord. = Vernonia bracteosa O. Hoffm. ■☆

46283 Baccharoides calvoana (Hook. f.) Isawumi = Vernonia calvoana (Hook. f.) Hook. f. ●☆

46284 Baccharoides calvoana (Hook. f.) Isawumi subsp. adolfi-friderici (Muschl.) Isawumi = Vernonia calvoana (Hook. f.) Hook. f. subsp. adolfi-friderici (Muschl.) C. Jeffrey ■☆

46285 Baccharoides calvoana (Hook. f.) Isawumi subsp. leucocalyx (O. Hoffm.) Isawumi et El-Ghazaly et B. Nord. = Vernonia calvoana (Hook. f.) Hook. f. subsp. leucocalyx (O. Hoffm.) C. Jeffrey ■☆

46286 Baccharoides calvoana (Hook. f.) Isawumi subsp. meridionalis (Wild) Isawumi,El-Ghazaly et B. Nord. = Vernonia calvoana (Hook. f.) Hook. f. subsp. meridionalis (Wild) C. Jeffrey ■☆

46287 Baccharoides calvoana (Hook. f.) Isawumi subsp. mokaensis (Mildbr. et Mattf.) Isawumi et El-Ghazaly et B. Nord. = Vernonia calvoana (Hook. f.) Hook. f. var. mokaensis (Mildbr. et Mattf.) C. Jeffrey ■☆

46288 Baccharoides calvoana (Hook. f.) Isawumi subsp. oehleri (Muschl.) Isawumi = Vernonia calvoana (Hook. f.) Hook. f. subsp. oehleri (Muschl.) C. Jeffrey ■☆

46289 Baccharoides calvoana (Hook. f.) Isawumi subsp. ruwenzoriensis (C. Jeffrey) Isawumi et El-Ghazaly et B. Nord. = Vernonia calvoana (Hook. f.) Hook. f. subsp. ruwenzoriensis C. Jeffrey ■☆

46290 Baccharoides calvoana (Hook. f.) Isawumi subsp. ulugurensis (O. Hoffm.) Isawumi et El-Ghazaly et B. Nord. = Vernonia calvoana (Hook. f.) Hook. f. subsp. ulugurensis (O. Hoffm.) C. Jeffrey ■☆

46291 Baccharoides calvoana (Hook. f.) Isawumi subsp. usambarensis (C. Jeffrey) Isawumi et El-Ghazaly et B. Nord. = Vernonia calvoana (Hook. f.) Hook. f. subsp. usambarensis C. Jeffrey ■☆

46292 Baccharoides calvoana (Hook. f.) Isawumi var. acuta (C. D. Adams) Isawumi = Vernonia calvoana (Hook. f.) Hook. f. var. acuta (C. D. Adams) C. Jeffrey ■☆

46293 Baccharoides calvoana (Hook. f.) Isawumi var. hymenolepis (A. Rich.) Isawumi = Vernonia hymenolepis A. Rich. ■☆

46294 Baccharoides calvoana (Hook. f.) Isawumi var. microcephala (C. D. Adams) Isawumi = Vernonia calvoana (Hook. f.) Hook. f. var. microcephala C. D. Adams ■☆

46295 Baccharoides cardiolepis (O. Hoffm.) Isawumi et El-Ghazaly et B. Nord. = Vernonia guineensis Benth. ■☆

46296 Baccharoides dumicola (S. Moore) Isawumi et El-Ghazaly et B. Nord. = Vernonia lasiopus O. Hoffm. ●☆

46297 Baccharoides filigera (Oliv. et Hiern) Isawumi et El-Ghazaly et B. Nord. = Vernonia filigera Oliv. et Hiern ■☆

46298 Baccharoides filipendula (Hiern) Isawumi et El-Ghazaly et B. Nord. = Vernonia filipendula Hiern ■☆

46299 Baccharoides guineensis (Benth.) H. Rob. = Vernonia guineensis Benth. ■☆

46300 Baccharoides guineensis (Benth.) H. Rob. var. cameroonica (C. D. Adams) Isawumi = Vernonia guineensis Benth. var. cameroonica C. D. Adams ●☆

46301 Baccharoides guineensis (Benth.) H. Rob. var. procera (O. Hoffm.) Isawumi = Vernonia procera O. Hoffm. ●☆

46302 Baccharoides hymenolepis (A. Rich.) Isawumi, B. Nord. et El-Ghazaly = Vernonia hymenolepis A. Rich. ■☆

46303 Baccharoides incompta (S. Moore) Isawumi et El-Ghazaly et B. Nord. = Vernonia incompta S. Moore ■☆

46304 Baccharoides kirungae (R. E. Fr.) Isawumi, El-Ghazaly et B. Nord. = Vernonia kirungae R. E. Fr. ●☆

46305 Baccharoides lasiopus (O. Hoffm.) H. Rob. = Vernonia lasiopus O. Hoffm. ●☆

46306 Baccharoides lasiopus (O. Hoffm.) H. Rob. var. acuta (C. Jeffrey) Isawumi, El-Ghazaly et B. Nord. = Vernonia lasiopus O. Hoffm. var. acuta C. Jeffrey ●☆

46307 Baccharoides lasiopus (O. Hoffm.) H. Rob. var. caudata (C. Jeffrey) Isawumi et El-Ghazaly et B. Nord. = Vernonia lasiopus O. Hoffm. var. caudata C. Jeffrey ●☆

46308 Baccharoides lasiopus (O. Hoffm.) H. Rob. var. grandiceps (C. Jeffrey) Isawumi et El-Ghazaly et B. Nord. = Vernonia lasiopus O. Hoffm. var. grandiceps C. Jeffrey ●☆

46309 Baccharoides lasiopus (O. Hoffm.) H. Rob. var. iodocalyx (O. Hoffm.) Isawumi et El-Ghazaly et B. Nord. = Vernonia lasiopus O. Hoffm. var. iodocalyx (O. Hoffm.) C. Jeffrey ●☆

46310 Baccharoides longipedunculata (De Wild.) Isawumi, El-Ghazaly et B. Nord. = Vernonia longipedunculata De Wild. ■☆

46311 Baccharoides longipedunculata (De Wild.) Isawumi, El-Ghazaly et B. Nord. var. retusa (R. E. Fr.) Isawumi, El-Ghazaly et B. Nord. = Vernonia longipedunculata De Wild. var. retusa (R. E. Fr.) G. V. Pope ■☆

46312 Baccharoides longipedunculata (De Wild.) Isawumi, El-Ghazaly et B. Nord. var. manikensis (De Wild.) Isawumi, El-Ghazaly et B. Nord. = Vernonia longipedunculata De Wild. var. manikensis (De Wild.) G. V. Pope ■☆

46313 Baccharoides nimbaensis (C. D. Adams) Isawumi et El-Ghazaly et B. Nord. = Vernonia nimbaensis C. D. Adams ■☆

46314 Baccharoides prolixa (S. Moore) Isawumi et El-Ghazaly et B. Nord. = Vernonia prolixa S. Moore ■☆

46315 Baccharoides pumila (Kotschy et Peyr.) Isawumi = Vernonia pumila Kotschy et Peyr. ●☆

46316 Baccharoides ringoetii (De Wild.) Isawumi, El-Ghazaly et B. Nord. = Vernonia ringoetii De Wild. ■☆

46317 Baccharoides schimperi (DC.) Isawumi et El-Ghazaly et B. Nord. = Vernonia schimperi DC. ■☆

46318 Baccharoides stenostegia (Stapf) Isawumi = Vernonia stenostegia (Stapf) Hutch. et Dalziel ■☆

46319 Baccharoides sunzuensis (Wild) Isawumi, El-Ghazaly et B. Nord. = Vernonia sunzuensis Wild ■☆

46320 Baccharoides tayloriana Isawumi = Vernonia hymenolepis A. Rich. ■☆

46321 Baccharoides tenoreana (O. Hoffm.) Isawumi = Vernonia tenoreana Oliv. ■☆

46322 Bachmannia Pax(1897);巴克曼山柑属●☆

46323 Bachmannia woodii (Oliv.) Gilg;巴克曼山柑●☆

46324 Backebergia Bravo = Mitrocereus (Backeb.) Backeb. ●☆

46325 Backebergia Bravo = Pachycereus (A. Berger) Britton et Rose ●

46326 Backebergia Bravo(1953);华装翁属■☆

46327 Backebergia militaris (Audot) Bravo;华装翁(冑状华装翁)■☆

46328 Backeria Bakh. f. = Anplectrum A. Gray ●■

46329 Backeria Bakh. f. = Diplectria (Blume) Rchb. ●■

46330 Backeria barbata (Wall. ex C. B. Clarke) Raizada = Diplectria barbata (Wall. ex C. B. Clarke) Franken et M. C. Roos ●■

46331 Backeria barbata Raizada = Diplectria barbata (Wall. ex C. B. Clarke) Franken et M. C. Roos ●■

46332 Backhousea Kuntze = Backhousia Hook. et Harv. ●☆

46333 Backhousia Hook. et Harv. (1845);巴克木属●☆

46334 Backhousia angustifolia F. Muell.;狭叶巴克木●☆

46335 Backhousia anisata Vickery;八角茴香巴克木;Aniseed Tree, Aniseed-tree, Ringwood ●☆

46336 Backhousia bancroftii Bailey;巴克木;Johnston River hardwood ●☆

46337 Backhousia citriodora F. Muell.;柠檬巴克木;Lemon-scented Myrtle, Sweet Verbena Tree ●☆

46338 Backhousia myrtifolia Hook. et Harv.;铁巴克木;Gray Myrtle, Ironwood ●☆

46339 Backhousia sciadophora F. Muell.;易裂巴克木;Shatterwood ●☆

46340 Baclea E. Fourn. (1877);巴氏萝藦属■☆

46341 Baclea Greene = Nemacladus Nutt. ■☆

46342 Baclea Greene = Pseudonemacladus McVaugh ■☆

46343 Baclea oppositifolia (B. L. Rob.) Greene;巴氏萝藦■☆

46344 Baclea oppositifolia Greene = Baclea oppositifolia (B. L. Rob.) Greene ■☆

46345 Baconia DC. = Pavetta L. ●

46346 Baconia corymbosa DC. = Pavetta corymbosa (DC.) F. N.

Williams ●☆

46347 Baconia montana Hook. f. = Pavetta hookeriana Hiern ●☆

46348 Bacopa Aubl. (1775)(保留属名);巴考婆婆纳属(过长沙属,假马齿苋属);Water Hyssop, Waterhissop, Water-hyssop ■

46349 Bacopa acuminata (Walter) B. L. Rob. = Mecardonia acuminata (Walter) Small ■☆

46350 Bacopa alternifolia Engl.;互叶巴考婆婆纳■☆

46351 Bacopa calycina (Benth.) Engl. ex De Wild. = Bacopa crenata (P. Beauv.) Hepper ■☆

46352 Bacopa caroliniana (Walter) B. L. Rob.;卡罗来纳巴考婆婆纳(卡罗来纳假马齿苋);Lemon Bacopa, Water Hyssop ■☆

46353 Bacopa chamaedryoides (Kunth) Wettst. = Bacopa procumbens (Mill.) Greenm. ■☆

46354 Bacopa crenata (P. Beauv.) Hepper;圆齿巴考婆婆纳■☆

46355 Bacopa decumbens (Fernald) F. N. Williams;外倾巴考婆婆纳■☆

46356 Bacopa egensis (Poepp. et Endl.) Pennell;巴西巴考婆婆纳(巴西假马齿苋);Brazilian Waterhyssop ■☆

46357 Bacopa egensis (Poepp.) Pennell = Bacopa egensis (Poepp. et Endl.) Pennell ■☆

46358 Bacopa erecta Hutch. et Dalziel = Bacopa decumbens (Fernald) F. N. Williams ■☆

46359 Bacopa floribunda (R. Br.) Wettst.;麦花草;Manyflower Waterhissop ■

46360 Bacopa hamiltoniana (Benth.) Wettst.;汉密尔顿巴考婆婆纳■☆

46361 Bacopa lisowskiana Mielcarek;利索巴考婆婆纳■☆

46362 Bacopa monnieri (L.) Pennell;假马齿苋(白花猪母菜,白线草,百克爬草,过长沙,蛇鳞菜);Brahmi, Coastal Waterhissop, Smooth Water Hyssop, Water Hyssop ■

46363 Bacopa monnieri (L.) Wettst. = Bacopa floribunda (R. Br.) Wettst. ■

46364 Bacopa nobsiana H. Mason = Bacopa rotundifolia (Michx.) Wettst. ■☆

46365 Bacopa occultans (Hiern) Hutch. et Dalziel;隐蔽巴考婆婆纳■☆

46366 Bacopa procumbens (Mill.) Greenm.;平卧假马齿苋(黄花过长沙舅)■☆

46367 Bacopa procumbens (Mill.) Greenm. = Mecardonia procumbens (Mill.) Small ■☆

46368 Bacopa pubescens (V. Naray.) Hutch. et Dalziel = Bacopa floribunda (R. Br.) Wettst. ■

46369 Bacopa punctata Engl.;斑点假马齿苋■☆

46370 Bacopa repens (Sw.) Wettst. = Bacopa monnieri (L.) Pennell ■

46371 Bacopa rotundifolia (Michx.) Wettst.;钝叶假马齿苋;Disc Water-hyssop, Water Hyssop ■☆

46372 Bacopa simulans Fernald = Bacopa rotundifolia (Michx.) Wettst. ■☆

46373 Bactris Jacq. = Bactris Jacq. ex Scop. ●

46374 Bactris Jacq. ex Scop. (1777);刺棒棕属(栗椰属,粮棕属,手杖椰子属,桃果椰子属,桃榈属,桃棕属);Gris Palm, Spiny Club Palm, Spiny Club-palm, Spiny-club Palm ●

46375 Bactris ciliata Mart. = Bactris gasipaes Kunth ●☆

46376 Bactris gasipaes Kunth;刺棒棕(桃果榈,桃榈);Peach Nut, Peach Palm, Pejibaye, Pejivalle, Pejivalle Pejibaye ●☆

46377 Bactris guineensis (L.) H. E. Moore;多巴哥刺棒棕;Prickly-pole, Tobago Cane ●☆

46378 Bactris horrida Oerst.;刺毛刺棒棕;Pristly Spiny-club Palm ●☆

46379 Bactris major Jacq.;手杖椰子(大刺棒棕,大桃榈);Beach Palm, Beach Spiny-clubpalm, Black Roseau, Black Roseau Palm, Prickly Palm ●

46380 Bactris mexican Mart.;墨西哥刺棒棕(墨西哥桃榈)●☆

46381 Bactris minor Jacq.;小刺棒棕;Small Spiny-club Palm ●☆

46382 Bactris pallidispina Mart.;白刺棒棕;White Spiny-club Palm ●☆

46383 Bactyrilobium Willd. = Cassia L.(保留属名)●■

46384 Bacularia F. Muell. = Linospadix H. Wendl. ●☆

46385 Bacularia F. Muell. ex Hook. f. = Linospadix H. Wendl. ●☆

46386 Badamia Gaertn. = Terminalia L.(保留属名)●

46387 Badamia commersonii Gaertn. = Terminalia catappa L. ●

46388 Badaroa Bert. ex Steud. = Sicyos L. ■

46389 Baderoa Bert. ex Hook. = Sicyos L. ■

46390 Badianifera Kuntze = Illicium L. ●

46391 Badianifera L. = Illicium L. ●

46392 Badianifera floridana (J. Ellis) Kuntze = Illicium floridanum J. Ellis ●☆

46393 Badianifera griffithii Kuntze = Illicium griffithii Hook. f. et Thomson ex Walp. ●

46394 Badianifera major Kuntze = Illicium majus Hook. f. et Thomson ●

46395 Badianifera parviflora (Michx. ex Vent.) Kuntze = Illicium parviflorum Michx. ex Vent. ●☆

46396 Badiera DC. (1824);巴迪远志属●■☆

46397 Badiera Hassk. = Polygala L. ●■

46398 Badiera acuminata DC.;渐尖巴迪远志■☆

46399 Badiera cubensis Britton;古巴巴迪远志■☆

46400 Badiera heterophylla Britton;异叶巴迪远志■☆

46401 Badiera montana Britton;山地巴迪远志■☆

46402 Badiera oblongata Britton;矩圆巴迪远志■☆

46403 Badiera punctata Britton;斑点巴迪远志■☆

46404 Badilloa R. M. King et H. Rob. (1975);点腺亮泽兰属●☆

46405 Badilloa salicina (Lam.) R. M. King et H. Rob.;柳叶点腺亮泽兰●☆

46406 Badula Juss. (1789);无梗药紫金牛属●☆

46407 Badula arborea Thouars ex A. DC. = Embelia arborea A. DC. ●☆

46408 Badula crassa A. DC.;粗无梗药紫金牛●☆

46409 Badula divaricata Thouars ex A. DC. = Oncostemum divaricatum A. DC. ●☆

46410 Badula laurifolia Bojer ex A. DC. = Oncostemum laurifolium (Bojer ex A. DC.) Mez ●☆

46411 Badula leandriana H. Perrier;利安无梗药紫金牛●☆

46412 Badula myrtifolia Thouars ex A. DC. = Oncostemum microphyllum (Roem. et Schult.) Mez ●☆

46413 Badula pauciflora Bojer ex A. DC. = Oncostemum pauciflorum A. DC. ●☆

46414 Badula pervilleana H. Perrier;佩尔无梗药紫金牛●☆

46415 Badula richardiana H. Perrier;理查德无梗药紫金牛●☆

46416 Badusa A. Gray(1859);白杜伞属●☆

46417 Badusa corymbifera (G. Forst.) A. Gray;白杜伞●☆

46418 Baea Comm. ex Juss. = Boea Comm. ex Lam. ■

46419 Baea Juss. = Boea Comm. ex Lam. ■

46420 Baechea Colla = Baeckea L. ●

46421 Baecka Cothen. = Baeckea L. ●

46422 Baeckea Burm. f. = Brunia Lam.(保留属名)●☆

46423 Baeckea L. (1753);岗松属;Baeckea ●

46424 Baeckea africana Burm. f. = Pseudobaeckea africana (Burm. f.) Pillans ●☆

46425 Baeckea brevifolia (Rudge) DC.;小叶岗松(短叶岗松);Heath Myrtle ●☆

46426 Baeckea chinensis Gaertn. = Baeckea frutescens L. ●

46427 Baeckea cochinchinensis Blume = Baeckea frutescens L. ●

46428 Baeckea cordata Burm. f. = Pseudobaeckea cordata (Burm. f.) Nied. ●☆

46429 Baeckea densifolia Sm.;密叶岗松;Dense-leaved Baeckea ●☆

46430 Baeckea frutescens L.;岗松(长松,鸡儿松,榕节花,扫把木,扫把枝,扫卡木,扫帚子,沙松,蛇虫草,石松草,松毛枝,铁扫把,香柴,羊脷木,羊脷叶,棕筛花);Shrubby Baeckea ●

46431 Baeckea frutescens L. var. brachyphylla Merr. et L. M. Perry;短叶岗松;Short-leaf Shrubby Baeckea ●

46432 Baeckea frutescens L. var. brachyphylla Merr. et L. M. Perry = Baeckea frutescens L. ●

46433 Baeckea gunniana Schauer ex Walp.;高山岗松;Alpine Baeckea ●☆

46434 Baeckea imbricata Druce;叠叶岗松●☆

46435 Baeckea lancifolia Eckl. et Zeyh. = Pseudobaeckea cordata (Burm. f.) Nied. ●☆

46436 Baeckea linifolia Rudge;线叶岗松;Swamp Baeckea ●☆

46437 Baeckea ramosissima A. Cunn.;玫瑰粉岗松;Rosy Baeckea, Rosy Heath Myrtle, Rosy Heath-myrtle ●☆

46438 Baeckea sumatrana Blume = Baeckea frutescens L. ●

46439 Baeckea virgata Andréws;高大岗松;Tall Baeckea ●☆

46440 Baeckeaceae Bercht. et J. Presl = Myrtaceae Juss.(保留科名)●

46441 Baeckia Andrews = Baeckea L. ●

46442 Baeckia R. Br. = Baeckea L. ●

46443 Baeica C. B. Clarke = Boeica T. Anderson ex C. B. Clarke ●■

46444 Baeobotrys J. Forst. et G. Forst. = Maesa Forssk. ●

46445 Baeobotrys argentea Wall. = Maesa argentea (Wall.) A. DC. ●

46446 Baeobotrys indica Roxb. = Maesa indica (Roxb.) A. DC. ●

46447 Baeobotrys japonica (Thunb.) Zipp. ex Scheffer = Maesa japonica (Thunb.) Moritzi ex Zoll. ●

46448 Baeobotrys japonica Zipp. ex Scheff. = Maesa japonica (Thunb.) Moritzi ex Zoll. ●

46449 Baeobotrys lanceolata (Forssk.) Vahl = Maesa lanceolata Forssk. ●☆

46450 Baeobotrys ramentacea Roxb. = Maesa ramentacea (Roxb.) A. DC. ●

46451 Baeobotrys rufescens E. Mey. = Choristylis rhamnoides Harv. ●☆

46452 Baeoehortus Ehrh. = Carex L. ■

46453 Baeolepis Decne. ex Moq. (1849);弱鳞萝藦属(南印度萝藦属)■☆

46454 Baeolepis nervosa (Wight et Arn.) Decne. = Baeolepis nervosa (Wight et Arn.) Decne. ex Moq. ■☆

46455 Baeolepis nervosa (Wight et Arn.) Decne. ex Moq.;弱鳞萝藦(南印度萝藦)■☆

46456 Baeometra Salisb. ex Endl. (1836);南非秋水仙属■☆

46457 Baeometra breyniana (L.) Baill.;南非秋水仙■☆

46458 Baeometra columellaris Salisb. = Baeometra uniflora (Jacq.) G. J. Lewis ■☆

46459 Baeometra uniflora (Jacq.) G. J. Lewis;单花南非秋水仙■☆

46460 Baeoterpe Salisb. = Baeometra Salisb. ex Endl. ■☆

46461 Baeoterpe Salisb. = Hyacinthus L. ■☆

46462 Baeothrion Pfeiff. = Baeothryon A. Dietr. ■

46463 Baeothryon A. Dietr. = Eleocharis R. Br. ■

46464 Baeothryon A. Dietr. = Scirpus L.(保留属名)■

46465 Baeothryon Ehrh. ex A. Dietr. = Eleocharis R. Br. ■

46466 Baeothryon Ehth. = Eleocharis R. Br. ■

46467 Baeothryon Ehth. = Scirpus L.(保留属名)■

46468 Baeothryon alpinum (L.) T. V. Egorova = Trichophorum alpinum (L.) Pers. ■☆

46469 Baeothryon alpinus Schleich. ex Gaudin = Scirpus pumilus Vahl ■

46470 Baeothryon alpinus Schleich. ex Gaudin = Trichophorum pumilum (Vahl) Schinz et Thell. ■

46471 Baeothryon cespitosum (L.) A. Dietr. = Scirpus caespitosus L. ■☆

46472 Baeothryon cespitosum (L.) A. Dietr. = Trichophorum cespitosum (L.) Schur ■☆

46473 Baeothryon distigmaticum (Kuk.) Y. C. Yang et M. Zhan = Trichophorum distigmaticum (Kuk.) Egorova ■

46474 Baeothryon planifolium (Spreng.) Soják = Trichophorum planifolium (Spreng.) Palla ■☆

46475 Baeothryon pumilum (Vahl) Á. Löve et D. Löve = Scirpus pumilus Vahl ■

46476 Baeothryon pumilum (Vahl) Á. Löve et D. Löve = Trichophorum pumilum (Vahl) Schinz et Thell. ■

46477 Baeothryon retroflexum A. Dietr. = Eleocharis retroflexa (Poir.) Urb. ■☆

46478 Baeothryon subcapitatum (Thwaites et Hook.) T. Koyama = Scirpus subcapitatus Thwaites et Hook. ■

46479 Baeothryon subcapitatum (Thwaites et Hook.) T. Koyama = Trichophorum subcapitatum (Thwaites et Hook.) D. A. Simpson ■

46480 Baeothryon verecundum (Fernald) Á. Löve et D. Löve = Trichophorum planifolium (Spreng.) Palla ■☆

46481 Baeria Fisch. et C. A. Mey. (1836);拜氏菊属■☆

46482 Baeria Fisch. et C. A. Mey. = Lasthenia Cass. ■☆

46483 Baeria bakeri J. T. Howell = Lasthenia californica DC. ex Lindl. subsp. bakeri (J. T. Howell) R. Chan ■☆

46484 Baeria burkei Greene = Lasthenia burkei (Greene) Greene ■☆

46485 Baeria californica (Hook.) K. L. Chambers = Lasthenia coronaria (Nutt.) Ornduff ■☆

46486 Baeria chrysostoma Fisch. et C. A. Mey. = Lasthenia californica DC. ex Lindl. ■☆

46487 Baeria chrysostoma Fisch. et C. A. Mey. subsp. gracilis (DC.) Ferris = Lasthenia gracilis (DC.) Greene ■☆

46488 Baeria coronaria (Nutt.) A. Gray = Lasthenia coronaria (Nutt.) Ornduff ■☆

46489 Baeria debilis Greene ex A. Gray = Lasthenia debilis (Greene ex A. Gray) Ornduff ■☆

46490 Baeria fremontii (Torr. ex A. Gray) A. Gray = Lasthenia fremontii (Torr. ex A. Gray) Greene ■☆

46491 Baeria fremontii (Torr. ex A. Gray) A. Gray var. conjugens (Greene) Ferris = Lasthenia conjugens Greene ■☆

46492 Baeria gracilis (DC.) A. Gray = Lasthenia gracilis (DC.) Greene ■☆

46493 Baeria macrantha (A. Gray) A. Gray = Lasthenia californica DC. ex Lindl. subsp. macrantha (A. Gray) R. Chan ■☆

46494 Baeria macrantha (A. Gray) A. Gray var. bakeri (J. T. Howell) D. D. Keck = Lasthenia californica DC. ex Lindl. subsp. bakeri (J. T. Howell) R. Chan ■☆

46495 Baeria macrantha (A. Gray) A. Gray var. pauciaristata A. Gray = Lasthenia californica DC. ex Lindl. subsp. macrantha (A. Gray) R. Chan ■☆

46496 Baeria macrantha (A. Gray) A. Gray var. thalassophila J. T. Howell = Lasthenia californica DC. ex Lindl. subsp. macrantha (A. Gray) R. Chan ■☆

46497 Baeria maritima (A. Gray) A. Gray = Lasthenia maritima (A. Gray) M. C. Vasey ■☆

46498 Baeria microglossa (DC.) Greene = Lasthenia microglossa (DC.) Greene ■☆

46499 Baeria minor (DC.) Ferris = Lasthenia minor (DC.) Ornduff ■☆

46500 Baeria minor (DC.) Ferris subsp. maritima (A. Gray) Ferris = Lasthenia maritima (A. Gray) M. C. Vasey ■☆

46501 Baeria platycarpha (A. Gray) A. Gray = Lasthenia platycarpha (A. Gray) Greene ■☆

46502 Baeriopsis J. T. Howell(1942);腺肉菊属(拟拜氏菊属)●☆

46503 Baeriopsis guadalupensis J. T. Howell;腺肉菊■☆

46504 Baeumerta P. Gaertn., B. Mey. et Scherb. = Cardaminum Moench (废弃属名)■

46505 Baeumerta P. Gaertn., B. Mey. et Scherb. = Nasturtium W. T. Aiton(保留属名)■

46506 Baeumerta P. Gaertn., B. Mey. et Scherb. = Rorippa Scop. ■

46507 Bafodeya Prance = Bafodeya Prance ex F. White ●☆

46508 Bafodeya Prance ex F. White(1976);西非金橡实属●☆

46509 Bafodeya benna (Scott-Elliot) Prance;西非金橡实●☆

46510 Bafutia C. D. Adams(1962);纤粉菊属■☆

46511 Bafutia tenuicaulis C. D. Adams;纤粉菊■☆

46512 Bafutia tenuicaulis C. D. Adams var. zapfackiana Beentje et B. J. Pollard;扎氏纤粉菊■☆

46513 Bagalatta Roxb. ex Rchb. = Tiliacora Colebr.(保留属名)●☆

46514 Bagassa Aubl.(1775);乳桑属;Bagasse,Tatajuba ●☆

46515 Bagassa guianensis Aubl.;圭亚那乳桑;Bagasse,Tatajuba ●☆

46516 Bagassa tiliifolia (Desv.) Benoist;椴叶乳桑●☆

46517 Bagnisia Becc. = Thismia Griff. ■

46518 Baguenaudiera Bubani = Colutea L. ●

46519 Bahamia Britton et Rose = Acacia Mill.(保留属名)●■

46520 Baharuia D. J. Middleton(1995);加岛夹竹桃属●☆

46521 Bahel Adans.(废弃属名) = Artanema D. Don(保留属名)■☆

46522 Bahelia Kuntze = Artanema D. Don(保留属名)■☆

46523 Bahelia Kuntze = Bahel Adans.(废弃属名)■☆

46524 Bahia Lag.(1816);黄羽菊属■☆

46525 Bahia Lag. = Picradeniopsis Rydb. ex Britton ■☆

46526 Bahia Nutt. = Trichophyllum Ehrh. ■

46527 Bahia absinthifolia Benth.;黄羽菊;Bahia, Desert Bahia, Hairyseed Bahia ■☆

46528 Bahia absinthifolia Benth. var. dealbata (A. Gray) A. Gray = Bahia absinthifolia Benth. ■☆

46529 Bahia achilleoides DC. = Eriophyllum lanatum (Pursh) J. Forbes var. achilleoides (DC.) Jeps. ■☆

46530 Bahia arachnoidea Fisch. et Avé-Lall. = Eriophyllum lanatum (Pursh) J. Forbes var. arachnoides (Fisch. et Avé-Lall.) Jeps. ■☆

46531 Bahia confertiflora DC. = Eriophyllum confertiflorum (DC.) A. Gray ■☆

46532 Bahia dealbata A. Gray = Bahia absinthifolia Benth. ■☆

46533 Bahia dissecta (A. Gray) Britton = Amauriopsis dissecta (A. Gray) Rydb. ■☆

46534 Bahia lanata DC. var. grandiflora A. Gray = Eriophyllum lanatum (Pursh) J. Forbes var. grandiflorum (A. Gray) Jeps. ■☆

46535 Bahia leucophylla DC. = Eriophyllum lanatum (Pursh) J. Forbes var. leucophyllum (DC.) W. R. Carter ■☆

46536 Bahia neomexicana (A. Gray) A. Gray = Schkuhria multiflora Hook. et Arn. ■☆

46537 Bahia nudicaulis A. Gray = Platyschkuhria integrifolia (A. Gray) Rydb. ■☆

46538 Bahia nudicaulis A. Gray var. desertorum (M. E. Jones) Cronquist = Platyschkuhria integrifolia (A. Gray) Rydb. ■☆

46539 Bahia nudicaulis A. Gray var. oblongifolia (A. Gray) Cronquist = Platyschkuhria integrifolia (A. Gray) Rydb. ■☆

46540 Bahia oppositifolia (Nutt.) DC. = Picradeniopsis oppositifolia (Nutt.) Rydb. ■☆

46541 Bahia schaffneri S. Watson;沙夫黄羽菊;Schaffner's Bahia ■☆

46542 Bahia wallacei A. Gray = Eriophyllum wallacei (A. Gray) A. Gray ■☆

46543 Bahia woodhousei (A. Gray) A. Gray = Picradeniopsis woodhousei (A. Gray) Rydb. ■☆

46544 Bahianthus R. M. King et H. Rob.(1972);胶黏柄泽兰属●☆

46545 Bahianthus viscosus (Spreng.) R. M. King et H. Rob.;胶黏柄泽兰●☆

46546 Bahiella J. F. Morales = Echites P. Browne ●☆

46547 Bahiella J. F. Morales(2006);巴西蛇木属●☆

46548 Bahiopsis Kellogg = Viguiera Kunth ●■☆

46549 Bahiopsis Kellogg(1863);黄目菊属■☆

46550 Bahiopsis laciniata (A. Gray) E. E. Schill. et Panero;黄目菊;San Diego County Viguiera,San Diego Sunflower ■☆

46551 Bahiopsis parishii (Greene) E. E. Schill. et Panero;帕里什黄目菊■☆

46552 Bahiopsis reticulata (S. Watson) E. E. Schill. et Panero;网状黄目菊;Death Valley Goldeneye ■☆

46553 Bahratherum echinatum Nees = Arthraxon lanceolatus (Roxb.) Hochst. var. echinatus (Nees) Hack. ■

46554 Baicalia Steller ex Grmel. = Astragalus L. + Oxytropis DC.(保留属名)●■

46555 Baijiania A. M. Lu et J. Q. Li(1993);白兼果属■

46556 Baijiania A. M. Lu et J. Q. Li = Siraitia Merr. ■

46557 Baijiania borneensis (Merr.) A. M. Lu et J. Q. Li = Siraitia borneensis (Merr.) C. Jeffrey ex A. M. Lu et Zhi Y. Zhang ■

46558 Baijiania taiwaniana (Hayata) A. M. Lu et J. Q. Li = Sinobaijiania taiwaniana (Hayata) C. Jeffrey et W. J. de Wilde ■

46559 Baijiania taiwaniana (Hayata) A. M. Lu et J. Q. Li = Siraitia taiwaniana (Hayata) C. Jeffrey ex A. M. Lu et Zhi Y. Zhang ■

46560 Baijiania yunnanensis (A. M. Lu et Zhi Y. Zhang) A. M. Lu et J. Q. Li = Sinobaijiania yunnanensis (A. M. Lu et Zhi Y. Zhang) C. Jeffrey et W. J. de Wilde ■

46561 Baijiania yunnanensis (A. M. Lu et Zhi Y. Zhang) A. M. Lu et J. Q. Li = Siraitia borneensis (Merr.) C. Jeffrey ex A. M. Lu et Zhi Y. Zhang var. yunnanensis A. M. Lu et Zhi Y. Zhang ■

46562 Baikiaea Benth.(1865);红苏木属●☆

46563 Baikiaea anomala Micheli = Tessmannia anomala (Micheli) Harms ●☆

46564 Baikiaea eminii Taub. = Baikiaea insignis Benth. subsp. minor (Oliv.) J. Léonard ●☆

46565 Baikiaea fragrantissima Baker f. = Baikiaea insignis Benth. subsp. minor (Oliv.) J. Léonard ●☆

46566 Baikiaea ghesquiereana J. Léonard;盖斯基埃红苏木●☆

46567 Baikiaea insignis Benth.;显著红苏木●☆

46568 Baikiaea insignis Benth. subsp. minor (Oliv.) J. Léonard;小显著红苏木●☆

46569 Baikiaea insignis Benth. var. fragrantissima (Baker f.) J. Léonard = Baikiaea insignis Benth. subsp. minor (Oliv.) J. Léonard ●☆

46570 Baikiaea lescrauwaetii De Wild. = Tessmannia lescrauwaetii (De Wild.) Harms ●☆

46571 Baikiaea minor Oliv. = Baikiaea insignis Benth. subsp. minor (Oliv.) J. Léonard ●☆

46572 Baikiaea minor Oliv. var. fragrantissima (Baker f.) J. Léonard = Baikiaea insignis Benth. subsp. minor (Oliv.) J. Léonard ●☆

46573 Baikiaea plurijuga Harms;红苏木(罗得西亚柚木);Rhodesian Chestnut,Rhodesian Teak,Zambesi Redwood ●☆

46574 Baikiaea robynsii Ghesq.;罗宾斯红苏木●☆

46575 Baikiaea suzannae Ghesq. = Baikiaea insignis Benth. subsp. minor (Oliv.) J. Léonard ●☆

46576 Baikiaea suzannae Ghesq. var. ripicola ? = Baikiaea insignis Benth. subsp. minor (Oliv.) J. Léonard ●☆

46577 Baikiaea zenkeri Harms = Baikiaea insignis Benth. subsp. minor (Oliv.) J. Léonard ●☆

46578 Baileya Harv. et A. Gray ex Torr. (1848);沙金盏属(白莱菊属,白莱氏菊属,贝利菊属);Desert Marigold ■☆

46579 Baileya australis Rydb. = Baileya multiradiata Harv. et A. Gray ex Torr. ■☆

46580 Baileya multiradiata Harv. et A. Gray ex Torr.;沙金盏(白莱氏菊,伯伊勒菊,多梗贝利菊,沙地白莱氏菊);Cloth of Gold,Desert Baileya,Desert Marigold ■☆

46581 Baileya multiradiata Harv. et A. Gray ex Torr. var. nudicaulis A. Gray = Baileya multiradiata Harv. et A. Gray ex Torr. ■☆

46582 Baileya multiradiata Harv. et A. Gray var. pleniradiata (Harv. et A. Gray) Coville = Baileya pleniradiata Harv. et A. Gray ■☆

46583 Baileya nervosa M. E. Jones = Baileya pleniradiata Harv. et A. Gray ■☆

46584 Baileya pauciradiata Harv. et A. Gray;少枝沙金盏(少梗白莱氏菊,少梗贝利菊)■☆

46585 Baileya perennis (A. Nelson) Rydb. = Baileya pleniradiata Harv. et A. Gray ■☆

46586 Baileya pleniradiata Harv. et A. Gray;多枝沙金盏(多梗白莱氏菊,富梗贝利菊)■☆

46587 Baileya pleniradiata Harv. et A. Gray var. multiradiata (Harv. et A. Gray) Kearney = Baileya multiradiata Harv. et A. Gray ex Torr. ■☆

46588 Baileya pleniradiata Harv. et A. Gray var. perennis A. Nelson = Baileya pleniradiata Harv. et A. Gray ■☆

46589 Baileya thurberi Rydb. = Baileya multiradiata Harv. et A. Gray ex Torr. ■☆

46590 Baileyoxylon C. T. White(1941);白氏木属●☆

46591 Baileyoxylon lanceolatum C. T. White;白氏木●☆

46592 Baillandea Roberty = Calycobolus Willd. ex Schult. ●☆

46593 Baillaudea mirabilis (Baker ex Oliv.) Roberty = Calycobolus campanulatus (K. Schum. ex Hallier f.) Heine ●☆

46594 Baillieria Aubl. = Clibadium F. Allam. ex L. ●■☆

46595 Baillonacanthus Kuntze = Solenoruellia Baill. ●☆

46596 Baillonella Pierre ex Dubard = Baillonella Pierre ●☆

46597 Baillonella Pierre(1890);毒籽山榄属●☆

46598 Baillonella africana (Pierre) Baehni = Tieghemella africana Pierre ●☆

46599 Baillonella dispar (N. E. Br.) Baehni = Vitellariopsis dispar (N. E. Br.) Aubrév. ●☆

46600 Baillonella djave (Engl.) Dubard = Baillonella toxisperma Pierre ●☆

46601 Baillonella heckelii (A. Chev.) Baehni = Tieghemella heckelii (A. Chev.) Pierre ex Dubard ●☆

46602 Baillonella marginata (N. E. Br.) Baehni = Vitellariopsis marginata (N. E. Br.) Aubrév. ●☆

46603 Baillonella obovata Pierre ex Dubard = Baillonella toxisperma Pierre var. obovata Aubrév. et Pellegr. ●☆

46604 Baillonella obovata Pierre ex Lecomte var. acuminata Lecomte ex Pellegr. = Baillonella toxisperma Pierre var. obovata Aubrév. et Pellegr. ●☆

46605 Baillonella pierreana (Engl.) A. Chev. = Baillonella toxisperma Pierre var. obovata Aubrév. et Pellegr. ●☆

46606 Baillonella sylvestris (S. Moore) Baehni = Vitellariopsis marginata (N. E. Br.) Aubrév. ●☆

46607 Baillonella toxisperma Pierre;毒籽山榄;Djave,Moabi ●☆

46608 Baillonella toxisperma Pierre var. obovata Aubrév. et Pellegr.;倒卵毒籽山榄●☆

46609 Baillonia Bocq. = Baillonia Bocq. ex Baill. ●☆

46610 Baillonia Bocq. ex Baill. (1862);白花不老树属(贝隆草属)●☆

46611 Baillonia amabilis Bocq. ex Baill.;白花不老树●☆

46612 Baillonodendron Heim = Dryobalanops C. F. Gaertn. ●☆

46613 Baimashania Al-Shehbaz(2000);白马芥属■★

46614 Baimashania pulvinata Al-Shehbaz;白马芥■

46615 Baimashania wangii Al-Shehbaz;王氏白马芥■

46616 Baimo Raf. = Fritillaria L. ■

46617 Baissea A. DC. (1844);白瑟木属●☆

46618 Baissea A. DC. = Cleghornia Wight ●

46619 Baissea acuminata (Wight) Benth. ex Hook. f. = Cleghornia acuminata Wight ●

46620 Baissea acuminata (Wight) Benth. ex Hook. f. = Cleghornia malaccensis (Hook. f.) King et Gamble ●

46621 Baissea aframensis Hutch. et Dalziel = Baissea zygodioides (K. Schum.) Stapf ●☆

46622 Baissea albo-rosea Gilg et Stapf = Baissea leonensis Benth. ●☆

46623 Baissea angolensis Stapf = Baissea multiflora A. DC. ●☆

46624 Baissea angolensis Stapf var. major Stapf = Baissea major (Stapf) Hiern ●☆

46625 Baissea axillaris (Benth.) Hua;腋花白瑟木●☆

46626 Baissea baillonii Hua;巴永白瑟木●☆

46627 Baissea brachyantha Stapf = Baissea leonensis Benth. ●☆

46628 Baissea breviloba Stapf = Baissea baillonii Hua ●☆

46629 Baissea calophylla (K. Schum.) Stapf = Baissea welwitschii (Baill.) Stapf ex Hiern ●☆

46630 Baissea campanulata (K. Schum.) de Kruif;风铃草状白瑟木●☆

46631 Baissea caudiloba Stapf = Baissea multiflora A. DC. ●☆

46632 Baissea dichotoma Stapf = Baissea leonensis Benth. ●☆

46633 Baissea elliptica Stapf = Baissea leonensis Benth. ●☆

46634 Baissea erythrosticta K. Schum. = Baissea welwitschii (Baill.) Stapf ex Hiern ●☆

46635 Baissea giorgii De Wild. = Baissea leonensis Benth. ●☆

46636 Baissea goossensii De Wild. = Baissea campanulata (K. Schum.) de Kruif ●☆

46637 Baissea gracillima (K. Schum.) Hua;细长白瑟木●☆

46638 Baissea guineensis (Thonn.) Roberty = Motandra guineensis (Thonn.) A. DC. ●☆

46639 Baissea haemantha Mildbr.;血红花白瑟木●☆

46640 Baissea heudelotii Hua = Baissea multiflora A. DC. ●☆

46641 Baissea hildebrandtii Vatke ex Markgr.;希尔德白瑟木●☆

46642 Baissea ivorensis A. Chev. = Baissea baillonii Hua ●☆

46643 Baissea kidengensis (K. Schum.) Pichon = Baissea myrtifolia

(Benth.) Pichon ●☆
46644 Baissea laurentii De Wild. = Baissea campanulata (K. Schum.) de Kruif ●☆
46645 Baissea laxiflora Stapf = Baissea multiflora A. DC. ●☆
46646 Baissea leonensis Benth.;短花白瑟木●☆
46647 Baissea likimiensis De Wild. = Baissea leonensis Benth. ●☆
46648 Baissea longipetiolata Van Dilst;长柄白瑟木●☆
46649 Baissea major (Stapf) Hiern;大白瑟木●☆
46650 Baissea malaccensis Hook. f. = Cleghornia malaccensis (Hook. f.) King et Gamble ●
46651 Baissea malchairii De Wild. = Baissea gracillima (K. Schum.) Hua ●☆
46652 Baissea melanocephala (K. Schum.) Pichon = Baissea myrtifolia (Benth.) Pichon ●☆
46653 Baissea micrantha Hua = Baissea gracillima (K. Schum.) Hua ●☆
46654 Baissea minutiflora (Benth.) Pichon;坛状白瑟木●☆
46655 Baissea mortehanii De Wild. = Baissea subrufa Stapf ●☆
46656 Baissea multiflora A. DC.;多花白瑟木●☆
46657 Baissea multiflora A. DC. var. caudiloba Stapf = Baissea multiflora A. DC. ●☆
46658 Baissea myrtifolia (Benth.) Pichon;香桃木叶白瑟木●☆
46659 Baissea odorata K. Schum. ex Stapf = Baissea leonensis Benth. ●☆
46660 Baissea ogowensis Hua = Baissea leonensis Benth. ●☆
46661 Baissea spectabilis Hua = Baissea wulfhorstii Schinz ●☆
46662 Baissea subrufa Stapf;浅红白瑟木●☆
46663 Baissea subsessilis (K. Schum.) Stapf ex Hutch. = Baissea campanulata (K. Schum.) de Kruif ●☆
46664 Baissea tenuiloba Stapf = Baissea leonensis Benth. ●☆
46665 Baissea thollonii Hua = Baissea multiflora A. DC. ●☆
46666 Baissea urceolata (Stapf) Pichon;壶状白瑟木●☆
46667 Baissea urceolata (Stapf) Pichon = Baissea minutiflora (Benth.) Pichon ●☆
46668 Baissea uropetala (K. Schum.) Hua = Baissea leonensis Benth. ●☆
46669 Baissea viridiflora (K. Schum.) de Kruif;绿花白瑟木●☆
46670 Baissea welwitschii (Baill.) Stapf ex Hiern;韦尔白瑟木●☆
46671 Baissea wulfhorstii Schinz;伍尔夫白瑟木●☆
46672 Baissea zygodioides (K. Schum.) Stapf;对称白瑟木●☆
46673 Baiswa Raf. = Paris L. ■
46674 Baitaria Ruiz. et Pav.(废弃属名) = Calandrinia Kunth(保留属名)■☆
46675 Bajacalia Loockerman, B. L. Turner et R. K. Jansen(2003);肉腺菊属●☆
46676 Bajacalia crassifolia (S. Watson) Loockerman, B. L. Turner et R. K. Jansen;厚叶肉腺菊●☆
46677 Bajacalia moranii B. L. Turner;莫朗肉腺菊●☆
46678 Bajacalia tridentata (Benth.) Loockerman, B. L. Turner et R. K. Jansen;三齿肉腺菊●☆
46679 Bajan Adans. = Amaranthus L. ■
46680 Bakera Post et Kuntze = Bakerantha L. B. Sm. ■☆
46681 Bakera Post et Kuntze = Bakeria André ■☆
46682 Bakera Post et Kuntze = Bakeria Seem. ●
46683 Bakera Post et Kuntze = Plerandra A. Gray ●
46684 Bakera Post et Kuntze = Rosa L. ●
46685 Bakerantha L. B. Sm. (1934);贝克凤梨属■☆
46686 Bakerantha L. B. Sm. = Hechtia Klotzsch ■☆
46687 Bakerantha tillandsioides (André) L. B. Sm.;哥伦比亚凤梨■☆
46688 Bakerantha tillandsioides L. B. Sm. = Bakerantha tillandsioides (André) L. B. Sm. ■☆
46689 Bakerella Tiegh. (1895);马岛寄生属●☆
46690 Bakerella analamerensis Balle;阿纳拉马岛寄生●☆
46691 Bakerella clavata (Desr.) Balle;棍棒马岛寄生●☆
46692 Bakerella collapsa (Lecomte) Balle;扁平马岛寄生●☆
46693 Bakerella diplocrater (Baker) Tiegh.;双杯马岛寄生●☆
46694 Bakerella gonoclada (Baker) Balle;棱枝马岛寄生●☆
46695 Bakerella grisea (Scott-Elliot) Balle;灰马岛寄生●☆
46696 Bakerella microcuspis (Baker) Tiegh.;小马岛寄生●☆
46697 Bakerella perrieri Balle;佩里耶马岛寄生●☆
46698 Bakerella poissonii (Lecomte) Balle;普瓦松马岛寄生●☆
46699 Bakerella tandrokensis (Lecomte) Balle;坦德马岛寄生●☆
46700 Bakerella tricostata (Lecomte) Balle;三脉马岛寄生●☆
46701 Bakerella viguieri (Lecomte) Balle;维基耶马岛寄生●☆
46702 Bakeria André = Bakerantha L. B. Sm. ■☆
46703 Bakeria André = Hechtia Klotzsch ■☆
46704 Bakeria Seem. = Plerandra A. Gray ●
46705 Bakeria Seem. = Schefflera J. R. Forst. et G. Forst.(保留属名)●
46706 Bakeridesia Hochr. (1913);巴伊锦葵属●☆
46707 Bakeridesia ferruginea (Martyn) Krapov.;锈色巴伊锦葵●☆
46708 Bakeridesia galeottii Hochr.;巴伊锦葵●☆
46709 Bakeridesia integerrima (Hook.) D. M. Bates;全缘巴伊锦葵●☆
46710 Bakeriella Dubard = Bakeriella Pierre ex Dubard ●☆
46711 Bakeriella Pierre ex Dubard = Synsepalum (A. DC.) Daniell ●☆
46712 Bakeriella Pierre ex Dubard = Vincentella Pierre ●☆
46713 Bakeriella brevipes (Baker) Dubard = Synsepalum brevipes (Baker) T. D. Penn. ●☆
46714 Bakeriella cerasifera (Welw.) Dubard = Synsepalum cerasiferum (Welw.) T. D. Penn. ●☆
46715 Bakeriella cinerea (Engl.) Dubard = Synsepalum brevipes (Baker) T. D. Penn. ●☆
46716 Bakeriella densiflora (Baker) Dubard = Synsepalum revolutum (Baker) T. D. Penn. ●☆
46717 Bakeriella disaco (Hiern) Dubard = Synsepalum cerasiferum (Welw.) T. D. Penn. ●☆
46718 Bakeriella dulcifica (Schumach. et Thonn.) Dubard = Synsepalum dulcificum (Schumach. et Thonn.) Daniell ●☆
46719 Bakeriella longistyla (Baker) Dubard = Synsepalum brevipes (Baker) T. D. Penn. ●☆
46720 Bakeriella pobeguiniana Pierre ex Dubard = Synsepalum pobeguinianum (Pierre ex Lecomte) Aké Assi et L. Gaut. ●☆
46721 Bakeriella revoluta (Baker) Dubard = Synsepalum revolutum (Baker) T. D. Penn. ●☆
46722 Bakerisideroxylon (Engl.) Engl. = Afrosersalisia A. Chev. ●☆
46723 Bakerisideroxylon (Engl.) Engl. = Synsepalum (A. DC.) Daniell ●☆
46724 Bakerisideroxylon (Engl.) Engl. = Vincentella Pierre ●☆
46725 Bakerisideroxylon Engl. = Afrosersalisia A. Chev. ●☆
46726 Bakerisideroxylon Engl. = Synsepalum (A. DC.) Daniell ●☆
46727 Bakerisideroxylon bruneelii De Wild. = Synsepalum revolutum (Baker) T. D. Penn. ●☆
46728 Bakerisideroxylon densiflorum (Baker) Engl. = Synsepalum revolutum (Baker) T. D. Penn. ●☆
46729 Bakerisideroxylon djalonense A. Chev. = Synsepalum cerasiferum (Welw.) T. D. Penn. ●☆
46730 Bakerisideroxylon passargei Engl. = Synsepalum passargei (Engl.) T. D. Penn. ●☆

46731 Bakerisideroxylon revolutum (Baker) Engl. = Synsepalum revolutum (Baker) T. D. Penn. ●☆
46732 Bakerisideroxylon revolutum (Baker) Engl. var. brevipetiolulatum Engl. = Synsepalum revolutum (Baker) T. D. Penn. ●☆
46733 Bakerisideroxylon sapinii De Wild. = Synsepalum passargei (Engl.) T. D. Penn. ●☆
46734 Bakerolimon Lincz. (1968);无叶补血草属(情人草属)●☆
46735 Bakerolimon peruvianum (Kuntze) Lincz.;无叶补血草●☆
46736 Bakerophyton (J. Léonard) Hutch. (1964);巴氏豆属●☆
46737 Bakerophyton (J. Léonard) Hutch. = Aeschynomene L. ●■
46738 Bakerophyton lateritium (Harms) Hutch. ex Maheshw.;砖红巴氏豆●☆
46739 Bakerophyton neglectum (Hepper) Maheshw.;忽视巴氏豆●☆
46740 Bakerophyton pulchellum (Planch. ex Baker) Maheshw.;美丽巴氏豆●☆
46741 Bakeros Raf. = Seseli L. ■
46742 Bakoa P. C. Boyce et S. Y. Wong(2008);巴科南星属■☆
46743 Balaka Becc. (1885);巴拉卡椰子属(巴拉卡榈属,巴拉卡棕属,仗椰属);Balaca, Balaca Palm ●☆
46744 Balaka seemannii Becc.;巴拉卡椰子(巴拉卡棕);Balaca, Seemann Acuminate Balaca ●☆
46745 Balakata Esser(1999);巴拉大戟属●☆
46746 Balanaulax Raf. = Pasania (Miq.) Oerst. ●
46747 Balanaulax Raf. = Quercus L. ●
46748 Balaneikon Setchell = Balanophora J. R. Forst. et G. Forst. ■
46749 Balaneikon tobiracola (Makino) Setch. = Balanophora tobiracola Makino ■
46750 Balaneikon tobiracola (Makino) Setch. = Balanophora wrightii Makino ■
46751 Balanghas Raf. = Southwellia Salisb. ●
46752 Balanghas Raf. = Sterculia L. ●
46753 Balania Noronha = Gnetum L. ●
46754 Balania Tiegh. = Balanophora J. R. Forst. et G. Forst. ■
46755 Balania harlandii (Hook. f.) Tiegh. = Balanophora harlandii Hook. f. ■
46756 Balania japonica (Makino) Tiegh. = Balanophora japonica Makino ■
46757 Balaniella Tiegh. = Balania Tiegh. ■
46758 Balaniella Tiegh. = Balanophora J. R. Forst. et G. Forst. ■
46759 Balaniella Tiegh. = Cynopsole Endl. ■
46760 Balaniella abbreviata (Blume) Tiegh. = Balanophora abbreviata Blume ■
46761 Balaniella distans Tiegh. = Balanophora abbreviata Blume ■
46762 Balaniella elongata (Blume) Tiegh. = Balanophora elongata Blume ■
46763 Balaniella hildebrandtii (Rchb. f.) Tiegh. = Balanophora abbreviata Blume ■
46764 Balanitaceae Endl. = Balanitaceae M. Roem. (保留科名)●☆
46765 Balanitaceae Endl. = Zygophyllaceae R. Br. (保留科名)●■
46766 Balanitaceae M. Roem. (1846)(保留科名);槲果科(卤水草科,龟头树科,翠蛋胚科)●☆
46767 Balanites Delile(1813)(保留属名);槲果属(翠蛋胚属,卤刺树属,卤水草属,橡形木属);Desert Dale ●☆
46768 Balanites aegyptiaca (L.) Delile;埃及槲果;Desert Date, Egyptian Myrobalan, Jericho Balsam, Soapberry Tree ●☆
46769 Balanites aegyptiaca (L.) Delile var. ferox (Poir.) DC.;多刺埃及槲果●☆
46770 Balanites aegyptiaca (L.) Delile var. pallida Sands;苍白埃及槲果●☆
46771 Balanites aegyptiaca (L.) Delile var. quarrei (De Wild.) G. C. C. Gilbert;卡雷槲果●☆
46772 Balanites aegyptiaca (L.) Delile var. tomentosa (Mildbr. et Schltr.) Sands;毛埃及槲果●☆
46773 Balanites angolensis (Welw.) Welw. ex Mildbr. et Schltr.;安哥拉槲果●☆
46774 Balanites angolensis (Welw.) Welw. ex Mildbr. et Schltr. subsp. welwitschii (Tiegh.) Sands;韦氏槲果●☆
46775 Balanites arabica (Tiegh.) Blatt. = Balanites aegyptiaca (L.) Delile ●☆
46776 Balanites australis Bremek. = Balanites pedicellaris Mildbr. et Schltr. ●☆
46777 Balanites fischeri Mildbr. et Schltr. = Balanites aegyptiaca (L.) Delile ●☆
46778 Balanites gillettii Cufod. = Balanites rotundifolia (Tiegh.) Blatt. ●☆
46779 Balanites gillettii Cufod. var. renifolia ? = Balanites rotundifolia (Tiegh.) Blatt. ●☆
46780 Balanites glabra Mildbr. et Schltr.;光滑槲果●☆
46781 Balanites horrida Mildbr. et Schltr. = Balanites pedicellaris Mildbr. et Schltr. ●☆
46782 Balanites latifolia (Tiegh.) Chiov. = Balanites aegyptiaca (L.) Delile ●☆
46783 Balanites mayumbensis Exell = Balanites wilsoniana Dawe et Sprague var. mayumbensis (Exell) Sands ●☆
46784 Balanites orbicularis Sprague;圆形槲果;Kullam Nut ●☆
46785 Balanites orbicularis Sprague = Balanites rotundifolia (Tiegh.) Blatt. ●☆
46786 Balanites patriziana Lusina = Balanites rotundifolia (Tiegh.) Blatt. ●☆
46787 Balanites pedicellaris Mildbr. et Schltr.;梗花槲果●☆
46788 Balanites pedicellaris Mildbr. et Schltr. subsp. somalensis (Mildbr. et Schltr.) Sands;索马里梗花槲果●☆
46789 Balanites quarrei De Wild. = Balanites aegyptiaca (L.) Delile var. quarrei (De Wild.) G. C. C. Gilbert ●☆
46790 Balanites racemosa Chiov. = Balanites aegyptiaca (L.) Delile var. pallida Sands ●☆
46791 Balanites rotundifolia (Tiegh.) Blatt.;圆叶槲果●☆
46792 Balanites rotundifolia (Tiegh.) Blatt. var. setulifera Sands;刚毛圆叶槲果●☆
46793 Balanites roxburghii Planch.;罗氏槲果●☆
46794 Balanites somalensis Mildbr. et Schltr. = Balanites pedicellaris Mildbr. et Schltr. subsp. somalensis (Mildbr. et Schltr.) Sands ●☆
46795 Balanites somalensis Mildbr. et Schltr. var. cinereocorticata Fiori ex Chiov. = Balanites pedicellaris Mildbr. et Schltr. subsp. somalensis (Mildbr. et Schltr.) Sands ●☆
46796 Balanites suckertii Chiov. = Balanites aegyptiaca (L.) Delile ●☆
46797 Balanites tieghemii A. Chev. = Balanites wilsoniana Dawe et Sprague ●☆
46798 Balanites tomentosa Mildbr. et Schltr. = Balanites aegyptiaca (L.) Delile var. tomentosa (Mildbr. et Schltr.) Sands ●☆
46799 Balanites welwitschii (Tiegh.) Exell et Mendonça = Balanites angolensis (Welw.) Welw. ex Mildbr. et Schltr. subsp. welwitschii (Tiegh.) Sands ●☆
46800 Balanites wilsoniana Dawe et Sprague;威氏槲果●☆

46801 Balanites wilsoniana Dawe et Sprague var. glabripetala Sands;光瓣威氏榭果●☆
46802 Balanites wilsoniana Dawe et Sprague var. mayumbensis (Exell) Sands;马永巴威氏榭果●☆
46803 Balanites zizyphoides Mildbr. et Schltr. = Balanites aegyptiaca (L.) Delile ●☆
46804 Balanocarpus Bedd.(1874);棒果香属●☆
46805 Balanocarpus Bedd. = Hopea Roxb.(保留属名)●
46806 Balanocarpus King = Neobalanocarpus P. S. Ashton ●☆
46807 Balanocarpus utilis Bedd.;棒果香●☆
46808 Balanopaceae Benth. = Balanopaceae Benth. et Hook. f.(保留科名)●☆
46809 Balanopaceae Benth. et Hook. f.(1880)(保留科名);橡子木科(假榭树科)●☆
46810 Balanophora J. R. Forst. et G. Forst. (1775);蛇菰属;Balanophora,Snakemushroom ■
46811 Balanophora abbreviata B. Hansen = Balanophora kainantensis Masam. ■
46812 Balanophora abbreviata Blume;短穗蛇菰(杯茎蛇菰);Cupstem Balanophora,Cupstem Snakemushroom ■
46813 Balanophora affinis Griff. = Balanophora dioica R. Br. ex Royle ■
46814 Balanophora cavaleriei H. Lév. = Balanophora abbreviata Blume ■
46815 Balanophora cryptocaudex S. Y. Chang et P. C. Tam;隐轴蛇菰;Hiddenaxle Balanophora,Hiddenaxle Snakemushroom ■
46816 Balanophora cryptocaudex S. Y. Chang et P. C. Tam = Balanophora indica (Arn.) Griff. ■
46817 Balanophora dioica R. Br. = Balanophora indica (Arn.) Griff. ■
46818 Balanophora dioica R. Br. ex Royle;粗穗蛇菰(鹿仙草,异株蛇菰)■
46819 Balanophora distans (Tiegh.) Harms = Balanophora abbreviata Blume ■
46820 Balanophora elongata Blume;长枝蛇菰;Longshoot Balanophora,Longshoot Snakemushroom ■
46821 Balanophora esquirolii H. Lév.;爱氏蛇菰(葛菌)■
46822 Balanophora esquirolii H. Lév. = Balanophora harlandii Hook. f. ■
46823 Balanophora fargesii (Tiegh.) Harms;川藏蛇菰;Farges Balanophora,Farges Snakemushroom ■
46824 Balanophora formosana Hayata;台湾蛇菰;Taiwan Balanophora,Taiwan Snakemushroom ■
46825 Balanophora formosana Hayata = Balanophora laxiflora Hemsl. ■
46826 Balanophora fungosa J. R. Forst. et G. Forst.;卵穗蛇菰(粗穗蛇菰,蛇菰,蕈状蛇菰);Dioecious Balanophora, Dioecious Snakemushroom ■
46827 Balanophora fungosa J. R. Forst. et G. Forst. subsp. indica (Arn.) B. Hansen = Balanophora indica (Arn.) Griff. ■
46828 Balanophora fungosa J. R. Forst. et G. Forst. var. kuroiwai (Makino) Makino = Balanophora fungosa J. R. Forst. et G. Forst. ■
46829 Balanophora harlandii Hook. f.;红冬蛇菰(笔头蛇菰,葛菌,蛇菰);Harland Balanophora,Harland Snakemushroom ■
46830 Balanophora harlandii Hook. f. = Balanophora multinoides Hayata ■
46831 Balanophora harlandii Hook. f. var. mutinoides (Hayata) F. W. Xing = Balanophora harlandii Hook. f. ■
46832 Balanophora harlandii Hook. f. var. spiralis P. C. Tam;旋生蛇菰■
46833 Balanophora harlandii Hook. f. var. spiralis P. C. Tam = Balanophora tobiracola Makino ■
46834 Balanophora harlandii Hook. f. var. spiralis P. C. Tam = Balanophora wrightii Makino ■
46835 Balanophora henryi Hemsl.;宜昌蛇菰(土菌子);Henry Balanophora,Henry Snakemushroom ■
46836 Balanophora henryi Hemsl. = Balanophora harlandii Hook. f. ■
46837 Balanophora hildebrandtii Rchb. f. = Balanophora abbreviata Blume ■
46838 Balanophora hongkongensis K. M. Lau et al. = Balanophora laxiflora Hemsl. ■
46839 Balanophora indica (Arn.) Griff.;印度蛇菰;India Snakemushroom,Indian Balanophora ■
46840 Balanophora involucrata Hook. f. = Balanophora involucrata Hook. f. et Thomson ■
46841 Balanophora involucrata Hook. f. et Thomson;筒鞘蛇菰(葛花,观音莲,红菌,黄药子,寄生黄,鹿仙草,蛇菇,文王一支笔);Involucrate Balanophora,Involucrate Snakemushroom ■
46842 Balanophora involucrata Hook. f. et Thomson var. cathcarti Hook. f. = Balanophora involucrata Hook. f. et Thomson ■
46843 Balanophora involucrata Hook. f. et Thomson var. flava Hook. f. = Balanophora involucrata Hook. f. et Thomson ■
46844 Balanophora involucrata Hook. f. et Thomson var. gracilis Hook. f. = Balanophora involucrata Hook. f. et Thomson ■
46845 Balanophora involucrata Hook. f. et Thomson var. rubra Hook. f.;红蛇菰■
46846 Balanophora involucrata Hook. f. et Thomson var. rubra Hook. f. = Balanophora involucrata Hook. f. et Thomson ■
46847 Balanophora involucrata Hook. f. var. rubra Hook. f. = Balanophora fargesii (Tiegh.) Harms ■
46848 Balanophora japonica Makino;日本蛇菰(葛花菜,葛菌,葛乳,葛蕈,红血莲,鸡心七,角菌,螺丝起,铺地开花,蛇菇,蛇菰,土心肝);Japan Snakemushroom,Japanese Balanophora ■
46849 Balanophora japonica Makino var. nipponica (Makino) Ohwi = Balanophora nipponica Makino ■
46850 Balanophora japonietz Makino = Balanophora japonica Makino ■
46851 Balanophora kainantensis Masam.;海南蛇菰;Hainan Balanophora,Hainan Snakemushroom ■
46852 Balanophora kainantensis Masam. = Balanophora abbreviata Blume ■
46853 Balanophora kawakamii Valeton = Balanophora harlandii Hook. f. ■
46854 Balanophora kawakamii Valeton = Balanophora multinoides Hayata ■
46855 Balanophora kiusiana Ohwi = Balanophora nipponica Makino ■
46856 Balanophora kudoi Yamam. = Balanophora harlandii Hook. f. ■
46857 Balanophora kuroiwai (Makino) Makino = Balanophora fungosa J. R. Forst. et G. Forst. ■
46858 Balanophora lancangensis Y. Y. Qian = Balanophora harlandii Hook. f. ■
46859 Balanophora laxiflora Hemsl.;疏花蛇菰(地荔枝,鹿仙草,山菠萝,石上莲,穗花蛇菰,太鲁阁蛇菰,通天蜡烛);Scatterflower Balanophora,Scatterflower Snakemushroom,Spike Balanophora,Spike Snakemushroom ■
46860 Balanophora minor Hemsl.;小蛇菰■
46861 Balanophora minor Hemsl. = Balanophora harlandii Hook. f. ■
46862 Balanophora morrisonicola Hayata = Balanophora laxiflora Hemsl. ■
46863 Balanophora multinoides Hayata;红烛蛇菰(仇人不见面,山狗球,深山不出头,石上莲,天麻公子);Redcandle Balanophora,Redcandle Snakemushroom ■
46864 Balanophora mutinoides Hayata = Balanophora harlandii Hook. f. ■
46865 Balanophora nipponica Makino = Balanophora japonica Makino ■

46866 Balanophora oshimae Yamam. = Balanophora laxiflora Hemsl. ■
46867 Balanophora parvior Hayata = Balanophora laxiflora Hemsl. ■
46868 Balanophora polyandra Griff.;多蕊蛇菰(木菌子,通天蜡烛,通天烛,土苁蓉); Manystamen Balanophora, Manystamen Snakemushroom ■
46869 Balanophora rugosa P. C. Tam;皱球蛇菰;Rugose Balanophora ■
46870 Balanophora rugosa P. C. Tam = Balanophora laxiflora Hemsl. ■
46871 Balanophora saxicola F. W. Xing et = Balanophora indica (Arn.) Griff. ■
46872 Balanophora simaoensis S. Y. Chang et P. C. Tam;思茅蛇菰(鹿仙草);Simao Balanophora, Simao Snakemushroom ■
46873 Balanophora simaoensis S. Y. Chang et P. C. Tam = Balanophora indica (Arn.) Griff. ■
46874 Balanophora spicata Hayata;穗花蛇菰■
46875 Balanophora spicata Hayata = Balanophora laxiflora Hemsl. ■
46876 Balanophora splendida P. C. Tam et D. Fang;彩丽蛇菰;Bright Balanophora, Bright Snakemushroom ■
46877 Balanophora splendida P. C. Tam et D. Fang. = Balanophora indica (Arn.) Griff. ■
46878 Balanophora subcupularis P. C. Tam;杯茎蛇菰■
46879 Balanophora subcupularis P. C. Tam = Balanophora abbreviata Blume ■
46880 Balanophora tobiracola Makino = Balanophora wrightii Makino ■
46881 Balanophora valida Diels = Balanophora multinoides Hayata ■
46882 Balanophora wrightii Makino;海桐生蛇菰(海桐蛇菰,鸟黐蛇菰,旋生蛇菰);Wright Balanophora ■
46883 Balanophora wrightii Makino = Balanophora tobiracola Makino ■
46884 Balanophora wrightii Makino ex Makino et Nemoto = Balanophora tobiracola Makino ■
46885 Balanophora yakushimensis Hatus. et Masam.;屋久岛蛇菰■☆
46886 Balanophora yuwanensis Agusawa et Sakuta = Balanophora yakushimensis Hatus. et Masam. ■☆
46887 Balanophoraceae L. C. A. Rich. et A. Rich. = Balanophoraceae Rich. (保留科名)●■
46888 Balanophoraceae Rich. (1822)(保留科名);蛇菰科(土鸟黐科);Balanophora Family, Snakemushroom Family ●■
46889 Balanophoraceae Rich. (保留科名) = Cynomoriaceae Endl. ex Lindl. (保留科名)■
46890 Balanophoraceae Rich. (保留科名) = Dactylanthaceae Takht. ■☆
46891 Balanoplis Raf. (废弃属名) = Castanopsis (D. Don) Spach(保留属名)●
46892 Balanops Baill. (1871);橡子木属(假槲树属,象子木属)●☆
46893 Balanops australiana F. Muell.;澳大利亚橡子木●☆
46894 Balanops microstachya Baill.;小穗橡子木●☆
46895 Balanops montana C. T. White;山地橡子木●☆
46896 Balanops reticulata S. Moore;网脉橡子木●☆
46897 Balanopseae Benth. = Balanopaceae Benth. et Hook. f. (保留科名)●☆
46898 Balanopsidaceae Engl. = Balanopseae Benth. ●☆
46899 Balanopsis Raf. = Ocotea Aubl. ●☆
46900 Balanopteris Gaertn. = Heritiera Aiton ●
46901 Balanostreblus Kurz = Sorocea A. St. -Hil. ●☆
46902 Balansaea Boiss. et Reut. = Geocaryum Coss. ■
46903 Balansaea fontanesii Boiss. et Reut. = Conopodium glaberrimum (Desf.) Engstrand ■☆
46904 Balansaea fontanesii Boiss. et Reut. var. maritima Batt. = Conopodium glaberrimum (Desf.) Engstrand ■☆
46905 Balansaea glaberrima (Desf.) Lange = Conopodium glaberrimum (Desf.) Engstrand ■☆
46906 Balansaephytum Drake = Poikilospermum Zipp. ex Miq. ●
46907 Balansaephytum tonkinense Drake = Poikilospermum suaveolens (Blume) Merr. ●
46908 Balansochloa Kuntze = Germainia Balansa et Poitr. ■
46909 Balantiaceae Dulac = Asclepiadaceae Borkh. (保留科名)●■
46910 Balantium Desv. ex Ham. = Parinari Aubl. ●☆
46911 Balardia Cambess. = Spergularia (Pers.) J. Presl et C. Presl(保留属名)■
46912 Balardia platensis Cambess. = Spergularia platensis (Cambess.) Fenzl ■☆
46913 Balaustion Hook. (1851);野石榴花属●☆
46914 Balaustion microphyllum C. A. Gardner;小叶野石榴花●☆
46915 Balaustion pulcherrimum Hook.;野石榴花●☆
46916 Balbisia Cav. (1833)(保留属名);巴尔果属(杜香果属)●☆
46917 Balbisia DC. = Rhetinodendron Meisn. ●☆
46918 Balbisia Willd. (废弃属名) = Balbisia Cav. (保留属名)●☆
46919 Balbisia Willd. (废弃属名) = Tridax L. ●■
46920 Balbisia gracilis (Meyen) Hunz. et Ariza;细巴尔果●☆
46921 Balbisia integrifolia R. Knuth;全缘巴尔果●☆
46922 Balbisia microphylla Reiche;小叶巴尔果●☆
46923 Balbisia miniata (I. M. Johnst.) Descole, O' Donell et Lourteig;小巴尔果●☆
46924 Balbisia verticillata Cav.;轮生巴尔果●☆
46925 Balboa Liebm. = Tephrosia Pers. (保留属名)●■
46926 Balboa Liebm. ex Didr. (废弃属名) = Balboa Planch. et Triana (保留属名)■☆
46927 Balboa Planch. et Triana = Chrysochlamys Poepp. ●☆
46928 Balboa Planch. et Triana(1860)(保留属名);哥伦比亚藤黄属■☆
46929 Balboa membranacea Planch. et Triana;哥伦比亚藤黄■☆
46930 Baldellia Parl. (1854);假泽泻属(圆果泻属);Baldellia, Lesser Water-plantain ■☆
46931 Baldellia ranunculoides (L.) Parl.;假泽泻;Lesser Water Plantain, Lesser Waterplantain, Lesser Water-plantain ■☆
46932 Baldellia ranunculoides Parl. = Baldellia ranunculoides (L.) Parl. ■☆
46933 Baldellia repens (Lam.) Lawalrée;匍匐假泽泻■☆
46934 Baldingera Dennst. = Premna L. (保留属名)●■
46935 Baldingera P. Gaertn., B. Mey. et Scherb. = Phalaris L. ■
46936 Baldingera P. Gaertn., B. Mey. et Scherb. = Typhoides Moench ■
46937 Baldingera arundinacea (L.) Dumort. = Phalaris arundinacea L. ■
46938 Baldingeria F. W. Schmidt = Leontodon L. (保留属名)■☆
46939 Baldingeria Neck. = Cotula L. ■
46940 Baldomiria Herter = Leptochloa P. Beauv. ■
46941 Balduina Nutt. (1818)(保留属名);蜂巢菊属(巴都菊属)■☆
46942 Balduina angustifolia (Pursh) B. L. Rob.;蜂巢菊(狭叶巴都菊)■☆
46943 Balduinia Raf. = Balduina Nutt. (保留属名)■☆
46944 Balduinia Raf. = Baldwinia Raf. ●■
46945 Baldwinia Raf. = Passiflora L. ●■
46946 Baldwinia Torr. et A. Gray = Balduina Nutt. (保留属名)■☆
46947 Balendasia Raf. = Passerina L. ●☆
46948 Balenerdia Comm. ex Steud. = Nanodea Banks ex C. F. Gaertn. ●☆
46949 Balessam Bruce = Balsamodendrum Kunth ●
46950 Balexerdia Comm. ex Endl. = Nanodea Banks ex C. F. Gaertn. ●☆

46951 Balfouria (H. Ohba) H. Ohba = Ohbaea Byalt et I. V. Sokolova ■★
46952 Balfouria R. Br. = Wrightia R. Br. ●
46953 Balfourina Kuntze = Didymaea Hook. f. ■☆
46954 Balfourodendron Corr. Mello ex Oliv. (1877);巴福芸香属●☆
46955 Balfourodendron eburneum Corr. Mello ex Oliv.;巴福芸香(象牙木);Guatambu Moroti, Ivorywood, Marfim, Pau Marfim ●☆
46956 Balfourodendron riedelianum (Engl.) Engl. = Balfourodendron eburneum Mello ex Oliv. ●☆
46957 Balfuria Rchb. = Balfouria R. Br. ●
46958 Balfuria Rchb. = Wrightia R. Br. ●
46959 Balgoya Morat et Meijden(1991);新喀远志属●☆
46960 Balgoya pacifica Morat et Meijden;新喀远志●☆
46961 Balingayum Blanco = Calogyne R. Br. ■
46962 Baliospermum Blume(1826);斑籽木属(斑籽属,微籽属);Baliosperm, Baliospermum ●
46963 Baliospermum angustifolium Y. T. Chang;狭叶斑籽木(狭叶斑籽);Narrowleaf Baliosperm, Narrow-leaved Baliospermum ●
46964 Baliospermum axillare Blume = Baliospermum montanum (Willd.) Müll. Arg. ●
46965 Baliospermum bilobatum T. L. Chin;西藏斑籽木(西藏斑籽);Bilobate Baliospermum, Xizang Baliosperm, Xizang Baliospermum ●
46966 Baliospermum calycinum Müll. Arg.; 微 籽; Calyx-shaped Baliospermum ●
46967 Baliospermum calycinum Müll. Arg. = Baliospermum angustifolium Y. T. Chang ●
46968 Baliospermum effusum Pax et K. Hoffm.;云南斑籽木(抱冬电,散微籽,微籽,云南斑籽); Effuse Baliospermum, Effusive Baliospermum, Yunnan Baliosperm ●
46969 Baliospermum micranthum Müll. Arg.;小花斑籽木(小花斑籽,小花微籽);Little Baliosperm, Small-flowered Baliospermum ●
46970 Baliospermum montanum (Willd.) Müll. Arg.;斑籽木(斑籽,山微籽,山微子); Montane Baliosperm, Montane Baliospermum, Montanous Baliospermum ●
46971 Baliospermum yui Y. T. Chang;心叶斑籽木(心叶斑籽);Hraerleaf Baliosperm, Yu Baliospermum ●
46972 Balisaea Taub. = Aeschynomene L. ●■
46973 Balizia Barneby et J. W. Grimes = Inga Mill. ●■☆
46974 Ballantinia Hook. f. ex E. A. Shaw(1974);牧人钱袋芥属■☆
46975 Ballantinia antipoda (F. Muell.) E. A. Shaw;牧人钱袋芥■☆
46976 Ballardia Montrouz. = Carpolepis (J. W. Dawson) J. W. Dawson ●☆
46977 Ballardia Montrouz. = Cloezia Brongn. et Gris ●☆
46978 Ballarion Raf. = Stellaria L. ■
46979 Ballela (Railn.) B. D. Jacks. = Merremia Dennst. ex Endl.(保留属名)●■
46980 Ballela Raf. = Campanula L. ●■
46981 Ballexerda Comm. ex A. DC. = Nanodea Banks ex C. F. Gaertn. ●☆
46982 Ballieria Juss. = Baillieria Aubl. ●■☆
46983 Ballieria Juss. = Clibadium F. Allam. ex L. ●■☆
46984 Ballimon Raf. = Daucus L. ■
46985 Ballochia Balf. f. (1884);脱被爵床属☆
46986 Ballochia atrovirgata Balf. f.;脱被爵床☆
46987 Ballochia rotundifolia Balf. f.;圆叶脱被爵床☆
46988 Ballosporum Salisb. = Gladiolus L. ■
46989 Ballota L. (1753);宽萼苏属(巴洛草属,巴娄塔属);Ballota, Black Horehound ●■☆
46990 Ballota acetabulosa Benth.;碟叶宽萼苏;False Dittany ●☆
46991 Ballota africana (L.) Benth.;非洲宽萼苏;Cat Herb ■☆
46992 Ballota africana Benth. = Ballota africana (L.) Benth. ■☆
46993 Ballota borealis Schweigg.;北方宽萼苏■☆
46994 Ballota bullata Pomel;泡状宽萼苏■☆
46995 Ballota damascena Boiss.;大马士革宽萼苏■☆
46996 Ballota deserti (Noë) Jury et al. = Marrubium deserti (Noë) Coss. ■☆
46997 Ballota foetida Lam. = Ballota nigra L. ■☆
46998 Ballota fruticosa Baker = Otostegia modesta S. Moore ■☆
46999 Ballota grisea Pojark.;灰宽萼苏■☆
47000 Ballota hildebrandtii Vatke et Kurtz = Otostegia hildebrandtii (Vatke et Kurtz) Sebald ●☆
47001 Ballota hirsuta Benth.;毛黑薄荷■☆
47002 Ballota hirsuta Benth. subsp. intermedia (Batt.) Patzak;间型宽萼苏■☆
47003 Ballota hirsuta Benth. subsp. maroccana (Murb.) Patzak;摩洛哥宽萼苏■☆
47004 Ballota hirsuta Benth. var. bullata (Pomel) Batt. = Ballota bullata Pomel ■☆
47005 Ballota hirsuta Benth. var. gracilis Maire = Ballota hirsuta Benth. ■☆
47006 Ballota hirsuta Benth. var. inermis (Emb. et Maire) Maire = Ballota hirsuta Benth. ■☆
47007 Ballota hirsuta Benth. var. saharica (Diels) Maire = Ballota hirsuta Benth. ■☆
47008 Ballota hirsuta Benth. var. tibestica Maire = Ballota hirsuta Benth. ■☆
47009 Ballota hirsuta Benth. var. vellerea (Maire et al.) Maire = Ballota vellerea Maire, Weiller et Wilczek ■☆
47010 Ballota lanata L. = Panzerina lanata (L.) Bunge ■
47011 Ballota microphylla (Desr.) Benth. = Otostegia fruticosa (Forssk.) Schweinf. ex Penz. subsp. schimperi (Benth.) Sebald ●☆
47012 Ballota microphylla Chiov. = Otostegia somala (Patzak) Sebald ●☆
47013 Ballota nigra L.;黑宽萼苏(臭巴洛草,黑巴洛草,黑夏至草);Black Archangel, Black Ballota, Black Horehound, Double Dumb Nettle, Dunny Nettle, Dunny-nettle, Foetid Horehound, Gypsywort, Hairhound, Henbit, Horehound, Madweed, Madwort, Marvel, Stinking Roger, Worral, Wurral ■☆
47014 Ballota nigra L. subsp. ruderalis (Sw.) Briq. = Ballota nigra L. ■☆
47015 Ballota nigra L. subsp. uncinata (Fiori et Bég.) Patzak;钩黑宽萼苏■☆
47016 Ballota nigra L. var. alba ?;白宽萼苏;Black Horehound ■☆
47017 Ballota nigra subsp. foetida (Lam.) Hayek;臭宽萼苏;Black Horehound ■☆
47018 Ballota pseudodictamnus (L.) Benth.;圆叶宽萼苏;False Dittany ■☆
47019 Ballota pseudodictamnus Benth. = Ballota pseudodictamnus (L.) Benth. ■☆
47020 Ballota ruderalis Sw. = Ballota nigra L. ■☆
47021 Ballota rupestris Vis.;岩地萼苏;Rock Ballota ■☆
47022 Ballota sagittata (Regel) Regel = Metastachydium sagittatum (Regel) C. Y. Wu et H. W. Li ■
47023 Ballota schimperi Benth. = Otostegia fruticosa (Forssk.) Schweinf. ex Penz. subsp. schimperi (Benth.) Sebald ●☆
47024 Ballota somala Patzak = Isoleucas somala (Patzak) Scheen ●☆
47025 Ballota stachydiformis (Hochst. ex Benth.) Jaub. et Spach = Leucas stachydiformis (Hochst. ex Benth.) Briq. ●☆

47026 Ballota suaveolens L. = Hyptis suaveolens (L.) Poit. ●■
47027 Ballota vellerea Maire, Weiller et Wilczek;羊毛宽萼苏■☆
47028 Ballote Mill. = Ballota L. ●■☆
47029 Balls-headleya F. Muell. ex F. M. Bailey(1886);澳洲吊片果属●☆
47030 Ballya Brenan = Aneilema R. Br. ■☆
47031 Ballya Brenan = Murdannia Royle(保留属名)■
47032 Ballya Brenan(1964);东非鸭跖草属■☆
47033 Ballya zebrina (Chiov.) Brenan;东非鸭跖草■☆
47034 Ballya zebrina (Chiov.) Brenan = Aneilema zebrinum Chiov. ■☆
47035 Ballyanthus Bruyns = Stapelia L.(保留属名)■
47036 Ballyanthus Bruyns(2000);索马里豹皮花属■☆
47037 Balmea Martinez(1942);巴尔木属(巴尔米木属,巴尔姆木属)●☆
47038 Balmea stormae Martinez;巴尔木●☆
47039 Balmeda Nocca = Grewia L. ●
47040 Balmisa Lag. = Arisarum Mill. ■☆
47041 Baloghia Endl. (1833);包洛格大戟属●■☆
47042 Baloghia lucida Endl.;包洛格大戟;Scrub Blood Wood, Scrub Bloodwood☆
47043 Baloghiaceae Baum. -Bod. = Euphorbiaceae Juss.(保留科名)●■
47044 Balonga Le Thomas = Uvaria L. ●
47045 Balonga Le Thomas(1968);巴朗木属(巴郎木属)●☆
47046 Balonga buchholzii (Engl. et Diels) Le Thomas;巴朗木●☆
47047 Baloskion Raf. = Restio Rottb.(保留属名)■☆
47048 Balowia paradoxa Bunge = Suaeda paradoxa Bunge ■
47049 Balsamaceae Dumort. = Burseraceae Kunth(保留科名)●
47050 Balsamaceae Lindl. = Altingiaceae Lindl.(保留科名)●
47051 Balsamaria Lour. = Calophyllum L. ●
47052 Balsamaria Lour. = Ponna Boehm. ●
47053 Balsamaria inophyllum (L.) Lour. = Calophyllum inophyllum L. ●
47054 Balsamaria inophyllum Lour. = Calophyllum inophyllum L. ●
47055 Balsamea Gled.(废弃属名) = Commiphora Jacq.(保留属名)●
47056 Balsamea africana (A. Rich.) Baill. = Commiphora africana (A. Rich.) Engl. ●☆
47057 Balsamea angolensis (Engl.) Hiern = Commiphora angolensis Engl. ●☆
47058 Balsamea capensis (Sond.) Engl. = Commiphora capensis (Sond.) Engl. ●☆
47059 Balsamea foliolosa Hiern = Haplocoelum foliolosum (Hiern) Bullock ●☆
47060 Balsamea fraxinoides Hiern = Zanha golungensis Hiern ●☆
47061 Balsamea harveyi Engl. = Commiphora harveyi (Engl.) Engl. ●☆
47062 Balsamea hildebrandtii Engl. = Commiphora hildebrandtii (Engl.) Engl. ●☆
47063 Balsamea kotschyi (O. Berg) Engl. = Commiphora africana (A. Rich.) Engl. ●☆
47064 Balsamea longebracteata (Engl.) Hiern = Commiphora angolensis Engl. ●☆
47065 Balsamea multijuga Hiern = Commiphora multijuga (Hiern) K. Schum. ●☆
47066 Balsamea pilosa Engl. = Commiphora africana (A. Rich.) Engl. ●☆
47067 Balsamea stocksiana Engl. = Commiphora gileadensis (L.) C. Chr. ●☆
47068 Balsamea stocksiana Engl. = Commiphora stocksiana (Engl.) Engl. ●☆
47069 Balsamea zanzibarica Baill. = Commiphora zanzibarica (Baill.) Engl. ●☆
47070 Balsameaceae Dumort. = Bursariaceae Kunth ●
47071 Balsamifiua Griff. = Populus L. ●
47072 Balsamiflua euphratica (Oliv.) Kimura = Populus euphratica Oliv. ●
47073 Balsamiflua pruinosa (Schrenk) Kimura = Populus pruinosa Schrenk ●◇
47074 Balsamina Mill. = Impatiens L. ■
47075 Balsamina Tourn. ex Scop. = Impatiens L. ■
47076 Balsamina hortensis Desp. = Impatiens balsamina L. ■
47077 Balsaminaceae A. Rich. (1822)(保留科名);凤仙花科;Balsam Family, Balsamina Family ■
47078 Balsaminaceae Bercht. et J. Presl = Balsaminaceae A. Rich.(保留科名)■
47079 Balsaminaceae DC. = Balsaminaceae A. Rich.(保留科名)■
47080 Balsamita Mill. = Chrysanthemum L.(保留属名)●■
47081 Balsamita Mill. = Tanacetum L. ●■
47082 Balsamita grandiflora Desf. = Plagius grandis (L.) Alavi et Heywood ■☆
47083 Balsamita major Desf. = Tanacetum balsamita L. ■☆
47084 Balsamita major Desf. var. tanacetoides (Boiss.) Moldenke = Balsamita major Desf. ■☆
47085 Balsamita vulgaris Willd. = Chrysanthemum balsamita L. ■☆
47086 Balsamita vulgaris Willd. = Tanacetum balsamita L. ■☆
47087 Balsamocarpon Clos = Caesalpinia L. ●
47088 Balsamocarpon Clos(1847);香果云实属■☆
47089 Balsamocarpon brevifolium Clos;香果云实;Algarobilla ■☆
47090 Balsamocitrus Stapf(1906);香胶橘属●☆
47091 Balsamocitrus camerunensis Letouzey;喀麦隆香胶橘●☆
47092 Balsamocitrus dawei Stapf;香胶橘●☆
47093 Balsamocitrus gabonensis Swingle = Afraegle gabonensis (Swingle) Engl. ●☆
47094 Balsamodendron DC. = Balsamodendron Kunth ●☆
47095 Balsamodendron Kunth = Commiphora Jacq.(保留属名)●
47096 Balsamodendron ehrenbergianum O. Berg = Commiphora gileadensis (L.) C. Chr. ●☆
47097 Balsamodendron mukul Hook. ex Stocks = Commiphora wightii (Arn.) Bhandari ●
47098 Balsamodendron pubescens Stocks = Commiphora stocksiana (Engl.) Engl. ●☆
47099 Balsamodendron roxburghii Stocks = Commiphora wightii (Arn.) Bhandari ●
47100 Balsamodendron wightii Arn. = Commiphora wightii (Arn.) Bhandari ●
47101 Balsamodendrum Kunth = Commiphora Jacq.(保留属名)●
47102 Balsamodendrum abyssinicum O. Berg = Commiphora kua (R. Br. ex Royle) Vollesen ●☆
47103 Balsamodendrum africanum (A. Rich.) Arn. = Commiphora africana (A. Rich.) Engl. ●☆
47104 Balsamodendrum africanum (A. Rich.) Arn. var. ramosissimus Oliv. = Commiphora africana (A. Rich.) Engl. var. ramosissima (Oliv.) Engl. ●☆
47105 Balsamodendrum capense Sond. = Commiphora capensis (Sond.) Engl. ●☆
47106 Balsamodendrum ehrenbergianum O. Berg = Commiphora gileadensis (L.) C. Chr. ●☆
47107 Balsamodendrum habessinica O. Berg = Commiphora kua (R. Br.

ex Royle) Vollesen ●☆
47108 Balsamodendrum kotschyi O. Berg = Commiphora africana (A. Rich.) Engl. ●☆
47109 Balsamodendrum kua R. Br. ex Royle = Commiphora kua (R. Br. ex Royle) Vollesen ●☆
47110 Balsamodendrum molle Oliv. = Commiphora mollis (Oliv.) Engl. ●☆
47111 Balsamodendrum myrrha T. Nees = Commiphora myrrha (T. Nees) Engl. ●
47112 Balsamodendrum pedunculatum Kotschy et Peyr. = Commiphora pedunculata (Kotschy et Peyr.) Engl. ●☆
47113 Balsamodendrum schimperi O. Berg = Commiphora schimperi (O. Berg) Engl. ●☆
47114 Balsamona Vand. = Cuphea Adans. ex P. Browne ●■
47115 Balsamophloeos O. Berg = Commiphora Jacq.(保留属名)●
47116 Balsamorhiza Hook. = Balsamorhiza Hook. ex Nutt. ■☆
47117 Balsamorhiza Hook. ex Nutt.(1840);香根属;Balsamroot ■☆
47118 Balsamorhiza Hook. ex Nutt. = Wyethia Nutt. ■☆
47119 Balsamorhiza Nutt. = Balsamorhiza Hook. ex Nutt. ■☆
47120 Balsamorhiza Nutt. = Wyethia Nutt. ■☆
47121 Balsamorhiza bolanderi A. Gray = Agnorhiza bolanderi (A. Gray) W. A. Weber ■☆
47122 Balsamorhiza careyana A. Gray;卡雷香根■☆
47123 Balsamorhiza careyana A. Gray var. intermedia Cronquist = Balsamorhiza careyana A. Gray ■☆
47124 Balsamorhiza deltoidea Nutt.;无毛香根;Deltoid Balsam-root ■☆
47125 Balsamorhiza glabrescens Benth. = Balsamorhiza deltoidea Nutt. ■☆
47126 Balsamorhiza helianthoides (Nutt.) Nutt. = Balsamorhiza sagittata (Pursh) Nutt. ■☆
47127 Balsamorhiza hirsuta Nutt. = Balsamorhiza hookeri Nutt. ■☆
47128 Balsamorhiza hirsuta Nutt. var. lagocephala W. M. Sharp = Balsamorhiza hookeri Nutt. ■☆
47129 Balsamorhiza hirsuta Nutt. var. neglecta W. M. Sharp = Balsamorhiza hookeri Nutt. ■☆
47130 Balsamorhiza hispidula W. M. Sharp;毛香根■☆
47131 Balsamorhiza hookeri Nutt.;胡克香根■☆
47132 Balsamorhiza hookeri Nutt. var. hirsuta (Nutt.) A. Nelson = Balsamorhiza hookeri Nutt. ■☆
47133 Balsamorhiza hookeri Nutt. var. hispidula (W. M. Sharp) Cronquist = Balsamorhiza hispidula W. M. Sharp ■☆
47134 Balsamorhiza hookeri Nutt. var. idahoensis (W. M. Sharp) Cronquist = Balsamorhiza macrophylla Nutt. ■☆
47135 Balsamorhiza hookeri Nutt. var. lagocephala (W. M. Sharp) Cronquist = Balsamorhiza hookeri Nutt. ■☆
47136 Balsamorhiza hookeri Nutt. var. lanata W. M. Sharp = Balsamorhiza lanata (W. M. Sharp) W. A. Weber ■☆
47137 Balsamorhiza hookeri Nutt. var. neglecta (W. M. Sharp) Cronquist = Balsamorhiza hookeri Nutt. ■☆
47138 Balsamorhiza hookeri Nutt. var. rosea (A. Nelson et J. F. Macbr.) W. M. Sharp = Balsamorhiza rosea A. Nelson et J. F. Macbr. ■☆
47139 Balsamorhiza incana Nutt.;香根;Balsam-root ■☆
47140 Balsamorhiza invenusta (Greene) Coville = Agnorhiza invenusta (Greene) W. A. Weber ■☆
47141 Balsamorhiza lanata (W. M. Sharp) W. A. Weber;绵毛胡克香根■☆
47142 Balsamorhiza macrolepis W. M. Sharp var. platylepis (W. M. Sharp) Ferris = Balsamorhiza hookeri Nutt. ■☆
47143 Balsamorhiza macrophylla Nutt.;大叶香根■☆
47144 Balsamorhiza macrophylla Nutt. var. idahoensis W. M. Sharp = Balsamorhiza macrophylla Nutt. ■☆
47145 Balsamorhiza platylepis W. M. Sharp = Balsamorhiza hookeri Nutt. ■☆
47146 Balsamorhiza rosea A. Nelson et J. F. Macbr.;粉香根■☆
47147 Balsamorhiza sagittata (Pursh) Nutt.;狭叶香根;Arrowleaf Balsam-root, Arrowroot, Balsam Root ■☆
47148 Balsamus Stackh. = Commiphora Jacq.(保留属名)●
47149 Balthasaria Verdc.(1969);长管山茶属(巴尔山茶属)●☆
47150 Balthasaria Verdc. = Melchiora Kobuski ●☆
47151 Balthasaria mannii (Oliv.) Verdc. = Melchiora mannii (Oliv.) Kobuski ●☆
47152 Balthasaria schliebenii (Melch.) Verdc.;长管山茶●☆
47153 Balthasaria schliebenii (Melch.) Verdc. var. intermedia (Boutique et Troupin) Verdc. = Melchiora schliebenii (Melch.) Kobuski var. intermedia (Boutique et Troupin) Kobuski ●☆
47154 Baltimora L.(1771)(保留属名);艳头菊属■☆
47155 Baltimora recta L.;艳头菊;Beautyhead ■☆
47156 Baltimorea Raf. = Baltimora L.(保留属名)■☆
47157 Bambekea Cogn.(1916);巴姆葫芦属(西非葫芦属)■☆
47158 Bambekea bequaertii (De Wild.) C. Jeffrey = Bambekea racemosa Cogn. ■☆
47159 Bambekea racemosa Cogn.;巴姆葫芦■☆
47160 Bamboga Baill. = Mamboga Blanco(废弃属名)●
47161 Bamboga Baill. = Mitragyna Korth.(保留属名)●
47162 Bambos Retz.(废弃属名) = Bambusa Schreb.(保留属名)●
47163 Bambos arundinacea Retz. = Bambusa arundinacea (Retz.) Willd. ●
47164 Bambos stricta Roxb. = Dendrocalamus strictus (Roxb.) Nees ●
47165 Bambos tootsik Siebold = Sinobambusa tootsik (Siebold) Makino ex Nakai ●
47166 Bamburanta L. Linden = Hybophrynium K. Schum. ■☆
47167 Bamburanta L. Linden = Trachyphrynium Benth. ■☆
47168 Bamburanta arnoldiana L. Linden = Trachyphrynium braunianum (K. Schum.) Baker ■☆
47169 Bambus Blanco = Bambusa Schreb.(保留属名)●
47170 Bambus J. F. Gmel. = Bambusa Schreb.(保留属名)●
47171 Bambusa Caldas = Bambusa Schreb.(保留属名)●
47172 Bambusa Mutis ex Caldas = Bambusa Schreb.(保留属名)●
47173 Bambusa Mutis ex Caldas = Guadua Kunth ●☆
47174 Bambusa Schreb.(1789)(保留属名);簕竹属(莿竹属,凤凰竹属,簕竹属,蓬莱竹属,山白竹属,孝顺竹属);Bamboo,Bambusa ●
47175 Bambusa abyssinica A. Rich. = Oxytenanthera abyssinica (A. Rich.) Munro ●☆
47176 Bambusa albofolia T. H. Wen et Hua = Bambusa multiplex (Lour.) Raeusch. ex Schult. et Schult. f. ●
47177 Bambusa albofolia T. H. Wen et S. C. Hua;皮竹●
47178 Bambusa albolineata L. C. Chia;花竹(火吹竹,火管竹,火广竹,绿篱竹);Colory Bamboo, Colory Bambusa, White-line Bamboo, White-lineate Bambusa ●
47179 Bambusa albo-striata (McClure) Ohrnb. = Bambusa albolineata L. C. Chia ●
47180 Bambusa alphonso-karri Mitford ex Satow = Bambusa multiplex (Lour.) Raeusch. ex Schult. et Schult. f. 'Alphonse-Karr' ●
47181 Bambusa alphonso-karrii Mitford ex Satow = Bambusa multiplex (Lour.) Raeusch. ex Schult. et Schult. f. ●

47182 Bambusa amplexicaulis W. T. Lin et Z. M. Wu;抱秆黄竹●

47183 Bambusa angulata Munro = Bambusa tuldoides Munro ●

47184 Bambusa angustiaurita W. T. Lin;狭耳坭竹(狭耳簕竹);Narrowear Bamboo,Narrow-eared Bambusa ●

47185 Bambusa angustissima L. C. Chia et H. L. Fung;狭耳簕竹(狭耳簕竹);Muchnarrow Bamboo,Narrowest-eared Bambusa ●

47186 Bambusa annulata (J. R. Xue et W. P. Zhang) D. Z. Li = Bambusa yunnanensis N. H. Xia ●

47187 Bambusa annulata W. T. Lin et Z. J. Feng;隆武竹;Annulate Bambusa ●

47188 Bambusa annulata W. T. Lin et Z. J. Feng = Bambusa textilis McClure ●

47189 Bambusa arnhemica F. Muell.;澳洲簕竹;Arnhem Land Bamboo,Australian Bamboo,Didgeridoo Bamboo ●☆

47190 Bambusa arundinacea (Retz.) Willd.;印度簕竹(茨竹,鸡儿竹,貌子竹,缅甸刺竹,芡竹,印度簕竹);Bamboo,Giant Thorny Bamboo,Indian Bamboo,Male Bamboo,Reed Bamboo,Spiny Bamboo,Thorny Bamboo,Thorny Bambusa ●

47191 Bambusa arundinacea (Retz.) Willd. = Bambusa vulgaris Schrad. ex J. C. Wendl. ●

47192 Bambusa arundinacea (Retz.) Willd. var. gigantea Bahadur;巨印度簕竹;Giant Reed-bamboo ●☆

47193 Bambusa aspera Schult. et Schult. f. = Dendrocalamus asper (Schult. et Schult. f.) Backer ex K. Heyne ●

47194 Bambusa aspera Schult. f. = Gigantochloa aspera (Schult.) Kurz ex Teijsm. et Binn. ●

47195 Bambusa atra Lindl.;茂物簕竹;Bogor Thin-walled Bamboo,New Guinea Thin-walled Bamboo ●☆

47196 Bambusa atrovirens T. H. Wen;光箨绿竹●

47197 Bambusa atrovirens T. H. Wen = Bambusa oldhamii Munro ●

47198 Bambusa atrovirens T. H. Wen = Dendrocalamopsis oldhamii (Munro) P. C. Keng ●

47199 Bambusa aureo-striata Regel = Shibataea chinensis Nakai 'Aureo-striata' ●

47200 Bambusa auriculata Kurz = Bambusa vulgaris Schrad. ex J. C. Wendl. ●

47201 Bambusa aurinuda McClure;裸耳竹;Bare-eared Bambusa,Nakeear Bamboo ●

47202 Bambusa baccifera Roxb. = Melocalamus baccifer (Roxb.) Kurz ●

47203 Bambusa balcooa Roxb. ex Roxb.;好望角簕竹;Balcooa Bamboo,Female Bamboo,Giant Bamboo ●☆

47204 Bambusa bambos (L.) Voss ex Vilm. = Bambusa arundinacea (Retz.) Willd. ●

47205 Bambusa basihirsuta McClure;扁竹●

47206 Bambusa basihirsuta McClure = Dendrocalamopsis basihirsuta (McClure) P. C. Keng et W. T. Lin ●

47207 Bambusa basihirsutoides N. H. Xia;凹箨绿竹●

47208 Bambusa beecheyana Munro = Dendrocalamopsis beecheyana (Munro) P. C. Keng ●

47209 Bambusa beecheyana Munro var. pubescens (P. F. Li) W. C. Lin = Dendrocalamopsis beecheyana (Munro) P. C. Keng var. pubescens (P. F. Li) P. C. Keng ●

47210 Bambusa beisitiku (Odash.) P. C. Keng = Bambusa pachinensis Hayata ●

47211 Bambusa bicicatricata (W. T. Lin) L. C. Chia et H. L. Fung;孟竹;Bicicatrical Dendrocalamopsis,Twinscar Greenbamboo ●

47212 Bambusa bicicatricata (W. T. Lin) L. C. Chia et H. L. Fung = Dendrocalamopsis bicicatricata (W. T. Lin) P. C. Keng ●

47213 Bambusa blumeana Schult. et Schult. f.;簕竹(刺竹,莿竹,大簕竹,菲律宾簕竹,簕竹,坭竹,郁竹,竹);Blume Bambusa,Blume Spine Bamboo,Kanayan-tinik,Narrow-spike Bambusa,Thorng Bamboo ●

47214 Bambusa blumeana Schult. et Schult. f. 'Blumeana';布氏簕竹●

47215 Bambusa blumeana Schult. et Schult. f. 'Wei-fang Lin';惠方簕竹(林氏簕竹)●

47216 Bambusa blumeana Schult. et Schult. f. f. weifang-lin (W. C. Lin) T. P. Yi = Bambusa stenostachya Hack. 'Wei-fang Lin' ●

47217 Bambusa blumeana Schult. et Schult. f. f. weifang-lin (W. C. Lin) T. P. Yi = Bambusa blumeana Schult. et Schult. f. 'Wei-fang Lin' ●

47218 Bambusa boniopsis McClure;妈竹;Mam Bamboo,Mather Bambusa ●

47219 Bambusa brandisii Munro = Dendrocalamus brandisii (Munro) Kurz ●

47220 Bambusa breviflora Munro;华南水竹;Short-flower Bambusa ●

47221 Bambusa breviflora Munro = Bambusa tuldoides Munro ●

47222 Bambusa breviflora Munro var. hainanensis G. A. Fu;海南水竹;Hainan Bamboo ●

47223 Bambusa breviligulata L. C. Chia et H. L. Fung = Bambusa duriuscula W. T. Lin ●

47224 Bambusa brunneoaciculia G. A. Fu;褐毛青皮竹●

47225 Bambusa burmanica Gamble;缅甸竹(缅孝顺竹);Burma Bamboo,Burma Bambusa,Burmese Weavers' Bamboo ●

47226 Bambusa caesia Siebold et Zucc. ex Munro = Bambusa multiplex (Lour.) Raeusch. ex Schult. et Schult. f. ●

47227 Bambusa calostachya Kurz = Dendrocalamus calostachyus (Kurz) Kurz ●

47228 Bambusa cantorii Munro = Pseudosasa cantorii (Munro) P. C. Keng ex S. L. Chen et al. ●

47229 Bambusa capensis Rupr. = Bambusa balcooa Roxb. ex Roxb. ●☆

47230 Bambusa cerosissima McClure;单竹(箪竹);Cerated Bamboo,Cerated Bambusa,Single Bamboo ●

47231 Bambusa chino Franch. et Sav. = Pleioblastus chino (Franch. et Sav.) Makino ●

47232 Bambusa chungii McClure;粉单竹(粉箪竹);Chung Bamboo,Chung Bambusa,Farinose Single Bamboo ●

47233 Bambusa chunii L. C. Chia et H. L. Fung;焕镛簕竹(焕镛簕竹);Chun Bambusa,Chun Spine Bamboo ●

47234 Bambusa contracta L. C. Chia et H. L. Fung;破篾黄竹;Compressed Bambusa,Contracte Bamboo,Splited Bambusa,Thinstrip Yellow Bamboo ●

47235 Bambusa corniculata L. C. Chia et H. L. Fung;东兴黄竹(东兴竹,黄竹);Corniculate Bamboo,Corniculate Bambusa,Dongxing Bamboo ●

47236 Bambusa cornigera McClure;牛角竹;Coriaceous Bambusa,Horny Bamboo ●

47237 Bambusa crispiaurita W. T. Lin et Z. M. Wu;皱耳石竹●

47238 Bambusa critica Kurz. = Bambusa pallida Munro ●

47239 Bambusa dendrocalamus ?;白藤竹●☆

47240 Bambusa diaoluoshanensis L. C. Chia et H. L. Fung;吊罗坭竹;Diaoluoshan Bamboo,Diaoluoshan Bambusa,Diaoluoshan Mud Bamboo ●

47241 Bambusa diffusa Blanco = Schizostachyum diffusum (Blanco) Merr. ●

47242 Bambusa dissemulator McClure;坭簕竹(坭刺竹,坭簕竹,坭

竹);Mud Spine Bamboo,Muddy Bamboo,Muddy Bambusa ●

47243 Bambusa dissemulator McClure var. albinodia McClure;白节簕竹(白节簕竹);Whitenode Bambusa,Whitenode Spine Bambusa ●

47244 Bambusa dissemulator McClure var. hispida McClure;毛簕竹(毛簕竹);Hair Spine Bamboo,Hispid Bambusa ●

47245 Bambusa dissimilis W. T. Lin = Bambusa indigena L. C. Chia et H. L. Fung ●

47246 Bambusa distegia (Keng et P. C. Keng) L. C. Chia et H. L. Fung;料慈竹;Doublestegine Bamboo,Upright Bamboo,Upright Bambusa ●

47247 Bambusa disticha Mitford = Pleioblastus argenteostriatus (Regel) Nakai 'Distichus' ●

47248 Bambusa disticha Mitford = Pleioblastus distichus (Mitford) Nakai ●

47249 Bambusa disticha Mitford = Sasa pygmaea (Miq.) E. G. Camus var. disticha (Mitford) C. S. Chao et G. G. Tang ●

47250 Bambusa dolichoclada Hayata;长枝竹(长枝仔竹,桶仔竹);Blowpipe Bamboo,Longbranch Bambusa,Long-branched Bamboo,Long-branched Bambusa,Long-shoot Bamboo,Longtwig Bamboo ●

47251 Bambusa dolichoclada Hayata 'Dolichoclada' = Bambusa dolichoclada Hayata ●

47252 Bambusa dolichoclada Hayata 'Stripe';条纹长枝竹(鼓节竹)●

47253 Bambusa dolichoclada Hayata f. stripe (W. C. Lin) T. P. Yi = Bambusa dolichoclada Hayata 'Stripe' ●

47254 Bambusa dolichomerithalla Hayata;火吹竹(火管竹,火广竹,火筒竹);Blow-pine Bamboo,Blowpipe Bamboo ●

47255 Bambusa dolichomerithalla Hayata 'Green Stripe-stem';金丝火广竹●

47256 Bambusa dolichomerithalla Hayata = Bambusa multiplex (Lour.) Raeusch. ex Schult. et Schult. f. ●

47257 Bambusa dumetorum Hance = Schizostachyum dumetorum (Hance) Munro ●

47258 Bambusa dumetorum Hance ex Walp. = Schizostachyum dumetorum (Hance) Munro ●

47259 Bambusa duriuscula W. T. Lin;蓬莱黄竹(石竹,黎庵高竹);Penglai Bambusa,Penglai Yellow Bamboo ●

47260 Bambusa duriuscula W. T. Lin = Bambusa insularis L. C. Chia et H. L. Fung ●

47261 Bambusa edulis (Odash.) P. C. Keng = Dendrocalamopsis edulis (Odash.) P. C. Keng ●

47262 Bambusa edulis (Odash.) W. C. Lin = Bambusa odashimae Hatus. ●

47263 Bambusa edulis Carrière = Phyllostachys edulis Rivière et C. Rivière ●

47264 Bambusa emeiensis L. C. Chia et H. L. Fung;慈竹(丛竹,钓鱼慈,酒米慈,绵竹,甜慈,孝竹,义竹,阴笋子,子母竹);Emei Bamboo,Emei Bambusa,Lovebamboo,Omei Mountain Bamboo,Similar Neosinocalamus ●

47265 Bambusa emeiensis L. C. Chia et H. L. Fung = Neosinocalamus affinis (Rendle) P. C. Keng ●

47266 Bambusa eutuldoides McClure;大眼竹;Bigeye Bamboo,Large-eye Bambusa,Large-eyed Bambusa ●

47267 Bambusa eutuldoides McClure var. basistriara McClure;银丝大眼竹(斑坭竹)●

47268 Bambusa eutuldoides McClure var. viridi-vittata (W. T. Lin) L. C. Chia;青丝大眼竹(青丝黄竹)●

47269 Bambusa excurrensa G. A. Fu;山凤凰竹●

47270 Bambusa fastuosa Mitford = Semiarundinaria fastuosa (Mitford) Makino ex Nakai ●

47271 Bambusa fauriei Hack. = Bambusa tuldoides Munro ●

47272 Bambusa fecunda McClure. = Bambusa boniopsis McClure ●

47273 Bambusa flavonoda W. T. Lin = Bambusa tuldoides Munro ●

47274 Bambusa flexuosa Carrière = Phyllostachys flexuosa (Carrière) Rivière et C. Rivière ●

47275 Bambusa flexuosa Munro;小簕竹(小勒竹,小簕竹,小坭竹);Flexuose Bamboo,Sinuate Bambusa,Small Spine Bamboo ●

47276 Bambusa floribunda (Büse) Zoll. et Mauro ex Steud. = Bambusa multiplex (Lour.) Raeusch. ex Schult. et Schult. f. 'Fernleaf' ●

47277 Bambusa floribunda (Büse) Zoll. et Mauro ex Steud. f. albo-variegata Nakai = Bambusa multiplex (Lour.) Raeusch. ex Schult. et Schult. f. 'Siverstripe' ●

47278 Bambusa floribunda (Büse) Zoll. et Mauro ex Steud. f. viridi-striata Nakai = Bambusa multiplex (Lour.) Raeusch. ex Schult. et Schult. f. 'Stripestem Fernleaf' ●

47279 Bambusa fortunei Van Houtte = Pleioblastus fortunei (Van Houtte) Nakai ●

47280 Bambusa fortunei Van Houtte = Sasa fortunei (Van Houtte) Fiori ●

47281 Bambusa funghomii McClure;鸡窦簕竹(鸡窦簕竹,鸡窦坭竹);Cocknest Spine Bamboo,Funghom Bamboo,Funghom Bambusa ●

47282 Bambusa galucescens var. riviereorum (Maire) L. C. Chia et H. L. Fung = Bambusa multiplex (Lour.) Raeusch. ex Schult. et Schult. f. var. riviereorum Maire ●

47283 Bambusa gibba McClure;坭竹(水黄竹);Gibbosity Bambusa,Gibbous Bamboo,Mud Bamboo ●

47284 Bambusa gibboides W. T. Lin;鱼肚腩竹(鱼肚腩竹);Fishbelly Bamboo,Gibbosity-like Bambusa,Like Gibbous Bamboo ●

47285 Bambusa glabro-vagina G. A. Fu;光鞘石竹;Glabrous-sheath Bamboo,Glabrous-sheath Bambusa,Smoothsheath Bamboo ●

47286 Bambusa glauca Lodd. ex Lindl. = Bambusa multiplex (Lour.) Raeusch. ex Schult. et Schult. f. ●

47287 Bambusa glaucescens (Lam.) Siebold ex Munro = Bambusa multiplex (Lour.) Raeusch. ex Schult. et Schult. f. ●

47288 Bambusa glaucescens (Willd.) Merr. = Bambusa multiplex (Lour.) Raeusch. ex Schult. et Schult. f. ●

47289 Bambusa glaucescens (Willd.) Merr. var. annulata (W. T. Lin et Z. J. Feng) N. H. Xia = Bambusa textilis McClure ●

47290 Bambusa glaucescens (Willd.) Siebold ex Munro 'Alphonse-Karr' = Bambusa multiplex (Lour.) Raeusch. ex Schult. et Schult. f. 'Alphonse-Karr' ●

47291 Bambusa glaucescens (Willd.) Siebold ex Munro 'Fernleaf' = Bambusa multiplex (Lour.) Raeusch. ex Schult. et Schult. f. 'Fernleaf' ●

47292 Bambusa glaucescens (Willd.) Siebold ex Munro 'Silverstripe' = Bambusa multiplex (Lour.) Raeusch. ex Schult. et Schult. f. 'Silverstripe' ●

47293 Bambusa glaucescens (Willd.) Siebold ex Munro 'Stripestem fernleaf' = Bambusa multiplex (Lour.) Raeusch. ex Schult. et Schult. f. 'Stripestem Fernleaf' ●

47294 Bambusa glaucescens (Willd.) Siebold ex Munro 'Yellowstripe' = Bambusa multiplex (Lour.) Raeusch. ex Schult. et Schult. f. var. 'Yellowstripe' ●

47295 Bambusa glaucescens (Willd.) Siebold ex Munro f. alphonsokarr (Mitford) T. H. Wen = Bambusa multiplex (Lour.) Raeusch. ex Schult. et Schult. f. 'Alphonse-Karr' ●

47296 Bambusa glaucescens (Willd.) Siebold ex Munro f. alphonsokarrii (Mitford ex Satow) Hatus. = Bambusa multiplex (Lour.) Raeusch. ex Schult. et Schult. f. ●

47297 Bambusa glaucescens (Willd.) Siebold ex Munro f. solida K. J. Mao et C. H. Zhao = Bambusa multiplex (Lour.) Raeusch. ex Schult. et Schult. f. ●

47298 Bambusa glaucescens (Willd.) Siebold ex Munro var. pubivagina (W. T. Lin et Z. J. Feng) N. H. Xia = Bambusa multiplex (Lour.) Raeusch. ex Schult. et Schult. f. var. incana B. M. Yang ●

47299 Bambusa glaucescens (Willd.) Siebold ex Munro var. riviereorum (Maire) L. C. Chia et H. L. Fung = Bambusa multiplex (Lour.) Raeusch. ex Schult. et Schult. f. var. riviereorum Maire ●

47300 Bambusa glaucescens (Willd.) Siebold ex Munro var. shimadai (Hayata) L. C. Chia et But = Bambusa multiplex (Lour.) Raeusch. ex Schult. et Schult. f. var. shimadae (Hayata) Sasaki ●

47301 Bambusa glaucescens (Willd.) Siebold ex Munro var. strigosa (T. H. Wen) L. C. Chia = Bambusa multiplex (Lour.) Raeusch. ex Schult. et Schult. f. var. incana B. M. Yang ●

47302 Bambusa grandis (Q. H. Dai et X. L. Tao) Ohrnb.;大绿竹;Big Greenbamboo,Dai Dendrocalamopsis ●

47303 Bambusa guangxiensis L. C. Chia et H. L. Fung;桂单竹(桂箪竹);Guangxi Bamboo,Guangxi Bambusa,Guangxi Single Bamboo ●

47304 Bambusa guangxiensis L. C. Chia et H. L. Fung = Lingnania funghomii McClure ●

47305 Bambusa hainanensis L. C. Chia et H. L. Fung;藤单竹(藤箪竹);Hainan Bambusa,Hainan Single Bamboo ●

47306 Bambusa hainanensis L. C. Chia et H. L. Fung = Lingnania hainanensis (L. C. Chia et H. L. Fung) T. P. Yi ●

47307 Bambusa heterocycla Carrière = Phyllostachys edulis Rivière et C. Rivière ●

47308 Bambusa heterocycla Carrière = Phyllostachys heterocycla (Carrière) Matsum. ●

47309 Bambusa humilis Rchb. ex Rupr.;苏里南竹;Surinam Dwarf Bamboo ●☆

47310 Bambusa humilis Rchb. ex Rupr. = Bambusa vulgaris Schrad. ex J. C. Wendl. ●

47311 Bambusa indigena L. C. Chia et H. L. Fung;乡土竹;Local Bambusa,Native Bamboo ●

47312 Bambusa insularis L. C. Chia et H. L. Fung;黎庵高竹;Insular Bambusa,Lian Bamboo ●

47313 Bambusa intermedia J. R. Xue et T. P. Yi;绵竹(凤尾竹,蛮竹,芒竹);Intermediate Bambusa,Soft Bambusa ●

47314 Bambusa intermedia J. R. Xue et T. P. Yi = Lingnania intermedia (J. R. Xue et T. P. Yi) T. P. Yi ●

47315 Bambusa kumasaca Zoll. ex Steud. = Shibataea kumasasa (Zoll. ex Steud.) Makino ex Nakai ●

47316 Bambusa kumasasa Zoll. = Shibataea kumasasa (Zoll. ex Steud.) Makino ex Nakai ●

47317 Bambusa lako Widjaja;帝汶黑簕竹;Timor Black Bamboo, Tropical Black Bamboo ●☆

47318 Bambusa lapidea McClure;油簕竹(橄榄竹,烂眼竹,马蹄竹,蛮竹,石竹,油簕竹);Horsehoof Bamboo, Oil Spine Bamboo, Stony Bambusa ●

47319 Bambusa latidellata W. T. Lin;软竹(软簕竹);Soft Bambusa ●

47320 Bambusa latiflora (Munro) Kurz = Dendrocalamus latiflorus Munro ●

47321 Bambusa lenta L. C. Chia;藤枝竹;Hard Bambusa, Rattan Bambusa,Vinetwig Bamboo ●

47322 Bambusa levis Blanco = Gigantochloa levis (Blanco) Merr. ●

47323 Bambusa lingnanioides W. T. Lin = Bambusa remotiflora (Kuntze) L. C. Chia et H. L. Fung ●

47324 Bambusa liukiuensis Hayata = Bambusa multiplex (Lour.) Raeusch. ex Schult. et Schult. f. ●

47325 Bambusa lixin J. R. Xue et T. P. Yi = Bambusa teres Buch.-Ham. ex Munro ●

47326 Bambusa lixin J. R. Xue et T. P. Yi = Bambusa tulda Roxb. ●

47327 Bambusa longiflora W. T. Lin;青秆●

47328 Bambusa longiflora W. T. Lin = Bambusa tuldoides Munro ●

47329 Bambusa longipalea W. T. Lin;紫斑竹(紫斑簕竹)●

47330 Bambusa longispiculata Gamble ex Brandis;花眉竹(长节竹);Longispiculate Bamboo,Long-spiculate Bambusa,Longspike Bamboo ●

47331 Bambusa macrotis L. C. Chia et H. L. Fung;大耳坭竹;Big-eared Bambusa,Largeear Bamboo ●

47332 Bambusa madagascariensis Rivière et C. Rivière = Bambusa vulgaris Schrad. ex J. C. Wendl. ●

47333 Bambusa malingensis McClure;马岭竹;Maling Bamboo, Maling Bambusa ●

47334 Bambusa marmorea Mitford = Chimonobambusa marmorea (Mitford) Makino ex Nakai ●◇

47335 Bambusa maxima (Lour.) Poir. = Gigantochloa verticillata (Willd.) Munro ●

47336 Bambusa membranacea (Munro) Stapleton et N. H. Xia = Dendrocalamus membranaceus Munro ●

47337 Bambusa metake Siebold ex Miq. = Arundinaria japonica A. Gray ●

47338 Bambusa metake Siebold ex Miq. = Pleioblastus simonii (Carrière) Nakai ●

47339 Bambusa metake Siebold ex Miq. = Pseudosasa japonica (Siebold et Zucc. ex Steud.) Makino ex Nakai ●

47340 Bambusa minutiligulata W. T. Lin et Z. M. Wu;微舌黄竹仔●

47341 Bambusa minutiligulata W. T. Lin et Z. M. Wu = Bambusa textilis McClure ●

47342 Bambusa miyiensis T. P. Yi;蛮竹;Miyi Bamboo ●

47343 Bambusa miyiensis T. P. Yi = Bambusa lapidea McClure ●

47344 Bambusa mollis L. C. Chia et H. L. Fung;拟黄竹;Pliant Bamboo,Soft Bambusa,Softhair Bamboo ●

47345 Bambusa multiplex (Lour.) Raeusch. ex Schult. et Schult. f.;孝顺竹(凤凰竹,凤尾竹,观音竹,蓬莱竹,扫把竹,石角竹);Hedge Bamboo,Hedge Bambusa,Karr Bamboo ●

47346 Bambusa multiplex (Lour.) Raeusch. ex Schult. et Schult. f. 'Alphonse-Karr';小琴丝竹(花茎竹,七弦竹,苏方竹,苏枋竹);Alphonse Karr Bamboo,Alphonse-karr Hedge Bamboo ●

47347 Bambusa multiplex (Lour.) Raeusch. ex Schult. et Schult. f. 'Fernleaf';凤尾竹(多花簕竹,凤凰竹,红凤凰竹,蕨叶竹,条纹凤尾竹); Abundant-flower Bambusa, Elegant Hedge Bamboo, Fernleaf Hedge Bamboo,Many-flowered Bamboo ●

47348 Bambusa multiplex (Lour.) Raeusch. ex Schult. et Schult. f. 'Floribunda';繁花孝顺竹;Chinese Dwarf Bamboo ●☆

47349 Bambusa multiplex (Lour.) Raeusch. ex Schult. et Schult. f. 'Golden Goddess';金色女神竹●☆

47350 Bambusa multiplex (Lour.) Raeusch. ex Schult. et Schult. f. 'Silverstripe';银丝竹(班入凤凰竹,红凤凰竹,条纹凤凰竹,银纹竹);Silverstripe Hedge Bamboo ●

47351 Bambusa multiplex (Lour.) Raeusch. ex Schult. et Schult. f. 'Stripestem Fernleaf';小叶琴丝竹;Stripestem Fernleaf Bamboo ●

47352 Bambusa multiplex (Lour.) Raeusch. ex Schult. et Schult. f. 'Willowy';垂柳竹;Willow Hedge Bamboo ●

47353 Bambusa multiplex (Lour.) Raeusch. ex Schult. et Schult. f. 'Yellowstripe';黄条竹;Yellowstripe Hedge Bamboo ●

47354 Bambusa multiplex (Lour.) Raeusch. ex Schult. et Schult. f. f. alphonsokarri (Mitford) Sasaki ex P. C. Keng = Bambusa multiplex (Lour.) Raeusch. ex Schult. et Schult. f. 'Alphonse-Karr' ●

47355 Bambusa multiplex (Lour.) Raeusch. ex Schult. et Schult. f. var. elagans (Koidz.) Muroi ex Sugim. f. viridi-striata (Makino ex Tsuboi) Muroi ex Sugim. = Bambusa multiplex (Lour.) Raeusch. ex Schult. et Schult. f. 'Stripestem Fernleaf' ●

47356 Bambusa multiplex (Lour.) Raeusch. ex Schult. et Schult. f. var. elegans (Koidz.) Muroi ex Sugim. = Bambusa multiplex (Lour.) Raeusch. ex Schult. et Schult. f. 'Fernleaf' ●

47357 Bambusa multiplex (Lour.) Raeusch. ex Schult. et Schult. f. var. elegans (Koidz.) Muroi ex Sugim. f. albo-varegata (Makino) Muroi ex Sugim. = Bambusa multiplex (Lour.) Raeusch. ex Schult. et Schult. f. 'Silverstripe' ●

47358 Bambusa multiplex (Lour.) Raeusch. ex Schult. et Schult. f. var. fernleaf R. A. Young = Bambusa multiplex (Lour.) Raeusch. ex Schult. et Schult. f. f. fernleaf (R. A. Young) T. P. Yi ●

47359 Bambusa multiplex (Lour.) Raeusch. ex Schult. et Schult. f. var. fernleaf R. A. Young = Bambusa multiplex (Lour.) Raeusch. ex Schult. et Schult. f. 'Fernleaf' ●

47360 Bambusa multiplex (Lour.) Raeusch. ex Schult. et Schult. f. var. gracillima (Makino ex E. G. Camus) Sad. Suzuki = Bambusa multiplex (Lour.) Raeusch. ex Schult. et Schult. f. 'Fernleaf' ●

47361 Bambusa multiplex (Lour.) Raeusch. ex Schult. et Schult. f. var. incana B. M. Yang;毛凤凰竹●

47362 Bambusa multiplex (Lour.) Raeusch. ex Schult. et Schult. f. var. lutea T. H. Wen;普陀孝顺竹●

47363 Bambusa multiplex (Lour.) Raeusch. ex Schult. et Schult. f. var. nana (Roxb.) P. C. Keng = Bambusa multiplex (Lour.) Raeusch. ex Schult. et Schult. f. ●

47364 Bambusa multiplex (Lour.) Raeusch. ex Schult. et Schult. f. var. nana (Roxb.) P. C. Keng = Bambusa multiplex (Lour.) Raeusch. ex Schult. et Schult. f. 'Fernleaf' ●

47365 Bambusa multiplex (Lour.) Raeusch. ex Schult. et Schult. f. var. nana (Roxb.) P. C. Keng = Bambusa multiplex (Lour.) Raeusch. ex Schult. et Schult. f. var. riviereorum Maire ●

47366 Bambusa multiplex (Lour.) Raeusch. ex Schult. et Schult. f. var. nortiplex f. alphonso-karri (Satow) Sasaki = Bambusa multiplex (Lour.) Raeusch. ex Schult. et Schult. f. 'Alphonse-Karr' ●

47367 Bambusa multiplex (Lour.) Raeusch. ex Schult. et Schult. f. var. pubivagina W. T. Lin et Z. J. Feng;毛鞘银丝竹;Hairy-sheathed Hedge Bamboo ●

47368 Bambusa multiplex (Lour.) Raeusch. ex Schult. et Schult. f. var. pubivagina W. T. Lin et Z. J. Feng = Bambusa multiplex (Lour.) Raeusch. ex Schult. et Schult. f. var. incana B. M. Yang ●

47369 Bambusa multiplex (Lour.) Raeusch. ex Schult. et Schult. f. var. riviereorum Maire;观音竹●

47370 Bambusa multiplex (Lour.) Raeusch. ex Schult. et Schult. f. var. shimadae (Hayata) Sasaki;石角竹;Shimada Hedge Bamboo ●

47371 Bambusa multiplex (Lour.) Raeusch. ex Schult. et Schult. f. var. silverstripe R. A. Young = Bambusa multiplex (Lour.) Raeusch. ex Schult. et Schult. f. f. silverstripe (R. A. Young) T. P. Yi ●

47372 Bambusa multiplex (Lour.) Raeusch. ex Schult. et Schult. f. var. silverstripe R. A. Young = Bambusa multiplex (Lour.) Raeusch. ex Schult. et Schult. f. 'Silverstripe' ●

47373 Bambusa multiplex (Lour.) Raeusch. ex Schult. et Schult. f. var. solida B. M. Yang = Bambusa multiplex (Lour.) Raeusch. ex Schult. et Schult. f. ●

47374 Bambusa multiplex (Lour.) Raeusch. ex Schult. et Schult. f. var. stripestem-fernleaf R. A. Young = Bambusa multiplex (Lour.) Raeusch. ex Schult. et Schult. f. f. stripestem-fernleaf (R. A. Young) T. P. Yi ●

47375 Bambusa multiplex (Lour.) Raeusch. ex Schult. et Schult. f. var. stripestem fernleaf R. A. Young = Bambusa multiplex (Lour.) Raeusch. ex Schult. et Schult. f. 'Stripestem Fernleaf' ●

47376 Bambusa mutabilis McClure;黄竹仔;Changeable Bamboo, Variable Bambusa ●

47377 Bambusa naibunensis (Hayata) Nakai = Ampelocalamus naibunensis (Hayata) T. H. Wen ●

47378 Bambusa nana Roxb. = Bambusa glaucescens (Lam.) Siebold ex Munro ●

47379 Bambusa nana Roxb. = Bambusa glaucescens (Willd.) Siebold ex Munro ●

47380 Bambusa nana Roxb. = Bambusa multiplex (Lour.) Raeusch. ex Schult. et Schult. f. ●

47381 Bambusa nana Roxb. f. albo-veriegata Makino = Bambusa multiplex (Lour.) Raeusch. ex Schult. et Schult. f. 'Silverstripe' ●

47382 Bambusa nana Roxb. f. viridi-striata Makino = Bambusa multiplex (Lour.) Raeusch. ex Schult. et Schult. f. 'Stripestem Fernleaf' ●

47383 Bambusa nana Roxb. var. albo-variegata E. G. Camus = Bambusa multiplex (Lour.) Raeusch. ex Schult. et Schult. f. 'Silverstripe' ●

47384 Bambusa nana Roxb. var. alphonso-karri (Satow) E. G. Camus = Bambusa multiplex (Lour.) Raeusch. ex Schult. et Schult. f. 'Alphonse-Karr' ●

47385 Bambusa nana Roxb. var. alphonso-karrii (Mitford ex Satow) Latour-Marliac ex E. G. Camus = Bambusa multiplex (Lour.) Raeusch. ex Schult. et Schult. f. ●

47386 Bambusa nana Roxb. var. gracillimma Makino ex E. G. Camus = Bambusa multiplex (Lour.) Raeusch. ex Schult. et Schult. f. 'Fernleaf' ●

47387 Bambusa nana Roxb. var. normalis ? f. alphonso-karri (Mitford ex Satow) Makino ex Shiras. = Bambusa multiplex (Lour.) Raeusch. ex Schult. et Schult. f. 'Alphonse-Karr' ●

47388 Bambusa nana Roxb. var. typica f. viridi-striata Makino ex Tsuboi = Bambusa multiplex (Lour.) Raeusch. ex Schult. et Schult. f. 'Stripestem Fernleaf' ●

47389 Bambusa nana Roxb. var. variegata E. G. Camus = Bambusa multiplex (Lour.) Raeusch. ex Schult. et Schult. f. ●

47390 Bambusa narihira (Bean) Makino = Semiarundinaria fastuosa (Mitford) Makino ex Nakai ●

47391 Bambusa nigra Lodd. ex Lindl. = Phyllostachys nigra (Lodd. ex Lindl.) Munro ●

47392 Bambusa nigrociliata Büse = Gigantochloa nigrociliata (Büse) Kurz ●

47393 Bambusa nutans Munro subsp. cupulata Stapleton. = Bambusa teres Buch. -Ham. ex Munro ●

47394 Bambusa nutans Wall. ex Munro;俯卧竹(俯竹);Burmese Timber Bamboo, Drooping Bambusa, Nutate Bamboo ●

47395 Bambusa odashimae Hatus. = Bambusa odashimae Hatus. ex Ohrnb. ●

47396 Bambusa odashimae Hatus. = Dendrocalamopsis edulis (Odash.) P. C. Keng ●
47397 Bambusa odashimae Hatus. ex Ohrnb.;乌脚绿竹●
47398 Bambusa odashimae Hatus. ex Ohrnb. = Dendrocalamopsis edulis (Odash.) P. C. Keng ●
47399 Bambusa oldhamii Munro = Dendrocalamopsis oldhamii (Munro) P. C. Keng ●
47400 Bambusa oldhamii Munro f. revoluta W. T. Lin et J. Y. Lin = Dendrocalamopsis oldhamii (Munro) P. C. Keng f. revoluta (W. T. Lin et J. Y. Lin) W. T. Lin ●
47401 Bambusa omeiensis L. C. Chia et H. L. Fung = Bambusa emeiensis L. C. Chia et H. L. Fung ●
47402 Bambusa pachinensis Hayata;米筛竹(八芝兰竹,空涵竹,有咸,矢竹,有咸);Pachilan Bamboo,Pan-chinese Bambusa,Ricesieve Bamboo ●
47403 Bambusa pachinensis Hayata var. hirsutissima (Odash.) W. C. Lin;长毛米筛竹(长毛八芝兰竹,黄竹);Hairy Pachilan Bamboo ●
47404 Bambusa pallida Munro;大薄竹;Pale Bamboo, Pallid-leaved Bambusa,Thin-walled Bambusa ●
47405 Bambusa papillata (Q. H. Dai) Q. H. Dai;水单竹(水箪竹);Papillate Bambusa,Water Single Bambusa ●
47406 Bambusa papillatoides Q. H. Dai et D. Y. Huang;细单竹(细箪竹);False Papillate Bambusa ●
47407 Bambusa parvifolia W. T. Lin = Bambusa tuldoides Munro ●
47408 Bambusa pervariabilis McClure;撑篙竹(白眉竹,花眉竹,硬头犁,油竹);Pole Bambusa, Punting Pole Bamboo, Punting Pole Bambusa,Puntingpole Bamboo,Puntong Bambusa ●
47409 Bambusa pervariabilis McClure var. viridis-striata Q. H. Dai et X. C. Liu;花撑篙竹;Colored Pole Bambusa ●
47410 Bambusa picta Siebold et Zucc. ex Munro = Pleioblastus fortunei (Van Houtte) Nakai ●
47411 Bambusa picta Siebold et Zucc. ex Munro = Sasa fortunei (Van Houtte) Fiori ●
47412 Bambusa piscaporum McClure;石竹仔;Fisher Bambusa, Fisherman Bambusa,Stone Bamboo ●
47413 Bambusa polymorpha Munro;灰秆竹(多型簕竹,灰竿竹);Greypole Bamboo,Maihop,Multiform Bambusa ●
47414 Bambusa prasina T. H. Wen = Bambusa basihirsuta McClure ●
47415 Bambusa prasina T. H. Wen = Dendrocalamopsis basihirsuta (McClure) P. C. Keng et W. T. Lin ●
47416 Bambusa prominens H. L. Fung et C. Y. Sia;牛耳竹(牛儿竹);Oxet Bamboo,Prominent Bamboo,Prominent Bambusa ●
47417 Bambusa pseudoarundinacea Steud. = Gigantochloa verticillata (Willd.) Munro ●
47418 Bambusa puberula Miq. = Phyllostachys nigra (Lodd. ex Lindl.) Munro var. henonis (Bean) Stapf ex Rendle ●
47419 Bambusa pubivaginata W. T. Lin et Z. M. Wu;毛鞘黄竹仔;Hair-sheath Bambusa ●
47420 Bambusa pubivaginata W. T. Lin et Z. M. Wu = Bambusa multiplex (Lour.) Raeusch. ex Schult. et Schult. f. var. incana B. M. Yang ●
47421 Bambusa purpure-vagian G. A. Fu;紫竹子;Purple-sheathed Bambusa ●
47422 Bambusa pygmaea Miq. = Pleioblastus fortunei (Van Houtte) Nakai ●
47423 Bambusa pygmaea Miq. = Sasa pygmaea (Miq.) E. G. Camus ●
47424 Bambusa quadrangularis Franceschi = Chimonobambusa quadrangularis (Franceschi) Makino ex Nakai ●
47425 Bambusa ramispinosa L. C. Chia et H. L. Fung;坭黄竹;Forkspine Bamboo,Rami-spined Bambusa,Ramispiny Bamboo,Spine Bambusa ●
47426 Bambusa regia Thomson ex Munro = Thyrsostachys siamensis (Kurz ex Munro) Gamble ●
47427 Bambusa remotiflora (Kuntze) L. C. Chia et H. L. Fung;甲竹(吊竹,麻竹,石竹,疏花单竹,水竹);Laxflower Bamboo, Remotiflower Bamboo, Remotiflower Bambusa, Remotiflowered Bambusa ●
47428 Bambusa remotiflora (Kuntze) L. C. Chia et H. L. Fung = Lingnania remotiflora (Kuntze) McClure ●
47429 Bambusa reticulata Rupr. = Phyllostachys reticulata (Rupr.) K. Koch ●
47430 Bambusa rigida Keng et P. C. Keng;硬头黄竹(硬头黄);Rigid Bamboo,Rigid Bambusa,Rigid Yellow Bamboo ●
47431 Bambusa rugata (W. T. Lin) Ohrnb.;皱纹单竹(皱纹箪竹)●
47432 Bambusa ruscifolia Siebold ex Munro = Shibataea kumasasa (Zoll. ex Steud.) Makino ex Nakai ●
47433 Bambusa rutila McClure;木竹(扁担竹);Glowing Bambusa, Shuangliu Bamboo, Shuangliu Bambusa, Wood Bamboo, Woody Bambusa ●
47434 Bambusa sanzaoensis W. T. Lin;三灶坭竹;Sanzao Bambusa ●
47435 Bambusa sanzaoensis W. T. Lin. = Bambusa xiashanensis L. C. Chia et H. L. Fung ●
47436 Bambusa scabriculma W. T. Lin. = Bambusa flexuosa Munro ●
47437 Bambusa sciptoria Dennst. = Bambusa multiplex (Lour.) Raeusch. ex Schult. et Schult. f. var. riviereorum Maire ●
47438 Bambusa semitecta W. T. Lin et Z. M. Wu;掩耳黄竹●
47439 Bambusa shimadae Hayata = Bambusa multiplex (Lour.) Raeusch. ex Schult. et Schult. f. var. shimadae (Hayata) Sasaki ●
47440 Bambusa shuangliuensis T. P. Yi;扁担竹;Shuangliu Bamboo, Shuangliu Bambusa ●
47441 Bambusa shuangliuensis T. P. Yi = Bambusa rutila McClure ●
47442 Bambusa siamensis Kurz ex Munro = Thyrsostachys siamensis (Kurz ex Munro) Gamble ●
47443 Bambusa sieberi Griseb. = Bambusa vulgaris Schrad. ex J. C. Wendl. ●
47444 Bambusa simonii Carrière = Pleioblastus simonii (Carrière) Nakai ●
47445 Bambusa sinospinosa McClure;车筒竹(簕楠竹,簕竹,车角竹,刺楠竹,刺竹,笒黎竹,大簕竹,大竻麻竹,大竹,莦竹,竻竹,槸竹,麻竹,水簕竹,水竻竹);China Thorny Bamboo,Chinese Spiny Bamboo,Chinese Thorny Bamboo,Chinese Thorny Bambusa ●
47446 Bambusa spinosa Blume ex Nees = Bambusa blumeana Schult. et Schult. f. ●
47447 Bambusa spinosa Merr. = Bambusa remotiflora (Kuntze) L. C. Chia et H. L. Fung ●
47448 Bambusa spinosa Roxb.;刺簕竹;Giant Thorny Bamboo, Spine Bambusa,Spiny Bamboo,Thorny Bamboo ●
47449 Bambusa spinosa Roxb. = Bambusa arundinacea (Retz.) Willd. ●
47450 Bambusa spinosa Roxb. ex Buch.-Ham. = Bambusa arundinacea (Retz.) Willd. ●
47451 Bambusa stenoaurita (W. T. Lin) T. H. Wen;黄麻竹;Jute Greenbamboo,Stenoauriculate Dendrocalamopsis ●
47452 Bambusa stenostachya Hack. ' Wei-fang Lin ' = Bambusa blumeana Schult. et Schult. f. ' Wei-fang Lin' ●
47453 Bambusa stenostachya Hack. = Bambusa blumeana Schult. et

Schult. f. ●

47454 Bambusa sterilis Kurz ex Miq. = Bambusa multiplex (Lour.) Raeusch. ex Schult. et Schult. f. ●

47455 Bambusa stipitata W. T. Lin. = Bambusa rigida Keng et P. C. Keng ●

47456 Bambusa striata Lodd. = Bambusa vulgaris Schrad. ex J. C. Wendl. 'Vittata' ●

47457 Bambusa striata Lodd. ex Lindl. = Bambusa vulgaris Schrad. ex J. C. Wendl. ●

47458 Bambusa striato-maculata G. A. Fu;海南斑竹;Hainan Bambusa ●

47459 Bambusa stricta (Roxb.) Roxb. = Dendrocalamus strictus (Roxb.) Nees ●

47460 Bambusa strigosa T. H. Wen;河边竹●

47461 Bambusa strigosa T. H. Wen = Bambusa multiplex (Lour.) Raeusch. ex Schult. et Schult. f. var. incana B. M. Yang ●

47462 Bambusa subaequalis H. L. Fung et C. Y. Sia; 锦竹; Brocade Bambusa,Near Equal Bamboo,Splendid Bambusa,Subequel Bamboo ●

47463 Bambusa subtruncata L. C. Chia et H. L. Fung; 信宜石竹; Subtruncate Bamboo,Subtruncate Stone Bamboo,Xinyi Bambusa ●

47464 Bambusa sulfurea Carrière = Phyllostachys sulphurea (Carrière) Rivière et C. Rivière ●

47465 Bambusa sulphurea Carrière = Phyllostachys sulphurea (Carrière) Rivière et C. Rivière ●

47466 Bambusa surinamensis Rupr. = Bambusa vulgaris Schrad. ex J. C. Wendl. ●

47467 Bambusa surrecta (Q. H. Dai) Q. H. Dai; 油竹; Erect Bamboo, Oily Bambusa ●

47468 Bambusa surrecta (Q. H. Dai) Q. H. Dai = Lingnania surrecta Q. H. Dai ●

47469 Bambusa taiwanensis L. C. Chia et H. L. Fung = Bambusa odashimae Hatus. ex Ohrnb. ●

47470 Bambusa taiwanensis L. C. Chia et H. L. Fung = Dendrocalamopsis edulis (Odash.) P. C. Keng ●

47471 Bambusa teba Miq. = Bambusa blumeana Schult. et Schult. f. ●

47472 Bambusa tengchongensis D. Z. Li et N. H. Xia = Bambusa xueana Ohrnb. ●

47473 Bambusa teres Buch. -Ham. ex Munro;马甲竹●

47474 Bambusa tessellata Munro = Indocalamus tessellatus (Munro) P. C. Keng ●

47475 Bambusa textilis McClure;青皮竹(广竹黄,土竹黄);Chinese Textile Bamboo, Chinese Textile Bambusa, Greenskin Bamboo, Weaver's Bamboo ●

47476 Bambusa textilis McClure 'Maculata';紫斑青皮竹(紫斑竹); Spotted Weavers' Bamboo ●

47477 Bambusa textilis McClure 'Purpurascens';紫秆青皮竹(紫竿竹,紫秆竹)●

47478 Bambusa textilis McClure f. maculata (McClure) T. P. Yi = Bambusa textilis McClure 'Maculata' ●

47479 Bambusa textilis McClure f. purpurascens (N. H. Xia) T. P. Yi = Bambusa textilis McClure 'Purpurascens' ●

47480 Bambusa textilis McClure var. albostriata McClure = Bambusa albolineata L. C. Chia ●

47481 Bambusa textilis McClure var. fusca McClure = Bambusa pachinensis Hayata var. hirsutissima (Odash.) W. C. Lin ●

47482 Bambusa textilis McClure var. glabra McClure;光秆青皮竹(光竿青皮竹);Slender Bamboo, Slender Weavers' Bamboo, Smooth Greenskin Bamboo ●

47483 Bambusa textilis McClure var. gracilis McClure;崖州竹;Slexder Bambusa ●

47484 Bambusa textilis McClure var. maculata McClure = Bambusa textilis McClure f. maculata (McClure) T. P. Yi ●

47485 Bambusa textilis McClure var. maculata McClure = Bambusa textilis McClure ●

47486 Bambusa textilis McClure var. maculata McClure = Bambusa textilis McClure 'Maculata' ●

47487 Bambusa textilis McClure var. persistens B. M. Yang;长沙青皮竹●

47488 Bambusa textilis McClure var. persistens B. M. Yang = Bambusa textilis McClure ●

47489 Bambusa textilis McClure var. purpurascens N. H. Xia = Bambusa textilis McClure 'Purpurascens' ●

47490 Bambusa thouarsii Kunth = Bambusa vulgaris Schrad. ex J. C. Wendl. ●

47491 Bambusa transvenula (W. T. Lin et Z. T. Feng) N. H. Xia;横脉单竹●

47492 Bambusa truncata B. M. Yang;平箨竹●

47493 Bambusa tulda Roxb.;俯竹(大耳竹,力新孝顺竹,马甲竹); Bengal Bamboo, Mandarin Jacker Bambusa, Spineless Indian Bamboo,Tuld Bamboo,Weskit Bamboo ●

47494 Bambusa tuldoides Munro;青秆竹(花眉竹,青竿竹,水竹,硬伞桃竹,硬生桃竹,硬头黄竹);Eyebrow Bambusa,Greenpole Bamboo, Punting Pole Bamboo,Puntingpole Bamboo,Verdant Bamboo ●

47495 Bambusa tuldoides Munro 'Swolleninternode';鼓节竹●

47496 Bambusa tuldoides Munro 'Swolleninternode' = Bambusa tuldoides Munro f. swolleninternode (N. H. Xia) T. P. Yi ●

47497 Bambusa tuldoides Munro 'Tuldoides' = Bambusa tuldoides Munro ●

47498 Bambusa tuldoides Munro 'Ventricosa'; 偏凸竹; Buddha Bamboo,Buddha's Belly Bamboo,Swollen-stemmed Bamboo ●☆

47499 Bambusa tuldoides Munro = Bambusa longispiculata Gamble ex Brandis ●

47500 Bambusa tuldoides Munro f. swolleninternode (N. H. Xia) T. P. Yi = Bambusa tuldoides Munro 'Swolleninternode' ●

47501 Bambusa utilis W. C. Lin; 乌叶竹; Blackleaf Bamboo, Useful Bamboo,Useful Bambusa ●

47502 Bambusa variegata Siebold ex Miq. = Pleioblastus fortunei (Van Houtte) Nakai ●

47503 Bambusa variegata Siebold ex Miq. = Sasa fortunei (Van Houtte) Fiori ●

47504 Bambusa varioaurita W. T. Lin et Z. J. Feng. = Bambusa textilis McClure ●

47505 Bambusa variostriata (W. T. Lin) L. C. Chia et H. L. Fung;吊丝单竹;Variable-striate Dendrocalamopsis,Versicolorline Greenbamboo ●

47506 Bambusa vario-striata (W. T. Lin) L. C. Chia et H. L. Fung = Dendrocalamopsis variostriata (W. T. Lin) P. C. Keng ●

47507 Bambusa veitchii Carrière = Sasa albomarginata (Miq.) Makino et Shibata ●

47508 Bambusa ventricosa McClure;佛肚竹(佛竹,葫芦竹,结头竹); Buddha Bamboo, Buddha Belly Bamboo, Buddha's Belly Bamboo, Buddhabelly Bamboo,Swollen Bamboo ●

47509 Bambusa ventricosa McClure 'Nana';小佛肚竹●

47510 Bambusa ventricosa McClure 'Nana' = Bambusa ventricosa McClure ●

47511 Bambusa verticillata Willd. = Gigantochloa verticillata (Willd.)

Munro ●

47512 Bambusa violascens Carrière = Phyllostachys violascens (Carrière) Rivière et C. Rivière ●

47513 Bambusa viridiglaucescens Carrière = Bambusa multiplex (Lour.) Raeusch. ex Schult. et Schult. f. ●

47514 Bambusa viridiglaucescens Carrière = Phyllostachys viridi-glaucescens (Carrière) Rivière et C. Rivière ●

47515 Bambusa viridi-vittata W. T. Lin = Bambusa eutuldoides McClure var. viridi-vittata (W. T. Lin) L. C. Chia ●

47516 Bambusa vulgaris Schrad. ex J. C. Wendl.;龙头竹(赤竹,泰山竹,银竹);Buddha's Common Bamboo, Common Bamboo, Dragonhead Bamboo, Dragon-head Bamboo, Feathery Bamboo, Golden Bamboo, Grand Bamboo, Thorny Bamboo, Unarmed Bamboo, Wamin Bamboo, Yellow-stemmed Bamboo ●

47517 Bambusa vulgaris Schrad. ex J. C. Wendl. 'Striata' = Bambusa vulgaris Schrad. ex J. C. Wendl. 'Vittata' ●

47518 Bambusa vulgaris Schrad. ex J. C. Wendl. 'Vittata';黄金间碧竹(桂绿竹,黄金间碧玉竹,金丝竹,青丝金竹);Giant Golden Bamboo, Giant Green-striped Bamboo, Golden Bamboo, Golden Common Bamboo, Green Stripe Common Bamboo, Green Striped Bamboo, Greenstripe Common Bamboo, Green-striped Bamboo, Ivory Bamboo, Ornamental Giant Bamboo, Painted Bamboo, Tiger Stripe Bamboo, Yellow Bamboo ●

47519 Bambusa vulgaris Schrad. ex J. C. Wendl. 'Vulgaris' = Bambusa vulgaris Schrad. ex J. C. Wendl. ●

47520 Bambusa vulgaris Schrad. ex J. C. Wendl. 'Wamin';大佛肚竹(短节泰山竹,葫芦龙头竹);Buddha Common Bamboo ●

47521 Bambusa vulgaris Schrad. ex J. C. Wendl. f. vittata (A. Riv. et C. Riv.) T. P. Yi = Bambusa vulgaris Schrad. ex J. C. Wendl. 'Vittata' ●

47522 Bambusa vulgaris Schrad. ex J. C. Wendl. f. wamin T. H. Wen = Bambusa vulgaris Schrad. ex J. C. Wendl. 'Wamin' ●

47523 Bambusa vulgaris Schrad. ex J. C. Wendl. var. striata (Lodd. ex Lindl.) Gamble = Bambusa vulgaris Schrad. ex J. C. Wendl. ●

47524 Bambusa vulgaris Schrad. ex J. C. Wendl. var. striata (Lodd.) Gamble = Bambusa vulgaris Schrad. ex J. C. Wendl. 'Vittata' ●

47525 Bambusa vulgaris Schrad. ex J. C. Wendl. var. vitatta Rivière et C. Rivière = Bambusa vulgaris Schrad. ex J. C. Wendl. 'Vittata' ●

47526 Bambusa vulgaris Schrad. ex J. C. Wendl. var. vittata Rivière et C. Rivière = Bambusa vulgaris Schrad. ex J. C. Wendl. f. vittata (A. Riv. et C. Riv.) T. P. Yi ●

47527 Bambusa vulgaris Schrad. ex J. C. Wendl. var. vittata Rivière et C. Rivière = Bambusa vulgaris Schrad. ex J. C. Wendl. ●

47528 Bambusa wamin Brandis ex E. G. Camus = Bambusa vulgaris Schrad. ex J. C. Wendl. 'Wamin' ●

47529 Bambusa wenchouensis (T. H. Wen) P. C. Keng ex Y. M. Lin et Q. F. Zheng = Bambusa wamin Brandis ex E. G. Camus ●

47530 Bambusa wenchouensis (T. H. Wen) Q. H. Dai;木毛单竹(大木竹,九层脑,毛单竹,王竹,温州箪竹);Wenzhou Bamboo, Wenzhou Bambusa ●

47531 Bambusa wenchouensis (T. H. Wen) Q. H. Dai = Lingnania wenchouensis T. H. Wen ●

47532 Bambusa xiashanensis L. C. Chia et H. L. Fung;霞山坭竹;Xiashan Bamboo, Xiashan Bambusa, Xiashan Mud Bamboo ●

47533 Bambusa xueana Ohrnb.;疙瘩竹;Yunnan Neosinocalamus ●

47534 Bambusa yunnanensis (J. R. Xue) D. Z. Li = Bambusa xueana Ohrnb. ●

47535 Bambusa yunnanensis N. H. Xia;毛环单竹;Yunnan Bamboo ●

47536 Bambusa yunnanensis N. H. Xia = Bambusa distegia (Keng et P. C. Keng) L. C. Chia et H. L. Fung ●

47537 Bambusaceae Bercht. et J. Presl = Gramineae Juss.(保留科名)●■

47538 Bambusaceae Bercht. et J. Presl = Poaceae Barnhart(保留科名)●■

47539 Bambusaceae Burnett = Gramineae Juss.(保留科名)●■

47540 Bambusaceae Burnett = Poaceae Barnhart(保留科名)●■

47541 Bambusaceae Nakai = Gramineae Juss.(保留科名)●■

47542 Bambusaceae Nakai = Poaceae Barnhart(保留科名)●■

47543 Bambusaceae Nakai;簕竹科●☆

47544 Bamia R. Br. ex Sims = Hibiscus L.(保留属名)●■

47545 Bamia R. Br. ex Wall. = Hibiscus L.(保留属名)●■

47546 Bamia cancellata Wall. = Abelmoschus crinitus Wall. ■

47547 Bamia chinensis Wall. = Abelmoschus moschatus (L.) Medik. ■

47548 Bamia crinita Wall. = Abelmoschus crinitus Wall. ■

47549 Bamiania Lincz. (1971);肉叶补血草属■☆

47550 Bamiania pachyeormum (Rech. f.) Lincz.;肉叶补血草■☆

47551 Bamlera K. Schum. et Lauterb. = Astronidium A. Gray(保留属名)●☆

47552 Bammia Rupp. = Bamia R. Br. ex Wall. ●■

47553 Bampsia Lisowski et Mielcarek(1983);邦氏婆婆纳属(巴氏玄参属)■☆

47554 Bampsia lawalreana Lisowski et Mielcarek;拉瓦尔邦氏婆婆纳■☆

47555 Bampsia symoensiana Lisowski et Mielcarek;西莫邦氏婆婆纳■☆

47556 Banalia Bubani = Hedysarum L.(保留属名)●■

47557 Banalia Moq. = Indobanalia A. N. Henry et B. Roy ■☆

47558 Banalia Raf. = Croton L. ●

47559 Banalia occidentalis Moq. = Nitrophila occidentalis (Moq.) S. Watson ■☆

47560 Banara Aubl. (1775);巴纳尔木属●☆

47561 Banara axilliflora Sleumer;腋花巴纳尔木●☆

47562 Banara cordifolia Urb. et Ekman;心叶巴纳尔木●☆

47563 Banara glandulosa Speg.;多腺巴纳尔木●☆

47564 Banara grandiflora Spruce ex Benth.;大花巴纳尔木●☆

47565 Banara guianensis Aubl.;巴纳尔木●☆

47566 Banara laxiflora Benth.;疏花巴纳尔木●☆

47567 Banara macrophylla Briq.;大叶巴纳尔木●☆

47568 Banara mexicana A. Gray;墨西哥巴纳尔木●☆

47569 Banara mollis (Poepp. et Endl.) Tul.;柔软巴纳尔木●☆

47570 Banara nitida Spruce ex Benth.;光亮巴纳尔木●☆

47571 Banara parviflora Benth.;小花巴纳尔木●☆

47572 Banara paucinervosa Steyerm.;寡脉巴纳尔木●☆

47573 Banara pubescens Spruce ex Benth.;毛巴纳尔木●☆

47574 Banava Juss. = Adamboe Adans. ●

47575 Banava Juss. = Lagerstroemia L. ●

47576 Bancalus Kuntze = Nauclea L. ●

47577 Bancalus Rumph. = Nauclea L. ●

47578 Bancalus Rumph. ex Kuntze = Nauclea L. ●

47579 Bancalus cuspidatus (Baker) Kuntze = Breonia cuspidata (Baker) Havil. ●☆

47580 Bancrofftia Steud. = Bancroftia Billb. ■☆

47581 Bancroftia Billb. = Arracacia Bancr. ■☆

47582 Bancroftia Macfad. = Tovaria Ruiz et Pav.(保留属名)●■

47583 Bandeiraea Benth. = Griffonia Baill. ■☆

47584 Bandeiraea Welw. = Griffonia Baill. ■☆

47585 Bandeiraea Welw. ex Benth. = Griffonia Baill. ■☆

47586 Bandeiraea Welw. ex Benth. et Hook. f. = Griffonia Baill. ■☆

47587 Bandeiraea simplicifolia (Vahl ex DC.) Benth. = Griffonia

simplicifolia (Vahl ex DC.) Baill. ■☆
47588 Bandeiraea speciosa Welw. ex Benth. = Griffonia speciosa (Welw. ex Benth.) Taub. ■☆
47589 Bandeiraea tenuiflora Benth. = Griffonia physocarpa Baill. ■☆
47590 Bandeiraea tessmannii De Wild. = Griffonia tessmannii (De Wild.) Compère ■☆
47591 Bandereia Baill. = Bandeiraea Welw. ex Benth. et Hook. f. ■☆
47592 Bandura Adans. = Nepenthes L. ●■
47593 Bandura Burm. = Nepenthes L. ●■
47594 Banffya Baumg. = Gypsophila L. ●■
47595 Banglium Buch. -Ham. ex Wall. = Boesenbergia Kuntze ■
47596 Bania Becc. = Carronia F. Muell. ●☆
47597 Banisteria L.(废弃属名) = Heteropterys Kunth(保留属名)●☆
47598 Banisteria benghalensis L. = Hiptage benghalensis (L.) Kurz ●
47599 Banisteria benghalensis L. f. cochinchinensis Pierre = Hiptage benghalensis (L.) Kurz ●
47600 Banisteria benghalensis L. f. typica Nied. = Hiptage benghalensis (L.) Kurz ●
47601 Banisteria kraussiana Hochst. = Acridocarpus natalitius A. Juss. ●☆
47602 Banisteria leona Cav. = Heteropterys leona (Cav.) Exell ●☆
47603 Banisteria ovata Cav. = Stigmaphyllon ovatum (Cav.) Nied. ●☆
47604 Banisteria timoriensis DC. = Rhyssopterys timoriensis (DC.) Blume ex A. Juss. ●
47605 Banisterioides Dubard et Dop = Sphedamnocarpus Planch. ex Benth. ●☆
47606 Banisteriopsis C. B. Rob. = Banisteriopsis C. B. Rob. ex Small ●☆
47607 Banisteriopsis C. B. Rob. ex Small(1910);槭果木属(巴尼金虎尾属,卡披木属)●☆
47608 Banisteriopsis argentea C. B. Rob.;阿根廷槭果木●☆
47609 Banisteriopsis caapi (Spruce ex Griseb.) Morton;卡披木●☆
47610 Banisteriopsis cornifolia C. B. Rob. ex Small;梾叶槭果木●☆
47611 Banisteriopsis inebrians Morton;酩酊槭果木●☆
47612 Banisterodes Kuntze = Xanthophyllum Roxb.(保留属名)●
47613 Banium Ces. ex Boiss. = Carum L. ■
47614 Banjolea Bowdich = Nelsonia R. Br. ■
47615 Bankesia Bruce = Banksia Bruce ●■☆
47616 Bankesia Bruce = Brayera Kunth ●■☆
47617 Banksea J. König = Costus L. ■
47618 Banksea J. König = Hellenia Retz. ■
47619 Banksea speciosa König = Costus speciosus (König) Sm. ■
47620 Banksia Bruce = Brayera Kunth ●■☆
47621 Banksia Bruce = Hagenia J. F. Gmel. ●☆
47622 Banksia Dombey ex DC. = Cuphea Adans. ex P. Browne ●■
47623 Banksia Gaertn. = Banksia L. f.(保留属名)●☆
47624 Banksia J. R. Forst. et G. Forst.(废弃属名) = Banksia L. f.(保留属名)●☆
47625 Banksia J. R. Forst. et G. Forst.(废弃属名) = Pimelea Banks ex Gaertn.(保留属名)●☆
47626 Banksia L. f.(1782)(保留属名);佛塔树属(澳洲山龙眼属,班克木属,贝克斯属);Australian Honeysuckle,Banksia ●☆
47627 Banksia abyssinica Bruce = Hagenia abyssinica (Bruce) J. F. Gmel. ●☆
47628 Banksia aemula R. Br.;花蜜班克木;Wallum,Wallum Banksia ●☆
47629 Banksia ashbyi Baker f.;爱须毕班克木;Ashby's Banksia ●☆
47630 Banksia australis R. Br. = Banksia marginata Cav. ●☆
47631 Banksia baueri R. Br.;绒毛班克木;Possum Banksia,Teddy-bear Banksia,Woolly ●☆
47632 Banksia baxteri R. Br.;贝克斯特班克木(巴氏佛塔树,贝克斯特贝克斯);Baxter Banksia,Baxter's Banksia ●☆
47633 Banksia blechnifolia F. Muell.;蕨叶班克木;Fern banksia ●☆
47634 Banksia caleyi R. Br.;勘莱班克木(勘莱贝克斯);Caley Banksia ●☆
47635 Banksia canei J. H. Willis;山班克木;Ountain Banksia ●☆
47636 Banksia coccinea R. Br.;红班克木(红贝克斯,红佛塔树);Scarlet Banksia ●☆
47637 Banksia collina R. Br.;丘生班克木(丘生贝克斯);Lowhill Banksia ●☆
47638 Banksia collina R. Br. = Banksia spinulosa Sm. ●☆
47639 Banksia dentata L. f.;齿叶班克木(齿叶贝克斯);Dentateleaf Banksia ●☆
47640 Banksia dryandroides Baxter ex Sweet;爪艾班克木;Dryandra-leaved banksia ●☆
47641 Banksia ericifolia L. f.;欧石楠叶班克木(欧石楠叶贝克斯,小叶佛塔树);Heath Banksia,Heathleaf Banksia,Heath-leaf Banksia,Heath-leaved Banksia ●☆
47642 Banksia gibbosa Sm. = Hakea gibbosa (Sm.) Cav. ●☆
47643 Banksia goodii R. Br.;古特班克木(古特贝克斯);Good Banksia ●☆
47644 Banksia grandis Willd.;大班克木(大形贝克斯);Bull Banksia,Large Banksia ●☆
47645 Banksia hookeriana Meisn.;橡子班克木;Acorn Banksia ●☆
47646 Banksia ilicifolia R. Br.;冬青叶班克木;Holly-leaf Banksia ●☆
47647 Banksia integrifolia L. f.;全缘叶班克木(全缘叶澳龙眼,全缘叶贝克斯);Beefsteak,Coast Banksia,Coastal Banksia,Entireleaf Banksia,Tree Honeysuckle,White Honeysuckle ●☆
47648 Banksia lehmanniana (Meisn.) Kuntze;梾檬班克木(梾檬贝克斯);Lehmann Banksia ●☆
47649 Banksia littoralis R. Br.;海滨班克木(海滨贝克斯);River Banksia,Shore Banksia,Swamp Banksia ●☆
47650 Banksia marginata Cav.;缘叶班克木(忍冬班克木,忍冬贝克斯,缘叶贝克斯);Honeysuckle Banksia,Silver Banksia ●☆
47651 Banksia media R. Br.;澳南平原班克木;Southern Plains Banksia ●☆
47652 Banksia meissneri Lehm.;美花班克木(梅斯纳贝克斯);Meissner Banksia ●☆
47653 Banksia menziesii R. Br.;梅兹班克木(梅兹贝克斯);Menzies Banksia ●☆
47654 Banksia nutans R. Br.;俯垂班克木(矮忍冬贝克斯,俯垂贝克斯);Scrubhoneysuckle Banksia ●☆
47655 Banksia oblongifolia Cav.;长圆叶班克木;Fern-leaf Banksia ●☆
47656 Banksia occidentalis R. Br.;西澳班克木(西方贝克斯);Red Awamp Banksia,Western Banksia ●☆
47657 Banksia paludosa R. Br.;喜湿班克木;Marsh banksia ●☆
47658 Banksia petiolaris F. Muell.;长叶柄班克木●☆
47659 Banksia pilostylis C. A. Gardner;毛柱班克木●☆
47660 Banksia praemorsa Dum. Cours.;啮蚀班克木;Banksia,Cut-leaf banksia ●☆
47661 Banksia prionotes Lindl.;锯齿叶班克木(锯齿叶贝克斯);Acorn Banksia,Serrateleaf Banksia ●☆
47662 Banksia pulchella R. Br.;美丽班克木(美丽贝克斯);Beautiful Banksia ●☆
47663 Banksia quercifolia R. Br.;栎叶班克木(栎叶贝克斯);Oakleaf Banksia ●☆
47664 Banksia repens Labill.;蔓生班克木(蔓生贝克斯);Creeping

Banksia ●☆
47665 Banksia robur Cav.;强力班克木(强力贝克斯);Strength Banksia,Swamp Banksia ●☆
47666 Banksia sceptrum Meisn.;笏班克木;Scepter Banksia ●☆
47667 Banksia serrata L. f.;红木班克木(红木贝克斯,锯叶班克木,锯叶佛塔树);Old Man Banksia, Red Honeysuckle, Redwood Banksia,Saw Banksia,Sawleaf banksia ●☆
47668 Banksia solandri R. Br.;索兰德班克木(索莱特班克木,索莱特贝克斯);Solander Banksia ●☆
47669 Banksia speciosa R. Br.;华艳班克木(华艳贝克斯);Showy Banksia ●☆
47670 Banksia sphaerocarpa R. Br.;球果班克木(球果贝克斯);Globular-fruit Banksia ●☆
47671 Banksia spinulosa Sm.;索发针班克木;Hairpin Banksia ●☆
47672 Banksia tenuifolia Salisb. = Hakea sericea Schrad. et J. C. Wendl. ●☆
47673 Banksia verticillata R. Br.;轮生叶班克木(轮生叶贝克斯);Whorlleaf Banksia ●☆
47674 Banksiaceae Bercht. et J. Presl = Proteaceae Juss.(保留科名)●■
47675 Baobab Adans. = Adansonia L. ●
47676 Baobabus Kuntze = Adansonia L. ●
47677 Baobabus Kuntze = Baobab Adans. ●
47678 Baobabus madagascariensis (Baill.) Kuntze = Adansonia madagascariensis Baill. ●☆
47679 Baolia H. W. Kung et G. L. Chu = Chenopodium L. ●■
47680 Baolia H. W. Kung et G. L. Chu(1978);苞藜属;Bractgoosefoot ■★
47681 Baolia bracteata H. W. Kung et G. L. Chu;苞藜;Bractgoosefoot ■
47682 Baoulia A. Chev. = Murdannia Royle(保留属名)■
47683 Baoulia tenuissima A. Chev. = Murdannia tenuissima (A. Chev.) Brenan ■☆
47684 Baphia Afzel. = Baphia Afzel. ex Lodd. ●☆
47685 Baphia Afzel. ex Lodd. (1820);杂色豆属(贝非属,非洲紫檀属);Camwood ●☆
47686 Baphia abyssinica Brummitt;阿比西尼亚杂色豆●☆
47687 Baphia acuminata De Wild. = Baphia pubescens Hook. f. ●☆
47688 Baphia albido-lenticellata De Wild. = Baphia capparidifolia Baker subsp. multiflora (Harms) Brummitt ●☆
47689 Baphia angolensis Welw. ex Baker;安哥拉杂色豆●☆
47690 Baphia bancoensis Aubrév. = Baphia pubescens Hook. f. ●☆
47691 Baphia bangweolensis R. E. Fr. = Baphia capparidifolia Baker subsp. bangweolensis (R. E. Fr.) Brummitt ●☆
47692 Baphia barombiensis Taub. = Baphia nitida Lodd. ●☆
47693 Baphia batangensis Harms = Baphia pilosa Baill. subsp. batangensis (Harms) Soladoye ●☆
47694 Baphia bequaertii De Wild.;贝卡尔杂色豆●☆
47695 Baphia bergeri De Wild.;贝格尔杂色豆●☆
47696 Baphia bipindensis Harms = Baphia buettneri Harms subsp. hylophila (Harms) Soladoye ●☆
47697 Baphia boonei De Wild. = Leucomphalos brachycarpus (Harms) Breteler ●☆
47698 Baphia brachybotrys Harms;短穗杂色豆●☆
47699 Baphia breteleriana Soladoye;布勒泰尔杂色豆●☆
47700 Baphia brevipedicellata De Wild. = Baphia wollastonii Baker f. ●☆
47701 Baphia buettneri Harms;比特纳杂色豆●☆
47702 Baphia buettneri Harms subsp. hylophila (Harms) Soladoye;喜盐杂色豆●☆
47703 Baphia burttii Baker f.;伯特杂色豆●☆
47704 Baphia busseana Harms = Baphia massaiensis Taub. subsp. busseana (Harms) Soladoye ●☆
47705 Baphia calophylla Harms = Baphia pilosa Baill. subsp. batangensis (Harms) Soladoye ●☆
47706 Baphia capparidifolia Baker;梨叶杂色豆●☆
47707 Baphia capparidifolia Baker subsp. bangweolensis (R. E. Fr.) Brummitt;班韦杂色豆●☆
47708 Baphia capparidifolia Baker subsp. multiflora (Harms) Brummitt;多花杂色豆●☆
47709 Baphia capparidifolia Baker subsp. polygalacea Brummitt;远志杂色豆●☆
47710 Baphia chrysophylla Taub.;金叶杂色豆●☆
47711 Baphia chrysophylla Taub. subsp. claessensii (De Wild.) Brummitt;克莱森斯杂色豆●☆
47712 Baphia claessensii De Wild. = Baphia chrysophylla Taub. subsp. claessensii (De Wild.) Brummitt ●☆
47713 Baphia compacta De Wild. = Baphia maxima Baker ●☆
47714 Baphia confusa Hutch. et Dalziel = Airyantha schweinfurthii (Taub.) Brummitt subsp. confusa (Hutch. et Dalziel) Brummitt ■☆
47715 Baphia conraui Harms = Baphia leptostemma Baill. var. conraui (Harms) Soladoye ●☆
47716 Baphia cordifolia Harms;心叶杂色豆●☆
47717 Baphia cornifolia Harms = Baphia massaiensis Taub. var. cornifolia (Harms) Soladoye ●☆
47718 Baphia crassifolia Harms;厚叶杂色豆●☆
47719 Baphia crassifolia Harms = Baphia laurifolia Baill. ●☆
47720 Baphia cuspidata Taub.;骤尖杂色豆●☆
47721 Baphia cymosa Breteler;聚伞杂色豆●☆
47722 Baphia densiflora Harms;密花杂色豆●☆
47723 Baphia densiflora Harms = Baphia laurifolia Baill. ●☆
47724 Baphia descampsii Vermoesen ex De Wild. = Baphia punctulata Harms subsp. descampsii (Vermoesen ex De Wild.) Soladoye ●☆
47725 Baphia dewevrei De Wild.;德韦杂色豆●☆
47726 Baphia dewevrei De Wild. var. inequalis Vermoesen = Baphia brachybotrys Harms ●☆
47727 Baphia dewildeana Soladoye;德维尔德杂色豆●☆
47728 Baphia dinklagei Harms = Baphia spathacea Hook. f. ●☆
47729 Baphia dubia De Wild.;疑惑杂色豆●☆
47730 Baphia elegans Lest.-Garl. = Baphia pilosa Baill. subsp. batangensis (Harms) Soladoye ●☆
47731 Baphia elegans Lest.-Garl. var. vestita ? = Baphia pilosa Baill. subsp. batangensis (Harms) Soladoye ●☆
47732 Baphia eriocalyx Harms;毛萼杂色豆●☆
47733 Baphia gabonensis De Wild. = Baphia leptostemma Baill. ●☆
47734 Baphia gilletii De Wild. = Baphia chrysophylla Taub. ●☆
47735 Baphia giorgii De Wild. = Baphia capparidifolia Baker subsp. bangweolensis (R. E. Fr.) Brummitt ●☆
47736 Baphia glabra De Wild. = Baphia wollastonii Baker f. ●☆
47737 Baphia glabra De Wild. var. oblongifolia ? = Baphia wollastonii Baker f. ●☆
47738 Baphia glauca A. Chev. = Baphia pubescens Hook. f. ●☆
47739 Baphia gomesii Baker f. = Baphia massaiensis Taub. subsp. gomesii (Baker f.) Brummitt ●☆
47740 Baphia goossensii De Wild. = Baphia capparidifolia Baker subsp. multiflora (Harms) Brummitt ●☆
47741 Baphia goossensii De Wild. var. grandifolia ? = Baphia dubia De

Wild. ●☆

47742 Baphia gossweileri Baker f.;戈斯杂色豆●☆

47743 Baphia gracilipedicellata De Wild. = Baphia longipedicellata De Wild. ●☆

47744 Baphia gracilipes Harms = Baphia leptostemma Baill. var. gracilipes (Harms) Soladoye ●☆

47745 Baphia haematoxylon (Schumach. et Thonn.) Hook. f. = Baphia nitida Lodd. ●☆

47746 Baphia henriquesiana Taub. = Baphia massaiensis Taub. var. obovata (Schinz) Brummitt ●☆

47747 Baphia heudelotiana Baill.;厄德杂色豆●☆

47748 Baphia hylophila Harms = Baphia buettneri Harms subsp. hylophila (Harms) Soladoye ●☆

47749 Baphia incerta De Wild.;可疑杂色豆●☆

47750 Baphia incerta De Wild. subsp. lebrunii (L. Touss.) Soladoye;勒布伦杂色豆●☆

47751 Baphia keniensis Brummitt = Baphia longipedicellata De Wild. subsp. keniensis (Brummitt) Soladoye ●☆

47752 Baphia kirkii Baker;柯克杂色豆●☆

47753 Baphia kirkii Baker subsp. ovata (Sim) Soladoye;卵形柯克杂色豆●☆

47754 Baphia klainei De Wild. = Baphia pilosa Baill. subsp. batangensis (Harms) Soladoye ●☆

47755 Baphia lancifolia Baill. ex Laness. = Baphia laurifolia Baill. ●☆

47756 Baphia laurentii De Wild.;洛朗杂色豆●☆

47757 Baphia laurifolia Baill.;桂叶杂色豆●☆

47758 Baphia lebrunii L. Touss. = Baphia incerta De Wild. subsp. lebrunii (L. Touss.) Soladoye ●☆

47759 Baphia leptobotrys Harms;细穗杂色豆●☆

47760 Baphia leptobotrys Harms subsp. silvatica (Harms) Soladoye;林地细穗杂色豆●☆

47761 Baphia leptostemma Baill.;细冠杂色豆●☆

47762 Baphia leptostemma Baill. var. conraui (Harms) Soladoye;康氏细冠杂色豆●☆

47763 Baphia leptostemma Baill. var. gracilipes (Harms) Soladoye;细梗细冠杂色豆●☆

47764 Baphia lescrauwaetii De Wild. = Baphia laurentii De Wild. ●☆

47765 Baphia letestui Pellegr.;莱泰斯图杂色豆●☆

47766 Baphia longepetiolata Taub.;长瓣杂色豆●☆

47767 Baphia longepetiolata Taub. = Baphia maxima Baker ●☆

47768 Baphia longipedicellata De Wild.;长柄杂色豆●☆

47769 Baphia longipedicellata De Wild. subsp. keniensis (Brummitt) Soladoye;肯尼亚长柄杂色豆●☆

47770 Baphia macrocalyx Harms;大萼杂色豆●☆

47771 Baphia madagascariensis C. H. Stirt. et Du Puy;马岛杂色豆●☆

47772 Baphia mambillensis Soladoye;曼比拉杂色豆●☆

47773 Baphia massaiensis Taub.;马萨杂色豆●☆

47774 Baphia massaiensis Taub. subsp. busseana (Harms) Soladoye;布瑟杂色豆●☆

47775 Baphia massaiensis Taub. subsp. cornifolia (Harms) Brummitt = Baphia massaiensis Taub. var. cornifolia (Harms) Soladoye ●☆

47776 Baphia massaiensis Taub. subsp. floribunda Brummitt;繁花杂色豆●☆

47777 Baphia massaiensis Taub. subsp. gomesii (Baker f.) Brummitt;格麦斯杂色豆●☆

47778 Baphia massaiensis Taub. var. cornifolia (Harms) Soladoye;角叶杂色豆●☆

47779 Baphia massaiensis Taub. var. obovata (Schinz) Brummitt;倒卵角叶杂色豆●☆

47780 Baphia massaiensis Taub. var. whitei (Brummitt) Soladoye;怀特杂色豆●☆

47781 Baphia maxima Baker;大杂色豆●☆

47782 Baphia megaphylla Breteler;大叶杂色豆●☆

47783 Baphia mildbraedii Harms = Baphia wollastonii Baker f. ●☆

47784 Baphia mocimboensis Pires de Lima = Baphia macrocalyx Harms ●☆

47785 Baphia multiflora Harms = Baphia capparidifolia Baker subsp. multiflora (Harms) Brummitt ●☆

47786 Baphia myrtifolia Lest. -Garl. = Baphia laurifolia Baill. ●☆

47787 Baphia nannanii Baker f. ex Lest. -Garl. = Baphia dewevrei De Wild. ●☆

47788 Baphia nitida Afzel. ex Lodd.;光亮杂色豆;Barwood Cam-wood, Camwood, Shiny Camwood ●☆

47789 Baphia nitida Lodd. = Baphia nitida Afzel. ex Lodd. ●☆

47790 Baphia nitida Lodd. var. pubescens A. Chev. = Baphia nitida Afzel. ex Lodd. ●☆

47791 Baphia obanensis Baker f.;奥班杂色豆●☆

47792 Baphia obovata Schinz = Baphia massaiensis Taub. var. obovata (Schinz) Brummitt ●☆

47793 Baphia odorata De Wild. = Baphia laurifolia Baill. ●☆

47794 Baphia orbiculata Baker f. = Baphia maxima Baker ●☆

47795 Baphia ovata Sim = Baphia kirkii Baker subsp. ovata (Sim) Soladoye ●☆

47796 Baphia ovato-acuminata De Wild. = Baphia dewevrei De Wild. ●☆

47797 Baphia pauloi Brummitt;保罗杂色豆●☆

47798 Baphia pierrei De Wild. = Baphia laurifolia Baill. ●☆

47799 Baphia pilosa Baill.;疏毛杂色豆●☆

47800 Baphia pilosa Baill. subsp. batangensis (Harms) Soladoye;巴唐杂色豆●☆

47801 Baphia polyantha Harms = Baphia spathacea Hook. f. subsp. polyantha (Harms) Soladoye ●☆

47802 Baphia polygalacea (Hook. f.) Baker = Baphia capparidifolia Baker subsp. polygalacea Brummitt ●☆

47803 Baphia polygalacea (Hook. f.) Baker var. hepperi Cavaco = Baphia capparidifolia Baker subsp. multiflora (Harms) Brummitt ●☆

47804 Baphia polygalacea Baker;多汁杂色豆;Walking-stick Camwood ●☆

47805 Baphia preussii Harms;普罗伊斯杂色豆●☆

47806 Baphia pubescens Hook. f.;柔毛杂色豆;Benin Camwood ●☆

47807 Baphia puguensis Brummitt;普古杂色豆●☆

47808 Baphia punctulata Harms;斑点杂色豆●☆

47809 Baphia punctulata Harms subsp. descampsii (Vermoesen ex De Wild.) Soladoye;德康杂色豆●☆

47810 Baphia punctulata Harms subsp. palmensis Soladoye;帕尔马杂色豆●☆

47811 Baphia pynaertii De Wild. = Baphia capparidifolia Baker subsp. multiflora (Harms) Brummitt ●☆

47812 Baphia pyrifolia (Desv.) Baill. = Baphia capparidifolia Baker ●☆

47813 Baphia racemosa (Hochst.) Baker;总花杂色豆(总花贝非)●☆

47814 Baphia racemosa Hochst. = Baphia racemosa (Hochst.) Baker ●☆

47815 Baphia radcliffei Baker f. = Baphiopsis parviflora Baker ●☆

47816 Baphia ringoetii De Wild. = Baphia bequaertii De Wild. ●☆

47817 Baphia schweinfurthii Taub. = Airyantha schweinfurthii (Taub.) Brummitt ■☆

47818 Baphia semseiana Brummitt;塞姆杂色豆●☆

47819 Baphia silvatica Harms = Baphia leptobotrys Harms subsp. silvatica (Harms) Soladoye ●☆

47820 Baphia solheidii De Wild. = Baphia pubescens Hook. f. ●☆

47821 Baphia spathacea Hook. f.;佛焰苞杂色豆●☆

47822 Baphia spathacea Hook. f. subsp. polyantha (Harms) Soladoye;多花佛焰苞杂色豆●☆

47823 Baphia speciosa J. B. Gillett et Brummitt;美丽杂色豆●☆

47824 Baphia sublucida De Wild. = Baphia brachybotrys Harms ●☆

47825 Baphia vermeulenii De Wild. = Baphia pilosa Baill. ●☆

47826 Baphia vermoesenii De Wild. = Baphia brachybotrys Harms ●☆

47827 Baphia verschuerenii De Wild. = Baphia angolensis Welw. ex Baker ●☆

47828 Baphia whitei Brummitt = Baphia massaiensis Taub. var. whitei (Brummitt) Soladoye ●☆

47829 Baphia wollastonii Baker f.;沃拉斯顿杂色豆●☆

47830 Baphia zenkeri Taub. = Baphia capparidifolia Baker subsp. multiflora (Harms) Brummitt ●☆

47831 Baphiastrum Harms(1913);小杂色豆属●☆

47832 Baphiastrum bequaertii De Wild. = Baphia pilosa Baill. ●☆

47833 Baphiastrum boonei (De Wild.) Vermoesen ex De Wild. = Leucomphalos brachycarpus (Harms) Breteler ●☆

47834 Baphiastrum brachycarpum Harms = Leucomphalos brachycarpus (Harms) Breteler ●☆

47835 Baphiastrum calophyllum (Harms) De Wild. = Baphia pilosa Baill. subsp. batangensis (Harms) Soladoye ●☆

47836 Baphiastrum claessensii De Wild. = Baphia pilosa Baill. ●☆

47837 Baphiastrum confusum (Hutch. et Dalziel) Pellegr. = Airyantha schweinfurthii (Taub.) Brummitt subsp. confusa (Hutch. et Dalziel) Brummitt ■☆

47838 Baphiastrum elegans (Lest. -Garl.) De Wild. = Baphia pilosa Baill. subsp. batangensis (Harms) Soladoye ●☆

47839 Baphiastrum pilosum (Baill.) De Wild. = Baphia pilosa Baill. ●☆

47840 Baphiastrum tisserantii Pellegr. = Airyantha schweinfurthii (Taub.) Brummitt ■☆

47841 Baphiastrum vermeulenii (De Wild.) De Wild. = Baphia pilosa Baill. ●☆

47842 Baphicacanthus Bremek. (1944);板蓝属(马蓝属);Baphicacanthus ●

47843 Baphicacanthus Bremek. = Strobilanthes Blume ●■

47844 Baphicacanthus cusia (Nees) Bremek. = Strobilanthes cusia (Nees) Kuntze ●

47845 Baphicacanthus cusia (Ness) Kuntze;板蓝(板蓝根,大菁,大蓝,大青,淀花,靛,靛花,靛沫,靛沫花,蓝靛,马蓝,南板蓝根,青黛,青缸花,软叶马蓝,土靛);Common Baphicacanthus, Common Conehead, Conehead, Flaccid Conehead ●

47846 Baphicacanthus multibracteolata Hung T. Chang et H. Chu;多小苞板蓝●

47847 Baphiopsis Benth. = Baphiopsis Benth. ex Baker ●☆

47848 Baphiopsis Benth. ex Baker(1871);拟杂色豆属(拟非洲紫檀属)●☆

47849 Baphiopsis parviflora Baker;拟杂色豆(拟非洲紫檀)●☆

47850 Baphiopsis stuhlmannii Taub. = Baphiopsis parviflora Baker ●☆

47851 Baphorhiza Link = Alkanna Tausch(保留属名)●☆

47852 Baphorhiza orientalis (L.) Font Quer & Rothm. = Alkanna orientalis (L.) Boiss. ●☆

47853 Baprea Pierre ex Pax et KHoffm. = Cladogynos Zipp. ex Span. ●

47854 Baptisia Vent. (1808);赛靛属(北美靛蓝属,赝靛属,野靛属);False Indigo, Wild Indigo ■☆

47855 Baptisia 'Purple Smoke';紫烟赛靛;Purple Smoke Indigo ■☆

47856 Baptisia alba (L.) Vent.;白赛靛;False Indigo, Milky White Indigo, White False Indigo, White Wild Indigo, Wild White Indigo ■☆

47857 Baptisia alba (L.) Vent. var. macrophylla (Larisey) Isely;大叶白赛靛;Large-leaved Wild Indigo, Milky White Indigo, White Wild Indigo ■☆

47858 Baptisia australis (L.) R. Br. = Baptisia australis (L.) R. Br. ex W. T. Aiton ■☆

47859 Baptisia australis (L.) R. Br. ex W. T. Aiton;赛靛花(北美南部靛蓝);Blue False Indigo, Blue Indigo, Blue Wild Indigo, False Indigo, Southern Wild Indigo, Wild Indigo ■☆

47860 Baptisia bicolor Greenm. et Larisey;二色赛靛■☆

47861 Baptisia bracteata Muhl. ex Elliott;草地赛靛;Black Rattlepod, Cream White Indigo, Long-bracted Wild Indigo, Plains Wild Indigo ■☆

47862 Baptisia bracteata Muhl. ex Elliott var. glabrescens (Larisey) Isely;变光长裂赛靛;Cream Wild Indigo, Long-bracted Wild Indigo, Plains Wild Indigo ■☆

47863 Baptisia bracteata Muhl. ex Elliott var. leucophaea (Nutt.) Kartesz et Gandhi = Baptisia bracteata Muhl. ex Elliott var. glabrescens (Larisey) Isely ■☆

47864 Baptisia gibbesii Small = Baptisia tinctoria (L.) R. Br. ex W. T. Aiton ■☆

47865 Baptisia lactea (Raf.) Thieret = Baptisia alba (L.) Vent. var. macrophylla (Larisey) Isely ■☆

47866 Baptisia lactea (Raf.) Thieret = Baptisia alba (L.) Vent. ■☆

47867 Baptisia laevicaulis (Canby) Small;光茎赛靛■☆

47868 Baptisia leucantha Torr. et A. Gray = Baptisia alba (L.) Vent. var. macrophylla (Larisey) Isely ■☆

47869 Baptisia leucantha Torr. et A. Gray = Baptisia alba (L.) Vent. ■☆

47870 Baptisia leucantha Torr. et A. Gray var. divaricata Larisey = Baptisia alba (L.) Vent. var. macrophylla (Larisey) Isely ■☆

47871 Baptisia leucantha Torr. et Gray;白花赛靛;Prairie False Indigo, White Wild Indigo, Wild False Indigo ■☆

47872 Baptisia leucophaea Nutt. = Baptisia bracteata Muhl. ex Elliott var. glabrescens (Larisey) Isely ■☆

47873 Baptisia leucophaea Nutt. = Baptisia bracteata Muhl. ex Elliott ■☆

47874 Baptisia leucophaea Nutt. var. glabrescens Larisey = Baptisia bracteata Muhl. ex Elliott ■☆

47875 Baptisia leucophaea Nutt. var. glabrescens Larisey = Baptisia bracteata Muhl. ex Elliott var. glabrescens (Larisey) Isely ■☆

47876 Baptisia nepalensis Hook. = Piptanthus nepalensis (Hook.) Sweet ●

47877 Baptisia pendula Larisey var. macrophylla Larisey = Baptisia alba (L.) Vent. var. macrophylla (Larisey) Isely ■☆

47878 Baptisia sphaerocarpa Nutt.;圆果赛靛;Round-fruited Yellow Wild Indigo, Yellow False Indigo, Yellow Wild Indigo ■☆

47879 Baptisia tinctoria (L.) R. Br. = Baptisia tinctoria (L.) R. Br. ex W. T. Aiton ■☆

47880 Baptisia tinctoria (L.) R. Br. ex W. T. Aiton;染料赛靛(北美靛蓝,赝靛);Horse-fly Weed, Horsefly-weed, Indigo Broom, Indigo-broom, Rattlebush, Rattleweed, Wild Indigo, Yellow Wild Indigo ■☆

47881 Baptisia tinctoria (L.) R. Br. var. crebra Fernald = Baptisia tinctoria (L.) R. Br. ex W. T. Aiton ■☆

47882 Baptisia tinctoria (L.) R. Br. var. projecta Fernald = Baptisia tinctoria (L.) R. Br. ex W. T. Aiton ■☆

47883 Baptislonia Barb. Rodr. = Oncidium Sw. (保留属名)■☆

47884 Baptistania Barb. Rodr. ex Pfltzer = Oncidium Sw. (保留属名)■☆
47885 Baptistania Pfitzer = Baptistonia Barb. Rodr. ■☆
47886 Baptistonia Barb. Rodr. = Oncidium Sw. (保留属名)■☆
47887 Baptorhachis Clayton et Renvoize(1986);染轴粟属■☆
47888 Baptorhachis foliacea (Clayton) Clayton;染轴粟■☆
47889 Baranda Llanos = Barringtonia J. R. Forst. et G. Forst. (保留属名)●
47890 Barathranthus (Korth.) Miq. = Baratranthus (Korth.) Miq. ●☆
47891 Barathranthus Danser = Baratranthus (Korth.) Miq. ●☆
47892 Barathranthus axanthus (Korth.) Miq. = Baratranthus axanthus (Korth.) Miq. ●☆
47893 Barathranthus axanthus Miq. = Barathranthus axanthus (Korth.) Miq. ●☆
47894 Baratostachys (Korth.) Kuntze = Phoradendron Nutt. ●☆
47895 Baratranthus (Korth.) Miq. (1856);凹花寄生属●☆
47896 Baratranthus Miq. = Baratranthus (Korth.) Miq. ●☆
47897 Baratranthus axanthus (Korth.) Miq.;凹花寄生●☆
47898 Baratranthus bicolor Tiegh.;二色凹花寄生●☆
47899 Baratranthus kingii Tiegh.;金氏凹花寄生●☆
47900 Barattia A. Gray et Engelm. = Encelia Adans. ●■☆
47901 Baraultia Spreng. = Barraldeia Thouars(废弃属名)●
47902 Baraultia Spreng. = Carallia Roxb. (保留属名)●
47903 Baraultia Steud. ex Spreng. = Barraldeia Thouars(废弃属名)●
47904 Baraultia Steud. ex Spreng. = Carallia Roxb. (保留属名)●
47905 Barbacenia Vand. (1788);巴尔翡若翠属■☆
47906 Barbacenia elegans (Balf.) Pax = Talbotia elegans Balf. ■☆
47907 Barbacenia goetzei Harms = Xerophyta goetzei (Harms) L. B. Sm. et Ayensu ■☆
47908 Barbacenia hereroensis Schinz = Xerophyta viscosa Baker ■☆
47909 Barbacenia humilis (Baker) Pax ex Burtt Davy = Xerophyta humilis (Baker) T. Durand et Schinz ■☆
47910 Barbacenia minuta (Baker) Dinter = Xerophyta humilis (Baker) T. Durand et Schinz ■☆
47911 Barbacenia naegelsbachii Dinter ex Friedr. -Holzh. = Xerophyta equisetoides Baker var. pauciramosa L. B. Sm. et Ayensu ●☆
47912 Barbacenia retinervis (Baker) Pax ex Burtt Davy = Xerophyta retinervis Baker ■☆
47913 Barbacenia rosea (Baker) Pax ex Burtt Davy = Xerophyta schlechteri (Baker) N. L. Menezes ■☆
47914 Barbacenia scabrida Pax = Xerophyta scabrida (Pax) T. Durand et Schinz ■☆
47915 Barbacenia villosa (Baker) Pax ex Burtt Davy = Xerophyta villosa (Baker) L. B. Sm. et Ayensu ■☆
47916 Barbacenia viscosa (Baker) Pax ex Burtt Davy = Xerophyta viscosa Baker ■☆
47917 Barbacenia wentzeliana Harms = Xerophyta equisetoides Baker var. pauciramosa L. B. Sm. et Ayensu ●☆
47918 Barbaceniaceae Arn. = Velloziaceae J. Agardh(保留科名)■
47919 Barbaceniopsis L. B. Sm. (1962);拟巴尔翡若翠属■☆
47920 Barbaceniopsis boliviensis (Baker) L. B. Sm.;玻利维亚拟巴尔翡若翠■☆
47921 Barbaceniopsis vargasiana (L. B. Sm.) L. B. Sm.;拟巴尔翡若翠■☆
47922 Barba-jovis Adans. = Anthyllis L. ■☆
47923 Barbajovis Mill. = Anthyllis L. ■☆
47924 Barba-jovis Ség. = Anthyllis L. ■☆
47925 Barbamine A. P. Khokhr. = Barbarea W. T. Aiton(保留属名)■
47926 Barbaraea Beckm. = Barbarea R. Br. ■
47927 Barbaraea vulgaris R. Br. = Barbarea vulgaris R. Br. ■☆
47928 Barbarea R. Br. = Barbarea W. T. Aiton(保留属名)■
47929 Barbarea Scop. (废弃属名) = Barbarea W. T. Aiton(保留属名)■
47930 Barbarea Scop. (废弃属名) = Dentaria L. ■☆
47931 Barbarea W. T. Aiton(1812)(保留属名);山芥属;American Cress, False Indigo, Upland Cress, Winter Cress, Wintercress ■
47932 Barbarea americana Rydb. = Barbarea orthoceras Ledeb. ■
47933 Barbarea arcuata (Opiz ex C. Presl) Rchb. = Barbaraea vulgaris R. Br. ■☆
47934 Barbarea arcuata (Opiz ex J. Presl et C. Presl) Rchb.;弧形山芥■☆
47935 Barbarea arcuata (Opiz ex J. Presl et C. Presl) Rchb. = Barbarea orthoceras Ledeb. ■
47936 Barbarea arcuata (Opiz ex J. Presl et C. Presl) Rchb. = Barbarea vulgaris R. Br. ■
47937 Barbarea arcuata Rchb. = Barbarea vulgaris R. Br. ■
47938 Barbarea arisanensis (Hayata) S. S. Ying;高山山芥■
47939 Barbarea arisanensis (Hayata) S. S. Ying = Cardamine flexuosa With. ■
47940 Barbarea cochlearifolia H. Boissieu = Barbarea orthoceras Ledeb. ■
47941 Barbarea derchiensis S. S. Ying = Brassica rapa L. ■
47942 Barbarea elata Hook. f. et Thomson = Rorippa dubia (Pers.) H. Hara ■
47943 Barbarea elata Hook. f. et Thomson = Rorippa elata (Hook. f. et Thomson) Hand. -Mazz. ■
47944 Barbarea grandiflora N. Busch;大花山芥■☆
47945 Barbarea hondoensis Nakai = Barbarea orthoceras Ledeb. ■
47946 Barbarea hongii Al-Shehbaz et G. Yang;洪氏山芥;Hong Wintercress ■
47947 Barbarea intermedia Boreau;羽裂山芥(间型山芥,羽裂叶山芥);Early Winter Cress, Intermediate Wintercress, Intermediate Yellow Rocket, Winter Cress, Yellow Rocket ■
47948 Barbarea minor K. Koch;小山芥■☆
47949 Barbarea orthoceras Ledeb.;山芥(山芥菜);American Yellow-rocket, Erecttop Wintercress, Northern Winter-cress ■
47950 Barbarea orthoceras Ledeb. var. dolichocarpa Fernald = Barbarea orthoceras Ledeb. ■
47951 Barbarea orthoceras Ledeb. var. formosana Kitam. = Barbarea taiwaniana Ohwi ■
47952 Barbarea patens Boiss. = Barbarea orthoceras Ledeb. ■
47953 Barbarea perennis Pomel = Barbarea vulgaris R. Br. ■
47954 Barbarea plantaginea DC.;扁山芥■☆
47955 Barbarea praecox (Sm.) R. Br. = Barbarea verna (Mill.) Asch. ■☆
47956 Barbarea sibirica Nakai;西伯利亚山芥;Siberia Crees ■☆
47957 Barbarea sibirica Nakai = Barbarea orthoceras Ledeb. ■
47958 Barbarea stolonifera Pomel = Barbarea vulgaris R. Br. ■
47959 Barbarea stricta Andrz.;直山芥;Small-flowered Land Cress, Small-flowered Winter Cress, Small-flowered Yellow Rocket ■☆
47960 Barbarea stricta Andrz. = Barbarea intermedia Boreau ■
47961 Barbarea stricta Andrz. = Barbarea vulgaris R. Br. ■
47962 Barbarea taiwaniana Ohwi;台湾山芥(山芥菜);Taiwan Wintercress ■
47963 Barbarea verna (Mill.) Asch.;春花山芥;American Crees, American Land Cress, American Winter Cress, Belle Isle Cress, Early Winter Cress, Early Yellowrocket, Early-flowering Yellow Rocket, Isle

Belle Cress, Land Cress, Normandy Cress, Spring Crees, Upland Cress, Winter Cress ■☆
47964 Barbarea vulgaris R. Br.;欧洲山芥; Bank Cress, Bitter Land Cress, Bitter Winter Cress, Cassabully, Common Winter Cress, French Cress, Garden Yellowrocket, Garden Yellow-rocket, Jack-by-the-hedge, Land Cress, Rocket Cress, Rocket Galant, St. Barbara's Cress, St. Barbara's Herb, Upland Cress, Upland Wintercress, Winter Cress, Winter Hedge Mustard, Winter Rocket, Wintercress, Winter-cress, Wound Rocket, Yellow Cress, Yellow Rocket, Yellowrocket, Yellow-rocket ■
47965 Barbarea vulgaris R. Br. = Barbarea orthoceras Ledeb. ■
47966 Barbarea vulgaris R. Br. 'Variegata';花叶欧洲山芥■
47967 Barbarea vulgaris R. Br. ex Aiton 'Flore-pleno';重瓣欧洲山芥; Double Yellow Rocket ■☆
47968 Barbarea vulgaris R. Br. ex Aiton subsp. intermedia (Boreau) Maire = Barbarea intermedia Boreau ■
47969 Barbarea vulgaris R. Br. ex Aiton var. arcuata (Opiz ex J. Presl et C. Presl) Fr. = Barbarea vulgaris R. Br. ■
47970 Barbarea vulgaris R. Br. ex Aiton var. arcuata (Rchb.) Coss. = Barbarea vulgaris R. Br. ■
47971 Barbarea vulgaris R. Br. ex Aiton var. brachycarpa Rouy et Foucaud = Barbarea vulgaris R. Br. ■
47972 Barbarea vulgaris R. Br. ex Aiton var. longisiliquosa Carion = Barbarea vulgaris R. Br. ■
47973 Barbarea vulgaris R. Br. ex Aiton var. orthoceras (Ledeb.) Regel = Barbarea orthoceras Ledeb. ■
47974 Barbarea vulgaris R. Br. ex Aiton var. perennis (Pomel) Batt. = Barbarea vulgaris R. Br. ■
47975 Barbarea vulgaris R. Br. ex Aiton var. sicula C. Presl = Barbarea vulgaris R. Br. ■
47976 Barbarea vulgaris R. Br. ex Aiton var. stricta (Andrz.) Coss. = Barbarea intermedia Boreau ■
47977 Barbarea vulgaris R. Br. ex Aiton var. sylvestris Fr. = Barbarea vulgaris R. Br. ■
47978 Barbarea vulgaris R. Br. ex Aiton var. transiens Font Quer et Maire = Barbarea intermedia Boreau ■
47979 Barbarea vulgaris R. Br. var. sicula Hook. f. et T. Anderson = Barbarea intermedia Boreau ■
47980 Barbarea vulgaris R. Br. var. taurica Hook. f. et T. Anderson = Barbarea vulgaris R. Br. ■
47981 Barbellina Cass. = Staehelina L. ●☆
47982 Barberetta Harv. (1868);南非血草属■☆
47983 Barberetta aurea Harv.;南非血草■☆
47984 Barberina Vell. = Symplocos Jacq. ●
47985 Barbeuia Thouars(1806);商陆藤属(节柄藤属,节柄属)●☆
47986 Barbeuia madagascariensis Steud.;商陆藤●☆
47987 Barbeuiaceae (H. Walter) Nakai = Phytolaccaceae R. Br.(保留科名)●■
47988 Barbeuiaceae Nakai = Phytolaccaceae R. Br.(保留科名)●■
47989 Barbeuiaceae Nakai(1942);商陆藤科●☆
47990 Barbeya Alboff = Amphoricarpus Vis. ●☆
47991 Barbeya Albov = Amphoricarpus Vis. ●☆
47992 Barbeya Schweinf. = Barbeya Schweinf. ex Penz. ●☆
47993 Barbeya Schweinf. ex Penz. (1892);钩毛树属(钩毛叶属)●☆
47994 Barbeya Schweinf. ex Penz. = Barbeya Schweinf. ●☆
47995 Barbeya oleoides Schweinf.;钩毛树(钩毛叶)●☆
47996 Barbeyaceae Rendle(1916)(保留科名);钩毛树科(钩毛叶科,合瓣莲科)●☆
47997 Barbeyastrum Cogn. = Dichaetanthera Endl. ●☆
47998 Barbeyastrum corymbosum Cogn. = Dichaetanthera corymbosa (Cogn.) Jacq. -Fél. ●☆
47999 Barbiera Spreng. = Barbieria DC. ●☆
48000 Barbieria DC. = Clitoria L. ●
48001 Barbieria Spreng. = Clitoria L. ●
48002 Barbilus P. Browne = Trichilia P. Browne(保留属名)●
48003 Barbosa Becc. (1887);东智利棕属(巴西金山葵属)●☆
48004 Barbosa Becc. = Syagrus Mart. ●
48005 Barbosa pseudococos (Raddi) Becc.;东智利棕●☆
48006 Barbosella Schltr. (1918);小棕兰属(巴波兰属)■☆
48007 Barbosella gardneri (Lindl.) Schltr.;小棕兰■☆
48008 Barbrodria Luer. (1981);米尔斯兰属■☆
48009 Barbrodria miersii (Lindl.) Luer;米尔斯兰■☆
48010 Barbula Lour. = Caryopteris Bunge ●
48011 Barbylus Juss. = Barbilus P. Browne ●
48012 Barbylus Juss. = Trichilia P. Browne(保留属名)●
48013 Barcella (Trail) Drude = Barcella (Trail) Trail ex Drude ●☆
48014 Barcella (Trail) Trail ex Drude(1881);亚马逊棕属(凹雌椰属,巴塞卢斯椰子属)●☆
48015 Barcella Drude = Barcella (Trail) Trail ex Drude ●☆
48016 Barcella odora (Trail) Trail ex Drude;亚马逊棕●☆
48017 Barcena Duges = Colubrina Rich. ex Brongn.(保留属名)●
48018 Barcenia Duges = Barcena Duges ●
48019 Barcenia Duges = Colubrina Rich. ex Brongn.(保留属名)●
48020 Barckhausenia Menke = Barckhausia DC. ■
48021 Barckhausia DC. = Barkhausia Moench ■
48022 Barckhausia DC. = Crepis L. ■
48023 Barclaya Wall. (1827)(保留属名);合瓣莲属■☆
48024 Barclaya longifolia Wall.;长叶合瓣莲■☆
48025 Barclaya rotundifolia Hotta;圆叶合瓣莲■☆
48026 Barclayaceae (Endl.) H. L. Li = Nymphaeaceae Salisb.(保留科名)■
48027 Barclayaceae H. L. Li = Nymphaeaceae Salisb.(保留科名)■
48028 Barclayaceae H. L. Li(1955);合瓣莲科■☆
48029 Bardana Hill = Arctium L. ■
48030 Bareria Juss. = Barreria Scop. ●☆
48031 Bareria Juss. = Poraqueiba Aubl. ●☆
48032 Baretia Comm. ex Cav. = Turraea L. ●
48033 Bargemontia Gaudich. = Nolana L. ex L. f. ■☆
48034 Barhamia Klotzsch = Croton L. ●
48035 Bariaea Rchb. f. = Cynorkis Thouars ■☆
48036 Barjonia Decne. (1844);巴尔萝藦属●■☆
48037 Barjonia chlorifolia Decne.;巴尔萝藦●■☆
48038 Barkania Ehrenb. = Halophila Thouars ■
48039 Barkeria Knowles et Westc. (1838);巴克兰属■☆
48040 Barkeria elegans Knowles et Westc.;巴克兰■☆
48041 Barkerwebbia Becc. (1905);巴克棕属(巴克伟榈属)●☆
48042 Barkerwebbia Becc. = Heterospathe Scheff. ●☆
48043 Barkerwebbia elegans Becc.;巴克棕●☆
48044 Barkhausenia Schur = Borckhausenia P. Gaertn., B. Mey. et Scherb. ■
48045 Barkhausenia Schur = Corydalis DC.(保留属名)■
48046 Barkhausia Moench = Crepis L. ■
48047 Barkhausia adenothrix Sch. Bip. ex A. Rich. = Crepis rueppellii Sch. Bip. ■☆

48048 Barkhausia amplexicaulis Coss. = Crepis amplexifolia (Godr.) Willk. ■☆

48049 Barkhausia arenaria Pomel = Crepis arenaria (Pomel) Pomel ■☆

48050 Barkhausia carbonaria (Sch. Bip.) A. Rich. = Crepis carbonaria Sch. Bip. ■☆

48051 Barkhausia clausonis Pomel = Crepis clausonis (Pomel) Batt. ■☆

48052 Barkhausia flexuosa (Ledeb.) DC. = Crepis flexuosa (Ledeb.) C. B. Clarke ■

48053 Barkhausia flexuosa (Ledeb.) DC. var. lyrata Schrenk = Crepis flexuosa (Ledeb.) C. B. Clarke ■

48054 Barkhausia floribunda Pomel = Crepis vesicaria L. subsp. myriocephala (Coss. et Durieu) Babc. ■☆

48055 Barkhausia foetida (L.) F. W. Schmidt = Crepis foetida L. ■☆

48056 Barkhausia grandiflora Nutt. = Pyrrhopappus grandiflorus (Nutt.) Nutt. ■☆

48057 Barkhausia hirsuta Pomel = Crepis vesicaria L. subsp. stellata (Ball) Babc. ■☆

48058 Barkhausia kralikii Pomel = Crepis nigricans Viv. ■☆

48059 Barkhausia nana (Rich.) DC. = Crepis nana Richardson ■

48060 Barkhausia numidica Pomel = Crepis vesicaria L. subsp. taraxacifolia (Thuill.) Schinz et Keller ■☆

48061 Barkhausia recognita DC. = Crepis vesicaria L. subsp. taraxacifolia (Thuill.) Schinz et Keller ■☆

48062 Barkhausia rubra Moench = Crepis rubra L. ■☆

48063 Barkhausia schimperi Sch. Bip. ex A. Rich. = Crepis foetida L. ■☆

48064 Barkhausia schultzii Hochst. ex A. Rich. = Crepis schultzii (Hochst. ex A. Richardson) Vatke ■☆

48065 Barkhausia senecioides (Delile) Spreng. = Crepis senecioides Delile ■☆

48066 Barkhausia taraxacifolia Thuill. = Crepis vesicaria L. subsp. taraxacifolia (Thuill.) Schinz et Keller ■☆

48067 Barkhausia taraxacifolia Thuill. var. atlantica Sennen et Mauricio = Crepis vesicaria L. subsp. taraxacifolia (Thuill.) Schinz et Keller ■☆

48068 Barkhausia taraxacifolia Thuill. var. macrophylla Sennen et Mauricio = Crepis vesicaria L. subsp. myriocephala (Coss. et Durieu) Babc. ■☆

48069 Barkhausia tenerrima Sch. Bip. = Crepis tenerrima (Sch. Bip.) R. E. Fr. ■☆

48070 Barkhausia versicolor (Fisch. ex Link) Spreng. = Ixeridium gramineum (Fisch.) Tzvelev ■

48071 Barkhausia versicolor (Fisch.) Spreng. = Ixeridium gramineum (Fisch.) Tzvelev ■

48072 Barkhausia vesicaria (L.) DC. = Crepis vesicaria L. ■

48073 Barkhusenia Hoppe = Barkhausia Moench ■

48074 Barkleyanthus H. Rob. et Brettell(1974);柳叶千里光属(巴克花属);Jarilla,Willow Ragwort ●☆

48075 Barkleyanthus salicifolius (Kunth) H. Rob. et Brettell;柳叶千里光(巴克花)●☆

48076 Barklya F. Muell. (1859);金花木属;Golden-blossom Tree, Leather Jacket ●☆

48077 Barklya F. Muell. = Bauhinia L. ●

48078 Barklya syringifolia F. Muell.;金花木;Golden-blossom Tree, Leather Jacket ●☆

48079 Barlaea Rchb. f. = Cynorkis Thouars ■☆

48080 Barleria L. (1753);假杜鹃属;Barleria,Falsecuckoo ●■

48081 Barleria acanthoides Oliv. = Barleria buxifolia L. ●☆

48082 Barleria acanthoides T. Anderson = Barleria lanceata (Forssk.) C. Chr. ●☆

48083 Barleria acanthoides Vahl = Barleria lanceolata (Schinz) Oberm. ■☆

48084 Barleria acanthoides Vahl f. lanceolata Schinz = Barleria lanceolata (Schinz) Oberm. ●☆

48085 Barleria acanthoides Vahl var. gracilispina Fiori = Barleria paolii Fiori ●☆

48086 Barleria acanthophora Nees;刺梗假杜鹃●☆

48087 Barleria adelensis Delile;阿代尔假杜鹃●☆

48088 Barleria affinis C. B. Clarke;近缘假杜鹃●☆

48089 Barleria alata S. Moore;具翅假杜鹃●☆

48090 Barleria alata S. Moore var. amoena Benoist;秀丽假杜鹃●☆

48091 Barleria albida Lindau = Barleria senensis Klotzsch ●☆

48092 Barleria albi-pilosa Hainz;白疏毛假杜鹃●☆

48093 Barleria albomarginata Hedrén;白边假杜鹃●☆

48094 Barleria albostellata C. B. Clarke;灰毛假杜鹃;Gray Barleria ●☆

48095 Barleria amanensis Lindau;阿曼假杜鹃●☆

48096 Barleria angustiloba Lindau = Barleria ventricosa Hochst. ex Nees ●☆

48097 Barleria antunesi Lindau;安图内思假杜鹃●☆

48098 Barleria arbuscula C. B. Clarke;小乔木假杜鹃●☆

48099 Barleria argentea Balf. f.;银白假杜鹃●☆

48100 Barleria argentea Balf. f. var. yemensis (Schweinf. ex Lindau) Fiori;也门假杜鹃●☆

48101 Barleria argillicola Oberm.;白土假杜鹃●☆

48102 Barleria aridicola Rendle ex Engl.;旱生假杜鹃●☆

48103 Barleria aromatica Oberm.;芳香假杜鹃●☆

48104 Barleria auriculata Schumach. = Hygrophila auriculata (Schumach.) Heine ●☆

48105 Barleria bagshawei S. Moore;巴格肖假杜鹃●☆

48106 Barleria barbata E. Mey. = Barleria gueinzii Sond. ●☆

48107 Barleria bechuanensis C. B. Clarke;贝专假杜鹃●☆

48108 Barleria benadirensis (Fiori) Chiov. = Barleria glandulifera Lindau ●☆

48109 Barleria benguellensis S. Moore;本格拉假杜鹃●☆

48110 Barleria bicolor Chiov. = Barleria proxima Lindau ●☆

48111 Barleria blepharoides Lindau;睫毛假杜鹃●☆

48112 Barleria boehmii Lindau;贝姆假杜鹃●☆

48113 Barleria boivinii Lindau = Barleria volkensii Lindau ●☆

48114 Barleria bolusii Oberm.;博卢斯假杜鹃●☆

48115 Barleria bonifacei Benoist = Barleria lancifolia T. Anderson subsp. charlesii (Benoist) J. -P. Lebrun et Monod ●☆

48116 Barleria boranensis Fiori;博兰假杜鹃●☆

48117 Barleria boranensis Fiori. f. leucosepala ? = Barleria boranensis Fiori ●☆

48118 Barleria brevispina (Fiori) Hedrén;短刺假杜鹃●☆

48119 Barleria brevispina R. Br. = Barleria setigera Rendle var. brevispina C. B. Clarke ●☆

48120 Barleria brevituba Benoist;短管假杜鹃●☆

48121 Barleria breyeri Oberm. = Barleria lugardii C. B. Clarke ●☆

48122 Barleria briartii De Wild. et T. Durand;布里亚特假杜鹃●☆

48123 Barleria brownei Juss. = Blechum pyramidatum (Lam.) Urb. ■

48124 Barleria brownii S. Moore;布朗假杜鹃●☆

48125 Barleria buddleioides S. Moore;醉鱼草假杜鹃●☆

48126 Barleria buxifolia L.;黄杨叶假杜鹃●☆

48127 Barleria calophylla Lindau;美叶假杜鹃●☆

48128 Barleria calophylloides Lindau;拟美叶假杜鹃●☆
48129 Barleria capitata Klotzsch;头状假杜鹃●☆
48130 Barleria cardiocalyx Solms = Barleria orbicularis Hochst. ex T. Anderson ●☆
48131 Barleria cavaleriei H. Lév. = Barleria cristata L. ●
48132 Barleria chlamydocalyx Lindau = Barleria orbicularis Hochst. ex T. Anderson ●☆
48133 Barleria ciliata Roxb. = Barleria cristata L. ●
48134 Barleria cinereicaulis C. B. Clarke = Barleria lancifolia T. Anderson ●☆
48135 Barleria clivorum C. B. Clarke;山岗假杜鹃●☆
48136 Barleria consanguinea Klotzsch;亲缘假杜鹃●☆
48137 Barleria cordata Oberm.;心形假杜鹃●☆
48138 Barleria cordifolia Hochst. ex T. Anderson = Barleria parviflora R. Br. ex T. Anderson ●☆
48139 Barleria crassa C. B. Clarke;粗假杜鹃●☆
48140 Barleria cristata L.;假杜鹃(大寒药,地狗胆,假红蓝,蓝花草,茗花,叶红草,紫靛);Bluebell Barleria, Bluebell Falsecuckoo, Crested Philippine Violet, Philippine Violet ●
48141 Barleria cristata L. var. alba Hort.;白花假杜鹃;White-flower Bluebell Falsecuckoo ●☆
48142 Barleria cristata L. var. mairei H. Lév.;禄劝假杜鹃;Maire Bluebell Falsecuckoo ●
48143 Barleria crossandriformis C. B. Clarke;粗蕊假杜鹃●☆
48144 Barleria crotalaria H. Lév. = Acanthocalyx nepalensis (D. Don) M. J. Cannon subsp. delavayi (Franch.) D. Y. Hong et F. Barrie ■
48145 Barleria cunenensis Benoist;库内内假杜鹃●☆
48146 Barleria cyanea S. Moore;蓝假杜鹃●☆
48147 Barleria damarensis T. Anderson;达马尔假杜鹃●☆
48148 Barleria decaisniana Nees;德凯纳假杜鹃●☆
48149 Barleria decaryi Benoist;德卡里假杜鹃●☆
48150 Barleria delagoensis Oberm.;迪拉果假杜鹃●☆
48151 Barleria delamerei S. Moore;德拉米尔假杜鹃●☆
48152 Barleria dentata Hedrén;具齿假杜鹃●☆
48153 Barleria descampsii Lindau;德康假杜鹃●☆
48154 Barleria diacantha Hochst. ex Nees = Barleria trispinosa (Forssk.) Vahl ●☆
48155 Barleria dichotoma Roxb. = Barleria cristata L. ●
48156 Barleria diffusa (Oliv.) Lindau;松散假杜鹃●☆
48157 Barleria dinteri Oberm.;丁特假杜鹃●☆
48158 Barleria dolomiticola M. Balkwill et K. Balkwill;多罗米蒂假杜鹃●☆
48159 Barleria dulcis Benoist;甜假杜鹃●☆
48160 Barleria eenii S. Moore = Barleria senensis Klotzsch ●☆
48161 Barleria elegans S. Moore;雅致假杜鹃●☆
48162 Barleria elliptica Benoist;椭圆假杜鹃●☆
48163 Barleria eranthemoides R. Br. ex C. B. Clarke;可爱花假杜鹃●☆
48164 Barleria exellii Benoist;埃克塞尔假杜鹃●☆
48165 Barleria eylesii S. Moore;艾尔斯假杜鹃●☆
48166 Barleria fissiflora Bojer ex Nees;半裂花假杜鹃●☆
48167 Barleria flava J. Jacq. = Barleria oenotheroides Dum. Cours. ●☆
48168 Barleria fulvostellata C. B. Clarke;黄褐星假杜鹃●☆
48169 Barleria galpinii C. B. Clarke;盖尔假杜鹃●☆
48170 Barleria glandulifera Lindau;具腺假杜鹃●☆
48171 Barleria glandulifera Lindau var. benadirensis Fiori = Barleria glandulifera Lindau ●☆
48172 Barleria glandulosa Hochst. ex Nees = Lepidagathis glandulosa Nees ex C. B. Clarke ■☆
48173 Barleria glaucobracteata Hedrén;灰苞假杜鹃●☆
48174 Barleria gossweileri S. Moore;戈斯假杜鹃●☆
48175 Barleria grandicalyx Lindau;大萼假杜鹃●☆
48176 Barleria grandicalyx Lindau var. vix-dentata C. B. Clarke;小齿大萼假杜鹃●☆
48177 Barleria grandiflora R. Br. = Barleria grandis Hochst. ex Nees ●☆
48178 Barleria grandipetala De Wild.;大瓣假杜鹃●☆
48179 Barleria grandis Hochst. ex Nees;大假杜鹃●☆
48180 Barleria grantii Oliv.;格兰特假杜鹃●☆
48181 Barleria greenii M. Balkwill et K. Balkwill;格林假杜鹃●☆
48182 Barleria gueinzii Sond.;吉内斯假杜鹃●☆
48183 Barleria halimoides Nees = Petalidium halimoides (Nees) S. Moore ■☆
48184 Barleria harnierii Solms = Barleria parviflora R. Br. ex T. Anderson ●☆
48185 Barleria hereroensis Engl. = Barleria lancifolia T. Anderson ●☆
48186 Barleria hereroensis Engl. var. charlesii Benoist = Barleria lancifolia T. Anderson subsp. charlesii (Benoist) J. -P. Lebrun et Monod ●☆
48187 Barleria heterotricha Lindau = Barleria affinis C. B. Clarke ●☆
48188 Barleria hildebrandtii S. Moore;希尔德假杜鹃●☆
48189 Barleria hirta Oberm.;多毛假杜鹃●☆
48190 Barleria hochstetteri Nees = Barleria hochstetteri Nees ex DC. ●☆
48191 Barleria hochstetteri Nees ex DC.;霍赫假杜鹃●☆
48192 Barleria holstii Lindau;霍尔假杜鹃●☆
48193 Barleria holubii C. B. Clarke;霍勒布假杜鹃●☆
48194 Barleria homoiotricha C. B. Clarke = Barleria mucronifolia Lindau ●☆
48195 Barleria horrida Buscal. et Muschl.;多刺假杜鹃●☆
48196 Barleria humbertii Benoist;亨伯特假杜鹃●☆
48197 Barleria humilis Benoist;低矮假杜鹃●☆
48198 Barleria hypocrateriformis Hochst. ex T. Anderson = Barleria eranthemoides R. Br. ex C. B. Clarke ●☆
48199 Barleria hystrix L. = Barleria prionitis L. ●
48200 Barleria ilicifolia Hedrén;冬青叶假杜鹃●☆
48201 Barleria induta C. B. Clarke = Barleria prionitis L. subsp. prionitoides (Engl.) Brummitt et J. R. I. Wood ●☆
48202 Barleria insericata Chiov. = Barleria glandulifera Lindau ●☆
48203 Barleria insolita Benoist;异常假杜鹃●☆
48204 Barleria integrisepala H. P. Tsui;全缘萼假杜鹃;Entire-sepal Barleria, Entire-sepal Falsecuckoo ●
48205 Barleria involucrata Nees;总苞假杜鹃●☆
48206 Barleria iodocephala Chiov. = Barleria mucronifolia Lindau ●☆
48207 Barleria irritans Nees var. rigida (Nees) C. B. Clarke = Barleria rigida Nees ●☆
48208 Barleria jasminiflora C. B. Clarke;茉莉花假杜鹃●☆
48209 Barleria jubata S. Moore;鬃毛假杜鹃●☆
48210 Barleria jucunda Benoist;愉悦假杜鹃●☆
48211 Barleria jucunda Benoist var. pilosula ?;疏毛愉悦假杜鹃●☆
48212 Barleria jucunda Lindau = Barleria diffusa (Oliv.) Lindau ●☆
48213 Barleria kaessneri S. Moore;卡斯纳假杜鹃●☆
48214 Barleria keniensis Mildbr.;肯尼亚假杜鹃■☆
48215 Barleria kilimandscharica Lindau = Barleria mucronata Lindau ●☆
48216 Barleria kirkii T. Anderson;柯克假杜鹃■☆
48217 Barleria kitchingii Baker;基钦假杜鹃●☆
48218 Barleria laceratiflora Lindau;裂花假杜鹃●☆

48219 Barleria laciniata Nees;刺苞假杜鹃;Laciniate Falsecuckoo ●
48220 Barleria laciniata Nees = Barleria cristata L. ●
48221 Barleria lactiflora Brummitt et Seyani;乳白花假杜鹃●☆
48222 Barleria lanceata (Forssk.) C. Chr.;披针形假杜鹃●☆
48223 Barleria lanceolata (Schinz) Oberm.;披针假杜鹃●☆
48224 Barleria lancifolia T. Anderson;披针叶假杜鹃●☆
48225 Barleria lancifolia T. Anderson subsp. charlesii (Benoist) J.-P. Lebrun et Monod;查尔斯假杜鹃●☆
48226 Barleria lancifolia T. Anderson var. charlesii (Benoist) Monod = Barleria lancifolia T. Anderson subsp. charlesii (Benoist) J.-P. Lebrun et Monod ●☆
48227 Barleria lateralis Oberm.;侧生假杜鹃●☆
48228 Barleria latiloba Engl. = Barleria lancifolia T. Anderson ●☆
48229 Barleria leandrii Benoist;利安假杜鹃●☆
48230 Barleria lichtensteiniana Nees;利希滕假杜鹃●☆
48231 Barleria linearifolia Rendle;线叶假杜鹃●☆
48232 Barleria linearifolia Rendle var. brevispina Fiori = Barleria brevispina (Fiori) Hedrén ●☆
48233 Barleria longifolia L. = Hygrophila auriculata (Schumach.) Heine ●☆
48234 Barleria longipes Benoist;长梗假杜鹃●☆
48235 Barleria longipes Benoist var. lutescens Benoist = Barleria puberula Benoist ●☆
48236 Barleria longipes Benoist var. puberula (Benoist) Benoist = Barleria puberula Benoist ●☆
48237 Barleria longissima Lindau;长假杜鹃●☆
48238 Barleria longistyla Lindau;长柱假杜鹃●☆
48239 Barleria lugardii C. B. Clarke;卢格德假杜鹃●☆
48240 Barleria lukafuensis De Wild.;卢卡夫假杜鹃●☆
48241 Barleria lupulina Lindl.;花叶假杜鹃(刺血红,七星剑,乌骨七星剑,血路草);Hophead Philippine Violet ●
48242 Barleria mackenii Hook. f.;梅肯假杜鹃●☆
48243 Barleria maclaudii Benoist;马克洛假杜鹃●☆
48244 Barleria macracantha R. Br. = Barleria trispinosa (Forssk.) Vahl ●☆
48245 Barleria macrostegia Nees;大盖假杜鹃●☆
48246 Barleria maculata S. Moore;斑点假杜鹃●☆
48247 Barleria marghilomaniae Volkens et Schweinf. = Barleria trispinosa (Forssk.) Vahl ●☆
48248 Barleria marginata Oliv.;具边假杜鹃●☆
48249 Barleria marlothii Engl. = Barleria damarensis T. Anderson ●☆
48250 Barleria massae Chiov.;马萨假杜鹃●☆
48251 Barleria matopensis S. Moore;马托假杜鹃●☆
48252 Barleria media C. B. Clarke;中间假杜鹃●☆
48253 Barleria megalosiphon Mildbr.;大管假杜鹃●☆
48254 Barleria merxmuelleri P. G. Mey.;梅尔假杜鹃●☆
48255 Barleria meyeriana Nees;迈尔假杜鹃●☆
48256 Barleria micans Nees;闪亮假杜鹃;Shiny Barleria, Shiny Falsecuckoo ●☆
48257 Barleria micrantha C. B. Clarke;热非小花假杜鹃●☆
48258 Barleria migiurtinorum Chiov. = Barleria hildebrandtii S. Moore ●☆
48259 Barleria molensis Wild;莫尔假杜鹃●☆
48260 Barleria mollis Lindau = Barleria stuhlmanni Lindau ●☆
48261 Barleria mollis R. Br. = Barleria ventricosa Hochst. ex Nees ●☆
48262 Barleria monticola Oberm.;山生假杜鹃●☆
48263 Barleria mosdenensis Oberm. = Barleria bolusii Oberm. ●☆
48264 Barleria mucronata Lindau;短尖假杜鹃●☆
48265 Barleria mucronifolia Lindau;钝尖假杜鹃●☆
48266 Barleria napalensis Nees = Barleria cristata L. ●
48267 Barleria natalensis Lindau;纳塔尔假杜鹃●☆
48268 Barleria neurophylla C. B. Clarke;脉叶假杜鹃●☆
48269 Barleria newtonii Lindau = Barleria calophylla Lindau ●☆
48270 Barleria norbertii Benoist;诺贝特假杜鹃●☆
48271 Barleria nyasensis C. B. Clarke;尼亚斯假杜鹃●☆
48272 Barleria obtusa Nees;钝形假杜鹃;Barleria, Obtuse Barleria, Obtuse Falsecuckoo ●☆
48273 Barleria obtusisepala C. B. Clarke;钝瓣假杜鹃●☆
48274 Barleria oenotheroides Dum. Cours.;月见草假杜鹃●☆
48275 Barleria opaca (Vahl) Nees;暗色假杜鹃●☆
48276 Barleria orbicularis Hochst. ex T. Anderson;圆形假杜鹃●☆
48277 Barleria ovata E. Mey. ex Nees;卵形假杜鹃●☆
48278 Barleria oxyphylla Lindau;尖叶假杜鹃●☆
48279 Barleria paludosa S. Moore;沼泽假杜鹃●☆
48280 Barleria pannosa Chiov.;毡状假杜鹃●☆
48281 Barleria paolii Fiori;保尔假杜鹃●☆
48282 Barleria papillosa T. Anderson;乳头假杜鹃■☆
48283 Barleria parviflora R. Br. ex T. Anderson;小花假杜鹃●☆
48284 Barleria parvispina Benoist;短小刺假杜鹃●☆
48285 Barleria paucidentata Benoist;疏齿假杜鹃●☆
48286 Barleria pauciflora Lindau = Barleria parviflora R. Br. ex T. Anderson ●☆
48287 Barleria perrieri Benoist;佩里耶假杜鹃●☆
48288 Barleria petrophila Lindau = Barleria senensis Klotzsch ●☆
48289 Barleria phillipseae Rendle = Barleria diffusa (Oliv.) Lindau ●☆
48290 Barleria phillyreifolia Baker;女贞叶假杜鹃●☆
48291 Barleria pirottae Lindau = Barleria hochstetteri Nees ex DC. ●☆
48292 Barleria polyneura S. Moore;多脉假杜鹃●☆
48293 Barleria polytricha Bojer = Crossandra stenandrium (Nees) Lindau ●☆
48294 Barleria pretoriensis C. B. Clarke;比勒陀利亚假杜鹃●☆
48295 Barleria prionitis L.;黄花假杜鹃(黄叶假杜鹃,锯假杜鹃);Porcupine Flower, Yellow Falsecuckoo, Yellowflower Barleria ●
48296 Barleria prionitis L. subsp. delagoensis (Oberm.) Brummitt et J. R. I. Wood;迪拉果黄花假杜鹃●☆
48297 Barleria prionitis L. subsp. prionitoides (Engl.) Brummitt et J. R. I. Wood;普通假杜鹃●☆
48298 Barleria prionitis L. subsp. pubiflora (Benth. ex Hohen.) Brummitt et J. R. I. Wood;短毛黄花假杜鹃●☆
48299 Barleria prionitis L. subsp. tanzaniana Brummitt et J. R. I. Wood;坦桑尼亚假杜鹃●☆
48300 Barleria prionitis L. var. setosa Klotzsch = Barleria eranthemoides R. Br. ex C. B. Clarke ●☆
48301 Barleria prionitoides Engl. = Barleria prionitis L. subsp. prionitoides (Engl.) Brummitt et J. R. I. Wood ●☆
48302 Barleria proxima Lindau;近基假杜鹃●☆
48303 Barleria pseudoprionitis Lindau = Barleria smithii Rendle ●☆
48304 Barleria puberula Benoist;微毛假杜鹃●☆
48305 Barleria pubiflora Benth. ex Hohen. = Barleria prionitis L. subsp. pubiflora (Benth. ex Hohen.) Brummitt et J. R. I. Wood ●☆
48306 Barleria puccionii Chiov.;普乔尼假杜鹃●☆
48307 Barleria pulchella Benoist;美丽假杜鹃●☆
48308 Barleria pulchra Lindau;马岛美丽假杜鹃●☆
48309 Barleria pumila Hochst. ex Nees = Justicia diffusa Willd. ■
48310 Barleria punctata Milne-Redh.;斑叶假杜鹃●☆

48311 Barleria pungens L. f. ;刺假杜鹃●☆
48312 Barleria pungens L. f. var. macrophylla Nees = Barleria elegans S. Moore ●☆
48313 Barleria purpureosepala H. P. Tsui;紫萼假杜鹃;Purple-sepaled Falsecuckoo ●
48314 Barleria pyramidata Lam. = Blechum pyramidatum (Lam.) Urb. ■
48315 Barleria quadriloba Oberm. = Barleria lugardii C. B. Clarke ●☆
48316 Barleria quadrispina Lindau;四刺假杜鹃●☆
48317 Barleria quadrispina Lindau var. linearifolia (Rendle) Chiov. = Barleria quadrispina Lindau ●☆
48318 Barleria querimbensis Klotzsch = Barleria repens Nees ●☆
48319 Barleria ramulosa C. B. Clarke;多枝假杜鹃●☆
48320 Barleria randii S. Moore;兰德假杜鹃●☆
48321 Barleria rautanenii Schinz = Barleria lancifolia T. Anderson ●☆
48322 Barleria rehmanni C. B. Clarke;雷曼假杜鹃●☆
48323 Barleria repens Nees;匍匐假杜鹃;Small Bush Violet ●☆
48324 Barleria rhodesiaca Oberm. ;罗得西亚假杜鹃●☆
48325 Barleria rhynchocarpa Klotzsch = Crossandra nilotica Oliv. ●☆
48326 Barleria richardiana Nees = Neuracanthus richardianus (Nees) Boivin ex Benoist ●☆
48327 Barleria rigida Nees;硬假杜鹃●☆
48328 Barleria rigida Nees var. ilicina (E. Mey. ex T. Anderson) Oberm. = Barleria rigida Nees ●☆
48329 Barleria rivae Lindau = Barleria hochstetteri Nees ex DC. ●☆
48330 Barleria rogersii S. Moore;罗杰斯假杜鹃●☆
48331 Barleria rotundifolia Oberm. ;圆叶假杜鹃●☆
48332 Barleria rotundisepala Rendle = Barleria orbicularis Hochst. ex T. Anderson ●☆
48333 Barleria ruellioides T. Anderson;芦莉草假杜鹃●☆
48334 Barleria ruspolii Lindau = Barleria parviflora R. Br. ex T. Anderson ●☆
48335 Barleria sacani Klotzsch ex Lindau = Neuracanthus africanus T. Anderson ex S. Moore ●☆
48336 Barleria sacleuxii Benoist;萨克勒假杜鹃●☆
48337 Barleria salicifolia S. Moore;柳叶假杜鹃●☆
48338 Barleria saxatilis Oberm. ;岩生假杜鹃●☆
48339 Barleria scassellatii Fiori = Barleria micrantha C. B. Clarke ●☆
48340 Barleria schenckii Schinz = Barleria rigida Nees ●☆
48341 Barleria schmittii Benoist;施米特假杜鹃●☆
48342 Barleria schweinfurthiana Lindau = Barleria diffusa (Oliv.) Lindau ●☆
48343 Barleria senegalensis Nees = Barleria oenotheroides Dum. Cours. ●☆
48344 Barleria senensis Klotzsch;塞纳假杜鹃●☆
48345 Barleria separata Benoist;分离假杜鹃●☆
48346 Barleria setigera Rendle;毛梗假杜鹃●☆
48347 Barleria setigera Rendle = Barleria fissiflora Bojer ex Nees ●☆
48348 Barleria setigera Rendle var. brevispina C. B. Clarke;短毛梗假杜鹃●☆
48349 Barleria setigera Rendle var. pumila ? = Barleria quadrispina Lindau ●☆
48350 Barleria seyrigii Benoist;塞里格假杜鹃●☆
48351 Barleria smithii Rendle;史密斯假杜鹃●☆
48352 Barleria somalensis Franch. = Asystasia guttata (Forssk.) Brummitt ●☆
48353 Barleria spathulata N. E. Br. = Barleria senensis Klotzsch ●☆
48354 Barleria spinisepala E. A. Bruce;刺萼假杜鹃●☆
48355 Barleria spinulosa Klotzsch;柄腺假杜鹃●☆
48356 Barleria splendens E. A. Bruce;光亮假杜鹃●☆
48357 Barleria squarrosa Klotzsch = Barleria spinulosa Klotzsch ●☆
48358 Barleria stellato-tomentosa S. Moore;星毛假杜鹃●☆
48359 Barleria stellato-tomentosa S. Moore var. ukambensis Lindau = Barleria salicifolia S. Moore ●☆
48360 Barleria stelligera Lindau;星状假杜鹃●☆
48361 Barleria steudneri C. B. Clarke;斯托德假杜鹃●☆
48362 Barleria stimulans E. Mey. ex Nees;具柄假杜鹃●☆
48363 Barleria stuhlmanni Lindau;斯图尔曼假杜鹃●☆
48364 Barleria subglobosa S. Moore;亚球形假杜鹃●☆
48365 Barleria subinermis Chiov. ;近无刺假杜鹃●☆
48366 Barleria submollis Lindau = Barleria stuhlmanni Lindau ●☆
48367 Barleria sudanica (Schweinf.) Lindau;苏丹假杜鹃●☆
48368 Barleria swynnertonii S. Moore;斯温纳顿假杜鹃●☆
48369 Barleria taitensis S. Moore;泰塔假杜鹃●☆
48370 Barleria talbotii S. Moore = Barleria brownii S. Moore ●☆
48371 Barleria tetraglochin Milne-Redh. ;四钩假杜鹃●☆
48372 Barleria tischeriana Chiov. ;蒂舍尔假杜鹃●☆
48373 Barleria transvaalensis Oberm. ;德兰士瓦假杜鹃●☆
48374 Barleria triacantha Hochst. ex Nees = Barleria lanceata (Forssk.) C. Chr. ●☆
48375 Barleria triacantha Nees = Barleria lanceata (Forssk.) C. Chr. ●☆
48376 Barleria trinervis Wall. = Lepidagathis trinervis Nees ■☆
48377 Barleria trispinosa (Forssk.) Vahl;三刺假杜鹃●☆
48378 Barleria ukamensis Lindau;乌卡姆假杜鹃●☆
48379 Barleria umbrosa Lindau = Barleria micrantha C. B. Clarke ●☆
48380 Barleria usambarica Lindau;乌桑巴拉假杜鹃●☆
48381 Barleria variabilis Oberm. ;易变假杜鹃●☆
48382 Barleria venosa Oberm. ;密脉假杜鹃●☆
48383 Barleria ventricosa Hochst. ex Nees;偏肿假杜鹃●☆
48384 Barleria verdickii De Wild. ;韦尔迪假杜鹃●☆
48385 Barleria villosa S. Moore;长柔毛假杜鹃●☆
48386 Barleria vincifolia Baker;野豌豆叶假杜鹃●☆
48387 Barleria violacea Hainz;堇色假杜鹃●☆
48388 Barleria violascens S. Moore;浅堇色假杜鹃●☆
48389 Barleria virgula C. B. Clarke;条纹假杜鹃●☆
48390 Barleria vix-dentata C. B. Clarke;齿假杜鹃●☆
48391 Barleria volkensii Lindau;福尔假杜鹃●☆
48392 Barleria waggana Rendle = Barleria quadrispina Lindau ●☆
48393 Barleria welwitschii S. Moore;韦尔假杜鹃●☆
48394 Barleria whytei S. Moore;怀特假杜鹃●☆
48395 Barleria wilmsiana Lindau;维尔姆斯假杜鹃●☆
48396 Barleria wilmsii Lindau ex C. B. Clarke = Barleria wilmsiana Lindau ●☆
48397 Barleria woodii C. B. Clarke = Barleria natalensis Lindau ●☆
48398 Barleria yemense Schweinf. ex Lindau = Barleria argentea Balf. f. var. yemensis (Schweinf. ex Lindau) Fiori ●☆
48399 Barleriacanthus Oerst. = Barleria L. ●■
48400 Barlerianthus Oerst. = Barleria L. ●■
48401 Barlerianthus grandis (Hochst. ex Nees) Oerst. = Barleria grandis Hochst. ex Nees ●☆
48402 Barleriola Oerst. (1855);小杜鹃花属●☆
48403 Barleriola solanifolia Oerst. ;小杜鹃花●☆
48404 Barleriopsis Oerst. = Barleria L. ●■
48405 Barleriosiphon Oerst. = Barleria L. ●■
48406 Barlerites Oerst. = Barleria L. ●■
48407 Barlerites hochstetteri (Nees) Oerst. = Barleria hochstetteri Nees

ex DC. ●☆
48408 Barlerites hochstetteri Oerst. = Barleria hochstetteri Nees ex DC. ●☆
48409 Barlia Parl. (1860);巴拉兰属■☆
48410 Barlia longibracteata Parl. = Himantoglossum robertianum (Loisel.) P. Delforge ■☆
48411 Barlia robertiana (Loisel.) Greuter;巴拉兰;Giant Orchid ■☆
48412 Barlia robertiana (Loisel.) Greuter = Himantoglossum robertianum (Loisel.) P. Delforge ■☆
48413 Barnadesia Mutis = Barnadesia Mutis ex L. f. ●☆
48414 Barnadesia Mutis ex L. f. (1782);刺菊木属●☆
48415 Barnadesia aculeata (Benth.) I. C. Chung;尖叶刺菊木●☆
48416 Barnadesia arborea Kunth;树刺菊木●☆
48417 Barnadesia jelskii Hieron.;耶尔刺菊木●☆
48418 Barnadesia kingii H. Rob.;金氏刺菊木●☆
48419 Barnadesia parviflora Benth. et Hook.;小花刺菊木●☆
48420 Barnadesia spinosa L. f.;刺菊木●☆
48421 Barnardia Lindl. (1826);类绵枣儿属■☆
48422 Barnardia Lindl. = Scilla L. ■
48423 Barnardia alboviridis (Hand. -Mazz.) Speta = Barnardia japonica (Thunb.) Schult. et Schult. f. ●
48424 Barnardia alboviridis Hand. -Mazz. = Barnardia japonica (Thunb.) Schult. et Schult. f. ●
48425 Barnardia bispatha (Hand. -Mazz.) Speta = Barnardia japonica (Thunb.) Schult. et Schult. f. ●
48426 Barnardia borealijaponica (M. Kikuchi) Speta = Barnardia japonica (Thunb.) Schult. et Schult. f. ●
48427 Barnardia japonica (Thunb.) Schult. et Schult. f. = Scilla scilloides (Lindl.) Druce ■
48428 Barnardia numidica (Poir.) Speta = Scilla numidica Poir. ■☆
48429 Barnardia pulchella (Kitag.) Speta = Barnardia japonica (Thunb.) Schult. et Schult. f. ●
48430 Barnardia pulchella (Kitag.) Speta = Scilla scilloides (Lindl.) Druce ■
48431 Barnardia scilloides Lindl. = Barnardia japonica (Thunb.) Schult. et Schult. f. ●
48432 Barnardia scilloides Lindl. = Scilla scilloides (Lindl.) Druce ■
48433 Barnardia sinensis (Lour.) Speta = Barnardia japonica (Thunb.) Schult. et Schult. f. ●
48434 Barnardia sinensis (Lour.) Speta = Scilla scilloides (Lindl.) Druce ■
48435 Barnardiella Goldblatt = Moraea Mill. (保留属名) ■
48436 Barnardiella Goldblatt(1976);肖绵枣儿属■☆
48437 Barnardiella spiralis (N. E. Br.) Goldblatt = Moraea herrei (L. Bolus) Goldblatt ■☆
48438 Barnebya W. R. Anderson et B. Gates(1981);巴恩木属●☆
48439 Barnebya dispar (Griseb.) W. R. Anderson et B. Gates;巴恩木●☆
48440 Barnebya harleyi W. R. Anderson et B. Gates;哈雷巴恩木●☆
48441 Barnebydendron J. H. Kirkbr. (1999);几内亚叶果豆属●☆
48442 Barnebydendron J. H. Kirkbr. = Phyllocarpus Riedel ex Endl. ●☆
48443 Barnebyella Podlech = Dorycnium Mill. ●■☆
48444 Barneoudia Gay(1845);肾果獐耳细辛属■☆
48445 Barneoudia balliana Britton;巴尔肾果獐耳细辛■☆
48446 Barneoudia chilensis Gay;肾果獐耳细辛■☆
48447 Barneoudia major Phil.;大肾果獐耳细辛■☆
48448 Barnettia Santisuk = Santisukia Brummitt ●☆
48449 Barnhartia Gleason(1926);繁花远志属●☆
48450 Barnhartia floribunda Schltr.;繁花远志■☆
48451 Barola Adans. = Barbilus P. Browne ●
48452 Barola Adans. = Trichilia P. Browne(保留属名) ●
48453 Barollaea Neck. = Caryocar F. Allam. ex L. ●☆
48454 Barombia Schltr. (1914);繁花兰属■☆
48455 Barombia Schltr. = Aerangis Rchb. f. ■☆
48456 Barombia gracillima (Kraenzl.) Schltr. = Aerangis gracillima (Kraenzl.) Arends et J. Stewart ■☆
48457 Barombia schliebenii (Mansf.) P. J. Cribb = Rangaeris schliebenii (Mansf.) P. J. Cribb ●☆
48458 Barombiella Szlach. = Barombia Schltr. ■☆
48459 Barongia Peter G. Wilson et B. Hyland(1988);巴隆木属●☆
48460 Barongia lophandra Peter G. Wilson et Hyland;巴隆木●☆
48461 Baronia Baker = Rhus L. ●
48462 Baronia Baker(1882);马达加斯加漆树属●☆
48463 Baronia taratana Baker;马达加斯加漆树●☆
48464 Baroniella Costantin et Gallaud(1907);巴龙萝藦属●☆
48465 Baroniella acuminata (Choux) Bullock;渐尖巴龙萝藦●☆
48466 Baroniella camptocarpoides Costantin et Gallaud;弯果巴龙萝藦●☆
48467 Baroniella capillacea Klack.;细毛巴龙萝藦●☆
48468 Baroniella ensifolia Klack.;剑叶巴龙萝藦●☆
48469 Baroniella linearis (Choux) Bullock;线状巴龙萝藦●☆
48470 Baroniella longicornis Klack.;长角巴龙萝藦●☆
48471 Baroniella multiflora (Choux) Bullock;多花巴龙萝藦●☆
48472 Barosma Willd. (1809) (保留属名);重香木属(重香属);Buchu ●☆
48473 Barosma Willd. (保留属名) = Agathosma Willd. (保留属名) ●☆
48474 Barosma acutata Sond. = Agathosma ovata (Thunb.) Pillans ●☆
48475 Barosma bathii Dümmer = Agathosma bathii (Dümmer) Pillans ●☆
48476 Barosma betulina (P. J. Bergius) Bartl. et H. L. Wendl. = Agathosma betulina (P. J. Bergius) Pillans ●☆
48477 Barosma betulina Bartl. et H. L. Wendl. = Agathosma betulina (P. J. Bergius) Pillans ●☆
48478 Barosma crenulata (L.) Hook. = Agathosma crenulata (L.) Pillans ●☆
48479 Barosma foetidissima Bartl. et H. L. Wendl. = Agathosma foetidissima (Bartl. et H. L. Wendl.) Steud. ●☆
48480 Barosma insignis Compton = Agathosma insignis (Compton) Pillans ●☆
48481 Barosma latifolia (L. f.) Roem. et Schult. = Agathosma odoratissima (Montin) Pillans ●☆
48482 Barosma microcarpa Sond. = Agathosma microcarpa (Sond.) Pillans ●☆
48483 Barosma oblonga (Thunb.) Bartl. et H. L. Wendl. = Agathosma ovata (Thunb.) Pillans ●☆
48484 Barosma ovata (Thunb.) Bartl. et H. L. Wendl. = Agathosma ovata (Thunb.) Pillans ●☆
48485 Barosma pulchella (L.) Bartl. = Agathosma pulchella (L.) Link ●☆
48486 Barosma pulchella (L.) Bartl. et H. L. Wendl. = Agathosma pulchella (L.) Link ●☆
48487 Barosma pulchella (L.) Bartl. et H. L. Wendl. var. tabularis Dümmer = Agathosma tabularis Sond. ●☆
48488 Barosma pulchra Cham. et Schltdl. = Agathosma odoratissima (Montin) Pillans ●☆
48489 Barosma pungens E. Mey. ex Sond. = Agathosma pungens (E. Mey. ex Sond.) Pillans ●☆

48490 Barosma scoparia Eckl. et Zeyh. = Agathosma ovata (Thunb.) Pillans ●☆
48491 Barosma serratifolia (Sims) Willd. = Agathosma serratifolia (Curtis) Spreeth ●☆
48492 Barosma serratifolia Willd. = Agathosma serratifolia (Curtis) Spreeth ●☆
48493 Barosma unicarpellata Fourc. = Agathosma unicarpellata (Fourc.) Pillans ●☆
48494 Barosma venusta Eckl. et Zeyh. = Agathosma venusta (Eckl. et Zeyh.) Pillans ●☆
48495 Barraldeia Thouars(废弃属名) = Carallia Roxb. (保留属名) ●
48496 Barrattia A. Gray = Barrattia A. Gray et Engelm. ●■☆
48497 Barrattia A. Gray = Encelia Adans. ●■☆
48498 Barrattia A. Gray et Engelm. (1847);巴勒特菊属●■☆
48499 Barrattia calva A. Gray et Engelm.;巴勒特菊●■☆
48500 Barrattia calva A. Gray et Engelm. = Simsia calva (A. Gray et Engelm.) A. Gray ●■☆
48501 Barreliera J. F. Gmel. = Barleria L. ●■
48502 Barreria L. = Diosma L. + Brunia Lam. (保留属名) ●☆
48503 Barreria Scop. = Poraqueiba Aubl. ●☆
48504 Barreria Willd. = Poraqueiba Aubl. ●☆
48505 Barreriaceae Mart. = Icacinaceae (Benth.) Miers ●■
48506 Barrettia Sim = Ricinodendron Müll. Arg. ●☆
48507 Barrettia umbrosa Sim = Ricinodendron heudelotii (Baill.) Pierre ex Heckel var. africanum (Müll. Arg.) J. Léonard ●☆
48508 Barringtonia J. R. Forst. et G. Forst. (1775)(保留属名);玉蕊属(金刀木属,棋盘脚树属);Barringtonia ●
48509 Barringtonia acutangula (L.) Gaertn.;锐角玉蕊(锐棱玉蕊);Chee Barringtonia, Freshwater Mangrove, Indian Oak ●☆
48510 Barringtonia asiatica (L.) Kurz;滨玉蕊(花玉蕊,基边脚,棋脚树,棋盘脚,棋盘脚树,台湾金刀木,台湾玉蕊,印度玉蕊);Asiatic Barringtonia, Fish-killer Tree, Indian Barringtonia, Sea Putat ●
48511 Barringtonia butonica J. Forst.;布敦玉蕊;Buton Barringtonia ●☆
48512 Barringtonia calophylla K. Schum. et Lauterb.;美叶玉蕊●☆
48513 Barringtonia calyptrata Benth.;有盖玉蕊●☆
48514 Barringtonia ceylanica (Miers) Gardner ex C. B. Clarke = Barringtonia racemosa (L.) Blume ex DC. ●
48515 Barringtonia fusicarpa Hu;云南玉蕊(金刀木,棱果玉蕊,马蛋果,疏果玉蕊,梭果玉蕊,小叶金刀木);Scatterfruit Barringtonia, Spindle-fruited Barringtonia, Yunnan Barringtonia ●
48516 Barringtonia macrostachya (Jack) Kurz;梭果玉蕊●
48517 Barringtonia pendula (Griff.) Kurz;垂穗金刀木●
48518 Barringtonia racemosa (L.) Blume ex DC.;玉蕊(海南金刀木,水茄冬,水茄苳,穗花棋盘脚树,细叶棋盘脚树);Barringtonia, Powderpuff Tree, Racemose Barringtonia, Smallleaved Barringtonia, Small-leaved Barringtonia ●
48519 Barringtonia racemosa (L.) Spreng. = Barringtonia racemosa (L.) Blume ex DC. ●
48520 Barringtonia scortechinii King;斯氏玉蕊●☆
48521 Barringtonia speciosa J. R. Forst. et G. Forst. = Barringtonia asiatica (L.) Kurz ●
48522 Barringtonia timorensis Blume = Barringtonia racemosa (L.) Blume ex DC. ●
48523 Barringtonia yunnanensis Hu = Barringtonia fusicarpa Hu ●
48524 Barringtoniaceae DC. ex F. Rudolphi (1830)(保留科名);翅玉蕊科(金刀木科);Barringtonia Family ●
48525 Barringtoniaceae F. Rudolphi = Barringtoniaceae DC. ex F. Rudolphi(保留科名) ●
48526 Barringtoniaceae F. Rudolphi = Lecythidaceae A. Rich. (保留科名) ●
48527 Barroetea A. Gray = Brickellia Elliott(保留属名) ●■
48528 Barroetea A. Gray(1880);刺叶修泽兰属■☆
48529 Barroetea brevipes B. L. Rob.;短梗刺叶修泽兰■☆
48530 Barroetea laxiflora Brandegee;疏花刺叶修泽兰■☆
48531 Barroetea setosa A. Gray;刺叶修泽兰■☆
48532 Barrosoa R. M. King et H. Rob. (1971);腺果柄泽兰属■☆
48533 Barrosoa apiculata (Gardn.) R. M. King et H. Rob.;腺果柄泽兰■☆
48534 Barrosoa viridiflora (Baker) R. M. King et H. Rob.;绿花腺果柄泽兰■☆
48535 Barrotia Brongn. = Pandanus Parkinson ex Du Roi ●■
48536 Barrotia Gaudich. = Pandanus Parkinson ex Du Roi ●■
48537 Barrowia Decne. = Orthanthera Wight ■☆
48538 Barrowia jasminiflora Decne. = Orthanthera jasminiflora (Decne.) Schinz ■☆
48539 Bartera Post et Kuntze = Barteria Hook. f. ●☆
48540 Barteria Hook. f. (1860);巴特西番莲属●☆
48541 Barteria Welw. = Brasenia Schreb. ■
48542 Barteria acuminata Baker f. = Barteria dewevrei De Wild. et T. Durand ●☆
48543 Barteria braunii Engl. = Barteria fistulosa Mast. ●☆
48544 Barteria dewevrei De Wild. et T. Durand;德韦巴特西番莲●☆
48545 Barteria fistulosa Mast.;管状巴特西番莲●☆
48546 Barteria nigritana Hook. f.;尼格里塔巴特西番莲●☆
48547 Barteria nigritana Hook. f. subsp. fistulosa (Mast.) Sleumer = Barteria fistulosa Mast. ●☆
48548 Barteria solida Breteler;污浊巴特西番莲●☆
48549 Barteria stuhlmannii Engl. et Gilg = Barteria dewevrei De Wild. et T. Durand ●☆
48550 Barteria urophylla Mildbr. = Barteria fistulosa Mast. ●☆
48551 Barthea Hook. f. (1867);棱果花属(刚毛药花属,毛药花属,深山野牡丹属);Barthea ●★
48552 Barthea barthei (Hance) Krasser;棱果花(芭茜,大野牡丹,刚毛药花,棱果木,毛药花,深山野牡丹,台湾毛花药);Formosan Barthea, S. China Barthea, South China Barthea, Taiwan Barthea ●
48553 Barthea barthei (Hance) Krasser var. valdealata C. Hansen;宽翅棱果花(宽翅毛药花);Broadwing South China Barthea ●
48554 Barthea blinii H. Lév. = Barthea esquirolii H. Lév. ●
48555 Barthea blinii H. Lév. = Plagiopetalum esquirolii (H. Lév.) Rehder ●
48556 Barthea cavaleriei H. Lév. = Phyllagathis longiradiosa (C. Chen) C. Chen ●■
48557 Barthea chinensis Hook. f. = Barthea barthei (Hance) Krasser ●
48558 Barthea esquirolii H. Lév. = Bredia esquirolii (H. Lév.) Lauener ●
48559 Barthea esquirolii H. Lév. = Plagiopetalum esquirolii (H. Lév.) Rehder ●
48560 Barthea formosana Hayata = Barthea barthei (Hance) Krasser ●
48561 Barthesia Comm. ex A. DC. = Ardisia Sw. (保留属名) ●■
48562 Barthlottia Eb. Fisch. (1996);巴氏玄参属●☆
48563 Barthlottia madagascariensis Eb. Fisch.;巴氏玄参●☆
48564 Bartholina R. Br. (1813);南非蜘蛛兰属(巴索兰属);Spider Orchid ■☆
48565 Bartholina burmanniana (L.) Ker Gawl.;布氏南非蜘蛛兰■☆
48566 Bartholina ethelae Bolus;南非蜘蛛兰;South African Spider

Orchid ■☆

48567 Bartholina lindleyana Rchb. f. ex Rolfe = Bartholina burmanniana (L.) Ker Gawl. ■☆

48568 Bartholina pectinata (Thunb.) R. Br. = Bartholina burmanniana (L.) Ker Gawl. ■☆

48569 Barthollesia Silva Manso = Bertholletia Bonpl. ●☆

48570 Bartholomaea Standl. et Steyerm. (1940);巴斯木属●☆

48571 Bartholomaea mollis Standl. et Steyerm.;柔软巴斯木●☆

48572 Bartholomaea paniculata Lundell;圆锥巴斯木●☆

48573 Bartholomaea sessiliflora (Standl.) Standl. et Steyerm.;巴斯木●☆

48574 Barthratherum Andersson = Arthraxon P. Beauv. ■

48575 Barthratherum Andersson = Bathratherum Nees ■

48576 Bartlettia A. Gray(1854);粗莛菊属(巴氏菊属)■☆

48577 Bartlettia scaposa A. Gray;粗莛菊(巴氏菊)■☆

48578 Bartlettina R. M. King et H. Rob. (1971);巴特菊属(柔冠毛泽兰属)●☆

48579 Bartlettina sordida (Less.) R. M. King et H. Rob.;暗色巴特菊●☆

48580 Bartlingia Brongn. = Pultenaea Sm. ●☆

48581 Bartlingia F. Muell. = Laxmannia R. Br. (保留属名)■☆

48582 Bartlingia F. Muell. ex Benth. = Laxmannia R. Br. (保留属名)■☆

48583 Bartlingia Rchb. = Plocama Aiton ●☆

48584 Bartolina Adans. = Tridax L. ●■

48585 Bartonia Muhl. ex Willd. (1801)(保留属名);巴顿龙胆属;Bartonia ■☆

48586 Bartonia Pursh = Mentzelia L. ●■☆

48587 Bartonia Pursh = Nuttallia Raf. ●■☆

48588 Bartonia Pursh ex Sims = Mentzelia L. ●■☆

48589 Bartonia Sims = Mentzelia L. ●■☆

48590 Bartonia aurea Lindl. = Mentzelia lindleyi Torr. et A. Gray ■☆

48591 Bartonia paniculata (Michx.) Muhl.;圆锥巴顿龙胆;Screw-stem,Twining Screw-stem ■☆

48592 Bartonia virginica (L.) Britton,Sterns et Poggenb.;维吉尼亚龙胆;Screwstem,Virginia Bartonia,Yellow Screw-stem ■☆

48593 Bartramia Bartram = Pinckneya Michx. ●☆

48594 Bartramia Ellis = Dodecatheon L. ■☆

48595 Bartramia L. = Triumfetta Plum. ex L. ●■

48596 Bartramia Salisb. = Penstemon Schmidel ●■

48597 Bartramia indica L. = Triumfetta rhomboidea Jacq. ●■

48598 Bartschella Britton et Rose = Mammillaria Haw. (保留属名)●

48599 Bartschia Dalla Torre et Harms = Bartsia L. (保留属名)●■☆

48600 Bartschia trixago (L.) Kuntze = Bartsia trixago L. ■☆

48601 Bartsia L. (1753)(保留属名);巴茨列当属(巴茨玄参属);Bartsia ●■☆

48602 Bartsia abyssinica Hochst. ex Benth. = Hedbergia abyssinica (Hochst. ex Benth.) Molau ■☆

48603 Bartsia abyssinica Hochst. ex Benth. var. nyikensis (R. E. Fr.) Hedberg et al. = Hedbergia abyssinica (Hochst. ex Benth.) Molau ■☆

48604 Bartsia alpina L. = Alicosta alpina (L.) Dulac ■☆

48605 Bartsia alpina L. = Staehelina alpina (L.) Crantz ■☆

48606 Bartsia aspera (Brot.) Lange var. caroli-paui Font Quer = Nothobartsia aspera (Brot.) Bolliger et Molau ■☆

48607 Bartsia asperrima (Link) Samp. = Nothobartsia aspera (Brot.) Bolliger et Molau ■☆

48608 Bartsia asperrima (Link) Samp. var. caroli-paui (Font Quer) Maire = Nothobartsia aspera (Brot.) Bolliger et Molau ■☆

48609 Bartsia bicolor DC. = Bellardia trixago (L.) All. ■☆

48610 Bartsia capensis (L.) Spreng. = Bartsia trixago L. ■☆

48611 Bartsia coccinea L. = Castilleja coccinea (L.) Spreng. ■☆

48612 Bartsia coccinea L. var. pallens Michx. = Castilleja coccinea (L.) Spreng. ■☆

48613 Bartsia decurva Hochst. ex Benth.;下延巴茨列当■☆

48614 Bartsia elgonensis R. E. Fr. = Hedbergia abyssinica (Hochst. ex Benth.) Molau ■☆

48615 Bartsia keniensis R. E. Fr. = Bartsia decurva Hochst. ex Benth. ■☆

48616 Bartsia keniensis Standl. = Bartsia longiflora Hochst. ex Benth. ■☆

48617 Bartsia kilimandscharica Engl. = Bartsia decurva Hochst. ex Benth. ■☆

48618 Bartsia latifolia (L.) Sibth. et Sm. = Parentucellia latifolia (L.) Caruel ■☆

48619 Bartsia longiflora Hochst. ex Benth.;长花巴茨列当■☆

48620 Bartsia longiflora Hochst. ex Benth. subsp. macrophylla (Hedberg) Hedberg;大叶巴茨列当■☆

48621 Bartsia macrocalyx R. E. Fr. = Bartsia decurva Hochst. ex Benth. ■☆

48622 Bartsia macrophylla Hedberg = Bartsia longiflora Hochst. ex Benth. subsp. macrophylla (Hedberg) Hedberg ■☆

48623 Bartsia macrophylla Hedberg subsp. gughensis Cufod. = Bartsia longiflora Hochst. ex Benth. ■☆

48624 Bartsia mannii Hemsl. = Hedbergia abyssinica (Hochst. ex Benth.) Molau ■☆

48625 Bartsia nyikensis R. E. Fr. = Hedbergia abyssinica (Hochst. ex Benth.) Molau ■☆

48626 Bartsia odontites Huds.;小齿巴茨列当;Red Bartsia,Twiny Legs ■☆

48627 Bartsia pallida L. = Castilleja pallida (L.) Kunth ■

48628 Bartsia petitiana (A. Rich.) Hemsl. = Hedbergia abyssinica (Hochst. ex Benth.) Molau ■☆

48629 Bartsia purpurea (Desf.) Ball = Odontites purpureus (Desf.) G. Don ■☆

48630 Bartsia scabra (Thunb.) Spreng. = Alectra sessiliflora (Vahl) Kuntze ■☆

48631 Bartsia similis Hemsl. = Bartsia longiflora Hochst. ex Benth. ■☆

48632 Bartsia trixago L.;非洲巴茨列当■☆

48633 Bartsia versicolor (Lam.) Pers. = Bellardia trixago (L.) All. ■☆

48634 Bartsia viscosa L. = Parentucellia viscosa (L.) Caruel ■☆

48635 Bartsiella Bolliger(1996);小巴茨列当属■☆

48636 Bartsiella rameauana (Emb.) Bolliger;小巴茨列当■☆

48637 Barya Klotzsch = Begonia L. ●■

48638 Barylucuma Ducke = Pouteria Aubl. ●

48639 Baryosma Gaertn. = Dipteryx Schreb. (保留属名)●☆

48640 Baryosma Roem. et Schult. = Barosma Willd. (保留属名)●☆

48641 Barysoma Bunge = Heracleum L. ■

48642 Baryxylum Lour. (废弃属名) = Peltophorum (Vogel) Benth. (保留属名)●

48643 Baryxylum Lour. (废弃属名) = Peltophorum Walp. + Gymnocladus Lam. (保留属名)●

48644 Baryxylum africanum Pierre = Peltophorum africanum Sond. ●☆

48645 Baryxylum tonkinense Pierre = Peltophorum tonkinense (Pierre) Gagnep. ●

48646 Basaltogeton Salisb. = Scilla L. ■

48647 Basanacantha Hook. f. = Randia L. ●

48648 Basananthe Peyr. (1859);基花莲属●■☆

48649 Basananthe Peyr. = Tryphostemma Harv. ●■☆

48650 Basananthe aciphylla Thulin;尖叶基花莲■☆
48651 Basananthe apetala (Baker f.) W. J. de Wilde;无瓣基花莲■☆
48652 Basananthe aristolochioides A. Robyns;马兜铃基花莲■☆
48653 Basananthe baumii (Harms) W. J. de Wilde;鲍姆基花莲■☆
48654 Basananthe baumii (Harms) W. J. de Wilde var. caerulescens (A. Fern. et R. Fern.) W. J. de Wilde;天蓝基花莲■☆
48655 Basananthe berberoides (Chiov.) W. J. de Wilde;小檗基花莲■☆
48656 Basananthe botryoidea A. Robyns;葡萄基花莲■☆
48657 Basananthe cupricola A. Robyns;喜铜基花莲■☆
48658 Basananthe gossweileri (Hutch. et K. Pearce) W. J. de Wilde;戈斯基花莲■☆
48659 Basananthe hanningtoniana (Mast.) W. J. de Wilde;汉宁顿基花莲■☆
48660 Basananthe hederae W. J. de Wilde;常春藤基花莲■☆
48661 Basananthe heterophylla Schinz;互叶基花莲■☆
48662 Basananthe hispidula W. J. de Wilde;细毛基花莲■☆
48663 Basananthe holmesii R. Fern. et A. Fern.;霍尔梅斯基花莲■☆
48664 Basananthe kottoensis W. J. de Wilde;科托基花莲■☆
48665 Basananthe kundelunguensis A. Robyns;昆德龙基花莲■☆
48666 Basananthe lanceolata (Engl.) W. J. de Wilde;披针形基花莲■☆
48667 Basananthe littoralis Peyr.;滨海基花莲■☆
48668 Basananthe longifolia (Harms) R. Fern. et A. Fern. = Basananthe sandersonii (Harv.) W. J. de Wilde ■☆
48669 Basananthe malaissei A. Robyns;马莱泽基花莲■☆
48670 Basananthe nummularia Welw.;铜钱基花莲■☆
48671 Basananthe papillosa (A. Fern. et R. Fern.) W. J. de Wilde;乳头基花莲■☆
48672 Basananthe parvifolia (Baker f.) W. J. de Wilde;小叶基花莲■☆
48673 Basananthe pedata (Baker f.) W. J. de Wilde;鸟足基花莲■☆
48674 Basananthe polygaloides (Hutch. et K. Pearce) W. J. de Wilde;远志基花莲■☆
48675 Basananthe pseudostipulata W. J. de Wilde;假托叶基花莲■☆
48676 Basananthe pubiflora W. J. de Wilde;短毛花基花莲■☆
48677 Basananthe reticulata (Baker f.) W. J. de Wilde;网状基花莲■☆
48678 Basananthe sandersonii (Harv.) W. J. de Wilde;桑德森基花莲■☆
48679 Basananthe scabrida A. Robyns;微糙基花莲■☆
48680 Basananthe scabrifolia (Dandy) W. J. de Wilde;糙叶基花莲■☆
48681 Basananthe spinosa W. J. de Wilde;具刺基花莲■☆
48682 Basananthe subsessilicarpa J. B. Gillett ex Verdc.;无果梗基花莲■☆
48683 Basananthe triloba (Bolus) W. J. de Wilde;三裂基花莲■☆
48684 Basananthe zanzibarica (Mast.) W. J. de Wilde;桑给巴尔基花莲■☆
48685 Basedowia E. Pritz. (1918);细弱金绒草属■☆
48686 Basedowia tenerrima (F. Muell. et Tate) J. M. Black;细弱金绒草■☆
48687 Basela L. = Basella L. ■
48688 Basella L. (1753);落葵属;Madeira-vine, Malabar Nightshade, Vine Spinach, Vinegreens, Vinespinach, Vine-spinach ■
48689 Basella alba L.;落葵(白虎下须,白花落葵,白落葵,潺菜,承露,寸金丹,豆腐菜,蘩露,红鸡尿藤,红鸡屎藤,红落葵,红藤菜,猴子七,胡燕脂,滑菜果,滑腹菜,滑藤,篱笆菜,木耳菜,蒲藤菜,染绛子,软姜子,软藤菜,藤菜,藤儿菜,藤葵,藤露,藤罗菜,藤七,天葵,西洋菜,胭脂菜,胭脂豆,胭脂藤,燕脂菜,燕脂豆,阳菜,御菜,粘藤,菸葵,紫草,紫豆藤,紫葵);Ceylon Spinach, Indian Spinach, Malabar Nightshade, Malabar Spinach, Poi, Red Malabar Nightshade, Red Vinespinach, Vine Spinach, Vine-spinach, White Malabar Nightshade, White Malabarnightshade, White Vine Spinach, White Vinespinach, White Vine-spinach ■
48690 Basella alba L. 'Rubra' = Basella rubra L. ■
48691 Basella cordifolia Lam. = Basella alba L. ■
48692 Basella cordifolia Lam. = Basella rubra L. ■
48693 Basella nigra Lour. = Basella alba L. ■
48694 Basella paniculata Volkens;圆锥落葵■☆
48695 Basella rubra L. = Basella alba L. ■
48696 Basella vesicaria Lam. = Anredera vesicaria Gaertn. f. ●☆
48697 Basellaceae Moq. = Basellaceae Raf. (保留科名)■
48698 Basellaceae Raf. (1837)(保留科名);落葵科;Basella Family, Madeira-Vine Family ■
48699 Baseonema Schltr. et Rendle(1896);基丝萝藦属■☆
48700 Baseonema acuminatum Choux = Baroniella acuminata (Choux) Bullock ●☆
48701 Baseonema camptocarpoides (Costantin et Gallaud) Choux = Baroniella camptocarpoides Costantin et Gallaud ●☆
48702 Baseonema gregorii Schltr. et Rendle;基丝萝藦■☆
48703 Baseonema lineare Choux = Baroniella linearis (Choux) Bullock ●☆
48704 Baseonema multiflorum Choux = Baroniella multiflora (Choux) Bullock ●☆
48705 Bashania P. C. Keng et T. P. Yi = Arundinaria Michx. ●
48706 Bashania P. C. Keng et T. P. Yi (1982);巴山木竹属(冷箭竹属);Bashanbamboo, Bashania ●★
48707 Bashania auctiaurita T. P. Yi;具耳巴山木竹(尖耳巴山木竹);Auriculate Bashania ●
48708 Bashania auctiaurita T. P. Yi = Indocalamus longiauritus Hand.-Mazz. ●
48709 Bashania faberi (Rendle) T. P. Yi = Arundinaria faberi Rendle ●
48710 Bashania fangiana (A. Camus) P. C. Keng et T. H. Wen = Arundinaria faberi Rendle ●
48711 Bashania fargesii (E. G. Camus) P. C. Keng et T. P. Yi;巴山木竹(法氏箬竹,秦岭箬竹,四川箬竹,镇巴木竹);Bashanbamboo, Farges Bashania, Farges Cane, Farges Canebrake ●
48712 Bashania fargesii (E. G. Camus) P. C. Keng et T. P. Yi = Arundinaria fargesii (E. G. Camus) P. C. Keng et T. P. Yi ●
48713 Bashania qingchengshanensis P. C. Keng et T. P. Yi = Arundinaria qingchengshanensis (P. C. Keng et T. P. Yi) D. Z. Li ●
48714 Bashania spanostachya T. P. Yi = Arundinaria spanostachya (T. P. Yi) D. Z. Li ●◇
48715 Bashania victorialis (P. C. Keng) T. P. Yi = Indocalamus victorialis P. C. Keng ●
48716 Basicarpus Tiegh. = Antidaphne Poepp. et Endl. ●☆
48717 Basicarpus Tiegh. = Loranthus Jacq. (保留属名)●
48718 Basigyne J. J. Sm. (1917);基蕊兰属■☆
48719 Basigyne J. J. Sm. = Dendrochilum Blume ■
48720 Basigyne muriculata J. J. Sm.;基蕊兰■☆
48721 Basilaea Juss. ex Lam. (废弃属名) = Eucomis L'Hér. (保留属名)■☆
48722 Basilaea coronata Lam. = Eucomis autumnalis (Mill.) Chitt. ■☆
48723 Basileophyta F. Muell. = Fieldia A. Cunn. ●☆
48724 Basilicum Moench(1802);小冠薰属;Basilicum ■
48725 Basilicum multiflorum (Benth.) Kuntze = Tetradenia riparia (Hochst.) Codd ●☆
48726 Basilicum myriostachyum (Benth.) Kuntze = Tetradenia riparia

(Hochst.) Codd ●☆
48727 Basilicum polystachyon (L.) Moench;小冠薰(假零陵香);Polystachyus Basilicum ■
48728 Basilicum polystachyon (L.) Moench var. flaccidum Briq. = Basilicum polystachyon (L.) Moench ■
48729 Basilicum polystachyon (L.) Moench var. stereocladum Briq. = Basilicum polystachyon (L.) Moench ■
48730 Basilicum riparium (Hochst.) Kuntze = Tetradenia riparia (Hochst.) Codd ●☆
48731 Basilima Raf. = Sorbaria (Ser. ex DC.) A. Braun(保留属名)●
48732 Basillaea R. Hedw. = Eucomis L'Hér.(保留属名)■☆
48733 Basiloxylon K. Schum. = Pterygota Schott et Endl. ●
48734 Basiphyllaea Schltr. (1921);基叶兰属■☆
48735 Basiphyllaea angustifolia Schltr. = Basiphyllaea corallicola (Small) Ames ■☆
48736 Basiphyllaea corallicola (Small) Ames;卡特基叶兰;Carter's Orchid ■☆
48737 Basisperma C. T. White(1942);基子金娘属●☆
48738 Basisperma lanceolata C. T. White;基子金娘●☆
48739 Basistelma Bartlett(1909);基冠萝藦属●☆
48740 Basistelma angustifolium Bartlett;基冠萝藦●☆
48741 Basistelma mexicanum (Brandegee) Bartlett;墨西哥基冠萝藦●☆
48742 Basistelma mexicanum Bartlett = Basistelma mexicanum (Brandegee) Bartlett ●☆
48743 Basistemon Turcz. (1863);基蕊玄参属●☆
48744 Basistemon bogotensis Turcz.;基蕊玄参●☆
48745 Basistemon intermedius Edwin;全缘基蕊玄参●☆
48746 Basistemon pulchellus (S. Moore) Barringer;美丽基蕊玄参●☆
48747 Baskervilla Lindl. (1840);秘鲁巴氏兰属■☆
48748 Baskervillea Lindl. (1840);巴斯克兰属■☆
48749 Basonca Raf. = Rogeria J. Gay ex Delile ■☆
48750 Bassecoia B. L. Burtt(1999);拟蓝盆花属■
48751 Basselinia Vieill. (1873);彩颈椰属(巴舍椰子属,巴西林椆属,喀里多尼亚椰属,新喀里多尼亚棕属)●☆
48752 Basselinia gracilis (Brongn. et Gris) Vieill.;彩颈椰●☆
48753 Bassia All. (1766);雾冰藜属(巴锡藜属,刺果藜属,肯诺藜属,雾冰草属);Bassia,Kochia,Summer-cypress ●■
48754 Bassia J. König = Madhuca Buch. -Ham. ex J. F. Gmel. ●
48755 Bassia J. König ex L. = Illipe J. König ex Gras ●
48756 Bassia J. König ex L. = Madhuca Buch. -Ham. ex J. F. Gmel. ●
48757 Bassia L. = Madhuca Buch. -Ham. ex J. F. Gmel. ●
48758 Bassia americana (S. Watson) A. J. Scott = Kochia americana S. Watson ■☆
48759 Bassia arabica (Boiss.) Maire et Weiller;阿拉伯雾冰藜■☆
48760 Bassia butyracea Roxb. = Diploknema butyraceum (Roxb.) H. J. Lam. ●
48761 Bassia californica (S. Watson) A. J. Scott = Kochia californica S. Watson ■☆
48762 Bassia dasyphylla (Fisch. et C. A. Mey.) Kuntze;雾冰藜(巴西藜,巴锡藜,肯诺藜,毛脊梁,毛叶费利菊,水蓬蒿,五星蒿,雾冰草,星状刺果藜);Divaricate Bassia ■
48763 Bassia diffusa (Thunb.) Kuntze;松散雾冰藜■☆
48764 Bassia dinteri (Botsch.) A. J. Scott;丁特雾冰藜■☆
48765 Bassia djave Laness. = Baillonella toxisperma Pierre ●☆
48766 Bassia eriophora (Schrad.) Asch.;毛梗雾冰藜■☆
48767 Bassia hirsuta (L.) Asch.;毛雾冰藜;Hairy Bassia, Hairy Seablite,Hairy Smotherweed,Summer-cypress ■☆
48768 Bassia hirsuta Asch. ex Nyman = Bassia hirsuta (L.) Asch. ■☆
48769 Bassia hyssopifolia (Pall.) Kuntze;钩刺雾冰藜(钩状刺果藜);Fivehook Bassia, Five-hook Bassia, Five-hooked Bassia, Fivehorn Smotherweed,Smotherweed ■
48770 Bassia indica (Wight) A. J. Scott;印度雾冰藜■☆
48771 Bassia iranica (Litv. ex Bornmüller) Bornmüller = Kochia stellaris Moq. ■
48772 Bassia iranica Bornm. = Kochia iranica Litv. ex Bornm. ■
48773 Bassia latifolia Roxb. = Madhuca indica J. F. Gmel. ●☆
48774 Bassia longifolia König = Madhuca longifolia (König) J. F. Macbr. ●☆
48775 Bassia longifolia L. = Madhuca longifolia (König) J. F. Macbr. ●☆
48776 Bassia longifolia W. Fitzg.;长叶雾冰藜●☆
48777 Bassia muricata (L.) Asch.;粗糙雾冰藜■☆
48778 Bassia parkii G. Don = Vitellaria paradoxa C. F. Gaertn. ●
48779 Bassia pasquieri Lecomte = Madhuca pasquieri (Dubard) H. J. Lam. ●◇
48780 Bassia prostrata (L.) A. J. Scott = Kochia prostrata (L.) Schrad. ●
48781 Bassia prostrata (L.) A. J. Scott = Salsola prostrata L. ●
48782 Bassia salsoloides (Fenzl) A. J. Scott;猪毛菜雾冰藜●☆
48783 Bassia scoparia (L.) A. J. Scott = Kochia scoparia (L.) Schrad. ■
48784 Bassia scoparia (L.) A. J. Scott var. trichophylla (Voss) S. L. Welsh = Kochia scoparia (L.) Schrad. var. trichophilla Schinz et Thell. ■
48785 Bassia scoparia (L.) Voss;帚状雾冰藜;Summer Cypress ■☆
48786 Bassia scoparia (L.) Voss 'Trichophylla';毛叶雾冰藜;Burning Bush ■☆
48787 Bassia scoparia (L.) Voss = Kochia scoparia (L.) Schrad. ■
48788 Bassia scoparia (L.) Voss subsp. densiflora (Turcz. ex Aellen) Cirujano et Velayos;密花雾冰藜■☆
48789 Bassia sedoides (Pall.) Asch. = Bassia sedoides (Schrad.) Asch. ■
48790 Bassia sedoides (Schrad.) Asch.;肉叶雾冰藜(短角雾冰藜);Fleshyleaf Bassia ■
48791 Bassia tomentosa (Lowe) Maire et Weiller;绒毛雾冰藜■☆
48792 Bassovia Aubl. = Solanum L. ●■
48793 Bastardia Kunth(1822);巴氏葵属●■☆
48794 Bastardia angulata Guillaumin et Perr. = Abutilon angulatum (Guillaumin et Perr.) Mast. ■☆
48795 Bastardia parvifolia Kunth;巴氏葵■☆
48796 Bastardiastrum (Rose) D. M. Bates(1978);小巴氏葵属●☆
48797 Bastardiastrum D. M. Bates = Bastardiastrum (Rose) D. M. Bates ●☆
48798 Bastardiastrum hirsutiflorum (C. Presl) D. M. Bates;小巴氏葵●☆
48799 Bastardiopsis (K. Schum.) Hassl. (1910);韧葵木属●☆
48800 Bastardiopsis Hassl. = Bastardiopsis (K. Schum.) Hassl. ●☆
48801 Bastardiopsis densiflora (Hook. et Arn.) Hassl.;丛花韧葵木●☆
48802 Bastera J. F. Gmel. = Basteria Houtt. ●■☆
48803 Basteria Houtt. = Berkheya Ehrh.(保留属名)●■☆
48804 Basteria Mill.(废弃属名) = Calycanthus L.(保留属名)●
48805 Basteria aculeata Houtt. = Cullumia aculeata (Houtt.) Rössler ■☆
48806 Bastia Steud. = Buphthalmum L. ■
48807 Bastia Steud. = Bustia Adans. ■
48808 Basutica E. Phillips = Gnidia L. ●☆
48809 Basutica aberrans (C. H. Wright) E. Phillips = Gnidia aberrans C. H. Wright ●☆

48810 Basutica propinqua Hilliard = Gnidia propinqua (Hilliard) B. Peterson ●☆
48811 Bataceae Mart. ex Meisn. = Bataceae Mart. ex Perleb(保留科名)●☆
48812 Bataceae Mart. ex Perleb(1838)(保留科名);肉穗果科(白楔科,藜木科)●☆
48813 Batania Hatus. = Pycnarrhena Miers ex Hook. f. et Thomson ●
48814 Batanthes Raf. = Gilia Ruiz et Pav. ●■☆
48815 Batanthes Raf. = Ipomopsis Michx. ■☆
48816 Bataprine Nieuwl. = Galium L. ●■
48817 Batatas Choisy = Ipomoea L. (保留属名)●■
48818 Batatas acetosifolia (Vahl) Choisy = Ipomoea imperati (Vahl) Griseb. ■
48819 Batatas crassicaulis Benth. = Ipomoea carnea Jacq. subsp. fistulosa (Mart. ex Choisy) D. F. Austin ●
48820 Batatas crassicaulis Benth. = Ipomoea fistulosa Mart. ex Choisy ●
48821 Batatas edulis (Thunb. ex Murray) Choisy = Ipomoea batatas (L.) Lam. ■
48822 Batatas edulis (Thunb.) Choisy = Ipomoea batatas (L.) Lam. ■
48823 Batatas littoralis (L.) Choisy = Ipomoea imperati (Vahl) Griseb. ■
48824 Batatas littoralis Choisy = Ipomoea imperati (Vahl) Griseb. ■
48825 Batatas paniculata (L.) Choisy = Ipomoea mauritiana Jacq. ■
48826 Batatas setosa (Ker Gawl.) Lindl. = Ipomoea setosa Ker Gawl. ■
48827 Batatas triloba (L.) Choisy = Ipomoea triloba L. ■
48828 Batemania Endl. = Batemannia Lindl. ■☆
48829 Batemannia Lindl. (1834);巴氏兰属(巴特兰属)■☆
48830 Batemannia armillata Rchb. f.;具环巴氏兰■☆
48831 Baterium Miers = Haematocarpus Miers ●☆
48832 Batesanthus N. E. Br. (1896);西非萝藦属●☆
48833 Batesanthus intrusus S. Moore = Batesanthus purpureus N. E. Br. ●☆
48834 Batesanthus mildbraedii Schltr. = Batesanthus purpureus N. E. Br. ●☆
48835 Batesanthus parviflorus C. Norman;小叶西非萝藦●☆
48836 Batesanthus purpureus N. E. Br.;紫西非萝藦●☆
48837 Batesanthus talbotii S. Moore = Batesanthus purpureus N. E. Br. ●☆
48838 Batesanthus talbotii S. Moore var. grandiflora ? = Batesanthus purpureus N. E. Br. ●☆
48839 Batesanthus talbotii S. Moore var. parviflora ? = Batesanthus purpureus N. E. Br. ●☆
48840 Batesia Spruce = Batesia Spruce ex Benth. ●☆
48841 Batesia Spruce ex Benth. (1865);巴北苏木属(贝茨苏木属)●☆
48842 Batesia floribunda Spruce ex Benth.;多花巴北苏木;Capurana-da-terra-firme ●☆
48843 Batesimalva Fryxell(1975);贝茨锦葵属●☆
48844 Batesimalva pulchella Fryxell;美丽贝茨锦葵●☆
48845 Batesimalva stipulata Fryxell;贝茨锦葵●☆
48846 Batesimalva violacea (Rose) Fryxell;堇色贝茨锦葵●☆
48847 Bathiaea Drake(1902);红花翼豆属●☆
48848 Bathiaea rubriflora Drake;红花翼豆●☆
48849 Bathiea Schltr. = Neobathiea Schltr. ■☆
48850 Bathiea perrieri (Schltr.) Schltr. = Neobathiea perrieri Schltr. ■☆
48851 Bathieaea Willis = Bathiaea Drake ●☆
48852 Bathieaea Willis = Rhododendron L. ●
48853 Bathiorchis Bosser et P. J. Cribb = Goodyera R. Br. ■
48854 Bathiorchis rosea (H. Perrier) Bosser et P. J. Cribb = Goodyera rosea (H. Perrier) Ormerod ■☆
48855 Bathiorhamnus Capuron(1966);巴斯鼠李属●☆
48856 Bathiorhamnus cryptophorus Capuron;隐梗巴斯鼠李●☆
48857 Bathiorhamnus louvelii (H. Perrier) Capuron;巴斯鼠李●☆
48858 Bathratherum Nees = Arthraxon P. Beauv. ■
48859 Bathratherum lanceolatum Nees = Arthraxon lanceolatus (Roxb.) Hochst. ■
48860 Bathratherum lancifolium W. Watson = Arthraxon lancifolius (Trin.) Hochst. ■
48861 Bathratherum micans Nees = Arthraxon micans (Nees) Hochst. ■
48862 Bathratherum molle Nees = Arthraxon lancifolius (Trin.) Hochst. ■
48863 Bathratherum serrulatum Hochst. ex Steud. = Arthraxon lanceolatus (Roxb.) Hochst. ■
48864 Bathratherum serrulatum Hochst. ex Steud. = Arthraxon prionodes (Steud.) Dandy ■
48865 Bathya Ravenna(2003);智利石蒜属■☆
48866 Bathysa C. Presl(1845);滨茜属■☆
48867 Bathysa australis K. Schum.;澳洲滨茜■☆
48868 Bathysa meridionalis L. B. Sm. et Downs;南滨茜■☆
48869 Bathysa multiflora L. O. Williams;多花滨茜■☆
48870 Bathysa pittieri (Standl.) Steyerm.;皮氏滨茜■☆
48871 Bathysograya Kuntze = Badusa A. Gray ●☆
48872 Batidacea (Dumort.) Greene = Rubus L. ●■
48873 Batidaceae Mart. ex Meisn. = Bataceae Mart. ex Perleb(保留科名)●☆
48874 Batidaea (Dumort.) Greene = Rubus L. ●■
48875 Batidaea Greene = Rubus L. ●■
48876 Batidaea Greene(1906);北美悬钩子属●☆
48877 Batidophaca Rydb. = Astragalus L. ●■
48878 Batindum Raf. = Oftia Adans. ●■☆
48879 Batis L. = Batis P. Browne ●☆
48880 Batis P. Browne(1756);肉穗果属(白楔属,藜木属);Saltwort ●☆
48881 Batis americana L.;北美肉穗果●☆
48882 Batis argillicola P. Royen;白土肉穗果●☆
48883 Batis fruticosa Roxb. = Cudrania fruticosa (Roxb.) Wight ex Kurz ●
48884 Batis fruticosa Roxb. = Maclura fruticosa (Roxb.) Corner ●
48885 Batis maritima L.;肉穗果(白楔);Beachwort,Saltwort ●☆
48886 Batis vermiculata Hook. = Sarcobatus vermiculatus (Hook.) Torr. ●☆
48887 Batocarpus H. Karst. (1863);荔枝桑属●☆
48888 Batocarpus amazonicus (Ducke) Fosberg;亚马逊荔枝桑●☆
48889 Batocarpus orinocensis H. Karst.;荔枝桑●☆
48890 Batocydia Mart. ex Britton et P. Wilson = Doxantha Miers ●☆
48891 Batocydia Mart. ex DC. = Bignonia L. (保留属名)●
48892 Batodendron Nutt. (1842);巴特杜鹃属●☆
48893 Batodendron Nutt. = Vaccinium L. ●
48894 Batodendron arboreum Nutt.;巴特杜鹃●☆
48895 Batopedina Verdc. (1953);平原茜属■☆
48896 Batopedina linearifolia (Bremek.) Verdc.;线叶平原茜■☆
48897 Batopedina linearifolia (Bremek.) Verdc. var. glabra E. M. Petit = Batopedina linearifolia (Bremek.) Verdc. ■☆
48898 Batopedina pulvinellata Robbr.;平原茜■☆
48899 Batopedina pulvinellata Robbr. subsp. glabrifolia Robbr.;光叶平原茜■☆

48900 Batopedina tenuis (A. Chev. ex Hutch. et Dalziel) Verdc.;纤细平原茜■☆
48901 Batopilasia G. L. Nesom et Noyes(2000);莲菀属■☆
48902 Batopilasia byei (S. D. Sundb. et G. L. Nesom) G. L. Nesom et Noyes;莲菀■☆
48903 Batrachium (DC.) Gray = Ranunculus L. ■
48904 Batrachium (DC.) Gray (1821);水毛茛属(梅花藻属);Batrachium ■
48905 Batrachium Gray = Batrachium (DC.) Gray ■
48906 Batrachium aquatile (L.) Dumort. = Ranunculus aquatilis L. ■☆
48907 Batrachium bungei (Steud.) L. Liou;水毛茛(扇叶水毛茛);Bunge Batrachium, Bunge Buttercup ■
48908 Batrachium bungei (Steud.) L. Liou var. flavidum (Hand.-Mazz.) L. Liou;黄花水毛茛;Yellow Bunge Batrachium, Yellowflower Batrachium ■
48909 Batrachium bungei (Steud.) L. Liou var. micranthum W. T. Wang;小花水毛茛;Smallflower Bunge Batrachium ■
48910 Batrachium carinatum Schur;龙骨水毛茛;Carinate Batrachium ■☆
48911 Batrachium circinatum (Sibth.) Spach subsp. subrigidum (W. B. Drew) Á. Löve et D. Löve = Ranunculus aquatilis L. var. diffusus With. ■☆
48912 Batrachium circinatum (Sibth.) Spach. = Batrachium foeniculaceum (Gilib.) Krecz. ■
48913 Batrachium dichotomum Schmalh. ex Trautv.;二枝水毛茛(二叉梅花藻);Dichotomous Batrachium ■☆
48914 Batrachium divaricatum (Schrank) Schur;歧裂水毛茛(广布梅花藻,散毛茛);Divaricate Batrachium ■
48915 Batrachium eradicatum (Laest.) Fr.;小水毛茛(浅生梅花藻,线状毛茛);Small Batrachium ■
48916 Batrachium flaccidum (Pers.) Rupr. = Ranunculus aquatilis L. var. diffusus With. ■☆
48917 Batrachium flavidum Hand.-Mazz. = Batrachium bungei (Steud.) L. Liou var. flavidum (Hand.-Mazz.) L. Liou ■
48918 Batrachium fluitans Wimm.;漂浮水毛茛;Eel-beds, Long-leaved Water Crowfoot, Naked Boys, River Crowfoot, River Water-crowfoot, Water Crowfoot ■☆
48919 Batrachium foeniculaceum (Gilib.) Krecz.;硬叶水毛茛(茴香梅花藻,卷叶毛茛);Fan-leaved Water-crowfoot, Rigidleaf Batrachium, Rigid-leaved Crowfoot ■
48920 Batrachium giliberti Krecz.;吉氏水毛茛(吉氏梅花藻)■☆
48921 Batrachium jingpoense G. Y. Zhang, Chen Wang et X. J. Liu = Batrachium trichophyllum (Chaix) Bosch var. jingpoense (G. Y. Chang, Chen Wang et X. J. Liu) W. T. Wang ■
48922 Batrachium kauffmannii (Clerc) V. I. Krecz.;长叶水毛茛(考氏梅花藻);Longleaf Batrachium ■
48923 Batrachium langei F. W. Schultz ex Nyman;郎氏水毛茛(郎氏梅花藻)■☆
48924 Batrachium lateriflorum (DC.) Ovcz.;侧花毛茛■☆
48925 Batrachium longirostre (Godr.) F. W. Schultz = Ranunculus aquatilis L. var. diffusus With. ■☆
48926 Batrachium marinum Fr.;海生水毛茛(海生梅花藻);Marine Batrachium ■☆
48927 Batrachium mongolicum (Krylov) V. I. Krecz.;蒙古水毛茛;Mongolian Batrachium ■☆
48928 Batrachium pachycaulon Nevski;厚茎水毛茛;Thick Stem Batrachium ■☆
48929 Batrachium pekinense L. Liou;北京水毛茛;Beijing Batrachium ■
48930 Batrachium porteri (Britton) Britton = Ranunculus aquatilis L. var. diffusus With. ■☆
48931 Batrachium rionii (Lagger) Nyman;钻托水毛茛(雷氏梅花藻)■
48932 Batrachium trichophyllum (Chaix ex Vill.) Bosch = Batrachium trichophyllum (Chaix) Bosch ■
48933 Batrachium trichophyllum (Chaix) Bosch;毛柄水毛茛(白水毛茛,毛叶毛茛,梅花藻);Dark Hair Crowfoot, Hairyleaf Batrachium, Thread-leaved Crowfoot, Thread-leaved Water-crowfoot, White Water-buttercup ■
48934 Batrachium trichophyllum (Chaix) Bosch = Batrachium bungei (Steud.) L. Liou ■
48935 Batrachium trichophyllum (Chaix) Bosch = Ranunculus aquatilis L. var. diffusus With. ■☆
48936 Batrachium trichophyllum (Chaix) Bosch = Ranunculus trichophyllus Chaix ex Vill. ■
48937 Batrachium trichophyllum (Chaix) Bosch subsp. lutulentum (Perrier et Songeon) Janch. ex V. V. Petrovsky = Ranunculus aquatilis L. var. diffusus With. ■☆
48938 Batrachium trichophyllum (Chaix) Bosch subsp. roinii (Lagger) C. D. K. Cook = Batrachium rionii (Lagger) Nyman ■
48939 Batrachium trichophyllum (Chaix) Bosch var. hirtellum L. Liou;多毛水毛茛;Hirtelous Hairyleaf Batrachium ■
48940 Batrachium trichophyllum (Chaix) Bosch var. jingpoense (G. Y. Chang, Chen Wang et X. J. Liu) W. T. Wang;镜泊水毛茛;Jingbohu Batrachium ■
48941 Batrachium trichophyllum (Chaix) Bosch var. paucistamineum (Tausch) Hand.-Mazz. = Batrachium eradicatum (Laest.) Fr. ■
48942 Batrachium trichophyllum (Chaix) Bosch var. paucistamineum (Tausch) Hand.-Mazz. f. terrestre (Gren. et Godr.) Hand.-Mazz. = Batrachium eradicatum (Laest.) Fr. ■
48943 Batrachium trichophyllum (Chaix) F. W. Schultz = Ranunculus aquatilis L. var. diffusus With. ■☆
48944 Batrachium triphyllum (Wallr.) Dumort.;三叶水毛茛■☆
48945 Batratherum Nees = Arthraxon P. Beauv. ■
48946 Batratherum echinatum Nees = Arthraxon echinatus Hochst. ■
48947 Batratherum lancifolium (Trin.) W. Watson = Arthraxon lanceolatus (Roxb.) Hochst. ■
48948 Batratherum micans Nees = Arthraxon hispidus (Thunb.) Makino ■
48949 Batratherum micans Nees = Arthraxon micans (Nees) Hochst. ■
48950 Batratherum molle Nees = Arthraxon lanceolatus (Roxb.) Hochst. ■
48951 Batratherum submuticum (Nees ex Steud.) W. Watson. = Arthraxon submuticus (Nees ex Steud.) Hochst. ■
48952 Batschia J. F. Gmel. = Lithospermum L. ■
48953 Batschia Moench = Eupatorium L. ●■
48954 Batschia Mutis ex Thunb. = Abuta Aubl. ●☆
48955 Batschia Mutis ex Thunb. = Trichoa Pars. ●☆
48956 Batschia Vahl = Humboldtia Vahl(保留属名)■☆
48957 Batschia canescens Michx. = Lithospermum canescens (Michx.) Lehm. ■☆
48958 Batschia gmelinii Michx. = Lithospermum caroliniense (Walter ex J. F. Gmel.) MacMill. subsp. croceum (Fernald) Cusick ■☆
48959 Batschia linearifolia (Goldie) Small = Lithospermum incisum Lehm. ■☆
48960 Battandiera Maire = Ornithogalum L. ■
48961 Battata Hill = Solanum L. ●■
48962 Bauchea E. Fourn. = Sporobolus R. Br. ■

48963 Bauchea E. Fourn. ex Benth. = Sporobolus R. Br. ■
48964 Baucis Phil. = Brachyclados Gillies ex D. Don ●☆
48965 Baudinia Lesch. ex DC. (1828) = Calothamnus Labill. ●☆
48966 Baudinia Lesch. ex DC. (1839) = Scaevola L.(保留属名)●■
48967 Baudouinia Baill. (1866);鲍德豆属●☆
48968 Baudouinia capuronii Du Puy et R. Rabev.;凯普伦鲍德豆●☆
48969 Baudouinia fluggeiformis Baill.;鲍德豆●☆
48970 Baudouinia louvelii R. Vig.;卢韦尔鲍德豆●☆
48971 Baudouinia orientalis R. Vig.;东方鲍德豆●☆
48972 Baudouinia rouxevillei H. Perrier;鲁氏鲍德豆●☆
48973 Baudouinia suarezensis Dumaz-le-Grand = Baudouinia fluggeiformis Baill. ●☆
48974 Bauera Banks = Bauera Banks ex Andréws ●☆
48975 Bauera Banks ex Andréws(1801);鲍氏木属(鲍耶尔属,常绿棱枝树属)●☆
48976 Bauera rubiifolia Salisb. = Bauera rubioides Andréws ●☆
48977 Bauera rubioides Andréws;鲍氏木(狗玫瑰);Dog Rose, River Rose ●☆
48978 Bauera sessiliflora F. Muell.;无梗鲍氏木(无梗花鲍耶尔);Grampians Bauera ●☆
48979 Baueraceae Lindl. (1830);鲍氏木科(常绿棱枝树科,常绿枝科,角瓣木科)●☆
48980 Baueraceae Lindl. = Cunoniaceae R. Br.(保留科名)●☆
48981 Bauerella Borzi = Acronychia J. R. Forst. et G. Forst.(保留属名)●
48982 Bauerella Borzi(1897);小鲍氏木属●☆
48983 Bauerella Schindl. = Baueropsis Hutch. ●☆
48984 Bauerella Schindl. = Psoralea L. ●■
48985 Bauerella australiana Borzi;小鲍氏木●☆
48986 Bauerella tomentosa Schindl. = Baueropsis tomentosa (Schindl.) Hutch. ●☆
48987 Baueropsis Hutch. (1964);拟鲍氏木属(绿棱枝树属)●☆
48988 Baueropsis Hutch. = Cullen Medik. ●■
48989 Baueropsis tomentosa (Schindl.) Hutch.;拟鲍氏木●☆
48990 Bauhinia L. (1753);羊蹄甲属;Bauhinia, Butter Tree, Camel's Foot, Mountain Ebony, Orchid Tree ●
48991 Bauhinia aculeata L.;针形羊蹄甲●☆
48992 Bauhinia aculeata L. subsp. grandiflora (Juss.) Wunderlin;大花马蹄豆●☆
48993 Bauhinia acuminata L.;白花羊蹄甲(马蹄豆,木椀树,木碗树,椀树);Acuminate Bauhinia, Orchid Bush, Snowy Bauhinia ●
48994 Bauhinia acuminata L. var. hirsuta (Weinm.) Craib = Bauhinia hirsuta Weinm. ●
48995 Bauhinia adansoniana Guillaumin et Perr. = Bauhinia rufescens Lam. ●☆
48996 Bauhinia alba Buch. -Ham. ex Wall. = Bauhinia variegata L. ●
48997 Bauhinia altefissa H. Lév. = Bauhinia brachycarpa Wall. ex Benth. ●
48998 Bauhinia altefissa H. Lév. = Bauhinia yunnanensis Franch. et Metcalf ●
48999 Bauhinia altifissa H. Lév. = Bauhinia brachycarpa Wall. ex Benth. ●
49000 Bauhinia altifissa H. Lév. = Bauhinia yunnanensis Franch. et Metcalf ●
49001 Bauhinia anguina Roxb. = Bauhinia scandens L. ●
49002 Bauhinia anguina Roxb. var. horsfieldii (Watt ex Prain) de Wit. = Bauhinia scandens L. var. horsfieldii (Watt ex Prain) K. Larsen et S. S. Larsen ●
49003 Bauhinia anguina Roxb. var. horsfieldii Watt ex Prain = Bauhinia scandens L. ●
49004 Bauhinia apertilobata Merr. et F. P. Metcalf;阔裂羊蹄甲(搭袋藤,阔裂叶羊蹄甲,亚那藤);Broadlobed Bauhinia, Broad-lobed Bauhinia ●
49005 Bauhinia argentea Chiov. = Tylosema argentea (Chiov.) Brenan ■☆
49006 Bauhinia aurantiaca Bojer;金黄羊蹄甲●☆
49007 Bauhinia aurea H. Lév.;红绒毛羊蹄甲(红毛藤,黄麻藤,火索麻,火索藤,金叶羊蹄甲,牛蹄藤,羊蹄藤);Firrope Bauhinia, Golden-yellow Bauhinia ●
49008 Bauhinia austrosinensis Ts. Tang et F. T. Wang = Bauhinia kerrii Gagnep. ●
49009 Bauhinia austrosinensis Ts. Tang et F. T. Wang = Bauhinia ornata Kurz var. austrosinensis (Gagnep.) K. Larsen et S. S. Larsen ●
49010 Bauhinia austrosinensis Ts. Tang et F. T. Wang = Bauhinia ornata Kurz var. kerrii (Gagnep.) K. Larsen et S. S. Larsen ●
49011 Bauhinia bainesii Schinz = Tylosema esculentum (Burch.) A. Schreib. ●☆
49012 Bauhinia bakeriana S. S. Larsen = Bauhinia ornata Kurz var. kerrii (Gagnep.) K. Larsen et S. S. Larsen ●
49013 Bauhinia balansae Gagnep. = Bauhinia ornata Kurz var. balansae (Gagnep.) K. Larsen et S. S. Larsen ●
49014 Bauhinia baviensis Drake = Bauhinia viridescens Desv. ●
49015 Bauhinia benzoin Kotschy = Piliostigma reticulatum (DC.) Hochst. ■☆
49016 Bauhinia bequaertii De Wild. = Tylosema fassoglensis (Schweinf.) Torre et Hillc. ●☆
49017 Bauhinia blakeana Dunn;红花羊蹄甲(羊蹄甲);Blake Bauhinia, Hong Kong Orchid, Hong Kong Orchidtree, Hong Kong Orchid-tree, Red Bauhinia, Red Flowered Camel's Foot, Red-flowered Bauhinia ●
49018 Bauhinia blakeana Dunn = Bauhinia variegata L. ●
49019 Bauhinia bohniana H. Y. Chen;丽江羊蹄甲;Lijiang Bauhinia ●
49020 Bauhinia bonatiana Pamp. = Bauhinia brachycarpa Wall. ex Benth. ●
49021 Bauhinia bowkeri Harv.;鲍克羊蹄甲●☆
49022 Bauhinia brachycarpa Wall. ex Benth.;鞍叶羊蹄甲(大飞扬,短果羊蹄甲,短荚羊蹄甲,蝴蝶风,蝴蝶凤,马鞍羊蹄甲,马鞍叶,马鞍叶羊蹄甲,桑角子,羊蹄藤,夜关门,夜合叶);Faber Bauhinia, Shortfruit Bauhinia ●
49023 Bauhinia brachycarpa Wall. ex Benth. var. cavaleriei (H. Lév.) T. C. Chen;刀果鞍叶羊蹄甲(刀果马鞍叶)●
49024 Bauhinia brachycarpa Wall. ex Benth. var. cavaleriei (H. Lév.) T. C. Chen = Bauhinia brachycarpa Wall. ex Benth. ●
49025 Bauhinia brachycarpa Wall. ex Benth. var. densiflora (Franch.) K. Larsen et S. S. Larsen;毛鞍叶羊蹄甲(密花羊蹄甲);Dense-flower Shortfruit Bauhinia ●
49026 Bauhinia brachycarpa Wall. ex Benth. var. densiflora (Franch.) K. Larsen et S. S. Larsen = Bauhinia brachycarpa Wall. ex Benth. ●
49027 Bauhinia brachycarpa Wall. ex Benth. var. microphylla (Oliv. ex Craib) K. Larsen et S. S. Larsen;小鞍叶羊蹄甲(小马鞍叶羊蹄甲,小叶马鞍叶);Small-leaf Shortfruit Bauhinia ●
49028 Bauhinia brachycarpa Wall. ex Benth. var. microphylla (Oliv. ex Craib) K. Larsen et S. S. Larsen = Bauhinia brachycarpa Wall. ex Benth. ●

49029 Bauhinia brevicalyx Du Puy et R. Rabev.;短萼羊蹄甲●☆

49030 Bauhinia bryoniflora Franch. = Bauhinia brachycarpa Wall. ex Benth. ●

49031 Bauhinia burkeana (Benth.) Harv. = Tylosema esculentum (Burch.) A. Schreib. ●☆

49032 Bauhinia calciphila D. X. Zhang et T. C. Chen;石山羊蹄甲●

49033 Bauhinia candida Roxb. = Bauhinia variegata L. 'Candida'●

49034 Bauhinia candida Roxb. = Bauhinia variegata L. var. candida (Roxb.) Voigt ●

49035 Bauhinia candida Roxb. = Bauhinia variegata L. ●

49036 Bauhinia capuronii Du Puy et R. Rabev.;凯普伦羊蹄甲●☆

49037 Bauhinia carcinophylla Merr.;蟹钳叶羊蹄甲●

49038 Bauhinia carronii F. Muell.;澳北羊蹄甲●☆

49039 Bauhinia caterviflora H. Y. Chen = Bauhinia glauca (Wall. ex Benth.) Benth. subsp. caterviflora (H. Y. Chen) T. C. Chen ●

49040 Bauhinia caterviflora H. Y. Chen = Bauhinia glauca (Wall. ex Benth.) Benth. subsp. tenuiflora (Watt ex C. B. Clarke) K. Larsen et S. S. Larsen ●

49041 Bauhinia caterviflora H. Y. Chen = Bauhinia tenuiflora Watt ex C. B. Clarke ●

49042 Bauhinia cavaleriei H. Lév. = Bauhinia brachycarpa Wall. ex Benth. ●

49043 Bauhinia cavaleriei H. Lév. = Bauhinia brachycarpa Wall. ex Benth. var. cavaleriei (H. Lév.) T. C. Chen ●

49044 Bauhinia cercidifolia D. X. Zhang;紫荆叶羊蹄甲●

49045 Bauhinia chalcophylla H. Y. Chen;多花羊蹄甲;Manyflower Bauhinia, Multiflorous Bauhinia, Multiflower Bauhinia ●

49046 Bauhinia championii (Benth.) Benth.;龙须藤(百代藤,串鼻藤,搭袋藤,蝶藤,飞扬藤,蛤叶,钩藤,过岗龙,过岗圆龙,黑皮藤,黄开口,九龙藤,九牛燥,菊花木,罗亚多藤,马脚藤,梅花入骨丹,木腰子,千打捶,双木鳖,双木蟹,田螺虎树,乌郎藤,乌皮藤,乌藤,乌鸦圪,五花血藤,五里藤,燕子尾,羊蹄叉,羊蹄风,羊蹄树,羊蹄藤,夜合草,圆过岗龙,圆龙,猪蹄叉,子燕藤);Champion Bauhinia, Chrysanthemum Wood ●

49047 Bauhinia championii (Benth.) Benth. var. acutifolia H. Y. Chen = Bauhinia championii (Benth.) Benth. ●

49048 Bauhinia championii (Benth.) Benth. var. apertilobata (Merr. et F. P. Metcalf) Hiroe = Bauhinia apertilobata Merr. et F. P. Metcalf ●

49049 Bauhinia championii (Benth.) Benth. var. yingtakensis (Merr. et F. P. Metcalf) T. C. Chen;英德羊蹄甲;Yingde Champion Bauhinia ●

49050 Bauhinia championii (Benth.) Benth. var. yingtakensis (Merr. et F. P. Metcalf) T. C. Chen = Bauhinia championii (Benth.) Benth. ●

49051 Bauhinia cissoides Oliv. = Tylosema fassoglensis (Schweinf.) Torre et Hillc. ●☆

49052 Bauhinia claviflora H. Y. Chen = Bauhinia nervosa (Wall. ex Benth.) Baker ●

49053 Bauhinia coccinea (Lour.) DC.;绯红羊蹄甲;Brightred Bauhinia, Scarlet Bauhinia ●

49054 Bauhinia coccinea (Lour.) DC. subsp. tonkinensis (Gagnep.) K. Larsen et S. S. Larsen;越南绯红羊蹄甲(绯红羊蹄甲,越南红花羊蹄甲);Tonkin Bauhinia tonkinensis ●

49055 Bauhinia commersonii Decne. = Bauhinia madagascariensis Desv. ●☆

49056 Bauhinia comosa Craib;川滇羊蹄甲(石山羊蹄甲);Comose Bauhinia, Tufted Bauhinia ●

49057 Bauhinia corymbosa Roxb. = Bauhinia corymbosa Roxb. ex DC. ●

49058 Bauhinia corymbosa Roxb. ex DC.;首冠藤(飞插藤,深裂羊蹄甲,深裂叶羊蹄甲);Corymbose Bauhinia, Phanera ●

49059 Bauhinia corymbosa Roxb. ex DC. var. longipes Hosok.;长序首冠藤●

49060 Bauhinia corymbosa Roxb. var. longipes Hosok. = Bauhinia corymbosa Roxb. ex DC. var. longipes Hosok. ●

49061 Bauhinia cunninghamii (Benth.) Benth.;澳东羊蹄甲●☆

49062 Bauhinia curyantha H. Y. Chen = Bauhinia chalcophylla H. Y. Chen ●

49063 Bauhinia damiaoshanensis T. C. Chen;大苗山羊蹄甲(飞蛾藤);Damiaoshan Bauhinia ●

49064 Bauhinia delavayi Franch.;薄荚羊蹄甲(川滇羊蹄甲,德氏羊蹄甲,迪氏羊蹄甲,羊蹄甲);Delavay Bauhinia ●

49065 Bauhinia densiflora Franch. = Bauhinia brachycarpa Wall. ex Benth. ●

49066 Bauhinia densiflora Franch. = Bauhinia brachycarpa Wall. ex Benth. var. densiflora (Franch.) K. Larsen et S. S. Larsen ●

49067 Bauhinia didyma L.;孪叶羊蹄甲(二裂片羊蹄甲,飞机藤,牛耳麻);Didymousleaf Bauhinia, Geminate Bauhinia ●

49068 Bauhinia dioscoreifolia H. Y. Chen = Bauhinia khasiana Baker ●

49069 Bauhinia dioscoreifolia L.;薯叶藤;Dioscorea-leaved Bauhinia, Yamleaf Bauhinia ●

49070 Bauhinia diptera Collett et Hemsl. = Bauhinia yunnanensis Franch. et Metcalf ●

49071 Bauhinia divaricata L.;叉分羊蹄甲●☆

49072 Bauhinia eberhardtii Gagnep. = Bauhinia ornata Kurz var. kerrii (Gagnep.) K. Larsen et S. S. Larsen ●

49073 Bauhinia ellenbeckii Harms;埃伦羊蹄甲●☆

49074 Bauhinia enigmatica Prain = Bauhinia brachycarpa Wall. ex Benth. ●

49075 Bauhinia erythropoda Hayata;锈荚藤;Rustpod Bauhinia, Rusty-podded Bauhinia ●

49076 Bauhinia erythropoda Hayata var. guangxiensis D. X. Zhang et T. C. Chen;广西锈荚藤●

49077 Bauhinia esculenta Burch.;食用羊蹄甲;Gemsbuck Bean, Morama Bean, Tsin Bean ●☆

49078 Bauhinia esculenta Burch. = Tylosema esculentum (Burch.) A. Schreib. ●☆

49079 Bauhinia esquirolii Gagnep.;元江羊蹄甲;Esquirol Bauhinia, Yuanjiang Bauhinia ●

49080 Bauhinia euryantha H. Y. Chen = Bauhinia chalcophylla H. Y. Chen ●

49081 Bauhinia euryantha L. Chen = Bauhinia chalcophylla H. Y. Chen ●

49082 Bauhinia excisa Hemsl.;缺刻羊蹄甲●☆

49083 Bauhinia exellii Torre et Hillc.;埃克塞尔羊蹄甲●☆

49084 Bauhinia faberi Oliv. = Bauhinia brachycarpa Wall. ex Benth. ●

49085 Bauhinia faberi Oliv. var. megaphylla Ts. Tang et F. T. Wang = Bauhinia brachycarpa Wall. ex Benth. var. densiflora (Franch.) K. Larsen et S. S. Larsen ●

49086 Bauhinia faberi Oliv. var. megaphylla Ts. Tang et F. T. Wang = Bauhinia brachycarpa Wall. ex Benth. var. cavaleriei (H. Lév.) T. C. Chen ●

49087 Bauhinia faberi Oliv. var. megaphylla Ts. Tang et F. T. Wang = Bauhinia brachycarpa Wall. ex Benth. ●

49088 Bauhinia faberi Oliv. var. microphylla Oliv. = Bauhinia brachycarpa Wall. ex Benth. var. microphylla (Oliv. ex Craib) K. Larsen et S. S. Larsen ●

49089 Bauhinia faberi Oliv. var. microphylla Oliv. = Bauhinia brachycarpa Wall. ex Benth. ●

49090 Bauhinia fassoglensis Schweinf. = Tylosema fassoglensis (Schweinf.) Torre et Hillc. ●☆

49091 Bauhinia ferruginea D. Dietr. var. tonkinensis Gagnep. = Bauhinia coccinea (Lour.) DC. subsp. tonkinensis (Gagnep.) K. Larsen et S. S. Larsen ●

49092 Bauhinia ferruginea Roxb. var. tonkinensis Gagnep. = Bauhinia coccinea (Lour.) DC. subsp. tonkinensis (Gagnep.) K. Larsen et S. S. Larsen ●

49093 Bauhinia forficata Link;剪形羊蹄甲(拱形羊蹄甲);Brazilian Orchid tree,Brazilian Orchid-tree ●☆

49094 Bauhinia galpinii N. E. Br.;南非羊蹄甲(斑点羊蹄甲,橙花羊蹄甲,非洲羊蹄甲,红花羊蹄甲);African Plume, Pride of De Kaap,Punctate Bauhinia,Red Bauhinia,South African Orchid Bush ●☆

49095 Bauhinia garipensis E. Mey. = Adenolobus garipensis (E. Mey.) Torre et Hillc. ■☆

49096 Bauhinia genuflexa Craib;囊萼羊蹄甲●

49097 Bauhinia genuflexa Craib = Bauhinia touranensis Gagnep. ●

49098 Bauhinia glauca (Wall. ex Benth.) Benth.;粉叶羊蹄甲(粉背羊蹄甲,过岗藤,过山龙,蝴蝶草,马蹄藤,拟粉羊蹄甲,拟粉叶羊蹄甲,缺药藤,缺叶藤,羊蹄甲,夜关门,猪腰子藤,竹药藤);Climbing Bauhinia,Glaucous Bauhinia ●

49099 Bauhinia glauca (Wall. ex Benth.) Benth. subsp. caterviflora (H. Y. Chen) T. C. Chen;簇花羊蹄甲(密花羊蹄甲);Denseflower Glaucous Bauhinia ●

49100 Bauhinia glauca (Wall. ex Benth.) Benth. subsp. caterviflora (H. Y. Chen) T. C. Chen = Bauhinia glauca (Wall. ex Benth.) Benth. subsp. tenuiflora (Watt ex C. B. Clarke) K. Larsen et S. S. Larsen ●

49101 Bauhinia glauca (Wall. ex Benth.) Benth. subsp. hupehana (Craib) T. C. Chen;鄂羊蹄甲(湖北羊蹄甲,马蹄,双肾藤,羊蹄藤);Hubei Glaucous Bauhinia,Hupeh Bauhinia ●

49102 Bauhinia glauca (Wall. ex Benth.) Benth. subsp. hupehana (Craib) T. C. Chen = Bauhinia glauca (Wall. ex Benth.) Benth. subsp. tenuiflora (Watt ex C. B. Clarke) K. Larsen et S. S. Larsen ●

49103 Bauhinia glauca (Wall. ex Benth.) Benth. subsp. pernervosa (H. Y. Chen) T. C. Chen;显脉羊蹄甲(大夜关门,多脉羊蹄甲,多脉叶羊蹄甲,关门草,羊蹄风,猪腰藤);Distinctvein Bauhinia, Manynerve Bauhinia ●

49104 Bauhinia glauca (Wall. ex Benth.) Benth. subsp. pernervosa (H. Y. Chen) T. C. Chen = Bauhinia glauca (Wall. ex Benth.) Benth. subsp. tenuiflora (Watt ex C. B. Clarke) K. Larsen et S. S. Larsen ●

49105 Bauhinia glauca (Wall. ex Benth.) Benth. subsp. tenuiflora (Watt ex C. B. Clarke) K. Larsen et S. S. Larsen;薄叶羊蹄甲(密花羊蹄甲);Thinleaf Bauhinia ●

49106 Bauhinia glauca (Wall. ex Benth.) Benth. var. caterviflora (H. Y. Chen) K. Larsen et S. S. Larsen ex D. X. Zhang = Bauhinia glauca (Wall. ex Benth.) Benth. subsp. tenuiflora (Watt ex C. B. Clarke) K. Larsen et S. S. Larsen ●

49107 Bauhinia glauca (Wall. ex Benth.) Benth. var. hupehana (Craib) K. Larsen et S. S. Larsen ex D. X. Zhang = Bauhinia glauca (Wall. ex Benth.) Benth. subsp. tenuiflora (Watt ex C. B. Clarke) K. Larsen et S. S. Larsen ●

49108 Bauhinia glauca (Wall. ex Benth.) Benth. var. pernervosa (H. Y. Chen) K. Larsen et S. S. Larsen ex D. X. Zhang = Bauhinia glauca (Wall. ex Benth.) Benth. subsp. tenuiflora (Watt ex C. B. Clarke) K. Larsen et S. S. Larsen ●

49109 Bauhinia glauca Wall. subsp. tenuiflora (Watt ex C. B. Clarke) K. et S. S. Larsen = Bauhinia glauca (Wall. ex Benth.) Benth. subsp. tenuiflora (Watt ex C. B. Clarke) K. Larsen et S. S. Larsen ●

49110 Bauhinia gossweileri Baker f.;戈斯羊蹄甲●☆

49111 Bauhinia grandidieri Baill.;格朗羊蹄甲●☆

49112 Bauhinia grandiglora Juss.;大花羊蹄甲●☆

49113 Bauhinia grevei Drake;格雷弗羊蹄甲●☆

49114 Bauhinia griffithiana (Benth.) Prain;勐龙羊蹄甲●

49115 Bauhinia guianensis Aubl.;圭亚那羊蹄甲●☆

49116 Bauhinia hainanensis Merr. et Chun ex H. Y. Chen;海南羊蹄甲;Hainan Bauhinia ●

49117 Bauhinia henryi Craib = Bauhinia genuflexa Craib ●

49118 Bauhinia henryi Craib = Bauhinia touranensis Gagnep. ●

49119 Bauhinia henryi Harms = Bauhinia comosa Craib ●

49120 Bauhinia herrerae (Britton et Rose) Standl. et Steyerm.;赫雷羊蹄甲●☆

49121 Bauhinia hildebrandtii Vatke;希尔德羊蹄甲●☆

49122 Bauhinia hirsuta Weinm.;粗毛羊蹄甲;Hirsute Bauhinia,Rough-haired Bauhinia ●

49123 Bauhinia hookeri F. Muell.;胡克羊蹄甲(虎氏包兴木);Mountain Ebony,Pegunny ●☆

49124 Bauhinia hookeri F. Muell. = Lysiphyllum hookeri (F. Muell.) Pedley ●☆

49125 Bauhinia horsfieldii (Miq.) J. F. Macbr. = Bauhinia scandens L. var. horsfieldii (Watt ex Prain) K. Larsen et S. S. Larsen ●

49126 Bauhinia horsfieldii (Miq.) J. F. Macbr. = Bauhinia scandens L. ●

49127 Bauhinia howii Merr. et Chun = Bauhinia khasiana Baker ●

49128 Bauhinia humblotiana Baill.;洪布羊蹄甲●☆

49129 Bauhinia humifusa Pic. Serm. et Roti Mich. = Tylosema humifusa (Pic. Serm. et Roti Mich.) Brenan ■☆

49130 Bauhinia hunanensis Hand.-Mazz. = Bauhinia championii (Benth.) Benth. ●

49131 Bauhinia hupehana Craib = Bauhinia glauca (Wall. ex Benth.) Benth. subsp. hupehana (Craib) T. C. Chen ●

49132 Bauhinia hupehana Craib = Bauhinia glauca (Wall. ex Benth.) Benth. subsp. tenuiflora (Watt ex C. B. Clarke) K. Larsen et S. S. Larsen ●

49133 Bauhinia hupehana Craib var. grandis Craib = Bauhinia glauca (Wall. ex Benth.) Benth. subsp. hupehana (Craib) T. C. Chen ●

49134 Bauhinia hupehana Craib var. grandis Craib = Bauhinia glauca (Wall. ex Benth.) Benth. subsp. tenuiflora (Watt ex C. B. Clarke) K. Larsen et S. S. Larsen ●

49135 Bauhinia hypochrysa T. C. Chen;绸缎藤(绸缎木);Golden-back Bauhinia,Satin Bauhinia,Silkfabric Bauhinia ●

49136 Bauhinia hypoglauca Ts. Tang et F. T. Wang ex T. C. Chen;滇南羊蹄甲;Hypoglaucous Bauhinia,S. Yunnan Bauhinia,South Yunnan Bauhinia ●

49137 Bauhinia inflexilobata Merr. = Bauhinia ornata Kurz var. kerrii (Gagnep.) K. Larsen et S. S. Larsen ●

49138 Bauhinia inflexilobata Merr. = Bauhinia ornata Kurz var. subumbellata (Pierre ex Gagnep.) K. Larsen et S. S. Larsen ●

49139 Bauhinia integrifolia Roxb. = Bauhinia ornata Kurz ●

49140 Bauhinia japonica Maxim.;日本羊蹄甲(广东羊蹄甲,粤羊蹄甲);Japan Bauhinia,Japanese Bauhinia,Kwangtung Bauhinia ●

49141 Bauhinia japonica Maxim. var. subrhombicarpa (Merr.) M. Hiroe = Bauhinia scandens L. ●

49142 Bauhinia japonica Maxim. var. subrhombicarpa (Merr.) M. Hiroe = Bauhinia scandens L. var. horsfieldii (Watt ex Prain) K. Larsen et S. S. Larsen ●
49143 Bauhinia kappleri Sagot = Bauhinia monandra Kurz ●☆
49144 Bauhinia kerrii Gagnep. = Bauhinia ornata Kurz var. kerrii (Gagnep.) K. Larsen et S. S. Larsen ●
49145 Bauhinia kerrii Gagnep. var. grandiflora Craib = Bauhinia ornata Kurz var. kerrii (Gagnep.) K. Larsen et S. S. Larsen ●
49146 Bauhinia khasiana Baker;牛蹄麻(侯氏羊蹄甲);Khasia Bauhinia,Khasya Bauhinia ●
49147 Bauhinia khasiana Baker var. gigalobia D. X. Zhang;巨荚牛蹄麻●
49148 Bauhinia khasiana Baker var. tomentella T. C. Chen;毛叶牛蹄麻(短毛牛蹄麻);Tomentose Khasia Bauhinia ●
49149 Bauhinia kirkii Oliv. = Tylosema fassoglensis (Schweinf.) Torre et Hillc. ●☆
49150 Bauhinia krugii Urb. = Bauhinia monandra Kurz ●☆
49151 Bauhinia kunthiana Vogel;孔氏羊蹄甲●☆
49152 Bauhinia kwangtungensis Merr. = Bauhinia japonica Maxim. ●
49153 Bauhinia laui Merr. = Bauhinia viridescens Desv. ●
49154 Bauhinia laui Merr. et H. Y. Chen = Bauhinia viridescens Desv. var. laui (Merr.) T. C. Chen ●
49155 Bauhinia lecomtei Gagnep.;黔羊蹄甲(尖叶龙须藤,马蹄叶);Guizhou Bauhinia ●
49156 Bauhinia leucantha Thulin;索马里白花羊蹄甲●☆
49157 Bauhinia lingyuenensis T. C. Chen;凌云羊蹄甲;Lingyun Bauhinia ●
49158 Bauhinia linnaei Ali;林奈羊蹄甲●☆
49159 Bauhinia loeseneriana Harms;勒泽纳羊蹄甲●☆
49160 Bauhinia longistipes T. C. Chen;长柄羊蹄甲;Longstalk Bauhinia,Long-stalked Bauhinia ●
49161 Bauhinia lunarioides A. Gray;得州羊蹄甲;Anacacho Orchid Tree,Texas Plume ●☆
49162 Bauhinia macrantha Oliv. = Bauhinia petersiana Bolle subsp. macrantha (Oliv.) Brummitt et J. H. Ross ●☆
49163 Bauhinia macrosiphon Harms;大管羊蹄甲●☆
49164 Bauhinia madagascariensis Desv.;马岛羊蹄甲●☆
49165 Bauhinia mairei Harms = Bauhinia comosa Craib ●
49166 Bauhinia malabarica Roxb. = Piliostigma malabaricum (Roxb.) Benth. ■☆
49167 Bauhinia manca Standl.;残缺羊蹄甲●☆
49168 Bauhinia marlothii Engl. = Adenolobus pechuelii (Kuntze) Torre et Hillc. ■☆
49169 Bauhinia megacarpa H. Y. Chen = Bauhinia hainanensis Merr. et Chun ex H. Y. Chen ●
49170 Bauhinia mendoncae Torre et Hillc.;门东萨羊蹄甲●☆
49171 Bauhinia mirabilis Gagnep. = Bauhinia rubrovillosa K. Larsen et S. S. Larsen ●
49172 Bauhinia mombassae Vatke;蒙巴萨羊蹄甲●☆
49173 Bauhinia monandra Kurz;蝴蝶花羊蹄甲(单体雄蕊羊蹄甲,蝶花羊蹄甲,孟南德洋紫荆);Butterfly Flower, Jerusalem Date, Napoleon's Plume,Orchid Tree ●☆
49174 Bauhinia moningerae Merr. = Bauhinia erythropoda Hayata ●
49175 Bauhinia mossamedensis (Torre et Hillc.) G. Cusset = Adenolobus pechuelii (Kuntze) Torre et Hillc. subsp. mossamedensis (Torre et Hillc.) Brummitt et J. H. Ross ■☆
49176 Bauhinia multinervia DC.;多脉羊蹄甲;Petite Flamboyant Bauhinia ●☆
49177 Bauhinia natalensis Oliv. = Bauhinia natalensis Oliv. ex Hook. ●☆
49178 Bauhinia natalensis Oliv. ex Hook.;纳塔尔羊蹄甲;Natal Bauhinia ●☆
49179 Bauhinia nervosa (Wall. ex Benth.) Baker;棒花羊蹄甲;Claviflower Bauhinia,Club-flowered Bauhinia ●
49180 Bauhinia ombrophila Du Puy et R. Rabev.;喜雨羊蹄甲●☆
49181 Bauhinia ornata Kurz;缅甸羊蹄甲;Burma Bauhinia,Decoratted Bauhinia ●
49182 Bauhinia ornata Kurz var. austrosinensis (Gagnep.) K. Larsen et S. S. Larsen;琼岛羊蹄甲●
49183 Bauhinia ornata Kurz var. austrosinensis (Ts. Tang et F. T. Wang) T. C. Chen = Bauhinia ornata Kurz var. kerrii (Gagnep.) K. Larsen et S. S. Larsen ●
49184 Bauhinia ornata Kurz var. balansae (Gagnep.) K. Larsen et S. S. Larsen;光叶羊蹄甲●
49185 Bauhinia ornata Kurz var. contigua T. C. Chen;叠片羊蹄甲●
49186 Bauhinia ornata Kurz var. contigua T. C. Chen = Bauhinia ornata Kurz var. kerrii (Gagnep.) K. Larsen et S. S. Larsen ●
49187 Bauhinia ornata Kurz var. kerrii (Gagnep.) K. Larsen et S. S. Larsen;褐毛羊蹄甲(绸缎木,琼岛羊蹄甲,羊蹄藤);Kerr Bauhinia ●
49188 Bauhinia ornata Kurz var. subumbellata (Pierre ex Gagnep.) K. et S. S. Larsen = Bauhinia ornata Kurz var. kerrii (Gagnep.) K. Larsen et S. S. Larsen ●
49189 Bauhinia ornata Kurz var. subumbellata (Pierre ex Gagnep.) K. Larsen et S. S. Larsen;伞序羊蹄甲(散序羊蹄甲);Subumbellate Bauhinia ●
49190 Bauhinia ovatifolia T. C. Chen;卵叶羊蹄甲;Ovateleaf Bauhinia, Ovate-leaved Bauhinia ●
49191 Bauhinia paraglauca Ts. Tang et F. T. Wang = Bauhinia glauca (Wall. ex Benth.) Benth. ●
49192 Bauhinia parviflora Vahl = Bauhinia racemosa Lam. ●
49193 Bauhinia paucinervata T. C. Chen;少脉羊蹄甲;Fewnerne Bauhinia,Few-veined Bauhinia,Few-veins Bauhinia ●
49194 Bauhinia pauletia Pers.;多刺羊蹄甲;Railroadfence, Railway Fence Bauhinia ●☆
49195 Bauhinia pechuelii Kuntze = Adenolobus pechuelii (Kuntze) Torre et Hillc. ■☆
49196 Bauhinia pernervosa H. Y. Chen = Bauhinia glauca (Wall. ex Benth.) Benth. subsp. tenuiflora (Watt ex C. B. Clarke) K. Larsen et S. S. Larsen ●
49197 Bauhinia pervernosa H. Y. Chen = Bauhinia glauca (Wall. ex Benth.) Benth. subsp. pernervosa (H. Y. Chen) T. C. Chen ●
49198 Bauhinia pervilleana Baill.;佩尔羊蹄甲●☆
49199 Bauhinia petelotii Merr. = Bauhinia ornata Kurz var. balansae (Gagnep.) K. Larsen et S. S. Larsen ●
49200 Bauhinia petersiana Bolle;非洲羊蹄甲;Orchid Bauhinia, Zambesi Coffee ●☆
49201 Bauhinia petersiana Bolle subsp. macrantha (Oliv.) Brummitt et J. H. Ross;大花佩尔羊蹄甲●☆
49202 Bauhinia petersiana Bolle subsp. serpae (Ficalho et Hiern) Brummitt et J. H. Ross = Bauhinia petersiana Bolle subsp. macrantha (Oliv.) Brummitt et J. H. Ross ●☆
49203 Bauhinia pierrei Gagnep. = Bauhinia khasiana Baker ●
49204 Bauhinia podopetala Baker;足瓣羊蹄甲●☆
49205 Bauhinia polycarpa Wall. ex Benth. = Bauhinia viridescens Desv. ●
49206 Bauhinia polystachya Gagnep. = Bauhinia khasiana Baker ●
49207 Bauhinia porosa Boivin ex Baill. = Bauhinia monandra Kurz ●☆

49208 Bauhinia punctata Bolle = Bauhinia galpinii N. E. Br. ●☆

49209 Bauhinia punctata Burch. ex Benth. = Bauhinia galpinii N. E. Br. ●☆

49210 Bauhinia punctiflora Baker = Bauhinia monandra Kurz ●☆

49211 Bauhinia purpurea L.;羊蹄甲(玲甲花,羊紫荆,洋紫荆,紫花羊蹄甲,紫羊蹄甲);Butterfly Tree,Butterfly-tree,Camel's Foot Tree,Orchid Tree,Purple Bauhinia,Purple Camel's-foot Tree,Purple Orchid Tree ●

49212 Bauhinia purpurea L. var. alba Bailey;粉白羊蹄甲●☆

49213 Bauhinia pyrrhoclada Drake;红毛羊蹄甲(红毛枝羊蹄甲);Redhair Bauhinia,Reddy Bauhinia ●

49214 Bauhinia quinanensis T. C. Chen;黔南羊蹄甲;Qiannan Bauhinia,S. Guizhou Bauhinia ●

49215 Bauhinia racemosa Lam.;硬叶羊蹄甲(穗状花洋紫荆,总状花羊蹄甲,总状羊蹄甲);Racemose Bauhinia ●

49216 Bauhinia racemosa Vahl = Bauhinia vahlii Wight et Arn. ●☆

49217 Bauhinia reticulata DC.;网纹羊蹄甲;Reticulate Bauhinia ●

49218 Bauhinia reticulata DC. = Piliostigma reticulatum (DC.) Hochst. ■☆

49219 Bauhinia retusa Roxb.;微凹羊蹄甲;Retuse Bauhinia ●☆

49220 Bauhinia richardiana DC.;理查德羊蹄甲●☆

49221 Bauhinia rocheri H. Lév. = Bauhinia touranensis Gagnep. ●

49222 Bauhinia rubrovillosa K. Larsen et S. S. Larsen;红背叶羊蹄甲;Redback Bauhinia,Red-backed Bauhinia,Red-villose Bauhinia ●

49223 Bauhinia rufa (Benth.) Baker = Bauhinia ornata Kurz var. kerrii (Gagnep.) K. Larsen et S. S. Larsen ●

49224 Bauhinia rufescens Lam.;浅红羊蹄甲●☆

49225 Bauhinia saxatilis Craib = Bauhinia comosa Craib ●

49226 Bauhinia scandens L.;攀缘羊蹄甲(鳞甲花);Climbing Bauhinia,Scandent Bauhinia ●

49227 Bauhinia scandens L. var. horsfieldii (Watt ex Prain) K. Larsen et S. S. Larsen;菱果羊蹄甲;Climbing Bauhinia ●

49228 Bauhinia scandens L. var. horsfieldii (Watt ex Prain) K. Larsen et S. S. Larsen = Bauhinia scandens L. ●

49229 Bauhinia serpae Ficalho et Hiern = Bauhinia petersiana Bolle subsp. macrantha (Oliv.) Brummitt et J. H. Ross ●☆

49230 Bauhinia somalensis Pic. Serm. et Roti Mich. = Bauhinia ellenbeckii Harms ●☆

49231 Bauhinia splendens Kunth;光泽羊蹄甲●☆

49232 Bauhinia subrhombicarpa Merr. = Bauhinia scandens L. var. horsfieldii (Watt ex Prain) K. Larsen et S. S. Larsen ●

49233 Bauhinia subrhombicarpa Merr. = Bauhinia scandens L. ●

49234 Bauhinia subumbellata (Pierre ex Gagnep.) K. Larsen et S. S. Larsen = Bauhinia ornata Kurz var. subumbellata (Pierre ex Gagnep.) K. Larsen et S. S. Larsen ●

49235 Bauhinia subumbellata Pierre ex Gagnep. = Bauhinia ornata Kurz var. kerrii (Gagnep.) K. Larsen et S. S. Larsen ●

49236 Bauhinia taitensis Taub.;泰塔羊蹄甲●☆

49237 Bauhinia tenuiflora Watt ex C. B. Clarke;密花羊蹄甲(薄叶羊蹄甲);Thin-flower Bauhinia ●

49238 Bauhinia tenuiflora Watt ex C. B. Clarke = Bauhinia glauca (Wall. ex Benth.) Benth. subsp. tenuiflora (Watt ex C. B. Clarke) K. Larsen et S. S. Larsen ●

49239 Bauhinia thonningii Schumach. = Piliostigma thonningii (Schumach.) Milne-Redh. ■☆

49240 Bauhinia timorana Decne. = Bauhinia viridescens Desv. ●

49241 Bauhinia tomentosa L.;黄花羊蹄甲(黄花洋紫荆);St. Thomas Bauhinia,St. Thomas Tree,Tomentose Bauhinia,Yellow Bauhinia,Yellow Bell Bauhinia ●

49242 Bauhinia tomentosa L. var. glabrata Hook. f. = Bauhinia tomentosa L. ●

49243 Bauhinia tomentosa Naves = Bauhinia linnaei Ali ●☆

49244 Bauhinia touranensis Gagnep.;囊托羊蹄甲(全缘羊蹄甲,越南羊蹄甲);Viet Nam Bauhinia ●

49245 Bauhinia urbaniana Schinz;乌尔巴尼亚羊蹄甲●☆

49246 Bauhinia vahli Wight et Arn.;瓦尔羊蹄甲●☆

49247 Bauhinia variegata L.;洋紫荆(白花羊蹄甲,白花洋紫荆,宫粉羊蹄甲,宫粉紫荆,红花羊蹄甲,红花紫荆,红紫荆,老白花,马蹄豆,埋修,双飞蝴蝶,弯叶树,羊蹄甲,洋荆树,洋蹄甲,猪迹树,猪迹羊蹄甲,紫荆树);Buddhist Bauhinia,Butterfly Bush,Hong Kong Orchid,Hong Kong Orchid Tree,Mountain Ebony,Orchid Tree,Orchid-tree,Purple Orchid-tree,White Bauhinia ●

49248 Bauhinia variegata L.'Candida';白花洋紫荆(白花羊蹄甲,白花紫荆,大白花);White-flower Bauhinia ●

49249 Bauhinia variegata L. var. alboflora de Wit = Bauhinia variegata L.'Candida' ●

49250 Bauhinia variegata L. var. candida (Roxb.) Baker = Bauhinia variegata L.'Candida' ●

49251 Bauhinia variegata L. var. candida (Roxb.) Voigt = Bauhinia variegata L.'Candida' ●

49252 Bauhinia variegata L. var. candida Buch.-Ham. = Bauhinia variegata L.'Candida' ●

49253 Bauhinia variegata L. var. chinensis DC. = Bauhinia variegata L. ●

49254 Bauhinia venustula T. C. Chen;小巧羊蹄甲;Beautiful Bauhinia,Enchanting Bauhinia ●

49255 Bauhinia viridescens Desv.;绿花羊蹄甲;Greenflower Bauhinia,Virescent Bauhinia ●

49256 Bauhinia viridescens Desv. var. baviensis (Drake) de Wit = Bauhinia viridescens Desv. ●

49257 Bauhinia viridescens Desv. var. laui (Merr.) T. C. Chen;白枝羊蹄甲;Lau Bauhinia ●

49258 Bauhinia viridescens Desv. var. laui (Merr.) T. C. Chen = Bauhinia viridescens Desv. ●

49259 Bauhinia viridiflora Blume ex Miq. = Bauhinia glauca (Wall. ex Benth.) Benth. ●

49260 Bauhinia volkensii Taub. = Bauhinia tomentosa L. ●

49261 Bauhinia wallichii J. F. Macbr.;圆叶羊蹄甲●

49262 Bauhinia welwitschii Oliv. = Tylosema fassoglensis (Schweinf.) Torre et Hillc. ●☆

49263 Bauhinia wituensis Harms = Bauhinia tomentosa L. ●

49264 Bauhinia wuzhengyii S. S. Larsen;征镒羊蹄甲●

49265 Bauhinia xerophyta Du Puy et R. Rabev.;旱生羊蹄甲●☆

49266 Bauhinia yingtakensis Merr. et F. P. Metcalf = Bauhinia championii (Benth.) Benth. ●

49267 Bauhinia yingtakensis Merr. et F. P. Metcalf = Bauhinia championii (Benth.) Benth. var. yingtakensis (Merr. et F. P. Metcalf) T. C. Chen ●

49268 Bauhinia yunnanensis Franch. et Metcalf;云南羊蹄甲;Yunnan Bauhinia ●

49269 Bauhiniaceae Martinov = Fabaceae Lindl.(保留科名)●■

49270 Bauhiniaceae Martinov = Leguminosae Juss.(保留科名)●■

49271 Bauhiniaceae Martinov;羊蹄甲科●

49272 Baukea Vatke = Rhynchosia Lour.(保留属名)●■

49273 Baukea Vatke(1881);显豆属■☆

49274 Baukea insignis Vatke = Rhynchosia baukea Du Puy et Labat ■☆
49275 Baukea maxima (Bojer ex Drake) Baill. = Rhynchosia baukea Du Puy et Labat ■☆
49276 Baumannia DC. = Damnacanthus C. F. Gaertn. ●
49277 Baumannia K. Schum. = Knoxia L. ■
49278 Baumannia K. Schum. = Neobaumannia Hutch. et Dalzell ■
49279 Baumannia Spach = Oenothera L. ●■
49280 Baumannia hedyotoidea K. Schum. = Knoxia hedyotoidea (K. Schum.) Puff et Robbr. ■☆
49281 Baumea Gaudich. (1829);鲍姆莎属■☆
49282 Baumea Gaudich. = Machaerina Vahl ■
49283 Baumea iridifolia Boeck. = Machaerina flexuosa (Boeck.) Kern subsp. polyanthemum (Kük.) Lye ■☆
49284 Baumea rubiginosa (Sol. ex G. Forst.) Boeck. 'Variegata';彩色剑叶莎;Variegated Rush ■☆
49285 Baumea rubiginosa (Sol. ex G. Forst.) Boeck. = Machaerina rubiginosa (Sol. ex G. Forst.) T. Koyama ■
49286 Baumgartenia Spreng. = Borya Labill. ■☆
49287 Baumgartia Moench(废弃属名) = Cocculus DC. (保留属名)●
49288 Baumia Engl. et Gilg(1903);鲍姆玄参属■☆
49289 Baumia angolensis Engl. et Gilg;安哥拉鲍姆玄参■☆
49290 Baumiella H. Wolff = Afrocarum Rauschert ■☆
49291 Baumiella imbricata (Schinz) H. Wolff = Afrocarum imbricatum (Schinz) Rauschert ■☆
49292 Baursea Hoffmanns. = Philodendron Schott(保留属名)●■
49293 Baursea Hort. ex Hoffmanns. = Philodendron Schott(保留属名)●■
49294 Baursia Schott = Baursea Hoffmanns. ●■
49295 Bauschia Seub. ex Warm. = Aneilema R. Br. ■☆
49296 Bauxia Neck. = Marica Ker Gawl. ■☆
49297 Bavera Poir. = Bauera Banks ex Andréws ●☆
49298 Baxtera Rchb. (废弃属名) = Baxteria R. Br. (保留属名)■☆
49299 Baxtera Rchb. (废弃属名) = Loniceroides Bullock ■☆
49300 Baxteria R. Br. (1843)(保留属名);西澳朱蕉属(无茎草属)■☆
49301 Baxteria R. Br. ex Hook. = Baxteria R. Br. (保留属名)■☆
49302 Baxteria R. Br. ex Hook. = Calectasia R. Br. ●☆
49303 Baxteria australis R. Br.;西澳朱蕉■☆
49304 Baxteriaceae Takht. = Baxteridaceae Takht. ■☆
49305 Baxteriaceae Takht. = Dasypogonaceae Dumort. ■☆
49306 Baxteridaceae Takht.;西澳朱蕉科(无茎草科)■☆
49307 Bayabusua W. J. de Wilde = Zanonia L. ●■
49308 Bayabusua W. J. de Wilde(1999);马来瓜属■☆
49309 Baynesia Bruyns(2000);纳米比亚萝藦属■☆
49310 Bayonia Dugand = Mansoa DC. ●☆
49311 Bayonia Dugand = Onohualcoa Lundell ●☆
49312 Baziasa Steud. = Sabazia Cass. ●■☆
49313 Bazina Raf. = Lindernia All. ■
49314 Bdallophyton Eichler = Bdallophytum Eichler ■☆
49315 Bdallophytum Eichler(1872);美洲簇花草属■☆
49316 Bdallophytum americanum Harms;美洲簇花草■☆
49317 Bdallophytum andrieuxii Eichler;安氏美洲簇花草■☆
49318 Bdallophytum oxylepis Harms;尖鳞美洲簇花草■☆
49319 Bdellium Baill. ex Laness. = Commiphora Jacq. (保留属名)●
49320 Bdellium Baill. ex Laness. = Heudelotia A. Rich. ●
49321 Bea C. B. Clarke = Boea Comm. ex Lam. ■
49322 Beadlea Small = Cyclopogon C. Presl ■☆
49323 Beadlea cranichoides (Griseb.) Small = Cyclopogon cranichoides (Griseb.) Schltr. ■☆
49324 Beadlea elata (Sw.) Small ex Britton = Cyclopogon elatus (Sw.) Schltr. ■☆
49325 Beadlea storeri (Chapm.) Small = Cyclopogon cranichoides (Griseb.) Schltr. ■☆
49326 Bealea Scribn. = Muhlenbergia Schreb. ■
49327 Bealia Scribn. = Muhlenbergia Schreb. ■
49328 Bealia Scribn. ex Vasey = Muhlenbergia Schreb. ■
49329 Beata O. F. Cook = Coccothrinax Sarg. ●☆
49330 Beata O. F. Cook(1941);贝塔棕属●☆
49331 Beata ekmani (Burret) O. F. Cook;贝塔棕●☆
49332 Beatonia Herb. = Tigridia Juss. ■
49333 Beatonia coelestina (W. Bartram) Klatt = Calydorea coelestina (W. Bartram) Goldblatt et Henrich ■☆
49334 Beatsonia Roxb. = Frankenia L. ●■
49335 Beaua Pourr. = Boea Comm. ex Lam. ■
49336 Beaucarnea Lem. (1861);酒瓶兰属●☆
49337 Beaucarnea Lem. = Nolina Michx. ●☆
49338 Beaucarnea bigelovii (Torr.) Baker = Nolina bigelovii S. Watson ■☆
49339 Beaucarnea erumpens (Torr.) Baker = Nolina erumpens (Torr.) S. Watson ■☆
49340 Beaucarnea gracilis Lem.;细酒瓶兰(纺缍百合)●☆
49341 Beaucarnea lindheimeriana (Scheele) Baker = Nolina lindheimeriana (Scheele) S. Watson ■☆
49342 Beaucarnea longifolia Baker = Nolina longifolia Hemsl. ●☆
49343 Beaucarnea recurvata Lem.;酒瓶兰(德利兰,德利木);Bottle Palm, Elephant's Foot, Elephant-foot Tree, Pony Tail Plant, Pony-tail, Ponytail Palm, Recurvate Nolina ●☆
49344 Beaucarnea recurvata Lem. = Nolina recurvata (Lem.) Hemsl. ●☆
49345 Beaucarnea stricta Lem.;劲直酒瓶兰;Bottle Palm, Ponytail Palm ●☆
49346 Beaufortia R. Br. = Beaufortia R. Br. ex Aiton ●☆
49347 Beaufortia R. Br. ex Aiton(1812);瓶刷树属●☆
49348 Beaufortia decussata R. Br.;交互瓶刷树●☆
49349 Beaufortia sparsa R. Br.;沼泽瓶刷树;Swamp Bottlebrush, Swamp Bottle-brush ●☆
49350 Beaufortia squarrosa Schauer;沙生瓶刷树;Sand Bottlebrush ●☆
49351 Beauharnoisia Ruiz et Pav. = Tovomita Aubl. ●☆
49352 Beauica Post et Kuntze = Boeica T. Anderson ex C. B. Clarke ●■
49353 Beaumaria Deless. = Aristotelia L'Hér. (保留属名)●☆
49354 Beaumaria Deless. ex Steud. = Aristotelia L'Hér. (保留属名)●☆
49355 Beaumontia Wall. (1824);清明花属(比蒙藤属);Herald-trumpet, Qingmingflower ●
49356 Beaumontia brevituba Oliv.;断肠花(大果夹竹桃,短管清明花);Shorttube Heraldtrumpet, Shorttube Qingmingflower, Short-tubed Herald-trumpet ●
49357 Beaumontia campanulata Pit. = Beaumontia pitardii Tsiang ●
49358 Beaumontia fragrans Pierre ex Pit. = Beaumontia murtonii Craib ●
49359 Beaumontia grandiflora (Roxb.) Wall.;清明花(比蒙藤,大花清明花,炮弹果,刹抱龙,刹枪龙);Easter Heraldtrumpet, Easter Herald-trumpet, Easter Lily Vine, Easter Qingmingflower, Herald's Trumpet ●◇
49360 Beaumontia grandiflora Wall. = Beaumontia grandiflora (Roxb.) Wall. ●◇
49361 Beaumontia indecora Baill. = Vallaris indecora (Baill.) Tsiang et P. T. Li ●

49362 Beaumontia khasiana Hook. f.; 云南清明花; Yunnan Heraldtrumpet, Yunnan Herald-trumpet, Yunnan Qingmingflower ●

49363 Beaumontia murtonii Craib; 思茅清明花(大萼清明花); Murton Heraldtrumpet, Murton Herald-trumpet, Murton Qingmingflower ●

49364 Beaumontia pitardii Tsiang; 锈毛清明花(广西清明花); Pitard Heraldtrumpet, Pitard Herald-trumpet, Pitard Qingmingflower ●

49365 Beaumontia yunnanensis Tsiang et W. C. Chen = Beaumontia khasiana Hook. f. ●

49366 Beaumulix Willd. ex Poir. = Reaumuria L. ●

49367 Beauprea Brongn. et Gris(1871); 新喀山龙眼属●☆

49368 Beauprea crassifolia Virot; 厚叶新喀山龙眼●☆

49369 Beauprea diversifolia Brongn. et Gris; 新喀山龙眼●☆

49370 Beauprea montana (Brongn. et Gris) Virot; 山地新喀山龙眼●☆

49371 Beauprea spathulifolia Brongn. et Gris; 匙叶新喀山龙眼●☆

49372 Beaupreopsis Virot(1968); 拟新喀山龙眼属●☆

49373 Beaupreopsis paniculata (Brongn. et Gris) Virot; 拟新喀山龙眼●☆

49374 Beautempsia Gaudich. = Capparis L. ●

49375 Beautia Comm. ex Poir. = Thilachium Lour. ●☆

49376 Beauverdia Herter = Ipheion Raf. ■☆

49377 Beauverdia Herter = Leucocoryne Lindl. ■☆

49378 Beauvisagea Pierre = Lucuma Molina ●

49379 Beauvisagea Pierre = Pouteria Aubl. ●

49380 Beauvisagea Pierre ex Baill. = Lucuma Molina ●

49381 Beauvisagea Pierre ex Baill. = Pouteria Aubl. ●

49382 Bebbia Greene(1885); 甜菊木属; Sweetbush ●☆

49383 Bebbia aspera (Greene) A. Nelson = Bebbia juncea (Benth.) Greene var. aspera Greene ●☆

49384 Bebbia juncea (Benth.) Greene; 甜菊木; Rush Bebbia ●☆

49385 Bebbia juncea (Benth.) Greene var. aspera Greene; 粗糙甜菊木●☆

49386 Beccabunga Fourr. = Veronica L. ■

49387 Beccabunga Hill = Veronica L. ■

49388 Beccarianthus Cogn. = Astronidium A. Gray(保留属名)●☆

49389 Beccariella Pierre = Planchonella Pierre(保留属名)●

49390 Beccariella Pierre = Pouteria Aubl. ●

49391 Beccarimnea Pierre ex Post et Kuntze = Beccariella Pierre ●

49392 Beccarimnia Pierre ex Koord. = Pouteria Aubl. ●

49393 Beccarina Tiegh. = Trithecanthera Tiegh. ●☆

49394 Beccarinda Kuntze(1891); 横蒴苣苔属; Beccarinda ■

49395 Beccarinda argentea (Anthony) B. L. Burtt; 岩饰横蒴苣苔(饰岩苣苔); Ellliptic Beccarinda, Ellliptic-leaf Beccarinda ■

49396 Beccarinda erythrotricha W. T. Wang; 红毛横蒴苣苔; Redhair Beccarinda, Red-haired Beccarinda ■

49397 Beccarinda minima K. Y. Pan; 小横蒴苣苔; Mini Beccarinda ■

49398 Beccarinda paucisetulosa C. Y. Wu ex H. W. Li; 少毛横蒴苣苔; Fewhair Beccarinda ■

49399 Beccarinda sinensis (Chun) B. L. Burtt = Beccarinda tonkinensis (Pellegr.) B. L. Burtt ■

49400 Beccarinda sinensis (Chun) Burtt = Beccarinda tonkinensis (Pellegr.) B. L. Burtt ■

49401 Beccarinda tonkinensis (Pellegr.) B. L. Burtt; 横蒴苣苔; Tonkin Beccarinda, Vietnam Beccarinda ■

49402 Beccariodendron Warb. (1891); 新几内亚番荔枝属●☆

49403 Beccariodendron Warb. = Goniothalamus (Blume) Hook. f. et Thomson ●

49404 Beccariodendron grandiflorum Warb.; 新几内亚番荔枝●☆

49405 Beccariophoenix Jum. et H. Perrier(1915); 贝加利椰子属(马岛窗孔椰属, 马岛刺葵属)●☆

49406 Beccariophoenix alfredii Rakotoarinivo, Ranarivelo et J. Dransf.; 阿尔贝加利椰子●☆

49407 Beccariophoenix madagascariensis Jum. et H. Perrier; 贝加利椰子(马达加斯加龟背棕, 马岛葵)●☆

49408 Becheria Ridl. = Ixora L. ●

49409 Bechium DC. (1836); 红腺尖鸠菊属■☆

49410 Bechium DC. = Vernonia Schreb. (保留属名)●■

49411 Bechium foliosum Klatt = Vernonia diversifolia Bojer ex DC. ■☆

49412 Bechium rubricaule DC. = Vernonia erythromarula Klatt ■☆

49413 Bechium scapiforme DC. = Vernonia nudicaulis Less. ■☆

49414 Bechonneria Hort. ex Carrière = Beschorneria Kunth ■☆

49415 Bechsteineria Muell. = Rechsteineria Regel(保留属名)■☆

49416 Becium Lindl. = Ocimum L. ●■

49417 Becium affine (Hochst. ex Benth.) Chiov. = Ocimum obovatum E. Mey. ex Benth. ■☆

49418 Becium affine (Hochst. ex Benth.) Chiov. var. cyclophyllum Chiov. = Ocimum filamentosum Forssk. ■☆

49419 Becium albostellatum Verdc. = Ocimum albostellatum (Verdc.) A. J. Paton ●☆

49420 Becium angustifolium (Benth.) N. E. Br. = Ocimum angustifolium Benth. ■☆

49421 Becium aureoviride P. A. Duvign. = Ocimum vanderystii (De Wild.) A. J. Paton ■☆

49422 Becium aureoviride P. A. Duvign. subsp. lupotoense (P. A. Duvign.) Ayob. = Ocimum vanderystii (De Wild.) A. J. Paton ■☆

49423 Becium aureoviride P. A. Duvign. var. lupotoense (P. A. Duvign.) Ayob. = Ocimum vanderystii (De Wild.) A. J. Paton ■☆

49424 Becium bicolor Lindl. = Ocimum grandiflorum Lam. ●☆

49425 Becium burchellianum (Benth.) N. E. Br. = Ocimum burchellianum Benth. ■☆

49426 Becium cameronii (Baker) P. J. Cribb = Ocimum fimbriatum Briq. ■☆

49427 Becium capitatum Agnew = Ocimum decumbens Gürke ■☆

49428 Becium centrali-africanum (R. E. Fr.) Sebald = Ocimum centrali-africanum R. E. Fr. ■☆

49429 Becium centrali-africanum (R. E. Fr.) Sebald var. linearifolium Ayob. = Ocimum centrali-africanum R. E. Fr. ■☆

49430 Becium choanum P. A. Duvign. et Plancke = Ocimum obovatum E. Mey. ex Benth. ■☆

49431 Becium choanum P. A. Duvign. et Plancke var. longifolium Ayob. = Ocimum obovatum E. Mey. ex Benth. ■☆

49432 Becium citriodorum S. D. Will. et K. Balkwill = Ocimum dolomiticola A. J. Paton ■☆

49433 Becium coddii S. D. Will. et K. Balkwill = Ocimum coddii (S. D. Will. et K. Balkwill) A. J. Paton ■☆

49434 Becium ctenodon Gilli = Ocimum fimbriatum Briq. var. ctenodon (Gilli) A. J. Paton ■☆

49435 Becium decumbens (Gürke) A. J. Paton = Ocimum decumbens Gürke ■☆

49436 Becium duvigneaudii Ayob. = Ocimum fimbriatum Briq. var. ctenodon (Gilli) A. J. Paton ■☆

49437 Becium ellenbeckii (Gürke) Cufod. = Ocimum ellenbeckii Gürke ■☆

49438 Becium ericoides P. A. Duvign. et Plancke = Ocimum ericoides (P. A. Duvign. et Plancke) A. J. Paton ●☆

49439 Becium fastigiatum A. J. Paton = Ocimum amicorum A. J. Paton ■☆
49440 Becium filamentosum (Forssk.) Chiov. = Ocimum filamentosum Forssk. ■☆
49441 Becium fimbriatum (Briq.) Sebald = Ocimum fimbriatum Briq. ■☆
49442 Becium fimbriatum (Briq.) Sebald var. angustilanceolatum (De Wild.) Sebald = Ocimum fimbriatum Briq. var. angustilanceolatum (De Wild.) A. J. Paton ■☆
49443 Becium fimbriatum (Briq.) Sebald var. bequaertii (De Wild.) Sebald = Ocimum fimbriatum Briq. var. bequaertii (De Wild.) A. J. Paton ■☆
49444 Becium fimbriatum (Briq.) Sebald var. ctenodon (Gilli) Sebald = Ocimum fimbriatum Briq. var. ctenodon (Gilli) A. J. Paton ■☆
49445 Becium fimbriatum (Briq.) Sebald var. microphyllum Sebald = Ocimum fimbriatum Briq. var. microphyllum (Sebald) A. J. Paton ■☆
49446 Becium formosum (Gürke) Chiov. ex Lanza = Ocimum formosum Gürke ●☆
49447 Becium frutescens (Sebald) A. J. Paton = Ocimum vihpyense A. J. Paton ■☆
49448 Becium grandiflorum (Lam.) Pic. Serm.;大花红腺尖鸠菊■☆
49449 Becium grandiflorum (Lam.) Pic. Serm. = Ocimum grandiflorum Lam. ●☆
49450 Becium grandiflorum (Lam.) Pic. Serm. subsp. densiflorum A. J. Paton = Ocimum grandiflorum Lam. subsp. densiflorum (A. J. Paton) A. J. Paton ●☆
49451 Becium grandiflorum (Lam.) Pic. Serm. subsp. turkanaense (Sebald) A. J. Paton = Ocimum grandiflorum Lam. subsp. turkanaense (Sebald) A. J. Paton ■☆
49452 Becium grandiflorum (Lam.) Pic. Serm. var. albostellatum (Verdc.) Sebald = Ocimum albostellatum (Verdc.) A. J. Paton ●☆
49453 Becium grandiflorum (Lam.) Pic. Serm. var. capitatum (Agnew) Sebald = Ocimum decumbens Gürke ■☆
49454 Becium grandiflorum (Lam.) Pic. Serm. var. cordatum (A. J. Paton) J. -P. Lebrun et Stork = Ocimum obovatum E. Mey. ex Benth. subsp. cordatum (A. J. Paton) A. J. Paton ■☆
49455 Becium grandiflorum (Lam.) Pic. Serm. var. crystallinum (A. J. Paton) J. -P. Lebrun et Stork = Ocimum obovatum E. Mey. ex Benth. subsp. crystallinum (A. J. Paton) A. J. Paton ■☆
49456 Becium grandiflorum (Lam.) Pic. Serm. var. decumbens (Gürke) Sebald = Ocimum decumbens Gürke ■☆
49457 Becium grandiflorum (Lam.) Pic. Serm. var. densiflorum (A. J. Paton) J. -P. Lebrun et Stork = Ocimum grandiflorum Lam. subsp. densiflorum (A. J. Paton) A. J. Paton ●☆
49458 Becium grandiflorum (Lam.) Pic. Serm. var. ericoides (P. A. Duvign. et Plancke) Sebald = Ocimum ericoides (P. A. Duvign. et Plancke) A. J. Paton ●☆
49459 Becium grandiflorum (Lam.) Pic. Serm. var. frutescens Sebald = Ocimum vihpyense A. J. Paton ■☆
49460 Becium grandiflorum (Lam.) Pic. Serm. var. galpinii (Gürke) Sebald = Ocimum dambicola A. J. Paton ■☆
49461 Becium grandiflorum (Lam.) Pic. Serm. var. latifolium Sebald;宽叶红腺尖鸠菊■☆
49462 Becium grandiflorum (Lam.) Pic. Serm. var. metallorum (P. A. Duvign.) Sebald = Ocimum metallorum (P. A. Duvign.) A. J. Paton ■☆
49463 Becium grandiflorum (Lam.) Pic. Serm. var. obovatum (E. Mey. ex Benth.) Sebald = Ocimum obovatum E. Mey. ex Benth. ■☆
49464 Becium grandiflorum (Lam.) Pic. Serm. var. punctatum (Baker) J. -P. Lebrun = Ocimum dambicola A. J. Paton ■☆
49465 Becium grandiflorum (Lam.) Pic. Serm. var. rubrocostatum (Robyns et Lebrun) Sebald = Ocimum decumbens Gürke ■☆
49466 Becium grandiflorum (Lam.) Pic. Serm. var. stuhlmannii (Gürke) Sebald = Ocimum obovatum E. Mey. ex Benth. ■☆
49467 Becium grandiflorum (Lam.) Pic. Serm. var. turkanaense Sebald = Ocimum grandiflorum Lam. subsp. turkanaense (Sebald) A. J. Paton ■☆
49468 Becium grandiflorum (Lam.) Pic. Serm. var. urundense (Robyns et Lebrun) Sebald = Ocimum urundense Robyns et Lebrun ■☆
49469 Becium grandiflorum (Lam.) Pic. Serm. var. vanderystii (De Wild.) Sebald = Ocimum vanderystii (De Wild.) A. J. Paton ■☆
49470 Becium hirsutissimum P. A. Duvign. = Ocimum hirsutissimum (P. A. Duvign.) A. J. Paton ■☆
49471 Becium hirtii Plancke ex Ayob. = Ocimum albostellatum (Verdc.) A. J. Paton ●☆
49472 Becium homblei (De Wild.) P. A. Duvign. et Plancke = Ocimum centrali-africanum R. E. Fr. ■☆
49473 Becium irvinei (J. K. Morton) Sebald = Ocimum irvinei J. K. Morton ■☆
49474 Becium knyanum (Vatke) N. E. Br. ex Broun et R. E. Massey = Ocimum filamentosum Forssk. ■☆
49475 Becium knyanum (Vatke) N. E. Br. ex Broun et R. E. Massey var. diffusum Ayob. = Ocimum filamentosum Forssk. ■☆
49476 Becium metallorum P. A. Duvign. = Ocimum metallorum (P. A. Duvign.) A. J. Paton ■☆
49477 Becium minutiflorum Sebald = Ocimum minutiflorum (Sebald) A. J. Paton ■☆
49478 Becium modestum (Briq.) G. Taylor = Ocimum obovatum E. Mey. ex Benth. ■☆
49479 Becium monocotyloides Ayob. = Ocimum monocotyloides (Ayob.) A. J. Paton ■☆
49480 Becium neumannii (Gürke) Cufod. = Ocimum obovatum E. Mey. ex Benth. ■☆
49481 Becium obovatum (E. Mey. ex Benth.) N. E. Br. = Ocimum obovatum E. Mey. ex Benth. ■☆
49482 Becium obovatum (E. Mey. ex Benth.) N. E. Br. subsp. cordatum A. J. Paton = Ocimum obovatum E. Mey. ex Benth. subsp. cordatum (A. J. Paton) A. J. Paton ■☆
49483 Becium obovatum (E. Mey. ex Benth.) N. E. Br. subsp. crystallinum A. J. Paton = Ocimum obovatum E. Mey. ex Benth. subsp. crystallinum (A. J. Paton) A. J. Paton ■☆
49484 Becium obovatum (E. Mey. ex Benth.) N. E. Br. subsp. punctatum (Baker) A. J. Paton = Ocimum dambicola A. J. Paton ■☆
49485 Becium obovatum (E. Mey. ex Benth.) N. E. Br. var. galpinii (Gürke) N. E. Br. = Ocimum obovatum E. Mey. ex Benth. var. galpinii (Gürke) A. J. Paton ■☆
49486 Becium obovatum (E. Mey. ex Benth.) N. E. Br. var. glabrius (Benth.) Cufod. = Ocimum obovatum E. Mey. ex Benth. ■☆
49487 Becium obovatum (E. Mey. ex Benth.) N. E. Br. var. hians (Benth.) N. E. Br. = Ocimum obovatum E. Mey. ex Benth. ■☆
49488 Becium obovatum (E. Mey. ex Benth.) N. E. Br. var. knyanum (Vatke) Cufod. = Ocimum filamentosum Forssk. ■☆
49489 Becium obovatum (E. Mey. ex Benth.) N. E. Br. var. latifolium Sebald = Ocimum obovatum E. Mey. ex Benth. ■☆
49490 Becium obovatum (E. Mey. ex Benth.) N. E. Br. var. macrocaulon (Briq.) Ayob. = Ocimum obovatum E. Mey. ex Benth. ■☆

49491 Becium obovatum (E. Mey. ex Benth.) N. E. Br. var. modestum (Briq.) Ayob. = Ocimum obovatum E. Mey. ex Benth. ■☆
49492 Becium peschianum P. A. Duvign. et Plancke = Ocimum metallorum (P. A. Duvign.) A. J. Paton ■☆
49493 Becium pumilum (Gürke) Chiov. ex Lanza = Ocimum obovatum E. Mey. ex Benth. ■☆
49494 Becium pyramidatum A. J. Paton = Ocimum pyramidatum (A. J. Paton) A. J. Paton ■☆
49495 Becium reclinatum S. D. Will. et K. Balkwill = Ocimum reclinatum (S. D. Will. et K. Balkwill) A. J. Paton ■☆
49496 Becium schweinfurthii (Briq.) N. E. Br. ex Broun et Massey = Ocimum obovatum E. Mey. ex Benth. ■☆
49497 Becium serpyllifolium (Forssk.) Wood = Ocimum serpyllifolium Forssk. ■☆
49498 Becium ternatum G. Taylor = Ocimum obovatum E. Mey. ex Benth. ■☆
49499 Becium thymifolium P. A. Duvign. = Ocimum metallorum (P. A. Duvign.) A. J. Paton ■☆
49500 Becium urundense (Robyns et Lebrun) A. J. Paton = Ocimum urundense Robyns et Lebrun ■☆
49501 Becium vandenbrandei P. A. Duvign. et Plancke ex Ayob. = Ocimum vandenbrandei (P. A. Duvign. et Plancke ex Ayob.) A. J. Paton ■☆
49502 Becium vanderystii De Wild. = Ocimum vanderystii (De Wild.) A. J. Paton ■☆
49503 Becium verticillifolium (Baker) Cufod. = Ocimum verticillifolium Baker ■☆
49504 Becium virgatum A. J. Paton = Ocimum canescens A. J. Paton ■☆
49505 Becium waterbergense S. D. Will. et K. Balkwill = Ocimum waterbergensis (S. D. Will. et K. Balkwill) A. J. Paton ■☆
49506 Beckea A. St. -Hil. = Baeckea L. ●
49507 Beckea Pers. = Baeckea Burm. f. ●
49508 Beckea Pers. = Brunia Lam. (保留属名)●☆
49509 Beckea St. Hil. = Baeckea L. ●
49510 Beckera Fresen = Snowdenia C. E. Hubb. ■☆
49511 Beckera glabrescens Steud. = Pennisetum unisetum (Nees) Benth. ■☆
49512 Beckera gracilis Hochst. = Snowdenia mutica (Hochst.) Pilg. ■☆
49513 Beckera mutica Hochst. = Snowdenia mutica (Hochst.) Pilg. ■☆
49514 Beckera nubica (Hochst.) Hochst. = Pennisetum nubicum (Hochst.) K. Schum. ex Engl. ■☆
49515 Beckera petiolaris (Hochst.) Hochst. = Pennisetum petiolare (Hochst.) Chiov. ■☆
49516 Beckera petitiana A. Rich. = Snowdenia petitiana (A. Rich.) C. E. Hubb. ■☆
49517 Beckera polystachia Fresen. = Snowdenia polystachya (Fresen.) Pilg. ■☆
49518 Beckera scabra Pilg. = Snowdenia petitiana (A. Rich.) C. E. Hubb. ■☆
49519 Beckera schimperi Hochst. = Snowdenia polystachya (Fresen.) Pilg. ■☆
49520 Beckera valida Fresen. = Snowdenia polystachya (Fresen.) Pilg. ■☆
49521 Beckeria Bernh. = Melica L. ■
49522 Beckeria Heynh. = Beckera Fresen ■☆
49523 Beckeria Heynh. = Snowdenia C. E. Hubb. ■☆
49524 Beckeropsis Fig. et De Not. = Pennisetum Rich. ■
49525 Beckeropsis laxior Clayton = Pennisetum laxior (Clayton) Clayton ■☆
49526 Beckeropsis nubica (Hochst.) Fig. et De Not. = Pennisetum nubicum (Hochst.) K. Schum. ex Engl. ■☆
49527 Beckeropsis petiolaris (Hochst.) Fig. et De Not. = Pennisetum petiolare (Hochst.) Chiov. ■☆
49528 Beckeropsis pirottae (Chiov.) Stapf et C. E. Hubb. = Pennisetum pirottae Chiov. ■☆
49529 Beckeropsis procera Stapf = Pennisetum procerum (Stapf) Clayton ■☆
49530 Beckeropsis uniseta (Nees) K. Schum. = Pennisetum unisetum (Nees) Benth. ■☆
49531 Beckia Raf. = Baeckea L. ●
49532 Beckmannia Host(1805);菵草属;Slough Grass,Sloughgrass ■
49533 Beckmannia baicalensis (I. V. Kusn.) Hultén = Beckmannia syzigachne (Steud.) Fernald ■
49534 Beckmannia eruciformis (L.) Host subsp. baicalensis (Kusn.) Hultén = Beckmannia syzigachne (Steud.) Fernald ■
49535 Beckmannia eruciformis (L.) Host var. uniflora Scribn. ex A. Gray = Beckmannia syzigachne (Steud.) Fernald ■
49536 Beckmannia eruciformis Host = Beckmannia syzigachne (Steud.) Fernald ■
49537 Beckmannia eruciformis Host subsp. baicalensis (I. V. Kusn.) Koyama et Kawano = Beckmannia syzigachne (Steud.) Fernald ■
49538 Beckmannia eruciformis Host var. baicalensis I. V. Kusn. = Beckmannia syzigachne (Steud.) Fernald ■
49539 Beckmannia hirsutiflora (Roshev.) Prob. = Beckmannia syzigachne (Steud.) Fernald var. hirsutiflora Roshev. ■
49540 Beckmannia syzigachne (Steud.) Fernald;菵草(水稗子,菵米);America Sloughgrass, American Slough Grass, American Sloughgrass, American Slough-grass, Beckmann's-grass, European Slough-grass ■
49541 Beckmannia syzigachne (Steud.) Fernald f. eriantha Kitag. = Beckmannia syzigachne (Steud.) Fernald var. hirsutiflora Roshev. ■
49542 Beckmannia syzigachne (Steud.) Fernald subsp. baicalensis (Kusn.) T. Koyama et Kawano = Beckmannia syzigachne (Steud.) Fernald ■
49543 Beckmannia syzigachne (Steud.) Fernald subsp. hirsutiflora (Roshev.) Tzvelev = Beckmannia syzigachne (Steud.) Fernald var. hirsutiflora Roshev. ■
49544 Beckmannia syzigachne (Steud.) Fernald var. hirsutiflora Roshev.;毛颖菵草;Hairyglume Sloughgrass ■
49545 Beckmannia syzigachne (Steud.) Fernald var. uniflora (Scribn. ex A. Gray) B. Boivin = Beckmannia syzigachne (Steud.) Fernald ■
49546 Beckwithia Jeps. = Ranunculus L. ■
49547 Beckwithia andersonii (A. Gray) Jeps. = Ranunculus andersonii A. Gray ■☆
49548 Beckwithia glacialis (L.) Á. Löve et D. Löve = Ranunculus glacialis L. ■☆
49549 Beclardia A. Rich. (1828);伯克兰属■☆
49550 Beclardia A. Rich. = Cryptopus Lindl. ■☆
49551 Beclardia brachystachya A. Rich. = Beclardia macrostachya (Thouars) A. Rich. ■☆
49552 Beclardia erostris Frapp. = Beclardia macrostachya (Thouars) A. Rich. ■☆
49553 Beclardia grandiflora Bosser;大花伯克兰■☆

49554 Beclardia humberti H. Perrier = Lemurella culicifera (Rchb. f.) H. Perrier ■☆
49555 Beclardia macrostachya (Thouars) A. Rich. ;大穗伯克兰■☆
49556 Becquerela Nees = Becquerelia Brongn. ■☆
49557 Becquerelia Brongn. (1833);贝克莎属■☆
49558 Becquerelia cymosa Brongn. ;贝克莎■☆
49559 Beddomea Hook. f. = Aglaia Lour. (保留属名)●
49560 Bedfordia DC. (1833);澳菊木属(线绒菊属)●☆
49561 Bedfordia arborescens Hochr. ;高大澳菊木●☆
49562 Bedfordia linearis (Labill.) DC. ;窄叶澳菊木●☆
49563 Bedfordia salicina (Labill.) DC. ;柳叶澳菊木●☆
49564 Bedousi Augier = Casearia Jacq. ●
49565 Bedousia Dennst. = Casearia Jacq. ●
49566 Bedusia Raf. = Casearia Jacq. ●
49567 Beehsa Endl. = Beesha Kunth ●
49568 Beehsa Endl. = Melocanna Trin. ●
49569 Been Schmidel = Limonium Mill. (保留属名)●■
49570 Beera P. Beauv. = Hypolytrum Rich. ex Pers. ■
49571 Beera P. Beauv. ex T. Lestib. = Hypolytrum Rich. ex Pers. ■
49572 Beesha Kunth = Melocanna Trin. ●
49573 Beesha Munro = Ochlandra Thwaites ●☆
49574 Beesha capitata (Kunth) Munro = Cathariostachys capitata (Kunth) S. Dransf. ●☆
49575 Beesia Balf. f. et W. W. Sm. (1915);铁破锣属;Beesia ■★
49576 Beesia calthifolia (Maxim. ex Oliv.) Ulbr. ;铁破锣(白细辛,贝茜花,贝西花,单叶升麻,滇豆根,定木香,葫芦七,花椒七,山豆根,土黄连,野大救驾);Marshmarigold-leaved Beesia ■
49577 Beesia calthifolia (Maxim.) Ulbr. = Beesia calthifolia (Maxim. ex Oliv.) Ulbr. ■
49578 Beesia cordata Balf. f. et W. W. Sm. = Beesia calthifolia (Maxim.) Ulbr. ■
49579 Beesia deltophylla C. Y. Wu;角叶铁破锣;Triangleleaf Beesia ■
49580 Beesia elongata Hand. -Mazz. = Beesia calthifolia (Maxim.) Ulbr. ■
49581 Beethovenia Engl. = Ceroxylon Bonpl. ex DC. ●☆
49582 Befaria Mutis ex L. (1771);贝氏杜鹃属●☆
49583 Befaria Mutis ex L. = Bejaria Mutis(保留属名)●☆
49584 Befaria Mutis ex L. = Brachysola Rye ●☆
49585 Befaria aestuans Mutis ex L. ;贝氏杜鹃●☆
49586 Befaria racemosa Vent. ;总花贝氏杜鹃;Andes Rose ●☆
49587 Begonia L. (1753);秋海棠属;Angel Wing Begonia, Beefsteak Geranium, Begonia, Elephant Ear, Elephant's Ear ●■
49588 Begonia 'Feastii' = Begonia erythrophylla Neuman ■☆
49589 Begonia 'Iron Cross' = Begonia masoniana Irmsch. ■
49590 Begonia 'Silver Jewel';银翠秋海棠■☆
49591 Begonia abyssinica Cufod. = Begonia wollastonii Baker f. ■☆
49592 Begonia acaulis Merr. et L. M. Perry;无茎秋海棠■☆
49593 Begonia acetosa Vell. ;微酸秋海棠(酸味秋海棠);Acid Begonia ■☆
49594 Begonia acetosella Craib;无翅秋海棠(红小姐,黄疸草,四棱秋海棠,酸味秋海棠);Wingless Begonia ■
49595 Begonia acetosella Craib var. hirtifolia Irmsch. ;粗毛无翅秋海棠(毛叶酸味秋海棠,毛叶无翅秋海棠);Hairyleaf Wingless Begonia ■
49596 Begonia acida Mart. ex A. DC. ;酸叶秋海棠;Acid-leaf Begonia ■☆
49597 Begonia aconitifolia A. DC. ;卷裂叶秋海棠;Aconite-leaved Begonia ■☆
49598 Begonia acutifolia Jacq. ;枸骨叶秋海棠;Holly-leaf Begonia ■☆
49599 Begonia acutitepala K. Y. Guan et D. K. Tian;尖被秋海棠;Sharppetal Begonia ■
49600 Begonia adolfi-friderici Gilg = Begonia poculifera Hook. f. ■☆
49601 Begonia adpressa Sosef;匍匐秋海棠■☆
49602 Begonia albo-coccinea Hook. ;印度秋海棠(白暗红秋海棠,白珊瑚秋海棠);Elephant's Ear Begonia ■
49603 Begonia albo-picta Bull. ;白彩秋海棠(星点秋海棠);Guinea-wing Begonia, Silverspot Begonia ■☆
49604 Begonia alepensis A. Chev. = Begonia fusialata Warb. ■☆
49605 Begonia algaia L. B. Sm. et Wassh. ;美丽秋海棠(虎爪龙,裂叶秋海棠);Pretty Begonia ■
49606 Begonia alnifolia A. DC. ;桤木叶秋海棠;Alder-leaf Begonia ■☆
49607 Begonia alveolata Te T. Yu;点叶秋海棠(蜂窝秋海棠);Alveolate Begonia ■
49608 Begonia amoena Wall. = Begonia tenella D. Don ■☆
49609 Begonia amoena Wall. ex A. DC. = Begonia tenella D. Don ■☆
49610 Begonia ampla Hook. f. ;膨大秋海棠■☆
49611 Begonia anceps Irmsch. ;二棱秋海棠;Two-edged Begonia ■
49612 Begonia anceps Irmsch. = Begonia morifolia Te T. Yu ■
49613 Begonia andersonii Hort. ;安氏秋海棠(印度秋海棠);Indian Begonia ■☆
49614 Begonia andina Rusby;安第斯山秋海棠;Andean Begonia ■☆
49615 Begonia androrangensis Humbert ex Rabenant. et Bosser;安德鲁兰加秋海棠■☆
49616 Begonia angolensis Irmsch. ;安哥拉秋海棠■☆
49617 Begonia angularis Raddi;有角秋海棠;Angular Begonia ■☆
49618 Begonia angularis Raddi = Begonia stipulacea Willd. ■☆
49619 Begonia anisosepala Hook. f. ;异萼秋海棠■☆
49620 Begonia ankaranensis Humbert ex Rabenant. et Bosser;安卡兰秋海棠■☆
49621 Begonia annobonensis A. DC. ;安诺本秋海棠■☆
49622 Begonia annulata K. Koch;轮纹秋海棠;Annular Begonia ■☆
49623 Begonia antongilensis Humbert ex Rabenant. et Bosser;安通吉尔秋海棠■☆
49624 Begonia antsingyensis Humbert ex Rabenant. et Bosser;安钦吉秋海棠■☆
49625 Begonia antsiranensis Aymonin et Bosser;安齐朗秋海棠■☆
49626 Begonia aptera Hayata;无翅果秋海棠(丹叶,红双通,红酸杆,红叶子,散血子,无翅秋海棠,野海棠,夜变红,圆果秋海棠,紫背天葵);Hayata Begonia ■
49627 Begonia aptera Hayata = Begonia longifolia Blume ■
49628 Begonia arboreta Y. M. Shui;树生秋海棠■
49629 Begonia argenteo-guttata L. H. Bailey;银星秋海棠(麻叶秋海棠,秋海棠);Anglewing Begonia, Silverstar Begonia, Trout Begonia, Trout-leaf Begonia ■
49630 Begonia argenteo-guttata Lemoine = Begonia argenteo-guttata L. H. Bailey ■
49631 Begonia argyrostigma Fisch. = Begonia maculata Raddi ■
49632 Begonia argyrostigma Fisch. ex Link et Otto;竹节叶秋海棠(斑叶竹节秋海棠);Spotted Begonia ■☆
49633 Begonia aridicaulis Ziesenh. ;干茎秋海棠;Arid-stem Begonia ■☆
49634 Begonia ascotiensis J. B. Weber;阿斯科特秋海棠;Ascotian Begonia ■☆
49635 Begonia asperifolia Irmsch. ;糙叶秋海棠;Rough-leaved Begonia ■
49636 Begonia asperifolia Irmsch. var. tomentosa Te T. Yu;俅江秋海棠(绒毛糙叶秋海棠,绒毛秋海棠);Tomentose Rough-leaved Begonia ■
49637 Begonia asperifolia Irmsch. var. unialata T. C. Ku;窄檐糙叶秋海

棠;Unialate Rough-leaved Begonia ■
49638 Begonia aspleniifolia Hook. f. ex A. DC.;铁线蕨叶秋海棠■☆
49639 Begonia asteropyrifolia Y. M. Shui et W. H. Chen;星果草叶秋海棠■
49640 Begonia atroglandulosa Sosef;黑腺秋海棠■
49641 Begonia augustinei Hemsl.;歪叶秋海棠(思茅秋海棠);Obliqueleaf Begonia ■
49642 Begonia auriculata Hook. f.;耳叶秋海棠;Auriculate-leaf Begonia,Cathedral Windows ■☆
49643 Begonia auritistipula Y. M. Shui et W. H. Chen;耳托秋海棠■
49644 Begonia austroguangxiensis Y. M. Shui et W. H. Chen;桂南秋海棠■
49645 Begonia austrotaiwanensis Y. K. Chen et C. I. Peng;南台湾秋海棠;S. Taiwan Begonia ■
49646 Begonia baccata Hook. f.;浆果秋海棠■
49647 Begonia bakeri C. DC.;巴克秋海棠(岜楞秋海棠);Baker Begonia ■☆
49648 Begonia balansana Gagnep.;北越秋海棠(香花秋海棠)■
49649 Begonia balansana Gagnep. var. rubropilosa S. H. Huang et Y. M. Shui = Begonia handelii Irmsch. var. rubropilosa (S. H. Huang et Y. M. Shui) C. I. Peng ■
49650 Begonia baronii Baker;巴龙秋海棠■☆
49651 Begonia batesii C. DC. = Begonia potamophila Gilg ■☆
49652 Begonia baumannii Lemoine;保曼秋海棠;Baumann Begonia ■☆
49653 Begonia baviensis Gagnep.;金平秋海棠;Jingping Begonia ■
49654 Begonia beddomei Hook. f.;白道木秋海棠;Boddome Begonia ■☆
49655 Begonia bellii H. Lév. = Begonia porteri H. Lév. et Vaniot ■
49656 Begonia bequaertii Robyns et Lawalrée;贝卡尔秋海棠■☆
49657 Begonia bernieri A. DC.;伯尼尔秋海棠■☆
49658 Begonia biflora T. C. Ku;双花秋海棠;Biflower Begonia ■
49659 Begonia binotii Hort.;比诺特秋海棠;Binot Begonia ■☆
49660 Begonia bipindensis Gilg ex Engl. = Begonia longipetiolata Gilg ■☆
49661 Begonia biserrata Lindl.;双锯齿秋海棠;Biserrate Begonia ■☆
49662 Begonia bismarckii Veitch;比斯马克秋海棠;Bismark Begonia ■☆
49663 Begonia bogneri Ziesenh.;博格纳秋海棠■☆
49664 Begonia boliviensis A. DC.;玻利维亚秋海棠(玻利秋海棠);Bolivian Begonia ■☆
49665 Begonia bonii Gagnep.;越南秋海棠■
49666 Begonia bonii Gagnep. var. remotisetulosa Y. M. Shui et W. H. Chen = Begonia debaoensis C. I. Peng et al. ■
49667 Begonia bosseri Rabenant.;博瑟秋海棠■☆
49668 Begonia bouffordii C. I. Peng;九九峰秋海棠■
49669 Begonia boweri Ziesenh.;鲍尔秋海棠(豹耳秋海棠,眉毛秋海棠);Eyelash Begonia ■☆
49670 Begonia boweri Ziesenh. 'Tiger';虎斑秋海棠■☆
49671 Begonia bowringiana Champ. ex Benth. = Begonia palmata D. Don var. bowringiana (Champ. ex Benth.) Golding et Kareg. ■
49672 Begonia bowringiana Hort. = Begonia cathayana Hemsl. ■
49673 Begonia brachyptera Hayata = Begonia aptera Hayata ■
49674 Begonia brachyptera Hayata = Begonia hayatae Gagnep. ■
49675 Begonia bradei Irmsch.;布拉得秋海棠;Brade Begonia ■☆
49676 Begonia brasiliensis Klotzsch;巴西秋海棠;Brazil Begonia ■☆
49677 Begonia bretschneideriana Hemsl. = Begonia leprosa Hance ■
49678 Begonia bretschneideriana Hemsl. et Irmsch. = Begonia leprosa Hance ■
49679 Begonia brevibracteata Kupicha;短苞秋海棠■☆
49680 Begonia brevicaulis T. C. Ku = Begonia gulinqingensis S. H. Huang et Y. M. Shui ■
49681 Begonia brevicaulis T. C. Ku = Begonia sinobrevicaulis T. C. Ku ■
49682 Begonia brevisetulosa C. Y. Wu;短刺秋海棠;Shortspine Begonia ■
49683 Begonia bruneelii De Wild. = Begonia macrocarpa Warb. ■☆
49684 Begonia buchholzii Gilg = Begonia preussii Warb. ■☆
49685 Begonia buddleiifolia A. DC.;醉鱼草叶秋海棠;Buddleia-leaved Begonia ■☆
49686 Begonia buimontana Yamam.;武威山秋海棠(武威秋海棠);Wuwei Begonia ■
49687 Begonia bulbosa H. Lév. = Begonia grandis Dryand. subsp. sinensis (A. DC.) Irmsch. ■
49688 Begonia buttonii Irmsch. = Begonia sutherlandii Hook. f. ■☆
49689 Begonia caffra Meisn. = Begonia dregei Otto et A. Dietr. ■☆
49690 Begonia caffra Meisn. = Begonia homonyma Steud. ■☆
49691 Begonia calabarica Stapf = Begonia quadrialata Warb. ■☆
49692 Begonia calophylla Gilg ex Engl. = Begonia anisosepala Hook. f. ■☆
49693 Begonia calophylla Irmsch. = Begonia algaia L. B. Sm. et Wassh. ■
49694 Begonia cameroonensis L. B. Sm. et Wassh. = Begonia ciliobracteata Warb. ■☆
49695 Begonia capillipes Gilg;纤毛秋海棠■☆
49696 Begonia caraguatatubensis Brade;卡拉秋秋海棠;Caraguatatuba Begonia ■☆
49697 Begonia carminata Veitch;洋红秋海棠;Carmine Begonia ■☆
49698 Begonia caroliniifolia Regel;喀罗林秋海棠;Carolinea-leaf Begonia ■☆
49699 Begonia carpinifolia Liebm.;鹅耳枥秋海棠;Beech-leaf Begonia ■☆
49700 Begonia cataractarum J. Braun et K. Schum. = Begonia polygonoides Hook. f. ■☆
49701 Begonia cathayana Hemsl.;花叶秋海棠(公鸡酸苔,花酸苔,花叶一口血,华秋海棠,苦酸苔,山海棠,中华秋海棠);Cathayan Begonia,Variegatedleaf Begonia ■
49702 Begonia cathcarti Hook. f. et Thomson;卡斯卡特秋海棠(花叶秋海棠);Cathcart Begonia ■☆
49703 Begonia cavaleriei H. Lév.;昌感秋海棠(昌感海棠,盾叶秋海棠,过山龙,红孩儿,金草箍,莲叶秋海棠,爬地龙,爬山猴,爬树龙,爬岩龙,麒麟叶,青竹标,岩蜈蚣,野海棠);Cavalerie Begonia,Peltateleaf Begonia ■
49704 Begonia cavaleriei H. Lév. = Begonia wangii Te T. Yu ■
49705 Begonia cavaleriei H. Lév. var. pinfaensis H. Lév. = Begonia cavaleriei H. Lév. ■
49706 Begonia cavaleriei H. Lév. var. pinfaensis H. Lév. = Begonia wangii Te T. Yu ■
49707 Begonia cavallyensis A. Chev.;卡瓦利耶秋海棠■☆
49708 Begonia cehengensis T. C. Ku;册亨秋海棠;Ceheng Begonia ■
49709 Begonia ceratocarpa S. H. Huang et Y. M. Shui;角果秋海棠;Corncarp Begonia ■
49710 Begonia cheimantha Everett = Begonia cheimantha Everett ex C. Weber ■☆
49711 Begonia cheimantha Everett ex C. Weber;圣诞秋海棠;Christmas Begonia ■☆
49712 Begonia cheimantha Everett ex C. Weber 'Gloire de Lorraine';洛林圣诞秋海棠;Christmas Begonia,Lorraine Begonia ■☆
49713 Begonia chevalieri Warb. ex A. Chev. = Begonia rostrata Welw. ex Hook. f. ■☆
49714 Begonia chingii Irmsch.;凤山秋海棠(广西秋海棠);Fengshan Begonia,Guangxi Begonia ■

49715 Begonia chishuiensis T. C. Ku;赤水秋海棠;Chishui Begonia ■
49716 Begonia chitoensis Ts. S. Liu et M. J. Lai;溪头秋海棠;Xitou Begonia ■
49717 Begonia chuniana C. Y. Wu;澄迈秋海棠;Chengmai Begonia ■
49718 Begonia chuniana C. Y. Wu = Begonia handelii Irmsch. var. prostrata (Irmsch.) Tebbitt ■
49719 Begonia chuyunshanensis C. I. Peng et Y. K. Chen;出云山秋海棠■
49720 Begonia ciliobracteata Warb.;毛苞秋海棠■☆
49721 Begonia cinnabarina Hook.;朱红秋海棠;Cinnabar Begonia ■☆
49722 Begonia circumlobata Hance et Irmsch.;周裂秋海棠(大麻酸杆,大麻酸汤杆,猴子酸,石酸苔,酸汤杆,野海棠,一口血);Circumlobed Begonia ■
49723 Begonia cirrosa L. B. Sm. et Wassh.;卷毛秋海棠(皱波秋海棠);Curly Begonia ■
49724 Begonia cladocarpa Baker;枝果秋海棠●☆
49725 Begonia cladocarpa Baker = Begonia oxyloba Welw. ex Hook. f. ■☆
49726 Begonia cladocarpoides Humbert ex Aymonin et Bosser;假枝果秋海棠■☆
49727 Begonia clarkei Hook. f.;克拉克秋海棠;C. B. Clarke Begonia ■☆
49728 Begonia clavicaulis Irmsch.;腾冲秋海棠;Tengchong Begonia ■
49729 Begonia coccinea Hook.;红花竹节秋海棠(巴西秋海棠,大红秋海棠,珊瑚秋海棠);Angel's Wing Begonia,Angel's-wing Begonia,Angelwing Begonia,Angel-wing Begonia,Scarlet Begonia ■☆
49730 Begonia compacta Bull.;密聚秋海棠;Compact Begonia ■☆
49731 Begonia comperei R. Wilczek = Begonia hirsutula Hook. f. ■
49732 Begonia conchifolia A. Dietr.;壳秋海棠(盘叶秋海棠);Shell Begonia ■☆
49733 Begonia concinna Schott;优雅秋海棠;Elegant Begonia ■☆
49734 Begonia conraui Gilg = Begonia oxyloba Welw. ex Hook. f. ■☆
49735 Begonia convolvulacea A. DC.;旋花秋海棠;Glorybind Begonia,Morning-glory Begonia ■☆
49736 Begonia cooperi C. DC.;库珀秋海棠;Cooper Begonia ■☆
49737 Begonia coptidifolia H. G. Ye et al.;阳春秋海棠■
49738 Begonia coptidi-montana C. Y. Wu;黄连山秋海棠;Huanglianshan Begonia ■
49739 Begonia corallina Carrière;珊瑚秋海棠;Coral Begonia ■☆
49740 Begonia cordifolia Thwaites;印度心叶秋海棠;Heart-leaved Begonia ■☆
49741 Begonia coriacea Hassk.;革质秋海棠;Leathery Begonia ■☆
49742 Begonia coursii Humbert ex Rabenant.;库尔斯秋海棠●☆
49743 Begonia crassipes Gilg ex Engl. = Begonia longipetiolata Gilg ■☆
49744 Begonia crassirostris Irmsch.;粗喙秋海棠(半边风,大半边莲,大海棠,鬼边榜,红半边莲,红莲,红小姐,老疸草,马酸通,肉半边莲,山蚂蝗,酸脚杆);Thickrostrate Begonia ■
49745 Begonia crassirostris Irmsch. = Begonia longifolia Blume ■
49746 Begonia crateris Exell = Begonia baccata Hook. f. ■
49747 Begonia crinita Oliv. ex Hook. f.;长毛秋海棠(具毛秋海棠);Longhair Begonia ■☆
49748 Begonia crispula Te T. Yu = Begonia cirrosa L. B. Sm. et Wassh. ■
49749 Begonia crispula Te T. Yu ex Irmsch. = Begonia cirrosa L. B. Sm. et Wassh. ■
49750 Begonia cristata Koord.;鸡冠秋海棠;Cristate Begonia ■☆
49751 Begonia crocea C. I. Peng;橙花秋海棠■
49752 Begonia crystallina Y. M. Shui et W. H. Chen;水晶秋海棠■
49753 Begonia cubensis Hassk.;冬青叶秋海棠;Cuban Holly,Hollyleaf Begonia,Holly-leaved Begonia ■☆
49754 Begonia cubincola A. DC. = Begonia cubensis Hassk. ■☆
49755 Begonia cucullata Willd.;兜状秋海棠;Clubed Begonia,Cucullate Begonia,Wax Begonia ■☆
49756 Begonia cucurbitifolia C. Y. Wu;瓜叶秋海棠;Melonleaf Begonia ■
49757 Begonia cultrata Irmsch. = Begonia capillipes Gilg ■☆
49758 Begonia curvicarpa S. M. Ku et al.;弯果秋海棠■
49759 Begonia cyclophylla Hook. f. = Begonia fimbristipula Hance ■
49760 Begonia cylindrica D. R. Liang et X. X. Chen;柱果秋海棠;Columcarp Begonia ■
49761 Begonia daveishanensis S. H. Huang et Y. M. Shui = Begonia dryadis Irmsch. ■
49762 Begonia davisii Veitch;戴维斯氏秋海棠;Davis Begonia ■☆
49763 Begonia daweishanensis S. H. Huang et Y. M. Shui;大围山秋海棠;Daweishan Begonia ■
49764 Begonia daweishanensis S. H. Huang et Y. M. Shui = Begonia dryadis Irmsch. ■
49765 Begonia daxinensis T. C. Ku;大新秋海棠;Daxin Begonia ■
49766 Begonia debaoensis C. I. Peng et al.;德保秋海棠■
49767 Begonia decandra Pav. ex A. DC.;十蕊秋海棠(十雄蕊秋海棠);Tenstamened Begonia ■☆
49768 Begonia decaryana Humbert ex Rabenant. et Bosser;德卡里秋海棠■
49769 Begonia decora Stapf;美爱秋海棠;Elegant Begonia ■☆
49770 Begonia delavayi Gagnep. = Begonia henryi Hemsl. ■
49771 Begonia deliciosa Lindl. ex Fotsch;爽快秋海棠■☆
49772 Begonia dentatobracteata C. Y. Wu;齿苞秋海棠;Toothbract Begonia ■
49773 Begonia dewildei Sosef;德维尔德秋海棠■☆
49774 Begonia diadema Lindau ex Rodigas;王冠秋海棠;Crown Begonia ■☆
49775 Begonia dichoroa Sprague;二色秋海棠(双色秋海棠,璎珞秋海棠);Two-coloured Begonia ■☆
49776 Begonia dichotoma Jacq.;叉叶秋海棠;Kidney Begonia ■☆
49777 Begonia dichrosa Sprague;璎珞秋海棠■
49778 Begonia dielsiana E. Pritz. ex Diels;南川秋海棠(鳞姜七);Nanchuan Begonia ■
49779 Begonia dielsiana Gilg = Begonia ciliobracteata Warb. ■☆
49780 Begonia dietrichiana Irmsch.;迪特秋海棠;Dietrich Begonia ■☆
49781 Begonia digitata Raddi;指状秋海棠;Digitate Begonia ■☆
49782 Begonia digyna Irmsch.;槭叶秋海棠(水八角,一口血);Mapleleaf Begonia,Two-styled Begonia ■
49783 Begonia dipetala Graham;二花秋海棠(二花瓣秋海棠);Two-petaled Begonia ■☆
49784 Begonia discolor R. Br. = Begonia grandis Dryand. ■
49785 Begonia discrepans Irmsch.;细茎秋海棠■
49786 Begonia discreta Craib;景洪秋海棠;Jinghong Begonia ■
49787 Begonia dissecta Irmsch. = Begonia sutherlandii Hook. f. ■☆
49788 Begonia domingensis A. DC.;道明秋海棠;Peanutbrittle Begonia ■☆
49789 Begonia dominicalis A. DC.;多米尼加秋海棠;Dominica Begonia ■☆
49790 Begonia dregei Otto et A. Dietr.;巴西葡萄叶秋海棠(葡叶秋海棠);Grape-leaf Begonia,Mapleleaf Begonia,Maple-leaf Begonia ■☆
49791 Begonia dryadis Irmsch.;厚叶秋海棠(红八角莲);Thickleaf Begonia ■
49792 Begonia duclouxii Gagnep. = Begonia henryi Hemsl. ■
49793 Begonia duclouxii Gagnep. et Irmsch.;川边秋海棠;Ducloux

Begonia ■
49794 Begonia duruensis De Wild. = Begonia ampla Hook. f. ■☆
49795 Begonia dusenii Warb. = Begonia quadrialata Warb. subsp. dusenii (Warb.) Sosef ■☆
49796 Begonia ealensis Irmsch. = Begonia eminii Warb. ■☆
49797 Begonia ebolowensis Engl.;埃博洛瓦秋海棠■☆
49798 Begonia echinata Royle = Begonia picta Sm. ■
49799 Begonia echinosepala Regel; 刺萼秋海棠; Prickly-sepaled Begonia ■☆
49800 Begonia ecuadoriensis Hort.;厄瓜多尔秋海棠;Ecuador Begonia ■☆
49801 Begonia edmundoi Brade;埃氏秋海棠;Edmundo Begonia ■☆
49802 Begonia edulis H. Lév.;食用秋海棠(南兰,葡萄叶秋海棠);Edible Begonia,Grapeleaf Begonia ■
49803 Begonia edulis H. Lév. var. henryi H. Lév. = Begonia palmata D. Don var. bowringiana (Champ. ex Benth.) Golding et Kareg. ■
49804 Begonia egregia N. E. Br.;非常秋海棠;Exellent Begonia ■☆
49805 Begonia elaeagnifolia Hook. f.;胡颓子秋海棠■☆
49806 Begonia elatostemmoides Hook. f.;楼梯草秋海棠■☆
49807 Begonia elliotii Gilg ex Engl. = Begonia rostrata Welw. ex Hook. f. ■☆
49808 Begonia emeiensis C. M. Hu ex C. Y. Wu et T. C. Ku;峨眉秋海棠;Emei Begonia ■
49809 Begonia eminii Warb.;爱妹秋海棠;Emin Begonia ■☆
49810 Begonia engleri Gilg;恩格勒秋海棠;Engler Begonia ■☆
49811 Begonia engleri Gilg var. nuda Irmsch. = Begonia engleri Gilg ■☆
49812 Begonia epilobioides Warb. = Begonia polygonoides Hook. f. ■☆
49813 Begonia epiphytica Hook. f. = Begonia mannii Hook. f. ■☆
49814 Begonia epipsila Brade;上裸秋海棠;Uponnaked Begonia ■☆
49815 Begonia erectocaulis Sosef;直立茎秋海棠■☆
49816 Begonia erectotricha Sosef;直立毛秋海棠■☆
49817 Begonia erubescens H. Lév. = Begonia grandis Dryand. ■
49818 Begonia erythrophylla Neuman;肾叶秋海棠(红叶秋海棠);Beefsteak,Kidney Begonia,Pond Lily ■☆
49819 Begonia esquirolii H. Lév. = Begonia cavaleriei H. Lév. ■
49820 Begonia estrellensis C. DC.;埃斯特雷亚秋海棠;Estrella Begonia ■☆
49821 Begonia evansiana Andréws = Begonia grandis Dryand. ■
49822 Begonia excelsa Hook. f. = Begonia mannii Hook. f. ■☆
49823 Begonia fangii Y. M. Shui et C. I. Peng;方氏秋海棠■
49824 Begonia favargeri Rech. = Begonia homonyma Steud. ■☆
49825 Begonia feastii Hort. = Begonia erythrophylla Neuman ■☆
49826 Begonia fengii T. C. Ku;矮小秋海棠;Feng Begonia ■
49827 Begonia fenicis Merr.;兰屿秋海棠;Lanyu Begonia ■
49828 Begonia fernandoi-costae Irmsch.;费尔南得秋海棠;Fernando Rib Begonia ■☆
49829 Begonia ferruginea Hayata = Begonia palmata D. Don var. bowringiana (Champ. ex Benth.) Golding et Kareg. ■
49830 Begonia ferruginea Hayata = Begonia randaiensis Sasaki ■
49831 Begonia ferruginea L. f.;锈叶秋海棠;Rusty-leaf Begonia ■☆
49832 Begonia ficicola Irmsch. = Begonia microsperma Warb. ■☆
49833 Begonia filicifolia Hallé = Begonia aspleniifolia Hook. f. ex A. DC. ■☆
49834 Begonia filiformis Irmsch.;丝形秋海棠;Filiform Begonia ■
49835 Begonia fimbriata Liebm.;流苏秋海棠;Fimbriate Begonia ■☆
49836 Begonia fimbribracteata Y. M. Shui et W. H. Chen;须苞秋海棠■
49837 Begonia fimbristipula Hance;紫背秋海棠(红水葵,红天葵,红叶,龙虎叶,散血子,天葵,夜渡红,圆叶秋海棠,紫背天葵);Fimbriate-stipulate Begonia,Roundleaf Begonia ■
49838 Begonia flava Marais = Begonia sutherlandii Hook. f. subsp. latior (Irmsch.) Kupicha ■☆
49839 Begonia flaviflora H. Hara;黄花秋海棠;Yellow Begonia ■
49840 Begonia flaviflora H. Hara var. gamblei (Irmsch.) Golding et Kareg.;浅裂黄花秋海棠■
49841 Begonia flaviflora H. Hara var. vivida Golding et Kareg.;乳黄秋海棠■
49842 Begonia floccifera Bedd.;绵毛秋海棠;Woolly Begonia ■☆
49843 Begonia floribunda T. C. Ku = Begonia sinofloribunda Dorr ■
49844 Begonia foliosa Kunth;多叶秋海棠;Fern-leaf Begonia,Fuchsia Begonia ■☆
49845 Begonia foliosa Kunth var. miniata L. B. Sm. et B. G. Schub.;倒挂金钟秋海棠(虎耳秋海棠,墨西哥秋海棠,小叶秋海棠);Corazon-de-jesua,Fuchsia Begonia ■
49846 Begonia fooningensis C. Y. Wu et al. = Begonia cirrosa L. B. Sm. et Wassh. ■
49847 Begonia fordii Irmsch.;西江秋海棠;Ford Begonia ■
49848 Begonia formosana (Hayata) Masam.;水鸭脚(台湾裂叶秋海棠);Taiwan Begonia ■
49849 Begonia formosana (Hayata) Masam. f. albomaculata Tang S. Liu et M. J. Lai;白斑水鸭脚;Whitespot Taiwan Begonia ■
49850 Begonia formosana (Hayata) Masam. f. albomaculata Tang S. Liu et M. J. Lai = Begonia formosana (Hayata) Masam. ■
49851 Begonia forrestii Irmsch.;陇川秋海棠;Forrest Begonia ■
49852 Begonia fragilis Baker = Begonia goudotii A. DC. ■☆
49853 Begonia francisia Ziesenh.;旱金莲秋海棠;Nastrutium-leaf Begonia ■☆
49854 Begonia francoisii Guillaumin;法兰西斯秋海棠■☆
49855 Begonia friburgensis Brade;弗里堡秋海棠;Fribourg Begonia ■☆
49856 Begonia froebelii A. DC.;圆心秋海棠;Froebel Begonia ■☆
49857 Begonia fruticosa A. DC.;灌木秋海棠;Shrubby Begonia ■☆
49858 Begonia fuchsioides Hook.;墨西哥秋海棠■
49859 Begonia fuchsioides Hook. = Begonia foliosa Kunth var. miniata L. B. Sm. et B. G. Schub. ■
49860 Begonia fulgens Lemoine;光亮秋海棠;Shining Begonia ■☆
49861 Begonia furfuracea Hook. f.;糠皮秋海棠■☆
49862 Begonia fusca Liebm. ex Klotzsch;红褐秋海棠;Fuscous Begonia ■☆
49863 Begonia fuscomaculata A. E. Lange;褐斑秋海棠(斑叶秋海棠,褐秋海棠);Fuscous-spotted Begonia ■☆
49864 Begonia fusialata Warb.;梭翅秋海棠■☆
49865 Begonia fusialata Warb. var. parviflora J. J. de Wilde;小花梭翅秋海棠■☆
49866 Begonia gabonensis J. J. de Wilde;加蓬秋海棠■☆
49867 Begonia gagnepainiana Irmsch.;昭通秋海棠;Gagnepain Begonia ■
49868 Begonia gentilii De Wild.;金提秋海棠;Gentil Begonia ■☆
49869 Begonia geranioides Hook. f.;老鹳草状秋海棠;Geranium-leaf Begonia,Geranium-like Begonia ■☆
49870 Begonia gesnerioides S. H. Huang et Y. M. Shui = Begonia hekouensis S. H. Huang ■
49871 Begonia gigantea Wall. = Begonia silletensis (A. DC.) C. B. Clarke ■
49872 Begonia gigaphylla Y. M. Shui et W. H. Chen = Begonia liuyanii C. I. Peng et al. ■
49873 Begonia gilgii Engl. = Begonia sessilifolia Hook. f. ■☆

49874 Begonia glabra Aubl.;光滑秋海棠(光秋海棠,无毛秋海棠);Glabrous Begonia ■☆

49875 Begonia glabra Aubl. var. cordifolia C. DC.;心叶光滑秋海棠■☆

49876 Begonia gladiifolia Engl. = Begonia longipetiolata Gilg ■☆

49877 Begonia glaucophylla Hook. f.;垂枝秋海棠;Ivy Begonia, Shrimp Begonia ■☆

49878 Begonia glechomifolia C. M. Hu ex C. Y. Wu et T. C. Ku;金秀秋海棠(心叶秋海棠);Jinxiu Begonia ■

49879 Begonia goegoensis N. E. Br.;火焰秋海棠;Fire-king Begonia ■☆

49880 Begonia gossweileri Irmsch.;戈斯秋海棠■☆

49881 Begonia goudotii A. DC.;马岛秋海棠■☆

49882 Begonia gouroana A. Chev. = Begonia macrocarpa Warb. ■☆

49883 Begonia gracilicaulis Irmsch. = Begonia macrocarpa Warb. ■☆

49884 Begonia gracilipetiolata De Wild. = Begonia longipetiolata Gilg ■☆

49885 Begonia gracilis Kunth;异叶秋海棠(细秋海棠);Hollyhock Begonia, Slender Begonia ■☆

49886 Begonia grandis Dryand.;秋海棠(八香,八月春,春海棠,大红袍,大花秋海棠,断肠草,断肠花,红白二丸,红白二元,红黑二丸,金线吊葫芦,山海棠,水八角,无名断肠草,无名相思草,相思草,小桃红,岩丸子,一点血,一口血,璎珞草,鸳鸯七,中华秋海棠);Evans Begonia, Hardy Begonia ■

49887 Begonia grandis Dryand. 'Alba';白色秋海棠;White Hardy Begonia ■☆

49888 Begonia grandis Dryand. subsp. evansiana (Andréws) Irmsch. = Begonia grandis Dryand. ■

49889 Begonia grandis Dryand. subsp. holostyla Irmsch.;全柱秋海棠;Holostyl ■

49890 Begonia grandis Dryand. subsp. sinensis (A. DC.) Irmsch.;中华秋海棠(红黑二丸,红黑二圆,扣子花,岩丸子,一点血,一口血,鸳鸯七,珠芽秋海棠);China Begonia ■

49891 Begonia gueinziana Irmsch. = Begonia sutherlandii Hook. f. ■☆

49892 Begonia guishanensis S. H. Huang et Y. M. Shui;圭山秋海棠;Guishan Begonia ■

49893 Begonia guishanensis S. H. Huang et Y. M. Shui = Begonia labordei H. Lév. ■

49894 Begonia gulinqingensis S. H. Huang et Y. M. Shui;古林箐秋海棠;Gulinqin Begonia ■

49895 Begonia gungshanensis C. Y. Wu;贡山秋海棠;Gonshan Begonia ■

49896 Begonia haageana W. Watson = Begonia scharffii Hook. f. ■☆

49897 Begonia hainanensis Chun et F. Chun;海南秋海棠;Hainan Begonia ■

49898 Begonia handelii Irmsch.;香花秋海棠(大香秋海棠,短茎秋海棠,铁木,香秋海棠,异株秋海棠);Handel Begonia ■

49899 Begonia handelii Irmsch. var. prostrata (Irmsch.) Tebbitt;铺地秋海棠(匍匐秋海棠,信宜秋海棠);Prostrate Begonia, Xinyi Begonia ■

49900 Begonia handelii Irmsch. var. rubropilosa (S. H. Huang et Y. M. Shui) C. I. Peng;红毛香花秋海棠■

49901 Begonia handroi Brade;韩氏秋海棠;Handro Begonia ■☆

49902 Begonia harrowiana Diels = Begonia labordei H. Lév. ■

49903 Begonia hatacoa Buch. -Ham. ex D. Don;墨脱秋海棠(赤脉秋海棠);Motuo Begonia ■

49904 Begonia haullevilleana De Wild. = Begonia poculifera Hook. f. ■☆

49905 Begonia hauttuynioides Te T. Yu = Begonia limprichtii Irmsch. ■

49906 Begonia hayatae Gagnep.;圆果秋海棠■

49907 Begonia hayatae Gagnep. = Begonia aptera Hayata ■

49908 Begonia hayatae Gagnep. = Begonia longifolia Blume ■

49909 Begonia heddei Warb. = Begonia oxyloba Welw. ex Hook. f. ■☆

49910 Begonia hekouensis S. H. Huang;河口秋海棠;Hekou Begonia ■

49911 Begonia hemsleyana Hook. f.;掌叶秋海棠(刺海棠,岩酸菇);Hemsley Begonia ■

49912 Begonia hemsleyana Hook. f. var. kwangsiensis Irmsch.;广西秋海棠(广西掌叶秋海棠);Guangxi Begonia ■

49913 Begonia henriquesii C. DC. = Begonia loranthoides Hook. f. ■

49914 Begonia henryi Hemsl.;独牛(红白二丸,活血,柔毛秋海棠,石鼓子,酸杆杆,岩酸,岩丸子,一口血,一面锣);Henry Begonia ■

49915 Begonia heracleifolia Cham. et Schltdl.;大苞秋海棠(裂叶秋海棠,星叶秋海棠);Star Begonia, Starleaf Begonia, Star-leaf Begonia ■

49916 Begonia heracleifolia Cham. et Schltdl. var. nigricans Hook.;黑大苞秋海棠■

49917 Begonia herbacea Vell.;草本秋海棠;Herbaceous Begonia ■☆

49918 Begonia heterochroma Sosef;异色秋海棠■

49919 Begonia heteropoda Baker;异足秋海棠■

49920 Begonia hidalgensis L. B. Sm. et B. G. Schub.;褐脉秋海棠■

49921 Begonia hiemalis Fotsch;冬花秋海棠;Elatior Begonia, Winter-flowering Begonia ■☆

49922 Begonia hirsuta Aubl.;硬毛秋海棠;Hirsute Begonia ■☆

49923 Begonia hirsutula Hook. f.;毛秋海棠■

49924 Begonia hirtella Link;微硬毛秋海棠;Brazilian Begonia, Hairy Begonia ■☆

49925 Begonia hispida Schott ex A. DC.;粗毛秋海棠;Hispid Begonia ■☆

49926 Begonia hispida Schott ex A. DC. var. cucullifera Irmsch.;肩背秋海棠;Piggyback Begonia, Piggy-back Begonia ■☆

49927 Begonia hispidivillosa Ziesenh.;柔毛秋海棠;Hispid-villous Begonia ■☆

49928 Begonia homblei De Wild. = Begonia princeae Gilg ■☆

49929 Begonia homonyma Steud.;全优秋海棠■☆

49930 Begonia hongkongensis F. W. Xing;香港秋海棠■

49931 Begonia hookeriana Gilg ex Engl. = Begonia ciliobracteata Warb. ■☆

49932 Begonia horticola Irmsch.;庭院秋海棠■☆

49933 Begonia houttuynioides Te T. Yu = Begonia limprichtii Irmsch. ■

49934 Begonia howii Merr. et Chun;侯氏秋海棠;How Begonia ■

49935 Begonia huangii Y. M. Shui et W. H. Chen;黄氏秋海棠■

49936 Begonia humbertii Rabenant.;亨伯特秋海棠■☆

49937 Begonia humilis Dryand.;矮秋海棠;Trinided Begonia ■☆

49938 Begonia hydrocotylifolia Otto ex Hook.;小睡莲秋海棠(铺地秋海棠);Miniature Pond-lily Begonia, Pennywort Begonia ■☆

49939 Begonia hymenocarpa C. Y. Wu;膜果秋海棠;Membranacarp Begonia ■

49940 Begonia imitans Irmsch.;鸡爪秋海棠■

49941 Begonia imperialis Lem.;壮丽秋海棠(绒毛秋海棠,毡毛秋海棠);Carpet Begonia, Imperial Begonia ■☆

49942 Begonia imperialis Lem. var. smaragdina Lem.;绿壮丽秋海棠;Green Imperial Begonia ■☆

49943 Begonia incana Lindl.;灰白毛秋海棠(莲叶秋海棠,绵毛秋海棠);Incanous Begonia, Lily-pad Begonia ■☆

49944 Begonia incarnata Link et Otto;肉红秋海棠;Incarnate Begonia ■☆

49945 Begonia inciso-serrata A. DC.;刻锯秋海棠;Incise-serrate Begonia ■☆

49946 Begonia inflata C. B. Clarke;膨胀秋海棠■☆

49947 Begonia injoloensis De Wild. = Begonia loranthoides Hook. f. subsp. rhopalocarpa (Warb.) J. J. de Wilde ■

49948 Begonia involucrata Liebm.;总苞秋海棠;Involucrate Begonia ■☆
49949 Begonia isalensis Humbert ex Rabenant. et Bosser;伊萨卢秋海棠■☆
49950 Begonia isoptera Dryand. ex Sm.;同翼秋海棠;Equal-winged Begonia ■☆
49951 Begonia jingxiensis D. Fang et Y. G. Wei;靖西秋海棠■
49952 Begonia johnstonii Oliv. ex Hook.;约翰司顿秋海棠;Johnston Begonia ■☆
49953 Begonia johnstonii Oliv. ex Hook. f. pilosa Irmsch. = Begonia johnstonii Oliv. ex Hook. ■☆
49954 Begonia josephii A. DC.;重齿秋海棠(钝叶秋海棠);Joseph Begonia ■
49955 Begonia jussiaeicarpa Warb. = Begonia oxyanthera Warb. ■☆
49956 Begonia karwinskyana A. DC.;卡尔温斯克秋海棠;Karwinsk Begonia ■☆
49957 Begonia kellermanii C. DC.;凯氏秋海棠■☆
49958 Begonia keniensis Gilg ex Engl. = Begonia wollastonii Baker f. ■☆
49959 Begonia kenworthyi Ziesenh.;肯沃奇秋海棠;Kenworthy Begonia ■☆
49960 Begonia keraudrenae Bosser;克罗德朗秋海棠■☆
49961 Begonia kewensis Gentil;邱园秋海棠;Kew Begonia ■☆
49962 Begonia klainei Pierre ex Pellegr. = Begonia hirsutula Hook. f. ■
49963 Begonia knowsleyana Hort.;努斯莱秋海棠;Knowsley Begonia ■☆
49964 Begonia komoensis Irmsch.;科莫秋海棠■☆
49965 Begonia kotoensis Hayata = Begonia fenicis Merr. ■
49966 Begonia kouy-tcheouensis Guillaumin;贵州秋海棠;Guizhou Begonia ■
49967 Begonia kouy-tcheouensis Guillaumin = Begonia palmata D. Don var. bowringiana (Champ. ex Benth.) Golding et Kareg. ■
49968 Begonia kribensis Engl. = Begonia longipetiolata Gilg ■☆
49969 Begonia kummeriae Gilg = Begonia oxyloba Welw. ex Hook. f. ■☆
49970 Begonia kunthiana Walp.;孔斯秋海棠;Kunth Begonia ■☆
49971 Begonia labordei H. Lév.;心叶秋海棠(大岩酸,红盘,丽江秋海棠,俅江秋海棠,一口血);Heartleaf Begonia,Laborde Begonia ■
49972 Begonia labordei H. Lév. var. unialata T. C. Ku;窄檐心叶秋海棠;Unialate Laborde Begonia ■
49973 Begonia lacerata Irmsch.;撕裂秋海棠(蒙自秋海棠);Lacerate Begonia ■
49974 Begonia laciniata Roxb.;峦大秋海棠(半边莲,红孩儿,红莲,红天葵,虎斑海棠,九齿莲,裂叶秋海棠,石莲,细裂秋海棠,血蜈蚣,岩红);Laciniate Begonia ■
49975 Begonia laciniata Roxb. = Begonia randaiensis Sasaki ■
49976 Begonia laciniata Roxb. ex Wall. = Begonia palmata D. Don ■
49977 Begonia laciniata Roxb. subsp. bowringiana Irmsch. = Begonia palmata D. Don var. bowringiana (Champ. ex Benth.) Golding et Kareg. ■
49978 Begonia laciniata Roxb. subsp. crassisetulosa Irmsch. = Begonia palmata D. Don var. crassisetulosa (Irmsch.) Golding et Kareg. ■
49979 Begonia laciniata Roxb. subsp. difformis Irmsch. = Begonia palmata D. Don var. difformis (Irmsch.) Golding et Kareg. ■
49980 Begonia laciniata Roxb. subsp. flava (C. B. Clarke) Irmsch. = Begonia flaviflora H. Hara ■
49981 Begonia laciniata Roxb. subsp. flaviflora Irmsch. = Begonia flaviflora H. Hara var. vivida Golding et Kareg. ■
49982 Begonia laciniata Roxb. subsp. gamblei Irmsch. = Begonia flaviflora H. Hara var. gamblei (Irmsch.) Golding et Kareg. ■
49983 Begonia laciniata Roxb. subsp. laevifolia Irmsch. = Begonia palmata D. Don var. laevifolia (Irmsch.) Golding et Kareg. ■
49984 Begonia laciniata Roxb. subsp. principalis Irmsch. = Begonia palmata D. Don var. bowringiana (Champ. ex Benth.) Golding et Kareg. ■
49985 Begonia laciniata Roxb. var. bowringiana (Champ. ex Benth.) A. DC. = Begonia palmata D. Don var. bowringiana (Champ. ex Benth.) Golding et Kareg. ■
49986 Begonia laciniata Roxb. var. bowringiana A. DC.;花晕细裂秋海棠;Bowring Begonia ■☆
49987 Begonia laciniata Roxb. var. flava Hook.;黄秋海棠;Yellow Begonia ■☆
49988 Begonia laciniata Roxb. var. formosana Hayata = Begonia formosana (Hayata) Masam. ■
49989 Begonia laciniata Roxb. var. laevifolia Irmsch. = Begonia palmata D. Don var. laevifolia (Irmsch.) Golding et Kareg. ■
49990 Begonia laciniata Roxb. var. nepalensis A. DC. = Begonia palmata D. Don ■
49991 Begonia laciniata Roxb. var. tuberculosa C. B. Clarke = Begonia palmata D. Don ■
49992 Begonia lacunosa Warb.;具腔秋海棠■☆
49993 Begonia laminariae Irmsch.;圆翅秋海棠(薄叶秋海棠,毛酸筒,酸汤竿);Roundwing Begonia ■
49994 Begonia lancangensis S. H. Huang;澜沧秋海棠;Lancang Begonia ■
49995 Begonia lanternaria Irmsch.;灯果秋海棠■
49996 Begonia laporteifolia Warb.;艾麻叶秋海棠■☆
49997 Begonia latipetiolata Irmsch. = Begonia subscutata De Wild. ■☆
49998 Begonia latistipula Engl. = Begonia macrocarpa Warb. ■☆
49999 Begonia leandrii Humbert ex Rabenant. et Bosser;利安秋海棠■☆
50000 Begonia lebrunii Robyns et Lawalrée = Begonia wollastonii Baker f. ■☆
50001 Begonia lehmbachii Warb. = Begonia oxyloba Welw. ex Hook. f. ■☆
50002 Begonia leichtiana Irmsch. ex Sosef = Begonia quadrialata Warb. subsp. nimbaensis Sosef ■☆
50003 Begonia lemurica Rabenant.;莱穆拉秋海棠■☆
50004 Begonia leprosa Hance;癞叶秋海棠(伯乐秋海棠,老虎耳,石上海棠,石上莲,石上莲秋海棠,石上秋海棠,团扇叶秋海棠);Scabby Begonia ■
50005 Begonia leprosa Hance = Begonia obsolescens Irmsch. ■
50006 Begonia leprosa Hance = Begonia versicolor Irmsch. ■
50007 Begonia leptotricha C. DC.;细毛秋海棠;Manda' s Woolly-bear Begonia ■☆
50008 Begonia letestui J. J. de Wilde;莱泰斯图秋海棠■☆
50009 Begonia lethomasiae R. Wilczek = Begonia ebolowensis Engl. ■☆
50010 Begonia letouzeyi Sosef;勒图秋海棠■☆
50011 Begonia liebmannii A. DC.;利氏秋海棠■☆
50012 Begonia limmingheana E. Morren;虾仔海棠(垂枝秋海棠);Ivy Begonia,Shrimp Begonia ■☆
50013 Begonia limprichtii Irmsch.;蕺叶秋海棠(七星花);Limbricht Begonia ■
50014 Begonia limprichtii Irmsch. = Begonia smithiana Te T. Yu ex Irmsch. ■
50015 Begonia lindleyana Walp.;林德利秋海棠(林得莱秋海棠);Lindley Begonia ■☆
50016 Begonia lipingensis Irmsch.;黎平秋海棠;Liping Begonia ■
50017 Begonia listida Brade;金线秋海棠■
50018 Begonia lithophila C. Y. Wu;石生秋海棠(石灰秋海棠);Cliff

Begonia ■
50019 Begonia liuyanii C. I. Peng et al.;刘演秋海棠■
50020 Begonia lobata Schott;裂叶秋海棠(分裂秋海棠);Honey-bear Begonia ■☆
50021 Begonia lokobeensis Humbert ex Rabenant. et Bosser = Begonia humilis Dryand. ■☆
50022 Begonia loloensis Gilg = Begonia elatostemmoides Hook. f. ■☆
50023 Begonia longanensis C. Y. Wu;隆安秋海棠;Long'an Begonia ■
50024 Begonia longialata K. Y. Guan et D. K. Tian;长翅秋海棠;Longwing Begonia ■
50025 Begonia longibarbata Brade;长髯毛秋海棠;Long-bard Begonia ■☆
50026 Begonia longicarpa K. Y. Guan et D. K. Tian;长果秋海棠;Longcarp Begonia ■
50027 Begonia longiciliata C. Y. Wu;长秋海棠;Longiciate Begonia ■
50028 Begonia longiciliata C. Y. Wu = Begonia rex Putz. ■
50029 Begonia longifolia Blume;长叶秋海棠(粗喙秋海棠)■
50030 Begonia longipetiolata Gilg;长梗秋海棠■☆
50031 Begonia longistyla Y. M. Shui et W. H. Chen;长柱秋海棠■
50032 Begonia loranthoides Hook. f.;桑寄生秋海棠■
50033 Begonia loranthoides Hook. f. subsp. rhopalocarpa (Warb.) J. J. de Wilde;棒果秋海棠■
50034 Begonia lubbersii E. Morren;双尖秋海棠(露伯秋海棠);Lubbers Begonia ■☆
50035 Begonia ludicra A. DC.;玩耍秋海棠;Playful Begonia ■☆
50036 Begonia ludwigii Irmsch.;路德维格秋海棠;Ludwig Begonia ■☆
50037 Begonia ludwigsii Gilg = Begonia longipetiolata Gilg ■☆
50038 Begonia lukuana Y. C. Liu et C. H. Ou;鹿谷秋海棠;Lugu Begonia ■
50039 Begonia luochengensis S. M. Ku et al.;罗城秋海棠■
50040 Begonia lutea L. B. Sm. et B. G. Schub.;纯黄秋海棠;Luteous Begonia ■☆
50041 Begonia luxurians Scheidw.;繁茂秋海棠;Palm-leaf Begonia ■☆
50042 Begonia luzhaiensis T. C. Ku;鹿寨秋海棠;Luzhai Begonia ■
50043 Begonia lyallii A. DC.;莱尔秋海棠■☆
50044 Begonia macambrarensis Exell = Begonia subalpestris A. Chev. ■☆
50045 Begonia macdougallii Ziesenh.;迈克刀秋海棠;Macdougall Begonia ■☆
50046 Begonia macrocarpa Warb.;大果秋海棠;Big-fruit Begonia ■☆
50047 Begonia macropoda Gilg = Begonia potamophila Gilg ■☆
50048 Begonia macrostyla Warb. = Begonia eminii Warb. ■☆
50049 Begonia macrotoma Irmsch.;大裂秋海棠■
50050 Begonia macrura Gilg = Begonia longipetiolata Gilg ■☆
50051 Begonia maculata Raddi;银点秋海棠(斑叶竹节秋海棠,银斑秋海棠,竹节秋海棠);Spotted Begonia ■
50052 Begonia madecassa Rabenant.;马德卡萨秋海棠●☆
50053 Begonia maguanensis S. H. Huang et Y. M. Shui = Begonia paucilobata C. Y. Wu var. maguanensis (S. H. Huang et Y. M. Shui) T. C. Ku ■
50054 Begonia mairei H. Lév. = Begonia henryi Hemsl. ■
50055 Begonia malabarica Lam. = Begonia roxburghii A. DC. ■☆
50056 Begonia malipoensis S. H. Huang et Y. M. Shui;麻栗坡秋海棠;Palipo Begonia ■
50057 Begonia mangorensis Humbert ex Rabenant.;曼古鲁秋海棠■☆
50058 Begonia manhaoensis S. H. Huang et Y. M. Shui;蛮耗秋海棠(蔓耗秋海棠);Manhao Begonia ■
50059 Begonia manicata Cels ex Vis.;长袖秋海棠;Long-sleeved Begonia ■☆
50060 Begonia manicata Cels ex Vis. 'Aurea-maculata';黄斑长袖秋海棠;Leopard Begonia ■☆
50061 Begonia manicata Cels ex Vis. 'Crispa';皱边长袖秋海棠■☆
50062 Begonia manicata Cels ex Vis. 'Cristata' = Begonia manicata Cels ex Vis. 'Crispa' ■☆
50063 Begonia mannii Hook. f.;玫瑰叶秋海棠;Rose-leaf Begonia ■☆
50064 Begonia marojejyensis Humbert;马罗秋海棠■☆
50065 Begonia martini H. Lév. = Begonia grandis Dryand. subsp. sinensis (A. DC.) Irmsch. ■
50066 Begonia mashanica D. Fang et D. H. Qin = Begonia jingxiensis D. Fang et Y. G. Wei ■
50067 Begonia masoniana Irmsch.;铁甲秋海棠(彩斑秋海棠,彩纹海棠,刺毛秋海棠,毛叶秋海棠,铁十字秋海棠);Iron Cross Begonia,Iron-cross Begonia ■
50068 Begonia masoniana Irmsch. var. maculata S. K. Chen,R. X. Zheng et D. Y. Xia = Begonia masoniana Irmsch. ■
50069 Begonia mayombensis Irmsch. = Begonia lacunosa Warb. ■☆
50070 Begonia mazae Ziesenh.;马兹秋海棠;Maza Begonia ■☆
50071 Begonia megalophyllaria C. Y. Wu;大叶秋海棠;Bigleaf Begonia ■
50072 Begonia megaptera A. DC.;大翅秋海棠;Large-winged Begonia ■☆
50073 Begonia membranifera Chun et F. Chun;红水葵(天葵海棠);Membraneous Begonia ■
50074 Begonia menglianensis Y. Y. Qian;孟连秋海棠;Menglian Begonia ■
50075 Begonia mengtzeana Irmsch.;肾托秋海棠(蒙自秋海棠)■
50076 Begonia metallica W. G. Sm.;金绿秋海棠(撒金秋海棠,洒金秋海棠);Metal-leaf Begonia,Metallic-leaf Begonia,Steel Begonia ■☆
50077 Begonia mexiae Standl.;墨氏秋海棠;Mex Begonia ■☆
50078 Begonia meyeri-johannis Engl.;迈尔约翰秋海棠■☆
50079 Begonia micranthera Griseb.;小花秋海棠(微花秋海棠);Small-flowered Begonia ■☆
50080 Begonia microsperma Warb.;小籽秋海棠■☆
50081 Begonia mildbraedii Gilg;米尔德秋海棠■☆
50082 Begonia minor Jacq.;细小秋海棠(小秋海棠);Minor Begonia ■☆
50083 Begonia minuta Sosef;微小秋海棠■☆
50084 Begonia minutifolia N. Hallé;微花秋海棠■☆
50085 Begonia miranda Irmsch.;截裂秋海棠(奇异秋海棠)■
50086 Begonia miranda Irmsch. = Begonia laminariae Irmsch. ■
50087 Begonia modestiflora Kurz;云南秋海棠(白棉胡,大麻酸汤杆,红耗儿,红耗子,化血丹,金蝉脱壳,老鸦枕头,山海棠,水八角,酸草果,酸苹果,小桃红,血当归,腰包花,野海棠,一口血);Yunnan Begonia ■
50088 Begonia modica Stapf = Begonia quadrialata Warb. ■☆
50089 Begonia molleri (C. DC.) Warb.;默勒秋海棠■☆
50090 Begonia monicae Aymonin et Bosser;念珠秋海棠■☆
50091 Begonia morifolia Te T. Yu;桑叶秋海棠;Mulberry-leaf Begonia ■
50092 Begonia morsei Irmsch.;龙州秋海棠;Longzhou Begonia ■
50093 Begonia morsei Irmsch. var. myriotricha Y. M. Shui et W. H. Chen;密毛龙州秋海棠■
50094 Begonia muliensis Te T. Yu;木里秋海棠;Muli Begonia ■
50095 Begonia multangula Blume;多角秋海棠;Many-anguled Begonia ■☆
50096 Begonia multiflora Benth.;繁花秋海棠;Many-flowered Begonia ■☆
50097 Begonia multinervia Liebm.;多脉秋海棠;Many-veined Begonia ■☆
50098 Begonia muricata Blume;粗糙秋海棠;Muricate Begonia ■☆

50099 Begonia nana L'Hér.;低矮秋海棠■☆
50100 Begonia nantoensis M. J. Lai et N. J. Chung;南投秋海棠;Nantou Begonia ■
50101 Begonia natalensis Hook.;纳塔尔秋海棠;Natal Begonia ■☆
50102 Begonia natalensis Hook. = Begonia dregei Otto et A. Dietr. ■☆
50103 Begonia ndongensis Engl.;恩东加秋海棠■☆
50104 Begonia nelumbiifolia Schltdl. et Cham.;莲叶秋海棠;Lilypad Begonia, Lily-pad Begonia, Pond-lily Begonia ■☆
50105 Begonia neoperrieri Humbert ex Rabenant.;佩里耶秋海棠■☆
50106 Begonia nepalensis Warb.;尼泊尔秋海棠■
50107 Begonia nicolai-hallei R. Wilczek = Begonia longipetiolata Gilg ■☆
50108 Begonia nigricans L. H. Bailey;黑色秋海棠;Black Begonia ■☆
50109 Begonia ningmingensis D. Fang et al.;宁明秋海棠■
50110 Begonia ningmingensis D. Fang et al. var. bella D. Fang et al.;丽叶秋海棠■
50111 Begonia nitida Dryand.;亮叶秋海棠;Glossy Begonia ■☆
50112 Begonia nossibea A. DC.;诺西波秋海棠■☆
50113 Begonia nyassensis Irmsch.;尼亚萨秋海棠■☆
50114 Begonia nymphaeafolia Te T. Yu;睡莲叶秋海棠;Waterlilyleaf Begonia ■☆
50115 Begonia nymphaeafolia Te T. Yu = Begonia cavaleriei H. Lév. ■
50116 Begonia obliqua Jacq. = Begonia grandis Dryand. ■
50117 Begonia obliquifolia S. H. Huang et Y. M. Shui;斜叶秋海棠■
50118 Begonia obscura Brade;不显秋海棠(不明显秋海棠);Obscure Begonia ■☆
50119 Begonia obsolescens Irmsch.;侧膜秋海棠(不显秋海棠,麻栗坡秋海棠);Obsolescent Begonia ■
50120 Begonia octopetala L'Hér.;八瓣秋海棠;Eightpetal Begonia ■☆
50121 Begonia odorata Willd.;香秋海棠;Fragrant Begonia ■☆
50122 Begonia olbia Kerch.;巴西槭叶秋海棠;Maple-leaf Begonia ■☆
50123 Begonia olsoniae L. B. Sm. et B. G. Schub.;奥氏秋海棠(奥尔逊尼亚秋海棠);Olson Begonia ■☆
50124 Begonia oreodoxa Chun et F. Chun ex C. Y. Wu et T. C. Ku;山地秋海棠■
50125 Begonia ornithophylla Irmsch.;鸟叶秋海棠■
50126 Begonia ovatifolia A. DC.;卵叶秋海棠;Ovateleaf Begonia ■☆
50127 Begonia oxyanthera Warb.;尖药秋海棠■☆
50128 Begonia oxyloba Welw. ex Hook. f.;尖裂叶秋海棠■☆
50129 Begonia oxyphylla A. DC.;尖叶秋海棠;Sharpleaf Begonia ■☆
50130 Begonia palmaris A. DC.;双齿秋海棠(红孩儿,细裂秋海棠,岩红);Biserrate Begonia ■☆
50131 Begonia palmata D. Don;九齿莲(八多酸,半边莲,红八角莲,红孩儿,红莲,裂叶秋海棠,石莲,岩红);Palmate Begonia ■
50132 Begonia palmata D. Don = Begonia randaiensis Sasaki ■
50133 Begonia palmata D. Don var. bowringiana (Champ. ex Benth.) Golding et Kareg.;红孩儿秋海棠(红孩儿,裂叶秋海棠,思茅食用秋海棠)■
50134 Begonia palmata D. Don var. crassisetulosa (Irmsch.) Golding et Kareg.;刺毛红孩儿秋海棠(刺毛红孩儿)■
50135 Begonia palmata D. Don var. difformis (Irmsch.) Golding et Kareg.;不齐红孩儿秋海棠(变形红孩儿)■
50136 Begonia palmata D. Don var. gamblei H. Hara = Begonia flaviflora H. Hara var. gamblei (Irmsch.) Golding et Kareg. ■
50137 Begonia palmata D. Don var. laevifolia (Irmsch.) Golding et Kareg.;光叶红孩儿秋海棠(光叶红孩儿,光叶秋海棠);Smooth-leaf Begonia ■
50138 Begonia parilis Irmsch.;均一秋海棠;Zigzag Begonia ■☆
50139 Begonia partita Irmsch. = Begonia dregei Otto et A. Dietr. ■☆
50140 Begonia parva Merr.;小秋海棠;Petty Begonia ■☆
50141 Begonia parva Sprague = Begonia horticola Irmsch. ■☆
50142 Begonia parvifolia Graham = Begonia dregei Otto et A. Dietr. ■☆
50143 Begonia parvula H. Lév. et Vaniot;小叶秋海棠(稍小秋海棠,小秋海棠);Small Begonia ■
50144 Begonia paucilobata C. Y. Wu;少裂秋海棠;Paucilobed Begonia ■
50145 Begonia paucilobata C. Y. Wu var. maguanensis (S. H. Huang et Y. M. Shui) T. C. Ku;马关秋海棠;Maguan Begonia ■
50146 Begonia paulensis A. DC.;保罗秋海棠(蜘蛛网秋海棠);Paul Begonia ■☆
50147 Begonia pearcei Hook. f.;皮尔斯氏秋海棠;Pearce Begonia ■☆
50148 Begonia pedatifida H. Lév.;掌裂秋海棠(风吹不动,枫香细辛,虎爪龙,花鸡公,裂叶秋海棠,水八角,水黄连,水蜈蚣,酸猴儿,一点血,一口血,掌裂叶秋海棠);Palmleaf Begonia, Pedatifid Begonia ■
50149 Begonia pedatifida H. Lév. var. kewensis H. Lév. = Begonia lipingensis Irmsch. ■
50150 Begonia pedatifida H. Lév. var. kewensis H. Lév. = Begonia pedatifida H. Lév. ■
50151 Begonia peii C. Y. Wu;裴氏秋海棠(小花秋海棠);Pei Begonia ■
50152 Begonia peltata Otto et Dietr. = Begonia incana Lindl. ■☆
50153 Begonia peltatifolia H. L. Li;盾叶秋海棠;Peltateleaf Begonia ■
50154 Begonia pendula Ridl.;悬垂秋海棠;Pendulous Begonia ■☆
50155 Begonia peperomioides Hook. f.;草胡椒秋海棠■☆
50156 Begonia perpusilla A. DC.;弱小秋海棠■☆
50157 Begonia petraea A. Chev. = Begonia prismatocarpa Hook. subsp. petraea (A. Chev.) Sosef ■☆
50158 Begonia petrophila Gilg = Begonia oxyloba Welw. ex Hook. f. ■☆
50159 Begonia philodendroides Ziesenh.;喜林芋状秋海棠;Philodendronlike Begonia ■☆
50160 Begonia phyllomaniaca Mart.;摇叶秋海棠;Crazy-leafpinetorum ■☆
50161 Begonia picta Sm.;樟木秋海棠(尼泊尔秋海棠);Nepal Begonia ■
50162 Begonia picturata Yan Liu et al.;一口血秋海棠■
50163 Begonia pinetorum A. DC.;松林秋海棠;Pineforest Begonia ■☆
50164 Begonia pingbienensis C. Y. Wu;睫毛秋海棠;Pingbian Begonia ■
50165 Begonia pinglinensis C. I. Peng;坪林秋海棠■
50166 Begonia plantanifolia Schott;悬铃木秋海棠;Plantree-leaf Begonia ■☆
50167 Begonia platycarpa Y. M. Shui et W. H. Chen;扁果秋海棠■
50168 Begonia plebeja Liebm.;普通秋海棠;Common Begonia ■☆
50169 Begonia poculifera Hook. f.;杯状秋海棠■☆
50170 Begonia poculifera Hook. f. var. teusziana (J. Braun et K. Schum.) J. J. de Wilde;托兹秋海棠■☆
50171 Begonia poggei Warb.;波哥秋海棠;Pogge Begonia ■☆
50172 Begonia poggei Warb. = Begonia eminii Warb. ■☆
50173 Begonia poggei Warb. var. albiflora T. Durand et H. Durand = Begonia eminii Warb. ■☆
50174 Begonia poikilantha Gilg = Begonia quadrialata Warb. ■☆
50175 Begonia polyantha H. Lév. = Begonia labordei H. Lév. ■
50176 Begonia polygonifolia A. DC.;蓼叶秋海棠;Knotweedleaf Begonia ■☆
50177 Begonia polygonoides Hook. f.;多加秋海棠■☆
50178 Begonia polypetala A. DC.;多瓣秋海棠;Manypetal Begonia ■☆
50179 Begonia polytricha C. Y. Wu;多毛秋海棠;Manyhair Begonia ■

50180 Begonia popenoei Standl.;波波诺秋海棠;Popenoe Begonia ■☆
50181 Begonia porteri H. Lév. et Vaniot;罗甸秋海棠(单花秋海棠);Luodian Begonia,Porter Begonia ■
50182 Begonia potamophila Gilg;河生秋海棠■☆
50183 Begonia preussii Warb.;普罗伊斯秋海棠■☆
50184 Begonia princeae Gilg;优越秋海棠■☆
50185 Begonia princeae Gilg f. grossidentata Irmsch. = Begonia princeae Gilg ■☆
50186 Begonia princeae Gilg f. racemigera Irmsch. = Begonia princeae Gilg ■☆
50187 Begonia princeae Gilg f. rhodesiana ? = Begonia princeae Gilg ■☆
50188 Begonia princeae Gilg f. vulgata Irmsch. = Begonia princeae Gilg ■☆
50189 Begonia princeae Gilg var. racemigera (Irmsch.) R. Wilczek = Begonia princeae Gilg ■☆
50190 Begonia princeps Klotzsch;帝王秋海棠;Princeps Begonia ■☆
50191 Begonia prismatocarpa Hook.;棱果秋海棠■☆
50192 Begonia prismatocarpa Hook. subsp. petraea (A. Chev.) Sosef;岩生棱果秋海棠■☆
50193 Begonia prostrata Irmsch. = Begonia handelii Irmsch. var. prostrata (Irmsch.) Tebbitt ■
50194 Begonia pruinata A. DC.;粉背秋海棠;Pruinate Begonia ■☆
50195 Begonia pseudimpatiens Gilg = Begonia macrocarpa Warb. ■☆
50196 Begonia pseudodaxinensis S. M. Ku et al.;假大新秋海棠■
50197 Begonia pseudodryadis C. Y. Wu;假厚叶秋海棠;Thickleaf-like Begonia ■
50198 Begonia pseudoleprosa C. I. Peng et al.;假癞叶秋海棠■
50199 Begonia pseudolubbersii Brade;假露伯秋海棠;False Lubbers Begonia ■☆
50200 Begonia pseudophyllomaniaca A. E. Lange;伪叶秋海棠■
50201 Begonia pseudoviola Gilg;假堇色秋海棠■
50202 Begonia psilophylla Irmsch.;光叶秋海棠(光滑秋海棠);Glabrousleaf Begonia ■
50203 Begonia pulcherrima Sosef;艳丽秋海棠■☆
50204 Begonia pulvinifera C. I. Peng et Yan Liu;肿柄秋海棠■
50205 Begonia purpurea A. Chev.;紫秋海棠;Purple Begonia ■☆
50206 Begonia purpurea Sw. = Begonia purpurea A. Chev. ■☆
50207 Begonia purpureofolia S. H. Huang et Y. M. Shui;紫叶秋海棠■
50208 Begonia purpureofolia S. H. Huang et Y. M. Shui = Begonia villifolia Irmsch. ■
50209 Begonia pustulata Liebm.;褐绒秋海棠(突点秋海棠);Blister Begonia ■☆
50210 Begonia pustulata Liebm. 'Argentata';银斑褐绒秋海棠■☆
50211 Begonia pustulata Liebm. 'Silver' = Begonia pustulata Liebm. 'Argentata' ■☆
50212 Begonia pycnocaulis Irmsch. = Begonia oxyloba Welw. ex Hook. f. ■☆
50213 Begonia pygmaea Irmsch.;极小秋海棠■☆
50214 Begonia quadrialata Warb.;四翅秋海棠■☆
50215 Begonia quadrialata Warb. subsp. dusenii (Warb.) Sosef;杜森秋海棠■☆
50216 Begonia quadrialata Warb. subsp. nimbaensis Sosef;尼恩巴秋海棠■☆
50217 Begonia quadrialata Warb. var. pilosa Sosef;疏毛秋海棠■☆
50218 Begonia quadrialata Warb. var. speciosa Irmsch. = Begonia quadrialata Warb. ■☆
50219 Begonia quintasii C. DC. = Begonia annobonensis A. DC. ■☆
50220 Begonia rajah Ridl.;王公秋海棠;Rajah Begonia ■☆
50221 Begonia ramosa Sosef = Begonia schaeferi Engl. ■☆
50222 Begonia randaiensis Sasaki = Begonia laciniata Roxb. ■
50223 Begonia randaiensis Sasaki = Begonia palmata D. Don var. bowringiana (Champ. ex Benth.) Golding et Kareg. ■
50224 Begonia ravenii C. I. Peng et Y. K. Chen;岩生秋海棠;Rocky Begonia ■
50225 Begonia raynaliorum R. Wilczek = Begonia ciliobracteata Warb. ■☆
50226 Begonia razafinjohanyi Aymonin et Bosser;拉扎芬秋海棠■☆
50227 Begonia reflexisquamosa C. Y. Wu;倒鳞秋海棠;Rreflexisquamose Begonia ■
50228 Begonia reniformis Berol. ex Klotzsch = Begonia A. dregei Otto et Dietr. ■☆
50229 Begonia reniformis Dryand.;肾形秋海棠;Grapeleaf Begonia,Kidney-form Begonia ■☆
50230 Begonia repenticaulis Irmsch.;匍茎秋海棠■
50231 Begonia retinervia D. Fang et al.;突脉秋海棠■
50232 Begonia rex Putz.;大王秋海棠(长纤秋海棠,大叶秋海棠,蟆叶秋海棠,毛叶秋海棠,紫叶秋海棠);Assam King Begonia,King Begonia,Painted Leaf Begonia,Painted-leaf Begonia,Rex Begonia ■
50233 Begonia rex Putz. = Begonia masoniana Irmsch. ■
50234 Begonia rex Putz. = Begonia picta Sm. ■
50235 Begonia rex-cultorum L. H. Bailey;虎耳秋海棠;Beefsteak Geranium,Rex Begonia ■☆
50236 Begonia rhipsaloides A. Chev. = Begonia polygonoides Hook. f. ■☆
50237 Begonia rhizocarpa Fisch. ex A. DC.;根果秋海棠■☆
50238 Begonia rhodophylla C. Y. Wu;红叶秋海棠;Redleaf Begonia ■
50239 Begonia rhodophylla C. Y. Wu. = Begonia guishanensis S. H. Huang et Y. M. Shui ■
50240 Begonia rhopalocarpa Warb. = Begonia loranthoides Hook. f. subsp. rhopalocarpa (Warb.) J. J. de Wilde ■
50241 Begonia rhynchocarpa Y. M. Shui et W. H. Chen;喙果秋海棠■
50242 Begonia richardsiana T. Moore = Begonia dregei Otto et A. Dietr. ■☆
50243 Begonia ricinifolia A. Dietr.;蓖麻叶秋海棠;Bronze-leaf Begonia,Castor-bean Begonia,Star Begonia ■☆
50244 Begonia rigida Regel ex A. DC.;坚硬秋海棠;Rigid Begonia ■☆
50245 Begonia riparia Irmsch.;溪畔秋海棠■☆
50246 Begonia robusta Blume;粗壮秋海棠;Robust Begonia ■☆
50247 Begonia rockii Irmsch.;滇缅秋海棠;Rok Begonia ■
50248 Begonia roezlii Regel;罗氏秋海棠;Roezl Begonia ■☆
50249 Begonia romeensis De Wild. = Begonia macrocarpa Warb. ■☆
50250 Begonia rongjiangensis T. C. Ku;榕江秋海棠;Rongjiang Begonia ■
50251 Begonia rosiflora Hook. f.;粉花秋海棠(维奇秋海棠);Veitch Begonia ■☆
50252 Begonia rostrata Welw. ex Hook. f.;喙状秋海棠■☆
50253 Begonia rostrata Welw. ex Hook. f. var. argutiserrata R. Fern.;尖齿秋海棠■☆
50254 Begonia rostrata Welw. ex Hook. f. var. brachyptera R. Fern.;短翅秋海棠■☆
50255 Begonia rotundifolia S. H. Huang et Y. M. Shui;圆叶秋海棠;Roundleaf Begonia ■
50256 Begonia roxburghii A. DC.;露氏秋海棠(马拉巴秋海棠);Malabar Begonia,Roxburgh Begonia ■☆
50257 Begonia rubella Buch. -Ham. ex D. Don;微红秋海棠;Reddish Begonia ■☆
50258 Begonia rubinea H. Z. Li et H. Ma;玉柄秋海棠■
50259 Begonia ruboides C. M. Hu ex C. Y. Wu et T. C. Ku;匍地秋海棠

(匍状秋海棠)■

50260 Begonia rubra Blume;粗糙红秋海棠(粗糙秋海棠);Muricate Begonia ■☆

50261 Begonia rubricaulis Hook.;红茎秋海棠■☆

50262 Begonia rubromarginata Gilg;红边秋海棠■☆

50263 Begonia rubronervata De Wild.;红脉秋海棠■☆

50264 Begonia rubropunctata S. H. Huang et Y. M. Shui;红斑秋海棠;Redspot Begonia ■

50265 Begonia rubropunctata S. H. Huang et Y. M. Shui = Begonia pedatifida H. Lév. ■

50266 Begonia rubro-venia Hook. = Begonia hatacoa Buch.-Ham. ex D. Don ■

50267 Begonia rudatisii Irmsch. = Begonia homonyma Steud. ■☆

50268 Begonia rumpiensis Kupicha;龙皮秋海棠■☆

50269 Begonia rupicola Miq.;岩地秋海棠;Rockliving Begonia ■☆

50270 Begonia rwandensis J. C. Arends;卢旺达秋海棠■☆

50271 Begonia salicifolia A. DC.;柳叶秋海棠;Willowleaf Begonia ■☆

50272 Begonia sambiranensis Humbert ex Rabenant. et Bosser;桑比朗秋海棠■☆

50273 Begonia sanguinea Raddi et Spreng.;牛耳海棠;Sanguine Begonia ■

50274 Begonia sanjeensis R. Wilczek = Begonia ebolowensis Engl. ■☆

50275 Begonia sartorii Liebm.;萨氏秋海棠;Grape-leaf Begonia ■☆

50276 Begonia sassandrensis A. Chev. = Begonia oxyloba Welw. ex Hook. f. ■☆

50277 Begonia scabrida A. DC.;略粗秋海棠(粗糙秋海棠);Scabrid Begonia ■☆

50278 Begonia scandens Klotzsch;欧洲旋花秋海棠;Glorybind Begonia ■☆

50279 Begonia scapigera Hook. f.;花茎秋海棠■☆

50280 Begonia scapigera Hook. f. subsp. australis Sosef;南方花茎秋海棠■☆

50281 Begonia sceptrum Bull.;深裂秋海棠;Scepter Begonia ■☆

50282 Begonia schaeferi Engl.;谢弗秋海棠■☆

50283 Begonia scharffiana Regel;沙弗秋海棠;Scharff Begonia ■☆

50284 Begonia scharffii Hook. f.;红秋海棠(红筋秋海棠,紫背秋海棠);Elephant's-ear Begonia ■☆

50285 Begonia schlechteri Gilg = Begonia laporteifolia Warb. ■☆

50286 Begonia schliebenii Irmsch.;施利本秋海棠■☆

50287 Begonia schmidtiana Regel;施密特秋海棠;Schmidt Begonia ■☆

50288 Begonia schultzei Engl. = Begonia elaeagnifolia Hook. f. ■☆

50289 Begonia schulziana Urb. et Ekman;舒尔茨秋海棠;Schulz Begonia ■☆

50290 Begonia sciaphila Gilg ex Engl.;荫生秋海棠■☆

50291 Begonia sciaphila Gilg ex Engl. var. longipedunculata R. Wilczek;长梗荫生秋海棠■☆

50292 Begonia scitifolia Irmsch.;成凤秋海棠■

50293 Begonia scutata Wall. = Begonia josephii A. DC. ■

50294 Begonia scutata Wall. = Begonia rubella Buch.-Ham. ex D. Don ■☆

50295 Begonia scutifolia Hook. f.;盾状秋海棠■☆

50296 Begonia scutulum Hook. f.;小盾秋海棠■☆

50297 Begonia semiparietalis Yan Liu et al.;半侧膜秋海棠■

50298 Begonia semperflorens Link et Otto;四季海秋棠(常花秋海棠,四季海棠,蚬肉海棠,蚬肉秋海棠,洋秋海棠);Bedding Begonia, Hooker Begonia, Perpetual Begonia, Tuberous Begonia, Wax Begonia, Wax Plant ■

50299 Begonia seretii De Wild. = Begonia oxyloba Welw. ex Hook. f. ■☆

50300 Begonia serratipetala Irmsch.;齿瓣秋海棠;Pink Spot Begonia ■☆

50301 Begonia sessilanthera Warb. = Begonia preussii Warb. ■☆

50302 Begonia sessilifolia Hook. f.;无柄秋海棠■☆

50303 Begonia setifolia Irmsch.;刚毛秋海棠;Setose Begonia, Setoseleaf Begonia ■

50304 Begonia setulosopeltata C. Y. Wu;刺盾叶秋海棠;Setuloso-peltate Begonia ■

50305 Begonia sikkimensis A. DC.;锡金秋海棠;Sikkim Begonia ■

50306 Begonia silletensis (A. DC.) C. B. Clarke;厚壁秋海棠;Thickwall Begonia ■

50307 Begonia silletensis (A. DC.) C. B. Clarke subsp. mengyangensis Tebbitt et K. Y. Guan;云南厚壁秋海棠■

50308 Begonia simii Stapf = Begonia macrocarpa Warb. ■☆

50309 Begonia sinensis A. DC. = Begonia grandis Dryand. subsp. sinensis (A. DC.) Irmsch. ■

50310 Begonia sinensis A. DC. = Begonia ravenii C. I. Peng et Y. K. Chen ■

50311 Begonia sinobrevicaulis T. C. Ku;短茎秋海棠;Shortstem Begonia ■

50312 Begonia sinobrevicaulis T. C. Ku. = Begonia gulinqingensis S. H. Huang et Y. M. Shui ■

50313 Begonia sinofloribunda Dorr;多花秋海棠;Manyflower Begonia ■

50314 Begonia sino-vietnamica C. Y. Wu;中越秋海棠;Cochin-China Begonia ■

50315 Begonia sinuata E. Mey. ex Otto et A. Dietr. = Begonia homonyma Steud. ■☆

50316 Begonia smithiana Te T. Yu ex Irmsch.;长柄秋海棠(红八角莲);Smith Begonia ■

50317 Begonia socotrana Hook. f.;索科特拉秋海棠;Socotra Begonia ■☆

50318 Begonia sonderiana Irmsch.;森诺秋海棠■☆

50319 Begonia sonderiana Irmsch. var. transgrediens ? = Begonia sonderiana Irmsch. ■☆

50320 Begonia sparsipila Baker;散生秋海棠(散生基粒秋海棠);Sparsepilum Begonia ■☆

50321 Begonia spinibarbis Irmsch.;刺髯秋海棠;Seersucker Begonia ■☆

50322 Begonia spraguei Weber = Begonia horticola Irmsch. ■☆

50323 Begonia squamulosa Hook. f.;小鳞秋海棠■☆

50324 Begonia squamulosa Hook. f. var. bipindensis (Gilg ex Engl.) N. Hallé = Begonia longipetiolata Gilg ■☆

50325 Begonia staudtii Gilg;施陶秋海棠■☆

50326 Begonia staudtii Gilg var. dispersipilosa Irmsch. = Begonia staudtii Gilg ■☆

50327 Begonia stellata Sosef;星状秋海棠■☆

50328 Begonia stigmosa Lindl.;小疤秋海棠;Stigmose Begonia ■☆

50329 Begonia stipulacea Willd.;托叶秋海棠;Stipulaceous Begonia ■☆

50330 Begonia stolzii Irmsch.;斯托尔兹秋海棠■☆

50331 Begonia strigillosa A. Dietr.;粗壮毛秋海棠;Strigillose Begonia ■☆

50332 Begonia subacuto-alata De Wild. = Begonia princeae Gilg ■☆

50333 Begonia subalpestris A. Chev.;亚高山秋海棠■☆

50334 Begonia subfalcata De Wild. = Begonia hirsutula Hook. f. ■

50335 Begonia subhowii S. H. Huang;粉叶秋海棠;How-like Begonia ■

50336 Begonia sublongipes Y. M. Shui;保亭秋海棠(长柄毛秋海棠)■

50337 Begonia suboblata D. Fang et D. H. Qin;都安秋海棠■

50338 Begonia subscutata De Wild.;长圆盾形秋海棠■☆

50339 Begonia subtilis Irmsch. = Begonia pseudoviola Gilg ■

50340 Begonia subvillosa Klotzsch;长柔毛秋海棠;Subvillose Begonia ■☆

50341 Begonia suffruticosa Meisn.;亚灌木秋海棠;Chinese Tree

Peony, Maple-leaf Begonia ■☆
50342 Begonia suffruticosa Meisn. = Begonia dregei Otto et A. Dietr. ■☆
50343 Begonia suffruticosa Meisn. var. gueinziana A. DC. = Begonia sutherlandii Hook. f. ■☆
50344 Begonia sulcata Scheidw.;有槽秋海棠;Sulcate Begonia ■☆
50345 Begonia summoglabra Te T. Yu = Begonia xishuiensis T. C. Ku ■
50346 Begonia susaniae Sosef;苏萨尼秋海棠■☆
50347 Begonia sutherlandii Hook. f.;萨瑟兰秋海棠;Sutherland Begonia ■☆
50348 Begonia sutherlandii Hook. f. f. densiserrata Irmsch. = Begonia sutherlandii Hook. f. subsp. latior (Irmsch.) Kupicha ■☆
50349 Begonia sutherlandii Hook. f. subsp. latior (Irmsch.) Kupicha;宽萨瑟兰秋海棠■☆
50350 Begonia sutherlandii Hook. f. var. latior Irmsch. = Begonia sutherlandii Hook. f. subsp. latior (Irmsch.) Kupicha ■☆
50351 Begonia taipeiensis C. I. Peng;台北秋海棠■
50352 Begonia taiwaniana Hayata;台湾秋海棠;Taiwan Begonia ■
50353 Begonia taiwaniana Hayata var. albomaculata S. S. Ying = Begonia taiwaniana Hayata ■
50354 Begonia taliensis Gagnep.;大理秋海棠;Dali Begonia ■
50355 Begonia tanala Humbert ex Rabenant. et Bosser;塔纳尔秋海棠■☆
50356 Begonia tarokoensis M. J. Lai;太鲁阁秋海棠;Taroko Begonia ■
50357 Begonia tarokoensis M. J. Lai = Begonia formosana (Hayata) Masam. ■
50358 Begonia tayloriana Irmsch.;泰勒秋海棠■☆
50359 Begonia tenella D. Don;可爱秋海棠;Charming Begonia ■☆
50360 Begonia tenera Dryand.;柔弱秋海棠;Tender Begonia ■☆
50361 Begonia tengchiana C. I. Peng et Y. K. Chen;藤枝秋海棠■
50362 Begonia tenuicaulis Irmsch. = Begonia discrepans Irmsch. ■
50363 Begonia tenuifolia Dryand.;细叶秋海棠;Dlender-leaf Begonia ■☆
50364 Begonia tessaricarpa C. B. Clarke;陀螺果秋海棠(特萨菊果秋海棠);Tessariumfruit Begonia ■☆
50365 Begonia tetrandra Irmsch.;四棱秋海棠(角果秋海棠);Fourangular Begonia ■
50366 Begonia teuscheri Linden ex André;图舍秋海棠;Teuscher Begonia ■☆
50367 Begonia teusziana J. Braun et K. Schum. = Begonia poculifera Hook. f. var. teusziana (J. Braun et K. Schum.) J. J. de Wilde ■☆
50368 Begonia thomeana C. DC.;汤姆秋海棠■☆
50369 Begonia togoensis Gilg = Begonia oxyloba Welw. ex Hook. f. ■☆
50370 Begonia tomentosa Schott;绒毛秋海棠;Tomentose Begonia ■☆
50371 Begonia triflora Irmsch. = Begonia scutifolia Hook. f. ■☆
50372 Begonia triflora Irmsch. var. caloskiadia N. Hallé = Begonia scutifolia Hook. f. ■☆
50373 Begonia truncatiloba Irmsch.;截叶秋海棠(截裂秋海棠);Truncate Begonia ■
50374 Begonia tsaii Irmsch.;屏边秋海棠;H. T. Tsai Begonia, Cai Begonia ■
50375 Begonia tsaii Irmsch. = Begonia setifolia Irmsch. ■
50376 Begonia tsaratananensis Aymonin et Bosser;察拉塔纳纳秋海棠■☆
50377 Begonia tsoongii C. Y. Wu;观光秋海棠;Tsoong Begonia ■
50378 Begonia tuberhydrida Voss;球茎秋海棠(球根海棠,球根秋海棠);Ballstem Begonia, Hybrid Tuberous Begonia, Tuberous Begonia, Tuberous-rooted Begonia ■
50379 Begonia ulmifolia Willd.;榆叶秋海棠;Elm-leaf Begonia ■☆
50380 Begonia umbraculifolia Y. Wan et B. N. Chang;伞叶秋海棠(龙虎山秋海棠);Longhushan Begonia ■
50381 Begonia umbraculifolia Y. Wan et B. N. Chang var. flocculosa Y. M. Shui et W. H. Chen;簇毛伞叶秋海棠■
50382 Begonia undulata Schott;波状秋海棠;Undulate Begonia ■☆
50383 Begonia urophylla Gilg ex De Wild.;尾叶秋海棠;Tailshapeleaf Begonia ■☆
50384 Begonia urophylla Hook. = Begonia urophylla Gilg ex De Wild. ■☆
50385 Begonia valdensium A. DC.;喜林芋叶秋海棠;Philodendron-leaf Begonia ■☆
50386 Begonia valerii Standl.;瓦勒秋海棠;Valeri Begonia ■☆
50387 Begonia variifolia Y. M. Shui et W. H. Chen;变异秋海棠■
50388 Begonia veitchii Hook. f.;维奇秋海棠;Veitch Begonia ■☆
50389 Begonia vellozoana Brade;韦洛秋海棠■☆
50390 Begonia venosa V. Naray. ex Hook. f.;有脉秋海棠;Veiny Begonia ■☆
50391 Begonia verdickii De Wild. = Begonia princeae Gilg ■☆
50392 Begonia versicolor Irmsch.;变色秋海棠(变叶秋海棠,花叶酸筒);Fairy-carpet Begonia ■
50393 Begonia vestita C. DC.;被覆秋海棠;Covered Begonia ■☆
50394 Begonia villifolia Irmsch.;毛叶秋海棠(长毛秋海棠,长柔毛秋海棠,紫叶秋海棠);Longhair Begonia, Purple-leaf Begonia, Villousleaf Begonia ■
50395 Begonia vitifolia Schott;葡萄叶秋海棠(葡叶秋海棠,小花秋海棠);Grape-leaf Begonia ■☆
50396 Begonia vittariifolia N. Hallé;圈叶秋海棠■☆
50397 Begonia wakefieldii Irmsch.;韦克菲尔德秋海棠■☆
50398 Begonia wakefieldii Irmsch. f. dentatiloba ? = Begonia wakefieldii Irmsch. ■☆
50399 Begonia wallichiana Lehm.;沃利克秋海棠;Wallich Begonia ■☆
50400 Begonia wangii Te T. Yu;少瓣秋海棠(富宁秋海棠,爬山猴,王氏秋海棠);Wang Begonia ■
50401 Begonia warburgii Gilg = Begonia eminii Warb. ■☆
50402 Begonia warpuri Hemsl. = Begonia nana L'Hér. ■☆
50403 Begonia wellmanii Gilg ex Engl. = Begonia princeae Gilg ■☆
50404 Begonia weltoniensis André;委东秋海棠;Grapevine Begonia, Mapleleaf Begonia, Maple-leaf Begonia ■☆
50405 Begonia wenshanensis C. M. Hu ex C. Y. Wu et T. C. Ku;文山秋海棠;Wenshan Begonia ■
50406 Begonia whytei Stapf = Begonia quadrialata Warb. ■☆
50407 Begonia wilczekiana N. Hallé = Begonia elaeagnifolia Hook. f. ■☆
50408 Begonia wilksii Sosef;维尔克斯秋海棠■☆
50409 Begonia wilsonii Gagnep.;一点血秋海棠(红砖草,网脉秋海棠,一点血);E. H. Wilson Begonia ■
50410 Begonia wollastonii Baker f.;沃拉斯顿秋海棠■☆
50411 Begonia wrightiana A. DC.;瑞特秋海棠;Wright Begonia ■☆
50412 Begonia wutaiana C. I. Peng et Y. K. Chen;雾台秋海棠■
50413 Begonia xanthina Hook. f.;黄瓣秋海棠(黄花秋海棠,金黄秋海棠);Golden Begonia ■
50414 Begonia xingyiensis T. C. Ku;兴义秋海棠;Xingyi Begonia ■
50415 Begonia xinyiensis T. C. Ku = Begonia handelii Irmsch. var. prostrata (Irmsch.) Tebbitt ■
50416 Begonia xishuiensis T. C. Ku;习水秋海棠(光叶秋海棠,最亮秋海棠);Glabrousleaf Begonia, Xishui Begonia ■
50417 Begonia yingjiangensis S. H. Huang;盈江秋海棠;Yingjiang Begonia ■
50418 Begonia yishanensis T. C. Ku;宜山秋海棠;Yishan Begonia ■
50419 Begonia yishanensis T. C. Ku = Begonia porteri H. Lév. et Vaniot ■
50420 Begonia yui Irmsch.;宿苞秋海棠(临沧秋海棠);Yu Begonia ■

50421 Begonia yunnanensis H. Lév. = Begonia modestiflora Kurz ■
50422 Begonia yunnanensis H. Lév. var. hypoleuca H. Lév. = Begonia modestiflora Kurz ■
50423 Begonia yunnanensis H. Lév. var. hypoleuca H. Lév. = Begonia yunnanensis H. Lév. ■
50424 Begonia zairensis Sosef;扎伊尔秋海棠■☆
50425 Begonia zairensis Sosef var. montana Sosef;山生秋海棠■☆
50426 Begonia zebrina Angl. ex Loudon = Begonia stipulacea Willd. ■☆
50427 Begonia zenkeri Irmsch. = Begonia zenkeriana L. B. Sm. et Wassh. ■☆
50428 Begonia zenkeri Warb. ex Exell = Begonia macrocarpa Warb. ■☆
50429 Begonia zenkeriana L. B. Sm. et Wassh.;岑克尔秋海棠■☆
50430 Begonia zhangii D. Fang et D. H. Qin = Begonia daxinensis T. C. Ku ■
50431 Begonia zhengyiana Y. M. Shui;吴氏秋海棠;Zhengyi Begonia ■
50432 Begonia zimmermannii Peter ex Irmsch.;齐默尔曼秋海棠■☆
50433 Begoniaceae C. Agardh(1824)(保留科名);秋海棠科;Begonia Family ●■
50434 Begoniella Oliv.(1872);小秋海棠属■☆
50435 Begoniella Oliv. = Begonia L. ●■
50436 Begoniella whitei Oliv.;瓦特小秋海棠■☆
50437 Beguea Capuron(1969);布格木属●☆
50438 Beguea apetala Capuron;布格木●☆
50439 Behaimia Griseb.(1866);古巴豆属●☆
50440 Behaimia cubensis Griseb.;古巴豆●☆
50441 Behen Hill = Centaurea L.(保留属名)●■
50442 Behen Hill = Jacea Mill. ●■
50443 Behen Hill = Vernonia Schreb.(保留属名)●■
50444 Behen Moench = Oberna Adans. ■
50445 Behen Moench = Silene L.(保留属名)■
50446 Behen vulgaris Moench = Silene cucubalus Wibel ■
50447 Behen vulgaris Moench = Silene vulgaris (Moench) Garcke ■
50448 Behenantha Schur = Behen Moench ■
50449 Behnia Didr.(1855);两型花属●☆
50450 Behnia reticulata (Thunb.) Didr.;两型花●☆
50451 Behniaceae Conran, M. W. Chase et Rudall = Agavaceae Dumort.(保留科名)●■
50452 Behniaceae Conran, M. W. Chase et Rudall(1997);两型花科(非拔契科)●☆
50453 Behniaceae R. Dahlgren ex Reveal = Behniaceae Conran, M. W. Chase et Rudall ●☆
50454 Behria Greene = Bessera Schult. f.(保留属名)■☆
50455 Behrinia Sieber = Berinia Brign. ■
50456 Behrinia Sieber = Crepis L. ■
50457 Behrinia Sieber ex Steud. = Berinia Brign. ■
50458 Behrinia Sieber ex Steud. = Crepis L. ■
50459 Behuria Cham.(1834);巴西野牡丹属■☆
50460 Behuria insignis Cham.;巴西野牡丹■☆
50461 Beilia (Baker) Eckl. ex Kuntze = Watsonia Mill.(保留属名)■☆
50462 Beilia Eckl. = Watsonia Mill.(保留属名)■☆
50463 Beilia Kuntze = Micranthus (Pers.) Eckl.(保留属名)■☆
50464 Beilschmidtia Rchb. = Beilschmiedia Nees ●
50465 Beilschmiedia Nees(1831);琼楠属;Slogwood, Slugwood ●
50466 Beilschmiedia acuta Kosterm.;尖琼楠●☆
50467 Beilschmiedia acutifolia (Engl. et K. Krause) Robyns et R. Wilczek = Beilschmiedia acuta Kosterm. ●☆
50468 Beilschmiedia alata Robyns et R. Wilczek;翅琼楠●☆
50469 Beilschmiedia ambigua Robyns et R. Wilczek;可疑琼楠●☆
50470 Beilschmiedia anacardioides (Engl. et K. Krause) Robyns et R. Wilczek;腰果琼楠●☆
50471 Beilschmiedia appendiculata (C. K. Allen) S. K. Lee et Y. T. Wei;山潺(槁琼楠,梅乐樟);Appendiculate Slugwood, Wildstickiness ●
50472 Beilschmiedia atrata C. K. Allen = Beilschmiedia pergamentacea C. K. Allen ●
50473 Beilschmiedia auriculata Robyns et R. Wilczek;耳形琼楠●☆
50474 Beilschmiedia bancroftii C. T. White;加那利琼楠;Canary Ash, Yellow Walnut ●☆
50475 Beilschmiedia baotingensis S. K. Lee et Y. T. Wei;保亭琼楠;Baoting Slugwood ●
50476 Beilschmiedia barensis (Engl. et K. Krause) Robyns et R. Wilczek;喀麦隆琼楠●☆
50477 Beilschmiedia batangensis (Engl.) Robyns et R. Wilczek;巴坦加琼楠●☆
50478 Beilschmiedia brachythyrsa H. W. Li;勐仑琼楠;Menglun Slugwood, Short Thyrse Slugwood, Short-thyrse Slugwood ●
50479 Beilschmiedia bracteata Robyns et R. Wilczek;具苞琼楠●☆
50480 Beilschmiedia brevifolia Y. T. Wei;短叶琼楠;Short-leaf Slugwood, Short-leaved Slugwood ●
50481 Beilschmiedia brevipaniculata C. K. Allen;短序琼楠;Short Panicle Slugwood ●
50482 Beilschmiedia calcitranthera Fouilloy;距药琼楠●☆
50483 Beilschmiedia caudata (Stapf) A. Chev.;尾琼楠●☆
50484 Beilschmiedia chevalieri Robyns et R. Wilczek;舍瓦利耶琼楠●☆
50485 Beilschmiedia chinensis Hance = Cryptocarya chinensis (Hance) Hemsl. ●
50486 Beilschmiedia cinnamomea (Stapf) Robyns et R. Wilczek;肉桂琼楠●☆
50487 Beilschmiedia conferta (S. Moore) Robyns et R. Wilczek = Beilschmiedia gaboonensis (Meisn.) Benth. et Hook. f. ●☆
50488 Beilschmiedia congestiflora (Engl. et K. Krause) Robyns et R. Wilczek;密花琼楠●☆
50489 Beilschmiedia congolana Robyns et R. Wilczek;刚果琼楠●☆
50490 Beilschmiedia corbisieri (Robyns) Robyns et R. Wilczek;科比西尔琼楠●☆
50491 Beilschmiedia corbisieri (Robyns) Robyns et R. Wilczek var. diversiflora (Pierre ex Robyns et R. Wilczek) Fouilloy = Beilschmiedia diversiflora Pierre ex Robyns et R. Wilczek ●☆
50492 Beilschmiedia crassifolia (Engl.) Robyns et R. Wilczek;非洲厚叶琼楠●☆
50493 Beilschmiedia crassipes (Engl. et K. Krause) Robyns et R. Wilczek;粗梗琼楠●☆
50494 Beilschmiedia cryptocaryoides Kosterm.;隐萼琼楠●☆
50495 Beilschmiedia cuspidata (Krause) Robyns et R. Wilczek;骤尖琼楠●☆
50496 Beilschmiedia cylindrica S. K. Lee et Y. T. Wei;柱果琼楠;Cylindric Fruit Slugwood, Cylindrical Slugwood ●
50497 Beilschmiedia dawei Robyns et R. Wilczek = Beilschmiedia ugandensis Rendle ●☆
50498 Beilschmiedia delicata S. K. Lee et Y. T. Wei;美脉琼楠;Prettynerved Slugwood, Pretty-nerved Slugwood ●
50499 Beilschmiedia descoingsii Fouilloy;德斯琼楠●☆
50500 Beilschmiedia dinklagei (Engl.) Robyns et R. Wilczek;丁克琼楠●☆

50501 Beilschmiedia discolor C. K. Allen = Beilschmiedia intermedia C. K. Allen ●

50502 Beilschmiedia discolor Robyns et R. Wilczek = Beilschmiedia versicolor Kosterm. ●☆

50503 Beilschmiedia diversiflora Pierre ex Robyns et R. Wilczek;异花琼楠●☆

50504 Beilschmiedia djalonensis A. Chev.;贾隆琼楠●☆

50505 Beilschmiedia donisii Robyns et R. Wilczek;多尼斯琼楠●☆

50506 Beilschmiedia elata Scott-Elliot = Beilschmiedia mannii (Meisn.) Benth. et Hook. f. ●☆

50507 Beilschmiedia erythrophloia Hayata;台琼楠(九芎舅,木耳树,琼楠);Red Bark Slugwood,Redbark Slugwood,Red-barked Slugwood ●

50508 Beilschmiedia erythrophloia Hayata var. tanakae (Hayata) Kaneh. = Beilschmiedia erythrophloia Hayata ●

50509 Beilschmiedia euryneura (Stapf) Robyns et R. Wilczek = Beilschmiedia caudata (Stapf) A. Chev. ●☆

50510 Beilschmiedia fagifolia Nees = Beilschmiedia roxburghiana Nees ●

50511 Beilschmiedia fasciata H. W. Li;白柴果(白琼楠);Fasciated Slugwood,Fasciatedpeduncle Slugwood ●

50512 Beilschmiedia foliosa (S. Moore) Robyns et R. Wilczek;多叶琼楠●☆

50513 Beilschmiedia fordii Dunn;广东琼楠;Ford Slugwood ●

50514 Beilschmiedia formosana C. E. Chang;台湾琼楠;Formosan Red Bark Slugwood,Taiwan Slugwood ●

50515 Beilschmiedia formosana C. E. Chang = Beilschmiedia tsangii Merr. ●

50516 Beilschmiedia foveolata Merr. = Cinnamomum foveolatum (Merr.) H. W. Li et J. Li ●

50517 Beilschmiedia fruticosa Engl.;灌木琼楠●☆

50518 Beilschmiedia fulva Robyns et R. Wilczek;黄褐琼楠●☆

50519 Beilschmiedia furfuracea Chun ex Hung T. Chang;糠秕琼楠;Furfuraceous Fruit Slugwood,Furfuraceous Slugwood ●

50520 Beilschmiedia gaboonensis (Meisn.) Benth. et Hook. f.;加蓬琼楠●☆

50521 Beilschmiedia gilbertii Robyns et R. Wilczek;吉尔伯特琼楠●☆

50522 Beilschmiedia gilbertii Robyns et R. Wilczek var. glabraHort.;光滑吉尔伯特琼楠●☆

50523 Beilschmiedia giorgii Robyns et R. Wilczek;乔治琼楠●☆

50524 Beilschmiedia glandulosa N. H. Xia et al.;香港琼楠●

50525 Beilschmiedia glauca S. K. Lee et L. Lau;粉背琼楠;Glaucous Leaf Slugwood,Glaucousback Slugwood,Grey-blue Slugwood ●

50526 Beilschmiedia glauca S. K. Lee et L. Lau var. glaucoides H. W. Li = Beilschmiedia glaucoides (H. W. Li) H. W. Li ●

50527 Beilschmiedia glaucoides (H. W. Li) H. W. Li;顶序琼楠;Glaucous-like Slugwood ●

50528 Beilschmiedia gradiosa C. K. Allen = Beilschmiedia intermedia C. K. Allen ●

50529 Beilschmiedia grandibracteata Robyns et R. Wilczek;大苞琼楠●☆

50530 Beilschmiedia grandiflora (Kosterm.) Kosterm. = Beilschmiedia velutina (Kosterm.) Kosterm. ●☆

50531 Beilschmiedia grandifolia (Stapf) Robyns et R. Wilczek;大叶琼楠●☆

50532 Beilschmiedia henghsienensis S. K. Lee et Y. T. Wei;横县琼楠;Hengxian Slugwood ●

50533 Beilschmiedia hermanii Robyns et R. Wilczek;赫尔曼琼楠●☆

50534 Beilschmiedia hutchinsoniana Robyns et R. Wilczek;哈钦森琼楠●☆

50535 Beilschmiedia insularum Robyns et R. Wilczek;海岛琼楠●☆

50536 Beilschmiedia intermedia C. K. Allen;琼楠(二色琼楠,番鬼榄,荔枝公);Intermediate Slugwood ●

50537 Beilschmiedia jabassensis (Engl. et K. Krause) Robyns et R. Wilczek;亚巴森琼楠●☆

50538 Beilschmiedia jacques-felixii Robyns et R. Wilczek;雅凯琼楠●☆

50539 Beilschmiedia klainei Robyns et R. Wilczek;克莱恩琼楠●☆

50540 Beilschmiedia kostermansiana Robyns et R. Wilczek;科斯特秋海棠■☆

50541 Beilschmiedia kwangsiensis Kosterm. = Syndiclis kwangsiensis (Kosterm.) H. W. Li ●

50542 Beilschmiedia kweichowensis W. C. Cheng;贵州琼楠;Guizhou Slugwood,Kweichou Slugwood ●

50543 Beilschmiedia laevis C. K. Allen;红枝琼楠(大叶槁,平滑琼楠);Redtwig Slugwood,Red-twigged Slugwood ●

50544 Beilschmiedia lancifolia (Engl. et K. Krause) Robyns et R. Wilczek = Beilschmiedia lancilimba Kosterm. ●☆

50545 Beilschmiedia lancilimba Kosterm.;剑片琼楠●☆

50546 Beilschmiedia lebrunii Robyns et R. Wilczek;勒布伦琼楠●☆

50547 Beilschmiedia ledermannii (Engl. et K. Krause) Robyns et R. Wilczek = Beilschmiedia robynsiana Kosterm. ●☆

50548 Beilschmiedia leemansii Robyns et R. Wilczek = Beilschmiedia zenkeri Engl. ●☆

50549 Beilschmiedia letouzeyi Robyns et R. Wilczek;勒图琼楠●☆

50550 Beilschmiedia linocieroides H. W. Li;李榄琼楠;Linociera-like Slugwood ●

50551 Beilschmiedia longicarpa Chun et S. K. Lee;长果琼楠●☆

50552 Beilschmiedia longipetiolata C. K. Allen;长柄琼楠;Longpetiole Slugwood,Long-petioled Slugwood ●

50553 Beilschmiedia louisii Robyns et R. Wilczek;路易斯琼楠●☆

50554 Beilschmiedia macrophylla (Hutch. et Dalziel) A. Chev. = Beilschmiedia hutchinsoniana Robyns et R. Wilczek ●☆

50555 Beilschmiedia macropoda C. K. Allen;肉柄琼楠;Fleshy-stalked Slugwood,Largestalk Slugwood,Succulent Stalk Slugwood ●

50556 Beilschmiedia madagascariensis (Baill.) Kosterm.;马岛琼楠●☆

50557 Beilschmiedia mannii (Meisn.) Benth. et Hook. f.;曼氏琼楠●☆

50558 Beilschmiedia mannioides Robyns et R. Wilczek;类苦木秋海棠■☆

50559 Beilschmiedia mayumbensis Robyns et R. Wilczek;马永巴琼楠●☆

50560 Beilschmiedia megaphylla Pierre ex Robyns et R. Wilczek = Beilschmiedia corbisieri (Robyns) Robyns et R. Wilczek ●☆

50561 Beilschmiedia membranacea (Stapf) Robyns et R. Wilczek = Beilschmiedia membranifolia Kosterm. ●☆

50562 Beilschmiedia membranifolia Kosterm.;膜叶琼楠●☆

50563 Beilschmiedia michelsonii Robyns et R. Wilczek;米歇尔松琼楠●☆

50564 Beilschmiedia microphylla (Kosterm.) Kosterm.;小叶琼楠●☆

50565 Beilschmiedia minutiflora (Meisn.) Benth. et Hook. f.;多花琼楠●☆

50566 Beilschmiedia moratii van der Werff;莫拉特琼楠●☆

50567 Beilschmiedia muricata Hung T. Chang;瘤果琼楠;Muricate Slugwood,Muricate-fruited Slugwood ●

50568 Beilschmiedia myrciifolia (S. Moore) Robyns et R. Wilczek;桂柳琼楠●☆

50569 Beilschmiedia natalensis J. H. Ross = Dahlgrenodendron natalense (J. H. Ross) J. J. M. van der Merwe et A. E. van Wyk ●☆

50570 Beilschmiedia ndongensis (Engl. et K. Krause) Robyns et R. Wilczek;恩东加琼楠●☆

50571 Beilschmiedia neoletestui Fouilloy et N. Hallé;新莱泰斯图琼楠●☆

50572 Beilschmiedia ngriki A. Chev. = Beilschmiedia jacques-felixii Robyns et R. Wilczek ●☆
50573 Beilschmiedia ningmingensis S. K. Lee et Y. T. Wei;宁明琼楠;Ningming Slugwood ●
50574 Beilschmiedia nitida Engl.;亮琼楠●☆
50575 Beilschmiedia obconica C. K. Allen;锈叶琼楠;Obconic Slugwood,Obconical Slugwood,Rustyleaf Slugwood ●
50576 Beilschmiedia oblongifolia Robyns et R. Wilczek;矩圆叶琼楠●☆
50577 Beilschmiedia obovata Kosterm.;倒卵叶叶琼楠●☆
50578 Beilschmiedia obscura (Stapf) Engl. ex A. Chev.;模糊琼楠●☆
50579 Beilschmiedia obscura Engl. ex Stapf = Beilschmiedia obscura (Stapf) Engl. ex A. Chev. ●☆
50580 Beilschmiedia obscurinervia Hung T. Chang;隐脉琼楠;Obscurenerve Slugwood,Obscure-nerved Slugwood ●
50581 Beilschmiedia olivacea Robyns et R. Wilczek;橄榄绿琼楠●☆
50582 Beilschmiedia opposita Kosterm.;对生琼楠●☆
50583 Beilschmiedia ovoidea F. N. Wei;卵果琼楠;Ovatefruit Slugwood,Ovate-fruited Slugwood ●
50584 Beilschmiedia papyracea (Stapf) Robyns et R. Wilczek;纸质琼楠●☆
50585 Beilschmiedia parvifolia Lecomte = Lindera communis Hemsl. ●
50586 Beilschmiedia pauciflora H. W. Li;少花琼楠;Fewflower Slugwood,Few-flowered Slugwood ●
50587 Beilschmiedia paulocordata Fouilloy et N. Hallé;心形琼楠●☆
50588 Beilschmiedia pedicellata van der Werff;梗花琼楠●☆
50589 Beilschmiedia pellegrinii Fouilloy et N. Hallé;佩尔格兰琼楠●☆
50590 Beilschmiedia percoriacea C. K. Allen;厚叶琼楠;Thickleaf Slugwood,Thick-leaved Slugwood ●
50591 Beilschmiedia percoriacea C. K. Allen var. ciliata H. W. Li;缘毛琼楠;Ciliate Thick-leaved Slugwood ●
50592 Beilschmiedia pergamentacea C. K. Allen;纸叶琼楠(黑叶琼楠);Perchment-like Leaf Slugwood,Pergamentaceous Slugwood ●
50593 Beilschmiedia pierreana Robyns et R. Wilczek;皮埃尔琼楠●☆
50594 Beilschmiedia piya A. Chev. = Beilschmiedia zenkeri Engl. ●☆
50595 Beilschmiedia preussii Engl.;普罗伊斯琼楠●☆
50596 Beilschmiedia preussioides Fouilloy et N. Hallé;拟普罗伊斯琼楠●☆
50597 Beilschmiedia punctilimba H. W. Li;点叶琼楠;Punctateleaf Slugwood,Punctate-leaved Slugwood ●
50598 Beilschmiedia purpurascens H. W. Li;紫叶琼楠;Purpleleaf Slugwood,Purple-leaved Slugwood ●
50599 Beilschmiedia robusta C. K. Allen;粗壮琼楠;Robust Slugwood ●
50600 Beilschmiedia robynsiana Kosterm.;罗宾斯琼楠●☆
50601 Beilschmiedia roxburghiana Nees;稠琼楠(琼楠);Roxburgh Slugwood ●
50602 Beilschmiedia rufohirtella H. W. Li;红毛琼楠;Redhair Slugwood,Red-hirtellous Slugwood ●
50603 Beilschmiedia rugosa van der Werff;皱纹琼楠●☆
50604 Beilschmiedia rwandensis R. Wilczek;卢旺达琼楠●☆
50605 Beilschmiedia sericans Kosterm.;绢毛琼楠●☆
50606 Beilschmiedia sessilifolia (Stapf) Engl. ex Fouilloy;无柄琼楠●☆
50607 Beilschmiedia sessilifolia Stapf ex Engl. = Beilschmiedia sessilifolia (Stapf) Engl. ex Fouilloy ●☆
50608 Beilschmiedia shangsiensis Y. T. Wei;上思琼楠;Shangsi Slugwood ●
50609 Beilschmiedia sichourensis H. W. Li;西畴琼楠;Xichou Slugwood ●◇
50610 Beilschmiedia stapfiana Robyns et R. Wilczek = Beilschmiedia mannii (Meisn.) Benth. et Hook. f. ●☆
50611 Beilschmiedia staudtii Engl.;施陶琼楠●☆
50612 Beilschmiedia supraglandulosa Y. K. Li;表腺琼楠;Supraglandulose Slugwood ●
50613 Beilschmiedia talbotiae (S. Moore) Robyns et R. Wilczek;塔尔博特琼楠●☆
50614 Beilschmiedia tanakae Hayata = Beilschmiedia erythrophloia Hayata ●
50615 Beilschmiedia tawa Kirk = Beilschnliedia tawa (Cunn.) Kirk ●☆
50616 Beilschmiedia thollonii Robyns et R. Wilczek;托伦琼楠●☆
50617 Beilschmiedia tisserantii A. Chev.;蒂斯朗特琼楠●☆
50618 Beilschmiedia troupinii R. Wilczek;特鲁皮尼琼楠●☆
50619 Beilschmiedia tsangii Merr.;网脉琼楠(广东琼楠,华河琼楠,牛奶奶果);Tsang Slugwood ●
50620 Beilschmiedia tsangii Merr. var. delicata (S. K. Lee et Y. T. Wei) J. Li et H. W. Li = Beilschmiedia delicata S. K. Lee et Y. T. Wei ●
50621 Beilschmiedia tungfangensis S. K. Lee et L. Lau;东方琼楠;Dongfang Slugwood,Oriental Slugwood ●
50622 Beilschmiedia ugandensis Rendle;乌干达琼楠●☆
50623 Beilschmiedia ugandensis Rendle var. katangensis Robyns et R. Wilczek;加丹加琼楠●☆
50624 Beilschmiedia variabilis Robyns et R. Wilczek;易变琼楠●☆
50625 Beilschmiedia velutina (Kosterm.) Kosterm.;绒毛琼楠●☆
50626 Beilschmiedia versicolor Kosterm.;变色琼楠●☆
50627 Beilschmiedia wangii C. K. Allen;海南琼楠(黄志琼楠);Hainan Slugwood,Wang Slugwood ●
50628 Beilschmiedia wilczekii Fouilloy;维尔切克琼楠●☆
50629 Beilschmiedia xizangensis H. P. Tsui;西藏琼楠;Xizang Slugwood ●
50630 Beilschmiedia xizangensis H. P. Tsui = Beilschmiedia robusta C. K. Allen ●
50631 Beilschmiedia yaanica (N. Chao ex H. W. Li et al.) N. Chao = Cryptocarya yaanica N. Chao ex H. W. Li et al. ●
50632 Beilschmiedia yangambiensis Robyns et R. Wilczek;扬甘比琼楠●☆
50633 Beilschmiedia yunnaneesis Hu;滇琼楠;Yunnan Slugwood ●
50634 Beilschmiedia zahnii (Krause) Robyns et R. Wilczek;察恩琼楠●☆
50635 Beilschmiedia zenkeri Engl.;岑克尔琼楠●☆
50636 Beilschnliedia tawa (Cunn.) Kirk;新西兰琼楠;New Zealand Walnut,Queensland Walnut,Tawa ●☆
50637 Beirnaertia Louis ex Troupin(1949);碧柰藤属(毕奈藤属)●☆
50638 Beirnaertia cabindensis (Exell et Mendonça) Troupin;碧柰藤●☆
50639 Beirnaertia yangambiensis Louis ex Troupin = Beirnaertia cabindensis (Exell et Mendonça) Troupin ●☆
50640 Beiselia Forman(1987);墨西哥橄榄属●☆
50641 Beiselia mexicana Forman;墨西哥橄榄●☆
50642 Bejaranoa R. M. King et H. Rob. (1978);寡花泽兰属(少花柄泽兰属)●☆
50643 Bejaranoa balansae (Hieron.) R. M. King et H. Rob.;寡花泽兰●☆
50644 Bejaria L. = Bejaria Mutis(保留属名)●☆
50645 Bejaria Mutis ex L. = Bejaria Mutis(保留属名)●☆
50646 Bejaria Mutis(1771)(保留属名);贝氏木属(贝亚利属,七瓣杜属);Andes Rose ●☆
50647 Bejaria Zea = Bejaria Mutis(保留属名)●☆
50648 Bejaria aestuans Mutis ex L.;安第斯山贝氏木●☆

50649 Bejaria boliviensis B. Fedtsch. et Basil. = Bejaria aestuans Mutis ex L. ●☆
50650 Bejaria coarctata Bonpl.;密集贝氏木(密集贝亚利)●☆
50651 Bejaria denticulata J. Rémy = Bejaria aestuans Mutis ex L. ●☆
50652 Bejaria glauca Bonpl. var. coarctata (Bonpl.) Mansf. et Sleumer = Bejaria aestuans Mutis ex L. ●☆
50653 Bejaria glauca Bonpl. var. glandulosa Mansf. et Sleumer = Bejaria aestuans Mutis ex L. ●☆
50654 Bejaria glauca Bonpl. var. setosa Mansf. et Sleumer = Bejaria aestuans Mutis ex L. ●☆
50655 Bejaria glauca Bonpl. var. tomentella Mansf. et Sleumer = Bejaria aestuans Mutis ex L. ●☆
50656 Bejaria hispida Poepp. et Endl. = Bejaria aestuans Mutis ex L. ●☆
50657 Bejaria pallens J. Rémy = Bejaria aestuans Mutis ex L. ●☆
50658 Bejaria parvifolia Rusby = Bejaria aestuans Mutis ex L. ●☆
50659 Bejaudia Gagnep. = Myrialepis Becc. ●☆
50660 Bejuco Loefl. = Hippocratea L. ●☆
50661 Beketovia tianschanica Krasn. = Braya scharnhorstii Regel et Schmalh. ■
50662 Beketowia Krassn. = Braya Sternb. et Hoppe ■
50663 Belairia A. Rich. (1846);加勒比豆属■☆
50664 Belairia angustifolia (Griseb.) Borhidi;狭叶加勒比豆■☆
50665 Belairia parvifoliola Britton;小叶加勒比豆■☆
50666 Belairia spinosa A. Rich.;加勒比豆■☆
50667 Belamcanda Adans. (1763)(保留属名);射干属;Blackberry Lily, Blackberrylily, Blackberry-lily, Leopard Flower ●■
50668 Belamcanda bulbifera (L.) Moench = Sparaxis bulbifera (L.) Ker Gawl. ■☆
50669 Belamcanda chinensis (L.) DC.;射干(萹竹,扁竹,扁竹兰,扁竹头,扁筑,草姜,寸干,寸干老鸦扇,地萹竹,地扁竹,凤凰草,凤翼,高搜山,高搜栽,鬼前,鬼扇,鬼仔扇,蝴蝶花,黄花扁蓄,黄远,黄知母,剪刀草,剪刀梏,交剪草,绞剪草,较剪草,较剪兰,金蝴蝶,金绞剪,金盏花,开喉箭,老鸹扇,老君扇,老鸦扇,冷水丹,冷水花,六甲花,蚂螂花,偏竹梁,秋蝴蝶,山蒲扇,扇把草,扇子草,麝干,铁扁担,乌吹,乌蒲,乌莲,乌翣,乌扇,仙人掌,萱草花,野萱花,夜干,鱼翅草,玉燕,玉燕鬼子扇,紫蝴蝶,紫金牛,紫良姜);Blackberry Lily, Blackberrylily, Blackberry-lily, Leopard Flower, Leopard Lily, Leopard-flower ■
50670 Belamcanda chinensis (L.) DC. = Iris domestica (L.) Goldblatt et Mabb. ■
50671 Belamcanda chinensis (L.) DC. var. taiwanensis S. S. Ying;台湾射干■
50672 Belamcanda chinensis (L.) DC. var. taiwanensis S. S. Ying = Belamcanda chinensis (L.) DC. ■
50673 Belamcanda pampaninii H. Lév. = Belamcanda chinensis (L.) DC. ■
50674 Belamcanda punctata Moench = Belamcanda chinensis (L.) DC. ■
50675 Belandra S. F. Blake = Prestonia R. Br. (保留属名)●☆
50676 Belangera Cambess. = Lamanonia Vell. ●☆
50677 Belangeraceae J. Agardh = Cunoniaceae R. Br. (保留科名)●☆
50678 Belanthera Post et Kuntze = Beloanthera Hassk. ■
50679 Belanthera Post et Kuntze = Hydrolea L. (保留属名)■
50680 Belantheria Nees = Brillantaisia P. Beauv. ●■☆
50681 Belemia J. M. Pires(1981);吊钟茉莉属●☆
50682 Belemia fucsioides J. M. Pires;吊钟茉莉■☆
50683 Belencita H. Karst. (1857);哥伦比亚山柑属(哥伦比亚白花菜属)●☆
50684 Belencita hagenii H. Karst.;哥伦比亚山柑●☆
50685 Belendenia Raf. = Tritonia Ker Gawl. ■
50686 Belenia Decne. = Physochlaina G. Don ■
50687 Belenia praealta Decne. = Physochlaina praealta (Decne.) Miers ■
50688 Belenidium Arn. = Hymenantherum Cass. ■☆
50689 Belenidium Arn. ex DC. = Hymenantherum Cass. ■☆
50690 Beleropone C. B. Clarke = Beloperone Nees ■☆
50691 Belharnnala Adans. = Sanguinaria L. ■☆
50692 Belia Steller = Claytonia Gronov. ex L. ■☆
50693 Belia Steller ex J. G. Gmel. = Claytonia Gronov. ex L. ■☆
50694 Belicea Lundell = Morinda L. ●■
50695 Beliceodendron Lundell = Lecointea Ducke ■☆
50696 Belicia Lundell = Morinda L. ●■
50697 Belilla Adans. = Mussaenda L. ●■
50698 Belingia Pierre = Zollingeria Kurz(保留属名)■☆
50699 Belis Salisb. (废弃属名) = Cunninghamia R. Br. (保留属名)●★
50700 Belis jaculifolia Salisb. = Cunninghamia lanceolata (Lamb.) Hook. ●
50701 Belis lanceolata (Lamb.) Hoffm. = Cunninghamia lanceolata (Lamb.) Hook. ●
50702 Belladona Adans. = Atropa L. ■
50703 Belladona Duhamel = Atropa L. ■
50704 Belladona Mill. = Atropa L. ■
50705 Belladonna (Sweet ex Endl.) Sweet ex Harv. = Amaryllis L. (保留属名)■☆
50706 Belladonna Boehm. = Belladona Mill. ■
50707 Belladonna Mill. = Atropa L. ■
50708 Belladonna Rupp. = Atropa L. ■
50709 Belladonna Sweet = Amaryllis L. (保留属名)■☆
50710 Bellardia All. (1785);伯氏玄参属;Mediterranean Lineseed ■☆
50711 Bellardia All. = Bartsia L. (保留属名)●■☆
50712 Bellardia Colla = Microseris D. Don ■☆
50713 Bellardia Schreb. = Coccocypselum P. Browne(保留属名)●☆
50714 Bellardia Schreb. = Salacia L. (保留属名)●
50715 Bellardia Schreb. = Tontelea Aubl. (废弃属名)●
50716 Bellardia Schreb. = Tontelea Miers(保留属名)●
50717 Bellardia trixago (L.) All.;伯氏玄参;Mediterranean Lineseed ■☆
50718 Bellardia trixago (L.) All. = Bartsia trixago L. ■☆
50719 Bellardia trixago (L.) All. var. flaviflora (Boiss.) Maire = Bartsia trixago L. ■☆
50720 Bellardia trixago (L.) All. var. versicolor (Lam.) Cout. = Bartsia trixago L. ■☆
50721 Bellardia trixago All. = Bellardia trixago (L.) All. ■☆
50722 Bellardiochloa Chiov. (1929);伯氏禾属■☆
50723 Bellardiochloa Chiov. = Poa L. ■
50724 Bellardiochloa polychloa (Trautv.) Roshev.;伯氏禾■☆
50725 Bellardiochloa polychloa (Trautv.) Roshev. = Festuca polychloa Trautv. ■☆
50726 Bellardiochloa variegata (Lam.) Kerguélen;斑叶伯氏禾■☆
50727 Bellardiochloa violacea Chiov.;堇色伯氏禾■☆
50728 Bellendena R. Br. (1810);塔岛山龙眼属●☆
50729 Bellendena montana R. Br.;塔岛山龙眼●☆
50730 Bellendenia Endl. = Bellendena R. Br. ●☆
50731 Bellendenia Raf. = Tritonia Ker Gawl. ■
50732 Bellendenia Raf. ex Endl. = Tritonia Ker Gawl. ■
50733 Bellermannia Klotzsch = ? Gonzalagunia Ruiz et Pav. ●☆

50734 Bellermannia Klotzsch ex H. Karst. (1846);阿氏茜属☆
50735 Bellevalia Delile = Althenia F. Petit ■☆
50736 Bellevalia Lapeyr. (1808)(保留属名);罗马风信子属■☆
50737 Bellevalia Montrouz. = Agatea A. Gray ■☆
50738 Bellevalia Montrouz. ex P. Beauvis. = Agatea A. Gray ■☆
50739 Bellevalia Roem. et Schult. = Richeria Vahl ●☆
50740 Bellevalia Scop. (废弃属名) = Bellevalia Lapeyr. (保留属名) ■☆
50741 Bellevalia Scop. (废弃属名) = Clerodendrum L. ●■
50742 Bellevalia Scop. (废弃属名) = Marurang Rumph. ex Adans. ●■
50743 Bellevalia acutifolia (Boiss.) Deloney;尖叶罗马风信子■☆
50744 Bellevalia albana Woronow;阿尔班罗马风信子■☆
50745 Bellevalia araxina Woronow;阿拉罗马风信子■☆
50746 Bellevalia atroviolacea Regel;暗紫罗马风信子■☆
50747 Bellevalia aucheri (Baker) Losinsk.;奥氏罗马风信子■☆
50748 Bellevalia ciliata (Cirillo) Nees;缘毛罗马风信子■☆
50749 Bellevalia ciliata (Cirillo) Nees var. glauca (Lindl.) Boiss. = Bellevalia ciliata (Cirillo) Nees ■☆
50750 Bellevalia cyrenaica Maire et Weiller;昔兰尼罗马风信子■☆
50751 Bellevalia dubia (Guss.) Rchb.;可疑罗马风信子■☆
50752 Bellevalia dubia (Guss.) Rchb. var. fallax (Pomel) Maire et Weiller = Bellevalia dubia (Guss.) Rchb. ■☆
50753 Bellevalia dubia (Guss.) Rchb. var. maura Braun-Blanq. et Maire = Bellevalia dubia (Guss.) Rchb. ■☆
50754 Bellevalia dubia (Guss.) Rchb. var. riphaena (Pau) Maire et Weiller = Bellevalia dubia (Guss.) Rchb. ■☆
50755 Bellevalia dubia (Guss.) Rchb. var. variabilis (Freyn) Maire = Bellevalia dubia (Guss.) Rchb. ■☆
50756 Bellevalia fallax Pomel = Bellevalia dubia (Guss.) Rchb. ■☆
50757 Bellevalia flexuosa Boiss.;曲折罗马风信子■☆
50758 Bellevalia flexuosa Boiss. var. galalensis Täckh. et Drar = Bellevalia flexuosa Boiss. ■☆
50759 Bellevalia fominii Woronow;佛敏罗马风信子■☆
50760 Bellevalia hyacinthoides (Bertol.) K. M. Perss. et Wendelbo;假风信子■☆
50761 Bellevalia lipskyi (Miscz.) Wulf;利普斯基罗马风信子■☆
50762 Bellevalia longistyla (Miscz.) Grossh.;长柱罗马风信子■☆
50763 Bellevalia lutea Bordz.;黄罗马风信子■☆
50764 Bellevalia macrobotrys Boiss.;大穗罗马风信子■☆
50765 Bellevalia mauritanica Pomel;毛里塔尼亚风信子■☆
50766 Bellevalia mauritanica Pomel var. tunetana Batt. = Bellevalia mauritanica Pomel ■☆
50767 Bellevalia montana (C. Koch) Boiss.;山地罗马风信子■☆
50768 Bellevalia pomelii Maire;波梅尔罗马风信子■☆
50769 Bellevalia pycnantha (K. Koch) Losinsk.;密花罗马风信子■☆
50770 Bellevalia romana (L.) Rchb.;罗马风信子;Roman Hyacinth ■☆
50771 Bellevalia romana (L.) Rchb. var. battandieri (Freyn) Durand et Batt. = Bellevalia mauritanica Pomel ■☆
50772 Bellevalia romana (L.) Rchb. var. mauritanica (Pomel) Bonnet et Barratte = Bellevalia mauritanica Pomel ■☆
50773 Bellevalia sarmatica (Pall.) Woronow;萨尔马特罗马风信子■☆
50774 Bellevalia saviczii Woronow;萨维罗马风信子■☆
50775 Bellevalia sessiliflora (Viv.) Kunth;无花梗罗马风信子■☆
50776 Bellevalia speciosa Woronow;美丽罗马风信子■☆
50777 Bellevalia trifoliata (Ten.) Kunth;三小叶罗马风信子■☆
50778 Bellevalia turkestanica Franck;土耳其四斯坦罗马风信子■☆
50779 Bellevalia wilhelmsii (Stev.) Woronow;维尔罗马风信子■☆
50780 Bellevalia zygomorpha Woronow;对称罗马风信子■☆
50781 Bellia Bubani = Chaerophyllum L. ■
50782 Bellida Ewart(1907);禾鼠麹属■☆
50783 Bellida graminea Ewart.;禾鼠麹■☆
50784 Bellidastrum (DC.) Scop. = Aster L. ●■
50785 Bellidastrum Scop. = Bellidastrum (DC.) Scop. ●■
50786 Bellidiaster Dumort. = Bellidastrum (DC.) Scop. ●■
50787 Bellidiastrum Cass. = Bellidastrum (DC.) Scop. ●■
50788 Bellidiastrum Less. = Osmites L. (废弃属名) ●☆
50789 Bellidiopsis Spach = Osmites L. (废弃属名) ●☆
50790 Bellidiopsis rotundifolia (Desf.) Pomel = Bellis rotundifolia (Desf.) Boiss. et Reut. ■☆
50791 Bellidiopsis rotundifolia (Desf.) Pomel var. parvifolia Pomel = Bellis rotundifolia (Desf.) Boiss. et Reut. ■☆
50792 Bellidistrum Rchb. = Bellidastrum Scop. ●■
50793 Bellidium Bertol. (1853);小白菊属■☆
50794 Bellidium Bertol. = Bellis L. ■
50795 Bellidium pappulosum Bertol.;小白菊■☆
50796 Bellidium rotundifolium Bertol.;圆叶小白菊■☆
50797 Bellilla Raf. = Belilla Adans. ●■
50798 Bellilla Raf. = Mussaenda L. ●■
50799 Bellingia Pierre = Zollingeria Kurz(保留属名) ■☆
50800 Bellinia Roem. et Schult. = Saracha Ruiz et Pav. ●☆
50801 Belliolum Tiegh. (1900);美林仙属●☆
50802 Belliolum Tiegh. = Zygogynum Baill. ●☆
50803 Belliolum crassifolium Tiegh.;厚叶美林仙●☆
50804 Belliolum gracile A. C. Sm.;纤细美林仙●☆
50805 Belliolum pancheri Tiegh.;美林仙●☆
50806 Belliopsis Pomel = Bellium L. ■☆
50807 Belliopsis rotundifolia (Desf.) Pomel = Bellis rotundifolia (Desf.) Boiss. et Reut. ■☆
50808 Bellis L. (1753);雏菊属;Daisy, English Daisy ■
50809 Bellis annua L.;西班牙雏菊(拟雏菊);Bastard Daisy, False Daisy, Spanish Daisy ■☆
50810 Bellis annua L. subsp. microcephala (Lange) Nyman;小头西班牙雏菊■☆
50811 Bellis annua L. subsp. minuta (DC.) Quézel et Santa = Bellis annua L. subsp. microcephala (Lange) Nyman ■☆
50812 Bellis annua L. var. minuta (DC.) Ball = Bellis annua L. subsp. microcephala (Lange) Nyman ■☆
50813 Bellis annua L. var. radicans Coss. = Bellis annua L. subsp. microcephala (Lange) Nyman ■☆
50814 Bellis annua L. var. vergens Pomel = Bellis annua L. ■☆
50815 Bellis atlantica Boiss. et Reut. = Bellis sylvestris Cirillo ■☆
50816 Bellis ciliata Raf. = Astranthium ciliatum (Raf.) G. L. Nesom ■☆
50817 Bellis coerulescens Coss.;天蓝雏菊■☆
50818 Bellis hyrcanica Woronow;希尔康雏菊■☆
50819 Bellis integrifolia Michx.;全缘叶雏菊■☆
50820 Bellis integrifolia Michx. = Astranthium integrifolium (Michx.) Nutt. ■☆
50821 Bellis jaculifolia Salisb. = Cunninghamia lanceolata (Lamb.) Hook. ●
50822 Bellis lanceolata (Salisb.) Sweet = Cunninghamia lanceolata (Lamb.) Hook. ●
50823 Bellis minuta (DC.) Bég. = Bellis annua L. subsp. microcephala (Lange) Nyman ■☆
50824 Bellis pappulosa Boiss. = Bellis sylvestris Cirillo ■☆
50825 Bellis perennis L.;雏菊(长寿菊,马兰头花,延命菊);Baby's

Pet, Bachelor's Buttons, Bairnwort, Baiyan-flower, Banebind, Bannergowan, Banwood, Banwort, Barnwort, Bennergowan, Bennert, Bessie Banwood, Bessy Bairnwort, Billy Buttons, Bone-flower, Bonwort, Brisewort, Bruisewort, Bryswort, Cat Posy, Children's Daisy, Cockiloorie, Comfrey, Common Daisy, Consound, Cumfirie, Curldoddy, Daisy, Day's Eye, Day's Eyes, Dazeg, Dicky Daisy, Dog Daisy, Double Daisy, Ducky Daisy, English Daisy, Ewe Gollan, Ewe Gowan, Ewe-gollan, Ewe-gowan, Eye of Day, Flower of Spring, Garden Daisy, Gowan, Gowen, Gowlan, Gracy Daisy, Hen-and-chickens, Herb Margaret, Innocent, Jack-an-apes-on-horseback, Lawn Daisy, Lawndaisy, Little Open Star, Little Star, March Daisy, Mary Gowan, May Gowan, Measure of Love, Miss Modesty, Nails, Noon-flower, Open Eyes, Primrose, Shepher's Daisy, Shepherd's Daisy, Silver Pennies, Silverpennies, Small Daisy, Sweep, True Daisy, Twelve Disciples, While Frills, White Yule ■

50826 Bellis perennis L. 'Prolifera';多育雏菊;Hen-and-chickens ■☆

50827 Bellis prostrata Pomel = Bellis repens Lam. ■☆

50828 Bellis repens Lam.;匍匐雏菊■☆

50829 Bellis rotundifolia (Desf.) Boiss. et Reut.;圆叶雏菊■☆

50830 Bellis stipitata Labill. = Lagenophora stipitata (Labill.) Druce ■

50831 Bellis sylvestris Cirillo;林间雏菊;Southern Daisy ■☆

50832 Bellis sylvestris Cirillo subsp. maroccana Sennen = Bellis sylvestris Cirillo ■☆

50833 Bellis sylvestris Cirillo subsp. natalis-Jesus Sennen et Mauricio = Bellis sylvestris Cirillo ■☆

50834 Bellis sylvestris Cirillo var. atlantica (Boiss. et Reut.) Batt. = Bellis sylvestris Cirillo ■☆

50835 Bellis sylvestris Cirillo var. cyrenaica Bég. = Bellis sylvestris Cirillo ■☆

50836 Bellis sylvestris Cirillo var. major Alleiz. = Bellis sylvestris Cirillo ■☆

50837 Bellis sylvestris Cirillo var. pappulosa (Boiss.) Lange = Bellis sylvestris Cirillo ■☆

50838 Bellis sylvestris Cirillo var. rotundifolia (Boiss. et Reut.) Batt. = Bellis sylvestris Cirillo ■☆

50839 Bellis velutina Pomel = Bellis sylvestris Cirillo ■☆

50840 Bellis vergens (Pomel) Bég. = Bellis annua L. ■☆

50841 Bellium L. (1771);拟雏菊属(丽菊属);Bastard Daisy ■☆

50842 Bellium bellidioides (L.) Desf. = Bellis annua L. ■☆

50843 Bellium minutum L.;小拟雏菊■☆

50844 Bellium rotundifolium (Desf.) DC. = Bellis rotundifolia (Desf.) Boiss. et Reut. ■☆

50845 Bellizinca Borhidi = Omiltemia Standl. ●☆

50846 Bellizinca Borhidi(2004);墨西哥茜草属●☆

50847 Belloa J. Rémy(1848);尖柱紫绒草属■☆

50848 Belloa chilensis J. Rémy;尖柱紫绒草■☆

50849 Belloa longifolia (Cuatrec. et Aristeg.) Sagást. et M. O. Dillon;长叶尖柱紫绒草■☆

50850 Bellonia L. (1753);贝隆苣苔属●☆

50851 Bellonia aspera L.;贝隆苣苔●☆

50852 Bellonia spinosa Sw.;刺贝隆苣苔●☆

50853 Belloniaceae Martinov = Gesneriaceae Rich. et Juss.(保留科名)●■

50854 Bellota A. Rich. ex Phil. = Boldu Adans.(废弃属名)●☆

50855 Bellota Gay = Beilschmiedia Nees ●

50856 Bellota Gay = Cryptocarya R. Br.(保留属名)●

50857 Bellota Gay = Ocotea Aubl. ●☆

50858 Belluccia Adans.(废弃属名) = Bellucia Neck. ex Raf.(保留属名)●☆

50859 Belluccia Adans.(废弃属名) = Ptelea L. ●

50860 Bellucia Neck. = Bellucia Neck. ex Raf.(保留属名)●☆

50861 Bellucia Neck. ex Raf. (1838)(保留属名);热美野牡丹属●☆

50862 Bellucia pentamera Naodin;五数热美野牡丹●☆

50863 Bellynkxia Müll. Arg. = Morinda L. ●■

50864 Belmontia E. Mey.(保留属名) = Sebaea Sol. ex R. Br. ■

50865 Belmontia baumiana Gilg = Sebaea baumiana (Gilg) Boutique ■☆

50866 Belmontia chevalieri Abbayes et Schnell = Sebaea teuszii (Schinz) P. Taylor ■☆

50867 Belmontia chionantha Gilg = Sebaea teuszii (Schinz) P. Taylor ■☆

50868 Belmontia cordata (L. f.) E. Mey. = Sebaea exacoides (L.) Schinz ■☆

50869 Belmontia cordata (L. f.) E. Mey. var. intermedia (Cham. et Schltdl.) Griseb. = Sebaea micrantha (Cham. et Schltdl.) Schinz var. intermedia (Cham. et Schltdl.) Marais ■☆

50870 Belmontia cordata (L. f.) E. Mey. var. micrantha (Cham. et Schltdl.) Griseb. = Sebaea micrantha (Cham. et Schltdl.) Schinz ■☆

50871 Belmontia debilis (Welw.) Benth. et Hook. = Sebaea debilis (Welw.) Schinz ■☆

50872 Belmontia debilis (Welw.) Schinz = Sebaea debilis (Welw.) Schinz ■☆

50873 Belmontia divaricata Baker = Exacum divaricatum (Baker) Schinz ●☆

50874 Belmontia flanaganii Schinz = Sebaea spathulata (E. Mey.) Steud. ■☆

50875 Belmontia gracilis Welw. = Sebaea gracilis (Welw.) Paiva et I. Nogueira ■☆

50876 Belmontia grandis E. Mey. = Sebaea grandis (E. Mey.) Steud. ■☆

50877 Belmontia hockii De Wild. = Sebaea hockii (De Wild.) Boutique ■☆

50878 Belmontia intermedia (Cham. et Schltdl.) Knobl. = Sebaea micrantha (Cham. et Schltdl.) Schinz var. intermedia (Cham. et Schltdl.) Marais ■☆

50879 Belmontia luteo-alba A. Chev. = Sebaea luteo-alba (A. Chev.) P. Taylor ■☆

50880 Belmontia micrantha (Cham. et Schltdl.) Gilg = Sebaea micrantha (Cham. et Schltdl.) Schinz ■☆

50881 Belmontia natalensis Schinz = Sebaea grandis (E. Mey.) Steud. ■☆

50882 Belmontia oligantha Gilg = Sebaea oligantha (Gilg) Schinz ■☆

50883 Belmontia platyptera Baker = Sebaea platyptera (Baker) Boutique ■☆

50884 Belmontia primuliflora (Welw.) Schinz = Sebaea primuliflora (Welw.) Sileshi ■☆

50885 Belmontia pumila Baker = Sebaea pumila (Baker) Schinz ■☆

50886 Belmontia spathulata E. Mey. = Sebaea spathulata (E. Mey.) Steud. ■☆

50887 Belmontia stricta Schinz = Sebaea madagascariensis Klack. ■☆

50888 Belmontia teuszii Schinz = Sebaea teuszii (Schinz) P. Taylor ■☆

50889 Belmontia teuszii Schinz var. angustifolia De Wild. = Sebaea teuszii (Schinz) P. Taylor ■☆

50890 Belmontia zambesiaca Baker = Sebaea grandis (E. Mey.) Steud. ■☆

50891 Beloakon Raf. = Phlomis L. ●■

50892 Beloanthera Hassk. = Hydrolea L.(保留属名)■

50893 Beloanthera oppositifolia Hassk. = Hydrolea zeylanica (L.) J.

Vahl ■
50894 Beloere Shuttlew. = Abutilaea F. Muell. ●■
50895 Beloere Shuttlew. = Abutilon Mill. ●■
50896 Beloere Shuttlew. = Beloere Shuttlew. ex A. Gray ●■
50897 Beloere Shuttlew. = Herissantia Medik. ●■
50898 Beloere Shuttlew. ex A. Gray = Abutilaea F. Muell. ●■
50899 Beloere Shuttlew. ex A. Gray = Abutilon Mill. ●■
50900 Beloere Shuttlew. ex A. Gray = Herissantia Medik. ●■
50901 Beloere crispa (L.) Shuttlew. ex A. Gray = Herissantia crispa (L.) Brizicky ■
50902 Beloglottis Schltr. (1920);刺舌兰属■☆
50903 Beloglottis Schltr. = Spiranthes Rich. (保留属名)■
50904 Beloglottis americana (C. Schweinf. et Garay) Szlach. = Manniella americana C. Schweinf. et Garay ■☆
50905 Beloglottis costaricensis (Rchb. f.) Schltr.;中脉肖绶草■☆
50906 Belonanthus Graebn. = Valeriana L. ●■
50907 Belonia Adans. = Bellonia L. ●☆
50908 Belonites B. Mey. = Pachypodium Lindl. ●☆
50909 Belonophora Hook. f. (1873);针茜属■☆
50910 Belonophora coffeoides Hook. f.;咖啡状针茜■☆
50911 Belonophora coffeoides Hook. f. subsp. hypoglauca (Welw. ex Hiern) S. E. Dawson et Cheek;里粉绿咖啡状针茜■☆
50912 Belonophora coriacea Hoyle;革质针茜■☆
50913 Belonophora gilletii A. Chev. = Belonophora coffeoides Hook. f. subsp. hypoglauca (Welw. ex Hiern) S. E. Dawson et Cheek ■☆
50914 Belonophora glomerata M. B. Moss = Belonophora coffeoides Hook. f. subsp. hypoglauca (Welw. ex Hiern) S. E. Dawson et Cheek ■☆
50915 Belonophora hypoglauca (Welw. ex Hiern) A. Chev. = Belonophora coffeoides Hook. f. subsp. hypoglauca (Welw. ex Hiern) S. E. Dawson et Cheek ■☆
50916 Belonophora lepidopoda (K. Schum.) Hutch. et Dalziel = Belonophora coffeoides Hook. f. subsp. hypoglauca (Welw. ex Hiern) S. E. Dawson et Cheek ■☆
50917 Belonophora morganae Hutch. = Belonophora coffeoides Hook. f. subsp. hypoglauca (Welw. ex Hiern) S. E. Dawson et Cheek ■☆
50918 Belonophora talbotii (Wernham) Keay;塔尔博特针茜■☆
50919 Belonophora uniflora K. Schum. = Aulacocalyx caudata (Hiern) Keay ●☆
50920 Belonophora wernhamii Hutch. et Dalziel;沃纳姆针茜■☆
50921 Beloperone Nees = Justicia L. ●■
50922 Beloperone Nees(1832);矢带爵床属(麒麟吐珠属)■☆
50923 Beloperone californica Benth. = Justicia californica (Benth.) D. N. Gibson ●☆
50924 Beloperone guttata Brandegee 'Variegata';花叶小虾花●☆
50925 Beloperone guttata Brandegee 'Yellow Queen';黄花小虾花●☆
50926 Beloperone guttata Brandegee = Calliaspidia guttata (Brandegee) Bremek. ●■
50927 Beloperone guttata Brandegee = Justicia brandegeana Wassh. et L. B. Sm. ●☆
50928 Beloperonidea Oerst. = Beloperone Nees ■☆
50929 Belospis Raf. = Salvia L. ●■
50930 Belostemma Wall. ex Wight = Tylophora R. Br. ●■
50931 Belostemma Wall. ex Wight(1834);箭药藤属;Belostemma ●★
50932 Belostemma cordifolium (Link, Klotzsch et Otto) M. G. Gilbert et P. T. Li;心叶箭药藤;Heart-leaf Belostemma ●
50933 Belostemma cordifolium (Link, Klotzsch et Otto) P. T. Li = Belostemma cordifolium (Link, Klotzsch et Otto) M. G. Gilbert et P. T. Li ●
50934 Belostemma hirsutum Wall. ex Wight;箭药藤(展冠藤);Hirsute Belostemma ●
50935 Belostemma hirsutum Wall. ex Wight = Tylophora belostemma Benth. ●
50936 Belostemma yunnanense Tsiang;镰药藤;Yunnan Belostemma ●
50937 Belosynapsis Hassk. (1871);假紫万年青属;Belosynapsis ■
50938 Belosynapsis capitata (Blume) Spreng. et C. E. C. Fisch. = Belosynapsis ciliata (Blume) R. S. Rao ■
50939 Belosynapsis ciliata (Blume) R. S. Rao;假紫万年青(毛绿蓝花草,毛叶鸭舌疝);Ciliate Belosynapsis ■
50940 Belosynapsis kawakamii (Hayata) C. I. Peng et Y. J. Chen;川上氏鸭舌疝■
50941 Belotia A. Rich. = Trichospermum Blume ●☆
50942 Belotia australis Little = Trichospermum australe (Little) Kosterm. ●☆
50943 Belotropis Raf. = Salicornia L. ●■
50944 Belou Adans(废弃属名) = Aegle Corrêa(保留属名)●
50945 Belovia Bunge = Suaeda Forssk. ex J. F. Gmel. (保留属名)●■
50946 Belovia Moq. = Suaeda Forssk. ex J. F. Gmel. (保留属名)●■
50947 Belovia paradoxa Bunge = Suaeda paradoxa Bunge ■
50948 Beltokon Raf. = Origanum L. ●■
50949 Beltrania Miranda = Enriquebeltrania Rzed. ☆
50950 Belutta Raf. = Allmania R. Br. ex Wight ■
50951 Belutta-kaka Adans. (废弃属名) = Chonemorpha G. Don(保留属名)●
50952 Beluttakaka Adans. = Chonemorpha G. Don(保留属名)●
50953 Beluttakaka Adans. ex Kuntze = Chonemorpha G. Don(保留属名)●
50954 Belvala Adans. (废弃属名) = Struthiola L. (保留属名)●☆
50955 Belvalla Delile = Althenia F. Petit ■☆
50956 Belvisia Desv. = Napoleona P. Beauv. ●☆
50957 Belvisiaceae R. Br. = Napoleonaenaceae P. Beauv. ●☆
50958 Bemarivea Choux = Tinopsis Radlk. ●☆
50959 Bemarivea dissitiflora (Baker) Choux = Tinopsis dissitiflora (Baker) Capuron ●☆
50960 Bembecodium Lindl. = Athanasia L. ●☆
50961 Bembicia Oliv. (1883);盾头木属●☆
50962 Bembicia axillaris Oliv.;盾头木●☆
50963 Bembiciaceae R. C. Keating et Takht. (1994);盾头木科●☆
50964 Bembiciaceae R. C. Keating et Takht. = Flacourtiaceae Rich. ex DC. (保留科名)●
50965 Bembiciaceae R. C. Keating et Takht. = Salicaceae Mirb. (保留科名)●
50966 Bembicidium Rydb. (1920);小盾头木属●☆
50967 Bembicidium Rydb. = Poitea Vent. ●☆
50968 Bembicina Kuntze = Bembicia Oliv. ●☆
50969 Bembiciopsis H. Perrier = Camellia L. ●
50970 Bembicium Mart. = Eupatorium L. ●■
50971 Bembicium Mart. ex Baker = Eupatorium L. ●■
50972 Bembicodium Post et Kuntze = Athanasia L. ●☆
50973 Bembicodium Post et Kuntze = Bembycodium Kuntze ●☆
50974 Bembix Lour. (废弃属名) = Ancistrocladus Wall. (保留属名)●
50975 Bembix tectoria Lour. = Ancistrocladus tectorius (Lour.) Merr. ●
50976 Bembycodium Kunze = Athanasia L. ●☆
50977 Bemsetia Raf. = Ixora L. ●

50978 Benaurea Raf. = Musschia Dumort. ●☆

50979 Bencomia Webb et Berthel. (1842);异地榆属●☆

50980 Bencomia brachystachya Nordborg;短穗异地榆●☆

50981 Bencomia caudata (Aiton) Webb et Berthel.;尾状异地榆●☆

50982 Bencomia maderensis Bornm.;梅德异地榆■☆

50983 Bencomia sphaerocarpa Svent.;球果异地榆●☆

50984 Benedicta Bernh. = Carbenia Adans. ●■

50985 Benedictaea Toledo = Ottelia Pers. ■

50986 Benedictella Maire = Lotus L. ■

50987 Benedictella benoistii Maire = Lotus benoistii (Maire) Lassen ■☆

50988 Beneditaea Toledo = Ottelia Pers. ■

50989 Benevidesia Saldanha et Cogn. (1888);贝内野牡丹属●☆

50990 Benevidesia Saldanha et Cogn. ex Cogn. = Benevidesia Saldanha et Cogn. ●☆

50991 Benevidesia organensis Saldanha et Cogn.;贝内野牡丹●☆

50992 Benguellia G. Taylor(1931);安哥拉草属■☆

50993 Benguellia lanceolata (Gürke) G. Taylor;安哥拉草■☆

50994 Benincasa Savi(1818);冬瓜属;Wax Gourd,Waxgourd ■

50995 Benincasa cerifera Savi = Benincasa hispida (Thunb.) Cogn. ■

50996 Benincasa cerifera Savi f. emarginata ?;微缺冬瓜■

50997 Benincasa hispida (Thunb. ex A. Murray) Cogn. = Benincasa hispida (Thunb.) Cogn. ■

50998 Benincasa hispida (Thunb.) Cogn.;冬瓜(白东瓜,白冬瓜,白瓜,白花,地芝,东瓜,濮瓜,水芝,枕瓜,猪子冬瓜);Ash Gourd, Ash Pumpkin, China Waxgourd, Chinese Preserving Melon, Chinese Pumpkin, Chinese Wax Gourd, Chinese Waxgourd, Petha, Tunka, Wax Gourd, Waxgourd, White Gourd, Zit-Kawa, Zit-Kwa ■

50999 Benincasa hispida (Thunb.) Cogn. = Benincasa hispida (Thunb. ex A. Murray) Cogn. ■

51000 Benincasa hispida (Thunb.) Cogn. var. chieh-qua F. C. How;节瓜;Chieh-gua ■

51001 Benincasa hispida (Thunb.) Cogn. var. chieh-qua F. C. How = Benincasa hispida (Thunb.) Cogn. ■

51002 Benitoa D. D. Keck = Lessingia Cham. ■☆

51003 Benitoa D. D. Keck(1956);光叶沙紫菀属■☆

51004 Benitoa occidentalis (H. M. Hall) D. D. Keck;光叶沙紫菀■☆

51005 Benitzta H. Karst. = Gymnosiphon Blume ■

51006 Benjamina Vell. = Dictyoloma A. Juss. (保留属名)●☆

51007 Benjaminia Mart. ex Benj. (1847);本氏婆婆纳属(本氏玄参属)■☆

51008 Benjaminia Mart. ex Benj. = Bacopa Aubl. (保留属名)■

51009 Benjaminia Mart. ex Benj. = Quinquelobus Benj. ■☆

51010 Benjaminia utriculariiformis Mart. ex Benj.;本氏婆婆纳■☆

51011 Benkara Adans. = Melastoma Burm. ●■

51012 Benkara Adans. = Randia L. ●

51013 Bennetia DC. = Bennettia Gray ●■

51014 Bennetia DC. = Saussurea DC. (保留属名)●■

51015 Bennetia Gray = Saussurea DC. (保留属名)●■

51016 Bennetia Raf. = Sporobolus R. Br. ■

51017 Bennettia Gray = Saussurea DC. (保留属名)●■

51018 Bennettia Miq. = Bennettiodendron Merr. ●

51019 Bennettia R. Br. = Galearia Zoll. et Moritzi(保留属名)●☆

51020 Bennettia leprosipes (Clos) Koord. = Bennettiodendron leprosipis (Clos) Merr. ●

51021 Bennettia longipes Oliv. = Bennettiodendron leprosipis (Clos) Merr. ●

51022 Bennettiaceae R. Br. = Scepaceae Lindl. + Pandaceae Engl. et Gilg(保留科名)●

51023 Bennettiaceae R. Br. ex Schnizl. = Scepaceae Lindl. + Pandaceae Engl. et Gilg(保留科名)●

51024 Bennettiodendron Merr. (1927);山桂花属;Bennettiodendron, Wildlaurel ●

51025 Bennettiodendron brevipes Merr.;短柄山桂花(短柄本勒木,假穗山桂花,山桂花,上思本勒木);False-raceme Bennettiodendron, Short-petioled Bennettiodendron, Short-stalk Bennettiodendron ●

51026 Bennettiodendron brevipes Merr. = Bennettiodendron leprosipis (Clos) Merr. ●

51027 Bennettiodendron brevipes Merr. var. margopatens S. S. Lai;延叶山桂花;Margo-spreading Bennettiodendron ●

51028 Bennettiodendron brevipes Merr. var. margopatens S. S. Lai = Bennettiodendron leprosipis (Clos) Merr. ●

51029 Bennettiodendron brevipes Merr. var. shangsiense (X. X. Chen et J. Y. Luo) S. S. Lai;上思山桂花;Shangsi Bennettiodendron ●

51030 Bennettiodendron brevipes Merr. var. shangsiense (X. X. Chen et J. Y. Luo) S. S. Lai = Bennettiodendron leprosipis (Clos) Merr. ●

51031 Bennettiodendron cordatum Merr.;心叶山桂花;Heart-leaved Bennettiodendron ●

51032 Bennettiodendron ellipticum C. Y. Wu;椭圆山桂花;Ellipticleaf Bennettiodendron ●

51033 Bennettiodendron lanceolatum H. L. Li;披针叶山桂花;Lanceleaf Bennettiodendron ●

51034 Bennettiodendron lanceolatum H. L. Li = Bennettiodendron leprosipis (Clos) Merr. ●

51035 Bennettiodendron lanceolatum H. L. Li var. obovatum S. S. Lai;倒卵叶山桂花●

51036 Bennettiodendron lanceolatum H. L. Li var. pilosum (G. S. Fan et Y. C. Hsu) S. S. Lai;毛叶山桂花(毛山桂花);Pilose Bigleaf Bennettiodendron ●

51037 Bennettiodendron leprosipes (Clos) Merr. var. pilosum G. S. Fan et Y. C. Hsu = Bennettiodendron lanceolatum H. L. Li var. pilosum (G. S. Fan et Y. C. Hsu) S. S. Lai ●

51038 Bennettiodendron leprosipes (Clos) Merr. var. stenophyllum S. S. Lai = Bennettiodendron brevipes Merr. var. shangsiense (X. X. Chen et J. Y. Luo) S. S. Lai ●

51039 Bennettiodendron leprosipis (Clos) Merr.;山桂花(本勒木,本勒木);Common Bennettiodendron, Wildlaurel ●

51040 Bennettiodendron leprosipis (Clos) Merr. var. ellipticum S. S. Lai;椭圆叶山桂花;Ellipticleaf Bennettiodendron ●

51041 Bennettiodendron leprosipis (Clos) Merr. var. ellipticum S. S. Lai = Bennettiodendron leprosipis (Clos) Merr. ●

51042 Bennettiodendron leprosipis (Clos) Merr. var. pilosum G. S. Fan et Y. C. Hsu = Bennettiodendron leprosipis (Clos) Merr. ●

51043 Bennettiodendron leprosipis (Clos) Merr. var. pilosum G. S. Fan et Y. C. Hsu = Bennettiodendron macrophyllim C. Y. Wu ex S. S. Lai var. pilosum (G. S. Fan et Y. C. Hsu) S. S. Lai ●

51044 Bennettiodendron leprosipis (Clos) Merr. var. rugosifolium S. S. Lai;皱叶山桂花;Wrinkledleaf Bennettiodendron ●

51045 Bennettiodendron leprosipis (Clos) Merr. var. rugosifolium S. S. Lai = Bennettiodendron leprosipis (Clos) Merr. ●

51046 Bennettiodendron longipes (Oliv.) Merr.;长柄山桂花(长柄本勒木);Long-stalk Bennettiodendron ●

51047 Bennettiodendron longipes (Oliv.) Merr. = Bennettiodendron leprosipis (Clos) Merr. ●

51048 Bennettiodendron macrophyllim C. Y. Wu ex S. S. Lai var.

obovatum S. S. Lai;倒卵大叶山桂花;Obovateleaf Bennettiodendron ●
51049 Bennettiodendron macrophyllum C. Y. Wu ex S. S. Lai;大叶山桂花(思茅山桂花);Bigleaf Bennettiodendron ●
51050 Bennettiodendron macrophyllum C. Y. Wu ex S. S. Lai = Bennettiodendron leprosipis (Clos) Merr. ●
51051 Bennettiodendron macrophyllum C. Y. Wu ex S. S. Lai var. pilosum (G. S. Fan et Y. C. Hsu) S. S. Lai = Bennettiodendron lanceolatum H. L. Li var. pilosum (G. S. Fan et Y. C. Hsu) S. S. Lai ●
51052 Bennettiodendron macrophyllum C. Y. Wu ex S. S. Lai var. pilosum (G. S. Fan et Y. C. Hsu) S. S. Lai = Bennettiodendron leprosipis (Clos) Merr. ●
51053 Bennettiodendron shangsiense X. X. Chen et J. Y. Luo = Bennettiodendron brevipes Merr. var. shangsiense (X. X. Chen et J. Y. Luo) S. S. Lai ●
51054 Bennettiodendron simaoense G. S. Fan = Bennettiodendron leprosipis (Clos) Merr. ●
51055 Bennettiodendron simaoense G. S. Fan = Bennettiodendron macrophyllum C. Y. Wu ex S. S. Lai ●
51056 Bennettiodendron subracemosum C. Y. Wu;假穗山桂花;False-raceme Bennettiodendron ●
51057 Bennettiodendron subracemosum C. Y. Wu = Bennettiodendron brevipes Merr. ●
51058 Bennettiodendron subracemosum C. Y. Wu = Bennettiodendron leprosipis (Clos) Merr. ●
51059 Bennettites Carruth. (1870);本内苏铁属(本勒苏铁属,赛凤尾蕉属)●☆
51060 Benoicanthus Heine et A. Raynal(1968);马岛芦莉草属■☆
51061 Benoicanthus tachiadenus Heine et A. Raynal;马岛芦莉草■☆
51062 Benoistia H. Perrier et Léandri(1938);伯努瓦大戟属☆
51063 Benoistia orientalis Radcl. -Sm.;东方伯努瓦大戟☆
51064 Benoistia sambiranensis H. Perrier et Léandri;伯努瓦大戟☆
51065 Bensonia Abrams et Bacig. = Bensoniella C. V. Morton ■☆
51066 Bensoniella C. V. Morton(1965);本森草属■☆
51067 Bensoniella oregona (Abrams et Bacig.) C. V. Morton;本森草■☆
51068 Benteca Adans. (废弃属名) = Hymenodictyon Wall. (保留属名)●
51069 Benthamantha Alef. = Coursetia DC. ●☆
51070 Benthamantha Alef. = Cracca Benth. (保留属名)●☆
51071 Benthamia A. Rich. (1828);本氏兰属●☆
51072 Benthamia Lindl. (1830) = Amsinckia Lehm. (保留属名)■☆
51073 Benthamia Lindl. (1833) = Dendrobenthamia Hutch. ●
51074 Benthamia Lindl. = Cornus L. ●
51075 Benthamia bathieana Schltr.;巴西本氏兰●☆
51076 Benthamia capitata (Wall.) Nakai = Cornus capitata Wall. ●
51077 Benthamia capitata (Wall.) Nakai = Dendrobenthamia capitata (Wall.) Hutch. ●
51078 Benthamia catatiana H. Perrier;卡他本氏兰●☆
51079 Benthamia chinensis Lavalee = Dendrobenthamia japonica (DC.) W. P. Fang var. chineneis (Osborn) W. P. Fang ●
51080 Benthamia cinnabarina (Rolfe) H. Perrier;朱红本氏兰●☆
51081 Benthamia cuspidata H. Perrier;骤尖本氏兰●☆
51082 Benthamia dauphinensis (Rolfe) Schltr.;多芬本氏兰●☆
51083 Benthamia elata Schltr.;高本氏兰●☆
51084 Benthamia exilis Schltr.;瘦小本氏兰●☆
51085 Benthamia flavida Schltr. = Benthamia cinnabarina (Rolfe) H. Perrier ●☆
51086 Benthamia fragifera Lindl. = Benthamidia capitata (Wall. ex Roxb.) H. Hara ●
51087 Benthamia fragifera Lindl. = Cornus capitata Wall. ●
51088 Benthamia fragifera Lindl. = Dendrobenthamia capitata (Wall.) Hutch. ●
51089 Benthamia glaberrima (Ridl.) H. Perrier;光滑本氏兰●☆
51090 Benthamia graminea Schltr. = Cynorkis graminea (Thouars) Schltr. ■☆
51091 Benthamia herminioides Schltr.;角盘兰本氏兰●☆
51092 Benthamia hongkongensis (Hemsl.) Nakai = Cornus hongkongensis Hemsl. ●
51093 Benthamia hongkongensis (Hemsl.) Nakai = Dendrobenthamia hongkongensis (Hemsl.) Hutch. ●
51094 Benthamia humbertii H. Perrier;亨伯特本氏兰●☆
51095 Benthamia japonica Siebold et Zucc. = Benthamidia japonica (Siebold et Zucc.) H. Hara ●
51096 Benthamia japonica Siebold et Zucc. = Cornus kousa Buerger ex Hance ●
51097 Benthamia japonica Siebold et Zucc. = Dendrobenthamia japonica (DC.) W. P. Fang ●
51098 Benthamia japonica Siebold et Zucc. var. sinensis Benth. = Benthamia hongkongensis (Hemsl.) Nakai ●
51099 Benthamia japonica Siebold et Zucc. var. sinensis Benth. = Cornus hongkongensis Hemsl. ●
51100 Benthamia japonica Siebold et Zucc. var. sinensis Benth. = Dendrobenthamia hongkongensis (Hemsl.) Hutch. ●
51101 Benthamia jiysa Bajau ? = Dendrobenthamia japonica (DC.) W. P. Fang ●
51102 Benthamia kousa Nakai = Benthamidia japonica (Siebold et Zucc.) H. Hara ●
51103 Benthamia latifolia Schltr. = Benthamia bathieana Schltr. ●☆
51104 Benthamia lycopsoides (Lehm.) Lindl. ex Druce = Amsinckia lycopsoides Lehm. ex Fisch. et C. A. Mey. ■☆
51105 Benthamia macra Schltr.;大本氏兰●☆
51106 Benthamia madagascariensis (Rolfe) Schltr.;马岛本氏兰●☆
51107 Benthamia melanopoda Schltr.;黑足本氏兰●☆
51108 Benthamia minutiflora (Ridl.) Schltr. = Benthamia perularioides Schltr. ●☆
51109 Benthamia misera (Ridl.) Schltr.;贫弱本氏兰●☆
51110 Benthamia monophylla Schltr.;单叶本氏兰●☆
51111 Benthamia nigrescens Schltr.;变黑本氏兰●☆
51112 Benthamia nigro-vaginata H. Perrier;黑鞘本氏兰●☆
51113 Benthamia nivea Schltr.;雪白本氏兰●☆
51114 Benthamia perrieri Schltr. = Benthamia cinnabarina (Rolfe) H. Perrier ●☆
51115 Benthamia perularioides Schltr.;微花本氏兰●☆
51116 Benthamia praecox Schltr.;早生本氏兰●☆
51117 Benthamia procera Schltr.;高大本氏兰●☆
51118 Benthamia rostrata Schltr.;喙本氏兰●☆
51119 Benthamia spiralis (Thouars) A. Rich.;螺旋本氏兰●☆
51120 Benthamia verecunda Schltr.;羞涩本氏兰●☆
51121 Benthamia viridis Nakai = Dendrobenthamia japonica (DC.) W. P. Fang ●
51122 Benthamidia Spach = Cornus L. ●
51123 Benthamidia Spach(1839);本氏茱萸属(四照花属,肖本氏兰属)●☆
51124 Benthamidia capitata (Wall. ex Roxb.) H. Hara = Cornus capitata Wall. ●

51125 Benthamidia capitata (Wall. ex Roxb.) H. Hara = Dendrobenthamia capitata (Wall.) Hutch. ●

51126 Benthamidia capitata (Wall. ex Roxb.) H. Hara var. mollis (Rehder) H. Hara = Cornus elliptica (Pojark.) Q. Y. Xiang et Boufford ●

51127 Benthamidia capitata (Wall. ex Roxb.) H. Hara var. mollis (Rehder) H. Hara = Dendrobenthamia angustata (Chun) W. P. Fang var. molis (Rehder) W. P. Fang ●

51128 Benthamidia capitata (Wall.) H. Hara = Cornus capitata Wall. ●

51129 Benthamidia fargifera ? = Dendrobenthamia capitata (Wall.) Hutch. ●

51130 Benthamidia ferruginea (Y. C. Wu) H. Hara = Cornus hongkongensis Hemsl. ●

51131 Benthamidia ferruginea (Y. C. Wu) H. Hara = Cornus hongkongensis Hemsl. subsp. ferruginea (Y. C. Wu) Q. Y. Xiang ●

51132 Benthamidia ferruginea (Y. C. Wu) H. Hara = Dendrobenthamia ferruginea (Y. C. Wu) W. P. Fang ●

51133 Benthamidia florida (L.) Spach = Cornus florida L. ●☆

51134 Benthamidia hongkongensis (Hemsl.) H. Hara = Cornus hongkongensis Hemsl. subsp. gigantea (Hand. -Mazz.) Q. Y. Xiang ●

51135 Benthamidia hongkongensis (Hemsl.) H. Hara = Cornus hongkongensis Hemsl. ●

51136 Benthamidia hongkongensis (Hemsl.) H. Hara = Dendrobenthamia hongkongensis (Hemsl.) Hutch. ●

51137 Benthamidia hongkongensis (Hemsl.) H. Hara var. gigantea (Hand. -Mazz.) H. Hara = Dendrobenthamia gigantea (Hand. -Mazz.) W. P. Fang ●

51138 Benthamidia hongkongensis (Hemsl.) H. Hara var. gigantea (Hand. -Mazz.) H. Hara = Cornus hongkongensis Hemsl. subsp. gigantea (Hand. -Mazz.) Q. Y. Xiang ●

51139 Benthamidia japonica (Siebold et Zucc.) H. Hara = Dendrobenthamia japonica (DC.) W. P. Fang ●

51140 Benthamidia japonica (Siebold et Zucc.) H. Hara f. magnifica (Nakai) H. Hara;华丽日本本氏茱萸●☆

51141 Benthamidia japonica (Siebold et Zucc.) H. Hara f. rosea (Honda) H. Hara;粉日本本氏茱萸●☆

51142 Benthamidia japonica (Siebold et Zucc.) H. Hara f. viridis (Nakai) Sugim.;绿日本本氏茱萸●☆

51143 Benthamidia japonica (Siebold et Zucc.) H. Hara var. angustata (Chun) H. Hara = Cornus elliptica (Pojark.) Q. Y. Xiang et Boufford ●

51144 Benthamidia japonica (Siebold et Zucc.) H. Hara var. chineneis (Osborn) H. Hara = Cornus kousa Buerger ex Hance subsp. chinensis (Osborn) Q. Y. Xiang ●

51145 Benthamidia japonica (Siebold et Zucc.) H. Hara var. chineneis (Osborn) H. Hara = Dendrobenthamia japonica (DC.) W. P. Fang var. chineneis (Osborn) W. P. Fang ●

51146 Benthamidia japonica (Siebold et Zucc.) H. Hara var. chinensis (Osborn) H. Hara = Cornus kousa Buerger ex Hance subsp. chinensis (Osborn) Q. Y. Xiang ●

51147 Benthamidia sinensis (Nakai) T. Yamaz. = Benthamidia japonica (Siebold et Zucc.) H. Hara var. chinensis (Osborn) H. Hara ●

51148 Benthamidia sinensis (Nakai) T. Yamaz. = Cornus hongkongensis Hemsl. ●

51149 Benthamidia sinensis (Nakai) T. Yamaz. = Cornus kousa Buerger ex Hance subsp. chinensis (Osborn) Q. Y. Xiang ●

51150 Benthamiella Speg. (1883);本氏茄属■☆

51151 Benthamiella acutifolia Speg.;本氏茄■☆

51152 Benthamina Tiegh. (1896);本氏寄生属●☆

51153 Benthamina alyxifolia (Benth.) Tiegh.;本氏寄生●☆

51154 Benthamistella Kuntze = Buchnera L. ■

51155 Benthamistella Kuntze = Stellularia Benth. ■

51156 Benthamistella nigricans (Benth.) Kuntze = Buchnera nigricans (Benth.) V. Naray. ■☆

51157 Benthamodendron Philipson = Dendrobenthamia Hutch. ●

51158 Bentheca Neck. = Benteca Adans. (废弃属名)●

51159 Bentheca Neck. = Hymenodictyon Wall. (保留属名)●

51160 Bentheka Neck. ex A. DC. = Willughbeia Roxb. (保留属名)●☆

51161 Bentia Rolfe = Justicia L. ●■

51162 Bentinckia Berry ex Roxb. = Bentinckia Berry ●

51163 Bentinckia Berry(1832);毛梗椰属(班氏椰子属,班秩克椰子属,本氏棕属,本亭琪亚棕属,边亭克榈属,尼科巴榈属,尼可巴椰属,尼可巴棕属);Bentinck Palm,Bentinckia ●

51164 Bentinckia nicobarica Becc.;毛梗椰(班秩克椰子,本亭琪亚棕);Nichobar Bentinckia,Nicobar Bentinckia,Orania ●

51165 Bentinckiopsis Becc. (1921);曲嘴椰子属(拟边亭榈属)●☆

51166 Bentinckiopsis Becc. = Clinostigma H. Wendl. ●☆

51167 Bentinckiopsis carolinensis (Becc.) Becc.;曲嘴椰子●☆

51168 Bentleya E. M. Benn. (1986);本特木属●☆

51169 Bentleya spinescens E. M. Benn.;本特木●☆

51170 Bentnickiopsis Becc. = Bentinckiopsis Becc. ●☆

51171 Benzingia Dodson(2010);本兰属■☆

51172 Benzingia estradae (Dodson) Dodson;本兰■☆

51173 Benzoe Fabr. = Lindera Thunb. (保留属名)●

51174 Benzoin Boerh. ex Schaeff. = Lindera Thunb. (保留属名)●

51175 Benzoin Hayne = Styrax L. ●

51176 Benzoin Nees = Lindera Thunb. (保留属名)●

51177 Benzoin Schaeff. (废弃属名) = Lindera Thunb. (保留属名)●

51178 Benzoin aestivale (L.) Nees = Lindera benzoin (L.) Blume ●☆

51179 Benzoin aestivale Nees = Lindera benzoin (L.) Blume ●☆

51180 Benzoin akoense (Hayata) Kamik. = Lindera akoensis Hayata ●

51181 Benzoin angusnfolium (W. C. Cheng) Nakai = Lindera angustifolia W. C. Cheng ●

51182 Benzoin bifarium (Nees) Chun = Lindera nacusua (D. Don) Merr. ●

51183 Benzoin caudatum (Nees) Kuntze = Iteadaphne candata (Nees) H. W. Li ●

51184 Benzoin caudatum (Nees) Kuntze = Lindera caudata Diels ●

51185 Benzoin cercidifolium (Hemsl.) Rehder = Lindera obtusiloba Blume ●

51186 Benzoin commune (Hemsl.) Rehder = Lindera communis Hemsl. ●

51187 Benzoin cubeba (Lour.) Hatus. = Litsea cubeba (Lour.) Pers. ●

51188 Benzoin cubeba Hatus. = Litsea cubeba (Lour.) Pers. ●

51189 Benzoin erythrocarpum (Makino) Rehder = Lindera erythrocarpa Makino ●

51190 Benzoin formosanum (Hayata) Kamikoti = Lindera communis Hemsl. ●

51191 Benzoin fragrans (Oliv.) Rehder = Lindera fragrans Oliv. ●

51192 Benzoin fruticosum (Hemsl.) Rehder = Lindera fruticosa Hemsl. ●

51193 Benzoin fruticosum (Hemsl.) Rehder = Lindera neesiana (Nees) Kurz ●

51194 Benzoin glaucum Siebold et Zucc. = Lindera glauca (Siebold et Zucc.) Blume ●

51195 Benzoin glaucum Siebold et Zucc. var. kawakamii (Hayata) Sasaki = Lindera glauca (Siebold et Zucc.) Blume ●
51196 Benzoin grandifolium Rehder = Lindera megaphylla Hemsl. ●
51197 Benzoin kariense (W. W. Sm.) Hand. -Mazz. = Lindera kariensis W. W. Sm. ●
51198 Benzoin kariensis (W. W. Sm.) Hand. -Mazz. = Lindera kariensis W. W. Sm. f. glabrescens H. W. Li ●
51199 Benzoin levinei (Merr.) Chun ex H. Liu = Neolitsea levinei Merr. ●
51200 Benzoin nacusua Kuntze = Lindera nacusua (D. Don) Merr. ●
51201 Benzoin nacusuum (D. Don) Kuntze = Lindera nacusua (D. Don) Merr. ●
51202 Benzoin neesianum Wall. ex Nees = Lindera neesiana (Nees) Kurz ●
51203 Benzoin obovatum (Franch.) Rehder = Litsea populifolia (Hemsl.) Gamble ●
51204 Benzoin obtusilobum (Blume) Kuntze = Lindera obtusiloba Blume ●
51205 Benzoin oldhamii (Hemsl.) Rehder = Lindera megaphylla Hemsl. ●
51206 Benzoin pedunculatum (Diels) Rehder = Litsea pedunculata (Diels) Yen C. Yang et P. H. Huang ●
51207 Benzoin praecox Siebold et Zucc. = Lindera praecox (Siebold et Zucc.) Blume ●
51208 Benzoin prattii (Gamble) Rehder = Lindera prattii Gamble ●
51209 Benzoin pricei (Hayata) Kamik. = Lindera megaphylla Hemsl. ●
51210 Benzoin pricei Kamik. = Lindera megaphylla Hemsl. ●
51211 Benzoin puberulum (Franch.) Rehder = Litsea moupinensis Lecomte ●
51212 Benzoin pulcherrimum (Nees) Kuntze = Lindera pulcherrima (Wall.) Benth. ex Hook. f. var. hemsleyana (Diels) H. P. Tsui f. velutina (Forrest) C. K. Allen ●
51213 Benzoin pulcherrimura Kuntze = Lindera pulcherrima (Wall.) Benth. ex Hook. f. ●
51214 Benzoin reflexum (Hemsl.) Rehder = Lindera reflexa Hemsl. ●
51215 Benzoin reflexum Rehder = Lindera reflexa Hemsl. ●
51216 Benzoin rubronervium (Gamble) Rehder = Lindera rubronervia Gamble ●
51217 Benzoin sericeum Siebold et Zucc. var. tenue Nakai = Lindera reflexa Hemsl. ●
51218 Benzoin setchuenense (Gamble) Rehder = Lindera setchuenensis Gamble ●
51219 Benzoin sikkimense (Meisn.) Kuntze = Lindera kariensis W. W. Sm. ●
51220 Benzoin strychnifolium (Siebold et Zucc.) Kuntze = Lindera aggregata (Sims) Kosterm. ●
51221 Benzoin strychnifolium (Siebold et Zucc.) Kuntze var. herrmleyanum Diels = Lindera pulcherrima (Wall.) Benth. ex Hook. f. var. hemsleyana (Diels) H. P. Tsui f. velutina (Forrest) C. K. Allen ●
51222 Benzoin subcaudatum (Merr.) Chun = Lindera pulcherrima (Wall.) Benth. ex Hook. f. var. attenuata C. K. Allen ●
51223 Benzoin supracostatum (Lecomte) Rehder = Lindera supracostata Lecomte ●
51224 Benzoin touyunense (H. Lév.) Rehder = Lindera megaphylla Hemsl. f. trichoclada (Rehder) W. C. Cheng ●
51225 Benzoin touyunense (H. Lév.) Rehder = Lindera megaphylla Hemsl. ●
51226 Benzoin touyunense (H. Lév.) Rehder f. megaphyllum (Hemsl.) Rehder = Lindera megaphylla Hemsl. ●
51227 Benzoin touyunense (H. Lév.) Rehder f. trichocladum Rehder = Lindera megaphylla Hemsl. ●
51228 Benzoin touyunense (H. Lév.) Rehder f. trichocladum Rehder = Lindera megaphylla Hemsl. f. trichoclada (Rehder) W. C. Cheng ●
51229 Benzoin umbellatum (Thunb.) Kuntze var. latifolium (Gamble) W. C. Cheng = Lindera reflexa Hemsl. ●
51230 Benzoin umbellatum Kuntze var. latifolium W. C. Cheng = Lindera reflexa Hemsl. ●
51231 Benzoin urophyllum Rehder = Lindera pulcherrima (Wall.) Benth. ex Hook. f. var. hemsleiana (Diels) H. P. Tsui ●
51232 Benzoin urophyllum Rehder = Lindera pulcherrima (Wall.) Benth. ex Hook. f. var. hemsleyana (Diels) H. P. Tsui f. velutina (Forrest) C. K. Allen ●
51233 Benzoina Raf. = Lindera Thunb. (保留属名) ●
51234 Benzonia Schumach. (1827);本索茜属☆
51235 Benzonia corymbosa Schumach.;本索茜☆
51236 Bequaertia R. Wilczek(1956);热非卫矛属●☆
51237 Bequaertia mucronata (Exell) R. Wilczek;热非卫矛●☆
51238 Bequaertiodendron De Wild. = Englerophytum K. Krause ●☆
51239 Bequaertiodendron congolense De Wild. = Englerophytum congolense (De Wild.) Aubrév. et Pellegr. ●☆
51240 Bequaertiodendron magalismontanum (Sond.) Heine et J. H. Hemsl. = Englerophytum magalismontanum (Sond.) T. D. Penn. ●☆
51241 Bequaertiodendron natalense (Sond.) Heine et J. H. Hemsl. = Englerophytum natalense (Sond.) T. D. Penn. ●☆
51242 Bequaertiodendron oblanceolatum (S. Moore) Heine et J. H. Hemsl. = Englerophytum oblanceolatum (S. Moore) T. D. Penn. ●☆
51243 Berardia Brongn. = Brunia Lam. (保留属名) ●☆
51244 Berardia Brongn. = Nebelia Neck. ex Sweet ●☆
51245 Berardia Vill. (1779);双绵菊属■☆
51246 Berardia affinis Brongn. = Nebelia fragarioides (Willd.) Kuntze ●☆
51247 Berardia affinis Sond. = Raspalia affinis Nied. ●☆
51248 Berardia angulata Sond. = Raspalia angulata (Sond.) Nied. ●☆
51249 Berardia aspera Sond. = Nebelia aspera (Sond.) Kuntze ●☆
51250 Berardia dregeana Sond. = Raspalia dregeana (Sond.) Nied. ■☆
51251 Berardia fragarioides (Willd.) Schltdl. = Nebelia fragarioides (Willd.) Kuntze ●☆
51252 Berardia globosa (Thunb.) Sond. = Nebelia fragarioides (Willd.) Kuntze ●☆
51253 Berardia laevis E. Mey. ex Sond. = Nebelia laevis (E. Mey. ex Sond.) Kuntze ●☆
51254 Berardia microphylla (Thunb.) Sond. = Raspalia microphylla (Thunb.) Brongn. ●☆
51255 Berardia paleacea (P. J. Bergius) Brongn. = Nebelia paleacea (P. J. Bergius) Sweet ●☆
51256 Berardia phyllicoides (Thunb.) Brongn. = Raspalia phylicoides (Thunb.) Arn. ●☆
51257 Berardia sphaerocephala Sond. = Nebelia sphaerocephala (Sond.) Kuntze ●☆
51258 Berardia subacaulis Vill.;双绵菊■☆
51259 Berardia trigyna Schltr. = Raspalia trigyna (Schltr.) Dümmer ●☆
51260 Berardia velutina Schltr. = Raspalia dregeana (Sond.) Nied. ■☆
51261 Berberidaceae Juss. (1789) (保留科名);小檗科;Barberry

Family ●■

51262 Berberidopsidaceae (Veldk.) Takht. = Flacourtiaceae Rich. ex DC. (保留科名) ●

51263 Berberidopsidaceae Takht. (1985);智利藤科●☆

51264 Berberidopsidaceae Takht. = Flacourtiaceae Rich. ex DC. (保留科名) ●

51265 Berberidopsis Hook. f. (1862);智利藤属●☆

51266 Berberidopsis beckleri (F. Muell.) Veldkamp;贝克智利藤●☆

51267 Berberidopsis corallina Hook. f.;珊瑚红智利藤(智利藤);Coral Plant ●☆

51268 Berberis L. (1753);小檗属;Barberry, Berberis, Oregon-grape ●

51269 Berberis acanthifolia (G. Don) Walp. = Mahonia napaulensis DC. ●

51270 Berberis acanthifolia (Wall. ex G. Don) Wall. ex Walp. = Mahonia napaulensis DC. ●

51271 Berberis actinacanta Mart. ex Roem. et Schult.;簇刺小檗;Actino-spiny Barberry ●☆

51272 Berberis actinacanta Mart. ex Roem. et Schult. f. var. crispa (Gay) Reiche = Berberis crispa Gay ●☆

51273 Berberis actinacanta Mart. ex Roem. et Schult. f. var. grevilleana (Gillies) C. K. Schneid. = Berberis grevilleana Gillies ex Hook. ●☆

51274 Berberis actinacanta Mart. ex Roem. et Schult. f. var. horrida (Gay) Reiche = Berberis horrida Gay ●☆

51275 Berberis acuminata Franch.;渐尖叶小檗(尖叶小檗,三颗针);Acuminate Barberry ●

51276 Berberis acuminata Stapf = Berberis gagnepainii C. K. Schneid. ●

51277 Berberis acuminata Veitch = Berberis veitchii C. K. Schneid. ●

51278 Berberis acutifolia Prantl = Berberis vulgaris L. var. acutifolia (Prantl) C. K. Schneid. ●☆

51279 Berberis aemulans C. K. Schneid.;康定黄花刺(峨眉小檗);Emei Barberry, Kangding Reddrop Barberry, Omei Mountain Barberry ●

51280 Berberis aetnensis C. Presl;埃得纳小檗;Aetna Barberry, Mount Etna Barberry ●☆

51281 Berberis affinis G. Don;近缘小檗(库芒小檗);Kumaon Barberry ●☆

51282 Berberis afghanica C. K. Schneid.;阿富汗小檗;Afghan Barberry, Afghanistan Barberry ●☆

51283 Berberis africana Hebenstr. et Ludw. ex Schult. f.;非洲小檗;African Barberry ●☆

51284 Berberis aggregata C. K. Schneid.;锥花小檗(刺黑珠,刺黄连,刺黄芩,短圆锥花序小檗,堆花小檗,黄檗刺,黄檗子,老鼠刺,猫儿刺,全缘叶锥花小檗,全缘锥花小檗,三颗针,小黄檗刺);Aggregate Barberry, Clustered Barberry, Entire-leaved Salmon Barberry, Salmon Barberry, Short-paniculate Barberry ●

51285 Berberis aggregata C. K. Schneid. var. intergrifolia Ahrendt = Berberis aggregata C. K. Schneid. ●

51286 Berberis aggregata C. K. Schneid. var. prattii (C. K. Schneid.) C. K. Schneid. = Berberis prattii C. K. Schneid. ●

51287 Berberis agricola Ahrendt;暗红小檗;Agricola Barberry, Dark-red Barberry ●

51288 Berberis aitchisonii Ahrendt;艾齐森小檗;Aitchison Barberry ●☆

51289 Berberis aldenhamensis Ahrendt;阿尔登海姆小檗;Aldenham Barberry ●☆

51290 Berberis alksuthiensis Ahrendt;阿尔克苏斯小檗;Alksuth Barberry ●☆

51291 Berberis alpicola C. K. Schneid. = Berberis kawakamii Hayata ●

51292 Berberis amabilis C. K. Schneid.;可爱小檗(全缘可爱小檗);Integrifolious Lovely Barberry, Lovely Barberry ●

51293 Berberis amabilis C. K. Schneid. var. holophylla C. Y. Wu et S. Y. Bao = Berberis amabilis C. K. Schneid. ●

51294 Berberis amabllis C. K. Schneid. var. holophylla C. Y. Wu = Berberis amabilis C. K. Schneid. ●

51295 Berberis ambigua Ahrendt;疑似小檗;Ambiguous Barberry, Doubtful Barberry ●

51296 Berberis ambrozyana C. K. Schneid. = Berberis dictyophylla Franch. var. epruinosa C. K. Schneid. ●

51297 Berberis amoena Dunn;美丽小檗(莫洛小檗,三颗针,伞花美丽小檗,伞花喜悦小檗,喜悦小檗);Beautiful Barberry, Pleasing Barberry, Umbelliferous Pleasing Barberry ●

51298 Berberis amoena Dunn var. moloensis Ahrendt = Berberis kongboensis Ahrendt ●

51299 Berberis amoena Dunn var. umbelliflora Ahrendt = Berberis amoena Dunn ●

51300 Berberis amplectens (Eastw.) L. C. Wheeler;环抱小檗●☆

51301 Berberis amurensis Rupr.;黄芦木(阿穆尔小檗,刺黄柏,刺黄檗,大叶小檗,刀口药,东北小檗,狗奶子,黄檗树,黄连,三颗针,山黄柏,山石榴,小檗,子檗);Amur Barberry ●

51302 Berberis amurensis Rupr. 'Dwarf Form';矮黄芦木;Barberry ●☆

51303 Berberis amurensis Rupr. f. bretschneideri (Rehder ex Sarg.) Ohwi;紫果小檗(伯乐小檗);Bretschneider Barberry, Purple Berry Barberry, Purpleberry Barberry, Purple-berry Barberry ●☆

51304 Berberis amurensis Rupr. f. brevifolia (Nakai) Ohwi = Berberis amurensis Rupr. f. bretschneideri (Rehder ex Sarg.) Ohwi ●☆

51305 Berberis amurensis Rupr. var. bretschneideri (Rehder ex Sarg.) H. Hara = Berberis amurensis Rupr. f. bretschneideri (Rehder ex Sarg.) Ohwi ●☆

51306 Berberis amurensis Rupr. var. japonica (Regel) Rehder = Berberis regeliana Koehne ex C. K. Schneid. ●☆

51307 Berberis amurensis Rupr. var. latifolia Nakai;阔叶黄芦木;Broad-leaved Amur Barberry ●☆

51308 Berberis amurensis Rupr. var. licentii Ahrendt = Berberis hersii Ahrendt ●

51309 Berberis andeana Job;安第斯小檗;Andean Barberry ●☆

51310 Berberis andreana Naudin;安德烈小檗;André Barberry ●☆

51311 Berberis angulosa Wall. ex Hook. f. et Thomson;有棱小檗(红珠小檗);Angular Barberry, Himalayan Barberry, Redbead Barberry, Red-bead Barberry ●

51312 Berberis angulosa Wall. ex Hook. f. et Thomson var. brevipes Franch. = Berberis minutiflora C. K. Schneid. ●

51313 Berberis angulosa Wall. ex Hook. f. et Thomson var. fasciculata Ahrendt;簇生小檗(簇生红珠小檗);Fascicled Redbead Barberry ●☆

51314 Berberis anhweiensis Ahrendt;安徽小檗(刺黄柏,黄柏,小檗);Anhui Barberry ●

51315 Berberis anniae Ahrendt;安小檗;Ann Barberry ●☆

51316 Berberis antoniana Ahrendt;安东尼小檗;Antony Barberry ●☆

51317 Berberis antucoana C. K. Schneid.;安托克小檗(细尖小檗);Antuco Barberry ●☆

51318 Berberis apiculata (Ahrendt) Ahrendt;细尖小檗(安托小檗);Apiculate Barberry ●☆

51319 Berberis approximata Sprague;接枝小檗(近似小檗,西南小檗,云南小檗);Approximate Barberry, Stiebritz Barberry ●

51320 Berberis aquifolium (Pursh) Nutt. = Mahonia aquifolium (Pursh) Nutt. ●☆

51321 Berberis aquifolium Nutt. = Berberis aquifolium (Pursh) Nutt. ●☆

51322 Berberis aquifolium Pursh = Berberis aquifolium (Pursh) Nutt. ●☆

51323 Berberis aquifolium Pursh var. dictyota (Jeps.) Jeps. = Berberis dictyota Jeps. ●☆

51324 Berberis aquifolium Pursh var. repens (Lindl.) Scoggan = Berberis repens Lindl. ●☆

51325 Berberis arguta (Franch.) C. K. Schneid.;锐齿小檗;Argute Barberry ●

51326 Berberis arido-calida Ahrendt;西固小檗;Dry-hot Barberry, Xigu Barberry ●☆

51327 Berberis aristata DC.;具芒小檗(刺齿小檗,芒檗);Aristate Barberry, Awned-leaved Barberry, Chitra, Darlahad Barberry, Indian Barberry, Nepal Barberry, Ophthalmic Barberry, Pepal Barberry, Spinetooth Barberry, Tanner's Barberry ●☆

51328 Berberis aristata DC. = Berberis chitria Buch.-Ham. ex Lindl. ●☆

51329 Berberis aristata DC. = Berberis glaucocarpa Stapf ●☆

51330 Berberis aristata DC. var. subintegra Engl. = Berberis holstii Engl. ●☆

51331 Berberis aristatoserrulata Hayata;密齿小檗(长叶小檗,齿状须小檗,刺尖锯齿小檗,芒齿小檗);Aristato-serrulate Barberry, Awned Serration Barberry ●

51332 Berberis armata Citerne;秘鲁长刺小檗;Armed Barberry ●☆

51333 Berberis asiatica Roxb. ex DC.;亚洲小檗;Asian Barberry, Asiatic Barberry, Indian Barberry, Raisin Barberry, Rusot ●☆

51334 Berberis asiatica Roxb. ex DC. var. clarkeana C. K. Schneid.;克拉克亚洲小檗;C. B. Clarke Asian Barberry ●☆

51335 Berberis asmyana C. K. Schneid.;直梗小檗(阿斯米小檗);Asmy Barberry ●

51336 Berberis atrocarpa C. K. Schneid.;黑果小檗(刺黄连,鸡脚刺,乐桂小檗,三颗针,深黑小檗,深黄小檗,铜叶刺,狭叶兴山小檗);Black-fruited Barberry, Dark-fruit Barberry, Jetbead Barberry, Narrow-leaved Forrest Barberry, Narrow-leaved Xingshan Barberry, Near-entire Dark-fruit Barberry ●

51337 Berberis atrocarpa C. K. Schneid. var. longipes Ahrendt;长柄黑果小檗●☆

51338 Berberis atrocarpa C. K. Schneid. var. subintegra Ahrendt = Berberis atrocarpa C. K. Schneid. ●

51339 Berberis atrocarpa C. K. Schneid. var. suijiangensis S. Y. Bao = Berberis insolita C. K. Schneid. ●

51340 Berberis atroprasina Ahrendt;深绿小檗;Dark-green Barberry, Dark-prasine Barberry ●

51341 Berberis atroviridiana T. S. Ying;那觉小檗;Najue Barberry ●

51342 Berberis atroviridis Steud. = Berberis atroviridiana T. S. Ying ●

51343 Berberis aurahuacensis Lem.;奥拉华库小檗;Aurahuaku-taquina Barberry ●☆

51344 Berberis baluchistanica Ahrendt;俾路支小檗;Baluchistan Barberry ●☆

51345 Berberis barandana Vidal;巴兰德小檗;Barand Barberry ●☆

51346 Berberis barbeyana C. K. Schneid.;巴比小檗;Barbey Barberry ●☆

51347 Berberis barilochensis Job;巴利罗切小檗;Bariloche Barberry ●☆

51348 Berberis batangensis T. S. Ying;巴塘小檗;Batang Barberry ●

51349 Berberis bealei Fortune = Mahonia bealei (Fortune) Carrière ●

51350 Berberis bealei Fortune var. planifolia Hook. f. = Mahonia bealei (Fortune) Carrière ●

51351 Berberis beaniana C. K. Schneid.;康松小檗(宾尼小檗);Bean Barberry ●

51352 Berberis beauverdiana C. K. Schneid.;博弗德小檗;Beauverd Barberry ●☆

51353 Berberis beesiana Ahrendt;比斯小檗;Bees Barberry ●☆

51354 Berberis beesiana Ahrendt var. glabra Ahrendt;无毛比斯小檗;Smooth Bees Barberry ●☆

51355 Berberis beijingensis T. S. Ying;北京小檗;Beijing Barberry ●

51356 Berberis bella C. K. Schneid.;美美小檗(美丽小檗);Handsome Barberry ●☆

51357 Berberis benoistiana J. F. Macbr.;伯努斯特小檗;Benoist Barberry ●☆

51358 Berberis berberidifolia (P. K. Hsiao et Y. S. Wang) Laferr. = Mahonia eurybracteata Fedde ●

51359 Berberis bergeriana C. K. Schneid.;伯杰小檗;Berger Barberry ●☆

51360 Berberis bergmanniae C. K. Schneid.;硬齿小檗(伯格曼小檗,刺黄连,汉源小檗,土黄柏);Bergmann Barberry, Chinese Barberry ●

51361 Berberis bergmanniae C. K. Schneid. var. acanthophylla C. K. Schneid.;刺叶硬齿小檗(刺叶伯格曼小檗,汶川小檗);Spiny Bergmann Barberry ●

51362 Berberis bhutanensis Ahrendt = Berberis griffithiana C. K. Schneid. var. pallida (Hook. f. et Thomson) D. F. Chamb. et C. M. Hu ●

51363 Berberis bicolor H. Lév.;二色小檗(薄叶小檗,长叶小檗);Bicolor Barberry, Thin-leaf Barberry ●

51364 Berberis bidentata Lechl.;二齿小檗;Two-toothed Barberry ●☆

51365 Berberis bodinieri (Gagnep.) Laferr. = Mahonia bodinieri Gagnep. ●

51366 Berberis boissieri C. K. Schneid.;布瓦西耶小檗(博西埃尔小檗);Boissier Barberry ●☆

51367 Berberis boliviana Lechl.;玻利维亚小檗;Bolivia Barberry ●☆

51368 Berberis bombycina Ahrendt;丝质小檗;Silky Barberry ●☆

51369 Berberis borealis (Takeda) Laferr. var. parryi (Ahrendt) Laferr. = Mahonia duclouxiana Gagnep. ●

51370 Berberis borealisinensis Nakai = Berberis sibirica Pall. ●

51371 Berberis boschanii C. K. Schneid. = Berberis mouillacana C. K. Schneid. ●

51372 Berberis brachyacantha Phil. ex Reiche;短刺小檗;Shortspiny Barberry ●☆

51373 Berberis brachybotria Gay;短总状花序小檗;Short-raceme Barberry ●☆

51374 Berberis brachypoda Maxim.;短柄小檗(刺黄连,毛叶小檗,三颗针,酸梨刺,小黄檗);Short-stalk Barberry, Short-stalked Barberry, Yellowspike Barberry ●

51375 Berberis brachypoda Maxim. var. salicaria (Fedde) C. K. Schneid. = Berberis salicaria Fedde ●

51376 Berberis brachystachys T. S. Ying;短穗小檗●

51377 Berberis brachystachys T. S. Ying = Berberis dictyoneura C. K. Schneid. ●

51378 Berberis bracteata (Ahrendt) Ahrendt;长苞小檗(长苞片小檗);Bracteate Barberry ●

51379 Berberis bracteolata (Takeda) Laferr. = Mahonia bracteolata Takeda ●

51380 Berberis bracteolata (Takeda) Laferr. var. zhongdianensis (S. Y. Bao) Laferr. = Mahonia bracteolata Takeda ●

51381 Berberis brandisiana Ahrendt;布兰迪斯小檗;Brandis Barberry ●☆

51382 Berberis bretschneideri Rehder = Berberis amurensis Rupr. f. bretschneideri (Rehder ex Sarg.) Ohwi ●☆

51383 Berberis bretschneideri Rehder ex Sarg. = Berberis amurensis Rupr. f. bretschneideri (Rehder ex Sarg.) Ohwi ●☆

51384 Berberis brevifolia Phil. ex Reiche;短叶小檗;Short-leaved

Barberry ●☆

51385 Berberis brevipaniculata C. K. Schneid. = Berberis aggregata C. K. Schneid. ●

51386 Berberis brevipes (Franch.) C. K. Schneid. = Berberis minutiflora C. K. Schneid. ●

51387 Berberis breviracema (Y. S. Wang et P. K. Hsiao) Laferr. = Mahonia breviracema Y. S. Wang et P. K. Hsiao ●

51388 Berberis breviscapa (Ahrendt) Ahrendt;短花茎小檗;Short-scape Barberry ●☆

51389 Berberis brevisepala Hayata;高山小檗●

51390 Berberis brevisepala Hayata = Berberis kawakamii Hayata ●

51391 Berberis brevisepala Hayata = Berberis micropetala T. S. Ying ●

51392 Berberis brevissima Jafri;短小檗●☆

51393 Berberis bristolensis Ahrendt;布里斯托尔小檗(刺齿小檗);Bristol Barberry ●☆

51394 Berberis brumalis J. F. Macbr.;冬至小檗;Winter Barberry ●☆

51395 Berberis buceronis J. F. Macbr.;弓茎小檗;Arching Barberry ●☆

51396 Berberis buchananii C. K. Schneid.;布查南小檗;Buchanan Barberry ●☆

51397 Berberis buchananii C. K. Schneid. var. tawangensis Ahrendt;达旺小檗;Tawang Buchanan Barberry ●☆

51398 Berberis bullata Ahrendt;水泡小檗;Blistered Barberry ●☆

51399 Berberis bumeliifolia C. K. Schneid.;布梅榄叶小檗;Bumelia-leaved Barberry ●☆

51400 Berberis burmanica Ahrendt;缅甸小檗;Burma Barberry ●☆

51401 Berberis buxifolia Lam.;黄杨叶小檗;Box-leaved Barberry, Magellan Barberry ●☆

51402 Berberis buxifolia Lam. var. antarctica C. K. Schneid.;南极黄杨叶小檗;Antarctic Magellan Barberry ●☆

51403 Berberis buxifolia Lam. var. inermis (Pers.) C. K. Schneid.;无刺黄杨叶小檗(无刺小檗);Unarmed Magellan Barberry ●☆

51404 Berberis buxifolia Lam. var. macracantha Phil. ex A. Usteri = Berberis spinosissima (Reiche) Ahrendt ●☆

51405 Berberis buxifolia Lam. var. nana A. Usteri;矮黄杨叶小檗●☆

51406 Berberis buxifolia Lam. var. nuda C. K. Schneid.;光茎黄杨叶小檗(矮黄杨叶小檗);Dwarf Magellan Barberry, Nude-stem Magellan Barberry ●☆

51407 Berberis buxifolia Lam. var. papilosa C. K. Schneid.;乳突黄杨叶小檗;Papilose Magellan Barberry ●☆

51408 Berberis buxifolia Lam. var. spinosissima Reiche = Berberis spinosissima (Reiche) Ahrendt ●☆

51409 Berberis bykoviana Pavlov;毕氏小檗●☆

51410 Berberis cabrerae Job;卡布雷拉小檗;Cabrera Barberry ●☆

51411 Berberis caelicolor (S. Y. Bao) Laferr. = Mahonia oiwakensis Hayata ●

51412 Berberis caesia (C. K. Schneid.) Laferr. = Mahonia bracteolata Takeda ●

51413 Berberis calcipratorum Ahrendt;钙原小檗;Calcisteppe Barberry ●

51414 Berberis californica Jeps. = Berberis dictyota Jeps. ●☆

51415 Berberis californica Jeps. = Mahonia californica (Jeps.) Ahrendt ●☆

51416 Berberis calliantha Mulligan;美花小檗;Beautiful-flower Barberry, Beautiful-flowered Barberry ●

51417 Berberis calliobotrys Bien. ex Aitch.;美穗小檗(加姆布小檗,瓦札里斯坦小檗);Beautiful-raceme Barberry, Gamble Barberry, Wazaristan Barberry ●☆

51418 Berberis campbellii Ahrendt;坎贝尔小檗;Campbell Barberry ●☆

51419 Berberis campos-portoi Brade;坎波斯-波尔图小檗;Campos-porto Barberry ●☆

51420 Berberis campylotropa T. S. Ying;弯果小檗;Campylotropous Barberry ●

51421 Berberis canadensis L. = Berberis canadensis Mill. ●☆

51422 Berberis canadensis Mill.;加拿大小檗(美国小檗,美洲小檗);Allegany Barberry, Alleghany Barberry, American Barberry, Canada Barberry ●☆

51423 Berberis candidula (C. K. Schneid.) C. K. Schneid. = Berberis candidula C. K. Schneid. ●

51424 Berberis candidula C. K. Schneid.;白背小檗(白叶小檗,单花小檗,单叶小檗);Paleleaf Barberry ●

51425 Berberis capillaris Cox = Berberis muliensis Ahrendt ●

51426 Berberis capillaris Cox ex Ahrendt = Berberis muliensis Ahrendt ●

51427 Berberis carinata Lechl.;龙骨小檗;Keeled Barberry ●☆

51428 Berberis carinata Lechl. var. echinata Diels;刺叶龙骨小檗(棘叶龙骨小檗);Echinate Keeled Barberry ●☆

51429 Berberis carminea Chitt. ex Ahrendt;洋红小檗(深红小檗);Carmine Barberry ●☆

51430 Berberis carminea Chitt. ex Ahrendt 'Barbarossa';巴巴洛沙洋红小檗●☆

51431 Berberis carminea Chitt. ex Ahrendt 'Pirate King';海盗王洋红小檗●☆

51432 Berberis caroli C. K. Schneid. var. hoanghensis C. K. Schneid. = Berberis vernae C. K. Schneid. ●

51433 Berberis carolii C. K. Schneid.;卡罗尔小檗(鄂尔多斯小檗,黄柏,三颗针);Carol Barberry ●

51434 Berberis carolii C. K. Schneid. var. hoanghensis C. K. Schneid. = Berberis vernae C. K. Schneid. ●

51435 Berberis caroliniana Loudon = Berberis canadensis L. ●☆

51436 Berberis caudatifolia S. Y. Bao = Berberis gagnepainii C. K. Schneid. ●

51437 Berberis cavaleriei H. Lév.;聚球滑叶小檗(长蕊小檗,刺黄连,大叶优秀小檗,贵州小檗,鸡脚刺,较大优秀小檗,卡瓦範利小檗,三颗针);Cavalerie Barberry, Crowd-ball Smooth-leaved Barberry, Long-stamen Barberry, Major Excellent Barberry ●

51438 Berberis cavaleriei H. Lév. var. pruinosa Bijh. = Berberis chingii W. C. Cheng ●

51439 Berberis cavaleriei H. Lév. var. pruinosa Byhouwer = Berberis chingii W. C. Cheng ●

51440 Berberis centiflora Diels;多花大黄连刺;Hundred-flower Barberry, Multiflorous Barberry ●

51441 Berberis cerasina Schrad.;樱桃红小檗;Cherry-red Barberry ●☆

51442 Berberis ceratophylla G. Don;角状叶小檗;Corneously-leaved Barberry ●☆

51443 Berberis ceylanica C. K. Schneid.;锡兰小檗;Ceylon Barberry ●☆

51444 Berberis chekiangensis Ahrendt = Berberis virgetorum C. K. Schneid. ●

51445 Berberis chenaultii Ahrendt;陈纳德小檗;Chennault Barberry ●☆

51446 Berberis chilensis Gillies ex Hook.;智利小檗;Chile Barberry ●☆

51447 Berberis chilensis Gillies ex Hook. var. diffusa (Gay) Reiche = Berberis diffusa Gay ●☆

51448 Berberis chillanensis (C. K. Schneid.) Sprague ex Bean;奇利安小檗;Chillan Barberry ●☆

51449 Berberis chillanensis (C. K. Schneid.) Sprague ex Bean var. hirsutipes Sprague;硬毛奇利安小檗(硬毛柄奇利安小檗);Hirsute-stalk Chillan Barberry ●☆

51450 Berberis chilternensis Ahrendt;奇尔特恩小檗;Chiltern Barberry ●☆

51451 Berberis chimboensis C. K. Schneid.;琴博小檗;Chimbo Barberry ●☆

51452 Berberis chinensis Poir.;中国小檗(大黄连,枸棘);Chinese Barberry ●

51453 Berberis chinensis Poir. var. paphlagonica (C. K. Schneid.) Ahrendt;帕夫拉哥尼亚小檗;Paphlagoni Barberrya ●☆

51454 Berberis chingii W. C. Cheng;华东小檗(安徽小檗,刺黄连,刺黄芩,刺小檗,鸡脚刺,江西小檗,秦氏小檗,三颗针,皖赣小檗,玉妹刺);Anhui Barberry,Ching Barberry,Jiangxi Barberry ●

51455 Berberis chingii W. C. Cheng subsp. subedentata C. M. Hu = Berberis chingii W. C. Cheng ●

51456 Berberis chingii W. C. Cheng subsp. wulingensis C. M. Hu = Berberis chingii W. C. Cheng ●

51457 Berberis chingshuiensis Shimizu;清水山小檗●

51458 Berberis chingshuiensis Shimizu = Berberis kawakamii Hayata ●

51459 Berberis chingshuiensis Shimizu = Berberis micropetala T. S. Ying ●

51460 Berberis chitria Buch. -Ham. ex Ker Gawl. var. sikkimensis C. K. Schneid. = Berberis sikkimensis (C. K. Schneid.) Ahrendt ●

51461 Berberis chitria Buch. -Ham. ex Lindl.;壶小檗(吉德小檗);Pot Barberry ●☆

51462 Berberis chitria Buch. -Ham. ex Lindl. var. occidentalis Ahrendt;西方壶小檗;Occidental Pot Barberry ●☆

51463 Berberis chitria Buch. -Ham. ex Lindl. var. occidentalis Ahrendt = Berberis chitria Buch. -Ham. ex Lindl. ●☆

51464 Berberis chitria Buch. -Ham. ex Lindl. var. sikkimensis C. K. Schneid. = Berberis sikkimensis Hemsl. ●

51465 Berberis chrysacantha C. K. Schneid.;黄刺小檗;Yellow-spiny Barberry ●☆

51466 Berberis chrysosphaera Mulligan;黄球小檗(金黄球小檗);Golden-sphaeroidal Barberry,Yellow-globular Barberry ●

51467 Berberis chunanensis T. S. Ying;淳安小檗;Chunan Barberry ●

51468 Berberis ciliaris Lindl.;缘毛小檗;Ciliate Barberry ●☆

51469 Berberis ciliaris Lindl. var. obtusara Ahrendt;钝圆缘毛小檗;Obtuse Ciliate Barberry ●☆

51470 Berberis circumserrata (C. K. Schneid.) C. K. Schneid.;秦岭小檗(刺黄檗,短刺秦岭小檗,黄檗,黄刺,老鼠刺);Cutleaf Barberry,Cut-leaved Barberry,Short-spiny Cutleaf Barberry ●

51471 Berberis circumserrata (C. K. Schneid.) C. K. Schneid. var. occidentalior Ahrendt;多萼秦岭小檗(多萼小檗,洮河小檗,多萼缘毛小檗);Taohe Cutleaf Barberry,Tao-river Cutleaf Barberry ●

51472 Berberis circumserrata (C. K. Schneid.) C. K. Schneid. var. subarmata Ahrendt = Berberis circumserrata (C. K. Schneid.) C. K. Schneid. ●

51473 Berberis circumserrata C. K. Schneid. = Berberis circumserrata (C. K. Schneid.) C. K. Schneid. ●

51474 Berberis circumserrata C. K. Schneid. var. subarmata Ahrendt = Berberis circumserrata (C. K. Schneid.) C. K. Schneid. ●

51475 Berberis citernei Ahrendt;西特恩小檗;Citerne Barberry ●☆

51476 Berberis clausenii Citerne;克劳森小檗;Claussen Barberry ●☆

51477 Berberis cliffortioides Diels;克利福特状小檗;Cliffortia-like Barberry ●☆

51478 Berberis coletioides Lechl.;拟科莱小檗;Collett-like Barberry ●☆

51479 Berberis collettii C. K. Schneid.;科莱小檗;Collett Barberry ●☆

51480 Berberis colombiana Ahrendt;哥伦比亚小檗;Colombian Barberry ●☆

51481 Berberis comberi Sprague et Sandwith;科姆伯小檗;Comber's Barberry ●☆

51482 Berberis commutata Fichler;变色小檗;Changed Barberry ●☆

51483 Berberis concinna Hook. f. et Thomson;雅洁小檗(美味小檗);Dainty Barberry,Elegant Barberry,Pretty Barberry ●

51484 Berberis concinna Hook. f. et Thomson var. brevior Ahrendt;短柄雅洁小檗;Short-pedicel Elegant Barberry ●

51485 Berberis concinna Hook. f. et Thomson var. extensiflora Ahrendt;伞花雅洁小檗;Extension-flowered Elegant Barberry ●

51486 Berberis concolor W. W. Sm.;同色小檗;Concolorous Barberry ●

51487 Berberis conferta Kunth;密集小檗;Crowded Barberry ●☆

51488 Berberis conferta Kunth var. karsteniana C. K. Schneid.;卡斯藤密集小檗;Karsten Crowded Barberry ●☆

51489 Berberis conferta Kunth var. lobbiana C. K. Schneid. = Berberis lobbiana C. K. Schneid. ●☆

51490 Berberis confusa (Sprague) Laferr. = Mahonia eurybracteata Fedde ●

51491 Berberis confusa (Sprague) Laferr. var. bournei (Ahrendt) Laferr. = Mahonia eurybracteata Fedde subsp. ganpinensis (H. Lév.) T. S. Ying et Boufford ●

51492 Berberis congestiflora Gay;智利密花小檗;Congest-flowered Barberry ●☆

51493 Berberis congestiflora Gay var. hakeoides Hook. f. = Berberis hakeoides (Hook. f.) C. K. Schneid. ●☆

51494 Berberis consimilis C. K. Schneid.;相似小檗;Similarest Barberry ●☆

51495 Berberis contracta T. S. Ying;德钦小檗;Deqin Barberry ●

51496 Berberis cooperi Ahrendt;库珀小檗;Cooper Barberry ●☆

51497 Berberis coquimbensis Munoz;科金博小檗;Coquimbo Barberry ●☆

51498 Berberis coriacea A. St. -Hil.;巴西革叶小檗;Brazil Leathern Barberry ●

51499 Berberis coriacea A. St. -Hil. var. oblanceifolia Ahrendt;倒披针叶巴西革叶小檗;Oblance-leaved Brazil Leathern Barberry ●

51500 Berberis coriaria Royle ex Lindl.;革叶小檗;Leathern Barberry ●☆

51501 Berberis coriaria Royle ex Lindl. = Berberis glaucocarpa Stapf ●☆

51502 Berberis coriaria Royle ex Lindl. var. patula Ahrendt;开展革叶小檗;Patulous Leathern Barberry ●☆

51503 Berberis coryi Veitch;贡山小檗;Cory Barberry ●

51504 Berberis corymbosa Hook. et Arn.;伞房花序小檗;Corymbose Barberry ●☆

51505 Berberis corymbosa Hook. et Arn. var. paniculata Phil.;圆锥状伞房花序小檗;Panicle-corymbose Barberry ●☆

51506 Berberis costulata Gand.;隆中脉小檗;Costulate Barberry ●☆

51507 Berberis coxii C. K. Schneid.;考克斯小檗;Cox Barberry ●☆

51508 Berberis crassilimba C. Y. Wu = Berberis crassilimba C. Y. Wu ex S. Y. Bao ●

51509 Berberis crassilimba C. Y. Wu ex S. Y. Bao;厚檐小檗;Crassilimbatus Barberry,Thick-limb Barberry ●

51510 Berberis crataegina DC.;山楂小檗;Hawthorn Barberry ●☆

51511 Berberis crataegina DC. var. armeniaca C. K. Schneid.;亚美尼亚山楂小檗;Armenian Hawthorn Barberry ●☆

51512 Berberis crataegina DC. var. cabulica ? = Berberis aitchisonii Ahrendt ●☆

51513 Berberis crataegina DC. var. lycica C. K. Schneid.;吕西亚山楂小檗;Lycian Hawthorn Barberry ●☆

51514 Berberis crenulata Schrad.;细圆齿小檗;Crenulate Barberry ●☆

51515 Berberis cretica L.;克里特小檗;Cretan Barberry ●☆

51516 Berberis crispa Gay;皱波小檗;Crisped Barberry ●☆

51517 Berberis cuneata DC.;楔叶小檗;Wedge-leaved Barberry ●☆

51518 Berberis daiana T. S. Ying;城口小檗;Chengkou Barberry ●

51519 Berberis daochengensis T. S. Ying;稻城小檗;Daocheng Barberry ●

51520 Berberis darwinii Hook.;达尔文小檗;Darwin Barberry,Darwin's Berberis ●☆

51521 Berberis darwinii Hook. 'Flame';火焰达尔文小檗●☆

51522 Berberis darwinii Hook. var. magellanica Ahrendt;麦哲伦达尔文小檗;Magellan Darwin Barberry ●☆

51523 Berberis dasyclada Ahrendt;粗枝小檗;Thick-branched Barberry ●☆

51524 Berberis dasystachya Maxim.;直穗小檗(长穗小檗,刺黄檗,刺黄皮,高甘肃小檗,黄檗,黄刺皮,黄三刺皮,吉朵尔,密穗小檗,三颗针,山黄檗,珊瑚刺,直序小檗);Longispiked Barberry,Long-raceme Barberry,Tall Gansu Barberry,Tangut Barberry ●

51525 Berberis dasystachya Maxim. var. pluriflora P. Y. Li;多花直穗小檗(多花密穗小檗);Many-flower Tangut Barberry ●

51526 Berberis davidii Ahrendt;密叶小檗(假小檗);David Barberry,Deceptive Barberry ●

51527 Berberis dawoensis K. Mey.;道孚小檗;Daofu Barberry,Dowo Barberry ●

51528 Berberis dealbata Lindl.;变白小檗;Whitened Barberry ●☆

51529 Berberis decandolleana Ahrendt;康多勒小檗(德坎多尔小檗)●☆

51530 Berberis decipiens (C. K. Schneid.) Laferr. = Mahonia decipiens C. K. Schneid. ●

51531 Berberis declinata Schrad.;垂枝小檗(下垂小檗);Drooping Barberry ●☆

51532 Berberis declinata Schrad. var. oxyphylla C. K. Schneid.;尖叶垂枝小檗(尖叶下垂小檗);Sharp-leaved Barberry ●☆

51533 Berberis deinacantha C. K. Schneid.;壮刺小檗(滇西小檗,多刺小檗,维西多刺小檗);Bigspiny Barberry,Multispinate Barberry ●

51534 Berberis deinacantha C. K. Schneid. var. valida C. K. Schneid. = Berberis valida (C. K. Schneid.) C. K. Schneid. ●

51535 Berberis delavayi C. K. Schneid. = Berberis phanera C. K. Schneid. ●

51536 Berberis delavayi C. K. Schneid. var. wachinensis Ahrendt = Berberis delavayi C. K. Schneid. ●

51537 Berberis delavayi C. K. Schneid. var. wachinensis Ahrendt = Berberis phanera C. K. Schneid. ●

51538 Berberis densa C. K. Schneid. = Berberis davidii Ahrendt ●

51539 Berberis densa Planch. et Linden ex Triana et Planch.;哥伦比亚密集小檗;Colombia Dense Barberry ●☆

51540 Berberis densiflora Boiss. et Buhse;密花小檗;Dense-flower Barberry ●☆

51541 Berberis densiflora Boiss. et Buhse = Berberis aitchisonii Ahrendt ●☆

51542 Berberis densiflora Boiss. et Buhse var. bungeana Ahrendt;邦奇密花小檗;Bunge Dense-flower Barberry ●☆

51543 Berberis densiflora Boiss. et Buhse var. macracantha Boiss.;大刺密花小檗;Large-spiny Dense-flower Barberry ●☆

51544 Berberis densiflora Boiss. et Buhse var. macrobotrys Ahrendt;长穗密花小檗;Elongate-raceme Dense-flower Barberry ●☆

51545 Berberis densiflora Boiss. et Buhse var. serratifolia Boiss.;锯齿叶密花小檗;Serrate-leaved Dense-flower Barberry ●☆

51546 Berberis densifolia Rusby;玻利维亚密叶小檗;Dense-leaved Barberry ●☆

51547 Berberis derongensis T. S. Ying;得荣小檗;Derong Barberry ●

51548 Berberis diaphana Maxim.;鲜黄小檗(单花黄花刺,黄檗,黄刺,黄花刺,三颗针,歪头小檗);Diaphonous Barberry,Reddrop Barberry,Red-drop Barberry,Solitary-flowered Reddrop Barberry ●

51549 Berberis diaphana Maxim. var. circumserrata C. K. Schneid. = Berberis circumserrata (C. K. Schneid.) C. K. Schneid. ●

51550 Berberis diaphana Maxim. var. tachiensis Ahrendt = Berberis aemulans C. K. Schneid. ●

51551 Berberis diaphana Maxim. var. tachiensis Ahrendt = Berberis tischleri C. K. Schneid. ●

51552 Berberis diaphana Maxim. var. uniflora Ahrendt = Berberis diaphana Maxim. ●

51553 Berberis dictyoneura C. K. Schneid.;细脉小檗(短穗小檗,松潘小檗);Netted Barberry,Net-veined Barberry,Short-spiked Barberry ●

51554 Berberis dictyoneura C. K. Schneid. var. bracteata Ahrendt = Berberis bracteata (Ahrendt) Ahrendt ●

51555 Berberis dictyophylla Franch.;刺红珠;Chalkleaf Barberry,Chalk-leaved Barberry ●

51556 Berberis dictyophylla Franch. var. albicaulis Hesse = Berberis dictyophylla Franch. ●

51557 Berberis dictyophylla Franch. var. approximata (Sprague) Rehder;刺齿刺红珠(刺齿刺红珠小檗);Spiny-toothed Chalk-leaved Barberry ●☆

51558 Berberis dictyophylla Franch. var. campyrogyna (Ahrendt) Ahrendt;弯心刺红珠(弯心刺红珠小檗)●

51559 Berberis dictyophylla Franch. var. epruinosa C. K. Schneid.;无粉刺红珠(安布罗齐小檗,黑石珠,拟刺红珠,三颗针);Ambrozy Barberry,Hoarless Chalk-leaved Barberry ●

51560 Berberis dictyophylla Hook. f. = Berberis approximata Sprague ●

51561 Berberis dictyophylla Hook. f. var. approximata (Sprague) Rehder = Berberis approximata Sprague ●

51562 Berberis dictyota Jeps. = Mahonia dictiota (Jeps.) Fedde ●☆

51563 Berberis dielsiana Fedde;首阳小檗(迪尔士小檗,黄柏刺,黄檗,黄檗刺,黄皮针刺);Diels Barberry ●

51564 Berberis diffusa Gay;披散小檗;Diffuse Barberry ●☆

51565 Berberis discolor Turcz.;异色小檗;Discolour Barberry ●☆

51566 Berberis discolorifolia (Ahrendt) Laferr. = Mahonia oiwakensis Hayata ●

51567 Berberis divaricata Rusby;分叉小檗;Divaricate Barberry ●☆

51568 Berberis dolichobotrys Fedde = Berberis dasystachya Maxim. ●

51569 Berberis dolichostemon Ahrendt = Berberis cavaleriei H. Lév. ●

51570 Berberis dolichostylis (Takeda) Laferr. = Mahonia duclouxiana Gagnep. ●

51571 Berberis dongchuanensis T. S. Ying;东川小檗;Dongchuan Barberry ●

51572 Berberis dubia C. K. Schneid.;置疑小檗(黄檗,吉尔雍);Doubtful Barberry,Dubious Barberry ●

51573 Berberis duclouxiana (Gagnep.) Laferr. = Mahonia duclouxiana Gagnep. ●

51574 Berberis duclouxiana (Gagnep.) Laferr. var. hilaica (Ahrendt) Laferr. = Mahonia duclouxiana Gagnep. ●

51575 Berberis dulcis Paxton = Berberis buxifolia Lam. var. papilosa C. K. Schneid. ●☆

51576 Berberis dulcis Sweet = Berberis buxifolia Lam. ●☆

51577 Berberis dumicola C. K. Schneid.;丛林小檗(灌丛小檗);Bush Barberry,Bush-wood Barberry ●

51578 Berberis durobrivensis C. K. Schneid.;杜洛布里小檗;Durobriv Barberry ●☆

51579 Berberis duthieana C. K. Schneid.;杜蒂小檗;Duthie Barberry ●☆
51580 Berberis edentata Rusby;无齿小檗;Edentate Barberry ●☆
51581 Berberis edgeworthiana C. K. Schneid.;埃奇沃斯小檗(埃季沃);Edgeworth Barberry ●☆
51582 Berberis elegans (Franch.) C. K. Schneid. = Berberis amoena Dunn ●
51583 Berberis elegans H. Lév. = Mahonia bodinieri Gagnep. ●
51584 Berberis elliotii Ahrendt;米林小檗●
51585 Berberis elliotii Ahrendt = Berberis tischleri C. K. Schneid. ●
51586 Berberis emarginata Willd.;凹叶小檗;Emarginate Barberry ●☆
51587 Berberis emarginata Willd. var. britzensis C. K. Schneid.;布里泽凹叶小檗;Britz Emarginate Barberry ●☆
51588 Berberis emilii C. K. Schneid.;埃米尔小檗;Emil Barberry ●
51589 Berberis emilii C. K. Schneid. = Berberis cavaleriei H. Lév. ●
51590 Berberis empetrifolia Lam.;岩高兰叶小檗(岩高兰小檗);Crowberry-leaved Barberry ●☆
51591 Berberis empetrifolia Lam. var. magellanica C. K. Schneid.;麦哲伦岩高兰叶小檗;Magellan Crowberry-leaved Barberry ●☆
51592 Berberis engleriana C. K. Schneid.;恩格勒小檗;Engler Barberry ●☆
51593 Berberis erythroclada Ahrendt;红枝小檗(迫隆红枝小檗);Redbranched Barberry, Red-branched Barberry, Trulung Redbranched Barberry ●
51594 Berberis erythroclada Ahrendt var. trulungensis Ahrendt = Berberis erythroclada Ahrendt ●
51595 Berberis esquirolii H. Lév. = Maytenus esquirolii (H. Lév.) C. Y. Cheng ●
51596 Berberis eurybracteata (Fedde) Laferr. = Mahonia eurybracteata Fedde ●
51597 Berberis everesiana Ahrendt;珠峰小檗;Everest Barberry ●
51598 Berberis everesiana Ahrendt var. nambuensis Ahrendt = Berberis parisepala Ahrendt ●
51599 Berberis everesiana Ahrendt var. ventosa Ahrendt;高地珠峰小檗;High-land Barberry ●
51600 Berberis everestiana Ahrendt var. nambuensis Ahrendt = Berberis parisepala Ahrendt ●
51601 Berberis faberi C. K. Schneid.;法贝尔小檗;Faber Barberry ●☆
51602 Berberis fallaciosa C. K. Schneid.;南川小檗;Nanchuan Barberry ●
51603 Berberis fallax C. K. Schneid.;假小檗;Deceptive Barberry ●
51604 Berberis fallax C. K. Schneid. = Berberis davidii Ahrendt ●
51605 Berberis fallax C. K. Schneid. var. latifolia C. Y. Wu et S. Y. Bao;阔叶假小檗;Broad-leaf Deceptive Barberry ●
51606 Berberis fargesii (Takeda) Laferr. = Mahonia sheridaniana C. K. Schneid. ●
51607 Berberis farinosa Benoist;粉背小檗;Farinose Barberry ●☆
51608 Berberis farreri Ahrendt;陇西小檗(法雷尔小檗);Farrer Barberry ●
51609 Berberis favosa W. W. Sm. = Berberis wilsonae Hemsl. var. favosa (W. W. Sm.) Ahrendt ●
51610 Berberis faxoniana C. K. Schneid.;法克昂小檗;Faxon Barberry ●☆
51611 Berberis feddeana C. K. Schneid.;异长穗小檗(异花穗小檗);Fedde Barberry ●
51612 Berberis fendleri A. Gray;芬得乐小檗(芬德勒小檗);Colorado Barberry, Fendler Barberry ●☆
51613 Berberis fengii S. Y. Bao;大果小檗;Feng Barberry ●
51614 Berberis ferdinandi-coburgii C. K. Schneid.;大叶小檗(刺黄莲,大黄檗,鸡脚刺,鸡脚黄连,昆明小檗,土黄连);Ferdinandi-coburg Barberry ●
51615 Berberis ferdinandi-coburgii C. K. Schneid. var. vernalis C. K. Schneid. = Berberis vernalis (C. K. Schneid.) Chamb. et C. M. Hu ●
51616 Berberis ferox Gay;智利多刺小檗;Frocious Barberry ●☆
51617 Berberis fiebrigii C. K. Schneid.;菲布里小檗;Fiebrig Barberry ●☆
51618 Berberis finetii C. K. Schneid. = Berberis papillifera (Franch.) Koehne ●
51619 Berberis flavida (C. K. Schneid.) Laferr. = Mahonia duclouxiana Gagnep. ●
51620 Berberis flavida (C. K. Schneid.) Laferr. var. integrifolia (Hand. -Mazz.) Laferr. = Mahonia duclouxiana Gagnep. ●
51621 Berberis flexuosa Ruiz et Pav.;蜿蜒小檗;Flexuose Barberry ●☆
51622 Berberis floribunda Wall. ex G. Don;繁花小檗;Free-flowering Barberry ●☆
51623 Berberis floribunda Wall. ex G. Don var. affinis Ahrendt = Berberis affinis G. Don ●☆
51624 Berberis florida Phil.;智利多花小檗;Chile Florifenoua Barberry ●☆
51625 Berberis fordii (C. K. Schneid.) Laferr. = Mahonia fordii C. K. Schneid. ●
51626 Berberis formosana Ahrendt = Berberis kawakamii Hayata ●
51627 Berberis formosana Ahrendt = Berberis micropetala T. S. Ying ●
51628 Berberis formosana H. L. Li = Berberis hayatana Mizush. ●
51629 Berberis forrestii Ahrendt;金江小檗;Forrest Barberry ●
51630 Berberis forsskaliana C. K. Schneid.;福斯卡尔小檗;Forsskal Barberry ●☆
51631 Berberis fortunei Lindl. = Mahonia fortunei (Lindl.) Fedde ex C. K. Schneid. ●
51632 Berberis fortunei Lindl. var. szechuanica (Ahrendt) Laferr. = Mahonia fortunei (Lindl.) Fedde ex C. K. Schneid. ●
51633 Berberis fragrans Phil. ex Reiche;芳香小檗;Fragrant Barberry ●☆
51634 Berberis franchetiana C. K. Schneid.;滇西北小檗(光柄滇西北小檗,光梗小檗);Franchet Barberry, Glabrous-stalk Franchet Barberry ●
51635 Berberis franchetiana C. K. Schneid. var. glabripes Ahrendt = Berberis franchetiana C. K. Schneid. ●
51636 Berberis franchetiana C. K. Schneid. var. glabripes Ahrendt = Mahonia eurybracteata Fedde subsp. ganpinensis (H. Lév.) T. S. Ying et Boufford ●
51637 Berberis franchetiana C. K. Schneid. var. gombalana C. Y. Wu et S. Y. Bao;贡巴拉小檗;Gongbala Franchet Barberry ●
51638 Berberis franchetiana C. K. Schneid. var. macrobotrys Ahrendt = Berberis lecomtei C. K. Schneid. ●
51639 Berberis francisci-ferdinandi C. K. Schneid.;大黄檗(三颗针);Pendant Barberry ●
51640 Berberis fremontii Torr.;弗氏小檗(沙漠小檗);Desert Barberry, Fremont Holly Grape, Fremont Holly-grape, Yellow-wood ●☆
51641 Berberis frikaltii C. K. Schneid. ex H. J. Van deLaar;弗里卡特小檗(富里卡小檗);Frikart Barberry ●☆
51642 Berberis fujianensis C. M. Hu;福建小檗;Fujian Barberry ●
51643 Berberis gagnepainii C. K. Schneid.;湖北小檗(黄花刺,加纳帕小檗,蓝果小檗,尾叶小檗,细柄湖北小檗);Blach Barberry, Caudate-leaf Barberry, Gagnepain Barberry, Gagnepain's Barberry, Thready-stalk Black Barberry Barberry ●
51644 Berberis gagnepainii C. K. Schneid. 'Chenault';彻瑙特湖北小檗●☆

51645 Berberis gagnepainii C. K. Schneid. 'Genty';更提湖北小檗●☆

51646 Berberis gagnepainii C. K. Schneid. 'Parkjuweed';帕克珠微湖北小檗●☆

51647 Berberis gagnepainii C. K. Schneid. 'Red Jewel';红珠宝湖北小檗●☆

51648 Berberis gagnepainii C. K. Schneid. 'Terra Nova';特拉诺瓦湖北小檗●☆

51649 Berberis gagnepainii C. K. Schneid. 'Tottenham';头腾汉湖北小檗●☆

51650 Berberis gagnepainii C. K. Schneid. var. filipes Ahrendt = Berberis gagnepainii C. K. Schneid. ●

51651 Berberis gagnepainii C. K. Schneid. var. lanceifolia Ahrendt = Berberis gagnepainii C. K. Schneid. ●

51652 Berberis gagnepainii C. K. Schneid. var. lanceofolia Ahrendt f. pluriflora Ahrendt = Berberis gagnepainii C. K. Schneid. ●

51653 Berberis gagnepainii C. K. Schneid. var. omeiensis Ahrendt;峨眉小檗(眉山小檗);Emwi Barberry ●

51654 Berberis gagnepainii C. K. Schneid. var. praestans Ahrendt;高雅湖小檗●☆

51655 Berberis gagnepainii C. K. Schneid. var. subovata C. K. Schneid.;瓦屋小檗●☆

51656 Berberis gambleana Ahrendt = Berberis calliobotrys Bien. ex Aitch. ●☆

51657 Berberis ganpinensis H. Lév. = Mahonia eurybracteata Fedde subsp. ganpinensis (H. Lév.) T. S. Ying et Boufford ●

51658 Berberis ganpinensis H. Lév. = Mahonia ganpinensis (H. Lév.) Fedde ●

51659 Berberis garciae Pau;加尔恰小檗;Garcia Barberry ●☆

51660 Berberis gautamae Laferr. = Mahonia napaulensis DC. ●

51661 Berberis gayi Citerne = Berberis citernei Ahrendt ●☆

51662 Berberis geraldii Veitch = Berberis aggregata C. K. Schneid. ●

51663 Berberis gibbsii Ahrendt;吉布斯小檗;Gibbs Barberry ●☆

51664 Berberis gilgiana C. K. Schneid. = Berberis pubescens Pamp. ●

51665 Berberis gilgiana Fedde;涝峪小檗(小叶黄檗);Gilg Barberry, Wildfire Barberry ●

51666 Berberis gilungensis T. S. Ying;吉隆小檗;Jilong Barberry ●

51667 Berberis giraldii Hesse = Berberis salicaria Fedde ●

51668 Berberis glauca Hort. ex K. Koch;粉绿小檗;Glaucous Barberry ●☆

51669 Berberis glaucescens A. St. -Hil.;变粉绿小檗;Glaucescent Barberry ●☆

51670 Berberis glaucocarpa Stapf;霜粉黑果小檗;Glauco-fruited Barberry, Great Barberry ●☆

51671 Berberis glazioviana Brade;格拉乔夫小檗;Glaziov Barberry ●☆

51672 Berberis globosa Benth.;球形小檗;Globose-fruited Barberry ●☆

51673 Berberis glomerate Hook. et Arn.;团集小檗;Glomerate Barberry ●☆

51674 Berberis glomerate Hook. et Arn. var. zahlbruckneriana (C. K. Schneid.) Ahrendt;札氏团集小檗;Zahlbruckner Glomerate Barberry ●☆

51675 Berberis goudotii Triana et Planch.;古多特小檗;Goudot Barberry ●☆

51676 Berberis gracilipes Oliv. = Mahonia gracilipes (Oliv.) Fedde ●

51677 Berberis graminea Ahrendt;草色小檗(狭叶小檗);Gramineous Barberry ●

51678 Berberis grandibracteata Ahrendt;大苞小檗;Large-bracteate Barberry ●☆

51679 Berberis grandiflora Turcz.;厄瓜多尔大花小檗;Ecuadoor Large-flowered Barberry ●☆

51680 Berberis grantii Ahrendt;格兰特小檗;Grant Barberry ●☆

51681 Berberis grantii Ahrendt = Berberis holstii Engl. ●☆

51682 Berberis grevilleana Gillies ex Hook.;格雷维尔小檗;Greville Barberry ●☆

51683 Berberis griffithiana C. K. Schneid.;印东北小檗(不等微翼枝小檗,错那小檗,微翼枝小檗);Griffith Barberry, Slightly Wing-branched Barberry, Variously Slightly Wing-branched Barberry ●

51684 Berberis griffithiana C. K. Schneid. var. pallida (Hook. f. et Thomson) D. F. Chamb. et C. M. Hu;灰叶小檗(不丹小檗,错那小檗,卷叶小檗,尼昔小檗);Bhutan Barberry, Slender-pedicelled Barberry, Trimo Barberry ●

51685 Berberis griffithii (Takeda) Laferr. = Mahonia napaulensis DC. ●

51686 Berberis grodtmannia C. K. Schneid.;安宁小檗(格罗德曼小檗,卷叶小檗);Grodtmann Barberry ●

51687 Berberis grodtmannia C. K. Schneid. var. flavoramea C. K. Schneid.;黄茎小檗(黄枝格罗德曼小檗);Yellow-branched Grodtmann Barberry Barberry ●

51688 Berberis guilache Triana et Planch.;吉拉策小檗;Guilache Barberry ●☆

51689 Berberis guipelii Koch et Bouché = Berberis chinensis Poir. ●

51690 Berberis guizhouensis T. S. Ying;贵州小檗(毕节小檗);Guizhou Barberry ●

51691 Berberis gyalaica Ahrendt ex F. Br.;波密小檗(黑龙小檗,西藏小檗,细梗小檗);Bomi Barberry, Gyala Barberry, Minute-stalk Gyala Barberry, Taylor Barberry, Tibetan Barberry ●

51692 Berberis gyalaica Ahrendt ex F. Br. var. maximiflora Ahrendt = Berberis gyalaica Ahrendt ex F. Br. ●

51693 Berberis gyalaica Ahrendt ex F. Br. var. minuate Ahrendt = Berberis gyalaica Ahrendt ex F. Br. ●

51694 Berberis haematocarpa Wooton;血果小檗;Algerita, Barberry, Desert Barberry, Mexican Barberry, Red Barberry, Red Holly Grape, Red-fruited Barberry ●☆

51695 Berberis haenkeana C. Presl ex Schult. f.;黑恩克小檗;Haenke Barberry ●☆

51696 Berberis hainesii Ahrendt;海恩斯小檗;Haines Barberry ●☆

51697 Berberis hainesii Ahrendt var. bravifilipes Ahrendt;短柄海恩斯小檗(短细柄海恩斯小檗);Short-thready-atalk Haines Barberry ●☆

51698 Berberis hakeoides (Hook. f.) C. K. Schneid.;墨水果小檗;Inkfruit Barberry ●☆

51699 Berberis hallii Hieron.;霍尔小檗;Hall Barberry ●☆

51700 Berberis hallii Hieron. var. wagneriana C. K. Schneid.;瓦格纳小檗;Wagner Barberry ●☆

51701 Berberis hamiltoniana Ahrendt;汉密尔顿小檗;Hamilton Barberry ●☆

51702 Berberis hancockiana (Takeda) Laferr. = Mahonia hancockiana Takeda ●

51703 Berberis haoi T. S. Ying;洮河小檗;Taohe Barberry ●

51704 Berberis harrisoniana Kearney et Peebles;哈氏小檗;Harrison's Barberry, Kofa Mountain Barberry ●☆

51705 Berberis hauanucensis (C. K. Schneid.) J. F. Macbr.;瓦努科小檗;Huanuco Barberry ●☆

51706 Berberis hauniensis C. K. Schneid.;豪恩小檗;Haun Barberry ●☆

51707 Berberis hayata Mizush. = Berberis mingetsensis Hayata ●

51708 Berberis hayatana Mizush.;南湖小檗(早田氏小檗);Hayata Barberry, Nanhu Barberry ●

51709 Berberis helenae Ahrendt;海伦小檗;Helen Barberry ●☆

51710 Berberis hemsleyana Ahrendt;拉萨小檗(三颗针);Hemsley Barberry ●

51711 Berberis henryana C. K. Schneid.;川鄂小檗(巴东小檗,亨利小檗,黄檗,三颗针);Henry Barberry ●

51712 Berberis henryi Laferr. = Mahonia conferta Takeda ●

51713 Berberis hersii Ahrendt;南阳小檗(赫斯小檗,利申特黄芦木);Hers Barberry,Licent Amur Barberry ●

51714 Berberis heteracantha Ahrendt;异形刺小檗;Various-spiny Barberry ●☆

51715 Berberis heterophylla Juss. ex Poir.;异叶小檗;Hedgehog Barberry ●☆

51716 Berberis heterophylla Juss. ex Poir. var. pluriflora Reiche = Berberis philippii Ahrendt ●☆

51717 Berberis heteropoda Schrenk;异柄小檗(刺黄柏,刺黄檗,黑果小檗,土耳其斯坦小檗,异果小檗);Turkestan Barberry ●

51718 Berberis heteropoda Schrenk var. oblonga Regel = Berberis oblonga (Regel) C. K. Schneid. ●☆

51719 Berberis heteropoda Schrenk var. sphaerocarpa (Kar. et Kir.) Ahrendt;球果土耳其斯坦小檗;Sphaerical Turkestan Barberry ●☆

51720 Berberis heteropsis Ahrendt;异形小檗;Different-appearance Barberry ●☆

51721 Berberis hibberdiana Ahrendt = Berberis pruinosa Franch. ●

51722 Berberis hieronymi C. K. Schneid.;海罗尼姆小檗;Hieronymus Barberry ●☆

51723 Berberis higginsiae Munz;希氏小檗●☆

51724 Berberis himalaica Ahrendt;喜马拉雅小檗;Himalayan Barberry ●☆

51725 Berberis hirtellipes Ahrendt;微硬毛小檗(微硬毛梗小檗);Hirtellous-pedicel Barberry ●☆

51726 Berberis hispanica Boiss. et Reut.;西班牙小檗;Spain Barberry ●☆

51727 Berberis hispanica Boiss. et Reut. var. hackeliana (C. K. Schneid.) Ahrendt;哈克尔西班牙小檗;Hackel Spain Barberry ●☆

51728 Berberis hobsonii Ahrendt;毛梗小檗(霍布森小檗,亚东小檗);Hobson Barberry,Yadong Barberry ●

51729 Berberis hochreutinerana J. F. Macbr.;霍赫鲁特纳小檗;Hochreutiner Barberry ●☆

51730 Berberis holocraspedon Ahrendt;全边小檗(滇西全边小檗,风庆小檗);Entire Barberry,Entire-marginal Barberry ●

51731 Berberis holstii Engl.;霍尔斯特小檗●☆

51732 Berberis honanensis Ahrendt;河南小檗;Henan Barberry ●

51733 Berberis hookeri Lem.;胡克小檗(虎克小檗);Hooker Barberry,Hooker's Barberry ●☆

51734 Berberis hookeri Lem. subsp. longipes D. F. Chamb. et C. M. Hu;长梗虎克小檗●

51735 Berberis hookeri Lem. var. candidula C. K. Schneid. = Berberis candidula (C. K. Schneid.) C. K. Schneid. ●

51736 Berberis hookeri Lem. var. microcarpa Ahrendt;小果虎克小檗;Small-fruited Hooker Barberry ●☆

51737 Berberis hookeri Lem. var. platyphylla Ahrendt;阔叶虎克小檗;Broad-leaved Hooker Barberry ●☆

51738 Berberis hookeri Lem. var. viridis C. K. Schneid.;绿虎克小檗;Green Hooker Barberry ●☆

51739 Berberis horrida Gay;南美刺小檗;South-american Prickly Barberry ●☆

51740 Berberis hsuyunensis P. K. Hsiao et W. C. Sung;叙永小檗;Hsuyun Barberry ●

51741 Berberis huegeliana C. K. Schneid.;许格小檗;Huegel Barberry ●☆

51742 Berberis huiliensis (Hand.-Mazz.) Laferr. = Mahonia sheridaniana C. K. Schneid. ●

51743 Berberis humbertiana J. F. Macbr.;汉伯特小檗;Humbert Barberry ●☆

51744 Berberis humido-umbrosa Ahrendt;阴湿小檗(不美阴湿小檗,湿荫小檗,荫湿小檗);Humid-shaped Barberry,Shady-humid Barberry ●

51745 Berberis humido-umbrosa Ahrendt var. dispersa Ahrendt;绿背小檗;Dispersed Humid-shaped Barberry ●☆

51746 Berberis humido-umbrosa Ahrendt var. inornata Ahrendt = Berberis lecomtei C. K. Schneid. ●

51747 Berberis hybrido-gagnepainii Suringar;假黑小檗;False Black Barberry ●☆

51748 Berberis hypericifolia T. S. Ying;金丝桃叶小檗(异叶小檗);Diversifoliaceous Barberry,St. Johnswort-leaved Barberry ●

51749 Berberis hyperythra Diels;厄瓜多尔微红小檗;Ecuador Slightly Barberry ●☆

51750 Berberis hypokerina Airy Shaw;紫点小檗;Violetbead Barberry ●☆

51751 Berberis hypoleuca Hort. = Berberis candidula C. K. Schneid. ●

51752 Berberis hypoleuca Lindl. = Berberis candidula C. K. Schneid. ●

51753 Berberis hypoxantha C. Y. Wu et S. Y. Bao;黄背小檗;Yellow-back Barberry,Yellow-backed Barberry ●

51754 Berberis ignorata C. K. Schneid.;藏黑果小檗(繁果小檗,黑果小檗,无知小檗);Ignorant Barberry ●

51755 Berberis ilicifolia J. Forst.;冬青叶小檗;Holly Barberry ●☆

51756 Berberis ilicina (Schltdl.) Hemsl. = Mahonia ilicina Schltdl. ●☆

51757 Berberis iliensis Popov;伊犁小檗(刺黄连,伊宁小檗);Ili Barberry,Yili Barberry ●

51758 Berberis iliensis Popov = Berberis nummularia Bunge ●

51759 Berberis impedita C. K. Schneid.;南岭小檗(敏山小檗);Minshan Barberry ●

51760 Berberis incrassata Ahrendt = Berberis insignis Hook. f. et Thomson subsp. incrassata (Ahrendt) D. F. Chamb. et C. M. Hu ●

51761 Berberis incrassata Ahrendt = Berberis insignis Hook. f. et Thomson ●

51762 Berberis incrassata Ahrendt var. bucahwangensis Ahrendt = Berberis insignis Hook. f. et Thomson subsp. incrassata (Ahrendt) D. F. Chamb. et C. M. Hu ●

51763 Berberis incrassata Ahrendt var. bucahwangensis Ahrendt = Berberis insignis Hook. f. et Thomson ●

51764 Berberis incrassata Ahrendt var. fugongensis S. Y. Bao = Berberis insignis Hook. f. et Thomson subsp. incrassata (Ahrendt) D. F. Chamb. et C. M. Hu ●

51765 Berberis inermis Pers. = Berberis buxifolia Lam. var. inermis (Pers.) C. K. Schneid. ●☆

51766 Berberis insignis Hook. f. et Thomson;显著小檗(大王刺);Remarkable Barberry ●

51767 Berberis insignis Hook. f. et Thomson subsp. incrassata (Ahrendt) D. F. Chamb. et C. M. Hu;球果小檗(福贡小檗,厚柄小檗,俅江厚柄小檗);Fugong Barberry,Qiujiang Barberry,Thick-pedicel Barberry,Thick-pediceled Barberry ●

51768 Berberis insignis Hook. f. et Thomson var. elegantifolia Ahrendt;美叶显著小檗;Elegant-leaved Remarkable Barberry ●☆

51769 Berberis insignis Hook. f. et Thomson var. gouldii Ahrendt;古尔德显著小檗;Gould Remarkable Barberry ●☆

51770 Berberis insignis Hook. f. et Thomson var. shergaonensis Ahrendt；舍岗显著小檗；Shergaon Remarkable Barberry ●☆

51771 Berberis insignis Hook. f. et Thomson var. tongloensis C. K. Schneid.；东洛显著小檗；Dongle Remarkable Barberry ●☆

51772 Berberis insignis Hook. f. et Thomson var. zelaica Ahrendt；泽拉显著小檗；Zela Remarkable Barberry ●☆

51773 Berberis insolita C. K. Schneid.；西昌小檗（绥江小檗，稀有小檗）；Suijiang Barberry，Uncommon Barberry ●

51774 Berberis integerrima Bunge = Berberis jamesiana Forrest et W. W. Sm. ●

51775 Berberis integripetala T. S. Ying；甘南小檗；Gannan Barberry ●

51776 Berberis interposia Ahrendt；居间小檗；Interpositional Barberry ●☆

51777 Berberis irwinii Bijh.；欧文小檗；Irwin Barberry ●☆

51778 Berberis iteophylla C. Y. Wu et S. Y. Bao；鼠刺叶小檗（柳叶小檗，鼠叶小檗）；Sweetspire Barberry，Sweetspire-leaved Barberry ●

51779 Berberis jaeschkeana C. K. Schneid.；贾施克小檗；Jaeschke Barberry ●☆

51780 Berberis jaeschkeana C. K. Schneid. var. bimbilacia Ahrendt；比巴小檗；Bimbila Barberry ●☆

51781 Berberis jaeschkeana C. K. Schneid. var. usteriana C. K. Schneid.；尤斯特小檗；Uster Barberry ●☆

51782 Berberis jamesiana Forrest et W. W. Sm.；川滇小檗（白果川滇小檗，篱笆川滇小檗，全缘小檗）；Hedge James Barberry，Integrifolious Barberry，James Barberry，White-fruited James Barberry ●

51783 Berberis jamesiana Forrest et W. W. Sm. var. leucocarpa（W. W. Sm.）Ahrendt = Berberis jamesiana Forrest et W. W. Sm. ●

51784 Berberis jamesiana Forrest et W. W. Sm. var. sepium Ahrendt = Berberis jamesiana Forrest et W. W. Sm. ●

51785 Berberis jamesonii Lindl.；詹姆森小檗；Jameson Barberry ●☆

51786 Berberis jamesonii Turcz. = Berberis grandiflora Turcz. ●☆

51787 Berberis japonica（Regel）C. K. Schneid. = Berberis regeliana Koehne ex C. K. Schneid. ●☆

51788 Berberis japonica（Thunb.）R. Br. = Mahonia japonica（Thunb.）DC. ●

51789 Berberis japonica（Thunb.）R. Br. var. gracillima（Fedde）Rehder = Mahonia japonica（Thunb.）DC. ●

51790 Berberis japonica（Thunb.）R. Br. var. trifurca（Lindl. et Paxton）Rehder = Mahonia bodinieri Gagnep. ●

51791 Berberis japonica Hort. = Berberis thunbergii DC. ●

51792 Berberis japonica Spreng. = Mahonia japonica（Thunb.）DC. ●

51793 Berberis javanica Miq. = Berberis xanthoxylon Hassk. ex C. K. Schneid. var. junghuhniana Ahrendt ●☆

51794 Berberis jelskiana Ahrendt；杰尔斯基小檗；Jelski Barberry ●☆

51795 Berberis jiangxiensis C. M. Hu；江西小檗；Jiangxi Barberry ●

51796 Berberis jiangxiensis C. M. Hu var. pulchella C. M. Hu；短叶江西小檗；Short-leaf Jiangxi Barberry ●

51797 Berberis jinfoshanensis T. S. Ying；金佛山小檗；Jinfoshan Barberry ●

51798 Berberis jinshajiangensis X. H. Li；小瓣小檗；Small-petal Barberry ●

51799 Berberis jiulongensis T. S. Ying；九龙小檗；Jiulong Barberry ●

51800 Berberis johannis Ahrendt；腰果小檗（西藏小檗，约翰小檗）；Johann Barberry ●

51801 Berberis johnstonii Standl. et Steyerm. = Mahonia johnstonii（Standl. et Steyerm.）Standl. et Steyerm. ●☆

51802 Berberis julianae C. K. Schneid.；豪猪刺（巴东豪猪刺，长叶豪猪刺，长圆叶豪猪刺，蚝猪刺，湖北小檗，鸡足黄连，九莲小檗，老鼠刺，三甲刺，三颗针，山黄连，石妹刺，土黄连）；Badong Wintergreen Barberry，Chinese Barberry，Dwarf Wintergreen Barberry，Oblong-leaved Wintergreen Barberry，Wintergreen Barberry ●

51803 Berberis julianae C. K. Schneid. var. oblongifolia Ahrendt = Berberis julianae C. K. Schneid. ●

51804 Berberis julianae C. K. Schneid. var. patungensis Ahrendt = Berberis julianae C. K. Schneid. ●

51805 Berberis kansuensis C. K. Schneid.；甘肃小檗（黄檗）；Gansu Barberry ●

51806 Berberis kansuensis C. K. Schneid. var. procera Ahrendt = Berberis dasystachya Maxim. ●

51807 Berberis karkaralensis Kornil. et Potapov；喀尔喀拉勒小檗●☆

51808 Berberis kartanica Ahrendt；卡达小檗；Karta Barberry ●☆

51809 Berberis kaschgarica Rupr.；喀什小檗；Kashi Barberry ●

51810 Berberis kashmirana Ahrendt；克什米尔小檗；Kashmir Barberry ●☆

51811 Berberis kawakamii Hayata；台湾小檗（阿里山小檗，薄叶小檗，川上氏小檗，高山小檗，黑果小檗，清水山小檗，土黄芩）；Alishan Barberry，Alpine Barberry，Kawakami Barberry，Qingshuishan Barberry，Short-sepal Barberry，Taiwan Barberry，Thin-ieaf Barberry ●

51812 Berberis kawakamii Hayata var. formosana（Ahrendt）Ahrendt = Berberis kawakamii Hayata ●

51813 Berberis kawakamii Hayata var. formosana（Ahrendt）Ahrendt = Berberis micropetala T. S. Ying ●

51814 Berberis keikoe Laferr. = Mahonia duclouxiana Gagnep. ●

51815 Berberis keissleriana C. K. Schneid.；基斯勒里小檗；Keissleri Barberry ●☆

51816 Berberis kerriana Ahrendt；克尔小檗（南方小檗）；Kerr Barberry ●

51817 Berberis kewensis C. K. Schneid.；邱园小檗；Kew Barberry ●☆

51818 Berberis khasiana Ahrendt；卡西亚小檗；Khasia Barberry ●☆

51819 Berberis knightii（Lindl.）C. Koch = Berberis knightii（Lindl.）K. Koch ●☆

51820 Berberis knightii（Lindl.）K. Koch；奈特小檗；Knight Barberry ●☆

51821 Berberis koehneana C. K. Schneid.；凯内小檗；Koehne Barberry ●☆

51822 Berberis koehneana C. K. Schneid. var. auramea Ahrendt；黄茎凯内小檗；Yellowstem Koehne Barberry ●☆

51823 Berberis kongboensis Ahrendt；工布小檗；Gongbogyanda Barberry，Gongbu Barberry ●

51824 Berberis koreana Palib.；朝鲜小檗（掌刺小檗）；Korea Barberry，Korean Barberry ●

51825 Berberis kumaonensis C. K. Schneid.；库芒小檗；Kumaon Barberry ●☆

51826 Berberis kunawurensis Royle；库纳乌尔小檗；Kunawur Barberry ●☆

51827 Berberis kunmingensis C. Y. Wu ex S. Y. Bao；昆明小檗（粉叶小檗，鸡脚刺，鸡脚黄连，宽叶鸡脚黄连，昆明鸡脚黄连，三根针，三颗针，土黄连）；Kunming Barberry ●

51828 Berberis kunmingensis C. Y. Wu ex S. Y. Bao = Berberis ferdinandi-coburgii C. K. Schneid. ●

51829 Berberis lambertii R. Parker；兰伯特小檗；Lambert Barberry ●☆

51830 Berberis lanceolata Benth. = Mahonia lanceolata（Benth.）Fedde ●☆

51831 Berberis laojunshanensis T. S. Ying；老君山小檗；Laojunshan Barberry ●

51832 Berberis lasioclema Ahrendt；毛枝小檗；Woolly-branched

Barberry ●☆

51833 Berberis latifolia Ruiz et Pav.;秘鲁阔叶小檗;Peru Broad-leaved Barberry ●☆

51834 Berberis laurina Billb.;月桂小檗;Laurel-like Barberry ●☆

51835 Berberis laxiflora Schrad.;疏花小檗;Loose-flowered Barberry ●☆

51836 Berberis laxiflora Schrad. var. langeana C. K. Schneid.;兰格疏花小檗;Lange Loose-flowered Barberry ●☆

51837 Berberis laxiflora Schrad. var. oblanceolata C. K. Schneid.;倒披针叶疏花小檗;Oblanceolata Loose-flowered Barberry ●☆

51838 Berberis leachiana Ahrendt;利奇小檗;Leachi Barberry ●☆

51839 Berberis leboensis T. S. Ying;雷波小檗;Leibo Barberry ●

51840 Berberis lechleriana C. K. Schneid.;勒克勒里小檗;Lechleri Barberry ●☆

51841 Berberis lecomtei C. K. Schneid.;光叶小檗(大球滇西北小檗,勒康特小檗,三颗针,土黄连,无粉小檗);Big-botrys Franchet Barberry,Inornate Humid-shaped Barberry,Lecomte Barberry ●

51842 Berberis lehmannii Hieron.;莱曼小檗;Lehmann Barberry ●☆

51843 Berberis leichtensteinii C. K. Schneid. = Berberis potaninii Maxim. ●

51844 Berberis lemoinei Ahrendt;勒莫奈小檗;Lemoine Barberry ●☆

51845 Berberis lempergiana Ahrendt;长柱小檗(黄檗,勒姆帕格小檗,天台小檗,土黄檗);Lemperg's Barberry ●

51846 Berberis lempergiana Ahrendt subsp. subedentata C. M. Wu;疏齿长柱小檗(疏齿小檗);Loose-tooth Lemperg's Barberry ●

51847 Berberis lempergiana Ahrendt subsp. wulingensis C. M. Hu;五岭长柱小檗(五岭小檗);Wuling Lemperg's Barberry ●

51848 Berberis lepidifolia Ahrendt;鳞叶小檗;Scaly-leaved Barberry ●

51849 Berberis leptoclada Diels = Berberis amoena Dunn ●

51850 Berberis leptodonta (Gagnep.) Laferr. = Mahonia leptodonta Gagnep. ●

51851 Berberis leptopoda Ahrendt;尼营小檗●

51852 Berberis leptopoda Ahrendt = Berberis griffithiana C. K. Schneid. var. pallida (Hook. f. et Thomson) D. F. Chamb. et C. M. Hu ●

51853 Berberis leschenaultii Wall. ex Wight et Arn. = Mahonia napaulensis DC. ●

51854 Berberis leucocarpa W. W. Sm. = Berberis jamesiana Forrest et W. W. Sm. ●

51855 Berberis leveilleana (C. K. Schneid.) Laferr. = Mahonia bodinieri Gagnep. ●

51856 Berberis levis Franch.;平滑小檗(短叶平滑小檗,洱源小檗,细齿洱源小檗);Eryuan Barberry,Serrulate Eryuan Barberry,Short-leaved Smooth Barberry,Smooth Barberry ●

51857 Berberis levis Franch. var. brachyphylla Ahrendt = Berberis levis Franch. ●

51858 Berberis libanotica Ehrenb. ex C. K. Schneid.;黎巴嫩小檗;Lebanon Barberry ●☆

51859 Berberis liechtensteninii C. K. Schneid. = Berberis potaninii Maxim. ●

51860 Berberis lijiangensis C. Y. Wu et S. Y. Bao;丽江小檗;Lijiang Barberry ●

51861 Berberis lilloana Job;利洛小檗;Lillo Barberry ●☆

51862 Berberis lindleyana Ahrendt;林德利小檗;Lindley Barberry ●☆

51863 Berberis linearifolia Phil.;线叶小檗;Jasperbells Barberry,Linear-leaved Barberry,Orange King ●☆

51864 Berberis linearifolia Phil. 'Orange King';橙王线叶小檗●☆

51865 Berberis linearifolia Phil. var. longifolia (Reiche) Ahrendt;长线叶小檗;Long Linear-leaved Barberry ●☆

51866 Berberis lintanensis Zh. G. Ma et Zh. Ma;临潭小檗●

51867 Berberis liophylla C. K. Schneid.;滑叶小檗;Smooth-leaved Barberry ●

51868 Berberis liophylla C. K. Schneid. var. conglobata Ahrendt = Berberis cavaleriei H. Lév. ●

51869 Berberis littoralis Phil.;滨海小檗;Seashore Barberry ●☆

51870 Berberis lobbiana C. K. Schneid.;罗布小檗;Lobb Barberry ●☆

51871 Berberis lologensis Sandwith;阿根廷小檗(罗罗格小檗,洛洛小檗);Lolog Barberry ●☆

51872 Berberis lologensis Sandwith 'Highdown';低矮阿根廷小檗●☆

51873 Berberis lologensis Sandwith 'Stapehill';思泰西尔阿根廷小檗(施塔珀希尔洛洛小檗)●☆

51874 Berberis lomariifolia (Takeda) Laferr. = Mahonia oiwakensis Hayata ●

51875 Berberis lomariifolia (Takeda) Laferr. var. estylis (C. Y. Wu ex S. Y. Bao) Laferr. = Mahonia oiwakensis Hayata ●

51876 Berberis longibracteata (Takeda) Laferr. = Mahonia longibracteata Takeda ●

51877 Berberis longispina T. S. Ying;长刺小檗;Longispiny Barberry,Long-spiny Barberry ●

51878 Berberis longlinensis (Y. S. Wang et P. K. Hsiao) Laferr. = Mahonia napaulensis DC. ●

51879 Berberis loudonii Ahrendt;路唐小檗;Loudon Barberry ●☆

51880 Berberis loxensis Benth.;洛哈小檗;Loja Barberry ●☆

51881 Berberis lubrica C. K. Schneid.;亮叶小檗(润滑小檗);Lubricant Barberry ●

51882 Berberis ludlowii Ahrendt;大花小檗●

51883 Berberis ludlowii Ahrendt = Berberis muliensis Ahrendt ●

51884 Berberis ludlowii Ahrendt var. capilaris (Cox ex Ahrendt) Ahrendt = Berberis muliensis Ahrendt ●

51885 Berberis ludlowii Ahrendt var. deleica (Ahrendt) Ahrendt = Berberis muliensis Ahrendt ●

51886 Berberis ludlowii Ahrendt var. sakdenensis Ahrendt = Berberis macrosepala Hook. f. et Thomson var. sakdenensis (Ahrendt) Ahrendt ●☆

51887 Berberis ludlowii Ahrendt var. saxiclivicola Ahrendt = Berberis muliensis Ahrendt var. atuntzeana Ahrendt ●

51888 Berberis luhuoensis T. S. Ying;炉霍小檗;Luhuo Barberry ●

51889 Berberis lutea Ruiz et Pav.;黄小檗;Yellow Barberry ●☆

51890 Berberis lycioides Stapf;枸杞状小檗;Box-thorn-like Barberry ●☆

51891 Berberis lycium Royle;枸杞小檗;Box-thorn Barberry ●☆

51892 Berberis lycium Royle var. simlensis Ahrendt;西姆拉枸杞小檗;Simla Box-thorn Barberry ●☆

51893 Berberis lycium Royle var. subfascicularis Ahrendt;近簇生枸杞小檗;Nearly-fascicled Box-thorn Barberry ●☆

51894 Berberis lycium Royle var. subvirescens Ahrendt;近缘枸杞小檗;Subvirescent Box-thorn Barberry ●☆

51895 Berberis macracantha Schrad.;大刺小檗;Large-spiny Barberry ●☆

51896 Berberis macracantha Schrad. var. pulchra Schrad.;美丽大刺小檗(丽大刺小檗);Beautiful Large-spiny Barberry ●☆

51897 Berberis macrosepala Hook. f. et Thomson;大萼小檗;Large-sepal Barberry ●☆

51898 Berberis macrosepala Hook. f. et Thomson var. deleica Ahrendt = Berberis muliensis Ahrendt ●

51899 Berberis macrosepala Hook. f. et Thomson var. sakdenensis (Ahrendt) Ahrendt;萨克丹大萼小檗;Sakden Large-sepal Barberry ●☆

51900 Berberis macrosepala Hook. f. et Thomson var. setifolia Ahrendt;刚毛叶大萼小檗;Setose-leaved Large-sepal Barberry ●☆
51901 Berberis maderensis Lowe;梅德小檗;Madeila Barberry ●☆
51902 Berberis magnifolia Ahrendt; 英国大叶小檗; Large-leaved Barberry ●☆
51903 Berberis mairei Ahrendt;迈氏小檗(东川小檗,麦氏小檗);Dongchuan Barberry,Maire Barberry ●
51904 Berberis malipoensis C. Y. Wu et S. Y. Bao;麻栗坡小檗;Malipo Barberry ●
51905 Berberis manipurana Ahrendt;曼尼普小檗;Manipur Barberry ●☆
51906 Berberis manipurensis (Takeda) Laferr. = Mahonia napaulensis DC. ●
51907 Berberis marginata Gay;白边小檗;White-margined Barberry ●☆
51908 Berberis masafuerana Skottsb.; 马沙弗尔小檗; Masafuer Barberry ●☆
51909 Berberis maximowiczii Regel = Berberis thunbergii DC. var. maximowiczii (Regel) Regel ●☆
51910 Berberis media Groot.;中间小檗;Medium Barberry ●☆
51911 Berberis medogensis T. S. Ying;矮生小檗;Medong Barberry ●
51912 Berberis meehanii C. K. Schneid. ex Rehder;梅汉小檗;Meehan Barberry ●☆
51913 Berberis mekongensis W. W. Sm.;湄公小檗;Mekong Barberry ●
51914 Berberis mentorensis L. M. Ames;门妥小檗(杂种日本小檗);Mentor Barberry ●☆
51915 Berberis metapolyantha Ahrendt; 万源小檗(变穗小檗);Metapolyanthous Barberry ●
51916 Berberis mianningensis T. S. Ying;冕宁小檗;Mianning Barberry ●
51917 Berberis miccia (Buch. -Ham. ex D. Don) Walp. = Mahonia napaulensis DC. ●
51918 Berberis miccia Walp. = Mahonia napaulensis DC. ●
51919 Berberis michayi Job;米夏小檗;Michay Barberry ●☆
51920 Berberis micrantha (Hook. f. et Thomson) Ahrendt;不丹小花小檗;Bhudan Small-flowered Barberry ●☆
51921 Berberis micropetala T. S. Ying = Berberis jinshajiangensis X. H. Li ●
51922 Berberis microphylla C. Koch = Berberis buxifolia Lam. ●☆
51923 Berberis microphylla G. Forst.; 智利小叶小檗; Small-leaved Barberry ●☆
51924 Berberis microtricha C. K. Schneid.;小毛花小檗(小毛小檗);Puberulent Barberry ●
51925 Berberis mikuna Job;米孔小檗;Mikun Barberry ●☆
51926 Berberis mingetsensis Hayata;薄叶小檗(阿里山小檗,眠月小檗,玉龙山小檗,早田氏小檗);Alishan Barberry,Hayata Barberry,Small-flowred Barberry,Yulongshan Barberry ●
51927 Berberis mingetsuensis Hayata = Berberis aristatoserrulata Hayata ●
51928 Berberis minutiflora C. K. Schneid.;小花小檗(光茎小檗);Small-flowred Barberry ●
51929 Berberis minutiflora C. K. Schneid. var. glabramea Ahrendt = Berberis minutiflora C. K. Schneid. ●
51930 Berberis minutiflora C. K. Schneid. var. yulongshanensis S. Y. Bao = Berberis minutiflora C. K. Schneid. ●
51931 Berberis miqueliana Ahrendt;米奎尔小檗●☆
51932 Berberis mitifolia Stapf = Berberis salicaria Fedde ●
51933 Berberis monosperma Ruiz et Pav.;独籽小檗●☆
51934 Berberis montana Gay;山地小檗●☆
51935 Berberis montevidensis C. K. Schneid.;蒙得维的亚小檗●☆
51936 Berberis monyulensis (Ahrendt) Laferr. = Mahonia monyulensis Ahrendt ●
51937 Berberis moranensis Schult. f.;墨西哥小檗●☆
51938 Berberis morenonis Kuntze;莫雷诺小檗●☆
51939 Berberis moritzii Hieron.;莫里茨小檗●☆
51940 Berberis morrisonensis Hayata; 玉山小檗(阿里山小檗);Morrison's Barberry,Mount Morrison Barberry,Yushan Barberry ●
51941 Berberis mouillacana C. K. Schneid.;变刺小檗(博斯钱小檗);Boschan Barberry,Mouillac Barberry ●
51942 Berberis mucrifolia Ahrendt;短尖叶小檗●☆
51943 Berberis muliensis Ahrendt;木里小檗(大萼大花小檗,大花小檗,得乐小檗,天宝山小檗,线柄大花小檗);Bigflower Barberry,Capillary-pedicel Barberry,Delei Ludlow Barberry,Ludlow Barberry,Muli Barberry,Tianbaoshan Barberry ●
51944 Berberis muliensis Ahrendt var. atuntzeana Ahrendt;阿墩小檗(阿顿小檗,白马小檗);Atuntze Barberry,Baima Barberry ●
51945 Berberis muliensis Ahrendt var. beimanica Ahrendt = Berberis muliensis Ahrendt var. atuntzeana Ahrendt ●
51946 Berberis multicaulis T. S. Ying; 多枝小檗; Many-stemmed Barberry,Multicauliforous Barberry ●
51947 Berberis multiflora Benth.;厄瓜多尔多花小檗●☆
51948 Berberis multiflora Benth. var. calvescens C. K. Schneid.;光秃厄瓜多尔多花小檗●☆
51949 Berberis multiovula T. S. Ying;多珠小檗;Many-ovele Barberry ●
51950 Berberis multiserrata T. S. Ying; 粗齿小檗; Manyserrate Barberry,Multiserrate Barberry ●
51951 Berberis mutabilis Phil.;易变小檗●☆
51952 Berberis nantoensis C. K. Schneid. = Berberis kawakamii Hayata ●
51953 Berberis nantoensis C. K. Schneid. = Berberis micropetala T. S. Ying ●
51954 Berberis napaulensis (DC.) Laferr. = Mahonia napaulensis DC. ●
51955 Berberis napaulensis sensu Hayata = Mahonia japonica (Thunb.) DC. ●
51956 Berberis napaulensis Spreng. var. leschenaultii (Wall. ex Wight et Arn.) Hook. f. et Thomson = Mahonia napaulensis DC. ●
51957 Berberis nemorosa C. K. Schneid.;林地小檗;Grovy Barberry ●
51958 Berberis nepalensis Spreng. = Mahonia japonica (Thunb.) DC. ●
51959 Berberis nervosa Pursh; 长叶小檗; Cascade Oregon Grape,Longleaf Holly Grape ●☆
51960 Berberis nervosa Pursh var. mendocinensis Roof = Berberis nervosa Pursh ●☆
51961 Berberis nevinii A. Gray;内文小檗●☆
51962 Berberis nevinii A. Gray var. haematocarpa (Wooton) L. D. Benson = Berberis haematocarpa Wooton ●☆
51963 Berberis nigricans Kuntze;变黑小檗;Blackish-fruited Barberry ●☆
51964 Berberis nilghiriensis Ahrendt;尼尔吉里小檗;Nilghiri Barberry ●☆
51965 Berberis nitens (C. K. Schneid.) Laferr. = Mahonia nitens C. K. Schneid. ●
51966 Berberis notabilis C. K. Schneid.; 杂种显著小檗; Notable Barberry ●☆
51967 Berberis nullinervis T. S. Ying; 无脉小檗; Enervious Barberry,Veinless Barberry ●
51968 Berberis nummularia Bunge;铜钱叶小檗(红果小檗);Money-leaved Barberry ●
51969 Berberis nummularia Bunge var. pyrocarpa C. K. Schneid.;红果铜钱叶小檗;Red-fruit Money-leaved Barberry ●☆
51970 Berberis nummularia Bunge var. schrenkiana C. K. Schneid. =

Berberis farreri Ahrendt ●
51971 Berberis nummularia Bunge var. schrenkiana C. K. Schneid. = Berberis iliensis Popov ●
51972 Berberis nummularia Bunge var. sinica C. K. Schneid. = Berberis jamesiana Forrest et W. W. Sm. ●
51973 Berberis nummularia Bunge var. szovitziana C. K. Schneid. = Berberis densiflora Boiss. et Buhse ●☆
51974 Berberis nutans Linden, Planch. et Sprague;俯垂小檗;Nodding Barberry ●☆
51975 Berberis nutanticarpa C. Y. Wu ex S. Y. Bao;垂果小檗;Dropin-fruit Barberry, Nutant-fruited Barberry ●
51976 Berberis oblanceifolia C. M. Hu;石门小檗;Shimen Barberry ●
51977 Berberis oblonceolata (C. K. Schneid.) Ahrendt = Berberis prattii C. K. Schneid. ●
51978 Berberis oblonga (Regel) C. K. Schneid.;长圆叶小檗(长圆果小檗);Oblong-leaved Barberry ●☆
51979 Berberis obovatifolia T. S. Ying;裂瓣小檗;Obovate-leaved Barberry ●
51980 Berberis oiwakensis (Hayata) Laferr. = Mahonia oiwakensis Hayata ●
51981 Berberis orientalis C. K. Schneid.;东方小檗;Oriental Barberry ●☆
51982 Berberis oritrepha C. K. Schneid.;血珠小檗(黄檗刺,三颗针,显脉小檗);Bloodbead Barberry, Distinct-veined Barberry ●
51983 Berberis orthobotrys Bien. ex Aitch.;直序小檗;Straight-raceme Barberry ●☆
51984 Berberis orthobotrys Bien. ex Aitch. subsp. capitata Jafri;头状直序小檗●☆
51985 Berberis orthobotrys Bien. ex Aitch. var. canescens Ahrendt;灰毛直总状花序小檗;Canescent Straight-raceme Barberry ●☆
51986 Berberis orthobotrys Bien. ex Aitch. var. conwayi Ahrendt;康威直总状花序小檗;Conway Straight-raceme Barberry ●☆
51987 Berberis orthobotrys Bien. ex Aitch. var. rubicunda Ahrendt;深红直总状花序小檗;Dark-red Straight-raceme Barberry ●☆
51988 Berberis orthobotrys Bien. ex Aitch. var. rupestris Ahrendt;岩生直总状花序小檗;Rock-loving Straight-raceme Barberry ●☆
51989 Berberis orthobotrys Bien. ex Aitch. var. sinthanensis Ahrendt;新坦直总状花序小檗;Sintan Straight-raceme Barberry ●☆
51990 Berberis osmastonii Dunn;奥斯马斯顿小檗;Osmaston Barberry ●☆
51991 Berberis ottawensis C. K. Schneid.;渥太华小檗(杂种紫叶小檗);Ottawa Barberry ●☆
51992 Berberis ottawensis C. K. Schneid. var. purpurea C. K. Schneid.;紫叶渥太华小檗;Purple-leaved Ottawa Barberry ●☆
51993 Berberis ovalifolia Rusby;广椭圆形叶小檗;Oval-leaved Barberry ●☆
51994 Berberis oxoniensis Ahrendt;牛津小檗;Oxford Barberry ●☆
51995 Berberis pachakshirensis (Ahrendt) Laferr. = Mahonia polyodonta Fedde ●
51996 Berberis pachyacantha Bien. ex Koehne;粗刺小檗;Thick-spiny Barberry ●☆
51997 Berberis pallens Franch.;淡色小檗(苍白小檗);Pallid Barberry ●
51998 Berberis pallida Hartw. = Mahonia pallida (Hartw.) Fedde ●☆
51999 Berberis paniculata Juss. ex DC.;圆锥花序小檗;Paniculate Barberry ●☆
52000 Berberis panlanensis Ahrendt = Berberis sanguinea Franch. ●
52001 Berberis panxianensis J. Y. Hsiao et S. Z. Ho;盘县小檗●
52002 Berberis papillifera (Franch.) Koehne;云南乳秃小檗(乳秃小檗);Yunnan Papilliferous Barberry ●
52003 Berberis papillosa Benoist;乳秃小檗(多乳秃小檗);Papilliferous Barberry, Papillose Barberry ●
52004 Berberis papillosa Benoist var. aequatorialis Ahrendt;赤道乳秃小檗;Equatorial Papillose Barberry ●☆
52005 Berberis papillosa Benoist var. opacifolia Ahrendt;暗淡乳秃小檗;Opaque-leaved Papillose Barberry ●☆
52006 Berberis parapruinosa T. S. Ying;拟粉叶小檗(似粉叶小檗);Parapruinose Barberry ●
52007 Berberis paraspecta Ahrendt;鸡脚连;Paraspectacular Barberry ●
52008 Berberis para-virescens Ahrendt;拟变绿小檗●☆
52009 Berberis parisepala Ahrendt;等萼小檗(南布小檗);Equal-sepal Barberry, Nambu Everest Barberry ●
52010 Berberis parkeriana C. K. Schneid.;帕克小檗;Parker Barberry ●☆
52011 Berberis parodii Job;帕罗德小檗;Parod Barberry ●☆
52012 Berberis parsonsii C. K. Schneid.;帕森斯小檗;Parsons Barberry ●☆
52013 Berberis parviflora Lindl.;南美小花小檗;Barberry ●☆
52014 Berberis parvifolia C. K. Schneid.;小叶小檗●
52015 Berberis parvifolia Sprague = Berberis wilsonae Hemsl. ●
52016 Berberis paucidentata Rusby;拟疏齿小檗●☆
52017 Berberis paucijuga (C. Y. Wu ex S. Y. Bao) Laferr. = Mahonia paucijuga C. Y. Wu ex S. Y. Bao ●
52018 Berberis pavoniana Ahrendt;帕丰小檗;Pavon Barberry ●☆
52019 Berberis pearcei Phil.;皮尔斯小檗;Pearce Barberry ●☆
52020 Berberis pectinata Hieron.;篦齿小檗;Pectinate Barberry ●☆
52021 Berberis pectinocraspedon C. Y. Wu ex S. Y. Bao;疏齿小檗(梳边小檗);Few-toothed Barberry, Pectinato-edge Barberry ●
52022 Berberis peruviana G. Schellenb.;秘鲁小檗;Peru Barberry ●☆
52023 Berberis petiolaris Wall. ex G. Don;具叶柄小檗;Petioled Barberry ●☆
52024 Berberis petiolaris Wall. ex G. Don = Berberis pachyacantha Bien. ex Koehne ●☆
52025 Berberis petiolaris Wall. ex G. Don var. garhwalana Ahrendt;加瓦尔具叶柄小檗;Garhwal Petioled Barberry ●☆
52026 Berberis petitiana C. K. Schneid.;帕蒂小檗;Petit Barberry ●☆
52027 Berberis petitiana C. K. Schneid. = Berberis holstii Engl. ●☆
52028 Berberis petrogena C. K. Schneid.;岩生小檗;Rocky Barberry ●
52029 Berberis phanera C. K. Schneid.;显脉小檗(近革叶小檗,瓦钦雪山小檗,雪山小檗);Delavay Barberry, Distinctvein Barberry, Obvious Barberry, Subcoriaceous Barberry, Wachin Delavy Barberry ●
52030 Berberis phanera C. K. Schneid. = Berberis delavayi C. K. Schneid. ●
52031 Berberis philippii Ahrendt;菲利普小檗;Philip Barberry ●☆
52032 Berberis photiniifolia C. M. Hu;石楠叶小檗;Photinia-leaved Barberry ●
52033 Berberis phyllacantha Rusby;刺叶小檗;Spiny-leaved Barberry ●☆
52034 Berberis pichinchensis Turcz.;皮钦查小檗;Pichincha Barberry ●☆
52035 Berberis pilosifolia Ahrendt;疏柔毛小檗;Pilose-leaved Barberry ●☆
52036 Berberis pindilicensis Hieron.;平迪利克小檗;Pindilic Barberry ●☆
52037 Berberis pingbienensis S. Y. Bao;屏边小檗;Pingbian Barberry ●
52038 Berberis pingjiangensis Q. L. Chen et B. M. Yang = Berberis virgetorum C. K. Schneid. ●
52039 Berberis pingshanensis W. C. Sung et P. K. Hsiao;屏山小檗(大

刺黄连,三颗针);Pingshan Barberry ●
52040 Berberis pingwuensis T. S. Ying;平武小檗;Pingwu Barberry ●
52041 Berberis pinnata Banks ex DC. = Berberis aquifolium Nutt. ●☆
52042 Berberis pinnata Buch ex DC. = Berberis nervosa Pursh ●☆
52043 Berberis pinnata Kunth = Berberis moranensis Schult. f. ●☆
52044 Berberis pinnata Lag. = Mahonia pinnata (Lag.) Fedde ●☆
52045 Berberis pinnata Roxb. = Mahonia japonica (Thunb.) DC. ●
52046 Berberis pinnata Sessé et Moc.;加州小檗;Californian Barberry, Cluster Holly Grape ●☆
52047 Berberis platyphylla (Ahrendt) Ahrendt;阔叶小檗;Broad-leaved Barberry ●
52048 Berberis podophylla C. K. Schneid.;秘鲁具柄叶小檗;Peru Petioled Barberry ●☆
52049 Berberis poiretii C. K. Schneid.;细叶小檗(北常山,波氏小檗,刺黄柏,刺溜溜,大叶小檗,刀口药,东北小檗,二籽针雀,狗奶子,红狗奶子,黄芦木,泡小檗,日本小檗,三颗针,山石榴,酸狗奶子,土常山,小檗,针雀,子檗);Biseminal Poiret Barberry, Poiret Barberry ●
52050 Berberis poiretii C. K. Schneid. var. biseminalis P. Y. Li = Berberis poiretii C. K. Schneid. ●
52051 Berberis poluninii Ahrendt;波卢宁小檗;Polunin Barberry ●☆
52052 Berberis polyantha Hemsl.;多花小檗(刺黄花,刺黄芩,三颗针);Chromeflower Barberry, Chromeflowered Barberry ●
52053 Berberis polyantha Hemsl. var. oblanceolata C. K. Schneid. = Berberis prattii C. K. Schneid. ●
52054 Berberis polymorpha Phil.;多型小檗;Polymorphous Barberry ●☆
52055 Berberis polyodonta (Fedde) Laferr. = Mahonia polyodonta Fedde ●
52056 Berberis polypetala Phil.;多瓣小檗;Many-petal Barberry ●☆
52057 Berberis pomensis (Ahrendt) Laferr. = Mahonia napaulensis DC. ●
52058 Berberis potaninii Maxim.;少齿小檗(岷江小檗);Liechtenstein Barberry, Minjiang Barberry, Potanin Barberry ●
52059 Berberis praecipua C. K. Schneid.;优秀小檗;Excellent Barberry ●
52060 Berberis praecipua C. K. Schneid. var. major Ahrendt = Berberis cavaleriei H. Lév. ●
52061 Berberis prainiana Schneid. ex Stapf = Berberis sublevis W. W. Sm. ●
52062 Berberis prattii C. K. Schneid.;短锥花小檗(垂花序普拉特小檗,刺黄连,倒披针叶小檗,拟变绿小檗,拟多花小檗,普拉特小檗,三颗针,弯梗普拉特小檗);Curved-stalk Pratt Barberry, Oblanceolate-leaved Barberry, Pendulous-panicle Pratt Barberry, Pratt Barberry ●
52063 Berberis prattii C. K. Schneid. var. laxipendula Ahrendt = Berberis prattii C. K. Schneid. ●
52064 Berberis prattii C. K. Schneid. var. recurvata C. K. Schneid. = Berberis prattii C. K. Schneid. ●
52065 Berberis prolifica Pittier;多育小檗;Prolific Barberry ●☆
52066 Berberis provincialis Audib. ex Schrad.;省区小檗;Provincial Barberry ●☆
52067 Berberis provincialis Audib. ex Schrad. var. serrata (Koehne) C. K. Schneid. = Berberis serrata Koehne ●☆
52068 Berberis pruinocarpa C. Y. Wu ex S. Y. Bao;粉果小檗;Pink-fruit Barberry, Pruinose-fruit Barberry ●
52069 Berberis pruinosa Franch.;粉叶小檗(大黄连,大黄连刺,点叶粉叶小檗,短柄粉叶小檗,黄连刺,混杂小檗,鸡脚刺,具点粉叶小檗,三颗针,石妹刺);Hibbered Barberry, Hollygreen Barberry, Holly-green Barberry, Pruinose Barberry, Punctate Pruinose Barberry, Short-pedicel Pruinose Barberry, Shortstalk Pruinose Barberry ●
52070 Berberis pruinosa Franch. var. barresiana Ahrendt;易江粉叶小檗(细柄粉叶小檗,易门小檗);Barres Pruinose Barberry, Slender-pedicel Pruinose Barberry ●
52071 Berberis pruinosa Franch. var. brevifolia Ahrendt;短叶粉叶小檗;Short-leaved Pruinose Barberry ●
52072 Berberis pruinosa Franch. var. brevipes Ahrendt = Berberis pruinosa Franch. ●
52073 Berberis pruinosa Franch. var. centiflora (Diels) Hand. -Mazz. = Berberis centiflora Diels ●
52074 Berberis pruinosa Franch. var. longifolia Ahrendt;长粉叶小檗;Long-leaved Pruinose Barberry ●
52075 Berberis pruinosa Franch. var. punctata Ahrendt = Berberis pruinosa Franch. ●
52076 Berberis pruinosa Franch. var. serratifolia Ahrendt;锯齿叶粉叶小檗;Serrate-leaved Pruinose Barberry ●
52077 Berberis pruinosa Franch. var. tenuipes Ahrendt = Berberis pruinosa Franch. var. barresiana Ahrendt ●
52078 Berberis pruinosa Franch. var. viridifolia C. K. Schneid. = Berberis wangii C. K. Schneid. ●
52079 Berberis prumosa Franch. var. centiflora (Diels) Hand. -Mazz. = Berberis centiflora Diels ●
52080 Berberis pseudoamoena T. S. Ying;假美丽小檗;False Pleasing Barberry ●
52081 Berberis pseudoilicifolia Skottsb.;假冬青叶小檗;False Holly-leaved Barberry ●☆
52082 Berberis pseudospinulosa Job;假小刺小檗;False Spinulose Barberry ●☆
52083 Berberis pseudotibetica C. Y. Wu ex S. Y. Bao;假藏小檗(假芷小檗);False Xizang Barberry ●
52084 Berberis pseudoumbellata R. Parker;假伞形花小檗;False Umbellate Barberry ●☆
52085 Berberis psilopoda Turcz.;秃柄小檗;Naked-pedicel Barberry ●☆
52086 Berberis pubescens Pamp.;柔毛小檗;Pubescent Barberry ●
52087 Berberis pulangensis T. S. Ying;普兰小檗;Pulan Barberry ●
52088 Berberis pumila Greene = Mahonia pumila (Greene) Fedde ●☆
52089 Berberis purdomii C. K. Schneid.;延安小檗;Purdom Barberry ●
52090 Berberis pycnophylla Bien. ex C. K. Schneid. = Berberis densiflora Boiss. et Buhse ●☆
52091 Berberis qiaojiaensis S. Y. Bao;巧家小檗;Qiaojia Barberry ●
52092 Berberis quelpaertensis Nakai;奎帕特小檗;Quelpaert Barberry ●☆
52093 Berberis quernseyensis Ahrendt;奎恩西小檗;Guernsey Barberry ●☆
52094 Berberis quindiuensis Kunth;金迪奥小檗;Quindio Barberry ●☆
52095 Berberis racemulosa T. S. Ying;短序小檗;Racemulose Barberry ●
52096 Berberis rariflora Lechl.;少花小檗;Rare-flowered Barberry ●☆
52097 Berberis rechingeri C. K. Schneid.;雷琴格小檗;Rechinger Barberry ●☆
52098 Berberis rectinervia Rusby;直脉小檗;Straight-veined Barberry ●☆
52099 Berberis recurvata Ahrendt = Berberis sargentiana C. K. Schneid. ●
52100 Berberis regeliana Koehne ex C. K. Schneid.;雷杰尔小檗;Regel Barberry ●☆
52101 Berberis regleriana Notcutt;雷格勒小檗;Regler Barberry ●☆
52102 Berberis rehderiana C. K. Schneid.;雷德小檗;Rehder Barberry ●☆
52103 Berberis reicheana C. K. Schneid.;赖歇小檗;Reiche Barberry ●☆

52104 Berberis repens Lindl.;匍匐小檗;Creeping Holly Grape, Creeping Oregon Grape ●☆

52105 Berberis replicata W. W. Sm.;卷叶小檗(折转小檗);Curlyleaf Barberry, Curly-leaved Barberry ●

52106 Berberis replicata W. W. Sm. var. dispar Ahrendt = Berberis griffithiana C. K. Schneid. var. pallida (Hook. f. et Thomson) D. F. Chamb. et C. M. Hu ●

52107 Berberis replicata W. W. Sm. var. dispar Ahrendt = Berberis griffithiana C. K. Schneid. ●

52108 Berberis reticulata Bijh.;网脉小檗;Net-veined Barberry ●

52109 Berberis reticulinervis (C. Y. Wu ex S. Y. Bao) Laferr. = Mahonia retinervis P. K. Hsiao et Y. S. Wang ●

52110 Berberis reticulinervis T. S. Ying;芒康小檗;Reticulinerve Barberry ●

52111 Berberis reticulinervis T. S. Ying var. brevipedicellata T. S. Ying;无梗小檗;Pedicelless Reticulinerve Barberry ●

52112 Berberis retinervis (P. K. Hsiao et Y. S. Wang) Laferr. = Mahonia retinervis P. K. Hsiao et Y. S. Wang ●

52113 Berberis retinervis Triana et Planch.;哥伦比亚网脉小檗;Colombia Net-veined Barberry ●☆

52114 Berberis retusa T. S. Ying;心叶小檗●

52115 Berberis rigida Hieron.;坚硬小檗;Rigid Barberry ●☆

52116 Berberis rigidifolia Kunth;硬叶小檗;Rigid-leaved Barberry ●☆

52117 Berberis rockii Ahrendt;摩顶山小檗;Modingshan Barberry, Rock Barberry ●

52118 Berberis rotundifolia Poepp. et Endl.;圆叶小檗;Round-leaved Barberry ●☆

52119 Berberis royleana Ahrendt;罗伊尔小檗;Royle Barberry ●☆

52120 Berberis rubrostilla Chitt.;威斯利红果小檗(红果小檗);Wisley Coral-red-fruited Barberry ●☆

52121 Berberis rubrostilla Chitt. var. chealii Cheal;奇尔小檗;Cheal's Barberry ●☆

52122 Berberis rubrostilla Chitt. var. crawleyensis Ahrendt;克劳利小檗;Crawley Barberry ●☆

52123 Berberis rubrostilla Chitt. var. erecta Hort. = Berberis suberecta Ahrendt ●

52124 Berberis rufescens Ahrendt;红茎小檗;Dark-red-stem Barberry ●☆

52125 Berberis rusbyana Ahrendt;鲁斯比小檗;Rusby Barberry ●☆

52126 Berberis ruscifolia Lam.;假叶树叶小檗;Ruscus-leaved Barberry ●☆

52127 Berberis sabulicola T. S. Ying;砂生小檗;Sabulicolous Barberry ●

52128 Berberis salicaria Fedde;柳叶小檗(季氏小檗,毛脉小檗,毛叶小檗);Girald Barberry, Mild-leaved Barberry, Shaanxi Barberry, Willow-leaved Barberry ●

52129 Berberis salweenensis (Ahrendt) Laferr. = Mahonia napaulensis DC. ●

52130 Berberis sanguinea Franch.;血红小檗(潘兰小檗);Panlanshan Barberry, Red-pedicel Barberry ●

52131 Berberis sargentiana C. K. Schneid.;刺黑珠(刺黄连,黑石珠,狼把针,毛叶小檗,三颗针,铜针刺,凸叶小檗,鲜黄小檗,相仿小檗,小黄柏,寻骨风);Convex-leaved Barberry, Sargent Barberry, Sargent's Barberry, Similar Barberry ●

52132 Berberis saxicola Lechl.;秘鲁岩生小檗;Peru Rocky Barberry ●☆

52133 Berberis saxorum Ahrendt;厄瓜多尔岩生小檗;Ecuador Rocky Barberry ●☆

52134 Berberis schneideri Rehder = Berberis amoena Dunn ●

52135 Berberis schneideriana Ahrendt = Berberis wangii C. K. Schneid. ●

52136 Berberis schochii (C. K. Schneid. ex Hand.-Mazz.) Laferr. = Mahonia nitens C. K. Schneid. ●

52137 Berberis schwerinii C. K. Schneid.;施威林小檗;Sehwerin Barberry ●☆

52138 Berberis sellowiana C. K. Schneid.;塞劳小檗;Sellow Barberry ●☆

52139 Berberis serotina Lange = Berberis chinensis Poir. ●

52140 Berberis serrata Koehne;锯齿小檗;Serrate Barberry ●☆

52141 Berberis serrato-dentata Lechl.;锯齿状齿缺小檗;Serrate-toothed Barberry ●☆

52142 Berberis setigrifolia Ahrendt;刺毛叶小檗;Bristly-leaved Barberry ●☆

52143 Berberis setosa (Gagnep.) Laferr. = Mahonia setosa Gagnep. ●

52144 Berberis shenii (Chun) Laferr. = Mahonia shenii Chun ●

52145 Berberis shenxiensis Ahrendt;陕西小檗;Shaanxi Barberry ●

52146 Berberis sheridaniana (C. K. Schneid.) Laferr. = Mahonia sheridaniana C. K. Schneid. ●

52147 Berberis sherriffii Ahrendt;短苞小檗;Sherriff Barberry ●

52148 Berberis siamensis (Takeda) Laferr. = Mahonia duclouxiana Gagnep. ●

52149 Berberis sibirica Pall.;西伯利亚小檗(刺叶小檗,西北小檗,小檗,小五台山小檗);Siberian Barberry, Xiaowutaishan Barberry ●

52150 Berberis sichuanica T. S. Ying;四川小檗;Sichuan Barberry ●

52151 Berberis sieboldii Miq.;西氏小檗(伏牛花,西保德小檗,席氏小檗);Siebold Barberry, Siebold's Barberry ●

52152 Berberis sikkimensis (C. K. Schneid.) Ahrendt = Berberis sikkimensis Hemsl. ●

52153 Berberis sikkimensis Hemsl.;锡金小檗(贝利锡金小檗,光滑锡金小檗,锡金无毛小檗);Bailey Sikkim Barberry, Sikkim Barberry, Smoothest Sikkim Barberry ●

52154 Berberis sikkimensis Hemsl. var. baileyi Ahrendt = Berberis sikkimensis Hemsl. ●

52155 Berberis sikkimensis Hemsl. var. glabramea Ahrendt = Berberis sikkimensis Hemsl. ●

52156 Berberis silva-taroucana C. K. Schneid.;华西小檗(大山黄刺,黄包刺棵子,三颗针);Taroucan Barberry, West China Barberry ●

52157 Berberis silvicola C. K. Schneid.;兴山小檗;Xingshan Barberry ●

52158 Berberis silvicola C. K. Schneid. var. angustata Ahrendt = Berberis atrocarpa C. K. Schneid. ●

52159 Berberis simonsii Ahrendt;西蒙斯小檗;Simons Barberry ●☆

52160 Berberis simulans C. K. Schneid. = Berberis sargentiana C. K. Schneid. ●

52161 Berberis sinensis DC. = Berberis chinensis Poir. ●

52162 Berberis sinensis Desf. = Berberis chinensis Poir. ●

52163 Berberis sinensis Desf. var. elegans Franch. = Berberis amoena Dunn ●

52164 Berberis sinensis Desf. var. paphlagonia C. K. Schneid. = Berberis chinensis Poir. var. paphlagonica (C. K. Schneid.) Ahrendt ●☆

52165 Berberis sinensis Desf. var. typica Franch. = Berberis lecomtei C. K. Schneid. ●

52166 Berberis smithiana Sprague ex Ahrendt;史密斯小檗;Smith Barberry ●☆

52167 Berberis solutiflora Ahrendt;离花小檗;Solut-flowered Barberry ●

52168 Berberis sonnei (Abrams) McMinn = Berberis repens Lindl. ●☆

52169 Berberis soulieana C. K. Schneid.;康定小檗(刺黄柏,刺黄檗,刺黄连,假蚝猪刺,假豪猪刺,老鼠刺,猫刺小檗,猫儿刺,拟豪猪刺,排骨筋,三颗针,铜针刺,土黄柏,狭叶小檗);Golden

Barberry, Hedge Barberry, Narrow-leaved Barberry, Procupine Barberry, Rosemary Barberry, Soulie Barberry ●

52170 Berberis soulieana C. K. Schneid. var. paucinervata Ahrendt;少脉康定小檗(少脉假豪猪刺);Few-veined Soulie Barberry ●

52171 Berberis soulieana C. K. Schneid. var. paucinervata Ahrendt = Berberis soulieana C. K. Schneid. ●

52172 Berberis spaethii C. K. Schneid. ;斯佩斯小檗;Spaeth Barberry ●☆

52173 Berberis spathulata C. K. Schneid. = Berberis chinensis Poir. ●

52174 Berberis sphalera Fedde = Berberis potaninii Maxim. ●

52175 Berberis spinosissima (Reiche) Ahrendt;多刺小檗;Most-spiny Barberry ●☆

52176 Berberis spinulosa A. St. -Hil. ;小刺小檗;Spinulose Barberry ●☆

52177 Berberis spinulosa Griseb. = Berberis pseudospinulosa Job ●☆

52178 Berberis spraguei Ahrendt;滇西小檗;Sprague Barberry ●

52179 Berberis spraguei Ahrendt var. pedunculata Ahrendt = Berberis virescens Hook. f. ●

52180 Berberis spruceana (C. K. Schneid.) Ahrendt;斯普鲁斯小檗;Spruce Barberry ●☆

52181 Berberis stapfiana C. K. Schneid. = Berberis wilsonae Hemsl. ●

52182 Berberis stearnii Ahrendt;中甸小檗(斯蒂恩小檗,斯泰恩小檗);Stearn Barberry ●

52183 Berberis stenophylla Hance = Berberis soulieana C. K. Schneid. ●

52184 Berberis stenophylla Lindl. ;狭叶小檗;Rosemary Barberry ●☆

52185 Berberis stenophylla Lindl. 'Corallina Compacta';致密珊瑚狭叶小檗●☆

52186 Berberis stenophylla Lindl. var. irwinii Hort. = Berberis irwinii Bijh. ●☆

52187 Berberis stenostachya Ahrendt; 短梗小檗; Narrow-spiked Barberry, Stwnostachyous Barberry ●

52188 Berberis stewartiana Jafri;斯图小檗●☆

52189 Berberis stiebritziana C. K. Schneid. ;西南小檗●

52190 Berberis stiebritziana C. K. Schneid. = Berberis approximata Sprague ●

52191 Berberis stolonifera Koehne et E. L. Wolf; 匍匐茎小檗; Stoloniferous Barberry ●☆

52192 Berberis stuebelii Hieron. ;斯图伯尔小檗;Stuebel Barberry ●☆

52193 Berberis subacuminata C. K. Schneid. ;亚尖小檗(薄叶小檗,尖叶小檗,近尖叶小檗,近渐尖小檗,亚尖叶小檗);Subacuminate Barberry, Subantarctic Barberry ●

52194 Berberis subantarctica Gand. ; 近南极小檗; Subantarctic Barberry ●☆

52195 Berberis subcaulialata C. K. Schneid. = Berberis wilsonae Hemsl. ●

52196 Berberis subcoriacea Ahrendt = Berberis phanera C. K. Schneid. ●

52197 Berberis subculialata C. K. Schneid. var. guhtzunica Ahrendt = Berberis wilsonae Hemsl. var. guhtzunica (Ahrendt) Ahrendt ●

52198 Berberis suberecta Ahrendt;近直立小檗;Suberect Barberry ●

52199 Berberis subholophylla C. Y. Wu ex S. Y. Bao;近全缘叶小檗(近缘小檗,近缘叶小檗);Subentire-leaved Barberry ●

52200 Berberis subimbricata (Chun et F. Chun) Laferr. = Mahonia subimbricata Chun et F. Chun ●

52201 Berberis sublevis W. W. Sm. ;近光滑小檗(大叶近光滑小檗,潞西近光滑小檗); Laege-leaved Nearsmooth Barberry, Luxi Nearsmooth Barberry, Nearsmooth Barberry, Subpsilate Barberry ●

52202 Berberis sublevis W. W. Sm. var. exquista Ahrendt = Berberis sublevis W. W. Sm. ●

52203 Berberis sublevis W. W. Sm. var. grandifolia C. K. Schneid. = Berberis sublevis W. W. Sm. ●

52204 Berberis sublevis W. W. Sm. var. macrocarpa (Hook. f. et Thomson) Ahrendt = Berberis sublevis W. W. Sm. ●

52205 Berberis sublevis W. W. Sm. var. microcarpa (Hook. f. et Thomson) Ahrendt; 小果近光滑小檗; Small-fruit Nearsmooth Barberry ●

52206 Berberis subpteroclada Ahrendt = Berberis griffithiana C. K. Schneid. ●

52207 Berberis subpteroclada Ahrendt var. impar Ahrendt = Berberis griffithiana C. K. Schneid. ●

52208 Berberis subpteroclada Ahrendt var. minoripes Ahrendt;短柄微翼枝小檗;Short-pedicel Slightly Wing-branched Barberry ●☆

52209 Berberis subscuminata C. K. Schneid. ;近见光小檗●

52210 Berberis subsessiliflora Pamp. ; 近无柄花小檗; Subsessile-flowered Barberry ●☆

52211 Berberis subtriplinervis Franch. = Mahonia gracilipes (Oliv.) Fedde ●

52212 Berberis sulcata Hort. = Berberis vulgaris L. var. sulcata Ahrendt ●☆

52213 Berberis swaseyi Buckley ex M. J. Young;斯韦齐小檗●☆

52214 Berberis taliensis C. K. Schneid. ;大理小檗;Dali Barberry ●

52215 Berberis tarokoensis S. Y. Lu et Yuen P. Yang;太鲁阁小檗●

52216 Berberis taronensis Ahrendt;独龙小檗;Dulong Barberry ●

52217 Berberis taronensis Ahrendt var. trimensis Ahrendt = Berberis griffithiana C. K. Schneid. var. pallida (Hook. f. et Thomson) D. F. Chamb. et C. M. Hu ●

52218 Berberis taylori Ahrendt = Berberis gyalaica Ahrendt ex F. Br. ●

52219 Berberis taylorii Ahrendt;黑龙小檗●

52220 Berberis temolaica Ahrendt;特莫拉小檗(粉叶小檗,林芝小檗);Temola Barberry ●☆

52221 Berberis temolaica Ahrendt var. artisepala Ahrendt = Berberis temolaica Ahrendt ●☆

52222 Berberis tenuipeticellata T. S. Ying; 细梗小檗; Slenderstalk Barberry ●

52223 Berberis thibetica C. K. Schneid. ;藏小檗(西藏小檗);Tibet Barberry, Xizang Barberry ●

52224 Berberis thomsoniana C. K. Schneid. ;汤姆森小檗;Thomson Barberry ●☆

52225 Berberis thunbergii DC. ;日本小檗(刺檗,刺黄柏,黄芦木,三颗针,山石榴,童氏小檗,小檗,子檗);Japan Barberry, Japanese Barberry, Red-leaf Japanese Barberry, Thunberg's Barberry ●

52226 Berberis thunbergii DC. 'Atropurpurea Nana';紫叶矮小檗(矮紫日本小檗);Purple Barberry ●☆

52227 Berberis thunbergii DC. 'Atropurpurea';紫叶小檗(深紫日本小檗); Crimson Pygmy Japanese Barberry, Dark-purple Japanese Barberry, Dwarf Red-leaved Barberry, Red Barberry ●☆

52228 Berberis thunbergii DC. 'Aurea';金叶小檗(黄叶日本小檗); Golden-leaved Japanese Barberry, Gold-leaved Barberry ●☆

52229 Berberis thunbergii DC. 'Bagatelle';巴特勒小檗●☆

52230 Berberis thunbergii DC. 'Bogozam';波沟山小檗●☆

52231 Berberis thunbergii DC. 'Concord';协和小檗;Dwarf Red-leaved Barberry ●☆

52232 Berberis thunbergii DC. 'Crimson Pygmy' = Berberis thunbergii DC. 'Atropurpurea Nana' ●☆

52233 Berberis thunbergii DC. 'Dart's Red Lady';达特红女小檗●☆

52234 Berberis thunbergii DC. 'Erecta'; 直立日本小檗; Erect Japanese Barberry ●☆

52235 Berberis thunbergii DC. 'Golden Ring';金环小檗(金环日本小

檗);Gold Ring Barberry,Japanese Barberry ●☆
52236 Berberis thunbergii DC. 'Helmond Pillar';荷孟柱小檗;Upright Red-leaved Barberry ●☆
52237 Berberis thunbergii DC. 'Red Chief';红首领小檗●☆
52238 Berberis thunbergii DC. 'Red Pillar';红柱小檗●☆
52239 Berberis thunbergii DC. 'Rose Glow';玫红小檗(艳紫日本小檗);Mottled Red-leaved Barberry,Rose Glow Japanese Barberry ●☆
52240 Berberis thunbergii DC. 'Sparkle';火花小檗;Japanese Barberry ●☆
52241 Berberis thunbergii DC. f. erecta Rehder = Berberis thunbergii DC. var. erecta (Rehder) Ahrendt ●☆
52242 Berberis thunbergii DC. var. argenteo-marginata C. K. Schneid.;银边日本小檗;Silver-margined Japanese Barberry ●☆
52243 Berberis thunbergii DC. var. atropurpurea Chenault = Berberis thunbergii DC. 'Atropurpurea' ●☆
52244 Berberis thunbergii DC. var. erecta (Rehder) Ahrendt = Berberis thunbergii DC. 'Erecta' ●☆
52245 Berberis thunbergii DC. var. glabra Franch. = Berberis lecomtei C. K. Schneid. ●
52246 Berberis thunbergii DC. var. maximowiczii (Regel) Regel;马氏日本小檗(马克西莫维奇小檗);Maximowicz Japanese Barberry ●☆
52247 Berberis thunbergii DC. var. minor Rehder;矮日本小檗;Low Japanese Barberry ●☆
52248 Berberis thunbergii DC. var. papillifera Franch. = Berberis papillifera (Franch.) Koehne ●
52249 Berberis thunbergii DC. var. pluriflora Koehne;多花日本小檗;Many-flowered Japanese Barberry ●☆
52250 Berberis thunbergii DC. var. rubrifolia Ahrendt;红叶日本小檗;Red-leaved Japanese Barberry ●☆
52251 Berberis thunbergii DC. var. uniflora Koehne;单花日本小檗;Uniflowered Japanese Barberry ●☆
52252 Berberis tianbaoshanensis S. Y. Bao = Berberis muliensis Ahrendt ●
52253 Berberis tianshuiensis T. S. Ying;天水小檗;Tianshui Barberry ●
52254 Berberis tibetensis Laferr. = Mahonia taronensis Hand. -Mazz. ●
52255 Berberis tibetica C. K. Schneid.;西藏小檗(藏小檗);Xizang Barberry ●
52256 Berberis tikushiensis (Hayata) Laferr. = Mahonia japonica (Thunb.) DC. ●
52257 Berberis tinctoria Lesch.;染用小檗;Dyeing Barberry ●☆
52258 Berberis tischleri C. K. Schneid.;川西小檗(蒂施籂小檗,康定小檗,米林小檗);Elliott Barberry,Tischler Barberry ●
52259 Berberis tischleri C. K. Schneid. var. abbreviata Ahrendt;短序蒂施籂小檗;Short-inflorescence Tischler Barberry ●
52260 Berberis tischleri C. K. Schneid. var. abbreviata Ahrendt = Berberis tischleri C. K. Schneid. ●
52261 Berberis tolimensis Planch. et Linden ex Triana et Planch. = Berberis psilopoda Turcz. ●☆
52262 Berberis tomentosa Ruiz et Pav.;绒毛小檗;Tomentose Barberry ●☆
52263 Berberis tomentulosa Ahrendt;微毛小檗(微绒花小檗);Tomentulose Barberry ●
52264 Berberis tragacanthoides DC.;羊角刺小檗●☆
52265 Berberis triacanthophora Fedde;芒齿小檗(刺齿小檗);Awntooth Barberry, Threespine Barberry, Three-spiny Barberry, Threethorned Barberry ●
52266 Berberis trichiata T. S. Ying;毛序小檗;Hairy Barberry,Trichiate Barberry ●
52267 Berberis trifoliolata Moric.;三小叶小檗;Agarito, Agritos, Algerita,Currant-of-texas ●☆
52268 Berberis trifurca Lindl. et Paxton = Mahonia bodinieri Gagnep. ●
52269 Berberis trifurca Loudon = Mahonia bodinieri Gagnep. ●
52270 Berberis trigona Kuntze ex Poepp. et Endl.;三角小檗;Trigonous Barberry ●☆
52271 Berberis trollii Diels;特罗尔小檗;Troll Barberry ●☆
52272 Berberis truxillensis Turcz.;特鲁希约小檗;Truxillo Barberry ●☆
52273 Berberis tsailunii Laferr. = Mahonia duclouxiana Gagnep. ●
52274 Berberis tsangpoensis Ahrendt;昌波小檗(藏布小檗);Tsangpo Barberry,Zangbu Barberry ●
52275 Berberis tsarica Ahrendt;隐脉小檗;Tsari Barberry ●
52276 Berberis tsarica Ahrendt var. ritangensis Ahrendt;里瑭小檗;Litang Barberry ●
52277 Berberis tsarongensis C. K. Schneid. = Berberis tsarongensis Stapf ●
52278 Berberis tsarongensis C. K. Schneid. var. breviscapa Ahrendt = Berberis breviscapa (Ahrendt) Ahrendt ●☆
52279 Berberis tsarongensis C. K. Schneid. var. megacarpa Ahrendt = Berberis tsarongensis Stapf ●
52280 Berberis tsarongensis C. K. Schneid. var. megacarpa Ahrendt = Berberis lecomtei C. K. Schneid. ●
52281 Berberis tsarongensis Stapf;察瓦龙小檗(大果金果小檗,金果小檗);Chawalong Barberry, Large-fruited Tsarong Barberry, Tsarong Barberry ●
52282 Berberis tschonoskyana Regel;柴氏小檗(柴科诺斯基小檗,黄小檗,四国大叶小檗);Tschonoski Barberry ●
52283 Berberis tsienii T. S. Ying;永思小檗;Tsien's Barberry ●
52284 Berberis turcomanica Kar. ex Ledeb.;土库曼小檗;Turcoman Barberry ●☆
52285 Berberis turcomanica Kar. ex Ledeb. var. buhseana (C. K. Schneid.) Ahrendt;布舍小檗;Bushe Turcoman Barberry ●☆
52286 Berberis turcomanica Kar. ex Ledeb. var. densiflora Rehder = Berberis densiflora Boiss. et Buhse ●☆
52287 Berberis ulcina Hook. f. et Thomson;尤里小檗(条叶小檗);Ulex-like Barberry ●
52288 Berberis umbellata Phil. = Berberis philippii Ahrendt ●☆
52289 Berberis umbellata Wall. ex G. Don;伞形花小檗;Umbellate Barberry ●☆
52290 Berberis umbellata Wall. ex G. Don var. brianii Ahrendt;布赖恩伞形花小檗;Brain Umbellate Barberry ●☆
52291 Berberis umbratica T. S. Ying;阴生小檗;Umbraticolor Barberry, Umbraticous Barberry ●
52292 Berberis uniflora F. N. Wei et Y. G. Wei;单花小檗;Singleflower Barberry ●
52293 Berberis usteriana (C. K. Schneid.) Parker = Berberis jaeschkeana C. K. Schneid. var. usteriana C. K. Schneid. ●☆
52294 Berberis usteriana (C. K. Schneid.) Parker var. apiculata Ahrendt = Berberis apiculata (Ahrendt) Ahrendt ●☆
52295 Berberis valdiviana Phil.;瓦尔蒂夫小檗(瓦尔的夫小檗,智利小檗);Valdiv Barberry ●☆
52296 Berberis valdiviana Phil. var. gracilifolia Ahrendt;狭叶瓦尔蒂夫小檗;Slender-leaved Valdiv Barberry ●☆
52297 Berberis valida (C. K. Schneid.) C. K. Schneid.;强枝小檗(宁远小檗);Strong Barberry,Strong-branched Barberry ●
52298 Berberis validisepala Ahrendt;强萼小檗;Strong-sepal Barberry, Strong-sepaled Barberry ●
52299 Berberis validisepala Ahrendt var. primoglauca Ahrendt;初粉强

萼小檗;Primoglauous Strong-sepal Barberry ●
52300 Berberis vanfleetii C. K. Schneid.;范弗利特小檗;Van Fleet Barberry ●☆
52301 Berberis variiflora C. K. Schneid.;异花小檗;Variable-flowered Barberry ●☆
52302 Berberis veitchii C. K. Schneid.;蓝果小檗(巴东小檗,三颗针);Lanceolate Barberry,Veitch Barberry ●
52303 Berberis veitchiorum Hemsl. et E. H. Wilson = Mahonia polyodonta Fedde ●
52304 Berberis veitchiorum Hemsl. et E. H. Wilson var. kingdon-wardiana (Ahrendt) Laferr. = Mahonia calamicaulis Sparre et C. E. C. Fisch. subsp. kingdon-wardiana (Ahrendt) T. S. Ying et Boufford ●
52305 Berberis venusta C. K. Schneid.;娇媚小檗;Charming Barberry ●☆
52306 Berberis vernae C. K. Schneid.;匙叶小檗(黄檗,黄刺,三颗针,婉娜小檗);Spoonleaf Barberry, Spoon-leaved Barberry, Verna Barberry,Vernal Barberry ●
52307 Berberis vernalis (C. K. Schneid.) Chamb. et C. M. Hu;春小檗;Vernal Barberry,Vernal Ferdinandi-coburg Barberry ●
52308 Berberis verruculosa Hemsl. et E. H. Wilson;疣枝小檗;Verruculose Barberry,Warted Barberry,Warty Barberry ●
52309 Berberis verschaffelti C. K. Schneid.;维尔沙费尔特小檗;Verschaffelt Barberry ●☆
52310 Berberis verticillata Turcz.;轮生小檗;Verticillate Barberry ●☆
52311 Berberis vilmorinii C. K. Schneid.;维尔莫林小檗;Vilmorin Barberry ●☆
52312 Berberis vinifera T. S. Ying;可食小檗;Wine-bearing Barberry ●
52313 Berberis virescens Hook. f.;变绿小檗(长叶小檗,刺黄连,具柄滇西小檗);Pedunculate Sprague Barberry,Virescent Barberry ●
52314 Berberis virescens Hook. f. et Thomson = Berberis virescens Hook. f. ●
52315 Berberis virescens Hook. f. et Thomson var. ignorata (C. K. Schneid.) Ahrendt = Berberis ignorata C. K. Schneid. ●
52316 Berberis virescens Hook. f. et Thomson var. macrocarpa Bean = Berberis para-virescens Ahrendt ●☆
52317 Berberis virescens Hook. f. var. ignorata Ahrendt = Berberis ignorata C. K. Schneid. ●
52318 Berberis virgata Ruiz et Pav.;细枝小檗;Twiggy Barberry ●☆
52319 Berberis virgata Ruiz et Pav. var. huanucensis C. K. Schneid. = Berberis provincialis Audib. ex Schrad. ●☆
52320 Berberis virgetorum C. K. Schneid.;庐山小檗(长叶小檗,刺黄连,黄刺柏,黄疸树,木黄连,平江小檗,三颗针,树黄连,铁篱笆,铁树黄连,土黄柏,土黄檗,土黄连,土水莲,浙江小檗);Lushan Barberry,Pingjiang Barberry,Zhejiang Barberry ●
52321 Berberis vitellina Hieron.;卵黄小檗;Vitelline Barberry ●☆
52322 Berberis vulgaris L.;欧洲小檗(檗木,刺檗,刺黄檗树,欧小檗,普通小檗,小檗);Barbareus, Barbary, Barberry, Barboranne, Berber, Berberis, Common Barberry, Europe Barberry, European Barberry, Guild, Guild Tree, Holy Thorn, Jaunders-berry, Jaunders-tree, Jaundice-berry, Jaundice-tree, Pipperidge Bush, Piprige Piprage, Sowberry, Woodsore, Woodsour ●
52323 Berberis vulgaris L. = Berberis baluchistanica Ahrendt ●☆
52324 Berberis vulgaris L. subsp. australis (Boiss.) Heywood = Berberis hispanica Boiss. et Reut. ●☆
52325 Berberis vulgaris L. var. acutifolia (Prantl) C. K. Schneid.;尖叶欧洲小檗;Acute-leaved Common Barberry ●☆
52326 Berberis vulgaris L. var. aetnensis (C. Presl) Fiori = Berberis aetnensis C. Presl ●☆
52327 Berberis vulgaris L. var. aetnensis (C. Presl) Hook. f. et Thomson = Berberis kunawurensis Royle ●☆
52328 Berberis vulgaris L. var. alba West.;白果欧洲小檗;White-fruied Common Barberry ●☆
52329 Berberis vulgaris L. var. albo-variegata Zabel;白彩欧洲小檗;White-variegated Common Barberry ●☆
52330 Berberis vulgaris L. var. amurensis (Rupr.) Regel = Berberis amurensis Rupr. ●
52331 Berberis vulgaris L. var. atropurpurea Regel;暗紫欧洲小檗;Dark-purple Common Barberry ●☆
52332 Berberis vulgaris L. var. aureo-marginata Regel;金边欧洲小檗;Golden-marginated Common Barberry ●☆
52333 Berberis vulgaris L. var. australis Boiss. = Berberis vulgaris L. subsp. australis (Boiss.) Heywood ●☆
52334 Berberis vulgaris L. var. brachybotrys Hook. f. = Berberis edgeworthiana C. K. Schneid. ●☆
52335 Berberis vulgaris L. var. brachybotrys Hook. f. = Berberis orthobotrys Bien. ex Aitch. ●☆
52336 Berberis vulgaris L. var. dulcis Loudon;甜实欧洲小檗;Sweet Common Barberry ●☆
52337 Berberis vulgaris L. var. emarginata (Willd.) Gordon = Berberis emarginata Willd. ●☆
52338 Berberis vulgaris L. var. enuclea West.;无籽欧洲小檗;Seedless Common Barberry ●☆
52339 Berberis vulgaris L. var. japonica Regel;日欧小檗;Japanese Common Barberry ●
52340 Berberis vulgaris L. var. japonica Regel = Berberis regeliana Koehne ex C. K. Schneid. ●☆
52341 Berberis vulgaris L. var. lutea L'Hér.;黄果欧洲小檗;Yellow-fruited Common Barberry ●☆
52342 Berberis vulgaris L. var. purpurea Hort. = Berberis vulgaris L. var. purpurifolia Ahrendt ●☆
52343 Berberis vulgaris L. var. purpurifolia Ahrendt;紫叶欧洲小檗;Purple-leaved Common Barberry ●☆
52344 Berberis vulgaris L. var. sulcata Ahrendt;具槽欧洲小檗;Sulcate-stem Common Barberry ●☆
52345 Berberis wallichiana DC.;瓦利赫小檗(三颗针,瓦科小檗,小檗)●☆
52346 Berberis wallichiana DC. f. arguta Franch. = Berberis arguta (Franch.) C. K. Schneid. ●
52347 Berberis wallichiana DC. f. parvifolia Franch. = Berberis davidii Ahrendt ●
52348 Berberis wallichiana DC. var. atroviridis ? = Berberis wallichiana DC. ●☆
52349 Berberis wallichiana DC. var. gracilipes Ahrendt = Berberis sublevis W. W. Sm. ●
52350 Berberis wallichiana DC. var. macrocarpa Hook. f. et Thunb. = Berberis sublevis W. W. Sm. ●
52351 Berberis wallichiana DC. var. microcarpa Hook. f. et Thomson = Berberis sublevis W. W. Sm. var. microcarpa (Hook. f. et Thomson) Ahrendt ●
52352 Berberis wallichiana DC. var. pallida Boiss. = Berberis candidula C. K. Schneid. ●
52353 Berberis wallichiana DC. var. pallida Hook. f. et Thomson = Berberis griffithiana C. K. Schneid. var. pallida (Hook. f. et Thomson) D. F. Chamb. et C. M. Hu ●
52354 Berberis walterana Ahrendt;沃尔特小檗;Walter Barberry ●☆

52355 Berberis wangii C. K. Schneid.;西山小檗(绿叶粉叶小檗,王氏小檗,维西小檗);Green-leaved Pruinose Barberry,Schneider Barberry,Wang Barberry,Weixi Barberry ●

52356 Berberis wanhuashanensis Y. J. Zhang;万花山小檗;Wanhuashan Barberry ●

52357 Berberis wardii C. K. Schneid.;沃尔德小檗;Ward Barberry ●☆

52358 Berberis warszewiczii Hieron.;瓦尔斯泽维奇小檗;Waszewicz Barberry ●☆

52359 Berberis watlingtonensis Ahrendt;沃特林顿小檗;Watlington Barberry ●☆

52360 Berberis wawrana C. K. Schneid.;沃拉小檗;Wawra Barberry ●☆

52361 Berberis wazaristanica Ahrendt = Berberis calliobotrys Bien. ex Aitch. ●☆

52362 Berberis weberbaueri C. K. Schneid.;韦伯鲍尔小檗;Weberbauer Barberry ●☆

52363 Berberis weddellii Lechl.;韦德尔小檗;Weddell Barberry ●☆

52364 Berberis weiningensis T. S. Ying;威宁小檗;Weining Barberry ●

52365 Berberis weisiensis C. Y. Wu ex S. Y. Bao;维西小檗;Weixi Barberry ●

52366 Berberis weixinensis C. Y. Bao;威信小檗;Weixin Barberry ●

52367 Berberis wettsteiniana C. K. Schneid.;韦特斯坦小檗;Wettstein Barberry ●☆

52368 Berberis wightiana C. K. Schneid.;怀特小檗;Wight Barberry ●☆

52369 Berberis willeana C. K. Schneid. = Berberis levis Franch. ●

52370 Berberis willeana C. K. Schneid. var. serrulata C. K. Schneid. = Berberis levis Franch. ●

52371 Berberis wilsonae Hemsl.;金花小檗(长果金花小檗,刺黄连,刺黄芩,近翅枝金花小檗,猫儿刺,木里小檗,三颗针,三爪黄连,酸咪咪,土黄连,小檗,小花小檗,小黄连,小黄连刺,小鸡脚黄刺,小三颗针,小叶金花小檗,小叶三颗针,小叶小檗);Gololen Barberry,Mrs. E. H. Wilson's Barberry,Mrs. Wilson's Barberry,Small-leaved Mrs. E. H. Wilson's Barberry,Stapf Barberry,Subcaulialate Mrs. E. H. Wilson's Barberry,Wilson Barberry,Wilson's Barberry ●

52372 Berberis wilsonae Hemsl. var. favosa (W. W. Sm.) Ahrendt;蜂房金花小檗;Favose Mrs. E. H. Wilson's Barberry ●

52373 Berberis wilsonae Hemsl. var. guhtzunica (Ahrendt) Ahrendt;古宗金花小檗(较宽金花小檗,宽叶金花小檗);Broad-leaved Mrs. E. H. Wilson's Barberry,Guhtzun Mrs. E. H. Wilson's Barberry ●

52374 Berberis wilsonae Hemsl. var. latior Ahrendt = Berberis wilsonae Hemsl. var. guhtzunica (Ahrendt) Ahrendt ●

52375 Berberis wilsonae Hemsl. var. parvifolia (Sprague) Ahrendt = Berberis wilsonae Hemsl. ●

52376 Berberis wilsonae Hemsl. var. stapfiana (C. K. Schneid.) C. K. Schneid. = Berberis wilsonae Hemsl. ●

52377 Berberis wilsonae Hemsl. var. subcaulialata (C. K. Schneid.) C. K. Schneid. = Berberis wilsonae Hemsl. ●

52378 Berberis wintonensis Ahrendt;温顿小檗;Winton Barberry ●☆

52379 Berberis wisleyensis Ahrendt;威斯利小檗;Wisley Barberry ●☆

52380 Berberis woomungensis C. Y. Wu et S. Y. Bao;乌蒙小檗;Woomung Barberry,Wumeng Barberry ●

52381 Berberis wuliangshanensis C. Y. Wu ex S. Y. Bao;无量山小檗;Wuliangshan Barberry ●

52382 Berberis wuyiensis C. M. Hu;武夷小檗;Wuyi Barberry ●

52383 Berberis xanthoclada C. K. Schneid.;梵净小檗(黄枝小檗);Yellow-branched Barberry ●

52384 Berberis xanthophloea Ahrendt;黄皮小檗;Yellow-bark Barberry,Yellow-barked Barberry ●

52385 Berberis xanthoxylon Hassk. ex C. K. Schneid.;黄木小檗;Yellow-wood Barberry ●☆

52386 Berberis xanthoxylon Hassk. ex C. K. Schneid. var. junghuhniana Ahrendt;容胡恩黄木小檗;Junghuhn Yellow-wood Barberry ●☆

52387 Berberis xanthoxylon Hassk. ex C. K. Schneid. var. sumatranica Ahrendt;苏门答腊黄木小檗;Sumatra Yellow-wood Barberry ●☆

52388 Berberis xinganensis G. H. Liu et S. Q. Zhou;兴安小檗;Xing'an Barberry ●

52389 Berberis xingwenensis T. S. Ying;兴文小檗;Xingwen Barberry ●

52390 Berberis yuii T. S. Ying;德浚小檗;Yu's Barberry ●

52391 Berberis yunnanensis Franch.;滇小檗(云南小檗);Yunnan Barberry ●

52392 Berberis yunnanensis Franch. var. platyphylla Ahrendt = Berberis platyphylla (Ahrendt) Ahrendt ●

52393 Berberis zabeliana C. K. Schneid.;札伯尔小檗;Zabel Barberry ●☆

52394 Berberis zabeliana C. K. Schneid. = Berberis pachyacantha Bien. ex Koehne ●☆

52395 Berberis zahlbruckneriana C. K. Schneid. = Berberis glomerate Hook. et Arn. var. zahlbruckneriana (C. K. Schneid.) Ahrendt ●☆

52396 Berberis zanlanscianensis Pamp.;沧浪山小檗(鄂西小檗,西南小檗);Canglangshan Barberry,Zanlanscian Barberry ●

52397 Berberis zayulana Ahrendt;察隅小檗;Chayu Barberry,Zayu Barberry ●

52398 Berberis ziyunensis P. K. Hsiao et Z. Y. Li;紫云小檗;Ziyun Barberry ●

52399 Berchemia Neck. = Berchemia Neck. ex DC. (保留属名) ●

52400 Berchemia Neck. ex DC. (1825)(保留属名);勾儿茶属(黄鳝藤属);Hooktea,Supplejack,Supple-jack ●

52401 Berchemia alnifolia H. Lév. = Corylopsis alnifolia (H. Lév.) C. K. Schneid. ●

52402 Berchemia annamensis Pit.;越南勾儿茶(柏子茶,柏子藤);Annam Supplejack,Annam Supple-jack,Vietnam Hooktea ●

52403 Berchemia arisanensis Y. C. Liu et F. Y. Lu;阿里山黄鳝藤;Alishan Supplejack ●

52404 Berchemia axilliflora W. C. Cheng = Berchemia edgeworthii M. A. Lawson ●

52405 Berchemia barbigera C. Y. Wu ex Y. L. Chen;腋毛勾儿茶;Barbiferous Hooktea,Barbiferous Supplejack,Barbiferous Supple-jack ●

52406 Berchemia berchemiifolia (Makino) Koidz. = Berchemiella berchemiifolia (Makino) Nakai ●

52407 Berchemia brachycarpa C. Y. Wu ex Y. L. Chen;短果勾儿茶;Shortfruit Hooktea,Shortfruited Supplejack,Short-fruited Supple-jack ●

52408 Berchemia cavaleriei H. Lév. = Sageretia henryi J. R. Drumm. et Sprague ●

52409 Berchemia chanetii H. Lév. = Sageretia thea (Osbeck) M. C. Johnst. ●

52410 Berchemia compressicarpa D. Fang et C. Z. Gao;扁果勾儿茶;Flat-fruited Supple-jack,Flattened-fruit Supplejack ●

52411 Berchemia congesta S. Moore = Rhamnella franguloides (Maxim.) Weberb. ●

52412 Berchemia discolor (Klotzsch) Hemsl.;异色勾儿茶●☆

52413 Berchemia edgeworthii M. A. Lawson;腋花勾儿茶(小叶勾儿茶,小叶铁包金);Axillaryflower Hooktea,Axillary-flowered Supplejack,Dwarf Hooktea,Dwarf Supplejack,Edgeworth Supple-jack ●

52414 Berchemia fagifolia Koidz. = Berchemia magna (Makino) Koidz. ●

52415 Berchemia fenchifuensis C. M. Wang et S. Y. Lu;奋起湖黄鳝藤●

52416 Berchemia flavescens (Wall. ex Roxb.) Brongn.;黄背勾儿茶(大鸭公藤,大叶甜果子,黄背勾儿藤,金背黄鳝藤,牛儿藤,牛儿膝,石罗藤,石萝藤,甜茶);Flavescet Hooktea, Flavescet Supplejack, Flavescet Supple-jack ●

52417 Berchemia flavescens Wall. = Berchemia flavescens (Wall. ex Roxb.) Brongn. ●

52418 Berchemia floribunda (Wall. ex Roxb.) Brongn.;多花勾儿茶(鼻朴子,扁担果,扁担藤,铳子藤,厝箕藤,大黄鳝藤,繁花枣,勾儿草,勾儿茶,勾勾茶,黑龙串筋,花眉跳架,黄鳝藤,金刚藤,老鼠藤,牛鼻角秧,牛鼻圈,牛鼻拳,牛儿藤,牛耳藤,牛脔子,铁包金,土黄耆,乌梢蛇,熊柳藤,羊母锁,皱皮草,皱纱皮);Girald's Supplejack, Manyflower Hooktea, Manyflower Supplejack, Multiflowered Supple-jack, Racemose Supplejack ●

52419 Berchemia floribunda (Wall. ex Roxb.) Brongn. var. megalophylla Schneid. = Berchemia floribunda (Wall. ex Roxb.) Brongn. ●

52420 Berchemia floribunda (Wall. ex Roxb.) Brongn. var. oblongifolia Y. L. Chen et P. K. Chou;矩叶勾儿茶;Oblong Hooktea, Oblong Supplejack ●

52421 Berchemia floribunda (Wall.) Brongn. = Berchemia floribunda (Wall. ex Roxb.) Brongn. ●

52422 Berchemia floribunda Wall. = Berchemia floribunda (Wall. ex Roxb.) Brongn. ●

52423 Berchemia formosana C. K. Schneid.;台湾勾儿茶(台湾黄鳝藤,台湾黄藤);Formosa Supple Jack, Formosan Supple-jack, Taiwan Hooktea, Taiwan Supplejack, Taiwan Supple-jack ●

52424 Berchemia giraldiana C. K. Schneid. = Berchemia floribunda (Wall. ex Roxb.) Brongn. ●

52425 Berchemia hirtella H. T. Tsai et K. M. Feng;大果勾儿茶(景东蛇藤);Bigfruit Hooktea, Bigfruit Supplejack, Big-fruited Supple-jack ●

52426 Berchemia hirtella H. T. Tsai et K. M. Feng var. glabrescens C. Y. Wu ex Y. L. Chen;大老鼠耳;Big-ear Supplejack ●

52427 Berchemia hispida (H. T. Tsai et K. M. Feng) Y. L. Chen et P. K. Chou;毛背勾儿茶(大鸭公藤,金背勾儿茶,牛儿藤,石萝藤);Hispid Hooktea, Hispid Supplejack, Hispid Supple-jack ●

52428 Berchemia hispida (H. T. Tsai et K. M. Feng) Y. L. Chen et P. K. Chou var. glabrata Y. L. Chen et P. K. Chou;光轴勾儿茶;Glabrous Hooktea, Glabrous Supplejack ●

52429 Berchemia huana Rehder;大叶勾儿茶(胡氏勾儿茶,毛勾儿茶);Bigleaf Hooktea, Bigleaf Supplejack, Big-leaved Supple-jack, Largeleaf Supplejack ●

52430 Berchemia huana Rehder = Berchemia magna (Makino) Koidz. ●

52431 Berchemia huana Rehder var. glabrescens W. C. Cheng ex Y. L. Chen;脱毛大叶勾儿茶;Glabrescent Hooktea, Glabrescent Supplejack ●

52432 Berchemia hypochrysa C. K. Schneid. = Berchemia flavescens (Wall. ex Roxb.) Brongn. ●

52433 Berchemia hypochrysa C. K. Schneid. = Berchemia hispida (H. T. Tsai et K. M. Feng) Y. L. Chen et P. K. Chou ●

52434 Berchemia hypochrysa C. K. Schneid. var. hispida H. T. Tsai et K. M. Feng = Berchemia hispida (H. T. Tsai et K. M. Feng) Y. L. Chen et P. K. Chou ●

52435 Berchemia kulingensis C. K. Schneid.;牯岭勾儿茶(常青藤,勾儿茶,画眉杠,青藤,山黄芪,铁箍散,铁骨散,小叶勾儿茶,小叶青,熊柳,紫青藤);Guling Hooktea, Guling Supplejack, Kuling Supple-jack ●

52436 Berchemia lineata (L.) DC.;铁包金(岗奸,勾儿茶,勾勾茶,狗脚刺,假榄仔,老鼠耳,老鼠乳,老鼠乌,绿末棵子,米拉藤,鼠米,鼠乳根,鼠乳头,提云草,乌金藤,乌口仔,乌李楝,乌龙根,乌痧头,细纹勾儿茶,细叶勾儿茶,小桃花,小叶黄鳝藤,小叶铁包金,鸭公青);Lineate Hooktea, Lineate Supplejack, Lineate Supple-jack, Supple Jack, Supple-jack ●

52437 Berchemia longipedicellata Y. L. Chen et P. K. Chou;细梗勾儿茶;Longipediceled Supple-jack, Slanderpedicel Hooktea, Slanderpedicel Supplejack, Slenderstalk Supplejack ●

52438 Berchemia longipes Y. L. Chen et P. K. Chou;长梗勾儿茶;Longpedicel Hooktea, Longpedicel Supplejack, Long-stalked Supple-jack ●

52439 Berchemia longiracemosa Okuyama;长序勾儿茶●☆

52440 Berchemia magna (Makino) Koidz. = Berchemia racemosa Siebold et Zucc. var. magna Makino ●

52441 Berchemia magna (Makino) Koidz. var. pubescens Ohwi = Berchemia magna (Makino) Koidz. ●

52442 Berchemia medogensis Y. L. Chen et Y. F. Du;墨脱勾儿茶;Motuo Hooktea, Motuo Supplejack ●

52443 Berchemia nana W. W. Sm. = Berchemia edgeworthii M. A. Lawson ●

52444 Berchemia ohwii Kaneh. = Berchemia formosana C. K. Schneid. ●

52445 Berchemia ohwii Kaneh. et Hatus. = Berchemia formosana C. K. Schneid. ●

52446 Berchemia omeiensis W. P. Fang ex Y. L. Chen;峨眉勾儿茶;Emei Hooktea, Emei Supplejack, Omei Supple-jack ●

52447 Berchemia pakistanica Browicz;巴基斯坦勾儿茶●☆

52448 Berchemia pauciflora Maxim.;小花勾儿茶●

52449 Berchemia pauciflora Maxim. var. longiracemosa (Okuyama) Satomi et Fukuoka = Berchemia longiracemosa Okuyama ●☆

52450 Berchemia polyphylla Wall. ex Lawson;多叶勾儿茶(光枝勾儿茶,金刚藤,水车藤,铁包金,小通花,鸭公头);Littleleaf Hooktea, Littleleaf Supplejack, Small-leaved Supple-jack ●

52451 Berchemia polyphylla Wall. ex Lawson var. leioclada Hand.-Mazz.;光枝勾儿茶(光枝水车藤);Smoothbrached Hooktea, Smoothbrached Supplejack ●

52452 Berchemia polyphylla Wall. ex Lawson var. trichophylla Hand.-Mazz.;毛叶勾儿茶;Hairy Littleleaf Supplejack ●

52453 Berchemia pycnantha C. K. Schneid. = Berchemia yunnanensis Franch. ●

52454 Berchemia racemosa Siebold et Zucc.;总状勾儿茶;Racemose Supplejack ●

52455 Berchemia racemosa Siebold et Zucc. = Berchemia floribunda (Wall. ex Roxb.) Brongn. ●

52456 Berchemia racemosa Siebold et Zucc. f. pubescens Nemoto;毛勾儿茶;Pubescent Racemose Supplejack ●☆

52457 Berchemia racemosa Siebold et Zucc. f. stenosperma Hatus.;瘦籽勾儿茶●☆

52458 Berchemia racemosa Siebold et Zucc. var. formosana (C. K. Schneid.) Kitam. = Berchemia formosana C. K. Schneid. ●

52459 Berchemia racemosa Siebold et Zucc. var. luxurians Hatus.;南国勾儿茶●

52460 Berchemia racemosa Siebold et Zucc. var. luxurians Hatus. = Berchemia racemosa Siebold et Zucc. ●

52461 Berchemia racemosa Siebold et Zucc. var. magna Makino;大勾儿茶(大黄鳝藤);Big Racemose Supplejack ●

52462 Berchemia racemosa Siebold et Zucc. var. magna Makino = Berchemia magna (Makino) Koidz. ●

52463 Berchemia racemosa Siebold et Zucc. var. magna Makino f. pubescens (Ohwi) Nemoto = Berchemia magna (Makino) Koidz. ●

52464 Berchemia racemosa Siebold et Zucc. var. pilosa Hatus.;毛脉勾儿茶●

52465 Berchemia racemosa Siebold et Zucc. var. pilosa Hatus. = Berchemia racemosa Siebold et Zucc. ●

52466 Berchemia scandens (Hill) K. Koch;攀缘勾儿茶(阿拉巴马勾儿茶);Alabama Supplejack, Rattan, Rattan Vine, Supplejack, Supple-Jack ●☆

52467 Berchemia scandens (Hill) Trel. = Berchemia scandens (Hill) K. Koch ●☆

52468 Berchemia sinica C. K. Schneid.;勾儿茶(牛鼻足秧,铁光棍);China Hooktea, Chinese Rattan Vine, Chinese Supplejack, Chinese Supple-jack ●

52469 Berchemia trichoclada (Rehder et E. H. Wilson) Hand.-Mazz. = Berchemia polyphylla Wall. ex Lawson ●

52470 Berchemia trichoclada (Rehder et E. H. Wilson) Hand.-Mazz. var. leioclada Hand.-Mazz. = Berchemia polyphylla Wall. ex Lawson var. leioclada Hand.-Mazz. ●

52471 Berchemia wilsonii (C. K. Schneid.) Koidz. = Berchemiella wilsonii (C. K. Schneid.) Nakai ●◇

52472 Berchemia wilsonii Koidz. = Berchemiella berchemiifolia (Makino) Nakai ●

52473 Berchemia yemensis Deflers = Sageretia thea (Osbeck) M. C. Johnst. ●

52474 Berchemia yunnanensis Franch.;云南勾儿茶(黑果子,女儿茶,青龙草,消黄散,鸭公青,鸭公藤,鸭公头);Yunnan Hooktea, Yunnan Supplejack, Yunnan Supple-jack ●

52475 Berchemia yunnanensis Franch. var. trichoclada Rehder et E. H. Wilson = Berchemia polyphylla Wall. ex Lawson ●

52476 Berchemia zeyheri (Sond.) Grubov;泽赫勾儿茶●☆

52477 Berchemiella Nakai (1923);小勾儿茶属;Berchemiella, Hookettea ●

52478 Berchemiella berchemiifolia (Makino) Nakai;日本小勾儿茶;Japan Hookettea, Japanese Berchemiella ●

52479 Berchemiella crenulata (Hand.-Mazz.) Hu = Chaydaia rubrinervis (H. Lév.) C. Y. Wu ●

52480 Berchemiella crenulata (Hand.-Mazz.) Hu = Rhamnella rubrinervis (H. Lév.) Rehder ●

52481 Berchemiella wilsonii (C. K. Schneid.) Nakai;小勾儿茶;E. H. Wilson Berchemiella, E. H. Wilson Hookettea, E. H. Wilson Supplejack, Wilson Berchemiella ●◇

52482 Berchemiella wilsonii (C. K. Schneid.) Nakai = Berchemiella berchemiifolia (Makino) Nakai ●

52483 Berchemiella wilsonii (C. K. Schneid.) Nakai var. pubipetiolata H. Qian;毛柄小勾儿茶●

52484 Berchemiella yunnanensis Y. L. Chen et P. K. Chou;云南小勾儿茶(滇小勾儿茶,女儿红,消黄散,鸭公青,鸭公藤,鸭公头);Yunnan Berchemiella, Yunnan Hookettea ●

52485 Berchtoldia C. Presl = Chaetium Nees ■☆

52486 Berchtoldia J. Presl = Chaetium Nees ■☆

52487 Berckheya C. Presl = Berkheya Ehrh.(保留属名)●■☆

52488 Berebera Baker = Berrebera Hochst. ●■

52489 Berebera Baker = Millettia Wight et Arn.(保留属名)●■

52490 Berendtia A. Gray = Hemichaena Benth. ■●☆

52491 Berendtia A. Gray(1868) = Berendtiella Wettst. et Harms ■●☆

52492 Berendtiella Wettst. et Harms = Hemichaena Benth. ■●☆

52493 Berendtiella Wettst. et Harms(1899);贝伦特玄参属■●☆

52494 Berendtiella levigata (B. L. Rob. et Greenm.) Thieret;平滑贝伦特玄参●☆

52495 Berendtiella rugosa (Benth.) Thieret;褶皱贝伦特玄参●☆

52496 Berenice Salisb. = Allium L. ■

52497 Berenice Salisb. = Loncostemon Raf. ■

52498 Berenice Tul. (1857);直药桔梗属●☆

52499 Berenice arguta Tul.;直药桔梗●☆

52500 Bergella Schnizl. = Bergia L. ●■

52501 Bergena Adans.(废弃属名) = Bergenia Moench(保留属名)■

52502 Bergena Adans.(废弃属名) = Lecythis Loefl. ●☆

52503 Bergenia Moench(1794)(保留属名);岩白菜属(朝鲜岩扇属);Bergenia, Elephant-ear, Elephant's Ears, Elephant's-ears, Saxifrage, Siberian Saxifrage ■

52504 Bergenia Moench(保留属名) = Saxifraga L. ■

52505 Bergenia Neck. = Lythrum L. ●■

52506 Bergenia Raf. = Bergenia Moench(保留属名)■

52507 Bergenia Raf. = Cuphea Adans. ex P. Browne ●■

52508 Bergenia bifolia Moench;二叶岩白菜;Two Leaves Bergenia ■☆

52509 Bergenia bifolia Moench = Bergenia crassifolia (L.) Fritsch ■

52510 Bergenia ciliata (Haw.) Sternb.;睫毛岩白菜(红岩七,腺毛岩白菜,缘毛虎耳草,缘毛岩白菜);Ciliate Bergenia ■☆

52511 Bergenia ciliata (Haw.) Sternb. f. ligulata (Engl.) P. F. Yeo = Bergenia pacumbis (Buch.-Ham. ex D. Don) C. Y. Wu et J. T. Pan ■

52512 Bergenia cordifolia (Haw.) Sternb. = Bergenia crassifolia (L.) Fritsch ■

52513 Bergenia coreana Nakai = Bergenia crassifolia (L.) Fritsch ■

52514 Bergenia crassifolia (L.) Fritsch;厚叶岩白菜(厚叶梅,盘龙七,心叶岩白菜);Elephant's Ear, Elephant's-ears, Heartleaf Bergenia, Heart-leaved Bergenia, Heartleaved Stonebreak, Leath Saxifrage, Leather Bergenia, Pig Squeak, Winter Blooming Bergenia ■

52515 Bergenia crassifolia (L.) Fritsch var. cordifolia (Haw.) Boriss. = Bergenia crassifolia (L.) Fritsch ■

52516 Bergenia crassifolia (L.) Fritsch var. elliptica Ledeb. = Bergenia crassifolia (L.) Fritsch ■

52517 Bergenia crassifolia (L.) Fritsch var. obovata Ser. = Bergenia crassifolia (L.) Fritsch ■

52518 Bergenia crassifolia (L.) Fritsch var. pacifica Nekr. = Bergenia crassifolia (L.) Fritsch ■

52519 Bergenia delavayi (Franck) Engl. = Bergenia purpurascens (Hook. f. et Thomson) Engl. ■

52520 Bergenia emeiensis C. Y. Wu ex J. T. Pan;峨眉岩白菜■

52521 Bergenia emeiensis C. Y. Wu var. rubellina J. T. Pan;淡红岩白菜;Reddish Emei Bergenia ■

52522 Bergenia himalaica Boriss. = Bergenia ciliata (Haw.) Sternb. f. ligulata (Engl.) P. F. Yeo ■

52523 Bergenia himalaica Boriss. = Bergenia pacumbis (Buch.-Ham. ex D. Don) C. Y. Wu et J. T. Pan ■

52524 Bergenia hissarica Boriss.;光托杯岩白菜■☆

52525 Bergenia ligulata (Wall.) Engl. = Bergenia ciliata (Haw.) Sternb. f. ligulata (Engl.) P. F. Yeo ■

52526 Bergenia ligulata (Wall.) Engl. = Bergenia pacumbis (Buch.-Ham. ex D. Don) C. Y. Wu et J. T. Pan ■

52527 Bergenia ligulata (Wall.) Engl. var. ciliata Haw. = Bergenia ciliata (Haw.) Sternb. ■☆

52528 Bergenia ligulata Engl. = Bergenia pacumbis (Buch. -Ham. ex D. Don) C. Y. Wu et J. T. Pan ■

52529 Bergenia ligulata sensu Blatt. = Bergenia ciliata (Haw.) Sternb. f. ligulata (Engl.) P. F. Yeo ■

52530 Bergenia pacifica Kom.;太平洋岩白菜;Pacific Bergenia ■☆

52531 Bergenia pacifica Stein = Bergenia crassifolia (L.) Fritsch ■

52532 Bergenia pacumbis (Buch. -Ham. ex D. Don) C. Y. Wu et J. T. Pan;舌岩白菜(舌状岩白菜,岩七);Elephant-leaved Saxifrage, Rocky Bergenia, Siberian Saxifrage, Strapleaf Bergenia, Tongue Saxifrage ■

52533 Bergenia purpurascens (Hook. f. et Thomson) Engl.;岩白菜(矮白菜,呆白菜,滇岩白菜,红岩七,厚叶岩白菜,兰花岩陀,蓝花岩陀,亮叶子,石白菜,石蚕,石山菖蒲,雪头开花,岩壁菜,岩菖蒲,岩七,云南岩白菜);Purple Bergenia ■

52534 Bergenia purpurascens (Hook. f. et Thomson) Engl. f. delavayi (Franch.) Hand. -Mazz. = Bergenia purpurascens (Hook. f. et Thomson) Engl. ■

52535 Bergenia purpurascens (Hook. f. et Thomson) Engl. var. delavayi (Franch.) Engl. et Irmsch.;云南岩白菜■

52536 Bergenia purpurascens (Hook. f. et Thomson) Engl. var. delavayi (Franch.) Engl. et Irmsch. = Bergenia purpurascens (Hook. f. et Thomson) Engl. ■

52537 Bergenia purpurascens (Hook. f. et Thomson) Engl. var. macrantha (Franch.) Diels = Bergenia purpurascens (Hook. f. et Thomson) Engl. ■

52538 Bergenia purpurascens (Hook. f. et Thomson) Engl. var. sessilis H. Chuang;短毛岩白菜;Shorthair Purple Bergenia ■

52539 Bergenia scopulosa T. P. Wang;秦岭岩白菜(地白菜,盘龙七,石白菜);Qinling Bergenia ■

52540 Bergenia sibirica?;西伯利亚岩白菜■☆

52541 Bergenia stracheyi (Hook. f. et Thomson) Engl.;短柄岩白菜(斯特拉彻岩白菜,喜马拉雅岩白菜);Strachey Bergenia ■

52542 Bergenia tianquanensis J. T. Pan;天全岩白菜;Tianquan Bergenia ■

52543 Bergenia ugamica V. N. Pavlov;塔什干岩白菜■☆

52544 Bergera J. König ex L. (废弃属名) = Murraya J. König ex L. (保留属名)●

52545 Bergera integerrima Buch. -Ham. ex Colebr. = Micromelum integerrimum (Buch. -Ham.) Wight et Arn. ex Roem. ●

52546 Bergera koenigii L. = Murraya koenigii (L.) Spreng. ●

52547 Bergera ternata Blanco = Melicope triphylla (Lam.) Merr. ●

52548 Bergeranthus Schwantes(1926);照波属■☆

52549 Bergeranthus addoensis L. Bolus = Bergeranthus multiceps (Salm-Dyck) Schwantes ■☆

52550 Bergeranthus artus L. Bolus = Bergeranthus multiceps (Salm-Dyck) Schwantes ■☆

52551 Bergeranthus caninus (Haw.) Schwantes = Carruanthus ringens (L.) Boom ■☆

52552 Bergeranthus concavus L. Bolus;凹照波■☆

52553 Bergeranthus derenbergianus (Dinter) Schwantes = Ebracteola derenbergiana (Dinter) Dinter et Schwantes ■☆

52554 Bergeranthus firmus L. Bolus = Bergeranthus multiceps (Salm-Dyck) Schwantes ■☆

52555 Bergeranthus glenensis N. E. Br. = Hereroa glenensis (N. E. Br.) L. Bolus ■☆

52556 Bergeranthus jamesii L. Bolus;濑波■☆

52557 Bergeranthus jamesii L. Bolus = Bergeranthus vespertinus (A. Berger) Schwantes ■☆

52558 Bergeranthus katbergensis L. Bolus = Bergeranthus concavus L. Bolus ■☆

52559 Bergeranthus leightoniae L. Bolus = Bergeranthus concavus L. Bolus ■☆

52560 Bergeranthus longisepalus L. Bolus;长萼照波■☆

52561 Bergeranthus multiceps (Salm-Dyck) Schwantes;照波(仙女花)■☆

52562 Bergeranthus rhomboideus (Salm-Dyck) Schwantes = Rhombophyllum rhomboideum (Salm-Dyck) Schwantes ●☆

52563 Bergeranthus scapiger (Haw.) N. E. Br.;翠锋■☆

52564 Bergeranthus scapiger (Haw.) Schwantes = Bergeranthus scapiger (Haw.) N. E. Br. ■☆

52565 Bergeranthus vespertinus (A. Berger) Schwantes;大群波■☆

52566 Bergeranthus vespertinus Schwantes = Bergeranthus vespertinus (A. Berger) Schwantes ■☆

52567 Bergeretia Bubani = Illecebrum L. ■☆

52568 Bergeretia Desv. = Clypeola L. ■☆

52569 Bergeria Koenig = Koenigia L. ■

52570 Bergeria Koenig ex Steud. = Koenigia L. ■

52571 Bergerocactus Britton et Rose(1909);碧彩柱属;Snakecactus ●☆

52572 Bergerocactus emoryi (Engelm.) Britton et Rose;碧彩柱;Golden Cereus, Golden Clubcactus, Golden Snakecactus ■☆

52573 Bergerocactus emoryi Britton et Rose = Bergerocactus emoryi (Engelm.) Britton et Rose ■☆

52574 Bergerocereus Britton et Rose = Bergerocactus Britton et Rose ●☆

52575 Bergerocereus Frič et Kreuz. = Bergerocactus Britton et Rose ●☆

52576 Bergeronia Micheli(1883);巴拉圭豆属(绢质豆属)■☆

52577 Bergeronia sericea Micheli;巴拉圭豆■☆

52578 Berghausia Endl. = Garnotia Brongn. ■

52579 Berghausia mutica Munro = Garnotia mutica (Munro) Druce ■

52580 Berghausia mutica Munro = Garnotia patula (Munro) Benth. var. mutica (Munro) Rendle ■

52581 Berghausia patula Munro = Garnotia patula (Munro) Benth. ■

52582 Berghausia tenella Arn. ex Miq. = Garnotia tenella (Arn. ex Miq.) Janowski ■

52583 Berghesia Nees(1847);伯格茜属☆

52584 Berghesia coccinea Nees;伯格茜☆

52585 Berghias Juss. = Bergkias Sonn. ●

52586 Berghias Juss. = Gardenia Ellis(保留属名)●

52587 Bergia L. (1771);田繁缕属(伯格草属);Bergia ●■

52588 Bergia abyssinica A. Rich. = Pollichia campestris Aiton ●☆

52589 Bergia aestivosa Wight et Arn.;夏田繁缕■☆

52590 Bergia alsinoides Friedr. -Holzh. = Bergia anagalloides E. Mey. ex Fenzl ■☆

52591 Bergia ammannioides Roth = Bergia ammannioides Roxb. ex Roth ●■

52592 Bergia ammannioides Roxb. ex Roth;田繁缕(伯格草,节节菜,密花菜,密花草);Common Bergia ●■

52593 Bergia ammannioides Roxb. ex Roth = Bergia serrata Blanco ●■

52594 Bergia ammannioides Roxb. ex Roth var. pentandra Wight = Bergia ammannioides Roxb. ex Roth ●■

52595 Bergia anagalloides E. Mey. ex Fenzl;石竹田繁缕■☆

52596 Bergia aquatica Roxb. = Bergia capensis L. ●■

52597 Bergia capensis L.;大叶田繁缕(水田繁缕);Bigleaf Bergia ●■

52598 Bergia decumbens Planch. ex Harv.;外倾田繁缕■☆

52599 Bergia erecta Guillaumin et Perr. = Bergia ammannioides Roxb.

ex Roth ●■
52600 Bergia erythroleuca Gilg = Bergia spathulata Schinz ■☆
52601 Bergia glandulosa Turcz. = Bergia ammannioides Roxb. ex Roth ●■
52602 Bergia glandulosa Turcz. = Bergia serrata Blanco ●■
52603 Bergia glomerata L. f.;团集田繁缕■☆
52604 Bergia glutinosa Dinter et Schulze-Menz;黏性田繁缕■☆
52605 Bergia guineensis Hutch. et Dalziel = Bergia suffruticosa (Delile) Fenzl ●☆
52606 Bergia herniarioides Range;治疝草田繁缕■☆
52607 Bergia integrifolia Dinter ex Friedr.-Holzh. = Bergia polyantha Sond. ■☆
52608 Bergia mairei Quézel;迈雷田繁缕■☆
52609 Bergia mossambicensis Wild = Bergia salaria Bremek. ■☆
52610 Bergia odorata Edgew. = Bergia suffruticosa (Delile) Fenzl. ●☆
52611 Bergia pentandra Cambess. ex Guillaumin et Perr. = Bergia ammannioides Roxb. ex Roth ●■
52612 Bergia pentheriana Keissl.;彭泰尔田繁缕■☆
52613 Bergia peploides Guillaumin et Perr. = Bergia ammannioides Roxb. ex Roth ●■
52614 Bergia polyantha Sond.;多花田繁缕■☆
52615 Bergia prostrata Schinz = Bergia pentheriana Keissl. ■☆
52616 Bergia salaria Bremek.;莫桑比克田繁缕■☆
52617 Bergia serrata Blanco;倍蕊田繁缕;Doublestamen Bergia ●■
52618 Bergia spathulata Schinz;匙状田繁缕■☆
52619 Bergia suffruticosa (Delile) Fenzl.;亚灌木田繁缕●☆
52620 Bergia texana (Hook.) Seub.;得州田繁缕■☆
52621 Bergia verticillata Willd. = Bergia capensis L. ●■
52622 Bergiera Neck. = Berginia Harv. ex Benth. et Hook. f. ■☆
52623 Berginia Harv. = Holographis Nees ■☆
52624 Berginia Harv. ex Benth. et Hook. f. = Holographis Nees ■☆
52625 Bergkias Sonn. = Gardenia Ellis(保留属名)●
52626 Bergsmia Blume = Ryparosa Blume ●☆
52627 Berhardia C. Muell. = Berardia Vill. ■☆
52628 Berhautia Balle(1956);伯氏寄生属●☆
52629 Berhautia senegalensis Balle;伯氏寄生●☆
52630 Beria Bubani = Sium L. ■
52631 Beriesa Steud. = Anredera Juss. ●■
52632 Beriesa Steud. = Siebera J. Gay(保留属名)■☆
52633 Beringeria (Neck.) Link = Ballota L. ●■☆
52634 Beringeria Link = Ballota L. ●■☆
52635 Beringeria Neck. = Ballota L. ●■☆
52636 Beringia R. A. Price, Al-Shehbaz et O'Kane(2001);白令芥属■☆
52637 Beringia bursifolia (DC.) R. A. Price, Al-Shehbaz et O'Kane;白令芥■☆
52638 Berinia Brignol. = Crepis L. ■
52639 Berinia chrysantha (Ledeb.) Sch. Bip. = Crepis chrysantha (Ledeb.) Turcz. ■
52640 Berinia crocea (Lam.) Sch. Bip. = Crepis crocea (Lam.) Babc. ■
52641 Berinia tenuifolia (Willd.) Sch. Bip. = Youngia tenuifolia (Willd.) Babc. et Stebbins ■
52642 Berkheya Ehrh. (1784)(保留属名);尖刺联苞菊属(贝克菊属)●■☆
52643 Berkheya acanthopoda (DC.) Rössler;刺梗尖刺联苞菊●☆
52644 Berkheya adlamii Hook. f. = Berkheya radula (Harv.) De Wild. ■☆
52645 Berkheya alba E. Phillips = Berkheya cirsiifolia (DC.) Rössler ■☆
52646 Berkheya amplexicaulis O. Hoffm. = Berkheya erysithales (DC.) Rössler ■☆
52647 Berkheya andongensis (Hiern) K. Schum. = Berkheya angolensis O. Hoffm. ■☆
52648 Berkheya angolensis O. Hoffm.;安哥拉尖刺联苞菊■☆
52649 Berkheya angusta Schltr.;狭尖刺联苞菊(窄叶贝克菊)■☆
52650 Berkheya angustifolia (Houtt.) Merr.;窄叶尖刺联苞菊(窄叶贝克菊)■☆
52651 Berkheya annectens Harv.;连合尖刺联苞菊■☆
52652 Berkheya antunesii O. Hoffm. = Hirpicium antunesii (O. Hoffm.) Rössler ■☆
52653 Berkheya arctiifolia O. Hoffm. = Berkheya montana J. M. Wood et M. S. Evans ■☆
52654 Berkheya armata (Vahl) Druce;尖刺联苞菊■☆
52655 Berkheya asteroides (L. f.) Druce = Berkheya fruticosa (L.) Ehrh. ■☆
52656 Berkheya barbata (L. f.) Hutch.;髯毛尖刺联苞菊■☆
52657 Berkheya bergiana Soderb.;贝格尖刺联苞菊■☆
52658 Berkheya bilabiata N. E. Br. = Berkheya montana J. M. Wood et M. S. Evans ■☆
52659 Berkheya bipinnatifida (Harv.) Rössler;双羽裂尖刺联苞菊■☆
52660 Berkheya bipinnatifida (Harv.) Rössler var. cordata G. V. Pope;心形双羽裂尖刺联苞菊■☆
52661 Berkheya bipinnatifida (Harv.) Rössler var. echinopsoides (Baker) Rössler;蓝刺头尖刺联苞菊■☆
52662 Berkheya bisulca (Thunb.) Willd. = Cullumia bisulca (Thunb.) Less. ■☆
52663 Berkheya buphthalmoides (DC.) Schltr.;牛眼菊状尖刺联苞菊■☆
52664 Berkheya caffra MacOwan;开菲尔尖刺联苞菊■☆
52665 Berkheya canescens DC.;灰尖刺联苞菊■☆
52666 Berkheya cardopatifolia (DC.) Rössler;蓝丝菊叶尖刺联苞菊■☆
52667 Berkheya carduiformis DC. = Berkheya carduoides (Less.) Hutch. ■☆
52668 Berkheya carduoides (Less.) Hutch.;飞廉尖刺联苞菊■☆
52669 Berkheya carlinifolia (DC.) Rössler;刺苞菊叶尖刺联苞菊■☆
52670 Berkheya carlinoides (Vahl) Willd.;刺苞菊状尖刺联苞菊■☆
52671 Berkheya carlinopsis Welw. ex O. Hoffm.;普通尖刺联苞菊■☆
52672 Berkheya carlinopsis Welw. ex O. Hoffm. subsp. sylvicola (S. Moore) Rössler;西尔维亚尖刺联苞菊■☆
52673 Berkheya carlinopsis Welw. ex O. Hoffm. var. sylvicola (S. Moore) Rössler = Berkheya carlinopsis Welw. ex O. Hoffm. subsp. sylvicola (S. Moore) Rössler ■☆
52674 Berkheya carthamoides (Thunb.) Willd. = Berkheya armata (Vahl) Druce ■☆
52675 Berkheya ciliaris (L.) Willd. = Cullumia ciliaris (L.) R. Br. ■☆
52676 Berkheya cirsiifolia (DC.) Rössler;蓟叶尖刺联苞菊■☆
52677 Berkheya coddii Rössler;科德尖刺联苞菊■☆
52678 Berkheya coriacea Harv.;革质尖刺联苞菊■☆
52679 Berkheya corymbosa DC. = Berkheya canescens DC. ■☆
52680 Berkheya cousinioides S. Moore = Berkheya mackenii (Harv.) Rössler ■☆
52681 Berkheya cruciata (Houtt.) Willd.;十字尖刺联苞菊■☆
52682 Berkheya cruciata (Houtt.) Willd. subsp. subintegra Rössler;亚全缘十字尖刺联苞菊■☆
52683 Berkheya cuneata (Thunb.) Willd.;楔形尖刺联苞菊■☆
52684 Berkheya cynaroides (Vahl) Willd. = Berkheya herbacea (L. f.) Druce ■☆
52685 Berkheya debilis MacOwan;弱小尖刺联苞菊■☆

52686 Berkheya decurrens (Thunb.) Willd.;下延尖刺联苞菊■☆
52687 Berkheya densifolia Bohnen ex Rössler;密叶尖刺联苞菊■☆
52688 Berkheya discolor (DC.) O. Hoffm. et Muschl.;异色尖刺联苞菊■☆
52689 Berkheya draco Rössler;龙尖刺联苞菊■☆
52690 Berkheya dregei Harv.;德雷联苞菊■☆
52691 Berkheya echinacea (Harv.) O. Hoffm. ex Burtt Davy;刺联苞菊■☆
52692 Berkheya echinacea (Harv.) O. Hoffm. ex Burtt Davy subsp. polyacantha (Baker) Rössler;多刺联苞菊■☆
52693 Berkheya echinopoda (DC.) Schönland = Berkheya decurrens (Thunb.) Willd. ■☆
52694 Berkheya echinopsoides Baker = Berkheya bipinnatifida (Harv.) Rössler var. echinopsoides (Baker) Rössler ■☆
52695 Berkheya ecklonis Harv. = Berkheya carlinoides (Vahl) Willd. ■☆
52696 Berkheya eriobasis (DC.) Rössler;红基尖刺联苞菊■☆
52697 Berkheya eryngiifolia Less. = Heterorhachis aculeata (Burm. f.) Rössler ●☆
52698 Berkheya erysithales (DC.) Rössler;热非尖刺联苞菊■☆
52699 Berkheya evansii Schltr. = Berkheya rhapontica (DC.) Hutch. et Burtt Davy var. exalata Rössler ■☆
52700 Berkheya excelsa Hutch. = Berkheya carlinifolia (DC.) Rössler ■☆
52701 Berkheya ferox O. Hoffm.;强刺贝克菊■☆
52702 Berkheya ferox O. Hoffm. var. glandulosa Rössler;具腺贝克菊■☆
52703 Berkheya ferox O. Hoffm. var. tomentosa Rössler;绒毛贝克菊■☆
52704 Berkheya francisci Bolus;弗朗西斯科尖刺联苞菊■☆
52705 Berkheya fruticosa (L.) Ehrh.;灌丛贝克菊■☆
52706 Berkheya gazanioides Harv. = Hirpicium gazanioides (Harv.) Rössler ■☆
52707 Berkheya glabrata (Thunb.) Fourc.;光滑贝克菊■☆
52708 Berkheya glabriuscula (DC.) Schönland = Berkheya decurrens (Thunb.) Willd. ■☆
52709 Berkheya gorterioides Oliv. et Hiern = Hirpicium gazanioides (Harv.) Rössler ■☆
52710 Berkheya gracilis O. Hoffm. = Hirpicium gracile (O. Hoffm.) Rössler ■☆
52711 Berkheya grandiflora (Thunb.) Willd. = Berkheya barbata (L. f.) Hutch. ■☆
52712 Berkheya grandiflora Willd. var. alternifolia E. Phillips = Berkheya rosulata Rössler ■☆
52713 Berkheya halenbergensis Dinter ex Range = Berkheya canescens DC. ■☆
52714 Berkheya herbacea (L. f.) Druce;草本贝克菊■☆
52715 Berkheya heterophylla (Thunb.) O. Hoffm.;异叶贝克菊;Prickly Gousblom ■☆
52716 Berkheya heterophylla (Thunb.) O. Hoffm. var. radiata (DC.) Rössler;放射异叶贝克菊■☆
52717 Berkheya heterophylla O. Hoffm. = Berkheya heterophylla (Thunb.) O. Hoffm. ■☆
52718 Berkheya hispida (L. f.) Willd. = Cullumia aculeata (Houtt.) Rössler ■☆
52719 Berkheya horrida Muschl. = Berkheya spinosissima (Thunb.) Willd. ■☆
52720 Berkheya ilicifolia (Vahl) Druce = Berkheya barbata (L. f.) Hutch. ■☆
52721 Berkheya incana (Thunb.) Willd. = Berkheya fruticosa (L.) Ehrh. ■☆
52722 Berkheya insignis (Harv.) Thell.;显著贝克菊■☆
52723 Berkheya johnstoniana Britten;约翰斯顿贝克菊■☆
52724 Berkheya kuntzei O. Hoffm. = Berkheya discolor (DC.) O. Hoffm. et Muschl. ■☆
52725 Berkheya lanceolata (Thunb.) Willd. = Berkheya angustifolia (Houtt.) Merr. ■☆
52726 Berkheya latifolia J. M. Wood et M. S. Evans;宽叶贝克菊■☆
52727 Berkheya lignosa Compton = Berkheya cardopatifolia (DC.) Rössler ■☆
52728 Berkheya mackenii (Harv.) Rössler;梅肯贝克菊■☆
52729 Berkheya macrocephala J. M. Wood;大头贝克菊■☆
52730 Berkheya maritima J. M. Wood et M. S. Evans;滨海贝克菊■☆
52731 Berkheya membranifolia (DC.) F. T. Hubb. = Berkheya decurrens (Thunb.) Willd. ■☆
52732 Berkheya microcephala (DC.) R. A. Dyer = Berkheya discolor (DC.) O. Hoffm. et Muschl. ■☆
52733 Berkheya milleriana Bolus;米勒贝克菊■☆
52734 Berkheya montana J. M. Wood et M. S. Evans;山地贝克菊■☆
52735 Berkheya multijuga (DC.) Rössler;多对贝克菊■☆
52736 Berkheya nivea N. E. Br.;雪白贝克菊■☆
52737 Berkheya obovata (Thunb.) Willd. = Berkheya spinosa (L. f.) Druce ■☆
52738 Berkheya onobromoides (DC.) O. Hoffm. et Muschl.;红花贝克菊■☆
52739 Berkheya onobromoides (DC.) O. Hoffm. et Muschl. var. carlinoides (Thunb.) Rössler;刺苞菊状贝克菊■☆
52740 Berkheya onopordifolia (DC.) O. Hoffm. ex Burtt Davy;大翅蓟贝克菊■☆
52741 Berkheya onopordifolia (DC.) O. Hoffm. ex Burtt Davy var. glabra Bohnen ex Rössler;光滑大翅蓟贝克菊■☆
52742 Berkheya oppositifolia (DC.) Hutch. = Berkheya spinosissima (Thunb.) Willd. var. namaensis Rössler ■☆
52743 Berkheya oppositifolia Range = Berkheya spinosissima (Thunb.) Willd. ■☆
52744 Berkheya palmata (Thunb.) Willd. = Heterorhachis aculeata (Burm. f.) Rössler ●☆
52745 Berkheya pannosa Hilliard;毡状贝克菊■☆
52746 Berkheya parvifolia Baker = Berkheya echinacea (Harv.) O. Hoffm. ex Burtt Davy subsp. polyacantha (Baker) Rössler ■☆
52747 Berkheya patula (Thunb.) Willd. = Cullumia patula (Thunb.) Less. ■☆
52748 Berkheya pauciflora Rössler;少花贝克菊■☆
52749 Berkheya pectinata (Thunb.) Willd. = Cullumia pectinata (Thunb.) Less. ■☆
52750 Berkheya petiolata (DC.) Schönland = Berkheya decurrens (Thunb.) Willd. ■☆
52751 Berkheya pinnata (Thunb.) Less. = Heterorhachis aculeata (Burm. f.) Rössler ●☆
52752 Berkheya pinnata (Thunb.) Less. var. minor Harv. = Berkheya cardopatifolia (DC.) Rössler ■☆
52753 Berkheya pinnatifida (Thunb.) Thell.;羽裂贝克菊■☆
52754 Berkheya pinnatifida (Thunb.) Thell. subsp. stobaeoides (Harv.) Rössler;尖刺贝克菊■☆
52755 Berkheya platyptera (Harv.) O. Hoffm. = Berkheya rhapontica (DC.) Hutch. et Burtt Davy subsp. platyptera (Harv.) Rössler ■☆
52756 Berkheya polyacantha (DC.) Schönland = Berkheya decurrens

(Thunb.) Willd. ■☆
52757 Berkheya polyacantha S. Moore = Berkheya discolor (DC.) O. Hoffm. et Muschl. ■☆
52758 Berkheya polycantha Baker = Berkheya echinacea (Harv.) O. Hoffm. ex Burtt Davy subsp. polyacantha (Baker) Rössler ■☆
52759 Berkheya pungens (Thunb.) Willd. = Berkheya carlinoides (Vahl) Willd. ■☆
52760 Berkheya purpurea (DC.) Mast.;紫贝克菊■☆
52761 Berkheya radula (Harv.) De Wild.;刮刀贝克菊■☆
52762 Berkheya radyeri Rössler;拉戴尔贝克菊■☆
52763 Berkheya rehmannii Thell. = Berkheya zeyheri Oliv. et Hiern var. rehmannii (Thell.) Rössler ■☆
52764 Berkheya rehmannii Thell. var. rogersiana? = Berkheya zeyheri Oliv. et Hiern var. rogersiana (Thell.) Rössler ■☆
52765 Berkheya rhapontica (DC.) Hutch. et Burtt Davy;缘膜菊贝克菊■☆
52766 Berkheya rhapontica (DC.) Hutch. et Burtt Davy subsp. platyptera (Harv.) Rössler;阔翅贝克菊■☆
52767 Berkheya rhapontica (DC.) Hutch. et Burtt Davy var. aristosa (DC.) Rössler;无翅贝克菊■☆
52768 Berkheya rhapontica (DC.) Hutch. et Burtt Davy var. exalata Rössler = Berkheya rhapontica (DC.) Hutch. et Burtt Davy var. aristosa (DC.) Rössler ■☆
52769 Berkheya rigida (Thunb.) Bolus et Wolley-Dod ex Adamson et T. M. Salter;硬贝克菊■☆
52770 Berkheya robusta Bohnen ex Rössler;粗壮贝克菊■☆
52771 Berkheya rosulata Rössler;莲座贝克菊■☆
52772 Berkheya schenckii O. Hoffm. = Berkheya spinosissima (Thunb.) Willd. ■☆
52773 Berkheya schinzii O. Hoffm.;欣兹贝克菊■☆
52774 Berkheya scolymoides DC. = Berkheya herbacea (L. f.) Druce ■☆
52775 Berkheya seminivea Harv. et Sond.;半白贝克菊■☆
52776 Berkheya setifera DC.;刚毛贝克菊■☆
52777 Berkheya setifera DC. var. tropica S. Moore = Berkheya setifera DC. ■☆
52778 Berkheya setosa (L.) Willd. = Cullumia setosa (L.) R. Br. ■☆
52779 Berkheya sonchifolia (Harv.) MacOwan = Berkheya erysithales (DC.) Rössler ■☆
52780 Berkheya speciosa (DC.) O. Hoffm.;美丽贝克菊■☆
52781 Berkheya speciosa (DC.) O. Hoffm. subsp. lanceolata Rössler;披针形贝克菊■☆
52782 Berkheya speciosa (DC.) O. Hoffm. subsp. ovata Rössler;卵形贝克菊■☆
52783 Berkheya spekeana Oliv.;斯皮克贝克菊■☆
52784 Berkheya spekeana Oliv. var. abyssinica Fiori = Berkheya spekeana Oliv. ■☆
52785 Berkheya sphaerocephala (DC.) Rössler;球头贝克菊■☆
52786 Berkheya spinosa (L. f.) Druce;具刺贝克菊■☆
52787 Berkheya spinosissima (Thunb.) Willd.;多刺贝克菊■☆
52788 Berkheya spinosissima (Thunb.) Willd. var. namaensis Rössler;对叶贝克菊■☆
52789 Berkheya spinulosa N. E. Br. = Berkheya discolor (DC.) O. Hoffm. et Muschl. ■☆
52790 Berkheya squarrosa (L.) Willd. = Cullumia squarrosa (L.) R. Br. ■☆
52791 Berkheya stobaeoides Harv. = Berkheya pinnatifida (Thunb.) Thell. subsp. stobaeoides (Harv.) Rössler ■☆
52792 Berkheya subteretifolia Thell. = Berkheya zeyheri Oliv. et Hiern ■☆
52793 Berkheya subulata Harv.;钻形贝克菊■☆
52794 Berkheya subulata Harv. var. wilmsiana Rössler;维尔姆斯贝克菊■☆
52795 Berkheya subulata J. M. Wood = Berkheya insignis (Harv.) Thell. ■☆
52796 Berkheya sulcata (Thunb.) Willd. = Cullumia sulcata (Thunb.) Less. ■☆
52797 Berkheya sylvicola S. Moore = Berkheya carlinopsis Welw. ex O. Hoffm. subsp. sylvicola (S. Moore) Rössler ■☆
52798 Berkheya tysonii Hutch.;泰森贝克菊■☆
52799 Berkheya umbellata DC.;小伞贝克菊■☆
52800 Berkheya viscosa (DC.) Hutch.;黏贝克菊■☆
52801 Berkheya welwitschii O. Hoffm.;韦尔贝克菊■☆
52802 Berkheya zeyheri Oliv. et Hiern;泽赫贝克菊■☆
52803 Berkheya zeyheri Oliv. et Hiern var. rehmannii (Thell.) Rössler;拉赫曼贝克菊■☆
52804 Berkheya zeyheri Oliv. et Hiern var. rogersiana (Thell.) Rössler;罗杰斯贝克菊■☆
52805 Berkheyopsis O. Hoffm. (1893);拟贝克菊属■☆
52806 Berkheyopsis O. Hoffm. = Hirpicium Cass. ■●☆
52807 Berkheyopsis aizoides O. Hoffm. = Hirpicium gazanioides (Harv.) Rössler ■☆
52808 Berkheyopsis angolensis O. Hoffm. = Hirpicium gazanioides (Harv.) Rössler ■☆
52809 Berkheyopsis bechuanensis S. Moore = Hirpicium bechuanense (S. Moore) Rössler ■☆
52810 Berkheyopsis brevisquama Mattf. = Hirpicium bechuanense (S. Moore) Rössler ■☆
52811 Berkheyopsis diffusa O. Hoffm. = Hirpicium diffusum (O. Hoffm.) Rössler ■☆
52812 Berkheyopsis echinus (Less.) O. Hoffm. = Hirpicium echinus Less. ■☆
52813 Berkheyopsis gorterioides (Oliv. et Hiern) Thell. = Hirpicium gazanioides (Harv.) Rössler ■☆
52814 Berkheyopsis gorterioides (Oliv. et Hiern) Thell. var. lobulata Thell. = Hirpicium gazanioides (Harv.) Rössler ■☆
52815 Berkheyopsis gossweileri S. Moore = Hirpicium gazanioides (Harv.) Rössler ■☆
52816 Berkheyopsis kuntzei O. Hoffm. = Hirpicium echinus Less. ■☆
52817 Berkheyopsis langii Bremek. et Oberm. = Hirpicium gazanioides (Harv.) Rössler ■☆
52818 Berkheyopsis linearifolia (Bolus) Burtt Davy = Hirpicium linearifolium (Bolus) Rössler ■☆
52819 Berkheyopsis pechuelii (Kuntze) O. Hoffm. = Hirpicium gazanioides (Harv.) Rössler ■☆
52820 Berkheyopsis rehmannii Thell. = Hirpicium bechuanense (S. Moore) Rössler ■☆
52821 Berkheyopsis schinzii O. Hoffm. = Hirpicium gorterioides (Oliv. et Hiern.) Rössler subsp. schinzii (O. Hoffm.) Rössler ■☆
52822 Berla Bubani = Sium L. ■
52823 Berlandiera DC. (1836);绿眼菊属(伯兰氏菊属);Green Eyes ■☆
52824 Berlandiera betonicifolia (Hook.) Small;药水苏叶绿眼菊■☆
52825 Berlandiera dealbata (Torr. et A. Gray) Small = Berlandiera pumila (Michx.) Nutt. ■☆
52826 Berlandiera incisa Torr. et A. Gray = Berlandiera lyrata Benth. ■☆

52827 Berlandiera longifolia Nutt. = Berlandiera texana DC. ■☆

52828 Berlandiera lyrata A. Gray var. monocephala B. L. Turner = Berlandiera monocephala (B. L. Turner) Pinkava ■☆

52829 Berlandiera lyrata Benth.;伯兰氏菊;Chocolate Daisy, Chocolate Flower ■☆

52830 Berlandiera lyrata Benth. var. macrophylla A. Gray = Berlandiera macrophylla (A. Gray) M. E. Jones ■☆

52831 Berlandiera macrophylla (A. Gray) M. E. Jones;大叶绿眼菊■☆

52832 Berlandiera monocephala (B. L. Turner) Pinkava;单头绿眼菊■☆

52833 Berlandiera pumila (Michx.) Nutt.;小绿眼菊■☆

52834 Berlandiera pumila (Michx.) Nutt. var. dealbata (Torr. et A. Gray) Trel. = Berlandiera pumila (Michx.) Nutt. ■☆

52835 Berlandiera pumila (Michx.) Nutt. var. scabrella G. L. Nesom et B. L. Turner = Berlandiera betonicifolia (Hook.) Small ■☆

52836 Berlandiera subacaulis (Nutt.) Nutt.;佛罗里达绿眼菊;Florida-dandelion, Green Eyes ■☆

52837 Berlandiera texana DC.;得州绿眼菊;Green Eyes ■☆

52838 Berlandiera texana DC. var. betonicifolia (Hook.) Torr. et A. Gray = Berlandiera betonicifolia (Hook.) Small ■☆

52839 Berlandiera tomentosa (Pursh) Nutt. = Berlandiera pumila (Michx.) Nutt. ■☆

52840 Berlandiera tomentosa (Pursh) Nutt. var. dealbata Torr. et A. Gray = Berlandiera pumila (Michx.) Nutt. ■☆

52841 Berliera Buch. -Ham. = Myrioneuron R. Br. ex Kurz ●

52842 Berliera Buch. -Ham. ex Wall. = Myrioneuron R. Br. ex Kurz ●

52843 Berlinia Sol. ex Hook. f. (1849)(保留属名);鞋木属;Abem, Berlinia, Ekpogoi ●☆

52844 Berlinia Sol. ex Hook. f. et Benth. = Berlinia Sol. ex Hook. f. (保留属名)●☆

52845 Berlinia acuminata Sol. ex Hook. f. et Benth.;尖叶鞋木●☆

52846 Berlinia acuminata Sol. ex Hook. f. et Benth. = Berlinia grandiflora (Vahl) Hutch. et Dalziel ●☆

52847 Berlinia acuminata Sol. ex Hook. f. et Benth. var. bruneelii De Wild. = Berlinia bruneelii (De Wild.) Torre et Hillc. ●☆

52848 Berlinia acuminata Sol. ex Hook. f. et Benth. var. pubescens De Wild. = Berlinia giorgii De Wild. var. pubescens (De Wild.) Hauman ●☆

52849 Berlinia angolensis Welw. ex Benth. = Isoberlinia angolensis (Welw. ex Benth.) Hoyle et Brenan ●☆

52850 Berlinia auriculata Benth.;耳状鞋木;Red Berlinia ●☆

52851 Berlinia baumii Harms = Julbernardia paniculata (Benth.) Troupin ●☆

52852 Berlinia bifoliolata Harms = Tetraberlinia bifoliolata (Harms) Hauman ●☆

52853 Berlinia bisulcata (A. Chev.) Troupin = Microberlinia bisulcata A. Chev. ●☆

52854 Berlinia bracteosa Benth. = Macroberlinia bracteosa (Benth.) Hauman ●☆

52855 Berlinia brazzavillensis (A. Chev.) Troupin = Microberlinia brazzavillensis A. Chev. ●☆

52856 Berlinia brieyi De Wild. = Julbernardia brieyi (De Wild.) Troupin ●☆

52857 Berlinia bruneelii (De Wild.) Torre et Hillc.;布吕内尔鞋木●☆

52858 Berlinia cabrae De Wild. = Berlinia giorgii De Wild. var. pubescens (De Wild.) Hauman ●☆

52859 Berlinia chevalieri De Wild. = Isoberlinia doka Craib et Stapf ●☆

52860 Berlinia confusa Hoyle;混乱鞋木●☆

52861 Berlinia congolensis (Baker f.) Keay;刚果鞋木●☆

52862 Berlinia coriacea Keay;革质鞋木●☆

52863 Berlinia craibiana Baker f.;克氏鞋木●☆

52864 Berlinia delevoyi De Wild. = Berlinia sapinii De Wild. ●☆

52865 Berlinia densiflora Baker = Isoberlinia angolensis (Welw. ex Benth.) Hoyle et Brenan var. lasiocalyx Hoyle et Brenan ●☆

52866 Berlinia eminii Taub. = Julbernardia globiflora (Benth.) Troupin ●☆

52867 Berlinia gilletii De Wild. = Berlinia giorgii De Wild. var. gilletii (De Wild.) Hauman ●☆

52868 Berlinia gilletii De Wild. var. gossweileri (Baker f.) Troupin = Berlinia giorgii De Wild. ●☆

52869 Berlinia giorgii De Wild.;吉奥鞋木●☆

52870 Berlinia giorgii De Wild. var. gilletii (De Wild.) Hauman;吉勒特鞋木●☆

52871 Berlinia giorgii De Wild. var. gossweileri Baker f. = Berlinia giorgii De Wild. ●☆

52872 Berlinia giorgii De Wild. var. pubescens (De Wild.) Hauman;短柔毛鞋木●☆

52873 Berlinia giorgii De Wild. var. vernicosa Hauman;光泽鞋木●☆

52874 Berlinia grandiflora (Vahl) Hutch. et Dalziel;大花鞋木●☆

52875 Berlinia grandiflora (Vahl) Hutch. et Dalziel var. bruneelii (De Wild.) Hauman = Berlinia bruneelii (De Wild.) Torre et Hillc. ●☆

52876 Berlinia grandiflora (Vahl) Hutch. et Dalziel var. pseudoauriculata Hauman;假耳鞋木●☆

52877 Berlinia grandiflora (Vahl) Hutch. et Dalziel var. smeathmannii Hauman = Berlinia congolensis (Baker f.) Keay ●☆

52878 Berlinia grandiflora Hutch. et Dalziel = Berlinia grandiflora (Vahl) Hutch. et Dalziel ●☆

52879 Berlinia heudelotiana Baill. = Berlinia grandiflora (Vahl) Hutch. et Dalziel ●☆

52880 Berlinia heudelotiana Baill. var. congolensis Baker f. = Berlinia congolensis (Baker f.) Keay ●☆

52881 Berlinia hollandii Hutch. et Dalziel;霍兰鞋木●☆

52882 Berlinia immaculata Mackinder et Wieringa;无斑鞋木●☆

52883 Berlinia ivorensis A. Chev. = Gilbertiodendron ivorense (A. Chev.) J. Léonard ●☆

52884 Berlinia kerstingii Harms = Isoberlinia doka Craib et Stapf ●☆

52885 Berlinia laurentii De Wild. = Berlinia grandiflora (Vahl) Hutch. et Dalziel ●☆

52886 Berlinia ledermannii Harms;莱氏鞋木●☆

52887 Berlinia ledermannii Harms = Julbernardia seretii (De Wild.) Troupin ●☆

52888 Berlinia lundensis Torre et Hillc.;隆德鞋木●☆

52889 Berlinia macrophylla Pierre ex Pellegr. = Berlinia bracteosa Benth. ●☆

52890 Berlinia magnistipulata Harms = Julbernardia magnistipulata (Harms) Troupin ●☆

52891 Berlinia mayumbensis De Wild. = Berlinia bracteosa Benth. ●☆

52892 Berlinia mengei De Wild. = Pseudomacrolobium mengei (De Wild.) Hauman ■☆

52893 Berlinia micrantha Harms = Oddoniodendron micranthum (Harms) Baker f. ■☆

52894 Berlinia niembaensis De Wild. = Isoberlinia angolensis (Welw. ex Benth.) Hoyle et Brenan var. niembaensis (De Wild.) Brenan ●☆

52895 Berlinia occidentalis Keay;西方鞋木●☆

52896 Berlinia orientalis Brenan;东方鞋木●☆

52897 Berlinia paniculata Benth. = Julbernardia paniculata (Benth.) Troupin ●☆

52898 Berlinia paniculata Benth. var. gossweileri Baker f. = Julbernardia gossweileri (Baker f.) Torre et Hillc. ●☆

52899 Berlinia platycarpa Pierre ex De Wild. = Berlinia bracteosa Benth. ●☆

52900 Berlinia polyphylla Harms = Tetraberlinia polyphylla (Harms) J. Léonard ●☆

52901 Berlinia preussii De Wild. = Berlinia craibiana Baker f. ●☆

52902 Berlinia rabiensis Mackinder;拉比鞋木●☆

52903 Berlinia sapinii De Wild.;萨潘鞋木●☆

52904 Berlinia scheffleri Harms = Isoberlinia scheffieri (Harms) Greenway ●☆

52905 Berlinia seretii De Wild. = Julbernardia seretii (De Wild.) Troupin ●☆

52906 Berlinia splendida A. Chev. ex Hutch. et Dalziel = Gilbertiodendron splendidum (A. Chev. ex Hutch. et Dalziel) J. Léonard ●☆

52907 Berlinia stipulacea Benth. = Gilbertiodendron stipulaceum (Benth.) J. Léonard ●☆

52908 Berlinia stolzii Harms = Isoberlinia angolensis (Welw. ex Benth.) Hoyle et Brenan var. lasiocalyx Hoyle et Brenan ●☆

52909 Berlinia tomentella Keay;短毛鞋木●☆

52910 Berlinia tomentosa Harms = Isoberlinia tomentosa (Harms) Craib et Stapf ●☆

52911 Berlinia verdickii De Wild. = Isoberlinia tomentosa (Harms) Craib et Stapf ●☆

52912 Berlinia viridicans Baker f.;绿色鞋木●☆

52913 Berlinianche (Harms) Vattimo(1955);云生花属■☆

52914 Berlinianche aethiopica (Welw.) Vattimo;云生花■☆

52915 Berlinianche holtzii (Engl.) Vattimo;豪尔云生花■☆

52916 Bermudiana Mill. = Sisyrinchium L. ■

52917 Bernarda Adans. = Bernardia L. ●

52918 Bernardia Endl. = Berardia Vill. ■☆

52919 Bernardia Houst. ex P. Browne = Adelia L. (保留属名)●☆

52920 Bernardia L. = Polyscias J. R. Forst. et G. Forst. ●

52921 Bernardia Mill. (废弃属名) = Adelia L. (保留属名)●☆

52922 Bernardia P. Browne = Adelia L. (保留属名)●☆

52923 Bernardina Baudo = Lysimachia L. ●■

52924 Bernardina pumila Baudo = Lysimachia pumila (Baudo) Franch. ■

52925 Bernardinia Planch. (1850);美洲牛栓藤属●☆

52926 Bernardinia Planch. = Rourea Aubl. (保留属名)●

52927 Bernardinia fluminensis Planch.;美洲牛栓藤●☆

52928 Berneuxia Decne. (1873);岩匙属(藏岩梅属);Berneuxine, Berneuxia ■★

52929 Berneuxia thibetica Decne.;岩匙(白奴花,藏岩梅,露寒草,石莲,小岩匙,岩菠菜,岩筋菜,云南岩匙);Tibet Berneuxia, Tibet Berneuxine, Yunnan Berneuxia, Yunnan Berneuxine ■

52930 Berneuxia yunnanensis H. L. Li = Berneuxia thibetica Decne. ■

52931 Bernhardia Post et Kuntze = Bernardia Mill. (废弃属名)●☆

52932 Berniera DC. (废弃属名) = Bernieria Baill. (保留属名)●

52933 Berniera DC. (废弃属名) = Gerbera L. (保留属名)■

52934 Berniera nepalensis DC. = Gerbera maxima (D. Don) Beauverd ■

52935 Bernieria Baill. (保留属名) = Beilschmiedia Nees ●

52936 Bernieria madagascariensis Baill. = Beilschmiedia madagascariensis (Baill.) Kosterm. ●☆

52937 Bernoullia Neck. = Geum L. ■

52938 Bernoullia Neck. ex Raf. (废弃属名) = Bernoullia Oliv. (保留属名)●☆

52939 Bernoullia Oliv. (1873)(保留属名);贝尔木棉属●☆

52940 Bernoullia flammea Oliv.;贝尔木棉●☆

52941 Bernullia Neck. ex Raf. = Bernoullia Oliv. (保留属名)●☆

52942 Bernullia Raf. = Bernoullia Neck. ■

52943 Bernullia Raf. = Geum L. ■

52944 Beroniella Zakirov et Nabiev = Heliotropium L. ●■

52945 Berrebera Hochst. = Millettia Wight et Arn. (保留属名)●■

52946 Berrebera ferruginea Hochst. = Millettia ferruginea (Hochst.) Baker ●☆

52947 Berresfordia L. Bolus(1932);紫锥花属■☆

52948 Berresfordia khamiesbergensis L. Bolus = Conophytum khamiesbergense (L. Bolus) Schwantes ■☆

52949 Berria Roxb. = Berrya Roxb. (保留属名)●

52950 Berroa Beauverd(1913);羽冠紫绒草属■☆

52951 Berroa gnaphalioides (Less.) Beauverd;羽冠紫绒草■☆

52952 Berrya DC. = Berrya Roxb. (保留属名)●

52953 Berrya Klein = Litsea Lam. (保留属名)●

52954 Berrya Roxb. (1820)(保留属名);六翅木属(浆果椴属);Berrya ●

52955 Berrya africana (Mast.) Kosterm. = Carpodiptera africana Mast. ●☆

52956 Berrya ammonilla Roxb. = Berrya cordifolia (Willd.) Burret ●

52957 Berrya cordifolia (Willd.) Burret;六翅木(浆果椴);Hamilla, Heartleaf Berrya, Heart-leaved Berrya, Trincomali Wood ●

52958 Berrya sansibarensis (Burret) Kosterm. = Carpodiptera africana Mast. ●☆

52959 Berryaceae Doweld = Malvaceae Juss. (保留科名)●■

52960 Bersama Fresen. (1837);伯萨木属(伯萨马属)●☆

52961 Bersama abyssinica Fresen.;伯萨木(伯萨马)●☆

52962 Bersama abyssinica Fresen. subsp. engleriana (Gürke) F. White = Bersama abyssinica Fresen. ●☆

52963 Bersama abyssinica Fresen. subsp. nyassae (Baker f.) White = Bersama abyssinica Fresen. ●☆

52964 Bersama abyssinica Fresen. subsp. paullinioides (Planch.) Verdc. = Bersama abyssinica Fresen. ●☆

52965 Bersama abyssinica Fresen. subsp. rosea (Hoyle) Mikkelsen;粉红伯萨木●☆

52966 Bersama abyssinica Fresen. var. engleriana (Gürke) Verdc. = Bersama abyssinica Fresen. ●☆

52967 Bersama abyssinica Fresen. var. gracilipes (Mildbr.) Verdc. = Bersama abyssinica Fresen. ●☆

52968 Bersama abyssinica Fresen. var. holstii (Gürke) Verdc. = Bersama abyssinica Fresen. ●☆

52969 Bersama abyssinica Fresen. var. kandtii (Gilg et Brehmer) Verdc. = Bersama abyssinica Fresen. ●☆

52970 Bersama abyssinica Fresen. var. nyassae (Baker f.) Verdc. = Bersama abyssinica Fresen. ●☆

52971 Bersama abyssinica Fresen. var. paullinioides (Planch.) Verdc. = Bersama abyssinica Fresen. ●☆

52972 Bersama abyssinica Fresen. var. ugandensis (Sprague) Verdc. = Bersama abyssinica Fresen. ●☆

52973 Bersama acutidens Welw. ex Hiern = Bersama abyssinica Fresen. ●☆

52974 Bersama andongensis Hiern = Bersama abyssinica Fresen. ●☆

52975 Bersama angolensis Baker f. = Bersama abyssinica Fresen. ●☆

52976 Bersama bolamensis Brehmer = Bersama abyssinica Fresen. ●☆
52977 Bersama chippii Sprague et Hutch. = Bersama abyssinica Fresen. ●☆
52978 Bersama chloroleuca Brehmer = Bersama abyssinica Fresen. ●☆
52979 Bersama coriacea Baker f. = Bersama abyssinica Fresen. ●☆
52980 Bersama deiningeri Brehmer = Bersama abyssinica Fresen. ●☆
52981 Bersama deneckeana Brehmer = Bersama abyssinica Fresen. ●☆
52982 Bersama engleri Gürke;恩格勒伯萨木●☆
52983 Bersama engleriana Gürke = Bersama abyssinica Fresen. ●☆
52984 Bersama erythrocarpa Brehmer = Bersama abyssinica Fresen. ●☆
52985 Bersama faucicola Gilg et Brehmer = Bersama abyssinica Fresen. ●☆
52986 Bersama gallensis Brehmer = Bersama abyssinica Fresen. ●☆
52987 Bersama goetzei Gürke = Bersama abyssinica Fresen. ●☆
52988 Bersama gossweileri Baker f. = Bersama abyssinica Fresen. ●☆
52989 Bersama gracilipes Mildbr. = Bersama abyssinica Fresen. ●☆
52990 Bersama hebecalyx Gilg et Brehmer = Bersama abyssinica Fresen. ●☆
52991 Bersama holstii Gürke = Bersama abyssinica Fresen. ●☆
52992 Bersama integrifolia A. Rich. = Bersama abyssinica Fresen. ●☆
52993 Bersama jaegeri Gilg et Brehmer = Bersama abyssinica Fresen. ●☆
52994 Bersama kandtii Gilg et Brehmer = Bersama abyssinica Fresen. ●☆
52995 Bersama kiwuensis Gürke = Bersama abyssinica Fresen. ●☆
52996 Bersama leiostegia Stapf = Bersama abyssinica Fresen. ●☆
52997 Bersama leucotricha Brehmer = Bersama abyssinica Fresen. ●☆
52998 Bersama lobulata Sprague et Hutch. = Bersama abyssinica Fresen. ●☆
52999 Bersama lucens (Hochst.) Szyszyl.;光亮伯萨木●☆
53000 Bersama magnifica A. Chev.;壮观伯萨木●☆
53001 Bersama maxima Baker = Bersama abyssinica Fresen. ●☆
53002 Bersama mildbraedii Gürke = Bersama abyssinica Fresen. ●☆
53003 Bersama mossambicensis Sim = Pseudobersama mossambicensis (Sim) Verdc. ●☆
53004 Bersama myriantha Gilg et Brehmer = Bersama abyssinica Fresen. ●☆
53005 Bersama ninagongensis Gürke = Bersama abyssinica Fresen. ●☆
53006 Bersama nyassae Baker f. = Bersama abyssinica Fresen. ●☆
53007 Bersama oligoneura Brehmer;寡脉伯萨木●☆
53008 Bersama pachythyrsa Brehmer = Bersama abyssinica Fresen. ●☆
53009 Bersama pallidinervia Brehmer;白脉伯萨木●☆
53010 Bersama palustris L. Touss. = Bersama abyssinica Fresen. ●☆
53011 Bersama paullinioides (Planch.) Baker = Bersama abyssinica Fresen. ●☆
53012 Bersama preussii Baker f. = Bersama abyssinica Fresen. ●☆
53013 Bersama rosea Hoyle = Bersama abyssinica Fresen. subsp. rosea (Hoyle) Mikkelsen ●☆
53014 Bersama schreberifolia Brehmer = Bersama abyssinica Fresen. ●☆
53015 Bersama schweinfurthii Brehmer = Bersama abyssinica Fresen. ●☆
53016 Bersama serrata A. Rich. = Bersama abyssinica Fresen. ●☆
53017 Bersama stayneri E. Phillips = Bersama tysoniana Oliv. ●☆
53018 Bersama suffruticosa Brehmer = Bersama abyssinica Fresen. ●☆
53019 Bersama swinnyi E. Phillips;斯温尼伯萨木●☆
53020 Bersama swynnertonii Baker f.;斯温纳顿伯萨木●☆
53021 Bersama tessmannii Brehmer = Bersama abyssinica Fresen. ●☆
53022 Bersama tysoniana Oliv.;泰森伯萨木●☆
53023 Bersama ugandensis Sprague = Bersama abyssinica Fresen. ●☆
53024 Bersama ugandensis Sprague var. serrata Baker f. = Bersama abyssinica Fresen. ●☆
53025 Bersama usambarica Gürke = Bersama abyssinica Fresen. ●☆
53026 Bersama ussanguensis Brehmer = Bersama abyssinica Fresen. ●☆
53027 Bersama volkensii Gürke = Bersama abyssinica Fresen. ●☆
53028 Bersama xanthotricha Gilg et Brehmer = Bersama abyssinica Fresen. ●☆
53029 Bersama yangambiensis L. Touss. = Bersama abyssinica Fresen. ●☆
53030 Bersama zombensis Dunkley = Bersama abyssinica Fresen. ●☆
53031 Bersamaceae Doweld = Melianthaceae Horan.(保留科名)●☆
53032 Bertauxia Szlach. = Habenaria Willd. ■
53033 Bertera Steud. = Gladiolus L. ■
53034 Berteroa DC.(1821);团扇荠属(波儿草属,团扇芥属);False Alyssum,Falsealyssum,Hoary Alison ■
53035 Berteroa Zipp. = Berteroa DC. ■
53036 Berteroa ascendens K. Koch;意大利团扇荠■☆
53037 Berteroa ascendens K. Koch = Alyssum mutabile Vent. ■
53038 Berteroa incana (L.) DC.;团扇荠(波儿草,团扇芥);Hoary Alison,Hoary Alyssum,Hoary False Alyssum,Hoary False Madwort,Hoary Falsealyssum,Hoary-alyssum ■
53039 Berteroa mutabilis DC.;路边团扇荠;Roadside False Madwort ■☆
53040 Berteroa potaninii Maxim. = Galitzkya potaninii (Maxim.) V. V. Botschantz. ■
53041 Berteroa potaninii Maxim. var. latifolia C. H. An = Berteroa potaninii Maxim. ■
53042 Berteroa potaninii Maxim. var. latifolia C. H. An = Galitzkya potaninii (Maxim.) V. V. Botschantz. ■
53043 Berteroa spathulata (Stephan ex Willd.) C. A. Mey. = Galitzkya spathulata (Steph. ex Willd.) V. V. Botschantz. ■
53044 Berteroa spathulata (Stephan) C. A. Mey. = Galitzkya spathulata (Steph. ex Willd.) V. V. Botschantz. ■
53045 Berteroella O. E. Schulz(1919);锥果芥属(北荠属,星毛芥属);Berteroella ■
53046 Berteroella maximowiczii (J. Palib.) O. E. Schulz = Berteroella maximowiczii (J. Palib.) O. E. Schulz ex Loes. ■
53047 Berteroella maximowiczii (J. Palib.) O. E. Schulz ex Loes.;锥果芥(北荠);Maximowicz Berteroella ■
53048 Berthelotia DC. = Pluchea Cass. ●■
53049 Berthelotia lanceolata DC. = Pluchea lanceolata (DC.) C. B. Clarke ■☆
53050 Berthelotia lanceolata DC. var. senegalensis? = Pluchea lanceolata (DC.) C. B. Clarke ■☆
53051 Berthiera Vent. = Bertiera Aubl. ■☆
53052 Bertholetia Brongn. = Bertholletia Bonpl. ●☆
53053 Bertholetia Rchb. = Berthelotia DC. ●■
53054 Bertholetia Rchb. = Pluchea Cass. ●■
53055 Bertholletia Bonpl.(1807);巴西果属(巴西坚果属,巴西栗属,栗油果属);Brazil-nut,Brazil-nut Tree,Butter-nut,Cream-nut,Para-nut ●☆
53056 Bertholletia Humb. et Bonpl. = Bertholletia Bonpl. ●☆
53057 Bertholletia excelsa Humb. et Bonpl.;巴西果(巴西坚果,巴西栗子,高大栗油果);Black Walnut,Brazil Nut,Brazil-nut,Brazil-nut Tree,Cream Nut,Juvia-nut,Nigger Toe,Niggertoe,Nigger-toe,Para Nut,Para-nut,Parfa Nut,Savoury-nut,Shoe Nut ●☆
53058 Bertiera Aubl.(1775);贝尔茜属■☆
53059 Bertiera Blume = Mycetia Reinw. ●
53060 Bertiera adamsii (Hepper) N. Hallé;亚当斯贝尔茜■☆
53061 Bertiera aequatorialis N. Hallé;赤道贝尔茜■☆

53062 Bertiera aethiopica Hiern;埃塞俄比亚贝尔茜■☆
53063 Bertiera africana A. Rich. = Bertiera spicata (C. F. Gaertn.) K. Schum. ■☆
53064 Bertiera annobonensis G. Taylor;安诺本贝尔茜■☆
53065 Bertiera arctistipula N. Hallé;直托叶贝尔茜■☆
53066 Bertiera batesii Wernham;贝茨贝尔茜■☆
53067 Bertiera bequaertii De Wild. = Bertiera iturensis K. Krause ■☆
53068 Bertiera bicarpellata (K. Schum.) N. Hallé;双小果贝尔茜■☆
53069 Bertiera bosscheana De Wild. = Bertiera subsessilis Hiern var. congolana (De Wild. et T. Durand) N. Hallé ■☆
53070 Bertiera bracteolata Hiern;小苞片贝尔茜■☆
53071 Bertiera breviflora Hiern;短花贝尔茜■☆
53072 Bertiera capitata De Wild. = Bertiera naucleoides (S. Moore) Bridson ■☆
53073 Bertiera chevalieri Hutch. et Dalziel;舍瓦利耶贝尔茜■☆
53074 Bertiera cinereoviridis K. Schum. = Bertiera aethiopica Hiern ■☆
53075 Bertiera coccinea (G. Don) G. Don = Mussaenda elegans Schumach. et Thonn. ●☆
53076 Bertiera congolana De Wild. et T. Durand = Bertiera subsessilis Hiern var. congolana (De Wild. et T. Durand) N. Hallé ■☆
53077 Bertiera dewevrei De Wild. et T. Durand = Bertiera racemosa (G. Don) K. Schum. var. dewevrei (De Wild. et T. Durand) N. Hallé ■☆
53078 Bertiera fasciculata Blume;簇生贝尔茜■☆
53079 Bertiera fimbriata (A. Chev. ex Hutch. et Dalziel) Hepper;流苏贝尔茜■☆
53080 Bertiera glabrata K. Schum. = Bertiera racemosa (G. Don) K. Schum. var. glabrata (K. Schum.) Hutch. et Dalziel ■☆
53081 Bertiera globiceps K. Schum.;球头贝尔茜■☆
53082 Bertiera gracilis De Wild. = Bertiera aethiopica Hiern ■☆
53083 Bertiera grandis Mildbr.;大贝尔茜■☆
53084 Bertiera iturensis K. Krause;伊图里贝尔茜■☆
53085 Bertiera laurentii De Wild.;洛朗贝尔茜■☆
53086 Bertiera laurentii De Wild. var. compacta? = Bertiera laurentii De Wild. ■☆
53087 Bertiera laxa Benth.;疏松贝尔茜■☆
53088 Bertiera laxa Benth. var. bamendae Hepper;巴门达贝尔茜■☆
53089 Bertiera laxa Benth. var. pedicellata Hiern = Bertiera pedicellata (Hiern) Wernham ■☆
53090 Bertiera laxissima K. Schum.;极松贝尔茜■☆
53091 Bertiera ledermannii K. Krause;莱德贝尔茜■☆
53092 Bertiera letouzeyi Hallé;勒图贝尔茜■☆
53093 Bertiera longiloba K. Krause;长裂贝尔茜■☆
53094 Bertiera longithyrsa Baker;长序贝尔茜■☆
53095 Bertiera lujae De Wild.;卢亚贝尔茜■☆
53096 Bertiera macrocarpa Benth. = Bertiera racemosa (G. Don) K. Schum. ■☆
53097 Bertiera maitlandii Hutch. et Dalziel = Bertiera laxa Benth. ■☆
53098 Bertiera mildbraedii K. Krause = Bertiera thonneri De Wild. et T. Durand ■☆
53099 Bertiera montana Hiern = Bertiera racemosa (G. Don) K. Schum. ■☆
53100 Bertiera naucleoides (S. Moore) Bridson;乌檀贝尔茜■☆
53101 Bertiera obversa K. Krause = Bertiera retrofracta K. Schum. ■☆
53102 Bertiera orthopetala (Hiern) N. Hallé;直瓣贝尔茜■☆
53103 Bertiera pauloi Verdc.;保罗贝尔茜■☆
53104 Bertiera pedicellata (Hiern) Wernham;梗花贝尔茜■☆
53105 Bertiera pomatium Benth. = Bertiera spicata (C. F. Gaertn.) K. Schum. ■☆
53106 Bertiera racemosa (G. Don) K. Schum.;总花贝尔茜■☆
53107 Bertiera racemosa (G. Don) K. Schum. var. dewevrei (De Wild. et T. Durand) N. Hallé;德韦贝尔茜■☆
53108 Bertiera racemosa (G. Don) K. Schum. var. elephantina N. Hallé;象贝尔茜■☆
53109 Bertiera racemosa (G. Don) K. Schum. var. glabrata (K. Schum.) Hutch. et Dalziel;光滑贝尔茜■☆
53110 Bertiera retrofracta K. Schum.;反曲贝尔茜■☆
53111 Bertiera rosseeliana Sonké et Esono et Nguembou;罗西尔贝尔茜■☆
53112 Bertiera simplicicaulis N. Hallé = Bertiera bicarpellata (K. Schum.) N. Hallé ■☆
53113 Bertiera sphaerica N. Hallé;球形贝尔茜■☆
53114 Bertiera spicata (C. F. Gaertn.) K. Schum.;穗状贝尔茜■☆
53115 Bertiera spicata (C. F. Gaertn.) K. Schum. var. minor K. Schum.;较小穗状贝尔茜■☆
53116 Bertiera stenothyrsus K. Schum. = Bertiera bracteolata Hiern ■☆
53117 Bertiera subsessilis Hiern;近无柄贝尔茜■☆
53118 Bertiera subsessilis Hiern var. congolana (De Wild. et T. Durand) N. Hallé;刚果贝尔茜■☆
53119 Bertiera tenuiflora Wernham = Bertiera iturensis K. Krause ■☆
53120 Bertiera tessmannii K. Krause;泰斯曼贝尔茜■☆
53121 Bertiera thollonii N. Hallé;托伦贝尔茜■☆
53122 Bertiera thonneri De Wild. et T. Durand;托内贝尔茜■☆
53123 Bertiera tisserantii N. Hallé = Bertiera iturensis K. Krause ■☆
53124 Bertiera troupinii N. Hallé;特鲁皮尼贝尔茜■☆
53125 Bertiera zenkeri Mildbr.;岑克尔贝尔茜■☆
53126 Bertolonia DC. = Bertolonia Raddi(保留属名)■☆
53127 Bertolonia DC. = Leucheria Lag. ■☆
53128 Bertolonia Moc. et Sessé = Cercocarpus Kunth ●☆
53129 Bertolonia Moc. et Sessé ex DC. = Cercocarpus Kunth ●☆
53130 Bertolonia Raddi(1820)(保留属名);华贵草属(拜氏野牡丹属)■☆
53131 Bertolonia Raf. = Phyla Lour. ■
53132 Bertolonia Spin(废弃属名) = Bertolonia Raddi(保留属名)■☆
53133 Bertolonia Spin(废弃属名) = Myoporum Banks et Sol. ex G. Forst. ●
53134 Bertolonia Spreng. = Tovomitopsis Planch. et Triana ●☆
53135 Bertolonia guttata Hook. = Gravesia guttata (Hook.) Triana ●☆
53136 Bertolonia maculata DC.;斑点华贵草■☆
53137 Bertolonia maculata DC. var. marmorata (Naudin) Planch. = Bertolonia marmorata (Naudin) Naudin ■☆
53138 Bertolonia marmorata (Naudin) Naudin;大理石状华贵草(斑纹华贵草)■☆
53139 Bertolonia marmorata (Naudin) Naudin var. aenea Cogn.;铜光大理石状华贵草■☆
53140 Bertolonia marmorata Naudin = Bertolonia marmorata (Naudin) Naudin ■☆
53141 Bertuchia Dennst. = Fagraea Thunb. ●
53142 Bertuchia Dennst. = Gardenia Ellis(保留属名)●
53143 Bertya Planch. (1845);贝梯大戟属●☆
53144 Bertya oblongifolia Müll. Arg.;矩圆叶贝梯大戟●☆
53145 Bertya oppositifolia F. Muell.;对叶贝梯大戟●☆
53146 Bertya pinifolia Planch.;松叶贝梯大戟●☆
53147 Bertya polymorpha Baill.;多形贝梯大戟●☆

53148 Bertya rotundifolia F. Muell.;圆叶贝梯大戟●☆
53149 Bertyaceae J. Agardh = Euphorbiaceae Juss.(保留科名)●■
53150 Berula Besser et W. D. J. Koch = Berula W. D. J. Koch ■
53151 Berula Hoffm. = Berula W. D. J. Koch ■
53152 Berula Hoffm. ex Besser = Berula W. D. J. Koch ■
53153 Berula Hoffm. ex Besser = Sium L. ■
53154 Berula W. D. J. Koch(1826);天山泽芹属(毕若拉属);Berula, Lesser Water-parsnip ■
53155 Berula angustifolia (L.) Mert. et Koch = Berula erecta (Huds.) Coville ■
53156 Berula angustifolia Mert. et W. D. J. Koch = Berula erecta (Huds.) Coville ■
53157 Berula erecta (Huds.) Coville;天山泽芹(窄叶毒人参);Cut-leaved Water-parsnip, Erect Berula, Lesser Water Parsnip, Lesser Water-parsnip, Low Water-parsnip, Narrowleaf Waterparsnip, Narrow-leaved Water Parsnip, Narrow-leaved Water-parsnip, Stalky Berula, Water Parsnip ■
53158 Berula erecta (Huds.) Coville subsp. thunbergii (DC.) B. L. Burtt;通贝里天山泽芹■☆
53159 Berula erecta (Huds.) Coville var. incisa (Torr.) Cronquist = Berula erecta (Huds.) Coville ■
53160 Berula incisa (Torr.) G. N. Jones = Berula erecta (Huds.) Coville ■
53161 Berula orientalis Woronow ex Schischk.;东方天山泽芹■☆
53162 Berula pusilla (Nutt. ex Torr. et A. Gray) Fernald = Berula erecta (Huds.) Coville ■
53163 Berula thunbergii (DC.) H. Wolff = Berula erecta (Huds.) Coville subsp. thunbergii (DC.) B. L. Burtt ■☆
53164 Beruniella Zakirov et Nabiev = Heliotropium L. ●■
53165 Beruniella micrantha (Pall.) Zakirov et Nabiev = Heliotropium micranthum (Pall.) Bunge ■
53166 Beryllis Salisb. = Ornithogalum L. ■
53167 Berylsimpsonia B. L. Turner(1993);弯刺钝柱菊属●☆
53168 Berylsimpsonia crassinervis (Urb.) B. L. Turner;粗脉弯刺钝柱菊●☆
53169 Berylsimpsonia vanillosma (C. Wright) B. L. Turner;弯刺钝柱菊●☆
53170 Berzelia Brongn.(1826);饰球花属(贝柘丽木属)●☆
53171 Berzelia Brongn. = Brunia L.(废弃属名)●☆
53172 Berzelia Brongn. = Brunia Lam.(保留属名)●☆
53173 Berzelia Mart. = Hermbstaedtia Rchb. ■●☆
53174 Berzelia abrotanoides (L.) Brongn.;青蒿饰球花●☆
53175 Berzelia alopecurioides (Thunb.) Sond. = Brunia alopecuroides Thunb. ●☆
53176 Berzelia arachnoidea (J. C. Wendl.) Eckl. et Zeyh. = Berzelia squarrosa (Thunb.) Sond. ●☆
53177 Berzelia burchellii Dümmer;伯切尔饰球花●☆
53178 Berzelia callunoides Oliv. = Mniothamnea callunoides (Oliv.) Nied. ●☆
53179 Berzelia commutata Sond.;变异饰球花●☆
53180 Berzelia comosa Eckl. et Zeyh. = Berzelia commutata Sond. ●☆
53181 Berzelia cordifolia Schltdl.;心叶饰球花●☆
53182 Berzelia dregeana Colozza;德雷饰球花●☆
53183 Berzelia ecklonii Pillans;埃氏饰球花●☆
53184 Berzelia galpinii Pillans;盖尔饰球花●☆
53185 Berzelia glauca (J. C. Wendl.) Mart. = Hermbstaedtia glauca (J. C. Wendl.) Rchb. ex Steud. ■☆
53186 Berzelia incurva Pillans;内折饰球花●☆
53187 Berzelia intermedia (D. Dietr.) Schltdl.;全叶饰球花●☆
53188 Berzelia lanuginosa (L.) Brongn.;饰球花(贝柘丽木)●☆
53189 Berzelia lanuginosa Brongn. = Berzelia lanuginosa (L.) Brongn. ●☆
53190 Berzelia rogersii N. E. Br. = Brunia albiflora E. Phillips ●☆
53191 Berzelia rubra Schltdl.;红饰球花●☆
53192 Berzelia squarrosa (Thunb.) Sond.;粗鳞饰球花●☆
53193 Berzelia squarrosa (Thunb.) Sond. var. reflexa Sond. = Berzelia rubra Schltdl. ●☆
53194 Berzelia superba (Donn) Eckl. et Zeyh. = Berzelia lanuginosa (L.) Brongn. ●☆
53195 Berzeliaceae Nakai = Bruniaceae R. Br. ex DC.(保留科名)●☆
53196 Berzeliaceae Nakai;饰球花科●
53197 Beschorneria Kunth(1850);龙舌草属■☆
53198 Beschorneria floribunda Hort. ex K. Koch;繁花龙舌草■☆
53199 Beschorneria multiflora Hort. ex K. Koch;多花龙舌草■☆
53200 Beschorneria yuccoides K. Koch;丝兰龙舌草■☆
53201 Besenna A. Rich. = Albizia Durazz. ●
53202 Besenna anthelmintica A. Rich. = Albizia anthelmintica Brongn. ●
53203 Besha D. Dietr. = Beesha Munro ●☆
53204 Besha D. Dietr. = Ochlandra Thwaites ●☆
53205 Besleria L.(1753);贝思乐苣苔属;Besleria ●■☆
53206 Besleria Plum. ex L. = Besleria L. ●■☆
53207 Besleria bivalvis L. f.;双门贝思乐苣苔■☆
53208 Besleria lutea L.;金黄贝思乐苣苔;Golden-yellow Besleria ■☆
53209 Besleriaceae Raf. = Gesneriaceae Rich. et Juss.(保留科名)■●
53210 Besoniaceae Bercht. et J. Presl;贝思乐苣苔科●■
53211 Bessera Schult.(废弃属名)= Bessera Schult. f.(保留属名)■☆
53212 Bessera Schult.(废弃属名)= Pulmonaria L. ■
53213 Bessera Schult. f.(1809)(保留属名);合丝韭属(白丝瑞属);Coral Drops, Coral-drops, Mexican Coral Drops ■☆
53214 Bessera Spreng. = Flueggea Willd. ●
53215 Bessera Spreng. = Xylosma G. Forst.(保留属名)●
53216 Bessera Vell. = Pisonia L. ●
53217 Bessera breviflora Jeps.;短脉合丝韭■☆
53218 Bessera elegans Schult. f.;合丝韭(白丝瑞);Coral Drops, Coral-drops ■☆
53219 Bessera tenuiflora J. F. Macbr.;细花合丝韭■☆
53220 Besseya Rydb.(1903);珊瑚参属■
53221 Besseya bullii (Eaton) Rydb.;布尔珊瑚参;Bull's Coral-drops, Kitten's-tails ■☆
53222 Bessia Raf. = Intsia Thouars ●☆
53223 Bestram Adans. = Antidesma L. ●
53224 Beta L.(1753);甜菜属(莙荙菜属);Beet, Chard ■
53225 Beta cicla L. = Beta vulgaris L. var. cicla L. ■
53226 Beta cicla L. = Beta vulgaris L. ■
53227 Beta cicla L. variegata?;智利甜菜;Chilean Beet ■☆
53228 Beta corolliflora L.;冠花甜菜■☆
53229 Beta diffusa Coss. = Patellifolia patellaris (Moq.) A. J. Scott, Ford-Lloyd et J. T. Williams ■☆
53230 Beta macrocarpa Guss.;大果甜菜■☆
53231 Beta macrorhiza Steven;大根甜菜;Sea Beet ■☆
53232 Beta maritima L. = Beta vulgaris L. subsp. maritima (L.) Arcang. ■☆
53233 Beta maritima L. = Beta vulgaris L. ■
53234 Beta monodiana Maire = Patellifolia patellaris (Moq.) A. J.

Scott, Ford-Lloyd et J. T. Williams ■☆
53235 Beta patellaris Moq.;碟状甜菜■☆
53236 Beta patellaris Moq. = Patellifolia patellaris (Moq.) A. J. Scott, Ford-Lloyd et J. T. Williams ■☆
53237 Beta patellaris Moq. f. luthereaui (Maire) Maire = Patellifolia patellaris (Moq.) A. J. Scott, Ford-Lloyd et J. T. Williams ■☆
53238 Beta patellaris Moq. var. diffusa (Coss.) Maire = Patellifolia patellaris (Moq.) A. J. Scott, Ford-Lloyd et J. T. Williams ■☆
53239 Beta patellaris Moq. var. luthereaui Maire = Patellifolia patellaris (Moq.) A. J. Scott, Ford-Lloyd et J. T. Williams ■☆
53240 Beta patellaris Moq. var. monodiana (Maire) Maire = Patellifolia patellaris (Moq.) A. J. Scott, Ford-Lloyd et J. T. Williams ■☆
53241 Beta patula Sol.;张开甜菜■☆
53242 Beta perennis (L.) Freyn;多年生甜菜■☆
53243 Beta procumbens C. Sm.;平铺甜菜;Cultivated Beet ■☆
53244 Beta procumbens C. Sm. = Patellifolia procumbens (C. Sm.) A. J. Scott et Ford-Lloyd et J. T. Williams ■☆
53245 Beta rapacea Hegetschw.;饲料甜菜;Mangel, Mangelwurzel, Mangold ■☆
53246 Beta rapacea Hegetschw. = Beta vulgaris L. ■
53247 Beta saccharifera? = Beta vulgaris L. var. saccharifera Alef. ■
53248 Beta trigina Waldst. et Kit.;三花柱甜菜;Caucasian Beet ■☆
53249 Beta vulgaris L.;甜菜(杓菜,出莙荙儿,根刀菜,光菜,海白菜,红牛皮菜,红头菜,厚皮菜,火焰菜,莙荙,莙荙菜,牛皮菜,茄茉菜,石菜,糖菜,糖萝卜,甜萝卜,菾菜,猪乸菜);Beet, Beetroot, Biennial Spinach Beet, Cattle-beet, Common Beet, Earthbeet, Fodder Beet, Mangelwurzel, Mangel-wurzel, Mangold, Perpetual Spinach, Root Beet, Sea-kale Beet, Spinach Beet, Spinach Chard, Sugar Beet, Swiss Chard ■
53250 Beta vulgaris L. hortensis? = Beta vulgaris L. var. esculenta Gürke ■☆
53251 Beta vulgaris L. hortensis? = Beta vulgaris L. var. rubra Moq. ■
53252 Beta vulgaris L. subsp. cicla (L.) Arcang. = Beta vulgaris L. var. cicla L. ■
53253 Beta vulgaris L. subsp. macrocarpa (Guss.) Thell. = Beta macrocarpa Guss. ■☆
53254 Beta vulgaris L. subsp. maritima (L.) Arcang.;海滨甜菜;Beet, Beetraw, Beetrie, Cattle Beet, Field Beet, Sea Beet, Sea Spinach ■☆
53255 Beta vulgaris L. var. altissima Döll;甜萝卜(极高甜菜,糖萝卜);Sugar Beet ■
53256 Beta vulgaris L. var. annua Asch. et Graebn. = Beta vulgaris L. subsp. maritima (L.) Arcang. ■☆
53257 Beta vulgaris L. var. cicla L.;莙荙菜(杓菜,观赏菾菜,光菜,海白菜,厚瓣菜,厚皮菜,牛皮菜,石菜,唐萵苣,甜菜,菾菜,猪乸菜);Beet Chard, Chard, Chilean Beet, Foliage Beet, Leaf Beet, Leaf-beet, Mangold, Ornamental Leaf Beet, Roman Jasmine, Roman Jessamine, Roman Kale, Sea-kale, Seakale Beet, Sicilian Beet, Silver Beet, Spanish Beet, Spinach Beet, Swiss Chard, White Beet ■
53258 Beta vulgaris L. var. debeauxii Clary = Beta vulgaris L. subsp. maritima (L.) Arcang. ■☆
53259 Beta vulgaris L. var. esculenta Gürke;食用甜菜;Garden Beet, Red Beet, Red Garden Beet ■☆
53260 Beta vulgaris L. var. flavescens DC. = Beta vulgaris L. var. cicla L. ■
53261 Beta vulgaris L. var. lutea DC.;饲用甜菜;Yellow Beet ■
53262 Beta vulgaris L. var. macrorrhiza?;大根莙荙菜;Fodder Beet ■☆
53263 Beta vulgaris L. var. maritima (L.) Boiss. = Beta vulgaris L. subsp. maritima (L.) Arcang. ■☆
53264 Beta vulgaris L. var. maritima (L.) K. Koch = Beta vulgaris L. subsp. maritima (L.) Arcang. ■☆
53265 Beta vulgaris L. var. perennis? = Beta vulgaris L. subsp. maritima (L.) Arcang. ■☆
53266 Beta vulgaris L. var. rapa Dumort.;砂糖莱菔(甜菜);Sugar Beet ■
53267 Beta vulgaris L. var. rosea Moq.;紫菜头(红菜头,紫萝卜);Beet, Beetroot, Rose Beet ■
53268 Beta vulgaris L. var. rosea Moq. = Beta vulgaris L. ■
53269 Beta vulgaris L. var. rubra Moq.;火焰菜;Beet-root, Garden Beet, Red Beet, Table Beet ■
53270 Beta vulgaris L. var. saccharifera Alef. = Beta vulgaris L. var. altissima Döll ■
53271 Beta vulgaris L. var. saccharifera Alef. = Beta vulgaris L. ■
53272 Beta vulgaris L. var. saccharifera Lange = Beta vulgaris L. var. rapa Dumort. ■
53273 Beta webbiana Moq. = Patellifolia webbiana (Moq.) A. J. Scott et al. ■☆
53274 Betaceae Burnett = Amaranthaceae Juss.(保留科名)●■
53275 Betaceae Burnett = Chenopodiaceae Vent.(保留科名)●■
53276 Betaceae Burnett = Chenopodiaecae + Amaranthaceae Juss.(保留科名)●■
53277 Betaceae Burnett;甜菜科■
53278 Betanthus melongenifolius A. Chev. = Atractogyne bracteata (Wernham) Hutch. et Dalziel ●☆
53279 Betchea Schltr. = Caldcluvia D. Don ●☆
53280 Betckea DC. (1830);贝才茜属■☆
53281 Betckea DC. = Valerianella Mill. ■
53282 Betckea caucasica Boiss.;高加索贝才茜■☆
53283 Betckea gilliesii Hook. et Arn.;吉尔贝才茜■☆
53284 Betckea heterophylla Phil.;互叶贝才茜■☆
53285 Betckea major Fisch. et E. Mey.;大贝才茜■☆
53286 Betela Raf. = Piper L. ●■
53287 Betenoourtia A. St. -Hil. = Galactia P. Browne ■
53288 Bethencourtia Choisy = Senecio L. ■●
53289 Bethencourtia Choisy(1825);小头尾药菊属●☆
53290 Bethencourtia palmensis (Nees) Choisy = Bethencourtia palmensis (Nees) Link ■☆
53291 Bethencourtia palmensis (Nees) Link;小头尾药菊■☆
53292 Bethencourtia rupicola (B. Nord.) B. Nord.;岩地小头尾药菊■☆
53293 Betonica L. (1753);药水苏属;Betonica, Woundwort ■
53294 Betonica L. = Stachys L. ●■
53295 Betonica abchasica (Bornm.) Chinth.;阿伯哈斯药水苏■☆
53296 Betonica algeriensis Noë = Stachys officinalis (L.) Trevis. subsp. algeriensis (Noë) Franco ■☆
53297 Betonica capensis Burm. f. = Stachys aethiopica L. ■☆
53298 Betonica foliosa Rupr.;多叶药水苏(藿香)■☆
53299 Betonica glabrata C. Koch = Betonica officinalis L. ■
53300 Betonica glabrata K. Koch = Betonica officinalis L. ■
53301 Betonica grandiflora Steph. ex Willd.;大花药水苏;Bigflower Betonica ■☆
53302 Betonica laevigata D. Don = Nepeta laevigata (D. Don) Hand. -Mazz. ■
53303 Betonica nivea Stev.;雪白药水苏■☆
53304 Betonica officinalis L.;药水苏(药用水苏);Beton, Betony, Bishop's Flower, Bishop's Wort, Bitny Bidny, Briny, Common Betony, Common Hedgenettle, Devil's Plaything, Harry Nettle, Hyssop,

Medicinal Betonica, St. Bride's Comb, Wild Hop, Wood Betony, Woolly Betony ■
53305 Betonica officinalis L. = Stachys officinalis (L.) Trevis. ■
53306 Betonica officinalis L. var. algeriensis (Noë) Ball = Stachys officinalis (L.) Trevis. subsp. algeriensis (Noë) Franco ■☆
53307 Betonica orientalis L.;东方药水苏■☆
53308 Betonica ossetica (Bornm.) Chinth.;骨质药水苏■☆
53309 Betula L. (1753);桦木属;Birch ●
53310 Betula × purpusii C. K. Schneid.;普氏桦;Purpus' Birch ●☆
53311 Betula × sandbergii Britton;桑氏桦;Sandberg's Birch ●☆
53312 Betula acuminata Wall. = Betula alnoides Buch. -Ham. ex D. Don ●
53313 Betula acuminata Wall. = Betula luminifera H. Winkl. ●
53314 Betula acuminata Wall. var. cylindrostachya (Lindl.) Regel = Betula cylindrostachya Lindl. ●
53315 Betula acuminata Wall. var. cylindrostachya Regel = Betula cylindrostachya Wall. ●
53316 Betula acuminata Wall. var. pyrifolia Franch. = Betula luminifera H. Winkl. ●
53317 Betula ajanensis Kom.;阿扬湾桦●☆
53318 Betula alajica Litv.;阿拉桦■☆
53319 Betula alaskana Sarg. = Betula neoalaskana Sarg. ●☆
53320 Betula alba L. = Betula pendula Roth ●
53321 Betula alba L. subsp. latifolia Regel = Betula platyphylla Sukaczev ●
53322 Betula alba L. subsp. mandshurica Regel = Betula platyphylla Sukaczev ●
53323 Betula alba L. subsp. populifolia (Marshall) Regel = Betula populifolia Marshall ●☆
53324 Betula alba L. subsp. soongarica var. microphylla Regel = Betula tianschanica Rupr. ●
53325 Betula alba L. subsp. tauschii Regel = Betula platyphylla Sukaczev ●
53326 Betula alba L. var. cordifolia (Regel) Fernald = Betula cordifolia Regel ●☆
53327 Betula alba L. var. cordifolia (Regel) Regel = Betula cordifolia Regel ●☆
53328 Betula alba L. var. dalecarlica L. f. = Betula pendula Roth ●
53329 Betula alba L. var. elobata Fernald = Betula papyrifera Marshall ●☆
53330 Betula alba L. var. fontqueri (Rothm.) Maire et Weiller = Betula pendula Roth subsp. fontqueri (Rothm.) Moreno et Peinado ●☆
53331 Betula alba L. var. japonica Miq. = Betula japonica Siebold ex H. Winkl. ●
53332 Betula alba L. var. papyrifera (Marshall) Spach = Betula papyrifera Marshall ●☆
53333 Betula alba L. var. populifolia (Marshall) Spach = Betula populifolia Marshall ●☆
53334 Betula alba L. var. pubescens (Ehrh.) Spach = Betula pubescens Ehrh. ●☆
53335 Betula albosinensis Burkill;红桦(风桦,红皮桦,鳞皮桦,纸皮桦);China Paper Birch, China-paper Birch, Chinese Birch, Chinese Red Birch, Chinese Red-barked Birch, White Chinese Birch ●
53336 Betula albosinensis Burkill var. septentrionalis C. K. Schneid.;牛皮桦(臭桦,毛桦);North Chinapaper Birch ●
53337 Betula albosinensis Burkill var. septentrionalis C. K. Schneid. = Betula utilis D. Don ●
53338 Betula alleghaniensis Britton;加拿大黄桦(黄桦,黄皮桦,魁北克桦,美国桦,银桦,硬桦,沼泽桦);American Birch, Canada Birch, Canadian Birch, Canadian Yellow Birch, Gray Birch, Hard Birch, Merisier, Quebec Birch, Silver Birch, Swamp Birch, Tall Birch, Yellow Birch ●☆
53339 Betula alleghaniensis Britton var. fallax (Fassett) Brayshaw = Betula alleghaniensis Britton ●☆
53340 Betula alleghaniensis Britton var. macrolepis (Fernald) Brayshaw = Betula alleghaniensis Britton ●☆
53341 Betula alnoides Buch. -Ham. = Betula luminifera H. Winkl. ●
53342 Betula alnoides Buch. -Ham. ex D. Don;西桦(西南桦,西南桦木);Alder Birch ●
53343 Betula alnoides Buch. -Ham. ex D. Don var. acuminata (Wall.) H. Winkl. = Betula alnoides Buch. -Ham. ex D. Don ●
53344 Betula alnoides Buch. -Ham. ex D. Don var. cylindrostachya (Lindl.) H. Winkl. = Betula alnoides Buch. -Ham. ex D. Don ●
53345 Betula alnoides Buch. -Ham. ex D. Don var. cylindrostachya Regel = Betula cylindrostachya Wall. ●
53346 Betula alnoides Buch. -Ham. ex D. Don var. pyrifolia (Franch.) Burkill = Betula luminifera H. Winkl. ●
53347 Betula alnoides Buch. -Ham. ex D. Don var. pyrifolia Franch.;蒙自桦;Mengzi Alder Birch ●
53348 Betula alnoides Buch. -Ham. var. cylindrostachya (Lindl.) H. Winkl. = Betula cylindrostachya Lindl. ●
53349 Betula alnus L. var. glutinosa L. = Alnus glutinosa (L.) Gaertn. ●
53350 Betula alnus L. var. incana L. = Alnus incana (L.) Moench ●☆
53351 Betula alnus L. var. rugosa Du Roi = Alnus incana (L.) Moench subsp. rugosa (Du Roi) R. T. Clausen ●☆
53352 Betula annae L.;安娜桦●☆
53353 Betula apoiensis Nakai;阿伯伊桦;Apoi Birch ●☆
53354 Betula austrosinensis Chun ex P. C. Li;华南桦;S. China Birch, South China Birch ●
53355 Betula aveczensis Kom.;库页白桦●☆
53356 Betula baeumkeri H. Winkl. = Betula luminifera H. Winkl. ●
53357 Betula baicalensis Sukaczev;拜卡尔桦●☆
53358 Betula bhojpattra Lindl. = Betula utilis D. Don ●
53359 Betula bhojpattra Lindl. ex Wall. = Betula utilis D. Don ●
53360 Betula bhojpattra Lindl. var. latifolia Regel = Betula utilis D. Don ●
53361 Betula bhojpattra Lindl. var. sinensis Franch. = Betula albosinensis Burkill ●
53362 Betula bhojpattra Wall.;印度纸桦;Indian Paper Birch ●☆
53363 Betula bhojpattra Wall. = Betula utilis D. Don ●
53364 Betula bhojpattra Wall. var. sinensis Franch. = Betula albosinensis Burkill ●
53365 Betula bhojpattra Wall. var. typica? = Betula utilis D. Don ●
53366 Betula bomiensis P. C. Li;波密桦;Bomi Birch ●
53367 Betula bomiensis P. C. Li. = Betula delavayi Pall. var. microstachys P. C. Li ●
53368 Betula borealis Spach = Betula pumila L. ●
53369 Betula caerulea Blanch.;蓝桦;Blue Birch ●☆
53370 Betula cajanderi Sukaczev;卡亚桦●☆
53371 Betula calcicola (W. W. Sm.) Hu ex P. C. Li;岩桦(毛枝高山桦);Calcicolous Birch ●
53372 Betula callosa Noto;硬皮桦●☆
53373 Betula candelae Koidz. = Betula maximowicziana Regel ●
53374 Betula carpinifolia Ehrh. = Betula lenta L. ●
53375 Betula celtiberica Rothm. et Vasc.;葡西桦●☆
53376 Betula ceratoptera G. H. Liu et Ma;角翅桦;Horned-wing Birch ●
53377 Betula ceratoptera G. H. Liu et Ma = Betula chinensis Maxim. ●

53378 Betula chichibuensis H. Hara;秩父桦●☆

53379 Betula chinensis Maxim.;坚桦(杵桦,杵榆,杵榆桦,垂榆,黑桦,桦,小桦木);China Birch,Chinese Birch ●

53380 Betula chinensis Maxim. f. linearisquama (Hatus.) S. L. Tung;线鳞坚桦(坚桦)●☆

53381 Betula chinensis Maxim. var. angusticarpa H. Winkl. = Betula chinensis Maxim. ●

53382 Betula chinensis Maxim. var. delavayi C. K. Schneid. = Betula delavayi Franch. ●

53383 Betula chinensis Maxim. var. fargesii (Franch.) Hu ex P. C. Li = Betula fargesii Franch. ●

53384 Betula chinensis Maxim. var. fargesii (Franch.) P. C. Li = Betula fargesii Franch. ●

53385 Betula chinensis Maxim. var. nana Liou = Betula chinensis Maxim. ●

53386 Betula concinna Gunnarsson;整洁桦●☆

53387 Betula cordifolia Regel;心叶桦;Heart-leaf Bbirch,Heartleaf Birch,Mountain White Birch ●☆

53388 Betula corylifolia Regel et Maxim.;日本榛叶桦(里白桦);Heart-leaf Birch,Japanese Birch,Mountain White Birch ●☆

53389 Betula costata Trautv.;硕桦(臭桦,风桦,枫桦,黄桦,驴脚桦,千层桦,日光桦);Ribbed Birch ●

53390 Betula costata Trautv. = Betula ermanii Champ. 'Grayswood Hill' ●

53391 Betula costata Trautv. var. pubescens Liou;柔毛硕桦;Pubescent Ribbed Birch ●

53392 Betula crispa Aiton = Alnus viridis (Vill.) DC. subsp. crispa (Aiton) Turrill ●☆

53393 Betula cylindrostachya Diels = Betula luminifera H. Winkl. ●

53394 Betula cylindrostachya Lindl.;长穗桦(桦,桦阁,桦槁树);Longspike Birch,Long-spiked Birch ●

53395 Betula cylindrostachya Lindl. var. resinosa Diels = Betula luminifera H. Winkl. ●

53396 Betula cylindrostachya Wall. = Betula alnoides Buch.-Ham. ex D. Don ●

53397 Betula cylindrostachya Wall. var. pilosa Regel = Betula alnoides Buch.-Ham. ex D. Don ●

53398 Betula cylindrostachya Wall. var. resinosa Diels = Betula luminifera H. Winkl. ●

53399 Betula cylindrostachya Wall. var. typica Regel = Betula cylindrostachya Wall. ●

53400 Betula dahurica Pall.;黑桦(长圆叶黑桦,臭桦,椴叶黑桦,棘皮桦,卵圆叶黑桦,马氏棘皮桦,千层桦,万昌桦);Asian Black Birch,Dahur Birch,Dahurian Birch,Maximowicz Birch,Oblong-leaf Dahurian Birch,Ovate-leaf Birch,Tilia-leaf Birch ●

53401 Betula dahurica Pall. var. oblongifolia Liou = Betula dahurica Pall. ●

53402 Betula dahurica Pall. var. okuboi Miyabe et Tatew.;大久桦●☆

53403 Betula dahurica Pall. var. ovalifolia Liou = Betula dahurica Pall. ●

53404 Betula davurica Pall. = Betula dahurica Pall. ●

53405 Betula delavayi Franch.;高山桦(大叶高山桦,崖桦);Delavay Birch ●

53406 Betula delavayi Franch. var. calcicola W. W. Sm. = Betula calcicola (W. W. Sm.) Hu ex P. C. Li ●

53407 Betula delavayi Franch. var. forrestii W. W. Sm. = Betula delavayi Franch. ●

53408 Betula delavayi Pall. var. calcicola W. W. Sm. = Betula calcicola (W. W. Sm.) Hu ex P. C. Li ●

53409 Betula delavayi Pall. var. microstachys P. C. Li;细穗高山桦;Smallspike Birch ●

53410 Betula delavayi Pall. var. polyneura L. C. Hu ex P. C. Li;多脉高山桦;Manyvein Delavay Birch ●

53411 Betula ermanii Champ.;岳桦;Erman's Birch,Gold Birch,Russian Rock Birch ●

53412 Betula ermanii Champ. 'Grayswood Hill';灰木山岳桦●

53413 Betula ermanii Champ. f. corticosa (Nakai) Sugim.;厚皮岳桦●☆

53414 Betula ermanii Champ. var. communis Koidz. = Betula ermanii Champ. ●

53415 Betula ermanii Champ. var. corticosa Nakai = Betula ermanii Champ. f. corticosa (Nakai) Sugim. ●☆

53416 Betula ermanii Champ. var. costata (Trautv.) Regel = Betula costata Trautv. ●

53417 Betula ermanii Champ. var. costata Regel = Betula costata Trautv. ●

53418 Betula ermanii Champ. var. ganjuensis Nakai = Betula ganjuensis Koidz. ●☆

53419 Betula ermanii Champ. var. genuina Regel = Betula ermanii Champ. ●

53420 Betula ermanii Champ. var. incisa Koidz.;缺刻岳桦●

53421 Betula ermanii Champ. var. japonica (Shirai) Koidz.;日本岳桦;Japonese Erman's Birch ●☆

53422 Betula ermanii Champ. var. lanata Regel = Betula ermanii Champ. ●

53423 Betula ermanii Champ. var. macrostrobila Liou;帽儿山岳桦;Maoershan Birch ●

53424 Betula ermanii Champ. var. nipponica Maxim. = Betula ermanii Champ. ●

53425 Betula ermanii Champ. var. parvifolia Koidz.;小叶岳桦●☆

53426 Betula ermanii Champ. var. subcordata (Regel) Koidz.;近心叶岳桦;Subcordate Ermans Birch ●

53427 Betula ermanii Champ. var. subglobosa Miyabe et Kudo;亚球叶岳桦●☆

53428 Betula ermanii Champ. var. typica Regel = Betula ermanii Champ. ●

53429 Betula ermanii Champ. var. yingkiliensis Liou et Z. Wang;英吉里岳桦;Yingkili Ermans Birch ●

53430 Betula exalata S. Moore = Betula chinensis Maxim. ●

53431 Betula excelsa Hook. = Betula lenta L. ●

53432 Betula excelsa L. ex B. D. Jacks.;黄桦;Yellow Birch ●☆

53433 Betula excelsa Pursh = Betula lutea Michx. ●☆

53434 Betula exilis Sukaczev;瘦桦●☆

53435 Betula exilis Sukaczev = Betula nana L. subsp. exilis (Sukaczev) Hultén ●☆

53436 Betula fargesii Franch.;狭翅桦(川鄂坚桦,窄翅桦);Farges Birch ●

53437 Betula fontinalis Sarg.;北美红桦;American Red Birch,Black Birch,Canyon Birch,Cherry Birch,Rocky Mountain Birch,Streamside Birch,Sweet Birch,Water Birch ●☆

53438 Betula fontinalis Sarg. = Betula occidentalis Hook. ●

53439 Betula fontinalis Sarg. var. inopina (Jeps.) Jeps. = Betula occidentalis Hook. ●

53440 Betula fontqueri Rothm. = Betula pendula Roth subsp. fontqueri (Rothm.) Moreno et Peinado ●☆

53441 Betula forrestii (W. W. Sm.) Hand.-Mazz.;佛氏桦;Forrest's Birch ●☆

53442 Betula forrestii (W. W. Sm.) Hand. -Mazz. = Betula delavayi Franch. ●

53443 Betula forrestii (W. W. Sm.) Hand. -Mazz. var. calcicola (W. W. Sm.) Hand. -Mazz. = Betula calcicola (W. W. Sm.) Hu ex P. C. Li ●

53444 Betula fruticosa Pall.;柴桦(柴桦条子,丛枝桦);Altai Birch, Fruticose Birch ●

53445 Betula fruticosa Pall. subsp. ruprechtiana (Trautv.) Kitag. = Betula fruticosa Pall. var. ruprechtiana Trautv. ●

53446 Betula fruticosa Pall. var. cuneifolia Regel = Betula microphylla Bunge ●

53447 Betula fruticosa Pall. var. gmelinii Regel = Betula gmelinii Bunge ●

53448 Betula fruticosa Pall. var. macrostachys S. L. Tung;长穗柴桦;Bigstachys Fruticose Birch ●

53449 Betula fruticosa Pall. var. ovalifolia (Rupr.) S. L. Tung = Betula fruticosa Pall. var. ruprechtiana Trautv. ●

53450 Betula fruticosa Pall. var. ovalifolia (Rupr.) S. L. Tung = Betula ovalifolia Rupr. ●

53451 Betula fruticosa Pall. var. ruprechtiana Trautv. = Betula ovalifolia Rupr. ●

53452 Betula ganjuensis Koidz.;小泉桦●☆

53453 Betula glandulifera (Regel) B. T. Butler = Betula pumila L. ●

53454 Betula glandulosa Michx.;腺桦;Bog Birch, Dwarf Birch, Dwarf Bog Birch, Glandular Birch, Hog Birch, Resin Birch ●☆

53455 Betula glandulosa Michx. var. glandulifera (Regel) Gleason = Betula pumila L. ●

53456 Betula glandulosa Michx. var. hallii (Howell) C. L. Hitchc. = Betula pumila L. ●

53457 Betula glandulosa Michx. var. rotundifolia (Spach) Regel = Betula rotundifolia Spach ●

53458 Betula glandulosa Michx. var. rotundifolia Regel = Betula rotundifolia Spach ●

53459 Betula glandulosa Michx. var. sibirica (Ledeb.) C. K. Schneid. = Betula nana L. subsp. exilis (Sukaczev) Hultén ●☆

53460 Betula globispica Shirai;地藏桦(球穗桦)●☆

53461 Betula gmelinii Bunge;砂生桦(圆叶桦);Gmelin Birch ●

53462 Betula gmelinii Bunge = Betula fruticosa Pall. ●

53463 Betula gmelinii Bunge var. zyzyphifolia (Z. Wang et S. L. Tung) G. H. Liu et E. W. Ma;枣叶桦;Jujube-leaved Birch ●

53464 Betula grandifolia Litv.;大叶桦●☆

53465 Betula grossa Siebold et Zucc.;日本樱桦(日本矮桦,日本香桦);Japanese Cherry Birch ●☆

53466 Betula gynoterminalis Y. C. Hsu et Ching J. Wang;贡山桦;Gongshan Birch ●

53467 Betula hallii Howell = Betula pumila L. ●

53468 Betula halophila R. C. Ching ex P. C. Li;盐桦;Halophilous Birch, Salt Birch, Salt-loving Birch ●◇

53469 Betula henanensis S. Y. Wang et C. L. Chang;豫白桦;Henan Birch ●

53470 Betula humilis Schrank;甸生桦(沼桦);Dwarf Birch ●

53471 Betula humilis Schrank var. genuina Regel = Betula humilis Schrank ●

53472 Betula humilis Schrank var. ovalifolia Regel = Betula ovalifolia Rupr. ●

53473 Betula humilis Schrank var. reticulata Regel = Betula ovalifolia Rupr. ●

53474 Betula humilis Schrank var. ruprechtii Regel = Betula ovalifolia Rupr. ●

53475 Betula humilis Schrank var. tatewakiana (M. Ohki et S. Watan.) Murata = Betula ovalifolia Rupr. ●

53476 Betula humilis Schrank var. vulgalis Perfiliev = Betula fruticosa Pall. ●

53477 Betula humilis Schrank var. vulgalis Perfiliev = Betula humilis Schrank ●

53478 Betula hupehensis C. K. Schneid. = Betula luminifera H. Winkl. ●

53479 Betula insignis Franch.;香桦;Frangrant Birch ●

53480 Betula irkutensis Sukaczev;伊尔库特桦●☆

53481 Betula jacquemontii Spach;喜马拉雅白桦;Jaquemont's Birch, West Himalayan Birch, Whitebarked Himalayan Birch, White-barked Himalayan Birch ●☆

53482 Betula jacquemontii Spach = Betula utilis D. Don ●

53483 Betula japonica (Miq.) Siebold ex H. Winkl.;日本白桦(白桦,木花,日本桦);Asian White Birch, Japan White Birch, Japanese White Birch ●

53484 Betula japonica (Miq.) Siebold ex H. Winkl. = Betula platyphylla Sukaczev var. japonica (Miq.) H. Hara ●

53485 Betula japonica (Miq.) Siebold ex H. Winkl. var. mandshurica (Regel) H. Winkl. = Betula platyphylla Sukaczev ●

53486 Betula japonica (Miq.) Siebold ex H. Winkl. var. rockii Rehder = Betula platyphylla Sukaczev ●

53487 Betula japonica (Miq.) Siebold ex H. Winkl. var. sachalinensis Koidz. = Betula platyphylla Sukaczev var. mandshurica (Regel) H. Hara ●

53488 Betula japonica (Miq.) Siebold ex H. Winkl. var. szechuanica C. K. Schneid. = Betula platyphylla Sukaczev ●

53489 Betula japonica Siebold ex H. Winkl. = Betula platyphylla Sukaczev ●

53490 Betula japonica Thunb. = Alnus japonica (Thunb.) Steud. ●

53491 Betula jarmolenkoana Golosk. = Betula tianschanica Rupr. ●

53492 Betula jiaodongensis S. B. Liang;胶东桦;Jiaodong Birch ●

53493 Betula jiaodongensis S. B. Liang = Betula chinensis Maxim. ●

53494 Betula jinpingensis P. C. Li;金平桦;Jinping Birch ●◇

53495 Betula jiulungensis L. C. Hu;九龙桦;Jiulong Birch ●

53496 Betula kamtschatica (Regel) V. N. Vassil. var. kenaica (W. H. Evans) C. -A. Jansson = Betula kenaica W. H. Evans ●☆

53497 Betula kelleriana Sukaczev;凯来尔桦●☆

53498 Betula kenaica W. H. Evans;基奈桦;Kenai Birch ●☆

53499 Betula kirghisorum Sawicz.;吉尔吉斯桦●☆

53500 Betula korshinskyi Litv.;科尔桦●☆

53501 Betula kusmisscheffii (Regel) Sukaczev;库斯米桦●☆

53502 Betula kweichowensis Hu = Betula insignis Franch. ●

53503 Betula latifolia Kom.;阔叶桦;Broad-leaved Birch ●☆

53504 Betula latifolia Kom. = Betula platyphylla Sukaczev ●

53505 Betula lenta L.;美加甜桦(甜桦);American Black Birch, American Cherry Birch, Black Birch, Cherry Birch, Mahogany Birch, Mountain Mahogany, Sweet Birch, Winter Green ●

53506 Betula lenta L. var. uber Ashe = Betula uber (Ashe) Fernald ●☆

53507 Betula liaotongensis A. I. Baranov;辽东桦;Liaodong Birch ●

53508 Betula liaotungensis A. I. Baranov = Betula chinensis Maxim. ●

53509 Betula litwinowii Doluch.;利特氏桦(列维诺夫桦)●☆

53510 Betula luminifera H. Winkl.;亮叶桦(狗啃木,光皮桦,光叶桦,红桦树,花胶树,桦角,桦树皮,桦桃树,亮皮桦,铁桦子);Bright Birch, Shining Leaf Birch, Shining-bark Birch, Shiningleaf Birch, Shining-leaved Birch ●

53511 Betula luminifera H. Winkl. var. baeumkeri (H. Winkl.) P. C. Kuo = Betula luminifera H. Winkl. ●

53512 Betula lutea Michx. = Betula alleghaniensis Britton ●☆

53513 Betula lutea Michx. f. fallax Fassett = Betula alleghaniensis Britton ●☆

53514 Betula lutea Michx. var. macrolepis Fernald = Betula alleghaniensis Britton ●☆

53515 Betula maackii Rupr. = Betula dahurica Pall. ●

53516 Betula mandshurica (Regel) Nakai = Betula platyphylla Sukaczev ●

53517 Betula mandshurica (Regel) Nakai var. japonica (Miq.) Rehder = Betula platyphylla Sukaczev var. japonica (Miq.) H. Hara ●

53518 Betula mandshurica (Regel) Nakai var. japonica (Miq.) Rehder = Betula japonica Siebold ex H. Winkl. ●

53519 Betula mandshurica (Regel) Nakai var. kamtschatica (Regel) Rehder = Betula platyphylla Sukaczev var. kamtschatica (Regel) H. Hara ●☆

53520 Betula maximowicziana Regel;大王桦(王桦);Japan Birch, Japanese Birch,Maximowicz's Birch,Monarch Birch ●

53521 Betula maximowicziana Regel = Betula dahurica Pall. ●

53522 Betula maximowiczii Rupr. = Betula dahurica Pall. ●

53523 Betula medwediewii Regel;高加索桦;Transcaucasian Birch ●☆

53524 Betula michauxii Spach;米氏桦;Michaux's Birch,Newfoundland Dwarf Birch ●☆

53525 Betula microphylla Bunge;小叶桦;Smallleaf Birch,Small-leaved Birch ●

53526 Betula microphylla Bunge var. fontinalis (Sarg.) M. E. Jones = Betula occidentalis Hook. ●

53527 Betula middendorffii Trautv. et C. A. Mey.;扇叶桦(小叶桦); Middendorff Birch ●

53528 Betula minor (Tuck.) Fernald;小矮桦;Dwarf Birch ●☆

53529 Betula murrayana B. V. Barnes et Dancik;默里桦;Murray's Birch ●☆

53530 Betula nana L.;小桦(矮北极桦,矮冠桦,矮小桦,高加索桦); Arctic Birch, Arctic Dwarf Birch, Arizona Walnut, Dwarf Arctic Birch,Dwarf Birch,Rock Birch ●☆

53531 Betula nana L. subsp. exilis (Sukaczev) Hultén;矮北极桦; Arctic Dwarf Birch ●☆

53532 Betula nana L. var. glandulifera (Regel) B. Boivin = Betula pumila L. ●

53533 Betula neoalaskana (Sarg.) Raup;新阿拉斯加桦;Paper Birch, Resin Birch ●☆

53534 Betula neoalaskana Sarg. = Betula neoalaskana (Sarg.) Raup ●☆

53535 Betula neoalaskana Sarg. var. kenaica (W. H. Evans) B. Boivin = Betula kenaica W. H. Evans ●☆

53536 Betula nigra L.;河岸黑桦(岸黑桦,河桦,美国黑桦);Black Birch,Fox Valley Dwarf River Birch,Red Birch,River Birch,Tropical Birch,Water Birch ●☆

53537 Betula nigra L. 'Heritage';传统河岸黑桦●☆

53538 Betula nigra L. 'Little King';矮王河岸黑桦●☆

53539 Betula nikoensis Koidz. = Betula costata Trautv. ●

53540 Betula occidentalis Hook.;西方桦(水桦);Black Birch, Red Birch,River Birch,Water Birch,Western Birch ●

53541 Betula occidentalis Hook. var. fecunda (Britton) Fernald = Betula occidentalis Hook. ●

53542 Betula occidentalis Hook. var. inopina (Jeps.) C. L. Hitchc. = Betula occidentalis Hook. ●

53543 Betula odorata Bechst. = Betula tortuosa L. ●☆

53544 Betula ovalifolia Rupr.;油桦(油桦条子);Ovalleaf Birch,Oval-leaved Birch ●

53545 Betula palustris Salisb. = Betula fruticosa Pall. ●

53546 Betula pamirica Litv.;帕米尔桦●☆

53547 Betula papyracea Aiton = Betula papyrifera Marshall ●☆

53548 Betula papyracea Aiton var. minor Tuck. = Betula minor (Tuck.) Fernald ●☆

53549 Betula papyrifera Marshall;纸皮桦(阿州桦,白桦,北美白桦,红桦,加拿大白桦,近心形纸桦,肯奈桦,美洲桦,山纸桦,西方纸桦,心形纸桦,银桦,纸桦);Alasja Birch,Canadian White Birch, Canoe Birch,Kenai Birch,Mountain Paper Birch,Northwestern Paper Birch, Paper Birch, Paper-bark Birch, Red Birch, Silver Birch, Western Paper Birch,White Birch ●☆

53550 Betula papyrifera Marshall subsp. humilis (Regel) Hultén = Betula neoalaskana Sarg. ●☆

53551 Betula papyrifera Marshall subsp. occidentalis (Hook.) Hultén = Betula occidentalis Hook. ●

53552 Betula papyrifera Marshall var. commutata (Regel) Fernald = Betula papyrifera Marshall ●☆

53553 Betula papyrifera Marshall var. cordifolia (Regel) Fernald = Betula cordifolia Regel ●☆

53554 Betula papyrifera Marshall var. elobata (Fernald) Sarg. = Betula papyrifera Marshall ●☆

53555 Betula papyrifera Marshall var. humilis (Regel) Fernald et Raup = Betula neoalaskana Sarg. ●☆

53556 Betula papyrifera Marshall var. kenaica (Evans) A. Henry;凯内卡纸皮桦●☆

53557 Betula papyrifera Marshall var. kenaica (W. H. Evans) A. Henry = Betula kenaica W. H. Evans ●☆

53558 Betula papyrifera Marshall var. macrostachya Fernald = Betula papyrifera Marshall ●☆

53559 Betula papyrifera Marshall var. neoalaskana (Sarg.) Raup = Betula neoalaskana Sarg. ●☆

53560 Betula papyrifera Marshall var. occidentalis (Hook.) Sarg. = Betula occidentalis Hook. ●

53561 Betula papyrifera Marshall var. pensilis Fernald = Betula papyrifera Marshall ●☆

53562 Betula papyrifera Marshall var. subcordata (Rydb.) Sarg. = Betula papyrifera Marshall ●☆

53563 Betula pendula Roth;垂枝桦(欧洲白桦,新疆白桦,疣桦,疣皮桦,疣枝桦); Bedwen Bedewen, Begh, Birch, Birchet, Birk, Bobbyns,Burk,Canoe Birch,Common Birch,Damsel of the Wood, Downy Birch,Drooping Birch,English Birch,Europe White Birch, European Aper Birch, European Birch, European Weeping Birch, European White Birch,Knotty Birch,Lady Birch,Lady of the Woods, Masur Wood,Paper Beech,Paper Birch,Paper-bark Birch,Pubescent Birch,Ribbon-tree, Silver Birch, Silver-leaved Tree, Swedish Birch, Warty Birch,Weeping Birch,White Birch ●

53564 Betula pendula Roth 'Dalecarlica';戴尔卡利垂枝桦;Cut-leaf Birch,Swedish Birch ●☆

53565 Betula pendula Roth 'Purpurea';紫垂枝桦;Purple-leaf Birch ●☆

53566 Betula pendula Roth 'Youngii';扬垂枝桦;Young's Weeping Birch ●☆

53567 Betula pendula Roth f. dalecarlica (L. f.) C. K. Schneid. = Betula pendula Roth ●

53568 Betula pendula Roth subsp. fontqueri (Rothm.) Moreno et

Peinado;丰特桦●☆
53569 Betula pendula Roth var. dalecarlica (L. f.) Rehder = Betula pendula Roth ●
53570 Betula pendula Roth var. japonica (Miq.) Rehder = Betula japonica Siebold ex H. Winkl. ●
53571 Betula pendula Roth var. japonica (Miq.) Rehder = Betula platyphylla Sukaczev var. japonica (Miq.) H. Hara ●
53572 Betula platyphylla Sukaczev;白桦(臭桦,川白桦,东北白桦,东北桦,粉桦,桦木,桦皮树,青海白桦,四川白桦);Asian White Birch, Japanese Birch, Manshurian Birch, Sichuan White Birch, Szechuan Birch, Szechuan White Birch, White Birch ●
53573 Betula platyphylla Sukaczev var. brunnea J. X. Huang;铁皮桦;Brown Asian White Birch ●
53574 Betula platyphylla Sukaczev var. cuneifolia (Nakai) H. Hara;楔叶白桦●☆
53575 Betula platyphylla Sukaczev var. japonica (Miq.) H. Hara 'Whitespire';白尖桦;Whitespire Japanese White Birch ●☆
53576 Betula platyphylla Sukaczev var. japonica (Miq.) H. Hara = Betula japonica (Miq.) Siebold ex H. Winkl. ●
53577 Betula platyphylla Sukaczev var. japonica (Miq.) H. Hara = Betula platyphylla Sukaczev ●
53578 Betula platyphylla Sukaczev var. japonica (Miq.) H. Hara f. laciniata (Miyabe et Tatew.) H. Hara;条裂白尖桦●☆
53579 Betula platyphylla Sukaczev var. japonica H. Hara = Betula platyphylla Sukaczev ●
53580 Betula platyphylla Sukaczev var. kamtschatica (Regel) H. Hara;勘察加白桦●☆
53581 Betula platyphylla Sukaczev var. mandshurica (Regel) H. Hara;东北白桦●
53582 Betula platyphylla Sukaczev var. mandshurica (Regel) H. Hara = Betula platyphylla Sukaczev ●
53583 Betula platyphylla Sukaczev var. phellodendroides S. L. Tung;栓皮白桦(栓皮桦)●
53584 Betula platyphylla Sukaczev var. pluricostata (Koidz.) Tatew.;多脉白桦●☆
53585 Betula platyphylla Sukaczev var. szechuanica (C. K. Schneid.) Rehder = Betula platyphylla Sukaczev ●
53586 Betula populifolia Marshall;杨叶桦;American White Birch, Fire Birch, Gray Birch, Grey Birch, Old Field Birch, Oldfield Birch, Poplar Birch, White Birch ●☆
53587 Betula potaninii Batalin;华西桦(矮桦,矮桦木);Potanin Birch ●
53588 Betula potaninii Batalin var. trichogemma Hu ex P. C. Li = Betula trichogemma (L. C. Hu) T. Hong ●
53589 Betula prochorowli Kuzen. et Litv.;普罗桦●☆
53590 Betula pubescens Ehrh.;欧洲桦(毛枝桦,柔毛桦);Brown Birch, Common Birch, Downy Birch, European Birch, European White Birch, Hairy Birch, Masur Wood, Pubescent Birch, Pubescet-branchlet Birch, Red Birch, Silver Birch, Weeping Birch, White Birch ●☆
53591 Betula pubescens Ehrh. subsp. borealis (Spach) Á. Löve et D. Löve = Betula pumila L. ●
53592 Betula pubescens Ehrh. subsp. minor (Tuck.) Á. Löve et D. Löve = Betula minor (Tuck.) Fernald ●☆
53593 Betula pubescens Ehrh. subsp. tortuosa (Ledeb.) Nyman = Betula tortuosa L. ●☆
53594 Betula pumila L.;矮桦(矮小桦);American Dwarf Birch, Bog Birch, Dwarf Birch, Low Birch, Swamp Birch, Western Bog Birch ●
53595 Betula pumila L. var. glabra Regel = Betula pumila L. ●
53596 Betula pumila L. var. glandulifera Regel = Betula pumila L. ●
53597 Betula pumila L. var. renifolia Fernald = Betula pumila L. ●
53598 Betula purpusii C. K. Schneid.;普尔桦;Purpus' Birch ●☆
53599 Betula purpusii C. K. Schneid. f. fallax (Fassett) B. Boivin = Betula alleghaniensis Britton ●☆
53600 Betula raddeana Trautv.;拉德桦●☆
53601 Betula resinifera (Regel) Britton = Betula neoalaskana Sarg. ●☆
53602 Betula reticulata Rupr. = Betula ovalifolia Rupr. ●
53603 Betula rezniczeckoana (Litv.) Schischk.;雷茨桦●☆
53604 Betula rhombibracteata Hu ex P. C. Li;菱苞桦;Rhomb-branched Birch, Rhomboidbract Birch ●
53605 Betula rotundifolia Spach;圆叶桦;Ground Birch ●
53606 Betula rubra Michx. = Betula nigra L. ●☆
53607 Betula sandbergii Britton;桑德桦;Sandberg's Birch ●☆
53608 Betula saposhnikovii Sukaczev;萨波桦●☆
53609 Betula saxophila Lepage = Betula minor (Tuck.) Fernald ●☆
53610 Betula schmidtii Regel;赛黑桦(辽东桦);Schmidt Birch, Schmidt's Birch ●
53611 Betula schmidtii Regel f. angustifolia (Shirai) Sugim. = Betula schmidtii Regel var. angustifolia (Shirai) Makino et Nemoto ●☆
53612 Betula schmidtii Regel var. angustifolia (Shirai) Makino et Nemoto;狭叶赛黑桦●☆
53613 Betula schmidtii Regel var. angustifolia Sugim. = Betula schmidtii Regel var. angustifolia (Shirai) Makino et Nemoto ●☆
53614 Betula shikokiana Nakai;四国桦;Sikoku Birch ●☆
53615 Betula sibirica Watson = Betula fruticosa Pall. ●
53616 Betula sibirica Watson = Betula humilis Schrank ●
53617 Betula subartica Orlova;亚北极桦●☆
53618 Betula subcordata Rydb. = Betula papyrifera Marshall ●☆
53619 Betula sukaczewii Soczava;苏氏桦;Sukaczew Birch ●☆
53620 Betula sunanensis Y. J. Zhang;肃南桦;South Gansu Birch ●
53621 Betula szechuanica (C. K. Schneid.) Jansen = Betula platyphylla Sukaczev ●
53622 Betula tatewakiana M. Ohki et S. Watan. = Betula ovalifolia Rupr. ●
53623 Betula tauschii (Regel) Koidz. = Betula japonica Siebold ex H. Winkl. ●
53624 Betula terra-novae Fernald = Betula michauxii Spach ●☆
53625 Betula tianschanica Rupr.;天山桦;Tianshan Birch ●
53626 Betula tortuosa L.;曲枝桦●☆
53627 Betula tortuosa Ledeb. = Betula tortuosa L. ●☆
53628 Betula trichogemma (L. C. Hu) T. Hong;峨眉矮桦(峨眉桦);Emei Birch, Emei Potanin Birch ●
53629 Betula turkestanica Litv.;中亚桦;Central-Asia Birch ●☆
53630 Betula uber (Ashe) Fernald;弗吉尼亚圆叶桦;Ashe Birch, Virginia Birch, Virginia Roundleaf Birch ●☆
53631 Betula ulmifolia Siebold et Zucc.;榆叶桦●☆
53632 Betula ulmifolia Siebold et Zucc. var. costata (Trautv.) Regel = Betula costata Trautv. ●
53633 Betula ulmifolia Siebold et Zucc. var. costata Regel = Betula costata Trautv. ●
53634 Betula ulmifolia Siebold et Zucc. var. glandulosa H. Winkl. = Betula ermanii Champ. ●
53635 Betula utilis D. Don;糙皮桦(臭桦,华西糙皮桦,西南糙皮桦,喜马拉雅银桦);Himalayan Birch, Himalayas Birch, Indian Birch, West-China Himalayan Birch, Whitebark ●
53636 Betula utilis D. Don subsp. intermedia sensu Kitam. = Betula

utils D. Don ●
53637 Betula utilis D. Don subsp. jacquemontii sensu Kitam. = Betula utilis D. Don ●
53638 Betula utilis D. Don subsp. occidentalis sensu Kitam. = Betula utilis D. Don ●
53639 Betula utilis D. Don var. jacquemontii (Spach) Winkl. = Betula jacquemontii Spach ●☆
53640 Betula utilis D. Don var. jacquemontii (Spach) Winkl. = Betula utilis D. Don ●
53641 Betula utilis D. Don var. prattii Burkill = Betula utilis D. Don ●
53642 Betula utilis D. Don var. sinensis (Franch.) H. Winkl. = Betula albosinensis Burkill ●
53643 Betula verrucosa Ehrh. = Betula pendula Roth ●
53644 Betula verrucosa Ehrh. var. platyphylla (Sukaczev) Lindq. ex R. K. Jansen = Betula platyphylla Sukaczev ●
53645 Betula viridis Chaix = Alnus viridis (Chaix) DC. ●☆
53646 Betula wilsoniana C. K. Schneid. = Betula luminifera H. Winkl. ●
53647 Betula wilsonii Bean = Betula potaninii Batalin ●
53648 Betula zyzyphifolia Z. Wang et S. L. Tung = Betula gmelinii Bunge var. zyzyphifolia (Z. Wang et S. L. Tung) G. H. Liu et E. W. Ma ●
53649 Betula-alnus Marshall = ? Alnus Mill. ●
53650 Betula-alnus maritima Marshall = Alnus maritima (Marshall) Muhl. ex Nutt. ●☆
53651 Betulaceae Gray(1822)(保留科名);桦木科;Birch Family ●
53652 Betulaster Spach = Betula L. ●
53653 Betulaster acuminata (Wall.) Spach. = Betula alnoides Buch.-Ham. ex D. Don ●
53654 Betulaster acuminata Spach = Betula alnoides Buch.-Ham. ex D. Don ●
53655 Betulaster cylindrostachya (Lindl.) Spach. = Betula cylindrostachya Lindl. ●
53656 Betulaster cylindrostachya Spach. = Betula cylindrostachya Wall. ●
53657 Beurera Kuntze = Bourreria P. Browne(保留属名)●☆
53658 Beureria Ehret(废弃属名) = Calycanthus L.(保留属名)●
53659 Beureria Spreng. = Bourreria P. Browne(保留属名)●☆
53660 Beurreria Jacq. = Bourreria P. Browne(保留属名)●☆
53661 Bevania Bridges ex Endl. = Desfontainia Ruiz et Pav. ●☆
53662 Beverin Collinson = Beureria Ehret(废弃属名)●
53663 Beverin Collinson = Calycanthus L.(保留属名)●
53664 Beverinekia Salisb. ex DC. = Azalea Desv. ●
53665 Beverinekia Salisb. ex DC. = Rhododendron L. ●
53666 Beverna Adans.(废弃属名) = Babiana Ker Gawl. ex Sims(保留属名)■☆
53667 Bewsia Gooss.(1941);非洲千金子属■☆
53668 Bewsia biflora (Hack.) Gooss.;非洲千金子■☆
53669 Beyeria Miq.(1844);拜尔大戟属;Turpentine Bursh☆
53670 Beyeria brevifolia (Müll. Arg.) Benth.;短叶拜尔大戟☆
53671 Beyeria lasiocarpa Müll. Arg.;毛果拜尔大戟☆
53672 Beyeria latifolia Baill.;宽叶拜尔大戟☆
53673 Beyeriopsis Müll. Arg. = Beyeria Miq. ☆
53674 Beyeriopsis brevifolia Müll. Arg. = Beyeria brevifolia (Müll. Arg.) Benth. ☆
53675 Beyrichia Cham. et Schltdl. = Achetaria Cham. et Schltdl. ■☆
53676 Beythea Endl. = Elaeocarpus L. ●
53677 Bezanilla J. Rémy = Psilocarphus Nutt. ■☆
53678 Bhesa Buch.-Ham. ex Arn.(1834);膝柄木属;Bhesa ●
53679 Bhesa robusta (Roxb.) Ding Hou;膝柄木(库林木);Robust Bhesa ●
53680 Bhesa sinica (Hung T. Chang et S. Ye Liang) Hung T. Chang et S. Ye Liang;中华膝柄木;China Bhesa ●◇
53681 Bhidea Stapf ex Bor(1949);印度禾属(印比草属)■☆
53682 Bhidea borii Deshp., V. Prakash et N. P. Singh;鲍尔印度禾■☆
53683 Bhidea burnsiana Bor;印度禾■☆
53684 Bhutanthera Renz(2001);高山兰属■
53685 Bhutanthera alpina (Hand.-Mazz.) Renz;高山兰■
53686 Bhutanthera humidicola (K. Y. Lang et D. S. Deng) Ormerod = Frigidorchis humidicola (K. Y. Lang et D. S. Deng) Z. J. Liu et S. C. Chen ■
53687 Bia Klotzsch = Tragia L. ●☆
53688 Biancaea Tod.(1860);多刺豆属●☆
53689 Biancaea Tod. = Caesalpinia L. ●
53690 Biancaea scandens Tod.;多刺豆●☆
53691 Biarum Schott(1832)(保留属名);袖珍南星属(双芋属)■☆
53692 Biarum arundanum Boiss. et Reut. = Biarum tenuifolium (L.) Schott subsp. arundanum (Boiss. et Reut.) Nyman ■☆
53693 Biarum bovei Blume subsp. dispar (Schott) Engl. = Biarum dispar (Schott) Talavera ■☆
53694 Biarum bovei Blume var. discolor Maire = Biarum tenuifolium (L.) Schott subsp. arundanum (Boiss. et Reut.) Nyman ■☆
53695 Biarum bovei Blume var. macroglossum (Pomel) Maire et Weiller = Biarum dispar (Schott) Talavera ■☆
53696 Biarum bovei Blume var. purpureum Engl. = Biarum dispar (Schott) Talavera ■☆
53697 Biarum bovei Blume var. rupestre (Pomel) Batt. et Trab. = Biarum dispar (Schott) Talavera ■☆
53698 Biarum bovei Blume var. viride Batt. et Trab. = Biarum dispar (Schott) Talavera ■☆
53699 Biarum bovei Blume var. zanonii Pamp. = Biarum dispar (Schott) Talavera ■☆
53700 Biarum dispar (Schott) Talavera;长叶袖珍南星■☆
53701 Biarum eximium Engl.;袖珍南星■☆
53702 Biarum longifolium Pomel = Biarum dispar (Schott) Talavera ■☆
53703 Biarum macroglossum Pomel = Biarum dispar (Schott) Talavera ■☆
53704 Biarum olivieri Blume;奥里维尔袖珍南星■☆
53705 Biarum rupestre Pomel = Biarum dispar (Schott) Talavera ■☆
53706 Biarum sewezowii Regel. = Arum korolkowii Regel ■
53707 Biarum tenuifolium (L.) Schott;细叶袖珍南星■☆
53708 Biarum tenuifolium (L.) Schott subsp. arundanum (Boiss. et Reut.) Nyman;芦苇细叶袖珍南星■☆
53709 Biaslia Vand. = Mayaca Aubl. ■☆
53710 Biasolettia Bertol. = Physocaulis (DC.) Tausch ■☆
53711 Biasolettia C. Presl = Hernandia L. ●
53712 Biasolettia Pohl ex Baker = Eupatorium L. ■●
53713 Biasolettia W. D. J. Koch = Freyera Rchb. ■☆
53714 Biasolettia W. D. J. Koch = Geocaryum Coss. ■
53715 Biasolettia nymphaeifolia C. Presl = Hernandia nymphaeifolia (C. Presl) Kubitzki ●
53716 Biassolettia Endl. = Biasolettia C. Presl ●
53717 Biatherium Desv. = Gymnopogon P. Beauv. ■☆
53718 Biaurieula Bubani = Iberis L. ●■
53719 Bicchia Parl. = Habenaria Willd. ■
53720 Bicchia Parl. = Leucorchis E. Mey. ■☆

53721 Bicchia Parl. = Pseudorchis Ség. ■☆
53722 Bichea Stokes(废弃属名) = Cola Schott et Endl.(保留属名)●☆
53723 Bichenta D. Don = Trichocline Cass. ■☆
53724 Bicornaceae Dulac = Saxifragaceae Juss.(保留科名)●■
53725 Bicornella Lindl. = Cynorkis Thouars ■☆
53726 Bicornella longifolia Lindl. = Cynorkis graminea (Thouars) Schltr. ■☆
53727 Bicornella parviflora Ridl. = Cynorkis graminea (Thouars) Schltr. ■☆
53728 Bicornella similis Schltr. = Cynorkis graminea (Thouars) Schltr. ■☆
53729 Bicornella stolonifera Schltr. = Cynorkis stolonifera (Schltr.) Schltr. ■☆
53730 Bicorona A. DC. = Melodinus J. R. Forst. et G. Forst. ●
53731 Bicuccula Adans. = Dicentra Bernh.(保留属名)■
53732 Bicuculla Borkh. = Adlumia Raf. ex DC.(保留属名)■
53733 Bicuculla canadensis (Goldie) Millsp. = Dicentra canadensis (Goldie) Walp. ■☆
53734 Bicuculla cucullaria (L.) Millsp. = Dicentra cucullaria (L.) Bernh. ■☆
53735 Bicuculla eximia (Ker Gawl.) Millsp. = Dicentra eximia (Ker Gawl.) Torr. ■☆
53736 Bicuculla occidentalis Rydb. = Dicentra cucullaria (L.) Bernh. ■☆
53737 Bicucullaria Juss. = Dicentra Bernh.(保留属名)■
53738 Bicucullaria Juss. ex Steud. = Dicentra Bernh.(保留属名)■
53739 Bicucullata Marchant ex Adans. = Bicucullaria Juss. ■
53740 Bicuiba W. J. de Wilde(1992);比蔻木属●☆
53741 Bicuiba oleifera W. J. de Wilde;比蔻木;Bicuhyba,Bicuhyba Fat ●☆
53742 Bicuspidaria (S. Watson) Rydb. = Mentzelia L. ●■☆
53743 Bicuspidaria Rydb. = Mentzelia L. ●■☆
53744 Bidaria (Endl.) Decne. = Gymnema R. Br. ●
53745 Bidaria Decne. = Gymnema R. Br. ●
53746 Bidaria Endl. = Gymnema R. Br. ●
53747 Bidaria foetida (Tsiang) P. T. Li = Gymnema foetidum Tsiang ●
53748 Bidaria foetida (Tsiang) P. T. Li var. mairei (Tsiang) P. T. Li = Gymnema foetidum Tsiang ●
53749 Bidaria hainanensis (Tsiang) P. T. Li = Gymnema hainanense Tsiang ●
53750 Bidaria inodora (Lour.) Decne. = Gymnema inodorum (Lour.) Decne. ●
53751 Bidaria latifolia (Wall. ex Wight) P. T. Li = Gymnema latifolium Wall. ex Wight ●
53752 Bidaria longiretinaculata (Tsiang) P. T. Li = Gymnema longiretinaculatum Tsiang ●
53753 Bidaria tingens (Roxb. ex Spreng.) Decne. = Gymnema inodorum (Lour.) Decne. ●
53754 Bidaria yunnanensis (Tsiang) P. T. Li = Gymnema yunnanense Tsiang ●
53755 Bidens L. (1753);鬼针草属(鬼针属,狼杷草属);Beggar Tick, Beggarticks, Beggar-ticks, Bur Marigold, Bur-marigold, Cuckold, Pitchfork, Sticktight, Stick-Tights, Tickseed ■●
53756 Bidens abyssinica Sch. Bip. ex Walp. = Bidens biternata (Lour.) Merr. et Sherff ex Sherff ■
53757 Bidens abyssinica Sch. Bip. ex Walp. var. glabrata Vatke = Bidens biternata (Lour.) Merr. et Sherff ex Sherff ■
53758 Bidens abyssinica Sch. Bip. ex Walp. var. quadriaristata Hochst. ex Schweinf. = Bidens biternata (Lour.) Merr. et Sherff ex Sherff ■
53759 Bidens acuta (Wiegand) Britton = Bidens comosa (A. Gray) Wiegand ■☆
53760 Bidens acuticaulis Sherff;锐茎鬼针草■☆
53761 Bidens acuticaulis Sherff var. filirostris (P. Taylor) T. G. J. Rayner;线喙鬼针草■☆
53762 Bidens acutiloba Sherff = Bidens schimperi Sch. Bip. ex Walp. ■☆
53763 Bidens alba (L.) DC. = Bidens pilosa L. var. radiata Sch. Bip. ■
53764 Bidens alba (L.) DC. = Bidens pilosa L. ■
53765 Bidens alba (L.) DC. var. radiata (Sch. Bip.) R. E. Ballard = Bidens pilosa L. ■
53766 Bidens ambacensis (Hiern) Sherff = Bidens steppia (Steetz) Sherff ■☆
53767 Bidens ambigua S. Moore = Bidens crocea Welw. ex O. Hoffm. ■☆
53768 Bidens amoena Sherff = Bidens elliotii (S. Moore) Sherff ■☆
53769 Bidens amplissima Greene;膨大鬼针草■☆
53770 Bidens andongensis Hiern;安东鬼针草■☆
53771 Bidens angustata (Sherff) Sherff = Bidens whytei Sherff ■☆
53772 Bidens arenicola (S. Moore) T. G. J. Rayner;沙生鬼针草■☆
53773 Bidens aristosa (Michx.) Britton;西部鬼针草;Bur Marigold, Midwestern Tickseed-sunflower, Swamp Marigold, Tickseed Sunflower, Western Tickseed ■☆
53774 Bidens aristosa (Michx.) Britton f. involucrata (Nutt.) Wunderlin = Bidens aristosa (Michx.) Britton ■☆
53775 Bidens aristosa (Michx.) Britton var. fritcheyi Fernald;弗里奇鬼针草■☆
53776 Bidens aristosa (Michx.) Britton var. fritcheyi Fernald = Bidens aristosa (Michx.) Britton ■☆
53777 Bidens aristosa (Michx.) Britton var. mutica (A. Gray) A. Gray ex Gatt. = Bidens aristosa (Michx.) Britton ■☆
53778 Bidens aristosa (Michx.) Britton var. mutica A. Gray = Bidens aristosa (Michx.) Britton ■☆
53779 Bidens aristosa (Michx.) Britton var. retrorsa (Sherff) Wunderlin = Bidens aristosa (Michx.) Britton ■☆
53780 Bidens artemisiifolia (Jacq.) Kuntze = Cosmos sulphureus Cav. ■
53781 Bidens articulata Sherff = Bidens rueppellii (Sch. Bip. ex Walp.) Sherff ■☆
53782 Bidens asperata (Hutch. et Dalziel) Sherff;粗糙鬼针草■☆
53783 Bidens aspiloides (Baker) Sherff;菊状鬼针草■☆
53784 Bidens atrosanguinea Lindl. = Cosmos atrosanguineus (Ortega) Voss ■☆
53785 Bidens atrosanguinea Lindl. = Cosmos diversifolius Otto ■☆
53786 Bidens aurea (Aiton) Sherff;金黄鬼针草■☆
53787 Bidens aurea (Aiton) Sherff var. wrightii (A. Gray) Sherff = Bidens aurea (Aiton) Sherff ■☆
53788 Bidens bampsiana Lisowski = Bidens moorei Sherff ■☆
53789 Bidens barteri (Oliv. et Hiern) T. G. J. Rayner;巴特鬼针草■☆
53790 Bidens baumii (O. Hoffm.) Sherff;鲍姆鬼针草■☆
53791 Bidens beckii Torr. = Bidens beckii Torr. ex Spreng. ■☆
53792 Bidens beckii Torr. ex Spreng. = Megalodonta beckii (Torr. ex Spreng.) Greene ■☆
53793 Bidens beguinotii (Chiov.) Cufod. = Bidens pachyloma (Oliv. et Hiern) Cufod. ■☆
53794 Bidens bequaertii De Wild. = Bidens crocea Welw. ex O. Hoffm. ■☆
53795 Bidens bidentoides (Nutt.) Brit;普通鬼针草■☆
53796 Bidens bidentoides var. mariana (S. F. Blake) Sherff = Bidens bidentoides (Nutt.) Brit ■☆

53797 Bidens bipinnata L.;婆婆针(叉婆子,刺儿鬼,刺针草,豆渣菜,钢叉草,跟人走,鬼钗草,鬼骨针,鬼黄花,鬼蒺藜,鬼菊,鬼针,鬼针草,家脱力草,金盏银盘,擂钻草,盲肠草,毛锥子草,清胃草,山东老鸦草,山虱母,索人衣,跳虱草,脱力草,乌藤菜,咸丰草,小鬼针,一把针,一包针,引线包,黏花衣,黏身草,针包草,止血草,锥叉菜);Beggarticks,Beggar-ticks,Black Jack,Bur Marigold,Cobbler's Pegs,Pitchfork,Spanish Needles,Spanishneedles,Sticktight,Sweethearts,Tickseed ■

53798 Bidens bipinnata L. var. biternatoides Sherff = Bidens bipinnata L. ■

53799 Bidens bipontina Sherff;大花鬼针草■☆

53800 Bidens biternata (Lour.) Merr. et Sherff = Bidens biternata (Lour.) Merr. et Sherff ex Sherff ■

53801 Bidens biternata (Lour.) Merr. et Sherff ex Sherff;金盏银盘(草鞋坪,刺针草,豆渣菜,方骨苦楝,感暑草,鬼针草,鬼针舅,黄花草,黄花母,黄花雾,金杯银盏,紧丝苦令,盲肠草,毛鬼针草,婆婆针,千条锇,铁筅箒,虾箝草,虾尾草,蟹钳草,一把针,一包针,引线包,玉盏载银杯,粘身草,针刺草);Biternate Beggarticks ■

53802 Bidens biternata (Lour.) Merr. et Sherff ex Sherff f abyssinica (Sch. Bip. ex Walp.) Sherff = Bidens biternata (Lour.) Merr. et Sherff ex Sherff ■

53803 Bidens biternata (Lour.) Merr. et Sherff ex Sherff f. lasiocarpa (Schultz) Sherff = Bidens biternata (Lour.) Merr. et Sherff ex Sherff ■

53804 Bidens biternata (Lour.) Merr. et Sherff ex Sherff var. glabrata (Vatke) Sherff = Bidens biternata (Lour.) Merr. et Sherff ex Sherff ■

53805 Bidens biternata (Lour.) Merr. et Sherff ex Sherff var. mayebarae (Kitam.) Kitam.;马屋原鬼针草■☆

53806 Bidens borianiana (Sch. Bip. ex Schweinf. et Asch.) Cufod.;博里鬼针草■☆

53807 Bidens bracteosa Sherff = Bidens ternata (Chiov.) Sherff ■☆

53808 Bidens bruceae Sherff = Bidens buchneri (Klatt) Sherff ■☆

53809 Bidens bruceae Sherff var. pubescentior? = Bidens buchneri (Klatt) Sherff ■☆

53810 Bidens buchneri (Klatt) Sherff;布赫纳鬼针草■☆

53811 Bidens burundiensis Mesfin;布隆迪鬼针草■☆

53812 Bidens camporum (Hutch.) Mesfin;弯鬼针草■☆

53813 Bidens campylotheca Sch. Bip.;夏威夷鬼针草■☆

53814 Bidens carinata (Hutch.) Cufod. = Bidens carinata Cufod. ex Mesfin ■☆

53815 Bidens carinata Cufod. ex Mesfin;龙骨鬼针草■☆

53816 Bidens cernua L.;柳叶鬼针草(俯垂鬼针草);Baclin,Dutch Agrimony,Hemp Agrimony,Nodding Agrimony,Nodding Beggarticks,Nodding Beggar-ticks,Nodding Bur Marigold,Nodding Bur-marigold,Nodding Water Hemp Agrimony,Small Bur Marigold,Sticktight,Water Agrimony,Water Marigold,Willowleaf Beggarticks ■

53817 Bidens cernua L. f. discoides (Wimm. et Grab.) Briq. et Cavill. = Bidens cernua L. ■

53818 Bidens cernua L. f. minima (Huds.) Larss. = Bidens cernua L. ■

53819 Bidens cernua L. var. dentata (Nutt.) B. Boivin = Bidens cernua L. ■

53820 Bidens cernua L. var. elata Torr. et A. Gray = Bidens amplissima Greene ■☆

53821 Bidens cernua L. var. elliptica Wiegand = Bidens cernua L. ■

53822 Bidens cernua L. var. integra Wiegand = Bidens cernua L. ■

53823 Bidens cernua L. var. limosa Nakai = Bidens cernua L. ■

53824 Bidens cernua L. var. minima (Huds.) Pursh = Bidens cernua L. ■

53825 Bidens cernua L. var. oligodonta Fernald et H. St. John = Bidens cernua L. ■

53826 Bidens cernua L. var. radiata DC. = Bidens cernua L. ■

53827 Bidens chaetodonta Sherff = Bidens camporum (Hutch.) Mesfin ■☆

53828 Bidens chaetodonta Sherff var. glabrior (Oliv. et Hiern) Sherff = Bidens camporum (Hutch.) Mesfin ■☆

53829 Bidens chandleri Sherff = Bidens ugandensis (S. Moore) Sherff ■☆

53830 Bidens chilensis DC.;智利鬼针草■

53831 Bidens chilensis Willd. = Bidens pilosa L. ■

53832 Bidens chinensis (L.) Willd. var. abyssinica (Sch. Bip. ex Walp.) O. E. Schulz = Bidens biternata (Lour.) Merr. et Sherff ex Sherff ■

53833 Bidens chinensis Willd. = Bidens biternata (Lour.) Merr. et Sherff ex Sherff ■

53834 Bidens chippii (M. B. Moss) Mesfin;奇普鬼针草■☆

53835 Bidens chrysanthemoides Michx. = Bidens laevis (L.) Britton, Stern et Poggenb. ■☆

53836 Bidens ciliata De Wild. = Bidens acuticaulis Sherff ■☆

53837 Bidens cinerea Sherff;灰鬼针草■☆

53838 Bidens cinerea Sherff var. tricuspidata? = Bidens cinerea Sherff ■☆

53839 Bidens cinereoides Sherff = Bidens cinerea Sherff ■☆

53840 Bidens cirsioides Sherff = Bidens camporum (Hutch.) Mesfin ■☆

53841 Bidens cochlearis Merxm. = Bidens diversa Sherff ■☆

53842 Bidens comosa (A. Gray) Wiegand;裂叶鬼针草;Leafy-bracted Beggarticks,Straw-stem Beggar-ticks,Swamp Marigold,Swamp Tickseed ■☆

53843 Bidens comosa (A. Gray) Wiegand = Bidens tripartita L. ■

53844 Bidens comosa (A. Gray) Wiegand var. acuta Wiegand = Bidens comosa (A. Gray) Wiegand ■☆

53845 Bidens connata Muhl. ex Willd.;伦敦鬼针草;Connate Beggarticks,London Bur-marigold,Purple-stem Beggar-ticks,Purple-stemmed Tickseed,Swamp Beggarticks ■☆

53846 Bidens connata Muhl. ex Willd. var. ambiversa Fassett = Bidens connata Muhl. ex Willd. ■☆

53847 Bidens connata Muhl. ex Willd. var. anomala Farw. = Bidens connata Muhl. ex Willd. ■☆

53848 Bidens connata Muhl. ex Willd. var. fallax (Warnst.) Sherff = Bidens connata Muhl. ex Willd. ■☆

53849 Bidens connata Muhl. ex Willd. var. gracilipes Fernald = Bidens connata Muhl. ex Willd. ■☆

53850 Bidens connata Muhl. ex Willd. var. inundata Fernald = Bidens connata Muhl. ex Willd. ■☆

53851 Bidens connata Muhl. ex Willd. var. petiolata (Nutt.) Farw. = Bidens connata Muhl. ex Willd. ■☆

53852 Bidens connata Muhl. ex Willd. var. petiolata (Nutt.) Farw. = Bidens tripartita L. ■

53853 Bidens connata Muhl. ex Willd. var. pinnata S. Watson = Bidens connata Muhl. ex Willd. ■☆

53854 Bidens connata Muhl. ex Willd. var. submutica Fassett = Bidens connata Muhl. ex Willd. ■☆

53855 Bidens connata Muhl. ex Willd. var. typica Fassett = Bidens connata Muhl. ex Willd. ■☆

53856 Bidens cooperi Sherff = Bidens camporum (Hutch.) Mesfin ■☆

53857 Bidens coriacea (O. Hoffm.) Sherff = Bidens buchneri (Klatt) Sherff ■☆

53858 Bidens coronata (L.) Britton;冠状鬼针草;Beggar Ticks, Northern Tickseed-sunflower, Swamp Beggar-ticks, Tall Swamp Marigold, Tick Sunflower, Tickseed Sunflower ■☆

53859 Bidens coronata (L.) Britton var. brachyodonta Fernald = Bidens coronata (L.) Britton ■☆

53860 Bidens coronata (L.) Britton var. brachyodonta Fernald = Bidens trichosperma (Michx.) Britton ■☆

53861 Bidens coronata (L.) Britton var. tenuiloba (A. Gray) Sherff = Bidens coronata (L.) Britton ■☆

53862 Bidens coronata (L.) Britton var. tenuiloba (A. Gray) Sherff = Bidens trichosperma (Michx.) Britton ■☆

53863 Bidens coronata (L.) Britton var. trichosperma (Michx.) Fernald = Bidens trichosperma (Michx.) Britton ■☆

53864 Bidens coronata (L.) Britton var. trichosperma (Michx.) Fernald = Bidens coronata (L.) Britton ■☆

53865 Bidens cosmoides (A. Gray) Sherff;秋英鬼针草■☆

53866 Bidens crataegifolia (O. Hoffm.) Sherff = Bidens kilimandscharica (O. Hoffm.) Sherff ■☆

53867 Bidens crataegifolia (O. Hoffm.) Sherff var. burttii Sherff = Bidens kilimandscharica (O. Hoffm.) Sherff ■☆

53868 Bidens crocea Welw. ex O. Hoffm.;镉黄鬼针草■☆

53869 Bidens crocea Welw. ex O. Hoffm. var. ornata Sherff = Bidens crocea Welw. ex O. Hoffm. ■☆

53870 Bidens crocea Welw. ex O. Hoffm. var. verrucifera S. Moore = Bidens crocea Welw. ex O. Hoffm. ■☆

53871 Bidens cuspidata Sherff = Bidens kilimandscharica (O. Hoffm.) Sherff ■☆

53872 Bidens cylindrica Sherff = Bidens biternata (Lour.) Merr. et Sherff ex Sherff ■

53873 Bidens cynopiifolia Kunth;桃叶鬼针草■☆

53874 Bidens denudata Turcz. = Glossocardia bidens (Retz.) Veldkamp ■

53875 Bidens dielsii Sherff = Bidens ternata (Chiov.) Sherff ■☆

53876 Bidens dielsii Sherff var. incisior? = Bidens ternata (Chiov.) Sherff ■☆

53877 Bidens dielsii Sherff var. intermedia? = Bidens oblonga (Sherff) Wild ■☆

53878 Bidens dielsii Sherff var. medusoides? = Bidens ternata (Chiov.) Sherff ■☆

53879 Bidens discoidea (Torr. et A. Gray) Britton;铁饼鬼针草;Beggar Ticks, Discoid Beggarticks, Few-bracted Beggar-ticks, Swamp Beggar-ticks ■☆

53880 Bidens diversa Sherff;异形鬼针草■☆

53881 Bidens diversa Sherff subsp. filiformis (Sherff) T. G. J. Rayner = Bidens diversa Sherff ■☆

53882 Bidens diversa Sherff var. megaglossa? = Bidens diversa Sherff ■☆

53883 Bidens diversa Sherff var. quilembana? = Bidens diversa Sherff ■☆

53884 Bidens dolosa Sherff = Bidens magnifolia Sherff ■☆

53885 Bidens dominicisaccardoi Fiori = Bidens somaliensis Sherff ■☆

53886 Bidens drummondii Wild = Bidens oligoflora (Klatt) Wild ■☆

53887 Bidens eatonii Fernald;伊顿鬼针草;Eaton's Bur-marigold ■☆

53888 Bidens eatonii Fernald var. fallax Fernald = Bidens eatonii Fernald ■☆

53889 Bidens eatonii Fernald var. illicita S. F. Blake = Bidens eatonii Fernald ■☆

53890 Bidens eatonii Fernald var. interstes (Fassett) Fassett = Bidens eatonii Fernald ■☆

53891 Bidens eatonii Fernald var. kennebecensis Fernald = Bidens eatonii Fernald ■☆

53892 Bidens eatonii Fernald var. major Fassett = Bidens eatonii Fernald ■☆

53893 Bidens eatonii Fernald var. mutabilis Fassett = Bidens eatonii Fernald ■☆

53894 Bidens eatonii Fernald var. simulans Fassett = Bidens eatonii Fernald ■☆

53895 Bidens elata (Torr. et A. Gray) Sherff = Bidens amplissima Greene ■☆

53896 Bidens elgonensis (Sherff) Agnew;埃尔贡鬼针草■☆

53897 Bidens elgonensis (Sherff) Agnew subsp. cheranganiensis T. G. J. Rayner = Bidens elgonensis (Sherff) Agnew ■☆

53898 Bidens elgonensis (Sherff) Agnew subsp. morotonensis (Sherff) T. G. J. Rayner = Bidens elgonensis (Sherff) Agnew ■☆

53899 Bidens ellenbeckii (O. Hoffm.) Cufod. = Bidens macroptera (Sch. Bip. ex Chiov.) Mesfin ■☆

53900 Bidens elliotii (S. Moore) Sherff;埃利鬼针草■☆

53901 Bidens engleri O. E. Schulz;恩格勒鬼针草■☆

53902 Bidens exilis (Sherff) Lisowski = Bidens oblonga (Sherff) Wild ■☆

53903 Bidens fengjiensis S. X. Liu et W. P. Li;奉节鬼针草;Fengjie Beggarticks ■

53904 Bidens ferulifolia (Jacq.) DC. = Bidens aurea (Aiton) Sherff ■☆

53905 Bidens ferulifolia (Jacq.) Sweet;大理石鬼针草;Apache Beggarticks, Gold Marble bidens, Toothache Plant ■☆

53906 Bidens filamentosa Rydb. = Bidens cernua L. ■

53907 Bidens filiformis Sherff = Bidens diversa Sherff ■☆

53908 Bidens fischeri (O. Hoffm.) Sherff;菲舍尔鬼针草■☆

53909 Bidens flabellata O. Hoffm.;扇状鬼针草■☆

53910 Bidens flagellata (Sherff) Mesfin;鞭状鬼针草■☆

53911 Bidens formosa (Bonato) Sch. Bip. = Cosmos bipinnatus Cav. ■

53912 Bidens fri-partita?;深裂鬼针草;Bur Marigold ■☆

53913 Bidens frondosa L.;大狼杷草(大狼巴草,接力草,外国脱力草);Beggar Ticks, Beggarticks, Beggar-ticks, Common Beggar-ticks, Devil's Beggar-ticks, Large Beggarticks, Large-leaved Beggarticks, Sticktight ■

53914 Bidens frondosa L. f. anomala (Porter ex Fernald) Fernald = Bidens frondosa L. ■

53915 Bidens frondosa L. var. anomala Porter ex Fernald = Bidens frondosa L. ■

53916 Bidens frondosa L. var. caudata Sherff = Bidens frondosa L. ■

53917 Bidens frondosa L. var. pallida (Wiegand) Wiegand = Bidens frondosa L. ■

53918 Bidens frondosa L. var. puberula Wiegand = Bidens vulgata Greene ■☆

53919 Bidens frondosa L. var. stenodonta Fernald et H. St. John = Bidens frondosa L. ■

53920 Bidens gardneri Baker;加德纳鬼针草;Ridge Beggartick ■☆

53921 Bidens gardullensis Cufod. = Bidens ternata (Chiov.) Sherff ■☆

53922 Bidens glaber O. Hoffm. = Bidens moorei Sherff ■☆

53923 Bidens glaucescens Greene = Bidens cernua L. ■

53924 Bidens gledhillii T. G. J. Rayner = Bidens sierra-leonensis Mesfin ■☆

53925 Bidens gossweileri Sherff = Bidens crocea Welw. ex O. Hoffm. ■☆

53926 Bidens gracilenta Greene = Bidens cernua L. ■

53927 Bidens gracilior (O. Hoffm.) Sherff = Bidens taylori (S. Moore)

Sherff ■☆

53928 Bidens gracilior (O. Hoffm.) Sherff var. ukerewensis Sherff = Bidens taylori (S. Moore) Sherff ■☆

53929 Bidens grandiflora Balb. = Bidens bipontina Sherff ■☆

53930 Bidens grandis Sherff = Bidens buchneri (Klatt) Sherff ■☆

53931 Bidens grantii (Oliv.) Sherff;格兰特鬼针草■☆

53932 Bidens grantii (Oliv.) Sherff var. dawei Sherff = Bidens grantii (Oliv.) Sherff ■☆

53933 Bidens grantii (Oliv.) Sherff var. scattae Sherff = Bidens grantii (Oliv.) Sherff ■☆

53934 Bidens grantii (Oliv.) Sherff var. stapfii Sherff = Bidens grantii (Oliv.) Sherff ■☆

53935 Bidens grantii (Oliv.) Sherff var. stapfioides Sherff = Bidens schimperi Sch. Bip. ex Walp. ■☆

53936 Bidens helianthoides Kunth = Bidens laevis (L.) Britton, Stern et Poggenb. ■☆

53937 Bidens heterodoxa (Fernald) Fernald et H. St. John;异型鬼针草■☆

53938 Bidens heterodoxa (Fernald) Fernald et H. St. John var. agnostica Fernald = Bidens heterodoxa (Fernald) Fernald et H. St. John ■☆

53939 Bidens heterodoxa (Fernald) Fernald et H. St. John var. monardifolia Fernald = Bidens heterodoxa (Fernald) Fernald et H. St. John ■☆

53940 Bidens heterodoxa (Fernald) Fernald et H. St. John var. orthodoxa Fernald et H. St. John = Bidens heterodoxa (Fernald) Fernald et H. St. John ■☆

53941 Bidens hildebrandtii O. Hoffm. ;希尔德鬼针草■☆

53942 Bidens hildebrandtii O. Hoffm. var. boranensis Lanza = Bidens hildebrandtii O. Hoffm. ■☆

53943 Bidens hoffmannii Sherff = Bidens whytei Sherff ■☆

53944 Bidens hoffmannii Sherff var. angustata? = Bidens whytei Sherff ■☆

53945 Bidens holstii (O. Hoffm.) Sherff;霍尔鬼针草■☆

53946 Bidens holstii (O. Hoffm.) Sherff var. rupestris (Sherff) Sherff = Bidens holstii (O. Hoffm.) Sherff ■☆

53947 Bidens hyperborea Greene;北方鬼针草;Northern Beggarticks ■☆

53948 Bidens hyperborea Greene var. arcuans Fernald = Bidens hyperborea Greene ■☆

53949 Bidens hyperborea Greene var. cathancensis Fernald = Bidens hyperborea Greene ■☆

53950 Bidens hyperborea Greene var. colpophila (Fernald et H. St. John) Fernald = Bidens hyperborea Greene ■☆

53951 Bidens hyperborea Greene var. gaspensis Fernald = Bidens hyperborea Greene ■☆

53952 Bidens hyperborea Greene var. svensonii Fassett = Bidens hyperborea Greene ■☆

53953 Bidens imatongensis Sherff = Bidens somaliensis Sherff ■☆

53954 Bidens incumbens Sherff = Bidens hildebrandtii O. Hoffm. ■☆

53955 Bidens incumbens Sherff var. muthicola? = Bidens hildebrandtii O. Hoffm. ■☆

53956 Bidens infirma Fernald = Bidens heterodoxa (Fernald) Fernald et H. St. John ■☆

53957 Bidens insecta (S. Moore) Sherff = Bidens kirkii (Oliv. et Hiern) Sherff ■☆

53958 Bidens insignis Sherff = Bidens kilimandscharica (O. Hoffm.) Sherff ■☆

53959 Bidens involucrata (Nutt.) Britton;总苞鬼针草;Tickseed ■☆

53960 Bidens jacksonii (S. Moore) Sherff = Guizotia jacksonii (S. Moore) J. Baagoe ■☆

53961 Bidens kamerunensis Sherff;喀麦隆鬼针草■☆

53962 Bidens kamtschatica Vassilcz. ;勘察加鬼针草■☆

53963 Bidens kasaiensis Lisowski = Bidens crocea Welw. ex O. Hoffm. ■☆

53964 Bidens kefensis T. G. J. Rayner = Bidens camporum (Hutch.) Mesfin ■☆

53965 Bidens kigeziensis Sherff = Bidens elliotii (S. Moore) Sherff ■☆

53966 Bidens kilimandscharica (O. Hoffm.) Sherff;基利鬼针草■☆

53967 Bidens kilimandscharica (O. Hoffm.) Sherff var. oxymera Sherff = Bidens kilimandscharica (O. Hoffm.) Sherff ■☆

53968 Bidens kilimandscharica (O. Hoffm.) Sherff var. retrorsa Sherff = Bidens kilimandscharica (O. Hoffm.) Sherff ■☆

53969 Bidens kirkii (Oliv. et Hiern) Sherff;柯克鬼针草■☆

53970 Bidens kirkii (Oliv. et Hiern) Sherff var. ciliato-vaginata Cufod. = Bidens flagellata (Sherff) Mesfin ■☆

53971 Bidens kirkii (Oliv. et Hiern) Sherff var. flagellata Sherff = Bidens flagellata (Sherff) Mesfin ■☆

53972 Bidens kivuensis Sherff = Bidens grantii (Oliv.) Sherff ■☆

53973 Bidens kivuensis Sherff var. armata? = Bidens grantii (Oliv.) Sherff ■☆

53974 Bidens kotschyi Sch. Bip. = Bidens biternata (Lour.) Merr. et Sherff ex Sherff ■

53975 Bidens laciniata Sch. Bip. ex Schweinf. et Asch. = Bidens biternata (Lour.) Merr. et Sherff ex Sherff ■

53976 Bidens laevis (L.) Britton, Stern et Poggenb. ;平滑鬼针草;Beggar Ticks, Begger's Tick, Bur-marigold, Larger Bur-marigold, Smooth Beggar-ticks ■☆

53977 Bidens lasiocarpa O. E. Schulz = Bidens biternata (Lour.) Merr. et Sherff ex Sherff ■

53978 Bidens lejolyana Lisowski;勒若利鬼针草■☆

53979 Bidens leptoglossa (Sherff) Lisowski = Bidens kilimandscharica (O. Hoffm.) Sherff ■☆

53980 Bidens leptolepis Sherff = Bidens urceolata De Wild. var. leptolepis (Sherff) T. G. J. Rayner ■☆

53981 Bidens leptolepis Sherff f. pallida? = Bidens urceolata De Wild. ■☆

53982 Bidens leptophylla C. H. An;薄叶鬼针草;Thinleaf Beggarticks ■

53983 Bidens leucantha (L.) Willd. = Bidens pilosa L. ■

53984 Bidens leucantha Meyen et Walp. = Bidens pilosa L. ■

53985 Bidens leucantha Willd. = Bidens pilosa L. ■

53986 Bidens leucantha Willd. var. sundaica (Blume) Hassk. = Bidens pilosa L. ■

53987 Bidens lindblomii Sherff = Bidens hildebrandtii O. Hoffm. ■☆

53988 Bidens linearifolia (Oliv. et Hiern) Sherff = Bidens ugandensis (S. Moore) Sherff ■☆

53989 Bidens lineariloba Oliv. ;线裂片鬼针草■☆

53990 Bidens lineariloba Oliv. var. deminuta Sherff = Bidens lineariloba Oliv. ■☆

53991 Bidens lineata Sherff = Bidens cinerea Sherff ■☆

53992 Bidens lineata Sherff var. tenuipes? = Bidens cinerea Sherff ■☆

53993 Bidens lynesii Sherff = Bidens magnifolia Sherff ■☆

53994 Bidens macrantha (Sch. Bip.) Cufod. = Bidens macroptera (Sch. Bip. ex Chiov.) Mesfin ■☆

53995 Bidens macroptera (Sch. Bip. ex Chiov.) Mesfin;大翅鬼针草■☆

53996 Bidens magnifolia Sherff;大叶鬼针草■☆

53997 Bidens magnifolia Sherff var. versuta? = Bidens magnifolia Sherff ■☆

53998 Bidens malawiense Mesfin;马拉维鬼针草■☆

53999 Bidens mannii T. G. J. Rayner;曼氏鬼针草■☆

54000 Bidens mariana S. F. Blake = Bidens bidentoides (Nutt.) Brit ■☆
54001 Bidens maximowicziana Oett.;羽叶鬼针草;Pinnate Beggarticks ■
54002 Bidens megapotamica Spreng. = Thelesperma longipes A. Gray ■☆
54003 Bidens melanocarpa Wiegand = Bidens frondosa L. ■
54004 Bidens meruensis Sherff = Bidens kilimandscharica (O. Hoffm.) Sherff ■☆
54005 Bidens meyeniana Walp. = Glossocardia bidens (Retz.) Veldkamp ■
54006 Bidens meyeniana Walp. = Glossogyne tenuifolia Cass. ■
54007 Bidens microcarpa Sherff = Bidens schimperi Sch. Bip. ex Walp. ■☆
54008 Bidens microphylla Sherff;小叶鬼针草■☆
54009 Bidens mildbraedii Sherff = Bidens ugandensis (S. Moore) Sherff ■☆
54010 Bidens minima Huds. = Bidens cernua L. ■
54011 Bidens minuta Miré et H. Gillet = Glossocardia bosvallia (L. f.) DC. ■☆
54012 Bidens mitis (Michx.) Sherff;沼泽鬼针草;Marsh Beggartick ■☆
54013 Bidens modesta Sherff = Bidens moorei Sherff ■☆
54014 Bidens moorei Sherff;穆尔鬼针草■☆
54015 Bidens moorei Sherff var. verrucosa? = Bidens moorei Sherff ■☆
54016 Bidens morotonensis (Sherff) Agnew = Bidens elgonensis (Sherff) Agnew ■☆
54017 Bidens mossiae Sherff = Bidens chippii (M. B. Moss) Mesfin ■☆
54018 Bidens napierae Sherff = Bidens holstii (O. Hoffm.) Sherff ■☆
54019 Bidens nashii Small = Bidens laevis (L.) Britton, Stern et Poggenb. ■☆
54020 Bidens natator Friis et Vollesen = Bidens negriana (Sherff) Cufod. ■☆
54021 Bidens navicularia Sherff = Bidens negriana (Sherff) Cufod. ■☆
54022 Bidens negriana (Sherff) Cufod.;内格里鬼针草■☆
54023 Bidens neumannii Sherff = Bidens ternata (Chiov.) Sherff ■☆
54024 Bidens nivea L. = Melanthera nivea (L.) Small ■☆
54025 Bidens nobilis Sherff = Bidens holstii (O. Hoffm.) Sherff ■☆
54026 Bidens nyikensis Sherff = Bidens oblonga (Sherff) Wild ■☆
54027 Bidens oblonga (Sherff) Wild;倒卵鬼针草■☆
54028 Bidens occidentalis (Hutch. et Dalziel) Mesfin;西方鬼针草■☆
54029 Bidens ochracea (O. Hoffm.) Sherff;淡黄褐鬼针草■☆
54030 Bidens ocymifolia Lam. = Spilanthes ocymifolia (Lam.) A. H. Moore ■☆
54031 Bidens odora (Sherff) T. G. J. Rayner;芳香鬼针草■☆
54032 Bidens odorata Cav. = Bidens pilosa L. ■
54033 Bidens oligoflora (Klatt) Wild;寡花鬼针草■☆
54034 Bidens oligoflora (Klatt) Wild var. robusta Sherff = Bidens oligoflora (Klatt) Wild ■☆
54035 Bidens onisciformis Sherff = Bidens grantii (Oliv.) Sherff ■☆
54036 Bidens overlaetii Sherff = Bidens urceolata De Wild. var. leptolepis (Sherff) T. G. J. Rayner ■☆
54037 Bidens pachyloma (Oliv. et Hiern) Cufod.;厚边鬼针草■☆
54038 Bidens pachyloma (Oliv. et Hiern) Cufod. var. inanis (Sherff) Cufod. = Bidens pachyloma (Oliv. et Hiern) Cufod. ■☆
54039 Bidens palustris Sherff = Bidens crocea Welw. ex O. Hoffm. ■☆
54040 Bidens palustris Sherff var. cubangora? = Bidens crocea Welw. ex O. Hoffm. ■☆
54041 Bidens palustris Sherff var. nematomera? = Bidens whytei Sherff ■☆
54042 Bidens parviflora Willd.;小花鬼针草(不怕日草,刺针草,鬼疙草,鬼疙针,鬼骨针,鬼针草,锅叉草,鹿角草,山黄连,桐花菜,土黄连,细叶刺针草,细叶鬼针草,小刺叉,小鬼叉,小鬼叉子,小花刺针草,一包针);Smallflower Beggarticks ■
54043 Bidens paupercula Sherff = Bidens acuticaulis Sherff ■☆
54044 Bidens paupercula Sherff var. filirostis P. Taylor = Bidens acuticaulis Sherff var. filirostris (P. Taylor) T. G. J. Rayner ■☆
54045 Bidens petiolata Nutt. = Bidens connata Muhl. ex Willd. ■☆
54046 Bidens phalangiphylla Sherff = Bidens crocea Welw. ex O. Hoffm. ■☆
54047 Bidens phelloptera Sherff = Bidens magnifolia Sherff ■☆
54048 Bidens pilosa L.;鬼针草(白花鬼针,豆渣菜,豆渣草,对叉草,鬼针,黄花雾,金杯银盏,金丝苦令,金盏银盘,盲肠草,美洲鬼针草,婆婆针,三叶鬼针草,虾钳草,蟹钳草,一把针,一包针,引线包,粘连子,粘人草);Beggar Tick, Beggar Ticks, Beggar-ticks, Black Jack, Blackfellows, Cobbler's Pegs, Duppy Needles, Hairy Beggarticks, Pitchfork, Railway Beggarticks, Railway Daisy, Spanish Needles, Sweethearts ■
54049 Bidens pilosa L. var. abyssinica (Sch. Bip. ex Walp.) Fiori = Bidens biternata (Lour.) Merr. et Sherff ex Sherff ■
54050 Bidens pilosa L. var. albiflora Maxim. = Bidens pilosa L. var. radiata Sch. Bip. ■
54051 Bidens pilosa L. var. bimucronata (Turcz.) Sch. Bip. = Bidens pilosa L. ■
54052 Bidens pilosa L. var. bipinnata (L.) Hook. = Bidens bipinnata L. ■
54053 Bidens pilosa L. var. glabrata (Vatke) Engl. = Bidens biternata (Lour.) Merr. et Sherff ex Sherff ■
54054 Bidens pilosa L. var. minor (Blume) Sherff;小白花鬼针(白花婆婆针,咸丰草)■
54055 Bidens pilosa L. var. minor (Blume) Sherff = Bidens pilosa L. var. radiata Sch. Bip. ■
54056 Bidens pilosa L. var. minor (Blume) Sherff = Bidens pilosa L. ■
54057 Bidens pilosa L. var. quadriseta (Hochst. ex Oliv. et Hiern) Engl. = Bidens biternata (Lour.) Merr. et Sherff ex Sherff ■
54058 Bidens pilosa L. var. radiata (Sch. Bip.) Sch. Bip. = Bidens pilosa L. ■
54059 Bidens pilosa L. var. radiata Sch. Bip.;白花鬼针草(大花咸丰草,金盏银盘,小白花鬼针,小白花鬼针草,小三叶鬼针草);Pilose Beggarticks ■
54060 Bidens pilosa L. var. radiata Sch. Bip. f. decumbens (Greenm.) Sherff;俯卧鬼针草■☆
54061 Bidens pilosa L. var. radiata Sch. Bip. f. indivisa Kayama;全裂鬼针草■☆
54062 Bidens pilosa L. var. rubiflorum S. S. Ying = Bidens pilosa L. ■
54063 Bidens pinnata Noronha;双羽鬼针草■☆
54064 Bidens pinnatipartita (O. Hoffm.) Wild;羽状深裂鬼针草■☆
54065 Bidens polylepis S. F. Blake;多鳞鬼针草■☆
54066 Bidens polylepis S. F. Blake = Bidens aristosa (Michx.) Britton ■☆
54067 Bidens polylepis S. F. Blake var. retrorsa Sherff = Bidens aristosa (Michx.) Britton ■☆
54068 Bidens polylepis S. F. Blake var. retrorsa Sherff = Bidens polylepis S. F. Blake ■☆
54069 Bidens praecox Sherff = Bidens taylori (S. Moore) Sherff ■☆
54070 Bidens prestinaria (Sch. Bip. ex Walp.) Cufod.;爆裂鬼针草■☆
54071 Bidens prestinariiformis (Vatke) Cufod. = Bidens prestinaria (Sch. Bip. ex Walp.) Cufod. ■☆
54072 Bidens prestinariiformis (Vatke) Cufod. var. incisa (Sherff)

Cufod. = Bidens prestinaria (Sch. Bip. ex Walp.) Cufod. ■☆
54073 Bidens prionophylla Greene = Bidens cernua L. ■
54074 Bidens prolixa S. Moore = Bidens schimperi Sch. Bip. ex Walp. ■☆
54075 Bidens puberula (Wiegand) Rydb. = Bidens vulgata Greene ■☆
54076 Bidens puberula Wiegand = Bidens vulgata Greene ■☆
54077 Bidens punctata Sherff = Bidens schimperi Sch. Bip. ex Walp. ■☆
54078 Bidens quadriaristata DC. = Bidens laevis (L.) Britton, Stern et Poggenb. ■☆
54079 Bidens quadriseta Hochst. ex Oliv. et Hiern = Bidens biternata (Lour.) Merr. et Sherff ex Sherff ■
54080 Bidens quadriseta Hochst. ex Oliv. et Hiern var. incisifolia Hochst. ex Chiov. = Bidens biternata (Lour.) Merr. et Sherff ex Sherff ■
54081 Bidens radiata Thuill.;大羽叶鬼针草(大羽鬼针草,辐鬼针草);Radiate Beggarticks ■
54082 Bidens radiata Thuill. var. microcephala C. H. An;小花大羽叶鬼针草(小花鬼针草)■
54083 Bidens radiata Thuill. var. microcephala C. H. An = Bidens radiata Thuill. ■
54084 Bidens radiata Thuill. var. pinnatifida (Turcz. ex DC.) Kitam. = Bidens maximowicziana Oett. ■
54085 Bidens repens D. Don = Bidens tripartita L. var. repens (D. Don) Sherff ■
54086 Bidens repens D. Don = Bidens tripartita L. ■
54087 Bidens rhodesiana Sherff = Bidens oblonga (Sherff) Wild ■☆
54088 Bidens riparia Kunth;溪岸鬼针草■☆
54089 Bidens robertianifolia H. Lév. = Bidens biternata (Lour.) Merr. et Sherff ex Sherff ■
54090 Bidens robertianifolia H. Lév. et Vaniot = Bidens biternata (Lour.) Merr. et Sherff ex Sherff ■
54091 Bidens robustior S. Moore = Bidens kilimandscharica (O. Hoffm.) Sherff ■☆
54092 Bidens rogersii Sherff = Bidens ugandensis (S. Moore) Sherff ■☆
54093 Bidens rotata Sherff = Bidens ternata (Chiov.) Sherff var. vatkei (Sherff) Mesfin ■☆
54094 Bidens ruandensis Sherff = Bidens baumii (O. Hoffm.) Sherff ■☆
54095 Bidens rubicundula Sherff;稍红鬼针草■☆
54096 Bidens rubicundula Sherff f. alba T. G. J. Rayner;白鬼针草■☆
54097 Bidens rubra De Wild. = Bidens urceolata De Wild. ■☆
54098 Bidens rueppellii (Sch. Bip. ex Walp.) Sherff;鲁氏鬼针草■☆
54099 Bidens rueppellioides Sherff = Bidens rueppellii (Sch. Bip. ex Walp.) Sherff ■☆
54100 Bidens rufovenosa Sherff = Bidens kirkii (Oliv. et Hiern) Sherff ■☆
54101 Bidens rupestris Sherff = Bidens holstii (O. Hoffm.) Sherff ■☆
54102 Bidens ruwenzoriensis (S. Moore) Sherff = Bidens buchneri (Klatt) Sherff ■☆
54103 Bidens schimperi Sch. Bip. ex Walp.;欣珀鬼针草■☆
54104 Bidens schimperi Sch. Bip. ex Walp. var. brachycera Sherff = Bidens schimperi Sch. Bip. ex Walp. ■☆
54105 Bidens schimperi Sch. Bip. ex Walp. var. brachyceroides Sherff = Bidens taylori (S. Moore) Sherff ■☆
54106 Bidens schimperi Sch. Bip. ex Walp. var. greenwayi Sherff = Bidens schimperi Sch. Bip. ex Walp. ■☆
54107 Bidens schimperi Sch. Bip. ex Walp. var. leiocera Sherff = Bidens schimperi Sch. Bip. ex Walp. ■☆
54108 Bidens schimperi Sch. Bip. ex Walp. var. leptocera Sherff = Bidens schimperi Sch. Bip. ex Walp. ■☆
54109 Bidens schimperi Sch. Bip. ex Walp. var. pilosa Sch. Bip. ex Schweinf. = Bidens schimperi Sch. Bip. ex Walp. ■☆
54110 Bidens schimperi Sch. Bip. ex Walp. var. punctata Sherff = Bidens schimperi Sch. Bip. ex Walp. ■☆
54111 Bidens schlechteri Sherff = Bidens kirkii (Oliv. et Hiern) Sherff ■☆
54112 Bidens schlechteri Sherff var. wildii? = Bidens kirkii (Oliv. et Hiern) Sherff ■☆
54113 Bidens schultzii Cufod. = Bidens macroptera (Sch. Bip. ex Chiov.) Mesfin ■☆
54114 Bidens schweinfurthii Sherff = Bidens ugandensis (S. Moore) Sherff ■☆
54115 Bidens seretii (De Wild.) Sherff = Bidens buchneri (Klatt) Sherff ■☆
54116 Bidens serrulata Sch. Bip. = Bidens bipontina Sherff ■☆
54117 Bidens setigera (Sch. Bip. ex Vatke) Sherff subsp. bipinnatopartita (Chiov.) Mesfin;羽状深裂刚毛鬼针草■☆
54118 Bidens setigera (Sch. Bip. ex Vatke) Sherff var. lobata Sherff = Bidens rueppellii (Sch. Bip. ex Walp.) Sherff ■☆
54119 Bidens setigera (Sch. Bip. ex Walp.) Sherff;刚毛鬼针草■☆
54120 Bidens setigera (Sch. Bip. ex Walp.) Sherff var. abyssinica (Sch. Bip.) Sherff = Bidens setigera (Sch. Bip. ex Walp.) Sherff ■☆
54121 Bidens setigeroides Sherff = Bidens negriana (Sherff) Cufod. ■☆
54122 Bidens setigeroides Sherff var. munita? = Bidens negriana (Sherff) Cufod. ■☆
54123 Bidens shimadai Hayata = Bidens tripartita L. ■
54124 Bidens sierra-leonensis Mesfin;塞拉里昂鬼针草■☆
54125 Bidens snowdenii Sherff = Bidens grantii (Oliv.) Sherff ■☆
54126 Bidens somaliensis Sherff;索马里鬼针草■☆
54127 Bidens somaliensis Sherff var. bukobensis? = Bidens baumii (O. Hoffm.) Sherff ■☆
54128 Bidens spathulata Sherff = Guizotia jacksonii (S. Moore) J. Baagoe ■☆
54129 Bidens squarrosa Less.;糙鬼针草■☆
54130 Bidens steppia (Steetz) Sherff;草原鬼针草■☆
54131 Bidens steppia (Steetz) Sherff var. ambaeensis (Hiern) Sherff = Bidens steppia (Steetz) Sherff ■☆
54132 Bidens steppia (Steetz) Sherff var. elskensis Sherff = Bidens steppia (Steetz) Sherff ■☆
54133 Bidens steppia (Steetz) Sherff var. garusonis Sherff = Bidens steppia (Steetz) Sherff ■☆
54134 Bidens steppia (Steetz) Sherff var. humbertii Sherff = Bidens grantii (Oliv.) Sherff ■☆
54135 Bidens steppia (Steetz) Sherff var. inarmata Sherff = Bidens steppia (Steetz) Sherff ■☆
54136 Bidens steppia (Steetz) Sherff var. kalamboensis Sherff = Bidens steppia (Steetz) Sherff ■☆
54137 Bidens steppia (Steetz) Sherff var. leptocarpa Sherff = Bidens steppia (Steetz) Sherff ■☆
54138 Bidens straminoides Sherff = Bidens grantii (Oliv.) Sherff ■☆
54139 Bidens stuhlmannii (O. Hoffm.) Sherff = Bidens buchneri (Klatt) Sherff ■☆
54140 Bidens sulphurea (Cav.) Sch. Bip. = Cosmos sulphureus Cav. ■
54141 Bidens sundaica Blume = Bidens pilosa L. ■
54142 Bidens sundaica Blume var. minor Blume = Bidens pilosa L. var. minor (Blume) Sherff ■

54143 Bidens sundaica Blume var. minor Blume = Bidens pilosa L. ■

54144 Bidens superba Sherff = Bidens ternata (Chiov.) Sherff ■☆

54145 Bidens superba Sherff var. angustifolia? = Bidens ternata (Chiov.) Sherff var. angustifolia (Sherff) Mesfin ■☆

54146 Bidens superba Sherff var. brachycarpa? = Bidens ternata (Chiov.) Sherff ■☆

54147 Bidens taitensis Sherff? = Bidens holstii (O. Hoffm.) Sherff ■☆

54148 Bidens taitensis Sherff var. aciculata? = Bidens holstii (O. Hoffm.) Sherff ■☆

54149 Bidens taylori (S. Moore) Sherff;泰勒鬼针草■☆

54150 Bidens tenuifolia Labill. = Glossocardia bidens (Retz.) Veldkamp ■

54151 Bidens tenuifolia Labill. = Glossogyne tenuifolia (Labill.) Cass. ■

54152 Bidens ternata (Chiov.) Sherff;三出鬼针草■☆

54153 Bidens ternata (Chiov.) Sherff var. angustifolia (Sherff) Mesfin;窄叶三出鬼针草■☆

54154 Bidens ternata (Chiov.) Sherff var. vatkei (Sherff) Mesfin;瓦特克鬼针草■☆

54155 Bidens trichosperma (Michx.) Britton;毛籽鬼针草;Tickseed Sunflower,Tickseed-sunflower ■☆

54156 Bidens trichosperma (Michx.) Britton = Bidens coronata (L.) Britton ■☆

54157 Bidens trifida Roxb. = Bidens tripartita L. ■

54158 Bidens tripartita L.;狼杷草(拔毒散,叉头,大狼把草,豆渣菜,豆渣草,鬼叉,鬼刺,鬼针,接力草,金盏银盘,欆,郎耶菜,郎耶草,狼把草,狼耶草,三裂鬼针草,王八叉,乌阶,乌杷,小鬼叉,小鬼钗,一包针,引线包,针包草,针线包);Bastard Agrimony,Beggar Ticks, Bur Beggarticks, Bur Beggar-ticks, Bur Marigold, Swamp Beggar Ticks,Trifid Bur Marigold,Trifid Bur-marigold,Trifid Hemp Agrimony,Tripartite Bur Marigold,Water Agrimony,Water Hemp ■

54159 Bidens tripartita L. = Bidens comosa (A. Gray) Wiegand ■☆

54160 Bidens tripartita L. f. limosa Kom. = Bidens tripartita L. ■

54161 Bidens tripartita L. f. pinnatifida Turcz. ex DC. = Bidens maximowicziana Oett. ■

54162 Bidens tripartita L. var. cenuifolia Sherff = Bidens tripartita L. ■

54163 Bidens tripartita L. var. fallax Warnst. = Bidens connata Muhl. ex Willd. ■☆

54164 Bidens tripartita L. var. heterodoxa Fernald = Bidens heterodoxa (Fernald) Fernald et H. St. John ■☆

54165 Bidens tripartita L. var. pinnatifida? = Bidens radiata Thuill. var. pinnatifida (Turcz. ex DC.) Kitam. ■

54166 Bidens tripartita L. var. quinqueloba C. H. An;五裂狼杷草;Fivelobed Bur Biebersteinia ■

54167 Bidens tripartita L. var. quinqueloba C. H. An = Bidens tripartita L. ■

54168 Bidens tripartita L. var. repens (D. Don) Sherff;矮狼杷草;Dwarf Bur Biebersteinia ■

54169 Bidens tripartita L. var. repens (D. Don) Sherff = Bidens tripartita L. ■

54170 Bidens tripartita L. var. shimadai (Hayata) Yamam. = Bidens tripartita L. ■

54171 Bidens tripartita L. var. shimadai Yamam. = Bidens tripartita L. ■

54172 Bidens ugandensis (S. Moore) Sherff;乌干达狼杷草■☆

54173 Bidens ugandensis (S. Moore) Sherff var. longisquama Sherff = Bidens ugandensis (S. Moore) Sherff ■☆

54174 Bidens ugandensis (S. Moore) Sherff var. rogersii (Sherff) Lisowski = Bidens ugandensis (S. Moore) Sherff ■☆

54175 Bidens ugandensis (S. Moore) Sherff var. schweinfurthii (Sherff) Lisowski = Bidens ugandensis (S. Moore) Sherff ■☆

54176 Bidens uhligii Sherff = Bidens steppia (Steetz) Sherff ■☆

54177 Bidens ukambensis S. Moore = Bidens kilimandscharica (O. Hoffm.) Sherff ■☆

54178 Bidens urceolata De Wild.;坛状鬼针草■☆

54179 Bidens urceolata De Wild. var. leptolepis (Sherff) T. G. J. Rayner;细鳞坛状鬼针草■☆

54180 Bidens vatkei Sherff = Bidens ternata (Chiov.) Sherff var. vatkei (Sherff) Mesfin ■☆

54181 Bidens volkensii O. Hoffm. = Bidens kilimandscharica (O. Hoffm.) Sherff ■☆

54182 Bidens vulgata Greene;普通大鬼针草;Beggar Ticks,Big Devil's Beggar-ticks,Common Beggarticks,Sticktight,Tall Beggarticks,Tall Beggar-ticks ■☆

54183 Bidens vulgata Greene f. puberula (Wiegand) Fernald = Bidens vulgata Greene ■☆

54184 Bidens vulgata Greene var. dissectior Sherff = Bidens vulgata Greene ■☆

54185 Bidens vulgata Greene var. puberula (Wiegand) Greene = Bidens vulgata Greene ■☆

54186 Bidens vulgata Greene var. schizantha Lunell = Bidens vulgata Greene ■☆

54187 Bidens whytei Sherff;怀特鬼针草■☆

54188 Bidens zairensis Lisowski;扎伊尔鬼针草■☆

54189 Bidens zavattarii Cufod.;扎瓦鬼针草■☆

54190 Bidens zavattarii Cufod. var. latisecta? = Bidens zavattarii Cufod. ■☆

54191 Biderdykia (L.) Dum. Cours. = Polygonum dumetorum L. ●

54192 Bidwellia Herb. = Bidwillia Herb. ■☆

54193 Bidwillia Herb. = ? Asphodelus L. ■☆

54194 Biebersteinia Stephan ex Fisch. (1806);熏倒牛属;Biebersteinia ■

54195 Biebersteinia emodii Jaub. et Spach = Biebersteinia odora Stephan ex Fisch. ●

54196 Biebersteinia heterostemon Maxim.;熏倒牛(臭花椒,臭婆娘,狼尾巴蒿);Biebersteinia,Heterostemonous Biebersteinia ●

54197 Biebersteinia multifida DC.;多裂熏倒牛;Multifid Biebersteinia ●

54198 Biebersteinia odora Stephan ex Fisch.;高山熏倒牛(西藏熏倒牛,香熏倒牛);Alp Biebersteinia ●

54199 Biebersteiniaceae Endl. (1841);熏倒牛科■

54200 Biebersteiniaceae Endl. = Geraniaceae Juss. (保留科名)■●

54201 Biebersteiniaceae Schnizl. = Geraniaceae Juss. (保留科名)■●

54202 Bielschmeidia Pancher et Sebert = Beilschmiedia Nees ●

54203 Bielzia Schur = Centaurea L. (保留属名)●■

54204 Bieneria Rchb. f. = Chloraea Lindl. ■☆

54205 Bienertia Bunge = Bienertia Bunge ex Boiss. ■☆

54206 Bienertia Bunge ex Boiss. (1879);翅果蓬属■☆

54207 Bienertia cycloptera Bunge = Bienertia cycloptera Bunge ex Boiss. ■☆

54208 Bienertia cycloptera Bunge ex Boiss.;翅果蓬■☆

54209 Bienertia kossinskyi (Iljin) Tzvelev = Suaeda kossinskyi Iljin ■

54210 Biermannia King et Pantl. (1897);胼胝兰属(尖囊兰属,胼胝体兰属);Biermannia ■

54211 Biermannia calcarata Aver.;胼胝体兰■

54212 Biermannia decumbens Griff. = Kingidium deliciosum (Rchb. f.) H. R. Sweet ■

54213 Biermannia navicularis (Z. H. Tsi ex Hashim.) Ts. Tang et F. T. Wang ex Gruss et Roellke = Phalaenopsis braceana (Hook. f.)

Christenson ■
54214 Biermannia taenialis (Lindl.) Ts. Tang et F. T. Wang = Kingidium taeniale (Lindl.) P. F. Hunt ■
54215 Biermannia taenialis (Lindl.) Ts. Tang et F. T. Wang = Phalaenopsis taenialis (Lindl.) Christenson et Pradhan ■
54216 Bifaria (Hack.) Kuntze = Mesosetum Steud. ■☆
54217 Bifaria Kuntze = Mesosetum Steud. ■☆
54218 Bifaria Tiegh. = Korthalsella Tiegh. ●
54219 Bifaria capensis Tiegh. = Korthalsella opuntia (Thunb.) Merr. var. gaudichaudii (Tiegh.) Danser ●☆
54220 Bifaria commersonii Tiegh. = Korthalsella commersonii (Tiegh.) Danser ●☆
54221 Bifaria davidiana Tiegh. = Korthalsella japonica (Thunb.) Engl. var. fasciculata (Tiegh.) H. S. Kiu ●
54222 Bifaria davidiana Tiegh. = Korthalsella japonica (Thunb.) Engl. ●
54223 Bifaria fasciculata Tiegh. = Korthalsella japonica (Thunb.) Engl. ●
54224 Bifaria gaudichaudii Tiegh. = Korthalsella opuntia (Thunb.) Merr. var. gaudichaudii (Tiegh.) Danser ●☆
54225 Bifaria japonica (Thunb.) Tiegh. = Korthalsella japonica (Thunb.) Engl. ●
54226 Bifaria opuntia (Thunb.) Merr. = Korthalsella japonica (Thunb.) Engl. ●
54227 Bifaria opuntia Merr. = Korthalsella japonica (Thunb.) Engl. ●
54228 Bifariaceae Nakai = Santalaceae R. Br. (保留科名)●■
54229 Bifariaceae Nakai = Viscaceae Miq. ●
54230 Bifolium Nieuwl. = Listera R. Br. (保留属名)■
54231 Bifolium P. Gaertn., B. Mey. et Scherb. = Maianthemum F. H. Wigg. (保留属名)■
54232 Bifolium Petiver = Listera R. Br. (保留属名)■
54233 Bifolium Petiver ex Nieuwl. = Listera R. Br. (保留属名)■
54234 Bifolium auriculatum (Wiegand) Nieuwl. = Listera auriculata Wiegand ■☆
54235 Bifolium convallarioides (Sw.) Nieuwl. = Listera convallarioides (Sw.) Elliott ■☆
54236 Bifolium cordatum (L.) Nieuwl. = Listera cordata (L.) R. Br. ■☆
54237 Bifora Hoffm. (1816)(保留属名);双孔芹属;Bifora ■☆
54238 Bifora dicocca Hoffm. = Bifora testiculata (L.) Spreng. ■☆
54239 Bifora radians M. Bieb.;辐射双孔芹;Wild Bishop ■☆
54240 Bifora testiculata (L.) DC.;双片双孔芹;European Bishop ■☆
54241 Bifora testiculata (L.) Spreng. = Bifora testiculata (L.) DC. ■☆
54242 Biforis Spreng. = Bifora Hoffm. (保留属名)■☆
54243 Bifrenaria Lindl. (1832);比佛瑞纳兰属(双柄兰属);Bifrenaria ■☆
54244 Bifrenaria atropurpurea Lindl.;紫花比佛瑞纳兰;Purpleflower Bifrenaria ■☆
54245 Bifrenaria aurantiaca Lindl.;黄花比佛瑞纳兰■☆
54246 Bifrenaria harrisoniae (Hook.) Rchb. f.;比佛瑞纳兰;Harrison Bifrenaria ■☆
54247 Bifrenaria inodora Lindl.;无味比佛瑞纳兰;Inodorous Bifrenaria ■☆
54248 Bifrenaria tetragona (Lindl.) Schltr.;四棱茎比佛瑞纳兰■☆
54249 Bifrenaria tyrianthina Rchb. f.;青紫花比佛瑞纳兰■☆
54250 Bigamea K. Koenig ex Endl. = Ancistrocladus Wall. (保留属名)●
54251 Bigelonia Raf. = Bigelowia Raf. (废弃属名)■
54252 Bigelonia Raf. = Stellaria L. ■
54253 Bigelovia Sm. = Forestiera Poir. (保留属名)●☆
54254 Bigelovia Spreng. (1821) = Casearia Jacq. ●
54255 Bigelovia Spreng. (1827) = Borreria G. Mey. (保留属名)●■
54256 Bigelovia Spreng. = Spermacoce L. ●■
54257 Bigelowia DC. (1836)(保留属名);暗黄花属;Rayless-goldenrod ●☆
54258 Bigelowia DC. (保留属名) = Spermacoce L. ●■
54259 Bigelowia DC. ex Ging. (1824) = Noisettia Kunth ■☆
54260 Bigelowia Raf. (废弃属名) = Bigelowia DC. (保留属名)●☆
54261 Bigelowia Raf. (废弃属名) = Stellaria L. ■
54262 Bigelowia acradenia Greene = Isocoma acradenia (Greene) Greene ■☆
54263 Bigelowia albida M. E. Jones ex A. Gray = Ericameria albida (M. E. Jones ex A. Gray) L. C. Anderson ●☆
54264 Bigelowia brachylepis A. Gray = Ericameria brachylepis (A. Gray) H. M. Hall ●☆
54265 Bigelowia cooperi A. Gray = Ericameria cooperi (A. Gray) H. M. Hall ●☆
54266 Bigelowia engelmannii A. Gray = Oönopsis engelmannii (A. Gray) Greene ■☆
54267 Bigelowia graveolens (Nutt.) A. Gray var. latisquamea A. Gray = Ericameria nauseosa (Pall. ex Pursh) G. L. Nesom et G. I. Baird var. latisquamea (A. Gray) G. L. Nesom et G. I. Baird ●☆
54268 Bigelowia graveolens Nutt. var. hololeuca A. Gray;烈味暗黄花;Rabbit Brush, Rabbitbrush, Rabbit-weed ●☆
54269 Bigelowia greenei A. Gray = Chrysothamnus greenei (A. Gray) Greene ■☆
54270 Bigelowia howardii (Parry ex A. Gray) A. Gray var. attenuata M. E. Jones = Ericameria parryi (A. Gray) G. L. Nesom et G. I. Baird var. attenuata (M. E. Jones) G. L. Nesom et G. I. Baird ●☆
54271 Bigelowia juncea Greene = Ericameria nauseosa (Pall. ex Pursh) G. L. Nesom et G. I. Baird var. juncea (Greene) G. L. Nesom et G. I. Baird ●☆
54272 Bigelowia leiosperma A. Gray = Ericameria nauseosa (Pall. ex Pursh) G. L. Nesom et G. I. Baird var. leiosperma (A. Gray) G. L. Nesom et G. I. Baird ●☆
54273 Bigelowia menziesii (Hook. et Arn.) A. Gray var. scopulorum M. E. Jones = Chrysothamnus scopulorum (M. E. Jones) Urbatsch, R. P. Roberts et Neubig ●☆
54274 Bigelowia mohavensis Greene = Ericameria nauseosa (Pall. ex Pursh) G. L. Nesom et G. I. Baird var. mohavensis (Greene) G. L. Nesom et G. I. Baird ●☆
54275 Bigelowia nudata (Michx.) DC.;裸暗黄花;Pineland Rayless-goldenrod ●☆
54276 Bigelowia nudata (Michx.) DC. subsp. australis L. C. Anderson = Bigelowia nudata (Michx.) DC. var. australis (L. C. Anderson) Shinners ●☆
54277 Bigelowia nudata (Michx.) DC. var. australis (L. C. Anderson) Shinners;南方裸暗黄花●☆
54278 Bigelowia nuttallii L. C. Anderson;纳托尔暗黄花;Nuttall's rayless-goldenrod ●☆
54279 Bigelowia paniculata A. Gray = Ericameria paniculata (A. Gray) Rydb. ●☆
54280 Bigelowia parishii Greene = Ericameria parishii (Greene) H. M. Hall ●☆
54281 Bigelowia pulchella A. Gray;美丽暗黄花●☆
54282 Bigelowia spathulata A. Gray = Ericameria cuneata (A. Gray) McClatchie var. spathulata (A. Gray) H. M. Hall ●☆

54283 Bigelowia turbinata M. E. Jones = Ericameria nauseosa (Pall. ex Pursh) G. L. Nesom et G. I. Baird var. turbinata (M. E. Jones) G. L. Nesom et G. I. Baird ●☆

54284 Bigelowia uniligulata DC. = Solidago uliginosa Nutt. ■☆

54285 Bigelowia vaseyi A. Gray = Chrysothamnus vaseyi (A. Gray) Greene ●☆

54286 Bigelowia veneta (Kunth) A. Gray var. sedoides Greene = Isocoma menziesii (Hook. et Arn.) G. L. Nesom var. sedoides (Greene) G. L. Nesom ■☆

54287 Bigelowia virgata (Nutt.) DC. = Bigelowia nudata (Michx.) DC. ●☆

54288 Biggina Raf. = Salix L.(保留属名)●

54289 Biglandularia H. Karst. = Leiphaimos Cham. et Schltdl. ■☆

54290 Biglandularia Seem. = Rosanowia Regel ●■☆

54291 Biglandularia Seem. = Sinningia Nees ●■☆

54292 Bignonia L.(1753)(保留属名);紫葳属(比格诺藤属,卷须紫葳属);Bignonia,Cross Vine,Trumpet-creeper ●

54293 Bignonia adenophylla Wall. ex G. Don = Fernandoa adenophylla (Wall. ex G. Don) Steenis ●☆

54294 Bignonia adenophylla Wall. ex G. Don = Haplophragma adenophyllum (Wall. ex G. Don) Dop ●☆

54295 Bignonia africana Lam. = Kigelia africana (Lam.) Benth. ●☆

54296 Bignonia anastomosans A. DC. = Phyllarthron ilicifolium (Pers.) H. Perrier ●☆

54297 Bignonia articulata Desf. ex Poir. = Phyllarthron articulatum (Desf. ex Poir.) K. Schum. ●☆

54298 Bignonia bojeri A. DC. = Ophiocolea floribunda (Bojer ex Lindl.) H. Perrier ●☆

54299 Bignonia bracteosa A. DC. = Rhodocolea involucrata (Bojer ex DC.) H. Perrier ●☆

54300 Bignonia callistegioides Cham. = Clytostoma callistegioides (Cham.) Bureau ex Griseb. ●

54301 Bignonia capensis Thunb. = Tecoma capensis (Thunb.) Lindl. ●

54302 Bignonia capensis Thunb. = Tecomaria capensis (Thunb.) Spach ●

54303 Bignonia capreolata L.;紫葳藤(缠绕猫爪藤,卷须紫葳);Cross Vine, Crossvine, Cross-vine, Garlic Vine, Monkey's-comb, Quarter Vine, Quartervine, Quarter-vine, Trumpet Flower, Trumpet-flower ●

54304 Bignonia catalpa Thunb. = Catalpa ovata G. Don ●

54305 Bignonia cauliflora (A. DC.) Sieber = Ophiocolea floribunda (Bojer ex Lindl.) H. Perrier ●☆

54306 Bignonia chamberlaynii Sims;张氏紫葳(蒜香藤);Chamberlayn Bignonia ●

54307 Bignonia chamberlaynii Sims = Anemopaegma chamberlaynei (Sims) Bureau et K. Schum. ●☆

54308 Bignonia chinensis Lam. = Campsis grandiflora (Thunb.) K. Schum. ●

54309 Bignonia chrysantha Jacq. = Tabebuia chrysantha (Jacq.) G. Nicholson ●☆

54310 Bignonia colais Buch. -Ham. ex Dillwyn = Stereospermum colais (Buch. -Ham. ex Dillwyn) Mabb. ●

54311 Bignonia colais Buch. -Ham. ex Wall. = Stereospermum colais (Buch. -Ham. ex Dillon) Mabb. ●

54312 Bignonia compressa Lam. = Rhodocolea racemosa (Lam.) H. Perrier ●☆

54313 Bignonia crucigera L. = Pithecoctenium crucigerum (L.) A. H. Gentry ●

54314 Bignonia discolor DC. = Ophiocolea floribunda (Bojer ex Lindl.) H. Perrier ●☆

54315 Bignonia discolor R. Br. = Stereospermum kunthianum Cham. ●☆

54316 Bignonia echinata Jacq. = Pithecoctenium crucigerum (L.) A. H. Gentry ●

54317 Bignonia euphorioides Bojer = Stereospermum euphorioides (Bojer) A. DC. ●☆

54318 Bignonia ferdinandi Welw. = Fernandoa ferdinandi (Welw.) K. Schum. ●☆

54319 Bignonia ghorta Buch. -Ham. = Pauldopia ghorta (Buch. -Ham. ex G. Don) Steenis ●

54320 Bignonia ghorta Buch. -Ham. ex G. Don = Pauldopia ghorta (Buch. -Ham. ex G. Don) Steenis ●

54321 Bignonia glandulosa Schumach. et Thonn. = Newbouldia laevis (P. Beauv.) Seem. ex Bureau ●☆

54322 Bignonia glauca Decne. = Tecomella undulata (Sm.) Seem. ●☆

54323 Bignonia grandiflora Thunb. = Campsis grandiflora (Thunb.) K. Schum. ●

54324 Bignonia ignea Vell. = Pyrostegia venusta (Ker Gawl.) Miers ●

54325 Bignonia ilicifolia Pers. = Phyllarthron ilicifolium (Pers.) H. Perrier ●☆

54326 Bignonia indica L. = Oroxylum indicum (L.) Vent. ●

54327 Bignonia lanata Fresen. = Stereospermum kunthianum Cham. ●☆

54328 Bignonia lindleyi DC. = Clytostoma callistegioides (Cham.) Bureau ex Griseb. ●

54329 Bignonia magnifica W. Bull;巨花紫葳●

54330 Bignonia marginata Cham. = Adenocalymma marginatum (Cham.) A. DC. ●☆

54331 Bignonia nematocarpa Bojer = Stereospermum nematocarpum A. DC. ●☆

54332 Bignonia pallida Lindl. = Tabebuia heterophylla (DC.) Britton ●☆

54333 Bignonia pandorana Andréws = Pandorea pandorana Steenis ●☆

54334 Bignonia pentandra Lour. = Oroxylum indicum (L.) Vent. ●

54335 Bignonia pentaphylla L. = Tabebuia heterophylla (DC.) Britton ●☆

54336 Bignonia picta Lindl. = Clytostoma callistegioides (Cham.) Bureau ex Griseb. ●

54337 Bignonia porteriana Wall. ex A. DC. = Radermachera glandulosa (Blume) Miq. ●

54338 Bignonia racemosa Lam. = Rhodocolea racemosa (Lam.) H. Perrier ●☆

54339 Bignonia radicans L. = Campsis radicans (L.) Seem. ex Bureau ●

54340 Bignonia selloi Spreng. = Arrabidaea selloi (Spreng.) Sandwith ●☆

54341 Bignonia speciosa Graham = Clytostoma callistegioides (Cham.) Bureau ex Griseb. ●

54342 Bignonia stans L. = Tecoma stans (L.) Juss. ex Kunth ●☆

54343 Bignonia stipulata (Wall.) Roxb. = Markhamia stipulata (Wall.) Seem. ex K. Schum. ●

54344 Bignonia stipulata Roxb. = Dolichandrone stipulata (Wall.) Benth. et Hook. f. ●

54345 Bignonia stipulata Roxb. = Markhamia stipulata (Wall.) Seem. ex K. Schum. ●

54346 Bignonia suaveolens Roxb. = Stereospermum colais (Buch. -Ham. ex Dillon) Mabb. ●

54347 Bignonia suberosa Roxb. = Millingtonia hortensis L. f. ●

54348 Bignonia tenuiflora DC. = Tecoma tenuiflora (DC.) Fabris ●☆

54349 Bignonia tomentosa Thunb. = Paulownia tomentosa (Thunb.)

Steud. ●

54350 Bignonia tulipifera Thonn. = Spathodea campanulata P. Beauv. ●

54351 Bignonia tweediana Lindl. = Macfadyena unguis-cati (L.) A. H. Gentry ●

54352 Bignonia undulata Roxb. = Tecomella undulata (Roxb.) Seem. ●☆

54353 Bignonia undulata Sm. = Tecomella undulata (Sm.) Seem. ●☆

54354 Bignonia unguis Mart. = Macfadyena unguis-cati (L.) A. H. Gentry ●

54355 Bignonia unguis-cati L. = Macfadyena unguis-cati (L.) A. H. Gentry ●

54356 Bignonia venusta Ker Gawl. = Pyrostegia venusta (Ker Gawl.) Miers ●

54357 Bignoniaceae Juss. (1789)(保留科名);紫葳科;Bignonia Family,Trumpet Creeper Family,Trumpet-creeper Family ●■

54358 Bihai Mill.(废弃属名) = Heliconia L.(保留属名)■

54359 Bihai metallica (Planch. et L. Linden ex Hook.) Kuntze = Heliconia metallica Planch. et Linden ex Hook. ■

54360 Bihaia Kuntze = Heliconia L.(保留属名)■

54361 Bihania Meisn. = Eusideroxylon Teijsm. et Binn.(保留属名)●☆

54362 Bijlia N. E. Br. (1928);碧波属■☆

54363 Bijlia cana N. E. Br.;碧波;Prince Albert Vygie ■☆

54364 Bijlia dilatata H. E. K. Hartmann;膨大碧波■☆

54365 Bijlia tugwelliae (L. Bolus) S. A. Hammer;特格碧波■☆

54366 Bikera Adans. = Tetragonotheca L. ■☆

54367 Bikinia Wieringa(1999);碧波豆属●☆

54368 Bikinia aciculifera Wieringa;针形碧波■☆

54369 Bikinia breynei (Bamps) Wieringa;布雷恩碧波■☆

54370 Bikinia congensis Wieringa;康格碧波■☆

54371 Bikinia coriacea (J. Morel ex Aubrév.) Wieringa;革质碧波(革质单瓣豆)■☆

54372 Bikinia durandii (F. Hallé et Normand) Wieringa;杜朗碧波■☆

54373 Bikinia evrardii (Bamps) Wieringa;埃夫拉尔碧波■☆

54374 Bikinia grisea Wieringa;灰碧波■☆

54375 Bikinia letestui (Pellegr.) Wieringa;莱泰斯图碧波■☆

54376 Bikinia letestui (Pellegr.) Wieringa subsp. mayumbensis Wieringa;马永巴碧波■☆

54377 Bikinia media Wieringa;中间碧波■☆

54378 Bikinia pellegrinii (A. Chev.) Wieringa;佩尔格兰碧波■☆

54379 Bikkia Reinw. (1825)(保留属名);比克茜属●☆

54380 Bikkia Reinw. ex Blume = Bikkia Reinw.(保留属名)●☆

54381 Bikkia australis DC. = Bikkia grandiflora Reinw. ex Blume ●☆

54382 Bikkia grandiflora Reinw. ex Blume;大花比克茜●☆

54383 Bikkia longicarpa Valeton;长果比克茜●☆

54384 Bikkia macrophylla K. Schum.;大叶比克茜●☆

54385 Bikkia pachyphylla Guillaumin;厚叶比克茜●☆

54386 Bikkia parviflora Schltr. et K. Krause;小花比克茜●☆

54387 Bikkiopsis Brongn. et Gris = Bikkia Reinw.(保留属名)●☆

54388 Bikukulla Adans.(废弃属名) = Dicentra Bernh.(保留属名)■

54389 Bilabium Miq. = Chirita Buch. -Ham. ex D. Don ●■

54390 Bilabium Miq. = Didymocarpus Wall.(保留属名)●■

54391 Bilabrella Lindl. = Habenaria Willd. ■

54392 Bilabrella falcicornis Burch. ex Lindl. = Habenaria falcicornis (Burch. ex Lindl.) Bolus ■☆

54393 Bilacunaria Pimenov et V. N. Tikhom. (1983);双沟芹属■☆

54394 Bilacunaria microcarpa (M. Bieb.) Pimenov et V. N. Tikhom.;双沟芹■☆

54395 Bilacus Kuntze = Aegle Corrêa(保留属名)●

54396 Bilacus Rumph. = Aegle Corrêa(保留属名)●

54397 Bilacus Rumph. ex Kuntze = Aegle Corrêa(保留属名)●

54398 Bilaeus Kuntze = Aegle Corrêa(保留属名)●

54399 Bilamista Raf. = Gentiana L. ■

54400 Bilderdykia Dumort. = Fallopia Adans. ●■

54401 Bilderdykia Dumort. = Tiniaria (Meisn.) Rchb. ●■

54402 Bilderdykia aubertii (L. Henry) Moldenke = Fallopia baldschuanica (Regel) Holub ■☆

54403 Bilderdykia aubertii (L. Henry) Moldenke = Polygonum aubertii (L. Henry) Holub ●

54404 Bilderdykia baldschuanica (Regel) D. A. Webb = Fallopia baldschuanica (Regel) Holub ■☆

54405 Bilderdykia baldschuanica (Regel) D. A. Webb = Polygonum aubertii (L. Henry) Holub ●

54406 Bilderdykia baldschuanica (Regel) D. A. Webb = Polygonum baldschuanicum Regel ■☆

54407 Bilderdykia cilinodis (Michx.) Greene = Fallopia cilinodis (Michx.) Holub ■☆

54408 Bilderdykia cilinodis (Michx.) Greene var. laevigata (Fernald) C. F. Reed = Fallopia cilinodis (Michx.) Holub ■☆

54409 Bilderdykia convolvulus (L.) Dumort. = Fallopia convolvulus (L.) Á. Löve ■

54410 Bilderdykia convolvulus (L.) Dumort. = Polygonum convolvulus L. ■

54411 Bilderdykia cristata (Engelm. et A. Gray) Greene = Fallopia scandens (L.) Holub ■☆

54412 Bilderdykia cristata (Engelm. et A. Gray) Greene = Polygonum scandens L. var. cristatum (Engelm. et A. Gray) Gleason ■☆

54413 Bilderdykia dentato-alata (F. Schmidt) Kitag. = Fallopia dentato-alata (F. Schmidt ex Maxim.) Holub ■

54414 Bilderdykia dumetora (L.) Dumort. = Fallopia dumetora (L.) Holub ■

54415 Bilderdykia multiflora (Thunb.) Roberty et Vautier = Fallopia multiflora (Thunb.) Haraldson ■

54416 Bilderdykia pauciflora (Maxim.) Nakai = Fallopia dumetora (L.) Holub var. pauciflora (Maxim.) A. J. Li ■

54417 Bilderdykia scandens (L.) Greene = Fallopia scandens (L.) Holub ■☆

54418 Bilderdykia scandens (L.) Greene var. cristata (Engelm. et A. Gray) C. F. Reed = Polygonum scandens L. var. cristatum (Engelm. et A. Gray) Gleason ■☆

54419 Bilderdykia scandens (L.) Greene var. cristata (Engelm. et A. Gray) C. F. Reed = Fallopia scandens (L.) Holub ■☆

54420 Bilderdykia scandens (L.) Greene var. dentatoalata (F. Schmidt) Nakai = Fallopia dentatoalata (F. Schmidt ex Maxim.) Holub ■

54421 Bilderdykia scandens (L.) Greene var. dumetorum (L.) Dumort. = Fallopia dumetora (L.) Holub ■

54422 Bilderkykia aubertii (L. Henry) Moldenke = Fallopia aubertii (L. Henry) Holub ●

54423 Bilegnum Brand = Rindera Pall. ■

54424 Bilegnum Brand(1915);肖翅果草属■☆

54425 Bilegnum bungei (Boiss.) Brand.;邦奇肖翅果草■☆

54426 Bileveillea Vaniot = Blumea DC.(保留属名)■●

54427 Bileveillea granulatifolia (Blume) H. Lév. = Blumea lanceolaria (Roxb.) Druce ■

54428 Bileveillea granulatifolia H. Lév. = Blumea lanceolaria (Roxb.)

Druce ■
54429 Bileveillea myriocephala DC. = Blumea lanceolaria (Roxb.) Druce ■
54430 Bileveillea spectabilis DC. = Blumea lanceolaria (Roxb.) Druce ■
54431 Billardiera Moench = Verbena L. ■●
54432 Billardiera Sm. (1793);比拉碟兰属(藤海桐属);Apple Berry, Billardiera ●■☆
54433 Billardiera Sm. = Verbena L. ■●
54434 Billardiera Vahl = Coussarea Aubl. ●☆
54435 Billardiera cymosa F. Muell.; 聚伞花比拉碟兰; Cymose Billardiera ■☆
54436 Billardiera longiflora Labill.;长花比拉碟兰(长花藤海桐,长叶比拉碟兰); Blue Appleberry, Blue Berry, Blueberry, Droopflower Billardiera, Purple Apple, Purple Apple Berry ●☆
54437 Billardiera scandens Sm.; 攀缘比拉碟兰; Apple Berry, Apple Dumpling, Dumplings, Twinebranch Billardiera ■☆
54438 Billardiera thyesoides Mart.; 密花比拉碟兰; Denseflower Billardiera ■☆
54439 Billbergia Thunb. (1821);水塔花属(比尔贝亚属,必尔褒奇属,凤兰属,红苞凤梨属,花尔贝属,水塔凤梨属,筒凤梨属,筒状凤梨属); Airbroom, Billbergia, Watertowerflower ■
54440 Billbergia × windii Baker;温氏水塔花;Angel's Tears ■☆
54441 Billbergia chantini Carrière = Aechmea chantinii (Carrière) Baker ■
54442 Billbergia decora Poepp. et Endl.;雅致水塔花■☆
54443 Billbergia distacaia Mez;双穗水塔花■☆
54444 Billbergia distacaia Mez var. concolor Reitz;同色双穗水塔花■☆
54445 Billbergia distacaia Mez var. maculata Reitz;斑点双穗水塔花■☆
54446 Billbergia euphemiae E. Morren;尤菲米亚水塔花■☆
54447 Billbergia fasciata Lindl. = Aechmea fasciata (Lindl.) Baker ■☆
54448 Billbergia granulosa Brongn.;颗粒水塔花■☆
54449 Billbergia horrida Regel;刺水塔花■☆
54450 Billbergia leopoldii Linden ex Houllet;利氏水塔花■☆
54451 Billbergia liboniana De Jonghe;里氏水塔花■☆
54452 Billbergia macrocalyx Hook.;大萼水塔花■☆
54453 Billbergia morelii Brongn.;莫氏水塔花■☆
54454 Billbergia nutans H. Wendl. ex Regel;垂花水塔花(比尔见亚,垂花菠萝,垂花凤梨,俯垂水塔花,狭叶水塔花); Angel's Tears, Friendship Plant, Nodding Billbergia, Nunate Watertowerflower, Queen's Tears ■
54455 Billbergia porteana Brongn. ex Beer;波氏水塔花■☆
54456 Billbergia pyramidalis (Sims) Lindl.;水塔花(红苞凤梨,红笔菠萝,水星波罗,香水塔花); Foolproof Vase Plant, Foolproofplant, Pyramidal Billbergia, Summer Torch, Watertowerflower ■
54457 Billbergia pyramidalis (Sims) Lindl. 'Concolor';火炬水塔花(火焰菠萝,火焰凤梨,同色水塔花)■☆
54458 Billbergia pyramidalis (Sims) Lindl. 'Kyoto';白条纹水塔花■☆
54459 Billbergia pyramidalis (Sims) Lindl. 'Variegata';黄纹水塔花■☆
54460 Billbergia pyramidalis (Sims) Lindl. var. concolor L. B. Sm. = Billbergia pyramidalis (Sims) Lindl. 'Concolor' ■☆
54461 Billbergia sanderiana E. Morren;桑氏水塔花;Sander's Billbergia ■☆
54462 Billbergia saundersii W. Bull;美叶水塔花(美叶菠萝,美叶凤梨)■☆
54463 Billbergia speciosa Thunb.;美丽水塔花■☆
54464 Billbergia thyrsoides Mart. ex Schult. f.;密花水塔花(密花比尔见亚)■☆
54465 Billbergia viridiflora H. Wendl.;绿水塔花■☆
54466 Billbergia vittata Brongn. ex Morel;粗线塔花■☆
54467 Billbergia zebrina (Herb.) Lindl.;斑马水塔花■☆
54468 Billburttia Magee et B.-E. van Wyk(2009);马岛草属☆
54469 Billia Peyr. (1858)(保留属名);三叶树属●☆
54470 Billia hippocastanum Peyr.;三叶树●☆
54471 Billieturnera Fryxell(1982);图尔锦葵属●☆
54472 Billieturnera helleri (Rose ex A. Heller) Fryxell;图尔锦葵●☆
54473 Billiotia Endl. = Billotta G. Don ●☆
54474 Billiotia G. Don = Billiottia DC. ●☆
54475 Billiotia G. Don = Melanopsidium Colla ●☆
54476 Billiotia Rchb. = Agonis (DC.) Sweet(保留属名)●☆
54477 Billiotia Rchb. = Billottia R. Br. ●☆
54478 Billiottia DC. = Melanopsidium Colla ●☆
54479 Billiottia Endl. = Billotta G. Don ●☆
54480 Billotia Colla = Calothamnus Labill. ●☆
54481 Billotia G. Don = Agonis (DC.) Sweet(保留属名)●☆
54482 Billotia R. Br. ex G. Don = Agonis (DC.) Sweet(保留属名)●☆
54483 Billotia Sch. Bip. = Barkhausia Moench ■
54484 Billotia Sch. Bip. = Crepis L. ■
54485 Billotta G. Don = Agonis (DC.) Sweet(保留属名)●☆
54486 Billotta G. Don = Billottia R. Br. ●☆
54487 Billottia Colla = Calothamnus Labill. ●☆
54488 Billottia R. Br. = Agonis (DC.) Sweet(保留属名)●☆
54489 Billya Cass.(废弃属名) = Billia Peyr.(保留属名)●☆
54490 Billya Cass.(废弃属名) = Helichrysum Mill.(保留属名)●■
54491 Billya Cass.(废弃属名) = Petalacte D. Don ●☆
54492 Biltia Small = Rhododendron L. ●
54493 Bima Noronha = Castanopsis (D. Don) Spach(保留属名) + Nephelium L. ●
54494 Bimcroftia Billb. = Arracacia Bancr. ■☆
54495 Binaria Raf. = Bauhinia L. ●
54496 Bindera Raf. = Aster L. ●■
54497 Binectaria Forssk. = Mimusops L. ●☆
54498 Binectaria laurifolia Forssk. = Mimusops laurifolia (Forssk.) Friis ●☆
54499 Bineetaria Forssk. = Imbricaria Comm. ex Juss. ●☆
54500 Bingeria A. Chev. = Turraeanthus Baill. ●
54501 Bingeria africana (Welw. ex C. DC.) A. Chev. = Turraeanthus africanus (Welw. ex C. DC.) Pellegr. ●☆
54502 Binghamia Backeb. = Borzicactus Riccob. ■☆
54503 Binghamia Britton et Rose = Espostoa Britton et Rose ●
54504 Binghamia Britton et Rose = Haageocereus Backeb. + Pseudoespostoa Backeb ■
54505 Binghamia Britton et Rose = Pseudoespostoa Backeb ■
54506 Binia Noronha ex Thouars = Noronhia Stadman ex Thouars ●☆
54507 Binnendijkia Kurz = Leptonychia Turcz. ●☆
54508 Binotia Rolfe(1905);比诺兰属■☆
54509 Binotia W. Watson = Hippeastrum Herb.(保留属名)■
54510 Binotia W. Watson = Worsleya (W. Watson ex Traub) Traub ■
54511 Binotia brasiliensis (Rolfe) Rolfe;比诺兰■☆
54512 Bioletta Greene = Trichocoronis A. Gray ■☆
54513 Biondea Usteri = Blondea Rich. ●
54514 Biondea Usteri = Sloanea L. ●
54515 Biondia Schltr. (1905);秦岭藤属;Biondia ●★
54516 Biondia chinensis Schltr.; 秦岭藤; China Biondia, Chinese Biondia ●◇

54517 Biondia crassipes M. G. Gilbert et P. T. Li;厚叶秦岭藤;Thick-leaf Biondia ●
54518 Biondia elliptica P. T. Li et Z. Y. Zhu = Biondia microcentra (Tsiang) P. T. Li ●
54519 Biondia hemsleyana (Warb. ex Schltr. et Diels) Tsiang et P. T. Li;宽叶秦岭藤(寸金藤);Hemsley Biondia ●
54520 Biondia henryi (Warb. ex Schltr. et Diels) Tsiang et P. T. Li;青龙藤(豆瓣绿,捆仙绳,捆仙丝,青龙筋,藤叶细辛,岩浆草);Henry Biondia ●
54521 Biondia henryi (Warb. ex Schltr. et Diels) Tsiang et P. T. Li var. longipedunculata M. Cheng et Z. J. Feng;长轴青龙藤●
54522 Biondia insignis Tsiang;黑水藤;Distinguished Biondia ●
54523 Biondia laxa M. G. Gilbert et P. T. Li;杯冠秦岭藤;Cupular Biondia ●
54524 Biondia longipes P. T. Li;长序梗秦岭藤;Long-stalk Biondia ●
54525 Biondia microcentra (Tsiang) P. T. Li;祛风藤(云南祛风藤,浙江乳突果);Elliptic Biondia,Zhejiang Adelostemma ●
54526 Biondia parviurnula M. G. Gilbert et P. T. Li;小花秦岭藤;Small-flower Biondia ●
54527 Biondia pilosa Tsiang et P. T. Li;宝兴藤;Pilose Biondia ●
54528 Biondia reveluta M. G. Gilbert et P. T. Li;卷冠秦岭藤;Revelute Biondia ●
54529 Biondia tsiukowensis M. G. Gilbert et P. T. Li;茨菇秦岭藤;Cigu Biondia ●
54530 Biondia yunnanensis (H. Lév.) Tsiang;短叶秦岭藤;Yunnan Biondia ●
54531 Bionia Mart. ex Benth. = Camptosema Hook. et Arn. ■☆
54532 Biophytum DC. (1824);感应草属(羞礼花属);Biophytum, Reactionggrass ■●
54533 Biophytum abyssinicum Steud. ex A. Rich.;阿比西尼亚感应草■☆
54534 Biophytum apodiscias (Turcz.) Edgew. et Hook. f. = Biophytum petersianum Klotzsch ●■
54535 Biophytum bequaertii De Wild. = Biophytum helenae Buscal. et Muschl. ■☆
54536 Biophytum bogoroense De Wild. = Biophytum helenae Buscal. et Muschl. ■☆
54537 Biophytum crassipes Engl.;粗梗感应草■☆
54538 Biophytum esquirolii H. Lév. = Biophytum fruticosum Blume ●■
54539 Biophytum fruticosum Blume;分枝感应草(大还魂草,筋骨草);Branch Reactionggrass,Esquirol Biophytum ●■
54540 Biophytum helenae Buscal. et Muschl.;海伦娜感应草■☆
54541 Biophytum homblei De Wild. = Biophytum helenae Buscal. et Muschl. ■☆
54542 Biophytum incrassatum Delhaye = Biophytum macrorrhizum R. E. Fr. ■☆
54543 Biophytum kamerunense Engl. = Biophytum talbotii (Baker f.) Hutch. et Dalziel ■☆
54544 Biophytum kassneri R. Knuth;卡斯纳感应草■☆
54545 Biophytum macrorrhizum R. E. Fr.;大根感应草■☆
54546 Biophytum nyikense Exell;尼卡感应草■☆
54547 Biophytum pedicellatum Delhaye = Biophytum talbotii (Baker f.) Hutch. et Dalziel ■☆
54548 Biophytum petersianum Klotzsch;无柄感应草(安胎药,降落伞,罗伞草,小感应草);Sessile Biophytum,Sessile Reactionggrass ●■
54549 Biophytum petersianum Klotzsch = Biophytum umbraculum Welw. ■☆
54550 Biophytum reinwardtii (Zucc.) Klotzsch = Biophytum abyssinicum Steud. ex A. Rich. ■☆
54551 Biophytum reinwardtii (Zucc.) Klotzsch subsp. abyssinicum (Steud. ex A. Rich.) Steenis = Biophytum abyssinicum Steud. ex A. Rich. ■☆
54552 Biophytum richardsiae Exell;理查兹感应草■☆
54553 Biophytum ringoetii De Wild. = Biophytum crassipes Engl. ■☆
54554 Biophytum rotundifolium Delhaye = Biophytum umbraculum Welw. ■☆
54555 Biophytum sensitivum (L.) DC.;感应草(降落草,罗伞草,吓唬草,小姑娘草,小礼花,羞礼花);Life Plant,Reactionggrass ●■
54556 Biophytum sessile (Buch.-Ham. ex Baill.) R. Knuth = Biophytum petersianum Klotzsch ●■
54557 Biophytum sessile Wall. = Biophytum petersianum Klotzsch ●■
54558 Biophytum talbotii (Baker f.) Hutch. et Dalziel;塔尔博特感应草■☆
54559 Biophytum thorelianum var. sinensis Guillaumin = Biophytum fruticosum Blume ●■
54560 Biophytum turianiense Kabuye;图里亚尼感应草■☆
54561 Biophytum umbraculum Welw.;伞状感应草■☆
54562 Biophytum uzungwaense Frim.-Moll.;乌尊季沃感应草■☆
54563 Biophytum zenkeri Guillaumin;岑克尔感应草■☆
54564 Biota (D. Don) Endl. = Platycladus Spach ●
54565 Biota (D. Don) Endl. = Thuja L. ●
54566 Biota D. Don = Platycladus Spach ●
54567 Biota D. Don ex Endl. = Platycladus Spach ●
54568 Biota D. Don ex Endl. = Thuja L. ●
54569 Biota occidentalis? = Thuja occidentalis L. ●
54570 Biota orientalis (L.) Endl. = Platycladus orientalis (L.) Franco ●
54571 Biota orientalis (L.) Endl. = Thuja orientalis L. ●
54572 Biota orientalis (L.) Endl. f. sieboldii (Endl.) W. C. Cheng et W. T. Wang = Platycladus orientalis (L.) Franco 'Sieboldii' ●
54573 Biota orientalis (L.) Endl. var. beverleyensis (Rehder) Hu = Platycladus orientalis (L.) Franco 'Beverleyensis' ●
54574 Biota orientalis (L.) Endl. var. nana Carrière = Platycladus orientalis (L.) Franco 'Sieboldii' ●
54575 Biota orientalis (L.) Endl. var. semperaurescens Lemoine ex Gordon = Platycladus orientalis (L.) Franco 'Semperaurescens' ●
54576 Biota orientalis (L.) Endl. var. sieboldii Endl. = Platycladus orientalis (L.) Franco 'Sieboldii' ●
54577 Biotia Cass. = Madia Molina ■☆
54578 Biotia DC. = Aster L. ●■
54579 Biotia commixta (Nees) DC. = Eurybia spectabilis (Aiton) G. L. Nesom ■☆
54580 Biotia corymbosa (Sol. ex Aiton) DC. var. discolor Regel = Doellingeria scabra (Thunb.) Nees ■
54581 Biotia corymbosa DC. = Eurybia divaricata (L.) G. L. Nesom ■☆
54582 Biotia corymbosa DC. var. alata (W. P. C. Barton) DC. = Eurybia divaricata (L.) G. L. Nesom ■☆
54583 Biotia corymbose DC. = Aster scaber Thunb. ■
54584 Biotia discolor Maxim. = Aster scaber Thunb. ■
54585 Biotia discolor Maxim. = Doellingeria scabra (Thunb.) Nees ■
54586 Biotia glomerata (Nees) DC. = Eurybia schreberi (Nees) Nees ■☆
54587 Biotia latifolia DC. = Eurybia macrophylla (L.) Cass. ■☆
54588 Biotia macrophylla (L.) DC. = Eurybia macrophylla (L.) Cass. ■☆
54589 Biotia macrophylla DC. var. divaricata (L.) DC. = Eurybia divaricata (L.) G. L. Nesom ■☆

54590 Biotia schreberi (Nees) DC. = Eurybia schreberi (Nees) Nees ■☆
54591 Biovularia Kamienski = Utricularia L. ■
54592 Biovularia cymbantha (Oliv.) Kamienski = Utricularia cymbantha Oliv. ■☆
54593 Bipinnula Comm. ex Juss. (1789);双羽兰属■☆
54594 Bipinnula montana Arechav.;山地双羽兰■☆
54595 Bipontia S. F. Blake = Soaresia Sch. Bip.(保留属名)●☆
54596 Bipontlnia Alef. = Psoralea L. ●■
54597 Biporeia Thouars = Quassia L. ●☆
54598 Biramella Tiegh. = Ochna L. ●
54599 Biramella acutifolia (Engl.) Tiegh. = Ochna holstii Engl. ●☆
54600 Biramella holstii (Engl.) Tiegh. = Ochna holstii Engl. ●☆
54601 Biramta Neraud = Macleania Hook. ●☆
54602 Birchea A. Rich. = Luisia Gaudich. ■
54603 Biris Medik. = Iris L. ■
54604 Birnbaumia Kostel. = Anisacanthus Nees ■☆
54605 Birolia Bellardi = Elatine L. ■
54606 Birolia Raf. = Clusia L. ●☆
54607 Birostula Raf. = Scandix L. ■
54608 Birostula Raf. = Tetraclis Hiern ●
54609 Bisasehersonla Kuntze = Diospyros L. ●
54610 Bisboeckelera Kuntze(1891);双伯莎属■☆
54611 Bisboeckelera angustifolia (Boeck.) Kuntze;狭叶双伯莎■☆
54612 Bisboeckelera angustifolia Kuntze = Bisboeckelera angustifolia (Boeck.) Kuntze ■☆
54613 Bisboeckelera bicolor Standl.;二色双伯莎■☆
54614 Bisboeckelera longifolia Kuntze;长叶双伯莎■☆
54615 Bisboeckelera microcephala (Boeck.) T. Koyama;小头双伯莎■☆
54616 Bischoffia Decne. = Bischofia Blume ●
54617 Bischoffia F. Muell. = Bischofia Blume ●
54618 Bischofia Blume(1827);重阳木属(别重阳木属,茄苳属,秋枫属);Bishopwood,Bishop-wood ●
54619 Bischofia cumingiana Decne. = Bischofia javanica Blume ●
54620 Bischofia javanica Blume;重阳木(常绿重阳木,赤木,大秋枫,丢了棒,高粱木,过冬梨,红桐,胡桐,胡杨,加当,加冬,千金不倒,茄冬,茄冬树,茄苳,茄苳树,秋风,秋风子,秋枫,秋枫木,秋枫树,三叶红,水梁木,水蚬,万年青树,乌杨,鸭脚枫,朱桐树,总花重阳木);Autumn Maple Tree, Bischofia, Bishopweed, Chinese Bishopwood, Java Bishopwood, Java Bishop-wood, Javanese Bishopwood,Javawood,Kainjul,Katang,Red Cedar,Toog ●
54621 Bischofia leptopoda Müll. Arg. = Bischofia javanica Blume ●
54622 Bischofia oblongifolia Decne. = Bischofia javanica Blume ●
54623 Bischofia polycarpa (H. Lév.) Airy Shaw;秋枫(多果重阳木,红桐,茄冬树,水枧木,乌杨,重阳木);Many-fruit Bishopwood, Multifruited Bishop-wood ●
54624 Bischofia racemosa W. C. Cheng et C. D. Chu = Bischofia polycarpa (H. Lév.) Airy Shaw ●
54625 Bischofia roeperiana (Wight et Arn.) Decne. = Bischofia javanica Blume ●
54626 Bischofia toui Decne. = Bischofia javanica Blume ●
54627 Bischofia trifoliata (Roxb.) Hook. = Bischofia javanica Blume ●
54628 Bischofiaceae (Müll. Arg.) Airy Shaw = Euphorbiaceae Juss.(保留科名)●■
54629 Bischofiaceae (Müll. Arg.) Airy Shaw = Phyllanthaceae J. Agardh ●■
54630 Bischofiaceae Airy Shaw = Euphorbiaceae Juss.(保留科名)●■
54631 Bischofiaceae Airy Shaw = Phyllanthaceae J. Agardh ●■
54632 Bischofiaceae Airy Shaw;重阳木科●
54633 Biscutela Raf. = Biscutella L.(保留属名)■☆
54634 Biscutella L. (1753)(保留属名);双碟荠属;Biscutella ■☆
54635 Biscutella algeriensis Jord. = Biscutella didyma L. ■☆
54636 Biscutella apula L. var. microcarpa (DC.) Ball = Biscutella baetica Boiss. et Reut. ■☆
54637 Biscutella atlantica (Maire) Greuter et Burdet;大西洋双碟荠■☆
54638 Biscutella auriculata L.;耳状双碟荠■☆
54639 Biscutella auriculata L. subsp. brevicalcarata Batt. = Biscutella brevicalcarata (Batt.) Batt. ■☆
54640 Biscutella auriculata L. var. bivillosa Maire = Biscutella auriculata L. ■☆
54641 Biscutella auriculata L. var. brevicalcarata Batt. = Biscutella auriculata L. ■☆
54642 Biscutella auriculata L. var. candollii (Jord.) Maire = Biscutella auriculata L. ■☆
54643 Biscutella auriculata L. var. coronata Maire = Biscutella auriculata L. ■☆
54644 Biscutella auriculata L. var. emarginata Gren. et Godr. = Biscutella auriculata L. ■☆
54645 Biscutella auriculata L. var. erigerifolia (DC.) Willk. = Biscutella auriculata L. ■☆
54646 Biscutella auriculata L. var. lamarckii (Jord.) Maire = Biscutella auriculata L. ■☆
54647 Biscutella auriculata L. var. maroccana (Murb.) Maire = Biscutella auriculata L. ■☆
54648 Biscutella auriculata L. var. mauritanica (Jord.) Batt. = Biscutella mauritanica Jord. ■☆
54649 Biscutella auriculata L. var. orcelitana Lag. = Biscutella auriculata L. ■☆
54650 Biscutella baetica Boiss. et Reut.;伯蒂卡双碟荠■☆
54651 Biscutella brevicalcarata (Batt.) Batt.;短距双碟荠■☆
54652 Biscutella brevicalcarata (Batt.) Batt. var. maroccana Murb. = Biscutella auriculata L. ■☆
54653 Biscutella candollii Jord. = Biscutella auriculata L. ■☆
54654 Biscutella chouletti Jord. = Biscutella didyma L. ■☆
54655 Biscutella ciliata DC. = Biscutella didyma L. ■☆
54656 Biscutella columnae Ten. = Biscutella didyma L. ■☆
54657 Biscutella confusa Pomel = Biscutella didyma L. ■☆
54658 Biscutella depressa Willd.;凹陷双碟荠■☆
54659 Biscutella didyma L.;对双碟荠■☆
54660 Biscutella didyma L. subsp. ciliata (DC.) Rouy et Foucaud = Biscutella didyma L. ■☆
54661 Biscutella didyma L. subsp. columnae (Ten.) Nyman;圆柱双碟荠■☆
54662 Biscutella didyma L. var. algeriensis (Jord.) Maire = Biscutella didyma L. ■☆
54663 Biscutella didyma L. var. chouletti (Jord.) Batt. = Biscutella didyma L. ■☆
54664 Biscutella didyma L. var. ciliata (DC.) Halácsy = Biscutella didyma L. ■☆
54665 Biscutella didyma L. var. columnae (Ten.) Halácsy = Biscutella didyma L. ■☆
54666 Biscutella didyma L. var. confusa (Pomel) Maire = Biscutella didyma L. ■☆
54667 Biscutella didyma L. var. coriophora Batt. = Biscutella didyma L. ■☆

54668 Biscutella didyma L. var. depressa (Willd.) El Naggar = Biscutella depressa Willd. ■☆
54669 Biscutella didyma L. var. elbensis (Chrtek) El Naggar = Biscutella elbensis Chrtek ■☆
54670 Biscutella didyma L. var. eriocarpa (DC.) Maire et Weiller = Biscutella maritima Ten. ■☆
54671 Biscutella didyma L. var. gymnocarpa Maire = Biscutella didyma L. ■☆
54672 Biscutella didyma L. var. haplotricha Maire = Biscutella didyma L. ■☆
54673 Biscutella didyma L. var. laxiflora (Spreng.) Maire = Biscutella didyma L. ■☆
54674 Biscutella didyma L. var. leiocarpa (DC.) Halácsy = Biscutella didyma L. ■☆
54675 Biscutella didyma L. var. lenticularis Pamp. = Biscutella didyma L. ■☆
54676 Biscutella didyma L. var. maritima (Ten.) Guss. = Biscutella maritima Ten. ■☆
54677 Biscutella didyma L. var. megacarpaea Boiss. = Biscutella depressa Willd. ■☆
54678 Biscutella didyma L. var. micraspis Maire = Biscutella didyma L. ■☆
54679 Biscutella didyma L. var. muscariodora Maire = Biscutella didyma L. ■☆
54680 Biscutella didyma L. var. orivilla Maire et Sam. = Biscutella didyma L. ■☆
54681 Biscutella didyma L. var. pseudoalgeriensis Maire = Biscutella didyma L. ■☆
54682 Biscutella didyma L. var. pseudomicrocarpa Maire = Biscutella didyma L. ■☆
54683 Biscutella didyma L. var. scabrida (Pau et Font Quer) Maire = Biscutella didyma L. ■☆
54684 Biscutella didyma L. var. taraxacifolia Kuntze = Biscutella didyma L. ■☆
54685 Biscutella elbensis Chrtek;厄尔巴双碟荠■☆
54686 Biscutella erigerifolia DC. = Biscutella auriculata L. ■☆
54687 Biscutella eriocarpa DC. = Biscutella maritima Ten. ■☆
54688 Biscutella frutescens Coss.;灌木双碟荠●☆
54689 Biscutella frutescens Coss. var. papillosa Maire = Biscutella frutescens Coss. ●☆
54690 Biscutella laevigata L.;双碟荠(李果荠);Biscutella, Buckler Mustard ■☆
54691 Biscutella laevigata L. subsp. atlantica (Maire) Maire = Biscutella atlantica (Maire) Greuter et Burdet ■☆
54692 Biscutella laevigata L. var. ajmasiana (Pau) Maire et Weiller = Biscutella atlantica (Maire) Greuter et Burdet ■☆
54693 Biscutella laevigata L. var. atlantica Maire = Biscutella atlantica (Maire) Greuter et Burdet ■☆
54694 Biscutella lamarckii Jord. = Biscutella auriculata L. ■☆
54695 Biscutella laxiflora C. Presl ex Spreng. = Biscutella didyma L. ■☆
54696 Biscutella leiocarpa DC. = Biscutella didyma L. ■☆
54697 Biscutella lyrata L.;大头羽裂双碟荠■☆
54698 Biscutella lyrata L. subsp. maritima (Ten.) Raffaelli = Biscutella maritima Ten. ■☆
54699 Biscutella lyrata L. var. algeriense Jord. = Biscutella maritima Ten. ■☆
54700 Biscutella maritima Ten.;滨海双碟荠■☆
54701 Biscutella mauritanica Jord.;毛里塔尼亚双碟荠■☆
54702 Biscutella megalocarpa Fisch. ex DC. = Megacarpaea megalocarpa (Fisch. ex DC.) Schischk. ex B. Fedtsch. ■
54703 Biscutella microcarpa DC.;小果双碟荠■☆
54704 Biscutella microcarpa DC. f. scabrida Pau et Font Quer = Biscutella didyma L. ■☆
54705 Biscutella montana Cav.;山地双碟荠■☆
54706 Biscutella montana Cav. var. ajmasiana Pau;艾马斯双碟荠■☆
54707 Biscutella radicata Coss. et Durieu = Biscutella raphanifolia Poir. ■☆
54708 Biscutella raphanifolia Poir.;萝卜叶双碟荠■☆
54709 Biscutella raphanifolia Poir. var. ditrichocarpa Maire = Biscutella raphanifolia Poir. ■☆
54710 Biscutella raphanifolia Poir. var. orivillosa Maire = Biscutella raphanifolia Poir. ■☆
54711 Biscutella sempervirens L.;常绿双碟荠■☆
54712 Biscutella valentina (L.) Heywood;强壮双碟荠■☆
54713 Bisedmondia Hutch. = Calycophysum H. Karst. et Triana ■☆
54714 Biserrula L. (1753);双齿黄耆属■☆
54715 Biserrula leiocarpa A. Rich. = Biserrula pelecinus L. subsp. leiocarpa (A. Rich.) J. B. Gillett ■☆
54716 Biserrula pelecinus L.;双齿黄耆■☆
54717 Biserrula pelecinus L. = Astragalus pelecinus (L.) Barneby ■☆
54718 Biserrula pelecinus L. subsp. leiocarpa (A. Rich.) J. B. Gillett;光果双齿黄耆■☆
54719 Biserrula pelecinus L. var. brevipes Murb. = Astragalus pelecinus (L.) Barneby ■☆
54720 Bisetaria Tiegh. = Campylospermum Tiegh. ●
54721 Bisetaria febrifuga (Engl. et Gilg) Tiegh. = Campylospermum lecomtei (Tiegh.) Farron ●☆
54722 Bisetaria lecomtei (Tiegh.) Tiegh. = Campylospermum lecomtei (Tiegh.) Farron ●☆
54723 Bisglaziovia Cogn. (1891);格拉野牡丹属☆
54724 Bisglaziovia behurioidesCogn.;格拉野牡丹☆
54725 Bisgoeppertia Kuntze(1891);双格佩龙胆属■☆
54726 Bisgoeppertia gracilis Kuntze;双格佩龙胆■☆
54727 Bishopalea H. Rob. (1981);毛瓣叉毛菊属■☆
54728 Bishopalea erecta H. Rob.;毛瓣叉毛菊■☆
54729 Bishopanthus H. Rob. (1983);单头黄安菊属●☆
54730 Bishopanthus soliceps H. Rob.;单头黄安菊■☆
54731 Bishopiella R. M. King et H. Rob. (1981);莲座柄泽兰属■●☆
54732 Bishopiella elegans R. M. King et H. Rob.;莲座柄泽兰■☆
54733 Bishovia R. M. King et H. Rob. (1978);繁花亮泽兰属■☆
54734 Bishovia boliviensis R. M. King et H. Rob.;玻利维亚繁花亮泽兰■☆
54735 Bishovia mikaniifolia R. M. King et H. Rob.;繁花亮泽兰■☆
54736 Bisluederitzia Kuntze = Neoluederitzia Schinz ●☆
54737 Bismalva Medik. = Malva L. ■
54738 Bismarckia Hildebr. et H. Wendl. (1881);霸王棕属(卑士麦棕属,卑斯麦榈属,卑斯麦棕属,比斯马棕属,俾氏榈属,贵榔属);Bismarck Palm, Bismarckia ●☆
54739 Bismarckia nobilis Hildebrandt et H. Wendl.;霸王棕(比斯马棕,俾斯麦椰子,俾斯麦棕);Bismarck Palm, Stately Bismarckia ●☆
54740 Bisnaga Orcutt = Ferocactus Britton et Rose ●
54741 Bisnicholsonia Kuntze = Neonicholsonia Dammer ●☆
54742 Bisphaeria Noronha = Poikilospermum Zipp. ex Miq. ●
54743 Bisquamaria Pichon = Laxoplumeria Markgr. ●☆

54744 Bisquamaria Pichon(1947);双鳞夹竹桃属(巴西夹竹桃属)●☆
54745 Bisquamaria macrophylla (Kuhlm.) Pichon;双鳞夹竹桃●☆
54746 Bisrautanenia Kuntze = Neorautanenia Schinz ■☆
54747 Bissea V. R. Fuentes = Henoonia Griseb. ●☆
54748 Bistania Noronha = Litsea Lam.(保留属名)●
54749 Bistella Adans.(废弃属名)= Vahlia Thunb.(保留属名)■☆
54750 Bistella capensis (L. f.) Bullock = Vahlia capensis (L. f.) Thunb. ■☆
54751 Bistella digyna (Retz.) Bullock = Vahlia digyna (Retz.) Kuntze ■☆
54752 Bistella geminiflora Delile = Vahlia geminiflora (Delile) Bridson ■☆
54753 Bistorta (L.) Adans. = Bistorta (L.) Scop. ■☆
54754 Bistorta (L.) Adans. = Persicaria (L.) Mill. ■
54755 Bistorta (L.) Mill. = Bistorta (L.) Adans. ■
54756 Bistorta (L.) Mill. = Persicaria (L.) Mill. ■
54757 Bistorta (L.) Scop. (1754);双曲蓼属(拳参属);Bistort ■☆
54758 Bistorta (L.) Scop. = Colubrina Rich. ex Brongn.(保留属名)●
54759 Bistorta (L.) Scop. = Persicaria (L.) Mill. ■
54760 Bistorta Adans. = Bistorta (L.) Adans. ■☆
54761 Bistorta L. = Bistorta (L.) Adans. ■☆
54762 Bistorta Scop. = Bistorta (L.) Scop. ■☆
54763 Bistorta abukumensis Yonek., Iketsu et H. Ohashi;阿武隈双曲蓼■☆
54764 Bistorta affinis (D. Don) Greene = Polygonum affine D. Don ●■
54765 Bistorta alopecuroides (Turcz. ex Besser) Kom. = Polygonum alopecuroides Turcz. ex Besser ■
54766 Bistorta alopecuroides (Turcz. ex Besser) Kom. f. pilosa (C. F. Fang) Kitag. = Polygonum alopecuroides Turcz. ex Besser ■
54767 Bistorta alopecuroides (Turcz.) Kom. = Polygonum alopecuroides Turcz. ex Besser ■
54768 Bistorta amplexicauis subsp. sinensis (Forbes et Hemsl.) Soják = Polygonum amplexicaule D. Don var. sinense Forbes et Hemsl. ex Stewart ■
54769 Bistorta amplexicaulis (D. Don) Greene = Polygonum amplexicaule D. Don ■
54770 Bistorta amplexicaulis (D. Don) Greene subsp. sinomontana (Sam.) Yonekura et H. Ohashi = Polygonum sinomontanum Sam. ■
54771 Bistorta amplexicaulis (D. Don) Greene var. speciosa (Meisn.) Munshi et Javeid;美丽双曲蓼■☆
54772 Bistorta bistortoides (Pursh) Small;西部双曲蓼;American Bistort,Smokeweed,Western Bistort ■☆
54773 Bistorta bistortoides (Pursh) Small var. oblongifolia (Meisn.) Moldenke = Bistorta bistortoides (Pursh) Small ■☆
54774 Bistorta chinensis H. Gross = Polygonum paleaceum Wall. ex Hook. f. ■
54775 Bistorta coriacea (Sam.) Yonek. et H. Ohashi = Polygonum coriaceum Sam. ■
54776 Bistorta emodi (Meisn.) Petr. = Polygonum emodii Meisn. ●■
54777 Bistorta emodi (Meisn.) Petr. subsp. dependens (Diels) Soják = Polygonum emodii Meisn. var. dependens Diels ■
54778 Bistorta emodi (Meisn.) Petr. var. dependens (Diels) Petr. = Polygonum emodii Meisn. var. dependens Diels ■
54779 Bistorta franchetiana Petr. = Polygonum suffultum Maxim. ■
54780 Bistorta griffithii (Hook. f.) Grierson = Polygonum griffithii Hook. f. ■
54781 Bistorta hayachinensis (Makino) H. Gross;早池峰双曲蓼■☆
54782 Bistorta henryi Yonek. et H. Ohashi = Polygonum amplexicaule D. Don var. sinense Forbes et Hemsl. ex Stewart ■
54783 Bistorta honanensis (H. W. Kung) Yonek. et H. Ohashi = Polygonum honanense H. W. Kung ■
54784 Bistorta lapidosa Kitag. = Polygonum bistorta L. ■
54785 Bistorta macrophyllum (D. Don) Soják = Polygonum macrophyllum D. Don ■
54786 Bistorta macrophyllum (D. Don) Soják var. stenophylla (Meisn.) Miyam. = Polygonum macrophyllum D. Don var. stenophyllum (Meisn.) A. J. Li ■
54787 Bistorta majanthemifolia Petr. = Polygonum suffultum Maxim. ■
54788 Bistorta majanthemifolium Petr. = Polygonum suffultum Maxim. ■
54789 Bistorta major Gray = Bistorta officinalis Delarbre ■☆
54790 Bistorta major Gray = Polygonum bistorta L. ■
54791 Bistorta major Gray subsp. elliptica (Spreng.) Á. Löve et D. Löve = Polygonum ellipticum Willd. ex Spreng. ■
54792 Bistorta major Gray subsp. plumosum (Small) H. Hara = Bistorta plumosa (Small) Greene ■☆
54793 Bistorta major Gray var. japonica H. Hara = Bistorta officinalis Delarbre subsp. japonica (H. Hara) Yonek. ■☆
54794 Bistorta manshuriensis Kom. = Polygonum manshuriense Petr. ex Kom. ■
54795 Bistorta milletii H. Lév. = Polygonum milletii (H. Lév.) H. Lév. ■
54796 Bistorta ochotensis (Petr. ex Kom.) Kom. = Polygonum ochotense Petr. ex Kom. ■
54797 Bistorta ochotensis (Petr.) Kom. = Polygonum ochotense Petr. ex Kom. ■
54798 Bistorta officinalis Delarbre;药用双曲蓼;European bistort ■☆
54799 Bistorta officinalis Delarbre subsp. japonica (H. Hara) Yonek.;日本药用双曲蓼■☆
54800 Bistorta officinalis Delarbre subsp. pacifica (Petr. ex Kom.) Yonek.;太平洋药用双曲蓼■☆
54801 Bistorta officinalis Raf. = Polygonum bistorta L. ■
54802 Bistorta pacifica (Petr. ex Kom.) Kom. = Polygonum pacificum Petr. ex Kom. ■
54803 Bistorta pacifica (Petr. ex Kom.) Kom. ex Kitag. = Bistorta officinalis Delarbre subsp. pacifica (Petr. ex Kom.) Yonek. ■☆
54804 Bistorta pacifica (Petr. ex Kom.) Kom. ex Kitag. = Polygonum pacificum Petr. ex Kom. ■
54805 Bistorta paleacea (Wall. ex Hook. f.) Yonek. et H. Ohashi = Polygonum paleaceum Wall. ex Hook. f. ■
54806 Bistorta pergracilis (Hemsl.) H. Gross = Polygonum suffultum Maxim. var. pergracile (Hemsl.) Sam. ■
54807 Bistorta perpusilla (Hook. f.) Greene;微小双曲蓼■☆
54808 Bistorta petiolata (D. Don) Petr. = Polygonum amplexicaule D. Don ■
54809 Bistorta plumosa (Small) Greene;羽状双曲蓼■☆
54810 Bistorta pseudosuffulta Petr. = Polygonum suffultum Maxim. var. pergracile (Hemsl.) Sam. ■
54811 Bistorta purpureonervosa (A. J. Li) Yonek. et H. Ohashi = Polygonum purpureonervosum A. J. Li ■
54812 Bistorta sinomontana (Sam.) Miyam. = Polygonum sinomontanum Sam. ■
54813 Bistorta speciosa (Meisn.) Greene = Polygonum amplexicaule D. Don ■
54814 Bistorta sphaerostachya (Meisn.) Greene = Polygonum macrophyllum D. Don ■

54815 Bistorta subscaposa (Diels) Petr. = Polygonum subscaposum Diels ■
54816 Bistorta suffulta (Maxim.) Greene ex H. Gross = Polygonum suffultum Maxim. ■
54817 Bistorta suffulta (Maxim.) H. Gross = Polygonum suffultum Maxim. ■
54818 Bistorta suffulta (Maxim.) H. Gross f. pubescens Hiyama;短毛双曲蓼■☆
54819 Bistorta suffulta (Maxim.) H. Gross subsp. pergracilis (Hemsl.) Sojάk = Polygonum suffultum Maxim. var. pergracile (Hemsl.) Sam. ■
54820 Bistorta taipaishanensis (H. W. Kung) Yonek. et H. Ohashi = Polygonum milletii (H. Lév.) H. Lév. ■
54821 Bistorta tenuicaulis (Bisset et S. Moore) Nakai;细茎双曲蓼■☆
54822 Bistorta tenuicaulis (Bisset et S. Moore) Nakai var. chionophila Yonek. et H. Ohashi;喜雪细茎双曲蓼■☆
54823 Bistorta vaccinifolium (Wall. ex Meisn.) Greene = Polygonum vaccinifolium Wall. ex Meisn. ■
54824 Bistorta vacciniifolia (Wall. ex Meisn.) Greene = Polygonum vaccinifolium Wall. ex Meisn. ■
54825 Bistorta vivipara (L.) Delarbre = Polygonum viviparum L. ■
54826 Bistorta vivipara (L.) Delarbre f. roessleri (Beck) Kitag.;洛氏珠芽蓼■☆
54827 Bistorta vivipara (L.) Delarbre var. roessleri (Beck) F. Maek. = Bistorta vivipara (L.) Delabre f. roessleri (Beck) Kitag. ■☆
54828 Bistorta vivipara (L.) Gray = Polygonum viviparum L. ■
54829 Bistorta vivipara (L.) Gray var. angustifolia Nakai = Polygonum viviparum L. ■
54830 Bistorta vulgaria Hill var. ovata Nakai ex H. Hara = Bistorta officinalis Delarbre subsp. pacifica (Petr. ex Kom.) Yonek. ■☆
54831 Bistorta yunnanense H. Gross. = Polygonum paleaceum Wall. ex Hook. f. ■
54832 Bistorta yunnanensis H. Gross = Polygonum macrophyllum D. Don ■
54833 Bistorta zigzag (H. Lév. et Vaniot) H. Gross = Polygonum emodii Meisn. var. dependens Diels ■
54834 Biswarea Cogn. (1882);三裂瓜属;Biswarea ■
54835 Biswarea tonglensis (C. B. Clarke) Cogn.;三裂瓜;Trilobe Biswarea ■
54836 Biteria Börner = Carex L. ■
54837 Bitteria Börner = Carex L. ■
54838 Bituminaria Fabr. = Psoralea L. ●■
54839 Bituminaria Heist. ex Fabr. (1759);沥青补骨脂属■☆
54840 Bituminaria acaulis (Stev.) C. H. Stirt.;无茎沥青补骨脂■☆
54841 Bituminaria bituminosa (L.) C. H. Stirt.;沥青补骨脂;Arabian Pea ■☆
54842 Biventraria Small = Asclepias L. ■
54843 Bivinia Jaub. ex Tul. (1857);比维木属●☆
54844 Bivinia Tul. = Calantica Jaub. ex Tul. ●☆
54845 Bivinia jalbertii Tul.;比维木●☆
54846 Bivolva Tiegh. = Balania Tiegh. ■
54847 Bivolva Tiegh. = Balanophora J. R. Forst. et G. Forst. ■
54848 Bivolva fargesii Tiegh. = Balanophora fargesii (Tiegh.) Harms ■
54849 Bivonaea DC. (1821)(保留属名);西地中海芥属■☆
54850 Bivonaea Moq. et Sessé = Cardionema DC. ■☆
54851 Bivonaea Moq. et Sessé ex DC. = Bivonaea DC. (保留属名)■☆
54852 Bivonaea lutea (Biv.) DC.;西地中海芥■☆
54853 Bivonea Raf. (废弃属名) = Bivonaea DC. (保留属名)■☆
54854 Bivonea Raf. (废弃属名) = Cnidoscolus Pohl ●☆
54855 Bivonea Raf. (废弃属名) = Jatropha L. (保留属名)●■
54856 Bivonia Raf. = Bivonea Raf. (废弃属名)●■
54857 Bivonia Spreng. = Bernardia L. ●
54858 Biwaldia Scop. = Garcinia L. ●
54859 Bixa L. (1753);红木属(胭脂树属);Anatto, Anatto Tree, Anattotree, Anatto-tree, Bixa ●
54860 Bixa arborea Huber;乔红木●☆
54861 Bixa katangensis Delep. = Bixa orellana L. ●
54862 Bixa orellana L.;红木(胭脂木,胭脂树);Anatto, Anatto Tree, Anatto-tree, Annatto, Lipstick Plant, Lipsticktree ●
54863 Bixaceae Kunth(1822)(保留科名);红木科(胭脂树科);Bixa Family ●■
54864 Bixaceae Link = Bixaceae Kunth(保留科名)●■
54865 Bixagrewia Kurz = Trichospermum Blume ●☆
54866 Bizanilla J. Rémy = Psilocarphus Nutt. ■☆
54867 Bizonula Pellegr. (1924);双带无患子属●☆
54868 Bizonula letestui Pellegr.;双带无患子●☆
54869 Blabea Baehni = Pouteria Aubl. ●
54870 Blabeia Baehni = Pouteria Aubl. ●
54871 Blaberopus A. DC. = Alstonia R. Br. (保留属名)●
54872 Blaberopus rupester Pichon = Alstonia rupestris Kerr ●
54873 Blaberopus rupestre (Kerr) Pichon = Alstonia rupestris Kerr ●
54874 Blachia Baill. (1858)(保留属名);留萼木属;Blachia ●
54875 Blachia andamanica (Kurz) Hook. f.;大果留萼木;Bigfruit Blachia ●
54876 Blachia chunii Y. T. Chang et P. T. Li;海南留萼木;Chun Blachia, Hainan Blachia ●
54877 Blachia longzhouensis X. X. Chen;龙州留萼木;Longzhou Blachia ●
54878 Blachia pentzii (Müll. Arg.) Benth.;留萼木;Pentz Blachia ●
54879 Blachia philippinensis Merr.;菲律宾留萼木;Philippine Blachia ●☆
54880 Blachia umbellata (Willd.) Baill.;印斯留萼木;Umbellate Blachia ●☆
54881 Blachia yaihsienensis F. W. Xing et;崖州留萼木;Yazhou Blachia ●
54882 Blackallia C. A. Gardner(1942);布莱鼠李属●☆
54883 Blackallia biloba C. A. Gardner;布莱鼠李●☆
54884 Blackbournea Kunth = Blackburnia J. R. Forst. et G. Forst. ●
54885 Blackburnia J. R. Forst. et G. Forst. = Zanthoxylum L. ●
54886 Blackia Schrank = Myriaspora DC. ●☆
54887 Blackiella Aellen = Atriplex L. ■●
54888 Blackiella inflata (F. Muell.) Aellen = Atriplex lindleyi Moq. subsp. inflata (F. Muell.) Paul G. Wilson ■☆
54889 Blackstonia A. Juss. = Blakstonia Scop. ●☆
54890 Blackstonia A. Juss. = Moronobea Aubl. ●☆
54891 Blackstonia Huds. (1762);布氏龙胆属;Yellow-wort ■☆
54892 Blackstonia acuminata (Koch et Ziz) Domin;渐尖布氏龙胆■☆
54893 Blackstonia grandiflora (Viv.) Pau;大花布氏龙胆■☆
54894 Blackstonia grandiflora (Viv.) Pau var. trimestris (Murb.) Zeltner = Blackstonia grandiflora (Viv.) Pau ■☆
54895 Blackstonia imperfoliata (L. f.) Samp.;不穿叶布氏龙胆■☆
54896 Blackstonia perfoliata (L.) Huds.;穿叶布氏龙胆;Common Yellow-wort, Earthgall, Greater Centaury, Greater Churmel, More Centaury, Yellow Gentian, Yellow Sanctuary, Yellow-wort ■☆
54897 Blackstonia perfoliata (L.) Huds. subsp. grandiflora (Viv.) Maire = Blackstonia grandiflora (Viv.) Pau ■☆
54898 Blackstonia perfoliata (L.) Huds. subsp. imperfoliata (L. f.)

Franco et Rocha Afonso = Blackstonia imperfoliata (L. f.) Samp. ■☆
54899 Blackstonia perfoliata (L.) Huds. subsp. intermedia (Ten.) Zeltner;间型穿叶布氏龙胆■☆
54900 Blackstonia perfoliata (L.) Huds. subsp. serotina (Rchb.) Vollm. = Blackstonia acuminata (Koch et Ziz) Domin ■☆
54901 Blackstonia perfoliata (L.) Huds. var. latibracteata Sennen et Mauricio = Blackstonia perfoliata (L.) Huds. ■☆
54902 Blackstonia perfoliata (L.) Huds. var. longidens (H. Lindb.) Maire = Blackstonia perfoliata (L.) Huds. ■☆
54903 Blackstonia serotina (Koch) Beck = Blackstonia acuminata (Koch et Ziz) Domin ■☆
54904 Blackwellia Comm. ex Juss. = Homalium Jacq. ●
54905 Blackwellia Gaertn. = Blakwellia Gaertn. ●■
54906 Blackwellia Gaertn. = Palladia Lam. ●■
54907 Blackwellia J. F. Gmel. = Blakwellia Comm. ex Juss. ●
54908 Blackwellia J. F. Gmel. = Homalium Jacq. ●
54909 Blackwellia Sieber ex Pax et K. Hoffm. = Claoxylon A. Juss. ●
54910 Blackwellia africana Hook. f. = Homalium africanum (Hook. f.) Benth. ●☆
54911 Blackwellia ceylanica Gardner = Homalium ceylanicum (Gardner) Bedd. ●
54912 Blackwellia dentata Harv. = Homalium dentatum (Harv.) Warb. ●☆
54913 Blackwellia fagifolia Lindl. = Homalium cochinchinense (Lour.) Druce ●
54914 Blackwellia padiflora Lindl. = Homalium cochinchinense (Lour.) Druce ●
54915 Blackwellia rufescens Arn. = Homalium rufescens Benth. ●☆
54916 Blackwelliaceae Sch. Bip. = Flacourtiaceae Rich. ex DC. (保留科名)●
54917 Bladhia Thunb.(废弃属名)= Ardisia Sw.(保留属名)●■
54918 Bladhia brevicaulis (Diels) Migo = Ardisia brevicaulis Diels ●
54919 Bladhia brevicaulis Migo = Ardisia brevicaulis Diels ●
54920 Bladhia chinensis (Benth.) Nakai = Ardisia chinensis Benth. ●
54921 Bladhia chinensis (Benth.) Nakai var. minor Nakai = Ardisia chinensis Benth. ●
54922 Bladhia citrifolia (Hayata) Nakai = Ardisia brevicaulis Diels ●
54923 Bladhia citrifolia Nakai = Ardisia brevicaulis Diels ●
54924 Bladhia cornudentata Nakai = Ardisia cornudentata Mez ●
54925 Bladhia crenata (Sims) H. Hara = Ardisia crenata Sims ●
54926 Bladhia crenata (Sims) H. Hara var. tequetii H. Hara = Ardisia crenata Sims ●
54927 Bladhia crispa Thunb. = Ardisia crispa (Thunb.) A. DC. ●
54928 Bladhia crispa Thunb. var. dielsii Nakai = Ardisia crispa (Thunb.) A. DC. ●
54929 Bladhia crispa Thunb. var. taguetii Nakai = Ardisia crenata Sims ●
54930 Bladhia elegans Koidz. = Ardisia elegans Andréws ●
54931 Bladhia glabra Thunb. = Sarcandra glabra (Thunb.) Nakai ●
54932 Bladhia humila (Vahl) Sasaki = Ardisia humilis Vahl ●
54933 Bladhia japonica Thunb. = Ardisia japonica (Thunb.) Blume ●
54934 Bladhia kotoensis (Hayata) Nakai = Ardisia elliptica Thunb. ●
54935 Bladhia lentiginosa (Ker Gawl.) Nakai var. lanceolata Masam. = Ardisia crenata Sims ●
54936 Bladhia lentiginosa Nakai = Ardisia crenata Sims ●
54937 Bladhia lentiginosa Nakai var. lanceolata Masam. = Ardisia crenata Sims ●
54938 Bladhia lentiginosa Nakai var. taquetii Nakai = Ardisia crenata Sims ●
54939 Bladhia montana (Miq.) Nakai = Ardisia japonica (Thunb.) Blume ●
54940 Bladhia montana (Miq.) Nakai = Ardisia walkeri Y. P. Yang ●
54941 Bladhia morrisonensis (Hayata) Nakai = Ardisia cornudentata Mez ●
54942 Bladhia morrisonensis Nakai = Ardisia cornudentata Mez ●
54943 Bladhia oldhamii (Mez) Masam. = Ardisia virens Kurz ●
54944 Bladhia oldhamii Masam. = Ardisia virens Kurz ●
54945 Bladhia primulifolia (Gardner et Champ.) Masam. = Ardisia primulifolia Gardner et Champ. ●
54946 Bladhia primulifolia Masam. = Ardisia primulifolia Gardner et Champ. ●
54947 Bladhia pseudoquinquegona Masam. = Ardisia quinquegona Blume ●
54948 Bladhia punctata (Lindl.) Nakai = Ardisia crispa (Thunb.) A. DC. ●
54949 Bladhia punctata (Lindl.) Nakai = Ardisia lindleyana D. Dietr. ●
54950 Bladhia punctata (Lindl.) Nakai = Ardisia punctata Lindl. ●
54951 Bladhia punctata Nakai = Ardisia lindleyana D. Dietr. ●
54952 Bladhia punctata Nakai = Bladhia punctata (Lindl.) Nakai ●
54953 Bladhia quinquegona (Blume) Nakai = Ardisia quinquegona Blume ●
54954 Bladhia quinquegona Nakai = Ardisia quinquegona Blume ●
54955 Bladhia radians (Hemsl. et Mez) Masam. = Ardisia virens Kurz ●
54956 Bladhia radians Masam. = Ardisia virens Kurz ●
54957 Bladhia recemosa Nakai = Ardisia squamulosa C. Presl ●
54958 Bladhia sciophila (T. Suzuki) Nakai. = Ardisia maclurei Merr. ●
54959 Bladhia sciophila Nakai = Ardisia maclurei Merr. ●
54960 Bladhia sieboldii (Miq.) Nakai = Ardisia sieboldii Miq. ●
54961 Bladhia sieboldii Nakai = Ardisia sieboldii Miq. ●
54962 Bladhia stenosepala (Hayata) Nakai. = Ardisia cornudentata Mez ●
54963 Bladhia villosa Thunb. = Ardisia pusilla A. DC. ●
54964 Bladhia villosa Thunb. = Ardisia violacea (T. Suzuki) W. Z. Fang et K. Yao ●
54965 Bladhia villosa Thunb. var. liukiuensis? = Ardisia pusilla A. DC. ●
54966 Bladhia violacea T. Suzuki = Ardisia brevicaulis Diels var. violacea (T. Suzuki) E. Walker ●
54967 Bladhia violacea T. Suzuki = Ardisia violacea (T. Suzuki) W. Z. Fang et K. Yao ●
54968 Blaeria L. = Erica L. ●☆
54969 Blaeria L. et E. Phillips = Erica L. ●☆
54970 Blaeria affinis N. E. Br. = Erica ericoides (L.) E. G. H. Oliv. ●☆
54971 Blaeria afromontana Alm et T. C. E. Fr. = Erica filago (Alm et T. C. E. Fr.) Beentje ●☆
54972 Blaeria albida Thunb. = Erica muscosa (Aiton) E. G. H. Oliv. ●☆
54973 Blaeria articulata L. = Erica similis (N. E. Br.) E. G. H. Oliv. ●☆
54974 Blaeria barbigera (Salisb.) G. Don = Erica barbigera Salisb. ●☆
54975 Blaeria bicolor Klotzsch = Erica inaequalis (N. E. Br.) E. G. H. Oliv. ●☆
54976 Blaeria bracteata J. C. Wendl. = Erica labialis Salisb. ●☆
54977 Blaeria breviflora Engl. = Erica silvatica (Engl.) Beentje ●☆
54978 Blaeria breviflora Engl. var. ulugurensis? = Erica silvatica (Engl.) Beentje ●☆
54979 Blaeria bugonii Welw. ex Engl. = Erica silvatica (Engl.) Beentje ●☆
54980 Blaeria caduca Thunb. = Erica polifolia Salisb. ex Benth. ●☆
54981 Blaeria campanulata Benth. = Erica equisetifolia Salisb. ●☆

54982 Blaeria carnea Klotzsch = Erica uberiflora E. G. H. Oliv. ●☆
54983 Blaeria ciliaris L. f. = Erica plumosa Thunb. ●☆
54984 Blaeria ciliciiflora (Salisb.) G. Don = Erica plumosa Thunb. ●☆
54985 Blaeria coccinea Klotzsch = Erica longimontana E. G. H. Oliv. ●☆
54986 Blaeria condensata Hochst. ex A. Rich. = Erica silvatica (Engl.) Beentje ●☆
54987 Blaeria depressa Licht. ex Roem. et Schult. = Erica glabella Thunb. ●☆
54988 Blaeria dumosa J. C. Wendl. = Erica equisetifolia Salisb. ●☆
54989 Blaeria dumosa J. C. Wendl. var. breviflora N. E. Br. = Erica equisetifolia Salisb. ●☆
54990 Blaeria elgonensis Alm et T. C. E. Fr. = Erica filago (Alm et T. C. E. Fr.) Beentje ●☆
54991 Blaeria equisetifolia (Salisb.) G. Don = Erica equisetifolia Salisb. ●☆
54992 Blaeria eriantha Willd. ex Steud. = Erica similis (N. E. Br.) E. G. H. Oliv. ●☆
54993 Blaeria ericoides L. = Erica ericoides (L.) E. G. H. Oliv. ●☆
54994 Blaeria fasciculata (Thunb.) Willd. = Erica glabella Thunb. ●☆
54995 Blaeria fastigiata Benth. = Erica longimontana E. G. H. Oliv. ●☆
54996 Blaeria filago Alm et T. C. E. Fr. = Erica filago (Alm et T. C. E. Fr.) Beentje ●☆
54997 Blaeria filago Alm et T. C. E. Fr. subsp. saxicola (Alm et T. C. E. Fr.) Hedberg = Erica filago (Alm et T. C. E. Fr.) Beentje ●☆
54998 Blaeria filago Alm et T. C. E. Fr. var. afromontana (Alm et T. C. E. Fr.) Alm et T. C. E. Fr. = Erica filago (Alm et T. C. E. Fr.) Beentje ●☆
54999 Blaeria filago Alm et T. C. E. Fr. var. elgonensis (Alm et T. C. E. Fr.) Alm et T. C. E. Fr. = Erica filago (Alm et T. C. E. Fr.) Beentje ●☆
55000 Blaeria flava Bolus = Erica equisetifolia Salisb. ●☆
55001 Blaeria flexuosa Benth. = Erica multiflexuosa E. G. H. Oliv. ●☆
55002 Blaeria friesii Weim. = Erica silvatica (Engl.) Beentje ●☆
55003 Blaeria fuscescens Klotzsch = Erica fuscescens (Klotzsch) E. G. H. Oliv. ●☆
55004 Blaeria glabella (Thunb.) Willd. = Erica glabella Thunb. ●☆
55005 Blaeria glabra (Thunb.) Thunb. = Erica inaequalis (N. E. Br.) E. G. H. Oliv. ●☆
55006 Blaeria glanduligera Engl. = Erica silvatica (Engl.) Beentje ●☆
55007 Blaeria glutinosa K. Schum. et Engl. = Erica silvatica (Engl.) Beentje ●☆
55008 Blaeria gracilis Bartl. = Erica benthamiana E. G. H. Oliv. ●☆
55009 Blaeria grandis N. E. Br. = Erica sagittata Klotzsch ex Benth. ●☆
55010 Blaeria granvikii Alm et T. C. E. Fr. = Erica silvatica (Engl.) Beentje ●☆
55011 Blaeria guguensis Pic. Serm. et Heiniger = Erica silvatica (Engl.) Beentje ●☆
55012 Blaeria hirsuta (Thunb.) Thunb. = Erica eriocephala Lam. ●☆
55013 Blaeria incana Bartl. = Erica plumosa Thunb. ●☆
55014 Blaeria johnstonii Engl. = Erica silvatica (Engl.) Beentje ●☆
55015 Blaeria johnstonii Engl. subsp. keniensis (Alm et T. C. E. Fr.) Hedberg = Erica silvatica (Engl.) Beentje ●☆
55016 Blaeria keilii Engl. = Erica silvatica (Engl.) Beentje ●☆
55017 Blaeria keniensis Alm et T. C. E. Fr. = Erica silvatica (Engl.) Beentje ●☆
55018 Blaeria kilimandjarica Alm et T. C. E. Fr. = Erica silvatica (Engl.) Beentje ●☆
55019 Blaeria kingaensis Engl. = Erica silvatica (Engl.) Beentje ●☆
55020 Blaeria klotzschii Alm et T. C. E. Fr. = Erica klotzschii (Alm et T. C. E. Fr.) E. G. H. Oliv. ●☆
55021 Blaeria kraussiana Klotzsch ex Walp. = Erica russakiana E. G. H. Oliv. ●☆
55022 Blaeria mannii (Engl.) Engl. = Erica silvatica (Engl.) Beentje ●☆
55023 Blaeria meyeri-johannis K. Schum. = Erica silvatica (Engl.) Beentje ●☆
55024 Blaeria microdonta C. H. Wright = Ericinella microdonta (C. H. Wright) Alm et T. C. E. Fr. ●☆
55025 Blaeria muirii L. Guthrie = Erica rosacea (L. Guthrie) E. G. H. Oliv. ●☆
55026 Blaeria multiflora Klotzsch = Erica uberiflora E. G. H. Oliv. ●☆
55027 Blaeria muscosa Aiton = Erica muscosa (Aiton) E. G. H. Oliv. ●☆
55028 Blaeria nudiflora (L.) Thunb. = Erica nudiflora L. ●☆
55029 Blaeria oppositifolia L. Guthrie = Erica equisetifolia Salisb. ●☆
55030 Blaeria paniculata (Thunb.) Thunb. = Erica quadrifida (Benth.) E. G. H. Oliv. ●☆
55031 Blaeria parviflora Klotzsch = Erica anguliger (N. E. Br.) E. G. H. Oliv. ●☆
55032 Blaeria patula (Engl.) Engl. = Erica silvatica (Engl.) Beentje ●☆
55033 Blaeria patula (Engl.) Engl. var. aberdarica Alm et T. C. E. Fr. = Erica silvatica (Engl.) Beentje ●☆
55034 Blaeria patula (Engl.) Engl. var. minima Brenan = Erica silvatica (Engl.) Beentje ●☆
55035 Blaeria patula Engl. var. tenuis Alm et T. C. E. Fr. = Erica silvatica (Engl.) Beentje ●☆
55036 Blaeria paucifolia J. C. Wendl. = Erica paucifolia (J. C. Wendl.) E. G. H. Oliv. ●☆
55037 Blaeria plumosa Thunb. = Erica plumosa Thunb. ●☆
55038 Blaeria ptilota E. Mey. ex Benth. = Erica plumosa Thunb. ●☆
55039 Blaeria puberula Klotzsch = Erica anguliger (N. E. Br.) E. G. H. Oliv. ●☆
55040 Blaeria purpurea L. f. = Erica equisetifolia Salisb. ●☆
55041 Blaeria purpurea P. J. Bergius = Erica glabella Thunb. ●☆
55042 Blaeria pusilla J. C. Wendl. = Erica muscosa (Aiton) E. G. H. Oliv. ●☆
55043 Blaeria pusilla Klotzsch = Erica klotzschii (Alm et T. C. E. Fr.) E. G. H. Oliv. ●☆
55044 Blaeria pusilla L. = Erica glabella Thunb. ●☆
55045 Blaeria revoluta Bartl. = Erica barbigeroides E. G. H. Oliv. ●☆
55046 Blaeria sagittata (Klotzsch ex Benth.) Alm et T. C. E. Fr. = Erica sagittata Klotzsch ex Benth. ●☆
55047 Blaeria saxicola Alm et T. C. E. Fr. = Erica filago (Alm et T. C. E. Fr.) Beentje ●☆
55048 Blaeria scabra (Thunb.) Willd. = Erica glabella Thunb. ●☆
55049 Blaeria serrata (Thunb.) Thunb. = Erica serrata Thunb. ●☆
55050 Blaeria silvatica Engl. = Erica silvatica (Engl.) Beentje ●☆
55051 Blaeria sphagnicola Sleumer = Erica silvatica (Engl.) Beentje ●☆
55052 Blaeria sphagnicola Sleumer. f. pseudobreviflora? = Erica silvatica (Engl.) Beentje ●☆
55053 Blaeria sphagnicola Sleumer. f. pubescens? = Erica silvatica (Engl.) Beentje ●☆
55054 Blaeria spicata Hochst. ex A. Rich. = Erica silvatica (Engl.) Beentje ●☆

55055 Blaeria spicata Hochst. ex A. Rich. subsp. mannii (Engl.) Wickens = Erica silvatica (Engl.) Beentje ●☆
55056 Blaeria spicata Hochst. ex A. Rich. var. mannii Engl. = Erica silvatica (Engl.) Beentje ●☆
55057 Blaeria spicata Hochst. ex A. Rich. var. patula Engl. = Erica silvatica (Engl.) Beentje ●☆
55058 Blaeria stolzii Alm et T. C. E. Fr. = Erica silvatica (Engl.) Beentje ●☆
55059 Blaeria subverticillata Engl. = Erica silvatica (Engl.) Beentje ●☆
55060 Blaeria tenuifolia Engl. = Erica silvatica (Engl.) Beentje ●☆
55061 Blaeria tenuipilosa Engl. ex Alm et T. C. E. Fr. = Erica silvatica (Engl.) Beentje ●☆
55062 Blaeria thunbergii G. Don = Erica eriocephala Lam. ●☆
55063 Blaeria viscosa Alm et T. C. E. Fr. = Erica filago (Alm et T. C. E. Fr.) Beentje ●☆
55064 Blaeria whyteana Engl. = Erica silvatica (Engl.) Beentje ●☆
55065 Blaeria xeranthemifolia (Salisb.) G. Don = Erica xeranthemifolia Salisb. ●☆
55066 Blainvillea Cass. (1823);异芒菊属(百能葳属);Blainvillea ■●
55067 Blainvillea acmella (L.) Philipson;异芒菊(百能葳,假麦菜草,鱼鳞菜);Common Blainvillea ■
55068 Blainvillea dalla-vedovae A. Terracc. = Blainvillea rhomboidea Cass. ■
55069 Blainvillea gayana Cass.;盖伊异芒菊■☆
55070 Blainvillea latifolia (L. f.) DC. = Blainvillea acmella (L.) Phillipson ■
55071 Blainvillea prieuriana DC. = Aspilia helianthoides (Schumach. et Thonn.) Oliv. et Hiern subsp. prieuriana (DC.) C. D. Adams ■☆
55072 Blainvillea rhomboidea Cass. = Blainvillea acmella (L.) Philipson ■
55073 Blainvillea rhomboidea sensu Dunn = Blainvillea acmella (L.) Phillipson ■
55074 Blairia Adans. = Priva Adans. ■☆
55075 Blairia Gled. = Blaeria L. ●☆
55076 Blairia Spreng. = Blaeria L. ●☆
55077 Blakburnia J. F. Gmel. = Blackburnia J. R. Forst. et G. Forst. ●
55078 Blakburnia J. F. Gmel. = Zanthoxylum L. ●
55079 Blakea P. Browne(1756);布氏野牡丹属■☆
55080 Blakea trinervia L.;三脉布氏野牡丹;Jamaican Rose ■☆
55081 Blakeaceae Rchb. ex Barnhart = Melastomataceae Juss.(保留科名)●■
55082 Blakeaceae Rchb. ex Barnhart;布氏野牡丹科■
55083 Blakeanthus R. M. King et H. Rob. (1972);杂腺菊属●☆
55084 Blakeanthus R. M. King et H. Rob. = Ageratum L. ■●
55085 Blakeanthus cordatus (S. F. Blake) R. M. King et H. Rob.;杂腺菊■☆
55086 Blakeochloa Veldkamp = Plinthanthesis Steud. ■☆
55087 Blakiella Cuatrec. (1968);卷边菀属●☆
55088 Blakiella bartsiifolia (S. F. Blake) Cuatrec.;卷边菀■☆
55089 Blakstonia Scop. = Moronobea Aubl. ●☆
55090 Blakwellia Comm. ex Juss. = Homalium Jacq. ●
55091 Blakwellia Gaertn. = Palladia Lam. ●■
55092 Blakwellia Scop. = Leea D. Royen ex L.(保留属名)●■
55093 Blakwellia Scop. = Nalagu Adans.(废弃属名)●■
55094 Blakwelliaceae T. Lestib. = Flacourtiaceae Rich. ex DC.(保留科名)●
55095 Blanca Hutch. = Blancoa Lindl. ■☆
55096 Blanchea Boiss. = Iphiona Cass.(保留属名)●■☆
55097 Blanchetia DC. (1836);黑毛落苞菊属■☆
55098 Blanchetia heterotricha DC.;黑毛落苞菊■☆
55099 Blanchetiastrum Hassl. (1910);小黑毛落苞菊属■☆
55100 Blanchetiastrum goetheoides Hassl.;小黑毛落苞菊■☆
55101 Blanchetiodendron Barneby et J. W. Grimes = Enterolobium Mart. ●
55102 Blanchetiodendron Barneby et J. W. Grimes(1996);巴西象耳豆属●
55103 Blanckia Neck. = Conobea Aubl. ■☆
55104 Blancoa Blume = Arenga Labill.(保留属名)●
55105 Blancoa Blume(1836) = Didymosperma H. Wendl. et Drude ex Benth. et Hook. f. ●
55106 Blancoa Blume(1847) = Harpullia Roxb. ●
55107 Blancoa Lindl. (1839) = Conostylis R. Br. ■☆
55108 Blancoa Lindl. (1840);布氏血草属■☆
55109 Blancoa canescens Lindl.;布氏血草;Red Bugle,Winter Bell ■☆
55110 Blandfordia Andréws(废弃属名) = Blandfordia Sm.(保留属名)■☆
55111 Blandfordia Andréws(废弃属名) = Galax Sims(保留属名)■☆
55112 Blandfordia Sm. (1804)(保留属名);香水花属(疣毛子属)■☆
55113 Blandfordia cordata Andréws;心形香水花(疣毛子)■☆
55114 Blandfordia grandiflora R. Br.;大花香水花(红钟百合,圣诞钟);Christmas Bell,Christmas Bells ■☆
55115 Blandfordia nobilis;名贵香水花;Christmas Bell ■☆
55116 Blandfordia punicea (Labill.) Sweet;深红香水花;Christmas Bell ■☆
55117 Blandfordiaceae R. Dahlgren et Clifford(1985);香水花科(疣毛子科)■☆
55118 Blandfortia Poir. = Galax Sims(保留属名)■☆
55119 Blandibractea Wernham(1917);光苞茜属☆
55120 Blandibractea brasiliensis Wernham;光苞茜☆
55121 Blandina Raf. = Leucas Burm. ex R. Br. ●■
55122 Blandowia Willd. = Apinagia Tul. emend. P. Royen ■☆
55123 Blanisia Pritz. = Cleome L. ●■
55124 Blanisia Pritz. = Polanisia Raf. ■
55125 Blastania Kotschy et Peyr. = Ctenolepis Hook. f. ■☆
55126 Blastania cerasiformis (Stocks) A. Meeuse = Ctenolepis cerasiformis (Stocks) Hook. f. ■☆
55127 Blastania fimbristipula Kotschy et Peyr. = Ctenolepis cerasiformis (Stocks) Hook. f. ■☆
55128 Blastania luederitziana Cogn. = Dactyliandra welwitschii Hook. f. ■☆
55129 Blastemanthus Planch. (1846);毛花金莲木属●☆
55130 Blastemanthus albidum Ruhland;白毛花金莲木●☆
55131 Blastemanthus densiflorus Hallier f.;密花毛花金莲木●☆
55132 Blastemanthus gemmiflorus Planch.;毛花金莲木●☆
55133 Blastocaulon Ruhland(1903);芽茎谷精草属■☆
55134 Blastotrophe Didr. = Alafia Thouars ●☆
55135 Blastus Lour. (1790);柏拉木属(伯拉木属);Blastus ●
55136 Blastus apricus (Hand. -Mazz.) H. L. Li;黄金梢(细黄金梢,线萼金花树,叶下红);Heliophilous Blastus ●
55137 Blastus apricus (Hand. -Mazz.) H. L. Li = Blastus pauciflorus (Benth.) Guillaumin ●
55138 Blastus apricus (Hand. -Mazz.) H. L. Li var. longiflorus (Hand. -Mazz.) C. Chen;长瓣黄金梢(长瓣金花树);Long Flower Heliophilous Blastus ●

55139 Blastus apricus (Hand. -Mazz.) H. L. Li var. longiflorus (Hand. -Mazz.) C. Chen = Blastus pauciflorus (Benth.) Guillaumin ●

55140 Blastus auriculatus Y. C. Huang; 耳基柏拉木; Auriculate Blastus, Earbase Blastus ●

55141 Blastus borneensis Cogn. ex Boerl.; 南亚柏拉木; Cogniaux Blastus, S. Asia Blastus ●

55142 Blastus brevissimus H. Lév.; 短柄柏拉木; Shortstalk Blastus, Short-stalked Blastus ●

55143 Blastus cavaleriei H. Lév. et Vaniot; 匙萼柏拉木(黔贵柏拉木, 黔贵野锦香); Cavalerie Blastus ●

55144 Blastus cavaleriei H. Lév. et Vaniot = Blastus pauciflorus (Benth.) Guillaumin ●

55145 Blastus cavaleriei H. Lév. et Vaniot var. tomentosus (H. L. Li) C. Chen; 腺毛柏拉木; Tomentose Cavalerie Blastus ●

55146 Blastus cavaleriei H. Lév. et Vaniot var. tomentosus (H. L. Li) C. Chen = Blastus pauciflorus (Benth.) Guillaumin ●

55147 Blastus cochinchinensis Lour.; 柏拉木(崩疮药, 伯拉木, 黄金梢, 山崩沙, 山甜娘, 野棉香); Cochinchina Blastus, Cochin-China Blastus, Greyblue Anplectrum ●

55148 Blastus cogniauxii Stapf = Blastus borneensis Cogn. ex Boerl. ●

55149 Blastus dunnianus H. Lév.; 金花树(谷皱草, 巨萼柏拉木, 六便狼, 木暗栅); Dunn Blastus ●

55150 Blastus dunnianus H. Lév. = Blastus pauciflorus (Benth.) Guillaumin ●

55151 Blastus dunnianus H. Lév. var. glandulosetosus C. Chen; 腺毛金花树; Glandular Setose Dunn Blastus ●

55152 Blastus dunnianus H. Lév. var. glandulosetosus C. Chen = Blastus pauciflorus (Benth.) Guillaumin ●

55153 Blastus ernae Hand. -Mazz.; 留行草(大莎药); Erna Blastus ●

55154 Blastus ernae Hand. -Mazz. = Blastus pauciflorus (Benth.) Guillaumin ●

55155 Blastus fengii S. Y. Hu = Sporoxeia sciadophila W. W. Sm. ●

55156 Blastus hindsii Hance = Blastus pauciflorus (Benth.) Guillaumin ●

55157 Blastus hirsutus H. L. Li = Sporoxeia hirsuta (H. L. Li) C. Y. Wu ●

55158 Blastus hirsutus H. L. Li = Sporoxeia sciadophila W. W. Sm. ●

55159 Blastus latifolius H. L. Li = Sporoxeia latifolia (H. L. Li) C. Y. Wu et Y. C. Huang ex C. Chen ●

55160 Blastus latifolius H. L. Li = Sporoxeia sciadophila W. W. Sm. ●

55161 Blastus lii M. P. Nayar = Blastus pauciflorus (Benth.) Guillaumin ●

55162 Blastus longiflorus Hand. -Mazz. = Blastus apricus (Hand. -Mazz.) H. L. Li var. longiflorus (Hand. -Mazz.) C. Chen ●

55163 Blastus longiflorus Hand. -Mazz. = Blastus pauciflorus (Benth.) Guillaumin ●

55164 Blastus longiflorus Hand. -Mazz. var. apricus (Hand. -Mazz.) Y. L. Zheng et N. H. Xia = Blastus pauciflorus (Benth.) Guillaumin ●

55165 Blastus lyi H. Lév. = Fordiophyton faberi Stapf ●■

55166 Blastus macrandii H. Lév. = Blastus cochinchinensis Lour. ●

55167 Blastus mairei H. Lév. = Bredia yunnanensis (H. Lév.) Diels ●■

55168 Blastus marchandii H. Lév. = Blastus cochinchinensis Lour. ●

55169 Blastus membranifolius H. L. Li = Neodriessenia membranifolia (H. L. Li) C. Hansen ●

55170 Blastus membranifolius H. L. Li = Stussenia membranifolia (H. L. Li) C. Hansen ●■

55171 Blastus mollissimus H. L. Li; 密毛柏拉木; Densehair Blastus, Dense-haired Blastus ●

55172 Blastus parviflorus (Benth.) Triana = Blastus cochinchinensis Lour. ●

55173 Blastus parviflorus Triana = Blastus cochinchinensis Lour. ●

55174 Blastus pauciflorus (Benth.) Guillaumin; 少花柏拉木; Few Flower Blastus, Fewflower Blastus, Pauciflorous Blastus ●

55175 Blastus setulosus Diels; 刺毛柏拉木; Finebristle Blastus, Setulose Blastus, Spinyhair Blastus ●

55176 Blastus spathulicalyx Hand. -Mazz. = Blastus cavaleriei H. Lév. et Vaniot ●

55177 Blastus spathulicalyx Hand. -Mazz. = Blastus pauciflorus (Benth.) Guillaumin ●

55178 Blastus spathulicalyx Hand. -Mazz. var. apricus Hand. -Mazz. = Blastus apricus (Hand. -Mazz.) H. L. Li ●

55179 Blastus spathulicalyx Hand. -Mazz. var. apricus Hand. -Mazz. = Blastus pauciflorus (Benth.) Guillaumin ●

55180 Blastus squamosus C. Y. Wu et Y. C. Huang = Blastus pauciflorus (Benth.) Guillaumin ●

55181 Blastus squamosus C. Y. Wu et Y. C. Huang ex C. Chen; 鳞毛柏拉木; Scalehair Blastus, Scaly Blastus ●

55182 Blastus tenuifolius Diels; 薄叶柏拉木; Thin Leaf Blastus, Thinleaf Blastus, Thin-leaved Blastus ●

55183 Blastus thaiyongii C. Hansen = Blastus pauciflorus (Benth.) Guillaumin ●

55184 Blastus tomentosus H. L. Li = Blastus cavaleriei H. Lév. et Vaniot var. tomentosus (H. L. Li) C. Chen ●

55185 Blastus tomentosus H. L. Li = Blastus pauciflorus (Benth.) Guillaumin ●

55186 Blastus tsaii H. L. Li; 云南柏拉木; H. T. Tsai Blastus, Tsai Blastus, Yunnan Blastus ●

55187 Blastus yunnanensis H. L. Li = Blastus tsaii H. L. Li ●

55188 Blastus yunnanensis H. Lév. = Bredia yunnanensis (H. Lév.) Diels ●■

55189 Blattaria Kuntze = Pentapetes L. ■●

55190 Blattaria Mill. = Verbascum L. ■●

55191 Blatti Adans. (废弃属名) = Sonneratia L. f. (保留属名) ●

55192 Blatti Rheede ex Adans. = Sonneratia L. f. (保留属名) ●

55193 Blatti acide Lam. = Sonneratia caseolaris (L.) Engl. ●

55194 Blattiaceae Engl. = Sonneratiaceae Engl. (保留科名) ●

55195 Blattiaceae Nied. = Lythraceae J. St. -Hil. (保留科名) ■●

55196 Blattiaceae Nied. = Sonneratiaceae Engl. (保留科名) ●

55197 Blaxium Cass. = Dimorphotheca Vaill. (保留属名) ■●☆

55198 Blaxium decumbens Cass. = Dimorphotheca fruticosa (L.) Less. ■☆

55199 Bleasdalea F. Muell. = Bleasdalea F. Muell. ex Domin ●

55200 Bleasdalea F. Muell. = Grevillea R. Br. ex Knight (保留属名) ●

55201 Bleasdalea F. Muell. ex Domin = Grevillea R. Br. ex Knight (保留属名) ●

55202 Blechum P. Browne (1756); 赛山蓝属(美爵床属); Blechum ■

55203 Blechum blechum (L.) Millsp. = Blechum pyramidatum (Lam.) Urb. ■

55204 Blechum blechum Millsp. = Blechum pyramidatum (Lam.) Urb. ■

55205 Blechum brownei Juss. = Blechum pyramidatum (Lam.) Urb. ■

55206 Blechum hamatum Klotzsch = Megalochlamys hamata (Klotzsch) Vollesen ●☆

55207 Blechum pyramidatum (Lam.) Urb.; 赛山蓝; Browne's

Blechum, Green Shrimp Plant, Pyramid Blechum ■
55208 Bleekeria Hassk. (1855);布拉克玫瑰树属(布利木属)●☆
55209 Bleekeria Hassk. = Ochrosia Juss. ●
55210 Bleekeria Miq. = Alchornea Sw. ●
55211 Bleekeria coccinea (Teijsm. et Binn.) Koidz. = Ochrosia coccinea (Teijsm. et Binn.) Miq. ●
55212 Bleekeria hexandra (Koidz.) Koidz.;六蕊布拉克玫瑰树●☆
55213 Bleekeria hexandra (Koidz.) Koidz. = Excavatia hexandra (Koidz.) Hatus. ●☆
55214 Bleekeria vitiensis (Markgr.) A. C. Sm.;布拉克玫瑰树(维提布利木)●☆
55215 Bleekrodea Blume = Streblus Lour. ●
55216 Bleekrodea Blume(1856);南鹊肾树属●☆
55217 Bleekrodea insignis Blume;南鹊肾树●☆
55218 Bleekrodea madagascariensis Blume;马岛南鹊肾树●☆
55219 Bleekrodea tonkinensis Eberh. et Dubard = Streblus tonkinensis (Eberh. et Dubard) Corner ●
55220 Blencocoes B. D. Jacks. = Blenocoes Raf. ●■
55221 Blencocoes B. D. Jacks. = Nicotiana L. ●■
55222 Blencocoes Raf. = Nierembergia Ruiz et Pav. ■☆
55223 Blennoderma Spach = Oenothera L. ●■
55224 Blennodia R. Br. (1849);黏液芥属■☆
55225 Blennodia canescens R. Br.;黏液芥■☆
55226 Blennosperma Less. (1832);黏子菊属;Stickyseed ■☆
55227 Blennosperma nanum (Hook.) S. F. Blake;黏子菊■☆
55228 Blennospora A. Gray = Calocephalus R. Br. ■●☆
55229 Blennospora A. Gray(1851);丝叶鼠麹草属■☆
55230 Blennospora drummondii A. Gray;丝叶鼠麹草■☆
55231 Blenocoes Raf. = Nicotiana L. ●■
55232 Blenocoes Raf. = Tabacus Moench ●■
55233 Blepetalon Raf. = Scutia (Comm. ex DC.) Brongn. (保留属名)●
55234 Blepetalon aculeatum Raf. = Scutia myrtina (Burm. f.) Kurz ●
55235 Blephanthera Raf. = Bulbine Wolf(保留属名)■☆
55236 Blepharacanthus Nees = Blepharis Juss. ●■
55237 Blepharacanthus Nees ex Lindl. = Blepharis Juss. ●■
55238 Blepharaden Dulac = Swertia L. ■
55239 Blepharandra Griseb. (1849);圭亚那金虎尾属●☆
55240 Blepharanthemum Klotzsch = Plagianthus J. R. Forst. et G. Forst. ●☆
55241 Blepharanthera Schltr. = Brachystelma R. Br. (保留属名)■
55242 Blepharanthera dinteri Schltr. = Brachystelma blepharanthera H. Huber ■☆
55243 Blepharanthera edulis Schltr. = Brachystelma blepharanthera H. Huber ■☆
55244 Blepharanthes Sm. = Adenia Forssk. ●
55245 Blepharanthes Sm. = Modecca Lam. ●
55246 Blepharidachne Hack. (1888);荒漠草属■☆
55247 Blepharidachne bigelovii Hack.;荒漠草■☆
55248 Blepharidium Standl. (1918);小毛茜属■☆
55249 Blepharidium guatemalense Standl.;小毛茜■☆
55250 Blepharidium mexicanum Standl.;墨西哥小毛茜■☆
55251 Blephariglottis Raf. = Platanthera Rich. (保留属名)■
55252 Blephariglottis lacera (Michx.) Farw. = Platanthera lacera (Michx.) G. Don ■☆
55253 Blephariglottis leucophaea (Nutt.) Farw. = Platanthera leucophaea (Nutt.) Lindl. ■☆
55254 Blephariglottis psycodes (L.) Rydb. = Platanthera psycodes (L.) Lindl. ■☆
55255 Blepharipappus Hook. (废弃属名) = Lebetanthus Endl. (保留属名)●☆
55256 Blepharipappus glandulosa Hook. = Layia glandulosa (Hook.) Hook. et Arn. ■☆
55257 Blepharis Juss. (1789);百簕花属●■
55258 Blepharis acanthodioides Klotzsch;普通百簕花■☆
55259 Blepharis acaulis Lindau = Acanthopsis disperma Nees ■☆
55260 Blepharis acuminata Oberm.;渐尖百簕花☆
55261 Blepharis aequisepala Vollesen;等萼百簕花☆
55262 Blepharis affinis Lindau;近缘百簕花■☆
55263 Blepharis angusta (Nees) T. Anderson;窄叶百簕花■☆
55264 Blepharis aspera Oberm.;粗糙百簕花■☆
55265 Blepharis asteracantha C. B. Clarke;星刺百簕花■☆
55266 Blepharis attenuata Napper;渐狭百簕花■☆
55267 Blepharis baguirmiensis A. Chev. ex Lindau = Blepharis involucrata Solms ■☆
55268 Blepharis bainesii S. Moore ex C. B. Clarke;贝恩斯百簕花■☆
55269 Blepharis bequaertii De Wild. = Blepharis asteracantha C. B. Clarke ■☆
55270 Blepharis boerhaviifolia Pers. = Blepharis maderaspatensis (L.) B. Heyne ex Roth ■☆
55271 Blepharis boerhaviifolia Pers. var. nigrovenulosa De Wild. et T. Durand = Blepharis maderaspatensis (L.) B. Heyne ex Roth ■☆
55272 Blepharis boranensis Vollesen;博兰百簕花■☆
55273 Blepharis bossii Oberm. = Blepharis obmitrata C. B. Clarke ■☆
55274 Blepharis breviciliata Fiori = Blepharis maderaspatensis (L.) B. Heyne ex Roth ■☆
55275 Blepharis breyeri Oberm.;布鲁尔百簕花■☆
55276 Blepharis buchneri Lindau;布赫纳百簕花■☆
55277 Blepharis burundiensis Vollesen;布隆迪百簕花■☆
55278 Blepharis calcitrapa Benoist;矢车菊百簕花■☆
55279 Blepharis calcitrapa Benoist var. decaryi Benoist = Blepharis calcitrapa Benoist ■☆
55280 Blepharis calcitrapa Benoist var. ilicifolia Benoist = Blepharis calcitrapa Benoist ■☆
55281 Blepharis calcitrapa Benoist var. sinuosa Benoist = Blepharis calcitrapa Benoist ■☆
55282 Blepharis calcitrapa Benoist var. velutina Benoist = Blepharis calcitrapa Benoist ■☆
55283 Blepharis caloneura S. Moore = Blepharis tenuiramea S. Moore ■☆
55284 Blepharis caloneura S. Moore var. angustifolia Oberm. = Blepharis tenuiramea S. Moore ■☆
55285 Blepharis capensis (L. f.) Pers.;好望角百簕花■☆
55286 Blepharis capensis (L. f.) Pers. var. latibracteata Oberm. = Blepharis capensis (L. f.) Pers. ■☆
55287 Blepharis capensis (L. f.) Pers. var. prostrata Oberm. = Blepharis capensis (L. f.) Pers. ■☆
55288 Blepharis carduacea Lindau = Blepharis grandis C. B. Clarke ■☆
55289 Blepharis carduifolia (L. f.) Nees var. glabra (Nees) T. Anderson = Acanthopsis carduifolia (L. f.) Schinz ■☆
55290 Blepharis carduifolia (L. f.) T. Anderson = Acanthopsis carduifolia (L. f.) Schinz ■☆
55291 Blepharis cataractae S. Moore = Blepharis bainesii S. Moore ex C. B. Clarke ■☆
55292 Blepharis chrysotricha Lindau;金毛百簕花■☆
55293 Blepharis ciliaris (L.) B. L. Burtt;缘毛百簕花■☆

55294 Blepharis clarkei Schinz = Blepharis integrifolia (L. f.) E. Mey. ex Schinz var. clarkei (Schinz) Oberm. ■☆
55295 Blepharis crinita Benoist;长软毛百簕花■☆
55296 Blepharis cristata S. Moore = Blepharis stuhlmanni Lindau ■☆
55297 Blepharis cuanzensis Welw. ex S. Moore;宽扎百簕花■☆
55298 Blepharis cuanzensis Welw. ex S. Moore subsp. tanganyikensis Napper = Blepharis tanganyikensis (Napper) Vollesen ■☆
55299 Blepharis cuanzensis Welw. ex S. Moore var. leptophylla S. Moore = Blepharis cuanzensis Welw. ex S. Moore ■☆
55300 Blepharis cuanzensis Welw. ex S. Moore var. parvispina Vollesen;小刺百簕花■☆
55301 Blepharis cuspidata Lindau;骤尖百簕花■☆
55302 Blepharis decussata S. Moore;交互对生百簕花■☆
55303 Blepharis dichotoma Engl. = Blepharis grossa (Nees) T. Anderson ■☆
55304 Blepharis dilatata C. B. Clarke;膨大百簕花■☆
55305 Blepharis diplodonta Vollesen;双齿百簕花■☆
55306 Blepharis diversispina (Nees) C. B. Clarke;异刺百簕花■☆
55307 Blepharis diversispina Eyles = Blepharis aspera Oberm. ■☆
55308 Blepharis drummondii Vollesen;德拉蒙德百簕花■☆
55309 Blepharis dunensis Vollesen;砂丘百簕花■☆
55310 Blepharis duvigneaudii Vollesen;迪维尼奥百簕花■☆
55311 Blepharis ecklonii C. B. Clarke = Blepharis hirtinervia (Nees) T. Anderson ■☆
55312 Blepharis edulis (Forssk.) Pers.;食用百簕花■☆
55313 Blepharis edulis (Forssk.) Pers. = Blepharis ciliaris (L.) B. L. Burtt ■☆
55314 Blepharis edulis Pers. = Blepharis edulis (Forssk.) Pers. ■☆
55315 Blepharis edulis Pers. f. hirta (Hochst. ex Nees) A. Terracc. = Blepharis linariifolia Pers. ■☆
55316 Blepharis edulis Pers. f. minima Chiov. = Blepharis edulis Pers. ■☆
55317 Blepharis edulis Pers. var. gracilis Maire = Blepharis ciliaris (L.) B. L. Burtt ■☆
55318 Blepharis edulis Pers. var. oblongata A. Terracc. = Blepharis edulis Pers. ■☆
55319 Blepharis espinosa E. Phillips;无刺百簕花■☆
55320 Blepharis evansii Turrill = Blepharis stuhlmanni Lindau ■☆
55321 Blepharis fenestralis Vollesen;窗孔百簕花■☆
55322 Blepharis ferox P. G. Mey.;多刺百簕花■☆
55323 Blepharis flava Vollesen;黄百簕花■☆
55324 Blepharis fleckii P. G. Mey.;弗莱克百簕花■☆
55325 Blepharis forgiarinii J.-P. Lebrun et Stork;福尔贾里尼百簕花■☆
55326 Blepharis frutescens Gilli = Blepharis grandis C. B. Clarke ■☆
55327 Blepharis fruticulosa C. B. Clarke = Blepharis hildebrandtii Lindau ■☆
55328 Blepharis furcata (L. f.) Pers.;叉分百簕花■☆
55329 Blepharis gazensis Vollesen;加兹百簕花■☆
55330 Blepharis gerlindae P. G. Mey. = Blepharis obmitrata C. B. Clarke ■☆
55331 Blepharis gigantea Oberm.;巨大百簕花■☆
55332 Blepharis glauca (Nees) T. Anderson = Acanthopsis glauca (Nees) Schinz ■☆
55333 Blepharis glomerans Benoist;团集百簕花■☆
55334 Blepharis glomerans Benoist var. brevifolia Benoist = Blepharis glomerans Benoist ■☆
55335 Blepharis glomerata (Lam.) Poir. = Blepharis procumbens (L. f.) Pers. ■☆
55336 Blepharis glumacea S. Moore;壳百簕花■☆
55337 Blepharis grandis C. B. Clarke;大百簕花■☆
55338 Blepharis grisea S. Moore = Blepharis diversispina (Nees) C. B. Clarke ■☆
55339 Blepharis grossa (Nees) T. Anderson;粗百簕花■☆
55340 Blepharis gueinzii T. Anderson = Blepharis maderaspatensis (L.) B. Heyne ex Roth ■☆
55341 Blepharis hildebrandtii Lindau;希尔德百簕花■☆
55342 Blepharis hildebrandtii Lindau subsp. phillipsiae (Rendle) Vollesen;菲利百簕花■☆
55343 Blepharis hirsuta Mildbr. = Blepharis affinis Lindau ■☆
55344 Blepharis hirta (Hochst. ex Nees) Martelli = Blepharis linariifolia Pers. ■☆
55345 Blepharis hirta (Hochst. ex Nees) Martelli var. latifolia Martelli = Blepharis linariifolia Pers. ■☆
55346 Blepharis hirtella Lindau = Blepharis cuanzensis Welw. ex S. Moore ■☆
55347 Blepharis hirtinervia (Nees) T. Anderson;毛脉百簕花■☆
55348 Blepharis homblei De Wild. = Blepharis cuanzensis Welw. ex S. Moore ■☆
55349 Blepharis hornbyae Milne-Redh. = Blepharis chrysotricha Lindau ■☆
55350 Blepharis huillensis Vollesen;威拉百簕花■☆
55351 Blepharis ilicifolia Napper;冬青叶百簕花■☆
55352 Blepharis ilicina Oberm.;冬青百簕花■☆
55353 Blepharis inaequalis C. B. Clarke;不对称百簕花■☆
55354 Blepharis inermis (Nees) C. B. Clarke;澳非无刺百簕花■☆
55355 Blepharis inflata Vollesen;膨胀百簕花■☆
55356 Blepharis inopinata Vollesen;意外百簕花■☆
55357 Blepharis integrifolia (L. f.) E. Mey. ex Schinz;全叶百簕花■☆
55358 Blepharis integrifolia (L. f.) E. Mey. ex Schinz var. clarkei (Schinz) Oberm.;克拉全叶百簕花■☆
55359 Blepharis integrifolia (L. f.) E. Mey. ex Schinz var. setosa (Nees) Oberm. = Blepharis integrifolia (L. f.) E. Mey. ex Schinz ■☆
55360 Blepharis involucrata Solms;总苞百簕花■☆
55361 Blepharis kassneri S. Moore = Blepharis glumacea S. Moore ■☆
55362 Blepharis katangensis De Wild.;加丹加百簕花■☆
55363 Blepharis kenyensis Vollesen;肯尼亚百簕花■☆
55364 Blepharis laevifolia Vollesen;光叶百簕花■☆
55365 Blepharis leendertziae Oberm.;伦德茨百簕花■☆
55366 Blepharis leptophylla (S. Moore) Hiern = Blepharis cuanzensis Welw. ex S. Moore ■☆
55367 Blepharis linariifolia Pers.;柳穿鱼叶百簕花■☆
55368 Blepharis longifolia Lindau;长叶百簕花■☆
55369 Blepharis longispica C. B. Clarke;长穗百簕花■☆
55370 Blepharis maculata Benoist;斑点百簕花■☆
55371 Blepharis madagascariensis Benoist = Blepharis glomerans Benoist ■☆
55372 Blepharis madagascariensis Benoist var. salinarum Benoist = Blepharis glomerans Benoist ■☆
55373 Blepharis madandensis S. Moore = Blepharis pungens Klotzsch ■☆
55374 Blepharis maderaspatensis (L.) B. Heyne ex Roth;马德拉斯百簕花■☆
55375 Blepharis maderaspatensis (L.) B. Heyne ex Roth subsp. rubiifolia (Schumach.) Napper;毛叶马德拉斯百簕花■☆
55376 Blepharis maderaspatensis (L.) B. Heyne ex Roth subsp. rubiifolia (Schumach.) Napper = Blepharis maderaspatensis (L.)

B. Heyne ex Roth ■☆
55377 Blepharis maderaspatensis (L.) B. Heyne ex Roth var. abyssinica Fiori = Blepharis maderaspatensis (L.) B. Heyne ex Roth ■☆
55378 Blepharis maderaspatensis (L.) Roth = Blepharis maderaspatensis (L.) B. Heyne ex Roth ■☆
55379 Blepharis malangensis S. Moore = Blepharis buchneri Lindau ■☆
55380 Blepharis marginata (Nees) C. B. Clarke;具边百簕花■☆
55381 Blepharis meyeri Vollesen;迈尔百簕花■☆
55382 Blepharis mitrata C. B. Clarke;僧帽百簕花■☆
55383 Blepharis molluginifolia Pers. = Blepharis integrifolia (L. f.) E. Mey. ex Schinz ■☆
55384 Blepharis montana Vollesen;山生百簕花■☆
55385 Blepharis naegelsbachii Oberm. = Blepharis obmitrata C. B. Clarke ■☆
55386 Blepharis natalensis Oberm.;纳塔尔百簕花■☆
55387 Blepharis obermeyerae Vollesen;奥伯迈尔百簕花■☆
55388 Blepharis obmitrata C. B. Clarke;倒僧帽百簕花■☆
55389 Blepharis obovata Chiov. = Blepharis edulis Pers. ■☆
55390 Blepharis obtusisepala Oberm.;钝萼百簕花■☆
55391 Blepharis ogadenensis Vollesen;欧加登百簕花■☆
55392 Blepharis panduriformis Lindau;琴形百簕花■☆
55393 Blepharis paradoxa Fritsch;奇异百簕花■☆
55394 Blepharis passargei Lindau = Blepharis linariifolia Pers. ■☆
55395 Blepharis persica (Burm. f.) Kuntze = Blepharis ciliaris (L.) B. L. Burtt ■☆
55396 Blepharis petalidioides Vollesen;瓣状百簕花■☆
55397 Blepharis petraea Vollesen;岩生百簕花■☆
55398 Blepharis phillipseae Rendle = Blepharis hildebrandtii Lindau subsp. phillipsiae (Rendle) Vollesen ■☆
55399 Blepharis pinguior C. B. Clarke = Blepharis involucrata Solms ■☆
55400 Blepharis pratensis S. Moore;草原百簕花■☆
55401 Blepharis procumbens (L. f.) Pers.;平铺百簕花■☆
55402 Blepharis procumbens T. Anderson = Blepharis serrulata (Nees) Ficalho et Hiern ■☆
55403 Blepharis pruinosa Engl.;粉百簕花■☆
55404 Blepharis pungens Klotzsch;刺百簕花■☆
55405 Blepharis pusilla Vollesen;微小百簕花■☆
55406 Blepharis quadrispina Lindau ex Pax = Barleria quadrispina Lindau ●☆
55407 Blepharis reekmansii Vollesen;里克曼斯百簕花■☆
55408 Blepharis refracta Mildbr.;反折百簕花■☆
55409 Blepharis repens (Vahl) Roth = Blepharis integrifolia (L. f.) E. Mey. ex Schinz ■☆
55410 Blepharis richardsiae Vollesen;理查兹百簕花■☆
55411 Blepharis rubiifolia Schumach. = Blepharis maderaspatensis (L.) B. Heyne ex Roth ■☆
55412 Blepharis rupicola Engl. = Blepharis integrifolia (L. f.) E. Mey. ex Schinz ■☆
55413 Blepharis ruwenzoriensis C. B. Clarke = Blepharis hildebrandtii Lindau ■☆
55414 Blepharis saturejifolia Pers. = Blepharis integrifolia (L. f.) E. Mey. ex Schinz ■☆
55415 Blepharis saxatilis Oberm. = Blepharis subvolubilis C. B. Clarke ■☆
55416 Blepharis scandens Vollesen;攀缘百簕花■☆
55417 Blepharis scullyi S. Moore = Acanthopsis scullyi (S. Moore) Oberm. ■☆
55418 Blepharis sericea Vollesen;绢毛百簕花■☆
55419 Blepharis serrulata (Nees) Ficalho et Hiern;细齿百簕花■☆
55420 Blepharis setosa Nees = Blepharis integrifolia (L. f.) E. Mey. ex Schinz ■☆
55421 Blepharis sinuata (Nees) C. B. Clarke;深波百簕花■☆
55422 Blepharis somaliensis Vollesen;索马里百簕花■☆
55423 Blepharis spathularis (Nees) T. Anderson = Acanthopsis spathularis (Nees) Schinz ■☆
55424 Blepharis spinescens Vollesen;细刺百簕花■☆
55425 Blepharis spinipes Vollesen;刺梗百簕花■☆
55426 Blepharis squarrosa (Nees) T. Anderson;粗鳞百簕花■☆
55427 Blepharis stuhlmanni Lindau;斯图尔曼百簕花■☆
55428 Blepharis subglabra Vollesen;近光百簕花■☆
55429 Blepharis subvolubilis C. B. Clarke;缠绕百簕花■☆
55430 Blepharis subvolubilis C. B. Clarke var. longifolia Oberm. = Blepharis subvolubilis C. B. Clarke ■☆
55431 Blepharis swaziensis Vollesen;斯威士百簕花■☆
55432 Blepharis tanae Napper;塔纳百簕花■☆
55433 Blepharis tanganyikensis (Napper) Vollesen;坦噶尼喀百簕花■☆
55434 Blepharis tanzaniensis Vollesen;坦桑尼亚百簕花■☆
55435 Blepharis teaguei Oberm. = Blepharis maderaspatensis (L.) B. Heyne ex Roth ■☆
55436 Blepharis tenuiramea S. Moore;细枝百簕花■☆
55437 Blepharis tetrasticha Lindau;四列百簕花■☆
55438 Blepharis thulinii Vollesen;图林百簕花■☆
55439 Blepharis togodelia Solms = Blepharis maderaspatensis (L.) B. Heyne ex Roth ■☆
55440 Blepharis torrei Vollesen;托雷百簕花■☆
55441 Blepharis transvaalensis Schinz;德兰士瓦百簕花■☆
55442 Blepharis trifida Vollesen;三裂百簕花■☆
55443 Blepharis trinervis Dewèvre = Blepharis stuhlmanni Lindau ■☆
55444 Blepharis trispina Napper;三刺百簕花■☆
55445 Blepharis trispinosa Hainz = Blepharis furcata (L. f.) Pers. ■☆
55446 Blepharis turkanae Vollesen;图尔卡纳百簕花■☆
55447 Blepharis uniflora C. B. Clarke;单花百簕花■☆
55448 Blepharis uzondoensis Vollesen;乌祖多伊百簕花■☆
55449 Blepharis verdickii De Wild. = Blepharis stuhlmanni Lindau ■☆
55450 Blepharis villosa C. B. Clarke = Blepharis mitrata C. B. Clarke ■☆
55451 Blepharis welwitschii S. Moore;韦尔百簕花■☆
55452 Blepharispermum Benth. = Blepharophyllum Klotzsch ●☆
55453 Blepharispermum Benth. = Scyphogyne Brongn. ●☆
55454 Blepharispermum DC. = Blepharispermum Wight ex DC. ■☆
55455 Blepharispermum Wight = Blepharispermum Wight ex DC. ■☆
55456 Blepharispermum Wight ex DC. (1834);睑子菊属■☆
55457 Blepharispermum arcuatum Erikss.;尖睑子菊■☆
55458 Blepharispermum brachycarphum Mattf.;短睑子菊■☆
55459 Blepharispermum canescens Erikss.;灰睑子菊■☆
55460 Blepharispermum ellenbeckii Cufod.;埃伦睑子菊■☆
55461 Blepharispermum fruticosum Klatt;灌丛睑子菊■☆
55462 Blepharispermum fruticosum Klatt var. lapathifolium Chiov. = Blepharispermum villosum O. Hoffm. ■☆
55463 Blepharispermum fruticosum Klatt var. typicum Chiov. = Blepharispermum fruticosum Klatt ■☆
55464 Blepharispermum lanceolatum Chiov. = Blepharispermum fruticosum Klatt ■☆
55465 Blepharispermum lobatum Klatt = Athroisma lobatum (Klatt) Mattf. ●☆
55466 Blepharispermum minus S. Moore;小睑子菊■☆

55467 Blepharispermum obovatum Chiov.;倒卵睑子菊■☆
55468 Blepharispermum pubescens S. Moore;短柔毛睑子菊■☆
55469 Blepharispermum spinulosum Oliv. et Hiern;细刺睑子菊■☆
55470 Blepharispermum villosum O. Hoffm.;长柔毛睑子菊■☆
55471 Blepharispermum xerothamnum Mattf.;干枝睑子菊■☆
55472 Blepharispermum yemense Deflers;也门睑子菊■☆
55473 Blepharispermum zanguebaricum Oliv. et Hiern;赞古睑子菊■☆
55474 Blepharistemma Benth. = Blepharistemma Wall. ex Benth. ●☆
55475 Blepharistemma Wall. ex Benth. (1858);睫毛树属●☆
55476 Blepharistemma corymbosum Wall. ex Benth.;睫毛树●☆
55477 Blepharitheca Pichon = Cuspidaria DC. (保留属名)●☆
55478 Blepharizonia (A. Gray) Greene(1885);睑菊属■☆
55479 Blepharizonia Greene = Blepharizonia (A. Gray) Greene ■☆
55480 Blepharizonia laxa Greene;疏松睑菊■☆
55481 Blepharizonia plumosa (Kellogg) Greene;羽状睑菊■☆
55482 Blepharizonia plumosa (Kellogg) Greene subsp. viscida D. D. Keck = Blepharizonia laxa Greene ■☆
55483 Blepharizonia plumosa (Kellogg) Greene var. subplumosa (A. Gray) Jeps. = Blepharizonia laxa Greene ■☆
55484 Blepharocalyx O. Berg(1856);毛萼金娘属●☆
55485 Blepharocalyx gigantea Lillo;巨毛萼金娘●☆
55486 Blepharocarya F. Muell. (1878);毛果漆属●☆
55487 Blepharocarya involucrigera F. Muell.;毛果漆●☆
55488 Blepharocaryaceae Airy Shaw = Anacardiaceae R. Br. (保留科名)●
55489 Blepharocaryaceae Airy Shaw;毛果漆科(毛萼金娘科)●
55490 Blepharochilum M. A. Clem. et D. L. Jones = Bulbophyllum Thouars(保留属名)■
55491 Blepharochlamys C. Presl = Mystropetalon Harv. ■☆
55492 Blepharochloa Endl. = Leersia Sw. (保留属名)■
55493 Blepharodachna Post et Kuntze = Blepharidachne Hack. ■☆
55494 Blepharodon Decne. (1844);毛齿萝藦属■☆
55495 Blepharodon amazonicus (Benth.) Fontella et Marquete;亚马逊毛齿萝藦■☆
55496 Blepharodon angustifolius Malme;窄叶毛齿萝藦■☆
55497 Blepharodon bicolor Decne.;二色毛齿萝藦■☆
55498 Blepharodon bidens Silveira;双毛齿萝藦■☆
55499 Blepharodon ciliatus Moldenke;睫毛毛齿萝藦■☆
55500 Blepharodon crassifolius Schltr.;厚叶毛齿萝藦■☆
55501 Blepharodon grandiflorus Benth.;大花毛齿萝藦■☆
55502 Blepharodon laurifolius Decne.;桂叶毛齿萝藦■☆
55503 Blepharodon minimus Woodson;小毛齿萝藦■☆
55504 Blepharodon mucronatus Decne.;短尖毛齿萝藦■☆
55505 Blepharodon pallidus Decne.;苍白毛齿萝藦■☆
55506 Blepharolepis Nees(1836) = Polpoda C. Presl ●☆
55507 Blepharolepis Nees(1843) = Scirpus L. (保留属名)■
55508 Blepharolepis eckloniana Nees = Polpoda capensis C. Presl ●☆
55509 Blepharoneuron Nash(1898);毛脉禾属(肋禾属)■☆
55510 Blepharoneuron Rydb. = Blepharoneuron Nash ■☆
55511 Blepharoneuron tricholepis (Torr.) Nash;毛脉禾■☆
55512 Blepharoneuron tricholepis Nash = Blepharoneuron tricholepis (Torr.) Nash ■☆
55513 Blepharopappua Post et Kuntze = Blepharipappus Hook. (废弃属名)●☆
55514 Blepharopappua Post et Kuntze = Lebetanthus Endl. (保留属名)●☆
55515 Blepharophyllum Klotzsch = Scyphogyne Brongn. ●☆
55516 Blepharophyllum divaricatum Klotzsch = Erica rigidula (N. E. Br.) E. G. H. Oliv. ●☆
55517 Blepharospermum Post et Kuntze = Blepharispermum Wight ex DC. ■☆
55518 Blepharostemma Fourr. = Asperula L. (保留属名)■
55519 Blepharostemrna Post et Kuntze = Blepharistemma Wall. ex Benth. ●☆
55520 Blepharozonia Post et Kuntze = Blepharizonia Greene ■☆
55521 Blepheuria Raf. = Campanula L. ■●
55522 Blephilia Raf. (1819);睫毛草属;Wood Mint ■☆
55523 Blephilia ciliata (L.) Benth.;俄亥俄睫毛草;Downy Pagoda-plant, Downy Wood Mint, Ohio Horse Mint, Ohio Horsemint ■☆
55524 Blephilia hirsuta (Pursh) Benth.;睫毛草;Hairy Pagoda-plant, Hairy Wood Mint, Wood Mint ■☆
55525 Blephiloma Raf. = Phlomis L. ●■
55526 Blephistelma Raf. = Disemma Labill. ●■
55527 Blephistelma Raf. = Passiflora L. ●■
55528 Blephixis Raf. = Chaerophyllum L. ■
55529 Bletia Ruiz et Pav. (1794);美洲白芨属(美洲白及属,拟白芨属,拟白及属)■☆
55530 Bletia acutipetala Hook. = Bletia purpurea (Lam.) DC. ■☆
55531 Bletia aphylla Nutt. = Hexalectris spicata (Walter) Barnhart ■☆
55532 Bletia bicallosa D. Don = Eulophia bicallosa (D. Don) P. F. Hunt et Summerh. ■
55533 Bletia bicallosa D. Don = Liparis nervosa (Thunb. ex A. Murray) Lindl. ■
55534 Bletia dabia D. Don = Eulophia dabia (D. Don) Hochr. ■
55535 Bletia flava (Blume) Wall. ex Lindl. = Phaius flavus (Blume) Lindl. ■
55536 Bletia formosana Hayata = Bletilla formosana (Hayata) Schltr. ■
55537 Bletia formosana Hayata f. kotoensis T. P. Li = Bletilla formosana (Hayata) Schltr. ■
55538 Bletia gebina Lindl. = Bletilla striata (Thunb. ex A. Murray) Rchb. f. ■
55539 Bletia graminifolia D. Don = Arundina graminifolia (D. Don) Hochr. ■
55540 Bletia havanensis Lindl. = Bletia purpurea (Lam.) DC. ■☆
55541 Bletia hyacinthina (Sm.) Aiton = Bletilla striata (Thunb. ex A. Murray) Rchb. f. ■
55542 Bletia hyacinthina (Sm.) Aiton var. gebina (Lindl.) Blume = Bletilla striata (Thunb. ex A. Murray) Rchb. f. ■
55543 Bletia hyacinthina (Sm.) R. Br. = Bletilla striata (Thunb. ex A. Murray) Rchb. f. ■
55544 Bletia kotoensis Hayata = Bletilla formosana (Hayata) Schltr. ■
55545 Bletia masuca D. Don = Calanthe sylvatica (Thouars) Lindl. ■
55546 Bletia morrisonicola Hayata = Bletilla formosana (Hayata) Schltr. ■
55547 Bletia obcordata Lindl. = Cephalantheropsis obcordata (Lindl.) Ormerod ■
55548 Bletia patula Graham;海地美洲白芨;Haitian Pine-pink ■☆
55549 Bletia purpurea (Lam.) DC.;紫花美洲白芨;Pine-pink, Purple Bletia, Sharp-petaled Bletia ■☆
55550 Bletia striata (Thunb. ex A. Murray) Druce = Bletilla striata (Thunb. ex A. Murray) Rchb. f. ■
55551 Bletia striata (Thunb.) Druce = Bletilla striata (Thunb. ex A. Murray) Rchb. f. ■
55552 Bletia sylvatica (Thouars) Bojer = Calanthe sylvatica (Thouars)

Lindl. ■
55553 Bletia tancarvilleae (L'Hér.) R. Br. = Phaius tankervilleae (Banks ex L'Hér.) Blume ■
55554 Bletia verecunda (Salisb.) R. Br. = Bletia purpurea (Lam.) DC. ■☆
55555 Bletia verecunda Chapm. = Pteroglossaspis ecristata (Fernald) Rolfe ■☆
55556 Bletia verecunda R. Br. = Bletia purpurea (Lam.) DC. ■☆
55557 Bletia woodfordii Hook. = Phaius flavus (Blume) Lindl. ■
55558 Bletiana Raf. = Bletia Ruiz et Pav. ■☆
55559 Bletilla Rchb. f. (1853)(保留属名);白芨属(白及属);Bletilla, Ground Orchid ●■
55560 Bletilla chinensis Schltr. = Bletilla sinensis (Rolfe) Schltr. ■
55561 Bletilla elegantula (Kraenzl.) Garay et Romero = Bletilla formosana (Hayata) Schltr. ■
55562 Bletilla elegantula (Kraenzl.) Garay et Romero = Bletilla striata (Thunb. ex A. Murray) Rchb. f. ■
55563 Bletilla formosana (Hayata) Schltr.;小白芨(白及,红头白及,兰屿白芨,兰屿白及,台湾白芨,台湾白及,小白及);Small Bletilla, Yunnan Bletilla ■
55564 Bletilla formosana (Hayata) Schltr. f. kotoensis (Hayata) T. P. Lin = Bletilla formosana (Hayata) Schltr. ■
55565 Bletilla formosana (Hayata) Schltr. f. rubrolabella S. S. Ying = Bletilla formosana (Hayata) Schltr. ■
55566 Bletilla formosana (Hayata) Schltr. var. limprichtii Schltdl.;白花小白芨;Whiteflower Small Bletilla ■
55567 Bletilla gebina (Lindl.) Rchb. f. = Bletilla striata (Thunb. ex A. Murray) Rchb. f. ■
55568 Bletilla hyacinthina (Sm.) Rchb. f. = Bletilla striata (Thunb. ex A. Murray) Rchb. f. ■
55569 Bletilla kotoensis (Hayata) Schltr. = Bletilla formosana (Hayata) Schltr. ■
55570 Bletilla morrisonensis (Hayata) Schltr. = Bletilla formosana (Hayata) Schltr. ■
55571 Bletilla morrisonicola (Hayata) Schltr. = Bletilla formosana (Hayata) Schltr. ■
55572 Bletilla ochracea Schltr.;黄花白芨(白及,黄花白及,黄术,狭叶白芨);Yellow Bletilla, Yellowflower Bletilla ■
55573 Bletilla scopulorum (W. W. Sm.) Schltr. = Pleione scopulorum W. W. Sm. ■
55574 Bletilla sinensis (Rolfe) Schltr.;华白芨(华白及,小白芨,中华白芨,中华白及);China Bletilla ■
55575 Bletilla striata (Thunb. ex A. Murray) Rchb. f.;白芨(白给,白根,白鸡,白鸡儿,白鸡娃,白及,白及子,白芷,白鸟儿头,百芨,百笠,冰球子,地螺丝,甘根,鸡头参,君球子,鞁口药,利知子,连及草,明白及,千年棕,箬兰,小白及,雪如米,羊角七,一兜棕,朱兰,竹粟胶,紫蕙,紫兰);Bletilla, Chinese Ground Orchid, Common Bletilla, Ground Orchid, Hardy Orchid, Hyacinth Bletilla, Urn Orchid ■
55576 Bletilla striata (Thunb. ex A. Murray) Rchb. f. f. gebina (Lindl.) Ohwi = Bletilla striata (Thunb. ex A. Murray) Rchb. f. ■
55577 Bletilla striata (Thunb. ex A. Murray) Rchb. f. var. albomarginata Makino;白边白芨■☆
55578 Bletilla striata (Thunb. ex A. Murray) Rchb. f. var. albomarginata Makino = Bletilla striata (Thunb. ex A. Murray) Rchb. f. ■
55579 Bletilla striata (Thunb. ex A. Murray) Rchb. f. var. gebina (Lindl.) Rchb. f.;大花白芨■
55580 Bletilla striata (Thunb. ex A. Murray) Rchb. f. var. gebina (Lindl.) Rchb. f. = Bletilla striata (Thunb. ex A. Murray) Rchb. f. ■
55581 Bletilla striata (Thunb. ex A. Murray) Rchb. f. var. kotoensis (Hayata) Masam.;红头白芨■
55582 Bletilla striata (Thunb. ex A. Murray) Rchb. f. var. kotoensis (Hayata) Masam. = Bletilla formosana (Hayata) Schltr. ■
55583 Bletilla striata (Thunb.) Rchb. f. var. kotoensis (Hayata) Masam. = Bletilla formosana (Hayata) Schltr. ■
55584 Bletilla szetschuanica Schltr. = Bletilla formosana (Hayata) Schltr. ■
55585 Bletilla yunnanensis Schltr. = Bletilla formosana (Hayata) Schltr. ■
55586 Bletilla yunnanensis Schltr. ex H. Limpr. = Bletilla formosana (Hayata) Schltr. ■
55587 Bletilla yunnanensis Schltr. ex H. Limpr. var. limprichtii Schltr. ex Limpr. = Bletilla formosana (Hayata) Schltr. ■
55588 Bletilla yunnanensis Schltr. var. limprichtii Schltr. = Bletilla formosana (Hayata) Schltr. ■
55589 Bletti Steud. = Blatti Adans. (废弃属名)●
55590 Bletti Steud. = Sonneratia L. f. (保留属名)●
55591 Blexum Raf. = Blechum P. Browne ■
55592 Blighia K. König(1806);阿开木属;Akee, Akee Apple ●☆
55593 Blighia kamerunensis Radlk. = Blighia welwitschii (Hiern) Radlk. ●☆
55594 Blighia laurentii De Wild. = Blighia welwitschii (Hiern) Radlk. ●☆
55595 Blighia mildbraedii Radlk. = Blighia welwitschii (Hiern) Radlk. ●☆
55596 Blighia sapida K. König;阿开木(库潘树,美味阿开木,西非荔枝果);Ackee, Akee, Akee Apple ●☆
55597 Blighia sapida K. König = Cupania sapida (König) Oken ●☆
55598 Blighia unijugata Baker;成双氏阿开木●☆
55599 Blighia welwitschii (Hiern) Radlk.;韦氏阿开木●☆
55600 Blighia wildemanniana Radlk. = Blighia welwitschii (Hiern) Radlk. ●☆
55601 Blighia zambesiaca Baker = Blighia unijugata Baker ●☆
55602 Blighiopsis Van de Veken(1960);拟阿开木属●☆
55603 Blighiopsis pseudostipularis Van der Veken;拟阿开木●☆
55604 Blinkworthia Choisy(1834);苞叶藤属;Blinkworthia ●
55605 Blinkworthia convolvuloides Prain;苞叶藤(鸡帐篷,马铃铛,旋花状苞叶藤);Bindweedlike Blinkworthia, Bindweed-like Blinkworthia, Blinkworthia, Discoid-stigma Blinkworthia, Glory-bind-like Blinkworthia ●
55606 Blinkworthia discostigma Hand.-Mazz. = Blinkworthia convolvuloides Prain ●
55607 Blismua Montand. = Blysmus Panz. ex Schult. (保留属名)■
55608 Blismus Friche-Joset et Montandon = Blysmus Panz. ex Schult. (保留属名)■
55609 Blismus Friche-Joset et Montandon = Scirpus L. (保留属名)■
55610 Blitaceae Adans. = Blitaceae Adans. ex T. Post et Kuntze ●■
55611 Blitaceae Adans. ex T. Post et Kuntze = Amaranthaceae Juss. (保留科名)●■
55612 Blitaceae Adans. ex T. Post et Kuntze = Chenopodiaceae Vent. (保留科名)●■
55613 Blitanthus Rchb. = Acroglochin Schrad. ■
55614 Blitoides Fabr. = Amaranthus L. ■
55615 Bliton Adans. = Blitum Fabr. ■
55616 Blitum Fabr. = Amaranthus L. ■

55617 Blitum Heist. ex Fabr. = Amaranthus L. ■
55618 Blitum Hill = Chenopodium L. ■●
55619 Blitum L. = Chenopodium L. ■●
55620 Blitum Scop. = Albersia Kunth ■
55621 Blitum Scop. = Amaranthus L. ■
55622 Blitum Scop. = Blitum L. ■●
55623 Blitum ambrosioides (L.) Beck = Dysphania ambrosioides (L.) Mosyakin et Clemants ■
55624 Blitum ambrosioides Beck = Chenopodium ambrosioides L. ■☆
55625 Blitum californicum S. Watson = Chenopodium californicum (S. Watson) S. Watson ■☆
55626 Blitum capitatum L. = Chenopodium capitatum Asch. ■☆
55627 Blitum chenopodioides L. = Chenopodium chenopodioides (L.) Aellen ■
55628 Blitum cristatum F. Muell. = Dysphania cristata (F. Muell.) Mosyakin et Clemants ■☆
55629 Blitum glaucum (L.) W. D. J. Koch = Chenopodium glaucum L. ■
55630 Blitum glaucum W. D. Koch = Chenopodium glaucum L. ■
55631 Blitum hastatum Rydb. = Chenopodium capitatum (L.) Ambrosi var. parvicapitatum S. L. Welsh ■☆
55632 Blitum nuttallianum Schult. = Monolepis nuttalliana (Schult. et Schult. f.) Greene ■☆
55633 Blitum polymorphum C. A. Mey. = Chenopodium chenopodioides (L.) Aellen ■
55634 Blitum polymorphum C. A. Mey. = Chenopodium rubrum L. ■
55635 Blitum rubrum (L.) Rchb. = Chenopodium rubrum L. ■
55636 Blitum rubrum Rchb. = Chenopodium rubrum L. ■
55637 Blitum virgatum L. = Chenopodium foliosum (Moench) Asch. ■
55638 Blitum virgatum L. var. minus Vahl = Chenopodium foliosum Asch. var. minus (Vahl) Asch. ■☆
55639 Blochmannia Rchb. = Triplaris Loefl. ex L. ●
55640 Blomia Miranda(1953);布氏无患子属●☆
55641 Blomia cupanioides Miranda;布氏无患子●☆
55642 Blondea Rich. = Sloanea L. ●
55643 Blondia Neck. = Tiarella L. ■
55644 Bloomeria Kellogg(1863);环丝韭属;Golden Star,Golden Stars ■☆
55645 Bloomeria aurea Kellogg = Bloomeria crocea (Torr.) Coville var. aurea (Kellogg) J. W. Ingram ■☆
55646 Bloomeria aurea Kellogg = Bloomeria crocea (Torr.) Coville ■☆
55647 Bloomeria clevelandii S. Watson;圣地亚哥环丝韭;San Diego Golden Star ■☆
55648 Bloomeria crocea (Torr.) Coville;环丝韭;Common Golden Star ■☆
55649 Bloomeria crocea (Torr.) Coville var. aurea (Kellogg) J. W. Ingram;黄环丝韭■☆
55650 Bloomeria crocea (Torr.) Coville var. aurea (Kellogg) J. W. Ingram = Bloomeria crocea (Torr.) Coville ■☆
55651 Bloomeria crocea (Torr.) Coville var. montana (Greene) J. W. Ingram;山地环丝韭;Inland Golden Star,Mountain Golden Star ■☆
55652 Bloomeria gracilis Borzí = Bloomeria crocea (Torr.) Coville ■☆
55653 Bloomeria humilis Hoover;小环丝韭;Dwarf Golden Star ■☆
55654 Bloomeria montana Greene = Bloomeria crocea (Torr.) Coville var. montana (Greene) J. W. Ingram ■☆
55655 Bloomeria transmontana (Greene) J. F. Macbr. = Muilla transmontana Greene ■☆
55656 Blossfeldia Werderm. (1937);松露玉属(松露球属);Blossfeldia ●☆
55657 Blossfeldia atroviridis F. Ritter;妖梦玉■☆
55658 Blossfeldia liliputana Werderm.;松露玉;Common Blossfeldia ■☆
55659 Blossfeldiana Megata = Frailea Britton et Rose ●
55660 Blotia Léandri(1957);布洛大戟属●■☆
55661 Blotia leandriana Petra Hoffm. et McPherson;利安布洛大戟●☆
55662 Blotia mimosoides (Baill.) Petra Hoffm. et McPherson;含羞草布洛大戟●☆
55663 Blotia oblongifolia (Baill.) Léandri;矩圆叶布洛大戟●☆
55664 Blotia tanalorum Léandri;塔纳尔布洛大戟●☆
55665 Bluffia Delile = Panicum L. ■
55666 Bluffia Nees = Alloteropsis J. Presl ex C. Presl ■
55667 Bluffia Nees = Panicum L. ■
55668 Bluffia eckloniana Nees = Alloteropsis semialata (R. Br.) Hitchc. var. eckloniana (Nees) Pilg. ■
55669 Blumea DC. (1833)(保留属名);艾纳香属(毛将军属,大艾属);Blumea ■●
55670 Blumea G. Don = Blumea DC. (保留属名)■●
55671 Blumea G. Don = Saurauia Willd. (保留属名)●
55672 Blumea Post et Kuntze = Blumea DC. (保留属名)■●
55673 Blumea Rchb. = Blumea DC. (保留属名)■●
55674 Blumea Rchb. = Neesia Blume(保留属名)●☆
55675 Blumea Zipp. ex Miq. = Didymosperma H. Wendl. et Drude ex Benth. et Hook. f. ●
55676 Blumea abyssinica Sch. Bip. ex A. Rich. = Blumea bovei (DC.) Vatke ■☆
55677 Blumea adamsii J. -P. Lebrun et Stork;亚当斯艾纳香■☆
55678 Blumea adenophora Franch.;具腺艾纳香;Glandular Blumea ■
55679 Blumea alata (D. Don) DC. = Blumea crispata (Vahl) Merxm. ■☆
55680 Blumea alata (D. Don) DC. = Laggera alata (D. Don) Sch. Bip. ex Oliv. ■
55681 Blumea alata (D. Don) Sch. Bip. ex Oliv. var. gracilis O. Hoffm. et Muschl. = Blumea adamsii J. -P. Lebrun et Stork ■☆
55682 Blumea alata (Roxb.) DC. = Laggera alata (D. Don) Sch. Bip. ex Oliv. ■
55683 Blumea amethystina Hance = Blumea fistulosa (Roxb.) Kurz ■
55684 Blumea arfakiana Martelli;阿法艾纳香■☆
55685 Blumea arnottiana Steud. = Duhaldea chinensis DC. ●■
55686 Blumea arnottiana Steud. = Inula cappa (Buch. -Ham.) DC. ●■
55687 Blumea aromatica DC.;馥芳艾纳香(薄叶艾纳香,大开门,黄药,山风,香艾,香艾纳);Aromatic Blumea ■
55688 Blumea aurita (L. f.) DC. = Pseudoconyza viscosa (Mill.) D'Arcy ■☆
55689 Blumea aurita (L. f.) DC. var. foliolosa (DC.) C. D. Adams = Pseudoconyza viscosa (Mill.) D'Arcy ■☆
55690 Blumea axillaris (Lam.) DC.;腋生艾纳香(柔毛艾纳香)■
55691 Blumea baccharoides Sch. Bip. = Pluchea dioscoridis (L.) DC. ●☆
55692 Blumea balansae Gagnep. = Blumea membranacea DC. ■
55693 Blumea balsamifera (L.) DC.;艾纳香(艾粉,艾脑香,冰片艾,冰片草,冰片叶,打蚊艾,大艾,大风艾,大风叶,大枫草,大骨风,大黄草,大毛药,牛耳艾,土冰片,学老麻,再风艾,紫再枫);Balmsamiferous Blumea Ai,Ngai,Ngai Camphor ■
55694 Blumea balsamifera (L.) DC. var. microcephala Kitam. = Blumea balsamifera (L.) DC. ■
55695 Blumea barbata (L.) DC. var. sericans Kurz = Blumea sericans (Kurz) Hook. f. ■
55696 Blumea bifoliata DC.;二叶艾纳香■☆

55697 Blumea bodinieri Vaniot = Blumea lacera (Burm. f.) DC. ■
55698 Blumea bovei (DC.) Vatke;博韦艾纳香■☆
55699 Blumea braunii (Vatke) J. -P. Lebrun et Stork;布劳恩艾纳香■☆
55700 Blumea brevipes (Oliv. et Hiern) Wild;短梗艾纳香■☆
55701 Blumea cafra (DC.) O. Hoffm.;纳塔尔艾纳香■☆
55702 Blumea cavaleriei H. Lév. et Vaniot = Blumea hamiltonii DC. ■
55703 Blumea cavaleriei H. Lév. et Vaniot = Blumea sericans (Kurz) Hook. f. ■
55704 Blumea chamissoniana DC. = Blumea dregeanoides Sch. Bip. ex A. Rich. ■☆
55705 Blumea chevalieri Gagnep. = Blumea lacera (Burm. f.) DC. ■
55706 Blumea chinensis DC. = Blumea megacephala (Randeria) C. T. Chang et C. H. Yu ■
55707 Blumea chinensis DC. = Blumea riparia (Blume) DC. ■
55708 Blumea chinensis Hook. et Arn. = Duhaldea chinensis DC. ●■
55709 Blumea chinensis Hook. et Arn. = Inula cappa (Buch. -Ham.) DC. ●■
55710 Blumea chinensis Walp. = Blumea hieraciifolia (D. Don) DC. ■
55711 Blumea clarkei Hook. f.;七里明(东风草);C. B. Clarke Blumea, Clarke Blumea ■
55712 Blumea conspicua Hayata;大花艾纳香(藤艾纳香)■
55713 Blumea conspicua Hayata = Blumea lanceolaria (Roxb.) Druce ■
55714 Blumea conyzoides H. Lév. et Vaniot = Conyza leucantha (D. Don) Ludlow et Raven ■
55715 Blumea conyzoides H. Lév. et Vaniot = Microglossa pyrifolia (Lam.) Kuntze ●
55716 Blumea crassifolia A. Rich. = Laggera crassifolia (A. Rich.) Oliv. et Hiern ■☆
55717 Blumea crispata (Vahl) Merxm.;皱波艾纳香■☆
55718 Blumea crispata (Vahl) Merxm. var. appendiculata (Robyns) Lisowski = Blumea brevipes (Oliv. et Hiern) Wild ■☆
55719 Blumea crispata (Vahl) Merxm. var. montana (C. D. Adams) J. -P. Lebrun et Stork;山地皱波艾纳香■☆
55720 Blumea dasycoma Boerl. var. pinnatifida (Miq.) Boerl. = Blumea densiflora DC. ■
55721 Blumea decurrens (Vahl) Merxm.;非洲艾纳香■☆
55722 Blumea densiflora DC.;密花艾纳香(大黑蒿,大升麻);Denseflower Blumea ■
55723 Blumea densiflora DC. var. hookeri (C. B. Clarke ex Hook. f.) C. C. Chang et Y. Q. Tseng = Blumea hookeri C. B. Clarke ex Hook. f. ■
55724 Blumea densiflora DC. var. hookeri Gagnep.;薄叶密花艾纳香;Hooker Denseflower Blumea ■
55725 Blumea densiflora DC. var. pinnatifida Miq. = Blumea densiflora DC. ■
55726 Blumea dregeanoides Sch. Bip. = Blumea dregeanoides Sch. Bip. ex A. Rich. ■☆
55727 Blumea dregeanoides Sch. Bip. ex A. Rich.;拟德雷艾纳香■☆
55728 Blumea dregeanoides Sch. Bip. ex A. Rich. = Blumea axillaris (Lam.) DC. ■
55729 Blumea eberhardtii Gagnep.;光叶艾纳香;Eberhardt Blumea ■●
55730 Blumea eberhardtii Gagnep. = Blumea repanda (Roxb.) Hand. -Mazz. ■
55731 Blumea elatior (R. E. Fr.) Lisowski;高大艾纳香■☆
55732 Blumea emeiensis Z. Y. Zhu = Blumea aromatica DC. ■
55733 Blumea esquirolii H. Lév. et Vaniot = Vernonia cinerea (L.) Less. ■
55734 Blumea excisa DC. = Blumea densiflora DC. ■
55735 Blumea fasciculata DC. = Blumea sessiliflora Decne. ■
55736 Blumea fistulosa (Roxb.) Kurz;聚花艾纳香(草骨黄,节节红,紫花大艾);Fistulose Blumea ■
55737 Blumea flava DC. = Blumeopsis flava (DC.) Gagnep. ■
55738 Blumea formasana Kitam.;台北艾纳香(里白艾纳香,台湾艾纳香);Taiwan Blumea ■
55739 Blumea fruticosa Koidz. = Blumea conspicua Hayata ■
55740 Blumea fruticulosa Hochst. = Phagnalon stenolepis Chiov. ■☆
55741 Blumea gariepina DC. = Blumea decurrens (Vahl) Merxm. ■☆
55742 Blumea glandulosa Benth. = Blumea laciniata (Roxb.) DC. ■
55743 Blumea glandulosa DC. = Blumea lacera (Burm. f.) DC. ■
55744 Blumea glomerata DC.;紫花大艾■
55745 Blumea glomerata DC. = Blumea fistulosa (Roxb.) Kurz ■
55746 Blumea gnaphalioides Hayata = Blumea sericans (Kurz) Hook. f. ■
55747 Blumea gracilis DC. = Blumea sessiliflora Decne. ■
55748 Blumea gracilis Dunn = Blumea tenuifolia C. Y. Wu ■
55749 Blumea hamiltonii DC.;少叶艾纳香(田芥菜仔);Hamilton Blumea ■
55750 Blumea hamiltonii DC. = Blumea sericans (Kurz) Hook. f. ■
55751 Blumea henryi Dunn;尖苞艾纳香;Henry Blumea ■
55752 Blumea henryi Dunn = Blumea martiniana Vaniot ■
55753 Blumea heudelotii (C. D. Adams) Lisowski;厄德艾纳香■☆
55754 Blumea hieraciifolia (D. Don) DC.;毛毡草(地菊,毛将军,坡艾草,丝毛毡草);Hawkweedleaf Blumea ■
55755 Blumea hieraciifolia Hayata = Blumea sericans (Kurz) Hook. f. ■
55756 Blumea hieraciifolia Hayata var. hamiltoni (DC.) C. B. Clarke = Blumea hamiltonii DC. ■
55757 Blumea hieraciifolia Hayata var. hamiltonii (DC.) C. B. Clarke = Blumea sericans (Kurz) Hook. f. ■
55758 Blumea hieraciifolia Hayata var. holosericea Benth. = Blumea sericans (Kurz) Hook. f. ■
55759 Blumea hieraciifolia Hayata var. macrostachya (DC.) Hook. f. = Blumea hieraciifolia (D. Don) DC. ■
55760 Blumea hongkongensis H. Lév. et Vaniot = Blumea clarkei Hook. f. ■
55761 Blumea hongkongensis Vaniot = Blumea clarkei Hook. f. ■
55762 Blumea hookeri C. B. Clarke = Blumea densiflora DC. var. hookeri Gagnep. ■
55763 Blumea hookeri C. B. Clarke ex Hook. f.;薄叶艾纳香■
55764 Blumea hymenophylla DC. = Blumea virens DC. ■
55765 Blumea kelleri Thell. = Pluchea kelleri (Thell.) Thulin ■☆
55766 Blumea lacera (Burm. f.) DC.;见霜黄(红根白毛倒提壶,红头草,黄花地胆头,甲冬仗,生毛将军);Lacerate-leaved Blumea, Malay Blumea, Malayan Blumea ■
55767 Blumea lacera (Burm. f.) DC. var. blumei DC. = Blumea laciniata (Roxb.) DC. ■
55768 Blumea laciniata (Roxb.) DC.;六耳铃(波缘艾纳香,吊钟黄,飞山虎,赶风茜,裂叶艾纳香,牛耳三稔,水马胎,羊耳三稔,走马风);Cutleaf False Oxtongue, Laciniate Blumea ■
55769 Blumea lanceolaria (Roxb.) Druce;千头艾纳香(大花艾纳香,大叶艾纳香,火油草,藤艾纳香,走马风,走马胎);Lanceolate Blumea, Taiwan Blumea ■
55770 Blumea lanceolaria (Roxb.) Druce var. spectabilis (DC.) Randeria = Blumea lanceolaria (Roxb.) Druce ■
55771 Blumea lapsanoides DC. = Blumea virens DC. ■

55772 Blumea lecomteana (O. Hoffm. et Muschl.) J.-P. Lebrun et Stork;勒孔特艾纳香■☆

55773 Blumea leptophylla Hayata = Blumea aromatica DC. ■

55774 Blumea lessingi Merr. = Blumea clarkei Hook. f. ■

55775 Blumea linearis C. I. Peng et W. P. Leu;狭叶艾纳香(条叶艾纳香)■●

55776 Blumea lyrata (Kunth) V. M. Badillo = Pseudoconyza viscosa (Mill.) D'Arcy ■☆

55777 Blumea macrostachya DC. = Blumea hieraciifolia (D. Don) DC. ■

55778 Blumea malabarica Hook. f. = Blumea clarkei Hook. f. ■

55779 Blumea malabarica Hook. f. = Blumea oblongifolia Kitam. ■

55780 Blumea martiniana Vaniot;裂苞艾纳香(大叶艾纳香,走马胎);Lobe-bracted Blumea, Lobed Bract Blumea, Lobedbract Blumea ■

55781 Blumea megacephala (Randeria) C. C. Chang et Y. Q. Tseng;东风草■●

55782 Blumea megacephala (Randeria) C. C. Chang et Y. Q. Tseng = Blumea riparia (Blume) DC. var. megacephala Randeria ■

55783 Blumea megacephala (Randeria) C. T. Chang et Y. Q. Tseng;大头艾纳香(白花九里明,东风草,管牙,华艾纳香,黄帽顶,九里明,青钓鱼杆,无毛大艾);Largehead Blumea ■

55784 Blumea membranacea DC.;长柄艾纳香(旗山艾纳香);Membranaceous Blumea ■

55785 Blumea membranacea Hayata = Blumea laciniata (Roxb.) DC. ■

55786 Blumea membranacea Hayata var. gracilis (DC.) Hook. f. = Blumea sessiliflora Decne. ■

55787 Blumea microphylla Chiov. = Pluchea somaliensis (Thell.) Thulin ■☆

55788 Blumea mollis (D. Don) Merr.;柔毛艾纳香(红头小仙,甲冬仗,毛艾纳香,毛干药,那猪草,紫背倒提壶);Pubescent Blumea ■

55789 Blumea mollis (D. Don) Merr. = Blumea axillaris (Lam.) DC. ■

55790 Blumea mollis (D. Don) Merr. = Blumea dregeanoides Sch. Bip. ex A. Rich. ■☆

55791 Blumea myriocephala DC. = Blumea lanceolaria (Roxb.) Druce ■

55792 Blumea napifolia DC.;芜菁叶艾纳香;Turnipleaf Blumea ■

55793 Blumea natalensis Sch. Bip. = Blumea cafra (DC.) O. Hoffm. ■☆

55794 Blumea nudiflora Hook. f. = Blumea fistulosa (Roxb.) Kurz ■

55795 Blumea obliqua (L.) Druce;斜艾纳香■

55796 Blumea oblongifolia Kitam.;长圆叶艾纳香(白叶,长叶艾纳香,大红草,大黄草,台湾艾纳香);Oblongleaf Blumea ■

55797 Blumea odorata?;芳香艾纳香■☆

55798 Blumea okinawensis Hayata = Blumea laciniata (Roxb.) DC. ■

55799 Blumea onnaensis Hayata = Blumea laciniata (Roxb.) DC. ■

55800 Blumea oxyodonta DC.;尖齿艾纳香;Sharptooth Blumea ■

55801 Blumea pappii Gand. = Pseudoconyza viscosa (Mill.) D'Arcy ■☆

55802 Blumea parvifolia DC. = Blumea mollis (D. Don) Merr. ■

55803 Blumea perrottetiana DC. = Blumea axillaris (Lam.) DC. ■

55804 Blumea petitiana A. Rich. = Laggera tomentosa (A. Rich.) Oliv. et Hiern ■☆

55805 Blumea phagnaloides Sch. Bip. ex A. Rich. = Phagnalon phagnaloides (Sch. Bip. ex A. Rich.) Cufod. ■☆

55806 Blumea procera DC. = Blumea repanda (Roxb.) Hand.-Mazz. ■

55807 Blumea pterodonta DC. = Blumea crispata (Vahl) Merxm. ■☆

55808 Blumea pterodonta DC. = Laggera pterodonta (DC.) Benth. ■

55809 Blumea pubigera (L.) Merr. = Blumea riparia (Blume) DC. ■

55810 Blumea purpurascens A. Rich. = Laggera crispata (Vahl) Hepper et J. R. I. Wood ■

55811 Blumea purpurea DC. = Blumea fistulosa (Roxb.) Kurz ■

55812 Blumea racemosa DC. = Blumea fistulosa (Roxb.) Kurz ■

55813 Blumea repanda (Roxb.) Hand.-Mazz.;高艾纳香;Repandous Blumea, Tall Blumea ■

55814 Blumea riparia (Blume) DC.;假东风草(管芽);Coastal Blumea ■

55815 Blumea riparia (Blume) DC. var. megacephala Randeria = Blumea megacephala (Randeria) C. T. Chang et Y. Q. Tseng ■

55816 Blumea runcinata DC. = Blumea laciniata (Roxb.) DC. ■

55817 Blumea sagittata Gagnep.;戟叶艾纳香;Sagittate Blumea ■

55818 Blumea salvifolia (Bory) DC. = Blumea crispata (Vahl) Merxm. ■☆

55819 Blumea saussureoides C. T. Chang et Y. Q. Tseng ex Y. Ling et Y. Q. Tseng;全裂艾纳香;Saussurea-like Blumea, Windhairlike Blumea ■

55820 Blumea sericans (Kurz) Hook. f.;丝毛艾纳香(毛将军,拟毛毡草,丝毛毛毡草);Silky Blumea ■

55821 Blumea sericans (Kurz) Hook. f. = Blumea hieraciifolia (D. Don) DC. ■

55822 Blumea serrata Chiov. = Pluchea kelleri (Thell.) Thulin ■☆

55823 Blumea sessiliflora Decne.;无梗艾纳香;Sessile Blumea, Sessileleaf False Oxtongue ■

55824 Blumea sinapifolia Gagnep. = Blumea laciniata (Roxb.) DC. ■

55825 Blumea solidaginoides (Poir.) DC. = Blumea axillaris (Lam.) DC. ■

55826 Blumea somaliensis Thell. = Pluchea somaliensis (Thell.) Thulin ■☆

55827 Blumea sonchifolia DC. = Blumea laciniata (Roxb.) DC. ■

55828 Blumea spectabilis DC. = Blumea lanceolaria (Roxb.) Druce ■

55829 Blumea squarrosa (Oliv. et Hiern) Wild;粗鳞艾纳香■☆

55830 Blumea subcapitata DC. = Blumea lacera (Burm. f.) DC. ■

55831 Blumea subcapitata Matsum. et Hayata = Conyza japonica (Thunb.) Less. ■

55832 Blumea suessenguthii Merxm. = Nicolasia heterophylla S. Moore ■☆

55833 Blumea tenuifolia C. Y. Wu;细叶艾纳香(狭叶艾纳香);Narrowleaf Blumea ■

55834 Blumea tomentosa A. Rich. = Laggera tomentosa (A. Rich.) Oliv. et Hiern ■☆

55835 Blumea tonkinensis Gagnep. = Blumea henryi Dunn ■

55836 Blumea tonkinensis Gagnep. = Blumea martiniana Vaniot ■

55837 Blumea trichophora DC. = Blumea mollis (D. Don) Merr. ■

55838 Blumea velutina H. Lév. et Vaniot = Blumea lacera (Burm. f.) DC. ■

55839 Blumea veronicifolia Franch.;纤枝艾纳香;Speedwell-leaf Blumea ■

55840 Blumea virens DC.;绿艾纳香;Green Blumea ■

55841 Blumea viscosa (Mill.) V. M. Badillo = Pseudoconyza viscosa (Mill.) D'Arcy ■☆

55842 Blumea volkensii (O. Hoffm.) J.-P. Lebrun et Stork;福尔艾纳香■☆

55843 Blumea wightiana DC. = Blumea axillaris (Lam.) DC. ■

55844 Blumea wightiana DC. = Blumea mollis (D. Don) Merr. ■

55845 Blumea wightiana Hook. f. = Blumea dregeanoides Sch. Bip. ex A. Rich. ■☆

55846 Blumella Tiegh. = Elytranthe (Blume) Blume ●

55847 Blumella Tiegh. = Elytranthe Blume + Macrosolen (Blume) Rchb. ●

55848 Blumella Tiegh. = Iticania Raf. ●

55849 Blumenbachia Koeler(废弃属名) = Blumenbachia Schrad.(保留属名)■☆

55850 Blumenbachia Koeler(废弃属名) = Sorghum Moench(保留属名)■

55851 Blumenbachia Schrad.(1825)(保留属名);布氏刺莲花属(南美刺莲花属)■☆

55852 Blumenbachia acaulis Phil.;无茎南美刺莲花■☆

55853 Blumenbachia aspera Urb.;粗糙南美刺莲花■☆

55854 Blumenbachia insignis Schrad.;南美刺莲花■☆

55855 Blumenbachia pterosperma G. Don;翅子南美刺莲花■☆

55856 Blumenbachia sylvestris Poepp.;林地南美刺莲花■☆

55857 Blumeodendron (Müll. Arg.) Kurz(1874);布氏木大戟属●☆

55858 Blumeodendron Kurz = Blumeodendron (Müll. Arg.) Kurz ●☆

55859 Blumeodendron tokbrai Kurz;布氏木大戟●☆

55860 Blumeopsis Gagnep.(1920);拟艾纳香属(假艾脑属,似艾脑属);Blumeopsis ■

55861 Blumeopsis falcata (D. Don) Merr. = Conyza leucantha (D. Don) Ludlow et Raven ■

55862 Blumeopsis flava (DC.) Gagnep.;拟艾纳香;Yellow Blumeopsis ■

55863 Blumeorchis Szlach.(2003);布氏兰属■☆

55864 Blumeorchis crochetii (Guillaumin) Szlach.;布氏兰■☆

55865 Blumia Meyen ex Endl. = ? Podochilus Blume ■

55866 Blumia Nees ex Blume = Blumea DC.(保留属名)■●

55867 Blumia Nees(废弃属名) = Blumea DC.(保留属名)■●

55868 Blumia Nees(废弃属名) = Talauma Juss. ●

55869 Blumia Spreng. = Reinwardtia Blume ex Nees ●

55870 Blumia Spreng. = Saurauia Willd.(保留属名)●

55871 Blutaparon Raf.(1838);银头苋属;Silverhead ■●☆

55872 Blutaparon Raf. = Philoxerus R. Br. ■

55873 Blutaparon vermiculare (L.) Mears;蠕虫状银头苋;Saltweed, Silverweed ■☆

55874 Blutaparon wrightii (Hook. f. ex Maxim.) Mears = Philoxerus wrightii Hook. f. ex Maxim. ■

55875 Blutaparon wrightii (Hook. f.) Mears = Philoxerus wrightii Hook. f. ex Maxim. ■

55876 Blynia Arn. = Cynanchum L. ●■

55877 Blysmocarex N. A. Ivanova = Kobresia Willd. ■

55878 Blysmocarex macrantha (Boeck.) N. A. Ivanova = Kobresia macrantha Boeck. ■

55879 Blysmocarex macrantha (Boeck.) N. A. Ivanova subsp. nudicarpa (Y. C. Yang) D. S. Deng = Kobresia macrantha Boeck. ■

55880 Blysmocarex macrantha (Boeck.) N. A. Ivanova subsp. stolonifera (Y. C. Tang ex P. C. Li) D. S. Deng = Kobresia hohxilensis R. F. Huang ■

55881 Blysmocarex nudicarpa Y. C. Yang = Kobresia macrantha Boeck. var. nudicarpa (Y. C. Yang) P. C. Li ■

55882 Blysmocarex nudicarpa Y. C. Yang = Kobresia macrantha Boeck. ■

55883 Blysmopsis Oteng-Yeb. = Blysmus Panz. ex Schult.(保留属名)■

55884 Blysmopsis rufa (Huds.) Oteng-Yeb. = Blysmus rufus (Huds.) Link ■

55885 Blysmoschoenus Palla(1910);泡箭莎属■☆

55886 Blysmoschoenus buchtienii Palla;泡箭莎■☆

55887 Blysmus Panz. = Blysmus Panz. ex Schult.(保留属名)■

55888 Blysmus Panz. ex Roem. et Schult. = Blysmus Panz. ex Schult.(保留属名)■

55889 Blysmus Panz. ex Schult.(1824)(保留属名);扁穗草属(扁穗莞属);Blysmus, Flat-sedge, Flatspike ■

55890 Blysmus compressus (L.) Link = Blysmus compressus (L.) Panz. ex Link ■

55891 Blysmus compressus (L.) Panz. = Blysmus compressus (L.) Panz. ex Link ■

55892 Blysmus compressus (L.) Panz. ex Link;扁穗草;Flat-sedge, Flatspike, Flattened Blysmus ■

55893 Blysmus compressus (L.) Panz. ex Link subsp. brevifolius (Decne.) Kukkonen;短叶扁穗草■☆

55894 Blysmus exilis (Printz) N. A. Ivanova = Blysmus rufus (Huds.) Link ■

55895 Blysmus rufus (Huds.) Link;大果扁穗草(内蒙古扁穗草);Red Bulrush, Saltmarsh Flat-sedge Red Clubrush ■

55896 Blysmus rufus (Huds.) Link subsp. exilis Printz = Blysmus rufus (Huds.) Link ■

55897 Blysmus sinocompressus Ts. Tang et F. T. Wang;华扁穗草(华扁穗苔);China Flatspike, Chinese Blysmus ■

55898 Blysmus sinocompressus Ts. Tang et F. T. Wang var. nodosus Ts. Tang et F. T. Wang;节秆扁穗草(节秆扁穗苔);Jointed-culm Blysmus, Node China Flatspike ■

55899 Blysmus sinocompressus Ts. Tang et F. T. Wang var. tenuifolius Ts. Tang et F. T. Wang;细叶扁穗草;Slender Leaf Blysmus, Thinleaf Flatspike ■

55900 Blysmus tenuifolius Ts. Tang et F. T. Wang = Blysmus sinocompressus Ts. Tang et F. T. Wang var. tenuifolius Ts. Tang et F. T. Wang ■

55901 Blyttia Arn.(1838);布吕特萝藦属■☆

55902 Blyttia Arn. = Vincetoxicum Wolf ●■

55903 Blyttia Fr. = Cinna L. ■

55904 Blyttia arabica Arn. = Blyttia fruticulosa (Decne.) D. V. Field ■☆

55905 Blyttia fruticulosa (Decne.) D. V. Field;灌木状布吕特萝藦■☆

55906 Blyttia spiralis (Forssk.) D. V. Field et J. R. I. Wood;螺旋布吕特萝藦■☆

55907 Blyxa Noronha = Blyxa Noronha ex Thouars ■

55908 Blyxa Noronha ex Thouars(1806);水筛属(簀藻属);Blyxa, Waterbolt ■

55909 Blyxa Thouars = Blyxa Noronha ex Thouars ■

55910 Blyxa Thouars ex Rich. = Blyxa Noronha ex Thouars ■

55911 Blyxa alternifolia (Miq.) Hartog;互叶水筛■☆

55912 Blyxa aubertii (Miq.) Hartog var. echinosperma (C. B. Clarke) Cook et L. = Blyxa echinosperma (C. B. Clarke) Hook. f. ■

55913 Blyxa aubertii Rich.;无尾水筛(瘤果簀藻);Aubert Blyxa, Roundfruit Blyxa, Tailless Waterbolt ■

55914 Blyxa bicaudata Nakai = Blyxa echinosperma (C. B. Clarke) Hook. f. ■

55915 Blyxa ceratosperma Maxim. ex Asch. et Gürke = Blyxa echinosperma (C. B. Clarke) Hook. f. ■

55916 Blyxa ceylanica Hook. f. = Blyxa aubertii Rich. ■

55917 Blyxa ecaudata Hayata = Blyxa aubertii Rich. ■

55918 Blyxa echinosperma (C. B. Clarke) Hook. f.;有尾水筛(杉毛藻,台湾簀藻,鸭舌草,有刺水筛);Tail Waterbolt ■

55919 Blyxa echinosperma (C. B. Clarke) Hook. f. var. ceratosperma (Maxim. ex Asch. et Gürke) Sugim. = Blyxa echinosperma (C. B. Clarke) Hook. f. ■

55920 Blyxa griffithii Planch. ex Hook. f. = Blyxa aubertii Rich. ■

55921 Blyxa hexandra C. D. K. Cook et Luond;六蕊水筛■☆

55922 Blyxa japonica (Miq.) Maxim. ex Asch. et Gurke;水筛(日本簀藻);Japanese Blyxa, Waterbolt ■

55923 Blyxa laevissima Hayata = Blyxa japonica (Miq.) Maxim. ex Asch. et Gurke ■
55924 Blyxa leiosperma Koidz.;光滑水筛;Velvetseed Waterbolt ●
55925 Blyxa malayana Ridl. = Blyxa aubertii Rich. ■
55926 Blyxa octandra (Roxb.) Planch. ex Thwaites;八药水筛■
55927 Blyxa oryzetorum (Decne.) Hook. f. = Blyxa aubertii Rich. ■
55928 Blyxa radicans Ridl.;具根水筛■☆
55929 Blyxa roxburghii Rich. = Blyxa octandra (Roxb.) Planch. ex Thwaites ■
55930 Blyxa senegalensis Dandy;塞内加尔水筛■☆
55931 Blyxa shimadai Hayata = Blyxa echinosperma (C. B. Clarke) Hook. f. ■
55932 Blyxa somai Hayata = Blyxa echinosperma (C. B. Clarke) Hook. f. ■
55933 Blyxa zeylanica Hook. f. = Blyxa aubertii Rich. ■
55934 Blyxaceae Nakai = Hydrocharitaceae Juss.(保留科名)■
55935 Blyxaceae Nakai;水筛科●■
55936 Blyxopsis Kuntze = Blyxa Noronha ex Thouars ■
55937 Blyxopsis Kuntze = Enhydrias Ridl. ■
55938 Boadschia All. = Bohadschia Crantz ■☆
55939 Boadschia All. = Peltaria Jacq. ■☆
55940 Boaria A. DC. = Maytenus Molina ●
55941 Bobaea A. Rich. = Bobea Gaudich. ●☆
55942 Bobartella Gaertn. = Mariscus Gaertn. ■
55943 Bobartia L. (1753)(保留属名);博巴鸢尾属■☆
55944 Bobartia L.(保留属名) = Cyperus L. ■
55945 Bobartia Salisb. = Bobartia L.(保留属名)■☆
55946 Bobartia anceps Baker = Bobartia macrospatha Baker subsp. anceps (Baker) Strid ■☆
55947 Bobartia aphylla (L. f.) Ker Gawl.;无叶博巴鸢尾■☆
55948 Bobartia burchellii Baker = Bobartia macrospatha Baker ■☆
55949 Bobartia fasciculata J. B. Gillett ex Strid;簇生博巴鸢尾■☆
55950 Bobartia filiformis (L. f.) Ker Gawl.;线形博巴鸢尾■☆
55951 Bobartia gladiata (L. f.) Ker Gawl.;剑形博巴鸢尾■☆
55952 Bobartia gladiata (L. f.) Ker Gawl. subsp. major (G. J. Lewis) Strid;大剑形博巴鸢尾■☆
55953 Bobartia gladiata (L. f.) Ker Gawl. subsp. teres Strid;圆柱博巴鸢尾■☆
55954 Bobartia gladiata (L. f.) Ker Gawl. var. major G. J. Lewis = Bobartia gladiata (L. f.) Ker Gawl. subsp. major (G. J. Lewis) Strid ■☆
55955 Bobartia gracilis Baker;纤细博巴鸢尾■☆
55956 Bobartia indica L.;印度博巴鸢尾■☆
55957 Bobartia keetii Phillips = Bobartia aphylla (L. f.) Ker Gawl. ■☆
55958 Bobartia lilacina G. J. Lewis;紫丁香博巴鸢尾■☆
55959 Bobartia longicyma J. B. Gillett;大芽博巴鸢尾■☆
55960 Bobartia longicyma J. B. Gillett subsp. microflora Strid;小花大芽博巴鸢尾■☆
55961 Bobartia macrocarpa Strid;大果博巴鸢尾■☆
55962 Bobartia macrospatha Baker;大匙博巴鸢尾■☆
55963 Bobartia macrospatha Baker subsp. anceps (Baker) Strid;二棱大匙博巴鸢尾■☆
55964 Bobartia orientalis J. B. Gillett;东方博巴鸢尾■☆
55965 Bobartia orientalis J. B. Gillett subsp. occidentalis Strid;西方博巴鸢尾■☆
55966 Bobartia paniculata G. J. Lewis;圆锥博巴鸢尾■☆
55967 Bobartia parva J. B. Gillett;较小博巴鸢尾■☆
55968 Bobartia purcellii J. B. Gillett = Bobartia indica L. ■☆
55969 Bobartia robusta Baker;粗壮博巴鸢尾■☆
55970 Bobartia rufa Strid;浅红博巴鸢尾■☆
55971 Bobartia spathacea (Thunb.) Ker Gawl. = Bobartia indica L. ■☆
55972 Bobartia tubata J. B. Gillett = Bobartia macrospatha Baker ■☆
55973 Bobartia umbellata (Thunb.) Ker Gawl. = Moraea umbellata Thunb. ■☆
55974 Bobea Gaudich. (1830);哈岛茜属●☆
55975 Boberella E. H. L. Krause = Atropa L. ■
55976 Boberella E. H. L. Krause = Lycium L. ●
55977 Boberella E. H. L. Krause = Nicandra Adans.(保留属名)■
55978 Boberella E. H. L. Krause = Physalis L. ■
55979 Bobgunnia J. H. Kirkbr. et Wiersema = Swartzia Schreb.(保留属名)●☆
55980 Bobgunnia fistuloides (Harms) J. H. Kirkbr. et Wiersema;哈岛茜●☆
55981 Bobgunnia madagascariensis (Desv.) J. H. Kirkbr. et Wiersema;马岛茜●☆
55982 Bobrovia A. P. Khokhr. = Trifolium L. ■
55983 Bobrovia A. P. Khokhr. = Ursifolium Doweld ■
55984 Bobu Adans. = Symplocos Jacq. ●
55985 Bobua Adans. = Symplocos Jacq. ●
55986 Bobua DC. = Bobu Adans. ●
55987 Bobua adinandrifolia (Hayata) Kaneh. et Sasald = Symplocos congesta Benth. ●
55988 Bobua austrosinensis Migo = Symplocos sumuntia Buch. -Ham. ex D. Don ●
55989 Bobua confusa (Brand) Kaneh. et Sasaki = Symplocos confusa Brand ●
55990 Bobua confusa (Brand) Kaneh. et Sasaki = Symplocos pendula Wight var. hirtistylus (C. B. Clarke) Noot. ●
55991 Bobua congesta (Bench.) Migo = Symplocos congesta Benth. ●
55992 Bobua crenatifolia Yamam. = Symplocos nokoensis (Hayata) Kaneh. ●
55993 Bobua divaricativena (Hayata) Kaneh. et Sasald = Symplocos cochinchinensis (Lour.) S. Moore var. laurina (Retz.) Raizada ●
55994 Bobua eriobotryifolia? = Symplocos stellaris Brand ●
55995 Bobua glauca (Thunb.) Nakai = Symplocos glauca (Thunb.) Koidz. ●
55996 Bobua groffii (Merr.) Migo = Symplocos groffii Merr. ●
55997 Bobua ilicifolia (Hayata) Kaneh. et Sasaki = Symplocos lucida (Thunb.) Siebold et Zucc ●
55998 Bobua ilicifolia (Hayata) Kaneh. et Sasaki = Symplocos setchuensis Brand ●
55999 Bobua kotoensis (Hayata) Yamam. = Symplocos cochinchinensis (Lour.) S. Moore var. philippinensis (Brand) Noot. ●
56000 Bobua modesta (Brand) Yamam. = Symplocos modesta Brand ●
56001 Bobua nakaii (Hayata) Kaneh. et Sasaki = Symplocos congesta Benth. ●
56002 Bobua neriifolia Miers = Symplocos glauca (Thunb.) Koidz. ●
56003 Bobua nokoensis (Hayata) Kaneh. et Sasaki = Symplocos nokoensis (Hayata) Kaneh. ●
56004 Bobua phaeophylla (Hayata) Kaneh. et Sasaki = Symplocos congesta Benth. ●
56005 Bobua pseudolancifolia Hatus. = Symplocos lancifolia Siebold et Zucc. ●
56006 Bobua stellaris (Brand) Migo = Symplocos stellaris Brand ●

56007 Bobua taiwaniana Hatus. = Symplocos congesta Benth. ●
56008 Bobua theifolia Kaneh. et Sasaki = Symplocos congesta Benth. ●
56009 Bobua wikstroemiifolia (Hayata) Kaneh. et Sasaki = Symplocos wikstroemiifolia Hayata ●
56010 Boca Vell. = Banara Aubl. ●☆
56011 Bocagea A. St. -Hil. (1825);花纹木属●☆
56012 Bocagea Blume = Sageraea Dalzell ●☆
56013 Bocagea alba A. St. -Hil.;花纹木●☆
56014 Bocagea canescens Spruce ex Benth.;灰花纹木●☆
56015 Bocageopsis R. E. Fr. (1931);类花纹木属●☆
56016 Bocageopsis canescens R. E. Fr.;灰类花纹木●☆
56017 Bocageopsis mattogrossensis (R. E. Fr.) R. E. Fr.;类花纹木●☆
56018 Bocageopsis mattogrossensis R. E. Fr. = Bocageopsis mattogrossensis (R. E. Fr.) R. E. Fr. ●☆
56019 Bocageopsis multiflora (Mart.) R. E. Fr.;多花类花纹木●☆
56020 Bocageopsis pleiosperma Maas;光果类花纹木●☆
56021 Bocco Steud. = Bocoa Aubl. ●☆
56022 Bocco Steud. = Swartzia Schreb. (保留属名)●☆
56023 Bocconia L. (1753);美罂粟属(博考尼属,博克尼属,博氏罂粟属,羽脉博落回属);False Nettle,Plume Poppy ●■☆
56024 Bocconia Plum. ex L. = Bocconia L. ●■☆
56025 Bocconia arborea S. Watson;树状美罂粟(博考尼,木状美罂粟,肖博落回)●☆
56026 Bocconia cordata Willd. = Macleaya cordata (Willd.) R. Br. ■
56027 Bocconia frutescens L.;灌木美罂粟(博克尼木,灌状美罂粟,树博考尼);Parrotweed,Parrot-weed,Tree Celandine,Tree-celandine ●☆
56028 Bocconia gracilis L.;细美罂粟●☆
56029 Bocconia microcarpa Maxim. = Macleaya microcarpa (Maxim.) Fedde ■
56030 Bockia Scop. = Mouriri Aubl. ●☆
56031 Bocoa Aubl. (1775);博科铁木豆属●☆
56032 Bocoa Aubl. = Swartzia Schreb. (保留属名)●☆
56033 Bocoa edulis Baill.;可食博科铁木豆(太平洋胡桃);Polynesian Chestnut,Tahiti Chestnut ●☆
56034 Bocoa prouacensis Aubl.;博科铁木豆●☆
56035 Bocquillonia Baill. (1862);新卡大戟属●☆
56036 Bocquillonia arborea Airy Shaw;树状新卡大戟●☆
56037 Bocquillonia brachypoda Baill.;短足新卡大戟●☆
56038 Bocquillonia grandidens Baill.;大齿新卡大戟●☆
56039 Bocquillonia lucidula Airy Shaw;光亮新卡大戟●☆
56040 Bodinierella cavaleriei H. Lév. = Enkianthus chinensis Franch. ●
56041 Bodinieria H. Lév. et Vaniot = Boenninghausenia Rchb. ex Meisn. (保留属名)●■
56042 Bodinteriella H. Lév. = Enkianthus Lour. ●
56043 Bodwichia Walp. = Bowdichia Kunth ●☆
56044 Boea Comm. ex Lam. (1785);旋蒴苣苔属(牛耳草属);Boea ■
56045 Boea Lam. = Boea Comm. ex Lam. ■
56046 Boea arachnoidea Diels = Ornithoboea arachnoidea (Diels) Craib ■
56047 Boea birmanica Craib = Trisepalum birmanicum (Craib) B. L. Burtt ■
56048 Boea cavaleriei H. Lév. et Vant = Rhabdothamnopsis sinensis Hemsl. ■
56049 Boea chaffanjoni H. Lév. = Paraboea sinensis (Oliv.) B. L. Burtt ●
56050 Boea clarkeana Hemsl.;大花旋蒴苣苔(散血草,旋蒴苣苔,岩巴绿);C. B. Clarke Boea ■
56051 Boea crassifolia Hemsl. = Paraboea crassifolia (Hemsl.) B. L. Burtt ■
56052 Boea darrisii H. Lév. = Ornithoboea feddei (H. Lév.) B. L. Burtt ■
56053 Boea densihispidula S. B. Zhou et X. H. Guo;安徽旋蒴苣苔;Anhui Boea ■
56054 Boea densihispidula S. B. Zhou et X. H. Guo = Boea clarkeana Hemsl. ■
56055 Boea dictyoneura Hance = Paraboea dictyoneura (Hance) B. L. Burtt ■
56056 Boea elephantopoides Chun = Boea philippensis C. B. Clarke ■
56057 Boea feddei H. Lév. = Ornithoboea feddei (H. Lév.) B. L. Burtt ■
56058 Boea glutinosa Hand. -Mazz. = Paraboea glutinosa (Hand. -Mazz.) K. Y. Pan ■
56059 Boea glutinosa Hand. -Mazz. = Paraboea martinii H. Lév. et Vaniot) B. L. Burtt ■
56060 Boea hainanensis Chun = Paraboea hainanensis (Chun) B. L. Burtt ■
56061 Boea hancei C. B. Clarke = Paraboea dictyoneura (Hance) B. L. Burtt ■
56062 Boea harroviana Craib = Paraboea dictyoneura (Hance) B. L. Burtt ■
56063 Boea hygrometrica (Bunge) R. Br.;旋蒴苣苔(八宝茶,地虎皮,翻魂草,蝴蝶草,还魂草,猫耳草,猫耳朵,绵还阳草,牛耳草,牛舌头,石胆草,石花子,崖青叶);Australian African Violet,Cas' ear Boea,Hygrometric Boea ■
56064 Boea macrophylla Drake = Paraboea sinensis (Oliv.) B. L. Burtt ●
56065 Boea mairei H. Lév. = Boea clarkeana Hemsl. ■
56066 Boea martinii (H. Lév. et Vaniot) H. Lév. = Paraboea martinii (H. Lév. et Vaniot) B. L. Burtt ■
56067 Boea martinii (H. Lév.) H. Lév. = Paraboea martinii (H. Lév. et Vaniot) B. L. Burtt ■
56068 Boea paniculata Hand. -Mazz. = Trisepalum birmanicum (Craib) B. L. Burtt ■
56069 Boea philippensis C. B. Clarke;地胆旋蒴苣苔(地胆苣苔);Philippine Boea ■
56070 Boea poilanei Pellegr. = Boea philippensis C. B. Clarke ■
56071 Boea rubicunda H. Lév. = Rhabdothamnopsis sinensis Hemsl. ■
56072 Boea rufescens Franch. = Paraboea rufescens (Franch.) B. L. Burtt ■
56073 Boea rufescens Franch. var. seguini (H. Lév. et Vaniot) H. Lév. = Paraboea rufescens (Franch.) B. L. Burtt ■
56074 Boea swinhoei Hance = Paraboea swinhoii (Hance) B. L. Burtt ●
56075 Boea thirionii H. Lév. = Paraboea thirionii (H. Lév.) B. L. Burtt ■
56076 Boea umbellata Drake = Paraboea rufescens (Franch.) B. L. Burtt var. umbellata (Drake) K. Y. Pan ■
56077 Boebera Willd. = Dyssodia Cav. ■☆
56078 Boebera papposa (Vent.) Rydb. = Dyssodia papposa (Vent.) Hitchc. ■☆
56079 Boeberastrum (A. Gray) Rydb. (1915);肉羽菊属■☆
56080 Boeberastrum Rydb. = Boeberastrum (A. Gray) Rydb. ■☆
56081 Boeberastrum anthemidifolium Rydb.;肉羽菊■☆
56082 Boeberoides (DC.) Strother(1986);多腺菊属■●☆
56083 Boeberoides grandiflora (DC.) Strother;多腺菊■☆
56084 Boechera Á. Löve et D. Löve = Arabis L. ●■
56085 Boecherarctica Á. Löve = Saxifraga L. ■
56086 Boecherarctica Á. Löve(1984);格陵兰虎耳草属■☆
56087 Boeckeleria T. Durand = Tetraria P. Beauv. ■☆
56088 Boeckeleria T. Durand(1888);伯克莎属■☆

56089 Boeckeleria nigrovaginata (Nees) Pfeiff. = Tetraria nigrovaginata (Nees) C. B. Clarke ■☆
56090 Boeckeleria nitrovaginata H. Pfeiff.;伯克莎■☆
56091 Boeckhia Kunth = Hypodiscus Nees(保留属名)■☆
56092 Boeckhia laevigata Kunth = Hypodiscus laevigatus (Kunth) H. P. Linder ■☆
56093 Boeckhia striata Kunth = Hypodiscus striatus (Kunth) Mast. ■☆
56094 Boehmeria Jacq. (1760);苎麻属;China Grass, False Nettle, Falsenettle, False-nettle, Ramie ●
56095 Boehmeria allophylla W. T. Wang;异叶苎麻;Diffferleaf Ramie ■
56096 Boehmeria arenicola Satake;沙丘苎麻■☆
56097 Boehmeria austrina Small = Boehmeria cylindrica (L.) Sw. ●☆
56098 Boehmeria bicuspis C. J. Chen;双尖苎麻;Bicuspidate Falsenettle, Doubletine Ramie ●
56099 Boehmeria bicuspis C. J. Chen = Boehmeria umbrosa (Hand.-Mazz.) W. T. Wang ■
56100 Boehmeria biloba Wedd.;双裂苎麻(二裂苎麻)●☆
56101 Boehmeria blinii H. Lév. = Boehmeria zollingeriana Wedd. var. blinii (H. Lév.) C. J. Chen ●
56102 Boehmeria blinii H. Lév. var. podocarpa W. T. Wang = Boehmeria wattersii (Hance) et Yuen P. Yang ●
56103 Boehmeria blinii H. Lév. var. podocarpa W. T. Wang = Boehmeria zollingeriana Wedd. var. podocarpa (W. T. Wang) W. T. Wang et C. J. Chen ●
56104 Boehmeria bodinieri H. Lév. = Laportea bulbifera (Siebold et Zucc.) Wedd. ■
56105 Boehmeria boninensis Nakai;小笠原苎麻●☆
56106 Boehmeria canescens Wedd. = Boehmeria macrophylla Hornem. var. canescens (Wedd.) D. G. Long ●
56107 Boehmeria caudata J. J. Sm.;尾状苎麻●☆
56108 Boehmeria chiangmaiensis Yahara = Boehmeria siamensis Craib ●
56109 Boehmeria chingshuishaniana S. S. Ying = Boehmeria hwaliensis Y. C. Liu et F. Y. Lu ●
56110 Boehmeria clidemioides Miq.;白面苎麻(水苎麻草,序叶苎麻);Clidemialike Ramie, White Falsenettle ■
56111 Boehmeria clidemioides Miq. var. cinerascens H. Hara = Boehmeria clidemioides Miq. ■
56112 Boehmeria clidemioides Miq. var. diffusa (Wedd.) Hand.-Mazz.;序叶苎麻(合麻仁,米麻,水苏麻,水苎麻,团水麻,野麻藤,叶序苎麻);Diffuse White Falsenettle ■
56113 Boehmeria clidemioides Miq. var. diffusa (Wedd.) Hand.-Mazz. = Boehmeria clidemioides Miq. ■
56114 Boehmeria clidemioides Miq. var. platyphylloides Yahara = Boehmeria clidemioides Miq. ■
56115 Boehmeria clidemioides Miq. var. umbrosa Hand.-Mazz. = Boehmeria umbrosa (Hand.-Mazz.) W. T. Wang ■
56116 Boehmeria comosa Wedd. = Boehmeria clidemioides Miq. var. diffusa (Wedd.) Hand.-Mazz. ■
56117 Boehmeria comosa Wedd. = Boehmeria clidemioides Miq. ■
56118 Boehmeria conica C. J. Chen;锥序苎麻●
56119 Boehmeria cylindrica (L.) Sw.;小穗苎麻(小苎麻);False Nettle, Small-spike False Nettle, Stingless Nettle ●☆
56120 Boehmeria cylindrica (L.) Sw. var. drummondiana (Wedd.) Wedd. = Boehmeria cylindrica (L.) Sw. ●☆
56121 Boehmeria cylindrica (L.) Sw. var. scabra Porter = Boehmeria cylindrica (L.) Sw. ●☆
56122 Boehmeria decurrens Small = Boehmeria cylindrica (L.) Sw. ●☆
56123 Boehmeria delavayi Gagnep. = Pouzolzia elegans Wedd. var. delavayi (Gagnep.) W. T. Wang ●
56124 Boehmeria delavayi Gagnep. = Pouzolzia elegans Wedd. ●
56125 Boehmeria densiflora Hook. et Arn.;密花苎麻(粗糠壳,木苎麻,山水柳,虾公须);Denseflower Falsenettle, Denseflowered False-nettle, Dense-flowered False-nettle ●
56126 Boehmeria densiglomerata W. T. Wang;密球苎麻(土麻仁,野紫苏);Denseball Ramie ■
56127 Boehmeria depauperata Wedd. = Boehmeria glomerulifera Miq. ●
56128 Boehmeria diffusa Wedd. = Boehmeria clidemioides Miq. var. diffusa (Wedd.) Hand.-Mazz. ■
56129 Boehmeria diffusa Wedd. = Boehmeria clidemioides Miq. ■
56130 Boehmeria diffusa Wedd. var. strigosa Wedd. = Boehmeria clidemioides Miq. var. diffusa (Wedd.) Hand.-Mazz. ■
56131 Boehmeria dolichostachya W. T. Wang;长序苎麻;Longhead Ramie, Long-stachys Falsenettle ●
56132 Boehmeria dolichostachya W. T. Wang var. mollis (W. T. Wang) W. T. Wang et C. J. Chen;绢毛长序苎麻(柔毛苎麻)●
56133 Boehmeria drummondiana Wedd. = Boehmeria cylindrica (L.) Sw. ●☆
56134 Boehmeria dura Satake;硬苎麻●☆
56135 Boehmeria egregia Satake;优秀苎麻●☆
56136 Boehmeria elegantula (W. W. Sm. et Jeffrey) Hand.-Mazz. = Pouzolzia elegans Wedd. ●
56137 Boehmeria elegantula Hand.-Mazz. = Pouzolzia elegans Wedd. ●
56138 Boehmeria erythropoda Miq. = Boehmeria macrophylla D. Don ●
56139 Boehmeria esquirolii H. Lév. = Maoutia puya (Hook.) Wedd. ●
56140 Boehmeria formosana Hayata;海岛苎麻(台湾野苎麻,台湾苎麻,西博苎麻,席氏苎麻);Formosa Falsenettle, Formosan False-nettle, Siebold Ramie, Taiwan Falsenettle, Taiwan Ramie ●
56141 Boehmeria formosana Hayata var. fuzhouensis W. T. Wang. = Boehmeria formosana Hayata var. stricta (C. H. Wright) C. J. Chen ●
56142 Boehmeria formosana Hayata var. stricta (C. H. Wright) C. J. Chen;福州苎麻;Fuzhou Falsenettle ●
56143 Boehmeria frondosa D. Don = Oreocnide frutescens (Thunb.) Miq. ●
56144 Boehmeria frondosa D. Don = Villebrunea frutescens (Thunb.) Blume ●
56145 Boehmeria frutescens (Thunb.) Thunb.;山苎麻(苴,青麻,青苎麻,纻);Shrubby Falsenettle, Shrubby False-nettle ●
56146 Boehmeria frutescens (Thunb.) Thunb. = Oreocnide frutescens (Thunb.) Miq. ●
56147 Boehmeria frutescens (Thunb.) Thunb. = Villebrunea frutescens (Thunb.) Blume ●
56148 Boehmeria frutescens Thunb. = Boehmeria frutescens (Thunb.) Thunb. ●
56149 Boehmeria frutescens Thunb. = Boehmeria nivea (L.) Gaudich. var. nipononivea (Koidz.) W. T. Wang ●
56150 Boehmeria frutescens Thunb. var. concolor (Makino) Nakai = Boehmeria nivea (L.) Gaudich. var. tenacissima (Gaudich.) Miq. ●
56151 Boehmeria frutescens Thunb. var. viridula (Yamam.) Suzuki = Boehmeria nivea (L.) Gaudich. var. nipononivea (Koidz.) W. T. Wang ●
56152 Boehmeria frutescens Thunb. var. viridula Yamam. = Boehmeria nivea (L.) Gaudich. var. viridula Yamam. ●
56153 Boehmeria fruticosa Gaudich. = Oreocnide frutescens (Thunb.) Miq. ●

56154 Boehmeria gigantea Satake = Boehmeria holosericea Blume ●

56155 Boehmeria glomerulifera Miq.;腋序苎麻(腋花苎麻,腋球苎麻);Glomerate Falsenettle ●

56156 Boehmeria glomerulifera Miq. = Boehmeria malabarica Wedd. ●

56157 Boehmeria glomerulifera Miq. var. leioclada W. T. Wang = Boehmeria glomerulifera Miq. ●

56158 Boehmeria glomerulifera Miq. var. leioclada W. T. Wang = Boehmeria malabarica Wedd. var. leioclada (W. T. Wang) W. T. Wang ●

56159 Boehmeria gracilis C. H. Wright = Boehmeria spicata (Thunb.) Thunb. ■

56160 Boehmeria grandifolia Wedd.;大叶苎麻●

56161 Boehmeria grandifolia Wedd. = Boehmeria japonica (L. f.) Miq. var. longispica (Steud.) Yahara ●

56162 Boehmeria grandifolia Wedd. = Boehmeria japonica (L. f.) Miq. ●■

56163 Boehmeria grandifolia Wedd. = Boehmeria longispica Steud. ●

56164 Boehmeria grandis A. Heller;夏威夷苎麻;Hawaiian False Nettle ●☆

56165 Boehmeria hamiltoniana Wedd.;细序苎麻;Hamilton Ramie, Thininflorescence Falsenettle ●

56166 Boehmeria hatusimae Satake;初岛苎麻●☆

56167 Boehmeria heteroidea Blume = Boehmeria zollingeriana Wedd. ●

56168 Boehmeria heteroidea Blume var. latifolia Gagnep. = Boehmeria zollingeriana Wedd. ●

56169 Boehmeria hirtella Satake;多毛苎麻●☆

56170 Boehmeria holosericea Blume = Boehmeria japonica (L. f.) Miq. ●

56171 Boehmeria holosericea Blume var. izuosimensis (Satake) Satake et M. Mizush. = Boehmeria izuosimensis Satake ●☆

56172 Boehmeria holosericea Blume var. strigosa W. T. Wang = Boehmeria dolichostachya W. T. Wang ●

56173 Boehmeria holosericea Blume var. strigosa W. T. Wang = Boehmeria strigosifolia W. T. Wang ●

56174 Boehmeria holosericea Blume var. strigosifolia W. T. Wang = Boehmeria strigosifolia W. T. Wang ●

56175 Boehmeria hwaliensis Y. C. Liu et F. Y. Lu;花莲苎麻;Hualian False-nettle, Hwalien False-nettle ●

56176 Boehmeria hwaliensis Y. C. Liu et F. Y. Lu = Boehmeria densiflora Hook. et Arn. ●

56177 Boehmeria hypoleuca (Steud.) Hochst. ex A. Rich. = Debregeasia saeneb (Forssk.) Hepper et J. R. I. Wood ●

56178 Boehmeria hypoleuca (Steud.) Hochst. ex A. Rich. = Debregeasia salicifolia (D. Don) Rendle ●

56179 Boehmeria ingjiangensis W. T. Wang;盈江苎麻;Yingjiang Falsenettle, Yingjiang Ramie ●

56180 Boehmeria izuosimensis Satake;佐竹苎麻●☆

56181 Boehmeria japonica (L. f.) Miq.;野线麻(日本苎麻,薮苎麻);Japan Ramie ●■

56182 Boehmeria japonica (L. f.) Miq. = Boehmeria longispica Steud. ●

56183 Boehmeria japonica (L. f.) Miq. var. appendiculata (Blume) Yahara = Boehmeria japonica (L. f.) Miq. ●■

56184 Boehmeria japonica (L. f.) Miq. var. appendiculata (Blume) Yahara = Boehmeria japonica (L. f.) Miq. var. longispica (Steud.) Yahara ●

56185 Boehmeria japonica (L. f.) Miq. var. longispica (Steud.) Yahara = Boehmeria longispica Steud. ●

56186 Boehmeria japonica (L. f.) Miq. var. longispica (Steud.) Yahara = Boehmeria japonica (L. f.) Miq. ●■

56187 Boehmeria japonica (L. f.) Miq. var. platanifolia Maxim. = Boehmeria tricuspis (Hance) Makino ●

56188 Boehmeria japonica Miq. var. platanifolia Maxim. = Boehmeria tricuspis (Hance) Makino ●

56189 Boehmeria kiusiana Satake;九州苎麻●☆

56190 Boehmeria kiyozumensis Satake;京泉氏苎麻●☆

56191 Boehmeria lanceolata Ridl.;北越苎麻●

56192 Boehmeria leiophylla W. T. Wang;光叶苎麻;Brightleaf Ramie, Shining-leaf Falsenettle ●

56193 Boehmeria leiophylla W. T. Wang = Boehmeria glomerulifera Miq. ●

56194 Boehmeria lohuiensis S. S. Chien;琼海苎麻;Qionghai Falsenettle, Qionghai Ramie ●

56195 Boehmeria longispica Steud.;长穗苎麻(长叶苎麻,大蛮婆草,大水麻,大叶苎麻,方麻,伏麻,火麻,火麻风,蒙自苎麻,山麻,山苎,水禾麻,水升麻,水苏麻,小赤麻,小苎麻,野线麻,野苎麻,重齿小赤麻);Bigleaf Ramie, Longleaf Falsenettle, Taiwan Falsenettle ●

56196 Boehmeria longispica Steud. = Boehmeria japonica (L. f.) Miq. var. longispica (Steud.) Yahara ●

56197 Boehmeria longispica Steud. = Boehmeria japonica (L. f.) Miq. ●■

56198 Boehmeria longispica Steud. var. dura (Satake) Satake = Boehmeria dura Satake ●☆

56199 Boehmeria longispica Steud. var. robusta (Nakai et Satake) Satake = Boehmeria robusta Nakai et Satake ●

56200 Boehmeria macrophylla D. Don = Boehmeria penduliflora Wedd. ex D. G. Long ●

56201 Boehmeria macrophylla D. Don var. canescens (Wedd.) D. G. Long = Boehmeria macrophylla Hornem. var. canescens (Wedd.) D. G. Long ●

56202 Boehmeria macrophylla D. Don var. dongtouensis W. T. Wang = Boehmeria macrophylla D. Don ●

56203 Boehmeria macrophylla D. Don var. loochooensis (Wedd.) W. T. Wang = Boehmeria penduliflora Wedd. var. loochoensis (Wedd.) W. T. Wang ●

56204 Boehmeria macrophylla D. Don var. rotundifolia (D. Don) W. T. Wang = Boehmeria macrophylla Hornem. var. rotundifolia (D. Don) W. T. Wang ●

56205 Boehmeria macrophylla D. Don var. scabrella (Roxb.) D. G. Long = Boehmeria macrophylla Hornem. var. scabrella (Roxb.) D. G. Long ●

56206 Boehmeria macrophylla Hornem.;水苎麻(八棱马,革芫,长叶苎麻,大接骨,大糯叶,大叶苎麻,鸽子枕头,癞蛤蟆果,米顶心,木苎麻,水麻,水细麻,折骨藤,折折藤,朱顶心);African Falsenettle, Bigleaf Falsenettle, Big-leaved False-nettle, Broad-leaved False-nettle, Largeleaf Falsenettle, Water Ramie ●■

56207 Boehmeria macrophylla Hornem. var. canescens (Wedd.) D. G. Long;灰绿水苎麻;Canescent Falsenettle, Greyish Largeleaf Falsenettle ●

56208 Boehmeria macrophylla Hornem. var. dongtouensis W. T. Wang;洞头水苎麻;Dongtou Falsenettle ●■

56209 Boehmeria macrophylla Hornem. var. dongtouensis W. T. Wang = Boehmeria macrophylla Hornem. ●■

56210 Boehmeria macrophylla Hornem. var. loochooensis (Wedd.) W. T. Wang;密花细辛(水苎麻)●

56211 Boehmeria macrophylla Hornem. var. rotundifolia (D. Don) W. T. Wang;圆叶水苎麻(圆叶苎麻);Round-leaf Largeleaf Falsenettle ●

56212 Boehmeria macrophylla Hornem. var. scabrella (Roxb.) D. G. Long;糙叶水苎麻(糙叶苎麻);Roughleaf Largeleaf Falsenettle, Scabrous Falsenettle ●

56213 Boehmeria macrostachya (Wight) F. M. Bailey = Boehmeria platyphylla D. Don ●

56214 Boehmeria malabarica Wedd.;腋球苎麻(腋序苎麻);Balabar Ramie, Glomerate Falsenettle ●

56215 Boehmeria malabarica Wedd. = Boehmeria glomerulifera Miq. ●

56216 Boehmeria malabarica Wedd. var. leioclada (W. T. Wang) W. T. Wang;光枝苎麻;Smoothbranch Glomerate Falsenettle, Smoothbranched Largeleaf Falsenettle ●

56217 Boehmeria malabarica Wedd. var. leioclada (W. T. Wang) W. T. Wang = Boehmeria glomerulifera Miq. ●

56218 Boehmeria martinii H. Lév. = Pilea martinii (H. Lév.) Hand.-Mazz. ■

56219 Boehmeria maximowiczii Nakai = Boehmeria tricuspis (Hance) Makino ●

56220 Boehmeria maximowiczii Nakai et Satake = Boehmeria tricuspis (Hance) Makino ●

56221 Boehmeria minor Satake;较小苎麻●☆

56222 Boehmeria nakashimae Yahara;长岛苎麻●☆

56223 Boehmeria nepalensis Wedd. = Pouzolzia sanguinea (Blume) Merr. var. nepalensis (Wedd.) H. Hara ●

56224 Boehmeria nepalensis Wedd. = Pouzolzia sanguinea (Blume) Merr. ●

56225 Boehmeria niponpremonivea Koidz. = Boehmeria nivea (L.) Gaudich. var. nipononivea (Koidz.) W. T. Wang ●

56226 Boehmeria nipononivea Koidz. = Boehmeria nivea (L.) Gaudich. var. tenacissima (Gaudich.) Miq. ●

56227 Boehmeria nipononivea Koidz. f. concolor (Makino) Kitam. ex Satake = Boehmeria nivea (L.) Gaudich. var. concolor Makino ●☆

56228 Boehmeria nipononivea Koidz. var. concolor (Makino) Ohwi = Boehmeria nivea (L.) Gaudich. var. concolor Makino ●☆

56229 Boehmeria nivea (L.) Gaudich.;苎麻(白麻,白苎麻,大麻,家麻,家苎麻,苴,青麻,天青地白,线麻,野麻,野苎麻,元麻,纻,苎仔,苧);Canton Linen, China Grass, China-grass, Chinese Grass, Chinese Silk Plant, Ramie, Silk-plant ●

56230 Boehmeria nivea (L.) Gaudich. = Urtica utilis L. ●

56231 Boehmeria nivea (L.) Gaudich. subsp. nipononivea (Koidz.) Kitam. = Boehmeria nivea (L.) Gaudich. var. nipononivea (Koidz.) W. T. Wang ●

56232 Boehmeria nivea (L.) Gaudich. subsp. nipononivea (Koidz.) Kitam. f. concolor (Makino) Kitam. = Boehmeria nivea (L.) Gaudich. var. concolor Makino ●☆

56233 Boehmeria nivea (L.) Gaudich. var. candicans Wedd.;楔基苎麻;Ramie, Silk-plant ●

56234 Boehmeria nivea (L.) Gaudich. var. candicans Wedd. = Boehmeria nivea (L.) Gaudich. var. tenacissima (Gaudich.) Miq. ●

56235 Boehmeria nivea (L.) Gaudich. var. concolor Makino = Boehmeria nivea (L.) Gaudich. var. tenacissima (Gaudich.) Miq. ●

56236 Boehmeria nivea (L.) Gaudich. var. crassifolia C. H. Wright.;云南苎麻;Chinese Silk Plant, Ramie, Snow-white False-nettle ●

56237 Boehmeria nivea (L.) Gaudich. var. crassifolia C. H. Wright. = Maoutia puya (Hook.) Wedd. ●

56238 Boehmeria nivea (L.) Gaudich. var. nipononivea (Koidz.) W. T. Wang;贴毛苎麻(伏毛苎麻);Strigose Ramie ●

56239 Boehmeria nivea (L.) Gaudich. var. nipononivea (Koidz.) W. T. Wang = Boehmeria nivea (L.) Gaudich. var. tenacissima (Gaudich.) Miq. ●

56240 Boehmeria nivea (L.) Gaudich. var. tenacissima (Gaudich.) Miq.;青叶苎麻(单色苎麻,光叶山苎麻,光叶苎麻,青苎麻);Concolorleaved Falsenettle ●

56241 Boehmeria nivea (L.) Gaudich. var. viridula Yamam.;微绿苎麻(绿背山苎麻,苎仔薯);Green Ramie, Greenish Ramie, Viridleaved Falsenettle ●

56242 Boehmeria nivea (L.) Gaudich. var. viridula Yamam. = Boehmeria nivea (L.) Gaudich. var. tenacissima (Gaudich.) Miq. ●

56243 Boehmeria nivea (L.) Hook. et Arn. = Boehmeria nivea (L.) Gaudich. ●

56244 Boehmeria oblongifolia W. T. Wang;长圆苎麻;Oblongleaf Falsenettle, Oblongleaf Ramie ●

56245 Boehmeria oblongifolia W. T. Wang. = Boehmeria glomerulifera Miq. ●

56246 Boehmeria ovalis Miq. = Pouzolzia sanguinea (Blume) Merr. ●

56247 Boehmeria pachystachya Satake;粗穗苎麻●☆

56248 Boehmeria pannosa Nakai et Satake = Boehmeria holosericea Blume ●

56249 Boehmeria paraspicata Nakai = Boehmeria gracilis C. H. Wright ■

56250 Boehmeria paraspicata Nakai = Boehmeria spicata (Thunb.) Thunb. ■

56251 Boehmeria parvifolia Wedd. = Droguetia iners (Forssk.) Schweinf. subsp. urticoides (Wight) Friis et Wilmot-Dear ■

56252 Boehmeria pauciflora (Steud.) Blume = Droguetia iners (Forssk.) Schweinf. ■☆

56253 Boehmeria penduliflora Wedd. = Boehmeria penduliflora Wedd. ex D. G. Long ●

56254 Boehmeria penduliflora Wedd. ex D. G. Long;长叶苎麻(米顶心,水麻,水细麻,折听藤);Longleaf Falsenettle, Longleaf Ramie ●

56255 Boehmeria penduliflora Wedd. ex D. G. Long var. loochooensis (Wedd.) W. T. Wang = Boehmeria densiflora Hook. et Arn. ●

56256 Boehmeria penduliflora Wedd. var. loochoensis (Wedd.) W. T. Wang = Boehmeria penduliflora Wedd. ex D. G. Long var. loochooensis (Wedd.) W. T. Wang ●

56257 Boehmeria pilosiuscula (Blume) Hassk.;疏毛苎麻(海南木苎麻,华南苎麻,疏毛水苎麻);Hainan Falsenettle, Hainan False-nettle, Laxhair Ramie, Laxhairy Falsenettle, Lax-hairy False-nettle ●■

56258 Boehmeria pilushanensis Y. C. Liu et F. Y. Lu;毕禄山苎麻;Bilushan False-nettle, Pilushan False-nettle ●

56259 Boehmeria pilushanensis Y. C. Liu et F. Y. Lu = Boehmeria japonica (L. f.) Miq. ●■

56260 Boehmeria platanifolia (Maxim.) Franch. et Sav. ex C. H. Wright;悬铃木叶苎麻(八角麻,大水庇,方麻,龟叶麻,火麻,山麻,水苎麻,透骨风,悬铃叶苎麻,野苎麻);Planetreeleaf Falsenettle ●

56261 Boehmeria platanifolia (Maxim.) Franch. et Sav. ex C. H. Wright = Boehmeria tricuspis (Hance) Makino ●

56262 Boehmeria platanifolia (Maxim.) Franch. et Sav. ex C. H. Wright var. silvestrii Pamp. = Boehmeria silvestrii (Pamp.) W. T. Wang ■

56263 Boehmeria platanifolia Franch. = Boehmeria tricuspis (Hance) Makino ●

56264 Boehmeria platanifolia Franch. et Sav. = Boehmeria tricuspis (Hance) Makino ●

56265 Boehmeria platanifolia Franch. et Sav. var. silvestrii Pamp. = Boehmeria silvestrii (Pamp.) W. T. Wang ■

56266 Boehmeria plataphylla (Maxim.) Franch. et Sav. ex C. H. Wright var. tricuspis Hance = Boehmeria tricuspis (Hance) Makino ●

56267 Boehmeria platyphylla Buch. -Ham. ex D. Don = Boehmeria macrophylla Hornem. ●■

56268 Boehmeria platyphylla Buch. -Ham. ex D. Don var. angolensis Rendle = Boehmeria macrophylla Hornem. ●■

56269 Boehmeria platyphylla Buch. -Ham. ex D. Don var. ugandensis Rendle = Boehmeria macrophylla Hornem. ●■

56270 Boehmeria platyphylla D. Don = Boehmeria macrophylla Hornem. ●■

56271 Boehmeria platyphylla D. Don var. canescens (Wedd. ex Blume) Wedd. = Boehmeria macrophylla Hornem. var. canescens (Wedd.) D. G. Long ●

56272 Boehmeria platyphylla D. Don var. canescens (Wedd.) Wedd. = Boehmeria macrophylla Hornem. var. canescens (Wedd.) D. G. Long ●

56273 Boehmeria platyphylla D. Don var. cinerascens? = Boehmeria clidemioides Miq. ■

56274 Boehmeria platyphylla D. Don var. clidemioides? = Boehmeria clidemioides Miq. ■

56275 Boehmeria platyphylla D. Don var. cuspidata Wedd. = Boehmeria platyphylla D. Don var. canescens (Wedd. ex Blume) Wedd. ●

56276 Boehmeria platyphylla D. Don var. hamiltoniana (Wedd.) Wedd. = Boehmeria hamiltoniana Wedd. ●

56277 Boehmeria platyphylla D. Don var. hamiltoniana Wedd. = Boehmeria hamiltoniana Wedd. ●

56278 Boehmeria platyphylla D. Don var. loochooensis Wedd. = Boehmeria densiflora Hook. et Arn. ●

56279 Boehmeria platyphylla D. Don var. loochooensis Wedd. = Boehmeria penduliflora Wedd. var. loochoensis (Wedd.) W. T. Wang ●

56280 Boehmeria platyphylla D. Don var. macrophylla Wedd. = Boehmeria japonica (L. f.) Miq. ●■

56281 Boehmeria platyphylla D. Don var. macrophylla Wedd. = Boehmeria longispica Steud. ●

56282 Boehmeria platyphylla D. Don var. macrostachya (Wight) Wedd. = Boehmeria macrophylla Hornem. ●■

56283 Boehmeria platyphylla D. Don var. macrostachya (Wight) Wedd. = Boehmeria macrophylla D. Don ●

56284 Boehmeria platyphylla D. Don var. pilosiuscula (Blume) Hand. - Mazz. = Boehmeria pilosiuscula (Blume) Hassk. ●■

56285 Boehmeria platyphylla D. Don var. rotundifolia (D. Don) Wedd. = Boehmeria macrophylla Hornem. var. rotundifolia (D. Don) W. T. Wang ●

56286 Boehmeria platyphylla D. Don var. rotundifolia (D. Don) Wedd. = Boehmeria macrophylla D. Don var. rotundifolia (D. Don) W. T. Wang ●

56287 Boehmeria platyphylla D. Don var. rotundifolia (D. Don) Wedd. = Boehmeria platyphylla D. Don ●

56288 Boehmeria platyphylla D. Don var. scabrella (Roxb.) Wedd. = Boehmeria macrophylla Hornem. var. scabrella (Roxb.) D. G. Long ●

56289 Boehmeria platyphylla D. Don var. scabrella (Roxb.) Wedd. = Boehmeria platyphylla D. Don ●

56290 Boehmeria platyphylla D. Don var. stricta C. H. Wright = Boehmeria formosana Hayata var. stricta (C. H. Wright) C. J. Chen ●

56291 Boehmeria platyphylla D. Don var. stricta C. H. Wright = Boehmeria formosana Hayata ●

56292 Boehmeria platyphylla D. Don var. tomentosa (Wedd.) Wedd. = Boehmeria tomentosa Wedd. ●

56293 Boehmeria platyphylla D. Don var. tricuspis Hance = Boehmeria tricuspis (Hance) Makino ●

56294 Boehmeria polystachya Wedd.;歧序苎麻(多穗苎麻);Manyfork Ramie,Polyspike Falsenettle ●■

56295 Boehmeria procridioides (Wedd.) Blume = Pouzolzia parasitica (Forssk.) Schweinf. ●☆

56296 Boehmeria pseudosieboldiana Honda = Boehmeria sieboldiana Blume ●

56297 Boehmeria pseudotricuspis W. T. Wang;滇黔苎麻;Dian-Qian Ramie,Yunnan-Kuichou Falsenettle ●

56298 Boehmeria pseudotricuspis W. T. Wang = Boehmeria umbrosa (Hand. -Mazz.) W. T. Wang ■

56299 Boehmeria puya Hook. = Maoutia puya (Hook.) Wedd. ●

56300 Boehmeria quelpaertensis Satake;朝鲜苎麻●☆

56301 Boehmeria rigida Benth. = Urera rigida (Benth.) Keay ■☆

56302 Boehmeria robusta Nakai et Satake;粗壮苎麻●

56303 Boehmeria rotundifolia D. Don = Boehmeria macrophylla D. Don var. rotundifolia (D. Don) W. T. Wang ●

56304 Boehmeria rotundifolia D. Don = Boehmeria platyphylla D. Don ●

56305 Boehmeria rugulosa Wedd.;略皱苎麻●

56306 Boehmeria rugulosa Wedd. var. tenuis = Pouzolzia sanguinea (Blume) Merr. var. nepalensis (Wedd.) H. Hara ●

56307 Boehmeria salicifolia D. Don = Debregeasia longifolia (Burm. f.) Wedd. ●

56308 Boehmeria salicifolia D. Don = Debregeasia saeneb (Forssk.) Hepper et J. R. I. Wood ●

56309 Boehmeria salicifolia D. Don = Debregeasia salicifolia (D. Don) Rendle ●

56310 Boehmeria sanguinea Hassk. = Pouzolzia sanguinea (Blume) Merr. ●

56311 Boehmeria scabra (Porter) Small = Boehmeria cylindrica (L.) Sw. ●☆

56312 Boehmeria siamensis Craib;束序苎麻(八棱麻,八楞麻,八楞马,大接骨,大糯叶,老母猪挂面,牛鼻子树,双合合,暹罗苎麻,野麻,野苎麻);Siam Falsenettle,Thailand Ramie ●

56313 Boehmeria sidifolia Wedd. = Boehmeria clidemioides Miq. ■

56314 Boehmeria sidifolia Wedd. = Leucosyke quadrinervia C. B. Rob. ●

56315 Boehmeria sieboldiana Blume = Boehmeria formosana Hayata ●

56316 Boehmeria sieboldiana Blume var. stenostachya (Satake) Kitam. = Boehmeria sieboldiana Blume ●

56317 Boehmeria silvestrii (Pamp.) W. T. Wang;赤麻(红芋,线麻);Red Ramie ■

56318 Boehmeria spicata (Thunb.) Thunb.;小赤麻(赤麻,东北苎麻,红锦麻,红线麻,麦麸草,水苎麻,细野麻,小红活麻,野线麻);Northeastern Falsenettle, Slender Falsenettle, Small Ramie, Small Red Ramie ■

56319 Boehmeria spicata (Thunb.) Thunb. = Boehmeria gracilis C. H. Wright ■

56320 Boehmeria spicata (Thunb.) Thunb. var. duploserrata C. H. Wright = Boehmeria japonica (L. f.) Miq. ●■

56321 Boehmeria spicata (Thunb.) Thunb. var. duploserrata Wright = Boehmeria longispica Steud. ●

56322 Boehmeria spicata (Thunb.) Thunb. var. microphylla Nakai ex Satake;小叶赤麻●☆

56323 Boehmeria spirei Gagnep. = Boehmeria siamensis Craib ●

56324 Boehmeria squamigera Wedd. = Chamabainia cuspidata Wight ■

56325 Boehmeria strigosifolia W. T. Wang; 伏毛苎麻; Prostratehair Ramie, Strigose-flower Falsenettle ●

56326 Boehmeria strigosifolia W. T. Wang var. mollis W. T. Wang; 柔毛苎麻; Hairy Falsenettle ●

56327 Boehmeria strigosifolia W. T. Wang var. mollis W. T. Wang = Boehmeria dolichostachya W. T. Wang var. mollis (W. T. Wang) W. T. Wang et C. J. Chen ●

56328 Boehmeria strigosifolia W. T. Wang. = Boehmeria dolichostachya W. T. Wang ●

56329 Boehmeria taiwaniana Nakai et Satake = Boehmeria japonica (L. f.) Miq. ●■

56330 Boehmeria taiwaniana Nakai et Satake = Boehmeria longispica Steud. ●

56331 Boehmeria taiwaniana Nakai et Satake ex Satake = Boehmeria japonica (L. f.) Miq. ●■

56332 Boehmeria taiwaniana Nakai et Satake ex Satake = Boehmeria longispica Steud. ●

56333 Boehmeria tenacissima Gaudich. = Boehmeria nivea (L.) Gaudich. var. tenacissima (Gaudich.) Miq. ●

56334 Boehmeria tenuifolia Satake = Boehmeria kiyozumensis Satake ●☆

56335 Boehmeria thailandica Yahara = Boehmeria nivea (L.) Gaudich. var. tenacissima (Gaudich.) Miq. ●

56336 Boehmeria tibetica C. J. Chen; 西藏苎麻; Xizang Falsenettle ●

56337 Boehmeria tibetica C. J. Chen = Boehmeria polystachya Wedd. ●■

56338 Boehmeria tiliifolia Satake; 椴叶苎麻●☆

56339 Boehmeria tomentosa Wedd.; 密毛苎麻(毛叶水苎麻); Hairy Ramie, Tomentose Falsenettle ●

56340 Boehmeria tonkinensis Gagnep.; 越南苎麻; Tonkin Falsenettle, Vietnam Ramie ●

56341 Boehmeria tonkinensis Gagnep. = Boehmeria lanceolata Ridl. ●

56342 Boehmeria tosaensis Miyazaki et H. Ohba; 土佐苎麻●☆

56343 Boehmeria tricuspis (Hance) Makino; 八角麻(白穗麻, 长白苎麻, 赤麻, 方麻, 龟叶麻, 山麻, 水苎麻, 透骨风, 悬铃叶苎麻, 野苎麻); Planeleaf Ramie, Tricuspidate Falsenettle ●

56344 Boehmeria tricuspis (Hance) Makino = Boehmeria silvestrii (Pamp.) W. T. Wang ■

56345 Boehmeria tricuspis (Hance) Makino subsp. paraspicata (Nakai ex H. Hara) Kitam. = Boehmeria gracilis C. H. Wright ■

56346 Boehmeria tricuspis (Hance) Makino var. unicuspis Makino = Boehmeria spicata (Thunb.) Thunb. ■

56347 Boehmeria tricuspis (Hance) Makino var. unicuspis Makino = Boehmeria gracilis C. H. Wright ■

56348 Boehmeria tricuspis (Hance) Makino var. unicuspis Makino ex Ohwi = Boehmeria spicata (Thunb.) Thunb. ■

56349 Boehmeria tricuspis (Hance) Makino var. unicuspis Makino ex Ohwi = Boehmeria gracilis C. H. Wright ■

56350 Boehmeria umbrosa (Hand. -Mazz.) W. T. Wang; 阴地苎麻; Shady Ramie ■

56351 Boehmeria utilis Blume = Boehmeria nivea (L.) Gaudich. var. candicans Wedd. ●

56352 Boehmeria vanioti H. Lév. = Pilea notata C. H. Wright ■

56353 Boehmeria villigera Satake; 长柔毛苎麻●☆

56354 Boehmeria viminea Blume = Pouzolzia sanguinea (Blume) Merr. ●

56355 Boehmeria wattersii (Hance) et Yuen P. Yang = Boehmeria zollingeriana Wedd. var. podocarpa (W. T. Wang) W. T. Wang et C. J. Chen ●

56356 Boehmeria yaeyamensis Hatus.; 八重山苎麻●☆

56357 Boehmeria zollingeriana Wedd.; 帚序苎麻(长叶苎麻, 金石榴); Broomhead Ramie, Zollinger Falsenettle, Zollinger False-nettle ●

56358 Boehmeria zollingeriana Wedd. var. blinii (H. Lév.) C. J. Chen; 黔桂苎麻; Blin Falsenettle, Qiangui Ramie ●

56359 Boehmeria zollingeriana Wedd. var. blinii (H. Lév.) C. J. Chen = Boehmeria blinii H. Lév. ●

56360 Boehmeria zollingeriana Wedd. var. podocarpa (W. T. Wang) W. T. Wang et C. J. Chen; 柄果苎麻(长叶苎麻); Stalkedfruit Falsenettle ●

56361 Boehmeriopsis Kom. (1901); 假苎麻属●☆

56362 Boehmeriopsis Kom. = Fatoua Gaudich. ●■

56363 Boehmeriopsis pallida Kom.; 假苎麻●☆

56364 Boeica C. B. Clarke = Boeica T. Anderson ex C. B. Clarke ●■

56365 Boeica T. Anderson ex C. B. Clarke (1874); 短筒苣苔属(比卡苣苔属); Boeica ●■

56366 Boeica ferruginea Drake; 锈毛短筒苣苔; Rusthair Boeica ■

56367 Boeica fulva C. B. Clarke; 短筒苣苔; Boeica ■

56368 Boeica guileana B. L. Burtt; 紫花短筒苣苔; Purpleflower Boeica ■

56369 Boeica multinervia K. Y. Pan; 多脉短筒苣苔; Veiny Boeica ■

56370 Boeica porosa C. B. Clarke; 孔药短筒苣苔(短筒苣苔); Holeanther Boeica, Porose Boeica ●

56371 Boeica stolonifera K. Y. Pan; 匍茎短筒苣苔; Creeping Boeica ■

56372 Boeica tonkinensis (Kraenzl.) B. L. Burtt = Boeica porosa C. B. Clarke ●

56373 Boeica yunnanensis (H. W. Li) K. Y. Pan; 翼柱短筒苣苔; Yunnan Boeica ●

56374 Boeicopsis H. W. Li = Boeica T. Anderson ex C. B. Clarke ●■

56375 Boeicopsis yunnanensis H. W. Li = Boeica yunnanensis (H. W. Li) K. Y. Pan ●

56376 Boelckea Rossow (1992); 伯尔克婆婆纳属■☆

56377 Boelckea beckii Rossow; 伯尔克婆婆纳■☆

56378 Boelia Webb = Genista L. ●

56379 Boenninghausenia Rchb. = Boenninghausenia Rchb. ex Meisn. (保留属名) ●■

56380 Boenninghausenia Rchb. ex Meisn. (1837) (保留属名); 石椒草属(臭节草属, 蛇皮草属, 松风草属); Chinaure ●■

56381 Boenninghausenia albiflora (Hook.) Rchb. ex Heynh. = Boenninghausenia albiflora (Hook.) Rchb. ex Meisn. ■

56382 Boenninghausenia albiflora (Hook.) Rchb. ex Meisn.; 蛇皮草(白虎草, 臭草, 臭虫草, 臭节草, 臭沙子, 大退癀, 大羊不食草, 大叶石椒, 地通花, 断根草, 二号黄药, 九牛二虎草, 苦黄草, 老蛇骚, 山羊草, 蛇根草, 蛇盘草, 生风草, 石胡椒, 石椒草, 松风草, 松气草, 烫伤草, 铜脚一枝蒿, 小黄药, 猩锈臭草, 岩椒草, 野椒); White Chinaure ■

56383 Boenninghausenia albiflora (Hook.) Rchb. ex Meisn. var. japonica (Nakai) Suzuki; 日本蛇皮草(日本松风草)■☆

56384 Boenninghausenia albiflora (Hook.) Rchb. ex Meisn. var. pilosa C. M. Tan; 毛蛇皮草(毛臭节草); Pilose White Chinaure ■

56385 Boenninghausenia albiflora Rchb. ex Meisn. var. brevipes Franch. = Boenninghausenia sessilicarpa H. Lév. ■

56386 Boenninghausenia albiflora Rchb. ex Meisn. var. japonica (Nakai ex Makino et Nemoto) H. Ohba; 日本松风草■☆

56387 Boenninghausenia brevipes H. Lév. = Boenninghausenia sessilicarpa H. Lév. ■

56388 Boenninghausenia japonica Nakai = Boenninghausenia albiflora (Hook.) Rchb. ex Meisn. var. japonica (Nakai) Suzuki ■☆

56389 Boenninghausenia japonica Nakai ex Makino et Nemoto = Boenninghausenia albiflora Rchb. ex Meisn. var. japonica (Nakai ex Makino et Nemoto) H. Ohba ■☆
56390 Boenninghausenia japonica var. livido-nitens? = Boenninghausenia albiflora (Hook.) Rchb. ex Meisn. var. japonica (Nakai) Suzuki ■☆
56391 Boenninghausenia schizocarpa S. Y. Hu = Boenninghausenia albiflora (Hook.) Rchb. ex Meisn. ■
56392 Boenninghausenia sessilicarpa H. Lév.;石椒草(白虎草,壁虱草,臭草,二号黄药,九牛二虎草,苦黄草,罗灶,千里马,蛇皮草,石胡椒,石交,石椒,铁扫把,铜脚一枝蒿,细叶石椒,小豆藤,小狼毒,羊不吃,羊不食草,羊山草,羊膻草);Sessilfruit Chinaure ■
56393 Boenninghausia Spreng.(废弃属名)= Boenninghausenia Rchb. ex Meisn.(保留属名)●■
56394 Boenninghausia Spreng.(废弃属名)= Chaetocalyx DC. ■☆
56395 Boerhaavia L. = Boerhavia L. ■
56396 Boerhaavia Mill. = Boerhavia L. ■
56397 Boerhavia L. (1753);黄细心属;Boerhavia,Spiderling ■
56398 Boerhavia adscendens Willd. = Boerhavia diffusa L. ■
56399 Boerhavia adscendens Willd. var. pubescens Choisy = Boerhavia coccinea Mill. var. pubescens (Choisy) Cufod. ■☆
56400 Boerhavia africana Lour. = Boerhavia diffusa L. ■
56401 Boerhavia agglutinans Batt. et Trab. = Boerhavia repens L. subsp. viscosa (Choisy) Maire ■
56402 Boerhavia ambigua (Meikle) Govaerts = Commicarpus ambiguus Meikle ●☆
56403 Boerhavia annulata Coville = Anulocaulis annulatus (Coville) Standl. ■☆
56404 Boerhavia boissieri Heimerl = Commicarpus boissieri (Heimerl) Cufod. ●☆
56405 Boerhavia bracteata T. Cooke = Boerhavia coccinea Mill. ■
56406 Boerhavia burchellii Choisy = Commicarpus pentandrus (Burch.) Heimerl ●☆
56407 Boerhavia caribaea Jacq. = Boerhavia coccinea Mill. ■
56408 Boerhavia chinensis (L.) Asch. et Schweinf. = Commicarpus chinensis (L.) Heimerl subsp. natalensis Meikle ●☆
56409 Boerhavia chinensis (L.) Asch. et Schweinf. = Commicarpus chinensis (L.) Heimerl ●
56410 Boerhavia chinensis (L.) Druce = Commicarpus chinensis (L.) Heimerl ●
56411 Boerhavia chinensis (L.) Rottb. = Commicarpus chinensis (L.) Heimerl ●
56412 Boerhavia ciliata Brandegee;缘毛黄细心■☆
56413 Boerhavia coccinea Mill.;红细心(猩红黄细心);Red Spiderling,Scarlet Spiderling ■
56414 Boerhavia coccinea Mill. var. pubescens (Choisy) Cufod.;毛红细心■☆
56415 Boerhavia coccinea Mill. var. viscosa (Lag. et Rodr.) Moscoso = Boerhavia coccinea Mill. ■
56416 Boerhavia coccinea sensu R. R. Stewart = Boerhavia procumbens Banks ex Roxb. ■☆
56417 Boerhavia commersonii Baill. = Commicarpus plumbagineus (Cav.) Standl. ●☆
56418 Boerhavia coulteri (Hook. f.) S. Watson;考特尔黄细心■☆
56419 Boerhavia coulteri (Hook. f.) S. Watson var. palmeri (S. Watson) Spellenb.;帕默黄细心■☆
56420 Boerhavia crispa K. Heyne;皱叶黄细心(缩叶黄细心);Crinkleleaf Spiderling ■
56421 Boerhavia deserticola Codd;荒漠黄细心■☆
56422 Boerhavia diandra L.;双蕊黄细心■☆
56423 Boerhavia diandra L. = Boerhavia repens L. subsp. diandra (L.) Maire et Weiller ■☆
56424 Boerhavia dichotoma Vahl = Commicarpus plumbagineus (Cav.) Standl. ●☆
56425 Boerhavia diffusa L.;黄细心(还少丹,黄寿丹,老来青,披散黄细心,匍匐黄细心,沙参,野兹菜);Diffuse Boerhavia,Diffuse Spiderling,Red Spiderling,Spreading Hogweed ■
56426 Boerhavia diffusa L. var. eudiffusa Helm ex Hand.-Mazz. = Boerhavia diffusa L. ■
56427 Boerhavia diffusa L. var. hirsuta Heimerl = Boerhavia coccinea Mill. ■
56428 Boerhavia diffusa L. var. hirta Balle;多毛黄细心■☆
56429 Boerhavia diffusa L. var. minor (Delile) Cufod. = Boerhavia repens L. ■
56430 Boerhavia diffusa L. var. mutabilis R. Br. = Boerhavia diffusa L. ■
56431 Boerhavia diffusa L. var. undulata (Asch. et Graebn.) Cufod.;波状黄细心■☆
56432 Boerhavia diffusa L. var. viscosa (Choisy) Cufod. = Boerhavia coccinea Mill. ■
56433 Boerhavia diffusa L. var. viscosa (Lag. et Rodr.) Heimerl = Boerhavia coccinea Mill. ■
56434 Boerhavia elegans Choisy;雅致黄细心■☆
56435 Boerhavia elegans Choisy = Boerhavia rubicunda Steud. ■☆
56436 Boerhavia elegans Choisy var. stenophylla Boiss. = Boerhavia elegans Choisy ■☆
56437 Boerhavia erecta L.;直立黄细心(西沙黄细心);Erect Spiderling,Spiderling ■
56438 Boerhavia erecta L. var. intermedia (M. E. Jones) Kearney et Peebles = Boerhavia intermedia M. E. Jones ■☆
56439 Boerhavia eriosolena A. Gray = Anulocaulis eriosolenus (A. Gray) Standl. ■☆
56440 Boerhavia fruticosa Dalzell = Commicarpus grandiflorus (A. Rich.) Standl. ●☆
56441 Boerhavia gracillima Heimerl;细长黄细心;Bush Spiderling ■☆
56442 Boerhavia graminicola Berhaut;草莺黄细心■☆
56443 Boerhavia grandiflora A. Rich. = Commicarpus grandiflorus (A. Rich.) Standl. ●☆
56444 Boerhavia greenwayi (Meikle) Govaerts = Commicarpus greenwayi Meikle ●☆
56445 Boerhavia gypsophiloides (M. Martens et Galeotti) J. M. Coult. = Cyphomeris gypsophiloides (M. Martens et Galeotti) Standl. ■☆
56446 Boerhavia helenae Roem. et Schult. = Commicarpus helenae (Roem. et Schult.) Meikle ●☆
56447 Boerhavia helenae Schult. = Commicarpus helenae (Schult.) Meikle ●☆
56448 Boerhavia hereroensis Heimerl;赫雷罗黄细心■☆
56449 Boerhavia hiranensis (Thulin) Govaerts = Commicarpus hiranensis Thulin ●☆
56450 Boerhavia hirsuta Jacq. = Boerhavia coccinea Mill. ■
56451 Boerhavia intermedia M. E. Jones;全叶黄细心■☆
56452 Boerhavia leiosolena Torr. = Anulocaulis leiosolenus (Torr.) Standl. ■☆
56453 Boerhavia lindheimeri Standl. = Boerhavia linearifolia A. Gray ■☆
56454 Boerhavia linearifolia A. Gray;线叶黄细心;Narrowleaf Spiderling ■☆

56455 Boerhavia marlothii Heimerl = Boerhavia coccinea Mill. ■
56456 Boerhavia maroccana Ball;摩洛哥黄细心■☆
56457 Boerhavia mathisiana F. B. Jones = Boerhavia ciliata Brandegee ■☆
56458 Boerhavia mista (Thulin) Govaerts = Commicarpus mistus Thulin ●☆
56459 Boerhavia montana (Miré et al.) Govaerts = Commicarpus montanus Miré et H. Gillet et Quézel ●☆
56460 Boerhavia paniculata Rich. = Boerhavia diffusa L. ■
56461 Boerhavia parviflora (Thulin) Govaerts = Commicarpus parviflorus Thulin ●☆
56462 Boerhavia pedunculosa A. Rich. = Commicarpus pedunculosus (A. Rich.) Cufod. ●☆
56463 Boerhavia pentandra Burch. = Commicarpus pentandrus (Burch.) Heimerl ●☆
56464 Boerhavia plumbaginea Cav. = Commicarpus plumbagineus (Cav.) Standl. ●☆
56465 Boerhavia plumbaginea Cav. var. grandiflora (A. Rich.) Asch. et Schweinf. = Commicarpus grandiflorus (A. Rich.) Standl. ●☆
56466 Boerhavia plumbaginea Cav. var. sinuato-lobata Chiov. = Commicarpus sinuatus Meikle ●☆
56467 Boerhavia plumbaginea Cav. var. trichocarpa Heimerl = Commicarpus plumbagineus (Cav.) Standl. var. trichocarpus (Heimerl) Meikle ■☆
56468 Boerhavia plumbaginea Cav. var. viscosa Boiss. = Commicarpus boissieri (Heimerl) Cufod. ●☆
56469 Boerhavia procumbens Banks ex Roxb.;平铺黄细心■☆
56470 Boerhavia ramosissima (Thulin) Govaerts = Commicarpus ramosissimus Thulin ●☆
56471 Boerhavia raynalii (J.-P. Lebrun et Meikle) Govaerts = Commicarpus raynalii J.-P. Lebrun et Meikle ●☆
56472 Boerhavia reboudiana Pomel = Boerhavia repens L. subsp. viscosa (Choisy) Maire ■
56473 Boerhavia reniformis Chiov. = Commicarpus reniformis (Chiov.) Cufod. ●☆
56474 Boerhavia repanda Willd. = Commicarpus chinensis (L.) Heimerl ●
56475 Boerhavia repens L. = Boerhavia diffusa L. ■
56476 Boerhavia repens L. subsp. viscosa (Choisy) Maire = Boerhavia coccinea Mill. ■
56477 Boerhavia repens L. var. annua (Batt. et Trab.) Maire = Boerhavia repens L. ■
56478 Boerhavia repens L. var. diffusa (L.) Heimerl ex Hook. f. = Boerhavia diffusa L. ■
56479 Boerhavia repens L. var. diffusa (L.) Hook. f. = Boerhavia diffusa L. ■
56480 Boerhavia repens L. var. elegans (Choisy) Asch. et Schweinf. = Boerhavia elegans Choisy ■☆
56481 Boerhavia repens L. var. glabra Choisy = Boerhavia diandra L. ■
56482 Boerhavia repens L. var. glabra Choisy = Boerhavia repens L. ■
56483 Boerhavia repens L. var. glutinosa (Vahl) Maire = Boerhavia repens L. ■
56484 Boerhavia repens L. var. minor Delile. = Boerhavia repens L. ■
56485 Boerhavia repens L. var. mollis Batt. et Trab. = Boerhavia repens L. ■
56486 Boerhavia repens L. var. pachypoda (Batt. et Trab.) Maire et Weiller = Boerhavia repens L. ■
56487 Boerhavia repens L. var. undulata (Ehrenb.) Asch. et Schweinf. = Boerhavia repens L. ■
56488 Boerhavia repens L. var. undulata Asch. et Graebn. = Boerhavia diffusa L. var. undulata (Asch. et Graebn.) Cufod. ■☆
56489 Boerhavia repens L. var. viscosa Choisy = Boerhavia coccinea Mill. ■
56490 Boerhavia repens L. var. vulvariifolia Boiss. = Boerhavia diandra L. ■
56491 Boerhavia rosei Standl. = Boerhavia coulteri (Hook. f.) S. Watson ■☆
56492 Boerhavia rubicunda Steud. = Boerhavia elegans Choisy ■☆
56493 Boerhavia rubicunda Steud. var. stenophylla (Boiss.) Fosberg = Boerhavia elegans Choisy ■☆
56494 Boerhavia scandens L. = Commicarpus scandens (L.) Standl. ●☆
56495 Boerhavia sinuata (Meikle) Greuter et Burdet = Commicarpus sinuatus Meikle ●☆
56496 Boerhavia somalensis (Chiov.) Govaerts = Acleisanthes somalensis (Chiov.) R. A. Levin ■☆
56497 Boerhavia spicata Choisy;匍匐黄细心;Creeping Spiderling ■☆
56498 Boerhavia spicata Choisy var. palmeri S. Watson = Boerhavia coulteri (Hook. f.) S. Watson var. palmeri (S. Watson) Spellenb. ■☆
56499 Boerhavia spicata Choisy var. torreyana S. Watson = Boerhavia torreyana (S. Watson) Standl. ■☆
56500 Boerhavia squarrosa Heimerl = Commicarpus squarrosus (Heimerl) Standl. ●☆
56501 Boerhavia stellata Wight = Commicarpus helenae (Roem. et Schult.) Meikle ●☆
56502 Boerhavia stenocarpa Chiov. = Commicarpus stenocarpus (Chiov.) Cufod. ●☆
56503 Boerhavia subumbellata Heimerl ex Engl.;小伞黄细心■☆
56504 Boerhavia tenuifolia A. Gray ex J. M. Coult. = Boerhavia linearifolia A. Gray ■☆
56505 Boerhavia torreyana (S. Watson) Standl.;托里黄细心■☆
56506 Boerhavia verticillata Poir. = Commicarpus plumbagineus (Cav.) Standl. ●☆
56507 Boerhavia verticillata Poir. var. fallacissima Heimerl = Commicarpus fallacissimus (Heimerl) Heimerl ex Oberm., Schweick. et I. Verd. ●☆
56508 Boerhavia verticillata Poir. var. glandulosa Franch. = Commicarpus plumbagineus (Cav.) Standl. ●☆
56509 Boerhavia viscosa Jacq. = Boerhavia coccinea Mill. ■
56510 Boerhavia viscosa Lag. et Rodr. = Boerhavia coccinea Mill. ■
56511 Boerhavia vulvariifolia Poir. = Boerhavia repens L. ■
56512 Boerlagea Cogn. (1890);加岛野牡丹属●☆
56513 Boerlagea Post et Kuntze = Boerlagella (Pierre ex Dubard) H. J. Lam ●☆
56514 Boerlagea Post et Kuntze = Boerlagia Pierre ●☆
56515 Boerlagea grandifolia Cogn.;加岛野牡丹●☆
56516 Boerlagella (Dubard) H. J. Lam = Boerlagella Pierre ex Cogn. ●☆
56517 Boerlagella (Pierre ex Dubard) H. J. Lam = Boerlagella Pierre ex Cogn. ●☆
56518 Boerlagella Cogn. = Boerlagella Pierre ex Cogn. ●☆
56519 Boerlagella Pierre ex Boerl. = Sideroxylon L. ●☆
56520 Boerlagella Pierre ex Cogn. (1891);苏门答腊山榄属●☆
56521 Boerlagella spectabilis (Dubard) H. J. Lam;苏门答腊山榄●☆
56522 Boerlagellaceae H. J. Lam = Sapotaceae Juss. (保留科名)●
56523 Boerlagellaceae H. J. Lam;苏门答腊山榄科●☆
56524 Boerlagia Pierre = Boerlagella Pierre ex Cogn. ●☆

56525 Boerlagiodendron Harms = Osmoxylon Miq. ●

56526 Boerlagiodendron Harms(1894);兰屿五加属(兰屿八角金盘属,兰屿加属,台湾五加属);Boerlagiodendron ●

56527 Boerlagiodendron kotoense (Hayata) Nakai = Boerlagiodendron pectinatum Merr. ●

56528 Boerlagiodendron kotoense (Hayata) Nakai = Osmoxylon pectinatum (Merr.) Philipson ●

56529 Boerlagiodendron palmatum (Zipp.) Harms;印尼兰屿五加(印度兰屿加,印尼兰屿加);Palmateleaf Boerlagiodendron ●

56530 Boerlagiodendron pectinatum Merr.;兰屿五加(兰屿八角金盘,兰屿加,台湾五加);Lanyu Boerlagiodendron, Pectinate Boerlagiodendron ●

56531 Boerlagiodendron pectinatum Merr. = Osmoxylon pectinatum (Merr.) Philipson ●

56532 Boesenbergia Kuntze(1891);凹唇姜属;Boesenbergia ■

56533 Boesenbergia albomaculata S. Q. Tong;白斑凹唇姜;Whitemaculate Boesenbergia ■

56534 Boesenbergia fallax Loes. = Boesenbergia longiflora (Wall.) Kuntze ■

56535 Boesenbergia longiflora (Wall.) Kuntze;心叶凹唇姜;Heartleaf Boesenbergia ■

56536 Boesenbergia pendulata (Roxb.) Schltr. = Boesenbergia rotunda (L.) Mansf. ■

56537 Boesenbergia rotunda (L.) Mansf.;凹唇姜(蓬莪术);Rotund Boesenbergia ■

56538 Bogenhardia Rchb. = Abutilon Mill. ●■

56539 Bogenhardia Rchb. = Herissantia Medik. ■●

56540 Bogenhardia crispa (L.) Kearney = Herissantia crispa (L.) Brizicky ■

56541 Bogenherdia crispa (L.) Kearney = Abutilon crispum (L.) Medik. ■

56542 Bognera Mayo et Nicolson(1984);鲍氏南星属■☆

56543 Bognera recondita (Madison) Mayo et Nicolson;鲍氏南星■☆

56544 Bogoria J. J. Sm. (1905);茂物兰属■☆

56545 Bogoria raciborskii J. J. Sm.;茂物兰■☆

56546 Bohadschia C. Presl = Turnera L. ●■☆

56547 Bohadschia Crantz = Peltaria Jacq. ■☆

56548 Bohadschia F. W. Schmidt = Hyoseris L. + Leontodon L.(保留属名)■☆

56549 Boheravia Parodi = Boerhavia L. ■

56550 Boholia Merr. (1926);菲律宾茜属☆

56551 Boholia nematostylis Merr.;菲律宾茜☆

56552 Boisalaea Lem. = Bossiaea Vent. ●☆

56553 Boisduvalia Spach = Epilobium L. ■

56554 Boisduvalia Spach(1835);穗报春属;Spike Primrose ■☆

56555 Boisduvalia densiflora (Lindl.) Bartl.;密花穗报春;Dense Spike Primrose ■☆

56556 Boissiaea Lem. = Bossiaea Vent. ●☆

56557 Boissiera Dombey ex DC. = Lardizabala Ruiz et Pav. ●☆

56558 Boissiera Haenseler ex Willd. = Gagea Salisb. ■

56559 Boissiera Hochst. ex Ledeb. = Bromus L.(保留属名)■

56560 Boissiera Hochst. ex Steud. (1840);糙雀麦属(布瓦氏草属);Boissiera ■

56561 Boissiera bromoides Hochst ex Steud. = Boissiera squarrosa (Banks et Sol.) Nevski ■☆

56562 Boissiera bromoides Hochst ex Steud. var. glabriflora Boiss. = Boissiera squarrosa (Banks et Sol.) Nevski ■☆

56563 Boissiera bromoides Hochst. ex Steud.;糙雀麦(布瓦氏草,粗糙布瓦氏草);Rough Boissiera ■

56564 Boissiera bromoides Hochst. ex Steud. = Boissiera pumilio Stapf ■☆

56565 Boissiera bromoides Hochst. ex Steud. var. glabriflora Boiss. = Boissiera squarrosa (Banks et Sol.) Nevski ■☆

56566 Boissiera danthoniae (Trin.) A. Braun = Bromus danthoniae Trin. ex C. A. Mey. ■

56567 Boissiera pumilio (Trin.) Hack. = Boissiera squarrosa (Banks et Sol.) Nevski ■☆

56568 Boissiera pumilio (Trin.) Hack. = Enneapogon borealis (Griseb.) Honda ■

56569 Boissiera pumilio Stapf = Boissiera bromoides Hochst. ex Steud. ■

56570 Boissiera pumilio Stapf = Enneapogon borealis (Griseb.) Honda ■

56571 Boissiera squarrosa (Banks et Sol.) Eig = Pappophorum squarrosum Banks et Sol. ■☆

56572 Boissiera squarrosa (Banks et Sol.) Nevski;粗鳞雀麦■☆

56573 Boissiera squarrosa (Banks et Sol.) Nevski = Boissiera pumilio Stapf ■☆

56574 Boissiera squarrosa (Banks et Sol.) Nevski = Pappophorum squarrosum Banks et Sol. ■☆

56575 Boita D. Don ex Endl. = Platycladus Spach ●

56576 Boita orientalis (L.) Endl. = Platycladus orientalis (L.) Franco ●

56577 Boita orientalis (L.) Endl. f. sieboldii (Endl.) W. C. Cheng et W. T. Wang = Platycladus orientalis (L.) Franco 'Sieboldii' ●

56578 Boita orientalis (L.) Endl. var. beverleyensis (Rehder) Hu = Platycladus orientalis (L.) Franco 'Beverleyensis' ●

56579 Boita orientalis (L.) Endl. var. nana Carrière = Platycladus orientalis (L.) Franco 'Sieboldii' ●

56580 Boita orientalis (L.) Endl. var. semperaurescens Lemoine ex Gordon = Platycladus orientalis (L.) Franco 'Semperaurescens' ●

56581 Boita orientalis (L.) Endl. var. sieboldii Endl. = Platycladus orientalis (L.) Franco 'Sieboldii' ●

56582 Boivinella (Pierre ex Baill.) Aubrév. et Pellegr. = Neoboivinella Aubrév. et Pellegr. ●☆

56583 Boivinella A. Camus = Cyphochlaena Hack. ■☆

56584 Boivinella Pierre ex Aubrév. et Pellegr. = Bequaertiodendron De Wild. ●☆

56585 Boivinella Pierre ex Aubrév. et Pellegr. = Englerophytum K. Krause ●☆

56586 Boivinella argyrophylla (Hiern) Aubrév. et Pellegr. = Englerophytum magalismontanum (Sond.) T. D. Penn. ●☆

56587 Boivinella glomerulifera (Hutch. et Dalziel) Aubrév. et Pellegr. = Englerophytum oblanceolatum (S. Moore) T. D. Penn. ●☆

56588 Boivinella kilimandscharica (G. M. Schulze) Aubrév. et Pellegr. = Englerophytum natalense (Sond.) T. D. Penn. ●☆

56589 Boivinella natalensis (Sond.) Pierre ex Aubrév. et Pellegr. = Englerophytum natalense (Sond.) T. D. Penn. ●☆

56590 Boivinella wilmsii (Engl.) Aubrév. et Pellegr. = Englerophytum magalismontanum (Sond.) T. D. Penn. ●☆

56591 Bojeria DC. = Inula L. ●■

56592 Bojeria Raf. = Doxanthes Raf. ■

56593 Bojeria Raf. = Euphorbia L. ●■

56594 Bojeria Raf. = Phaeomeria Lindl. ex K. Schum. ■☆

56595 Bojeria glabra Klatt = Aedesia glabra (Klatt) O. Hoffm. ■☆

56596 Bojeria lanceolata (Harv.) Benth. ex B. D. Jacks. = Pegolettia lanceolata Harv. ■☆

56597 Bojeria nutans Bolus = Printzia nutans (Bolus) Leins ■☆

56598 Bojeria perrieri Humbert = Inula perrieri (Humbert) Mattf. ■☆
56599 Bojeria speciosa DC. = Inula speciosa (DC.) O. Hoffm. ■☆
56600 Bojeria vestita Baker = Inula shirensis Oliv. ■☆
56601 Bokkeveldia D. Müll. -Doblies et U. Müll. -Doblies(1985);南非石蒜属■☆
56602 Bokkeveldia aestivalis (Snijman) D. Müll. -Doblies et U. Müll. -Doblies = Strumaria aestivalis Snijman ■☆
56603 Bokkeveldia perryae (Snijman) D. Müll. -Doblies et U. Müll. -Doblies = Strumaria perryae Snijman ■☆
56604 Bokkeveldia picta (W. F. Barker) D. Müll. -Doblies et U. Müll. -Doblies = Strumaria picta W. F. Barker ■☆
56605 Bokkeveldia pubescens (W. F. Barker) D. Müll. -Doblies et U. Müll. -Doblies = Strumaria pubescens W. F. Barker ■☆
56606 Bokkeveldia salteri (W. F. Barker) D. Müll. -Doblies et U. Müll. -Doblies = Strumaria salteri W. F. Barker ■☆
56607 Bokkeveldia watermeyeri (L. Bolus) D. Müll. -Doblies et U. Müll. -Doblies;沃特迈耶南非石蒜■☆
56608 Bokkeveldia watermeyeri (L. Bolus) D. Müll. -Doblies et U. Müll. -Doblies = Strumaria watermeyeri L. Bolus ■☆
56609 Bolandia Cron(2006);异果千里光属■☆
56610 Bolandia argillacea (Cron) Cron;白土异果千里光■☆
56611 Bolandia pedunculosa (DC.) Cron;梗花异果千里光■☆
56612 Bolandra A. Gray(1868);节柱菊属■☆
56613 Bolandra californica A. Gray;节柱菊■☆
56614 Bolanosa A. Gray(1852);博拉菊属■☆
56615 Bolanosa coulteri A. Gray;博拉菊■☆
56616 Bolanthus (Ser.) Rchb. (1841);爪翅花属■☆
56617 Bolanthus chelmicus Phitos;爪翅花■☆
56618 Bolax Comm. ex Juss. (1789);垫芹属■☆
56619 Bolax Comm. ex Juss. = Azorella Lam. ■☆
56620 Bolax gummifer Spreng.;垫芹;Azorella ■☆
56621 Bolbidium (Lindl.) Lindl. = Dendrobium Sw. (保留属名)■
56622 Bolbidium Brieger = Dendrobium Sw. (保留属名)■
56623 Bolbidium Lindl. = Dendrobium Sw. (保留属名)■
56624 Bolbophyllaria Rchb. f. = Bulbophyllum Thouars(保留属名)■
56625 Bolbophyllopsis Rchb. = Cirrhopetalum Lindl. (保留属名)■
56626 Bolbophyllopsis Rchb. f. = Cirrhopetalum Lindl. (保留属名)■
56627 Bolbophyllum Spreng. = Bulbophyllum Thouars(保留属名)■
56628 Bolborchis Lindl. = Coelogyne Lindl. ■
56629 Bolborchis Zoll. et Moritzi = Nervilia Comm. ex Gaudich. (保留属名)■
56630 Bolborchis crociformis Zoll. et Moritzi = Nervilia crociformis (Zoll. et Moritzi) Seidenf. ■
56631 Bolborchis crociformis Zoll. et Moritzi = Nervilia simplex (Thouars) Schltr. ■☆
56632 Bolbosaponaria Bondarenko = Gypsophila L. ■●
56633 Bolboschoenus (Asch.) Palla(1905);块茎藨草属(荆三棱属,球茎藨草属);Bolax,Tuberous Bulrush ■
56634 Bolboschoenus Palla = Bolboschoenus (Asch.) Palla ■
56635 Bolboschoenus affinis (Roth) Drobow = Scirpus strobilinus Roxb. ■
56636 Bolboschoenus fluviatilis (Torr.) Soják;河岸块茎藨草;River Bulrush ■☆
56637 Bolboschoenus fluviatilis (Torr.) Soják subsp. yagara (Ohwi) T. Koyama;大井藨草(灯芯草,湖三棱,江囊果,京三棱,荆三棱,老母拐子,马胡须,三棱,三棱草,三楞果,铁荸荠,野荸荠);Yagara Bulrush ■
56638 Bolboschoenus glaucus (Lam.) S. G. Sm.;灰蓝块茎藨草■☆
56639 Bolboschoenus grandispicus (Steud.) Lewej. et Lobin;大穗块茎藨草■☆
56640 Bolboschoenus koshevnikovii (Litv.) A. E. Kozhevn. = Scirpus planiculmis F. Schmidt ■
56641 Bolboschoenus maritimus (L.) Palla;沼泽块茎藨草;Alkali Bulrush,Bayonet Grass,Bayonet-grass,Prairie Bulrush,Salt-marsh Bulrush ■☆
56642 Bolboschoenus maritimus (L.) Palla = Scirpus maritimus L. ■
56643 Bolboschoenus maritimus (L.) Palla subsp. paludosus (A. Nelson) T. Koyama;沼生块茎藨草■☆
56644 Bolboschoenus nipponicus (Makino) T. Koyama = Schoenoplectus nipponicus (Makino) Soják ■☆
56645 Bolboschoenus nobilis (Ridl.) Goetgh. et D. A. Simpson;名贵藨草■☆
56646 Bolboschoenus novae-angliae (Britton) S. G. Sm.;新英格兰藨草■☆
56647 Bolboschoenus paludosus (A. Nelson) Soó = Bolboschoenus maritimus (L.) Palla ■☆
56648 Bolboschoenus planiculmis (F. Schmidt) T. V. Egorova;扁杆荆三棱■
56649 Bolboschoenus planiculmis (F. Schmidt) T. V. Egorova = Scirpus planiculmis F. Schmidt ■
56650 Bolboschoenus popovii T. V. Egorova = Bolboshoenus strobilinus (Roxb.) V. I. Krecz. ■
56651 Bolboschoenus popovii T. V. Egorova = Scirpus strobilinus Roxb. ■
56652 Bolboschoenus robustus (Pursh) Soják;粗壮块茎藨草;Seacoast Bulrush ■☆
56653 Bolboschoenus yagara (Ohwi) A. E. Kozhevn. = Bolboschoenus fluviatilis (Torr.) Soják subsp. yagara (Ohwi) T. Koyama ■
56654 Bolboschoenus yagara (Ohwi) Y. C. Yang et M. Zhan = Bolboschoenus fluviatilis (Torr.) Soják subsp. yagara (Ohwi) T. Koyama ■
56655 Bolboshoenus strobilinus (Roxb.) V. I. Krecz.;球穗三棱草■
56656 Bolbostemma Franquet(1930);假贝母属;Bolbostemma ●★
56657 Bolbostemma biglandulosum (Hemsl.) Franquet;刺儿瓜(拉拉藤);Biglangulose Bolbostemma ■
56658 Bolbostemma biglandulosum (Hemsl.) Franquet var. sinusto-lobulatum C. Y. Wu;波裂叶刺儿瓜;Wavylobulate Bolbostemma ■
56659 Bolbostemma paniculatum (Maxim.) Franquet;假贝母(草贝,大贝母,地苦胆,苦地胆,土贝,土贝母);Paniculate Bolbostemma ■
56660 Bolbostylis Gardner = Eupatorium L. ■●
56661 Bolboxalis Small = Oxalis L. ■●
56662 Boldea Juss. = Boldu Adans. (废弃属名)●
56663 Boldea Juss. = Peumus Molina(保留属名)●☆
56664 Boldoa Cav. = Boldoa Cav. ex Lag. ■☆
56665 Boldoa Cav. ex Lag. (1816);钩毛茉莉属■☆
56666 Boldoa Endl. = Boldu Adans. (废弃属名)●
56667 Boldoa purpurascens Cav. ex Lag.;钩毛茉莉■☆
56668 Boldu Adans. (废弃属名) = Peumus Molina(保留属名)●☆
56669 Boldu Feuill. ex Adans. = Peumus Molina(保留属名)●☆
56670 Boldu Nees = Beilschmiedia Nees ●
56671 Boldus Neck. = Taralea Aubl. (废弃属名)●☆
56672 Boldus Kuntze = Boldu Adans. (废弃属名)●
56673 Boldus Kuntze = Peumus Molina(保留属名)●☆
56674 Boldus Schult. et Schult. f. = Peumus Molina(保留属名)●☆
56675 Boldus Schult. f. = Peumus Molina(保留属名)●☆

56676 Bolelia Raf. (废弃属名) = Downingia Torr. (保留属名)■☆
56677 Boleum Desv. (1815);西班牙芥属■☆
56678 Boleum Desv. = Vella L. ●☆
56679 Boleum asperum (Pers.) Desv.;西班牙芥■☆
56680 Bolina Raf. = Bertolonia Raddi(保留属名)■☆
56681 Bolivaraceae Griseb. = Oleaceae Hoffmanns. et Link(保留科名)●
56682 Bolivaria Cham. et Schltdl. = Menodora Humb. et Bonpl. ●☆
56683 Bolivariaceae Griseb. = Oleaceae Hoffmanns. et Link(保留科名)●
56684 Bolivicactus Doweld = Echinocactus Link et Otto ●
56685 Bolivicereus Cárdenas = Borzicactus Riccob. ■☆
56686 Bolivicereus Cárdenas = Cleistocactus Lem. ●☆
56687 Bollaea Parl. = Pancratium L. ■
56688 Bollea Rchb. f. (1852);宝丽兰属;Bollea ■☆
56689 Bollea coelestis (Rchb. f.) Rchb. f.;宝丽兰;Glorious Bollea ■☆
56690 Bollea lanindei (Rchb. f.) Rchb. f.;拉氏宝丽兰;Lalinde Bollea ■☆
56691 Bollea violacea (Lindl.) Rchb. f.;紫红宝丽兰;Lalinde Bollea ■☆
56692 Bollwilleria Zabel = Pyrus L. ●
56693 Bolocephalus Hand. -Mazz. (1938);球菊属(丝苞菊属);Balldaisy, Bolocephalus ●★
56694 Bolocephalus Hand. -Mazz. = Dolomiaea DC. ■
56695 Bolocephalus saussureoides Hand. -Mazz.;球菊(丝苞菊);Common Balldaisy, Common Bolocephalus ■
56696 Bolophyta Nutt. (1840);掷菊属■☆
56697 Bolophyta Nutt. = Parthenium L. ■●
56698 Bolophyta alpina Nutt. = Parthenium alpinum (Nutt.) Torr. et A. Gray ■☆
56699 Bolosia Pourr. ex Willd. et Lange = Hispidella Barnadez ex Lam. ■☆
56700 Boltonia L'Hér. (1789);偶雏菊属(北美马兰属);Boltonia, Doll's-daisy, False Chamomile ■☆
56701 Boltonia apalachicolensis L. C. Anderson;阿帕拉契科拉偶雏菊;Apalachicola Doll's-daisy ■☆
56702 Boltonia asteroides (L.) L'Hér. var. decurrens (Torr. et A. Gray) Engelm. ex A. Gray = Boltonia decurrens (Torr. et A. Gray) A. W. Wood ■☆
56703 Boltonia asteroides (L.) L'Hér. var. glastifolia (Hill) Fernald = Boltonia asteroides (L.) L'Hér. ■☆
56704 Boltonia asteroides (L.) L'Hér. var. latisquama (A. Gray) Cronquist;宽鳞偶雏菊;False Aster, White Doll's-daisy ■☆
56705 Boltonia asteroides (L.) L'Hér. var. recognita (Fernald et Griscom) Cronquist;平滑偶雏菊;False Aster, White Doll's-daisy ■☆
56706 Boltonia asteroides (L.) L'Hér.;北美偶雏菊(北美马兰);Asterlike Boltonia, False Aster, False Starwort, White Boltonia, White Doll's-daisy, White Doll's Daisy ■☆
56707 Boltonia asteroides (L.) L'Hér. var. decurrens (Torr. et A. Gray) Fernald et Griscom = Boltonia decurrens (Torr. et A. Gray) A. W. Wood ■☆
56708 Boltonia asteroides (L.) L'Hér. var. microcephala Fernald et Griscom = Boltonia asteroides (L.) L'Hér. ■☆
56709 Boltonia asteroides Sims = Boltonia diffusa Elliott ■☆
56710 Boltonia cantonensis (Lour.) Franch. et Sav. = Kalimeris indica (L.) Sch. Bip. ■
56711 Boltonia caroliniana (Walter) Fernald;卡罗来纳偶雏菊;Carolina Doll's-daisy ■☆
56712 Boltonia decurrens (Torr. et A. Gray) A. W. Wood;下延偶雏菊;Clasping-leaf Doll's-daisy, Decurrent False Aster ■☆
56713 Boltonia diffusa Elliott;小头偶雏菊;False Aster, False Starwort, Smallhead Doll's-daisy ■☆
56714 Boltonia glastifolia (Hill) L'Hér. = Boltonia asteroides (L.) L'Hér. ■☆
56715 Boltonia glastifolia Michx. var. decurrens Torr. et A. Gray = Boltonia decurrens (Torr. et A. Gray) A. W. Wood ■☆
56716 Boltonia incisa (Fisch.) Benth. = Kalimeris incisa (Fisch.) DC. ■
56717 Boltonia indica Benth. = Aster indicus L. ■
56718 Boltonia integrifolia (Turcz.) Benth. et Hook. f. = Kalimeris integrifolia Turcz. ex DC. ■
56719 Boltonia latisquama A. Gray = Boltonia asteroides (L.) L'Hér. var. latisquama (A. Gray) Cronquist ■☆
56720 Boltonia latisquama A. Gray var. decurrens (Torr. et A. Gray) Fernald et Griscom = Boltonia decurrens (Torr. et A. Gray) A. W. Wood ■☆
56721 Boltonia latisquama A. Gray var. microcephala Fernald et Griscom = Boltonia asteroides (L.) L'Hér. var. recognita (Fernald et Griscom) Cronquist ■☆
56722 Boltonia latisquama A. Gray var. occidentalis A. Gray = Boltonia asteroides (L.) L'Hér. var. recognita (Fernald et Griscom) Cronquist ■☆
56723 Boltonia latisquama A. Gray var. recognita Fernald et Griscom = Boltonia asteroides (L.) L'Hér. var. recognita (Fernald et Griscom) Cronquist ■☆
56724 Boltonia lautureana Debeaux;劳氏北美马兰■☆
56725 Boltonia lautureana Debeaux = Kalimeris lautureana (Debeaux) Kitam. ■
56726 Boltonia lautureana Debeaux var. holophylla Chen = Kalimeris lautureana (Debeaux) Kitam. ■
56727 Boltonia pekinensis Hance = Kalimeris integrifolia Turcz. ex DC. ■
56728 Boltonia ravenelii Fernald et Griscom = Chrysanthemum caroliniana Walter ■☆
56729 Boltonia recognita (Fernald et Griscom) G. N. Jones = Boltonia asteroides (L.) L'Hér. var. recognita (Fernald et Griscom) Cronquist ■☆
56730 Bolusafra Kuntze(1891);沥青豆属■☆
56731 Bolusafra bituminosa (L.) Kuntze;沥青豆■☆
56732 Bolusanthemum Schwantes = Bijlia N. E. Br. ■☆
56733 Bolusanthemum tugwelliae Schwantes = Bijlia dilatata H. E. K. Hartmann ■☆
56734 Bolusanthus Harms(1906);树紫藤属●☆
56735 Bolusanthus speciosus (Bolus) Harms;树紫藤;Elephant Wood, Rhodesian Wistaria, South African Wisteria, Wild Wistaria ●☆
56736 Bolusanthus speciosus (Bolus) Harms. f. albescens Yakovlev;白树紫藤●☆
56737 Bolusia Benth. (1873);托叶齿豆属■☆
56738 Bolusia acuminata (DC.) Polhill;尖托叶齿豆■☆
56739 Bolusia amboensis (Schinz) Harms;安博托叶齿豆■☆
56740 Bolusia capensis Benth. = Bolusia acuminata (DC.) Polhill ■☆
56741 Bolusia ervoides (Welw. ex Baker) Torre;野豌豆托叶齿豆■☆
56742 Bolusia grandis B. -E. van Wyk;大托叶齿豆■☆
56743 Bolusia polhilliana Lisowski;普尔齿豆■☆
56744 Bolusia resupinata Milne-Redh.;倒置托叶齿豆■☆
56745 Bolusia rhodesiana Corbishley = Bolusia amboensis (Schinz)

Harms ■☆

56746 Bolusiella Schltr. (1918);波鲁兰属■☆

56747 Bolusiella batesii (Rolfe) Schltr.;贝茨波鲁兰■☆

56748 Bolusiella imbricata (Rolfe) Schltr. = Bolusiella maudiae (Bolus) Schltr. ■☆

56749 Bolusiella iridifolia (Rolfe) Schltr.;鸢尾叶托叶齿豆■☆

56750 Bolusiella iridifolia (Rolfe) Schltr. subsp. picea P. J. Cribb;沥青波鲁兰■☆

56751 Bolusiella maudiae (Bolus) Schltr.;莫迪波鲁兰■☆

56752 Bolusiella talbotii (Rendle) Summerh.;塔尔博特波鲁兰■☆

56753 Bolusiella zenkeri (Kraenzl.) Schltr.;岑克尔波鲁兰■☆

56754 Bolvicereus Cardenas = Borzicactus Riccob. ■☆

56755 Bomarea Mirb. (1804);竹叶吊钟属(玻玛莉属,藤本百合水仙属);Climbing Amaryllis ■☆

56756 Bomarea andimarcana (Herb.) Baker;彩花竹叶吊钟■☆

56757 Bomarea caldasii (Kunth) Asch. et Graebn.;橘红竹叶吊钟■☆

56758 Bomarea edulis Herb.;食用竹叶吊钟(食用玻玛莉);Salsilla ■☆

56759 Bomarea salsilla Herb.;藤本竹叶吊钟(藤本水百合)■☆

56760 Bomaria Kunth = Bomarea Mirb. ■☆

56761 Bombacaceae Kunth (1822)(保留科名);木棉科;Bombax Family,Silk-Cotton Family ●

56762 Bombacaceae Kunth(保留科名) = Malvaceae Juss.(保留科名)●■

56763 Bombacopsis Pittier = Pachira Aubl. ●

56764 Bombacopsis Pittier (1916)(保留属名);类木棉属;Mahot Coton ●☆

56765 Bombacopsis amazonica A. Robyns;亚马逊类木棉●☆

56766 Bombacopsis cubensis A. Robyns;古巴类木棉●☆

56767 Bombacopsis glabra (Pasq.) A. Robyns;光滑类木棉●☆

56768 Bombacopsis glabra (Pasq.) A. Robyns = Pachira glabra Pasq. ●☆

56769 Bombacopsis macrocalyx (Ducke) A. Robyns;大萼类木棉●☆

56770 Bombax L. (1753)(保留属名);木棉属;Bombax, Kapok-tree, Silk Cotton Tree,Silk-cotton Tree ●

56771 Bombax aculaetnm L. = Bombax ceiba L. ●

56772 Bombax anceps Pierre var. cambodiense (Pierre) A. Robyns = Bombax cambodiense Pierre ●

56773 Bombax andrieui Pellegr. et Vuill. = Bombax costatum Pellegr. et Vuill. ●☆

56774 Bombax angulicarpum Ulbr. = Bombax buonopozense P. Beauv. ●☆

56775 Bombax aquaticum (Aubl.) K. Schum. = Pachira aquatica Aubl. ●

56776 Bombax bouunopozense P. Beauv.;布诺木棉●☆

56777 Bombax brevicuspe Sprague;短尖木棉●☆

56778 Bombax buesgenii Ulbr. = Bombax buonopozense P. Beauv. ●☆

56779 Bombax buonopozense P. Beauv.;火焰色木棉;East African Bombax ●☆

56780 Bombax buonopozense P. Beauv. subsp. reflexum (Sprague) A. Robyns = Bombax buonopozense P. Beauv. ●☆

56781 Bombax buonopozense P. Beauv. var. cristata A. Chev. = Bombax buonopozense P. Beauv. ●☆

56782 Bombax buonopozense P. Beauv. var. vuilletii Pellegr. = Bombax costatum Pellegr. et Vuill. ●☆

56783 Bombax cambodiense Pierre;澜沧木棉●

56784 Bombax ceiba L.;印度木棉(木棉);Devil's Tree, Indian Silk Cotton Tree,Jumbie Tree,Malabar Simal Tree,Red Silkcotton Tree, Red Silk-cotton Tree,Silk Cotton Tree,Simal,Simul ●

56785 Bombax ceiba L. = Bombax malabaricum DC. ●

56786 Bombax chevalieri Pellegr.;加蓬木棉;Gabon Bombax ●☆

56787 Bombax chevalieri Pellegr. = Bombax brevicuspe Sprague ●☆

56788 Bombax costatum Pellegr. et Vuill.;红花木棉;Red-flowering Silk-cotton Tree,Sunset-tree ●☆

56789 Bombax ellipticum Kunth;椭圆木棉;White Shaving Brush Tree ●☆

56790 Bombax flammeum Ulbr. = Bombax buonopozense P. Beauv. ●☆

56791 Bombax globosum Aubl.;球形木棉●☆

56792 Bombax guineensis Schum. et Thonn. = Ceiba pentandra (L.) Gaertn. ●

56793 Bombax heptaphyllum L. = Bombax ceiba L. ●

56794 Bombax houardii Pellegr. et Vuill. = Bombax costatum Pellegr. et Vuill. ●☆

56795 Bombax insigne Wall.;长果木棉;Longfruit Bombax, Long-fruited Bombax ●

56796 Bombax insigne Wall. var. cambodiense (Pierre) Prain = Bombax cambodiense Pierre ●

56797 Bombax insigne Wall. var. tenebrosum (Dunn) A. Robyns;荫生长果木棉●

56798 Bombax kerrii Prain = Bombax cambodiense Pierre ●

56799 Bombax kimuenzae De Wild. et T. Durand = Pachira glabra Pasq. ●☆

56800 Bombax lukayense De Wild. et T. Durand;卢卡亚木棉●☆

56801 Bombax macrocarpum (Schltdl. et Cham.) K. Schum. = Pachira aquatica Aubl. ●

56802 Bombax malabaricum DC.;木棉(斑芝花,斑芝棉,斑芝树,广东海桐,槁木,海桐,红棉,吉贝,棉,莫连,木棉树,攀支棉,攀枝,攀枝花,塞瓦木棉,英雄树);Bombax, Common Bombax, Cotton Tree,Indian Cottowood,Malabar Bombax,Malabar Simal Tree,Red Silk-cotton Tree, Silk Cotton Tree, Silk-cotton Tree, Simal, Wood Cotton-tree ●

56803 Bombax malabaricum DC. = Bombax ceiba L. ●

56804 Bombax mossambicense A. Robyns = Bombax rhodognaphalon K. Schum. var. tomentosum A. Robyns ●☆

56805 Bombax munguba Mart. et Zucc.;莽木棉;Munguba ●☆

56806 Bombax oleagineum (Decne.) A. Robyns = Pachira glabra Pasq. ●☆

56807 Bombax orientate L. = Ceiba pentandra (L.) Gaertn. ●

56808 Bombax orientate Spreng. = Ceiba pentandra (L.) Gaertn. ●

56809 Bombax pentandrum L. = Ceiba pentandra (L.) Gaertn. ●

56810 Bombax pyramidale Cav. ex Lam. = Ochroma pyramidale (Cav. ex Lam.) Urb. ●

56811 Bombax reflexum Sprague = Bombax buonopozense P. Beauv. ●☆

56812 Bombax religiosum L. = Cochlospermum religiosum (L.) Alston ●☆

56813 Bombax rhodognaphalon K. Schum.;红鼠麴木棉●☆

56814 Bombax rhodognaphalon K. Schum. var. tomentosum A. Robyns;绒毛木棉●☆

56815 Bombax sessile (Benth.) Bakh. = Pachira sessilis Benth. ●☆

56816 Bombax stolzii Ulbr. = Bombax rhodognaphalon K. Schum. var. tomentosum A. Robyns ●☆

56817 Bombax tenebrosum Dunn = Bombax insigne Wall. var. tenebrosum (Dunn) A. Robyns ●

56818 Bombax tenebrosum Dunn = Bombax insigne Wall. ●

56819 Bombax vitifolium Willd. = Cochlospermum vitifolium (Willd.) Spreng. ●☆

56820 Bombax vuilletii Pellegr. = Bombax costatum Pellegr. et Vuill. ●☆

56821 Bombix Medik. = Hibiscus L.(保留属名)●■
56822 Bombusa gigantea Wall. ex Munro = Dendrocalamus giganteus (Wall.) Munro ●
56823 Bombycella (DC.) Lindl. = Bombax L.(保留属名)●
56824 Bombycella Lindl. = Bombax L.(保留属名)●
56825 Bombycidendron Zoll. et Moritzi = Hibiscus L.(保留属名)●■
56826 Bombycidendron grewiifolium (Hassk.) Zoll. et Moritzi = Hibiscus grewiifolius Hassk. ●
56827 Bombycilaena (DC.) Smoljan. (1955);光果紫绒草属■☆
56828 Bombycilaena (DC.) Smoljan. = Micropus L. ■☆
56829 Bombycilaena californica (Fisch. et C. A. Mey.) Holub;加州光果紫绒草■☆
56830 Bombycilaena californica (Fisch. et C. A. Mey.) Holub = Micropus californicus Fisch. et C. A. Mey. ■☆
56831 Bombycilaena discolor (Pers.) M. Lainz;异色光果紫绒草■☆
56832 Bombycilaena erecta (L.) Smoljan.;直立光果紫绒草■☆
56833 Bombycodendrum Post et Kuntze = Bombycidendron Zoll. et Moritzi ●■
56834 Bombycospermum C. Presl = Ipomoea L.(保留属名)●■
56835 Bombynia Noronha = Elaeagnus L. ●
56836 Bombyx Moench = Bombix Medik. ●■
56837 Bombyx Moench = Hibiscus L.(保留属名)●■
56838 Bona Medik. = Vicia L. ■
56839 Bonafidia Neck. = Amorpha L. ●
56840 Bonafousia A. DC. = Tabernaemontana L. ●
56841 Bonaga Medik. = Ononis L. ■●
56842 Bonamia A. Gray = Breweria R. Br. ●☆
56843 Bonamia Thouars(1804)(保留属名);伯纳旋花属●☆
56844 Bonamia alternifolia J. St. -Hil.;互叶伯纳旋花●☆
56845 Bonamia althoffiana Dammer = Convolvulus kilimandschari Engl. ■☆
56846 Bonamia ankaranensis Deroin;安卡兰伯纳旋花●☆
56847 Bonamia boivinii Hallier f.;博伊文伯纳旋花●☆
56848 Bonamia capensis (E. Mey. ex Choisy) Burtt Davy = Seddera capensis (E. Mey. ex Choisy) Hallier f. ●☆
56849 Bonamia cordata (Hallier f.) Hallier f. = Bonamia semidigyna (Roxb.) Hallier f. ●☆
56850 Bonamia cymosa (Roem. et Schult.) Hallier f. = Bonamia thunbergiana (Roem. et Schult.) F. N. Williams ●☆
56851 Bonamia gabonensis Breteler;加蓬伯纳旋花●☆
56852 Bonamia glomerata (Balf. f.) Hallier f. = Seddera glomerata (Balf. f.) O. Schwartz ●☆
56853 Bonamia hildebrandtii (Vatke) Hallier f. = Bonamia spectabilis (Choisy) Hallier f. ●☆
56854 Bonamia lebrunii Petit ex Evrard = Bonamia longitubulosa Breteler ●☆
56855 Bonamia longitubulosa Breteler;长管伯纳旋花●☆
56856 Bonamia madagascariensis Poir. = Bonamia alternifolia J. St. -Hil. ●☆
56857 Bonamia minor Hallier f. = Bonamia spectabilis (Choisy) Hallier f. ●☆
56858 Bonamia minor Hallier f. var. argentea R. E. Fr. = Bonamia spectabilis (Choisy) Hallier f. ●☆
56859 Bonamia mossambicensis (Klotzsch) Hallier f.;莫桑比克伯纳旋花●☆
56860 Bonamia poranoides Hallier f. = Metaporana densiflora (Hallier f.) N. E. Br. ●☆
56861 Bonamia schizantha (Hallier f.) A. Meeuse = Seddera schizantha Hallier f. ●☆
56862 Bonamia sedderoides Rendle;赛德旋花伯纳旋花●☆
56863 Bonamia semidigyna (Roxb.) Hallier f.;半双蕊伯纳旋花●☆
56864 Bonamia spectabilis (Choisy) Hallier f.;壮观伯纳旋花●☆
56865 Bonamia suffruticosa (Schinz) Burtt Davy et R. Pott = Seddera suffruticosa (Schinz) Hallier f. ●☆
56866 Bonamia thouarsii Scott-Elliot = Bonamia alternifolia J. St. -Hil. ●☆
56867 Bonamia thunbergiana (Roem. et Schult.) F. N. Williams;通贝里伯纳旋花●☆
56868 Bonamia velutina Verdc.;短绒毛伯纳旋花●☆
56869 Bonamia vignei Hoyle;维涅伯纳旋花●☆
56870 Bonamia volkensii Dammer = Hewittia malabarica (L.) Suresh ■
56871 Bonamica Vell. = Chionanthus L. ●
56872 Bonamica Vell. = Linociera Sw. ex Schreb.(保留属名)●
56873 Bonamica Vell. = Mayepea Aubl.(废弃属名)●
56874 Bonamiopsis (Boberty) Roberty = Bonamia Thouars(保留属名) ●☆
56875 Bonamiopsis Roberty = Bonamia Thouars(保留属名)●☆
56876 Bonamya Neck. = Stachys L. ●■
56877 Bonania A. Rich. (1850);伯南大戟属☆
56878 Bonania cubana A. Rich.;伯南大戟☆
56879 Bonania elliptica Urb.;椭圆伯南大戟☆
56880 Bonannia C. Presl = Brassica L. ■●
56881 Bonannia Guss. (1843)(保留属名);西西里草属■☆
56882 Bonannia Raf.(废弃属名)= Blighia K. König ●☆
56883 Bonannia Raf.(废弃属名)= Bonannia Guss.(保留属名)■☆
56884 Bonannia resinifera Guss.;西西里草■☆
56885 Bonanox Raf. = Ipomoea L.(保留属名)●■
56886 Bonapa Larranaga = Tillandsia L. ■☆
56887 Bonapartea Haw. = Agave L. ■
56888 Bonapartea Ruiz et Pav. (1802);刺子凤梨属■☆
56889 Bonapartea Ruiz et Pav. = Tillandsia L. ■☆
56890 Bonapartea glaucum Hort. = Dasylirion glaucophyllum Hook. ●☆
56891 Bonapartea juncea Ruiz et Pav.;刺子凤梨■☆
56892 Bonarota Adans. = Paederota L. ■☆
56893 Bonatea Willd. (1805);长须兰属■☆
56894 Bonatea bilabrella Lindl. = Habenaria falcicornis (Burch. ex Lindl.) Bolus ■☆
56895 Bonatea boltonii (Harv.) Bolus;博尔顿长须兰■☆
56896 Bonatea bracteata G. McDonald et McMurtry = Habenaria transvaalensis Schltr. ■☆
56897 Bonatea cassidea Sond.;决明长须兰■☆
56898 Bonatea cirrhata Lindl. = Habenaria cirrhata (Lindl.) Rchb. f. ■☆
56899 Bonatea clavata Lindl. = Habenaria clavata (Lindl.) Rchb. f. ■☆
56900 Bonatea densiflora Sond. = Bonatea speciosa (L. f.) Willd. ■☆
56901 Bonatea eminii (Kraenzl.) Rolfe;埃明长须兰■☆
56902 Bonatea foliosa (Sw.) Lindl. = Habenaria epipactidea Rchb. f. ■☆
56903 Bonatea foliosa (Sw.) Lindl. var. pauciflora Sond. = Habenaria epipactidea Rchb. f. ■☆
56904 Bonatea incarnata Lindl. = Habenaria incarnata Rchb. f. ■☆
56905 Bonatea insignis (Schltr.) Summerh. = Bonatea polypodantha (Rchb. f.) L. Bolus ■☆
56906 Bonatea kayseri (Kraenzl.) Rolfe = Bonatea steudneri (Rchb. f.) T. Durand et Schinz ■☆
56907 Bonatea lamprophylla J. L. Stewart;亮叶长须兰■☆

56908 Bonatea liparophylla Schelpe = Habenaria transvaalensis Schltr. ■☆
56909 Bonatea micrantha Lindl. = Habenaria arenaria Lindl. ■☆
56910 Bonatea phillipsii (Rolfe) Rolfe = Bonatea steudneri (Rchb. f.) T. Durand et Schinz ■☆
56911 Bonatea pirottae Cortesi = Bonatea steudneri (Rchb. f.) T. Durand et Schinz ■☆
56912 Bonatea polypodantha (Rchb. f.) L. Bolus;多花长须兰■☆
56913 Bonatea porrecta (Bolus) Summerh.;外伸长须兰■☆
56914 Bonatea pulchella Summerh.;美丽长须兰■☆
56915 Bonatea punduana Lindl. ex Wall. = Habenaria digitata Lindl. ■☆
56916 Bonatea rabaiensis (Rendle) Rolfe;拉巴伊长须兰■☆
56917 Bonatea saundersiae (Harv.) T. Durand et Schinz;桑德斯长须兰■☆
56918 Bonatea saundersioides (Kraenzl. et Schltr.) Cortesi;拟桑德斯长须兰■☆
56919 Bonatea speciosa (L. f.) Willd.;雅丽长须兰■☆
56920 Bonatea stereophylla (Kraenzl.) Summerh.;硬叶长须兰■☆
56921 Bonatea steudneri (Rchb. f.) T. Durand et Schinz;斯托德长须兰■☆
56922 Bonatea sudanensis Rolfe = Bonatea steudneri (Rchb. f.) T. Durand et Schinz ■☆
56923 Bonatea tentaculifera Summerh. = Habenaria bonateoides Ponsie ■☆
56924 Bonatea tetrapetala Lindl. = Habenaria falcicornis (Burch. ex Lindl.) Bolus ■☆
56925 Bonatea ugandae Rolfe ex Summerh. = Bonatea steudneri (Rchb. f.) T. Durand et Schinz ■☆
56926 Bonatea verdickii De Wild. = Habenaria verdickii (De Wild.) Schltr. ■☆
56927 Bonatea volkensiana (Kraenzl.) Rolfe;福尔长须兰■☆
56928 Bonatia Schltr. et Krause = Tarenna Gaertn. ●
56929 Bonatoa Post et Kuntze = Bonatea Willd. ■☆
56930 Bonavera securidaca (L.) Desv. = Securigera securidaca (L.) Degen. et Dorf. ■☆
56931 Bonaveria Scop.(废弃属名) = Coronilla L.(保留属名)●■
56932 Bonaveria Scop.(废弃属名) = Securigera DC.(保留属名)●■
56933 Bondtia Kuntze = Bontia L. + Eremophila R. Br. + Pholidia R. Br. ●☆
56934 Bonduc Adans. = Caesalpinia L. ●
56935 Bonduc Mill. = Caesalpinia L. ●
56936 Bonduc Mill. = Guilandina L. ●
56937 Bonduc majus Medik. = Caesalpinia major (Medik.) Dandy et Exell ●
56938 Bonellia Bertero ex Colla = Jacquinia L.(保留属名)●☆
56939 Bonetiella Rzed.(1957);墨西哥漆树属●☆
56940 Bonetiella anomala (I. M. Johnst.) Rzed.;墨西哥漆树●☆
56941 Bongardia C. A. Mey.(1831);长瓣囊果草属(邦加草属)■☆
56942 Bongardia chrysogenum (L.) Boiss. = Bongardia chrysogonum (L.) Spach ●☆
56943 Bongardia chrysogonum (L.) Griseb. = Bongardia chrysogenum (L.) Boiss. ●☆
56944 Bongardia chrysogonum (L.) Spach;长瓣囊果草(黄邦加草)●☆
56945 Bongardia chrysogonum (L.) Spach = Bongardia chrysogenum (L.) Boiss. ●☆
56946 Bongardia margalla R. R. Stewart = Bongardia chrysogonum (L.) Spach ●☆
56947 Bongardia olivieri C. A. Mey. = Bongardia chrysogonum (L.) Spach ●☆
56948 Bongardia rauwolfii C. A. Mey. = Bongardia chrysogonum (L.) Spach ●☆
56949 Bonia Balansa = Bambusa Schreb.(保留属名)●
56950 Bonia Balansa = Monocladus H. C. Chia, H. L. Fung et Y. L. Yang ●★
56951 Bonia Balansa(1890);异箣竹属(单枝竹属,异簕竹属);Monocladus, Singlebamboo ●★
56952 Bonia amplexicaulis (L. C. Chia et al.) N. H. Xia;芸香竹;Amplexicaul Monocladus, Rue Singlebamboo ●◇
56953 Bonia levigata (L. C. Chia et al.) N. H. Xia;响子竹(散穗弓果黍);Smooth Monocladus, Smooth Singlebamboo ●
56954 Bonia parvifloscula (W. T. Lin) N. H. Xia;小花单枝竹;Smallflower Singlebamboo ●
56955 Bonia saxatilis (L. C. Chia et al.) N. H. Xia;单枝竹;Saxitil Monocladus, Singlebamboo ●
56956 Bonia saxatilis (L. C. Chia et al.) N. H. Xia var. solida (C. D. Chu et C. S. Chao) D. Z. Li;箭杆竹;Arrowshaft Singlebamboo ●
56957 Bonia solida (C. D. Chu et C. S. Chao) N. H. Xia = Bonia saxatilis (L. C. Chia et al.) N. H. Xia var. solida (C. D. Chu et C. S. Chao) D. Z. Li ●
56958 Bonifacia Silva Manso ex Steud. = Augusta Pohl(保留属名)■☆
56959 Bonifazia Standl. et Steyerm. = Disocactus Lindl. ●☆
56960 Boninia Planch.(1872);博南芸香属●☆
56961 Boninia glabra Planch. = Melicope quadrilocularis (Hook. et Arn.) T. G. Hartley ●☆
56962 Boninia glabra Planch. f. macrophylla Nakai = Melicope grisea (Planch.) T. G. Hartley ●☆
56963 Boninia glabra Planch. var. crassifolia Nakai = Melicope grisea (Planch.) T. G. Hartley var. crassifolia (Nakai) Yonek. ●☆
56964 Boninia grisea Planch. = Melicope grisea (Planch.) T. G. Hartley ●☆
56965 Boninia grisea Planch. var. crassifolia (Nakai) T. Yamaz. ex H. Ohba = Melicope grisea (Planch.) T. G. Hartley var. crassifolia (Nakai) Yonek. ●☆
56966 Boninofatsia Nakai = Fatsia Decne. et Planch. ●
56967 Boninofatsia Nakai(1924);小笠原五加属●☆
56968 Boninofatsia oligocarpella (Koidz.) Nakai = Fatsia oligocarpella Koidz. ●☆
56969 Boninofatsia wilsonii Nakai = Fatsia oligocarpella Koidz. ●☆
56970 Boniodendron Gagnep.(1946);黄梨木属;Boniodendron ●
56971 Boniodendron minus (Hemsl.) T. C. Chen;黄梨木(采木树,黄达木,米琼);Small Boniodendron ●
56972 Boniodendron parviflorum (Lecomte) Gagnep.;小花黄梨木;Smallflower Boniodendron ●☆
56973 Bonjeanea Rchb. = Bonjeania Rchb. ●■☆
56974 Bonjeania Rchb. = Dorycnium Mill. ●■☆
56975 Bonjeania hirsuta (L.) Rchb. = Dorycnium hirsutum (L.) Ser. ■☆
56976 Bonjeania recta (L.) Rchb. = Dorycnium rectum (L.) Ser. ■☆
56977 Bonnaya Link et Otto = Lindernia All. ■
56978 Bonnaya antipoda (L.) Druce = Lindernia antipoda (L.) Alston ■
56979 Bonnaya brachiata Link et Otto = Lindernia ciliata (Colsm.) Pennell ■
56980 Bonnaya hyssopioides (L.) Benth. = Lindernia hyssopioides (L.) Haines ■
56981 Bonnaya parviflora (Roxb.) Benth. = Lindernia parviflora

(Roxb.) Haines ■☆
56982 Bonnaya pumila (D. Don) Spreng. = Chirita pumila D. Don ■
56983 Bonnaya reptans (Roxb.) Spreng. = Lindernia ruellioides (Colsm.) Pennell ■
56984 Bonnaya tenuifolia (Colsm.) Spreng. = Lindernia tenuifolia (Vahl) Alston ■
56985 Bonnaya trichotoma Oliv. = Lindernia madiensis Dandy ■☆
56986 Bonnaya veronicifolia (Retz.) Spreng. = Lindernia antipoda (L.) Alston ■
56987 Bonnaya veronicifolia (Retz.) Spreng. var. grandiflora? = Lindernia antipoda (L.) Alston ■
56988 Bonnaya veronicifolia (Retz.) Spreng. var. verbenifolia? = Lindernia antipoda (L.) Alston ■
56989 Bonnayodes Blatt. et Hallb. = Limnophila R. Br. (保留属名) ■
56990 Bonneria B. D. Jacks. = Bonniera Cordem. ■☆
56991 Bonnetia Mart. (1826)(保留属名);多籽树属(多子树属)●☆
56992 Bonnetia Mart. et Zucc. = Bonnetia Mart. (保留属名)●☆
56993 Bonnetia Neck. = Buchnera L. ■
56994 Bonnetia Schreb. (废弃属名) = Bonnetia Mart. (保留属名)●☆
56995 Bonnetia Schreb. (废弃属名) = Mahurea Aubl. ●☆
56996 Bonnetia anceps Mart.;多籽树●☆
56997 Bonnetiaceae L. Beauvis. = Clusiaceae Lindl. (保留科名)●■
56998 Bonnetiaceae L. Beauvis. = Guttiferae Juss. (保留科名)●■
56999 Bonnetiaceae L. Beauvis. ex Nakai(1948);多籽树科(多子科)●☆
57000 Bonnetiaceae Nakai = Bonnetiaceae P. Beauv. ●☆
57001 Bonnetiaceae P. Beauv. ex Nakai = Clusiaceae Lindl. (保留科名)●■
57002 Bonnetiaceae P. Beauv. ex Nakai = Guttiferae Juss. (保留科名)●■
57003 Bonniera Cordem. (1899);留岛兰属■☆
57004 Bonniera appendiculata Cordem.;留岛兰■☆
57005 Bonnierella R. Vig. = Polyscias J. R. Forst. et G. Forst. ●
57006 Bonplandia Cav. (1800);邦普花葱属(墨西哥花葱属)●☆
57007 Bonplandia Willd. = Angostura Roem. et Schult. ●☆
57008 Bonplandia geminiflora Cav.;邦普花葱●☆
57009 Bonplandia linearis B. L. Rob.;线形邦普花葱●☆
57010 Bontia L. (1753);假瑞香属(美槛蓝属)●☆
57011 Bontia P. Br. = Avicennia L. ●
57012 Bontia daphnoides L.;假瑞香(美槛蓝);Mangle Bobo ●☆
57013 Bontia germinans L. = Avicennia germinans (L.) L. ●☆
57014 Bontiaceae Horan.;假瑞香科●
57015 Bontiaceae Horan. = Myoporaceae R. Br. (保留科名)●
57016 Bontiaceae Horan. = Scrophulariaceae Juss. (保留科名)●■
57017 Bonyunia M. R. Schomb. = Bonyunia M. R. Schomb. ex Progel. ●☆
57018 Bonyunia M. R. Schomb. ex Progel. (1868);热美马钱属●☆
57019 Bonyunia minor N. E. Br.;小热美马钱●☆
57020 Bonzetia Post et Kuntze = Bouzetia Montrouz. ●☆
57021 Boophone Herb. (1821);非洲箭毒草属(非洲石蒜属)■☆
57022 Boophone ciliaris (L.) Herb. = Crossyne guttata (L.) D. Müll. -Doblies et U. Müll. -Doblies ■☆
57023 Boophone disticha (L. f.) Herb.;二列非洲箭毒草(二列非洲石蒜)■☆
57024 Boophone disticha (L. f.) Herb. var. ernesti-ruschii Dinter et G. M. Schulze = Boophone haemanthoides F. M. Leight. ■☆
57025 Boophone fischeri Baker = Sansevieria fischeri (Baker) Marais ■☆
57026 Boophone flava W. F. Barker ex Snijman = Crossyne flava (W. F. Barker ex Snijman) D. Müll. -Doblies et U. Müll. -Doblies ■☆
57027 Boophone guttata (L.) Herb. = Crossyne guttata (L.) D. Müll. -Doblies et U. Müll. -Doblies ■☆
57028 Boophone haemanthoides F. M. Leight.;血红非洲箭毒草(血红非洲石蒜)■☆
57029 Boophone longepedicellata Pax = Boophone disticha (L. f.) Herb. ■☆
57030 Boophone pulchra W. F. Barker = Brunsvigia pulchra (W. F. Barker) D. Müll. -Doblies et U. Müll. -Doblies ■☆
57031 Boophone toxicaria Herb.;非洲箭毒草■☆
57032 Boopidaceae Cass. = Calyceraceae R. Br. ex Rich. (保留科名)●■☆
57033 Boopis Juss. (1803);牛眼萼角花属(南美萼角花属)■☆
57034 Boopis acaulis Phil.;无茎牛眼萼角花(无茎南美萼角花)■☆
57035 Boopis alpina Poepp. ex Less.;高山牛眼萼角花(高山南美萼角花)■☆
57036 Boopis australis Phil.;南方牛眼萼角花(南方南美萼角花)■☆
57037 Boopis filifolia Speg.;线叶牛眼萼角花(线叶南美萼角花)■☆
57038 Boopis integrifolia Phil.;全缘牛眼萼角花(全缘南美萼角花)■☆
57039 Boopis leptophylla Speg.;细叶牛眼萼角花(细叶南美萼角花)■☆
57040 Boopis monocephala Phil.;单头牛眼萼角花(单头南美萼角花)■☆
57041 Boosia Speta = Urginea Steinh. ■☆
57042 Boosia Speta(2001);澳非海葱属■☆
57043 Boothia Douglas ex Benth. = Platystemon Benth. ■☆
57044 Bootia Adans. = Aspalathus L. ●☆
57045 Bootia Adans. = Borbonia L. ●☆
57046 Bootia Bagel. = Potentilla L. ■●
57047 Bootia Neck. = Saponaria L. ■
57048 Bootrophis Steud. = Botrophis Raf. ●■
57049 Bootrophis Steud. = Cimicifuga L. ●■
57050 Boottia Ayres ex Baker = Pleurostylia Wight et Arn. ●
57051 Boottia Wall. = Ottelia Pers. ■
57052 Boottia abyssinica Ridl. = Ottelia ulvifolia (Planch.) Walp. ■☆
57053 Boottia acuminata Gagnep. = Ottelia acuminata (Gagnep.) Dandy ■
57054 Boottia aschersoniana Gürke = Ottelia muricata (C. H. Wright) Dandy ■☆
57055 Boottia brachyphylla Gürke = Ottelia brachyphylla (Gürke) Dandy ■☆
57056 Boottia cordata Wall. = Ottelia cordata (Wall.) Dandy ■
57057 Boottia crassifolia Ridl. = Ottelia ulvifolia (Planch.) Walp. ■☆
57058 Boottia crispa Hand. -Mazz. = Ottelia acuminata (Gagnep.) Dandy var. crispa (Hand. -Mazz.) H. Li ■
57059 Boottia cylindrica T. C. E. Fr. = Ottelia cylindrica (T. C. E. Fr.) Dandy ■☆
57060 Boottia echinata W. W. Sm. = Ottelia acuminata (Gagnep.) Dandy ■
57061 Boottia esquirolii H. Lév. et Vaniot = Ottelia acuminata (Gagnep.) Dandy ■
57062 Boottia exserta Ridl. = Ottelia exserta (Ridl.) Dandy ■☆
57063 Boottia fischeri Gürke = Ottelia fischeri (Gürke) Dandy ■☆
57064 Boottia heterophylla Merr. et F. P. Metcalf = Ottelia cordata (Wall.) Dandy ■
57065 Boottia kunenensis Gürke = Ottelia kunenensis (Gürke) Dandy ■☆
57066 Boottia macrantha C. H. Wright = Ottelia ulvifolia (Planch.) Walp. ■☆

57067 Boottia mairei H. Lév. = Monochoria vaginalis (Burm. f.) C. Presl ex Kunth ■
57068 Boottia muricata C. H. Wright = Ottelia muricata (C. H. Wright) Dandy ■☆
57069 Boottia polygonifolia Gagnep. = Ottelia acuminata (Gagnep.) Dandy ■
57070 Boottia rautanenii Gürke = Ottelia exserta (Ridl.) Dandy ■☆
57071 Boottia rohrbachiana Asch. et Gürke = Ottelia ulvifolia (Planch.) Walp. ■☆
57072 Boottia scabra (Baker) Benth. et Hook. f. = Ottelia scabra Baker ■☆
57073 Boottia schinziana Asch. et Gürke = Ottelia exserta (Ridl.) Dandy ■☆
57074 Boottia sinensis H. Lév. et Vaniot = Ottelia sinensis (H. Lév. et Vaniot) H. Lév. ex Dandy ■
57075 Boottia yunnanensis Gagnep. = Ottelia acuminata (Gagnep.) Dandy ■
57076 Bopusia C. Presl = Graderia Benth. ■●☆
57077 Bopusia scabra (L. f.) C. Presl = Graderia scabra (L. f.) Benth. ■☆
57078 Bopusia scabra Hiern = Alectra aurantiaca Hemsl. ■☆
57079 Boquila Decne. (1837);南美木通属●☆
57080 Boquila trifoliolata (DC.) Decne.;南美木通●☆
57081 Borabora Steud. = Mariscus Gaertn. ■
57082 Boraeva Boiss. = Boreava Jaub. et Spach ■☆
57083 Boraginaceae Adans. = Boraginaceae Juss.(保留科名)■●
57084 Boraginaceae Juss. (1789)(保留科名);紫草科;Borage Family,Forget-me-not Family ■●
57085 Boraginella Kuntze = Borraginoides Boehm.(废弃属名)●■
57086 Boraginella Kuntze = Trichodesma R. Br.(保留属名)●■
57087 Boraginella Siegesb. = Trichodesma R. Br.(保留属名)●■
57088 Boraginella Siegesb. ex Kuntze = Trichodesma R. Br.(保留属名)●■
57089 Boraginella africana (L.) Kuntze = Trichodesma africanum (L.) Sm. ■☆
57090 Boraginella angustifolia (Harv.) Kuntze = Trichodesma angustifolium Harv. ■☆
57091 Boraginodes Post et Kuntze = Boraginella Kuntze ●■
57092 Boraginodes Post et Kuntze = Borraginoides Boehm.(废弃属名)●■
57093 Boraginodes Post et Kuntze = Trichodesma R. Br.(保留属名)●■
57094 Boraginoides Moench = Trichodesma R. Br.(保留属名)●■
57095 Borago L. (1753);琉璃苣属(玻璃苣属);Borage ■☆
57096 Borago africana L. = Trichodesma africanum (L.) R. Br. ■☆
57097 Borago crassifolia Vent. = Caccinia macranthera var. crassifolia (Vent.) Brand ■☆
57098 Borago indica L. = Trichodesma indicum (L.) Lehm. ■☆
57099 Borago laxiflora DC.;疏花琉璃苣;Slender Borage ■☆
57100 Borago longifolia Poir.;长叶琉璃苣■☆
57101 Borago officinalis L.;琉璃苣;Bee Bread, Blue Robin, Borage, Burridge Burrage, Common Borage, Common Bugloss, Cool Tankard, Devil-in-church, Granny's Nightcap, Ox Tongue, Ox-tongue, Starflower, Starflower Oil, Tailwort, Talewort, Virgin's Robe ■☆
57102 Borago officinalis L. var. stenopetala Ducell. et Maire = Borago officinalis L. ■☆
57103 Borago orientalis L.;东方琉璃苣;Abraham-isaac-jacob ■☆
57104 Borago pygmaea (DC.) Chater et Greuter = Borago laxiflora DC. ■☆
57105 Borago trabutii Maire;特拉布特琉璃苣■☆
57106 Borago tristis G. Forst. = Trichodesma africanum (L.) Sm. ■☆
57107 Borago verrucosa Forssk. = Trichodesma africanum (L.) R. Br. ■☆
57108 Borago zeylanica Burm. f. = Trichodesma zeylanicum (Burm. f.) R. Br. ■☆
57109 Borassaceae O. F. Cook = Arecaceae Bercht. et J. Presl(保留科名)●
57110 Borassaceae O. F. Cook = Palmae Juss.(保留科名)●
57111 Borassaceae O. F. Cook;糖棕科●
57112 Borassaceae Schultz Sch. = Arecaceae Bercht. et J. Presl(保留科名)●
57113 Borassaceae Schultz Sch. = Palmae Juss.(保留科名)●
57114 Borassodendron Becc. (1914);垂裂棕属(毛果榈属,树头榈属,树头木榈属,树头木属);Borassodendron ●☆
57115 Borassodendron machadonis (Ridl.) Becc.;垂裂棕(树头木);Borassodendron ●☆
57116 Borassus L. (1753);糖棕属(贝叶棕属,扁叶槟榔属,扇榈属,扇椰子属,扇叶糖棕属,树头榈属,树头棕属,糖椰属);Borassus, Sweetpalm ●
57117 Borassus aethiopium Mart.;非洲糖棕(阿比尼亚扇椰子,非洲扇棕,树头榈);Rhum Palm ●☆
57118 Borassus aethiopium Mart. var. senegalensis Becc.;非洲棕;African Rhun Palm ●☆
57119 Borassus caudatus Lour. = Arenga caudata (Lour.) H. E. Moore ●
57120 Borassus caudatus Lour. = Didymosperma caudatum (Lour.) H. Wendl. et Drude ●
57121 Borassus deleb Becc. = Borassus aethiopium Mart. ●☆
57122 Borassus flabellifer L.;糖棕(桹多,多罗,扇椰子,扇叶树头榈,扇叶树头棕,扇棕);Bassine Fibre, Brabtree, Lontar, Meelalla, Odiyal Flour, Palmyra, Palmyra Palm, Sweetpalm, Toddy Palm, Wine Palm ●
57123 Borassus flabellifer L. var. aethiopum (Mart.) Warb. = Borassus aethiopium Mart. ●☆
57124 Borassus flabellifer L. var. madagascariensis Jum. et H. Perrier = Borassus madagascariensis (Jum. et H. Perrier) Jum. et H. Perrier ●☆
57125 Borassus flabelliformis Murray = Borassus flabellifer L. ●
57126 Borassus heineana Becc.;海涅糖棕●☆
57127 Borassus madagascariensis (Jum. et H. Perrier) Jum. et H. Perrier;马岛糖棕●☆
57128 Borbasia Gand. = Dianthus L. ■
57129 Borbonia Adans. = Ocotea Aubl. ●☆
57130 Borbonia L. = Aspalathus L. ●☆
57131 Borbonia Mill. = Persea Mill.(保留属名)●
57132 Borbonia alata Willd. ex Spreng. = Aspalathus crenata (L.) R. Dahlgren ●☆
57133 Borbonia alpestris Benth. = Aspalathus alpestris (Benth.) R. Dahlgren ●☆
57134 Borbonia angustifolia Lam. = Aspalathus angustifolia (Lam.) R. Dahlgren ●☆
57135 Borbonia barbata Lam. = Aspalathus barbata (Lam.) R. Dahlgren ●☆
57136 Borbonia candolleana Eckl. et Zeyh. = Aspalathus perforata (Thunb.) R. Dahlgren ●☆
57137 Borbonia capitata (Thunb.) Poir. = Liparia capitata Thunb. ■☆
57138 Borbonia ciliata Willd. = Aspalathus perforata (Thunb.) R. Dahlgren ●☆

57139 Borbonia commutata Vogel = Aspalathus commutata (Vogel) R. Dahlgren ●☆

57140 Borbonia complicata Benth. = Aspalathus complicata (Benth.) R. Dahlgren ●☆

57141 Borbonia cordata L. = Aspalathus cordata (L.) R. Dahlgren ●☆

57142 Borbonia cordifolia Lam. = Aspalathus cordata (L.) R. Dahlgren ●☆

57143 Borbonia crenata L. = Aspalathus crenata (L.) R. Dahlgren ●☆

57144 Borbonia decipiens E. Mey. = Aspalathus angustifolia (Lam.) R. Dahlgren ●☆

57145 Borbonia elliptica E. Phillips = Aspalathus elliptica (E. Phillips) R. Dahlgren ●☆

57146 Borbonia ericifolia L. = Amphithalea ericifolia (L.) Eckl. et Zeyh. ■☆

57147 Borbonia graminifolia (L.) Lam. = Liparia graminifolia L. ■☆

57148 Borbonia hirsuta (Thunb.) Poir. = Liparia hirsuta Thunb. ■☆

57149 Borbonia laevigata L. = Liparia laevigata (L.) Thunb. ■☆

57150 Borbonia lanceolata L. = Aspalathus angustifolia (Lam.) R. Dahlgren ●☆

57151 Borbonia lanceolata L. f. angustifolia (Lam.) Walp. = Aspalathus angustifolia (Lam.) R. Dahlgren ●☆

57152 Borbonia lanceolata L. var. robusta E. Phillips = Aspalathus angustifolia (Lam.) R. Dahlgren subsp. robusta (E. Phillips) R. Dahlgren ●☆

57153 Borbonia lanceolata L. var. villosa E. Phillips = Aspalathus angustifolia (Lam.) R. Dahlgren ●☆

57154 Borbonia latifolia Benth. = Aspalathus elliptica (E. Phillips) R. Dahlgren ●☆

57155 Borbonia leiantha E. Phillips = Aspalathus crenata (L.) R. Dahlgren ●☆

57156 Borbonia monosperma DC. = Aspalathus monosperma (DC.) R. Dahlgren ●☆

57157 Borbonia multiflora (Harv.) E. Phillips = Aspalathus perfoliata (Lam.) R. Dahlgren subsp. phillipsii R. Dahlgren ●☆

57158 Borbonia myrtifolia (Thunb.) Poir. = Liparia myrtifolia Thunb. ■☆

57159 Borbonia parviflora Lam. = Aspalathus crenata (L.) R. Dahlgren ●☆

57160 Borbonia perfoliata Lam. = Aspalathus perfoliata (Lam.) R. Dahlgren ●☆

57161 Borbonia perfoliata Thunb. = Rafnia acuminata (E. Mey.) G. J. Campb. et B. -E. van Wyk ■☆

57162 Borbonia perforata Eckl. et Zeyh. = Aspalathus perforata (Thunb.) R. Dahlgren ●☆

57163 Borbonia perforata Thunb. = Aspalathus perforata (Thunb.) R. Dahlgren ●☆

57164 Borbonia perforata Thunb. var. breviflora Walp. = Aspalathus perforata (Thunb.) R. Dahlgren ●☆

57165 Borbonia perforata Thunb. var. pauciflora Harv. = Aspalathus perforata (Thunb.) R. Dahlgren ●☆

57166 Borbonia perforata Thunb. var. pluriflora Harv. = Aspalathus perforata (Thunb.) R. Dahlgren ●☆

57167 Borbonia pinifolia Marloth = Aspalathus linearis (Burm. f.) R. Dahlgren ●☆

57168 Borbonia pungens Mundt ex Benth. = Aspalathus alpestris (Benth.) R. Dahlgren ●☆

57169 Borbonia ruscifolia Sims = Aspalathus crenata (L.) R. Dahlgren ●☆

57170 Borbonia serrulata Thunb. = Aspalathus crenata (L.) R. Dahlgren ●☆

57171 Borbonia sphaerica (L.) Lam. = Liparia splendens (Burm. f.) Bos et de Wit ■☆

57172 Borbonia tomentosa L. = Liparia vestita Thunb. ■☆

57173 Borbonia trinervia L. = Cliffortia ruscifolia L. ●☆

57174 Borbonia umbellifera (Thunb.) Poir. = Liparia umbellifera Thunb. ■☆

57175 Borbonia undulata Thunb. = Aspalathus commutata (Vogel) R. Dahlgren ●☆

57176 Borbonia undulata Thunb. var. ciliata E. Phillips = Aspalathus commutata (Vogel) R. Dahlgren ●☆

57177 Borbonia undulata Thunb. var. multiflora Harv. = Aspalathus perfoliata (Lam.) R. Dahlgren subsp. phillipsii R. Dahlgren ●☆

57178 Borbonia vestita (Thunb.) Poir. = Liparia vestita Thunb. ■☆

57179 Borbonia villosa Harv. = Aspalathus lanifera R. Dahlgren ●☆

57180 Borbonia villosa Thunb. = Liparia angustifolia (Eckl. et Zeyh.) A. L. Schutte ■☆

57181 Borboraceae Dulac = Scheuchzeriaceae F. Rudolphi(保留科名) + Juncaginaceae Rich.(保留科名) + Alismataceae Vent.(保留科名)■

57182 Borboya Raf. = Hyacinthus L. ■☆

57183 Borckhausenia P. Gaertn., B. Mey. et Scherb. = Corydalis DC.(保留属名)■

57184 Borckhausenia P. Gaertn., B. Mey. et Scherb. = Pseudo-fumaria Medik. ■

57185 Borckhausenia Roth = Teedia Rudolphi ■●☆

57186 Bordasia Krapov. (2003);巴拉圭锦葵属☆

57187 Borderea Miégev. (1866);无翅薯蓣属■☆

57188 Borderea Miégev. = Dioscorea L.(保留属名)■

57189 Borderea pyrenaica Miégev.;无翅薯蓣■☆

57190 Borea Meisn. = Bovea Decne. ■

57191 Borea Meisn. = Lindenbergia Lehm. ■

57192 Borealluma Plowes = Caralluma R. Br. ■

57193 Borealluma munbyana (Decne. ex Munby) Plowes = Apteranthes munbyana (Decne.) Meve et Liede ■☆

57194 Boreava Jaub. et Spach(1841);钩喙芥属(博里花属)■☆

57195 Boreava aptera Boiss. et Heldr.;无翅钩喙芥■☆

57196 Boreava orientalis Jaub. et Spach;东方钩喙芥(东方博里花)■☆

57197 Borellia Neck. = Cordia L.(保留属名)●

57198 Boretta Kuntze = Daboecia D. Don(保留属名)●☆

57199 Boretta Neck. = Daboecia D. Don(保留属名)●☆

57200 Boretta Neck. ex Baill. = Daboecia D. Don(保留属名)●☆

57201 Borinda Stapleton = Fargesia Franch. ●

57202 Borinda Stapleton(1994);北方箭竹属(北风箭竹属)■☆

57203 Borinda albocerea (J. R. Xue et T. P. Yi) Stapleton = Fargesia albo-cerea J. R. Xue et T. P. Yi ●

57204 Borinda edulis (J. R. Xue et T. P. Yi) Stapleton = Fargesia edulis J. R. Xue et T. P. Yi ●

57205 Borinda extensa (T. P. Yi) Stapleton = Fargesia extensa T. P. Yi ●

57206 Borinda farcta (T. P. Yi) Stapleton = Fargesia farcta T. P. Yi ●

57207 Borinda frigidis (T. P. Yi) Stapleton = Fargesia frigida T. P. Yi ●

57208 Borinda glabrifolia (T. P. Yi) Stapleton = Fargesia glabrifolia T. P. Yi ●

57209 Borinda grossa (T. P. Yi) Stapleton = Fargesia grossa T. P. Yi ●

57210 Borinda lushuiensis (J. R. Xue et T. P. Yi) Stapleton = Fargesia lushuiensis J. R. Xue et T. P. Yi ●

57211 Borinda macclureana (Bor) Stapleton = Fargesia macclureana (Bor) Stapleton ●
57212 Borinda papyrifera (T. P. Yi) Stapleton = Fargesia papyrifera T. P. Yi ●
57213 Borinda perlonga (J. R. Xue et T. P. Yi) Stapleton = Fargesia perlonga J. R. Xue et T. P. Yi ●
57214 Borinda setosa (T. P. Yi) Stapleton = Fargesia macclureana (Bor) Stapleton ●
57215 Boriskellera Terechov = Eragrostis Wolf ■
57216 Borismene Barneby(1972);月牛藤属●☆
57217 Borismene japutvnsis (C. Mart.) Barneby;月牛藤●☆
57218 Borissa Raf. = Asterolinon Hoffmanns. et Link ■☆
57219 Borissa Raf. = Lysimachia L. ●■
57220 Borissa Raf. ex Steud. = Lysimachia L. ●■
57221 Borith Adans. = Anabasis L. ●■
57222 Borkhausia Nutt. = Barkhausia Moench ■
57223 Borkhausia Nutt. = Crepis L. ■
57224 Borkonstia Ignatov = Aster L. ●■
57225 Borkonstia Ignatov = Krylovia Schischk. ■
57226 Borkonstia Ignatov = Rhinactinidia Novopokr. ■
57227 Bormiera Post et Kuntze = Bonniera Cordem. ■☆
57228 Borneacanthus Bremek. (1960);婆罗刺属●■☆
57229 Borneacanthus angustifolius Bremek.;窄叶婆罗刺■☆
57230 Borneacanthus grandifolius Bremek.;大叶婆罗刺■☆
57231 Borneacanthus mesargyreus (Hallier f.) Bremek.;婆罗刺●■☆
57232 Borneacanthus paniculatus Bremek.;圆锥婆罗刺■☆
57233 Borneacanthus parvus Bremek.;小管婆罗刺■☆
57234 Borneacanthus stenothyrsus Bremek.;细穗婆罗刺●■☆
57235 Borneodendron Airy Shaw(1963);三数大戟属●☆
57236 Borneodendron aenigmaticum Airy Shaw;三数大戟●☆
57237 Borneosicyos W. J. de Wilde(1998);加岛瓜属●☆
57238 Bornmuellera Hausskn. (1897);岩园葶苈属■●☆
57239 Bornmuellera angustifolia (Hausskn. ex Bornm.) Cullen et T. R. Dudley;狭叶岩园葶苈●☆
57240 Bornmuellera tymphaea Hausskn.;岩园葶苈●☆
57241 Bornmuellerantha Rothm. (1943);博恩列当属■☆
57242 Bornmuellerantha Rothm. = Odontites Ludw. ■
57243 Bornmuellerantha aucheri (Boiss.) Rothm.;博恩列当■☆
57244 Bornoa O. F. Cook = Attalea Kunth ●☆
57245 Borodinia N. Busch(1921);贝加尔芥属■☆
57246 Borodinia baicalensis N. Busch;贝加尔芥■☆
57247 Borojoa Cuatrec. (1949);博罗茜属●☆
57248 Borojoa claviflora (K. Schum.) Cuatrec.;棒花博罗茜●☆
57249 Borojoa lanceolata (Cham.) Cuatrec.;剑叶博罗茜●☆
57250 Borojoa patinoi Cuatrec.;博罗茜●☆
57251 Boronella Baill. (1872);小博龙香木属●☆
57252 Boronella Baill. = Boronia Sm. ●☆
57253 Boronella crassifolia Guillaumin;厚叶小博龙香木●☆
57254 Boronella pancheri Baill.;小博龙香木●☆
57255 Boronella parvifolia Baker f.;小叶小博龙香木●☆
57256 Boronella verticillata Baill. ex Guillaumin;轮生小博龙香木●☆
57257 Boronia Sm. (1798);博龙香木属(宝容木属,香波龙属);Boronia ●☆
57258 Boronia alata Sm.;翅博龙香木●☆
57259 Boronia anemonifolia A. Cunn.;黏博龙香木;Sticky Boronia ●☆
57260 Boronia caerulescens F. Muell.;蓝博龙香木;Blue Boronia ●☆
57261 Boronia clavata Paul G. Wilson;棍棒博龙香木;Boronia ●☆
57262 Boronia crenulata Sm.;细圆齿博龙香木;Boronia ●☆
57263 Boronia denticulata Sm.;细齿博龙香木;Boronia ●☆
57264 Boronia floribunda Sieber ex Spreng.;粉花博龙香木;Pale Pink Boronia ●☆
57265 Boronia heterophylla F. Muell.;红花博龙香木;Kalgan Boronia, Red Boronia ●☆
57266 Boronia ledifolia J. Gay;西尼博龙香木;Ledum Boronia, Sydny Boronia ●☆
57267 Boronia megastigma Nees ex Bartl.;褐色博龙香木(大柱宝容木,大柱香波龙);Brown Boronia, Scented Boronia ●☆
57268 Boronia megastigma Nees ex Bartl. 'Harlequin';哈乐奎褐色博龙香木●☆
57269 Boronia megastigma Nees ex Bartl. 'Heaven Scent';天香褐色博龙香木●☆
57270 Boronia megastigma Nees ex Bartl. 'Lutea';黄绿褐色博龙香木●☆
57271 Boronia mollis A. Cunn. ex Lindl.;软博龙香木;Soft Boronia ●☆
57272 Boronia molloyae J. R. Drumm.;高博龙香木;Tall Boronia ●☆
57273 Boronia muelleri (Benth.) Cheel;默勒博龙香木;Boronia ●☆
57274 Boronia pilosa Labill.;毛博龙香木;Hairy Boronia ●☆
57275 Boronia pinnata Sm.;羽叶博龙香木(粉花宝容木);Pinnate Boronia ●☆
57276 Boronia polygalifolia Sm.;远志叶龙香木;Milkwort Boronia ●☆
57277 Boronia serrulata Sm.;博龙香木;Native Boronia, Native Rose, Sydney Rock-rose ●☆
57278 Boroniaceae J. Agardh = Rutaceae Juss. (保留科名)●■
57279 Boroniaceae J. Agardh;博龙香木科●
57280 Borrachinea Lavy = Borago L. ■☆
57281 Borraginoides Boehm. (废弃属名) = Trichodesma R. Br. (保留属名)●■
57282 Borraginoides Moench = Trichodesma R. Br. (保留属名)●■
57283 Borraginoides aculeata Moench = Trichodesma africanum (L.) Sm. ■☆
57284 Borraginoides africana (L.) Hiern = Trichodesma africanum (L.) Sm. ■☆
57285 Borrago Mill. = Borago L. ■☆
57286 Borrera Ach. (废弃属名) = Borreria G. Mey. (保留属名)●■
57287 Borrera Spreng. = Borreria G. Mey. (保留属名)●■
57288 Borrera Spreng. = Spermacoce L. ●■
57289 Borreria G. Mey. (1818)(保留属名);丰花草属(半丰草属,糙叶丰花草属);Borreria ●■
57290 Borreria G. Mey. (保留属名) = Spermacoce L. ●■
57291 Borreria alata (Aubl.) DC = Spermacoce latifolia Aubl. ■
57292 Borreria alata (Aubl.) DC. = Borreria latifolia (Aubl.) K. Schum. ■
57293 Borreria andongensis (Hiern) K. Schum. = Spermacoce andongensis (Hiern) R. D. Good ■☆
57294 Borreria articularis (L. f.) F. N. Williams = Spermacoce articularis (L. f.) G. Mey. ■
57295 Borreria articularis (L. f.) F. N. Williams = Spermacoce hispida L. ■☆
57296 Borreria articularis (L. f.) G. Mey.;糙叶丰花草(捕鱼草,粗叶丰花草,铺地毡草,鸭舌癀,鸭舌癀舅);Scabrousleaf Borreria ■
57297 Borreria articularis (L. f.) G. Mey. = Spermacoce articularis (L. f.) G. Mey. ■
57298 Borreria arvensis (Hiern) K. Schum. = Spermacoce arvensis (Hiern) R. D. Good ■☆

57299 Borreria bambusicola Berhaut = Spermacoce bambusicola (Berhaut) J. -P. Lebrun et Stork ■☆

57300 Borreria bangweolensis R. E. Fr. = Spermacoce bangweolensis (R. E. Fr.) Verdc. ■☆

57301 Borreria bequaertii De Wild. = Spermacoce bequaertii (De Wild.) Verdc. ■☆

57302 Borreria chaetocephala (DC.) Hepper = Spermacoce chaetocephala DC. ■☆

57303 Borreria chaetocephala (DC.) Hepper var. minor Hepper = Spermacoce chaetocephala DC. var. minor (Hepper) Puff ■☆

57304 Borreria compacta (Hochst. ex Hiern) K. Schum. = Spermacoce chaetocephala DC. ■☆

57305 Borreria compressa Hutch. et Dalziel = Spermacoce hepperana Verdc. ■☆

57306 Borreria dibrachiata (Oliv.) K. Schum. = Spermacoce dibrachiata Oliv. ■☆

57307 Borreria diodon K. Schum. = Spermacoce kirkii (Hiern) Verdc. ■☆

57308 Borreria exilis L. O. Williams;瘦小丰花草■☆

57309 Borreria exilis L. O. Williams = Spermacoce exilis (L. O. Williams) C. D. Adams ex W. C. Burger et C. M. Taylor ■☆

57310 Borreria ferruginea (A. St. -Hil.) DC. = Mitracarpus hirtus (L.) DC. ■

57311 Borreria filifolia (Schumach. et Thonn.) K. Schum. = Spermacoce filifolia (Schumach. et Thonn.) J. -P. Lebrun et Stork ●☆

57312 Borreria filiformis (Hiern) Hutch. et Dalziel = Spermacoce filiformis Hiern ■☆

57313 Borreria filituba K. Schum. = Spermacoce filituba (K. Schum.) Verdc. ■☆

57314 Borreria galeopsidis (DC.) Berhaut;热非丰花草■☆

57315 Borreria garuensis K. Krause;加鲁丰花草■☆

57316 Borreria globularioides Cham. et Schltdl. = Spermacoce verticillata L. ●☆

57317 Borreria gracilis L. O. Williams = Spermacoce exilis (L. O. Williams) C. D. Adams ex W. C. Burger et C. M. Taylor ■☆

57318 Borreria graminifolia M. Martens et Galeotti = Spermacoce verticillata L. ●☆

57319 Borreria hebecarpa Hochst. ex A. Rich. = Spermacoce chaetocephala DC. var. minor (Hepper) Puff ■☆

57320 Borreria hedraeanthoides Chiov.;草钟丰花草■☆

57321 Borreria hispida (L.) K. Schum.;刚毛丰花草;Hispid Borreria ■

57322 Borreria hispida (L.) K. Schum. = Borreria articularis (L. f.) G. Mey. ■

57323 Borreria hispida (L.) K. Schum. = Spermacoce hispida L. ■☆

57324 Borreria hispida K. Schum. = Borreria hispida (L.) K. Schum. ■

57325 Borreria hockii De Wild. = Spermacoce hockii (De Wild.) Dessein ●☆

57326 Borreria intricans Hepper = Spermacoce intricans (Hepper) H. M. Burkill ●☆

57327 Borreria kohautiana Cham. et Schltdl. = Spermacoce verticillata L. ●☆

57328 Borreria kotschyana (Oliv.) K. Schum. = Spermacoce chaetocephala DC. ■☆

57329 Borreria laevia (Lamkey) Griseb.;平滑丰花草(台湾丰花草,小破得力);Small Borreria, White Broom ■

57330 Borreria laevigata M. Martens et Galeotti = Spermacoce verticillata L. ●☆

57331 Borreria laevis (Lam.) Griseb. = Spermacoce laevis Lam. ■☆

57332 Borreria laevis (Lam.) Griseb. = Spermacoce tenuior L. ■☆

57333 Borreria latifolia (Aubl.) K. Schum.;阔叶丰花草(阔叶破得力,阔叶鸭舌癀舅,阔叶鸭舌舅,翼丰花草);Broad-leaved Borreria ■

57334 Borreria latifolia (Aubl.) K. Schum. = Spermacoce latifolia Aubl. ■

57335 Borreria latituba K. Schum. = Spermacoce latituba (K. Schum.) Verdc. ■☆

57336 Borreria leucadea (Hochst. ex Hiern) K. Schum. = Spermacoce stachydea DC. ●☆

57337 Borreria macrantha Hepper = Spermacoce ivorensis Govaerts ●☆

57338 Borreria macrantha Hepper var. glabra? = Spermacoce ivorensis Govaerts ●☆

57339 Borreria miegei Assemien = Spermacoce octodon (Hepper) J. -P. Lebrun et Stork ●☆

57340 Borreria minutiflora K. Schum. = Spermacoce minutiflora (K. Schum.) Verdc. ●☆

57341 Borreria molleri Gand. = Spermacoce verticillata L. ●☆

57342 Borreria monticola Mildbr. ex Hutch. et Dalziel = Spermacoce ruelliae DC. ■☆

57343 Borreria natalensis (Hochst.) S. Moore = Spermacoce natalensis Hochst. ●☆

57344 Borreria neglecta A. Rich. = Spermacoce abyssinica Kuntze ●☆

57345 Borreria oaxacana M. Martens et Galeotti = Spermacoce verticillata L. ●☆

57346 Borreria octodon Hepper = Spermacoce octodon (Hepper) J. -P. Lebrun et Stork ●☆

57347 Borreria ocymoides (Burm. f.) DC.;罗勒丰花草●☆

57348 Borreria ocymoides DC. = Borreria ocymoides (Burm. f.) DC. ●☆

57349 Borreria oligantha K. Schum.;寡花丰花草●☆

57350 Borreria paludosa Hepper = Spermacoce quadrisulcata (Bremek.) Verdc. ■☆

57351 Borreria paolii Chiov. = Spermacoce paolii (Chiov.) Verdc. ●☆

57352 Borreria phyteuma (Schweinf. ex Hiern) Dandy = Spermacoce phyteuma Schweinf. ex Hiern ■☆

57353 Borreria podocephala DC. = Spermacoce verticillata L. ●☆

57354 Borreria princeae K. Schum. var. pubescens Hepper = Spermacoce princeae (K. Schum.) Verdc. var. pubescens (Hepper) Verdc. ●☆

57355 Borreria pusilla (Wall.) DC. = Borreria stricta (L. f.) G. Mey. ●

57356 Borreria pusilla (Wall.) DC. = Spermacoce pusilla Wall. ■

57357 Borreria quadrisulcata Bremek. = Spermacoce quadrisulcata (Bremek.) Verdc. ■☆

57358 Borreria radiata DC. = Spermacoce radiata (DC.) Hiern ■☆

57359 Borreria repens DC.;二萼丰花草(蔓鸭舌癀舅)●

57360 Borreria repens DC. = Spermacoce exilis (L. O. Williams) C. D. Adams ex W. C. Burger et C. M. Taylor ■☆

57361 Borreria repens DC. = Spermacoce mauritiana Gideon ■

57362 Borreria rhodesica Suess. = Spermacoce senensis (Klotzsch) Hiern ■☆

57363 Borreria ruelliae (DC.) Thoms = Spermacoce ruelliae DC. ■☆

57364 Borreria scabra (Schumach. et Thonn.) K. Schum. = Spermacoce ruelliae DC. ■☆

57365 Borreria senensis (Klotzsch) K. Schum. = Spermacoce senensis (Klotzsch) Hiern ■☆

57366 Borreria setosa (Hiern) K. Schum. = Spermacoce octodon (Hepper) J. -P. Lebrun et Stork ●☆

57367 Borreria shandongensis F. Z. Li et X. D. Chen = Spermacoce

shangdongensis (F. Z. Li et X. D. Chen) X. D. Chen et al. ●
57368 Borreria somalica K. Schum. = Spermacoce sphaerostigma (A. Rich.) Vatke ■☆
57369 Borreria squarrosa Schinz = Spermacoce senensis (Klotzsch) Hiern ■☆
57370 Borreria stachydea (DC.) Hutch. et Dalziel var. phyllocephala (DC.) Hepper = Spermacoce stachydea DC. var. phyllocephala (DC.) J. -P. Lebrun et Stork ■☆
57371 Borreria stolzii K. Krause = Spermacoce senensis (Klotzsch) Hiern ■☆
57372 Borreria stricta (L. f.) G. Mey. = Spermacoce stricta L. f. ■
57373 Borreria stricta (L. f.) G. Mey. = Spermacoce verticillata L. ●☆
57374 Borreria subvulgata K. Schum. = Spermacoce subvulgata (K. Schum.) J. G. Garcia ●☆
57375 Borreria tenuiflora Chiov. = Spermacoce filituba (K. Schum.) Verdc. ■☆
57376 Borreria tenuissima (Hiern) K. Schum. = Spermacoce tenuissima Hiern ●☆
57377 Borreria tetraodon K. Schum. = Spermacoce bequaertii (De Wild.) Verdc. ■☆
57378 Borreria thymoidea (Hiern) K. Schum. = Spermacoce thymoidea (Hiern) Verdc. ■☆
57379 Borreria velorensis Berhaut = Spermacoce hepperana Verdc. ■☆
57380 Borreria verticillata (L.) G. Mey.;轮生丰花草●☆
57381 Borreria verticillata (L.) G. Mey. = Spermacoce verticillata L. ●☆
57382 Borreria verticillata (L.) G. Mey. var. thymiformis B. L. Rob. = Spermacoce verticillata L. ●☆
57383 Borrichia Adans. (1763);滨菊蒿属;Sea Oxeye Daisy ●■☆
57384 Borrichia arborescens (L.) DC.;匍匐滨菊蒿●☆
57385 Borrichia frutescens (L.) DC.;灌木滨菊蒿●☆
57386 Borsczowia Bunge(1877);异子蓬属(浆果蓬属);Borsczowia ■
57387 Borsczowia aralocaspica Bunge;异子蓬;Common Borszczowia ■
57388 Borszczowia Bunge = Borsczowia Bunge ■
57389 Borthwickia W. W. Sm. (1912);节蒴木属;Borthwickia ●
57390 Borthwickia trifoliana Sm.;节蒴木;Trifoliolate Borthwickia ●
57391 Bortyodendraceae J. Agarth = Araliaceae Juss. (保留科名)●■
57392 Borya Labill. (1805);耐旱草属■☆
57393 Borya Montrouz. ex P. Beauv. = Oxera Labill. ●☆
57394 Borya Willd. = Adelia L. (保留属名)●☆
57395 Borya Willd. = Forestiera Poir. (保留属名)●☆
57396 Borya nitida Labill.;耐旱草■☆
57397 Boryaceae Rudall, M. W. Chase et Conran(1997);耐旱草科(澳韭兰科)■☆
57398 Borzicactella F. Ritter = Cleistocactus Lem. ●☆
57399 Borzicactella Johnson = Borzicactus Riccob. ■☆
57400 Borzicactella Johnson ex F. Ritter = Cleistocactus Lem. ●☆
57401 Borzicactus Riccob. (1909);花冠柱属■☆
57402 Borzicactus Riccob. = Cleistocactus Lem. ●☆
57403 Borzicactus acanthurus (Vaupel) Britton et Rose = Submatucana aurantiaca (Vaupel) Backeb. ■☆
57404 Borzicactus celsianus (Lem. ex Salm-Dyck) Kimnach = Oreocereus celsianus Riccob. ●☆
57405 Borzicactus formosus (F. Ritter) Donald;美丽黄仙玉■☆
57406 Borzicactus formosus (F. Ritter) Donald = Submatucana formosa (F. Ritter) Backeb. ■☆
57407 Borzicactus icosagonus (Kunth) Britton et Rose;金玉兔■☆
57408 Borzicactus roezlii (W. Haage) Backeb.;伟冠柱■☆
57409 Borzicactus samaipatanus (Cárdenas) Kimnach;彩舞柱■☆
57410 Borzicactus samnensis F. Ritter;雨树花冠柱■☆
57411 Borzicactus sepium Britton et Rose;篱笆花冠柱■☆
57412 Borzicactus serpens (Kunth) Kimnach;匍匐花冠柱■☆
57413 Borzicactus sextonianus (Backeb.) Kimnach;六强花冠柱■☆
57414 Borzicactus trollii (Kupper) Kimnach = Oreocereus trollii (Kupper) Kupper ●
57415 Borzicereus Frič et Kreuz. = Borzicactus Riccob. ■☆
57416 Borzicereus Frič et Kreuz. = Cleistocactus Lem. ●☆
57417 Bosca Vell. = Daphnopsis Mart. ●☆
57418 Boscheria Carrière = Bosscheria de Vriese et Teijsm. ●
57419 Boscheria Carrière = Ficus L. ●
57420 Boschia Korth. (1844);博什木棉属●☆
57421 Boschia Korth. = Durio Adans. ●
57422 Boschia acutifolia Mast.;尖叶博什木棉●☆
57423 Boschia excelsa Korth.;博什木棉●☆
57424 Boschia grandiflora Mast.;大花博什木棉●☆
57425 Boschia griffithii Mast.;格氏博什木棉●☆
57426 Boschia mansonii Gamble;曼森博什木棉●☆
57427 Boschia oblongifolia Ridl.;矩圆博什木棉●☆
57428 Boschniakia C. A. Mey. = Boschniakia C. A. Mey. ex Bong. ■
57429 Boschniakia C. A. Mey. ex Bong. (1832);草苁蓉属;Boschniakia, Cistancheherb ■
57430 Boschniakia glabra C. A. Mey. = Boschniakia rossica (Cham. et Schltdl.) B. Fedtsch. et Flerow ■
57431 Boschniakia glabra C. A. Mey. ex Bong. = Boschniakia rossica (Cham. et Schltdl.) B. Fedtsch. et Flerow ■
57432 Boschniakia handelii Beck = Boschniakia himalaica Hook. f. et Thomson ■
57433 Boschniakia handelii Beck f. minor Beck = Boschniakia himalaica Hook. f. et Thomson ■
57434 Boschniakia himalaica Hook. f. et Thomson;丁座草(半夏,川上氏肉苁蓉,枇杷玉,枇杷芋,千斤坠,千金重,西域丁座草,喜马拉雅草苁蓉);Himalayan Boschniakia, Himalayas Cistancheherb, Himalayas Xylanche ■
57435 Boschniakia himalaica Hook. f. et Thomson = Xylanche himalaica (Hook. f. et Thomson) Beck ■
57436 Boschniakia kawakamii Hayata = Boschniakia himalaica Hook. f. et Thomson ■
57437 Boschniakia rossica (Cham. et Schltdl.) B. Fedtsch. et Flerow;草苁蓉(不老草,苁蓉,列当,肉苁蓉);Russia Cistancheherb, Russian Boschniakia ■
57438 Boschniakia rossica (Cham. et Schltdl.) B. Fedtsch. et Flerow var. flavida Y. Zhang et J. Y. Ma;黄色草苁蓉;Yellow Russian Boschniakia ■
57439 Boscia Lam. = Boscia Lam. ex J. St. -Hil. (保留属名)■☆
57440 Boscia Lam. ex J. St. -Hil. (1793)(保留属名);非洲白花菜属■☆
57441 Boscia Thunb. (废弃属名) = Asaphes DC. ●☆
57442 Boscia Thunb. (废弃属名) = Boscia Lam. ex J. St. -Hil. (保留属名)■☆
57443 Boscia albitrunca (Burch.) Gilg et Gilg-Ben.;咖啡非洲白花菜;Coffee Tree ●☆
57444 Boscia longifolia Hadj-Moust.;长叶非洲白花菜●☆
57445 Boscia madagascariensis (DC.) Hadj-Moust.;马岛非洲白花菜●☆
57446 Boscia oleoides (Burch. ex DC.) Toelken;油非洲白花菜●☆
57447 Boscia plantefolii Hadj-Moust.;扁白花菜●☆

57448 Boscia senegalensis (Pers.) Lam.;塞内加尔白花菜●☆
57449 Boscia welwitschii Gilg;韦尔白花菜●☆
57450 Bosciopsis B. C. Sun = Hypselandra Pax et K. Hoffm. ■☆
57451 Boscoa Post et Kuntze = Bosca Vell. ●☆
57452 Boscoa Post et Kuntze = Daphnopsis Mart. ●☆
57453 Bosea L. (1753);木苋属(浆苋藤属)●☆
57454 Bosea amherstiana (Moq.) Hook. f.;喜马拉雅木苋●☆
57455 Bosea yervamora L.;加那利木苋;Goldenrod Tree, Goldenrod-tree, Hediondo, Hierbamora ●☆
57456 Bosia Mill. = Bosea L. ●☆
57457 Bosistoa F. Muell. = Bosistoa F. Muell. ex Benth. ●☆
57458 Bosistoa F. Muell. ex Benth. (1863);巴博芸香属●☆
57459 Bosistoa euodiformis F. Muell.;巴博芸香●☆
57460 Bosleria A. Nelson = Solanum L. ●■
57461 Bosqueia Thouars = Bosqueia Thouars ex Baill. ●☆
57462 Bosqueia Thouars ex Baill. (1863);热非桑属(鳞桑属)●☆
57463 Bosqueia Thouars ex Baill. = Trilepisium Thouars ●☆
57464 Bosqueia angolensis Ficalho = Trilepisium madagascariense DC. ●☆
57465 Bosqueia calcicola Léandri = Trilepisium madagascariense DC. ●☆
57466 Bosqueia carvalhoana Engl. = Trilepisium madagascariense DC. ●☆
57467 Bosqueia cerasiflora Volkens ex Engl. = Trilepisium madagascariense DC. ●☆
57468 Bosqueia danguyana Léandri = Trilepisium madagascariense DC. ●☆
57469 Bosqueia manongarivensis Léandri = Trilepisium madagascariense DC. ●☆
57470 Bosqueia occidentalis Léandri = Trilepisium madagascariense DC. ●☆
57471 Bosqueia orientalis Léandri = Trilepisium madagascariense DC. ●☆
57472 Bosqueia phoberos Baill.;热非桑;Bosqueia ●☆
57473 Bosqueia phoberos Baill. = Trilepisium madagascariense DC. ●☆
57474 Bosqueia spinosa Engl. = Chaetachme aristata Planch. ●☆
57475 Bosqueia welwitschii Engl. = Trilepisium madagascariense DC. ●☆
57476 Bosqueiopsis De Wild. et T. Durand (1901);拟热非桑属(假鳞桑属)●☆
57477 Bosqueiopsis carvalhoana Engl.;拟热非桑●☆
57478 Bosqueiopsis gilletii De Wild. et T. Durand;吉勒特拟热非桑●☆
57479 Bosqueiopsis lujae De Wild.;卢氏拟热非桑●☆
57480 Bosqueiopsis parvifolia Engl.;小叶拟热非桑●☆
57481 Bosquiea B. D. Jacks. = Bosqueia Thouars ex Baill. ●☆
57482 Bosscheria de Vriese et Teijsm. = Ficus L. ●
57483 Bossea (DC.) Rchb. = Cynosbata (DC.) Rchb. ●■
57484 Bossekia Neck. = Rubus L. ●■
57485 Bossekia Neck. ex Greene = Rubacer Rydb. ●■☆
57486 Bossekia Neck. ex Greene = Rubus L. ●■
57487 Bossekia Raf. = Waldsteinia Willd. ■
57488 Bossera Léandri = Alchornea Sw. ●
57489 Bossiaea Vent. (1800);波思豆属●☆
57490 Bossiaea aquifolium Benth.;冬青叶波思豆●☆
57491 Bossiaea cinerea R. Br.;灰毛波思豆;Showy Bossiaea ●☆
57492 Bossiaea cordifolia Sweet;心叶波思豆●☆
57493 Bossiaea foliosa A. Cunn.;多叶波思豆;Leafy Bossiaea ●☆
57494 Bossiaea heterophylla Vent.;变叶波思豆;Variable Bossiaea ●☆
57495 Bossiaea kiamensis Benth.;佳木波思豆●☆
57496 Bossiaea lenticularis DC.;双凸镜波思豆●☆
57497 Bossiaea leptacantha E. Pritz.;细花波思豆●☆
57498 Bossiaea linophylla R. Br.;窄叶波思豆●☆
57499 Bossiaea rhombifolia DC.;菱叶波思豆●☆
57500 Bossiaea scolopendria (Andréws) Sm.;蜈蚣波思豆;Centipede Bossiaea, Leafless Bossiaea ●☆
57501 Bossiaea walkeri F. Muell.;仙人掌波思豆;Cactus Bossiaea ●☆
57502 Bossiena B. D. Jacks. = Bossiaea Vent. ●☆
57503 Bostrychanthera Benth. (1876);毛药花属(环药花属);Bostrychanthera ●★
57504 Bostrychanthera deflexa Benth.;毛药花(垂花铃子香,华麝香草,环药花);Deflexed Bostrychanthera ■
57505 Bostrychanthera yaoshanensis S. L. Mo et F. N. Wei;瑶山毛药花;Yaoshan Bostrychanthera ■
57506 Bostrychode Miq. ex O. Berg = Syzygium R. Br. ex Gaertn. (保留属名)●
57507 Boswellia Roxb. = Boswellia Roxb. ex Colebr. ●☆
57508 Boswellia Roxb. ex Colebr. (1807);乳香树属;Frankincense, Incense, Incense Tree, Incense-tree, Olibanum ●☆
57509 Boswellia bhar-dajiana Birdw.;鲍达乳香树●☆
57510 Boswellia bhar-dajiana Birdw. = Boswellia sacra Flueck. ●☆
57511 Boswellia bhar-dajiana Birdw. var. serrulata Engl. = Boswellia sacra Flueck. ●☆
57512 Boswellia boranensis Engl. = Boswellia rivae Engl. ●☆
57513 Boswellia brichettii (Chiov.) Chiov. = Lannea obovata (Hook. f. ex Oliv.) Engl. ●☆
57514 Boswellia bullata Thulin;泡状乳香树●☆
57515 Boswellia campestris Engl. = Boswellia neglecta S. Moore ●☆
57516 Boswellia carteri Birdw.;乳香树(阿拉伯乳香树,卡氏乳香树);Bible Frankincense ●☆
57517 Boswellia carteri Birdw. var. undulato-crenata Engl. = Boswellia sacra Flueck. ●☆
57518 Boswellia chariensis Guillaumin = Boswellia papyrifera (Delile) Hochst. ●☆
57519 Boswellia dalzielii Hutch.;达尔齐尔乳香●☆
57520 Boswellia elegans Engl. = Boswellia neglecta S. Moore ●☆
57521 Boswellia frereana Birdw.;弗里乳香树;African Elemi, E. African Elemi, Elemi Frankincense ●☆
57522 Boswellia globosa Thulin;球形乳香树●☆
57523 Boswellia hildebrandtii Engl. = Boswellia neglecta S. Moore ●☆
57524 Boswellia holstii Engl. = Boswellia neglecta S. Moore ●☆
57525 Boswellia madagascariensis Capuron;马岛乳香树●☆
57526 Boswellia microphylla Chiov. = Boswellia neglecta S. Moore ●☆
57527 Boswellia multifoliolata Engl. = Boswellia neglecta S. Moore ●☆
57528 Boswellia neglecta S. Moore;野乳香树●☆
57529 Boswellia occidentalis Engl. = Boswellia papyrifera (Delile) Hochst. ●☆
57530 Boswellia odorata Hutch. = Boswellia papyrifera (Delile) Hochst. ●☆
57531 Boswellia ogadensis Vollesen;欧加登乳香树●☆
57532 Boswellia papyrifera (Delile) Hochst.;纸乳香树;Elephant Tree ●☆
57533 Boswellia papyrifera Hochst. = Boswellia papyrifera (Delile) Hochst. ●☆
57534 Boswellia pirottae Chiov.;皮罗特乳香树●☆
57535 Boswellia rivae Engl.;沟乳香树●☆
57536 Boswellia ruspoliana Engl. = Boswellia rivae Engl. ●☆
57537 Boswellia sacra Flueck.;荐骨乳香树;Frankincense, Frankincense Tree, Somalia ●☆
57538 Boswellia sahariensis A. Chev.;萨哈里乳香树●☆

57539 Boswellia serrata Roxb. ex Colebr.;齿叶乳香树(返魂香);Indian Frankincense,Indian Olibanum,Salai Tree ●☆
57540 Boswellia socotrana Balf. f.;索科特拉乳香树●☆
57541 Boswellia undulato-crenata (Engl.) Engl. = Boswellia sacra Flueck. ●☆
57542 Botelua Lag. = Bouteloua Lag.(保留属名)■
57543 Botherbe Steud. ex Klatt = Calydorea Herb. ■☆
57544 Bothriochilus Lem. = Coelia Lindl. ■☆
57545 Bothriochloa Kuntze(1891);孔颖草属(臭根子草属,白羊草属);Bothriochloa,Holeglumegrass ■
57546 Bothriochloa anamitica Kuntze = Bothriochloa bladhii (Retz.) S. T. Blake ■
57547 Bothriochloa assimilis (Steud.) Ohwi = Capillipedium assimile (Steud.) A. Camus ■
57548 Bothriochloa bladhii (Retz.) S. T. Blake;臭根子草;Australian Bluestem,Caucasian Bluestem,Intermediate Bothriochloa,Stinkroot Holeglumegrass ■
57549 Bothriochloa bladhii (Retz.) S. T. Blake var. punctata (Roxb.) R. R. Stewart;孔颖臭根子草;Punctate Bothriochloa,Punctate Stinkroot Holeglumegrass ■
57550 Bothriochloa caucasica (Trin.) C. E. Hubb. = Bothriochloa bladhii (Retz.) S. T. Blake ■
57551 Bothriochloa caucasica (Trin.) Henrard;高加索孔颖草■
57552 Bothriochloa caucasica Trin. = Bothriochloa bladhii (Retz.) S. T. Blake ■
57553 Bothriochloa glabra (Roxb.) A. Camus;光孔颖草(歧穗臭根子草);Nake Bothriochloa,Nake Holeglumegrass ■
57554 Bothriochloa glabra (Roxb.) A. Camus = Bothriochloa bladhii (Retz.) S. T. Blake ■
57555 Bothriochloa glabra (Roxb.) A. Camus subsp. haenkei (J. Presl) Henrard = Bothriochloa bladhii (Retz.) S. T. Blake ■
57556 Bothriochloa glabra (Roxb.) A. Camus var. epunctata Jacks. = Bothriochloa bladhii (Retz.) S. T. Blake ■
57557 Bothriochloa gracilis W. Z. Fang;细瘦孔颖草;Weak Bothriochloa,Weak Holeglumegrass ■
57558 Bothriochloa gracilis W. Z. Fang = Pseudosorghum fasciculare (Roxb.) A. Camus ■
57559 Bothriochloa haenkei (J. Presl) Ohwi = Bothriochloa bladhii (Retz.) S. T. Blake ■
57560 Bothriochloa insculpta (Hochst. ex A. Rich.) A. Camus;雕刻孔颖草■☆
57561 Bothriochloa insculpta (Hochst. ex A. Rich.) A. Camus var. hirta (Chiov.) Cufod. = Bothriochloa insculpta (Hochst. ex A. Rich.) A. Camus ■☆
57562 Bothriochloa insculpta (Hochst. ex A. Rich.) A. Camus var. vegetior (Hack.) C. E. Hubb. = Bothriochloa bladhii (Retz.) S. T. Blake ■
57563 Bothriochloa intermedia (R. Br.) A. Camus = Bothriochloa bladhii (Retz.) S. T. Blake ■
57564 Bothriochloa intermedia (R. Br.) A. Camus var. acidula (Stapf) C. E. Hubb. = Bothriochloa bladhii (Retz.) S. T. Blake ■
57565 Bothriochloa intermedia (R. Br.) A. Camus var. punctata (Roxb.) Keng = Bothriochloa bladhii (Retz.) S. T. Blake ■
57566 Bothriochloa intermedia (R. Br.) A. Camus var. punctata (Roxb.) Keng = Bothriochloa bladhii (Retz.) S. T. Blake var. punctata (Roxb.) R. R. Stewart ■
57567 Bothriochloa ischaemum (L.) Keng;白羊草(白草,东印度须芒草);Digitate Bothriochloa,East Indies Bluestem,Whitesheep Holeglumegrass,Yellow Bluestem ■
57568 Bothriochloa ischaemum (L.) Keng f. songarica (Rupr.) Kitag.;准噶尔白羊草;Turkestan Bluestem,Yellow Bluestem ■☆
57569 Bothriochloa ischaemum (L.) Keng var. songarica (Rupr. ex Fisch. et C. A. Mey.) Celarier et J. R. Harlan = Bothriochloa ischaemum (L.) Keng ■
57570 Bothriochloa ischaemum (L.) Mansf. = Dichanthium ischaemum (L.) Roberty ■
57571 Bothriochloa kwashotensis (Hayata) Ohwi = Capillipedium kwashotense (Hayata) C. C. Hsu ■
57572 Bothriochloa laguroides (DC.) Herter;银色孔颖草;Silver Beardgrass,Silver Bluestem ■☆
57573 Bothriochloa nana W. Z. Fang;小孔颖草;Dwarf Bothriochloa,Dwarf Holeglumegrass ■
57574 Bothriochloa nana W. Z. Fang = Bothriochloa pertusa (L.) A. Camus ■
57575 Bothriochloa odorata (Lisboa) A. Camus = Bothriochloa bladhii (Retz.) S. T. Blake ■
57576 Bothriochloa parviflora (R. Br.) Ohwi = Capillipedium parviflorum (R. Br.) Stapf ■
57577 Bothriochloa parviflora (R. Br.) Ohwi var. specigera (Benth.) Ohwi = Capillipedium parviflorum (R. Br.) Stapf var. spicigerum (Benth.) C. C. Hsu ■
57578 Bothriochloa parviflora (R. Br.) Ohwi var. spicigera (Benth.) Ohwi = Capillipedium spicigerum (Benth.) S. T. Blake ■
57579 Bothriochloa pertusa (L.) A. Camus;孔颖草;Holeglumegrass,Perforated Bothriochloa,Pitted Beardgrass ■
57580 Bothriochloa pertusa (L.) A. Camus var. decipiens (Hack.) Maire et Weiller = Dichanthium insculptum (Hochst. ex A. Rich.) Clayton ■☆
57581 Bothriochloa pertusa (L.) A. Camus var. maroccana Maire = Dichanthium insculptum (Hochst. ex A. Rich.) Clayton ■☆
57582 Bothriochloa pertusa (L.) A. Camus var. panormitana (Parl.) Maire et Weiller = Dichanthium insculptum (Hochst. ex A. Rich.) Clayton ■☆
57583 Bothriochloa pertusa (L.) A. Camus var. tunetana Dubuis et Faurel = Dichanthium insculptum (Hochst. ex A. Rich.) Clayton ■☆
57584 Bothriochloa picta Ohwi = Capillipedium assimile (Steud.) A. Camus ■
57585 Bothriochloa pseudoischaemum Sultan et Stewart = Bothriochloa ischaemum (L.) Keng ■
57586 Bothriochloa punctata (Roxb.) L. Liou = Bothriochloa bladhii (Retz.) S. T. Blake var. punctata (Roxb.) R. R. Stewart ■
57587 Bothriochloa punctata Roxb. = Bothriochloa bladhii (Retz.) S. T. Blake var. punctata (Roxb.) R. R. Stewart ■
57588 Bothriochloa radicans (Lehm.) A. Camus;索马里孔颖草■☆
57589 Bothriochloa saccharoides (Sw.) Rydb.;甘蔗孔颖草;Silver Bluestem ■☆
57590 Bothriochloa saccharoides (Sw.) Rydb. = Bothriochloa laguroides (DC.) Herter ■☆
57591 Bothriochloa saccharoides (Sw.) Rydb. var. torreyana (Steud.) Gould = Bothriochloa laguroides (DC.) Herter ■☆
57592 Bothriochloa spicigera (Benth.) T. Koyama = Capillipedium spicigerum (Benth.) S. T. Blake ■
57593 Bothriochloa tuberculata W. Z. Fang;疣毛孔颖草;Tuberoculate Holeglumegrass ■

57594 Bothriochloa tuberculata W. Z. Fang = Dichanthium annulatum (Forssk.) Stapf ■

57595 Bothriochloa yunnanensis W. Z. Fang;云南孔颖草;Yunnan Bothriochloa, Yunnan Holeglumegrass ■

57596 Bothriochloa yunnanensis W. Z. Fang = Pseudosorghum fasciculare (Roxb.) A. Camus ■

57597 Bothriocline Oliv. ex Benth. (1873);宽肋瘦片菊属■☆

57598 Bothriocline aggregata (Hutch.) Wild et G. V. Pope;聚集宽肋瘦片菊■☆

57599 Bothriocline alternifolia O. Hoffm. = Erlangea alternifolia (O. Hoffm.) S. Moore ■☆

57600 Bothriocline amplexicaulis (Muschl.) Wild et G. V. Pope = Bothriocline inyangana N. E. Br. var. amplexicaulis (Muschl.) C. Jeffrey ■☆

57601 Bothriocline amplifolia (O. Hoffm. et Muschl.) M. G. Gilbert;大叶宽肋瘦片菊■☆

57602 Bothriocline angelinii Fiori = Bothriocline schimperi Oliv. et Hiern ex Benth. ■☆

57603 Bothriocline angolensis (Hiern) Wild et G. V. Pope;安哥拉宽肋瘦片菊■☆

57604 Bothriocline argentea (O. Hoffm.) Wild et G. V. Pope;银白宽肋瘦片菊■☆

57605 Bothriocline atroviolacea Wech.;暗堇色宽肋瘦片菊■☆

57606 Bothriocline attenuata (Muschl.) Lisowski;渐狭宽肋瘦片菊■☆

57607 Bothriocline auriculata (M. Taylor) C. Jeffrey;耳形宽肋瘦片菊■☆

57608 Bothriocline bagshawei (S. Moore) C. Jeffrey;巴格肖宽肋瘦片菊■☆

57609 Bothriocline bampsii Lisowski;邦氏宽肋瘦片菊■☆

57610 Bothriocline calycina (S. Moore) M. G. Gilbert = Erlangea calycina S. Moore ■☆

57611 Bothriocline carrissoi Wech.;卡里索宽肋瘦片菊■☆

57612 Bothriocline centauroides (S. Moore) M. G. Gilbert = Erlangea centauroides (S. Moore) S. Moore ■☆

57613 Bothriocline concinna (S. Moore) Wech.;整洁宽肋瘦片菊■☆

57614 Bothriocline congesta (M. Taylor) Wech.;密集宽肋瘦片菊■☆

57615 Bothriocline cuneifolia Lisowski;楔叶宽肋瘦片菊■☆

57616 Bothriocline diversifolia O. Hoffm. = Gutenbergia cordifolia Benth. ex Oliv. ■☆

57617 Bothriocline duemmeri (S. Moore) Wild = Bothriocline bagshawei (S. Moore) C. Jeffrey ■☆

57618 Bothriocline emilioides C. Jeffrey;一点红宽肋瘦片菊■☆

57619 Bothriocline ethulioides C. Jeffrey;都丽菊状宽肋瘦片菊■☆

57620 Bothriocline eupatorioides (Hutch. et B. L. Burtt) Wild et G. V. Pope = Bothriocline longipes (Oliv. et Hiern) N. E. Br. ■☆

57621 Bothriocline fusca (S. Moore) M. G. Gilbert;棕色宽肋瘦片菊■☆

57622 Bothriocline glabrescens C. Jeffrey;光滑宽肋瘦片菊■☆

57623 Bothriocline globosa (Robyns) C. Jeffrey;球形宽肋瘦片菊■☆

57624 Bothriocline glomerata (O. Hoffm. et Muschl.) C. Jeffrey;团集宽肋瘦片菊■☆

57625 Bothriocline grandicapitulata Lisowski;大头宽肋瘦片菊■☆

57626 Bothriocline hispida (S. Moore) Wild et G. V. Pope;硬毛宽肋瘦片菊■☆

57627 Bothriocline hoyoensis Lisowski;丰予宽肋瘦片菊■☆

57628 Bothriocline huillensis (Hiern) Wild et G. V. Pope;威拉宽肋瘦片菊■☆

57629 Bothriocline inyangana N. E. Br.;伊尼扬加宽肋瘦片菊■☆

57630 Bothriocline inyangana N. E. Br. var. amplexicaulis (Muschl.) C. Jeffrey;抱茎宽肋瘦片菊■☆

57631 Bothriocline ituriensis Lisowski;伊图里宽肋瘦片菊■☆

57632 Bothriocline katangensis Lisowski;加丹加宽肋瘦片菊■☆

57633 Bothriocline kundelungensis Lisowski;昆德龙宽肋瘦片菊■☆

57634 Bothriocline kungwensis C. Jeffrey;昆圭宽肋瘦片菊■☆

57635 Bothriocline laxa N. E. Br.;疏松宽肋瘦片菊■☆

57636 Bothriocline laxa N. E. Br. subsp. mbalensis Wild et G. V. Pope = Bothriocline mbalensis (Wild et G. V. Pope) C. Jeffrey ■☆

57637 Bothriocline leonardiana Lisowski;莱奥宽肋瘦片菊■☆

57638 Bothriocline linearifolia O. Hoffm. = Erlangea linearifolia (O. Hoffm.) S. Moore ■☆

57639 Bothriocline longipes (Oliv. et Hiern) N. E. Br.;长梗宽肋瘦片菊■☆

57640 Bothriocline madagascariensis (DC.) C. Jeffrey;马岛宽肋瘦片菊■☆

57641 Bothriocline malaissei Lisowski;马莱泽宽肋瘦片菊■☆

57642 Bothriocline marginata O. Hoffm. = Gutenbergia cordifolia Benth. ex Oliv. ■☆

57643 Bothriocline marungensis Lisowski;马龙加宽肋瘦片菊■☆

57644 Bothriocline mbalensis (Wild et G. V. Pope) C. Jeffrey;姆巴莱宽肋瘦片菊■☆

57645 Bothriocline microcephala (S. Moore) Wild et G. V. Pope;小头宽肋瘦片菊■☆

57646 Bothriocline milanjiensis (S. Moore) Wild et G. V. Pope;米兰吉宽肋瘦片菊■☆

57647 Bothriocline misera (Oliv. et Hiern) O. Hoffm. = Erlangea misera (Oliv. et Hiern) S. Moore ■☆

57648 Bothriocline monocephala (Hiern) Wild et G. V. Pope;单头宽肋瘦片菊■☆

57649 Bothriocline monticola (M. Taylor) Wech.;山地宽肋瘦片菊■☆

57650 Bothriocline mooreana (Alston) Wild et G. V. Pope = Bothriocline trifoliata (De Wild. et Muschl.) Wild et G. V. Pope ■☆

57651 Bothriocline moramballae (Oliv. et Hiern) O. Hoffm.;莫拉宽肋瘦片菊■☆

57652 Bothriocline nyungwensis Wech.;尼永圭宽肋瘦片菊■☆

57653 Bothriocline pauciseta O. Hoffm.;少刚毛宽肋瘦片菊■☆

57654 Bothriocline pauwelsii Lisowski;保韦尔斯宽肋瘦片菊■☆

57655 Bothriocline pectinata (O. Hoffm.) Wild et G. V. Pope;篦状宽肋瘦片菊■☆

57656 Bothriocline quercifolia C. Jeffrey;栎叶宽肋瘦片菊■☆

57657 Bothriocline richardsiae Wech. = Erlangea richardsiae (Wech.) C. Jeffrey ■☆

57658 Bothriocline ruwenzoriensis (S. Moore) C. Jeffrey;鲁文佐里宽肋瘦片菊■☆

57659 Bothriocline schimperi Oliv. et Hiern ex Benth.;欣珀宽肋瘦片菊■☆

57660 Bothriocline schimperi Oliv. et Hiern ex Benth. var. angolensis Hiern = Bothriocline angolensis (Hiern) Wild et G. V. Pope ■☆

57661 Bothriocline schimperi Oliv. et Hiern ex Benth. var. huillensis Hiern = Bothriocline huillensis (Hiern) Wild et G. V. Pope ■☆

57662 Bothriocline schimperi Oliv. et Hiern ex Benth. var. longipes Oliv. et Hiern = Bothriocline longipes (Oliv. et Hiern) N. E. Br. ■☆

57663 Bothriocline schimperi Oliv. et Hiern ex Benth. var. tomentosa Oliv. et Hiern = Bothriocline longipes (Oliv. et Hiern) N. E. Br. ■☆

57664 Bothriocline schinzii (O. Hoffm.) O. Hoffm. = Erlangea misera (Oliv. et Hiern) S. Moore ■☆

57665 Bothriocline smithii (S. Moore) M. G. Gilbert = Erlangea smithii S. Moore ■☆

57666 Bothriocline somalensis (O. Hoffm.) M. G. Gilbert = Erlangea smithii S. Moore ■☆

57667 Bothriocline steetziana Wild et G. V. Pope;斯蒂兹宽肋瘦片菊■☆

57668 Bothriocline subcordata (De Wild.) Wech.;近心形宽肋瘦片菊■☆

57669 Bothriocline tomentosa (Oliv. et Hiern) Wild et G. V. Pope = Bothriocline longipes (Oliv. et Hiern) N. E. Br. ■☆

57670 Bothriocline trifoliata (De Wild. et Muschl.) Wild et G. V. Pope;三小叶宽肋瘦片菊■☆

57671 Bothriocline ugandensis (S. Moore) M. G. Gilbert;乌干达宽肋瘦片菊■☆

57672 Bothriocline upembensis Lisowski;乌彭贝宽肋瘦片菊■☆

57673 Bothriocline wittei Lisowski;维特宽肋瘦片菊■☆

57674 Bothriopodium Rizzini = Urbanolophium Melch. ●☆

57675 Bothriospermum Bunge (1833);斑种草属(细叠子草属);Bothriospermum, Spotseed ■

57676 Bothriospermum asperugoides Siebold et Zucc. = Bothriospermum zeylanicum (J. Jacq.) Druce ■

57677 Bothriospermum bicarunculatum Fisch. et C. A. Mey. = Bothriospermum chinense Bunge ■

57678 Bothriospermum chinense Bunge;斑种草(蛤蟆草,细叠子草);China Spotseed, Chinese Bothriospermum ■

57679 Bothriospermum decumbens Kitag. = Bothriospermum kusnetzowii Bunge ex DC. ■

57680 Bothriospermum hispidissimum Hand.-Mazz.;云南斑种草(刚毛斑种菜,云南斑种菜);Hispid Bothriospermum, Yunnan Spotseed ■

57681 Bothriospermum kusnetzowii Bunge ex DC.;狭苞斑种草(细叠子草);Kusnezow Bothriospermum, Kusnezow Spotseed ■

57682 Bothriospermum majasculum (Hayata) Suzuki = Thyrocarpus sampsonii Hance ■

57683 Bothriospermum secundum Maxim.;多苞斑种草(蛤蟆草,毛萝菜,山蚂蝗,山蚂癀,野山蚂蝗);Manybract Bothriospermum, Manybract Spotseed ■

57684 Bothriospermum secundum Maxim. = Bothriospermum hispidissimum Hand.-Mazz. ■

57685 Bothriospermum tenellum (Hornem.) Fisch. et C. A. Mey.;细弱斑种草(柔弱斑种草);Leaf Between Flower ■

57686 Bothriospermum tenellum (Hornem.) Fisch. et C. A. Mey. = Bothriospermum zeylanicum (J. Jacq.) Druce ■

57687 Bothriospermum tenellum (Hornem.) Fisch. et C. A. Mey. var. asperugoides (Siebold et Zucc.) Maxim. = Bothriospermum zeylanicum (J. Jacq.) Druce ■

57688 Bothriospermum tenellum (Hornem.) Fisch. et C. A. Mey. var. majasculum Hayata = Thyrocarpus sampsonii Hance ■

57689 Bothriospermum zeylanicum (J. Jacq.) Druce;柔弱斑种草(鬼点灯,细迭子草,细叠子草,细茎斑种草,细累子草,小马耳朵);Slender Pitred-seed, Tender Bothriospermum, Tender Spotseed ■

57690 Bothriospermum zollingeri (A. DC.) Johnst. = Lithospermum zollingeri A. DC. ■

57691 Bothriospora Hook. f. (1870);孔子茜属☆

57692 Bothriospora corymbosa (Benth.) Hook. f.;孔子茜☆

57693 Bothrocaryum (Koehne) Pojark. (1950);灯台树属;Bothrocaryum, Lampstandtree ●

57694 Bothrocaryum (Koehne) Pojark. = Cornus L. ●

57695 Bothrocaryum controversum (Hemsl. ex Prain) Pojark.;灯台树(灯台木,鸡肫皮,六角树,猫猫头,瑞木,伞柄树,乌牙树);Giant Bothrocaryum, Giant Dogwood, Lampstandtree, Lantern Dogwood, Table Dogwood, Tabletop Dogwood, Wedding-cake Tree ●

57696 Bothrocaryum controversum (Hemsl. ex Prain) Pojark. = Cornus controversa Hemsl. ex Prain ●

57697 Bothrocaryum controversum (Hemsl. ex Prain) Pojark. = Swida controversa (Hemsl. ex Prain) Soják ●

57698 Bothrocaryum controversum (Hemsl.) Pojark. = Cornus controversa Hemsl. ex Prain ●

57699 Bothrocaryum longipetiolatum (Hayata) Pojark. = Bothrocaryum controversum (Hemsl. ex Prain) Pojark. ●

57700 Bothrocaryum longipetiolatum (Hayata) Pojark. = Cornus macrophylla Wall. ●

57701 Bothrocaryum longipetiolatum (Hayata) Pojark. = Swida macrophylla (Wall.) Soják ●

57702 Botor Adans. (废弃属名) = Psophocarpus Neck. ex DC. (保留属名) ■

57703 Botor tertragonolobus (L.) Kuntze = Psophocarpus tetragonolobus (L.) DC. ■

57704 Botria Lour. (废弃属名) = Ampelocissus Planch. (保留属名) ●

57705 Botria africana Lour. = Ampelocissus africana (Lour.) Merr. ●☆

57706 Botrophis Raf. = Cimicifuga L. ●■

57707 Botrophis Raf. = Megotrys Raf. ●■

57708 Botrya Juss. = Ampelocissus Planch. (保留属名) ●

57709 Botryadenia Fisch. et C. A. Mey. = Myriactis Less. ■

57710 Botryanthe Klotzsch = Fragariopsis A. St.-Hil. ●☆

57711 Botryanthe Klotzsch = Plukenetia L. ●☆

57712 Botryanthus Kunth = Muscari Mill. ■☆

57713 Botryarrhena Ducke (1932);串雄茜属☆

57714 Botryarrhena pendula Ducke;串雄茜☆

57715 Botrycarpum (A. Rich.) Opiz = Ribes L. ●

57716 Botrycarpum A. Rich. = Ribes L. ●

57717 Botrycarpum nigrum (L.) A. Rich. = Ribes nigrum L. ●

57718 Botrycarpum nigrum A. Rich. = Ribes nigrum L. ●

57719 Botryceras Willd. = Laurophyllus Thunb. ●☆

57720 Botryceras laurinum Willd. = Laurophyllus capensis Thunb. ●☆

57721 Botrycomus Fourr. = Leopoldia Parl. (保留属名) ■☆

57722 Botrycomus Fourr. = Muscari Mill. ■☆

57723 Botrydendrum Post et Kuntze = Botryodendrum Endl. ●☆

57724 Botrydendrum Post et Kuntze = Meryta J. R. Forst. et G. Forst. ●☆

57725 Botrydium Spach = Chenopodium L. ■●

57726 Botrydium Spach = Neobotrydium Moldenke ■●

57727 Botrydium botrys (L.) Small = Dysphania botrys (L.) Mosyakin et Clemants ■

57728 Botrylotus Post et Kuntze = Botryolotus Jaub. et Spach ■

57729 Botrylotus Post et Kuntze = Trigonella L. ■

57730 Botrymorus Miq. = Pipturus Wedd. ●

57731 Botryocarpium (A. Rich.) Spach = Ribes L. ●

57732 Botryocarpium Spach = Ribes L. ●

57733 Botryocytinus (Baker f.) Watan. = Cytinus L. (保留属名) ■☆

57734 Botryocytinus baroni (Baker f.) Watan. = Cytinus baroni Baker f. ■☆

57735 Botryodendraceae J. Agardh = Araliaceae Juss. (保留科名) ●■

57736 Botryodendrum Endl. = Meryta J. R. Forst. et G. Forst. ●☆

57737 Botryolides Wolf = Muscari Mill. ■☆

57738 Botryoloranthus (Engl. et K. Krause) Balle = Oedina Tiegh. ●☆

57739 Botryoloranthus (Engl. et K. Krause) Balle (1954);总状桑寄生

属●☆

57740 Botryoloranthus pendens (Engl. et K. Krause) Balle = Oedina pendens (Engl. et K. Krause) Polhill et Wiens ●☆

57741 Botryolotus Jaub. et Spach = Trigonella L. ■

57742 Botryolotus cachemirianus (Cambess.) Jaub. et Spach = Trigonella cachemiriana Cambess. ■

57743 Botryomeryta R. Vig. = Meryta J. R. Forst. et G. Forst. ●☆

57744 Botryopanax Miq. = Polyscias J. R. Forst. et G. Forst. ●

57745 Botryopanax fulvus (Hiern) Hutch. = Polyscias fulva (Hiern) Harms ●☆

57746 Botryophora Bompard(废弃属名) = Botryophora Hook. f. (保留属名)●☆

57747 Botryophora Hook. f. (1867)(保留属名);总状大戟属●☆

57748 Botryophora Post et Kuntze = Botryophora Hook. f. (保留属名)●☆

57749 Botryophora geniculata (Miq.) Beumee ex Airy Shaw;总状大戟●☆

57750 Botryopleuron Hemsl. = Veronicastrum Heist. ex Fabr. ■

57751 Botryopleuron axillare (Siebold et Zucc.) Hemsl. = Veronicastrum axillare (Siebold et Zucc.) T. Yamaz. ■

57752 Botryopleuron caulopterum (Hance) Airy Shaw = Veronicastrum caulopterum (Hance) T. Yamaz. ■

57753 Botryopleuron formosanum Masam. = Veronicastrum axillare (Siebold et Zucc.) T. Yamaz. ■

57754 Botryopleuron kitamurae (Ohwi) Ohwi = Veronicastrum formosanum (Masam.) T. Yamaz. ■

57755 Botryopleuron kitamurae Ohwi = Veronicastrum kitamurae (Ohwi) T. Yamaz. ■

57756 Botryopleuron latifolium Hemsl. = Veronicastrum latifolium (Hemsl.) T. Yamaz. ■

57757 Botryopleuron liukiuense Ohwi = Veronicastrum liukiuense (Ohwi) T. Yamaz. ■

57758 Botryopleuron longispicatum Merr. = Veronicastrum longispicatum (Merr.) T. Yamaz. ■

57759 Botryopleuron macrophyllum H. L. Li = Veronicastrum villosulum (Miq.) T. Yamaz. ■

57760 Botryopleuron plukenetii T. Yamaz. = Veronicastrum stenostachyum (Hemsl.) T. Yamaz. subsp. plukenetii (T. Yamaz.) D. Y. Hong ■

57761 Botryopleuron rhombifolium Hand.-Mazz. = Veronicastrum rhombifolium (Hand.-Mazz.) P. C. Tsoong ■

57762 Botryopleuron stenostachyum Hemsl. = Veronicastrum stenostachyum (Hemsl.) T. Yamaz. ■

57763 Botryopleuron tagawae Ohwi = Veronicastrum tagawae (Ohwi) T. Yamaz. ■☆

57764 Botryopleuron villosulum (Miq.) Makino = Veronicastrum villosulum (Miq.) T. Yamaz. ■

57765 Botryopleuron yamatsutae T. Yamaz. = Veronicastrum stenostachyum (Hemsl.) T. Yamaz. ■

57766 Botryopleuron yunnanense W. W. Sm. = Veronicastrum yunnanense (W. W. Sm.) T. Yamaz. ■

57767 Botryopsis Miers = Chondrodendron Ruiz et Pav. ●☆

57768 Botryoropis C. Presl = Barringtonia J. R. Forst. et G. Forst. (保留属名)●

57769 Botryosicyos Hochst. = Dioscorea L. (保留属名)■

57770 Botryosicyos pentaphyllus Hochst. = Dioscorea quartiniana A. Rich. ■☆

57771 Botryospora B. D. Jacks. = Botryophora Hook. f. (保留属名)●☆

57772 Botryostege Stapf = Cladothamnus Bong. ●☆

57773 Botryostege Stapf(1934);串盖杜鹃属●☆

57774 Botryostege bracteata (Maxim.) Stapf = Tripetaleia bracteata Maxim. ●☆

57775 Botrypanax Post et Kuntze = Botryopanax Miq. ●

57776 Botrypanax Post et Kuntze = Polyscias J. R. Forst. et G. Forst. ●

57777 Botryphile Salisb. = Muscari Mill. ■☆

57778 Botryphora Post et Kuntze = Botryophora Hook. f. (保留属名)●☆

57779 Botryropis Post et Kuntze = Barringtonia J. R. Forst. et G. Forst. (保留属名)●

57780 Botryropis Post et Kuntze = Botryoropis C. Presl ●

57781 Botrys Fourr. = Teucrium L. ●■

57782 Botrys Nieuwl. = Chenopodium L. ■●

57783 Botrys Rchb. ex Nieuwl. = Chenopodium L. ■●

57784 Botschantzevia Nabiev = Erysimum L. ■●

57785 Botschantzevia Nabiev(1972);哈萨克芥属■☆

57786 Botschantzevia karatavica (Lipsch.) Nabiev;哈萨克芥■☆

57787 Bottegoa Chiov. (1916);博特无患子属●☆

57788 Bottegoa insignis Chiov.;博特无患子●☆

57789 Bottionea Colla = Trichopetalum Lindl. ■☆

57790 Bottionea Colla(1834);智利吊兰属■☆

57791 Bottionea thysanthoides Colla;智利吊兰■☆

57792 Boucerosia Wight et Arn. = Caralluma R. Br. ■

57793 Boucerosia acutangula (Decne.) Decne. = Caralluma acutangula (Decne.) N. E. Br. ■☆

57794 Boucerosia cylindrica Brongn. = Echidnopsis cereiformis Hook. f. ■☆

57795 Boucerosia decaisneana Lem. = Orbea decaisneana (Lehm.) Bruyns ■☆

57796 Boucerosia edulis Edgew. = Caralluma edulis (Edgew.) Benth. ■☆

57797 Boucerosia incarnata (L. f.) N. E. Br. = Quaqua incarnata (L. f.) Bruyns ■☆

57798 Boucerosia mammillaris (L.) N. E. Br. = Quaqua mammillaris (L.) Bruyns ■☆

57799 Boucerosia maroccana Hook. f. = Apteranthes europaea (Guss.) Plowes ■☆

57800 Boucerosia munbyana Decne. ex Munby = Apteranthes munbyana (Decne.) Meve et Liede ■☆

57801 Boucerosia munbyana Decne. ex Munby var. hispanica Coincy = Apteranthes munbyana (Decne.) Meve et Liede ■☆

57802 Boucerosia russeliana Courbon ex Brongn. = Caralluma acutangula (Decne.) N. E. Br. ■☆

57803 Boucerosia socotrana Balf. f. = Caralluma socotrana (Balf. f.) N. E. Br. ■☆

57804 Boucerosia stocksiana Boiss. = Caralluma edulis (Edgew.) Benth. ■☆

57805 Boucerosia tombuctuensis A. Chev. = Caralluma acutangula (Decne.) N. E. Br. ■☆

57806 Bouchardatia Baill. (1867);布沙芸香属●☆

57807 Bouchardatia Baill. = Melicope J. R. Forst. et G. Forst. ●

57808 Bouchardatia australia Baill.;澳大利亚布沙芸香●☆

57809 Bouchea Cham. (1832)(保留属名);布谢草属●☆

57810 Bouchea Cham. (保留属名) = Chascanum E. Mey. (保留属名)●☆

57811 Bouchea adenostachya Schauer = Chascanum adenostachyum (Schauer) Moldenke ●☆

57812 Bouchea caespitosa H. Pearson = Chascanum caespitosum (H.

Pearson) Moldenke ●☆
57813 Bouchea cernua (L.) Schauer = Chascanum cernuum (L.) E. Mey. ●☆
57814 Bouchea cuneifolia (L. f.) Schauer = Chascanum cuneifolium (L. f.) E. Mey. ●☆
57815 Bouchea garepensis (E. Mey.) Schauer = Chascanum garipense E. Mey. ●☆
57816 Bouchea glandulifera H. Pearson = Chascanum garipense E. Mey. ●☆
57817 Bouchea guerkeana Loes. ex Dinter = Chascanum pumilum E. Mey. ●☆
57818 Bouchea hanningtonii Oliv. = Chascanum hanningtonii (Oliv.) Moldenke ●☆
57819 Bouchea hederacea Sond. = Chascanum hederaceum (Sond.) Moldenke ●☆
57820 Bouchea hederacea Sond. var. natalensis H. Pearson = Chascanum hederaceum (Sond.) Moldenke var. natalense (H. Pearson) Moldenke ●☆
57821 Bouchea incisa H. Pearson = Chascanum incisum (H. Pearson) Moldenke ●☆
57822 Bouchea integrifolia H. Pearson = Chascanum integrifolium (H. Pearson) Moldenke ●☆
57823 Bouchea krookii Gürke ex Zahlbr. = Chascanum krookii (Gürke ex Zahlbr.) Moldenke ●☆
57824 Bouchea latifolia Harv. = Chascanum latifolium (Harv.) Moldenke ●☆
57825 Bouchea latifolia Harv. var. glabrescens H. Pearson = Chascanum latifolium (Harv.) Moldenke var. glabrescens (H. Pearson) Moldenke ●☆
57826 Bouchea lignosa Dinter = Chascanum pumilum E. Mey. ●☆
57827 Bouchea linifolia A. Gray;亚麻叶布谢草;Flaxleaf Buchea ●☆
57828 Bouchea longipetala H. Pearson = Chascanum adenostachyum (Schauer) Moldenke ●☆
57829 Bouchea marrubiifolia (Fenzl ex Walp.) Schauer = Chascanum marrubiifolium Fenzl ex Walp. ●☆
57830 Bouchea namaquana Bolus ex H. Pearson = Chascanum namaquanum (Bolus ex H. Pearson) Moldenke ●☆
57831 Bouchea pinnatifida (L. f.) Schauer = Chascanum pinnatifidum (L. f.) E. Mey. ●☆
57832 Bouchea pterygocarpa Schauer = Chascanum laetum Walp. ●☆
57833 Bouchea pumila Schauer = Chascanum pinnatifidum (L. f.) E. Mey. ●☆
57834 Bouchea rariflora (A. Terracc.) Chiov. = Chascanum rariflorum (A. Terracc.) Moldenke ●☆
57835 Bouchea schlechteri Gürke = Chascanum schlechteri (Gürke) Moldenkei ●☆
57836 Bouchea sessilifolia Vatke = Chascanum sessilifolium (Vatke) Moldenke ●☆
57837 Bouchea spathulata Torr.;匙形布谢草;Spoon-leaf Bouchea ●☆
57838 Bouchea wilmsii Gürke = Chascanum hederaceum (Sond.) Moldenke var. natalense (H. Pearson) Moldenke ●☆
57839 Bouchetia DC. = Bouchetia DC. ex Dunal ■☆
57840 Bouchetia DC. ex Dunal(1852);布谢茄属■☆
57841 Bouchetia Dunal = Bouchetia DC. ex Dunal ■☆
57842 Bouchetia anomala (Miers) Loes.;布谢茄■☆
57843 Bouchetia erecta DC. ex Dunal;直立布谢茄■☆
57844 Bouchetia erecta Dunal = Bouchetia erecta DC. ex Dunal ■☆
57845 Bouchetia procumbens DC. ex Dunal;匍匐布谢茄■☆
57846 Bouea Meisn. (1837);波漆属(对叶杜属)●
57847 Bouea macrophylla Griff.;大叶波漆(庚大利,羹大利);Gandaria, Plum Mango ●
57848 Bouea oppositifolia (Roxb.) Meisn.;波漆●
57849 Bouetia A. Chev. = Hemizygia (Benth.) Briq. ●■☆
57850 Bouetia ocymoides A. Chev. = Hemizygia bracteosa (Benth.) Briq. ●☆
57851 Bougainvillea Comm. = Bougainvillea Comm. ex Juss. (保留属名)●
57852 Bougainvillea Comm. ex Juss. (1789)(保留属名)('Buginvillaea');叶子花属(宝巾属,九重葛属,南美紫茉莉属,三角花属);Bougainvillea, Leafyflower ●
57853 Bougainvillea Spach = Bougainvillea Comm. ex Juss. (保留属名)●
57854 Bougainvillea × buttiana Williams;黄色九重葛(红苞九重葛);Butt Bougainvillea ●
57855 Bougainvillea brasiliensis Raeusch.;九重葛(龟花,南美紫茉莉);Hairy Bougainvillea ●
57856 Bougainvillea fastuosa Hérincq;娇人叶子花●☆
57857 Bougainvillea glabra Choisy;光叶子花(宝巾,宝巾花,光九重葛,光三角花,光叶龟花,光叶九重葛,红苞紫茉莉,角花,九重葛,簕杜鹃,绿红苞紫茉莉,三角花,三角梅,桃红紫茉莉,小叶九重葛,叶子花,紫三角,紫亚兰);Bougainvillea, Glabrous Bougainvillea, Lesser Bougainvillea, Naked Leafyflower, Paper Flower, Paperflower, Rosy-red Bracted Bougainvillea ●
57858 Bougainvillea glabra Choisy 'Brazil';茄色宝巾●☆
57859 Bougainvillea glabra Choisy 'Crimson Lake';红湖宝巾●☆
57860 Bougainvillea glabra Choisy 'Elizabeth Doxey';白色宝巾●☆
57861 Bougainvillea glabra Choisy 'Sanderiana' = Bougainvillea glabra Choisy 'Variegata'●☆
57862 Bougainvillea glabra Choisy 'Snow White';白雪光叶子花●☆
57863 Bougainvillea glabra Choisy 'Variegata';斑叶光叶子花●☆
57864 Bougainvillea glabra Choisy var. sanderiana Bosschere;玫瑰宝巾花;Paper-flower ●☆
57865 Bougainvillea glabra Choisy var. sanderiana Bosschere = Bougainvillea glabra Choisy 'Sanderiana'●☆
57866 Bougainvillea peruviana Nees et Mart. = Bougainvillea spectabilis Willd. ●
57867 Bougainvillea speciosa Schnizl. = Bougainvillea spectabilis Willd. ●
57868 Bougainvillea spectabilis Willd.;叶子花(红花九重葛,九重葛,毛宝巾,南美紫茉莉,三角花);Beautiful Bougainvillea, Bougainvillea, Brazil Bougainvillea, Great Bougainvillea, Leafyflower, Mary Palmer, Paper Flower ●
57869 Bougainvillea spectabilis Willd. var. glabra (Choisy) Hook. = Bougainvillea glabra Choisy ●
57870 Bougainvillea spectabilis Willd. var. lateritia Lem.;砖红宝巾●☆
57871 Bougainvilleaceae J. Agardh = Nyctaginaceae Juss. (保留科名)●■
57872 Bougainvilleaceae J. Agardh;叶子花科●
57873 Bougueria Decne. (1836);单蕊车前属■☆
57874 Bougueria Decne. = Plantago L. ■●
57875 Bougueria nubicola Decne.;单蕊车前■☆
57876 Boulardia F. Schultz = Orobanche L. ■
57877 Boulaya Gand. = Rubus L. ●■
57878 Bouletia M. A. Clem. et D. L. Jones = Dendrobium Sw. (保留属名)■
57879 Bouletia M. A. Clem. et D. L. Jones(2002);新喀石斛属■☆

57880 Boulia A. Chev. = Murdannia Royle(保留属名)■
57881 Bouphon Lem. = Bouphone Lem. ■☆
57882 Bouphone Lem. = Boophone Herb. ■☆
57883 Bourasaha Thouars. = Burasaia Thouars ●☆
57884 Bourasaia Thouars. = Burasaia Thouars ●☆
57885 Bourdaria A. Chev. = Cincinnobotrys Gilg ■☆
57886 Bourdaria felicis A. Chev. = Cincinnobotrys felicis (A. Chev.) Jacq. -Fél. ■☆
57887 Bourdonia Greene = Chaetopappa DC. ■☆
57888 Bourdonia Greene = Keerlia A. Gray et Engelm. ■☆
57889 Bourgaea Coss. = Cynara L. ■
57890 Bourgaea humilis (L.) Coss. = Cynara humilis L. ■☆
57891 Bourgaea humilis (L.) Coss. var. leucantha Coss. = Cynara humilis L. ■☆
57892 Bourgia Scop. = Cordia L. (保留属名)●
57893 Bourgia Scop. = Salimori Adans. ●
57894 Bourjotia Pomel = Heliotropium L. ●■
57895 Bourjotia erosa (Lehm.) Pomel = Heliotropium bacciferum Forssk. subsp. erosum (Lehm.) Riedl ■☆
57896 Bourjotia kralikii Pomel = Heliotropium bacciferum Forssk. subsp. erosum (Lehm.) Riedl ■☆
57897 Bourlageodendron K. Schum. = Boerlagiodendron Harms ●
57898 Bournea Oliv. (1893);四数苣苔属;Bournea ■★
57899 Bournea leiophylla (W. T. Wang) W. T. Wang et K. Y. Pan = Bournea leiophylla (W. T. Wang) W. T. Wang et K. Y. Pan ex W. T. Wang ■
57900 Bournea leiophylla (W. T. Wang) W. T. Wang et K. Y. Pan ex W. T. Wang; 五数苣苔(光叶石上莲); Smoothleaf Bournea, Smoothleaf Oreocharis ■
57901 Bournea leiophylla W. T. Wang = Bournea leiophylla (W. T. Wang) W. T. Wang et K. Y. Pan ex W. T. Wang ■
57902 Bournea sinensis Oliv.; 四数苣苔; China Bournea, Chinese Bournea ■
57903 Bourreria P. Browne(1756)(保留属名);鲍雷木属;Strongbark ●☆
57904 Bourreria axiphylla Standl.;轴叶鲍雷木●☆
57905 Bourreria huanita (La Llave et Lex.) Hemsl.;墨西哥鲍雷木; Huanita ●☆
57906 Bourreria lyciacea Thulin = Hilsenbergia lyciacea (Thulin) J. S. Mill. ●☆
57907 Bourreria nemoralis (Gürke) Thulin = Hilsenbergia nemoralis (Gürke) J. S. Mill. ●☆
57908 Bourreria orbicularis (Hutch. et E. A. Bruce) Thulin = Hilsenbergia orbicularis (Hutch. et E. A. Bruce) J. S. Mill. ●☆
57909 Bourreria petiolaris (Lam.) Thulin = Hilsenbergia petiolaris (Lam.) J. S. Mill. ●☆
57910 Bourreria teitensis (Gürke) Thulin = Hilsenbergia teitensis (Gürke) J. S. Mill. ●☆
57911 Bousigonia Pierre(1898);奶子藤属(菠锡岗属);Bousigonia ●
57912 Bousigonia angustifolia Pierre;闷奶果;Angustifoliate Bousigonia, Narrowleaf Bousigonia ●
57913 Bousigonia mekongensis Pierre;奶子藤;Mekong Bousigonia ●
57914 Boussingaultia Kunth = Anredera Juss. ●■
57915 Boussingaultia baselloides Kunth = Anredera cordifolia (Ten.) Steenis ●
57916 Boussingaultia cordifolia Ten. = Anredera cordifolia (Ten.) Steenis ●
57917 Boussingaultia gracilis Miers = Anredera cordifolia (Ten.) Steenis ●
57918 Boussingaultia gracilis Miers f. pseudobaselloides Hauman = Anredera cordifolia (Ten.) Steenis ●
57919 Boussingaultia gracilis Miers var. pseudobaseelloides (Hauman) L. H. Bailey = Anredera cordifolia (Ten.) Steenis ●
57920 Boussingaultia leptostachys Moq. = Anredera vesicaria Gaertn. f. ●☆
57921 Bouteloua Hornem. ex P. Beauv. = Atheropogon Muhlenb. ex Willd. ■
57922 Bouteloua Lag. (1805)(保留属名)('Botelua');格兰马草属(垂穗草属); Gama-grass, Gamma-grass, Grama, Gramma, Gramma Grass, Sideoats Grama ■
57923 Bouteloua acuminata Griffiths;渐尖格兰马草■☆
57924 Bouteloua curtipendula (Michx.) Torr.; 垂穗草; Sideoats Grama, Side-oats Grama ■
57925 Bouteloua dimorpha Columbus;二型格兰马草;Acapulco Grass ■☆
57926 Bouteloua eriopoda (Torr.) Torr.;毛梗格兰马草;Black Grama ■☆
57927 Bouteloua filiformis Griffiths;线形格兰马草■☆
57928 Bouteloua gracilis (Kunth) Lag. ex Griffiths = Bouteloua gracilis (Kunth) Lag. ex Steud. ■
57929 Bouteloua gracilis (Kunth) Lag. ex Griffiths var. stricta (Vasey) Hitchc. = Bouteloua gracilis (Kunth) Lag. ex Steud. ■
57930 Bouteloua gracilis (Kunth) Lag. ex Steud.; 格兰马草; Blue Grama, Eyelash Grass, Graceful Grama Grass, Grama, Mosquito-grass ■
57931 Bouteloua hirsuta Lag.;硬毛垂穗草;Hairy Grama Grass, Hirsute Grama ■
57932 Bouteloua multifida (Griffiths) Columbus;多裂格兰马草■☆
57933 Bouteloua oligostachya (Nutt.) Torr. ex A. Gray = Bouteloua gracilis (Kunth) Lag. ex Steud. ■
57934 Bouteloua procumbens (P. Durand) Griffiths;匍匐格兰马草■☆
57935 Bouteloua procumbens Griffiths = Bouteloua procumbens (P. Durand) Griffiths ■☆
57936 Boutiquea Le Thomas = Neostenanthera Exell ●☆
57937 Boutiquea Le Thomas(1966);包氏木属●☆
57938 Boutiquea platypetala (Engl. et Diels) Le Thomas;包氏木●☆
57939 Boutonia Bojer = Boutonia Bojer ex Baill. ●
57940 Boutonia Bojer = Cordemoya Baill. ●☆
57941 Boutonia Bojer ex Baill. = Cordemoya Baill. ●☆
57942 Boutonia Bojer ex Baill. = Mallotus Lour. ●
57943 Boutonia DC. (1838);马岛爵床属●☆
57944 Boutonia Erfurt. ex Steud. = Goodenia Sm. ●■☆
57945 Boutonia Hort. Erfurt. ex Steud. = Goodenia Sm. ●■☆
57946 Boutonia Steud. = Goodenia Sm. ●■☆
57947 Boutonia cuspidata DC.;马岛爵床●☆
57948 Bouvardia Salisb. (1807);寒丁子属(鲍伐茜属,波华丽属); Bouvardia ●■☆
57949 Bouvardia angustifolia Kunth;窄叶寒丁子●☆
57950 Bouvardia humboldtii Hort.;洪堡寒丁子;Sweet Bouvardia ●☆
57951 Bouvardia hybrida Hort.;杂种寒丁子●☆
57952 Bouvardia jacquini Kunth = Bouvardia ternifolia Schltdl. ●☆
57953 Bouvardia laevis M. Martens et Galeotti;平滑寒丁子(光波华丽)●☆
57954 Bouvardia leiantha Benth.;光花寒丁子;Scarlet Bouvardia ●☆
57955 Bouvardia longiflora (Cav.) Kunth;寒丁子花(白玉冠,香波华丽);Scented Bouvardia ●☆

57956 Bouvardia multiflora Schult.;多花寒丁子(多花波华丽)●☆
57957 Bouvardia tenuifolia Standl.;三叶寒丁子(三叶波华丽,三叶寒丁子花);Scarlet Bourvardia,Trumpetilla ●☆
57958 Bouvardia ternifolia Schltdl.;细茎寒丁子(红玉冠,三叶鲍伐茜,细茎波华丽);Scarlet Bouvardia,Scarlet Trompetilla ●☆
57959 Bouvardia triphylla Salisb. = Bouvardia tenuifolia Schltdl. ●☆
57960 Bouvardia versicolor Ker Gawl.;变色寒丁子(变色波华丽)●☆
57961 Bouzetia Montrouz.(1860);新喀芸香属●☆
57962 Bouzetia maritima Montrouz.;新喀芸香●☆
57963 Bovea Decne. = Lindenbergia Lehm. ■
57964 Bovea sinaica Decne. = Lindenbergia indica (L.) Vatke ■☆
57965 Bovonia Chiov.(1923);鲍温草属■☆
57966 Bovonia Chiov. = Aeollanthus Mart. ex Spreng. ■☆
57967 Bovonia diphylla Chiov.;鲍温草■☆
57968 Bowdichia Kunth(1824);鲍迪木属(鲍迪豆属,博递奇亚木属,褐心木属);Sucupira ●☆
57969 Bowdichia nitida Spruce ex Benth.;鲍迪木(博递奇亚木,光鲍迪豆,光亮鲍迪豆)●☆
57970 Bowdichia virgilioides Kunth;小枝鲍迪木(鲍迪豆,维吉豆状鲍迪豆)●☆
57971 Bowenia Hook. = Bowenia Hook. ex Hook. f. ●☆
57972 Bowenia Hook. ex Hook. f.(1863);波温苏铁属(波温铁属,莲铁属)●☆
57973 Bowenia serrulata Chamb.;齿叶波温苏铁(齿叶莲铁,裂叶苏铁)●☆
57974 Bowenia spectabilis Hook.;波温苏铁(波温铁,全缘莲铁);Byfield Fern ●☆
57975 Boweniaceae D. W. Stev.(1981);波温铁科●☆
57976 Boweniaceae D. W. Stev. = Zamiaceae Rchb. ●☆
57977 Bowiea Harv. ex Hook. f.(1824)(保留属名);苍角殿属(仙鞭草属)■☆
57978 Bowiea Haw.(废弃属名) = Aloe L. ●■
57979 Bowiea Haw.(废弃属名) = Bowiea Harv. ex Hook. f.(保留属名)■☆
57980 Bowiea Hook. f. et Haw. = Bowiea Harv. ex Hook. f.(保留属名)■☆
57981 Bowiea africana Haw. = Aloe bowiea Schult. et Schult. f. ■☆
57982 Bowiea kilimandscharica Mildbr. = Bowiea volubilis Harv. ex Hook. f. ■☆
57983 Bowiea myriacantha Haw. = Aloe myriacantha (Haw.) Schult. et Schult. f. ●☆
57984 Bowiea volubilis Harv. ex Hook. f.;苍角殿(仙鞭草);Climbing Onion,Sea Onion ■☆
57985 Bowkeria Harv.(1859);布克木属;Shellflower Bush ●☆
57986 Bowkeria calceolarioides Diels = Bowkeria cymosa MacOwan ●☆
57987 Bowkeria citrina Thode;黄花布克木;Yellow Shellflower Bush ●☆
57988 Bowkeria cymosa MacOwan;伞花布克木●☆
57989 Bowkeria gerrardiana Harv. ex Hiern = Bowkeria verticillata (Eckl. et Zeyh.) Schinz ●☆
57990 Bowkeria natalensis Schinz = Bowkeria verticillata (Eckl. et Zeyh.) Schinz ●☆
57991 Bowkeria simpliciflora MacOwan = Bowkeria verticillata (Eckl. et Zeyh.) Schinz ●☆
57992 Bowkeria triphylla Harv. = Bowkeria verticillata (Eckl. et Zeyh.) Schinz ●☆
57993 Bowkeria triphylla Harv. var. pubescens Kuntze = Bowkeria verticillata (Eckl. et Zeyh.) Schinz ●☆
57994 Bowkeria triphylla Harv. var. subglabra Kuntze = Bowkeria verticillata (Eckl. et Zeyh.) Schinz ●☆
57995 Bowkeria velutina Harv. ex Hiern = Bowkeria verticillata (Eckl. et Zeyh.) Schinz ●☆
57996 Bowkeria verticillata (Eckl. et Zeyh.) Schinz;轮叶布克木(轮生布克木);Natal Shellflower Bush ●☆
57997 Bowkeria verticillata Druce = Bowkeria verticillata (Eckl. et Zeyh.) Schinz ●☆
57998 Bowlesia Ruiz et Pav.(1794);鲍尔斯草属(鲍尔斯属)■☆
57999 Bowlesia asiatica Nasir = Bowlesia incana Ruiz et Pav. ■☆
58000 Bowlesia glandulosa (Poir.) Kuntze;腺点鲍尔斯草■☆
58001 Bowlesia incana Ruiz et Pav.;鲍尔斯草■☆
58002 Bowlesia septentrionalis Coult. et Rose = Bowlesia incana Ruiz et Pav. ■☆
58003 Bowlesia tenera Spreng. = Bowlesia incana Ruiz et Pav. ■☆
58004 Bowmania Garda. = Trixis P. Browne ■●☆
58005 Bowringia Champ. ex Benth.(1852);藤槐属(鲍氏槐属);Bowringia ●
58006 Bowringia callicarpa Champ. ex Benth.;藤槐(包金豆,赤竹子,放屁藤,鸡公合藤,石岩风);Bowringia,Common Bowringia ●
58007 Bowringia discolor J. B. Hall = Leucomphalos discolor (J. B. Hall) Breteler ●☆
58008 Bowringia madagascariensis R. Vig. = Leucomphalos mildbraedii (Harms) Breteler ●☆
58009 Bowringia mildbraedii Harms;米氏藤槐●☆
58010 Bowringia mildbraedii Harms = Leucomphalos mildbraedii (Harms) Breteler ●☆
58011 Boyania Wurdack(1964);博延野牡丹属■☆
58012 Boyania ayangannae Wurdack;博延野牡丹■☆
58013 Boykiana Raf.(废弃属名) = Boykinia Nutt.(保留属名)●■☆
58014 Boykiana Raf.(废弃属名) = Rotala L. ■
58015 Boykinia Nutt.(1834)(保留属名);八幡草属;Boykinia ●■☆
58016 Boykinia Nutt.(保留属名) = Cayaponia Silva Manso(保留属名)●■☆
58017 Boykinia Nutt. ex Raf. = Cayaponia Silva Manso(保留属名)■☆
58018 Boykinia Raf. = Rotala L. ■
58019 Boykinia aconitifolia Nutt.;乌头叶八幡草(八幡草);Monkshoodleaf Boykinia ●☆
58020 Boykinia elata Greene;高八幡草;Tall Boykinia ●☆
58021 Boykinia jamesii Engl.;詹氏八幡草■☆
58022 Boykinia lycoctonifolia (Maxim.) Engl.;牛扁叶八幡草●☆
58023 Boykinia major A. Gray;大八幡草;Sierra Boykinia ●☆
58024 Boykinia rotundifolia A. Gray;圆叶八幡草;San Gabriel Boykinia ●☆
58025 Boykinia tellimoides (Maxim.) Engl. = Peltoboykinia tellimoides (Maxim.) H. Hara ■
58026 Boymia A. Juss. = Evodia J. R. Forst. et G. Forst. ●
58027 Boymia glabrifolia Champ. ex Benth. = Evodia glabrifolia (Champ. ex Benth.) C. C. Huang ●
58028 Boymia rutaecarpa Juss. = Evodia ruticarpa (Juss.) Benth. ●
58029 Bozea Raf. = Bosea L. ●☆
58030 Braasiella Braem,Lückel et Russmann = Oncidium Sw.(保留属名)■☆
58031 Brabejaria Burm. f. = Brabejum L. ●☆
58032 Brabejum L.(1753);南非野杏属;South African Wild Almond,Wild Almond,Wild Chestnut ●☆
58033 Brabejum stellatifolium L.;南非野杏;South African Wild

Almond, Wild Almond, Wild Chestnut ●☆
58034 Brabyla L. = Brabejum L. ●☆
58035 Bracea Britton = Neobracea Britton ●☆
58036 Braced King = Sarcosperma Hook. f. ●
58037 Bracera Engelm. = Brayera Kunth ■●☆
58038 Brachanthemum DC. (1838); 短舌菊属; Brachanthemum, Shorttonguedaisy ●■
58039 Brachanthemum arrectum (Durieu et Schinz) Stent; 非洲短舌菊; Tanner Grass ■☆
58040 Brachanthemum deflexum (Schum.) Robyns; 外折短舌菊●☆
58041 Brachanthemum fruticulosum (Ledeb.) DC.; 短舌菊(灌木短舌菊, 新疆短舌菊); Shorttonguedaisy ●
58042 Brachanthemum gobicum Krasch.; 戈壁短舌菊; Gobi Brachanthemum ■
58043 Brachanthemum kirghisorum Krasch.; 吉尔吉斯短舌菊(天山短舌菊) ■
58044 Brachanthemum mongolicum Krasch.; 蒙古短舌菊; Mongol Shorttonguedaisy, Mongolian Brachanthemum ■
58045 Brachanthemum muticum (Forssk.) Stapf; 无尖短舌菊; Para Grass, Parit Grass ●☆
58046 Brachanthemum nanschanicum Krasch. = Brachanthemum pulvinatum (Hand. -Mazz.) C. Shih ■
58047 Brachanthemum pulvinatum (Hand. -Mazz.) C. Shih; 星毛短舌菊(南山短舌菊); Starhair Shorttonguedaisy, Stellatehair Brachanthemum ■
58048 Brachanthemum titovii Krasch.; 无毛短舌菊(准噶尔短舌菊); Titov Brachanthemum ■
58049 Brachatera Desv. = Danthonia DC. (保留属名) ■
58050 Brachatera Desv. = Sieglingia Bernh. (废弃属名) ■
58051 Bracheilema R. Br. = Vernonia Schreb. (保留属名) ●■
58052 Brachiaria (Trin.) Griseb. (1853); 臂形草属; Armgrass, Signalgrass, Singnalgrass ■
58053 Brachiaria Griseb. = Brachiaria (Trin.) Griseb. ■
58054 Brachiaria advena Vickery; 外来臂形草■☆
58055 Brachiaria ambigua (Trin.) A. Camus = Urochloa paspaloides J. Presl ex C. Presl ■
58056 Brachiaria andongensis (Rendle) Stapf; 安东臂形草■☆
58057 Brachiaria arida (Mez) Stapf; 旱生臂形草■☆
58058 Brachiaria arrecta (Hack. ex T. Durand et Schinz) Stent; 直立臂形草; Tanner Grass ■☆
58059 Brachiaria bemarivensis A. Camus; 贝马里武臂形草■☆
58060 Brachiaria bemarivensis A. Camus subsp. ankarafantsikaensis A. Camus = Brachiaria bemarivensis A. Camus ■☆
58061 Brachiaria benoistii A. Camus = Brachiaria bemarivensis A. Camus ■☆
58062 Brachiaria bequaertii Robyns = Brachiaria decumbens Stapf ■☆
58063 Brachiaria bomaensis Vanderyst = Brachiaria jubata (Fig. et De Not.) Stapf ■☆
58064 Brachiaria bovonei (Chiov.) Robyns; 博奥臂形草■☆
58065 Brachiaria brachylopha Stapf = Brachiaria serrata (Thunb.) Stapf ■☆
58066 Brachiaria breviglumis Clayton; 短颖臂形草■☆
58067 Brachiaria brevis Stapf = Brachiaria jubata (Fig. et De Not.) Stapf ■☆
58068 Brachiaria brevispicata (Rendle) Stapf; 短穗臂形草■☆
58069 Brachiaria bulawayensis (Hack.) Henrard = Urochloa oligotricha (Fig. et De Not.) Henrard ■☆
58070 Brachiaria callopus (Pilg.) Stapf = Echinochloa callopus (Pilg.) Clayton ■☆
58071 Brachiaria chusqueoides (Hack.) Clayton; 楚氏竹臂形草■☆
58072 Brachiaria ciliaris Vanderyst; 缘毛臂形草■☆
58073 Brachiaria clavipila (Chiov.) Robyns; 棒臂形草■☆
58074 Brachiaria clavuliseta Chiov. = Brachiaria deflexa (Schumach.) C. E. Hubb. ex Robyns ■☆
58075 Brachiaria comata (Hochst. ex A. Rich.) Stapf; 束毛臂形草■☆
58076 Brachiaria comorensis (Mez) A. Camus = Panicum comorense Mez ■☆
58077 Brachiaria coronifera Pilg.; 冠生臂形草■☆
58078 Brachiaria decumbens Stapf; 俯卧尾稃草; Spreading Liverseed Grass ■☆
58079 Brachiaria deflexa (Schumach.) C. E. Hubb. ex Robyns; 外折臂形草; Deflexed Brachiaria ■☆
58080 Brachiaria dictyoneura (Fig. et De Not.) Stapf; 网脉臂形草■☆
58081 Brachiaria distachya (L.) Stapf = Urochloa distachya (L.) T. Q. Nguyen ■☆
58082 Brachiaria distachyoides Stapf; 双穗臂形草■☆
58083 Brachiaria distichophylla (Trin.) Stapf = Brachiaria villosa (Lam.) A. Camus ■
58084 Brachiaria dura Stapf; 硬臂形草■☆
58085 Brachiaria dura Stapf var. pilosa J. G. Anderson; 疏毛臂形草■☆
58086 Brachiaria eminii (Mez) Robyns; 埃明臂形草■☆
58087 Brachiaria epaleata Stapf = Brachiaria comata (Hochst. ex A. Rich.) Stapf ■☆
58088 Brachiaria eruciformis (Sm.) Griseb.; 臂形草; Armgrass, Signal Grass, Singnalgrass, Sweet Signalgrass ■
58089 Brachiaria extensa Chase = Urochloa platyphylla (Munro ex C. Wright) R. D. Webster ■☆
58090 Brachiaria falcifera (Trin.) Stapf; 镰臂形草■☆
58091 Brachiaria filifolia Stapf = Brachiaria subulifolia (Mez) Clayton ■☆
58092 Brachiaria fulva Stapf = Brachiaria jubata (Fig. et De Not.) Stapf ■☆
58093 Brachiaria fusiformis Reeder; 细毛臂形草■☆
58094 Brachiaria glauca Stapf = Brachiaria ovalis Stapf ■☆
58095 Brachiaria glomerata (Stapf) A. Camus; 团集臂形草■☆
58096 Brachiaria glycerioides Chiov.; 甜茅臂形草■☆
58097 Brachiaria grossa Stapf; 粗臂形草■☆
58098 Brachiaria hagerupii Hitchc. = Brachiaria orthostachys (Mez) Clayton ■☆
58099 Brachiaria heterocraspeda (Peter) Pilg. = Brachiaria scalaris Pilg. ■☆
58100 Brachiaria hians Stapf = Brachiaria bovonei (Chiov.) Robyns ■☆
58101 Brachiaria holosericea (R. Br.) Hughes; 绢毛臂形草■
58102 Brachiaria hubbardii A. Camus = Acroceras hubbardii (A. Camus) Clayton ■☆
58103 Brachiaria humbertiana A. Camus; 亨伯特臂形草■☆
58104 Brachiaria humidicola (Rendle) Schweick.; 湿地臂形草■☆
58105 Brachiaria interstipitata Stapf = Brachiaria oligobrachiata (Pilg.) Henrard ■☆
58106 Brachiaria isachne (Roem. et Schult.) Stapf = Brachiaria eruciformis (Sm.) Griseb. ■
58107 Brachiaria isachne (Roth ex Roem. et Schult.) Stapf = Brachiaria eruciformis (Sm.) Griseb. ■
58108 Brachiaria isachne (Roth. ex Roem.) Stapf = Brachiaria eruciformis (Sm.) Griseb. ■

58109 Brachiaria jubata (Fig. et De Not.) Stapf;鬃毛臂形草■☆

58110 Brachiaria keniensis Henrard = Brachiaria dictyoneura (Fig. et De Not.) Stapf ■☆

58111 Brachiaria kotschyana (Steud.) Stapf = Brachiaria comata (Hochst. ex A. Rich.) Stapf ■☆

58112 Brachiaria kurzii (Hook. f.) A. Camus;无名臂形草■

58113 Brachiaria lachnantha (Hochst.) Stapf;白毛臂形草■

58114 Brachiaria lata (Schumach.) C. E. Hubb.;阔臂形草(宽臂形草)■☆

58115 Brachiaria lata (Schumach.) C. E. Hubb. var. pubescens C. E. Hubb.;柔毛阔臂形草■☆

58116 Brachiaria latifolia Stapf = Brachiaria arrecta (Hack. ex T. Durand et Schinz) Stent ■☆

58117 Brachiaria leersioides (Hochst.) Stapf;假稻臂形草■☆

58118 Brachiaria leucacrantha (K. Schum.) Stapf;白臂形草■☆

58119 Brachiaria lindiensis (Pilg.) Clayton;林迪臂形草■☆

58120 Brachiaria longiflora Clayton;长花臂形草■☆

58121 Brachiaria longifolia Gilli = Echinochloa colona (L.) Link ■

58122 Brachiaria malacodes (Mez et K. Schum.) H. Scholz;软臂形草■☆

58123 Brachiaria marlothii (Hack.) Stent;马洛斯臂形草■☆

58124 Brachiaria melanotyla (Hack.) Henrard = Brachiaria nigropedata (Munro ex Ficalho et Hiern) Stapf ■☆

58125 Brachiaria miliiformis (C. Presl) Chase = Brachiaria subquadripara (Trin.) Hitchc. var. miliiformis (C. Presl) S. L. Chen et Y. X. Jin ■

58126 Brachiaria miliiformis (J. Presl) Chase = Urochloa subquadripara (Trin.) R. D. Webster ■

58127 Brachiaria multispiculata Scholz;多细刺臂形草■☆

58128 Brachiaria mutica (Forssk.) Stapf;巴拉草(荷兰尾稃草,紫黍草);Buffalo Grass, Dutch Grass, Obtuse Armgrass, Para Grass, Paragrass, Purple Panicgrass ■

58129 Brachiaria mutica (Forssk.) Stapf = Urochloa mutica (Forssk.) T. Q. Nguyen ■

58130 Brachiaria nana Stapf;矮小臂形草■☆

58131 Brachiaria nigropedata (Munro ex Ficalho et Hiern) Stapf;黑鸟足状臂形草■☆

58132 Brachiaria numidiana (Lam.) Henrard = Brachiaria mutica (Forssk.) Stapf ■

58133 Brachiaria obvoluta Stapf = Brachiaria dictyoneura (Fig. et De Not.) Stapf ■☆

58134 Brachiaria oligobrachiata (Pilg.) Henrard;寡枝臂形草■☆

58135 Brachiaria orthostachys (Mez) Clayton;直穗臂形草■☆

58136 Brachiaria ovalis Stapf;卵臂形草■☆

58137 Brachiaria paspaloides (C. Presl) C. E. Hubb. = Urochloa paspaloides J. Presl ex C. Presl ■

58138 Brachiaria perrieri A. Camus;佩里耶臂形草■☆

58139 Brachiaria plantaginea (Link) Hitchc.;车前状臂形草;Plantago Singnalgrass ■

58140 Brachiaria platynota (K. Schum.) Robyns;阔背臂形草■☆

58141 Brachiaria platyphylla (Griseb.) Nash = Urochloa platyphylla (Griseb.) R. D. Webster ■☆

58142 Brachiaria platyphylla (Munro ex C. Wright) Nash = Urochloa platyphylla (Munro ex C. Wright) R. D. Webster ■☆

58143 Brachiaria platyrhachis Chiov.;阔轴臂形草■☆

58144 Brachiaria platytaenia Stapf = Brachiaria oligobrachiata (Pilg.) Henrard ■☆

58145 Brachiaria poaeoides Stapf = Brachiaria malacodes (Mez et K. Schum.) H. Scholz ■☆

58146 Brachiaria prostrata (Lam.) Griseb. = Brachiaria reptans (L.) C. A. Gardner et C. E. Hubb. ■

58147 Brachiaria prostrata (Lam.) Griseb. = Urochloa repans (L.) Stapf ■

58148 Brachiaria psammophila (Welw. ex Rendle) Launert;喜沙臂形草■☆

58149 Brachiaria pseudodichotoma Bosser;假二歧臂形草■☆

58150 Brachiaria pubescens (Chiov.) S. M. Phillips;短柔毛臂形草■☆

58151 Brachiaria pubifolia Stapf = Brachiaria xantholeuca (Hack.) Stapf ■☆

58152 Brachiaria purpurascens (Raddi) Henrard = Brachiaria mutica (Forssk.) Stapf ■

58153 Brachiaria purpurascens (Raddi) Henrard = Urochloa mutica (Forssk.) T. Q. Nguyen ■

58154 Brachiaria radicans Napper = Brachiaria arrecta (Hack. ex T. Durand et Schinz) Stent ■☆

58155 Brachiaria ramosa (L.) Stapf;多枝臂形草;Brown-top Millet, Dixie Signalgrass, Manybranched Armgrass, Manybranched Singnalgrass ■

58156 Brachiaria rautanenii (Hack.) Stapf = Brachiaria humidicola (Rendle) Schweick. ■☆

58157 Brachiaria regularis (Nees) Stapf = Brachiaria deflexa (Schumach.) C. E. Hubb. ex Robyns ■☆

58158 Brachiaria regularis (Nees) Stapf var. nidulans (Mez) Tackh. = Brachiaria ramosa (L.) Stapf ■

58159 Brachiaria reptans (L.) C. A. Gardner et C. E. Hubb. = Urochloa reptans (L.) Stapf ■

58160 Brachiaria reticulata Stapf;网状臂形草■☆

58161 Brachiaria rovumensis (Pilg.) Pilg. = Eriochloa rovumensis (Pilg.) Clayton ■☆

58162 Brachiaria rugulosa Stapf;稍皱臂形草■☆

58163 Brachiaria ruziziensis R. Germ. et C. M. Evrard = Urochloa ruziziensis (R. Germ. et C. M. Evrard) Crins ■☆

58164 Brachiaria sadinii Vanderyst = Panicum sadinii (Vanderyst) Renvoize ■☆

58165 Brachiaria scalaris Pilg.;梯臂形草■☆

58166 Brachiaria schoenfelderi C. E. Hubb. et Schweick.;舍恩臂形草■☆

58167 Brachiaria secernenda (Hochst. ex Mez) Henrard = Brachiaria comata (Hochst. ex A. Rich.) Stapf ■☆

58168 Brachiaria semiundulata (Hitchc. ex A. Rich.) Stapf;半波臂形草(短颖臂形草);Shortglume Armgrass ■

58169 Brachiaria semiundulata (Hitchc.) Stapf = Brachiaria semiundulata (Hitchc. ex A. Rich.) Stapf ■

58170 Brachiaria serpens (Kunth) C. E. Hubb.;蛇形臂形草■☆

58171 Brachiaria serrata (Thunb.) Stapf;具齿臂形草■☆

58172 Brachiaria serrata (Thunb.) Stapf var. gossypina (A. Rich.) Stapf = Brachiaria serrata (Thunb.) Stapf ■☆

58173 Brachiaria serrifolia (Hochst.) Stapf;齿叶臂形草■☆

58174 Brachiaria serrifolia (Hochst.) Stapf var. pubescens Chiov. = Brachiaria pubescens (Chiov.) S. M. Phillips ■☆

58175 Brachiaria setigera (Retz.) C. E. Hubb. = Urochloa setigera (Retz.) Stapf ■

58176 Brachiaria soluta Stapf = Brachiaria jubata (Fig. et De Not.) Stapf ■☆

58177 Brachiaria somalensis C. E. Hubb. = Brachiaria ovalis Stapf ■☆

58178 Brachiaria squarrosa (Peter) Clayton = Panicum peteri Pilg. ■☆

58179 Brachiaria stefaninii Chiov.;斯特臂形草■☆
58180 Brachiaria stigmatisata (Mez) Stapf;柱头臂形草■☆
58181 Brachiaria stipitata C. E. Hubb. = Echinochloa callopus (Pilg.) Clayton ■☆
58182 Brachiaria stolonifera Gooss. = Urochloa stolonifera (Gooss.) Chippind. ■☆
58183 Brachiaria subquadripara (Trin.) Hitchc.;四生臂形草(两穗臂形草);Four-armed Grass, Fourbearing Singnalgrass ■
58184 Brachiaria subquadripara (Trin.) Hitchc. = Urochloa subquadripara (Trin.) R. D. Webster ■
58185 Brachiaria subquadripara (Trin.) Hitchc. var. miliiformis (C. Presl) S. L. Chen et Y. X. Jin;锐头臂形草(臂形草);Milletformis Singnalgrass ■
58186 Brachiaria subquadripara (Trin.) Hitchc. var. setulosa S. L. Chen et Y. X. Jin;刺毛臂形草■
58187 Brachiaria subquadripara (Trin.) Hitchc. var. setulosa S. L. Chen et Y. X. Jin = Brachiaria fusiformis Reeder ■☆
58188 Brachiaria subrostrata A. Camus;喙状臂形草■☆
58189 Brachiaria subulifolia (Mez) Clayton;钻叶臂形草■☆
58190 Brachiaria texana (Buckley) S. T. Blake = Urochloa texana (Buckley) R. D. Webster ■☆
58191 Brachiaria turbinata Van der Veken;陀螺形臂形草■☆
58192 Brachiaria ukambensis Henrard = Brachiaria xantholeuca (Hack.) Stapf ■☆
58193 Brachiaria umbellata (Trin.) Clayton;小伞臂形草■☆
58194 Brachiaria umboensis Stent et J. M. Rattray = Brachiaria rugulosa Stapf ■☆
58195 Brachiaria umbratilis Napper;荫蔽臂形草■☆
58196 Brachiaria urochlooides S. L. Chen et Y. X. Jin;尾稃臂形草;Taillemma Armgrass ■
58197 Brachiaria uzondoiensis Sánchez-Ken;乌祖多伊臂形草■☆
58198 Brachiaria verdickii Robyns = Echinochloa callopus (Pilg.) Clayton ■☆
58199 Brachiaria villosa (Lam.) A. Camus;毛臂形草;Hairy Signalgrass, Villous Armgrass, Villous Singnalgrass ■
58200 Brachiaria villosa (Lam.) A. Camus = Urochloa villosa (Lam.) T. Q. Nguyen ■
58201 Brachiaria villosa (Lam.) A. Camus f. glabriglumis (Ohwi) Ohwi;光皮臂形草■☆
58202 Brachiaria villosa (Lam.) A. Camus var. barbata Bor;髯毛臂形草;Barbate Singnalgrass ■
58203 Brachiaria villosa (Lam.) A. Camus var. barbata Bor = Brachiaria villosa (Lam.) A. Camus ■
58204 Brachiaria villosa (Lam.) A. Camus var. glabrata S. L. Chen et Y. X. Jin;无毛臂形草;Glabrous Singnalgrass ■
58205 Brachiaria villosa (Lam.) A. Camus var. glabriglumis Ohwi = Brachiaria villosa (Lam.) A. Camus f. glabriglumis (Ohwi) Ohwi ■☆
58206 Brachiaria villosa Vanderyst = Brachiaria villosa (Lam.) A. Camus ■
58207 Brachiaria viridula Stapf = Brachiaria bovonei (Chiov.) Robyns ■☆
58208 Brachiaria vittata Stapf = Brachiaria oligobrachiata (Pilg.) Henrard ■☆
58209 Brachiaria wittei Robyns;维特臂形草■☆
58210 Brachiaria xantholeuca (Hack.) Stapf;黄白臂形草■☆
58211 Brachilobos Desv. = Brachiolobos All. ■
58212 Brachilobus Desv. = Brachiolobos All. ■
58213 Brachiolobos All. = Radicula Hill ■
58214 Brachiolobos All. = Rorippa Scop. ■
58215 Brachiolobos hispidus Desv. = Rorippa palustris (L.) Besser subsp. hispida (Desv.) Jonsell ■☆
58216 Brachiolobus Bernh. = Brachiolobos All. ■
58217 Brachionidium Lindl. (1859);臂兰属■☆
58218 Brachionidium brachycladum Luer et R. Escobar;短枝臂兰■☆
58219 Brachionidium concolor Lindl.;臂兰■☆
58220 Brachionidium elegans Luer et Hirtz;雅致臂兰■☆
58221 Brachionidium floribundum Garay;多花臂兰■☆
58222 Brachionidium parvifolium Lindl.;小叶臂兰■☆
58223 Brachionidium parvum Cogn.;小管臂兰■☆
58224 Brachionostylum Mattf. (1932);齿叶蟹甲木属●☆
58225 Brachionostylum pullet Mattf.;齿叶蟹甲木●☆
58226 Brachiostemon Hand. -Mazz. = Ornithoboea Parish ex C. B. Clarke ■
58227 Brachiostemon macrocalyx Hand. -Mazz. = Ornithoboea wildeana Craib ■
58228 Brachistus Miers(1849);短茄属●☆
58229 Brachistus actinocalyx Winkler;星萼短茄●☆
58230 Brachistus affinis (C. V. Morton) D'Arcy, J. L. Gentry et Averett;近缘短茄●☆
58231 Brachistus ciliatus Miers;缘毛短茄●☆
58232 Brachistus dimorphus Miers;二型短茄●☆
58233 Brachistus fasciculatus Rusby;簇生短茄●☆
58234 Brachlstepis Thouars = Beclardia A. Rich. ■☆
58235 Brachlstepis Thouars = Epidendrum L. (保留属名)■☆
58236 Brachoneuron Post et Kuntze = Brochoneura Warb. ●☆
58237 Brachramphus DC. = Launaea Cass. ■
58238 Brachtia Rchb. f. (1850)(保留属名);布拉兰属(勃拉兰属)■☆
58239 Brachtia Trevis. (废弃属名) = Brachtia Rchb. f. (保留属名)■☆
58240 Brachtia andina Rchb. f.;布拉兰■☆
58241 Brachtia brevis Kraenzl.;短布拉兰■☆
58242 Brachtia diphylla H. R. Sweet;二叶布拉兰■☆
58243 Brachtia minutiflora Kraenzl.;小花布拉兰■☆
58244 Brachtia sulphurea Rchb. f.;硫色布拉兰■☆
58245 Brachyachaenium Baker = Dicoma Cass. ●☆
58246 Brachyachenium Baker = Dicoma Cass. ●☆
58247 Brachyachenium incanum Baker = Dicoma incana (Baker) O. Hoffm. ●☆
58248 Brachyachne (Benth.) Stapf(1922);短毛草属■☆
58249 Brachyachne Stapf = Brachyachne (Benth.) Stapf ■☆
58250 Brachyachne chrysolepis C. E. Hubb. = Brachyachne patentiflora (Stent et J. M. Rattray) C. E. Hubb. ■☆
58251 Brachyachne fibrosa C. E. Hubb. = Brachyachne fulva Stapf ■☆
58252 Brachyachne fulva Stapf;黄褐短毛草■☆
58253 Brachyachne kundelungensis Van der Veken = Brachyachne fulva Stapf ■☆
58254 Brachyachne obtusiflora (Benth.) C. E. Hubb.;钝花短毛草■☆
58255 Brachyachne patentiflora (Stent et J. M. Rattray) C. E. Hubb.;展花短毛草■☆
58256 Brachyachne pilosa Van der Veken;疏毛短毛草■☆
58257 Brachyachne simonii Kupicha et Cope;西蒙短毛草■☆
58258 Brachyachne upembaensis Van der Veken = Brachyachne patentiflora (Stent et J. M. Rattray) C. E. Hubb. ■☆
58259 Brachyachyris Spreng. = Brachyris Nutt. ■●☆
58260 Brachyachyris Spreng. = Gutierrezia Lag. ■●☆

58261 Brachyactis Ledeb.(1845);短星菊属;Brachyactis ■
58262 Brachyactis angusta (Lindl.) Britton = Brachyactis ciliata (Ledeb.) Ledeb. subsp. angusta (Lindl.) A. G. Jones ■☆
58263 Brachyactis angusta (Lindl.) Britton = Symphyotrichum ciliatum (Ledeb.) G. L. Nesom ■☆
58264 Brachyactis anomala (DC.) Kitam.;香短星菊;Fragrant Brachyactis,Mintodorous Brachyactis ■
58265 Brachyactis ciliata (Ledeb.) Ledeb.;短星菊;Alkali Rayless Aster,Ciliate Brachyactis,Narrow Aster ■
58266 Brachyactis ciliata (Ledeb.) Ledeb. subsp. angusta (Lindl.) A. G. Jones;窄短星菊;Alkali Rayless Aster ■☆
58267 Brachyactis ciliata (Ledeb.) Ledeb. subsp. angusta (Lindl.) A. G. Jones = Aster brachyactis S. F. Blake ■☆
58268 Brachyactis ciliata Ledeb. = Brachyactis ciliata (Ledeb.) Ledeb. ■
58269 Brachyactis ciliata Ledeb. subsp. angusta (Lindl.) A. G. Jones = Symphyotrichum ciliatum (Ledeb.) G. L. Nesom ■☆
58270 Brachyactis frondosa (Nutt.) A. Gray = Symphyotrichum frondosum (Nutt.) G. L. Nesom ■☆
58271 Brachyactis iliensis Rupr. = Psychrogeton nigromontanus (Boiss. et Buhse) Grierson ■
58272 Brachyactis indica C. B. Clarke = Brachyactis anomalum (DC.) Kitam. ■
58273 Brachyactis latisquama (Maxim.) Kitag. = Brachyactis ciliata Ledeb. ■
58274 Brachyactis latisquamata (Maxim.) Kitag. = Brachyactis ciliata Ledeb. ■
58275 Brachyactis linearifolia C. Winkl.;线叶短星菊■
58276 Brachyactis menthodora Benth. = Brachyactis anomalum (DC.) Kitam. ■
58277 Brachyactis menthora Benth. = Brachyactis anomalum (DC.) Kitam. ■
58278 Brachyactis pubescens (DC.) Aitch. et C. B. Clarke;腺毛短星菊;Pubescent Brachyactis ■
58279 Brachyactis robusta Benth. = Brachyactis pubescens (DC.) Aitch. et C. B. Clarke ■
58280 Brachyactis roylei (DC.) Wendelbo;西疆短星菊(西藏短星菊);Royle Brachyactis ■
58281 Brachyactis umbrosa (Kar. et Kir.) Benth. = Brachyactis roylei (DC.) Wendelbo ■
58282 Brachyandra Naudin(废弃属名) = Brachyandra Phil.(保留属名)●☆
58283 Brachyandra Naudin(废弃属名) = Pterolepis (DC.) Miq.(保留属名)●■☆
58284 Brachyandra Phil.(1860)(保留属名);短蕊修泽兰属●☆
58285 Brachyandra Phil.(保留属名) = Helogyne Nutt. ●☆
58286 Brachyanthes Chem. ex Dunal = Petunia Juss.(保留属名)■
58287 Brachyapium (Baill.) Maire = Stoibrax Raf. ■☆
58288 Brachyapium dichotomum (L.) Maire = Stoibrax dichotomum (L.) Raf. ■☆
58289 Brachyapium hanotei (Braun-Blanq. et Maire) Maire = Stoibrax hanotei (Maire) B. L. Burtt ■☆
58290 Brachyapium involucratum Maire = Stoibrax involucratum (Braun-Blanq. et Maire) B. L. Burtt ■☆
58291 Brachyapium pomelianum Maire = Stoibrax pomelianum (Maire) B. L. Burtt ■☆
58292 Brachyapium pomelianum Maire var. vegetum (Pau et Font Quer) Maire = Stoibrax pomelianum (Maire) B. L. Burtt ■☆
58293 Brachyaster Ambrosi = Aster L. ●■
58294 Brachyaster Ambrosi = Bellidastrum Scop. ●■
58295 Brachyathera Kuntze = Danthonia DC.(保留属名)■
58296 Brachyathera Post et Kuntze = Brachatera Desv. ■
58297 Brachyathera Post et Kuntze = Danthonia DC.(保留属名)■
58298 Brachybotrys Maxim. = Brachybotrys Maxim. ex Oliv. ■
58299 Brachybotrys Maxim. ex Oliv.(1878);山茄子属(短穗草属,短穗花属,短序花属,山茄属);Brachybotrys,Wildeggplant ■
58300 Brachybotrys paridiformis Maxim. ex Oliv.;山茄子(假王孙,帕利草,人参幌子);Common Brachybotrys,Wildeggplant ■
58301 Brachycalycium Backeb.(1942);新世界属■☆
58302 Brachycalycium Backeb. = Gymnocalycium Pfeiff. ex Mittler ●
58303 Brachycalycium tilcarense Backeb.;新世界(白云阁)■☆
58304 Brachycalyx Sweet ex DC. = Rhododendron L. ●
58305 Brachycarpaea DC.(1821);南非短果芥属■☆
58306 Brachycarpaea flava (L. f.) Druce = Brachycarpaea juncea (P. J. Bergius) Marais ■☆
58307 Brachycarpaea juncea (P. J. Bergius) Marais;南非短果芥■☆
58308 Brachycarpaea laxa (Thunb.) Sond. = Brachycarpaea juncea (P. J. Bergius) Marais ■☆
58309 Brachycarpaea laxa (Thunb.) Sond. var. stricta Sond. = Brachycarpaea juncea (P. J. Bergius) Marais ■☆
58310 Brachycarpaea linifolia Eckl. et Zeyh. = Brachycarpaea juncea (P. J. Bergius) Marais ■☆
58311 Brachycarpaea polygaloides Eckl. et Zeyh. = Brachycarpaea juncea (P. J. Bergius) Marais ■☆
58312 Brachycarpaea varians DC. = Brachycarpaea juncea (P. J. Bergius) Marais ■☆
58313 Brachycarpaea varians DC. var. purpurascens = Brachycarpaea juncea (P. J. Bergius) Marais ■☆
58314 Brachycaulaceae Panigrahi et Dikshit = Saxifragaceae Juss.(保留科名)●■
58315 Brachycaulaceae Panigrahi et Dikshit;短茎蔷薇科●☆
58316 Brachycaulos Dikshit et Panigrahi = Chamaerhodos Bunge ■●
58317 Brachycaulos Dikshit et Panigrahi(1981);短茎蔷薇属●☆
58318 Brachycaulos simplicifolius Dikshit et Panigrahi;短茎蔷薇●☆
58319 Brachycentrum Meisn. = Centronia D. Don ●☆
58320 Brachycereus Britton et Rose(1920);飞龙柱属●☆
58321 Brachycereus nesieticus (Rob.) Backeb.;飞龙柱●☆
58322 Brachycereus nesioticus (K. Schum.) Backeb. = Brachycereus nesieticus (Rob.) Backeb. ●☆
58323 Brachychaeta Torr. et A. Gray = Solidago L. ■
58324 Brachychaeta cordata (Short et R. Peter) Torr. et A. Gray = Solidago sphacelata Raf. ■☆
58325 Brachychaeta sphacelata (Raf.) Britton = Solidago sphacelata Raf. ■☆
58326 Brachycheila Harv. ex Eckl. et Zeyh. = Diospyros L. ●
58327 Brachycheila Harv. ex Eckl. et Zeyh. = Euclea L. ●☆
58328 Brachychilum (R. Br. ex Wall.) Petersen = Hedychium J. König ■
58329 Brachychilum (Wall.) Petersen = Hedychium J. König ■
58330 Brachychilum Petersen = Hedychium J. König ■
58331 Brachychilus Post et Kuntze = Brachycheila Harv. ex Eckl. et Zeyh. ●
58332 Brachychilus Post et Kuntze = Brachychilum (R. Br. ex Wall.) Petersen ■
58333 Brachychilus Post et Kuntze = Diospyros L. ●

58334 Brachychiton Schott et Endl. (1832);瓶木属(澳梧桐属,澳洲瓶子树属,瓶子木属);Bottle Tree,Bottle-tree,Brachychiton ●☆

58335 Brachychiton acerifolium (A. Cunn.) F. Muell.;槭叶苹婆(濠州梧桐,火焰瓶木,槭叶瓶木,槭叶桐);Brachiton,Flame Bottle Tree,Flame Bottle-tree,Flame Kurrajong,Flame Tree,Flame-tree,Illawara Flame-tree,Illawarra Flame Tree,Maple-leaf Sterculia ●

58336 Brachychiton acerifolium Macarthur et C. Moore = Brachychiton acerifolium (A. Cunn.) F. Muell. ●

58337 Brachychiton australe (Schott et Endl.) A. Terracc.;北方瓶木(澳洲桐);Broad-leaf Bottle-tree,Northern Bottle Tree ●☆

58338 Brachychiton bidwillii Hook.;毕氏瓶木;Dwarf Kurrajong,Little Kurrajong ●☆

58339 Brachychiton discolor F. Muell.;异色瓶木;Lacebark Kurrajong,Pink Flame Tree,Scrub Bottle-tree ●☆

58340 Brachychiton gregorii F. Muell.;沙漠瓶木(三裂酒瓶树);Desert Kurrajong ●☆

58341 Brachychiton luridus F. Muell. = Brachychiton discolor F. Muell. ●☆

58342 Brachychiton paradoxus Schott et Endl.;奇异瓶木●☆

58343 Brachychiton platanoides R. Br. = Brachychiton australis (Schott et Endl.) A. Terracc. ●☆

58344 Brachychiton populneum (Cav.) R. Br. = Brachychiton populneum (Schott et Endl.) R. Br. ●☆

58345 Brachychiton populneum (Cav.) R. Br. = Sterculia diversifolia G. Don ●☆

58346 Brachychiton populneum (Schott et Endl.) R. Br.;掌叶酒瓶树●☆

58347 Brachychiton roseum Guymer;杂种瓶木;Hybrid Kurrajong ●☆

58348 Brachychiton rupestris (Lindl.) K. Schum.;昆士兰瓶木(酒瓶树,瓶树);Barrel Bottle Tree,Barrel Bottle-tree,Queensland Bottle Tree,Queensland Bottle-tree,Queensland Kurrajong ●☆

58349 Brachychiton trichosiphon (Benth.) Audas = Brachychiton australis (Schott et Endl.) A. Terracc. ●☆

58350 Brachychlaena Post et Kuntze = Brachylaena R. Br. ●☆

58351 Brachychloa S. M. Phillips(1982);非洲矮草属■☆

58352 Brachychloa fragilis S. M. Phillips;非洲矮草■☆

58353 Brachycladium (Luer) Luer = Lepanthes Sw. ■☆

58354 Brachyclados D. Don = Brachyclados Gillies ex D. Don ●☆

58355 Brachyclados Gillies ex D. Don(1832);短枝菊属●☆

58356 Brachyclados lycioides D. Don;短枝菊●☆

58357 Brachyclados megalanthus Speg.;大花短枝菊●☆

58358 Brachyclados obtusifolius Kuntze;钝叶短枝菊●☆

58359 Brachycladus Post et Kuntze = Brachyclados D. Don ●☆

58360 Brachycodon (Benth.) Progel = Pagaea Griseb. ■☆

58361 Brachycodon Fed. = Brachycodonia Fed. ex Kolak. ■●

58362 Brachycodon Fed. = Campanula L. ■●

58363 Brachycodon Progel = Pagaea Griseb. ■☆

58364 Brachycodonia Fed. = Campanula L. ■●

58365 Brachycodonia Fed. ex Kolak. = Campanula L. ■●

58366 Brachycome Cass. = Brachyscome Cass. ●■☆

58367 Brachycome Gaudich. = Vittadinia A. Rich. ■☆

58368 Brachycome hispida (Vatke) Klatt = Gyrodoma hispida (Vatke) Wild ■☆

58369 Brachycome iberidifolia Benth. = Brachyscome iberidifolia Benth. ■☆

58370 Brachycome mossambicensis Oliv. et Hiern = Gyrodoma hispida (Vatke) Wild ■☆

58371 Brachycome palustris O. Hoffm. = Jeffreya palustris (O. Hoffm.) Wild ■☆

58372 Brachycorys Schrad. = Lindenbergia Lehm. ■

58373 Brachycorythis Lindl. (1838);苞叶兰属;Brachycorythis ■

58374 Brachycorythis acutiloba Rendle = Brachycorythis ovata Lindl. subsp. welwitschii (Rchb. f.) Summerh. ■☆

58375 Brachycorythis allisoni Rolfe = Brachycorythis ovata Lindl. ■☆

58376 Brachycorythis angolensis (Schltr.) Schltr.;安哥拉苞叶兰■☆

58377 Brachycorythis basifoliata Summerh.;基小叶苞叶兰■☆

58378 Brachycorythis briantiana Kraenzl. ex De Wild. et T. Durand = Brachycorythis pleistophylla Rchb. f. ■

58379 Brachycorythis briartiana Kraenzl. = Brachycorythis pleistophylla Rchb. f. ■

58380 Brachycorythis buchananii (Schltr.) Rolfe;布坎南苞叶兰■☆

58381 Brachycorythis bulbinella Rchb. f. = Schizochilus bulbinellus (Rchb. f.) Bolus ■☆

58382 Brachycorythis congoensis Kraenzl.;刚果苞叶兰■☆

58383 Brachycorythis conica (Summerh.) Summerh.;圆锥苞叶兰■☆

58384 Brachycorythis conica (Summerh.) Summerh. subsp. longilabris Summerh.;长唇圆锥苞叶兰■☆

58385 Brachycorythis conica (Summerh.) Summerh. subsp. transvaalensis Summerh.;德兰士瓦苞叶兰■☆

58386 Brachycorythis crassicornis Kraenzl. = Brachycorythis tenuior Rchb. f. ■☆

58387 Brachycorythis engleriana Kraenzl. = Brachycorythis tenuior Rchb. f. ■☆

58388 Brachycorythis friesii (Schltr.) Summerh.;弗里斯苞叶兰■☆

58389 Brachycorythis galeandra (Rchb. f.) Summerh.;短距苞叶兰(宽唇苞叶兰,犸骝草,拟粉蝶兰,肾草);Short-spur Brachycorythis ■

58390 Brachycorythis gerrardii Rchb. f. = Schizochilus gerrardii (Rchb. f.) Bolus ■☆

58391 Brachycorythis goetzeana Kraenzl. = Brachycorythis pubescens Harv. ■☆

58392 Brachycorythis grandis Kraenzl. var. ugandensis Braid = Brachycorythis ovata Lindl. subsp. schweinfurthii (Rchb. f.) Summerh. ■☆

58393 Brachycorythis henryi (Schltr.) Summerh.;长叶苞叶兰;Henry Brachycorythis ■

58394 Brachycorythis hirschbergii Braid = Brachycorythis congoensis Kraenzl. ■☆

58395 Brachycorythis inhambanensis (Schltr.) Schltr.;伊尼扬巴内苞叶兰■☆

58396 Brachycorythis kalbreyeri Rchb. f.;卡尔苞叶兰■☆

58397 Brachycorythis kalbreyeri Rchb. f. var. glandulosa Braid = Brachycorythis kalbreyeri Rchb. f. ■☆

58398 Brachycorythis kassneriana Kraenzl. = Brachycorythis pubescens Harv. ■☆

58399 Brachycorythis lastii Rolfe = Schwartzkopffia lastii (Rolfe) Schltr. ■☆

58400 Brachycorythis lisowskiana Szlach. et Olszewski;利索苞叶兰■☆

58401 Brachycorythis macclouniei Braid = Brachycorythis pleistophylla Rchb. f. ■

58402 Brachycorythis macowaniana Rchb. f.;麦克欧文苞叶兰■☆

58403 Brachycorythis macrantha (Lindl.) Summerh.;大花苞叶兰■☆

58404 Brachycorythis menglianensis Y. Y. Qian;孟连苞叶兰;Menglian Brachycorythis ■

58405 Brachycorythis mixta Summerh.;混杂苞叶兰■☆

58406 Brachycorythis oligophylla Kraenzl. = Brachycorythis angolensis (Schltr.) Schltr. ■☆
58407 Brachycorythis ovata Lindl.;卵形苞叶兰■☆
58408 Brachycorythis ovata Lindl. subsp. schweinfurthii (Rchb. f.) Summerh.;施韦苞叶兰■☆
58409 Brachycorythis ovata Lindl. subsp. welwitschii (Rchb. f.) Summerh.;韦尔苞叶兰■☆
58410 Brachycorythis parviflora Rolfe = Brachycorythis buchananii (Schltr.) Rolfe ■☆
58411 Brachycorythis paucifolia Summerh.;少叶苞叶兰■☆
58412 Brachycorythis perrieri Schltr. = Brachycorythis pleistophylla Rchb. f. ■
58413 Brachycorythis pilosa Summerh.;疏毛苞叶兰■☆
58414 Brachycorythis pleistophylla Rchb. f.;多叶苞叶兰;Manyleaf Brachycorythis ■
58415 Brachycorythis pubescens Harv.;短柔毛苞叶兰■☆
58416 Brachycorythis pulchra Schltr. = Brachycorythis pleistophylla Rchb. f. ■
58417 Brachycorythis pumilio (Lindl.) Rchb. f. = Schwartzkopffia pumilio (Lindl.) Schltr. ■☆
58418 Brachycorythis rhodostachys (Schltr.) Summerh.;粉红穗苞叶兰■☆
58419 Brachycorythis rhomboglossa Kraenzl. = Brachycorythis tenuior Rchb. f. ■☆
58420 Brachycorythis rosea A. Chev. = Schwartzkopffia pumilio (Lindl.) Schltr. ■☆
58421 Brachycorythis sceptrum Schltr.;王杖苞叶兰■☆
58422 Brachycorythis schlechteri Geerinck = Schwartzkopffia lastii (Rolfe) Schltr. ■☆
58423 Brachycorythis schweinfurthii Rchb. f. = Brachycorythis ovata Lindl. subsp. schweinfurthii (Rchb. f.) Summerh. ■☆
58424 Brachycorythis stolzii Schltr. = Brachycorythis pubescens Harv. ■☆
58425 Brachycorythis sudanica Schltr. = Brachycorythis pubescens Harv. ■☆
58426 Brachycorythis tanganyikensis Summerh.;坦噶尼喀苞叶兰■☆
58427 Brachycorythis tenuior Rchb. f.;细苞叶兰■☆
58428 Brachycorythis truncatolabellata (Hayata) S. S. Ying = Brachycorythis galeandra (Rchb. f.) Summerh. ■
58429 Brachycorythis tysonii Bolus = Neobolusia tysonii (Bolus) Schltr. ■☆
58430 Brachycorythis ugandensis Schltr. = Brachycorythis ovata Lindl. subsp. schweinfurthii (Rchb. f.) Summerh. ■☆
58431 Brachycorythis velutina Schltr.;短绒毛苞叶兰■☆
58432 Brachycorythis virginea (Bolus) Rolfe = Dracomonticola virginea (Bolus) H. P. Linder et Kurzweil ■☆
58433 Brachycorythis welwitschii Rchb. f. = Brachycorythis ovata Lindl. subsp. welwitschii (Rchb. f.) Summerh. ■☆
58434 Brachycorythis zeyheri (Sond.) Rchb. f. = Schizochilus zeyheri Sond. ■☆
58435 Brachycylix (Harms) R. S. Cowan (1975);艳花短杯豆属■☆
58436 Brachycylix vageleri (Harms) R. S. Cowan;艳花短杯豆■☆
58437 Brachycyrtis Koidz. (1924);日本铃兰属■☆
58438 Brachycyrtis Koidz. = Tricyrtis Wall. (保留属名)■
58439 Brachycyrtis macrantha Koidz.;日本铃兰■☆
58440 Brachyderea Cass. = Crepis L. ■
58441 Brachyderea abyssinica (Sch. Bip.) Kuntze = Crepis rueppellii Sch. Bip. ■☆
58442 Brachyderea carbonaria (Sch. Bip.) Sch. Bip. ex Schweinf. = Crepis carbonaria Sch. Bip. ■☆
58443 Brachyderea rueppellii (Sch. Bip.) Schweinf. = Crepis rueppellii Sch. Bip. ■☆
58444 Brachyderea schultzii (Hochst. ex A. Rich.) Schweinf. = Crepis schultzii (Hochst. ex A. Richardson) Vatke ■☆
58445 Brachyderea tenerrima (Sch. Bip.) Schweinf. = Crepis tenerrima (Sch. Bip.) R. E. Fr. ■☆
58446 Brachyderea xylorrhiza (Sch. Bip.) Schweinf. = Crepis xylorrhiza Sch. Bip. ■☆
58447 Brachyelytrum P. Beauv. (1812);短颖草属;Shorthusk ■
58448 Brachyelytrum africanum Hack. = Festuca africana (Hack.) Clayton ■☆
58449 Brachyelytrum aristosum (Michx.) P. Beauv. ex Branner et Coville var. glabratum Vasey ex Millsp. = Brachyelytrum aristosum (Michx.) Trel. ■☆
58450 Brachyelytrum aristosum (Michx.) Trel.;长芒短颖草;Bearded Shorthusk, Long-awned Wood Grass, Northern Shorthusk ■☆
58451 Brachyelytrum aristosum (Michx.) Trel. var. glabratum Vasey ex Millsp. = Brachyelytrum aristosum (Michx.) Trel. ■☆
58452 Brachyelytrum erectum (Schreb. ex Spreng.) P. Beauv.;短颖草;Bearded Shorthusk, Erect Dilepyrum, Long-awned Wood Grass ■
58453 Brachyelytrum erectum (Schreb. ex Spreng.) P. Beauv. subsp. japonicum (Hack.) T. Koyama et Kawano = Brachyelytrum japonicum (Hack.) Matsum. ex Honda ■
58454 Brachyelytrum erectum (Schreb. ex Spreng.) P. Beauv. var. glabratum (Vasey ex Millsp.) T. Koyama et Kawano = Brachyelytrum aristosum (Michx.) Trel. ■☆
58455 Brachyelytrum erectum (Schreb. ex Spreng.) P. Beauv. var. japonicum Hack. = Brachyelytrum japonicum (Hack.) Matsum. ex Honda ■
58456 Brachyelytrum erectum (Schreb. ex Spreng.) P. Beauv. var. septentrionale Babel = Brachyelytrum aristosum (Michx.) Trel. ■☆
58457 Brachyelytrum erectum (Schreb.) P. Beauv. subsp. japonicum (Hack.) T. Koyama et Kawano = Brachyelytrum erectum (Schreb.) P. Beauv. var. japonicum Hack. ■
58458 Brachyelytrum erectum (Schreb.) P. Beauv. subsp. japonicum (Hack.) T. Koyama et Kawano = Brachyelytrum japonicum (Hack.) Matsum. ex Honda ■
58459 Brachyelytrum erectum (Schreb.) P. Beauv. var. japonicum Hack. = Brachyelytrum japonicum (Hack.) Matsum. ex Honda ■
58460 Brachyelytrum erectum P. Beauv. = Brachyelytrum erectum (Schreb. ex Spreng.) P. Beauv. ■
58461 Brachyelytrum japonicum (Hack.) Hack. ex Honda = Brachyelytrum erectum (Schreb.) P. Beauv. var. japonicum Hack. ■
58462 Brachyelytrum japonicum (Hack.) Matsum. ex Honda;日本短颖草;Japan Wildeggplant, Japanese Shorthusk ■
58463 Brachyelytrum japonicum (Hack.) Matsum. ex Honda = Brachyelytrum erectum (Schreb.) P. Beauv. var. japonicum Hack. ■
58464 Brachyelytrum japonicum Hack. ex Matsum. = Brachyelytrum erectum (Schreb.) P. Beauv. var. japonicum Hack. ■
58465 Brachyelytrum septentrionale (Babel) G. C. Tucker = Brachyelytrum aristosum (Michx.) Trel. ■☆
58466 Brachyelytrum silvaticum (K. Schum.) Hack. = Festuca africana (Hack.) Clayton ■☆
58467 Brachyeorys Schrad. = Lindenbergia Lehm. ■
58468 Brachyglottis J. R. Forst. et G. Forst. (1775);短喉木属(常春菊

属);Shrub Ragwort ●■☆
58469 Brachyglottis bidwillii (Hook. f.) B. Nord.;苹叶短喉木●☆
58470 Brachyglottis compacta (Kirk) B. Nord.;密生短喉木(密丛常春菊)●☆
58471 Brachyglottis elaeagnifolia (Hook. f.) B. Nord.;胡颓子叶短喉木●☆
58472 Brachyglottis greyi (Hook. f.) B. Nord.;灰叶短喉木●☆
58473 Brachyglottis huntii (F. Muell.) B. Nord.;湿生短喉木;Rautini ●☆
58474 Brachyglottis laxifolia (Buchanan) B. Nord.;疏叶短喉木(疏叶常春菊)●☆
58475 Brachyglottis monroi (Hook. f.) B. Nord.;健壮短喉木(沿海常春菊);Monro's Ragwort ●☆
58476 Brachyglottis perdicioides (Hook. f.) B. Nord.;薄叶短喉木;Raukumara ●☆
58477 Brachyglottis repanda J. R. Forst. et G. Forst.;波叶短喉木(常春菊);Hedge Ragwort,Pukapuka,Rangiora ●☆
58478 Brachyglottis rotundifolia J. R. Forst. et G. Forst.;圆叶短喉木(圆叶常春菊)●☆
58479 Brachygyne (Benth.) Small = Dasistoma Raf. ■●☆
58480 Brachygyne Cass. = Eriocephalus L. ●☆
58481 Brachygyne Small = Seymeria Pursh(保留属名)■☆
58482 Brachyhelus (Benth.) Post et Kuntze = Schwenckia L. ■●☆
58483 Brachyilema Post et Kuntze = Bracheilema R. Br. ●■
58484 Brachyilema Post et Kuntze = Vernonia Schreb.(保留属名)●■
58485 Brachylaena R. Br. (1817);非洲木菊属(短被菊属,短衣菊属)●☆
58486 Brachylaena dentata (Thunb.) Harv. = Brachylaena glabra (L. f.) Druce ●☆
58487 Brachylaena dentata DC. = Brachylaena elliptica (Thunb.) DC. ●☆
58488 Brachylaena dentata DC. var. salicina? = Brachylaena elliptica (Thunb.) DC. ●☆
58489 Brachylaena discolor DC.;异色非洲木菊(非洲木菊);Coast Silver Oak ●☆
58490 Brachylaena discolor DC. subsp. transvaalensis (E. Phillips et Schweick.) Paiva = Brachylaena transvaalensis E. Phillips et Schweick. ●☆
58491 Brachylaena discolor DC. var. mossambicensis Paiva = Brachylaena discolor DC. ●☆
58492 Brachylaena discolor DC. var. rotundata (S. Moore) Beentje;圆叶异色非洲木菊●☆
58493 Brachylaena elliptica (Thunb.) DC.;椭圆非洲木菊(椭圆短衣菊)●☆
58494 Brachylaena elliptica (Thunb.) DC. var. salicina (DC.) Harv. = Brachylaena elliptica (Thunb.) DC. ●☆
58495 Brachylaena glabra (L. f.) Druce;全缘非洲木菊●☆
58496 Brachylaena glabra Druce = Brachylaena glabra (L. f.) Druce ●☆
58497 Brachylaena grandifolia DC. = Brachylaena glabra (L. f.) Druce ●☆
58498 Brachylaena huillensis O. Hoffm.;威拉非洲木菊●☆
58499 Brachylaena hutchinsii Hutch. = Brachylaena huillensis O. Hoffm. ●☆
58500 Brachylaena ilicifolia (Lam.) E. Phillips et Schweick.;冬青叶非洲木菊●☆
58501 Brachylaena microphylla Humbert;小叶非洲木菊●☆
58502 Brachylaena mullensis O. Hoffm. = Brachylaena huillensis O. Hoffm. ●☆
58503 Brachylaena natalensis Sch. Bip. ex Walp. = Brachylaena discolor DC. ●☆
58504 Brachylaena neriifolia (L.) R. Br.;夹竹桃叶非洲木菊●☆
58505 Brachylaena perrieri (Drake) Humbert;佩里耶非洲木菊●☆
58506 Brachylaena racemosa (Thunb.) DC. = Brachylaena ilicifolia (Lam.) E. Phillips et Schweick. ●☆
58507 Brachylaena ramiflora (DC.) Humbert;枝花非洲木菊●☆
58508 Brachylaena rotundata S. Moore = Brachylaena discolor DC. var. rotundata (S. Moore) Beentje ●☆
58509 Brachylaena stellulifera Humbert;星状非洲木菊●☆
58510 Brachylaena transvaalensis E. Phillips et Schweick.;德兰士瓦非洲木菊●☆
58511 Brachylaena trinervia Sond.;三脉非洲木菊●☆
58512 Brachylaena uniflora Harv.;单花非洲木菊●☆
58513 Brachylepis C. A. Mey. = Brachylepis C. A. Mey. ex Ledeb. ■
58514 Brachylepis C. A. Mey. ex Ledeb. (1829);短鳞藜属(短鳞木贼属)■
58515 Brachylepis Hook. et Arn. = Melinia Decne. ■☆
58516 Brachylepis Wight et Arn. = Baeolepis Decne. ex Moq. ■☆
58517 Brachylepis elatior C. A. Mey. = Anabasis elatior (C. A. Mey.) Schischk. ●
58518 Brachylepis eriopoda Schrenk = Anabasis eriopoda (Schrenk) Benth. ex Volkens ●
58519 Brachylepis salsa C. A. Mey. = Anabasis salsa (C. A. Mey.) Benth. ex Volkens ●
58520 Brachylepis truncata Schrenk = Anabasis truncata (Schrenk) Bunge ●
58521 Brachylobos Desv. = Brachylobus Link ■
58522 Brachylobus Dulac = Melilotus (L.) Mill. ■
58523 Brachylobus Link = Brachiolobos All. ■
58524 Brachylobus Link = Rorippa Scop. ■
58525 Brachyloma Hanst. = Isoloma Decne. ●■☆
58526 Brachyloma Hanst. = Kohleria Regel ●■☆
58527 Brachyloma Sond. (1854);瑞香石南属●☆
58528 Brachyloma daphnoides (Sm.) Benth.;瑞香石南;Daphne Heath ●☆
58529 Brachylophon Oliv. (1887);短脊木属●☆
58530 Brachylophon curtisii Oliv.;短脊木●☆
58531 Brachylophon niedenzuianum Engl. = Flabellariopsis acuminata (Engl.) R. Wilczek ●☆
58532 Brachymeris DC. = Phymaspermum Less. ●☆
58533 Brachymeris athanasioides (S. Moore) Hutch. = Phymaspermum athanasioides (S. Moore) Källersjö ■☆
58534 Brachymeris bolusii Hutch. = Phymaspermum bolusii (Hutch.) Källersjö ●☆
58535 Brachymeris erubescens Hutch. = Phymaspermum erubescens (Hutch.) Källersjö ●☆
58536 Brachymeris montana Hutch. = Phymaspermum montanum (Hutch.) Källersjö ■☆
58537 Brachymeris peglerae Hutch. = Phymaspermum peglerae (Hutch.) Källersjö ■☆
58538 Brachymeris scoparia DC. = Phymaspermum scoparium (DC.) Källersjö ■☆
58539 Brachynema Benth. (1857)(保留属名);短丝铁青树属●☆
58540 Brachynema F. Muell. = Abrophyllum Hook. f. ex Benth. ●☆
58541 Brachynema Griff.(废弃属名) = Brachynema Benth.(保留属

名)●☆
58542 Brachynema Griff.(废弃属名)= Sphenodesme Jack ●
58543 Brachynema ramiflorum Benth.;短丝铁青树●☆
58544 Brachyoglotis Lam. = Brachyglottis J. R. Forst. et G. Forst. ●■☆
58545 Brachyolobos DC. = Brachiolobos All. ■
58546 Brachyolobos DC. = Rorippa Scop. ■
58547 Brachyonostylum Mattf. = Brachionostylum Mattf. ●☆
58548 Brachyotum (DC.) Triana = Brachyotum (DC.) Triana ex Benth. ●☆
58549 Brachyotum (DC.) Triana ex Benth.(1867);短野牡丹属●☆
58550 Brachyotum Triana = Alifana Raf. ●☆
58551 Brachyotum Triana = Brachyotum (DC.) Triana ex Benth. ●☆
58552 Brachyotum alpinum Cogn.;高山短野牡丹●☆
58553 Brachyotum angustifolium Wurdack;狭叶短野牡丹●☆
58554 Brachyotum asperum Cogn.;粗糙短野牡丹●☆
58555 Brachyotum floribundum Triana;多花短野牡丹●☆
58556 Brachyotum intermedium Wurdack;全缘短野牡丹●☆
58557 Brachyotum longisepalum Wurdack;长瓣短野牡丹●☆
58558 Brachyotum maximowiczii Cogn.;马氏短野牡丹●☆
58559 Brachyotum minimum Markgr.;小短野牡丹●☆
58560 Brachyotum multinervium Wurdack;多脉短野牡丹●☆
58561 Brachyotum parvifolium Cogn.;小叶短野牡丹●☆
58562 Brachyotum rotundifolium Cogn.;圆叶短野牡丹●☆
58563 Brachypappus Sch. Bip. = Senecio L. ■●
58564 Brachypetalum Nutt. ex Lindl. = Smilacina Desf.(保留属名)■
58565 Brachypeza Garay(1972);短足兰属■☆
58566 Brachypeza Schltr. ex Garay. = Brachypeza Garay ■☆
58567 Brachypeza archytas (Ridl.) Garay;短足兰■☆
58568 Brachypeza stenoglottis (Hook. f.) Garay;窄舌短足兰■☆
58569 Brachyphragma Rydb. = Astragalus L. ●■
58570 Brachypoda Raf. = Eclipta L.(保留属名)■
58571 Brachypodandra Gagnep.(1948);短蕊椴属●☆
58572 Brachypodium P. Beauv.(1812);短柄草属;False Brome, False Brome Grass, Falsebrome, False-brome, False-brome Grass, Slender False-brome ■
58573 Brachypodium arbuscula Knoche;木本短柄草●☆
58574 Brachypodium bolusii Stapf;博卢斯短柄草■☆
58575 Brachypodium ciliare (Trin. ex Bunge) Maxim. = Elymus ciliaris (Trin. ex Bunge) Tzvelev ■
58576 Brachypodium commutatum (Schrad.) P. Beauv. = Bromus racemosus L. ■
58577 Brachypodium diaphanum Cufod. = Brachypodium flexum Nees ■☆
58578 Brachypodium distachyon (L.) P. Beauv.;二穗短柄草(紫短柄草);Purple False Brome, Purple Falsebrome, Stiff Brome ■
58579 Brachypodium distachyon (L.) P. Beauv. = Trachynia distachya (L.) Link ■
58580 Brachypodium distachyon (L.) P. Beauv. var. asperum (DC.) Parl. = Trachynia distachya (L.) Link ■☆
58581 Brachypodium distachyon (L.) P. Beauv. var. hispidum Pamp. = Brachypodium distachyon (L.) P. Beauv. ■
58582 Brachypodium distachyon (L.) P. Beauv. var. platystachyon Coss. et Durieu = Trachynia platystachya (Coss.) H. Scholz ■☆
58583 Brachypodium distachyon (L.) P. Beauv. var. velutinum Pamp. = Trachynia distachya (L.) Link ■☆
58584 Brachypodium durum Keng = Elymus durus (Keng) S. L. Chen ■
58585 Brachypodium durum Keng = Roegneria dura (Keng) Keng ex Keng et S. L. Chen ■
58586 Brachypodium flexum Nees;弯曲短柄草■☆
58587 Brachypodium flexum Nees var. abyssinicum Hochst. = Brachypodium flexum Nees ■☆
58588 Brachypodium formosanum Hayata;台湾短柄草;Taiwan Falsebrome ■
58589 Brachypodium formosanum Hayata = Brachypodium luzoniense Hack. ■
58590 Brachypodium formosanum Hayata = Brachypodium sylvaticum (Huds.) P. Beauv. ■
58591 Brachypodium hayatanum Honda = Brachypodium sylvaticum (Huds.) P. Beauv. ■
58592 Brachypodium kawakamii Hayata;川上氏短柄草;Kawakami Falsebrome ■
58593 Brachypodium kelungense Honda = Brachypodium sylvaticum (Huds.) P. Beauv. ■
58594 Brachypodium luzoniense Hack. = Brachypodium sylvaticum (Huds.) P. Beauv. ■
58595 Brachypodium madagascariensis A. Camus et Perr.;马岛短柄草■☆
58596 Brachypodium manshuricum Kitag.;东北短柄草;Northeast Falsebrome ■
58597 Brachypodium manshuricum Kitag. = Brachypodium sylvaticum (Huds.) P. Beauv. ■
58598 Brachypodium miserum Koidz. = Brachypodium luzoniense Hack. ■
58599 Brachypodium nepalense Nees ex Steud. = Brachypodium sylvaticum (Huds.) P. Beauv. ■
58600 Brachypodium perrieri A. Camus;佩里耶短柄草■☆
58601 Brachypodium phoenicoides (L.) Roem. et Schult.;凤凰短柄草;Thinleaf False Brome ■☆
58602 Brachypodium phoenicoides (L.) Roem. et Schult. var. brevisetum (DC.) St. -Yves = Brachypodium phoenicoides (L.) Roem. et Schult. ■☆
58603 Brachypodium phoenicoides (L.) Roem. et Schult. var. villiglume Emb. et Maire = Brachypodium phoenicoides (L.) Roem. et Schult. ■☆
58604 Brachypodium pinnatum (L.) P. Beauv.;羽状短柄草(短芒短柄草,兴安短柄草);Chalk False-brome, Heath False Brome, Japanese False Brome, Japanese Falsebrome, Japanese False-brome, Tor-grass ■
58605 Brachypodium pinnatum L. var. australe Gren. et Godr. = Brachypodium phoenicoides (L.) Roem. et Schult. ■☆
58606 Brachypodium pinnatum L. var. phoenicoides (L.) Trab. = Brachypodium phoenicoides (L.) Roem. et Schult. ■☆
58607 Brachypodium pinnatum P. Beauv. = Brachypodium pinnatum (L.) P. Beauv. ■
58608 Brachypodium pratense Keng ex P. C. Keng;草地短柄草;Meadow Falsebrome ■
58609 Brachypodium quartinianum (A. Rich.) Hack. ex Engl. = Brachypodium flexum Nees ■☆
58610 Brachypodium ramosum (L.) Roem. et Schult. = Brachypodium retusum (Pers.) P. Beauv. ■☆
58611 Brachypodium ramosum (L.) Roem. et Schult. var. gigas St. -Yves = Brachypodium phoenicoides (L.) Roem. et Schult. ■☆
58612 Brachypodium ramosum (L.) Roem. et Schult. var. roemeri St. -Yves = Brachypodium retusum (Pers.) P. Beauv. ■☆
58613 Brachypodium ramosum (L.) Roem. et Schult. var. scabriculme Maire = Brachypodium retusum (Pers.) P. Beauv. ■☆

58614 Brachypodium retusum (Pers.) P. Beauv.;微凹短柄草■☆
58615 Brachypodium retusum P. Beauv. = Brachypodium retusum (Pers.) P. Beauv. ■☆
58616 Brachypodium rupestre Roem. et Schult.;沼泽短柄草■☆
58617 Brachypodium schimperi (Hochst. ex A. Rich.) Chiov.;欣珀短柄草■☆
58618 Brachypodium schumannianum Pilg. = Brachypodium flexum Nees ■☆
58619 Brachypodium sylvaticum (Huds.) P. Beauv.;短柄草(基隆短柄草);Slender False Brome, Slender False-brome, Woods Falsebrome ■
58620 Brachypodium sylvaticum (Huds.) P. Beauv. subsp. glaucovirens Murb.;灰绿短柄草■☆
58621 Brachypodium sylvaticum (Huds.) P. Beauv. subsp. luzoniense Hack. = Brachypodium sylvaticum (Huds.) P. Beauv. ■
58622 Brachypodium sylvaticum (Huds.) P. Beauv. var. breviglume Keng ex P. C. Keng;小颖短柄草;Shortglume Falsebrome ■
58623 Brachypodium sylvaticum (Huds.) P. Beauv. var. breviglume Keng ex P. C. Keng = Brachypodium sylvaticum (Huds.) P. Beauv. ■
58624 Brachypodium sylvaticum (Huds.) P. Beauv. var. glabrescens Coss. et Germ. = Brachypodium sylvaticum (Huds.) P. Beauv. ■
58625 Brachypodium sylvaticum (Huds.) P. Beauv. var. gracile (Weigel) Keng;细株短柄草(细珠短柄草);Slender Falsebrome ■
58626 Brachypodium sylvaticum (Huds.) P. Beauv. var. gracile (Weigel) Keng = Brachypodium sylvaticum (Huds.) P. Beauv. ■
58627 Brachypodium sylvaticum (Huds.) P. Beauv. var. kelungense (Honda) C. C. Hsu = Brachypodium sylvaticum (Huds.) P. Beauv. ■
58628 Brachypodium sylvaticum (Huds.) P. Beauv. var. kelungense (Honda) C. C. Hsu = Brachypodium kelungense Honda ■
58629 Brachypodium sylvaticum (Huds.) P. Beauv. var. luzoniense (Hack.) Hara;吕宋短柄草(基隆短柄草)■
58630 Brachypodium sylvaticum (Huds.) P. Beauv. var. luzoniense (Hack.) Hara = Brachypodium sylvaticum (Huds.) P. Beauv. ■
58631 Brachypodium sylvaticum (Huds.) P. Beauv. var. luzoniense (Hack.) Hara = Brachypodium luzoniense Hack. ■
58632 Brachypodium sylvaticum (Huds.) P. Beauv. var. miserum (Thunb.) Koidz. = Brachypodium luzoniense Hack. ■
58633 Brachypodium sylvaticum (Huds.) P. Beauv. var. pseudodistachyon (C. B. Clarke) Hook. f. = Brachypodium sylvaticum (Huds.) P. Beauv. ■
58634 Brachypodium sylvaticum (Huds.) P. Beauv. var. pseudodistachyon Hook. f. = Brachypodium sylvaticum (Huds.) P. Beauv. ■
58635 Brachypodium sylvaticum (Huds.) P. Beauv. var. villosum Lej. et Courtois = Brachypodium sylvaticum (Huds.) P. Beauv. ■
58636 Brachypodium sylvaticum (Huds.) P. Beauv. var. wattii (C. B. Clarke) Hook. f. = Brachypodium sylvaticum (Huds.) P. Beauv. ■
58637 Brachypodium tataricum Munro ex Aitch. = Elymus canaliculatus (Nevski) Tzvelev ■
58638 Brachypodium villosum Drobow;毛短柄草■☆
58639 Brachypodium wattii C. B. Clarke = Brachypodium sylvaticum (Huds.) P. Beauv. ■
58640 Brachypremna Gleason = Ernestia DC. ☆
58641 Brachypteris Griseb. = Brachypterys A. Juss. ●☆
58642 Brachypterum (Wight et Arn.) Benth. = Derris Lour.(保留属名)●
58643 Brachypterum Benth. = Derris Lour.(保留属名)●
58644 Brachypterum microphyllum Miq. = Derris microphylla (Miq.) B. D. Jacks. ●☆
58645 Brachypterum robustum (Roxb. ex DC.) Dalzell et Gibson = Derris robusta (Roxb. ex DC.) Benth. ●
58646 Brachypterys A. Juss. (1838);短翼金虎尾属●☆
58647 Brachypteryx Dalla Torre et Harms = Brachypterys A. Juss. ●☆
58648 Brachypus Ledeb. = Fibigia Medik. ■☆
58649 Brachypus Ledeb. = Lunaria L. ■☆
58650 Brachyramphus DC. = Lactuca L. ■
58651 Brachyramphus DC. = Launaea Cass. ■
58652 Brachyramphus goraeensis (Lam.) DC. = Launaea intybacea (Jacq.) Beauverd ■☆
58653 Brachyramphus goraeensis DC. = Launaea intybacea (Jacq.) Beauverd ■☆
58654 Brachyramphus intybacea (Jacq.) DC. = Launaea intybacea (Jacq.) Beauverd ■☆
58655 Brachyramphus nudicaulis Klatt = Launaea rarifolia (Oliv. et Hiern) Boulos ■☆
58656 Brachyramphus ramosissimus Benth. = Paraixeris denticulata (Houtt.) Nakai ■
58657 Brachyramphus sinicus Miq. = Pterocypsela indica (L.) C. Shih ■
58658 Brachyrhynchos Less. = Senecio L. ■●
58659 Brachyrhynchos eupatorioides Less. = Senecio othonniflorus DC. ■☆
58660 Brachyrhynchos junceus DC. = Senecio junceus (DC.) Harv. ■☆
58661 Brachyrhynchos tuberosus DC. = Senecio tuberosus (DC.) Harv. ■☆
58662 Brachyridium Meisn. = Lepidophyllum Cass. ●☆
58663 Brachyris Nutt. = Gutierrezia Lag. ■●☆
58664 Brachyris californica DC. = Gutierrezia californica (DC.) Torr. et A. Gray ■☆
58665 Brachyris dracunculoides DC. = Amphiachyris dracunculoides (DC.) Nutt. ■☆
58666 Brachyris microcephala DC. = Gutierrezia microcephaia (DC.) A. Gray ■☆
58667 Brachyris ovatifolia DC. = Solidago sphacelata Raf. ■☆
58668 Brachyscias J. M. Hart et Henwood(1999);澳洲短伞芹属■☆
58669 Brachyscome Cass. (1816);五色菊属(短毛菊属,鹅河菊属);Swan River Daisy, Swanriver Daisy, Swan-river Daisy ●■☆
58670 Brachyscome Cass. = Brachycome Cass. ●■☆
58671 Brachyscome iberidifolia Benth.;五色菊(鹅河菊);Swan River Daisy, Swan-river Daisy ■☆
58672 Brachyscome multifida DC.;多裂五色菊;Swan River Daisy ●☆
58673 Brachyscome nivalis F. Muell.;雪白五色菊●☆
58674 Brachyscypha Baker = Lachenalia J. Jacq. ex Murray ■☆
58675 Brachysema R. Br. (1811);西澳木属●☆
58676 Brachysema aphyllum Hook.;无叶西澳木●☆
58677 Brachysema celsianum Lem.;湿生西澳木;Swan River Pae ●☆
58678 Brachysema lanceolatum Meisn.;剑叶西澳木;Scimitar Shrub, Swan River Pea Shrub, Swan River Peabush ●☆
58679 Brachysiphon A. Juss. (1846);南非管萼木属●☆
58680 Brachysiphon acutus (Thunb.) A. Juss.;尖南非管萼木●☆
58681 Brachysiphon ericifolius A. Juss. = Stylapterus ericifolius (A. Juss.) R. Dahlgren ●☆
58682 Brachysiphon fucatus (L.) Gilg;着色南非管萼木●☆
58683 Brachysiphon imbricatus (Graham) A. Juss. = Brachysiphon fucatus (L.) Gilg ●☆
58684 Brachysiphon microphyllus Rourke;小叶南非管萼木●☆

58685 Brachysiphon mundii Sond. ;蒙德南非管萼木●☆
58686 Brachysiphon petraeus W. F. Barker = Sonderothamnus petraeus (W. F. Barker) R. Dahlgren ●☆
58687 Brachysiphon rupestris Sond. ;岩生南非管萼木●☆
58688 Brachysiphon speciosus Sond. = Sonderothamnus speciosus (Sond.) R. Dahlgren ●☆
58689 Brachysola Rye(2000);星毛灌属●☆
58690 Brachysola coerulea (F. Muell. et Tate) Rye;蓝星毛灌●☆
58691 Brachysola halganiacea (F. Muell.) Rye;星毛灌●☆
58692 Brachyspatha Schott = Amorphophallus Blume ex Decne.(保留属名)■●
58693 Brachyspatha konjac Schott ex Miq. = Amorphophallus rivieri Durieu ex Carrière ■
58694 Brachyspatha variabilis Benth. = Amorphophallus variabilis Blume ■
58695 Brachystachys Klotzsch = Croton L. ●
58696 Brachystachyum Keng = Semiarundinaria Makino ex Nakai ●
58697 Brachystachyum Keng(1940);短穗竹属;Shortspikebamboo, Shortspikilet Bamboo, Short-spikileted Bamboo ●★
58698 Brachystachyum densiflorum (Rendle) Keng;短穗竹;Dense-flowered Bamboo, Short Spikelet Bamboo, Shortspikebamboo, Shortspikilet Bamboo, Short-tassled Bamboo ●
58699 Brachystachyum densiflorum (Rendle) Keng = Semiarundinaria densiflora (Rendle) T. H. Wen ●
58700 Brachystachyum densiflorum (Rendle) Keng var. villosum S. L. Chen et C. Y. Yao;毛环短穗竹;Villose Shortspikilet Bamboo ●
58701 Brachystegia Benth. (1865);短盖豆属●☆
58702 Brachystegia allenii Hutch. et Burtt Davy;阿伦短盖豆●☆
58703 Brachystegia allenii Hutch. et Burtt Davy var. giorgii (De Wild.) Hoyle = Brachystegia allenii Hutch. et Burtt Davy ●☆
58704 Brachystegia angustistipulata De Wild. ;窄托叶短盖豆●☆
58705 Brachystegia apertifolia Hutch. et Burtt Davy = Brachystegia longifolia Benth. ●☆
58706 Brachystegia appendiculata Benth. = Brachystegia spiciformis Benth. ●☆
58707 Brachystegia astlei Hoyle = Brachystegia michelmorei Hoyle ●☆
58708 Brachystegia bakeriana Hutch. et Burtt Davy;贝克短盖豆●☆
58709 Brachystegia boehmii Taub. ;贝姆短盖豆;Prince of Wales' Feathers Tree ●☆
58710 Brachystegia boehmii Taub. var. katangensis (De Wild.) Hoyle = Brachystegia boehmii Taub. ●☆
58711 Brachystegia bournei Greenway = Brachystegia longifolia Benth. ●☆
58712 Brachystegia bragaei Harms = Brachystegia spiciformis Benth. ●☆
58713 Brachystegia bussei Harms;布瑟短盖豆●☆
58714 Brachystegia cynometroides Harms;茎花豆短盖豆●☆
58715 Brachystegia diloloensis De Wild. = Brachystegia utilis Hutch. et Burtt Davy ●☆
58716 Brachystegia edulis Hutch. et Burtt Davy = Brachystegia spiciformis Benth. ●☆
58717 Brachystegia eurycoma Harms;宽短盖豆●☆
58718 Brachystegia euryphylla Harms = Brachystegia spiciformis Benth. ●☆
58719 Brachystegia falcato-appendiculata De Wild. = Brachystegia longifolia Benth. ●☆
58720 Brachystegia ferruginea De Wild. = Brachystegia boehmii Taub. ●☆
58721 Brachystegia ferruginea De Wild. var. angustifoliotata? = Brachystegia boehmii Taub. ●☆
58722 Brachystegia ferruginea De Wild. var. interrupta? = Brachystegia boehmii Taub. ●☆
58723 Brachystegia ferruginea De Wild. var. querrei? = Brachystegia boehmii Taub. ●☆
58724 Brachystegia ferruginea De Wild. var. robynsii? = Brachystegia boehmii Taub. ●☆
58725 Brachystegia filiformis Hutch. et Burtt Davy = Brachystegia boehmii Taub. ●☆
58726 Brachystegia fischeri Taub. = Brachystegia microphylla Harms ●☆
58727 Brachystegia flagristipulata Taub. = Brachystegia boehmii Taub. ●☆
58728 Brachystegia fleuryana A. Chev. = Didelotia minutiflora (A. Chev.) J. Léonard ●☆
58729 Brachystegia floribunda Benth. ;多花短盖豆●☆
58730 Brachystegia gairdnerae Hutch. et Burtt Davy = Brachystegia bakeriana Hutch. et Burtt Davy ●☆
58731 Brachystegia giorgii De Wild. = Brachystegia allenii Hutch. et Burtt Davy ●☆
58732 Brachystegia glaberrima R. E. Fr. = Brachystegia longifolia Benth. ●☆
58733 Brachystegia glaucescens Hutch. et Burtt Davy = Brachystegia microphylla Harms ●☆
58734 Brachystegia globiflora Benth. = Julbernardia globiflora (Benth.) Troupin ●☆
58735 Brachystegia goetzei Harms = Brachystegia longifolia Benth. ●☆
58736 Brachystegia gossweileri Hutch. et Burtt Davy;戈斯短盖豆●☆
58737 Brachystegia hockii De Wild. = Brachystegia spiciformis Benth. ●☆
58738 Brachystegia holtzii Harms = Brachystegia longifolia Benth. ●☆
58739 Brachystegia homblei De Wild. = Brachystegia longifolia Benth. ●☆
58740 Brachystegia hopkinsii Suess. = Brachystegia boehmii Taub. ●☆
58741 Brachystegia itoliensis Taub. = Brachystegia spiciformis Benth. ●☆
58742 Brachystegia kalongoensis De Wild. = Brachystegia wangermeeana De Wild. ●☆
58743 Brachystegia kassneri Baker f. = Brachystegia stipulata De Wild. ●☆
58744 Brachystegia katangensis De Wild. = Brachystegia boehmii Taub. ●☆
58745 Brachystegia kennedyi Hoyle;肯尼迪短盖豆●☆
58746 Brachystegia klainei Pierre ex Harms = Librevillea klainei (Pierre ex Harms) Hoyle ■☆
58747 Brachystegia laurentii (De Wild.) Louis ex Hoyle;洛朗短盖豆●☆
58748 Brachystegia leonensis Hutch. et Burtt Davy;莱昂短盖豆●☆
58749 Brachystegia letestui De Wild. ;莱泰斯图盖豆●☆
58750 Brachystegia longifolia Benth. ;长叶短盖豆●☆
58751 Brachystegia longifolia Benth. var. parvifolia (Benth.) Topham = Brachystegia longifolia Benth. ●☆
58752 Brachystegia longifoliolata De Wild. = Brachystegia longifolia Benth. ●☆
58753 Brachystegia lufirensis De Wild. = Brachystegia stipulata De Wild. ●☆
58754 Brachystegia luishiensis De Wild. = Brachystegia longifolia Benth. ●☆
58755 Brachystegia lujae De Wild. = Brachystegia spiciformis Benth. ●☆
58756 Brachystegia malengensis De Wild. = Brachystegia boehmii Taub. ●☆
58757 Brachystegia michelmorei Hoyle;米氏短盖豆●☆
58758 Brachystegia microphylla Harms;小叶短盖豆●☆
58759 Brachystegia mildbraedii Harms;米尔德短盖豆●☆

58760 Brachystegia mimosifolia Hutch. et Burtt Davy = Brachystegia taxifolia Harms ●☆
58761 Brachystegia mpalensis Micheli = Brachystegia spiciformis Benth. ●☆
58762 Brachystegia mpalensis Micheli var. latifoliolata De Wild. = Brachystegia spiciformis Benth. ●☆
58763 Brachystegia nchangensis Greenway = Brachystegia floribunda Benth. ●☆
58764 Brachystegia nigerica Hoyle et A. P. D. Jones;尼日利亚短盖豆●☆
58765 Brachystegia nzang Pellegr. = Brachystegia mildbraedii Harms ●☆
58766 Brachystegia obliqua Hutch. et Burtt Davy = Brachystegia bakeriana Hutch. et Burtt Davy ●☆
58767 Brachystegia oblonga Sims;矩圆短盖豆●☆
58768 Brachystegia oliveri Taub. = Brachystegia spiciformis Benth. ●☆
58769 Brachystegia pectinata Sim = Brachystegia microphylla Harms ●☆
58770 Brachystegia polyantha Harms = Brachystegia floribunda Benth. ●☆
58771 Brachystegia pruinosa De Wild. = Brachystegia allenii Hutch. et Burtt Davy ●☆
58772 Brachystegia puberula Hutch. et Burtt Davy;微毛短盖豆●☆
58773 Brachystegia randii Baker f. = Brachystegia spiciformis Benth. ●☆
58774 Brachystegia reticulata Hutch. et Burtt Davy = Brachystegia microphylla Harms ●☆
58775 Brachystegia robynsii De Wild. = Brachystegia microphylla Harms ●☆
58776 Brachystegia russelliae I. M. Johnst.;鲁塞尔短盖豆●☆
58777 Brachystegia sapinii De Wild. = Brachystegia wangermeeana De Wild. ●☆
58778 Brachystegia schliebenii Harms = Brachystegia allenii Hutch. et Burtt Davy ●☆
58779 Brachystegia spiciformis Benth.;美丽短盖豆●☆
58780 Brachystegia spiciformis Benth. var. kwangensis Hoyle = Brachystegia spiciformis Benth. ●☆
58781 Brachystegia spiciformis Benth. var. latifoliolata (De Wild.) Hoyle = Brachystegia spiciformis Benth. ●☆
58782 Brachystegia spiciformis Benth. var. mpalensis (Micheli) Hoyle = Brachystegia spiciformis Benth. ●☆
58783 Brachystegia spiciformis Benth. var. schmitzii Hoyle = Brachystegia spiciformis Benth. ●☆
58784 Brachystegia stipulacea Taub.;热非短盖豆●☆
58785 Brachystegia stipulata De Wild.;托叶短盖豆●☆
58786 Brachystegia stipulata De Wild. var. lufirensis (De Wild.) Hoyle = Brachystegia stipulata De Wild. ●☆
58787 Brachystegia stipulata De Wild. var. velutina (De Wild.) Hoyle = Brachystegia stipulata De Wild. ●☆
58788 Brachystegia subfalcato-foliolata De Wild. = Brachystegia taxifolia Harms ●☆
58789 Brachystegia tamarindoides Welw. ex Benth.;酸豆短盖豆●☆
58790 Brachystegia tamarindoides Welw. ex Benth. subsp. microphylla (Welw. ex Benth.) Chikuni = Brachystegia microphylla Harms ●☆
58791 Brachystegia tamarindoides Welw. ex Benth. subsp. torrei (Hoyle) Chikuni = Brachystegia torrei Hoyle ●☆
58792 Brachystegia taubertiana Hutch. et Burtt Davy = Brachystegia spiciformis Benth. ●☆
58793 Brachystegia taxifolia Harms;紫杉叶短盖豆●☆
58794 Brachystegia thomasii De Wild. = Brachystegia stipulata De Wild. ●☆
58795 Brachystegia torrei Hoyle;托雷短盖豆●☆
58796 Brachystegia trijuga R. E. Fr. = Brachystegia spiciformis Benth. ●☆
58797 Brachystegia utilis Hutch. et Burtt Davy;有用短盖豆●☆
58798 Brachystegia velutina De Wild. = Brachystegia stipulata De Wild. ●☆
58799 Brachystegia venosa Hutch. et Burtt Davy = Brachystegia spiciformis Benth. ●☆
58800 Brachystegia wangermeeana De Wild.;旺格短盖豆●☆
58801 Brachystegia wildemaniana R. E. Fr. = Brachystegia wangermeeana De Wild. ●☆
58802 Brachystegia woodiana Harms = Brachystegia boehmii Taub. ●☆
58803 Brachystegia zenkeri Harms;岑克尔短盖豆●☆
58804 Brachystele Schltr. (1920);短柱兰属■☆
58805 Brachystele affinis (C. Schweinf.) Burns-Bal.;近缘短柱兰■☆
58806 Brachystele aguacatensis Schltr.;短柱兰■☆
58807 Brachystele guyanensis (Lindl.) Schltr.;圭亚那短柱兰■☆
58808 Brachystele longiflora Schltr.;长花短柱兰■☆
58809 Brachystele minutiflora (A. Rich. et Galeotti) Burns-Bal.;小花短柱兰■☆
58810 Brachystele oxyanthos Szlach.;尖花短柱兰■☆
58811 Brachystele polyantha (Kuntze) Burns-Bal.;多花短柱兰■☆
58812 Brachystele tenuissima (L. O. Williams) Burns-Bal.;细短柱兰■☆
58813 Brachystelma R. Br. (1822)(保留属名);润肺草属(短梗藤属,球萝藦属);Brachystelma ■
58814 Brachystelma albipilosum Peckover;白疏毛润肺草■☆
58815 Brachystelma alpinum R. A. Dyer;高山润肺草■☆
58816 Brachystelma angustum Peckover;窄润肺草■☆
58817 Brachystelma arachnoideum Masinde;蛛毛润肺草■☆
58818 Brachystelma arenarium S. Moore;沙地润肺草■☆
58819 Brachystelma arnotii Baker;阿诺特润肺草■☆
58820 Brachystelma asmarense Chiov. = Brachystelma lineare A. Rich. ■☆
58821 Brachystelma atacorense A. Chev. = Brachystelma togoense Schltr. ■☆
58822 Brachystelma australe R. A. Dyer;南方润肺草■☆
58823 Brachystelma bagshawei S. Moore = Brachystelma johnstonii N. E. Br. ■☆
58824 Brachystelma barberae Harv. ex Hook. f.;巴尔巴拉润肺草■☆
58825 Brachystelma bingeri A. Chev. = Raphionacme bingeri (A. Chev.) Lebrun et Stork ■☆
58826 Brachystelma blepharanthera H. Huber;毛药润肺草■☆
58827 Brachystelma bolusii N. E. Br. = Brachystelma circinatum E. Mey. ■☆
58828 Brachystelma bracteolatum Meve;小苞片润肺草■☆
58829 Brachystelma brevipedicellatum Turrill;短小梗润肺草■☆
58830 Brachystelma bruceae R. A. Dyer;布鲁斯润肺草■☆
58831 Brachystelma bruceae R. A. Dyer subsp. hirsutum R. A. Dyer;粗毛润肺草■☆
58832 Brachystelma buchananii N. E. Br.;布坎南润肺草■☆
58833 Brachystelma burchellii (Decne.) Peckover;伯切尔润肺草■☆
58834 Brachystelma burchellii (Decne.) Peckover var. grandiflora (N. E. Br.) Meve = Brachystelma burchellii (Decne.) Peckover ■☆
58835 Brachystelma caffrum (Schltr.) N. E. Br.;开菲尔润肺草■☆
58836 Brachystelma campanulatum N. E. Br.;风铃草状润肺草■☆
58837 Brachystelma canum R. A. Dyer;灰色润肺草■☆
58838 Brachystelma cathcartense R. A. Dyer;卡斯卡特润肺草■☆
58839 Brachystelma caudatum (Thunb.) N. E. Br. = Brachystelma tuberosum (Meerb.) R. Br. ex Sims ■☆
58840 Brachystelma chloranthum (Schltr.) Peckover;绿花润肺草■☆

58841 Brachystelma chlorozonum E. A. Bruce;绿润肺草■☆
58842 Brachystelma circinatum E. Mey.;小叶润肺草■☆
58843 Brachystelma coddii R. A. Dyer;科德润肺草■☆
58844 Brachystelma commixtum N. E. Br. = Brachystelma circinatum E. Mey. ■☆
58845 Brachystelma comptum N. E. Br.;装饰润肺草■☆
58846 Brachystelma constrictum J. B. Hall;缢缩润肺草■☆
58847 Brachystelma crispum Graham = Brachystelma tuberosum (Meerb.) R. Br. ex Sims ■☆
58848 Brachystelma cupulatum R. A. Dyer;杯状润肺草■☆
58849 Brachystelma decipiens N. E. Br.;迷惑润肺草■☆
58850 Brachystelma delicatum R. A. Dyer;姣美润肺草■☆
58851 Brachystelma dimorphum R. A. Dyer;二型润肺草■☆
58852 Brachystelma dimorphum R. A. Dyer subsp. gratum R. A. Dyer;可爱润肺草■☆
58853 Brachystelma dinteri Schltr.;丁特润肺草■☆
58854 Brachystelma discoideum R. A. Dyer;盘状润肺草■☆
58855 Brachystelma distinctum N. E. Br. = Brachystelma elongatum (Schltr.) N. E. Br. ■☆
58856 Brachystelma duplicatum R. A. Dyer;双润肺草■☆
58857 Brachystelma dyeri K. Balkwill et M. Balkwill;戴尔润肺草■☆
58858 Brachystelma edulis Collett et Hemsl.;润肺草(地饼,短梗藤);Edible Brachystelma ■
58859 Brachystelma elegantulum S. Moore;雅致润肺草■☆
58860 Brachystelma ellipticum A. Rich. = Brachystelma lineare A. Rich. ■☆
58861 Brachystelma elongatum (Schltr.) N. E. Br.;伸长润肺草■☆
58862 Brachystelma exile Bullock;瘦小润肺草■☆
58863 Brachystelma festucifolium E. A. Bruce;羊茅叶润肺草■☆
58864 Brachystelma filifolium (N. E. Br.) Peckover;线叶润肺草■☆
58865 Brachystelma filiforme Harv. = Brachystelma circinatum E. Mey. ■☆
58866 Brachystelma flavidum Schltr. = Brachystelma pygmaeum (Schltr.) N. E. Br. subsp. flavidum (Schltr.) R. A. Dyer ■☆
58867 Brachystelma floribundum R. A. Dyer = Brachystelma duplicatum R. A. Dyer ■☆
58868 Brachystelma floribundum Turrill;繁花润肺草■☆
58869 Brachystelma foetidum Schltr.;臭润肺草■☆
58870 Brachystelma franksiae N. E. Br.;弗兰克斯润肺草■☆
58871 Brachystelma franksiae N. E. Br. subsp. grandiflorum A. P. Dold et Bruyns;大花润肺草■☆
58872 Brachystelma furcatum Boele;叉分润肺草■☆
58873 Brachystelma galpinii (Schltr.) N. E. Br. = Brachystelma circinatum E. Mey. ■☆
58874 Brachystelma gerrardii Harv.;杰勒德润肺草■☆
58875 Brachystelma gracile E. A. Bruce;纤细润肺草■☆
58876 Brachystelma gracillimum R. A. Dyer;细长润肺草■☆
58877 Brachystelma grossartii Dinter = Brachystelma arnotii Baker ■☆
58878 Brachystelma gymnopodum (Schltr.) Bruyns;裸梗润肺草■☆
58879 Brachystelma hirsutum E. Mey. = Raphionacme hirsuta (E. Mey.) R. A. Dyer ■☆
58880 Brachystelma hirtellum Weim.;多毛润肺草■☆
58881 Brachystelma huttonii (Harv.) N. E. Br.;赫顿润肺草■☆
58882 Brachystelma incanum R. A. Dyer;灰毛润肺草■☆
58883 Brachystelma inconspicuum S. Venter;显著润肺草■☆
58884 Brachystelma johnstonii N. E. Br.;约翰斯顿润肺草■☆
58885 Brachystelma keniense Schweinf.;凯尼润肺草■☆
58886 Brachystelma kerrii Craib;长节润肺草;Kerr Brachystelma ■
58887 Brachystelma lancasteri Boele;兰开斯特润肺草■☆
58888 Brachystelma lanceolatum Turrill = Brachystelma johnstonii N. E. Br. ■☆
58889 Brachystelma letestui Pellegr.;莱泰斯图润肺草■☆
58890 Brachystelma lineare A. Rich.;线形润肺草■☆
58891 Brachystelma linearifolium Turrill = Brachystelma plocamoides Oliv. ■☆
58892 Brachystelma longifolium (Schltr.) N. E. Br.;长叶润肺草■☆
58893 Brachystelma luteum Peckover;黄润肺草■☆
58894 Brachystelma macropetalum (Schltr.) N. E. Br.;大瓣润肺草■☆
58895 Brachystelma macrorrhiza E. Mey.;大根润肺草■☆
58896 Brachystelma mafekingense N. E. Br. = Ceropegia mafekingensis (N. E. Br.) R. A. Dyer ■☆
58897 Brachystelma magicum N. E. Br. = Brachystelma buchananii N. E. Br. ■☆
58898 Brachystelma megasepalum Peckover;大萼润肺草■☆
58899 Brachystelma meyerianum Schltr.;迈尔润肺草■☆
58900 Brachystelma micranthum E. Mey.;小花润肺草■☆
58901 Brachystelma minimum R. A. Dyer;极小润肺草■☆
58902 Brachystelma minor E. A. Bruce;小润肺草■☆
58903 Brachystelma modestum R. A. Dyer;适度润肺草■☆
58904 Brachystelma montanum R. A. Dyer;山地润肺草■☆
58905 Brachystelma mortonii Walker;莫顿润肺草■☆
58906 Brachystelma nanum (Schltr.) N. E. Br.;矮小润肺草■☆
58907 Brachystelma natalense (Schltr.) N. E. Br.;纳塔尔润肺草■☆
58908 Brachystelma nauseosum De Wild. = Brachystelma buchananii N. E. Br. ■☆
58909 Brachystelma ngomense R. A. Dyer;恩戈姆润肺草■☆
58910 Brachystelma nigrum R. A. Dyer = Brachystelma gerrardii Harv. ■☆
58911 Brachystelma occidentale Schltr.;西方润肺草■☆
58912 Brachystelma omissum Bullock;奥米萨润肺草■☆
58913 Brachystelma ovatum Oliv. = Brachystelma circinatum E. Mey. ■☆
58914 Brachystelma pachypodium R. A. Dyer;粗足润肺草■☆
58915 Brachystelma pallidum (Schltr.) N. E. Br. = Brachystelma circinatum E. Mey. ■☆
58916 Brachystelma parviflora Morton = Brachystelma mortonii Walker ■☆
58917 Brachystelma parvulum R. A. Dyer;较小润肺草■☆
58918 Brachystelma pellacibellum L. E. Newton = Brachystelma lineare A. Rich. ■☆
58919 Brachystelma petraeum R. A. Dyer;岩生润肺草■☆
58920 Brachystelma phyteumoides K. Schum. = Brachystelma lineare A. Rich. ■☆
58921 Brachystelma pilosum R. A. Dyer = Brachystelma hirtellum Weim. ■☆
58922 Brachystelma plocamoides Oliv.;卷毛茜润肺草■☆
58923 Brachystelma prostratum E. A. Bruce;平卧润肺草■☆
58924 Brachystelma pulchellum (Harv.) Schltr.;美丽润肺草■☆
58925 Brachystelma punctatum Boele;斑点润肺草■☆
58926 Brachystelma pygmaeum (Schltr.) N. E. Br.;短花润肺草■☆
58927 Brachystelma pygmaeum (Schltr.) N. E. Br. subsp. flavidum (Schltr.) R. A. Dyer;黄短花润肺草■☆
58928 Brachystelma pygmaeum (Schltr.) N. E. Br. var. breviflorum? = Brachystelma pygmaeum (Schltr.) N. E. Br. ■☆
58929 Brachystelma ramosissimum (Schltr.) N. E. Br.;多分枝润肺草■☆
58930 Brachystelma recurvatum Bruyns;反折润肺草■☆

58931　Brachystelma rehmannii Schltr. = Brachystelma foetidum Schltr. ■☆
58932　Brachystelma remotum R. A. Dyer;稀疏润肺草■☆
58933　Brachystelma richardsii Peckover;理查兹润肺草■☆
58934　Brachystelma ringens E. A. Bruce = Brachystelma brevipedicellatum Turrill ■☆
58935　Brachystelma rubellum (E. Mey.) Peckover;微红润肺草■☆
58936　Brachystelma sandersonii (Oliv.) N. E. Br.;桑德森润肺草■☆
58937　Brachystelma schinzii (K. Schum.) N. E. Br.;欣兹润肺草■☆
58938　Brachystelma schizoglossoides (Schltr.) N. E. Br.;裂舌润肺草■☆
58939　Brachystelma schoenlandianum Schltr.;舍恩润肺草■☆
58940　Brachystelma schultzei (Schltr.) Bruyns;舒尔茨润肺草■☆
58941　Brachystelma setosum Peckover;刚毛润肺草■☆
58942　Brachystelma shirense Schltr. = Brachystelma buchananii N. E. Br. ■☆
58943　Brachystelma simplex Schltr.;简单润肺草■☆
58944　Brachystelma sinuatum E. Mey. = Fockea sinuata (E. Mey.) Druce ●☆
58945　Brachystelma spathulatum Lindl. = Brachystelma tuberosum (Meerb.) R. Br. ex Sims ■☆
58946　Brachystelma stellatum E. A. Bruce et R. A. Dyer;星状润肺草■☆
58947　Brachystelma stenophyllum (Schltr.) R. A. Dyer;窄叶润肺草■☆
58948　Brachystelma subaphyllum K. Schum. = Caralluma edulis (Edgew.) Benth. ■☆
58949　Brachystelma swazicum R. A. Dyer;斯威士润肺草■☆
58950　Brachystelma tabularium R. A. Dyer;扁平润肺草■☆
58951　Brachystelma tenellum R. A. Dyer;柔弱润肺草■☆
58952　Brachystelma tenue R. A. Dyer;细润肺草■☆
58953　Brachystelma thunbergii N. E. Br.;通贝里润肺草■☆
58954　Brachystelma togoense Schltr.;多哥润肺草■☆
58955　Brachystelma tuberosum (Meerb.) R. Br. ex Sims;块状润肺草■☆
58956　Brachystelma villosum (Schltr.) N. E. Br.;长柔毛润肺草■☆
58957　Brachystelma virgatum Dieter. = Sisyranthus virgatus E. Mey. ■☆
58958　Brachystelma viridiflorum Turrill = Raphionacme velutina Schltr. ■☆
58959　Brachystelmaria Schltr. = Brachystelma R. Br. (保留属名)■
58960　Brachystelmaria gerrardii (Harv.) Schltr. = Brachystelma gerrardii Harv. ■☆
58961　Brachystelmaria longifolia Schltr. = Brachystelma longifolium (Schltr.) N. E. Br. ■☆
58962　Brachystelmaria macropetalum Schltr. = Brachystelma macropetalum (Schltr.) N. E. Br. ■☆
58963　Brachystelmaria natalensis (Schltr.) Schltr. = Brachystelma sandersonii (Oliv.) N. E. Br. ■☆
58964　Brachystelmaria occidentalis (Schltr.) Schltr. = Brachystelma occidentale Schltr. ■☆
58965　Brachystemma D. Don(1825);短瓣花属(短瓣石竹属,短瓣藤属);Brachystemma,Shortpetalflower ■
58966　Brachystemma calycinum D. Don;短瓣花(抽筋藤,短瓣石竹,短瓣藤,活抽筋,生筋藤,生烟叶,松筋藤,土牛七,土牛膝);Calyx Brachystemma,Shortpetalflower ■
58967　Brachystemma ovatifolium Mizush. = Stellaria ovatifolia (Mizush.) Mizush. ■
58968　Brachystemum Michx. = Pycnanthemum Michx. (保留属名)■☆
58969　Brachystephanus Nees(1847);短冠爵床属■☆
58970　Brachystephanus africanus S. Moore;非洲短绒毛短冠爵床■☆
58971　Brachystephanus africanus S. Moore var. brevicuspis Mildbr. = Brachystephanus africanus S. Moore ■☆
58972　Brachystephanus africanus S. Moore var. velutinus (De Wild.) Figueiredo;短绒毛短冠爵床■☆
58973　Brachystephanus bequaertii De Wild. = Brachystephanus africanus S. Moore ■☆
58974　Brachystephanus coeruleus S. Moore = Oreacanthus coeruleus (S. Moore) Champl. et Figueiredo ■☆
58975　Brachystephanus cuspidatus Baker = Ecbolium syringifolium (Vahl) Vollesen ●☆
58976　Brachystephanus glaberrimus Champl.;光滑短冠爵床■☆
58977　Brachystephanus longiflorus Lindau;长花短冠爵床■☆
58978　Brachystephanus lyallii Nees;莱尔短冠爵床■☆
58979　Brachystephanus myrmecophilus Champl.;蚂蚁短冠爵床■☆
58980　Brachystephanus nemoralis S. Moore;森林短冠爵床■☆
58981　Brachystephanus nimbae Heine;尼恩巴短冠爵床■☆
58982　Brachystephanus occidentalis Lindau;西方短冠爵床■☆
58983　Brachystephanus velutinus De Wild. = Brachystephanus africanus S. Moore var. velutinus (De Wild.) Figueiredo ■☆
58984　Brachystephium Less. = Brachycome Cass. ●■☆
58985　Brachystepis Pritz. = Beclardia A. Rich. ■☆
58986　Brachystepis Pritz. = Oeonia Lindl. (保留属名)■☆
58987　Brachystigma Pennell = Agalinis Raf. (保留属名)■☆
58988　Brachystigma Pennell(1928);短柱头列当属■☆
58989　Brachystigma wrightii (A. Gray) Pennell;短柱头列当■☆
58990　Brachystigma wrightii Pennell = Brachystigma wrightii (A. Gray) Pennell ■☆
58991　Brachystylis E. Mey. ex DC. = Marasmodes DC. ■☆
58992　Brachystylus Dulac = Koeleria Pers. ■
58993　Brachythalamus Gilg = Gyrinops Gaertn. ●☆
58994　Brachythrix Wild et G. V. Pope(1978);短毛瘦片菊属■☆
58995　Brachythrix brevipapposa Wild et G. V. Pope = Brachythrix glomerata (Mattf.) C. Jeffrey ■☆
58996　Brachythrix brevipapposa Wild et G. V. Pope subsp. malawiensis? = Brachythrix malawiensis (Wild et G. V. Pope) G. V. Pope ■☆
58997　Brachythrix glomerata (Mattf.) C. Jeffrey;团集短毛瘦片菊■☆
58998　Brachythrix malawiensis (Wild et G. V. Pope) G. V. Pope;马拉维短毛瘦片菊■☆
58999　Brachythrix pawekiae Wild et G. V. Pope;帕维基短毛瘦片菊■☆
59000　Brachythrix sonchoides Wild et G. V. Pope;苦苣菜短毛瘦片菊■☆
59001　Brachythrix stolzii (S. Moore) Wild et G. V. Pope;斯托尔兹瘦片菊■☆
59002　Brachytome Hook. f. (1871);短萼齿木属(短口木属);Brachytome ●
59003　Brachytome hainanensis C. Y. Wu ex W. C. Chen;海南短萼齿木;Hainan Brachytome ●
59004　Brachytome hirtellata Hu;滇短萼齿木(滇短口树);Yunnan Brachytome ●
59005　Brachytome hirtellata Hu var. glabrescens W. C. Chen;疏毛短萼齿木;Glabrescent Yunnan Brachytome ●
59006　Brachytome wallichii Hook. f.;短萼齿木(短口树);Wallich Brachytome ●
59007　Brachytome wallichii Hook. f. = Brachytome hainanensis C. Y. Wu ex W. C. Chen ●
59008　Brachytophora T. Durand = Brachylophon Oliv. ●☆
59009　Brachytropis Rchb. = Polygala L. ●■
59010　Bracisepalum J. J. Sm. (1933);短萼兰属(马来西亚兰属)■☆
59011　Bracisepalum selebicum J. J. Sm.;短萼兰■☆

59012 Brackenridgea A. Gray(1853);布氏木属●☆
59013 Brackenridgea alboserrata (Engl.) Tiegh. = Brackenridgea zanguebarica Oliv. ●☆
59014 Brackenridgea arenaria (De Wild. et T. Durand) N. Robson;沙地布氏木●☆
59015 Brackenridgea bussei Gilg = Brackenridgea zanguebarica Oliv. ●☆
59016 Brackenridgea denticulata Furtado;齿布氏木●☆
59017 Brackenridgea fascicularis (Blanco) Fern. -Vill.;簇生布氏木●☆
59018 Brackenridgea ferruginea (Engl.) Tiegh. = Brackenridgea arenaria (De Wild. et T. Durand) N. Robson ●☆
59019 Brackenridgea hookeri Gilg;胡克布氏木●☆
59020 Brackenridgea kingii Tiegh.;金氏布氏木●☆
59021 Brackenridgea nitida A. Gray;光亮布氏木●☆
59022 Brackenridgea rubescens Tiegh.;红布氏木●☆
59023 Brackenridgea zanguebarica Oliv.;桑岛布氏木●☆
59024 Braconotia Godr. = Agropyron Gaertn. ■
59025 Braconotia Godr. = Elymus L. ■
59026 Braconotia Godr. = Elytrigia Desv. ■
59027 Braconotia canina (L.) Fourr. = Elymus caninus (L.) L. ■
59028 Braconotia elymoides Godr. = Elymus caninus (L.) L. ■
59029 Braconotia officinarum Godr. = Elytrigia repens (L.) Desv. ex B. D. Jacks. ■
59030 Bracteantha Anderb. = Bracteantha Anderb. et L. Haegi ■☆
59031 Bracteantha Anderb. et L. Haegi = Xerochrysum Tzvelev ■☆
59032 Bracteantha Anderb. et L. Haegi(1991);小蜡菊属■☆
59033 Bracteantha acuminata (DC.) Anderb. et Haegi;小蜡菊■☆
59034 Bracteantha bicolor (Lindl.) Anderb. et Haegi;二色小蜡菊■☆
59035 Bracteantha bracteata (Vent.) Anderb. et Haegi = Helichrysum bracteatum (Vent.) Andréws ■
59036 Bracteantha bracteata (Vent.) Anderb. et Haegi = Xerochrysum bracteatum (Vent.) Tzvelev ■
59037 Bracteanthus Ducke = Siparuna Aubl. ●☆
59038 Bracteanthus Ducke(1930);巴西香材树属●☆
59039 Bracteanthus glycycarpus Ducke;巴西香材树●☆
59040 Bractearia DC. = Tibouchina Aubl. ●■☆
59041 Bracteocarpaceae Melikian et A. V. Bobrov = Podocarpaceae Endl. (保留科名)●
59042 Bracteocarpus A. V. Bobrov et Melikyan = Dacrycarpus (Endl.) de Laub. ●
59043 Bracteocarpus Melikian et A. V. Bobrov = Dacrycarpus (Endl.) de Laub. ●
59044 Bracteocarpus kawaii (Hayata) A. V. Bobrov et Melikyan = Dacrycarpus imbricatus (Blume) de Laub. var. patulus de Laub. ●
59045 Bracteola Swallen = Chrysochloa Swallen ■☆
59046 Bracteola lucida Swallen = Chrysochloa subaequigluma (Rendle) Swallen ■☆
59047 Bracteola orientalis C. E. Hubb. = Chrysochloa orientalis (C. E. Hubb.) Swallen ■☆
59048 Bracteolanthus de Wit = Bauhinia L. ●
59049 Bracteolaria Hochst. = Baphia Afzel. ex Lodd. ●☆
59050 Bracteolaria polygalacea Hook. f. = Baphia capparidifolia Baker subsp. polygalacea Brummitt ●☆
59051 Bracteolaria racemosa Hochst. = Baphia racemosa (Hochst.) Baker ●☆
59052 Bractillaceae Dulac = Amaryllidaceae J. St. -Hil. (保留科名)●■
59053 Bradburia Torr. et A. Gray = Chrysopsis (Nutt.) Elliott(保留属名)■☆
59054 Bradburia Torr. et A. Gray(1842)(保留属名);软金菀属;Goldenaster ■☆
59055 Bradburia hirtella Torr. et A. Gray;软金菀■☆
59056 Bradburia pilosa (Nutt.) Semple;毛软金菀;Golden Aster,Soft Goldenaster ■☆
59057 Bradburya Raf. (废弃属名) = Bradburia Torr. et A. Gray(保留属名)■☆
59058 Bradburya Raf. (废弃属名) = Centrosema (DC.) Benth. (保留属名)●■☆
59059 Braddleya Vell. = Amphirrhox Spreng. (保留属名)■☆
59060 Bradea Standl. = Bradea Standl. ex Brade ●☆
59061 Bradea Standl. ex Brade(1932);布雷德茜属☆
59062 Bradea brasiliensis Standl.;巴西布雷德茜●☆
59063 Bradea montana Brade;山地布雷德茜●☆
59064 Bradlaeia Neck. = Siler Mill. ●☆
59065 Bradlea Adans. = Apios Fabr. (保留属名) + Wisteria Nutt. (保留属名)●
59066 Bradlea Adans. = Apios Fabr. (保留属名)●
59067 Bradleia Banks ex Gaertn. = Glochidion J. R. Forst. et G. Forst. (保留属名)●
59068 Bradleia Cav. = Glochidion J. R. Forst. et G. Forst. (保留属名)●
59069 Bradleia Raf. = Bradlaeia Neck. ●☆
59070 Bradleia Raf. = Siler Mill. ●☆
59071 Bradleia hirsuta Roxb. = Glochidion hirsutum (Roxb.) Voigt ●
59072 Bradleia lanceolaria Roxb. = Glochidion lanceolarium (Roxb.) Voigt ●
59073 Bradleia philippensis Willd. = Glochidion philippicum (Cav.) C. B. Rob. ●
59074 Bradleia philippica Cav. = Glochidion philippicum (Cav.) C. B. Rob. ●
59075 Bradleia sinica Gaertn. = Glochidion puberum (L.) Hutch. ●
59076 Bradleia zeylanica Gaertn. = Glochidion zeylanicum (Gaertn.) A. Juss. ●
59077 Bradleja Banks ex Gaertn. = Glochidion J. R. Forst. et G. Forst. (保留属名)●
59078 Bradleya Kuntze = Braddleya Vell. ■☆
59079 Bradleya Post et Kuntze = Amphirrhox Spreng. (保留属名)■☆
59080 Bradleya Post et Kuntze = Apios Fabr. (保留属名)●
59081 Bradleya Post et Kuntze = Braddleya Vell. ■☆
59082 Bradleya Post et Kuntze = Bradlaeia Neck. ●☆
59083 Bradleya Post et Kuntze = Bradlea Adans. ●
59084 Bradleya Post et Kuntze = Bradleia Banks ex Gaertn. ●
59085 Bradleya Post et Kuntze = Glochidion J. R. Forst. et G. Forst. (保留属名)●
59086 Bradleya Post et Kuntze = Siler Mill. ●☆
59087 Bradshawia F. Muell. = Rhamphicarpa Benth. ■☆
59088 Braemea Jenny = Houlletia Brongn. ■☆
59089 Braemia Jenny = Houlletia Brongn. ■☆
59090 Bragaia Esteves,Hofacker et P. J. Braun(2009);巴西掌属■☆
59091 Bragantia Lour. = Apama Lam. ●
59092 Bragantia Lour. = Thottea Rottb. ●
59093 Bragantia Vand. = Gomphrena L. ●■
59094 Brahea Mart. = Brahea Mart. ex Endl. ●☆
59095 Brahea Mart. ex Endl. (1837);石棕属(巴夏榈属,长穗棕属,短茎棕属,岩榈属,中美石棕属);Blue Rock Palm,Brahea Palm,Erythea,Hesper Palm,Rock Palm ●☆
59096 Brahea armata S. Watson;刺石棕(石榈,石棕);Blue Fan Palm,

Blue Hesper Palm, Blue Palm, Gray Goddess, Hesper Palm, Mexican Blue Palm ●☆
59097 Brahea brandegeei (Purpus) H. E. Moore; 圣何塞石棕; Brandegee Hesper Palm, San Jose Hesper Palm ●☆
59098 Brahea dulcis Mart.; 甜石棕; Palm Dulce, Rock Palm, Sweet Brahea Palm ●☆
59099 Brahea edulis H. Wendl. ex S. Watson; 可食石棕(加州棕,食用石棕); Guadalupe Erythea, Guadalupe Palm ●☆
59100 Brahea elegans (Becc.) H. E. Moore; 优雅石棕; Franceschi Palm ●☆
59101 Brahea filamentosa Watson = Washingtonia filifera (Linden ex André) H. Wendl. ex de Bary ●
59102 Brahea filifera Watson = Washingtonia filifera (Linden ex André) H. Wendl. ex de Bary ●
59103 Brahea pimo Becc.; 皮谋石棕; Pimo Rock Palm ●☆
59104 Brahea serrulata (Michx.) H. Wendl. = Serenoa repens (Bartram) Small ●☆
59105 Brami Adans. (废弃属名) = Bacopa Aubl. (保留属名) ■
59106 Bramia Lam. = Bacopa Aubl. (保留属名) ■
59107 Bramia Lam. = Brami Adans. (废弃属名) ■
59108 Bramia indica Lam. = Bacopa monnieri (L.) Pennell ■
59109 Bramia monniera (L.) Drake = Bacopa monnieri (L.) Pennell ■
59110 Bramia monnieri (L.) Pennell = Bacopa monnieri (L.) Pennell ■
59111 Bramia rotundifolia (Michx.) Britton = Bacopa rotundifolia (Michx.) Wettst. ■☆
59112 Branciona Salisb. = Albuca L. ■☆
59113 Brandegea Cogn. (1890); 伯兰得瓜属(布兰德瓜属) ■☆
59114 Brandegea bigelovii (S. Watson) Cogn.; 伯兰得瓜(布兰德瓜) ■☆
59115 Brandella R. R. Mill(1986); 布雷德草属●☆
59116 Brandella erythraea (Brand) R. R. Mill.; 布雷德草●☆
59117 Brandella erythraea (Brand) R. R. Mill. = Paracaryum erythraeum Schweinf. ex Brand ●☆
59118 Brandella erythraea (Brand) R. R. Mill. f. subexalata Riedl = Paracaryum erythraeum Schweinf. ex Brand ●☆
59119 Brandesia Mart. = Alternanthera Forssk. ■
59120 Brandesia Mart. = Telanthera R. Br. ■
59121 Brandisia Hook. f. et Thomson(1865); 来江藤属; Brandisia ●
59122 Brandisia cauliflora P. C. Tsoong et A. M. Lu; 茎花来江藤(红花金银藤); Cauliflorous Brandisia, Stemflower Brandisia ●
59123 Brandisia discolor Hook. f. et Thomson; 异色来江藤; Discolor Brandisia ●
59124 Brandisia discolor Hook. f. et Thomson = Brandisia hancei Hook. f. ●
59125 Brandisia glabrescens Rehder; 退毛来江藤; Glabrous Brandisia ●
59126 Brandisia glabrescens Rehder var. hypochrysa P. C. Tsoong; 黄背退毛来江藤●
59127 Brandisia hancei Hook. f.; 来江藤(蜂蜜果,蜂糖罐,蜂糖花,猫花,猫咪花,蜜通花,蜜桶花,蜜札札,铁林杆,小白叶,小叶来江藤,野连翘,叶上花); Brandisia, Hance Brandisia ●
59128 Brandisia kwangsiensis H. L. Li; 广西来江藤; Guangxi Brandisia, Kwangsi Brandisia ●
59129 Brandisia laetevirens Rehder = Brandisia hancei Hook. f. ●
59130 Brandisia racemosa Hemsl.; 总花来江藤; Racemose Brandisia ●
59131 Brandisia rosea W. W. Sm.; 红花来江藤(黄花红花来江藤); Roseflower Brandisia, Rose-flowered Brandisia ●
59132 Brandisia rosea W. W. Sm. var. flava C. E. C. Fisch.; 黄花来江藤; Yellow-flowered Roseflower Brandisia ●
59133 Brandisia souliei Bonati = Chelonopsis souliei (Bonati) Merr. ●
59134 Brandisia swinglei Merr.; 岭南来江藤; Swingle Brandisia ●
59135 Brandonia Rchb. = Pinguicula L. ■
59136 Brandtia Kunth = Arundinella Raddi ■
59137 Brandzeia Baill. = Bathiaea Drake + Albizia Durazz. ●
59138 Brandzeia filicifolia Baill. = Bathiaea rubriflora Drake ●☆
59139 Branica Endl. = Bacopa Aubl. (保留属名) ■
59140 Branica Endl. = Bramia Lam. ■
59141 Branicia Andrz. = Senecio L. ■●
59142 Branicia Andrz. ex Trautv. = Senecio L. ■●
59143 Brasea Voss = Senecio L. ■●
59144 Brasenia Schreb. (1789); 莼菜属(莼属); Brasenia, Target, Water Shield, Watershield, Water-shield ■
59145 Brasenia peltata Pursh = Brasenia schreberi J. F. Gmel. ■
59146 Brasenia purpurea (Michx.) Casp. = Brasenia schreberi J. F. Gmel. ■
59147 Brasenia purpurea Casp. = Brasenia schreberi J. F. Gmel. ■
59148 Brasenia schreberi J. F. Gmel.; 莼菜(莼,蓴菜,凫葵,锦带,露葵,马粟草,马蹄草,茆,屏风,缺盆草,水案板,水葵,水芹,丝莼); Common Watershield, Purple Wen-dock, Target, Water Shield, Water Target, Watershield, Water-shield ■
59149 Brasilaelia Campacci = Cattleya Lindl. ■
59150 Brasilettia (DC.) Kuntze = Caesalpinia L. ●
59151 Brasilettia (DC.) Kuntze = Peltophorum (Vogel) Benth. (保留属名) ●
59152 Brasilettia Kuntze = Peltophorum (Vogel) Benth. (保留属名) ●
59153 Brasilettia africana (Sond.) Kuntze = Peltophorum africanum Sond. ●☆
59154 Brasilia G. M. Barroso = Calea L. ●■☆
59155 Brasiliastrum Lam. = Comocladia P. Browne + Picramnia Sw. (保留属名) ●☆
59156 Brasiliastrum Lam. = Picramnia Sw. (保留属名) ●☆
59157 Brasiliastrum Lam. = Pseudobrasilium Plum. ex Adans. ●☆
59158 Brasilicactus Backeb. (1942); 雪晃属■☆
59159 Brasilicactus Backeb. = Acanthocephala Backeb. ●
59160 Brasilicactus Backeb. = Notocactus (K. Schum.) A. Berger et Backeb. ■
59161 Brasilicactus Backeb. = Parodia Speg. (保留属名) ●
59162 Brasilicactus graessneri (K. Schum. ex Rümpler) Backeb. ex Schallert = Notocactus graessneri (K. Schum.) A. Berger ■
59163 Brasilicactus graessneri (K. Schum.) Backeb. ex Schallert; 黄雪晃(白绢仙人球,黄雪光,肖雪晃); Graessner Parodia ■
59164 Brasilicactus haselbergii (Rümpler) Backeb. ex Schaffnit; 雪晃(白绢仙人球,雪光); Haselberg Ballcactus, Scarlet Ball Cactus, Scarlet Ball-cactus ■
59165 Brasilicactus haselbergii (Rümpler) Backeb. ex Schaffnit = Notocactus haselbergii (Rümpler) A. Berger ■
59166 Brasilicactus haselbergii Backeb. ex Schaffnit = Notocactus haselbergii (Rümpler) A. Berger ■
59167 Brasilicereus Backeb. (1938); 巴西柱属●☆
59168 Brasilicereus phaeacanthus (Gürke) Backeb.; 巴西柱●☆
59169 Brasilidium Campacci = Oncidium Sw. (保留属名) ■☆
59170 Brasiliocroton P. E. Berry et Cordeiro(2005); 巴西巴豆属●☆
59171 Brasiliopuntia (K. Schum.) A. Berger = Opuntia Mill. ●
59172 Brasiliopuntia A. Berger = Opuntia Mill. ●
59173 Brasiliopuntia brasiliensis A. Berger = Opuntia brasiliensis

(Willd.) Haw. ■

59174 Brasiliorchis R. B. Singer, S. Koehler et Carnevali (2007); 巴西兰属■☆

59175 Brasiliparodia F. Ritter = Parodia Speg. (保留属名)●

59176 Brasilium J. F. Gmel. = Brasiliastrum Lam. ●☆

59177 Brasilium J. F. Gmel. = Tariri Aubl. (废弃属名)●☆

59178 Brasilocactus Frič = Brasilicactus Backeb. ■☆

59179 Brasilocactus Frič = Notocactus (K. Schum.) A. Berger et Backeb. ■

59180 Brasilocactus Frič = Parodia Speg. (保留属名)●

59181 Brasilocalamus Nakai = Merostachys Spreng. ●☆

59182 Brasilocycnis G. Gerlach et M. W. Whitten (1999); 巴西天鹅兰属■☆

59183 Brassaia Endl. = Schefflera J. R. Forst. et G. Forst. (保留属名)●

59184 Brassaia actinophylla Endl. = Schefflera actinophylla (Endl.) Harms ●

59185 Brassaia mannii (Hook. f.) Hutch. = Schefflera mannii (Hook. f.) Harms ●☆

59186 Brassaia volkensii (Harms) Hutch. = Schefflera volkensii (Harms) Harms ●☆

59187 Brassaiopsis Decne. et Planch. (1854); 罗伞属(柏拉参属,柏那参属,柏氏参属,阴阳枫属,掌叶树属); Bigumbrella, Brassaiopsis, Euaraliopsis, Palmleaftree ●

59188 Brassaiopsis acuminata H. L. Li; 尖叶罗伞(披针叶柏那参); Acuminateleaf Brassaiopsis, Sharpleaf Bigumbrella ●

59189 Brassaiopsis acuminata H. L. Li = Brassaiopsis glomerulata (Blume) Regel ●

59190 Brassaiopsis acuminata H. L. Li var. multiflora G. Hoo = Brassaiopsis glomerulata (Blume) Regel var. longipedicellata H. L. Li ●

59191 Brassaiopsis acuminata H. L. Li var. multiflora G. Hoo = Brassaiopsis producta (Dunn) C. B. Shang ●

59192 Brassaiopsis alpina C. B. Clarke; 高山罗伞; Alpine Brassaiopsis ●

59193 Brassaiopsis alpina C. B. Clarke = Merrilliopanax alpinus (C. B. Clarke) C. B. Shang ●

59194 Brassaiopsis angustifolia K. M. Feng; 狭叶罗伞(狭叶柏那参); Narrowleaf Bigumbrella, Narrowleaf Brassaiopsis, Narrow-leaved Brassaiopsis ●

59195 Brassaiopsis bodinieri (H. Lév.) J. Wen et Lowry; 直序罗伞●

59196 Brassaiopsis chengkangensis Hu; 镇康罗伞(镇康柏那参); Zhenkang Bigumbrella, Zhenkang Brassaiopsis ●

59197 Brassaiopsis ciliata Seem.; 纤齿罗伞(刺龙桐,假通草,睫毛掌叶树,纤齿柏那参,纤细罗伞); Ciliate Brassaiopsis, Ciliate Euaraliopsis, Ciliate Palmleaftree ●

59198 Brassaiopsis coriacea W. W. Sm. = Brassaiopsis glomerulata (Blume) Regel var. coriacea (W. W. Sm.) H. L. Li ●

59199 Brassaiopsis dumicola W. W. Sm.; 翅叶罗伞(翅叶掌叶树,狭翅柏那参,狭翅罗伞); Wingleaf Euaraliopsis, Wingleaf Palmleaftree, Wing-leaved Brassaiopsis ●

59200 Brassaiopsis fatsioides Buch. -Ham.; 盘叶罗伞(八角金盘,柏氏参,盘叶柏那参,盘叶掌叶树); Basinleaf Palmleaftree, Fatsialike Brassaiopsis, Fatsia-like Brassaiopsis, Fatsialike Euaraliopsis ●

59201 Brassaiopsis ferruginea (H. L. Li) G. Hoo; 锈毛罗伞(锈毛柏那参,锈毛树参,锈毛掌叶树,阴阳枫); Rusty-haired Brassaiopsis, Rustyhairy Brassaiopsis, Rustyhairy Euaraliopsis, Rustyhairy Palmleaftree ●

59202 Brassaiopsis ficifolia Dunn; 榕叶罗伞(榕叶柏那参,榕叶掌叶树); Figleaf Brassaiopsis, Figleaf Euaraliopsis, Figleaf Palmleaftree, Fig-leaved Brassaiopsis ●

59203 Brassaiopsis floribunda (Miq.) Seem. = Macropanax dispermus (Blume) Kuntze ●

59204 Brassaiopsis floribunda Seem. = Brassaiopsis glomerulata (Blume) Regel ●

59205 Brassaiopsis gaussenii Bui = Brassaiopsis bodinieri (H. Lév.) J. Wen et Lowry ●

59206 Brassaiopsis gaussenii Bui = Brassaiopsis dumicola W. W. Sm. ●

59207 Brassaiopsis glomerulata (Blume) Regel; 鸭脚罗伞(柏那参,刺鸭脚,刺鸭脚木,华丽柏那参,空壳洞,罗伞,七加皮,掌叶木,掌叶树); Glomerulate Bigumbrella, Glomerulate Brassaiopsis ●

59208 Brassaiopsis glomerulata (Blume) Regel var. angustifolia Y. R. Li; 狭叶柏那参; Narrowleaf Glomerulate Brassaiopsis ●

59209 Brassaiopsis glomerulata (Blume) Regel var. angustifolia Y. R. Li = Brassaiopsis glomerulata (Blume) Regel ●

59210 Brassaiopsis glomerulata (Blume) Regel var. brevipedicellata H. L. Li; 短梗罗伞; Shortpedicel Glomerulate Brassaiopsis ●

59211 Brassaiopsis glomerulata (Blume) Regel var. brevipedicellata H. L. Li = Brassaiopsis glomerulata (Blume) Regel ●

59212 Brassaiopsis glomerulata (Blume) Regel var. coriacea (W. W. Sm.) H. L. Li; 厚叶罗伞(厚叶柏那参); Coriaceousleaf Glomerulate Brassaiopsis ●

59213 Brassaiopsis glomerulata (Blume) Regel var. coriacea (W. W. Sm.) H. L. Li = Brassaiopsis glomerulata (Blume) Regel ●

59214 Brassaiopsis glomerulata (Blume) Regel var. longipedicellata H. L. Li; 长梗罗伞(长梗柏那参,长花柄掌叶树); Longpedicel Brassaiopsis ●

59215 Brassaiopsis glomerulata (Blume) Regel var. longipedicellata H. L. Li = Brassaiopsis glomerulata (Blume) Regel ●

59216 Brassaiopsis gracilis Hand. -Mazz.; 细梗罗伞(细梗柏那参); Slender Bigumbrella, Slender Brassaiopsis ●

59217 Brassaiopsis grushvitzkyi J. Wen et al.; 南星毛罗伞●

59218 Brassaiopsis hainla (Buch. -Ham.) Seem.; 浅裂罗伞(浅裂掌叶树,掌裂柏那参); Lobed Brassaiopsis, Lobed Euaraliopsis, Lobed Palmleaftree ●

59219 Brassaiopsis hispida Seem.; 粗毛罗伞(粗毛柏那参,粗毛掌叶树); Hispid Brassaiopsis, Hispid Euaraliopsis, Hispid Palmleaftree ●

59220 Brassaiopsis karmalaica Philipson = Brassaiopsis shweliensis W. W. Sm. ●

59221 Brassaiopsis kwangsiensis G. Hoo; 广西罗伞(广西柏那参,广西掌叶树); Guangxi Bigumbrella, Guangxi Brassaiopsis, Kwangsi Brassaiopsis ●

59222 Brassaiopsis lepidota K. M. Feng et Y. R. Li = Brassaiopsis producta (Dunn) C. B. Shang ●

59223 Brassaiopsis lepitoda K. M. Feng et Y. R. Li; 鳞片罗伞(豆豉杆,鳞片柏那参); Lepidoid Brassaiopsis ●

59224 Brassaiopsis liana Y. F. Deng = Brassaiopsis glomerulata (Blume) Regel ●

59225 Brassaiopsis moumingeneis (Y. R. Ling) C. B. Shang = Brassaiopsis moumingensis C. B. Shang ●

59226 Brassaiopsis moumingensis C. B. Shang; 茂名罗伞(茂名掌叶树); Maoming Bigumbrella, Maoming Brassaiopsis, Maoming Euaraliopsis, Maoming Palmleaftree ●

59227 Brassaiopsis palmata (Roxb.) Kurz = Brassaiopsis polyacantha (Wall.) R. N. Banerjee ●

59228 Brassaiopsis palmata Kurz = Brassaiopsis polyacantha (Wall.) R. N. Banerjee ●

59229 Brassaiopsis palmipes Forrest ex W. W. Sm.;阔翅罗伞(假柄掌叶树,阔翅柏那参,阔翅掌叶树);Palmstipe Brassaiopsis, Palmstipe Euaraliopsis, Palmstipe Palmleaftree, Palm-stiped Brassaiopsis ●

59230 Brassaiopsis palmipes Forrest ex W. W. Sm. = Brassaiopsis fatsioides Buch. -Ham. ●

59231 Brassaiopsis papayoides Hand. -Mazz.;瓜叶掌叶树;Papaya Like Bigumbrella, Papaya Like Brassaiopsis ●

59232 Brassaiopsis pentalocula G. Hoo;五室罗伞(五室柏那参);Fivelocular Brassaiopsis, Fiveloculous Bigumbrella, Pentalocule Brassaiopsis ●

59233 Brassaiopsis pentalocula G. Hoo = Brassaiopsis producta (Dunn) C. B. Shang ●

59234 Brassaiopsis phanerophlebia (Merr. et Chun) C. N. Ho = Brassaiopsis tripteris (H. Lév.) Rehder ●

59235 Brassaiopsis polyacantha (Wall.) R. N. Banerjee;掌叶柏那参;Palmateleaf Bigumbrella ●

59236 Brassaiopsis polyacantha (Wall.) R. N. Banerjee = Brassaiopsis hainla (Buch. -Ham.) Seem. ●

59237 Brassaiopsis producta (Dunn) C. B. Shang;尖苞罗伞(尖苞柏那参,尖苞掌叶树);Spinybract Bigumbrella, Spinybract Brassaiopsis, Stretched Brassaiopsis ●

59238 Brassaiopsis pseudoficifolia Lowry et C. B. Shang;假榕叶罗伞●

59239 Brassaiopsis quaircifolia G. Hoo;栎叶罗伞(栎叶柏那参);Oakleaf Bigumbrella, Oakleaf Brassaiopsis, Oak-leaved Brassaiopsis ●

59240 Brassaiopsis shweliensis W. W. Sm.;瑞丽罗伞(瑞丽柏那参);Ruili Bigumbrella, Ruili Brassaiopsis ●

59241 Brassaiopsis simplicifolia C. B. Clarke;单叶罗伞●

59242 Brassaiopsis speciosa Decne. et Planch. = Brassaiopsis glomerulata (Blume) Regel ●

59243 Brassaiopsis spinibracteata G. Hoo = Brassaiopsis producta (Dunn) C. B. Shang ●

59244 Brassaiopsis stellata K. M. Feng;星毛罗伞(星毛柏那参);Stellar Brassaiopsis, Stellatehair Bigumbrella, Stellatehair Brassaiopsis ●

59245 Brassaiopsis suberipetala K. M. Feng et Y. R. Li;栓瓣罗伞(栓瓣柏那参);Cork-petaled Brassaiopsis ●

59246 Brassaiopsis suberipetala K. M. Feng et Y. R. Li. = Brassaiopsis shweliensis W. W. Sm. ●

59247 Brassaiopsis tibetanus C. B. Shang;西藏罗伞(樟木柏那参);Xizang Bigumbrella, Xizang Brassaiopsis ●

59248 Brassaiopsis trevesioides W. W. Sm. = Brassaiopsis fatsioides Buch. -Ham. ●

59249 Brassaiopsis triloba K. M. Feng;三裂罗伞(三裂柏那参,三叶罗伞);Trilobate Brassaiopsis, Trilobe Bigumbrella, Trilobe Brassaiopsis ●

59250 Brassaiopsis tripteris (H. Lév.) Rehder;显脉罗伞(三叶莲,三叶罗伞,显脉掌叶树);Manifestnerved Bigumbrella, Manifest-nerved Bigumbrella, Manifestnerved Brassaiopsis, Trifoliolate Bigumbrella, Trifoliolate Brassaiopsis ●

59251 Brassaiopsis zhangmuensis Y. R. Li = Brassaiopsis tibetanus C. B. Shang ●

59252 Brassavola Adans. (废弃属名) = Brassavola R. Br. (保留属名) ■☆

59253 Brassavola Adans. (废弃属名) = Helenium L. ■

59254 Brassavola R. Br. (1813) (保留属名);巴拉索兰属(柏拉兰属);Brassavola ■☆

59255 Brassavola acaulis Lindl. et Paxton;无茎巴索拉兰■☆

59256 Brassavola digbyana Lindl. = Rhyncholaelia digbyana Schltr. ■☆

59257 Brassavola fragrans Barb. Rodr.;芳香巴索拉兰;Fragrans Brassavola ■☆

59258 Brassavola glauca Lindl.;青灰巴索拉兰■☆

59259 Brassavola nodosa Lindl.;大花巴索拉兰(柏拉兰,夜夫人);Common Brassavola, Lady-of-the-night ■☆

59260 Brassavolaea Poepp. et Endl. = Brassavola R. Br. (保留属名) ■☆

59261 Brassavolaea Post et Kuntze = Brassavola R. Br. (保留属名) ■☆

59262 Brassavolaea Post et Kuntze = Helenium L. ■

59263 Brassavolea Spreng. = Brassavola R. Br. (保留属名) ■☆

59264 Brassenia Heynh. = Brasenia Schreb. ■

59265 Brassia R. Br. (1813);长萼兰属(巴西亚兰属,蜘蛛兰属);Spider Orchid, Spider-orchid ■☆

59266 Brassia angusta Lindl. = Brassia longissima (Rchb. f.) Schltr. ■☆

59267 Brassia brachiata Lindl.;双枝长萼兰■☆

59268 Brassia caudata (L.) Lindl.;尾状长萼兰(褐斑蜘蛛兰)■☆

59269 Brassia caudata Lindl. = Brassia caudata (L.) Lindl. ■☆

59270 Brassia gireoudiana Rchb. f. et Warsz.;长萼兰(巴西亚兰,香蜘蛛兰);Gireoud Spider Orchid, Spider Orchid ■☆

59271 Brassia guttata Lindl. = Brassia maculata R. Br. ■☆

59272 Brassia lawrenceana Lindl. = Brassia longissima (Rchb. f.) Schltr. ■☆

59273 Brassia longissima (Rchb. f.) Schltr.;大长萼兰(长叶蜘蛛兰)■☆

59274 Brassia maculata R. Br.;斑花长萼兰;Spotted Spider Orchid ■☆

59275 Brassia verrucosa Lindl.;具疣长萼兰■☆

59276 Brassiantha A. C. Sm. (1941);新几内亚卫矛属●☆

59277 Brassiantha pentamera A. C. Sm.;新几内亚卫矛●☆

59278 Brassica L. (1753);芸苔属(甘蓝属,芥属,芸薹属);Brassica, Cabbage, Cole, Greens, Mustard, Rape ■●

59279 Brassica adpressa (Moench) Ball = Hirschfeldia incana (L.) Lagr. -Foss. ■☆

59280 Brassica adpressa (Moench) Ball var. lasiocarpa Ball = Hirschfeldia incana (L.) Lagr. -Foss. ■☆

59281 Brassica alba (L.) Boiss. = Sinapis alba L. ■

59282 Brassica alba (L.) Rabenh. = Brassica hirta Moench ■

59283 Brassica alba (L.) Rabenh. = Sinapis alba L. ■

59284 Brassica alba Rabenh.;白花芥蓝(白芥);White Mustard ■

59285 Brassica alboglabra L. H. Bailey = Brassica oleracea L. var. albiflora Kuntze ■

59286 Brassica alboglabra L. H. Bailey var. acephala DC.;芥蓝菜(芥蓝)■

59287 Brassica amplexicaulis (Desf.) Pomel = Guenthera amplexicaulis (Desf.) Gómez-Campo ■☆

59288 Brassica amplexicaulis (Desf.) Pomel subsp. souliei (Batt.) Maire et Weiller = Guenthera amplexicaulis (Desf.) Gómez-Campo subsp. souliei (Batt.) Gómez-Campo ■☆

59289 Brassica amplexicaulis (Desf.) Pomel var. hirtula Pau = Guenthera amplexicaulis (Desf.) Gómez-Campo ■☆

59290 Brassica amplexicaulis (Desf.) Pomel var. maroccana O. E. Schulz = Guenthera amplexicaulis (Desf.) Gómez-Campo ■☆

59291 Brassica amplexicaulis (Desf.) Pomel var. occidentalis Pau = Guenthera amplexicaulis (Desf.) Gómez-Campo ■☆

59292 Brassica amplexicaulis Hochst. ex A. Rich. = Brassica rapa L. ■

59293 Brassica antiquorum H. Lév. = Brassica rapa L. var. chinensis (L.) Kitam. ■

59294 Brassica arabica (Fisch. et C. A. Mey.) Fiori = Erucastrum arabicum Fisch. et C. A. Mey. ■☆

59295 Brassica argyi H. Lév. = Brassica juncea (L.) Czern. ■

59296 Brassica arvensis (L.) Kuntze = Brassica tornefortii Gouan ■☆

59297 Brassica arvensis (L.) Kuntze = Sinapis arvensis L. ■

59298 Brassica arvensis (L.) Rabenh. = Sinapis arvensis L. ■

59299 Brassica arvensis L. = Moricandia arvensis (L.) DC. ■

59300 Brassica asperifolia Lam. = Brassica rapa L. var. oleifera DC. ■

59301 Brassica aurasiaca Coss. et Kralik = Guenthera loncholoma (Pomel) Gómez-Campo ■☆

59302 Brassica barrelieri (L.) Janka;巴雷芸苔■☆

59303 Brassica barrelieri (L.) Janka subsp. sabularia Maire = Brassica barrelieri (L.) Janka ■☆

59304 Brassica barrelieri (L.) Janka var. laevigata (Lag.) Maire et Weiller = Brassica barrelieri (L.) Janka ■☆

59305 Brassica barrelieri (L.) Janka var. papillaris (Boiss.) O. E. Schulz = Brassica barrelieri (L.) Janka ■☆

59306 Brassica barrelieri (L.) Janka var. psammophila (Pomel) Batt. = Brassica barrelieri (L.) Janka ■☆

59307 Brassica blancoana Boiss. = Guenthera repanda (Willd.) Gómez-Campo subsp. blancoana (Boiss.) Gómez-Campo ■☆

59308 Brassica boissieri Munby = Guenthera gravinae (Ten.) Gómez-Campo ■☆

59309 Brassica bourgeaui (H. Christ) Kuntze;布瓦芸苔■☆

59310 Brassica brachyloma Boiss. et Reut. = Guenthera gravinae (Ten.) Gómez-Campo ■☆

59311 Brassica brevirostrata C. H. An = Brassica elongata Ehrh. ■

59312 Brassica caespitosa Pomel = Guenthera gravinae (Ten.) Gómez-Campo ■☆

59313 Brassica campestris L. = Brassica rapa L. var. oleifera DC. ■

59314 Brassica campestris L. = Brassica rapa L. ■

59315 Brassica campestris L. subsp. chinensis (L.) Makino = Brassica rapa L. var. chinensis (L.) Kitam. ■

59316 Brassica campestris L. subsp. chinensis (L.) Makino var. communis M. Tsen et S. H. Lee = Brassica chinensis L. ■

59317 Brassica campestris L. subsp. chinensis var. amplexicaulis (Tanaka et Ono) Makino = Brassica rapa L. var. chinensis (L.) Kitam. ■

59318 Brassica campestris L. subsp. napus (L.) Hook. f. = Brassica napus L. ■

59319 Brassica campestris L. subsp. napus (L.) Hook. f. et T. Anderson = Brassica napus L. ■

59320 Brassica campestris L. subsp. napus Hook. f. et T. Anderson var. nippo-oleifera Makino = Brassica napus L. ■

59321 Brassica campestris L. subsp. narinosa (L. H. Bailey) G. Olsson = Brassica rapa L. var. chinensis (L.) Kitam. ■

59322 Brassica campestris L. subsp. nipposinica (L. H. Bailey) G. Olsson = Brassica rapa L. var. oleifera DC. ■

59323 Brassica campestris L. subsp. oleifera (DC.) Schübl. et Mart. = Brassica rapa L. var. oleifera DC. ■

59324 Brassica campestris L. subsp. pekinensis (Lour.) G. Olsson = Brassica rapa L. var. glabra Regel ■

59325 Brassica campestris L. subsp. rapa (L.) Hook. f. = Brassica rapa L. ■

59326 Brassica campestris L. subsp. rapifera (Metzg.) Sinskaya = Brassica rapa L. ■

59327 Brassica campestris L. subvar. rapa Hook. f. et Anderson = Brassica rapa L. ■

59328 Brassica campestris L. var. amplexicaulis (Tanaka et Ono) Makino = Brassica chinensis L. ■

59329 Brassica campestris L. var. biennis Schübl. et Mart. = Brassica napus L. ■

59330 Brassica campestris L. var. chinensis (L.) T. Ito = Brassica chinensis L. ■

59331 Brassica campestris L. var. chinensis (L.) T. Ito = Brassica rapa L. var. chinensis (L.) Kitam. ■

59332 Brassica campestris L. var. chinoleifera Viehoever = Brassica rapa L. var. oleifera DC. ■

59333 Brassica campestris L. var. dichotoma? = Brassica rapa L. var. dichotoma Kitam. ■☆

59334 Brassica campestris L. var. komatsuna Matsum. et Nakai = Brassica rapa L. var. perviridis L. H. Bailey ■

59335 Brassica campestris L. var. napobriassica (L.) DC. = Brassica napus L. var. napobrassica (L.) Rchb. ■

59336 Brassica campestris L. var. napus (L.) Bab. = Brassica napus L. ■

59337 Brassica campestris L. var. narinosa (L. H. Bailey) Kitam. = Brassica rapa L. var. chinensis (L.) Kitam. ■

59338 Brassica campestris L. var. nippoleifera Makino = Brassica rapa L. var. oleifera DC. ■

59339 Brassica campestris L. var. oleifera DC. = Brassica rapa L. var. oleifera DC. ■

59340 Brassica campestris L. var. parachinensis (L. H. Bailey) Makino = Brassica rapa L. var. chinensis (L.) Kitam. ■

59341 Brassica campestris L. var. pekinensis (Lour.) Viehoever = Brassica rapa L. var. glabra Regel ■

59342 Brassica campestris L. var. purpuraria L.; 紫菜苔; Zicaitai Mustard ■

59343 Brassica campestris L. var. rapa (L.) Hartm. = Brassica rapa L. ■

59344 Brassica campestris L. var. sarson Prain;印度白菜;Indian Colza ■☆

59345 Brassica campestris L. var. toona Makino;唐白菜■

59346 Brassica capitata (L.) H. Lév. = Brassica oleracea L. var. capitata L. ■

59347 Brassica capitata H. Lév. = Brassica oleracea L. var. capitata L. ■

59348 Brassica carinata A. Braun;背棱芸苔(阿比西尼亚芥菜);Abyssinian Mustard, Chinese Mustard, Ethiopian Rape, Indian Mustard, Texsel Greens, Yellow Mustard ■☆

59349 Brassica caulorapa (DC.) Pasq. = Brassica oleracea L. var. gongylodes L. ■

59350 Brassica caulorapa Pasq. = Brassica oleracea L. var. gongylodes L. ■

59351 Brassica celerifolia (M. Tsen et S. H. Lee) Y. C. Lan et T. Y. Cheo;鸡冠菜■

59352 Brassica cernua (Thunb.) F. B. Forbes et Hemsl. = Brassica juncea (L.) Czern. ■

59353 Brassica cernua (Thunb.) Forbes et Hemsl. var. chirimenna Makino = Brassica juncea (L.) Czern. ■

59354 Brassica cheiranthos Vill. = Coincya monensis (L.) Greuter et Burdet subsp. cheiranthos (Vill.) Aedo, Leadlay et Munoz Garm. ■☆

59355 Brassica chinensis L. = Brassica rapa L. var. chinensis (L.) Kitam. ■

59356 Brassica chinensis L. var. angustifolia V. G. Sun = Brassica rapa L. var. oleifera DC. ■

59357 Brassica chinensis L. var. communis M. Tsen et S. H. Lee = Brassica rapa L. var. chinensis (L.) Kitam. ■

59358 Brassica chinensis L. var. oleifera Makino et Nemoto;油白菜(白

菜型油菜,油菜,芸苔);Yubaicai Pakchoi ■

59359 Brassica chinensis L. var. pandurata V. G. Sun = Brassica rapa L. var. glabra Regel ■

59360 Brassica chinensis L. var. parachinensis (L. H. Bailey) Sinskaya = Brassica rapa L. var. chinensis (L.) Kitam. ■

59361 Brassica chinensis L. var. pekinensis (Lour.) V. G. Sun = Brassica rapa L. var. glabra Regel ■

59362 Brassica chinensis L. var. rosularis M. Tsen et S. H. Lee = Brassica napus L. var. chinensis (L.) O. E. Schulz ■

59363 Brassica chinensis L. var. rosularis M. Tsen et S. H. Lee = Brassica narinosa L. H. Bailey ■

59364 Brassica chinensis L. var. rosularis M. Tsen et S. H. Lee = Brassica rapa L. var. chinensis (L.) Kitam. ■

59365 Brassica chinensis L. var. utilis M. Tsen et S. H. Lee = Brassica rapa L. var. oleifera DC. ■

59366 Brassica cossoniana (Reut.) Reut. et Boiss. = Brassica fruticulosa Cirillo subsp. cossoniana (Reut.) Maire ■☆

59367 Brassica cossoniana Boiss. et Reut. var. rifana (Emb. et Maire) Pau et Font Quer = Brassica fruticulosa Cirillo subsp. cossoniana (Reut.) Maire ■☆

59368 Brassica cossoniana Boiss. et Reut. var. subacaulis Maire = Brassica fruticulosa Cirillo subsp. cossoniana (Reut.) Maire ■☆

59369 Brassica crassifolia Forssk. = Erucaria crassifolia (Forssk.) Delile ■☆

59370 Brassica cretica Lam.;地中海甘蓝■☆

59371 Brassica cretica Lam. subsp. atlantica (Coss.) Onno = Brassica insularis Moris ■☆

59372 Brassica crispa Raf.;皱波芸苔;Ornamental Cabbage ■☆

59373 Brassica cyrenaica Spreng. = Brassica rapa L. ■

59374 Brassica deflexa Boiss.;外折芸苔■☆

59375 Brassica deflexa Boiss. var. lasiocalycina (Boiss. et Hausskn.) O. E. Schulz = Brassica deflexa Boiss. ■☆

59376 Brassica deflexa Boiss. var. tigridis (Boiss.) Boiss. = Brassica deflexa Boiss. ■☆

59377 Brassica dimorpha Coss. et Durieu = Guenthera dimorpha (Coss. et Durieu) Gómez-Campo ■☆

59378 Brassica dubiosa L. H. Bailey = Brassica rapa L. var. oleifera DC. ■

59379 Brassica elata Ball = Erucastrum elatum (Ball) O. E. Schulz ■☆

59380 Brassica elongata Ehrh.;短喙芥;Elongated Mustard, Long-stalked Rape, Short-beaked Brassica ■

59381 Brassica elongata Ehrh. = Guenthera elongata (Ehrh.) Andr. ■☆

59382 Brassica elongata Ehrh. subsp. subscaposa (Maire et Weiller) Maire = Guenthera elongata (Ehrh.) Andr. subsp. subscaposa (Maire et Weiller) Gómez-Campo ■☆

59383 Brassica erosa Turcz. = Erucastrum strigosum (Thunb.) O. E. Schulz ■☆

59384 Brassica eruca L. = Eruca sativa Mill. ■

59385 Brassica eruca L. = Eruca vesicaria (L.) Cav. ■☆

59386 Brassica erucastrum L. = Erucastrum gallicum (Willd.) O. E. Schulz ■☆

59387 Brassica fimbricata?;流苏芸苔;Curdly-greens, Curled Kale, Winter Greens ■☆

59388 Brassica foliata Pau et Font Quer;多叶芸苔■☆

59389 Brassica fruticulosa Cirillo;拟地中海甘蓝;Mediterranean Cabbage ■☆

59390 Brassica fruticulosa Cirillo subsp. cossoniana (Reut.) Maire = Diplotaxis erucoides (L.) DC. subsp. cossoniana (Reut.) Mart.-Laborde ■☆

59391 Brassica fruticulosa Cirillo subsp. dolichocarpa Emb. et Maire = Erucastrum rifanum (Emb. et Maire) Gómez-Campo ■☆

59392 Brassica fruticulosa Cirillo subsp. glaberrima (Pomel) Batt.;光滑芸苔■☆

59393 Brassica fruticulosa Cirillo subsp. mauritanica (Coss.) Maire;毛里塔尼亚芸苔■☆

59394 Brassica fruticulosa Cirillo subsp. numidica (Coss.) Maire;努米底亚芸苔■☆

59395 Brassica fruticulosa Cirillo subsp. radicata (Desf.) Batt.;大根芸苔■☆

59396 Brassica fruticulosa Cirillo subsp. rifana (Emb. et Maire) Maire = Erucastrum rifanum (Emb. et Maire) Gómez-Campo ■☆

59397 Brassica fruticulosa Cirillo var. christobalii Maire et Sennen = Brassica fruticulosa Cirillo ■☆

59398 Brassica fruticulosa Cirillo var. cossoniana (Reut.) Coss. = Diplotaxis erucoides (L.) DC. subsp. cossoniana (Reut.) Mart.-Laborde ■☆

59399 Brassica fruticulosa Cirillo var. hispida Faure et Maire = Brassica fruticulosa Cirillo ■☆

59400 Brassica fruticulosa Cirillo var. leucantha Coss. = Brassica fruticulosa Cirillo subsp. glaberrima (Pomel) Batt. ■☆

59401 Brassica fruticulosa Cirillo var. mauritanica Coss. = Brassica fruticulosa Cirillo ■☆

59402 Brassica fruticulosa Cirillo var. mauritanica Coss. = Brassica fruticulosa Cirillo subsp. mauritanica (Coss.) Maire ■☆

59403 Brassica fruticulosa Cirillo var. numidica Coss. = Brassica fruticulosa Cirillo subsp. numidica (Coss.) Maire ■☆

59404 Brassica fruticulosa Cirillo var. radicata (Desf.) Coss. = Brassica fruticulosa Cirillo subsp. radicata (Desf.) Batt. ■☆

59405 Brassica fruticulosa Cirillo var. subacaulis Maire = Diplotaxis erucoides (L.) DC. subsp. cossoniana (Reut.) Mart.-Laborde ■☆

59406 Brassica fruticulosa Cirillo var. tenuisiliqua Faure et Maire = Brassica fruticulosa Cirillo ■☆

59407 Brassica gallica Druce;法国甘蓝;Hairy Rocket ■☆

59408 Brassica gemmifera (DC.) H. Lév. = Brassica oleracea L. var. gemmifera (DC.) Zenker ■

59409 Brassica gemmifera H. Lév. = Brassica oleracea L. var. gemmifera (DC.) Zenker ■

59410 Brassica glaberrima Pomel = Brassica fruticulosa Cirillo subsp. glaberrima (Pomel) Batt. ■☆

59411 Brassica glauca Kuntze;粉绿芥■☆

59412 Brassica gracilis Pomel = Erucastrum varium (Durieu) Durieu ■☆

59413 Brassica gravinae Ten. = Guenthera gravinae (Ten.) Gómez-Campo ■☆

59414 Brassica gravinae Ten. var. atlantica Batt. = Guenthera repanda (Willd.) Gómez-Campo subsp. confusa (Emb. et Maire) Gómez-Campo ■☆

59415 Brassica gravinae Ten. var. brachyloma (Boiss. et Reut.) Batt. = Guenthera gravinae (Ten.) Gómez-Campo ■☆

59416 Brassica gravinae Ten. var. caespitosa (Pomel) Batt. = Guenthera gravinae (Ten.) Gómez-Campo ■☆

59417 Brassica gravinae Ten. var. djurdjurae Batt. = Guenthera gravinae (Ten.) Gómez-Campo ■☆

59418 Brassica gravinae Ten. var. italica Maire et Weiller = Guenthera gravinae (Ten.) Gómez-Campo ■☆

59419 Brassica gravinae Ten. var. rupicola (Pomel) Batt. = Guenthera

gravinae (Ten.) Gómez-Campo ■☆
59420 Brassica griffithii Hook. f. et Thomson = Diplotaxis griffithii (Hook. f. et Thomson) Boiss ■☆
59421 Brassica griquensis N. E. Br. = Erucastrum griquense (N. E. Br.) O. E. Schulz ■☆
59422 Brassica havardi Pomel = Erucastrum varium (Durieu) Durieu ■☆
59423 Brassica hirta Moench = Sinapis alba L. ■
59424 Brassica humilis DC. = Guenthera repanda (Willd.) Gómez-Campo ■☆
59425 Brassica humilis DC. var. atlantica Batt. = Guenthera repanda (Willd.) Gómez-Campo ■☆
59426 Brassica humilis DC. var. nudicaulis (Lag.) Coss. = Guenthera repanda (Willd.) Gómez-Campo subsp. africana (Maire) Gómez-Campo ■☆
59427 Brassica insularis Moris;撒丁白菜■☆
59428 Brassica integrifolia (H. West) O. E. Schulz = Brassica juncea (L.) Czern. ■
59429 Brassica integrifolia (H. West) Rupr. = Brassica carinata A. Braun ■☆
59430 Brassica integrifolia (H. Willd.) Rupr. = Brassica juncea (L.) Czern. ■
59431 Brassica integrifolia O. E. Schulz = Brassica integrifolia (H. West) O. E. Schulz ■
59432 Brassica iranica Rech. f. et al. = Brassica deflexa Boiss. ■☆
59433 Brassica japonica (Thunb.) Siebold ex Miq. = Brassica juncea (L.) Czern. ■
59434 Brassica japonica (Thunb.) Siebold ex Miq. var. indivisa Makino;米布那白菜■
59435 Brassica japonica (Thunb.) Siebold ex Miq. var. isena Makino;伊塞那白菜■
59436 Brassica japonica (Thunb.) Siebold ex Miq. var. suigukina Makino;梵菜■
59437 Brassica japonica Thunb. = Brassica juncea (L.) Czern. ■
59438 Brassica juncea (L.) Czern.;芥菜(白芥子,菜芽,冲菜,大芥,大叶芥菜,多裂芥,多裂叶芥,高油菜,花芥,黄芥,黄芥菜,家芥,芥,芥菜型油菜,芥蓝,芥蓝菜,芥叶,金丝菜,京菜,苦菜,苦芥,腊菜,辣芥,莨,南芥,千筋菜,青菜,全缘菜,日本白菜,霜不老,水菜,苔菜,夏芥,夏台,辛菜,辛芳,辛芥,银丝菜,银丝芥,幽芥,油菜,油芥菜,皱叶芥,紫芥);Bitter Mustard, Brown Mustard, Chinese Cabbage, Chinese Mustard, Curled Mustard, Dijon Mustard, Entireleaf Mustard, Ethiopian Rape, India Mustard, Indian Mustard, Largeleaf Mustard, Leaf Mustard, Leaf-mustard, Manysect Leaf Mustard, Mustard Greens, Pot Herb Mustard, Rai, Sarepta Mustard, Texsel Greens, Yellow Mustard, Yujiecai Leaf-mustard ■
59439 Brassica juncea (L.) Czern. 'Crispifolia' = Brassica juncea (L.) Czern. var. crispifolia L. ■
59440 Brassica juncea (L.) Czern. = Brassica carinata A. Braun ■☆
59441 Brassica juncea (L.) Czern. et Coss. = Brassica juncea (L.) Czern. ■
59442 Brassica juncea (L.) Czern. subsp. integrifolia (West.) Thell. = Brassica juncea (L.) Czern. ■
59443 Brassica juncea (L.) Czern. subsp. napiformis (Pailleux et Bois) Gladis = Brassica integrifolia (H. West) O. E. Schulz ■
59444 Brassica juncea (L.) Czern. var. crispifolia L.;皱叶芥菜;China Mustard, Chinese Mustard, Crisped-leaf Mustard, Japan Greens, Japanese Greens ■
59445 Brassica juncea (L.) Czern. var. crispifolia L. = Brassica juncea (L.) Czern. 'Crispifolia' ■
59446 Brassica juncea (L.) Czern. var. crispifolia L. = Brassica juncea (L.) Czern. ■
59447 Brassica juncea (L.) Czern. var. crispifolia L. H. Bailey = Brassica juncea (L.) Czern. ■
59448 Brassica juncea (L.) Czern. var. foliosa L. = Brassica juncea (L.) Czern. ■
59449 Brassica juncea (L.) Czern. var. foliosa L. H. Bailey = Brassica juncea (L.) Czern. ■
59450 Brassica juncea (L.) Czern. var. gracilis M. Tsen et S. H. Lee = Brassica juncea (L.) Czern. ■
59451 Brassica juncea (L.) Czern. var. integrifolia (Stokes) Sinskaya = Brassica juncea (L.) Czern. ■
59452 Brassica juncea (L.) Czern. var. japonica (Thunb.) L. H. Bailey = Brassica juncea (L.) Czern. ■
59453 Brassica juncea (L.) Czern. var. longidens L. H. Bailey = Brassica juncea (L.) Czern. ■
59454 Brassica juncea (L.) Czern. var. longipes M. Tsen et S. H. Lee = Brassica juncea (L.) Czern. ■
59455 Brassica juncea (L.) Czern. var. megarrhiza M. Tsen et S. H. Lee;大头菜■
59456 Brassica juncea (L.) Czern. var. megarrhiza M. Tsen et S. H. Lee = Brassica juncea (L.) Czern. var. napiformis (Pailleux et Bois) Kitam. ■
59457 Brassica juncea (L.) Czern. var. megarrhiza M. Tsen et S. H. Lee = Brassica napiformis (Pailleux et Bois) L. H. Bailey ■
59458 Brassica juncea (L.) Czern. var. multiceps M. Tsen et S. H. Lee;雪里蕻(雪菜,雪里红);Xuelihong Leaf-mustard ■
59459 Brassica juncea (L.) Czern. var. multiceps M. Tsen et S. H. Lee = Brassica juncea (L.) Czern. ■
59460 Brassica juncea (L.) Czern. var. multisecta L. H. Bailey = Brassica juncea (L.) Czern. ■
59461 Brassica juncea (L.) Czern. var. napiformis (Pailleux et Bois) Kitam.;芥菜疙瘩(大头菜,根用芥,芥菜,芥疙瘩,辣疙瘩,玉根,诸葛菜);Datoucai Leaf-mustard, Rapeform Mustard ■
59462 Brassica juncea (L.) Czern. var. napiformis (Pailleux et Bois) Kitam. = Brassica napiformis (Pailleux et Bois) L. H. Bailey ■
59463 Brassica juncea (L.) Czern. var. napiformis Pailleux et Bois = Brassica napiformis (Pailleux et Bois) L. H. Bailey ■
59464 Brassica juncea (L.) Czern. var. oleifera Makino;油芥菜■☆
59465 Brassica juncea (L.) Czern. var. rugosa (Roxb.) Kitam. = Brassica juncea (L.) Czern. ■
59466 Brassica juncea (L.) Czern. var. strumata M. Tsen et S. H. Lee = Brassica juncea (L.) Czern. ■
59467 Brassica juncea (L.) Czern. var. subintegrifolia Sinskaya = Brassica juncea (L.) Czern. ■
59468 Brassica juncea (L.) Czern. var. tsatsai Mao = Brassica juncea (L.) Czern. var. tumida M. Tsen et S. H. Lee ■
59469 Brassica juncea (L.) Czern. var. tumida M. Tsen et S. H. Lee;榨菜(菱角菜,羊角儿菜);Zhacai Leaf-mustard ■
59470 Brassica juncea Coss. = Brassica juncea (L.) Czern. ■
59471 Brassica juncea L. subsp. napiformis (Pailleux et Bois) Gladis = Brassica napiformis (Pailleux et Bois) L. H. Bailey ■
59472 Brassica juncea L. var. megarrhiza M. Tsen et S. H. Lee = Brassica napiformis (Pailleux et Bois) L. H. Bailey ■
59473 Brassica kaber (DC.) L. C. Wheeler = Sinapis arvensis L. ■
59474 Brassica kaber (DC.) L. C. Wheeler var. pinnatifida (Stokes)

L. C. Wheeler = Sinapis arvensis L. ■

59475 Brassica kaber (DC.) L. C. Wheeler var. schkuhriana (Rchb.) L. C. Wheeler = Sinapis arvensis L. ■

59476 Brassica laevigata Lag. = Brassica barrelieri (L.) Janka ■☆

59477 Brassica lanceolata (DC.) Lange = Brassica juncea (L.) Czern. ■

59478 Brassica lasiocalycina (Boiss. et Hausskn.) Boiss. = Brassica deflexa Boiss. ■☆

59479 Brassica leptocarpa Boiss. = Brassica deflexa Boiss. ■☆

59480 Brassica leptopetala (DC.) Sond. = Erucastrum strigosum (Thunb.) O. E. Schulz ■☆

59481 Brassica loncholoma Pomel = Guenthera loncholoma (Pomel) Gómez-Campo ■☆

59482 Brassica lyrata Desf. = Enarthrocarpus clavatus Delile ex Godr. ■☆

59483 Brassica maurorum Durieu;莫尔芸苔■☆

59484 Brassica monensis (L.) Huds. = Coincya monensis (L.) Greuter et Burdet ■☆

59485 Brassica monensis (L.) Huds. var. maroccana Pau et Font Quer = Coincya monensis (L.) Greuter et Burdet ■☆

59486 Brassica muralis (L.) Boiss. = Diplotaxis muralis (L.) DC. ■

59487 Brassica napiformis (Pailleux et Bois) L. H. Bailey = Brassica juncea (L.) Czern. var. napiformis (Pailleux et Bois) Kitam. ■

59488 Brassica napiformis (Pailleux et Bois) L. H. Bailey var. multisecta (L. H. Bailey) A. I. Baranov = Brassica juncea (L.) Czern. ■

59489 Brassica napiformis (Pailleux et Bois) L. H. Bailey var. multisecta Baranov = Brassica juncea (L.) Czern. ■

59490 Brassica napobrassica (L.) Mill. = Brassica napus L. subsp. napobrassica (L.) Jafri ■

59491 Brassica napobrassica (L.) Mill. = Brassica napus L. var. napobrassica (L.) Rchb. ■

59492 Brassica napobrassica (L.) Mill. = Brassica napus L. ■

59493 Brassica napus L.;欧洲油菜(菜子,芜菁芥,油菜,芸台);Borecole Cole, Colesat, Coleseed, Colewort, Collard, Collards, Colza, Curled-leaved Kale, Kale, Knolles, Knolls, Long Rape, Midden Mylies, Nape, Navet, Navet-gentle, Navew, Oil-seed Rape, Pottage Iterb, Rape, Rape Kale, Rapeseed, Rawp, Rutabaga, Shirt, Swede, Swedish Turnip, Turnip, Winter Rape, Yellow-flower ■

59494 Brassica napus L. = Brassica oleracea L. var. acephala DC. ■

59495 Brassica napus L. subsp. napobrassica (L.) Hanelt = Brassica napobrassica (L.) Mill. ■

59496 Brassica napus L. subsp. napobrassica (L.) Hanelt = Brassica napus L. var. napobrassica (L.) Rchb. ■

59497 Brassica napus L. subsp. napobrassica (L.) Jafri = Brassica napobrassica (L.) Mill. ■

59498 Brassica napus L. subsp. napobrassica (L.) Jafri = Brassica napus L. var. napobrassica (L.) Rchb. ■

59499 Brassica napus L. subsp. oleifera (DC.) Metzg. = Brassica napus L. ■

59500 Brassica napus L. var. arvensis (Duch.) Thell. = Brassica napus L. ■

59501 Brassica napus L. var. biennis (Schübl. et Mart.) Rchb. = Brassica napus L. ■

59502 Brassica napus L. var. chinensis (L.) O. E. Schulz = Brassica chinensis L. ■

59503 Brassica napus L. var. chinensis (L.) O. E. Schulz = Brassica rapa L. var. chinensis (L.) Kitam. ■

59504 Brassica napus L. var. dichotoma = Brassica rapa L. var. dichotoma Kitam. ■☆

59505 Brassica napus L. var. edulis Delile = Brassica napus L. var. napobrassica (L.) Rchb. ■

59506 Brassica napus L. var. leptorrhiza Spach. = Brassica napus L. ■

59507 Brassica napus L. var. napobrassica (L.) Rchb.;蔓菁甘蓝(布留克,大头菜,瑞典芜菁,芜菁甘蓝,洋大头,洋大头菜,洋蔓菁);Russian Turnip, Rutabaga, Swede, Swede Turnip, Swedish Turnip, Yellow Turnip ■

59508 Brassica napus L. var. napobrassica (L.) Rchb. = Brassica napobrassica (L.) Mill. ■

59509 Brassica napus L. var. oleifera DC. = Brassica napus L. ■

59510 Brassica napus L. var. rapifera Metzg. = Brassica napus L. var. napobrassica (L.) Rchb. ■

59511 Brassica napus L. var. sahariensis A. Chev. = Brassica napus L. ■

59512 Brassica narinosa L. H. Bailey;塌棵菜(飘儿菜,瓢儿菜,塌菜,塌地白菜,塌古菜,蹋菜,蹋地白菜,蹋棵菜,乌鸡白,乌塌菜,乌蹋菜,雪里青);Broadbeaked Mustard ■

59513 Brassica narinosa L. H. Bailey = Brassica chinensis L. ■

59514 Brassica narinosa L. H. Bailey = Brassica rapa L. var. chinensis (L.) Kitam. ■

59515 Brassica nigra (L.) W. D. J. Koch;黑芥(黑芥子,排菜);Black Mustard, Brown Mustard, Red Mustard, Senvie, Senvy ■

59516 Brassica nipposinica L. H. Bailey = Brassica rapa L. var. oleifera DC. ■

59517 Brassica nudicaulis (Lag.) Gonz. Albo = Guenthera repanda (Willd.) Gómez-Campo subsp. africana (Maire) Gómez-Campo ■☆

59518 Brassica nudicaulis Pomel = Guenthera repanda (Willd.) Gómez-Campo subsp. africana (Maire) Gómez-Campo ■☆

59519 Brassica nudicaulis Pomel var. africana O. E. Schulz = Guenthera repanda (Willd.) Gómez-Campo subsp. africana (Maire) Gómez-Campo ■☆

59520 Brassica oleracea L.;野甘蓝(包包菜,番芥蓝,番牡丹,甘蓝,高丽菜,卷心菜,葵花白菜,莲花白,蔬食芥);Cabbage, Chinese Kale, Coleslaw, Head Cabbage, Ornamemtal Cabbage, Ornamental Brassicas, Ornamental Kale, Sauerkraut, Wild Cabbage, Wild Kale Sea Cabbage ■

59521 Brassica oleracea L. f. atlantica Coss. = Brassica insularis Moris ■☆

59522 Brassica oleracea L. f. costata?;中脉甘蓝;Couve Tronchuda, Portuguese Cabbage ■☆

59523 Brassica oleracea L. subsp. gemmifera (DC.) Schwarz. = Brassica oleracea L. var. gemmifera (DC.) Zenker ■

59524 Brassica oleracea L. var. acephala DC.;羽衣甘蓝(芥蓝菜,绿叶甘蓝);Black Cabbage, Borecole, Collard, Collards, Crilly-greens, Curled-leaved Kale, Flowering Cabbage, Kale, Kales, Marrow Cabbage, Mustard, Rape, Rape Kale ■

59525 Brassica oleracea L. var. acephala DC. f. tricolor Hort.;三色羽衣甘蓝(三色芥蓝菜,羽衣甘蓝)■☆

59526 Brassica oleracea L. var. albiflora Kuntze;白花野甘蓝(白花甘蓝,芥蓝,芥蓝菜);Chinese Kale, Whiteglabrous Mustard ■

59527 Brassica oleracea L. var. arvensis Duch. = Brassica napus L. ■

59528 Brassica oleracea L. var. asparagoides DC. = Brassica oleracea L. var. italica Planch. ■

59529 Brassica oleracea L. var. botrytis L.;花椰菜(菜花,花菜,球花甘蓝);Broccoli, Cauliflower ■

59530 Brassica oleracea L. var. bullata DC.;皱叶甘蓝;Savoy, Savoy Cabbage ■☆

59531 Brassica oleracea L. var. bullata DC. subvar. gemmifera DC. = Brassica oleracea L. var. gemmifera (DC.) Zenker ■

59532 Brassica oleracea L. var. capitata L.;甘蓝(包包菜,包菜,包心菜,玻璃菜,大头菜,高丽菜,疙瘩白,结球甘蓝莲花白,卷心菜,葵花白菜,蓝菜,苤菜,西土蓝,洋白菜,椰菜,圆白菜);Bow Kail, Bow-kail, Brock, Cabbage, Cabbish, Callards, Caskets, Castocks, Caul, Cave, Cole, Colewort, Collets, Colluts, Headed Cabbage, Keele, Red Cole, Redwort, Savoy Cabbage, Sea Colewort, Slick Greens ■

59533 Brassica oleracea L. var. capitata L. f. alba?;白花甘蓝(白花莲花白);White Cabbage ■☆

59534 Brassica oleracea L. var. capitata L. rubra?;红花甘蓝(红花莲花白);Red Cabbage ■☆

59535 Brassica oleracea L. var. caulorapa DC. = Brassica caulorapa Pasq. ■

59536 Brassica oleracea L. var. caulorapa DC. = Brassica oleracea L. var. gongylodes L. ■

59537 Brassica oleracea L. var. chinensis (L.) Prain = Brassica rapa L. var. chinensis (L.) Kitam. ■

59538 Brassica oleracea L. var. fimbriata Mill.;流苏甘蓝■☆

59539 Brassica oleracea L. var. gemmifera (DC.) Zenker;抱子甘蓝(姬甘蓝,球芽甘蓝,汤菜);Baby Cabbage, Brussel's Sprouts, Brussels Sprout, Brussels Sprouts ■

59540 Brassica oleracea L. var. glauca A. Chev. = Brassica oleracea L. ■

59541 Brassica oleracea L. var. gongylodes L.;擘蓝(掰蓝,菜折,甘蓝,芥兰头,芥蓝,芥蓝头,撇蓝,苤蓝,茄连,茄莲,球茎甘蓝,芫甘蓝,玉蔓菁,玉头);Kohlrabi, Kohl-rabi, Portuguese Cabbage, Turnip-cabbage ■

59542 Brassica oleracea L. var. gongylodes L. = Brassica caulorapa Pasq. ■

59543 Brassica oleracea L. var. gongylodes L. = Brassica oleracea L. var. caulorapa DC. ■

59544 Brassica oleracea L. var. hongnoensis H. Lév. = Brassica napus L. ■

59545 Brassica oleracea L. var. italica Plenck;绿花椰菜(茎用花椰菜,绿花菜,伊太利甘蓝);Asparagus Broccoli, Broccilo, Broccoli, Italian Broccoli, Purple Sprouting, Sprouting Broccoli ■

59546 Brassica oleracea L. var. napobrassica L. = Brassica napus L. subsp. napobrassica (L.) Jafri ■

59547 Brassica oleracea L. var. napobrassica L. = Brassica napus L. var. napobrassica (L.) Rchb. ■

59548 Brassica oleracea L. var. pseudocolza H. Lév. = Brassica napus L. ■

59549 Brassica oleracea L. var. tricolor Hort. = Brassica oleracea L. var. acephala DC. f. tricolor Hort. ■☆

59550 Brassica oleracea L. var. tsiekentsiensis H. Lév. = Brassica rapa L. var. chinensis (L.) Kitam. ■

59551 Brassica oleracea L. viridis?;绿羽衣甘蓝;Kale ■☆

59552 Brassica orientalis L. = Conringia orientalis (L.) Dum. Cours. ■☆

59553 Brassica pachypoda Thell. = Erucastrum austroafricanum Al-Shehbaz et Warwick ■☆

59554 Brassica palmensis Kuntze = Sinapis pubescens L. ■☆

59555 Brassica papillaris Boiss. = Brassica barrelieri (L.) Janka ■☆

59556 Brassica parachinensis L. H. Bailey;菜苔;False Pakchoi ■

59557 Brassica parachinensis L. H. Bailey = Brassica chinensis L. ■

59558 Brassica parachinensis L. H. Bailey = Brassica rapa L. var. chinensis (L.) Kitam. ■

59559 Brassica pekinensis (Lour.) Rupr. = Brassica rapa L. var. glabra Regel ■

59560 Brassica pekinensis (Lour.) Rupr. = Brassica rapa L. var. glabra Regel 'Pe-tsai' ■

59561 Brassica pekinensis (Lour.) Rupr. var. cephalata M. Tsen et S. H. Lee = Brassica rapa L. var. glabra Regel ■

59562 Brassica pekinensis (Lour.) Rupr. var. cylindrica M. Tsen et S. H. Lee = Brassica rapa L. var. glabra Regel ■

59563 Brassica pekinensis (Lour.) Rupr. var. dentata Matsum. et Nakai = Brassica rapa L. var. pekinensis (Lour.) Kitam. 'Dentata' ■

59564 Brassica pekinensis (Lour.) Rupr. var. laxa M. Tsen et S. H. Lee = Brassica rapa L. var. glabra Regel ■

59565 Brassica pekinensis (Lour.) Rupr. var. petsai Lour. = Brassica rapa L. var. glabra Regel ■

59566 Brassica persica Boiss. et Hohen. = Brassica elongata Ehrh. ■

59567 Brassica perviridis (L. H. Bailey) L. H. Bailey;西班牙白菜;Spanish Mustard, Tendergreens ■☆

59568 Brassica perviridis (L. H. Bailey) L. H. Bailey = Brassica rapa L. var. oleifera DC. ■

59569 Brassica petsai (Lour.) L. H. Bailey = Brassica rapa L. var. glabra Regel ■

59570 Brassica petsai L. H. Bailey = Brassica rapa L. var. glabra Regel ■

59571 Brassica pinnatifida Desf. = Eruca pinnatifida (Desf.) Pomel ■☆

59572 Brassica polymorpha Murray = Sisymbrium polymorphum (Murray) Roth ■

59573 Brassica procumbens (Poir.) O. E. Schulz;平铺蔓菁■☆

59574 Brassica psammophila Pomel = Brassica barrelieri (L.) Janka ■☆

59575 Brassica rapa L.;蔓菁(扁萝卜,大芥,大头菜,地蔓菁,葑,葑苁,狗头芥,鸡毛菜,鸡头菜,芥,芥荛,九英蔓菁,九英菘,盘菜,荛,蕵芜,台菜,温菘,芜,芜菁,须,圆菜头,圆菜头菜,圆根,诸葛菜);Bird's Rape, Chonnocks, Field Mustard, Kai Choy, Keblock, Knoll, Knolles, Knolls, Lukes, Mip, Mit, Moor, Navew, Neaps Neeps, Nenufar, Oil-seed Rape, Pak-choi, Rape Mustard, Sanson, Swedish Tummit, Turmet, Turmit, Turnip, Wild Mustard ■

59576 Brassica rapa L. subsp. campestris (L. H. Bailey) L. H. Bailey = Brassica rapa L. var. chinensis (L.) Kitam. ■

59577 Brassica rapa L. subsp. campestris (L.) Clapham = Brassica rapa L. var. oleifera DC. ■

59578 Brassica rapa L. subsp. chinensis (L.) Hanelt = Brassica rapa L. var. chinensis (L.) Kitam. ■

59579 Brassica rapa L. subsp. chinensis (L.) Hanelt var. parachinensis (L. H. Bailey) Hanelt = Brassica rapa L. var. chinensis (L.) Kitam. ■

59580 Brassica rapa L. subsp. chinensis (L.) Hanelt var. rosularis (M. Tsen et S. H. Lee) Hanelt = Brassica rapa L. var. chinensis (L.) Kitam. ■

59581 Brassica rapa L. subsp. narinosa (L. H. Bailey) Hanelt = Brassica rapa L. var. chinensis (L.) Kitam. ■

59582 Brassica rapa L. subsp. nipposinica (L. H. Bailey) Hanelt = Brassica rapa L. var. oleifera DC. ■

59583 Brassica rapa L. subsp. oleifera (DC.) Metzg. = Brassica rapa L. var. oleifera DC. ■

59584 Brassica rapa L. subsp. pekinensis (Lour.) Hanelt = Brassica rapa L. var. glabra Regel ■

59585 Brassica rapa L. subsp. pekinensis (Lour.) Hanelt var. laxa (M. Tsen et S. H. Lee) Hanelt = Brassica rapa L. var. glabra Regel ■

59586 Brassica rapa L. subsp. pekinensis (Lour.) Hanelt var. pandurata (V. G. Sun) Gladis = Brassica rapa L. var. glabra Regel ■

59587 Brassica rapa L. subsp. rapifera Metzg. = Brassica rapa L. ■

59588 Brassica rapa L. subsp. sylvestris (L.) Janch. = Brassica rapa L. ■

59589 Brassica rapa L. subvar. dentata (Matsum. et Nakai) Kitam. = Brassica rapa L. var. pekinensis (Lour.) Kitam. 'Dentata' ■

59590 Brassica rapa L. var. amplexicaulis Tanaka et Ono = Brassica rapa L. var. chinensis (L.) Kitam. ■

59591 Brassica rapa L. var. amplexicaulis Tanaka et Ono subvar. pe-tsai (L. H. Bailey) Kitam. = Brassica rapa L. var. glabra Regel ■

59592 Brassica rapa L. var. campestris (L.) Clapham = Brassica rapa L. var. oleifera DC. ■

59593 Brassica rapa L. var. campestris (L.) Peterm. = Brassica rapa L. subsp. campestris (L.) Clapham ■

59594 Brassica rapa L. var. campestris (L.) Petermann = Brassica rapa L. var. oleifera DC. ■

59595 Brassica rapa L. var. chinensis (L.) Kitam.;青菜(白菜,白菘,不结球白菜,菜子,黄芽白,江门白菜,牛肚菘,菘,菘菜,体菜,土白菜,无心菜,夏菘,小白菜,小青菜,小油菜,油白菜,油菜,芸薹菜,中国芜菁);Buk Choy,Chinese Cabbage,Chinese Mustard,Choi Sum, Choy Sum, Pakchoi, Pak-choi Cabbege, Pe-tsai, Shantung Cabbage ■

59596 Brassica rapa L. var. chinensis (L.) Kitam. = Brassica chinensis L. ■

59597 Brassica rapa L. var. chinoleifera (Viehoever) Kitam. = Brassica rapa L. var. oleifera DC. ■

59598 Brassica rapa L. var. dichotoma Kitam.;二叉芥■☆

59599 Brassica rapa L. var. glabra (Sinskaya) Kitam. = Brassica rapa L. ■

59600 Brassica rapa L. var. glabra Regel;白菜(包心白菜,长白菜,大白菜,光蔓菁,花交菜,黄矮菜,黄芽白,黄芽白菜,黄芽白绍菜,黄芽菜,结球白菜,京白菜,卷心白,卷心白菜,绍菜,菘);Beijing Cabbage,Celery Cabbage,Chihli,Chinese Cabbage,Peking Cabbage,Pe-tsai ■

59601 Brassica rapa L. var. glabra Regel 'Pe-tsai' = Brassica rapa L. var. glabra Regel ■

59602 Brassica rapa L. var. nippoleifera (Makino) Kitam. = Brassica rapa L. var. oleifera DC. ■

59603 Brassica rapa L. var. oleifera DC.;芸苔(白菜,菜苔,菜子,春菜,梵菜,甘蓝,寒菜,红油菜,胡菜,花芥,蓝菜,蔓菁,盘科菜,青菜,台菜,苔芥,芜菁,油菜,芸苔菜,紫菘);Bargeman's Cabbage, Bird Rape, Colza, Field Cabbage, Field Mustard, Naven, Rape, Rapeseed,Rutabaga,Turnip-rape,Wild Turnip,Yellow Mustard ■

59604 Brassica rapa L. var. pekinensis (Lour.) Kitam. 'Dentata';尖齿芸苔■

59605 Brassica rapa L. var. pekinensis (Lour.) Kitam. = Brassica rapa L. var. glabra Regel ■

59606 Brassica rapa L. var. pekinensis (Lour.) Kitam. = Brassica rapa L. var. glabra Regel 'Pe-tsai' ■

59607 Brassica rapa L. var. perviridis L. H. Bailey;小松菜■

59608 Brassica rapa L. var. perviridis L. H. Bailey = Brassica rapa L. var. oleifera DC. ■

59609 Brassica repanda (Willd.) DC. = Guenthera repanda (Willd.) Gómez-Campo ■☆

59610 Brassica repanda (Willd.) DC. subsp. africana (Maire) Greuter et Burdet = Guenthera repanda (Willd.) Gómez-Campo subsp. africana (Maire) Gómez-Campo ■☆

59611 Brassica repanda (Willd.) DC. subsp. confusa (Emb. et Maire) Heywood = Guenthera repanda (Willd.) Gómez-Campo subsp. confusa (Emb. et Maire) Gómez-Campo ■☆

59612 Brassica repanda (Willd.) DC. subsp. nudicaulis (Lag.) Heywood = Guenthera repanda (Willd.) Gómez-Campo subsp. africana (Maire) Gómez-Campo ■☆

59613 Brassica repanda (Willd.) DC. subsp. silenifolia (Emb.) Greuter et Burdet = Guenthera repanda (Willd.) Gómez-Campo subsp. silenifolia (Emb.) Gómez-Campo ■☆

59614 Brassica rerayensis Ball = Erucastrum elatum (Ball) O. E. Schulz ■☆

59615 Brassica rifana Emb. et Maire = Brassica fruticulosa Cirillo subsp. cossoniana (Reut.) Maire ■☆

59616 Brassica rugosa (Roxb.) L. H. Bailey = Brassica juncea (L.) Czern. ■

59617 Brassica rupicola Pomel = Guenthera gravinae (Ten.) Gómez-Campo ■☆

59618 Brassica rutabaga DC. ex H. Lév. = Brassica napus L. var. napobrassica (L.) Rchb. ■

59619 Brassica ruvo L. H. Bailey;意大利芸苔;Italian Turnip Broccoli, Italian Turnipbroccoli,Ruvo Kale ■☆

59620 Brassica sabularia Brot. = Brassica barrelieri (L.) Janka ■☆

59621 Brassica sabularia Brot. var. psammophila (Pomel) Batt. = Brassica barrelieri (L.) Janka ■☆

59622 Brassica saxatilis (Lam.) Amo = Guenthera repanda (Willd.) Gómez-Campo ■☆

59623 Brassica saxatilis (Lam.) Amo subsp. africana Maire = Guenthera repanda (Willd.) Gómez-Campo subsp. africana (Maire) Gómez-Campo ■☆

59624 Brassica saxatilis (Lam.) Amo subsp. confusa Emb. et Maire = Guenthera repanda (Willd.) Gómez-Campo subsp. confusa (Emb. et Maire) Gómez-Campo ■☆

59625 Brassica saxatilis (Lam.) Amo subsp. nudicaulis (Lag.) Maire = Guenthera repanda (Willd.) Gómez-Campo subsp. africana (Maire) Gómez-Campo ■☆

59626 Brassica saxatilis (Lam.) Amo subsp. silenifolia Emb. = Guenthera repanda (Willd.) Gómez-Campo subsp. silenifolia (Emb.) Gómez-Campo ■☆

59627 Brassica saxatilis (Lam.) Amo var. berberica Litard. et Maire = Guenthera repanda (Willd.) Gómez-Campo subsp. confusa (Emb. et Maire) Gómez-Campo ■☆

59628 Brassica saxatilis (Lam.) Amo var. brevisiliqua Emb. et Maire = Guenthera repanda (Willd.) Gómez-Campo subsp. confusa (Emb. et Maire) Gómez-Campo ■☆

59629 Brassica saxatilis (Lam.) Amo var. diplotaxiformis Maire = Guenthera repanda (Willd.) Gómez-Campo subsp. diplotaxiformis (Maire) Gómez-Campo ■☆

59630 Brassica saxatilis (Lam.) Amo var. embergeri Maire = Guenthera repanda (Willd.) Gómez-Campo subsp. confusa (Emb. et Maire) Gómez-Campo ■☆

59631 Brassica saxatilis (Lam.) Amo var. purpurascens Humbert et Maire = Guenthera repanda (Willd.) Gómez-Campo ■☆

59632 Brassica saxatilis (Lam.) Amo var. silenifolia (Emb.) Maire = Guenthera repanda (Willd.) Gómez-Campo subsp. silenifolia (Emb.) Gómez-Campo ■☆

59633 Brassica schimperi Boiss. = Erucastrum arabicum Fisch. et C. A. Mey. ■☆

59634 Brassica scopulorum Coss. et Durieu = Brassica spinescens Pomel ■☆

59635 Brassica setulosa (Boiss. et Reut.) Coss. = Guenthera setulosa

(Boiss. et Reut.) Gómez-Campo ■☆
59636 Brassica sinaica Boiss. = Moricandia sinaica (Boiss.) Boiss. ■☆
59637 Brassica sinapis Vis. = Sinapis arvensis L. ■
59638 Brassica sinapistrum Boiss. = Sinapis arvensis L. ■
59639 Brassica somalensis Hedge et A. G. Mill.;索马里芸苔■☆
59640 Brassica souliei (Batt.) Batt. = Guenthera amplexicaulis (Desf.) Gómez-Campo subsp. souliei (Batt.) Gómez-Campo ■☆
59641 Brassica souliei (Batt.) Batt. subsp. amplexicaulis (Desf.) Greuter et Burdet = Guenthera amplexicaulis (Desf.) Gómez-Campo ■☆
59642 Brassica spinescens Pomel;细刺芸苔■☆
59643 Brassica stocksii Hook. f. et Thomson = Brassica tornefortii Gouan ■☆
59644 Brassica strigosa (Thunb.) DC. var. glabrata Sond. = Erucastrum strigosum (Thunb.) O. E. Schulz ■☆
59645 Brassica subhastata Willd. = Sisymbrium orientale L. ■
59646 Brassica subscaposa Maire et Weiller = Guenthera elongata (Ehrh.) Andr. subsp. subscaposa (Maire et Weiller) Gómez-Campo ■☆
59647 Brassica suffruticosa Desf. = Moricandia suffruticosa (Desf.) Coss. et Durieu ●☆
59648 Brassica taquetii H. Lév. = Brassica juncea (L.) Czern. ■
59649 Brassica teretifolia Desf. = Pseuderucaria teretifolia (Desf.) O. E. Schulz ■☆
59650 Brassica tigridis Boiss. = Brassica deflexa Boiss. ■☆
59651 Brassica tornefortii Gouan;地中海白菜;African Mustard, Asian Mustard, Mediterranean Turnip, Pale Cabbage, Wild Turnip ■☆
59652 Brassica torulosa Durieu = Diplotaxis siifolia Kuntze ■☆
59653 Brassica tournefortii Gouan;微花芸苔■☆
59654 Brassica tournefortii Gouan var. dasycarpa O. E. Schulz = Brassica tournefortii Gouan ■☆
59655 Brassica tournefortii Gouan var. leiocarpa Maire et Weiller = Brassica tournefortii Gouan ■☆
59656 Brassica varia Durieu = Erucastrum varium (Durieu) Durieu ■☆
59657 Brassica vesicaria L. = Eruca vesicaria (L.) Cav. ■☆
59658 Brassica violacea L. = Orychophragmus violaceus (L.) O. E. Schulz ■
59659 Brassica willdenowii Boiss. = Brassica juncea (L.) Czern. ■
59660 Brassica xinjiangensis Y. C. Lan et T. Y. Cheo = Sinapis arvensis L. ■
59661 Brassicaceae Burnett (1835) (保留科名);十字花科■●
59662 Brassicaceae Burnett (保留科名) = Cruciferae Juss. (保留科名) ■●
59663 Brassicaria (Gren. et Godron) Pomel = Brassica L. ■●
59664 Brassicaria Pomel = Brassica L. ■●
59665 Brassicastrum Link = Brassica L. ■●
59666 Brassicastrum Link = Guenthera Andrz. ex Besser ■☆
59667 Brassicella Fourr. = Brassica L. ■●
59668 Brassicella Fourr. ex O. E. Schnlz = Rhynchosinapis Hayek ■☆
59669 Brassicella Fourr. ex O. E. Schulz = Coincya Rouy ■☆
59670 Brassicella coincyoides Humbert et Maire = Coincya monensis (L.) Greuter et Burdet ■☆
59671 Brassicella coincyoides Humbert et Maire var. leptocarpa Maire = Coincya monensis (L.) Greuter et Burdet ■☆
59672 Brassicella monensis (L.) O. E. Schulz = Coincya monensis (L.) Greuter et Burdet ■☆
59673 Brassicella monensis (L.) O. E. Schulz subsp. coincyoides (Humbert et Maire) Maire = Coincya monensis (L.) Greuter et Burdet ■☆
59674 Brassicella monensis (L.) O. E. Schulz var. leptocarpa (Maire) Maire = Coincya monensis (L.) Greuter et Burdet ■☆
59675 Brassicella monensis (L.) O. E. Schulz var. maroccana Pau et Font Quer = Coincya monensis (L.) Greuter et Burdet ■☆
59676 Brassidium × hybridium Hort.;蜘蛛文心兰■
59677 Brassiodendron C. K. Allen = Endiandra R. Br. ●
59678 Brassiodendron C. K. Allen (1942);肖土楠属●☆
59679 Brassiodendron fragrans C. K. Allen;肖土楠●☆
59680 Brassiophoenix Burret (1935);三叉羽椰属(布拉索椰属)●☆
59681 Brassiophoenix drymophloeoides Burret;三叉羽椰●☆
59682 Brassiophoenix schumannii (Becc.) Essig;舒曼三叉羽椰●☆
59683 Brathydium Spach = Hypericum L. ■●
59684 Brathys L. f. = Hypericum L. ■●
59685 Brathys Mutis ex L. f. = Hypericum L. ■●
59686 Brathys japonica (Thunb. ex Murray) Wight = Hypericum japonicum Thunb. ex Murray ■
59687 Brathys japonica (Thunb.) Wight = Hypericum japonicum Thunb. ■
59688 Brathys laxa Blume = Hypericum japonicum Thunb. ■
59689 Braunblanquetia Eskuche = Fonkia Phil. ■
59690 Braunblanquetia Eskuche (1974);阿根廷婆婆纳属(阿根廷玄参属)■☆
59691 Braunblanquetia littoralis Eskuche;阿根廷婆婆纳■☆
59692 Braunea Willd. (废弃属名) = Tiliacora Colebr. (保留属名)●☆
59693 Brauneria Neck. = Echinacea Moench ■☆
59694 Brauneria Neck. ex Britton = Echinacea Moench ■☆
59695 Brauneria Neck. ex Porter et Britton = Echinacea Moench ■☆
59696 Brauneria laevigata C. L. Boynton et Beadle = Echinacea laevigata (C. L. Boynton et Beadle) S. F. Blake ■☆
59697 Brauneria pallida (Nutt.) Britton = Echinacea pallida (Nutt.) Nutt. ■☆
59698 Brauneria paradoxa Norton = Echinacea paradoxa (Norton) Britton ■☆
59699 Brauneria purpurea (L.) Britton = Echinacea purpurea (L.) Moench ■☆
59700 Brauneria tennesseensis Beadle = Echinacea tennesseensis (Beadle) Small ■☆
59701 Braunlowia A. DC. = Brownlowia Roxb. (保留属名)●☆
59702 Braunsia Schwantes (1928);碧玉莲属●☆
59703 Braunsia apiculata (Kensit) L. Bolus;细尖碧玉莲●☆
59704 Braunsia bina (N. E. Br.) Schwantes;碧玉莲●☆
59705 Braunsia edentula (Haw.) N. E. Br. = Ruschia edentula (Haw.) L. Bolus ●☆
59706 Braunsia geminata (Haw.) L. Bolus;双生碧玉莲●☆
59707 Braunsia maximilianii (Schltr. et A. Berger) Schwantes;马氏碧玉莲●☆
59708 Braunsia nelii Schwantes;尼尔碧玉莲●☆
59709 Braunsia stayneri (L. Bolus) L. Bolus;斯泰纳碧玉莲●☆
59710 Braunsia vanrensburgii (L. Bolus) L. Bolus;范伦碧玉莲●☆
59711 Bravaisia DC. = Braunsia Schwantes ●☆
59712 Bravoa La Llave et Lex. = Bravoa Lex. ■☆
59713 Bravoa La Llave et Lex. = Polianthes L. ■
59714 Bravoa Lex. (1824);布拉沃兰属(红花月下香属)■☆
59715 Bravoa geminiflora Lex.;布拉沃兰(红花月下香,双花晚香玉);Mexican Twinbloom ■☆

59716 Bravoa geminiflora Lex. = Polianthes geminiflora (Lex.) Rose ■☆
59717 Bravocactus Doweld(1998);布拉沃仙人掌属■☆
59718 Bravocactus horripilus (Lem.) Doweld;布拉沃仙人掌■☆
59719 Braxilia Raf. = Pyrola L. ●■
59720 Braxilia minor (L.) House = Pyrola minor L. ●
59721 Braxilia parvifolia Raf. = Pyrola minor L. ●
59722 Braxipis Raf. = Cola Schott et Endl.(保留属名)●☆
59723 Braxireon Raf. = Narcissus L. ■
59724 Braxireon Raf. = Tapeinanthus Herb.(废弃属名)●☆
59725 Braxireon humile (Cav.) Raf. = Narcissus cavanillesii Barra et G. López ■☆
59726 Braxylis Raf. = Ilex L. ●
59727 Braya Sternb. et Hoppe(1815);肉叶荠属(柏蕾荠属,肉叶芥属);Braya ■
59728 Braya Vell. = Hirtella L. ●☆
59729 Braya aenea Bunge = Braya rosea (Turcz.) Bunge ■
59730 Braya aenea Bunge subsp. pseudoaenia Petr. = Braya rosea (Turcz.) Bunge ■
59731 Braya alpine Sternb. et Hoppe = Aphragmus oxycarpus (Hook. f. et Thomson) Jafri ■
59732 Braya alpine Sternb. et Hoppe = Braya thomsonii Hook. f. ■
59733 Braya angustifolia (N. Busch) Vassilcz. = Braya rosea (Turcz.) Bunge ■
59734 Braya brachycarpa Vassilcz. = Braya rosea (Turcz.) Bunge ■
59735 Braya brevicaulis Schmid = Braya rosea (Turcz.) Bunge ■
59736 Braya densiflora Muschl. = Weberbauera densiflora (Muschl.) Gilg et Muschl. ■☆
59737 Braya foliosa Pamp. = Aphragmus oxycarpus (Hook. f. et Thomson) Jafri ■
59738 Braya forrestii W. W. Sm.;细肉叶芥(弗氏肉叶芥);Forrest Braya ■
59739 Braya forrestii W. W. Sm. = Braya rosea (Turcz.) Bunge ■
59740 Braya forrestii W. W. Sm. var. puberula W. T. Wang;毛细肉叶芥;Hairy Forrest Braya ■
59741 Braya forrestii W. W. Sm. var. puberula W. T. Wang = Braya forrestii W. W. Sm. ■
59742 Braya glacialis Korsh. = Draba altaica (C. A. Mey.) Bunge ■
59743 Braya heterophylla W. W. Sm. = Eutrema heterophylla (W. W. Sm.) H. Hara ■
59744 Braya humilis (C. A. Mey.) B. L. Rob. = Neotorularia humilis (C. A. Mey.) Hedge et J. Léonard ■
59745 Braya kokonorica O. E. Schulz;青海肉叶荠;Qinghai Braya ■
59746 Braya kokonorica O. E. Schulz = Phaeonychium villosum (Maxim.) Al-Shehbaz ■
59747 Braya limosella Bunge = Braya rosea (Turcz.) Bunge ■
59748 Braya limoselloides Bunge ex Ledeb. = Braya rosea (Turcz.) Bunge ■
59749 Braya marinellii Pamp. = Eurycarpus marinellii (Pamp.) Al-Shehbaz et G. Yan ■
59750 Braya oxycarpa Hook. f. et Thomson = Aphragmus oxycarpus (Hook. f. et Thomson) Jafri ■
59751 Braya oxycarpa Hook. f. et Thomson f. glabra Vassilcz. = Aphragmus oxycarpus (Hook. f. et Thomson) Jafri ■
59752 Braya oxycarpa Hook. f. et Thomson var. scharnhorstii (Regel et Schmalh.) O. E. Schulz = Braya scharnhorstii Regel et Schmalh. ■
59753 Braya oxycarpa Hook. f. et Thomson var. stenocarpa O. E. Schulz = Aphragmus oxycarpus (Hook. f. et Thomson) Jafri ■
59754 Braya pamirica (Korsh.) O. Fedtsch.;帕米尔肉叶荠■☆
59755 Braya pamirica (Korsh.) O. Fedtsch. = Braya scharnhorstii Regel et Schmalh. ■
59756 Braya pamirica (Korsh.) O. Fedtsch. var. glabra O. Fedtsch. = Braya scharnhorstii Regel et Schmalh. ■
59757 Braya purpurascens (R. Br.) Bunge;紫色肉叶荠■☆
59758 Braya rosea (Turcz.) Bunge;红花肉叶荠(短果肉叶荠,狭叶肉叶荠);Redflower Braya ■
59759 Braya rosea (Turcz.) Bunge var. aenea (Bunge) Malyschev;古铜色肉叶荠;Bronzecolour Braya ■
59760 Braya rosea (Turcz.) Bunge var. aenea (Bunge) Malyschev = Braya rosea (Turcz.) Bunge ■
59761 Braya rosea (Turcz.) Bunge var. brachycarpa (Vassilcz.) Malyschev = Braya rosea (Turcz.) Bunge ■
59762 Braya rosea (Turcz.) Bunge var. glabra Regel et Schmalh. = Braya rosea (Turcz.) Bunge ■
59763 Braya rosea (Turcz.) Bunge var. glabrata Regel et Schmalh.;无毛肉叶荠;Glabrous Braya ■
59764 Braya rosea (Turcz.) Bunge var. glabrata Regel et Schmalh. = Braya rosea (Turcz.) Bunge ■
59765 Braya rosea (Turcz.) Bunge var. leiocarpa O. E. Schulz = Braya rosea (Turcz.) Bunge ■
59766 Braya rosea (Turcz.) Bunge var. multicaulis B. Fedtsch. = Braya rosea (Turcz.) Bunge ■
59767 Braya rosea (Turcz.) Bunge var. simplicior B. Fedtsch. = Braya rosea (Turcz.) Bunge ■
59768 Braya rubicunda Franch. = Aphragmus oxycarpus (Hook. f. et Thomson) Jafri ■
59769 Braya scharnhorstii Regel et Schmalh.;黄花肉叶荠(线叶丛菔);Linearifolious Solms-laubachia ■
59770 Braya siliquosa Bunge;长角肉叶荠;Silique Braya ■
59771 Braya sinensis Hemsl. = Pegaeophyton scapiflorum (Hook. f. et Thomson) C. Marquand et Airy Shaw ■
59772 Braya sinensis Hemsl. = Pegaeophyton scapiflorum (Hook. f. et Thomson) C. Marquand et Airy Shaw var. robustum (O. E. Schulz) R. L. Guo et T. Y. Cheo ■
59773 Braya sinuata Maxim. = Braya rosea (Turcz.) Bunge ■
59774 Braya sternbergii Krasser = Braya scharnhorstii Regel et Schmalh. ■
59775 Braya thomsonii Hook. f. = Braya rosea (Turcz.) Bunge ■
59776 Braya thomsonii Hook. f. var. pamirica (Korsh.) O. E. Schulz = Braya scharnhorstii Regel et Schmalh. ■
59777 Braya thomsonii Hook. f. var. pilosa Schulz = Braya thomsonii Hook. f. ■
59778 Braya tibetica Hook. f. et Thomson;西藏肉叶荠;Tibet Braya, Xizang Braya ■
59779 Braya tibetica Hook. f. et Thomson = Braya rosea (Turcz.) Bunge ■
59780 Braya tibetica Hook. f. et Thomson f. breviscapa Pamp.;短葶肉叶荠;Shortscape Braya ■
59781 Braya tibetica Hook. f. et Thomson f. breviscapa Pamp. = Braya rosea (Turcz.) Bunge ■
59782 Braya tibetica Hook. f. et Thomson f. linearifolia C. H. An;条叶肉叶荠;Linearleaf Xizang Braya ■
59783 Braya tibetica Hook. f. et Thomson f. linearifolia C. H. An = Braya rosea (Turcz.) Bunge ■
59784 Braya tibetica Hook. f. et Thomson f. sinuata (Maxim.) O. E.

Schulz;羽叶肉叶荠;Pinnate Braya ■
59785 Braya tibetica Hook. f. et Thomson f. sinuata (Maxim.) O. E. Schulz = Braya rosea (Turcz.) Bunge ■
59786 Braya tinkleri Schmid = Braya rosea (Turcz.) Bunge ■
59787 Braya uniflora Hook. f. et Thomson = Pycnoplinthus uniflorus (Hook. f. et Thomson) O. E. Schulz ■
59788 Braya versicolor Turcz. = Braya siliquosa Bunge ■
59789 Braya verticillata (Jeffrey et W. W. Sm.) W. W. Sm. = Staintoniella verticillata (Jeffrey et W. W. Sm.) H. Hara ■
59790 Braya verticillata (Jeffrey et W. W. Sm.) W. W. Sm. = Taphrospermum verticillatum (Jeffrey et W. W. Sm.) Al-Shehbaz ■
59791 Brayera Kunth = Hagenia J. F. Gmel. ●☆
59792 Brayera Kunth ex A. Rich. = Hagenia J. F. Gmel. ●☆
59793 Brayera Kunth(1822);苦苏属(哈根花属)■●☆
59794 Brayera anthelmintica Kunth;苦苏;Kousso ●☆
59795 Brayodendron Small = Diospyros L. ●
59796 Brayopsis Gilg et Muschl. (1909);假肉叶芥属■☆
59797 Brayopsis Gilg et Muschl. = Englerocharis Muschl. ■☆
59798 Brayopsis alpaminae Gilg et Muschl.;假肉叶芥■☆
59799 Brayopsis grandiflora Gilg et Muschl.;大花假肉叶芥■☆
59800 Brayopsis pycnophylla Gilg et Muschl.;密叶假肉叶芥■☆
59801 Brayopsis trichocarpa Gilg et Muschl.;毛果假肉叶芥■☆
59802 Brayulinea Small = Guilleminea Kunth ■☆
59803 Brayulinea Small(1903);厄瓜多尔苋属■☆
59804 Brayulinea australis (Griseb.) Schinz;南方厄瓜多尔苋■☆
59805 Brayulinea densa (Willd. ex Roem. et Schult.) Small = Guilleminea densa (Humb. et Bonpl. ex Schult.) Moq. ■☆
59806 Brayulinea densa (Willd.) Small = Guilleminea densa (Willd.) Moq. ■☆
59807 Brayulinea gracilis (Fr.) Schinz;细厄瓜多尔苋■☆
59808 Brazocactus A. Frič = Brasilicactus Backeb. ■☆
59809 Brazocactus A. Frič = Notocactus (K. Schum.) A. Berger et Backeb. ■
59810 Brazoria Engelm. et A. Gray(1845);响尾花属(布拉梭属)■☆
59811 Brazoria Engelm. ex A. Gray = Brazoria Engelm. et A. Gray ■☆
59812 Brazoria truncata Engelm. et A. Gray;响尾花;Rattlesnake Flower ■☆
59813 Brazzeia Baill. (1886);梭果革瓣花属●☆
59814 Brazzeia acuminata Tiegh. = Brazzeia soyauxii (Oliv.) Tiegh. var. acuminata (Tiegh.) Letouzey ●☆
59815 Brazzeia congoensis Baill.;刚果梭果革瓣花●☆
59816 Brazzeia longipedicellata Verdc.;长梗梭果革瓣花●☆
59817 Brazzeia soyauxii (Oliv.) Tiegh.;索亚梭果革瓣花●☆
59818 Brazzeia soyauxii (Oliv.) Tiegh. var. acuminata (Tiegh.) Letouzey;渐尖梭果革瓣花●☆
59819 Brebissonia Spach = Fuchsia L. ●■
59820 Bredemeyera Willd. (1801);布雷木属●☆
59821 Bredemeyera floribunda Willd.;多花布雷木●☆
59822 Bredia Blume(1849);野海棠属(布勒德木属,金石榴属);Bredia ●■
59823 Bredia amoena Diels;秀丽野海棠(白矮茶,大号狗卵,大山落苏,大叶活血,高脚山茄,华东野海棠,活血丹,活血藤,金石榴,老鼠柴,落地山落乌,山糖浆,水杨树,野靛,野海棠);Splended Bredia ●■
59824 Bredia amoena Diels = Bredia quadrangularis Cogn. ●
59825 Bredia amoena Diels var. eglandulata B. Y. Ding;无腺野海棠●
59826 Bredia amoena Diels var. eglandulata B. Y. Ding = Bredia quadrangularis Cogn. ●
59827 Bredia amoena Diels var. serrata H. L. Li = Bredia amoena Diels ●■
59828 Bredia amoena Diels var. serrata H. L. Li = Bredia quadrangularis Cogn. ●
59829 Bredia amoena Diels var. trimera C. Chen;三数野海棠;Trimerous Splended Bredia ●
59830 Bredia amoena Diels var. trimera C. Chen = Bredia quadrangularis Cogn. ●
59831 Bredia biglandularis C. Chen;双腺野海棠;Biglandular Bredia, Dualgland Bredia, Two-glandules Bredia ●
59832 Bredia bodinieri H. Lév. = Blastus cavaleriei H. Lév. et Vaniot ●
59833 Bredia cavaleriei (H. Lév.) Diels = Phyllagathis longiradiosa (C. Chen) C. Chen ●■
59834 Bredia cavaleriei H. Lév. et Vaniot = Fordiophyton faberi Stapf ●■
59835 Bredia chinensis Merr. = Bredia amoena Diels ●■
59836 Bredia chinensis Merr. = Bredia quadrangularis Cogn. ●
59837 Bredia cordata H. L. Li = Bredia esquirolii (H. Lév.) Lauener var. cordata (H. L. Li) C. Chen ●
59838 Bredia cordata H. L. Li = Bredia esquirolii (H. Lév.) Lauener ●
59839 Bredia esquirolii (H. Lév.) Lauener;赤水野海棠(小猫子草);Chishui Bredia, Esquirol Bredia ●
59840 Bredia esquirolii (H. Lév.) Lauener = Phyllagathis longiradiosa (C. Chen) C. Chen ●■
59841 Bredia esquirolii (H. Lév.) Lauener = Plagiopetalum esquirolii (H. Lév.) Rehder ●
59842 Bredia esquirolii (H. Lév.) Lauener var. cordata (H. L. Li) C. Chen;心叶野海棠(罐罐草,红水麻叶,鸡窝红麻,千地红,山红活麻,向天胡芦);Cordate Esquirol Bredia, Cordate-leaf Bredia ●
59843 Bredia esquirolii (H. Lév.) Lauener var. cordata (H. L. Li) C. Chen = Bredia esquirolii (H. Lév.) Lauener ●
59844 Bredia fordii (Hance) Diels;叶底红(白还魂,大毛蛇,还魂红,假紫苏,江南野海棠,沙崩草,血还魂,野海棠,叶下红);Ford Metalleaf ●
59845 Bredia fordii (Hance) Diels = Phyllagathis fordii (Hance) C. Chen ●
59846 Bredia gibba Ohwi;尖瓣野海棠(垂花布勒德木,恒春布勒德木,小金石榴);Gibbous Bredia, Sharp-petal Bredia ●
59847 Bredia gibba Ohwi = Bredia oldhamii Hook. f. ●
59848 Bredia glabra Merr. = Bredia sinensis (Diels) H. L. Li ●
59849 Bredia gracilis (Hand.-Mazz.) Diels = Phyllagathis gracilis (Hand.-Mazz.) C. Chen ●■
59850 Bredia hainanensis Merr. et Chun = Phyllagathis hainanensis (Merr. et Chun) C. Chen ●
59851 Bredia hirsuta Blume;布勒德藤●
59852 Bredia hirsuta Blume var. rotundifolia (Tang S. Liu et F. Y. Lu ex F. Y. Lu) S. F. Huang et T. C. Huang = Bredia hirsuta Blume var. rotundifolia (Tang S. Liu et F. Y. Lu) S. F. Huang et T. C. Huang ●
59853 Bredia hirsuta Blume var. rotundifolia (Tang S. Liu et F. Y. Lu) S. F. Huang et T. C. Huang;圆叶布勒德藤●
59854 Bredia hirsuta Blume var. rotundifolia (Y. C. Liu et C. H. Ou) S. F. Huang et T. C. Huang = Bredia hirsuta Blume var. scandens Ito et Matsum. ●
59855 Bredia hirsuta Blume var. scandens Ito et Matsum.;野海棠(布勒德藤);Climbing Bredia, Climbing Hirsute Bredia, Common Bredia, Scandent Bredia ●
59856 Bredia hispidissima C. Chen = Phyllagathis hispidissima (C. Chen) C. Chen ●

59857 Bredia longiloba (Hand. -Mazz.) Diels;长萼野海棠(露水葛,女儿红,天青地红,血经草,叶下红,紫背红);Long Calyx Bredia, Longcalyx Bredia ●

59858 Bredia longiradiosa C. Chen = Phyllagathis longiradiosa (C. Chen) C. Chen ●■

59859 Bredia longiradiosa C. Chen ex Govaerts = Phyllagathis longiradiosa (C. Chen) C. Chen ●■

59860 Bredia mairei H. Lév. = Bredia yunnanensis (H. Lév.) Diels ●■

59861 Bredia mairei H. Lév. = Fordiophyton faberi Stapf ●■

59862 Bredia microphylla H. L. Li;小叶野海棠;Littleleaf Bredia, Smallleaf Bredia ●

59863 Bredia okinawensis (Matsum.) H. L. Li;冲绳野海棠●☆

59864 Bredia oldhamii Hook. f.;金石榴(垂花布勒德木,地丁,俄氏布勒德木);Oldham Bredia ●

59865 Bredia oldhamii Hook. f. var. ovata Ohwi;台湾金石榴(卵叶金石榴)●

59866 Bredia oldhamii Hook. f. var. ovata Ohwi = Bredia oldhamii Hook. f. ●

59867 Bredia omeiensis H. L. Li = Bredia fordii (Hance) Diels ●■

59868 Bredia omeiensis H. L. Li = Bredia tuberculata (Guillaumin) Diels ●

59869 Bredia penduliflora S. S. Ying = Bredia gibba Ohwi ●

59870 Bredia penduliflora S. S. Ying = Bredia oldhamii Hook. f. ●

59871 Bredia pricei F. P. Metcalf = Bredia amoena Diels ●■

59872 Bredia pricei F. P. Metcalf. = Bredia quadrangularis Cogn. ●

59873 Bredia quadrangularis Cogn.;过路惊;Four-anglar Bredia, Fourangled Bredia ●

59874 Bredia rotundifolia Y. C. Liu et C. H. Ou;圆叶野海棠(圆叶布勒德藤);Roundleaf Bredia ●

59875 Bredia rotundifolia Y. C. Liu et C. H. Ou = Bredia hirsuta Blume var. rotundifolia (Y. C. Liu et C. H. Ou) S. F. Huang et T. C. Huang ●

59876 Bredia rotundifolia Y. C. Liu et C. H. Ou = Bredia hirsuta Blume var. scandens Ito et Matsum. ●

59877 Bredia scandens (Ito et Matsum.) Hayata = Bredia hirsuta Blume var. scandens Ito et Matsum. ●

59878 Bredia sepalosa Diels = Bredia fordii (Hance) Diels ●■

59879 Bredia sepalosa Diels = Phyllagathis fordii (Hance) C. Chen ●

59880 Bredia sessilifolia H. L. Li;短柄野海棠(水牡丹);Sessile Bredia, Short-stalk Bredia ●

59881 Bredia sinensis (Diels) H. L. Li;鸭脚茶(九节兰,山落茄,鸭海棠,雨伞子,中华野海棠);Chinese Bredia, Duckpalm Bredia ●

59882 Bredia soneriloides H. Lév. = Oxyspora paniculata (D. Don) DC. ●

59883 Bredia stenophylla Merr. et Chun = Phyllagathis stenophylla (Merr. et Chun) H. L. Li ●

59884 Bredia tuberculata (Guillaumin) Diels;瘤药鸭脚茶(红毛野海棠,山红活麻);Tuberculate Bredia, Tuberculate-anthered Bredia ●

59885 Bredia tuberculata (Guillaumin) Diels = Bredia fordii (Hance) Diels ●■

59886 Bredia tuberculata (Guillaumin) Diels = Phyllagathis fordii (Hance) C. Chen ●

59887 Bredia velutina Diels = Phyllagathis velutina (Diels) C. Chen ●■

59888 Bredia yaeyamensis (Matsum.) H. L. Li;八重山野海棠●☆

59889 Bredia yunnanensis (H. Lév.) Diels;云南野海棠;Yunnan Bredia ●■

59890 Breea Less. = Cirsium Mill. ■

59891 Breea arvensis (L.) Less. = Cirsium arvense (L.) Scop. ■

59892 Breea arvensis Less. = Cirsium arvense (L.) Scop. ■

59893 Breea segetum (Bunge) Kitam. = Cirsium segetum Bunge ■

59894 Breea setosa (Willd.) Kitam. = Cirsium setosum (Willd.) M. Bieb. ■

59895 Brehmia Harv. = Strychnos L. ●

59896 Brehmia Schrank = Pavonia Cav. (保留属名)●■☆

59897 Brehnia Baker = Behnia Didr. ●☆

59898 Breitungia Á. Löve et D. Löve = Sedum L. ●■

59899 Bremekampia Sreem. = Haplanthodes Kuntze ■

59900 Bremeria Razafim. et Alejandro(2005);布雷默茜属■☆

59901 Bremeria asperula (Wernham) Razafim. et Alejandro;粗糙布雷默茜■☆

59902 Bremeria erectiloba (Wernham) Razafim. et Alejandro;直立布雷默茜■☆

59903 Bremeria fuscopilosa (Baker) Razafim. et Alejandro;褐毛布雷默茜■☆

59904 Bremeria gerrardii (Homolle) Razafim. et Alejandro;杰氏布雷默茜■☆

59905 Bremeria humblotii (Wernham) Razafim. et Alejandro;洪布布雷默茜■☆

59906 Bremeria hymenopogonoides (Baker) Razafim. et Alejandro;血毛布雷默茜■☆

59907 Bremeria lantziana (Homolle) Razafim. et Alejandro;兰兹布雷默茜■☆

59908 Bremeria latisepala (Homolle) Razafim. et Alejandro;宽萼布雷默茜■☆

59909 Bremeria monantha (Wernham) Razafim. et Alejandro;山地布雷默茜■☆

59910 Bremeria perrieri (Homolle) Razafim. et Alejandro;佩里耶布雷默茜■☆

59911 Bremeria pervillei (Wernham) Razafim. et Alejandro;佩尔布雷默茜■☆

59912 Bremeria pilosa (Baker) Razafim. et Alejandro;疏毛布雷默茜■☆

59913 Bremeria ramosissima (Wernham) Razafim. et Alejandro, Erasmo;多枝布雷默茜■☆

59914 Bremeria scabridior (Wernham) Razafim. et Alejandro;微糙布雷默茜■☆

59915 Bremeria trichophlebia (Baker) Razafim. et Alejandro;毛脉布雷默茜■☆

59916 Bremeria vestita (Baker) Razafim. et Alejandro;包被布雷默茜■☆

59917 Bremontiera DC. (1825);布雷豆属☆

59918 Bremontiera ammoxylon DC.;布雷豆☆

59919 Brenandendron H. Rob. (1999);蕨序鸡菊花属●☆

59920 Brenandendron donianum (DC.) H. Rob.;蕨序鸡菊花●☆

59921 Brenania Keay(1958);布氏茜属●☆

59922 Brenania brieyi (De Wild.) E. M. Petit;布里布氏茜●☆

59923 Brenania rhomboideifolia E. M. Petit;菱叶布氏茜●☆

59924 Brenania spathulifolia (R. D. Good) Keay = Brenania brieyi (De Wild.) E. M. Petit ●☆

59925 Brenaniodendron J. Léonard = Micklethwaitia G. P. Lewis et Schrire ●☆

59926 Brenaniodendron J. Léonard(1999);非洲茎花豆属●☆

59927 Brenesia Schltr. = Pleurothallis R. Br. ■☆

59928 Brenierea Humbert(1959);拟羊蹄甲属●☆

59929 Brenierea insignis Humbert;拟羊蹄甲●☆

59930 Breonadia Ridsdale(1975);大苞风箱树属(布雷那茜属)●☆

59931 Breonadia microcephala (Delile) Ridsdale = Breonadia salicina (Vahl) Hepper et J. R. I. Wood ●☆

59932 Breonadia salicina (Vahl) Hepper et J. R. I. Wood;大苞风箱树(布雷那茜)●☆
59933 Breonadia salicina (Vahl) Hepper et J. R. I. Wood = Cephalanthus spathelliferus Baker ●☆
59934 Breonadia salicina (Vahl) Hepper et J. R. I. Wood var. galpinii (Oliv.) Hepper et J. R. I. Wood = Breonadia salicina (Vahl) Hepper et J. R. I. Wood ●☆
59935 Breonia A. Rich. = Breonia A. Rich. ex DC. ●☆
59936 Breonia A. Rich. ex DC. (1830);黄梁木属(团花属)●☆
59937 Breonia boivinii Havil.;博伊文黄梁木●☆
59938 Breonia capuronii Razafim.;凯普伦黄梁木●☆
59939 Breonia chinensis (Lam.) Capuron;支那黄梁木●☆
59940 Breonia citrifolia (Poir.) Ridsdale = Breonia chinensis (Lam.) Capuron ●☆
59941 Breonia coriacea Havil. = Breonia chinensis (Lam.) Capuron ●☆
59942 Breonia cuspidata (Baker) Havil.;骤尖黄梁木●☆
59943 Breonia decaryana Homolle;德卡里黄梁木●☆
59944 Breonia keliravina Homolle = Breonia decaryana Homolle ●☆
59945 Breonia louvelii Homolle;卢韦尔黄梁木●☆
59946 Breonia lowryi Razafim.;劳里黄梁木●☆
59947 Breonia macrocarpa Homolle;大果黄梁木●☆
59948 Breonia madagascariensis A. Rich. ex DC.;马岛黄梁木●☆
59949 Breonia mauritiana Havil. = Breonia chinensis (Lam.) Capuron ●☆
59950 Breonia membranacea Havil.;膜质黄梁木●☆
59951 Breonia parviflora Havil. = Breonia sphaerantha (Baill.) Homolle ex Ridsdale ●☆
59952 Breonia perrieri Homolle;佩里耶黄梁木●☆
59953 Breonia richardiana (Baill.) Havil. = Breonia chinensis (Lam.) Capuron ●☆
59954 Breonia richardsonii Razafim.;理查黄梁木●☆
59955 Breonia sambiranensis Razafim.;桑比朗黄梁木●☆
59956 Breonia sphaerantha (Baill.) Homolle ex Ridsdale;球花黄梁木●☆
59957 Breonia stipulata Havil.;托叶黄梁木●☆
59958 Breonia taolagnaroensis Razafim.;陶拉纳鲁黄梁木●☆
59959 Breonia tayloriana Razafim.;泰勒黄梁木●☆
59960 Breonia tsaratananensis Razafim.;察拉塔纳纳黄梁木●☆
59961 Brephocton Raf. = Baccharis L.(保留属名)●■☆
59962 Breraontiera DC. = Indigofera L. ●■
59963 Breteuillia Buc'hoz ex DC. = Didelta L'Hér.(保留属名)■☆
59964 Breteuillia Buc'hoz(废弃属名) = Didelta L'Hér.(保留属名)■☆
59965 Bretschneidera Hemsl. (1901);伯乐树属(钟萼木属);Bretschneidera ●★
59966 Bretschneidera sinensis Hemsl.;伯乐树(山桃树,钟萼木);China Bretschneidera, Chinese Bretschneidera ●◇
59967 Bretschneidera yunshanensis Chun et F. C. How = Bretschneidera sinensis Hemsl. ●◇
59968 Bretschneideraceae Engl. et Gilg(1924)(保留科名);伯乐树科(钟萼木科);Bretschneidera Family ●
59969 Bretschneideraceae Engl. et Gilg(保留科名) = Akaniaceae Stapf(保留科名)●☆
59970 Bretschneideraceae Engl. et Gilg(保留科名) = Brexiaceae Lindl. ●☆
59971 Breueria R. Br. = Bonamia Thouars(保留属名)●☆
59972 Breueriopsis Roberty = Bonamia Thouars(保留属名)●☆
59973 Brevidens Miq. ex C. B. Clarke = Cyrtandromoea Zoll. ■
59974 Breviea Aubrév. et Pellegr. (1935);长籽山榄属●☆
59975 Breviea leptosperma (Baehni) Heine = Breviea sericea Aubrév. et Pellegr. ●☆
59976 Breviea sericea Aubrév. et Pellegr.;长籽山榄木●☆
59977 Breviglandium Dulac = Hottonia L. ■☆
59978 Brevilongium Christenson = Oncidium Sw.(保留属名)■☆
59979 Brevipodium Á. Löve et D. Löve = Brachypodium P. Beauv. ■
59980 Brevipodium sylvaticum (Huds.) Á. Löve et D. Löve = Brachypodium sylvaticum (Huds.) P. Beauv. ■
59981 Brevoortia A. Wood = Dichelostemma Kunth ■☆
59982 Brevoortia A. Wood (1867);缘檐丽韭属;Firecracker Flower, Fire-cracker Flower ■☆
59983 Brevoortia coccinea (A. Gray) S. Watson = Brevoortia idamaia A. Wood ■☆
59984 Brevoortia coccinea (A. Gray) S. Watson = Dichelostemma idamaia Greene ■☆
59985 Brevoortia idamaia A. Wood;缘檐丽韭;Firecracker Flower, Fire-cracker Flower ■☆
59986 Brevoortia idamaia A. Wood = Dichelostemma idamaia (A. Wood) Greene ■☆
59987 Brevoortia venusta Greene = Dichelostemma idamaia Greene ■☆
59988 Brewcaria L. B. Sm., Steyerm. et H. Rob. (1984);布鲁凤梨属■☆
59989 Brewcaria duidensis L. B. Sm., Steyerm. et Robinson;布鲁凤梨■☆
59990 Breweria R. Br. = Bonamia Thouars(保留属名)●☆
59991 Breweria africana (G. Don) Benth. et Hook. f. = Calycobolus africanus (G. Don) Heine ●☆
59992 Breweria alternifolia (Planch.) Radlk. = Calycobolus africanus (G. Don) Heine ●☆
59993 Breweria argentea A. Terracc. = Seddera latifolia Hochst. et Steud. var. argentea (A. Terracc.) Capua ●☆
59994 Breweria baccharoides Baker = Seddera suffruticosa (Schinz) Hallier f. ●☆
59995 Breweria buddleoides Baker = Bonamia mossambicensis (Klotzsch) Hallier f. ●☆
59996 Breweria campanulata (K. Schum. ex Hallier f.) Baker = Calycobolus campanulatus (K. Schum. ex Hallier f.) Heine ●☆
59997 Breweria capensis (E. Mey. ex Choisy) Baker = Seddera capensis (E. Mey. ex Choisy) Hallier f. ●☆
59998 Breweria codonanthus Baker ex Oliv. = Calycobolus africanus (G. Don) Heine ●☆
59999 Breweria conglomerata Baker = Seddera suffruticosa (Schinz) Hallier f. var. hirsutissima Hallier f. ●☆
60000 Breweria glomerata Balf. f. = Seddera glomerata (Balf. f.) O. Schwartz ●☆
60001 Breweria heudelotii Baker ex Oliv. = Calycobolus heudelotii (Baker ex Oliv.) Heine ●☆
60002 Breweria hildebrandtii Vatke = Bonamia spectabilis (Choisy) Hallier f. ●☆
60003 Breweria hispida Franch. = Seddera arabica (Forssk.) Choisy ●☆
60004 Breweria malvacea Klotzsch = Astripomoea malvacea (Klotzsch) A. Meeuse ■☆
60005 Breweria microcephala Baker = Seddera suffruticosa (Schinz) Hallier f. ●☆
60006 Breweria mirabilis Baker ex Oliv. = Calycobolus campanulatus (K. Schum. ex Hallier f.) Heine ●☆
60007 Breweria oxycarpa A. Rich. = Seddera arabica (Forssk.)

Choisy ●☆

60008 Breweria pedunculata Balf. f. = Seddera pedunculata (Balf. f.) Hallier f. ●☆

60009 Breweria pickeringii (Torr.) A. Gray = Stylisma pickeringii (Torr.) A. Gray ■☆

60010 Breweria secunda (G. Don) Benth. = Bonamia thunbergiana (Roem. et Schult.) F. N. Williams ●☆

60011 Breweria sessiliflora Baker = Seddera suffruticosa (Schinz) Hallier f. ●☆

60012 Breweria somalensis Vatke = Seddera arabica (Forssk.) Choisy ●☆

60013 Breweria spectabilis Choisy = Bonamia spectabilis (Choisy) Hallier f. ●☆

60014 Breweria suffruticosa Schinz = Seddera suffruticosa (Schinz) Hallier f. ●☆

60015 Breweria tiliifolia Baker = Rapona tiliifolia (Baker) Verdc. ●☆

60016 Breweria virgata (Hochst. et Steud.) Vatke = Seddera virgata Hochst. et Steud. ●☆

60017 Brewerina A. Gray = Arenaria L. ■

60018 Brewerina suffrutescens A. Gray = Arenaria suffrutescens (A. Gray) A. Heller ■☆

60019 Brewerina suffrutescens A. Gray = Eremogone congesta (Nutt.) Ikonn. var. suffrutescens (A. Gray) R. L. Hartm. et Rabeler ■☆

60020 Breweriopsis Roberty = Bonamia Thouars(保留属名)●☆

60021 Brewertna A. Gray = Arenaria L. ■

60022 Brewiera Roberty = Bonamia Thouars(保留属名)●☆

60023 Brewiera Roberty = Breweria R. Br. ●☆

60024 Brewieropsis Roberty = Bonamia Thouars(保留属名)●☆

60025 Brewieropsis Roberty = Breweriopsis Roberty ●☆

60026 Brewstera M. Poem. = Ixonanthes Jack ●

60027 Brexia Noronha ex Thouars(1806)(保留属名);雨湿木属(流苏边脉属)●☆

60028 Brexia apoda H. Perrier;无梗雨湿木●☆

60029 Brexia arborea H. Perrier;树状雨湿木●☆

60030 Brexia cauliflora Tul.;茎花雨湿木●☆

60031 Brexia coursiana H. Perrier;库尔斯雨湿木●☆

60032 Brexia decurrens H. Perrier;下延雨湿木●☆

60033 Brexia horombensis J. -F. Leroy = Brexia humbertii H. Perrier ●☆

60034 Brexia humbertii H. Perrier;亨伯特雨湿木●☆

60035 Brexia madagascariensis (Lam.) Ker Gawl.;雨湿木(伯力木);Mfukufuku ●☆

60036 Brexia madagascariensis Thouars = Brexia madagascariensis (Lam.) Ker Gawl. ●☆

60037 Brexia montana H. Perrier;山地雨湿木●☆

60038 Brexia montana H. Perrier var. bracteata H. Perrier = Brexia humbertii H. Perrier ●☆

60039 Brexiaceae Lindl. = Celastraceae R. Br.(保留科名)●

60040 Brexiaceae Lindl. = Grossulariaceae DC.(保留科名)●

60041 Brexiaceae Loudon = Grossulariaceae DC.(保留科名)●

60042 Brexiaceae Loudon(1830);雨湿木科(流苏边脉科)●☆

60043 Brexiella H. Perrier(1933);小雨湿木属(小流苏边脉属)●☆

60044 Brexiella illicifolia H. Perrier;小流苏边脉●☆

60045 Brexiopsis H. Perrier = Drypetes Vahl ●

60046 Brexiopsis H. Perrier(1942);拟雨湿木属(拟流苏边脉属)●☆

60047 Brexiopsis aquifolia H. Perrier;拟雨湿木(拟流苏边脉)●☆

60048 Brexiopsis aquifolia H. Perrier = Drypetes bathiei Capuron et Léandri ●☆

60049 Breynia J. R. Forst. et G. Forst.(1775)(保留属名);黑面神属(山漆茎属);Breynia ●

60050 Breynia L.(废弃属名) = Breynia J. R. Forst. et G. Forst.(保留属名)●

60051 Breynia L.(废弃属名) = Capparis L. ●

60052 Breynia L.(废弃属名) = Linnaeobreynia Hutch. ●

60053 Breynia accrescens Hayata;宽萼山漆茎(小红仔珠,小山漆茎);Largecalyx Breynia ●

60054 Breynia accrescens Hayata = Breynia officinalis Hemsl. ●

60055 Breynia accrescens Hayata = Breynia vitis-idaea (Burm. f.) C. E. C. Fisch. ●

60056 Breynia cernua (Poir.) Müll. Arg.;垂头黑面神●☆

60057 Breynia disticha J. R. Forst. et G. Forst.;二列黑面神(雪丛黑面神);Bilow Breynia, Foliage Flower, Otaheite Gooseberry, Snow Bush, Snowbush ●

60058 Breynia disticha J. R. Forst. et G. Forst. 'Roseopicta';红斑雪丛黑面神●☆

60059 Breynia disticha J. R. Forst. et G. Forst. f. nivosa (W. Bull) Croizat ex Radcl. -Sm.;雪白二列黑面神●☆

60060 Breynia fordii Hemsl.;华南逼迫子;Ford Breynia ●

60061 Breynia formosana (Hayata) Hayata;台湾山漆茎;Formosan Breynia ●

60062 Breynia formosana (Hayata) Hayata = Breynia officinalis Hemsl. ●

60063 Breynia formosana (Hayata) Hayata = Breynia vitis-idaea (Burm. f.) C. E. C. Fisch. ●

60064 Breynia fruticosa (L.) Hook. f.;黑面神(暗鬼木,狗脚刺,鬼划符,鬼画符,锅盖木,锅盖仔,黑面树,黑面叶,鸡肾叶,老鸦写字,庙公仔,漆鼓,漆舅,青凡木,青漆,青丸木,山树兰,山夜兰,四眼草,四眼叶,田中逵,铁甲将军,乌漆臼,乌漆血,细清漆树,野甜菜,夜兰,夜兰茶,蚁惊树,钟馗草);Fruticose Breynia ●

60065 Breynia glauca Craib;苍白黑面神●☆

60066 Breynia hyposauropa Croizat;广西黑面神(红子仔,节节红花,小黑面神,小叶黑面神);Guangxi Breynia ●

60067 Breynia keithii Ridl. = Breynia vitis-idaea (Burm. f.) C. E. C. Fisch. ●

60068 Breynia microcalyx Ridl. = Breynia vitis-idaea (Burm. f.) C. E. C. Fisch. ●

60069 Breynia nivosa (W. Bull) Small = Breynia disticha J. R. Forst. et G. Forst. f. nivosa (W. Bull) Croizat ex Radcl. -Sm. ●☆

60070 Breynia nivosa Small;白叶黑面神(山漆茎);Snow Bush, Snowbush, Snow-bush ●

60071 Breynia nivosa Small = Breynia disticha J. R. Forst. et G. Forst. ●

60072 Breynia oblongifolia (Müll. Arg.) Müll. Arg.;小黑面神;Dwarf Apples ●☆

60073 Breynia officinalis Hemsl.;红仔珠(黑面神,红心仔,红薏仔,红珠,红珠仔,七日晕,山漆茎,药用黑面神);Medicinal Breynia, Official Breynia ●

60074 Breynia officinalis Hemsl. = Breynia vitis-idaea (Burm. f.) C. E. C. Fisch. ●

60075 Breynia officinalis Hemsl. var. accrescens (Hayata) M. J. Deng et J. C. Wang;小红仔珠●

60076 Breynia officinalis Hemsl. var. accrescens (Hayata) M. J. Deng et J. C. Wang = Breynia vitis-idaea (Burm. f.) C. E. C. Fisch. ●

60077 Breynia patens (Roxb.) Benth. = Breynia retusa (Dennst.) Alston ●

60078 Breynia retusa (Dennst.) Alston;钝叶黑面神(地石榴,黑面叶,跳八丈,小面瓜,小柿子,小叶黑面神,小叶黑面叶,小叶山漆

茎,枝展黑面神);Obuse Breynia,Retuse Breynia ●
60079 Breynia rhamnoides (Willd.) Müll. Arg. = Breynia vitis-idaea (Burm. f.) C. E. C. Fisch. ●
60080 Breynia rostrata Merr.;喙果黑面神(尾叶黑面叶,小面瓜,小柿子);Beak-shaped Breynia,Rostratefruit Breynia ●
60081 Breynia stipitata Müll. Arg.;具柄黑面神●☆
60082 Breynia stipitata Müll. Arg. var. formosana Hayata = Breynia vitis-idaea (Burm. f.) C. E. C. Fisch. ●
60083 Breynia vitis-idaea (Burm. f.) C. E. C. Fisch.;小叶黑面神(红子仔,宽萼山漆茎,葡萄黑面神,山漆茎,鼠李状山漆茎,小红仔珠,小山漆茎,药用黑面神,一叶一枝花);Small-leaved Breynia ●
60084 Breyniopsis Beille = Sauropus Blume ●■
60085 Breyniopsis pierrei Beille = Sauropus pierrei (Beille) Croizat ●
60086 Breza heterophylla (Kar. et Kir.) Moq. = Suaeda heterophylla (Kar. et Kir.) Bunge ■
60087 Breza heterophylla Moq. = Suaeda heterophylla (Kar. et Kir.) Bunge ■
60088 Brezia Moq. (1849);横翼碱蓬属■☆
60089 Brezia Moq. = Suaeda Forssk. ex J. F. Gmel. (保留属名)●■
60090 Brezia heterophylla Moq.;横翼碱蓬■☆
60091 Brianhuntleya Chess., S. A. Hammer et I. Oliv. (2003);澳非舟叶花属●☆
60092 Brianhuntleya Chess., S. A. Hammer et I. Oliv. = Mesembryanthemum L. (保留属名)■●
60093 Brianhuntleya intrusa (Kensit) Chess., S. A. Hammer et I. M. Oliv.;澳非舟叶花●☆
60094 Bricchettia Pax = Cocculus DC. (保留属名)●
60095 Brickellia Elliott(1823)(保留属名);肋泽兰属(布氏菊属,鞘冠菊属);Coleostephus ■●
60096 Brickellia Raf. (废弃属名) = Brickellia Elliott(保留属名)■●
60097 Brickellia Raf. (废弃属名) = Gilia Ruiz et Pav. ■●☆
60098 Brickellia Raf. (废弃属名) = Ipomopsis Michx. ■☆
60099 Brickellia adenocarpa B. L. Rob.;腺果肋泽兰(腺果布氏菊)■☆
60100 Brickellia arguta B. L. Rob. = Brickellia atractyloides A. Gray var. arguta (B. L. Rob.) Jeps. ■☆
60101 Brickellia arguta B. L. Rob. var. odontolepis B. L. Rob. = Brickellia atractyloides A. Gray var. odontolepis (B. L. Rob.) Jeps. ■☆
60102 Brickellia atractyloides A. Gray var. arguta (B. L. Rob.) Jeps.;光亮肋泽兰■☆
60103 Brickellia atractyloides A. Gray var. odontolepis (B. L. Rob.) Jeps.;齿鳞肋泽兰■☆
60104 Brickellia betonicifolia A. Gray var. conduplicata B. L. Rob. = Brickellia lemmonii A. Gray var. conduplicata (B. L. Rob.) B. L. Turner ■☆
60105 Brickellia brachiata A. Gray = Brickellia coulteri A. Gray var. brachiata (A. Gray) B. L. Turner ■☆
60106 Brickellia brachyphylla (A. Gray) A. Gray;短叶肋泽兰■☆
60107 Brickellia brachyphylla (A. Gray) A. Gray var. hinckleyi (Standl.) Flyr = Brickellia hinckleyi Standl. ■☆
60108 Brickellia californica (Torr. et Gray) A. Gray;加州肋泽兰(加州布氏菊);California Brickellia ■☆
60109 Brickellia cavanillesii Gray;卡布氏菊■☆
60110 Brickellia chenopodina (Greene) B. L. Rob.;藜肋泽兰■☆
60111 Brickellia conduplicata (B. L. Rob.) B. L. Rob. = Brickellia lemmonii A. Gray var. conduplicata (B. L. Rob.) B. L. Turner ■☆
60112 Brickellia cordifolia Elliott;心叶肋泽兰■☆
60113 Brickellia coulteri A. Gray;库尔特肋泽兰;Coulter's Brickellia ■☆
60114 Brickellia coulteri A. Gray var. brachiata (A. Gray) B. L. Turner;双枝肋泽兰■☆
60115 Brickellia dentata (DC.) Sch. Bip.;齿叶肋泽兰■☆
60116 Brickellia eupatorioides (L.) Shinners;泽兰库恩菊;False Boneset ■☆
60117 Brickellia eupatorioides (L.) Shinners = Kuhnia eupatorioides L. ■☆
60118 Brickellia eupatorioides (L.) Shinners var. chlorolepis (Wooton et Standl.) B. L. Turner;绿鳞泽兰库恩菊■☆
60119 Brickellia eupatorioides (L.) Shinners var. corymbulosa (Torr. et A. Gray) Shinners;小伞序泽兰库恩菊;False Boneset ■☆
60120 Brickellia eupatorioides (L.) Shinners var. floridana (R. W. Long) B. L. Turner;佛罗里达库恩菊■☆
60121 Brickellia eupatorioides (L.) Shinners var. gracillima (A. Gray) B. L. Turner;纤细泽兰库恩菊■☆
60122 Brickellia eupatorioides (L.) Shinners var. ozarkana (Shinners) Shinners = Brickellia eupatorioides (L.) Shinners var. texana (Shinners) Shinners ■☆
60123 Brickellia eupatorioides (L.) Shinners var. texana (Shinners) Shinners;得州泽兰库恩菊■☆
60124 Brickellia eupatorioides L. var. chlorolepis (Wooton et Standl.) Cronquist = Brickellia eupatorioides (L.) Shinners var. chlorolepis (Wooton et Standl.) B. L. Turner ■☆
60125 Brickellia fendleri A. Gray = Brickelliastrum fendleri (A. Gray) R. M. King et H. Rob. ■☆
60126 Brickellia grandiflora (Hook.) Nutt.;大花布氏菊;Tassel Flower ■☆
60127 Brickellia hinckleyi Standl.;欣克利布氏菊■☆
60128 Brickellia laciniata A. Gray;裂叶布氏菊;Cutleaf Brickellia ■☆
60129 Brickellia lemmonii A. Gray;莱蒙布氏菊■☆
60130 Brickellia lemmonii A. Gray var. conduplicata (B. L. Rob.) B. L. Turner;对折布氏菊■☆
60131 Brickellia lemmonii A. Gray var. wootonii (Greene) B. L. Rob. = Brickellia lemmonii A. Gray ■☆
60132 Brickellia leptophylla (Scheele) Shinners = Brickellia eupatorioides (L.) Shinners var. gracillima (A. Gray) B. L. Turner ■☆
60133 Brickellia longifolia S. Watson;长叶布氏菊■☆
60134 Brickellia longifolia S. Watson var. multiflora (Kellogg) Cronquist;多花长叶布氏菊■☆
60135 Brickellia microphylla (Nutt.) A. Gray;小叶布氏菊■☆
60136 Brickellia microphylla (Nutt.) A. Gray subsp. scabra (A. Gray) W. A. Webe = Brickellia microphylla (Nutt.) A. Gray var. scabra A. Gray ■☆
60137 Brickellia microphylla (Nutt.) A. Gray var. scabra A. Gray;粗糙布氏菊■☆
60138 Brickellia microphylla (Nutt.) A. Gray var. watsonii (B. L. Rob.) S. L. Welsh = Brickellia microphylla (Nutt.) A. Gray ■☆
60139 Brickellia monocephala Robins.;单头布氏菊■☆
60140 Brickellia mosieri (Small) Shinners;佛罗里达肋泽兰■☆
60141 Brickellia mosieri (Small) Shinners = Kuhnia eupatorioides L. var. floridana R. W. Long ■☆
60142 Brickellia multiflora Kellogg = Brickellia longifolia S. Watson var. multiflora (Kellogg) Cronquist ■☆
60143 Brickellia oblongifolia Nutt.;矩圆肋泽兰■☆
60144 Brickellia oblongifolia Nutt. linifolia D. C. Eaton = Brickellia oblongifolia Nutt. var. linifolia (D. C. Eaton) B. L. Rob. ■☆

60145 Brickellia oblongifolia Nutt. subsp. linifolia (D. C. Eaton) Cronquist = Brickellia oblongifolia Nutt. var. linifolia (D. C. Eaton) B. L. Rob. ■☆

60146 Brickellia oblongifolia Nutt. var. linifolia (D. C. Eaton) B. L. Rob. ;亚麻叶矩圆肋泽兰■☆

60147 Brickellia paniculata B. L. Rob. ;锥花布氏菊■☆

60148 Brickellia rosmarinifolia (Vent.) W. A. Weber subsp. chlorolepis (Wooton et Standl.) W. A. Weber = Brickellia eupatorioides (L.) Shinners var. chlorolepis (Wooton et Standl.) B. L. Turner ■☆

60149 Brickellia scabra (A. Gray) A. Nelson ex B. L. Rob. = Brickellia microphylla (Nutt.) A. Gray var. scabra A. Gray ■☆

60150 Brickellia secundiflora A. Gray;偏花布氏菊■☆

60151 Brickellia squamulosa A. Gray = Asanthus squamulosus (A. Gray) R. M. King et H. Rob. ■☆

60152 Brickellia tenuiflora (DC.) D. J. Keil et Pinkava = Carminatia tenuiflora DC. ■☆

60153 Brickellia venosa (Wooton et Standl.) B. L. Rob. ;多脉肋泽兰■☆

60154 Brickellia veronicifolia (Kunth) A. Gray;婆婆纳叶布氏菊■☆

60155 Brickellia veronicifolia (Kunth) A. Gray var. petrophila (B. L. Rob.) B. L. Rob. = Brickellia veronicifolia (Kunth) A. Gray ■☆

60156 Brickellia viejensis Flyr = Brickellia lemmonii A. Gray var. conduplicata (B. L. Rob.) B. L. Turner ■☆

60157 Brickellia watsonii B. L. Rob. = Brickellia microphylla (Nutt.) A. Gray ■☆

60158 Brickelliastrum R. M. King et H. Rob. (1972);落冠肋泽兰属(小肋泽兰属)●☆

60159 Brickelliastrum fendleri (A. Gray) R. M. King et H. Rob. ;落冠肋泽兰;Fendler's Brickellbush ■☆

60160 Bricour Adans. = Myagrum L. ■☆

60161 Bridelia Spreng. = Bridelia Willd. (保留属名)●

60162 Bridelia Willd. (1806)(保留属名)('Briedelia');土密树属(土蜜树属);Bridelia, Prikly Bridelia ●

60163 Bridelia abyssinica Pax = Bridelia cathartica G. Bertol. f. lingelsheimii (Gehrm.) Radcl. -Sm. ●☆

60164 Bridelia angolensis Welw. ex Müll. Arg. = Bridelia scleroneura Müll. Arg. subsp. angolensis (Welw. ex Müll. Arg.) Radcl. -Sm. ●

60165 Bridelia atroviridis Müll. Arg. ;墨绿土密树●☆

60166 Bridelia aubrevillei Pellegr. ;奥氏土密树●☆

60167 Bridelia balansae Tutcher;禾串树(大叶逼迫子,刺杜密,橡土蜜树);Bridelia ●

60168 Bridelia balansae Tutcher = Bridelia insulana Hance ●

60169 Bridelia brideliifolia (Pax) Fedde;单室土密树●☆

60170 Bridelia brideliifolia (Pax) Fedde subsp. pubescentifolia J. Léonard = Bridelia brideliifolia (Pax) Fedde ●☆

60171 Bridelia cathartica G. Bertol. ;泻下土密树●☆

60172 Bridelia cathartica G. Bertol. f. fischeri (Pax) Radcl. -Sm. ;菲舍尔土密树●☆

60173 Bridelia cathartica G. Bertol. f. lingelsheimii (Gehrm.) Radcl. -Sm. ;林格土密树●☆

60174 Bridelia cathartica G. Bertol. f. melanthesoides (Baill.) Radcl. -Sm. ;黑面神土密树●☆

60175 Bridelia cathartica G. Bertol. f. pubescens Radcl. -Sm. ;短柔毛泻下土密树●☆

60176 Bridelia cathartica G. Bertol. var. melanthesoides (Baill.) Radcl. -Sm. = Bridelia cathartica G. Bertol. f. melanthesoides (Baill.) Radcl. -Sm. ●☆

60177 Bridelia duvigneaudii J. Léonard;迪维尼奥土密树●☆

60178 Bridelia elegans Müll. Arg. = Bridelia tenuifolia Müll. Arg. var. elegans (Müll. Arg.) Hutch. ●☆

60179 Bridelia ferruginea Benth. ;锈色土密树(非洲土密树)●☆

60180 Bridelia fischeri Pax = Bridelia cathartica G. Bertol. f. fischeri (Pax) Radcl. -Sm. ●☆

60181 Bridelia fischeri Pax var. lingelsheimii (Gehrm.) Hutch. = Bridelia cathartica G. Bertol. f. lingelsheimii (Gehrm.) Radcl. -Sm. ●☆

60182 Bridelia fordii Hemsl. ;大叶土密树(华南逼迫子,虾公木);Ford Bridelia ●

60183 Bridelia gambecola Baill. = Bridelia micrantha (Hochst.) Baill. ●

60184 Bridelia glauca Blume f. balansae (Tutcher) Hatus. = Bridelia insulata Hance ●

60185 Bridelia grandis Pierre ex Hutch. ;大土密树●☆

60186 Bridelia grandis Pierre ex Hutch. subsp. puberula J. Léonard;微毛土密树●☆

60187 Bridelia griffithii Hook. f. var. penangiana (Hook. f.) Gehrm. = Bridelia insulana Hance ●

60188 Bridelia henryana Jabl. = Bridelia tomentosa Blume ●

60189 Bridelia insulana Hance;禾串土密树(禾串树,大叶逼迫子,大叶土密树,刺杜密);Insular Bridelia, Prickly Bridelia ●☆

60190 Bridelia lingelsheimii Gehrm. = Bridelia cathartica G. Bertol. f. lingelsheimii (Gehrm.) Radcl. -Sm. ●☆

60191 Bridelia melanthesoides (Baill.) Klotzsch = Bridelia cathartica G. Bertol. f. melanthesoides (Baill.) Radcl. -Sm. ●☆

60192 Bridelia micrantha (Hochst.) Baill. ;小花土密树●

60193 Bridelia microphylla Chiov. ;小叶土密树●

60194 Bridelia mildbraedii Gehrm. = Bridelia micrantha (Hochst.) Baill. ●

60195 Bridelia minutiflora Hook. f. = Bridelia insulana Hance ●

60196 Bridelia mollis Hutch. ;绢毛土密树●

60197 Bridelia monoica Merr. = Bridelia tomentosa Blume ●

60198 Bridelia montana (Roxb.) Willd. ;波叶土密树(山土密树);Wave-leaved Bridelia ●

60199 Bridelia ndellensis Beille;恩代尔土密树●☆

60200 Bridelia niedenzui Gehrm. var. pilosa? = Bridelia cathartica G. Bertol. f. fischeri (Pax) Radcl. -Sm. ●☆

60201 Bridelia nigricans Gehrm. = Bridelia taitensis Vatke et Pax ●☆

60202 Bridelia ovata Decne. ;卵叶土密树●☆

60203 Bridelia ovata Decne. = Bridelia insulana Hance ●

60204 Bridelia pachynensis Hayata ex Matsum. et Hayata = Bridelia insulana Hance ●

60205 Bridelia paxii Gehrm. = Bridelia scleroneura Müll. Arg. ●

60206 Bridelia penangiana Hook. f. = Bridelia insulana Hance ●

60207 Bridelia perrotii Beille = Bridelia speciosa Müll. Arg. ●☆

60208 Bridelia pierrei Gagnep. ;贵州土密树;Guizhou Bridelia, Pierre Bridelia ●

60209 Bridelia platyphylla Merr. = Bridelia insulana Hance ●

60210 Bridelia poilanei Gagnep. ;圆叶土密树;Poilane Bridelia, Round-leaved Bridelia ●

60211 Bridelia pubescens Kurz;膜叶土密树(毛土密树);Duwny Bridelia, Pubescent Bridelia ●

60212 Bridelia retusa (L.) Spreng. ;微凹土密树●

60213 Bridelia retusa (L.) Spreng. = Bridelia spinosa (Roxb.) Willd. ●

60214 Bridelia retusa A. Juss. = Bridelia stipularis (L.) Blume ●

60215 Bridelia ripicola J. Léonard;岩地土密树●☆

60216 Bridelia scandens (Roxb.) Willd. = Bridelia stipularis (L.)

Blume ●
60217 Bridelia schlechteri Hutch. = Bridelia cathartica G. Bertol. ●☆
60218 Bridelia scleroneura Müll. Arg.;硬脉土密树●
60219 Bridelia scleroneura Müll. Arg. subsp. angolensis (Welw. ex Müll. Arg.) Radcl. -Sm.;安哥拉土密树●
60220 Bridelia scleroneuroides Pax = Bridelia scleroneura Müll. Arg. ●
60221 Bridelia somalensis Hutch.;索马里土密树●
60222 Bridelia speciosa Müll. Arg.;美丽土密树●☆
60223 Bridelia spinosa (Roxb.) Willd.;密脉土密树(微凹土密树);Dense-veins Bridelia, Thorny Bridelia ●
60224 Bridelia stenocarpa Müll. Arg. = Bridelia micrantha (Hochst.) Baill. ●
60225 Bridelia stipularis (L.) Blume;土密藤(大串连果,狗舌果,托叶土密树);Pikpoktsai, Spitular Bridelia, Spitulate Bridelia ●
60226 Bridelia stipularis Blume = Bridelia mollis Hutch. ●
60227 Bridelia taitensis Vatke et Pax;泰塔土密树●☆
60228 Bridelia tenuifolia Müll. Arg.;瘦叶土密树●☆
60229 Bridelia tenuifolia Müll. Arg. var. elegans (Müll. Arg.) Hutch.;雅致土密树●☆
60230 Bridelia tomentosa Blume;土密树(逼迫子,补锅树,补脑根,夹骨木,土蜜树,土知母,猪牙木);Bigbagdzi, Pikpoktsai, Pikpoktsai Bridelia ●
60231 Bridelia tomentosa Blume var. chinensis (Müll. Arg.) Gehrm. = Bridelia tomentosa Blume ●
60232 Bridelia tomentosa Blume var. glabrata Schweinf. = Bridelia scleroneura Müll. Arg. ●
60233 Bridelia zanzibarensis Vatke et Pax = Bridelia micrantha (Hochst.) Baill. ●
60234 Bridelia zenkeri Pax = Bridelia atroviridis Müll. Arg. ●☆
60235 Bridgesia Backeb. = Gymnocalycium Sweet ex Mittler ●
60236 Bridgesia Backeb. = Neoporteria Britton et Rose ●■
60237 Bridgesia Backeb. = Rebutia K. Schum. ●
60238 Bridgesia Bertero ex Cambess. (1834)(保留属名);布里无患子属●☆
60239 Bridgesia Hook.(废弃属名) = Bridgesia Bertero ex Cambess.(保留属名)●☆
60240 Bridgesia Hook.(废弃属名) = Polyachyrus Lag. ●■☆
60241 Bridgesia Hook. et Arn. = Bridgesia Bertero ex Cambess.(保留属名)●☆
60242 Bridgesia Hook. et Arn. = Ercilla A. Juss. ●☆
60243 Bridgesia incisifolia Bertero ex Cambess.;布里无患子●☆
60244 Briedelia Willd. = Bridelia Willd.(保留属名)●
60245 Briegeria Senghas = Epidendrum L.(保留属名)■☆
60246 Briegeria Senghas(1980);布里兰属■☆
60247 Briegeria teretifolia (Sw.) Senghas;布里兰■☆
60248 Brieya De Wild. = Piptostigma Oliv. ●☆
60249 Brieya fasciculatum De Wild. = Piptostigma fasciculatum (De Wild.) Boutique ●☆
60250 Brieya latipetala Exell = Piptostigma exellii R. E. Fr. ●☆
60251 Briggsia Craib(1919);粗筒苣苔属(佛肚苣苔属);Briggsia ■
60252 Briggsia acutiloba K. Y. Pan;尖瓣粗筒苣苔;Sharplobe Briggsia ■
60253 Briggsia agnesiae (Forrest) Craib;灰毛粗筒苣苔;Greyhair Briggsia ■
60254 Briggsia amabilis (Diels) Craib = Briggsia kurzii (C. B. Clarke) W. E. Evans ■
60255 Briggsia amabilis (Diels) Craib var. taliensis Craib = Briggsia kurzii (C. B. Clarke) W. E. Evans ■
60256 Briggsia aurantiaca B. L. Burtt;黄花粗筒苣苔;Yellowflower Briggsia ■
60257 Briggsia beanuverdiana (H. Lév.) Craib = Briggsiopsis delavayi (Franch.) K. Y. Pan ■
60258 Briggsia cavaleriei (H. Lév. et Vaniot) Craib = Loxostigma cavaleriei (H. Lév. et Vaniot) B. L. Burtt ■
60259 Briggsia chienii Chun;浙皖粗筒苣苔(佛肚花,虎皮,秦氏佛肚苣苔,小荷草,岩白菜,岩青菜);Chien Briggsia ■
60260 Briggsia crenulata Hand. -Mazz. = Briggsia rosthornii (Diels) B. L. Burtt var. crenulata (Hand. -Mazz.) K. Y. Pan ■
60261 Briggsia delavayi (Franch.) Chun = Briggsiopsis delavayi (Franch.) K. Y. Pan ■
60262 Briggsia dongxingensis Chun ex K. Y. Pan;东兴粗筒苣苔;Dongxing Briggsia ■
60263 Briggsia elegantissima (H. Lév. et Vaniot) Craib;紫花粗筒苣苔;Mostbeautiful Briggsia ■
60264 Briggsia forrestii Craib;云南粗筒苣苔;Forrest Briggsia ■
60265 Briggsia fritschii (H. Lév. et Vaniot) Craib = Briggsia mihieri (Franch.) Craib ■
60266 Briggsia fritschii (H. Lév.) Craib = Briggsia mairei Craib ■
60267 Briggsia hians Chun = Briggsia rosthornii (Diels) B. L. Burtt ■
60268 Briggsia humilis K. Y. Pan;小粗筒苣苔;Small Briggsia ■
60269 Briggsia kurzii (C. B. Clarke) W. E. Evans;粗筒苣苔(苦苣苔);Amaible Briggsia, Lovely Briggsia ■
60270 Briggsia kurzii (C. B. Clarke) W. E. Evans = Loxostigma kurzii (C. B. Clarke) B. L. Burtt ■
60271 Briggsia latisepala Chun ex K. Y. Pan;宽萼粗筒苣苔;Broadsepal Briggsia ■
60272 Briggsia longicaulis W. T. Wang et K. Y. Pan;长茎粗筒苣苔;Longstem Briggsia ■
60273 Briggsia longifolia Craib;长叶粗筒苣苔;Longleaf Briggsia ■
60274 Briggsia longifolia Craib var. multiflora S. Y. Chen ex K. Y. Pan;多花粗筒苣苔;Multiflower Longleaf Briggsia ■
60275 Briggsia longipes (Hemsl. ex Oliv.) Craib;盾叶粗筒苣苔(盾叶佛肚苣苔);Longstalk Brigssia, Peltate Briggsia ■
60276 Briggsia macrosiphon (Hance) Chun;大嘴粗筒苣苔■
60277 Briggsia mairei Craib;东川粗筒苣苔;Maire Briggsia ■
60278 Briggsia mihieri (Franch.) Craib;革叶粗筒苣苔(小岩青菜,锈草,岩枇杷,岩莴苣);Coriaceous Briggsia, Coriaceousleaf Briggsia, Fritsch Briggsia ■
60279 Briggsia muscicola (Diels) Craib;藓丛粗筒苣苔;Mossliving Brigssia ■
60280 Briggsia parvifolia K. Y. Pan;小叶粗筒苣苔;Smallleaf Briggsia ■
60281 Briggsia penlopi C. E. C. Fisch. = Briggsia muscicola (Diels) Craib ■
60282 Briggsia pinfaensis (H. Lév.) Craib;平伐粗筒苣苔;Pingfa Briggsia ■
60283 Briggsia rosthornii (Diels) B. L. Burtt;川鄂粗筒苣苔;Rosthorn Briggsia ■
60284 Briggsia rosthornii (Diels) B. L. Burtt var. crenulata (Hand.-Mazz.) K. Y. Pan;贞丰粗筒苣苔■
60285 Briggsia rosthornii (Diels) B. L. Burtt var. wenshanensis K. Y. Pan;文山粗筒苣苔;Wenshan Briggsia ■
60286 Briggsia rosthornii (Diels) B. L. Burtt var. xingrenensis K. Y. Pan;锈毛粗筒苣苔;Xingren Briggsia ■
60287 Briggsia speciosa (Hemsl.) Craib;鄂西粗筒苣苔(丫头还阳,雅头还阳);Beautiful Briggsia ■

60288 Briggsia stewardii Chun；广西粗筒苣苔(蟾蜍草)；Guangxi Briggsia ■

60289 Briggsiopsis K. Y. Pan(1985)；筒花苣苔属；Briggsiopsis ●★

60290 Briggsiopsis delavayi (Franch.) K. Y. Pan；筒花苣苔；Briggsiopsis ■

60291 Brighamia A. Gray(1866)；布里桔梗属●☆

60292 Brighamia insignis A. Gray；布里桔梗；Olulu ●☆

60293 Brignolia Bertol. = Kundmannia Scop. ■☆

60294 Brignolia DC. = Isertia Schreb. ●☆

60295 Brillantaisia P. Beauv. (1818)；伴帕爵床属●■☆

60296 Brillantaisia alata T. Anderson ex Oliv. = Brillantaisia patula T. Anderson ■☆

60297 Brillantaisia anomala Lindau = Brillantaisia pubescens T. Anderson ex Oliv. ■☆

60298 Brillantaisia bagshawei S. Moore；巴格肖伴帕爵床■☆

60299 Brillantaisia borellii Lindau = Hygrophila borellii (Lindau) Heine ■☆

60300 Brillantaisia cicatricosa Lindau；疤痕伴帕爵床■☆

60301 Brillantaisia debilis Burkill = Brillantaisia soyauxii Lindau ■☆

60302 Brillantaisia dewevrei De Wild. et T. Durand；德韦伴帕爵床■☆

60303 Brillantaisia eminii Lindau = Brillantaisia lamium (Nees) Benth. ■☆

60304 Brillantaisia grandidentata S. Moore；大齿伴帕爵床■☆

60305 Brillantaisia grottanellii Pic. Serm.；格罗塔伴帕爵床■☆

60306 Brillantaisia lamium (Nees) Benth.；喉花伴帕爵床■☆

60307 Brillantaisia lancifolia Lindau；披针叶伴帕爵床■☆

60308 Brillantaisia leonensis Burkill = Brillantaisia owariensis P. Beauv. ■☆

60309 Brillantaisia madagascariensis T. Anderson ex Lindau；马岛伴帕爵床■☆

60310 Brillantaisia mahonii C. B. Clarke；马洪伴帕爵床■☆

60311 Brillantaisia majestica Wernham = Justicia preussii (Lindau) C. B. Clarke ■☆

60312 Brillantaisia molleri Lindau = Brillantaisia vogeliana (Nees) Benth. ■☆

60313 Brillantaisia nitens Lindau = Brillantaisia owariensis P. Beauv. ■☆

60314 Brillantaisia nyanzarum Burkill = Brillantaisia owariensis P. Beauv. ■☆

60315 Brillantaisia oligantha Milne-Redh.；寡花伴帕爵床■☆

60316 Brillantaisia owariensis P. Beauv.；尾张伴帕爵床■☆

60317 Brillantaisia palisotii Lindau = Brillantaisia lamium (Nees) Benth. ■☆

60318 Brillantaisia patula T. Anderson；伸展伴帕爵床■☆

60319 Brillantaisia patula T. Anderson var. welwitschii Burkill；韦尔伴帕爵床■☆

60320 Brillantaisia preussii Lindau = Brillantaisia vogeliana (Nees) Benth. ■☆

60321 Brillantaisia pubescens T. Anderson ex Oliv.；短柔毛伴帕爵床■☆

60322 Brillantaisia pubescens T. Anderson ex Oliv. var. riparia Vollesen et Brummitt；溪畔伴帕爵床■☆

60323 Brillantaisia pubescens T. Anderson ex Oliv. var. rutenbergiana (Vatke) Benoist = Brillantaisia pubescens T. Anderson ex Oliv. ■☆

60324 Brillantaisia rutenbergiana Vatke = Brillantaisia pubescens T. Anderson ex Oliv. ■☆

60325 Brillantaisia salviiflora Lindau = Brillantaisia owariensis P. Beauv. ■☆

60326 Brillantaisia schumanniana Lindau；舒曼伴帕爵床■☆

60327 Brillantaisia soyauxii Lindau；索亚伴帕爵床■☆

60328 Brillantaisia spicata Lindau；穗状伴帕爵床■☆

60329 Brillantaisia stenopteris Sidwell；狭翅伴帕爵床■☆

60330 Brillantaisia subcordata De Wild. et T. Durand；亚心形伴帕爵床■☆

60331 Brillantaisia subcordata De Wild. et T. Durand var. macrophylla；大叶伴帕爵床■☆

60332 Brillantaisia subulugurica Burkill；拟乌卢古尔伴帕爵床■☆

60333 Brillantaisia talbotii S. Moore = Brillantaisia lancifolia Lindau ■☆

60334 Brillantaisia ulugurica Lindau；乌卢古尔伴帕爵床■☆

60335 Brillantaisia verruculosa Lindau；小疣伴帕爵床■☆

60336 Brillantaisia vogeliana (Nees) Benth.；沃格尔伴帕爵床■☆

60337 Brimeura Salisb. (1866)；钟花风信子属■☆

60338 Brimeura Salisb. = Hyacinthus L. ■☆

60339 Brimeura amethystina (L.) Chouard；钟花风信子；Alpine Hyacinth, Pyrenean Hyacinth, Spanish-hyacinth ■☆

60340 Brimys Scop. = Drimys J. R. Forst. et G. Forst. (保留属名)●☆

60341 Brindonia Thouars = Garcinia L. ●

60342 Brintonia Greene = Solidago L. ■

60343 Brintonia Greene(1895)；一枝白花属■☆

60344 Brintonia discoidea (Elliott) Greene；一枝白花；Rayless Mock Goldenrod ■☆

60345 Briquetastrum Robyns et Lebrun = Leocus A. Chev. ■●☆

60346 Briquetastrum africanum (Baker ex Scott-Elliot) Robyns et Lebrun = Leocus africanus (Baker ex Scott-Elliot) J. K. Morton ■●☆

60347 Briquetia Hochr. (1902)；布里锦葵属■●☆

60348 Briquetia ancylocarpa Hochr.；布里锦葵●☆

60349 Briquetia brasiliensis Fryxell；巴西布里锦葵●☆

60350 Briquetina J. F. Macbr. (1931)；秘鲁茶茱萸属●☆

60351 Briquetina J. F. Macbr. = Citronella D. Don ●☆

60352 Briquetina incarum J. F. Macbr.；秘鲁茶茱萸●☆

60353 Briquetina mollis Sleumer；柔软秘鲁茶茱萸●☆

60354 Brisegnoa J. Rémy = Oxytheca Nutt. ■☆

60355 Briseis Salisb. = Allium L. ■

60356 Brissonia Neck. = Brissonia Neck. ex Desv. ●■

60357 Brissonia Neck. ex Desv. = Indigofera L. + Tephrosia Pers. (保留属名)●■

60358 Brissonia Neck. ex Desv. = Tephrosia Pers. (保留属名)●■

60359 Britoa O. Berg = Campomanesia Ruiz et Pav. ●☆

60360 Brittenia Cogn. (1890)；巴拉圭野牡丹属■☆

60361 Brittenia subacaulis Cogn. ex Boerl.；巴拉圭野牡丹■☆

60362 Brittonamra Kuntze = Cracca Benth. (保留属名)●☆

60363 Brittonastrum Briq. = Agastache J. Clayton ex Gronov. ■

60364 Brittonastrum mexicanum (Kunth) Briq. = Agastache mexicana (Kunth) Lint et Epling ■☆

60365 Brittonella Rusby = Mionandra Griseb. ●☆

60366 Brittonia C. A. Armstr. = Ferocactus Britton et Rose ●

60367 Brittonia Houghton ex C. A. Armstr. = Hamatocactus Britton et Rose +? Thelocactus (K. Schum.) Britton et Rose ●

60368 Brittonia Kuntze = Brissonia Neck. ●■

60369 Brittonia Kuntze = Indigofera L. + Tephrosia Pers. (保留属名)●■

60370 Brittonrosea Speg. = Echinofossulocactus Lawr. ■

60371 Brittonrosea Speg. = Melocactus Link et Otto(保留属名)●

60372 Briza L. (1753)；凌风草属(铃茅属, 银鳞茅属)；Quake Grass, Quakegrass, Quaking Grass, Quaking-grass, Shaking Grass ■

60373 Briza bipinnata L. = Desmostachya bipinnata (L.) Stapf ■

60374 Briza brizoides (Lam.) Kuntze;普通凌风草■☆
60375 Briza capensis Thunb. = Eragrostis capensis (Thunb.) Trin. ■☆
60376 Briza deltoidea Burm. f. = Briza minor L. ■
60377 Briza eragrostis L. = Eragrostis cilianensis (All.) Vignolo ex Janch. ■
60378 Briza geniculata Thunb. = Eragrostis obtusa Munro ex Ficalho et Hiern ■☆
60379 Briza maxima L.;大凌风草(大银鳞草,凌风草,小判草);Big Quakegrass, Big Quaking Grass, Big Quaking-grass, Great Quaking Grass, Greater Quaking Grass, Greater Quaking-grass, Large Quaking Grass, Large Quaking-grass, Large-eaered Quaking Grass, Pearl-grass, Quaking Oats, Rattlesnake Grass ■
60380 Briza maxima L. var. glabriflora Röhl. = Briza maxima L. ■
60381 Briza maxima L. var. pubescens Nicotra = Briza maxima L. ■
60382 Briza media L.;凌风草;Bloody Thumbs, Common Quaking Grass, Common Quaking-grass, Cow Quakers, Cow Quakes, Dawdle-grass, Didder Grass, Diddery Dock, Dillies, Dithering Grass, Dithery Dother, Dodder Grass, Doddering Dickies, Doddering Dillies, Doddering-dillies, Doddle Grass, Dother Grass, Dothering, Dothering Dick, Dothering Dickies, Dothering Dillies, Dothering Dock, Dothering Grass, Dothering Nancy, Dothery, Earth Quakes, Fairy Grass, Hay Shakers, Horses-and-chariot, Horses-and-chariots, Jiggle-joggles, Jockies, Lady's Hair, Lady's Hand, Lady's Hands, Lady's Shakes, Lady's Tresses, Maiden's Hair, Nodding Isabel, Perennial Quakegrass, Perennial Quaking Grass, Perennial Quaking-grass, Quake Grass, Quaker Grass, Quakers, Quaking Grass, Quaking-grass, Rattle-basket, Rattle-grass, Seller Tassels, Shackle Basket, Shackle Box, Shackle-basket, Shackle-box, Shackle-grass, Shadow, Shaeklle Box, Shakers, Shaking Grass, Shaking Shadow, Shaky Grass, Shander-grass, Shekel Basket, Shekel Box, Shickle Shacklers, Shickle-shacklers, Shiver-grass, Shivering Jimmy, Shiver-shakes, Siller Tassel, Silver Ginglers, Silver Shackle, Silver Shackles, Silver Shakers, Silver Shekels, Toddling Grass, Totter Grass, Totter-grass, Tottering Grass, Totty Grass, Trembling Grass, Trembling Jockies, Trembling Jocks, Trembling Shadow, Virgin's Hair, Wag Wafers, Wag Wantons, Waggering Grass, Wagtails, Wag-wafers, Wag-wams, Wag-wanting, Wag-wantons, Wagwants, Wegwants, Wig Wagons, Wiggle Waggles, Wiggle-waggle Grass, Wiggle-waggle Wantons, Wigglewaggles, Wiggle-waggle-wantons, Wiggle-wants, Wigwag Wantons, Wig-wagons, Wigwag-wantons, Wigwams, Wigwants, Wing-wangs, Woman's Tongue ■
60383 Briza minor L.;银鳞茅(小凌风草,银鳞草);Lesser Quaking-grass, Little Quaking Grass, Little Quakinggrass, Little Quaking-grass, Small Quaking-grass ■
60384 Briza multiflora Forst. ex P. Beauv.;多花银鳞草■☆
60385 Briza nigra Burch. ex Steud. = Eragrostis obtusa Munro ex Ficalho et Hiern ■☆
60386 Briza rubella Steud. = Eragrostis turgida (Schumach.) De Wild. ■☆
60387 Briza rufa (C. Presl) Steud.;红凌风草;Red Quakegrass ■☆
60388 Briza spicata Burm. f.;穗状凌风草■☆
60389 Briza spicata Sibth. et Sm. = Briza spicata Burm. f. ■☆
60390 Briza subaristata Lam.;近无芒凌风草;Halfawn Quakegrass ■☆
60391 Briza triloba Nees = Briza subaristata Lam. ■☆
60392 Briza virens L. = Briza minor L. ■
60393 Brizochloa V. Jirásek et Chrtek = Briza L. ■
60394 Brizophila Salisb. = Honorius Gray ■☆
60395 Brizophila Salisb. = Ornithogalum L. ■
60396 Brizopyrum J. Presl = Distichlis Raf. ■☆
60397 Brizopyrum Link = Desmazeria Dumort. ■☆
60398 Brizopyrum Stapf = Tribolium Desv. ■☆
60399 Brizopyrum acutiflorum Nees = Tribolium acutiflorum (Nees) Renvoize ■☆
60400 Brizopyrum acutiflorum Nees var. capillaris = Tribolium acutiflorum (Nees) Renvoize ■☆
60401 Brizopyrum alternans Nees = Tribolium uniolae (L. f.) Renvoize ■☆
60402 Brizopyrum brachystachyum (Nees) Stapf = Tribolium brachystachyum (Nees) Renvoize ■☆
60403 Brizopyrum capense (Trin.) Trin. var. villosum Stapf = Tribolium uniolae (L. f.) Renvoize ■☆
60404 Brizopyrum capense Trin. var. brachystachyum Nees = Tribolium brachystachyum (Nees) Renvoize ■☆
60405 Brizopyrum ciliare Stapf = Tribolium ciliare (Stapf) Renvoize ■☆
60406 Brizopyrum glomeratum (Thunb.) Stapf = Tribolium obtusifolium (Nees) Renvoize ■☆
60407 Brizopyrum mucronatum (L.) Nees = Halopyrum mucronatum (L.) Stapf ■☆
60408 Brizopyrum obliterum (Hemsl.) Stapf = Tribolium obliterum (Hemsl.) Renvoize ■☆
60409 Brizula Hieron. = Aphelia R. Br. ●☆
60410 Brocchia Mauri ex Ten. = Simmondsia Nutt. ●☆
60411 Brocchia Vis. (1836);黏周菊属■☆
60412 Brocchia Vis. = Cotula L. ■
60413 Brocchia cinerea (Delile) Vis.;黏周菊■☆
60414 Brocchia cinerea (Delile) Vis. = Cotula cinerea Delile ■☆
60415 Brocchinia Schult. et Schult. f. = Brocchinia Schult. f. ■☆
60416 Brocchinia Schult. f. (1830);布洛凤梨属(布蕊金属)■☆
60417 Brocchinia acuminata L. B. Sm.;渐尖布洛凤梨■☆
60418 Brocchinia amazonica L. B. Sm.;亚马孙布洛凤梨■☆
60419 Brocchinia cryptantha L. B. Sm.;隐花布洛凤梨■☆
60420 Brocchinia paniculata Schult. f.;布洛凤梨■☆
60421 Brocchoneura Warb. = Brochoneura Warb. ●☆
60422 Brochoneura Warb. (1897);显脉木属●☆
60423 Brochoneura acuminata (Lam.) Warb.;尖显脉木●☆
60424 Brochoneura madagascariensis (Lam.) Warb.;马岛显脉木●☆
60425 Brochoneura usambarensis Warb. = Cephalosphaera usambarensis (Warb.) Warb. ●☆
60426 Brochoneura vouri (Baill.) Warb.;显脉木●☆
60427 Brochosiphon Nees = Dicliptera Juss. (保留属名)■
60428 Brockmania W. Fitzg. (1918);西澳木槿属●☆
60429 Brockmania W. Fitzg. = Hibiscus L. (保留属名)●■
60430 Brockmania membranacea W. Fitzg.;西澳锦葵●☆
60431 Brodiaea Sm. (1810)(保留属名);花韭属(布罗地属,布若地属);Brodiaea, California Hyacinth, Cluster-lily, Pretty Face, Starflower, Wild Hyacinth ■☆
60432 Brodiaea bicolor Suksd. = Triteleia grandiflora Lindl. ■☆
60433 Brodiaea bridgesii S. Watson = Triteleia bridgesii (S. Watson) Greene ■☆
60434 Brodiaea californica (Torr.) Jeps. = Dichelostemma volubile (Kellogg) A. Heller ■☆
60435 Brodiaea californica Jeps. = Brodiaea californica Lindl. ■☆
60436 Brodiaea californica Lindl.;布罗地花韭(布罗地,布罗地石蒜);Brodiaea ■☆

60437 Brodiaea candida (Greene) Baker = Triteleia laxa Benth. ■☆

60438 Brodiaea capitata Benth. = Dichelostemma capitatum (Benth.) A. W. Wood ■☆

60439 Brodiaea capitata Benth. = Dichelostemma pulchellum A. Heller ■☆

60440 Brodiaea capitata Benth. var. insularis (Greene) J. F. Macbr. = Dichelostemma capitatum (Benth.) A. W. Wood ■☆

60441 Brodiaea capitata Benth. var. pauciflora Torr. = Dichelostemma capitatum subsp. pauciflorum (Torr.) Keator ■☆

60442 Brodiaea clementina (Hoover) Munz = Triteleia clementina Hoover ■☆

60443 Brodiaea coccinea A. Gray = Brevoortia idamaia A. Wood ■☆

60444 Brodiaea coccinea A. Gray = Dichelostemma idamaia Greene ■☆

60445 Brodiaea congesta Sm.;密集布罗地;Long-leaved Grass Nut ■☆

60446 Brodiaea congesta Sm. = Dichelostemma congestum Kunth ■☆

60447 Brodiaea coronaria (Salisb.) Engl.;冠花韭;Harvest Brodiaea, Triplet Lily ■☆

60448 Brodiaea coronaria (Salisb.) Engl. subsp. rosea (Greene) T. F. Niehaus;印度冠花韭;Indian valley brodiaea ■☆

60449 Brodiaea coronaria (Salisb.) Jeps. = Brodiaea coronaria (Salisb.) Engl. ■☆

60450 Brodiaea coronaria (Salisb.) Jeps. var. mundula Jeps. = Brodiaea elegans Hoover ■☆

60451 Brodiaea coronaria Engl. = Brodiaea coronaria (Salisb.) Engl. ■☆

60452 Brodiaea coronaria Engl. var. macropoda (Torr.) Hoover = Brodiaea terrestris Kellogg ■☆

60453 Brodiaea crocea (A. W. Wood) S. Watson = Triteleia crocea (A. W. Wood) Greene ■☆

60454 Brodiaea crocea (A. W. Wood) S. Watson var. modesta (H. M. Hall) Munz = Triteleia crocea (A. W. Wood) Greene ■☆

60455 Brodiaea dissimulata M. Peck = Triteleia hyacinthina (Lindl.) Greene ■☆

60456 Brodiaea douglasii S. Watson;道格拉斯氏布罗地■☆

60457 Brodiaea douglasii S. Watson = Triteleia grandiflora Lindl. ■☆

60458 Brodiaea douglasii S. Watson var. howellii (S. Watson) M. Peck = Triteleia grandiflora Lindl. ■☆

60459 Brodiaea dudleyi (Hoover) Munz = Triteleia dudleyi Hoover ■☆

60460 Brodiaea elegans Hoover;雅致布罗地;Elegant Brodiaea, Elegant Cluster-lily, Harvest Brodiaea ■☆

60461 Brodiaea elegans Hoover var. mundula (Jeps.) Hoover = Brodiaea elegans Hoover ■☆

60462 Brodiaea filifolia S. Watson;线叶布罗地■☆

60463 Brodiaea filifolia S. Watson var. orcuttii (Greene) Jeps. = Brodiaea orcuttii (Greene) Baker ■☆

60464 Brodiaea gracilis S. Watson = Triteleia montana Hoover ■☆

60465 Brodiaea grandiflora (Lindl.) J. F. Macbr. = Triteleia grandiflora Lindl. ■☆

60466 Brodiaea grandiflora Sm. = Brodiaea coronaria (Salisb.) Engl. ■☆

60467 Brodiaea grandiflora Sm. = Brodiaea elegans Hoover ■☆

60468 Brodiaea grandiflora Sm. var. brachypoda Torr. = Dichelostemma multiflorum (Benth.) A. Heller ■☆

60469 Brodiaea grandiflora Sm. var. elatior Baker = Brodiaea californica Lindl. ■☆

60470 Brodiaea grandiflora Sm. var. macropoda Torr. = Brodiaea terrestris Kellogg ■☆

60471 Brodiaea grandiflora Sm. var. minor Benth. = Brodiaea minor (Benth.) S. Watson ■☆

60472 Brodiaea hendersonii (Greene) S. Watson = Triteleia hendersonii Greene ■☆

60473 Brodiaea hendersonii Watson;亨氏布罗地■☆

60474 Brodiaea howellii S. Watson = Brodiaea coronaria (Salisb.) Engl. ■☆

60475 Brodiaea howellii S. Watson = Triteleia grandiflora Lindl. ■☆

60476 Brodiaea hyacinthina (Lindl.) Baker = Triteleia hyacinthina (Lindl.) Greene ■☆

60477 Brodiaea hyacinthina (Lindl.) Baker var. greenei (Hoover) Munz = Triteleia lilacina Greene ■☆

60478 Brodiaea hyacinthina (Lindl.) Baker var. lactea (Lindl.) Baker = Triteleia hyacinthina (Lindl.) Greene ■☆

60479 Brodiaea hyacinthina (Lindl.) Baker var. lilacina (S. Watson) Jeps. = Triteleia hyacinthina (Lindl.) Greene ■☆

60480 Brodiaea ida-maia (A. W. Wood) Greene = Dichelostemma idamaia Greene ■☆

60481 Brodiaea insignis (Jeps.) T. F. Niehaus;显著布罗地;Firecracker Plant, Kaweah Brodiaea ■☆

60482 Brodiaea insularis Greene = Dichelostemma capitatum (Benth.) A. W. Wood ■☆

60483 Brodiaea ixioides (W. T. Aiton) S. Watson = Triteleia ixioides (W. T. Aiton) Greene ■☆

60484 Brodiaea ixioides (W. T. Aiton) S. Watson var. lugens (Greene) Jeps. = Triteleia lugens Greene ■☆

60485 Brodiaea ixioides (W. T. Aiton) S. Watson var. scabra (Greene) Smiley = Triteleia ixioides (W. T. Aiton) Greene subsp. scabra (Greene) L. W. Lenz ■☆

60486 Brodiaea jolonensis Eastw.;丛林布罗地;Chaparral Cluster-lily, Mesa Brodiaea ■☆

60487 Brodiaea kinkiensis T. F. Niehaus;近畿布罗地;San Clemente Island Brodiaea ■☆

60488 Brodiaea lactea (Lindl.) S. Watson = Triteleia hyacinthina (Lindl.) Greene ■☆

60489 Brodiaea lactea (Lindl.) S. Watson var. lilacina S. Watson = Triteleia hyacinthina (Lindl.) Greene ■☆

60490 Brodiaea lactea S. Watson = Triteleia hyacinthina (Lindl.) Greene ■☆

60491 Brodiaea laxa (Benth.) S. Watson = Triteleia laxa Benth. ■☆

60492 Brodiaea laxa (Benth.) S. Watson var. candida (Greene) Jeps. = Triteleia laxa Benth. ■☆

60493 Brodiaea laxa (Benth.) S. Watson var. nimia Jeps. = Triteleia laxa Benth. ■☆

60494 Brodiaea laxa (Benth.) S. Watson var. tracyi Jeps. = Triteleia laxa Benth. ■☆

60495 Brodiaea laxa S. Watson = Triteleia laxa Benth. ■☆

60496 Brodiaea leachiae M. Peck = Triteleia hendersonii Greene ■☆

60497 Brodiaea lemmoniae S. Watson = Triteleia lemmoniae (S. Watson) Greene ■☆

60498 Brodiaea lilacina (Greene) Baker = Triteleia lilacina Greene ■☆

60499 Brodiaea lugens (Greene) Baker = Triteleia lugens Greene ■☆

60500 Brodiaea lutea (Lindl.) C. V. Morton = Triteleia ixioides (W. T. Aiton) Greene ■☆

60501 Brodiaea lutea (Lindl.) C. V. Morton var. anilina (Greene) Munz = Triteleia ixioides (W. T. Aiton) Greene subsp. anilina (Greene) L. W. Lenz ■☆

60502 Brodiaea lutea (Lindl.) C. V. Morton var. cookii (Hoover) Munz = Triteleia ixioides (W. T. Aiton) Greene subsp. cookii (Hoover) L. W. Lenz ■☆

60503 Brodiaea lutea (Lindl.) C. V. Morton var. lugens (Greene) C. V. Morton = Triteleia lugens Greene ■☆
60504 Brodiaea lutea (Lindl.) C. V. Morton var. scabra (Greene) Munz = Triteleia ixioides (W. T. Aiton) Greene subsp. scabra (Greene) L. W. Lenz ■☆
60505 Brodiaea minor (Benth.) S. Watson;矮小布罗地;Dwarf Brodiaea ■☆
60506 Brodiaea minor (Benth.) S. Watson var. nana (Hoover) Hoover = Brodiaea minor (Benth.) S. Watson ■☆
60507 Brodiaea modesta H. M. Hall = Triteleia crocea (A. W. Wood) Greene ■☆
60508 Brodiaea multiflora Benth.;多花布罗地■☆
60509 Brodiaea multiflora Benth. = Dichelostemma multiflorum (Benth.) A. Heller ■☆
60510 Brodiaea nana Hoover = Brodiaea minor (Benth.) S. Watson ■☆
60511 Brodiaea orcuttii (Greene) Baker;奥科特布罗地;Orcutt's Brodiaea ■☆
60512 Brodiaea pallida Hoover;苍白布罗地;Chinese Camp Brodiaea ■☆
60513 Brodiaea palmeri S. Watson = Triteleiopsis palmeri (S. Watson) Hoover ■☆
60514 Brodiaea parviflora Torr. et A. Gray = Dichelostemma multiflorum (Benth.) A. Heller ■☆
60515 Brodiaea peduncularis (Lindl.) S. Watson = Triteleia peduncularis Lindl. ■☆
60516 Brodiaea pulchella (Salisb.) Greene;美丽布罗地;Ookow ■☆
60517 Brodiaea pulchella (Salisb.) Greene var. pauciflora (Torr.) C. V. Morton = Dichelostemma capitatum subsp. pauciflorum (Torr.) Keator ■☆
60518 Brodiaea purdyi Eastw. = Brodiaea purdyi S. Watson ■☆
60519 Brodiaea purdyi S. Watson;珀迪布罗地;Purdy's Brodiaea, Sierran Cluster-lily ■☆
60520 Brodiaea scabra (Greene) Baker = Triteleia ixioides (W. T. Aiton) Greene subsp. scabra (Greene) L. W. Lenz ■☆
60521 Brodiaea scabra (Greene) Baker var. anilina (Greene) Baker = Triteleia ixioides (W. T. Aiton) Greene subsp. anilina (Greene) L. W. Lenz ■☆
60522 Brodiaea stellaris S. Watson;星状布罗地;Star Brodiaea, Star-flower Cluster-lily ■☆
60523 Brodiaea synandra (A. Heller) Jeps. = Brodiaea coronaria (Salisb.) Engl. ■☆
60524 Brodiaea synandra (A. Heller) Jeps. var. insignis Jeps. = Brodiaea insignis (Jeps.) T. F. Niehaus ■☆
60525 Brodiaea terrestris Kellogg;陆生布罗地;Dwarf Brodiaea ■☆
60526 Brodiaea uniflora Engl.;单花布罗地;Oneflower Brodiaea, Spring Star-flower ■☆
60527 Brodiaea uniflora Engl. = Tristagma uniflorum (Lindl.) Traub ■☆
60528 Brodiaea venusta (Greene) Greene = Dichelostemma idamaia Greene ■☆
60529 Brodiaea volubilis (Kellogg) Baker = Dichelostemma volubile (Kellogg) A. Heller ■☆
60530 Brodriguesia R. S. Cowan(1981);飘柔丝蕊豆属■☆
60531 Brodriguesia santosii R. S. Cowan;飘柔丝蕊豆■☆
60532 Bromaceae Bercht. et J. Presl = Gramineae Juss.(保留科名)■●
60533 Bromaceae Bercht. et J. Presl = Poaceae Barnhart(保留科名)■●
60534 Bromaceae Burnett = Gramineae Juss.(保留科名)■●
60535 Bromaceae Burnett = Poaceae Barnhart(保留科名)■●
60536 Bromaceae Burnett = Sterculiaceae Vent.(保留科名)●■
60537 Bromaceae K. Koch = Gramineae Juss.(保留科名)■●
60538 Bromaceae K. Koch = Poaceae Barnhart(保留科名)■●
60539 Brombya F. Muell.(1865);澳东北芸香属●☆
60540 Brombya F. Muell. = Melicope J. R. Forst. et G. Forst. ●
60541 Brombya platynema F. Muell.;澳东北芸香●☆
60542 Bromelia Adans. = Bromelia L. ■☆
60543 Bromelia Adans. = Pitcairnia L'Hér.(保留属名)■☆
60544 Bromelia L.(1753);红心凤梨属(菠萝属,布洛美属,布诺美丽亚属,凤梨属,观赏凤梨属,美凤梨属,强刺凤梨属,强刺属,野凤梨属,真凤梨属);Bromelia ■☆
60545 Bromelia ananas L. = Ananas comosus (L.) Merr. ■
60546 Bromelia balansae Mez;红心凤梨;Heart of Flame, Heart-of-flame ■☆
60547 Bromelia carolinae Beer = Neoregelia carolinae (Beer) L. B. Sm. ■☆
60548 Bromelia chrysantha Jacq.;黄花红心凤梨■☆
60549 Bromelia comosa L. = Ananas comosus (L.) Merr. ■
60550 Bromelia humilis Jacq.;矮红心凤梨■☆
60551 Bromelia karatas L.;野凤梨■☆
60552 Bromelia pauciflora K. Koch;疏花心凤梨■☆
60553 Bromelia pinguin L.;肥美凤梨;Pinguin, Pinguin Fibre ■☆
60554 Bromelia plumieri (E. Morren) L. B. Sm.;普吕米凤梨■☆
60555 Bromelia serra Griseb. 'Variegata';杂色凤梨■☆
60556 Bromelia striata? = Schizachne purpurascens (Torr.) Swallen ■☆
60557 Bromeliaceae Juss.(1789)(保留科名);凤梨科(菠萝科);Bromelia Family, Bromeliad Family, Pineapple Family, Rhodostachys Family ■
60558 Bromelica (Thurb.) Farw. = Melica L. ■
60559 Bromelica Farw. = Melica L. ■
60560 Bromelica smithii (Porter ex A. Gray) Farw. = Melica smithii (Porter ex A. Gray) Vasey ■☆
60561 Bromfeldia Neck. = Jatropha L.(保留属名)●■
60562 Bromheadia Lindl.(1841);布隆兰属(布氏兰属)■☆
60563 Bromheadia finlaysoniana Rchb. f.;芬莱布隆兰(芬莱布氏兰)■☆
60564 Bromidium Nees et Meyen = Agrostis L.(保留属名)■
60565 Bromidium Nees et Meyen = Deyeuxia Clarion ■
60566 Bromopsis (Dumort.) Fourr.(1869);小雀麦属;Brome, Brome Grass ■
60567 Bromopsis (Dumort.) Fourr. = Bromus L.(保留属名)■
60568 Bromopsis Fourr. = Bromus L.(保留属名)■
60569 Bromopsis angrenica (Drobow) Holub = Bromus paulsenii Hack. ex Paulsen ■
60570 Bromopsis atlanticus (H. Lindb.) Holub = Bromopsis benekenii (Lange) Holub ■
60571 Bromopsis benekenii (Lange) Holub = Bromus benekenii (Lange) Trimen ■
60572 Bromopsis canadensis (Michx.) Holub = Bromus canadensis Michx. ■
60573 Bromopsis canadensis (Michx.) Holub = Bromus ciliatus L. ■
60574 Bromopsis ciliata (L.) Holub = Bromus ciliatus L. ■
60575 Bromopsis confinis (Nees ex Steud.) Holub = Bromus confinis Nees ex Steud. ■
60576 Bromopsis erecta (Huds.) Fourr. = Bromus erectus Huds. ■
60577 Bromopsis erecta (Huds.) Fourr. subsp. permixta (H. Lindb.) H. Scholz et Valdés;混乱小雀麦■☆
60578 Bromopsis erecta Fourr. = Bromus erectus Huds. ■
60579 Bromopsis erecta Fourr. subsp. microchaeta (Font Quer) H.

Scholz et Valdés;微毛小雀麦■☆

60580 Bromopsis erecta Fourr. subsp. permixta (H. Lindb.) H. Scholz et Valdés = Bromopsis erecta (Huds.) Fourr. subsp. permixta (H. Lindb.) H. Scholz et Valdés ■☆

60581 Bromopsis himalaica (Stapf) Holub = Bromus himalaicus Stapf ■

60582 Bromopsis inermis (Leyss.) Holub = Bromus inermis Leyss. ■

60583 Bromopsis inermis (Leyss.) Holub var. malzevii (Drobow) Tzvelev = Bromus inermis Leyss. var. malzevii Drobow ■

60584 Bromopsis kalmii (A. Gray) Holub = Bromus kalmii A. Gray ■☆

60585 Bromopsis korotkiji (Drobow) Holub = Bromus korotkiji Drobow ■

60586 Bromopsis maroccana (Pau et Font Quer) Holub;摩洛哥小雀麦■☆

60587 Bromopsis microchaeta (Font Quer) Holub = Bromopsis erecta Fourr. subsp. microchaeta (Font Quer) H. Scholz et Valdés ■☆

60588 Bromopsis pamirica (Drobow) Holub = Bromus paulsenii Hack. ex Paulsen ■

60589 Bromopsis paulsenii (Hack. ex Paulsen) Holub = Bromus paulsenii Hack. ex Paulsen ■

60590 Bromopsis paulsenii (Hack. ex Paulsen) Holub subsp. angrenica (Drobow) Tzvelev = Bromus angrenicus Drobow ■

60591 Bromopsis paulsenii (Hack. ex Paulsen) Holub subsp. angrenica (Drobow) Tzvelev = Bromus paulsenii Hack. ex Paulsen ■

60592 Bromopsis paulsenii (Hack. ex Paulsen) Holub subsp. pamirica (Drobow) Tzvelev = Bromus paulsenii Hack. ex Paulsen ■

60593 Bromopsis paulsenii (Hack. ex Paulsen) Holub subsp. turkestanica (Drobow) Tzvelev = Bromus paulsenii Hack. ex Paulsen ■

60594 Bromopsis permixta (H. Lindb.) Holub = Bromopsis erecta Fourr. subsp. permixta (H. Lindb.) H. Scholz et Valdés ■☆

60595 Bromopsis pubescens (Muhl. ex Willd.) Holub = Bromus pubescens Muhl. ex Willd. ■☆

60596 Bromopsis pumpelliana (Scribn.) Holub = Bromopsis pumpelliana Scribn. ■

60597 Bromopsis pumpelliana (Scribn.) Holub = Bromus pumpellianus Scribn. ■

60598 Bromopsis pumpelliana (Scribn.) Holub subsp. arctica (Shear ex Scribn. et Merr.) Tzvelev = Bromus arcticus Shear ex Scribn. et Merr. ■☆

60599 Bromopsis pumpelliana Scribn.;耐酸草(缘毛雀麦);Fringed Brome,Richardson Fringed Brome ■

60600 Bromopsis pumpelliana Scribn. subsp. korotkiji (Drobow) Tzvelev = Bromus korotkiji Drobow ■

60601 Bromopsis pumpelliana Scribn. subsp. korotkiji (Drobow) Tzvelev var. ircutensis (Kom.) Tzvelev = Bromus irkutensis Kom. ■

60602 Bromopsis ramosa (Huds.) Holub = Bromus ramosus Huds. ■

60603 Bromopsis riparia (Rehmann) Holub = Bromus riparius Rehmann ■

60604 Bromopsis stenostachya (Boiss.) Holub = Bromus stenostachyus Boiss. ■

60605 Bromopsis tomentella (Boiss.) Holub subsp. cappadocica (Boiss. et Balansa) Tzvelev = Bromus cappadocicus Boiss. et Balansa ■

60606 Bromopsis turkestanica (Drobow) Holub = Bromus paulsenii Hack. ex Paulsen ■

60607 Bromopsis tyttholepis (Nevski) Holub = Bromus tyttholepis Nevski ■

60608 Bromopsis variegata (M. Bieb.) Holub = Bromus variegatus M. Bieb. ■

60609 Bromuniola Stapf et C. E. Hubb. (1926);安哥拉禾属(雀麦草属)■☆

60610 Bromuniola gossweileri Stapf et C. E. Hubb.;安哥拉禾■☆

60611 Bromus Dill. ex L. = Bromus L.(保留属名)■

60612 Bromus L.(1753)(保留属名);雀麦属;Brome,Brome Grass,Bromegrass,Brome-grass,Bromus,Cheat,Chess ■

60613 Bromus Scop. = Triticum L. ■

60614 Bromus abolinii Drobow = Bromus japonicus Thunb. ■

60615 Bromus abortiflorus St. -Amans = Bromus tectorum L. ■

60616 Bromus adoensis Steud. = Bromus pectinatus Thunb. ■

60617 Bromus aegyptiacus Tausch;埃及雀麦■☆

60618 Bromus alaicus Korsh. = Littledalea alaica (Korsh.) Petr. ex Kom. ■

60619 Bromus alaicus Korsh. = Littledalea alaica (Korsh.) Petr. ex Nevski ■

60620 Bromus aleutensis Trin.;阿留申雀麦;Aleutian Brome ■

60621 Bromus alopecuros Poir.;看麦娘雀麦;Alopecurus Brome,Weedy Brome ■

60622 Bromus alopecuros Poir. var. calvus Halácsy = Bromus alopecuros Poir. ■

60623 Bromus alopecuros Poir. var. poiretianus Maire et Weiller = Bromus alopecuros Poir. ■

60624 Bromus altissimus Pursh;耳叶雀麦;Ear-leaved Brome ■☆

60625 Bromus altissimus Pursh = Bromus latiglumis (Shear) Hitchc. ■☆

60626 Bromus anatolicus Boiss. et Heldr.;小亚细亚雀麦■☆

60627 Bromus anatolicus Boiss. et Heldr. = Bromus japonicus Thunb. ■

60628 Bromus angrenicus Drobow;安格雀麦■

60629 Bromus angrenicus Drobow = Bromus paulsenii Hack. ex Paulsen ■

60630 Bromus annuus Jacq. ex Stapf = Bromus japonicus Thunb. ■

60631 Bromus arcticus Shear ex Scribn. et Merr.;北极雀麦■☆

60632 Bromus arenarius Labill.;澳大利亚雀麦;Australian Brome ■☆

60633 Bromus arundinaceus (Schreb.) Roth. = Festuca arundinacea Schreb. ■

60634 Bromus arvensis L.;野雀麦(田雀麦);Field Brome,Field Brome Gras,Field Brome-grass ■

60635 Bromus arvensis L. var. phragmitoides (A. Nyar.) Borza = Bromus arvensis L. ■

60636 Bromus arvensis L. var. phragmitoides (A. Nyar.) Borza = Bromus phragmitoides A. Nyar. ■

60637 Bromus arvensis L. var. racemosus (L.) Neilr. = Bromus racemosus L. ■

60638 Bromus asper Murray = Bromus ramosus Huds. ■

60639 Bromus asper Murray var. angustifolia = Bromus ramosus Huds. ■

60640 Bromus asper Murray var. benekenii (Lange) Syme = Bromus benekenii (Lange) Trimen ■

60641 Bromus asper Murray var. depauperata? = Bromus ramosus Huds. ■

60642 Bromus atlanticus H. Lindb. = Bromopsis benekenii (Lange) Holub ■

60643 Bromus benekenii (Lange) Trimen;本氏雀麦(毕尼氏雀麦,密丛雀麦);Beneken Brome,Lesser Hairy-brome ■

60644 Bromus benekenii (Lange) Trimen = Bromopsis benekenii (Lange) Holub ■

60645 Bromus berterianus Colla;智利雀麦;Chilean Chess ■☆

60646 Bromus biebersteinii Roem. et Schult.;毕氏雀麦■☆

60647 Bromus bifidus Thunb. = Trisetum bifidum (Thunb.) Ohwi ■

60648 Bromus bornmulleri Hausskn. = Bromus gracillimus Bunge ■

60649 Bromus brachystachys Hornung;短轴雀麦;Shortspike Brome ■

60650 Bromus braunii Sennen = Bromus hordeaceus L. ■
60651 Bromus breviaristatus Buckley = Bromus marginatus Nees ex Steud. var. breviaristatus (Buckley) Beetle ■☆
60652 Bromus brevis Nees ex Steud. = Ceratochloa brevis (Nees ex Steud.) B. D. Jacks. ■☆
60653 Bromus briziformis Fisch. et C. A. Mey.;凌风草叶雀麦;Quake-grass, Rattlegrass, Rattlesnake Brome, Rattlesnake Chess ■☆
60654 Bromus canadensis Michx.;加拿大雀麦;Canada Brome ■
60655 Bromus canadensis Michx. = Bromus ciliatus L. ■
60656 Bromus canadensis Michx. subsp. yezoensis (Ohwi) Vorosch. = Bromus canadensis Michx. ■
60657 Bromus canescens Viv. = Anisantha rubens (L.) Nevski ■
60658 Bromus cappadocicus Boiss. et Balansa;卡帕雀麦(卡帕多细亚雀麦)■
60659 Bromus carinatus Hook. et Arn.;龙骨状雀麦(显脊雀麦);Californian Brome, Keel Brome ■
60660 Bromus caroli-henrici Greuter;卡罗尔雀麦■☆
60661 Bromus catharticus Vahl;扁穗雀麦(大扁雀麦);Rescue Grass ■
60662 Bromus catharticus Vahl = Bromus unioloides (Willd.) Kunth ■
60663 Bromus catharticus Vahl = Bromus willdenowii Kunth ■
60664 Bromus catharticus Vahl = Ceratochloa cathartica (Vahl) Herter ■☆
60665 Bromus chitralensis Melderis = Bromus ramosus Huds. ■
60666 Bromus chrysopogon Viv.;金芒雀麦■☆
60667 Bromus ciliatus L.;缘毛雀麦(加拿大雀麦);Ciliate Brome-grass, Fringed Brome ■
60668 Bromus ciliatus L. = Bromopsis pumpelliana Scribn. ■
60669 Bromus ciliatus L. = Bromus inermis Leyss. ■
60670 Bromus ciliatus L. f. denudatus Wiegand = Bromus ciliatus L. ■
60671 Bromus ciliatus L. sensu Baum = Bromus altissimus Pursh ■☆
60672 Bromus ciliatus L. var. denudatus (Wiegand) Fernald = Bromus ciliatus L. ■
60673 Bromus ciliatus L. var. genuinus Fernald = Bromus ciliatus L. ■
60674 Bromus ciliatus L. var. intonsus Fernald = Bromus ciliatus L. ■
60675 Bromus ciliatus L. var. laeviglumis Scribn. ex Shear = Bromus kalmii A. Gray ■☆
60676 Bromus ciliatus L. var. richardsonii (Link) Y. C. Jiang = Bromopsis pumpelliana Scribn. ■
60677 Bromus cincinnatus Ten. = Helictotrichon cincinnatum (Ten.) Röser ■☆
60678 Bromus cognatus Steud. = Bromus leptoclados Nees ■☆
60679 Bromus coloratus Steud.;有色雀麦;Color Brome ■
60680 Bromus commutatus Schrad.;变雀麦;European Brome, Hairy Brome, Hairy Chess, Meadow Brome, Varied Brome ■
60681 Bromus commutatus Schrad. = Bromus racemosus L. ■
60682 Bromus commutatus Schrad. subsp. neglectus (Parl.) P. M. Sm.;忽视雀麦■☆
60683 Bromus commutatus Schrad. var. apricorum Simonk. = Bromus commutatus Schrad. ■
60684 Bromus commutatus Schrad. var. neglectus (Parl.) Trab. = Bromus commutatus Schrad. subsp. neglectus (Parl.) P. M. Sm. ■☆
60685 Bromus commutatus Schrad. var. villosus Batt. et Trab. = Bromus commutatus Schrad. ■
60686 Bromus confertus M. Bieb. = Bromus scoparius L. ■
60687 Bromus confinis Nees ex Steud.;毗邻雀麦;Confined Brome ■
60688 Bromus contortus Desf. = Bromus alopecuros Poir. ■
60689 Bromus crinitus Boiss. et Hohen. = Bromus gracillimus Bunge ■
60690 Bromus cristatum L. = Agropyron cristatum (L.) Gaertn. ■
60691 Bromus danthoniae Trin. = Bromus danthoniae Trin. ex C. A. Mey. ■
60692 Bromus danthoniae Trin. ex C. A. Mey.;三芒雀麦(但陶雀麦)■
60693 Bromus danthoniae Trin. var. lanuginosus Roshev. = Bromus danthoniae Trin. ex C. A. Mey. ■
60694 Bromus degenii Pénzes = Bromus scoparius L. ■
60695 Bromus dertonensis All. = Vulpia bromoides (L.) Gray ■☆
60696 Bromus diandrus Roth;两雄雀麦;Great Brome, Ripgut Brome, Ripgut Grass, Twostamens Brome ■
60697 Bromus diandrus Roth = Anisantha diandra (Roth) Tzvelev ■
60698 Bromus diandrus Roth subsp. maximus (Desf.) Soó = Anisantha diandra (Roth) Tzvelev ■
60699 Bromus diandrus Roth subsp. rigidus (Roth) O. Boiss., Masalles et Vigo = Bromus rigidus Roth ■
60700 Bromus diandrus Roth subsp. rigidus (Roth) Sales;硬两雄雀麦;Ripgut Brome, Ripgut Grass ■☆
60701 Bromus diandrus Roth var. rigidus (Roth) Sales = Bromus rigidus Roth ■
60702 Bromus dilatatus Poir. = Anisantha rubens (L.) Nevski ■
60703 Bromus distachyos L. = Brachypodium distachyon (L.) P. Beauv. ■
60704 Bromus dudleyi Fernald;北美雀麦;Dudley's Brome-grass ■☆
60705 Bromus dudleyi Fernald = Bromus ciliatus L. ■
60706 Bromus epilis P. C. Keng;光稃雀麦;Smooth Brome ■
60707 Bromus erectus Huds.;直立雀麦(直梗雀麦);Erect Brome, European Brome, Meadow Brome, Upright Brome, Upright Brome Grass, Upright Brome-grass ■
60708 Bromus erectus Huds. = Bromopsis erecta Fourr. ■
60709 Bromus erectus Huds. = Bromopsis riparia (Rehmann) Holub ■
60710 Bromus erectus Huds. = Bromus riparius Rehmann ■
60711 Bromus erectus Huds. subsp. microchaetus (Font Quer) Maire et Weiller = Bromopsis erecta Fourr. subsp. microchaeta (Font Quer) H. Scholz et Valdés ■☆
60712 Bromus erectus Huds. subsp. permixtus H. Lindb. = Bromopsis erecta Fourr. subsp. permixta (H. Lindb.) H. Scholz et Valdés ■☆
60713 Bromus erectus Huds. var. arvensis (L.) Huds. = Bromus arvensis L. ■
60714 Bromus erectus Huds. var. embergeri Maire = Bromopsis erecta Fourr. ■
60715 Bromus erectus Huds. var. glabratus Maire et Weiller = Bromopsis erecta Fourr. ■
60716 Bromus erectus Huds. var. longispiculatus Maire = Bromopsis erecta Fourr. ■
60717 Bromus erectus Huds. var. pubescens Maire et Weiller = Bromopsis erecta Fourr. ■
60718 Bromus fasciculatus C. Presl;束生雀麦;Fasciculate Brome ■
60719 Bromus fasciculatus C. Presl = Anisantha fasciculata (C. Presl) Nevski ■
60720 Bromus fasciculatus C. Presl subsp. delilei (Boiss.) H. Scholz = Anisantha fasciculata (C. Presl) Nevski subsp. delilei (Boiss.) H. Scholz et Valdés ■☆
60721 Bromus fasciculatus C. Presl var. tenuiflorus (Viv.) Bég. et Vacc. = Anisantha fasciculata (C. Presl) Nevski ■
60722 Bromus firmior (Nees) Stapf;坚实雀麦■☆
60723 Bromus firmior (Nees) Stapf var. leiorhachis Stapf = Bromus firmior (Nees) Stapf ■☆

60724 Bromus formosanus Honda;台湾雀麦;Taiwan Brome, Taiwan Bromegrass ■
60725 Bromus garamas Maire = Bromus pectinatus Thunb. ■
60726 Bromus gedrosianus Pénzes;直芒雀麦■
60727 Bromus gedrosianus Pénzes = Bromus pectinatus Thunb. ■
60728 Bromus giganteus L. = Festuca gigantea (L.) Vill. ■
60729 Bromus glomeratus Tausch. = Bromus hordeaceus L. ■
60730 Bromus gracilis Krösche = Bromus lepidus E. Holmb. ■
60731 Bromus gracilis Popov;纤细雀麦■
60732 Bromus gracilis Popov = Bromus tytthanthus Nevski ■
60733 Bromus gracilis Weigel = Brachypodium sylvaticum (Huds.) P. Beauv. ■
60734 Bromus gracilis Weigel = Brachypodium sylvaticum (Huds.) P. Beauv. var. gracile (Weigel) Keng ■
60735 Bromus gracillimus Bunge;细雀麦;Thin Brome ■
60736 Bromus gracillimus Bunge = Nevskiella gracillima (Bunge) V. I. Krecz. et Vved. ■
60737 Bromus grandis (Stapf) Melderis;大花雀麦;Big-flowered Brome ■
60738 Bromus grandis (Stapf) Melderis = Bromus porphyranthos Cope ■
60739 Bromus grossus Desf. ex DC. et A. Camus;粗大雀麦(粗雀麦);Gross Brome, Whiskered Brome ■
60740 Bromus gussonii Parl. var. rigidus (Roth) H. Lindb. = Bromus rigidus Roth ■
60741 Bromus himalaicus Stapf;喜马拉雅雀麦(藏雀麦);Himalayan Brome, Himalayas Bromegrass ■
60742 Bromus himalaicus Stapf var. grandis Stapf = Bromus grandis (Stapf) Melderis ■
60743 Bromus himalaicus Stapf var. grandis Stapf = Bromus porphyranthos Cope ■
60744 Bromus hirtus Licht. ex Roem. et Schult.;多毛雀麦■☆
60745 Bromus hordeaceus L.;毛雀麦(大麦状雀麦);Bald Brome, Barleyshaped Brome, Lopgrass, Soft Brome, Soft Chess ■
60746 Bromus hordeaceus L. subsp. divaricatus (Bonnier et Layens) Kerguélen = Bromus hordeaceus L. subsp. molliformis (J. Lloyd) Maire et A. Weiller ■☆
60747 Bromus hordeaceus L. subsp. lepidus (Holmb.) A. Pedersen = Bromus lepidus E. Holmb. ■
60748 Bromus hordeaceus L. subsp. molliformis (J. Lloyd) Maire et A. Weiller;柔软雀麦;Soft Brome ■☆
60749 Bromus hordeaceus L. subsp. mollis (L.) Hyl. = Bromus hordeaceus L. ■
60750 Bromus hordeaceus L. subsp. mollis (L.) Maire et Weiller = Bromus hordeaceus L. ■
60751 Bromus hordeaceus L. var. contractus (Lange) Asch. et Graebn. = Bromus hordeaceus L. ■
60752 Bromus hordeaceus L. var. intermedius (Guss.) Shear = Bromus intermedius Guss. ■
60753 Bromus hordeaceus L. var. laeviculmis Maire = Bromus hordeaceus L. ■
60754 Bromus hordeaceus L. var. leiostachys Hartm. = Bromus hordeaceus L. ■
60755 Bromus hordeaceus L. var. molliformis (Lloyd) Halácsy = Bromus hordeaceus L. subsp. molliformis (J. Lloyd) Maire et A. Weiller ■☆
60756 Bromus hordeaceus L. var. ramosus (Ball) Maire = Bromus hordeaceus L. ■
60757 Bromus hordeaceus L. var. tunetanus Hack. = Bromus hordeaceus L. subsp. molliformis (J. Lloyd) Maire et A. Weiller ■☆
60758 Bromus incrassatus Lam. = Vulpia geniculata (L.) Link ■☆
60759 Bromus inermis (Leyss.) Holub subsp. pumpellianus (Scribn.) Wagnon = Bromus pumpellianus Scribn. ■
60760 Bromus inermis Leyss.;无芒雀麦;Awnless Brome, Awnless Bromegrass, Awnless Brome-grass, Hungarian Brome, Smooth Brome, Smooth Brome Grass ■
60761 Bromus inermis Leyss. f. aristatus (Schur) Fernald = Bromus inermis Leyss. ■
60762 Bromus inermis Leyss. f. villosus (Mert. et W. D. J. Koch) Fernald = Bromus inermis Leyss. ■
60763 Bromus inermis Leyss. subsp. pumpellianus (Scribn.) Wagnon = Bromopsis pumpelliana Scribn. ■
60764 Bromus inermis Leyss. subsp. pumpellianus (Scribn.) Wagnon = Bromus pumpellianus Scribn. ■
60765 Bromus inermis Leyss. var. confinis (Nees ex Steud.) Stapf = Bromus confinis Nees ex Steud. ■
60766 Bromus inermis Leyss. var. divaricatus?;光滑雀麦;Smooth Brome ■☆
60767 Bromus inermis Leyss. var. longiflorus Keng;长花雀麦;Longflower Awnless Bromegrass, Longflower Brome ■
60768 Bromus inermis Leyss. var. malzevii Drobow;短枝雀麦;Malzev Awnless Bromegrass ■
60769 Bromus inermis Leyss. var. sibiricus (Drobow) Krylov. = Bromus sibiricus Drobow ■
60770 Bromus inermis Leyss. var. sibiricus Krylov. = Bromus sibiricus Drobow ■
60771 Bromus inermis Leyss. var. villosus (Mert. et Koch) Beck = Bromus inermis Leyss. ■
60772 Bromus inopinatus C. T. Brues et B. B. Brues = Bromus inermis Leyss. ■
60773 Bromus intermedius Guss.;中间雀麦;Intermediate Brome ■
60774 Bromus interruptus (Hack.) Druce;间断雀麦;Interrupted Brome ■☆
60775 Bromus ircutensis Kom. = Bromus korotkiji Drobow ■
60776 Bromus irkutensis Kom.;沙地雀麦(伊尔库特雀麦);Irkut Brome ■
60777 Bromus japonicus Thunb.;雀麦(杜姥草,爵麦,瞌睡草,蘥,牡姓草,牛星草,山大麦,山稷子,午星草,燕麦,野大麦,野麦,野梅签,野小麦,野燕麦,蘥);Japan Bromegrass, Japanese Brome, Japanese Chess ■
60778 Bromus japonicus Thunb. ex Murray = Bromus japonicus Thunb. ■
60779 Bromus japonicus Thunb. ex Murray var. porrectus Hack. = Bromus japonicus Thunb. ■
60780 Bromus japonicus Thunb. subsp. anatolicus (Boiss. et Heldr.) Pénzes = Bromus japonicus Thunb. ■
60781 Bromus japonicus Thunb. subsp. sinaicus Hack. = Bromus pectinatus Thunb. ■
60782 Bromus japonicus Thunb. var. acutidens Melderis = Bromus gedrosianus Pénzes ■
60783 Bromus japonicus Thunb. var. acutidens Melderis = Bromus pectinatus Thunb. ■
60784 Bromus japonicus Thunb. var. acutidens Melderis = Bromus rechingeri Melderis ex Bor ■
60785 Bromus japonicus Thunb. var. falconeri (Stapf) R. R. Stewart = Bromus pectinatus Thunb. ■
60786 Bromus japonicus Thunb. var. pectinatus (Thunb.) Asch. =

Bromus pectinatus Thunb. ■

60787 Bromus japonicus Thunb. var. pectinatus (Thunb.) Asch. et Graebn. = Bromus pectinatus Thunb. ■

60788 Bromus japonicus Thunb. var. porrectus Hack. = Bromus japonicus Thunb. ■

60789 Bromus japonicus Thunb. var. sinaicus Hack. = Bromus pectinatus Thunb. ■

60790 Bromus japonicus Thunb. var. subsquarrosus (Borbás) Savul. et Rayss = Bromus japonicus Thunb. ■

60791 Bromus japonicus Thunb. var. velutinus (Koch) Bornm. = Bromus japonicus Thunb. ■

60792 Bromus japonicus Thunb. var. velutinus Asch. et Graebn. = Bromus pectinatus Thunb. ■

60793 Bromus kalmii A. Gray;耳雀麦叶;Arctic Brome, Ear-leaved Brome, Kalm's Brome, Prairie Brome ■☆

60794 Bromus kerkeranus Sennen et Mauricio = Anisantha rubens (L.) Nevski ■

60795 Bromus kopetdagensis Drobow;科佩特雀麦■☆

60796 Bromus korotkiji Drobow;甘蒙雀麦■

60797 Bromus lanceolatus Roth;大穗雀麦(披针形雀麦);Lanceolate Brome, Lanceolate Bromegrass, Mediterranean Brome ■

60798 Bromus lanceolatus Roth subsp. intermedius (Guss.) Lloret = Bromus intermedius Guss. ■

60799 Bromus lanceolatus Roth subsp. macrostachys (Desf.) Maire = Bromus lanceolatus Roth ■

60800 Bromus lanceolatus Roth subsp. oxyodon (Schrenk) Tzvelev = Bromus oxyodon Schrenk ■

60801 Bromus lanceolatus Roth var. danthoniae (Trin.) Dinsm. = Bromus danthoniae Trin. ex C. A. Mey. ■

60802 Bromus lanceolatus Roth var. dasystachys Maire = Bromus lanceolatus Roth ■

60803 Bromus lanceolatus Roth var. lanuginosus (Poir.) Dinsm. = Bromus lanceolatus Roth ■

60804 Bromus lanceolatus Roth var. leiostachys Maire = Bromus lanceolatus Roth ■

60805 Bromus lanuginosus Poir. = Bromus lanceolatus Roth var. lanuginosus (Poir.) Dinsm. ■☆

60806 Bromus lanuginosus Poir. = Bromus lanceolatus Roth ■

60807 Bromus latiglumis (Shear) Hitchc.;凸边雀麦;Flanged Brome, Hairy Woodbrome ■☆

60808 Bromus latiglumis (Shear) Hitchc. = Bromus altissimus Pursh ■☆

60809 Bromus latiglumis (Shear) Hitchc. f. incanus (Shear) Fernald = Bromus altissimus Pursh ■☆

60810 Bromus laxiflorus Spreng.;稀花雀麦■☆

60811 Bromus lepidus E. Holmb.;鳞稃雀麦;Slender Brome, Slender Soft Brome ■

60812 Bromus leptoclados Nees;细枝雀麦■☆

60813 Bromus longifolius Schousb. = Bromopsis benekenii (Lange) Holub ■

60814 Bromus macrantherus Hack. ex Henriq. = Anisantha macranthera (Hack. ex Henriq.) P. Silva ■☆

60815 Bromus macrostachys Desf. = Bromus lanceolatus Roth ■

60816 Bromus macrostachys Desf. var. danthoniae (Trin.) Asch. et Graebn. = Bromus danthoniae Trin. ex C. A. Mey. ■

60817 Bromus macrostachys Desf. var. lanuginosus (Poir.) Coss. et Durieu = Bromus lanceolatus Roth ■

60818 Bromus macrostachys Desf. var. oxyodon (Schrenk) Griseb. = Bromus oxyodon Schrenk ■

60819 Bromus macrostachys Desf. var. triaristatus Hack. = Bromus danthoniae Trin. ex C. A. Mey. ■

60820 Bromus madritensis L.;马德里雀麦(马德雀麦);Compact Brome, Madrit Brome, Spanish Brome ■

60821 Bromus madritensis L. naturalised = Anisantha madritensis (L.) Nevski ■

60822 Bromus madritensis L. subsp. delilei (Boiss.) Maire et Weiller = Anisantha fasciculata (C. Presl) Nevski subsp. delilei (Boiss.) H. Scholz et Valdés ■☆

60823 Bromus madritensis L. subsp. kunkelii H. Scholz = Anisantha rubens (L.) Nevski subsp. kunkelii (H. Scholz) H. Scholz ■☆

60824 Bromus madritensis L. subsp. rubens (L.) Husn. = Bromus rubens L. ■

60825 Bromus madritensis L. var. ambiguus Coss. et Durieu = Anisantha madritensis (L.) Nevski ■

60826 Bromus madritensis L. var. ciliatus Guss. = Anisantha madritensis (L.) Nevski ■

60827 Bromus madritensis L. var. delilei Boiss. = Anisantha fasciculata (C. Presl) Nevski subsp. delilei (Boiss.) H. Scholz et Valdés ■☆

60828 Bromus madritensis L. var. glabriculmis Maire et Weiller = Anisantha madritensis (L.) Nevski ■

60829 Bromus madritensis L. var. pubiculmis Maire et Weiller = Anisantha madritensis (L.) Nevski ■

60830 Bromus madritensis L. var. rigidus (Roth) Bab. ex Syme = Bromus rigidus Roth ■

60831 Bromus madritensis L. var. victorini (Sennen et Mauricio) Maire = Anisantha madritensis (L.) Nevski ■

60832 Bromus madritensis L. var. villiglumis Maire et Weiller = Anisantha madritensis (L.) Nevski ■

60833 Bromus madritensis L. var. villosissimus Maire = Anisantha madritensis (L.) Nevski ■

60834 Bromus magnus Keng;大雀麦;Big Brome ■

60835 Bromus mairei Hack.;梅氏雀麦(云南雀麦);Maire Brome, Maire Bromegrass ■

60836 Bromus mairei Hack. ex Hand.-Mazz. = Bromus mairei Hack. ■

60837 Bromus mairei Sennen et Mauricio = Anisantha tectorum (L.) Nevski ■

60838 Bromus manroi Boiss. = Bromus confinis Nees ex Steud. ■

60839 Bromus marcostachys Desf. var. lanuginosus (Poir.) Coss. et Durieu = Bromus lanceolatus Roth ■

60840 Bromus marginatus Nees ex Steud.;山地雀麦(边雀麦);Margin Brome, Western Brome ■

60841 Bromus marginatus Nees ex Steud. = Ceratochloa marginata (Nees ex Steud.) W. A. Weber ■

60842 Bromus marginatus Nees ex Steud. var. breviaristatus (Buckley) Beetle;短芒山地雀麦;Mountain Brome ■☆

60843 Bromus maroccanus Pau et Font Quer = Bromopsis maroccana (Pau et Font Quer) Holub ■☆

60844 Bromus maximus Desf. = Anisantha rigida (Roth) Hyl. ■

60845 Bromus maximus Desf. subsp. macrantherus (Hack. ex Henriq.) Trab. = Anisantha macranthera (Hack. ex Henriq.) P. Silva ■☆

60846 Bromus maximus Desf. var. gussonei Parl. = Anisantha diandra (Roth) Tzvelev ■

60847 Bromus maximus Desf. var. minor Boiss. = Anisantha rigida (Roth) Hyl. ■

60848 Bromus membranaceus Jacquem. ex Griseb. = Duthiea bromoides

Hack. ■☆

60849 Bromus microchaetus Font Quer = Bromopsis erecta Fourr. subsp. microchaeta (Font Quer) H. Scholz et Valdés ■☆

60850 Bromus milanjianus Rendle = Helictotrichon milanjianum (Rendle) C. E. Hubb. ■☆

60851 Bromus moeszii Pénzes = Bromus sericeus Drobow ■

60852 Bromus molliformis J. Lloyd = Bromus hordeaceus L. subsp. molliformis (J. Lloyd) Maire et A. Weiller ■☆

60853 Bromus mollis L.;绢毛雀麦(软雀麦);Bob-grass,Lob-grass,Lop-grass,Soft Brome,Soft Chess,Son Brome ■

60854 Bromus mollis L. = Bromus hordeaceus L. ■

60855 Bromus mollis L. var. commutatus (Schrad.) Sanio = Bromus racemosus L. ■

60856 Bromus mollis L. var. lloydianus (Gren. et Godr.) Trab. = Bromus hordeaceus L. subsp. molliformis (J. Lloyd) Maire et A. Weiller ■☆

60857 Bromus mollis L. var. major Trab. = Bromus hordeaceus L. ■

60858 Bromus mollis L. var. ramosus Ball = Bromus hordeaceus L. ■

60859 Bromus mollis L. var. secalinus (L.) Huds. = Bromus secalinus L. ■

60860 Bromus morrisonensis Honda;玉山雀麦;Yushan Brome ■

60861 Bromus munroi Boiss. = Bromus confinis Nees ex Steud. ■

60862 Bromus mutabilis F. W. Schultz var. com-mutatus (Schrad.) F. W. Schultz = Bromus racemosus L. ■

60863 Bromus natalensis Stapf;纳塔尔雀麦■☆

60864 Bromus natalensis Stapf var. lasiophilus? = Bromus natalensis Stapf ■☆

60865 Bromus nepalensis Melderis;尼泊尔雀麦;Nepal Brome ■

60866 Bromus nototrophus Rupr. = Bromus oxyodon Schrenk ■

60867 Bromus nottowayanus Fernald;弗吉尼亚雀麦(维吉尼亚雀麦);Satin Brome,Virginia Brome ■☆

60868 Bromus occidentalis (Nevski) Pavlov = Bromopsis pumpelliana Scribn. ■

60869 Bromus ornans Kom.;装饰雀麦■☆

60870 Bromus ovatus Gaertn. = Bromus scoparius L. ■

60871 Bromus oxyodon Schrenk;尖齿雀麦;Oxytoothed Brome ■

60872 Bromus oxyodon Schrenk var. lanuginosus Rozhev = Bromus oxyodon Schrenk ■

60873 Bromus pamiricus Drobow;帕米尔雀麦;Pamir Brome ■

60874 Bromus pamiricus Drobow = Bromus paulsenii Hack. ex Paulsen ■

60875 Bromus patulus Merr. et W. D. J. Koch var. falconers Stapf = Bromus pectinatus Thunb. ■

60876 Bromus patulus Merr. et W. D. J. Koch var. microstachya Stapf = Bromus japonicus Thunb. ■

60877 Bromus patulus Merr. et W. D. J. Koch var. pectinatus (Thunb.) Stapf = Bromus pectinatus Thunb. ■

60878 Bromus patulus Merr. et W. D. J. Koch var. velutinus Koch = Bromus japonicus Thunb. ■

60879 Bromus patulus Merr. et W. D. J. Koch var. vestitus (Schrad.) Stapf = Bromus pectinatus Thunb. ■

60880 Bromus patulus Mert. et W. D. J. Koch = Bromus japonicus Thunb. ■

60881 Bromus pauciflorum (Thunb.) Hack. = Bromus remotiflorus (Steud.) Ohwi ■

60882 Bromus paulsenii Hack. ex Paulsen;波申雀麦;Paulsen Brome ■

60883 Bromus pectinatus Thunb.;篦齿雀麦;Pectinate Brome ■

60884 Bromus persicus Boiss. = Bromus tomentosus Trin. ■☆

60885 Bromus petitianus A. Rich. = Bromus leptoclados Nees ■☆

60886 Bromus phragmitoides A. Nyar.;苇状雀麦■

60887 Bromus phragmitoides A. Nyar. = Bromus arvensis L. ■

60888 Bromus phrygius Boiss. = Bromus japonicus Thunb. ■

60889 Bromus pinanensis (Ohwi) L. Liou;卑南雀麦;Pinan Brome ■

60890 Bromus pinnatus L. = Brachypodium pinnatum (L.) P. Beauv. ■

60891 Bromus plurinodis Keng;多节雀麦;Manynodes Brome ■

60892 Bromus popovii Drobow;波陂雀麦;Popov Brome ■

60893 Bromus popovii Drobow = Bromus racemosus L. ■

60894 Bromus porphyranthos Cope;大药雀麦;Big-flowered Brome ■

60895 Bromus pseudodanthoniae Drobow = Bromus lanceolatus Roth ■

60896 Bromus pseudoramosus P. C. Keng; 假枝雀麦; Falsebranch Brome ■

60897 Bromus pseudoramosus P. C. Keng var. sedgioides B. X. Sun et H. Peng;莎叶雀麦■

60898 Bromus pskemensis Pavlov;普康雀麦;Pukang Brome ■

60899 Bromus pskemensis Pavlov = Bromus inermis Leyss. ■

60900 Bromus pubescens Muhl. ex Willd.;柔毛雀麦;Canada Brome,Canadian Brome,Hairy Woodland Brome,Laxative Brome-grass ■☆

60901 Bromus pulchellus Fig. et De Not. = Bromus pectinatus Thunb. ■

60902 Bromus pumpellianus (Scribn.) Holub subsp. korotkiji (Drobow) Tzvelev = Bromus korotkiji Drobow ■

60903 Bromus pumpellianus (Scribn.) Holub var. ircutensis (Kom.) Tzvelev = Bromus korotkiji Drobow ■

60904 Bromus pumpellianus Scribn.;紧穗雀麦■

60905 Bromus purgans L. = Bromus kalmii A. Gray ■☆

60906 Bromus purgans L. = Bromus pubescens Muhl. ex Willd. ■☆

60907 Bromus purgans L. f. glabriflorus Wiegand = Bromus pubescens Muhl. ex Willd. ■☆

60908 Bromus purgans L. f. laevivaginatus Wiegand = Bromus pubescens Muhl. ex Willd. ■☆

60909 Bromus purgans L. var. laeviglumis (Scribn. ex Shear) Swallen = Bromus kalmii A. Gray ■☆

60910 Bromus purgans L. var. latiglumis Shear = Bromus latiglumis (Shear) Hitchc. ■☆

60911 Bromus purgans sensu Wagnon = Bromus altissimus Pursh ■☆

60912 Bromus racemosus L.; 总状雀麦; Bald Brome, Cheat, Hairy Chess,Racemose Brome,Smooth Brome,Smooth Chess ■

60913 Bromus racemosus L. subsp. commutatus (Schrad.) Tourlet = Bromus commutatus Schrad. ■

60914 Bromus racemosus L. var. commutatus (Schrad.) Coss. et Durieu = Bromus racemosus L. ■

60915 Bromus racemosus L. var. commutatus (Schrad.) Coss. et Durieu = Bromus commutatus Schrad. ■

60916 Bromus racemosus L. var. glabriglumis Maire et Weiller = Bromus commutatus Schrad. ■

60917 Bromus racemosus L. var. villosus (Trab.) Maire et Weiller = Bromus commutatus Schrad. subsp. neglectus (Parl.) P. M. Sm. ■☆

60918 Bromus ramosus Huds.;多枝雀麦(糙雀麦,粗糙雀麦,分枝雀麦,类雀麦);Hairy Brome,Hairy Brome Grass,Hairy Brome-grass,Manybranches Brome,Rough Brome ■

60919 Bromus ramosus Huds. = Bromopsis ramosa (Huds.) Holub ■

60920 Bromus ramosus Huds. subsp. benekenii (Lange) Schinz et Thell. = Bromopsis benekenii (Lange) Holub ■

60921 Bromus ramosus Huds. subsp. benekenii (Lange) Tzvelev = Bromus benekenii (Lange) Trimen ■

60922 Bromus ramosus Huds. var. algeriensis Maire et Weiller =

Bromopsis benekenii (Lange) Holub ■

60923 Bromus ramosus Huds. var. benekenii (Lange) Asch. et Graebn. = Bromus benekenii (Lange) Trimen ■

60924 Bromus ramosus Huds. var. macrostachys Litard. et Maire = Bromopsis benekenii (Lange) Holub ■

60925 Bromus rechingeri Melderis = Bromus gedrosianus Pénzes ■

60926 Bromus rechingeri Melderis = Bromus pectinatus Thunb. ■

60927 Bromus rechingeri Melderis ex Bor;丽庆雀麦;Liqing Brome ■

60928 Bromus remotiflorus (Steud.) Ohwi;疏花雀麦(狐茅);Laxflower Brome,Laxflower Bromegrass ■

60929 Bromus remotiflorus (Steud.) Ohwi var. pinanensis Ohwi = Bromus pinanensis (Ohwi) L. Liou ■

60930 Bromus retusa Pers. = Brachypodium retusum (Pers.) P. Beauv. ■☆

60931 Bromus richardsonii Link;理查德雀麦(耐酸草)■

60932 Bromus richardsonii Link = Bromus ciliatus L. var. richardsonii (Link) Y. C. Jiang ■

60933 Bromus richardsonii Link = Bromus pumpellianus Scribn. ■

60934 Bromus rigens L. = Bromus scoparius L. ■

60935 Bromus rigidus Roth;硬雀麦;Rigid Brome, Rigid Bromegrass, Ripgut,Ripgut Brome ■

60936 Bromus rigidus Roth subsp. gussonei (Parl.) Maire = Anisantha diandra (Roth) Tzvelev ■

60937 Bromus rigidus Roth subsp. maximus (Desf.) Rothm. et = Anisantha rigida (Roth) Hyl. ■

60938 Bromus rigidus Roth var. ambigens (Jord.) Maire et Weiller = Anisantha rigida (Roth) Hyl. ■

60939 Bromus rigidus Roth var. gussonei (Parl.) Coss. et Durieu = Anisantha diandra (Roth) Tzvelev ■

60940 Bromus rigidus Roth var. macrantherus (Hack. ex Henriq.) Maire = Anisantha macranthera (Hack. ex Henriq.) P. Silva ■☆

60941 Bromus rigidus Roth var. minor (Boiss.) Maire = Anisantha rigida (Roth) Hyl. ■

60942 Bromus riparius Rehmann;山丹雀麦(河边雀麦)■

60943 Bromus rubens L.;红雀麦;Foxtail Brome, Foxtail Chess, Red Brome,Red Bromegrass ■

60944 Bromus rubens L. = Anisantha rubens (L.) Nevski ■

60945 Bromus rubens L. subsp. fasciculatus (C. Presl) Trab. = Anisantha fasciculata (C. Presl) Nevski ■

60946 Bromus rubens L. subsp. fasciculatus (C. Presl) Trab. = Bromus fasciculatus C. Presl ■

60947 Bromus rubens L. subsp. kunkelii (H. Scholz) H. Scholz = Anisantha rubens (L.) Nevski subsp. kunkelii (H. Scholz) H. Scholz ■☆

60948 Bromus rubens L. var. alexandrinus (Thell.) Maire et Weiller = Anisantha rubens (L.) Nevski ■

60949 Bromus rubens L. var. ambiguus Maire = Anisantha rubens (L.) Nevski ■

60950 Bromus rubens L. var. canescens (Viv.) Coss. = Anisantha rubens (L.) Nevski ■

60951 Bromus rubens L. var. fallax Maire = Anisantha rubens (L.) Nevski ■

60952 Bromus rubens L. var. glabriglumis Maire = Anisantha rubens (L.) Nevski ■

60953 Bromus rubens L. var. puberulus Maire et Weiller = Anisantha rubens (L.) Nevski ■

60954 Bromus rubens L. var. rigidus (Roth) Mutel = Bromus rigidus Roth ■

60955 Bromus runssoroensis K. Schum. = Bromus leptoclados Nees ■☆

60956 Bromus scabridus Hook. f. = Bromus leptoclados Nees ■☆

60957 Bromus scoparius L.;帚雀麦(刷雀麦);Broom Brome ■

60958 Bromus scoparius L. subsp. chrysopogon (Viv.) Chrtek et Slavík = Bromus chrysopogon Viv. ■☆

60959 Bromus scoparius L. var. psilostachys Halácsy = Bromus scoparius L. ■

60960 Bromus scoparius L. var. rubens (L.) St. -Amans = Bromus rubens L. ■

60961 Bromus scoparius L. var. stenantha Stapf = Bromus scoparius L. ■

60962 Bromus scoparius L. var. villiglumis Maire et Weiller = Bromus scoparius L. ■

60963 Bromus secalinus L.;黑麦状雀麦(黑麦雀麦);Brome Grass, Cheat, Chess, Chess Brome, Rye Brome, Ryelike Brome, Rye-like Brome,Secale-like Brome ■

60964 Bromus secalinus L. var. hirsutus Kindb. = Bromus secalinus L. ■

60965 Bromus secalinus L. var. hirtus (F. W. Schultz) Hegi = Bromus secalinus L. ■

60966 Bromus secalinus L. var. hordeaceus (L.) L. = Bromus hordeaceus L. ■

60967 Bromus sericeus Drobow = Bromus tectorum L. subsp. lucidus Sales ■

60968 Bromus sewertzovii Regel;密穗雀麦(北疆雀麦);Sewertzov Brome ■

60969 Bromus sibiricus Drobow;西伯利亚雀麦(西伯利亚无芒雀麦);Siberia Bromegrass,Siberian Brome ■

60970 Bromus sibiricus Drobow = Bromus pumpellianus Scribn. ■

60971 Bromus sinaicus (Hack.) Tackh. = Bromus pectinatus Thunb. ■

60972 Bromus sinensis P. C. Keng;华雀麦;China Bromegrass, Chinese Brome ■

60973 Bromus sinensis P. C. Keng var. minor L. Liou;小华雀麦;Small Chinese Brome ■

60974 Bromus sitchensis Trin.;锡珍雀麦;Sitchen Brome, Sitchen Bromegrass ■

60975 Bromus speciosus Nees;美丽雀麦■☆

60976 Bromus speciosus Nees var. firmior = Bromus firmior (Nees) Stapf ■☆

60977 Bromus squarrosus L.;偏穗雀麦(糙雀麦,单列雀麦);Corn Brome, Downy Brome, Nodding Brome, One-way Brome, One-way Chess,Rough Brome ■

60978 Bromus squarrosus L. var. racemosus (L.) Regel = Bromus racemosus L. ■

60979 Bromus squarrosus L. var. villosus (C. C. Gmel.) Koch = Bromus squarrosus L. ■

60980 Bromus staintonii Melderis;大序雀麦;Bigspike Brome ■

60981 Bromus staintonii Melderis var. guoxunianus N. X. Zhao et M. F. Li;国勋雀麦;Guoxun Brome ■

60982 Bromus stamineus E. Desv.;雄蕊雀麦; Roadside Brome, Southern Brome,Stamenlike Brome ■

60983 Bromus stamineus E. Desv. = Ceratochloa staminea (E. Desv.) Stace ■

60984 Bromus stenostachyus Boiss.;窄序雀麦;Narrowspike Brome ■

60985 Bromus sterilis L.;不实雀麦(贫育雀麦); Barren Brome, Poverty Brome ■

60986 Bromus sterilis L. = Anisantha sterilis (L.) Nevski ■

60987 Bromus stipoides L. = Vulpiella stipoides (L.) Maire ■☆

60988 Bromus sylvaticus (Huds.) Lyons = Brachypodium sylvaticum (Huds.) P. Beauv. ■

60989 Bromus tectorum L.;旱雀麦;Cheat Grass,Cheatgrass,Cheatgrass Brome,Downy Brome,Downy Chess,Drooping Brome,June Grass,Roof Brome Grass,Xeric Bromegrass ■

60990 Bromus tectorum L. = Anisantha tectorum (L.) Nevski ■

60991 Bromus tectorum L. subsp. lucidus Sales;绢雀麦(绢毛雀麦)■

60992 Bromus tectorum L. var. glabratus Spenn.;无毛旱雀麦;Cheat Grass,Downy Brome,Downy Chess,June Grass ■☆

60993 Bromus tectorum L. var. grandiflorus Hack. ex Fedtsch. = Bromus sericeus Drobow ■

60994 Bromus tectorum L. var. hirsutus Regel = Bromus tectorum L. ■

60995 Bromus tectorum L. var. nudus Klett et Richt. = Bromus tectorum L. var. glabratus Spenn. ■☆

60996 Bromus tenuiflorus Viv. = Anisantha fasciculata (C. Presl) Nevski ■

60997 Bromus tenuis Tineo = Vulpiella tenuis (Tineo) Kerguélen ■☆

60998 Bromus tibesticus Maire = Bromus pectinatus Thunb. ■

60999 Bromus tomentellus Boiss.;微毛雀麦■☆

61000 Bromus tomentellus Boiss. subsp. cappadocicus (Boiss. et Balansa) Tzvelev = Bromus cappadocicus Boiss. et Balansa ■

61001 Bromus tomentosus Trin.;绒毛雀麦■☆

61002 Bromus trichopodus A. Rich. = Streblochaete longiarista (A. Rich.) Pilg. ■☆

61003 Bromus turkestanicus Drobow;土耳其雀麦;Turkestan Brome ■

61004 Bromus turkestanicus Drobow = Bromus paulsenii Hack. ex Paulsen ■

61005 Bromus tytthanthus Nevski;裂稃雀麦■

61006 Bromus tyttholepis Nevski;土沙雀麦■

61007 Bromus ugamicus Drobow = Bromus japonicus Thunb. ■

61008 Bromus unioloides (Willd.) Kunth;单穗雀麦(扁穗雀麦,大扁雀麦);Oneflower Brome,Rescue Brome,Rescue Grass,Rescuegrass ■

61009 Bromus unioloides (Willd.) Kunth = Bromus catharticus Vahl ■

61010 Bromus unioloides (Willd.) Kunth = Ceratochloa cathartica (Vahl) Herter ■☆

61011 Bromus unioloides (Willd.) Raspail = Bromus catharticus Vahl ■

61012 Bromus unioloides Kunth = Bromus catharticus Vahl ■

61013 Bromus unioloides Kunth = Bromus willdenowii Kunth ■

61014 Bromus uralensis Govor. = Bromus pumpellianus Scribn. ■

61015 Bromus valdivianus Phil. = Ceratochloa staminea (E. Desv.) Stace ■

61016 Bromus variegatus M. Bieb.;变色雀麦(斑雀麦)■

61017 Bromus vestitus Schrad. = Bromus pectinatus Thunb. ■

61018 Bromus viciosoi Sennen = Anisantha madritensis (L.) Nevski ■

61019 Bromus villosus Forssk.;长柔毛雀麦;Villose Brome ■

61020 Bromus villosus Forssk. = Anisantha madritensis (L.) Nevski ■

61021 Bromus villosus Forssk. = Bromus rigidus Roth ■

61022 Bromus villosus Forssk. subsp. gussonei (Parl.) Holmb. = Anisantha diandra (Roth) Tzvelev ■

61023 Bromus villosus Forssk. subsp. rigidus (Roth) Braun-Blanq. = Anisantha rigida (Roth) Hyl. ■

61024 Bromus villosus Forssk. var. gussonei (Parl.) Bonnet et Barratte = Anisantha diandra (Roth) Tzvelev ■

61025 Bromus villosus Forssk. var. minor (Boiss.) Briq. = Anisantha diandra (Roth) Tzvelev ■

61026 Bromus villosus Forssk. var. rigidus (Roth) Asch. et Graebn. = Anisantha diandra (Roth) Tzvelev ■

61027 Bromus villosus Scop. var. rigidus (Roth) Asch. et Graebn. = Bromus rigidus Roth ■

61028 Bromus willdenowii Kunth = Bromus catharticus Vahl ■

61029 Bromus willdenowii Kunth = Ceratochloa cathartica (Vahl) Herter ■☆

61030 Bromus willdenowii Kunth = Ceratochloa unioloides (Willd.) P. Beauv. ■

61031 Bromus wolgensis Fisch. ex J. Jacq. = Bromus squarrosus L. ■

61032 Bromus wolgensis Fisch. ex Willd.;沃京雀麦■

61033 Bromus yezoensis Ohwi = Bromus canadensis Michx. ■

61034 Bromus yezoensis Ohwi = Bromus ciliatus L. ■

61035 Brongniartia Blume = Kibara Endl. ●☆

61036 Brongniartia Kunth(1824);豌豆树属●☆

61037 Brongniartia Walp. = Brongniartia Kunth ●☆

61038 Brongniartia alamosana Rydb.;豌豆树●☆

61039 Brongniartia minutifolia S. Watson;加州豌豆树;California Brickellia ●☆

61040 Brongniartikentia Becc. (1921);裂鞘椰属(邦铁榈属,布朗尼亚椰属)●☆

61041 Brongniartikentia lanuginosa H. E. Moore;多毛裂鞘椰●☆

61042 Brongniartikentia vaginata Becc.;裂鞘椰●☆

61043 Bronnia Kunth = Fouquieria Kunth ●☆

61044 Bronwenia W. R. Anderson et C. Davis = Banisteria L. (废弃属名)●☆

61045 Bronwenia W. R. Anderson et C. Davis = Heteropterys Kunth(保留属名)●☆

61046 Bronwenia W. R. Anderson et C. Davis(2007);布龙异翅藤属●☆

61047 Brookea Benth. (1876);布鲁草属●☆

61048 Brookea dasyantha Benth.;粗毛布鲁草■☆

61049 Brookea tomentosa Benth.;绒毛布鲁草■☆

61050 Brosimopsis S. Moore = Brosimum Sw. (保留属名)●☆

61051 Brosimopsis S. Moore(1895);类饱食桑属●☆

61052 Brosimopsis oblongifolia Ducke;长圆叶类饱食桑●☆

61053 Brosimum Sw. (1788)(保留属名);饱食桑属(饱食木属);Breadfruit-tree,Breadnut,Breadnut Tree ●☆

61054 Brosimum acutifolium Huber;尖叶饱食桑●☆

61055 Brosimum aubletii Poepp. et Endl. = Piratinera guianensis Aubl. ●☆

61056 Brosimum columbianum S. F. Blake;哥伦比亚饱食桑●☆

61057 Brosimum galactodendron D. Don ex Sweet;乳汁饱食桑;Cow Breadnut Tree,Cow Breadnut-tree,Cow Tree ●☆

61058 Brosimum gaudichaudii Trécul;巴西饱食桑(巴西饱食木)●☆

61059 Brosimum guianense (Aubl.) Huber;圭亚那饱食桑(圭亚那饱食木,圭亚那蛇桑);Amourette,Leopard Wood,Letterwood,Snakewood ●☆

61060 Brosimum guyanense Huber = Piratinera guianensis Aubl. ●☆

61061 Brosimum paraense Huber = Brosimum rubescens Taub. ●☆

61062 Brosimum rubescens Taub.;红饱食桑(红饱食木,红变饱食桑);Brazil Redwood,Brazilian Redwood,Brazilwood,Cardinal Wood,Satine,Snake Wood,Snakewood ●☆

61063 Brosimum utile (Kunth) Oken = Brosimum galactodendron D. Don ex Sweet ●☆

61064 Brosimum utile H. Karst. = Brosimum galactodendron D. Don ex Sweet ●☆

61065 Brossaea L. = Gaultheria L. ●

61066 Brossardia Boiss. (1841);两伊芥属■☆

61067 Brossardia papyracea Boiss.;两伊芥■☆

61068 Brossea Kuntze = Gaultheria L. ●
61069 Brossea trichophylla (Royle) Kuntze = Gaultheria trichophylla Royle ●
61070 Brotera Cav. = Melhania Forssk. ●■
61071 Brotera Spreng. (1800) = Flaveria Juss. ■●
61072 Brotera Spreng. (1800) = Nauenburgia Willd. ■●
61073 Brotera Spreng. (1801) = Hyptis Jacq. (保留属名)●■
61074 Brotera Vell. = Luehea Willd. (保留属名)●☆
61075 Brotera Willd. = Cardopatium Juss. ■☆
61076 Brotera bracteosa Guillaumin et Perr. = Melhania denhamii R. Br. ●☆
61077 Brotera ovata Cav. = Melhania ovata (Cav.) Spreng. ●☆
61078 Brotera sprengelii Cass. = Flaveria trinervia (Spreng.) C. Mohr ■☆
61079 Brotera trinervata (Willd.) Pers. = Flaveria trinervia (Spreng.) C. Mohr ■☆
61080 Broteroa DC. = Brotera Spreng. (1800)■●
61081 Broteroa DC. = Flaveria Juss. ■●
61082 Broteroa Kuntze = Brotera Willd. ■☆
61083 Broteroa Kuntze = Cardopatium Juss. ■☆
61084 Broteroa amethystina (Spach) Kit. = Cardopatum amethystinum Spach ■☆
61085 Brotobroma H. Karst. et Triana = Herrania Goudot ●☆
61086 Broughtonia R. Br. = Broughtonia Wall. ex Lindl. ■
61087 Broughtonia Wall. ex Lindl. (1813);宝通兰属(波东兰属,布劳顿兰属,西印第安兰属);Broughtonia ■
61088 Broughtonia Wall. ex Lindl. = Otochilus Lindl. ■
61089 Broughtonia fusca (Lindl.) Wall. ex Hook. f. = Otochilus fuscus Lindl. ■
61090 Broughtonia linearis Wall. ex Lindl. = Coelogyne fimbriata Lindl. ■
61091 Broughtonia sanguinea (Sw.) R. Br.; 宝 通 兰; Bloodred Broughtonia ■☆
61092 Brousemichea Balansa = Zoysia Willd. (保留属名)■
61093 Brousemichea Willis = Brousemichea Balansa ■
61094 Broussaisia Gaudich. (1830);夏威夷绣球属●☆
61095 Broussaisia arguta Gaudich.;哈瓦利绣球●☆
61096 Broussonetia L'Hér. ex Vent. (1799)(保留属名);构树属(阿里桑属,楮属,构属,落叶花桑属);Allaeanthus, Paper Mulberry, Paper-mulberry ●
61097 Broussonetia Ortega(废弃属名) = Broussonetia L'Hér. ex Vent. (保留属名)●
61098 Broussonetia Ortega(废弃属名) = Sophora L. ●■
61099 Broussonetia brasiliensis C. Mart. = Maclura brasiliensis Endl. ●☆
61100 Broussonetia jiangxiensis Z. X. Yu; 江 西 构; Jiangxi Paper-mulberry ●
61101 Broussonetia kaempferi Siebold;藤葡蟠(楮树,葡蟠,藤构,小构树);Kaempfer Paper-mulberry, Small-paper Mulberry ●
61102 Broussonetia kaempferi Siebold var. australis T. Suzuki;藤构(构桑,谷皮藤,尖叶楮皮,蔓构,南方藤构,藤葡蟠);South Paper-mulberry ●
61103 Broussonetia kazinoki Siebold;小构树(楮,楮树,丁字黄心羌,飞天拢,构皮麻,谷皮树,谷皮藤,谷树,过里丹,黄疸藤,尖叶楮皮,酱叶树,蜡皮树,冷丹藤,女谷,葡蟠,生藤,野构树,野构桃,纸皮);Kazinoki Paper-mulberry, Kozo ●
61104 Broussonetia kazinoki Siebold var. ruyangensis P. H. Ling et X. W. Wei = Broussonetia kazinoki Siebold ●
61105 Broussonetia kurzii (Hook. f.) Corner; 落 叶 花 桑; Kurz Allaeanthus, Kurz Paper-mulberry ●
61106 Broussonetia monoica Hance = Broussonetia kazinoki Siebold ●
61107 Broussonetia papyrifera (L.) L'Hér. ex Vent.;构树(楮,楮桑,楮树,楮桃,楮桃树,大构树,大骨皮,构,构泡,谷,谷浆树,谷木,谷皮树,谷皮树子,谷桑,谷沙树,谷树,瓜槌草,假杨梅,酱黄木,角树,柯树,壳树,鹿仔树,毛桃儿,奶树,沙皮树,纱纸树,商庭树,野杨梅,造纸树);Common Paper Mulberry, Common Paper-mulberry, Kapa, Kou-shuipaper Mulberry, Paper Mulberry, Paper-mulberry, Tapa Cloth, Tapa Cloth Tree, Tapa-tree ●
61108 Broussonetia papyrifera (L.) L'Hér. ex Vent. f. leucocarpa (Ser.) H. Wei Jen;白果构树;White-fruit Paper-mulberry ●
61109 Broussonetia plumerii Spreng. = Maclura tinctoria (L.) D. Don ex Steud. ●☆
61110 Broussonetia rupicola F. T. Wang et Ts. Tang;南川构;Nanchuan Paper-mulberry ●
61111 Broussonetia secundiflora Ortega = Sophora secundiflora (Ortega) Lag. ex DC. ●☆
61112 Broussonetia sieboldii Blume = Broussonetia kaempferi Siebold var. australis T. Suzuki ●
61113 Broussonetia tinctoria (L.) Kunth = Maclura tinctoria (L.) D. Don ex Steud. ●☆
61114 Broussonetia tricolor Hort. ex K. Koch;三色构●☆
61115 Broussonetia zanthoxylon (L.) C. Mart. = Maclura tinctoria (L.) D. Don ex Steud. ●☆
61116 Brouvalea Adans. = Browallia L. ■☆
61117 Brovallia L. = Browallia L. ■☆
61118 Browallia L. (1753); 歪 头 花 属 (布 洛 华 丽 属); Amethyst Flower, Browallia ■☆
61119 Browallia americana L.;美洲歪头花(歪头花,小布洛华丽);Amethyst Flower, Jamaican Forget-me-not ■☆
61120 Browallia czerniakowskiana Hort. = Browallia viscosa Kunth ■☆
61121 Browallia czerniakowskiana Warsz. = Browallia viscosa Kunth ■☆
61122 Browallia demissa L.;下垂歪头花■☆
61123 Browallia demissa L. = Browallia americana L. ■☆
61124 Browallia elata L. = Browallia americana L. ■☆
61125 Browallia elongata Kunth = Browallia americana L. ■☆
61126 Browallia grandiflora Graham;大花歪头花(大花布洛华丽)■☆
61127 Browallia humifusa Forssk. = Cycniopsis humifusa (Forssk.) Engl. ■☆
61128 Browallia nervosa Miers = Browallia americana L. ■☆
61129 Browallia pulchella Vilm. = Browallia viscosa Kunth ■☆
61130 Browallia roezlii Nicholson = Browallia grandiflora Graham ■☆
61131 Browallia speciosa Hook.;长管歪头花(布洛华丽,蓝宝花);Browallia, Bush Violet, Methyst Flower, Sappire Flower ■☆
61132 Browallia viscosa Kunth; 黏 萼 歪 头 花 (粘 萼 布 洛 华 丽); Browallia ■☆
61133 Browalliaceae Bercht. et J. Presl = Solanaceae Juss. (保留科名)●■
61134 Brownaea Jacq. = Brownea Jacq. (保留属名)●☆
61135 Brownanthus Schwantes(1927);褐花属(露花树属)●☆
61136 Brownanthus arenosus (Schinz) Ihlenf. et Bittrich;砂褐花●☆
61137 Brownanthus ciliatus (Aiton) Schwantes = Brownanthus vaginatus (Lam.) Chess. et M. Pignal ●☆
61138 Brownanthus ciliatus (Aiton) Schwantes subsp. schenkii (Schinz) Ihlenf. et Bittrich = Brownanthus vaginatus (Lam.) Chess. et M. Pignal subsp. schenckii (Schinz) Chess. et M. Pignal ●☆
61139 Brownanthus corallinus (Thunb.) Ihlenf. et Bittrich;珊瑚色褐花●☆

61140 Brownanthus fraternus Klak;兄弟褐花●☆
61141 Brownanthus glareicola Klak;石砾褐花●☆
61142 Brownanthus kuntzei (Schinz) Ihlenf. et Bittrich;孔策褐花●☆
61143 Brownanthus lignescens Klak;木质褐花●☆
61144 Brownanthus marlothii (Pax) Schwantes;马尔褐花●☆
61145 Brownanthus namibensis (Marloth) Bullock;纳米布褐花●☆
61146 Brownanthus nucifer (Ihlenf. et Bittrich) S. M. Pierce et Gerbaulet;坚果褐花●☆
61147 Brownanthus pubescens (N. E. Br. ex C. A. Maass) Bullock;毛褐花●☆
61148 Brownanthus schenkii (Schinz) Schwantes = Brownanthus vaginatus (Lam.) Chess. et M. Pignal subsp. schenckii (Schinz) Chess. et M. Pignal ●☆
61149 Brownanthus schlichtianus (Sond.) Ihlenf. et Bittrich = Brownanthus arenosus (Schinz) Ihlenf. et Bittrich ●☆
61150 Brownanthus simplex (N. E. Br. ex C. A. Maass) Bullock = Brownanthus vaginatus (Lam.) Chess. et M. Pignal subsp. schenckii (Schinz) Chess. et M. Pignal ●☆
61151 Brownanthus solutifolius (A. Berger) H. Jacobsen = Brownanthus marlothii (Pax) Schwantes ●☆
61152 Brownanthus vaginatus (Lam.) Chess. et M. Pignal;普通褐花●☆
61153 Brownanthus vaginatus (Lam.) Chess. et M. Pignal subsp. schenckii (Schinz) Chess. et M. Pignal;申克褐花●☆
61154 Brownea Jacq. (1760)(保留属名)('Brownaea');热木豆属(宝冠木属)●☆
61155 Brownea ariza Benth.;红苞热木豆;Rose of Venezuela ●☆
61156 Brownea capitella Jacq.;美丽热木豆●☆
61157 Brownea coccinea Jacq.;绯红热木豆(宝冠木)●☆
61158 Brownea grandiceps Jacq.;大花热木豆;Rose of Venezuela, Rose-of-venezuela, Scarlet Flame Bean ●☆
61159 Brownea latifolia Jacq.;宽叶热木豆;Guaramaco ●☆
61160 Browneopsis Huber(1906);拟热木豆属●☆
61161 Browneopsis ucayalina Huber;拟热木豆●☆
61162 Brownetara Rich. ex Tratt. = Phyllocladus Rich. ex Mirb.(保留属名)●☆
61163 Brownetara Rich. ex Tratt. = Podocarpus Labill. ●
61164 Brownetera Rich. = Phyllocladus Rich. ex Mirb.(保留属名)●☆
61165 Browningia Britton et Rose(1920);群蛇柱属(青铜龙属)●☆
61166 Browningia candelaris (Meyen) Britton et Rose;群蛇柱(青铜龙);Eastern Redbud ●☆
61167 Browningia hertlingiana (Backeb.) Buxb.;佛塔(佛头)●☆
61168 Brownleea Harv. = Brownleea Harv. ex Lindl. ■☆
61169 Brownleea Harv. ex Lindl. (1842);布郎兰属;Brownleea ■☆
61170 Brownleea alpina (Hook. f.) N. E. Br.;高山布郎兰;Alpine Brownleea ■☆
61171 Brownleea alpina (Hook. f.) N. E. Br. = Brownleea parviflora Harv. ex Lindl. ■☆
61172 Brownleea apetala (Kraenzl.) N. E. Br. = Brownleea parviflora Harv. ex Lindl. ■☆
61173 Brownleea coerulea Harv. = Brownleea coerulea Harv. ex Lindl. ■☆
61174 Brownleea coerulea Harv. ex Lindl.;天蓝布郎兰;Sky-blue Brownleea ■☆
61175 Brownleea fanniniae Rolfe = Brownleea galpinii Bolus subsp. major (Bolus) H. P. Linder ■☆
61176 Brownleea flavescens Schltr. = Brownleea galpinii Bolus ■☆
61177 Brownleea galpinii Bolus;盖尔布郎兰■☆
61178 Brownleea galpinii Bolus subsp. major (Bolus) H. P. Linder;大盖尔布郎兰■☆
61179 Brownleea galpinii Bolus var. major? = Brownleea galpinii Bolus subsp. major (Bolus) H. P. Linder ■☆
61180 Brownleea gracilis Schltr. = Brownleea parviflora Harv. ex Lindl. ■☆
61181 Brownleea leucantha Schltr. = Brownleea galpinii Bolus subsp. major (Bolus) H. P. Linder ■☆
61182 Brownleea macroceras Sond.;大角布郎兰■☆
61183 Brownleea maculata P. J. Cribb;斑点布郎兰■☆
61184 Brownleea madagascarica Ridl. = Brownleea coerulea Harv. ex Lindl. ■☆
61185 Brownleea monophylla Schltr. = Brownleea macroceras Sond. ■☆
61186 Brownleea natalensis Rolfe = Brownleea recurvata Sond. ■☆
61187 Brownleea nelsonii Rolfe = Brownleea coerulea Harv. ex Lindl. ■☆
61188 Brownleea parviflora Harv. ex Lindl.;小花布郎兰■☆
61189 Brownleea pentheriana Kraenzl. ex Zahlbr. = Disa ophrydea (Lindl.) Bolus ■☆
61190 Brownleea perrieri Schltr. = Brownleea parviflora Harv. ex Lindl. ■☆
61191 Brownleea recurvata Sond.;布郎兰;Recurved Brownleea ■☆
61192 Brownleea transvaalensis Schltr. = Brownleea parviflora Harv. ex Lindl. ■☆
61193 Brownleea woodii Rolfe = Brownleea coerulea Harv. ex Lindl. ■☆
61194 Brownlowia Roxb. (1820)(保留属名);布朗椴属●☆
61195 Brownlowia argentata Kurz;银叶布朗椴●☆
61196 Brownlowia elliptica Ridl.;椭圆布朗椴●☆
61197 Brownlowia glabrata Stapf ex Ridl.;无毛布朗椴●☆
61198 Brownlowiaceae Cheek = Malvaceae Juss.(保留科名)●■
61199 Brucea J. F. Mill. (1979-1780)(保留属名);鸦胆子属(鸭胆子属,雅胆子属);Brucea ●
61200 Brucea acuminata H. L. Li = Brucea mollis Wall. ex Kurz ●
61201 Brucea amarissima (Lour.) Desv. ex Gomez = Brucea javanica (L.) Merr. ●
61202 Brucea amarissima (Lour.) Merr.;阿麻利士鸦胆子●☆
61203 Brucea amarissima Desv. ex Gomes = Brucea javanica (L.) Merr. ●
61204 Brucea antidysenterica J. F. Mill.;抗痢鸦胆子●☆
61205 Brucea erythraeae Chiov. = Brucea antidysenterica J. F. Mill. ●☆
61206 Brucea ferruginea L'Hér. = Brucea antidysenterica J. F. Mill. ●☆
61207 Brucea guineensis G. Don;几内亚鸦胆子●☆
61208 Brucea javanica (L.) Merr.;鸦胆子(苦参子,苦榛子,老鸦胆,鲜苦楝,小苦楝,丫蛋子,鸦胆,鸦胆树,鸦旦子,鸦蛋子,鸭胆子,鸭旦子,鸭蛋子,雅胆子,欲槐麦);Brucea, Java Brucea, Javan Brucea ●
61209 Brucea macrocarpa Stannard;大果鸦胆子●☆
61210 Brucea macrophylla Oliv.;大叶鸦胆子●☆
61211 Brucea mollis Wall. ex Kurz;毛鸦胆子(大果鸦胆子,柔毛鸦胆子);Hairy Brucea, Softhair Brucea ●
61212 Brucea mollis Wall. ex Kurz var. tonkinensis Lecomte = Brucea mollis Wall. ex Kurz ●
61213 Brucea paniculata Lam. = Trichoscypha smythei Hutch. et Dalziel ●☆
61214 Brucea quinensis G. Don;奎尼鸦胆子●☆
61215 Brucea salutaris A. Chev. = Brucea antidysenterica J. F. Mill. ●☆
61216 Brucea sumatrana Roxb. = Bischofia javanica Blume ●
61217 Brucea sumatrana Roxb. = Brucea javanica (L.) Merr. ●
61218 Brucea tenuifolia Engl.;细叶鸦胆子●☆

61219 Brucea trichotoma (Lour.) Spreng. = Evodia trichotoma (Lour.) Pierre ●
61220 Bruchmannia Nutt. = Beckmannia Host ■
61221 Bruckenthalia Rchb. (1831);布鲁杜鹃属;Spike Heath ●☆
61222 Bruckenthalia Rchb. = Erica L. ●☆
61223 Bruckenthalia spiculifolia (Salisb.) Rchb.;布鲁杜鹃(细尖叶石南);Spike Heath, Spike-heath ●☆
61224 Bruckenthalia spiculifolia (Salisb.) Rchb. = Erica spiculifolia Salisb. ●☆
61225 Bruea Gaudich. (1830);孟加拉大戟属■☆
61226 Bruea bengalensis Gaudich.;孟加拉大戟■☆
61227 Brueckea Klotzsch et H. Karst. = Aegiphila Jacq. ●■☆
61228 Bruennichia Willd. = Brunnichia Banks ex Gaertn. ●☆
61229 Brugmansia Blume = Rhizanthes Dumort. ■☆
61230 Brugmansia Pers. (1805);曼陀罗木属(花曼陀罗属,木本曼陀罗属,木曼陀罗属);Angel's Trumpet, Brugmansia ●
61231 Brugmansia arborea (L.) Steud.;曼陀罗木(橙花曼陀罗,大花曼陀罗,木本曼陀罗,木曼陀罗,乔木状曼陀罗,树曼陀罗);Angel's Trumpet, Angels-trumpet, Arboreous Brugmansia, Floripondio Datura, Maikoa ●
61232 Brugmansia arborea (L.) Steud. = Datura arborea L. ●
61233 Brugmansia aurea Lagerh.;金曼陀罗木(金黄木曼陀罗);Golden Angel's Trumpet ●☆
61234 Brugmansia aurea Lagerh. = Brugmansia candida Pers. ●☆
61235 Brugmansia candida Pers.;南美曼陀罗木(白花曼陀罗,白花木曼陀罗,白木曼陀罗);Angel's Trumpet, Angel's-trumpet, Thorn Apple Datur, Yellow Angel's Trumpet ●☆
61236 Brugmansia insignis (Barb. Rodr.) Lockwood ex R. E. Schult.;白花曼陀罗木;Angel's Trumpet ●☆
61237 Brugmansia knightii Dun = Brugmansia arborea (L.) Steud. ●
61238 Brugmansia knightii Dun = Datura arborea L. ●
61239 Brugmansia metaloides?;芹叶曼陀罗木;Angel's Trumpet ●☆
61240 Brugmansia sanguinea (Ruiz et Pav.) D. Don;红曼陀罗木(红花木曼陀罗);Red Angel's Trumpet ●☆
61241 Brugmansia suaveolens (Humb. et Bonpl. ex Wild.) Bercht. et Presl = Datura suaveolens Humb. et Bonpl. ex Willd. ●
61242 Brugmansia suaveolens (Humb. et Bonpl. ex Willd.) Bercht. et C. Presl;大花曼陀罗(墨西哥曼陀萝,香甜曼陀罗木);Angel's Trumpet, Angel's Trumpets, Angel's-tears, Angel's-trumpet, Angel-tears Datura, Brazilian Angel's Trumpets ●
61243 Brugmansia suaveolens (Humb. et Bonpl. ex Willd.) Sweet = Brugmansia suaveolens (Humb. et Bonpl. ex Willd.) Bercht. et C. Presl ●
61244 Brugmansia suaveolens (Humb. et Bonpl. ex Willd.) Sweet = Datura suaveolens Humb. et Bonpl. ex Willd. ●
61245 Brugmansia versicolor Lagerh.;变色曼陀罗木;Arborescent Angel's-tears ●☆
61246 Bruguiera Lam. = Bruguiera Sav. ●
61247 Bruguiera Pfeiff. = Bruquieria Pourr. ex Ortega ■
61248 Bruguiera Pfeiff. = Mirabilis L. ■
61249 Bruguiera Rich. ex DC. = Conostegia D. Don ■☆
61250 Bruguiera Sav. (1798);木榄属(红树属);Bruguiera ●
61251 Bruguiera Thouars = Lumnitzera Willd. ●
61252 Bruguiera capensis Blume = Bruguiera gymnorhiza (L.) Lam. ●
61253 Bruguiera caryophylloides (Burm. f.) Blume = Bruguiera cylindrica (L.) Blume ●
61254 Bruguiera caryophylloides Blume;石竹木榄;Rui, Thua Khao Rui ●
61255 Bruguiera caryophylloides Blume = Bruguiera cylindrica (L.) Blume ●
61256 Bruguiera conjugata (L.) Merr. = Bruguiera gymnorhiza (L.) Savigny ●
61257 Bruguiera conjugata Merr. = Bruguiera gymnorhiza (L.) Savigny ●
61258 Bruguiera cylindrica (L.) Blume;柱果木榄(五脚里,五梨跤);Cylinder-fruit Bruguiera, Cylindric Bruguiera ●
61259 Bruguiera cylindrica Blume = Bruguiera gymnorhiza (L.) Lam. ●
61260 Bruguiera eriopetala Wight et Arn.;绵毛瓣木榄●☆
61261 Bruguiera eriopetala Wight et Arn. = Bruguiera sexangula (Lour.) Poir. ●
61262 Bruguiera gymnorhiza (L.) Lam. = Bruguiera gymnorhiza (L.) Savigny ●
61263 Bruguiera gymnorhiza (L.) Savigny;木榄(包罗剪定,大头榄,红茄冬,红树,鸡爪榄,鸡爪浪,枷定,剪定,裸根木榄,缅甸木榄,铁榄,五跤梨,五脚里);Burma Mangrove, Common Bruguiera, Manypetaled Mangrove, Multipetalous Bruguiera, Oriental Mangrove ●
61264 Bruguiera littorea (Jack) Steud. = Lumnitzera littorea (Jacq.) Voigt ●◇
61265 Bruguiera madagascariensis DC. = Lumnitzera racemosa Willd. ●
61266 Bruguiera parviflora (Roxb.) Griff.;小花木榄木;Smallflower Bruguiera ●☆
61267 Bruguiera sexangula (Lour.) Poir.;海莲(剪定树,罗古,小叶格拿稍);Oriental Mangrove, Six-angled Bruguiera ●
61268 Bruguiera sexangula (Lour.) Poir. var. rhynchopetala W. C. Ko;尖瓣海莲;Beakpetal Bruguiera ●
61269 Bruguiera sexangula Wight et Arn. var. rhynchopetala W. C. Ko = Bruguiera sexangula (Lour.) Poir. ●
61270 Bruinsmania Miq. = Isertia Schreb. ●☆
61271 Bruinsmia Boerl. et Koord. (1893);歧序安息香属(歧序野茉莉属)●
61272 Bruinsmia polysperma (C. B. Clarke) Steenis;歧序安息香(歧序野茉莉)●
61273 Brunella L. = Prunella L. ■
61274 Brunella Mill. = Prunella L. ■
61275 Brunella Moench = Prunella L. ■
61276 Brunella alba M. Bieb. = Prunella laciniata (L.) L. ■☆
61277 Brunella algeriensis Noë = Prunella vulgaris L. ■
61278 Brunella grandiflora Moench = Prunella grandiflora (L.) Jacq. ■
61279 Brunella ovata Wall. = Prunella hispida Benth. ■
61280 Brunella vulgaris L. = Prunella vulgaris L. ■
61281 Brunella vulgaris L. var. elongata Benth. = Prunella vulgaris L. var. lanceolata (W. P. C. Barton) Fernald ■
61282 Brunella vulgaris L. var. hispida Benth. = Prunella hispida Benth. ■
61283 Brunellia Ruiz et Pav. (1794);槽柱花属(西印度黄栌属);West Indian Sumach ●☆
61284 Brunellia boliviana Britton;玻利维亚槽柱花●☆
61285 Brunellia brittonii Rusby = Brunellia boliviana Britton ●☆
61286 Brunellia crenata Engl. = Brunellia rhoides Rusby ●☆
61287 Brunellia cutervensis Cuatrec. = Brunellia oliveri Britton ●☆
61288 Brunellia oliveri Britton;奥里弗槽柱花●☆
61289 Brunellia pinnata (Pax) Cuatrec. = Brunellia boliviana Britton ●☆
61290 Brunellia rhoides Rusby;盐肤木槽柱花●☆
61291 Brunelliaceae Engl. (1897)(保留科名);槽柱花科(瓣裂果科,西印度黄栌科)●☆
61292 Brunelliaceae Engl. (保留科名) = Cunoniaceae R. Br. (保留科

名)●☆

61293 Brunfelsia L.(1753)(保留属名);番茉莉属(鸳鸯茉莉属);Rain Tree,Raintree,Rain-tree ●

61294 Brunfelsia acuminata Benth.;尖番茉莉(鸳鸯茉莉)●☆

61295 Brunfelsia americana L.;番茉莉(尖番茉莉,美洲番茉莉,夜香花);Franciscan Raintree,Lady of the Night,Lady-of-the-night ●

61296 Brunfelsia australis Benth.;南方番茉莉;Today and Tomorrow,Yesterday ●☆

61297 Brunfelsia calycina Benth.;大鸳鸯茉莉(大花番茉莉,大鸳鸯番茉莉);Large-calyxed Brunfelsia,Yesterday Today and Tomorrow ●

61298 Brunfelsia calycina Benth. = Brunfelsia pauciflora (Cham. et Schltdl.) Benth. ●☆

61299 Brunfelsia calycina Benth. var. eximia (Scheidw. ex Moore et Ayers) L. H. Bailey et Raffill;卓越大鸳鸯茉莉●☆

61300 Brunfelsia calycina Benth. var. macrantha (Lem.) L. H. Bailey et Raffill = Brunfelsia grandiflora D. Don ●☆

61301 Brunfelsia eximia (Scheidw. ex Moore et Ayers) Bosse = Brunfelsia calycina Benth. var. eximia (Scheidw. ex Moore et Ayers) L. H. Bailey et Raffill ●☆

61302 Brunfelsia eximia Blume = Brunfelsia calycina Benth. var. eximia (Scheidw. ex Moore et Ayers) L. H. Bailey et Raffill ●☆

61303 Brunfelsia grandiflora D. Don;大花番茉莉●☆

61304 Brunfelsia hopeana Benth.;二色茉莉(番茉莉,香素馨,紫夜香花);Brazil Raintree,Manaca Raintree,Vegetable Mercury ●☆

61305 Brunfelsia hopeana Benth. = Brunfelsia uniflora D. Don ●☆

61306 Brunfelsia latifolia Benth.;鸳鸯茉莉;Broadleaf Rain Tree,Broadleaf Raintree,Kiss-me-quick,Large-leaved Brunfelsia ●

61307 Brunfelsia macrantha Lem. = Brunfelsia calycina Benth. var. macrantha (Lem.) L. H. Bailey et Raffill ●☆

61308 Brunfelsia macrantha Lem. = Brunfelsia grandiflora D. Don ●☆

61309 Brunfelsia nitida Benth.;古巴茉莉;Cuban Raintree ●☆

61310 Brunfelsia pauciflora (Cham. et Schltdl.) Benth.;少花番茉莉(疏花鸳鸯茉莉);Large-calyxed Brunfelsia,Yesterday Today and Tomorrow,Yesterday-today-and-tomorrow,Yesterday-today-tomorrow ●☆

61311 Brunfelsia pauciflora (Cham. et Schltdl.) Benth. 'Floribunda Compacta';丰花紧凑番茉莉●☆

61312 Brunfelsia pauciflora (Cham. et Schltdl.) Benth. 'Floribunda';丰花番茉莉●☆

61313 Brunfelsia pauciflora (Cham. et Schltdl.) Benth. 'Macrantha';寡大花番茉莉(大花疏花鸳鸯茉莉)●☆

61314 Brunfelsia pauciflora Benth. = Brunfelsia pauciflora (Cham. et Schltdl.) Benth. ●☆

61315 Brunfelsia undulata Sw.;白花番茉莉;Rain Tree,White Raintree ●☆

61316 Brunfelsia uniflora D. Don;单花番茉莉(变色茉莉);One-flower Raintree ●☆

61317 Brunfelsiopsis (Urb.) Kuntze = Brunfelsia L.(保留属名)●

61318 Brunfelsiopsis Urb. = Brunfelsia L.(保留属名)●

61319 Brunia L.(废弃属名) = Brunia Lam.(保留属名)●☆

61320 Brunia Lam.(1785)(保留属名);鳞叶树属(布鲁尼木属)●☆

61321 Brunia albiflora E. Phillips;白花鳞叶树(白花布鲁尼木)●☆

61322 Brunia alopecuroides Thunb.;看麦娘鳞叶树●☆

61323 Brunia candicans Steud.;纯白鳞叶树●☆

61324 Brunia capitella Thunb. = Staavia capitella (Thunb.) Sond. ●☆

61325 Brunia capitellata E. Mey. = Raspalia staavioides (Sond.) Pillans ●☆

61326 Brunia dregeana C. Presl = Staavia capitella (Thunb.) Sond. ●☆

61327 Brunia fragarioides Willd. = Nebelia fragarioides (Willd.) Kuntze ●☆

61328 Brunia globosa Thunb. = Nebelia fragarioides (Willd.) Kuntze ●☆

61329 Brunia glutinosa P. J. Bergius = Staavia glutinosa (P. J. Bergius) Dahl ●☆

61330 Brunia laevis Thunb.;平滑鳞叶树●☆

61331 Brunia lancifolia (Eckl. et Zeyh.) Walp. = Pseudobaeckea cordata (Burm. f.) Nied. ●☆

61332 Brunia laxa Thunb. = Tittmannia laxa (Thunb.) C. Presl ●☆

61333 Brunia levisanus L. = Leucadendron levisanus (L.) P. J. Bergius ●☆

61334 Brunia macrocephala Willd.;大头鳞叶树●☆

61335 Brunia marlothii Schltr. = Brunia macrocephala Willd. ●☆

61336 Brunia microcephala E. Mey. ex Sond. = Nebelia sphaerocephala (Sond.) Kuntze ●☆

61337 Brunia microphylla Thunb. = Raspalia microphylla (Thunb.) Brongn. ●☆

61338 Brunia neglecta Schltr.;忽视鳞叶树●☆

61339 Brunia nodiflora L.;湿生鳞叶树(高山维林图柏,湿生布鲁尼木);Mlanje Cedar,Mountain Cedar,Mountain Cypress,Sapree Wood ●☆

61340 Brunia noduliflora Goldblatt et J. C. Manning;节花鳞叶树●☆

61341 Brunia paleacea P. J. Bergius = Nebelia paleacea (P. J. Bergius) Sweet ●☆

61342 Brunia palustris Schltr. ex Kinscher = Raspalia palustris (Schltr. ex Kirchn.) Pillans ●☆

61343 Brunia passerinoides Schltdl. = Raspalia phylicoides (Thunb.) Arn. ●☆

61344 Brunia phylicoides E. Mey. = Raspalia dregeana (Sond.) Nied. ●☆

61345 Brunia phylicoides Thunb. = Raspalia phylicoides (Thunb.) Arn. ●☆

61346 Brunia pinifolia (L. f.) Brongn. = Pseudobaeckea africana (Burm. f.) Pillans ●☆

61347 Brunia racemosa Brongn. = Pseudobaeckea cordata (Burm. f.) Nied. ●☆

61348 Brunia sacculata Bolus ex Kirchn. = Raspalia sacculata (Bolus ex Kirchn.) Pillans ●☆

61349 Brunia squalida Sond. = Raspalia phylicoides (Thunb.) Arn. ●☆

61350 Brunia staavioides Sond. = Raspalia staavioides (Sond.) Pillans ●☆

61351 Brunia stokoei E. Phillips;斯托克鳞叶树●☆

61352 Brunia teres Oliv. = Pseudobaeckea teres (Oliv.) Dümmer ●☆

61353 Brunia thyrsophora Walp. = Pseudobaeckea africana (Burm. f.) Pillans ●☆

61354 Brunia verticillata Eckl. et Zeyh. = Raspalia virgata (Brongn.) Pillans ●☆

61355 Brunia verticillata L. f. = Staavia verticillata (L. f.) Pillans ●☆

61356 Brunia villosa (C. Presl) Sond. = Raspalia villosa C. Presl ●☆

61357 Brunia virgata Brongn. = Raspalia virgata (Brongn.) Pillans ●☆

61358 Bruniaceae Bercht. et J. Presl = Bruniaceae R. Br. ex DC.(保留科名)●☆

61359 Bruniaceae DC. = Bruniaceae R. Br. ex DC.(保留科名)●☆

61360 Bruniaceae R. Br. ex DC.(1825)(保留科名);鳞叶树科(不路尼亚科,布鲁尼科,假石楠科,鳞叶木科,小叶树科,小叶树木科)●☆

61361 Bruniera Franch. = Wolffia Horkel ex Schleid.(保留属名)■

61362 Bruniera columbiana (H. Karst.) Nieuwl. = Wolffia columbiana H. Karst. ■☆
61363 Bruniera punctata (Griseb.) Nieuwl. = Wolffia brasiliensis Wedd. ■☆
61364 Brunnera Steven(1851);蓝珠草属;Brunnera, Great Forget-me-not ■☆
61365 Brunnera macrophylla (Adams) I. M. Johnst. = Brunnera macrophylla (M. Bieb.) I. M. Johnst. ■☆
61366 Brunnera macrophylla (M. Bieb.) I. M. Johnst.;大叶蓝珠草(西伯利亚牛舌草);Brunnera, Great Forget-me-not, Jack Frost Siberian Bugloss, Largeleaf Brunnera, Siberian Bugloss ■☆
61367 Brunnera macrophylla I. M. Johnst. 'Dawson's White';乳斑大叶蓝珠草■☆
61368 Brunnera orientalis I. M. Johnst.;东方蓝珠草■☆
61369 Brunnera sibirica Steven;西伯利亚蓝珠草■☆
61370 Brunnichia Banks = Brunnichia Banks ex Gaertn. ●☆
61371 Brunnichia Banks ex Gaertn. (1788);黄珊瑚藤属(北美蓼属);American Buckwheat-vine, Anserine Liana, Eardrop Vine, Ladies'-eardrops ●☆
61372 Brunnichia africana Welw. = Afrobrunnichia africana (Welw.) Hutch. et Dalziel ■☆
61373 Brunnichia africana Welw. var. erecta (Asch.) Büttner = Afrobrunnichia erecta (Asch.) Hutch. et Dalziel ■☆
61374 Brunnichia africana Welw. var. glabra Dammer = Afrobrunnichia erecta (Asch.) Hutch. et Dalziel ■☆
61375 Brunnichia cirrhosa Gaertn. = Brunnichia ovata (Walter) Shinners ■☆
61376 Brunnichia congoensis Dammer = Afrobrunnichia erecta (Asch.) Hutch. et Dalziel ■☆
61377 Brunnichia erecta Asch. = Afrobrunnichia erecta (Asch.) Hutch. et Dalziel ■☆
61378 Brunnichia ovata (Walter) Shinners;黄珊瑚藤(北美蓼);Buckwheat Vine, Ladies' Eardrops ■☆
61379 Brunonia R. Br. = Brunonia Sm. ex R. Br. ●■☆
61380 Brunonia Sm. = Brunonia Sm. ex R. Br. ●■☆
61381 Brunonia Sm. ex R. Br. (1810);蓝针花属(兰针垫花属,兰针花属,蓝花根叶草属,留粉花属)●■☆
61382 Brunonia ausfralis Sm. ex R. Br.;蓝针花(兰针花)●■☆
61383 Brunonia australis R. Br. = Brunonia ausfralis Sm. ex R. Br. ●■☆
61384 Brunoniaceae Dumort. (1829)(保留科名);蓝针花科(兰针垫花科,兰针花科,蓝花根叶草科,留粉花科)●☆
61385 Brunoniaceae Dumort. (保留科名) = Goodeniaceae R. Br. (保留科名)●■
61386 Brunoniella Bremek. (1964);小蓝针花属■☆
61387 Brunoniella acaulis (R. Br.) Bremek.;小蓝针花■☆
61388 Brunsfelsia L. = Brunfelsia L. (保留属名)●
61389 Brunsvia Neck. = Croton L. ●
61390 Brunsvigia Heist. (1755);花盏属;Candelabra Flower, Umbrella Lily ■☆
61391 Brunsvigia comptonii W. F. Barker;康普顿花盏■☆
61392 Brunsvigia cooperi Baker;库珀氏花盏■☆
61393 Brunsvigia gigantea Heist. = Brunsvigia josephinae Ker Gawl. ■☆
61394 Brunsvigia grandiflora Lindl.;大花花盏■☆
61395 Brunsvigia gregaria R. A. Dyer;聚生花盏■☆
61396 Brunsvigia gydobergensis D. Müll. -Doblies et U. Müll. -Doblies = Brunsvigia josephinae (Redouté) Ker Gawl. ■☆
61397 Brunsvigia herrei F. M. Leight. ex W. F. Barker;赫勒花盏■☆
61398 Brunsvigia josephinae (Redouté) Ker Gawl.;大花盏(红花盏);Giant Brunsvigia, Josephine's Lily ■☆
61399 Brunsvigia josephinae Ker Gawl. = Brunsvigia josephinae (Redouté) Ker Gawl. ■☆
61400 Brunsvigia litoralis R. A. Dyer;滨海花盏■☆
61401 Brunsvigia lucida Herb. = Nerine laticoma (Ker Gawl.) T. Durand et Schinz ■☆
61402 Brunsvigia marginata (Jacq.) Aiton;具边花盏■☆
61403 Brunsvigia massaiana L. Linden et Rodigas = Crinum kirkii Baker ■☆
61404 Brunsvigia minor Lindl. = Brunsvigia striata (Jacq.) Aiton ■☆
61405 Brunsvigia namaquana D. Müll. -Doblies et U. Müll. -Doblies;纳马夸花盏■☆
61406 Brunsvigia natalensis Baker;纳塔尔花盏■☆
61407 Brunsvigia orientalis (L.) Aiton ex Eckl.;东方花盏;Candelabra Flower ■☆
61408 Brunsvigia orientalis Aiton ex Eckl. = Brunsvigia orientalis (L.) Aiton ex Eckl. ■☆
61409 Brunsvigia pulchra (W. F. Barker) D. Müll. -Doblies et U. Müll. -Doblies;美丽花盏■☆
61410 Brunsvigia radula (Jacq.) Aiton;刮刀花盏■☆
61411 Brunsvigia radulosa Herb.;多刮刀花盏■☆
61412 Brunsvigia striata (Jacq.) Aiton;条纹花盏■☆
61413 Brunsvigia undulata F. M. Leight.;波状花盏■☆
61414 Brunsvigiaceae Horan.;花盏科■
61415 Brunsvigiaceae Horan. = Amaryllidaceae J. St. -Hil. (保留科名)●■
61416 Brunswigiaceae Horan. = Amaryllidaceae J. St. -Hil. (保留科名)●■
61417 Brunyera Bubani = Oenothera L. ●■
61418 Bruquieria Pourr. ex Ortega = Calyxhymenia Ortega ■
61419 Bruquieria Pourr. ex Ortega = Mirabilis L. ■
61420 Bruschia Bertol. = Nyctanthes L. ●
61421 Bruxanelia Dennst. = Blachia Baill. (保留属名)●
61422 Bruxanellia Dennst. ex Kostel. (废弃属名) = Blachia Baill. (保留属名)●
61423 Brya P. Browne = Aldina Endl. (保留属名)●☆
61424 Brya P. Browne(1756);柏雷木属(椰豆木属)●☆
61425 Brya Vell. = Hirtella L. ●☆
61426 Brya buxifolia Urb.;黄杨叶柏雷木(黄杨叶椰豆木);Cocoswood, Cocuswood ●☆
61427 Brya ebenus DC.;牙买加柏雷木(黑椰豆木);American Ebony, Brown Ebony, Coccus Wood, Cocos Wood, Cocus Wood, Cocuswood, Ebony Coccuswood, Granadilla, Jamaica Ebony, Jamaican Ebony, West Indian Ebony, West Indies Ebony ●☆
61428 Brya leonensis Lodd. ex Loudon;莱昂柏雷木●☆
61429 Bryantea Raf. (废弃属名) = Neolitsea (Benth. et Hook. f.) Merr. (保留属名)●
61430 Bryanthus J. G. Gmel. (1769);繁花鹃属(布利安木属,布利杜鹃属,线香石南属)●☆
61431 Bryanthus coeruleus Dippel = Phyllodoce caerulea (L.) Bab. ●◇
61432 Bryanthus gmelinii D. Don;线香石南(布利安木,布利杜鹃)●☆
61433 Bryanthus musciformis (Poir.) Nakai = Bryanthus gmelinii D. Don ●☆
61434 Bryanthus taxifolius Gray = Phyllodoce caerulea (L.) Bab. ●◇
61435 Bryantia Webb ex Gaudich. = Pandanus Parkinson ex Du Roi ●■
61436 Bryantiella J. M. Porter = Gilia Ruiz et Pav. ■●☆
61437 Bryaspis P. A. Duvign. (1954);满盾豆属■☆

61438 Bryaspis humularioides Gledhill?;葎草满盾豆■☆
61439 Bryaspis humularioides Gledhill subsp. falcistipulata?;镰托叶满盾豆■☆
61440 Bryaspis lupulina (Planch.) P. A. Duvign.;满盾豆■☆
61441 Brylkinia F. Schmidt(1868);曲柄草属(白列金氏草属,扁穗草属,扁穗属);Brylkinia ■
61442 Brylkinia caudata (Munro ex A. Gray) F. Schmidt;曲柄草(扁穗,扁穗草);Caudate Brylkinia ■
61443 Brylkinia caudata (Munro) F. Schmidt = Brylkinia caudata (Munro ex A. Gray) F. Schmidt ■
61444 Brylkinia caudata (Thunb.) F. Schmidt = Brylkinia caudata (Munro ex A. Gray) F. Schmidt ■
61445 Brylkinia schmidtii Ohwi = Brylkinia caudata (Munro ex A. Gray) F. Schmidt ■
61446 Bryobium Lindl. (1838);藓兰属■
61447 Bryobium Lindl. = Eria Lindl.(保留属名)■
61448 Bryobium pudicum (Ridl.) Y. P. Ng et P. J. Cribb;藓兰(版纳毛兰);Xishuangbanna Eria, Xishuangbanna Hairorchis ■
61449 Bryocarpum Hook. f. et Thomson(1857);长果报春属(藓果草属,藓蒴报春属);Bryocarpum ■☆
61450 Bryocarpum himalaicum Hook. f. et Thomson;长果报春(藓果草,藓蒴报春);Common Bryocarpum ■
61451 Bryocles Salisb. = Funkia Spreng. ■
61452 Bryocles ventricosa Salisb. = Hosta ventricosa (Salisb.) Stearn ■
61453 Bryodes Benth. (1846);马斯克林透骨草属■☆
61454 Bryodes madagascariensis Bonati = Psammetes madagascariensis (Bonati) Eb. Fisch. et Hepper ■☆
61455 Bryodes micrantha Benth.;马岛透骨草■☆
61456 Bryodes perrieri Bonati = Lindernia nummulariifolia (D. Don) Wettst. ■
61457 Bryomorpha Kar. et Kir. = Thylacospermum Fenzl ■
61458 Bryomorpha rupifraga Kar. et Kir. = Thylacospermum caespitosum (Cambess.) Schischk. ■
61459 Bryomorphe Harv. (1863);帚藓菊属●☆
61460 Bryomorphe aretioides (Turcz.) Druce = Bryomorphe lycopodioides (Sch. Bip. ex Walp.) Levyns ■☆
61461 Bryomorphe lycopodioides (Sch. Bip. ex Walp.) Levyns;帚藓菊■☆
61462 Bryomorphe lycopodioides (Walp.) Levyns. = Bryomorphe lycopodioides (Sch. Bip. ex Walp.) Levyns ■☆
61463 Bryomorphe zeyheri Harv. = Bryomorphe lycopodioides (Sch. Bip. ex Walp.) Levyns ■☆
61464 Bryonia L. (1753);泻根属(欧薯蓣属,欧洲甜瓜属);Bryony, White Bryony ■☆
61465 Bryonia L. = Zehneria Endl. ■
61466 Bryonia abyssinica Lam. = Coccinia abyssinica (Lam.) Cogn. ■☆
61467 Bryonia acuta Desf.;尖泻根■☆
61468 Bryonia affinis Endl. = Diplocyclos palmatus (L.) C. Jeffrey ■
61469 Bryonia afghanica Podlech = Bryonia aspera Stev. ex Ledeb. ■☆
61470 Bryonia africana L. = Kedrostis africana (L.) Cogn. ■☆
61471 Bryonia alba L.;白泻根;White Bryonia, White Bryony ■☆
61472 Bryonia alba sensu Boiss. = Bryonia aspera Stev. ex Ledeb. ■☆
61473 Bryonia alba sensu Vassilcz. = Bryonia monoica Aitch. et Hemsl. ■☆
61474 Bryonia althaeoides Ser. = Mukia maderaspatana (L.) M. Roem. ■
61475 Bryonia amplexicaulis Lam. = Solena amplexicaulis (Lam.) Gandhi ■
61476 Bryonia aspera Stev. ex Ledeb.;粗糙泻根■☆
61477 Bryonia callosa Rottler = Cucumis melo L. var. agrestis Naudin ■
61478 Bryonia capillacea Schumach. = Zehneria capillacea (Schumach.) C. Jeffrey ■☆
61479 Bryonia cheirophylla Wall. = Luffa cylindrica (L.) M. Roem. ■☆
61480 Bryonia ciliata Moench = Kedrostis africana (L.) Cogn. ■☆
61481 Bryonia cochinchinensis Lour. = Gymnopetalum chinense (Lour.) Merr. ■
61482 Bryonia collosa Rottler = Cucumis melo L. subsp. agrestis (Naudin) Pangalo ■
61483 Bryonia cordata Thunb. = Zehneria scabra (L. f.) Sond. ●☆
61484 Bryonia cordifolia L. = Mukia maderaspatana (L.) M. Roem. ■
61485 Bryonia cretica L.;泻根;Cretan Bryony ■☆
61486 Bryonia cretica L. subsp. dioica (Jacq.) Tutin = Bryonia dioica Jacq. ■☆
61487 Bryonia cretica L. subsp. dioica Jacq. = Bryonia dioica Jacq. ■☆
61488 Bryonia cucumeroides Ser. = Trichosanthes cucumeroides (Ser.) Maxim. ex Franch. et Sav. ■
61489 Bryonia deltoidea Schumach. et Thonn. = Zehneria hallii C. Jeffrey ■☆
61490 Bryonia digyna Pomel = Bryonia dioica Jacq. ■☆
61491 Bryonia dioica Jacq. = Bryonia dioica Sessé et Moc. ■☆
61492 Bryonia dioica Jacq. subsp. acuta (Desf.) Batt. = Bryonia acuta Desf. ■☆
61493 Bryonia dioica Jacq. var. acuta (Desf.) Coss. = Bryonia acuta Desf. ■☆
61494 Bryonia dioica Jacq. var. digyna (Pomel) Batt. = Bryonia dioica Jacq. ■☆
61495 Bryonia dioica Jacq. var. lavifrons Pau = Bryonia dioica Jacq. ■☆
61496 Bryonia dioica sensu Ledeb. = Bryonia aspera Stev. ex Ledeb. ■☆
61497 Bryonia dioica Sessé et Moc.;异株泻根(欧洲甜瓜,泻根);Actaea Rubra, Big-root, Bryouy, Canterbury Jack, Cowbind, Cow's Lick, Cretan Bryony, Dead Creepers, Death Warrant, Devil's Cherries, Devil's Cherry, Devil's Turnip, Dog's Cherries, Dog's Cherry, Elphamy, English Bryony, English Mandrake, Fellon-berry, Grapewort, Hedge Grape, Hedge Vine, Hop, Isle of Wight Vine, Jack-in-the-hedge, Lady's Seal, Mad Nip, Mandrake, Mednip, Mmurrain Berries, Murren, Murren-berries, Nigh' r-bonnets, Night Bonnets, Our Lady's Seal, Poisoning Berries, Red Bryony, Red-berried Bryony, Red-berry Bryony, Rowberry, Snakeberry, Tetter-berries, Tetterberry, Vine, White Bryony, White Vine, White Wild Vine, Wild Cucumber, Wild Hop, Wild Navew, Wild Nep, Wild Nept, Wild Vine, Woman-drake, Wood Vine ■☆
61498 Bryonia dissecta Thunb. = Kedrostis africana (L.) Cogn. ■☆
61499 Bryonia epigaea Rottler = Corallocarpus epigaeus (Rottler) Hook. f. ex C. B. Clarke ■☆
61500 Bryonia grandis L. = Coccinia grandis (L.) Voigt ■
61501 Bryonia grossulariifolia E. Mey. = Kedrostis africana (L.) Cogn. ■☆
61502 Bryonia guineensis G. Don = Lagenaria guineensis (G. Don) C. Jeffrey ■☆
61503 Bryonia hastata Lour. = Solena amplexicaulis (Lam.) Gandhi ■
61504 Bryonia hastata Lour. = Solena heterophylla Lour. ■
61505 Bryonia hausknechtiana Bornm. = Bryonia aspera Stev. ex Ledeb. ■☆
61506 Bryonia indica (Lour.) Rabenant. = Zehneria japonica (Thunb. ex A. Murray) H. Y. Liu ■

61507 Bryonia japonica Thunb. = Neoachmandra japonica (Thunb.) W. J. de Wilde et Duyfjes ■

61508 Bryonia japonica Thunb. ex A. Murray = Zehneria japonica (Thunb. ex A. Murray) H. Y. Liu ■

61509 Bryonia jatrophifolia A. Rich. = Coccinia adoensis (A. Rich.) Cogn. ■☆

61510 Bryonia laciniosa L. = Bryonopsis laciniosa (L.) Naudin ■

61511 Bryonia laciniosa L. = Diplocyclos palmatus (L.) C. Jeffrey ■

61512 Bryonia laevis Thunb. = Ceratiosicyos laevis (Thunb.) A. Meeuse ●☆

61513 Bryonia leucocarpa Blume = Zehneria indica (Lour.) Rabenant. ■

61514 Bryonia macrostylis Heilbr. et Bilge = Bryonia aspera Stev. ex Ledeb. ■☆

61515 Bryonia marginata Blume = Scopellaria marginata (Blume) W. J. de Wilde et Duyfjes ■

61516 Bryonia marginata Blume = Zehneria marginata (Blume) Rabenant. ■

61517 Bryonia maysorensis Wight et Arn. = Zehneria maysorensis (Wight et Arn.) Arn. ■

61518 Bryonia monoica Aitch. et Hemsl.;同株泻根■☆

61519 Bryonia mucronata Blume = Zehneria mucronata (Blume) Miq. ■

61520 Bryonia multifida E. Mey. = Kedrostis africana (L.) Cogn. ■☆

61521 Bryonia multiflora sensu Chakr. = Bryonia monoica Aitch. et Hemsl. ■☆

61522 Bryonia nana Lam. = Kedrostis nana (Lam.) Cogn. ■☆

61523 Bryonia napaulensis Ser. = Solena heterophylla Lour. subsp. napaulensis (Ser.) W. J. de Wilde et Duyfjes ■

61524 Bryonia obtusa A. Rich. = Cucumis ficifolius A. Rich. ■☆

61525 Bryonia palmata L. = Diplocyclos palmatus (L.) C. Jeffrey ■

61526 Bryonia pectinata E. Mey. = Trochomeria hookeri Harv. ■☆

61527 Bryonia pedunculosa Ser. = Herpetospermum pedunculosum (Ser.) C. B. Clarke ■

61528 Bryonia perrottetiana Ser. = Kedrostis foetidissima (Jacq.) Cogn. ■☆

61529 Bryonia pinnatifida Burch. = Kedrostis africana (L.) Cogn. ■☆

61530 Bryonia punctata Thunb.;斑点泻根■☆

61531 Bryonia quinqueloba Thunb. = Coccinia quinqueloba (Thunb.) Cogn. ■☆

61532 Bryonia rostrata Rottler = Kedrostis foetidissima (Jacq.) Cogn. ■☆

61533 Bryonia scabra L. f. = Zehneria scabra (L. f.) Sond. ●☆

61534 Bryonia scabrella L. = Mukia maderaspatana (L.) M. Roem. ■

61535 Bryonia scabrella L. f. = Cucumis maderaspatanus L. ■

61536 Bryonia scrobiculata Hochst. ex A. Rich. = Zehneria scabra (L. f.) Sond. ●☆

61537 Bryonia tenuis Klotzsch = Diplocyclos tenuis (Klotzsch) C. Jeffrey ■☆

61538 Bryonia transoxana Vassilcz.;外阿穆达尔泻根■☆

61539 Bryonia transoxana Vassilcz. = Bryonia monoica Aitch. et Hemsl. ■☆

61540 Bryonia triloba Thunb. = Kedrostis nana (Lam.) Cogn. ■☆

61541 Bryonia umbellata (Klein ex Willd.) Roxb. = Solena amplexicaulis (Lam.) Gandhi ■

61542 Bryonia verrucosa Dryand.;多疣泻根■☆

61543 Bryoniaceae Adans. ex Post et Kuntze = Cucurbitaceae Juss.(保留科名)●■

61544 Bryoniaceae G. Mey. = Cucurbitaceae Juss.(保留科名)●■

61545 Bryoniaceae Post et Kuntze = Cucurbitaceae Juss.(保留科名)●■

61546 Bryoniaceae Post et Kuntze;泻根科■

61547 Bryoniastrum Fabr. = Sicyos L. ■

61548 Bryoniastrum Heist. ex Fabr. = Sicyos L. ■

61549 Bryonopsis Arn. (1840);拟泻根属;Bryonopsis ■

61550 Bryonopsis Arn. = Kedrostis Medik. ■☆

61551 Bryonopsis affinis (Endl.) Cogn. = Diplocyclos palmatus (L.) C. Jeffrey ■

61552 Bryonopsis laciniosa (L.) Naudin;拟泻根(毒瓜);Marble Vine ■

61553 Bryonopsis laciniosa (L.) Naudin = Diplocyclos palmatus (L.) C. Jeffrey ■

61554 Bryonopsis laciniosa (L.) Naudin var. erythrocarpa (Naudin) Naudin = Diplocyclos palmatus (L.) C. Jeffrey ■

61555 Bryonopsis laciniosa (L.) Naudin var. erythrocarpa Naudin = Diplocyclos palmatus (L.) C. Jeffrey ■

61556 Bryonopsis laciniosa (L.) Naudin var. walkeri Chakr. = Diplocyclos palmatus (L.) C. Jeffrey ■

61557 Bryonopsis laciniosa Naudin = Diplocyclos palmatus (L.) C. Jeffrey ■

61558 Bryonopsis laciniosa Naudin var. walkeri Chakr. = Diplocyclos palmatus (L.) C. Jeffrey ■

61559 Bryophthalmum E. Mey. = Moneses Salisb. ex Gray ■

61560 Bryophthalmum uniflorum (L.) E. Mey. = Moneses uniflora (L.) A. Gray ■

61561 Bryophyllum Salisb. (1805);落地生根属;Kalanhoe ■

61562 Bryophyllum Salisb. = Kalanchoe Adans. ●■

61563 Bryophyllum beauverdi (Raym. -Hamet) A. Berger = Kalanchoe beauverdii Raym. -Hamet ■☆

61564 Bryophyllum beauverdii (Raym. -Hamet) A. Berger;线叶落地生根(纤叶伽蓝菜);Slender Kalanchoe, Sotre-sotry ■☆

61565 Bryophyllum calycinum Salisb. = Bryophyllum pinnatum (Lam.) Oken ■

61566 Bryophyllum calycinum Salisb. = Kalanchoe pinnata (Lam.) Pers. ■

61567 Bryophyllum crenatum Baker = Kalanchoe laxiflora Baker ■☆

61568 Bryophyllum daigremontianum (Raym. -Hamet et H. Perrier) A. Berger;大叶落地生根(戴氏伽蓝菜,锐叶掌上珠,纹叶伽蓝菜);Airplant, Devil's Backbone, Devil's-backbone, Flopper, Largeleaf Kalanchoe, Maternity Plant, Mexican Hat Plant ■

61569 Bryophyllum daigremontianum (Raym. -Hamet et H. Perrier) A. Berger = Kalanchoe daigremontiana Raym. -Hamet et H. Perrier ■

61570 Bryophyllum fedtschenkoi (Raym. -Hamet et H. Perrier) C. Y. Cheng;璎珞洋吊钟;Kalanchoe Stonecrop, Lavender Scallops, Lavender-scallops, South American Air Plant, South American Air-plant ■☆

61571 Bryophyllum fedtschenkoi (Raym. -Hamet et H. Perrier) Lauz. -March. = Kalanchoe fedtschenkoi Raym. -Hamet et H. Perrier ■☆

61572 Bryophyllum pinnatum (L. f.) Oken;落地生根(打不死,大疔癀,大还魂,古仔灯,厚面皮,火炼丹,脚目草,接骨草,接骨丹,落叶生根,枪刀草,枪刀叶,晒不死,伤药,天灯笼,土三七,新娘灯,叶爆芽,叶生,叶生根,着生药);Air Plant, Air-plant, Cathedral Bells, Curtain Plant, Floppers, Good Luck Leaf, Life-plant, Never-die, Oliwa-ku-kahakai, Pinnate Kalanchoe ■

61573 Bryophyllum pinnatum (Lam.) Oken = Bryophyllum pinnatum (L. f.) Oken ■

61574 Bryophyllum pinnatum (Lam.) Oken = Kalanchoe pinnata (Lam.) Pers. ■

61575 Bryophyllum proliferum Bowie ex Hook.;多育落地生根;

Blooming Boxes ■☆
61576 Bryophyllum proliferum Bowie ex Hook. = Kalanchoe prolifera (Bowie ex Hook.) Raym. -Hamet ■☆
61577 Bryophyllum scandens (H. Perrier) A. Berger = Kalanchoe beauverdii Raym. -Hamet ■☆
61578 Bryophyllum schizophyllum (Baker) A. Berger = Kalanchoe schizophylla (Baker) Baill. ■☆
61579 Bryophyllum tubiflorum Harv. = Kalanchoe delagoensis Eckl. et Zeyh. ■
61580 Bryophyllum tubiflorum Harv. = Kalanchoe tubiflora (Harv.) Raym. -Hamet ■
61581 Bryophyllum verticillatum (Scott-Elliot) A. Berger = Kalanchoe delagoensis Eckl. et Zeyh. ■
61582 Bryopsis Reiche = Reicheella Pax ■☆
61583 Bubalina Raf. = Burchellia R. Br. ●☆
61584 Bubania Girard = Limoniastrum Fabr. ●☆
61585 Bubania migiurtina Chiov. = Ceratolimon migiurtinum (Chiov.) M. B. Crespo et Lledò ■☆
61586 Bubbia Tiegh. (1900); 巴布林仙属(布比林仙属,布波林仙属)●☆
61587 Bubbia Tiegh. = Takhtajania Baranova et J. -F. Leroy ●☆
61588 Bubbia Tiegh. = Zygogynum Baill. ●☆
61589 Bubbia perrieri Capuron = Takhtajania perrieri (Capuron) Baranova et J. -F. Leroy ●☆
61590 Bubon L. = Athamanta L. ■☆
61591 Bubon aphyllum Cham. et Schltdl. = Deverra denudata (Viv.) Pfisterer et Podlech subsp. aphylla (Cham. et Schltdl.) Pfisterer et Podlech ■☆
61592 Bubon buchtormensis Fisch. = Libanotis buchtormensis (Fisch.) DC. ■
61593 Bubon capense (Eckl. et Zeyh.) Sond. = Peucedanum polyactinum B. L. Burtt ■☆
61594 Bubon eriocephalus Pall. ex Spreng. = Seseli eriocephalum (Pall. ex Spreng.) Schischk. ■
61595 Bubon galbanum L. = Peucedanum galbanum (L.) Drude ■☆
61596 Bubon gummiferum L. = Peucedanum gummiferum (L.) Wijnands ■☆
61597 Bubon hypoleucum Meisn. = Peucedanum tenuifolium Thunb. ■☆
61598 Bubon laevigatum Aiton = Peucedanum capense (Thunb.) Sond. var. lanceolatum Sond. ●■☆
61599 Bubon macedonicum L. = Athamanta macedonica (L.) Spreng. ■☆
61600 Bubon montanum Sond. = Peucedanum tenuifolium Thunb. ■☆
61601 Bubon proliferum Burm. f. = Glia prolifera (Burm. f.) B. L. Burtt ■☆
61602 Bubon tenuifolium (Thunb.) Sond. = Peucedanum tenuifolium Thunb. ■☆
61603 Bubon tortuosus Desf. = Deverra tortuosa (Desf.) DC. ■☆
61604 Bubonium Hill = Asteriscus Mill. ●■☆
61605 Bubonium aquaticum (L.) Hill = Asteriscus aquaticus (L.) Less. ■☆
61606 Bubonium graveolens (Forssk.) Maire = Asteriscus graveolens (Forssk.) Less. ■☆
61607 Bubonium graveolens (Forssk.) Maire subsp. graveolens = Asteriscus graveolens (Forssk.) Less. ■☆
61608 Bubonium graveolens (Forssk.) Maire subsp. odorum (Schousb.) Wiklund = Asteriscus graveolens (Forssk.) Less. subsp. odorus (Schousb.) Greuter ■☆
61609 Bubonium graveolens (Forssk.) Maire subsp. stenophyllum (Link) Halvorsen = Asteriscus graveolens (Forssk.) Less. subsp. stenophyllus (Link) Greuter ■☆
61610 Bubonium graveolens (Forssk.) Maire var. ambiguum Maire = Asteriscus graveolens (Forssk.) Less. ■☆
61611 Bubonium imbricatum (Cav.) Litard. = Asteriscus imbricatus (Cav.) DC. ■☆
61612 Bubonium intermedium (DC.) Halvorsen et Wiklund = Asteriscus intermedius (DC.) Pit. et Proust ■☆
61613 Bubonium longiradiatum Maire = Asteriscus schultzii (Bolle) Pit. et Proust ■☆
61614 Bubonium odorum (Schousb.) Maire = Asteriscus graveolens (Forssk.) Less. subsp. odorus (Schousb.) Greuter ■☆
61615 Bubonium odorum (Schousb.) Maire var. cavanillesi (Caball.) Maire = Asteriscus graveolens (Forssk.) Less. subsp. odorus (Schousb.) Greuter ■☆
61616 Bubonium odorum (Schousb.) Maire var. cavanillesii (Caball.) Maire = Asteriscus schultzii (Bolle) Pit. et Proust ■☆
61617 Bubonium odorum (Schousb.) Maire var. eriactinum Maire = Asteriscus graveolens (Forssk.) Less. subsp. odorus (Schousb.) Greuter ■☆
61618 Bubonium odorum (Schousb.) Maire var. fruticosum Maire = Asteriscus schultzii (Bolle) Pit. et Proust ■☆
61619 Bubonium odorum (Schousb.) Maire var. fruticosus Maire = Asteriscus graveolens (Forssk.) Less. subsp. odorus (Schousb.) Greuter ■☆
61620 Bubonium odorum (Schousb.) Maire var. paui (Caball.) Maire = Asteriscus graveolens (Forssk.) Less. subsp. odorus (Schousb.) Greuter ■☆
61621 Bubonium odorum (Schousb.) Maire. f. condensatum Maire = Asteriscus graveolens (Forssk.) Less. subsp. odorus (Schousb.) Greuter ■☆
61622 Bubonium schultzii (Bolle) Svent. = Asteriscus schultzii (Bolle) Pit. et Proust ■☆
61623 Bubonium sericeum (L. f.) Halvorsen et Wiklund = Asteriscus sericeus (L. f.) DC. ■☆
61624 Bubonium sericeum (L. f.) Maire = Asteriscus sericeus (L. f.) DC. ■☆
61625 Bubonium smithii (Webb) Halvorsen = Asteriscus smithii (Webb) Walp. ■☆
61626 Bubroma Ehrh. = Trifolium L. ■
61627 Bubroma Schreb. = Guazuma Mill. ●☆
61628 Bucafer Adans. = Ruppia L. ■
61629 Bucanion Steven = Heliotropium L. ●■
61630 Buccafarrea Bubani = Potamogeton L. ■
61631 Buccaferrea Mich. ex Petagna = Bucafer Adans. ■
61632 Buccaferrea Petagna = Bucafer Adans. ■
61633 Buccaferrea Petagna = Ruppia L. ■
61634 Buccaferrea cirrhosa Petagna = Ruppia cirrhosa (Petagna) Grande ■☆
61635 Buccella Luer = Masdevallia Ruiz et Pav. ■☆
61636 Buccella Luer(2006); 布克兰属■☆
61637 Bucco J. C. Wendl(废弃属名) = Agathosma Willd. (保留属名) ●☆
61638 Bucco acuminata J. C. Wendl. = Agathosma imbricata (L.) Willd. ●☆
61639 Bucco cuspidata J. C. Wendl. = Agathosma serpyllacea Licht. ex

Roem. et Schult. ●☆

61640 Bucco erecta J. C. Wendl. = Agathosma capensis (L.) Dümmer ●☆

61641 Bucco linifolia Roem. et Schult. = Agathosma linifolia (Roem. et Schult.) Licht. ex Bartl. et H. L. Wendl. ●☆

61642 Bucco ovata (Thunb.) J. C. Wendl. = Agathosma ovata (Thunb.) Pillans ●☆

61643 Bucco ventenatiana Roem. et Schult. = Agathosma corymbosa (Montin) G. Don ●☆

61644 Bucculina Lindl. = Holothrix Rich. ex Lindl.(保留属名)■☆

61645 Bucculina aspera Lindl. = Holothrix aspera (Lindl.) Rchb. f. ■☆

61646 Bucephalandra Schott(1858);加岛南星属■☆

61647 Bucephalandra motleyana Schott;加岛南星●☆

61648 Bucephalon L.(废弃属名)= Trophis P. Browne(保留属名)●☆

61649 Bucephalophora Pau(1887);地中海蓼属■☆

61650 Bucephalophora aculeata Pau = Rumex bucephalophora L. ■☆

61651 Bucephalora Pau = Rumex L. ■●

61652 Bucera P. Browne = Bucida L.(保留属名)●☆

61653 Buceragenia Greenm.(1897);闭壳骨属☆

61654 Buceragenia glandulosa Léonard;多腺闭壳骨☆

61655 Buceragenia hirsuta Léonard;毛闭壳骨☆

61656 Buceragenia minutiflora Greenm.;闭壳骨☆

61657 Buceras Haller = Trigonella L. ■

61658 Buceras Haller ex All. = Trigonella L. ■

61659 Buceras P. Browne(废弃属名)= Bucida L.(保留属名)●☆

61660 Buceras P. Browne(废弃属名)= Terminalia L.(保留属名)●

61661 Bucerosia Endl. = Boucerosia Wight et Arn. ■☆

61662 Bucetum Parn. = Festuca L. ■

61663 Buch'osia Vell. = Heteranthera Ruiz et Pav.(保留属名)■☆

61664 Buchanania Sm. = Colebrookea Sm. ●

61665 Buchanania Spreng.(1802);山様子属(山羡子属,天干果属);Buchanania ●

61666 Buchanania arborescens (Blume) Blume;山様子(乔木山様子,山様木,山様仔,山羡子);Arborescent Buchanania, Buchanania ●

61667 Buchanania florida Schauer = Buchanania arborescens (Blume) Blume ●

61668 Buchanania florida Schauer var. arborescens Pierre = Buchanania arborescens (Blume) Blume ●

61669 Buchanania florida Schauer var. dongnaiensis Pierre = Buchanania arborescens (Blume) Blume ●

61670 Buchanania lanzan Spreng.;兰桑山様子;Almondette, Cheronjee, Cuddapah Almond ●☆

61671 Buchanania latifolia Roxb.;豆腐果(山様子,天干果,云南山様子);Broadleaf Buchanania, Broad-leaved Buchanania ●

61672 Buchanania microphylla Engl.;小叶山様子(赤南,山马耳);Smallleaf Buchanania, Small-leaved Buchanania ●

61673 Buchanania obovata Engl.;倒卵山様子●☆

61674 Buchanania spinosa?;刺山様子(有刺山様子)●☆

61675 Buchanania yunnanensis C. Y. Wu;云南山様子;Yunnan Buchanania ●◇

61676 Buchaniana Pierre = Buchanania Spreng. ●

61677 Bucharea Raf. = Convolvulus L. ■●

61678 Buchena Heynh. = Thryptomene Endl.(保留属名)●☆

61679 Buchenavia Eichler(1866)(保留属名);布切木属●☆

61680 Buchenavia capitata (Vahl) Eichler;头状布切木●☆

61681 Buchenavia capitata Eichler = Buchenavia capitata (Vahl) Eichler ●☆

61682 Buchenroedera Eckl. et Zeyh. = Lotononis (DC.) Eckl. et Zeyh.(保留属名)■

61683 Buchenroedera alpina Eckl. et Zeyh. = Lotononis alpina (Eckl. et Zeyh.) B. -E. van Wyk ■☆

61684 Buchenroedera biflora Bolus = Lotononis eriocarpa (E. Mey.) B. -E. van Wyk ■☆

61685 Buchenroedera caerulescens (E. Mey.) C. Presl = Lotononis caerulescens (E. Mey.) B. -E. van Wyk ■☆

61686 Buchenroedera glabrescens Dümmer = Lotononis glabrescens (Dümmer) B. -E. van Wyk ■☆

61687 Buchenroedera glabriflora N. E. Br. = Lotononis caerulescens (E. Mey.) B. -E. van Wyk ■☆

61688 Buchenroedera gracilis Eckl. et Zeyh. = Lotononis alpina (Eckl. et Zeyh.) B. -E. van Wyk subsp. multiflora (Eckl. et Zeyh.) B. -E. van Wyk ■☆

61689 Buchenroedera griquana Schltr. = Lotononis stricta (Eckl. et Zeyh.) B. -E. van Wyk ■☆

61690 Buchenroedera holosericea (E. Mey.) Benth. = Lotononis holosericea (E. Mey.) B. -E. van Wyk ■☆

61691 Buchenroedera jacottetii Schinz = Lotononis jacottetii (Schinz) B. -E. van Wyk ■☆

61692 Buchenroedera lanceolata (E. Mey.) C. Presl = Lotononis lanceolata (E. Mey.) Benth. ■☆

61693 Buchenroedera lotononoides Scott-Elliot = Lotononis lotononoides (Scott-Elliot) B. -E. van Wyk ■☆

61694 Buchenroedera macowanii Dümmer = Lotononis pulchella (E. Mey.) B. -E. van Wyk ■☆

61695 Buchenroedera meyeri C. Presl = Lotononis meyeri (C. Presl) B. -E. van Wyk ■☆

61696 Buchenroedera multiflora Eckl. et Zeyh. = Lotononis alpina (Eckl. et Zeyh.) B. -E. van Wyk subsp. multiflora (Eckl. et Zeyh.) B. -E. van Wyk ■☆

61697 Buchenroedera pauciflora Schltr. = Lotononis carnosa (Eckl. et Zeyh.) Benth. ■☆

61698 Buchenroedera sparsiflora J. M. Wood et M. S. Evans = Lotononis galpinii Dümmer ■☆

61699 Buchenroedera spicata Harv. = Lotononis harveyi B. -E. van Wyk ●☆

61700 Buchenroedera tenuifolia Eckl. et Zeyh.;细叶布切木●☆

61701 Buchenroedera tenuifolia Eckl. et Zeyh. = Lotononis pulchella (E. Mey.) B. -E. van Wyk ■☆

61702 Buchenroedera tenuifolia Eckl. et Zeyh. var. pulchella (E. Mey.) Harv. = Lotononis pulchella (E. Mey.) B. -E. van Wyk ■☆

61703 Buchenroedera teretifolia Eckl. et Zeyh. = Aspalathus albens L. ●☆

61704 Buchenroedera trichodes (E. Mey.) C. Presl = Lotononis trichodes (E. Mey.) B. -E. van Wyk ■☆

61705 Buchenroedera umbellata Harv. = Lotononis meyeri (C. Presl) B. -E. van Wyk ■☆

61706 Buchenroedera uniflora Dümmer = Lotononis caerulescens (E. Mey.) B. -E. van Wyk ■☆

61707 Buchenroedera viminea (E. Mey.) C. Presl = Lotononis viminea (E. Mey.) B. -E. van Wyk ■☆

61708 Buchera Rchb. = Hornungia Rchb. ■

61709 Bucheria Heynh. = Gomphotis Raf.(废弃属名)●☆

61710 Bucheria Heynh. = Thryptomene Endl.(保留属名)●☆

61711 Buchholzia Engl.(1886);西非白花菜属●☆

61712 Buchholzia coriacea Engl.;革质西非白花菜●☆

61713 Buchholzia engleri Gilg = Buchholzia coriacea Engl. ●☆
61714 Buchholzia macrophylla Pax = Buchholzia coriacea Engl. ●☆
61715 Buchholzia macrothyrsa Gilg et Gilg-Ben. = Buchholzia tholloniana Hua ●☆
61716 Buchholzia polyantha Gilg et Gilg-Ben. = Buchholzia tholloniana Hua ●☆
61717 Buchholzia tholloniana Hua;托尔西非白花菜●☆
61718 Buchia D. Dietr. = Bouchea Cham.(保留属名)●☆
61719 Buchia Kunth = Perama Aubl. ●☆
61720 Buchingera Boiss. et Hohen. = Asperuginoides Rauschert ■☆
61721 Buchingera F. Schultz = Cuscuta L. ■
61722 Buchingera axillaris Boiss. et Hohen. = Asperuginoides axillaris (Boisset Hohen.) Rauschert ■☆
61723 Buchloe Engelm.(1859)(保留属名);野牛草属;Buffalo Grass,Buffalograss ■
61724 Buchloe dactyloides (Nutt.) Engelm.;野牛草;Buffalo Grass, Buffalograss,Buffalo-grass,Common Buffalograss,Texoka Buffalograss ■
61725 Buchlomimus Reeder,C. Reeder et Rzed.(1965);拟野牛草属■☆
61726 Buchlomimus nervatus (Swallen) Reeder,Reeder et Rzed.;拟野牛草■☆
61727 Buchnera L.(1753);黑草属(鬼羽箭属);Blackgrass,Buchnera ■
61728 Buchnera aethiopica L. = Sutera aethiopica (L.) Kuntze ■☆
61729 Buchnera affinis De Wild.;近缘黑草■☆
61730 Buchnera africana L. = Bartsia trixago L. ■☆
61731 Buchnera albiflora V. Naray.;白花黑草■☆
61732 Buchnera americana L.;美洲黑草;Blue Hearts ■☆
61733 Buchnera andongensis Hiern;安东黑草■☆
61734 Buchnera androsacea Merxm.;点地梅黑草■☆
61735 Buchnera angolensis Engl.;安哥拉黑草■☆
61736 Buchnera angustifolia D. Don = Striga angustifolia (D. Don) C. J. Saldanha ■
61737 Buchnera arenicola R. E. Fr.;沙生黑草■☆
61738 Buchnera asiatica L. = Striga asiatica (L.) Kuntze ■
61739 Buchnera attenuata V. Naray.;渐狭黑草■☆
61740 Buchnera aurantiaca Burch. = Jamesbrittenia aurantiaca (Burch.) Hilliard ■☆
61741 Buchnera bampsiana Mielcarek;邦氏黑草■☆
61742 Buchnera bangweolensis R. E. Fr.;班韦黑草■☆
61743 Buchnera baumii Engl. et Gilg;鲍姆黑草■☆
61744 Buchnera benthamiana V. Naray. = Buchnera nigricans (Benth.) V. Naray. ■☆
61745 Buchnera bequaertii De Wild.;贝卡尔黑草■☆
61746 Buchnera bilabiata Thunb. = Striga bilabiata (Thunb.) Kuntze ■☆
61747 Buchnera brevibractealis Hiern = Buchnera longespicata Schinz ■☆
61748 Buchnera browniana Schinz = Buchnera hispida Buch.-Ham. ex D. Don ■
61749 Buchnera buchneroides (S. Moore) Brenan;普通黑草■☆
61750 Buchnera buettneri Engl. = Striga macrantha (Benth.) Benth. ■☆
61751 Buchnera bukamensis De Wild.;布卡姆黑草■☆
61752 Buchnera butayei De Wild.;布塔耶黑草■☆
61753 Buchnera candida P. A. Duvign. et Van Bockstal = Buchnera candida S. Moore ■☆
61754 Buchnera candida S. Moore;纯白黑草■☆
61755 Buchnera capensis L. = Polycarena capensis (L.) Benth. ■☆
61756 Buchnera capitata Benth.;头状黑草■☆
61757 Buchnera cernua L. = Chascanum cernuum (L.) E. Mey. ●☆
61758 Buchnera chimanimaniensis Philcox;奇马尼马尼黑草■☆
61759 Buchnera ciliolata Engl.;睫毛黑草■☆
61760 Buchnera coccinea (Hook.) Benth. = Striga asiatica (L.) Kuntze ■
61761 Buchnera congoensis S. Moore;刚果黑草■☆
61762 Buchnera convallicola S. Moore = Buchnera cryptocephala (Baker) Philcox ■☆
61763 Buchnera cordifolia L. f. = Priva cordifolia (L. f.) Druce ■☆
61764 Buchnera crassifolia Engl.;厚叶黑草■☆
61765 Buchnera cruciata Buch.-Ham. = Buchnera cruciata Buch.-Ham. ex D. Don ■
61766 Buchnera cruciata Buch.-Ham. ex D. Don;黑草(鬼羽箭,黑骨草,黑鬼草,克草,坡饼,幼克草,羽箭,羽箭草);Cruciate Blackgrass,Cruciate Buchnera ■
61767 Buchnera cryptocephala (Baker) Philcox;隐头黑草■☆
61768 Buchnera cryptocephala (Baker) Philcox var. mwinilungensis Philcox;穆维尼黑草■☆
61769 Buchnera cuneifolia L. f. = Chascanum cuneifolium (L. f.) E. Mey. ●☆
61770 Buchnera cupricola Robyns = Buchnera henriquesii Engl. ■☆
61771 Buchnera densiflora Benth. = Striga densiflora Benth. ■
61772 Buchnera densiflora Hook. et Arn. = Buchnera cruciata Buch.-Ham. ex D. Don ■
61773 Buchnera descampsii De Wild. et Ledoux = Buchnera quadrifaria Baker ■☆
61774 Buchnera dilungensis Mielcarek;迪龙黑草■☆
61775 Buchnera dundensis Cavaco;敦达黑草■☆
61776 Buchnera dura Benth.;硬黑草■☆
61777 Buchnera ebracteolata Philcox;无苞黑草■☆
61778 Buchnera ensifolia Engl.;剑叶黑草■☆
61779 Buchnera ensifolia Engl. var. andongensis?;安东剑叶黑草■☆
61780 Buchnera erinoides Jaroscz;刺黑草■☆
61781 Buchnera euphrasioides Vahl = Striga bilabiata (Thunb.) Kuntze subsp. linearifolia (Schumach. et Thonn.) Mohamed ■☆
61782 Buchnera eylesii S. Moore;艾尔斯黑草■☆
61783 Buchnera foetida Andréws = Sutera foetida Roth ■☆
61784 Buchnera foliosa V. Naray.;多叶黑草■☆
61785 Buchnera fulgens Engl. = Striga fulgens (Engl.) Hepper ■☆
61786 Buchnera garuensis Pilg.;加鲁黑草■☆
61787 Buchnera geminiflora Philcox;对花黑草■☆
61788 Buchnera gesnerioides Willd. = Striga gesnerioides (Willd.) Vatke ■☆
61789 Buchnera glabrata Benth. = Buchnera simplex (Thunb.) Druce ■☆
61790 Buchnera gossweileri S. Moore;戈斯黑草■☆
61791 Buchnera granitica S. Moore;花岗岩黑草■☆
61792 Buchnera henriquesii Engl.;亨利克斯黑草■☆
61793 Buchnera hermonthica Delile = Striga hermonthica (Delile) Benth. ■☆
61794 Buchnera hispida Buch.-Ham. = Buchnera hispida Buch.-Ham. ex D. Don ■
61795 Buchnera hispida Buch.-Ham. ex D. Don;粗硬毛黑草(刚毛黑草);Hispid Buchnera ■
61796 Buchnera hockii De Wild. = Buchnera multicaulis Engl. ■☆
61797 Buchnera humifusa (Forssk.) Vahl = Cycniopsis humifusa (Forssk.) Engl. ■☆
61798 Buchnera humilis V. Naray.;低矮黑草■☆
61799 Buchnera humpatensis Hiern;洪帕塔黑草■☆
61800 Buchnera inflata (De Wild.) V. Naray.;膨胀黑草■☆

61801 Buchnera kassneri S. Moore = Buchnera lastii Engl. ■☆
61802 Buchnera keilii Mildbr.;凯尔黑草■☆
61803 Buchnera klingii Engl. = Striga klingii (Engl.) V. Naray. ■☆
61804 Buchnera lastii Engl.;拉斯特黑草■☆
61805 Buchnera lastii Engl. subsp. pubiflora Philcox;短毛花黑草■☆
61806 Buchnera latibracteata V. Naray.;宽苞黑草■☆
61807 Buchnera laxiflora Philcox;疏花黑草■☆
61808 Buchnera ledermannii Pilg.;莱德黑草■☆
61809 Buchnera leptostachya Benth.;细穗黑草■☆
61810 Buchnera libenii Mielcarek;利本黑草■☆
61811 Buchnera linearifolia Schumach. et Thonn. = Striga bilabiata (Thunb.) Kuntze subsp. linearifolia (Schumach. et Thonn.) Mohamed ■☆
61812 Buchnera lippioides Vatke ex Engl.;三脉黑草■☆
61813 Buchnera lisowskiana Mielcarek;利索黑草■☆
61814 Buchnera lithospermifolia Kunth;紫草叶黑草■☆
61815 Buchnera longespicata Schinz;长穗黑草■☆
61816 Buchnera longifolia Klotzsch = Buchnera hispida Buch.-Ham. ex D. Don ■
61817 Buchnera lundensis Cavaco;隆德黑草■☆
61818 Buchnera macrantha Benth. = Striga macrantha (Benth.) Benth. ■☆
61819 Buchnera macrocarpa Hochst. = Buchnera hispida Buch.-Ham. ex D. Don ■
61820 Buchnera masuria Buch.-Ham. ex Benth. = Striga masuria (Buch.-Ham. ex Benth.) Benth. ■
61821 Buchnera masuria Ham. ex Benth. = Striga masuria (Buch.-Ham. ex Benth.) Benth. ■
61822 Buchnera metallorum P. A. Duvign. et Van Bockstal;光泽黑草■☆
61823 Buchnera minutiflora Engl.;小花黑草■☆
61824 Buchnera mossambicensis Klotzsch = Buchnera leptostachya Benth. ■☆
61825 Buchnera mossambicensis Klotzsch var. usafuensis Engl. = Buchnera eylesii S. Moore ■☆
61826 Buchnera multicaulis Engl.;多茎黑草■☆
61827 Buchnera multicaulis Engl. var. grandifolia Norl. = Buchnera wildii Philcox ■☆
61828 Buchnera namuliensis V. Naray.;纳木里黑草■☆
61829 Buchnera nervosa Philcox;多脉黑草■☆
61830 Buchnera nigricans (Benth.) V. Naray.;变黑黑草■☆
61831 Buchnera nitida V. Naray.;光亮黑草■☆
61832 Buchnera nuttii V. Naray.;纳特黑草■☆
61833 Buchnera nyassica Gilli;尼亚萨黑草■☆
61834 Buchnera oppositifolia Steud. = Sutera hispida (Thunb.) Druce ■☆
61835 Buchnera orobanchoides R. Br. = Striga gesnerioides (Willd.) Vatke ■☆
61836 Buchnera pallescens Engl.;变苍白黑草■☆
61837 Buchnera paucidentata Engl. ex Hemsl. et V. Naray.;少齿黑草■☆
61838 Buchnera peduncularis Brenan;梗花黑草■☆
61839 Buchnera pedunculata Andrέws = Jamesbrittenia argentea (L. f.) Hilliard ■☆
61840 Buchnera pinnatifida L. f. = Chascanum pinnatifidum (L. f.) E. Mey. ●☆
61841 Buchnera poggei Engl.;波格黑草■☆
61842 Buchnera prorepens Engl. et Gilg;匍匐黑草■☆
61843 Buchnera pruinosa Gilli;白粉黑草■☆
61844 Buchnera pulcherrima R. E. Fr.;艳丽黑草■☆
61845 Buchnera pulchra V. Naray. ex S. Moore = Buchnera cryptocephala (Baker) Philcox ■☆
61846 Buchnera pusilla De Wild. = Buchnera henriquesii Engl. ■☆
61847 Buchnera pusilla Kunth;微小黑草;Pygmy Bluehearts ■☆
61848 Buchnera pusilliflora S. Moore;微花黑草■☆
61849 Buchnera quadrangularis S. Moore = Buchnera foliosa V. Naray. ■☆
61850 Buchnera quadrifaria Baker;四出黑草■☆
61851 Buchnera randii S. Moore;兰德黑草■☆
61852 Buchnera reducta Hiern;退缩黑草■☆
61853 Buchnera remotiflora Schinz;稀花黑草■☆
61854 Buchnera robynsii Mielcarek;罗宾斯黑草■☆
61855 Buchnera rubriflora P. A. Duvign. et Van Bockstal;红花黑草■☆
61856 Buchnera rungwensis Engl.;伦圭黑草■☆
61857 Buchnera ruwenzoriensis V. Naray.;鲁文佐里黑草■☆
61858 Buchnera scabridula E. A. Bruce;微糙黑草■☆
61859 Buchnera schliebenii Melch.;施利本黑草■☆
61860 Buchnera similis V. Naray. = Buchnera lastii Engl. ■☆
61861 Buchnera simplex (Thunb.) Druce;简单黑草■☆
61862 Buchnera speciosa V. Naray.;美丽黑草■☆
61863 Buchnera splendens Engl.;闪烁黑草■☆
61864 Buchnera stachytarphetoides Mildbr. et Melch. = Buchnera usuiensis Oliv. ■☆
61865 Buchnera stricta Benth. = Buchnera cruciata Buch.-Ham. ex D. Don ■
61866 Buchnera strictissima Engl. et Gilg;刚直黑草■☆
61867 Buchnera subcapitata Engl.;亚头状黑草■☆
61868 Buchnera subglabra Philcox;近光黑草■☆
61869 Buchnera symoensiana Mielcarek;西莫黑草■☆
61870 Buchnera tenuifolia Philcox;细叶黑草■☆
61871 Buchnera tetrasticha Wall.;四纵列黑草;Fourverticalrank Buchnera ■☆
61872 Buchnera thunbergii D. Dietr. = Striga bilabiata (Thunb.) Kuntze ■☆
61873 Buchnera trilobata V. Naray.;三裂黑草■☆
61874 Buchnera trinervia Engl. = Buchnera lippioides Vatke ex Engl. ■☆
61875 Buchnera tuberosa V. Naray. = Buchnera lastii Engl. ■☆
61876 Buchnera usafuensis (Engl.) Melch. = Buchnera eylesii S. Moore ■☆
61877 Buchnera usuiensis Oliv.;乌苏黑草■☆
61878 Buchnera vandenberghenii Mielcarek;范登黑草■☆
61879 Buchnera verbenoides Klotzsch;马鞭草状黑草■☆
61880 Buchnera verdickii V. Naray. = Buchnera henriquesii Engl. ■☆
61881 Buchnera viscosa Aiton = Sutera caerulea (L. f.) Hiern ■☆
61882 Buchnera welwitschii Engl.;韦尔黑草■☆
61883 Buchnera wildii Philcox;维尔德黑草■☆
61884 Buchneraceae Benth. = Orobanchaceae Vent.(保留科名)●■
61885 Buchneraceae Lilja = Orobanchaceae Vent.(保留科名)●■
61886 Buchnerodendron Gürke(1893);比希纳木属●☆
61887 Buchnerodendron bussei Gilg = Buchnerodendron lasiocalyx (Oliv.) Gilg ●☆
61888 Buchnerodendron eximium (Gilg) Engl. = Buchnerodendron lasiocalyx (Oliv.) Gilg ●☆
61889 Buchnerodendron lasiocalyx (Oliv.) Gilg;毛萼比希纳木●☆
61890 Buchnerodendron laurentii De Wild. = Buchnerodendron speciosum Gürke ●☆
61891 Buchnerodendron nanum Gilg = Buchnerodendron lasiocalyx

(Oliv.) Gilg ●☆

61892 Buchnerodendron speciosum Gürke;美丽比希纳木●☆

61893 Buchnerodendron stipulatum (Oliv.) Bullock = Oncoba stipulata Oliv. ●☆

61894 Bucholtzia Meisn. = Bucholzia Mart. ■

61895 Bucholzia Mart. = Alternanthera Forssk. ■

61896 Bucholzia Stadtm. ex Wiliem. = Combretum Loefl.(保留属名)●

61897 Bucholzia maritima Mart. = Alternanthera littoralis P. Beauv. var. maritima (Mart.) Pedersen ■☆

61898 Bucholzia maritima Mart. = Alternanthera maritima (Mart.) A. St. -Hil. ■☆

61899 Bucholzia philoxeroides C. Mart. = Alternanthera philoxeroides (Mart.) Griseb. ■

61900 Bucholzia philoxeroides Mart. = Alternanthera philoxeroides (Mart.) Griseb. ■

61901 Buchozia L'Hér. ex Juss. = Serissa Comm. ex Juss. ●

61902 Buchozia Pfeiffer = Buch'osia Vell. ■☆

61903 Buchozia Pfeiffer = Heteranthera Ruiz et Pav.(保留属名)■☆

61904 Buchtienia Schltr.(1929);布枯兰属■☆

61905 Buchtienia boliviensis Schltr.;玻利维亚布枯兰■☆

61906 Buchtienia rosea Garay;粉红布枯兰■☆

61907 Bucida L.(1759)(保留属名);拉美使君子属(布希达属)●☆

61908 Bucida L.(保留属名) = Terminalia L.(保留属名)●

61909 Bucida buceras;牛角拉美使君子木;Black Olive,Jucaro Oxhorn Bucida,Pucte ●☆

61910 Bucidaceae Spreng. = Combretaceae R. Br.(保留科名)●

61911 Bucindia (Wiehler) Wiehler = Columnea L. ●■☆

61912 Bucindia Wiehler = Columnea L. ●■☆

61913 Bucinella Wiehler = Bucinellina Wiehler ●☆

61914 Bucinellina Wiehler = Columnea L. ●■☆

61915 Bucinellina Wiehler(1981);布奇苣苔属●☆

61916 Bucinellina nariniana (Wiehler) Wiehler;布奇苣苔●☆

61917 Buckinghamia F. Muell.(1868);象牙弯木属(布根海秘属,布根海密属);Buckinghamia ●☆

61918 Buckinghamia celsissima F. Muell.;象牙弯木(布根海秘,高耸布根海密);Celsus Buckinghamia,Ivory Curl Tree ●☆

61919 Bucklandia R. Br. = Bucklandia R. Br. ex Griff. ●

61920 Bucklandia R. Br. ex Griff.(1836);拟马蹄荷属(白克木属)●

61921 Bucklandia R. Br. ex Griff. = Exbucklandia R. W. Br. ●

61922 Bucklandia populifolia Hook. f. et Thomson = Exbucklandia populnea (R. Br.) R. W. Br. ●

61923 Bucklandia populnea R. Br. = Exbucklandia populnea (R. Br.) R. W. Br. ●

61924 Bucklandia populnea R. Br. ex Griff. = Exbucklandia populnea (R. Br. ex Griff.) R. W. Br. ●

61925 Bucklandia populnea sensu Merr. = Bucklandia tonkinensis Lecomte ●

61926 Bucklandia tonkinensis Lecomte = Exbucklandia tonkinensis (Lecomte) Hung T. Chang ●

61927 Bucklandiaceae J. Agardh = Hamamelidaceae R. Br.(保留科名)●

61928 Bucklandiaceae J. Agardh;马蹄荷科●

61929 Buckleya Torr.(1843)(保留属名);米面蓊属;Piratebush ●

61930 Buckleya angulosa S. B. Zhou et X. H. Guo;狭叶米面蓊●

61931 Buckleya distichophylla (Nutt.) Torr.;北美米面翁;N. America Piratebush ●☆

61932 Buckleya graebneriana Diels;秦岭米面翁(面牛,面翁,面瓮,线苞,线苞米面蓊,痒痒树);Graebner Piratebush,Qinling Piratebush ●

61933 Buckleya henryi Diels = Buckleya lanceolata (Siebold et Zucc.) Miq. ●

61934 Buckleya joan Siebold = Buckleya lanceolata (Siebold et Zucc.) Miq. ●

61935 Buckleya lanceolata (Siebold et Zucc.) Miq.;米面蓊(柴骨皮,都念子,凤凰草,九层皮,尿尿皮,羽毛球树,撞羽,籽米驼);Henry Piratebush,Lanceolate Piratebush ●

61936 Buckleya lanceolata (Siebold et Zucc.) Miq. f. tanigawaensis (Ohki) Sugim.;谷川米面蓊■☆

61937 Buckleya lanceolata (Siebold et Zucc.) Miq. var. tanigawaensis (Ohki) Honda = Buckleya lanceolata (Siebold et Zucc.) Miq. f. tanigawaensis (Ohki) Sugim. ■☆

61938 Buckleya quadriala Benth. et Hook. f. = Buckleya lanceolata (Siebold et Zucc.) Miq. ●

61939 Bucknera Michx. = Buchnera L. ■

61940 Buckollia Venter et R. L. Verh.(1994);热非萝藦属☆

61941 Buckollia tomentosa (E. A. Bruce) Venter et R. L. Verh.;绒毛米面蓊●☆

61942 Buckollia volubilis (Schltr.) Venter et R. L. Verh.;缠绕米面蓊●☆

61943 Bucquetia DC.(1828);比凯野牡丹属●☆

61944 Bucquetia nigritella (Naudin) Triana;比凯野牡丹●☆

61945 Bucranion Raf. = Utricularia L. ■

61946 Bucranion capense (Spreng.) Raf = Utricularia bisquamata Schrank ■☆

61947 Buda Adans.(废弃属名) = Spergularia (Pers.) J. Presl et C. Presl(保留属名)■

61948 Budawangia I. Telford(1992);杉石南属(澳洲石南属)●☆

61949 Budawangia gnidioides (Summerh.) Telford.;杉石南(澳洲石南)●☆

61950 Buddleia L. = Buddleja L. ●■

61951 Buddleia virgata L. f. = Gomphostigma virgatum (L. f.) Baill. ●☆

61952 Buddleiaceae K. Wilh. = Buddlejaceae K. Wilh.(保留科名)●

61953 Buddleja L.(1753);醉鱼草属(白埔姜属,扬波属);Buddleia,Butterfly Bush,Butterflybush,Butterfly-bush,Butter-fly-bush,Summer Lilac,Summer-lilac,Summerlilic ●■

61954 Buddleja × intermedia Carrière;间型醉鱼草●☆

61955 Buddleja × intermedia Carrière var. insignis (Carrière) Rehder;显著醉鱼草●☆

61956 Buddleja × pikei H. R. Fletcher;皮克醉鱼草●☆

61957 Buddleja × whiteana R. J. Moore;威氏醉鱼草;Butterfly Bush,Weyer Hybrid Butterfly Bush,Weyeriana Butterfly Bush,Yellow Butterfly Bush ●☆

61958 Buddleja acosma C. Marquand = Buddleja crispa Benth. ●

61959 Buddleja acuminata Poir.;尖醉鱼草●☆

61960 Buddleja acuminata R. Br. = Buddleja polystachya Fresen. ●☆

61961 Buddleja acuminatissima Blume = Buddleja asiatica Lour. ●

61962 Buddleja acutifolia C. H. Wright = Buddleja paniculata Wall. ●

61963 Buddleja adenantha Diels;腺冠醉鱼草(暗蓝花醉鱼草);Glandflower Butterflybush,Glandular-colored Butterfly-bush,Glandularflower Butterflybush,Glandularflower Summerlilic ●

61964 Buddleja adenantha Diels = Buddleja myriantha Diels ●

61965 Buddleja agathosma Diels = Buddleja crispa Benth. ●

61966 Buddleja agathosma Diels var. glandulifera C. Marquand = Buddleja crispa Benth. ●

61967 Buddleja alata Rehder et E. H. Wilson;翅枝醉鱼草;Wingbranch

Summerlilic, Winged Butterflybush, Winged Butterfly-bush ●

61968 Buddleja albiflora Hemsl.;巴东醉鱼草(白花醉鱼草);Badong Summerlilic, Whiteflower Butterflybush, White-flowered Butterfly-bush ●

61969 Buddleja albiflora Hemsl. var. giraldii (Diels) Rehder et E. H. Wilson = Buddleja albiflora Hemsl. ●

61970 Buddleja albiflora Hemsl. var. hemsleyana (Koehne) C. K. Schneid. = Buddleja albiflora Hemsl. ●

61971 Buddleja alternifolia Maxim.;互叶醉鱼草(白芨,白芨梢,白积梢,白箕稍,互生叶醉鱼草,小叶醉鱼草,泽当醉鱼草);Alternateleaf Butterfly Bush, Alternate-leaved Butterfly-bush, Fountain Buddleja, Fountain Butterflybush, Fountain Butterfly-bush, Fountain Summerlilic ●

61972 Buddleja amentacea Kraenzl. = Buddleja asiatica Lour. ●

61973 Buddleja americana L.;美洲醉鱼草;American Butterflybush ●

61974 Buddleja arfakensis Kaneh. et Hatus. = Buddleja asiatica Lour. ●

61975 Buddleja asiatica Lour.;白背枫(白花洋泡,白花醉鱼草,白埔姜,白鱼号,白鱼尾,驳骨丹,驳骨醉鱼草,独叶埔姜,独叶埔羌,黄合叶,蒲羌癀,七里香,山埔姜,水黄花,王记叶,溪桃,狭叶醉鱼草,扬波,野桃,醉鱼草);Asian Butterflybush, Asian Butterfly-bush, Asian Summerlilic, Asiatic Butter-fly-bush, Dogtail, White Butterfly-bush ●

61976 Buddleja asiatica Lour. var. brevicuspe Koord. = Buddleja asiatica Lour. ●

61977 Buddleja asiatica Lour. var. densiflora (Blume) Koord. et Valeton = Buddleja asiatica Lour. ●

61978 Buddleja asiatica Lour. var. salicina (Lam.) Koord. et Valeton = Buddleja asiatica Lour. ●

61979 Buddleja asiatica Lour. var. stipulata Gagnep. = Buddleja myriantha Diels ●

61980 Buddleja asiatica Lour. var. sundaica (Blume) Koord. et Valeton = Buddleja asiatica Lour. ●

61981 Buddleja aurantiaco-maculata Gilg = Buddleja salviifolia (L.) Lam. ●☆

61982 Buddleja auriculata Benth.;卷叶醉鱼草;Weeping Sage ●☆

61983 Buddleja auriculata Benth. var. euryifolia Prain et Cummins = Buddleja auriculata Benth. ●☆

61984 Buddleja australis Vell.;澳大利亚醉鱼草(南方醉鱼草);Australian Butterflybush ●

61985 Buddleja axillaris Willd. ex Roem. et Schult.;腋花醉鱼草●☆

61986 Buddleja bangii Kraenzl. = Buddleja tucumanensis Griseb. ●☆

61987 Buddleja brachystachya Diels;短序醉鱼草;Shortin Florescens Butterflybush, Shortinflorescens Summerlilic, Short-racemed Butterfly-bush ●

61988 Buddleja candelabrum Kraenzl. = Buddleja acuminata Poir. ●☆

61989 Buddleja candida Dunn;密香醉鱼草(密香树,喜马拉雅醉鱼草);Pure-white Butterfly-bush, White Butterflybush ●

61990 Buddleja canescens Rusby = Buddleja tucumanensis Griseb. ●☆

61991 Buddleja capitata Jacq. = Buddleja globosa Hope ●☆

61992 Buddleja caryopteridifolia W. W. Sm.;莸叶醉鱼草;Blue-beard Butterfly-bush, Bluebeardleaf Butterflybush, Bluebeardleaf Summerlilic ●

61993 Buddleja caryopteridifolia W. W. Sm. = Buddleja crispa Benth. ●

61994 Buddleja caryopteridifolia W. W. Sm. var. eremophila (W. W. Sm.) C. Marquand;簇花醉鱼草;Dense-flower Bluebeardleaf Butterflybush ●

61995 Buddleja caryopteridifolia W. W. Sm. var. eremophila (W. W. Sm.) C. Marquand = Buddleja crispa Benth. ●

61996 Buddleja caryopteridifolia W. W. Sm. var. fasciculiflora Z. Ying Zhang = Buddleja crispa Benth. ●

61997 Buddleja caryopteridifolia W. W. Sm. var. fasciculiflora Z. Ying Zhang = Buddleja caryopteridifolia W. W. Sm. var. eremophila (W. W. Sm.) C. Marquand ●

61998 Buddleja caryopteridifolia W. W. Sm. var. lanuginosa C. Marquand = Buddleja crispa Benth. ●

61999 Buddleja cochabambensis Rusby = Buddleja tucumanensis Griseb. ●☆

62000 Buddleja colvilei Hook. f. et Thomson;大花醉鱼草(尼泊尔醉鱼草);Bigflower Butterflybush, Bigflower Summerlilic, Largeflower Butterflybush, Large-flowered Butterfly-bush, Summer Lilic ●

62001 Buddleja colvilei Hook. f. et Thomson 'Kewensis';邱园大花醉鱼草●☆

62002 Buddleja comorensis Baker = Buddleja axillaris Willd. ex Roem. et Schult. ●☆

62003 Buddleja cooperi W. W. Sm. = Buddleja forrestii Diels ●

62004 Buddleja cordata Kunth;心形醉鱼草●☆

62005 Buddleja coriacea Remy;革质醉鱼草●☆

62006 Buddleja coriacea Remy var. beta Wedd. = Buddleja montana Britton ●☆

62007 Buddleja coroicense Rusby = Buddleja diffusa Ruiz et Pav. ●☆

62008 Buddleja corrugata (Benth.) E. Phillips = Buddleja loricata Leeuwenb. ●☆

62009 Buddleja crispa Benth.;皱叶醉鱼草(染饭花);Curly Butterflybush, Curly Butterfly-bush, Curly Summerlilic ●

62010 Buddleja crispa Benth. var. amplexicaulis Z. Ying Zhang;抱茎醉鱼草;Stem-clasping Butterflybush ●

62011 Buddleja crispa Benth. var. amplexicaulis Z. Ying Zhang = Buddleja crispa Benth. ●

62012 Buddleja crispa Benth. var. dicipiens Schmidt = Buddleja crispa Benth. ●

62013 Buddleja crispa Benth. var. farreri (Balf. f. et W. W. Sm.) Hand. -Mazz. = Buddleja crispa Benth. ●

62014 Buddleja crispa Benth. var. glandulifera (C. Marquand) S. Y. Pao = Buddleja crispa Benth. ●

62015 Buddleja curviflora Hook. et Arn.;台湾醉鱼草(弯花醉鱼草,弯花醉鱼木);Formosan Butter-fly-bush, Taiwan Butterflybush, Taiwan Butterfly-bush, Taiwan Summerlilic ●

62016 Buddleja curviflora Hook. et Arn. f. venenifera (Makino) T. Yamaz.;毒台湾醉鱼草●☆

62017 Buddleja curviflora Hook. et Arn. var. venenifera (Makino) Makino = Buddleja curviflora Hook. et Arn. f. venenifera (Makino) T. Yamaz. ●☆

62018 Buddleja cylindrostachya Kraenzl. = Buddleja macrostachya Wall. ex Benth. ●

62019 Buddleja davidii Franch.;大叶醉鱼草(白背叶醉鱼草,白壶子,大蒙花,绛花醉鱼草,酒曲花,酒药花,蒙花,穆坪醉鱼草,兴山醉鱼草,紫花醉鱼草);Butterfly Bush, Butterfly-bush, Common Butterfly Bush, Orange Eye, Orange Eye Butterflybush, Orange-eye Butterfly Bush, Orangeeye Butterflybush, Orange-eye Butterfly-bush, Orangeeye Summerlilic, Summer Lilac, Summer Lilic, Summer-lilac ●

62020 Buddleja davidii Franch. 'African Queen';非洲皇后大叶醉鱼草;Butterfly Bush ●☆

62021 Buddleja davidii Franch. 'Black Knight';黑骑士大叶醉鱼草;Butterfly Bush ●☆

62022 Buddleja davidii Franch. 'Dart's Ornamental White';达特美白大叶醉鱼草●☆

62023 Buddleja davidii Franch. 'Dartmoor';达特姆尔大叶醉鱼草●☆

62024 Buddleja davidii Franch. 'Empire Blue';帝国蓝大叶醉鱼草(品蓝大叶醉鱼草);Butterfly Bush ●☆

62025 Buddleja davidii Franch. 'Harlequin';小丑大叶醉鱼草;Harlequin Butterfly Bush, Variegated Butterfly Bush ●☆

62026 Buddleja davidii Franch. 'Nanho Blue';南波蓝大叶醉鱼草●☆

62027 Buddleja davidii Franch. 'Nanho Purple';南波紫大叶醉鱼草●☆

62028 Buddleja davidii Franch. 'Orchid Beauty';紫美人大叶醉鱼草;Butterfly Bush ●☆

62029 Buddleja davidii Franch. 'Peace';和平大叶醉鱼草●☆

62030 Buddleja davidii Franch. 'Pink Pearl';粉珍珠大叶醉鱼草●☆

62031 Buddleja davidii Franch. 'Royal Red';皇家红大叶醉鱼草●☆

62032 Buddleja davidii Franch. 'Superba' = Buddleja davidii Franch. ●

62033 Buddleja davidii Franch. 'Veitchiana' = Buddleja davidii Franch. ●

62034 Buddleja davidii Franch. 'White Cloud';白云大叶醉鱼草●☆

62035 Buddleja davidii Franch. 'White Profusion';夏丁香醉鱼草;Butterfly-bush, Summer Lilac ●☆

62036 Buddleja davidii Franch. var. alba Rehder et E. H. Wilson = Buddleja davidii Franch. ●

62037 Buddleja davidii Franch. var. glabrescens Gagnep. = Buddleja davidii Franch. ●

62038 Buddleja davidii Franch. var. magnifica (E. H. Wilson) Rehder et E. H. Wilson = Buddleja davidii Franch. ●

62039 Buddleja davidii Franch. var. nanhoensis (Chitt.) Rehder = Buddleja davidii Franch. ●

62040 Buddleja davidii Franch. var. superba (Veitch) Rehder et E. H. Wilson = Buddleja davidii Franch. ●

62041 Buddleja davidii Franch. var. superba Rehder et E. H. Wilson = Buddleja davidii Franch. ●

62042 Buddleja davidii Franch. var. veitchiana (Veitch) Rehder et Bailey = Buddleja davidii Franch. ●

62043 Buddleja davidii Franch. var. vetichiana (Veitch) Rehder = Buddleja davidii Franch. ●

62044 Buddleja davidii Franch. var. wilsonii (E. H. Wilson) Rehder et E. H. Wilson = Buddleja davidii Franch. ●

62045 Buddleja delavayi Gagnep.;腺叶醉鱼草;Delavay Butterflybush, Delavay Butterfly-bush, Delavay Summerlilic ●

62046 Buddleja delavayi Gagnep. var. tomentosa Comber = Buddleja delavayi Gagnep. ●

62047 Buddleja densiflora Blume = Buddleja asiatica Lour. ●

62048 Buddleja diffusa Ruiz et Pav.;松散醉鱼草●☆

62049 Buddleja discolor Roth = Buddleja asiatica Lour. ●

62050 Buddleja duclouxii C. Marquand = Buddleja myriantha Diels ●

62051 Buddleja dysophylla (Benth.) Radlk.;斜叶醉鱼草●☆

62052 Buddleja eremophila W. W. Sm. = Buddleja caryopteridifolia W. W. Sm. var. eremophila (W. W. Sm.) C. Marquand ●

62053 Buddleja eremophila W. W. Sm. = Buddleja crispa Benth. ●

62054 Buddleja fallowiana Balf. f. et W. W. Sm.;紫花醉鱼草(白叶花,蓝花密蒙花);Butterfly Bush, Fallows Buddleja, Purpleflower Butterflybush, Purpleflower Summerlilic, Purple-flowered Butterfly-bush ●

62055 Buddleja fallowiana Balf. f. et W. W. Sm. 'Lochinch';罗奇科紫花醉鱼草●☆

62056 Buddleja fallowiana Balf. f. et W. W. Sm. var. alba Sabour. = Buddleja fallowiana Balf. f. et W. W. Sm. ●

62057 Buddleja farreri Balf. f. et W. W. Sm. = Buddleja crispa Benth. ●

62058 Buddleja formosana Hatus. = Buddleja curviflora Hook. et Arn. ●

62059 Buddleja forrestii Diels;滇川醉鱼草(苍山醉鱼草,瑞丽醉鱼草);Forrest Butterflybush, Forrest Butterfly-bush, Forrest Summerlilic ●

62060 Buddleja forrestii Diels var. gracilis Lingelsh. = Buddleja forrestii Diels ●

62061 Buddleja giraldii Diels = Buddleja albiflora Hemsl. ●

62062 Buddleja glaberrima Loisel. = Freylinia lanceolata (L. f.) G. Don ●☆

62063 Buddleja glabrescens W. W. Sm. = Buddleja delavayi Gagnep. ●

62064 Buddleja globosa Hope;球花醉鱼草(球序醉鱼草);Globe Butterfly-bush, Golden Ball Butterfly Bush, Honeyball, Orange Ball Tree, Orange Ball-tree, Orange Butterfly Bush, Orange-ball-tree, Pincushion Tree ●☆

62065 Buddleja glomerata H. L. Wendl.;团集醉鱼草●☆

62066 Buddleja gracilis Lingelsh. = Buddleja forrestii Diels ●

62067 Buddleja griffithii (C. B. Clarke) Marquart = Buddleja macrostachya Wall. ex Benth. var. griffithii C. B. Clarke ●

62068 Buddleja gynandra C. Marquand = Buddleja paniculata Wall. ●

62069 Buddleja hancockii Kraenzl. = Buddleja macrostachya Wall. ex Benth. ●

62070 Buddleja harrowiana Balf. f. et W. W. Sm.;蓝花密蒙花;Harrow's Butterflybush ●

62071 Buddleja hastata Prain ex C. Marquand;戟叶醉鱼草;Hastate Butterflybush, Hastate Butterfly-bush, Hastileleaf Summerlilic ●

62072 Buddleja hastata Prain ex C. Marquand = Buddleja crispa Benth. ●

62073 Buddleja heliophila W. W. Sm.;全缘叶醉鱼草(羊耳枝,缘叶醉鱼草);Entire Butterflybush, Entire Butterfly-bush, Entire Summerlilic ●

62074 Buddleja heliophila W. W. Sm. = Buddleja davidii Franch. ●

62075 Buddleja heliophila W. W. Sm. = Buddleja delavayi Gagnep. ●

62076 Buddleja heliophila W. W. Sm. var. adenophora Hand. -Mazz. = Buddleja delavayi Gagnep. ●

62077 Buddleja heliophila W. W. Sm. var. angustifolia C. Marquand = Buddleja delavayi Gagnep. ●

62078 Buddleja heliophila W. W. Sm. var. pubescens C. Marquand = Buddleja delavayi Gagnep. ●

62079 Buddleja hemsleyana Koehne. = Buddleja albiflora Hemsl. ●

62080 Buddleja henryi H. Lév. = Buddleja forrestii Diels ●

62081 Buddleja henryi Rehder et E. H. Wilson = Buddleja forrestii Diels ●

62082 Buddleja henryi Rehder et E. H. Wilson var. glabrescens C. Marquand = Buddleja forrestii Diels ●

62083 Buddleja henryi Rehder et E. H. Wilson var. hancockii (Kraenzl.) C. Marquand = Buddleja macrostachya Wall. ex Benth. ●

62084 Buddleja heterophylla Lindl. = Buddleja madagascariensis Lam. ●

62085 Buddleja hookeri C. Marquand = Buddleja macrostachya Wall. ex Benth. ●

62086 Buddleja hosseusiana Kraenzl. = Buddleja macrostachya Wall. ex Benth. ●

62087 Buddleja hypoleuca Kraenzl. = Buddleja tucumanensis Griseb. ●☆

62088 Buddleja ignea Kraenzl. = Buddleja tucumanensis Griseb. ●☆

62089 Buddleja incompta L. f. = Gomphostigma incomptum (L. f.) N. E. Br. ●☆

62090 Buddleja incompta W. W. Sm. = Buddleja crispa Benth. ●

62091 Buddleja inconspicua Kraenzl. = Buddleja tucumanensis

Griseb. ●☆

62092 Buddleja indica Lam.;蔓生醉鱼草●☆

62093 Buddleja insignis Carrière;日本溪畔醉鱼草;Japanese Butterfly-bush ●☆

62094 Buddleja insignis Carrière = Buddleja japonica Hemsl. ●

62095 Buddleja insignis Carrière = Buddleja lindleyana Fortune ●

62096 Buddleja insignis Dippel = Buddleja lindleyana Fortune ●

62097 Buddleja intermedia Carrière = Buddleja lindleyana Fortune ●

62098 Buddleja intermedia Carrière var. insignis (Carrière) Rehder = Buddleja lindleyana Fortune ●

62099 Buddleja japonica Hemsl.;日本醉鱼草(醉鱼草);Butterfly Bush, Japan Summerlilic, Japanese Butterfly Bush, Japanese Butterflybush, Japanese Butter-flybush, Japanese Butterfly-bush ●

62100 Buddleja japonica Hemsl. = Buddleja insignis Carrière ●☆

62101 Buddleja japonica Hemsl. = Buddleja lindleyana Fortune ●

62102 Buddleja japonica Hemsl. f. albiflora Akasawa;白花日本醉鱼草●☆

62103 Buddleja japonica Hemsl. var. insignis (Carrière) E. H. Wilson = Buddleja lindleyana Fortune ●

62104 Buddleja latiflora S. Y. Pao;宽管醉鱼草;Broad-flower Butterflybush ●

62105 Buddleja latiflora S. Y. Pao = Buddleja forrestii Diels ●

62106 Buddleja lavandulacea Kraenzl. = Buddleja paniculata Wall. ●

62107 Buddleja legendrei Gagnep. = Buddleja alternifolia Maxim. ●

62108 Buddleja limitanea W. W. Sm.;扁脉醉鱼草(有梗醉鱼草);Borderline Summerlilic, Limit Butterflybush, Pedunculate Butterfly-bush ●

62109 Buddleja limitanea W. W. Sm. = Buddleja forrestii Diels ●

62110 Buddleja lindleyana Fortune;醉鱼草(白袍花,白皮消,闭鱼花,洞庭草,毒鱼草,毒鱼藤,防痛树,光子,红鱼波,红鱼皂,花玉成,鸡公尾,金鸡尾,老阳花,鲤鱼花草,楼梅草,樚木,驴尾草,萝卜树,满山香,闹鱼花,闹鱼子,七里香,钱线尾,水泡木,四方麻,四季青,四棱麻,四楞麻,糖茶,铁线尾,铁帚尾,土蒙花,五霸蔷,羊白婆,羊饱药,羊脑髓,羊尾巴,阳包树,洋波,痒见消,药杆子,药鳗老醋,药鱼子,野巴豆,鱼鳞子,鱼泡草,鱼藤草,鱼尾草,雨背子花,雉尾花,醉鱼儿草);Butterfly Bush, Lindley Butterfly Bush, Lindley Butterflybush, Lindley Butterfly-bush, Lindley Summerlilic, Lindley's Butterfly Bush, Lindley's Butterflybush ●

62111 Buddleja lindleyana Fortune var. sinuatodentata Hemsl. = Buddleja lindleyana Fortune ●

62112 Buddleja longifolia Gagnep. = Buddleja forrestii Diels ●

62113 Buddleja loricata Leeuwenb.;甲醉鱼草●☆

62114 Buddleja macrostachya Wall. ex Benth.;大序醉鱼草(白叶子,长穗醉鱼草,锡金醉鱼草,羊巴巴叶);Biginflorescens Summerlilic, Big-spiked Butterfly-bush, Largeinflorescence Butterflybush ●

62115 Buddleja macrostachya Wall. ex Benth. var. griffithii C. B. Clarke;不丹醉鱼草;Griffith's Butterflybush ●

62116 Buddleja macrostachya Wall. ex Benth. var. yunnanensis Diels = Buddleja nivea Duthie ●

62117 Buddleja macrostachya Wall. ex Benth. var. yunnanensis Dop = Buddleja nivea Duthie ●

62118 Buddleja madagascariensis Lam.;浆果醉鱼草(非洲醉鱼草,假黄花,假蒙花,马达加斯加醉鱼草);Butterfly Bush, Madegascar Butterflybush, Madegascar Butterfly-bush, Madegascar Summerlilic, Nicodemia, Smokebush ●

62119 Buddleja madagascariensis Lam. = Nicodemia madagascariensis (Lam.) R. Parker ●

62120 Buddleja mairei H. Lév. = Buddleja paniculata Wall. ●

62121 Buddleja mairei H. Lév. f. albiflora H. Lév. = Buddleja paniculata Wall. ●

62122 Buddleja manei H. Lév. f. albiflora H. Lév. = Buddleja albiflora Hemsl. ●

62123 Buddleja marrubiifolia Benth.;毛醉鱼草(夏至草叶醉鱼草);Wooly Butterfly Bush ●☆

62124 Buddleja martii Schmidt. = Buddleja macrostachya Wall. ex Benth. ●

62125 Buddleja minima S. Y. Pao = Buddleja alternifolia Maxim. ●

62126 Buddleja montana Britton;山地醉鱼草●☆

62127 Buddleja myriantha Diels;酒药花醉鱼草(多花醉鱼草);Flowery Summerlilic, Manyflower Butterflybush, Multiflorous Butterfly-bush ●

62128 Buddleja nana W. W. Sm. = Buddleja brachystachya Diels ●

62129 Buddleja neemda Buch.-Ham. ex Roxb. = Buddleja asiatica Lour. ●

62130 Buddleja neemda Buch.-Ham. ex Roxb. var. philippensis Cham. et Schltdl. = Buddleja asiatica Lour. ●

62131 Buddleja nivea Duthie;金沙江醉鱼草(雪白醉鱼草);Snowwhite Butterflybush, Snow-white Butterfly-bush, Snowwhite Summerlilic ●

62132 Buddleja nivea Duthie var. yunnanensis (Dop) Rehder et E. H. Wilson = Buddleja nivea Duthie ●

62133 Buddleja oblongifolia Rusby = Buddleja montana Britton ●☆

62134 Buddleja officinalis Maxim.;蜜蒙花(草春条,虫见死草,疙瘩皮树花,寒不调,黄饭花,黄花树,黄花醉鱼草,鸡骨头花,老蒙花,莽草,莽爪,蒙花,蒙花树,米汤花,密蒙花,绵糊条子,绵条子,染饭花,水锦花,小锦花,羊春条,羊耳朵,羊耳朵朵尖);Mimeng Summerlilic, Pale Butterflybush, Pale Butterfly-bush ●

62135 Buddleja officinalis Maxim. 'Spring Primise';春日乳白蜜蒙花●☆

62136 Buddleja officinalis Maxim. var. macrantha Lingelsh. = Buddleja officinalis Maxim. ●

62137 Buddleja officinalis Maxim. var. sinuato-dentata Hemsl.;波叶醉鱼草;Undulate Butterflybush, Undulate Summerlilic ●

62138 Buddleja paniculata Wall.;喉药醉鱼草(羊耳朵,圆锥醉鱼草);Laryngitis Summerlilic, Paniculate Butterflybush, Paniculate Butterfly-bush ●

62139 Buddleja paniculata Wall. = Buddleja crispa Benth. ●

62140 Buddleja perfoliata Kunth;穿叶醉鱼草●☆

62141 Buddleja plectranthoidea H. Lév. = Elsholtzia fruticosa (D. Don) Rehder ●

62142 Buddleja poiretii Spreng. = Buddleja acuminata Poir. ●☆

62143 Buddleja polystachya Fresen.;多穗醉鱼草●☆

62144 Buddleja polystachya Fresen. var. parvifolia Marquand = Buddleja polystachya Fresen. ●☆

62145 Buddleja powellii Kraenzl. = Buddleja polystachya Fresen. ●☆

62146 Buddleja praecox Lingelsh. = Buddleja crispa Benth. ●

62147 Buddleja pterocaulis A. B. Jacks. = Buddleja forrestii Diels ●

62148 Buddleja pulchella N. E. Br.;美丽醉鱼草●☆

62149 Buddleja purdomii W. W. Sm.;甘肃醉鱼草(白胡子花);Gansu Butterflybush, Gansu Summerlilic, Purdom Butterflybush, Purdom Butterfly-bush ●◇

62150 Buddleja purdomii W. W. Sm. = Buddleja brachystachya Diels ●

62151 Buddleja purdomii W. W. Sm. var. fulvotomentosa Z. Ying Zhang;黄毛醉鱼草;Yellow-hairs Gansu Butterflybush ●

62152 Buddleja purdomii W. W. Sm. var. fulvotomentosa Z. Ying Zhang = Buddleja brachystachya Diels ●
62153 Buddleja purdomii W. W. Sm. var. fulvotomentosa Z. Ying Zhang = Buddleja purdomii W. W. Sm. ●◇
62154 Buddleja rufa Fresen. = Buddleja polystachya Fresen. ●☆
62155 Buddleja salicifolia Jacq. = Buddleja saligna Willd. ●☆
62156 Buddleja salicina Lam. = Buddleja asiatica Lour. ●
62157 Buddleja saligna Willd.；橄榄叶醉鱼草；Bastard Olive, Squarestem Butterflybush ●☆
62158 Buddleja saltiana Steud. = Buddleja polystachya Fresen. ●☆
62159 Buddleja salvifolia Lam. = Buddleja salviifolia（L.）Lam. ●☆
62160 Buddleja salviifolia（L.）Lam.；南非醉鱼草；South African Sagewood, Winter Buddleja ●☆
62161 Buddleja serrulata Roth = Buddleja asiatica Lour. ●
62162 Buddleja sessilifolia B. S. Sun et S. Y. Pao；无柄醉鱼草；Stalkless Butterflybush ●
62163 Buddleja sessilifolia B. S. Sun ex S. Y. Pao = Buddleja colvilei Hook. f. et Thomson ●
62164 Buddleja shaanxiensis Z. Ying Zhang；陕西醉鱼草；Shaanxi Butterflybush ●
62165 Buddleja shaanxiensis Z. Ying Zhang = Buddleja davidii Franch. ●
62166 Buddleja shimidzuana Nakai = Buddleja davidii Franch. ●
62167 Buddleja sinuata Willd. ex Roem. et Schult. = Buddleja acuminata Poir. ●☆
62168 Buddleja stenostachya Rehder et E. H. Wilson = Buddleja nivea Duthie ●
62169 Buddleja sterniana Cotton = Buddleja crispa Benth. ●
62170 Buddleja stirata Z. Ying Zhang var. zhouquensis Z. Ying Zhang = Buddleja davidii Franch. ●
62171 Buddleja striata Z. Ying Zhang；多纹醉鱼草；Striate Butterflybush ●
62172 Buddleja striata Z. Ying Zhang = Buddleja davidii Franch. ●
62173 Buddleja striata Z. Ying Zhang var. zhouquensis Z. Ying Zhang；舟曲醉鱼草；Zhouqu Striate Butterflybush ●
62174 Buddleja subherbacea Keenan = Buddleja forrestii Diels ●
62175 Buddleja subserrata Buch. -Ham. ex D. Don = Buddleja asiatica Lour. ●
62176 Buddleja subserrata Ham. ex D. Don = Buddleja asiatica Lour. ●
62177 Buddleja sundaica Blume = Buddleja asiatica Lour. ●
62178 Buddleja taliensis W. W. Sm.；大理醉鱼草；Dali Butterflybush, Dali Butterfly-bush, Dali Summerlilic ●
62179 Buddleja taliensis W. W. Sm. = Buddleja forrestii Diels ●
62180 Buddleja tibetica W. W. Sm. = Buddleja crispa Benth. ●
62181 Buddleja tibetica W. W. Sm. var. farreri（Balf. f. et W. W. Sm.）C. Marquand = Buddleja crispa Benth. ●
62182 Buddleja tibetica W. W. Sm. var. glandulifera C. Marquand = Buddleja crispa Benth. ●
62183 Buddleja tibetica W. W. Sm. var. grandiflora C. Marquand = Buddleja crispa Benth. ●
62184 Buddleja tibetica W. W. Sm. var. truncatifolia（H. Lév.）C. Marquand = Buddleja crispa Benth. ●
62185 Buddleja truncata Gagnep. = Buddleja crispa Benth. ●
62186 Buddleja truncatifolia H. Lév. = Buddleja crispa Benth. ●
62187 Buddleja tsetangensis C. Marquand = Buddleja wardii C. Marquand ●
62188 Buddleja tucumanensis Griseb.；土库曼醉鱼草●☆
62189 Buddleja usambarensis Gilg = Buddleja pulchella N. E. Br. ●☆
62190 Buddleja variabilis Hemsl. = Buddleja davidii Franch. ●
62191 Buddleja variabilis Hemsl. var. magnifica E. H. Wilson = Buddleja davidii Franch. ●
62192 Buddleja variabilis Hemsl. var. nanhoensis Chitt. = Buddleja davidii Franch. ●
62193 Buddleja variabilis Hemsl. var. prostrata C. K. Schneid. = Buddleja davidii Franch. ●
62194 Buddleja variabilis Hemsl. var. superba Veitch = Buddleja davidii Franch. ●
62195 Buddleja variabilis Hemsl. var. veitchiana Veitch = Buddleja davidii Franch. ●
62196 Buddleja variabilis Hemsl. var. wilsonii E. H. Wilson. = Buddleja davidii Franch. ●
62197 Buddleja variabilis Veitch = Buddleja davidii Franch. ●
62198 Buddleja venenifera Makino = Buddleja curviflora Hook. et Arn. f. venenifera（Makino）T. Yamaz. ●☆
62199 Buddleja venenifera Makino = Buddleja curviflora Hook. et Arn. ●
62200 Buddleja venenifera Makino f. calvescens Ohwi = Buddleja curviflora Hook. et Arn. ●
62201 Buddleja verbascifolia Kunth = Buddleja americana L. ●
62202 Buddleja virgata Blanco = Buddleja asiatica Lour. ●
62203 Buddleja wardii C. Marquand；互对醉鱼草（高山醉鱼草）；Ward Butterfly-bush, Ward Summerlilic, Ward's Butterflybush ●
62204 Buddleja whitei Kraenzl. = Buddleja crispa Benth. ●
62205 Buddleja woodii Gilg = Buddleja pulchella N. E. Br. ●☆
62206 Buddleja yunnanensis Gagnep.；云南醉鱼草（滇醉鱼草，猫屎树）；Yunnan Butterflybush, Yunnan Butterfly-bush, Yunnan Summerlilic ●
62207 Buddlejaceae Bartl. = Buddleja L. + Scrophulariaceae Juss.（保留科名）●■
62208 Buddlejaceae K. Wilh.（1910）（保留科名）；醉鱼草科；Butterfly-bush Family ●
62209 Buddlejaceae K. Wilh.（保留科名）= Scrophulariaceae Juss.（保留科名）●■
62210 Buechnera Roth = Buchnera L. ■
62211 Buechnera Wettst. = Buchnera L. ■
62212 Buechneria Roth = Buechnera Wettst. ■
62213 Bueckia A. Rich. = Buekia Nees ■☆
62214 Bueckia A. Rich. = Neesenbeckia Levyns ■☆
62215 Buekia Giseke（废弃属名）= Alpinia Roxb.（保留属名）■
62216 Buekia Nees = Neesenbeckia Levyns ■☆
62217 Buellia Raf. = Ruellia L. ■●
62218 Buelowia Schumach. = Smeathmannia Sol. ex R. Br. ●☆
62219 Buelowia Schumach. et Thonn. = Smeathmannia Sol. ex R. Br. ●☆
62220 Buena Cav. = Gonzalagunia Ruiz et Pav. ●☆
62221 Buena Pohl = Cosmibuena Ruiz et Pav.（保留属名）●☆
62222 Buergeria Miq. = Cladrastis Raf. ●
62223 Buergeria Miq. = Maackia Rupr. et Maxim. ●
62224 Buergeria Siebold et Zucc. = Magnolia L. ●
62225 Buergeria foribunda Miq. = Maackia floribunda（Miq.）Takeda ●
62226 Buergersiochloa Pilg.（1914）；伊里安禾属■☆
62227 Buergersiochloa bambusoides Pilg.；伊里安禾■☆
62228 Buesiella C. Schweinf.（1952）；布埃兰属■☆
62229 Buesiella C. Schweinf. = Rusbyella Rolfe ex Rusby ■☆
62230 Buesiella pusilla C. Schweinf.；布埃兰■☆
62231 Buettnera J. F. Gmel. = Byttneria Loefl.（保留属名）●
62232 Buettneria L. = Byttneria Loefl.（保留属名）●

62233 Buettneria Murray = Byttneria Loefl.(保留属名)●
62234 Buettneria aspera Colebr. = Byttneria aspera Colebr. ●
62235 Buettneria grandifolia DC. = Byttneria grandifolia DC. ●
62236 Buettneria nepalensis Turcz. = Byttneria grandifolia DC. ●
62237 Buettneriaceae Barnhart = Calycanthaceae Lindl.(保留科名)●
62238 Buffonea Koch. = Bufonia L. ■☆
62239 Buffonia Batsch = Bufonia L. ■☆
62240 Buffonia L. = Bufonia L. ■☆
62241 Buffonia macrocarpa Ser. = Bufonia macrocarpa Ser. ■☆
62242 Buffonia oliveriana Ser. = Bufonia oliveriana Ser. ■☆
62243 Buffonia parviflora Griseb. = Bufonia parviflora Griseb. ■☆
62244 Bufonia L.(1753);蟾蜍草属■☆
62245 Bufonia battandieri Batt. = Bufonia duvaljouvei Batt. et Trab. subsp. battandieri (Batt.) Maire ■☆
62246 Bufonia chevallieri Batt.;舍瓦利耶蟾蜍草■☆
62247 Bufonia duvaljouvei Batt. et Trab.;杜瓦蟾蜍草■☆
62248 Bufonia duvaljouvei Batt. et Trab. subsp. battandieri (Batt.) Maire;巴坦蟾蜍草■☆
62249 Bufonia duvaljouvei Batt. et Trab. var. aurasiaca Maire = Bufonia duvaljouvei Batt. et Trab. ■☆
62250 Bufonia duvaljouvei Batt. et Trab. var. chabertiana (Maire) Maire = Bufonia duvaljouvei Batt. et Trab. ■☆
62251 Bufonia duvaljouvei Batt. et Trab. var. clausonii Maire = Bufonia paniculata Dubois ■☆
62252 Bufonia macrocarpa Ser.;大果蟾蜍草■☆
62253 Bufonia macropetala Willk.;大瓣蟾蜍草■☆
62254 Bufonia macropetala Willk. var. strohlii (Emb. et Maire) Pau et Font Quer = Bufonia macropetala Willk. ■☆
62255 Bufonia macrosperma J. Gay subsp. parviflora Batt. = Bufonia tenuifolia L. ■☆
62256 Bufonia macrosperma J. Gay var. parviflora (Batt.) Batt. = Bufonia tenuifolia L. ■☆
62257 Bufonia macrosperma J. Gay var. trinervia Coss. ex Rouy et Foucaud = Bufonia tenuifolia L. ■☆
62258 Bufonia mauritanica Murb.;毛里塔尼亚蟾蜍草■☆
62259 Bufonia mauritanica Murb. var. longipetala Maire = Bufonia mauritanica Murb. ■☆
62260 Bufonia murbeckii Emb.;穆尔拜克蟾蜍草■☆
62261 Bufonia oliveriana Ser.;奥里弗蟾蜍草■☆
62262 Bufonia paniculata Dubois;圆锥蟾蜍草■☆
62263 Bufonia paniculata Dubois subsp. parviflora (Batt.) Maire = Bufonia tenuifolia L. ■☆
62264 Bufonia parviflora Griseb.;小花蟾蜍草■☆
62265 Bufonia perennis Pourr. subsp. mauritanica (Murb.) Pau et Font Quer = Bufonia mauritanica Murb. ■☆
62266 Bufonia strohlii Emb. et Maire = Bufonia macropetala Willk. ■☆
62267 Bufonia strohlii Emb. et Maire var. laevis Sauvage = Bufonia macropetala Willk. ■☆
62268 Bufonia teneriiae Christ = Bufonia paniculata Dubois ■☆
62269 Bufonia tenuifolia L.;细叶蟾蜍草■☆
62270 Bufonia tenuifolia L. subsp. battandieriana Munoz Garm. et Pedrol = Bufonia tenuifolia L. ■☆
62271 Bufonia tenuifolia L. subsp. parviflora (Batt.) Maire = Bufonia tenuifolia L. ■☆
62272 Bufonia tenuifolia L. var. parviflora (Batt.) Maire = Bufonia tenuifolia L. ■☆
62273 Bufonia tenuifolia L. var. trinervia (Coss. ex Rouy et Foucaud) Maire et Weiller = Bufonia tenuifolia L. ■☆
62274 Bufonia willkommiana Boiss. var. chabertiana Maire = Bufonia duvaljouvei Batt. et Trab. ■☆
62275 Buforrestia C. B. Clarke(1881);透鞘花属■☆
62276 Buforrestia brachycarpa Gilg et Lederm. ex Mildbr. = Stanfieldiella brachycarpa (Gilg et Ledermann ex Mildbr.) Brenan ■☆
62277 Buforrestia brachycarpa Gilg et Lederm. ex Mildbr. var. hirsuta Brenan = Stanfieldiella brachycarpa (Gilg et Ledermann ex Mildbr.) Brenan var. hirsuta (Brenan) Brenan ■☆
62278 Buforrestia glabrisepala De Wild. = Stanfieldiella imperforata (C. B. Clarke) Brenan var. glabrisepala (De Wild.) Brenan ■☆
62279 Buforrestia imperforata C. B. Clarke = Stanfieldiella imperforata (C. B. Clarke) Brenan ■☆
62280 Buforrestia imperforata C. B. Clarke var. glabrisepala (De Wild.) Brenan = Stanfieldiella imperforata (C. B. Clarke) Brenan var. glabrisepala (De Wild.) Brenan ■☆
62281 Buforrestia mannii C. B. Clarke;曼透鞘花■☆
62282 Buforrestia minor K. Schum. = Stanfieldiella imperforata (C. B. Clarke) Brenan var. glabrisepala (De Wild.) Brenan ■☆
62283 Buforrestia obovata Brenan;倒卵透鞘花■☆
62284 Buforrestia oligantha Mildbr. = Stanfieldiella oligantha (Mildbr.) Brenan ■☆
62285 Buforrestia tenuis C. B. Clarke = Amischotolype tenuis (C. B. Clarke) R. S. Rao ■☆
62286 Bugenvillea Endl. = Buginvillaea Comm. ex Juss. ●
62287 Buginvillaea Comm. ex Juss. = Bougainvillea Comm. ex Juss.(保留属名)●
62288 Buglossa Gray = Lycopsis L. ■
62289 Buglossaceae Hoffmanns. et Link = Boraginaceae Juss.(保留科名)■●
62290 Buglossites Bubani = Lycopsis L. ■
62291 Buglossites Moris = Borago L. ■☆
62292 Buglossoides I. M. Johnst. = Lithospermum L. ■
62293 Buglossoides Moench = Lithospermum L. ■
62294 Buglossoides Moench(1794);麦家公属(地仙桃属,拟紫草属);Buglossoides ■
62295 Buglossoides arvense (L.) I. M. Johnst. = Lithospermum arvense L. ■
62296 Buglossoides arvensis (L.) I. M. Johnst. subsp. gasparrinii (Guss.) R. Fern. = Lithospermum incrassatum Guss. ■☆
62297 Buglossoides arvensis (L.) I. M. Johnst. subsp. permixta (F. W. Schultz) R. Fern. = Lithospermum incrassatum Guss. ■☆
62298 Buglossoides arvensis (L.) I. M. Johnst. subsp. sibthorpiana (Griseb.) R. Fern. = Lithospermum sibthorpianum Griseb. ■☆
62299 Buglossoides arvensis (L.) I. M. Johnst. var. coerulescens (DC.) A. Hansen et Sunding = Lithospermum incrassatum Guss. ■☆
62300 Buglossoides incrassata (Guss.) I. M. Johnst. = Lithospermum incrassatum Guss. ■☆
62301 Buglossoides tenuiflora (L. f.) I. M. Johnst.;细花麦家公■☆
62302 Buglossoides tenuiflora (L. f.) I. M. Johnst. = Lithospermum tenuiflorum L. f. ■☆
62303 Buglossoides zollingeri (A. DC.) I. M. Johnst. = Lithospermum zollingeri A. DC. ■
62304 Buglossum Mill. = Anchusa L. ■
62305 Bugranopsis Pomel = Ononis L. ■●
62306 Bugranopsis alopecuroides (L.) Pomel = Ononis alopecuroides L. ●☆

62307 Bugranopsis cephalantha (Pomel) Pomel = Ononis cephalantha Pomel ●☆

62308 Bugranopsis crinita (Pomel) Pomel = Ononis crinita Pomel ●☆

62309 Bugranopsis euphrasiifolia (Desf.) Pomel = Ononis euphrasiifolia Desf. ●☆

62310 Bugranopsis megalostachys (Munby) Pomel = Ononis megalostachys Munby ●☆

62311 Bugranopsis mitissima (L.) Pomel = Ononis mitissima L. ●☆

62312 Bugranopsis rosea (Durieu) Pomel = Ononis rosea Durieu ●☆

62313 Bugranopsis salzmanniana (Boiss. et Reut.) Pomel = Ononis alopecuroides L. ●☆

62314 Bugranopsis stricta (Pomel) Pomel = Ononis stricta Pomel ●☆

62315 Bugranopsis tournefortii (Coss.) Pomel = Ononis tournefortii Coss. ■☆

62316 Bugranopsis variegata (L.) Pomel = Ononis variegata L. ■☆

62317 Buguinvillaea Humb. et Bonpl. = Bougainvillea Comm. ex Juss. (保留属名)●

62318 Bugula Mill. = Ajuga L. ■●

62319 Bugula Tourn. ex Mill. = Ajuga L. ■●

62320 Buhsia Bunge(1859);布赫山柑属●☆

62321 Buhsia coluteoides Bunge;布赫山柑●☆

62322 Buinalis Raf. = Siphonychia Torr. et A. Gray(保留属名)■☆

62323 Buiningia Buxb. = Coleocephalocereus Backeb. ●☆

62324 Bujacia E. Mey. = Glycine Willd.(保留属名)■

62325 Bukiniczia Lincz.(1971);裂萼补血草属(阿富汗白花丹属)■☆

62326 Bukiniczia cabulica (Boiss.) Lincz.;裂萼补血草■☆

62327 Bulbedulis Raf. = Camassia Lindl.(保留属名)■☆

62328 Bulbedulis Raf. = Quamasia Raf. ■☆

62329 Bulbilis Kuntze = Buchloe Engelm.(保留属名)■

62330 Bulbilis Raf. = Buchloe Engelm.(保留属名)■

62331 Bulbilis dactyloides (Nutt.) Raf. ex Kuntze = Buchloe dactyloides (Nutt.) Engelm. ■

62332 Bulbillaria Zucc. = Gagea Salisb. ■

62333 Bulbine Gaertn. = Bulbine Wolf(保留属名)■☆

62334 Bulbine Wolf(1788)(保留属名);球百合属;Bulbine ■☆

62335 Bulbine abyssinica A. Rich.;阿比西尼亚球百合■☆

62336 Bulbine alba Van Jaarsv.;白花球百合■☆

62337 Bulbine alooides (L.) Willd.;芦荟球百合;Renosterveld Bulbine ■☆

62338 Bulbine alooides Willd. = Bulbine alooides (L.) Willd. ■☆

62339 Bulbine altissima (Mill.) Fourc. = Bulbine asphodeloides (L.) Spreng. ■☆

62340 Bulbine alveolata S. A. Hammer;蜂窝球百合■☆

62341 Bulbine angustifolia Poelln.;窄叶球百合■☆

62342 Bulbine annua (L.) Willd.;一年生球百合■☆

62343 Bulbine asphodeloides (L.) Spreng.;阿福花状球百合■☆

62344 Bulbine asphodeloides (L.) Spreng. var. denticulifera Poelln. = Bulbine lagopus (Thunb.) N. E. Br. ■☆

62345 Bulbine asphodeloides (L.) Spreng. var. filifolioides De Wild. = Bulbine abyssinica A. Rich. ■☆

62346 Bulbine asphodeloides (L.) Spreng. var. monticola Poelln. = Bulbine abyssinica A. Rich. ■☆

62347 Bulbine asphodeloides (L.) Spreng. var. otaviensis Poelln. = Bulbine capitata Poelln. ■☆

62348 Bulbine asphodeloides (L.) Spreng. var. xanthobotrys (Engl. et Gilg) Weim. = Bulbine abyssinica A. Rich. ■☆

62349 Bulbine bachmanii Baker = Bulbine nutans (Jacq.) Spreng. ■☆

62350 Bulbine bachmanniana Schinz = Bulbine nutans (Jacq.) Spreng. ■☆

62351 Bulbine bisulcata Haw. = Bulbine cepacea (Burm. f.) Wijnands ■☆

62352 Bulbine brevifolia (Thunb.) Roem. et Schult. f. = Caesia contorta (L. f.) T. Durand et Schinz ■☆

62353 Bulbine breviracemosa Poelln. = Jodrellia fistulosa (Chiov.) Baijnath ■☆

62354 Bulbine brunsvigiifolia Baker = Bulbine latifolia (L. f.) Spreng. ■☆

62355 Bulbine bruynsii S. A. Hammer;布勒伊斯球百合■☆

62356 Bulbine bulbosa Haw.;澳洲球百合;Australian Native Leek ■☆

62357 Bulbine caespitosa Baker = Bulbine lagopus (Thunb.) N. E. Br. ■☆

62358 Bulbine canaliculata (Aiton) Spreng. = Trachyandra ciliata (L. f.) Kunth ■☆

62359 Bulbine canaliculata G. Will. = Bulbine erectipilosa G. Will. ■☆

62360 Bulbine capitata Poelln.;头状球百合■☆

62361 Bulbine cataphyllata Poelln. = Bulbine cepacea (Burm. f.) Wijnands ■☆

62362 Bulbine cauda-felis (L. f.) Schult. et Schult. f. = Bulbinella cauda-felis (L. f.) T. Durand et Schinz ■☆

62363 Bulbine caulescens L. = Bulbine frutescens (L.) Willd. ■☆

62364 Bulbine cepacea (Burm. f.) Wijnands;小花球百合■☆

62365 Bulbine ciliata (L. f.) Link = Trachyandra ciliata (L. f.) Kunth ■☆

62366 Bulbine circinata Schltr. ex Poelln. = Bulbine torta N. E. Br. ■☆

62367 Bulbine circinata Schltr. ex Poelln. var. minor Poelln. = Bulbine torta N. E. Br. ■☆

62368 Bulbine concinna Baker = Bulbine favosa (Thunb.) Schult. et Schult. f. ■☆

62369 Bulbine cremnophila Van Jaarsv.;悬崖球百合■☆

62370 Bulbine crispa (Thunb.) Roem. et Schult. = Chlorophytum crispum (Thunb.) Baker ■☆

62371 Bulbine crocea L. Guthrie = Bulbine asphodeloides (L.) Spreng. ■☆

62372 Bulbine dactylopsoides G. Will.;指状球百合■☆

62373 Bulbine decurvata Peter ex Poelln. = Bulbine abyssinica A. Rich. ■☆

62374 Bulbine densiflora Baker = Bulbine narcissifolia Salm-Dyck ■☆

62375 Bulbine dielsii Poelln. = Bulbine asphodeloides (L.) Spreng. ■☆

62376 Bulbine diphylla Schltr. ex Poelln.;二叶球百合■☆

62377 Bulbine dubia Schult. et Schult. f. = Bulbine favosa (Thunb.) Schult. et Schult. f. ■☆

62378 Bulbine ensifolia Baker = Bulbine latifolia (L. f.) Spreng. ■☆

62379 Bulbine erectipilosa G. Will.;直立毛球百合■☆

62380 Bulbine esterhuyseniae Baijnath;埃斯特球百合■☆

62381 Bulbine falcata (L. f.) Schult. et Schult. f. = Trachyandra falcata (L. f.) Kunth ■☆

62382 Bulbine fallax Poelln.;迷惑球百合■☆

62383 Bulbine favosa (Thunb.) Schult. et Schult. f.;泡状球百合■☆

62384 Bulbine filifolia Baker = Bulbine favosa (Thunb.) Schult. et Schult. f. ■☆

62385 Bulbine fistulosa Chiov. = Jodrellia fistulosa (Chiov.) Baijnath ■☆

62386 Bulbine flexicaulis Baker;曲茎球百合■☆

62387 Bulbine flexuosa Schltr.;曲折球百合■☆

62388 Bulbine foleyi E. Phillips;福莱球百合■☆

62389 Bulbine fragilis G. Will.;脆球百合■☆
62390 Bulbine frutescens (L.) Willd.;具柄球百合;Orange Bulbine, Rankkopieva, Stalked Bulbine ■☆
62391 Bulbine frutescens Willd. = Bulbine frutescens (L.) Willd. ■☆
62392 Bulbine graminea Haw. = Bulbine lagopus (Thunb.) N. E. Br. ■☆
62393 Bulbine hallii G. Will.;霍尔球百合■☆
62394 Bulbine hamata Peter ex Poelln. = Bulbine abyssinica A. Rich. ■☆
62395 Bulbine hantamensis Poelln. = Bulbine succulenta Compton ■☆
62396 Bulbine haworthioides B. Nord.;霍沃斯球百合■☆
62397 Bulbine huilensis Poelln. = Bulbine abyssinica A. Rich. ■☆
62398 Bulbine inexpecta Poelln. = Bulbine cepacea (Burm. f.) Wijnands ■☆
62399 Bulbine inflata Oberm.;膨胀球百合■☆
62400 Bulbine inops N. E. Br. = Bulbine flexicaulis Baker ■☆
62401 Bulbine lagopus (Thunb.) N. E. Br.;兔足球百合■☆
62402 Bulbine lamprophylla G. Will.;亮叶球百合■☆
62403 Bulbine latibracteata Poelln. = Bulbine narcissifolia Salm-Dyck ■☆
62404 Bulbine latifolia (L. f.) Spreng.;宽叶球百合;Bulbine, Rooiwortel ■☆
62405 Bulbine latifolia Spreng. = Bulbine latifolia (L. f.) Spreng. ■☆
62406 Bulbine latitepala Poelln. = Bulbine abyssinica A. Rich. ■☆
62407 Bulbine lavranii G. Will. et Baijnath;拉夫连球百合■☆
62408 Bulbine laxiflora Baker = Bulbine praemorsa (Jacq.) Spreng. ■☆
62409 Bulbine longifolia Schinz;长叶球百合■☆
62410 Bulbine longiscapa (Jacq.) Willd. = Bulbine asphodeloides (L.) Spreng. ■☆
62411 Bulbine lydenburgensis Poelln. = Bulbine capitata Poelln. ■☆
62412 Bulbine mackenii Hook. f. = Eriospermum mackenii (Hook. f.) Baker ■☆
62413 Bulbine macrophylla Salm-Dyck;大叶球百合■☆
62414 Bulbine macrophylla Salm-Dyck = Bulbine alooides (L.) Willd. ■☆
62415 Bulbine mallyana Schltr. ex Poelln. = Bulbine torta N. E. Br. ■☆
62416 Bulbine mayori Beauverd = Bulbine favosa (Thunb.) Schult. et Schult. f. ■☆
62417 Bulbine meiringii Van Jaarsv.;迈林球百合■☆
62418 Bulbine melanovaginata G. Will. = Bulbine foleyi E. Phillips ■☆
62419 Bulbine mesembryanthoides Haw.;日中花球百合■☆
62420 Bulbine mesembryanthoides Haw. subsp. namaquensis G. Will.;纳马夸球百合■☆
62421 Bulbine mettinghi Ten. = Bulbine asphodeloides (L.) Spreng. ■☆
62422 Bulbine migiurtina Chiov. = Jodrellia migiurtina (Chiov.) Baijnath ■☆
62423 Bulbine minima Baker;极小球百合■☆
62424 Bulbine monophylla Poelln.;单叶球百合■☆
62425 Bulbine muscicola G. Will.;苔地球百合■☆
62426 Bulbine namaensis Schinz;纳马球百合■☆
62427 Bulbine narcissifolia Salm-Dyck;水仙叶球百合■☆
62428 Bulbine natalensis Baker = Bulbine latifolia (L. f.) Spreng. ■☆
62429 Bulbine nutans (Jacq.) Spreng.;低垂球百合■☆
62430 Bulbine ophiophylla G. Will.;蛇叶球百合■☆
62431 Bulbine otaviensis (Poelln.) Sölch = Bulbine capitata Poelln. ■☆
62432 Bulbine pallida Baker = Bulbine asphodeloides (L.) Spreng. ■☆
62433 Bulbine parviflora Baker = Bulbine cepacea (Burm. f.) Wijnands ■☆
62434 Bulbine pendens G. Will. et Baijnath;彭达球百合■☆
62435 Bulbine platyphylla Baker = Bulbine alooides (L.) Willd. ■☆
62436 Bulbine praemorsa (Jacq.) Spreng.;啮蚀球百合■☆
62437 Bulbine pugioniformis (Jacq.) Link = Bulbine cepacea (Burm. f.) Wijnands ■☆
62438 Bulbine quartzicola G. Will.;阔茨球百合■☆
62439 Bulbine ramosa Van Jaarsv.;分枝球百合■☆
62440 Bulbine retinens Van Jaarsv. et S. A. Hammer;密着球百合■☆
62441 Bulbine rhopalophylla Dinter;棒叶球百合■☆
62442 Bulbine rigidula Schltr. ex Poelln. = Bulbine favosa (Thunb.) Schult. et Schult. f. ■☆
62443 Bulbine rostrata Willd. = Bulbine frutescens (L.) Willd. ■☆
62444 Bulbine rupicola G. Will.;岩生球百合■☆
62445 Bulbine scabra (L. f.) Schult. et Schult. f. = Trachyandra scabra (L. f.) Kunth ■☆
62446 Bulbine seineri Engl. et K. Krause = Ornithogalum seineri (Engl. et K. Krause) Oberm. ■☆
62447 Bulbine setifera Poelln. = Bulbine favosa (Thunb.) Schult. et Schult. f. ■☆
62448 Bulbine stenophylla I. Verd. = Bulbine capitata Poelln. ■☆
62449 Bulbine striata Baijnath et Van Jaarsv.;条纹球百合■☆
62450 Bulbine succulenta Compton;多汁球百合■☆
62451 Bulbine suurbergensis Van Jaarsv. et A. E. van Wyk;叙尔贝赫球百合■☆
62452 Bulbine tetraphylla Dinter = Bulbine praemorsa (Jacq.) Spreng. ■☆
62453 Bulbine thomasiae Van Jaarsv.;托马斯球百合■☆
62454 Bulbine torsiva G. Will.;扭转球百合■☆
62455 Bulbine torta N. E. Br.;缠扭球百合■☆
62456 Bulbine tortifolia I. Verd. = Bulbine angustifolia Poelln. ■☆
62457 Bulbine transvaalensis Baker = Bulbine latifolia (L. f.) Spreng. ■☆
62458 Bulbine trichophylla Baker = Bulbine favosa (Thunb.) Schult. et Schult. f. ■☆
62459 Bulbine triebneri Dinter = Bulbine frutescens (L.) Willd. ■☆
62460 Bulbine triquetra (L. f.) Schult. et Schult. f. = Bulbinella triquetra (L. f.) Kunth ■☆
62461 Bulbine truncata G. Will.;平截球百合■☆
62462 Bulbine tuberosa (Mill.) Oberm. = Bulbine cepacea (Burm. f.) Wijnands ■☆
62463 Bulbine unifolia Schult. f. ex Baker;单花球百合■☆
62464 Bulbine urgineoides Baker = Bulbine praemorsa (Jacq.) Spreng. ■☆
62465 Bulbine vesicularis Dinter = Bulbine namaensis Schinz ■☆
62466 Bulbine vitrea G. Will. et Baijnath;透明球百合■☆
62467 Bulbine vittatifolia G. Will.;粗线叶球百合■☆
62468 Bulbine wiesei L. I. Hall;维塞球百合■☆
62469 Bulbine xanthobotrys Engl. et Gilg = Bulbine abyssinica A. Rich. ■☆
62470 Bulbine zeyheri Baker = Bulbine praemorsa (Jacq.) Spreng. ■☆
62471 Bulbinella Kunth(1843);猫尾花属;Cat's Tail ■☆
62472 Bulbinella aitonii (Baker) T. Durand et Schinz = Trachyandra filiformis (Aiton) Oberm. ■☆
62473 Bulbinella brevifolia (Thunb.) Kunth = Caesia contorta (L. f.) T. Durand et Schinz ■☆
62474 Bulbinella brevifolia (Thunb.) Roem. et Schult. = Caesia contorta (L. f.) T. Durand et Schinz ■☆
62475 Bulbinella burkei (Baker) Benth. et Hook. f. = Trachyandra burkei (Baker) Oberm. ■☆

62476 Bulbinella capillaris (Poir.) Kunth = Bulbinella triquetra (L. f.) Kunth ■☆
62477 Bulbinella carnosum (Baker) Baker = Ornithogalum paludosum Baker ■☆
62478 Bulbinella cauda-felis (L. f.) T. Durand et Schinz;猫尾花■☆
62479 Bulbinella caudata (Thunb.) Kunth = Bulbinella cauda-felis (L. f.) T. Durand et Schinz ■☆
62480 Bulbinella caudata (Thunb.) Kunth var. ciliolata (Kunth) Baker = Bulbinella ciliolata Kunth ■☆
62481 Bulbinella chartacea P. L. Perry;纸质猫尾花■☆
62482 Bulbinella ciliolata Kunth;睫毛猫尾花■☆
62483 Bulbinella divaginata P. L. Perry;双鞘猫尾花■☆
62484 Bulbinella elata P. L. Perry;高猫尾花■☆
62485 Bulbinella elegans P. L. Perry;雅致猫尾花■☆
62486 Bulbinella filiformis (Aiton) Kunth = Trachyandra filiformis (Aiton) Oberm. ■☆
62487 Bulbinella floribunda (Aiton) T. Durand et Schinz;繁花猫尾花■☆
62488 Bulbinella gracilis Kunth;纤细猫尾花■☆
62489 Bulbinella graminifolia P. L. Perry;禾叶猫尾花■☆
62490 Bulbinella latifolia Kunth;宽叶猫尾花■☆
62491 Bulbinella latifolia Kunth subsp. denticulata P. L. Perry;细齿宽叶猫尾花■☆
62492 Bulbinella nana P. L. Perry;矮小猫尾花■☆
62493 Bulbinella nutans (Thunb.) T. Durand et Schinz;低垂猫尾花■☆
62494 Bulbinella nutans (Thunb.) T. Durand et Schinz var. nutans = Bulbinella nutans (Thunb.) T. Durand et Schinz ■☆
62495 Bulbinella ornithogaloides Kunth = Ornithogalum flexuosum (Thunb.) U. Müll. -Doblies et D. Müll. -Doblies ■☆
62496 Bulbinella peronata Kunth = Bulbinella triquetra (L. f.) Kunth ■☆
62497 Bulbinella potbergensis P. L. Perry;贝尔热猫尾花■☆
62498 Bulbinella robusta Kunth = Bulbinella nutans (Thunb.) T. Durand et Schinz ■☆
62499 Bulbinella robusta Kunth var. latifolia (Kunth) Baker = Bulbinella latifolia Kunth ■☆
62500 Bulbinella setifolia Kunth = Bulbinella triquetra (L. f.) Kunth ■☆
62501 Bulbinella setosa (Willd.) T. Durand et Schinz = Bulbinella nutans (Thunb.) T. Durand et Schinz ■☆
62502 Bulbinella squamea (L. f.) Kunth = Trachyandra hispida (L.) Kunth ■☆
62503 Bulbinella trinervis (Baker) P. L. Perry;三脉猫尾花■☆
62504 Bulbinella triquetra (L. f.) Kunth;三棱猫尾花■☆
62505 Bulbinella triquetra (L. f.) Kunth var. trinervis Baker = Bulbinella trinervis (Baker) P. L. Perry ■☆
62506 Bulbinopsis Borzi = Bulbine Wolf(保留属名)■☆
62507 Bulbinopsis bulbosa (R. Br.) Borzi = Bulbine bulbosa Haw. ■☆
62508 Bulbisperma Reinw. ex Blume = Peliosanthes Andrews ■
62509 Bulbocapnos Bernh. = Capnites (DC.) Dumort. ■
62510 Bulbocapnos Bernh. = Corydalis DC. (保留属名)■
62511 Bulbocastanum Lag. = Conopodium W. D. J. Koch(保留属名)■☆
62512 Bulbocastanum Mill. = Bunium L. ■☆
62513 Bulbocastanum Schur = Carum L. ■
62514 Bulbocodiaceae Salisb.;春水仙科■
62515 Bulbocodiaceae Salisb. = Colchicaceae DC. (保留科名)■
62516 Bulbocodiaceae Salisb. = Melanthiaceae Batsch ex Borkh. (保留科名)■
62517 Bulbocodium Gronov. = Romulea Maratti(保留属名)■☆
62518 Bulbocodium L. (1753);春水仙属(欧洲鸢尾属);Meadow Saffron ■☆
62519 Bulbocodium L. = Colchicum L. ■
62520 Bulbocodium Ludw. = Romulea Maratti(保留属名)■☆
62521 Bulbocodium Ludw. ex Kuntze = Romulea Maratti(保留属名)■☆
62522 Bulbocodium cameroonianum (Baker) Kuntze = Romulea camerooniana Baker ■☆
62523 Bulbocodium serotinum L. = Lloydia serotina (L.) Salisb. ex Rchb. ■
62524 Bulbocodium vernum Desf. = Merendera filifolia Cambess. ■☆
62525 Bulbocodium vernum L.;春水仙(春花秋水仙);Spring Colchicum, Spring Meadow Saffron ■☆
62526 Bulbocodium versicolor (Ker. Gawl.) Spreng.;彩色春水仙■☆
62527 Bulbophyllaria Rchb. f. = Bulbophyllum Thouars(保留属名)■
62528 Bulbophyllaria S. Moore = Bulbophyllum Thouars(保留属名)■
62529 Bulbophyllopsis Rchb. f. = Bulbophyllum Thouars(保留属名)■
62530 Bulbophyllum Thouars (1822) (保留属名);石豆兰属(豆兰属,卷瓣兰属);Bulbophyllum, Curlylip-orchis, Stonebean-orchis ■
62531 Bulbophyllum acutebracteatum De Wild.;尖苞石豆兰■☆
62532 Bulbophyllum acutebracteatum De Wild. subsp. fuscoides (Petersen) W. Sanford = Bulbophyllum acutebracteatum De Wild. var. rubrobrunneopapillosum (De Wild.) J. J. Verm. ■☆
62533 Bulbophyllum acutebracteatum De Wild. var. rubrobrunneopapillosum (De Wild.) J. J. Verm.;红褐乳突石豆兰■☆
62534 Bulbophyllum acutisepalum De Wild. = Bulbophyllum schimperianum Kraenzl. ■☆
62535 Bulbophyllum acutispicatum H. Perrier;尖穗石豆兰■☆
62536 Bulbophyllum affine Lindl.;赤唇石豆兰(高士佛豆兰,恒春石豆兰,纹星兰);Redlip Bulbophyllum, Redlip Stonebean-orchis ■
62537 Bulbophyllum africanum Hawkes = Bulbophyllum nigritianum Rendle ■☆
62538 Bulbophyllum afzelii Schltr.;阿芙泽尔石豆兰■☆
62539 Bulbophyllum aggregatum Bosser;聚集石豆兰■☆
62540 Bulbophyllum albidum De Wild. = Bulbophyllum nigritianum Rendle ■☆
62541 Bulbophyllum albociliatum (Tang S. Liu et H. J. Su) Nackej.;白毛卷瓣兰(白缘石豆兰);Whitehair Bulbophyllum, Whitehair Curlylip-orchis ■
62542 Bulbophyllum albociliatum (Tang S. Liu et H. J. Su) Nackej. var. weiminianum T. P. Lin et Kuo Huang = Bulbophyllum albociliatum (Tang S. Liu et H. J. Su) Nackej. ■
62543 Bulbophyllum albociliatum (Tang S. Liu et H. J. Su) Seidenf. = Bulbophyllum albociliatum (Tang S. Liu et H. J. Su) Nackej. ■
62544 Bulbophyllum album Jum. et Perrier = Bulbophyllum humblotii Rolfe ■☆
62545 Bulbophyllum alexandrae Schltr.;亚历山大石豆兰■☆
62546 Bulbophyllum alleizettei Schltr.;阿雷卷瓣兰■☆
62547 Bulbophyllum amanicum Kraenzl. = Bulbophyllum josephii (Kuntze) Summerh. ■☆
62548 Bulbophyllum amauryae Rendle = Bulbophyllum intertextum Lindl. ■☆
62549 Bulbophyllum ambongense Schltr. = Bulbophyllum rubrum Jum. et Perrier ■☆
62550 Bulbophyllum ambreae H. Perrier = Bulbophyllum septatum Schltr. ■☆
62551 Bulbophyllum ambrense H. Perrier;昂布尔卷瓣兰■☆
62552 Bulbophyllum ambrosium (Hance) Schltr.;芳香石豆兰(肥猪

草);Fragrant Bulbophyllum,Fragrant Stonebean-orchis ■
62553 Bulbophyllum ambrosium (Hance) Schltr. subsp. nepalense J. J. Wood;西南石豆兰■
62554 Bulbophyllum amoenum Bosser;秀丽石豆兰■☆
62555 Bulbophyllum amplifolium (Rolfe) N. P. Balakr. et S. Chowdhury;大叶卷瓣兰(抱茎卷瓣兰,大叶卷唇兰,一匹草);Largeleaf Bulbophyllum,Largeleaf Curlylip-orchis ■
62556 Bulbophyllum amplum (Lindl.) Rchb. f. = Epigeneium amplum (Lindl. ex Wall.) Summerh. ■
62557 Bulbophyllum andersonii (Hook. f.) J. J. Sm.;梳帽卷瓣兰(果上叶,卷瓣兰,梳帽卷唇兰,梳帽石豆兰,一匹草,一匹叶);Anderson Bulbophyllum,Anderson Cirrhopetalum,Anderson Curlylip-orchis ■
62558 Bulbophyllum andohahelense H. Perrier;安杜哈赫尔石豆兰■☆
62559 Bulbophyllum andongense Rchb. f. = Bulbophyllum cocoinum Bateman ex Lindl. ■☆
62560 Bulbophyllum andringitranum Schltr. = Bulbophyllum nutans Thouars ■☆
62561 Bulbophyllum anguste-ellipticum Seidenf. = Bulbophyllum nigrescens Rolfe ■
62562 Bulbophyllum ankaizinense (Jum. et H. Perrier) Schltr.;安凯济纳石豆兰■☆
62563 Bulbophyllum ankaratranum Schltr.;安卡拉特拉石豆兰■☆
62564 Bulbophyllum antongilense Schltr.;安通吉尔石豆兰■☆
62565 Bulbophyllum apetalum Lindl. = Genyorchis apetala (Lindl.) J. J. Verm. ■☆
62566 Bulbophyllum apodum Hook. f.;无柄石豆兰■
62567 Bulbophyllum approximatum Ridl.;相似石豆兰■☆
62568 Bulbophyllum arnoldianum (De Wild.) De Wild. = Bulbophyllum velutinum (Lindl.) Rchb. f. ■☆
62569 Bulbophyllum aubrevillei Bosser;奥布石豆兰■☆
62570 Bulbophyllum aurantiacum Hook. f. = Bulbophyllum josephii (Kuntze) Summerh. ■☆
62571 Bulbophyllum aureolabellum T. P. Lin;台湾石豆兰(耳唇石豆兰,细豆兰,小豆兰);Taiwan Bulbophyllum,Taiwan Stonebean-orchis ■
62572 Bulbophyllum aureolabellum T. P. Lin = Bulbophyllum drymoglossum Maxim. ex Okubo ■
62573 Bulbophyllum bakossorum Schltr. = Bulbophyllum bufo (Lindl.) Rchb. f. ■☆
62574 Bulbophyllum ballii P. J. Cribb;鲍尔石豆兰■☆
62575 Bulbophyllum bambiliense De Wild. = Bulbophyllum scaberulum (Rolfe) Bolus ■☆
62576 Bulbophyllum barbigerum Lindl.;毛唇石豆兰■☆
62577 Bulbophyllum baronii Ridl.;巴龙石豆兰■☆
62578 Bulbophyllum bathieanum Schltr.;巴西石豆兰■☆
62579 Bulbophyllum bequaertii De Wild. = Bulbophyllum cochleatum Lindl. var. bequaertii (De Wild.) J. J. Verm. ■☆
62580 Bulbophyllum bequaertii De Wild. var. brachyanthum Summerh. = Bulbophyllum cochleatum Lindl. var. tenuicaule (Lindl.) J. J. Verm. ■☆
62581 Bulbophyllum bibundiense Schltr. = Bulbophyllum sandersonii (Hook. f.) Rchb. f. ■☆
62582 Bulbophyllum bicolor (Lindl.) Hook. f. = Sunipia bicolor Lindl. ■
62583 Bulbophyllum bicolor Jum. et H. Perrier = Bulbophyllum bicoloratum Schltr. ■☆
62584 Bulbophyllum bicoloratum Schltr.;双色石豆兰■☆
62585 Bulbophyllum bidenticulatum J. J. Verm.;双齿石豆兰■☆
62586 Bulbophyllum bifarium Hook. f.;二列石豆兰■☆
62587 Bulbophyllum biflorum Teijsm. et Binn.;双花石豆兰■☆
62588 Bulbophyllum bittnerianum Schltr.;团花石豆兰;Bittner Bulbophyllum,Bittner Stonebean-orchis ■
62589 Bulbophyllum boiteaui H. Perrier;博特■☆
62590 Bulbophyllum bomiense Z. H. Tsi;波密卷瓣兰(波密卷唇兰);Bomi Bulbophyllum,Bomi Curlylip-orchis ■
62591 Bulbophyllum boninense (Schltr.) J. J. Sm.;小笠原石豆兰■☆
62592 Bulbophyllum bosseri K. Lemcke = Bulbophyllum reflexiflorum H. Perrier ■☆
62593 Bulbophyllum brachyphyton Schltr.;短石豆兰■☆
62594 Bulbophyllum brachystachyum Schltr.;短穗石豆兰■☆
62595 Bulbophyllum braunii Kraenzl. = Bulbophyllum falcipetalum Lindl. ■
62596 Bulbophyllum brevidenticulatum De Wild. = Bulbophyllum cocoinum Bateman ex Lindl. ■☆
62597 Bulbophyllum brevipedunculatum T. C. Hsu et S. W. Chung;短葶卷瓣兰■
62598 Bulbophyllum brevipetalum H. Perrier;短瓣石豆兰■☆
62599 Bulbophyllum brevipetalum H. Perrier subsp. majus H. Perrier = Bulbophyllum brevipetalum H. Perrier ■☆
62600 Bulbophyllum brevipetalum H. Perrier subsp. speculiferum H. Perrier = Bulbophyllum brevipetalum H. Perrier ■☆
62601 Bulbophyllum brevispicatum Z. H. Tsi et S. C. Chen;短序石豆兰;Shortspike Bulbophyllum,Shortspike Stonebean-orchis ■
62602 Bulbophyllum brixhei De Wild. = Bulbophyllum velutinum (Lindl.) Rchb. f. ■☆
62603 Bulbophyllum buchenauianum (Kraenzl.) De Wild. = Bulbophyllum calyptratum Kraenzl. ■☆
62604 Bulbophyllum bufo (Lindl.) Rchb. f.;蟾蜍石豆兰■☆
62605 Bulbophyllum buntingii Rendle = Bulbophyllum oxychilum Schltr. ■☆
62606 Bulbophyllum burttii Summerh.;伯特石豆兰■☆
62607 Bulbophyllum calabaricum Rolfe = Bulbophyllum pumilum (Sw.) Lindl. ■☆
62608 Bulbophyllum calamarioides Schltr. = Bulbophyllum erectum Thouars ■☆
62609 Bulbophyllum calamarium Lindl. var. albociliatum Finet = Bulbophyllum finetii Szlach. et Olszewski ■☆
62610 Bulbophyllum callosum Bosser;硬皮石豆兰■☆
62611 Bulbophyllum calodictyon Schltr. = Bulbophyllum griffithii (Lindl.) Rchb. f. ■
62612 Bulbophyllum calvum Summerh.;光秃石豆兰■☆
62613 Bulbophyllum calyptratum Kraenzl.;帽状石豆兰■☆
62614 Bulbophyllum calyptratum Kraenzl. var. graminifolium (Summerh.) J. J. Verm.;禾叶帽状石豆兰■☆
62615 Bulbophyllum candidum (Lindl.) Hook. f. = Sunipia candida (Lindl.) P. F. Hunt ■
62616 Bulbophyllum capituliflorum Rolfe;头花石豆兰■☆
62617 Bulbophyllum capuronii Bosser;凯普伦石豆兰■☆
62618 Bulbophyllum cardiobulbum Bosser;心球石豆兰■☆
62619 Bulbophyllum careyanum (Hook.) Spreng. = Bulbophyllum orientale Seidenf. ■
62620 Bulbophyllum carinatum G. Will. = Bulbophyllum unifoliatum De Wild. var. infracarinatum (G. Will.) J. J. Verm. ■☆
62621 Bulbophyllum cariniflorum Rchb. f.;尖叶石豆兰;Sharpleaf

Bulbophyllum, Sharpleaf Stonebean-orchis ■

62622 Bulbophyllum carnosilabium Summerh. ;肉质石豆兰■☆

62623 Bulbophyllum carnosisepalum J. J. Verm. ;肉萼石豆兰■☆

62624 Bulbophyllum cataractarum Schltr. ;瀑布群高原石豆兰■☆

62625 Bulbophyllum catenarium Ridl. ;链状石豆兰■

62626 Bulbophyllum caudatum Lindl. ;尾萼卷瓣兰(尾萼卷唇兰); Tailcalyx Bulbophyllum, Tailcalyx Curlylip-orchis ■

62627 Bulbophyllum cauliflorum Hook. f. ;茎花石豆兰; Cauliflorous Bulbophyllum, Cauliflorous Stonebean-orchis ■

62628 Bulbophyllum chevalieri De Wild. = Bulbophyllum scaberulum (Rolfe) Bolus ■☆

62629 Bulbophyllum chinense (Lindl.) Rchb. f. ;中华卷瓣兰; China Curlylip-orchis, Chinese Bulbophyllum ■

62630 Bulbophyllum chitouense S. S. Ying; 溪头石豆兰; Xitou Bulbophyllum, Xitou Stonebean-orchis ■

62631 Bulbophyllum chitouense S. S. Ying = Bulbophyllum griffithii (Lindl.) Rchb. f. ■

62632 Bulbophyllum chrondriophorum (Gagnep.) Seidenf. ;城口卷瓣兰(城口卷唇兰); Chengkou Bulbophyllum, Chengkou Curlylip-orchis ■

62633 Bulbophyllum chrysobulbum H. Perrier = Bulbophyllum nutans Thouars ■☆

62634 Bulbophyllum ciliatilabrum H. Perrier;缘毛卷瓣兰■☆

62635 Bulbophyllum ciliatum Schltr. = Bulbophyllum maximum (Lindl.) Rchb. f. ■☆

62636 Bulbophyllum ciliisepalum T. C. Hsu et S. W. Chung = Bulbophyllum setaceum T. P. Lin ■

62637 Bulbophyllum clarkeanum King et Pantl. = Bulbophyllum stenobulbon Parl. et Rchb. f. ■

62638 Bulbophyllum clarkei (Rolfe) Schltr. = Bulbophyllum scaberulum (Rolfe) Bolus ■☆

62639 Bulbophyllum clavigerum H. Perrier = Bulbophyllum lemurense Bosser et P. J. Cribb ■☆

62640 Bulbophyllum coccinatum H. Perrier;浆果卷瓣兰■☆

62641 Bulbophyllum cochleatum Lindl. ;螺状卷瓣兰■☆

62642 Bulbophyllum cochleatum Lindl. var. bequaertii (De Wild.) J. J. Verm. ;贝卡尔石豆兰■☆

62643 Bulbophyllum cochleatum Lindl. var. brachyanthum (Summerh.) J. J. Verm. ;短花螺状卷瓣兰■☆

62644 Bulbophyllum cochleatum Lindl. var. gravidum (Lindl.) J. J. Verm. ;重螺状卷瓣兰■☆

62645 Bulbophyllum cochleatum Lindl. var. tenuicaule (Lindl.) J. J. Verm. ;细茎螺状卷瓣兰■☆

62646 Bulbophyllum cocoinum Bateman ex Lindl. ;短梗卷瓣兰■☆

62647 Bulbophyllum coeruleum H. Perrier = Bulbophyllum bicoloratum Schltr. ■☆

62648 Bulbophyllum colomaculosum Z. H. Tsi et S. C. Chen;豹斑石豆兰; Leopard Bulbophyllum, Leopard Stonebean-orchis ■

62649 Bulbophyllum colomaculosum Z. H. Tsi et S. C. Chen = Bulbophyllum leopardinum (Wall.) Lindl. ■

62650 Bulbophyllum colubrinum (Rchb. f.) Rchb. f. ;蛇状卷瓣兰■☆

62651 Bulbophyllum comatum Lindl. ;束毛卷瓣兰■☆

62652 Bulbophyllum comatum Lindl. var. inflatum (Rolfe) J. J. Verm. ;膨胀卷瓣兰■☆

62653 Bulbophyllum comosum Collett et Hemsl. ;冠毛石豆兰■☆

62654 Bulbophyllum compactum Kraenzl. = Bulbophyllum coriophorum Ridl. ■☆

62655 Bulbophyllum complanatum H. Perrier;扁平卷瓣兰■☆

62656 Bulbophyllum concatenatum P. J. Cribb et P. Taylor;链状卷瓣兰■☆

62657 Bulbophyllum congestum Rolfe = Bulbophyllum odoratissimum (Sm.) Lindl. ■

62658 Bulbophyllum congolanum Schltr. = Bulbophyllum scaberulum (Rolfe) Bolus ■☆

62659 Bulbophyllum congolanum Schltr. subsp. leucorrhachis Perez-Vera;白轴石豆兰■☆

62660 Bulbophyllum congolense (De Wild.) De Wild. = Bulbophyllum imbricatum Lindl. var. purpureum W. Sanford ■☆

62661 Bulbophyllum corallinum Tixier et Guillaumin; 环唇石豆兰; Coral Bulbophyllum, Coral Stonebean-orchis ■

62662 Bulbophyllum coriophorum Ridl. ;革梗卷瓣兰■☆

62663 Bulbophyllum coursianum H. Perrier = Bulbophyllum rutenbergianum Schltr. ■☆

62664 Bulbophyllum craibianum Kerr = Bulbophyllum shweliense W. W. Sm. ■

62665 Bulbophyllum crassipes Hook. f. ;短耳石豆兰(粗柄石豆兰); Shortear Bulbophyllum, Shortear Stonebean-orchis ■

62666 Bulbophyllum crassipetalum H. Perrier;厚瓣石豆兰■☆

62667 Bulbophyllum crenulatum Rolfe = Bulbophyllum coriophorum Ridl. ■☆

62668 Bulbophyllum cryptostachyum Schltr. ;隐穗石豆兰■☆

62669 Bulbophyllum cupuligerum Kraenzl. = Stolzia cupuligera (Kraenzl.) Summerh. ■☆

62670 Bulbophyllum curvifolium Schltr. ;折叶石豆兰■☆

62671 Bulbophyllum cyclanthum Schltr. ;环花石豆兰■☆

62672 Bulbophyllum cylindraceum Lindl. ;大苞石豆兰(长苞石豆兰,石灵芝,石枣子,石藻); Bigbract Bulbophyllum, Bigbract Stonebean-orchis ■

62673 Bulbophyllum cylindraceum Lindl. var. khasyanum (Griff.) Hook. f. = Bulbophyllum khasyanum Griff. ■

62674 Bulbophyllum cylindraceum Lindl. var. khasyanum Hook. f. = Bulbophyllum khasyanum Griff. ■

62675 Bulbophyllum cyrtopetalum Schltr. = Bulbophyllum maximum (Lindl.) Rchb. f. ■☆

62676 Bulbophyllum dahlemense Schltr. = Bulbophyllum falcatum (Lindl.) Rchb. f. ■

62677 Bulbophyllum daloaense P. J. Cribb et Pérez-Vera = Bulbophyllum resupinatum Ridl. var. filiforme (Kraenzl.) J. J. Verm. ■☆

62678 Bulbophyllum dearei Rchb. f. ;戴氏石豆兰■☆

62679 Bulbophyllum decaryanum H. Perrier;德卡里石豆兰■☆

62680 Bulbophyllum decipiens Schltr. = Bulbophyllum colubrinum (Rchb. f.) Rchb. f. ■☆

62681 Bulbophyllum deistelianum (Kraenzl.) Schltr. = Bulbophyllum bufo (Lindl.) Rchb. f. ■☆

62682 Bulbophyllum delitescens Hance;直唇卷瓣兰(直唇卷唇兰); Straightlip Bulbophyllum, Straightlip Cirrhopetalum, Straightlip Curlylip-orchis ■

62683 Bulbophyllum densiflorum Rolfe = Bulbophyllum cariniflorum Rchb. f. ■

62684 Bulbophyllum denticulatum Rolfe;细齿卷瓣兰■☆

62685 Bulbophyllum depressum King et Pantl. ;戟唇石豆兰(箭唇石豆兰); Harberdlip Stonebean-orchis, Hastatelip Bulbophyllum ■

62686 Bulbophyllum derchianum S. S. Ying = Bulbophyllum tokioi

Fukuy. ■

62687 Bulbophyllum devangiriense N. P. Balakr. = Bulbophyllum pteroglossum Schltr. ■

62688 Bulbophyllum discilabium H. Perrier;盘状卷瓣兰■☆

62689 Bulbophyllum distans Lindl. = Bulbophyllum finetii Szlach. et Olszewski ■☆

62690 Bulbophyllum divaricatum H. Perrier;叉开卷瓣兰■☆

62691 Bulbophyllum djumaense (De Wild.) De Wild. = Bulbophyllum maximum (Lindl.) Rchb. f. ■☆

62692 Bulbophyllum djumaense (De Wild.) De Wild. var. grandifolium De Wild. = Bulbophyllum maximum (Lindl.) Rchb. f. ■☆

62693 Bulbophyllum dolabriforme J. J. Verm.;斧形卷瓣兰■☆

62694 Bulbophyllum dorotheae Rendle = Bulbophyllum pumilum (Sw.) Lindl. ■☆

62695 Bulbophyllum drallei Rchb. f. = Bulbophyllum pumilum (Sw.) Lindl. ■☆

62696 Bulbophyllum drymoglossum Maxim. ex Okubo;圆叶石豆兰(果上叶,破石珠,狭萼豆兰,狭萼石豆兰,小石豆兰);Roundleaf Bulbophyllum,Roundleaf Stonebean-orchis ■

62697 Bulbophyllum drymoglossum Maxim. ex Okubo f. atrosanguiflorum Masam. et Satomi;暗血红圆叶石豆兰■☆

62698 Bulbophyllum dulongjiangense X. H. Jin;独龙江石豆兰■

62699 Bulbophyllum dyerianum (King et Pantl.) Seidenf. = Bulbophyllum rolfei (Kuntze) Seidenf. ■

62700 Bulbophyllum ealaense De Wild. = Bulbophyllum scaberulum (Rolfe) Bolus ■☆

62701 Bulbophyllum ebulbum King et Pantl. = Bulbophyllum apodum Hook. f. ■

62702 Bulbophyllum ebulbum King et Pantl. = Bulbophyllum spathaceum Rolfe ■

62703 Bulbophyllum eburneum (Pfitzer ex Kraenzl.) De Wild. = Bulbophyllum scaberulum (Rolfe) Bolus ■☆

62704 Bulbophyllum edentatum H. Perrier;无齿卷瓣兰■☆

62705 Bulbophyllum elachon J. J. Verm. = Bulbophyllum pumilum (Sw.) Lindl. ■☆

62706 Bulbophyllum elatum (Hook. f.) J. J. Sm.;高茎卷瓣兰(高茎卷唇兰);Tall Bulbophyllum,Tall Curlylip-orchis ■

62707 Bulbophyllum electrinum Seidenf.;长轴卷瓣兰■

62708 Bulbophyllum electrinum Seidenf. = Bulbophyllum hirundinis (Gagnep.) Seidenf. ■

62709 Bulbophyllum elliotii Rolfe;埃利石豆兰■☆

62710 Bulbophyllum ellipticum De Wild. = Bulbophyllum oxychilum Schltr. ■☆

62711 Bulbophyllum elongatum (De Wild.) De Wild. = Bulbophyllum ivorense P. J. Cribb et Pérez-Vera ■☆

62712 Bulbophyllum emarginatum (Finet) J. J. Sm.;匍茎卷瓣兰(匍茎卷唇兰);Emaginate Bulbophyllum, Emaginate Cirrhopetalum, Emaginate Curlylip-orchis ■

62713 Bulbophyllum erectum Thouars;挺立卷瓣兰■☆

62714 Bulbophyllum erythroglossum Bosser;红舌卷瓣兰■☆

62715 Bulbophyllum erythrostachyum Rolfe;红穗卷瓣兰■☆

62716 Bulbophyllum eublepharum Rchb. f.;墨脱石豆兰;Motuo Bulbophyllum,Motuo Stonebean-orchis ■

62717 Bulbophyllum eurhachis Schltr.;良轴石豆兰■☆

62718 Bulbophyllum falcatum (Lindl.) Rchb. f.;镰刀卷瓣兰■

62719 Bulbophyllum falcatum (Lindl.) Rchb. f. var. bufo (Lindl.) J. J. Verm. = Bulbophyllum bufo (Lindl.) Rchb. f. ■☆

62720 Bulbophyllum falcatum (Lindl.) Rchb. f. var. velutinum (Lindl.) J. J. Verm. = Bulbophyllum velutinum (Lindl.) Rchb. f. ■☆

62721 Bulbophyllum falcipetalum Lindl.;镰瓣卷瓣兰■

62722 Bulbophyllum farreri (W. W. Sm.) Seidenf.;麻栗坡卷瓣兰■

62723 Bulbophyllum fascinator (Rolfe) Rolfe;粗柄石豆兰■

62724 Bulbophyllum fayi J. J. Verm.;费伊卷瓣兰■☆

62725 Bulbophyllum fenghuangshanianum S. S. Ying;凤凰山石豆兰(红心豆兰);Fenghuangshan Bulbophyllum, Fenghuangshan Stonebean-orchis ■

62726 Bulbophyllum fenghuangshanianum S. S. Ying = Bulbophyllum rubrolabellum T. P. Lin ■

62727 Bulbophyllum filiforme Kraenzl. = Bulbophyllum resupinatum Ridl. var. filiforme (Kraenzl.) J. J. Verm. ■☆

62728 Bulbophyllum fimbriatum H. Perrier = Bulbophyllum peyrotii Bosser ■☆

62729 Bulbophyllum fimbriperianthium W. M. Lin;钝萼卷瓣兰■

62730 Bulbophyllum finetii Szlach. et Olszewski;菲内石豆兰■☆

62731 Bulbophyllum flavidum Lindl. = Bulbophyllum pumilum (Sw.) Lindl. ■☆

62732 Bulbophyllum flavidum Lindl. var. elongatum De Wild. = Bulbophyllum ivorense P. J. Cribb et Pérez-Vera ■☆

62733 Bulbophyllum flaviflorum (Tang S. Liu et H. J. Su) Seidenf. = Bulbophyllum pectenveneris (Gagnep.) Seidenf. ■

62734 Bulbophyllum flavisepalum Hayata = Bulbophyllum retusiusculum Rchb. f. ■

62735 Bulbophyllum flectens P. J. Cribb et P. Taylor = Bulbophyllum unifoliatum De Wild. var. flectens (P. J. Cribb et P. Taylor) J. J. Verm. ■☆

62736 Bulbophyllum flickingeranum A. D. Hawkes = Bulbophyllum peyrotii Bosser ■☆

62737 Bulbophyllum florulentum Schltr.;多花石豆兰■☆

62738 Bulbophyllum fordii (Rolfe) J. J. Sm.;狭唇卷瓣兰(狭唇卷唇兰);Narroelip Bulbophyllum,Narroelip Curlylip-orchis ■

62739 Bulbophyllum formosanum (Rolfe) Seidenf.;台湾豆兰(伞苞石豆兰,台湾卷瓣兰)■

62740 Bulbophyllum forrestii Seidenf.;尖角卷瓣兰(尖角卷唇兰);Forrest Bulbophyllum,Forrest Curlylip-orchis ■

62741 Bulbophyllum forsythianum Kraenzl.;福赛斯石豆兰■☆

62742 Bulbophyllum fractiflexum Kraenzl. = Bulbophyllum velutinum (Lindl.) Rchb. f. ■☆

62743 Bulbophyllum francoisii H. Perrier;弗朗卷瓣兰■☆

62744 Bulbophyllum fuerstenbergianum (De Wild.) De Wild. = Bulbophyllum scaberulum (Rolfe) Bolus var. fuerstenbergianum (De Wild.) J. J. Verm. ■☆

62745 Bulbophyllum funingense Z. H. Tsi et H. C. Chen;富宁卷瓣兰;Funing Bulbophyllum,Funing Curlylip-orchis ■

62746 Bulbophyllum fuscescens (Griff.) Rchb. f. = Epigeneium fuscescens (Griff.) Summerh. ■

62747 Bulbophyllum fuscoides Petersen = Bulbophyllum acutebracteatum De Wild. var. rubrobrunneopapillosum (De Wild.) J. J. Verm. ■☆

62748 Bulbophyllum fuscum Lindl. = Bulbophyllum acutebracteatum De Wild. var. rubrobrunneopapillosum (De Wild.) J. J. Verm. ■☆

62749 Bulbophyllum fuscum Lindl. var. melinostachyum (Schltr.) J. J. Verm. = Bulbophyllum melinostachyum Schltr. ■☆

62750 Bulbophyllum gabonis Lindl. et Rchb. f. = Bulbophyllum pumilum (Sw.) Lindl. ■☆

62751 Bulbophyllum gabunense Schltr. = Bulbophyllum colubrinum

(Rchb. f.) Rchb. f. ■☆

62752 Bulbophyllum galeatum (Sw.) Lindl. = Polystachya galeata (Sw.) Rchb. f. ■☆

62753 Bulbophyllum gentilii Rolfe = Bulbophyllum phaeopogon Schltr. ■☆

62754 Bulbophyllum gilgianum Kraenzl.;吉尔格石豆兰■☆

62755 Bulbophyllum gilletii (De Wild.) De Wild. = Bulbophyllum imbricatum Lindl. var. purpureum W. Sanford ■☆

62756 Bulbophyllum gongshanense Z. H. Tsi;贡山卷瓣兰(贡山卷唇兰);Gongshan Bulbophyllum,Gongshan Curlylip-orchis ■

62757 Bulbophyllum graciliscapum Summerh. = Bulbophyllum finetii Szlach. et Olszewski ■☆

62758 Bulbophyllum gracillimum Hayata = Bulbophyllum aureolabellum T. P. Lin ■

62759 Bulbophyllum gracillimum Hayata = Bulbophyllum drymoglossum Maxim. ex Okubo ■

62760 Bulbophyllum graminifolium Summerh. = Bulbophyllum calyptratum Kraenzl. var. graminifolium (Summerh.) J. J. Verm. ■☆

62761 Bulbophyllum grandiflorum Blume;大花卷瓣兰■☆

62762 Bulbophyllum gravidum Lindl. = Bulbophyllum cochleatum Lindl. var. gravidum (Lindl.) J. J. Verm. ■☆

62763 Bulbophyllum griffithii (Lindl.) Rchb. f.;短齿石豆兰(果上叶,美网石豆兰,石串莲,石斛,小果上叶,小绿芳,小绿石豆兰);Shorttooth Bulbophyllum,Shorttooth Stonebean-orchis ■

62764 Bulbophyllum gustavii Schltr. = Bulbophyllum josephii (Kuntze) Summerh. ■☆

62765 Bulbophyllum guttulatum (Hook. f.) N. P. Balakr.;钻齿卷瓣兰(钻齿卷唇兰);Awltooth Bulbophyllum,Awltooth Curlylip-orchis ■

62766 Bulbophyllum gymnopus Hook. f.;线瓣石豆兰;Linearpetal Bulbophyllum,Linearpetal Stonebean-orchis ■

62767 Bulbophyllum hainanense Z. H. Tsi;海南石豆兰;Hainan Bulbophyllum,Hainan Stonebean-orchis ■

62768 Bulbophyllum hamelinii W. Watson;哈梅林石豆兰■☆

62769 Bulbophyllum haniffii Carrière;飘带石豆兰;Streamer Bulbophyllum,Streamer Stonebean-orchis ■

62770 Bulbophyllum hastatum Ts. Tang et F. T. Wang = Bulbophyllum depressum King et Pantl. ■

62771 Bulbophyllum helenae (Kuntze) J. J. Sm.;角萼卷瓣兰(角萼卷唇兰);Helena Bulbophyllum,Helena Curlylip-orchis ■

62772 Bulbophyllum hemirhachis (Pfitzer) De Wild. = Bulbophyllum falcatum (Lindl.) Rchb. f. ■

62773 Bulbophyllum henanense J. L. Lu;河南卷瓣兰(河南卷唇兰);Henan Bulbophyllum,Henan Curlylip-orchis ■

62774 Bulbophyllum henrici Schltr.;昂里克石豆兰■☆

62775 Bulbophyllum henryi (Rolfe) J. J. Sm. = Bulbophyllum andersonii (Hook. f.) J. J. Sm. ■

62776 Bulbophyllum herminiostachys (Rchb. f.) Rchb. f. = Bulbophyllum pumilum (Sw.) Lindl. ■☆

62777 Bulbophyllum hildebrandtii Rchb. f.;希尔德卷瓣兰■☆

62778 Bulbophyllum hirsutissimum Kraenzl. = Bulbophyllum comatum Lindl. ■☆

62779 Bulbophyllum hirtum (Sm.) Lindl.;落叶石豆兰;Fallingleaf Bulbophyllum,Fallingleaf Stonebean-orchis ■

62780 Bulbophyllum hirundinis (Gagnep.) Seidenf.;花莲卷瓣兰(红花缘石豆兰,莲花卷唇兰,疏花石豆兰,朱红冠毛兰);Lotus Bulbophyllum,Lotus Curlylip-orchis ■

62781 Bulbophyllum hirundinis (Gagnep.) Seidenf. var. electrinum (Seidenf.) S. S. Ying = Bulbophyllum electrinum Seidenf. ■

62782 Bulbophyllum hirundinis (Gagnep.) Seidenf. var. electrinum (Seidenf.) S. S. Ying = Bulbophyllum hirundinis (Gagnep.) Seidenf. ■

62783 Bulbophyllum horizontale Bosser;平展石豆兰■☆

62784 Bulbophyllum horridulum J. J. Verm.;小刺石豆兰■☆

62785 Bulbophyllum humbertii Schltr.;亨伯特石豆兰■☆

62786 Bulbophyllum humblotii Rolfe;洪布石豆兰■☆

62787 Bulbophyllum hyacinthiodorum W. W. Sm. = Bulbophyllum odoratissimum (Sm.) Lindl. ■

62788 Bulbophyllum hyalinum Schltr.;透明石豆兰■☆

62789 Bulbophyllum ikongoense H. Perrier;伊孔古卷瓣兰■☆

62790 Bulbophyllum imbricatum Lindl.;覆瓦石豆兰■☆

62791 Bulbophyllum imbricatum Lindl. var. luteum W. Sanford;黄覆瓦石豆兰■☆

62792 Bulbophyllum imbricatum Lindl. var. purpureum W. Sanford;紫覆瓦石豆兰■☆

62793 Bulbophyllum imogeniae K. Hamillon = Bulbophyllum pumilum (Sw.) Lindl. ■☆

62794 Bulbophyllum implexum Jum. et Perrier = Bulbophyllum minutum Thouars ■☆

62795 Bulbophyllum imschootianum (Rolfe) De Wild. = Bulbophyllum colubrinum (Rchb. f.) Rchb. f. ■☆

62796 Bulbophyllum inabae (Hayata) Hayata = Bulbophyllum japonicum (Makino) Makino ■

62797 Bulbophyllum inabae Hayata = Bulbophyllum japonicum (Makino) Makino ■

62798 Bulbophyllum inaequale Rchb. f. = Bulbophyllum colubrinum (Rchb. f.) Rchb. f. ■☆

62799 Bulbophyllum inconspicuum Maxim.;麦斛(单叶石枣,根上子,瓜子莲,褂子连,果上叶,黄豆鞭,灵芝角,楼上楼,七仙桃,青兰,石豆,石莲子,石龙石尾,石蚊虫,石仙桃,石杨梅,石荚,石枣子,万年桃,小扣子兰,鸭雀嘴,雅雀嘴,羊奶草,一挂鱼,子上叶)■

62800 Bulbophyllum inflatum Rolfe = Bulbophyllum comatum Lindl. var. inflatum (Rolfe) J. J. Verm. ■☆

62801 Bulbophyllum infracarinatum G. Will. = Bulbophyllum unifoliatum De Wild. var. infracarinatum (G. Will.) J. J. Verm. ■☆

62802 Bulbophyllum inopinatum W. W. Sm.;意外石豆兰■☆

62803 Bulbophyllum inornatum J. J. Verm.;无饰石豆兰■☆

62804 Bulbophyllum insolitum Bosser;异常石豆兰■☆

62805 Bulbophyllum insulsoides Seidenf. = Bulbophyllum insulsum (Gagnep.) Seidenf. ■

62806 Bulbophyllum insulsum (Gagnep.) Seidenf.;穗花卷瓣兰(瓶壶卷瓣兰,穗花卷唇兰);Insipid Bulbophyllum,Insipid Curlylip-orchis,Spike Bulbophyllum,Spike Curlylip-orchis ■

62807 Bulbophyllum insulsum (Gagnep.) Seidenf. = Bulbophyllum levinei Schltr. ■

62808 Bulbophyllum intermedium De Wild. = Bulbophyllum calyptratum Kraenzl. var. graminifolium (Summerh.) J. J. Verm. ■☆

62809 Bulbophyllum intertextum Lindl.;络合石豆兰■☆

62810 Bulbophyllum intertextum Lindl. var. parvilabium G. Will. = Bulbophyllum intertextum Lindl. ■☆

62811 Bulbophyllum ituriense De Wild. = Bulbophyllum lupulinum Lindl. ■☆

62812 Bulbophyllum ivorense P. J. Cribb et Pérez-Vera;伊沃里石豆兰■☆

62813 Bulbophyllum japonicum (Makino) Makino;瘤唇卷瓣兰(瘤唇卷唇兰,日本红花石豆兰,日本卷瓣兰);Japan Bulbophyllum,Japan Curlylip-orchis ■

62814 Bulbophyllum japonicum (Makino) Makino f. lutescens (Murata) Masam. et Satomi;黄瘤唇卷瓣兰■☆

62815 Bulbophyllum jespersenii De Wild. = Bulbophyllum scaberulum (Rolfe) Bolus ■☆

62816 Bulbophyllum johannis H. Wendl. et Kraenzl.;约翰石豆兰■☆

62817 Bulbophyllum johannum H. Perrier = Bulbophyllum hildebrandtii Rchb. f. ■☆

62818 Bulbophyllum josephii (Kuntze) Summerh.;约瑟夫石豆兰■☆

62819 Bulbophyllum josephii (Kuntze) Summerh. var. mahonii (Rolfe) J. J. Verm.;马洪石豆兰■☆

62820 Bulbophyllum jumelleanum Schltr.;朱迈尔石豆兰■☆

62821 Bulbophyllum jungwirthianum Schltr. = Bulbophyllum cochleatum Lindl. ■☆

62822 Bulbophyllum kamerunense Schltr. = Bulbophyllum imbricatum Lindl. var. luteum W. Sanford ■☆

62823 Bulbophyllum kewense Schltr. = Bulbophyllum velutinum (Lindl.) Rchb. f. ■☆

62824 Bulbophyllum khaoyaiense Seidenf.;白花卷瓣兰(白花卷唇兰);White Bulbophyllum, White Curlylip-orchis ■

62825 Bulbophyllum khasyanum Griff.;卷苞石豆兰;Khasy Bulbophyllum, Khasy Stonebean-orchis ■

62826 Bulbophyllum kindtianum De Wild. = Bulbophyllum finetii Szlach. et Olszewski ■☆

62827 Bulbophyllum kivuense J. J. Verm.;基伍石豆兰■☆

62828 Bulbophyllum kraenzlianum De Wild. = Bulbophyllum comatum Lindl. ■☆

62829 Bulbophyllum kuanwuense S. W. Chung et T. C. Hsu;台南卷瓣兰■

62830 Bulbophyllum kupense P. J. Cribb et B. J. Pollard;库普石豆兰■☆

62831 Bulbophyllum kusukusense Hayata = Bulbophyllum affine Lindl. ■

62832 Bulbophyllum kwangtungense Schltr.;广东石豆兰(单叶岩珠,广石豆兰,扣子兰,石枣子,岩枣);Guangdong Stonebean-orchis, Kwangtung Bulbophyllum ■

62833 Bulbophyllum labatii Bosser;拉巴卷瓣兰■☆

62834 Bulbophyllum laggiarae Schltr. = Bulbophyllum humblotii Rolfe ■☆

62835 Bulbophyllum lancisepalum H. Perrier;剑萼卷瓣兰■☆

62836 Bulbophyllum lanuriense De Wild. = Bulbophyllum velutinum (Lindl.) Rchb. f. ■☆

62837 Bulbophyllum latipetalum H. Perrier;宽瓣卷瓣兰■☆

62838 Bulbophyllum laurentianum Kraenzl. = Bulbophyllum imbricatum Lindl. var. purpureum W. Sanford ■☆

62839 Bulbophyllum leandrianum H. Perrier;利安石豆兰■☆

62840 Bulbophyllum ledermannii (Kraenzl.) De Wild. = Bulbophyllum imbricatum Lindl. var. purpureum W. Sanford ■☆

62841 Bulbophyllum ledungense Ts. Tang et F. T. Wang;乐东石豆兰;Ledong Bulbophyllum, Ledong Stonebean-orchis, Letung Bulbophyllum ■

62842 Bulbophyllum lemniscatum Parl.;垂花石豆兰■☆

62843 Bulbophyllum lemuraeoides H. Perrier;拟莱穆拉石豆兰■☆

62844 Bulbophyllum lemurense Bosser et P. J. Cribb;莱穆拉石豆兰■☆

62845 Bulbophyllum leopardinum (Wall.) Lindl.;短葶石豆兰;Shortscap Bulbophyllum, Shortscap Stonebean-orchis ■

62846 Bulbophyllum lepidum (Blume) J. J. Sm;南方卷瓣兰■

62847 Bulbophyllum leptochlamys Schltr.;细被石豆兰■☆

62848 Bulbophyllum leptorrhachis Schltr. = Bulbophyllum falcatum (Lindl.) Rchb. f. ■

62849 Bulbophyllum leptostachyum Schltr.;细穗石豆兰■☆

62850 Bulbophyllum leucopogon Kraenzl. = Bulbophyllum pumilum (Sw.) Lindl. ■☆

62851 Bulbophyllum leucorhachis (Rolfe) Schltr. = Bulbophyllum imbricatum Lindl. var. luteum W. Sanford ■☆

62852 Bulbophyllum levinei Schltr.;齿瓣石豆兰(密珠石豆兰,印度石豆兰);Levine Bulbophyllum, Levine Stonebean-orchis ■

62853 Bulbophyllum linchianum S. S. Ying = Bulbophyllum melanoglossum Hayata ■

62854 Bulbophyllum linderi Summerh. = Bulbophyllum imbricatum Lindl. var. purpureum W. Sanford ■☆

62855 Bulbophyllum lindleyi (Rolfe) Schltr. = Bulbophyllum calyptratum Kraenzl. ■☆

62856 Bulbophyllum leariligulatum Schltr.;线舌石豆兰■☆

62857 Bulbophyllum linguiforme P. J. Cribb = Bulbophyllum humblotii Rolfe ■☆

62858 Bulbophyllum lizae J. J. Verm.;利扎石豆兰■☆

62859 Bulbophyllum lobbii Lindl.;劳氏石豆兰■☆

62860 Bulbophyllum lobulatum Schltr. = Bulbophyllum erectum Thouars ■☆

62861 Bulbophyllum longibrachiatum Z. H. Tsi;长臂卷瓣兰(长臂卷唇兰);Longarm Bulbophyllum, Longarm Curlylip-orchis ■

62862 Bulbophyllum longibulbum Schltr. = Bulbophyllum bufo (Lindl.) Rchb. f. ■☆

62863 Bulbophyllum longiflorum Thouars;长花石豆兰■☆

62864 Bulbophyllum longisepalum Rolfe;长萼石豆兰■☆

62865 Bulbophyllum longispicatum Kraenzl. et Schltr. = Bulbophyllum resupinatum Ridl. var. filiforme (Kraenzl.) J. J. Verm. ■☆

62866 Bulbophyllum longivaginans H. Perrier;长鞘石豆兰■☆

62867 Bulbophyllum lubiense De Wild. = Bulbophyllum bufo (Lindl.) Rchb. f. ■☆

62868 Bulbophyllum lucidum Schltr.;亮石豆兰■☆

62869 Bulbophyllum lupulinum Lindl.;狼石豆兰■☆

62870 Bulbophyllum luteobracteatum Jum. et Perrier;黄苞石豆兰■☆

62871 Bulbophyllum luteolabium H. Perrier = Bulbophyllum humblotii Rolfe ■☆

62872 Bulbophyllum lutescens (Rolfe) De Wild. = Bulbophyllum falcipetalum Lindl. ■

62873 Bulbophyllum macraei (Lindl.) Rchb. f.;乌来卷瓣兰(牧野氏卷瓣兰,乌来卷唇兰,乌来石豆兰,一枝瘤,紫花石豆兰);Macrae Bulbophyllum, Macrae Curlylip-orchis ■

62874 Bulbophyllum macraei (Lindl.) Rchb. f. var. autumale (Fukuy.) S. S. Ying = Bulbophyllum macraei (Lindl.) Rchb. f. ■

62875 Bulbophyllum macraei (Lindl.) Rchb. f. var. tanegashimense (Masam.) F. Maek.;种子岛卷瓣兰■☆

62876 Bulbophyllum macraei var. autumnale (Fukuy.) S. S. Ying = Bulbophyllum macraei (Lindl.) Rchb. f. ■

62877 Bulbophyllum maculatum Jum. et Perrier = Bulbophyllum hildebrandtii Rchb. f. ■☆

62878 Bulbophyllum madagascariense Schltr. = Bulbophyllum hildebrandtii Rchb. f. ■☆

62879 Bulbophyllum magnibracteatum Summerh.;大苞卷瓣兰■☆

62880 Bulbophyllum mahonii Rolfe = Bulbophyllum josephii (Kuntze) Summerh. var. mahonii (Rolfe) J. J. Verm. ■☆

62881 Bulbophyllum makakense Hansen = Bulbophyllum colubrinum (Rchb. f.) Rchb. f. ■☆

62882 Bulbophyllum makinoanum (Schltr.) Masam. = Bulbophyllum macraei (Lindl.) Rchb. f. ■

62883 Bulbophyllum makoyanum Rchb. f. = Bulbophyllum pectenveneris (Gagnep.) Seidenf. ■

62884 Bulbophyllum malawiense B. Morris = Bulbophyllum elliotii Rolfe ■☆

62885 Bulbophyllum malawiense Morris = Bulbophyllum elliotii Rolfe ■☆

62886 Bulbophyllum maleolens Kraenzl.;小锤石豆兰■☆

62887 Bulbophyllum malipoense Z. J. Liu, S. C. Chen et S. P. Lei = Bulbophyllum farreri (W. W. Sm.) Seidenf. ■

62888 Bulbophyllum mananjarense Poiss.;马南扎里石豆兰■☆

62889 Bulbophyllum mandrakanum Schltr. = Bulbophyllum coriophorum Ridl. ■☆

62890 Bulbophyllum mangenotii Bosser;芒热诺石豆兰■☆

62891 Bulbophyllum mannii Hook. f. = Bulbophyllum cochleatum Lindl. ■☆

62892 Bulbophyllum marojejiense H. Perrier;马鲁杰石豆兰■☆

62893 Bulbophyllum masoalanum Schltr.;马苏阿拉石豆兰■☆

62894 Bulbophyllum maudeae A. D. Hawkes;黑花石豆兰■☆

62895 Bulbophyllum maximum (Lindl.) Rchb. f.;高大石豆兰■☆

62896 Bulbophyllum maximum (Lindl.) Rchb. f. var. oxypterum (Lindl.) Pérez-Vera;尖翅高大石豆兰■☆

62897 Bulbophyllum mayae A. D. Hawkes = Bulbophyllum peyrotii Bosser ■☆

62898 Bulbophyllum mayombeense Garay;马永贝石豆兰■☆

62899 Bulbophyllum mediocre Summerh.;中位石豆兰■☆

62900 Bulbophyllum melanoglossum Hayata;紫纹卷瓣兰(红斑石豆兰,邵氏卷瓣兰,邵氏卷唇兰,邵氏石豆兰,紫纹石豆兰);Blacktongue Bulbophyllum, Blacktongue Curlylip-orchis, Linch Bulbophyllum, Linch Curlylip-orchis, Redspoted Bulbophyllum ■

62901 Bulbophyllum melanoglossum Hayata var. rubropunctatum (S. S. Ying) S. S. Ying = Bulbophyllum melanoglossum Hayata ■

62902 Bulbophyllum melanoglossum Hayata var. rubropunctatum S. S. Ying = Bulbophyllum melanoglossum Hayata ■

62903 Bulbophyllum melanopogon Schltr. = Bulbophyllum hildebrandtii Rchb. f. ■☆

62904 Bulbophyllum melanorrhachis (Rchb. f.) De Wild. = Bulbophyllum velutinum (Lindl.) Rchb. f. ■☆

62905 Bulbophyllum melinostachyum Schltr.;苹果穗石豆兰■☆

62906 Bulbophyllum melleri (Hook. f.) Rchb. f. = Bulbophyllum sandersonii (Hook. f.) Rchb. f. ■☆

62907 Bulbophyllum melleum H. Perrier;蜜色石豆兰■☆

62908 Bulbophyllum menghaiense Z. H. Tsi;勐海石豆兰;Menghai Bulbophyllum, Menghai Stonebean-orchis ■

62909 Bulbophyllum menglunense Z. H. Tsi et Y. Z. Ma;勐仑石豆兰;Menglun Bulbophyllum, Menglun Stonebean-orchis ■

62910 Bulbophyllum microglossum Perrier = Bulbophyllum moldenkeanum A. D. Hawkes ■☆

62911 Bulbophyllum micropetalum Lindl. = Genyorchis micropetala (Lindl.) Schltr. ■☆

62912 Bulbophyllum mildbraedii Kraenzl. = Bulbophyllum finetii Szlach. et Olszewski ■☆

62913 Bulbophyllum millenii (Rolfe) Schltr. = Bulbophyllum velutinum (Lindl.) Rchb. f. ■☆

62914 Bulbophyllum minax Schltr.;突出石豆兰■☆

62915 Bulbophyllum minutiflorum W. Sanford;小花石豆兰■☆

62916 Bulbophyllum minutilabrum H. Perrier;微小石豆兰■☆

62917 Bulbophyllum minutum (Rolfe) Engl. = Bulbophyllum velutinum (Lindl.) Rchb. f. ■☆

62918 Bulbophyllum minutum (Rolfe) Engl. var. purpureum (De Wild.) De Wild. = Bulbophyllum velutinum (Lindl.) Rchb. f. ■☆

62919 Bulbophyllum minutum Thouars;小石豆兰■☆

62920 Bulbophyllum modicum Summerh. = Bulbophyllum josephii (Kuntze) Summerh. var. mahonii (Rolfe) J. J. Verm. ■☆

62921 Bulbophyllum moireanum Hawkes = Bulbophyllum maximum (Lindl.) Rchb. f. ■☆

62922 Bulbophyllum moldenkeanum A. D. Hawkes;莫尔石豆兰■☆

62923 Bulbophyllum moliwense Schltr. = Bulbophyllum pumilum (Sw.) Lindl. ■☆

62924 Bulbophyllum monanthum (Kuntze) J. J. Sm. = Bulbophyllum pteroglossum Schltr. ■

62925 Bulbophyllum monticola Hook. f. = Bulbophyllum cochleatum Lindl. var. gravidum (Lindl.) J. J. Verm. ■☆

62926 Bulbophyllum mooreanum Robyns et Tournay = Bulbophyllum sandersonii (Hook. f.) Rchb. f. ■☆

62927 Bulbophyllum moramanganum Schltr.;莫拉芒石豆兰■☆

62928 Bulbophyllum moratii Bosser;莫拉特石豆兰■☆

62929 Bulbophyllum multiflorum Ridl.;繁花石豆兰■☆

62930 Bulbophyllum multiligulatum H. Perrier;多舌石豆兰■☆

62931 Bulbophyllum multivaginatum Jum. et Perrier;多鞘石豆兰■☆

62932 Bulbophyllum myrmecochilum Schltr.;蚂蚁石豆兰■☆

62933 Bulbophyllum nanum De Wild. = Bulbophyllum pumilum (Sw.) Lindl. ■☆

62934 Bulbophyllum neglectum Bosser;忽视石豆兰■☆

62935 Bulbophyllum nigericum Summerh.;尼日利亚石豆兰■☆

62936 Bulbophyllum nigrescens Rolfe;钩梗石豆兰;Black Bulbophyllum, Blacken Stonebean-orchis ■

62937 Bulbophyllum nigrilabium H. Perrier = Bulbophyllum maudeae A. D. Hawkes ■☆

62938 Bulbophyllum nigripetalum Rolfe;黑瓣石豆兰■☆

62939 Bulbophyllum nigritianum Rendle;尼格里塔石豆兰■☆

62940 Bulbophyllum nitens Jum. et Perrier;光亮石豆兰■☆

62941 Bulbophyllum nitens Jum. et Perrier var. minus H. Perrier = Bulbophyllum nitens Jum. et Perrier ■☆

62942 Bulbophyllum nudiscapum Rolfe = Bulbophyllum finetii Szlach. et Olszewski ■☆

62943 Bulbophyllum nummularium (Kraenzl.) Rolfe;铜钱石豆兰■☆

62944 Bulbophyllum nutans Thouars;低垂石豆兰■☆

62945 Bulbophyllum nyassanum Schltr. = Bulbophyllum maximum (Lindl.) Rchb. f. ■☆

62946 Bulbophyllum obanense Rendle = Bulbophyllum melinostachyum Schltr. ■☆

62947 Bulbophyllum obscuriflorum H. Perrier;暗花石豆兰■☆

62948 Bulbophyllum obtusatum Schltr.;钝石豆兰■☆

62949 Bulbophyllum obtusiangulum Z. H. Tsi;黄花卷瓣兰(黄花卷唇兰,紫花卷瓣兰);Yellow Bulbophyllum, Yellow Curlylip-orchis ■

62950 Bulbophyllum obtusiangulum Z. H. Tsi = Bulbophyllum lepidum (Blume) J. J. Sm ■

62951 Bulbophyllum obtusilabium W. Kittr.;钝唇石豆兰■☆

62952 Bulbophyllum obtusum Jum. et Perrier = Bulbophyllum obtusatum Schltr. ■☆

62953 Bulbophyllum occlusum Ridl.;闭合石豆兰■☆

62954 Bulbophyllum occultum Thouars;隐蔽石豆兰■☆

62955 Bulbophyllum odoratissimum (Sm.) Lindl.;密花石豆兰(果上叶,极香石豆兰,石串莲,石橄榄,石米,石枣子,万年桃,小果上叶,羊奶果,一匹草);Denseflower Bulbophyllum, Flowery

Stonebean-orchis ■
62956 Bulbophyllum odoratissimum (Sm.) Lindl. var. rubrolabellum (T. P. Lin) S. S. Ying = Bulbophyllum rubrolabellum T. P. Lin ■
62957 Bulbophyllum ogoouense Guillaumin = Bulbophyllum fuscum Lindl. ■☆
62958 Bulbophyllum omerandrum Hayata;毛药卷瓣兰(黄唇卷瓣兰,溪头卷瓣兰);Hairanther Bulbophyllum, Hairanther Curlylip-orchis ■
62959 Bulbophyllum ophiuchus Ridl. var. ankaizinensis Jum. et H. Perrier = Bulbophyllum ankaizinense (Jum. et H. Perrier) Schltr. ■☆
62960 Bulbophyllum oreogenes (W. W. Sm.) Seidenf. = Bulbophyllum retusiusculum Rchb. f. ■
62961 Bulbophyllum orientale Seidenf.;麦穗石豆兰(穗花卷瓣兰);Oriental Bulbophyllum, Oriental Stonebean-orchis, Yunnan Bulbophyllum ■
62962 Bulbophyllum ornatissimum J. J. Sm.;华丽石豆兰■☆
62963 Bulbophyllum otoglossum Tuyama;德钦石豆兰;Deqin Bulbophyllum, Deqin Stonebean-orchis ■
62964 Bulbophyllum otoglossum Tuyama = Bulbophyllum yunanense Rolfe ■
62965 Bulbophyllum ovalifolium (Blume) Lindl.;卵叶石豆兰(卵唇石豆兰)■
62966 Bulbophyllum ovatilabellum Seidenf. = Bulbophyllum ovalifolium (Blume) Lindl. ■
62967 Bulbophyllum oxycalyx Schltr.;尖萼石豆兰■☆
62968 Bulbophyllum oxychilum Schltr.;尖唇石豆兰■☆
62969 Bulbophyllum oxychilum Schltr. var. roseum W. Sanford;粉红尖唇石豆兰■☆
62970 Bulbophyllum oxyodon Rchb. f. = Bulbophyllum falcatum (Lindl.) Rchb. f. ■
62971 Bulbophyllum oxypterum (Lindl.) Rchb. f. = Bulbophyllum maximum (Lindl.) Rchb. f. var. oxypterum (Lindl.) Pérez-Vera ■☆
62972 Bulbophyllum oxypterum (Lindl.) Rchb. f. var. mozambicense (Finet) De Wild. = Bulbophyllum maximum (Lindl.) Rchb. f. ■☆
62973 Bulbophyllum pachypus Schltr.;粗足石豆兰■☆
62974 Bulbophyllum pachyrachis (A. Rich.) Griseb.;粗轴石豆兰■☆
62975 Bulbophyllum pallens Schltr.;变苍白石豆兰■☆
62976 Bulbophyllum pallescens Kraenzl. = Bulbophyllum bifarium Hook. f. ■☆
62977 Bulbophyllum pallidiflorum Schltr.;苍白花石豆兰■☆
62978 Bulbophyllum pandurella Schltr.;琴形石豆兰■☆
62979 Bulbophyllum papillosum Finet = Bulbophyllum pumilum (Sw.) Lindl. ■☆
62980 Bulbophyllum parvulum (Hook. f.) J. J. Sm. = Bulbophyllum rolfei (Kuntze) Seidenf. ■
62981 Bulbophyllum parvum Summerh.;较小石豆兰■☆
62982 Bulbophyllum pauciflorum Ames;寡白花石豆兰(白花石豆兰)■
62983 Bulbophyllum pavimentatum Lindl. = Bulbophyllum pumilum (Sw.) Lindl. ■☆
62984 Bulbophyllum pectenveneris (Gagnep.) Seidenf.;斑唇卷瓣兰(斑唇卷唇兰,翠花卷瓣兰,黄花卷瓣兰,黄花石豆兰,毛边卷瓣兰,米谷还阳,石上桃,石枣);Bulbophyllum, Spotlip Curlylip-orchis ■
62985 Bulbophyllum pectinatum Finet;长足石豆兰(阿里山豆兰,瓜子莲,石三棱,石山莲,石仙桃);Alishan Bulbophyllum, Alishan Stonebean-orchis, Pectinate Bulbophyllum, Pectinate Stonebean-orchis ■
62986 Bulbophyllum pectinatum Finet var. transarisanense (Hayata) S. S. Ying;阿里山石豆兰■
62987 Bulbophyllum peniculus Schltr. = Bulbophyllum rutenbergianum Schltr. ■☆
62988 Bulbophyllum peperomioides Kraenzl. = Stolzia peperomioides (Kraenzl.) Summerh. ■☆
62989 Bulbophyllum perpusillum H. Wendl. et Kraenzl.;微小卷瓣兰■☆
62990 Bulbophyllum perreflexum Bosser et P. J. Cribb;反折卷瓣兰■☆
62991 Bulbophyllum perrieri Schltr.;佩里耶卷瓣兰■☆
62992 Bulbophyllum pertenue Kraenzl. = Bulbophyllum intertextum Lindl. ■☆
62993 Bulbophyllum pervillei Rolfe;佩尔石豆兰■☆
62994 Bulbophyllum petrae G. A. Fisch., Sieder et P. J. Cribb;岩生石豆兰■☆
62995 Bulbophyllum peyrotii Bosser;佩罗石豆兰■☆
62996 Bulbophyllum phaeopogon Schltr.;褐毛石豆兰■☆
62997 Bulbophyllum pholidotoides Kraenzl. = Bulbophyllum cochleatum Lindl. ■☆
62998 Bulbophyllum picturatum (Lodd.) Rchb. f.;彩色卷瓣兰■
62999 Bulbophyllum pingtungense S. S. Ying et C. Chen ex S. S. Ying;屏东石豆兰(屏东豆兰)■
63000 Bulbophyllum platirachis De Wild. = Bulbophyllum acutebracteatum De Wild. ■☆
63001 Bulbophyllum platypodum H. Perrier;阔足石豆兰■☆
63002 Bulbophyllum platyrhachis (Rolfe) Schltr. = Bulbophyllum maximum (Lindl.) Rchb. f. ■☆
63003 Bulbophyllum pleiopterum Schltr.;多翅石豆兰■☆
63004 Bulbophyllum pobeguinii (Finet) De Wild. = Bulbophyllum scaberulum (Rolfe) Bolus ■☆
63005 Bulbophyllum poilanei Gagnep. = Bulbophyllum repens Griff. ■
63006 Bulbophyllum polyrrhizum Lindl.;锥茎石豆兰;Awlstem Bulbophyllum, Awlstem Stonebean-orchis ■
63007 Bulbophyllum porphyroglossum Kraenzl. = Bulbophyllum pumilum (Sw.) Lindl. ■☆
63008 Bulbophyllum porphyrostachys Summerh.;紫穗石豆兰■☆
63009 Bulbophyllum prorepens Summerh.;匍匐石豆兰■☆
63010 Bulbophyllum protectum H. Perrier;易变石豆兰■☆
63011 Bulbophyllum pseudonutans H. Perrier = Bulbophyllum brachystachyum Schltr. ■☆
63012 Bulbophyllum psittacoglossum Rchb. f.;滇南石豆兰;S. Yunnan Bulbophyllum, S. Yunnan Stonebean-orchis ■
63013 Bulbophyllum psychoon Rchb. f. = Bulbophyllum levinei Schltr. ■
63014 Bulbophyllum pteroglossum Schltr.;曲萼石豆兰;Curvecalyx Bulbophyllum, Curvecalyx Stonebean-orchis ■
63015 Bulbophyllum pumilum (Sw.) Lindl.;矮小卷瓣兰■☆
63016 Bulbophyllum purpureifolium Aver. = Bulbophyllum longibrachiatum Z. H. Tsi ■
63017 Bulbophyllum purpureorhachis (De Wild.) Schltr.;紫石豆兰■☆
63018 Bulbophyllum pusillum (Rolfe) De Wild. = Bulbophyllum sandersonii (Hook. f.) Rchb. f. ■☆
63019 Bulbophyllum quadrangulum Z. H. Tsi;浙杭卷瓣兰(四棱卷瓣兰,浙杭卷唇兰);Zhejiang Bulbophyllum, Zhejiang Curlylip-orchis ■
63020 Bulbophyllum quadrangulum Z. H. Tsi = Bulbophyllum chrondriophorum (Gagnep.) Seidenf. ■
63021 Bulbophyllum quadrialatum H. Perrier;四翅石豆兰■☆
63022 Bulbophyllum quintasii Rolfe = Bulbophyllum intertextum Lindl. ■☆
63023 Bulbophyllum racemosum Hayata = Bulbophyllum insulsum

(Gagnep.) Seidenf. ■
63024 Bulbophyllum radiatum Lindl.;辐射石豆兰■☆
63025 Bulbophyllum ranomafanae Bosser et P. J. Cribb;拉努马法纳■☆
63026 Bulbophyllum rauhii Toill. -Gen. et Bosser;劳氏卷瓣兰■☆
63027 Bulbophyllum recurvum Lindl. = Bulbophyllum pumilum (Sw.) Lindl. ■☆
63028 Bulbophyllum reflexiflorum H. Perrier;曲花卷瓣兰■☆
63029 Bulbophyllum refractoides Seidenf. = Bulbophyllum wallichii Rchb. f. ■
63030 Bulbophyllum refractum Rchb. f.;膝曲卷瓣兰;Windmill Orchid ■☆
63031 Bulbophyllum repens Griff.;球花石豆兰;Ballflower Bulbophyllum,Ballflower Stonebean-orchis ■
63032 Bulbophyllum reptans (Lindl.) Lindl.;伏生石豆兰(果上叶,麦斛,牛虱子,匍匐石豆兰,石链子,小绿芨);Creep Stonebean-orchis,Creeping Bulbophyllum ■
63033 Bulbophyllum resupinatum Ridl.;倒置卷瓣兰■☆
63034 Bulbophyllum resupinatum Ridl. var. filiforme (Kraenzl.) J. J. Verm.;线形倒置卷瓣兰■☆
63035 Bulbophyllum retusiusculum Rchb. f.;藓叶卷瓣兰(黄萼卷瓣兰);Bryoleaf Bulbophyllum,Bryoleaf Curlylip-orchis ■
63036 Bulbophyllum retusiusculum Rchb. f. var. oreogenes (W. W. Sm.) Z. H. Tsi = Bulbophyllum retusiusculum Rchb. f. ■
63037 Bulbophyllum retusiusculum Rchb. f. var. tigridum (Hance) Z. H. Tsi = Bulbophyllum tigridum Hance ■
63038 Bulbophyllum rhizomatosum Schltr. = Bulbophyllum obtusilabium W. Kittr. ■☆
63039 Bulbophyllum rhizophorae Lindl. = Bulbophyllum velutinum (Lindl.) Rchb. f. ■☆
63040 Bulbophyllum rhodopetalum Kraenzl. = Bulbophyllum sandersonii (Hook. f.) Rchb. f. subsp. stenopetalum (Kraenzl.) J. J. Verm. ■☆
63041 Bulbophyllum rhodostachys Schltr.;粉红穗石豆兰■☆
63042 Bulbophyllum rictorium Schltr. = Bulbophyllum lucidum Schltr. ■☆
63043 Bulbophyllum ridleyi Kraenzl. = Bulbophyllum multiflorum Ridl. ■☆
63044 Bulbophyllum riyanum Fukuy.;白花石豆兰(白花豆兰,非豆兰,假豆兰);White Bulbophyllum,White Stonebean-orchis ■
63045 Bulbophyllum riyanum Fukuy. = Bulbophyllum pauciflorum Ames ■
63046 Bulbophyllum robustum Rolfe = Bulbophyllum coriophorum Ridl. ■☆
63047 Bulbophyllum rolfei (Kuntze) Seidenf.;高山卷瓣兰(罗尔夫卷瓣兰,若氏卷瓣兰)■
63048 Bulbophyllum rothschildianum (O'Brien) J. J. Sm.;美花卷瓣兰(美花卷唇兰);Beautyflower Bulbophyllum,Beautyflower Curlylip-orchis,Rothschild Cirrhopetalum ■
63049 Bulbophyllum rotundatum (Lindl.) Rchb. f. = Epigeneium rotundatum (Lindl.) Summerh. ■
63050 Bulbophyllum rubescens Schltr. var. meizobulbon Schltr. = Bulbophyllum oxycalyx Schltr. ■☆
63051 Bulbophyllum rubiginosum Schltr.;锈红石豆兰■☆
63052 Bulbophyllum rubrilabium Schltr.;红唇石豆兰■☆
63053 Bulbophyllum rubrobrunneopapillosum De Wild. = Bulbophyllum acutebracteatum De Wild. var. rubrobrunneopapillosum (De Wild.) J. J. Verm. ■☆
63054 Bulbophyllum rubrolabellum T. P. Lin;红心石豆兰(红心豆兰);Redheart Bulbophyllum,Redheart Stonebean-orchis ■
63055 Bulbophyllum rubropunctatum S. S. Ying = Bulbophyllum melanoglossum Hayata ■
63056 Bulbophyllum rubroviolaceum De Wild. = Bulbophyllum resupinatum Ridl. var. filiforme (Kraenzl.) J. J. Verm. ■☆
63057 Bulbophyllum rubrum Jum. et Perrier;红石豆兰■☆
63058 Bulbophyllum rufinum Rchb. f.;窄苞石豆兰;Narrowbract Bulbophyllum,Narrowbract Stonebean-orchis ■
63059 Bulbophyllum ruginosum H. Perrier;褶皱石豆兰■☆
63060 Bulbophyllum rugosibulbum Summerh.;皱纹卷瓣兰■☆
63061 Bulbophyllum rutenbergianum Schltr.;鲁藤贝格石豆兰■☆
63062 Bulbophyllum saltatorium Lindl. var. albociliatum (Finet) J. J. Verm. = Bulbophyllum finetii Szlach. et Olszewski ■☆
63063 Bulbophyllum sambiranense Jum. et Perrier;桑比朗石豆兰■☆
63064 Bulbophyllum sandersonii (Hook. f.) Rchb. f.;桑德森石豆兰■☆
63065 Bulbophyllum sandersonii (Hook. f.) Rchb. f. subsp. stenopetalum (Kraenzl.) J. J. Verm.;窄瓣桑德森石豆兰■☆
63066 Bulbophyllum sanguineum H. Perrier;血红石豆兰■☆
63067 Bulbophyllum sarcorhachis Schltr.;肉轴石豆兰■☆
63068 Bulbophyllum saruwatarii Hayata = Bulbophyllum umbellatum Lindl. ■
63069 Bulbophyllum scaberulum (Rolfe) Bolus;略糙石豆兰■☆
63070 Bulbophyllum scaberulum (Rolfe) Bolus var. album Perez-Vera;白略糙石豆兰■☆
63071 Bulbophyllum scaberulum (Rolfe) Bolus var. crotalicaudatum J. J. Verm.;尾状石豆兰■☆
63072 Bulbophyllum scaberulum (Rolfe) Bolus var. fuerstenbergianum (De Wild.) J. J. Verm.;富尔石豆兰■☆
63073 Bulbophyllum scaphiforme J. J. Verm.;囊唇石豆兰■
63074 Bulbophyllum scariosum Summerh.;干膜质石豆兰■☆
63075 Bulbophyllum schimperianum Kraenzl.;欣珀石豆兰■☆
63076 Bulbophyllum schinzianum Kraenzl.;欣兹石豆兰■☆
63077 Bulbophyllum schinzianum Kraenzl. var. phaeopogon (Schltr.) J. J. Verm. = Bulbophyllum phaeopogon Schltr. ■☆
63078 Bulbophyllum schlechteri De Wild. = Bulbophyllum josephii (Kuntze) Summerh. ■☆
63079 Bulbophyllum secundum Hook. f.;少花石豆兰(下垂石豆兰);Poorflower Bulbophyllum,Poorflower Stonebean-orchis ■
63080 Bulbophyllum sennii Chiov. = Bulbophyllum josephii (Kuntze) Summerh. ■☆
63081 Bulbophyllum septatum Schltr.;具隔石豆兰■☆
63082 Bulbophyllum seretii De Wild. = Bulbophyllum bufo (Lindl.) Rchb. f. ■☆
63083 Bulbophyllum serratum H. Perrier = Bulbophyllum septatum Schltr. ■☆
63084 Bulbophyllum setaceum T. P. Lin;鹤冠卷瓣兰(鹤冠兰,台中豆兰,台中石豆兰);Marabou-coronal Bulbophyllum,Marabou-coronal Curlylip-orchis ■
63085 Bulbophyllum seychellarum Rchb. f. = Bulbophyllum intertextum Lindl. ■☆
63086 Bulbophyllum shanicum King et Pantl.;二叶石豆兰;Shan Bulbophyllum,Shan Stonebean-orchis ■
63087 Bulbophyllum shweliense W. W. Sm.;伞花石豆兰(白冰球);Ruili Stonebean-orchis,Umbellate Bulbophyllum ■
63088 Bulbophyllum sigilliforme H. Perrier = Bulbophyllum complanatum H. Perrier ■☆
63089 Bulbophyllum simonii Summerh. = Bulbophyllum velutinum (Lindl.) Rchb. f. ■☆
63090 Bulbophyllum solheidii De Wild. = Bulbophyllum velutinum

(Lindl.) Rchb. f. ■☆

63091 Bulbophyllum somae Hayata = Bulbophyllum drymoglossum Maxim. ex Okubo ■

63092 Bulbophyllum somai Hayata;狭萼豆兰■

63093 Bulbophyllum somai Hayata = Bulbophyllum drymoglossum Maxim. ex Okubo ■

63094 Bulbophyllum spathaceum Rolfe; 柄叶石豆兰; Spadix Bulbophyllum, Spadix Stonebean-orchis ■

63095 Bulbophyllum spathaceum Rolfe = Bulbophyllum apodum Hook. f. ■

63096 Bulbophyllum spathulatum (Rolfe ex E. Cooper) Seidenf.;匙萼卷瓣兰;Spooncalyx Bulbophyllum, Spooncalyx Curlylip-orchis ■

63097 Bulbophyllum spathulifolium H. Perrier = Bulbophyllum rutenbergianum Schltr. ■☆

63098 Bulbophyllum sphaericum Z. H. Tsi et H. Li;球茎卷瓣兰(球茎卷唇兰);Sphaerical Bulbophyllum, Sphaerical Curlylip-orchis ■

63099 Bulbophyllum sphaerobulbum H. Perrier;球状石豆兰■☆

63100 Bulbophyllum stenobulbon Parl. et Rchb. f.; 短足石豆兰; Smallbulb Bulbophyllum, Smallbulb Stonebean-orchis ■

63101 Bulbophyllum stenopetalum Kraenzl. = Bulbophyllum sandersonii (Hook. f.) Rchb. f. subsp. stenopetalum (Kraenzl.) J. J. Verm. ■☆

63102 Bulbophyllum stenorhachis Kraenzl. = Bulbophyllum imbricatum Lindl. var. purpureum W. Sanford ■☆

63103 Bulbophyllum stolzii Schltr.;斯托尔兹石豆兰■☆

63104 Bulbophyllum striatum (Griff.) Rchb. f.;细柄石豆兰;Thinstalk Bulbophyllum, Thinstalk Stonebean-orchis ■

63105 Bulbophyllum strobiliferum Kraenzl. = Bulbophyllum imbricatum Lindl. var. luteum W. Sanford ■☆

63106 Bulbophyllum suavissimum Rolfe; 直莛石豆兰; Straghtscape Bulbophyllum, Straghtscape Stonebean-orchis ■

63107 Bulbophyllum subapproximatum H. Perrier;相似卷瓣兰■☆

63108 Bulbophyllum subclavatum Schltr.;棍棒卷瓣兰■☆

63109 Bulbophyllum subcoriaceum De Wild. = Bulbophyllum maximum (Lindl.) Rchb. f. ■☆

63110 Bulbophyllum subcrenulatum Schltr.;微圆齿卷瓣兰■☆

63111 Bulbophyllum subparviflorum Z. H. Tsi et S. C. Chen = Bulbophyllum secundum Hook. f. ■

63112 Bulbophyllum subsecundum Schltr.;单侧卷瓣兰■☆

63113 Bulbophyllum subsessile Schltr.;近无柄卷瓣兰■☆

63114 Bulbophyllum sulfureum Schltr.;硫黄卷瓣兰■☆

63115 Bulbophyllum summerhayesii Hawkes = Bulbophyllum scaberulum (Rolfe) Bolus ■☆

63116 Bulbophyllum sutepense (Rolfe ex Downie) Seidenf.;聚株石豆兰;Clump Bulbophyllum, Clump Stonebean-orchis ■

63117 Bulbophyllum taeniophyllum Parl. et Rchb. f.;带叶卷瓣兰(带叶卷唇兰);Beltleaf Bulbophyllum, Beltleaf Curlylip-orchis ■

63118 Bulbophyllum taichungianum S. S. Ying = Bulbophyllum albociliatum (Tang S. Liu et H. J. Su) Nackej. ■

63119 Bulbophyllum taichungianum S. S. Ying = Bulbophyllum setaceum T. P. Lin ■

63120 Bulbophyllum taitungianum S. S. Ying;台东石豆兰■

63121 Bulbophyllum taitungianum S. S. Ying = Bulbophyllum setaceum T. P. Lin ■

63122 Bulbophyllum taiwanense (Fukuy.) Nackej.;台湾卷瓣兰(台湾卷唇兰,台湾石豆兰);Taiwan Bulbophyllum, Taiwan Curlylip-orchis ■

63123 Bulbophyllum taiwanense (Fukuy.) Seidenf. = Bulbophyllum taiwanense (Fukuy.) Nackej. ■

63124 Bulbophyllum talbotii Rendle = Bulbophyllum cochleatum Lindl. ■☆

63125 Bulbophyllum tampoketsense H. Perrier;唐波凯茨■☆

63126 Bulbophyllum tengchongense Z. H. Tsi;云北石豆兰;N. Yunnan Bulbophyllum, N. Yunnan Stonebean-orchis ■

63127 Bulbophyllum tentaculigerum Rchb. f. = Bulbophyllum sandersonii (Hook. f.) Rchb. f. ■☆

63128 Bulbophyllum tenuicaule Lindl. = Bulbophyllum cochleatum Lindl. var. tenuicaule (Lindl.) J. J. Verm. ■☆

63129 Bulbophyllum teretifolium Schltr.;圆柱叶石豆兰■☆

63130 Bulbophyllum tetragonum Lindl.;四角石豆兰■☆

63131 Bulbophyllum theiochlamys Schltr. = Bulbophyllum bicoloratum Schltr. ■☆

63132 Bulbophyllum thomense Summerh.;爱岛石豆兰■☆

63133 Bulbophyllum thompsonii Ridl.;汤普森石豆兰■☆

63134 Bulbophyllum tiagii A. S. Chauhan = Bulbophyllum pteroglossum Schltr. ■

63135 Bulbophyllum tianguii K. Y. Lang et D. Luo;天贵卷瓣兰■

63136 Bulbophyllum tibeticum Rolfe = Bulbophyllum umbellatum Lindl. ■

63137 Bulbophyllum tigridum Hance;虎斑卷瓣兰■

63138 Bulbophyllum tigridum Hance = Bulbophyllum retusiusculum Rchb. f. var. tigridum (Hance) Z. H. Tsi ■

63139 Bulbophyllum tokioi Fukuy.;小叶石豆兰(白花小石豆兰,小叶豆兰);Littleleaf Bulbophyllum, Littleleaf Stonebean-orchis ■

63140 Bulbophyllum tokioi Fukuy. f. alboviride Fukuy. = Bulbophyllum tokioi Fukuy. ■

63141 Bulbophyllum transarisanense Hayata = Bulbophyllum pectinatum Finet ■

63142 Bulbophyllum transarisanense Hayata = Bulbophyllum pectinatum Finet var. transarisanense (Hayata) S. S. Ying ■

63143 Bulbophyllum transarisanense Hayata f. alboviride Fukuy. = Bulbophyllum pectinatum Finet ■

63144 Bulbophyllum transarisanense Hayata f. alboviride Fukuy. = Bulbophyllum pectinatum Finet var. transarisanense (Hayata) S. S. Ying ■

63145 Bulbophyllum triaristellum Kraenzl. et Schltr. = Bulbophyllum intertextum Lindl. ■☆

63146 Bulbophyllum trichocephalum (Schltr.) Ts. Tang et F. T. Wang = Bulbophyllum odoratissimum (Sm.) Lindl. ■

63147 Bulbophyllum trichocephalum (Schltr.) Ts. Tang et F. T. Wang var. wallongense Agrawala, Sabapathy et H. J. Chowdhery = Bulbophyllum odoratissimum (Sm.) Lindl. ■

63148 Bulbophyllum trichochlamys H. Perrier;毛被石豆兰■☆

63149 Bulbophyllum trifarium Rolfe;三列石豆兰■☆

63150 Bulbophyllum trilineatum H. Perrier;三线石豆兰■☆

63151 Bulbophyllum tripudians E. C. Parish et Rchb. f. var. pumilum Seidenf. et Smitinand = Bulbophyllum khaoyaiense Seidenf. ■

63152 Bulbophyllum tripudians H. Perrier var. pumilum Seidenf. et Smitinand = Bulbophyllum khaoyaiense Seidenf. ■

63153 Bulbophyllum triste (Rolfe) Schltr. = Bulbophyllum imbricatum Lindl. var. purpureum W. Sanford ■☆

63154 Bulbophyllum triste Rchb. f.; 球茎石豆兰; Sphaerical Bulbophyllum, Sphaerical Stonebean-orchis ■

63155 Bulbophyllum tsanum (S. Y. Hu et Barretto) Z. H. Tsi;香港卷瓣兰;Hongkong Bulbophyllum, Hongkong Curlylip-orchis ■

63156 Bulbophyllum tsinjoarivense H. Perrier = Bulbophyllum nutans Thouars ■☆
63157 Bulbophyllum turkii Bosser et P. J. Cribb;图尔克石豆兰■☆
63158 Bulbophyllum ugandae (Rolfe) De Wild. = Bulbophyllum falcatum (Lindl.) Rchb. f. ■
63159 Bulbophyllum umbellatum Lindl.;伞花卷瓣兰(伞花卷唇兰,伞形卷瓣兰);Umbell Bulbophyllum, Umbell Curlylip-orchis ■
63160 Bulbophyllum unciniferum Seidenf.;直立卷瓣兰;Erect Bulbophyllum, Erect Curlylip-orchis ■
63161 Bulbophyllum uniflorum Griff. = Bulbophyllum pteroglossum Schltr. ■
63162 Bulbophyllum unifoliatum De Wild.;单小叶石豆兰■☆
63163 Bulbophyllum unifoliatum De Wild. var. flectens (P. J. Cribb et P. Taylor) J. J. Verm.;弯曲石豆兰■☆
63164 Bulbophyllum unifoliatum De Wild. var. infracarinatum (G. Will.) J. J. Verm.;下龙骨石豆兰■☆
63165 Bulbophyllum uraiense Hayata = Bulbophyllum macraei (Lindl.) Rchb. f. ■
63166 Bulbophyllum urbanianum Kraenzl. = Bulbophyllum lupulinum Lindl. ■☆
63167 Bulbophyllum usambarae Kraenzl. = Bulbophyllum intertextum Lindl. ■☆
63168 Bulbophyllum variegatum Thouars;杂色石豆兰■☆
63169 Bulbophyllum velutinum (Lindl.) Rchb. f.;短绒毛卷瓣兰■☆
63170 Bulbophyllum verecundum Summerh. = Bulbophyllum pumilum (Sw.) Lindl. ■☆
63171 Bulbophyllum verruculiferum H. Perrier;小疣卷瓣兰■☆
63172 Bulbophyllum vestitum Bosser;包被卷瓣兰■☆
63173 Bulbophyllum victoris P. J. Cribb et Pérez-Vera = Bulbophyllum resupinatum Ridl. var. filiforme (Kraenzl.) J. J. Verm. ■☆
63174 Bulbophyllum viguieri Schltr.;维基耶石豆兰■☆
63175 Bulbophyllum violaceolabellum Seidenf.;等萼卷瓣兰(等萼卷唇兰);Equalsepal Bulbophyllum, Equalsepal Curlylip-orchis ■
63176 Bulbophyllum viride Rolfe = Bulbophyllum intertextum Lindl. ■☆
63177 Bulbophyllum viridiflorum Hayata = Bulbophyllum pectinatum Finet var. transarisanense (Hayata) S. S. Ying ■
63178 Bulbophyllum viridiflorum Hayata = Bulbophyllum pectinatum Finet ■
63179 Bulbophyllum vitiense Rolfe = Bulbophyllum cocoinum Bateman ex Lindl. ■☆
63180 Bulbophyllum vulcanicum Kraenzl.;火山卷瓣兰■☆
63181 Bulbophyllum vulcanorum H. Perrier;火卷瓣兰■☆
63182 Bulbophyllum wallichii Rchb. f.;双叶卷瓣兰(双叶卷唇兰);Wallich Bulbophyllum, Wallich Cirrhopetalum, Wallich Curlylip-orchis ■
63183 Bulbophyllum watsonianum Rchb. f. = Bulbophyllum ambrosium (Hance) Schltr. ■
63184 Bulbophyllum wendlandianum J. J. Sm.;温氏石豆兰■☆
63185 Bulbophyllum wightii Rchb. f.;睫毛卷瓣兰(大花豆兰,睫毛卷唇兰);Eyelash Bulbophyllum, Eyelash Curlylip-orchis ■
63186 Bulbophyllum winkleri Schltr. = Bulbophyllum pumilum (Sw.) Lindl. ■☆
63187 Bulbophyllum wrightii Summerh. = Bulbophyllum tetragonum Lindl. ■☆
63188 Bulbophyllum wuzhishanense X. H. Jin;五指山石豆兰■
63189 Bulbophyllum xanthobulbum Schltr.;黄球石豆兰■☆
63190 Bulbophyllum xanthoglossum Schltr. = Bulbophyllum schimperianum Kraenzl. ■☆
63191 Bulbophyllum yangambiense Louis et Mullend. ex Geerinck = Bulbophyllum pumilum (Sw.) Lindl. ■☆
63192 Bulbophyllum yoksunense J. J. Sm. = Bulbophyllum emarginatum (Finet) J. J. Sm. ■
63193 Bulbophyllum youngsayeanum S. Y. Hu = Bulbophyllum stenobulbon Parl. et Rchb. f. ■
63194 Bulbophyllum youngsayeanum S. Y. Hu et Barretto = Bulbophyllum stenobulbon Parl. et Rchb. f. ■
63195 Bulbophyllum yuanyangense Z. H. Tsi;元阳石豆兰;Yuanyang Bulbophyllum, Yuanyang Stonebean-orchis ■
63196 Bulbophyllum yuanyangense Z. H. Tsi = Bulbophyllum eublepharum Rchb. f. ■
63197 Bulbophyllum yunanense Rolfe;蒙自石豆兰;Yunnan Bulbophyllum, Yunnan Stonebean-orchis ■
63198 Bulbophyllum zobiaense De Wild. = Bulbophyllum scaberulum (Rolfe) Bolus ■☆
63199 Bulboscodium serotinum L. = Lloydia serotina (L.) Rchb. ■
63200 Bulbospermum Blume = Peliosanthes Andréws ■
63201 Bulbostylis DC. = Brickellia Elliott(保留属名)■●
63202 Bulbostylis DC. = Coleosanthus Cass.(废弃属名)■●
63203 Bulbostylis Gardner = Eupatorium L. ■●
63204 Bulbostylis Kunth(1837)(保留属名);球柱草属;Bulbostyle, Bulbostylis, Sallstyle ■
63205 Bulbostylis Steven(废弃属名) = Bulbostylis Kunth(保留属名)■
63206 Bulbostylis Steven(废弃属名) = Eleocharis R. Br. ■
63207 Bulbostylis abortiva (Steud.) C. B. Clarke = Abildgaardia abortiva (Steud.) Lye ■☆
63208 Bulbostylis afroorientalis (Lye) R. W. Haines = Abildgaardia afroorientalis Lye ■☆
63209 Bulbostylis andongensis (Ridl.) C. B. Clarke;安东球柱草■☆
63210 Bulbostylis annua Nutt. = Psathyrotes annua (Nutt.) A. Gray ■☆
63211 Bulbostylis aphyllanthoides (Ridl.) C. B. Clarke = Abildgaardia pilosa (Willd.) Nees ■☆
63212 Bulbostylis arenaria (Nees) Lindm. = Bulbostylis juncoides (Vahl) Kük. ex Osten ■☆
63213 Bulbostylis argentina Palla = Bulbostylis juncoides (Vahl) Kük. ex Osten ■☆
63214 Bulbostylis argentobrunnea C. B. Clarke = Abildgaardia argentobrunnea (C. B. Clarke) Lye ■☆
63215 Bulbostylis atrosanguinea (Boeck.) C. B. Clarke = Abildgaardia setifolia (Hochst. ex A. Rich.) Lye ■☆
63216 Bulbostylis barbata (Rottb.) C. B. Clarke;球柱草(大毛草,高雄球柱草,龙爪草,毛球柱草,牛毛草,畦莎,旗茅,畎莎,秧草,油麻草);Bearbed Bulbostylis, Bearbed Sallstyle, Sandy Bulbostyl, Watergrass ■
63217 Bulbostylis barbata (Rottb.) Kunth = Bulbostylis barbata (Rottb.) C. B. Clarke ■
63218 Bulbostylis bodardii S. S. Hooper;博达尔球柱草■☆
63219 Bulbostylis boeckeleriana (Schweinf.) Beetle;伯氏球柱草■☆
63220 Bulbostylis breviculmis Kunth = Bulbostylis humilis (Kunth) C. B. Clarke ■☆
63221 Bulbostylis buchananii C. B. Clarke = Abildgaardia buchananii (C. B. Clarke) Lye ■☆
63222 Bulbostylis burchellii (Ficalho et Hiern) C. B. Clarke = Abildgaardia burchellii (Ficalho et Hiern) Lye ■☆

63223 Bulbostylis burkei C. B. Clarke;伯克球柱草■☆
63224 Bulbostylis caespitosa Peter = Abildgaardia oritrephes (Ridl.) Lye ■☆
63225 Bulbostylis californica Torr. et A. Gray = Brickellia california (Torr. et Gray) A. Gray ■☆
63226 Bulbostylis capillaris (L.) C. B. Clarke = Bulbostylis capillaris (L.) Kunth ex C. B. Clarke ■☆
63227 Bulbostylis capillaris (L.) C. B. Clarke var. abortiva (Steud.) Pfeiff. = Abildgaardia abortiva (Steud.) Lye ■☆
63228 Bulbostylis capillaris (L.) C. B. Clarke var. coarctata (Elliott) J. F. Macbr. = Bulbostylis ciliatifolia (Elliott) Torr. var. coarctata (Elliott) Král ■☆
63229 Bulbostylis capillaris (L.) Kunth ex C. B. Clarke;三叶球柱草; Hair Sedge, Hair-like Stenophyllus, Thread-leaved Beak-seed ■☆
63230 Bulbostylis capillaris (L.) Kunth ex C. B. Clarke var. capitata (Miq.) Makino = Bulbostylis densa (Wall.) Hand. -Mazz. var. capitata (Miq.) Ohwi ■☆
63231 Bulbostylis capillaris (L.) Kunth ex C. B. Clarke var. crebra Fernald = Bulbostylis capillaris (L.) Kunth ex C. B. Clarke ■☆
63232 Bulbostylis capillaris (L.) Kunth ex C. B. Clarke var. trifida (Nees) C. B. Clarke = Bulbostylis densa (Wall.) Hand. -Mazz. ■
63233 Bulbostylis capillaris (L.) Kunth var. trifida (Nees) C. B. Clarke = Abildgaardia densa (Wall.) Lye ■☆
63234 Bulbostylis cardiocarpa (Ridl.) C. B. Clarke var. holubii C. B. Clarke = Abildgaardia collina (Ridl.) Lye ■☆
63235 Bulbostylis ciliatifolia (Elliott) Fernald = Bulbostylis ciliatifolia (Elliott) Torr. ■☆
63236 Bulbostylis ciliatifolia (Elliott) Torr.;缘毛叶球柱草■☆
63237 Bulbostylis ciliatifolia (Elliott) Torr. var. coarctata (Elliott) Král;狭缘毛叶球柱草■☆
63238 Bulbostylis cinnamomea (Boeck.) C. B. Clarke;肉桂色球柱草■☆
63239 Bulbostylis claessensii De Wild. = Abildgaardia abortiva (Steud.) Lye ■☆
63240 Bulbostylis clarkeana Hutch. ex M. Bodard = Abildgaardia clarkeana (Hutch. ex M. Bodard) Lye ■☆
63241 Bulbostylis coarctata (Elliott) Fernald = Bulbostylis ciliatifolia (Elliott) Torr. var. coarctata (Elliott) Král ■☆
63242 Bulbostylis coleotricha (Hochst. ex A. Rich.) C. B. Clarke = Abildgaardia coleotricha (Hochst. ex A. Rich.) Lye ■☆
63243 Bulbostylis coleotricha (Hochst. ex A. Rich.) C. B. Clarke var. lanifera (Boeck.) C. B. Clarke = Abildgaardia lanifera (Boeck.) Lye ■☆
63244 Bulbostylis coleotricha (Hochst. ex A. Rich.) C. B. Clarke var. miegei (M. Bodard) R. W. Haines = Abildgaardia coleotricha (Hochst. ex A. Rich.) Lye var. miegei (Bodard) Lye ■☆
63245 Bulbostylis collina (Kunth) C. B. Clarke = Abildgaardia contexta (Nees) Lye ■☆
63246 Bulbostylis congolensis De Wild. = Abildgaardia congolensis (De Wild.) Lye ■☆
63247 Bulbostylis contexta (Nees) M. Bodard = Abildgaardia contexta (Nees) Lye ■☆
63248 Bulbostylis cruciformis (Lye) R. W. Haines = Abildgaardia cruciformis Lye ■☆
63249 Bulbostylis cupricola Goetgh.;喜铜球柱草■☆
63250 Bulbostylis cylindrica C. B. Clarke;柱形球柱草■☆
63251 Bulbostylis cyrtathera Cherm. = Bulbostylis fimbristyloides C. B. Clarke ■☆
63252 Bulbostylis densa (Wall. ex Roxb.) Hand. -Mazz. = Bulbostylis densa (Wall.) Hand. -Mazz. ■
63253 Bulbostylis densa (Wall.) Hand. -Mazz.;丝叶球柱草(黄毛草,龙须草,球柱草,细黄毛草,羊胡须); Filamentous-leaf Bulbostylis, Filiformleaf Sallstyle ■
63254 Bulbostylis densa (Wall.) Hand. -Mazz. = Abildgaardia densa (Wall.) Lye ■☆
63255 Bulbostylis densa (Wall.) Hand. -Mazz. subsp. afromontana (Lye) R. W. Haines;非洲山生球柱草■☆
63256 Bulbostylis densa (Wall.) Hand. -Mazz. subsp. afromontana (Lye) R. W. Haines = Abildgaardia densa (Wall.) Lye subsp. afromontana Lye ■☆
63257 Bulbostylis densa (Wall.) Hand. -Mazz. var. cameroonensis (C. B. Clarke) S. S. Hooper;喀麦隆球柱草■☆
63258 Bulbostylis densa (Wall.) Hand. -Mazz. var. capitata (Miq.) Ohwi;头状丝叶球柱草■☆
63259 Bulbostylis disticha Ohwi = Bulbostylis barbata (Rottb.) C. B. Clarke ■
63260 Bulbostylis disticha Ohwi et T. Koyama = Bulbostylis barbata (Rottb.) C. B. Clarke ■
63261 Bulbostylis elegans Cherm.;雅致球柱草■☆
63262 Bulbostylis elegantissima (Lye) R. W. Haines = Abildgaardia elegantissima Lye ■☆
63263 Bulbostylis equitans (Kük.) Raymond = Nemum equitans (Kük.) J. Raynal ■☆
63264 Bulbostylis erratica (Hook. f.) C. B. Clarke = Abildgaardia erratica (Hook. f.) Lye ■☆
63265 Bulbostylis exilis (Kunth) Lye = Abildgaardia hispidula (Vahl) Lye ■☆
63266 Bulbostylis fasciculata Cherm.;簇生球柱草■☆
63267 Bulbostylis filamentosa (Vahl) C. B. Clarke = Abildgaardia filamentosa (Vahl) Lye ■☆
63268 Bulbostylis filamentosa (Vahl) C. B. Clarke var. barbata C. B. Clarke;髯毛球柱草■☆
63269 Bulbostylis filamentosa (Vahl) C. B. Clarke var. metralis (Cherm.) R. W. Haines = Abildgaardia filamentosa (Vahl) Lye ■☆
63270 Bulbostylis filamentosa (Vahl) C. B. Clarke var. scabricaulis (Cherm.) M. Bodard = Abildgaardia collina (Ridl.) Lye ■☆
63271 Bulbostylis filiformis C. B. Clarke = Abildgaardia hispidula (Vahl) Lye subsp. filiformis (C. B. Clarke) Lye ■☆
63272 Bulbostylis fimbristyloides C. B. Clarke;流苏柱球柱草■☆
63273 Bulbostylis flexuosa (Ridl.) Goetgh.;曲折球柱草■☆
63274 Bulbostylis floridana (Britton) Fernald = Bulbostylis barbata (Rottb.) C. B. Clarke ■
63275 Bulbostylis fusiformis Goetgh.;梭形球柱草■☆
63276 Bulbostylis glaberrima Kük. = Abildgaardia glaberrima (Kük.) Lye ■☆
63277 Bulbostylis grandibulbosa Kük. = Abildgaardia scleropus (C. B. Clarke) Lye ■☆
63278 Bulbostylis guineensis Cherm. ex Bodard;几内亚球柱草■☆
63279 Bulbostylis hensii (C. B. Clarke) Haines = Abildgaardia hispidula (Vahl) Lye subsp. brachyphylla (Cherm.) Lye ■☆
63280 Bulbostylis hirta (Thunb.) Svenson; 粗 糙 球 柱 草; Rough Hairsedge ■☆
63281 Bulbostylis hispidula (Vahl) R. W. Haines = Abildgaardia hispidula (Vahl) Lye ■☆
63282 Bulbostylis hispidula (Vahl) R. W. Haines subsp. brachyphylla

(Cherm.) R. W. Haines = Abildgaardia hispidula (Vahl) Lye subsp. brachyphylla (Cherm.) Lye ■☆
63283 Bulbostylis hispidula (Vahl) R. W. Haines subsp. filiformis (C. B. Clarke) Haines = Abildgaardia hispidula (Vahl) Lye subsp. filiformis (C. B. Clarke) Lye ■☆
63284 Bulbostylis hispidula (Vahl) R. W. Haines subsp. halophila (Lye) R. W. Haines = Abildgaardia hispidula (Vahl) Lye subsp. halophila Lye ■☆
63285 Bulbostylis hispidula (Vahl) R. W. Haines subsp. intermedia (Lye) R. W. Haines = Abildgaardia hispidula (Vahl) Lye subsp. intermedia Lye ■☆
63286 Bulbostylis hispidula (Vahl) R. W. Haines subsp. pyriformis (Lye) R. W. Haines;梨形球柱草■☆
63287 Bulbostylis hispidula (Vahl) R. W. Haines subsp. pyriformis (Lye) R. W. Haines = Abildgaardia hispidula (Vahl) Lye subsp. pyriformis (Lye) Lye ■☆
63288 Bulbostylis holotricha Peter = Abildgaardia congolensis (De Wild.) Lye ■☆
63289 Bulbostylis humilis (Kunth) C. B. Clarke;低矮球柱草■☆
63290 Bulbostylis humpatensis Meneses;洪帕塔球柱草■☆
63291 Bulbostylis igneotonsa Raymond = Abildgaardia igneotonsa (Raymond) Kornas ■☆
63292 Bulbostylis johnstonii C. B. Clarke = Abildgaardia johnstonii (C. B. Clarke) Lye ■☆
63293 Bulbostylis juncoides (Vahl) Kük. ex Osten;灯心草状球柱草■☆
63294 Bulbostylis kirkii C. B. Clarke = Bulbostylis contexta (Nees) M. Bodard ■☆
63295 Bulbostylis langsdorffiana (Kunth) C. B. Clarke = Bulbostylis juncoides (Vahl) Kük. ex Osten ■☆
63296 Bulbostylis laniceps (K. Schum.) C. B. Clarke ex T. Durand et Schinz;绵毛梗球柱草■☆
63297 Bulbostylis lanifera (Boeck.) Beetle = Abildgaardia lanifera (Boeck.) Lye ■☆
63298 Bulbostylis leiolepis (Kük.) R. W. Haines = Abildgaardia leiolepis (Kük.) Lye ■☆
63299 Bulbostylis lineolata Goetgh.;细线球柱草■☆
63300 Bulbostylis longiradiata Goetgh.;长射线球柱草■☆
63301 Bulbostylis macra (Ridl.) C. B. Clarke = Abildgaardia macra (Ridl.) Lye ■☆
63302 Bulbostylis macrostachya (Lye) R. W. Haines = Abildgaardia macrostachya Lye ■☆
63303 Bulbostylis megastachys (Ridl.) C. B. Clarke;大穗球柱草■☆
63304 Bulbostylis megastachys (Ridl.) C. B. Clarke = Abildgaardia megastachys (Ridl.) Lye ■☆
63305 Bulbostylis melanocephala (Ridl.) C. B. Clarke;黑头球柱草■☆
63306 Bulbostylis metralis Cherm. = Abildgaardia filamentosa (Vahl) Lye ■☆
63307 Bulbostylis microcarpa (Lye) R. W. Haines = Abildgaardia microcarpa Lye ■☆
63308 Bulbostylis microelegans (Lye) R. W. Haines = Abildgaardia microelegans Lye ■☆
63309 Bulbostylis micromucronata Goetgh.;小短尖球柱草■☆
63310 Bulbostylis microphylla Nutt. = Brickellia microphylla (Nutt.) A. Gray ■☆
63311 Bulbostylis miegei M. Bodard = Abildgaardia coleotricha (Hochst. ex A. Rich.) Lye var. miegei (Bodard) Lye ■☆
63312 Bulbostylis moggii Schönland et Turrill = Bulbostylis pusilla (Hochst. ex A. Rich.) C. B. Clarke ■☆
63313 Bulbostylis mozambica Raymond;莫桑比克球柱草■☆
63314 Bulbostylis mucronata C. B. Clarke;短尖球柱草■☆
63315 Bulbostylis multispiculata Cherm.;多细刺球柱草■☆
63316 Bulbostylis nemoides Goetgh.;丝状球柱草■☆
63317 Bulbostylis oligostachya (Boeck.) C. B. Clarke;寡穗球柱草■☆
63318 Bulbostylis oligostachys (Hochst. ex A. Rich.) Lye = Abildgaardia oligostachys (Hochst. ex A. Rich.) Lye ■☆
63319 Bulbostylis oritrephes (Ridl.) C. B. Clarke;山地球柱草■☆
63320 Bulbostylis oritrephes (Ridl.) C. B. Clarke = Abildgaardia oritrephes (Ridl.) Lye ■☆
63321 Bulbostylis oritrephes (Ridl.) C. B. Clarke subsp. australis B. L. Burtt = Bulbostylis oritrephes (Ridl.) C. B. Clarke ■☆
63322 Bulbostylis pallescens (Lye) R. W. Haines = Abildgaardia pallescens Lye ■☆
63323 Bulbostylis parva (Ridl.) C. B. Clarke = Abildgaardia pusilla (Hochst. ex A. Rich.) Lye ■☆
63324 Bulbostylis parvinux C. B. Clarke;小果球柱草■☆
63325 Bulbostylis pilosa (Willd.) Cherm. = Abildgaardia pilosa (Willd.) Nees ■☆
63326 Bulbostylis polytricha Cherm. = Abildgaardia congolensis (De Wild.) Lye ■☆
63327 Bulbostylis puberula (Poir.) C. B. Clarke;毛鳞球柱草;Hairy-scale Bulbostylis, Haiscale Sallstyle ■
63328 Bulbostylis puberula (Poir.) C. B. Clarke var. cameroonensis C. B. Clarke = Bulbostylis densa (Wall.) Hand.-Mazz. var. cameroonensis (C. B. Clarke) S. S. Hooper ■
63329 Bulbostylis puberula (Poir.) Kunth = Bulbostylis puberula (Poir.) C. B. Clarke ■
63330 Bulbostylis pusilla (Hochst. ex A. Rich.) C. B. Clarke;微小球柱草■☆
63331 Bulbostylis pusilla (Hochst. ex A. Rich.) C. B. Clarke = Abildgaardia pusilla (Hochst. ex A. Rich.) Lye ■☆
63332 Bulbostylis pusilla (Hochst. ex A. Rich.) C. B. Clarke subsp. congolensis (De Wild.) R. W. Haines = Abildgaardia congolensis (De Wild.) Lye ■☆
63333 Bulbostylis pusilla (Hochst. ex A. Rich.) C. B. Clarke subsp. yalingensis (Cherm.) R. W. Haines et Lye = Abildgaardia pusilla (Hochst. ex A. Rich.) Lye ■☆
63334 Bulbostylis rarissima (Steud.) C. B. Clarke;罕见球柱草■☆
63335 Bulbostylis rehmanni C. B. Clarke = Bulbostylis filamentosa (Vahl) C. B. Clarke ■☆
63336 Bulbostylis rhizomatosa (Lye) R. W. Haines = Abildgaardia rhizomatosa Lye ■☆
63337 Bulbostylis rotundata (Kük.) R. W. Haines = Abildgaardia rotundata (Kük.) Lye ■☆
63338 Bulbostylis rumokensis Cherm. ex Goetgh.;卢莫克球柱草■☆
63339 Bulbostylis scabricaulis Cherm.;茎球柱草■☆
63340 Bulbostylis scabricaulis Cherm. = Abildgaardia collina (Ridl.) Lye ■☆
63341 Bulbostylis schaffneri (Boeck.) C. B. Clarke;沙氏球柱草■☆
63342 Bulbostylis schimperiana (A. Rich.) C. B. Clarke var. leiolepis Kük. = Abildgaardia leiolepis (Kük.) Lye ■☆
63343 Bulbostylis schimperiana (Hochst. ex A. Rich.) C. B. Clarke = Abildgaardia schimperiana (A. Rich.) Lye ■☆
63344 Bulbostylis schlechteri C. B. Clarke;施莱球柱草■☆
63345 Bulbostylis schoenoides (Kunth) C. B. Clarke;拟舍恩球柱草■☆

63346 Bulbostylis schoenoides (Kunth) C. B. Clarke = Abildgaardia erratica (Hook. f.) Lye subsp. schoenoides (Kunth) Lye ■☆

63347 Bulbostylis scleropus C. B. Clarke = Abildgaardia scleropus (C. B. Clarke) Lye ■☆

63348 Bulbostylis seretii De Wild. = Abildgaardia coleotricha (Hochst. ex A. Rich.) Lye ■☆

63349 Bulbostylis setifolia (Hochst. ex A. Rich.) M. Bodard = Abildgaardia setifolia (Hochst. ex A. Rich.) Lye ■☆

63350 Bulbostylis somaliensis Lye;索马里球柱草■☆

63351 Bulbostylis somaliensis Lye subsp. confusa Lye;混乱球柱草■☆

63352 Bulbostylis sphaerocarpa (Boeck.) C. B. Clarke = Abildgaardia sphaerocarpa (Boeck.) Lye ■☆

63353 Bulbostylis stenophylla (Elliott) C. B. Clarke;窄叶球柱草■☆

63354 Bulbostylis striatella C. B. Clarke = Abildgaardia striatella (C. B. Clarke) Lye ■☆

63355 Bulbostylis stricta Turrill = Abildgaardia megastachys (Ridl.) Lye ■☆

63356 Bulbostylis subumbellata (Lye) R. W. Haines = Abildgaardia subumbellata Lye ■☆

63357 Bulbostylis tanzaniae (Lye) R. W. Haines = Abildgaardia tanzaniae Lye ■☆

63358 Bulbostylis taylorii (K. Schum.) C. B. Clarke = Abildgaardia taylorii (K. Schum.) Lye ■☆

63359 Bulbostylis tenerrima (Fisch. et C. A. Mey.) Palla;柔弱球柱草■☆

63360 Bulbostylis tisserantii (Cherm.) Lye = Bulbostylis viridecarinata (De Wild.) Goetgh. ■☆

63361 Bulbostylis togoensis Cherm. = Abildgaardia lanifera (Boeck.) Lye ■☆

63362 Bulbostylis trabeculata C. B. Clarke = Abildgaardia trabeculata (C. B. Clarke) Lye ■☆

63363 Bulbostylis trabeculata C. B. Clarke var. microglumis (Lye) R. W. Haines = Abildgaardia trabeculata (C. B. Clarke) Lye var. microglumis Lye ■☆

63364 Bulbostylis transiens (K. Schum.) C. B. Clarke = Abildgaardia boeckeleriana (Schweinf.) Lye var. transiens (K. Schum.) Lye ■☆

63365 Bulbostylis trichobasis (Baker) C. B. Clarke var. caespitosa (Peter) Kük. = Abildgaardia oritrephes (Ridl.) Lye ■☆

63366 Bulbostylis trichobasis (Baker) C. B. Clarke var. leptocaulis C. B. Clarke = Abildgaardia oritrephes (Ridl.) Lye ■☆

63367 Bulbostylis trichobasis (Baker) C. B. Clarke var. uniseriata C. B. Clarke = Abildgaardia oritrephes (Ridl.) Lye ■☆

63368 Bulbostylis trifida (Nees) C. B. Clarke = Bulbostylis densa (Wall.) Hand. -Mazz. ■

63369 Bulbostylis trifida (Nees) Kunth = Bulbostylis densa (Wall.) Hand. -Mazz. ■

63370 Bulbostylis trifida (Nees) Nelmes = Abildgaardia densa (Wall.) Lye ■☆

63371 Bulbostylis trifida (Nees) Nelmes var. biegensis Cherm. = Abildgaardia densa (Wall.) Lye ■☆

63372 Bulbostylis trullata Goetgh.;杓球柱草■☆

63373 Bulbostylis ugandensis (Lye) R. W. Haines = Abildgaardia ugandensis Lye ■☆

63374 Bulbostylis vaginosa Kük.;多鞘球柱草■☆

63375 Bulbostylis vanderystii Cherm. = Abildgaardia vanderystii (Cherm.) Lye ■☆

63376 Bulbostylis viridecarinata (De Wild.) Goetgh.;绿棱球柱草■☆

63377 Bulbostylis wallichiana (Schult.) Beetle = Abildgaardia wallichiana (Schult.) Lye ■☆

63378 Bulbostylis warei (Torr.) C. B. Clarke;瓦氏球柱草■☆

63379 Bulbostylis wittei Cherm. = Bulbostylis laniceps (K. Schum.) C. B. Clarke ex T. Durand et Schinz ■☆

63380 Bulbostylis woronowii Palla;沃氏球柱草■☆

63381 Bulbostylis yalingensis Cherm. = Abildgaardia pusilla (Hochst. ex A. Rich.) Lye ■☆

63382 Bulbostylis zambesica C. B. Clarke = Abildgaardia macra (Ridl.) Lye ■☆

63383 Bulbostylis zambesica C. B. Clarke var. occidentalis M. Bodard = Bulbostylis bodardii S. S. Hooper ■☆

63384 Bulbostylis zeyheri (Boeck.) C. B. Clarke = Abildgaardia contexta (Nees) Lye ■☆

63385 Bulbulus Swallen = Rehia Fijten ■☆

63386 Bulga Kuntze = Ajuga L. ■●

63387 Bulga Kuntze = Bugula Mill. ■●

63388 Bulleyia Schltr. (1912);蜂腰兰属;Bulleyia ■★

63389 Bulleyia yunnanensis Schltr.;蜂腰兰(云南蜂腰兰);Bulleyia, Yunnan Bulleyia ■

63390 Bulliarda DC. = Crassula L. ●■☆

63391 Bulliarda DC. = Tillaea L. ■

63392 Bulliarda Neck. = Annona L. + Xylopia L.(保留属名)●

63393 Bulliarda abyssinica A. Rich. = Crassula hedbergii Wickens et M. Bywater ■☆

63394 Bulliarda alpina (Eckl. et Zeyh.) Harv. = Crassula umbellata Thunb. ■☆

63395 Bulliarda brevifolia Eckl. et Zeyh. = Crassula decumbens Thunb. var. brachyphylla (Adamson) Toelken ●☆

63396 Bulliarda capensis (L. f.) E. Mey. ex Drège = Crassula natans Thunb. ■☆

63397 Bulliarda dregei Harv. = Crassula pellucida L. subsp. brachypetala (Drège ex Harv.) Toelken ●☆

63398 Bulliarda elatinoides Eckl. et Zeyh. = Crassula elatinoides (Eckl. et Zeyh.) Friedrich ●☆

63399 Bulliarda filifomis Eckl. et Zeyh. = Crassula natans Thunb. var. minus (Eckl. et Zeyh.) G. D. Rowley ■☆

63400 Bulliarda perfoliata (L. f.) DC. = Crassula inanis Thunb. ■☆

63401 Bulliarda trichotoma Eckl. et Zeyh. = Crassula decumbens Thunb. ●☆

63402 Bulliarda vaillantii (Willd.) DC. = Crassula vaillantii (Willd.) Roth ●☆

63403 Bulliarda vaillantii (Willd.) DC. var. subulata Harv. = Crassula vaillantii (Willd.) Roth ●☆

63404 Bullockia (Bridson) Razafim., Lantz et B. Bremer = Canthium Lam. ●

63405 Bulnesia Gay(1846);布奈木属(布藜属,南美洲蒺藜木属);Bulnesia ●☆

63406 Bulnesia arborea (Jacq.) Engl.;乔状布奈木(乔状南美洲蒺藜木);Maracaibo Lignum, Verawood ●☆

63407 Bulnesia arborea Engl. = Bulnesia arborea (Jacq.) Engl. ●☆

63408 Bulnesia retama Griseb.;布奈木(勒塔木);Retamo, Retamo Wax ●☆

63409 Bulnesia sarmientoi Lorentz ex Griseb.;萨米布奈木(萨米南美洲蒺藜木,萨氏布藜);Guaiac, Palo Santo, Paraguaylignum, Pau-santo ●☆

63410 Bulowia Hook. = Buelowia Schumach. ●☆

63411 Bulowia Hook. = Smeathmannia Sol. ex R. Br. ●☆
63412 Bulwera Post et Kuntze = Bulweria F. Muell. ●☆
63413 Bulweria F. Muell. = Deplanchea Vieill. ●☆
63414 Bumalda Thunb. (1783);布马无患子属●☆
63415 Bumalda Thunb. = Staphylea L. ●
63416 Bumalda trifolia Thunb. = Staphylea bumalda (Thunb.) DC. ●
63417 Bumelia Sw. (1788)(保留属名);刺李山榄属(布玛利木属)●☆
63418 Bumelia Sw. (保留属名) = Sideroxylon L. ●☆
63419 Bumelia dulcifica Schumach. et Thonn. = Synsepalum dulcificum (Schumach. et Thonn.) Daniell ●☆
63420 Bumelia lanuginosa (Michx.) Pers. subsp. oblongifolia (Nutt.) Cronquist = Sideroxylon lanuginosa Michx. ●☆
63421 Bumelia lanuginosa (Michx.) Pers. subsp. oblongifolia (Nutt.) Cronquist var. albicans Sarg. = Sideroxylon lanuginosa Michx. ●☆
63422 Bumelia lanuginosa Pers.;多毛刺李山榄(多毛布玛利木);Gum Bumelia ●☆
63423 Bumelia lucida Small;亮刺李山榄(光亮布玛利木)●☆
63424 Bumelia lycioides (L.) Pers. = Sideroxylon lycioides L. ●☆
63425 Bumelia lycioides Willd.;光叶刺李山榄(光叶布玛利木);Buckthorn Bumelia ●☆
63426 Bumelia obtusifolia Roem. et Schult.;钝叶刺李山榄木●☆
63427 Bumeliaceae Barnhart = Sapotaceae Juss. (保留科名)●
63428 Bumeliaceae Barnhart;刺李山榄科●
63429 Bunburia Harv. = Vincetoxicum Wolf ●■
63430 Bunburia elliptica Harv. = Cynanchum ellipticum (Harv.) R. A. Dyer ●☆
63431 Bunburya Meisn. ex Hochst. = Natalanthe Sond. ●
63432 Bunburya Meisn. ex Hochst. = Tricalysia A. Rich. ex DC. ●
63433 Bunburya capensis Meisn. ex Hochst. = Tricalysia capensis (Meisn. ex Hochst.) Sim ●☆
63434 Bunchosia Rich. ex Juss. = Bunchosia Rich. ex Kunth ●☆
63435 Bunchosia Rich. ex Kunth(1822);邦乔木属●☆
63436 Bunchosia armeniaca DC.;文雀邦乔木(文雀西亚木)●☆
63437 Bunchosia glandulosa (Cav.) DC.;腺质邦乔木●☆
63438 Bunchosia swartziana Griseb.;斯氏邦乔木●☆
63439 Bungarimba K. M. Wong = Porterandia Ridl. ●
63440 Bungea C. A. Mey. (1831);斑沼草属(本格草属)■☆
63441 Bungea sheareri S. Moore = Monochasma sheareri (S. Moore) Maxim. ■
63442 Bungea trifida (Vahl) C. A. Mey.;斑沼草■☆
63443 Bunias L. (1753);匙荠属(班倪属);Bunias, Corn Rocket, Spooncress, Warty-cabbage ■
63444 Bunias aegyptiaca L. = Ochthodium aegyptiacum (L.) DC. ■☆
63445 Bunias balearica L. = Succowia balearica (L.) Medik. ■☆
63446 Bunias cochlearioides Murray;匙荠(草原女真荠);Cochlear-like Bunias, Spooncress ■
63447 Bunias cornuta L. = Pugionium cornutum (L.) Gaertn. ■
63448 Bunias edentula Bigelow = Cakile edentula (Bigelow) Hook. ■☆
63449 Bunias edentula Bigelow = Cakile maritima Scop. ■☆
63450 Bunias erucago L.;冠毛匙荠;Corn Rocket, Crested Bunias, Crested Wartycabbage, Southern Warty-cabbage ■☆
63451 Bunias myagroides L. = Erucaria hispanica (L.) Druce ■☆
63452 Bunias orientalis L.;疣果匙荠;Hill Mustard, Oriental Bunias, Oriental Spooncress, Turkish Rocket, Turkish Wartycabbage, Turkish Warty-cabbage, Warted Bunias, Warty Cabbage, Warty-cabbage ■
63453 Bunias prostrata Desf. = Muricaria prostrata (Desf.) Desv. ■☆
63454 Bunias spinosa L. = Zilla spinosa (L.) Prantl ■☆
63455 Bunias syriaca (L.) M. Bieb. = Euclidium syriacum (L.) R. Br. ■
63456 Bunias syriaca Gaertn. = Euclidium syriacum (L.) R. Br. ■
63457 Bunias tatarica Willd. = Euclidium tenuissimum (Pall.) B. Fedtsch. ■
63458 Bunias tatarica Willd. = Litwinowia tenuissima (Pall.) Woronow ex Pavlov ■
63459 Bunias tenuifolia Sibth. et Sm. = Didesmus aegyptius (L.) Desv. ■☆
63460 Bunias tscheliensis Debeaux = Bunias cochlearioides Murray ■
63461 Buniella Schischk. = Bunium L. ■☆
63462 Bunion St. -Lag. = Bunium L. ■☆
63463 Bunioseris Jord. = Lactuca L. ■
63464 Buniotrinia Stapf et Wettst. = Ferula L. ■
63465 Buniotrinia Stapf et Wettst. ex Stapf = Ferula L. ■
63466 Bunium Koch = Pimpinella L. ■
63467 Bunium L. (1753);布留芹属(布尼芹属);Great Pignut, Hawk Nut, Hawknut ■☆
63468 Bunium W. D. J. Koch = Pimpinella L. ■
63469 Bunium alpinum Waldst. et Kit.;高山布留芹■☆
63470 Bunium alpinum Waldst. et Kit. subsp. atlanticum Maire;大西洋布留芹■☆
63471 Bunium badghysii Korovin;巴德布留芹■☆
63472 Bunium bourgaei (Boiss.) Freyn et Sint.;布尔布留芹■☆
63473 Bunium brevifolium Lowe;短叶布留芹■☆
63474 Bunium bubocastanum L. = Bunium bulbocastanum Bertero ex DC. ■☆
63475 Bunium bulbocastanum Bertero ex DC.;球布留芹(球栗布尼芹);Arnut, Awnut, Ciper Nut, Earth Chestnut, Earth-nut, Ernut, Great Earth Nut, Great Earthnut, Great Pignut, Ground Nut, Kippernut, Lousy Arnot, Lousy Arnut, Lucy Arnut, Pignut, Square Parsley ■☆
63476 Bunium bulbocastanum L. var. elatum Batt. = Bunium fontanesii (Pers.) Maire ■☆
63477 Bunium bulbocastanum L. var. peucedanoides (Desf.) J. M. Monts. = Bunium fontanesii (Pers.) Maire ■☆
63478 Bunium buriaticum (Turcz.) Drude = Carum buriaticum Turcz. ■
63479 Bunium buriaticum Drude = Carum buriaticum Turcz. ■
63480 Bunium capusii (Franch.) Korovin;卡普布留芹■☆
63481 Bunium carvi (L.) M. Bieb. = Carum carvi L. ■
63482 Bunium chaerophylloides (Regel et Schmalh.) Drude;细叶布留芹(细叶芹)■☆
63483 Bunium copticum (L.) Spreng. = Trachyspermum ammi (L.) Sprague ■
63484 Bunium crassifolium Batt.;厚叶布留芹■☆
63485 Bunium cylindricum (Boiss. et Hausskn.) Drude;布留芹(布尼芹,圆筒布尼芹)■
63486 Bunium cylindricum Grossh. = Bunium cylindricum (Boiss. et Hausskn.) Drude ■
63487 Bunium elatum Batt.;高布留芹■☆
63488 Bunium elegans (Fenzl.) Freyn;雅致布留芹■☆
63489 Bunium flexuosum Brot. = Conopodium majus (Gouan) Loret ■☆
63490 Bunium fontanesii (Pers.) Maire;丰塔纳布留芹■☆
63491 Bunium fontanesii (Pers.) Maire var. aphyllum Nègre = Bunium fontanesii (Pers.) Maire ■☆
63492 Bunium fontanesii (Pers.) Maire var. glaucum (Maire) Emb. et Maire = Bunium fontanesii (Pers.) Maire ■☆

63493 Bunium fontanesii (Pers.) Maire var. litorale Maire, Weiller et Wilczek = Bunium fontanesii (Pers.) Maire ■☆
63494 Bunium fontanesii (Pers.) Maire var. mauritanicum (Boiss. et Reut.) Maire = Bunium fontanesii (Pers.) Maire ■☆
63495 Bunium fontanesii (Pers.) Maire var. perrotii (Braun-Blanq. et Maire) Maire = Bunium fontanesii (Pers.) Maire ■☆
63496 Bunium gypsaceum Korovin;钙布留芹■☆
63497 Bunium hissaricum Korovin;希萨尔■☆
63498 Bunium imbricatum (Schinz) Drude = Afrocarum imbricatum (Schinz) Rauschert ■☆
63499 Bunium incrassatum (Boiss.) Batt. = Bunium pachypodum P. W. Ball ■☆
63500 Bunium intermedium Korovin;间型布留芹■☆
63501 Bunium kuhitangi Nevski;库希塘布留芹■☆
63502 Bunium longipes Freyn;长梗布留芹■☆
63503 Bunium majus Gouan;大布留芹■☆
63504 Bunium mauritanicum (Boiss. et Reut.) Batt. = Bunium fontanesii (Pers.) Maire ■☆
63505 Bunium pachypodum P. W. Ball;粗足布留芹■☆
63506 Bunium paucifolium DC.;少叶布留芹■☆
63507 Bunium perrotii Braun-Blanq. et Maire = Bunium fontanesii (Pers.) Maire ■☆
63508 Bunium persicum (Boiss.) B. Fedtsch.;波斯布留芹■☆
63509 Bunium scabrellum Korovin;粗糙布留芹■☆
63510 Bunium seravschanicum Korovin;塞拉夫布留芹■☆
63511 Bunium setaceum (Schrenk) H. Wolff = Scaligeria setacea (Schrenk) Korovin ■
63512 Bunium trichophyllum (Schrenk) H. Wolff = Hyalolaena trichophylla (Schrenk) Pimenov et Kljuykov ■
63513 Bunium vaginatum Korovin;具鞘布留芹■☆
63514 Bunnya F. Muell. = Cyanostegia Turcz. ●☆
63515 Bunochilus D. L. Jones et M. A. Clem. (2002);丘舌兰属■☆
63516 Bunophila Willd. ex Roem. et Schult. = Machaonia Bonpl. ■☆
63517 Buonapartea G. Don = Bonapartea Ruiz et Pav. ■☆
63518 Buonapartea G. Don = Tillandsia L. ■☆
63519 Bupariti Duhamel(废弃属名) = Thespesia Sol. ex Corrêa(保留属名)●
63520 Bupariti lampas (Cav.) Rothm. = Thespesia lampas (Cav.) Dalzell et A. Gibson ●
63521 Bupariti populnea (L.) Rothm. = Thespesia populnea (L.) Sol. ex Corrêa ●
63522 Buphane Herb. = Boophone Herb. ■☆
63523 Buphane angolensis Baker = Ammocharis angolensis (Baker) Milne-Redh. et Schweick. ■☆
63524 Buphthalmum L. (1753);牛眼菊属;Oxeye, Ox-eye, Oxeyedaisy, Sunwheel, Yellow Oxeye ■
63525 Buphthalmum Mill. = Anthemis L. ■
63526 Buphthalmum angustifolium Pursh = Balduina angustifolia (Pursh) B. L. Rob. ■☆
63527 Buphthalmum arborescens L. = Borrichia arborescens (L.) DC. ●☆
63528 Buphthalmum capense L. = Oedera capensis (L.) Druce ●☆
63529 Buphthalmum frutescens L. = Borrichia frutescens (L.) DC. ●☆
63530 Buphthalmum garcinii Burm. f. = Anvillea garcinii (Burm. f.) DC. ●☆
63531 Buphthalmum grandiflorum L.;大花牛眼菊;Oxeye, Showy Oxeye ■☆
63532 Buphthalmum graveolens Forssk. = Asteriscus graveolens (Forssk.) Less. ■☆
63533 Buphthalmum helianthoides L. = Heliopsis helianthoides (L.) Sweet ■☆
63534 Buphthalmum imbricatum Cav. = Asteriscus imbricatus (Cav.) DC. ■☆
63535 Buphthalmum laevigatum Willd. = Vieraea laevigata (Willd.) Webb et Berthel. ■☆
63536 Buphthalmum maritimum L. = Pallenis maritima (L.) Greuter ■☆
63537 Buphthalmum odorum Schousb. = Asteriscus graveolens (Forssk.) Less. subsp. odorus (Schousb.) Greuter ■☆
63538 Buphthalmum sagittatum Pursh = Balsamorhiza sagittata (Pursh) Nutt. ■☆
63539 Buphthalmum salicifolium L.;牛眼菊(黄柳菊,黄牛眼菊,柳叶牛眼菊);Willowleaf Oxeye, Willow-leaf Oxeye, Willowleaf Oxeyedaisy, Willow-leaf Sunwheel, Yellow Oxeye ■
63540 Buphthalmum scandens Schumach. et Thonn. = Melanthera scandens (Schumach. et Thonn.) Roberty ■☆
63541 Buphthalmum sericeum L. f. = Asteriscus sericeus (L. f.) DC. ■☆
63542 Buphthalmum speciossimum Ard.;高牛眼菊;Tall Oxeye ■☆
63543 Buphthalmum speciosum Schreb. = Telekia speciosa (Schreb.) Baumg. ■☆
63544 Buphthalmum spinosum L. = Pallenis spinosa (L.) Cass. ■☆
63545 Buplerum Raf. = Bupleurum L. ●■
63546 Bupleuraceae Bercht. et J. Presl = Apiaceae Lindl. (保留科名)●■
63547 Bupleuraceae Bercht. et J. Presl = Umbelliferae Juss. (保留科名)■●
63548 Bupleuraceae Martinov = Apiaceae Lindl. (保留科名)●■
63549 Bupleuraceae Martinov = Bupleuraceae Bercht. et J. Presl ●■
63550 Bupleuraceae Martinov = Umbelliferae Juss. (保留科名)■●
63551 Bupleuroides Moench = Phyllis L. ●☆
63552 Bupleurum Ehrh. = Bupleurum falcatum L. ●■
63553 Bupleurum L. (1753);柴胡属;Hare's Ear, Hare's-ear, Thoroughwax, Thorowax ●■
63554 Bupleurum abchasicum Manden.;阿伯哈斯柴胡■☆
63555 Bupleurum acerosum E. Mey. = Anginon swellendamense (Eckl. et Zeyh.) B. L. Burtt ■☆
63556 Bupleurum aciphyllum Webb et Berthel. = Bupleurum salicifolium R. Br. subsp. aciphyllum (Parl.) Sunding et G. Kunkel ■☆
63557 Bupleurum affine Sadler;亲缘柴胡■☆
63558 Bupleurum ajanense (Regel) Krasnob. ex T. Yamaz.;阿扬湾柴胡■☆
63559 Bupleurum alatum R. H. Shan et M. L. Sheh;翅果柴胡;Wingfruit Thorowax ■
63560 Bupleurum album Maire;白柴胡■☆
63561 Bupleurum angolense C. Norman = Heteromorpha gossweileri (C. Norman) C. Norman ●☆
63562 Bupleurum angustissimum (Franch.) Kitag.;线叶柴胡(筲柴胡,细叶柴胡);Linearleaf Thorowax ■
63563 Bupleurum antonii Maire;安东柴胡■☆
63564 Bupleurum asperuloides Heldr.;车叶草柴胡■☆
63565 Bupleurum atlanticum Murb.;大西洋柴胡■☆
63566 Bupleurum atlanticum Murb. subsp. algeriense Cauwet et Carb.;阿尔及利亚柴胡■☆
63567 Bupleurum atlanticum Murb. subsp. mairei (Panel. et Vindt) Cauwet et Carb.;迈雷柴胡■☆

63568 Bupleurum atropurpureum (C. Y. Wu) C. Y. Wu. = Bupleurum candollei Wall. ex DC. var. atropurpureum C. Y. Wu ■

63569 Bupleurum atropurpureum C. Y. Wu = Bupleurum candollei Wall. ex DC. var. atropurpureum C. Y. Wu ■

63570 Bupleurum atroviolaceum (O. E. Schulz) Nasir;暗堇色柴胡■☆

63571 Bupleurum aureum (Hoffm.) Fisch. ex Spreng.;金黄柴胡(穿叶柴胡,贯叶柴胡,黄金柴胡,金色柴胡);Goldenyellow Thorowax ■

63572 Bupleurum aureum Fisch. var. breviivolucratum Trautv.;短苞金黄柴胡(短苞黄金柴胡);Shortbract Thorowax, Shortinvolucre Goldenyellow Thorowax ■

63573 Bupleurum balansae Boiss. et Reut.;巴拉柴胡;Small Hare's-ear ■☆

63574 Bupleurum balansae Boiss. et Reut. var. lazari Maire et Sennen = Bupleurum balansae Boiss. et Reut. ■☆

63575 Bupleurum balansae Boiss. et Reut. var. longiradiatum Faure et Maire = Bupleurum balansae Boiss. et Reut. ■☆

63576 Bupleurum balansae Boiss. et Reut. var. mauritii Maire et Sennen = Bupleurum balansae Boiss. et Reut. ■☆

63577 Bupleurum balansae Boiss. et Reut. var. sessile Clary = Bupleurum balansae Boiss. et Reut. ■☆

63578 Bupleurum benoistii Litard. et Maire;本诺柴胡■☆

63579 Bupleurum bicaule Helm;锥叶柴胡(红柴胡,双茎柴胡);Acicular Thorowax ■

63580 Bupleurum bicaule Helm f. latifolium (Y. C. Chu) Y. C. Chu = Bupleurum bicaule Helm var. latifolium Y. C. Chu ■

63581 Bupleurum bicaule Helm var. latifolium Y. C. Chu;呼玛柴胡■

63582 Bupleurum boissieri Post;布瓦西耶柴胡■☆

63583 Bupleurum boissieuanum H. Wolff;紫花阔叶柴胡■

63584 Bupleurum brachiatum K. Koch ex Boiss.;枝状柴胡■☆

63585 Bupleurum breviradiatum Regel = Bupleurum komarovianum Lincz. ■

63586 Bupleurum canaliculatum Diels.;具沟柴胡■☆

63587 Bupleurum candollei Franch. = Bupleurum petiolatum Franch. var. frachetii H. Boissieu ■

63588 Bupleurum candollei Wall. ex DC.;川滇柴胡(麦门冬叶柴胡,飘带草,窄叶柴胡,窄叶飘带草);Candolle Thorowax ■

63589 Bupleurum candollei Wall. ex DC. var. atropurpureum C. Y. Wu;紫红川滇柴胡;Deeppurple Candolle Thorowax ■

63590 Bupleurum candollei Wall. ex DC. var. virgatissimum C. Y. Wu;多枝川滇柴胡;Manybranch Candolle Thorowax ■

63591 Bupleurum canescens Schousb.;灰白柴胡■☆

63592 Bupleurum captatum (L. f.) Thunb. = Hermas capitata L. f. ■☆

63593 Bupleurum chaishoui R. H. Shan et M. L. Sheh;柴首;Chaishou Thorowax ■

63594 Bupleurum chinense DC.;北柴胡(柴草,柴胡,茈胡,地熏,黑柴胡,韭叶柴胡,蚂蚱腿,茹草,山菜,山柴胡,山根菜,铁苗柴胡,硬苗柴胡,竹叶柴胡);China Thorowax, Chinese Thorowax ■

63595 Bupleurum chinense DC. f. chiliosciadium (H. Wolff) R. H. Shan et Yin Li;多伞北柴胡;Manyumbell Thorowax ■

63596 Bupleurum chinense DC. f. octoradiatum (Bunge) R. H. Shan et M. L. Sheh;百花山柴胡(八伞柴胡);Whiteflower Thorowax ■

63597 Bupleurum chinense DC. f. pekinense (Franch.) R. H. Shan et Yin Li;北京柴胡;Beijing Thorowax, Peking Thorowax ■

63598 Bupleurum chinense DC. f. vanheurckii (Müll. Arg.) R. H. Shan et Yin Li;烟台柴胡;Vanheurck Thorowax, Yantai Thorowax ■

63599 Bupleurum chinense DC. f. vanheurckii (Müll. Arg.) R. H. Shan et Yin Li = Bupleurum chinense DC. ■

63600 Bupleurum chinense DC. var. komarovianum (O. A. Lincz.) S. L. Liou et Y. Huei Huang = Bupleurum komarovianum Lincz. ■

63601 Bupleurum chinense Franch. = Bupleurum chinense DC. f. pekinense (Franch.) R. H. Shan et Yin Li ■

63602 Bupleurum chinense Franch. = Bupleurum chinense DC. ■

63603 Bupleurum choulettei Pomel = Bupleurum oligactis Boiss. ■☆

63604 Bupleurum ciliatum (L. f.) Thunb. = Hermas ciliata L. f. ■☆

63605 Bupleurum clarkeanum (H. Wolff) Nasir;克拉克柴胡■☆

63606 Bupleurum collinum (Eckl. et Zeyh.) D. Dietr. = Heteromorpha arborescens (Spreng.) Cham. et Schltdl. var. collina (Eckl. et Zeyh.) Sond. ●☆

63607 Bupleurum columnae Guss. = Bupleurum lancifolium Hornem. ■☆

63608 Bupleurum commelynoideum H. Boissieu;紫花鸭跖柴胡(宽苞柴胡,小柴胡,鸭跖草叶柴胡);Purple Dayflowerlike Thorowax, Purpleflower Dayflowerlike Thorowax ■

63609 Bupleurum commelynoideum H. Boissieu var. flaviflorum R. H. Shan et Yin Li;黄花鸭跖柴胡;Yellow Dayflowerlike Thorowax, Yellowflower Dayflowerlike Thorowax ■

63610 Bupleurum commutatum Boiss. et Balansa;多变柴胡■☆

63611 Bupleurum condensatum R. H. Shan et Yin Li;簇生柴胡;Clustered Thorowax, Condensed Thorowax ■

63612 Bupleurum dahuricum Fisch. et C. A. Mey. ex Turcz. = Bupleurum sibiricum Vest ex Roem. et Schult. ■

63613 Bupleurum dalhousianum (C. B. Clarke) Koso-Pol.;匍枝柴胡;Stoloniferous Thorowax ■

63614 Bupleurum dauricum Fisch. et C. A. Mey. ex Turcz. = Bupleurum sibiricum Vest ex Roem. et Schult. ■

63615 Bupleurum densiflorum Rupr.;密花柴胡;Denseflower Thorowax ■

63616 Bupleurum dielsianum H. Wolff;太白柴胡;Diels Thorowax ■

63617 Bupleurum difforme L. = Anginon difforme (L.) B. L. Burtt ■☆

63618 Bupleurum distichophyllum Wight et Arn.;两列叶柴胡;Tworow Thorowax ■

63619 Bupleurum diversifolium Rchb.;叉叶柴胡■☆

63620 Bupleurum dumosum Coss. et Balansa;棘丛柴胡■☆

63621 Bupleurum ecklonianum Gand. = Bupleurum mundii Cham. et Schltdl. ■☆

63622 Bupleurum euphorbioides Nakai;大苞柴胡;Bigbract Thorowax, Largebract Thorowax ■

63623 Bupleurum exaltatum M. Bieb.;新疆柴胡;Sinkiang Thorowax, Xinjiang Thorowax ■

63624 Bupleurum exaltatum M. Bieb. var. linearifolium? = Bupleurum linearifolium DC. ■☆

63625 Bupleurum falcatum L.;镰形柴胡(镰刀叶柴胡);Falcate Hare's-ear, Sickle Hare's Ear, Sickle-leaved Hare's-ear ■☆

63626 Bupleurum falcatum L. f. angustissimum (Franch.) C. P'ei et R. H. Shan = Bupleurum angustissimum (Franch.) Kitag. ■

63627 Bupleurum falcatum L. f. ensifolium H. Wolff = Bupleurum chinense DC. ■

63628 Bupleurum falcatum L. f. stenophyllum (H. Wolff) P. K. Mukh. et B. D. Naithani = Bupleurum marginatum Wall. ex DC. var. stenophyllum (H. Wolff) R. H. Shan et Yin Li ■

63629 Bupleurum falcatum L. subf. angustissimum (Franch.) H. Wolff = Bupleurum angustissimum (Franch.) Kitag. ■

63630 Bupleurum falcatum L. subsp. eufalcatum? = Bupleurum gracillimum Klotzsch ■

63631 Bupleurum falcatum L. subsp. eufalcatum? var. gracillimum H. Wolff? = Bupleurum gracillimum Klotzsch ■

63632 Bupleurum falcatum L. subsp. eufalcatum? var. scorzonerifolium (Willd.) H. Wolff = Bupleurum scorzonerifolium Willd. ■

63633 Bupleurum falcatum L. subsp. eufalcatum? var. scorzonerifolium (Willd.) H. Wolff f. ensifolium H. Wolff = Bupleurum chinense DC. ■

63634 Bupleurum falcatum L. subsp. eufalcatum? var. scorzonerifolium (Willd.) H. Wolff f. ensifolium H. Wolff subf. angustissimum H. Wolff = Bupleurum angustissimum (Franch.) Kitag. ■

63635 Bupleurum falcatum L. subsp. exaltatum? = Bupleurum exaltatum M. Bieb. ■

63636 Bupleurum falcatum L. subsp. exaltatum? var. euexaltatum H. Wolff = Bupleurum exaltatum M. Bieb. ■

63637 Bupleurum falcatum L. subsp. flexuosum Koso-Pol. = Bupleurum krylovianum Schischk. ex Krylov ■

63638 Bupleurum falcatum L. subsp. komarovianum (O. A. Lincz.) Vorosch. = Bupleurum komarovianum Lincz. ■

63639 Bupleurum falcatum L. subsp. marginatum (Wall. ex DC.) Clarke ex H. Wolff = Bupleurum marginatum Wall. ex DC. ■

63640 Bupleurum falcatum L. subsp. marginatum (Wall. ex DC.) H. Wolff;神荞(膜缘柴胡,竹叶柴胡,竹叶防风,紫柴胡)■

63641 Bupleurum falcatum L. subsp. marginatum (Wall. ex DC.) H. Wolff = Bupleurum marginatum Wall. ex DC. ■

63642 Bupleurum falcatum L. subsp. scorzonerifolium (Willd.) Koso-Pol. = Bupleurum scorzonerifolium Willd. ■

63643 Bupleurum falcatum L. var. africanum P. J. Bergius = Itasina filifolia (Thunb.) Raf. ■☆

63644 Bupleurum falcatum L. var. angustissimum Franch. = Bupleurum angustissimum (Franch.) Kitag. ■

63645 Bupleurum falcatum L. var. bicaule (Helm) H. Wolff = Bupleurum bicaule Helm ■

63646 Bupleurum falcatum L. var. bicaule H. Wolff = Bupleurum bicaule Helm ■

63647 Bupleurum falcatum L. var. chilioschiadium H. Wolff = Bupleurum chinense DC. f. chiliosciadium (H. Wolff) R. H. Shan et Yin Li ■

63648 Bupleurum falcatum L. var. euexaltatum H. Wolff = Bupleurum exaltatum M. Bieb. ■

63649 Bupleurum falcatum L. var. exaltatum = Bupleurum exaltatum M. Bieb. ■

63650 Bupleurum falcatum L. var. gracillimum (Klotzsch) H. Wolff = Bupleurum gracillimum Klotzsch ■

63651 Bupleurum falcatum L. var. komarowii Koso-Pol. = Bupleurum scorzonerifolium Willd. ■

63652 Bupleurum falcatum L. var. linearifolium? = Bupleurum exaltatum M. Bieb. ■

63653 Bupleurum falcatum L. var. marginatum (Wall. ex DC.) C. B. Clarke = Bupleurum marginatum Wall. ex DC. ■

63654 Bupleurum falcatum L. var. marginatum (Wall.) Clarke = Bupleurum marginatum Wall. ex DC. ■

63655 Bupleurum falcatum L. var. scorzonerifolium (Willd.) Ledeb. = Bupleurum scorzonerifolium Willd. ■

63656 Bupleurum falcatum L. var. scorzonerifolium (Willd.) Ledeb. f. angustissimum (Franch.) H. Wolff = Bupleurum scorzonerifolium Willd. ■

63657 Bupleurum falcatum L. var. stenophyllum H. Wolff = Bupleurum marginatum Wall. ex DC. var. stenophyllum (H. Wolff) R. H. Shan et Yin Li ■

63658 Bupleurum falcatum Ledeb. = Bupleurum krylovianum Schischk. ex Krylov ■

63659 Bupleurum falcatum R. H. Shan = Bupleurum chinense DC. ■

63660 Bupleurum faurelii Maire;福雷尔柴胡■☆

63661 Bupleurum faurelii Maire var. tazzertense Quézel = Bupleurum faurelii Maire ■☆

63662 Bupleurum flexuosum Wall. = Bupleurum tenue Buch. -Ham. ex D. Don ■

63663 Bupleurum foliosum DC.;密叶柴胡■☆

63664 Bupleurum fontanesii Guss. = Bupleurum odontites L. ■☆

63665 Bupleurum fontannesii Guss. ex Caruel;方氏柴胡■☆

63666 Bupleurum fruticescens L.;灌木状柴胡■☆

63667 Bupleurum fruticescens L. subsp. spinosum (Gouan) O. Bolòs et Vigo = Bupleurum spinosum Gouan ■☆

63668 Bupleurum fruticosum L.;垂柴胡(灌丛柴胡,灌木柴胡);Shrub Thoroughwax,Shrubby Hare's Ear,Shrubby Hare's-ear ■☆

63669 Bupleurum gerardii All.;吉氏柴胡■☆

63670 Bupleurum gerardii All. var. patens Rchb. = Bupleurum gerardii All. ■☆

63671 Bupleurum gibraltaricum Lam.;直布罗陀柴胡■☆

63672 Bupleurum giganteum (L. f.) Thunb. = Hermas gigantea L. f. ■☆

63673 Bupleurum giraldii (H. Wolff) Koso-Pol. = Bupleurum longicaule Wall. ex DC. var. giraldii H. Wolff ■

63674 Bupleurum giraldii (H. Wolff) Koso-Pol. = Bupleurum petiolatum Franch. var. giraldii H. Wolff ■

63675 Bupleurum glaucum Robill. et Castagne;蓝灰柴胡■☆

63676 Bupleurum glaucum Robill. et Castagne = Bupleurum semicompositum L. ■☆

63677 Bupleurum gracile DC.;美柴胡■☆

63678 Bupleurum gracilescens Rech. = Bupleurum mundii Cham. et Schltdl. ■☆

63679 Bupleurum gracilipes Diels;细柄柴胡;Slenderstalk Thorowax, Smallstipe Thorowax ■

63680 Bupleurum gracillimum Klotzsch;纤细柴胡;Slenderest Thorowax ■

63681 Bupleurum hamiltonii N. P. Balakr.;海氏柴胡(小柴胡)■☆

63682 Bupleurum hamiltonii N. P. Balakr. = Bupleurum tenue Buch. -Ham. ex D. Don ■

63683 Bupleurum hamiltonii N. P. Balakr. var. humile (Franch.) R. H. Shan et M. L. Sheh;矮海氏柴胡(小柴胡)■☆

63684 Bupleurum handelii H. Wolff = Bupleurum rockii H. Wolff ■

63685 Bupleurum heterophyllum Link = Bupleurum lancifolium Hornem. ■☆

63686 Bupleurum himalayense G. Klotz;喜马拉雅柴胡(喜山柴胡);Himalayas Thorowax ■☆

63687 Bupleurum intermedium (Loisel.) Steud. = Bupleurum lancifolium Hornem. ■☆

63688 Bupleurum intermedium (Loisel.) Steud. = Bupleurum subovatum Spreng. ■☆

63689 Bupleurum jeholense Nakai = Bupleurum sibiricum Vest ex Roem. et Schult. var. jeholense (Nakai) Y. C. Chu ex R. H. Shan et Yin Li ■

63690 Bupleurum jeholense Nakai var. latifolium Nakai = Bupleurum sibiricum Vest ex Roem. et Schult. var. jeholense (Nakai) Y. C. Chu ex R. H. Shan et Yin Li ■

63691 Bupleurum jucundum Kurz;异叶柴胡(西藏柴胡)■

63692 Bupleurum kaoi Tang S. Liu, C. Y. Chao et T. I. Chuang;台湾柴胡(高氏柴胡)■

63693 Bupleurum komarovianum Lincz.;长白柴胡(柞柴胡);Komarov Thorowax ■
63694 Bupleurum koso-poljanskyi Grossh.;科索柴胡■☆
63695 Bupleurum krylovianum Schischk. ex Krylov;阿尔泰柴胡(新疆柴胡);Altai Mountain Thorowax,Altai Thorowax ■
63696 Bupleurum kunmingense Y. Li et S. L. Pan;韭叶柴胡(竹叶柴胡);Kunming Thorowax ■
63697 Bupleurum kweichowense R. H. Shan;贵州柴胡;Guizhou Thorowax,Kweichow Thorowax ■
63698 Bupleurum lanceolatum Wall.;披针形柴胡■☆
63699 Bupleurum lancifolium Hornem.;剑叶柴胡;Lanceleaf Thorow Wax ■☆
63700 Bupleurum lancifolium Hornem. = Bupleurum subovatum Spreng. ■☆
63701 Bupleurum lancifolium Hornem. var. heterophyllum (Link) Maire = Bupleurum lancifolium Hornem. ■☆
63702 Bupleurum lancifolium Hornem. var. intermedium Loisel. = Bupleurum lancifolium Hornem. ■☆
63703 Bupleurum lateriflorum Batt.;侧花柴胡■☆
63704 Bupleurum leveillei H. Boissieu = Bupleurum longiradiatum Turcz. ■
63705 Bupleurum linearifolium DC.;丝叶柴胡■☆
63706 Bupleurum longicaule Diels = Bupleurum petiolatum Franch. var. giraldii H. Wolff ■
63707 Bupleurum longicaule Wall. ex DC.;长茎柴胡(大柴胡,金柴胡,飘带草);Longstem Thorowax ■
63708 Bupleurum longicaule Wall. ex DC. = Bupleurum petiolatum Franch. var. frachetii H. Boissieu ■
63709 Bupleurum longicaule Wall. ex DC. = Bupleurum petiolatum Franch. ■
63710 Bupleurum longicaule Wall. ex DC. = Bupleurum smithii H. Wolff ■
63711 Bupleurum longicaule Wall. ex DC. var. amplexicaule C. Y. Wu;抱茎柴胡;Amplexicaul Thorowax,Holdstem Thorowax ■
63712 Bupleurum longicaule Wall. ex DC. var. dalhousieanum C. B. Clarke = Bupleurum dalhousianum (C. B. Clarke) Koso-Pol. ■
63713 Bupleurum longicaule Wall. ex DC. var. franchetii H. Boissieu;空心长茎柴胡(长茎柴胡,空心柴胡,银柴胡,竹叶柴胡);Hollowstem Thorowax ■
63714 Bupleurum longicaule Wall. ex DC. var. giraldii H. Wolff;秦岭柴胡(竹叶柴胡);Qinling Thorowax,Tsinling Mountain Thorowax ■
63715 Bupleurum longicaule Wall. ex DC. var. himalayense (G. Klotz) C. B. Clarke = Bupleurum himalayense G. Klotz ■☆
63716 Bupleurum longicaule Wall. ex DC. var. himalayense?;喜马拉雅长茎柴胡■☆
63717 Bupleurum longicaule Wall. ex DC. var. ramosum?;分枝长茎柴胡■☆
63718 Bupleurum longicaule Wall. ex DC. var. strictum C. B. Clarke;坚挺柴胡;Strict Thorowax ■
63719 Bupleurum longicaule Wall. ex DC. var. strictum C. B. Clarke = Bupleurum longicaule Wall. ex DC. ■
63720 Bupleurum longicaule Wall. ex DC. var. tibetanicum H. Wolff = Bupleurum petiolatum Franch. ■
63721 Bupleurum longicaule Wall. ex DC. var. tibetanicum H. Wolff f. stenophyllum (H. Wolff) H. J. Chowdhery et Wadhwa = Bupleurum petiolatum Franch. ■
63722 Bupleurum longifolium L.;长叶柴胡■☆
63723 Bupleurum longifolium L. subvar. breviinvolucratum Trautv. ex H. Wolff = Bupleurum aureum Fisch. var. breviivolucratum Trautv. ■
63724 Bupleurum longifolium Turcz. var. aureum (Fisch. ex Hoffm.) H. Wolff = Bupleurum aureum (Hoffm.) Fisch. ex Spreng. ■
63725 Bupleurum longifolium Turcz. var. aureum (Fisch.) H. Wolff = Bupleurum aureum (Hoffm.) Fisch. ex Spreng. ■
63726 Bupleurum longifolium Turcz. var. aureum (Fisch.) H. Wolff subvar. breviinvolucratum Trautv. ex H. Wolff = Bupleurum aureum Fisch. var. breviivolucratum Trautv. ■
63727 Bupleurum longiinvolucratum Krylov;长总苞柴胡■☆
63728 Bupleurum longiradiatum Turcz.;大叶柴胡;Bigleaf Thorowax ■
63729 Bupleurum longiradiatum Turcz. f. australe R. H. Shan et Yin Li;南方大叶柴胡;S. China Bigleaf Thorowax,South Bigleaf Thorowax ■
63730 Bupleurum longiradiatum Turcz. f. elatius Koso-Pol. = Bupleurum longiradiatum Turcz. var. elatius (Koso-Pol.) Kitag. ■☆
63731 Bupleurum longiradiatum Turcz. f. leveillei (Boiss.) Kitag. = Bupleurum longiradiatum Turcz. ■
63732 Bupleurum longiradiatum Turcz. f. sachalinense (F. Schmidt) Nakai = Bupleurum longiradiatum Turcz. var. sachalinense (F. Schmidt) H. Boissieu ■☆
63733 Bupleurum longiradiatum Turcz. subsp. sachalinense (F. Schmidt) Kitag. = Bupleurum longiradiatum Turcz. var. sachalinense (F. Schmidt) H. Boissieu ■☆
63734 Bupleurum longiradiatum Turcz. subsp. shikotanense (M. Hiroe) Vorosch. = Bupleurum longiradiatum Turcz. var. shikotanense (M. Hiroe) Ohwi ■☆
63735 Bupleurum longiradiatum Turcz. var. breviradiatum F. Schmidt;短伞大叶柴胡;Shortradiate Bigleaf Thorowax ■
63736 Bupleurum longiradiatum Turcz. var. elatius (Koso-Pol.) Kitag.;挺举柴胡■☆
63737 Bupleurum longiradiatum Turcz. var. porphyranthum R. H. Shan et Yin Li;紫花大叶柴胡;Purpleflower Bigleaf Thorowax ■
63738 Bupleurum longiradiatum Turcz. var. porphyranthum R. H. Shan et Yin Li = Bupleurum boissieuanum H. Wolff ■
63739 Bupleurum longiradiatum Turcz. var. pseudonipponicum Kitag.;假日本柴胡■☆
63740 Bupleurum longiradiatum Turcz. var. sachalinense (F. Schmidt) H. Boissieu;库页大叶柴胡■☆
63741 Bupleurum longiradiatum Turcz. var. shikotanense (M. Hiroe) Ohwi;色丹柴胡■☆
63742 Bupleurum luxiense Y. Li et S. L. Pan;泸西柴胡(竹叶柴胡);Luxi Thorowax ■
63743 Bupleurum mairei Panelatti et Vindt = Bupleurum atlanticum Murb. subsp. mairei (Panel. et Vindt) Cauwet et Carb. ■☆
63744 Bupleurum malconense R. H. Shan et Yin Li;马尔康柴胡(马尾柴胡,竹叶柴胡);Maerkang Thorowax,Malcon Thorowax ■
63745 Bupleurum marginatum Wall. ex DC.;竹叶柴胡(膜缘柴胡,神荠,竹叶防风,紫柴胡);Bambooleaf Thorowax ■
63746 Bupleurum marginatum Wall. ex DC. = Bupleurum falcatum L. subsp. marginatum (Wall. ex DC.) H. Wolff ■
63747 Bupleurum marginatum Wall. ex DC. var. minutum X. F. Zhang;小竹叶柴胡;Small Bambooleaf Thorowax ■
63748 Bupleurum marginatum Wall. ex DC. var. stenophyllum (H. Wolff) R. H. Shan et Yin Li;窄竹叶柴胡(西藏柴胡);Narrow Bambooleaf Thorowax,Narrowleaf Thorowax ■
63749 Bupleurum marschallianum C. A. Mey.;马尔沙柴胡■☆
63750 Bupleurum martjanovii Krylov;马尔特柴胡■☆
63751 Bupleurum mauritanicum Batt. = Bupleurum oligactis Boiss. ■☆

63752 Bupleurum melilense Pau = Bupleurum balansae Boiss. et Reut. ■☆

63753 Bupleurum mesatlanticum Litard. et Maire;梅萨柴胡■☆

63754 Bupleurum mesatlanticum Litard. et Maire var. plurivittatum Maire = Bupleurum mesatlanticum Litard. et Maire ■☆

63755 Bupleurum mesatlanticum Litard. et Maire var. univittatum Maire = Bupleurum mesatlanticum Litard. et Maire ■☆

63756 Bupleurum microcephalum Diels;马尾柴胡(线柴胡,线叶柴胡,竹叶柴胡);Horsetail Thorowax ■

63757 Bupleurum montanum Coss.;山地柴胡■☆

63758 Bupleurum montanum Coss. var. atlanticum (Murb.) Pau et Font Quer = Bupleurum atlanticum Murb. ■☆

63759 Bupleurum montanum Coss. var. baboranum (Debeaux et E. Rev.) Maire = Bupleurum montanum Coss. ■☆

63760 Bupleurum montanum Coss. var. oblongifolium (Ball) Maire = Bupleurum montanum Coss. ■☆

63761 Bupleurum multinerve DC.;多脉柴胡;Multinerved Thorowax ■☆

63762 Bupleurum mundii Cham. et Schltdl.;蒙德柴胡■☆

63763 Bupleurum nanum Poir.;低矮柴胡■☆

63764 Bupleurum nipponicum Koso-pol.;本州柴胡(白山柴胡)■☆

63765 Bupleurum nipponicum Koso-Pol. f. stenolepis Kitag.;窄鳞本州柴胡■☆

63766 Bupleurum nipponicum Koso-Pol. var. yesoense (Nakai ex H. Hara) H. Hara;北海道柴胡■☆

63767 Bupleurum nodiflorum Sibth. et Sm.;节花柴胡■☆

63768 Bupleurum nodiflorum Sibth. et Sm. subsp. nanum (Poir.) Jafri = Bupleurum nanum Poir. ■☆

63769 Bupleurum nordmannianum Ledeb.;诺德柴胡■☆

63770 Bupleurum oblongifolium Ball = Bupleurum montanum Coss. ■☆

63771 Bupleurum octoradiatum Bunge = Bupleurum chinense DC. f. octoradiatum (Bunge) R. H. Shan et M. L. Sheh ■

63772 Bupleurum odontites L.;狭叶柴胡;Narrowleaf Thorow Wax ■☆

63773 Bupleurum oligactis Boiss.;寡柴胡■☆

63774 Bupleurum oligactis Boiss. var. choulettei (Pomel) Panelatti = Bupleurum oligactis Boiss. ■☆

63775 Bupleurum opacum Lange;狭柴胡;Narrow Hare's Ear ■☆

63776 Bupleurum pauciradiatum Fenzl ex Boiss.;灯心柴胡■☆

63777 Bupleurum pekinense Franch. = Bupleurum chinense DC. f. pekinense (Franch.) R. H. Shan et Yin Li ■

63778 Bupleurum petiolatum Franch.;有柄柴胡;Petiolulate Thorowax ■

63779 Bupleurum petiolatum Franch. var. amplexicaule C. Y. Wu;抱茎有柄柴胡■

63780 Bupleurum petiolatum Franch. var. frachetii H. Boissieu;空心有柄柴胡(空心柴胡)■

63781 Bupleurum petiolatum Franch. var. giraldii H. Wolff;秦岭有柄柴胡(秦岭柴胡)■

63782 Bupleurum petiolatum Franch. var. strictum C. B. Clarke = Bupleurum longicaule Wall. ex DC. var. strictum C. B. Clarke ■

63783 Bupleurum petiolatum Franch. var. tenerum R. H. Shan et Yin Li;细茎有柄柴胡;Slenderstem Thorowax ■

63784 Bupleurum plantagineum Desf.;车前柴胡■☆

63785 Bupleurum polyclonum Y. Li et S. L. Pan;多枝柴胡;Multibranch Thorowax ■

63786 Bupleurum polymorphum Albov;多形柴胡■☆

63787 Bupleurum polyphyllum Ledeb.;多叶柴胡■☆

63788 Bupleurum procumbens Desf. = Bupleurum lancifolium Hornem. ■☆

63789 Bupleurum protractum Link et Hoffm. = Bupleurum rotundifolium L. ■☆

63790 Bupleurum protractum Link et Hoffm. subsp. heterophyllum (Link) Murb. = Bupleurum lancifolium Hornem. ■☆

63791 Bupleurum protractum Link et Hoffm. var. heterophyllum (Link) Boiss. = Bupleurum lancifolium Hornem. ■☆

63792 Bupleurum pusillum Krylov;短茎柴胡(矮小柴胡);Shortstem Thorowax ■

63793 Bupleurum qinghaiense Yin Li et J. X. Guo;青海柴胡■

63794 Bupleurum quadriradiatum Kitag. = Bupleurum longiradiatum Turcz. ■

63795 Bupleurum quinquedentatum (L. f.) Thunb. = Hermas quinquedentata L. f. ■☆

63796 Bupleurum ranunculoides L.;毛茛柴胡■☆

63797 Bupleurum ranunculoides L. var. triradiatum (Adams ex Hoffm.) Regel = Bupleurum triradiatum Adams ex Hoffm. ■

63798 Bupleurum ranunculoides L. var. triradiatum Regel = Bupleurum triradiatum Adams ex Hoffm. ■

63799 Bupleurum rigescens Maire et Sennen = Bupleurum balansae Boiss. et Reut. ■☆

63800 Bupleurum rigidum L.;硬柴胡■☆

63801 Bupleurum rigidum L. subsp. paniculatum (Brot.) H. Wolff;圆锥柴胡■☆

63802 Bupleurum rigidum L. var. angustifolium Lange = Bupleurum rigidum L. ■☆

63803 Bupleurum rockii H. Wolff;丽江柴胡(丽江小柴胡,银柴胡);Handel Thorowax, Rock Thorowax ■

63804 Bupleurum rossicum Woronow;俄罗斯柴胡;Russia Thorowax ■☆

63805 Bupleurum rotundifolium L.;圆叶柴胡;Book Leaf, Hare's Ear, Roundleaf Thorowax, Round-leaved Hare's-ear, Round-leaved Thoroughwax, Thoroughwax, Thorough-wax, Thorowax, Thorow-wax, Throw-wax ■☆

63806 Bupleurum rupestre Edgew. = Bupleurum longicaule Wall. ex DC. ■

63807 Bupleurum sachalinense F. Schmidt;库页柴胡;Sachalin Thorowax ■☆

63808 Bupleurum sachalinense F. Schmidt = Bupleurum longiradiatum Turcz. var. breviradiatum F. Schmidt ■

63809 Bupleurum salicifolium R. Br.;柳叶柴胡;Hinojo ■☆

63810 Bupleurum salicifolium R. Br. subsp. aciphyllum (Parl.) Sunding et G. Kunkel;尖柳叶柴胡■☆

63811 Bupleurum salicifolium R. Br. var. robustum (Burch.) Cauwet et Sunding = Bupleurum salicifolium R. Br. ■☆

63812 Bupleurum salicifolium Sol. ex Lowe = Bupleurum salicifolium R. Br.

63813 Bupleurum scorzonerifolium Willd.;红柴胡(柴草,此胡,地熏,韭叶柴胡,南柴胡,茹草,软柴胡,软苗柴胡,山菜,细叶柴胡,狭叶柴胡,香柴胡,小柴胡,芽胡);Red Thorowax ■

63814 Bupleurum scorzonerifolium Willd. f. ensifolium (H. Wolff) Nakai = Bupleurum chinense DC. ■

63815 Bupleurum scorzonerifolium Willd. f. longiradiatam R. H. Shan et Yin Li;长伞红柴胡;Longradiate Red Thorowax, Longumbrella Red Thorowax ■

63816 Bupleurum scorzonerifolium Willd. f. pauciflorum R. H. Shan et Yin Li;少花红柴胡;Few Flower Thorowax, Fewflower Red Thorowax ■

63817 Bupleurum scorzonerifolium Willd. subsp. angustissimum (Franch.) Kitag. = Bupleurum angustissimum (Franch.) Kitag. ■

63818 Bupleurum scorzonerifolium Willd. subsp. angustissimum Kitag.

= Bupleurum angustissimum (Franch.) Kitag. ■
63819 Bupleurum scorzonerifolium Willd. var. angustissimum (Franch.) Y. Huei Huang = Bupleurum angustissimum (Franch.) Kitag. ■
63820 Bupleurum scorzonerifolium Willd. var. obvallatum Nakai;被包红柴胡■☆
63821 Bupleurum scorzonerifolium Willd. var. stenophyllum Nakai;曲茎柴胡■☆
63822 Bupleurum scorzonerifolium Willd. var. stenophyllum Nakai = Bupleurum scorzonerifolium Willd. ■
63823 Bupleurum scorzonerifolium Willd. var. stenophyllum Nakai f. kiusianum (Kitag.) Kitag. = Bupleurum chinense DC. ■
63824 Bupleurum semicompositum L. var. glaucum (Robill. et Castagne) Batt. = Bupleurum semicompositum L. ■☆
63825 Bupleurum semicompositum L. var. pseudodontites (Rouy) Batt. = Bupleurum semicompositum L. ■☆
63826 Bupleurum semicompostium L.;半混合柴胡;Narrow-leaved Hare's-ear ■☆
63827 Bupleurum sibiricum Vest ex Roem. et Schult.;兴安柴胡;Siberia Thorowax,Siberian Thorowax ■
63828 Bupleurum sibiricum Vest ex Roem. et Schult. var. jeholense (Nakai) Y. C. Chu ex R. H. Shan et Yin Li;雾灵柴胡;Johol Siberian Thorowax,Wuling Thorowax ■
63829 Bupleurum sichuanense S. L. Pan et P. S. Hsu = Buplenurum malconense R. H. Shan et Yin Li ■
63830 Bupleurum sinensium Gand. = Bupleurum scorzonerifolium Willd. ■
63831 Bupleurum smithii H. Wolff;黑柴胡;Black Thorowax ■
63832 Bupleurum smithii H. Wolff var. auriculatum R. H. Shan et Yin Li;耳叶黑柴胡(黑耳叶柴胡);Earleaf Thorowax, Earshapeleaf Thorowax ■
63833 Bupleurum smithii H. Wolff var. parvifolium R. H. Shan et Yin Li;小叶黑柴胡;Littleleaf Black Thorowax ■
63834 Bupleurum sosnovskyi Manden.;索斯诺夫柴胡■☆
63835 Bupleurum spinosum Gouan;具刺柴胡■☆
63836 Bupleurum spinosum Gouan var. lucidum Batt. = Bupleurum spinosum Gouan ■☆
63837 Bupleurum spinosum Gouan var. mauritanicum Cauwet = Bupleurum spinosum Gouan ■☆
63838 Bupleurum stellatum Lapeyr.;星形柴胡;Starry Hare's-ear ■☆
63839 Bupleurum stenophyllum (Nakai) Kitag. var. kiusianum Kitag. = Bupleurum chinense DC. ■
63840 Bupleurum stenophyllum (Nakai) Kitag. var. obvallatum (Nakai) Kitag. = Bupleurum scorzonerifolium Willd. var. obvallatum Nakai ■☆
63841 Bupleurum stewartianum Nasir;喜西柴胡(喜马拉雅柴胡);Stewart Thorowax ■
63842 Bupleurum subovatum Spreng.;亚卵形柴胡;False Thorow-wax ■☆
63843 Bupleurum subovatum Spreng. = Bupleurum lancifolium Hornem. ■☆
63844 Bupleurum subovatum Spreng. var. heterophyllum (Link) W. Wight = Bupleurum subovatum Spreng. ■☆
63845 Bupleurum subovatum Spreng. var. longifolium Pamp. = Bupleurum subovatum Spreng. ■☆
63846 Bupleurum subspinosum Maire et Weiller;小刺柴胡■☆
63847 Bupleurum subuniflorum Boiss. et Heldr.;单花柴胡■☆
63848 Bupleurum swatianum Nasir;斯瓦特柴胡■☆
63849 Bupleurum tatudinense A. I. Baranov = Bupleurum euphorbioides Nakai ■
63850 Bupleurum tenue Buch.-Ham. ex D. Don;小柴胡(滇银柴胡,金柴胡,芫荽柴胡,窄叶飘带草,竹叶柴胡);Small Thorowax ■
63851 Bupleurum tenue Buch.-Ham. ex D. Don = Bupleurum hamiltonii N. P. Balakr. ■☆
63852 Bupleurum tenue Buch.-Ham. ex D. Don var. humile Franch.;矮小柴胡;Dwarf Thorowax ■
63853 Bupleurum tenue Buch.-Ham. ex D. Don var. humile Franch. = Bupleurum hamiltonii N. P. Balakr. var. humile (Franch.) R. H. Shan et M. L. Sheh ■☆
63854 Bupleurum tenue Buch.-Ham. ex D. Don var. paucefulcrans C. Y. Wu ex R. H. Shan et Yin Li;三苞小柴胡;Threebract Small Thorowax,Threebract Thorowax ■
63855 Bupleurum tenuissimum L.;细柴胡;Slender Hare's Ear,Slender Hare's-ear,Smallest Hare's Ear ■☆
63856 Bupleurum tenuissimum L. var. columnae (Guss.) Gren. et Godr. = Bupleurum lancifolium Hornem. ■☆
63857 Bupleurum tenuissimum L. var. procumbens (Desf.) K. Richt. = Bupleurum lancifolium Hornem. ■☆
63858 Bupleurum thomsonii C. B. Clarke;托马森柴胡■☆
63859 Bupleurum tianschanicum Freyn;天山柴胡;Tianshan Mountain Thorowax,Tianshan Thorowax ■
63860 Bupleurum togasii Kitag. = Bupleurum chinense DC. ■
63861 Bupleurum trichopodum Boiss. et Spruner;毛足柴胡●☆
63862 Bupleurum trichopodum Boiss. et Spruner var. depauperatum Boiss. = Bupleurum trichopodum Boiss. et Spruner ●☆
63863 Bupleurum trifoliatum H. L. Wendl. = Heteromorpha arborescens (Spreng.) Cham. et Schltdl. var. abyssinica (Hochst. ex A. Rich.) H. Wolff ●☆
63864 Bupleurum triradiatum Adams ex Hoffm.;三辐柴胡;Triradiate Thorowax ■
63865 Bupleurum vankeurckii Müll. Arg. = Bupleurum chinense DC. f. vanheurckii (Müll. Arg.) R. H. Shan et Yin Li ■
63866 Bupleurum vankeurckii Müll. Arg. = Bupleurum chinense DC. ■
63867 Bupleurum villosum L. = Hermas villosa (L.) Thunb. ■☆
63868 Bupleurum wenchuanense R. H. Shan et Yin Li;汶川柴胡(马尾柴胡);Wenchuan Thorowax ■
63869 Bupleurum wittmannii Stev.;维特曼柴胡■☆
63870 Bupleurum woronowii Manden.;沃氏柴胡■☆
63871 Bupleurum yesoense Nakai ex H. Hara = Bupleurum nipponicum Koso-Pol. var. yesoense (Nakai ex H. Hara) H. Hara ■☆
63872 Bupleurum yinchowense R. H. Shan et Yin Li;银州柴胡(红柴胡,红软柴胡,软柴胡,卧银花);Yinchow Thorowax, Yinzhou Thorowax ■
63873 Bupleurum yunnanense Franch.;云南柴胡(滇柴胡,金柴胡,飘带草,竹柴胡);Yunnan Thorowax ■
63874 Buplevrum Raf. = Bupleurum L. ●■
63875 Buprestis Spreng. = Bupleurum L. ●■
63876 Buprestis arborescens Spreng. = Heteromorpha arborescens (Spreng.) Cham. et Schltdl. ●☆
63877 Bupthalmum Neck. = Buphthalmum L. ■
63878 Buraeavia Baill. (1873)(保留属名);布拉大戟属●☆
63879 Buraeavia Baill. (保留属名) = Austrobuxus Miq. ●☆
63880 Buraeavia Baill. (保留属名) = Baloghia Endl. ●■☆
63881 Burasaia Thouars(1806);马岛啤酒藤属(马岛防己属)●☆
63882 Burasaia australis Scott-Elliot;南方马岛啤酒藤●☆
63883 Burasaia congesta Decne.;密集马岛啤酒藤●☆
63884 Burasaia gracilis Decne.;细马岛啤酒藤●☆

63885 Burasaia madagascariensis DC.;马岛啤酒藤●☆
63886 Burbidgea Hook. f. (1879);大萼姜属■☆
63887 Burbidgea schizochaila Hort.;大萼姜■☆
63888 Burbonia Fabr. = Persea Mill. (保留属名)●
63889 Burcarda J. F. Gmel. = Burcardia Schreb. ■●☆
63890 Burcardia Duhamel = Callicarpa L. ●
63891 Burcardia Heist. ex Duhamel(废弃属名) = Burchardia R. Br. (保留属名)■☆
63892 Burcardia Heist. ex Duhamel(废弃属名) = Callicarpa L. ●
63893 Burcardia Neck. ex Raf. = Campomanesia Ruiz et Pav. ●☆
63894 Burcardia Raf. = Campomanesia Ruiz et Pav. ●☆
63895 Burcardia Schreb. = Piriqueta Aubl. ■●☆
63896 Burchardia B. D. Jacks. = Burcardia Duhamel ●
63897 Burchardia B. D. Jacks. = Callicarpa L. ●
63898 Burchardia Neck. = Psidium L. ●
63899 Burchardia R. Br. (1810)(保留属名);球茎草属■☆
63900 Burchardia umbellata R. Br.;球茎草■☆
63901 Burchardiaceae Takht.;球茎草科■☆
63902 Burchardiaceae Takht. = Colchicaceae DC. (保留科名)■
63903 Burchellia R. Br. (1820);布切尔木属;Buffalo Wood, South African Pomegranate ●☆
63904 Burchellia bubalina (L. f.) Sims;布切尔木;Buffalo Wood, South African Pomegranate, Wild Granaat ●☆
63905 Burchellia bubalina Sims = Burchellia bubalina (L. f.) Sims ●☆
63906 Burchellia capensis R. Br. = Burchellia bubalina (L. f.) Sims ●☆
63907 Burckella Pierre(1890);布克榄属●☆
63908 Burckella obovata (G. Forst.) Pierre;布克榄●☆
63909 Burdachia Juss. ex Endl. (1840);巴北木属●☆
63910 Burdachia Mart. ex A. Juss. = Burdachia Juss. ex Endl. ●☆
63911 Burdachia prismatocarpa Mart. ex A. Juss.;巴北木●☆
63912 Bureaua Kuntze = Austrobuxus Miq. ●☆
63913 Bureaua Kuntze = Buraeavia Baill. (保留属名)●☆
63914 Bureava Baill. (废弃属名) = Buraeavia Baill. (保留属名)●☆
63915 Bureava Baill. (废弃属名) = Combretum Loefl. (保留属名)●
63916 Bureava crotonoides Baill. = Combretum micranthum G. Don ●☆
63917 Bureavella Pierre = Lucuma Molina ●
63918 Bureavella Pierre = Pouteria Aubl. ●
63919 Burgesia F. Muell. (1859);伯吉斯豆属●☆
63920 Burgesia F. Muell. = Brachysema R. Br. ●☆
63921 Burgesia homalocIada F. Muell.;伯吉斯豆●☆
63922 Burghartia Scop. = Piriqueta Aubl. ■●☆
63923 Burglaria Wendl. = Ilex L. ●
63924 Burglaria Wendl. ex Steud. = Ilex L. ●
63925 Burgsdorfia Moench = Cunila Mill. ■●
63926 Burgsdorfia Moench = Sideritis L. ■●
63927 Burkartia Crisci(1976);木垫钝柱菊属●☆
63928 Burkartia lanigera (Hook. et Arn.) Crisci;木垫钝柱菊●☆
63929 Burkea Benth. = Burkea Hook. ●☆
63930 Burkea Hook. (1843);伯克豆属(白奇木属,伯克苏木属,布克豆属)●☆
63931 Burkea africana Hook.;非洲伯克豆(伯克苏木,布克豆,非洲布克豆);Rhodeslan Ash, Wild Syringa ●☆
63932 Burkea africana Hook. var. andongensis Oliv. = Burkea africana Hook. ●☆
63933 Burkea africana Hook. var. cordata Welw. ex Oliv. = Burkea africana Hook. ●☆
63934 Burkea caperangau Baill.;伯克豆(布克豆)●☆
63935 Burkhardia Benth. et Hook. f. = Burghartia Scop. ■●☆
63936 Burkhardia Benth. et Hook. f. = Piriqueta Aubl. ■●☆
63937 Burkillanthus Swingle(1939);布尔芸香属●☆
63938 Burkillanthus malaccensis (Ridl.) Swingle;布尔芸香●☆
63939 Burkillia Ridl. = Alloburkillia Whitmore ●☆
63940 Burkillia Ridl. = Burkilliodendron Sastry ●☆
63941 Burkilliodendron Sastry(1969);马来布豆属●☆
63942 Burkilliodendron album (Ridl.) Sastry;马来布豆●☆
63943 Burlemarxia N. L. Menezes et Semir(1991);岩地翡若翠属■☆
63944 Burlemarxia pungens N. L. Menezes et Semir;岩地翡若翠■☆
63945 Burlingtonia Lindl. = Rodriguezia Ruiz et Pav. ■☆
63946 Burmabambus P. C. Keng = Sinarundinaria Nakai ●
63947 Burmabambus P. C. Keng = Yushania P. C. Keng ●
63948 Burmabambus P. C. Keng(1982);缅竹属●
63949 Burmannia L. (1753);水玉簪属;Burmannia ■
63950 Burmannia aphylla Blume;小水玉簪■
63951 Burmannia aptera Schltr. = Burmannia congesta (C. H. Wright) Jonker ■☆
63952 Burmannia bakeri Hochr. = Burmannia madagascariensis Mart. ■☆
63953 Burmannia bicolor Mart. var. africana Ridl. = Burmannia madagascariensis Mart. ■☆
63954 Burmannia bicolor Mart. var. micrantha Engl. et Gilg = Burmannia madagascariensis Mart. ■☆
63955 Burmannia bifida Gagnep. = Burmannia oblonga Ridl. ■
63956 Burmannia biflora L.;二花水玉簪;Twoflower Burmannia ■
63957 Burmannia blanda Gilg = Burmannia madagascariensis Mart. ■☆
63958 Burmannia caillei A. Chev. = Burmannia madagascariensis Mart. ■☆
63959 Burmannia capitata (Walter ex J. F. Gmel.) Mart.;头状水玉簪■☆
63960 Burmannia championii Thwaites;头花水玉簪(日本水玉簪);Champion Burmannia ■
63961 Burmannia chariensis Schltr. = Burmannia madagascariensis Mart. ■☆
63962 Burmannia chinensis Gand.;香港水玉簪;Hongkong Burmannia ■
63963 Burmannia chinensis Gand. = Burmannia coelestis D. Don ■
63964 Burmannia coelestis D. Don;三品一枝花(地沙,米洋参,蛆儿草,少花木,玉簪);Skyblue Burmannia ■
63965 Burmannia congesta (C. H. Wright) Jonker;密集水玉簪■
63966 Burmannia cryptopetala Makino;透明水玉簪;Cryptopetalous Burmannia ■
63967 Burmannia cryptopetala Makino var. daxikangensis Z. Wei et Y. B. Chang;大西坑水玉簪;Daxikeng Burmannia ■
63968 Burmannia cryptopetala Makino var. daxikangensis Z. Wei et Y. B. Chang = Burmannia cryptopetala Makino ■
63969 Burmannia dalzieli Rendle = Burmannia championii Thwaites ■
63970 Burmannia densiflora Schltr.;密花水玉簪■
63971 Burmannia disticha L.;水玉簪(苍山贝母);Common Burmannia ■
63972 Burmannia fadouensis H. Li;石山水玉簪;Fadou Burmannia ■
63973 Burmannia fadouensis H. Li = Burmannia nepalensis (Miers) Hook. f. ■
63974 Burmannia filamentosa D. X. Zhang et R. M. K. Saunders;粤东水玉簪■
63975 Burmannia hunanensis K. M. Liu et C. L. Long = Burmannia championii Thwaites ■
63976 Burmannia inhambanensis Schltr. = Burmannia madagascariensis Mart. ■☆

63977 Burmannia itoana Makino;纤草水玉簪(纤草,紫水玉簪);Ito Burmannia ■
63978 Burmannia latialata Hua = Burmannia madagascariensis Mart. ■☆
63979 Burmannia letestui Schltr. = Burmannia madagascariensis Mart. ■☆
63980 Burmannia liberica Engl. = Burmannia madagascariensis Mart. ■☆
63981 Burmannia liukiuensis Hayata;琉球水玉簪■
63982 Burmannia liukiuensis Hayata = Burmannia nepalensis (Miers) Hook. f. ■
63983 Burmannia madagascariensis Baker = Burmannia madagascariensis Mart. ■☆
63984 Burmannia madagascariensis Mart.;马岛水玉簪■☆
63985 Burmannia nana (Fukuy. et T. Suzuki) Tuyama = Burmannia aphylla Blume ■
63986 Burmannia nana Fukuy. et T. Suzuki = Gymnosiphon aphyllus Blume ■
63987 Burmannia nepalensis (Miers) Hook. f.;宽翅水玉簪;Nepal Burmannia ■
63988 Burmannia oblonga Ridl.;裂萼水玉簪;Oblong Burmannia ■
63989 Burmannia obscurata Schltr.;隐涩水玉簪■☆
63990 Burmannia paniculata Schult. f. = Burmannia madagascariensis Mart. ■☆
63991 Burmannia pingbianensis H. Li;屏边水玉簪;Pingbian Burmannia ■
63992 Burmannia pingbienensis H. Li = Burmannia itoana Makino ■
63993 Burmannia pulcherrima A. Chev.;美丽水玉簪■
63994 Burmannia pusilla (Wall. ex Miers) Thwaites;微小水玉簪■
63995 Burmannia pusilla (Wall. ex Miers) Thwaites var. hongkongensis Jonker = Burmannia chinensis Gand. ■
63996 Burmannia takeoi Hayata = Burmannia itoana Makino ■
63997 Burmannia tisserantii Schltr.;蒂斯朗特水玉簪■
63998 Burmannia wallichii (Miers) Hook. f.;亭立水玉簪(亭立);Wallich Burmannia ■
63999 Burmannia welwitschii Schltr. = Burmannia madagascariensis Mart. ■☆
64000 Burmanniaceae Blume (1827)(保留科名);水玉簪科;Burmannia Family ■
64001 Burmeistera H. Karst. et Triana = Burmeistera Triana ●■☆
64002 Burmeistera Triana(1855);南美桔梗属●■☆
64003 Burmeistera andersonii Jeppesen;南美桔梗■☆
64004 Burmeistera aspera Wimm.;粗糙南美桔梗■☆
64005 Burmeistera loejtnantii Jeppesen;劳氏南美桔梗■☆
64006 Burmeistera oblongifolia Wimm.;矩圆叶南美桔梗■☆
64007 Burmeistera rubrosepala (Wimm.) Wimm.;红萼南美桔梗■☆
64008 Burnatastrum Briq. = Plectranthus L'Hér.(保留属名)●■
64009 Burnatastrum lanceolatum (Bojer ex Benth.) Briq. = Plectranthus lanceolatus Bojer ex Benth. ●☆
64010 Burnatastrum lavanduloides (Baker) Briq. = Plectranthus lanceolatus Bojer ex Benth. ●☆
64011 Burnatia Micheli(1881);东非泽泻属(柏那特泽泻属,比尔纳泽泻属)■☆
64012 Burnatia alismatoides Peter = Burnatia enneandra Micheli ■☆
64013 Burnatia alismatoides Peter var. elliptica? = Burnatia enneandra Micheli ■☆
64014 Burnatia enneandra Micheli;九蕊东非泽泻(比尔纳泽泻,九蕊比尔纳泽泻)■☆
64015 Burnatia enneandra Micheli var. linearis Peter = Burnatia enneandra Micheli ■☆
64016 Burnatia oblonga Peter = Burnatia enneandra Micheli ■☆
64017 Burnettia Lindl.(1840);塔斯马尼亚兰属■☆
64018 Burnettia cuneata Lindl.;塔斯马尼亚兰;Lizard Orchid ■☆
64019 Burneya Cham. et Schltdl.(废弃属名) = Timonius DC.(保留属名) + Bobea Gaudich. ●
64020 Burneya Cham. et Schltdl.(废弃属名) = Timonius DC.(保留属名)●
64021 Burnsbaloghia Szlach.(1991);伯恩兰属■☆
64022 Burnsbaloghia diaphana (Lindl.) Szlach.;伯恩兰■☆
64023 Burragea Donn. Sm. et Rose = Gongylocarpus Schltdl. et Cham. ■☆
64024 Burretiodendron Rehder(1936);柄翅果属;Burretiodendron ●
64025 Burretiodendron combretoides Chun et F. C. How = Craigia yunnanensis W. W. Sm. et Evans ●◇
64026 Burretiodendron esquirolii (H. Lév.) Rehder;柄翅果;Esquirol Burretiodendron ●◇
64027 Burretiodendron hsienmu Chun et F. C. How = Excentrodendron hsienmu (Chun et F. C. How) Hung T. Chang et R. H. Miao ●◇
64028 Burretiodendron hsienmu Chun et F. C. How = Excentrodendron tonkinense (A. Chev.) Hung T. Chang et R. H. Miao ●
64029 Burretiodendron kydiifolium Y. C. Hsu et Zhuge;元江柄翅果●
64030 Burretiodendron longistipitatum R. H. Miao = Burretiodendron esquirolii (H. Lév.) Rehder ●◇
64031 Burretiodendron longistipitatum R. H. Miao ex Hung T. Chang;长柄翅果;Longistiped Burretiodendron, Longstipe Burretiodendron ●
64032 Burretiodendron obconicum Chun et F. C. How = Excentrodendron obconicum (Chun et F. C. How) Hung T. Chang et R. H. Miao ●
64033 Burretiodendron tonkinense (A. Chev.) Kosterm. = Excentrodendron tonkinense (A. Chev.) Hung T. Chang et R. H. Miao ●
64034 Burretiodendron tonkinense Kosterm. = Excentrodendron tonkinense (A. Chev.) Hung T. Chang et R. H. Miao ●
64035 Burretiodendron yunnanense (Sm. et Evans) Kosterm. = Craigia yunnanensis W. W. Sm. et Evans ●◇
64036 Burretiodendron yunnanense Kosterm. = Craigia yunnanensis W. W. Sm. et Evans ●◇
64037 Burretiokentia Pic. Serm.(1955);裂柄椰属(棱籽椰属,裂柄棕属)●☆
64038 Burretiokentia vieillardi (Brongn. et Gris) Pic. Serm.;裂柄椰●☆
64039 Burriela Baill. = Burrielia DC. ■☆
64040 Burrielia DC. = Lasthenia Cass. ■☆
64041 Burrielia chrysostoma (Fisch. et C. A. Mey.) Torr. et A. Gray var. macrantha A. Gray = Lasthenia californica DC. ex Lindl. subsp. macrantha (A. Gray) R. Chan ■☆
64042 Burrielia gracilis DC. = Lasthenia gracilis (DC.) Greene ■☆
64043 Burrielia lanosa A. Gray = Eriophyllum lanosum (A. Gray) A. Gray ■☆
64044 Burrielia leptalea A. Gray = Lasthenia leptalea (A. Gray) Ornduff ■☆
64045 Burrielia maritima A. Gray = Lasthenia maritima (A. Gray) M. C. Vasey ■☆
64046 Burrielia microglossa DC. = Lasthenia microglossa (DC.) Greene ■☆
64047 Burrielia nivea D. C. Eaton = Eatonella nivea (D. C. Eaton) A. Gray ■☆
64048 Burrielia platycarpha A. Gray = Lasthenia platycarpha (A. Gray)

Greene ■☆

64049 Burriellia Engl. = Burrielia DC. ■☆

64050 Burroughsia Moldenke = Lippia L. ●■☆

64051 Bursa Boehm. = Capsella Medik.(保留属名)■

64052 Bursa Weber = Capsella Medik.(保留属名)■

64053 Bursa bursa-pastoris (L.) Britton = Capsella bursa-pastoris (L.) Medik. ■

64054 Bursa bursa-pastoris (L.) Britton var. bifida? = Capsella bursa-pastoris (L.) Medik. ■

64055 Bursa djurdjurae Shull = Capsella bursa-pastoris (L.) Medik. ■

64056 Bursa occidentalis Shull = Capsella bursa-pastoris (L.) Medik. ■

64057 Bursa occidentalis Shull subsp. mairei? = Capsella bursa-pastoris (L.) Medik. ■

64058 Bursa rubella (Reut.) Druce = Capsella bursa-pastoris (L.) Medik. ■

64059 Bursaia Steud. = Burasaia Thouars ●☆

64060 Bursapastoris Quer = Capsella Medik.(保留属名)■

64061 Bursa-pastoris Rupp. = Capsella Medik.(保留属名)■

64062 Bursa-pastoris Ség.(废弃属名)= Capsella Medik.(保留属名)■

64063 Bursaria Cav.(1797);少子果属(囊花属);Bursaria ●☆

64064 Bursaria spinosa Cav.;少子果(多刺囊花);Australian Boxthorn, Blackthorn, Box Thorn, Prickly Box, Sweet Bursaria ●☆

64065 Bursaria spinosa Cav. subsp. lasiophylla (E. M. Benn.) L. W. Cayzer, Crisp et I. Telford;毛叶少子果●☆

64066 Bursaria tenuifolia F. M. Bailey;薄叶少子果●☆

64067 Bursariaceae Kunth = Pittosporaceae R. Br.(保留科名)●

64068 Bursera Jacq. ex L.(1762)(保留属名);裂榄属;American Elemi, Bursera, Linaloa Oil, Linaloe ●☆

64069 Bursera aloexylon Engl.;伽罗木●☆

64070 Bursera arida (Rose) Standl.;旱裂榄●☆

64071 Bursera bipinnata Engl.;二羽裂榄●☆

64072 Bursera delpechiana Poiss. ex Engl.;香裂榄(伽罗木,香榄木)●☆

64073 Bursera fagaroides Engl.;崖椒裂榄;Elephant Tree, Fragrant Bursera ●☆

64074 Bursera glabrifolia Engl.;光叶裂榄●☆

64075 Bursera graveolens (Kunth) Triana et Planch.;烈味裂榄●☆

64076 Bursera gummifera L. = Bursera simaruba (L.) Sarg. ●☆

64077 Bursera hindsiana Engl.;海因兹裂榄;Copal, Torote Prieto ●☆

64078 Bursera jorullensis Engl.;乔鲁裂榄●☆

64079 Bursera laxiflora S. Watson;疏花裂榄●☆

64080 Bursera microphylla A. Gray;小叶裂榄(大象树);Copal, Elephant Tree, Elephant-tree, Torote, Torote Colorado ●☆

64081 Bursera odorata Brandegee;齿裂榄;Elephant Tree, Torote Blanco ●☆

64082 Bursera penicillata (DC.) Engl.;画笔裂榄;Indian Lavender ●☆

64083 Bursera penicillata Engl. = Bursera penicillata (DC.) Engl. ●☆

64084 Bursera serrata Wall. ex Colebr. = Protium serratum (Wall. ex Colebr.) Engl. ●

64085 Bursera simaruba (L.) Sarg.;苦木裂榄(胶裂榄,苦樗裂榄);American Elemi, Budge Gum, Chibou, Gum Elemi, Gumbo Limbo Tree, Gumbolimbo, Gumbo-limbo, Hog Doctor Tree, Incense Tree, Incense-tree, Naked Indian Tree, West Indian Birch, Westindian Birch ●☆

64086 Bursera simplicifolia DC.;单叶裂榄;Bushman Candle, Torote Prieto ●☆

64087 Bursera tecomaca Standl.;黄钟花裂榄●☆

64088 Burseraceae Kunth (1824)(保留科名);橄榄科;Bursera Family, Thochwood Family ●

64089 Burseranthe Rizzini(1974);东巴楝属●☆

64090 Burseranthe pinnata Rizzini;东巴楝●☆

64091 Burseria Jacq. = Bursera Jacq. ex L.(保留属名)●☆

64092 Burseria Loefl. = Verbena L. ■●

64093 Burshia Raf. = Myriophyllum L. ■

64094 Burshia humile Raf. = Myriophyllum humile Morong ■

64095 Bursinopetalum Wight = Mastixia Blume ●

64096 Burtinia Buc'hoz = Magnolia L. ●

64097 Burtonia R. Br.(1811)(保留属名);澳洲水龙骨豆属●☆

64098 Burtonia R. Br.(保留属名)= Gompholobium Sm. ●☆

64099 Burtonia R. Br. ex Aiton = Burtonia R. Br.(保留属名)●☆

64100 Burtonia Salisb.(废弃属名)= Burtonia R. Br.(保留属名)●☆

64101 Burtonia Salisb.(废弃属名)= Hibbertia Andréws ●☆

64102 Burttdavya Hoyle(1936);布尔茜属☆

64103 Burttdavya nyasica Hoyle;布尔茜☆

64104 Burttia Baker f. et Exell(1931);伯特藤属●☆

64105 Burttia prunoides Baker f. et Exell.;伯特藤●☆

64106 Busbeckea Endl. = Capparis L. ●

64107 Busbeckea Mart. = Salpichroa Miers ●☆

64108 Busbeckia Hecart = Syringa L. ●

64109 Busbeckia Rchb. = Busbeckea Endl. ●

64110 Busbequia Salisb. = Hyacinthus L. ■☆

64111 Buschia Ovcz.(1940);布施毛茛属■☆

64112 Buschia Ovcz. = Ranunculus L. ■

64113 Buschia lateriflora (DC.) Ovcz.;侧花布施毛茛■☆

64114 Busea Miq. = Cyrtandromoea Zoll. ■

64115 Buseria T. Durand = Coffea L. ●

64116 Bushia Nieuwl. = Kochia Roth ●■

64117 Bushiola Nieuwl. = Kochia Roth ●■

64118 Busipho Salisb. = Aloe L. ●■

64119 Bussea Harms(1902);布瑟苏木属(奥契瑟苏木属,巴瑟苏木属)●☆

64120 Bussea eggelingii Verdc.;埃格林布瑟苏木●☆

64121 Bussea gossweileri Baker f.;戈斯布瑟苏木●☆

64122 Bussea massaiensis (Taub.) Harms;马萨布瑟苏木●☆

64123 Bussea massaiensis (Taub.) Harms subsp. rhodesica Brenan;罗得西亚布瑟苏木●☆

64124 Bussea occidentalis Hutch.;西方布瑟苏木(西方巴瑟苏木)●☆

64125 Bussea perrieri R. Vig.;佩里耶布瑟苏木(佩里耶巴瑟苏木)●☆

64126 Bussea sakalava Du Puy et R. Rabev.;萨卡拉瓦●☆

64127 Bussea xylocarpa (Sprague) Sprague et Craib;尖果布瑟苏木(尖果巴瑟苏木)●☆

64128 Busseria Cramer = Bursera Jacq. ex L.(保留属名)●☆

64129 Busseria Cramer = Priva Adans. ■☆

64130 Busseuillia Lesson = Eriocaulon L. ■

64131 Bustamenta Alaman ex DC. = Eupatorium L. ■●

64132 Bustelina B. D. Jacks. = Bustelma E. Fourn. ●■

64133 Bustelma E. Fourn. = Oxystelma R. Br. ●■

64134 Bustia Adans. = Buphthalmum L. ■

64135 Bustillosia Clos = Asteriscium Cham. et Schltdl. ■☆

64136 Butania P. C. Keng = Arundinaria Michx. ●

64137 Butania P. C. Keng = Sinarundinaria Nakai ●

64138 Butania P. C. Keng = Yushania P. C. Keng ●

64139 Butayea De Wild. = Sclerochiton Harv. ●☆

64140 Butayea congolana De Wild. = Sclerochiton vogelii (Nees) T. Anderson subsp. congolanus (De Wild.) Vollesen ●☆

64141 Butea K. König et Blatt. = Butea Roxb. ex Willd.(保留属名)●
64142 Butea K. König ex Roxb. = Butea Roxb. ex Willd.(保留属名)●
64143 Butea Roxb. = Butea Roxb. ex Willd.(保留属名)●
64144 Butea Roxb. ex Willd.(1802)(保留属名);紫铆树属(紫矿树属,紫矿属,紫铆属);Butea ●
64145 Butea braamiana DC.;绒毛紫矿;Braam Butea, Floss Butea, Velvet Atylosia, Velvet Butea ●
64146 Butea frondosa K. König ex Roxb. = Butea monosperma (Lam.) Taub. ●
64147 Butea frondosa Roxb. = Butea monosperma (Lam.) Taub. ●
64148 Butea monosperma (Lam.) Taub.;紫铆(单籽紫铆,单子紫铆,胶虫树,麻路子,紫矿,紫矿树,紫鑛);Bastard Teak, Bengal Kino, Bustard Teak, Dhak, Dhak Tree, Dhak-tree, Flame of the Forest, Flame-of-the-forest, Leafy Butea, Muduga Oil, Oneseed Butea, Palas, Palas Tree, Palasa, Palas-tree, Pulas-tree, Tisso Flowers ●
64149 Butea parviflora Roxb.;小花紫铆●☆
64150 Butea sericophylla Wall. = Spatholobus roxburghii Benth. ●
64151 Butea suberecta (Dunn) Blatt. = Spatholobus suberectus Dunn ●
64152 Butea suberecta Blatt. = Spatholobus suberectus Dunn ●
64153 Butea superba Roxb.;华丽紫铆●☆
64154 Buteraea Nees = Strobilanthes Blume ●■
64155 Buteraea Nees(1832);缅甸爵床属■☆
64156 Buteraea parvifolia Bremek.;小叶缅甸爵床■☆
64157 Buteraea rhamnifolia Nees;缅甸爵床■☆
64158 Buteraea ulmifolia Nees;榆叶缅甸爵床■☆
64159 Butia (Becc.) Becc.(1916);果冻棕属(贝蒂棕属,波蒂亚棕属,布迪椰子属,布帝亚椰子属,布齐亚椰子属,布提棕属,冻椰属,冻子椰子属,弓葵属,果冻椰子属,菩提棕属,普提榈属);Butia Palm, Jelly Palm, Palm, Yatay Palm ●☆
64160 Butia Becc. = Butia (Becc.) Becc. ●☆
64161 Butia capitata (Mart.) Becc.;果冻棕(布迪椰子,布帝亚椰子,冻子椰子,果冻椰子);Butia Palm, Jelly Palm, Jetty Palm, Pindo Palm, South American Jelly Palm ●☆
64162 Butia eriospatha (Mart. ex Drude) Becc.;毛果冻棕(棉苞椰);Wooly Jelly Palm ●☆
64163 Butia yatay (Mart.) Becc.;亚泰果冻棕(雅塔椰);Jelly Palm, Yatay Palm ●☆
64164 Butinia Boiss. = Conopodium W. D. J. Koch(保留属名)■☆
64165 Butinia capnoides Decne. = Chaerophyllum capnoides (Decne.) Benth. ■☆
64166 Butneria Duhamel(废弃属名) = Byttneria Loefl.(保留属名)●
64167 Butneria Duhamel(废弃属名) = Calycanthus L.(保留属名)●
64168 Butneria P. Browne = Buttneria P. Browne ●☆
64169 Butneria P. Browne = Casasia A. Rich. ●☆
64170 Butneria occidentalis (Hook. et Arn.) Greene = Calycanthus occidentalis Hook. et Arn. ●☆
64171 Butneria praecox (L.) C. K. Schneid. = Chimonanthus praecox (L.) Link ●
64172 Butneriaceae Barnhart = Calycanthaceae Lindl.(保留科名)●
64173 Butomaceae Mirb.(1804)(保留科名);花蔺科;Butomus Family, Flowering Rush Family, Floweringrush Family, Flowering-Rush Family ■
64174 Butomaceae Rich. = Butomaceae Mirb.(保留科名)■
64175 Butomissa Salisb. = Allium L. ■
64176 Butomopsis Kunth = Tenagocharis Hochst. ■☆
64177 Butomopsis Kunth (1841);拟花蔺属(假花蔺属);Bloomingrush, Tenagocharis ■☆
64178 Butomopsis lanceolata Kunth = Butomopsis latifolia (D. Don) Kunth ■
64179 Butomopsis latifolia (D. Don) Kunth;拟花蔺(假花蔺);Bloomingrush, Broadleaf Butomopsis, Broadleaf Tenagocharis ■
64180 Butomus L.(1753);花蔺属(蔛草属);Florrush, Flowering Rush, Flowering-rush ■
64181 Butomus junceus Turcz.;灯心草花蔺■☆
64182 Butomus lanceolatus Roxb. = Butomopsis latifolia (D. Don) Kunth ■
64183 Butomus latifolius D. Don = Butomopsis latifolia (D. Don) Kunth ■
64184 Butomus umbellatus L.;花蔺(花蔺草,面碌礴,蔛草);Florrush, Flowering Rush, Flowering-rush, Grassy Rush, Hen-and-chickens, Lotus Flower, Pride of the Thames, Raxen, Water Gladiole, Water Gladiolus ■
64185 Butonica Lam. = Barringtonia J. R. Forst. et G. Forst.(保留属名)●
64186 Butonica ceylanica Miers = Barringtonia racemosa (L.) Blume ex DC. ●
64187 Butonica inclyta Miers = Barringtonia racemosa (L.) Blume ex DC. ●
64188 Butonica terrestris Miers = Barringtonia racemosa (L.) Blume ex DC. ●
64189 Butonicoides R. Br. = Planchonia Blume ●☆
64190 Butonicoides R. Br. ex T. Durand = Planchonia Blume ●☆
64191 Buttneria Duhamel = Planchonia Blume ●☆
64192 Buttneria P. Browne = Casasia A. Rich. ●☆
64193 Buttneria Schreb. = Byttneria Loefl.(保留属名)●
64194 Buttonia McKen ex Benth.(1871);巴顿列当属(巴顿玄参属)●☆
64195 Buttonia hildebrandtii Engl. = Buttonia natalensis McKen ex Benth. ●☆
64196 Buttonia latifolia (L.) Griseb. var. flaviflora Boiss. = Parentucellia latifolia (L.) Caruel ■☆
64197 Buttonia natalensis McKen ex Benth.;纳塔尔巴顿列当●☆
64198 Buttonia superba Oberm.;巴顿列当●☆
64199 Butumia G. Taylor = Saxicolella Engl. ■☆
64200 Butumia G. Taylor(1953);尼日利亚苔草属■☆
64201 Butumia marginalis G. Taylor = Saxicolella marginalis (G. Taylor) C. Cusset ex Cheek ■☆
64202 Butyrospermum Kotschy = Vitellaria C. F. Gaertn. ●
64203 Butyrospermum Kotschy(1865);牛油果属(黄油树属);Butter Seed, Butterfruit, Butyrospermum ●
64204 Butyrospermum kirkii Baker = Vitellariopsis kirkii (Baker) Dubard ●☆
64205 Butyrospermum mangifolium (Pierre ex A. Chev.) A. Chev. = Vitellaria paradoxa C. F. Gaertn. ●
64206 Butyrospermum niloticum Kotschy = Vitellaria paradoxa C. F. Gaertn. subsp. nilotica (Kotschy) A. N. Henry, Chithra et N. C. Nair ●☆
64207 Butyrospermum paradoxum (C. F. Gaertn.) Hepper;奇异牛油果;Shea Butter, Shea Nut, Shea-nut Tree ●☆
64208 Butyrospermum paradoxum (C. F. Gaertn.) Hepper = Butyrospermum parkii Kotschy ●
64209 Butyrospermum paradoxum (C. F. Gaertn.) Hepper = Vitellaria paradoxa C. F. Gaertn. ●
64210 Butyrospermum paradoxum (C. F. Gaertn.) Hepper subsp. niloticum (Kotschy) Hepper = Vitellaria paradoxa C. F. Gaertn.

subsp. nilotica (Kotschy) A. N. Henry, Chithra et N. C. Nair ●☆

64211 Butyrospermum paradoxum (C. F. Gaertn.) Hepper subsp. parkii (G. Don) Hepper = Vitellaria paradoxa C. F. Gaertn. ●

64212 Butyrospermum parkii (G. Don) Kotschy = Vitellaria paradoxa C. F. Gaertn. ●

64213 Butyrospermum parkii (G. Don) Kotschy subsp. niloticum (Kotschy) J. H. Hemsl. = Vitellaria paradoxa C. F. Gaertn. subsp. nilotica (Kotschy) A. N. Henry, Chithra et N. C. Nair ●☆

64214 Butyrospermum parkii (G. Don) Kotschy var. cuneata A. Chev. = Vitellaria paradoxa C. F. Gaertn. ●

64215 Butyrospermum parkii (G. Don) Kotschy var. ferruginea A. Chev. = Vitellaria paradoxa C. F. Gaertn. ●

64216 Butyrospermum parkii (G. Don) Kotschy var. floccosa A. Chev. = Vitellaria paradoxa C. F. Gaertn. ●

64217 Butyrospermum parkii (G. Don) Kotschy var. mangifolium Pierre ex A. Chev. = Vitellaria paradoxa C. F. Gaertn. ●

64218 Butyrospermum parkii (G. Don) Kotschy var. niloticum (Kotschy) Pierre ex Engl.;尼洛牛油果●☆

64219 Butyrospermum parkii (G. Don) Kotschy var. niloticum (Kotschy) Pierre ex Engl. = Vitellaria paradoxa C. F. Gaertn. subsp. nilotica (Kotschy) A. N. Henry, Chithra et N. C. Nair ●☆

64220 Butyrospermum parkii (G. Don) Kotschy var. parvifolia A. Chev. = Vitellaria paradoxa C. F. Gaertn. ●

64221 Butyrospermum parkii (G. Don) Kotschy var. poissoni A. Chev. = Vitellaria paradoxa C. F. Gaertn. ●

64222 Butyrospermum parkii (G. Don) Kotschy var. serotina A. Chev. = Vitellaria paradoxa C. F. Gaertn. ●

64223 Butyrospermum parkii Kotschy;牛油果(黄油木);Butterfruit, Butter-tree, Shea Butter, Shea Butter Tree, Shea Butternut, Shea Butter-seed, Shea Butter-tree, Shea Nut, Sheabutter Tree, Shes Butter Seed ●

64224 Buxaceae Dumort. (1822)(保留科名);黄杨科;Box Family, Boxwood Family, Box-wood Family ●■

64225 Buxaceae Loisel = Buxaceae Dumort. (保留科名)●■

64226 Buxanthus Tiegh. = Buxus L. ●

64227 Buxanthus hildebrantii (Baill.) Tiegh. = Buxus hildebrandtii Baill. ●☆

64228 Buxella Small = Gaylussacia Kunth(保留属名)●☆

64229 Buxella Tiegh. = Buxus L. ●

64230 Buxella macowanii (Oliv.) Tiegh. = Buxus macowanii Oliv. ●☆

64231 Buxella madagascarica (Baill.) Tiegh. = Buxus madagascarica Baill. ●☆

64232 Buxiphyllum W. T. Wang = Paraboea (C. B. Clarke) Ridl. ■

64233 Buxiphyllum W. T. Wang et C. Z. Gao = Paraboea (C. B. Clarke) Ridl. ■

64234 Buxiphyllum velutinum W. T. Wang et C. Z. Gao = Paraboea velutina (W. T. Wang et C. Z. Gao) B. L. Burtt ●

64235 Buxus L. (1753);黄杨属;Box, Box Tree, Boxwood, Box-wood ●

64236 Buxus acuminata (Gilg) Hutch. = Buxus acutata Friis ●☆

64237 Buxus acutata Friis;锐黄杨●☆

64238 Buxus aemulans (Rehder et E. H. Wilson) S. C. Li et S. H. Wu = Buxus sinica (Rehder et E. H. Wilson) W. C. Cheng ex M. Cheng subsp. aemulans (Rehder et E. H. Wilson) M. Cheng ●

64239 Buxus austro-yunnanensis Hatus.;滇南黄杨(河滩黄杨);Austral-yunnan Box, South Yunnan Box ●

64240 Buxus balearica Lam.;西班牙黄杨;Balearic Box, Minorca Box, Spanish Box ●☆

64241 Buxus benguellensis Gilg;本格拉黄杨●☆

64242 Buxus benguellensis Gilg var. hirta Hutch.;多毛黄杨●☆

64243 Buxus bodinieri H. Lév.;雀舌黄杨(匙叶黄杨,大样满天星,黄秧树,黄杨木,千年矮,万年青,细叶黄杨,小叶黄杨);Bodinier Box ●

64244 Buxus calcarea G. E. Schatz et Lowry;石灰黄杨●☆

64245 Buxus calophylla Pax = Buxus hildebrandtii Baill. ●☆

64246 Buxus capuronii G. E. Schatz et Lowry;凯普伦黄杨●☆

64247 Buxus caucasica Hort. ex K. Koch;高加索黄杨●☆

64248 Buxus cephalantha H. Lév. et Vaniot;头花黄杨(千年矮,万年青,细叶黄杨);Capitate Box ●

64249 Buxus cephalantha H. Lév. et Vaniot var. shantouensis M. Cheng;汕头黄杨;Shantou Box, Shantou Capitate Box ●

64250 Buxus chinensis Link = Simmondsia chinensis (Link) C. K. Schneid. ●☆

64251 Buxus colchica Pojark.;考卡黄杨●☆

64252 Buxus cordata (Radcl. -Sm.) Friis;心形黄杨●☆

64253 Buxus hainanensis Merr.;海南黄杨;Hainan Box ●

64254 Buxus harlandii Hance;匙叶黄杨(华南黄杨,雀舌黄杨,细叶黄杨);Harland Box, Harland Boxwood ●

64255 Buxus harlandii Hance = Buxus bodinieri H. Lév. ●

64256 Buxus harlandii Hance var. cephalantha (H. Lév. et Vaniot) Rehder = Buxus cephalantha H. Lév. et Vaniot ●

64257 Buxus harlandii Hance var. linearis Hand. -Mazz. = Buxus cephalantha H. Lév. et Vaniot ●

64258 Buxus hebecarpa Hatus.;毛果黄杨;Downy-fruited Box, Hairyfruit Box ●

64259 Buxus henryi Mayr et Dümmer;大花黄杨(桃叶黄杨);Bigflower Box, Big-flowered Box ●

64260 Buxus hildebrandtii Baill.;希尔黄杨●☆

64261 Buxus hirta (Hutch.) Mathou = Buxus benguellensis Gilg var. hirta Hutch. ●☆

64262 Buxus humbertii G. E. Schatz et Lowry;亨伯特黄杨●☆

64263 Buxus hyrcana Pojark.;希尔康黄杨●☆

64264 Buxus ichangensis Hatus.;宜昌黄杨;Yichang Box ●

64265 Buxus intermedia Kaneh. = Buxus sinica (Rehder et E. H. Wilson) W. C. Cheng ex M. Cheng var. intermedia (Kaneh.) M. Cheng ●

64266 Buxus itremoensis G. E. Schatz et Lowry;伊特雷穆黄杨●☆

64267 Buxus japonica Müll. Arg. = Buxus microphylla Siebold et Zucc. var. japonica (Müll. Arg. ex Miq.) Rehder et E. H. Wilson ●☆

64268 Buxus japonica Müll. Arg. f. rubra? = Buxus microphylla Siebold et Zucc. var. japonica (Müll. Arg. ex Miq.) Rehder et E. H. Wilson 'Rubra' ●

64269 Buxus japonica Müll. Arg. var. microphylla? = Buxus microphylla Siebold et Zucc. ●

64270 Buxus latistyla Gagnep.;阔柱黄杨(假黄杨,假山枝子);Broadstyle Box, Broad-styled Box ●

64271 Buxus linearifolia M. Cheng;线叶黄杨;Linearleaf Box, Linear-leaved Box ●

64272 Buxus lisowskii Bamps et Malaisse;利索黄杨●☆

64273 Buxus liukiuensis (Makino) Makino;琉球黄杨●

64274 Buxus liukiuensis (Makino) Makino = Buxus sinica (Rehder et E. H. Wilson) W. C. Cheng ex M. Cheng var. intermedia (Kaneh.) M. Cheng ●

64275 Buxus liukiuensis (Makino) Makino f. glabra Hiyama = Buxus liukiuensis (Makino) Makino ●

64276 Buxus longifolia Boiss. ;长叶黄杨;Long-leaved Box ●☆

64277 Buxus macowanii Oliv. ;海角黄杨(南非黄杨);Cape Box,E. London Box,East London Box,East London Boxwood ●☆

64278 Buxus macowanii Oliv. var. benguellensis (Gilg) Mathou = Buxus benguellensis Gilg ●☆

64279 Buxus macrocarpa Capuron;大果黄杨●☆

64280 Buxus madagascarica Baill. ;马岛黄杨●☆

64281 Buxus madagascarica Baill. subsp. sambiranensis H. Perrier = Buxus monticola G. E. Schatz et Lowry ●☆

64282 Buxus madagascarica Baill. subsp. tropophila H. Perrier = Buxus moratii G. E. Schatz et Lowry ●☆

64283 Buxus madagascarica Baill. subsp. xerophila H. Perrier = Buxus humbertii G. E. Schatz et Lowry ●☆

64284 Buxus megistophylla H. Lév. ;大叶黄杨(长叶黄杨);Big-leaf Box,Big-leaved Box,Largeleaf Box ●

64285 Buxus microphylla Siebold et Zucc. ;小叶黄杨(黄杨);Chinese Box,Japanese Box,Korean Box,Small-leaved Box ●

64286 Buxus microphylla Siebold et Zucc. 'Compacta';紧密小叶黄杨●☆

64287 Buxus microphylla Siebold et Zucc. 'Curly Locks';弯洛克小叶黄杨●☆

64288 Buxus microphylla Siebold et Zucc. 'Cushion';垫子小叶黄杨●☆

64289 Buxus microphylla Siebold et Zucc. 'Faulkner';福克纳小叶黄杨●☆

64290 Buxus microphylla Siebold et Zucc. 'Green Gem';绿宝石小叶黄杨●☆

64291 Buxus microphylla Siebold et Zucc. 'Green Jade';绿玉小叶黄杨●☆

64292 Buxus microphylla Siebold et Zucc. 'Green Pillow';绿枕小叶黄杨●☆

64293 Buxus microphylla Siebold et Zucc. 'Green Velvet';绿丝绒小叶黄杨●☆

64294 Buxus microphylla Siebold et Zucc. 'John Baldwin';直亮小叶黄杨●☆

64295 Buxus microphylla Siebold et Zucc. 'Morris Midget';小莫里斯小叶黄杨●☆

64296 Buxus microphylla Siebold et Zucc. 'Tide Hill';潮汐山小叶黄杨●☆

64297 Buxus microphylla Siebold et Zucc. 'Winter Beauty';冬美人小叶黄杨●☆

64298 Buxus microphylla Siebold et Zucc. subsp. sinica (Rehder et E. H. Wilson) Hatus. = Buxus sinica (Rehder et E. H. Wilson) W. C. Cheng ex M. Cheng ●

64299 Buxus microphylla Siebold et Zucc. subsp. sinica (Rehder et E. H. Wilson) Hatus. var. aemulans (Rehder et E. H. Wilson) Hatus. = Buxus sinica (Rehder et E. H. Wilson) W. C. Cheng ex M. Cheng subsp. aemulans (Rehder et E. H. Wilson) M. Cheng ●

64300 Buxus microphylla Siebold et Zucc. var. arborescens? = Buxus microphylla Siebold et Zucc. ●

64301 Buxus microphylla Siebold et Zucc. var. insularis Nakai;朝鲜小叶黄杨●☆

64302 Buxus microphylla Siebold et Zucc. var. intermedia (Kaneh.) H. L. Li = Buxus sinica (Rehder et E. H. Wilson) W. C. Cheng ex M. Cheng var. intermedia (Kaneh.) M. Cheng ●

64303 Buxus microphylla Siebold et Zucc. var. japonica (Müll. Arg. ex Miq.) Rehder et E. H. Wilson;日本小叶黄杨;Japanese Box, Japanese Little-leaf Box,Japanese Small-leaved Box ●☆

64304 Buxus microphylla Siebold et Zucc. var. japonica (Müll. Arg. ex Miq.) Rehder et E. H. Wilson f. tenuis Makino;纤细日本小叶黄杨●☆

64305 Buxus microphylla Siebold et Zucc. var. japonica (Müll. Arg. ex Miq.) Rehder et E. H. Wilson f. minutissima Makino;极小日本小叶黄杨●☆

64306 Buxus microphylla Siebold et Zucc. var. japonica (Müll. Arg. ex Miq.) Rehder et E. H. Wilson 'Rubra';红色日本小叶黄杨●

64307 Buxus microphylla Siebold et Zucc. var. japonica (Müll. Arg. ex Miq.) Rehder et E. H. Wilson f. riparia (Makino) Makino = Buxus microphylla Siebold et Zucc. var. riparia (Makino) Makino ●☆

64308 Buxus microphylla Siebold et Zucc. var. japonica (Müll. Arg. ex Miq.) Rehder et E. H. Wilson f. major Hatus. = Buxus microphylla Siebold et Zucc. var. kitashimae (Yanagita) H. Ohba ●☆

64309 Buxus microphylla Siebold et Zucc. var. japonica (Müll. Arg. ex Miq.) Rehder et E. H. Wilson f. rubra? = Buxus microphylla Siebold et Zucc. ●

64310 Buxus microphylla Siebold et Zucc. var. japonica (Müll. Arg. ex Miq.) Rehder et E. H. Wilson f. rubra? = Buxus microphylla Siebold et Zucc. var. japonica (Müll. Arg. ex Miq.) Rehder et E. H. Wilson 'Rubra' ●

64311 Buxus microphylla Siebold et Zucc. var. kiangsiensis Hu et F. H. Chen = Buxus sinica (Rehder et E. H. Wilson) W. C. Cheng ex M. Cheng subsp. aemulans (Rehder et E. H. Wilson) M. Cheng ●

64312 Buxus microphylla Siebold et Zucc. var. kitashimae (Yanagita) H. Ohba;北岛黄杨●☆

64313 Buxus microphylla Siebold et Zucc. var. koreana? = Buxus microphylla Siebold et Zucc. var. insularis Nakai ●☆

64314 Buxus microphylla Siebold et Zucc. var. platyphylla (Schneid.) Hand. -Mazz. = Buxus bodinieri H. Lév. ●

64315 Buxus microphylla Siebold et Zucc. var. prostrata W. W. Sm. = Buxus rugulosa Hatus. var. prostrata (W. W. Sm.) M. Cheng ●

64316 Buxus microphylla Siebold et Zucc. var. riparia (Makino) Makino;溪畔黄杨●☆

64317 Buxus microphylla Siebold et Zucc. var. rotundifolia? = Buxus microphylla Siebold et Zucc. ●

64318 Buxus microphylla Siebold et Zucc. var. rupicola W. W. Sm. = Buxus rugulosa Hatus. ●

64319 Buxus microphylla Siebold et Zucc. var. sinica Rehder et E. H. Wilson = Buxus sinica (Rehder et E. H. Wilson) W. C. Cheng ex M. Cheng ●

64320 Buxus microphylla Siebold et Zucc. var. suffruticosa (Siebold) Makino f. minutissima? = Buxus microphylla Siebold et Zucc. ●

64321 Buxus microphylla Siebold et Zucc. var. suffruticosa (Siebold) Makino f. tenuis? = Buxus microphylla Siebold et Zucc. ●

64322 Buxus microphylla Siebold et Zucc. var. suffruticosa (Siebold) Makino = Buxus microphylla Siebold et Zucc. ●

64323 Buxus microphylla Siebold et Zucc. var. tarokoensis S. Y. Lu et Yuen P. Yang;太鲁阁黄杨;Taroko Small-leaved Box ●

64324 Buxus mollicula W. W. Sm. ;软毛黄杨(毛黄杨);Pubescent Box,Soft-hairy Box ●

64325 Buxus mollicula W. W. Sm. var. glabra Hand. -Mazz. ;光叶黄杨(变光软毛黄杨);Glabrous Box,Glabrous Pubescent Box ●

64326 Buxus monticola G. E. Schatz et Lowry;山地黄杨●☆

64327 Buxus moratii G. E. Schatz et Lowry;莫拉特黄杨●☆

64328 Buxus myrica H. Lév. ;杨梅黄杨(结青树);Bayberry Box, Myrica Box ●

64329 Buxus myrica H. Lév. var. angustifolia Gagnep.;狭叶杨梅黄杨;Narrowleaf Bayberry Box,Narrowleaf Box ●

64330 Buxus natalensis (Oliv.) Hutch.;纳塔尔黄杨●☆

64331 Buxus nyasica Hutch.;尼亚斯黄杨●☆

64332 Buxus obcordata-variegata Fortune = Buxus microphylla Siebold et Zucc. ●

64333 Buxus obtusifolia (Mildbr.) Hutch.;钝叶黄杨●☆

64334 Buxus ovalifolia Siebold ex K. Koch;卵叶黄杨●☆

64335 Buxus papillosa C. K. Schneid.;乳头黄杨●☆

64336 Buxus pubiramea Merr. et Chun;毛枝黄杨;Hairy-twig Box,Pubescentbranch Box ●

64337 Buxus rotundifolia Hort. ex K. Koch;圆叶黄杨●☆

64338 Buxus rugulosa Hatus.;皱叶黄杨(高山黄杨);Wrinkledleaf Box,Wrinkle-leaved Box ●

64339 Buxus rugulosa Hatus. subsp. prostrata (W. W. Sm.) Hatus. = Buxus rugulosa Hatus. var. prostrata (W. W. Sm.) M. Cheng ●

64340 Buxus rugulosa Hatus. var. intermedia Hatus. = Buxus rugulosa Hatus. var. prostrata (W. W. Sm.) M. Cheng ●

64341 Buxus rugulosa Hatus. var. prostrata (W. W. Sm.) M. Cheng;平卧皱叶黄杨(铺地黄杨);Prostrate Box,Prostrate Wrinkledleaf Box ●

64342 Buxus rugulosa Hatus. var. rupicola (W. W. Sm.) S. C. Li et S. H. Wu;岩生皱叶黄杨(石生黄杨,岩生黄杨);Rupicolous Winkledleaf Box ●

64343 Buxus salicifolia Hort. ex K. Koch;柳叶黄杨●☆

64344 Buxus saligna D. Don = Sarcococca saligna (D. Don) Müll. Arg. ●

64345 Buxus sempervirens L.;锦熟黄杨(黄梨树,黄杨,黄杨木,土耳其黄杨,窄叶黄杨);Abassian Box,Abassian Boxwood,Boxwood,Burying Box,Busch,Bush Tree,Chairs-and-tables,Common Box,Common Boxwood,Crocks-and-kettles,Dudgeon,English Box,English Boxwood,Europe Box,European Box,European Boxwood,European Box-wood,European Buckeye,Iranian Box,Kettles-and Crocks,Milkstools,Palm,Persian Box,Pots-and-kettles,Tables-and-chairs,Turkey Box,Turkey Boxwood ●

64346 Buxus sempervirens L. 'Arborescens';乔木状锦熟黄杨;Tree Boxwood ●☆

64347 Buxus sempervirens L. 'Argentea' = Buxus sempervirens L. 'Argenteovariegata' ●☆

64348 Buxus sempervirens L. 'Argenteovariegata';黄边锦熟黄杨;Variegated Boxwood ●☆

64349 Buxus sempervirens L. 'Edgar Anderson';埃德加安德森锦熟黄杨●☆

64350 Buxus sempervirens L. 'Elegantissima';极美锦熟黄杨;Elegans Boxwood,Variegated Boxwood ●☆

64351 Buxus sempervirens L. 'Handsworthiensis';汉兹沃斯锦熟黄杨(汉德·沃斯锦熟黄杨)●☆

64352 Buxus sempervirens L. 'Latifolia Maculata';阔叶变色锦熟黄杨●☆

64353 Buxus sempervirens L. 'Marginata';黄斑锦熟黄杨●☆

64354 Buxus sempervirens L. 'Memorial';纪念锦熟黄杨●☆

64355 Buxus sempervirens L. 'Ponteyi';庞特伊锦熟黄杨●☆

64356 Buxus sempervirens L. 'Subfruticosa';亚灌木锦熟黄杨(矮灌锦熟黄杨);Dutch Box,Dwarf Box,Edging Box,English Boxwood,French Box ●☆

64357 Buxus sempervirens L. 'Vardar Valley';瓦尔德山谷锦熟黄杨;Vardar Valley Boxwood ●☆

64358 Buxus sempervirens L. = Buxus harlandii Hance ●

64359 Buxus sempervirens L. = Buxus microphylla Siebold et Zucc. ●

64360 Buxus sempervirens L. = Buxus sinica (Rehder et E. H. Wilson) W. C. Cheng ex M. Cheng ●

64361 Buxus sempervirens L. sensu Stewart et Brandis = Buxus papillosa C. K. Schneid. ●☆

64362 Buxus sempervirens L. var. angustifolia Loudon = Buxus sempervirens L. ●

64363 Buxus sempervirens L. var. japonica? = Buxus microphylla Siebold et Zucc. ●

64364 Buxus sempervirens L. var. microphylla H. Lév. = Buxus cephalantha H. Lév. et Vaniot ●

64365 Buxus sempervirens L. var. microphylla H. Lév. = Buxus microphylla Siebold et Zucc. ●

64366 Buxus sempervirens L. var. subfruticosa = Buxus microphylla Siebold et Zucc. ●

64367 Buxus sempervirens L. var. subfruticosa? = Buxus sempervirens L. 'Subfruticosa' ●☆

64368 Buxus sempervirens Thunb. = Buxus japonica Müll. Arg. ●

64369 Buxus sinica (Rehder et E. H. Wilson) W. C. Cheng ex M. Cheng;黄杨(百日红,豆板黄杨,豆瓣黄杨,瓜子黄杨,黄杨木,锦熟黄杨,千年矮,山黄杨,万年青,乌龙木,细叶黄杨,小黄杨,小叶黄杨);Chinese Box,Japanese Boxwood,Little Leaf Boxwood ●

64370 Buxus sinica (Rehder et E. H. Wilson) W. C. Cheng ex M. Cheng = Buxus microphylla Siebold et Zucc. subsp. sinica (Rehder et E. H. Wilson) Hatus. ●

64371 Buxus sinica (Rehder et E. H. Wilson) W. C. Cheng ex M. Cheng subsp. aemulans (Rehder et E. H. Wilson) M. Cheng;尖叶黄杨(长叶黄杨,黄杨木,江西黄杨);Acuteleaf Box,Acute-leaved Box,Jiangxi Smallleaved Box ●

64372 Buxus sinica (Rehder et E. H. Wilson) W. C. Cheng ex M. Cheng var. insularis (Nakai) M. Cheng = Buxus microphylla Siebold et Zucc. var. insularis Nakai ●☆

64373 Buxus sinica (Rehder et E. H. Wilson) W. C. Cheng ex M. Cheng var. intermedia (Kaneh.) M. Cheng;中间黄杨(琉球黄杨,台湾黄杨);Intermediate Box,Liuqiu Box,Taiwan Littleleaf Box ●

64374 Buxus sinica (Rehder et E. H. Wilson) W. C. Cheng ex M. Cheng var. koreana (Nakai ex Rehder) Q. L. Wang;朝鲜黄杨●☆

64375 Buxus sinica (Rehder et E. H. Wilson) W. C. Cheng ex M. Cheng var. parvifolia M. Cheng;鱼鳞黄杨(小叶黄杨,鱼鳞木);Littleleaf Box ●

64376 Buxus sinica (Rehder et E. H. Wilson) W. C. Cheng ex M. Cheng var. pumila M. Cheng;矮生黄杨;Low Box,Low Chinese Box ●

64377 Buxus sinica (Rehder et E. H. Wilson) W. C. Cheng ex M. Cheng var. vaccinifolia M. Cheng;越橘叶黄杨;Blueberryleaf Box ●

64378 Buxus stenophylla Hance;狭叶黄杨;Narrowleaf Box,Narrow-leaved Box ●

64379 Buxus terminalis Siebold et Zucc.;顶蕊黄杨●

64380 Buxus wallichiana Baill.;喜马拉雅黄杨;Himalayan Box ●☆

64381 Buxus wallichiana Baill. var. velutina Franch. = Buxus mollicula W. W. Sm. ●

64382 Bwusemichea Balansa = Zoysia Willd.(保留属名)■

64383 Byblidaceae (Engl. et Gilg) Domin = Byblidaceae Domin ●☆

64384 Byblidaceae Domin(1922)(保留科名);二型腺毛科(捕虫纸草科,腺毛草科)●☆

64385 Byblis Salisb.(1808);二型腺毛属(白布莉斯属,捕虫纸草属,腺毛草属)●■☆

64386 Byblis filifolia Planch.;线叶二型腺毛●☆

64387 Byblis gigantea Lindl.;大二型腺毛●☆
64388 Bygnonia Barcena = Bignonia L.(保留属名)●
64389 Byrnesia Rose = Graptopetalum Rose ■●☆
64390 Byronia Endl. = Ilex L. ●
64391 Byrsa Noronha = Stephania Lour. ●■
64392 Byrsanthes C. Presl(废弃属名) = Byrsanthus Guill.(保留属名)●☆
64393 Byrsanthes C. Presl(废弃属名) = Siphocampylus Pohl ■●☆
64394 Byrsanthus Guill.(1838)(保留属名);西非大风子属●☆
64395 Byrsanthus brownii Guillaumin;西非大风子●☆
64396 Byrsanthus brownii Guillaumin var. latifolius A. Fern.;宽叶西非大风子●☆
64397 Byrsella Luer(2006);袋兰属■☆
64398 Byrsocarpus Schumach. = Rourea Aubl.(保留属名)●
64399 Byrsocarpus Schumach. et Thonn. = Rourea Aubl.(保留属名)●
64400 Byrsocarpus albidoflavescens (Gilg) Greenway ex Burtt Davy = Rourea thomsonii (Baker) Jongkind ●☆
64401 Byrsocarpus astragalifolius A. Chev. = Rourea coccinea (Thonn. ex Schumach.) Benth. ●☆
64402 Byrsocarpus boivinianus (Baill.) G. Schellenb. = Rourea coccinea (Thonn. ex Schumach.) Benth. subsp. boiviniana (Baill.) Jongkind ●☆
64403 Byrsocarpus caillei A. Chev. = Dalbergia boehmii Taub ●☆
64404 Byrsocarpus cassioides (Hiern) G. Schellenb. = Rourea cassioides Hiern ●☆
64405 Byrsocarpus coccineus (Thonn. ex Schumach.) Benth. var. parviflorus Planch. ex G. Schellenb. = Rourea coccinea (Thonn. ex Schumach.) Benth. ●☆
64406 Byrsocarpus coccineus Thonn. ex Schumach. = Rourea coccinea (Thonn. ex Schumach.) Benth. ●☆
64407 Byrsocarpus dinklagei (Gilg) G. Schellenb. = Rourea coccinea (Thonn. ex Schumach.) Benth. var. viridis (Gilg) Jongkind ●☆
64408 Byrsocarpus goetzei (Gilg) Greenway = Rourea coccinea (Thonn. ex Schumach.) Benth. subsp. boiviniana (Baill.) Jongkind ●☆
64409 Byrsocarpus ledermannii G. Schellenb. = Rourea coccinea (Thonn. ex Schumach.) Benth. ●☆
64410 Byrsocarpus maximus Baker = Rourea coccinea (Thonn. ex Schumach.) Benth. subsp. boiviniana (Baill.) Jongkind ●☆
64411 Byrsocarpus orientalis (Baill.) Baker = Rourea orientalis Baill. ●☆
64412 Byrsocarpus orientalis (Baill.) Baker var. hirtella Rabenant. = Rourea orientalis Baill. ●☆
64413 Byrsocarpus orientalis (Baill.) Baker var. pubescens Rabenant. = Rourea orientalis Baill. ●☆
64414 Byrsocarpus ovatifolius Baker = Rourea coccinea (Thonn. ex Schumach.) Benth. subsp. boiviniana (Baill.) Jongkind ●☆
64415 Byrsocarpus papillosus G. Schellenb. = Rourea coccinea (Thonn. ex Schumach.) Benth. var. viridis (Gilg) Jongkind ●☆
64416 Byrsocarpus parviflorus (Gilg) G. Schellenb. = Rourea parviflora Gilg ●☆
64417 Byrsocarpus poggeanus (Gilg) G. Schellenb. = Rourea coccinea (Thonn. ex Schumach.) Benth. var. viridis (Gilg) Jongkind ●☆
64418 Byrsocarpus puberulus G. Schellenb. = Rourea coccinea (Thonn. ex Schumach.) Benth. ●☆
64419 Byrsocarpus puniceus Thonn. = Rourea coccinea (Thonn. ex Schumach.) Benth. ●☆
64420 Byrsocarpus tisserantii Aubrév. et Pellegr. = Rourea coccinea (Thonn. ex Schumach.) Benth. ●☆
64421 Byrsocarpus tomentosus G. Schellenb. = Rourea orientalis Baill. ●☆
64422 Byrsocarpus usambaricus G. Schellenb. = Rourea coccinea (Thonn. ex Schumach.) Benth. var. viridis (Gilg) Jongkind ●☆
64423 Byrsonima Juss. = Byrsonima Rich. ex Juss. ●☆
64424 Byrsonima Rich. = Byrsonima Rich. ex Juss. ●☆
64425 Byrsonima Rich. ex Juss.(1822);金匙树属(糅皮木属);Byrsonima, Locust, Locust Bean, Surette ●☆
64426 Byrsonima Rich. ex Kunth = Byrsonima Rich. ex Juss. ●☆
64427 Byrsonima aerugo Sagot;铜绿金匙树(铜绿糅皮木)●☆
64428 Byrsonima bucidifolia Standl.;使君子金匙树(糅皮木)●☆
64429 Byrsonima coriacea DC.;革金匙树(革糅皮木);Locust Berry ●☆
64430 Byrsonima crassifolia Lunan ex Griseb.;厚叶金匙树(厚叶糅皮木);Golden Spoon, Nance ●☆
64431 Byrsonima cuneata P. Wilson;楔形金匙树(楔形糅皮木)●☆
64432 Byrsonima spicata Rich. ex Juss.;穗状金匙木;Maricao ●☆
64433 Byrsophyllum Hook. f.(1873);袋叶茜属●■☆
64434 Byrsophyllum ellipticum Hook. f.;椭圆袋叶茜●■☆
64435 Byrsophyllum tetrandrum Hook. f.;袋叶茜●■☆
64436 Bystropogon L'Hér.(1789)(保留属名);绒萼木属●☆
64437 Bystropogon canariensis (L.) L'Hér.;加那利绒萼木(加那利毕斯特罗木)●☆
64438 Bystropogon canariensis (L.) L'Hér. var. smithianus Christ = Bystropogon canariensis (L.) L'Hér. ●☆
64439 Bystropogon coarctatus Schumach. et Thonn. = Hyptis pectinata (L.) Poit. ■☆
64440 Bystropogon graveolens Blume = Hyptis suaveolens (L.) Poit. ●■
64441 Bystropogon maderensis Webb;梅德绒萼木●☆
64442 Bystropogon odoratissimus Bolle;芳香绒萼木●☆
64443 Bystropogon origanifolius L'Hér.;牛至叶绒萼木●☆
64444 Bystropogon origanifolius L'Hér. var. canariae La Serna = Bystropogon origanifolius L'Hér. ●☆
64445 Bystropogon origanifolius L'Hér. var. ferrensis (Ceballos et Ort.) La Serna = Bystropogon origanifolius L'Hér. ●☆
64446 Bystropogon origanifolius L'Hér. var. palmensis Bornm. = Bystropogon origanifolius L'Hér. ●☆
64447 Bystropogon plumosus L'Hér.;羽萼绒萼木(羽萼毕斯特罗木)●☆
64448 Bystropogon punctatus L'Hér.;斑点绒萼木(斑点毕斯特罗木)●☆
64449 Bystropogon suaveolens (L.) Blume = Hyptis suaveolens (L.) Poit. ●■
64450 Bystropogon wildpretii La Serna;维尔德绒萼木●☆
64451 Bythophyton Hook. f.(1884);水中透骨草属■☆
64452 Bythophyton indicum Hook. f.;水中透骨草■☆
64453 Bytneria Jacq. = Byttneria Loefl.(保留属名)●
64454 Byttneria Loefl.(1758)(保留属名);刺果藤属;Burvine, Byttneria ●
64455 Byttneria Steud. = Butneria Duhamel(废弃属名)●
64456 Byttneria Steud. = Calycanthus L.(保留属名)●
64457 Byttneria africana Mast. = Byttneria catalpifolia Jacq. subsp. africana (Mast.) Exell et Mendonça ●☆
64458 Byttneria aspera Colebr. = Byttneria grandifolia DC. ●
64459 Byttneria aspera Colebr. ex Wall. = Byttneria grandifolia DC. ●
64460 Byttneria catalpifolia Jacq. subsp. africana (Mast.) Exell et Mendonça;非洲刺果藤●☆
64461 Byttneria dahomensis N. Hallé;达荷姆刺果藤●☆
64462 Byttneria devredii R. Germ. = Megatritheca devredii (R. Germ.)

Cristóbal ●☆
64463 Byttneria elegans Ridl. = Byttneria pilosa Roxb. ●
64464 Byttneria fruticosa K. Schum. ex Engl.;灌木刺果藤●☆
64465 Byttneria glabra K. Schum. et Engl.;光刺果藤●☆
64466 Byttneria grandifolia DC.;刺果藤(大胶藤);Burvine,Scabrous Byttneria ●
64467 Byttneria grossedenticulata Bodard et Pellegr. = Megatritheca grossedenticulata (M. Bodard et Pellegr.) Cristóbal ●☆
64468 Byttneria guineensis Keay et Milne-Redh.;几内亚刺果藤●☆
64469 Byttneria integrifolia Lace;全缘刺果藤;Entire Byttneria,Entireleaf Burvine,Entire-leaved Burvine ●
64470 Byttneria ivorensis N. Hallé;伊沃里刺果藤●☆
64471 Byttneria pilosa Roxb.;粗毛刺果藤(毛刺果藤,野枇杷藤);Pilose Byttneria,Pilous Byttneria,Scabrous Burvine ●
64472 Byttneria pilosa Roxb. var. pellita Gagnep. = Byttneria pilosa Roxb. ●
64473 Byttneria siamensis Craib. = Byttneria grandifolia DC. ●
64474 Byttneriaceae R. Br. (1814)(保留科名);刺果藤科(利末花科)●■☆
64475 Byttneriaceae R. Br.(保留科名) = Malvaceae Juss.(保留科名)●■
64476 Byttneriaceae R. Br.(保留科名)= Sterculiaceae Vent.(保留科名)●■
64477 Caampyloa Post et Kuntze = Campuloa Desv. ■☆
64478 Caampyloa Post et Kuntze = Ctenium Panz.(保留属名)■☆
64479 Caapeba Mill. = Cissampelos L. ●
64480 Caapeba Plum. ex Adans. = Cissampelos L. ●
64481 Caatinganthus H. Rob. (1999);短命地胆草属■☆
64482 Caatinganthus harleyi H. Rob.;短命地胆草■☆
64483 Caballeria Ruiz et Pav. = Myrsine L. ●
64484 Caballeroa Font Quer(1935);非洲合柱补血草属■☆
64485 Caballeroa ifniensis (Caball.) Font Quer = Saharanthus ifniensis (Caball.) M. B. Crespo et Lledó ●☆
64486 Cabanisia Klotzsch ex Schltdl. = Eichhornia Kunth(保留属名)■
64487 Cabi Ducke = Callaeum Small ●☆
64488 Cabobanthus H. Rob. (1999);聚花瘦片菊属■☆
64489 Cabobanthus bullulatus (S. Moore) H. Rob. = Vernonia bullulata S. Moore ●☆
64490 Cabobanthus polysphaerus (Baker) H. Rob. = Vernonia polysphaera Baker ●☆
64491 Cabomba Aubl. (1775);竹节水松属(水盾草属);Fanwort,Fish Grass ■
64492 Cabomba aquctica Aubl.;黄竹节水松(黄菊花草);Giant Cabomba,Golden Cabomba ■☆
64493 Cabomba aquctica DC. = Cabomba caroliniana A. Gray ■
64494 Cabomba aubletii Michx. = Cabomba caroliniana A. Gray ■
64495 Cabomba australis Speg.;金竹节水松(金菊花草);Fanwort,Washington Grass ■☆
64496 Cabomba caroliniana A. Gray;竹节水松(绿菊花草,水盾草,羽衣藻,紫菊花草);Carolina Fanwort,Carolina Water Shield,Fanwort,Fish Grass,Fish-grass,Washington Fanwort,Washington Grass,Washington Plant,Washington-grass ■
64497 Cabomba caroliniana A. Gray var. pulcherrima R. M. Harper = Cabomba caroliniana A. Gray ■
64498 Cabomba caroliniana A. Gray var. pulcherrima R. M. Harper = Cabomba pulcherrima (R. M. Harper) Fassett ■
64499 Cabomba furcata Schult. f.;叉状竹节水松;Forked Fanwort ■☆
64500 Cabomba pulcherrima (R. M. Harper) Fassett = Cabomba caroliniana A. Gray ■
64501 Cabomba viridiflora Hort. = Cabomba caroliniana A. Gray ■
64502 Cabombaceae A. Rich. = Cabombaceae Rich. ex A. Rich.(保留科名)■
64503 Cabombaceae Rich. ex A. Rich. (1822)(保留科名);竹节水松科(莼菜科,莼科);Cabomba Family,Water-shield Family ■
64504 Cabombaceae Rich. ex A. Rich.(保留科名) = Nymphaeaceae Salisb.(保留科名)■
64505 Cabralea A. Juss. (1830);南美楝属(卡氏楝属,南美洲楝属)●☆
64506 Cabralea cangerana Saldanha = Cabralea canjerana (Vell.) Mart. ●☆
64507 Cabralea canjerana (Vell.) Mart.;南美楝(南美洲楝);Cancharana Cabralea ●☆
64508 Cabralea eichleriana DC.;艾勒南美楝(艾勒卡氏楝)●☆
64509 Cabralea laevis C. DC.;平滑南美楝(平滑南美洲楝)●☆
64510 Cabralea oblongifoliola C. DC.;长圆叶南美楝(长圆叶南美洲楝)●☆
64511 Cabralia Schrank = Centratherum Cass. ■☆
64512 Cabralia Schrank = Spixia Schrank ■☆
64513 Cabrera Lag. = Axonopus P. Beauv. ■
64514 Cabrera Lag. = Paspalum L. ■
64515 Cabreraea Bonif. (2009);阿安菊属●☆
64516 Cabreraea Bonif. = Chiliophyllum Phil.(保留属名)●☆
64517 Cabreriella Cuatrec. (1980);光藤菊属●☆
64518 Cabreriella oppositicordia (Cuatrec.) Cuatrec.;对心光藤菊●☆
64519 Cabreriella sanctae-martae (Greenm.) Cuatrec.;光藤菊●☆
64520 Cabucala Pichon = Petchia Livera ●☆
64521 Cabucala Pichon(1948);卡布木属●☆
64522 Cabucala brachyantha Pichon = Carissa spinarum L. ●
64523 Cabucala caudata Markgr. = Petchia erythrocarpa (Vatke) Leeuwenb. ●☆
64524 Cabucala crassifolia Pichon = Petchia erythrocarpa (Vatke) Leeuwenb. ●☆
64525 Cabucala cryptophlebia (Baker) Pichon = Petchia cryptophlebia (Baker) Leeuwenb. ●☆
64526 Cabucala erythrocarpa (Vatke) Markgr. = Petchia erythrocarpa (Vatke) Leeuwenb. ●☆
64527 Cabucala erythrocarpa (Vatke) Markgr. var. angustifolia (Pichon) Markgr. = Petchia erythrocarpa (Vatke) Leeuwenb. ●☆
64528 Cabucala erythrocarpa (Vatke) Markgr. var. intermedia (Pichon) Markgr. = Petchia erythrocarpa (Vatke) Leeuwenb. ●☆
64529 Cabucala fasciculata Pichon;簇生卡布木●☆
64530 Cabucala fasciculata Pichon = Petchia erythrocarpa (Vatke) Leeuwenb. ●☆
64531 Cabucala glauca Pichon = Petchia erythrocarpa (Vatke) Leeuwenb. ●☆
64532 Cabucala humbertii Markgr. = Petchia humbertii (Markgr.) Leeuwenb. ●☆
64533 Cabucala intermedia Pichon = Petchia erythrocarpa (Vatke) Leeuwenb. ●☆
64534 Cabucala longipes Pichon = Petchia erythrocarpa (Vatke) Leeuwenb. ●☆
64535 Cabucala macrophylla Pichon = Petchia erythrocarpa (Vatke) Leeuwenb. ●☆
64536 Cabucala macrophylla Pichon var. acuta Markgr. = Petchia erythrocarpa (Vatke) Leeuwenb. ●☆

64537 Cabucala macrophylla Pichon var. oxyphylla Markgr. = Petchia madagascariensis (A. DC.) Leeuwenb. ●☆

64538 Cabucala madagascariensis (A. DC.) Pichon = Petchia erythrocarpa (Vatke) Leeuwenb. ●☆

64539 Cabucala madagascariensis (A. DC.) Pichon var. acuta (Markgr.) Markgr. = Petchia madagascariensis (A. DC.) Leeuwenb. ●☆

64540 Cabucala madagascariensis (A. DC.) Pichon var. intermedia Pichon = Petchia erythrocarpa (Vatke) Leeuwenb. ●☆

64541 Cabucala madagascariensis (A. DC.) Pichon var. latifolia Pichon = Petchia erythrocarpa (Vatke) Leeuwenb. ●☆

64542 Cabucala madagascariensis (A. DC.) Pichon var. longipes (Pichon) Markgr. ex Boiteau = Petchia erythrocarpa (Vatke) Leeuwenb. ●☆

64543 Cabucala monarthron Pichon = Petchia erythrocarpa (Vatke) Leeuwenb. ●☆

64544 Cabucala montana Pichon = Petchia montana (Pichon) Leeuwenb. ●☆

64545 Cabucala multiflora Pichon = Petchia erythrocarpa (Vatke) Leeuwenb. ●☆

64546 Cabucala oblongo-ovata Markgr. = Petchia erythrocarpa (Vatke) Leeuwenb. ●☆

64547 Cabucala penduliflora Markgr. = Petchia madagascariensis (A. DC.) Leeuwenb. ●☆

64548 Cabucala plectaneiifolia Pichon = Petchia plectaneiifolia (Pichon) Leeuwenb. ●☆

64549 Cabucala polysperma (Scott-Elliot) Pichon = Petchia erythrocarpa (Vatke) Leeuwenb. ●☆

64550 Cabucala striolata Pichon = Petchia erythrocarpa (Vatke) Leeuwenb. ●☆

64551 Cacabus Bernh. = Exodeconus Raf. ■☆

64552 Cacalia DC. = Psacalium Cass. ■☆

64553 Cacalia Kuntze = Adenostyles Cass. ■☆

64554 Cacalia Kuntze = Behen Hill ●■

64555 Cacalia L. = Parasenecio W. W. Sm. et J. Small ■

64556 Cacalia L. = Senecio L. ■●

64557 Cacalia Lour. = Crassocephalum Moench(废弃属名)■

64558 Cacalia × cuneata (Honda) Kitam. = Parasenecio × cuneatus (Honda) H. Koyama ■☆

64559 Cacalia × koidzumiana Kitam. = Parasenecio koidzumianus (Kitam.) H. Koyama ■☆

64560 Cacalia × shiroumense Shizuo Ito et H. Koyama = Parasenecio × shiroumensis (Shizuo Ito et H. Koyama) H. Koyama ■☆

64561 Cacalia acaulis L. f. = Senecio acaulis (L. f.) Sch. Bip. ■☆

64562 Cacalia achyrotricha (Diels) Y. Ling = Ligularia achyrotricha (Diels) Y. Ling ■

64563 Cacalia aconitifolia Bunge = Syneilesis aconitifolia (Bunge) Maxim. ■

64564 Cacalia adenocauloides Hand. -Mazz. = Parasenecio roborowskii (Maxim.) Y. L. Chen ■

64565 Cacalia adenostyloides (Franch. et Sav. ex Maxim.) Matsum. = Parasenecio adenostyloides (Franch. et Sav. ex Maxim.) H. Koyama ■

64566 Cacalia ainsiliaeflora (Franch.) Hand. -Mazz. = Parasenecio ainsliiflorus (Franch.) Y. L. Chen ■

64567 Cacalia amagiensis Kitam. = Parasenecio amagiensis (Kitam.) H. Koyama ●☆

64568 Cacalia ambigua Y. Ling = Parasenecio ambiguus (Y. Ling) Y. L. Chen ■

64569 Cacalia ambigua Y. Ling var. wangiana Y. Ling = Parasenecio ambiguus (Y. Ling) Y. L. Chen var. wangianus (Y. Ling) Y. L. Chen ■

64570 Cacalia angulata Vahl = Solanecio angulatus (Vahl) C. Jeffrey ■☆

64571 Cacalia angulosa (Wall.) DC. = Gynura cusimbua (D. Don) S. Moore ■

64572 Cacalia angulosa Wall. = Gynura cusimbua (D. Don) S. Moore ■

64573 Cacalia anteuphorbia L. = Kleinia anteuphorbium (L.) Haw. ■☆

64574 Cacalia appendiculata L. f. = Pericallis appendiculata (L. f.) B. Nord. ■☆

64575 Cacalia arachnantha (Franch.) Hand. -Mazz. = Senecio arachnanthus Franch. ■

64576 Cacalia arbuscula Thunb. = Othonna arbuscula (Thunb.) Sch. Bip. ●☆

64577 Cacalia aristata (DC.) Kuntze = Vernonia natalensis Sch. Bip. ex Walp. ■☆

64578 Cacalia articulata L. f. = Senecio articulatus (L. f.) Sch. Bip. ■

64579 Cacalia atriplicifolia L. = Arnoglossum atriplicifolium (L.) H. Rob. ■☆

64580 Cacalia aurantiaca Blume = Gynura aurantiaca (Blume) DC. ■☆

64581 Cacalia auriculata DC. = Parasenecio auriculatus (DC.) J. R. Grant ■

64582 Cacalia auriculata DC. = Parasenecio gansuensis Y. L. Chen ■

64583 Cacalia auriculata DC. var. bulbifera Koidz. = Parasenecio auriculatus (DC.) J. R. Grant var. bulbifer (Koidz.) H. Koyama ■☆

64584 Cacalia auriculata DC. var. kamtschatica (Maxim.) Matsum. = Parasenecio auriculatus (DC.) H. Koyama ■

64585 Cacalia auriculata DC. var. ochotensis Kom. = Parasenecio auriculatus (DC.) H. Koyama ■

64586 Cacalia begoniifolia (Franch.) Hand. -Mazz. = Parasenecio begoniifolius (Franch.) Y. L. Chen ■

64587 Cacalia bicolor Roxb. ex Willd. = Gynura bicolor (Roxb. ex Willd.) DC. ■

64588 Cacalia bipinnata Thunb. = Senecio bipinnatus (Thunb.) Less. ■☆

64589 Cacalia bulbifera (Maxim.) Matsumura var. piligera Y. Ling = Parasenecio hwangshanicus (Y. Ling) C. I. Peng et S. W. Ching ■

64590 Cacalia bulbifera Maxim. var. acerina Makino = Parasenecio farfarifolius (Siebold et Zucc.) H. Koyama var. acerinus (Makino) H. Koyama ■☆

64591 Cacalia bulbifera Maxim. var. piligera Y. Ling = Parasenecio hwangshanicus (Y. Ling) Y. L. Chen ■

64592 Cacalia bulbiferoides Hand. -Mazz. = Parasenecio bulbiferoides (Hand. -Mazz.) Y. L. Chen ■

64593 Cacalia bulbosa Lour. = Gynura pseudochina (L.) DC. ■

64594 Cacalia capensis (Houtt.) Kuntze = Vernonia capensis (Houtt.) Druce ■☆

64595 Cacalia caroli (C. Winkl.) C. C. Chang = Sinacalia caroli (C. Winkl.) C. Jeffrey et Y. L. Chen ■

64596 Cacalia coccinea Sims = Emilia coccinea (Sims) G. Don ■

64597 Cacalia cordifolia L. f. = Mikania cordifolia (L. f.) Willd. ■☆

64598 Cacalia corymbosa (L. f.) Kuntze = Vernonia tigna Klatt ●☆

64599 Cacalia crepidifolia Nakai = Dendrocacalia crepidifolia (Nakai) Nakai ●☆

64600 Cacalia cusimbua D. Don = Gynura cusimbua (D. Don) S. Moore ■

64601 Cacalia cyclota (Bureau et Franch.) Hand. -Mazz. = Parasenecio cyclotus (Bureau et Franch.) Y. L. Chen ■

64602 Cacalia cylindrica Lam. = Othonna cylindrica (Lam.) DC. ●☆
64603 Cacalia dasythyrsa Hand.-Mazz. = Parasenecio dasythyrsus (Hand.-Mazz.) Y. L. Chen ■
64604 Cacalia davidii (Franch.) Hand.-Mazz. = Sinacalia davidii (Franch.) H. Koyama ■
64605 Cacalia decomposita A. Gray = Psacalium decompositum (A. Gray) H. Rob. et Brettell ■☆
64606 Cacalia delphiniifolia Siebold et Zucc. = Parasenecio delphiniifolius (Siebold et Zucc.) H. Koyama ■☆
64607 Cacalia delphiniifolia Siebold et Zucc. var. breviloba Sugim. et Sugino = Parasenecio delphiniifolius (Siebold et Zucc.) H. Koyama var. brevilobus (Sugim. et Sugino) Yonek. ■☆
64608 Cacalia delphyniphylla (H. Lév.) Hand.-Mazz. = Parasenecio delphiniphyllus (H. Lév.) Y. L. Chen ■
64609 Cacalia deltophylla (Maxim.) Mattf. ex Rehder et Kobuski = Parasenecio deltophyllus (Maxim.) Y. L. Chen ■
64610 Cacalia dephiniphylla (H. Lév.) Hand.-Mazz. = Parasenecio delphiniphyllus (H. Lév.) Y. L. Chen ■
64611 Cacalia diantha (Franch.) Hand.-Mazz. = Synotis erythropappa (Bureau et Franch.) C. Jeffrey et Y. L. Chen ■
64612 Cacalia didymantha (Dunn) Hand.-Mazz. = Sinacalia davidii (Franch.) H. Koyama ■
64613 Cacalia diversifolia Torr. et A. Gray = Arnoglossum diversifolium (Torr. et A. Gray) H. Rob. ■☆
64614 Cacalia elaeagnoides (DC.) Kuntze = Vernonia oligocephala (DC.) Sch. Bip. ex Walp. ■☆
64615 Cacalia elliottii (R. M. Harper) Shinners = Arnoglossum ovatum (Walter) H. Rob. ■☆
64616 Cacalia farfarifolia Kuntze subsp. petasitoides (H. Lév.) H. Koyama = Parasenecio petasitoides (H. Lév.) Y. L. Chen ■
64617 Cacalia farfarifolia Siebold et Zucc. = Parasenecio farfarifolius (Siebold et Zucc.) H. Koyama ■☆
64618 Cacalia farfarifolia Siebold et Zucc. subsp. petasitoides (H. Lév.) Koyama = Parasenecio petasitoides (H. Lév.) Y. L. Chen ■
64619 Cacalia farfarifolia Siebold et Zucc. var. acerina (Makino) Kitam. = Parasenecio farfarifolius (Siebold et Zucc.) H. Koyama var. acerinus (Makino) H. Koyama ■☆
64620 Cacalia ficoides L. = Senecio ficoides (L.) Sch. Bip. ■☆
64621 Cacalia firma Kom. = Parasenecio firmus (Kom.) Y. L. Chen ■
64622 Cacalia floridana A. Gray = Arnoglossum floridanum (A. Gray) H. Rob. ■☆
64623 Cacalia forrestii (W. W. Sm. et J. Small.) Hand.-Mazz. = Parasenecio forrestii W. W. Sm. et J. Small ■
64624 Cacalia gerrardii (Harv.) Kuntze = Vernonia gerrardii Harv. ■☆
64625 Cacalia harveyi Kuntze = Vernonia galpinii Klatt ■☆
64626 Cacalia hastata L. = Parasenecio hastatus (L.) H. Koyama ■
64627 Cacalia hastata L. f. glabra (Ledeb.) Kitag. = Parasenecio hastatus (L.) H. Koyama var. glaber (Ledeb.) Y. L. Chen ■
64628 Cacalia hastata L. subsp. farfarifolia (Koidz.) Kitam. = Parasenecio maximowiczianus (Nakai et F. Maek. ex H. Hara) H. Koyama ■☆
64629 Cacalia hastata L. subsp. farfarifolia (Koidz.) Kitam. var. alata (F. Maek.) Kitam. = Parasenecio maximowiczianus (Nakai et F. Maek. ex H. Hara) H. Koyama var. alatus (F. Maek.) H. Koyama ■☆
64630 Cacalia hastata L. subsp. komaroviana (Pojark.) Kitag. = Parasenecio komarovianus (Poljakov) Y. L. Chen ■
64631 Cacalia hastata L. subsp. lancifolia (Franch.) H. Koyama = Parasenecio lancifolius (Franch.) Y. L. Chen ■
64632 Cacalia hastata L. subsp. orientalis Kitam. = Parasenecio hastatus (L.) H. Koyama subsp. orientalis (Kitam.) H. Koyama ■☆
64633 Cacalia hastata L. subsp. orientalis Kitam. var. chokaiensis (Kudo) Kitam. = Calamagrostis adpressiramea Ohwi ■☆
64634 Cacalia hastata L. subsp. orientalis Kitam. var. hayachinensis Kitam. = Parasenecio hayachinensis (Kitam.) Kadota ■☆
64635 Cacalia hastata L. subsp. orientalis Kitam. var. ramosa (Maxim.) Kitam. = Parasenecio hastatus (L.) H. Koyama subsp. orientalis (Kitam.) H. Koyama var. ramosus (Maxim.) H. Koyama ■☆
64636 Cacalia hastata L. subsp. orientalis Kitam. var. tanakae (Franch. et Sav.) Kitam. = Parasenecio hastatus (L.) H. Koyama subsp. tanakae (Franch. et Sav.) H. Koyama ■☆
64637 Cacalia hastata L. subsp. tanakae (Franch. et Sav.) H. Koyama = Parasenecio hastatus (L.) H. Koyama subsp. tanakae (Franch. et Sav.) H. Koyama ■☆
64638 Cacalia hastata L. var. farfarifolia (Koidz.) Ohwi = Parasenecio maximowiczianus (Nakai et F. Maek. ex H. Hara) H. Koyama ■☆
64639 Cacalia hastata L. var. glabra Ledeb. = Parasenecio hastatus (L.) H. Koyama var. glaber (Ledeb.) Y. L. Chen ■
64640 Cacalia hastata L. var. lancifolia (Franch.) Koyama = Parasenecio lancifolius (Franch.) Y. L. Chen ■
64641 Cacalia hastata L. var. orientalis (Kitam.) Ohwi = Parasenecio hastatus (L.) H. Koyama subsp. orientalis (Kitam.) H. Koyama ■☆
64642 Cacalia hastata L. var. pubescens Ledeb. = Parasenecio hastatus (L.) H. Koyama ■
64643 Cacalia haworthii Sweet = Senecio haworthii (Sweet) Sch. Bip. ■
64644 Cacalia heterophylla W. Bartram = Garberia heterophylla (W. Bartram) Merr. et F. Harper ■☆
64645 Cacalia hupehensis Hand.-Mazz. = Parasenecio phyllolepis (Franch.) Y. L. Chen ■
64646 Cacalia hwangshanica Y. Ling = Parasenecio hwangshanicus (Y. Ling) C. I. Peng et S. W. Ching ■
64647 Cacalia hwangshanica Y. Ling = Parasenecio hwangshanicus (Y. Ling) Y. L. Chen ■
64648 Cacalia ianthophylla (Franch.) Hand.-Mazz. = Parasenecio ianthophyllus (Franch.) Y. L. Chen ■
64649 Cacalia incana L. = Gynura divaricata (L.) DC. ■
64650 Cacalia intermedia Hayata = Syneilesis intermedia (Hayata) Kitam. ■
64651 Cacalia intermedia Hayata var. subglabrata Gogeleinet Sasaki = Syneilesis subglabrata (Gogeleinet Sasaki) Kitam. ■
64652 Cacalia intermedia Hayata var. subglabrata Yamam. et Sasaki = Syneilesis subglabrata (Yamam. et Sasaki) Kitam. ■
64653 Cacalia kamtschatica (Maxim.) Kudo = Parasenecio auriculatus (DC.) J. R. Grant var. kamtschaticus (Maxim.) H. Koyama ■☆
64654 Cacalia kangxianensis Z. Ying Zhang et Y. H. Guo = Parasenecio kangxianensis (Z. Y. Zhang et Y. H. Gou) Y. L. Chen ■
64655 Cacalia kitamurana Nakai = Parasenecio hastatus (L.) H. Koyama subsp. orientalis (Kitam.) H. Koyama ■☆
64656 Cacalia komaroviana (Pojark.) Pojark. = Parasenecio komarovianus (Poljakov) Y. L. Chen ■
64657 Cacalia koualapensis (Franch.) Hand.-Mazz. = Parasenecio koualapensis (Franch.) Y. L. Chen ■
64658 Cacalia krameri Matsum. = Syneilesis australis Y. Ling ■
64659 Cacalia lanceolata Nutt. = Arnoglossum ovatum (Walter) H. Rob. ■☆

64660 Cacalia lanceolata Nutt. var. elliottii (R. M. Harper) Král et R. K. Godfrey = Arnoglossum ovatum (Walter) H. Rob. ■☆

64661 Cacalia latipes (Franch.) Hand.-Mazz. = Parasenecio latipes (Franch.) Y. L. Chen ■

64662 Cacalia leucanthema (Dunn) Y. Ling = Parasenecio ainsliiflorus (Franch.) Y. L. Chen ■

64663 Cacalia leucocephala (Franch.) Hand.-Mazz. = Parasenecio leucocephalus (Franch.) Y. L. Chen ■

64664 Cacalia lidjiangensis Hand.-Mazz. = Parasenecio lijiangensis (Hand.-Mazz.) Y. L. Chen ■

64665 Cacalia lidjiangensis Hand.-Mazz. var. acerina Koyama = Parasenecio rockianus (Hand.-Mazz.) Y. L. Chen ■

64666 Cacalia longispica Hand.-Mazz. = Parasenecio longispicus (Hand.-Mazz.) Y. L. Chen ■

64667 Cacalia longispica Z. Y. Zhang et Y. H. Guo = Ligulariopsis shichuana Y. L. Chen ■

64668 Cacalia macrocephala Hand.-Mazz. = Sinacalia macrocephala (H. Rob. et Brettell) C. Jeffrey et Y. L. Chen ■

64669 Cacalia makinoana (Yatabe) Makino = Miricacalia makinoana (Yatabe) Kitam. ■☆

64670 Cacalia matsudae Kitam. = Parasenecio matsudai (Kitam.) Y. L. Chen ■

64671 Cacalia matsumurana Kudo = Parasenecio auriculatus (DC.) J. R. Grant var. bulbifer (Koidz.) H. Koyama ●☆

64672 Cacalia maximowicziana Nakai et F. Maek. ex H. Hara = Parasenecio maximowiczianus (Nakai et F. Maek. ex H. Hara) H. Koyama ■☆

64673 Cacalia maximowicziana Nakai et F. Maek. ex H. Hara var. alata F. Maek. = Parasenecio maximowiczianus (Nakai et F. Maek. ex H. Hara) H. Koyama var. alatus (F. Maek.) H. Koyama ■☆

64674 Cacalia monantha (Diels) Hayata = Parasenecio morrisonensis Y. L. Chen ■

64675 Cacalia muhlenbergii (Sch. Bip.) Fernald = Arnoglossum reniforme (Hook.) H. Rob. ■☆

64676 Cacalia nardosmia A. Gray = Cacaliopsis nardosmia (A. Gray) A. Gray ■☆

64677 Cacalia nardosmia A. Gray var. glabrata (Piper) B. Boivin = Cacaliopsis nardosmia (A. Gray) A. Gray ■☆

64678 Cacalia nokoensis Masam. et Suzuki = Parasenecio nokoensis (Masam. et Suzuki) C. I. Peng et S. W. Chung ■

64679 Cacalia nokoensis Masam. et Suzuki = Parasenecio nokoensis (Masam. et Suzuki) Y. L. Chen ■

64680 Cacalia nudicaulis (DC.) Kuntze = Vernonia dregeana Sch. Bip. ■☆

64681 Cacalia odora Forssk. = Kleinia odora (Forssk.) DC. ■☆

64682 Cacalia otopteryx Hand.-Mazz. = Parasenecio otopteryx (Hand.-Mazz.) Y. L. Chen ■

64683 Cacalia ovalis Ker Gawl. = Gynura divaricata (L.) DC. ■

64684 Cacalia ovata Walter = Arnoglossum ovatum (Walter) H. Rob. ■☆

64685 Cacalia palmatisecta (Jeffrey) Hand.-Mazz. = Parasenecio otopteryx (Hand.-Mazz.) Y. L. Chen ■

64686 Cacalia palmatisecta (Jeffrey) Hand.-Mazz. = Parasenecio palmatisectus (Jeffrey) Y. L. Chen ■

64687 Cacalia palmatisecta (Jeffrey) Hand.-Mazz. var. moupingensis (Franch.) Koyama = Parasenecio palmatisectus (Jeffrey) Y. L. Chen var. moupingensis (Franch.) Y. L. Chen ■

64688 Cacalia palmatisecta (Jeffrey) Hand.-Mazz. var. moupingensis (Franch.) Koyama f. pilipes Koyama = Parasenecio palmatisectus (Jeffrey) Y. L. Chen var. moupingensis (Franch.) Y. L. Chen ■

64689 Cacalia palmatisecta (Jeffrey) Hand.-Mazz. var. pubescens (Jeffrey) C. Y. Wu = Parasenecio palmatisectus (Jeffrey) Y. L. Chen var. moupingensis (Franch.) Y. L. Chen ■

64690 Cacalia paniculata Raf. = Arnoglossum atriplicifolium (L.) H. Rob. ■☆

64691 Cacalia paniculata Raf. = Arnoglossum plantagineum Raf. ■☆

64692 Cacalia papillaris L. = Tylecodon papillaris (L.) G. D. Rowley ■☆

64693 Cacalia pendula Forssk. = Kleinia pendula (Forssk.) DC. ■☆

64694 Cacalia penicillata Cass. = Senecio penicillatus (Cass.) Sch. Bip. ■☆

64695 Cacalia penninervis H. Koyama = Senecio kumaonensis Duthie ex C. Jeffrey et Y. L. Chen ■

64696 Cacalia pentaloba Hand.-Mazz. = Parasenecio quinquelobus (Wall. ex DC.) Y. L. Chen ■

64697 Cacalia pentaloba Hand.-Mazz. var. moupinensis (Franch.) Hand.-Mazz. = Parasenecio palmatisectus (Jeffrey) Y. L. Chen var. moupingensis (Franch.) Y. L. Chen ■

64698 Cacalia pentaloba Hand.-Mazz. var. sinuata H. Koyama = Parasenecio quinquelobus (Wall. ex DC.) Y. L. Chen var. sinuatus (Koyama) Y. L. Chen ■

64699 Cacalia phyllolepis (Franch.) Hand.-Mazz. = Parasenecio phyllolepis (Franch.) Y. L. Chen ■

64700 Cacalia pilgeriana (Diels) Y. Ling = Parasenecio pilgerianus (Diels) Y. L. Chen ■

64701 Cacalia pilgeriana (Diels) Y. Ling subsp. dephiniphyllus (H. Lév.) H. Koyama = Parasenecio delphiniphyllus (H. Lév.) Y. L. Chen ■

64702 Cacalia pinnatifida Lour. = Gynura japonica (Thunb.) Juel ■

64703 Cacalia pinnatifida P. J. Bergius = Senecio pinnatifidus (P. J. Bergius) Less. ■☆

64704 Cacalia plantaginea (Raf.) Shinners; 印度蟹甲草; Indian Plantain ■☆

64705 Cacalia plantaginea (Raf.) Shinners = Arnoglossum plantagineum Raf. ■☆

64706 Cacalia potaninii (C. Winkl.) Mattf. = Ligularia potaninii (C. Winkl.) Y. Ling ■

64707 Cacalia praetermissa (Poljakov) Poljakov = Parasenecio praetermissus (Poljakov) Y. L. Chen ■

64708 Cacalia procumbens Lour. = Gynura procumbens (Lour.) Merr. ■

64709 Cacalia profundora (Dunn) Hand.-Mazz. = Parasenecio profundorum (Dunn) Y. L. Chen ■

64710 Cacalia pteranthes Raf. = Arnoglossum plantagineum Raf. ■☆

64711 Cacalia quinqueloba (Wall. ex DC.) Kitam. = Parasenecio quinquelobus (Wall. ex DC.) Y. L. Chen ■

64712 Cacalia quinqueloba Thunb. = Senecio quinquelobus (Thunb.) DC. ■☆

64713 Cacalia radicans L. f. = Senecio radicans (L. f.) Sch. Bip. ■

64714 Cacalia reniformis Muhl. ex Willd. = Arnoglossum reniforme (Hook.) H. Rob. ■☆

64715 Cacalia roborowskii (Maxim.) Y. Ling = Parasenecio roborowskii (Maxim.) Y. L. Chen ■

64716 Cacalia robusta Tolm. = Parasenecio hastatus (L.) H. Koyama subsp. orientalis (Kitam.) H. Koyama ■☆

64717 Cacalia rockiana Hand.-Mazz. = Parasenecio rockianus (Hand.-Mazz.) Y. L. Chen ■

64718 Cacalia rotundifolia (Raf.) House = Arnoglossum atriplicifolium (L.) H. Rob. ■☆
64719 Cacalia rubescens (S. Moore) Matsuda = Parasenecio rubescens (S. Moore) Y. L. Chen ■
64720 Cacalia rufipilis (Franch.) Y. Ling = Parasenecio rufipilis (Franch.) Y. L. Chen ■
64721 Cacalia rugelia (A. Gray) T. M. Barkley et Cronquist = Rugelia nudicaulis Shuttlew. ex Chapm. ■☆
64722 Cacalia sagittata Willd. = Emilia coccinea (Sims) G. Don ■
64723 Cacalia sarmentosa Blume = Gynura procumbens (Lour.) Merr. ■
64724 Cacalia scandens Aiton = Senecio deltoideus Less. ■☆
64725 Cacalia segetum Lour. = Gynura japonica (Thunb.) Juel ■
64726 Cacalia sinicus Y. Ling = Parasenecio sinicus (Y. Ling) Y. L. Chen ■
64727 Cacalia sonchifolia L. = Emilia sonchifolia (L.) DC. ex Wight ■
64728 Cacalia souliei (Franch.) Hand. -Mazz. = Parasenecio souliei (Franch.) Y. L. Chen ■
64729 Cacalia suaveolens (Kunth) Kuntze = Cacalia suaveolens L. ■☆
64730 Cacalia suaveolens Kuntze = Cacalia suaveolens L. ■☆
64731 Cacalia suaveolens L.;北美芳香蟹甲草;Indian Plantain, Sweet-scented Indian Plantain ■☆
64732 Cacalia suaveolens L. = Hasteola suaveolens (L.) Pojark. ■☆
64733 Cacalia suaveolens Zeakert? = Cacalia suaveolens L. ■☆
64734 Cacalia subglabra C. C. Chang = Parasenecio subglaber (C. C. Chang) Y. L. Chen ■
64735 Cacalia subglabrata (Gogeleinet Sasaki) Kitam. = Syneilesis subglabrata (Gogeleinet Sasaki) Kitam. ■
64736 Cacalia sulcata Fernald = Arnoglossum sulcatum (Fernald) H. Rob. ■☆
64737 Cacalia taliensis (Franch.) Hand. -Mazz. = Parasenecio taliensis (Franch.) Y. L. Chen ■
64738 Cacalia tangutica (Franch.) Hand. -Mazz. = Sinacalia tangutica (Maxim.) B. Nord. ■
64739 Cacalia tatsienensis (Bureau et Franch.) Hand. -Mazz. = Parasenecio roborowskii (Maxim.) Y. L. Chen ■
64740 Cacalia tebakoensis (Makino) Makino = Parasenecio tebakoensis (Makino) H. Koyama ■☆
64741 Cacalia teniana Hand. -Mazz. = Parasenecio tenianus (Hand. -Mazz.) Y. L. Chen ■
64742 Cacalia thunbergii Nakai = Syneilesis palmata (Thunb.) Maxim. ■☆
64743 Cacalia tomentosa Thunb. = Senecio thunbergii Harv. ■☆
64744 Cacalia tricuspis (Franch.) Hand. -Mazz. = Senecio tricuspis Franch. ■
64745 Cacalia tripteria Hand. -Mazz. = Parasenecio tripteris (Hand. -Mazz.) Y. L. Chen ■
64746 Cacalia tsinlingensis Hand. -Mazz. = Parasenecio tsinlingensis (Hand. -Mazz.) Y. L. Chen ■
64747 Cacalia tuberosa Nutt. = Arnoglossum plantagineum Raf. ■☆
64748 Cacalia tuberosa Nutt. = Cacalia plantaginea (Raf.) Shinners ■☆
64749 Cacalia tuberosa Nutt.;块状蟹甲草;Tuberous Indian Plantain ■☆
64750 Cacalia vespertilio (Franch.) Hand. -Mazz. = Parasenecio vespertilo (Franch.) Y. L. Chen ■
64751 Cacalia volubilis Blume = Cissampelopsis volubilis (Blume) Miq. ●
64752 Cacalia wangiana Y. Ling = Parasenecio ambiguus (Y. Ling) Y. L. Chen var. wangianus (Y. Ling) Y. L. Chen ■
64753 Cacalia xanthotricha Grüning = Ligularia xanthotricha (Grüning) Y. Ling ■
64754 Cacalia xinjiashanensis Z. Ying Zhang et Y. H. Guo = Parasenecio xinjiashanensis (Z. Ying Zhang et Y. H. Guo) Y. L. Chen ■
64755 Cacalia yatabei Matsum. et Koidz. = Parasenecio yatabei (Matsum. et Koidz.) H. Koyama ■☆
64756 Cacalia yatabei Matsum. et Koidz. var. occidentalis F. Maek. ex Kitam. = Parasenecio yatabei (Matsum. et Koidz.) H. Koyama var. occidentalis (F. Maek. ex Kitam.) H. Koyama ■☆
64757 Cacaliopsis A. Gray(1883);类蟹甲属■
64758 Cacaliopsis nardosmia (A. Gray) A. Gray;类蟹甲■☆
64759 Cacao Mill. = Theobroma L. ●
64760 Cacao Tourn. ex Mill. = Theobroma L. ●
64761 Cacaoaceae Augier ex T. Post et Kuntze = Malvaceae Juss.(保留科名)●■
64762 Cacaoaceae T. Post et Kuntze = Malvaceae Juss.(保留科名)●■
64763 Cacara Rumph. ex Thouars = Pachyrhizus Rich. ex DC.(保留属名)■
64764 Cacara Thouars(废弃属名) = Pachyrhizus Rich. ex DC.(保留属名)■
64765 Cacara orbicularis (Welw. ex Baker) Hiern = Neorautanenia mitis (A. Rich.) Verdc. ■☆
64766 Cacatali Adans. = Pedalium D. Royen ex L. ■☆
64767 Caccinia Savi(1832);卡克草属■☆
64768 Caccinia crassifolia (Vent.) K. Koch = Caccinia macranthera (Banks et Sol.) Brand var. crassifolia (Vent.) Brand ■☆
64769 Caccinia dubia Bunge;可疑卡克草■☆
64770 Caccinia glauca Savi = Caccinia macranthera (Banks et Sol.) Brand var. crassifolia (Vent.) Brand ■☆
64771 Caccinia macranthera (Banks et Sol.) Brand;大药卡克草■☆
64772 Caccinia macranthera (Banks et Sol.) Brand var. crassifolia (Vent.) Brand;厚叶卡克草■☆
64773 Cachris D. Dietr. = Cachrys L. ■
64774 Cachrydium Link = Cachrys L. ■
64775 Cachrydium Link = Hippomarathrum Link ■☆
64776 Cachrys L. (1753);绵果芹属;Cachris, Cachrys ■
64777 Cachrys abyssinica Hochst. ex A. Rich. = Diplolophium africanum Turcz. ■☆
64778 Cachrys alpina M. Bieb.;高山绵果芹■☆
64779 Cachrys athamanthoides M. Bieb. = Stenocoelium athamantoides (M. Bieb.) Ledeb. ■
64780 Cachrys cristata DC.;冠毛绵果芹■☆
64781 Cachrys didyma Regel = Cryptodiscus didymus (Regel) Korovin ■
64782 Cachrys didyma Regel = Prangos didyma (Regel) Pimenov et V. N. Tikhom. ■
64783 Cachrys herderi Regel;海尔绵果芹■☆
64784 Cachrys humilis Schousb. = Cachrys libanotis L. ■☆
64785 Cachrys libanotis L.;香料绵果芹■☆
64786 Cachrys libanotis L. var. leiocarpa (Boiss. et Reut.) Pau = Cachrys libanotis L. ■☆
64787 Cachrys libanotis L. var. pterochlaena DC. = Cachrys sicula L. ■☆
64788 Cachrys macrocarpa Ledeb.;大果绵果芹;Bigfruit Cachrys ■
64789 Cachrys macrocarpa Ledeb. = Prangos ledebourii Herrnst. et Heyn ■
64790 Cachrys peucedanoides Desf.;前胡绵果芹■☆
64791 Cachrys pubescens (Pall.) Schischk.;毛绵果芹■☆
64792 Cachrys pungens Guss.;刚毛绵果芹■☆

64793 Cachrys seseloides (Hoffm.) M. Bieb. = Saposhnikovia divaricata (Turcz.) Schischk. ■

64794 Cachrys sibirica Fisch. ex Spreng. = Phlojodicarpus sibiricus (Fisch. ex Spreng.) Koso-Pol. ■

64795 Cachrys sibirica Steph. ex Fisch. = Phlojodicarpus sibiricus (Stephan ex Spreng.) Koso-Pol. ■

64796 Cachrys sicula L.;西西里绵果芹■☆

64797 Cachrys tomentosa Desf. = Magydaris pastinacea (Lam.) Fiori ■☆

64798 Cachrys vaginata Ledeb. = Schrenkia vaginata (Ledeb.) Fisch. et C. A. Mey. ■

64799 Cachyris Zumagl. = Cachrys L. ■

64800 Caconapea Cham. = Bacopa Aubl.(保留属名)■

64801 Caconapea Cham. = Mella Vand. ■

64802 Caconobea Walp. = Caconapea Cham. ■

64803 Cacosmanthus de Vriese = Madhuca Buch.-Ham. ex J. F. Gmel. ●

64804 Cacosmanthus Miq. = Kakosmanthus Hassk. ●

64805 Cacosmanthus Miq. = Payena A. DC. ●☆

64806 Cacosmia Kunth(1820);无冠黄安菊属●☆

64807 Cacosmia harlingii B. Nord.;哈氏无冠黄安菊●☆

64808 Cacosmia hieronymi H. Rob.;希氏无冠黄安菊●☆

64809 Cacosmia rugosa Kunth;无冠黄安菊●☆

64810 Cacotanis Raf. = Boltonia L'Hér. ■☆

64811 Cacoucia Aubl. = Combretum Loefl.(保留属名)●

64812 Cacoucia barteri Hemsl. = Combretum platypterum (Welw.) Hutch. et Dalziel ●☆

64813 Cacoucia bracteata M. A. Lawson = Combretum bracteatum (M. A. Lawson) Engl. et Diels ●☆

64814 Cacoucia littorea Engl. = Combretum falcatum (Welw. ex Hiern) Jongkind ●☆

64815 Cacoucia longispicata Engl. = Combretum longispicatum (Engl.) Engl. et Diels ●☆

64816 Cacoucia paniculata M. A. Lawson = Combretum platypterum (Welw.) Hutch. et Dalziel ●☆

64817 Cacoucia platyptera Hemsl. = Combretum bracteatum (M. A. Lawson) Engl. et Diels ●☆

64818 Cacoucia platyptera Welw. = Combretum platypterum (Welw.) Hutch. et Dalziel ●☆

64819 Cacoucia splendens Hemsl. = Combretum bracteatum (M. A. Lawson) Engl. et Diels ●☆

64820 Cacoucia velutina S. Moore = Combretum mooreanum Exell ●☆

64821 Cacoucia villosa Lawson = Combretum dolichopetalum Engl. et Diels ●☆

64822 Cactaceae Juss.(1789)(保留科名);仙人掌科;Cactus Family ●■

64823 Cactodendron Bigelow = Opuntia Mill. ●

64824 Cactus Britton et Rose = Melocactus Link et Otto(保留属名)●

64825 Cactus Kuntze = Mammillaria Haw.(保留属名)●

64826 Cactus L.(废弃属名)= Mammillaria Haw.(保留属名)●

64827 Cactus L.(废弃属名) = Melocactus Link et Otto + Mammallaria Haw. + Cerena Mill. + Opuntia Mill. + Pereakia Mill. 等●■

64828 Cactus Lem. = Opuntia Mill. ●

64829 Cactus sensu Britton et Rose = Melocactus Link et Otto(保留属名)●

64830 Cactus bicolor Terán et Berland. = Hamatocactus bicolor (Terán et Berland.) I. M. Johnst. ■☆

64831 Cactus chinensis Roxb. = Opuntia ficus-indica (L.) Mill. ●

64832 Cactus clavatus (Engelm.) Lem. = Grusonia clavata (Engelm.) H. Rob. ■☆

64833 Cactus cochenillifer L. = Nopalea cochenillifera (L.) Salm-Dyck ●☆

64834 Cactus cochenillifer L. = Opuntia cochinellifera (L.) Mill. ■

64835 Cactus cylindricus James = Cylindropuntia imbricata (Haw.) F. M. Knuth ■☆

64836 Cactus cylindricus Lam. = Austrocylindropuntia cylindrica (Lam.) Backeb. ■☆

64837 Cactus decumanus Willd. = Opuntia ficus-indica (L.) Mill. ●

64838 Cactus dillenii Ker Gawl. = Opuntia dillenii (Ker Gawl.) Haw. ●

64839 Cactus dillenii Ker Gawl. = Opuntia stricta (Haw.) Haw. var. dillenii (Ker Gawl.) L. D. Benson ■

64840 Cactus echinocarpus (Engelm. et Bigelow) Lem. = Cylindropuntia echinocarpa (Engelm. et J. M. Bigelow) F. M. Knuth ■☆

64841 Cactus ferox Nutt. = Opuntia polyacantha Haw. ●☆

64842 Cactus ficus-indica L. = Opuntia ficus-indica (L.) Mill. ●

64843 Cactus fragilis Nutt. = Opuntia fragilis (Nutt.) Haw. ■☆

64844 Cactus grandiflorus L. = Cereus grandiflorus (L.) Mill. ●

64845 Cactus humifusus Raf. = Opuntia humifusa (Raf.) Raf. ●☆

64846 Cactus indicus Roxb. = Opuntia monacantha (Willd.) Haw. ■

64847 Cactus microdasys Lehm. = Opuntia microdasys (Lehm.) Lehm. ex Pfeiff. ■

64848 Cactus monacanthos Willd. = Opuntia monacantha (Willd.) Haw. ■

64849 Cactus opuntia L. = Opuntia ficus-indica (L.) Mill. ●

64850 Cactus opuntia L. var. inermis DC. = Opuntia inermis (DC.) DC. ■☆

64851 Cactus pereskia L. = Pereskia aculeata Mill. ●

64852 Cactus peruvianus L. = Cereus peruvianus (L.) Mill. ●

64853 Cactus phyllanthus L. = Epiphyllum phyllanthus (L.) Haw. ■☆

64854 Cactus proliferus Mill. = Mammillaria prolifera (Mill.) Haw. ■☆

64855 Cactus pusillus Haw. = Opuntia pusilla (Haw.) Haw. ■☆

64856 Cactus radiosus (Engelm.) J. M. Coult. var. alversonii J. M. Coult. = Coryphantha alversonii (J. M. Coult.) Orcutt ■☆

64857 Cactus scopa Link et Otto = Cereus scopa Salm-Dyck ex DC. ■☆

64858 Cactus scopa Spreng. = Cereus scopa Salm-Dyck ex DC. ■☆

64859 Cactus scopa Spreng. = Echinocactus scopa Link et Otto ■

64860 Cactus scopa Spreng. = Malacocarpus scopa (Spreng.) Britton et Rose ■

64861 Cactus scopa Spreng. = Notocactus scopa (Spreng.) Backeb. ■

64862 Cactus scopa Spreng. = Peronocactus scopa (Spreng.) Doweld ■

64863 Cactus strictus Haw. = Opuntia santa-rita (Griffiths et Hare) Rose ■☆

64864 Cactus strictus Haw. = Opuntia stricta (Haw.) Haw. ■

64865 Cactus triacanthos Willd. = Opuntia triacantha (Willd.) Sweet ■☆

64866 Cactus triangularis L. = Hylocereus triangularis (L.) Britton et Rose ●

64867 Cactus tuna L. = Opuntia tuna (L.) Mill. ■☆

64868 Cactus tunicatus Lehm. = Cylindropuntia tunicata (Lehm.) F. M. Knuth ■☆

64869 Cactus viviparus Nutt. = Coryphantha vivipara (Nutt.) Britton et Rose ■☆

64870 Cacucia J. F. Gmel. = Cacoucia Aubl. ●

64871 Cacucia J. F. Gmel. = Combretum Loefl.(保留属名)●

64872 Cacuvallum Medik. = Mucuna Adans.(保留属名)●■

64873 Cadaba Forssk.(1775);腺花山柑属(热带白花菜属)●☆
64874 Cadaba adenotricha Gilg et Gilg-Ben. = Cadaba farinosa Forssk. subsp. adenotricha (Gilg et Gilg-Ben.) R. A. Graham ●☆
64875 Cadaba aphylla (Thunb.) Wild;无叶腺花山柑●☆
64876 Cadaba apiculata Gilg et Gilg-Ben. = Cadaba farinosa Forssk. subsp. adenotricha (Gilg et Gilg-Ben.) R. A. Graham ●☆
64877 Cadaba baccarinii Chiov.;巴卡林腺花山柑●☆
64878 Cadaba barbigera Gilg;髯毛腺花山柑●☆
64879 Cadaba benguellensis Mendes;本格拉腺花山柑●☆
64880 Cadaba carneo-viridis Gilg et Gilg-Ben.;肉绿色腺花山柑●☆
64881 Cadaba dasyantha Gilg et Gilg-Ben. = Cadaba carneo-viridis Gilg et Gilg-Ben. ●☆
64882 Cadaba divaricata Gilg;叉开腺花山柑●☆
64883 Cadaba dubia DC. = Cadaba farinosa Forssk. ●☆
64884 Cadaba farinosa Forssk.;被粉腺花山柑●☆
64885 Cadaba farinosa Forssk. subsp. adenotricha (Gilg et Gilg-Ben.) R. A. Graham;腺毛山柑●☆
64886 Cadaba farinosa Forssk. subsp. rariflora Jafri;疏花被粉腺花山柑●☆
64887 Cadaba gillettii R. A. Graham;吉莱特腺花山柑●☆
64888 Cadaba glaberrima Gilg et Gilg-Ben.;光滑腺花山柑●☆
64889 Cadaba glandulosa Forssk.;具腺腺花山柑●☆
64890 Cadaba heterotricha Stocks ex Hook.;异毛山柑●☆
64891 Cadaba juncea (Sparrm.) Harv. ex Hook. f. = Cadaba aphylla (Thunb.) Wild ●☆
64892 Cadaba kassasii Chrtek;卡萨斯腺花山柑●☆
64893 Cadaba kirkii Oliv.;柯克腺花山柑●☆
64894 Cadaba longifolia DC.;长叶腺花山柑●☆
64895 Cadaba longifolia DC. var. frutescens Chiov. = Cadaba longifolia DC. ●☆
64896 Cadaba longifolia DC. var. scandens (Pax) Chiov. = Cadaba longifolia DC. ●☆
64897 Cadaba macropoda Gilg = Cadaba termitaria N. E. Br. ●☆
64898 Cadaba madagascariensis Baill. = Cadaba virgata Bojer ●☆
64899 Cadaba mirabilis Gilg;奇异腺花山柑●☆
64900 Cadaba mombassana Gilg et Gilg-Ben. = Cadaba farinosa Forssk. ●☆
64901 Cadaba nakakope Gilg et Gilg-Ben. = Cadaba carneo-viridis Gilg et Gilg-Ben. ●☆
64902 Cadaba natalensis Sond.;纳塔尔腺花山柑●☆
64903 Cadaba obovata E. A. Bruce = Cadaba stenopoda Gilg et Gilg-Ben. ●☆
64904 Cadaba parvula Polhill;较小腺花山柑●☆
64905 Cadaba rotundifolia Forssk.;圆叶腺花山柑●☆
64906 Cadaba ruspolii Gilg;鲁斯波利腺花山柑●☆
64907 Cadaba scandens Pax = Cadaba longifolia DC. ●☆
64908 Cadaba somalensis Franch.;索马里腺花山柑●☆
64909 Cadaba stenopoda Gilg et Gilg-Ben.;细梗腺花山柑●☆
64910 Cadaba termitaria N. E. Br.;顶生腺花山柑●☆
64911 Cadaba virgata Bojer;条纹腺花山柑●☆
64912 Cadacya Raf. = Kadakia Raf. ■
64913 Cadacya Raf. = Monochoria C. Presl ■
64914 Cadalvena Fenzl = Costus L. ■
64915 Cadalvena dalzielii C. H. Wright = Costus spectabilis (Fenzl) K. Schum. ■☆
64916 Cadalvena pistiifolia (K. Schum.) Baker = Costus spectabilis (Fenzl) K. Schum. ■☆
64917 Cadalvena spectabilis Fenzl = Costus spectabilis (Fenzl) K. Schum. ■☆
64918 Cadamba Sonn. = Guettarda L. ●
64919 Cadelari Adans. = Achyranthes L.(保留属名)■
64920 Cadelari Medik. = Pupalia Juss.(保留属名)■☆
64921 Cadelaria Raf. = Cadelari Adans. ■
64922 Cadelium Medik. = Phaseolus L. ■
64923 Cadellia F. Muell.(1860);澳洲海人树属●☆
64924 Cadellia pentastylis F. Muell.;澳洲海人树●☆
64925 Cadetia Gaudich.(1829);卡德兰属■☆
64926 Cadetia adenantha Schltr.;腺花卡德兰■☆
64927 Cadetia albiflora Schltr.;白花卡德兰■☆
64928 Cadetia angustifolia Blume;窄叶卡德兰■☆
64929 Cadetia lactiflora Schltr.;乳花卡德兰■☆
64930 Cadetia lucida Schltr.;光亮卡德兰■☆
64931 Cadetia wariana Schltr.;新几内亚卡德兰■☆
64932 Cadia Forssk.(1775);卡迪豆属(卡迪亚豆属);Cadia ●■☆
64933 Cadia anomala Vatke = Xanthocercis madagascariensis Baill. ●☆
64934 Cadia baronii Drake = Neoharmsia baronii (Drake) R. Vig. ●☆
64935 Cadia catatii Drake = Cadia commersoniana Baill. ●☆
64936 Cadia commersoniana Baill.;科梅逊卡迪豆●☆
64937 Cadia ellisiana Baker;卡迪豆(卡迪亚豆)●☆
64938 Cadia emarginatior M. Pelt.;微缺卡迪豆●☆
64939 Cadia pedicellata Baker;梗花卡迪豆●☆
64940 Cadia pubescens Bojer ex Baker;短柔毛卡迪豆●☆
64941 Cadia purpurea (G. Piccioli) Aiton;紫卡迪豆(紫卡迪亚豆)●☆
64942 Cadia purpurea Forssk. = Cadia purpurea (G. Piccioli) Aiton ●☆
64943 Cadia rubra R. Vig.;红卡迪豆●☆
64944 Cadiscus E. Mey. ex DC.(1838);水漂菊属■☆
64945 Cadiscus aquaticus E. Mey. ex DC.;水漂菊■☆
64946 Cadsura Spreng. = Kadsura Kaempf. ex Juss. ●
64947 Caela Adans. = Torenia L. ■
64948 Caelebogyne J. Sm. = Coelebogyne J. Sm. ●
64949 Caelebogyne Rchb. = Coelebogyne J. Sm. ●
64950 Caelestina Cass. = Ageratum L. ■●
64951 Caelia G. Don = Coelia Lindl. ■☆
64952 Caelobogyne N. T. Burb. = Coelebogyne J. Sm. ●
64953 Caelodepas Benth. = Koilodepas Hassk. ●
64954 Caeloglossum Steud. = Coeloglossum Hartm. ■
64955 Caelogyne Wall. = Coelogyne Lindl. ■
64956 Caelogyne Wall. ex Steud. = Coelogyne Lindl. ■
64957 Caelospermum Blume = Coelospermum Blume ●
64958 Caelospermum decipiens Merr. = Coelospermum decipiens Merr. ●☆
64959 Caelospermum kanehire Merr. = Coelospermum kanehire Merr. ●
64960 Caelospermum morindiforme Pierre ex Pit. = Coelospermum morindiforme Pierre ex Pit. ●
64961 Caenotus (Nutt.) Raf. = Conyza Less.(保留属名)■
64962 Caenotus Raf. = Erigeron L. ■●
64963 Caepha Leschen. ex Rchb. = Platysace Bunge ■☆
64964 Caesalpina Plum. ex L. = Caesalpinia L. ●
64965 Caesalpinia L.(1753);云实属(苏木属);Brasiletto, Caesalpinia, Nicaragua Wood, Poinciana ●
64966 Caesalpinia aestivalis Chun et F. C. How;夏花云实(夏花,夏苏木);Summer-flower Caesalpinia ●
64967 Caesalpinia aestivalis Chun et F. C. How = Pterolobium punctatum Hemsl. ex Forbes et Hemsl. ●

64968 Caesalpinia angolensis (Welw. ex Oliv.) Herend. et Zarucchi = Mezoneuron angolense Welw. ex Oliv. ●☆

64969 Caesalpinia antsiranensis Capuron = Caesalpinia madagascariensis (R. Vig.) Senesse ●☆

64970 Caesalpinia benthamiana (Baill.) Herend. et Zarucchi = Mezoneuron benthamianum Baill. ●☆

64971 Caesalpinia bonduc (L.) Roxb.;大托叶云实(刺果苏木,老虎心,杧果钉);Bonduc,Bonduc Nut,Brazilian Redwood,Brazilwood,Farnambuck,Fever Nut,Grey Nicker,Horse Nicker,Nicker Bean,Nicker Nut,Nickernut Caesalpinia,Pernambuco Redwood,Pernambuco Wood,Spinyfruit Caesalpinia,Stipulete Caesalpinia,Yellow Nicker Bean ●

64972 Caesalpinia bonducella (L.) Fleming = Caesalpinia bonduc (L.) Roxb. ●

64973 Caesalpinia bonducella Fleming;鹰叶刺(答肉刺,老虎心,蛇药,蚁蛔);Bonduc Seed,Nicker Nut,Nicker-nut Caesalpinia ●

64974 Caesalpinia bracteata Germish.;具苞苏木●☆

64975 Caesalpinia brasiliensis L.;巴西苏木(巴西云实);Brazilian Redwood,Brazilwood,Brazil-wood,Lima Wood ●☆

64976 Caesalpinia brevifolia Baill.;短叶苏木●☆

64977 Caesalpinia brevifolia Baill. = Balsamocarpon brevifolium Clos ■☆

64978 Caesalpinia cacalaco Humb. et Bonpl.;卡卡木;Cascalote Caesalpinia ●☆

64979 Caesalpinia caesia Hand. -Mazz.;粉叶云实(粉白苏木,粉叶苏木,广西云实);Caesious Caesalpinia,Grey-blue-leaved Caesalpinia ●

64980 Caesalpinia californica (A. Gray) Standl.;加州云实;Californian Bird of Paradise ●☆

64981 Caesalpinia coriaria (Jacq.) Willd. = Caesalpinia coriaria (Jacq.) Willd. ex Kunth ●

64982 Caesalpinia coriaria (Jacq.) Willd. ex Kunth;鞣料云实(狄薇豆,卷果云实);American Sumach,Cascalote,Dividivi,Divi-divi ●

64983 Caesalpinia crista L.;华南云实(川云实,刺果苏木,搭肉刺,大托叶云实,虎耳藤,假老虎簕,见血飞,老虎心,杧果钉,双角龙,台湾云实,鹰叶刺);Cristate Caesalpinia,Guangdong Caesalpinia,Nicker Nut Caesalpinia,Nickernut Caesalpinia,Nicker-nut Caesalpinia,Sichuan Caesalpinia,Wood Gossip Caesalpinia,Woodgossip Caesalpinia ●

64984 Caesalpinia crista L. = Caesalpinia bonduc (L.) Roxb. ●

64985 Caesalpinia crista L. = Caesalpinia bonducella Fleming ●

64986 Caesalpinia crista L. = Caesalpinia decapetala (Roth) Alston ●

64987 Caesalpinia cucullata Roxb. = Mezoneuron cuculatum (Roxb.) Wight et Arn. ●

64988 Caesalpinia dalei Brenan et J. B. Gillett = Stuhlmannia moavi Taub. ●☆

64989 Caesalpinia dauensis Thulin;达瓦云实●☆

64990 Caesalpinia decapetala (Roth) Alston;云实(百鸟不宿,百鸟不停,草云母,臭草,臭草子,刺皂角,倒钩刺,倒挂刺,翻天云,粉刺,红总管,虎头刺,黄牛刺,拦蛇刺,老虎刺尖,马豆,猫爪刺,鸟不落,鸟不栖,牛王茨,牛王刺,爬墙刺,杉刺,蛇不过,十瓣云实,水皂荚,水皂角,四时青,天豆,铁场豆,小霸王,芽皮刀,阎王刺,羊石子草,药王子,野皂荚,野皂角,员实,云英,粘刺);Decapetalous Caesalpinia,Mauritius Thorn,Mysore Thorn,Mysorethorn,Nicker-nut Caesalpinia,Shoofly ●

64991 Caesalpinia decapetala (Roth) Alston var. japonica (Siebold et Zucc.) H. Ohashi;日本云实●

64992 Caesalpinia decapetala (Roth) Alston var. japonica (Siebold et Zucc.) H. Ohashi = Caesalpinia decapetala (Roth) Alston ●

64993 Caesalpinia decapetala (Roth) Alston var. pubescens (Ts. Tang et F. T. Wang) P. C. Huang;毛云实(朝天子,大牛昂,倒挂牛,多毛叶云实,多毛云实,毛叶云实,牛王刺,日本云实,云实);Japanese Caesalpinia,Mysore Thorn,Pubescent Mysorethorn ●

64994 Caesalpinia decapetala (Roth) Alston var. pubescens (Ts. Tang et F. T. Wang) P. C. Huang = Caesalpinia decapetala (Roth) Alston ●

64995 Caesalpinia delphinensis Du Puy et R. Rabev.;德尔芬云实●☆

64996 Caesalpinia digyna Rottler = Caesalpinia digyna Rottler ex Willd. ●

64997 Caesalpinia digyna Rottler ex Willd.;肉荚云实(二蕊苏木);Digynian Caesalpinia,Towri ●

64998 Caesalpinia dinteri Harms = Haematoxylum dinteri (Harms) Harms ●☆

64999 Caesalpinia drepanocarpa (A. Gray) Fisher;镰果云实;Rush Pea ●☆

65000 Caesalpinia drepanocarpa Fisher = Caesalpinia drepanocarpa (A. Gray) Fisher ●☆

65001 Caesalpinia dubia Spreng. = Peltophorum dubium (Spreng.) Taub. ●☆

65002 Caesalpinia echinata Lam.;巴西云实(巴西木,巴西苏枋木,巴西铁木,刺云实,帕拉木,猬毛云实);Ara Wood,Bahia Wood,Brazil Redwood,Brazil Wood,Braziletto,Brazilian Ironwood,Brazilian Redwood,Brazilredwood,Brazilwood,Brazil-wood,Peach Wood,Peachwood,Pernambuco,Pernambuco Wood,Prickly Brazilwood ●☆

65003 Caesalpinia elliptifolia S. J. Li,Z. Y. Chen et D. X. Zhang;椭叶云实●

65004 Caesalpinia enneaphylla Roxb. = Mezoneuron enneaphyllum (Roxb.) Wight et Arn. ●

65005 Caesalpinia erianthera Chiov.;毛药云实●☆

65006 Caesalpinia erianthera Chiov. var. pubescens Brenan;短柔毛药云实●☆

65007 Caesalpinia erlangeri Harms = Caesalpinia trothae Harms subsp. erlangeri (Harms) Brenan ●☆

65008 Caesalpinia ferrea C. Mart. ex Tul.;豹木(铁云实);Brazilian Ironwood,Leopard Tree,Leopardtree ●☆

65009 Caesalpinia ferrea Mart. = Caesalpinia ferrea C. Mart. ex Tul. ●☆

65010 Caesalpinia ferrea Mart. var. leiostachya Benth. = Caesalpinia leiostachya (Benth.) Ducke ●☆

65011 Caesalpinia ferruginea Decne. = Peltophorum pterocarpum (DC.) K. Heyne ●

65012 Caesalpinia gillettii Hutch. et E. A. Bruce = Parkinsonia scioana (Chiov.) Brenan ●☆

65013 Caesalpinia gilliesii (Hook.) D. Dietr. = Caesalpinia gilliesii (Wall. ex Hook.) Benth. ●☆

65014 Caesalpinia gilliesii (Wall. ex Hook.) Benth.;红蝴蝶(红蕊蝴蝶,红蕊云实,极乐鸟云实);Bird of Paradise Bush,Bird-of-paradise,Bird-of-paradise Shrub,Desert Bird of Paradise,Gillies Caesalpinia,Yellow Bird of Paradise ●☆

65015 Caesalpinia gilliesii (Wall. ex Hook.) D. Dietr. = Caesalpinia gilliesii (Wall. ex Hook.) Benth. ●☆

65016 Caesalpinia glandulosopedicellata R. Wilczek;腺梗云实●☆

65017 Caesalpinia globulorum Bakh. f. et P. Royen = Caesalpinia major (Medik.) Dandy et Exell ●

65018 Caesalpinia hildebrandtii (Vatke) Baill.;马岛见血飞●☆

65019 Caesalpinia homblei R. Wilczek;洪布勒云实●☆

65020 Caesalpinia hymenocarpa (Prain) Hattink;膜荚见血飞;Hymenocarpous Caesalpinia,Membranaceous-podded Caesalpinia ●

65021 Caesalpinia hypoglauca Chun et F. C. How = Caesalpinia caesia

Hand. -Mazz. ●

65022 Caesalpinia inermis Roxb. = Peltophorum pterocarpum (DC.) Backer ex K. Heyne ●

65023 Caesalpinia insolita (Harms) Brenan et J. B. Gillett = Stuhlmannia moavi Taub. ●☆

65024 Caesalpinia jamesii (Torr. et A. Gray) Fisher;詹姆斯云实;Rush Pea ●☆

65025 Caesalpinia japonica Siebold et Zucc. = Caesalpinia decapetala (Roth) Alston var. japonica (Siebold et Zucc.) H. Ohashi ●

65026 Caesalpinia japonica Siebold et Zucc. = Caesalpinia decapetala (Roth) Alston ●

65027 Caesalpinia japonica Siebold et Zucc. = Caesalpinia sepiaria Roxb. ●

65028 Caesalpinia jayabo M. Gómez = Caesalpinia globulorum Bakh. f. et P. Royen ●

65029 Caesalpinia jayabo M. Gómez = Caesalpinia major (Medik.) Dandy et Exell ●

65030 Caesalpinia kwangtungensis Merr. = Caesalpinia crista L. ●

65031 Caesalpinia leiostachya (Benth.) Ducke;光穗苏木●☆

65032 Caesalpinia madagascariensis (R. Vig.) Senesse;马岛苏木●☆

65033 Caesalpinia magnifoliolata F. P. Metcalf;大叶云实(刺藤,铁藤);Bigleaf Caesalpinia, Big-leaved Caesalpinia, Largeleaf Caesalpinia ●

65034 Caesalpinia major (Medik.) Dandy et Exell;莲实藤(大云实)●

65035 Caesalpinia melanosticta Spreng. = Hoffmannseggia burchellii (DC.) Benth. ex Oliv. ■☆

65036 Caesalpinia merxmuellerana A. Schreib.;梅尔苏木●☆

65037 Caesalpinia mexicana A. Gray;墨西哥云实;Mexican Bird of Paradise, Mexican Holdback ●☆

65038 Caesalpinia millettii Hook. et Arn.;小叶云实(假南蛇簕,假楠);Smallleaf Caesalpinia, Small-leaved Caesalpinia ●

65039 Caesalpinia mimosoides Lam.;含羞云实(草云实,臭菜,印度含羞草);Mimosalike Caesalpinia, Mimoselike Caesalpinia ●

65040 Caesalpinia minax Hance;喙荚云实(飞天龙,广石莲,喙荚苏木,苦石莲,老鸦枕头,莲实藤,莲子簕,猫爪簕,南蛇艻,南蛇勒,南蛇簕,南蛇茸子,青蛇簕,雀不站,蚺蛇艻,石莲,石莲簕,石莲藤,烫耙苗,土石莲,阎王刺);Whiteflower Cacalia, White-flowered Cacalia, White-flowered Caesalpinia ●

65041 Caesalpinia minax Hance = Caesalpinia major (Medik.) Dandy et Exell ●

65042 Caesalpinia morsei Dunn = Caesalpinia minax Hance ●

65043 Caesalpinia nuga (L.) W. T. Aiton = Caesalpinia crista L. ●

65044 Caesalpinia nuga W. T. Aiton = Caesalpinia crista L. ●

65045 Caesalpinia obovata Schinz = Haematoxylum dinteri (Harms) Harms ●☆

65046 Caesalpinia oleosperma Roxb. = Caesalpinia digyna Rottler ex Willd. ●

65047 Caesalpinia oligophylla Harms;寡叶云实●☆

65048 Caesalpinia parvifolia Steud. = Caesalpinia sinensis (Hemsl.) J. E. Vidal ●

65049 Caesalpinia paucijuga Oliv. = Caesalpinia punctata Willd. ●☆

65050 Caesalpinia pearsonii L. Bolus;皮尔逊云实●☆

65051 Caesalpinia pectinata Cav. = Caesalpinia spinosa (Molina) Kuntze ●☆

65052 Caesalpinia peltophoroides Benth.;盾梗云实●☆

65053 Caesalpinia platyloba S. Watson;弯云实;Curly Paela ●☆

65054 Caesalpinia praecox Ruiz et Pav. ex Hook. et Arn.;早发云实●☆

65055 Caesalpinia pulcherrima (L.) Sw.;金凤花(蝶花,番蝴蝶,红紫,黄蝴蝶,蛱蝶花,洋金凤);Barbados Flower-fence, Barbados Pride, Barbadospride, Bird of Paradise, Doctor Doodles, Dwarf Poinciana, Flower Fence, Flowerfence Poinciana, Flower-fence Poinciana, Paradise Flower, Peacock Flower, Prettiest Caesalpinia, Pride of Barbados, Pride-of-barbados, Red Bird of Paradise ●

65056 Caesalpinia punctata Willd.;斑点苏木●☆

65057 Caesalpinia rhombifolia J. E. Vidal;菱叶云实●

65058 Caesalpinia rostrata N. E. Br.;喙苏木●☆

65059 Caesalpinia rubra (Engl.) Brenan;红色苏木●☆

65060 Caesalpinia sappan L.;苏木(赤木,戈方,红柴,红木,红苏木,苏方,苏方木,苏枋,苏枋木,窊木,紫纳,棕木);Brasiletto, Brazil Wood, Brazilwood, Bukkum Wood, Bukkum-wood, Peachwood, Red Wood, Sappan, Sappan Caesalpinia, Sappan Wood ●

65061 Caesalpinia sepiaria Roxb. = Caesalpinia decapetala (Roth) Alston ●

65062 Caesalpinia sepiaria Roxb. var. japonica (Siebold et Zucc.) Makino = Caesalpinia decapetala (Roth) Alston ●

65063 Caesalpinia sepiaria Roxb. var. pubescens Ts. Tang et F. T. Wang = Caesalpinia decapetala (Roth) Alston var. pubescens (Ts. Tang et F. T. Wang) P. C. Huang ●

65064 Caesalpinia sepiaria Roxb. var. pubescens Ts. Tang et F. T. Wang = Caesalpinia decapetala (Roth) Alston ●

65065 Caesalpinia sinense (Hemsl. ex Forbes et Hemsl.) J. E. Vidal = Mezoneuron sinense Hemsl. ex Forbes et Hemsl. ●■

65066 Caesalpinia sinensis (Hemsl.) J. E. Vidal = Mezoneuron sinense Hemsl. ex Forbes et Hemsl. ●■

65067 Caesalpinia spicata Dalziel;穗云实●☆

65068 Caesalpinia spinosa (Molina) Kuntze;刺云实(多刺云实);Spiny Holdback, Tara ●☆

65069 Caesalpinia stenoptera Merr. = Caesalpinia sinense (Hemsl. ex Forbes et Hemsl.) J. E. Vidal ●

65070 Caesalpinia szechuenensis Craib = Caesalpinia crista L. ●

65071 Caesalpinia tinctoria (Kunth) Benth. = Caesalpinia spinosa (Molina) Kuntze ●☆

65072 Caesalpinia tinctoria Dombey ex DC.;南美苏木(染用云实)●☆

65073 Caesalpinia tortuosa Roxb.;扭果云实(扭果苏木);Tortuous Caesalpinia, Tortuousfruit Caesalpinia ●

65074 Caesalpinia trothae Harms;特罗塔云实●☆

65075 Caesalpinia trothae Harms subsp. erlangeri (Harms) Brenan;厄兰格云实●☆

65076 Caesalpinia tsoongii Merr. = Caesalpinia sinense (Hemsl. ex Forbes et Hemsl.) J. E. Vidal ●

65077 Caesalpinia vernalis Champ.;春云实(南蛇簕,乌爪簕藤);Vernal Caesalpinia ●

65078 Caesalpinia versicolor Hort.;彩色云实;Nicker-nut Caesalpinia ●☆

65079 Caesalpinia violacea (Mill.) Standl.;青紫苏木;Violet Caesalpinia ●

65080 Caesalpinia volkensii Harms;福尔云实●☆

65081 Caesalpinia welwitschiana (Oliv.) Brenan;韦尔云实●☆

65082 Caesalpinia yaoshanensis Chun = Caesalpinia magnifoliolata F. P. Metcalf ●

65083 Caesalpinia yunnanensis S. J. Li, D. X. Zhang et Z. Y. Chen;云南云实●

65084 Caesalpiniaceae R. Br. (1814)(保留科名);云实科(苏木科);Caesalpinia Family, Senna Family ●■

65085 Caesalpiniaceae R. Br. (保留科名) = Fabaceae Lindl. (保留科

名)●■
65086 Caesalpiniaceae R. Br.(保留科名)= Leguminosae Juss.(保留科名)●■
65087 Caesalpiniodes Kuntze = Gleditsia L. ●
65088 Caesalpinioides africanum (Welw. ex Benth.) Kuntze = Erythrophleum africanum (Welw. ex Benth.) Harms ●☆
65089 Caesalpinioides triacanthum (L.) Kuntze = Gleditsia triacanthos L. ●
65090 Caesarea Cambess. = Viviania Cav. ■☆
65091 Caesia R. Br.(1810);蔡斯吊兰属■☆
65092 Caesia Vell. = Cormonema Reissek ex Endl. ●
65093 Caesia africana Baker = Chlorophytum africanum (Baker) Engl. ■☆
65094 Caesia brevicaulis (Baker) T. Durand et Schinz = Caesia contorta (L. f.) T. Durand et Schinz ■☆
65095 Caesia brevifolia (Thunb.) T. Durand et Schinz = Caesia contorta (L. f.) T. Durand et Schinz ■☆
65096 Caesia capensis (Bolus) Oberm.;好望角蔡斯吊兰■☆
65097 Caesia comosa (Thunb.) Spreng. = Chlorophytum comosum (Thunb.) Baker ●■
65098 Caesia contorta (L. f.) T. Durand et Schinz;缠绕蔡斯吊兰■☆
65099 Caesia dregeana Kunth;德雷蔡斯吊兰■☆
65100 Caesia eckloniana Schult. et Schult. f.;埃氏蔡斯吊兰■☆
65101 Caesia subulata Baker;钻形蔡斯吊兰■☆
65102 Caesia thunbergii Roem. et Schult. f. = Caesia contorta (L. f.) T. Durand et Schinz ■☆
65103 Caesulia Roxb.(1798);腋序菊属■☆
65104 Caesulia axillaris Roxb.;腋序菊■☆
65105 Caesulia radicans Willd. = Enydra radicans (Willd.) Lack ■☆
65106 Caeta Steud. = Caela Adans. ■
65107 Caeta Steud. = Torenia L. ■
65108 Caetocapnia Endl. = Bravoa Lex. ■☆
65109 Caetocapnia Endl. = Coetocapnia Link et Otto ■
65110 Caex variabilis L. H. Bailey var. elatior L. H. Bailey = Carex emoryi Dewey ■☆
65111 Caex virginiana Woods var. elongata Boeck. = Carex emoryi Dewey ■☆
65112 Cafe Adans. = Coffea L. ●
65113 Caffea Noronha = Coffea L. ●
65114 Cahota H. Karst. = Clusia L. ●☆
65115 Caidbeja Forssk. = Forsskaolea L. ■☆
65116 Caidbeja adhaerens Forssk. = Forsskaolea tenacissima L. ■☆
65117 Caihetus Lour. = Phyllanthus L. ●■
65118 Cailliea Guill. et Perr.(废弃属名)= Dichrostachys (A. DC.) Wight et Arn.(保留属名)●
65119 Cailliea cinerea (L.) Roberty = Dichrostachys cinerea (L.) Wight et Arn. ●
65120 Cailliea dichrostachys Guillaumin et Perr. = Dichrostachys cinerea (L.) Wight et Arn. var. africana Brenan et Brummitt ●☆
65121 Cailliea dichrostachys Guillaumin et Perr. var. leptostachys (DC.) Guillaumin et Perr. = Dichrostachys cinerea (L.) Wight et Arn. var. africana Brenan et Brummitt ●☆
65122 Cailliea glomerata (Forssk.) J. F. Macbr. = Dichrostachys cinerea (L.) Wight et Arn. ●
65123 Cailliella Jacq. -Fél.(1939);西非野牡丹属●☆
65124 Cailliella praerupticola Jacq. -Fél.;西非野牡丹●☆
65125 Caina Panch. ex Baill. = Neuburgia Blume ●☆
65126 Cainito Adans. = Chrysophyllum L. ●
65127 Cainito Plum. ex Adans. = Chrysophyllum L. ●
65128 Caiophora C. Presl(1831);南美刺莲花属(烧莲属)■☆
65129 Caiophora albiflora (Griseb.) Urb. et Gilg;白花南美刺莲花■☆
65130 Caiophora boliviana Urb. et Gilg;玻利维亚南美刺莲花■☆
65131 Caiophora clavata Urb. et Gilg;棒状南美刺莲花■☆
65132 Caiophora coronata (Gillies ex Arn.) Hook. et Arn.;冠状南美刺莲花■☆
65133 Caiophora kurtzii Urb. et Gilg ex Kurtz;孔策南美刺莲花■☆
65134 Caiophora lateritia Benth.;瓦红南美刺莲花■☆
65135 Caiophora macrocarpa Urb. et Gilg;大果南美刺莲花■☆
65136 Caiophora nivalis Lillo;雪白南美刺莲花■☆
65137 Caiophora tenuis Killip;纤细南美刺莲花■☆
65138 Cajalbania Urb. = Gliricidia Kunth ●☆
65139 Cajalbania Urb. = Poitea Vent. ●☆
65140 Cajalbania Urb. = Sauvallella Rydb. ●☆
65141 Cajan Adans. = Cajanus Adans.(保留属名)●
65142 Cajanum Raf. = Cajanus Adans.(保留属名)●
65143 Cajanus Adans.(1763)(保留属名);木豆属(虫豆属);Cajan, Pigeonpea, Pigeon-pea ●
65144 Cajanus DC. = Cajanus Adans.(保留属名)●
65145 Cajanus bicolor DC. = Cajanus cajan (L.) Millsp. ●
65146 Cajanus cajan (L.) Millsp.;木豆(阿瓦豆,豆蓉,鸽豆,观音豆,花豆,豇豆,柳豆,米豆,面豆,扭豆,三叶豆,山豆根,树豆,野黄豆,印度虫豆,猪屎豆);Ambrevade, Angola Pea, Arhar, Ca Jan Bean, Cajan, Cajun Bean, Catjang, Common Dholl, Congo Bean, Congo Pea, Congopea, Dhal, Indian Cajan, No-eye Pea, Pigeon Pea, Pigeonpea, Red Grain ●
65147 Cajanus crassus (Prain ex King) Maesen;虫豆(木虫豆);Thick Cajan ●
65148 Cajanus flavus DC. = Cajanus cajan (L.) Millsp. ●
65149 Cajanus goensis Dalzell;硬毛虫豆;Bearded Atylosia, Hardhair Cajan, Hirsute Cajan ●
65150 Cajanus grandiflorus (Benth. ex Baker) Maesen;大花虫豆(大花木豆);Bigflower Cajan, Big-flowered Cajan, Large Flower Atylosia ●
65151 Cajanus indicus Spreng. = Cajanus cajan (L.) Millsp. ●
65152 Cajanus kerstingii Harms;克斯廷木豆●☆
65153 Cajanus mollis (Benth.) Maesen;长叶虫豆(蝙蝠豆,缠绕虫豆,虫豆,软木豆);Longleaf Cajan, Pubescent Atylosia, Pubescent Cajan, Twining Atylosia ●
65154 Cajanus niveus (Benth.) Maesen;白虫豆(白毛虫豆);Snow-white Cajan, White Pigeonpea, Whitehair Atylosia ●
65155 Cajanus niveus Graham ex Wall. = Cajanus niveus (Benth.) Maesen ●
65156 Cajanus scarabaeoides (L.) Thouars;蔓草虫豆(虫豆,蔓草木豆,蔓虫豆);Deciduous Atylosia, Dungbeetle Cajan, Showy Pigeonpea, Spread Atylosia ●
65157 Cajanus scarabaeoides (L.) Thouars var. argyrophyllus (Y. T. Wei et S. K. Lee) Y. T. Wei et S. K. Lee;白蔓草虫豆;Silver-leaf Atylosia, White Dungbeetle Cajan, White Spread Atylosia ●
65158 Cajophora C. Presl = Caiophora C. Presl ■☆
65159 Cajophora Endl. = Caiophora C. Presl ■☆
65160 Caju Kuntze = Pongamia Adans.(保留属名)●
65161 Cajum Kuntze = Pongamia Adans.(保留属名)●
65162 Cajuputi Adans. = Melaleuca L.(保留属名)●
65163 Caju-puti Adans. = Melaleuca L.(保留属名)●
65164 Cakile Mill.(1754);海滨芥属(海凯菜属);Sea Rocket ■☆

65165 Cakile aegyptia (L.) Maire et Weiller = Cakile maritima Scop. ■☆
65166 Cakile aegyptia (L.) Maire et Weiller var. australis Coss. = Cakile maritima Scop. ■☆
65167 Cakile aegyptia (L.) Maire et Weiller var. edentula (Bigelow) Loret et Barratte = Cakile maritima Scop. ■☆
65168 Cakile aegyptia (L.) Maire et Weiller var. hispanica (Jord.) Maire = Cakile maritima Scop. ■☆
65169 Cakile aegyptia (L.) Maire et Weiller var. latifolia Desf. = Cakile maritima Scop. ■☆
65170 Cakile aegyptia (L.) Maire et Weiller var. litoralis (Jord.) Rouy = Cakile maritima Scop. ■☆
65171 Cakile aegyptia (L.) Maire et Weiller var. susica Maire et Wilczek = Cakile maritima Scop. ■☆
65172 Cakile americana Nutt. = Cakile edentula (Bigelow) Hook. ■☆
65173 Cakile arctica Pobed.;北极海滨芥■☆
65174 Cakile californica A. Heller = Cakile edentula (Bigelow) Hook. ■☆
65175 Cakile edentula (Bigelow) Hook.;无齿海滨芥;American Sea-rocket ■☆
65176 Cakile edentula (Bigelow) Hook. = Cakile maritima Scop. ■☆
65177 Cakile edentula (Bigelow) Hook. subsp. californica (A. Heller) Hultén = Cakile edentula (Bigelow) Hook. ■☆
65178 Cakile edentula (Bigelow) Hook. subsp. lacustris (Fernald) Hultén = Cakile lacustris (Fernald) Pobed. ■☆
65179 Cakile edentula (Bigelow) Hook. var. californica (A. Heller) Fernald = Cakile edentula (Bigelow) Hook. ■☆
65180 Cakile edentula (Bigelow) Hook. var. lacustris Fernald = Cakile lacustris (Fernald) Pobed. ■☆
65181 Cakile hispanica Jord. = Cakile maritima Scop. ■☆
65182 Cakile lacustris (Fernald) Pobed.;湖中海滨芥;Sea-rocket ■☆
65183 Cakile latifolia Poir.;宽叶海滨芥■☆
65184 Cakile litoralis Jord. = Cakile maritima Scop. ■☆
65185 Cakile maritima Scop.;海滨芥;American Sea Rocket, American Sea-rocket, European Searocket, Sea Rocket ■☆
65186 Cakile maritima Scop. subsp. integrifolia (Hornem.) Greuter et Burdet;全叶海滨芥■☆
65187 Cakile maritima Scop. var. aegyptia (L.) Coss. = Cakile maritima Scop. ■☆
65188 Cakile maritima Scop. var. hispanica (Jord.) Rouy = Cakile maritima Scop. ■☆
65189 Cakile maritima Scop. var. integrifolia Hornem. = Cakile maritima Scop. subsp. integrifolia (Hornem.) Greuter et Burdet ■☆
65190 Cakile maritima Scop. var. littoralis (Jord.) Rouy = Cakile maritima Scop. ■☆
65191 Cakpethia Britton = Anemone L.(保留属名)■
65192 Calaba Mill. = Calophyllum L. ●
65193 Calacanthus Kuntze = Calacanthus T. Anderson ex Benth. et Hook. f. ■☆
65194 Calacanthus T. Anderson ex Benth. = Calacanthus T. Anderson ex Benth. et Hook. f. ■☆
65195 Calacanthus T. Anderson ex Benth. et Hook. f. (1876);丽刺爵床属■☆
65196 Calacanthus dalzellianus T. Anderson ex Benth. et Hook. f.;丽刺爵床■☆
65197 Calacanthus grandiflorus (Dalzell) Kuntze = Calacanthus grandiflorus (Dalzell) Radlk. ■☆
65198 Calacanthus grandiflorus (Dalzell) Radlk.;大花丽刺爵床■☆
65199 Calachyris Post et Kuntze = Calliachyris Torr. et A. Gray ■☆
65200 Calachyris Post et Kuntze = Layia Hook. et Arn. ex DC.(保留属名)■☆
65201 Calacinum Raf.(废弃属名) = Muehlenbeckia Meisn.(保留属名)●☆
65202 Caladenia R. Br. (1810);裂缘兰属(卡拉迪兰属);Caladenia ■☆
65203 Caladenia alba R. Br.;白花裂缘兰(白花卡拉迪兰);White Caladenia ■☆
65204 Caladenia caerulea R. Br.;蓝裂缘兰;Blue Caladenia ■☆
65205 Caladenia carnea R. Br.;裂缘兰(卡拉迪兰);Flesh-coloured Caladenia ■☆
65206 Caladenia catenata (Sm.) Druce;链状裂缘兰;White Fingers ■☆
65207 Caladenia dimorpha R. Br.;二型裂缘兰;Two-formes Caladenia ■☆
65208 Caladeniastrum (Szlach.) Szlach. (2003);小裂缘兰属■☆
65209 Caladeniastrum (Szlach.) Szlach. = Caladenia R. Br. ■☆
65210 Caladiaceae Salisb.;五彩芋科■
65211 Caladiaceae Salisb. = Araceae Juss.(保留科名)■●
65212 Caladiopsis Engl. = Chlorospatha Engl. ■☆
65213 Caladium Raf. = Caladium Vent. ■
65214 Caladium Vent. (1801);五彩芋属(花叶芋属,叶芋属,彩叶芋属);Angel Wings, Angel-wings, Caladium, Elephat's-ear, Garishtaro, Mother-in-law Plant ■
65215 Caladium 'Candidum' = Caladium bicolor (Aiton) Vaniot 'Candidum' ■☆
65216 Caladium 'Crimson Wave';红浪花叶芋■☆
65217 Caladium 'Hok Long';孔雀花叶芋■☆
65218 Caladium bicolor (Aiton) Vaniot;五彩芋(独角芋,红半夏,红水芋,红芋头,花水芋,花叶杯芋,花叶芋,石芋头,珍珠莫玉散);Angel Wings, Angels' Wings, Caladium, Cocoa-root, Common Caladium, Common Garishtaro, Fancy-leafed Caladium, Fancy-leaved Caladium, Heart-of-Jesus ■
65219 Caladium bicolor (Aiton) Vaniot 'Candidum';白亮花叶芋(白鹭花叶芋)■☆
65220 Caladium bicolor (Aiton) Vaniot 'John Peed';约翰·皮德花叶芋■☆
65221 Caladium bicolor (Aiton) Vaniot 'Pink Beauty';粉丽花叶芋■☆
65222 Caladium bicolor (Aiton) Vaniot 'Pink Gloud';粉云花叶芋■☆
65223 Caladium bicolor (Aiton) Vaniot 'Splenden';嫣红花叶芋;Angel Wings, Fancy-leaved Caladium ■☆
65224 Caladium colocasia (L.) W. Wight ex Safford = Colocasia esculenta (L.) Schott ■
65225 Caladium cucullata Pers. = Alocasia cucullata (Lour.) Schott et Endl. ●■
65226 Caladium esculentum Vaniot = Colocasia esculenta (L.) Schott ■
65227 Caladium giganteum Blume ex Hassk = Colocasia gigantea (Blume) Hook. f. ■
65228 Caladium hortulanum Birdsey;彩叶芋(杂种花叶芋);Angels' Wings, Fancy Leaved Caladium ■
65229 Caladium humboldtii (Raf.) Schott;银斑芋(小锦芋);Caladium Star Light, Miniature Caladium ■☆
65230 Caladium lindenii (André) Madison;林登五彩芋(玲殿黄肉芋);Angel's Wing, Indian Kale, Yautia ■☆
65231 Caladium lindenii Madison = Caladium lindenii (André) Madison ■☆
65232 Caladium maculatum Lodd. = Dieffenbachia picta (Lodd.) Schott ●■

65233 Caladium maculatum Lodd. = Dieffenbachia seguine (Jacq.) Schott ●■

65234 Caladium nymphaeifolium Vent. = Colocasia esculenta (L.) Schott ■

65235 Caladium odorum Lindl. = Alocasia macrorrhiza Schott ■

65236 Caladium pictum Lodd. = Dieffenbachia picta (Lodd.) Schott ●■

65237 Caladium picturatum K. Koch et C. D. Bouché;画叶芋(狭叶洋芋);Painted Caladium, Strap-leaf Caladium ■☆

65238 Caladium pumilum D. Don = Gonatanthus pumilus (D. Don) Engl. et Krause ■

65239 Caladium zamiifolium Lodd.;龙凤木(金钱树,雪铁芋,泽米叶天南星);Aroid Palm, Arum Fern ■☆

65240 Caladium zamiifolium Lodd. = Zamioculcas zamiifolia (Lodd.) Engl. ■☆

65241 Calaena Schltdl. = Caleana R. Br. ■☆

65242 Calais DC. = Microseris D. Don ■☆

65243 Calais aphantocarpha A. Gray = Microseris douglasii (DC.) Sch. Bip. subsp. tenella (A. Gray) K. L. Chambers ■☆

65244 Calais douglasii DC. = Microseris douglasii (DC.) Sch. Bip. ■☆

65245 Calais glauca (Hook.) A. Gray var. procera A. Gray = Microseris laciniata (Hook.) Sch. Bip. ■☆

65246 Calais lindleyi DC. = Uropappus lindleyi (DC.) Nutt. ■☆

65247 Calais platycarpha A. Gray = Microseris douglasii (DC.) Sch. Bip. subsp. platycarpha (A. Gray) K. L. Chambers ■☆

65248 Calais tenella A. Gray = Microseris douglasii (DC.) Sch. Bip. subsp. tenella (A. Gray) K. L. Chambers ■☆

65249 Calamaceae Kunth ex Perleb = Arecaceae Bercht. et J. Presl(保留科名)●

65250 Calamaceae Kunth ex Perleb = Palmae Juss.(保留科名)●

65251 Calamaceae Y. R. Ling = Arecaceae Bercht. et J. Presl(保留科名)●

65252 Calamaceae Y. R. Ling = Palmae Juss.(保留科名)●

65253 Calamagristis kokonorica (Keng) Tzvelev = Deyeuxia kokonorica Keng ■

65254 Calamagrostis Adans. (1763);拂子茅属;Feather Grass, Feather Reed Grass, Reedbentgrass, Small-reed, Woodreed ■

65255 Calamagrostis × aristata Ohwi;具芒拂子茅■☆

65256 Calamagrostis × goyozanensis M. Kikuchi;五叶山拂子茅■☆

65257 Calamagrostis × muramatsui Ohwi;村松拂子茅■☆

65258 Calamagrostis × yatabei Maxim.;谷田拂子茅■☆

65259 Calamagrostis aculeolata (Hack.) Ohwi = Calamagrostis neglecta (Ehrh.) Gaertn., B. Mey. et Scherb. subsp. inexpansa (A. Gray) Tzvelev ■☆

65260 Calamagrostis acutiflora DC.;尖花拂子茅■☆

65261 Calamagrostis acutiflora DC. 'Karl Foerster';卡尔福斯特尖花拂子茅;Karl Foerster Feather Reed Grass ■☆

65262 Calamagrostis acutiflora DC. 'Overdam';彩叶尖花拂子茅;Variegated Feather Reed Grass ■☆

65263 Calamagrostis acutiflora DC. nothovar. yatabei (Maxim.) Sugim. = Calamagrostis × yatabei Maxim. ■☆

65264 Calamagrostis acutifolia Rchb. f.;尖叶拂子茅(尖叶野青茅);Feather Reed Grass ■☆

65265 Calamagrostis adpressiramea Ohwi;日本小花野青茅■☆

65266 Calamagrostis alajica Litv.;阿拉拂子茅■☆

65267 Calamagrostis alexeenkoana Litv.;阿氏拂子茅■☆

65268 Calamagrostis alopecuroides Roshev. = Calamagrostis turkestanica Hack. ■

65269 Calamagrostis altaica Tzvelev;阿尔泰野青茅;Altai Caladium ■

65270 Calamagrostis angustifolia Kom.;狭叶野青茅■

65271 Calamagrostis angustifolia Kom. = Deyeuxia angustifolia (Kom.) Y. L. Chang ■

65272 Calamagrostis aniselytron Govaerts = Aniselytron agrostoides Merr. ■

65273 Calamagrostis anomala Suksd. = Calamagrostis canadensis (Michx.) P. Beauv. ■☆

65274 Calamagrostis anthoxanthoides (Munro) Regel;黄花茅状野青茅(短毛野青茅)■

65275 Calamagrostis argentea DC. = Achnatherum calamagrostis (L.) P. Beauv. ■

65276 Calamagrostis argentea DC. subsp. mesatlantica Quézel = Achnatherum calamagrostis (L.) P. Beauv. subsp. mesatlanticum (Quézel) Dobignard ■☆

65277 Calamagrostis arundinacea (L.) Roth;苇状拂子茅■☆

65278 Calamagrostis arundinacea (L.) Roth = Deyeuxia arundinacea (L.) P. Beauv. ■

65279 Calamagrostis arundinacea (L.) Roth subsp. adpressiramea (Ohwi) T. Koyama = Calamagrostis adpressiramea Ohwi ■☆

65280 Calamagrostis arundinacea (L.) Roth subsp. brachytricha (Steud.) Tzvelev = Calamagrostis brachytricha Steud. ■

65281 Calamagrostis arundinacea (L.) Roth subsp. sugawarae (Ohwi) Tzvelev = Calamagrostis brachytricha Steud. ■

65282 Calamagrostis arundinacea (L.) Roth var. adpressiramea (Ohwi) Ohwi = Calamagrostis adpressiramea Ohwi ■☆

65283 Calamagrostis arundinacea (L.) Roth var. brachytricha (Steud.) Hack. = Calamagrostis brachytricha Steud. ■

65284 Calamagrostis arundinacea (L.) Roth var. brachytricha (Steud.) Hack. = Deyeuxia arundinacea (L.) P. Beauv. var. brachytricha (Steud.) P. C. Kuo et S. L. Lu ■

65285 Calamagrostis arundinacea (L.) Roth var. ciliata Honda = Deyeuxia arundinacea (L.) P. Beauv. var. ciliata (Honda) P. C. Kuo et S. L. Lu ■

65286 Calamagrostis arundinacea (L.) Roth var. inaequata Hack. ex Honda = Calamagrostis brachytricha Steud. var. inaequata (Hack. ex Honda) Yonek. ■☆

65287 Calamagrostis arundinacea (L.) Roth var. latifolia (Rendle) Kitag. = Deyeuxia arundinacea (L.) P. Beauv. var. latifolia (Rendle) P. C. Kuo et S. L. Lu ■

65288 Calamagrostis arundinacea (L.) Roth var. robusta (Franch. et Sav.) Nakai ex Honda = Deyeuxia arundinacea (L.) P. Beauv. var. robusta (Franch. et Sav.) P. C. Kuo et S. L. Lu ■

65289 Calamagrostis arundinacea (L.) Roth var. sciuroides (Franch. et Sav.) Hack. = Deyeuxia arundinacea (L.) P. Beauv. var. sciuroides (Franch. et Sav.) P. C. Kuo et S. L. Lu ■

65290 Calamagrostis atropurpurea Nash = Calamagrostis canadensis (Michx.) P. Beauv. ■☆

65291 Calamagrostis autumnalis Koidz.;秋拂子茅■☆

65292 Calamagrostis autumnalis Koidz. subsp. insularis (Honda) Tateoka;海岛秋拂子茅■☆

65293 Calamagrostis autumnalis Koidz. subsp. microtis (Ohwi) T. Koyama = Calamagrostis autumnalis Koidz. var. microtis Ohwi ■☆

65294 Calamagrostis autumnalis Koidz. var. microtis Ohwi;小秋拂子茅■☆

65295 Calamagrostis balansae Boiss.;巴拉萨拂子茅■☆

65296 Calamagrostis borii Tzvelev = Deyeuxia rosea Bor ■

65297 Calamagrostis brachytricha Steud. = Calamagrostis arundinacea (L.) Roth var. brachytricha (Steud.) Hack. ■

65298 Calamagrostis brachytricha Steud. = Deyeuxia brachytricha (Steud.) Y. L. Chang ■
65299 Calamagrostis brachytricha Steud. var. ciliata (Honda) Ibaragi et H. Ohashi;短睫毛拂子茅■☆
65300 Calamagrostis brachytricha Steud. var. inaequata (Hack. ex Honda) Yonek.;不等拂子茅■☆
65301 Calamagrostis bungeana Petr.;布氏拂子茅■☆
65302 Calamagrostis californica Kearney = Calamagrostis stricta (Timm) Koeler subsp. inexpansa (A. Gray) C. W. Greene ■☆
65303 Calamagrostis canadensis (Michx.) P. Beauv.;加拿大拂子茅;Big Reed Grass, Bluejoint, Blue-joint Grass, Bluejoint Reedgrass, Canada Reed-grass, Canadian Reedgrass, Marsh Pinegrass, Marsh Reedgrass, Meadow Pinegrass ■☆
65304 Calamagrostis canadensis (Michx.) P. Beauv. subsp. langsdorfii (Link) Hultén = Calamagrostis purpurea (Trin.) Trin. subsp. langsdorfii (Link) Tzvelev ■
65305 Calamagrostis canadensis (Michx.) P. Beauv. var. acuminata Vasey ex Shear et Rydb. = Calamagrostis canadensis (Michx.) P. Beauv. ■☆
65306 Calamagrostis canadensis (Michx.) P. Beauv. var. langsdorfii (Link) Inman = Calamagrostis purpurea (Trin.) Trin. subsp. langsdorfii (Link) Tzvelev ■
65307 Calamagrostis canadensis (Michx.) P. Beauv. var. macouniana (Vasey) Stebbins;马昆拂子茅;Macoun's Blue-joint, Macoun's Reed Grass ■☆
65308 Calamagrostis canadensis (Michx.) P. Beauv. var. macouniana (Vasey) Stebbins = Calamagrostis canadensis (Michx.) P. Beauv. ■☆
65309 Calamagrostis canadensis (Michx.) P. Beauv. var. pallida (Vasey et Scribn.) Stebbins = Calamagrostis canadensis (Michx.) P. Beauv. ■☆
65310 Calamagrostis canadensis (Michx.) P. Beauv. var. robusta Vasey = Calamagrostis canadensis (Michx.) P. Beauv. ■☆
65311 Calamagrostis canadensis (Michx.) P. Beauv. var. typica Stebbins = Calamagrostis canadensis (Michx.) P. Beauv. ■☆
65312 Calamagrostis canescens (F. H. Wigg.) Roth = Calamagrostis canescens (Webb) Roth ■☆
65313 Calamagrostis canescens (Webb) Roth;灰白拂子茅;Purple Small-reed ■☆
65314 Calamagrostis capensis Stapf = Calamagrostis epigejos (L.) Roth var. capensis Stapf ■☆
65315 Calamagrostis caucasica Trin.;高加索拂子茅■☆
65316 Calamagrostis chordorrhiza Porsild = Calamagrostis stricta (Timm) Koeler subsp. inexpansa (A. Gray) C. W. Greene ■☆
65317 Calamagrostis compacta (Munro ex Hook. f.) Hack. ex Paulsen = Calamagrostis holciformis Jaub. et Spach ■
65318 Calamagrostis compacta (Munro ex Hook. f.) Hack. ex Paulsen = Deyeuxia compacta Munro ex Duthie ■
65319 Calamagrostis compacta (Munro ex Hook. f.) Hack. ex Paulsen = Deyeuxia holciformis (Jaub. et Spach) Bor ■
65320 Calamagrostis decora Hook. f.;饰拂子茅■☆
65321 Calamagrostis deschampsioides Trin.;发草拂子茅■☆
65322 Calamagrostis deschampsioides Trin. var. hayachinensis Ohwi = Calamagrostis nana Takeda subsp. hayachinensis (Ohwi) Tateoka ■☆
65323 Calamagrostis dubia Bunge = Calamagrostis pseudophragmites (Hallier f.) Koeler subsp. dubia (Bunge) Tzvelev ■
65324 Calamagrostis elata Blytt;高拂子茅■☆
65325 Calamagrostis emodensis Griseb.;单蕊拂子茅(单穗拂子茅);Singlespikelet Reedbentgrass, Singlespikelet Woodreed ■
65326 Calamagrostis epigeios (L.) Roth = Calamagrostis epigejos (L.) Roth ■
65327 Calamagrostis epigeios (L.) Roth subsp. macrolepis (Litv.) Tzvelev = Calamagrostis macrolepis Litv. ■
65328 Calamagrostis epigeios (L.) Roth var. densiflora Griseb. = Calamagrostis epigejos (L.) Roth ■
65329 Calamagrostis epigeios (L.) Roth var. sylvatica T. F. Wang = Calamagrostis epigejos (L.) Roth ■
65330 Calamagrostis epigejos (L.) Roth;拂子茅(怀绒草,狼尾草,山拂草,水茅草,羽顶拂子茅);Chee Reed Grass, Chee Reedbentgrass, Chee Reed-grass, Chee Woodreed, Feathertop, Wood Small-reed ■
65331 Calamagrostis epigejos (L.) Roth subsp. extremiorientalis Tzvelev = Calamagrostis epigeios (L.) Roth ■
65332 Calamagrostis epigejos (L.) Roth var. capensis Stapf;好望角拂子茅■☆
65333 Calamagrostis epigejos (L.) Roth var. densiflora Griseb.;密花拂子茅;Denseflower Chee Woodreed, Denseflower Reedbentgrass ■
65334 Calamagrostis epigejos (L.) Roth var. densiflora Griseb. = Calamagrostis epigejos (L.) Roth ■
65335 Calamagrostis epigejos (L.) Roth var. extremiorientalis (Tzvelev) Kitag. = Calamagrostis epigeios (L.) Roth ■
65336 Calamagrostis epigejos (L.) Roth var. georgica?;乔治拂子茅;Chee Reedgrass ■☆
65337 Calamagrostis epigejos (L.) Roth var. parviflora Keng ex T. F. Wang;小花拂子茅;Smallflower Chee Woodreed, Smallflower Reedbentgrass ■
65338 Calamagrostis epigejos (L.) Roth var. sylvatica T. F. Wang;林中拂子茅;Woodland Chee Woodreed, Woods Reedbentgrass ■
65339 Calamagrostis epigejos (L.) Roth var. sylvatica T. F. Wang. = Calamagrostis epigejos (L.) Roth ■
65340 Calamagrostis expansa Rickett et Gilly = Calamagrostis stricta (Timm) Koeler subsp. inexpansa (A. Gray) C. W. Greene ■☆
65341 Calamagrostis expansa Rickett et Gilly var. robusta (Vasey) Stebbins = Calamagrostis canadensis (Michx.) P. Beauv. ■☆
65342 Calamagrostis extremiorientalis (Tzvelev) Prob. = Calamagrostis epigejos (L.) Roth ■
65343 Calamagrostis fauriei Hack.;法氏拂子茅■☆
65344 Calamagrostis fernaldii Louis-Marie = Calamagrostis stricta (Timm) Koeler subsp. inexpansa (A. Gray) C. W. Greene ■☆
65345 Calamagrostis filiformis Griseb. = Calamagrostis scabrescens Griseb. ■
65346 Calamagrostis flaccida (Keng) P. C. Keng = Deyeuxia flaccida Keng ■
65347 Calamagrostis flexuosa Rupr.;曲折拂子茅■☆
65348 Calamagrostis formosana Hayata = Deyeuxia brachytricha (Steud.) Chang ■
65349 Calamagrostis formosana Hayata = Deyeuxia formosana (Hayata) C. C. Hsu ■
65350 Calamagrostis fusca Kom.;棕色拂子茅■☆
65351 Calamagrostis garwhalensis C. E. Hubb. et Bor = Calamagrostis emodensis Griseb. ■
65352 Calamagrostis gigantea (Roshev.) Roshev. = Calamagrostis macrolepis Litv. ■
65353 Calamagrostis gigantea Roshev. = Calamagrostis macrolepis Litv. ■

65354 Calamagrostis gigas Takeda;大拂子茅■☆

65355 Calamagrostis gigas Takeda var. aspera (Honda) Ohwi = Calamagrostis gigas Takeda ■☆

65356 Calamagrostis glauca (M. Bieb.) Rchb. = Calamagrostis pseudophragmites (Hallier f.) Koeler ■

65357 Calamagrostis glauca (M. Bieb.) Trin.;灰绿拂子茅■☆

65358 Calamagrostis grandiflora Hack. et Roshev.;大花拂子茅■☆

65359 Calamagrostis grandiseta Takeda;巨毛拂子茅■☆

65360 Calamagrostis hakonensis Franch. et Sav. = Deyeuxia hakonensis (Franch. et Sav.) Keng ■

65361 Calamagrostis hedbergii Melderis;赫德拂子茅■☆

65362 Calamagrostis hedinii Pilg.;短芒拂子茅;Hedin Reedbentgrass, Hedin Woodreed, Tatar Reedbentgrass ■

65363 Calamagrostis heterogluma Honda;朝鲜拂子茅■

65364 Calamagrostis hissarica Nevski;希萨尔拂子茅■☆

65365 Calamagrostis holciformis Jaub. et Spach = Deyeuxia holciformis (Jaub. et Spach) Bor ■

65366 Calamagrostis holmii Lange;豪氏拂子茅■☆

65367 Calamagrostis huttoniae Hack. = Lachnagrostis lachnantha (Nees) Rúgolo et A. M. Molina ■☆

65368 Calamagrostis hyperborea Lange var. americana (Vasey) Kearney = Calamagrostis stricta (Timm) Koeler subsp. inexpansa (A. Gray) C. W. Greene ■☆

65369 Calamagrostis hyperborea Lange var. elongata Kearney = Calamagrostis stricta (Timm) Koeler subsp. inexpansa (A. Gray) C. W. Greene ■☆

65370 Calamagrostis iberica Little;伊比利亚拂子茅■☆

65371 Calamagrostis inexpansa A. Gray = Calamagrostis neglecta (Ehrh.) Gaertn., B. Mey. et Scherb. subsp. inexpansa (A. Gray) Tzvelev ■☆

65372 Calamagrostis inexpansa A. Gray = Calamagrostis stricta (Timm) Koeler ■☆

65373 Calamagrostis inexpansa A. Gray = Calamagrostis stricta (Timm) Koeler subsp. inexpansa (A. Gray) C. W. Greene ■☆

65374 Calamagrostis inexpansa A. Gray var. barbulata Kearney = Calamagrostis stricta (Timm) Koeler subsp. inexpansa (A. Gray) C. W. Greene ■☆

65375 Calamagrostis inexpansa A. Gray var. brevior (Vasey) Stebbins = Calamagrostis stricta (Timm) Koeler subsp. inexpansa (A. Gray) C. W. Greene ■☆

65376 Calamagrostis inexpansa A. Gray var. brevior (Vasey) Stebbins = Calamagrostis stricta (Timm) Koeler ■☆

65377 Calamagrostis inexpansa A. Gray var. cuprea Kearney = Calamagrostis canadensis (Michx.) P. Beauv. ■☆

65378 Calamagrostis inexpansa A. Gray var. novae-angliae Stebbins = Calamagrostis stricta (Timm) Koeler subsp. inexpansa (A. Gray) C. W. Greene ■☆

65379 Calamagrostis inexpansa A. Gray var. robusta (Vasey) Stebbins = Calamagrostis canadensis (Michx.) P. Beauv. ■☆

65380 Calamagrostis insperata Swallen = Calamagrostis porteri A. Gray ■☆

65381 Calamagrostis insularis Honda = Calamagrostis autumnalis Koidz. subsp. insularis (Honda) Tateoka ■☆

65382 Calamagrostis jacquemontii Hook. f. = Agrostis pilosula Trin. ■

65383 Calamagrostis japonica (Hack.) Govaerts = Aniselytron treutleri (Kuntze) Soják ■

65384 Calamagrostis kengii T. F. Wang;东北拂子茅(耿氏拂子茅);Keng Reedbentgrass, Keng Woodreed ■

65385 Calamagrostis kirishimensis Honda = Calamagrostis autumnalis Koidz. ■☆

65386 Calamagrostis kokonorica (Keng) Tzvelev;青海拂子茅;Qinghai Reedbentgrass ■

65387 Calamagrostis kolymaensis Kom.;科雷马野青茅■☆

65388 Calamagrostis korotkyi Litv.;考劳氏拂子茅■☆

65389 Calamagrostis korotkyi Litv. subsp. turczaninowii (Litv.) Tzvelev = Deyeuxia turczaninowii (Litv.) Y. L. Chang ■

65390 Calamagrostis korshinskyi Litv.;科尔拂子茅■☆

65391 Calamagrostis labradorica Kearney = Calamagrostis stricta (Timm) Koeler subsp. inexpansa (A. Gray) C. W. Greene ■☆

65392 Calamagrostis lacustris (Kearney) Nash = Calamagrostis stricta (Timm) Koeler subsp. inexpansa (A. Gray) C. W. Greene ■☆

65393 Calamagrostis laguroides Regel;兔尾禾拂子茅■☆

65394 Calamagrostis lagurus Koeler = Imperata cylindrica (L.) P. Beauv. ■

65395 Calamagrostis lahulensis Singh = Deyeuxia pulchella (Griseb.) Hook. f. ■

65396 Calamagrostis lanceolata Aitch. = Calamagrostis pseudophragmites (Hallier f.) Koeler ■

65397 Calamagrostis lanceolata Roth;披针拂子茅■☆

65398 Calamagrostis lanceolata Roth = Calamagrostis pseudophragmites (Hallier f.) Koeler ■

65399 Calamagrostis lanceolata Roth var. somalensis Chiov.;索马里拂子茅■☆

65400 Calamagrostis langsdorffii (Link) Trin. = Deyeuxia langsdorffii (Link) Kunth ■

65401 Calamagrostis langsdorffii (Link) Trin. var. punctulata Ohwi = Calamagrostis purpurea (Trin.) Trin. subsp. langsdorfii (Link) Tzvelev ■

65402 Calamagrostis langsdorfii (Link) Trin. = Calamagrostis purpurea (Trin.) Trin. subsp. langsdorfii (Link) Tzvelev ■

65403 Calamagrostis lapponica (Wahlenb.) Hartm. = Deyeuxia lapponica (Wahlenb.) Kunth ■

65404 Calamagrostis lapponica (Wahlenb.) Hartm. var. brevipilis Stebbins = Calamagrostis stricta (Timm) Koeler subsp. inexpansa (A. Gray) C. W. Greene ■☆

65405 Calamagrostis laxa Host = Calamagrostis pseudophragmites (Hallier f.) Koeler ■

65406 Calamagrostis littorea (Schrad.) P. Beauv. = Calamagrostis pseudophragmites (Hallier f.) Koeler ■

65407 Calamagrostis littorea (Schrad.) P. Beauv. var. tartarica Hook. f. = Calamagrostis pseudophragmites (Hallier f.) Koeler subsp. tatarica (Hook. f.) Tzvelev ■

65408 Calamagrostis littorea (Schrad.) P. Beauv. var. tartarica Hook. f. = Calamagrostis hedinii Pilg. ■

65409 Calamagrostis littorea DC. var. tatarica Hook. f. = Calamagrostis pseudophragmites (Hallier f.) Koeler subsp. tatarica (Hook. f.) Tzvelev ■

65410 Calamagrostis littorea DC. var. tatarica Hook. f. = Calamagrostis tatarica (Hook. f.) D. F. Cui ■

65411 Calamagrostis longiflora (Keng) P. C. Keng = Deyeuxia longiflora Keng ■

65412 Calamagrostis longiflora Keng ex P. C. Keng = Deyeuxia flavens Keng ■

65413 Calamagrostis longiseta Hack.;长毛拂子茅■☆

65414 Calamagrostis longiseta Hack. subsp. longearistata (Takeda) T.

Koyama = Calamagrostis grandiseta Takeda ■☆

65415 Calamagrostis longiseta Hack. var. longearistata (Takeda) Ohwi = Calamagrostis grandiseta Takeda ■☆

65416 Calamagrostis longiseta Hack. var. masamunei (Honda) T. Koyama = Calamagrostis masamunei Honda ■☆

65417 Calamagrostis macilenta (Griseb.) Litv. = Deyeuxia macilenta (Griseb.) Keng ■

65418 Calamagrostis macouniana (Vasey) Vasey = Calamagrostis canadensis (Michx.) P. Beauv. var. macouniana (Vasey) Stebbins ■☆

65419 Calamagrostis macrolepis Litv.;大稃拂子茅(大拂子茅);Large Reedbentgrass, Large Woodreed ■

65420 Calamagrostis macrolepis Litv. var. rigidula T. F. Wang;刺稃拂子茅;Rigid Large Woodreed, Stiff Reedbentgrass ■

65421 Calamagrostis macrolepis Litv. var. rigidula T. F. Wang. = Calamagrostis macrolepis Litv. ■

65422 Calamagrostis mannii (Hook. f.) Engl. = Agrostis mannii (Hook. f.) Stapf ■☆

65423 Calamagrostis masamunei Honda;正宗氏拂子茅■☆

65424 Calamagrostis matsudanae Honda = Deyeuxia matsudanae (Honda) Keng ■

65425 Calamagrostis matsumurae Maxim.;松村氏拂子茅■☆

65426 Calamagrostis megalantha (Keng ex P. C. Keng) P. C. Keng = Deyeuxia pulchella (Griseb.) Hook. f. ■

65427 Calamagrostis megalantha (Keng) P. C. Keng = Deyeuxia megalantha Keng ■

65428 Calamagrostis megalantha Keng et P. C. Keng = Deyeuxia pulchella (Griseb.) Hook. f. ■

65429 Calamagrostis mongolicola Kitag.;蒙古拂子茅■☆

65430 Calamagrostis montana DC.;山地拂子茅■☆

65431 Calamagrostis monticola Petr. ex Kom.;山生拂子茅■☆

65432 Calamagrostis moupinensis Franch. = Deyeuxia moupinensis (Franch.) Pilg. ■

65433 Calamagrostis munroana (Aitch. et Hemsl.) Boiss. = Agrostis munroana Aitch. et Hemsl. ■

65434 Calamagrostis nana Takeda;矮小拂子茅■☆

65435 Calamagrostis nana Takeda subsp. hayachinensis (Ohwi) Tateoka;早池峰山拂子茅■☆

65436 Calamagrostis nana Takeda subsp. ohminensis Tateoka;大峰拂子茅■☆

65437 Calamagrostis neglecta (Ehrh.) Gaertn., B. Mey. et Scherb. subsp. aculeolata (Hack.) T. Koyama = Calamagrostis neglecta (Ehrh.) Gaertn., B. Mey. et Scherb. subsp. inexpansa (A. Gray) Tzvelev ■☆

65438 Calamagrostis neglecta (Ehrh.) Gaertn., B. Mey. et Scherb. subsp. inexpansa (A. Gray) Tzvelev;北美拂子茅;Contracted Reedgrass, New England Northern Reed Grass, Northern Reed Grass, Slimstem Reed Grass ■☆

65439 Calamagrostis neglecta (Ehrh.) Gaertn., B. Mey. et Scherb. var. aculeolata (Hack.) Miyabe et Kudo = Calamagrostis neglecta (Ehrh.) Gaertn., B. Mey. et Scherb. subsp. inexpansa (A. Gray) Tzvelev ■☆

65440 Calamagrostis neglecta (Ehrh.) Gaertn., Mey. et Scherb. = Deyeuxia neglecta (Ehrh.) Kunth ■

65441 Calamagrostis neglecta Gaertn., B. Mey. et Scherb. = Calamagrostis stricta (Timm) Koeler ■☆

65442 Calamagrostis neglecta Gaertn., B. Mey. et Scherb. subsp. stricta (Timm) Tzvelev = Calamagrostis stricta (Timm) Koeler ■☆

65443 Calamagrostis neglecta Gaertn., B. Mey. et Scherb. var. gracilis (Scribn.) Scribn. = Calamagrostis stricta (Timm) Koeler ■☆

65444 Calamagrostis neglecta Gaertn., B. Mey. et Scherb. var. micrantha (Kearney) Stebbins = Calamagrostis stricta (Timm) Koeler ■☆

65445 Calamagrostis nepalensis Nees ex Steud. = Calamagrostis pseudophragmites (Hallier f.) Koeler ■

65446 Calamagrostis niitakayamensis Honda subsp. masamunei (Honda) T. Koyama = Calamagrostis masamunei Honda ■☆

65447 Calamagrostis ningxiaensis D. Z. Ma et J. N. Li;宁夏拂子茅;Ningxia Reedbentgrass ■

65448 Calamagrostis obtusata Trin.;钝叶拂子茅■☆

65449 Calamagrostis olympica Boiss.;奥林匹克拂子茅■☆

65450 Calamagrostis onibitoana Tateoka;日本拂子茅■☆

65451 Calamagrostis pappophorea Hack. = Stephanachne pappophorea (Hack.) Keng ■

65452 Calamagrostis paradoxa Lipsky;奇异拂子茅■☆

65453 Calamagrostis pavlovi (Roshev.) Roshev.;帕氏拂子茅■

65454 Calamagrostis persica Boiss.;波斯拂子茅■☆

65455 Calamagrostis petelotii (Hitchc.) Govaerts = Deyeuxia petelotii (Hitchc.) S. M. Phillips et Wen L. Chen ■

65456 Calamagrostis pickeringii A. Gray var. lacustris (Kearney) Hitchc. = Calamagrostis stricta (Timm) Koeler subsp. inexpansa (A. Gray) C. W. Greene ■☆

65457 Calamagrostis pilosula (Trin.) Hook. f. = Agrostis pilosula Trin. ■

65458 Calamagrostis pilosula (Trin.) Hook. f. var. ciliata (Trin.) Hook. f. = Agrostis pilosula Trin. ■

65459 Calamagrostis pilosula (Trin.) Hook. f. var. scabra Hook. f. = Agrostis pilosula Trin. ■

65460 Calamagrostis pilosula (Trin.) Hook. f. var. wallichiana (Steud.) Hook. f. = Agrostis pilosula Trin. ■

65461 Calamagrostis poplawskae Roshev.;波普拉夫斯基拂子茅■☆

65462 Calamagrostis porteri A. Gray;波尔特拂子茅;Ofer Hollow Reedgrass ■☆

65463 Calamagrostis przevalskyi Tzvelev = Deyeuxia tibetica Bor var. przevalskyi (Tzvelev) P. C. Kuo et S. L. Lu ■

65464 Calamagrostis pseudophragmites (Hallier f.) Koeler;假苇拂子茅;Falsereed Reedbentgrass, Falsereed Woodreed ■

65465 Calamagrostis pseudophragmites (Hallier f.) Koeler subsp. dubia (Bunge) Tzvelev;可疑拂子茅■

65466 Calamagrostis pseudophragmites (Hallier f.) Koeler subsp. tatarica (Hook. f.) Tzvelev = Calamagrostis hedinii Pilg. ■

65467 Calamagrostis pseudophragmites (Hallier f.) Koeler subsp. tatarica (Hook. f.) Tzvelev = Calamagrostis tatarica (Hook. f.) D. F. Cui ■

65468 Calamagrostis pseudophragmites (Hallier f.) Koeler var. tatarica (Hook. f.) R. R. Stewart = Calamagrostis hedinii Pilg. ■

65469 Calamagrostis pulchella Griseb. = Deyeuxia pulchella (Griseb.) Hook. f. ■

65470 Calamagrostis purpurascens R. Br.;浅紫拂子茅■☆

65471 Calamagrostis purpurea (Trin.) Trin.;紫拂子茅(大叶章);Purple Reedbentgrass, Scandinavian Small-reed ■

65472 Calamagrostis purpurea (Trin.) Trin. subsp. langsdorffii (Link) Tzvelev = Calamagrostis purpurea (Trin.) Trin. ■

65473 Calamagrostis purpurea (Trin.) Trin. subsp. langsdorfii (Link) Tzvelev;大叶章;Langsdorff Small Reed, Langsdorff Smallreed ■

65474 Calamagrostis rigidula A. I. Baranov et Skvortsov = Calamagrostis macrolepis Litv. var. rigidula T. F. Wang ■

65475 Calamagrostis robusta Franch. et Sav. = Deyeuxia arundinacea (L.) P. Beauv. var. robusta (Franch. et Sav.) P. C. Kuo et S. L. Lu ■

65476 Calamagrostis ruprechtii Nevski;鲁氏拂子茅■☆

65477 Calamagrostis sachalinensis Eastw.;库页拂子茅■☆

65478 Calamagrostis sachalinensis F. Schmidt = Calamagrostis sachalinensis Eastw. ■☆

65479 Calamagrostis salicina Tzvelev = Calamagrostis macilenta (Griseb.) Litv. ■

65480 Calamagrostis scabrescens Griseb. = Deyeuxia scabrescens (Griseb.) Munro ex Duthie ■

65481 Calamagrostis scabrescens Griseb. var. humilis Griseb. = Deyeuxia scabrescens (Griseb.) Munro ex Duthie var. humilis (Griseb.) Hook. f. ■

65482 Calamagrostis schimperiana Hochst. = Leptagrostis schimperiana (Hochst.) C. E. Hubb. ■☆

65483 Calamagrostis schugnanica Litv.;舒格南拂子茅■☆

65484 Calamagrostis sciuroides Franch. et Sav. = Deyeuxia arundinacea (L.) P. Beauv. var. sciuroides (Franch. et Sav.) P. C. Kuo et S. L. Lu ■

65485 Calamagrostis scopulorum M. E. Jones var. bakeri Stebbins = Calamagrostis stricta (Timm) Koeler subsp. inexpansa (A. Gray) C. W. Greene ■☆

65486 Calamagrostis sesquiflora (Trin.) Tzvelev;异花拂子茅■☆

65487 Calamagrostis sesquiflora (Trin.) Tzvelev subsp. urelytra (Hack.) Prob.;日本异花拂子茅■☆

65488 Calamagrostis sichuanensis J. L. Yang;四川野青茅;Sichuan Reedbentgrass,Sichuan Woodreed ■

65489 Calamagrostis stenophylla Hand.-Mazz. = Deyeuxia stenophylla (Hand.-Mazz.) P. C. Kuo et S. L. Lu ■

65490 Calamagrostis stricta (Timm) Koeler;劲直拂子茅(北方拂子茅);Narrow Small-reed, Northern Reed Grass, Slim-stem Reed Grass,Slim-stem Small Reed Grass ■☆

65491 Calamagrostis stricta (Timm) Koeler subsp. inexpansa (A. Gray) C. W. Greene;新英格兰拂子茅;New England Northern Reed Grass,Northern Reed Grass,Slim-stem Reed Grass ■☆

65492 Calamagrostis stricta (Timm) Koeler var. aculeolata Hack. = Calamagrostis neglecta (Ehrh.) Gaertn., B. Mey. et Scherb. subsp. inexpansa (A. Gray) Tzvelev ■☆

65493 Calamagrostis stricta (Timm) Koeler var. brevior Vasey = Calamagrostis stricta (Timm) Koeler subsp. inexpansa (A. Gray) C. W. Greene ■☆

65494 Calamagrostis stricta (Timm) Koeler var. brevior Vasey = Calamagrostis stricta (Timm) Koeler ■☆

65495 Calamagrostis stricta (Timm) Koeler var. lacustris (Kearney) C. W. Greene = Calamagrostis stricta (Timm) Koeler subsp. inexpansa (A. Gray) C. W. Greene ■☆

65496 Calamagrostis suizanensis (Hayata) Honda = Deyeuxia suizanensis (Hayata) Ohwi ■

65497 Calamagrostis sylvatica (Schrad.) Besser = Calamagrostis arundinacea (L.) ■

65498 Calamagrostis tartarica (Hook. f.) D. F. Cui = Calamagrostis hedinii Pilg. ■

65499 Calamagrostis tashiroi Ohwi;田代拂子茅■☆

65500 Calamagrostis tashiroi Ohwi subsp. sikokiana (Ohwi) Tateoka;四国拂子茅■☆

65501 Calamagrostis tatarica (Hook. f.) D. F. Cui = Calamagrostis hedinii Pilg. ■

65502 Calamagrostis teberdensis Litv.;捷别尔达拂子茅■☆

65503 Calamagrostis tianschanica Rupr. = Deyeuxia tianschanica (Rupr.) Bor ■

65504 Calamagrostis tibetica (Bor) Tzvelev = Deyeuxia tibetica Bor ■

65505 Calamagrostis treutleri (Kuntze) Shukla = Aniselytron treutleri (Kuntze) Soják ■

65506 Calamagrostis trichantha Schischk.;毛花拂子茅■☆

65507 Calamagrostis turczaninowii Litv.;兴安短毛野青茅;Turczaninov Reedbentgrass ■

65508 Calamagrostis turczaninowii Litv. = Deyeuxia turczaninowii (Litv.) Y. L. Chang ■

65509 Calamagrostis turkestanica Hack.;突厥野青茅;Turkestan Reedbentgrass ■

65510 Calamagrostis turkestanica Hack. = Calamagrostis pseudophragmites (Hallier f.) Koeler subsp. tatarica (Hook. f.) Tzvelev ■

65511 Calamagrostis uralensis Litv.;乌拉尔拂子茅■☆

65512 Calamagrostis urelytra Hack. = Calamagrostis sesquiflora (Trin.) Tzvelev subsp. urelytra (Hack.) Prob. ■☆

65513 Calamagrostis urelytra Hack. = Calamagrostis sesquiflora (Trin.) Tzvelev ■☆

65514 Calamagrostis varia Bol. = Deyeuxia bolanderi Vasey ■☆

65515 Calamagrostis varia Gren. et Godr. = Calamagrostis montana DC. ■☆

65516 Calamagrostis varia Host = Deyeuxia halleriana Vasey ■☆

65517 Calamagrostis varia P. Beauv.;欧亚拂子茅■☆

65518 Calamagrostis varia P. Beauv. var. macilenta Griseb. = Deyeuxia macilenta (Griseb.) Keng ■

65519 Calamagrostis variegata With. = Phalaris arundinacea L. ■

65520 Calamagrostis villosa (Chaix) J. F. Gmel.;柔毛拂子茅■☆

65521 Calamagrostis vilnensis Besser ex Roem. et Schult.;维尔纳拂子茅■☆

65522 Calamagrostis welwitschii Rendle = Lachnagrostis lachnantha (Nees) Rúgolo et A. M. Molina ■☆

65523 Calamagrostis wiluica Litv. ex Petr.;韦卢拂子茅■☆

65524 Calamagrostis yanyuanensis J. L. Yang;盐源野青茅;Yanyuan Reedbentgrass,Yanyuan Woodreed ■

65525 Calamina P. Beauv. = Apluda L. ■

65526 Calamintha Adans. = Glecoma L. ■

65527 Calamintha Adans. = Nepeta L. ■●

65528 Calamintha Mill. (1754);新风轮菜属(新风轮属);Calamint, Calamintha,Savory ■

65529 Calamintha Mill. = Clinopodium L. ■●

65530 Calamintha abyssinica (Hochst. ex Benth.) A. Rich. = Clinopodium abyssinicum (Hochst. ex Benth.) Kuntze ■☆

65531 Calamintha acinos (L.) Clairv.;葡萄新风轮菜■☆

65532 Calamintha acinos (L.) Clairv. = Acinos arvensis (Lam.) Dandy ■☆

65533 Calamintha acinos (L.) Clairv. ex Gaudin = Calamintha arvensis Lam. ■☆

65534 Calamintha albiflora Vaniot = Nepeta cataria L. ■

65535 Calamintha alpina Lam.;高山新风轮菜;Alpine Savory ■☆

65536 Calamintha annua Schrenk = Calamintha debilis (Bunge) Benth. ■

65537 Calamintha argyi H. Lév. = Clinopodium confine (Hance) Kuntze ■

65538 Calamintha arkansana (Nutt.) Shinners;矮新风轮;Arkansas

Calamint, Calamint, Limestone Calamint, Low Calamint ■☆
65539 Calamintha arvensis Lam.;田野新风轮■☆
65540 Calamintha ascendens Jord.;山地新风轮;Basil Thyme, Calamint, Calamint Balm, Hore Calamint, Mountain Balm, Mountain Calamint, Mountain Mint, Nespite ■☆
65541 Calamintha ascendens Jord. = Calamintha sylvatica Bromf. subsp. ascendens (Jord.) P. W. Ball ■☆
65542 Calamintha atlantica (Ball) Ball = Clinopodium atlanticum (Ball) N. Galland ■☆
65543 Calamintha baborensis Batt. = Calamintha grandiflora (L.) Moench subsp. baborensis (Batt.) N. Galland ■☆
65544 Calamintha baetica Boiss. et Reut. = Calamintha sylvatica Bromf. subsp. ascendens (Jord.) P. W. Ball ■☆
65545 Calamintha barosma W. W. Sm. = Micromeria barosma (W. W. Sm.) Hand. -Mazz. ■
65546 Calamintha cacondensis G. Taylor = Clinopodium myrianthum (Baker) Ryding ■☆
65547 Calamintha candidissima (Munby) Benth.;极白新风轮■☆
65548 Calamintha cavaleriei H. Lév. et Vaniot = Melissa axillaris (Benth.) Bakh. f. ■
65549 Calamintha chinensis Benth. = Clinopodium chinense (Benth.) Kuntze ■
65550 Calamintha chinensis Benth. var. grandiflora Maxim. = Clinopodium urticifolium (Hance) C. Y. Wu et S. J. Hsuan ex H. W. Li ■
65551 Calamintha chinensis Benth. var. megalantha Diels = Clinopodium megalanthum (Diels) C. Y. Wu et S. J. Hsuan ex H. W. Li ■
65552 Calamintha clinopodium Benth. = Clinopodium urticifolium (Hance) C. Y. Wu et S. J. Hsuan ex H. W. Li ■
65553 Calamintha clinopodium Benth. = Clinopodium vulgare L. ■☆
65554 Calamintha clinopodium Benth. subsp. atlantica Ball = Clinopodium atlanticum (Ball) N. Galland ■☆
65555 Calamintha clinopodium Benth. subsp. villosum (Noë) Maire = Clinopodium vulgare L. subsp. arundanum (Boiss.) Nyman ■☆
65556 Calamintha clinopodium Benth. var. abbreviata Maire = Clinopodium vulgare L. ■☆
65557 Calamintha clinopodium Benth. var. chinensis (Benth.) Miq. = Clinopodium chinense (Benth.) Kuntze ■
65558 Calamintha clinopodium Benth. var. discolor (Diels) Dunn = Clinopodium discolor (Diels) C. Y. Wu et S. J. Hsuan ex H. W. Li ■
65559 Calamintha clinopodium Benth. var. glabrescens (Pomel) Batt. = Clinopodium vulgare L. ■☆
65560 Calamintha clinopodium Benth. var. megalantha (Diels) Dunn = Clinopodium megalanthum (Diels) C. Y. Wu et S. J. Hsuan ex H. W. Li ■
65561 Calamintha clinopodium Benth. var. nepalensis Dunn = Clinopodium polycephalum (Vaniot) C. Y. Wu et S. J. Hsuan ex H. W. Li ■
65562 Calamintha clinopodium Benth. var. polycephala (Vaniot) Dunn = Clinopodium polycephalum (Vaniot) C. Y. Wu et S. J. Hsuan ex H. W. Li ■
65563 Calamintha clinopodium Benth. var. pratensis Dunn = Clinopodium polycephalum (Vaniot) C. Y. Wu et S. J. Hsuan ex H. W. Li ■
65564 Calamintha clinopodium Benth. var. repens (Buch. -Ham. ex D. Don) Dunn = Clinopodium repens (D. Don) Vell. ■
65565 Calamintha clinopodium Benth. var. urticifolia Hance = Clinopodium urticifolium (Hance) C. Y. Wu et S. J. Hsuan ex H. W. Li ■
65566 Calamintha clipeata Vaniot = Mosla chinensis Maxim. ■
65567 Calamintha confinis Hance = Clinopodium confine (Hance) Kuntze ■
65568 Calamintha confinis Hance = Clinopodium gracile (Benth.) Matsum. ■
65569 Calamintha coreana H. Lév. = Clinopodium urticifolium (Hance) C. Y. Wu et S. J. Hsuan ex H. W. Li ■
65570 Calamintha cryptantha Vatke = Clinopodium simense (Benth.) Kuntze ■☆
65571 Calamintha cryptantha Vatke var. filiformis Chiov. = Clinopodium simense (Benth.) Kuntze ■☆
65572 Calamintha cryptantha Vatke var. mildbraedii Perkins = Clinopodium simense (Benth.) Kuntze ■☆
65573 Calamintha debilis (Bunge) Benth.;新风轮菜(新风轮);Slender Calamintha ■
65574 Calamintha discolor Diels = Clinopodium discolor (Diels) C. Y. Wu et S. J. Hsuan ex H. W. Li ■
65575 Calamintha elgonensis Bullock = Clinopodium uhligii (Gürke) Ryding ■☆
65576 Calamintha esquirolii H. Lév. = Coleus esquirolii (H. Lév.) Dunn ■
65577 Calamintha euosma W. Sm. = Micromeria euosma (W. W. Sm.) C. Y. Wu ■
65578 Calamintha glabella (Michx.) Benth. var. angustifolia (Torr.) DeWolf = Calamintha arkansana (Nutt.) Shinners ■☆
65579 Calamintha gracilis Benth. = Clinopodium gracile (Benth.) Matsum. ■
65580 Calamintha granatensis Boiss. et Reut. = Acinos alpinus (L.) Moench subsp. meridionalis (Nyman) P. W. Ball ■☆
65581 Calamintha grandiflora (L.) Moench;大花新风轮菜(大花香草);Beautiful Mint, Large-flowered Calamint ■☆
65582 Calamintha grandiflora (L.) Moench = Satureja grandiflora Scheele ■☆
65583 Calamintha grandiflora (L.) Moench subsp. baborensis (Batt.) N. Galland;巴布尔新风轮菜■☆
65584 Calamintha grandiflora (L.) Moench var. parviflora Coss. = Calamintha grandiflora (L.) Moench subsp. baborensis (Batt.) N. Galland ■☆
65585 Calamintha heterotricha Boiss. et Reut.;异毛新风轮菜■☆
65586 Calamintha hispidula Boiss. et Reut.;细毛新风轮菜■☆
65587 Calamintha incana (Sm.) Boiss.;灰毛新风轮菜■☆
65588 Calamintha kilimandschari Gürke = Clinopodium kilimandschari (Gürke) Ryding ■☆
65589 Calamintha laxiflora Hayata = Clinopodium laxiflorum (Hayata) Mori ■
65590 Calamintha longicaulis Benth.;长柄新风轮菜;Longstalk Calamintha ■☆
65591 Calamintha macrostema (Moc. et Sessé ex Benth.) Benth.;大蕊新风轮■☆
65592 Calamintha masukuensis (Baker) S. Moore = Clinopodium myrianthum (Baker) Ryding ■☆
65593 Calamintha megalantha (Diels) Hand. -Mazz. = Clinopodium megalanthum (Diels) C. Y. Wu et S. J. Hsuan ex H. W. Li ■
65594 Calamintha menthifolia Host var. baetica (Boiss. et Reut.) Ball = Calamintha sylvatica Bromf. subsp. ascendens (Jord.) P. W. Ball ■☆
65595 Calamintha multicaulis Maxim. = Clinopodium multicaule (Maxim.) Kuntze ■☆

65596 Calamintha nepeta (L.) Savi = Clinopodium calamintha Kuntze ■☆
65597 Calamintha nepeta (L.) Savi = Satureja nepeta Scheele ■☆
65598 Calamintha nepeta (L.) Savi subsp. glandulosa (Req.) P. W. Ball = Calamintha nepeta (L.) Savi subsp. spuneri (Boiss.) Nyman ■☆
65599 Calamintha nepeta (L.) Savi subsp. glandulosa (Req.) P. W. Ball = Clinopodium calamintha Kuntze ■☆
65600 Calamintha nepeta (L.) Savi subsp. spuneri (Boiss.) Nyman;斯普奈新风轮菜■☆
65601 Calamintha nepeta (L.) Savi subsp. sylvatica (Bromf.) R. Morales = Calamintha sylvatica Bromf. ■☆
65602 Calamintha nervosa Pomel;多脉新风轮菜■☆
65603 Calamintha officinalis Moench;药用新风轮菜(新风轮塔花);Calamint Savory ■☆
65604 Calamintha paradoxa Vatke = Clinopodium paradoxum (Vatke) Ryding ■☆
65605 Calamintha parvula S. Moore = Clinopodium simense (Benth.) Kuntze ■☆
65606 Calamintha polycephala Vaniot = Clinopodium polycephalum (Vaniot) C. Y. Wu et S. J. Hsuan ex H. W. Li ■
65607 Calamintha radicans Vaniot = Clinopodium gracile (Benth.) Matsum. ■
65608 Calamintha repens (Buch. -Ham. ex D. Don) Benth. = Clinopodium repens (D. Don) Vell. ■
65609 Calamintha repens (D. Don) Benth. = Clinopodium repens (D. Don) Vell. ■
65610 Calamintha rotundifolia (Pers.) Benth. = Acinos rotundifolius Pers. ■☆
65611 Calamintha rotundifolia (Pers.) Benth. var. micrantha Murb. = Acinos rotundifolius Pers. ■☆
65612 Calamintha simensis Benth. = Clinopodium simense (Benth.) Kuntze ■☆
65613 Calamintha simensis Benth. f. flaccida Vatke = Clinopodium uhligii (Gürke) Ryding var. obtusifolium (Avetta) Ryding ■☆
65614 Calamintha simensis Benth. var. obtusifolia Avetta = Clinopodium uhligii (Gürke) Ryding var. obtusifolium (Avetta) Ryding ■☆
65615 Calamintha sylvatica Bromf. = Clinopodium ascendens Samp. ■☆
65616 Calamintha sylvatica Bromf. subsp. ascendens (Jord.) P. W. Ball = Clinopodium ascendens Samp. ■☆
65617 Calamintha sylvatica Bromf. subsp. ascendens P. W. Ball = Clinopodium ascendens Samp. ■☆
65618 Calamintha tsacapanensis H. Lév. = Clinopodium polycephalum (Vaniot) C. Y. Wu et S. J. Hsuan ex H. W. Li ■
65619 Calamintha umbrosa (M. Bieb.) Benth. = Clinopodium polycephalum (Vaniot) C. Y. Wu et S. J. Hsuan ex H. W. Li ■
65620 Calamintha umbrosa Rchb.;耐荫新风轮菜;Shadeloving Calamintha ■☆
65621 Calamintha urticifolia (Hance) Hand. -Mazz. = Clinopodium urticifolium (Hance) C. Y. Wu et S. J. Hsuan ex H. W. Li ■
65622 Calamochloa E. Fourn. = Sohnsia Airy Shaw ■☆
65623 Calamochloe Rchb. = Arundinella Raddi ■
65624 Calamochloe Rchb. = Goldbachia Trin. (废弃属名) ■
65625 Calamophyllum Schwantes = Cylindrophyllum Schwantes ●☆
65626 Calamophyllum Schwantes(1927);苇叶番杏属■☆
65627 Calamophyllum cylindricum (Haw.) Schwantes;柱形苇叶番杏■☆
65628 Calamophyllum teretifolium (Haw.) Schwantes;圆柱叶苇叶番杏■☆
65629 Calamophyllum teretiusculum (Haw.) Schwantes;圆柱苇叶番杏■☆
65630 Calamosagus Griff. = Korthalsia Blume ●☆
65631 Calamovilfa (A. Gray) Hack. (1890);沙茅属■☆
65632 Calamovilfa Hack. = Calamovilfa (A. Gray) Hack. ■☆
65633 Calamovilfa longifolia (Hook.) Scribn.;长叶沙茅;Long-leaved Reed Grass, Prairie Sand Reed, Prairie Sandreed, Prairie Sand-reed, Sand-reed Grass ■☆
65634 Calampelis D. Don = Eccremocarpus Ruiz et Pav. ●☆
65635 Calamphoreus Chinnock = Eremophila R. Br. ●☆
65636 Calamphoreus Chinnock(2007);澳洲沙漠木属●☆
65637 Calamus L. (1753);省藤属(白藤属,水藤属,藤属);Cane Palms, Malacca Cane, Rattan, Rattan Palm, Rattan Palms, Rattanpalm, Rattan-palm, Reed Palm, White-awhile Vine ●
65638 Calamus Pall. = Acorus L. ●
65639 Calamus akimensis Becc. = Calamus deerratus G. Mann et H. Wendl. ●☆
65640 Calamus angustifolia Griff. = Daemonorops angustifolia Mart. ●☆
65641 Calamus aquatilis Ridl.;浮游省藤(南方省藤);Aquatic Rattan Palm ●☆
65642 Calamus austroguangxiensis S. J. Pei et S. Y. Chen;桂南省藤;S. Guangxi Rattanpalm, South Guangxi Rattan Palm ●
65643 Calamus axillaris Becc.;腋生省藤;Axillary Rattan Palm ●☆
65644 Calamus balansaeanus Becc.;小白藤;Balanse Rattan Palm, Small Rattanpalm ●
65645 Calamus balansaeanus Becc. var. castaneolepis (C. F. Wei) S. J. Pei et S. Y. Chen;褐鳞省藤;Brownscale Small Rattanpalm ●
65646 Calamus balingensis Furtado;巴玲省藤;Baling Rattan Palm ●☆
65647 Calamus barteri Drude = Calamus deerratus G. Mann et H. Wendl. ●☆
65648 Calamus belumutensis Furtado;百禄省藤;Belumut Rattan Palm ●☆
65649 Calamus benomensis Furtado;白农省藤;Benom Rattan Palm ●☆
65650 Calamus blumei Becc.;布鲁墨省藤;Blume Rattan Palm ●☆
65651 Calamus bonianus Becc.;多穗白藤;Boni Rattan Palm, Manyspike Rattanpalm ●
65652 Calamus brevispadix Ridl.;短佛焰省藤;Shortspadix Rattan Palm ●☆
65653 Calamus bubuensis Becc.;布布省藤;Bubu Rattan Palm ●☆
65654 Calamus burkillianus Becc.;布尔基利省藤;Burkili Rattan Palm ●☆
65655 Calamus cabrae De Wild. et T. Durand = Eremospatha cabrae (T. Durand et Schinz) De Wild. ●☆
65656 Calamus caesius Blume;竹藤(蓝灰省藤,蓝灰藤);Blueish-grey Rattanpalm, Bluish-grey Rattan Palm, Rotan Segar ●☆
65657 Calamus caryotoides A. Cunn. ex Mart.;鱼尾葵省藤●☆
65658 Calamus castaneus Griff.;褐色省藤;Chestnut Rattan Palm ●☆
65659 Calamus chibehensis Furtado;其本省藤;Chibenh Rattan Palm ●☆
65660 Calamus ciliaris Blume;缘毛省藤(睫毛藤);Ciliate Rattan Palm ●☆
65661 Calamus collinus Griff. = Calamus erectus Roxb. ●◇
65662 Calamus collinus Griff. = Salacca secunda Griff. ●◇
65663 Calamus compsostachys Burret;短轴省藤;Beautiful-spiked Rattan Palm, Shortaxis Rattanpalm ●
65664 Calamus concinnus Mart.;优雅省藤;Elegant Rattan Palm ●☆
65665 Calamus corneri Furtado;考奈省藤;Corner Rattan Palm ●☆
65666 Calamus cuspidatus G. Mann et H. Wendl. = Eremospatha

cuspidata (G. Mann et H. Wendl.) G. Mann et H. Wendl. ●☆

65667 Calamus deerratus G. Mann et H. Wendl.;迷惑省藤●☆

65668 Calamus densiflorus Becc.;稠密花省藤(密花省藤);Denseflower Rattan Palm ●☆

65669 Calamus dianbaiensis C. F. Wei;电白省藤;Dianbai Rattan Palm, Dianbai Rattanpalm,Yunnan White Beautyberry ●

65670 Calamus dichotoma (Lour.) K. Koch;白棠子树(大叶毛将军,尖尾枫,毛毛茶,梅灯狗散,青含条,细亚锡饭,小米干饭,小叶鸦鹊饭,小叶鸭鹊饭,小叶紫珠,小紫珠,珍珠风,子条,紫珠,紫珠草);Chinese Beautyberry,Korean Beautyberry,Purple Beauty Berry,Purple Beauty Bush,Purple Beautyberry,Purple Purplepearl ●

65671 Calamus diepenhorstii Miq.;蝶盆豪氏省藤;Diepenhorst Rattan Palm ●☆

65672 Calamus distichoideus Furtado;两列状省藤;Tworows Rattan Palm ●☆

65673 Calamus distichus Ridl.;二列省藤;Bilow Rattanpalm,Distinchous Rattan Palm,Distinchous Rattanpalm ●

65674 Calamus distichus Ridl. var. shangsiensis S. J. Pei et S. Y. Chen;上思省藤;Shangsi Distinchous Rattan Palm ●

65675 Calamus egregius Burret;短叶省藤(厘藤);Shortleaf Rattan Palm,Shortleaf Rattanpalm ●

65676 Calamus elegans Becc. ex Ridl.;雅致省藤;Elegant Rattan Palm ●☆

65677 Calamus erectus Roxb.;直立省藤;Erect Rattanpalm, Upright Rattan Palm ●◇

65678 Calamus erectus Roxb. var. birmanicus Becc.;滇缅省藤;Yunnan-Burma Rattanpalm ●

65679 Calamus erectus Roxb. var. collina Becc. = Calamus erectus Roxb. ●◇

65680 Calamus erectus Roxb. var. macrocarpus Becc. = Calamus erectus Roxb. ●◇

65681 Calamus exilis Griff. ex Mart.;瘦直省藤;Thin Rattan Palm ●☆

65682 Calamus faberi Becc.;大白藤(多果省藤);Big Rattanpalm, Faber Rattan Palm,Faber Rattanpalm ●

65683 Calamus faberi Becc. = Calamus walkerii Hance ●

65684 Calamus faberi Becc. var. brevispicatus (C. F. Wei) S. J. Pei et S. Y. Chen;短穗省藤;Shortspike Faber Rattanpalm, Shortspike Rattanpalm ●

65685 Calamus falabensis Becc. = Calamus deerratus G. Mann et H. Wendl. ●☆

65686 Calamus fasciculatus Roxb. = Calamus viminalis Willd. var. fasciculatus (Roxb.) Becc. ●

65687 Calamus feanus Becc.;缅甸省藤;Burma Rattan Palm, Burma Rattanpalm ●

65688 Calamus feanus Becc. var. medogensis S. J. Pei et S. Y. Chen;墨脱省藤;Motuo Rattanpalm ●

65689 Calamus filipendulus Becc.;丝悬省藤;Threadpendulous Rattan Palm ●☆

65690 Calamus fissus Miq. = Daemonorops fissa Blume ●☆

65691 Calamus flabellatus Becc.;扇形省藤;Fanshape Rattan Palm ●☆

65692 Calamus flabelloides Furtado;扇叶省藤;Flabellate Rattan Palm ●☆

65693 Calamus flagellum Griff. = Calamus flagellum Griff. ex Mart. ●

65694 Calamus flagellum Griff. ex Mart.;长鞭藤(长鞭省藤);Flagellum Rattan Palm,Longscourge Rattanpalm ●

65695 Calamus flagellum Griff. var. furvifurfuraceus S. J. Pei et S. Y. Chen;黑鳞秕藤;Black-scaled Longscourge Rattanpalm ●

65696 Calamus flagellum Griff. var. karinensis Becc. = Calamus karinensis (Becc.) S. J. Pei et S. Y. Chen ●

65697 Calamus formosanus Becc.;台湾省藤(水藤,台湾水藤,土藤);Formosan Cans, Formosan Rattan Palm, Formosan Rattanpalm, Taiwan Rattanpalm ●

65698 Calamus giganteus Becc.;巨藤(巨大省藤);Giant Rattanpalm, Gigant Rattan Palm ●

65699 Calamus giganteus Becc. var. robustus S. J. Pei et S. Y. Chen;粗壮省藤;Robust Rattanpalm ●

65700 Calamus gracilis Roxb.;小省藤(海南省藤,细茎省藤,纤细省藤);Little Rattanpalm,Thin Rattan Palm ●

65701 Calamus grandis Griff. = Daemonorops grandis (Griff.) Mart. ●☆

65702 Calamus guangxiensis C. F. Wei;广西省藤;Guangxi Rattan Palm,Guangxi Rattanpalm ●

65703 Calamus hainanensis Hung T. Chang et Z. R. Xu ex R. H. Miao;海南省藤;Hainan Rattanpalm ●

65704 Calamus hainanensis Hung T. Chang et Z. R. Xu ex R. H. Miao = Calamus gracilis Roxb. ●

65705 Calamus hendersonii Furtado;亨氏省藤;Henderson Rattan Palm ●☆

65706 Calamus henryanus Becc.;滇南省藤(白藤,山省藤);Henry Rattan Palm,S. Yunnan Rattanpalm ●

65707 Calamus henryanus Becc. var. castaneolepis C. F. Wei = Calamus balansaeanus Becc. var. castaneolepis (C. F. Wei) S. J. Pei et S. Y. Chen ●

65708 Calamus heudelotii Drude = Calamus deerratus G. Mann et H. Wendl. ●☆

65709 Calamus holttumii Furtado;郝氏省藤;Holttum Rattan Palm ●☆

65710 Calamus hookeri G. Mann et H. Wendl. = Eremospatha hookeri (G. Mann et H. Wendl.) H. Wendl. ●☆

65711 Calamus hoplites Dunn;高毛鳞省藤(高毛省藤);Higher-hairy-scaled Rattan Palm,Tall Rattanpalm ●

65712 Calamus insignis Griff.;明显省藤;Distinct Rattan Palm ●☆

65713 Calamus javensis Blume;爪哇省藤(灰藤);Java Rattan Palm, Rotan Lilin ●☆

65714 Calamus jenkinsianus Griff. = Calamus flagellum Griff. ex Mart. ●

65715 Calamus karinensis (Becc.) S. J. Pei et S. Y. Chen;勐腊鞭藤;Mengla Rattan Palm,Mengla Rattanpalm ●

65716 Calamus kemamanensis Furtado;克马曼省藤;Kemaman Rattan Palm ●☆

65717 Calamus koribanus Furtado;克里巴省藤;Koriba Rattan Palm ●☆

65718 Calamus laevis G. Mann et H. Wendl. = Laccosperma laeve (G. Mann et H. Wendl.) H. Wendl. ●☆

65719 Calamus latifolius Roxb. = Calamus palustris Griff. var. cochinchinensis Becc. ●

65720 Calamus laurentii De Wild. = Calamus deerratus G. Mann et H. Wendl. ●☆

65721 Calamus laxiflorus Becc.;疏花省藤;Loose-flowered Rattan Palm ●☆

65722 Calamus laxissimus Ridl.;极疏省藤;Most Loose Rattan Palm ●☆

65723 Calamus leprieurii Becc. = Calamus deerratus G. Mann et H. Wendl. ●☆

65724 Calamus litoralis Blume;水藤;Rotan Ajer ●☆

65725 Calamus longisetus Griff.;长刚毛省藤;Longsetose Rattan Palm ●☆

65726 Calamus longispathus Ridl.;长佛焰苞省藤;Longspathe Rattan Palm ●☆

65727 Calamus luridus Becc.;褐黄省藤;Pale Yellow Rattan Palm ●☆

65728 Calamus macrocarpus G. Mann et H. Wendl. = Eremospatha

macrocarpa (G. Mann et H. Wendl.) H. Wendl. ●☆
65729 Calamus macrocarpus Griff. = Calamus erectus Roxb. ●◇
65730 Calamus macrorrhynchus Burret;大喙省藤(喙尖黄藤,白藤);Big-beak Rattan Palm,Bigbeak Rattanpalm ●
65731 Calamus manan Miq.;马南省藤;Manan Rattan Palm ●☆
65732 Calamus mannii H. Wendl. = Oncocalamus mannii (H. Wendl.) H. Wendl. ●☆
65733 Calamus margaritae Hance = Calamus orientalis C. E. Chang ●
65734 Calamus margaritae Hance = Daemonorops margaritae (Hance) Becc. ●
65735 Calamus mawaiensis Furtado;马尾省藤;Mawai Rattan Palm ●☆
65736 Calamus melanochrous Burret;瑶山省藤(长果省藤);Yaoshan Rattan Palm,Yaoshan Rattanpalm ●
65737 Calamus melanoloma Mart. = Calamus distichus Ridl. var. shangsiensis S. J. Pei et S. Y. Chen ●
65738 Calamus moorhousei Furtado;木豪斯省藤;Moorhouse Rattan Palm ●☆
65739 Calamus multirameus Ridl.;多枝省藤;Manybranch Rattan Palm ●☆
65740 Calamus multispicatus Burret;裂苞省藤;Multispiked Rattan Palm,Splitbract Rattanpalm ●
65741 Calamus nambariensis Becc.;南巴省藤;Nanba Rattanpalm,Nanbar Rattan Palm ●
65742 Calamus nambariensis Becc. var. alpinus S. J. Pei et S. Y. Chen;高地省藤;Alpine Nanbar Rattan Palm ●
65743 Calamus nambariensis Becc. var. furfuraceus S. J. Pei et S. Y. Chen;鳞秕省藤●
65744 Calamus nambariensis Becc. var. furfuraceus S. J. Pei et S. Y. Chen = Calamus nambariensis Becc. var. yingjiangensis S. J. Pei et S. Y. Chen ●
65745 Calamus nambariensis Becc. var. menglongensis S. J. Pei et S. Y. Chen;勐龙省藤;Menglong Nanbar Rattan Palm ●
65746 Calamus nambariensis Becc. var. xishuangbannanensis S. J. Pei et S. Y. Chen;版纳省藤;Xishuangbanna Rattanpalm ●
65747 Calamus nambariensis Becc. var. yingjiangensis S. J. Pei et S. Y. Chen;盈江省藤;Yingjiang Rattanpalm ●
65748 Calamus obovoideus S. J. Pei et S. Y. Chen;倒卵果省藤;Obovoid-fruited Rattan Palm,Obvatefruit Rattanpalm ●
65749 Calamus opacus G. Mann et H. Wendl. = Laccosperma opacum (G. Mann et H. Wendl.) Drude ●☆
65750 Calamus oreophilus Furtado;山地省藤;Montane Rattan Palm ●☆
65751 Calamus orientalis C. E. Chang;阔叶省藤;Broadleaf Rattanpalm,Oriental Rattan Palm ●
65752 Calamus orientalis C. E. Chang = Calamus quiquesetinervius Burret ●
65753 Calamus ornatus Blume;装饰省藤;Decorative Rattan Palm ●☆
65754 Calamus oxleyanus Teijsm. et Binn. ex Miq.;欧克莱省藤;Oxley Rattan Palm ●☆
65755 Calamus oxycarpus Becc.;尖果省藤;Acutefruit Rattanpalm,Sharped-fruit Rattanpalm,Sharped-fruited Rattan Palm ●
65756 Calamus oxycarpus Becc. var. angustifolius S. Y. Chen ex K. L. Wang;窄叶尖果省藤●
65757 Calamus padangensis Furtado;巴当省藤;Padang Rattan Palm ●☆
65758 Calamus pallidulus Becc.;苍白省藤;Pallid Rattan Palm ●☆
65759 Calamus palustris Griff.;泽生藤(沼泽省藤);Bog Rattanpalm,Marshy Rattan Palm,Marshy Rattanpalm ●
65760 Calamus palustris Griff. var. cochinchinensis Becc.;滇越省藤;Cochichina Bog Rattanpalm ●
65761 Calamus palustris Griff. var. longistachys S. J. Pei et S. Y. Chen;长穗泽生藤(长穗省藤);Longspike Bog Rattanpalm ●
65762 Calamus palustris Griff. var. mediostachys S. J. Pei et S. Y. Chen;中穗泽生藤(中穗省藤);Medewspike Bog Rattanpalm ●
65763 Calamus pandanosmus Furtado;露兜树省藤;Screwpine Rattan Palm ●☆
65764 Calamus parakensis Becc.;波拉克省藤;Perak Rattan Palm ●☆
65765 Calamus paspalanthus Becc.;雀稗花省藤;Paspalum-flower Rattan Palm ●☆
65766 Calamus penicillatus Roxb.;画笔状省藤;Penicillate Rattan Palm ●☆
65767 Calamus peregrinus Furtado;外来省藤;Foreign Rattan Palm ●☆
65768 Calamus perrottetii Becc. = Calamus deerratus G. Mann et H. Wendl. ●☆
65769 Calamus platyacanthoides Merr.;阔刺省藤(省藤);Broadspine Rattan Palm,Common Rattan Palm ●☆
65770 Calamus platyacanthus Warb. ex Becc.;宽刺藤(宽藤);Broad-spine Rattan Palm,Broadspine Rattanpalm ●
65771 Calamus platycanthus Warb. ex Becc. var. mediostachys S. J. Pei et S. Y. Chen;中穗宽刺藤(中穗省藤);Mediostachys Broad-spine Rattan Palm ●
65772 Calamus polystachys Becc.;多穗省藤;Manyspike Rattan Palm ●☆
65773 Calamus pulaiensis Becc.;布莱省藤;Pulai Rattan Palm ●☆
65774 Calamus pulchellus Burret;阔叶鸡藤(阔叶省藤,猫藤);Broadleaf Rattanpalm,Broad-leaved Rattan Palm ●
65775 Calamus quinquesetinervius Burret;五脉刚毛省藤(黄藤);Fiveveine Rattanpalm,Five-veined Rattan Palm ●
65776 Calamus radulosus Becc.;粗糙省藤;Scabrous Rattan Palm ●☆
65777 Calamus ramosissimus Griff.;多分枝省藤;Manybranched Rattan Palm ●☆
65778 Calamus rhabdocladus Burret;华南省藤(白藤,弹弓藤,弓藤,木藤,手杖藤,杖藤,棕藤);South China Rattan Palm,South China Rattanpalm,Stick Rattanpalm ●
65779 Calamus rhabdocladus Burret var. glabulosus S. J. Pei et S. Y. Chen;弓弦藤●
65780 Calamus ridleyanus Becc.;瑞得莱省藤;Ridley Rattan Palm ●☆
65781 Calamus riparius Furtado;河岸省藤;Riverbank Rattan Palm ●☆
65782 Calamus rotang L.;省藤(罗丹滕);Common Rattan Palm,Ratan,Rattan,Rattan Cane,Rattan Palm,Rotang Rattan Palm,Rotang Rattan-palm ●☆
65783 Calamus roxburghii Griff. = Calamus rotang L. ●☆
65784 Calamus rugosus Becc.;皱纹省藤;Rugose Rattan Palm ●☆
65785 Calamus scabridulus Becc.;略粗糙省藤;Slightly-scabrous Rattan Palm ●☆
65786 Calamus schweinfurthii Becc. = Calamus deerratus G. Mann et H. Wendl. ●☆
65787 Calamus scipionum Lour.;天竺藤(杖省藤);Malacca Cane,Rotan Semaboe,Stick Rattan Palm,Stick Rattanpalm ●☆
65788 Calamus secundiflorus P. Beauv. = Laccosperma secundiflorum (P. Beauv.) Kuntze ●☆
65789 Calamus siamensis Becc.;泰国省藤;Siam Rattan Palm ●☆
65790 Calamus simplex Becc.;单式省藤;Simple Rattan Palm ●☆
65791 Calamus simplicifolius C. F. Wei;单叶省藤(省藤);Singleleaf Rattanpalm,Unifolious Rattan Palm ●
65792 Calamus siphonospathus Mart.;管苞省藤;Pipebract Rattanpalm,Tubular-bract Rattan Palm ●

65793 Calamus siphonospathus Mart. var. sublaevis Becc. ;兰屿省藤;Lanyu Rattan Palm ●
65794 Calamus spathulatus Becc. ;匙形省藤;Spatulate Rattan Palm ●☆
65795 Calamus speciocissimus Furtado;极美丽省藤;Mostbeautiful Rattan Palm ●☆
65796 Calamus spectatissimus Furtado;非寻常省藤;Wonderful Rattan Palm ●☆
65797 Calamus tanakadatei Furtado;田中氏省藤;Tanaka Rattan Palm ●☆
65798 Calamus tenuis Roxb. = Calanthe plantaginea Wall. ●☆
65799 Calamus tetradactyloides Burret;多刺鸡藤(高山鸡藤);Four-fingers-like Rattanpalm, Manyspine Rattan Palm, Manyspine Rattanpalm ●
65800 Calamus tetradactylus Hance;白省藤(白藤,鸡藤,山甘蔗);Fourfinger Rattan Palm, Fourfingers Rattan Palm, Fourfingers Rattanpalm, Rattan Palm, White Rattanpalm ●
65801 Calamus tetradactylus Hance var. bonianus (Becc.) Conrard = Calamus bonianus Becc. ●
65802 Calamus thysanolepis Hance;毛鳞省藤;Hairscale Rattanpalm, Hairy-scaled Rattan Palm ●
65803 Calamus thysanolepis Hance = Calamus simplicifolius C. F. Wei ●
65804 Calamus thysanolepis Hance var. polylepis C. F. Wei;多鳞省藤(毛鳞省藤);Hairy-scaled Rattan Palm, Many-hairy-scale Rattanpalm, Poly-hairscale Rattanpalm ●
65805 Calamus tomentosus Becc. ;绒毛省藤;Tomentose Rattan Palm ●☆
65806 Calamus tonkinensis Becc. = Calamus walkerii Hance ●
65807 Calamus tonkinensis Becc. var. brevispicatus C. F. Wei = Calamus faberi Becc. var. brevispicatus (C. F. Wei) S. J. Pei et S. Y. Chen ●
65808 Calamus tumidus Furtado;肿胀省藤;Swollen Rattan Palm ●☆
65809 Calamus viminalis Willd. ;柳条省藤(柳枝状省藤,勐捧省藤);Osier-like Rattan Palm, Osier-like Rattanpalm, Willowtwig Rattanpalm ●
65810 Calamus viminalis Willd. var. fasciculatus (Roxb.) Becc. ;勐捧省藤;Fasciculate Osier-like Rattan Palm ●
65811 Calamus viminalis Willd. var. fasciculatus (Roxb.) Becc. = Calamus viminalis Willd. ●
65812 Calamus viridispinus Becc. ;绿刺省藤;Greenspine Rattan Palm ●☆
65813 Calamus wailong S. J. Pei et S. Y. Chen;大藤;Wailong Rattan Palm, Wailong Rattanpalm ●
65814 Calamus walkerii Hance;多果省藤;Fruitful Rattanpalm, Walker Rattan Palm ●
65815 Calamus yangchunensis C. F. Wei;阳春省藤;Yangchun Rattan Palm, Yangchun Rattanpalm ●
65816 Calamus yunnanensis S. J. Pei et S. Y. Chen;云南省藤;Yunnan Rattan Palm, Yunnan Rattanpalm ●
65817 Calamus yunnanensis S. J. Pei et S. Y. Chen var. densiflorus S. J. Pei et S. Y. Chen;密花省藤;Denseflower Yunnan Rattanpalm ●
65818 Calamus yunnanensis S. J. Pei et S. Y. Chen var. intermedius S. J. Pei et S. Y. Chen;屏边省藤;Pingbian Rattanpalm ●
65819 Calanassa Post et Kuntze = Callianassa Webb et Berthel. ●☆
65820 Calanassa Post et Kuntze = Isoplexis (Lindl.) Loudon ●☆
65821 Calanchoe Pers. = Kalanchoe Adans. ●■
65822 Calanda K. Schum. (1903);热非紫草属■☆
65823 Calanda rubricaulis K. Schum. ;热非紫草■☆
65824 Calandarium Juss. ex Steud. = Calandrinia Kunth(保留属名)■☆
65825 Calandra Post et Kuntze = Calliandra Benth. (保留属名)●
65826 Calandra Post et Kuntze = Inga Mill. ●■☆
65827 Calandrinia Kunth (1823) (保留属名);岩马齿苋属;Parakeelya, Rock Pueslane, Rockpueslane ■☆
65828 Calandrinia Ruiz. et Pav. = Calandrinia Kunth(保留属名)■☆
65829 Calandrinia acaulis Kunth;无茎马齿苋■☆
65830 Calandrinia ambigua (S. Watson) Howell;沙地岩马齿苋;Desert Pot-herb ■☆
65831 Calandrinia ambigua (S. Watson) Howell = Cistanthe ambigua (S. Watson) Carolin ex Hershk. ■☆
65832 Calandrinia balonensis Lindl. ;宽叶岩马齿苋;Broad-leaved Parakeelya, Parakedya ■☆
65833 Calandrinia breweri S. Watson;布鲁尔岩马齿苋■☆
65834 Calandrinia caulescens Kunth = Calandrinia ciliata DC. ■☆
65835 Calandrinia ciliata (Ruiz et Pav.) DC. ;缘毛岩马齿苋;Red Maids ■☆
65836 Calandrinia ciliata (Ruiz et Pav.) DC. var. menziesii (Hook.) J. F. Macbr. = Calandrinia ciliata (Ruiz et Pav.) DC. ■☆
65837 Calandrinia ciliata DC. = Calandrinia ciliata (Ruiz et Pav.) DC. ■☆
65838 Calandrinia ciliata DC. var. menziesii (Hook.) J. F. Macbr. = Calandrinia ciliata DC. ■☆
65839 Calandrinia columbiana Howell ex A. Gray;哥伦比亚岩马齿苋■☆
65840 Calandrinia cotyledon S. Watson = Lewisia cotyledon (S. Watson) B. L. Rob. ■☆
65841 Calandrinia discolor Schrad. ;异色岩马齿苋■☆
65842 Calandrinia elegans Hort. = Calandrinia discolor Schrad. ■☆
65843 Calandrinia grandiflora Lindl. ;大花岩马齿苋;Common Rock Pueslane ■☆
65844 Calandrinia grayi Britton = Lewisia pygmaea (A. Gray) B. L. Rob. ■☆
65845 Calandrinia howellii S. Watson = Lewisia cotyledon (S. Watson) B. L. Rob. var. howellii (S. Watson) Jeps. ■☆
65846 Calandrinia leeana Porter = Lewisia leeana (Porter) B. L. Rob. ■☆
65847 Calandrinia maritima Nutt. = Cistanthe maritima (Nutt.) Carolin ex Hershk. ■☆
65848 Calandrinia menziesii (Hook.) Torr. et A. Gray var. macrocarpa A. Gray = Calandrinia breweri S. Watson ■☆
65849 Calandrinia micrantha Schltdl. = Calandrinia ciliata (Ruiz et Pav.) DC. ■☆
65850 Calandrinia nevadensis A. Gray = Lewisia nevadensis (A. Gray) B. L. Rob. ■☆
65851 Calandrinia oppositifolia S. Watson = Lewisia oppositifolia (S. Watson) B. L. Rob. ■☆
65852 Calandrinia pygmaea (A. Gray) A. Gray = Lewisia pygmaea (A. Gray) B. L. Rob. ■☆
65853 Calandrinia spectabilis Otto et A. Dietr. ;岩马齿苋;Rock Purslane ■☆
65854 Calandrinia tweedyi A. Gray = Cistanthe tweedyi (A. Gray) Hershk. ■☆
65855 Calandrinia umbellata DC. ;秘鲁岩马齿苋;Peru Rock Purslane, Peruvian Rock Pueslane ■☆
65856 Calandriniopsis E. Franz = Calandrinia Kunth(保留属名)■☆
65857 Calandriniopsis E. Franz(1908);拟岩马齿苋属■☆
65858 Calandriniopsis montana (Phil.) E. Franz;拟岩马齿苋■☆
65859 Calanira Post et Kuntze = Callianira Miq. ●■
65860 Calanira Post et Kuntze = Piper L. ●■

65861 Calanthe Ker Gawl. = Calanthe R. Br.(保留属名)■
65862 Calanthe R. Br.(1821)(保留属名);虾脊兰属(根节兰属);Calanthe ■
65863 Calanthe × dominii Lindl.;道明虾脊兰(白花长距虾脊兰)■
65864 Calanthe × matsumurana Schltr. = Calanthe × dominii Lindl. ■
65865 Calanthe × okinawensis Hayata;冲绳虾脊兰■☆
65866 Calanthe actinomorpha Fukuy.;辐射虾脊兰(辐射鹤顶兰,辐形根节兰);Radiative Calanthe ■
65867 Calanthe albolongicalcarata S. S. Ying;白花长距虾脊兰(白花长距根节兰);White-longspur Calanthe ■
65868 Calanthe albolongicalcarata S. S. Ying = Calanthe dominii Lindl. ■☆
65869 Calanthe alismifolia Lindl.;泽泻虾脊兰(八仙草,长青九龙盘,九子连环草,克马七,山卡拉,山蜘蛛,细点根节兰,棕叶七);Waterplantainleaf Calanthe ■
65870 Calanthe alleizettei Gagnep.;长柄虾脊兰■
65871 Calanthe alpina Hook. f. ex Lindl.;流苏虾脊兰(大仙茅,高山虾脊兰,九子连,九子连环草,马牙七,肉连环,铁牛杆子,羽唇根节兰);Tassel Calanthe ■
65872 Calanthe alpina Hook. f. ex Lindl. var. fimbriatomarginata (Fukuy.) F. Maek. = Calanthe alpina Hook. f. ex Lindl. ■
65873 Calanthe alpina Hook. f. ex Lindl. var. schlechteri (H. Hara) F. Maek. = Calanthe alpina Hook. f. ex Lindl. ■
65874 Calanthe amamiana Fukuy.;奄美虾脊兰■
65875 Calanthe amamiana Fukuy. = Calanthe aristulifera Rchb. f. ■
65876 Calanthe amoena W. W. Sm. = Calanthe puberulla Lindl. ■
65877 Calanthe angusta Lindl. = Calanthe odora Griff. ■
65878 Calanthe angusta Lindl. var. laeta Hand.-Mazz. = Calanthe odora Griff. ■
65879 Calanthe angustifolia (Blume) Lindl.;狭叶虾脊兰(矮根节兰,小根节兰);Narrowleaf Calanthe ■
65880 Calanthe anjanii Lucksom = Calanthe griffithii Lindl. ■
65881 Calanthe arcuata Rolfe;弧距虾脊兰(尾唇根节兰);Arcuate Calanthe ■
65882 Calanthe arcuata Rolfe = Calanthe brevicornu Lindl. ■☆
65883 Calanthe arcuata Rolfe var. brevifolia Z. H. Tsi;短叶虾脊兰;Shortleaf Arcuate Calanthe ■
65884 Calanthe arcuata Rolfe var. brevifolia Z. H. Tsi = Calanthe arcuata Rolfe ■
65885 Calanthe argenteo-striata C. Z. Tang et S. J. Cheng;银带虾脊兰;Silverline Calanthe ■
65886 Calanthe arisanensis Hayata;台湾虾脊兰(阿里山根节兰);Alishan Calanthe ■
65887 Calanthe aristulifera Rchb. f.;翘距虾脊兰(垂花根节兰,翘距根节兰);Raisespur Calanthe ■
65888 Calanthe aristulifera Rchb. f. var. amamiana (Fukuy.) Hatus. = Calanthe amamiana Fukuy. ■
65889 Calanthe aristulifera Rchb. f. var. kirishimensis (Yatabe) Honda = Calanthe aristulifera Rchb. f. ■
65890 Calanthe austrokiusiuensis Ohwi = Calanthe alismifolia Lindl. ■
65891 Calanthe biloba Lindl.;二裂虾脊兰;Bilobe Calanthe ■
65892 Calanthe brachychila Gagnep. = Calanthe mannii Hook. f. ■
65893 Calanthe brevicolumna Hayata = Calanthe herbacea Lindl. ■
65894 Calanthe brevicolumna Hayata = Calanthe triplicata (Willemet) Ames ■
65895 Calanthe brevicornu Lindl.;肾唇虾脊兰(地棕,短角虾脊兰,九子连环草,青野棕,细抓);Reniformlip Calanthe Calanthe, Shorthorn Calanthe ■☆
65896 Calanthe buccinifera Rolfe = Calanthe alpina Hook. f. ex Lindl. ■
65897 Calanthe buccinifera Rolfe ex Hemsl. = Calanthe alpina Hook. f. ex Lindl. ■
65898 Calanthe bungoana Ohwi = Calanthe davidii Franch. var. bungoana (Ohwi) T. Hashim. ■
65899 Calanthe bungoana Ohwi = Calanthe davidii Franch. ■
65900 Calanthe bursicola Gagnep. = Cephalantheropsis obcordata (Lindl.) Ormerod ■
65901 Calanthe cardioglossa Schltr.;心唇虾脊兰;Heartshaped Calanthe ■☆
65902 Calanthe caudatilabella Hayata;尾唇根节兰■
65903 Calanthe caudatilabella Hayata = Calanthe arcuata Rolfe ■
65904 Calanthe caudatilabella Hayata var. latiloba F. Maek. ex Gogelein = Calanthe arcuata Rolfe ■
65905 Calanthe cheniana Hand.-Mazz. = Calanthe discolor Lindl. ■
65906 Calanthe clavata Lindl.;棒距虾脊兰(棒距根节兰);Clavate Calanthe, Stickspur Calanthe ■
65907 Calanthe clavata Lindl. = Calanthe formosana Rolfe ■
65908 Calanthe clavata Lindl. var. malipoensis Z. H. Tsi;麻栗坡虾脊兰;Malipo Calanthe ■
65909 Calanthe clavata Lindl. var. malipoensis Z. H. Tsi = Calanthe clavata Lindl. ■
65910 Calanthe coelogyniformis Kraenzl. = Calanthe delavayi Finet ■
65911 Calanthe coelogyniformis Kraenzl. = Phaius delavayi (Finet) P. J. Cribb et Perner ■
65912 Calanthe corymbosa Lindl. = Calanthe sylvatica (Thouars) Lindl. ■
65913 Calanthe crinita Gagnep. = Phaius mishmensis (Lindl. et Paxton) Rchb. f. ■
65914 Calanthe curculigoides Lindl. = Calanthe formosana Rolfe ■
65915 Calanthe davidii Franch.;剑叶虾脊兰(长叶根节兰,九龙草,九子连,兰草,螺丝七,马齿七,铁梳子,羊角七);David Calanthe, Swordleaf Calanthe ■
65916 Calanthe davidii Franch. var. bungoana (Ohwi) T. Hashim. = Calanthe davidii Franch. ■
65917 Calanthe delavayi Finet;少花虾脊兰(九子连环草,肉连环);Delavay Calanthe ■
65918 Calanthe delavayi Finet = Phaius delavayi (Finet) P. J. Cribb et Perner ■
65919 Calanthe delphinioides Kraenzl. = Calanthe sylvatica (Thouars) Lindl. ■
65920 Calanthe densiflora Lindl.;密花虾脊兰(密花根节兰,竹叶根节兰);Denseflower Calanthe ■
65921 Calanthe discolor Lindl.;虾脊兰(串白鸡,九节虫,九子连环草,连环草,肉连环,三棱虾脊兰,夜白鸡,一串纽子,异色虾脊兰,硬九头狮子草,珠串珠);Calanthe, Common Calanthe ■
65922 Calanthe discolor Lindl. f. kanashiroi (Fukuy.) K. Nakaj. = Calanthe discolor Lindl. var. kanashiroi Fukuy. ■☆
65923 Calanthe discolor Lindl. f. rosea (Hatus.) Honda;粉红虾脊兰■☆
65924 Calanthe discolor Lindl. f. rufoaurantiaca (Iwata) Honda;浅橙虾脊兰■☆
65925 Calanthe discolor Lindl. f. sieboldii (Decne. ex Regel) Ohwi = Calanthe sieboldii Decne. ex Regel ■
65926 Calanthe discolor Lindl. f. trilabellata F. Maek.;三唇虾脊兰■☆
65927 Calanthe discolor Lindl. f. tyo-harae (Makino) Honda;原宽虾脊兰■☆

65928 Calanthe discolor Lindl. f. viridialba (Maxim.) Honda;绿白虾脊兰■☆

65929 Calanthe discolor Lindl. var. amamiana (Fukuy.) Masam. = Calanthe amamiana Fukuy. ■

65930 Calanthe discolor Lindl. var. bicolor (Lindl.) Makino;二色虾脊兰■☆

65931 Calanthe discolor Lindl. var. divaricatipetala Ida;叉瓣虾脊兰■☆

65932 Calanthe discolor Lindl. var. flava Yatabe = Calanthe sieboldii Decne. ex Regel ■

65933 Calanthe discolor Lindl. var. flava Yatabe = Calanthe striata R. Br. ■

65934 Calanthe discolor Lindl. var. kanashiroi Fukuy.;金城虾脊兰■☆

65935 Calanthe discolor Lindl. var. sieboldii (Decne. ex Regel) Maxim. = Calanthe sieboldii Decne. ex Regel ■

65936 Calanthe disticha Ts. Tang et F. T. Wang = Calanthe formosana Rolfe ■

65937 Calanthe disticha Ts. Tang et F. T. Wang = Calanthe speciosa (Blume) Lindl. ■

65938 Calanthe dolichopoda Fukuy. = Cephalantheropsis calanthoides (Ames) Tang S. Liu et H. J. Su ■

65939 Calanthe dominii Lindl.;多明虾脊兰■☆

65940 Calanthe dulongensis H. Li, R. Li et Z. L. Dao;独龙虾脊兰;Dulong Calanthe ■

65941 Calanthe ecarinata Rolfe ex Hemsl.;天全虾脊兰;Tianquan Calanthe ■

65942 Calanthe elliptica Hayata = Calanthe aristulifera Rchb. f. ■

65943 Calanthe emeishanica K. Y. Lang et Z. H. Tsi;峨眉虾脊兰;Emei Calanthe ■

65944 Calanthe ensifolia Rolfe = Calanthe davidii Franch. ■

65945 Calanthe esquirolei Schltr. = Calanthe discolor Lindl. ■

65946 Calanthe fargesii Finet;天府虾脊兰;Farges Calanthe ■

65947 Calanthe fauriei Schltr. = Calanthe alismifolia Lindl. ■

65948 Calanthe fimbriata Franch. = Calanthe alpina Hook. f. ex Lindl. ■

65949 Calanthe fimbriatomarginata Fukuy. = Calanthe alpina Hook. f. ex Lindl. ■

65950 Calanthe foerstermannii Rchb. f. = Calanthe lyroglossa Rchb. f. ■

65951 Calanthe foesythiiflora Hayata = Calanthe lyroglossa Rchb. f. ■

65952 Calanthe formosana Rolfe;二列叶虾脊兰(两列虾脊兰,两列叶虾脊兰,台湾根节兰);Distichous Calanthe, Taiwan Calanthe, Two-row Calanthe ■

65953 Calanthe formosana Rolfe = Calanthe speciosa (Blume) Lindl. ■

65954 Calanthe forsythiiflora Hayata = Calanthe lyroglossa Rchb. f. ■

65955 Calanthe fugongensis X. H. Jin et S. C. Chen;福贡虾脊兰■

65956 Calanthe furcata Bateman = Calanthe triplicata (Willemet) Ames ■

65957 Calanthe furcata Bateman ex Lindl. = Calanthe triplicata (Willemet) Ames ■

65958 Calanthe furcata Bateman ex Lindl. = Dactylorhiza umbrosa (Kar. et Kir.) Nevski ■

65959 Calanthe furcata Bateman ex Lindl. f. fauriei (Schltr.) M. Hiroe = Calanthe alismifolia Lindl. ■

65960 Calanthe furcata Bateman ex Lindl. f. masuca (D. Don) M. Hiroe = Calanthe sylvatica (Thouars) Lindl. ■

65961 Calanthe furcata Bateman ex Lindl. f. matsudae (Hayata) M. Hiroe = Calanthe davidii Franch. ■

65962 Calanthe furcata Bateman ex Lindl. f. raishaensis (Hayata) M. Hiroe = Calanthe aristulifera Rchb. f. ■

65963 Calanthe furcata Bateman ex Lindl. f. textorii (Miq.) M. Hiroe = Calanthe sylvatica (Thouars) Lindl. ■

65964 Calanthe furcata Bateman ex Lindl. var. alismatifolia (Lindl.) M. Hiroe = Calanthe alismifolia Lindl. ■

65965 Calanthe gebina (Lindl.) Lindl. = Bletilla striata (Thunb. ex A. Murray) Rchb. f. ■

65966 Calanthe graciliflora Hayata;钩距虾脊兰(千斤桩,四里麻,细花根节兰,纤花根节兰,一莞棕);Hamate Calanthe, Hookspur Calanthe ■

65967 Calanthe graciliflora Hayata var. xuafengensis Z. H. Tsi;雪峰虾脊兰;Xuefeng Calanthe ■

65968 Calanthe gracilis Lindl. = Cephalantheropsis gracilis (Lindl.) S. Y. Hu ■

65969 Calanthe gracilis Lindl. = Cephalantheropsis obcordata (Lindl.) Ormerod ■

65970 Calanthe gracilis Lindl. var. venusta (Schltr.) F. Maek. = Cephalantheropsis obcordata (Lindl.) Ormerod ■

65971 Calanthe gracilis Lindl. var. venusta (Schltr.) F. Maek. = Cephalantheropsis gracilis (Lindl.) S. Y. Hu ■

65972 Calanthe griffithii Lindl.;通麦虾脊兰;Griffith Calanthe ■

65973 Calanthe hamata Hand.-Mazz. = Calanthe graciliflora Hayata ■

65974 Calanthe hancockii Rolfe;叉唇虾脊兰(九子连环草);Hancock Calanthe ■

65975 Calanthe hattorii Schltr.;服部虾脊兰■☆

65976 Calanthe henryi Rolfe;疏花虾脊兰;Poorflower Calanthe ■

65977 Calanthe herbacea Lindl.;西南虾脊兰;SW. China Calanthe ■

65978 Calanthe hoshii S. Kobay = Calanthe triplicata (Willemet) Ames ■

65979 Calanthe humbertii H. Perrier;亨伯特虾脊兰■☆

65980 Calanthe izu-insularis Ohwi et Satomi;伊豆虾脊兰■☆

65981 Calanthe japonica Blume ex Miq. = Calanthe alismifolia Lindl. ■

65982 Calanthe kawakamii Hayata = Calanthe sieboldii Decne. ex Regel ■

65983 Calanthe kazuoi Gogelein = Calanthe densiflora Lindl. ■

65984 Calanthe kintaroi Gogelein = Calanthe sylvatica (Thouars) Lindl. ■

65985 Calanthe kintaroi Yamam. = Calanthe sylvatica (Thouars) Lindl. ■

65986 Calanthe kirishimensis Yatabe = Calanthe aristulifera Rchb. f. ■

65987 Calanthe kooshunensis Fukuy. = Cephalantheropsis calanthoides (Ames) Tang S. Liu et H. J. Su ■

65988 Calanthe kooshunensis Fukuy. = Cephalantheropsis halconensis (Ames) S. S. Ying ■

65989 Calanthe koshunensis Fukuy. = Cephalantheropsis calanthoides (Ames) Tang S. Liu et H. J. Su ■

65990 Calanthe labrosa (Rchb. f.) Hook. f.;葫芦茎虾脊兰;Lipped Calanthe ■

65991 Calanthe lamellata Hayata = Calanthe tricarinata Lindl. ■

65992 Calanthe lamellosa Rolfe = Calanthe brevicornu Lindl. ■☆

65993 Calanthe lechangensis Z. H. Tsi et Ts. Tang;乐昌虾脊兰;Lechang Calanthe ■

65994 Calanthe lepida W. W. Sm. = Calanthe puberulla Lindl. ■

65995 Calanthe limprichtii Schltr.;开唇虾脊兰;Limpricht Calanthe ■

65996 Calanthe liukiuensis Schltr. = Calanthe lyroglossa Rchb. f. ■

65997 Calanthe longicalcarata Hayata ex Gogelein = Calanthe masuca (D. Don) Lindl. ■

65998 Calanthe longicalcarata Hayata ex Gogelein = Calanthe sylvatica (Thouars) Lindl. ■

65999 Calanthe longipes Hook. f. = Cephalantheropsis gracilis (Lindl.) S. Y. Hu ■

66000 Calanthe longipes Hook. f. = Cephalantheropsis longipes (Hook.

f.) Ormerod ■

66001 Calanthe lyroglossa Rchb. f. ;南方虾脊兰(黄花根节兰,连翘根节兰);Foerstermann Calanthe,Southern Calanthe ■

66002 Calanthe lyroglossa Rchb. f. var. foesythiiflora (Hayata) S. S. Ying = Calanthe lyroglossa Rchb. f. ■

66003 Calanthe madacasgariensis Rolfe ex Hook. f. ;马达加斯加虾脊兰;Madagascar Calanthe ■☆

66004 Calanthe madagascariensis Rolfe = Calanthe madacasgariensis Rolfe ex Hook. f. ■☆

66005 Calanthe mannii Hook. f. ;细花虾脊兰(九子连环草,马里虾脊兰,肉连环,铁连环);Mann Calanthe ■

66006 Calanthe masuca (D. Don) Lindl. = Calanthe sylvatica (Thouars) Lindl. ■

66007 Calanthe masuca (D. Don) Lindl. ex Wall. = Calanthe sylvatica (Thouars) Lindl. ■

66008 Calanthe masuca (D. Don) Lindl. ex Wall. f. albiflora (Ida) Nackej. ;白细花长距虾脊兰■☆

66009 Calanthe masuca (D. Don) Lindl. ex Wall. var. sinensis Rendle = Calanthe sylvatica (Thouars) Lindl. ■

66010 Calanthe masuca (D. Don) Lindl. var. sinensis Rendle = Calanthe sylvatica (Thouars) Lindl. ■

66011 Calanthe matsudae Hayata;长叶根节兰■

66012 Calanthe matsudae Hayata = Calanthe davidii Franch. ■

66013 Calanthe matsumurana Schltr. = Calanthe × dominii Lindl. ■

66014 Calanthe megalopha Franch. = Calanthe tricarinata Lindl. ■

66015 Calanthe metoensis Z. H. Tsi et K. Y. Lang;墨脱虾脊兰;Motuo Calanthe ■

66016 Calanthe nankunensis Z. H. Tsi;南昆虾脊兰;Nankun Calanthe ■

66017 Calanthe natalensis (Rchb. f.) Rchb. f. = Calanthe sylvatica (Thouars) Lindl. ■

66018 Calanthe neglecta Schltr. = Calanthe sylvatica (Thouars) Lindl. ■

66019 Calanthe nigropuncticulata Fukuy. = Calanthe alismifolia Lindl. ■

66020 Calanthe nipponica Makino;戟形虾脊兰;Hastate Calanthe ■

66021 Calanthe oblanceolata Ohwi et T. Koyama = Calanthe mannii Hook. f. ■

66022 Calanthe odora Griff. ;香花虾脊兰(大仙茅,大线毛);Fragrant Calanthe ■

66023 Calanthe okinawensis Hayata = Calanthe sylvatica (Thouars) Lindl. ■

66024 Calanthe pachystalix Rchb. f. ex Hook. f. = Calanthe davidii Franch. ■

66025 Calanthe pantlingii Schltr. = Calanthe tricarinata Lindl. ■

66026 Calanthe patsinensis S. Y. Hu = Calanthe formosana Rolfe ■

66027 Calanthe patsinensis S. Y. Hu = Calanthe speciosa (Blume) Lindl. ■

66028 Calanthe petelotiana Gagnep. ;圆唇虾脊兰;Roundlip Calanthe ■

66029 Calanthe phajoides Rchb. f. = Calanthe angustifolia (Blume) Lindl. ■

66030 Calanthe pilchra (Blume) Lindl. var. formosana (Rolfe) S. S. Ying = Calanthe speciosa (Blume) Lindl. ■

66031 Calanthe plantaginea Lindl. ;车前虾脊兰;Plantain Calanthe ■

66032 Calanthe plantaginea Lindl. var. lushuiensis K. Y. Lang et Z. H. Tsi;泸水车前虾脊兰;Lushui Plantain Calanthe ■

66033 Calanthe plantaginea Wall. = Calanthe plantaginea Lindl. ■

66034 Calanthe puberula Lindl. var. okushirensis (Miyabe et Tatew.) Hiroe;奥尻虾脊兰■☆

66035 Calanthe puberula Lindl. var. reflexa (Maxim.) M. Hiroe;卷萼根节兰■

66036 Calanthe puberula Lindl. var. reflexa (Maxim.) M. Hiroe = Calanthe reflexa (Kuntze) Maxim. ■

66037 Calanthe puberulla Lindl. ;镰萼虾脊兰(反卷根节兰,卷萼根节兰,镰叶虾脊兰);Sicklecalyx Calanthe ■

66038 Calanthe pulchra (Blume) Lindl. var. formosana (Rolfe) S. S. Ying = Calanthe speciosa (Blume) Lindl. ■

66039 Calanthe pulchra (Blume) Lindl. var. formosana (Rolfe) S. S. Ying = Calanthe formosana Rolfe ■

66040 Calanthe pumila Fukuy. ;矮根节兰■

66041 Calanthe pumila Fukuy. = Calanthe angustifolia (Blume) Lindl. ■

66042 Calanthe pusilla Finet = Calanthe mannii Hook. f. ■

66043 Calanthe raishaensis Hayata = Calanthe aristulifera Rchb. f. ■

66044 Calanthe ramosa Gagnep. = Phaius mishmensis (Lindl. et Paxton) Rchb. f. ■

66045 Calanthe ramosii Ames = Cephalantheropsis obcordata (Lindl.) Ormerod ■

66046 Calanthe reflexa (Kuntze) Maxim. ;反瓣虾脊兰(反卷根节兰,假虾脊兰,卷萼根节兰,相似虾脊兰);Reflex Calanthe,Reflexpetal Calanthe ■

66047 Calanthe reflexa (Kuntze) Maxim. = Calanthe puberulla Lindl. ■

66048 Calanthe reflexa Maxim. = Calanthe puberula Lindl. var. reflexa (Maxim.) M. Hiroe ■

66049 Calanthe reflexa Maxim. var. okushirensis (Miyabe et Tatew.) Ohwi = Calanthe puberula Lindl. var. okushirensis (Miyabe et Tatew.) Hiroe ■☆

66050 Calanthe repens Schltr. ;匍匐虾脊兰■☆

66051 Calanthe rosea (Lindl.) Benth. ;红花虾脊兰;Redflower Calanthe ■☆

66052 Calanthe rosea Benth. = Calanthe rosea (Lindl.) Benth. ■☆

66053 Calanthe rubens Ridl. ;红唇虾脊兰;Redlip Calanthe ■☆

66054 Calanthe rubicallosa Masam. = Calanthe triplicata (Willemet) Ames ■

66055 Calanthe rubicallosa Masam. = Dactylorhiza umbrosa (Kar. et Kir.) Nevski ■

66056 Calanthe sacculata Schltr. ;囊爪虾脊兰;Sacculate Calanthe ■

66057 Calanthe sacculata Schltr. var. tchengkeoutinensis Ts. Tang et F. T. Wang;城口虾脊兰;Chengkou Calanthe ■

66058 Calanthe sacculata Schltr. var. tchengkeoutinensis Ts. Tang et F. T. Wang = Calanthe sacculata Schltr. ■

66059 Calanthe sanderiana Rolfe = Calanthe sylvatica (Thouars) Lindl. ■

66060 Calanthe sasakii Hayata = Calanthe arisanensis Hayata ■

66061 Calanthe scaposa Z. H. Tsi et K. Y. Lang;西藏虾脊兰;Xizang Calanthe ■

66062 Calanthe scaposa Z. H. Tsi et K. Y. Lang = Calanthe brevicornu Lindl. ■☆

66063 Calanthe schlechteri H. Hara;羽唇根节兰■

66064 Calanthe schlechteri H. Hara = Calanthe alpina Hook. f. ex Lindl. ■

66065 Calanthe schliebenii Mansf. = Calanthe sylvatica (Thouars) Lindl. ■

66066 Calanthe seikooensis Gogelein = Calanthe sylvatica (Thouars) Lindl. ■

66067 Calanthe shweliensis W. W. Sm. = Calanthe odora Griff. ■

66068 Calanthe sieboldii Decne. ex Regel;大黄花虾脊兰(红纹唇虾脊兰,黄根节兰);Redstriatelip Calanthe,Siebold Calanthe ■

66069 Calanthe sieboldii Decne. ex Regel = Calanthe striata R. Br. ■

66070 Calanthe similis Schltr. = Calanthe puberulla Lindl. ■
66071 Calanthe similis Schltr. = Calanthe reflexa (Kuntze) Maxim. ■
66072 Calanthe simplex Seidenf.;匙瓣虾脊兰;Spoonpetal Calanthe ■
66073 Calanthe sinica Z. H. Tsi;中华虾脊兰;China Calanthe ■
66074 Calanthe speciosa (Blume) Lindl.;台湾根节兰(二列叶虾脊兰)■
66075 Calanthe stolzii Schltr. = Calanthe sylvatica (Thouars) Lindl. ■
66076 Calanthe striata (Banks) R. Br. = Calanthe sieboldii Decne. ex Regel ■
66077 Calanthe striata (Sw.) R. Br. = Calanthe graciliflora Hayata ■
66078 Calanthe striata (Sw.) R. Br. = Calanthe sieboldii Decne. ex Regel ■
66079 Calanthe striata (Sw.) R. Br. f. sieboldii (Decne. ex Regel) Ohwi = Calanthe sieboldii Decne. ex Regel ■
66080 Calanthe striata (Sw.) R. Br. var. pumila (Fukuy.) S. S. Ying = Calanthe angustifolia (Blume) Lindl. ■
66081 Calanthe striata R. Br. f. sieboldii (Decne. ex Regel) Ohwi = Calanthe striata R. Br. ■
66082 Calanthe striata R. Br. var. pumila (Fukuy.) S. S. Ying = Calanthe angustifolia (Blume) Lindl. ■
66083 Calanthe striata R. Br. var. sieboldii (Decne. ex Regel) Maxim. = Calanthe sieboldii Decne. ex Regel ■
66084 Calanthe striata R. Br. var. sieboldii (Decne.) Maxim. = Calanthe sieboldii Decne. ex Regel ■
66085 Calanthe striata R. Br. var. sieboldii Maxim.;黄根节兰■
66086 Calanthe sylvatica (Thouars) Lindl.;长距虾脊兰(长距根节兰,林生虾脊兰);Forest Calanthe, Longspur Calanthe ■
66087 Calanthe sylvatica (Thouars) Lindl. var. natalensis Rchb. f. = Calanthe sylvatica (Thouars) Lindl. ■
66088 Calanthe takeoi Hayata = Calanthe sieboldii Decne. ex Regel ■
66089 Calanthe takeoi Hayata = Phaius takeoi (Hayata) H. J. Su ■
66090 Calanthe tangmaiensis K. Y. Lang et Y. Tateishi = Calanthe griffithii Lindl. ■
66091 Calanthe textori Miq. = Calanthe masuca (D. Don) Lindl. ■
66092 Calanthe textori Miq. = Calanthe sylvatica (Thouars) Lindl. ■
66093 Calanthe textori Miq. var. longicalcarata (Hayata ex Gogelein) Garay et H. R. Sweet = Calanthe sylvatica (Thouars) Lindl. ■
66094 Calanthe textori Miq. var. longicalcarata (Hayata ex Gogelein) Garay et H. R. Sweet = Calanthe masuca (D. Don) Lindl. ■
66095 Calanthe tokunoshimensis Hatus. et Ida;德之岛虾脊兰■☆
66096 Calanthe tokunoshimensis Hatus. et Ida = Calanthe aristulifera Rchb. f. ■
66097 Calanthe tongmaiensis K. Y. Lang et Tateishi = Calanthe griffithii Lindl. ■
66098 Calanthe torifera Schltr. = Calanthe tricarinata Lindl. ■
66099 Calanthe tricarinata Lindl.;三棱虾脊兰(九子连环草,三板根节兰,铁连环,竹叶石凤丹);Triangular Calanthe, Tricarinate Calanthe ■
66100 Calanthe trifida Ts. Tang et F. T. Wang;裂距虾脊兰(三裂虾脊兰);Trifid Calanthe ■
66101 Calanthe triplicata (Willemet) Ames;三褶虾脊兰(白鹤兰,藜芦叶虾脊兰,裂唇虾脊兰,芦叶虾脊兰,肉连环,石上蕉);Falsehelleboreleaf Calanthe, Forkedlip Calanthe, Triplicate Calanthe ■
66102 Calanthe triplicata (Willemet) Ames f. purpureoflora S. S. Ying = Calanthe triplicata (Willemet) Ames ■
66103 Calanthe triplicata (Willemet) Ames f. purpureoflora S. S. Ying = Dactylorhiza umbrosa (Kar. et Kir.) Nevski ■
66104 Calanthe trulliformis King et Pantl. ex King var. hastata Finet = Calanthe nipponica Makino ■
66105 Calanthe trulliformis King et Pantl. var. hastata Finet = Calanthe nipponica Makino ■
66106 Calanthe tsoongiana Ts. Tang et F. T. Wang;无距虾脊兰(观光虾脊兰);Spurless Calanthe ■
66107 Calanthe tsoongiana Ts. Tang et F. T. Wang var. guizhouensis Z. H. Dai;贵州虾脊兰;Guizhou Calanthe ■
66108 Calanthe tubifera Hook. f. = Cephalantheropsis obcordata (Lindl.) Ormerod ■
66109 Calanthe undulata Schltr. = Calanthe tricarinata Lindl. ■
66110 Calanthe veitchii Lindl.;维奇虾脊兰■☆
66111 Calanthe venusta Schltr. = Cephalantheropsis gracilis (Lindl.) S. Y. Hu ■
66112 Calanthe venusta Schltr. = Cephalantheropsis obcordata (Lindl.) Ormerod ■
66113 Calanthe veratrifolia Ker Gawl. = Dactylorhiza umbrosa (Kar. et Kir.) Nevski ■
66114 Calanthe veratrifolia R. Br. = Calanthe triplicata (Willemet) Ames ■
66115 Calanthe vestita Wall. ex Lindl.;乳白虾脊兰■☆
66116 Calanthe violacea Rolfe = Calanthe sylvatica (Thouars) Lindl. ■
66117 Calanthe viridifusca Hook. = Tainia ruybarrettoi (S. Y. Hu et Barretto) Z. H. Tsi ■
66118 Calanthe viridifusca Hook. = Tainia viridifusca (Hook.) Benth. et Hook. f. ■
66119 Calanthe volkensii Rolfe = Calanthe sylvatica (Thouars) Lindl. ■
66120 Calanthe wardii W. W. Sm. = Calanthe whiteana King et Pantl. ■
66121 Calanthe warpuri Rolfe = Calanthe madagascariensis Rolfe ■☆
66122 Calanthe whiteana King et Pantl.;四川虾脊兰;Sichuan Calanthe ■
66123 Calanthe yuana Ts. Tang et F. T. Wang;峨边虾脊兰(截帽虾脊兰,石鸭子);Yuan Calanthe ■
66124 Calanthe yunnanensis Rolfe = Calanthe brevicornu Lindl. ■☆
66125 Calanthe yushunii K. Mori et Gogelein = Calanthe speciosa (Blume) Lindl. ■
66126 Calanthe yushunii K. Mori et Yamam. = Calanthe speciosa (Blume) Lindl. ■
66127 Calanthea (DC.) Miers(1865);箭羽芭蕉属(热美白花菜属)■☆
66128 Calanthea Miers = Calanthea (DC.) Miers ■☆
66129 Calanthea pulcherrima (Jacq.) Miers;箭羽芭蕉■☆
66130 Calanthea pulcherrima Miers = Calanthea pulcherrima (Jacq.) Miers ■☆
66131 Calanthemum Post et Kuntze = Callianthemum C. A. Mey. ■
66132 Calanthera Hook. = Buchloe Engelm. (保留属名)■
66133 Calanthera Kunth ex Hook. = Buchloe Engelm. (保留属名)■
66134 Calanthera dactyloides (Nutt.) Kunth ex Hook. = Buchloe dactyloides (Nutt.) Engelm. ■
66135 Calanthidium Pfitzer = Calanthe R. Br. (保留属名)■
66136 Calanthidium labrosum (Rchb. f.) Pfitzer = Calanthe labrosa (Rchb. f.) Hook. f. ■
66137 Calanthus Oerst. = Calanthus Oerst. ex Hanst. ●■☆
66138 Calanthus Oerst. ex Hanst. = Alloplectus Mart. (保留属名)●■☆
66139 Calanthus Oerst. ex Hanst. = Drymonia Mart. ●☆
66140 Calanthus Post et Kuntze = Callanthus Rchb. ■☆
66141 Calanthus Post et Kuntze = Watsonia Mill. (保留属名)■☆
66142 Calantica Jaub. ex Tul. (1857);东非大风子属●☆

66143 Calantica cerasifolia Tul.;东非大风子●☆

66144 Calantica grandiflora Jaub. ex Tul.;大花东非大风子●☆

66145 Calantica jalbertii (Tul.) Baill. = Bivinia jalbertii Tul. ●☆

66146 Calantica lucida Scott-Elliot;亮东非大风子●☆

66147 Calanticaria (B. L. Rob. et Greenm.) E. E. Schill. et Panero (2002);少花葵属●☆

66148 Calanticaria bicolor (S. F. Blake) E. E. Schill. et Panero;二色少花葵●☆

66149 Calanticaria brevifolia (Greenm.) E. E. Schill. et Panero;短叶少花葵●☆

66150 Calanticaria oligantha (S. González, M. González et Rzed.) E. E. Schill. et Panero;少花葵●☆

66151 Calappa Kuntze = Cocos L. ●

66152 Calappa Rumph. = Cocos L. ●

66153 Calappa Rumph. ex Kuntze = Cocos L. ●

66154 Calappa Steck = Cocos L. ●

66155 Calarnagrostis tianschanica Rupr. = Deyeuxia tianschanica (Rupr.) Bor ■

66156 Calasias Raf. (废弃属名) = Anisotes Nees(保留属名)●☆

66157 Calathea G. Mey. (1818);肖竹芋属;Calathea ■

66158 Calathea allovia Lindl.;热美肖竹芋;Sweet Root-corn, Topeetambo, Topinambour, Topi-tamboo ■☆

66159 Calathea argyraea Körn.;银色肖竹芋;Calathea ■☆

66160 Calathea bachemiana E. Morren;巴氏肖竹芋■☆

66161 Calathea bicolor Ker.;豹斑竹芋■☆

66162 Calathea conferta Benth. = Ataenidia conferta (Benth.) Milne-Redh. ■☆

66163 Calathea gymnocarpa H. Kenn.;裸果肖竹芋■☆

66164 Calathea illustris Nicholson;白边肖竹芋;Whitemargin Calathea ■☆

66165 Calathea insignis Petersen et Cufod.;明显肖竹芋(箭羽叶竹芋,箭羽竹芋,响尾蛇竹竽,紫背葛郁金);Oliveblotch Calathea, Rattlesnake Plant ■☆

66166 Calathea lancifolia Boom;剑叶肖竹芋;Rattlesnake Plant ■☆

66167 Calathea leopardina (W. Bull) Regel;翠锦竹芋■☆

66168 Calathea leopardina Regel = Calathea leopardina (W. Bull) Regel ■☆

66169 Calathea lietzei E. Morren;列兹肖竹芋(节根竹芋,小竹芋);Lietze Calathea ■☆

66170 Calathea lindeniana Wallis;林登肖竹芋■☆

66171 Calathea louisae Gagnep.;白竹芋(龟壳纹竹芋)■☆

66172 Calathea luciani (Linden) N. E. Br.;变叶肖竹芋■☆

66173 Calathea lutea (Aubl.) G. Mey.;黄肖竹芋;Balasier, Cachibou, Cauassu ■☆

66174 Calathea majestica (Linden) H. A. Kenn. 'Roseolineata';红羽肖竹芋■☆

66175 Calathea makoyana (E. Morren) E. Morren;孔雀竹芋(马克肖竹芋);Makoy Calathea, Peacock Plant ■

66176 Calathea makoyana E. Morren = Calathea makoyana (E. Morren) E. Morren ■

66177 Calathea mannii Benth. = Marantochloa mannii (Benth.) Milne-Redh. ■☆

66178 Calathea mediopicta E. Morren;白肋肖竹芋(银道竹芋);Whiterib Calathea ■☆

66179 Calathea metalica Hort. = Calathea picturata K. Koch et Linden ■☆

66180 Calathea micans Körn.;闪亮肖竹芋(光亮肖竹芋);Shiny Calathea, Whitefeather Calathea ■☆

66181 Calathea musaica (Bull.) Bailey;撒金竹芋■☆

66182 Calathea musaica Hort. = Calathea musaica (Bull.) Bailey ■☆

66183 Calathea oppenheimiana E. Morren;三色叶竹芋■☆

66184 Calathea oppenheimiana E. Morren = Ctenanthe oppenheimiana K. Schum. ■☆

66185 Calathea ornata (Lem.) Körn.;肖竹芋(双线竹芋);Bigleaf Calathea, Calathea ■

66186 Calathea ornata (Lem.) Körn. 'Koseolineata';红羽竹竽■☆

66187 Calathea ornata (Lem.) Körn. 'Roseolineata' = Calathea majestica (Linden) H. A. Kenn. 'Roseolineata' ■☆

66188 Calathea ornata (Lem.) Körn. 'Sanderiana';桃红双线竹芋■☆

66189 Calathea ovandensis Matuda;墨西哥肖竹芋■☆

66190 Calathea pavonii Körn.;帕冯肖竹芋■☆

66191 Calathea picta Hook. f.;彩色肖竹芋■☆

66192 Calathea picturata K. Koch = Calathea picturata K. Koch et Linden ■☆

66193 Calathea picturata K. Koch et Linden;彩纹肖竹芋■☆

66194 Calathea picturata K. Koch et Linden 'Argentea';丽白竹芋■☆

66195 Calathea picturata K. Koch et Linden ex K. Koch = Calathea picturata K. Koch et Linden ■☆

66196 Calathea picturata K. Koch et Linden var. vandenheckei (Regel) Shimizu;范氏彩纹肖竹芋(花纹竹芋);Vandenhecke Calathea ■☆

66197 Calathea picturata K. Koch et Linden var. vandenheckei Regel = Calathea picturata K. Koch et Linden var. vandenheckei (Regel) Shimizu ■☆

66198 Calathea princeps (Lindl.) Regel;绿道竹芋(绿羽竹芋)■☆

66199 Calathea rhizantha K. Schum. = Afrocalathea rhizantha (K. Schum.) K. Schum. ■☆

66200 Calathea roseopicta (Lindl.) Regel;红边肖竹芋(彩虹竹芋,玫瑰竹芋,肖竹芋);Prayer Plant, Red Margin Calathea, Redmargin Calathea, Rosy Arrowroot ■

66201 Calathea roseopicta (Lindl.) Regel 'Asian Beauty';红玉肖竹芋■

66202 Calathea rotundifolia (Lindl.) Regel 'Fasciata';圆叶竹芋■☆

66203 Calathea rufibarba (Lindl.) Regel 'Wavestar';波浪竹竽;Furry Feather Calathea ■☆

66204 Calathea sanderiana (Sander) Gentil;桑德肖竹芋■☆

66205 Calathea splendida (Lem.) Corrêa;凤尾竹芋■☆

66206 Calathea tubispatha Hook. = Calathea pavonii Körn. ■☆

66207 Calathea undulata Regel;绣边竹芋■☆

66208 Calathea utilis H. Kenn.;厄瓜多尔肖竹芋■☆

66209 Calathea vaginata A. Chev. = Halopegia azurea (K. Schum.) K. Schum. ■☆

66210 Calathea vandenheckei Regel = Calathea picturata K. Koch et Linden var. vandenheckei (Regel) Shimizu ■☆

66211 Calathea veitchiana Hook. f.;维奇肖竹芋(美丽竹芋,维氏肖竹芋);Veitch Calathea ■☆

66212 Calathea violacea (Roscoe) Lindl.;堇色肖竹芋■☆

66213 Calathea vittata Körn.;纵带肖竹芋(美丽竹芋);Ribboned Maranta, Vittate Calathea ■☆

66214 Calathea warscewiczii Körn.;瓦尔肖竹芋■☆

66215 Calathea zebrina (Sims) Lindl.;绒叶肖竹芋(斑叶肖竹芋,斑叶竹芋);Zebra Calathea, Zebra Plant, Zebraplant ■

66216 Calathea zebrina (Sims) Lindl. 'Humilior';绿背天鹅绒竹芋■☆

66217 Calathiana Delarbre = Gentiana L. ■

66218 Calathlnus Raf. = Narcissus L. ■

66219 Calathodes Hook. f. et Thomson (1855);鸡爪草属;Calathodes,

Cockclawflower ■★

66220 Calathodes oxycarpa Sprague; 鸡爪草(虎脚板); Cockclawflower, Sharpfruit Calathodes ■

66221 Calathodes palmata Hook. f. et Thomson; 黄花鸡爪草(掌叶鸡爪草); Yellowflower Calathodes, Yellowflower Cockclawflower ■

66222 Calathodes palmata Hook. f. et Thomson = Calathodes oxycarpa Sprague ■

66223 Calathodes palmata Hook. f. et Thomson var. appendiculata Brühl = Calathodes oxycarpa Sprague ■

66224 Calathodes polycarpa Ohwi; 多果鸡爪草(台湾鸡爪草, 铁血子); Manyfruit Calathodes, Manyfruit Cockclawflower ■

66225 Calathodes unciformis W. T. Wang; 钩突鸡爪草; Unciform Calathodes ■

66226 Calathostelma E. Fourn. (1885); 篮冠萝藦属☆

66227 Calathostelma ditassoides E. Fourn.; 篮冠萝藦☆

66228 Calatola Standl. (1923); 热美茶茱萸属☆

66229 Calatola colombiana Sleumer; 哥伦比亚热美茶茱萸☆

66230 Calatola laevigata Standl.; 平滑热美茶茱萸☆

66231 Calatola mollis Standl.; 软热美茶茱萸☆

66232 Calaunia Grudz. = Streblus Lour. ●

66233 Calboa Cav. = Ipomoea L. (保留属名) ●■

66234 Calcalia Krock. = Adenostyles Cass. ■☆

66235 Calcalia Krock. = Cacalia L. ●■

66236 Calcaratolobelia Wilbur = Lobelia L. ●■

66237 Calcareoboea C. Y. Wu = Calcareoboea C. Y. Wu ex H. W. Li ●★

66238 Calcareoboea C. Y. Wu = Platyadenia B. L. Burtt ■☆

66239 Calcareoboea C. Y. Wu ex H. W. Li (1982); 朱红苣苔属■★

66240 Calcareoboea H. W. Li = Calcareoboea C. Y. Wu ex H. W. Li ■★

66241 Calcareoboea coccinea C. Y. Wu ex H. W. Li; 朱红苣苔■

66242 Calcareophilum gillettii Meve et Liede = Calciphila gillettii Liede et Meve ●☆

66243 Calcarunia Raf. = Monochoria C. Presl ■

66244 Calcatrippa Heist. = Delphinium L. ■

66245 Calcearia Blume = Corybas Salisb. ■

66246 Calcearia sinii (Ts. Tang et F. T. Wang) M. A. Clem. et D. L. Jones = Corybas sinii Ts. Tang et F. T. Wang ■

66247 Calcearia taiwanensis (T. P. Lin et S. Y. Leu) M. A. Clem. et D. L. Jones = Corybas taiwanensis T. P. Lin et S. Y. Leu ■

66248 Calcearia taliensis (Tang et F. T. Wang) M. A. Clem. et D. L. Jones = Corybas taiwanensis T. P. Lin et S. Y. Leu ■

66249 Calceolangis Thouars = Angraecum Bory ■

66250 Calceolaria Fabr. = Cypripedium L. ■

66251 Calceolaria Heist. ex Fabr. = Cypripedium L. ■

66252 Calceolaria L. (1770) (保留属名); 蒲包花属(风帽草属, 荷包花属, 鞋形草属); Calceolaria, Slipper Flower, Slipperwort ■●☆

66253 Calceolaria Loefl. (废弃属名) = Calceolaria L. (保留属名) ■●☆

66254 Calceolaria Loefl. (废弃属名) = Hybanthus Jacq. (保留属名) ●■

66255 Calceolaria Loefl. ex Kuntze = Calceolaria L. (保留属名) ■●☆

66256 Calceolaria Loefl. ex Kuntze = Cypripedium L. ■

66257 Calceolaria acutifolia Witasek = Calceolaria polyrrhiza Cav. ■☆

66258 Calceolaria angustifolia Sweet = Calceolaria integrifolia L. var. angustifolia Lindl. ●☆

66259 Calceolaria arachnoidea Graham; 蛛毛蒲包花■☆

66260 Calceolaria biflora Lam.; 智利荷包花●☆

66261 Calceolaria chelidonioides Kunth; 白屈菜状荷包花; Slipperwort ■☆

66262 Calceolaria corymbosa Ruiz et Pav.; 伞房荷包花(伞花鞋形草) ■☆

66263 Calceolaria crenatiflora Cav.; 荷包花(蒲包花); Crenateflower Calceolaria, Pocketbook Plant, Slipper Flower, Slipperwort ●☆

66264 Calceolaria crenatiflora Cav. 'Grandiflora Pumila Compacta'; 大花矮生荷包花●☆

66265 Calceolaria crenatiflora Cav. 'Grandiflora'; 大花荷包花●☆

66266 Calceolaria crenatiflora Cav. 'Multiflora Nana'; 矮生多花荷包花●☆

66267 Calceolaria crenatiflora Cav. 'Multiflora'; 多花荷包花●☆

66268 Calceolaria darwinii Benth. = Calceolaria uniflora Ruiz et Pav. var. darwinii (Benth.) Witasek ■☆

66269 Calceolaria fothergillii Sol.; 红斑蒲包花■☆

66270 Calceolaria herbacea Hort. ex Vilmorin = Calceolaria herbeohybrida Voss ■☆

66271 Calceolaria herbeohybrida Voss; 杂种荷包花; Calceolaria, Slipper Flower, Voss Calceolaria ■☆

66272 Calceolaria hybrida Hort. ex Vilmorin = Calceolaria herbeohybrida Voss ■☆

66273 Calceolaria integrifolia L.; 灌木荷包花(全缘叶荷包花); Bush Calceolaria, Pouch Flower, Slipper Flower, Slipperwort ●☆

66274 Calceolaria integrifolia L. f. robusta (A. Dietr.) Voss = Calceolaria integrifolia L. ●☆

66275 Calceolaria integrifolia L. subvar. rugosa (Ruiz et Pav.) Clos = Calceolaria integrifolia L. ●☆

66276 Calceolaria integrifolia L. var. angustifolia Lindl.; 窄叶灌木荷包花●☆

66277 Calceolaria integrifolia L. var. ferruginea Linde; 锈色灌木荷包花●☆

66278 Calceolaria integrifolia L. var. robusta A. Dietr. = Calceolaria integrifolia L. ●☆

66279 Calceolaria integrifolia L. var. viscosissima Hook.; 黏灌木荷包花●☆

66280 Calceolaria integrifolia Murray = Calceolaria integrifolia L. ●☆

66281 Calceolaria mexicana Benth.; 墨西哥荷包花; Mexican Calceolaria ■☆

66282 Calceolaria paralia Hook. = Calceolaria corymbosa Ruiz et Pav. ■☆

66283 Calceolaria pavonii Benth.; 攀缘蒲包花■☆

66284 Calceolaria pinnata L.; 羽状蒲包花■☆

66285 Calceolaria polyrrhiza Cav.; 多根蒲包花■☆

66286 Calceolaria rugosa Ruiz et Pav. = Calceolaria integrifolia L. ●☆

66287 Calceolaria scabiosifolia Sieber ex Sims; 松虫草叶荷包花●☆

66288 Calceolaria tenella Poepp. et Endl.; 细茎蒲包花■☆

66289 Calceolaria thyrsiflora Graham; 聚花蒲包花■☆

66290 Calceolaria uniflora Ruiz et Pav. var. darwinii (Benth.) Witasek; 皱叶蒲包花■☆

66291 Calceolaria wheeleri Sweet = Calceolaria herbeohybrida Voss ■☆

66292 Calceolariaceae Olmstead = Calceolariaceae Raf. ex Olmstead ■☆

66293 Calceolariaceae Raf. ex Olmstead (2001); 蒲包花科(荷包花科) ■☆

66294 Calceolus Adans. = Cypripedium L. ■

66295 Calceolus Mill. = Cypripedium L. ■

66296 Calceolus Nieuwl. = Cypripedium L. ■

66297 Calchas P. V. Heath = Plectranthus L'Hér. (保留属名) ●■

66298 Calcicola W. R. Anderson et C. Davis = Malpighia L. ●

66299 Calcicola W. R. Anderson et C. Davis (2007); 墨西哥金虎尾属●☆

66300 Calciphila Liede et Meve (2006); 钙竹桃属●☆

66301 Calciphila galgalensis (Liede) Liede et Meve; 钙竹桃●☆

66302 Calciphila gillettii Liede et Meve;吉莱特钙竹桃●☆
66303 Calcitrapa Adans. = Centaurea L.(保留属名)●■
66304 Calcitrapa Haller = Centaurea L.(保留属名)●■
66305 Calcitrapa Heist. ex Fabr. = Centaurea L.(保留属名)●■
66306 Calcitrapa aegyptiaca (L.) Sweet = Centaurea aegyptiaca L. ■☆
66307 Calcitrapa alexandrina (Delile) Soják = Centaurea alexandrina Delile ■☆
66308 Calcitrapa aspera (L.) Cass. = Centaurea aspera L. ■☆
66309 Calcitrapa calcitrapa (L.) Hill = Centaurea calcitrapa L. ■☆
66310 Calcitrapa diluta (Aiton) Holub = Centaurea diluta Aiton ■☆
66311 Calcitrapa eriophora (L.) Moench = Centaurea eriophora L. ■☆
66312 Calcitrapa ferox (Desf.) Moench = Centaurea ferox Desf. ■☆
66313 Calcitrapa furfuracea (Coss. et Durieu) Holub = Centaurea furfuracea Coss. et Durieu ■☆
66314 Calcitrapa hyalolepis (Boiss.) Holub = Centaurea hyalolepis Boiss. ■☆
66315 Calcitrapa iberica (Trevis.) Schur. = Centaurea iberica Trevir. ex Spreng. ■
66316 Calcitrapa maroccana (Ball) Holub = Centaurea hyalolepis Boiss. ■☆
66317 Calcitrapa melitensis (L.) Soják = Centaurea melitensis L. ■☆
66318 Calcitrapa napifolia (L.) Moench = Centaurea napifolia L. ■☆
66319 Calcitrapa nicaeeensis (L.) Holub = Centaurea nicaeensis All. ■☆
66320 Calcitrapa pallescens (Delile) Soják = Centaurea pallescens Delile ■☆
66321 Calcitrapa pungens (Pomel) Holub = Centaurea pungens Pomel ■☆
66322 Calcitrapa sulphurea (Willd.) Soják = Centaurea sulphurea Willd. ■☆
66323 Calcitrapoides Fabr. = Centaurea L.(保留属名)●■
66324 Calcitrapoides praecox (Oliv. et Hiern) Holub = Centaurea praecox Oliv. et Hiern ■☆
66325 Calcoa Salisb. = Luzuriaga Ruiz et Pav.(保留属名)■☆
66326 Caldasia Humb. ex Willd. = Bonplandia Cav. ●☆
66327 Caldasia Lag. = Oreomyrrhis Endl. ■
66328 Caldasia Mutis ex Caldas = Helosis Rich.(保留属名)■☆
66329 Caldasia Willd. = Bonplandia Cav. ●☆
66330 Caldcluvia D. Don = Ackama A. Cunn. ●☆
66331 Caldcluvia D. Don(1830);圆锥火把树属●☆
66332 Caldcluvia paniculata D. Don.;圆锥火把树●☆
66333 Caldenbachia Pohl ex Nees = Stenandrium Nees(保留属名)■☆
66334 Calderonella Soderstr. et H. F. Decker(1974);丝柄穗顶草属■☆
66335 Calderonella sylvatica Soderstr. et H. F. Decker;丝柄穗顶草■☆
66336 Calderonia Standl. = Simira Aubl. ■☆
66337 Caldesia Parl.(1860);泽苔草属(圆叶泽泻属);Caldesia, Dampsedge ■
66338 Caldesia acanthocarpa (F. Muell.) Buchenau = Caldesia oligococca (F. Muell.) Buchenau ■☆
66339 Caldesia grandis Samuel;宽叶泽苔草■
66340 Caldesia grandis Samuel = Caldesia parnassifolia (Bassi. ex L.) Parl. ■
66341 Caldesia oligococca (F. Muell.) Buchenau;广布泽苔草■☆
66342 Caldesia oligococca (F. Muell.) Buchenau var. echinata Hartog;刺广布泽苔草■☆
66343 Caldesia parnassifolia (Bassi ex L.) Parl.;泽苔草(宽叶泽苔草,梅花草叶泽苔草,圆叶泽泻);Broadleaf Caldesia, Broadleaf Dampsedge, Dampsedge, Kidney-shaped Caldesia, Parnassis-leaf Caldesia ■
66344 Caldesia parnassifolia (Bassi ex L.) Parl. f. natans Glück;浮游泽苔草■☆
66345 Caldesia parnassifolia (Bassi ex L.) Parl. var. major (Micheli) Buchenau = Caldesia reniformis (D. Don) Makino ■
66346 Caldesia parnassifolia (Bassi ex L.) Parl. var. major Micheli = Caldesia parnassifolia (Bassi ex L.) Parl. ■
66347 Caldesia parnassifolia (Bassi ex L.) Parl. var. nilotica Buchenau = Caldesia reniformis (D. Don) Makino ■
66348 Caldesia reniformis (D. Don) Makino = Caldesia parnassifolia (Bassi ex L.) Parl. ■
66349 Caldesia reniformis (D. Don) Makino f. natans H. Hara = Caldesia parnassifolia (Bassi. ex L.) Parl. f. natans Glück ■☆
66350 Calea L.(1763);多鳞菊属(美菊属)●■☆
66351 Calea Sw. = Neurolaena R. Br. ■●☆
66352 Calea anomala Hassl.;异常多鳞菊■☆
66353 Calea aspera Jacq. = Melanthera nivea (L.) Small ■☆
66354 Calea brevifolia Rusby;短叶多鳞菊■☆
66355 Calea coriacea DC.;革质多鳞菊■☆
66356 Calea cymosa Less.;聚伞多鳞菊■☆
66357 Calea jamaicensis (L.) L.;牙买加多鳞菊(美菊,牙买加美菊)■☆
66358 Calea lantanoides Gardner;马缨丹多鳞菊■☆
66359 Calea lutea Pruski et Urbatsch;黄多鳞菊■☆
66360 Calea nematophylla Pruski;虫叶多鳞菊■☆
66361 Calea pinnatifida (R. Br.) Banks ex Steud.;羽裂多鳞菊(羽裂美菊)■☆
66362 Calea rhombifolia S. F. Blake;菱叶多鳞菊■☆
66363 Calea serrata Less.;锯齿多鳞菊(锯齿美菊)■☆
66364 Calea solidaginea Kunth;坚实多鳞菊■☆
66365 Calea urticifolia (Mill.) DC.;荨麻叶多鳞菊■☆
66366 Calea urticifolia (Mill.) DC. var. yucatanesis Wussow, Urb. et Sullivan;尤卡多鳞菊(尤卡美菊)■☆
66367 Caleacte R. Br. = Calea L. ●■☆
66368 Caleana R. Br.(1810);卡丽娜兰属;Caleana ■☆
66369 Caleana major R. Br.;卡丽娜兰;Common Caleana ■☆
66370 Caleana minor R. Br.;小卡丽娜兰;Duck Orchid, Minor Caleana ■☆
66371 Caleatia Mart. ex Steud. = Lucuma Molina ●
66372 Calebrachys Cass. = Calea L. ●■☆
66373 Calectasia R. Br.(1810);澳丽花属●☆
66374 Calectasia cyanea R. Br.;澳丽花;Blue Tinsel Lily ●☆
66375 Calectasiaceae Endl.;澳丽花科(篮花木科)●☆
66376 Calectasiaceae Endl. = Dasypogonaceae Dumort. ■☆
66377 Calectasiaceae Endl. = Xanthorrhoeaceae Dumort.(保留科名)●■☆
66378 Calectasiaceae Schnizl. = Calectasiaceae Endl. ●☆
66379 Calendelia Kuntze = Calendula L. ●■
66380 Calendula L.(1753);金盏花属(金盏菊属);Calendula, Marigold, Pot Marigold ●■
66381 Calendula aegyptiaca Pers. = Calendula sancta L. ■☆
66382 Calendula aegyptiaca Pers. subsp. ceratosperma (Viv.) Murb. = Calendula sancta L. ■☆
66383 Calendula aegyptiaca Pers. subsp. tripterocarpa (Rupr.) Lanza = Calendula tripterocarpa Rupr. ■☆
66384 Calendula aegyptiaca Pers. var. ceratosperma (Viv.) Pamp. = Calendula sancta L. ■☆

66385 Calendula aegyptiaca Pers. var. exalata-longirostris Lanza = Calendula sancta L. ■☆
66386 Calendula aegyptiaca Pers. var. hymenocarpa (DC.) Pamp. = Calendula sancta L. ■☆
66387 Calendula aegyptiaca Pers. var. intermedia (Coss. et Kralik) Pamp. = Calendula sancta L. ■☆
66388 Calendula aegyptiaca Pers. var. microcephala Boiss. = Calendula sancta L. ■☆
66389 Calendula aegyptiaca Pers. var. platycarpa (Coss.) Batt. = Calendula sancta L. ■☆
66390 Calendula aegyptiaca Pers. var. suberostris Boiss. = Calendula sancta L. ■☆
66391 Calendula algeriensis Boiss. et Reut. = Calendula stellata Cav. ■☆
66392 Calendula amplexicaulis Thunb. = Tripteris amplexicaulis (Thunb.) Less. ●☆
66393 Calendula arborescens Jacq. = Tripteris dentata (Burm. f.) Harv. ●☆
66394 Calendula arvensis Batt. = Calendula sancta L. ■☆
66395 Calendula arvensis L.;小金盏花(长春草,长春花,长春菊,金仙花,金盏草,金盏儿花,金盏花,欧洲金盏花,田野金盏菊,醒酒花);Field Marigold,Marigold,Wild Marigold ■
66396 Calendula arvensis L. subsp. hydruntina (Fiori) Lanza = Calendula arvensis L. ■
66397 Calendula arvensis L. subsp. macroptera Rouy = Calendula arvensis L. ■
66398 Calendula arvensis L. var. bicolor (Raf.) DC. = Calendula sancta L. ■☆
66399 Calendula arvensis L. var. echinata Ball = Calendula arvensis L. ■
66400 Calendula arvensis L. var. micrantha Ball = Calendula arvensis L. ■
66401 Calendula arvensis L. var. parviflora (Raf.) Batt. = Calendula arvensis L. ■
66402 Calendula arvensis L. var. sicula (Willd.) Quézel et Santa = Calendula arvensis L. ■
66403 Calendula arvensis L. var. stellata Cav. = Calendula stellata Cav. ■☆
66404 Calendula aspera Thunb. = Osteospermum rigidum Aiton ■☆
66405 Calendula balansae Boiss. et Reut. = Calendula suffruticosa Vahl subsp. balansae (Boiss. et Reut.) Ohle ■☆
66406 Calendula bicolor Raf. = Calendula sancta L. ■☆
66407 Calendula bicolor Raf. var. cossonii Quézel et Santa = Calendula stellata Cav. ■☆
66408 Calendula bicolor Raf. var. faurelii Quézel et Santa = Calendula stellata Cav. ■☆
66409 Calendula bicolor Raf. var. odettei Quézel et Santa = Calendula stellata Cav. ■☆
66410 Calendula chrysanthemifolia Vent. = Dimorphotheca chrysanthemifolia (Vent.) DC. ■☆
66411 Calendula cuneata Thunb. = Dimorphotheca cuneata (Thunb.) Less. ■☆
66412 Calendula decurrens Thunb. = Dimorphotheca pluvialis (L.) Moench ■☆
66413 Calendula diffusa Salisb. = Dimorphotheca fruticosa (L.) Less. ■☆
66414 Calendula echinata DC.;具刺金盏花■☆
66415 Calendula echinata DC. subsp. lanzae Maire = Calendula lanzae Maire ■☆
66416 Calendula echinata DC. subsp. murbeckii (Lanza) Maire = Calendula maroccana Ball subsp. murbeckii (Lanza) Ohle ■☆
66417 Calendula echinata DC. var. pinnatiloba Maire = Calendula maroccana Ball ■☆
66418 Calendula flaccida Vent. = Dimorphotheca tragus (Aiton) B. Nord. ■☆
66419 Calendula foliosa Batt. = Calendula suffruticosa Vahl subsp. tlemcensis Ohle ■☆
66420 Calendula fruticosa L. = Dimorphotheca fruticosa (L.) Less. ■☆
66421 Calendula glabrata Thunb. = Tripteris oppositifolia (Aiton) B. Nord. ●☆
66422 Calendula gracilis DC.;纤细金盏花■☆
66423 Calendula graminifolia L. = Dimorphotheca nudicaulis (L.) DC. var. graminifolia (L.) Harv. ■☆
66424 Calendula hispida Thunb. = Aster bakerianus Burtt Davy ex C. A. Sm. ■☆
66425 Calendula hybrida L. = Dimorphotheca pluvialis (L.) Moench ■☆
66426 Calendula karakalensis Vassilcz.;卡拉卡利金盏花■☆
66427 Calendula kewensis?;冬金盏花;Scotch Marigold,Winter-flowering Wallflower ■☆
66428 Calendula lanzae Maire;兰扎金盏花■☆
66429 Calendula lusitanica Boiss. = Calendula suffruticosa Vahl subsp. lusitanica (Boiss.) Ohle ■☆
66430 Calendula maderensis DC.;梅德金盏花■☆
66431 Calendula malacitana Boiss. et Reut. = Calendula arvensis L. ■
66432 Calendula malvaecarpa Pomel = Calendula sancta L. ■☆
66433 Calendula marginata Willd. = Calendula suffruticosa Vahl subsp. tomentosa (Desf.) Murb. ■☆
66434 Calendula maritima Guss.;西西里金盏花(欧洲金盏花);Sicily Calendula ■☆
66435 Calendula maroccana Ball;摩洛哥金盏花■☆
66436 Calendula maroccana Ball subsp. murbeckii (Lanza) Ohle;穆尔拜克金盏花■☆
66437 Calendula meuselii Ohle;穆塞尔金盏花■☆
66438 Calendula monardii Boiss. et Reut. = Calendula suffruticosa Vahl subsp. monardii (Boiss. et Reut.) Ohle ■☆
66439 Calendula monstruosa Burm. f. = Monoculus monstruosus (Burm. f.) B. Nord. ■☆
66440 Calendula murbeckii Lanza = Calendula maroccana Ball subsp. murbeckii (Lanza) Ohle ■☆
66441 Calendula murbeckii Lanza subsp. lanzae (Maire) Maire = Calendula lanzae Maire ■☆
66442 Calendula murbeckii Lanza var. pinnatiloba (Coss.) Maire = Calendula maroccana Ball subsp. murbeckii (Lanza) Ohle ■☆
66443 Calendula muricata Thunb. = Osteospermum grandiflorum DC. ●☆
66444 Calendula nudicaulis L. = Dimorphotheca nudicaulis (L.) DC. ■☆
66445 Calendula officinalis L.;金盏花(棒红,大金盏花,甘菊花,黄花秋,黄金盏,金盏儿花,金盏菊,山金菊,水涨菊);Calendula, Common Marigold, English Earigold, Garden Gold, Gold, Gold Bloom, Goldings, Goldwort, Gools, Goulan, Gowan, Gowlan, Gowles, Guild, Gules, Hardhow, Holigold, Husbandman's Dial, Jack-an-apes-on-horseback, Mally-gowl, Marigold, Mary Gold, Mary Gooles, Mary Gowlan, Marybud, Measle-flower, Merry-go-rounds, Nobody's-flower, Pot Marigold, Pot-marigold, Potmarigold Calendula, Rodewort, Rods Gold, Rodsgold, Rud, Ruddes, Ruddles, Scotch Marigold, Soucique, Summer's Bride, Sunflower, Sun's Bride, Sun's Herb ■
66446 Calendula officinalis L. 'Fiesta Gitana';吉普赛节日金盏花■☆
66447 Calendula officinalis L. 'Geisha Girl';歌舞女郎金盏花■☆

66448 Calendula officinalis L. var. chrysantha Hort. ;黄花金盏花■
66449 Calendula officinalis L. var. subspathulata Miq. ;亚匙金盏花■
66450 Calendula officinalis L. var. subspathulata Miq. = Calendula arvensis L. ■
66451 Calendula oppositifolia Aiton = Tripteris oppositifolia (Aiton) B. Nord. ●☆
66452 Calendula parviflora Thunb. = Oligocarpus calendulaceus (L. f.) Less. ■☆
66453 Calendula persica C. A. Mey. ;波斯金盏花■☆
66454 Calendula persica C. A. Mey. = Calendula arvensis Batt. ■
66455 Calendula pinnata Thunb. = Osteospermum pinnatum (Thunb.) Norl. ■☆
66456 Calendula platycarpa Coss. = Calendula sancta L. ■☆
66457 Calendula platycarpa Coss. var. malvaecarpa (Pomel) Batt. = Calendula sancta L. ■☆
66458 Calendula pluvialis L. = Dimorphotheca pluvialis (L.) Moench ■☆
66459 Calendula rigida Aiton = Osteospermum grandiflorum DC. ●☆
66460 Calendula rosmarinifolia Houtt. = Osteospermum polygaloides L. ■☆
66461 Calendula sancta L. ;神圣金盏花■☆
66462 Calendula scabra P. J. Bergius = Dimorphotheca acutifolia Hutch. ■☆
66463 Calendula stellata Cav. ;星形金盏花■☆
66464 Calendula stellata Cav. var. hymenocarpa Coss. = Calendula sancta L. ■☆
66465 Calendula stellata Cav. var. intermedia Coss. = Calendula sancta L. ■☆
66466 Calendula subinermis Pomel = Calendula sancta L. ■☆
66467 Calendula suffruticosa Vahl;灌木金盏花;Shrubby Calendula ●☆
66468 Calendula suffruticosa Vahl subsp. balansae (Boiss. et Reut.) Ohle;巴兰萨金盏花■☆
66469 Calendula suffruticosa Vahl subsp. boissieri Lanza;布瓦西耶金盏花■☆
66470 Calendula suffruticosa Vahl subsp. ifniensis Font Quer = Calendula lanzae Maire ■☆
66471 Calendula suffruticosa Vahl subsp. lusitanica (Boiss.) Ohle = Calendula suffruticosa Vahl subsp. tomentosa (Desf.) Murb. ■☆
66472 Calendula suffruticosa Vahl subsp. marginata (Willd.) Maire = Calendula suffruticosa Vahl subsp. tomentosa (Desf.) Murb. ■☆
66473 Calendula suffruticosa Vahl subsp. maroccana Ball = Calendula maroccana Ball ■☆
66474 Calendula suffruticosa Vahl subsp. monardii (Boiss. et Reut.) Ohle;莫纳尔金盏花■☆
66475 Calendula suffruticosa Vahl subsp. tlemcensis Ohle;阿尔及利亚金盏花■☆
66476 Calendula suffruticosa Vahl subsp. tomentosa (Desf.) Maire = Calendula suffruticosa Vahl subsp. tomentosa (Desf.) Murb. ■☆
66477 Calendula suffruticosa Vahl subsp. tomentosa (Desf.) Murb. ;绒毛金盏花■☆
66478 Calendula suffruticosa Vahl subsp. tunetana (Cuénod) Pott.-Alap. = Calendula suffruticosa Vahl ●☆
66479 Calendula suffruticosa Vahl var. balansae (Boiss. et Reut.) Maire = Calendula meuselii Ohle ■☆
66480 Calendula suffruticosa Vahl var. dichroa Batt. = Calendula suffruticosa Vahl ●☆
66481 Calendula suffruticosa Vahl var. maroccana (Ball) Maire = Calendula maroccana Ball ■☆
66482 Calendula suffruticosa Vahl var. obtusifolia Maire = Calendula suffruticosa Vahl ●☆
66483 Calendula suffruticosa Vahl var. polymorphocarpa Quézel et Santa = Calendula suffruticosa Vahl ●☆
66484 Calendula suffruticosa Vahl var. trimorphocarpa Lanza = Calendula suffruticosa Vahl subsp. boissieri Lanza ■☆
66485 Calendula suffruticosa Vahl var. tunetana (Cuénod) Ohle = Calendula suffruticosa Vahl ●☆
66486 Calendula thapsiaecarpa Pomel = Calendula tripterocarpa Rupr. ■☆
66487 Calendula tomentosa Desf. = Calendula suffruticosa Vahl ●☆
66488 Calendula tomentosa Desf. subsp. marginata (Willd.) Lanza = Calendula suffruticosa Vahl ●☆
66489 Calendula tomentosa Desf. var. balansae (Boiss. et Reut.) Quézel et Santa = Calendula suffruticosa Vahl subsp. tomentosa (Desf.) Murb. ■☆
66490 Calendula tomentosa L. f. = Inuloides tomentosa (L. f.) B. Nord. ■☆
66491 Calendula tragus Aiton = Dimorphotheca tragus (Aiton) B. Nord. ■☆
66492 Calendula tragus Aiton var. flaccida (Vent.) Pers. = Dimorphotheca tragus (Aiton) B. Nord. ■☆
66493 Calendula tripterocarpa Rupr. ;三翅果金盏花■☆
66494 Calendula tunetana Cuénod = Calendula suffruticosa Vahl subsp. tunetana (Cuénod) Pott.-Alap. ●☆
66495 Calendula versicolor Salisb. = Dimorphotheca pluvialis (L.) Moench ■☆
66496 Calendula vidalii Pau = Calendula stellata Cav. ■☆
66497 Calendula viscosa Andréws = Dimorphotheca cuneata (Thunb.) Less. ■☆
66498 Calendulaceae Bercht. et J. Presl = Asteraceae Bercht. et J. Presl (保留科名)●■
66499 Calendulaceae Bercht. et J. Presl = Compositae Giseke(保留科名)●■
66500 Calendulaceae Link = Asteraceae Bercht. et J. Presl(保留科名)●■
66501 Calendulaceae Link = Compositae Giseke(保留科名)●■
66502 Calendulaceae Link;金盏花科●■
66503 Caleopsis Fedde = Goldmanella Greenm. ■☆
66504 Calepina Adans. (1763);卡来荠属■☆
66505 Calepina corvini Desv. = Calepina irregularis (Asso) Thell. ■☆
66506 Calepina irregularis (Asso) Thell. ;卡来荠;White Ball Mustard, White Ballmustard ■☆
66507 Calesia Raf. = Lannea A. Rich. (保留属名)●
66508 Calesiam Adans. (废弃属名) = Lannea A. Rich. (保留属名)●
66509 Calesiam alatum Engl. = Lannea alata (Engl.) Engl. ●☆
66510 Calesiam ambacensis Hiern = Lannea ambacensis (Hiern) Engl. ●☆
66511 Calesiam rubra Hiern = Lannea rubra (Hiern) Engl. ●☆
66512 Calesiam somalensis Chiov. = Lannea somalensis (Chiov.) Cufod. ●☆
66513 Calesiam tomentosum Engl. = Lannea humilis (Oliv.) Engl. ●☆
66514 Calesiam welwitschii Hiern = Lannea welwitschii (Hiern) Engl. ●☆
66515 Calesium Kuntze = Lannea A. Rich. (保留属名)●
66516 Calesium cinereum Engl. = Lannea cinerea (Engl.) Engl. ●☆
66517 Calesium fulvum Engl. = Lannea fulva (Engl.) Engl. ●☆
66518 Calesium grandis (Dennst.) Kuntze = Lannea coromandelica (Houtt.) Merr. ●

66519 Calesium obcordatum Engl. = Lannea obcordata (Engl.) Engl. ●☆
66520 Calestania Koso-Pol. = Peucedanum L. ■
66521 Calestania Koso-Pol. = Thyselium Raf. ■☆
66522 Caletia Baill. = Micrantheum Desf. ●☆
66523 Caleya R. Br. = Caleana R. Br. ■☆
66524 Caleyana T. Post et Kuntze = Caleya R. Br. ■☆
66525 Calhounia A. Nels. = Lagascea Cav. (保留属名)●■☆
66526 Calia Berland. = Sophora L. ●■
66527 Calia Teton et Beriand. = Sophora L. ●■
66528 Calia secundiflora (Ortega) Yakovlev = Sophora secundiflora (Ortega) Lag. ex DC. ●☆
66529 Calibanus Rose(1906);墨西哥龙血树属●☆
66530 Calibanus caespitesus (Scheidw.) Rose;墨西哥龙血树●☆
66531 Calibrachoa Cerv. (1825);卡利茄属;Million Bells ■☆
66532 Calibrachoa Cerv. ex La Llave et Lex. = Petunia Juss. (保留属名)■
66533 Calibrachoa La Llave et Lex. = Petunia Juss. (保留属名)■
66534 Calibrachoa elegans (Miers) Stehmann et Semir;雅致卡利茄■☆
66535 Calibrachoa hybrida Hort.;杂种卡利茄;Million Bells, Trailing Petunia ■☆
66536 Calibrachoa ovalifolia (Miers) Stehmann et J. Semir;卵叶卡利茄■☆
66537 Calibrachoa parviflora (Jussieu) D'Arcy;小花卡利茄■☆
66538 Calibrachoa procumbens Cerv.;卡利茄■☆
66539 Calibrachoa pubescens (Spreng.) Stehmann;毛卡利茄■☆
66540 Calicanthus Raf. = Calycanthus L. (保留属名)●
66541 Calicera Cav. = Calycera Cav. (保留属名)■☆
66542 Calicoca Raf. = Callicocca Schreb. ●
66543 Calicoca Raf. = Cephaelis Sw. (保留属名)●
66544 Calicorema Hook. f. (1880);亮红苋属●☆
66545 Calicorema capitata (Moq.) Hook. f.;亮红苋■☆
66546 Calicorema squarrosa (Schinz) Schinz;粗鳞亮红苋■☆
66547 Calicotome Link = Calycotome Link ●☆
66548 California Aldasoro, C. Navarro, P. Vargas, L. Sáez et Aedo = Erodium L'Hér. ex Aiton ■●
66549 California Aldasoro, C. Navarro, P. Vargas, L. Sáez et Aedo (2002);加州牻牛儿苗属■☆
66550 Caligula Klotzsch = Agapetes D. Don ex G. Don ●
66551 Calimaris canescens Nees = Heteropappus altaicus (Willd.) Novopokr. var. canescens (Nees) Serg. ■
66552 Calimeris Nees = Aster L. ●■
66553 Calimeris Nees = Kalimeris (Cass.) Cass. ■
66554 Calimeris alberitii Regel = Heteropappus altaicus (Willd.) Novopokr. var. canescens (Nees) Serg. ■
66555 Calimeris altaica (Willd.) Nees = Heteropappus altaicus (Willd.) Novopokr. ■
66556 Calimeris altaica (Willd.) Nees var. scabra Avé-Lall. = Heteropappus altaicus (Willd.) Novopokr. var. scaber (Avé-Lall.) Wang-Wei ■
66557 Calimeris altaica (Willd.) Nees var. subincana Avé-Lall. = Heteropappus altaicus (Willd.) Novopokr. ■
66558 Calimeris altaica (Willd.) Nees var. subincana Lanunem. = Heteropappus altaicus (Willd.) Novopokr. ■
66559 Calimeris altaicus Nees var. scabra Avé-Lall. = Heteropappus altaicus (Willd.) Novopokr. var. scaber (Avé-Lall.) Wang-Wei ■
66560 Calimeris alyssoides DC. = Asterothamnus alyssoides (Turcz.) Novopokr. ●
66561 Calimeris biennis (Lindl.) Ledeb. = Heteropappus tataricus (Lindl. ex DC.) Tamamsch. ■
66562 Calimeris canescens Nees = Heteropappus altaicus (Willd.) Novopokr. var. canescens (Nees) Serg. ■
66563 Calimeris ciliata A. Gray = Heteropappus ciliosus (Turcz.) Y. Ling ■
66564 Calimeris ciliosa Turcz. = Heteropappus ciliosus (Turcz.) Y. Ling ■
66565 Calimeris exilis DC. = Heteropappus altaicus (Willd.) Novopokr. ■
66566 Calimeris fruticosus C. Winkl. = Asterothamnus fruticosus (C. Winkl.) Novopokr. ●
66567 Calimeris hispida (Thunb.) Nees = Heteropappus hispidus (Thunb.) Less. ■
66568 Calimeris incisa DC. var. holophylla Maxim. = Kalimeris lautureana (Debeaux) Kitam. ■
66569 Calimeris japonica Sch. Bip. = Turczaninovia fastigiata (Fisch.) DC. ■
66570 Calimeris japonica Sch. Bip. var. achilleiformes H. Lév. = Turczaninovia fastigiata (Fisch.) DC. ■
66571 Calimeris tatarica Lindl. ex DC. = Heteropappus tataricus (Lindl. ex DC.) Tamamsch. ■
66572 Calinea Aubl. = Doliocarpus Rol. ●☆
66573 Calinux Raf. = Pyrularia Michx. ●
66574 Caliphruria Herb. (废弃属名) = Eucharis Planch. et Linden(保留属名)■☆
66575 Calipogon Raf. = Calopogon R. Br. (保留属名)■☆
66576 Calirhoe Raf. = Callirhoe Nutt. ■●☆
66577 Calisaya Hort. ex Pav. = Cinchona L. ■●
66578 Calispepla Vved. (1952);中亚银豆属■☆
66579 Calispepla aegacanthoides Vved.;中亚银豆■☆
66580 Calispermum Lour. = Embelia Burm. f. (保留属名)●■
66581 Calispermum oblongifolium (Hemsl.) Nakai = Embelia vestita Roxb. ●
66582 Calispermum oblongifolium Nakai = Embelia vestita Roxb. ●
66583 Calispermum rude (Hand.-Mazz.) Nakai = Embelia vestita Roxb. ●
66584 Calispermum scandens Lour. = Embelia scandens (Lour.) Mez ●
66585 Calista Ritg. = Callista Lour. (废弃属名)■
66586 Calista Ritg. = Dendrobium Sw. (保留属名)■
66587 Calistaehya Raf. = Veronicastrum Heist. ex Fabr. ■
66588 Calistegia Raf. = Calystegia R. Br. (保留属名)■
66589 Calisto Neraud. (1826);毛里求斯莎草属■☆
66590 Calius Blanco = Streblus Lour. ●
66591 Calixnos Raf. = Crawfurdia Wall. ■
66592 Calixnos Raf. = Gentiana L. ■
66593 Calla L. (1753);水芋属;Bog Arum, Calla, Calla Lily, Water Arum, Wild Calla ■
66594 Calla aethiopica L. = Zantedeschia aethiopica (L.) Spreng. ■
66595 Calla calyptrata Roxb. = Schismatoglottis calyptrata (Roxb.) Zoll. et Moritzi ■
66596 Calla elliotiana (W. Watson) W. Watson = Zantedeschia elliottiana (W. Watson) Engl. ■☆
66597 Calla montana Blume = Anadendrum montanum (Blume) Schott ●
66598 Calla occulta Lour. = Homalomena occulta (Lour.) Schott ■
66599 Calla oculata Lindl. = Zantedeschia albomaculata (Hook.) Baill. ■

66600 Calla palustris L. ;水芋(水浮莲,水葫芦);Bog Arum, Calla, Water Arum, Water-arum, Wild Calla ■

66601 Calla picta Roxb. = Aglaonema pictum (Roxb.) Kunth ■

66602 Calla sagittifolia Michx. = Peltandra sagittifolia (Michx.) Morong ■☆

66603 Callaceae Baetl. = Araceae Juss.(保留科名)■●

66604 Callaceae Rchb. ex Bartl. ;水芋科■

66605 Callaceae Rchb. ex Bartl. = Araceae Juss.(保留科名)●■

66606 Calladium R. Br. = Caladium Vent. ■

66607 Calladium Raf. = Caladium Vent. ■

66608 Callaeocarpus Miq. = Castanopsis (D. Don) Spach(保留属名)●

66609 Callaeolepium H. Karst. = Fimbristemma Turcz. ☆

66610 Callaeum Small(1910);冠虎尾属●☆

66611 Callaeum macropterum (DC.) D. M. Johnson;冠虎尾;Female Dragons, Water Dragons ●☆

66612 Callaion Raf. = Calla L. ■

66613 Callanthus Rchb. = Watsonia Mill.(保留属名)■☆

66614 Callaria Raf. = Calla L. ■

66615 Callerya Endl.(1843);鸡血藤属(昆明鸡血藤属,崖豆藤属)●■

66616 Callerya Endl. = Millettia Wight et Arn.(保留属名)●■

66617 Callerya bonatiana (Pamp.) P. K. Lôc;滇桂崖豆藤(白花藤,白藤,大发汗,大毛豆,断肠叶);Bonat Cliffbean, Bonat Millettia ●

66618 Callerya bonatiana (Pamp.) P. K. Lôc = Millettia bonatiana Pamp. ●

66619 Callerya championii (Benth.) P. K. Lôc = Millettia championii Benth. ●

66620 Callerya championii (Benth.) X. Y. Zhu;绿花崖豆藤(白跌打,老京藤,老荆藤,水苦楝,羊药头,硬骨藤);Champion Millettia, Greenflower Cliffbean ●

66621 Callerya cinerea (Benth.) Schot;灰毛崖豆藤;Greyhair Cliffbean, Greyhair Millettia, Grey-haired Millettia ●■

66622 Callerya cinerea (Benth.) Schot = Millettia cinerea Benth. ●■

66623 Callerya congestiflora (T. C. Chen) Z. Wei et Pedley;密花崖豆藤(青叶烂麻藤);Denseflower Cliffbean, Denseflower Millettia, Dense-flowered Millettia ■●

66624 Callerya dielsiana (Harms) P. K. Lôc;香花崖豆藤(大巴豆,大血藤,丰城鸡血藤,鸡血藤,昆明鸡血藤,老人根,山鸡血藤,香花鸡血藤,血风藤,崖豆藤,岩豆根,岩豆藤,银瓣崖豆藤);Diels Millettia, Spice Cliffbean ●

66625 Callerya dielsiana (Harms) P. K. Lôc = Millettia dielsiana Harms ex Diels ●

66626 Callerya dielsiana (Harms) P. K. Lôc var. heterocarpa (Chun ex T. C. Chen) X. Y. Zhu ex Z. Wei et Pedley;异果崖豆藤;Cliffbean, Differentfruit Millettia ●

66627 Callerya dielsiana (Harms) P. K. Lôc var. solida (T. C. Chen ex Z. Wei) X. Y. Zhu ex Z. Wei et Pedley;雪峰山崖豆藤(雪峰崖豆藤);Solid Cliffbean, Solid Millettia ●

66628 Callerya dorwardii (Collett et Hemsl.) Z. Wei et Pedley;滇缅崖豆藤;Dorward Cliffbean, Dorward Millettia ●

66629 Callerya eurybotrya (Drake) Schot;宽序崖豆藤;Broad-inflorescence Millettia, Broadraceme Cliffbean, Broad-spiked Millettia ●

66630 Callerya eurybotrya (Drake) Schot = Millettia eurybotrya Drake ●

66631 Callerya fordii (Dunn) Schot;广东崖豆藤(披针叶崖豆藤,苦牛大力);Ford Millettia, Guangdong Cliffbean, Kwangtung Cliffbean ■●

66632 Callerya fordii (Dunn) Schot = Millettia fordii Dunn ■●

66633 Callerya gentiliana (H. Lév.) Z. Wei et Pedley;黔滇崖豆藤;Gentil Cliffbean, Gentil Millettia, Yunnan-Guizhou Millettia, Yunnan-Kweichow Millettia ●

66634 Callerya kiangsiensis (Z. Wei) Z. Wei et Pedley;江西崖豆藤;Jiangxi Cliffbean, Jiangxi Millettia, Kiangsi Millettia ●

66635 Callerya lantsangensis (Z. Wei) H. Sun = Millettia lantsangensis Z. Wei ●■

66636 Callerya longipedunculata (Z. Wei) X. Y. Zhu;长梗崖豆藤;Long-pediceled Millettia, Long-peduncle Millettia, Longpedunculate Cliffbean ●

66637 Callerya nitida (Benth.) R. Geesink;光叶崖豆藤(光叶刈藤,光叶鱼藤,贵州崖豆藤,贵州岩石藤,鸡血藤,老京滕,老人根,亮叶鸡血藤,亮叶崖豆藤,血风藤,血筋藤,血藤,硬根藤);Cliffbean, Shiningleaf Millettia, Shining-leaved Millettia ●

66638 Callerya nitida (Benth.) R. Geesink = Millettia nitida Benth. ●

66639 Callerya nitida (Benth.) R. Geesink var. hirsutissima (Z. Wei) X. Y. Zhu;丰城崖豆藤(丰城鸡血藤);Fengcheng Cliffbean, Fengcheng Millettia, Hirsute Cliffbean, Hirsute Millettia ●

66640 Callerya nitida (Benth.) R. Geesink var. minor (Z. Wei) X. Y. Zhu;峨眉崖豆藤;Emei Cliffbean, Emei Millettia, Omei Millettia ●

66641 Callerya oosperma (Dunn) Z. Wei et Pedley;皱果崖豆藤(大叶京老藤,大种崖豆藤,蛋果崖豆藤,豆藤);Wrinkled Fruit Millettia, Wrinkled-fruited Millettia, Wrinklepod Cliffbean ●

66642 Callerya pachyloba (Drake) H. Sun = Millettia pachyloba Drake ●■

66643 Callerya reticulata (Benth.) Schot;网络崖豆藤(白骨藤,白血藤,柴徼藤,蟾蜍藤,冲天子,酢甲,醋甲藤,大肠藤,光叶朱藤,过山龙,红藤,黄藤,黄昭藤,鸡血藤,苦檀子,昆明鸡血藤,蓝藤,老荆藤,老凉藤,老鼠豆,芦藤,马尿血豆,青皮活血,石柱藤,石桩藤,松藤,土鸡血,血防藤,血灌皮,血藤,崖儿藤,苅藤,硬壳藤,紫藤);Evergreen Wisteria, Leatherleaf Millettia, Leather-leaved Millettia, Millettia Vine, Net Cliffbean ■

66644 Callerya reticulata (Benth.) Schot = Millettia reticulata Benth. ■

66645 Callerya reticulata (Benth.) Schot var. championii (Benth.) H. Sun = Callerya championii (Benth.) X. Y. Zhu ●

66646 Callerya reticulata (Benth.) Schot var. stenophylla (Merr. et Chun) X. Y. Zhu;线叶崖豆藤(红叶老凉藤,绒叶崖豆藤,狭叶鸡血藤);Narrowleaf Cliffbean, Narrowleaf Millettia ●

66647 Callerya sericosema (Hance) Z. Wei et Pedley;锈毛崖豆藤(崖豆花);Embroiderhair Cliffbean, Rustyhair Millettia, Rusty-haired Millettia ●

66648 Callerya speciosa (Champ. ex Benth.) Schot;美丽崖豆藤(扒山虎,扳山虎,大口唇,大力牛,大力薯,大莲藕,倒吊金钟,地藕,金钟根,九龙串珠,老惊藤,牛大力,牛大力藤,牛古大力,牛牯大力藤,坡莲藕,山葛,山莲藕,甜牛大力,血藤,猪脚笠);Beautiful Cliffbean, Beautiful Millettia ■●

66649 Callerya speciosa (Champ. ex Benth.) Schot = Millettia speciosa Champ. ex Benth. ■●

66650 Callerya speciosa (Champ.) Schot = Millettia speciosa Champ. ex Benth. ■●

66651 Callerya sphaerosperma (Z. Wei) Z. Wei et Pedley;球子崖豆藤;Ballseed Cliffbean, Sphereseed Millettia, Sphere-seeded Millettia ●

66652 Callerya sphaerosperma (Z. Wei) Z. Wei et Pedley = Millettia sphaerosperma Z. Wei ●

66653 Callerya tsui (F. P. Metcalf) Z. Wei et Pedley;喙果崖豆藤(虎崖豆藤,老虎豆,连珠豆);Beakfruit Millettia, Beakpod Cliffbean, Tsu Millettia ■●

66654 Callerya tsui (F. P. Metcalf) Z. Wei et Pedley = Millettia tsui F.

P. Metcalf ■●
66655 Callerya unijuga (Gagnep.) H. Sun = Craspedolobium unijugum (Gagnep.) Z. Wei et Pedley ●
66656 Calliachyris Torr. et A. Gray = Layia Hook. et Arn. ex DC. (保留属名) ■☆
66657 Calliachyris fremontii Torr. et A. Gray = Layia fremontii (Torr. et A. Gray) A. Gray ■☆
66658 Calliagrostis Ehrh. = Bromus L. (保留属名) ■
66659 Callianassa Webb et Beethel. = Isoplexis (Lindl.) Loudon ●☆
66660 Calliandra Benth. (1840) (保留属名); 朱缨花属(美洲合欢属); Calliandra, Pauderpuff, Pauderpuff Tree ●
66661 Calliandra alternans Benth. = Viguieranthus alternans (Benth.) Villiers ●☆
66662 Calliandra amblyphylla Harms = Calliandra falcata Benth. ●☆
66663 Calliandra anomala (Kunth) J. F. Macbr.; 异形朱缨花; Diverse Mesquitilla ●☆
66664 Calliandra boliviana Britton = Calliandra haematocephala Hassk. ●
66665 Calliandra californica Benth.; 墨西哥朱缨花; Baja Fairy Duster, Red Fairy Duster ●☆
66666 Calliandra calothyrsus Meisn.; 中美朱缨花(危地马拉朱樱花) ●☆
66667 Calliandra conferta Benth.; 密刺朱缨花●☆
66668 Calliandra emarginata Benth.; 微缺朱缨花; Powder Puff ●☆
66669 Calliandra eriophylla Benth.; 绵毛叶朱缨花(绯合欢, 红合欢, 毛叶朱缨花); Fairy Duster, Fairy-duster, False Mesquite, Hairy-leaved Calliandra, Mesquitilla, Mock Mesquite, Pink Fairy Duster ●☆
66670 Calliandra falcata Benth.; 镰形朱缨花●☆
66671 Calliandra gilbertii Thulin et Hunde; 吉尔朱缨花●☆
66672 Calliandra grandiflora (L'Hér.) Benth. = Calliandra anomala (Kunth) J. F. Macbr. ●☆
66673 Calliandra haematocephala Hassk.; 朱缨花(红合欢, 红花刷芾豆, 红头朱缨花, 美蕊花, 美洲合欢, 血头朱缨花); Blood Red Tassel Flower, Blood-red Tassel Flower, Pink Powder Puff, Powderpuff Tree, Red Powderpuff, Red-headed Calliandra, Tweed Calliandra ●
66674 Calliandra harrisii (Lindl.) Benth.; 哈里斯朱缨花●☆
66675 Calliandra houstoniana (Mill.) Standl.; 休斯敦朱缨花●☆
66676 Calliandra houstonii (L'Hér.) Benth. = Calliandra inermis (L.) Druce ●☆
66677 Calliandra inaequilatera Rusby = Calliandra haematocephala Hassk. ●
66678 Calliandra inermis (L.) Druce = Gleditsia triacanthos L. ●
66679 Calliandra megalophylla R. Vig. = Viguieranthus megalophyllus (R. Vig.) Villiers ●☆
66680 Calliandra mollissima (Humb. et Bonpl. ex Willd.) Benth.; 柔软朱缨花●☆
66681 Calliandra perrieri R. Vig. = Viguieranthus perrieri (R. Vig.) Villiers ●☆
66682 Calliandra portoricensis (Jacq.) Benth.; 白朱缨花(波多朱缨花); White Powderpuff, White Tassel Flower ●☆
66683 Calliandra redacta (J. H. Ross) Thulin et Asfaw; 回复朱缨花●☆
66684 Calliandra scottiana R. Vig. = Viguieranthus scottianus (R. Vig.) Villiers ●☆
66685 Calliandra simulans R. Vig. = Viguieranthus simulans (R. Vig.) Villiers ●☆
66686 Calliandra surinamensis Benth.; 苏里南朱缨花(粉扑花, 美蕊花, 苏利南合欢, 小叶合欢, 小朱缨花, 朱缨花); Pink-and-white Powderpuff, Surinam Calliandra, Surinamese Stickpea ●
66687 Calliandra tetragona Benth.; 四角朱樱花●☆
66688 Calliandra thouarsiana Baill. = Viguieranthus alternans (Benth.) Villiers ●☆
66689 Calliandra tweedii Benth.; 巴西朱缨花(墨西哥朱缨花, 朱缨花); Brazilian Flame Bush, Cunure, Mexican Flamebush, Red Tassel Flower, Tweed Calliandra ●☆
66690 Calliandra umbrosa (Wall.) Benth.; 云南朱缨花; Yunnan Calliandra ●
66691 Calliandra xylocarpa Sprague = Bussea xylocarpa (Sprague) Sprague et Craib ●☆
66692 Calliandropsis H. M. Hern. et P. Guinet (1990); 拟朱缨花属(多脉合欢草属) ●☆
66693 Calliandropsis nervosus (Britton et Rose) H. M. Hern. et P. Guinet; 拟朱缨花●☆
66694 Callianira Miq. = Piper L. ●■
66695 Callianthemoides Tamura (1992); 美花毛茛属■☆
66696 Callianthemoides semiverticillatus (Phil.) Tamura; 美花毛茛■☆
66697 Callianthemum C. A. Mey. (1830); 美花草属; Callianthemum ■
66698 Callianthemum alatavicum Freyn; 厚叶美花草■
66699 Callianthemum anemonoides (Zahlbr.) Endl.; 银莲花状美花草■☆
66700 Callianthemum anemonoides (Zahlbr.) Schott = Callianthemum anemonoides (Zahlbr.) Endl. ■☆
66701 Callianthemum angustifolium Witasek; 薄叶美花草; Narrowleaf Callianthemum ■
66702 Callianthemum cashmirianum Cambess. = Callianthemum pimpinlloides (D. Don) Hook. f. et Thomson ■
66703 Callianthemum coriandrifolium Rchb.; 芫荽叶美花草■☆
66704 Callianthemum cuneilobum Hand.-Mazz. = Callianthemum farreri W. W. Sm. ■
66705 Callianthemum endlicheri Walp. = Oxygraphis endicheri (Walp.) Bennet et S. Chandra ■
66706 Callianthemum farreri W. W. Sm.; 川甘美花草; Farrer Callianthemum ■
66707 Callianthemum imbricatum Hand.-Mazz. = Callianthemum pimpinlloides (D. Don) Hook. f. et Thomson ■
66708 Callianthemum insigne (Nakai) Nakai; 显著美花草■☆
66709 Callianthemum insigne (Nakai) Nakai var. hondoense Ohwi; 本州美花草■☆
66710 Callianthemum isopyroides (DC.) Witasek; 扁果草状美花草■☆
66711 Callianthemum kernerianum Freyn ex Kern.; 欧岩荠状美花草■☆
66712 Callianthemum miyabeanum Tatew.; 宫部美花草■☆
66713 Callianthemum pimpinlloides (D. Don) Hook. f. et Thomson; 美花草; Common Callianthemum ■
66714 Callianthemum rutifolium (L.) Rchb. = Callianthemum anemonoides (Zahlbr.) Endl. ■☆
66715 Callianthemum rutifolium C. A. Mey. = Callianthemum coriandrifolium Rchb. ■☆
66716 Callianthemum rutifolium Rchb. = Callianthemum anemonoides (Zahlbr.) Endl. ■☆
66717 Callianthemum sajanense (Regel) Witasek; 萨因美花草■☆
66718 Callianthemum taipaicum W. T. Wang; 太白美花草(重叶莲); Taibaishan Callianthemum ■
66719 Callianthemum tibeticum Witasek = Callianthemum pimpinlloides (D. Don) Hook. f. et Thomson ■
66720 Callias Cass. = Heliopsis Pers. (保留属名) ■☆
66721 Callias Cass. = Kallias Cass. ■☆

66722 Calliaspidia Bremek.(1948);虾衣花属(麒麟吐珠属,虾衣草属)■

66723 Calliaspidia Bremek. = Drejerella Lindau ■

66724 Calliaspidia Bremek. = Justicia L. ●■

66725 Calliaspidia guttata (Brandegee) Bremek.;虾衣花(狐尾木,麒麟塔,麒麟吐珠,虾衣草,虾夷花,小海老草,小虾花);Guttate Drejerella, Shrimp Plant, Shrimp-plant, Yellow Shrimp Plant ■●

66726 Calliaspidia guttata (Brandegee) Bremek. = Drejerella guttata (Brandegee) Bremek. ■●

66727 Callicarpa L.(1753);紫珠属;Beauty Berry, Beauty Bush, Beautyberry, Beauty-berry, French Mulberry, Purplepearl ●

66728 Callicarpa × shirasawana Makino;白泽紫珠●☆

66729 Callicarpa acuminata Roxb. = Callicarpa nudiflora Hook. et Arn. ●

66730 Callicarpa acuminata Roxb. var. angustifolia F. P. Metcalf = Callicarpa nudiflora Hook. et Arn. ●

66731 Callicarpa acuminatissima Tang S. Liu et C. J. Tseng = Callicarpa ticusikensis Masam. ●

66732 Callicarpa acutifolia Hung T. Chang;尖叶紫珠;Acuteleaf Beautyberry, Acuteleaf Purplepearl, Acute-leaved Beautyberry ●

66733 Callicarpa americana L.;美洲紫珠(白毛紫珠,北美紫珠);America Purplepearl, American Beauty Berry, American Beauty Bush, American Beautyberry, American Beauty-berry, Beauty Berry, Bird Eye, Fox Berry, French Mulberry, Indian Currant ●

66734 Callicarpa americana Lour. = Callicarpa candicans (Burm. f.) Hochr. ●

66735 Callicarpa anisophylla C. Y. Wu ex W. Z. Fang;异叶紫珠;Anisophyllous Beautyberry, Anisophyllous Purplepearl ●

66736 Callicarpa antaoensis Hayata = Callicarpa japonica Thunb. var. luxurians Rehder ●

66737 Callicarpa antaoensis Hayata = Callicarpa kotoensis Hayata ●

66738 Callicarpa arborea Roxb.;木紫珠(白叶子树,大树紫珠,马踏皮,梅发破,南洋紫珠,乔木紫珠,紫珠);Arborous Beautyberry, Tree Beautyberry, Tree Purplepearl ●

66739 Callicarpa aspera Hand. -Mazz. = Callicarpa formosana Rolfe ●

66740 Callicarpa ausrealis Koidz. = Callicarpa japonica Thunb. var. luxurians Rehder ●

66741 Callicarpa australis Koidz.;澳洲紫珠●

66742 Callicarpa basitruncata Merr.;平基紫珠(基截紫珠)●

66743 Callicarpa bodinieri H. Lév.;紫珠(白木姜,白棠子树,爆竹柴,大叶斑鸠米,大叶鸦雀饭,大叶鸦鹊饭,鲤鱼下子,漆大白,漆大伯,鱼子,珍珠风,珍珠枫,珍珠柳,珠子树);Beauty Berry, Beautyberry, Bodinier Beautyberry, Purplepearl ●

66744 Callicarpa bodinieri H. Lév. var. giraldii (Hesse) Rehder = Callicarpa giraldii Hesse ex Rehder ●

66745 Callicarpa bodinieri H. Lév. var. iteophylla C. Y. Wu;鼠刺叶紫珠(柳叶紫珠);Sweetspire-leaf Bodinier Beautyberry ●

66746 Callicarpa bodinieri H. Lév. var. lyi (H. Lév.) Rehder = Callicarpa giraldiana Hesse ex Rehder var. subcanescens Rehder ●

66747 Callicarpa bodinieri H. Lév. var. lyi (H. Lév.) Rehder = Callicarpa giraldii Hesse ex Rehder var. lyi (H. Lév.) C. Y. Wu ●

66748 Callicarpa bodinieri H. Lév. var. rosthornii (Diels) Rehder;南川紫珠(斑鸠米,罗桑氏紫珠,罗斯托紫珠);Rosthorn Bodinier Beautyberry ●

66749 Callicarpa bodinierioides R. H. Miao;拟紫珠;Bodinier-like Beautyberry ●

66750 Callicarpa breviceps (Benth.) Hance f. serrulata C. P'ei = Callicarpa brevipes (Benth.) Hance ●

66751 Callicarpa breviceps (Benth.) Hance f. yingtakensis C. P'ei = Callicarpa collina Diels ●

66752 Callicarpa breviceps (Benth.) Hance f. yingtakensis Hung T. Chang = Callicarpa dentosa (Hung T. Chang) W. Z. Fang ●

66753 Callicarpa breviceps (Benth.) Hance var. dentosa Hung T. Chang = Callicarpa dentosa (Hung T. Chang) W. Z. Fang ●

66754 Callicarpa breviceps (Benth.) Hance var. obovata Hung T. Chang = Callicarpa brevipes (Benth.) Hance var. obovata Hung T. Chang ●

66755 Callicarpa brevipes (Benth.) Hance;短柄紫珠(红米碎木);Shortstipe Beautyberry, Shortstipe Purplepearl, Short-stiped Beautyberry ●

66756 Callicarpa brevipes (Benth.) Hance f. serrulata C. P'ei = Callicarpa brevipes (Benth.) Hance ●

66757 Callicarpa brevipes (Benth.) Hance f. yingtakensis C. P'ei = Callicarpa collina Diels ●

66758 Callicarpa brevipes (Benth.) Hance var. dentosa Hung T. Chang = Callicarpa dentosa (Hung T. Chang) W. Z. Fang ●

66759 Callicarpa brevipes (Benth.) Hance var. obovata Hung T. Chang;倒卵叶短柄紫珠;Obovate Shortstipe Beautyberry, Obovateleaf Shortstipe Beautyberry ●

66760 Callicarpa cana L. = Callicarpa candicans (Burm. f.) Hochr. ●

66761 Callicarpa candicans (Burm. f.) Hochr.;白毛紫珠(紫珠草);White-haired Beautyberry, Whitehairy Beautyberry, Whitehairy Purplepearl ●

66762 Callicarpa cathayana Hung T. Chang;华紫珠(创伤草,鲤鱼显子,米筛子,小叶珍珠风,鱼显子,鱼泻子,珍珠风,止血草,紫红鞭);China Purplepearl, Chinese Beautyberry ●

66763 Callicarpa cavaleriei H. Lév. = Ilex chinensis Sims ●

66764 Callicarpa chinyunensis C. P'ei et W. Z. Fang = Callicarpa giraldii Hesse ex Rehder var. chinyunensis (C. P'ei et W. Z. Fang) S. L. Chen ●

66765 Callicarpa collina Diels;丘陵紫珠;Hill Beautyberry, Hill-growing Beautyberry, Hill-growing Purplepearl ●

66766 Callicarpa cuspidata Lam. et Bakh. = Callicarpa longipes Dunn ●

66767 Callicarpa dentosa (Hung T. Chang) W. Z. Fang;多齿紫珠(有齿紫珠);Dentate Beautyberry, Manytooth Purplepearl, Toothed Beautyberry ●

66768 Callicarpa dichotoma (Lour.) K. Koch = Calamus dichotoma (Lour.) K. Koch ●

66769 Callicarpa dichotoma (Lour.) K. Koch f. albifructa Okuyama;白果白棠子树●☆

66770 Callicarpa dielsii (H. Lév.) C. P'ei = Callicarpa rubella Lindl. ●

66771 Callicarpa dunniana H. Lév. = Callicarpa macrophylla Vahl ●

66772 Callicarpa eaquirolli H. Lév. = Caryopteris paniculata C. B. Clarke ●

66773 Callicarpa eriochona Schauer;绵毛紫珠(滇南紫珠);Woolly Beautyberry ●

66774 Callicarpa eriochona Schauer = Callicarpa yunnanensis W. Z. Fang ●

66775 Callicarpa erioclona C. Y. Wu = Callicarpa yunnanensis W. Z. Fang ●

66776 Callicarpa erythrosticta Merr. et Chun;红腺紫珠(红点紫珠);Redgland Purplepearl, Red-glandular Purplepearl, Redspot Beautyberry ●

66777 Callicarpa esquirolii H. Lév. = Caryopteris paniculata C. B. Clarke ●

66778 Callicarpa feddei H. Lév. = Callicarpa bodinieri H. Lév. ●

66779 Callicarpa formosana Rolfe;杜虹花(白毛柴,白奶雪草,白棠子树,粗糠柴,粗糠仔,大丁黄,灯黄,华紫珠,老蟹眼,毛将军,螃蟹目,山埔姜,台湾紫珠,鸦雀饭,鸦鹊板,雅木草,雅目草,贼子草,止血草,紫荆,紫珠,紫珠草);Formosan Beauty-berry,Taiwan Beautyberry,Taiwan Purplepearl ●

66780 Callicarpa formosana Rolfe f. albiflora Sawada et Gogelein;白粗糠仔;White Taiwan Beautyberry ●

66781 Callicarpa formosana Rolfe var. chinensis C. P'ei = Callicarpa integerrima Champ. var. chinensis (C. P'ei) S. L. Chen ●

66782 Callicarpa formosana Rolfe var. chinensis C. P'ei = Callicarpa peii Hung T. Chang ●

66783 Callicarpa formosana Rolfe var. glabra T. T. Chen, Chaw et Yuen P. Yang;六龟粗糠树●

66784 Callicarpa formosana Rolfe var. longifolia Suzuki;长叶杜虹花(长叶粗糠树);Longleaf Beautyberry ●

66785 Callicarpa giraldiana Hesse = Callicarpa giraldii Hesse ex Rehder ●

66786 Callicarpa giraldiana Hesse ex Rehder var. rosthornii (Diels) Rehder = Callicarpa bodinieri H. Lév. var. rosthornii (Diels) Rehder ●

66787 Callicarpa giraldiana Hesse ex Rehder var. subcanescens Rehder;毛叶老鸦糊(珍珠树,紫珠树);Ly Purplepearl,Ly's Beautyberry ●

66788 Callicarpa giraldiana Hesse ex Rehder var. subcanescens Rehder = Callicarpa bodinieri H. Lév. ●

66789 Callicarpa giraldii Hesse ex Rehder;老鸦糊(百子爆,爆竹子,菜子木,粗糠草,大麻雀米,大叶鲤鱼泻子,丰果紫珠,红泡果,猴草,鸡米树,尖叶蜘蛛,路金子,毛风,没翻叶,米筛花,米筛子,牛舌癀,舌癀,万年青,小米团花,鱼胆,珍珠子,粥香,紫珠);Girald Beautyberry,Girald Bodinier Beautyberry,Girald Purplepearl ●

66790 Callicarpa giraldii Hesse ex Rehder var. chinyunensis (C. P'ei et W. Z. Fang) S. L. Chen;缙云紫珠;Jinyun Beautyberry, Jinyun Purplepearl ●

66791 Callicarpa giraldii Hesse ex Rehder var. lyi (H. Lév.) C. Y. Wu = Callicarpa giraldiana Hesse ex Rehder var. subcanescens Rehder ●

66792 Callicarpa giraldii Hesse ex Rehder var. rosthornii (Diels) Rehder = Callicarpa bodinieri H. Lév. var. rosthornii (Diels) Rehder ●

66793 Callicarpa girsea Hand. -Mazz. = Callicarpa giraldiana Hesse ex Rehder var. subcanescens Rehder ●

66794 Callicarpa glabra Koidz.;光紫珠●☆

66795 Callicarpa gracilipes Rehder;湖北紫珠;Hubei Beautyberry, Hubei Purplepearl,Hupeh Beautyberry ●

66796 Callicarpa gracilis Siebold et Zucc. = Callicarpa dichotoma (Lour.) K. Koch ●

66797 Callicarpa grisea Hand. -Mazz. = Callicarpa giraldiana Hesse ex Rehder var. subcanescens Rehder ●

66798 Callicarpa grisea Hand. -Mazz. = Callicarpa giraldii Hesse ex Rehder var. lyi (H. Lév.) C. Y. Wu ●

66799 Callicarpa guizhouensis H. T. Chang et Z. R. Xu;贵州紫珠;Guizhou Beautyberry,Guizhou Purplepearl ●

66800 Callicarpa hungtaii C. P'ei et S. L. Chen;厚萼紫珠(浑氏紫珠);Hungta Beautyberry,Thickcalyx Purplepearl ●

66801 Callicarpa hypoleucophylla Wei-Fang;灰背叶紫珠(里白杜虹花)●

66802 Callicarpa inamoena C. Y. Wu = Callicarpa giraldiana Hesse ex Rehder var. subcanescens Rehder ●

66803 Callicarpa incana Roxb. = Callicarpa macrophylla Vahl ●

66804 Callicarpa integerrima Champ.;全缘叶紫珠(全缘紫珠,月中风);Entire Beautyberry,Entire Purplepearl ●

66805 Callicarpa integerrima Champ. var. chinensis (C. P'ei) S. L. Chen;藤紫珠(裴氏紫珠,三爪风,粤赣紫珠);Pei Beautyberry,Pei Purplepearl ●

66806 Callicarpa integerrima Champ. var. serrulata H. L. Li = Callicarpa formosana Rolfe ●

66807 Callicarpa integrifolia Forbes et Hemsl. = Callicarpa integerrima Champ. ●

66808 Callicarpa japonica Thunb.;日本紫珠(红头紫珠,吉隆紫珠,女儿茶,紫珠);Japan Purplepearl, Japanese Beauty Berry, Japanese Beauty Bush,Japanese Beautyberry,Japanese Beauty-berry,Japanese Callicarpa,Jilong Beautyberry ●

66809 Callicarpa japonica Thunb. f. albibacca H. Hara;白果日本紫珠;Whitefruit Japan Purplepearl ●☆

66810 Callicarpa japonica Thunb. f. albiflora Moldenke;白花日本紫珠●☆

66811 Callicarpa japonica Thunb. f. angustata (Rehder) Ohwi = Callicarpa japonica Thunb. var. angustata Rehder ●

66812 Callicarpa japonica Thunb. f. glabra C. P'ei = Callicarpa siongsaiensis F. P. Metcalf ●

66813 Callicarpa japonica Thunb. f. kiruninsularis Masam. = Callicarpa japonica Thunb. ●

66814 Callicarpa japonica Thunb. f. microcarpa (Nakai) Ohwi = Callicarpa japonica Thunb. var. microcarpa Nakai ●☆

66815 Callicarpa japonica Thunb. f. taquetii (H. Lév.) Ohwi = Callicarpa japonica Thunb. var. taquetii (H. Lév.) Nakai ●☆

66816 Callicarpa japonica Thunb. subsp. luxurians (Rehder) Masam. et Yanagih. = Callicarpa japonica Thunb. var. luxurians Rehder ●

66817 Callicarpa japonica Thunb. var. angustata Rehder;窄叶紫珠(峦大紫珠,狭叶女儿茶,狭叶紫珠,止血草);Narrowleaf Beautyberry, Narrowleaf Purplepearl,Narrow-leaved Beauty-berry ●

66818 Callicarpa japonica Thunb. var. angustata Rehder = Callicarpa membranacea Hung T. Chang ●

66819 Callicarpa japonica Thunb. var. angustata sensu Rehder = Callicarpa membranacea Hung T. Chang ●

66820 Callicarpa japonica Thunb. var. dichotoma Bakh. = Callicarpa oligantha Merr. ●

66821 Callicarpa japonica Thunb. var. kotoensis (Hayata) Masam. = Callicarpa kotoensis Hayata ●

66822 Callicarpa japonica Thunb. var. leucocarpa Nakai = Callicarpa japonica Thunb. f. albibacca H. Hara ●☆

66823 Callicarpa japonica Thunb. var. luxurians Masam. = Callicarpa japonica Thunb. f. kiruninsularis Masam. ●

66824 Callicarpa japonica Thunb. var. luxurians Rehder;朝鲜紫珠(大序日本紫珠,兰屿女儿茶);Korea Purplepearl,Korean Beautyberry ●

66825 Callicarpa japonica Thunb. var. luxurians Rehder f. albifructa H. Hara;白果朝鲜紫珠●☆

66826 Callicarpa japonica Thunb. var. microcarpa Nakai;小果日本紫珠;Smallfruit Japan Purplepearl ●☆

66827 Callicarpa japonica Thunb. var. taquetii (H. Lév.) Nakai;小叶日本紫珠;Smallleaf Japan Purplepearl ●☆

66828 Callicarpa japonica Thunb. var. typica Lam. et Bakjh. = Callicarpa japonica Thunb. ●

66829 Callicarpa kochiana Makino;枇杷叶紫珠(长叶紫珠,鬼紫珠,黄毛紫珠,黄紫珠,劳莱氏紫珠,牛舌癀,枇杷紫珠,山枇杷,野枇杷);Koch Beautyberry, Loquatleaf Beautyberry, Loquatleaf Purplepearl,Loureir Beauty-berry,Wild Purplepearl ●

66830 Callicarpa kochiana Makino var. laxiflora (Hung T. Chang) W. Z. Fang;散花紫珠(有梗劳莱氏紫珠);Laxflower Beautyberry ●

66831 Callicarpa kotoensis Hayata;红头紫珠;Hongtou Beautyberry, Koto Purplepearl,Red-head Beautyberry ●

66832 Callicarpa kotoensis Hayata = Callicarpa japonica Thunb. var. luxurians Rehder ●

66833 Callicarpa kwangdungensis Chun;广东紫珠(臭常山,金刀菜,老鸦饭,万年青,小叶紫珠菜,珍珠风,止血柴);Guangdong Beautyberry,Guangdong Purplepearl,Kwangtung Beautyberry ●

66834 Callicarpa kwangdungensis Chun var. trichocarpa L. Q. Li;毛果广东紫珠;Hairy-fruited Guangdong Beautyberry ●

66835 Callicarpa lanata L.;印度粗糠树;Indian Beautyberry ●☆

66836 Callicarpa lanceolaria Roxb. = Callicarpa longifolia Lam. var. laceolaria (Roxb.) C. B. Clarke ●

66837 Callicarpa lingii Merr.;光叶紫珠(江西紫珠,绿荚柴,无柄紫珠);Glabrousleaf Beautyberry, Glabrousleaf Purplepearl, Glabrous-leaved Beautyberry ●

66838 Callicarpa lobo-apiculata F. P. Metcalf;尖萼紫珠(尖裂紫珠);Acutesepal Beautyberry, Apiculate-lobed Beautyberry, Lobeapiculate Beautyberry,Sharpcalyx Purplepearl ●

66839 Callicarpa longibracteata Hung T. Chang;长苞紫珠;Longbract Beautyberry,Longbract Purplepearl,Long-bracted Beautyberry ●

66840 Callicarpa longifolia Lam.;长叶紫珠(兰屿长叶紫珠,兰屿女儿茶,老哈眼,万年青);Lanyu Long-leaved Beauty-berry, Longleaf Beautyberry,Longleaf Purplepearl,Long-leaved Beautyberry ●

66841 Callicarpa longifolia Lam. f. floccosa Schauer = Callicarpa longifolia Lam. var. floccosa Schauer ●

66842 Callicarpa longifolia Lam. var. brevipes Benth. = Callicarpa brevipes (Benth.) Hance ●

66843 Callicarpa longifolia Lam. var. floccosa Schauer;白毛长叶紫珠;Floccose Beautyberry,Floccose Purplepearl ●

66844 Callicarpa longifolia Lam. var. laceolaria (Roxb.) C. B. Clarke;披针叶紫珠;Lanceolate Beautyberry,Lanceolate Purplepearl ●

66845 Callicarpa longifolia Lam. var. longissima Hemsl. = Callicarpa longissima (Hemsl.) Merr. ●

66846 Callicarpa longifolia Lam. var. rosthornii Diels = Callicarpa bodinieri H. Lév. var. rosthornii (Diels) Rehder ●

66847 Callicarpa longifolia Lam. var. subglabrata Schauer = Callicarpa japonica Thunb. ●

66848 Callicarpa longifolia P. Court = Callicarpa lingii Merr. ●

66849 Callicarpa longiloba Merr. = Callicarpa kochiana Makino ●

66850 Callicarpa longiloba Merr. = Callicarpa loureiri Hook. et Arn. ●

66851 Callicarpa longipes Dunn;长柄紫珠;Longstipe Beautyberry, Longstipe Purplepearl,Long-stiped Beautyberry ●

66852 Callicarpa longipes Dunn var. laui Moldenke = Callicarpa longipes Dunn ●

66853 Callicarpa longissima (Hemsl.) Merr.;尖尾枫(长叶紫珠,赤药子,穿骨风,穿骨枫,大风药,大样尖尾枫,赶风柴,赶风晒,赶风帅,赶疯晒,尖尾风,尖尾峰,拿手风,牛舌癀,起疯晒,雪突,黏手风,黏手枫);Longestapex Beautyberry, Longestapex Purplepearl, Longest-apexed Beautyberry,Long-leaved Beauty-berry ●

66854 Callicarpa longissima (Hemsl.) Merr. f. subglabra C. P'ei;秃尖尾枫;Subglabrous Longestapex Beautyberry, Subglabrous Longestapex Purplepearl ●

66855 Callicarpa longituba? = Callicarpa kochiana Makino ●

66856 Callicarpa loureiri Hook. et Arn.;野枇杷(长叶紫珠,鬼紫珠,黄毛紫珠,牛舌癀,枇杷叶紫珠,山枇杷);Loureir Beauty-berry, Wild Purplepearl ●

66857 Callicarpa loureiri Hook. et Arn. = Callicarpa kochiana Makino ●

66858 Callicarpa loureiri Hook. et Arn. ex Merr. = Callicarpa kochiana Makino ●

66859 Callicarpa loureiri Hook. et Arn. ex Merr. var. laxiflora Hung T. Chang = Callicarpa kochiana Makino var. laxiflora (Hung T. Chang) W. Z. Fang ●

66860 Callicarpa loureiri Hook. et Arn. var. laxiflora Hung T. Chang = Callicarpa kochiana Makino var. laxiflora (Hung T. Chang) W. Z. Fang ●

66861 Callicarpa luteopunctata Hung T. Chang;黄腺紫珠;Yellowdotted Beautyberry,Yellow-dotted Beautyberry,Yellowdotted Purplepearl ●

66862 Callicarpa lyi H. Lév. = Callicarpa giraldiana Hesse ex Rehder var. subcanescens Rehder ●

66863 Callicarpa lyi H. Lév. = Callicarpa giraldii Hesse ex Rehder var. lyi (H. Lév.) C. Y. Wu ●

66864 Callicarpa macrophylla Vahl;大叶紫珠(白骨风,白骨枫,大风叶,大叶粗糠树,赶风柴,赶风紫,红大白,假大艾,尖裂紫珠,羊耳朵,贼子叶,止血草,紫珠草);Big-leaf Beauty Berry, Bigleaf Purplepearl,Big-leaved Beautyberry,Largeleaf Beautyberry ●

66865 Callicarpa macrophylla Vahl var. kouytchensis H. Lév. = Callicarpa macrophylla Vahl ●

66866 Callicarpa macrophylla Vahl var. sinensis C. B. Clarke = Callicarpa nudiflora Hook. et Arn. ●

66867 Callicarpa mairei H. Lév. = Callicarpa giraldii Hesse ex Rehder ●

66868 Callicarpa martinii H. Lév. = Caryopteris paniculata C. B. Clarke ●

66869 Callicarpa membranacea Hung T. Chang;膜叶紫珠(窄叶紫珠);Membranaleaf Purplepearl ●

66870 Callicarpa mimuraskai Hassk. = Callicarpa japonica Thunb. ●

66871 Callicarpa minutiflora Y. Y. Qian;细花紫珠;Thinflower Purplepearl ●

66872 Callicarpa mixiensis Z. X. Yu;密溪紫珠;Mixi Beautyberry, Mixi Purplepearl ●

66873 Callicarpa mollis Siebold et Zucc.;毛紫珠(白棠子树,高山紫珠,女儿茶);Hairy Purplepearl ●

66874 Callicarpa mollis Siebold et Zucc. f. albifructa Nanko;白果毛紫珠●☆

66875 Callicarpa mollis Siebold et Zucc. var. microphylla Siebold et Zucc.;小叶毛紫珠●☆

66876 Callicarpa murasaki Siebold = Callicarpa japonica Thunb. ●

66877 Callicarpa ningpoensis Matsuda = Callicarpa formosana Rolfe ●

66878 Callicarpa nishimurae Koidz. = Callicarpa parvifolia Hook. et Arn. ●☆

66879 Callicarpa nudiflora Hook. et Arn.;裸花紫珠(白花茶,大斑鸠米,大贼仔,饭汤叶,赶风柴,节节红,蟹目周,鸦鹊饭,贼公叶,贼佬药,贼仔叶);Bare-flowered Beautyberry, Nakedflower Beautyberry,Nakedflower Purplepearl ●

66880 Callicarpa okinawensis Nakai = Callicarpa oshimensis Hayata var. okinawensis (Nakai) Hatus. ●☆

66881 Callicarpa oligantha Merr.;罗浮紫珠;Lessflower Beautyberry, Luofu Purplepearl,Oliganthous Beautyberry ●

66882 Callicarpa oshimensis Hayata;奄美紫珠●☆

66883 Callicarpa oshimensis Hayata var. iriomotensis (Masam.) Hatus.;西表紫珠●☆

66884 Callicarpa oshimensis Hayata var. okinawensis (Nakai) Hatus.;冲绳紫珠●☆

66885 Callicarpa panduriformis H. Lév. = Callicarpa rubella Lindl. ●

66886 Callicarpa paniculata Lam. = Buddleja saligna Willd. ●☆

66887 Callicarpa parvifolia Hayata = Callicarpa randaiensis Hayata ●

66888 Callicarpa parvifolia Hook. et Arn.;小叶紫珠●☆

66889 Callicarpa pauciflora Chun ex Hung T. Chang;少花紫珠(疏花紫珠);Fewflower Beautyberry, Fewflower Purplepearl, Pauciflorous Beautyberry ●

66890 Callicarpa pedunculata Lam. et Bakh. = Callicarpa formosana Rolfe ●

66891 Callicarpa pedunculata R. Br. = Callicarpa formosana Rolfe ●

66892 Callicarpa pedunculata R. Br. var. chinensis (C. P'ei) F. P. Metcalf = Callicarpa peii Hung T. Chang ●

66893 Callicarpa pedunculata R. Br. var. chinensis (C. P'ei) F. P. Metcalf = Callicarpa integerrima Champ. var. chinensis (C. P'ei) S. L. Chen ●

66894 Callicarpa pedunculata R. Br. var. longifolia (Suzuki) Hung T. Chang = Callicarpa formosana Rolfe var. longifolia Suzuki ●

66895 Callicarpa peichieniana Chun et S. L. Chen;钩毛紫珠(红斑鸠米);Hookedhairy Beautyberry, Hookedhairy Purplepearl, Hook-haired Beautyberry ●

66896 Callicarpa peii Hung T. Chang = Callicarpa integerrima Champ. var. chinensis (C. P'ei) S. L. Chen ●

66897 Callicarpa pilosissima Maxim.;长毛紫珠(红面将军,锐叶紫珠,细叶紫珠);Longpilose Beautyberry, Longpilose Purplepearl, Long-pilosed Beautyberry, Narrowleaf Beauty-berry ●

66898 Callicarpa pilosissima Maxim. var. henryi Gogelein = Callicarpa pilosissima Maxim. ●

66899 Callicarpa pingshanensis C. Y. Wu ex W. Z. Fang;屏山紫珠;Pingshan Beautyberry, Pingshan Purplepearl ●

66900 Callicarpa poilanei Dop;白背紫珠;Whiteback Beautyberry, Whiteback Purplepearl, White-backed Beautyberry ●

66901 Callicarpa prolifera C. Y. Wu;抽芽紫珠;Proliferous Beautyberry, Proliferous Purplepearl ●

66902 Callicarpa pseudorubella Hung T. Chang;拟红紫珠(假红紫珠);False Reddish Beautyberry ●

66903 Callicarpa purpurea Juss. = Callicarpa dichotoma (Lour.) K. Koch ●

66904 Callicarpa randaiensis Hayata;峦大紫珠(大叶紫珠);Luanda Purplepearl ●

66905 Callicarpa reevesii Wall. = Callicarpa nudiflora Hook. et Arn. ●

66906 Callicarpa reevesii Wall. ex Schauer = Callicarpa nudiflora Hook. et Arn. ●

66907 Callicarpa remotiflora Wei-Fang Lin et Jen-Li Wang;疏花紫珠;Remoteflower Beauty-berry ●

66908 Callicarpa remotiflora Wei-Fang Lin et Jen-Li Wang = Callicarpa remotiserrulata Hayata ●

66909 Callicarpa remotiserrulata Hayata;疏齿紫珠(恒春紫珠,疏细齿紫珠);Remoteserrate Beauty-berry, Remote-serrulate Beautyberry, Remotiserrulate Beautyberry, Scattertooth Purplepearl ●

66910 Callicarpa rhandaiensis Hayata;峦代紫珠;Beautyberry ●☆

66911 Callicarpa roxburghii C. P'ei = Callicarpa lobo-apiculata F. P. Metcalf ●

66912 Callicarpa roxburghii Wall. = Callicarpa macrophylla Vahl ●

66913 Callicarpa rubella Lindl.;红紫珠(白金子风,对节树,复生药,红叶紫珠,空壳树,空壳铁砂子,漆大伯,山霸王,山槟榔,细米油珠,小红米果,野蓝靛,贼仔树);Beauty Berry, Chinese Beauty Bush, Chinese Beauty-berry, Reddish Beautyberry, Reddish Purplepearl ●

66914 Callicarpa rubella Lindl. f. angustata C. P'ei;狭叶红紫珠(斑鸠钻,节节风,狭叶紫珠);Narrowleaf Reddish Beautyberry, Narrowleaf Reddish Purplepearl ●

66915 Callicarpa rubella Lindl. f. crenata C. P'ei;钝齿红紫珠(大刀药,钝齿紫珠,毛跌打,沙药草);Crenate Reddish Beautyberry, Crenate Reddish Purplepearl ●

66916 Callicarpa rubella Lindl. f. robusta C. P'ei = Callicarpa formosana Rolfe ●

66917 Callicarpa rubella Lindl. f. villosa M. Cheng et Z. J. Feng;长毛红紫珠;Villose Reddish Beautyberry ●

66918 Callicarpa rubella Lindl. var. dielsii (H. Lév.) H. L. Li = Callicarpa rubella Lindl. var. subglabra (C. P'ei) Hung T. Chang ●

66919 Callicarpa rubella Lindl. var. dielsii (H. Lév.) H. L. Li = Callicarpa rubella Lindl. ●

66920 Callicarpa rubella Lindl. var. hemsleyana Diels = Callicarpa rubella Lindl. ●

66921 Callicarpa rubella Lindl. var. hemsleyana Diels f. subglabra C. P'ei = Callicarpa rubella Lindl. var. subglabra (C. P'ei) Hung T. Chang ●

66922 Callicarpa rubella Lindl. var. subglabra (C. P'ei) Hung T. Chang;秃红紫珠;Subglabrous Reddish Beautyberry, Subglabrous Reddish Purplepearl ●

66923 Callicarpa rubella Rehder = Callicarpa rubella Lindl. f. angustata C. P'ei ●

66924 Callicarpa ruptofoliata R. H. Miao;裂叶紫珠;Ruptofoliate Purplepearl ●

66925 Callicarpa salicifolia C. P'ei et W. Z. Fang;水金花;Willowleaf Beautyberry, Willowleaf Purplepearl, Willow-leaved Beautyberry ●

66926 Callicarpa seguinii H. Lév. = Callicarpa bodinieri H. Lév. ●

66927 Callicarpa shikokiana Makino;四国紫珠;Sikoku Purplepearl ●☆

66928 Callicarpa shikokiana Makino f. albiflora S. Toyama;白花四国紫珠●☆

66929 Callicarpa sinensis Steud. = Callicarpa candicans (Burm. f.) Hochr. ●

66930 Callicarpa siong-saiensis F. P. Metcalf;上狮紫珠;Shangshi Beautyberry, Shangshi Purplepearl ●

66931 Callicarpa subpubescens Hook. et Arn.;亚光紫珠●☆

66932 Callicarpa taiwaniana Suzuki = Callicarpa formosana Rolfe var. longifolia Suzuki ●

66933 Callicarpa takakumensis Hatus. = Callicarpa longissima (Hemsl.) Merr. ●

66934 Callicarpa tangutica Maxim. var. brachyodonta Hand.-Mazz. = Caryopteris trichosphaera W. W. Sm. ●

66935 Callicarpa taquetii H. Lév. = Callicarpa japonica Thunb. ●

66936 Callicarpa tenuiflora Champ. = Callicarpa rubella Lindl. ●

66937 Callicarpa tenuiflora Champ. ex Benth. = Callicarpa rubella Lindl. ●

66938 Callicarpa ticusikensis Masam.;锐叶紫珠●

66939 Callicarpa tingwuensis Hung T. Chang;鼎湖紫珠;Dinghu Beautyberry, Dinghu Purplepearl ●

66940 Callicarpa tomentosa Lam. = Callicarpa candicans (Burm. f.) Hochr. ●

66941 Callicarpa tomentosa Lam. et Bakh. = Callicarpa arborea Roxb. ●

66942 Callicarpa tonkinensis Dop = Callicarpa bodinieri H. Lév. ●

66943 Callicarpa tosaensis Makino;绒毛紫珠●☆

66944 Callicarpa triloba Lour. = Cissus triloba (Lour.) Merr. ●

66945 Callicarpa tsiangii Moldenke = Callicarpa bodinieri H. Lév. ●

66946 Callicarpa vastifolia Diels = Viburnum rhytidophyllum Hemsl. ex Forbes et Hemsl. ●

66947 Callicarpa yakusimensis Koidz. = Callicarpa shikokiana Makino ●☆

66948 Callicarpa yunnanensis W. Z. Fang;云南紫珠;Yunnan Beautyberry,Yunnan Purplepearl ●
66949 Callicephalus C. A. Mey. (1831);丽头菊属(肖美头菊属)■☆
66950 Callicephalus C. A. Mey. = Centaurea L. (保留属名)●■
66951 Callicephalus niteus (M. Bieb.) C. A. Mey.;丽头菊■☆
66952 Callichilia Stapf(1902);丽唇夹竹桃属●☆
66953 Callichilia barteri (Hook. f.) Stapf;巴特丽唇夹竹桃●☆
66954 Callichilia basileis Beentje;基丽唇夹竹桃●☆
66955 Callichilia bequaertii De Wild.;大萼丽唇夹竹桃●☆
66956 Callichilia inaequalis Stapf;不等丽唇夹竹桃●☆
66957 Callichilia macrocalyx Schellenb. ex Markgr. = Callichilia bequaertii De Wild. ●☆
66958 Callichilia magnifica Good = Callichilia bequaertii De Wild. ●☆
66959 Callichilia mannii Stapf = Callichilia inaequalis Stapf ●☆
66960 Callichilia monopodialis (K. Schum.) Stapf;单足夹竹桃●☆
66961 Callichilia orientalis S. Moore;东方丽唇夹竹桃●☆
66962 Callichilia stenosepala Stapf = Callichilia subsessilis (Benth.) Stapf ●☆
66963 Callichilia subsessilis (Benth.) Stapf;无梗丽唇夹竹桃●☆
66964 Callichlamys Miq. (1845);美苞紫葳属●☆
66965 Callichlamys riparia Miq.;美苞紫葳●☆
66966 Callichloe Pfeiff. = Callichloe Willd. ex Steud. ■☆
66967 Callichloe Willd. ex Steud. = Elionurus Humb. et Bonpl. ex Willd. (保留属名)■☆
66968 Callichloea Spreng. ex Steud. = Callichloe Willd. ex Steud. ■☆
66969 Callichloea Steud. = Elionurus Humb. et Bonpl. ex Willd. (保留属名)■☆
66970 Callichroa Fisch. et C. A. Mey. = Layia Hook. et Arn. ex DC. (保留属名)■☆
66971 Callichroa nutans Greene = Harmonia nutans (Greene) B. G. Baldwin ■☆
66972 Callichroa platyglossa Fisch. et C. A. Mey. = Layia platyglossa (Fisch. et C. A. Mey.) A. Gray ■☆
66973 Callicocca Schreb. = Cephaelis Sw. (保留属名)●
66974 Callicocca Schreb. = Psychotria L. (保留属名)●
66975 Callicoma Andréws(1809);美毛木属(卡利寇马属,瓦特木属);Black Wattle ●☆
66976 Callicoma serratifolia Andréws;瓦特木;Black Wattle ●☆
66977 Callicomaceae J. Agardh = Cunoniaceae R. Br. (保留科名)●☆
66978 Callicomis Wittst. = Callicoma Andréws ●☆
66979 Callicore Link = Amaryllis L. (保留属名)■☆
66980 Callicornia Burm. f. = Asteropterus Adans. ●☆
66981 Callicysthus Endl. = Vigna Savi(保留属名)■
66982 Callidrynos Neraud = Molinaea Comm. ex Juss. ●☆
66983 Calliglossa Hook. et Arn. (1839);美舌菊属■☆
66984 Calliglossa Hook. et Arn. = Layia Hook. et Arn. ex DC. (保留属名)■☆
66985 Calliglossa douglasii Hook. et Arn.;美舌菊■☆
66986 Calligonaceae Khalk.;沙拐枣科●
66987 Calligonaceae Khalk. = Polygonaceae Juss. (保留科名)●■
66988 Calligonum L. (1753);沙拐枣属;Calligonum,Kneejujube ●
66989 Calligonum Lour. = Tetracera L. ●
66990 Calligonum acanthopterum I. G. Borshch.;尖翅沙拐枣●☆
66991 Calligonum affine Popov;近缘沙拐枣●☆
66992 Calligonum affine Popov = Calligonum rubicundum Bunge ●
66993 Calligonum alaschanicum Losinsk.;阿拉善沙拐枣;Alashan Calligonum,Alashan Kneejujube ●
66994 Calligonum alatiforme Pavlov;翅形沙拐枣●☆
66995 Calligonum alatum Litv.;高沙拐枣●☆
66996 Calligonum androssowii Litv.;安德罗索夫沙拐枣●☆
66997 Calligonum anfractuosum Bunge = Calligonum leucocladum (Schrenk) Bunge ●
66998 Calligonum aphyllum (Pall.) Gürke;无叶沙拐枣;Leafless Calligonum,Leafless Kneejujube ●
66999 Calligonum aphyllus Kar. et Kir. = Calligonum leucocladum (Schrenk) Bunge ●
67000 Calligonum aralense I. G. Borshch.;阿拉尔沙拐枣●☆
67001 Calligonum arborescens Litv.;乔木沙拐枣(乔木状沙拐枣);Arborescent Calligonum,Arborescent Kneejujube ●
67002 Calligonum borszczowii Litv.;鲍尔沙拐枣●☆
67003 Calligonum bubyri B. Fedtsch. ex Pavlov;布贝沙拐枣●☆
67004 Calligonum calcareum Pavlov;石灰沙拐枣●☆
67005 Calligonum calliphysa Bunge;泡果沙拐枣●
67006 Calligonum calvescens Maire;光秃沙拐枣●☆
67007 Calligonum canescens Pursh = Atriplex canescens (Pursh) Nutt. ●
67008 Calligonum caput-medusae Schrenk;头状沙拐枣;Capitate Calligonum,Headshape Kneejujube ●
67009 Calligonum cartilagineum Pavlov;软骨质沙拐枣●☆
67010 Calligonum chinense Losinsk.;中华沙拐枣(甘肃沙拐枣,中国沙拐枣);Chinese Calligonum,Gansu Kneejujube ●
67011 Calligonum colubrinum I. G. Borshch.;褐色沙拐枣;Brown Kneejujube,Snake-like Calligonum ●
67012 Calligonum comosum L'Hér. = Calligonum polygonoides L. subsp. comosum (L'Hér.) Soskov ●☆
67013 Calligonum cordatum Korovin ex Pavlov;心形沙拐枣;Heart Calligonum,Heartshape Kneejujube ●
67014 Calligonum coriaceum Pavlov;革质沙拐枣●☆
67015 Calligonum crinitum Boiss.;长软毛沙拐枣●☆
67016 Calligonum crispum Bunge = Calligonum rubicundum Bunge ●
67017 Calligonum cristatum Pavlov;冠状沙拐枣●☆
67018 Calligonum densum I. G. Borshch.;密刺沙拐枣;Dense Calligonum,Densethorn Kneejujube ●
67019 Calligonum dielsianum K. S. Hao;美节蓼;Diels Calligonum,Diels Kneejujube ●
67020 Calligonum dielsianum K. S. Hao = Calligonum mongolicum Turcz. ●
67021 Calligonum dissectum Popov;深裂沙拐枣●☆
67022 Calligonum ebi-nurcum Ivanova ex Soskov;艾比湖沙拐枣(精河沙拐枣);Ebi-lake Calligonum,Ebilake Kneejujube ●
67023 Calligonum elatum Litv.;高大沙拐枣●☆
67024 Calligonum erinaceum I. G. Borshch.;刺沙拐枣●☆
67025 Calligonum eugenii-korovinii Pavlov;番樱桃沙拐枣●☆
67026 Calligonum ferganense Pavlov;费尔干沙拐枣●☆
67027 Calligonum flavidum Bunge = Calligonum rubicundum Bunge ●
67028 Calligonum gobicum (Bunge) Losinsk.;戈壁沙拐枣;Gobi Calligonum,Gobi Kneejujube ●
67029 Calligonum gracile Litv.;纤细沙拐枣●☆
67030 Calligonum griseum Korovin ex Pavlov;灰沙拐枣●☆
67031 Calligonum horridum I. G. Borshch. = Calligonum calliphysa Bunge ●
67032 Calligonum humile Litv.;低矮沙拐枣●☆
67033 Calligonum involutum Pavlov;内卷沙拐枣●☆
67034 Calligonum jimunaicum Z. M. Mao;吉木乃沙拐枣;Jimunai Calligonum,Jimunai Kneejujube ●

67035 Calligonum junceum (Fisch. et C. A. Mey.) Litv.;灯心草沙拐枣(泡果沙拐枣);Dunkfruit Kneejujube,Junceus Calligonum ●
67036 Calligonum juochiangense Y. X. Liou;若羌沙拐枣;Ruoqiang Calligonum,Ruoqiang Kneejujube ●
67037 Calligonum juochiangense Y. X. Liou = Calligonum pumilum Losinsk. ●
67038 Calligonum klementzii Losinsk.;奇台沙拐枣(新疆沙拐枣);Klementz Calligonum,Qitai Kneejujube ●
67039 Calligonum korlaense Z. M. Mao;库尔勒沙拐枣;Kuerle Calligonum,Kuerle Kneejujube ●
67040 Calligonum koslovii Losinsk. = Calligonum gobicum (Bunge) Losinsk. ●
67041 Calligonum koslovii Losinsk. = Calligonum zaidamense Losinsk. ●
67042 Calligonum lanciculatum Pavlov;披针沙拐枣●☆
67043 Calligonum leucocladum (Schrenk) Bunge;白皮沙拐枣(淡枝沙拐枣);Whitebark Calligonum,White-barked Calligonum ●
67044 Calligonum lipskyi Litv.;利普斯基沙拐枣●☆
67045 Calligonum litwinowii Drobow;利特氏沙拐枣●☆
67046 Calligonum macrocarpum I. G. Borshch.;大果沙拐枣●
67047 Calligonum membranaceum (I. G. Borshch.) Litv.;膜质沙拐枣●☆
67048 Calligonum microcarpum I. G. Borshch.;小果沙拐枣●☆
67049 Calligonum molle Litv.;柔软沙拐枣●☆
67050 Calligonum mongolicum Turcz.;沙拐枣(蒙古沙拐枣,砂拐枣,头发草);Mongolian Calligonum,Mongolian Kneejujube ●
67051 Calligonum mongolicum Turcz. var. gobicum Bunge ex Meisn. = Calligonum gobicum (Bunge) Losinsk. ●
67052 Calligonum muravljanskyi Pavlov;穆拉沙拐枣●☆
67053 Calligonum obtusum Litv.;钝沙拐枣●☆
67054 Calligonum orthotrichum Pavlov;直毛沙拐枣●☆
67055 Calligonum paletzldanum Litv.;帕来沙拐枣●☆
67056 Calligonum pallasia L'Hér. = Calligonum aphyllum (Pall.) Gürke ●
67057 Calligonum patens Litv.;铺展沙拐枣●☆
67058 Calligonum pellucidum Pavlov;透明沙拐枣●☆
67059 Calligonum persicum Boiss.;伊朗沙拐枣;Persian Calligonum ●
67060 Calligonum petunnikowii Litv.;佩通尼科夫沙拐枣●☆
67061 Calligonum physopterum Pavlov;囊翅沙拐枣●☆
67062 Calligonum platyacanthum I. G. Borshch.;阔刺沙拐枣●☆
67063 Calligonum plicatum Pavlov;折扇沙拐枣●☆
67064 Calligonum polygonoides L.;多节沙拐枣●☆
67065 Calligonum polygonoides L. subsp. comosum (L'Hér.) Soskov;簇毛沙拐枣●☆
67066 Calligonum polygonoides Pall. = Calligonum aphyllum (Pall.) Gürke ●
67067 Calligonum potaninii Losinsk. = Calligonum mongolicum Turcz. ●
67068 Calligonum przewalskii Losinsk. = Calligonum alaschanicum Losinsk. ●
67069 Calligonum pulcherrimum Korovin ex Pavlov;美丽沙拐枣●☆
67070 Calligonum pumilum Losinsk.;小沙拐枣(小果沙拐枣);Dwarf Calligonum,Small Kneejujube ●
67071 Calligonum quadraepterum Korovin = Calligonum quadraepterum Korovin ex Pavlov ●☆
67072 Calligonum quadraepterum Korovin ex Pavlov;土耳其斯坦四翅沙拐枣●☆
67073 Calligonum rigidum Litv. = Calligonum aphyllum (Pall.) Gürke ●
67074 Calligonum roborovskii Losinsk.;南疆沙拐枣(昆仑沙拐枣,塔里木沙拐枣);Roborovsk Calligonum,South Xinjiang Calligonum,Talimu Kneejujube ●
67075 Calligonum rubicundum Bunge;红果沙拐枣(红皮沙拐枣);Red-barked Calligonum,Redfruit Kneejujube ●
67076 Calligonum russanovii Pavlov;路萨沙拐枣●☆
67077 Calligonum setosum Litv.;刚毛沙拐枣●☆
67078 Calligonum songaricum Endl. = Calligonum rubicundum Bunge ●
67079 Calligonum spinosissimum Pavlov;具刺沙拐枣●☆
67080 Calligonum squarrosum Pavlov;粗糙沙拐枣;Roughness Kneejujube,Squarrose Calligonum ●
67081 Calligonum tenue Pavlov;细小沙拐枣●☆
67082 Calligonum tetrapterum Jaub. et Spach;四翅沙拐枣●☆
67083 Calligonum trifarium Z. M. Mao;三列沙拐枣(三裂沙拐枣);Trifarious Calligonum,Trirow Kneejujube ●
67084 Calligonum triste Litv.;暗淡沙拐枣●☆
67085 Calligonum turkestanicum (Korovin) Pavlov;土耳其斯坦沙拐枣●☆
67086 Calligonum yengisaricum Z. M. Mao;英吉沙沙拐枣;Yigisar Kneejujube ●
67087 Calligonum zaidamense Losinsk.;柴达木沙拐枣;Chadamu Kneejujube,Zaidam Calligonum ●
67088 Callilepis DC.(1836);美鳞鼠麹木属■●☆
67089 Callilepis caerulea (Hutch.) Leins;蓝美鳞鼠麹木●☆
67090 Callilepis glabra DC. = Callilepis laureola DC. ●☆
67091 Callilepis hispida DC. = Callilepis laureola DC. ●☆
67092 Callilepis lancifolia Burtt Davy;剑叶美鳞鼠麹木●☆
67093 Callilepis laureola DC.;美鳞鼠麹木●☆
67094 Callilepis laureola DC. var. glabra (DC.) Harv. = Callilepis laureola DC. ●☆
67095 Callilepis laureola DC. var. hispida (DC.) Harv. = Callilepis laureola DC. ●☆
67096 Callilepis leptophylla Harv.;细叶美鳞鼠麹木●☆
67097 Callilepis salicifolia Oliv.;柳叶美鳞鼠麹木●☆
67098 Callionia Greene = Potentilla L. ■●
67099 Calliopea D. Don = Crepis L. ■
67100 Calliopsis Rchb. = Coreopsis L. ●■
67101 Calliopsis basalis A. Dietr. = Coreopsis basalis (A. Dietr.) S. F. Blake ■
67102 Calliopsis tinctoria (Nutt.) DC. = Coreopsis tinctoria Nutt. ■
67103 Calliopsis wrightii (A. Gray) H. M. Parker ex E. B. Sm. = Coreopsis basalis (A. Dietr.) S. F. Blake ■
67104 Calliparion (Link) Rchb. ex Wittst. = Aconitum L. ■
67105 Callipeltis Steven(1829);美盾茜属■☆
67106 Callipeltis aperta Boiss. et Buhse = Callipeltis cucullaris (L.) Steven ■☆
67107 Callipeltis cucullaris (L.) Steven;美盾茜■☆
67108 Calliphruria Lindl. = Eucharis Planch. et Linden(保留属名)■☆
67109 Calliphyllon Bubani = Epipactis Zinn(保留属名)■
67110 Calliphyllon Bubani et Penz. = Epipactis Zinn(保留属名)■
67111 Calliphysa Fisch. et C. A. Mey. = Calligonum L. ●
67112 Calliphysa juncea Fisch. et C. A. Mey. = Calligonum junceum (Fisch. et C. A. Mey.) Litv. ●
67113 Calliprena Salisb. = Allium L. ■
67114 Calliprora Lindl. = Brodiaea Sm.(保留属名)■☆
67115 Calliprora Lindl. = Triteleia Douglas ex Lindl. ■☆
67116 Calliprora anilina (Greene) A. Heller = Triteleia ixioides (W. T. Aiton) Greene subsp. anilina (Greene) L. W. Lenz ■☆

67117 Calliprora aurantea Kellogg = Triteleia ixioides (W. T. Aiton) Greene subsp. scabra (Greene) L. W. Lenz ■☆

67118 Calliprora ixioides (W. T. Aiton) Greene = Triteleia ixioides (W. T. Aiton) Greene ■☆

67119 Calliprora ixioides (W. T. Aiton) Greene var. lugens (Greene) Abrams = Triteleia lugens Greene ■☆

67120 Calliprora lugens (Greene) Greene = Triteleia lugens Greene ■☆

67121 Calliprora lutea Lindl. = Triteleia ixioides (W. T. Aiton) Greene ■☆

67122 Calliprora scabra Greene var. anilina Greene = Triteleia ixioides (W. T. Aiton) Greene subsp. anilina (Greene) L. W. Lenz ■☆

67123 Callipsyche Herb. = Eucrosia Ker Gawl. ■☆

67124 Callirhoe Nutt. (1821); 罂粟葵属; Poppy Mallow ■●☆

67125 Callirhoe alcaeoides (Michx.) A. Gray; 苍白罂粟葵(光亮罂粟葵); Light Poppy Mallow, Pale Poppy Mallow, Pink Poppy Mallow, Purple Poppy Mallow ●☆

67126 Callirhoe bushii Fernald; 布什罂粟葵; Bush's Poppy Mallow ●☆

67127 Callirhoe digitata Nutt.; 指状罂粟葵; Finger Poppy Mallow ●☆

67128 Callirhoe involucrata (Torr. et A. Gray) A. Gray; 矮罂粟葵(矮粟葵); Low Poppy Mallow, Poppy Mallow, Purple Mallow, Purple Poppy Mallow, Wine Cup, Wine Cups ●☆

67129 Callirhoe papaver (Cav.) A. Gray; 罂粟葵; Poppy Mallow ●☆

67130 Callirhoe papaver (Cav.) A. Gray var. bushii (Fernald) Waterf. = Callirhoe bushii Fernald ●☆

67131 Callirhoe triangulata (Leavenw.) A. Gray; 三角罂粟葵; Clustered Poppy Mallow ●☆

67132 Callirrhoe A. Gray = Callirhoe Nutt. ■●☆

67133 Callisace Fisch. = Angelica L. ■

67134 Callisace Fisch. ex Hoffm. = Angelica L. ■

67135 Callisace dahurica Fisch. ex Hoffm. = Angelica dahurica (Fisch. ex Hoffm.) Benth. et Hook. f. ex Franch. et Sav. ■

67136 Callisace ternata (Regel et Schmalh.) Koso-Pol. = Angelica ternata Regel et Schmalh. ■

67137 Callisace ternata Koso-Pol. = Angelica ternata Regel et Schmalh. ■

67138 Callisema Steud. = Callisemaea Benth. ■☆

67139 Callisemaea Benth. = Platypodium Vogel ■☆

67140 Callisia L. = Asteropterus Adans. ●☆

67141 Callisia L. = Leysera L. ■●☆

67142 Callisia Loefl. (1758); 锦竹草属(卡利草属, 洋竹草属) ■☆

67143 Callisia cordifolia (Sw.) E. S. Anderson et Woodson; 心叶锦竹草(心叶洋竹草) ■☆

67144 Callisia elegans Alexander ex H. E. Moore; 斑纹锦竹草(斑纹鸭跖草); Striped Inch Plant ■☆

67145 Callisia fragrans (Lindl.) Woodson; 芳香锦竹草(芳香洋竹草); Basketplant, Inch Plant, Spironema ■☆

67146 Callisia gnaphalodes L. = Leysera gnaphalodes (L.) L. ●☆

67147 Callisia graminea (Small) G. C. Tucker; 禾叶锦竹草(禾叶洋竹草) ■☆

67148 Callisia micrantha (Torr.) D. R. Hunt; 小花锦竹草(小花洋竹草) ■☆

67149 Callisia monandra (Sw.) Schult. f.; 美洲锦竹草(美洲洋竹草) ■☆

67150 Callisia navicularis (Ortgies) D. R. Hunt; 叠叶草(重扇, 小舟卡利草) ■☆

67151 Callisia ornata (Small) G. C. Tucker; 装饰锦竹草 ■☆

67152 Callisia repens L.; 铺地锦竹草(垂锦竹草, 翠玲珑, 洋竹草) ■

67153 Callisia rosea (Vent.) D. R. Hunt; 粉红锦竹草(粉红洋竹草) ■☆

67154 Callista D. Don = Erica L. ●☆

67155 Callista Lour. (废弃属名) = Dendrobium Sw. (保留属名) ■

67156 Callista adunca (Wall. ex Lindl.) Kuntze = Dendrobium aduncum Wall. ex Lindl. ■

67157 Callista aggregata Kuntze = Dendrobium lindleyi Steud. ■

67158 Callista aggregata Kuntze var. jenkinsii (Wall. ex Lindl.) Brieger = Dendrobium jenkinsii Lindl. ■

67159 Callista alpestris Kuntze = Dendrobium monticola P. F. Hunt et Summerh. ■

67160 Callista amabile Lour. = Dendrobium thyrsiflorum Rchb. f. ■

67161 Callista ampla (Lindl.) Kuntze = Epigeneium amplum (Lindl. ex Wall.) Summerh. ■

67162 Callista angustifolia (Blume) Kuntze = Flickingeria angustifolia (Blume) A. D. Hawkes ■

67163 Callista annamensis Kraenzl. = Dendrobium hercoglossum Rchb. f. ■

67164 Callista aurantiaca Kuntze = Dendrobium chryseum Rolfe ■

67165 Callista aurea (Lindl.) Kuntze = Dendrobium heterocarpum Wall. ex Lindl. ■

67166 Callista binocularis (Rchb. f.) Kuntze = Dendrobium gibsonii Lindl. ■

67167 Callista boxallii (Rchb. f.) Kuntze = Dendrobium gratiosissimum Rchb. f. ■

67168 Callista brymeriana (Rchb. f.) Kuntze = Dendrobium brymerianum Rchb. f. ■

67169 Callista calceolaria (Carey ex Hook.) Kuntze = Dendrobium moschatum (Buch. -Ham.) Sw. ■

67170 Callista candida (Wall. ex Lindl.) Kuntze = Dendrobium moniliforme (L.) Sw. ■

67171 Callista capillipes (Rchb. f.) Kuntze = Dendrobium capillipes Rchb. f. ■

67172 Callista carinifera (Rchb. f.) Kuntze = Dendrobium cariniferum Rchb. f. ■

67173 Callista chrysantha (Wall. ex Lindl.) Kuntze = Dendrobium chrysanthum Wall. ex Lindl. ■

67174 Callista chrysotoxa (Lindl.) Kuntze = Dendrobium chrysotoxum Lindl. ■

67175 Callista clavata Kuntze = Dendrobium aurantiacum Rchb. f. var. denneanum (Kerr) Z. H. Tsi ■

67176 Callista coelogyne (Rchb. f.) Kuntze = Epigeneium amplum (Lindl. ex Wall.) Summerh. ■

67177 Callista comata (Blume) Kuntze = Flickingeria comata (Blume) A. D. Hawkes ■

67178 Callista crassinodis (Benson ex Rchb. f.) Kuntze = Dendrobium pendulum Roxb. ■

67179 Callista crepidata (Lindl. et Paxton) Kuntze = Dendrobium crepidatum Lindl. et Paxton ■

67180 Callista cretacea (Lindl.) Kuntze = Dendrobium primulinum Lindl. ■

67181 Callista crumenata (Sw.) Kuntze = Dendrobium crumenatum Sw. ■

67182 Callista crystallina (Rchb. f.) Kuntze = Dendrobium crystallinum Rchb. f. ■

67183 Callista densiflora (Wall.) Kuntze = Dendrobium densiflorum Lindl. ex Wall. ■

67184 Callista devoniana (Paxton) Kuntze = Dendrobium devonianun Paxton ■

67185 Callista dixantha (Rchb. f.) Kuntze = Dendrobium dixanthum Rchb. f. ■

67186 Callista falconeri (Hook.) Kuntze = Dendrobium falconeri Hook. ■

67187 Callista fimbriata (Hook.) Kuntze = Dendrobium fimbriatum Hook. ■

67188 Callista findlayana (E. C. Parish et Rchb. f.) Kuntze = Dendrobium findlayanum Parl. et Rchb. f. ■

67189 Callista flavescens (Lindl.) Kuntze = Polystachya concreta (Jacq.) Garay et H. R. Sweet ■

67190 Callista flavidula (Ridl. ex Hook. f.) Kuntze = Dendrobium stuposum Lindl. ■

67191 Callista floribunda (D. Don) Kuntze = Mycaranthes floribunda (D. Don) S. C. Chen et J. J. Wood ■

67192 Callista fuscescens (Griff.) Kuntze = Epigeneium fuscescens (Griff.) Summerh. ■

67193 Callista gibsonii (Lindl.) Kuntze = Dendrobium gibsonii Lindl. ■

67194 Callista gratiosissima (Rchb. f.) Kuntze = Dendrobium gratiosissimum Rchb. f. ■

67195 Callista harveyana (Rchb. f.) Kuntze = Dendrobium harveyanum Rchb. f. ■

67196 Callista hercoglossa (Rchb. f.) Kuntze = Dendrobium hercoglossum Rchb. f. ■

67197 Callista heterocarpa (Wall. ex Lindl.) Kuntze = Dendrobium heterocarpum Wall. ex Lindl. ■

67198 Callista hookeriana (Lindl.) Kuntze = Dendrobium hookerianum Lindl. ■

67199 Callista intermedia (Teijsm. et Binn.) Kuntze = Dendrobium salaccense (Blume) Lindl. ■

67200 Callista japonica (Blume) Kuntze = Dendrobium moniliforme (L.) Sw. ■

67201 Callista jenkinsii (Griff.) Kuntze = Dendrobium parciflorum Rchb. f. ex Lindl. ■

67202 Callista lawiana (Lindl.) Kuntze = Dendrobium crepidatum Lindl. et Paxton ■

67203 Callista linawiana (Rchb. f.) Kuntze = Dendrobium linawianum Rchb. f. ■

67204 Callista lituiflora (Lindl.) Kuntze = Dendrobium lituiflorum Lindl. ■

67205 Callista loddigesii (Rolfe) Kuntze = Dendrobium loddigesii Rolfe ■

67206 Callista longicornu (Lindl.) Kuntze = Dendrobium longicornu Lindl. ■

67207 Callista moniliforme (Lindl.) Kuntze = Dendrobium moniliforme (L.) Sw. ■

67208 Callista moniliformis (L.) Kuntze = Dendrobium moniliforme (L.) Sw. ■

67209 Callista moschata (Buch. -Ham.) Kuntze = Dendrobium moschatum (Buch. -Ham.) Sw. ■

67210 Callista moulmeinensis (E. C. Parish ex Hook. f.) Kuntze = Dendrobium devonianun Paxton ■

67211 Callista nobilis (Lindl.) Kuntze = Dendrobium nobile Lindl. ●■

67212 Callista oculata (Hook.) Kuntze = Dendrobium fimbriatum Hook. ■

67213 Callista parishii (Rchb. f.) Kuntze = Dendrobium parishii Rchb. f. ■

67214 Callista pendula (Roxb.) Kuntze = Dendrobium pendulum Roxb. ■

67215 Callista porphyrochila (Lindl.) Kuntze = Dendrobium prophyrochilum Lindl. ■

67216 Callista primulina (Lindl.) Kuntze = Dendrobium primulinum Lindl. ■

67217 Callista reptans Kuntze = Conchidium japonicum (Maxim.) S. C. Chen et J. J. Wood ■

67218 Callista rotundata (Lindl.) Kuntze = Epigeneium rotundatum (Lindl.) Summerh. ■

67219 Callista salaccense (Blume) Kuntze = Dendrobium salaccense (Blume) Lindl. ■

67220 Callista spatella (Rchb. f.) Kuntze = Dendrobium spatella Rchb. f. ■

67221 Callista spathacea (Lindl.) Kuntze = Dendrobium moniliforme (L.) Sw. ■

67222 Callista stricklandiana (Rchb. f.) Kuntze = Dendrobium catenatum Lindl. ■

67223 Callista strongylantha (Rchb. f.) Kuntze = Dendrobium strongylanthum Rchb. f. ■

67224 Callista stuposa (Lindl.) Kuntze = Dendrobium stuposum Lindl. ■

67225 Callista sulcata (Lindl.) Kuntze = Dendrobium sulcatum Lindl. ■

67226 Callista terminalis (E. C. Parish et Rchb. f.) Kuntze = Dendrobium terminale Parl. et Rchb. f. ■

67227 Callista thyrsiflora (Rchb. f. ex André) M. A. Clem. = Dendrobium thyrsiflorum Rchb. f. ■

67228 Callista trogonopus (Rchb. f.) Kuntze = Dendrobium trigonopus Rchb. f. ■

67229 Callista vexans (Dammer) Kraenzl. = Dendrobium hercoglossum Rchb. f. ■

67230 Callista wardiana (Warner) Kuntze = Dendrobium wardianum Warner ■

67231 Callista wattii (Hook. f.) Kuntze = Dendrobium wattii (Hook. f.) Rchb. f. ■

67232 Callista williamsonii (J. Day et Rchb. f.) Kuntze = Dendrobium williamsonii J. Day et Rchb. f. ■

67233 Callistachya Raf. = Veronicastrum Heist. ex Fabr. ■

67234 Callistachya Sm. = Callistachys Vent. (废弃属名) ■☆

67235 Callistachya Sm. = Oxylobium Andréws (保留属名) ■☆

67236 Callistachys Heuffel = Carex L. ■

67237 Callistachys Heuffel = Heuffelia Opiz ■

67238 Callistachys Raf. = Veronica L. ■

67239 Callistachys Vent. (废弃属名) = Oxylobium Andréws (保留属名) ■☆

67240 Callistanthos Szlach. (2008); 粉红肥根兰属 ■☆

67241 Callistanthos Szlach. = Pelexia Poit. ex Lindl. (保留属名) ■

67242 Callistema Cass. = Callistephus Cass. (保留属名) ■

67243 Callistemma Boiss. = Tremastelma Raf. ■☆

67244 Callistemma Cass. (废弃属名) = Callistephus Cass. (保留属名) ■

67245 Callistemma chinensis (L.) Skeels = Callistephus chinensis (L.) Nees ■

67246 Callistemma honense Cass. = Callistephus chinensis (L.) Nees ■

67247 Callistemma sinensis (L.) Broth. = Callistephus chinensis (L.) Nees ■

67248 Callistemon R. Br. (1814); 红千层属 (瓶刷树属, 瓶子刷树属); Bottle Brush, Bottlebrush, Bottle-brush ●

67249 Callistemon R. Br. = Melaleuca L. (保留属名) ●

67250 Callistemon acuminatus Cheel; 细叶红千层; Thin-leaved

Bottlebrush ●☆

67251 Callistemon acuminatus Cheel 'Nabiac Red';纳比红细叶红千层●☆

67252 Callistemon brachyandrus Lindl.; 短叶红千层; Prickly Bottlebrush ●☆

67253 Callistemon citrinus (Curtis) Skeels;柠檬红千层(橙红红千层,红花红千层,红千层);Australian Bottlcbrush, Bottlebrush, Crimson Bottlebrush, Lemon Bottlebrush, Lemon Bottle-brush, Lemon-scent Bottlebrush ●

67254 Callistemon citrinus (Curtis) Skeels 'Angela';安哥拉红花红千层●☆

67255 Callistemon citrinus (Curtis) Skeels 'Austraflora Firebrand';火焰红花红千层●☆

67256 Callistemon citrinus (Curtis) Skeels 'Burgundy';暗红红花红千层●☆

67257 Callistemon citrinus (Curtis) Skeels 'Endeavour' = Callistemon citrinus (Curtis) Skeels 'Splendens' ●☆

67258 Callistemon citrinus (Curtis) Skeels 'Jeffersii';杰弗斯红花红千层●☆

67259 Callistemon citrinus (Curtis) Skeels 'Reeves Pink';李维斯粉红花红千层●☆

67260 Callistemon citrinus (Curtis) Skeels 'Splendens';光彩照人红花红千层(华美红千层,艳红红千层)●☆

67261 Callistemon citrinus (Curtis) Skeels 'White Anzac';白澳斯红花红千层●☆

67262 Callistemon coccineus F. Muell.;绯红红千层(红牙刷,杉叶榴);Scarlet Bottlebrush, Scarlet Bottle-brush ●☆

67263 Callistemon comboynensis Cheel;悬崖红千层;Cliff Bottlebrush ●☆

67264 Callistemon formosus S. T. Blake;美丽红千层;Cliff Bottlebrush ●☆

67265 Callistemon formosus S. T. Blake 'Carmina';胭脂红美丽红千层●☆

67266 Callistemon glaucus (Bonpl.) Sweet = Callistemon speciosus (Sims) DC. ●

67267 Callistemon hybridus (Otto) DC.;杂种红千层;Bottlebrush ●☆

67268 Callistemon hybridus (Otto) DC. 'Harkness';聆听红千层●☆

67269 Callistemon hybridus (Otto) DC. 'King's Park Special';皇园特选红千层●☆

67270 Callistemon hybridus (Otto) DC. 'Little John';小约翰红千层●☆

67271 Callistemon hybridus (Otto) DC. 'Mauve Mist';紫红红千层●☆

67272 Callistemon hybridus (Otto) DC. 'Old Duninald';顿翁红千层●☆

67273 Callistemon hybridus (Otto) DC. 'Perth Pink';佩思粉红千层●☆

67274 Callistemon hybridus (Otto) DC. 'Western Glory';西荣红千层●☆

67275 Callistemon lanceolatus (Sm.) Sweet;剑叶红千层(红千层,花槙,金宝树);Lemon Bottle-brush ●☆

67276 Callistemon lanceolatus (Sm.) Sweet = Callistemon citrinus (Curtis) Skeels ●

67277 Callistemon lanceolatus Sweet = Callistemon lanceolatus (Sm.) Sweet ●☆

67278 Callistemon linearis (Sm.) DC.;狭叶红千层(狭叶串钱柳);Narrow-leaf Bottlebrush, Narrow-leaved Bottlebrush ●☆

67279 Callistemon linearis (Sm.) DC. 'Pumila';矮生狭叶红千层●☆

67280 Callistemon lophanthus Lodd. = Callistemon salignus DC. ●

67281 Callistemon pachyphyllus Cheel; 厚叶红千层; Wallum Bottlebrush ●☆

67282 Callistemon pallidus (Bonpl.) DC.;黄花红千层(淡白瓶刷树);Lemon Bottlebrush ●☆

67283 Callistemon pallidus (Bonpl.) DC. = Callistemon salignus DC. ●

67284 Callistemon pallidus DC. = Callistemon salignus DC. ●

67285 Callistemon pearsonii R. D. Spencer et Lumley;尖叶红千层;Blackdown Bottlebrush ●☆

67286 Callistemon phoeniceus Lindl.;猩红红千层;Lesser Bottlebrush ●☆

67287 Callistemon pinifolius (Wendl.) Sweet;松针红千层;Pine-leaved Bottlebrush ●☆

67288 Callistemon pithyoides Miq.;高山红千层(黄花瓶刷树);Mountain Bottlebrush ●☆

67289 Callistemon polandii F. M. Bailey;波蓝迪红千层●☆

67290 Callistemon recurvus R. D. Spencer et Lumley;弯红千层;Tinaroo Bottlebrush ●☆

67291 Callistemon rigidus R. Br.;红千层(红花瓶刷子树,红瓶子刷树,瓶刷树,瓶子刷树);Rigid Bottlebrush, Stiff Bottle Brush, Stiff Bottlebrush ●

67292 Callistemon rugulosus (Link) DC.; 略皱红千层; Scarlet Bottlebrush ●☆

67293 Callistemon salignus (Sm.) Sweet;柳叶红千层(黄花瓶刷子树);Pink Tips, White Bottlebrush, Willow Bottlebrush, Willowleaf Bottlebrush, Willow-leaved Bottlebrush ●

67294 Callistemon salignus (Sm.) Sweet f. viridiflorus (Sims) Siebert et Voss;绿花柳叶红千层;Green Bottle-brush, Green-flower Willowleaf Bottlebrush, Weidenblatt Callistemon ●

67295 Callistemon salignus DC. = Callistemon salignus (Sm.) Sweet ●

67296 Callistemon salignus DC. var. viridiflorus F. Muell. = Callistemon salignus (Sm.) Sweet f. viridiflorus (Sims) Siebert et Voss ●

67297 Callistemon sieberi DC.;湿生红千层(硬叶瓶刷树);River Bottlebrush ●☆

67298 Callistemon speciosus (Sims) DC.;杉叶红千层(灰叶红千层,美丽红千层,杉叶榴);Albany Bottlebrush, Albany Bottle-brush, Beautiful Bottlebrush, Showy Bottle-brush ●

67299 Callistemon speciosus DC. = Callistemon speciosus (Sims) DC. ●

67300 Callistemon subulatus Cheel;钻叶红千层(亮叶红千层)●☆

67301 Callistemon teretifolius F. Muell.;针叶红千层;Needle-leaved Bottlebrush ●☆

67302 Callistemon viminalis Cheel;垂枝红千层(串钱柳,垂花红千层,垂枝花槙,轮枝红千层);Pendulous Bottlebrush, Weeping Bottlebrush ●☆

67303 Callistemon viminalis Cheel 'Captain Cook';库克船长垂枝红千层●☆

67304 Callistemon viminalis Cheel 'Dawson River';道森河垂枝红千层●☆

67305 Callistemon viminalis Cheel 'Hannah Ray';汉纳雷垂枝红千层●☆

67306 Callistemon viminalis Cheel 'Little John';小约翰垂枝红千层;Callistemon 'Little John', Dwarf Bottlebrush ●☆

67307 Callistemon viminalis Cheel 'Wild River';急流垂枝红千层●☆

67308 Callistemon viridiflorus (Sims) Sweet;绿花红千层;Green Bottlebrush ●☆

67309 Callistemon viridiflorus (Sims) Sweet = Callistemon salignus DC. var. viridiflorus F. Muell. ●

67310 Callistemon viridiflorus Sweet = Callistemon viridiflorus (Sims) Sweet ●☆

67311 Callistephana Fourr. = Coronilla L.(保留属名)●■

67312 Callistephus Cass.(1825)(保留属名);翠菊属;China Aster,

China-aster, Chinese Aster ■

67313 Callistephus biennis Lindl. ex DC. = Heteropappus tataricus (Lindl.) Tamamsch. ■

67314 Callistephus chinensis (L.) Nees; 翠菊(佛螺,江南腊,江西腊,蓝菊,六月菊,宿根翠菊,五月菊); Annual Aster, China Aster, Chinese Aster, Common China Aster, Common China-aster, Garden Aster ■

67315 Callistephus hortensis Cass. = Callistephus chinensis (L.) Nees ■

67316 Callisteris Greene = Batanthes Raf. ■☆

67317 Callisteris Greene = Gilia Ruiz et Pav. ■●☆

67318 Callisteris Greene = Ipomopsis Michx. ■☆

67319 Callisthene Mart. (1826); 美丽囊萼花属 ☆

67320 Callisthene fasciculata Mart.; 美丽囊萼花 ☆

67321 Callisthene major Mart.; 大美丽囊萼花 ☆

67322 Callisthene microphylla Warm.; 小叶美丽囊萼花 ☆

67323 Callisthene minor Mart.; 小美丽囊萼花 ☆

67324 Callisthene robusta Briq. et Glaz.; 粗壮美丽囊萼花 ☆

67325 Callisthenia Spreng. = Callisthene Mart. ☆

67326 Callistigma Dinter et Schwantes = Mesembryanthemum L. (保留属名) ■●

67327 Callistroma Fenzl = Oliveria Vent. ■☆

67328 Callistylon Pittier = Coursetia DC. ●☆

67329 Callithauma Herb. = Stenomesson Herb. ■☆

67330 Callithronum Ehrh. = Cephalanthera Rich. ■

67331 Callithronum Ehrh. = Serapias L. (保留属名) ■☆

67332 Callitraceae Seward = Cupressaceae Gray (保留科名) ●

67333 Callitrichaceae Bercht. et J. Presl = Callitrichaceae Link (保留科名) ■

67334 Callitrichaceae Link (1821) (保留科名); 水马齿科; Water Starwort Family, Waterstarwort Family, Water-starwort Family ■

67335 Callitrichaceae Link (保留科名) = Plantaginaceae Juss. (保留科名) ■

67336 Callitriche L. (1753); 水马齿属; Starwort, Water Starwort, Waterstarwort, Water-starwort ■

67337 Callitriche anceps Fernald = Callitriche heterophylla Pursh ■☆

67338 Callitriche anisoptera Schotsman; 异翅水马齿 ■☆

67339 Callitriche austinii Engelm. = Callitriche terrestris Raf. ■☆

67340 Callitriche autumnalis L. = Callitriche hermaphroditica L. ■

67341 Callitriche autumnalis L. f. macrocarpa Hegelm. = Callitriche hermaphroditica L. subsp. macrocarpa (Hegelm.) Lansdown ■

67342 Callitriche bengalensis Petr. = Callitriche palustris L. var. elegans (Petr.) Y. L. Chang ■

67343 Callitriche bifida (L.) Morong. = Callitriche hermaphroditica L. ■

67344 Callitriche bolusii Schönland et Pax ex Marloth; 博卢斯水马齿 ■☆

67345 Callitriche brutia Petagna; 布吕特水马齿; Pedunculate Water-starwort ■☆

67346 Callitriche clausonis Hegelm. = Callitriche hermaphroditica L. subsp. clausonis (Hegelm.) Maire ■☆

67347 Callitriche compressa N. E. Br.; 扁水马齿 ■☆

67348 Callitriche deflexa A. Br. = Callitriche deflexa A. Br. ex Hegelm. ■

67349 Callitriche deflexa A. Br. ex Hegelm.; 弯茎水马齿; Deflex Waterstarwort ■

67350 Callitriche deflexa A. Braun = Callitriche terrestris Raf. ■☆

67351 Callitriche deflexa A. Braun var. austinii (Engelm.) Hegelm. = Callitriche terrestris Raf. ■☆

67352 Callitriche deflexa A. Braun var. subsessilis Fassett = Callitriche terrestris Raf. ■☆

67353 Callitriche elegans Petr. = Callitriche palustris L. var. elegans (Petr.) Y. L. Chang ■

67354 Callitriche fallax Petr. = Callitriche palustris L. ■

67355 Callitriche favargeri Schotsman; 法瓦尔热水马齿 ■☆

67356 Callitriche fehmedianii Majeed Kak et Javeid; 西南水马齿 ■

67357 Callitriche fuscicarpa Lansdown; 褐果水马齿 ■

67358 Callitriche glareosa Lansdown; 西藏水马齿 ■

67359 Callitriche hamulata W. D. J. Koch; 具沟水马齿; Intermediate Water-starwort ■☆

67360 Callitriche hedbergiorum Schotsman; 赫德水马齿 ■☆

67361 Callitriche hermaphroditica L.; 线叶水马齿(秋水马齿); Autumn Water Starwort, Autumn Waterstarwort, Autumn Water-starwort, Autumnal Starwort, Autumnal Water-starwort, Linearleaf Waterstarwort, Narrow Water Starwort, Northern Water-starwort ■

67362 Callitriche hermaphroditica L. subsp. bifida (L.) Schinz et Hell. = Callitriche truncata Guss. ■☆

67363 Callitriche hermaphroditica L. subsp. clausonis (Hegelm.) Maire = Callitriche lusitanica Schotsman ■☆

67364 Callitriche hermaphroditica L. subsp. macrocarpa (Hegelm.) Lansdown; 大果线叶水马齿(大果水马齿) ■

67365 Callitriche hermaphroditica L. subsp. truncata (Guss.) Jahand. et Maire = Callitriche truncata Guss. ■☆

67366 Callitriche heterophylla Pursh; 异叶水马齿; Large Water Starwort, Large Water-starwort ■☆

67367 Callitriche japonica Engelm. ex Hegelm.; 日本水马齿; Japan Waterstarwort ■

67368 Callitriche keniensis Schotsman; 肯尼亚水马齿 ■☆

67369 Callitriche lusitanica Schotsman; 葡萄牙水马齿 ■☆

67370 Callitriche marginata Torr.; 花边水马齿; Winged Water-starwort ■☆

67371 Callitriche mathezii Schotsman; 马泰水马齿 ■☆

67372 Callitriche nana B. C. Ho et G. T. P. Vo = Callitriche japonica Engelm. ex Hegelm. ■

67373 Callitriche obtusangula Le Gall; 钝果水马齿; Blunt-fruited Water-starwort ■☆

67374 Callitriche oreophila Schotsman; 喜山水马齿 ■☆

67375 Callitriche oryzetorum Petr. = Callitriche palustris L. var. oryzetorum (Petr.) Lansdown ■

67376 Callitriche palustris L.; 沼生水马齿(春水马齿,迷惑水马齿,水马齿,雅致水马齿); Common Waterstarwort, Common Water-starwort, Star-grass, Vernal Water Starwort, Vernal Water-starwort, Water Fennel, Water Starwort, Waterwort ■

67377 Callitriche palustris L. = Callitriche platycarpa Kütz. ■

67378 Callitriche palustris L. subsp. obtusangula (Le Gall) Maire = Callitriche obtusangula Le Gall ■☆

67379 Callitriche palustris L. subsp. pedunculata (DC.) Maire = Callitriche brutia Petagna ■☆

67380 Callitriche palustris L. subsp. stagnalis (Scop.) Schinz et Thell. = Callitriche stagnalis Scop. ■

67381 Callitriche palustris L. var. bifida L. = Callitriche hermaphroditica L. subsp. bifida (L.) Schinz et Hell. ■☆

67382 Callitriche palustris L. var. elegans (Petr.) Y. L. Chang; 东北水马齿(雅致水马齿); Elegant Waterstarwort ■

67383 Callitriche palustris L. var. elegans (Petr.) Y. L. Chang = Callitriche palustris L. ■

67384 Callitriche palustris L. var. oryzetorum (Petr.) Lansdown; 广东水马齿; Guangdong Waterstarwort, Kwangtung Waterstarwort ■

67385 Callitriche palustris L. var. verna (L.) Fassett = Callitriche palustris L. ■
67386 Callitriche palustris L. var. verna (L.) Fenley ex Jeps. = Callitriche palustris L. ■
67387 Callitriche pedunculata DC. = Callitriche brutia Petagna ■☆
67388 Callitriche peploides Nutt.;凹果水马齿(台湾水马齿)■
67389 Callitriche platycarpa Kütz.;宽果水马齿(多变水马齿);Common Water-starwort,Various-leaved Water-starwort ■
67390 Callitriche polymorpha E. J. Lönnr. = Callitriche platycarpa Kütz. ■
67391 Callitriche pulchra Schotsman;美丽水马齿■☆
67392 Callitriche raveniana Lansdown;细苞水马齿■
67393 Callitriche stagnalis Scop.;水马齿;Common Starwort, Common Water-starwort, Pond Water Starwort, Pond Waterstarwort, Pond Water-starwort,Starwort,Water Starwort,Water-starwort ■
67394 Callitriche subanceps Petr.;两栖水马齿■☆
67395 Callitriche terrestris Raf.;陆生水马齿;Terrestrial Starwort, Terrestrial Water-starwort ■☆
67396 Callitriche truncata Guss.;短叶水马齿;Short-leaved Water-starwort ■☆
67397 Callitriche verna L. = Callitriche palustris L. ■
67398 Callitriche verna L. subsp. elegans Petr. ex Kom. et Aliss. = Callitriche palustris L. var. elegans (Petr.) Y. L. Chang ■
67399 Callitriche verna L. var. elegans (Petr.) Kitag. = Callitriche palustris L. var. elegans (Petr.) Y. L. Chang ■
67400 Callitriche verna L. var. fallax (Petr.) H. Hara = Callitriche palustris L. ■
67401 Callitriche vernalis Koch;春水马齿;Spring Waterstarwort ■
67402 Callitriche vulcanicola Schotsman;火山水马齿■☆
67403 Callitriche wightiana Wall. = Callitriche stagnalis Scop. ■
67404 Callitriche wightiana Wall. ex Wight et Arn. = Callitriche stagnalis Scop. ■
67405 Callitris Vent. (1808);澳洲柏属(澳柏属,美丽柏属);Australian Cypress Pine,Callitris,Cypress Pine,Cypress-pine ●
67406 Callitris arborea Schrad. ex E. Mey.;乔木澳洲柏●☆
67407 Callitris arenosa Sweet;多沙澳柏;Queenslamd Cypress-pine ●☆
67408 Callitris articulata (Vahl) Link = Tetraclinis articulata (Vahl) Mast. ●☆
67409 Callitris articulata (Vahl) Link var. glaucostrobilea Sennen = Tetraclinis articulata (Vahl) Mast. ●☆
67410 Callitris baileyi C. T. White;贝利澳洲柏(贝利美丽柏)●☆
67411 Callitris calcarata R. Br. ex Miers;马刺苞澳洲柏(有距澳柏);Black Pine ●☆
67412 Callitris calcarata R. Br. ex Miers = Callitris endlicheri (Parl.) F. M. Bailey ●☆
67413 Callitris canescens (Parl.) S. T. Blake;灰叶澳洲柏(灰叶美丽柏)●☆
67414 Callitris columellaris F. Muell.;柱状澳洲柏(粉叶澳洲柏,小柱美丽柏,柱状美丽柏);Bribie Island Pine,Sand Cypress Pine,White Cypress Pine,White Cypress-pine ●☆
67415 Callitris columellaris F. Muell. var. campestris Silba = Callitris glaucophylla J. Thomps. et L. A. S. Johnson ●☆
67416 Callitris drummondii Benth. et Hook. f.;德拉蒙德澳洲柏(德鲁美丽柏);Drummond's Cypress Pine ●☆
67417 Callitris endlicheri (Parl.) F. M. Bailey;恩氏澳洲柏(恩得利美丽柏,恩氏美丽柏,南威尔士柏);Australian Sandarac,Black Cypress,Black Cypress Pine,Endlicher Callitris,Red Cypress Pine ●☆
67418 Callitris glaucophylla J. Thomps. et L. A. S. Johnson;粉绿叶美丽柏;White Cypress Pine,White Cypress-pine ●☆
67419 Callitris gracilis R. T. Baker;山地澳洲柏(纤细澳柏,纤细澳洲柏);Moutain Cypress Pine ●☆
67420 Callitris hugelii (Carrière) Franco = Callitris glaucophylla J. Thomps. et L. A. S. Johnson ●☆
67421 Callitris intratropica R. T. Baker et H. G. Sm.;热带澳洲柏(热带美丽柏);Northern Cypress Pine ●☆
67422 Callitris macleayana (F. Muell.) F. Muell.;麦夸里澳洲柏(麦加利澳柏,条裂美丽柏);Port Macquarie Pine,Stringybark Cypress Pine ●☆
67423 Callitris morrisoni R. T. Baker;西南澳洲柏;Morrison's Cypress Pine ●☆
67424 Callitris muelleri (Parl.) F. Muell.;昆士兰澳洲柏(昆士兰柏,马勒美丽柏);Mueller Callitris,Mueller's Cypress Pine ●☆
67425 Callitris neocaledonica Dümmer;新喀澳洲柏(新喀里多尼亚柏,新卡利登柏);New Caledonia Cypress Pine ●☆
67426 Callitris oblonga Rich.;矩圆澳洲柏(塔斯马尼亚柏,圆柱美丽柏);South Esk Pine,Tasmanian Cypress Pine,Tasmanian Cypress-pine ●☆
67427 Callitris preissii Miq.;普氏澳洲柏(布勒斯美丽柏);Preiss Callitris,Preiss Cypress Pine ●☆
67428 Callitris preissii Miq. subsp. verrucosa (A. Cunn. ex Endl.) J. Garden = Callitris verrucosa (Endl.) F. Muell. ●☆
67429 Callitris preissii Miq. subsp. verrucosa (Endl.) J. Garden = Callitris verrucosa (Endl.) F. Muell. ●☆
67430 Callitris preissii Miq. var. verrucosa (A. Cunn. ex Endl.) Silba = Callitris verrucosa (Endl.) F. Muell. ●☆
67431 Callitris quadrivalvis Vent. = Tetraclinis articulata Mast. ●☆
67432 Callitris rhomboidea R. Br. ex A. Rich. et Rich.;菱苞澳洲柏(菱形美丽柏);Cypress-pine, Drooping Cypress-pine, Illawara Pine, Oyster Bay Pine,Oyster Flay Pine,Port Jackson Cypress,Port Jackson Cypress Pine,Port Jackson Pine ●
67433 Callitris robusta R. Br. ex Mirb.;粗枝澳洲柏(粗壮澳柏);Common Cypress-pine, Cypress Pine, Great Murray Pine, Rottnest Island Pine,Sturdy Cypress-pine ●
67434 Callitris verrucosa (Endl.) F. Muell.;多疣澳洲柏(多疣美丽柏);Mallee Cypress Pine,Scrub Cypress-pine ●☆
67435 Callitropsis Compton = Chamaecyparis Spach ●
67436 Callitropsis Compton = Neocallitropsis Florin ●☆
67437 Callitropsis Oerst. = Chamaecyparis Spach ●
67438 Callitropsis benthamii (Endl.) D. P. Little = Cupressus benthamii Endl. ●
67439 Callixene Comm. ex Juss.(废弃属名) = Luzuriaga Ruiz et Pav.(保留属名)■☆
67440 Callobuxus Panch. ex Brongn. et Gris = Tristania R. Br. ●
67441 Calloglossum Schltr. = Cymbidiella Rolfe ■☆
67442 Callopisma Mart. = Deianira Cham. et Schltdl. ■☆
67443 Callopsis Engl. (1895);拟水芋属■☆
67444 Callopsis hallaei Bogner = Nephthytis hallaei (Bogner) Bogner ■☆
67445 Callopsis volkensii Engl.;拟水芋■☆
67446 Callosmia C. Presl = Anneslea Wall.(保留属名)●
67447 Callosmia fragrans (Wall.) Presl = Anneslea fragrans Wall. ●
67448 Callostylis Blume = Eria Lindl.(保留属名)■
67449 Callostylis Blume(1825);美柱兰属;Beautystyle ■
67450 Callostylis bambusifolia (Lindl.) S. C. Chen et J. J. Wood;竹叶美柱兰■
67451 Callostylis rigida Blume;美柱兰;Beautystyle ■

67452 Callostylis rigida Blume subsp. discolor (Lindl.) Brieger = Callostylis rigida Blume ■
67453 Callothlaspi F. K. Mey. = Thlaspi L. ■
67454 Callotropis G. Don = Galega L. ■
67455 Calluna Salisb. (1802); 帚石南属(佳萝属); Heather, Ling, Summer Heather ●☆
67456 Calluna vulgaris (L.) Hill; 帚石南(佳萝,四方柏); Basam, Bassam, Bassom, Bend, Bent, Besom, Broom, Common Heather, Crow Ling, Dog Heather, Grig, Griglans, Griglings, Griglum, Hadder, Haddyr, Hather, He Heather, Heath, Heather, Hedder, He-heather, Honeybottle, Ling, Ling Heather, Long Heath, Moor, Mountain Mist, Red Heath, Red Ling, Satin-balls, Scotch Heather, Scottish Heather, Small Heath, White Heather, Yeth ●☆
67457 Calluna vulgaris (L.) Hill 'Alba Plena'; 白花重瓣帚石南(白花瓣帚石南)●☆
67458 Calluna vulgaris (L.) Hill 'Alexandra'; 亚历山大帚石南●☆
67459 Calluna vulgaris (L.) Hill 'Alicia'; 阿莉西亚帚石南●☆
67460 Calluna vulgaris (L.) Hill 'Allegro'; 阿莱格罗帚石南●☆
67461 Calluna vulgaris (L.) Hill 'Alportii'; 阿伯特帚石南●☆
67462 Calluna vulgaris (L.) Hill 'Anette'; 阿内塔帚石南●☆
67463 Calluna vulgaris (L.) Hill 'Annemarie'; 安妮玛丽帚石南●☆
67464 Calluna vulgaris (L.) Hill 'Anthony Davis'; 安东尼·戴维斯帚石南●☆
67465 Calluna vulgaris (L.) Hill 'Beoly Gold'; 倍莱金叶帚石南(贝奥尼金叶帚石南)●☆
67466 Calluna vulgaris (L.) Hill 'Beoly Silver'; 倍莱银叶帚石南(贝奥尼银叶帚石南)●☆
67467 Calluna vulgaris (L.) Hill 'Blazeaway'; 火焰帚石南●☆
67468 Calluna vulgaris (L.) Hill 'Bonfire Brilliance'; 篝火辉煌帚石南●☆
67469 Calluna vulgaris (L.) Hill 'Boskoop'; 博丝库普帚石南●☆
67470 Calluna vulgaris (L.) Hill 'County Wicklow'; 半匍匐帚石南(威克罗锦帚石南)●☆
67471 Calluna vulgaris (L.) Hill 'Dark Beauty'; 黑美人帚石南●☆
67472 Calluna vulgaris (L.) Hill 'Darkness'; 黑红帚石南(暗色帚石南)●☆
67473 Calluna vulgaris (L.) Hill 'Elsie Purnell'; 爱尔西·普勒帚石南●☆
67474 Calluna vulgaris (L.) Hill 'Finale'; 终曲帚石南●☆
67475 Calluna vulgaris (L.) Hill 'Firefly'; 萤火虫帚石南●☆
67476 Calluna vulgaris (L.) Hill 'Foxii Nana'; 福克斯矮种帚石南●☆
67477 Calluna vulgaris (L.) Hill 'Fred J. Chapple'; 弗雷德·查普尔帚石南●☆
67478 Calluna vulgaris (L.) Hill 'Gold Haze'; 金叶帚石南(金雾帚石南)●☆
67479 Calluna vulgaris (L.) Hill 'Golden Feather'; 金羽帚石南●☆
67480 Calluna vulgaris (L.) Hill 'H. E. Beele'; 比尔帚石南●☆
67481 Calluna vulgaris (L.) Hill 'Hammondii Aureifolia'; 哈蒙德黄叶帚石南●☆
67482 Calluna vulgaris (L.) Hill 'Hibernica'; 爱尔兰帚石南●☆
67483 Calluna vulgaris (L.) Hill 'J. H. Hamilton'; 哈密尔顿帚石南●☆
67484 Calluna vulgaris (L.) Hill 'Joy Vanstone'; 乔伊·范斯通帚石南●☆
67485 Calluna vulgaris (L.) Hill 'Kerstin'; 克斯汀帚石南●☆
67486 Calluna vulgaris (L.) Hill 'Kinlochruel'; 加勒克鲁尔帚石南(金诺赫瑞帚石南)●☆
67487 Calluna vulgaris (L.) Hill 'Loch Turret'; 海湾炮塔帚石南●☆
67488 Calluna vulgaris (L.) Hill 'Mair's Variety'; 迈尔品种帚石南●☆
67489 Calluna vulgaris (L.) Hill 'Marleen'; 马尔里恩帚石南●☆
67490 Calluna vulgaris (L.) Hill 'Mullion'; 放射帚石南●☆
67491 Calluna vulgaris (L.) Hill 'Multicolor'; 多色帚石南(多彩帚石南)●☆
67492 Calluna vulgaris (L.) Hill 'My Dream'; 梦境帚石南●☆
67493 Calluna vulgaris (L.) Hill 'My Dream' = Calluna vulgaris (L.) Hill 'Snowball'●☆
67494 Calluna vulgaris (L.) Hill 'Peter Sparkes'; 彼得·斯巴克斯帚石南●☆
67495 Calluna vulgaris (L.) Hill 'Rica'; 里卡帚石南●☆
67496 Calluna vulgaris (L.) Hill 'Robert Chapman'; 金叶紫花帚石南(罗伯特·查普曼帚石南)●☆
67497 Calluna vulgaris (L.) Hill 'Ruth Sparkes'; 路斯·斯巴克斯帚石南●☆
67498 Calluna vulgaris (L.) Hill 'Silver Knight'; 银骑士帚石南●☆
67499 Calluna vulgaris (L.) Hill 'Silver Queen'; 银后帚石南; Summer Heather ●☆
67500 Calluna vulgaris (L.) Hill 'Sir John Charrington'; 约翰·查林顿帚石南●☆
67501 Calluna vulgaris (L.) Hill 'Sister Anne'; 安尼姐帚石南●☆
67502 Calluna vulgaris (L.) Hill 'Snowball'; 雪球帚石南●☆
67503 Calluna vulgaris (L.) Hill 'Snowkall' = Calluna vulgaris (L.) Hill 'My Dream'●☆
67504 Calluna vulgaris (L.) Hill 'Spring Cream'; 春乳白帚石南●☆
67505 Calluna vulgaris (L.) Hill 'Spring Torch'; 春火把帚石南(火焰帚石南)●☆
67506 Calluna vulgaris (L.) Hill 'Sunset'; 夕阳帚石南●☆
67507 Calluna vulgaris (L.) Hill 'Tib'; 王牌帚石南●☆
67508 Calluna vulgaris (L.) Hill 'White Lawn'; 白草地帚石南●☆
67509 Calluna vulgaris (L.) Hill 'Wickwar Flame'; 威克沃火焰帚石南●☆
67510 Calluna vulgaris Salisb. = Calluna vulgaris (L.) Hill ●☆
67511 Callyntranthele Nied. = Byrsonima Rich. ex Juss. ●☆
67512 Calobota Eckl. et Zeyh. = Lebeckia Thunb. ■☆
67513 Calobotrya Spach = Ribes L. ●
67514 Calobuxus T. Post et Kuntze = Callobuxus Panch. ex Brongn. et Gris ●
67515 Calobuxus T. Post et Kuntze = Tristania R. Br. ●
67516 Calocapnos Spach = Corydalis DC. (保留属名)■
67517 Calocapnos nobilis (L.) Spach = Corydalis nobilis (L.) Pers. ■
67518 Calocapnos nobilis (L.) Spach = Corydalis solida (L.) Clairv. ■
67519 Calocapnos nobilis Spach = Corydalis nobilis (L.) Pers. ■
67520 Calocarpum Pierre = Calospermum Pierre ●
67521 Calocarpum Pierre = Pouteria Aubl. ●
67522 Calocarpum Pierre ex Engl. (1904); 美果榄属(美果山榄属)●☆
67523 Calocarpum Pierre ex Engl. = Pouteria Aubl. ●
67524 Calocarpum sapota (Jacq.) Merr.; 美果榄(美果山榄木); Mamee-sapote, Mammee Sapote, Mammey, Mammey Sapote, Marmalade Plum, Sapodilla, Sapote ●☆
67525 Calocarpus Post et Kuntze = Callicarpa L. ●
67526 Calocedrus Kurz (1873); 翠柏属(肖楠属); Incense Cedar, Nothern Incense Cedar ●
67527 Calocedrus decurrens (Torr.) Florin = Libocedrus decurrens Torr. ●☆
67528 Calocedrus formosana (Florin) Florin = Calocedrus macrolepis

Kurz var. formosana (Florin) W. C. Cheng et L. K. Fu ●◇
67529 Calocedrus macrolepis Kurz;翠柏(长柄翠柏,大鳞肖楠,黄肉树,梢楠,肖楠);China Incense Cedar,Chinese Incense Cedar ●◇
67530 Calocedrus macrolepis Kurz var. formosana (Florin) W. C. Cheng et L. K. Fu;台湾翠柏(黄肉树,黄肉仔,台湾肖楠,肖楠,肖楠木);Taiwan Incense Cedar,Taiwan Incense-cedar ●◇
67531 Calocedrus macrolepis Kurz var. longipes W. C. Cheng et L. K. Fu = Calocedrus macrolepis Kurz ●◇
67532 Calocephalus R. Br. (1817);美头菊属■●☆
67533 Calocephalus brownii (Cass.) F. Muell.;布氏美头菊;Cushion Bush ●☆
67534 Calocephalus brownii F. Muell. = Calocephalus brownii (Cass.) F. Muell. ●☆
67535 Calocephalus brownii F. Muell. = Leucophyta brownii Cass. ●☆
67536 Calocephalus drummondii (A. Gray) Benth.;美头菊●☆
67537 Calocephalus drummondii Benth. = Calocephalus drummondii (A. Gray) Benth. ●☆
67538 Calocephalus multiflorus (Turcz.) Benth.;多花美头菊●☆
67539 Calocephalus multiflorus Benth. = Calocephalus multiflorus (Turcz.) Benth. ●☆
67540 Calochilus Post et Kuntze = Callichilia Stapf ●☆
67541 Calochilus R. Br. (1810);卡洛基兰属;Calochilus ■☆
67542 Calochilus campestris R. Br.;卡洛基兰;Common Calochilus, Copper Beard ■☆
67543 Calochilus robertsonii Benth.;罗氏卡洛基兰;Beard Orchid ■☆
67544 Calochlamys C. Presl = Congea Roxb. ●
67545 Calochlamys Post et Kuntze = Callichlamys Miq. ●☆
67546 Calochloa Kunze = Elionurus Humb. et Bonpl. ex Willd.(保留属名)■☆
67547 Calochloa Post et Kuntze = Callichloe Willd. ex Steud. ■☆
67548 Calochloa Post et Kuntze = Elionurus Humb. et Bonpl. ex Willd. (保留属名)■☆
67549 Calochone Keay(1958);丽蔓属●☆
67550 Calochone acuminata Keay;尖叶丽蔓●☆
67551 Calochone redingii (De Wild.) Keay;丽蔓●☆
67552 Calochortaceae Dumort. (1829);美莲草科(油点草科,裂果草科)■
67553 Calochortaceae Dumort. = Liliaceae Juss.(保留科名)■●
67554 Calochortus Pursh(1814);美莲草属(蝶花百合属,丽草属,仙灯属,油点草属);Butterfly Tulip, Butterfly-lily, Cat's-ears, Fairy Lantern, Globe Tulip, Globe-tulip, Mariposa, Mariposa Lily, Mariposa Tulip, Mariposa-lily, Star Tulip ■☆
67555 Calochortus albus (Benth.) Douglas ex Benth.;仙灯(白花仙灯);Fairy Lantern, Fairy Lanterns, Globe Lily, White Globe-lily, White Globe-tulip ■☆
67556 Calochortus albus (Benth.) Douglas ex Benth. var. paniculatus (Lindl.) Baker;圆锥仙灯■☆
67557 Calochortus albus (Benth.) Douglas ex Benth. var. rubellus Greene = Calochortus albus (Benth.) Douglas ex Benth. ■☆
67558 Calochortus albus Douglas = Calochortus albus (Benth.) Douglas ex Benth. ■☆
67559 Calochortus albus Douglas ex Benth. = Calochortus albus (Benth.) Douglas ex Benth. ■☆
67560 Calochortus amabilis Purdy;金灯(爱丽草,金仙灯);Diogenes' Lantern, Golden Fairy Lantern, Golden Fairy Lanterns, Golden Globe Tulip, Golden Globe-tulip ■☆
67561 Calochortus ambiguus (M. E. Jones) Ownbey;可疑仙灯■☆
67562 Calochortus amoenus Greene;紫花仙灯;Purple Globe-tulip ■☆
67563 Calochortus argillosus (Hoover) Zebell et P. L. Fiedler;白土仙灯■☆
67564 Calochortus aureus S. Watson;黄仙灯■☆
67565 Calochortus barbatus (Kunth) J. H. Painter;须毛仙灯■☆
67566 Calochortus benthamii Baker;本瑟姆仙灯(拜氏仙灯);Yellow Star-tulip ■☆
67567 Calochortus bruneaunis A. Nelson et J. F. Macbr.;布朗蝶花百合■☆
67568 Calochortus caeruleus S. Watson;毛瓣仙灯;Hairypetal Fairy Lantern ■☆
67569 Calochortus catalinae S. Watson;卡他里那蝶花百合;Catalina Mariposa-lily ■☆
67570 Calochortus clavatus S. Watson;棒毛仙灯;Clavatehair Fairy Lantern, Yellow Mariposa Lily ■☆
67571 Calochortus clavatus S. Watson var. avius Jeps.;愉悦棒毛仙灯;Pleasant Valley Mariposa-lily ■☆
67572 Calochortus clavatus S. Watson var. gracilis Ownbey;纤细棒毛仙灯;Slender Mariposa-lily ■☆
67573 Calochortus clavatus S. Watson var. pallidus (Hoover) P. L. Fiedler et Zebell;苍白棒毛仙灯■☆
67574 Calochortus clavatus S. Watson var. recurvifolius (Hoover) P. L. Fiedler et Zebell;卷叶棒毛仙灯■☆
67575 Calochortus coeruleus (Kellogg) S. Watson;蓝蝶花百合;Beavertail-grass ■☆
67576 Calochortus coeruleus (Kellogg) S. Watson var. fimbriatus Ownbey = Calochortus coeruleus (Kellogg) S. Watson ■☆
67577 Calochortus coeruleus (Kellogg) S. Watson var. westonii (Eastw.) Ownbey = Calochortus westonii Eastw. ■☆
67578 Calochortus concolor (Baker) Purdy;同色蝶花百合;Goldenbowl Mariposa-lily ■☆
67579 Calochortus coxii M. R. Godfrey et Callahan;考克斯蝶花百合;Cox's Mariposa-lily ■☆
67580 Calochortus davidsonianus Abrams = Calochortus splendens Douglas ex Benth. ■☆
67581 Calochortus dunnii Purdy;邓恩蝶花百合;Dunn's Mariposa-lily ■☆
67582 Calochortus elegans Pursh;猫耳蝶花百合;Cat's Ear ■☆
67583 Calochortus elegans Pursh var. nanus A. W. Wood;小猫耳蝶花百合■☆
67584 Calochortus elegans Pursh var. oreophilis Ownbey = Calochortus elegans Pursh var. nanus A. W. Wood ■☆
67585 Calochortus elegans Pursh var. selwayensis (H. St. John) Ownbey;塞地猫耳蝶花百合■☆
67586 Calochortus eurycarpus S. Watson;蝶花百合■☆
67587 Calochortus flexuosus S. Watson;卷曲仙灯;Straggling Mariposa ■☆
67588 Calochortus greenei S. Watson;格林蝶花百合;Greene's Mariposa-lily ■☆
67589 Calochortus kennedyi Porter;肯氏仙灯;Desert Mariposa, Desert Mariposa Lily ■☆
67590 Calochortus leichtlinii Hook. f.;莱氏蝶花百合■☆
67591 Calochortus lilacinus Kellogg = Calochortus uniflorus Hook. et Arn. ■☆
67592 Calochortus lobbii (Baker) Purdy = Calochortus subalpinus Piper ■☆
67593 Calochortus longebarbatus S. Watson;长毛蝶花百合;Long-haired Star-tulip ■☆
67594 Calochortus luteus Douglas ex Kunth = Calochortus luteus

Douglas ex Lindl. ■☆

67595 Calochortus luteus Douglas ex Lindl.;纤毛仙灯(仙灯);Slender Fairy Lantern, Yellow Mariposa Lily, Yellow Mariposa Tulip ■☆

67596 Calochortus luteus Douglas ex Lindl. var. citrinus (Baker) S. Watson;橘色纤毛仙灯■☆

67597 Calochortus luteus Douglas ex Lindl. var. concolor Baker = Calochortus concolor (Baker) Purdy ■☆

67598 Calochortus luteus Douglas ex Lindl. var. oculatus S. Watson;斑纹纤毛仙灯■☆

67599 Calochortus luteus Douglas ex Lindl. var. robustus Purdy;粗壮纤毛仙灯■☆

67600 Calochortus luteus Douglas ex Lindl. var. weedii (A. W. Wood) Baker = Calochortus weedii A. W. Wood ■☆

67601 Calochortus macrocarpus Douglas;大果仙灯;Green-banded Mariposa-lily, Largefruit Fairy Lantern ■☆

67602 Calochortus macrocarpus Douglas var. maculosus (A. Nelson et J. F. Macbr.) A. Nelson et J. F. Macbr.;斑点大果仙灯■☆

67603 Calochortus maculosus A. Nelson et J. F. Macbr. = Calochortus macrocarpus Douglas var. maculosus (A. Nelson et J. F. Macbr.) A. Nelson et J. F. Macbr. ■☆

67604 Calochortus maweanus Leichtlin ex Baker = Calochortus coeruleus (Kellogg) S. Watson ■☆

67605 Calochortus monanthus Ownbey;单花蝶花百合;Single-flowered Mariposa-lily ■☆

67606 Calochortus monophyllus (Lindl.) Lem.;单叶蝶花百合(单叶仙灯);Yellow Star-tulip ■☆

67607 Calochortus monophyllus Lem. = Calochortus monophyllus (Lindl.) Lem. ■☆

67608 Calochortus nitidus Douglas;无毛仙灯;Hairless Fairy Lantern, Shasta Star-tulip ■☆

67609 Calochortus nitidus Douglas var. eurycarpus L. F. Hend. = Calochortus eurycarpus S. Watson ■☆

67610 Calochortus nudus Douglas var. shastensis (Purdy) Jeps. = Calochortus nitidus Douglas ■☆

67611 Calochortus nuttallii Torr.;纳托尔蝶花百合;Sego-lily ■☆

67612 Calochortus nuttallii Torr. var. aureus (S. Watson) Ownbey = Calochortus aureus S. Watson ■☆

67613 Calochortus nuttallii Torr. var. bruneaunis (A. Nelson et J. F. Macbr.) Ownbey = Calochortus bruneaunis A. Nelson et J. F. Macbr. ■☆

67614 Calochortus nuttallii Torr. var. panamintensis Ownbey = Calochortus panamintensis (Ownbey) Reveal ■☆

67615 Calochortus obispoensis Lemmon;奥维斯波苔草;San Luis Mariposa-lily ■☆

67616 Calochortus palmeri S. Watson;帕默蝶花百合;Palmer's Mariposa-lily ■☆

67617 Calochortus palmeri S. Watson var. dunnii (Purdy) Jeps. et Ames = Calochortus dunnii Purdy ■☆

67618 Calochortus palmeri S. Watson var. munzii Ownbey;蒙茨蝶花百合;Munz's Mariposa-lily ■☆

67619 Calochortus panamintensis (Ownbey) Reveal;帕地苔草;Panamint Mariposa-lily ■☆

67620 Calochortus persistens Ownbey;锡斯基尤蝶花百合;Siskiyou Mariposa-lily ■☆

67621 Calochortus plummerae Greene;普卢默蝶花百合;Plummer's Mariposa-lily ■☆

67622 Calochortus pulchellus (Benth.) A. W. Wood;星状蝶花百合;Star Tulip ■☆

67623 Calochortus raichei Farwig et V. Girard;拉氏蝶花百合;Cedars Fairy Lantern ■☆

67624 Calochortus selwayensis H. St. John = Calochortus elegans Pursh var. selwayensis (H. St. John) Ownbey ■☆

67625 Calochortus shastensis Purdy = Calochortus nitidus Douglas ■☆

67626 Calochortus simulans (Hoover) Munz;相似蝶花百合;San Luis Obispo Mariposa-lily ■☆

67627 Calochortus splendens Douglas ex Benth.;辉花仙灯;Splendid mariposa Lily ■☆

67628 Calochortus striatus Parish;碱地蝶花百合;Alkali Mariposa-lily ■☆

67629 Calochortus subalpinus Piper;亚高山蝶花百合■☆

67630 Calochortus superbus Purdy ex J. T. Howell;华丽仙灯;Superb Fairy Lantern ■☆

67631 Calochortus tolmiei Hook. et Arn.;托尔蝶花百合;Pussy Ears ■☆

67632 Calochortus umbellatus A. W. Wood;奥克兰蝶花百合;Oakland Star-tulip ■☆

67633 Calochortus umpquaensis Fredr.;尤地蝶花百合;Umpqua Mariposa-lily ■☆

67634 Calochortus uniflorus Hook. et Arn.;单花仙灯;Butterfly Tulip, Large-flowered Star-tulip, Mariposa Lily ■☆

67635 Calochortus venustus Douglas ex Benth.;异色仙灯(挺拔仙灯);Butterfly Mariposa, Butterfly Tulip, Mariposa Lily, Pretty Fairy Lantern, White Mariposa Lily, White Mariposa-lily ■☆

67636 Calochortus venustus Douglas ex Benth. var. brachysepalus Regel;短瓣异色仙灯■☆

67637 Calochortus venustus Douglas ex Benth. var. citrinus Baker;橘色异色仙灯■☆

67638 Calochortus venustus Douglas ex Benth. var. eldorado Purdy;黄金乡异色仙灯■☆

67639 Calochortus venustus Douglas ex Benth. var. purpurascens Purdy;紫色异色仙灯■☆

67640 Calochortus venustus Douglas ex Benth. var. robustus Hort;粗壮异色仙灯■☆

67641 Calochortus venustus Douglas ex Benth. var. roseus Reuthe;粉红异色仙灯■☆

67642 Calochortus venustus Douglas ex Benth. var. rubra Purdy;红异色仙灯■☆

67643 Calochortus venustus Douglas ex Benth. var. sulphureus Purdy;斑花异色仙灯■☆

67644 Calochortus vesta Purdy;大华丽仙灯(锈点仙灯);Vesta Fairy Lantern ■☆

67645 Calochortus watsonii M. E. Jones var. ambiguus M. E. Jones = Calochortus ambiguus (M. E. Jones) Ownbey ■☆

67646 Calochortus weedii A. W. Wood;橙花仙灯■☆

67647 Calochortus weedii A. W. Wood var. purpurascens S. Watson = Calochortus plummerae Greene ■☆

67648 Calochortus westonii Eastw.;雪莉仙灯;Mariposa-lily, Shirley Meadows Star-tulip ■☆

67649 Calochroa Post et Kuntze = Callichroa Fisch. et C. A. Mey. ■☆

67650 Calochroa Post et Kuntze = Layia Hook. et Arn. ex DC. (保留属名) ■☆

67651 Calococca Post et Kuntze = Callicocca Schreb. ●

67652 Calococca Post et Kuntze = Cephaelis Sw. (保留属名) ●

67653 Calococcus Kurz ex Teijsm. = Margaritaria L. f. ●

67654 Calococcus Kurz ex Teijsm. et Binnend. = Margaritaria L. f. ●

67655 Calocoma T. Post et Kuntze = Callicoma Andréws ●☆

67656 Calocorema Post et Kuntze = Calicorema Hook. f. ●☆
67657 Calocornia Post et Kuntze = Callicornia Burm. f. ●☆
67658 Calocornia Post et Kuntze = Leysera L. ■●☆
67659 Calocrater K. Schum. (1895);丽杯夹竹桃属●☆
67660 Calocrater preussii K. Schum.;丽杯夹竹桃●☆
67661 Calocysthus Post et Kuntze = Callicysthus Endl. ■
67662 Calocysthus Post et Kuntze = Vigna Savi(保留属名)■
67663 Calodecaryia J. -F. Leroy(1960);马达加斯加楝属●☆
67664 Calodecaryia crassifolia J. -F. Leroy;厚叶马达加斯加楝●☆
67665 Calodecaryia pauciflora J. -F. Leroy;少花马达加斯加楝●☆
67666 Calodendrum Thunb. (1782)(保留属名);丽芸木属(好望角美树属,卡罗树属,美木芸香属);Cape Chestnut ●☆
67667 Calodendrum capensis Thunb.;丽芸木(好望角卡罗树,好望角美木芸香,卡罗树);Cape Chestnut,Wild Chestnut ●☆
67668 Calodendrum eickii Engl.;爱克丽芸木●☆
67669 Calodium Lour. = Cassytha L. ■●
67670 Calodonta Nutt. = Tolpis Adans. ●■☆
67671 Calodracon Planch. = Cordyline Comm. ex R. Br. (保留属名)●
67672 Calodryum Desv. = Quivisia Comm. ex Juss. ●
67673 Calodryum Desv. = Turraea L. ●
67674 Caloglossa Post et Kuntze = Calliglossa Hook. et Arn. ■☆
67675 Caloglossa Post et Kuntze = Layia Hook. et Arn. ex DC. (保留属名)■☆
67676 Caloglossum Schltr. = Cymbidiella Rolfe ■☆
67677 Caloglossum flabellatum (Thouars) Schltr. = Cymbidiella flabellata (Thouars) Rolfe ■☆
67678 Caloglossum humblotii (Rolfe) Schltr. = Cymbidiella falcigera (Rchb. f.) Garay ■☆
67679 Caloglossum magnificum Schltr. = Cymbidiella falcigera (Rchb. f.) Garay ■☆
67680 Caloglossum rhodochilum (Rolfe) Schltr. = Cymbidiella pardalina (Rchb. f.) Garay ■☆
67681 Calogonum Post et Kuntze = Calligonum L. ●
67682 Calogonum Post et Kuntze = Tetracera L. ●
67683 Calographis Thouars = Eulophidium Pfitzer ■☆
67684 Calographis Thouars = Limodorum Boehm. (保留属名)■☆
67685 Calogyna T. Post et Kuntze = Calogyne R. Br. ■
67686 Calogyne R. Br. (1810);离根香属(离根菜属,美柱草属,美柱兰属);Calogyne ■
67687 Calogyne R. Br. = Goodenia Sm. ●■☆
67688 Calogyne chinensis Benth. = Calogyne pilosa R. Br. subsp. chinensis (Benth.) H. S. Kiu ■
67689 Calogyne pilosa R. Br.;离根香(根风藤,利根香,毛离根香,美花草,美柱草,肉桂草,山茼蒿,山茵蒿);Pilose Calogyne ■
67690 Calogyne pilosa R. Br. subsp. chinensis (Benth.) H. S. Kiu;华离根香(离根香,美柱草);Chinese Calogyne ■
67691 Calolepis T. Post et Kuntze = Callilepis DC. ■●☆
67692 Calolinea macrocarpa Cham. et Schltdl. = Pachira macrocarpa (Schltdl. et Champ.) Walp. ●
67693 Calolisianthus Gilg = Irlbachia Mart. ■☆
67694 Calomecon Spach = Papaver L. ■
67695 Calomeria Vent. (1804);香木菊属(苋菊属,香木属)■●☆
67696 Calomeria Vent. = Humea Sm. ●☆
67697 Calomeria africana (S. Moore) Heine;非洲香木菊(非洲香木)●☆
67698 Calomeria amaranthoides Vent.;香木菊(苋菊,香木);Fountain Humea,Humea,Incense Bush,Incense Plant ●☆
67699 Calomeria elegans? = Calomeria amaranthoides Vent. ●☆
67700 Calomicta Post et Kuntze = Actinidia Lindl. ●
67701 Calomicta Post et Kuntze = Kolomikta Regel ●
67702 Calomorphe Kuntze ex Walp. = Lennea Klotzsch ■☆
67703 Calomyrtus Blume = Myrtus L. ●
67704 Caloncoba Gilg(1908);卡洛木属;Caloncoba ●☆
67705 Caloncoba angolensis Exell et Sleumer;安哥拉卡洛木(安哥拉大风子)●☆
67706 Caloncoba angolensis Exell et Sleumer = Oncoba suffruticosa (Milne-Redh.) Hul et Breteler ●☆
67707 Caloncoba aristata (Oliv.) Gilg = Oncoba mannii Oliv. ●☆
67708 Caloncoba brevipes (Stapf) Gilg = Oncoba brevipes Stapf ●☆
67709 Caloncoba brevipes Gilg = Oncoba brevipes Stapf ●☆
67710 Caloncoba cauliflora Sleumer = Oncoba welwitschii Oliv. ●☆
67711 Caloncoba crepiniana (De Wild. et T. Durand) Gilg = Oncoba crepiniana De Wild. et T. Durand ●☆
67712 Caloncoba dusenii Gilg = Oncoba glauca (P. Beauv.) Planch. ●☆
67713 Caloncoba echinata (Oliv.) Gilg = Oncoba echinata Oliv. ●☆
67714 Caloncoba echinata Gilg = Oncoba echinata Oliv. ●☆
67715 Caloncoba flagelliflora (Mildbr.) Gilg ex Pellegr. = Oncoba flagelliflora (Mildbr.) Hul ●☆
67716 Caloncoba gigantocarpa Perkins et Gilg = Oncoba welwitschii Oliv. ●☆
67717 Caloncoba gilgiana (Sprague) Gilg = Oncoba gilgiana Sprague ●☆
67718 Caloncoba glauca (P. Beauv.) Gilg = Oncoba glauca (P. Beauv.) Planch. ●☆
67719 Caloncoba glauca Gilg = Oncoba glauca (P. Beauv.) Planch. ●☆
67720 Caloncoba grotei Gilg ex Engl. = Oncoba welwitschii Oliv. ●☆
67721 Caloncoba longipetiolata Gilg = Oncoba crepiniana De Wild. et T. Durand ●☆
67722 Caloncoba lophocarpa (Oliv.) Gilg = Oncoba lophocarpa Oliv. ●☆
67723 Caloncoba mannii (Oliv.) Gilg = Oncoba mannii Oliv. ●☆
67724 Caloncoba schweinfurthii Gilg = Oncoba crepiniana De Wild. et T. Durand ●☆
67725 Caloncoba subtomentosa Gilg = Oncoba subtomentosa (Gilg) Hul et Breteler ●☆
67726 Caloncoba suffruticosa (Milne-Redh.) Exell et Sleumer = Oncoba suffruticosa (Milne-Redh.) Hul et Breteler ●☆
67727 Caloncoba welwitschii (Oliv.) Gilg = Oncoba welwitschii Oliv. ●☆
67728 Calonema (Lindl.) D. L. Jones et M. A. Clem. = Caladenia R. Br. ■☆
67729 Calonema (Lindl.) Szlach. = Caladenia R. Br. ■☆
67730 Calonemorchis Szlach. = Caladenia R. Br. ■☆
67731 Calonnea Buc'hoz = Gaillardia Foug. ■
67732 Calonyction Choisy = Bonanox Raf. ●■
67733 Calonyction Choisy = Ipomoea L. (保留属名)●■
67734 Calonyction Choisy(1834);月光花属(夜喇叭花属);Moon Flower,Moonflower ■
67735 Calonyction acanthocarpum Choisy = Ipomoea obscura (L.) Ker Gawl. ■
67736 Calonyction aculeatum (L.) House;月光花(嫦娥奔月);Common Moonflower,Large Moonflower,Moonflower,Moonvine ■
67737 Calonyction aculeatum (L.) House = Ipomoea alba L. ■
67738 Calonyction aculeatum (L.) House var. lobatum (Hallier f.) C. Y. Wu;裂叶月光花;Lobed Large Moonflower ■
67739 Calonyction aculeatum (L.) House var. lobatum (Hallier f.) C. Y. Wu = Ipomoea alba L. ■

67740 Calonyction aculeatum House = Calonyction aculeatum (L.) House ■
67741 Calonyction album (L.) House = Ipomoea alba L. ■
67742 Calonyction bona-nox (L.) Bojer = Ipomoea alba L. ■
67743 Calonyction bona-nox (L.) Bojer var. lobata Hallier f. = Ipomoea alba L. ■
67744 Calonyction grandiflorum (Jacq.) Choisy = Ipomoea violacea L. ■
67745 Calonyction grandiflorum Choisy = Ipomoea violacea L. ■
67746 Calonyction jacquinii G. Don = Ipomoea violacea L. ■
67747 Calonyction longiflorum Hassk. = Ipomoea turbinata Lag. ■
67748 Calonyction mollissimum Zoll. = Ipomoea aculeata (L.) Kuntze var. mollissima (Zoll.) Hallier f. ex V. Oost. ■
67749 Calonyction mollissimum Zoll. var. galbrior Miq. = Ipomoea alba L. ■
67750 Calonyction mollissimum Zoll. var. glabrior Miq. = Ipomoea aculeata (L.) Kuntze var. mollissima (Zoll.) Hallier f. ex V. Oost. ■
67751 Calonyction muricatum (L.) G. Don = Ipomoea turbinata Lag. ■
67752 Calonyction pavonii (Choisy) Hallier f. = Ipomoea setosa Ker Gawl. ■
67753 Calonyction pavonii Hallier f.;刺毛月光花;Pavon Moonflower ■
67754 Calonyction pteripes G. Don;翅梗盒果藤(翅盒果藤)■☆
67755 Calonyction setosum (Ker Gawl.) Hallier f. = Ipomoea setosa Ker Gawl. ■
67756 Calonyction speciosum Choisy = Calonyction aculeatum (L.) House ■
67757 Calonyction speciosum Choisy = Ipomoea alba L. ■
67758 Calonyction speciosum Choisy var. muricatum Choisy = Ipomoea turbinata Lag. ■
67759 Calonyction tuba (Schltdl.) Colla = Ipomoea violacea L. ■
67760 Calonyction tuba Schltdl. = Ipomoea violacea L. ■
67761 Calopanax Post et Kuntze = Kalopanax Miq. ●
67762 Calopappus Meyen = Nassauvia Comm. ex Juss. ●☆
67763 Calopappus Meyen(1834);智利网菊属●☆
67764 Calopappus aceresus Meyen;智利网菊●☆
67765 Caloparion Post et Kuntze = Aconitum L. ■
67766 Caloparion Post et Kuntze = Calliparion (Link) Rchb. ex Wittst. ■
67767 Calopeltis Post et Kuntze = Callipeltis Steven ■☆
67768 Calopetalon Harv. = Marianthus Hügel ex Endl. ●☆
67769 Calopetalon J. Drumm. ex Harv. = Marianthus Hügel ex Endl. ●☆
67770 Calophaca Fisch. = Calophaca Fisch. ex DC. ●
67771 Calophaca Fisch. ex DC. (1825);丽豆属;Calophaca, Prettybean ●
67772 Calophaca acaulis (Baker) Kom.;无茎丽豆●☆
67773 Calophaca acaulis (Baker) Kom. = Chesneya acaulis (Baker) Popov ●
67774 Calophaca chinensis Boriss.;华丽豆(塔城丽豆);China Prettybean, Chinese Calophaca ●
67775 Calophaca crassicaulis (Benth. ex Baker) Kom. = Chesneya nubigena (D. Don) Ali ●☆
67776 Calophaca cuneata (Benth.) Kom. = Caragana cuneata (Benth.) Baker ●
67777 Calophaca cuneata (Benth.) Kom. = Chesneya cuneata (Benth.) Ali ●
67778 Calophaca depressa Oliv. = Chesneya depressa (Oliv.) Popov ●☆
67779 Calophaca ferganensis B. Fedtsch.;费尔干丽豆●☆
67780 Calophaca grandiflora Regel;大花丽豆●☆
67781 Calophaca hovenii Schrenk = Calophaca soongorica Kar. et Kir. ●
67782 Calophaca nigricans B. Fedtsch.;黑丽豆●☆
67783 Calophaca polystichoides Hand.-Mazz. = Chesneya polystichoides (Hand.-Mazz.) Ali ●
67784 Calophaca sericea B. Fedtsch. ex Boriss.;绢毛丽豆●☆
67785 Calophaca sinica Rehder;丽豆;Chinese Calophaca, Common Chinese Calophaca ●◇
67786 Calophaca soongorica Kar. et Kir.;新疆丽豆;Sinkiang Calophaca, Soógar Calophaca, Xinjiang Calophaca, Xinjiang Prettybean ●
67787 Calophaca tianschanica (B. Fedtsch.) Boriss.;天山丽豆;Tianshan Calophaca ●
67788 Calophaca tomentosa Blatt. et Hallb.;毛丽豆●☆
67789 Calophaca wolgarica Fisch.;伏尔加丽豆●☆
67790 Calophanes D. Don = Dyschoriste Nees ■●
67791 Calophanes adscendens Hochst. ex Nees = Dyschoriste adscendens (Hochst. ex Nees) Kuntze ■☆
67792 Calophanes burkei T. Anderson = Dyschoriste radicans Nees ■☆
67793 Calophanes clarkei Vatke = Dyschoriste clarkei (Vatke) Benoist ■☆
67794 Calophanes costatus (Nees) T. Anderson = Chaetacanthus costatus Nees ■☆
67795 Calophanes crenata Schinz = Dyschoriste depressa (L.) Nees ■
67796 Calophanes fasciculiflora (Fenzl ex Sond.) Martelli = Dyschoriste perrottetii (Nees) Kuntze ■☆
67797 Calophanes gracilis Nees = Dyschoriste gracilis (Nees) Kuntze ■
67798 Calophanes heudelotianus Nees = Dyschoriste heudelotiana (Nees) Kuntze ■☆
67799 Calophanes hildebrandtii S. Moore = Dyschoriste hildebrandtii (S. Moore) Lindau ■☆
67800 Calophanes hyssopifolius Nees = Dyschoriste hyssopifolia (Nees) Kuntze ■☆
67801 Calophanes linifolius T. Anderson ex C. B. Clarke = Strobilanthopsis linifolia (T. Anderson ex C. B. Clarke) Milne-Redh. ●☆
67802 Calophanes madagascariensis Nees = Dyschoriste madagascariensis (Nees) Kuntze ■☆
67803 Calophanes multicaulis T. Anderson = Dyschoriste multicaulis (T. Anderson) Kuntze ■☆
67804 Calophanes nagchana Nees = Dyschoriste erecta (Burm.) Kuntze ■
67805 Calophanes natalensis T. Anderson = Dyschoriste depressa (L.) Nees ■
67806 Calophanes perrottetii Nees = Dyschoriste perrottetii (Nees) Kuntze ■☆
67807 Calophanes persoonii (Nees) T. Anderson = Chaetacanthus setiger (Pers.) Lindl. ■☆
67808 Calophanes radicans (Nees) T. Anderson = Dyschoriste radicans Nees ■☆
67809 Calophanes radicans T. Anderson var. mutica S. Moore = Dyschoriste mutica (S. Moore) C. B. Clarke ■☆
67810 Calophanes setosus Nees = Ruelliopsis setosa (Nees) C. B. Clarke ■☆
67811 Calophanes siphonanthus Nees = Dyschoriste siphonantha (Nees) Kuntze ●☆
67812 Calophanes thunbergiiflora S. Moore = Dyschoriste thunbergiiflora (S. Moore) Lindau ■☆
67813 Calophanes trichocalyx Oliv. = Dyschoriste trichocalyx (Oliv.) Lindau ■☆
67814 Calophanes verticillaris T. Anderson ex Oliv. = Dyschoriste

verticillaris (T. Anderson ex Oliv.) C. B. Clarke ■☆
67815 Calophanoides (C. B. Clarke) Ridl. = Justicia L. ●■
67816 Calophanoides Ridl. (1923);杜根藤属(赛爵床属);Calophanoides ●■
67817 Calophanoides Ridl. = Justicia L. ●■
67818 Calophanoides albovelata (W. W. Sm.) C. Y. Wu ex C. C. Hu;绵毛杜根藤;Lanatous Calophanoides ●
67819 Calophanoides alboviridis (Benoist) C. Y. Wu et H. S. Lo;大叶赛爵床■
67820 Calophanoides buxifolia (H. S. Lo et D. Fang) C. Y. Wu ex C. C. Hu;黄杨叶赛爵床●
67821 Calophanoides chinensis (Benth.) C. Y. Wu et H. S. Lo;圆苞杜根藤(杜根藤,中华赛爵床);China Calophanoides, Chinese Calophanoides ●
67822 Calophanoides chinensis (Champ.) C. Y. Wu et H. S. Lo ex Y. C. Tang = Calophanoides chinensis (Benth.) C. Y. Wu et H. S. Lo ●
67823 Calophanoides hainanensis C. Y. Wu et H. S. Lo;海南赛爵床;Hainan Calophanoides ●
67824 Calophanoides kouytcheensis (H. Lév.) H. S. Lo;贵州赛爵床;Guizhou Calophanoides ●
67825 Calophanoides kwangxiensis H. S. Lo;广西杜根藤(广西赛爵床);Guangxi Calophanoides ●
67826 Calophanoides loberi (C. B. Clarke) Bremek.;狭叶赛爵床●
67827 Calophanoides multinodis (Benoist) C. Y. Wu et H. S. Lo;白节赛爵床;Multinode Calophanoides ●
67828 Calophanoides quadrifaria (Nees) Ridl.;杜根藤(大青草)●
67829 Calophanoides salicifolia (T. Atders.) Bremek.;柳叶杜根藤(柳叶赛爵床)●☆
67830 Calophanoides siccanea (W. W. Sm.) C. Y. Wu et al. = Calophanoides siccanea (W. W. Sm.) C. Y. Wu ex C. C. Hu ●
67831 Calophanoides siccanea (W. W. Sm.) C. Y. Wu ex C. C. Hu;旱杜根藤;Dried Calophanoides ●
67832 Calophanoides wardii (W. W. Sm.) C. Y. Wu ex C. C. Hu;高山杜根藤;Ward Calophanoides ●
67833 Calophanoides xantholeuca (W. W. Sm.) C. Y. Wu ex C. C. Hu;黄白杜根藤;Yellow-white Calophanoides ●
67834 Calophanoides xerobatica (W. W. Sm.) C. Y. Wu ex C. C. Hu;滇东杜根藤;East Yunnan Calophanoides ●
67835 Calophanoides xerophila (W. W. Sm.) C. Y. Wu ex C. C. Hu;干地杜根藤;Xerophilous Calophanoides ●
67836 Calophanoides xylopoda (W. W. Sm.) C. Y. Wu ex C. C. Hu;木柄杜根藤;Wood-stalk Calophanoides ●
67837 Calophanoides yunnanensis (W. W. Sm.) C. Y. Wu ex H. P. Tsui;滇杜根藤;Yunnan Calophanoides ●
67838 Calophruria Post et Kuntze = Eucharis Planch. et Linden(保留属名)■☆
67839 Calophthalmum Rchb. = Blainvillea Cass. ■●
67840 Calophylica C. Presl = Phylica L. ●☆
67841 Calophyllaceae J. Agardh = Clusiaceae Lindl. (保留科名)●■
67842 Calophyllaceae J. Agardh = Guttiferae Juss. (保留科名)●■
67843 Calophyllaceae J. Agardh;红厚壳科●
67844 Calophylloides Smeathman = Eugenia L. ●
67845 Calophylloides Smeathman ex DC. = Eugenia L. ●
67846 Calophyllum L. (1753);红厚壳属(胡桐属,琼崖海棠属);Beauty Leaf, Beautyleaf, Beauty-leaf, Bintangor ●
67847 Calophyllum Post et Kuntze = Calliphyllon Bubani et Penz. ■
67848 Calophyllum Post et Kuntze = Epipactis Zinn(保留属名)■
67849 Calophyllum Post et Kuntze = Kallophyllon Pohl ●☆
67850 Calophyllum Post et Kuntze = Symphyopappus Turcz. ●☆
67851 Calophyllum antillanum Britton;拉美胡桐;Santa Maria ●
67852 Calophyllum apetalum Willd.;无瓣红厚壳;Petalless Beautyleaf ●☆
67853 Calophyllum blancoi Planch. et Triana;兰屿红厚壳(兰屿胡桐);Blanco Beautyleaf, Lanyu Beautyleaf ●
67854 Calophyllum bracteatum Thwaites;苞片红厚壳;Bracteate Beautyleaf ●☆
67855 Calophyllum brasiliense Cambess.;巴西红厚壳(巴西胡桐);Brazil Calaba, Galba, Jacareuba, Santa Maria ●☆
67856 Calophyllum calaba L.;卡拉红厚壳;Calaba, Ceylon Beauty Leaf, Ceylon Beauty-leaf ●☆
67857 Calophyllum changii N. Robson = Calophyllum blancoi Planch. et Triana ●
67858 Calophyllum chapelieri Drake;马达加斯加胡桐●☆
67859 Calophyllum glabrum Merr.;兰屿海棠(小叶红厚壳);Lanyu Indiapoon Beautyleaf ●
67860 Calophyllum inophyllum L.;红厚壳(海棠,海棠果,海棠壳,海棠木,海棠树,呵喇菩,胡桐,君子树,南印胡桐,琼崖海棠,琼崖海棠树);Alexandria Laurel, Alexandrian Laurel, Beauty Leaf, Beauty Leaf Mastwood, Beautyleaf, Beauty-leaf Mastwood, Borneo Mahogany, India Poon, Indian Laurel, Indiapoon Beautyleaf, Kalofilum, Kalofilum Kathing, Kamani, Kathing, Laurelwood, Oionut Tree, Poon, Punnai-nutus, Tacamahac, Tongan Oil ●
67861 Calophyllum lanigerum Miq.;绵毛胡桐(绵毛红厚壳)●☆
67862 Calophyllum lanigerum Miq. var. austrocoriaceum (Whitmore) P. F. Stevens;革质绵毛胡桐(革质绵毛红厚壳)●☆
67863 Calophyllum lucidum Benth.;光胡桐●☆
67864 Calophyllum membranaceum Gardner et Champ.;薄叶红厚壳(篦子王,薄叶胡桐,跌打将军,独角风,独筋猪尾,横经席,绞枫,皮子黄,梳篦木,梳篦王,碎骨莲,铁将军,小果海棠木);Membranaceous Beautyleaf, Membranaceous Beauty-leaf ●
67865 Calophyllum nagassarium Burm. f. = Mesua ferrea L. ●
67866 Calophyllum polyanthum Wall. ex Choisy;滇南红厚壳(滇南红壳桂,泰国红厚壳,云南红厚壳,云南胡桐);S. Yunnan Beautyleaf, South Yunnan Beauty-leaf, Yunnan Beautyleaf ●◇
67867 Calophyllum saigonense Pierre;西贡胡桐(西贡海棠果);Cong ●
67868 Calophyllum smilesianum Craib = Calophyllum polyanthum Wall. ex Choisy ●◇
67869 Calophyllum smilesianum Craib var. luteum Craib = Calophyllum polyanthum Wall. ex Choisy ●◇
67870 Calophyllum spectabile Hook. et Arn. = Calophyllum membranaceum Gardner et Champ. ●
67871 Calophyllum tacamahaca Willd.;马达加斯加红厚壳;Calaba Balsam, Madagascar Beautyleaf, Tacamahac ●☆
67872 Calophyllum thorelii Pierre = Calophyllum polyanthum Wall. ex Choisy ●◇
67873 Calophyllum tomentosum Wight;茸毛胡桐(毛红厚壳,茸毛海棠果);Poon, Tomentose Beautyleaf, Tomentose Calaba ●☆
67874 Calophyllum williamsianum Craib = Calophyllum polyanthum Wall. ex Choisy ●◇
67875 Calophysa DC. = Clidemia D. Don ●☆
67876 Calophysa Post et Kuntze = Calligonum L. ●
67877 Calophysa Post et Kuntze = Calliphysa Fisch. et C. A. Mey. ●
67878 Calopisma Post et Kuntze = Callopisma Mart. ■☆
67879 Calopisma Post et Kuntze = Deianira Cham. et Schltdl. ■☆

67880 Caloplectus Oerst. = Alloplectus Mart.(保留属名)●■☆
67881 Caloplectus Oerst. = Drymonia Mart. ●☆
67882 Calopogon R. Br. (1813)(保留属名);北美毛唇兰属(毛唇兰属);Calopogon, Grass pink, Grass-pink Orchid ■☆
67883 Calopogon barbatus (Walter) Ames;北美毛唇兰(北美兰);Bearded Grass-pink ■☆
67884 Calopogon barbatus (Walter) Ames var. multiflorus (Lindl.) Correll = Calopogon multiflorus Lindl. ■☆
67885 Calopogon multiflorus Lindl.;多花北美毛唇兰(多花北美兰);Many-flowered grass-pink ■☆
67886 Calopogon oklahomensis D. H. Goldman;俄克拉何马北美毛唇兰(俄克拉何马北美兰);Oklahoma Grass Pink ■☆
67887 Calopogon pallidus Chapm.;苍白北美毛唇兰;Pale grass-pink ■☆
67888 Calopogon parviflorus Lindl. = Calopogon barbatus (Walter) Ames ■☆
67889 Calopogon parviflorus Raf. = Calopogon barbatus (Walter) Ames ■☆
67890 Calopogon pulchellus (Salisb.) R. Br. = Calopogon tuberosus (L.) Britton, Sterns et Poggenb. ■☆
67891 Calopogon pulchellus (Salisb.) R. Br. var. graminifolius Elliott = Calopogon barbatus (Walter) Ames ■☆
67892 Calopogon pulchellus (Salisb.) R. Br. var. latifolius (H. St. John) Fernald = Calopogon tuberosus (L.) Britton, Sterns et Poggenb. ■☆
67893 Calopogon pulchellus (Salisb.) R. Br. var. simpsonii (Small) Ames = Calopogon tuberosus (L.) Britton, Sterns et Poggenb. var. simpsonii (Small) Magrath ■☆
67894 Calopogon tuberosus (L.) Britton, Sterns et Poggenb.;块茎北美毛唇兰(块茎北美兰);Grass Pink, Grass Pink Orchid, Grass-pink Orchid, Tuberous Grass Pink ■☆
67895 Calopogon tuberosus (L.) Britton, Sterns et Poggenb. var. latifolius (H. St. John) B. Boivin = Calopogon tuberosus (L.) Britton, Sterns et Poggenb. ■☆
67896 Calopogon tuberosus (L.) Britton, Sterns et Poggenb. var. simpsonii (Small) Magrath;西氏北美毛唇兰■☆
67897 Calopogonium Desv. (1826);毛蔓豆属(拟大豆属);Hairvinebean ●
67898 Calopogonium mucunoides Desv.;毛蔓豆(拟大豆);Hairvinebean ●
67899 Caloprena Post et Kuntze = Allium L. ■
67900 Caloprena Post et Kuntze = Calliprena Salisb. ■
67901 Caloprora Post et Kuntze = Brodiaea Sm.(保留属名)■☆
67902 Caloprora Post et Kuntze = Calliprora Lindl. ■☆
67903 Calopsis P. Beauv. (1828);南非帚灯草属■☆
67904 Calopsis P. Beauv. ex Desv. = Calopsis P. Beauv. ■☆
67905 Calopsis P. Beauv. ex Juss. = Calopsis P. Beauv. ■☆
67906 Calopsis P. Beauv. ex Juss. = Leptocarpus R. Br.(保留属名)■
67907 Calopsis adpressa Esterh.;匍匐南非帚灯草■☆
67908 Calopsis andreaeana (Pillans) H. P. Linder;安氏南非帚灯草■☆
67909 Calopsis aspera (Mast.) H. P. Linder;粗糙南非帚灯草■☆
67910 Calopsis burchellii (Mast.) H. P. Linder;布尔南非帚灯草■☆
67911 Calopsis clandestina Esterh.;隐匿南非帚灯草■☆
67912 Calopsis dura Esterh.;杜拉南非帚灯草■☆
67913 Calopsis esterhuyseniae (Pillans) H. P. Linder = Askidiosperma esterhuyseniae (Pillans) H. P. Linder ■☆
67914 Calopsis festucacea Kunth = Calopsis viminea (Rottb.) H. P. Linder ■☆
67915 Calopsis filiformis (Mast.) H. P. Linder;线形南非帚灯草■☆
67916 Calopsis fruticosa (Mast.) H. P. Linder;灌丛南非帚灯草■☆
67917 Calopsis gracilis (Mast.) H. P. Linder;纤细南非帚灯草■☆
67918 Calopsis hirtella Kunth = Calopsis viminea (Rottb.) H. P. Linder ■☆
67919 Calopsis hyalina (Mast.) H. P. Linder;无色南非帚灯草■☆
67920 Calopsis impolita (Kunth) H. P. Linder;暗色南非帚灯草■☆
67921 Calopsis incurvata (Thunb.) Kunth = Willdenowia incurvata (Thunb.) H. P. Linder ■☆
67922 Calopsis incurvata Pillans = Calopsis viminea (Rottb.) H. P. Linder ■☆
67923 Calopsis levynsiae (Pillans) H. P. Linder;勒温斯南非帚灯草■☆
67924 Calopsis marlothii (Pillans) H. P. Linder;马洛斯南非帚灯草■☆
67925 Calopsis membranacea (Pillans) H. P. Linder;膜质南非帚灯草■☆
67926 Calopsis monostylis (Pillans) H. P. Linder;单柱南非帚灯草■☆
67927 Calopsis muirii (Pillans) H. P. Linder;缪里南非帚灯草■☆
67928 Calopsis neglecta Hochst. = Restio multiflorus Spreng. ■☆
67929 Calopsis nudiflora (Pillans) H. P. Linder;裸花南非帚灯草■☆
67930 Calopsis oxylepis Kunth = Calopsis viminea (Rottb.) H. P. Linder ■☆
67931 Calopsis paniculata (Rottb.) Desv.;圆锥南非帚灯草■☆
67932 Calopsis peronata Kunth = Calopsis viminea (Rottb.) H. P. Linder ■☆
67933 Calopsis pulchra Esterh.;美丽南非帚灯草■☆
67934 Calopsis ramiflora (Nees) Kunth = Calopsis paniculata (Rottb.) Desv. ■☆
67935 Calopsis rigida (Mast.) H. P. Linder;硬南非帚灯草■☆
67936 Calopsis sparsa Esterh.;散生南非帚灯草■☆
67937 Calopsis triticea (Rottb.) Kunth = Restio triticeus Rottb. ■☆
67938 Calopsis viminea (Rottb.) H. P. Linder;软枝南非帚灯草■☆
67939 Calopsyche Post et Kuntze = Callipsyche Herb. ■☆
67940 Calopsyche Post et Kuntze = Eucrosia Ker Gawl. ■☆
67941 Calopteryx A. C. Sm. (1946);南美杜鹃属●☆
67942 Calopteryx A. C. Sm. = Thibaudia Ruiz et Pav. ●☆
67943 Calopteryx insignis A. C. Sm.;南美杜鹃●☆
67944 Calopteryx sessiliflora A. C. Sm.;无梗南美杜鹃●☆
67945 Caloptilium Lag. = Nassauvia Comm. ex Juss. ●☆
67946 Calopyxis Tul. (1856);美果使君子属●☆
67947 Calopyxis Tul. = Combretum Loefl.(保留属名)●
67948 Calopyxis alata Tul.;高美果使君子●☆
67949 Calopyxis brevistyla H. Perrier;短柱美果使君子●☆
67950 Calopyxis macrocalyx (Tul.) H. Perrier;大萼美果使君子●☆
67951 Calopyxis malifolia Baker;苹果叶美果使君子●☆
67952 Calopyxis trichophylla Baker;毛叶美果使君子●☆
67953 Calopyxis velutina Tul.;黏美果使君子●☆
67954 Calorchis Barb. Rodr. = Ponthieva R. Br. ■☆
67955 Calorezia Panero = Perezia Lag. ■☆
67956 Calorezia Panero(2007);智利钝柱菊属■☆
67957 Calorhabdos Benth. (1835);四方麻属■
67958 Calorhabdos Benth. = Veronicastrum Heist. ex Fabr. ■
67959 Calorhabdos arisanensis Masam. = Veronicastrum kitamurae (Ohwi) T. Yamaz. ■
67960 Calorhabdos brunoniana Benth. = Veronicastrum brunonianum (Benth.) D. Y. Hong ■
67961 Calorhabdos cauloptera Hance = Veronicastrum caulopterum (Hance) T. Yamaz. ■

67962 Calorhabdos chinensis (Maxim.) Franch. = Scrofella chinensis Maxim. ■
67963 Calorhabdos fargesii Franch. = Veronicastrum stenostachyum (Hemsl.) T. Yamaz. ■
67964 Calorhabdos formosana (Masam.) Ohwi = Veronicastrum formosanum (Masam.) T. Yamaz. ■
67965 Calorhabdos formosana (Masam.) Ohwi var. latifolia Masam. = Veronicastrum kitamurae (Ohwi) T. Yamaz. ■
67966 Calorhabdos kitamurae Ohwi = Veronicastrum formosanum (Masam.) T. Yamaz. ■
67967 Calorhabdos kitamurae Ohwi = Veronicastrum kitamurae (Ohwi) T. Yamaz. ■
67968 Calorhabdos latifolia Hemsl. = Veronicastrum latifolium (Hemsl.) T. Yamaz. ■
67969 Calorhabdos longispicata (Merr.) Masam. = Veronicastrum longispicatum (Merr.) T. Yamaz. ■
67970 Calorhabdos robusta Diels = Veronicastrum robustum (Diels) D. Y. Hong ■
67971 Calorhabdos simadai Masam. = Veronicastrum axillare (Siebold et Zucc.) T. Yamaz. ■
67972 Calorhabdos simadai Masam. = Veronicastrum simadai (Masam.) T. Yamaz. ■
67973 Calorhabdos stenostachya Hemsl. = Veronicastrum stenostachyum (Hemsl.) T. Yamaz. ■
67974 Calorhabdos sutchuenensis Franch. = Veronicastrum brunonianum (Benth.) D. Y. Hong ■
67975 Calorhabdos sutchuenensis Franch. = Veronicastrum brunonianum (Benth.) D. Y. Hong subsp. sutchunense (Franch.) D. Y. Hong ■
67976 Calorhabdos vennosa Hemsl. = Veronicastrum stenostachyum (Hemsl.) T. Yamaz. ■
67977 Calorhabdos villosula (Miq.) Benth. et Hook. f. ex Najubi = Veronicastrum villosulum (Miq.) T. Yamaz. ■
67978 Calorhabdos yunnanensis (W. W. Sm.) Masam. = Veronicastrum yunnanense (W. W. Sm.) T. Yamaz. ■
67979 Calorhobdos axillaris (Siebold et Zucc.) Benth. et Hook. f. ex et Moore = Veronicastrum axillare (Siebold et Zucc.) T. Yamaz. ■
67980 Calorhoe Post et Kuntze = Callirhoe Nutt. ■●☆
67981 Calorophus Labill. (废弃属名) = Hypolaena R. Br. (保留属名) ■☆
67982 Calorophus anceps (Mast.) Kuntze = Platycaulos anceps (Mast.) H. P. Linder ■☆
67983 Calorophus asper Kuntze = Calopsis aspera (Mast.) H. P. Linder ■☆
67984 Calorophus burchellii (Mast.) Kuntze = Calopsis membranacea (Pillans) H. P. Linder ■☆
67985 Calorophus digitatus (Thunb.) Kuntze = Mastersiella digitata (Thunb.) Gilg-Ben. ●☆
67986 Calorophus filiformis (Mast.) Kuntze = Calopsis filiformis (Mast.) H. P. Linder ■☆
67987 Calorophus gracilis (Mast.) Kuntze = Calopsis gracilis (Mast.) H. P. Linder ■☆
67988 Calorophus laxiflorus (Nees) Kuntze = Anthochortus laxiflorus (Nees) H. P. Linder ■☆
67989 Calorophus tenuis (Mast.) Kuntze = Anthochortus ecklonii Nees ■☆
67990 Calorophus virgatus (Mast.) Kuntze = Restio debilis Nees ■☆
67991 Calosace Post et Kuntze = Angelica L. ■
67992 Calosace Post et Kuntze = Callisace Fisch. ex Hoffm. ■
67993 Calosacme Wall. = Chirita Buch. -Ham. ex D. Don ●■
67994 Calosanthes Blume = Oroxylum Vent. ●
67995 Calosanthes indica (L.) Blume = Oroxylum indicum (L.) Vent. ●
67996 Calosanthes indica Blume = Oroxylum indicum (L.) Vent. ●
67997 Calosanthos Rchb. = Kalosanthes Haw. ●■☆
67998 Calosciadium Endl. = Aciphylla J. R. Forst. et G. Forst. ■☆
67999 Caloscilla Jord. et Fourr. = Scilla L. ■
68000 Caloscordum Herb. (1844); 合被韭属■☆
68001 Caloscordum Herb. = Allium L. ■
68002 Caloscordum Herb. = Nothoscordum Kunth(保留属名)■☆
68003 Caloscordum exsertum Lindl. = Allium chinense G. Don ■
68004 Caloscordum inutile (Makino) Okuyama et Kitag. = Allium inutile Makino ■
68005 Caloscordum neriniflorum Herb. = Allium neriniflorum (Herb.) Baker ■
68006 Caloscordum neriniflorum Herb. = Nothoscordum nerinifolium Benth. et Hook. f. ■☆
68007 Caloscordum tubiflorum (Rendle) Traub. = Allium tubiflorum Rendle ■
68008 Calosemaea Post et Kuntze = Callisemaea Benth. ■☆
68009 Calosemaea Post et Kuntze = Platypodium Vogel ■☆
68010 Caloserts Benth. = Onoseris Willd. ●■☆
68011 Calosmon Bercht. et J. Presl = Benzoin Boerh. ex Schaeff. ●
68012 Calosmon Bercht. et J. Presl = Lindera Thunb. (保留属名)●
68013 Calosmon J. Presl = Lindera Thunb. (保留属名)●
68014 Calospatha Becc. (1911); 美苞棕属; Calospatha ●☆
68015 Calospatha confusa Furtado; 紊乱美苞棕; Confused Calospatha ●☆
68016 Calospatha scortechinii Becc.; 斯考氏美苞棕; Scortechin Calospatha ●☆
68017 Calospermum Pierre = Lucuma Molina ●
68018 Calospermum Pierre = Pouteria Aubl. ●
68019 Calosphace (Benth.) Raf. = Salvia L. ●■
68020 Calosphace Raf. = Salvia L. ●■
68021 Calostachya T. Post et Kuntze = Callistachys Raf. ■
68022 Calostachys T. Post et Kuntze = Callistachys Heuffel ■
68023 Calostachys T. Post et Kuntze = Callistachys Vent. (废弃属名)■☆
68024 Calostachys T. Post et Kuntze = Carex L. ■
68025 Calostachys T. Post et Kuntze = Oxylobium Andréws(保留属名) ■☆
68026 Calosteca Desv. = Briza L. ■
68027 Calosteca Desv. = Calotheca Desv. ■
68028 Calostelma D. Don = Liatris Gaertn. ex Schreb. (保留属名)■☆
68029 Calostemma Post et Kuntze = Callistemma Boiss. ■☆
68030 Calostemma Post et Kuntze = Tremastelma Raf. ■☆
68031 Calostemma R. Br. (1810); 东澳石蒜属; Gariand-lily ■☆
68032 Calostemma luteum Sims; 黄东澳石蒜; Yellow Gariand-lily ■☆
68033 Calostemma purpureum R. Br.; 紫东澳石蒜; Purple Gariand-lily ■☆
68034 Calostemon T. Post et Kuntze = Callistemon R. Br. ●
68035 Calostephana Post et Kuntze = Callistephana Fourr. ●■
68036 Calostephana Post et Kuntze = Coronilla L. (保留属名)●■
68037 Calostephane Benth. (1872); 丽冠菊属■☆
68038 Calostephane angolensis (O. Hoffm.) Anderb.; 安哥拉丽冠菊■☆
68039 Calostephane divaricata Benth.; 叉丽冠菊■☆
68040 Calostephane divaricata Benth. var. schinzii (O. Hoffm.) Thell.

= Calostephane divaricata Benth. ■☆
68041 Calostephane eyelesii Thell. = Calostephane divaricata Benth. ■☆
68042 Calostephane foliosa Klatt = Calostephane divaricata Benth. ■☆
68043 Calostephane huillensis (Hiern) Anderb.;威拉丽冠菊■☆
68044 Calostephane madagascariensis (Humbert) Anderb.;马岛丽冠菊■☆
68045 Calostephane marlothiana O. Hoffm.;马洛斯丽冠菊■☆
68046 Calostephane punctulata (Hiern) Anderb.;斑点丽冠菊■☆
68047 Calostephane schinzii O. Hoffm. = Calostephane divaricata Benth. ■☆
68048 Calostephane setosa Alston = Calostephane divaricata Benth. ■☆
68049 Calostephus Post et Kuntze = Callistephus Cass.(保留属名)■
68050 Calostigma Decne.(1838);丽柱萝藦属■☆
68051 Calostigma Schott = Philodendron Schott(保留属名)■●
68052 Calostigma Schott ex B. D. Jacks. = Calostigma Decne. ■☆
68053 Calostigma multiflorum Malme;多花丽柱萝藦☆
68054 Calostima Raf. = Urera Gaudich. ●☆
68055 Calostroma Post et Kuntze = Callistroma Fenzl ■☆
68056 Calostroma Post et Kuntze = Oliveria Vent. ■☆
68057 Calostrophus F. Muell. = Hypolaena R. Br.(保留属名)■☆
68058 Calostylis Kuntze = Callostylis Blume ■
68059 Calostylis Kuntze = Eria Lindl.(保留属名)■
68060 Calota Hare. ex Lindl. = Ceratandra Eckl. ex F. A. Bauer ■☆
68061 Calotesta P. O. Karis(1990);白苞鼠麹木属●☆
68062 Calotesta alba P. O. Karis;白苞鼠麹木●☆
68063 Calothamnus Labill.(1806);网木属(半边花属);Net Bush, Woolly Netbush ●☆
68064 Calothamnus quadrifidus R. Br.;网木(四裂半边花);Common Net Brush, One-sided Bottle-brush ●☆
68065 Calothamnus rupestris Schauer;岩生网木;Cilff Net Bush ●☆
68066 Calothamnus sanguineus Labill.;血花网木;Blood Flower ●☆
68067 Calothamnus validus S. Moore;健壮网木;Barrens Claw Flower ●☆
68068 Calothamnus villosus R. Br.;毛网木;Silky Net Bush, Woolly Net Bush, Woolly Netbush ●☆
68069 Calothauma Post et Kuntze = Callithauma Herb. ■☆
68070 Calothauma Post et Kuntze = Stenomesson Herb. ■☆
68071 Calotheca Desv. = Briza L. ■
68072 Calotheca P. Beauv. = Briza L. ■
68073 Calotheca Spreng. = Aeluropus Trin. ■
68074 Calotheca sabulosa Steud. = Eragrostis sabulosa (Steud.) Schweick. ■☆
68075 Calotheria Steud. = Enneapogon Desv. ex P. Beauv. ■
68076 Calotheria Wight et Arn. = Pappophorum Schreb. ■☆
68077 Calotheria Wight et Arn. ex Steud. = Pappophorum Schreb. ■☆
68078 Calothyrsus Spach = Aesculus L. ●
68079 Calotis R. Br.(1820);刺冠菊属;Bur Daisy, Calotis ■
68080 Calotis anamitica (Kuntze) Merr. = Calotis caespitosa C. C. Chang ■
68081 Calotis caespitosa C. C. Chang;刺冠菊;Common Calotis ■
68082 Calotis cuneifolia R. Br.;楔叶刺冠菊;Bur Daisy, Wedgeleaf Calotis ■☆
68083 Calotriche Post et Kuntze = Callitriche L. ■
68084 Calotropis Post et Kuntze = Callotropis G. Don ■
68085 Calotropis Post et Kuntze = Galega L. ■
68086 Calotropis R. Br.(1810);牛角瓜属;Calotrope, Madar, Mudar, Mudar Fibre ●
68087 Calotropis busseana K. Schum. = Calotropis procera (Aiton) R. Br. ●
68088 Calotropis gigantea (L.) Dryand. ex W. T. Aiton;牛角瓜(断肠草,曼陀罗花,五狗卧花心,羊浸树);Akund Calotrope, Bowstring Hemp, Crownplant, Giant Milkweed, Madar, Mudar, Wara, Yercum ●
68089 Calotropis gigantea (L.) R. Br. = Calotropis gigantea (L.) Dryand. ex W. T. Aiton ●
68090 Calotropis gigantea (L.) W. T. Aiton = Calotropis gigantea (L.) Dryand. ex W. T. Aiton ●
68091 Calotropis hamiltonii Wight = Calotropis procera (L.) Dryand. ex W. T. Aiton subsp. hamiltonii (Wight) Ali ●☆
68092 Calotropis inflexa Chiov. = Calotropis procera (Aiton) R. Br. ●
68093 Calotropis procera (Aiton) R. Br. = Calotropis procera (L.) Dryand. ex W. T. Aiton ●
68094 Calotropis procera (Aiton) W. T. Aiton = Calotropis procera (L.) Dryand. ex W. T. Aiton ●
68095 Calotropis procera (L.) Dryand. ex W. T. Aiton;白花牛角瓜;Akund Fibre, Auricula Tree, French Cotton, Roostertree, Swallow-wort, White-flowered Calotrope ●
68096 Calotropis procera (L.) Dryand. ex W. T. Aiton subsp. hamiltonii (Wight) Ali;哈氏牛角瓜●☆
68097 Caloxene Post et Kuntze = Callixene Comm. ex Juss.(废弃属名)■☆
68098 Caloxene Post et Kuntze = Luzuriaga Ruiz et Pav.(保留属名)■☆
68099 Calpandria Blume = Camellia L. ●
68100 Calpicarpum G. Don = Kopsia Blume(保留属名)●
68101 Calpicarpum G. Don = Ochrosia Juss. ●
68102 Calpidia Thouars = Ceodes J. R. Forst. et G. Forst. ●
68103 Calpidia Thouars = Pisonia L. ●
68104 Calpidia excelsa (Blume) Heim. = Ceodes umbellifera J. R. Forst. et G. Forst. ●
68105 Calpidia excelsa (Blume) Heim. = Pisonia umbellifera (J. R. Forst. et G. Forst.) Seem. ●
68106 Calpidisca Barnhart = Utricularia L. ■
68107 Calpidisca capensis (Spreng.) Barnhart = Utricularia bisquamata Schrank ■☆
68108 Calpidochlamys Diels = Trophis P. Browne(保留属名)●☆
68109 Calpidosicyos Harms = Momordica L. ■
68110 Calpidosicyos friesiorum Harms = Momordica friesiorum (Harms) C. Jeffrey ■☆
68111 Calpigyne Blume = Koilodepas Hassk. ●
68112 Calpigyne hainanensis Merr. = Koilodepas hainanense (Merr.) Airy Shaw ●
68113 Calpocalyx Harms(1897);瓮萼豆属;Calpocalyx ●☆
68114 Calpocalyx atlanticus Villiers;大西洋瓮萼豆●☆
68115 Calpocalyx aubrevillei Pellegr.;奥布瓮萼豆●☆
68116 Calpocalyx brevibracteatus Harms;短苞瓮萼豆●☆
68117 Calpocalyx brevifolius Villiers;短叶瓮萼豆●☆
68118 Calpocalyx cauliflorus Hoyle;茎花瓮萼豆●☆
68119 Calpocalyx crawfordianus Mendes = Calpocalyx dinklagei Harms ●☆
68120 Calpocalyx dinklagei Harms;丁克瓮萼豆木●☆
68121 Calpocalyx heitzii Pellegr.;海氏瓮萼豆木●☆
68122 Calpocalyx klainei Pierre ex Harms;克莱恩瓮萼豆木●☆
68123 Calpocalyx letestui Pellegr.;莱泰斯图瓮萼豆木●☆
68124 Calpocalyx macrostachys Harms = Calpocalyx brevibracteatus Harms ●☆
68125 Calpocalyx ngouniensis Pellegr.;恩古涅瓮萼豆木●☆

68126 Calpocalyx sericeus Hutch. et Dalziel = Pseudoprosopis sericea (Hutch. et Dalziel) Brenan ●☆

68127 Calpocalyx winkleri (Harms) Harms;温克勒瓮萼豆木●☆

68128 Calpocarpus Post et Kuntze = Calpicarpum G. Don ●

68129 Calpocarpus Post et Kuntze = Kopsia Blume(保留属名)●

68130 Calpurnia E. Mey. (1836);翼荚豆属■☆

68131 Calpurnia antunesii Taub. = Pterocarpus lucens Lepr. ex Guillaumin et Perr. subsp. antunesii (Taub.) Rojo ●☆

68132 Calpurnia aurea (Aiton) Benth.;黄翼荚豆;African Laburnum ■☆

68133 Calpurnia aurea (Aiton) Benth. subsp. sylvatica (Burch.) Brummitt = Calpurnia aurea (Aiton) Benth. ■☆

68134 Calpurnia aurea (Aiton) Benth. var. major Baker f. = Calpurnia aurea (Aiton) Benth. ■☆

68135 Calpurnia australiana F. Muell.;翼荚豆■☆

68136 Calpurnia capensis (Burm. f.) Druce;好望角翼荚豆■☆

68137 Calpurnia floribunda Harv.;多花翼荚豆■☆

68138 Calpurnia glabrata Brummitt;光滑翼荚豆■☆

68139 Calpurnia intrusa (R. Br.) E. Mey.;长柔毛翼荚豆■☆

68140 Calpurnia intrusa (R. Br.) E. Mey. var. glabrata Yakovlev = Calpurnia intrusa (R. Br.) E. Mey. ■☆

68141 Calpurnia lasiogyna E. Mey.;毛柱翼荚豆■☆

68142 Calpurnia lasiogyne E. Mey. = Calpurnia aurea (Aiton) Benth. ■☆

68143 Calpurnia mucronulata Harms ex Kuntze = Calpurnia sericea Harv. ■☆

68144 Calpurnia obovata Schinz = Calpurnia sericea Harv. ■☆

68145 Calpurnia obovata Schinz var. pubescens Yakovlev = Calpurnia sericea Harv. ■☆

68146 Calpurnia reflexa A. J. Beaumont;反折翼荚豆■☆

68147 Calpurnia sericea Harv.;绢毛翼荚豆■☆

68148 Calpurnia subdecandra (L'Hér.) Schweick.;狭翼荚豆■☆

68149 Calpurnia subdecandra (L'Hér.) Schweick. = Calpurnia aurea (Aiton) Benth. ■☆

68150 Calpurnia sylvatica (Burch.) E. Mey. = Calpurnia aurea (Aiton) Benth. ■☆

68151 Calpurnia uarandensis Chiov. = Dalbergia uarandensis (Chiov.) Thulin ●☆

68152 Calpurnia villosa Harv.;毛翼荚豆■☆

68153 Calpurnia villosa Harv. = Calpurnia intrusa (R. Br.) E. Mey. ■☆

68154 Calpurnia woodii Schinz;伍得翼荚豆■☆

68155 Calsiama Raf. = Calesiam Adans.(废弃属名)●

68156 Calsiama Raf. = Lannea A. Rich.(保留属名)●

68157 Caltha L. (1753);驴蹄草属;Kingcup, Marsh Marigold, Marshmarigold, Marsh-marigold, Populage ■

68158 Caltha Mill. = Calendula L. ●■

68159 Caltha Tourn. ex Adans. = Calendula L. ●■

68160 Caltha arctica R. Br.;北极驴蹄草;Arctic Marshmarigold ■☆

68161 Caltha arctica R. Br. = Caltha palustris L. ■

68162 Caltha asarifolia DC. = Caltha palustris L. ■

68163 Caltha biflora DC.;双花驴蹄草;Twin-flowered Marsh Marigold ■☆

68164 Caltha biflora DC. = Caltha leptosepala DC. ■☆

68165 Caltha biflora DC. subsp. howellii (Huth) Abrams = Caltha leptosepala DC. ■☆

68166 Caltha biflora DC. var. rotundifolia (Huth) C. L. Hitchc. = Caltha leptosepala DC. ■☆

68167 Caltha caespitosa Schipcz.;丛生驴蹄草(簇生驴蹄草);Clustered Marshmarigold, Tufted Marshmarigold ■

68168 Caltha camtschatica Spreng. = Oxygraphis glacialis (Fisch.) Bunge ■

68169 Caltha elata Duthie;高驴蹄草;Tall Marshmarigold ■

68170 Caltha fistulosa Schipcz. = Caltha natans Pall. ■

68171 Caltha fistulosa Schipcz. = Caltha palustris L. var. barthei Hance ■

68172 Caltha fistulosa Schipcz. f. atrorubra W. T. Wang = Caltha natans Pall. ■

68173 Caltha fistulosa Schipcz. f. atrorubra W. T. Wang = Caltha palustris L. var. barthei Hance ■

68174 Caltha flabellifolia Pursh = Caltha palustris L. ■

68175 Caltha glacialis (Fisch. ex DC.) Spreng. = Oxygraphis glacialis (Fisch. ex DC.) Bunge ■

68176 Caltha glacialis Spreng. = Oxygraphis glacialis (Fisch.) Bunge ■

68177 Caltha gracilis Hand.-Mazz. = Caltha sinogracilis W. T. Wang ■

68178 Caltha howellii (Huth) Greene = Caltha leptosepala DC. ■☆

68179 Caltha introloba F. Muell.;内裂驴蹄草■☆

68180 Caltha kamchatica (DC.) Spreng. = Oxygraphis glacialis (Fisch.) Bunge ■

68181 Caltha leptosepala DC.;细萼驴蹄草(白花驴蹄草);Elk's Lip, Marsh Marigold ■☆

68182 Caltha leptosepala DC. var. rotundifolia Huth = Caltha leptosepala DC. ■☆

68183 Caltha leptosepala DC. var. sulfurea C. L. Hitchc. = Caltha leptosepala DC. ■☆

68184 Caltha membranacea (Turcz.) Schipcz. = Caltha palustris L. var. membranacea Turcz. ■

68185 Caltha membranacea (Turcz.) Schipcz. var. grandiflora S. H. Li et Y. H. Huang = Caltha palustris L. var. membranacea Turcz. ■

68186 Caltha natans Pall.;白花驴蹄草;Floating Marsh Marigold, Floating Marshmarigold, Floating Marsh-marigold ■

68187 Caltha natans Pall. var. arctica (R. Br.) Hultén = Caltha palustris L. ■

68188 Caltha natans Pall. var. asarifolia (DC.) Huth = Caltha palustris L. ■

68189 Caltha palustris L.;驴蹄草(立金花,驴蹄菜,马蹄草,马蹄叶,水八角,水葫芦,小马蹄当归);Bachelor's Buttons, Bee's Rest, Bel Buttons, Bel-buttons, Bells Ivy, Big Buttercup, Billy Buttons, Billy-o'-buttons, Blib Blob, Blib-blob, Blogga, Blugga Bludda, Blughtyns, Bobby's Butions, Bobby's Buttons, Bog Daisy, Boots, Bout, Brave Bassinet, Brave Bassinets, Brave Celandine, Bright Meadow, Bull Buttercup, Bull Cup, Bull Flower, Bull Rush, Bulldogs, Bullrushes, Bull's Eyes, Butter Blob, Butter Flower, Butterblob, Butter-blob, Buttercup, Caltrop, Carlicups, Carlock Cups, Chirms, Christ's Eyes, Common Marsh Marigold, Common Marshmarigold, Cow Crane, Cow Cranes, Cow Lily, Cowflock, Cowslip, Cowslop, Crane Johnny, Crazy, Crazy Bess, Crazy Bet, Crazy Bets, Crazy Betsy, Crazy Betty, Crazy Lily, Crazy-pates, Crow Crane, Crow Flower, Cuckoo-pint, Cup-and-saucer, Dale Cup, Dale-cup, Dandelion, Darmell Goddard, Dill-cup, Downscwob, Drunkards, Fiddle, Fire-o'-gold, Gilcup, Gildcup, Gilty Cup, Gilty-cup, Go-cup, Golden Buttercup, Golden Cup, Golden Kingcup, Golden Knob, Goldicup, Goldilocks, Golland, Gollin, Gools, Goulan, Gowan, Gowles, Grandfather's Buttons, Guild, Gules, Gypsy's Money, Halcup, Hobble-gobbles, Horse Blob, Horse Blobs, Horse Buttercup, Horse Hoof, Horse's Hooves, Ivy Bells, Janet-flower, John George, John Georges, Johnny Cranes, Jonette, King Cob, King Cup, King Kong, King-cob, Kingcup, Kingcups, King's Cob, Knob, Liver, Lucken Golland, Mare Blob, Marigold, Marsh Horsegowl, Marsh Lily,

Marsh Marigold, Marsh-marigold, Marybout, Marybud, Mary's Gold, May Blob, May Blub, May Bubble, May Gollen, May-blob, May-blub, May-bubbles, May-flower, Meadow Bout, Meadow Gowan, Meadow-bout, Meadowbright, Mire Blob, Moll Blob, Moll-blob, Molly Blob, Mollyblobs, Monkey Bells, Old Man's Buttons, Open Gowan, Open-gowan, Paddock Flower, Pates Crazy, Policeman's Buttons, Publicans, Saligo, Shrub of Beltaine, Soldier Buttons, Soldier's Buttons, Water Babies, Water Baby, Water Bleb, Water Blob, Water Blubbers, Water Boats, Water Bubbles, Water Buttercup, Water Caltrop, Water Goggles, Water Golland, Water Gowan, Water Gowland, Water Lily, Water Nut, Waterbleb, Yellow Blob, Yellow Boots, Yellow Crazy, Yellow Gowlan, Yellow Marsh Marigold, Yellow Marsh-marigold ■

68190 Caltha palustris L. subsp. arctica (R. Br.) Hultén = Caltha palustris L. ■

68191 Caltha palustris L. subsp. asarifolia (DC.) Hultén = Caltha palustris L. ■

68192 Caltha palustris L. subsp. radicans (J. R. Forst.) Beck = Caltha palustris L. ■

68193 Caltha palustris L. subsp. sibirica (Regel) Hultén = Caltha palustris L. var. sibirica (Regel) Hultén ■

68194 Caltha palustris L. var. arctica (R. Br.) Huth = Caltha palustris L. ■

68195 Caltha palustris L. var. asarifolia (DC.) Huth = Caltha palustris L. ■

68196 Caltha palustris L. var. barthei Hance;空茎驴蹄草(管茎驴蹄草,红花空茎驴蹄草,空管状驴蹄草);Fistulous Marshmarigold, Hollowstem Marsh Marigold, Hollowstem Marshmarigold, Redflower Fistulous Marshmarigold ■

68197 Caltha palustris L. var. barthei Hance f. atrorubra (W. T. Wang) W. T. Wang = Caltha natans Pall. ■

68198 Caltha palustris L. var. barthei Hance f. atrorubra (W. T. Wang) W. T. Wang = Caltha palustris L. var. barthei Hance ■

68199 Caltha palustris L. var. flabellifolia (Pursh) Torr. et A. Gray = Caltha palustris L. ■

68200 Caltha palustris L. var. himalaica Tamura;长柱驴蹄草;Longstyle Marshmarigold ■

68201 Caltha palustris L. var. membranacea Turcz.;膜叶驴蹄草(薄叶驴蹄草,马蹄草立金花,膜质驴蹄草);Membranaceous Marshmarigold ■

68202 Caltha palustris L. var. multiflora Kom. ex Schipcz. = Caltha natans Pall. ■

68203 Caltha palustris L. var. multiflora Kom. ex Schipcz. = Caltha palustris L. var. barthei Hance ■

68204 Caltha palustris L. var. orientali-sinensis X. H. Guo;华东驴蹄草;E. China Marshmarigold ■

68205 Caltha palustris L. var. orientali-sinensis X. H. Guo = Caltha palustris L. ■

68206 Caltha palustris L. var. pygmaea Makino;矮驴蹄草;Dwarf Marshmarigold ■

68207 Caltha palustris L. var. radicans (J. R. Forst.) Huth = Caltha palustris L. ■

68208 Caltha palustris L. var. scaposa (Hook. f. et Thomson) Maxim. = Caltha scaposa Hook. f. et Thomson ■

68209 Caltha palustris L. var. scaposa Maxim. = Caltha scaposa Hook. f. et Thomson ■

68210 Caltha palustris L. var. sibirica (Regel) Hultén;三角叶驴蹄草(驴蹄草);Siberian Marshmarigold ■

68211 Caltha palustris L. var. sibirica Regel subvar. palmata Takeda. = Caltha palustris L. var. umbrosa Diels ■

68212 Caltha palustris L. var. umbrosa Diels;掌叶驴蹄草(掌裂驴蹄草);Shady Marshmarigold ■

68213 Caltha polypetala (Hochst.) Boiss.;多瓣驴蹄草;Manypetal Marshmarigold ■

68214 Caltha radicans J. R. Forst. = Caltha palustris L. ■

68215 Caltha rubriflora B. L. Burtt et Lauener = Caltha sinogracilis W. T. Wang ■

68216 Caltha rubriflora B. L. Burtt et Lauener = Caltha sinogracilis W. T. Wang f. rubriflora (B. L. Burtt et Lauener) W. T. Wang ■

68217 Caltha scaposa Hook. f. et Thomson;花葶驴蹄草(立金花,水八角,水葫芦,小马蹄当归);Scapose Marshmarigold ■

68218 Caltha scaposa Hook. f. et Thomson var. parnassioides Ulbr. = Caltha scaposa Hook. f. et Thomson ■

68219 Caltha scaposa Hook. f. et Thomson var. smithii Ulbr. = Caltha scaposa Hook. f. et Thomson ■

68220 Caltha sibirica (Regel) Tolm. = Caltha palustris L. var. sibirica (Regel) Hultén ■

68221 Caltha silvestris Vorosch.;野生驴蹄草;Wild Marshmarigold ■☆

68222 Caltha sinogracilis W. T. Wang;细茎驴蹄草;Slenderstem Marshmarigold ■

68223 Caltha sinogracilis W. T. Wang f. rubriflora (B. L. Burtt et Lauener) W. T. Wang = Caltha sinogracilis W. T. Wang ■

68224 Caltha sinogracilis W. T. Wang f. rubriflora (B. L. Burtt Lauener) W. T. Wang;红花细茎驴蹄草;Redflower Slenderstem Marshmarigold ■

68225 Caltha uniflora Rydb. = Caltha leptosepala DC. ■☆

68226 Calthaceae Martinov = Ranunculaceae Juss.(保留科名)●■

68227 Calthoides B. Juss. ex DC. = Othonna L. ●■☆

68228 Calucechinus Hombr. et Jacquinot ex Decne. = Nothofagus Blume (保留属名)●☆

68229 Calucechinus Hombr. et Jacquinot(废弃属名) = Nothofagus Blume(保留属名)●☆

68230 Caluera Dodson et Determann(1983);卡卢兰属■☆

68231 Caluera surinamensis Dodson et Determann;卡卢兰■☆

68232 Calusia Bert. ex Steud. = Myrospermum Jacq. ●☆

68233 Calusparassus Hombr. et Jacquinot ex Decne. = Nothofagus Blume(保留属名)●☆

68234 Calusparassus Hombr. et Jacquinot(废弃属名) = Nothofagus Blume(保留属名)●☆

68235 Calvaria C. F. Gaertn. = Sideroxylon L. ●☆

68236 Calvaria Comm. ex C. F. Gaertn. = Sideroxylon L. ●☆

68237 Calvaria diospyroides (Baker) Dubard = Sideroxylon inerme L. subsp. diospyroides (Baker) J. H. Hemsl. ●☆

68238 Calvaria inermis (L.) Dubard = Sideroxylon inerme L. ●☆

68239 Calvaria inermis (L.) Dubard var. zanzibarensis Pierre ex Dubard = Sideroxylon inerme L. subsp. diospyroides (Baker) J. H. Hemsl. ●☆

68240 Calvelia Moq. = Suaeda Forssk. ex J. F. Gmel.(保留属名)●■

68241 Calvoa Hook. f. (1867);非洲野牡丹属■☆

68242 Calvoa angolensis A. Fern. et R. Fern.;安哥拉野牡丹■☆

68243 Calvoa bequaertii De Wild. = Calvoa hirsuta Hook. f. ■☆

68244 Calvoa calliantha Jacq.-Fél. = Calvoa zenkeri Gilg ex Engl. ■☆

68245 Calvoa confertifolia Exell;密集非洲野牡丹■☆

68246 Calvoa crassinoda Hook. f.;粗节非洲野牡丹■☆

68247 Calvoa grandifolia Cogn.;大叶非洲野牡丹☆

68248 Calvoa henriquesii Cogn. = Calvoa hirsuta Hook. f. ■☆
68249 Calvoa hirsuta Hook. f.;毛非洲野牡丹■☆
68250 Calvoa integrifolia Cogn.;全叶非洲野牡丹■☆
68251 Calvoa jacques-felixii Figueiredo;雅凯非洲野牡丹■☆
68252 Calvoa leonardii Jacq. -Fél.;莱奥非洲野牡丹■☆
68253 Calvoa maculata M. E. Leal;斑点非洲野牡丹■☆
68254 Calvoa molleri Gilg = Calvoa crassinoda Hook. f. ■☆
68255 Calvoa monticola A. Chev. ex Hutch. et Dalziel;山地非洲野牡丹■☆
68256 Calvoa orientalis Taub.;东方非洲野牡丹■☆
68257 Calvoa polychaeta Guinea = Calvoa trochainii Jacq. -Fél. ■☆
68258 Calvoa pulcherrima Gilg ex Engl.;美丽非洲野牡丹■☆
68259 Calvoa robusta Cogn. = Calvoa crassinoda Hook. f. ■☆
68260 Calvoa rosularis Gilg ex Engl. = Calvoa hirsuta Hook. f. ■☆
68261 Calvoa sapinii De Wild.;萨潘非洲野牡丹■☆
68262 Calvoa sapinii De Wild. var. angolensis (A. Fern. et R. Fern.) Cavaco = Calvoa angolensis A. Fern. et R. Fern. ■☆
68263 Calvoa seretii De Wild.;赛雷非洲野牡丹■☆
68264 Calvoa seretii De Wild. subsp. wildemaniana (Exell) Figueiredo;怀尔德曼非洲野牡丹■☆
68265 Calvoa sessiliflora Cogn. = Calvoa orientalis Taub. ■☆
68266 Calvoa sinuata Hook. f.;深波非洲野牡丹■☆
68267 Calvoa stenophylla Jacq. -Fél.;窄叶非洲野牡丹■☆
68268 Calvoa subquinquenervia De Wild.;亚五脉非洲野牡丹■☆
68269 Calvoa superba A. Chev. = Amphiblemma cymosum (Schrad. et J. C. Wendl.) Naudin ●☆
68270 Calvoa trochainii Jacq. -Fél.;特罗尚非洲野牡丹■☆
68271 Calvoa uropetala Mildbr. = Calvoa crassinoda Hook. f. ■☆
68272 Calvoa wildemaniana Exell = Calvoa seretii De Wild. subsp. wildemaniana (Exell) Figueiredo ■☆
68273 Calvoa zenkeri Gilg ex Engl.;岑克尔非洲野牡丹■☆
68274 Calycacanthus K. Schum. (1889);新几内亚爵床属■☆
68275 Calycacanthus magnusianus K. Schum.;新几内亚爵床■☆
68276 Calycadenia DC. (1836);腺萼菊属■☆
68277 Calycadenia DC. = Hemizonia DC. ■☆
68278 Calycadenia ciliosa Greene = Calycadenia fremontii A. Gray ■☆
68279 Calycadenia elegans Greene = Calycadenia fremontii A. Gray ■☆
68280 Calycadenia fremontii A. Gray;弗氏腺萼菊■☆
68281 Calycadenia hispida (Greene) Greene = Calycadenia multiglandulosa DC. ■☆
68282 Calycadenia hispida (Greene) Greene subsp. reducta D. D. Keck = Calycadenia multiglandulosa DC. ■☆
68283 Calycadenia micrantha R. L. Carr et G. D. Carr;小花腺萼菊■☆
68284 Calycadenia mollis A. Gray;柔软腺萼菊■☆
68285 Calycadenia multiglandulosa DC.;多腺腺萼菊■☆
68286 Calycadenia multiglandulosa DC. subsp. bicolor (Greene) D. D. Keck = Calycadenia multiglandulosa DC. ■☆
68287 Calycadenia multiglandulosa DC. subsp. cephalotes (DC.) D. D. Keck = Calycadenia multiglandulosa DC. ■☆
68288 Calycadenia multiglandulosa DC. subsp. robusta D. D. Keck = Calycadenia multiglandulosa DC. ■☆
68289 Calycadenia oppositifolia (Greene) Greene;对叶腺萼菊■☆
68290 Calycadenia pauciflora A. Gray;少花腺萼菊■☆
68291 Calycadenia pauciflora A. Gray var. elegans Jeps. = Calycadenia fremontii A. Gray ■☆
68292 Calycadenia plumosa Kellogg = Blepharizonia plumosa (Kellogg) Greene ■☆
68293 Calycadenia spicata (Greene) Greene;穗状腺萼菊■☆
68294 Calycadenia tenella (Nutt.) Torr. et A. Gray = Osmadenia tenella Nutt. ■☆
68295 Calycadenia truncata DC.;平截腺萼菊;Rosinweed ■☆
68296 Calycadenia truncata DC. subsp. scabrella (Drew) D. D. Keck = Calycadenia truncata DC. ■☆
68297 Calycampe O. Berg = Myrcia DC. ex Guill. ●☆
68298 Calycandra Lepr. ex A. Rich. = Cordyla Lour. ●☆
68299 Calycandra pinnata Lepr. ex A. Rich. = Cordyla pinnata (Lepr. ex A. Rich.) Milne-Redh. ●☆
68300 Calycanthaceae Lindl. (1819)(保留科名);蜡梅科;Allspice Family, Calycanthus Family, Strawbcrryshrub Family, Strawberry Shrub Family, Strawberry-Shrub Family ●
68301 Calycanthemeae [L.] Vent. = Lythraceae J. St. -Hil. (保留科名)■●
68302 Calycanthemeae Vent. = Lythraceae J. St. -Hil. (保留科名)■●
68303 Calycanthemum Klotzsch = Ipomoea L. (保留属名)●■
68304 Calycanthemum leucanthemum Klotzsch = Ipomoea leucanthemum (Klotzsch) Hallier f. ●☆
68305 Calycantherum Klotzsch = Ipomoea L. (保留属名)●■
68306 Calycanthus L. (1759)(保留属名);夏蜡梅属(美国蜡梅属,夏腊梅属,洋蜡梅属,泽蜡梅属);Allspice, Calycanthus, Carolina Allspice, Spicebush, Strawberry Shrub, Strawberryshrub, Strawberry-shrub, Sweet Shrub, Sweet-scented Bush, Sweet-scented Shrub, Sweetshrub ●
68307 Calycanthus brockianus Ferry et Ferry f. = Calycanthus floridus L. ●
68308 Calycanthus chinensis (W. C. Cheng et S. Y. Chang) W. C. Cheng et S. Y. Chang ex P. T. Li;夏蜡梅(大叶柴,黄梅花,蜡木,牡丹木,夏梅);China Allspice, Chinese Allspice, Chinese Calycanthus, Chinese Sweet Shrub, Chinese Sweetshrub ●◇
68309 Calycanthus chinensis W. C. Cheng et S. Y. Chang = Calycanthus chinensis (W. C. Cheng et S. Y. Chang) W. C. Cheng et S. Y. Chang ex P. T. Li ●◇
68310 Calycanthus fertilis Walter;东南夏蜡梅(东南美国蜡梅);Allspice, Carolina Allspice, Pale sweetshrnb Pale Sweet Shrub, Smooth Sweet Shrub, Sweetshrub ●
68311 Calycanthus fertilis Walter = Calycanthus floridus L. var. glaucus (Willd.) Torr. et A. Gray ●☆
68312 Calycanthus fertilis Walter = Calycanthus glaucus Willd. ●☆
68313 Calycanthus floridus L.;美国夏蜡梅(美国蜡梅);Allspice, Atllspice, Carolina, Carolina Allspice, Carolina-allspice, Common Sweet Shrub, Common Sweetshrub, Common Sweet-shrub, Pale Sweetshrub, Pineapple Shrub, Spicebush, Strawberry Shrub, Sweet Betsy, Sweet Bubby Bush, Sweet Bush, Sweet Shrub, Sweetshrub ●
68314 Calycanthus floridus L. 'Athens';加州夏蜡梅;Carolina Allspice, Common Sweetshrub ●☆
68315 Calycanthus floridus L. var. glaucus (Willd.) Torr. et A. Gray = Calycanthus glaucus Willd. ●☆
68316 Calycanthus floridus L. var. laevigatus (Willd.) Torr. et A. Gray = Calycanthus floridus L. var. glaucus (Willd.) Torr. et A. Gray ●☆
68317 Calycanthus floridus L. var. laevigatus (Willd.) Torr. et A. Gray = Calycanthus glaucus Willd. ●☆
68318 Calycanthus floridus L. var. oblongifolius (Nutt.) Boufford et Spongberg = Calycanthus floridus L. var. glaucus (Willd.) Torr. et A. Gray ●☆
68319 Calycanthus floridus L. var. oblongifolius (Nutt.) Boufford et

Spongberg = Calycanthus glaucus Willd. ●☆
68320 Calycanthus fragrans = Chimonanthus praecox (L.) Link ●
68321 Calycanthus glaucus Willd.;粉绿夏蜡梅;Glaucous Allspice, Pale Sweet-shrub ●☆
68322 Calycanthus glaucus Willd. = Calycanthus floridus L. var. glaucus (Willd.) Torr. et A. Gray ●☆
68323 Calycanthus laevigatus Willd.;光叶红;Smooth Carolina Allspice ●
68324 Calycanthus laevigatus Willd. = Calycanthus floridus L. var. laevigatus (Willd.) Torr. et A. Gray ●☆
68325 Calycanthus mohrii (Small) Pollard = Calycanthus floridus L. ●
68326 Calycanthus nanus (Loisel.) Small = Calycanthus floridus L. var. glaucus (Willd.) Torr. et A. Gray ●☆
68327 Calycanthus nanus (Loisel.) Small = Calycanthus glaucus Willd. ●☆
68328 Calycanthus nitens (Oliv.) Rehder = Chimonanthus nitens Oliv. ●
68329 Calycanthus occidentalis Hook. et Arn.;西美夏蜡梅(加州夏蜡梅,西美蜡梅);California spicebush, Californian Allspice, Spice Bush, Sweetshrub ●☆
68330 Calycanthus praecox L. = Chimonanthus praecox (L.) Link ●
68331 Calycera Cav. (1797)(保留属名);萼角花属(萼角属,头花草属)■☆
68332 Calycera balsamitifolia Rich.;萼角花;Balsam-leaf Calycera ●☆
68333 Calyceraceae R. Br. ex Rich. (1820)(保留科名);萼角花科(萼角科,头花草科)■●☆
68334 Calycinum alyssoides L. = Alyssum alyssoides (L.) L. ■
68335 Calycium Elliott = Heterotheca Cass. ■☆
68336 Calycobolus Schult. = Calycobolus Willd. ex Schult. ●☆
68337 Calycobolus Willd. ex Roem. et Schult. = Calycobolus Willd. ex Schult. ●☆
68338 Calycobolus Willd. ex Roem. et Schult. = Prevostea Choisy ●☆
68339 Calycobolus Willd. ex Schult. (1819);落萼旋花属●☆
68340 Calycobolus acuminatus (Pilg.) Heine;渐尖落萼旋花●☆
68341 Calycobolus acutus (Pilg.) Heine;尖落萼旋花●☆
68342 Calycobolus africanus (G. Don) Heine;非洲落萼旋花●☆
68343 Calycobolus bampsianus Lejoly et Lisowski;邦氏落萼旋花●☆
68344 Calycobolus breviflorus (De Wild.) Heine = Calycobolus heudelotii (Baker ex Oliv.) Heine ●☆
68345 Calycobolus cabrae (De Wild. et T. Durand) Heine = Calycobolus heudelotii (Baker ex Oliv.) Heine ●☆
68346 Calycobolus campanulatus (K. Schum. ex Hallier f.) Heine;风铃草状落萼旋花●☆
68347 Calycobolus campanulatus (K. Schum. ex Hallier f.) Heine subsp. oddonii (De Wild.) Lejoly et Lisowski;奥氏落萼旋花●☆
68348 Calycobolus claessensii (De Wild.) Heine;克莱森斯落萼旋花●☆
68349 Calycobolus cordatus (Hallier f.) Heine;心形落萼旋花●☆
68350 Calycobolus gilgianus (Pilg.) Heine;吉尔格落萼旋花●☆
68351 Calycobolus goodii Heine;古得落萼旋花●☆
68352 Calycobolus heudelotii (Baker ex Oliv.) Heine;短花落萼旋花●☆
68353 Calycobolus insignis (Rendle) Heine;显著落萼旋花●☆
68354 Calycobolus kasaiensis Lejoly et Lisowski;开赛落萼旋花●☆
68355 Calycobolus klaineanus (Pierre ex Pellegr.) Heine;克莱恩落萼旋花●☆
68356 Calycobolus letouzeyanus Lejoly et Lisowski;勒图落萼旋花●☆
68357 Calycobolus longiracemosus Lejoly et Lisowski;长序落萼旋花●☆
68358 Calycobolus mayombensis (Pellegr.) Heine;马永贝落萼旋花●☆
68359 Calycobolus mortehanii (De Wild.) Heine;莫特汉落萼旋花●☆
68360 Calycobolus oddonii (De Wild.) Heine = Calycobolus campanulatus (K. Schum. ex Hallier f.) Heine subsp. oddonii (De Wild.) Lejoly et Lisowski ●☆
68361 Calycobolus parviflorus (Mangenot) Heine;小花落萼旋花●☆
68362 Calycobolus petitianus Lejoly et Lisowski;佩蒂蒂落萼旋花●☆
68363 Calycobolus racemosus (R. D. Good) Heine;总花落萼旋花●☆
68364 Calycobolus robynsianus Lejoly et Lisowski;罗宾斯落萼旋花●☆
68365 Calycobolus thollonii Lejoly et Lisowski;托伦落萼旋花●☆
68366 Calycobolus upembaensis Lejoly et Lisowski;乌彭贝落萼旋花●☆
68367 Calycobolus zairensis Lejoly;扎伊尔落萼旋花●☆
68368 Calycocarpum (Nutt.) Spach = Calycocarpum Nutt. ex Torr. et A. Gray ●☆
68369 Calycocarpum Nutt. ex Spach = Calycocarpum Nutt. ex Torr. et A. Gray ●☆
68370 Calycocarpum Nutt. ex Torr. et A. Gray (1838);杯子藤属;Cupseed ●☆
68371 Calycocarpum lyonii (Pursh) A. Gray;杯子藤;Cupseed ●☆
68372 Calycocorsus F. W. Schmidt = Chondrilla L. ■
68373 Calycocorsus F. W. Schmidt = Willemetia Neck. ex Cass. ■☆
68374 Calycocorsus F. W. Schmidt(1795);肖鳞果苣属■☆
68375 Calycocorsus hieracioides F. W. Schmidt;肖鳞果苣■☆
68376 Calycocorsus tuberosus (Fisch. et Mey. ex DC.) Rauschert;块状肖鳞果苣■☆
68377 Calycodaphne Bojer = Ocotea Aubl. ●☆
68378 Calycodendron A. C. Sm. (1936);萼木属●☆
68379 Calycodendron A. C. Sm. = Psychotria L. (保留属名)●
68380 Calycodendron milneri (A. Gray) A. C. Sm.;米奈萼木●☆
68381 Calycodon Nutt. = Muhlenbergia Schreb. ■
68382 Calycodon Wendl. = Hyospathe Mart. ●☆
68383 Calycogonium DC. (1828);萼叶茜属●☆
68384 Calycogonium angulatum Griseb.;窄萼叶茜●☆
68385 Calycogonium biflorum Cogn.;双花萼叶茜●☆
68386 Calycogonium brevifolium Urb. et Ekman;短叶萼叶茜●☆
68387 Calycogonium calycopteris Urb.;美翅萼叶茜●☆
68388 Calycogonium glabratum DC.;无毛萼叶茜●☆
68389 Calycogonium lanceolatum Griseb.;剑叶萼叶茜●☆
68390 Calycogonium maculatum Urb. et Ekman;斑点萼叶茜●☆
68391 Calycogonium microphyllum Wright;小叶萼叶茜●☆
68392 Calycogonium pauciflorum Wright;少花萼叶茜●☆
68393 Calycogonium reticulatum (Cogn.) Judd et Skean;网萼叶茜●☆
68394 Calycogonium rubens Borhidi;红萼叶茜●☆
68395 Calycogonium saxicola Britton et P. Wilson;岩生萼叶茜●☆
68396 Calycolpus O. Berg(1856);沟萼桃金娘属●☆
68397 Calycolpus alternifolius (Gleason) Landrum;异花沟萼桃金娘●☆
68398 Calycolpus angustifolius L. Riley;窄叶沟萼桃金娘●☆
68399 Calycolpus australis Landrum;澳洲沟萼桃金娘●☆
68400 Calycolpus calophyllus O. Berg;美叶沟萼桃金娘●☆
68401 Calycolpus cordatus L. Riley;心形沟萼桃金娘●☆
68402 Calycolpus glaber O. Berg;光沟萼桃金娘●☆
68403 Calycolpus gracilis (O. Berg) L. Riley;纤细沟萼桃金娘●☆
68404 Calycolpus ovalifolius O. Berg;卵叶沟萼桃金娘●☆
68405 Calycomelia Kostel. = Fraxinus L. ●
68406 Calycomis D. Don = Acrophyllum Benth. ●☆
68407 Calycomis R. Br. = Callicoma Andréws ●☆
68408 Calycomis R. Br. ex T. Nees et Sinning = Callicoma Andréws ●☆
68409 Calycomis T. Nees = Callicoma Andréws ●☆
68410 Calycomorphum C. Presl = Trifolium L. ■

68411 Calycomorphum subterraneum C. Presl = Trifolium subterraneum L. ■☆

68412 Calycopeplus Planch. (1861);萼被大戟属☆

68413 Calycopeplus ephedroides Planch.;萼被大戟☆

68414 Calycophisum H. Karst. et Triana = Calycophysum H. Karst. et Triana ■☆

68415 Calycophyllum DC. (1830);萼叶木属●☆

68416 Calycophyllum candidissimum (Vahl) DC.;白萼叶木(萼叶茜草);Degame,Degame Lancewood,Degami ●☆

68417 Calycophyllum megistocaulum (K. Krause) C. M. Taylor;萼叶木●☆

68418 Calycophyllum multiflorum Griseb.;多花萼叶木;Palo Amarillo ●☆

68419 Calycophyllum spruceanum (Benth.) Hook. f. ex K. Schum.;云杉萼叶木●☆

68420 Calycophysum H. Karst. et Triana(1855);南美葫芦属■☆

68421 Calycophysum Triana = Calycophysum H. Karst. et Triana ■☆

68422 Calycophysum brevipes Pittier;短梗南美葫芦■☆

68423 Calycophysum pedunculatum H. karst. et Triana;南美葫芦■☆

68424 Calycoplectus Oerst. = Alloplectus Mart. (保留属名)●■☆

68425 Calycopteris Lam. (1811);翅萼使君子属●

68426 Calycopteris Lam. = Getonia Roxb. ●

68427 Calycopteris Lam. ex Poir. = Getonia Roxb. ●

68428 Calycopteris Poir. = Calycopteris Lam. ●

68429 Calycopteris Poir. = Getonia Roxb. ●

68430 Calycopteris Rich. ex DC. = Calycogonium DC. ●☆

68431 Calycopteris floribunda (Roxb.) Lam. = Getonia floribunda Roxb. ●

68432 Calycopteris floribunda (Roxb.) Lam. ex Poir. = Getonia floribunda Roxb. ●

68433 Calycopteris joan Siebold et Zucc. = Buckleya lanceolata (Siebold et Zucc.) Miq. ●

68434 Calycopteris nutans (Roxb.) Kurz = Calycopteris floribunda (Roxb.) Lam. ●

68435 Calycopteris nutans (Roxb.) Kurz var. glabriuscula Kurz = Calycopteris floribunda (Roxb.) Lam. ●

68436 Calycopteris nutans (Roxb.) Kurz var. roxburghii Kurz = Calycopteris floribunda (Roxb.) Lam. ●

68437 Calycopterls Siebold = Buckleya Torr. (保留属名)●

68438 Calycorectes O. Berg(1856);直萼木属●☆

68439 Calycorectes australis D. Legrand;南方直萼木●☆

68440 Calycorectes densiflorus Nied.;密花直萼木●☆

68441 Calycorectes ferrugineus Mattos;锈色直萼木●☆

68442 Calycorectes macrocalyx Rusby;大萼直萼木●☆

68443 Calycorectes maximus McVaugh;大直萼木●☆

68444 Calycoseris A. Gray(1853);杯苣属;Tackstem,Tack-stem ■☆

68445 Calycoseris parryi A. Gray;黄杯苣;Tackstem,Yellow tack-stem ■☆

68446 Calycoseris wrightii A. Gray;白杯苣;Tackstem,White Tack-stem ■☆

68447 Calycosia A. Gray(1858);索岛茜属☆

68448 Calycosia fragrans Gillespie;香索岛茜☆

68449 Calycosia glabra Turrill;光滑索岛茜☆

68450 Calycosia laxiflora Gillespie;疏花索岛茜☆

68451 Calycosia trichocalyx Drake;毛萼索岛茜☆

68452 Calycosiphonia (Pierre) Lebrun = Coffea L. ●

68453 Calycosiphonia Pierre ex Robbr. (1981);管萼茜属●☆

68454 Calycosiphonia macrochlamys (K. Schum.) Leroy = Calycosiphonia macrochlamys (K. Schum.) Robbr. ●☆

68455 Calycosiphonia macrochlamys (K. Schum.) Robbr.;大被管萼茜●☆

68456 Calycosiphonia spathicalyx (K. Schum.) Lebrun = Calycosiphonia spathicalyx (K. Schum.) Robbr. ●☆

68457 Calycosiphonia spathicalyx (K. Schum.) Robbr.;匙萼管萼茜●☆

68458 Calycosorus Endl. = Calycocorsus F. W. Schmidt ■☆

68459 Calycosorus Endl. = Chondrilla L. ■

68460 Calycostegia Lem. = Calystegia R. Br. (保留属名)■

68461 Calycostemma Hanst. = Isoloma Decne. ●■☆

68462 Calycostemma Hanst. = Kohleria Regel ●■☆

68463 Calycostylis Hort. = Beloperone Nees ■☆

68464 Calycostylis Hort. ex Viim. = Beloperone Nees ■☆

68465 Calycothrix Meisn. = Calytrix Labill. ●☆

68466 Calycotome E. Mey. = Dichilus DC. ■☆

68467 Calycotome E. Mey. = Melinospermum Walp. ■☆

68468 Calycotome Link = Calicotome Link ●☆

68469 Calycotome Link(1822);刺桂豆属(刺桂属);Thorny Broom ●☆

68470 Calycotome grosii Pau et Font Quer = Calycotome intermedia (Salzm.) C. Presl ●☆

68471 Calycotome infesta (C. Presl) Guss.;多刺刺桂豆;Thorny Broom ●☆

68472 Calycotome infesta (C. Presl) Guss. subsp. intermedia (C. Presl) Greuter = Calycotome intermedia (Salzm.) C. Presl ●☆

68473 Calycotome intermedia (Salzm.) C. Presl;间型刺桂豆●☆

68474 Calycotome rigida (Viv.) Maire et Weiller;硬刺桂豆●☆

68475 Calycotome spinosa (L.) Link;刺桂豆(刺桂)●☆

68476 Calycotome spinosa (L.) Link = Calycotome infesta (C. Presl) Guss. ●☆

68477 Calycotome spinosa (L.) Link subsp. rigida (Viv.) Maire = Calycotome rigida (Viv.) Maire et Weiller ●☆

68478 Calycotome villosa (Poir.) Link;长柔毛刺桂豆●☆

68479 Calycotome villosa (Poir.) Link subsp. intermedia (Salzm.) Maire = Calycotome intermedia (Salzm.) C. Presl ●☆

68480 Calycotome villosa (Poir.) Link var. grosii (Pau et Font Quer) Maire = Calycotome grosii Pau et Font Quer ●☆

68481 Calycotome villosa (Poir.) Link var. intermedia (Salzm.) Ball = Calycotome intermedia (Salzm.) C. Presl ●☆

68482 Calycotome villosa (Poir.) Link var. rigida (Viv.) Bég. et Vacc. = Calycotome rigida (Viv.) Maire et Weiller ●☆

68483 Calycotomon Hoffmanns. = Calicotome Link ●☆

68484 Calycotomus Rich. = Conostegia D. Don ■☆

68485 Calycotomus Rich. ex DC. = Conostegia D. Don ■☆

68486 Calycotropis Turcz. (1862);墨西哥白鼓钉属■☆

68487 Calycotropis Turcz. = Polycarpaea Lam. (保留属名)■●

68488 Calycotropis minuartioides Turcz.;墨西哥白鼓钉■☆

68489 Calyctenium Greene = Rubus L. ●■

68490 Calyculogygas Krapov. (1960);手萼锦葵属●☆

68491 Calyculogygas uruguayensis Krapov.;手萼锦葵●☆

68492 Calydermos Lag. = Calea L. ●■☆

68493 Calydermos Ruiz et Pav. = Nicandra Adans. (保留属名)■

68494 Calydorea Herb. (1843);矛鞘鸢尾属■☆

68495 Calydorea coelestina (W. Bartram) Goldblatt et Henrich;北美矛鞘鸢尾;Bartram's Ixia,Celestial-lily ■☆

68496 Calydorea pallens Griseb. = Tamia pallens (Griseb.) Ravenna ■☆

68497 Calydorea punctata (Herb.) Baker = Alophia drummondii (Graham) R. C. Foster ■☆

68498 Calydorea texana (Herb.) Baker = Nemastylis geminiflora

Nutt. ■☆
68499 Calygogonium G. Don = Calycogonium DC. ●☆
68500 Calygonium D. Dietr. = Calycogonium DC. ●☆
68501 Calylophis Spach = Calylophus Spach ■☆
68502 Calylophus Spach(1835);北美夜来香属;Evening-primrose ■☆
68503 Calylophus australis Towner et P. H. Raven = Calylophus serrulatus (Nutt.) P. H. Raven ■☆
68504 Calylophus drummondianus Spach;德拉蒙德北美夜来香;Texas Primrose ■☆
68505 Calylophus drumondii Spach ex Steud.;德氏北美夜来香;Sun Drops ■☆
68506 Calylophus hartwegii (Benth.) P. H. Raven;哈氏北美夜来香;Sierra Sundrop, Sundrop ■☆
68507 Calylophus hartwegii (Benth.) P. H. Raven subsp. fendleri (A. Gray) Towner et P. H. Raven;芬氏北美夜来香;Fendler's Sundrops ■☆
68508 Calylophus serrulatus (Nutt.) P. H. Raven;黄北美夜来香;Plains Evening Primrose, Plains Yellow Primrose, Toothed Evening Primrose, Yellow Evening-primrose, Yellow Sundrops ■☆
68509 Calymenia Pers. = Calyxhymenia Ortega ■
68510 Calymenia Pers. = Mirabilis L. ■
68511 Calymenia Pers. = Oxybaphus L'Hér. ex Willd. ■
68512 Calymeris Post et Kuntze = Aster L. ●■
68513 Calymeris Post et Kuntze = Kalimeris (Cass.) Cass. ■
68514 Calymmandra Torr. et A. Gray = Evax Gaertn. ■☆
68515 Calymmandra candida Torr. et A. Gray = Diaperia candida (Torr. et A. Gray) Benth. et Hook. f. ■☆
68516 Calymmanthera Schltr. (1913);纱药兰属■☆
68517 Calymmanthera filiformis Schltr.;线形纱药兰■☆
68518 Calymmanthera major Schltr.;大纱药兰■☆
68519 Calymmanthera montana Schltr.;山地纱药兰■☆
68520 Calymmanthera paniculata Schltr.;圆锥纱药兰■☆
68521 Calymmanthera tenuis Schltr.;细纱药兰■☆
68522 Calymmanthium F. Ritter(1962);灌木柱属●☆
68523 Calymmanthium substerile F. Ritter;灌木柱■☆
68524 Calymmatium O. E. Schulz(1933);面纱芥属■☆
68525 Calymmatium draboides (Korsh.) O. L. Schulz;面纱芥■☆
68526 Calymmostachya Bremek. = Justicia L. ●■
68527 Calyntranthele Nied. = Byrsonima Rich. ex Juss. ●☆
68528 Calynux Raf. = Calinux Raf. ●
68529 Calynux Raf. = Pyrularia Michx. ●
68530 Calyplectus Ruiz et Pav. = Lafoensia Vand. ●
68531 Calypso Salisb. (1807)(保留属名);布袋兰属(匙唇兰属);Calypso, Fairy-slipper ■
68532 Calypso Thouars(废弃属名) = Calypso Salisb. (保留属名)■
68533 Calypso Thouars(废弃属名) = Johnia Roxb. ●
68534 Calypso Thouars(废弃属名) = Salacia L. (保留属名)●
68535 Calypso africana (Willd.) G. Don = Loeseneriella africana (Willd.) N. Hallé ●☆
68536 Calypso americana R. Br. = Calypso bulbosa (L.) Oakes var. americana (R. Br.) Luer ■☆
68537 Calypso borealis (Sw.) Salisb. = Calypso bulbosa (L.) Oakes ■
68538 Calypso bulbosa (L.) Oakes;布袋兰(鳞茎匙唇兰);Calypso, Calypso Orchid, Common Calypso, Fairy-slipper Orchid, Northern Calypso ■
68539 Calypso bulbosa (L.) Oakes f. albiflora Satomi;白花布袋兰■☆
68540 Calypso bulbosa (L.) Oakes f. occidentalis Holz. = Calypso bulbosa (L.) Oakes var. occidentalis (Holz.) B. Boivin ■☆
68541 Calypso bulbosa (L.) Oakes var. americana (R. Br.) Luer;美国布袋兰;Calypso Orchid, Fairy-slipper Orchid ■☆
68542 Calypso bulbosa (L.) Oakes var. japonica Makino;日本布袋兰■☆
68543 Calypso bulbosa (L.) Oakes var. occidentalis (Holz.) B. Boivin;西方布袋兰■☆
68544 Calypso bulbosa (L.) Oakes var. speciosa (Schltr.) Makino;美丽布袋兰(布袋兰)■
68545 Calypso debilis G. Don = Salacia debilis (G. Don) Walp. ●☆
68546 Calypso sativa L.;栽培布袋兰■☆
68547 Calypso speciosa Schltr. = Calypso bulbosa (L.) Oakes var. speciosa (Schltr.) Makino ■
68548 Calypsodium Link = Calypso Salisb. (保留属名)■
68549 Calypteriopetalon Hassk. = Croton L. ●
68550 Calypthrantes Raeusch. = Calyptranthes Sw. (保留属名)●☆
68551 Calyptocarpus Less. (1832);金腰箭舅属(隐果菊属)■●
68552 Calyptocarpus blepharolepis B. L. Rob. = Calyptocarpus vialis Less. ●
68553 Calyptocarpus vialis Less.;金腰箭舅;Straggler Daisy ●
68554 Calyptochloa C. E. Hubb. (1933);昆士兰隐草属■☆
68555 Calyptochloa gracillima C. E. Hubb.;昆士兰隐草■☆
68556 Calyptosepalum S. Moore = Drypetes Vahl ●
68557 Calyptosepalum S. Moore(1925);隐萼大戟属●☆
68558 Calyptosepalum sumatranum S. Moore;隐萼大戟●☆
68559 Calyptospermum A. Dietr. = Bolivaria Cham. et Schltdl. ●☆
68560 Calyptostylis Arènes = Rhynchophora Arènes ●☆
68561 Calyptostylis Arènes(1946);隐柱金虎尾属●☆
68562 Calyptostylis humbertii Arènes = Rhynchophora phillipsonii W. R. Anderson ●☆
68563 Calyptracordia Britton = Cordia L. (保留属名)●
68564 Calyptracordia Britton = Varronia P. Browne ●☆
68565 Calyptracordia abyssinica (R. Br.) Friesen = Cordia africana Lam. ●☆
68566 Calyptraemalva Krapov. (1965);异锦葵属●☆
68567 Calyptraemalva catharinensis Krapov.;异锦葵●☆
68568 Calyptranthe (Maxim.) Nakai = Hydrangea L. ●
68569 Calyptranthera Klack. (1996);隐药萝藦属☆
68570 Calyptranthera baronii Klack.;巴龙隐药萝藦●☆
68571 Calyptranthera brevicaudata Klack.;短尾隐药萝藦●☆
68572 Calyptranthera caudiclava (Choux) Klack.;棒尾隐药萝藦●☆
68573 Calyptranthera filifera Klack.;丝隐药萝藦●☆
68574 Calyptranthera gautieri Klack.;戈捷隐药萝藦●☆
68575 Calyptranthera grandiflora Klack.;大花隐药萝藦●☆
68576 Calyptranthera pubipetala Klack.;短毛瓣隐药萝藦●☆
68577 Calyptranthera schatziana Klack.;沙茨隐药萝藦●☆
68578 Calyptranthera sulphurea Klack.;硫色隐药萝藦●☆
68579 Calyptranthera villosa Klack.;长柔毛隐药萝藦●☆
68580 Calyptranthes Sw. (1788)(保留属名);冠花树属●☆
68581 Calyptranthes chytraculis (L.) Sw. var. americana McVaugh;美国冠花树●☆
68582 Calyptranthes grandiflora O. Berg;大花冠花树●☆
68583 Calyptranthes guineensis Willd. = Syzygium guineense (Willd.) DC. ●☆
68584 Calyptranthes longifolia O. Berg;长叶冠花树●☆
68585 Calyptranthes lucida Mart. ex DC.;光亮冠花树●☆
68586 Calyptranthes mangiferifolia Hance ex Walp. = Cleistocalyx operculatus (Roxb.) Merr. et L. M. Perry ●

68587 Calyptranthes mangiferifolia Hance ex Walp. = Syzygium nervosum DC. ●
68588 Calyptranthes micrantha Wright ex Griseb.;小花冠花树●☆
68589 Calyptranthus Blume = Syzygium R. Br. ex Gaertn.(保留属名)●
68590 Calyptranthus Juss. = Calyptranthes Sw.(保留属名)●☆
68591 Calyptranthus Thouars = Capparis L. ●
68592 Calyptraria Naudin = Centronia D. Don ●☆
68593 Calyptrella Naudin = Graffenrieda DC. ●☆
68594 Calyptridium Nutt. = Calyptridium Nutt. ex Torr. et A. Gray ■☆
68595 Calyptridium Nutt. = Cistanthe Spach ■☆
68596 Calyptridium Nutt. ex Torr. et A. Gray(1838);裂果猫爪苋属■☆
68597 Calyptridium monandrum Nutt. = Cistanthe monandra (Nutt.) Hershk. ■☆
68598 Calyptridium monospermum Greene = Cistanthe monosperma (Greene) Hershk. ■☆
68599 Calyptridium parryi A. Gray = Cistanthe parryi (A. Gray) Hershk. ■☆
68600 Calyptridium parryi A. Gray subsp. nevadense (J. T. Howell) Munz = Cistanthe parryi (A. Gray) Hershk. ■☆
68601 Calyptridium parryi A. Gray var. arizonicum J. T. Howell = Cistanthe parryi (A. Gray) Hershk. ■☆
68602 Calyptridium parryi A. Gray var. nevadense J. T. Howell = Cistanthe parryi (A. Gray) Hershk. ■☆
68603 Calyptridium pulchellum (Eastw.) Hoover = Cistanthe pulchella (Eastw.) Hershk. ■☆
68604 Calyptridium pygmaeum Parish ex Rydb. = Cistanthe pygmaea (Parish ex Rydb.) Hershk. ■☆
68605 Calyptridium quadripetalum S. Watson = Cistanthe quadripetala (S. Watson) Hershk. ■☆
68606 Calyptridium roseum S. Watson = Cistanthe rosea (S. Watson) Hershk. ■☆
68607 Calyptridium umbellatum (Torr.) Greene;裂果猫爪苋;Pussy-paws ■☆
68608 Calyptridium umbellatum (Torr.) Greene = Cistanthe umbellata (Torr.) Hershk. ■☆
68609 Calyptridium umbellatum (Torr.) Greene var. caudiciferum (A. Gray) Jeps. = Cistanthe umbellata (Torr.) Hershk. ■☆
68610 Calyptrimalva Krapov. = Calyptraemalva Krapov. ●☆
68611 Calyptrion Ging. = Corynostylis Mart. ■☆
68612 Calyptrion Ging. ex DC. = Corynostylis Mart. ■☆
68613 Calyptriopetalum Hassk. ex Müll. Arg. = Croton L. ●
68614 Calyptrocalyx Blume(1838);隐萼椰子属(被萼榈属,盖萼棕属,隐萼榈属,隐萼椰属);Calyptrocalyx,Henahena Palm ●
68615 Calyptrocalyx angustifrons Becc.;窄花隐萼椰子(窄花盖萼棕)●☆
68616 Calyptrocalyx laxiflorus Becc.;疏花隐萼椰子(疏花盖萼棕)●☆
68617 Calyptrocalyx spicatus Blume;隐萼椰子;Nibung,Pinang Utan ●
68618 Calyptrocarpus Rchb. = Calyptocarpus Less. ■●
68619 Calyptrocarya Nees(1834);隐果莎草属■☆
68620 Calyptrocarya angustifolia Nees;狭叶隐果莎草■☆
68621 Calyptrocarya bicolor (H. Pfeiff.) Koyama;二色隐果莎草■☆
68622 Calyptrocarya bicolor Nees = Calyptrocarya bicolor (H. Pfeiff.) Koyama ■☆
68623 Calyptrocarya brevicaulis Nees;短茎隐果莎草■☆
68624 Calyptrocarya monocephala Steud.;单头隐果莎草■☆
68625 Calyptrochilum Kraenzl. (1895);帽唇兰属■☆
68626 Calyptrochilum christyanum (Rchb. f.) Summerh.;帽唇兰■☆
68627 Calyptrochilum emarginatum (Sw.) Schltr.;无边帽唇兰■☆
68628 Calyptrochilum orientale Schltr. = Calyptrochilum christyanum (Rchb. f.) Summerh. ■☆
68629 Calyptrochilum preussii Kraenzl. = Calyptrochilum emarginatum (Sw.) Schltr. ■☆
68630 Calyptrocoryne Schott = Theriophonum Blume ■☆
68631 Calyptrogenia Burret(1941);热美桃金娘属●☆
68632 Calyptrogenia biflora Alain;双花热美桃金娘●☆
68633 Calyptrogenia grandiflora Burret;大花热美桃金娘●☆
68634 Calyptrogenia riedeliana (O. Berg ex Mart.) Burret;热美桃金娘●☆
68635 Calyptrogyne H. Wendl. (1859);草椰属(被蕊榈属,盖雌棕属,隐雌椰属,隐蕊椰子属)●☆
68636 Calyptrogyne ghiesbreghtiana (Linden et H. Wendl.) H. Wendl.;草椰●☆
68637 Calyptrolepis Steud. = Rhynchospora Vahl(保留属名)■
68638 Calyptromyrcia O. Berg = Myrcia DC. ex Guill. ●☆
68639 Calyptronoma Griseb. (1864);肖椰子属●☆
68640 Calyptronoma dulcis H. Wendl.;肖椰子●☆
68641 Calyptronoma rivalis (O. F. Cook) L. H. Bailey;溪肖椰子●☆
68642 Calyptroon Miq. = Baccaurea Lour. ●
68643 Calyptropetalum Post et Kuntze = Calyptriopetalum Hassk. ex Müll. Arg. ●
68644 Calyptropetalum Post et Kuntze = Croton L. ●
68645 Calyptropsidium O. Berg = Psidium L. ●
68646 Calyptropsidium friedrichsthalianum O. Berg = Psidium friedrichsthalianum (O. Berg) Nied. ●☆
68647 Calyptrosciadium Rech. f. et Kuber(1964);隐伞芹属■☆
68648 Calyptrosciadium polycladum Rech. f. et Kuber;隐伞芹■☆
68649 Calyptrosicyos Rabenant. = Corallocarpus Welw. ex Benth. et Hook. f. ■☆
68650 Calyptrosicyos grevei Rabenant. = Corallocarpus grevei (Rabenant.) Rabenant. ■☆
68651 Calyptrosicyos perrieri Rabenant. = Corallocarpus perrieri (Rabenant.) Rabenant. ■☆
68652 Calyptrospatha Klotzsch ex Baill. = Acalypha L. ●■
68653 Calyptrospatha pubiflora (Baill.) Klotzsch = Acalypha pubiflora Baill. ■☆
68654 Calyptrospermum A. Dietr. = Bolivaria Cham. et Schltdl. ●☆
68655 Calyptrospermum A. Dietr. = Menodora Humb. et Bonpl. ●☆
68656 Calyptrostegia C. A. Mey. = Pimelea Banks ex Gaertn.(保留属名)●☆
68657 Calyptrostigma Klotzsch = Beyeria Miq. ☆
68658 Calyptrostigma Trautv. et C. A. Mey. = Macrodiervilla Nakai ●☆
68659 Calyptrostigma Trautv. et C. A. Mey. = Wagneria Lem. ●☆
68660 Calyptrostigma Trautv. et C. A. Mey. = Weigela Thunb. ●
68661 Calyptrostylis Nees = Rhynchospora Vahl(保留属名)■
68662 Calyptrotheca Gilg(1897);冠盖树属●☆
68663 Calyptrotheca somalensis Gilg;索马里冠盖树●☆
68664 Calyptrotheca taitense (Pax et Vatke) Brenan;冠盖树●☆
68665 Calysaccion Wight = Mammea L. ●
68666 Calysaccion Wight = Ochrocarpos Thouars ●
68667 Calysericos Eckl. et Zeyh. = Cryptadenia Meisn. ●☆
68668 Calysericos Eckl. et Zeyh. ex Meisn. = Cryptadenia Meisn. ●☆
68669 Calysphyrum Bunge = Weigela Thunb. ●
68670 Calysphyrum floridum Bunge = Weigela florida (Bunge) A. DC. ●
68671 Calystegia R. Br. (1810)(保留属名);打碗花属(滨旋花属);

Bindweed, Calystegia, Glorybind ■

68672 Calystegia abyssinica Engl. = Calystegia hederacea Wall. ex Roxb. ■

68673 Calystegia acetosifolia (Turcz.) Turcz. = Calystegia hederacea Wall. ex Roxb. ■

68674 Calystegia acetosifolia Turcz. = Calystegia hederacea Wall. ex Roxb. ■

68675 Calystegia calystegioides Choisy = Calystegia hederacea Wall. ex Roxb. ■

68676 Calystegia dahurica (Herb.) Choisy; 毛打碗花(打碗花,大收旧花,夫儿妙,狗狗秧,狗娃秧,马刺楷); Dahuria Calystegia, Dahuria Glorybind, Hairy Bindweed ■

68677 Calystegia dahurica (Herb.) Choisy = Calystegia pellita (Ledeb.) G. Don ■

68678 Calystegia dahurica (Herb.) Choisy f. anestia (Fernald) H. Hara; 缠枝牡丹■

68679 Calystegia dahurica (Herb.) Choisy f. anestia (Fernald) H. Hara = Calystegia pubescens Lindl. ■

68680 Calystegia dahurica (Herb.) Choisy var. pellita (Ledeb.) Choisy = Calystegia pellita (Ledeb.) G. Don ■

68681 Calystegia dahurica (Herb.) Choisy var. pellitus Choisy = Calystegia pellita (Ledeb.) G. Don ■

68682 Calystegia dahurica (Herb.) G. Don = Calystegia pulchra Brummitt et Heywood ■☆

68683 Calystegia dahurica (Herb.) G. Don = Calystegia sepium (L.) R. Br. subsp. spectabilis Brummitt ■

68684 Calystegia dahurica Herb. f. anestia (Fernald) H. Hara = Calystegia pubescens Lindl. ■

68685 Calystegia glabrata (Hallier f.) Chiov. = Convolvulus kilimandschari Engl. ■☆

68686 Calystegia hederacea Wall. ex Roxb.; 打碗花(常春藤叶打碗花,常春藤叶天剑,打破碗花,大碗花,扶苗,扶七秧子,扶秧,扶子苗,葍,葍子根,富苗秧,富斯劳草,钩耳藤,狗儿蔓,狗儿完,狗儿秧,狗耳苗,狗耳丸,胶旋花,喇叭花,来江藤儿,老母猪草,面根,面根草,面根藤,面根藤儿,奶浆藤,南面根,盘肠参,铺地参,蒲地参,矢叶旋花,兔儿苗,兔耳草,小旋花,旋花苦,燕覆子,秧子根,栅栏打碗花,走丝牡丹); Ivy Glorybind, Ivy-like Calystegia, Japanese Bindweed, Japanese False Bindweed, Syrian Bindweed ■

68687 Calystegia hederacea Wall. ex Roxb. var. elongata Liou et Y. Ling = Calystegia hederacea Wall. ex Roxb. ■

68688 Calystegia hederacea Wall. f. pentapetala (Makino) H. Hara; 五瓣打碗花■☆

68689 Calystegia hederacea Wall. f. sakuraii Hiyama; 白花打碗花■☆

68690 Calystegia hederacea Wall. var. elongata Liou et Y. Ling = Calystegia hederacea Wall. ex Roxb. ■

68691 Calystegia japonica (Thunb.) Koidz. = Calystegia hederacea Wall. ex Roxb. ■

68692 Calystegia japonica Choisy; 日本打碗花(长裂篱天,长裂旋花,打碗花,夫儿苗,狗狗秧,狗娃秧,日本天剑,旋花); Japan Glorybind, Japan Hedge Glorybind ■

68693 Calystegia japonica Choisy = Calystegia pellita (Ledeb.) G. Don ■

68694 Calystegia japonica Choisy = Calystegia pubescens Lindl. ■

68695 Calystegia japonica Choisy = Calystegia sepium (L.) R. Br. var. japonica (Choisy) Makino ■

68696 Calystegia japonica Choisy f. albiflora (Makino) H. Hara = Calystegia pubescens Lindl. f. albiflora (Makino) Yonek. ■☆

68697 Calystegia japonica Choisy f. albiflora (Makino) H. Hara = Calystegia pubescens Lindl. ■

68698 Calystegia japonica Choisy var. albiflora Makino = Calystegia pubescens Lindl. ■

68699 Calystegia macounii (Greene) Brummitt; 篱笆打碗花; Hedge Bindweed ■☆

68700 Calystegia maritimus Lam. = Calystegia soldanella (L.) R. Br. ■

68701 Calystegia pellita (Ledeb.) G. Don; 藤长苗(缠绕天剑,打碗花,大土拉苗,辐儿苗,狗儿秧,狗藤花,毛胡弯,兔耳苗,脱毛天剑,野山药,野兔子苗); Densepubescent Glorybind, Skinned Calystegia ■

68702 Calystegia pellita (Ledeb.) G. Don subsp. longifolia Brummitt; 长叶藤长苗■

68703 Calystegia pellita (Ledeb.) G. Don subsp. stricta Brummitt; 直立藤长苗■

68704 Calystegia physoides Pomel = Calystegia sylvatica (Kit.) Griseb. ■

68705 Calystegia pubescens Lindl.; 柔毛打碗花; Hedge Bindweed ■

68706 Calystegia pubescens Lindl. 'Flore Pleno'; 重瓣柔毛打碗花■☆

68707 Calystegia pubescens Lindl. f. albiflora (Makino) Yonek.; 白花柔毛打碗花■☆

68708 Calystegia pubescens Lindl. f. major (Makino) Yonek.; 大柔毛打碗花■☆

68709 Calystegia pulchra Brummitt et Heywood; 美丽打碗花■☆

68710 Calystegia reniformis (R. Br.) Poir. = Calystegia soldanella (L.) R. Br. ■

68711 Calystegia reniformis R. Br. = Calystegia soldanella (L.) R. Br. ■

68712 Calystegia sepium (L.) Pursh var. pubescens A. Gray = Calystegia sepium (L.) R. Br. ■

68713 Calystegia sepium (L.) R. Br.; 旋花(苞颈草,串枝莲,打破碗花花,打碗花,扽肠草,吊茄子,饭豆藤,饭藤,饭藤子,葍花,葍旋花,葍子根,狗儿释花,狗儿弯藤,鼓子花,挂金灯,鸡屎条,筋根,筋根花,宽叶打碗花,篱打碗花,篱天剑,美草,美花草,面根藤,面根藤儿,牵枝牡丹,葬花,天剑草,兔儿苗,乌鮎花,续筋根,旋葍花,旋花苗,燕葍花,野苕,肫肠草); Bearbind, Bedwind, Bell Woodbind, Bellbind, Bellbinder, Bellbine, Bellflower, Bellwind, Bellwinder, Beswind, Beswlne, Bethroot, Bind, Bine Lily, Binf-lily, Bryony, Bunks, Campanelle, Chemise, Corn Lily, Creep Ivy, Creeper, Creeping Jenny, Creep-ivy, Daddy's White Shirt, Devil's Garter, Devil's Garters, Devil's Guts, Devil's Nightcap, Devil's Vine, Fairy Trumpet, Fairy Trumpets, Flags, Grandmother's Nightcap, Granny's Bonnet, Granny's Bonnets, Granny's Night Bonnet, Granny's Night-bonnets, Granny's Nightcap, Great Bind, Great Bindweed, Ground Ivy, Harvest Lily, Hedge Bells, Hedge Bindweed, Hedge Calystegia, Hedge False Bindweed, Hedge Glorybind, Hedge Lily, Hellweed, Holland Smock, Honeysuckle, Hooded Bindweed, Jack-run-in-the-country, Jack-run-in-the-hedge, Knutsford Devil, Lady's Chemise, Lady's Nightcap, Lady's Shimmey, Lady's Shimmy, Lady's Smock, Lady's Umbrella, Larger Bindweed, Larger Sunshade, Lily, Lily-bind, Lily-flower, London Bells, Milkmaids, Morning Glory, Mother's Nightcap, Nightcap, Nightshirt, Nit Clickers, Old Lady's Smock, Old Man's Nightcap, Old Man's Shirt, Old Woman's Nightcap, Our Lady's Nightcap, Our Lady's Smock, Pisspot, Robin-run-the-hedge, Rutland Beauty, Shimmies, Shimmies-and-shirt, Shimmies-and-shirts, Shimmy-and-buttons, Shimmy-shirt, Shimmy-shirts, Smock, Snake's Flower, Strangleweed, Tare, Trumpet, Trumpet Flower, Umbrella, Vicar's Tresses, Wavewind, Wavewine, Waywind, Weatherwind, Weedbind, White Bellflower, White Lily, White Shirt, White Smock, Wild Lily,

Wild Morning Glory, Winding Lily, Wireweed, Witherwine, Withewind, With-vine, Withwind, Withywind, Woodbine ■
68714 Calystegia sepium (L.) R. Br. subsp. americana (Sims) Brummitt = Calystegia sepium (L.) R. Br. ■
68715 Calystegia sepium (L.) R. Br. subsp. angulata Brummitt = Calystegia sepium (L.) R. Br. ■
68716 Calystegia sepium (L.) R. Br. subsp. spectabilis Brummitt;欧旋花■
68717 Calystegia sepium (L.) R. Br. subsp. sylvatica (Kit.) Batt. = Calystegia sylvatica (Kit.) Griseb. ■
68718 Calystegia sepium (L.) R. Br. var. americana Matsuda = Calystegia sepium (L.) R. Br. var. communis (R. M. Tryon) H. Hara ■
68719 Calystegia sepium (L.) R. Br. var. angulata (Brummitt) N. H. Holmgren = Calystegia sepium (L.) R. Br. ■
68720 Calystegia sepium (L.) R. Br. var. barbara (Pomel) Batt. = Calystegia sylvatica (Kit.) Griseb. ■
68721 Calystegia sepium (L.) R. Br. var. communis (R. M. Tryon) H. Hara;阔叶打碗花(宽叶打碗花);Broadleaf Hedge Glorybind ■
68722 Calystegia sepium (L.) R. Br. var. hirsuta K. T. Fu;硬毛打碗花;Hirsute Hedge Calystegia ■
68723 Calystegia sepium (L.) R. Br. var. integrifolia Liou et Y. Ling = Calystegia sepium (L.) R. Br. subsp. spectabilis Brummitt ■
68724 Calystegia sepium (L.) R. Br. var. japonica (Choisy) Makino;长裂旋花(长裂篱天,打碗花,夫儿苗,狗狗秧,狗娃秧,日本天剑,旋花);Japan Hedge Glorybind ■
68725 Calystegia sepium (L.) R. Br. var. japonica (Choisy) Makino = Calystegia pubescens Lindl. ■
68726 Calystegia sepium (L.) R. Br. var. japonica (Choisy) Makino = Calystegia japonica Choisy ■
68727 Calystegia sepium (L.) R. Br. var. japonica (Choisy) Makino f. albiflora (Makino) T. Yamaz. = Calystegia pubescens Lindl. f. albiflora (Makino) Yonek. ■☆
68728 Calystegia sepium (L.) R. Br. var. maritima (Gouan) Choisy = Calystegia sepium (L.) R. Br. ■
68729 Calystegia sepium (L.) R. Br. var. physoides (Pomel) Sauvage et Vindt = Calystegia sylvatica (Kit.) Griseb. ■
68730 Calystegia sepium (L.) R. Br. var. repens (L.) A. Gray = Calystegia sepium (L.) R. Br. ■
68731 Calystegia sepium (L.) R. Br. var. tangerina Pau = Calystegia soldanella (L.) R. Br. ex Roem. et Schult. ■
68732 Calystegia silvatica (Kit.) Griseb.;林生打碗花;Hedge Bindweed, Large Bindweed, Shortstalk False Bindweed ■☆
68733 Calystegia silvatica (Kit.) Griseb. subsp. orientalis Brummitt;鼓子花■
68734 Calystegia silvatica Choisy = Calystegia silvatica (Kit.) Griseb. ■☆
68735 Calystegia silvatica Choisy subsp. orientalis Brummitt = Calystegia silvatica (Kit.) Griseb. subsp. orientalis Brummitt ■
68736 Calystegia soldanella (L.) R. Br. = Calystegia soldanella (L.) R. Br. ex Roem. et Schult. ■
68737 Calystegia soldanella (L.) R. Br. ex Roem. et Schult.;肾叶打碗花(滨打碗花,滨旋花,扶子苗,李扇草,马鞍藤,沙马藤,肾叶天剑);Beach Morning Glory, Kidneyshaped Leaf Calystegia, Prince's Flower, Red Cole, Scottish Scurvy-grass, Sea Bells, Sea Bindweed, Sea Cawle, Sea Colewort, Sea Foalfoot, Seashore Glorybind ■
68738 Calystegia soldanella (L.) Roem. et Schult. = Calystegia soldanella (L.) R. Br. ex Roem. et Schult. ■
68739 Calystegia soldanella (L.) Roem. et Schult. f. albiflora H. Hara;白花肾叶打碗花■☆
68740 Calystegia soldanella (L.) Roem. et Schult. f. rubriflora Asai;红花肾叶打碗花■☆
68741 Calystegia soldanelloides Makino = Calystegia soldanella (L.) R. Br. ■
68742 Calystegia spithamaea (L.) Pursh;小打碗花;Dwarf Morning Glory, Low Bindweed, Low False Bindweed, Upright Bindweed ■☆
68743 Calystegia spithamaea (L.) Pursh subsp. stans (Michx.) Brummitt = Calystegia spithamaea (L.) Pursh ■☆
68744 Calystegia subvolubilis (Ledeb.) G. Don = Calystegia pellita (Ledeb.) G. Don ■
68745 Calystegia sylvatica (Kit.) Griseb. = Calystegia sylvestris (Willd.) Roem. et Schult. ■☆
68746 Calystegia sylvestris (Willd.) Roem. et Schult.;林地打碗花;American Bellbine, Great Bindweed ■☆
68747 Calystegia wallichianus Spreng. = Calystegia hederacea Wall. ex Roxb. ■
68748 Calythrix DC. = Calytrix Labill. ●☆
68749 Calythrix Labill. = Calytrix Labill. ●☆
68750 Calythropsis C. A. Gardner = Calytrix Labill. ●☆
68751 Calythropsis C. A. Gardner(1942);萼红木属●☆
68752 Calythropsis aurea C. A. Gardner;萼红木●☆
68753 Calytriplex Ruiz et Pav. = Bacopa Aubl.(保留属名)■
68754 Calytriplex Ruiz et Pav. = Brami Adans.(废弃属名)■
68755 Calytrix Labill.(1806);星花木属;Fringe Myrtle, Fringe-myrtle, Fringe-myrtles, Star Flower ●☆
68756 Calytrix alpestris (Lindl.) Court;高山星花木;Grampians Fringe Myrtle, Snow Myrtle ●☆
68757 Calytrix aurea Lindl.;金星花木;Golden Fringe Myrtle ●☆
68758 Calytrix depressa (Turcz.) Benth.;细叶星花木●☆
68759 Calytrix exstipulata DC.;北部星花木;Kimberley Heath, Northern Fringe Myrtle, Turkey Bush ●☆
68760 Calytrix sullivanii (F. Muell.) B. D. Jacks.;澳洲星花木;Grampians Fringe-myrtle ●☆
68761 Calytrix tetragona Labill.;四棱星花木;Common Fringe-myrtle, Fringe Myrtle ●☆
68762 Calyxhymenia Ortega = Mirabilis L. ■
68763 Camacum Adans. ex Steud. = Camacum Steud. ●
68764 Camacum Steud. = Comacum Adans. ●
68765 Camacum Steud. = Myristica Gronov.(保留属名)●
68766 Camaion Raf. = Helicteres L. ●
68767 Camara Adans. = Lantana L.(保留属名)●
68768 Camara vulgaris Benth. = Lantana camara L. ●
68769 Camarandraceae Dulac = Rhamnaceae Juss.(保留科名)●
68770 Camarea A. St. -Hil.(1823);拱顶金虎尾属●☆
68771 Camarea affinis A. St. -Hil.;近缘拱顶金虎尾●☆
68772 Camarea axillaris A. St. -Hil.;腋生拱顶金虎尾●☆
68773 Camarea linearifolia A. St. -Hil.;线叶拱顶金虎尾●☆
68774 Camarea pulchella Griseb.;美丽拱顶金虎尾●☆
68775 Camarea salicifolia Chodat;柳叶拱顶金虎尾●☆
68776 Camarea sericea A. St. -Hil.;绢毛拱顶金虎尾●☆
68777 Camarea triphylla Mart.;三叶拱顶金虎尾●☆
68778 Camaridium Lindl. = Maxillaria Ruiz et Pav. ■☆
68779 Camarilla Salisb. = Allium L. ■
68780 Camarinnea Bubani et Penz. = Empetrum L. ●

68781 Camarotea Scott-Elliot(1891);拱顶爵床属☆
68782 Camarotea souiensis Scott-Elliot;拱顶爵床●☆
68783 Camarotis Lindl.(1833);拱顶兰属(卡马洛兰属);Camarotis ■☆
68784 Camarotis Lindl. = Micropera Lindl. ■
68785 Camarotis philippinensis Lindl.;菲律宾拱顶兰(卡马洛兰);Philippine Camarotis ■☆
68786 Camarotis poilanei (Guillaumin) Seidenf. et Smitinand = Micropera poilanei (Guillaumin) Garay ■
68787 Camarotis rostrata (Roxb.) Rchb. f.;卡马洛兰;Common Camarotis ■☆
68788 Camassia Eckl. ex Pfeiff. = Gonioma E. Mey. ●☆
68789 Camassia Lindl.(1832)(保留属名);克美莲属(雏百合属,卡马莲属,卡玛百合属);Camas,Camash,Camass,Camassia,Camus Lily,Quamash,Wild Hyacinth ■☆
68790 Camassia angusta (Engelm. et A. Gray) Blank.;草原克美莲;Prairie Camas,Wild Hyacinth ■☆
68791 Camassia cusickii S. Watson;克美莲;Cusick Camas, Cusick Camass,Indian Quamash,Wild Hyacinth ■☆
68792 Camassia esculenta (Ker Gawl.) B. L. Rob. = Camassia scilloides (Raf.) Cory ■☆
68793 Camassia esculenta Lindl. var. leichtlinii (Baker) Baker = Camassia leichtlinii (Baker) S. Watson ■☆
68794 Camassia esculenta Rob.;印度克美莲;Bear Grass, Camass, Common Camas, Common Camass, Cusickii Camass, Indian Quamash,Quamash,Wild Hyacinth ■☆
68795 Camassia esculenta Rob. = Camassia fraseri Torr. ■☆
68796 Camassia esculenta Rob. = Camassia quamash (Pursh) Greene ■☆
68797 Camassia fraseri Torr.;弗氏克美莲■☆
68798 Camassia fraseri Torr. var. angusta (Engelm. et A. Gray) Torr. = Camassia angusta (Engelm. et A. Gray) Blank. ■☆
68799 Camassia howellii S. Watson;豪氏克美莲;Howell Cmas ■☆
68800 Camassia leichtlinii (Baker) S. Watson;卡玛百合(克美莲);Leichtlin Camass,Small Camas ■☆
68801 Camassia leichtlinii (Baker) S. Watson 'Semiplena';半重瓣克美莲■☆
68802 Camassia leichtlinii (Baker) S. Watson subsp. suksdorfii (Greenm.) Gould;苏克卡玛百合■☆
68803 Camassia quamash (Pursh) Greene;大克美莲(可食卡马莲,蓝克美莲);Blue Camassia, Camass, Common Camas, Common Camash, Common Camass, Common Camassia, Large Camas, Quamash ■☆
68804 Camassia quamash (Pursh) Greene subsp. azurea (A. Heller) Gould;天蓝大克美莲■☆
68805 Camassia quamash (Pursh) Greene subsp. walpolei (Piper) Gould;瓦氏大克美莲■☆
68806 Camassia scilloides (Raf.) Cory;大西洋克美莲;Atlantic Camas,Eastern Camas,Eastern Camass,Wild Hyacinth ■☆
68807 Camassia scilloides (Raf.) Cory f. petersenii Steyerm. = Camassia scilloides (Raf.) Cory ■☆
68808 Camassia scilloides (Raf.) Cory f. variegata Steyerm. = Camassia scilloides (Raf.) Cory ■☆
68809 Camassia suksdorfii Greenm. = Camassia leichtlinii (Baker) S. Watson subsp. suksdorfii (Greenm.) Gould ■☆
68810 Camassia walpolei (Piper) J. F. Macbr. = Camassia quamash (Pursh) Greene subsp. walpolei (Piper) Gould ■☆
68811 Camax Schreb. = Diospyros L. ●
68812 Camax Schreb. = Ropourea Aubl. ●
68813 Cambania Comm. ex M. Roem. = Dysoxylum Blume ●
68814 Cambderia Steud. = Campderia A. Rich. ■☆
68815 Cambderia Steud. = Vellozia Vand. ■☆
68816 Cambea Endl. = Careya Roxb.(保留属名)●☆
68817 Cambea Endl. = Cumbia Buch. -Ham. ●☆
68818 Cambessedea Kunth(废弃属名) = Buchanania Spreng. ●
68819 Cambessedea Kunth(废弃属名) = Cambessedesia DC.(保留属名)●■☆
68820 Cambessedea Wight et Arn. = Bouea Meisn. ●
68821 Cambessedesia DC.(1828)(保留属名);巴南野牡丹属●■☆
68822 Cambessedesia gracilis Wurdack;纤细巴南野牡丹☆
68823 Cambessedesia latevenosa DC.;巴南野牡丹☆
68824 Cambessedesia purpurata DC.;紫巴南野牡丹☆
68825 Cambessedesia rugosa Cogn.;皱巴南野牡丹☆
68826 Cambessedesia tenuis Markgr.;细巴南野牡丹☆
68827 Cambogia L. = Garcinia L. ●
68828 Cambogiaceae Horan. = Clusiaceae Lindl.(保留科名)●■
68829 Cambogiaceae Horan. = Guttiferae Juss.(保留科名)●■
68830 Camchaya Gagnep.(1920);凋缨菊属;Camchaya ■
68831 Camchaya calcarea Kitam. = Koyamasia calcarea (Kitam.) H. Rob. ■☆
68832 Camchaya kampotensis Gagnep.;柬凋缨菊■☆
68833 Camchaya loloana Kerr;凋缨菊(罗罗菊);Common Camchaya ■
68834 Camdenia Scop. = Evolvulus L. ●■
68835 Camdenia Scop. = Vistnu Adans. ●■
68836 Camderia Dumort. = Heritiera J. F. Gmel. ■☆
68837 Camderia Dumort. = Lachnanthes Elliott(保留属名)■☆
68838 Camelia Raf. = Camellia L. ●
68839 Camelina Crantz(1762);亚麻荠属;Camelina, False Flax, Falseflax,Gold of Pleasure,Gold-of-pleasure ■
68840 Camelina albiflora Kotschy ex Boiss.;白花亚麻荠■
68841 Camelina alyssum (Mill.) Thell.;庭荠状亚麻荠;Gold-of-pleasure ■☆
68842 Camelina barbareaedolia DC. = Rorippa barbareifolia (DC.) Kitag. ■
68843 Camelina barbareifolia DC. = Rorippa barbareifolia (DC.) Kitag. ■
68844 Camelina caucasica (Sinskaya) Vassilcz.;高加索亚麻荠■☆
68845 Camelina caucasica (Sinskaya) Vassilcz. = Camelina sativa (L.) Crantz ■
68846 Camelina glabrata (DC.) Fritsch = Camelina sativa (L.) Crantz ■
68847 Camelina hispida Boiss.;硬毛亚麻荠■☆
68848 Camelina laxa C. A. Mey.;疏松亚麻荠■☆
68849 Camelina longistyla Bordz. = Camelina microcarpa Andrz. ex DC. ■
68850 Camelina microcarpa Andrz. ex DC.;小果亚麻荠(长梗亚麻荠);Dutch Flax, Flaxweed, Gold-of-pleasure, Lesser Gold-of-pleasure, Littlepod False Flax, Little-pod False Flax, Longstalk Camelina, Longstalk Smallleaf Falseflax, Smallfruit Camelina, Smallfruit Falseflax,Small-fruited False Flax ■
68851 Camelina microcarpa Andrz. ex DC. f. longistipata C. H. An = Camelina microcarpa Andrz. ex DC. ■
68852 Camelina microcarpa DC. = Camelina microcarpa Andrz. ex DC. ■
68853 Camelina microcarpa DC. subsp. sylvestris (Wallr.) Hiitonen = Camelina microcarpa DC. ■
68854 Camelina microphylla C. H. An;小叶亚麻荠;Smallleaf Camelina,Smallleaf Falseflax ■
68855 Camelina microphylla C. H. An = Camelina microcarpa Andrz. ex

DC. ■
68856 Camelina parodii Ibarra et La Porte;帕罗德亚麻荠■☆
68857 Camelina pilosa (DC.) N. W. Zinger;纤毛亚麻荠■☆
68858 Camelina pilosa (DC.) N. W. Zinger = Camelina sativa (L.) Crantz ■
68859 Camelina rumelica Velen.;优美亚麻荠;Graceful False Flax ■☆
68860 Camelina sativa (L.) Crantz;亚麻荠(无毛亚麻荠);Big-seeded False Flax, Common False Flax, Cultivated Camelina, Cultivated Falseflax, Dodder-cake Plant, False Flax, Gold of Pleasure, Gold-of-pleasure, Large-seed False Flax, Oilseed ■
68861 Camelina sativa (L.) Crantz subsp. alyssum (Mill.) Thell. = Camelina alyssum (Mill.) Thell. ■☆
68862 Camelina sativa (L.) Crantz subsp. microcarpa (Andrz. ex DC.) Schmid = Camelina microcarpa Andrz. ex DC. ■
68863 Camelina sativa (L.) Crantz subsp. microcarpa (DC.) Hegi et Schmid = Camelina microcarpa DC. ■
68864 Camelina sativa (L.) Crantz subsp. pilosa (DC.) N. W. Zinger;毛亚麻荠■☆
68865 Camelina sativa (L.) Crantz var. caucasica Sinskaya = Camelina sativa (L.) Crantz ■
68866 Camelina sativa (L.) Crantz var. glabrata DC. = Camelina sativa (L.) Crantz ■
68867 Camelina sativa (L.) Crantz var. pilosa DC. = Camelina sativa (L.) Crantz subsp. pilosa (DC.) N. W. Zinger ■☆
68868 Camelina sativa (L.) Crantz var. pilosa DC. = Camelina sativa (L.) Crantz ■
68869 Camelina sativa (L.) Crantz var. sylvestris (Wallr.) Coss. et Germ. = Camelina microcarpa DC. ■
68870 Camelina souliei Batt. = Guenthera amplexicaulis (Desf.) Gómez-Campo subsp. souliei (Batt.) Gómez-Campo ■☆
68871 Camelina sylvestris Wallr.;野亚麻荠;Wild Camelina, Wild Falseflax ■
68872 Camelina sylvestris Wallr. = Camelina microcarpa Andrz. ex DC. ■
68873 Camelina yunnanensis W. W. Sm.;云南亚麻荠;Yunnan Camelina, Yunnan Falseflax ■
68874 Camelina yunnanensis W. W. Sm. = Rorippa globosa (Turcz. ex Fisch. et C. A. Mey.) Hayek ■
68875 Camelinopsis A. G. Mill. (1978);曲柄荠属■☆
68876 Camelinopsis campylopoda (Bornm. et Gauba) A. G. Mill.;曲柄荠■☆
68877 Camellia L. (1753);山茶属(茶属);Calla Lily, Camellia, Tea, Tea Bush, Tea Plant, Trumpet Lily ●
68878 Camellia 'Julie';尤利山茶;Julie Camellia ●☆
68879 Camellia 'Winter's Rose';冬蔷薇山茶;Winter's Rose Camellia ●☆
68880 Camellia achrysantha Hung T. Chang et S. Ye Liang;中东金花茶;Middle-east Tea ●
68881 Camellia achrysantha Hung T. Chang et S. Ye Liang = Camellia petelotii (Merr.) Sealy ●
68882 Camellia acuticalyx Hung T. Chang;尖萼瘤果茶;Acutesepal Camellia, Sharpcalyx Camellia ●
68883 Camellia acuticalyx Hung T. Chang = Camellia anlungensis Hung T. Chang var. acutiperulata (Hung T. Chang et C. X. Ye) T. L. Ming ●
68884 Camellia acutiperulata Hung T. Chang et C. X. Ye = Camellia anlungensis Hung T. Chang var. acutiperulata (Hung T. Chang et C. X. Ye) T. L. Ming ●
68885 Camellia acutisepala H. T. Tsai et K. M. Feng = Camellia forrestii (Diels) Cohen-Stuart var. acutisepala (H. T. Tsai et K. M. Feng) Hung T. Chang ●
68886 Camellia acutisepala Hung T. Chang = Camellia anlungensis Hung T. Chang var. acutiperulata (Hung T. Chang et C. X. Ye) T. L. Ming ●
68887 Camellia acutisepala Sealy = Camellia longicalyx Hung T. Chang ●
68888 Camellia acutiserrata Hung T. Chang;尖齿离蕊茶;Acute-leaved Camellia, Acutiserrate Camellia, Sharptooth Camellia ●
68889 Camellia acutiserrata Hung T. Chang = Camellia anlungensis Hung T. Chang var. acutiperulata (Hung T. Chang et C. X. Ye) T. L. Ming ●
68890 Camellia acutiserrata Hung T. Chang = Camellia yunnanensis (Pit. ex Diels) Cohen-Stuart ●
68891 Camellia acutissima Hung T. Chang = Camellia cuspidata (Kochs) Wright ex Garden var. grandiflora Sealy ●
68892 Camellia albescens Hung T. Chang;褪色红山茶;Albescent Camellia ●
68893 Camellia albescens Hung T. Chang = Camellia reticulata Lindl. ●◇
68894 Camellia albogigas Hu;大白山茶;White-giant Camellia, White-large Camellia ●
68895 Camellia albogigas Hu = Camellia granthamiana Sealy ●◇
68896 Camellia albo-sericea Hung T. Chang;白丝毛红山茶(白丝红山茶);White-sericeous Camellia, Whitesilk Camellia ●
68897 Camellia albosericea Hung T. Chang = Camellia reticulata Lindl. ●◇
68898 Camellia albovillosa Hu ex Hung T. Chang;白毛红山茶;White-hairs Camellia, Whitevillose Camellia, White-villosed Camellia ●
68899 Camellia albovillosa Hu ex Hung T. Chang = Camellia pitardii Cohen-Stuart ●
68900 Camellia albovillosa Hung T. Chang = Camellia reticulata Lindl. ●◇
68901 Camellia aluberculata Hung T. Chang;赤水红山茶;Chishui Camellia ●
68902 Camellia amplexicaulis (Pit.) Cohen;越南抱茎山茶;Viet Nam Camellia ●☆
68903 Camellia amplexifolia Merr. et Chun;抱茎短蕊茶;Amplexicaul Camellia, Amplexicaul Tea, Amplexifoliate Camellia ●
68904 Camellia angustifolia Hung T. Chang;狭叶茶(狭叶山茶);Angustifoliate Camellia, Narrowleaf Camellia, Narrowleaf Tea ●
68905 Camellia angustifolia Hung T. Chang = Camellia sinensis (L.) Kuntze var. pubilimba Hung T. Chang ●
68906 Camellia anlungensis Hung T. Chang;安龙瘤果茶;Anlong Camellia ●
68907 Camellia anlungensis Hung T. Chang var. acutiperulata (Hung T. Chang et C. X. Ye) T. L. Ming;尖苞瘤果茶;Acutiperulate Anlong Camellia, Acutiperulate Camellia, Sharpperula Camellia ●
68908 Camellia apolyodonta Hung T. Chang et Q. M. Chen;假多齿红山茶;False Many-tooth Camellia, Shammanytooth Camellia ●
68909 Camellia apolyodonta Hung T. Chang et Q. M. Chen = Camellia polyodonta F. C. How ex Hu var. longicaudata (Hung T. Chang et S. Ye Liang ex Hung T. Chang) T. L. Ming ●
68910 Camellia arborescens Hung T. Chang et F. L. Yu = Camellia sinensis (L.) Kuntze ●
68911 Camellia arborescens Hung T. Chang, F. L. Yu et P. S. Wang;大树茶;Big-tree Camellia, Tree Tea ●
68912 Camellia arborescens Hung T. Chang, F. L. Yu et P. S. Wang = Camellia sinensis (L.) Kuntze ●
68913 Camellia assamica (J. W. Mast.) Hung T. Chang = Camellia

sinensis (L.) Kuntze var. assamica (Mast.) Kitam. ●◇

68914 Camellia assamica (J. W. Mast.) Hung T. Chang var. kucha (Hung T. Chang et H. S. Wang) Hung T. Chang et H. S. Wang = Camellia sinensis (L.) Kuntze var. assamica (Mast.) Kitam. ●◇

68915 Camellia assamica (J. W. Mast.) Hung T. Chang var. polyneura (Hung T. Chang et Y. J. Tang) Hung T. Chang = Camellia sinensis (L.) Kuntze var. assamica (Mast.) Kitam. ●◇

68916 Camellia assamica (Mast.) Hung T. Chang = Camellia sinensis (L.) Kuntze var. assamica (Mast.) Kitam. ●◇

68917 Camellia assamica (Mast.) Hung T. Chang var. kucha (Hung T. Chang, H. S. Wang et B. H. Chen) Hung T. Chang et H. S. Wang;苦茶;Kucha Camellia ●

68918 Camellia assamica (Mast.) Hung T. Chang var. polyneura (Hung T. Chang et Ts. Tang) Hung T. Chang;多脉普洱茶;Many-nernes Camellia ●

68919 Camellia assimilis Champ. = Camellia caudata Wall. ●

68920 Camellia assimilis Champ. ex Benth.;香港毛蕊茶;Hongkong Camellia, Hongkong Tea ●

68921 Camellia assimiloides Sealy;大萼毛蕊茶;Big-calyxed Camellia, Largecalyx Tea, Like Hongkong Camellia ●

68922 Camellia atrothea Hung T. Chang et H. S. Wang = Camellia crassicolumna Hung T. Chang ●

68923 Camellia atrothea Hung T. Chang, H. S. Wang et B. H. Chen;老黑茶;Black Camellia, Dark Tea ●

68924 Camellia atrothea Hung T. Chang, H. S. Wang et B. H. Chen = Camellia crassicolumna Hung T. Chang ●

68925 Camellia atuberculata Hung T. Chang;直脉瘤果茶;Atuberculate Camellia, Tumorless Camellia ●

68926 Camellia atuberculata Hung T. Chang = Camellia tuberculata S. S. Chien var. atuberculata (Hung T. Chang) T. L. Ming ●

68927 Camellia aurea Hung T. Chang;五室金花茶(多变金花茶,五室茶,中华五室金花茶);Five-locular Camellia, Five-locules Camellia, Goldyellow Tea ●

68928 Camellia austrosinica Hung T. Chang;华南茶;South China Camellia ●

68929 Camellia austroyunnanensis Hung T. Chang;滇南茶;South Yunnan Camellia ●

68930 Camellia axillaris Griff. = Camellia caudata Wall. ●

68931 Camellia axillaris Roxb. ex Ker Gawl. = Gordonia axillaris (Roxb. ex Ker Gawl.) Endl. ●

68932 Camellia axillaris Roxb. ex Ker Gawl. = Polyspora axillaris (Roxb. ex Ker Gawl.) Sweet ex G. Don ●

68933 Camellia azalea C. F. Wei;杜鹃叶山茶(杜鹃红山茶,假大头茶,阳春红山茶,张氏红山茶);Chang Camellia, Yangchun Camellia ●

68934 Camellia azelea Wei = Camellia changii C. X. Ye ●

68935 Camellia bailinshanica Hung T. Chang, H. S. Liu et G. X. Xiang;白灵山红山茶;Bailingshan Camellia ●

68936 Camellia bailinshanica Hung T. Chang, H. S. Liu et G. X. Xiang = Camellia reticulata Lindl. ●◇

68937 Camellia bambusifolia Hung T. Chang, H. S. Liu et Y. Z. Zhang;竹叶红山茶;Bambooleaf Camellia ●

68938 Camellia bambusifolia Hung T. Chang, H. S. Liu et Y. Z. Zhang = Camellia reticulata Lindl. ●◇

68939 Camellia biflora Makino;双花山茶;Biflorous Camellia ●

68940 Camellia bonnardii Berl. = Camellia japonica L. ●

68941 Camellia boreali-yunnanica Hung T. Chang;滇北红山茶;N. Yunnan Camellia, North Yunnan Camellia ●

68942 Camellia boreali-yunnanica Hung T. Chang = Camellia reticulata Lindl. ●◇

68943 Camellia brachyandra Hung T. Chang;短蕊茶;Shortanther Tea, Short-stamen Camellia, Short-stamened Camellia ●

68944 Camellia brachyandra Hung T. Chang = Camellia xanthochroma K. M. Feng et L. S. Xie ●

68945 Camellia brachygyna Hung T. Chang;短蕊红山茶;Shortpistil Camellia ●

68946 Camellia brevicolumna Hung T. Chang, H. S. Liu et Y. Z. Zhang;短轴红山茶;Short-column Camellia ●

68947 Camellia brevicolumna Hung T. Chang, H. S. Liu et Y. Z. Zhang = Camellia reticulata Lindl. ●◇

68948 Camellia brevigyna Hung T. Chang = Camellia reticulata Lindl. ●◇

68949 Camellia brevipetiolata Hung T. Chang;短柄红山茶;Shortpetiole Camellia, Shortstalk Camellia ●

68950 Camellia brevipetiolata Hung T. Chang = Camellia reticulata Lindl. ●◇

68951 Camellia brevissima Hung T. Chang = Camellia brevistyla (Hayata) Cohen-Stuart ●

68952 Camellia brevissima Hung T. Chang et S. Ye Liang;红紫花短柱茶;Short-stalked Camellia ●

68953 Camellia brevistyla (Hayata) Cohen-Stuart;短柱油茶(短柱茶,短柱山茶);Shortstyle Camellia, Short-style Camellia, Short-styled Camellia ●

68954 Camellia brevistyla (Hayata) Cohen-Stuart f. rubida P. L. Chiu = Camellia brevistyla (Hayata) Cohen-Stuart ●

68955 Camellia brevistyla (Hayata) Cohen-Stuart var. microphylla (Merr.) T. L. Ming;细叶短柱油茶●

68956 Camellia brevistyla (Hayata) Cohen-Stuart var. microphylla (Merr.) T. L. Ming = Camellia microphylla (Merr.) S. S. Chien ●

68957 Camellia buisanensis Sasaki;武威山茶●

68958 Camellia buisanensis Sasaki = Camellia caudata Wall. ●

68959 Camellia buisanensis Sasaki = Pyrenaria microcarpa (Dunn) H. Keng var. ovalifolia (H. L. Li) T. L. Ming et S. X. Yang ●

68960 Camellia buxifolia Hung T. Chang;黄杨叶连蕊茶;Boxleaf Camellia, Boxleaf Tea, Box-leaved Camellia ●

68961 Camellia buxifolia Hung T. Chang = Camellia polyodonta F. C. How ex Hu ●

68962 Camellia buxifolia Hung T. Chang = Camellia rosthorniana Hand. -Mazz. ●

68963 Camellia calcicola T. L. Ming;长叶山茶;Long-leaf Camellia ●

68964 Camellia callidonta Hung T. Chang;美齿连蕊茶;Beautiful-tooth Camellia, Beautiful-toothed Camellia, Finetooth Tea ●

68965 Camellia callidonta Hung T. Chang = Camellia tsaii Hu ●

68966 Camellia campanisepala Hung T. Chang;钟萼连蕊茶;Bellcalyx Camellia, Bellsepal Tea, Campanulate-sepaled Camellia ●

68967 Camellia campanisepala Hung T. Chang = Camellia cuspidata (Kochs) Wright ex Garden var. chekiangensis Sealy ●

68968 Camellia candida Hung T. Chang;白毛蕊茶(白毛毛蕊茶);Whitehair Camellia, Whitehaired Camellia ●

68969 Camellia cantonensis Lour. = Camellia sinensis (L.) Kuntze ●

68970 Camellia caudata Wall.;长尾毛蕊茶(尾叶山茶,武威山茶,云南大叶茶);Buwi Camellia, Caudate Camellia, Longtail Tea, Tail-leaf Camellia ●

68971 Camellia caudata Wall. var. gracilis (Hemsl.) Gogelein ex H. Keng;小长尾毛蕊茶●

68972 Camellia caudata Wall. var. gracilis (Hemsl.) Gogelein ex H. Keng = Camellia caudata Wall. ●
68973 Camellia caudata Wall. var. gracilis (Hemsl.) Yamam. ex H. Keng = Camellia caudata Wall. ●
68974 Camellia caudata Wall. var. gracilis Gogeleinex Keng = Camellia caudata Wall. ●
68975 Camellia caudatum (Wall.) Nakai = Camellia caudata Wall. ●
68976 Camellia cavaleriana (H. Lév.) Nakai = Camellia pitardii Cohen-Stuart ●
68977 Camellia changii C. X. Ye = Camellia azalea C. F. Wei ●
68978 Camellia changningensis F. C. Zhang, W. R. Ding et Y. Huang;昌宁茶;Changning Camellia ●
68979 Camellia changningensis F. C. Zhang, W. R. Ding et Y. Huang = Camellia taliensis (W. W. Sm.) Melch. ●
68980 Camellia chekiangoleosa Hu;浙江山茶(红花油茶,厚叶红山茶,离蕊红山茶,闪光红山茶,浙江红花油茶,浙江红山茶);Redflower Camellia, Red-flowered Camellia, Zhejiang Camellia ●
68981 Camellia chekiangoleosa Hu f. tangii P. L. Chiu;白花浙江山茶;Tang's Zhejiang Camellia ●
68982 Camellia chekiangoleosa Hu f. tanglii P. L. Chiu = Camellia chekiangoleosa Hu ●
68983 Camellia chiupeiensis Hu;邱北山茶;Chiupei Camellia, Qiubei Camellia ●
68984 Camellia chrysantha (Hu) Tuyama = Camellia nitidissima C. W. Chi ●
68985 Camellia chrysantha (Hu) Tuyama = Camellia petelotii (Merr.) Sealy ●
68986 Camellia chrysantha (Hu) Tuyama f. longistyla S. L. Mo et Y. C. Zhong = Camellia petelotii (Merr.) Sealy ●
68987 Camellia chrysantha (Hu) Tuyama var. longistyla S. L. Mo et Y. C. Zhong = Camellia nitidissima C. W. Chi ●
68988 Camellia chrysantha (Hu) Tuyama var. macrophylla S. L. Mo et S. Z. Huang = Camellia euphlebia Merr. ex Sealy ●◇
68989 Camellia chrysantha (Hu) Tuyama var. microcarpa S. L. Mo et S. Z. Huang = Camellia microcarpa (S. L. Mo et S. Z. Huang) S. L. Mo ●
68990 Camellia chrysantha (Hu) Tuyama var. microcarpa S. L. Mo et S. Z. Huang = Camellia nitidissima C. W. Chi var. microcarpa (S. L. Mo et S. Z. Huang) Hung T. Chang et C. X. Ye ●
68991 Camellia chrysantha (Hu) Tuyama var. microcarpa S. L. Mo et S. Z. Huang = Camellia petelotii (Merr.) Sealy var. microcarpa (S. L. Mo et S. Z. Huang) T. L. Ming et W. J. Zhang ●
68992 Camellia chrysanthoides Hung T. Chang = Camellia tonkinensis (Pit.) Cohen-Stuart ●
68993 Camellia chungkingensis Hung T. Chang;重庆山茶;Chongqing Camellia ●
68994 Camellia chungkingensis Hung T. Chang = Camellia tuberculata S. S. Chien ●
68995 Camellia chunii Hung T. Chang;陈氏红山茶;Chun Camellia ●
68996 Camellia chunii Hung T. Chang = Camellia reticulata Lindl. ●◇
68997 Camellia chunii Hung T. Chang var. pentaphylax (Hung T. Chang) Hung T. Chang = Camellia reticulata Lindl. ●◇
68998 Camellia chunii Hung T. Chang var. pentaphylax Hung T. Chang;五列木红山茶;Pentaphylax-like Camellia ●
68999 Camellia cochinchinensis Lour.;越南茶(普洱茶);Cochinchinese Camellia ●☆
69000 Camellia compressa Hung T. Chang et X. K. Wen = Camellia compressa Hung T. Chang et X. K. Wen ex Hung T. Chang ●
69001 Camellia compressa Hung T. Chang et X. K. Wen = Camellia pitardii Cohen-Stuart var. variabilis (Hung T. Chang et X. K. Wen) T. L. Ming ●
69002 Camellia compressa Hung T. Chang et X. K. Wen ex Hung T. Chang;扁果红山茶;Flatfruit Camellia ●
69003 Camellia compressa Hung T. Chang et X. K. Wen ex Hung T. Chang = Camellia pitardii Cohen-Stuart var. variabilis Hung T. Chang et X. K. Wen ●
69004 Camellia compressa Hung T. Chang et X. K. Wen ex Hung T. Chang var. varibilis Hung T. Chang et X. K. Wen;多变扁果红山茶;Variable Flatfruit Camellia ●
69005 Camellia compressa Hung T. Chang et X. K. Wen var. variabilis Hung T. Chang et X. K. Wen = Camellia pitardii Cohen-Stuart var. variabilis (Hung T. Chang et X. K. Wen) T. L. Ming ●
69006 Camellia confusa Craib;小果短柱茶;Smallfruit Camellia, Small-fruited Camellia ●
69007 Camellia connatistyla S. L. Mo et Y. C. Zhong;合柱糙果茶;Connatestyle Camellia, Unistyle Camellia ●
69008 Camellia connatistyla S. L. Mo et Y. C. Zhong = Camellia lapidea Y. C. Wu ●
69009 Camellia connatistyla S. L. Mo et Y. C. Zhong = Camellia mairei (H. Lév.) Melch. var. lapidea (W. C. Wu) Sealy ●
69010 Camellia cordifolia (F. P. Metcalf) Nakai;心叶毛蕊茶(野山茶);Cordate Camellia, Heartleaf Tea ●
69011 Camellia cordifolia (F. P. Metcalf) Nakai var. glabrisepala T. L. Ming;光萼心叶毛蕊茶●
69012 Camellia costata Hung T. Chang;突肋茶(榕江茶);Costate Camellia, Costate Tea, Ribbed Camellia, Yungkiang Camellia ●
69013 Camellia costata Hung T. Chang et S. Ye Liang = Camellia costata Hung T. Chang ●
69014 Camellia costei H. Lév.;贵州连蕊茶;Coste Tea, Guizhou Camellia ●
69015 Camellia crapnelliana Tutcher;红皮糙山茶(八瓣糙果茶,博白大果油茶,博白油茶,茶梨,红皮糙果茶,红皮茶,红皮山茶,梨茶,小叶金花茶,窄叶短柱茶);Bigfruit Oiltea Camellia, Big-fruited Camellia, Eightpetal Camellia, Eight-petaled Camellia, Redbark Camellia, Red-barked Camellia ●◇
69016 Camellia crassicolumna Hung T. Chang;厚轴茶(厚轴山茶);Thick Rachis Camellia, Thickcolumn Tea, Thick-columned Camellia ●
69017 Camellia crassicolumna Hung T. Chang var. multiplex (Hung T. Chang et C. M. Tan) T. L. Ming;光萼厚轴茶●
69018 Camellia crassicolumna Hung T. Chang var. shangbaensis F. C. Zhang = Camellia crassicolumna Hung T. Chang ●
69019 Camellia crassipes Sealy;粗梗连蕊茶(粗柄山茶,厚柄连蕊茶,厚柄毛蕊茶);Thickstalk Tea, Thick-stalked Camellia ●
69020 Camellia crassipetala Hung T. Chang;厚瓣短蕊茶;Thickpeta Tea, Thickpetal Camellia, Thick-petaled Camellia ●
69021 Camellia crassipetala Hung T. Chang = Camellia tsaii Hu ●
69022 Camellia crassissima Hung T. Chang et W. J. Shi = Camellia chekiangoleosa Hu ●
69023 Camellia crassissima Hung T. Chang et W. J. Shi ex Hung T. Chang;厚叶红山茶(厚壳红山茶);Thickleaf Camellia, Thick-shelled Camellia ●
69024 Camellia crassissima Hung T. Chang et W. J. Shi ex Hung T. Chang = Camellia chekiangoleosa Hu ●
69025 Camellia cratera Hung T. Chang;杯萼毛蕊茶;Cupcalyx Tea, Cupularcalyx Camellia, Cupular-calyxed Camellia ●

69026 Camellia cratera Hung T. Chang = Camellia assimiloides Sealy ●
69027 Camellia crispula Hung T. Chang;皱叶茶;Creaseleaf Camellia, Creaseleaf Tea, Crispate Camellia, Winkledleaf Camellia ●
69028 Camellia crispula Hung T. Chang = Camellia crassicolumna Hung T. Chang ●
69029 Camellia cryptoneura Hung T. Chang;隐脉红山茶;Crypto-nerved Camellia, Cryptonervius Camellia, Latentvein Camellia ●
69030 Camellia cryptoneura Hung T. Chang = Camellia pitardii Cohen-Stuart var. cryptoneura (Hung T. Chang) T. L. Ming ●
69031 Camellia cryptoneura Hung T. Chang = Camellia pitardii Cohen-Stuart ●
69032 Camellia cupiformis T. L. Ming;滇南连蕊茶●
69033 Camellia cuspidata (Kochs) Wright ex Garden;连蕊茶(火烟子,尖连蕊茶,尖毛蕊茶,尖叶山茶);Cuspidate Tea, Taperleaf Camellia, Taper-leaved Camellia ●
69034 Camellia cuspidata (Kochs) Wright ex Garden var. chekiangensis Sealy;浙江连蕊茶(浙江尖连蕊茶);Zhejiang Taperleaf Camellia ●
69035 Camellia cuspidata (Kochs) Wright ex Garden var. grandiflora Sealy;大花连蕊茶(长尖连蕊茶,长尖毛蕊茶,大花尖连蕊茶);Acutest Camellia, Bigflower Taperleaf Camellia, Longtine Tea ●
69036 Camellia cuspidata (Kochs) Wright ex Garden var. trichandra (Hung T. Chang) T. L. Ming;毛丝连蕊茶(毛丝毛蕊茶);Hairander Tea, Hairyfilament Camellia, Hairy-filamented Camellia ●
69037 Camellia cuspidata (Kochs) Wright ex Garden var. trichandra (Hung T. Chang) T. L. Ming = Camellia trichandra Hung T. Chang ●
69038 Camellia cylindracea T. L. Ming;柱蕊茶;Cylindrical Camellia ●
69039 Camellia danzaiensis Hung T. Chang et K. M. Lan;丹寨秃茶(丹寨茶);Danzhai Camellia ●
69040 Camellia danzaiensis Hung T. Chang et K. M. Lan = Camellia costata Hung T. Chang ●
69041 Camellia danzaiensis K. M. Lan = Camellia costata Hung T. Chang ●
69042 Camellia dehungensis Hung T. Chang et B. H. Chen = Camellia sinensis (L.) Kuntze var. dehungensis (Hung T. Chang et B. H. Chen) T. L. Ming ●
69043 Camellia dehungensis Hung T. Chang, H. S. Wang et B. H. Chen = Camellia sinensis (L.) Kuntze var. dehungensis (Hung T. Chang et B. H. Chen) T. L. Ming ●
69044 Camellia delicata Y. K. Li = Camellia mairei (H. Lév.) Melch. var. lapidea (W. C. Wu) Sealy ●
69045 Camellia delicata Y. K. Li = Camellia mairei (H. Lév.) Melch. ●
69046 Camellia dishiensis F. C. Zhang, X. Y. Chen et G. B. Chen;底圩茶;Diyu Camellia ●
69047 Camellia dishiensis F. C. Zhang, X. Y. Chen et G. B. Chen = Camellia assamica (Mast.) Hung T. Chang ●
69048 Camellia dishiensis F. C. Zhang, X. Y. Chen et G. B. Chen = Camellia sinensis (L.) Kuntze var. pubilimba Hung T. Chang ●
69049 Camellia drupifera Lour.;越南油茶(红皮糙果茶);Himalayan Camellia ●
69050 Camellia drupifera Lour. = Camellia oleifera Abel ●
69051 Camellia drupifera Lour. f. biflora (Hayata) S. S. Ying = Camellia oleifera Abel ●
69052 Camellia dubia Sealy;秃梗连蕊茶;Doubtful Camellia, Nakestalk Tea ●
69053 Camellia dubia Sealy = Camellia costei H. Lév. ●
69054 Camellia edentata Hung T. Chang;无齿毛蕊茶;Toothless Camellia, Toothless Tea ●
69055 Camellia edentata Hung T. Chang = Camellia caudata Wall. var. gracilis (Hemsl.) Gogelein ex H. Keng ●
69056 Camellia edithae Hance;东南山茶(尖萼红山茶);Acutesepal Camellia, Acute-sepaled Camellia, Tinecalyx Camellia ●
69057 Camellia elongata (Rehder et E. H. Wilson) Rehder;长管连蕊茶(长管毛蕊茶);Graceful Camellia, Longtube Camellia, Longtube Tea ●
69058 Camellia euonymifolia (Hu) C. Y. Wu;卫矛叶连蕊茶;Euonymusleaf Camellia ●
69059 Camellia euphlebia Merr. ex Sealy;显脉金花茶(大叶金花茶);Distinct Camellia, Obviousvein Tea ●◇
69060 Camellia euphlebia Merr. ex Sealy var. macrophylla (S. L. Mo et S. Z. Huang) C. X. Ye et J. Y. Liang;大叶金花茶;Bigleaf Obviousvein Tea ●
69061 Camellia euphlebia Merr. ex Sealy var. macrophylla (S. L. Mo et S. Z. Huang) C. X. Ye et J. Y. Liang = Camellia euphlebia Merr. ex Sealy ●◇
69062 Camellia euphlebia Merr. ex Sealy var. yunnanensis Chen J. Wang et G. S. Fang;云南显脉金花茶;Yunnan Obviousvein Tea ●
69063 Camellia euphlebia Merr. ex Sealy var. yunnanensis Chen J. Wang et G. S. Fang = Camellia fascicularis Hung T. Chang ●
69064 Camellia euryoides Lindl.;柃叶连蕊茶(绿果青);Euryaleaf Tea, Euryalike Camellia, Eurya-like Camellia ●
69065 Camellia euryoides Lindl. var. nokoensis (Hayata) T. L. Ming;毛蕊柃叶连蕊茶●
69066 Camellia euryoides Lindl. var. nokoensis (Hayata) T. L. Ming = Camellia nokoensis Hayata ●
69067 Camellia fangchengensis S. Ye. Liang et Y. C. Zhong;防城茶(陆川油茶);Fangcheng Camellia, Fangcheng Tea ●◇
69068 Camellia fascicularis Hung T. Chang;云南金花茶(簇蕊金花茶);Clustered Camellia, Fasciculate Tea ●
69069 Camellia fengkingensis Hung T. Chang;风庆连蕊茶;Fenqing Camellia ●
69070 Camellia flava (Pit.) Sealy;黄花茶;Yellow Camellia ●☆
69071 Camellia flavida Hung T. Chang;淡黄金花茶(淡黄离蕊茶);Thinyellow Tea, Yellowish Camellia ●◇
69072 Camellia flavida Hung T. Chang f. polypetala R. G. Li et S. Q. He = Camellia flavida Hung T. Chang ●◇
69073 Camellia flavida Hung T. Chang var. patens (S. L. Mo et Y. C. Zhong) T. L. Ming;多变淡黄金花茶●
69074 Camellia flavida Hung T. Chang var. patens (S. L. Mo et Y. C. Zhong) T. L. Ming = Camellia petelotii (Merr.) Sealy var. microcarpa (S. L. Mo et S. Z. Huang) T. L. Ming et W. J. Zhang ●
69075 Camellia fleuryi (Chev.) Sealy;屏边山茶;Fleury Camellia ●
69076 Camellia florida Salisb. = Camellia japonica L. ●
69077 Camellia fluviatilis Hand.-Mazz.;窄叶油茶(防城茶,窄叶短柱茶);Angustifoliate Camellia, Narrowleaf Camellia ●
69078 Camellia fluviatilis Hand.-Mazz. var. megalantha (Hung T. Chang) T. L. Ming = Camellia kissi Wall. var. megalantha Hung T. Chang ●
69079 Camellia forrestii (Diels) Cohen-Stuart;云南连蕊茶(蒙自连蕊茶,小花山茶);Forrest Camellia, Forrest Tea ●
69080 Camellia forrestii (Diels) Cohen-Stuart var. acutisepala (H. T. Tsai et K. M. Feng) Hung T. Chang;尖萼云南连蕊茶(尖萼连蕊茶,尖萼蒙自连蕊茶,尖萼山茶);Acute-sepal Camellia ●
69081 Camellia forrestii (Diels) Cohen-Stuart var. pentamera (Hung T. Chang) T. L. Ming = Camellia pentamera Hung T. Chang ●

69082 Camellia fraterna Hance;毛花连蕊茶(连蕊茶,毛柄连蕊茶,毛柄毛蕊茶);Ellipticleaf Camellia,Elliptic-leaved Camellia,Hairstalk Tea ●

69083 Camellia fukienensis Hung T. Chang;福建红山茶;Fujian Camellia ●

69084 Camellia furfuracea (Merr.) Cohen-Stuart;糙果茶(垢果山茶);Furfuraceousfruit Camellia,Furfuraceous-fruited Camellia ●

69085 Camellia furfuracea (Merr.) Cohen-Stuart var. latipetiolata (C. W. Chi) T. L. Ming = Camellia latipetiolata C. M. Chi ●

69086 Camellia furfuracea (Merr.) Cohen-Stuart var. lutea Hu = Camellia furfuracea (Merr.) Cohen-Stuart ●

69087 Camellia furfuracea (Merr.) Cohen-Stuart var. melanosticta Hung T. Chang;黑腺糙果茶●

69088 Camellia furfuracea (Merr.) Cohen-Stuart var. shanglinensis T. L. Ming;上林糙果茶;Shanglin Furfuraceousfruit Camellia ●

69089 Camellia furfuracea (Merr.) Cohen-Stuart var. yaoshanica S. Ye Liang et Y. C. Zhong = Camellia pubifurfuracea Y. C. Zhong ●

69090 Camellia fusuiensis S. Ye Liang et X. J. Dong = Camellia indochinensis Merr. ●

69091 Camellia fusuiensis S. Ye Liang et X. J. Dong = Camellia limonia C. F. Liang et S. L. Mo ●

69092 Camellia gauchowensis Hung T. Chang;高州山茶(高州油茶);Gaozhou Oil Camellia ●

69093 Camellia gauchowensis Hung T. Chang = Camellia drupifera Lour. ●

69094 Camellia gaudichaudii (Gagnep.) Sealy;硬叶糙果茶(山油茶);Gaudichaud Camellia,Hardleaf Camellia ●

69095 Camellia gigantocarpa Hu et T. C. Huang = Camellia crapnelliana Tutcher ●◇

69096 Camellia gilbertii (A. Chev.) Sealy;中越短蕊茶●

69097 Camellia glaberrima Hung T. Chang;秃山茶(秃茶);Glabrous Camellia,Smooth Tea ●

69098 Camellia glaberrima Hung T. Chang = Camellia gymnogyna Hung T. Chang ●

69099 Camellia glabricostata T. L. Ming;秃肋连蕊茶;Glabrous-midrib Camellia ●

69100 Camellia glabriperulata Hung T. Chang = Camellia saluenensis Stapf ex Bean ●

69101 Camellia gracilipes Merr. ex Sealy;狭叶长梗茶●

69102 Camellia gracilis Hemsl. = Camellia caudata Wall. var. gracilis (Hemsl.) Gogelein ex H. Keng ●

69103 Camellia gracilis Hemsl. = Camellia caudata Wall. ●

69104 Camellia grandibracteata Hung T. Chang, Y. J. Tan, F. L. Yu et P. S. Wang;大苞茶;Bigbract Camellia ●

69105 Camellia grandis (C. F. Liang et S. L. Mo) Hung T. Chang et S. Ye Liang;弄岗金花茶;Nonggang Camellia ●

69106 Camellia grandis (C. F. Liang et S. L. Mo) Hung T. Chang et S. Ye Liang = Camellia flavida Hung T. Chang ●◇

69107 Camellia grandis (C. F. Liang et S. L. Mo) Hung T. Chang et S. Ye Liang = Camellia parvisepala Hung T. Chang ●

69108 Camellia granthamiana Sealy;大苞白山茶(大苞茶,大苞山茶);Big-bracted Camellia ●◇

69109 Camellia grijsii Hance;长瓣短柱茶(芳香短柱茶,闽萼山茶);Bilobedpetal Camellia,Bilobed-petaled Camellia,Shortpetal Camellia ●◇

69110 Camellia grijsii Hance var. shensiensis (Hung T. Chang) T. L. Ming;小叶短柱茶●

69111 Camellia gymnogyna Hung T. Chang;秃房茶(秃山茶);Glabrous Camellia, Glabrouslocular Camellia, Glabrous-locular Camellia, Nakepistil Tea ●

69112 Camellia gymnogyna Hung T. Chang var. remotiserrata (Hung T. Chang, H. S. Wang et P. S. Wang) T. L. Ming;疏齿秃房茶●

69113 Camellia gymnogyna Hung T. Chang var. remotiserrata (Hung T. Chang, H. S. Wang et P. S. Wang) T. L. Ming = Camellia tachangensis F. S. Zhang var. remotiserrata (Hung T. Chang, H. S. Wang et B. H. Wang) T. L. Ming ●

69114 Camellia gymnogyna Hung T. Chang var. remotiserrata (Hung T. Chang, H. S. Wang et P. S. Wang) T. L. Ming = Camellia remotiserrata Hung T. Chang, H. S. Wang et P. S. Wang ●

69115 Camellia gymnogynoides Hung T. Chang et B. H. Chen = Camellia tachangensis F. S. Zhang var. remotiserrata (Hung T. Chang, H. S. Wang et B. H. Wang) T. L. Ming ●

69116 Camellia gymnogynoides Hung T. Chang, H. S. Wang et B. H. Chen = Camellia tachangensis F. S. Zhang var. remotiserrata (Hung T. Chang, H. S. Wang et B. H. Wang) T. L. Ming ●

69117 Camellia gymnogynoides Hung T. Chang, H. S. Wang et F. H. Chen = Camellia dehungensis Hung T. Chang, H. S. Wang et B. H. Chen ●

69118 Camellia gymnogynoides Hung T. Chang, H. S. Wang et F. H. Chen = Camellia gymnogyna Hung T. Chang var. remotiserrata (Hung T. Chang, H. S. Wang et P. S. Wang) T. L. Ming ●

69119 Camellia haaniensis Hung T. Chang et F. L. Yu = Camellia crassicolumna Hung T. Chang ●

69120 Camellia hainanensis Hung T. Chang, H. S. Wang et B. H. Chen = Camellia crassicolumna Hung T. Chang ●

69121 Camellia handelii Sealy;岳麓连蕊茶;Handel Camellia, Handel Tea ●

69122 Camellia handelii Sealy = Camellia transarisanensis (Hayata) Cohen-Stuart ●

69123 Camellia hekouensis Ching J. Wang et G. S. Fan;河口长梗茶(河口长柄茶,河口超长柄茶);Hekou Camellia,Hekou Tea ●

69124 Camellia hengchunensis C. E. Chang;恒春山茶●

69125 Camellia hengchunensis C. E. Chang = Camellia brevistyla (Hayata) Cohen-Stuart ●

69126 Camellia hengchunensis C. E. Chang = Camellia tenuifolia (Hayata) Cohen-Stuart ●

69127 Camellia henryana Cohen-Stuart;光果山茶(蒙自山茶);Henry Camellia ●

69128 Camellia henryana Cohen-Stuart = Camellia yunnanensis (Pit. ex Diels) Cohen-Stuart ●

69129 Camellia henryana Cohen-Stuart var. pilocarpa T. L. Ming = Camellia yunnanensis (Pit. ex Diels) Cohen-Stuart var. camellioides (Hu) T. L. Ming ●

69130 Camellia henryana Cohen-Stuart var. trichocarpa (Hung T. Chang) T. L. Ming = Camellia trichocarpa Hung T. Chang ●

69131 Camellia henryana Cohen-Stuart var. trichocarpa (Hung T. Chang) T. L. Ming = Camellia yunnanensis (Pit. ex Diels) Cohen-Stuart var. camellioides (Hu) T. L. Ming ●

69132 Camellia heterophylla Hu = Camellia reticulata Lindl. ●◇

69133 Camellia hiemalis Nakai;冬红短柱茶(冬茶梅,冬红山茶,冬山茶);Snow Camellia,Winter Camellia ●

69134 Camellia hiemalis Nakai = Camellia sasanqua Thunb. 'Shishigashira' ●

69135 Camellia hongkongensis Seem.;香港山茶(广东山茶,香港椿,

香港红山茶);Hongkong Camellia ●
69136 Camellia hozanensis (Hayata) Hayata = Camellia japonica L. ●
69137 Camellia huana T. L. Ming et W. T. Zhang;贵州金花茶●
69138 Camellia huiliensis Hung T. Chang = Camellia parvimuricata Hung T. Chang ●
69139 Camellia huiliensis Hung T. Chang = Camellia pitardii Cohen-Stuart ●
69140 Camellia huiliensis Hung T. Chang = Camellia reticulata Lindl. ●◇
69141 Camellia hunanica Hung T. Chang et L. L. Qi = Camellia pitardii Cohen-Stuart ●
69142 Camellia hupehensis Hung T. Chang;湖北瘤果茶(湖北茶);Hubei Camellia ●
69143 Camellia hupehensis Hung T. Chang = Camellia parvimuricata Hung T. Chang var. hupehensis (Hung T. Chang) T. L. Ming ●
69144 Camellia ilicifolia H. L. Li;冬青叶瘤果茶(冬青叶茶,冬青叶山茶);Hollyleaf Camellia,Holly-leaved Camellia ●
69145 Camellia ilicifolia H. L. Li f. rubimuricata (Hung T. Chang et Z. R. Xu) T. L. Ming = Camellia ilicifolia H. L. Li ●
69146 Camellia ilicifolia H. L. Li var. nerifolia (Hung T. Chang) T. L. Ming;狭叶瘤果茶;Narrow Hollyleaf Camellia, Narrow-leaves Camellia,Oleanderleaf Camellia ●
69147 Camellia impressinervis Hung T. Chang et S. Ye Liang;凹脉金花茶(大苞白山茶);Concavevein Tea,Sunkenvein Camellia,Sunken-veined Camellia ●◇
69148 Camellia indochinensis Merr.;柠檬金花茶(长叶越南油茶,中越山茶);Indo-China Camellia,Indochinese Camellia,Lemon Tea,Lemon-like Camellia,Sino-Vietnam Tea ●
69149 Camellia indochinensis Merr. var. tunghinensis (Hung T. Chang) T. L. Ming et W. J. Zhang;东兴金花茶;Dongxing Camellia ●
69150 Camellia infraferruginosa Hung T. Chang;褐背山茶;Brown-backed Camellia ●
69151 Camellia integerrima Hung T. Chang;全缘糙果茶;Absolute Entire Camellia,Entire Camellia ●
69152 Camellia integerrima Hung T. Chang = Camellia hongkongensis Seem. ●
69153 Camellia irrawadiensis P. K. Barua;滇缅茶;Irrawadi Camellia,Yunnan-Burma Camellia ●
69154 Camellia irrawadiensis P. K. Barua = Camellia taliensis (W. W. Sm.) Melch. ●
69155 Camellia japonica H. Lév. = Camellia pitardii Cohen-Stuart ●
69156 Camellia japonica L.;山茶(宝珠茶,宝珠山茶,茶花,海榴茶,海石榴,鹤顶山茶,红茶,红茶花,红山茶,曼陀罗,耐冬花,千叶白,千叶红,日本红山茶,日本山茶,山茶花,石榴茶,晚山茶,洋茶,一捻红,踯躅茶);Camellia,Camellia Flower,Common Camellia,Japan Camellia,Japanese Camellia ●
69157 Camellia japonica L. 'Alba Plena';白洋茶山茶●☆
69158 Camellia japonica L. 'Big Red';大红山茶●☆
69159 Camellia japonica L. 'Bob Hope';希望山茶●☆
69160 Camellia japonica L. 'Bob's Tinsie';廷西山茶●☆
69161 Camellia japonica L. 'Bokuhan';波库汉山茶●☆
69162 Camellia japonica L. 'Brushfield's Yellow';黄金山茶●☆
69163 Camellia japonica L. 'Chandleri Elegans' = Camellia japonica L. 'Elegans'●☆
69164 Camellia japonica L. 'Deptante';出现山茶●☆
69165 Camellia japonica L. 'Desire';渴望山茶●☆
69166 Camellia japonica L. 'Dona Herzilia de Frietas Magalhaes';麦哲伦山茶●☆
69167 Camellia japonica L. 'Elegans Champagne';香槟秀美山茶●☆
69168 Camellia japonica L. 'Elegans Supreme';至高秀美山茶●☆
69169 Camellia japonica L. 'Elegans Variegated';白斑秀美山茶●☆
69170 Camellia japonica L. 'Elegans';秀美山茶●☆
69171 Camellia japonica L. 'Fimbriata';毛边山茶●☆
69172 Camellia japonica L. 'Gongfen';宫粉茶●☆
69173 Camellia japonica L. 'Great Eastern';东部之光山茶●☆
69174 Camellia japonica L. 'Higo Camellias';希高山茶●☆
69175 Camellia japonica L. 'Janet Waterhouse';珍妮特山茶●☆
69176 Camellia japonica L. 'Lady Loch';圣女湖山茶●☆
69177 Camellia japonica L. 'Magellan' = Camellia japonica L. 'Dona Herzilia de Frietas Magalhaes'●☆
69178 Camellia japonica L. 'Magnoliiflora';玉兰粉花山茶●☆
69179 Camellia japonica L. 'Nuccio's Cameo';浮雕山茶●☆
69180 Camellia japonica L. 'Nuccio's Carousel';传送带山茶●☆
69181 Camellia japonica L. 'Nuccio's Gem';珠宝山茶●☆
69182 Camellia japonica L. 'Nuccio's Jewel';宝石山茶●☆
69183 Camellia japonica L. 'Nuccio's Pearl';牛西奥宝珠山茶●☆
69184 Camellia japonica L. 'Roma Risorta';罗马之宴山茶●☆
69185 Camellia japonica L. 'Rubescens Major';红色大调山茶●☆
69186 Camellia japonica L. 'Tama-no-ura';白边山茶●☆
69187 Camellia japonica L. 'The Czar';独裁者山茶●☆
69188 Camellia japonica L. 'Tinsie' = Camellia japonica L. 'Bokuhan'●☆
69189 Camellia japonica L. 'Tomorrow';明日山茶●☆
69190 Camellia japonica L. 'Tomorrow's Dawn';次日黎明山茶●☆
69191 Camellia japonica L. 'Twilight';曙光山茶●☆
69192 Camellia japonica L. f. grosseserrata Uyeki = Camellia japonica L. ●
69193 Camellia japonica L. f. hexapetala H. Hara;六瓣山茶●☆
69194 Camellia japonica L. f. ilicifolia Makino;八角叶山茶●☆
69195 Camellia japonica L. f. incarnata Lodd.;无雄山茶●☆
69196 Camellia japonica L. f. lancifolia H. Hara;尖叶山茶●☆
69197 Camellia japonica L. f. lancifolia H. Hara = Camellia japonica L. ●
69198 Camellia japonica L. f. leucantha Makino;白花山茶●☆
69199 Camellia japonica L. f. lilifolia Makino;狭披针叶山茶●☆
69200 Camellia japonica L. f. macrocarpa (Masam.) Tuyama = Camellia japonica L. var. macrocarpa Masam. ●☆
69201 Camellia japonica L. f. parviflora Makino;小花山茶●☆
69202 Camellia japonica L. f. parviflora Makino = Camellia japonica L. ●
69203 Camellia japonica L. f. polypetala Makino;多瓣山茶●☆
69204 Camellia japonica L. f. trifida Makino;三裂山茶●☆
69205 Camellia japonica L. subsp. hozanensis (Hayata) Kitam. = Camellia japonica L. ●
69206 Camellia japonica L. subsp. hozanensis (Hayata) Kitam. = Camellia japonica L. var. hozanensis (Hayata) Gogeleinet Mori ●
69207 Camellia japonica L. subsp. rusticana (Honda) Kitam.;短柄山茶(短柄红山茶,雪山茶);Short-stalk Japanese Camellia ●
69208 Camellia japonica L. subsp. rusticana (Honda) Kitam. = Camellia japonica L. var. rusticana (Honda) T. L. Ming ●
69209 Camellia japonica L. subsp. rusticana (Honda) Kitam. = Camellia japonica L. ●
69210 Camellia japonica L. subsp. rusticana (Honda) Kitam. f. albiflora (Ikegami) Tuyama;白花短柄山茶●☆
69211 Camellia japonica L. subsp. rusticana (Honda) Kitam. f. kagamontana Tuyama;加贺山茶●☆
69212 Camellia japonica L. subsp. rusticana (Honda) Kitam. f.

leucantha (Honda) Tuyama = Camellia japonica L. subsp. rusticana (Honda) Kitam. f. albiflora (Ikegami) Tuyama ●☆

69213 Camellia japonica L. subsp. rusticana (Honda) Kitam. f. plena (Honda) Tuyama;重瓣短柄山茶●☆

69214 Camellia japonica L. var. decumbens Sugim.;伏毛山茶●☆

69215 Camellia japonica L. var. decumbens Sugim. = Camellia japonica L. subsp. rusticana (Honda) Kitam. ●

69216 Camellia japonica L. var. hozanensis (Hayata) Gogeleinet Mori;台湾山茶花(台湾山茶);Taiwan Camellia ●

69217 Camellia japonica L. var. hozanensis (Hayata) Gogeleinet Mori = Camellia japonica L. ●

69218 Camellia japonica L. var. macrocarpa Masam.;大果山茶●☆

69219 Camellia japonica L. var. macrocarpa Masam. = Camellia japonica L. ●

69220 Camellia japonica L. var. rusticana (Honda) T. L. Ming = Camellia japonica L. subsp. rusticana (Honda) Kitam. ●

69221 Camellia japonica L. var. rusticana (Honda) Tuyama = Camellia japonica L. subsp. rusticana (Honda) Kitam. ●

69222 Camellia japonica var. hozanensis (Hayata) Gogelein = Camellia japonica L. ●

69223 Camellia japonica var. macrocarpa Masam. = Camellia japonica L. ●

69224 Camellia jingyunshanica Hung T. Chang et G. X. Xiang;缙云山茶;Jinyun Camellia ●

69225 Camellia jinshajiangica Hung T. Chang = Camellia reticulata Lindl. ●◇

69226 Camellia jinshajiangica Hung T. Chang et S. L. Lee;金沙江红山茶;Jinshajiang Camellia ●

69227 Camellia jinshajiangica Hung T. Chang et S. L. Lee = Camellia reticulata Lindl. ●◇

69228 Camellia jinyunshanica Hung T. Chang et J. H. Xiong = Camellia tachangensis F. S. Zhang var. remotiserrata (Hung T. Chang, H. S. Wang et B. H. Wang) T. L. Ming ●

69229 Camellia jiuyishanica Hung T. Chang et L. L. Qi = Camellia cuspidata (Kochs) Wright ex Garden var. chekiangensis Sealy ●

69230 Camellia juiyishanica Hung T. Chang et L. L. Qi;九嶷山连蕊茶(九嶷山毛蕊茶);Jiuyishan Camellia, Jiuyishan Tea ●

69231 Camellia kaempferia Reboul = Camellia japonica L. ●

69232 Camellia kangdianica Hung T. Chang, H. S. Liu et G. X. Xiang = Camellia bailinshanica Hung T. Chang, H. S. Liu et G. X. Xiang ●

69233 Camellia kangdianica Hung T. Chang, H. S. Liu et G. X. Xiang = Camellia reticulata Lindl. ●◇

69234 Camellia keina Buch. -Ham. ex D. Don = Camellia kissi Wall. ●

69235 Camellia kissi Wall.;落瓣油茶(绿瓣油茶,落瓣短柱茶,南皮山茶);Caducouspetal Camellia, Caducous-petaled Camellia, Kiss Camellia ●

69236 Camellia kissi Wall. var. megalantha Hung T. Chang;大花短柱茶;Big-flower Caducouspetal Camellia ●

69237 Camellia kissi Wall. var. stenophylla (Kobuski) Sealy = Camellia fluviatilis Hand. -Mazz. ●

69238 Camellia kissii Wall. var. confusa (Craib) T. L. Ming;大叶落瓣油茶●

69239 Camellia kissii Wall. var. megalantha Hung T. Chang = Camellia fluviatilis Hand. -Mazz. var. megalantha (Hung T. Chang) T. L. Ming ●

69240 Camellia kissii Wall. var. stenophylla (Kobuski) Sealy = Camellia fluviatilis Hand. -Mazz. ●

69241 Camellia krempfii (Gagnep.) Sealy;越南长叶山茶●☆

69242 Camellia kwangnanica Hung T. Chang et B. H. Chen = Camellia kwangsiensis Hung T. Chang var. kwangnanica (Hung T. Chang et B. H. Chen) T. L. Ming ●

69243 Camellia kwangnanica Hung T. Chang, H. S. Wang et B. H. Chen;广南茶;Guangnan Camellia, Guangnan Tea ●

69244 Camellia kwangsiensis Hung T. Chang;广西茶;Guangxi Camellia, Guangxi Tea, Kwangsi Camellia ●

69245 Camellia kwangsiensis Hung T. Chang var. kwangnanica (Hung T. Chang et B. H. Chen) T. L. Ming = Camellia kwangnanica Hung T. Chang, H. S. Wang et B. H. Chen ●

69246 Camellia kwangtungensis Hung T. Chang;广东山茶(广东秃茶);Guangdong Camellia, Guangdong Tea, Kwangtung Camellia ●

69247 Camellia kwangtungensis Hung T. Chang = Camellia costata Hung T. Chang ●

69248 Camellia kweichouensis Hung T. Chang;贵州红山茶;Guizhou Camellia ●

69249 Camellia kweichouensis Hung T. Chang = Camellia reticulata Lindl. ●◇

69250 Camellia lanceisepala L. K. Ling = Camellia longicalyx Hung T. Chang ●

69251 Camellia lanceoleosa Hung T. Chang = Camellia fluviatilis Hand. -Mazz. var. megalantha (Hung T. Chang) T. L. Ming ●

69252 Camellia lanceoleosa Hung T. Chang et P. S. Chiu ex Hung T. Chang et S. X. Ren;狭叶油茶;Lanceoleaf Camellia, Narrowleaf Oil Camellia ●

69253 Camellia lancicalyx Hung T. Chang;披针萼连蕊茶;Lanceolatecalyx Camellia, Lanceolatecalyx Tea, Lanceolateleaf Camellia ●

69254 Camellia lancicalyx Hung T. Chang = Camellia tsingpienensis Hu var. pubisepala Hung T. Chang ●

69255 Camellia lancilimba Hung T. Chang;披针叶连蕊茶;Lanceolate-leaved Camellia ●

69256 Camellia lancilimba Hung T. Chang = Camellia cuspidata (Kochs) Wright ex Garden var. chekiangensis Sealy ●

69257 Camellia lanosituba Hung T. Chang;绵管红山茶;Cottontube Camellia, Woolly-tube Camellia ●

69258 Camellia lanosituba Hung T. Chang = Camellia mairei (H. Lév.) Melch. var. lapidea (W. C. Wu) Sealy ●

69259 Camellia lanosituba Hung T. Chang = Camellia reticulata Lindl. ●◇

69260 Camellia lapidea Y. C. Wu;石果红山茶;Rocky Camellia, Stonefruit Camellia, Woodfruit Camellia ●

69261 Camellia lapidea Y. C. Wu = Camellia mairei (H. Lév.) Melch. var. lapidea (W. C. Wu) Sealy ●

69262 Camellia latilimba Hu;梨茶(大油茶,山桐茶);Broadleaf Camellia ●

69263 Camellia latilimba Hu = Camellia crapnelliana Tutcher ●◇

69264 Camellia latipetiolata C. W. Chi;阔柄糙果茶;Broad-petioled Camellia, Latipetiolate Camellia ●

69265 Camellia latipetiolata C. W. Chi = Camellia furfuracea (Merr.) Cohen-Stuart var. latipetiolata (C. W. Chi) T. L. Ming ●

69266 Camellia lawii Sealy;四川毛蕊茶;Law Camellia, Sichuan Tea ●

69267 Camellia leptophylla S. Ye Liang et Hung T. Chang;膜叶茶;Membranousleaf Camellia, Thinleaf Tea, Thin-leaved Camellia ●

69268 Camellia leyeensis Hung T. Chang et Y. C. Zhong;乐业瘤果茶;Leye Camellia ●

69269 Camellia leyeensis Hung T. Chang et Y. C. Zhong = Camellia anlungensis Hung T. Chang ●

69270 Camellia liberistamina Hung T. Chang et J. S. Chiu;离蕊红山茶;Discretestamen Camellia, Free-stamen Camellia ●

69271 Camellia liberistamina Hung T. Chang et J. S. Chiu = Camellia chekiangoleosa Hu ●

69272 Camellia liberistyla Hung T. Chang;散柱茶;Free-style Camellia, Separatestyle Camellia, Spread-styled Camellia ●

69273 Camellia liberistyla Hung T. Chang = Camellia yunnanensis (Pit. ex Diels) Cohen-Stuart var. camellioides (Hu) T. L. Ming ●

69274 Camellia liberistyloides Hung T. Chang;肖散柱茶;Free-style-like Camellia, Shamseparatestyle Camellia, Spread-style-like Camellia ●

69275 Camellia liberistyloides Hung T. Chang = Camellia yunnanensis (Pit. ex Diels) Cohen-Stuart var. camellioides (Hu) T. L. Ming ●

69276 Camellia liberofilamenta Hung T. Chang et C. H. Yang = Camellia huana T. L. Ming et W. T. Zhang ●

69277 Camellia lienshanensis Hung T. Chang;连山红山茶(连山离蕊茶);Lianshan Camellia ●

69278 Camellia lienshanensis Hung T. Chang = Camellia subintegra P. C. Huang ex Hung T. Chang ●

69279 Camellia limonia C. F. Liang et S. L. Mo = Camellia indochinensis Merr. ●

69280 Camellia limonia C. F. Liang et S. L. Mo f. obovata S. L. Mo et Y. C. Zhong = Camellia indochinensis Merr. ●

69281 Camellia limonia C. F. Liang et S. L. Mo f. obovata S. L. Mo et Y. C. Zhong = Camellia limonia C. F. Liang et S. L. Mo ●

69282 Camellia limonia C. F. Liang et S. L. Mo var. obovata S. L. Mo et Y. C. Zhong = Camellia flavida Hung T. Chang ●◇

69283 Camellia lioui H. T. Tsai et K. M. Feng = Camellia forrestii (Diels) Cohen-Stuart ●

69284 Camellia lipingensis Hung T. Chang;黎平瘤果茶;Liping Camellia ●

69285 Camellia lipingensis Hung T. Chang = Camellia rhytidocarpa Hung T. Chang et S. Ye Liang ●

69286 Camellia lipoensis Hung T. Chang et Z. R. Xu;荔波连蕊茶;Libo Camellia, Libo Tea ●

69287 Camellia lipoensis Hung T. Chang et Z. R. Xu = Camellia polyodonta F. C. How ex Hu ●

69288 Camellia lipoensis Hung T. Chang et Z. R. Xu = Camellia rosthorniana Hand. -Mazz. ●

69289 Camellia litchi Hung T. Chang = Camellia ilicifolia H. L. Li ●

69290 Camellia litchi Hung T. Chang = Camellia obovatifolia Hung T. Chang ●

69291 Camellia liui H. T. Tsai et K. M. Feng = Camellia forrestii (Diels) Cohen-Stuart ●

69292 Camellia longgangensis C. F. Liang et S. L. Mo = Camellia flavida Hung T. Chang ●◇

69293 Camellia longgangensis C. F. Liang et S. L. Mo var. grandis C. F. Liang et S. L. Mo;大样弄岗茶●

69294 Camellia longgangensis C. F. Liang et S. L. Mo var. grandis C. F. Liang et S. L. Mo = Camellia flavida Hung T. Chang ●◇

69295 Camellia longgangensis C. F. Liang et S. L. Mo var. patens S. L. Mo et Y. C. Zhong = Camellia petelotii (Merr.) Sealy var. microcarpa (S. L. Mo et S. Z. Huang) T. L. Ming et W. J. Zhang ●

69296 Camellia longgangensis C. F. Liang et S. L. Mo var. patens S. L. Mo et Y. C. Zhong = Camellia flavida Hung T. Chang var. patens (S. L. Mo et Y. C. Zhong) T. L. Ming ●

69297 Camellia longicalyx Hung T. Chang;长萼连蕊茶(长萼毛蕊茶,披针萼连蕊茶);Lance-sepal Camellia, Lance-sepaled Camellia, Longcalyx Camellia, Longcalyx Tea, Longcalyxed Camellia ●

69298 Camellia longicarpa Hung T. Chang;长果连蕊茶;Longfruit Camellia, Longfruit Tea, Long-fruited Camellia ●

69299 Camellia longicarpa Hung T. Chang = Camellia synaptica Sealy ●

69300 Camellia longicaudata Hung T. Chang et S. Ye Liang = Camellia polyodonta F. C. How ex Hu var. longicaudata (Hung T. Chang et S. Ye Liang ex Hung T. Chang) T. L. Ming ●

69301 Camellia longicaudata Hung T. Chang et S. Ye Liang ex Hung T. Chang;长尾红山茶;Longcaudate Camellia, Long-caudated Camellia, Longtail Camellia ●

69302 Camellia longicaudata Hung T. Chang et S. Ye Liang ex Hung T. Chang = Camellia polyodonta F. C. How ex Hu var. longicaudata (Hung T. Chang et S. Ye Liang ex Hung T. Chang) T. L. Ming ●

69303 Camellia longicuspis S. Ye Liang;长尖连蕊茶;Longcusp Camellia, Long-cusp Camellia, Longcusp Tea ●

69304 Camellia longicuspis S. Ye Liang ex Hung T. Chang = Camellia cuspidata (Kochs) Wright ex Garden var. grandiflora Sealy ●

69305 Camellia longigyna Hung T. Chang;长蕊红山茶;Long-famele Camellia, Longpistil Camellia ●

69306 Camellia longigyna Hung T. Chang = Camellia mairei (H. Lév.) Melch. var. lapidea (W. C. Wu) Sealy ●

69307 Camellia longipedicellata (Hu) Hung T. Chang et D. Fang;长梗茶(长柄山茶,长梗山茶,山茶花);Long-pedicele Camellia, Long-pediceled Camellia, Longpetiole Tea ●

69308 Camellia longipetiolata (Hu) Hung T. Chang et D. Fang;长柄山茶●

69309 Camellia longissima Hung T. Chang et S. Ye Liang;超长梗茶(超长柄茶);Extralongstalk Tea, Longest Camellia, Longest-pediceled Camellia ●

69310 Camellia longistyla Hung T. Chang ex F. A. Zeng et H. Zhou = Camellia mairei (H. Lév.) Melch. var. lapidea (W. C. Wu) Sealy ●

69311 Camellia longituba Hung T. Chang;长管红山茶;Long-tube Camellia ●

69312 Camellia longituba Hung T. Chang = Camellia pitardii Cohen-Stuart var. variabilis (Hung T. Chang et X. K. Wen) T. L. Ming ●

69313 Camellia longlingensis F. C. Zhang, G. B. Chen et M. D. Tang;龙陵茶;Longling Camellia ●

69314 Camellia longlingensis F. C. Zhang, G. B. Chen et M. D. Tang = Camellia sinensis (L.) Kuntze ●

69315 Camellia longruiensis S. Ye Liang et X. J. Dong = Camellia flavida Hung T. Chang ●◇

69316 Camellia longruiensis S. Ye Liang et X. J. Dong = Camellia limonia C. F. Liang et S. L. Mo ●

69317 Camellia longzhouensis J. Y. Luo = Camellia chrysanthoides Hung T. Chang ●

69318 Camellia longzhouensis J. Y. Luo = Camellia tonkinensis (Pit.) Cohen-Stuart ●

69319 Camellia lucidissima Hung T. Chang;闪光红山茶;Flash Camellia, Lucid Camellia ●

69320 Camellia lucidissima Hung T. Chang = Camellia chekiangoleosa Hu ●

69321 Camellia lungshenensis Hung T. Chang;龙胜红山茶;Longsheng Camellia ●

69322 Camellia lungshenensis Hung T. Chang = Camellia pitardii Cohen-Stuart var. cryptoneura (Hung T. Chang) T. L. Ming ●

69323 Camellia lungshenensis Hung T. Chang = Camellia pitardii

Cohen-Stuart ●

69324 Camellia lungyaiensis (Hu) Tuyama = Camellia brevistyla (Hayata) Cohen-Stuart ●

69325 Camellia lungzhouensis J. Y. Luo;龙州金花茶;Longzhou Camellia, Longzhou Tea ●

69326 Camellia lutchuensis Ito = Camellia lutchuensis Ito ex Ito et Matsum. ●

69327 Camellia lutchuensis Ito ex Ito et Matsum.;台湾连蕊茶(琉球连蕊茶)●

69328 Camellia lutchuensis Ito ex Ito et Matsum. var. minutiflora (Hung T. Chang) T. L. Ming = Camellia minutiflora Hung T. Chang ●

69329 Camellia luteoflora Y. K. Li ex H. T. Chang et F. A. Zeng;小黄花茶;Yellowflower Camellia ●

69330 Camellia lutescens Dyer = Camellia kissi Wall. ●

69331 Camellia macrosepala Hung T. Chang;大萼连蕊茶;Big-calyxed Camellia, Bigsepal Camellia, Bigsepal Tea ●

69332 Camellia macrosepala Hung T. Chang = Camellia cuspidata (Kochs) Wright ex Garden var. grandiflora Sealy ●

69333 Camellia magniflora Hung T. Chang;大花红山茶;Bigflower Camellia ●

69334 Camellia magniflora Hung T. Chang = Camellia pitardii Cohen-Stuart var. variabilis (Hung T. Chang et X. K. Wen) T. L. Ming ●

69335 Camellia magnocarpa (Hu et T. C. Huang) Hung T. Chang = Camellia semiserrata C. W. Chi var. magnocarpa Hu et Hung T. Chang ex Hu ●

69336 Camellia magnocarpa Hung T. Chang;大果红山茶;Bigfruit Camellia, Big-fruited Camellia, Largefruit Camellia ●

69337 Camellia magnocarpa Hung T. Chang = Camellia semiserrata C. W. Chi var. magnocarpa Hu et Hung T. Chang ex Hu ●

69338 Camellia mairei (H. Lév.) Melch.;毛蕊山茶(成凤山茶,峨眉红山茶,毛蕊红山茶);Maire Camellia ●

69339 Camellia mairei (H. Lév.) Melch. f. alba (Hung T. Chang) T. L. Ming = Camellia mairei (H. Lév.) Melch. var. alba Hung T. Chang ●

69340 Camellia mairei (H. Lév.) Melch. f. alba (Hung T. Chang) T. L. Ming = Camellia mairei (H. Lév.) Melch. ●

69341 Camellia mairei (H. Lév.) Melch. var. alba Hung T. Chang;白花毛蕊山茶(白花毛蕊茶);Whiteflower Maire Camellia ●

69342 Camellia mairei (H. Lév.) Melch. var. alba Hung T. Chang = Camellia mairei (H. Lév.) Melch. ●

69343 Camellia mairei (H. Lév.) Melch. var. lapidea (W. C. Wu) Sealy = Camellia lapidea Y. C. Wu ●

69344 Camellia mairei (H. Lév.) Melch. var. velutina Sealy;滇南毛蕊山茶(绒毛成凤山茶);S. Yunnan Maire Camellia, Velvety Maire Camellia ●

69345 Camellia makuanica Hung T. Chang et Y. J. Tang = Camellia crassicolumna Hung T. Chang ●

69346 Camellia makuanica Hung T. Chang, Y. J. Tang et P. S. Tang;马关茶;Maguan Camellia, Maguan Tea ●

69347 Camellia makuanica Hung T. Chang, Y. J. Tang et P. S. Tang = Camellia crassicolumna Hung T. Chang ●

69348 Camellia maliflora Lindl.;樱花短柱茶(苹果花山茶);Cherryflower Camellia ●

69349 Camellia manglaensis Hung T. Chang, Y. J. Tan et P. S. Tang = Camellia dehungensis Hung T. Chang, H. S. Wang et B. H. Chen ●

69350 Camellia mastersii Griff. = Camellia kissi Wall. ●

69351 Camellia meiocarpa Hu = Camellia oleifera Abel var. monosperma Hung T. Chang ●

69352 Camellia melliana Hand.-Mazz.;广东毛蕊茶;Guangdong Camellia, Guangdong Tea ●

69353 Camellia membranacea Hung T. Chang;膜叶连蕊茶;Membranaceous Camellia, Membrana-leaf Camellia, Membraneleaf Tea ●

69354 Camellia membranacea Hung T. Chang = Camellia caudata Wall. ●

69355 Camellia micrantha S. Ye Liang et Y. C. Zhong;小花金花茶(龙州金花茶);Smallanther Tea, Small-flower Camellia ●

69356 Camellia micrantha S. Ye Liang et Y. C. Zhong = Camellia parvisepala Hung T. Chang ●

69357 Camellia microcarpa (S. L. Mo et S. Z. Huang) S. L. Mo;小金花茶(小果金花茶)●

69358 Camellia microcarpa (S. L. Mo et S. Z. Huang) S. L. Mo = Camellia nitidissima C. W. Chi var. microcarpa (S. L. Mo et S. Z. Huang) Hung T. Chang et C. X. Ye ●

69359 Camellia microcarpa (S. L. Mo et S. Z. Huang) S. L. Mo = Camellia petelotii (Merr.) Sealy var. microcarpa (S. L. Mo et S. Z. Huang) T. L. Ming et W. J. Zhang ●

69360 Camellia microdonta Hung T. Chang = Camellia pitardii Cohen-Stuart ●

69361 Camellia microdonta Hung T. Chang et C. S. Ye = Camellia oviformis Hung T. Chang ●

69362 Camellia microdonta Hung T. Chang et C. S. Ye = Camellia pitardii Cohen-Stuart ●

69363 Camellia microphylla (Merr.) S. S. Chien;细叶短柱茶;Minileaf Camellia, Small-leaf Shortstyle Camellia, Small-leaved Camellia ●

69364 Camellia microphylla (Merr.) S. S. Chien = Camellia brevistyla (Hayata) Cohen-Stuart var. microphylla (Merr.) T. L. Ming ●

69365 Camellia mileensis T. L. Ming;弥勒糙果茶●

69366 Camellia mileensis T. L. Ming var. microphylla T. L. Ming;小叶弥勒糙果茶●

69367 Camellia minor Hung T. Chang = Camellia saluenensis Stapf ex Bean ●

69368 Camellia minutiflora Hung T. Chang;微花连蕊茶;Miniflower Tea, Verylittle-flower Camellia ●

69369 Camellia minutiflora Hung T. Chang = Camellia lutchuensis Ito ex Ito et Matsum. var. minutiflora (Hung T. Chang) T. L. Ming ●

69370 Camellia miyagii (Koidz.) Makino et Nemoto;冲绳山茶;Miyagi Camellia ●☆

69371 Camellia mongshanica Hung T. Chang et C. S. Ye;莽山红山茶;Mangshan Camellia ●

69372 Camellia mongshanica Hung T. Chang et C. S. Ye = Camellia semiserrata C. W. Chi var. magnocarpa Hu et Hung T. Chang ex Hu ●

69373 Camellia monodelphia Hu;单体连蕊茶;Monosomic Camellia ●

69374 Camellia multibracteata Hung T. Chang = Camellia crapnelliana Tutcher ●◇

69375 Camellia multibracteata Hung T. Chang et Z. Q. Mo;多苞糙果茶;Manybract Camellia, Multibracteate Camellia ●

69376 Camellia multiperulata Hung T. Chang = Camellia semiserrata C. W. Chi ●

69377 Camellia multipetala S. Ye Liang et C. Z. Deng = Camellia flavida Hung T. Chang var. patens (S. L. Mo et Y. C. Zhong) T. L. Ming ●

69378 Camellia multipetala S. Ye Liang et C. Z. Deng = Camellia petelotii (Merr.) Sealy var. microcarpa (S. L. Mo et S. Z. Huang) T. L. Ming et W. J. Zhang ●

69379 Camellia multiplex Hung T. Chang et C. M. Tan = Camellia

makuanica Hung T. Chang, Y. J. Tang et P. S. Tang ●
69380 Camellia multiplex Hung T. Chang et Y. J. Tang = Camellia crassicolumna Hung T. Chang var. multiplex (Hung T. Chang et C. M. Tan) T. L. Ming ●
69381 Camellia multiplex Hung T. Chang, Y. J. Tan et P. S. Wang = Camellia crassicolumna Hung T. Chang var. multiplex (Hung T. Chang et C. M. Tan) T. L. Ming ●
69382 Camellia multisepala Hung T. Chang = Camellia sinensis (L.) Kuntze var. assamica (Mast.) Kitam. ●◇
69383 Camellia multisepala Hung T. Chang et Y. J. Tang = Camellia sinensis (L.) Kuntze var. assamica (Mast.) Kitam. ●◇
69384 Camellia multisepala Hung T. Chang, Y. J. Tan et P. S. Wang;多萼茶;Manysepal Camellia, Manysepal Tea ●
69385 Camellia muricatula Hung T. Chang;瘤叶短蕊茶(瘤果短蕊茶);Muricate Camellia, Tuberculate-fruited Camellia, Tumorleaf Tea ●
69386 Camellia muricatula Hung T. Chang = Camellia wardii Kobuski var. muricatula (Hung T. Chang) T. L. Ming ●
69387 Camellia mutabilis Paxton = Camellia japonica L. ●
69388 Camellia nakaii (Hayata) Hayata = Camellia japonica L. ●
69389 Camellia nanchuanica Hung T. Chang et J. H. Xiong;南川茶;Nanchuan Camellia, Nanchuan Tea ●
69390 Camellia nanchuanica Hung T. Chang et J. H. Xiong = Camellia gymnogyna Hung T. Chang var. remotiserrata (Hung T. Chang, H. S. Wang et P. S. Wang) T. L. Ming ●
69391 Camellia nanchuanica Hung T. Chang et J. H. Xiong = Camellia tachangensis F. S. Zhang var. remotiserrata (Hung T. Chang, H. S. Wang et B. H. Wang) T. L. Ming ●
69392 Camellia neriifolia Hung T. Chang = Camellia ilicifolia H. L. Li var. nerifolia (Hung T. Chang) T. L. Ming ●
69393 Camellia nitidissima C. W. Chi = Camellia petelotii (Merr.) Sealy ●
69394 Camellia nitidissima C. W. Chi var. microcarpa (S. L. Mo et S. Z. Huang) Hung T. Chang et C. X. Ye = Camellia petelotii (Merr.) Sealy var. microcarpa (S. L. Mo et S. Z. Huang) T. L. Ming et W. J. Zhang ●
69395 Camellia nitidissima C. W. Chi var. microcarpa Hung T. Chang et C. X. Ye = Camellia petelotii (Merr.) Sealy var. microcarpa (S. L. Mo et S. Z. Huang) T. L. Ming et W. J. Zhang ●
69396 Camellia nitidissima C. W. Chi var. phaeopubisperma S. Ye Liang et Z. H. Tang = Camellia petelotii (Merr.) Sealy var. microcarpa (S. L. Mo et S. Z. Huang) T. L. Ming et W. J. Zhang ●
69397 Camellia nitidissima C. W. Chi var. phaeopubisperma S. Ye Liang et Z. H. Tang = Camellia petelotii (Merr.) Sealy ●
69398 Camellia nokoensis Hayata;能高山茶(柃木叶山茶,南投连蕊茶,能高连蕊茶,生毛胡桃);Nantou Camellia, Nenggaoshan Camellia, Nengka Camellia, Noko Tea ●
69399 Camellia nokoensis Hayata = Camellia euryoides Lindl. var. nokoensis (Hayata) T. L. Ming ●
69400 Camellia oblata Hung T. Chang;扁糙果茶;Flat Camellia, Oblate Camellia ●
69401 Camellia oblata Hung T. Chang = Camellia furfuracea (Merr.) Cohen-Stuart ●
69402 Camellia obovatifolia Hung T. Chang;倒卵叶瘤果茶(倒卵瘤果茶);Obovateleaf Camellia, Obovate-leaved Camellia ●
69403 Camellia obovatifolia Hung T. Chang = Camellia anlungensis Hung T. Chang ●
69404 Camellia obscurinervis H. T. Tsai et K. M. Feng;无脉山茶;Obscurevein Camellia ●
69405 Camellia obscurinervis H. T. Tsai et K. M. Feng = Camellia villicarpa S. S. Chien ●
69406 Camellia obtusifolia Hung T. Chang;钝叶短柱茶;Obtuseleaf Camellia, Obtuseleaf Shortstyle Camellia, Obtuse-leaved Camellia ●
69407 Camellia obtusifolia Hung T. Chang = Camellia brevistyla (Hayata) Cohen-Stuart ●
69408 Camellia obtusifolia Hung T. Chang f. rubella Z. H. Cheng;粉红钝叶短柱茶;Reddish Obtuseleaf Camellia ●
69409 Camellia octopetala Hu = Camellia crapnelliana Tutcher ●◇
69410 Camellia odorata L. S. Xie et Zhan Y. Zhang = Camellia grijsii Hance ●◇
69411 Camellia oleifera Abel;油茶(白花茶,茶麸,茶油树,茶子木,茶子树,长柱金花茶,建茶,苦茶,柃木,桃茶,野茶籽,油子树,楂);Himalayan Camellia, Oil Camellia, Oil Tea, Oiltea Camellia, Oil-tea Camellia, Tea-oil Camellia, Tea-oil Plant ●
69412 Camellia oleifera Abel 'Lushan Snow';庐山雪油茶●
69413 Camellia oleifera Abel = Camellia oleosa (Lour.) Rehder ●
69414 Camellia oleifera Abel var. confusa (Craib) Sealy;澜沧油茶;Confused Oiltea Camellia ●
69415 Camellia oleifera Abel var. confusa (Craib) Sealy = Camellia confusa Craib ●
69416 Camellia oleifera Abel var. monosperma Hung T. Chang;单籽油茶(单籽红山茶,小叶油茶);One-seed Oiltea Camellia ●
69417 Camellia oleifera Abel var. monosperma Hung T. Chang = Camellia oleifera Abel ●
69418 Camellia oleifera C. Abel var. confusa (Craib) Sealy = Camellia kissii Wall. var. confusa (Craib) T. L. Ming ●
69419 Camellia oleosa (Lour.) C. Y. Wu = Camellia oleifera Abel ●
69420 Camellia oleosa (Lour.) Rehder = Camellia sinensis (L.) Kuntze ●
69421 Camellia oligophlebia Hung T. Chang;寡脉红山茶;Fewnerve Camellia, Fewvein Camellia ●
69422 Camellia oligophlebia Hung T. Chang = Camellia reticulata Lindl. ●◇
69423 Camellia omeiensis Hung T. Chang;峨眉红山茶;Emei Camellia ●
69424 Camellia omeiensis Hung T. Chang = Camellia mairei (H. Lév.) Melch. ●
69425 Camellia ovatifolia Hung T. Chang;卵叶短尾茶;Ovate-leaved Camellia ●
69426 Camellia oviformis Hung T. Chang;卵果红山茶;Eggfruit Camellia, Ovate-fruit Camellia, Ovate-fruited Camellia ●
69427 Camellia oviformis Hung T. Chang = Camellia polyodonta F. C. How ex Hu ●
69428 Camellia pachyandra Hu;滇南离蕊茶(沧源山茶,厚短蕊茶);Short-thicked Camellia, Thickanther Tea, Thick-stamened Camellia ●
69429 Camellia pankiangensis Hung T. Chang;盘江连蕊茶;Panjiang Camellia ●
69430 Camellia parafurfuracea S. Ye Liang ex Hung T. Chang;肖糙果茶;Furfuraceous-leaf-like Camellia, Like Furfuraceousfruit Camellia, Sham Chaff Camellia ●
69431 Camellia parafurfuracea S. Ye Liang ex Hung T. Chang = Camellia furfuracea (Merr.) Cohen-Stuart ●
69432 Camellia parapolyodonta Hung T. Chang;多齿红山茶;Wantien-like Camellia ●
69433 Camellia parvicaudata Hung T. Chang;小长尾连蕊茶;Parvicaudate Camellia, Smalltail Tea ●

69434 Camellia parvicaudata Hung T. Chang = Camellia tsingpienensis Hu var. pubisepala Hung T. Chang ●

69435 Camellia parvicuspidata Hung T. Chang;细尖连蕊茶;Smallcuspidate Camellia,Smallcuspidate Tea,Smallwoodyfruit Camellia ●

69436 Camellia parvicuspidata Hung T. Chang = Camellia cuspidata (Kochs) Wright ex Garden ●

69437 Camellia parviflora Merr. et Chun ex Sealy;细花短蕊茶;Littleflower Camellia,Small-flowered Camellia ●

69438 Camellia parvifolia (Hayata) Cohen-Stuart = Camellia transarisanensis (Hayata) Cohen-Stuart ●

69439 Camellia parvifolia (Hayata) Nakai = Camellia transarisanensis (Hayata) Cohen-Stuart ●

69440 Camellia parvifolia Cohen-Stuart = Camellia transarisanensis (Hayata) Cohen-Stuart ●

69441 Camellia parvilapidea Hung T. Chang;小石果连蕊茶;Smallstonefruit Camellia, Smallstonefruit Tea, Small-woody-fruited Camellia ●

69442 Camellia parvilapidea Hung T. Chang = Camellia caudata Wall. ●

69443 Camellia parvilimba Merr. et F. P. Metcalf;细叶连蕊茶(小叶山茶);Littleleaf Tea,Ovalleaf Camellia,Small-leaved Camellia ●

69444 Camellia parvilimba Merr. et F. P. Metcalf = Camellia euryoides Lindl. ●

69445 Camellia parvilimba Merr. et F. P. Metcalf var. brevipes Hung T. Chang;短柄细叶连蕊茶;Short-stalk Ovalleaf Camellia ●

69446 Camellia parvilimba Merr. et F. P. Metcalf var. brevipes Hung T. Chang = Camellia euryoides Lindl. ●

69447 Camellia parvimuricata Hung T. Chang;小瘤果茶;Smallmuricate Camellia, Smallmuricatefruit Camellia, Small-muricate-fruited Camellia ●

69448 Camellia parvimuricata Hung T. Chang var. hupehensis (Hung T. Chang) T. L. Ming;大萼小瘤果茶●

69449 Camellia parvimuricata Hung T. Chang var. songtaoensis K. M. Lan et H. H. Zhang ex T. L. Ming et Y. C. Zhong;光枝小瘤果茶;Smooth-branch Smallmuricatefruit Camellia ●

69450 Camellia parviovata Hung T. Chang et S. S. Wang = Camellia synaptica Sealy var. parviovata (Hung T. Chang et S. S. Wang) T. L. Ming ●

69451 Camellia parviovata Hung T. Chang et S. S. Wang ex Hung T. Chang;小卵叶连蕊茶;Smallegg Tea,Small-ovate-leaf Camellia ●

69452 Camellia parvipetala J. Y. Liang et Z. M. Su;小瓣金花茶;Smallpetal Camellia,Smallpetal Tea ●

69453 Camellia parvipetala J. Y. Liang et Z. M. Su = Camellia grandis (C. F. Liang et S. L. Mo) Hung T. Chang et S. Ye Liang ●

69454 Camellia parvipetala J. Y. Liang et Z. M. Su = Camellia indochinensis Merr. ●

69455 Camellia parvipetala J. Y. Liang et Z. M. Su = Camellia limonia C. F. Liang et S. L. Mo ●

69456 Camellia parvisepala Hung T. Chang;细萼茶(白毛茶);Smallsepal Camellia,Smallsepal Tea,Small-sepaled Camellia ●

69457 Camellia parvisepala Hung T. Chang = Camellia sinensis (L.) Kuntze var. pubilimba Hung T. Chang ●

69458 Camellia parvisepaloides Hung T. Chang et H. S. Wang = Camellia sinensis (L.) Kuntze var. dehungensis (Hung T. Chang et B. H. Chen) T. L. Ming ●

69459 Camellia parvisepaloides Hung T. Chang, H. S. Wang et B. H. Chen;拟细萼茶;Like Smallsepal Camellia,Near-smallsepal Tea ●

69460 Camellia paucigyrata Hung T. Chang;寡节短柱茶;Pauciannulated Camellia ●

69461 Camellia paucigyrata Hung T. Chang var. yaoshanensis Hung T. Chang;瑶山短柱茶;Yaoshan Pauciannulated Camellia ●

69462 Camellia pauciperulata Hung T. Chang;贫萼红山茶;Paucisepaled Camellia ●

69463 Camellia paucipetala Hung T. Chang;寡瓣红山茶;Fewpetal Camellia ●

69464 Camellia paucipetala Hung T. Chang = Camellia reticulata Lindl. ●◇

69465 Camellia paucipunctata (Merr. et Chun) Chun;腺叶离蕊茶(海南山茶);Glandleaf Tea,Paucipunctated Camellia ●

69466 Camellia pentamera Hung T. Chang;五数离蕊茶;Fived Camellia,Pentamerous Camellia,Pentamerous Tea ●

69467 Camellia pentamera Hung T. Chang = Camellia forrestii (Diels) Cohen-Stuart var. pentamera (Hung T. Chang) T. L. Ming ●

69468 Camellia pentapetala Hung T. Chang;五瓣红山茶;Five-petal Camellia ●

69469 Camellia pentapetala Hung T. Chang = Camellia reticulata Lindl. ●◇

69470 Camellia pentaphylacoides Hung T. Chang = Camellia oligophlebia Hung T. Chang ●

69471 Camellia pentaphylacoides Hung T. Chang = Camellia reticulata Lindl. ●◇

69472 Camellia pentaphylax Hung T. Chang = Camellia chunii Hung T. Chang var. pentaphylax Hung T. Chang ●

69473 Camellia pentaphylax Hung T. Chang = Camellia reticulata Lindl. ●◇

69474 Camellia pentastyla Hung T. Chang;五柱茶;Fivestyle Camellia, Fivestyle Tea,Five-styled Camellia ●

69475 Camellia pentastyla Hung T. Chang = Camellia reticulata Lindl. ●◇

69476 Camellia pentastyla Hung T. Chang = Camellia taliensis (W. W. Sm.) Melch. ●

69477 Camellia percuspidata Hung T. Chang;超尖连蕊茶;Extracuspidate Tea,Percuspidate Camellia ●

69478 Camellia percuspidata Hung T. Chang = Camellia caudata Wall. ●

69479 Camellia petaphylax Hung T. Chang = Camellia chunii Hung T. Chang var. pentaphylax Hung T. Chang ●

69480 Camellia petelotii (Merr.) Sealy;金花茶(大叶金花茶,多瓣山茶,亮叶离蕊茶,亮叶连蕊茶,小花金花茶,张氏红山茶);Golden Camellia, Golden-flowered Camellia, Goldflower Tea, Petelot's Camellia,Shining Camellia,Shinyleaf Camellia ●

69481 Camellia petelotii (Merr.) Sealy var. microcarpa (S. L. Mo et S. Z. Huang) T. L. Ming et W. J. Zhang;小果金花茶(小金花茶);Little Petelot's Camellia ●

69482 Camellia phaeoclada Hung T. Chang;褐枝短柱茶;Brown-branch Camellia,Brown-branched Camellia,Browntwig Tea ●

69483 Camellia phaeoclada Hung T. Chang = Camellia saluenensis Stapf ex Bean ●

69484 Camellia phellocapsa Hung T. Chang et B. K. Lee;栓壳红山茶;Kork-fruited Camellia,Korkhull Camellia,Phellemy Camellia ●

69485 Camellia phellocapsa Hung T. Chang et B. K. Lee = Camellia semiserrata C. W. Chi ●

69486 Camellia phelloderma Hung T. Chang, H. S. Liu et Y. Z. Zhang;栓皮红山茶;Cork Camellia,Phellem-bark Camellia ●

69487 Camellia phelloderma Hung T. Chang, H. S. Liu et Y. Z. Zhang = Camellia mairei (H. Lév.) Melch. var. lapidea (W. C. Wu) Sealy ●

69488 Camellia phelloderma Hung T. Chang, H. S. Liu et Y. Z. Zhang =

Camellia reticulata Lindl. ●◇

69489 Camellia pilistyla Hung T. Chang; 毛柱红山茶; Hairy-styled Camellia ●

69490 Camellia pilosperma S. Ye Liang; 毛籽短蕊茶(毛籽离蕊茶); Hairseed Tea, Hairy-seed Camellia, Hairy-seeded Camellia ●

69491 Camellia pingguoensis D. Fang; 平果金花茶(小瓣金花茶); Pingguo Camellia, Pingguo Tea ●◇

69492 Camellia pingguoensis D. Fang var. terminalis (J. Y. Liang et Z. M. Su) T. L. Ming et W. J. Zhang; 顶生金花茶; Terminal Camellia ●◇

69493 Camellia pingguoensis D. Fang var. terminalis (J. Y. Liang et Z. M. Su) S. Ye. Liang = Camellia pingguoensis D. Fang var. terminalis (J. Y. Liang et Z. M. Su) T. L. Ming et W. J. Zhang ●◇

69494 Camellia pitardii Cohen-Stuart; 西南山茶(滇野山茶,匹他山茶,西南红山茶,野山茶); Cavalerie Camellia, Pitard Camellia, SW. China Camellia ●

69495 Camellia pitardii Cohen-Stuart f. alba (Hung T. Chang) T. L. Ming = Camellia pitardii Cohen-Stuart ●

69496 Camellia pitardii Cohen-Stuart f. cavaleriana (H. Lév.) Sealy = Camellia pitardii Cohen-Stuart ●

69497 Camellia pitardii Cohen-Stuart var. alba Hung T. Chang; 西南白山茶; White Pitard Camellia ●

69498 Camellia pitardii Cohen-Stuart var. alba Hung T. Chang = Camellia pitardii Cohen-Stuart ●

69499 Camellia pitardii Cohen-Stuart var. cryptoneura (Hung T. Chang) T. L. Ming = Camellia cryptoneura Hung T. Chang ●

69500 Camellia pitardii Cohen-Stuart var. longistaminata J. L. Liu et Q. Luo = Camellia pitardii Cohen-Stuart ●

69501 Camellia pitardii Cohen-Stuart var. lucidissima (H. Lév.) Rehder = Camellia saluenensis Stapf ex Bean ●

69502 Camellia pitardii Cohen-Stuart var. panxiensis J. L. Liu = Camellia pitardii Cohen-Stuart ●

69503 Camellia pitardii Cohen-Stuart var. variabilis (Hung T. Chang et X. K. Wen) T. L. Ming; 多变西南山茶; Variable Pitard Camellia ●

69504 Camellia pitardii Cohen-Stuart var. yunnanica Sealy; 窄叶西南红山茶(狭叶西南红山茶,野山茶,云南野山茶); Yunnan Pitard Camellia ●

69505 Camellia pitardii Cohen-Stuart var. yunnanica Sealy = Camellia reticulata Lindl. ●◇

69506 Camellia planipetala Lem. = Camellia japonica L. ●

69507 Camellia pleurocarpa (Gagnep.) Sealy; 簕果茶●☆

69508 Camellia polygama (Hu) Hu = Camellia forrestii (Diels) Cohen-Stuart ●

69509 Camellia polyneura Hung T. Chang et Y. J. Tang = Camellia sinensis (L.) Kuntze var. assamica (Mast.) Kitam. ●◇

69510 Camellia polyneura Hung T. Chang, Y. J. Tan et P. S. Wang = Camellia assamica (Mast.) Hung T. Chang var. polyneura (Hung T. Chang, Y. J. Tan et P. S. Wang) Hung T. Chang ●

69511 Camellia polyneura Hung T. Chang, Y. J. Tan et P. S. Wang = Camellia sinensis (L.) Kuntze var. assamica (Mast.) Kitam. ●◇

69512 Camellia polyneura Hung T. Chang, Y. J. Tan et P. S. Wang var. kucha Hung T. Chang et H. S. Wang = Camellia assamica (Mast.) Hung T. Chang var. kucha (Hung T. Chang, H. S. Wang et B. H. Chen) Hung T. Chang et H. S. Wang ●

69513 Camellia polyodonta F. C. How ex Hu; 多齿山茶(长毛红山茶,多齿红山茶,卵果红山茶,宛田红花油茶); Manytooth Camellia, Wantian Camellia, Wantien Camellia ●

69514 Camellia polyodonta F. C. How ex Hu f. alba T. L. Ming = Camellia polyodonta F. C. How ex Hu ●

69515 Camellia polyodonta F. C. How ex Hu var. longicaudata (Hung T. Chang et S. Ye Liang ex Hung T. Chang) T. L. Ming; 长尾多齿山茶(长尾红山茶,桂峰山油茶,假多齿红山茶); Longtail Wantien Camellia ●

69516 Camellia polypetala Hung T. Chang; 多瓣糙果茶; Manypetal Camellia ●

69517 Camellia polypetala Hung T. Chang = Camellia furfuracea (Merr.) Cohen-Stuart ●

69518 Camellia polystyla Hung T. Chang; 多柱茶; Polystylous Camellia ●

69519 Camellia pseudoelongata Hung T. Chang et S. X. Ren = Camellia tsaii Hu ●

69520 Camellia ptilophylla Hung T. Chang; 毛叶茶(毛茶); Downy-leaved Camellia, Hairleaf Tea, Hairy-leaf Camellia ●

69521 Camellia ptilosperma S. Ye Liang = Camellia flavida Hung T. Chang ●◇

69522 Camellia ptilosperma S. Ye Liang et Q. D. Chen = Camellia flavida Hung T. Chang ●◇

69523 Camellia ptilosperma S. Ye Liang et Q. D. Chen = Camellia grandis (C. F. Liang et S. L. Mo) Hung T. Chang et S. Ye Liang ●

69524 Camellia pubescens Hung T. Chang et C. X. Ye = Camellia ptilophylla Hung T. Chang ●

69525 Camellia pubescens Hung T. Chang et S. Ye Liang; 汝城毛叶茶; Pubescent Tea, Rucheng Camellia ●

69526 Camellia pubescens Hung T. Chang et S. Ye Liang = Camellia ptilophylla Hung T. Chang ●

69527 Camellia pubifurfuracea Y. C. Zhong; 毛糙果茶; Hairy Camellia ●

69528 Camellia pubipetala Y. Wan et S. Z. Huang; 毛瓣金花茶(毛籽金花茶); Hairpetal Tea, Hairy-petal Camellia ●◇

69529 Camellia pubisepala D. Fang = Camellia furfuracea (Merr.) Cohen-Stuart ●

69530 Camellia pubisepula D. Fang; 毛萼金花茶; Hairsepal Camellia ●

69531 Camellia punctata (Kochs) Cohen-Stuart; 斑枝毛蕊茶; Dotted Camellia, Punctate Camellia, Punctate Tea ●

69532 Camellia puniceiflora Hung T. Chang; 粉红短柱茶; Pink Camellia, Pink-flower Camellia, Scarlet-flowered Camellia ●

69533 Camellia puniceiflora Hung T. Chang = Camellia brevistyla (Hayata) Cohen-Stuart ●

69534 Camellia purpurea Hung T. Chang et B. H. Chen = Camellia crassicolumna Hung T. Chang ●

69535 Camellia purpurea Hung T. Chang, H. S. Wang et B. H. Chen; 紫果茶; Purple Camellia, Purplefruit Tea ●

69536 Camellia pyxidiacea Hung T. Chang et F. H. Chen var. rubituberculata (Hung T. Chang) T. L. Ming; 红花三江瘤果茶●

69537 Camellia pyxidiacea Hung T. Chang et F. H. Chen var. rubituberculata (Hung T. Chang) T. L. Ming = Camellia rubituberculata Hung T. Chang ●◇

69538 Camellia pyxidiacea Z. R. Xu, F. P. Chen et C. Y. Deng; 三江瘤果茶●

69539 Camellia pyxidiacea Z. R. Xu, F. P. Chen et C. Y. Deng = Camellia anlungensis Hung T. Chang ●

69540 Camellia quinquebracteata Hung T. Chang et C. X. Ye = Camellia taliensis (W. W. Sm.) Melch. ●

69541 Camellia quinquelocularis Hung T. Chang et S. Ye Liang = Camellia tachangensis F. S. Zhang ●

69542 Camellia quinqueloculosa S. L. Mo et Y. C. Zhong = Camellia aurea Hung T. Chang ●

69543 Camellia quinqueloculosa S. L. Mo et Y. C. Zhong = Camellia flavida Hung T. Chang var. patens (S. L. Mo et Y. C. Zhong) T. L. Ming ●

69544 Camellia quiquelocularis Hung T. Chang et S. Ye Liang = Camellia tachangensis F. S. Zhang ●

69545 Camellia remotiserrata Hung T. Chang, F. L. Yu et P. S. Wang = Camellia remotiserrata Hung T. Chang, H. S. Wang et P. S. Wang ●

69546 Camellia remotiserrata Hung T. Chang, F. L. Yu et P. S. Wang = Camellia tachangensis F. S. Zhang var. remotiserrata (Hung T. Chang, H. S. Wang et B. H. Wang) T. L. Ming ●

69547 Camellia remotiserrata Hung T. Chang, H. S. Wang et P. S. Wang; 疏齿茶; Distanttooth Tea, Loose-toothed Camellia ●

69548 Camellia renshanxiangiae C. X. Ye et X. Q. Zheng; 毛药山茶; Hairanther Camellia ●

69549 Camellia reriifolia Hung T. Chang; 窄叶瘤果茶(狭叶瘤果茶); Reriifoliate Camellia ●

69550 Camellia reticulata Benth. = Tutcheria championi Nakai ●

69551 Camellia reticulata Benth. = Tutcheria spectabilis Dunn ●

69552 Camellia reticulata Lindl.; 滇山茶(白丝毛红山茶, 百灵山红山茶, 陈氏红山茶, 单瓣滇山茶, 滇北红山茶, 短柄糙果茶, 短柄红山茶, 短蕊红山茶, 短轴红山茶, 寡瓣红山茶, 寡脉红山茶, 贵州红山茶, 红花山茶, 红花油茶, 会理红山茶, 假五列木红山茶, 金沙江红山茶, 康滇红山茶, 离瓣红山茶, 毛瓣金花茶, 绵管红山茶, 木果红山茶, 南山茶, 栓皮红山茶, 唐椿, 唐山茶, 腾冲红花油茶, 五瓣红山茶, 五列木红山茶, 五柱茶, 野茶花, 野山茶, 异叶山茶, 云南红花油茶, 云南红山茶, 云南山茶, 云南野山茶, 窄叶西南红山茶, 竹叶红山茶); Heterophyllous Camellia, Jinshajiang Camellia, Netted Camellia, Netvein Camellia, Reticulate Camellia, Short-petioled Camellia, Simple Petal Camellia ●◇

69553 Camellia reticulata Lindl. 'Arch of Triumph'; 凯旋门滇山茶(凯旋门南山茶) ●

69554 Camellia reticulata Lindl. 'Butterfly Wings' = Camellia reticulata Lindl. 'Houye Diechi' ●

69555 Camellia reticulata Lindl. 'Captain Rawés'; 船长滇山茶(拉维斯船长南山茶) ●

69556 Camellia reticulata Lindl. 'Change of Day'; 岁月变幻滇山茶●

69557 Camellia reticulata Lindl. 'Cornelian'; 大玛瑙滇山茶●

69558 Camellia reticulata Lindl. 'Highlight'; 聚焦点滇山茶●

69559 Camellia reticulata Lindl. 'Houye Diechi'; 蝶翅南山茶●

69560 Camellia reticulata Lindl. 'Narrow-leaved Short Silk'; 狭叶射线滇山茶●

69561 Camellia reticulata Lindl. 'Nuccio's Ruby'; 牛西奥红宝石滇山茶●

69562 Camellia reticulata Lindl. 'Otto Hopfer'; 大奥托滇山茶●

69563 Camellia reticulata Lindl. 'Purple Gown'; 紫袍滇山茶●

69564 Camellia reticulata Lindl. 'Red Crystal'; 红水晶滇山茶●

69565 Camellia reticulata Lindl. 'Short Silk'; 射丝线滇山茶●

69566 Camellia reticulata Lindl. f. albescens (Hung T. Chang) T. L. Ming = Camellia reticulata Lindl. ●◇

69567 Camellia reticulata Lindl. f. simplex Sealy = Camellia reticulata Lindl. ●◇

69568 Camellia reticulata Lindl. var. rosea Makino = Camellia uraku Kitam. ●

69569 Camellia rhytidocarpa Hung T. Chang et S. Ye Liang; 皱果茶; Crinklefruit Camellia, Wrinkledfruit Camellia, Wrinkle-fruited Camellia ●

69570 Camellia rhytidocarpa Hung T. Chang et S. Ye Liang var. microphylla Y. C. Zhong; 小叶皱果茶(小皱果茶); Little-leaf Wrinkledfruit Camellia ●

69571 Camellia rhytidophylla Y. K. Li et M. Z. Yang = Camellia tuberculata S. S. Chien ●

69572 Camellia rosaeflora Hook.; 玫瑰连蕊茶(玫瑰茶, 玫瑰毛蕊茶); Roseflower Camellia, Roseflower Tea, Rosy-flowered Camellia ●

69573 Camellia rosthorniana Hand. -Mazz.; 川鄂连蕊茶; Rosthorn Camellia, Sichuan-Hubei Tea ●

69574 Camellia rotundata Hung T. Chang et F. L. Yu = Camellia crassicolumna Hung T. Chang ●

69575 Camellia rotundata Hung T. Chang et Te T. Yu; 圆基茶; Rotunbase Tea, Rotundate Camellia ●

69576 Camellia rotundata Hung T. Chang, Y. J. Tan et P. S. Wang = Camellia crassicolumna Hung T. Chang ●

69577 Camellia rubimuricata Hung T. Chang et Z. R. Xu; 荔波红瘤果茶(荔波瘤果茶); Libo Camellia, Rubimuricate Camellia ●◇

69578 Camellia rubimuricata Hung T. Chang et Z. R. Xu = Camellia ilicifolia H. L. Li ●

69579 Camellia rubituberculata Hung T. Chang; 厚壳红瘤果茶(红花瘤果茶); Rubituberculate Camellia, Thickhull Camellia ●◇

69580 Camellia rubituberculata Hung T. Chang ex M. J. Lin et Q. M. Lu = Camellia pyxidiacea Hung T. Chang et F. H. Chen var. rubituberculata (Hung T. Chang) T. L. Ming ●

69581 Camellia rubro-anthera Hung T. Chang = Camellia pitardii Cohen-Stuart ●

69582 Camellia rubro-anthera Hung T. Chang ex M. J. Lin et Q. M. Lu = Camellia pitardii Cohen-Stuart ●

69583 Camellia rusticana Honda = Camellia japonica L. subsp. rusticana (Honda) Kitam. ●

69584 Camellia rusticana Honda = Camellia japonica L. var. rusticana (Honda) Tuyama ●

69585 Camellia rusticana Honda f. plena (Honda) H. Hara = Camellia japonica L. subsp. rusticana (Honda) Kitam. f. plena (Honda) Tuyama ●☆

69586 Camellia salicifolia Champ. ex Benth.; 柳叶毛蕊茶(金毛山茶, 柳叶山茶, 毛叶山茶, 生毛胡桃); Willowleaf Camellia, Willowleaf Tea, Willow-leaved Camellia ●

69587 Camellia salicifolia Champ. ex Benth. var. longisepala Keng = Camellia salicifolia Champ. ex Benth. ●

69588 Camellia salicifolia Champ. var. longisepala H. Keng = Camellia salicifolia Champ. ex Benth. ●

69589 Camellia saluenensis Stapf ex Bean; 怒江山茶(狗爪爪, 怒江红山茶, 威宁短柱茶, 小怒江红山茶, 小怒江山茶, 野茶花); Nujiang Camellia, Small Nujiang Camellia, Weining Camellia ●

69590 Camellia saluenensis Stapf ex Bean f. minor Sealy = Camellia saluenensis Stapf ex Bean ●

69591 Camellia saluenensis Stapf ex Bean var. minor (Sealy) H. T. Chang = Camellia saluenensis Stapf ex Bean ●

69592 Camellia sasanqua Sims = Camellia maliflora Lindl. ●

69593 Camellia sasanqua Thunb. = Camellia sasanqua Thunb. ex A. Murray ●

69594 Camellia sasanqua Thunb. ex A. Murray; 茶梅(腊至, 山茶, 山茶花, 小叶油茶, 油茶, 云南山茶); Chansonette Camellia, Christmas Camellia, Sasanqua, Sasanqua Camellia, Sasanqua Oil Camellia, Sasanqua Tea ●

69595 Camellia sasanqua Thunb. ex A. Murray 'Agnes O. Solomon'; 所罗门茶梅●

69596 Camellia sasanqua Thunb. ex A. Murray 'Cotton Candy';棉花糖茶梅●
69597 Camellia sasanqua Thunb. ex A. Murray 'Gulf Glory';海湾之光茶梅●
69598 Camellia sasanqua Thunb. ex A. Murray 'Mine-no-yuki';山顶之雪茶梅●
69599 Camellia sasanqua Thunb. ex A. Murray 'Misty Moon';月光茶梅●
69600 Camellia sasanqua Thunb. ex A. Murray 'Narumigata';奇异茶梅●
69601 Camellia sasanqua Thunb. ex A. Murray 'Paradise Belinda';天堂茶梅●
69602 Camellia sasanqua Thunb. ex A. Murray 'Plantation Pink';粉红茶梅●
69603 Camellia sasanqua Thunb. ex A. Murray 'Red Willow';红柳茶梅●
69604 Camellia sasanqua Thunb. ex A. Murray 'Satan's Robe';礼炮茶梅●
69605 Camellia sasanqua Thunb. ex A. Murray 'Setsugekka';波白茶梅●
69606 Camellia sasanqua Thunb. ex A. Murray 'Shishigashira';狮子头茶梅●
69607 Camellia sasanqua Thunb. ex A. Murray 'Snow on the Peak' = Camellia sasanqua Thunb. ex A. Murray 'Mine-no-yuki'●
69608 Camellia sasanqua Thunb. ex A. Murray 'Wavy White' = Camellia sasanqua Thunb. ex A. Murray 'Setsugekka'●
69609 Camellia sasanqua Thunb. var. fujikoana Makino = Camellia sasanqua Thunb. 'Shishigashira'●
69610 Camellia scariosisepala Hung T. Chang;膜萼离蕊茶;Filmsepal Tea, Membrana-sepal Camellia, Scarious-sepaled Camellia●
69611 Camellia scariosisepala Hung T. Chang = Camellia yunnanensis (Pit. ex Diels) Cohen-Stuart●
69612 Camellia sealyana T. L. Ming;老挝茶;Laos Camellia●☆
69613 Camellia semiserrata C. W. Chi;南山茶(茶梅,多苞红山茶,多萼红山茶,广宁油茶,华南红花油茶,毛籽红山茶,牛牯茶,栓壳红山茶);Halftooth Camellia, Multiperulated Camellia, Semiserrate Camellia, Semiserrate Oil Camellia●
69614 Camellia semiserrata C. W. Chi f. albiflora (Hu et T. C. Huang) T. L. Ming = Camellia semiserrata C. W. Chi var. albiflora Hu et T. C. Huang●
69615 Camellia semiserrata C. W. Chi f. albiflora (Hu et T. C. Huang) T. L. Ming = Camellia semiserrata C. W. Chi●
69616 Camellia semiserrata C. W. Chi var. albiflora Hu et T. C. Huang;白花南山茶(白毛南山茶);White-flowered Semiserrate Camellia●
69617 Camellia semiserrata C. W. Chi var. albiflora Hu et T. C. Huang = Camellia semiserrata C. W. Chi●
69618 Camellia semiserrata C. W. Chi var. magnocarpa Hu et Hung T. Chang ex Hu;大果南山茶(大果红山茶,莽山红山茶);Bigfruit Semiserrate Camellia●
69619 Camellia semiserrata C. W. Chi var. magnocarpa Hu et Hung T. Chang ex Hu = Camellia mongshanica Hung T. Chang et C. S. Ye●
69620 Camellia septempetala Hung T. Chang et L. L. Qi;七瓣连蕊茶;Sevenpetal Tea●
69621 Camellia septempetala Hung T. Chang et L. L. Qi = Camellia cuspidata (Kochs) Wright ex Garden var. grandiflora Sealy●
69622 Camellia septempetala Hung T. Chang et L. L. Qi var. rubra Hung T. Chang et L. L. Qi;红花七瓣连蕊茶;Red-flower Seve-petal Camellia●
69623 Camellia septempetala Hung T. Chang et L. L. Qi var. rubra Hung T. Chang et L. L. Qi. = Camellia cuspidata (Kochs) Wright ex Garden var. grandiflora Sealy●
69624 Camellia setiperulata Hung T. Chang = Camellia pitardii Cohen-Stuart var. cryptoneura (Hung T. Chang) T. L. Ming●
69625 Camellia setiperulata Hung T. Chang et B. K. Lee = Camellia pitardii Cohen-Stuart●
69626 Camellia setiperulata Hung T. Chang et B. K. Lee = Camellia setiperulata Hung T. Chang et B. K. Lee ex Hung T. Chang●
69627 Camellia setiperulata Hung T. Chang et B. K. Lee ex Hung T. Chang;粗毛红山茶(硬毛红山茶);Hirsute Camellia, Setiperulate Camellia, Shag Camellia●
69628 Camellia shensiensis Hung T. Chang;陕西短柱茶;Shaanxi Camellia●
69629 Camellia shensiensis Hung T. Chang = Camellia grijsii Hance var. shensiensis (Hung T. Chang) T. L. Ming●
69630 Camellia shinkoensis (Hayata) Cohen-Stuart = Tutcheria shinkoensis (Hayata) Nakai●
69631 Camellia shinkoensis (Hayata) Makino = Pyrenaria microcarpa (Dunn) H. Keng●
69632 Camellia shinkoensis Makino = Tutcheria shinkoensis (Hayata) Nakai●
69633 Camellia simplicifolia Griff. = Camellia kissi Wall.●
69634 Camellia sinensis (L.) Kuntze;茶(茶树,长叶茶,荈,高树茶,广宁油茶,槚,苦茶,酪奴,龙陵茶,茗,蔎,乌龙茶,小叶种茶,芽菜);Chinese Tea, Common Tea, Tea, Tea Bush, Tea Camellia, Tea Plant●
69635 Camellia sinensis (L.) Kuntze 'Blushing Bride';红新娘茶●
69636 Camellia sinensis (L.) Kuntze f. formosensis Kitam.;台湾山茶;Taiwan Tea●
69637 Camellia sinensis (L.) Kuntze f. formosensis Kitam. = Camellia sinensis (L.) Kuntze●
69638 Camellia sinensis (L.) Kuntze f. macrophylla (Siebold ex Miq.) Kitam. = Camellia sinensis (L.) Kuntze var. macrophylla Siebold ex Miq.●
69639 Camellia sinensis (L.) Kuntze f. macrophylla (Siebold) Kitam. = Camellia sinensis (L.) Kuntze●
69640 Camellia sinensis (L.) Kuntze f. parvifolia (Miq.) Sealy = Camellia sinensis (L.) Kuntze●
69641 Camellia sinensis (L.) Kuntze f. rosea (Makino) Kitam.;粉红花茶●☆
69642 Camellia sinensis (L.) Kuntze subsp. buisanensis (Sasaki) S. Y. Lu et Y. P. Yang = Pyrenaria microcarpa (Dunn) H. Keng var. ovalifolia (H. L. Li) T. L. Ming et S. X. Yang●
69643 Camellia sinensis (L.) Kuntze subsp. buisanensis (Sasaki) S. Y. Lu et Y. P. Yang = Camellia sinensis (L.) Kuntze f. formosensis Kitam.●
69644 Camellia sinensis (L.) Kuntze var. assamica (Mast.) Kitam.;普洱茶(阿萨姆茶,博白油茶,茶叶,川茶,大叶茶,普雨茶,人头茶,野茶树);Assam Tea, Puer Tea, Tea Plant●◇
69645 Camellia sinensis (L.) Kuntze var. assamica (Mast.) Kitam. = Camellia assamica (Mast.) Hung T. Chang●
69646 Camellia sinensis (L.) Kuntze var. dehungensis (Hung T. Chang et B. H. Chen) T. L. Ming;德宏茶;Dehong Camellia, Dehong Tea●
69647 Camellia sinensis (L.) Kuntze var. dehungensis (Hung T. Chang, H. S. Wang et B. H. Chen) T. L. Ming = Camellia sinensis (L.) Kuntze var. dehungensis (Hung T. Chang et B. H. Chen) T.

L. Ming ●

69648 Camellia sinensis (L.) Kuntze var. kucha Hung T. Chang et H. S. Wang = Camellia assamica (Mast.) Hung T. Chang var. kucha (Hung T. Chang, H. S. Wang et B. H. Chen) Hung T. Chang et H. S. Wang ●

69649 Camellia sinensis (L.) Kuntze var. kucha Hung T. Chang et H. S. Wang = Camellia sinensis (L.) Kuntze var. assamica (Mast.) Kitam. ●◇

69650 Camellia sinensis (L.) Kuntze var. macrophylla Siebold = Camellia sinensis (L.) Kuntze var. macrophylla Siebold ex Miq. ●

69651 Camellia sinensis (L.) Kuntze var. macrophylla Siebold ex Miq.;皋芦(大叶茶,皋芦,瓜芦,过罗,拘罗,苦登,苦苧,物罗);Largeleaf Tea ●

69652 Camellia sinensis (L.) Kuntze var. pubilimba Hung T. Chang;白毛茶(白花茶);Hairy Tea ●

69653 Camellia sinensis (L.) Kuntze var. waldeniae (S. Y. Hu) Hung T. Chang = Camellia sinensis (L.) Kuntze ●

69654 Camellia sinensis (L.) Kuntze var. waldensae (S. Y. Hu) Hung T. Chang;香花茶(长叶茶);Fragrant Tea ●

69655 Camellia skogiana C. X. Ye = Camellia yunnanensis (Pit. ex Diels) Cohen-Stuart var. camellioides (Hu) T. L. Ming ●

69656 Camellia sophiae Hu = Pyrenaria sophiae (Hu) S. X. Yang et T. L. Ming ●

69657 Camellia sophiae Hu = Tutcheria sophiae (Hu) Hung T. Chang ●

69658 Camellia speciosa (Kochs) Cohen-Stuart = Gordonia szechuanensis Hung T. Chang ●

69659 Camellia speciosa (Kochs) Cohen-Stuart = Polyspora speciosa (Kochs) B. M. Barthol. et T. L. Ming ●

69660 Camellia speciosa (Pit. ex Diels) Melch. = Camellia pitardii Cohen-Stuart ●

69661 Camellia speciosa Hort. = Camellia saluenensis Stapf ex Bean ●

69662 Camellia spectabilis Champ. = Pyrenaria spectabilis (Champ.) C. Y. Wu et S. X. Yang ●

69663 Camellia spectabilis Champ. = Tutcheria championi Nakai ●

69664 Camellia spectabilis Champ. = Tutcheria spectabilis Dunn ●

69665 Camellia spectabilis Champ. var. florepleno Seem. = Camellia reticulata Lindl. ●◇

69666 Camellia stenophylla Kobuski = Camellia fluviatilis Hand. -Mazz. ●

69667 Camellia stichoclada Hung T. Chang = Camellia reticulata Lindl. ●◇

69668 Camellia stictoclada Hung T. Chang;斑枝红山茶;Spot-branch Camellia, Spottwig Camellia ●

69669 Camellia stuartiana Sealy;五室连蕊茶(元江山茶);Fiveroom Tea, Stuart Camellia ●

69670 Camellia suaveolens C. X. Ye et al. = Camellia furfuracea (Merr.) Cohen-Stuart ●

69671 Camellia subacutissima Hung T. Chang;肖长尖连蕊茶(肖长尖毛蕊茶);Acutestlike Camellia, Acutest-like Camellia, Near-muchsharp Tea ●

69672 Camellia subacutissima Hung T. Chang = Camellia costei H. Lév. ●

69673 Camellia subglabra Hung T. Chang;半秃连蕊茶(半秃毛蕊茶);Semiglabrous Tea, Subglabrous Camellia ●

69674 Camellia subglabra Hung T. Chang = Camellia caudata Wall. var. gracilis (Hemsl.) Gogelein ex H. Keng ●

69675 Camellia subintegra P. C. Huang ex Hung T. Chang;全缘山茶(连山红山茶,连山离蕊茶,全缘叶红山茶);Entire Camellia, Subentire Camellia ●

69676 Camellia subliberipetala Hung T. Chang = Camellia pitardii Cohen-Stuart ●

69677 Camellia subliberipetala Hung T. Chang = Camellia reticulata Lindl. ●◇

69678 Camellia sunmingensis Hu;嵩明山茶;Songming Camellia, Sungming Camellia ●

69679 Camellia sylvestris Berl. = Camellia japonica L. ●

69680 Camellia symplocifolia Griff. = Camellia kissi Wall. ●

69681 Camellia synaptica Sealy = Camellia tsaii Hu var. synaptica (Sealy) Hung T. Chang ●

69682 Camellia synaptica Sealy var. parviovata (Hung T. Chang et S. S. Wang) T. L. Ming;毛蕊川滇连蕊茶●

69683 Camellia synaptica Sealy var. parviovata (Hung T. Chang) T. L. Ming = Camellia parviovata Hung T. Chang et S. S. Wang ex Hung T. Chang ●

69684 Camellia szechuanensis C. W. Chi;四川离蕊茶(半宿萼茶,四川山茶);Sichuan Camellia ●

69685 Camellia szemaoensis Hung T. Chang;斑叶离蕊茶(思茅短蕊茶);Simao Camellia, Simao Tea ●

69686 Camellia tachangensis F. S. Zhang;大厂茶(缙云山茶,四球茶,五室茶);Dachang Camellia, Dachang Tea, Fivelocular Camellia, Jinyunshan Tea, Tetracoccous Camellia ●

69687 Camellia tachangensis F. S. Zhang var. remotiserrata (Hung T. Chang, H. S. Wang et B. H. Wang) T. L. Ming = Camellia remotiserrata Hung T. Chang, F. L. Yu et P. S. Wang ●

69688 Camellia taheishanensis F. S. Zhang;大黑山野山茶;Daheishan Camellia ●

69689 Camellia taheishanensis F. S. Zhang = Camellia pachyandra Hu ●

69690 Camellia taimiaushanica Hung T. Chang;大苗山茶;Damiaoshan Camellia ●

69691 Camellia taliensis (W. W. Sm.) Melch.;大理茶(感通寺茶);Dali Camellia, Dali Tea, Tali Camellia ●

69692 Camellia tenii Sealy;小糙果茶(大姚短柱茶,盐丰山茶);Dayao Camellia, Ten Camellia ●

69693 Camellia tenuiflora (Hayata) Cohen-Stuart = Camellia brevistyla (Hayata) Cohen-Stuart ●

69694 Camellia tenuifolia (Hayata) Cohen-Stuart;细叶山茶(细叶茶);Slender-leaved Camellia ●

69695 Camellia tenuivalvis Hung T. Chang;薄壳红山茶;Thinhull Camellia, Thinvalve Camellia ●

69696 Camellia tenuivalvis Hung T. Chang = Camellia saluenensis Stapf ex Bean ●

69697 Camellia terminalis J. Y. Liang et Z. M. Su = Camellia pingguoensis D. Fang var. terminalis (J. Y. Liang et Z. M. Su) T. L. Ming et W. J. Zhang ●◇

69698 Camellia tetracocca Hung T. Chang = Camellia tachangensis F. S. Zhang ●

69699 Camellia thea L. = Camellia sinensis (L.) Kuntze ●

69700 Camellia thea Link = Camellia sinensis (L.) Kuntze ●

69701 Camellia theifera Griff. = Camellia sinensis (L.) Kuntze var. assamica (Mast.) Kitam. ●◇

69702 Camellia theiformis Hance = Camellia euryoides Lindl. ●

69703 Camellia tianeensis S. Ye Liang et Y. T. Luo = Camellia huana T. L. Ming et W. T. Zhang ●

69704 Camellia tonkinensis (Pit.) Cohen-Stuart;薄叶金花茶;Golden-flowered-like Camellia, Thinleaf Camellia, Thinleaf Tea, Tonkin Camellia ●

69705 Camellia transarisanensis (Hayata) Cohen-Stuart;毛萼连蕊茶(阿里山茶,阿里山茶梅,阿里山连蕊茶,阿里山毛蕊茶,小叶茶梅,小叶山茶);Alishan Camellia, Alishan Tea, Small-leaved Camellia, Smooth-leaved Camellia ●

69706 Camellia transnokoensis Hayata;南投秃连蕊茶(泛能高山茶,光叶茶梅,光叶山茶,南投连蕊茶,南投秃毛蕊茶,赛能高山茶,台湾秃连蕊茶);Nantou Camellia, Nantou Tea, Smooth-leaved Camellia ●

69707 Camellia transnokoensis Hayata = Camellia lutchuensis Ito ex Ito et Matsum. ●

69708 Camellia triantha Hung T. Chang;三花连蕊茶(三花毛蕊茶,银花茶);Trianther Tea, Triflorous Camellia, Triflower Camellia ●

69709 Camellia triantha Hung T. Chang = Camellia caudata Wall. ●

69710 Camellia trichandra Hung T. Chang = Camellia cuspidata (Kochs) Wright ex Garden var. trichandra (Hung T. Chang) T. L. Ming ●

69711 Camellia trichocarpa Hung T. Chang;毛果山茶;Hairyfruit Camellia, Hairy-fruited Camellia ●

69712 Camellia trichocarpa Hung T. Chang = Camellia yunnanensis (Pit. ex Diels) Cohen-Stuart var. camellioides (Hu) T. L. Ming ●

69713 Camellia trichoclada (Rehder) S. S. Chien;毛枝连蕊茶(毛枝毛蕊茶);Hairtwig Tea, Hairybranch Camellia, Hairy-branched Camellia ●

69714 Camellia trichoclada (Rehder) S. S. Chien f. leucantha P. L. Chiu = Camellia trichoclada (Rehder) S. S. Chien ●

69715 Camellia trichosperma Hung T. Chang;毛籽红山茶;Hairseed Camellia, Hairy-seed Camellia, Hairy-seeded Camellia, Trigonous-fruited Camellia ●

69716 Camellia trichosperma Hung T. Chang = Camellia semiserrata C. W. Chi ●

69717 Camellia trigonocarpa Hung T. Chang;棱果毛蕊茶;Threeangularfruit CamelliaTriangleaf Tea ●

69718 Camellia trigonocarpa Hung T. Chang = Camellia assimiloides Sealy ●

69719 Camellia truncata Hung T. Chang et C. S. Ye;截叶连蕊茶;Truncate Camellia, Truncate Tea ●

69720 Camellia truncata Hung T. Chang et C. S. Ye = Camellia forrestii (Diels) Cohen-Stuart ●

69721 Camellia tsaii Hu;窄叶连蕊茶(泸西山茶,云南连蕊茶);Tsai Camellia, Tsai Tea ●

69722 Camellia tsaii Hu var. synaptica (Sealy) Hung T. Chang;川滇连蕊茶;Southwest Camellia, Yellowpistil Camellia ●

69723 Camellia tsaii Hu var. synaptica (Sealy) Hung T. Chang = Camellia synaptica Sealy ●

69724 Camellia tsingpienensis Hu;金屏连蕊茶(屏边连蕊茶);Jinping Camellia, Jinping Tea, Tsingpien Camellia ●

69725 Camellia tsingpienensis Hu var. macrophylla T. L. Ming;大叶金屏连蕊茶(大叶屏边连蕊茶);Bigleaf Jinping Camellia ●

69726 Camellia tsingpienensis Hu var. pubisepala Hung T. Chang;毛萼金屏连蕊茶(毛萼金屏茶,毛萼屏边连蕊茶);Hairy-sepal Jinping Camellia, Hairy-sepal Jinping Tea ●

69727 Camellia tsofui S. S. Chien;细萼连蕊茶(细萼毛蕊茶);Thin-sepal Camellia, Tsofu Camellia, Tsofu Tea ●

69728 Camellia tsofui S. S. Chien = Camellia euryoides Lindl. var. nokoensis (Hayata) T. L. Ming ●

69729 Camellia tsubakki Crantz = Camellia japonica L. ●

69730 Camellia tuberculata S. S. Chien;瘤果茶;Tuberculate Camellia ●

69731 Camellia tuberculata S. S. Chien var. atuberculata (Hung T. Chang) T. L. Ming;秃蕊瘤果茶;Atuberculate Camellia ●

69732 Camellia tubiformis Hung T. Chang et S. X. Ren = Camellia caudata Wall. var. gracilis (Hemsl.) Gogelein ex H. Keng ●

69733 Camellia tuboleifera Hung T. Chang et S. X. Ren;管蕊山茶;Tuber Camellia ●

69734 Camellia tuboleifera Hung T. Chang et S. X. Ren = Camellia brevistyla (Hayata) Cohen-Stuart ●

69735 Camellia tunganica Hung T. Chang et B. K. Lee;东安红山茶;Dong'an Camellia ●

69736 Camellia tunganica Hung T. Chang et B. K. Lee = Camellia pitardii Cohen-Stuart ●

69737 Camellia tunghinensis Hung T. Chang = Camellia indochinensis Merr. var. tunghinensis (Hung T. Chang) T. L. Ming et W. J. Zhang ●

69738 Camellia tzegingensis Y. K. Li;织金山茶;Zhijin Camellia ●

69739 Camellia uraku Kitam.;单体红山茶;Uraku Camellia ●

69740 Camellia vernalis (Makino) Makino;春山茶●☆

69741 Camellia vietnamensis T. C. Huang ex Hu;越南山茶(越南油茶);Viet Nam Camellia, Vietnam Camellia ●

69742 Camellia vietnamensis T. C. Huang ex Hu = Camellia drupifera Lour. ●

69743 Camellia villicarpa S. S. Chien;小果毛蕊茶;Smallfruit Camellia, Small-fruited Camellia, Villform Tea ●

69744 Camellia villosa Hung T. Chang = Camellia polyodonta F. C. How ex Hu ●

69745 Camellia villosa Hung T. Chang et S. Ye Liang ex Hung T. Chang;长毛红山茶;Villose Camellia ●

69746 Camellia villosa Hung T. Chang et S. Ye Liang ex Hung T. Chang = Camellia polyodonta F. C. How ex Hu ●

69747 Camellia villosa Hung T. Chang et S. Ye Liang ex Hung T. Chang = Camellia mairei (H. Lév.) Melch. var. lapidea (W. C. Wu) Sealy ●

69748 Camellia virgata (Koidz.) Makino et Nemoto = Pyrenaria microcarpa (Dunn) H. Keng ●

69749 Camellia viridicalyx Hung T. Chang et S. Ye Liang ex Hung T. Chang;绿萼连蕊茶(绿萼毛蕊茶);Greencalyx Camellia, Greencalyx Tea, Green-calyxed Camellia ●

69750 Camellia viridicalyx Hung T. Chang et S. Ye Liang ex Hung T. Chang var. linearifolia T. L. Ming;线叶连蕊茶;Line-leaf Greencalyx Camellia ●

69751 Camellia wabisuke (Makino) Kitam. f. uraku (Kitam.) Kitam. = Camellia uraku Kitam. ●

69752 Camellia wabisuke (Makino) Kitam. var. bicolor (Makino) Tuyama = Camellia wabisuke (Makino) Kitam. var. campanulata (Makino) Kitam. f. bicolor (Makino) Kitam. ●☆

69753 Camellia wabisuke (Makino) Kitam. var. campanulata (Makino) Kitam.;风铃草状茶●☆

69754 Camellia wabisuke (Makino) Kitam. var. campanulata (Makino) Kitam. f. bicolor (Makino) Kitam.;二色风铃草状茶●☆

69755 Camellia waldeniae S. Y. Hu = Camellia sinensis (L.) Kuntze ●

69756 Camellia waldensae S. Y. Hu = Camellia sinensis (L.) Kuntze var. waldensae (S. Y. Hu) Hung T. Chang ●

69757 Camellia wardii Kobuski;滇缅离蕊茶(陇川山茶);Sino-Burma Tea, Ward Camellia ●

69758 Camellia wardii Kobuski var. muricatula (Hung T. Chang) T. L. Ming;毛滇缅离蕊茶●

69759 Camellia weiningensis Y. K. Li ex Hung T. Chang = Camellia saluenensis Stapf ex Bean ●

69760 Camellia wenshanensis Hu;文山毛蕊茶(文山山茶);Wenshan Camellia,Wenshan Tea ●

69761 Camellia wenshanensis Hu = Camellia cordifolia (F. P. Metcalf) Nakai ●

69762 Camellia wumingensis S. Ye Liang et C. R. Fu = Camellia flavida Hung T. Chang ●◇

69763 Camellia wumingensis S. Ye Liang et C. R. Fu = Camellia flavida Hung T. Chang var. patens (S. L. Mo et Y. C. Zhong) T. L. Ming ●

69764 Camellia wumingensis S. Ye Liang et C. R. Fu = Camellia petelotii (Merr.) Sealy var. microcarpa (S. L. Mo et S. Z. Huang) T. L. Ming et W. J. Zhang ●

69765 Camellia xanthochroma K. M. Feng et L. S. Xie;黄花短蕊茶(海南黄山茶);Yellow Tea, Yellow-flower Camellia, Yellow-flowered Camellia ●

69766 Camellia xiashiensis S. Ye Liang et C. Z. Deng = Camellia chrysanthoides Hung T. Chang ●

69767 Camellia xiashiensis S. Ye Liang et C. Z. Deng = Camellia limonia C. F. Liang et S. L. Mo ●

69768 Camellia xiashiensis S. Ye Liang et C. Z. Deng = Camellia parvipetala J. Y. Liang et Z. M. Su ●

69769 Camellia xiashiensis S. Ye Liang et C. Z. Deng = Camellia tonkinensis (Pit.) Cohen-Stuart ●

69770 Camellia xichangensis Hung T. Chang = Camellia albo-sericea Hung T. Chang ●

69771 Camellia xichangensis Hung T. Chang = Camellia reticulata Lindl. ●◇

69772 Camellia xifongensis Y. K. Li ex X. C. Chen et F. Z. Zheng = Camellia pitardii Cohen-Stuart ●

69773 Camellia xylocarpa (Hu) Hung T. Chang;木果红山茶;Woodfruit Camellia,Xylocarpous Camellia ●

69774 Camellia xylocarpa (Hu) Hung T. Chang = Camellia reticulata Lindl. ●◇

69775 Camellia yangkiangensis Hung T. Chang = Camellia crassipes Sealy ●

69776 Camellia yankiangensis Hung T. Chang;元江短蕊茶;Yuanjiang Camellia,Yuanjiang Tea ●

69777 Camellia yankiangensis Hung T. Chang = Camellia sinensis (L.) Kuntze var. pubilimba Hung T. Chang ●

69778 Camellia yuhsienensis Hu;攸县油茶;Youxian Camellia ●

69779 Camellia yuhsienensis Hu = Camellia grijsii Hance ●◇

69780 Camellia yungkiangensis Hung T. Chang = Camellia costata Hung T. Chang ●

69781 Camellia yungkiangensis Hung T. Chang, H. S. Wang et B. H. Chen;榕江茶;Rongjiang Camellia,Rongjiang Tea ●

69782 Camellia yunnanensis (Pit. ex Diels) Cohen-Stuart;猴子木(五柱滇山茶);Yunnan Camellia ●

69783 Camellia yunnanensis (Pit. ex Diels) Cohen-Stuart var. camellioides (Hu) T. L. Ming;毛果猴子木(假核果茶);Falseyunnan Camellia ●

69784 Camellia yunnanensis (Pit. ex Diels) Cohen-Stuart var. trichocarpa (Hung T. Chang) T. L. Ming = Camellia yunnanensis (Pit. ex Diels) Cohen-Stuart var. camellioides (Hu) T. L. Ming ●

69785 Camellia yuyang Hung T. Chang;乳阳红山茶;Ruyang Camellia ●

69786 Camellia zengii Hung T. Chang = Camellia rhytidocarpa Hung T. Chang et S. Ye Liang ●

69787 Camelliaceae DC. = Theaceae Mirb. (保留科名) ●

69788 Camelliaceae Dumort. = Theaceae Mirb. (保留科名) ●

69789 Camelliaceae Mirb. = Theaceae Mirb. (保留科名) ●

69790 Camelliastrum Nakai = Camellia L. ●

69791 Camelliastrum assimile (Champ.) Nakai = Camellia assimilis Champ. ex Benth. ●

69792 Camelliastrum assimile (Champ.) Nakai = Camellia caudata Wall. ●

69793 Camelliastrum buisanense (Sasaki) Nakai = Pyrenaria microcarpa (Dunn) H. Keng var. ovalifolia (H. L. Li) T. L. Ming et S. X. Yang ●

69794 Camelliastrum busisanense (Sasaki) Nakai = Camellia caudata Wall. ●

69795 Camelliastrum caudatum (Wall.) Nakai = Camellia caudata Wall. ●

69796 Camelliastrum gracile (Hemsl.) Nakai = Camellia caudata Wall. var. gracilis (Hemsl.) Gogelein ex H. Keng ●

69797 Camelliastrum gracile (Hemsl.) Nakai = Camellia caudata Wall. ●

69798 Camelliastrum mairei (H. Lév.) Nakai = Camellia mairei (H. Lév.) Melch. ●

69799 Camelliastrum salicifolium (Champ.) Nakai = Camellia salicifolia Champ. ex Benth. ●

69800 Camelostalix Pfitzer = Pholidota Lindl. ex Hook. ■

69801 Cameraria Boehm. = Hemerocallis L. ■

69802 Cameraria Dill. ex Moench = Montia L. ■☆

69803 Cameraria Fabr. = Montia L. ■☆

69804 Cameraria L. (1753);鸭蛋花属●

69805 Cameraria latifolia L.;鸭蛋花●

69806 Cameraria zeylanica Retz. = Hunteria zeylanica (Retz.) Garden ex Thwaites ●

69807 Cameridium Rchb. = Camaridium Lindl. ■☆

69808 Cameridium Rchb. f. = Camaridium Lindl. ■☆

69809 Camerunia (Pichon) Boiteau = Tabernaemontana L. ●

69810 Camerunia bouquetii Boiteau = Tabernaemontana bouquetii (Boiteau) Leeuwenb. ●☆

69811 Camerunia penduliflora (K. Schum.) Boiteau = Tabernaemontana penduliflora K. Schum. ●☆

69812 Camforosma Spreng. = Camphorosma L. ●■

69813 Camilleugenia Frapp. ex Cordem. = Cynorkis Thouars ■☆

69814 Camilleugenia coccinelloies Frapp. = Cynorkis coccinelloides Schltr. ■☆

69815 Camirium Gaertn. = Aleurites J. R. Forst. et G. Forst. ●

69816 Camirium Rumph. ex Gaertn. = Aleurites J. R. Forst. et G. Forst. ●

69817 Camissonia Link = Oenothera L. ●■

69818 Camissonia Link(1818);卡密柳叶菜属■☆

69819 Camissonia boothii (Douglas) P. H. Raven;布斯卡密柳叶菜■☆

69820 Camissonia brevipes (A. Gray) P. H. Raven = Oenothera brevipes A. Gray ex Torr. ■☆

69821 Camissonia californica (Nutt. ex Torr. et A. Gray) P. H. Raven;加州卡密柳叶菜;Mustard Evening Primrose ■☆

69822 Camissonia cheiranthifolia Raim. = Oenothera cheiranthifolia Hornem. ex Spreng. ■☆

69823 Camissoniopsis W. L. Wagner et Hoch = Agassizia Spach ■☆

69824 Camissoniopsis W. L. Wagner et Hoch(2007);拟卡密柳叶菜属■☆

69825 Cammarum Fourr. = Aconitum L. ■

69826 Cammarum Hill(废弃属名) = Eranthis Salisb. (保留属名) ■

69827 Cammarum hyemale (L.) Greene = Eranthis hyemalis (L.)

Salisb. ■☆
69828 Camocladia L. = Comocladia P. Browne ●☆
69829 Camoensia Welw. = Camoensia Welw. ex Benth. et Hook. f. (保留属名)●☆
69830 Camoensia Welw. ex Benth. et Hook. f. (1865)(保留属名);西非豆藤属(西非豆属)●☆
69831 Camoensia brevicalyx Benth.;短萼西非豆●☆
69832 Camoensia laurentii De Wild. = Camoensia brevicalyx Benth. ●☆
69833 Camoensia maxima Welw. = Camoensia scandens (Welw.) J. B. Gillett ●☆
69834 Camoensia maxima Welw. ex Benth. = Camoensia scandens (Welw.) J. B. Gillett ●☆
69835 Camoensia scandens (Welw.) J. B. Gillett;西非豆藤●☆
69836 Camolenga Post et Kuntze = Benincasa Savi ■
69837 Camomilla Gilib. = Matricaria L. ■
69838 Camonea Raf. (废弃属名) = Merremia Dennst. ex Endl. (保留属名)●■
69839 Camontagnea Pujals = Acanthonema J. Agardh(废弃属名)■☆
69840 Campana Post et Kuntze = Tecomanthe Baill. ■☆
69841 Campanea Decne. = Capanea Decne. ex Planch. ●■☆
69842 Campanemia Post et Kuntze = Capanemia Barb. Rodr. ■☆
69843 Campanocalyx Valeton = Keenania Hook. f. ●
69844 Campanolea Gilg et Schellenb. = Chionanthus L. ●
69845 Campanolea Gilg et Schellenb. = Linociera Sw. ex Schreb. (保留属名)●
69846 Campanolea mildbraedii Gilg et G. Schellenb. = Chionanthus mildbraedii (Gilg et G. Schellenb.) Stearn ●☆
69847 Campanopsis (R. Br.) Kuntze = Wahlenbergia Schrad. ex Roth (保留属名)■●
69848 Campanopsis Kuntze = Wahlenbergia Schrad. ex Roth(保留属名)■●
69849 Campanopsis androsacea (A. DC.) Kuntze = Wahlenbergia androsacea A. DC. ■☆
69850 Campanopsis arenaria (A. DC.) Kuntze = Wahlenbergia androsacea A. DC. ■☆
69851 Campanopsis arguta (Hook. f.) Kuntze = Wahlenbergia krebsii Cham. subsp. arguta (Hook. f.) Thulin ■☆
69852 Campanopsis banksiana (A. DC.) Kuntze = Wahlenbergia banksiana A. DC. ■☆
69853 Campanopsis bojeri (A. DC.) Kuntze = Wahlenbergia undulata (L. f.) A. DC. ■☆
69854 Campanopsis caledonica (Sond.) Kuntze = Wahlenbergia undulata (L. f.) A. DC. ■☆
69855 Campanopsis campanuloides (Delile) Kuntze = Wahlenbergia campanuloides (Delile) Vatke ■☆
69856 Campanopsis capillacea (L. f.) Kuntze = Wahlenbergia capillacea (L. f.) A. DC. ■☆
69857 Campanopsis etbaica (Schweinf.) Kuntze = Wahlenbergia lobelioides (L. f.) Link subsp. nutabunda (Guss.) Murb. ■☆
69858 Campanopsis hilsenbergii (A. DC.) Kuntze = Wahlenbergia madagascariensis A. DC. ■☆
69859 Campanopsis humilis (A. DC.) Kuntze = Wahlenbergia lobelioides (L. f.) Link subsp. riparia (A. DC.) Thulin ■☆
69860 Campanopsis inhambanensis (Klotzsch) Kuntze = Wahlenbergia androsacea A. DC. ■☆
69861 Campanopsis lobelioides (L. f.) Kuntze = Wahlenbergia lobelioides (L. f.) Link ■☆
69862 Campanopsis madagascariensis (A. DC.) Kuntze = Wahlenbergia madagascariensis A. DC. ■☆
69863 Campanopsis mannii (Vatke) Kuntze = Wahlenbergia silenoides Hochst. ex A. Rich. ■☆
69864 Campanopsis oppositifolia (A. DC.) Kuntze = Wahlenbergia madagascariensis A. DC. ■☆
69865 Campanopsis pusilla (Hochst. ex A. Rich.) Kuntze = Wahlenbergia pusilla Hochst. ex A. Rich. ■☆
69866 Campanopsis riparia (A. DC.) Kuntze = Wahlenbergia lobelioides (L. f.) Link subsp. riparia (A. DC.) Thulin ■☆
69867 Campanopsis silenoides (Hochst. ex A. Rich.) Kuntze = Wahlenbergia silenoides Hochst. ex A. Rich. ■☆
69868 Campanopsis undulata (L. f.) Kuntze = Wahlenbergia undulata (L. f.) A. DC. ■☆
69869 Campanopsis zeyheri (H. Buek) Kuntze = Wahlenbergia krebsii Cham. ■☆
69870 Campanula L. (1753);风铃草属(桔梗属);Bell Flower, Bellflower, Bell-flower, Blue Bells, Bluebell, Campanula, Harebell ■●
69871 Campanula abietina Griseb. et Schenk;冷杉风铃草■☆
69872 Campanula adpressa Thunb. = Wahlenbergia adpressa (Thunb.) Sond. ■☆
69873 Campanula adscendens Vest. ex Roem. et Schult.;上举风铃草■☆
69874 Campanula afganica Pomel;阿富汗风铃草■☆
69875 Campanula afra Cav.;非洲风铃草■☆
69876 Campanula afra Cav. subsp. hypocrateriformis (Dobignard) Fennane = Campanula hypocrateriformis Dobignard ■☆
69877 Campanula afra Cav. var. leptosiphon (Pau et Sennen) Maire = Campanula afra Cav. ■☆
69878 Campanula afra Cav. var. pallida (Maire) Dobignard = Campanula afra Cav. ■☆
69879 Campanula alaskana (A. Gray) W. Wight ex J. P. Anderson = Campanula rotundifolia L. ■
69880 Campanula alata Desf.;高大风铃草■☆
69881 Campanula albertii Trautv.;新疆风铃草;Albert Bellflower ■
69882 Campanula albovii Kolak.;阿氏风铃草■☆
69883 Campanula aldanensis Fed. et Karav.;阿尔丹风铃草■☆
69884 Campanula algida Fisch. ex A. DC. = Campanula lasiocarpa Cham. ■☆
69885 Campanula alliariifolia Willd.;葱芥叶风铃草;Cornish Bellflower, Spurred Bellflower ■☆
69886 Campanula allionii Vill.;阿尔卑斯风铃草■☆
69887 Campanula alpestris All. = Campanula allionii Vill. ■☆
69888 Campanula alpigena K. Koch;山地风铃草■☆
69889 Campanula alpina Jacq.;高山风铃草;Alpine Bellflower ■☆
69890 Campanula alpini L. = Adenophora liliifolia (L.) Ledeb. ex A. DC. ■
69891 Campanula altaica Ledeb.;阿尔泰风铃草■☆
69892 Campanula americana L.;美洲风铃草;American Bellflower, American Bluebell, Bellflower, Tall Bellflower ■☆
69893 Campanula americana L. var. illinoensis (Fresen.) Farw. = Campanula americana L. ■☆
69894 Campanula androsacea (A. DC.) A. Dietr. = Wahlenbergia androsacea A. DC. ■☆
69895 Campanula angustifolia Lam. = Campanula rotundifolia L. ■
69896 Campanula annae Kolak.;安娜风铃草■☆
69897 Campanula anomala Fomin;异常风铃草■☆
69898 Campanula antiatlantica Maire et al. = Campanula filicaulis

Durieu ■☆
69899 Campanula aparinoides Pursh;沼泽风铃草;Bedstraw Bellflower, Marsh Bellflower, Swamp Bluebell ■☆
69900 Campanula aparinoides Pursh var. grandiflora Holz.;大花沼泽风铃草;Marsh Bellflower ■☆
69901 Campanula aparinoides Pursh var. uliginosa (Rydb.) Gleason = Campanula aparinoides Pursh var. grandiflora Holz. ■☆
69902 Campanula aprica Nannf.;田风铃草■
69903 Campanula aprica Nannf. = Campanula cana Wall. ■
69904 Campanula argyrotricha Wall.;银毛风铃草■☆
69905 Campanula aristata Wall.;钻裂风铃草(针叶风铃草);Aristate Bellflower ■
69906 Campanula aristata Wall. var. longisepala C. Marquand = Campanula aristata Wall. ■
69907 Campanula atlantica Batt. = Campanula afganica Pomel ■☆
69908 Campanula atlantica Batt. var. glabra Bonnet = Campanula afganica Pomel ■☆
69909 Campanula atlantica Coss. et Durieu var. guergourensis Batt. = Campanula afganica Pomel ■☆
69910 Campanula atlantis Gatt., Maire et Weiler;亚特兰大风铃草■☆
69911 Campanula aucheri A. DC.;奥氏风铃草■☆
69912 Campanula aurasiaca (Batt. et Trab.) Batt. et Trab. = Asyneuma rigidum (Willd.) Grossh. subsp. aurasiacum (Batt. et Trab.) Damboldt ■☆
69913 Campanula austroxinjiangensis Y. K. Yang, J. K. Wu et J. Z. Li;南疆风铃草;S. Xingjiang Bellflower ■
69914 Campanula baborensis Quézel;巴布尔风铃草■☆
69915 Campanula banksiana (A. DC.) A. Dietr. = Wahlenbergia banksiana A. DC. ■☆
69916 Campanula barbata L.;髯毛风铃草(流苏风铃草);Bearded Bellflower ■☆
69917 Campanula bellidifolia Adam;雅叶风铃草■☆
69918 Campanula benthamii Wall. ex Kitam. = Campanula dimorphantha Schweinf. ■
69919 Campanula betulifolia K. Koch;桦叶风铃草■☆
69920 Campanula biflora Ruiz et Pavon = Triodanis biflora (Ruiz et Pav.) Greene ■
69921 Campanula bononiensis L.;波伦风铃草;European Bellflower ■☆
69922 Campanula bordesiana Maire;博尔德斯风铃草■☆
69923 Campanula bordesiana Maire subsp. tibestica Quézel;提贝斯提博尔德风铃草■☆
69924 Campanula bordesiana Maire var. minuta Quézel;微小风铃草■☆
69925 Campanula brotheri Sommier et H. Lév.;布拉泽风铃草■☆
69926 Campanula broussonnetiana Roem. et Schult. = Campanula lusitanica Loefl. ■☆
69927 Campanula caespitosa Scop.;簇生风铃草■☆
69928 Campanula calcarata Sommier et H. Lév.;距风铃草■☆
69929 Campanula calcarea Albov ex Charadze;石灰风铃草■☆
69930 Campanula calcicola W. W. Sm.;灰岩风铃草;Calcicolous Bellflower ■
69931 Campanula calycanthema Turcz. ex Herder;萼花风铃草;Cup-and-saucer Canterbury Bell, Cup-and-saucer Canterbury Bells ■☆
69932 Campanula calycina Boeber ex Roem. et Schult.;萼状风铃草■☆
69933 Campanula camtschatica Pall. ex Roem. et Schult.;勘察加风铃草■☆
69934 Campanula cana Wall.;灰毛风铃草(曲茎风铃草,着色风铃草);Bendstem Bellflower, Grayhairy Bellflower, Greyhair Bellflower ■
69935 Campanula canescens Wall. ex A. DC. = Campanula benthamii Wall. ex Kitam. ■
69936 Campanula canescens Wall. ex A. DC. = Campanula dimorphantha Schweinf. ■
69937 Campanula capensis L. = Wahlenbergia capensis (L.) A. DC. ■☆
69938 Campanula caperonoides Klotzsch = Campanula sylvatica Wall. ■☆
69939 Campanula capillacea L. f. = Wahlenbergia capillacea (L. f.) A. DC. ■☆
69940 Campanula capusii (Franch.) Fed.;卡氏风铃草■☆
69941 Campanula carnosa Wall. = Peracarpa carnosa (Wall.) Hook. f. et Thomson ■
69942 Campanula carpatica Jacq.;喀尔巴阡山风铃草(丛生风铃草,广口风铃草);Bellflower, Carolina Hickory, Carpatian Bellflower, Tussock Bellflower, Tussock Bluebell ■
69943 Campanula carpatica Jacq. 'Blue Clips';大花蓝风铃草(大花蓝)■
69944 Campanula carpatica Jacq. 'Bressingham White';纯白广口风铃草■
69945 Campanula carpatica Jacq. 'Carpatica Blue';喀尔巴蓝风铃草■
69946 Campanula carpatica Jacq. 'Jewel';宝石广口风铃草■
69947 Campanula carpatica Jacq. 'Turbinata';螺旋广口风铃草■
69948 Campanula carpatica Jacq. 'White Clips';大花白风铃草(大花白)■
69949 Campanula caucasica M. Bieb.;高加索风铃草■☆
69950 Campanula cenisia L.;采尼斯风铃草■☆
69951 Campanula cephalotes Fisch. ex Nakai = Campanula glomerata L. var. dahurica Fisch. ex Ker Gawl. ■
69952 Campanula cephalotes Nakai = Campanula glomerata L. subsp. cephalotes (Fisch. ex Nakai) D. Y. Hong ■
69953 Campanula cernua Thunb. = Wahlenbergia cernua (Thunb.) A. DC. ■☆
69954 Campanula cervicaria L.;硬毛风铃草■☆
69955 Campanula cervicina (A. DC.) A. Dietr. = Wahlenbergia campanuloides (Delile) Vatke ■☆
69956 Campanula chamissonis Al. Fed.;查米森风铃草■☆
69957 Campanula chamissonis Al. Fed. f. albiflora (Miyabe et Tatew.) T. Shimizu;白花查米森风铃草■☆
69958 Campanula charkeviczii Fed.;哈氏风铃草■☆
69959 Campanula chinensis D. Y. Hong;长柱风铃草;Chinese Bellflower ■
69960 Campanula chinganensis Baranov = Campanula langsdorffiana Fisch. ex Trautv. et C. A. Mey. ■
69961 Campanula choziatowskyi Fomin;考氏风铃草■☆
69962 Campanula chrysoaplenifolia Franch.;丝茎风铃草(白毛风铃草);Goldsaxifrageleaf Bellflower, Silkstem Bellflower ■
69963 Campanula ciliata Steven;缘毛风铃草■☆
69964 Campanula ciliata Thunb. = Wahlenbergia uitenhagensis (H. Buek) Lammers ■☆
69965 Campanula cinerea Hegetschw. = Campanula rotundifolia L. ■
69966 Campanula cinerea L. f. = Wahlenbergia cinerea (L. f.) Lammers ■☆
69967 Campanula circaeoides Schmidt = Peracarpa carnosa (Wall.) Hook. f. et Thomson ■
69968 Campanula circaeoides Schmidt ex Miq. = Peracarpa carnosa (Wall.) Hook. f. et Thomson ■
69969 Campanula circassica Fomin;切尔卡西亚风铃草■☆
69970 Campanula cochlearifolia Lam.;岩荠叶风铃草(仙女风铃草);

Fairy Thimbles, Fairy's Thimbles ■☆
69971 Campanula collina M. Bieb.;山丘风铃草■☆
69972 Campanula colorata Wall.;西南风铃草(鸡肉参,兰花石参,蓝花石参,山鹅儿肠,土桔梗,土人参,土沙参,小果风铃草,小石参,岩兰花,岩蓝花,着色风铃草);Coloured Bellflower, Smallcarp Bellflower ■
69973 Campanula colorata Wall. = Campanula pallida Wall. ■☆
69974 Campanula colorata Wall. var. moorcroftiana? = Campanula pallida Wall. ■☆
69975 Campanula colorata Wall. var. tibetica Hook. f. et Thomson = Campanula colorata Wall. ■
69976 Campanula colorata Wall. var. tibetica Hook. f. et Thomson = Campanula pallida Wall. var. tibetica (Hook. f. et Thomson) H. Hara ■☆
69977 Campanula cordifolia K. Koch;心叶风铃草■☆
69978 Campanula coronopifolia Fisch. ex Roem. et Schult. = Adenophora gmelinii (Spreng.) Fisch. ■
69979 Campanula crenulata Franch.;流石风铃草(补肺参);Crenulate Bellflower ■
69980 Campanula crispa Lam.;皱波风铃草■☆
69981 Campanula cylindrica (Pax et K. Hoffm.) Nannf. = Campanula aristata Wall. ■
69982 Campanula daghestanica Fomin;达赫斯坦风铃草■☆
69983 Campanula dasyantha M. Bieb.;毛花风铃草(密花风铃草)■☆
69984 Campanula dasyantha M. Bieb. = Campanula chamissonis Al. Fed. ■☆
69985 Campanula dasyantha M. Bieb. subsp. chamissonis (Al. Fed.) Victorov = Campanula chamissonis Al. Fed. ■☆
69986 Campanula dehiscens Roxb. = Wahlenbergia marginata (Thunb.) A. DC. ■
69987 Campanula delavayi Franch.;丽江风铃草;Delavay Bellflower ■
69988 Campanula denticulata (Fisch.) Spreng. = Adenophora tricuspidata (Fisch. ex Roem. et Schult.) A. DC. ■
69989 Campanula denticulata Burch. = Wahlenbergia denticulata (Burch.) A. DC. ■☆
69990 Campanula denticulata Spreng. = Adenophora tricuspidata (Fisch. ex Roem. et Schult.) A. DC. ■
69991 Campanula dichotoma L.;二歧风铃草■☆
69992 Campanula dichotoma L. subsp. afra (Cav.) Maire = Campanula afra Cav. ■☆
69993 Campanula dichotoma L. subsp. kremeri (Boiss. et Reut.) Batt. = Campanula kremeri Boiss. et Reut. ■☆
69994 Campanula dichotoma L. var. glabrescens Maire = Campanula afra Cav. ■☆
69995 Campanula dichotoma L. var. kebdanensis (Sennen) Maire = Campanula afra Cav. ■☆
69996 Campanula dichotoma L. var. leptosiphon (Pau et Sennen) Maire = Campanula afra Cav. ■☆
69997 Campanula dichotoma L. var. pallida Maire = Campanula afra Cav. ■☆
69998 Campanula dichotoma L. var. parviflora Maire et M. Peltier = Campanula kremeri Boiss. et Reut. ■☆
69999 Campanula dichotoma L. var. peltieri Maire = Campanula dichotoma L. ■☆
70000 Campanula diffusa Vahl = Campanula fragilis Cirillo ■☆
70001 Campanula dimorphantha Schweinf.;一年风铃草(桔梗);Annual Bellflower ■
70002 Campanula divaricata Michx.;南方风铃草;Panicled Bellflower, Southern Harebell ■☆
70003 Campanula divergens Willd.;稍叉风铃草■☆
70004 Campanula dolomitica E. A. Busch;多罗米蒂风铃草■☆
70005 Campanula drabifolia Sibth. et Sm.;葶苈叶风铃草;Greek Bellflower ■☆
70006 Campanula dubia A. DC. = Campanula rotundifolia L. ■
70007 Campanula edulis Forssk.;可食风铃草■☆
70008 Campanula elatinoides Moretti;高风铃草■☆
70009 Campanula elatior (Fomin) Grossh. ex Fed.;较高风铃草■☆
70010 Campanula elegans Roem. et Schult.;雅致风铃草■☆
70011 Campanula elegantissima Grossh.;极雅风铃草■☆
70012 Campanula embergeri Litard. et Maire = Campanula filicaulis Durieu subsp. embergeri (Litard. et Maire) Dobignard ■☆
70013 Campanula embergeri Litard. et Maire subsp. schotteri Quézel = Campanula filicaulis Durieu ■☆
70014 Campanula erinus L.;绵毛风铃草■☆
70015 Campanula erinus L. var. hernandezii Sennen et Mauricio = Campanula erinus L. ■☆
70016 Campanula erysimoides Roem. et Schult. = Adenophora gmelinii (Spreng.) Fisch. ■
70017 Campanula erysimoides Vest ex Roem. et Schult. = Adenophora gmelinii (Spreng.) Fisch. ■
70018 Campanula esculenta A. Rich. = Campanula edulis Forssk. ■☆
70019 Campanula eugeniae Fed.;番樱桃风铃草■☆
70020 Campanula excisa Schleich. ex Murith;缺刻风铃草■☆
70021 Campanula farinosa Andrz. ex Besser;被粉风铃草■☆
70022 Campanula fasciculata L. f. = Wahlenbergia desmantha Lammers ■☆
70023 Campanula fastigiata A. DC.;帚状风铃草■☆
70024 Campanula fenestrellata Feer.;光泽风铃草■☆
70025 Campanula filicaulis Durieu;线茎风铃草■☆
70026 Campanula filicaulis Durieu subsp. embergeri (Litard. et Maire) Dobignard;恩贝格尔风铃草■☆
70027 Campanula filicaulis Durieu subsp. reboudiana (Pomel) Maire = Campanula filicaulis Durieu ■☆
70028 Campanula filicaulis Durieu var. antiatlantica (Maire et al.) Quézel = Campanula filicaulis Durieu ■☆
70029 Campanula filicaulis Durieu var. cedretorum H. Lindb. = Campanula filicaulis Durieu ■☆
70030 Campanula filicaulis Durieu var. elata (Faure et Maire) Quézel = Campanula filicaulis Durieu ■☆
70031 Campanula filicaulis Durieu var. gattefossei (Maire et Weiller) Quézel = Campanula filicaulis Durieu ■☆
70032 Campanula filicaulis Durieu var. genuina Maire = Campanula filicaulis Durieu ■☆
70033 Campanula filicaulis Durieu var. gomarica Font Quer = Campanula filicaulis Durieu ■☆
70034 Campanula filicaulis Durieu var. intermedia Vindt = Campanula filicaulis Durieu ■☆
70035 Campanula filicaulis Durieu var. mairei Quézel = Campanula filicaulis Durieu ■☆
70036 Campanula filicaulis Durieu var. maroccana (Ball) Dobignard = Campanula filicaulis Durieu ■☆
70037 Campanula filicaulis Durieu var. maroccana Pau = Campanula filicaulis Durieu ■☆
70038 Campanula filicaulis Durieu var. parielii Quézel = Campanula

filicaulis Durieu ■☆

70039 Campanula filicaulis Durieu var. pseudoantiatlantica Quézel = Campanula filicaulis Durieu ■☆

70040 Campanula filicaulis Durieu var. pseudoradicosa Litard. et Maire = Campanula filicaulis Durieu ■☆

70041 Campanula filicaulis Durieu var. reboudiana (Pomel) Maire = Campanula filicaulis Durieu ■☆

70042 Campanula filicaulis Durieu var. schotteri (Quézel) Dobignard = Campanula filicaulis Durieu ■☆

70043 Campanula filicaulis Durieu var. tibestica Quézel;西藏线茎风铃草(提贝斯提线茎风铃草)■☆

70044 Campanula fischeriana Spreng. = Adenophora gmelinii (Spreng.) Fisch. ■

70045 Campanula fominii Grossh.;福明风铃草■☆

70046 Campanula fondervisii Albov.;丰德韦斯风铃草■☆

70047 Campanula fragilis Cirillo;脆弱风铃草■☆

70048 Campanula froedinii Rech. f.;弗勒丁风铃草■☆

70049 Campanula fulgens Wall. = Asyneuma fulgens (Wall.) Briq. ●

70050 Campanula garganica Ten.;常春藤叶风铃草■☆

70051 Campanula garganica Ten. 'W. H. Paine';佩音常春藤叶风铃草■☆

70052 Campanula gieseckiana Vest ex Roem. et Schult. = Campanula rotundifolia L. ■

70053 Campanula glauca Thunb. = Platycodon grandiflorus (Jacq.) A. DC. ■

70054 Campanula glomerata L.;北疆风铃草(灯笼花,聚花风铃草,聚铃花,球花风铃草);Campanula glomerata, Canterbury Bells, Clustered Bellflower, Dane's Blood, Dane's-blood, Danesblood Bellflower, Dane's-blood Bellflower ■

70055 Campanula glomerata L. 'Snow';雪白北疆风铃草;Clustered Bellflower, Dane's Blood, Danesblood, White Clustered Bellflower ■

70056 Campanula glomerata L. 'Superba';华丽聚花风铃草■☆

70057 Campanula glomerata L. f. canescens (Maxim.) Kitag.;白毛风铃草■☆

70058 Campanula glomerata L. subsp. cephalotes (Fisch. ex Nakai) D. Y. Hong;聚花风铃草(头状风铃草,新疆风铃草);N. Xingjiang Bellflower ■

70059 Campanula glomerata L. subsp. cephalotes (Fisch. ex Nakai) D. Y. Hong = Campanula glomerata L. var. dahurica Fisch. ex Ker Gawl. ■

70060 Campanula glomerata L. subsp. daqingshanica D. Y. Hong et Y. Z. Zhao;大青山风铃草;Daqingshan Bellflower ■

70061 Campanula glomerata L. var. dahurica Fisch. ex Ker Gawl.;达呼里风铃草(聚花风铃草)■

70062 Campanula glomerata L. var. dahurica Fisch. ex Ker Gawl. f. alba (Nakai) Nakai ex T. B. Lee;白花达呼里风铃草■☆

70063 Campanula glomerata L. var. salviifolia Kom. = Campanula glomerata L. subsp. cephalotes (Fisch. ex Nakai) D. Y. Hong ■

70064 Campanula glomerata L. var. speciosa DC. = Campanula speciosa Hornem. ■☆

70065 Campanula glomeratoides D. Y. Hong;头花风铃草;Glomeratelike Bellflower, Headflower Bellflower ■

70066 Campanula gmelinii Spreng. = Adenophora gmelinii (Spreng.) Fisch. ■

70067 Campanula grandiflora Jacq. = Platycodon grandiflorus (Jacq.) A. DC. ■

70068 Campanula grandiflora Lam. = Campanula medium L. ■

70069 Campanula grandis Fisch. et C. A. Mey.;奥林匹克风铃草;Olympic Bellflower ■☆

70070 Campanula grandis Fisch. et C. A. Mey. = Campanula latiloba A. DC. ■☆

70071 Campanula grestis Wall. = Wahlenbergia marginata (Thunb.) A. DC. ■

70072 Campanula griffithii Hook. f. et Thomson = Campanula leucoclada Boiss. ■☆

70073 Campanula grossekii Heuff.;格氏风铃草;Grossek Bellflower ■☆

70074 Campanula grossheimii Charadze;格罗风铃草■☆

70075 Campanula hemschinica K. Koch;海姆什风铃草■☆

70076 Campanula herminii Hoffmanns. et Link var. atlantica Jahand. et Maire = Campanula mairei Maire ■☆

70077 Campanula heterodoxa Bong. = Campanula rotundifolia L. ■

70078 Campanula heterophylla Gray = Campanula rotundifolia L. ■

70079 Campanula himalayensis Klotzsch = Campanula pallida Wall. ■☆

70080 Campanula hirsutissima Sennen et Mauricio = Campanula afra Cav. ■☆

70081 Campanula hispidula Thunb. = Wahlenbergia hispidula (Thunb.) A. DC. ■☆

70082 Campanula hoffineisteri Klotzsch = Campanula pallida Wall. ■☆

70083 Campanula humilis (A. DC.) A. Dietr. = Wahlenbergia lobelioides (L. f.) Link subsp. riparia (A. DC.) Thulin ■☆

70084 Campanula hybida L. = Legousia hybrida (L.) Delarbre ■☆

70085 Campanula hypocrateriformis Dobignard;杯状风铃草■☆

70086 Campanula hypopolia Trautv.;背白风铃草■☆

70087 Campanula imeretina Rupr.;小亚细亚风铃草■☆

70088 Campanula incanescens Boiss.;浅灰风铃草■☆

70089 Campanula infundibulum Rchb.;漏斗状风铃草■☆

70090 Campanula integerrima Buch.-Ham. ex D. Don = Campanula sylvatica Wall. ■☆

70091 Campanula intercedens Witasek = Campanula rotundifolia L. ■

70092 Campanula irinae A. I. Kuth.;伊里娜风铃草■☆

70093 Campanula isophylla Moretti;同叶风铃草;Italian Bellflower, Star of Bethlehem ■☆

70094 Campanula jacobaea Webb;雅各菊风铃草■☆

70095 Campanula jacobaea Webb var. hispida Bolle = Campanula jacobaea Webb ■☆

70096 Campanula jacobaea Webb var. humilis Bolle = Campanula jacobaea Webb ■☆

70097 Campanula japonica Vatke = Asyneuma japonicum (Miq.) Briq. ●

70098 Campanula jurjurensis Pomel;北非风铃草■☆

70099 Campanula karakuschensis Grossh.;卡拉库风铃草■☆

70100 Campanula kemulariae Fomin;凯穆拉利亚风铃草■☆

70101 Campanula keniensis Thulin;肯尼亚风铃草■☆

70102 Campanula khasiana Hook. f. et Thomson = Adenophora khasiana (Hook. f. et Thomson) Collett et Hemsl. ■

70103 Campanula kolakovskyi Charadze;科拉科夫斯基风铃草■☆

70104 Campanula komarovii Maleev;科马罗夫风铃草;Komarov Bellflower ■☆

70105 Campanula krebsii (Cham.) D. Dietr. = Wahlenbergia krebsii Cham. ■☆

70106 Campanula kremeri Boiss. et Reut.;小花风铃草■☆

70107 Campanula lactiflora M. Bieb.;神钟花(白花风铃草);Milky Bellflower ■☆

70108 Campanula lactiflora M. Bieb. 'Brantwood';布兰特伍德神钟花

(布兰特伍德阔叶风铃草)■☆
70109 Campanula lactiflora M. Bieb. 'Loddon Anna';劳顿·安娜神钟花■☆
70110 Campanula lactiflora M. Bieb. 'Prichard's Variety';布利查德神钟花■☆
70111 Campanula lamarkii D. Dietr. = Adenophora lamarkii Fisch. ■
70112 Campanula lamiifolia M. Bieb. = Campanula alliariifolia Willd. ■☆
70113 Campanula lanata Friv.;软毛风铃草;Woolly Bellflower ■☆
70114 Campanula lancifolia (Roxb.) Merr. = Cyclocodon lancifolius (Roxb.) Kurz ■
70115 Campanula lancifolia Roxb. = Cyclocodon lancifolius (Roxb.) Kurz ■
70116 Campanula langsdorffiana Fisch. ex Trautv. et C. A. Mey.;石生风铃草;Longsdorff Bellflower ■
70117 Campanula lasiocarpa Cham.;毛果风铃草■☆
70118 Campanula lasiocarpa Cham. f. albiflora Tatew.;白花毛果风铃草■☆
70119 Campanula latifolia L.;宽叶风铃草(阔叶风铃草);Foxglove, Giant Bellflower, Giant Throatwort, Gowk's Hose, Great Bellflower, Haskwort, Polypodium, Sea Spinach, White Foxglove ■☆
70120 Campanula latifolia L. var. macrantha Sims;大花宽叶风铃草;Royal Bellflower ■☆
70121 Campanula latiloba A. DC.;宽裂风铃草(宽叶风铃草,阔叶风铃草);Olympic Bellflower ■☆
70122 Campanula latiloba A. DC. 'Percy Piper';珀西花边宽裂风铃草■☆
70123 Campanula latisepala Hultén = Campanula lasiocarpa Cham. ■☆
70124 Campanula lavandulifolia Reinw. ex Blume = Wahlenbergia marginata (Thunb.) A. DC. ■
70125 Campanula leptosiphon Pau et Sennen = Campanula afra Cav. ■☆
70126 Campanula leucantha Gilli;白花风铃草■☆
70127 Campanula leucoclada Boiss.;白枝风铃草■☆
70128 Campanula leucosiphon Boisset Heldr.;白管风铃草■☆
70129 Campanula leucotricha C. Y. Wu = Campanula chrysoaplenifolia Franch. ■
70130 Campanula liliifolia Hegetschw. = Adenophora liliifolia (L.) Besser ■
70131 Campanula liliifolia Hegetschw. = Campanula liliifolia L. ■
70132 Campanula liliifolia L. = Adenophora communis Fisch. ■
70133 Campanula liliifolia L. = Adenophora liliifolia (L.) Besser ■
70134 Campanula linearis L. f. = Microcodon lineare (L. f.) H. Buek ■☆
70135 Campanula linifolia Scop.;亚麻叶风铃草■☆
70136 Campanula lobelioides L. f. = Wahlenbergia lobelioides (L. f.) Link ■☆
70137 Campanula loeflingii Brot. = Campanula lusitanica Loefl. ■☆
70138 Campanula loeflingii Brot. var. maura Murb. = Campanula lusitanica Loefl. ■☆
70139 Campanula lusitanica Loefl.;葡萄牙风铃草■☆
70140 Campanula lusitanica Loefl. subsp. specularioides (Coss.) Aldasoro et L. Sáenz;桔梗状风铃草■☆
70141 Campanula lusitanica Loefl. var. broussonetiana (Roem. et Schult.) Pau = Campanula lusitanica Loefl. ■☆
70142 Campanula lusitanica Loefl. var. maura (Murb.) Quézel = Campanula lusitanica Loefl. ■☆
70143 Campanula lyrata Lam.;大头羽裂风铃草■☆
70144 Campanula macrophylla Sims = Campanula alliariifolia Willd. ■☆
70145 Campanula macrorhiza Gay;长根风铃草■☆
70146 Campanula macrorrhiza J. Gay var. jurjurensis (Pomel) Chabert = Campanula jurjurensis Pomel ■☆
70147 Campanula macrorrhiza J. Gay var. rotundata Chabert = Campanula jurjurensis Pomel ■☆
70148 Campanula madagascariensis (A. DC.) A. Dietr. = Wahlenbergia madagascariensis A. DC. ■☆
70149 Campanula mairei Maire;迈氏风铃草■☆
70150 Campanula mairei Maire var. anremerica Litard. et Maire = Campanula mairei Maire ■☆
70151 Campanula mairei Maire var. atlantica (Jahand. et Maire) Maire = Campanula mairei Maire ■☆
70152 Campanula mairei Maire var. flahaultiana Emb. = Campanula mairei Maire ■☆
70153 Campanula mairei Maire var. tenera Emb. et Maire = Campanula mairei Maire ■☆
70154 Campanula mairei Pau = Campanula mairei Maire ■☆
70155 Campanula makaschvilii E. A. Busch;马卡什夫风铃草■☆
70156 Campanula marginata Thunb. = Wahlenbergia marginata (Thunb.) A. DC. ■
70157 Campanula maroccana (Ball) Batt. var. pseudoradicosa Litard. et Maire = Campanula filicaulis Durieu ■☆
70158 Campanula maroccana Ball = Campanula filicaulis Durieu ■☆
70159 Campanula marsupiiflora Fisch. ex Roem. et Schult. = Adenophora stenanthina (Ledeb.) Kitag. ■
70160 Campanula marsupiiflora Roem. et Schult. = Adenophora stenanthina (Ledeb.) Kitag. ■
70161 Campanula massalskyi Fomin;马萨尔斯基风铃草■☆
70162 Campanula mauritanica Pomel = Campanula trachelium L. subsp. mauritanica (Pomel) Quézel ■☆
70163 Campanula mauritanica Pomel var. parviflora Batt. = Campanula trachelium L. subsp. mauritanica (Pomel) Quézel ■☆
70164 Campanula medium L.;风铃草(彩萼钟花,银叶菊,钟花);Canterbury Bell, Mariet, Mercury's Violet, Mugs. Without-handles, Old Woman's Bonnets, Old Woman's Nightcap, Parachutes, Wineglasses ■
70165 Campanula medium L. 'Bell of Holland';荷兰钟风铃草■☆
70166 Campanula medium L. var. calycanthema Nichols;萼花钟风铃草;Cup and Saucer, Cup-and-saucer ■☆
70167 Campanula mekongensis Diels ex C. Y. Wu;澜沧风铃草;Lancang Bellflower ■
70168 Campanula meyeriana Rupr.;迈尔风铃草■☆
70169 Campanula microcarpa C. Y. Wu = Campanula colorata Wall. ■
70170 Campanula microdonta Koidz.;小齿风铃草■☆
70171 Campanula microphylla Cav. = Campanula mollis L. ■☆
70172 Campanula mirabilis Albov.;奇异风铃草■☆
70173 Campanula modesta Hook. f. et Thomson;藏滇风铃草;Moderate Bellflower ■
70174 Campanula mollis L.;绢毛风铃草;Soft Bellflower ■☆
70175 Campanula mollis L. var. canescens Quézel = Campanula mollis L. ■☆
70176 Campanula mollis L. var. hosmarensis (Maire) Quézel = Campanula mollis L. ■☆
70177 Campanula mollis L. var. longicornis Maire = Campanula mollis L. ■☆
70178 Campanula mollis L. var. mesatlantica Quézel = Campanula mollis L. ■☆
70179 Campanula mollis L. var. microphylla (Cav.) A. DC. =

Campanula mollis L. ■☆

70180 Campanula mollis L. var. oranensis Maire = Campanula mollis L. ■☆

70181 Campanula mollis L. var. pseudovelata Maire = Campanula mollis L. ■☆

70182 Campanula mollis L. var. rifana Emb. et Maire = Campanula mollis L. ■☆

70183 Campanula mollis L. var. tlemcenensis Quézel = Campanula mollis L. ■☆

70184 Campanula mollis L. var. umbricola (Font Quer) Quézel = Campanula mollis L. ■☆

70185 Campanula monodiana Maire;莫诺风铃草■☆

70186 Campanula montevidensis Spreng. = Triodanis biflora (Ruiz et Pav.) Greene ■

70187 Campanula morettiana Rchb.;垫状风铃草■☆

70188 Campanula multiflora A. DC. = Campanula bononiensis L. ■☆

70189 Campanula multiflora Waldst. et Kit.;欧洲多花风铃草■☆

70190 Campanula multiflora Willd. ex Roem. et Schult. = Campanula polyantha Roem. et Schult. ■☆

70191 Campanula nakaoi Kitam.;藏南风铃草;S. Xizang Bellflower, South Tibet Bellflower ■

70192 Campanula nana Lam. = Campanula allionii Vill. ■☆

70193 Campanula nepetifolia H. Lév. et Vaniot = Campanula colorata Wall. ■

70194 Campanula nephrophylla C. Y. Wu;肾叶风铃草;Kidneyleaf Bellflower ■

70195 Campanula nicaeensis Risso ex A. DC. = Campanula macrorhiza Gay ■☆

70196 Campanula nobilis Lindl. = Campanula punctata Lam. ■

70197 Campanula numidica Durieu;努米底亚风铃草■☆

70198 Campanula oblongifolia (K. Koch) Kharadze;矩圆叶风铃草■☆

70199 Campanula occidentalis Nyman;西方风铃草■☆

70200 Campanula ochroleuca (Kem.-Nath.) Kem.-Nath.;绿白风铃草■☆

70201 Campanula odontosepala Boiss.;齿瓣风铃草■☆

70202 Campanula omeiensis (Z. Y. Zhu) D. Y. Hong et Z. Y. Li;峨眉风铃草(峨眉沙参);Emei Ladybell ■

70203 Campanula ossetica M. Bieb.;骨质风铃草■☆

70204 Campanula pallida Wall.;苍白风铃草■☆

70205 Campanula pallida Wall. var. tibetica (Hook. f. et Thomson) H. Hara;西藏苍白风铃草■☆

70206 Campanula paniculata Thunb. = Wahlenbergia paniculata (Thunb.) A. DC. ■☆

70207 Campanula parviflora Lam. = Campanula sibirica L. ■

70208 Campanula parviflora Salisb. = Wahlenbergia lobelioides (L. f.) Link ■☆

70209 Campanula pasumensis Marquart = Campanula cana Wall. ■

70210 Campanula patula L.;多枝风铃草;Rambling Bellflower, Spreading Bellflower ■☆

70211 Campanula pereskiifolia Fisch. ex Roem. et Schult. = Adenophora pereskiifolia (Fisch. ex Roem. et Schult.) G. Don ■

70212 Campanula pereskiifolia Fisch. ex Roem. et Schult. = Adenophora pereskiifolia (Fisch. ex Roem. et Schult.) Fisch. ex Loudon ■

70213 Campanula perfoliata L. = Triodanis perfoliata (L.) Nieuwl. ■

70214 Campanula persicifolia L.;桃叶风铃草;Narrow-leaved Bellflower, Peach Bells, Peachleaf Bellflower, Peach-leaved Bellflower, Peachleaved Campanula ■☆

70215 Campanula persicifolia L. 'Fleur de Neige';雪花桃叶风铃草■☆

70216 Campanula persicifolia L. 'Kelly's Gold';金色桃叶风铃草;Gold-leaved Bellflower ■☆

70217 Campanula persicifolia L. 'Pride of Exmouth';无口丽桃叶风铃草■☆

70218 Campanula persicifolia L. 'Telham Beauty';大花桃叶风铃草(泰尔汉姆丽桃叶风铃草);Large-flowered Bellflower ■☆

70219 Campanula petiolata A. DC. = Campanula rotundifolia L. ■

70220 Campanula petrophila Rupr.;喜岩风铃草■☆

70221 Campanula pilosa Pall. = Campanula pilosa Pall. ex Roem. et Schult. ■☆

70222 Campanula pilosa Pall. ex Roem. et Schult.;茸毛风铃草■☆

70223 Campanula pilosa Pall. ex Roem. et Schult. var. dasyantha Herd = Campanula dasyantha M. Bieb. ■☆

70224 Campanula polyantha Roem. et Schult.;西伯利亚多花风铃草■☆

70225 Campanula polyclada Rech. f. et Schimann-Czeika;多茎风铃草■☆

70226 Campanula polymorpha Witasek;多型风铃草■☆

70227 Campanula pontica Albov.;蓬特风铃草■☆

70228 Campanula portenschlagiana Roem. et Schult.;波氏风铃草(普氏风铃草);Adria Bellflower, Delmatian Bellflower ■☆

70229 Campanula poscharskyana Degen;肋瓣风铃草;Serbian Bellflower, Trailing Bellflower ■☆

70230 Campanula prenanthoides Durand;加州风铃草;Californian Harebell ■☆

70231 Campanula procumbens Thunb. = Wahlenbergia procumbens (Thunb.) A. DC. ■☆

70232 Campanula propinqua Fisch. et C. A. Mey.;邻近风铃草■☆

70233 Campanula pulla L.;暗钟花;Solitary Harebell ■☆

70234 Campanula punctata Lam.;紫斑风铃草(灯笼花,吊钟花,山小菜);Bellflower, Spotted Bellflower ■

70235 Campanula punctata Lam. f. inpunctata N. Yonez.;无斑风铃草■☆

70236 Campanula punctata Lam. f. lucida Sugim.;光泽紫斑风铃草■☆

70237 Campanula punctata Lam. f. rubriflora (Makino) T. Shimizu;红花紫斑风铃草■

70238 Campanula punctata Lam. f. rubriflora Makino = Campanula punctata Lam. f. rubriflora (Makino) T. Shimizu ■

70239 Campanula punctata Lam. subsp. hondoensis (Kitam.) Kitam. = Campanula punctata Lam. var. hondoensis (Kitam.) Ohwi ■☆

70240 Campanula punctata Lam. subsp. microdonta (Koidz.) Kitam. = Campanula microdonta Koidz. ■☆

70241 Campanula punctata Lam. var. hondoensis (Kitam.) Ohwi;本州风铃草■☆

70242 Campanula punctata Lam. var. hondoensis (Kitam.) Ohwi f. albiflora T. Shimizu;白花本州风铃草■☆

70243 Campanula purpurea (Wall.) Spreng. = Codonopsis purpurea Wall. ■

70244 Campanula purpurea Spreng. = Codonopsis purpurea Wall. ■

70245 Campanula pusilla Haenke = Campanula caespitosa Scop. ■☆

70246 Campanula pusilla Hegetschw. = Campanula cochlearifolia Lam. ■☆

70247 Campanula pyramidalis L.;火箭风铃草(塔钟花);Bridal Wreath, Chimney Bellflower, Chimney Plant, Perrymedoll, Steeple Bellflower, Steeple Bells ■☆

70248 Campanula quartiniana A. Rich. = Campanula edulis Forssk. ■☆

70249 Campanula rabelaisiana Roem. et Schult. = Adenophora gmelinii (Spreng.) Fisch. ■

70250 Campanula raddeana Trautv.;拉德风铃草■☆

70251 Campanula radula Fisch. ex Tchih.;刮刀风铃草■☆

70252 Campanula rainerii Perp. ;擎钟花■☆
70253 Campanula ramulosa Wall. = Campanula pallida Wall. ■☆
70254 Campanula rapuncula L. ; 芜 菁 风 铃 草; Rampion, Rampion Bellflower, Ramps ■☆
70255 Campanula rapunculoides L. ; 钓 钟 花; Bellflower, Creeping Bellflower, European Bellflower, False Rampion, Rampion Bellflower, Rover Bellflower, Taurian Bellflower ■☆
70256 Campanula rapunculoides L. var. ucranica (Besser) K. Koch = Campanula rapunculoides L. ■☆
70257 Campanula rapunculus L. ;匍匐风铃草;Rampion ■☆
70258 Campanula rapunculus L. var. grandiflora Font Quer = Campanula rapunculus L. ■☆
70259 Campanula rapunculus L. var. hirta Peterm. = Campanula rapunculus L. ■☆
70260 Campanula rapunculus L. var. spiciformis Boiss. = Campanula rapunculus L. ■☆
70261 Campanula rapunculus L. var. strigulosa Batt. = Campanula rapunculus L. var. verruculosa (Hoffmanns. et Link) H. Lindb. ■☆
70262 Campanula rapunculus L. var. verruculosa (Hoffmanns. et Link) H. Lindb. ;小疣匍匐风铃草■☆
70263 Campanula reboudiana Pomel = Campanula filicaulis Durieu ■☆
70264 Campanula regelii Trautv. = Sergia regelii (Trautv.) Fed. ■☆
70265 Campanula remotiflora Siebold et Zucc. = Adenophora remotiflora (Siebold et Zucc.) Miq. ■
70266 Campanula rhomboidea Borbás = Adenophora pereskiifolia (Fisch. ex Roem. et Schult.) Fisch. ex Loudon ■
70267 Campanula richteri Borbás = Adenophora tricuspidata (Fisch. ex Roem. et Schult.) A. DC. ■
70268 Campanula rigescens Pall. ex Roem. et Schult. ;硬直风铃草■☆
70269 Campanula rigidipila Steud. et Hochst. ex A. Rich. = Campanula edulis Forssk. ■☆
70270 Campanula riparia Bojer ex A. DC. = Wahlenbergia lobelioides (L. f.) Link subsp. riparia (A. DC.) Thulin ■☆
70271 Campanula rotundifolia L. ;圆叶风铃草(兴安风铃草);Airbell, Airbell Hairbell, Blaver, Blawort, Blue Bells, Blue Bells of Scotland, Blue Blaver, Bluebell, Bluebell Bellflower, Bluebells-of-scotland, Bluebottle, Cuckoo, Cuckoo's Stockings, Ding-dong, Fairy Bells, Fairy Ringers, Fairy Thimbles, Fairy's Cap, Fairy's Cup, Fairy's Thimbles, Gowk's Hose, Gowk's Thumb, Granny's Tears, Harebell, Harvest Bells, Heath Bellflower, Heathbell, Heather Bell, Lady's Thimble, Lady's Thimbles, Milkwort, Ofthe Field, Old Man's Beard, Old Man's Bells, Our Lady's Thimble, Our Lady's Thimbles, Roundleaf Bluebell, Round-leaved Bellflower, School Bells, Scottish Bluebell, Scttish Bluebell, Sheep Bells, Thimbles, Thumble, Witch Bells, Witches' Thimbles ■
70272 Campanula rotundifolia L. var. alaskana A. Gray = Campanula rotundifolia L. ■
70273 Campanula rotundifolia L. var. albiflora G. Don;白花圆叶风铃草■☆
70274 Campanula rotundifolia L. var. alpina Tuck. ;高山圆叶风铃草■☆
70275 Campanula rotundifolia L. var. arctica Lange;北极圆叶风铃草■☆
70276 Campanula rotundifolia L. var. intercedens (Witasek) Farw. = Campanula rotundifolia L. ■
70277 Campanula rotundifolia L. var. jurjurensis (Pomel) Quézel = Campanula jurjurensis Pomel ■☆
70278 Campanula rotundifolia L. var. lancifolia Mert. et W. D. J. Koch = Campanula rotundifolia L. ■
70279 Campanula rotundifolia L. var. langsdorffiana (A. DC.) Britton; 朗氏风铃草■☆
70280 Campanula rotundifolia L. var. petiolata (A. DC.) J. K. Henry = Campanula rotundifolia L. ■
70281 Campanula rotundifolia L. var. velutina A. DC. = Campanula rotundifolia L. ■
70282 Campanula rupestris Risso ex A. DC. = Campanula macrorhiza Gay ■☆
70283 Campanula ruprechtii Boiss. ;鲁普雷希特风铃草■☆
70284 Campanula sacajaweana M. Peck = Campanula rotundifolia L. ■
70285 Campanula sarmentosa Hochst. ex A. Rich. = Campanula edulis Forssk. ■☆
70286 Campanula sauvagei Quézel;索瓦热风铃草■☆
70287 Campanula saxatilis L. ;岩地风铃草■☆
70288 Campanula saxifraga M. Bieb. ;岩生风铃草■☆
70289 Campanula saxifragoides Doum. ;虎耳草状风铃草■☆
70290 Campanula saxifragoides Doum. var. gatteffossei Maire et Weiller = Campanula saxifragoides Doum. ■☆
70291 Campanula saxifragoides Doum. var. guinetii Quézel = Campanula saxifragoides Doum. ■☆
70292 Campanula saxifragoides Doum. var. latifolia (Litard. et Maire) Emb. = Campanula saxifragoides Doum. ■☆
70293 Campanula scandens Pall. ex Roem. et Schult. ;攀缘风铃草■☆
70294 Campanula scheuchzeri Vill. ; 苏 氏 风 铃 草; Scheuchzer's Bellflower ■☆
70295 Campanula schimperi Vatke = Campanula edulis Forssk. ■☆
70296 Campanula schischkinii Kolak. et Sachokia;希施风铃草■☆
70297 Campanula sclerotricha Boiss. ;粗毛风铃草■☆
70298 Campanula seminuda Vest;半裸风铃草■☆
70299 Campanula semisecta Murb. ;半裂风铃草■☆
70300 Campanula semisecta Murb. var. basiclada = Campanula semisecta Murb. ■☆
70301 Campanula serpylliformis Batt. et Trab. = Campanula velata Pomel subsp. serpylliformis (Batt. et Trab.) Quézel ■☆
70302 Campanula sessiliflora K. Koch = Campanula latiloba A. DC. ■☆
70303 Campanula sessiliflora L. f. = Wahlenbergia subulata (L'Hér.) Lammers ■☆
70304 Campanula sewerzowii Regel = Sergia sewerzowii (Regel) Fed. ■☆
70305 Campanula sibirica L. ;刺毛风铃草;Siberian Bellflower ■
70306 Campanula simplex Lam. ex DC. ;简单风铃草■☆
70307 Campanula sommieri Charadze;索米尔风铃草■☆
70308 Campanula sosnowskyi Charadze;锁斯诺夫斯基风铃草■☆
70309 Campanula speciosa Hornem. ;美丽风铃草;Pyrenean Bellflower ■☆
70310 Campanula speculum L. = Legousia speculum-veneris (L.) Chaix ●☆
70311 Campanula staintonii Rech. f. et Schiman-Czeika;斯坦顿风铃草 ■☆
70312 Campanula stellata Thunb. ;星状风铃草■☆
70313 Campanula stenanthina Ledeb. = Adenophora stenanthina (Ledeb.) Kitag. ■
70314 Campanula stevenii M. Bieb. ;史蒂文风铃草■☆
70315 Campanula striata Kitam. ;条纹风铃草■☆
70316 Campanula strigosa Vahl;硬尖风铃草■☆
70317 Campanula subcapitata Popov;亚头状风铃草■☆
70318 Campanula sulphurea Boiss. ;硫色风铃草■☆
70319 Campanula sylvatica Wall. ;林地风铃草■☆

70320 Campanula takhtadzhianii Fed. ;塔氏风铃草■☆
70321 Campanula talievii Juz. ;塔里风铃草■☆
70322 Campanula taurica Juz. ;克里木风铃草■☆
70323 Campanula tenella L. f. = Wahlenbergia tenella (L. f.) Lammers ■☆
70324 Campanula tenuissima Dunn;极细风铃草■☆
70325 Campanula tetraphylla Thunb. = Adenophora triphylla (Thunb.) A. DC. ■
70326 Campanula thalictrifolia Spreng. = Codonopsis thalictrifolia Wall. ■
70327 Campanula thomsonii Hook. f. = Asyneuma thomsonii (Hook. f.) Bornm. ■☆
70328 Campanula thyrsoides L. ;密花序风铃草;Yellow Bellflower ■☆
70329 Campanula tortuosa C. Y. Wu = Campanula cana Wall. ■
70330 Campanula trachelioides Munby = Campanula trachelium L. subsp. mauritanica (Pomel) Quézel ■☆
70331 Campanula trachelium L. ;荨麻叶风铃草(宽钟风铃草);Bat-in-the-belfry, Bats in the Belfry, Bats-in-the-belfry, Blue Foxglove, Canterbury Bells, Coral Bells, Coventry Bells, Haskwort, Mercury's Violet, Neckwort, Nettle-leaved Bellflower, Throatwort, Throat-wort, Uvula-wort ■☆
70332 Campanula trachelium L. 'Bernice';重瓣宽钟风铃草■☆
70333 Campanula trachelium L. subsp. genuina (Maire) Quézel = Campanula trachelium L. ■☆
70334 Campanula trachelium L. subsp. mauritanica (Pomel) Quézel;毛里塔尼亚风铃草■☆
70335 Campanula trachelium L. var. hirta Quézel = Campanula trachelium L. ■☆
70336 Campanula trachelium L. var. munbyana Quézel = Campanula trachelium L. ■☆
70337 Campanula trachelium L. var. parviflora Batt. = Campanula trachelium L. ■☆
70338 Campanula trautvetteri Grossh. ex Fed. ;特劳特风铃草■☆
70339 Campanula trichocalycina Ten. ;毛萼风铃草■☆
70340 Campanula tricuspidata Fisch. ex Roem. et Schult. = Adenophora tricuspidata (Fisch. ex Roem. et Schult.) A. DC. ■
70341 Campanula tridentata Schreb. ;三齿风铃草■☆
70342 Campanula triphylla Thunb. = Adenophora triphylla (Thunb.) A. DC. ■
70343 Campanula trista Kitam. ;暗淡风铃草■☆
70344 Campanula turbinata Schott = Campanula turbinata Schott, Nyman et Kotschy ■☆
70345 Campanula turbinata Schott, Nyman et Kotschy;陀螺风铃草■☆
70346 Campanula turczaninovii Fed. ;屠氏风铃草■☆
70347 Campanula uliginosa Rydb. ; 湿地蓝风铃草; Blue Marsh Bellflower, Blue Marsh-bellflower ■☆
70348 Campanula uliginosa Rydb. = Campanula aparinoides Pursh var. grandiflora Holz. ■☆
70349 Campanula undulata L. f. = Wahlenbergia undulata (L. f.) A. DC. ■☆
70350 Campanula unidentata L. f. = Wahlenbergia unidentata (L. f.) Lammers ■☆
70351 Campanula uniflora L. ; 单花风铃草; Arctic Bellflower, One-flowered Bellflower, One-flowered Bluebell, One-flowered Harebell ■☆
70352 Campanula uralensis Nevski ex B. Fedtsch. = Campanula wolgensis C. C. Davis ■☆
70353 Campanula vaidae Pénzes;瓦氏风铃草■☆
70354 Campanula vaillantii Quézel;瓦扬风铃草■☆
70355 Campanula velata Pomel;缘膜风铃草■☆
70356 Campanula velata Pomel subsp. serpylliformis (Batt. et Trab.) Quézel;百里香风铃草■☆
70357 Campanula velata Pomel var. rifana Maire = Campanula velata Pomel ■☆
70358 Campanula velutina Desf. = Campanula mollis L. ■☆
70359 Campanula veronicifolia Hance = Campanula canescens Wall. ex A. DC. ■
70360 Campanula veronicifolia Hance = Campanula dimorphantha Schweinf. ■
70361 Campanula verticillata Pall. = Adenophora tetraphylla (Thunb.) A. DC. ■
70362 Campanula vidalii H. C. Watson; 维达尔风铃草; Azores Bellflower ■☆
70363 Campanula vincaeflora Pau = Campanula lusitanica Loefl. ■☆
70364 Campanula volubilis Willd. ex Roem. et Schult. ;缠绕风铃草■☆
70365 Campanula waldsteiniana Roem. et Schult. ;瓦尔德风铃草■☆
70366 Campanula wolgensis C. C. Davis;乌拉尔风铃草;Ural Bellflower ■☆
70367 Campanula woronovii Charadze;沃氏风铃草■☆
70368 Campanula yunnanensis D. Y. Hong; 云南风铃草; Yunnan Bellflower ■
70369 Campanula zeyheri (H. Buek) D. Dietr. = Wahlenbergia krebsii Cham. ■☆
70370 Campanula zoysii Wulfen;瓶花风铃草;Crimped Bellflower ■☆
70371 Campanulaceae Adans. = Campanulaceae Juss. (保留科名)■●
70372 Campanulaceae Juss. (1789)(保留科名);桔梗科;Bellflower Family ■●
70373 Campanulastrum Small = Campanula L. ■●
70374 Campanulastrum americanum (L.) Small = Campanula americana L. ■☆
70375 Campanuloides A. DC. (1830);拟风铃草属■
70376 Campanuloides A. DC. = Lightfootia L'Hér. ■●
70377 Campanuloides A. DC. = Wahlenbergia Schrad. ex Roth(保留属名)■●
70378 Campanuloides Hort. Kew. ex A. DC. = Lightfootia L'Hér. ■●
70379 Campanuloides subulata A. DC. ;拟风铃草■☆
70380 Campanulopsis (Roberty) Roberty = Convolvulus L. ■●
70381 Campanulopsis Zoll. et Moritzi = Wahlenbergia Schrad. ex Roth (保留属名)■●
70382 Campanulorchis Brieger = Eria Lindl. (保留属名)■
70383 Campanulorchis Brieger(1981);钟兰属■
70384 Campanulorchis thao (Gagnep.) S. C. Chen et J. J. Wood;钟兰(石豆毛兰);Stonebean Eria, Stonebean Hairorchis ■
70385 Campanumoea Blume = Codonopsis Wall. ex Roxb. ■
70386 Campanumoea Blume(1826);金钱豹属;Campanumoea, Leopard ■
70387 Campanumoea axillaris Oliv. = Campanumoea lancifolia (Roxb.) Merr. ■
70388 Campanumoea axillaris Oliv. = Cyclocodon lancifolius (Roxb.) Kurz ■
70389 Campanumoea celebica Blume = Cyclocodon celebicus (Blume) D. Y. Hong ■
70390 Campanumoea cordata Maxim. = Campanumoea javanica Blume ■
70391 Campanumoea cordata Miq. = Campanumoea javanica Blume ■
70392 Campanumoea cordifolia Kom. = Campanumoea javanica Blume ■
70393 Campanumoea inflata (Hook. f.) C. B. Clarke;藏南金钱豹;Inflated Campanumoea, Inflated Leopard ●

70394 Campanumoea japonica Maxim. = Campanumoea javanica Blume ■

70395 Campanumoea javanica (Blume) Hook. f. = Campanumoea javanica Blume ■

70396 Campanumoea javanica Blume;大花金钱豹(白人参,白洋参,柴党参,川人参,对月参,浮瓶子,浮萍参,桂党参,孩儿葛,假丽参,金钱豹,金线吊葫芦,蔓桔梗,蔓人参,模登果,奶参,奶浆藤,南人参,人参薯,算盘果,土参,土党参,土人参,土沙参,土羊乳,土洋参,香浮参,香浮萍,野党参);Java Campanumoea,Java Leopard ■

70397 Campanumoea javanica Blume subsp. japonica (Makino) D. Y. Hong = Campanumoea javanica Blume var. japonica Makino ■

70398 Campanumoea javanica Blume var. japonica Makino;金钱豹(日本金钱豹,小花金钱豹,小花土党参);Japan Leopard ■

70399 Campanumoea javanica Blume var. japonica Makino = Campanumoea javanica Blume subsp. japonica (Makino) D. Y. Hong ■

70400 Campanumoea labordei H. Lév. = Campanumoea javanica Blume ■

70401 Campanumoea lanceolata Siebold et Zucc. = Codonopsis lanceolata (Siebold et Zucc.) Benth. et Hook. f. ■●

70402 Campanumoea lanceolata Siebold et Zucc. = Codonopsis lanceolata (Siebold et Zucc.) Trautv. ■

70403 Campanumoea lancifolia (Roxb.) Merr.;长叶轮钟草(红果参,披针叶金钱豹,肉算盘,沙参,山荸荠,台湾土党参,蜘蛛果);Longleaf Campanumoea,Longleaf Leopard ■

70404 Campanumoea lancifolia (Roxb.) Merr. = Cyclocodon lancifolius (Roxb.) Kurz ■

70405 Campanumoea lancifolia Roxb. = Campanumoea celebica Blume ■

70406 Campanumoea maximowiczii Honda = Campanumoea javanica Blume subsp. japonica (Makino) D. Y. Hong ■

70407 Campanumoea maximowiczii Honda = Campanumoea javanica Blume var. japonica Makino ■

70408 Campanumoea parviflora (Wall. ex A. DC.) Benth. et Hook. f.;小花轮钟草;Smallflower Campanumoea,Smallflower Leopard ■

70409 Campanumoea parviflora (Wall. ex A. DC.) Benth. et Hook. f. = Cyclocodon parviflorus (Wall. ex A. DC.) Hook. f. et Thomson ■

70410 Campanumoea pilosula Franch. = Codonopsis pilosula (Franch.) Nannf. ■

70411 Campanumoea truncata (Wall. ex A. DC.) Diels = Cyclocodon lancifolius (Roxb.) Kurz ■

70412 Campanumoea truncata (Wall.) Diels = Campanumoea lancifolia (Roxb.) Merr. ■

70413 Campanumoea violifolia H. Lév. = Codonopsis micrantha Chipp ■

70414 Campbellia Wight = Christisonia Gardner ■

70415 Campderia A. Rich. = Vellozia Vand. ■☆

70416 Campderia Benth. = Coccoloba P. Browne(保留属名)●

70417 Campderia Lag. = Kundmannia Scop. ■☆

70418 Campe Dulac = Barbarea W. T. Aiton(保留属名)■

70419 Campe barbarea (L.) W. Wight ex Piper = Barbarea vulgaris R. Br. ■

70420 Campe stricta (Andrz.) W. Wight ex Piper = Barbarea vulgaris R. Br. ■

70421 Campecarpus H. Wendl. (1921);曲果椰属(坎佩卡普椰属,密根柱椰属,曲果属)●☆

70422 Campecarpus H. Wendl. ex Becc. = Campecarpus H. Wendl. ●☆

70423 Campecarpus H. Wendl. ex Benth. et Hook. f. = Cyphophoenix H. Wendl. ex Benth. et Hook. f. ●☆

70424 Campecarpus fulcitus (Brongn.) Becc.;曲果椰(密根柱椰)●☆

70425 Campecia Adans. = Caesalpinia L. ●

70426 Campeiostachys Drobow = Elymus L. ■

70427 Campeiostachys schrenkiana (Fisch. et C. A. Mey.) Drobow = Elymus schrenkianus (Fisch. et C. A. Mey.) Tzvelev ■

70428 Campelepis Falc. = Periploca L. ●

70429 Campelia Kunth = Campella Link ■

70430 Campelia Kunth = Deschampsia P. Beauv. ■

70431 Campelia Rich. = Tradescantia L. ■

70432 Campella Link = Deschampsia P. Beauv. ■

70433 Campereia Engl. = Champereia Griff. ●

70434 Campesia Wight et Arn. ex Steud. = Galactia P. Browne ■

70435 Campestigma Pierre ex Costantin(1912);曲柱萝藦属☆

70436 Campestigma purpureum Pierre ex Costantin;曲柱萝藦☆

70437 Camphora Fabr. (废弃属名) = Cinnamomum Schaeff. (保留属名)●

70438 Camphora camphora (L.) H. Karst. = Cinnamomum camphora (L.) J. Presl ●

70439 Camphora glandulifera (Wall.) Nees = Cinnamomum glanduliferum (Wall.) Nees ●

70440 Camphora glandulifera Nees = Cinnamomum glanduliferum (Wall.) Nees ●

70441 Camphora officinarum Fabr. = Cinnamomum camphora (L.) J. Presl ●

70442 Camphora officinarum Nees = Cinnamomum camphora (L.) T. Nees et C. H. Eberm. ●

70443 Camphora officinarum Nees var. glaucescens A. Braun = Cinnamomum camphora (L.) T. Nees et C. H. Eberm. ●

70444 Camphora parthenoxylon (Jack) Nees = Cinnamomum parthenoxylum (Jack) Meisn. ●

70445 Camphora parthenoxylon Nees = Cinnamomum parthenoxylon Nees ●

70446 Camphora porrecta (Roxb.) Voigt = Cinnamomum parthenoxylum (Jack) Meisn. ●

70447 Camphora porrecta (Roxb.) Voigt = Cinnamomum porrectum (Roxb.) Kosterm. ●

70448 Camphora porrecta Voigt = Cinnamomum porrectum (Roxb.) Kosterm. ●

70449 Camphora porrecta Voigt = Cinnamomum parthenoxylum (Jack) Meisn. ●

70450 Camphorata Fabr. = Selago L. ●☆

70451 Camphorata Mill. = Camphorosma L. ●■

70452 Camphorata Tourn. ex Crantz = Camphorosma L. ●■

70453 Camphorata Zinn = Camphorosma L. ●■

70454 Camphorina Noronha = Desmos Lour. ●

70455 Camphoromoea Nees = Ocotea Aubl. ●☆

70456 Camphoromoea Nees ex Meisn. = Ocotea Aubl. ●☆

70457 Camphoromyrtus Schauer = Baeckea L. ●

70458 Camphoromyrtus Schauer = Triplarina Raf. ●

70459 Camphoropsis Moq. ex Pfeiff. = Nanophyton Less. ●■

70460 Camphorosma L. (1753);樟味藜属;Camphorfume,Camphorfume,Stink Groundpine ●■

70461 Camphorosma annuua Pall.;一年生樟味藜;Annual Stink Groundpine ●☆

70462 Camphorosma lessingii Litv. = Camphorosma monspeliaca L. subsp. lessingii (Litv.) Aellen ●

70463 Camphorosma monspeliaca L.;樟味藜(蒙山樟味藜,樟臭草);Eurasia Stink Groundpine,Mediterranean Camphorfume ●

70464 Camphorosma monspeliaca L. subsp. lessingii (Litv.) Aellen;同

齿樟味藜(累氏樟味藜,李氏樟味藜);Lessing Mediterranean Camphorfume, Mediterranea Stink Groundpine ●

70465 Camphorosma ruthenica M. Bieb. = Camphorosma monspeliaca L. ●

70466 Camphorosma songorica Bunge;准噶尔樟味藜;Dzungar Stink Groundpine ●☆

70467 Camphusia de Vriese = Scaevola L.(保留属名)●■

70468 Camphyleia Spreng. = Campuleia Thouars ■

70469 Camphyleia Spreng. = Striga Lour. ■

70470 Campia Dombey ex Endl. = Capia Dombey ex Juss. ●☆

70471 Campia Dombey ex Endl. = Lapageria Ruiz et Pav. ●☆

70472 Campilostachys A. Juss. = Campylostachys Kunth ●☆

70473 Campimia Ridl.(1911);南洋野牡丹属●☆

70474 Campimia wrayi (King) Ridl.;南洋野牡丹●☆

70475 Campnosperma Thwaites(1854)(保留属名);曲籽漆属;Tigasco Oil ●☆

70476 Campnosperma auriculatum Hook. f.;耳状籽漆●☆

70477 Campnosperma brevipetiolatum Volk;短柄籽漆●☆

70478 Campnosperma coriaceum (Jack) Steenis;革质籽漆●☆

70479 Campnosperma gummiferum Marchand;树胶籽漆●☆

70480 Campnosperma lepidotum Capuron ex Randrian. et J. S. Mill.;鳞籽漆●☆

70481 Campnosperma micranteium Marchand;小籽漆●☆

70482 Campnosperma panamense Standl.;巴拿马籽漆●☆

70483 Campnosperma parvifolium Capuron ex J. S. Mill. et Randrian.;小花籽漆●☆

70484 Campnosperma schatzii Randrian. et J. S. Mill.;沙茨籽漆●☆

70485 Campnosperma zacharyi Randrian. et Lowry;扎卡里籽漆●☆

70486 Campnosperma zeylanicum Thwaites;斯里兰卡籽漆●☆

70487 Campoearpus Post et Kuntze = Campecarpus H. Wendl. ●☆

70488 Campolepis Post et Kuntze = Campelepis Falc. ●

70489 Campolepis Post et Kuntze = Periploca L. ●

70490 Campomanesia Ruiz et Pav.(1794);坎波木属;Para Guava ●☆

70491 Campomanesia aromatica (Aubl.) Griseb.;芳香坎波木●☆

70492 Campomanesia grandiflora Sagot;大花坎波木●☆

70493 Campovassouria R. M. King et H. Rob.(1971);显脉泽兰属■☆

70494 Campovassouria bupleurifolia (DC.) R. M. King et H. Rob.;显脉泽兰■☆

70495 Campsanthus Steud. = Compsanthus Spreng. ■

70496 Campsanthus Steud. = Tricyrtis Wall.(保留属名)■

70497 Campsiandra Benth.(1840);弯蕊豆属(卡姆苏木属,弯花属)●☆

70498 Campsiandra comosa Benth.;丛毛弯蕊豆(丛毛弯花)●☆

70499 Campsiandra laurifolia Benth.;月桂叶弯蕊豆(月桂叶卡姆苏木,月桂叶弯花)●☆

70500 Campsidium Seem.(1862);小凌霄花属●☆

70501 Campsidium valdivianum (Phil.) Bull;小凌霄花●☆

70502 Campsis Lour.(1790)(保留属名);凌霄花属(凌霄属,紫葳属);Trumpet Creeper, Trumpet Vine, Trumpetcreeper, Trumpet-creeper ●

70503 Campsis × tagliabuana (Vis.) Rehder;塔利亚布凌霄花;Rumpet Creeper ●☆

70504 Campsis adrepens Lour. = Campsis grandiflora (Thunb.) K. Schum. ●

70505 Campsis chinensis (Lam.) Voss = Campsis grandiflora (Thunb.) K. Schum. ●

70506 Campsis fortunei Seem. = Paulownia fortunei (Seem.) Hemsl. ●

70507 Campsis grandiflora (Thunb.) K. Schum.;凌霄花(白狗肠,白华芰,傍墙花,吹风亭,此葳花,倒挂金钟,吊墙花,杜灵霄,堕胎花,鬼目,过路蜈蚣,红花倒水莲,黄花蕈,黄花凌霄,接骨丹,接骨风,九龙下海,九重藤,瞿陵,凌霄,凌霄藤,陵居腹,陵苕,陵苕,陵时花,陵霄,陵召,马捕花,女葳,女葳花,上树龙,上树蜈蚣,苕花,苕华,搜骨风,碎骨风,藤罗,藤罗草,藤五加,五爪龙,武威,芰华,争墙风,追风箭,追罗,紫葳,钻方风);China Trumpetcreeper, Chinese Trumpet Creeper, Chinese Trumpet Flower, Chinese Trumpet Vine, Chinese Trumpetcreeper, Chinese Trumpet-creeper, Chinese Trumpet-vine, Trumpet Vine ●

70508 Campsis grandiflora (Thunb.) K. Schum. = Campsis grandiflora (Thunb.) Loisel. ●

70509 Campsis grandiflora (Thunb.) K. Schum. f. thunbergii (Carrière) Rehder;通贝里凌霄(通氏凌霄,童氏凌霄)●☆

70510 Campsis grandiflora (Thunb.) Loisel. = Campsis grandiflora (Thunb.) K. Schum. ●

70511 Campsis radicans (L.) Bureau = Campsis radicans (L.) Seem. ●

70512 Campsis radicans (L.) Seem.;厚萼凌霄(杜凌霄,辐射黄钟花,凌霄花,美国凌霄,美国紫葳,美洲凌霄,美洲凌霄花,硬骨凌霄);America Trumpetcreeper, American Trumpet Creeper, Common Trumpet Creeper, Common Trumpetcreeper, Common Trumpet-creeper, Cow-Itch, Devil's Shoelaces, Hell Vine, Rooting, Shoestrings, Trumpet Climber, Trumpet Creeper, Trumpet Honeysuckle, Trumpet Vine, Trumpet-climber, Trumpetcreeper, Trumpet-creeper, Trumpet-flower, Trumpet-honeysuckle, Trumpet-vine ●

70513 Campsis radicans (L.) Seem. 'Flava';黄厚萼凌霄;Yellow Trumpet ●☆

70514 Campsis radicans (L.) Seem. ex Bureau = Campsis radicans (L.) Seem. ●

70515 Campsoneura T. Durand et Jacks. = Compsoneura (A. DC.) Warb. ●☆

70516 Camptacra N. T. Burb.(1982);根茎层菀属■☆

70517 Camptacra barbata N. T. Burb.;根茎层菀■☆

70518 Camptandra Ridl.(1899);曲蕊姜属(弯蕊花属)■

70519 Camptandra angustifolia Ridl.;窄叶曲蕊姜■☆

70520 Camptandra fongyuensis (Gagnep.) K. Schum. = Pyrgophyllum yunnanense (Gagnep.) T. L. Wu et Z. Y. Chen ■

70521 Camptandra latifolia Ridl.;宽叶曲蕊姜■☆

70522 Camptandra ovata Ridl.;卵叶曲蕊姜■☆

70523 Camptederia Steud. = Campderia A. Rich. ■☆

70524 Camptederia Steud. = Vellozia Vand. ■☆

70525 Camptocarpus Decne.(1844)(保留属名);弯果萝藦属●■☆

70526 Camptocarpus K. Koch(废弃属名) = Alkanna Tausch(保留属名)●☆

70527 Camptocarpus K. Koch(废弃属名) = Camptocarpus Decne.(保留属名)●■☆

70528 Camptocarpus K. Koch(废弃属名) = Oskampia Moench ●■☆

70529 Camptocarpus acuminatus (Choux) Venter;渐尖弯果萝藦●☆

70530 Camptocarpus bojeri Jum. et H. Perrier = Camptocarpus mauritianus (Lam.) Decne. ●☆

70531 Camptocarpus bojerianus Decne. = Camptocarpus mauritianus (Lam.) Decne. ●☆

70532 Camptocarpus cornutus Klack.;角状弯果萝藦●☆

70533 Camptocarpus crassifolius Decne.;厚叶弯果萝藦●☆

70534 Camptocarpus decaryi (Choux) Venter;德卡里弯果萝藦●☆

70535 Camptocarpus lanceolatus Klack.;剑叶弯果萝藦●☆

70536 Camptocarpus linearis Decne.;线叶弯果萝藦●☆

70537 Camptocarpus madagascariensis (Schltr.) Venter =

Camptocarpus crassifolius Decne. ●☆

70538 Camptocarpus mauritianus (Lam.) Decne.;毛里求斯弯果萝藦●☆

70539 Camptocarpus mauritianus (Lam.) Decne. var. madagascariensis Costantin et Gallaud = Camptocarpus mauritianus (Lam.) Decne. ●☆

70540 Camptocarpus semihastatus Klack.;半戟弯果萝藦●☆

70541 Camptolepis Radlk. (1907);弯鳞无患子属●☆

70542 Camptolepis crassifolia Capuron;糙叶弯鳞无患子●☆

70543 Camptolepis grandiflora Capuron;大花弯鳞无患子●☆

70544 Camptolepis hygrophila Capuron;喜水弯鳞无患子●☆

70545 Camptolepis ramiflora (Taub.) Radlk.;非洲弯鳞无患子●☆

70546 Camptoloma Benth. (1846);弯边玄参属■●☆

70547 Camptoloma lyperiiflorum (Vatke) Hilliard;弯边玄参■●☆

70548 Camptoloma oxypetalum Wagner et Vierh. = Camptoloma lyperiiflorum (Vatke) Hilliard ■●☆

70549 Camptoloma rotundifolia Benth.;圆叶弯边玄参■☆

70550 Camptoloma villosa Balf. f. = Camptoloma lyperiiflorum (Vatke) Hilliard ■●☆

70551 Camptophytum Pierre ex A. Chev. = Tarenna Gaertn. ●

70552 Camptophytum klaineanum Pierre ex A. Chev. = Tarenna eketensis Wernham ●☆

70553 Camptopus Hook. f. = Psychotria L. (保留属名)●

70554 Camptopus densinervia K. Krause = Psychotria densinervia (K. Krause) Verdc. ●☆

70555 Camptopus goetzei (K. Schum.) K. Krause = Psychotria megalopus Verdc. ●☆

70556 Camptopus mannii Hook. f. = Psychotria camptopus Verdc. ●☆

70557 Camptorrhiza Hutch. (1934);弯根秋水仙属■☆

70558 Camptorrhiza flexuosa (Baker) Sterling;弯根秋水仙■☆

70559 Camptorrhiza junodii (Schinz) Sterling;朱诺德弯根秋水仙■☆

70560 Camptorrhiza strumosa (Baker) Oberm.;多疣秋水仙■☆

70561 Camptosema Hook. et Arn. (1833);曲藤豆属■☆

70562 Camptosema rubicundum Hook. et Arn.;曲藤豆■☆

70563 Camptostemon Mast. (1872);曲蕊木棉属●☆

70564 Camptostemon schultzii Mast.;曲蕊木棉●☆

70565 Camptostylus Gilg(1898);弯柱大风子属●☆

70566 Camptostylus aristatus (Oliv.) Gilg = Oncoba mannii Oliv. ●☆

70567 Camptostylus caudatus Gilg;弯柱大风子●☆

70568 Camptostylus caudatus Gilg = Oncoba ovalis Oliv. ●☆

70569 Camptostylus kivuensis Bamps = Oncoba kivuensis (Bamps) Hul et Breteler ●☆

70570 Camptostylus litoralis Gilg = Oncoba mannii Oliv. ●☆

70571 Camptostylus mannii (Oliv.) Gilg = Oncoba mannii Oliv. ●☆

70572 Camptostylus ovalis (Oliv.) Chipp = Oncoba ovalis Oliv. ●☆

70573 Camptostylus petiolaris (Pierre) Gilg = Oncoba mannii Oliv. ●☆

70574 Camptotheca Decne. (1873);喜树属(旱莲木属,旱莲属);Camptotheca ●★

70575 Camptotheca acuminata Decne.;喜树(旱莲,旱莲木,南京梧桐,千张树,千丈树,水白杂,水栗,水栗子,水漠子,水桐树,忑芭蕉,天梓树,野芭蕉);Camptotheca,Common Camptotheca ●

70576 Camptotheca acuminata Decne. var. rotundifolia B. M. Yang et L. D. Duan = Camptotheca acuminata Decne. ●

70577 Camptotheca acuminata Decne. var. tenuifolia W. P. Fang et Soong;薄叶喜树;Thinleaf Camptotheca ●

70578 Camptotheca acuminata Decne. var. tenuifolia W. P. Fang et Soong = Camptotheca acuminata Decne. ●

70579 Camptotheca lowreyana S. Y. Li;洛氏喜树●

70580 Camptotheca yunnanensis Dode = Camptotheca acuminata Decne. ●

70581 Camptouratea Tiegh. = Ouratea Aubl. (保留属名)●

70582 Campuleia Thouars = Striga Lour. ■

70583 Campuleia coccinea Hook. = Striga asiatica (L.) Kuntze ■

70584 Campuloa Desv. = Ctenium Panz. (保留属名)■☆

70585 Campuloa hirsuta Desv. = Harpochloa falx (L. f.) Kuntze ■☆

70586 Campuloclinium DC. (1836);大头柄泽兰属■●☆

70587 Campuloclinium DC. = Eupatorium L. ■●

70588 Campuloclinium macrocephalum (Less.) DC.;大头柄泽兰■☆

70589 Campulosus Desv. (废弃属名) = Ctenium Panz. (保留属名)■☆

70590 Campulosus falcatus (L. f.) P. Beauv. = Harpochloa falx (L. f.) Kuntze ■☆

70591 Campydorum Salisb. = Polygonatum Mill. ■

70592 Campylandra Baker = Tupistra Ker Gawl. ■

70593 Campylandra Baker(1875);开口箭属(扁竹枝属);Tupistra ■

70594 Campylandra annulata (H. Li et J. L. Huang) M. N. Tamura, S. Yun Liang et N. J. Turland;环花开口箭;Annulate Tupistra ■

70595 Campylandra annulata (H. Li et J. L. Huang) M. N. Tamura, S. Yun Liang et N. J. Turland = Tupistra annulata H. Li et J. L. Huang ■

70596 Campylandra aurantiaca Wall. ex Baker;橙花开口箭(黄花开口箭);Orangeflower Tupistra ■

70597 Campylandra aurantiaca Wall. ex Baker = Tupistra aurantiaca Wall. ex Baker ■

70598 Campylandra cauliflora Chun = Campylandra wattii C. B. Clarke ■

70599 Campylandra cauliflora Chun = Tupistra wattii (C. B. Clarke) Hook. f. ■

70600 Campylandra chinensis (Baker) M. N. Tamura, S. Yun Liang et N. J. Turland;开口箭(贺县开口箭,开喉箭,老蛇莲,牛尾七,山萝卜,疏花开口箭,万年攀,万年青,心不干,岩七,斩蛇剑,竹根参,竹根七,竹节七);China Tupistra, Chinese Tupistra, Hexian Tupistra, Laxflower Tupistra ■

70601 Campylandra chinensis (Baker) M. N. Tamura, S. Yun Liang et N. J. Turland = Tupistra chinensis Baker ■

70602 Campylandra delavayi (Franch.) M. N. Tamura, S. Yun Liang et N. J. Turland;筒花开口箭;Delavay Tupistra, Tubeflower Tupistra ■

70603 Campylandra delavayi (Franch.) M. N. Tamura, S. Yun Liang et N. J. Turland = Tupistra delavayi Franch. ■

70604 Campylandra emeiensis (Z. Y. Zhu) M. N. Tamura, S. Yun Liang et N. J. Turland;峨眉开口箭;Emei Tupistra ■

70605 Campylandra ensifolia (F. T. Wang et Ts. Tang) M. N. Tamura, S. Yun Liang et N. J. Turland;剑叶开口箭(白马分宗,搜山虎,小万年青,岩七,云南铁扁担,竹节七);Sword Tupistra, Swordleaf Tupistra ■

70606 Campylandra ensifolia (F. T. Wang et Ts. Tang) M. N. Tamura, S. Yun Liang et N. J. Turland = Tupistra ensifolia F. T. Wang et Ts. Tang ■

70607 Campylandra fimbriata (Hand. -Mazz.) M. N. Tamura, S. Yun Liang et N. J. Turland;齿瓣开口箭(短瓣开口箭,开喉箭,流苏开口箭,万年青,须瓣开口箭,云南铁扁担,竹根七);Fimbriate Tupistra, Shortlobed Fimbriate Tupistra ■

70608 Campylandra fimbriata (Hand. -Mazz.) M. N. Tamura, S. Yun Liang et N. J. Turland = Tupistra fimbriata Hand. -Mazz. ■

70609 Campylandra jinshanensis (Z. L. Yang et X. G. Luo) M. N. Tamura, S. Yun Liang et N. J. Turland;金山开口箭;Jinshan Tupistra ■

70610 Campylandra jinshanensis (Z. L. Yang et X. G. Luo) M. N. Tamura, S. Yun Liang et N. J. Turland = Tupistra jinshanensis Z. L. Yang et X. G. Luo ■

70611 Campylandra kwangtungensis Dandy = Campylandra chinensis (Baker) M. N. Tamura, S. Yun Liang et N. J. Turland ■

70612 Campylandra liangshanensis (Z. Y. Zhu) M. N. Tamura, S. Yun Liang et N. J. Turland;凉山开口箭;Liangshan Tupistra ■

70613 Campylandra liangshanensis (Z. Y. Zhu) M. N. Tamura, S. Yun Liang et N. J. Turland = Tupistra liangshanensis Z. Y. Zhu ■

70614 Campylandra lichuanensis (Y. K. Yang, J. K. Wu et D. T. Peng) M. N. Tamura, S. Yun Liang et N. J. Turland;利川开口箭;Lichuan Tupistra ■

70615 Campylandra lichuanensis (Y. K. Yang, J. K. Wu et D. T. Peng) M. N. Tamura, S. Yun Liang et N. J. Turland = Tupistra lichuanensis Y. K. Yang, J. K. Wu et D. T. Peng ■

70616 Campylandra longibracteata F. T. Wang et Ts. Tang = Campylandra wattii C. B. Clarke ■

70617 Campylandra longibracteata F. T. Wang et Ts. Tang = Tupistra wattii (C. B. Clarke) Hook. f. ■

70618 Campylandra longipedunculata (F. T. Wang et S. Ye Liang) M. N. Tamura, S. Yun Liang et N. J. Turland;长梗开口箭;Longpedicel Tupistra ■

70619 Campylandra longipedunculata (F. T. Wang et S. Ye Liang) M. N. Tamura, S. Yun Liang et N. J. Turland = Tupistra longipedunculata F. T. Wang et S. Yun Liang ■

70620 Campylandra longispica Y. T. Wan et X. H. Lu = Tupistra longispica Y. Wan et X. H. Lu ■

70621 Campylandra lovangtungneis Dandy = Campylandra chinensis (Baker) M. N. Tamura, S. Yun Liang et N. J. Turland ■

70622 Campylandra lovangtungneis Dandy = Tupistra chinensis Baker ■

70623 Campylandra pachynema F. T. Wang et Ts. Tang = Tupistra chinensis Baker ■

70624 Campylandra tui (F. T. Wang et Ts. Tang) M. N. Tamura, S. Yun Liang et N. J. Turland;碟花开口箭;Dishflower Tupistra ■

70625 Campylandra tui (F. T. Wang et Ts. Tang) M. N. Tamura, S. Yun Liang et N. J. Turland = Tupistra tui (F. T. Wang et Ts. Tang) F. T. Wang et S. Yun Liang ■

70626 Campylandra urotepala (Hand. -Mazz.) M. N. Tamura, S. Yun Liang et N. J. Turland;尾萼开口箭;Tailsepal Tupistra ■

70627 Campylandra urotepala (Hand. -Mazz.) M. N. Tamura, S. Yun Liang et N. J. Turland = Tupistra urotepala (Hand. -Mazz.) F. T. Wang et Ts. Tang ■

70628 Campylandra verruculosa (Q. H. Chen) M. N. Tamura, S. Yun Liang et N. J. Turland;疣点开口箭;Verruculose Tupistra ■

70629 Campylandra verruculosa (Q. H. Chen) M. N. Tamura, S. Yun Liang et N. J. Turland = Tupistra verruculosa Q. H. Chen ■

70630 Campylandra viridiflora (Franch.) Hand. -Mazz. = Campylandra chinensis (Baker) M. N. Tamura, S. Yun Liang et N. J. Turland ■

70631 Campylandra viridiflora (Franch.) Hand. -Mazz. = Tupistra chinensis Baker ■

70632 Campylandra watanabei (Hayata) Dandy = Campylandra chinensis (Baker) M. N. Tamura, S. Yun Liang et N. J. Turland ■

70633 Campylandra watanabei (Hayata) Dandy = Tupistra chinensis Baker ■

70634 Campylandra watanabei (Hayata) Dandy = Tupistra watanabei (Hayata) F. T. Wang et C. F. Liang ■

70635 Campylandra wattii C. B. Clarke;弯蕊开口箭(白跌打,扁竹兰,柄叶开口箭,柄叶竹根七,见血封口,兰花草,老蛇莲,石风丹,石枫丹,岩七,竹叶兰);Watt Tupistra ■

70636 Campylandra wattii C. B. Clarke = Tupistra wattii (C. B. Clarke) Hook. f. ■

70637 Campylandra yunnanensis (F. T. Wang et S. Yun Liang) M. N. Tamura, S. Yun Liang et N. J. Turland;云南开口箭;Yunnan Tupistra ■

70638 Campylandra yunnanensis (F. T. Wang et S. Yun Liang) M. N. Tamura, S. Yun Liang et N. J. Turland = Tupistra yunnanensis F. T. Wang et S. Yun Liang ■

70639 Campylanthera Hook. (1837);弯药海桐属●☆

70640 Campylanthera Hook. = Pronaya Hügel ex Endl. ●☆

70641 Campylanthera Hook. = Spiranthera Hook. ●☆

70642 Campylanthera Schott et Endl. = Ceiba Mill. ●

70643 Campylanthera Schott et Endl. = Eriodendron DC. ●

70644 Campylanthus Roth(1821);弯花婆婆纳属(弯花玄参属)●☆

70645 Campylanthus anisotrichus (A. G. Mill.) Hjertson et A. G. Mill.;异毛弯花婆婆纳●☆

70646 Campylanthus benthamii Webb var. hirsutus = Campylanthus glaber Benth. ●☆

70647 Campylanthus glaber Benth.;光弯花婆婆纳●☆

70648 Campylanthus glaber Benth. subsp. spathulatus (A. Chev.) Brochmann et al.;匙光弯花婆婆纳●☆

70649 Campylanthus glaber Benth. var. puberulus Cout. = Campylanthus glaber Benth. subsp. spathulatus (A. Chev.) Brochmann et al. ●☆

70650 Campylanthus glaber Benth. var. pumilus Pett. = Campylanthus glaber Benth. ●☆

70651 Campylanthus incanus A. G. Mill.;灰毛弯花婆婆纳●☆

70652 Campylanthus incanus A. G. Mill. var. anisotrichus A. G. Mill. = Campylanthus anisotrichus (A. G. Mill.) Hjertson et A. G. Mill. ●☆

70653 Campylanthus junceus Edgew.;灯心草弯花婆婆纳●☆

70654 Campylanthus monicae Dobignard;念珠弯花婆婆纳●☆

70655 Campylanthus parviflorus Hjertson et A. G. Mill.;小弯花婆婆纳●☆

70656 Campylanthus reconditus Hjertson et Thulin;隐蔽弯花婆婆纳●☆

70657 Campylanthus salsoloides (L. f.) Roth;猪毛菜婆婆纳●☆

70658 Campylanthus salsoloides (L. f.) Roth var. leucantha Svent. = Campylanthus salsoloides (L. f.) Roth ●☆

70659 Campylanthus somaliensis A. G. Mill.;索马里婆婆纳●☆

70660 Campylanthus spathulatus A. Chev. = Campylanthus glaber Benth. subsp. spathulatus (A. Chev.) Brochmann et al. ●☆

70661 Campylanthus spinosus Balf. f.;多刺婆婆纳●☆

70662 Campyleia Spreng. = Campuleia Thouars ■

70663 Campyleia Spreng. = Striga Lour. ■

70664 Campyleja Post et Kuntze = Campyleia Spreng. ■

70665 Campylia Lindl. ex Sweet = Pelargonium L'Hér. ex Aiton ●■

70666 Campylia veronicifolia Eckl. et Zeyh. = Pelargonium ovale (Burm. f.) L'Hér. subsp. veronicifolium (Eckl. et Zeyh.) Hugo ■☆

70667 Campylobotrys Lem. = Hoffmannia Sw. ●■☆

70668 Campylocaryum DC. ex A. DC. = Alkanna Tausch(保留属名)●☆

70669 Campylocaryum DC. ex Meisn. = Alkanna Tausch(保留属名)●☆

70670 Campylocentron Benth. et Hook. f. = Campylocentrum Benth. ■☆

70671 Campylocentrum Benth. (1881);弯唇兰属■☆

70672 Campylocentrum pachyrrhizum (Rchb. f.) Rolfe;粗根弯唇兰;Leafless Bentspur Orchid ■☆

70673 Campylocentrum porrectum (Rchb. f.) Rolfe = Harrisella porrecta (Rchb. f.) Fawc. et Rendle ■☆

70674 Campylocera Nutt. = Legousia T. Durand ●■☆

70675 Campylocera Nutt. = Triodanis Raf. ■

70676 Campylocercum Tiegh. = Campylospermum Tiegh. ●

70677 Campylocercum Tiegh. = Ouratea Aubl.(保留属名)●
70678 Campylocercum striatum Tiegh. = Campylospermum striatum (Tiegh.) M. C. E. Amaral ●
70679 Campylochinium B. D. Jacks. = Campyloclinium Endl. ■●
70680 Campylochinium B. D. Jacks. = Eupatorium L. ■●
70681 Campylochinium Endl. = Campuloclinium DC. ■●☆
70682 Campylochinium Endl. = Eupatorium L. ■●
70683 Campylochiton Welw. ex Hiern = Combretum Loefl.(保留属名)●
70684 Campylochiton platypterus (Welw.) Hiern = Combretum platypterum (Welw.) Hutch. et Dalziel ●☆
70685 Campylochnella Tiegh. = Ochna L. ●
70686 Campylochnella angustifolia (Engl. et Gilg) Tiegh. = Brackenridgea arenaria (De Wild. et T. Durand) N. Robson ●☆
70687 Campylochnella arenaria (De Wild. et T. Durand) Tiegh. = Brackenridgea arenaria (De Wild. et T. Durand) N. Robson ●☆
70688 Campylochnella katangensis (De Wild.) Tiegh. = Ochna katangensis De Wild. ●☆
70689 Campylochnella pungens Tiegh. = Brackenridgea arenaria (De Wild. et T. Durand) N. Robson ●☆
70690 Campylochnella roseiflora (Engl. et Gilg) Tiegh. = Brackenridgea arenaria (De Wild. et T. Durand) N. Robson ●☆
70691 Campylochnella thollonii Tiegh. = Brackenridgea arenaria (De Wild. et T. Durand) N. Robson ●☆
70692 Campyloclinium Endl. = Campuloclinium DC. ■●☆
70693 Campyloclinium Endl. = Eupatorium L. ■●
70694 Campylogyne Welw. ex Hemsl. = Combretum Loefl.(保留属名)●
70695 Campylogyne Welw. ex Hemsl. = Quisqualis L. ●
70696 Campylonema Poir. = Campynema Labill. ■☆
70697 Campylonema Schult. et Schult. f. = Campynema Labill. ■☆
70698 Campylopelma Rchb. = Hypericum L. ■●
70699 Campylopetalum Forman(1954);弯瓣木属●☆
70700 Campylopetalum siamense Forman;弯瓣木●☆
70701 Campylopora Tiegh. = Brackenridgea A. Gray ●☆
70702 Campyloptera Boiss. (1841);弯翅芥属■☆
70703 Campyloptera Boiss. = Aethionema R. Br. ■☆
70704 Campyloptera carnea (Banks et Sol.) Botsch. et Vved.;肉色弯翅芥■☆
70705 Campyloptera heterocarpa Baill.;异果弯翅芥■☆
70706 Campyloptera syriaca Boiss.;叙利亚弯翅芥■☆
70707 Campylopus Spach(1836) = Campylopelma Rchb. ■●
70708 Campylopus Spach(1836) = Hypericum L. ■●
70709 Campylosiphon Benth. (1882);弯管水玉簪属■☆
70710 Campylosiphon St. -Lag. = Siphocampylus Pohl ■●☆
70711 Campylosiphon purpurascens Benth.;弯管水玉簪■☆
70712 Campylospermum Tiegh. (1902);赛金莲木属(奥里木属);Gomphia ●
70713 Campylospermum Tiegh. = Gomphia Schreb. ●
70714 Campylospermum affine (Hook. f.) Tiegh. = Rhabdophyllum affine (Hook. f.) Tiegh. ●☆
70715 Campylospermum amplectens (Stapf) Farron;环抱赛金莲木●☆
70716 Campylospermum bukobense (Gilg) Farron;布科巴赛金莲木●☆
70717 Campylospermum cabrae (Gilg) Farron;卡布拉赛金莲木●☆
70718 Campylospermum calanthum (Gilg) Farron;美花赛金莲木●☆
70719 Campylospermum calophyllum (Hook. f.) Tiegh. = Rhabdophyllum calophyllum (Hook. f.) Tiegh. ●☆
70720 Campylospermum claessensii (De Wild.) Farron;克莱森斯赛金莲木●☆
70721 Campylospermum congestum (Oliv.) Farron;密集赛金莲木●☆
70722 Campylospermum densiflorum (De Wild. et T. Durand) Farron;密花赛金莲木●☆
70723 Campylospermum descoingsii Farron;德斯赛金莲木●☆
70724 Campylospermum duparquetianum (Baill.) Tiegh.;迪帕赛金莲木●☆
70725 Campylospermum dybowskii Tiegh.;迪布赛金莲木●☆
70726 Campylospermum elongatum (Oliv.) Tiegh.;伸长赛金莲木●☆
70727 Campylospermum excavatum (Tiegh.) Farron;凹陷赛金莲木●☆
70728 Campylospermum flavum (Schumach. et Thonn.) Farron;黄赛金莲木●☆
70729 Campylospermum glaberrimum (P. Beauv.) Farron;光滑赛金莲木●☆
70730 Campylospermum glaucum (Tiegh.) Farron;灰绿赛金莲木●☆
70731 Campylospermum hiernii (Tiegh.) Exell;希尔恩赛金莲木●☆
70732 Campylospermum katangense Farron;加丹加赛金莲木●☆
70733 Campylospermum klainei (Tiegh.) Farron;克莱恩赛金莲木●☆
70734 Campylospermum laeve (De Wild. et T. Durand) Farron;平滑赛金莲木●☆
70735 Campylospermum laxiflorum (De Wild. et T. Durand) Tiegh.;疏花赛金莲木●☆
70736 Campylospermum lecomtei (Tiegh.) Farron;勒孔特赛金莲木●☆
70737 Campylospermum letouzeyi Farron;勒图赛金莲木●☆
70738 Campylospermum mannii (Oliv.) Tiegh.;曼氏赛金莲木●☆
70739 Campylospermum monticola (Gilg) Cheek;山生赛金莲木●☆
70740 Campylospermum oliveri (Tiegh.) Farron;奥氏赛金莲木●☆
70741 Campylospermum oliverianum (Gilg) Farron;奥里弗赛金莲木●☆
70742 Campylospermum paucinervatum Sosef;少脉赛金莲木●☆
70743 Campylospermum reticulatum (P. Beauv.) Farron;网状赛金莲木●☆
70744 Campylospermum reticulatum (P. Beauv.) Farron var. turnerae (Hook. f.) Farron;图尔网状赛金莲木●☆
70745 Campylospermum sacleuxii (Tiegh.) Farron;萨克勒赛金莲木●☆
70746 Campylospermum scheffleri (Engl. et Gilg) Farron;谢夫勒赛金莲木●☆
70747 Campylospermum schoenleinianum (Klotzsch) Farron;舍恩赛金莲木●☆
70748 Campylospermum serratum (Gaertn.) Bittrich et M. C. E. Amaral;齿叶赛金莲木(裂瓣奥里木,裂瓣赛金莲木);Serrate Gomphia,Toothleaf Gomphia ●
70749 Campylospermum squamosum (DC.) Farron;多鳞赛金莲木●☆
70750 Campylospermum striatum (Tiegh.) M. C. E. Amaral;赛金莲木(奥里木);Gomphia,Striped Gomphia ●
70751 Campylospermum strictum (Tiegh.) Farron;刚直赛金莲木●☆
70752 Campylospermum subcordatum (Stapf) Farron;亚心形赛金莲木●☆
70753 Campylospermum sulcatum (Tiegh.) Farron;纵沟赛金莲木●☆
70754 Campylospermum umbricola (Tiegh.) Farron;荫地赛金莲木●☆
70755 Campylospermum vogelii (Hook. f.) Farron;沃氏赛金莲木●☆
70756 Campylospermum vogelii (Hook. f.) Farron var. angustifolium (Engl.) Farron;窄叶沃氏赛金莲木●■☆
70757 Campylospermum vogelii (Hook. f.) Farron var. costatum (Tiegh.) Farron;单脉沃氏赛金莲木●☆
70758 Campylospermum vogelii (Hook. f.) Farron var. molleri (Tiegh.) Farron;柔软沃氏赛金莲木●☆
70759 Campylospermum vogelii (Hook. f.) Farron var. poggei (Engl.) Farron;波格赛金莲木●☆

70760 Campylospermum zenkeri (Engl. ex Tiegh.) Farron;岑克尔赛金莲木●☆
70761 Campylosporua Spach = Hypericum L. ■●
70762 Campylostachys E. Mey. = Fimbristylis Vahl(保留属名)■
70763 Campylostachys Kunth(1832);弯穗木属●☆
70764 Campylostachys abbreviata E. Mey. = Euthystachys abbreviata (E. Mey.) A. DC. ●☆
70765 Campylostachys cernua (L. f.) Kunth;弯穗木●☆
70766 Campylostachys phylicoides Sond. = Kogelbergia phylicoides (A. DC.) Rourke ●☆
70767 Campylostemon E. Mey. = Justicia L. ●■
70768 Campylostemon Erdtman = Campylopetalum Forman ●☆
70769 Campylostemon Welw. = Campylostemon Welw. ex Benth. et Hook. f. ●☆
70770 Campylostemon Welw. ex Benth. et Hook. f. (1867);曲蕊卫矛属●☆
70771 Campylostemon angolensis Welw. ex Oliv.;曲蕊卫矛●☆
70772 Campylostemon bequaertii De Wild.;贝卡尔曲蕊卫矛●☆
70773 Campylostemon kennedyi (Hoyle) Hoyle et Brenan = Campylostemon laurentii De Wild. ●☆
70774 Campylostemon laurentii De Wild.;洛朗曲蕊卫矛●☆
70775 Campylostemon mitophorum Loes.;线梗曲蕊卫矛●☆
70776 Campylostemon nigrisilvae N. Hallé = Tristemonanthus nigrisilvae (N. Hallé) N. Hallé ●☆
70777 Campylostemon warneckeanum Loes. ex Fritsch;沃内克曲蕊卫矛●☆
70778 Campylosus Post et Kuntze = Campulosus Desv.(废弃属名)■☆
70779 Campylosus Post et Kuntze = Ctenium Panz.(保留属名)■☆
70780 Campylotheca Cass. = Bidens L. ■●
70781 Campylotropis Bunge(1835);杭子梢属(弯龙骨属);Clover Shrub,Clovershrub ●
70782 Campylotropis alata Schindl. = Campylotropis trigonoclada (Franch.) Schindl. ●
70783 Campylotropis alba Iokawa et H. Ohashi;白花杭子梢●
70784 Campylotropis alopochroa H. Ohashi;西藏杭子梢(察隅杭子梢);Chayu Clovershrub ●
70785 Campylotropis argentea Schindl.;银叶杭子梢(白叶花,白叶花柴,百叶花,银背叶杭子梢);Silverleaf Clovershrub, Silveryleaf Clovershrub,Silvery-leaved Clovershrub ●
70786 Campylotropis balfouriana (Diels ex Schindl.) Schindl. = Campylotropis trigonoclada (Franch.) Schindl. ●
70787 Campylotropis bodinieri Schindl. = Campylotropis macrocarpa (Bunge) Rehder var. giraldii (Schindl.) K. T. Fu ex P. Y. Fu ●
70788 Campylotropis bonatiana (Pamp.) Schindl.;毛三棱枝杭子梢(马尿鞭,马尿藤,毛棱枝杭子梢,毛三穗枝杭子梢,三棱梢,三麦梢);Bonat Clovershrub ●
70789 Campylotropis bonatiana (Pamp.) Schindl. = Campylotropis trigonoclada (Franch.) Schindl. var. bonatiana (Pamp.) Iokawa et H. Ohashi ●
70790 Campylotropis bonii Schindl.;密脉杭子梢(密花杭子梢);Bon Clovershrub,Densevein Clovershrub,Veiny Clovershrub ●
70791 Campylotropis bonii Schindl. var. stipellata Iokawa et H. Ohashi;密花杭子梢●
70792 Campylotropis brevifolia Ricker;短序杭子梢(德荣杭子梢);Derong Clovershrub,Shortleaf Clovershrub,Short-leaved Clovershrub, Shortraceme Clovershrub ●
70793 Campylotropis capillipes (Franch.) Schindl.;细花梗杭子梢(细柄杭子梢,细梗杭子梢);Capillary-pedicel Clovershrub,Hairy-stalked Clovershrub,Thinpedicel Clovershrub ●
70794 Campylotropis capillipes (Franch.) Schindl. subsp. prainii (Collett et Hemsl.) Iokawa et H. Ohashi;草山杭子梢(滇南杭子梢);Prain Clovershrub ●
70795 Campylotropis cavaleriei (H. Lév.) C. Y. Wu;绒毛杭子梢(三匹叶);Hispid Clovershrub ●
70796 Campylotropis chinensis Bunge = Campylotropis macrocarpa (Bunge) Rehder ●
70797 Campylotropis cytisoides Miq.;小花杭子梢;Littleflower Clovershrub,Small-flower Clovershrub,Small-flowered Clovershrub ●
70798 Campylotropis decora (Kurz) Schindl.;华美杭子梢●
70799 Campylotropis delavayi (Franch.) Schindl.;西南杭子梢(豆角柴);Delavay Clovershrub ●
70800 Campylotropis diversifolia (Franch.) Schindl.;异叶杭子梢;Abnormalleaf Clovershrub,Differentleaf Clovershrub ●
70801 Campylotropis eriocarpa Schindl. = Campylotropis speciosa (Royle ex Schindl.) Schindl. subsp. eriocarpa (Schindl.) Iokawa et H. Ohashi ●
70802 Campylotropis esquirolii Schindl.;滇黔杭子梢●
70803 Campylotropis esquirolii Schindl. = Campylotropis henryi (Schindl.) Schindl. ●
70804 Campylotropis falconeri (Prain) Schindl. = Campylotropis meeboldii (Schindl.) Schindl. ●☆
70805 Campylotropis filipes Ricker = Campylotropis yunnanensis (Franch.) Schindl. subsp. filipes (Ricker) Iokawa et H. Ohashi ●
70806 Campylotropis filipes Ricker = Campylotropis yunnanensis (Franch.) Schindl. var. filipes (Ricker) P. Y. Fu ●
70807 Campylotropis franchetiana Lingelsh. et Borza = Campylotropis bonatiana (Pamp.) Schindl. ●
70808 Campylotropis franchetiana Lingelsh. et Borza = Campylotropis trigonoclada (Franch.) Schindl. var. bonatiana (Pamp.) Iokawa et H. Ohashi ●
70809 Campylotropis fulva Schindl.;暗黄杭子梢;Dimyellow Clovershrub,Reddish-brown Clovershrub,Yellow Clovershrub ●
70810 Campylotropis giraldii (Schindl.) Schindl.;弯龙骨●
70811 Campylotropis giraldii (Schindl.) Schindl. = Campylotropis macrocarpa (Bunge) Rehder var. giraldii (Schindl.) K. T. Fu ex P. Y. Fu ●
70812 Campylotropis giraldii (Schindl.) Schindl. = Campylotropis macrocarpa (Bunge) Rehder f. giraldii (Schindl.) P. Y. Fu ●
70813 Campylotropis giraldii Schindl. = Campylotropis macrocarpa (Bunge) Rehder var. giraldii (Schindl.) K. T. Fu ex P. Y. Fu ●
70814 Campylotropis giraldii Schindl. = Campylotropis macrocarpa (Bunge) Rehder var. hupehensis (Pamp.) Iokawa et H. Ohashi ●
70815 Campylotropis glauca (Schindl.) Schindl. = Campylotropis macrocarpa (Bunge) Rehder var. giraldii (Schindl.) K. T. Fu ex P. Y. Fu ●
70816 Campylotropis glauca (Schindl.) Schindl. = Campylotropis macrocarpa (Bunge) Rehder var. hupehensis (Pamp.) Iokawa et H. Ohashi ●
70817 Campylotropis gracilis Ricker = Campylotropis macrocarpa (Bunge) Rehder ●
70818 Campylotropis grandifolia Schindl.;大叶杭子梢(弥勒杭子梢);Largeleaf Clovershrub ●
70819 Campylotropis harmsii Schindl.;思茅杭子梢(滇南杭子梢,干枝柳,化食草,米过穴,三叶豆);Harms Clovershrub ●

70820 Campylotropis henryi (Schindl.) Schindl.;元江杭子梢;Henry Clovershrub ●

70821 Campylotropis hersii Ricker = Campylotropis macrocarpa (Bunge) Rehder ●

70822 Campylotropis hirtella (Franch.) Schindl.;毛杭子梢(白蓝地花,扁皂角,大和红,大红袍,地油根,地油花,山黄豆,山皮条,锈钉子,野黄豆,硬毛杭子梢);Hair Clovershrub,Hairy Clovershrub ●

70823 Campylotropis howellii Schindl.;腾冲杭子梢;Howell Clovershrub ●

70824 Campylotropis huberi Ricker = Campylotropis macrocarpa (Bunge) Rehder ●

70825 Campylotropis ichangensis (Schindl.) Cheng f. et al. = Campylotropis macrocarpa (Bunge) Rehder ●

70826 Campylotropis ichangensis Schindl. = Campylotropis macrocarpa (Bunge) Rehder ●

70827 Campylotropis ichangensis Schindl. ex S. H. Cheng et al. = Campylotropis macrocarpa (Bunge) Rehder ●

70828 Campylotropis kingdonii H. Ohashi = Campylotropis thomsonii (Benth. ex Baker) Schindl. ●

70829 Campylotropis latifolia (Dunn) Schindl.;阔叶杭子梢;Broadleaf Clovershrub,Broad-leaved Clovershrub ●

70830 Campylotropis longepedunculata Ricker = Campylotropis macrocarpa (Bunge) Rehder var. hupehensis (Pamp.) Iokawa et H. Ohashi ●

70831 Campylotropis longepedunculata Ricker = Campylotropis macrocarpa (Bunge) Rehder f. longepedunculata (Ricker) P. Y. Fu ●

70832 Campylotropis luhitensis H. Ohashi;藏东杭子梢●

70833 Campylotropis macrocarpa (Bunge) Rehder;杭子梢(胡枝子,灰靛花,假大红袍,见肿消,落豆花,马胡烧,马料梢,排兜根,万年梢,小叶乌梢,宜昌杭子梢,壮筋草);Chinese Clovershrub,Clovershrub,Ichang Clovershrub,Yichang Clovershrub ●

70834 Campylotropis macrocarpa (Bunge) Rehder f. giraldii (Schindl.) P. Y. Fu = Campylotropis macrocarpa (Bunge) Rehder var. giraldii (Schindl.) K. T. Fu ex P. Y. Fu ●

70835 Campylotropis macrocarpa (Bunge) Rehder f. hupehensis (Pamp.) P. Y. Fu = Campylotropis macrocarpa (Bunge) Rehder var. hupehensis (Pamp.) Iokawa et H. Ohashi ●

70836 Campylotropis macrocarpa (Bunge) Rehder f. lanceolata P. Y. Fu;披针叶杭子梢;Lanceolate Clovershrub ●

70837 Campylotropis macrocarpa (Bunge) Rehder f. lanceolata P. Y. Fu = Campylotropis macrocarpa (Bunge) Rehder ●

70838 Campylotropis macrocarpa (Bunge) Rehder f. longepedunculata (Ricker) P. Y. Fu;长梗杭子梢(长叶杭子梢);Long-peduncle Clovershrub ●

70839 Campylotropis macrocarpa (Bunge) Rehder f. micriphylla K. T. Fu ex P. Y. Fu;小叶杭子梢;Small-leaf Clovershrub ●

70840 Campylotropis macrocarpa (Bunge) Rehder subsp. hengduanshanensis C. J. Chen;横断山杭子梢;Hengduanshan Clovershrub ●

70841 Campylotropis macrocarpa (Bunge) Rehder subsp. hengduanshanensis C. J. Chen = Campylotropis macrocarpa (Bunge) Rehder ●

70842 Campylotropis macrocarpa (Bunge) Rehder var. giraldii (Schindl.) K. T. Fu ex P. Y. Fu;太白杭子梢(蜻蛉胡枝子,太白山杭子梢,弯龙骨);Chinese Clover Shrub,Taibaishan Clovershrub ●

70843 Campylotropis macrocarpa (Bunge) Rehder var. giraldii (Schindl.) K. T. Fu ex P. Y. Fu = Campylotropis macrocarpa (Bunge) Rehder var. hupehensis (Pamp.) Iokawa et H. Ohashi ●

70844 Campylotropis macrocarpa (Bunge) Rehder var. giraldii (Schindl.) K. T. Fu ex P. Y. Fu f. hupehensis (Pamp.) P. Y. Fu = Campylotropis macrocarpa (Bunge) Rehder var. hupehensis (Pamp.) Iokawa et H. Ohashi ●

70845 Campylotropis macrocarpa (Bunge) Rehder var. giraldii (Schindl.) K. T. Fu ex P. Y. Fu f. longepedunculata (Ricker) P. Y. Fu = Campylotropis macrocarpa (Bunge) Rehder var. hupehensis (Pamp.) Iokawa et H. Ohashi ●

70846 Campylotropis macrocarpa (Bunge) Rehder var. giraldii (Schindl.) K. T. Fu ex P. Y. Fu f. microphylla K. T. Fu ex P. Y. Fu = Campylotropis macrocarpa (Bunge) Rehder var. hupehensis (Pamp.) Iokawa et H. Ohashi ●

70847 Campylotropis macrocarpa (Bunge) Rehder var. hupehensis (Pamp.) Iokawa et H. Ohashi;太白山杭子梢(丝苞杭子梢);Hubei Clovershrub,Hupeh Clovershrub ●

70848 Campylotropis macrocarpa (Bunge) Rehder var. hupehensis (Pamp.) Iokawa et H. Ohashi = Campylotropis macrocarpa (Bunge) Rehder f. hupehensis (Pamp.) P. Y. Fu ●

70849 Campylotropis macrocarpa (Bunge) Rehder var. hupehensis Pamp. = Campylotropis macrocarpa (Bunge) Rehder f. hupehensis (Pamp.) P. Y. Fu ●

70850 Campylotropis macrostyla (D. Don) Schindl.;大柱杭子梢●☆

70851 Campylotropis macrostyla (D. Don) Schindl. var. eriocarpa (Maxim.) H. Ohashi = Campylotropis speciosa (Royle ex Schindl.) Schindl. subsp. eriocarpa (Schindl.) Iokawa et H. Ohashi ●

70852 Campylotropis macrostyla (D. Don) Schindl. var. stenocarpa (Klotz.) Ohashi;窄果大柱杭子梢●☆

70853 Campylotropis meeboldii (Schindl.) Schindl.;米氏大柱杭子梢●☆

70854 Campylotropis mortolana Ricker = Campylotropis macrocarpa (Bunge) Rehder ●

70855 Campylotropis muehleana (Schindl.) Schindl.;蜀杭子梢;Sichuan Clovershrub,Szechuan Clovershrub ●

70856 Campylotropis muehleana (Schindl.) Schindl. = Campylotropis polyantha (Franch.) Schindl. ●

70857 Campylotropis neglecta Schindl. = Campylotropis polyantha (Franch.) Schindl. var. neglecta (Schindl.) Iokawa et H. Ohashi ●

70858 Campylotropis paniculata Schindl.;圆锥杭子梢●

70859 Campylotropis parviflora (Kurz) Schindl. ex Gagnep. = Campylotropis cytisoides Miq. ●

70860 Campylotropis pauciflora C. J. Chen;少花杭子梢;Poorflower Clovershrub ●

70861 Campylotropis pinetorum (Kurz) Schindl.;缅南杭子梢(松林杭子梢);S. Burma Clovershrub,South-Burma Clovershrub ●☆

70862 Campylotropis pinetorum (Kurz) Schindl. subsp. albopubescens (Iokawa et H. Ohashi) Iokawa et H. Ohashi;白柔毛杭子梢●

70863 Campylotropis pinetorum (Kurz) Schindl. subsp. velutina (Dunn) H. Ohashi;绒毛叶杭子梢(绒毛杭子梢);Velvety Clovershrub ●

70864 Campylotropis pinetorum (Kurz) Schindl. subsp. velutina (Dunn) H. Ohashi = Campylotropis velutina (Dunn) Schindl. ●

70865 Campylotropis pinetorum (Kurz) Schindl. var. albopubescens Iokawa et H. Ohashi = Campylotropis pinetorum (Kurz) Schindl. subsp. albopubescens (Iokawa et H. Ohashi) Iokawa et H. Ohashi ●

70866 Campylotropis polyantha (Franch.) Schindl.;多花杭子梢(小雀花,多花胡枝子);Flowery Clovershrub, Many Flowered

Clovershrub, Manyflower Clovershrub, Multiflorous Clovershrub ●

70867 Campylotropis polyantha (Franch.) Schindl. f. macrophylla P. Y. Fu;大叶小雀花;Largeleaf Clovershrub ●

70868 Campylotropis polyantha (Franch.) Schindl. f. macrophylla P. Y. Fu = Campylotropis polyantha (Franch.) Schindl. ●

70869 Campylotropis polyantha (Franch.) Schindl. f. souliei (Schindl.) P. Y. Fu;狭叶小雀花;Narrowleaf Clovershrub ●

70870 Campylotropis polyantha (Franch.) Schindl. f. souliei (Schindl.) P. Y. Fu = Campylotropis polyantha (Franch.) Schindl. ●

70871 Campylotropis polyantha (Franch.) Schindl. var. leiocarpa (Pamp.) E. Peter;光果小雀花;Smooth-fruit Clovershrub ●

70872 Campylotropis polyantha (Franch.) Schindl. var. neglecta (Schindl.) Iokawa et H. Ohashi;蒙自杭子梢;Mengtze Clovershrub, Mengzi Clovershrub, Neglect Clovershrub ●

70873 Campylotropis polyantha (Franch.) Schindl. var. tomentosa P. Y. Fu;密毛小雀花;Tomentose Clovershrub ●

70874 Campylotropis polyantha (Franch.) Schindl. var. tomentosa P. Y. Fu = Campylotropis polyantha (Franch.) Schindl. ●

70875 Campylotropis prainii (Collett et Hemsl.) Schindl. = Campylotropis capillipes (Franch.) Schindl. subsp. prainii (Collett et Hemsl.) Iokawa et H. Ohashi ●

70876 Campylotropis purpurascens Ricker = Campylotropis sulcata Schindl. ●

70877 Campylotropis reticulata Ricker;多网杭子梢●

70878 Campylotropis reticulata Ricker = Campylotropis polyantha (Franch.) Schindl. ●

70879 Campylotropis reticulata S. S. Chien = Campylotropis polyantha (Franch.) Schindl. ●

70880 Campylotropis reticulinervis C. Y. Wu;网脉杭子梢●

70881 Campylotropis reticulinervis C. Y. Wu = Campylotropis polyantha (Franch.) Schindl. ●

70882 Campylotropis rockii Schindl.;滇南杭子梢(橄榄坝杭子梢);Rock Clovershrub ●

70883 Campylotropis rockii Schindl. = Campylotropis sulcata Schindl. ●

70884 Campylotropis rogersii Schindl. = Campylotropis thomsonii (Benth. ex Baker) Schindl. ●

70885 Campylotropis sargentiana (Schindl.) H. Lév. = Campylotropis polyantha (Franch.) Schindl. ●

70886 Campylotropis sargentiana Schindl. = Campylotropis polyantha (Franch.) Schindl. ●

70887 Campylotropis schneideri Schindl. = Campylotropis polyantha (Franch.) Schindl. var. leiocarpa (Pamp.) E. Peter ●

70888 Campylotropis sericophylla (Collett et Hemsl.) Schindl. = Campylotropis decora (Kurz) Schindl. ●

70889 Campylotropis sessilifolia Schindl. = Campylotropis decora (Kurz) Schindl. ●

70890 Campylotropis smithii Ricker = Campylotropis macrocarpa (Bunge) Rehder ●

70891 Campylotropis smithii Ricker = Campylotropis polyantha (Franch.) Schindl. ●

70892 Campylotropis souliei Schindl. = Campylotropis polyantha (Franch.) Schindl. f. souliei (Schindl.) P. Y. Fu ●

70893 Campylotropis souliei Schindl. = Campylotropis polyantha (Franch.) Schindl. ●

70894 Campylotropis speciosa (Royle ex Schindl.) Schindl. subsp. eriocarpa (Schindl.) Iokawa et H. Ohashi;绵毛果杭子梢●

70895 Campylotropis speciosa Schindl.;美丽杭子梢●☆

70896 Campylotropis stenocarpa (Klotzsch) Schindl.;窄果杭子梢●☆

70897 Campylotropis stenocarpa (Klotzsch) Schindl. = Campylotropis macrostyla (D. Don) Schindl. var. stenocarpa (Klotz.) Ohashi ●☆

70898 Campylotropis sulcata Schindl.;槽茎杭子梢;Sulcate Clovershrub, Sulcatestem Clovershrub, Sulcate-stem Clovershrub ●

70899 Campylotropis tenuiramea P. Y. Fu;细枝杭子梢;Slendertwig Clovershrub, Tenuous Clovershrub, Tenuous-branch Clovershrub ●

70900 Campylotropis teretiracemosa P. C. Li et C. J. Chen;柱序杭子梢;Teretiraceme Clovershrub ●

70901 Campylotropis thomsonii (Benth. ex Baker) Schindl.;汤姆逊杭子梢●

70902 Campylotropis tomentosipetiolata P. Y. Fu;绒柄杭子梢;Flosspetiole Clovershrub, Tomentose Clovershrub, Tomentose Pedicel Clovershrub ●

70903 Campylotropis tomentosipetiolata P. Y. Fu = Campylotropis polyantha (Franch.) Schindl. ●

70904 Campylotropis trigonoclada (Franch.) Schindl.;三棱枝杭子梢(大发表,黄花马尿藤,马尿藤,爬山豆,三股筋,三棱草,三棱草佳不齐,三棱梢爬山豆,三楞草,山罗松,山落花生,小落花生,野蚕豆);Threeanglesbranch Clovershrub, Trigonous Clovershrub, Tririb Clovershrub ●

70905 Campylotropis trigonoclada (Franch.) Schindl. var. bonatiana (Pamp.) Iokawa et H. Ohashi;马尿藤●

70906 Campylotropis velutina (Dunn) Schindl. = Campylotropis pinetorum (Kurz) Schindl. subsp. velutina (Dunn) H. Ohashi ●

70907 Campylotropis velutina Schindl. = Campylotropis pinetorum (Kurz) Schindl. subsp. velutina (Dunn) H. Ohashi ●

70908 Campylotropis velutina Schindl. = Campylotropis velutina (Dunn) Schindl. ●

70909 Campylotropis wangii Ricker = Campylotropis polyantha (Franch.) Schindl. ●

70910 Campylotropis wenshanica P. Y. Fu;秋杭子梢;Autumn Clovershrub, Wenshan Clovershrub ●

70911 Campylotropis wilsonii Schindl.;四川杭子梢(秋杭子梢,小叶杭子梢,雅江杭子梢);E. H. Wilson Clovershrub, Wilson Clovershrub, Yajiang Clovershrub ●

70912 Campylotropis yajiangensis P. Y. Fu = Campylotropis wilsonii Schindl. ●

70913 Campylotropis yajiangensis P. Y. Fu var. deronica P. Y. Fu = Campylotropis brevifolia Ricker ●

70914 Campylotropis yunnanensis (Franch.) Schindl.;滇杭子梢;Yunnan Clovershrub ●

70915 Campylotropis yunnanensis (Franch.) Schindl. subsp. filipes (Ricker) Iokawa et H. Ohashi = Campylotropis yunnanensis (Franch.) Schindl. var. filipes (Ricker) P. Y. Fu ●

70916 Campylotropis yunnanensis (Franch.) Schindl. var. filipes (Ricker) P. Y. Fu;丝梗杭子梢;Thread-pedicel Clovershrub ●

70917 Campylotropis yunnanensis (Franch.) Schindl. var. filipes (Ricker) P. Y. Fu = Campylotropis yunnanensis (Franch.) Schindl. subsp. filipes (Ricker) Iokawa et H. Ohashi ●

70918 Campylotropis yunnanensis (Franch.) Schindl. var. zhongdianensis P. Y. Fu;中甸杭子梢;Zhongdian Clovershrub ●

70919 Campylotropis yunnanensis (Franch.) Schindl. var. zhongdianensis P. Y. Fu = Campylotropis yunnanensis (Franch.) Schindl. ●

70920 Campylus Lour.(废弃属名) = Tinospora Miers(保留属名)●■

70921 Campylus sinensis Lour. = Tinospora sinensis (Lour.) Merr. ■
70922 Campynema Labill. (1805);金梅草属(弯丝草属)■☆
70923 Campynema lineare Labill.;金梅草■☆
70924 Campynemaceae Dumort. = Campynemataceae Dumort. ■☆
70925 Campynemanthe Baill. (1893);曲丝花属■☆
70926 Campynemanthe parva Goldblatt;小管曲丝花■☆
70927 Campynemanthe viridiflora Baill.;曲丝花■☆
70928 Campynemataceae Dumort. (1829);金梅草科■☆
70929 Campynemataceae Dumort. = Hypoxidaceae R. Br. (保留科名)■☆
70930 Camunium Adans. = Trichogamila P. Browne ●
70931 Camunium Kuntze = Chalcas L. ●
70932 Camunium Kuntze = Murraya J. König ex L. (保留属名)●
70933 Camunium Roxb. = Aglaia Lour. (保留属名)●
70934 Camunium exoticum (L.) Kuntze = Murraya paniculata (L.) Jack ●
70935 Camusia Lorch = Acrachne Wight et Arn. ex Chiov. ■
70936 Camusia Lorch = Dactyloctenium Willd. ■
70937 Camusiella Bosser = Setaria P. Beauv. (保留属名)■
70938 Camutia Bonat. ex Steud. = Melampodium L. ■●
70939 Canabis Roth = Cannabis L. ■
70940 Canaca Guillaumin = Austrobuxus Miq. ●☆
70941 Canacomyrica Guillaumin = Myrica L. ●
70942 Canacomyrica Guillaumin(1940);高山杨梅属●☆
70943 Canacomyrica monticola Guillaumin;高山杨梅●☆
70944 Canacomyricaceae Baum. -Bod. = Myricaceae Rich. ex Kunth(保留科名)●
70945 Canacomyricaceae Baum. -Bod. ex Doweld = Myricaceae Rich. ex Kunth(保留科名)●
70946 Canacorchis Guillaumin = Bulbophyllum Thouars(保留属名)■
70947 Canadaea Gand. = Campanula L. ■●
70948 Canadanthus G. L. Nesom(1994);沼菀属;Aster ■☆
70949 Canadanthus modestus (Lindl.) G. L. Nesom;沼菀;Great Northern Aster, Western Bog Aster ■☆
70950 Canahia Steud. = Kanahia R. Br. ■☆
70951 Canahia laniflora (Forssk.) Steud. = Kanahia laniflora (Forssk.) R. Br. ■☆
70952 Canala Pohl = Spigelia L. ■☆
70953 Canalia F. W. Schmidt = Gnidia L. ●☆
70954 Cananga (DC.) Hook. f. et Thomson(1855)(保留属名);依兰属(加拿楷属,夷兰属);Cananga ●
70955 Cananga Aubl. (废弃属名) = Cananga (DC.) Hook. f. et Thomson(保留属名)●
70956 Cananga Aubl. (废弃属名) = Guatteria Ruiz et Pav. (保留属名)●☆
70957 Cananga Hook. f. et Thomson = Cananga (DC.) Hook. f. et Thomson(保留属名)●
70958 Cananga Raf. = Cananga (DC.) Hook. f. et Thomson(保留属名)●
70959 Cananga Rumph. ex Hook. f. et Thomson = Cananga (DC.) Hook. f. et Thomson(保留属名)●
70960 Cananga odorata (Lam.) Hook. f. et Thomson;依兰(加拿楷,香水树,香依兰,依兰香,夷兰);Fragrant Cananga, Ilang-Ilang, Ilan-ilan, Kenanga, Macassar Oil Tree, Ylang Ylang, Ylang-ylang, Ylan-ylan ●
70961 Cananga odorata (Lam.) Hook. f. et Thomson var. fruticosa (Craib) J. Sinclair;小依兰(小夷兰);Shrubby Cananga ●
70962 Canangium Baill. = Cananga (DC.) Hook. f. et Thomson(保留属名)●
70963 Canangium fruticosum Craib = Cananga odorata (Lam.) Hook. f. et Thomson var. fruticosa (Craib) J. Sinclair ●
70964 Canangium odoratum (Lam.) King = Cananga odorata (Lam.) Hook. f. et Thomson ●
70965 Canangium odoratum Baill. = Cananga odorata (Lam.) Hook. f. et Thomson ●
70966 Canangium odoratum King = Cananga odorata (Lam.) Hook. f. et Thomson ●
70967 Canangium odoratum King var. fruticosum (Craib) Comer = Cananga odorata (Lam.) Hook. f. et Thomson var. fruticosa (Craib) J. Sinclair ●
70968 Canaria L. = Canarina L. (保留属名)■☆
70969 Canariastrum Engl. = Uapaca Baill. ■☆
70970 Canariellum Engl. = Canarium L. ●
70971 Canarina L. (1771)(保留属名);加那利参属■☆
70972 Canarina abyssinica Engl.;阿比西尼亚加那利参■☆
70973 Canarina campanulata L. = Canarina canariensis Kuntze ■☆
70974 Canarina canariensis (L.) Vatke;加那利参;Canary Bellflower, Canary Island Bellflower ■☆
70975 Canarina canariensis Kuntze = Canarina canariensis (L.) Vatke ■☆
70976 Canarina elegantissima T. C. E. Fr. = Canarina eminii Asch. et Schweinf. ■☆
70977 Canarina eminii Asch. et Schweinf.;埃敏加那利参■☆
70978 Canarina eminii Asch. et Schweinf. var. elgonensis T. C. E. Fr. = Canarina eminii Asch. et Schweinf. ■☆
70979 Canarina zanguebar Lour.;赞古加那利参■☆
70980 Canarion St. -Lag. = Canarina L. (保留属名)■☆
70981 Canariopsis Hochr. = Canarina L. (保留属名)■☆
70982 Canariopsis Miq. = Canarina L. (保留属名)■☆
70983 Canariothamnus B. Nord. (2006);加那利泽菊属■☆
70984 Canariothamnus B. Nord. = Bethencourtia Choisy ●☆
70985 Canariothamnus B. Nord. = Cineraria L. ■●☆
70986 Canariothamnus palmensis (Nees) B. Nord. = Bethencourtia palmensis (Nees) Link ■☆
70987 Canariothamnus rupicola B. Nord. = Bethencourtia rupicola (B. Nord.) B. Nord. ■☆
70988 Canarium L. (1759);橄榄属;Black Dammar, Canarium, Canary Tree, Canarytree, Canary-tree, China Olive, Chinese Olive, Kedondong, Olive, Pili Nut ●
70989 Canarium album (Lour.) Raeusch.;橄榄(白榄,波斯橄榄,方榄,甘榄,橄棪,广青果,红榄,黄榄,黄榔果,谏果,榄子,绿榄,青橄榄,青果,青子,山榄,忠果);Chinese Olive, White Canary Tree, White Canarytree, White Chinese Olive, White Olive ●
70990 Canarium album Leenh. = Canarium tokinense Engl. ●
70991 Canarium auriculatum H. Winkl. = Canarium zeylanicum (Retz.) Blume ●☆
70992 Canarium bengalense Roxb.;方榄(三角榄);Bengal Canary Tree, Bengal Canarytree, Bengal Olive ●
70993 Canarium boivini Engl.;博氏橄榄木●☆
70994 Canarium buettneri Engl. = Dacryodes buettneri (Engl.) H. J. Lam ●☆
70995 Canarium chevalieri Guillaumin = Canarium schweinfurthii Engl. ●☆
70996 Canarium commune L.;普通橄榄(爪哇橄榄);Canari-nut-tree, Common Canarytree, Java Almond Canary-tree, Java Almond Tree,

Java-almond, Java-almond Canary Tree, Java-almond Tree, Kanari-nut-tree ●
70997 Canarium edule (G. Don) Hook. f. = Dacryodes edulis (G. Don) H. J. Lam ●☆
70998 Canarium euphyllum Kurz; 美叶橄榄; Dhup, Indian White Mahogany, Pill Nut ●☆
70999 Canarium harveyi Seem.; 波利尼西亚橄榄●☆
71000 Canarium indicum L.; 爪哇橄榄; Ngali Nut, Solomon Nut Oil ●
71001 Canarium kerrii Craib = Canarium subulatum Guillaumin ●◇
71002 Canarium khiala A. Chev. = Canarium schweinfurthii Engl. ●☆
71003 Canarium liebertianum Engl. = Canarium madagascariense Engl. ●☆
71004 Canarium luzonicum (Blume) A. Gray; 吕宋橄榄; Java Almond, Luzon Canarytree, Manila Elemi ●☆
71005 Canarium luzonicum Miq. = Canarium luzonicum (Blume) A. Gray ●☆
71006 Canarium macrophyllum Oliv. = Dacryodes macrophylla (Oliv.) H. J. Lam ●☆
71007 Canarium madagascariense Engl.; 马达加斯加橄榄●☆
71008 Canarium muelleri F. M. Bailey; 昆士兰橄榄树(米勒橄榄); Queensland Elemi Tree ●☆
71009 Canarium nigrum (Lour.) Engl. = Canarium pimela K. D. Koenig ●
71010 Canarium nigrum Roxb. = Canarium pimela Leenh. ●
71011 Canarium nigrum Roxb. = Canarium tramdenum C. D. Dai et Yakovlev ●
71012 Canarium occidentale A. Chev. = Canarium schweinfurthii Engl. ●☆
71013 Canarium ovatum Engl.; 卵果橄榄(卵橄榄树); Pili Nut ●☆
71014 Canarium parvum Leenh.; 小叶榄; Littleleaf Canarytree, Littleleaf Olive, Small-leaved Canary Tree ●
71015 Canarium pimela K. D. Koenig; 乌榄(黑橄榄,黑榄,榄,木威子,乌橄榄); Black Canary Tree, Black Canarytree, Black Canary-tree, Black Chinese Olive, Black Olive, Chinese Black Olive, Chinese Olive ●
71016 Canarium pimela Leenh. = Canarium pimela K. D. Koenig ●
71017 Canarium pimeloides Govaerts = Canarium pimela K. D. Koenig ●
71018 Canarium resiniferum Bruce ex King = Canarium strictum Roxb. ●
71019 Canarium rotundifolium Guillaumin = Canarium subulatum Guillaumin ●◇
71020 Canarium saphu Engl. = Dacryodes edulis (G. Don) H. J. Lam ●☆
71021 Canarium schweinfurthii Engl.; 非洲橄榄; African Canarium, African Elemi, Incense Tree ●☆
71022 Canarium sikkimense King = Canarium strictum Roxb. ●
71023 Canarium strictum Roxb.; 滇榄(劲直橄榄); Black Dhup, Blackdamar, Canary Tree, Strict Canarium, Yunnan Olive ●
71024 Canarium subulatum Guillaumin; 毛叶榄; Awl-shape-leaved Canary Tree, Hairleaf Olive, Subulate Canarytree ●◇
71025 Canarium thollonicum Guillaumin = Canarium schweinfurthii Engl. ●☆
71026 Canarium thorelianum Guillaumin = Canarium subulatum Guillaumin ●◇
71027 Canarium tokinense Engl.; 越榄(黄榄果); Tonkin Canary Tree, Tonkin Canarytree, Vietnam Olive, Vietnamese Canary Tree ●
71028 Canarium tonkinense Guillaumin = Canarium parvum Leenh. ●
71029 Canarium tramdenum C. D. Dai et Yakovlev = Canarium pimela K. D. Koenig ●
71030 Canarium velutinum Guillaumin = Canarium schweinfurthii Engl. ●☆
71031 Canarium venosum Craib = Canarium subulatum Guillaumin ●◇
71032 Canarium vittatistipulatum Guillaumin = Canarium subulatum Guillaumin ●◇
71033 Canarium williamsii C. B. Rob.; 威氏榄●☆
71034 Canarium zeylanicum (Retz.) Blume; 锡兰橄榄●☆
71035 Canavali Adans. = Canavalia Adans. (保留属名) ●■
71036 Canavalia Adans. (1763) (保留属名) ('Canavali'); 刀豆属; Jack Bean, Jackbean, Jack-bean, Knifebean ●■
71037 Canavalia DC. = Canavalia Adans. (保留属名) ●■
71038 Canavalia africana Dunn; 非洲刀豆■☆
71039 Canavalia bonariensis Lindl.; 博纳利刀豆■☆
71040 Canavalia cathartica Thouars; 小果刀豆(野刀板豆,小刀豆); Maunaloa, Small Knifebean, Small Sword Jackbean ■
71041 Canavalia cryptodon Meisn. = Canavalia bonariensis Lindl. ■☆
71042 Canavalia ensiformis (L.) DC.; 直生刀豆(大刀豆,刀豆,关刀豆,龙爪豆,马眼镰扁豆,水流豆,洋刀豆,直立刀豆); Broad Bean, Common Jack Bean, Common Jack-bean, Cut-eye Bean, Ensiform Knifebean, Horse Bean, Horse-eye Bean, Jack Bean, Jamaica Horse Bean, Jamaican Horse Bean, One-eye Bean, Overlook Bean, Patagonian Bean, Sabre Bean, Sword Bean, Sword Jack-bean, Swordform Jackbean, Wonder Bean, Wonderbean ■
71043 Canavalia ensiformis (L.) DC. var. mucunoides Baill. = Canavalia madagascariensis J. D. Sauer ●☆
71044 Canavalia ensiformis (L.) DC. var. turgida Baker = Canavalia cathartica Thouars ■
71045 Canavalia ensiformis (L.) DC. var. turgida Graham ex Baker = Canavalia cathartica Thouars ■
71046 Canavalia ensiformis (L.) DC. var. virosa (Roxb.) Baker = Canavalia virosa (Roxb.) Wight et Arn. ■
71047 Canavalia ferruginea Piper = Canavalia africana Dunn ■☆
71048 Canavalia gladiata (Jacq.) DC. = Canavalia gladiata (Savi) DC. ■
71049 Canavalia gladiata (Jacq.) DC. f. alba (Makino) H. Ohashi = Canavalia gladiata (Savi) DC. f. alba (Makino) H. Ohashi ■☆
71050 Canavalia gladiata (Savi) DC.; 刀豆(白凤豆,炒刀豆,大刀豆,大弋豆,刀巴豆,刀把豆,刀坝豆,刀豆角,刀豆子,刀培豆,刀鞘豆,刀侠豆,刀挟豆,葛豆,关刀豆,马刀豆,泥鳅豆,四季豆,挟剑豆,洋刀豆,野刀板藤,野刀板藤豆,皂荚豆); Jack Bean, Knifebean, Parang Bean, Sword Bean, Sword Jack Bean, Sword Jackbean, Swordbean ■
71051 Canavalia gladiata (Savi) DC. = Canavalia ensiformis (L.) DC. ■
71052 Canavalia gladiata (Savi) DC. f. alba (Makino) H. Ohashi; 白直生刀豆■☆
71053 Canavalia gladiolata Sauer; 尖萼刀豆; Sword Jackbean, Tinesepal Knifebean ■
71054 Canavalia grandis (Wall. ex Benth.) Kurz = Dysolobium grande (Benth.) Prain ●
71055 Canavalia lineata (Thunb.) DC.; 狭刀豆(滨刀豆,肥猪豆); Lineate Jackbean, Narrow Knifebean ■
71056 Canavalia lineata (Thunb.) DC. f. albiflora Y. Kimura; 白花狭刀豆■☆
71057 Canavalia loureirii G. Don = Canavalia gladiata (Savi) DC. ■
71058 Canavalia madagascariensis J. D. Sauer; 马岛刀豆●☆
71059 Canavalia maritima (Aubl.) Thouars = Canavalia rosea (Sw.) DC. ■
71060 Canavalia maritima Thouars = Canavalia rosea (Sw.) DC. ■

71061 Canavalia microcarpa (DC.) Piper = Canavalia cathartica Thouars ■
71062 Canavalia moneta Welw. = Canavalia rosea (Sw.) DC. ■
71063 Canavalia monodon E. Mey. = Canavalia bonariensis Lindl. ■☆
71064 Canavalia obcordata (Roxb.) Voigt = Canavalia maritima (Aubl.) Thouars ■
71065 Canavalia obcordata (Roxb.) Voigt = Canavalia rosea (Sw.) DC. ■
71066 Canavalia obtusifolia (Lam.) DC.;钝叶刀豆;Seaside Sword Bean ■
71067 Canavalia obtusifolia (Lam.) DC. = Canavalia maritima (Aubl.) Thouars ■
71068 Canavalia obtusifolia (Lam.) DC. = Canavalia rosea (Sw.) DC. ■
71069 Canavalia obtusifolia DC. = Canavalia rosea (Sw.) DC. ■
71070 Canavalia polystachya Schweinf. = Canavalia cathartica Thouars ■
71071 Canavalia regalis Piper et Dunn;王刀豆■☆
71072 Canavalia rosea (Sw.) DC.;海刀豆(滨刀豆);Rose Jackbean, Sea Knifebean, Seahore Jackbean ■
71073 Canavalia rosea (Sw.) DC. = Canavalia maritima (Aubl.) Thouars ■
71074 Canavalia sericea A. Gray;绢毛刀豆;Silky Jackbean ■☆
71075 Canavalia turgida Graham ex A. Gray = Canavalia cathartica Thouars ■
71076 Canavalia villosa Benth.;柔毛刀豆■☆
71077 Canavalia virosa (Roxb.) Wight et Arn.;野刀豆■
71078 Canbya Parry ex A. Gray = Canbya Parry ■☆
71079 Canbya Parry(1877);矮罂粟属;Pygmy-poppy ■☆
71080 Canbya aurea S. Watson;矮罂粟;Yellow Pygmy-poppy ■☆
71081 Canbya candida Parry ex A. Gray;白矮罂粟;White pygmy-poppy ■☆
71082 Cancellaria (DC.) Mattei = Pavonia Cav.(保留属名)●■☆
71083 Cancellaria Mattei = Pavonia Cav.(保留属名)●■☆
71084 Cancellaria Sch. Bip. ex Oliver = Adelostigma Steetz ■☆
71085 Cancrinia Kar. et Kir. (1842);小甘菊属;Cancrinia, Sweetdaisy ●■
71086 Cancrinia botschantzevii (Kovalevsk.) Tzvelev;包沙小甘菊■☆
71087 Cancrinia brachypappus C. Winkl. = Cancrinia discoides (Ledeb.) Poljakov ex Tzvelev ■
71088 Cancrinia chrysocephala Kar. et Kir.;黄头小甘菊;Common Cancrinia, Yellowhead Sweetdaisy ■
71089 Cancrinia chrysocephala Kar. et Kir. subsp. tianschanica Krasch. = Cancrinia tianschanica (Krasch.) Tzvelev ●■
71090 Cancrinia discoides (Ledeb.) Poljakov ex Tzvelev;小甘菊(草甘菊,甘菊,金扭扣);Discoid Cancrinia, Straw Sweetdaisy, Sweetdaisy ■
71091 Cancrinia ferganensis (Kovalevsk.) Tzvelev;费尔干小甘菊■☆
71092 Cancrinia goloskokovii (Poljakov) Tzvelev;戈罗斯小甘菊■☆
71093 Cancrinia karatavica Tzvelev;卡拉塔夫小甘菊●☆
71094 Cancrinia komarovii (Krasch. et N. I. Rubtzov) S. Y. Hu = Kaschgarica komarovii (Krasch. et N. I. Rubtzov) Poljakov ●
71095 Cancrinia lasiocarpa C. Winkl.;毛果小甘菊(甘菊,金纽扣);Hairyfruit Cancrinia, Hairyfruit Sweetdaisy ■
71096 Cancrinia maximowiczii C. Winkl.;灌木小甘菊(灌木状小甘菊);Maximowicz Cancrinia, Maximowicz Sweetdaisy, Maximowicz's Cancrinia ●
71097 Cancrinia mucronata (Regel et Schmalh.) Tzvelev;短尖小甘菊●☆
71098 Cancrinia nevskii Tzvelev;涅夫小甘菊●☆
71099 Cancrinia pamiralaica (Kovalevsk.) Tzvelev;帕米尔小甘菊■☆
71100 Cancrinia paucicephala Y. Ling = Cancrinia maximowiczii C. Winkl. ●
71101 Cancrinia setacea (Regel et Schmalh.) Tzvelev;刚毛小甘菊■☆
71102 Cancrinia submarginata (Kovalevsk.) Tzvelev;亚边小甘菊■☆
71103 Cancrinia subsimilis (Rech. f.) Tzvelev;相似小甘菊■☆
71104 Cancrinia tadshikorum (Kudrjanzev) Tzvelev;塔什克小甘菊■☆
71105 Cancrinia tianschanica (Krasch.) Tzvelev;天山小甘菊;Tianshan Cancrinia, Tianshan Sweetdaisy ●■
71106 Cancriniella Tzvelev(1961);木甘菊属;Hairy Lady's Smock ●☆
71107 Cancriniella krascheninnikovii (Rubtzov) Tzvelev;木甘菊;Hairy Lady's Smock ●☆
71108 Candarum Reichenb. ex Schott et Endl. = Amorphophallus Blume ex Decne.(保留属名)■●
71109 Candarum Schott = Amorphophallus Blume ex Decne.(保留属名)■●
71110 Candelabria Hochst. = Bridelia Willd.(保留属名)●
71111 Candelabria micrantha Hochst. = Bridelia micrantha (Hochst.) Baill. ●
71112 Candelium Medik. = Vigna Savi(保留属名)■
71113 Candidea Ten. = Baccharoides Moench ●■
71114 Candidea Ten. = Vernonia Schreb.(保留属名)●■
71115 Candidea stenostegia Stapf = Vernonia stenostegia (Stapf) Hutch. et Dalziel ■☆
71116 Candjera Decne. = Cansjera Juss.(保留属名)●
71117 Candollea Baumg. = Menziesia Sm. ●☆
71118 Candollea Labill. (1805) = Stylidium Sw. ex Willd.(保留属名)■
71119 Candollea Labill. (1806) = Eeldea T. Durand ●☆
71120 Candollea Labill. (1806) = Hibbertia Andréws ●☆
71121 Candollea Steud. = Agrostis L.(保留属名)■
71122 Candollea Steud. = Decandolia Bastard ■
71123 Candolleaceae F. Muell. = Stylidiaceae R. Br.(保留科名)●■
71124 Candolleaceae Schonl. = Stylidiaceae R. Br.(保留科名)●■
71125 Candolleodendron R. S. Cowan(1966);巴西坎多豆属●☆
71126 Candolleodendron brachystachyum R. S. Cowan;巴西坎多豆●☆
71127 Candollina Tiegh. = Amyema Tiegh. ●☆
71128 Canella Dombey ex Endl. = Drimys J. R. Forst. et G. Forst.(保留属名)●☆
71129 Canella P. Browne(1756)(保留属名);白桂皮属(白樟属,假樟属);Canella, Wild-cinnamon ●☆
71130 Canella Post et Kuntze = Cannella Schott ex Meisn. ●☆
71131 Canella Post et Kuntze = Ocotea Aubl. ●☆
71132 Canella alba Murray;白桂皮;Bahama Whitewood, Canella, Cinnamon-bark, West Indian Whitewood, Whitewood, Wild Cinnamon ●☆
71133 Canella canella Sudw. = Canella alba Murray ●☆
71134 Canella jamaica;牙买加白桂皮●☆
71135 Canella laurifolia Lodd. ex Sweet = Canella alba Murray ●☆
71136 Canella winterana Gaertn. = Canella alba Murray ●☆
71137 Canella winteriana (L.) Gaertn. = Canella alba Murray ●☆
71138 Canellaceae Mart. (1832)(保留科名);白桂皮科(白樟科,假樟科);Canella Family, Wild-cinnamon Family ●☆
71139 Canephora Juss. (1789);苇梗茜属■☆
71140 Canephora angustifolia Wernham;窄叶苇梗茜■☆
71141 Canephora goudotii Wernham;古氏苇梗茜■☆

71142 Canephora humblotii Drake;胡氏苇梗茜■☆
71143 Canephora madagascariensis J. F. Gmel.;马岛苇梗茜■☆
71144 Canephora maroana Aug. DC.;马洛苇梗茜■☆
71145 Canhamo Perini = Hibiscus L.(保留属名)●■
71146 Canicidia Vell. = Connarus L. ●
71147 Canidia Salisb. = Allium L. ■
71148 Caniram Thouars ex Steud. = Strychnos L. ●
71149 Canistrum E. Morren(1873);筒凤梨属(卡尼斯楚属,笼凤梨属,心花凤梨属,心花属);Canistrum ■☆
71150 Canistrum amazonicum Mez;筒凤梨;Canistrum ■☆
71151 Canistrum fosterianum L. B. Sm.;福斯特筒凤梨;Foster Canistrum ■☆
71152 Canistrum fragrans (Linden) Mabb.;林氏筒凤梨(彪纹菠萝,玲典筒凤梨);Linden Canistrum ■☆
71153 Canistrum lindenii (Regel) Mez = Canistrum fragrans (Linden) Mabb. ■☆
71154 Canizaresia Britton = Piscidia L.(保留属名)■☆
71155 Cankrienia de Vriese = Primula L. ■
71156 Canna L.(1753);美人蕉属(美人焦属,昙华属);Canna, Flowering Reed, Indian Shot, Indian Shot Plant ■
71157 Canna Noronha = Calamus L. ●
71158 Canna Noronha = Daemonorops Blume ●
71159 Canna Noronha = Plectocomia Mart. et Blume ●
71160 Canna achiras Gillies;智利美人蕉■☆
71161 Canna angustifolia L.;狭叶美人蕉■☆
71162 Canna angustifolia Walter = Canna flaccida Salisb. ■
71163 Canna bidentata Bertol.;非洲美人蕉;African Arrowroot ■☆
71164 Canna bidentata Bertol. = Canna indica L. ■
71165 Canna chinensis Willd. = Canna indica L. ■
71166 Canna coccinea Aiton;绯红美人蕉(红花美人蕉)■☆
71167 Canna coccinea Mill. = Canna indica L. ■
71168 Canna discolor Lindl. = Canna indica L. ■
71169 Canna edulis Ker Gawl.;蕉芋(芭蕉芋,飞天子,姜芽,姜芋,蕉藕,藕芋,羌芋,食用美人蕉,香珠);Achira, Edible Canna, Queensland Arrowroot, Tous-les-mois ■
71170 Canna edulis Ker Gawl. = Canna indica L. ■
71171 Canna flaccida Salisb.;柔瓣美人蕉(黄花美人蕉,黄花昙华);Bandana-of-the-everglades, Flaccid Canna, Golden Canna, Schlaffes Blumenrohr ■
71172 Canna flavescens Lindk = Canna indica L. var. flava (Roscoe) Baker ■
71173 Canna generalis L. H. Bailey;大花美人蕉(美人蕉,艳蕉);Canna Lily, Common Garden Canna, Largeflower Canna ■
71174 Canna generalis L. H. Bailey 'Doc';博士美人蕉■☆
71175 Canna generalis L. H. Bailey 'Eureca';尤里卡美人蕉■☆
71176 Canna generalis L. H. Bailey 'Pfitzer's Salmon Pink';鲑粉美人蕉■☆
71177 Canna generalis L. H. Bailey 'Striatus';条纹美人蕉■☆
71178 Canna generalis L. H. Bailey 'The President';总统美人蕉■☆
71179 Canna glauca L.;粉美人蕉(白花美人蕉);Brazilian Arrowroot, Canna, Glaucous Canna ■
71180 Canna indica Hell. subsp. orientalis (Roscoe) Baker = Canna indica L. ■
71181 Canna indica L.;美人蕉(白芭蕉,白花昙花,白莲蕉花,凤尾花,观音姜,红蕉,红玉簪花,虎头蕉,兰蕉,莲蕉,美人焦,破血红,水蕉,五筋草,小芭蕉,小芭蕉头,洋芭蕉);Cana, Edible Cana, India Canna, Indian Cane, Indian Canna, Indian Flowering Reed, Indian Shot, Indian-shot, Indian-shot Plant, Queensland Arrowroot, Tous-les-mois ■
71182 Canna indica L. var. flava (Roscoe) Baker;黄花美人蕉;Yellowflower Canna ■
71183 Canna indica L. var. flava Roscoe = Canna indica L. var. flava (Roscoe) Baker ■
71184 Canna indica L. var. nepalensis = Canna indica L. ■
71185 Canna indica L. var. orientalis (Roscoe) Baker = Canna indica L. ■
71186 Canna indica L. var. orientalis Hook. f.;东方美人蕉(美人蕉,昙华)■
71187 Canna indica L. var. orientalis Roscoe = Canna indica L. ■
71188 Canna indica L. var. rubra Aiton = Canna indica L. ■
71189 Canna iridiflora Ruiz et Pav.;垂花美人蕉(鸢尾花美人蕉)■☆
71190 Canna liliflora Warsz.;百合花美人蕉■☆
71191 Canna lutea Mill. = Canna indica L. ■
71192 Canna nepalensis D. Dietr. = Canna indica L. ■
71193 Canna orchioides L. H. Bailey;兰花美人蕉(黄花美人蕉,杂色美人蕉);Orchid Canna, Orchidflowered Canna ■
71194 Canna orientalis Roscoe = Canna indica L. ■
71195 Canna orientalis Roscoe var. flava Roscoe = Canna indica L. var. flava (Roscoe) Baker ■
71196 Canna patens Roscoe;兰昙华(和兰昙花,荷兰莲蕉,荷兰昙花)■
71197 Canna reevesii Lindl.;中国昙花■
71198 Canna speciosa Roscoe ex Sims;雅致美人蕉■☆
71199 Canna warscewiczii A. Dietr.;紫叶美人蕉;Purpleleaf Canna, Warscewicz Canna ■
71200 Canna warscewiczii A. Dietr. = Canna indica L. ■
71201 Cannabaceae Endl. = Cannabaceae Martinov(保留科名)■
71202 Cannabaceae Martinov (1820)(保留科名);大麻科;Hemp Family, Hop Family ■
71203 Cannabidaceae Endl. = Cannabaceae Martinov(保留科名)■
71204 Cannabina Mill. = Datisca L. ●■☆
71205 Cannabina Tourn. ex Medik. = Datisca L. ●■☆
71206 Cannabinaceae Lindl. = Cannabaceae Martinov(保留科名)■
71207 Cannabinastrum Fabr. = Galeopsis L. ■
71208 Cannabinastrum Heist. ex Fabr. = Galeopsis L. ■
71209 Cannabis L.(1753);大麻属;Hemp, Indian Hemp, Marihuana, Marijuana ■
71210 Cannabis australis L.;澳洲大麻■☆
71211 Cannabis chinensis Delile = Cannabis sativa L. ■
71212 Cannabis indica Lam. = Cannabis sativa L. var. indica (Lam.) Wehmer ■
71213 Cannabis indica Lam. = Cannabis sativa L. ■
71214 Cannabis indica Lam. var. kafiristanica Vavilov = Cannabis sativa L. ■
71215 Cannabis ruderalis Janisch. = Cannabis sativa L. f. ruderalis (Janisch.) S. Z. Liou ■
71216 Cannabis ruderalis Janisch. = Cannabis sativa L. ■
71217 Cannabis saliva L. subvar. indica (Lam.) Asch. et Graebn. = Cannabis sativa L. ■
71218 Cannabis sativa L.;大麻(荸,汉麻,胡麻,黄麻,火麻,火麻草,苴,麻,麻勃,麻蓝,麻母,麻线,绵麻子,青葛,青羊,山丝苗,唐麻,乌麻花,枲,枲实,线麻,宪麻子,小麻,野麻);Barren Hemp, Cannabis, Carl Hemp, Churl Hemp, Common Hemp, Femble, Fimble, Gallow-grass, Grass, Hashish, Hemp, Indian Hay, Indian Hemp,

Marihuana, Marijuana, Maryjane, Muggles, Neckweed, Nogs, Pot, Redroot, Reefers, Russian Hemp, Skunk, St. Audre's Lace, Staudre's Lace, Thistle Hemp, Tristram's Knot, Welsh Parsley ■

71219 Cannabis sativa L. f. ruderalis (Janisch.) S. Z. Liou = Cannabis sativa L. var. ruderalis (Janisch.) S. Z. Liou ■

71220 Cannabis sativa L. subsp. indica (Lam.) E. Small et Cronquist; 印度大麻(火麻); Bhang, Churras, Dagga, Ganja, Hashish, Indian Hemp, Kif, Marijuana ■

71221 Cannabis sativa L. subsp. indica (Lam.) E. Small et Cronquist = Cannabis sativa L. ■

71222 Cannabis sativa L. subsp. indica (Lam.) E. Small et Cronquist var. kafiristanica (Vavilov) E. Small et Cronquist = Cannabis sativa L. ■

71223 Cannabis sativa L. var. chinensis (Delile) Asch. et Graebn. = Cannabis sativa L. ■

71224 Cannabis sativa L. var. indica (Lam.) E. Small et Cronquist = Cannabis sativa L. subsp. indica (Lam.) E. Small et Cronquist ■

71225 Cannabis sativa L. var. indica (Lam.) E. Small et Cronquist = Cannabis sativa L. ■

71226 Cannabis sativa L. var. indica (Lam.) Wehmer = Cannabis sativa L. subsp. indica (Lam.) E. Small et Cronquist ■

71227 Cannabis sativa L. var. kif DC. = Cannabis sativa L. ■

71228 Cannabis sativa L. var. ruderalis (Janisch.) S. Z. Liou; 野大麻; Wild Hemp ■

71229 Cannabis sativa L. var. spontanea Vavilov = Cannabis sativa L. ■

71230 Cannaboides B. -E. van Wyk (1999); 拟大麻属■☆

71231 Cannaboides andohahelensis (Humbert) B. -E. van Wyk; 拟大麻 ■☆

71232 Cannaboides andohahelensis (Humbert) B. -E. van Wyk = Heteromorpha andohahelensis Humbert ■☆

71233 Cannaboides andohahelensis (Humbert) B. -E. van Wyk var. denudata (Humbert) B. -E. van Wyk = Cannaboides andohahelensis (Humbert) B. -E. van Wyk ■☆

71234 Cannaboides betsileensis (Humbert) B. -E. van Wyk; 马岛拟大麻■☆

71235 Cannaceae Juss. (1789)(保留科名); 美人蕉科(昙花科,昙华科); Canna Family ■

71236 Cannacorus Mill. = Canna L. ■

71237 Cannacorus Tourn. ex Medik. = Canna L. ■

71238 Cannaeorchis M. A. Clem. et D. L. Jones = Dendrobium Sw. (保留属名) ■

71239 Cannaeovus Mill. = Canna L. ■

71240 Cannella Schott ex Meisn. = Ocotea Aubl. ●☆

71241 Cannomois P. Beauv. ex Desv. (1828); 歪果帚灯草属■☆

71242 Cannomois acuminata (Kunth) Pillans = Cannomois parviflora (Thunb.) Pillans ■☆

71243 Cannomois aristata Mast.; 具芒歪果帚灯草■☆

71244 Cannomois cephalotes Desv. = Cannomois virgata (Rottb.) Steud. ■☆

71245 Cannomois complanatus Mast. = Cannomois parviflora (Thunb.) Pillans ■☆

71246 Cannomois congesta Mast.; 密集歪果帚灯草■☆

71247 Cannomois dregei Pillans = Cannomois scirpoides (Kunth) Mast. ■☆

71248 Cannomois nitida (Mast.) Pillans; 光亮歪果帚灯草■☆

71249 Cannomois parviflora (Thunb.) Pillans; 小花歪果帚灯草■☆

71250 Cannomois schlechteri Mast. = Cannomois parviflora (Thunb.) Pillans ■☆

71251 Cannomois scirpoides (Kunth) Mast.; 藨草歪果帚灯草■☆

71252 Cannomois scirpoides (Kunth) Mast. var. minor Pillans = Cannomois congesta Mast. ■☆

71253 Cannomois scirpoides (Kunth) Mast. var. primosii Pillans = Cannomois congesta Mast. ■☆

71254 Cannomois simplex Kunth = Cannomois parviflora (Thunb.) Pillans ■☆

71255 Cannomois taylorii H. P. Linder; 泰勒歪果帚灯草■☆

71256 Cannomois virgata (Rottb.) Steud.; 条纹歪果帚灯草■☆

71257 Canonanthus G. Don = Siphocampylus Pohl ■●☆

71258 Canophollis G. Don = Conopholis Walk. ■☆

71259 Canophora Post et Kuntze = Canephora Juss. ■☆

71260 Canopodaceae C. Presl = Santalaceae R. Br. (保留科名) ●■

71261 Canopus C. Presl = Exocarpos Labill. (保留属名) ●☆

71262 Canothus Rain. = Ceanothus L. ●☆

71263 Canotia Torr. (1857); 墨西哥卫矛属(卡诺希属) ●☆

71264 Canotia holacantha Torr.; 墨西哥卫矛(卡诺希); Canotia, Crucifixion Thorn ●☆

71265 Canotiaceae Airy Shaw = Celastraceae R. Br. (保留科名) ●

71266 Canotiaceae Britton = Celastraceae R. Br. (保留科名) ●

71267 Canotiaceae Britton; 墨西哥卫矛科(卡诺希科) ●☆

71268 Canschi Adans. = Trewia L. ●

71269 Canscora Griseb. = Canscora Lam. ■

71270 Canscora Lam. (1785); 穿心草属(贯叶草属,堪司哥拉属); Canscora ■

71271 Canscora alata (Roth ex Roem. et Schult.) Wall.; 对生穿心草■☆

71272 Canscora andrographioides Griff. = Centaurium pulchellum (Sw.) Druce var. altaicum (Griseb.) Kitag. et H. Hara ■

71273 Canscora andrographioides Griff. ex C. B. Clarke; 罗星草(白花田草,糖果草); Melastomalike Canscora, Whiteflower Canscora ■

71274 Canscora carinata Dop = Cracosna carinata (Dop) Thiv ■☆

71275 Canscora decussata (Roxb.) Schult. et Schult. f. = Canscora alata (Roth ex Roem. et Schult.) Wall. ■☆

71276 Canscora diffusa (Vahl) R. Br. ex Roem. et Schult.; 铺地穿心草■

71277 Canscora kirkii N. E. Br. = Canscora diffusa (Vahl) R. Br. ex Roem. et Schult. ■

71278 Canscora lawii Wight = Canscora diffusa (Vahl) R. Br. ex Roem. et Schult. ■

71279 Canscora lucidissima (H. Lév. et Vaniot) Hand. -Mazz.; 穿心草(穿钱草,穿心莲,串钱草,顶心风,狮子草); Common Canscora ■

71280 Canscora melastomacea Hand. -Mazz. = Canscora andrographioides Griff. ex C. B. Clarke ■

71281 Canscora perfoliata Lam.; 贯叶草; Penetrateleaf Canscora ■

71282 Canscora ramosissima Baker = Crepidorhopalon spicatus (Engl.) Eb. Fisch. ■☆

71283 Canscora rubiflora X. X. Chen; 红花罗星草; Redflower Canscora ■

71284 Canscora rubiflora X. X. Chen = Canscora diffusa (Vahl) R. Br. ex Roem. et Schuldt ■

71285 Canscora tenella Wall. = Canscora diffusa (Vahl) R. Br. ex Roem. et Schult. ■

71286 Canscora tetragona Schinz = Schinziella tetragona (Schinz) Gilg ■☆

71287 Cansenia Raf. = Bauhinia L. ●

71288 Cansiera Spreng. = Potameia Thouars ●☆

71289 Cansjera Juss. (1789)(保留属名); 山柑藤属; Cansjera ●

71290 Cansjera manillana Blume = Champereia manillana (Blume) Merr. ●

71291 Cansjera pentandra Blanco = Antidesma pentandrum (Blanco) Merr. ●

71292 Cansjera rheedii J. F. Gmel.;山柑藤(捞饺藤,山柑,山柚藤,牙刷树);Rheed Cansjera ●

71293 Cansjera scandens Roxb. = Cansjera rheedii J. F. Gmel. ●

71294 Cansjeraceae J. Agardh = Opiliaceae Valeton(保留科名)●

71295 Cansjeraceae J. Agardh;山柑藤科●

71296 Cantalea Raf. = Lycium L. ●

71297 Cantalea Raf. = Trozelia Raf.(废弃属名)●☆

71298 Cantamine muliensis W. T. Wang = Cardamine yunnanensis Franch. ■

71299 Cantamine multijuga Franch. = Cardamine griffithii Hook. f. et Thomson ■

71300 Cantamine polyphylla D. Don = Cardamine macrophylla Willd. ■

71301 Cantamine pratensis subsp. chinensis O. E. Schulz = Cardamine stenoloba Hemsl. ■

71302 Cantamine pratensis subsp. chinensis O. E. Schulz = Loxostemon stenolobus (Hemsl.) Y. C. Lan et T. Y. Cheo ■

71303 Cantamine scutata var. regeliana (Miq.) H. Hara = Cardamine scutata Thunb. ■

71304 Cantharospermum Wight et Arn. = Atylosia Wight et Arn. ●■

71305 Cantharospermum Wight et Arn. = Cajanus Adans.(保留属名)●

71306 Cantharospermum molle (Benth.) Taub. = Cajanus mollis (Benth.) Maesen ●

71307 Cantharospermum niveum (Benth.) Raizada = Cajanus niveus (Benth.) Maesen ●

71308 Cantharospermum scarabaeoides (L.) Baill. = Cajanus scarabaeoides (L.) Thouars ●

71309 Cantharospermum volubilis (Blanco) Merr. = Cajanus crassus (Prain ex King) Maesen ●

71310 Cantharosperrnum scarabaeoides (L.) Baill. = Cajanus scarabaeoides (L.) Thouars ●

71311 Canthiopsis Seem. = Randia L. ●

71312 Canthiopsis Seem. = Tarenna Gaertn. ●

71313 Canthium Lam.(1785);鱼骨木属(步散属);Canthium ●

71314 Canthium abbreviatum (K. Schum.) S. Moore = Pygmaeothamnus zeyheri (Sond.) Robyns ●☆

71315 Canthium acarophytum (De Wild.) Evrard = Psydrax acutiflora (Hiern) Bridson ●☆

71316 Canthium acutiflorum Hiern = Psydrax acutiflora (Hiern) Bridson ●☆

71317 Canthium affine (Robyns) Hepper = Pyrostria affinis (Robyns) Bridson ●☆

71318 Canthium afzelianum Hiern;阿芙泽尔鱼骨木●☆

71319 Canthium amplum S. Moore = Multidentia crassa (Hiern) Bridson et Verdc. var. ampla (S. Moore) Bridson et Verdc. ●☆

71320 Canthium andringitrense Cavaco;安德林吉特拉山鱼骨木●☆

71321 Canthium ankaranense Arènes ex Cavaco;安卡兰鱼骨木●☆

71322 Canthium anomocarpum DC. = Psydrax horizontalis (Schumach.) Bridson ●☆

71323 Canthium arnoldianum (De Wild. et T. Durand) Hepper = Psydrax arnoldiana (De Wild. et T. Durand) Bridson ●☆

71324 Canthium barteri Hiern = Keetia venosa (Oliv.) Bridson ●☆

71325 Canthium benthamianum Baill. = Rytigynia canthioides (Benth.) Robyns ■☆

71326 Canthium bibracteatum (Baker) Hiern = Pyrostria bibracteata (Baker) Cavaco ●☆

71327 Canthium bogosense (Martelli) Penz. = Pyrostria phyllanthoidea (Baill.) Bridson ●☆

71328 Canthium boinense Arènes ex Cavaco;博伊纳鱼骨木●☆

71329 Canthium bosseri Cavaco;博瑟鱼骨木●☆

71330 Canthium brownii Bullock = Keetia tenuiflora (Hiern) Bridson ●☆

71331 Canthium burttii Bullock;伯特鱼骨木●☆

71332 Canthium burttii Bullock subsp. glabrum Bridson;无毛伯特鱼骨木●☆

71333 Canthium burttii Bullock var. nanguanum Tennant = Canthium racemulosum S. Moore var. nanguanum (Tennant) Bridson ●☆

71334 Canthium captum Bullock = Canthium oligocarpum Hiern subsp. captum (Bullock) Bridson ●☆

71335 Canthium caudatiflorum Hiern = Psydrax horizontalis (Schumach.) Bridson ●☆

71336 Canthium celastroides Baill. = Rytigynia celastroides (Baill.) Verdc. ●☆

71337 Canthium charadrophilum (K. Krause) Bullock = Keetia gueinzii (Sond.) Bridson ●☆

71338 Canthium cienkowskii (Schweinf.) Roberty = Fadogia cienkowskii Schweinf. ●☆

71339 Canthium ciliatum (Klotzsch ex Eckl. et Zeyh.) Kuntze;缘毛鱼骨木●☆

71340 Canthium clityophilum Bullock = Psydrax livida (Hiern) Bridson ●☆

71341 Canthium congensis Hiern = Keetia gracilis (Hiern) Bridson ●☆

71342 Canthium cornelia Cham. et Schltdl. = Keetia cornelia (Cham. et Schltdl.) Bridson ●☆

71343 Canthium cornelianum St. -Lag. = Keetia cornelia (Cham. et Schltdl.) Bridson ●☆

71344 Canthium crassum Hiern = Multidentia crassa (Hiern) Bridson et Verdc. ●☆

71345 Canthium decaryi Homolle ex Cavaco;德卡里鱼骨木●☆

71346 Canthium dewildemanianum E. M. Petit et Evrard = Psydrax splendens (K. Schum.) Bridson ●☆

71347 Canthium dicoccum (Gaertn.) Merr. = Canthium dicoccum (Gaertn.) Teijsm. et Binn. ●

71348 Canthium dicoccum (Gaertn.) Teijsm. et Binn.;鱼骨木(白骨木,步散,双鱼骨木,铁屎米);Butulang Canthium ●

71349 Canthium dicoccum (Gaertn.) Teijsm. et Binn. var. obovatifolium G. A. Fu;倒卵叶鱼骨木;Obovateleaf Canthium ●

71350 Canthium dictyophlebum S. Moore = Multidentia crassa (Hiern) Bridson et Verdc. ●☆

71351 Canthium didymum C. F. Gaertn. = Canthium dicoccum (Gaertn.) Teijsm. et Binn. ●

71352 Canthium discolor Benth. = Vangueriella discolor (Benth.) Verdc. ●☆

71353 Canthium dubium Lindl. = Diplospora dubia (Lindl.) Masam. ●

71354 Canthium dundusanense (De Wild.) Evrard = Keetia venosa (Oliv.) Bridson ●☆

71355 Canthium dunnianum H. Lév. = Lasianthus japonicus Miq. ●

71356 Canthium egregium Bullock = Psydrax kraussioides (Hiern) Bridson ●☆

71357 Canthium euonymoides (Schweinf. ex Hiern) Baill. = Rytigynia senegalensis Blume ■●☆

71358 Canthium euryoides Bullock ex Hutch. et Dalziel;柃状鱼骨木●☆

71359 Canthium euryoides Bullock ex Hutch. et Dalziel = Psydrax splendens (K. Schum.) Bridson ●☆

71360 Canthium fadenii Bridson;法登鱼骨木●☆

71361 Canthium fanshawei Tennant = Multidentia fanshawei (Tennant) Bridson ●☆

71362 Canthium favosum Hutch. et Dalziel = Keetia hispida (Benth.) Bridson ●☆

71363 Canthium foetidum Hiern = Keetia foetida (Hiern) Bridson ●☆

71364 Canthium foliosum (Burtt Davy) Burtt Davy = Pavetta eylesii S. Moore ●☆

71365 Canthium fragrantissimum (K. Schum.) Cavaco = Psydrax fragrantissima (K. Schum.) Bridson ●☆

71366 Canthium frangula S. Moore = Canthium glaucum Hiern subsp. frangula (S. Moore) Bridson ●☆

71367 Canthium gilfillanii (N. E. Br.) O. B. Mill.;吉尔菲兰鱼骨木●☆

71368 Canthium giordanii Chiov. = Psydrax parviflora (Afzel.) Bridson ●☆

71369 Canthium glabriflorum Hiern = Psydrax subcordata (DC.) Bridson ●☆

71370 Canthium glaucum Hiern;灰绿毛鱼骨木●☆

71371 Canthium glaucum Hiern subsp. frangula (S. Moore) Bridson;脆鱼骨木●☆

71372 Canthium glaucum Hiern var. frangula (S. Moore) Bridson = Canthium glaucum Hiern subsp. frangula (S. Moore) Bridson ●☆

71373 Canthium glaucum Hiern var. pubescens Bridson;短柔毛鱼骨木●☆

71374 Canthium golungense Hiern = Psydrax parviflora (Afzel.) Bridson ●☆

71375 Canthium golungense Hiern var. parviflorum S. Moore = Psydrax parviflora (Afzel.) Bridson ●☆

71376 Canthium gracile Hiern = Keetia gracilis (Hiern) Bridson ●☆

71377 Canthium greenwayi Bullock = Canthium mombazense Baill. ●☆

71378 Canthium gueinzii Sond. = Keetia gueinzii (Sond.) Bridson ●☆

71379 Canthium gymnosporioides Launert = Psydrax livida (Hiern) Bridson ●☆

71380 Canthium gynochodes Baill.;拟狗骨●

71381 Canthium gynochthodes Baill.;兰屿鱼骨木(兰屿观吉木,朴莱木,台湾猪肚木);Lanyu Canthium ●

71382 Canthium henriquesianum (K. Schum.) G. Taylor = Psydrax acutiflora (Hiern) Bridson ●☆

71383 Canthium henryi H. Lév. = Damnacanthus henryi (H. Lév.) H. S. Lo ●

71384 Canthium hirtellum Ridl.;粗毛鱼骨木●☆

71385 Canthium hispidum Benth. = Keetia hispida (Benth.) Bridson ●☆

71386 Canthium horizontale (Schumach.) Hiern = Psydrax horizontalis (Schumach.) Bridson ●☆

71387 Canthium horridum Blume;猪肚木(刺鱼骨木,跌掌随,山石榴,铁拳随,猪肚簕,猪腹簕);Bristly Canthium ●

71388 Canthium huillense Hiern = Psydrax livida (Hiern) Bridson ●☆

71389 Canthium humbertianum Cavaco;亨伯特鱼骨木●☆

71390 Canthium impressinervium Bridson;陷脉鱼骨木●☆

71391 Canthium inaequilaterum Hutch. et Dalziel = Keetia inaequilatera (Hutch. et Dalziel) Bridson ●☆

71392 Canthium inerme (L. f.) Kuntze;无刺鱼骨木●☆

71393 Canthium infaustum (Burch.) Baill. = Vangueria infausta Burch. ■☆

71394 Canthium inflatum Cavaco;膨胀鱼骨木●☆

71395 Canthium inopinatum Bullock = Canthium mombazense Baill. ●☆

71396 Canthium italyense Cavaco;意大利鱼骨木●☆

71397 Canthium junodii (Burtt Davy) Burtt Davy = Psydrax livida (Hiern) Bridson ●☆

71398 Canthium kaessneri S. Moore = Psydrax kaessneri (S. Moore) Bridson ●☆

71399 Canthium keniense Bullock;肯尼亚鱼骨木●☆

71400 Canthium kilifiense Bridson;基利菲鱼骨木●☆

71401 Canthium kitsoni S. Moore = Keetia multiflora (Schumach. et Thonn.) Bridson ●☆

71402 Canthium kraussioides Hiern = Psydrax kraussioides (Hiern) Bridson ●☆

71403 Canthium kuntzeanum Bridson;孔策鱼骨木●☆

71404 Canthium labordei H. Lév. = Damnacanthus labordei (H. Lév.) H. S. Lo ●

71405 Canthium lactescens Hiern;乳白鱼骨木●☆

71406 Canthium lactescens Hiern var. grandifolium S. Moore = Canthium lactescens Hiern ●☆

71407 Canthium lacusvictoriae Bullock = Psydrax acutiflora (Hiern) Bridson ●☆

71408 Canthium lagoensis Baill. = Rytigynia umbellulata (Hiern) Robyns ■☆

71409 Canthium lanciflorum Hiern = Vangueriopsis lanciflora (Hiern) Robyns ●☆

71410 Canthium latiflorum Homolle ex Cavaco;宽花鱼骨木●☆

71411 Canthium latifolium F. Muell. ex Benth.;宽叶鱼骨木;Plum-bush ●☆

71412 Canthium laurinum (Poir.) Roberty = Craterispermum laurinum (Poir.) Benth. ●☆

71413 Canthium libericum Dinkl.;离生鱼骨木●☆

71414 Canthium lividum Hiern = Psydrax livida (Hiern) Bridson ●☆

71415 Canthium loandensis S. Moore = Pavetta loandensis (S. Moore) Bremek. ●☆

71416 Canthium locuples (K. Schum.) Codd = Psydrax locuples (K. Schum.) Bridson ●☆

71417 Canthium lucidum R. Br. = Psydrax schimperiana (A. Rich.) Bridson ●☆

71418 Canthium macrocalyx (Sond.) Baill. = Pachystigma macrocalyx (Sond.) Robyns ●☆

71419 Canthium macrostipulatum (De Wild.) Evrard = Keetia molundensis (K. Krause) Bridson var. macrostipulata (De Wild.) Bridson ●☆

71420 Canthium malacocarpum (K. Schum. et K. Krause) Bullock = Psydrax kraussioides (Hiern) Bridson ●☆

71421 Canthium maleolens Chiov. = Vangueria madagascariensis J. F. Gmel. ■☆

71422 Canthium mandrarense Cavaco;曼德拉鱼骨木●☆

71423 Canthium manense Aubrév. et Pellegr. = Psydrax manensis (Aubrév. et Pellegr.) Bridson ●☆

71424 Canthium mannii Hiern = Keetia mannii (Hiern) Bridson ●☆

71425 Canthium marojejyense Cavaco;马罗鱼骨木●☆

71426 Canthium martinii Dunkley = Psydrax martinii (Dunkley) Bridson ●☆

71427 Canthium medusulum Hiern = Keetia hispida (Benth.) Bridson ●☆

71428 Canthium melanophengos Bullock = Psydrax parviflora (Afzel.) Bridson subsp. melanophengos (Bullock) Bridson ●☆

71429 Canthium micans Bullock = Psydrax micans (Bullock) Bridson ●☆

71430 Canthium microdon S. Moore = Canthium setiflorum Hiern ●☆

71431 Canthium mombazense Baill.;蒙巴兹鱼骨木●☆

71432 Canthium mortehanii (De Wild.) Evrard = Keetia molundensis (K. Krause) Bridson var. macrostipulata (De Wild.) Bridson ●☆

71433 Canthium multiflorum (Schumach. et Thonn.) Hiern = Keetia multiflora (Schumach. et Thonn.) Bridson ●☆

71434 Canthium mundianum Cham. et Schltdl.;蒙德鱼骨木●☆

71435 Canthium myrtifolium S. Moore = Psydrax schimperiana (A. Rich.) Bridson ●☆

71436 Canthium neglectum Hiern = Rytigynia neglecta (Hiern) Robyns ■☆

71437 Canthium nervosum Hiern;多脉鱼骨木●☆

71438 Canthium ngonii Bridson;恩贡鱼骨木●☆

71439 Canthium nitens Hiern = Psydrax splendens (K. Schum.) Bridson ●☆

71440 Canthium oatesii Rolfe = Pygmaeothamnus zeyheri (Sond.) Robyns ●☆

71441 Canthium obovatum Klotzsch ex Eckl. et Zeyh. = Psydrax obovata (Klotzsch ex Eckl. et Zeyh.) Bridson ●☆

71442 Canthium obovatum Klotzsch ex Eckl. et Zeyh. var. pyrifolium (Klotzsch ex Eckl. et Zeyh.) Sond. = Psydrax obovata (Klotzsch ex Eckl. et Zeyh.) Bridson ●☆

71443 Canthium oligocarpum Hiern;寡果鱼骨木●☆

71444 Canthium oligocarpum Hiern subsp. angustifolium Bridson;窄叶寡果鱼骨木●☆

71445 Canthium oligocarpum Hiern subsp. captum (Bullock) Bridson;头状寡果鱼骨木●☆

71446 Canthium oligocarpum Hiern subsp. friesiorum (Robyns) Bridson;弗里斯鱼骨木●☆

71447 Canthium oligocarpum Hiern subsp. intermedium Bridson;间型鱼骨木●☆

71448 Canthium opimum S. Moore = Multidentia crassa (Hiern) Bridson et Verdc. ●☆

71449 Canthium orbiculare (K. Schum.) R. D. Good = Rytigynia orbicularis (K. Schum.) Robyns ■☆

71450 Canthium orthacanthum (Mildbr.) Robyns = Vangueriella orthacantha (Mildbr.) Bridson et Verdc. ●☆

71451 Canthium ovatum (Burtt Davy) Burtt Davy = Plectroniella armata (K. Schum.) Robyns ●☆

71452 Canthium pallidum (K. Schum.) Bullock = Canthium mombazense Baill. ●☆

71453 Canthium palma (K. Schum.) R. D. Good = Psydrax palma (K. Schum.) Bridson ●☆

71454 Canthium parviflorum Bartl. ex DC.;小花鱼骨木●☆

71455 Canthium parvifolium Roxb.;小叶鱼骨木(铁屎米,小叶铁屎米);Small-leaved Canthium ●

71456 Canthium parvifolium Roxb. = Canthium horridum Blume ●

71457 Canthium pauciflorum (Klotzsch ex Eckl. et Zeyh.) Kuntze = Canthium kuntzeanum Bridson ●☆

71458 Canthium pauciflorum Baill. = Rytigynia pauciflora (Schweinf. ex Hiern) Robyns ■☆

71459 Canthium perrieri Cavaco;佩里耶鱼骨木●☆

71460 Canthium peteri Bridson;彼得鱼骨木●☆

71461 Canthium phyllanthoideum Baill. = Pyrostria phyllanthoidea (Baill.) Bridson ●☆

71462 Canthium platyphyllum Hiern = Vangueriopsis lanciflora (Hiern) Robyns ●☆

71463 Canthium pobeguinii Hutch. et Dalziel = Multidentia pobeguinii (Hutch. et Dalziel) Bridson ●☆

71464 Canthium polycarpum Schweinf. ex Hiern = Psydrax subcordata (DC.) Bridson ●☆

71465 Canthium pseudosetiflorum Bridson;假毛花鱼骨木●☆

71466 Canthium pseudoverticillatum S. Moore;拟轮生鱼骨木●☆

71467 Canthium pseudoverticillatum S. Moore subsp. somaliense Bridson;索马里鱼骨木●☆

71468 Canthium pubipes S. Moore = Chazaliella abrupta (Hiern) E. M. Petit et Verdc. ●☆

71469 Canthium purpurascens Bullock = Keetia purpurascens (Bullock) Bridson ●☆

71470 Canthium pynaertii (De Wild.) C. M. Evrard = Keetia gueinzii (Sond.) Bridson ●☆

71471 Canthium pyrifolium Klotzsch ex Eckl. et Zeyh. = Psydrax obovata (Klotzsch ex Eckl. et Zeyh.) Bridson ●☆

71472 Canthium racemulosum S. Moore;小总花鱼骨木●☆

71473 Canthium racemulosum S. Moore var. nanguanum (Tennant) Bridson;南古鱼骨木●☆

71474 Canthium randii S. Moore = Canthium lactescens Hiern ●☆

71475 Canthium recurvifolium Bullock = Psydrax recurvifolia (Bullock) Bridson ●☆

71476 Canthium rhamnifolium (Chiov.) Cufod. = Coffea rhamnifolia (Chiov.) Bridson ●☆

71477 Canthium rhamnoides Hiern = Vangueriella rhamnoides (Hiern) Verdc. ●☆

71478 Canthium ripae (De Wild.) Evrard = Keetia ripae (De Wild.) Bridson ●☆

71479 Canthium robynsianum Bullock = Canthium pseudoverticillatum S. Moore ●☆

71480 Canthium robynsianum C. M. Evrard = Keetia acuminata (De Wild.) Bridson ●☆

71481 Canthium rondoense Bridson;龙多鱼骨木●☆

71482 Canthium rubens Hiern = Keetia rubens (Hiern) Bridson ●☆

71483 Canthium rubrocostatum Robyns = Psydrax parviflora (Afzel.) Bridson subsp. rubrocostata (Robyns) Bridson ●☆

71484 Canthium rufivillosum Robyns ex Hutch. et Dalziel = Keetia rufivillosa (Robyns ex Hutch. et Dalziel) Bridson ●☆

71485 Canthium ruwenzoriense Bullock = Canthium oligocarpum Hiern ●☆

71486 Canthium sahafaryense Cavaco;萨哈法里鱼骨木●☆

71487 Canthium salubenii Bridson;萨卢本鱼骨木●☆

71488 Canthium sapini (De Wild.) Robyns = Vangueriella sapini (De Wild.) Verdc. ●☆

71489 Canthium sarogliae Chiov. = Rytigynia uhligii (K. Schum. et K. Krause) Verdc. ■☆

71490 Canthium scabrosum Bullock = Keetia gueinzii (Sond.) Bridson ●☆

71491 Canthium schimperianum A. Rich. = Psydrax schimperiana (A. Rich.) Bridson ●☆

71492 Canthium sclerocarpum (K. Schum.) Bullock = Multidentia sclerocarpa (K. Schum.) Bridson ●☆

71493 Canthium senegalense A. Rich. = Keetia cornelia (Cham. et Schltdl.) Bridson ●☆

71494 Canthium sennii (Chiov.) Cufod. = Pavetta uniflora Bremek. ●☆

71495 Canthium setiflorum Hiern;毛花鱼骨木●☆

71496 Canthium shabanii Bridson;沙邦鱼骨木●☆

71497 Canthium sidamense Cufod. = Canthium oligocarpum Hiern ●☆

71498 Canthium simile Merr. et Chun;大叶鱼骨木(六大天王);Bigleaf Canthium,Largeleaf Canthium,Large-leaved Canthium ●

71499 Canthium sordidum (K. Schum.) Bullock;污浊鱼骨木●☆

71500 Canthium spinosissimum Merr. = Fagerlindia depauperata (Drake) Tirveng. ●

71501 Canthium spinosum (Klotzsch ex Eckl. et Zeyh.) Kuntze;多刺鱼骨木●☆

71502 Canthium stenosepalum Lantz;窄萼鱼骨木●☆

71503 Canthium stuhlmannii Bullock = Multidentia castaneae (Robyns) Bridson et Verdc. ●☆

71504 Canthium subcordatum DC.;亚心形鱼骨木●☆

71505 Canthium subcordatum DC. = Psydrax subcordata (DC.) Bridson ●☆

71506 Canthium suberosum Codd;木栓质鱼骨木●☆

71507 Canthium subopacum (K. Schum. et K. Krause) Bullock = Tricalysia ovalifolia Hiern ●☆

71508 Canthium swynnertonii S. Moore = Canthium inerme (L. f.) Kuntze ●☆

71509 Canthium sylvaticum Hiern = Keetia venosa (Oliv.) Bridson ●☆

71510 Canthium syringodorum (K. Schum.) Bullock = Psydrax livida (Hiern) Bridson ●☆

71511 Canthium tamatavense Cavaco;塔马塔夫鱼骨木●☆

71512 Canthium tekbe Aubrév. et Pellegr. = Psydrax arnoldiana (De Wild. et T. Durand) Bridson ●☆

71513 Canthium tenuiflorum Hiern = Keetia tenuiflora (Hiern) Bridson ●☆

71514 Canthium tetraphyllum Baill. = Meyna tetraphylla (Schweinf. ex Hiern) Robyns ●☆

71515 Canthium thonningii Benth.;通宁鱼骨木●☆

71516 Canthium thunbergianum Cham. et Schltdl. = Canthium inerme (L. f.) Kuntze ●☆

71517 Canthium tophami Bullock et Dunkley = Keetia foetida (Hiern) Bridson ●☆

71518 Canthium transvaalensis S. Moore = Pygmaeothamnus zeyheri (Sond.) Robyns ●☆

71519 Canthium tsaratananense Homolle ex Cavaco;察拉塔纳纳鱼骨木●☆

71520 Canthium umbrosum Hiern = Canthium lactescens Hiern ●☆

71521 Canthium urophyllum Chiov. = Rytigynia bugoyensis (K. Krause) Verdc. ■☆

71522 Canthium vanguerioides Hiern = Vangueriella nigerica (Robyns) Verdc. ●☆

71523 Canthium vanwykii Tilney et Kok;万维鱼骨木●☆

71524 Canthium vatkeanum Hiern = Rytigynia neglecta (Hiern) Robyns var. vatkeana (Hiern) Verdc. ●☆

71525 Canthium venosissimum Hutch. et Dalziel = Vangueriella discolor (Benth.) Verdc. ●☆

71526 Canthium venosum (Oliv.) Hiern = Keetia venosa (Oliv.) Bridson ●☆

71527 Canthium venosum (Oliv.) Hiern var. pubescens Hiern = Keetia venosa (Oliv.) Bridson ●☆

71528 Canthium ventosum (L.) Kuntze = Olinia ventosa (L.) Cufod. ●☆

71529 Canthium viguieri Homolle ex Cavaco;维基耶鱼骨木●☆

71530 Canthium virgatum Hiern = Psydrax virgata (Hiern) Bridson ●☆

71531 Canthium viridissimum Wernham = Rytigynia membranacea (Hiern) Robyns ■☆

71532 Canthium vollesenii Bridson;福勒森鱼骨木●☆

71533 Canthium vulgare (K. Schum.) Bullock = Psydrax parviflora (Afzel.) Bridson ●☆

71534 Canthium welwitschii Hiern = Psydrax subcordata (DC.) Bridson ●☆

71535 Canthium whitei (Bridson) White = Psydrax whitei Bridson ●☆

71536 Canthium wildii (Suess.) Codd = Psydrax livida (Hiern) Bridson ●☆

71537 Canthium zanzibaricum Klotzsch = Keetia zanzibarica (Klotzsch) Bridson ●☆

71538 Canthium zanzibaricum Klotzsch var. glabristyla Hiern = Keetia zanzibarica (Klotzsch) Bridson subsp. cornelioides (De Wild.) Bridson ●☆

71539 Canthopsis Miq. = Catunaregam Wolf ●

71540 Canthopsis Miq. = Randia L. ●

71541 Cantleya Ridl. (1922);香茶茱萸属●☆

71542 Cantleya corniculata (Becc.) R. A. Howard;角香茶茱萸●☆

71543 Cantua Juss. = Cantua Juss. ex Lam. ●☆

71544 Cantua Juss. ex Lam. (1785);坎图木属(坎吐阿木属,坎吐阿属,魔力花属);Magic Flower of the Incas,Sacred Flower of the Incas ●☆

71545 Cantua Lam. = Cantua Juss. ex Lam. ●☆

71546 Cantua bicolor Lem.;二色坎图木●☆

71547 Cantua buxifolia Lam.;魔花坎图木(黄杨叶坎吐阿,魔力花);Flower-of-the-Incas,Magic Flower,Sacret Flower of the Incas,Scarlet Gilia,Skyrocket ●☆

71548 Cantua buxifolia Lam. 'Hot Pants';紧身裤魔花坎图木●☆

71549 Cantua dependens Pers. = Cantua buxifolia Lam. ●☆

71550 Cantua pyrifolia Juss. ex Lam.;梨叶坎图木(梨叶坎吐阿木)●☆

71551 Cantua quercifolia Juss.;栎叶坎图木(栎叶坎吐阿木)●☆

71552 Cantuffa J. F. Gmel. (废弃属名) = Pterolobium R. Br. ex Wight et Arn. (保留属名)●

71553 Cantuffa exosa J. F. Gmel. = Pterolobium stellatum (Forssk.) Brenan ●☆

71554 Caonabo Turpin ex Raf. = Columnea L. ●■☆

71555 Caopia Adans. (废弃属名) = Vismia Vand. (保留属名)●☆

71556 Caoutchoua J. F. Gmel. = Hevea Aubl. ●

71557 Capanea Decne. = Capanea Decne. ex Planch. ●■☆

71558 Capanea Decne. ex Planch. (1849);卡帕苣苔属●■☆

71559 Capanea Planch. = Capanea Decne. ex Planch. ●■☆

71560 Capanea affinis Fritsch;近缘卡帕苣苔●■☆

71561 Capanemia Barb. Rodr. (1877);卡氏兰属(卡班兰属)■☆

71562 Capanemia uliginosa Barb. Rodr.;卡氏兰■☆

71563 Capassa Klotzsch = Lonchocarpus Kunth(保留属名)●■☆

71564 Capassa Klotzsch = Philenoptera Fenzl ex A. Rich. ●■☆

71565 Capassa violacea Klotzsch = Philenoptera violacea (Klotzsch) Schrire ●☆

71566 Capelio B. Nord. (2002);多绒菊属■☆

71567 Capelio caledonica B. Nord.;卡利登多绒菊■☆

71568 Capelio tabularis (Thunb.) B. Nord.;扁平多绒菊■☆

71569 Capelio tomentosa (Burm. f.) B. Nord.;毛多绒菊■☆

71570 Capellenia Hassk. = Capellia Blume ■☆

71571 Capellenia Hassk. = Dillenia L. ●

71572 Capellenia Teijsm. et Binn. = Endospermum Benth. (保留属名)●

71573 Capellia Blume = Dillenia Heist. ex Fabr. ■☆

71574 Capellia Blume = Sherardia L. ■☆

71575 Capeobolus J. Browning = Costularia C. B. Clarke ex Dyer ■☆
71576 Capeobolus brevicaulis (C. B. Clarke) Browning = Costularia brevicaulis C. B. Clarke ■☆
71577 Caperonia A. St. -Hil. (1826);卡普龙大戟属(羊大戟属)■☆
71578 Caperonia angusta S. F. Blake;巴拿马羊大戟■☆
71579 Caperonia buchananii Baker = Caperonia fistulosa Beille ■☆
71580 Caperonia capiibariensis Eskuche;羊大戟■☆
71581 Caperonia castaneifolia A. St. -Hil.;栗叶羊大戟;Chestnutleaf False Croton ■☆
71582 Caperonia chevalieri Beille = Caperonia serrata (Turcz.) C. Presl ■☆
71583 Caperonia corchoroides;圭亚那羊大戟■☆
71584 Caperonia cubana Pax et K. Hoffm.;古巴羊大戟■☆
71585 Caperonia fistulosa Beille;管状卡普龙大戟■☆
71586 Caperonia fistulosa Beille var. pubescens J. Léonard;短柔毛卡普龙大戟■☆
71587 Caperonia gallabatensis Pax et K. Hoffm. = Caperonia serrata (Turcz.) C. Presl ■☆
71588 Caperonia hirtella Beille = Caperonia fistulosa Beille ■☆
71589 Caperonia latifolia Pax;宽叶羊大戟■☆
71590 Caperonia macrocarpa Pax et K. Hoffm.;大果卡普龙大戟■☆
71591 Caperonia senegalensis Müll. Arg. = Caperonia serrata (Turcz.) C. Presl ■☆
71592 Caperonia serrata (Turcz.) C. Presl;塞内加尔羊大戟■☆
71593 Caperonia stuhlmannii Pax;斯图尔曼羊大戟■☆
71594 Caperonia subrotunda Chiov.;近圆形卡普龙大戟■☆
71595 Capethia Britton = Anemone L. (保留属名)■
71596 Capethia Britton = Oreithales Schltdl. ■☆
71597 Capethia Britton(1891);全叶獐耳细辛属■☆
71598 Capethia integrifolia (DC.) Britton;全叶獐耳细辛■☆
71599 Capethia integrifolia (DC.) Britton = Anemone integrifolia Humb. et Bonpl ■☆
71600 Capethia weddellii Britton;韦德尔全叶獐耳细辛■☆
71601 Capia Dombey ex Juss. = Lapageria Ruiz et Pav. ●☆
71602 Capillipedium Stapf (1917);细柄草属;Capillipedium, Ministalkgrass ■
71603 Capillipedium assimile (Steud.) A. Camus;硬秆细柄草(硬秆子草,竹枝细柄草);Firmculm Capillipedium, Harsculm Ministalkgrass ■
71604 Capillipedium cinctum (Steud.) A. Camus;具穗细柄草;Spiked Capillipedium ■
71605 Capillipedium glaucopsis (Steud.) Stapf = Capillipedium assimile (Steud.) A. Camus ■
71606 Capillipedium kuoi L. B. Cai;郭氏细柄草■
71607 Capillipedium kwashotense (Hayata) C. C. Hsu;绿岛细柄草;Greenisland Ministalkgrass ■
71608 Capillipedium parviflorum (R. Br.) Stapf;细柄草(吊丝草);Ministalkgrass, Smallflower Capillipedium ■
71609 Capillipedium parviflorum (R. Br.) Stapf f. villosulus (Nees) Kitag. = Capillipedium parviflorum (R. Br.) Stapf var. villosulus (Hack.) C. Y. Wu ■
71610 Capillipedium parviflorum (R. Br.) Stapf f. violascens (Trin.) Kitag.;堇色细柄草■☆
71611 Capillipedium parviflorum (R. Br.) Stapf f. viviparum Kitag.;珠芽细柄草■☆
71612 Capillipedium parviflorum (R. Br.) Stapf var. spicigerum (Benth.) C. C. Hsu = Capillipedium spicigerum (Benth.) S. T. Blake ■
71613 Capillipedium parviflorum (R. Br.) Stapf var. villosulus (Hack.) C. Y. Wu;柔毛吊丝草;Villose Manynode Capillipedium ■
71614 Capillipedium spicigerum (Benth.) S. T. Blake;多节细柄草;Manynode Capillipedium ■
71615 Capillipedium spicigerum S. T. Blake = Capillipedium parviflorum (R. Br.) Stapf var. spicigerum (Benth.) C. C. Hsu ■
71616 Capillipedium subrepens (Steud.) Henrard = Capillipedium assimile (Steud.) A. Camus ■
71617 Capirona Spruce(1859);卡比茜属■☆
71618 Capirona decorticans Spruce;脱皮卡比茜■☆
71619 Capitanopsis S. Moore(1916);马岛香茶菜属●☆
71620 Capitanopsis albida (Baker) Hedge;白花马岛香茶菜●☆
71621 Capitanopsis angustifolia (Moldenke) Capuron;窄叶马岛香茶菜●☆
71622 Capitanopsis cloiselii S. Moore;克卢塞尔马岛香茶菜●☆
71623 Capitanya Gürke = Capitanya Schweinf. ex Gürke ●☆
71624 Capitanya Schweinf. ex Gürke = Plectranthus L'Hér. (保留属名)●■
71625 Capitanya Schweinf. ex Gürke(1895);卡匹塔草属●☆
71626 Capitanya Schweinf. ex Penz. = Plectranthus L'Hér. (保留属名)●■
71627 Capitanya otostegioides Gürke = Plectranthus otostegioides (Gürke) Ryding ■☆
71628 Capitanya rogleoides Schweinf. ex Penz. = Plectranthus otostegioides (Gürke) Ryding ■☆
71629 Capitellaria Naudin = Clidemia D. Don ●☆
71630 Capitellaria Naudin = Sagraea DC. ●☆
71631 Capitularia J. V. Suringar = Capitularina Kern ■☆
71632 Capitularina Kern(1974);五棱莎草属■☆
71633 Capitularina involucrata (J. V. Suringar) J. Kern;五棱莎草■☆
71634 Capnites (DC.) Dumort. = Corydalis DC. (保留属名)■
71635 Capnites Dumort. = Corydalis DC. (保留属名)■
71636 Capnitis E. Mey. = Lotononis (DC.) Eckl. et Zeyh. (保留属名)■
71637 Capnitis clandestina E. Mey. = Lotononis platycarpa (Viv.) Pic. Serm. ■☆
71638 Capnitis porrecta E. Mey. = Lotononis platycarpa (Viv.) Pic. Serm. ■☆
71639 Capnocystis Juss. = Corydalis DC. (保留属名)■
71640 Capnodes Kuntze = Capnoides Mill. (废弃属名)■
71641 Capnodes Kuntze = Corydalis DC. (保留属名)■
71642 Capnodes aureum (Willd.) Kuntze = Corydalis aurea Willd. ■☆
71643 Capnodes flavulum (Raf.) Kuntze = Corydalis flavula (Raf.) DC. ■☆
71644 Capnodes flavulum (Raf.) Kuntze = Fumaria flavula Raf. ■☆
71645 Capnogonium Benth. ex Endl. = Corydalis DC. (保留属名)■
71646 Capnogorium Bernh. = Corydalis DC. (保留属名)■
71647 Capnogorium nobile (L.) Bernh. = Corydalis nobilis (L.) Pers. ■
71648 Capnoides Mill. (废弃属名) = Corydalis DC. (保留属名)■
71649 Capnoides Tourn. ex Adans. = Corydalis DC. (保留属名)■
71650 Capnoides aureum (Willd.) Kuntze = Corydalis aurea Willd. ■☆
71651 Capnoides brachycarpum Rydb. = Corydalis caseana A. Gray subsp. brachycarpa (Rydb.) G. B. Ownbey ■☆
71652 Capnoides cusickii (S. Watson) A. Heller = Corydalis caseana A. Gray subsp. cusickii (S. Watson) G. B. Ownbey ■☆
71653 Capnoides halei Small = Corydalis micrantha (Engelm. ex A. Gray) A. Gray subsp. australis (Chapm.) G. B. Ownbey ■☆

71654 Capnoides hastatum Rydb. = Corydalis caseana A. Gray subsp. hastata (Rydb.) G. B. Ownbey ■☆
71655 Capnoides micranthum (Engelm. ex A. Gray) Britton = Corydalis micrantha (Engelm. ex A. Gray) A. Gray ■☆
71656 Capnoides sempervirens (L.) Borkh. = Corydalis sempervirens (L.) Pers. ■☆
71657 Capnophyllum Gaertn. (1790);烟叶草属■☆
71658 Capnophyllum africanum (L.) W. D. J. Koch;非洲烟叶草■☆
71659 Capnophyllum africanum (L.) W. D. J. Koch var. leiocarpon Sond. = Capnophyllum leiocarpon (Sond.) Manning et Goldblatt ■☆
71660 Capnophyllum dichotomum (Desf.) Ball = Krubera peregrinum (L.) Hoffm. ■☆
71661 Capnophyllum jacquinii DC. = Dasispermum suffruticosum (P. J. Bergius) B. L. Burtt ■☆
71662 Capnophyllum leiocarpon (Sond.) Manning et Goldblatt;光果烟叶草■☆
71663 Capnophyllum nodiflorum (Coss.) Drude = Sclerosciadium nodiflorum (Schousb.) Ball ■☆
71664 Capnophyllum peregrinum (L.) Lag. = Krubera peregrinum (L.) Hoffm. ■☆
71665 Capnorchis Borkh. = Eucapnos Bernh. ■
71666 Capnorchis Borkh. = Lamprocapnos Endl. ■
71667 Capnorchis Mill. (废弃属名) = Dicentra Bernh. (保留属名)■
71668 Capnorchis spectabilis (L.) Borkh. = Lamprocapnos spectabilis (L.) Fukuhara ■
71669 Capnorea Raf. (废弃属名) = Hesperochiron S. Watson(保留属名)■☆
71670 Capparaceae Adans. = Capparaceae Juss. (保留科名)●■
71671 Capparaceae Juss. (1789)(保留科名);山柑科(白花菜科,醉蝶花科);Caper Family ●■
71672 Capparicordis Iltis et Cornejo = Capparis L. ●
71673 Capparicordis Iltis et Cornejo(2007);美洲山柑属●☆
71674 Capparidaceae Juss. = Capparaceae Juss. (保留科名)●■
71675 Capparidastrum Hutch. = Capparis L. ●
71676 Capparis L. (1753);山柑属(槌果藤属,马槟榔属,山柑仔属);Caper,Caper Bush,Caperbush ●
71677 Capparis acuminata De Wild. = Capparis erythrocarpos Isert var. acuminata (De Wild.) Hauman ●☆
71678 Capparis acuminata Lindl. = Capparis acutifolia Sweet ●
71679 Capparis acutifolia Sweet;独行千里(单兵救主,独虎龙,黑钩榕,黑皮蛇,尖破石,尖叶槌果藤,扣钮子,落地金鸡,落杆薯,膜叶槌果藤,锐叶山柑,石钻子,狭叶山柑,下洞底);Acuteleaf Caper,Acute-leaved Caper,Longwalk Alone,Narrow-leaved Caper ●
71680 Capparis acutifolia Sweet subsp. bodinieri (H. Lév.) Jacobs = Capparis bodinieri H. Lév. ●
71681 Capparis acutifolia Sweet subsp. sabiifolia (Hook. f. et Thomson) Jacobs = Capparis sabiifolia Hook. f. et Thomson ●
71682 Capparis acutifolia Sweet subsp. viminea Jacobs = Capparis membranifolia Kurz ●
71683 Capparis acutissima Gilg et Gilg-Ben. = Capparis erythrocarpos Isert var. rosea (Klotzsch) DeWolf ●☆
71684 Capparis aegyptia Lam.;埃及山柑●☆
71685 Capparis afzelii DC. = Capparis erythrocarpos Isert ●☆
71686 Capparis afzelii Pax = Maerua duchesnei (De Wild.) F. White ●☆
71687 Capparis albitrunca Burch. = Boscia albitrunca (Burch.) Gilg et Gilg-Ben. ●☆
71688 Capparis alexandrae Chiov. = Capparis tomentosa Lam. ●☆
71689 Capparis aphylla Hayne ex Roth;无叶山柑;Leafless Caper ●
71690 Capparis aphylla Hayne ex Roth = Capparis decidua (Forssk.) Edgew. ●
71691 Capparis aphylla Roth = Capparis aphylla Hayne ex Roth ●
71692 Capparis aphylla Roth = Capparis decidua (Forssk.) Edgew. ●
71693 Capparis arborea (F. Muell.) Maiden;树山柑;Australian Native Quince,Bush Caper Berry ●☆
71694 Capparis assamica Hook. f. et Thunb.;总序山柑;Assam Caper, Racemose Caper ●
71695 Capparis atlantica Inocencio et D. Rivera et al.;大西洋山柑●☆
71696 Capparis bangweolensis R. E. Fr. = Capparis fascicularis DC. var. elaeagnoides (Gilg) DeWolf ●☆
71697 Capparis bequaertii De Wild. = Capparis erythrocarpos Isert ●☆
71698 Capparis bhamoensis Raizada = Capparis yunnanensis Craib et W. W. Sm. ●
71699 Capparis bodinieri H. Lév.;野香橼花(刺珠,猫胡子花,黔桂槌果藤,青刺尖,小毛毛花,叶上花);Bodinier Caper, Wild Citron flower ●
71700 Capparis boscioides Pax = Capparis sepiaria L. var. boscioides (Pax) Kers ●☆
71701 Capparis brachyandra Pax = Capparis fascicularis DC. var. elaeagnoides (Gilg) DeWolf ●☆
71702 Capparis brassii DC.;布拉斯山柑●☆
71703 Capparis brevis Spreng. = Glyphaea brevis (Spreng.) Monach. ●☆
71704 Capparis bussei Gilg et Gilg-Ben. = Maerua bussei (Gilg et Gilg-Ben.) R. Wilczek ●☆
71705 Capparis calvescens Gilg et Gilg-Ben. = Capparis fascicularis DC. ●☆
71706 Capparis canescens Banks ex DC.;灰山柑;Dog Caper ●☆
71707 Capparis cantoniensis Lour.;广州山柑(广州槌果藤,老虎须,山柑子);Canton Caper,Guangzhou Caper ●
71708 Capparis capensis Thunb. = Capparis sepiaria L. var. citrifolia (Lam.) Toelken ●☆
71709 Capparis cartilaginea Decne.;软骨山柑●☆
71710 Capparis carvalhoana Gilg = Capparis erythrocarpos Isert var. rosea (Klotzsch) DeWolf ●☆
71711 Capparis cathcartii Hemsl. ex Gamble = Capparis sikiimensis Kurz ●
71712 Capparis cerasifera Gilg = Capparis sepiaria L. var. fischeri (Pax) DeWolf ●☆
71713 Capparis cerasifolia A. Gray = Capparis pubiflora DC. ●
71714 Capparis chinensis Don = Capparis acutifolia Sweet ●
71715 Capparis chingiana B. S. Sun;野槟榔(山水槟榔,水槟榔,子农山柑);Ching Caper ●
71716 Capparis chionantha Gilg et Gilg-Ben. = Capparis sepiaria L. var. boscioides (Pax) Kers ●☆
71717 Capparis chrysomeia Bojer;金山柑●☆
71718 Capparis chrysomeia Bojer var. richardii (Baill.) Hadj.-Moust. = Capparis chrysomeia Bojer ●☆
71719 Capparis citrifolia Lam. = Capparis sepiaria L. var. citrifolia (Lam.) Toelken ●☆
71720 Capparis clutiifolia Burch. ex DC. = Boscia oleoides (Burch. ex DC.) Toelken ●☆
71721 Capparis coriacea Burch. ex DC. = Boscia oleoides (Burch. ex DC.) Toelken ●☆
71722 Capparis corymbifera E. Mey. ex Sond. = Capparis tomentosa Lam. ●☆

71723 Capparis corymbosa Lam. var. sansibarensis Pax = Capparis sepiaria L. var. subglabra (Oliv.) DeWolf ●☆
71724 Capparis corymbosa Lam. var. subglabra Oliv. = Capparis sepiaria L. var. subglabra (Oliv.) DeWolf ●☆
71725 Capparis cuspidata B. S. Sun = Capparis urophylla F. Chun ●
71726 Capparis cynophallophora L.;牙买加山柑;Jamaica Caper Tree ●☆
71727 Capparis dasyphylla Merr. et F. P. Metcalf;多毛山柑(厚叶槌果藤);Hairy Caper,Manyhair Caper,Wooly-leaved Caper ●
71728 Capparis decidua (Forssk.) Edgew.;脱落山柑●
71729 Capparis decidua Edgew. = Capparis aphylla Roth ●
71730 Capparis decidua Pax = Capparis sodada R. Br. ●☆
71731 Capparis deckenii Chiov. = Maerua denhardtiorum Gilg ●☆
71732 Capparis deserti (Zohary) Täckh. et Boulos = Capparis spinosa L. var. deserti Zohary ●☆
71733 Capparis djurica Gilg et Gilg-Ben. = Capparis sepiaria L. var. fischeri (Pax) DeWolf ●☆
71734 Capparis duchesnei De Wild. = Maerua duchesnei (De Wild.) F. White ●☆
71735 Capparis elaeagnoides Gilg = Capparis fascicularis DC. var. elaeagnoides (Gilg) DeWolf ●☆
71736 Capparis elaeagnoides Gilg var. zizyphoides (Gilg) Hauman = Capparis fascicularis DC. var. elaeagnoides (Gilg) DeWolf ●☆
71737 Capparis erythrocarpos Isert;红果山柑●☆
71738 Capparis erythrocarpos Isert var. acuminata (De Wild.) Hauman;渐尖山柑●☆
71739 Capparis erythrocarpos Isert var. rosea (Klotzsch) DeWolf;粉红山柑●☆
71740 Capparis fascicularis DC.;扁山柑●☆
71741 Capparis fascicularis DC. var. elaeagnoides (Gilg) DeWolf;胡颓子山柑●☆
71742 Capparis fascicularis DC. var. scheffleri (Gilg et Gilg-Ben.) DeWolf;谢夫勒山柑●☆
71743 Capparis fascicularis DC. var. zeyheri (Turcz.) Toelken;泽赫山柑●☆
71744 Capparis fengii B. S. Sun;文山山柑(国楣山柑);Feng Caper,Wenshan Caper ●
71745 Capparis fischeri Pax = Capparis sepiaria L. var. fischeri (Pax) DeWolf ●☆
71746 Capparis flacata Lour. = Crateva falcata (Lour.) DC. ●
71747 Capparis flanaganii Gilg et Gilg-Ben. = Capparis fascicularis DC. ●☆
71748 Capparis flexicaulis Hance = Capparis sepiaria L. ●
71749 Capparis floribunda Wight;少蕊山柑(多花山柑);Few-flower Caper,Flowery Caper ●
71750 Capparis fohaiensis B. S. Sun;勐海山柑(佛海山柑);Menghai Caper ●
71751 Capparis formosana Hemsl.;台湾山柑(山柑,山柑仔,台湾槌果藤,台湾马槟榔);Formosa Caper,Taiwan Caper ●
71752 Capparis formosana Hemsl. = Capparis henryi Matsum. et Hayata ●
71753 Capparis galeata Fresen. = Capparis cartilaginea Decne. ●☆
71754 Capparis glauca Wall. ex Hook. f. et Thomson = Capparis sepiaria L. ●
71755 Capparis globifera Delile ex Rochet;球山柑●☆
71756 Capparis gueinzii Sond. = Capparis brassii DC. ●☆
71757 Capparis hainanensis Oliv.;海南山柑(海南槌果藤,山柑);Hainan Caper ●
71758 Capparis hastigera Hance = Capparis zeylanica L. ●
71759 Capparis hastigera Hance var. obcordata Merr. et F. P. Metcalf = Capparis zeylanica L. ●
71760 Capparis henryi Matsum. et Hayata;长刺山柑(山柑仔);Henry Smallspine Caper,Henry's Caper,Longspine Caper ●
71761 Capparis hereroensis Schinz;赫雷罗山柑●☆
71762 Capparis himalayensis Jafri;爪瓣山柑(菠里克果,槌果藤,刺马槟榔,刺山柑,勾刺槌果藤,抗旱草,狼西瓜,老鼠瓜,山柑,续随子,野西瓜);Caper,Caper Bush,Caper-bush,Caper-tree,Clawpetal Caper,Common Caper,Himalayan Caper ●
71763 Capparis holliensis A. Chev. = Capparis brassii DC. ●☆
71764 Capparis horrida L. f.;刺蝴蝶木(马来西亚山柑)●
71765 Capparis horrida L. f. = Capparis zeylanica L. ●
71766 Capparis humblotii Baill. = Crateva humblotii (Baill.) Hadj-Moust. ●☆
71767 Capparis hypericoides Hochst. = Capparis tomentosa Lam. ●☆
71768 Capparis hypovellerea Gilg et Gilg-Ben. = Capparis sepiaria L. var. fischeri (Pax) DeWolf ●☆
71769 Capparis incana Kunth;灰白山柑;Hoary Caper ●☆
71770 Capparis incanescens DC. = Capparis sepiaria L. ●
71771 Capparis inermis Forssk. = Capparis cartilaginea Decne. ●☆
71772 Capparis jodotricha Gilg et Gilg-Ben. = Capparis fascicularis DC. ●☆
71773 Capparis kanehirai Hayata ex Kaneh. = Capparis formosana Hemsl. ●
71774 Capparis khuamak Gagnep.;屏边山柑(毛柄山柑);Pingbian Caper ●
71775 Capparis kikuchii Hayata;毛瓣蝴蝶木(菊池氏山柑);Kikuchi Caper ●
71776 Capparis kikuchii Hayata = Capparis acutifolia Sweet ●
71777 Capparis kirkii Oliv. = Maerua kirkii (Oliv.) F. White ●☆
71778 Capparis koi Merr. et Chun = Capparis versicolor Griff. ●
71779 Capparis lanceolaris DC.;兰屿山柑;Lanceolate Caper,Lanyu Caper ●
71780 Capparis lasiantha R. Br. ex DC.;毛花山柑●☆
71781 Capparis laurifolia Gilg et Gilg-Ben. = Capparis sepiaria L. var. citrifolia (Lam.) Toelken ●☆
71782 Capparis leptophylla Hayata;细叶山柑;Thin-leaved Caper ●
71783 Capparis leptophylla Hayata = Capparis acutifolia Sweet ●
71784 Capparis leptophylla Hayata = Capparis membranacea Gardner et Champ. var. angustissima Hemsl. ●
71785 Capparis leucophylla Collett = Capparis himalayensis Jafri ●
71786 Capparis leucophylla DC. = Capparis sicula Veill. subsp. leucophylla (DC.) Inocencio et D. Rivera et al. ●☆
71787 Capparis leucophylla Hayata = Capparis himalayensis Jafri ●
71788 Capparis liangii Merr. et Chun = Capparis micracantha DC. ●
71789 Capparis lilacina Gilg = Capparis viminea Hook. f. et Thomson ex Oliv. ●
71790 Capparis lucens Hauman;光亮山柑●☆
71791 Capparis lucida (DC.) Benth.;明亮山柑●☆
71792 Capparis lutaoensis C. C. Chang = Capparis pubiflora DC. ●
71793 Capparis macrosperma Delile ex Rochet;大籽山柑●☆
71794 Capparis magna Lour. = Crateva magna (Lour.) DC. ●
71795 Capparis marlothii Gilg et Gilg-Ben. = Capparis fascicularis DC. ●☆
71796 Capparis masaikai H. Lév.;马槟榔(马大白,马金南,马金囊,屈头鸡,山槟榔,水槟榔,太极子,紫槟榔);Masaika Caper ●
71797 Capparis masaikai H. Lév. var. elabra (Gagnep.) Rehder;紫槟

榔(马槟榔,马金囊,山槟榔,太极子)●

71798 Capparis membranacea Gardner et Champ.;膜叶槌果藤(独虎龙,独行千里,扣钮子,勒儿根,落地金鸡,落杆薯,下洞底);Membranaceous Caper ●

71799 Capparis membranacea Gardner et Champ. = Capparis acutifolia Sweet ●

71800 Capparis membranacea Gardner et Champ. var. angustissima Hemsl. = Capparis acutifolia Sweet ●

71801 Capparis membranacea Gardner et Champ. var. puberula B. S. Sun = Capparis acutifolia Sweet ●

71802 Capparis membranifolia Kurz;雷公橘(老虎木,纤枝槌果藤,一扫光);Membranaceousleaf Caper, Membranaceous-leaved Caper, Thundergod Caper ●

71803 Capparis micracantha DC.;小刺山柑(牛眼睛);Smallspine Caper, Small-spined Caper ●

71804 Capparis micracantha DC. var. henryi (Matsum.) Jacobs = Capparis henryi Matsum. et Hayata ●

71805 Capparis micrantha A. Rich.;小花山柑●☆

71806 Capparis mildbraedii Gilg = Capparis sepiaria L. var. rivae (Gilg) DeWolf ●☆

71807 Capparis mildbraedii Gilg var. roseiflora (Gilg et Gilg-Ben.) Hauman = Capparis sepiaria L. var. rivae (Gilg) DeWolf ●☆

71808 Capparis mitchellii Lindl.;野山柑;Bumbil, Bumble Tree, Mitchell's Caper, Native Orange, Wild Orange ●☆

71809 Capparis multiflora Hook. f. et Thomson;多花山柑;Flowery Caper, Manyflower Caper, Multiflorous Caper ●

71810 Capparis murrayana Graham = Capparis spinosa L. ●

71811 Capparis nepaulensis DC. = Capparis spinosa L. ●

71812 Capparis oblongifolia Forssk. = Maerua oblongifolia (Forssk.) A. Rich. ●☆

71813 Capparis olacifolia Hook. f. et Thomson;藏东南山柑●

71814 Capparis oleoides Burch. ex DC. = Boscia oleoides (Burch. ex DC.) Toelken ●☆

71815 Capparis oligantha Gilg et Gilg-Ben. = Capparis fascicularis DC. var. elaeagnoides (Gilg) DeWolf ●☆

71816 Capparis oligostema Hayata = Capparis floribunda Wight ●

71817 Capparis oliveriana Gilg = Capparis fascicularis DC. ●☆

71818 Capparis orientalis Veill. = Capparis spinosa L. subsp. rupestris (Sibth. et Sm.) Nyman ●☆

71819 Capparis orthacantha Gilg et Gilg-Ben. = Capparis viminea Hook. f. et Thomson ex Oliv. var. orthacantha (Gilg et Gilg-Ben.) DeWolf ●☆

71820 Capparis ovata Desf.;卵叶山柑●☆

71821 Capparis ovata Desf. subsp. myrtifolia Inocencio et D. Rivera et al.;香桃木叶山柑●☆

71822 Capparis ovata Desf. var. sicula (Veill.) Zohary = Capparis sicula Veill. ●☆

71823 Capparis pachyphylla Jacobs;厚叶山柑●

71824 Capparis persicifolia A. Rich. = Capparis tomentosa Lam. ●☆

71825 Capparis persicoides A. Chev. = Maerua duchesnei (De Wild.) F. White ●☆

71826 Capparis pittieri Standl.;哥斯达黎加山柑●☆

71827 Capparis poggei Pax;波格山柑●☆

71828 Capparis polymorpha A. Rich. = Capparis tomentosa Lam. ●☆

71829 Capparis pterocarpa Chun = Capparis masaikai H. Lév. ●

71830 Capparis pubiflora DC.;毛蕊山柑(毛花山柑,毛叶山柑);Downy-flowered Caper, Hairpistil Caper, Hairy-flower Caper ●

71831 Capparis pubifolia B. S. Sun;毛叶山柑(毛蕊山柑,桃叶槌果藤);Hairyleaf Caper, Hairy-leaved Caper ●

71832 Capparis pumila Champ. ex Benth. = Capparis cantoniensis Lour. ●

71833 Capparis punctata Burch. = Boscia albitrunca (Burch.) Gilg et Gilg-Ben. ●☆

71834 Capparis racemulosa A. DC. = Maerua racemulosa (A. DC.) Gilg et Gilg-Ben. ●☆

71835 Capparis reflexa Thonn. = Ritchiea reflexa (Thonn.) Gilg et Gilg-Ben. ●☆

71836 Capparis richardii Baill.;理查德山柑●☆

71837 Capparis rivae Gilg = Capparis sepiaria L. var. rivae (Gilg) DeWolf ●☆

71838 Capparis rosanowiana B. Fedtsch.;罗萨山柑●☆

71839 Capparis rosea (Klotzsch) Oliv. = Capparis erythrocarpos Isert var. rosea (Klotzsch) DeWolf ●☆

71840 Capparis roseiflora Gilg et Gilg-Ben. = Capparis sepiaria L. var. rivae (Gilg) DeWolf ●☆

71841 Capparis rothii Oliv. = Capparis fascicularis DC. ●☆

71842 Capparis roxburghii DC. = Capparis yunnanensis Craib et W. W. Sm. ●

71843 Capparis rudatisii Gilg et Gilg-Ben. = Capparis fascicularis DC. ●☆

71844 Capparis rupestris Sibth. et Sm. = Capparis spinosa L. subsp. rupestris (Sibth. et Sm.) Nyman ●☆

71845 Capparis rupestris Sibth. et Sm. = Capparis spinosa L. ●

71846 Capparis sabiifolia Hook. f. et Thomson;黑叶山柑;Blackleaf Caper, Black-leaved Caper ●

71847 Capparis sansibarensis (Pax) Gilg = Capparis sepiaria L. var. subglabra (Oliv.) DeWolf ●☆

71848 Capparis scheffleri Gilg et Gilg-Ben. = Capparis fascicularis DC. var. scheffleri (Gilg et Gilg-Ben.) DeWolf ●☆

71849 Capparis schlechteri Schinz = Capparis fascicularis DC. ●☆

71850 Capparis sciaphila Hance = Capparis cantoniensis Lour. ●

71851 Capparis sepiaria L.;青皮刺(公须花,篱边生山柑,曲枝槌果藤);Greenpeel Caper, Hedge Caper ●

71852 Capparis sepiaria L. var. boscioides (Pax) Kers;鸭山柑●☆

71853 Capparis sepiaria L. var. citrifolia (Lam.) Toelken;桔叶青皮刺●☆

71854 Capparis sepiaria L. var. fischeri (Pax) DeWolf;菲氏青皮刺●☆

71855 Capparis sepiaria L. var. rivae (Gilg) DeWolf;沟山柑●☆

71856 Capparis sepiaria L. var. stuhlmannii (Gilg) DeWolf;斯图尔曼山柑●☆

71857 Capparis sepiaria L. var. subglabra (Oliv.) DeWolf;近光青皮刺●☆

71858 Capparis sicula Veill.;西西里山柑●☆

71859 Capparis sicula Veill. subsp. leucophylla (DC.) Inocencio et D. Rivera et al.;白叶西西里山柑●☆

71860 Capparis sikiimensis Kurz;锡金山柑●

71861 Capparis sikiimensis Kurz subsp. yunnanensis (Craib et W. W. Sm.) Jacobs = Capparis yunnanensis Craib et W. W. Sm. ●

71862 Capparis sikkimensis Kurz subsp. formosana (Hemsl.) Jacobs = Capparis formosana Hemsl. ●

71863 Capparis sikkimensis Kurz subsp. masaikai (H. Lév.) Jacobs = Capparis masaikai H. Lév. ●

71864 Capparis sikkimensis Kurz subsp. yunnanensis (Craib et W. W. Smith) Jacobs = Capparis yunnanensis Craib et W. W. Sm. ●

71865 Capparis sinaica Veill. = Capparis aegyptia Lam. ●☆

71866 Capparis sodada R. Br.;凋落山柑●☆

71867 Capparis sodada R. Br. = Capparis decidua (Forssk.) Edgew. ●

71868 Capparis solanoides Bojer = Capparis chrysomeia Bojer ●☆

71869 Capparis solanoides Gilg et Gilg-Ben. = Capparis fascicularis DC. ●☆

71870 Capparis somalensis Gilg = Capparis fascicularis DC. ●☆

71871 Capparis spinosa L.；刺山柑（山柑）；Caper，Caper Bush，Capper，Common Caper，Common Capers，Puapilo，Spineless Caper ●

71872 Capparis spinosa L. = Capparis himalayensis Jafri ●

71873 Capparis spinosa L. f. coriacea（Coss.）Maire；革质山柑●☆

71874 Capparis spinosa L. subsp. cartilaginea（Decne.）Maire et Weiller = Capparis cartilaginea Decne. ●☆

71875 Capparis spinosa L. subsp. orientalis（Veill.）Jafri = Capparis spinosa L. subsp. rupestris（Sibth. et Sm.）Nyman ●☆

71876 Capparis spinosa L. subsp. rupestris（Sibth. et Sm.）Nyman；岩生山柑●☆

71877 Capparis spinosa L. var. aegyptia（Lam.）Boiss. = Capparis aegyptia Lam. ●☆

71878 Capparis spinosa L. var. canescens Coss. = Capparis sicula Veill. ●☆

71879 Capparis spinosa L. var. coriacea（Coss.）Maire = Capparis ovata Desf. ●☆

71880 Capparis spinosa L. var. coriacea Coss. ex Maire = Capparis spinosa L. f. coriacea（Coss.）Maire ●☆

71881 Capparis spinosa L. var. deserti Zohary = Capparis aegyptia Lam. ●☆

71882 Capparis spinosa L. var. galeata（Fresen.）Hook. f. et Thomson = Capparis cartilaginea Decne. ●☆

71883 Capparis spinosa L. var. himalayensis（Jafri）Jacobs = Capparis himalayensis Jafri ●

71884 Capparis spinosa L. var. inermis Turra = Capparis spinosa L. subsp. rupestris（Sibth. et Sm.）Nyman ●☆

71885 Capparis spinosa L. var. kruegeriana（Pamp.）Jafri = Capparis ovata Desf. ●☆

71886 Capparis spinosa L. var. ovata（Desf.）Batt. = Capparis ovata Desf. ●☆

71887 Capparis spinosa L. var. parviflora Boiss. = Capparis spinosa L. ●

71888 Capparis spinosa L. var. pubescens Zohary = Capparis sicula Veill. subsp. leucophylla（DC.）Inocencio et D. Rivera et al. ●☆

71889 Capparis spinosa L. var. rupestris（Sibth. et Sm.）Boiss. = Capparis ovata Desf. ●☆

71890 Capparis spinosa L. var. sicula（Veill.）Hausskn. = Capparis ovata Desf. ●☆

71891 Capparis stuhlmannii Gilg = Capparis sepiaria L. var. stuhlmannii（Gilg）DeWolf ●☆

71892 Capparis subglabra（Oliv.）Gilg et Gilg-Ben. = Capparis sepiaria L. var. subglabra（Oliv.）DeWolf ●☆

71893 Capparis subsessilis B. S. Sun；无柄山柑；Sessile-leaved Caper，Stallkless Caper ●

71894 Capparis subtenera Craib et W. W. Sm. = Capparis bodinieri H. Lév. ●

71895 Capparis subtomentosa De Wild.；微绒毛山柑●☆

71896 Capparis sulphurea Gilg et Gilg-Ben. = Capparis erythrocarpos Isert ●☆

71897 Capparis sunbisiniana M. L. Zhang et G. C. Tucker；倒卵叶山柑●

71898 Capparis swinhoei Hance = Capparis zeylanica L. ●

71899 Capparis tenera Dalzell；薄叶山柑；Filmleaf Caper，Thinleaf Caper，Thin-leaved Caper ●

71900 Capparis tenera Dalzell = Capparis bodinieri H. Lév. ●

71901 Capparis tenera Dalzell var. caudata B. S. Sun = Capparis urophylla F. Chun ●

71902 Capparis tenera Dalzell var. dalzellii Hook. f. et Thomson = Capparis tenera Dalzell ●

71903 Capparis tenuifolia Hayata = Capparis acutifolia Sweet ●

71904 Capparis thonningii Schumach. = Capparis brassii DC. ●☆

71905 Capparis tomentosa Lam.；绒毛山柑●☆

71906 Capparis transvaalensis Schinz = Capparis fascicularis DC. ●☆

71907 Capparis transvaalensis Schinz var. calvescens（Gilg et Gilg-Ben.）Marais = Capparis fascicularis DC. ●☆

71908 Capparis trichocarpa B. S. Sun；毛果山柑；Hairfruit Caper，Hairy-fruited Cape ●

71909 Capparis trichopoda B. S. Sun = Capparis khuamak Gagnep. ●

71910 Capparis trifoliata Roxb. = Crateva trifoliata（Roxb.）B. S. Sun ●

71911 Capparis triphylla Thunb. = Maerua cafra（DC.）Pax ●☆

71912 Capparis uberiflora F. Muell.；乳头花山柑●☆

71913 Capparis umbonata R. Br. ex DC.；鳞脐山柑●☆

71914 Capparis uncinata Edgew. = Capparis cartilaginea Decne. ●☆

71915 Capparis undulata Zeyh. ex Eckl. et Zeyh. = Maerua racemulosa（A. DC.）Gilg et Gilg-Ben. ●☆

71916 Capparis urophylla F. Chun；小绿刺（尖叶山柑，尾叶槌果藤，尾叶马槟榔，尾叶山柑）；Small Greenspine，Urophyllous Caper ●

71917 Capparis venenata Schinz = Capparis tomentosa Lam. ●☆

71918 Capparis verdickii De Wild. = Capparis tomentosa Lam. ●☆

71919 Capparis versicolor Griff.；屈头鸡（保亭槌果藤，马槟榔，山木通，树屈头鸡，锡朋槌果藤，圆头鸡）；Bowhead Cock，Changingcolour Caper，Versicolor Caper，Versicolous Caper ●

71920 Capparis viburnifolia Gagnep.；荚蒾叶山柑；Viburnum-leaf Caper，Viburnum-leaved Caper ●

71921 Capparis victoriae-nyanzae Brenan = Capparis sepiaria L. var. rivae（Gilg）DeWolf ●☆

71922 Capparis vientianensis Gagnep. = Capparis sabiifolia Hook. f. et Thomson ●

71923 Capparis viminea Hook. f. et Thomson = Capparis acutifolia Sweet ●

71924 Capparis viminea Hook. f. et Thomson = Capparis membranifolia Kurz ●

71925 Capparis viminea Hook. f. et Thomson ex Oliv.；软枝山柑（纤枝山柑）；Slender-branch Caper ●

71926 Capparis viminea Hook. f. et Thomson ex Oliv. var. orthacantha（Gilg et Gilg-Ben.）DeWolf；直刺山柑●☆

71927 Capparis viminea Hook. f. et Thomson var. ferruginea B. S. Sun = Capparis membranifolia Kurz ●

71928 Capparis volkensii Gilg = Capparis tomentosa Lam. ●☆

71929 Capparis warneckei Gilg et Gilg-Ben. = Capparis erythrocarpos Isert ●☆

71930 Capparis welwitschii Pax et Gilg = Capparis viminea Hook. f. et Thomson ex Oliv. ●

71931 Capparis woodii Gilg et Gilg-Ben. = Capparis sepiaria L. var. citrifolia（Lam.）Toelken ●☆

71932 Capparis wui B. S. Sun；元江山柑（征镒山柑）；Wu's Caper，Yuanjiang Caper ●

71933 Capparis yunnanensis Craib et W. W. Sm.；苦子马槟榔（马槟榔）；Bitterseed Caper，Yunnan Caper ●

71934 Capparis zeyheri Turcz. = Capparis fascicularis DC. var. zeyheri（Turcz.）Toelken ●☆

71935 Capparis zeylanica L.；牛眼睛（槌果藤，锡兰蝴蝶木，锡兰山柑）；Ceylon Caper，Oxeye Caper ●

71936 Capparis zizyphoides Gilg = Capparis fascicularis DC. var. elaeagnoides (Gilg) DeWolf ●☆
71937 Capparis zoharyi Inocencio et D. Rivera et al.;佐哈里山柑●☆
71938 Cappidastrum Hutch. = Capparidastrum Hutch. ●
71939 Capraea Opiz = Salix L.(保留属名)●
71940 Capraria L.(1753);羊玄参属■☆
71941 Capraria arabica (Endl.) Steud. et Hochst. ex Benth. = Anticharis arabica Endl. ■☆
71942 Capraria biflora L.;双花羊玄参■☆
71943 Capraria crustacea L. = Lindernia crustacea (L.) F. Muell. ■
71944 Capraria dissecta Delile = Jamesbrittenia dissecta (Delile) Kuntze ■☆
71945 Capraria lanceolata L. f. = Freylinia lanceolata (L. f.) G. Don ●☆
71946 Capraria longiflora Thunb. = Freylinia longiflora Benth. ●☆
71947 Capraria lucida Sol. = Teedia lucida (Sol.) Rudolphi ●☆
71948 Capraria rigida Thunb. = Ehretia rigida (Thunb.) Druce ●☆
71949 Capraria salicifolia Salisb. = Freylinia lanceolata (L. f.) G. Don ●☆
71950 Capraria undulata L. f. = Freylinia undulata (L. f.) Benth. ●☆
71951 Caprariaceae Martinov = Scrophulariaceae Juss.(保留科名)●■
71952 Caprella Raf. = Capsella Medik.(保留属名)■
71953 Caprificus Gasp. = Ficus L. ●
71954 Caprifoliaceae Adans. = Caprifoliaceae Juss.(保留科名)●■
71955 Caprifoliaceae Juss.(1789)(保留科名);忍冬科;Honeysuckle Family ●■
71956 Caprifolium Mill. = Lonicera L. ●■
71957 Caprifolium altmannii (Regel et Schmalh.) Kuntze = Lonicera altmannii Regel et Schmalh. ex Regel ●
71958 Caprifolium altmannii Kuntze = Lonicera altmannii Regel et Schmalh. ex Regel ●
71959 Caprifolium angustifolium (Wall. ex DC.) Kuntze = Lonicera angustifolia Wall. ex DC. ●
71960 Caprifolium angustifolium Kuntze = Lonicera angustifolia Wall. ex DC. ●
71961 Caprifolium bournei (Hemsl.) Kuntze = Lonicera bournei Hemsl. ●
71962 Caprifolium bowrnei Kuntze = Lonicera bournei Hemsl. ex Forbes et Hemsl. ●■
71963 Caprifolium chrysanthum (Turcz. ex Ledeb.) Kuntze = Lonicera chrysantha Turcz. ex Ledeb. ●
71964 Caprifolium chrysanthum Kuntze = Lonicera chrysantha Turcz. ex Ledeb. ●
71965 Caprifolium confusum (Sweet) Spach = Lonicera confusa (Sweet) DC. ●■
71966 Caprifolium confusum Spach = Lonicera confusa (Sweet) DC. ●■
71967 Caprifolium decipiens Kuntze = Lonicera lanceolata Wall. ●
71968 Caprifolium elisae (Franch.) Kuntze = Lonicera elisae Franch. ●
71969 Caprifolium ferdinandii (Franch.) Kuntze = Lonicera ferdinandi Franch. ●
71970 Caprifolium ferdinandii Kuntze = Lonicera ferdinandii Franch. ●
71971 Caprifolium fragrantissimum (Lindl. et Paxton) Kuntze = Lonicera fragrantissima Lindl. et Paxton ●
71972 Caprifolium fragrantissimum Kuntze = Lonicera fragrantissima Lindl. et Paxton ●
71973 Caprifolium fuchsioides (Hemsl.) Kuntze = Lonicera acuminata Wall. ●■
71974 Caprifolium fuchsioides Kuntze = Lonicera acuminata Wall. ●■
71975 Caprifolium gynochlamydeum (Hemsl.) Kuntze = Lonicera gynochlamydea Hemsl. ex Forbes et Hemsl. ●
71976 Caprifolium gynochlamydeum Kuntze = Lonicera gynochlamydea Hemsl. ex Forbes et Hemsl. ●
71977 Caprifolium hemsleyanum Kuntze = Lonicera hemsleyana (Kuntze) Rehder ●
71978 Caprifolium henryi (Hemsl.) Kuntze = Lonicera acuminata Wall. ●■
71979 Caprifolium henryi Kuntze = Lonicera acuminata Wall. ●■
71980 Caprifolium hispidum (Pall. ex Roem. et Schult.) Kuntze = Lonicera hispida Pall. ex Roem. et Schult. ●
71981 Caprifolium hispidum Kuntze = Lonicera hispida Pall. ex Roem. et Schult. ●
71982 Caprifolium humile (Kar. et Kir.) Kuntze = Lonicera humilis Kar. et Kir. ●
71983 Caprifolium humile Kuntze = Lonicera humilis Kar. et Kir. ●
71984 Caprifolium hypoglaucum (Miq.) Otto = Lonicera hypoglauca Miq. ●
71985 Caprifolium hypoglaucum Otto = Lonicera hypoglauca Miq. ●
71986 Caprifolium japonicum (Thunb.) Dum. Cours. = Lonicera japonica Thunb. ●
71987 Caprifolium japonicum Dum. Cours. = Lonicera japonica Thunb. ●
71988 Caprifolium karelinii Kuntze = Lonicera heterophylla Decne. ●
71989 Caprifolium ligustrinum (Wall.) Kuntze = Lonicera ligustrina Wall. ●
71990 Caprifolium longiflorum Lindl. = Lonicera longiflora (Lindl.) DC. ●
71991 Caprifolium maackii (Rupr.) Kuntze = Lonicera maackii (Rupr.) Maxim. ●
71992 Caprifolium maackii Kuntze = Lonicera maackii (Rupr.) Maxim. ●
71993 Caprifolium macranthum D. Don = Lonicera macrantha (D. Don) Spreng. ●
71994 Caprifolium maximowiczii (Rupr.) Kuntze = Lonicera maximowiczii (Rupr. ex Maxim.) Rupr. ex Maxim. ●
71995 Caprifolium maximowiczii Kuntze = Lonicera maximowiczii (Rupr. ex Maxim.) Rupr. ex Maxim. ●
71996 Caprifolium micranthum (Trautv.) Kuntze = Lonicera tatarica L. var. micrantha Trautv. ●
71997 Caprifolium micranthum Kuntze = Lonicera tatarica L. var. micrantha Trautv. ex Regel ●
71998 Caprifolium microphyllum (Willd. ex Roem. et Schult.) Kuntze = Lonicera microphylla Willd. ex Roem. et Schult. ●
71999 Caprifolium microphyllum Kuntze = Lonicera microphylla Willd. ex Roem. et Schult. ●
72000 Caprifolium mollissimum Otto = Lonicera hypoglauca Miq. ●
72001 Caprifolium nervosum (Maxim.) Kuntze = Lonicera nervosa Maxim. ●
72002 Caprifolium nervosum Kuntze = Lonicera nervosa Maxim. ●
72003 Caprifolium nigrum (L.) Kuntze = Lonicera nigra L. ●
72004 Caprifolium nigrum Kuntze = Lonicera nigra L. ●
72005 Caprifolium parvifolium Kuntze = Lonicera myrtillus Hook. f. et Thomson ●
72006 Caprifolium phyllocarpum (Maxim.) Kuntze = Lonicera fragrantissima Lindl. et Paxton subsp. phyllocarpa (Maxim.) P. S. Hsu et H. J. Wang ●
72007 Caprifolium phyllocarpum Kuntze = Lonicera fragrantissima

Lindl. et Paxton subsp. phyllocarpa (Maxim.) P. S. Hsu et H. J. Wang ●

72008 Caprifolium pileatum (Oliv.) Kuntze = Lonicera pileata Oliv. ●

72009 Caprifolium pileatum Kuntze = Lonicera pileata Oliv. ●

72010 Caprifolium reticulamm Kuntze = Lonicera rhytidophylla Hand. -Mazz. ●■

72011 Caprifolium rupicola Kuntze = Lonicera rupicola Hook. f. et Thomson ●

72012 Caprifolium ruprechtianum (Regel) Kuntze = Lonicera ruprechtiana Regel ●

72013 Caprifolium ruprechtianum Kuntze = Lonicera ruprechtiana Regel ●

72014 Caprifolium semenovii (Regel) Kuntze = Lonicera semenovii Regel ●

72015 Caprifolium semnovii Kuntze = Lonicera semenovii Regel ●

72016 Caprifolium sempervirens (L.) Moench = Lonicera sempervirens L. ●■

72017 Caprifolium simile (Hemsl.) Kuntze = Lonicera similis Hemsl. ●

72018 Caprifolium simile Kuntze = Lonicera similis Hemsl. ex Forbes et Hemsl. ●■

72019 Caprifolium spinosum (Jacquem. ex Walp.) Kuntze = Lonicera spinosa Jacquem. ex Walp. ●

72020 Caprifolium spinosum Kuntze = Lonicera spinosa Jacquem. ex Walp. ●

72021 Caprifolium standishii (Carrière) Kuntze = Lonicera fragrantissima Lindl. et Paxton subsp. standishii (Carrière) P. S. Hsu et H. J. Wang ●

72022 Caprifolium standishii Kuntze = Lonicera fragrantissima Lindl. et Paxton subsp. standishii (Carrière) P. S. Hsu et H. J. Wang ●

72023 Caprifolium tanguticum (Maxim.) Kuntze = Lonicera tangutica Maxim. ●

72024 Caprifolium tanguticum Kuntze = Lonicera tangutica Maxim. ●

72025 Caprifolium tataricum (L.) Kuntze = Lonicera tatarica L. ●

72026 Caprifolium tataricum Kuntze = Lonicera tatarica L. ●

72027 Caprifolium tatarinovii (Maxim.) Kuntze = Lonicera tatarinowii Maxim. ●

72028 Caprifolium tatarinovii Kuntze = Lonicera tatarinowii Maxim. ●

72029 Caprifolium thomsonii Kuntze = Lonicera semenovii Regel ●

72030 Caprifolium tomentellum Kuntze = Lonicera tomentella Hook. f. et Thomson ●

72031 Caprifolium tragophyllum Kuntze = Lonicera tragophylla Hemsl. ex Forbes et Hemsl. ●■

72032 Capriola Adans.(废弃属名)= Cynodon Rich.(保留属名)■

72033 Capriola dactylon (L.) Kuntze = Cynodon dactylon (L.) Pers. ■

72034 Caprosma G. Don = Coprosma J. R. Forst. et G. Forst. ●☆

72035 Caproxylon Tussac = Hedwigia Sw. ●☆

72036 Caproxylon Tussac = Tetragastris Gaertn. ●☆

72037 Capsella Medik. (1792)(保留属名);荠属(荠菜属);Shepherd's Purse, Shepherdspurse, Shepherd's-purse ■

72038 Capsella bursa-pastoris (L.) Medik.;荠(荎,大蕺,大荠,地菜,地地菜,地米菜,地米花,耳勾草,饭锹头草,护生草,鸡脚菜,鸡心菜,鸡翼菜,假水菜,净肠草,榄豉菜,老荠,菱角菜,马驹,马辛,蒲蝇花,荠菜,荠菜儿,荠荠菜,荠实,荠只菜,芊菜,芊草,清明草,三角菜,三角草,沙荠,上已菜,菥蓂,香料娘,香荠菜,香芹娘,香善菜,香田荠,烟盒菜,烟盒草,枕头草,正花,粽子菜);Bad Man's Oatmeal, Blindweed, Caseweed, Casewort, Casse-weed, Churchyard Elder, Clappedepouch, Cocowort, Crow Peck, Fat Hen, Gentleman's Purse, Hen Cress, Hen-and-chickens, Hencress, Lady's Purse, Man's Purse, Money Bags, Mother's Heart, Naughty Man's Plaything, Old Woman's Bonnets, Our Lady's Pincushion, Pepper-and-salt, Pepper-grass, Pepper-plant, Pick Purse, Pickpocket, Pick-purse, Pick-your-mother's-heart-out, Poor Man's Parmacetty, Poor Man's Purse, Poverty Purse, Poverty-purse, Purse Flower, Rattle-pouch, Rifle-the-Ladies'-purse, Sanguinary, Schepherospurse, Shepherd's Bag, Shepherd's Bags, Shepherd's Pedlar, Shepherd's Pocket, Shepherd's Pouch, Shepherd's Pounce, Shepherds Purse, Shepherd's Purse, Shepherd's Scrip, Shepherd's-purse, Shovelweed, Snakeflower, Stanche, Staunch, St. James' Wort, Stony-in-the-well, Tacker-weed, Toothwort, Toy-wort, Wardseed, Whoreman's Permacetty, Whoreman's Permacity, Wild Purse, Witches' Pouch ■

72039 Capsella bursa-pastoris (L.) Medik. subsp. occidentalis (Shull) Maire = Capsella bursa-pastoris (L.) Medik. ■

72040 Capsella bursa-pastoris (L.) Medik. subsp. rubella (Reut.) Hook. = Capsella bursa-pastoris (L.) Medik. ■

72041 Capsella bursa-pastoris (L.) Medik. var. concava Almq. = Capsella bursa-pastoris (L.) Medik. ■

72042 Capsella bursa-pastoris (L.) Medik. var. djurdjurae (Shull) Maire et Weiller = Capsella bursa-pastoris (L.) Medik. ■

72043 Capsella bursa-pastoris (L.) Medik. var. gracilis (Gren.) Batt. = Capsella bursa-pastoris (L.) Medik. ■

72044 Capsella bursa-pastoris (L.) Medik. var. macrocarpa Albert = Capsella bursa-pastoris (L.) Medik. ■

72045 Capsella bursa-pastoris (L.) Medik. var. mairei (Shull) Maire = Capsella bursa-pastoris (L.) Medik. ■

72046 Capsella bursa-pastoris (L.) Medik. var. rubella (Reut.) Batt. = Capsella bursa-pastoris (L.) Medik. ■

72047 Capsella bursa-pastoris (L.) Medik. var. triangularis Gruner;日本荠菜■☆

72048 Capsella bursa-pastoris Medik. = Capsella bursa-pastoris (L.) Medik. ■

72049 Capsella elliptica C. A. Mey. = Hymenolobus procumbens (L.) Nutt. ■

72050 Capsella gracilis Gren. = Capsella bursa-pastoris (L.) Medik. ■

72051 Capsella grandiflora (Fauche et Chaub.) Boiss.;大花荠■☆

72052 Capsella hirsuta L. = Cardamine hirsuta L. ■

72053 Capsella hyrcana Grossh.;希尔康荠菜■☆

72054 Capsella orientalis Klokov;东方荠;Oriental Shepherd's Purse ■☆

72055 Capsella pauciflora Koch = Hornungia procumbens (L.) Hayek ■

72056 Capsella procumbens (L.) Fr. = Hornungia procumbens (L.) Hayek ■

72057 Capsella procumbens (L.) Fr. = Hymenolobus procumbens (L.) Nutt. ■

72058 Capsella procumbens (L.) Fr. var. diffusa (Jord.) Maire = Hymenolobus procumbens (L.) Nutt. ■

72059 Capsella rubella Reut.;红荠;Pink Shepherd's-purse ■☆

72060 Capsella rubella Reut. = Capsella bursa-pastoris (L.) Medik. subsp. rubella (Reut.) Hook. ■

72061 Capsella rubella Reut. = Capsella bursa-pastoris (L.) Medik. ■

72062 Capsella thomsonii Hook. f. = Hedinia tibetica (Thomson) Ostenf. ■

72063 Capsicodendron Hoehne = Cinnamodendron Endl. ●

72064 Capsicodendron Hoehne(1933);辣樟属●☆

72065 Capsicodendron pimenteira Hoehne;辣樟●☆

72066 Capsicophysalis (Bitter) Averett et M. Martínez = Chamaesaracha (A. Gray) Benth. et Hook. f. ■☆

72067 Capsicophysalis (Bitter) Averett et M. Martínez(2009);墨西哥刺酸浆属■☆
72068 Capsicum L. (1753);辣椒属;Bell Pepper, Capsicum, Cyaennepepper, Green Pepper, Pimento, Red Pepper, Redpepper, Red-pepper, Sweet Pepper ●■
72069 Capsicum abyssinicum A. Rich. = Capsicum annuum L. ■●
72070 Capsicum annuum L.;辣椒(长辣椒,大椒,番椒,海椒,红海椒,鸡嘴椒,辣虎,辣角,辣茄,辣子,牛角椒,秦椒);Bell Pepper, Bird Pepper, Bush Red Pepper, Bush Redpepper, Cayenne Pepper, Chile, Chile Pepper, Chili, Chilli, Christmas Pepper, Goat Pepper, Green Pepper, Guinea Pepper, Guinea-pods, Hot Pepper, Indish Pepper, Natal Chili, Natal Chilli, Ornamental Pepper, Paprika, Pepper, Pepper Chili, Pimento, Red Pepper, Redpepper, Scarlet Pepper, Spanish Pepper, Spur Pepper, Sweet Pepper, Tabasco Pepper ■●
72071 Capsicum annuum L. 'Acuminatum';尖辣椒(辣椒);Chili Pepper ●
72072 Capsicum annuum L. 'Cerasiforme';樱桃椒(五彩椒,五色椒);Cherry Pepper ●
72073 Capsicum annuum L. 'Conoides';朝天椒(长柄椒,小辣椒,指天椒);Cone Redpepper, Conelike Redpepper ●
72074 Capsicum annuum L. 'Fasciculatum';簇生椒(朝天椒,五色椒);Cluster Peppers, Cluster Redpepper, Fascicled Redpepper, Red None Pepper ●
72075 Capsicum annuum L. 'Grossum';灯笼椒(菜椒,生番姜);Bell Pepper, Cherry Pepper, Cone Pepper, Green Pepper, Pimento, Red Pepper, Sweet Pepper, Thick Redpepper ●
72076 Capsicum annuum L. 'Holiday Cheer';假日情辣椒●☆
72077 Capsicum annuum L. 'Longum';长果辣椒(长辣椒,牛角椒);Bird Pepper, Cayenne Pepper, Chilli Pepper, Paprika ■●☆
72078 Capsicum annuum L. f. bicolor Makino;两色辣椒●
72079 Capsicum annuum L. f. incrassatum (Fingerh.) Makino;薄叶辣椒■☆
72080 Capsicum annuum L. var. acuminatum Fingerh. = Capsicum annuum L. 'Acuminatum' ●
72081 Capsicum annuum L. var. angulosum Mill.;柿子椒(灯笼椒,狮头椒)●
72082 Capsicum annuum L. var. aviculare (Dierb.) D'Arcy et Eshbaugh = Capsicum annuum L. var. glabriusculum (Dunal) Heiser et Pickersgill ●☆
72083 Capsicum annuum L. var. cerasiforum Irish = Capsicum annuum L. 'Cerasiforme' ●
72084 Capsicum annuum L. var. conoides (Mill.) Irish = Capsicum annuum L. ■●
72085 Capsicum annuum L. var. conoides (Mill.) Irish = Capsicum annuum L. 'Conoides' ●
72086 Capsicum annuum L. var. fasciculatum (Sturtev.) Irish = Capsicum annuum L. 'Fasciculatum' ●
72087 Capsicum annuum L. var. fasciculatum (Sturtev.) Irish = Capsicum annuum L. ■●
72088 Capsicum annuum L. var. glabriusculum (Dunal) Heiser et Pickersgill;非洲鸟椒;Bird Pepper ●☆
72089 Capsicum annuum L. var. grossum (L.) Sendtn. = Capsicum annuum L. 'Grossum' ●
72090 Capsicum annuum L. var. grossum (L.) Sendtn. = Capsicum annuum L. ■●
72091 Capsicum annuum L. var. longum Sendtn. = Capsicum annuum L. 'Longum' ■●☆
72092 Capsicum annuum L. var. minimum = Capsicum annuum L. var. glabriusculum (Dunal) Heiser et Pickersgill ●☆
72093 Capsicum annuum L. var. palvoacuminatum Makino;小指天椒●
72094 Capsicum anomalum Franch. et Sav. = Tubocapsicum anomalum (Franch. et Sav.) Makino ■
72095 Capsicum baccatum L.;鸟椒;Bird Pepper ■☆
72096 Capsicum boninense Koidz. = Tubocapsicum boninense (Koidz.) Koidz. ex H. Hara ■☆
72097 Capsicum chinensis Jacq.;中国椒■☆
72098 Capsicum conoides Mill. = Capsicum annuum L. ■●
72099 Capsicum fasciculatum Sturtev. = Capsicum annuum L. ■●
72100 Capsicum frutescens L.;小米椒(番姜,非洲辣椒,红海椒,辣椒,辣椒树,辣子,米辣,牛角椒,涮涮辣,小米辣,野辣子);Bush Red Pepper, Bush Redpepper, Bush Red-pepper, Spanish Pepper, Tabasco ●
72101 Capsicum frutescens L. = Capsicum annuum L. ■●
72102 Capsicum frutescens L. var. conoides L. Bailey = Capsicum annuum L. ●
72103 Capsicum frutescens L. var. fasciculatum L. Bailey = Capsicum annuum L. ■●
72104 Capsicum frutescens L. var. grossum L. Bailey = Capsicum annuum L. ■●
72105 Capsicum frutescens L. var. longum L. Bailey = Capsicum annuum L. ■●
72106 Capsicum fuscoviolaceum (Cufod.) C. V. Morton et Standl.;褐紫辣椒●☆
72107 Capsicum grossum L. = Capsicum annuum L. var. grossum (L.) Sendtn. ●
72108 Capsicum grossum L. = Capsicum annuum L. ■●
72109 Capsicum hispidum Dunal subsp. glabriusculum Dunal = Capsicum annuum L. var. glabriusculum (Dunal) Heiser et Pickersgill ●☆
72110 Capsicum indicum Dierb. var. aviculare Dierb. = Capsicum annuum L. var. glabriusculum (Dunal) Heiser et Pickersgill ●☆
72111 Capsicum longum DC. = Capsicum annuum L. 'Longum' ●■
72112 Capsicum minimum Mill. = Capsicum annuum L. var. glabriusculum (Dunal) Heiser et Pickersgill ●☆
72113 Capsicum minimum Roxb.;小辣椒(番姜,山新尔);Smallest Redpepper ●
72114 Capsicum minimum Walker = Tubocapsicum anomalum (Franch. et Sav.) Makino ■
72115 Capsicum pubescens Ruiz. et Pav.;紫花椒●
72116 Captaincookia N. Hallé(1973);新喀茜属☆
72117 Captaincookia margaretae N. Hallé;新喀茜☆
72118 Capura Blanco = Otophora Blume ●
72119 Capura L.(废弃属名) = Wikstroemia Endl.(保留属名)●
72120 Capura purpurata L. = Wikstroemia indica (L.) C. A. Mey. ●
72121 Capurodendron Aubrrév.(1962);卡普山榄属●☆
72122 Capurodendron androyense Aubrév.;安德罗卡普山榄●☆
72123 Capurodendron ankaranense Aubrév.;安卡兰卡普山榄●☆
72124 Capurodendron antongiliense Aubrév.;安通吉尔卡普山榄●☆
72125 Capurodendron apollonioides Aubrév.;太阳楠卡普山榄●☆
72126 Capurodendron bakeri (Scott-Elliot) Aubrév.;贝克卡普山榄●☆
72127 Capurodendron costatum Aubrév.;单脉卡普山榄●☆
72128 Capurodendron delphinense Aubrév.;德尔芬卡普山榄●☆
72129 Capurodendron gracilifolium Aubrév.;细叶卡普山榄●☆

72130 Capurodendron greveanum Aubrév.;格雷弗卡普山榄●☆

72131 Capurodendron madagascariense (Lecomte) Aubrév.;马岛卡普山榄●☆

72132 Capurodendron mandrarense Aubrév.;曼德拉卡普山榄●☆

72133 Capurodendron microlobum (Baker) Aubrév. = Capurodendron pervillei (Engl.) Aubrév. ●☆

72134 Capurodendron microphyllum (Scott-Elliot) Aubrév.;小叶卡普山榄●☆

72135 Capurodendron nodosum Aubrév.;多节卡普山榄●☆

72136 Capurodendron perrieri (Lecomte) Aubrév.;佩里耶卡普山榄●☆

72137 Capurodendron pervillei (Engl.) Aubrév.;佩尔卡普山榄●☆

72138 Capurodendron pseudoterminalia Aubrév.;顶生卡普山榄●☆

72139 Capurodendron rubrocostatum (Jum. et H. Perrier) Aubrév.;红单脉卡普山榄●☆

72140 Capurodendron rufescens Aubrév.;浅红卡普山榄●☆

72141 Capurodendron sakalavum Aubrév.;萨卡拉瓦卡普山榄●☆

72142 Capurodendron suarezense Aubrév.;苏亚雷斯卡普山榄●☆

72143 Capurodendron tampinense (Lecomte) Aubrév.;淡边卡普山榄●☆

72144 Capurodendron terminalioides Aubrév.;拟顶生卡普山榄●☆

72145 Capuronetta Markgr. = Tabernaemontana L. ●

72146 Capuronetta elegans Markgr. = Tabernaemontana capuronii Leeuwenb. ●☆

72147 Capuronia Lourteig(1960);卡普莱属●☆

72148 Capuronia madagascariensis Lourteig;卡普莱●☆

72149 Capuronianthus J.-F. Leroy(1958);卡普楝属●☆

72150 Capuronianthus mahafalensis J.-F. Leroy;卡普楝●☆

72151 Capusia Lecomte = Siphonodon Griff. ●☆

72152 Capusiaceae Gagnep. = Celastraceae R. Br.(保留科名)●

72153 Capusiaceae Gagnep. = Siphonodontaceae Gagnep. et Tardieu(保留科名)●

72154 Caquepiria J. F. Gmel. = Gardenia Ellis(保留属名)●

72155 Caquepiria bergkia J. F. Gmel. = Gardenia thunbergia Thunb. ●☆

72156 Carabichea Post et Kuntze = Cephaelis Sw.(保留属名)●

72157 Caracalla Tod. = Phaseolus L. ■

72158 Caracasia Szyszyl.(1894);加拉加斯藤属●☆

72159 Caracasia tremadena (Ernst) Szyszył.;加拉加斯藤●☆

72160 Caracasia tremadena Szyszył. = Caracasia tremadena (Ernst) Szyszył. ●☆

72161 Caracasia viridiflora (Ernst) Szyszył.;绿花加拉加斯藤●☆

72162 Caracasia viridiflora Szyszył. = Caracasia viridiflora (Ernst) Szyszył. ●☆

72163 Carachera Juss. = Charachera Forssk. ●

72164 Carachera Juss. = Lantana L.(保留属名)●

72165 Caradesia Raf. = Eupatorium L. ■●

72166 Caraea Hochst. = Euryops (Cass.) Cass. ●■☆

72167 Caraea pinifolia Hochst. ex Steud. = Euryops arabicus Steud. ■●☆

72168 Caragana Fabr.(1763);锦鸡儿属;Pea Shrub,Pea Tree,Peashrub,Pea-shrub,Peatree,Pea-tree ●

72169 Caragana Lam. = Caragana Fabr. ●

72170 Caragana acanthophylla Kom.;刺叶锦鸡儿;Spinyleaf Peashrub,Spiny-leaved Pea-shrub ●

72171 Caragana acanthophylla Kom. subsp. macrocalyx Yakovlev = Caragana acanthophylla Kom. ●

72172 Caragana acualis Baker = Chesneya acaulis (Baker) Popov ●

72173 Caragana aegacanthoides (R. Parker) L. B. Chaudhary et S. K. Srivast.;萨迦锦鸡儿;Sajia Peashrub,Sajia Pea-shrub ●

72174 Caragana aitchisonii Prain = Caragana decorticans Hemsl. ●☆

72175 Caragana alaica Pojark.;阿赖锦鸡儿●☆

72176 Caragana alpina Y. X. Liou = Caragana chumbica Prain ●

72177 Caragana altagana Poir. = Caragana microphylla Lam. ●

72178 Caragana altaica (Kom.) Pojark.;阿尔泰锦鸡儿●

72179 Caragana ambigua Stocks;可疑锦鸡儿●☆

72180 Caragana arborescens Lam.;树锦鸡儿(锦鸡儿,蒙古锦鸡儿,小黄刺条);Caragana,Pea Tree,Pea-tree,Siberian Pea Shrub,Siberian Pea Tree,Siberian Peashrub,Siberian Pea-shrub,Siberian Pea-tree,Treelike Peashrub ●

72181 Caragana arborescens Lam. 'Lorbergii';洛伯格树锦鸡儿●☆

72182 Caragana arborescens Lam. 'Nana';矮生树锦鸡儿●☆

72183 Caragana arborescens Lam. 'Walker';步行者树锦鸡儿●☆

72184 Caragana arcuata Y. X. Liou;弯枝锦鸡儿;Arcuate Peashrub,Arcuate Pea-shrub,Bendtwig Peashrub ●

72185 Caragana arcuata Y. X. Liou = Caragana acanthophylla Kom. ●

72186 Caragana argenta Lam. = Halimodendron halodendron (Pall.) C. K. Schneid. ●

72187 Caragana aurantiaca Koehne;镰叶锦鸡儿;Dwarf Peashrub,Orange-coloured Peashrub,Orange-coloured Pea-shrub,Sickleleaf Peashrub ●

72188 Caragana aurantiaca Koehne var. deserticola Kom. = Caragana leucophloea Poljakov ●

72189 Caragana aurantiaca var. conferta Kom. = Caragana aurantiaca Koehne ●

72190 Caragana bicolor Kom.;二色锦鸡儿;Bicolor Peashrub,Two-coloured Peashrub,Two-coloured Pea-shrub ●

72191 Caragana boisii C. K. Schneid.;扁刺锦鸡儿(野皂角);Bois Pea-shrub,Flatspine Peashrub ●

72192 Caragana boisii C. K. Schneid. var. platycarpa Yakovlev = Caragana boisii C. K. Schneid. ●

72193 Caragana bongardiana (Fisch. et C. A. Mey.) Pojark.;边塞锦鸡儿;Bongard Peashrub,Bongard Pea-shrub,Frontier Peashrub ●

72194 Caragana brachyantha Rech. f.;短花锦鸡儿●☆

72195 Caragana brachypoda Pojark.;矮脚锦鸡儿;Shortfoot Peashrub,Short-stalked Peashrub,Short-stalked Pea-shrub ●

72196 Caragana brevifolia Kom.;短叶锦鸡儿(猪儿刺);Shortleaf Peashrub,Short-leaved Pea-shrub ●

72197 Caragana brevispina Benth.;短刺锦鸡儿●☆

72198 Caragana brevispina Benth. var. catenata = Caragana catenata Kom. ●☆

72199 Caragana bungei Ledeb.;邦奇锦鸡儿;Bunge Pea-shrub ●☆

72200 Caragana camilli-schneideri Kom.;北疆锦鸡儿(库车锦鸡儿,新疆锦鸡儿);Sinkiang Peashrub,Sinkiang Pea-shrub,Xingjiang Peashrub ●

72201 Caragana catenata Kom.;链状短刺锦鸡儿●☆

72202 Caragana chamlago B. Mey. = Sophora davidii (Franch.) Pavol. ●

72203 Caragana chamlagu Lam. = Caragana sinica (Buc'hoz) Rehder ●

72204 Caragana changduensis Y. X. Liou;昌都锦鸡儿;Changdu Peashrub,Changdu Pea-shrub ●

72205 Caragana chinensis Turcz. ex Maxim. = Caragana sinica (Buc'hoz) Rehder ●

72206 Caragana chinghaiensis Y. X. Liou;青海锦鸡儿;Qinghai Peashrub,Qinghai Pea-shrub ●

72207 Caragana chinghaiensis Y. X. Liou var. minima Y. X. Liou;小青海锦鸡儿;Small Qinghai Peashrub,Small Qinghai Pea-shrub ●

72208 Caragana chumbica Prain;高山锦鸡儿;Alp Peashrub,Alpine Peashrub,Alpine Pea-shrub ●

72209 Caragana conferta Benth. ex Baker;密集锦鸡儿●☆
72210 Caragana crassicaulis Benth. ex Baker = Chesneya nubigena (D. Don) Ali ●☆
72211 Caragana crassispina C. Marquand;粗枝锦鸡儿●
72212 Caragana cuneata (Benth.) Baker = Chesneya cuneata (Benth.) Ali ●
72213 Caragana cuneato-alata Y. X. Liou;楔翼锦鸡儿;Cuneate-wing Peashrub,Cuneate-winged Caunealwing Peashrub,Pea-shrub ●
72214 Caragana czetyrkinii Sanchir = Caragana jubata (Pall.) Poir. var. czetyrkininii (Sanchir) Y. X. Liou ●
72215 Caragana dasyphylla Pojark.;粗毛锦鸡儿;Shag Peashrub, Thickleaf Peashrub,Woolly-leaved Pea-shrub ●
72216 Caragana davazamcii Sanchir;沙地锦鸡儿;Davazamc Pea-shrub, Dene Peashrub,Sandy Peashrub ●
72217 Caragana davazamcii Sanchir = Caragana korshinskii Kom. ●
72218 Caragana davazamcii Sanchir var. viridis Y. X. Liou;绿沙地锦鸡儿;Green Sandy Peashrub,Green Sandy Pea-shrub ●
72219 Caragana davazamcii Sanchir var. viridis Y. X. Liou = Caragana davazamcii Sanchir ●
72220 Caragana decorticans Hemsl.;脱皮锦鸡儿●☆
72221 Caragana densa Kom.;密叶锦鸡儿;Denseleaf Peashrub,Dense-leaved Pea-shrub ●
72222 Caragana erenensis Y. X. Liou;二连锦鸡儿;Erlian Peashrub, Erlian Pea-shrub ●
72223 Caragana erenensis Y. X. Liou = Caragana davazamcii Sanchir ●
72224 Caragana erenensis Y. X. Liou = Caragana intermedia Kuang et H. C. Fu ●
72225 Caragana erinacea Kom.;川西锦鸡儿;W. Sichuan Peashrub, West Sichuan Peashrub,West Sichuan Pea-shrub ●
72226 Caragana forrestii Sanchir = Caragana jubata (Pall.) Poir. var. czetyrkininii (Sanchir) Y. X. Liou ●
72227 Caragana franchetiana Kom.;云南锦鸡儿(川青锦鸡儿,阳雀花);Franchet Peashrub,Yunnan Peashrub,Franchet Pea-shrub ●
72228 Caragana franchetiana Kom. var. gyrongensis (C. C. Ni) Y. X. Liou;吉隆锦鸡儿;Jilong Peashrub,Jilong Pea-shrub ●
72229 Caragana frutescens Medik. = Caragana frutex (L.) K. Koch ●
72230 Caragana frutescens Medik. var. ussuriensis Regel = Caragana ussuriensis (Regel) Pojark. ●
72231 Caragana frutescnes Medik. var. turfanensis Krasn. = Caragana turfanensis (Krasn.) Kom. ●
72232 Caragana frutex (L.) C. Koch = Caragana frutex (L.) K. Koch ●
72233 Caragana frutex (L.) K. Koch;黄刺条锦鸡儿(黄刺条,黄荆条,金雀花,金雀锦鸡儿,木锦鸡儿);Bush Pea-shrub, Huangjingtiao Peashrub, Russian Pea Shrub, Russian Peashrub, Russian Pea-tree,Shrubby Peashrub ●
72234 Caragana frutex (L.) K. Koch 'Globosa';圆球黄荆条●☆
72235 Caragana frutex (L.) K. Koch var. latifolia C. K. Schneid.;宽叶黄荆条(宽叶黄刺条);Broad-leaf Pea-shrub, Broad-leaf Shrubby Peashrub,Russian Pea Shrub ●
72236 Caragana frutex (L.) K. Koch var. latifolia C. K. Schneid. = Caragana frutex (L.) K. Koch ●
72237 Caragana fruticosa (Pall.) Besser;极东锦鸡儿;Extremeeast Peashrub,Shrub Peashrub,Shrubby Pea-shrub ●
72238 Caragana fruticosa (Pall.) Besser var. multiflora H. Xie et Y. T. Zhao = Caragana arborescens Lam. ●
72239 Caragana gerardiana Benth.;印度锦鸡儿;Gerard Pea-shrub, Indian Peashrub,Indo Peashrub ●
72240 Caragana gerardiana Benth. var. glabrescens Franch. = Caragana franchetiana Kom. ●
72241 Caragana grandiflora DC.;大花锦鸡儿;Big-flowered Pea-shrub ●☆
72242 Caragana grandiflora DC. = Caragana aurantiaca Koehne ●
72243 Caragana gyirongensis C. C. Ni = Caragana franchetiane Kom. var. gyrongensis (C. C. Ni) Y. X. Liou ●
72244 Caragana hololeuca Bunge ex Kom.;绢毛锦鸡儿;Sericeous Peashrub,Sericeous Pea-shrub,Silky Peashrub ●
72245 Caragana hololeuca Bunge ex Kom. = Caragana tragacanthoides (Pall.) Poir. ●
72246 Caragana hopeiensis Yakovlev = Caragana pekinensis Kom. ●
72247 Caragana intermedia Kuang et H. C. Fu = Caragana korshinskii Kom. ●
72248 Caragana jilungensis C. C. Ni = Caragana franchetiane Kom. var. gyrongensis (C. C. Ni) Y. X. Liou ●
72249 Caragana jubata (Pall.) Poir.;鬼箭锦鸡儿(藏锦鸡儿,冠毛锦鸡儿,鬼见愁,浪麻,着母香);Ghost-arrow Peashrub, Sharpspine Peashrub,Sharp-spined Pea-shrub ●
72250 Caragana jubata (Pall.) Poir. f. seczuanica Kom. = Caragana jubata (Pall.) Poir. ●
72251 Caragana jubata (Pall.) Poir. f. szechuanica Kom.;四川鬼箭锦鸡儿(四川鬼箭);Sichuan Peashrub,Sichuan Pea-shrub ●
72252 Caragana jubata (Pall.) Poir. var. biaurita Y. X. Liou;两耳鬼箭;Two-ear Peashrub,Two-ear Pea-shrub ●
72253 Caragana jubata (Pall.) Poir. var. czetyrkininii (Sanchir) Y. X. Liou;浪麻鬼箭;Czetyrkinin Peashrub,Czetyrkinin Pea-shrub ●
72254 Caragana jubata (Pall.) Poir. var. recurva Y. X. Liou;弯耳鬼箭;Bent-ear Peashrub,Bent-ear Pea-shrub ●
72255 Caragana junatovii Gorbunova;通天河锦鸡儿●
72256 Caragana kansuensis Pojark.;甘肃锦鸡儿(臭柴,甘青锦鸡儿,母猪刺,牛筋条);Gansu Peashrub,Gansu Pea-shrub ●
72257 Caragana kirghisorum Pojark.;囊萼锦鸡儿(刺萼锦鸡儿); Bagcalyx Peashrub,Pea-shrub,Saccate-calyx Peashrub ●
72258 Caragana korshinskii Kom.;柠条锦鸡儿(白柠条,老虎刺,马集柴,毛条,柠条,牛筋条,中间锦鸡儿);Intermediate Peashrub, Intermediate Pea-shrub, Korshinsk Peashrub, Korshinsk Pea-shrub, Middle Peashrub ●
72259 Caragana korshinskii Kom. f. brachypoda Y. X. Liou = Caragana korshinskii Kom. ●
72260 Caragana korshinskii Kom. var. brachypoda Y. X. Liou;短荚柠条;Shortpod Korshinsk Peashrub,Shortpod Korshinsk Pea-shrub ●
72261 Caragana korshinskii Kom. var. brachypoda Y. X. Liou = Caragana korshinskii Kom. ●
72262 Caragana korshinskii Kom. var. davazamcii (Sanchir) Yakovlev = Caragana davazamcii Sanchir ●
72263 Caragana korshinskii Kom. var. intermedia (Kuang et H. C. Fu) M. L. Zhang et G. H. Zhu = Caragana intermedia Kuang et H. C. Fu ●
72264 Caragana korshinskii Kom. var. intermedia (Kuang et H. C. Fu) M. L. Zhang et G. H. Zhu = Caragana korshinskii Kom. ●
72265 Caragana korshinskii Kom. var. ordosica Yakovlev = Caragana intermedia Kuang et H. C. Fu ●
72266 Caragana kozlowii Kom.;沧江锦鸡儿;Cangjiang Peashrub, Kozlov Peashrub,Kozlov Pea-shrub ●
72267 Caragana laeta Kom.;阿拉套锦鸡儿●
72268 Caragana laetevirens Poljakov;鲜绿锦鸡儿●☆
72269 Caragana leduensis Y. Z. Zhao et al = Caragana tangutica Maxim. ●

72270 Caragana leucophloea Poljakov;白皮锦鸡儿(金雀花,锦鸡儿);Whitebark Peashrub,Whitebark Pea-shrub ●

72271 Caragana leucospina Kom.;白刺锦鸡儿;Whitespine Peashrub,White-spine Pea-shrub ●

72272 Caragana leveillei Kom.;毛掌叶锦鸡儿(母猪鬃);Hairpalmleaf Peashrub,Leveille Peashrub,Leveille Pea-shrub ●

72273 Caragana licentiana Hand.-Mazz.;白毛锦鸡儿;Licent Pea-shrub,Whitehair Peashrub ●

72274 Caragana limprichtii Harms = Caragana bicolor Kom. ●

72275 Caragana liouana Zhao Y. Chang et Yakovlev;中间锦鸡儿●

72276 Caragana litwinowii Kom.;金州锦鸡儿;Jinzhou Peashrub,Litwinow Peashrub,Litwinow Pea-shrub ●

72277 Caragana longiunguiculata C. W. Chang = Caragana rosea Turcz. var. longiunguicaulata (C. W. Chang) Y. X. Liou ●

72278 Caragana longiunguiculata C. W. Chang = Caragana sinica (Buc'hoz) Rehder ●

72279 Caragana manshurica (Kom.) Kom.;东北锦鸡儿(骨担草);Manchurian Peashrub,Manchurian Pea-shrub,NE. China Peashrub ●

72280 Caragana manshurica Kom. = Caragana manshurica (Kom.) Kom. ●

72281 Caragana maximoviziana Kom. = Caragana erinacea Kom. ●

72282 Caragana maximoviziana Kom. = Caragana maximowicziana Kom. ●

72283 Caragana maximowicziana Kom.;繁花锦鸡儿;Maximowicz Peashrub,Maximowicz Pea-shrub ●

72284 Caragana maximowiziana Kom. = Caragana erinacea Kom. ●

72285 Caragana microphylla Lam.;小叶锦鸡儿(黑柠条,猴獠刺,连针,柠条,牛筋条,小叶金雀花,雪里洼);Littleleaf Pea Shrub,Littleleaf Peashrub, Little-leaved Pea Shrub, Microphyllous Pea-shrub,Small-leaved Pea-shrub ●

72286 Caragana microphylla Lam. f. cinerea Kom.;灰叶锦鸡儿(灰色小叶锦鸡儿);Grey-leaf Peashrub,Grey-leaf Pea-shrub ●

72287 Caragana microphylla Lam. f. cinerea Kom. = Caragana microphylla Lam. ●

72288 Caragana microphylla Lam. f. daurica Kom.;兴安锦鸡儿;Daurian Peashrub,Daurian Pea-shrub ●

72289 Caragana microphylla Lam. f. daurica Kom. = Caragana microphylla Lam. ●

72290 Caragana microphylla Lam. f. manshurica Kom. = Caragana manshurica (Kom.) Kom. ●

72291 Caragana microphylla Lam. f. pallasiana Kom.;毛序锦鸡儿;Hairy-inflorescens Peashrub,Hairy-inflorescens Pea-shrub ●

72292 Caragana microphylla Lam. f. pallasiana Kom. = Caragana microphylla Lam. ●

72293 Caragana microphylla Lam. f. tomentosa Kom.;毛枝锦鸡儿;Tomentose Peashrub,Tomentose Pea-shrub ●

72294 Caragana microphylla Lam. f. viridis Kom.;绿叶锦鸡儿(绿色小叶锦鸡儿);Green-leaf Peashrub,Green-leaf Pea-shrub ●

72295 Caragana microphylla Lam. f. viridis Kom. = Caragana microphylla Lam. ●

72296 Caragana microphylla Lam. var. crasseaculeata Bois = Caragana boisii C. K. Schneid. ●

72297 Caragana microphylla Lam. var. potaninii (Kom.) Y. X. Liou = Caragana potaninii Kom. ●

72298 Caragana microphylla Lam. var. potaninii (Kom.) Y. X. Liou ex L. Z. Shue = Caragana potaninii Kom. ●

72299 Caragana microphylla Lam. var. tomentosa Kom. = Caragana korshinskii Kom. ●

72300 Caragana mollis Besser;柔毛锦鸡儿;Woolly Pea-shrub ●☆

72301 Caragana moorcroftiana Benth. = Sophora moocroftiana (Graham) Benth. ex Baker ●

72302 Caragana nepalensis Kitam.;尼泊尔锦鸡儿●☆

72303 Caragana opulens Kom.;甘蒙锦鸡儿;Gansu-Mongol Peashrub,Gansu-Mongol Pea-shrub,Kansu-mongolian Peashrub ●

72304 Caragana opulens Kom. var. licentiana (Hand.-Mazz.) Yakovlev = Caragana licentiana Hand.-Mazz. ●

72305 Caragana opulens Kom. var. perforata Merrgen et Y. Q. Ma = Caragana opulens Kom. ●

72306 Caragana opulens Kom. var. trichophylla Z. H. Gao et S. C. Zhang;毛叶锦鸡儿;Hairyleaf Gansu-Mongol Peashrub,Hairyleaf Gansu-Mongol Pea-shrub ●

72307 Caragana opulens Kom. var. trichophylla Z. H. Gao et S. C. Zhang = Caragana opulens Kom. ●

72308 Caragana ordosica Y. Z. Zhao et al. = Caragana tibetica (Maxim. ex C. K. Schneid.) Kom. ●

72309 Caragana oreophila W. W. Sm. = Caragana franchetiana Kom. ●

72310 Caragana pekinensis Kom.;北京锦鸡儿(灰叶黄刺条);Beijing Peashrub,Peking Pea-shrub ●

72311 Caragana pleiophylla (Regel) Pojark.;多叶锦鸡儿;Leafy Peashrub,Manyleaves Peashrub,Pleiophyllous Pea-shrub ●

72312 Caragana polourensis Franch.;昆仑锦鸡儿;Polour Peashrub,Polour Pea-shrub ●

72313 Caragana potaninii Kom.;五台锦鸡儿;Potanin Peashrub,Potanin Pea-shrub ●

72314 Caragana potaninii Kom. = Caragana microphylla Lam. var. potaninii (Kom.) Y. X. Liou ●

72315 Caragana prainii C. K. Schneid. = Caragana decorticans Hemsl. ●☆

72316 Caragana prainii Kom.;阿富汗锦鸡儿●☆

72317 Caragana pruinosa Kom.;粉刺锦鸡儿;Pale Peashrub,Pruinose Pea-shrub,Waxy-powdery Peashrub ●

72318 Caragana przewalskii Pojark. = Caragana roborovskyi Kom. ●

72319 Caragana pulcherrima Vassilcz. = Caragana gerardiana Benth. ●

72320 Caragana pumila Pojark.;草原锦鸡儿●

72321 Caragana purdomii Rehder;秦晋锦鸡儿(马柠条,朴氏锦鸡儿);Purdom Peashrub,Purdom Pea-shrub ●

72322 Caragana pygmaea (L.) DC.;矮锦鸡儿(线叶锦鸡儿);Dwarf Peashrub,Dwarf Pea-shrub ●

72323 Caragana pygmaea (L.) DC. = Caragana versicolor Benth. ●

72324 Caragana pygmaea (L.) DC. f. longifolia Kom.;长叶矮锦鸡儿;Long-leaf Dwarf Peashrub,Pea-shrub ●

72325 Caragana pygmaea (L.) DC. f. longifolia Kom. = Caragana pygmaea (L.) DC. ●

72326 Caragana pygmaea (L.) DC. subsp. altaica (Kom.) Bondareva = Caragana altaica (Kom.) Pojark. ●

72327 Caragana pygmaea (L.) DC. var. altaica Kom. = Caragana altaica (Kom.) Pojark. ●

72328 Caragana pygmaea (L.) DC. var. angustissima C. K. Schneid.;窄叶矮锦鸡儿;Narrow-leaf Dwarf Peashrub,Narrow-leaf Dwarf Pea-shrub ●

72329 Caragana pygmaea (L.) DC. var. pallasiana Kom. = Caragana stenophylla Pojark. ●

72330 Caragana pygmaea (L.) DC. var. parviflora H. C. Fu;小花矮锦鸡儿;Small-flower Dwarf Peashrub,Small-flower Dwarf Pea-shrub ●

72331 Caragana qingheensis Z. Y. Chang et al.;青河锦鸡儿●

72332 Caragana reticulate Rehder = Caragana stipitata Kom. ●

72333 Caragana roborovskyi Kom.;荒漠锦鸡儿(阿拉善锦鸡儿,洛氏锦鸡儿,猫耳刺,猫耳锦鸡儿,通天河锦鸡儿);Desert Peashrub, Przewalski Peashrub, Przewalski Pea-shrub, Roborovsky Pea-shrub ●

72334 Caragana rosea Turcz.;红花锦鸡儿(黄枝条,金雀儿,金雀花);Dwarf Pea Shrub, Red Peashrub, Redflower Peashrub, Red-flowered Pea-shrub ●

72335 Caragana rosea Turcz. ex Maxim. var. longiunguiculata (C. W. Chang) Y. X. Liou = Caragana sinica (Buc'hoz) Rehder ●

72336 Caragana rosea Turcz. var. longiunguicaulata (C. W. Chang) Y. X. Liou;长爪红花锦鸡儿(长爪锦鸡儿);Long-chawed Peashrub, Long-chawed Pea-shrub ●

72337 Caragana sajaensis Z. C. Ni = Caragana aegacanthoides (R. Parker) L. B. Chaudhary et S. K. Srivast. ●

72338 Caragana sericea Pamp. ex Kom. = Caragana stipitata Kom. ●

72339 Caragana shensiensis C. W. Chang;秦岭锦鸡儿(陕西锦鸡儿);Shaanxi Peashrub, Shaanxi Pea-shrub ●

72340 Caragana sibirica Fabr.;西伯利亚锦鸡儿(骨担草,蒙古锦鸡儿,树锦鸡儿);Siberian Peashrub, Siberian Pea-shrub ●

72341 Caragana sibirica Medik. = Caragana arborescens Lam. ●

72342 Caragana sinica (Buc'hoz) Rehder;锦鸡儿(坝齿花,白心皮,板参,斧头花,黄棘,黄雀花,黄雀梅,甲鱼嘴,甲鱼嘴花,江南金凤,酱瓣子,金雀花,金鹊花,娘娘袜,群雀花,铁扫帚,土黄芪,阳雀花,阳鹊花,野黄芪,粘粘袜,猪蹄花);China Peashrub, Chinese Pea Tree, Chinese Peashrub, Chinese Pea-shrub ●

72343 Caragana sinica (Buc'hoz) Rehder var. longipedunculata C. W. Chang;长柄锦鸡儿;Long-pedicel Chinese Peashrub, Long-pedicel Chinese Pea-shrub ●

72344 Caragana sinica (Buc'hoz) Rehder var. longipedunculata C. W. Chang = Caragana leveillei Kom. ●

72345 Caragana soongorica Grubov;准噶尔锦鸡儿;Dzungar Peashrub, Junggar Pea-shrub, Songar Peashrub, Soógorian Pea-shrub ●

72346 Caragana spinifera Kom.;西藏锦鸡儿(多刺锦鸡儿);Thornbearing Peashrub, Thorn-bearing Pea-shrub, Xizang Peashrub ●

72347 Caragana spinifera Kom. = Caragana erinacea Kom. ●

72348 Caragana spinosa (L.) DC.;多刺锦鸡儿;Spiny Peashrub, Spiny Pea-shrub, Thorny Peashrub ●

72349 Caragana spinosa (L.) Hornem. = Caragana spinosa (L.) DC. ●

72350 Caragana spinosissima Benth. = Caragana gerardiana Benth. ●

72351 Caragana stenophylla Pojark.;狭叶锦鸡儿(红刺,红柠条,母猪刺,柠角,皮溜刺,羊柠角);Narrowleaf Peashrub, Narrow-leaved Pea-shrub, Stenophyllous Pea-shrub ●

72352 Caragana stenophylla Pojark. var. parviflora Zhan Wang et H. C. Fu;细叶小花锦鸡儿;Small-flower Narrowleaf Peashrub, Small-flower Narrowleaf Pea-shrub ●

72353 Caragana stenophylla Pojark. var. parviflora Zhan Wang et H. C. Fu = Caragana stenophylla Pojark. ●

72354 Caragana stenophylla Pojark. var. sericea H. C. Fu;绢毛狭叶锦鸡儿(狭叶锦鸡儿);Silky Naroowleaf Peashrub, Silky Naroowleaf Pea-shrub ●

72355 Caragana stenophylla Pojark. var. sericea H. C. Fu = Caragana pygmaea (L.) DC. var. angustissima C. K. Schneid. ●

72356 Caragana stipitata Kom.;柄荚锦鸡儿;Stipitate Peashrub, Stipitate Pea-shrub ●

72357 Caragana sukiensis C. K. Schneid.;锡金锦鸡儿●☆

72358 Caragana tangutica Maxim.;青甘锦鸡儿(甘青锦鸡儿);Tangut Peashrub, Tangut Pea-shrub ●

72359 Caragana tangutica Maxim. ex Kom. = Caragana tangutica Maxim. ●

72360 Caragana tangutica Maxim. var. yushuensis Y. H. Wu = Caragana kozlowii Kom. ●

72361 Caragana tekesiensis Y. Z. Zhao et D. W. Zhou;特克斯锦鸡儿;Tekesi Peashrub, Tekesi Pea-shrub ●

72362 Caragana tibetica (Maxim. ex C. K. Schneid.) Kom.;川青锦鸡儿(藏锦鸡儿,垫状锦鸡儿,黑毛头刺,康青锦鸡儿,毛刺锦鸡儿,头刺);Hairspine Peashrub, Tibet Peashrub, Tibet Pea-shrub ●

72363 Caragana tibetica Kom. = Caragana tibetica (Maxim. ex C. K. Schneid.) Kom. ●

72364 Caragana tragacanthoides (Pall.) Poir.;中亚锦鸡儿;Central Asia Peashrub, Central Asian Peashrub, Central Asian Pea-shrub ●

72365 Caragana tragacanthoides (Pall.) Poir. var. pallasiana Fisch. et C. A. Mey. = Caragana przewalskii Pojark. ●

72366 Caragana tragacanthoides (Pall.) Poir. var. pleiophylla Regel = Caragana pleiophylla (Regel) Pojark. ●

72367 Caragana tragacanthoides (Pall.) Poir. var. tibetica Maxim. ex C. K. Schneid. = Caragana tibetica (Maxim. ex C. K. Schneid.) Kom. ●

72368 Caragana tragacanthoides (Pall.) Poir. var. villosa Regel = Caragana przewalskii Pojark. ●

72369 Caragana triflora Lindl. = Caragana brevispina Benth. ●☆

72370 Caragana turfanensis (Krasn.) Kom.;吐鲁番锦鸡儿(伊犁锦鸡儿);Tulufan Peashrub, Tulufan Pea-shrub, Turfan Peashrub ●

72371 Caragana turkestanica Kom.;新疆锦鸡儿;Xinjiang Peashrub, Xinjiang Pea-shrub ●

72372 Caragana ulicina Stocks;荆豆锦鸡儿●☆

72373 Caragana ussuriensis (Regel) Pojark.;乌苏里锦鸡儿;Ussuri Peashrub, Ussuri Pea-shrub, Wusuli Peashrub ●

72374 Caragana versicolor Benth.;变色锦鸡儿;Variedcolor Peashrub, Variously Coloured Peashrub, Versicolorous Pea-shrub ●

72375 Caragana wenhsienensis C. W. Chang;文县锦鸡儿;Wenxian Peashrub, Wenxian Pea-shrub ●

72376 Caragana wenhsienensis C. W. Chang = Caragana shensiensis C. W. Chang ●

72377 Caragana wenhsienensis C. W. Chang var. inermis C. W. Chang;无刺锦鸡儿;Spineless Wenxian Peashrub, Spineless Wenxian Pea-shrub ●

72378 Caragana wenhsienensis C. W. Chang var. inermis C. W. Chang = Caragana shensiensis C. W. Chang ●

72379 Caragana williamsii Vassilcz. = Caragana brevispina Benth. ●☆

72380 Caragana zahlbruckneri C. K. Schneid.;南口锦鸡儿;Nankou Peashrub, Zahlbruckner Peashrub, Zahlbruckner Pea-shrub ●

72381 Caragana zahlbruckneri C. K. Schneid. subsp. litwinowii (Kom.) Yakovelev = Caragana litwinowii Kom. ●

72382 Caragana zahlbruckneri C. K. Schneid. var. pekinensis (Kom.) Yakovlev = Caragana pekinensis Kom. ●

72383 Caragana zahlbruckneri C. K. Schneid. var. pilosa Yakovlev = Caragana pekinensis Kom. ●

72384 Caragna Medik. = Caragana Fabr. ●

72385 Caraguata Adans. = Tillandsia L. ■☆

72386 Caraguata Lindl. = Guzmania Ruiz et Pav. ■☆

72387 Caraipa Aubl. (1775);南美洲藤黄属(卡瑞藤黄属)●☆

72388 Caraipa densiflora Mart.;密花南美洲藤黄●☆

72389 Carajaea (Tul.) Wedd. = Castelnavia Tul. et Wedd. ■☆

72390 Carajaea Wedd. = Castelnavia Tul. et Wedd. ■☆

72391 Carallia Roxb. (1811)(保留属名);竹节树属(鹅山木属);Carallia ●

72392 Carallia Roxb. ex R. Br. = Carallia Roxb. (保留属名)●

72393 Carallia brachiata (Lour.) Merr.;竹节树(鹅件树,鹅肾木,气管木,山竹公,山竹犁,竹节木,竹球);India Carallia, Indian Carallia ●

72394 Carallia diplopetala Hand.-Mazz.;锯叶竹节树(铁巴掌,叶上花,鱼骨木,鱼钻姆);Serrateleaf Carallia, Serrate-leaved Carallia ●◇

72395 Carallia garciniifolia F. C. How et C. N. Ho;大叶竹节树;Bigleaf Carallia, Big-leaved Carallia ●

72396 Carallia integerrima DC. = Carallia brachiata (Lour.) Merr. ●

72397 Carallia longipes Chun ex W. C. Ko = Carallia pectinifolia W. C. Ko ●

72398 Carallia lucida Roxb. = Carallia brachiata (Lour.) Merr. ●

72399 Carallia madagascariensis (DC.) Tul. = Carallia brachiata (Lour.) Merr. ●

72400 Carallia pectinifolia W. C. Ko;旁杞木(齿叶竹节树,锯齿王,锯叶竹节树,旁栏木,旁杞树);Longstipe Carallia, Long-stiped Carallia ●

72401 Carallia sinensis Arn. = Carallia brachiata (Lour.) Merr. ●

72402 Caralluma R. Br. (1810);水牛角属(龙角属,水牛掌属);Caralluma ■

72403 Caralluma acutangula (Decne.) N. E. Br.;锐棱水牛角■☆

72404 Caralluma acutiloba N. E. Br. = Quaqua acutiloba (N. E. Br.) Bruyns ■☆

72405 Caralluma adscendens (Roxb.) Haw.;上举水牛角■☆

72406 Caralluma adscendens (Roxb.) R. Br. = Caralluma adscendens (Roxb.) Haw. ■☆

72407 Caralluma affinis De Wild. = Caralluma europaea (Guss.) N. E. Br. ■☆

72408 Caralluma albocastanea (Marloth) L. C. Leach = Orbea albocastanea (Marloth) Bruyns ■☆

72409 Caralluma aperta (Masson) N. E. Br. = Tromotriche aperta (Masson) Bruyns ■☆

72410 Caralluma arachnoidea (P. R. O. Bally) M. G. Gilbert;蛛网水牛角■☆

72411 Caralluma arachnoidea (P. R. O. Bally) M. G. Gilbert var. breviloba (Bally) M. G. Gilbert;短裂蛛网水牛角■☆

72412 Caralluma arenicola N. E. Br. = Quaqua arenicola (N. E. Br.) Plowes ■☆

72413 Caralluma arida (Masson) N. E. Br. = Quaqua arida (Masson) Bruyns ■☆

72414 Caralluma armata N. E. Br. = Quaqua armata (N. E. Br.) Bruyns ■☆

72415 Caralluma atrosanguinea (N. E. Br.) N. E. Br. = Piaranthus atrosanguineus (N. E. Br.) Bruyns ■☆

72416 Caralluma aucheriana (Decne.) N. E. Br.;奥切尔水牛角■☆

72417 Caralluma aurea C. A. Lückh. = Quaqua aurea (C. A. Lückh.) Plowes ■☆

72418 Caralluma ausana Dinter et A. Berger = Quaqua incarnata (L. f.) Bruyns subsp. hottentotorum (N. E. Br.) Bruyns ■☆

72419 Caralluma australis Nel = Orbea melanantha (Schltr.) Bruyns ■☆

72420 Caralluma baldratii A. C. White et B. Sloane = Orbea baldratii (A. C. White et B. Sloane) Bruyns ■☆

72421 Caralluma baradii Lavranos;巴拉德水牛角■☆

72422 Caralluma bredae R. A. Dyer = Orbea miscella (N. E. Br.) Meve ■☆

72423 Caralluma bredae R. A. Dyer var. thomallae? = Orbea miscella (N. E. Br.) Meve ■☆

72424 Caralluma brownii Dinter et A. Berger = Orbea lutea (N. E. Br.) Bruyns subsp. vaga (N. E. Br.) Bruyns ■☆

72425 Caralluma burchardii N. E. Br.;龙角;Burchard Caralluma ■☆

72426 Caralluma burchardii N. E. Br. subsp. maura (Maire) Meve et F. Albers = Caralluma burchardii N. E. Br. var. maura Maire ■☆

72427 Caralluma burchardii N. E. Br. var. maura Maire;莫尔龙角■☆

72428 Caralluma burchardii N. E. Br. var. purpurascens Gatt. et Maire = Apteranthes burchardii (N. E. Br.) Plowes ■☆

72429 Caralluma burchardii N. E. Br. var. sventenii E. Lamb et B. M. Lamb = Apteranthes burchardii (N. E. Br.) Plowes ■☆

72430 Caralluma carnosa Stent = Orbea carnosa (Stent) Bruyns ■☆

72431 Caralluma caudata N. E. Br.;尾瓣水牛角■☆

72432 Caralluma caudata N. E. Br. = Orbea caudata (N. E. Br.) Bruyns ■☆

72433 Caralluma caudata N. E. Br. subsp. rhodesiaca L. C. Leach = Orbea caudata (N. E. Br.) Bruyns subsp. rhodesica (L. C. Leach) Bruyns ■☆

72434 Caralluma caudata N. E. Br. var. chibensis (C. A. Lückh.) C. A. Lückh. = Orbea caudata (N. E. Br.) Bruyns subsp. rhodesica (L. C. Leach) Bruyns ■☆

72435 Caralluma caudata N. E. Br. var. fusca C. A. Lückh. = Orbea caudata (N. E. Br.) Bruyns ■☆

72436 Caralluma caudata N. E. Br. var. milleri Nel = Orbea caudata (N. E. Br.) Bruyns subsp. rhodesica (L. C. Leach) Bruyns ■☆

72437 Caralluma caudata N. E. Br. var. rhodesiaca (L. C. Leach) L. C. Leach = Orbea caudata (N. E. Br.) Bruyns subsp. rhodesica (L. C. Leach) Bruyns ■☆

72438 Caralluma caudata N. E. Br. var. stevensonii Oberm. = Orbea caudata (N. E. Br.) Bruyns subsp. rhodesica (L. C. Leach) Bruyns ■☆

72439 Caralluma chibensis C. A. Lückh. = Orbea caudata (N. E. Br.) Bruyns subsp. rhodesica (L. C. Leach) Bruyns ■☆

72440 Caralluma chlorantha Schltr.;绿花水牛角■☆

72441 Caralluma cincta C. A. Lückh. = Quaqua cincta (C. A. Lückh.) Bruyns ■☆

72442 Caralluma circes M. G. Gilbert = Orbea circes (M. G. Gilbert) Bruyns ■☆

72443 Caralluma codonoides K. Schum. = Caralluma speciosa (N. E. Br.) N. E. Br. ■☆

72444 Caralluma commutata A. Berger = Orbea sprengeri (Schweinf.) Bruyns subsp. commutata (A. Berger) Bruyns ■☆

72445 Caralluma commutata A. Berger subsp. hesperidum (Maire) Maire = Orbea decaisneana (Lehm.) Bruyns ■☆

72446 Caralluma compta (N. E. Br.) Schltr. = Piaranthus comptus N. E. Br. ■☆

72447 Caralluma congestiflora P. R. O. Bally;密花水牛角■☆

72448 Caralluma corrugata N. E. Br. = Caralluma socotrana (Balf. f.) N. E. Br. ■☆

72449 Caralluma crassa N. E. Br. = White-Sloanea crassa (N. E. Br.) Chiov. ■☆

72450 Caralluma dalzielii N. E. Br.;达尔齐尔水牛角■☆

72451 Caralluma dalzielii N. E. Br. = Caralluma adscendens (Roxb.) Haw. ■☆

72452 Caralluma decaisneana (Lem.) N. E. Br. = Orbea decaisneana (Lehm.) Bruyns ■☆

72453 Caralluma decora (Masson) Schltr. = Piaranthus geminatus (Masson) N. E. Br. subsp. decorus (Masson) Bruyns ■☆

72454 Caralluma denboefii Lavranos = Orbea denboefii (Lavranos)

Bruyns ■☆
72455 Caralluma dependens N. E. Br. = Quaqua parviflora (Masson) Bruyns subsp. dependens (N. E. Br.) Bruyns ■☆
72456 Caralluma dicapuae (Chiov.) A. C. White et B. Sloane subsp. turneri (E. A. Bruce) P. R. O. Bally = Caralluma turneri E. A. Bruce ■☆
72457 Caralluma dicapuae (Chiov.) A. C. White et B. Sloane subsp. ukambensis P. R. O. Bally;乌卡水牛角■☆
72458 Caralluma distincta E. A. Bruce = Orbea distincta (E. A. Bruce) Bruyns ■☆
72459 Caralluma dodsoniana Lavranos = Pseudolithos dodsonianus (Lavranos) Bruyns et Meve ■☆
72460 Caralluma dummeri (N. E. Br.) A. C. White et B. Sloane = Orbea dummeri (N. E. Br.) Bruyns ■☆
72461 Caralluma edulis (Edgew.) Benth.;可食水牛掌■☆
72462 Caralluma edulis (Edgew.) Hook. f. = Caralluma edulis (Edgew.) Benth. ■☆
72463 Caralluma edulis A. Chev. = Caralluma edulis (Edgew.) Benth. ■☆
72464 Caralluma edwardsiae (M. G. Gilbert) M. G. Gilbert;爱德华兹水牛角■☆
72465 Caralluma elata Chiov. = Caralluma priogonium K. Schum. ■☆
72466 Caralluma ericeta Nel = Quaqua parviflora (Masson) Bruyns subsp. gracilis (C. A. Lückh.) Bruyns ■☆
72467 Caralluma europaea (Guss.) N. E. Br.;欧洲水牛角(赤缟龙角)■☆
72468 Caralluma europaea (Guss.) N. E. Br. subsp. gussoneana (J. C. Mikan) Maire = Caralluma europaea (Guss.) N. E. Br. ■☆
72469 Caralluma europaea (Guss.) N. E. Br. subsp. maroccana (Hook. f.) Maire = Caralluma europaea (Guss.) N. E. Br. ■☆
72470 Caralluma europaea (Guss.) N. E. Br. var. affinis (De Wild.) Berger = Caralluma europaea (Guss.) N. E. Br. ■☆
72471 Caralluma europaea (Guss.) N. E. Br. var. albotigrina Maire = Caralluma europaea (Guss.) N. E. Br. ■☆
72472 Caralluma europaea (Guss.) N. E. Br. var. barrueliana Maire = Caralluma europaea (Guss.) N. E. Br. ■☆
72473 Caralluma europaea (Guss.) N. E. Br. var. decipiens Maire = Caralluma europaea (Guss.) N. E. Br. ■☆
72474 Caralluma europaea (Guss.) N. E. Br. var. gattefossei Maire = Caralluma europaea (Guss.) N. E. Br. ■☆
72475 Caralluma europaea (Guss.) N. E. Br. var. marmaricensis Berger = Caralluma europaea (Guss.) N. E. Br. ■☆
72476 Caralluma europaea (Guss.) N. E. Br. var. micrantha Maire = Caralluma europaea (Guss.) N. E. Br. ■☆
72477 Caralluma europaea (Guss.) N. E. Br. var. simonis Berger = Caralluma europaea (Guss.) N. E. Br. ■☆
72478 Caralluma europaea (Guss.) N. E. Br. var. smuckiana Gatt. et Maire = Caralluma europaea (Guss.) N. E. Br. ■☆
72479 Caralluma europaea (Guss.) N. E. Br. var. tristis Maire = Caralluma europaea (Guss.) N. E. Br. ■☆
72480 Caralluma europaea (Guss.) N. E. Br. var. typica Berger = Caralluma europaea (Guss.) N. E. Br. ■☆
72481 Caralluma europaea N. E. Br. = Caralluma europaea (Guss.) N. E. Br. ■☆
72482 Caralluma flavovirens L. E. Newton;黄绿水牛角■☆
72483 Caralluma foetida E. A. Bruce;臭水牛角■☆
72484 Caralluma fosteri Pillans = Orbea carnosa (Stent) Bruyns subsp. keithii (R. A. Dyer) Bruyns ■☆
72485 Caralluma framesii Pillans = Quaqua framesii (Pillans) Bruyns ■☆
72486 Caralluma furta P. R. O. Bally;蜥蜴角■☆
72487 Caralluma geminata (Masson) Schltr. = Piaranthus geminatus (Masson) N. E. Br. ■☆
72488 Caralluma gemugofana M. G. Gilbert = Orbea gemugofana (M. G. Gilbert) Bruyns ■☆
72489 Caralluma gerstneri Letty = Orbea gerstneri (Letty) Bruyns ■☆
72490 Caralluma gerstneri Letty subsp. elongata R. A. Dyer = Orbea gerstneri (Letty) Bruyns subsp. elongata (R. A. Dyer) Bruyns ■☆
72491 Caralluma gossweileri S. Moore = Orbea huillensis (Hiern) Bruyns ■☆
72492 Caralluma gracilipes K. Schum.;细梗水牛角■☆
72493 Caralluma gracilipes K. Schum. subsp. arachnoidea P. R. O. Bally = Caralluma arachnoidea (P. R. O. Bally) M. G. Gilbert ■☆
72494 Caralluma gracilipes K. Schum. subsp. breviloba P. R. O. Bally = Caralluma arachnoidea (P. R. O. Bally) M. G. Gilbert var. breviloba (Bally) M. G. Gilbert ■☆
72495 Caralluma gracilipes K. Schum. subsp. edwardsiae M. G. Gilbert = Caralluma edwardsiae (M. G. Gilbert) M. G. Gilbert ■☆
72496 Caralluma gracilis C. A. Lückh. = Quaqua parviflora (Masson) Bruyns subsp. gracilis (C. A. Lückh.) Bruyns ■☆
72497 Caralluma grandidens I. Verd. = Orbea maculata (N. E. Br.) L. C. Leach ■☆
72498 Caralluma grivana (N. E. Br.) Schltr. = Piaranthus decipiens (N. E. Br.) Bruyns ■☆
72499 Caralluma hahnii Nel = Orbea lutea (N. E. Br.) Bruyns subsp. vaga (N. E. Br.) Bruyns ■☆
72500 Caralluma hesperidum Maire = Orbea decaisneana (Lehm.) Bruyns ■☆
72501 Caralluma hirtiflora N. E. Br. = Caralluma acutangula (Decne.) N. E. Br. ■☆
72502 Caralluma hottentotorum (N. E. Br.) N. E. Br. = Quaqua incarnata (L. f.) Bruyns subsp. hottentotorum (N. E. Br.) Bruyns ■☆
72503 Caralluma hottentotorum (N. E. Br.) N. E. Br. var. major N. E. Br. = Quaqua incarnata (L. f.) Bruyns subsp. hottentotorum (N. E. Br.) Bruyns ■☆
72504 Caralluma hottentotorum (N. E. Br.) N. E. Br. var. minor C. A. Lückh. = Quaqua incarnata (L. f.) Bruyns subsp. hottentotorum (N. E. Br.) Bruyns ■☆
72505 Caralluma hottentotorum (N. E. Br.) N. E. Br. var. tubata C. A. Lückh. = Quaqua incarnata (L. f.) Bruyns subsp. hottentotorum (N. E. Br.) Bruyns ■☆
72506 Caralluma huernioides P. R. O. Bally = Orbea huernioides (P. R. O. Bally) Bruyns ■☆
72507 Caralluma huillensis Hiern = Orbea huillensis (Hiern) Bruyns ■☆
72508 Caralluma incarnata (L. f.) N. E. Br. = Quaqua incarnata (L. f.) Bruyns ■☆
72509 Caralluma incarnata (L. f.) N. E. Br. var. alba (G. Don) N. E. Br. = Quaqua incarnata (L. f.) Bruyns ■☆
72510 Caralluma intermedia (N. E. Br.) Schltr.;间型水牛角■☆
72511 Caralluma inversa N. E. Br. = Quaqua inversa (N. E. Br.) Bruyns ■☆
72512 Caralluma joannis Maire;蓝灰龙角■☆
72513 Caralluma kalaharica Nel = Orbea knobelii (E. Phillips) Bruyns ■☆
72514 Caralluma keithii R. A. Dyer = Orbea carnosa (Stent) Bruyns

subsp. keithii (R. A. Dyer) Bruyns ■☆
72515 Caralluma knobelii (E. Phillips) E. Phillips = Orbea knobelii (E. Phillips) Bruyns ■☆
72516 Caralluma kochii Lavranos = Orbea sacculata (N. E. Br.) Bruyns ■☆
72517 Caralluma lamellosa M. G. Gilbert et Thulin;片状水牛角■☆
72518 Caralluma langii A. C. White et B. Sloane = Orbea knobelii (E. Phillips) Bruyns ■☆
72519 Caralluma lateritia N. E. Br. = Orbea lutea (N. E. Br.) Bruyns ■☆
72520 Caralluma lateritia N. E. Br. var. stevensonii A. C. White et B. Sloane = Orbea lutea (N. E. Br.) Bruyns ■☆
72521 Caralluma lateritia N. E. Br. var. vansonii (Bremek. et Oberm.) C. A. Lückh. = Orbea lutea (N. E. Br.) Bruyns ■☆
72522 Caralluma leendertziae N. E. Br. = Orbea melanantha (Schltr.) Bruyns ■☆
72523 Caralluma linearis N. E. Br. = Quaqua linearis (N. E. Br.) Bruyns ■☆
72524 Caralluma longecornuta Croizat ex Gomes et Sousa = Orbea caudata (N. E. Br.) Bruyns ■☆
72525 Caralluma longicuspis N. E. Br. = Orbea lugardii (N. E. Br.) Bruyns ■☆
72526 Caralluma longidens N. E. Br. = Caralluma edulis (Edgew.) Benth. ■☆
72527 Caralluma longiflora M. G. Gilbert;长花水牛角■☆
72528 Caralluma longipes N. E. Br. = Pectinaria longipes (N. E. Br.) Bruyns ■☆
72529 Caralluma lugardii N. E. Br. = Orbea lugardii (N. E. Br.) Bruyns ■☆
72530 Caralluma lutea N. E. Br.;大龙角;Yellow Caralluma ■☆
72531 Caralluma lutea N. E. Br. = Orbea lutea (N. E. Br.) Bruyns ■☆
72532 Caralluma lutea N. E. Br. var. lateritia (N. E. Br.) Nel = Orbea lutea (N. E. Br.) Bruyns ■☆
72533 Caralluma maculata N. E. Br. = Orbea maculata (N. E. Br.) L. C. Leach ■☆
72534 Caralluma maculata N. E. Br. var. brevidens H. Huber = Orbea maculata (N. E. Br.) L. C. Leach subsp. rangeana (Dinter et A. Berger) Bruyns ■☆
72535 Caralluma mammillaris (L.) N. E. Br. = Quaqua mammillaris (L.) Bruyns ■☆
72536 Caralluma marlothii N. E. Br. = Quaqua marlothii (N. E. Br.) Bruyns ■☆
72537 Caralluma marlothii N. E. Br. var. viridis E. Lamb = Quaqua marlothii (N. E. Br.) Bruyns ■☆
72538 Caralluma maroccana (Hook. f.) N. E. Br. = Caralluma europaea (Guss.) N. E. Br. ■☆
72539 Caralluma melanantha (Schltr.) N. E. Br. = Orbea melanantha (Schltr.) Bruyns ■☆
72540 Caralluma melanantha (Schltr.) N. E. Br. var. sousae A. C. White ex Gomes = Orbea melanantha (Schltr.) Bruyns ■☆
72541 Caralluma moniliformis P. R. O. Bally;串珠水牛角■☆
72542 Caralluma montana R. A. Dyer et E. A. Bruce = Echidnopsis montana (R. A. Dyer et E. A. Bruce) P. R. O. Bally ■☆
72543 Caralluma mouretii A. Chev. = Caralluma edulis (Edgew.) Benth. ■☆
72544 Caralluma mouretii A. Chev. = Caudanthera edulis (Edgew.) Meve et Liede ■☆
72545 Caralluma multiflora R. A. Dyer = Quaqua multiflora (R. A. Dyer) Bruyns ■☆
72546 Caralluma munbyana (Decne. ex Munby) N. E. Br. = Apteranthes munbyana (Decne.) Meve et Liede ■☆
72547 Caralluma munbyana (Decne. ex Munby) N. E. Br. var. hispanica (Coincy) Maire = Apteranthes munbyana (Decne.) Meve et Liede ■☆
72548 Caralluma nebrownii A. Berger;水牛角(水牛掌);Caralluma, Nebrown Caralluma ■
72549 Caralluma ortholoba Lavranos = Quaqua acutiloba (N. E. Br.) Bruyns ■☆
72550 Caralluma oxyodonta Chiov. = Caralluma speciosa (N. E. Br.) N. E. Br. ■☆
72551 Caralluma parviflora (Masson) N. E. Br. = Quaqua parviflora (Masson) Bruyns ■☆
72552 Caralluma peckii P. R. O. Bally;佩克水牛角■☆
72553 Caralluma penicillata (Deflers) N. E. Br.;帚状水牛角■☆
72554 Caralluma penicillata (Deflers) N. E. Br. var. robusta (N. E. Br.) A. C. White et B. Sloane;粗壮帚状水牛角■☆
72555 Caralluma peschii Nel = Australluma peschii (Nel) Plowes ■☆
72556 Caralluma piaranthoides Oberm. = Orbea schweinfurthii (A. Berger) Bruyns ■☆
72557 Caralluma pillansii N. E. Br. = Quaqua pillansii (N. E. Br.) Bruyns ■☆
72558 Caralluma plurifasciculata P. R. O. Bally = Caralluma congestiflora P. R. O. Bally ■☆
72559 Caralluma praegracilis Oberm. = Orbea caudata (N. E. Br.) Bruyns ■☆
72560 Caralluma priogonium K. Schum.;热非水牛角■☆
72561 Caralluma pruinosa (Masson) N. E. Br. = Quaqua pruinosa (Masson) Bruyns ■☆
72562 Caralluma pruinosa (Masson) N. E. Br. var. nigra C. A. Lückh. = Quaqua pruinosa (Masson) Bruyns ■☆
72563 Caralluma punctata (Masson) Schltr. = Piaranthus punctatus (Masson) R. Br. ■☆
72564 Caralluma ramosa (Masson) N. E. Br. = Quaqua ramosa (Masson) Bruyns ■☆
72565 Caralluma rangeana Dinter et A. Berger = Orbea maculata (N. E. Br.) L. C. Leach subsp. rangeana (Dinter et A. Berger) Bruyns ■☆
72566 Caralluma reflexa C. A. Lückh. = Quaqua parviflora (Masson) Bruyns subsp. dependens (N. E. Br.) Bruyns ■☆
72567 Caralluma retrospiciens (Ehrenb.) N. E. Br. subsp. tombuctuensis (A. Chev.) A. Chev. = Caralluma acutangula (Decne.) N. E. Br. ■☆
72568 Caralluma retrospiciens (Ehrenb.) N. E. Br. var. acutangula (Decne.) A. C. White et B. Sloane = Caralluma acutangula (Decne.) N. E. Br. ■☆
72569 Caralluma retrospiciens (Ehrenb.) N. E. Br. var. acutangula (Decne.) A. Chev. = Caralluma acutangula (Decne.) N. E. Br. ■☆
72570 Caralluma retrospiciens (Ehrenb.) N. E. Br. var. glabra N. E. Br. = Caralluma acutangula (Decne.) N. E. Br. ■☆
72571 Caralluma retrospiciens (Ehrenb.) N. E. Br. var. hirtiflora (N. E. Br.) A. Berger = Caralluma acutangula (Decne.) N. E. Br. ■☆
72572 Caralluma retrospiciens (Ehrenb.) N. E. Br. var. laxiflora Maire = Caralluma acutangula (Decne.) N. E. Br. ■☆
72573 Caralluma retrospiciens (Ehrenb.) N. E. Br. var. tombuctuensis (A. Chev.) A. C. White et B. Sloane = Caralluma acutangula (Decne.) N. E. Br. ■☆

72574 Caralluma rivae Chiov. = Caralluma socotrana (Balf. f.) N. E. Br. ■☆
72575 Caralluma robusta N. E. Br. = Caralluma penicillata (Deflers) N. E. Br. var. robusta (N. E. Br.) A. C. White et B. Sloane ■☆
72576 Caralluma rogersii (L. Bolus) E. A. Bruce et R. A. Dyer = Orbea rogersii (L. Bolus) Bruyns ■☆
72577 Caralluma rosengrenii Vierh. = Caralluma socotrana (Balf. f.) N. E. Br. ■☆
72578 Caralluma rubiginosa Werderm. = Orbea melanantha (Schltr.) Bruyns ■☆
72579 Caralluma russeliana (Courbon ex Brongn.) Cufod. = Caralluma acutangula (Decne.) N. E. Br. ■☆
72580 Caralluma russelliana (Courbon ex Brongn.) Cufod. = Desmidorchis acutangula Decne. ■☆
72581 Caralluma sacculata N. E. Br. = Orbea sacculata (N. E. Br.) Bruyns ■☆
72582 Caralluma schweickerdtii Oberm. = Orbea carnosa (Stent) Bruyns subsp. keithii (R. A. Dyer) Bruyns ■☆
72583 Caralluma schweinfurthii A. Berger = Orbea schweinfurthii (A. Berger) Bruyns ■☆
72584 Caralluma serpentina Nel = Notechidnopsis tessellata (Pillans) Lavranos et Bleck ■☆
72585 Caralluma serrulata (Jacq.) Schltr. = Piaranthus geminatus (Masson) N. E. Br. subsp. decorus (Masson) Bruyns ■☆
72586 Caralluma simulans N. E. Br. = Quaqua marlothii (N. E. Br.) Bruyns ■☆
72587 Caralluma socotrana (Balf. f.) N. E. Br.;索科特拉水牛角■☆
72588 Caralluma somalica N. E. Br.;索马里水牛角■☆
72589 Caralluma speciosa (N. E. Br.) N. E. Br.;美丽水牛角■☆
72590 Caralluma sprengeri (Schweinf.) N. E. Br.;紫花水牛角■☆
72591 Caralluma sprengeri (Schweinf.) N. E. Br. = Orbea sprengeri (Schweinf.) Bruyns ■☆
72592 Caralluma sprengeri N. E. Br. = Caralluma sprengeri (Schweinf.) N. E. Br. ■☆
72593 Caralluma sprengeri N. E. Br. subsp. foetida M. G. Gilbert = Orbea sprengeri (Schweinf.) Bruyns subsp. foetida (M. G. Gilbert) Bruyns ■☆
72594 Caralluma sprengeri N. E. Br. subsp. laticorona M. G. Gilbert = Orbea laticorona (M. G. Gilbert) Bruyns ■☆
72595 Caralluma sprengeri N. E. Br. subsp. ogadensis M. G. Gilbert = Orbea sprengeri (Schweinf.) Bruyns subsp. ogadensis (M. G. Gilbert) Bruyns ■☆
72596 Caralluma stalagmifera C. E. C. Fisch.;石笋水牛掌■☆
72597 Caralluma subterranea E. A. Bruce et P. R. O. Bally = Orbea subterranea (E. A. Bruce et P. R. O. Bally) Bruyns ■☆
72598 Caralluma subterranea E. A. Bruce et P. R. O. Bally var. minutiflora? = Orbea subterranea (E. A. Bruce et P. R. O. Bally) Bruyns ■☆
72599 Caralluma subulata (Forssk.) Decne. = Caralluma adscendens (Roxb.) Haw. ■☆
72600 Caralluma swanepoelii Lavranos = Quaqua parviflora (Masson) Bruyns subsp. swanepoelii (Lavranos) Bruyns ■☆
72601 Caralluma tessellata Pillans = Notechidnopsis tessellata (Pillans) Lavranos et Bleck ■☆
72602 Caralluma tombuctuensis (A. Chev.) N. E. Br. = Caralluma acutangula (Decne.) N. E. Br. ■☆
72603 Caralluma tsumebensis Oberm. = Orbea huillensis (Hiern) Bruyns ■☆
72604 Caralluma tuberculata N. E. Br.;多疣水牛角■☆
72605 Caralluma tubiformis E. A. Bruce et P. R. O. Bally = Orbea tubiformis (E. A. Bruce et P. R. O. Bally) Bruyns ■☆
72606 Caralluma turneri E. A. Bruce;特纳水牛角■☆
72607 Caralluma ubomboensis I. Verd. = Australluma ubomboensis (I. Verd.) Bruyns ■☆
72608 Caralluma umdausensis Nel = Tromotriche umdausensis (Nel) Bruyns ■☆
72609 Caralluma vaduliae Lavranos;瓦杜尔水牛角■☆
72610 Caralluma vaga (N. E. Br.) A. C. White et B. Sloane = Orbea lutea (N. E. Br.) Bruyns subsp. vaga (N. E. Br.) Bruyns ■☆
72611 Caralluma valida N. E. Br. = Orbea valida (N. E. Br.) Bruyns ■☆
72612 Caralluma vansonii Bremek. et Oberm. = Orbea lutea (N. E. Br.) Bruyns ■☆
72613 Caralluma venenosa Maire = Orbea decaisneana (Lehm.) Bruyns ■☆
72614 Caralluma vibratilis E. A. Bruce et P. R. O. Bally = Orbea vibratilis (E. A. Bruce et P. R. O. Bally) Bruyns ■☆
72615 Caralluma villetii C. A. Lückh. = Quaqua inversa (N. E. Br.) Bruyns ■☆
72616 Caralluma virescens C. A. Lückh. = Quaqua parviflora (Masson) Bruyns ■☆
72617 Caralluma vittata N. E. Br. = Caralluma edulis (Edgew.) Benth. ■☆
72618 Caralluma wilfriedii Dinter = Quaqua acutiloba (N. E. Br.) Bruyns ■☆
72619 Caralluma wilsonii P. R. O. Bally = Orbea wilsonii (P. R. O. Bally) Bruyns ■☆
72620 Caralluma winkleriana (Dinter) A. C. White et B. Sloane = Quaqua mammillaris (L.) Bruyns ■☆
72621 Caramanica Tineo = Taraxacum F. H. Wigg.(保留属名)■
72622 Carambola Adans. = Averrhoa L. ●
72623 Caramuri Aubrév. et Pellegr. = Pouteria Aubl. ●
72624 Caranda Gaertn. = Canthium Lam. ●
72625 Carandas Adans.(废弃属名) = Carissa L.(保留属名)●
72626 Carandas Rumph. ex Adans. = Carissa L.(保留属名)●
72627 Carandas edulis (Forssk.) Hiern = Carissa edulis Vahl ●
72628 Carandra Gaertn. = Psydrax Gaertn. ●☆
72629 Caranga Juss. = Curanga Juss. ■☆
72630 Caranga Vahl = Curanga Juss. ■☆
72631 Carania Chiov. = Basananthe Peyr. ■●☆
72632 Carania Chiov. = Tryphostemma Harv. ■●☆
72633 Carania berberoides Chiov. = Basananthe berberoides (Chiov.) W. J. de Wilde ■☆
72634 Carapa Aubl. (1775);酸渣树属(苦油楝属,苦油树属);Crab Wood ●☆
72635 Carapa angustifolia Harms;窄叶酸渣树●☆
72636 Carapa batesii C. DC.;贝茨酸渣树●☆
72637 Carapa dinklagei Harms;丁克酸渣树●☆
72638 Carapa grandiflora Sprague;大花酸渣树(大花苦油楝)●☆
72639 Carapa grandiflora Sprague = Carapa procera DC. ●☆
72640 Carapa guianensis Aubl.;酸渣树(圭亚那苦油楝,圭亚那苦油树);Andiroba, Bastard Mahogany, Coondi, Crab Oil, Crab Wood, Guiana Crabwood, Tallicona ●
72641 Carapa gummiflua C. DC. = Carapa procera DC. ●☆
72642 Carapa hygrophila Harms;喜水酸渣树●☆

72643 Carapa macrantha Harms = Carapa procera DC. ●☆

72644 Carapa microcarpa A. Chev.;小果酸渣树(小果苦油楝)●☆

72645 Carapa microcarpa A. Chev. = Carapa procera DC. ●☆

72646 Carapa moluccensis Lam.;马六甲酸渣树(马六甲苦油树,马鲁古苦油楝)●☆

72647 Carapa moluccensis Lam. = Xylocarpus moluccensis (Lam.) M. Roem. ●☆

72648 Carapa nicaraguensis C. DC.;尼加拉瓜酸渣树(尼加拉瓜苦油楝)●☆

72649 Carapa parviflora Harms = Carapa procera DC. ●☆

72650 Carapa procera DC.;高大酸渣树(高大苦油楝);African Crabwood,Kunda-oll Tree ●☆

72651 Carapa procera DC. var. palustre G. C. C. Gilbert;沼泽酸渣树●☆

72652 Carapa surinamensis Miq.;苏里南酸渣树(苏里南苦油楝)●☆

72653 Carapa velutina C. DC. = Carapa procera DC. ●☆

72654 Carapichea Aubl.(废弃属名)= Cephaelis Sw.(保留属名)●

72655 Carapichea Aubl.(废弃属名)= Psychotria L.(保留属名)●

72656 Carara Medik. = Coronopus Zinn(保留属名)■

72657 Carara didyma (L.) Britton = Coronopus didymus (L.) Sm. ■

72658 Caratas Raf. = Bromelia L. ■☆

72659 Caratas Raf. = Karatas Mill. ■☆

72660 Caraxeron Raf. = Iresine P. Browne(保留属名)●■

72661 Carbeni Adans. = Cnicus L.(保留属名)■●

72662 Carbenia Adans. = Cnicus L.(保留属名)■●

72663 Carbenia Benth. = Carbeni Adans. ■●

72664 Carbenis benedicta (L.) Benth. = Cnicus benedictus L. ■

72665 Carcerulaceae Dulac = Tiliaceae Juss.(保留科名)●■

72666 Carcia Raeusch. = Garcia Rohr ●☆

72667 Carcinetrum Post et Kuntze = Karkinetron Raf.(废弃属名)●☆

72668 Carcinetrum Post et Kuntze = Muehlenbeckia Meisn.(保留属名)●☆

72669 Carcinetrum Post et Kuntze = Polygonum L.(保留属名)■●

72670 Carda Noronha = Aleurites J. R. Forst. et G. Forst. ●

72671 Cardamindaceae Link = Tropaeolaceae Juss. ex DC.(保留科名)■

72672 Cardamindum Adans. = Tropaeolum L. ■

72673 Cardamindum Tourn. ex Adans. = Tropaeolum L. ■

72674 Cardamine L. (1753);碎米荠属;Bitter Cress,Bittercress,Bitter-cress,Cress,Cuckoo Flower ■

72675 Cardamine africana L.;非洲碎米荠■☆

72676 Cardamine agyokumontana Hayata;阿玉碎米荠■

72677 Cardamine agyokumontana Hayata = Cardamine circaeoides Hook. f. et Thomson ■

72678 Cardamine akitensis Mochizuki;秋田碎米荠■☆

72679 Cardamine amara L.;苦味碎米荠;Bitter Lady's Smock,Land Cress,Large Bittercress,Large Bitter-cress,Water Cress,Watercress ■☆

72680 Cardamine anemonoides O. E. Schulz;银莲花碎米荠■☆

72681 Cardamine anemonoides O. E. Schulz f. major Hiyama;大银莲花碎米荠■☆

72682 Cardamine anemonoides O. E. Schulz f. suavis (O. E. Schulz) Hiyama;芳香银莲花碎米荠■☆

72683 Cardamine angulata Hook. var. kamtschatica Regel = Cardamine scutata Thunb. ■

72684 Cardamine anhuiensis D. C. Zhang et C. Z. Shao;安徽碎米荠;Anhui Bittercress ■

72685 Cardamine appendiculata Franch. et Sav.;附属物碎米荠■☆

72686 Cardamine arakiana Koidz.;荒木碎米荠■☆

72687 Cardamine arenicola Britton = Cardamine parviflora L. var. arenicola (Britton) O. E. Schulz ■☆

72688 Cardamine argyi H. Lév. = Cardamine lyrata Bunge ■

72689 Cardamine arisanensis Hayata;阿里山荠(高山山芥,高山碎米荠);Alpine Bittercress ■

72690 Cardamine arisanensis Hayata = Barbarea arisanense (Hayata) S. S. Ying ■

72691 Cardamine arisanensis Hayata = Cardamine flexuosa With. ■

72692 Cardamine atlantica (Ball) Maire = Roripella atlantica (Ball) Greuter et Burdet ■☆

72693 Cardamine autumnalis Koidz. = Cardamine scutata Thunb. ■

72694 Cardamine baishanensis P. Y. Fu;长白山碎米荠;Baishan Bittercress ■

72695 Cardamine baishanensis P. Y. Fu = Cardamine scutata Thunb. ■

72696 Cardamine basisagittata W. T. Wang;箭基碎米荠;Basisagittate Bittercress ■

72697 Cardamine basisagittata W. T. Wang = Cardamine impatiens L. ■

72698 Cardamine bellidiflora All.;菊叶碎米荠(皱菊碎米荠);Daisy-leaved Bittercress ■☆

72699 Cardamine bijiangensis W. T. Wang;碧江碎米荠;Bijiang Bittercress ■

72700 Cardamine bijiangensis W. T. Wang = Cardamine yunnanensis Franch. ■

72701 Cardamine bodinieri (H. Lév.) Lauener;博氏碎米荠;Bodinier Bittercress ■

72702 Cardamine borbonica Pers. = Cardamine africana L. ■☆

72703 Cardamine borealis Andrz. ex DC. = Cardamine prorepens Fisch. ex DC. ■

72704 Cardamine boryi Boiss. = Murbeckiella boryi (Boiss.) Rothm. ■☆

72705 Cardamine brachycarpa Franch. = Cardamine parviflora L. ■

72706 Cardamine bracteata S. Moore = Eutrema tenue (Miq.) Makino ■

72707 Cardamine bulbifera (L.) Crantz = Dentaria bulbifera L. ■☆

72708 Cardamine bulbosa (Schreb. ex Muhl.) Britton, Sterns et Poggenb.;北美碎米荠;Bulbous Bitter-cress, Spring Cress, Springcress,Spring-cress ■☆

72709 Cardamine bulbosa (Schreb. ex Muhl.) Britton, Sterns et Poggenb. var. purpurea (Torr.) Britton, Sterns et Poggenb. = Cardamine douglassii (Torr.) Britton ■☆

72710 Cardamine bulbosa (Schreb. ex) Britton, Sterns et Poggenb. f. fontinalis E. J. Palmer et Steyerm. = Cardamine bulbosa (Schreb. ex Muhl.) Britton,Sterns et Poggenb. ■☆

72711 Cardamine burchellii Spreng. = Cardamine africana L. ■☆

72712 Cardamine calcicola W. W. Sm.;岩生碎米荠;Kalkliving Bittercress,Rocky Bittercress ■

72713 Cardamine caledonica (Sond.) Kuntze = Rorippa fluviatilis (E. Mey. ex Sond.) Thell. var. caledonica (Sond.) Marais ■☆

72714 Cardamine calthifolia H. Lév.;驴蹄碎米荠■

72715 Cardamine caroides C. Y. Wu;细裂碎米荠;Thinsplit Bittercress ■

72716 Cardamine cathayensis Migo = Cardamine leucantha (Tausch) O. E. Schulz ■

72717 Cardamine changbaiana Al-Shehbaz;天池碎米荠;Resedaleaf Bittercress ■

72718 Cardamine chejiangensis var. huangshanensis D. C. Zhang;黄山碎米荠;Huangshan Bittercress ■

72719 Cardamine chenopodiifolia Pers.;南美碎米荠■☆

72720 Cardamine cheotaiyienii Al-Shehbaz et G. Yang;周氏碎米荠;Zhou's Bittercress ■

72721 Cardamine circaeoides Hook. f. et Thomson;露珠碎米荠(阿玉

碎米荠,肾叶碎米荠);Agyoku Bittercress,Circaea-like Bittercress,Dewdrop Bittercress,Kidneyleaf Bittercress ■

72722 Cardamine circaeoides Hook. f. et Thomson var. diversifolia O. E. Schulz = Cardamine circaeoides Hook. f. et Thomson ■

72723 Cardamine concatenata (Michx.) O. Schwarz;裂叶碎米荠;Cut-leaved Toothwort,Five-parted Toothwort,Toothwort ■☆

72724 Cardamine cordifolia A. Gray;心叶碎米荠;Heart-leaved Bitter-cress ■☆

72725 Cardamine crassifolia Opiz = Cardamine pratensis L. ■

72726 Cardamine cryptantha (A. Rich.) Kuntze var. pinnatodentata Kuntze = Rorippa cantoniensis (Lour.) Ohwi ■

72727 Cardamine dasycarpa M. Bieb. = Cardamine impatiens L. var. dasycarpa (M. Bieb.) T. Y. Cheo et R. C. Fang ■

72728 Cardamine dasycarpa M. Bieb. = Cardamine impatiens L. ■

72729 Cardamine dasyloba (Turcz.) Miq. = Cardamine leucantha (Tausch) O. E. Schulz ■

72730 Cardamine dasyloba Miq. = Cardamine leucantha (Tausch) O. E. Schulz ■

72731 Cardamine debilis D. Don = Cardamine flexuosa With. ■

72732 Cardamine delavayi Franch.;洱源碎米荠;Delavay Bittercress ■

72733 Cardamine densiflora Gontsch.;密花碎米荠■☆

72734 Cardamine dentata Schult.;锯齿碎米荠■☆

72735 Cardamine dentipetala Matsum.;齿瓣碎米荠■☆

72736 Cardamine dentipetala Matsum. = Cardamine scutata Thunb. ■

72737 Cardamine dentipetala Matsum. var. longifructa (Ohwi) Hiyama;长果齿瓣碎米荠■☆

72738 Cardamine denudata O. E. Schulz = Cardamine scaposa Franch. ■

72739 Cardamine diphylla (Michx.) A. W. Wood;阔叶碎米荠;Broad-leaved Toothwort,Crinkleroot,Crinkle-root ■☆

72740 Cardamine douglassii (Torr.) Britton;道氏碎米荠;Limestone Bitter-cress,Northern Bitter Cress,Purple Cress,Purple Spring-cress ■☆

72741 Cardamine drakeana H. Boissieu = Cardamine scutata Thunb. ■

72742 Cardamine engleriana O. E. Schulz;光头山碎米荠(光头碎米荠);Engler Bittercress,Guangtoushan Bittercress ■

72743 Cardamine enneaphyllos Crantz;九叶碎米荠;Nine-leaved Bittercress ■☆

72744 Cardamine fallax (O. E. Schulz) Nakai;迷惑碎米荠■☆

72745 Cardamine fallax (O. E. Schulz) Nakai = Cardamine parviflora L. ■

72746 Cardamine fargesiana Al-Shehbaz;佛欧里碎米荠;Farges Bittercress ■

72747 Cardamine fauriei Franch.;法氏碎米荠■☆

72748 Cardamine flexuoides W. T. Wang var. glabricaulis W. T. Wang;汶川碎米荠;Wenchuan Bittercress ■

72749 Cardamine flexuosa With.;弯曲碎米荠(白带草,高山山芥,高山碎米荠,焊菜,萝目草,米花香荠菜,曲枝碎米荠,碎米荠,小叶地豇豆,野荠菜);Alpine Bittercress,Bending Bitter-cress,Flexuose Bittercress,Greater Bittercress,Roadside Bittercress,Wavy Bittercress,Wavy Bitter-cress,Wood Bittercress,Woodland Bittercress ■

72750 Cardamine flexuosa With. = Cardamine scutata Thunb. ■

72751 Cardamine flexuosa With. subsp. debilis (D. Don) O. E. Schulz = Cardamine flexuosa With. ■

72752 Cardamine flexuosa With. subsp. debilis O. E. Schulz = Cardamine flexuosa With. ■

72753 Cardamine flexuosa With. subsp. debilis O. E. Schulz var. occulata (Hornem.) O. E. Schulz = Cardamine flexuosa With. ■

72754 Cardamine flexuosa With. subsp. debilis O. E. Schulz var. occulata O. E. Schulz = Cardamine flexuosa With. ■

72755 Cardamine flexuosa With. subsp. fallax O. E. Schulz = Cardamine fallax (O. E. Schulz) Nakai ■☆

72756 Cardamine flexuosa With. subsp. fallax O. E. Schulz = Cardamine parviflora L. ■

72757 Cardamine flexuosa With. subsp. fallax O. E. Schulz f. microphylla O. E. Schulz = Cardamine parviflora L. ■

72758 Cardamine flexuosa With. subsp. regeliana (Miq.) O. E. Schulz = Cardamine scutata Thunb. ■

72759 Cardamine flexuosa With. subsp. regeliana (Miq.) O. E. Schulz = Cardamine regeliana Miq. ■☆

72760 Cardamine flexuosa With. subsp. regeliana (Miq.) O. E. Schulz var. scutata (Thunb.) O. E. Schulz = Cardamine scutata Thunb. ■

72761 Cardamine flexuosa With. var. debilis (D. Don) T. Y. Cheo et R. C. Fang;柔弱弯曲碎米荠;Soft Flexuose Bittercress,Weak Flexuose Bittercress ■

72762 Cardamine flexuosa With. var. debilis (D. Don) T. Y. Cheo et R. C. Fang = Cardamine flexuosa With. ■

72763 Cardamine flexuosa With. var. debilis (O. E. Schulz) T. Y. Cheo et R. C. Fang = Cardamine flexuosa With. ■

72764 Cardamine flexuosa With. var. fallax (O. E. Schulz) T. Y. Cheo et R. C. Fang;假弯曲碎米荠;False Flexuose Bittercress ■

72765 Cardamine flexuosa With. var. fallax (O. E. Schulz) T. Y. Cheo et R. C. Fang = Cardamine parviflora L. ■

72766 Cardamine flexuosa With. var. fallax (O. E. Schulz) T. Y. Cheo et R. C. Fang = Cardamine fallax (O. E. Schulz) Nakai ■☆

72767 Cardamine flexuosa With. var. kamtschatica (Regel) Matsum. = Cardamine scutata Thunb. ■

72768 Cardamine flexuosa With. var. latifolia (Maxim.) Makino = Cardamine scutata Thunb. var. latifolia (Maxim.) H. Hara ■☆

72769 Cardamine flexuosa With. var. manshurica Kom. = Cardamine scutata Thunb. ■

72770 Cardamine flexuosa With. var. ovatifolia T. Y. Cheo et R. C. Fang;卵叶弯曲碎米荠;Ovateleaf Flexuose Bittercress ■

72771 Cardamine flexuosa With. var. ovatifolia T. Y. Cheo et R. C. Fang = Cardamine flexuosa With. ■

72772 Cardamine flexuosa With. var. regeliana (Miq.) Kom. = Cardamine scutata Thunb. ■

72773 Cardamine flexuosa With. var. regeliana (Miq.) Matsum. = Cardamine regeliana Miq. ■☆

72774 Cardamine flexuosoides W. T. Wang = Cardamine trifoliolata Hook. f. et Thomson ■

72775 Cardamine flexuosoides W. T. Wang var. glabricaulis W. T. Wang = Cardamine trifoliolata Hook. f. et Thomson ■

72776 Cardamine fragarifolia O. E. Schulz;莓叶碎米荠(阿拉泰阴山荠,翅柄岩荠)■

72777 Cardamine fragarifolia O. E. Schulz = Cardamine trifoliolata Hook. f. et Thomson ■

72778 Cardamine franchetiana Diels;宽翅碎米荠(白花弯蕊芥);Smith Curvedstamencress,White Curvedstamencress ■

72779 Cardamine gemmifera Matsum. = Arabidopsis halleri (L.) O'Kane et Al-Shehbaz subsp. gemmifera (Matsum.) O'Kane et Al-Shehbaz ■

72780 Cardamine gemmifera Matsum. = Arabis gemmifera (Matsum.) Makino ■

72781 Cardamine glandulosa Blanco = Rorippa indica (L.) Hiern ■

72782 Cardamine glaphyropoda O. E. Schulz = Cardamine impatiens L. ■

72783 Cardamine glaphyropoda O. E. Schulz var. crenata T. Y. Cheo et R. C. Fang = Cardamine impatiens L. ■

72784 Cardamine gracilis (O. E. Schulz) T. Y. Cheo et R. C. Fang;纤细碎米荠;Slender Bittercress ■

72785 Cardamine graeca L. ;希腊碎米荠■☆

72786 Cardamine granulifera (Franch.) Diels;颗粒碎米荠(三叶弯蕊芥,弯蕊石蕊芥);Threeleaves Curvedstamencress, Trifolious Curvedstamencress ■

72787 Cardamine granulifera (Franch.) Diels = Cardamine simplex Hand. -Mazz. ■

72788 Cardamine granulifera Diels = Cardamine granulifera (Franch.) Diels ■

72789 Cardamine greatrexii Miyabe et Kudo = Arabidopsis halleri (L.) O'Kane et Al-Shehbaz subsp. gemmifera (Matsum.) O'Kane et Al-Shehbaz ■

72790 Cardamine griffithii Hook. f. et Thomson;山芥碎米荠(山芥菜);Griffith Bittercress ■

72791 Cardamine griffithii Hook. f. et Thomson subsp. multijuga (Franch.) O. E. Schulz = Cardamine multijuga Franch. ■

72792 Cardamine griffithii Hook. f. et Thomson var. grandifolia T. Y. Cheo et R. C. Fang;大叶山芥碎米荠;Largeleaf Griffith Bittercress ■

72793 Cardamine griffithii Hook. f. et Thomson var. grandifolia T. Y. Cheo et R. C. Fang = Cardamine engleriana O. E. Schulz ■

72794 Cardamine griffithii Hook. f. et Thomson var. pentaloba W. T. Wang;五裂碎米荠;Fivelobed Griffith Bittercress ■

72795 Cardamine griffithii Hook. f. et Thomson var. pentaloba W. T. Wang = Cardamine griffithii Hook. f. et Thomson ■

72796 Cardamine heishuiensis W. T. Wang;黑水碎米荠(黑水华羽芥);Heishui Bittercress ■

72797 Cardamine heishuiensis W. T. Wang = Sinosophiopsis heishuiensis (W. T. Wang) Al-Shehbaz ■

72798 Cardamine heterandra J. Z. Sun et K. L. Chang = Cardamine circaeoides Hook. f. et Thomson ■

72799 Cardamine heterophylla T. Y. Cheo et R. C. Fang;异叶碎米荠;Different Leaf Bittercress, Heterophyllous Bittercress, Pinnate Coral-root ■

72800 Cardamine heterophylla T. Y. Cheo et R. C. Fang = Cardamine yunnanensis Franch. ■

72801 Cardamine hickinii O. E. Schulz = Orychophragmus limprichtianus (Pax) Al-Shehbaz et G. Yang ■

72802 Cardamine hirsuta L. ;碎米荠(白带草,蔊菜,辣米子,毛碎米荠,雀儿菜,野芹菜,硬毛碎米荠);Hairy Bitter Cress, Hairy Bittercress, Hairy Bitter-cress, Hairy Lady's Smock, Hoary Bitter Cress, Lamb's Cress, Land Cress, Pennsylvania Bittercress, Pennsylvania Hitter Cress, Spiky-flowers, Touch-me-not ■

72803 Cardamine hirsuta L. subsp. flaxuosa With. ex Forbes et Hemsl. = Cardamine flexuosa With. ■

72804 Cardamine hirsuta L. subsp. flexuosa (With.) F. B. Forbes et Hemsl. = Cardamine flexuosa With. ■

72805 Cardamine hirsuta L. subsp. flexuosa (With.) Hook. f. = Cardamine flexuosa With. ■

72806 Cardamine hirsuta L. subsp. sylvatica (Link) Syme = Cardamine flexuosa With. ■

72807 Cardamine hirsuta L. var. flaccida Franch. = Cardamine flexuosa With. ■

72808 Cardamine hirsuta L. var. flaccida Franch. = Cardamine hirsuta L. ■

72809 Cardamine hirsuta L. var. formosana Hayata;宝岛碎米荠;Taiwan Pennsylvania Bittercress ■

72810 Cardamine hirsuta L. var. formosana Hayata = Cardamine hirsuta L. ■

72811 Cardamine hirsuta L. var. latifolia Maxim. = Cardamine scutata Thunb. ■

72812 Cardamine hirsuta L. var. omeiensis T. Y. Cheo et R. C. Fang;峨眉碎米荠;Emei Pennsylvania Bittercress, Omei Bittercress ■

72813 Cardamine hirsuta L. var. omeiensis T. Y. Cheo et R. C. Fang = Cardamine flexuosa With. ■

72814 Cardamine hirsuta L. var. oxycarpa Hook. f. et T. Anderson = Cardamine yunnanensis Franch. ■

72815 Cardamine hirsuta L. var. pensylvanica Muhl. ex Willd. = Cardamine pensylvanica Muhl. ex Willd. ■☆

72816 Cardamine hirsuta L. var. pilosa O. E. Schulz = Cardamine hirsuta L. ■

72817 Cardamine hirsuta L. var. regeliana (Miq.) Maxim. = Cardamine scutata Thunb. ■

72818 Cardamine hirsuta L. var. rotundiloba Hayata;圆裂碎米荠;Rotundilobe Pennsylvania Bittercress, Rotundilobed Bittercress ■

72819 Cardamine hirsuta L. var. rotundiloba Hayata = Cardamine scutata Thunb. ■

72820 Cardamine hirsuta L. var. rotundiloba Hayata = Cardamine scutata Thunb. var. rotundiloba (Hayata) Tang S. Liu et S. S. Ying ■

72821 Cardamine hirsuta L. var. rotundiloba Hayata = Cardamine scutata Thunb ■

72822 Cardamine hirsuta L. var. sylvatica (Link) Ball = Cardamine flexuosa With. ■

72823 Cardamine hirsuta L. var. sylvatica (Link) Hook. f. et T. Anderson = Cardamine flexuosa With. ■

72824 Cardamine hirsuta L. var. sylvatica (Link) Syme = Cardamine flexuosa With. ■

72825 Cardamine hirsuta Pall. = Cardamine prorepens Fisch. ex DC. ■

72826 Cardamine holtziana Engl. et O. E. Schulz = Cardamine africana L. ■☆

72827 Cardamine hydrocotyloides W. T. Wang;德钦碎米荠(肾叶碎米荠);Kidneyleaf Bittercress ■

72828 Cardamine hygrophila T. Y. Cheo et R. C. Fang;湿生碎米荠;Hygrophilous Bittercress, Moistureloving Bittercress ■

72829 Cardamine impatiens L. ;弹裂碎米荠(钝齿四川碎米荠,水菜花,水花菜,四川碎米荠,松潘碎米荠);Crenate Sichuan Bittercress, Crenate Szechuan Bittercress, Impatient Bittercress, Impatient Lady's Smock, Narrowleaf Bittercress, Narrow-leaved Bittercress, Narrow-leaved Bitter-cress, Sichuan Bittercress, Spring Bittercress ■

72830 Cardamine impatiens L. subsp. elongata O. E. Schulz = Cardamine impatiens L. ■

72831 Cardamine impatiens L. var. angustifolia O. E. Schulz;窄叶碎米荠(狭叶弹裂碎米荠);Narrowleaf Impatient Bittercress, Narrowleaf Spring Bittercress ■

72832 Cardamine impatiens L. var. angustifolia O. E. Schulz = Cardamine impatiens L. ■

72833 Cardamine impatiens L. var. dasycarpa (M. Bieb.) T. Y. Cheo et R. C. Fang = Cardamine impatiens L. var. eriocarpa DC. ■

72834 Cardamine impatiens L. var. dasycarpa (M. Bieb.) T. Y. Cheo et R. C. Fang = Cardamine impatiens L. ■

72835 Cardamine impatiens L. var. eriocarpa DC.;毛果弹裂碎米荠(毛果碎米荠);Hairfruit Spring Bittercress, Hairyfruit Impatient Bittercress ■

72836 Cardamine impatiens L. var. eriocarpa DC. = Cardamine impatiens L. ■

72837 Cardamine impatiens L. var. fumaria H. Lév. = Cardamine impatiens L. ■

72838 Cardamine impatiens L. var. longipes Hatus.;长梗弹裂碎米荠■☆

72839 Cardamine impatiens L. var. microphylla O. E. Schulz = Cardamine impatiens L. ■

72840 Cardamine impatiens L. var. obtusifolia (Knaf) O. E. Schulz;钝叶碎米荠;Obtuseleaf Impatient Bittercress, Obtuseleaf Spring Bittercress ■

72841 Cardamine impatiens L. var. obtusifolia (Knaf) O. E. Schulz = Cardamine impatiens L. ■

72842 Cardamine impatiens L. var. obtusifolia Knaf = Cardamine impatiens L. ■

72843 Cardamine impatiens L. var. pilosa O. E. Schulz = Cardamine impatiens L. ■

72844 Cardamine inayatii O. E. Schulz = Cardamine flexuosa With. ■

72845 Cardamine inayatii O. E. Schulz = Cardamine yunnanensis Franch. ■

72846 Cardamine insignis O. E. Schulz = Cardamine circaeoides Hook. f. et Thomson ■

72847 Cardamine jinshaensis Q. H. Chen et T. L. Xu;金沙碎米荠;Jinsha Bittercress ■

72848 Cardamine jinshaensis Q. H. Chen et T. L. Xu = Cardamine anhuiensis D. C. Zhang et C. Z. Shao ■

72849 Cardamine johnstonii Oliv. = Cardamine obliqua Hochst. ex A. Rich. ■☆

72850 Cardamine kiusiana H. Hara = Cardamine yezoensis Maxim. var. kiusiana (H. Hara) Ohwi ■☆

72851 Cardamine komarovii Nakai;翼柄碎米荠;Komarov Bittercress ■

72852 Cardamine konaensis H. St. John = Cardamine flexuosa With. ■

72853 Cardamine koshiensis Koidz. = Cardamine parviflora L. ■

72854 Cardamine laciniata (Muhl. ex Willd.) A. W. Wood = Cardamine concatenata (Michx.) O. Schwarz ■☆

72855 Cardamine lamontii Hance = Rorippa indica (L.) Hiern ■

72856 Cardamine latifolia Lej. = Cardamine raphanifolia Pourr. ■☆

72857 Cardamine lazica Boiss. et Balansa;拉扎碎米荠■☆

72858 Cardamine leucantha (Tausch) O. E. Schulz;白花碎米荠(白花石芥菜,菜子七,假芹菜,角蒿,山芥菜);White Bittercress, Whiteflowered Bittercress ■

72859 Cardamine leucantha (Tausch) O. E. Schulz var. crenata D. C. Zhang;圆齿白花碎米荠;Rotundtooth White Bittercress ■

72860 Cardamine leucantha (Tausch) O. E. Schulz var. crenata D. C. Zhang = Cardamine leucantha (Tausch) O. E. Schulz ■

72861 Cardamine leucantha (Tausch) O. E. Schulz var. glaberrima F. Maek. ex H. Hara;光白花碎米荠■

72862 Cardamine levicaulis W. T. Wang;光茎碎米荠;Smoothstem Bittercress ■

72863 Cardamine levicaulis W. T. Wang = Cardamine yunnanensis Franch. ■

72864 Cardamine lihengiana Al-Shehbaz;李恒碎米荠;Liheng Bittercress ■

72865 Cardamine limprichtiana Pax = Orychophragmus limprichtianus (Pax) Al-Shehbaz et G. Yang ■

72866 Cardamine longifructa Ohwi = Cardamine dentipetala Matsum. var. longifructa (Ohwi) Hiyama ■☆

72867 Cardamine longipedicellata Z. M. Tan et G. H. Chen 长梗碎米荠;Longpedicel Bittercress ■

72868 Cardamine longipedicellata Z. M. Tan et G. H. Chen = Cardamine yunnanensis Franch. ■

72869 Cardamine longisrostris Janka;长喙碎米荠■☆

72870 Cardamine longistyla W. T. Wang;长柱碎米荠;Longstyle Bittercress ■

72871 Cardamine longistyla W. T. Wang = Cardamine yunnanensis Franch. ■

72872 Cardamine loxostemonoides O. E. Schulz = Loxostemon loxostemonoides (O. E. Schulz) Y. C. Lan et T. Y. Cheo ■

72873 Cardamine lyrata Bunge;水田碎米荠(水田芥,水田荠,小水田荠);Lyrate Bittercress ■

72874 Cardamine mecrocarpa Brandegee;大果碎米荠■

72875 Cardamine macrocepala Z. M. Tan et S. C. Zhou = Cardamine circaeoides Hook. f. et Thomson ■

72876 Cardamine macrophylla Willd.;大叶碎米荠(假芹菜,普贤菜,石格菜,石芹菜,蜈蚣七);Largeleaf Bittercress ■

72877 Cardamine macrophylla Willd. subsp. polyphylla (D. Don) O. E. Schulz = Cardamine macrophylla Willd. ■

72878 Cardamine macrophylla Willd. var. crenata Trautv. = Cardamine macrophylla Willd. ■

72879 Cardamine macrophylla Willd. var. crenata Trautv. et C. A. Mey.;钝圆齿碎米荠;Crenate Bittercress, Crenate Largeleaf Bittercress ■

72880 Cardamine macrophylla Willd. var. crenata Trautv. et C. A. Mey. = Cardamine macrophylla Willd. ■

72881 Cardamine macrophylla Willd. var. dentariifolia Hook. f. et T. Anderson = Cardamine macrophylla Willd. ■

72882 Cardamine macrophylla Willd. var. diplodonta T. Y. Cheo;重齿碎米荠;Doubleteethed Bittercress ■

72883 Cardamine macrophylla Willd. var. diplodonta T. Y. Cheo = Cardamine macrophylla Willd. ■

72884 Cardamine macrophylla Willd. var. foliosa Hook. f. et T. Anderson = Cardamine macrophylla Willd. ■

72885 Cardamine macrophylla Willd. var. lobata Hook. f. et T. Anderson = Cardamine macrophylla Willd. ■

72886 Cardamine macrophylla Willd. var. moupinensis Franch. = Cardamine macrophylla Willd. ■

72887 Cardamine macrophylla Willd. var. parviflora Trautv. = Cardamine leucantha (Tausch) O. E. Schulz ■

72888 Cardamine macrophylla Willd. var. polyphylla (D. Don) R. C. Fang;多叶碎米荠;Manyleaves Bittercress ■

72889 Cardamine macrophylla Willd. var. polyphylla (D. Don) R. C. Fang = Cardamine macrophylla Willd. ■

72890 Cardamine macrophylla Willd. var. polyphylla (D. Don) T. Y. Cheo et R. C. Fang = Cardamine macrophylla Willd. ■

72891 Cardamine macrophylla Willd. var. sikkimensis Hook. f. et T. Anderson = Cardamine macrophylla Willd. ■

72892 Cardamine manshurica (Kom.) Nakai = Cardamine parviflora L. ■

72893 Cardamine matsumurana Nemoto = Cardamine anemonoides O. E. Schulz ■☆

72894 Cardamine matthioli Moretti ex Comolli = Cardamine pratensis L. ■

72895 Cardamine maxima (Nutt.) A. W. Wood;大碎米荠;Large Toothwort, Three-leaved Toothwort ■☆

72896 Cardamine microsperma (DC.) Kuntze = Rorippa cantoniensis (Lour.) Ohwi ■
72897 Cardamine microzyga O. E. Schulz;小叶碎米荠;Smallleaf Bittercress ■
72898 Cardamine microzyga O. E. Schulz var. duplolobata C. Y. Wu ex T. Y. Cheo;重齿小叶碎米荠;Doublelobed Smallleaf Bittercress ■
72899 Cardamine microzyga O. E. Schulz var. purpurascens O. E. Schulz = Cardamine purpurascens (O. E. Schulz) Al-Shehbaz, T. Y. Cheo, L. L. Lou et G. Yang ■
72900 Cardamine minuta Willd.;微小碎米荠■☆
72901 Cardamine muliensis W. T. Wang;木里碎米荠;Muli Bittercress ■
72902 Cardamine muliensis W. T. Wang = Cardamine yunnanensis Franch. ■
72903 Cardamine multiflora T. Y. Cheo et R. C. Fang;多花碎米荠;Manyflowers Bittercress, Multiflower Bittercress ■
72904 Cardamine multijuga Franch.;多裂碎米荠■
72905 Cardamine multijuga Franch. = Cardamine gracilis (O. E. Schulz) T. Y. Cheo et R. C. Fang ■
72906 Cardamine multijuga Franch. = Cardamine griffithii Hook. f. et Thomson ■
72907 Cardamine multijuga Franch. var. gracilis O. E. Schulz = Cardamine gracilis (O. E. Schulz) T. Y. Cheo et R. C. Fang ■
72908 Cardamine nakaiana H. Lév. = Cardamine impatiens L. ■
72909 Cardamine niigatensis H. Hara;新泻碎米荠■☆
72910 Cardamine nipponica Franch. et Sav.;日本碎米荠(日本焊菜,峰芥)■
72911 Cardamine nivalis Pall. = Macropodium nivale (Pall.) R. Br. ■
72912 Cardamine nudicaulis L. = Parrya nudicaulis (L.) Regel ■
72913 Cardamine nymanii Gand. = Cardamine pratensis L. ■
72914 Cardamine obliqua Hochst. ex A. Rich.;偏斜碎米荠■☆
72915 Cardamine occulata Hornem. = Cardamine flexuosa With. ■
72916 Cardamine ovata Benth. = Cardamine macrophylla Willd. ■
72917 Cardamine palustre (L.) Kuntze = Rorippa palustris (L.) Besser ■
72918 Cardamine palustris Bubani = Cardamine pratensis L. ■
72919 Cardamine paradoxa Hance = Yinshania paradoxa (Hance) Y. Z. Zhao ■
72920 Cardamine parviflora L.;小花碎米荠(小叶碎米荠);Dry-land Bitter-cress, Parviflorous Bittercress, Sand Bitter-cress, Smallflower Bittercress, Small-flowered Bitter Cress, Small-flowered Bitter-cress, Small-leaved Bittercress ■
72921 Cardamine parviflora L. f. hispida Franch. = Cardamine parviflora L. ■
72922 Cardamine parviflora L. var. arenicola (Britton) O. E. Schulz;沙地小花碎米荠;Dry-land Bitter-cress, Sand Bitter-cress, Small-flowered Bitter-cress ■☆
72923 Cardamine parviflora L. var. manshurica Kom. = Cardamine parviflora L. ■
72924 Cardamine paucifolia Hand. -Mazz.;少叶碎米荠■
72925 Cardamine pectinata Pall.;篦状碎米荠■☆
72926 Cardamine pedata Regel et Tiling;鸟足碎米荠■☆
72927 Cardamine pensylvanica Muhl. ex Willd.;宾州碎米荠;Bitter Cress, Brook Cress, Pennsylvania Bittercress, Pennsylvania Bitter-cress ■☆
72928 Cardamine pensylvanica Muhl. ex Willd. var. brittoniana Farw. = Cardamine pensylvanica Muhl. ex Willd. ■☆
72929 Cardamine pentaphyllos Crantz;五小叶碎米荠■☆
72930 Cardamine petraea Townson;岩碎米荠■☆
72931 Cardamine pilosa Willd. = Cardamine prorepens Fisch. ex DC. ■
72932 Cardamine polyphylla D. Don = Cardamine macrophylla Willd. ■
72933 Cardamine potentillifolia H. Lév. = Orychophragmus violaceus (L.) O. E. Schulz ■
72934 Cardamine pratensis L.;草甸碎米荠(草原碎米荠,紫花蔊菜);Apple Pie, Biddy's Eyes, Bird's Eye, Bitter-cress, Bog-flower, Bogspinks, Bonny Bird's Eye, Bonny Bird's Eyes, Bread-and-butter, Bread-and-milk, Canterbury Bells, Cuckoo, Cuckoo Bitter-cress, Cuckoo Flower, Cuckoo Hitter Cress, Cuckoo Pint, Cuckoo Pintel, Cuckoo Spice, Cuckoo Spit, Cuckoo-bread, Cuckoo-buds, Cuckooflower, Cuckoo-flower, Cuckoo-pint, Cuckoo-pintle, Cuckoo's Shoes-and-stockings, Cuckoo-spice, Darmell Goddard, Gilloflower, Goocoo-flower, Gookoo Buttons, Headache, Ladies' Flower, Ladies' Smock, Lady Smock, Lady's Cloak, Lady's Flock, Lady's Glove, Lady's Gloves, Lady's Mantle, Lady's Milksile, Lady's Pride, Lady's Shoes, Lady's Smock, Lady's-smock, Lamb's Lakens, Laylock, Lilac, Lonesome Lady, Lucy Locket, May Blob, May-blob, Mayflower, May-flower, Meadow, Meadow Bittercress, Meadow Bitter-cress, Meadow Cress, Meadow Pink, Meadow-cresses, Meadow-cuckoo, Meadowflower, Meadow-flower, Milkies Milkgirl, Milking Maids, Milkmaids, Milksile, Milky Maidens, Milky Malden, Moll Blob, Moll-blob, Mylady's Smock, Naked Ladies, Naked Lady, Nightingale Flower, Paigle, Pick-folly, Pickpocket, Pigeon's Eye, Pigeon's Eyes, Pig's Eye, Pig's Eyes, Pink, Pleasant-in-sight, Shoes-and-stockings, Smell Smock, Smell-smock, Smick Smock, Smick-smock, Spink, Spinks, Swamp's Companion, Water Cuckoo, Water Lily, Wet Cuckoo, Whitsuntide Gillif Ower, Wild Rocket ■
72935 Cardamine pratensis L. 'Flore Pleno';重瓣草甸碎米荠■☆
72936 Cardamine pratensis L. subsp. atlantica (Emb. et Maire) Greuter et Burdet;大西洋碎米荠■☆
72937 Cardamine pratensis L. subsp. chinensis O. E. Schulz = Cardamine stenoloba Hemsl. ■
72938 Cardamine pratensis L. var. atlantica Emb. et Maire = Cardamine pratensis L. subsp. atlantica (Emb. et Maire) Greuter et Burdet ■☆
72939 Cardamine pratensis L. var. palustris Wimm. et Grab.;沼泽草甸碎米荠;Cuckoo-flower ■☆
72940 Cardamine pratensis L. var. prorepens (Fisch. ex DC.) Maxim. = Cardamine prorepens Fisch. ex DC. ■
72941 Cardamine prattii Hemsl. et E. H. Wilson = Cardamine microzyga O. E. Schulz ■
72942 Cardamine prorepens Fisch. ex DC.;浮水碎米荠(伏水碎米荠);Repent Bittercress ■
72943 Cardamine pubescens Steven = Cardamine prorepens Fisch. ex DC. ■
72944 Cardamine pulchella (Hook. f. et Thomson) Al-Shehbaz et G. Yan;细巧碎米荠(弯蕊芥);Curvedstamencress, Showy Curvedstamencress ■
72945 Cardamine purpurascens (O. E. Schulz) Al-Shehbaz, T. Y. Cheo, L. L. Lou et G. Yang;紫花碎米荠(紫花弯蕊芥,紫花小叶碎米荠);Purple Curvedstamencress ■
72946 Cardamine purpurea Cham. et Schltdl.;紫碎米荠■☆
72947 Cardamine pusilla Hochst. ex A. Rich. = Arabidopsis thaliana (L.) Heynh. ■
72948 Cardamine raphanifolia Pourr.;萝卜叶碎米荠;Greater Cuckoo Flower, Radish-leaved Bittercress ■☆
72949 Cardamine regeliana Miq. = Cardamine scutata Thunb. ■

72950 Cardamine regeliana Miq. var. manshurica (Kom.) Kitag. = Cardamine scutata Thunb. ■

72951 Cardamine reniformis Hayata;肾叶碎米荠■

72952 Cardamine reniformis Hayata = Cardamine circaeoides Hook. f. et Thomson ■

72953 Cardamine repens (Franch.) Diels;匍匐碎米荠(匍匐弯蕊芥);Creeping Curvedstamencress ■

72954 Cardamine repens Diels = Cardamine repens (Franch.) Diels ■

72955 Cardamine resedifolia L. var. longisiliqua Font Quer = Murbeckiella boryi (Boiss.) Rothm. ■☆

72956 Cardamine resedifolia L. var. morii Nakai. = Cardamine changbaiana Al-Shehbaz ■

72957 Cardamine rhomboidea (Pers.) DC. = Cardamine bulbosa (Schreb. ex Muhl.) Britton, Sterns et Poggenb. ■☆

72958 Cardamine rhomboidea (Pers.) DC. var. purpurea Torr. = Cardamine douglassii (Torr.) Britton ■☆

72959 Cardamine rivularis Schur = Cardamine pratensis L. ■

72960 Cardamine rockii O. E. Schulz;鞭枝碎米荠;Rock Bittercress, Whiptwig Bittercress ■

72961 Cardamine rotundifolia Michx.;圆叶碎米荠;Mountain Watercress ■☆

72962 Cardamine sachalinensis Miyabe et Miyake = Cardamine macrophylla Willd. ■

72963 Cardamine sachokiana N. Busch;萨氏碎米荠■☆

72964 Cardamine scaposa Franch.;裸茎碎米荠(落叶梅);Nakedstem Bittercress ■

72965 Cardamine schinziana O. E. Schulz;欣兹碎米荠■☆

72966 Cardamine schinziana O. E. Schulz f. lasiocarpa (H. Hara) H. Hara;毛果欣兹碎米荠■☆

72967 Cardamine schinziana O. E. Schulz var. lasiocarpa (H. Hara) Koidz. = Cardamine schinziana O. E. Schulz f. lasiocarpa (H. Hara) H. Hara ■☆

72968 Cardamine schulziana Baehni = Cardamine trifida (Lam. ex Poir.) B. M. G. Jones ■

72969 Cardamine scoriarum W. W. Sm. = Cardamine fragarifolia O. E. Schulz ■

72970 Cardamine scoriarum W. W. Sm. = Cardamine trifoliolata Hook. f. et Thomson ■

72971 Cardamine scutata Thunb.;圆齿碎米荠(雷氏碎米荠,碎米荠);Bucklershaped Bittercress, Scutate-dentate Bittercress ■

72972 Cardamine scutata Thunb. subsp. fallax (O. E. Schulz) H. Hara = Cardamine parviflora L. ■

72973 Cardamine scutata Thunb. subsp. fallax (O. E. Schulz) H. Hara = Cardamine fallax (O. E. Schulz) Nakai ■☆

72974 Cardamine scutata Thunb. subsp. flexuosa (With.) H. Hara = Cardamine flexuosa With. ■

72975 Cardamine scutata Thunb. subsp. flexuosa (With.) H. Hara = Cardamine scutata Thunb. ■

72976 Cardamine scutata Thunb. subsp. regeliana (Miq.) H. Hara = Cardamine regeliana Miq. ■☆

72977 Cardamine scutata Thunb. var. formosana (Hayata) Tang S. Liu et S. S. Ying = Cardamine hirsuta L. ■

72978 Cardamine scutata Thunb. var. koshiensis Ohwi et Okuyama = Cardamine niigatensis H. Hara ■☆

72979 Cardamine scutata Thunb. var. latifolia (Maxim.) H. Hara;宽叶圆齿碎米荠■☆

72980 Cardamine scutata Thunb. var. longiloba P. Y. Fu;大顶叶碎米荠;Longlobe Bucklershaped Bittercress ■

72981 Cardamine scutata Thunb. var. longiloba P. Y. Fu = Cardamine scutata Thunb. ■

72982 Cardamine scutata Thunb. var. regeliana (Miq.) H. Hara = Cardamine scutata Thunb. ■

72983 Cardamine scutata Thunb. var. rotundiloba (Hayata) Tang S. Liu et S. S. Ying;台湾碎米荠■

72984 Cardamine scutata Thunb. var. rotundiloba (Hayata) Tang S. Liu et S. S. Ying = Cardamine scutata Thunb. ■

72985 Cardamine seidlitziana Albov;塞德碎米荠■☆

72986 Cardamine senanensis Franch. et Sav. = Cardamine impatiens L. ■

72987 Cardamine sikkimensis H. Hara = Cardamine yunnanensis Franch. ■

72988 Cardamine simensis Hochst. ex Oliv. = Cardamine hirsuta L. ■

72989 Cardamine simplex Hand.-Mazz.;单茎碎米荠;Simplestem Bittercress ■

72990 Cardamine sinica Rashida et H. Ohba = Cardamine yunnanensis Franch. ■

72991 Cardamine sinomanshurica (Kitag.) Kitag. = Cardamine macrophylla Willd. ■

72992 Cardamine smithiana Biswas;腺萼碎米荠;Smith Bittercress ■

72993 Cardamine smithiana Biswas = Cardamine fragarifolia O. E. Schulz ■

72994 Cardamine stenoloba Hemsl.;狭叶碎米荠■

72995 Cardamine stenoloba Hemsl. = Loxostemon stenolobus (Hemsl.) Y. C. Lan et T. Y. Cheo ■

72996 Cardamine sublyrata Miq. = Rorippa dubia (Pers.) H. Hara ■

72997 Cardamine sylvatica Link = Cardamine flexuosa With. ■

72998 Cardamine sylvatica Link var. regeliana (Miq.) Franch. et Sav. = Cardamine scutata Thunb. ■

72999 Cardamine talamontiana Chiov. = Cardamine trichocarpa Hochst. ex A. Rich. ■☆

73000 Cardamine tanakae Franch. et Sav. ex Maxim.;田中氏碎米荠■☆

73001 Cardamine tangutora O. E. Schulz;唐古特碎米荠(石芥菜,石芥花,石荠菜,唐古碎米荠,紫花碎米荠);Tangut Bittercress ■

73002 Cardamine taquetii H. Lév. = Cardamine scutata Thunb. ■

73003 Cardamine tenera S. G. Gmel. ex C. A. Mey.;柔碎米荠■☆

73004 Cardamine tenuifoha (Ledeb.) Turcz. = Cardamine trifida (Lam. ex Poir.) B. M. G. Jones ■

73005 Cardamine tenuifolia (Ledeb.) Turcz. var. granulifera Franch. = Cardamine granulifera (Franch.) Diels ■

73006 Cardamine tenuifolia (Ledeb.) Turcz. var. repens (Franch.) Franch. = Cardamine repens (Franch.) Diels ■

73007 Cardamine tenuifolia Ledeb. = Cardamine schulziana Baehni ■

73008 Cardamine tenuifolia Turcz. = Cardamine schulziana Baehni ■

73009 Cardamine tenuifolia Turcz. var. granulifera Franch. = Cardamine granulifera (Franch.) Diels ■

73010 Cardamine tenuifolia Turcz. var. repens (Franch.) Franch. = Cardamine repens (Franch.) Diels ■

73011 Cardamine tibetana Rashid et H. Ohba = Cardamine loxostemonoides O. E. Schulz ■

73012 Cardamine tibetana Rashid et H. Ohba = Loxostemon loxostemonoides (O. E. Schulz) Y. C. Lan et T. Y. Cheo ■

73013 Cardamine torrentis Nakai;急流碎米荠■☆

73014 Cardamine trichocarpa A. Rich. var. elegans Engl. = Cardamine trichocarpa Hochst. ex A. Rich. ■☆

73015 Cardamine trichocarpa Hochst. ex A. Rich.;毛果碎米荠■☆

73016 Cardamine trichocarpa Hochst. ex A. Rich. f. leiocarpa O. E. Schulz = Cardamine trichocarpa Hochst. ex A. Rich. ■☆

73017 Cardamine trichocarpa Hochst. ex A. Rich. subsp. elegans (Engl.) O. E. Schulz = Cardamine trichocarpa Hochst. ex A. Rich. ■☆

73018 Cardamine trifida (Lam. ex Poir.) B. M. G. Jones;细叶碎米荠(细叶石芥菜);Fineleaf Bittercress,Tenuifolious Bittercress ■

73019 Cardamine trifolia L.;车轴草叶碎米荠;Trefoil Cress,Trefoil Milkmaid ■☆

73020 Cardamine trifolia Pall. = Cardamine amara L. ■☆

73021 Cardamine trifolia Thunb. = Cardamine scutata Thunb. ■

73022 Cardamine trifolia Wahlenb. = Cardamine amara L. ■☆

73023 Cardamine trifoliolata Hook. f. et Thomson;三小叶碎米荠(菱叶碎米荠,拟弯曲碎米荠,三叶碎米荠,汶川碎米荠,小三叶碎米荠);Flexuose-like Bittercress,Three Leaflets Bittercress,Trifoliolate Bittercress ■

73024 Cardamine truncatolobata W. T. Wang;截裂碎米荠;Trucatelobed Bittercress ■

73025 Cardamine truncatolobata W. T. Wang = Cardamine simplex Hand. -Mazz. ■

73026 Cardamine uliginosa M. Bieb.;沼泽碎米荠■☆

73027 Cardamine umbellata Greene;小伞碎米荠■☆

73028 Cardamine umbrosa Andrz. = Cardamine flexuosa With. ■

73029 Cardamine urbaniana O. E. Schulz;华中碎米荠(半边菜,菜子七,妇人参,普贤菜);Central China Bittercress ■

73030 Cardamine urbaniana O. E. Schulz = Cardamine macrophylla Willd. ■

73031 Cardamine verticillata Jeffrey et W. W. Sm. = Staintoniella verticillata (Jeffrey et W. W. Sm.) H. Hara ■

73032 Cardamine verticillata Jeffrey et W. W. Sm. = Taphrospermum verticillatum (Jeffrey et W. W. Sm.) Al-Shehbaz ■

73033 Cardamine victoris N. Busch;维多利亚碎米荠■☆

73034 Cardamine violacea (D. Don) Wall. ex Hook. f. et Thomson;堇色碎米荠(紫花山芥);Violet Bittercress ■

73035 Cardamine violacea (D. Don) Wall. ex Hook. f. et Thomson subsp. bhutanica Grierson = Cardamine violacea (D. Don) Wall. ex Hook. f. et Thomson ■

73036 Cardamine violifolia O. E. Schulz;堇叶碎米荠;Violetleaf Bittercress ■

73037 Cardamine violifolia O. E. Schulz = Cardamine circaeoides Hook. f. et Thomson ■

73038 Cardamine violifolia O. E. Schulz var. diversifolia O. E. Schulz;异堇叶碎米荠;Different Violetleaf Bittercress ■

73039 Cardamine violifolia O. E. Schulz var. diversifolia O. E. Schulz = Cardamine circaeoides Hook. f. et Thomson ■

73040 Cardamine violifolia O. E. Schulz var. pilosa K. L. Chang et H. L. Huang = Cardamine circaeoides Hook. f. et Thomson ■

73041 Cardamine weixiensis W. T. Wang;维西碎米荠;Weixi Bittercress ■

73042 Cardamine weixiensis W. T. Wang = Cardamine yunnanensis Franch. ■

73043 Cardamine wightiana Wall. = Cardamine africana L. ■☆

73044 Cardamine yezoensis Maxim.;北海道碎米荠■☆

73045 Cardamine yezoensis Maxim. var. kiusiana (H. Hara) Ohwi;九州碎米荠■☆

73046 Cardamine yezoensis Maxim. var. schinziana (O. E. Schulz) Ohwi = Cardamine schinziana O. E. Schulz ■☆

73047 Cardamine yungshunensis W. T. Wang = Neomartinella yungshunensis (W. T. Wang) Al-Shehbaz ■

73048 Cardamine yunnanensis Franch.;云南碎米荠(云南石芥菜);Yunnan Bittercress ■

73049 Cardamine yunnanensis Franch. var. obtusata C. Y. Wu ex T. Y. Cheo et R. C. Fang;钝叶云南碎米荠;Obtuse Yunnan Bittercress,Obtuseleaf Yunnan Bittercress ■

73050 Cardamine yunnanensis Franch. var. obtusata C. Y. Wu ex T. Y. Cheo et R. C. Fang = Cardamine paucifolia Hand. -Mazz. ■

73051 Cardamine yunshunensis W. T. Wang;永顺碎米荠;Yongshun Bittercress ■

73052 Cardamine zheiiangensis T. Y. Cheo et R. C. Fang;浙江碎米荠(大叶水尖);Zhejiang Bittercress ■

73053 Cardamine zhejiangensis T. Y. Cheo et R. C. Fang = Cardamine scutata Thunb. ■

73054 Cardamine zhejiangensis T. Y. Cheo et R. C. Fang var. huangshanensis D. C. Zhang = Cardamine scutata Thunb. ■

73055 Cardamine zollingeri Turcz. = Cardamine flexuosa With. ■

73056 Cardaminopsis (C. A. Mey.) Hayek = Arabidopsis Heynh.(保留属名)■

73057 Cardaminopsis (C. A. Mey.) Hayek = Arabis L. ●■

73058 Cardaminopsis (C. A. Mey.) Hayek = Cardamine L. ■

73059 Cardaminopsis (C. A. Mey.) Hayek(1908);假碎米荠属■

73060 Cardaminopsis Hayek = Cardaminopsis (C. A. Mey.) Hayek ■

73061 Cardaminopsis arenosa L. = Arabis arenosa (L.) Scop. ■☆

73062 Cardaminopsis gemmifera (Matsum.) Berk. = Arabidopsis halleri (L.) O' Kane et Al-Shehbaz subsp. gemmifera (Matsum.) O' Kane et Al-Shehbaz ■

73063 Cardaminopsis kamchatica (DC.) O. E. Schulz = Arabidopsis kamchatica (DC.) K. Shimizu et Kudoh ■☆

73064 Cardaminopsis kamchatica (Fisch. ex DC.) O. E. Schulz = Arabidopsis lyrata (L.) O'Kane et Al-Shehbaz subsp. kamchatica (Fisch. ex DC.) O'Kane et Al-Shehbaz ■

73065 Cardaminum Moench(废弃属名) = Nasturtium W. T. Aiton(保留属名)■

73066 Cardamomum Kuntze = Amomum Roxb.(保留属名)■

73067 Cardamomum Rumph. = Amomum Roxb.(保留属名)■

73068 Cardamomum Rumph. ex Kuntze = Amomum Roxb.(保留属名)■

73069 Cardamomum Salisb. = Elettaria Maton ■

73070 Cardamon Beck = Lepidium L. ■

73071 Cardamon Fourr. = Lepidium L. ■

73072 Cardanoglyphus Post et Kuntze = Kardanoglyphos Schltdl. ■☆

73073 Cardanthera Buch. -Ham. = Synnema Benth. ●■☆

73074 Cardanthera Buch. -Ham. ex Benth. = Hygrophila R. Br. ●■

73075 Cardanthera Buch. -Ham. ex Nees = Synnema Benth. ●■☆

73076 Cardanthera Buch. -Ham. ex Voigt = Synnema Benth. ●■☆

73077 Cardanthera africana (T. Anderson) Benth. = Hygrophila africana (T. Anderson) Heine ■☆

73078 Cardanthera africana (T. Anderson) Benth. var. schweinfurthii S. Moore = Hygrophila abyssinica (Hochst. ex Nees) T. Anderson ●☆

73079 Cardanthera breviflora (Burkill) Turrill = Hygrophila brevituba (Burkill) Heine ■☆

73080 Cardanthera justicioides S. Moore = Hygrophila abyssinica (Hochst. ex Nees) T. Anderson ●☆

73081 Cardanthera parviflora Turrill = Hygrophila brevituba (Burkill) Heine ■☆

73082 Cardanthera triflora (Nees) Buch. -Ham. ex C. B. Clarke = Hygrophila difformis (L. f.) Blume ■☆

73083 Cardanthera triflora Buch. -Ham. = Hygrophila difformis (L. f.) Blume ■☆

73084 Cardanthera triflora Buch. -Ham. ex Nees = Hygrophila difformis (L. f.) Blume ■☆

73085 Cardaria Desv. (1815);群心菜属;Cardaria,Whitetop ■

73086 Cardaria Desv. = Lepidium L. ■

73087 Cardaria boissieri (N. Busch) Soó = Cardaria draba (L.) Desv. subsp. chalepensis (L.) O. E. Schulz ■

73088 Cardaria chalepensis (L.) Hand. -Mazz. = Cardaria draba (L.) Desv. subsp. chalepensis (L.) O. E. Schulz ■

73089 Cardaria chalepensis (L.) Hand. -Mazz. = Cardaria draba (L.) Desv. ■

73090 Cardaria draba (L.) Desv.;群心菜(毛独行菜,细粒独行菜);Common Cardaria, Hairy Cress, Heart-podded Hoary Cress, Heart-podded Hoary-cress, Hoary Cress, Hoary Pepperwort, Lenspod Whitetop, Lens-podded Hoary Cress, Pepper Cress, Pepperweed Whitetop, Pepper-weed White-top, Perennial Pepper-grass, Thanet Cress, Thanet Weed, Thompson's Curse, White Top, Whitetop, White-top, Whitlow Pepperwort ■

73091 Cardaria draba (L.) Desv. = Lepidium draba L. ■

73092 Cardaria draba (L.) Desv. subsp. chalepensis (L.) O. E. Schulz;球果群心菜;Sphaeroidalfruit Cardaria ■

73093 Cardaria draba (L.) Desv. subsp. chalepensis (L.) O. E. Schulz var. repens (Schrenk) O. E. Schulz = Cardaria draba (L.) Desv. subsp. chalepensis (L.) O. E. Schulz ■

73094 Cardaria draba (L.) Desv. subsp. chalepensis var. repens (Schrenk) O. E. Schulz = Cardaria draba (L.) Desv. subsp. chalepensis (L.) O. E. Schulz ■

73095 Cardaria fenestrata (Boiss.) Rollins = Cardaria draba (L.) Desv. subsp. chalepensis (L.) O. E. Schulz ■

73096 Cardaria macrocarpa (Franch.) Rollins = Cardaria draba (L.) Desv. subsp. chalepensis (L.) O. E. Schulz ■

73097 Cardaria propinqua (Fisch. et C. A. Mey.) N. Busch. = Cardaria draba (L.) Desv. subsp. chalepensis (L.) O. E. Schulz ■

73098 Cardaria pubescens (C. A. Mey.) Jarm.;毛果群心菜(泡果芥,甜萝卜缨子);Globe-podded Hoary Cress, Globe-podded Hoary-cress, Hairy Whitetop, Hairy White-top, Hoary Cress, Pubescent Cardaria, White Top ■

73099 Cardaria pubescens (C. A. Mey.) Jarm. var. elongata Rollins = Cardaria pubescens (C. A. Mey.) Jarm. ■

73100 Cardaria pubescens (C. A. Mey.) Rollins = Cardaria pubescens (C. A. Mey.) Jarm. ■

73101 Cardaria repens (Schrenk) Boiss. = Cardaria chalepensis (L.) Hand. -Mazz. ■

73102 Cardaria repens (Schrenk) Jarm. = Cardaria draba (L.) Desv. subsp. chalepensis (L.) O. E. Schulz ■

73103 Cardarninum Moench = Rorippa Scop. ■

73104 Cardenanthus R. C. Foster(1945);基管鸢尾属■☆

73105 Cardenanthus boliviensis R. C. Foster;玻利维亚基管鸢尾■☆

73106 Cardenanthus longitubus R. C. Foster;长管基管鸢尾■☆

73107 Cardenasia Rusby = Bauhinia L. ●

73108 Cardenasiodendron F. A. Barkley(1954);卡尔漆属●☆

73109 Cardenasiodendron brachypterum (Loes.) F. A. Barkley;卡尔漆●☆

73110 Carderina Cass. = Senecio L. ■●

73111 Cardia Dulac = Veronica L. ■

73112 Cardiaca L. = Leonurus L. ■

73113 Cardiaca Mill. = Leonurus L. ■

73114 Cardiaca leonuroides Willd. = Chaiturus marrubiastrum (L.) Spenn. ■

73115 Cardiaca marrubiastrum (L.) Medik. = Chaiturus marrubiastrum (L.) Spenn. ■

73116 Cardiaca marrubiastrum Medik. = Chaiturus marrubiastrum (L.) Spenn. ■

73117 Cardiacanthus Nees et Schauer(废弃属名) = Carlowrightia A. Gray(保留属名)☆

73118 Cardiacanthus Schauer = Jacobinia Nees ex Moric.(保留属名)●■☆

73119 Cardiandra Siebold et Zucc. (1839);草绣球属(草八仙花属,草紫阳花属,人心药属);Cardiandra ●

73120 Cardiandra alternifolia (Siebold) Siebold et Zucc.;互叶草绣球(台湾草紫阳花)●☆

73121 Cardiandra alternifolia Siebold et Zucc. f. formosa Honda;美丽互叶草绣球●☆

73122 Cardiandra alternifolia Siebold et Zucc. f. mirabilis (Takeda) Sugim. ex H. Ohba;奇异互叶草绣球●☆

73123 Cardiandra alternifolia Siebold et Zucc. f. multiplex Hayashi;多倍互叶草绣球●☆

73124 Cardiandra alternifolia Siebold et Zucc. f. oppositifolia (Honda) F. Maek.;对叶草绣球●☆

73125 Cardiandra alternifolia Siebold et Zucc. subsp. moellendorffii (Hance) H. Hara et H. Ohba var. binata F. Maek. = Cardiandra formosana Hayata ●

73126 Cardiandra alternifolia Siebold et Zucc. subsp. moellendorffii (Hance) H. Hara et H. Ohba = Cardiandra moellendorffii (Hance) Migo ●

73127 Cardiandra alternifolia Siebold et Zucc. var. amamiohsimensis (Koidz.) Masam. = Cardiandra amamiohsimensis Koidz. ●☆

73128 Cardiandra alternifolia Siebold et Zucc. var. hakonensis Ohba ex H. Ohba;箱根草绣球●☆

73129 Cardiandra alternifolia Siebold et Zucc. var. moellendorffii (Hance) Engl. = Cardiandra moellendorffii (Hance) Migo ●

73130 Cardiandra alternifolia Siebold et Zucc. var. oppositifolia Honda = Cardiandra alternifolia Siebold et Zucc. f. oppositifolia (Honda) F. Maek. ●☆

73131 Cardiandra amamiohsimensis Koidz.;奄美草绣球●☆

73132 Cardiandra densifolia C. F. Wei;密叶草绣球;Denseleaf Cardiandra ●

73133 Cardiandra densifolia C. F. Wei = Cardiandra formosana Hayata ●

73134 Cardiandra formosana Hayata;台湾草绣球(台湾草紫阳花,台湾人心药);Taiwan Cardiandra ●

73135 Cardiandra laxiflora H. L. Li = Cardiandra moellendorffii (Hance) Migo var. laxiflora (H. L. Li) C. F. Wei ●

73136 Cardiandra moellendorffii (Hance) Migo;草绣球(草紫阳花,牡丹三七,人心药,紫阳花);Moellendorff Cardiandra ●

73137 Cardiandra moellendorffii (Hance) Migo var. laxiflora (H. L. Li) C. F. Wei;疏花草绣球;Loose-flower Moellendorff Cardiandra ●

73138 Cardiandra sinensis Hemsl. = Cardiandra formosana Hayata ●

73139 Cardiandra sinensis Hemsl. = Cardiandra moellendorffii (Hance) Migo ●

73140 Cardianthera Hance = Cardanthera Buch. -Ham. ex Voigt ●■☆

73141 Cardinalis Fabr. = Lobelia L. ●■

73142 Cardinalis Rupp. = Lobelia L. ●■

73143 Cardiobatus Greene = Rubus L. ●■

73144 Cardiocarpus Reinw. = Soulamea Lam. ●☆

73145 Cardiochilos P. J. Cribb(1977);心唇兰属■☆

73146 Cardiochilos williamsonii P. J. Cribb;心唇兰■☆

73147 Cardiochlamys Oliv. (1883);心被旋花属●☆

73148 Cardiochlamys discifera (C. K. Schneid.) C. Y. Wu = Porana discifera C. K. Schneid. ●

73149 Cardiochlamys discifera (C. K. Schneid.) C. Y. Wu = Poranopsis discifera (C. K. Schneid.) Staples ●

73150 Cardiochlamys madagascariensis Oliv.;马岛心被旋花●☆

73151 Cardiochlamys sinensis Hand. -Mazz. = Porana discifera C. K. Schneid. ●

73152 Cardiochlamys sinensis Hand. -Mazz. = Poranopsis sinensis (Hemsl.) Staples ●

73153 Cardiochlamys velutina Hallier f.;短绒毛心被旋花●☆

73154 Cardiocrinum (Endl.) Lindl. (1846);大百合属(荞麦叶贝母属);Cardiocrinum,Giant Lily,Largelily ■

73155 Cardiocrinum Endl. = Cardiocrinum (Endl.) Lindl. ■

73156 Cardiocrinum Lindl. = Cardiocrinum (Endl.) Lindl. ■

73157 Cardiocrinum cathayanum (E. H. Wilson) Stearn;荞麦叶大百合(百合莲,大百合,广兜铃,号筒花,喇叭,荞麦叶贝母,山丹,水百合); Cardio Crinum, Cardiocrinum, Cathay Lily, China Largelily, Chinese Cardiocrinum ■

73158 Cardiocrinum cordatum (Thunb.) Makino; 心叶大百合; Heartleaf Lily ■

73159 Cardiocrinum giganteum (Wall.) Makino;大百合(海百合,号筒花,马兜铃,荞麦叶,荞麦叶贝母,荞麦叶大百合,山菠萝根,山芋头,水草蒙,心叶百合,心叶大百合,云南大百合); Giant Cardiocrinum,Giant Lily,Largelily ■

73160 Cardiocrinum giganteum (Wall.) Makino var. yunnanense (Leichtlin ex Elwes) Stearn;云南大百合■

73161 Cardiocrinum glehni Makino;本州大百合■☆

73162 Cardiodaphnopsis Hutch. = Caryodaphnopsis Airy Shaw ●

73163 Cardiogyne Bureau = Maclura Nutt. (保留属名)●

73164 Cardiogyne africana Bureau = Maclura africana (Bureau) Corner ●☆

73165 Cardiolepis Raf. = Endotropis Raf. ●

73166 Cardiolepis Raf. = Rhamnus L. ●

73167 Cardiolepis Wallr. = Cardaria Desv. ■

73168 Cardiolepis Wallr. = Lepidium L. ■

73169 Cardiolochia Raf. = Aristolochia L. ■●

73170 Cardiolochia Raf. ex Rchb. = Aristolochia L. ■●

73171 Cardiolophus Griff. = Bacopa Aubl. (保留属名)■

73172 Cardionema DC. (1828);沙垫花属;Sandcarpet ■☆

73173 Cardionema ramosissima (Weinm.) A. Nelson et J. F. Macbr.;沙垫花■☆

73174 Cardiopetalum Schltdl. (1834-1835);心瓣花属●☆

73175 Cardiopetalum calophyllum Schltdl.;心瓣花●☆

73176 Cardiophora Benth. = Soulamea Lam. ●☆

73177 Cardiophyllarium Choux = Doratoxylon Thouars ex Hook. f. ●☆

73178 Cardiophyllum Ehrh. = Listera R. Br. (保留属名)■

73179 Cardiophyllum Ehrh. = Ophrys L. ■☆

73180 Cardiopleryx Wall. ex Blume = Peripterygium Hassk. ●■

73181 Cardiopteridaceae Blume (1847)(保留科名);心翼果科; Cardiopteris Family ■

73182 Cardiopteris Wall. ex Blume = Peripterygium Hassk. ●■

73183 Cardiopteris Wall. ex Royle = Peripterygium Hassk. ●■

73184 Cardiopteris Wall. ex Royle(1834);心翼果属;Peripterygium ●■

73185 Cardiopteris lobata R. Br. = Peripterygium quinquelobum Hassk. ■

73186 Cardiopteris lobata Wall. ex Benn. et R. Br.;裂叶心翼果●■☆

73187 Cardiopteris moluccana Blume = Peripterygium platycarpum (Gagnep.) Sleumer ●■

73188 Cardiopteris platycarpa Gagnep.;大心翼果;Big Peripterygium ●■

73189 Cardiopteris platycarpa Gagnep. = Peripterygium platycarpum (Gagnep.) Sleumer ●■

73190 Cardiopteris quinqueloba Hassk.;心翼果(裂叶心翼果); Peripterygium ■

73191 Cardiopteris quinqueloba Hassk. = Peripterygium quinquelobum Hassk. ■

73192 Cardiopterygaceae Blume = Cardiopteridaceae Blume(保留科名)●■

73193 Cardiopterygaceae Tiegh. = Cardiopteridaceae Blume(保留科名)●■

73194 Cardiospermum L. (1753);倒地铃属(灯笼藤属);Heartseed, Heart-seed ■

73195 Cardiospermum alatum Bremek. et Oberm. = Cardiospermum corindum L. ■

73196 Cardiospermum canescens L.;灰色倒地铃■☆

73197 Cardiospermum clematideum A. Rich. = Cardiospermum corindum L. ■

73198 Cardiospermum corindum L. = Cardiospermum halicacabum L. ■

73199 Cardiospermum corindum L. f. clematideum (A. Rich.) Radlk. = Cardiospermum corindum L. ■

73200 Cardiospermum elegans Kunth = Cardiospermum grandiflorum Sw. ■☆

73201 Cardiospermum grandiflorum Sw.;大花倒地铃(宿根风船葛); Heart's Dea,Heart's Seed,Showy Balloonvine ■☆

73202 Cardiospermum grandiflorum Sw. f. elegans (Kunth) Radlk. = Cardiospermum grandiflorum Sw. ■☆

73203 Cardiospermum grandiflorum Sw. f. genuinum Radlk. = Cardiospermum grandiflorum Sw. ■☆

73204 Cardiospermum grandiflorum Sw. f. hirsutum (Willd.) Radlk. = Cardiospermum grandiflorum Sw. ■☆

73205 Cardiospermum grandiflorum Sw. var. elegans (Kunth) Hiern = Cardiospermum grandiflorum Sw. ■☆

73206 Cardiospermum grandiflorum Sw. var. hirsutum (Willd.) Hiern = Cardiospermum grandiflorum Sw. ■☆

73207 Cardiospermum halicacabum L.;倒地铃(白花草,包袱草,带藤苦楝,灯笼草,风船葛,鬼灯笼,假苦瓜,假苦楝,假蒲达,金丝苦楝,金丝苦楝藤,炮掌果,三角灯笼,三角泡,小果倒地铃,眼睛草,野苦瓜,一胎三子,粽子草); Balloon Vine, Balloon-vine, Balloonvine Heartseed,Blister Creeper,Common Balloon Vine,Heart Pea,Heart Seed,Heart-pea,Heartseed,Heart-seed,Love in a Puff, Love-in-a-puff,Palsy-curer,Pigeon's Knee,Winter Cherry ■

73208 Cardiospermum halicacabum L. var. microcarpum (Kunth) Blume;小果倒地铃(包袱草,倒地铃,风船葛,金丝苦楝,炮卜草,软枝苦楝,三角包,三角泡,三角藤,粽子草);Smallfruit Heartseed ■

73209 Cardiospermum halicacabum L. var. microcarpum (Kunth) Blume = Cardiospermum halicacabum L. ■

73210 Cardiospermum hirsutum Willd. = Cardiospermum grandiflorum Sw. ■☆

73211 Cardiospermum integerrimum Radlk.;全缘倒地铃■☆

73212 Cardiospermum microcarpum Kunth = Cardiospermum halicacabum L. var. microcarpum (Kunth) Blume ■

73213 Cardiospermum microcarpum Kunth = Cardiospermum

halicacabum L. ■
73214 Cardiospermum pechuelii Kuntze;佩休倒地铃■☆
73215 Cardiostegia C. Presl = Melhania Forssk. ●■
73216 Cardiostigma Baker = Calydorea Herb. ■☆
73217 Cardiostigma Baker = Sphenostigma Baker ■☆
73218 Cardioteucris C. Y. Wu = Caryopteris Bunge ●
73219 Cardioteucris C. Y. Wu = Rubiteucris Kudo ■
73220 Cardioteucris C. Y. Wu(1962);心叶石蚕属(腺香莸属);Cardioteucris ●★
73221 Cardioteucris cordifolia C. Y. Wu;心叶石蚕(野苏麻);Cordateleaf Cardioteucris ■
73222 Cardioteucris cordifolia C. Y. Wu = Caryopteris siccanea W. W. Sm. ■
73223 Cardiotheca Ehrenb. ex Steud. = Anarrhinum Desf.(保留属名)■●☆
73224 Cardlocarpus Reinw. = Soulamea Lam. ●☆
73225 Cardonaea Aristeg., Maguire et Steyerm. = Gongylolepis R. H. Schomb. ●☆
73226 Cardopatium Juss.(1805);蓝丝菊属■☆
73227 Cardopatium amethystinum Spach;紫水晶蓝丝菊■☆
73228 Cardopatium corymbosum Pers.;蓝丝菊■☆
73229 Cardosanctus Bubani = Cnicus L.(保留属名)■●
73230 Cardosoa S. Ortiz et Paiva = Anisopappus Hook. et Arn. ■
73231 Cardosoa S. Ortiz et Paiva(2010);安哥拉山黄菊属■☆
73232 Carduaceae Bercht. et J. Presl = Asteraceae Bercht. et J. Presl(保留科名)●■
73233 Carduaceae Bercht. et J. Presl = Compositae Giseke(保留科名)●■
73234 Carduaceae Dumort. = Asteraceae Bercht. et J. Presl(保留科名)●■
73235 Carduaceae Dumort. = Compositae Giseke(保留科名)●■
73236 Carduaceae Small = Asteraceae Bercht. et J. Presl(保留科名)●■
73237 Carduaceae Small;飞廉科■
73238 Carduncellus Adans.(1763);小飞廉属(类飞廉属)■☆
73239 Carduncellus atlanticus Bonnet et Barratte = Carthamus plumosus (Pomel) Greuter ■☆
73240 Carduncellus atractyloides Batt.;纺锤菊小飞廉■☆
73241 Carduncellus atractyloides Pomel = Carthamus atractyloides (Pomel) Greuter ■☆
73242 Carduncellus atractyloides Pomel var. elatus Chabert = Carthamus atractyloides (Pomel) Greuter ■☆
73243 Carduncellus battandieri Barratte et Chevall. = Carthamus atractyloides (Pomel) Greuter ■☆
73244 Carduncellus caeruleus (L.) A. DC. = Carthamus caeruleus L. ■☆
73245 Carduncellus caeruleus (L.) A. DC. subsp. tingitanus (L.) Rivas Goday et Rivas Mart. = Carthamus caeruleus L. subsp. tingitanus (L.) Batt. ■☆
73246 Carduncellus calvus Boiss. et Reut. = Carthamus calvus (Boiss. et Reut.) Batt. ■☆
73247 Carduncellus calvus Boiss. et Reut. var. glaucescens (Faure et Maire) Hanelt = Carthamus calvus (Boiss. et Reut.) Batt. ■☆
73248 Carduncellus carthamoides (Pomel) Hanelt = Carthamus carthamoides (Pomel) Batt. ■☆
73249 Carduncellus chouletteanus (Pomel) Batt. = Carthamus chouletteanus (Pomel) Greuter ■☆
73250 Carduncellus chouletteanus (Pomel) Batt. var. gracilis Maire = Carthamus pomelianus (Batt.) Batt. ■☆
73251 Carduncellus cryptocephalus Baker = Atractylis cryptocephalus (Baker) F. G. Davies ■☆
73252 Carduncellus eriocephalus Boiss. = Carthamus eriocephalus (Boiss.) Greuter ■☆
73253 Carduncellus eriocephalus Boiss. var. albiflora Gauba = Carthamus eriocephalus (Boiss.) Greuter ■☆
73254 Carduncellus eriocephalus Boiss. var. glaucescens Cavara = Carthamus eriocephalus (Boiss.) Greuter ■☆
73255 Carduncellus eriocephalus Boiss. var. leucanthus Cavara et Trotter = Carthamus eriocephalus (Boiss.) Greuter ■☆
73256 Carduncellus eriocephalus Boiss. var. sulphureus Andr. = Carthamus eriocephalus (Boiss.) Greuter ■☆
73257 Carduncellus fruticosus (Maire) Hanelt = Carthamus fruticosus Maire ■☆
73258 Carduncellus helenoides (Desf.) Hanelt = Carthamus helenoides Desf. ■☆
73259 Carduncellus ilicifolius Pomel = Carthamus ilicifolius (Pomel) Greuter ■☆
73260 Carduncellus lucens (Ball) Ball = Carthamus pinnatus Desf. subsp. lucens (Ball) Dobignard ■☆
73261 Carduncellus mareoticus (Delile) Hanelt = Carthamus mareoticus Delile ■☆
73262 Carduncellus mareoticus (Delile) Hanelt subsp. deserticola Chrtek = Carthamus mareoticus Delile ■☆
73263 Carduncellus multifidus (Desf.) Coss. = Carthamus multifidus Desf. ■☆
73264 Carduncellus pectinatus (Desf.) DC. = Carthamus pectinatus Desf. ■☆
73265 Carduncellus pinnatus (Desf.) DC. = Carthamus pinnatus Desf. ■☆
73266 Carduncellus pinnatus (Desf.) DC. subsp. lucens Ball = Carthamus pinnatus Desf. subsp. lucens (Ball) Dobignard ■☆
73267 Carduncellus pinnatus (Desf.) DC. var. acaulos (C. Presl) Guss. = Carthamus pinnatus Desf. ■☆
73268 Carduncellus pinnatus (Desf.) DC. var. purpureus Maire = Carthamus pinnatus Desf. ■☆
73269 Carduncellus plumosus Pomel = Carthamus plumosus (Pomel) Greuter ■☆
73270 Carduncellus plumosus Pomel var. ilicifolius (Pomel) Batt. = Carthamus plumosus (Pomel) Greuter ■☆
73271 Carduncellus pomelianus (Batt.) Batt. var. gracilis (Maire) Maire = Carthamus pomelianus (Batt.) Batt. ■☆
73272 Carduncellus pomelianus (Batt.) Batt. var. mesatlanticus Maire = Carthamus pomelianus (Batt.) Batt. ■☆
73273 Carduncellus pomelianus (Batt.) Batt. var. rifanus Emb. et Maire = Carthamus pomelianus (Batt.) Batt. ■☆
73274 Carduncellus pomelianus Batt. = Carthamus pomelianus (Batt.) Batt. ■☆
73275 Carduncellus reboudianus Batt. = Carthamus reboudianus Batt. ■☆
73276 Carduncellus rhaponticoides Pomel = Carthamus rhaponticoides (Pomel) Greuter ■☆
73277 Carduncellus strictus (Pomel) Hanelt = Carthamus strictus (Pomel) Batt. ■☆
73278 Carduus L.(1753);飞廉属;Bristlethistle, Bristle-thistle, Chardon, Plumeless Thistle, Thistle ■
73279 Carduus abyssinicus Sch. Bip. = Carduus leptacanthus Fresen. ■☆
73280 Carduus acanthocephalus C. A. Mey.;尖头飞廉■☆
73281 Carduus acanthoides L.;节毛飞廉(藏飞廉,刺飞廉,飞廉,利

刺飞廉); Acanthus Bristle Thistle, Acanthuslike Bristlethistle, Plumeless Thistle, Spiny Plumeless Thistle ■

73282 Carduus acanthoides L. = Carduus crispus L. ■

73283 Carduus acicularis Bertol.;针形飞廉■☆

73284 Carduus adpressus C. A. Mey.;匍匐飞廉■☆

73285 Carduus afromontanus R. E. Fr.;非洲山生飞廉■☆

73286 Carduus afromontanus R. E. Fr. var. breviflorus? = Carduus afromontanus R. E. Fr. ■☆

73287 Carduus ahaicus Patrin ex DC. = Serratula marginata Tausch ■

73288 Carduus albidus M. Bieb.;白飞廉■☆

73289 Carduus algeriensis Munby;阿尔及利亚飞廉■☆

73290 Carduus altissimus L. = Cirsium altissimum (L.) Spreng. ■☆

73291 Carduus altissimus L. = Cirsium altissimum Hill ■☆

73292 Carduus ammophilus Hoffmanns. et Link = Carduus meonanthus Hoffmanns. et Link ■☆

73293 Carduus arabicus Jacq.;阿拉伯飞廉■☆

73294 Carduus arabicus Murray = Carduus arabicus Jacq. ■☆

73295 Carduus arabicus Murray = Carduus pycnocephalus L. subsp. arabicus (Murray) Nyman ■☆

73296 Carduus arctioides Willd.;北方飞廉■☆

73297 Carduus arenarius Desf. = Onopordum arenarium Pomel ■☆

73298 Carduus argentatus L.;银白飞廉■☆

73299 Carduus argentatus L. var. polycephalus Post = Carduus argentatus L. ■☆

73300 Carduus argyrancanthus Wall. = Cirsium argyracanthum DC. ■

73301 Carduus argyroa Biv.;银色飞廉■☆

73302 Carduus armenus Boiss. = Carduus nutans L. ■

73303 Carduus arvensis (L.) Robson = Cirsium arvense (L.) Scop. ■

73304 Carduus arvensis Robins. = Cirsium arvense (L.) Scop. ■

73305 Carduus atlanticus Pomel = Carduus macrocephalus Desf. ■☆

73306 Carduus atlantis Humbert et Maire = Carduus spachianus Durieu subsp. atlantis (Humbert et Maire) Greuter ■☆

73307 Carduus atlantis Humbert et Maire = Carduus spachianus Durieu ■☆

73308 Carduus atlantis Humbert et Maire subsp. megalatlanticus (Emb.) Maire = Carduus spachianus Durieu subsp. megalatlanticus (Emb.) Greuter ■☆

73309 Carduus atriplicifolius Trevis. = Synurus deltoides (Aiton) Nakai ■

73310 Carduus atropatanicus Sosn. ex Grossh.;暗飞廉■☆

73311 Carduus australis L. f. = Carduus pycnocephalus L. subsp. arabicus (Murray) Nyman ■☆

73312 Carduus baeocephalus Webb et Berthel.;小头飞廉■☆

73313 Carduus baeocephalus Webb et Berthel. var. glabra Pit. = Carduus baeocephalus Webb et Berthel. ■☆

73314 Carduus baeocephalus Webb et Berthel. var. tomentosa Pit. = Carduus baeocephalus Webb et Berthel. ■☆

73315 Carduus balansae Boiss. et Reut. = Carduus myriacanthus DC. ■☆

73316 Carduus ballii Hook. f.;鲍尔飞廉■☆

73317 Carduus beckerianus Tamamsch.;贝克尔飞廉■☆

73318 Carduus benedictus Steud.;洋飞廉; Bitter Thistle, Blessed Thistle, Cursed Thistle, Hitter Thistle, Holy Thistle, Ladles' Thistle, Our Lady's Thistle, Spatted Thistle, St. Benedict's Thistle, Virgin Mary's Thistle ■☆

73319 Carduus benedictus Steud. = Cnicus benedictus L. ■

73320 Carduus blepharolepis Chiov. = Carduus ruwenzoriensis S. Moore ■☆

73321 Carduus bourgeanus Boiss. et Reut.;布尔热飞廉■☆

73322 Carduus bourgeaui Kazmi;布尔飞廉■☆

73323 Carduus butagensis De Wild. = Carduus ruwenzoriensis S. Moore ■☆

73324 Carduus candidissimus Greene = Cirsium occidentale (Nutt.) Jeps. var. candidissimum (Greene) J. F. Macbr. ■☆

73325 Carduus canovirens Rydb. = Cirsium cymosum (Greene) J. T. Howell var. canovirens (Rydb.) D. J. Keil ■☆

73326 Carduus carolinianus Walter = Cirsium carolinianum (Walter) Fernald et B. G. Schub. ■☆

73327 Carduus cephalanthus Viv.;头花飞廉■☆

73328 Carduus cernuus Patrin ex Ledeb. = Alfredia cernua (L.) Cass. ■

73329 Carduus cernuus Steud.;灰飞廉■☆

73330 Carduus chamaecephalus (Vatke) Oliv. et Hiern = Carduus schimperi Sch. Bip. ■☆

73331 Carduus chevallieri Barratte;舍瓦飞廉■☆

73332 Carduus cinereus M. Bieb. = Carduus cernuus Steud. ■☆

73333 Carduus collinus Waldst. et Kit.;山飞廉■☆

73334 Carduus coloradensis Rydb. = Cirsium scariosum Nutt. var. coloradense (Rydb.) D. J. Keil ■☆

73335 Carduus coloratus Tamamsch. = Carduus nutans L. ■

73336 Carduus crassicaulis Greene = Cirsium crassicaule (Greene) Jeps. ■☆

73337 Carduus crispus L.;丝毛飞廉(刺打草,刺飞廉,刺盖,大蓟,大力王,飞帘,飞廉,飞廉蒿,飞轻,飞雉,枫头棵,伏兔,伏猪,红花草,卷飞廉,老牛锉,老牛错,雷公菜,漏芦蒿,木禾,天荠,小蓟); Curled Thistle, Curly Bristle Thistle, Curly Bristlethistle, Curly Plumeless Thistle, Welted Thistle ■

73338 Carduus crispus L. = Carduus acanthoides L. ■

73339 Carduus crispus L. subsp. agrestis (Kern.) Vollm.;田野丝毛飞廉■☆

73340 Carduus crispus L. subsp. agrestis (Kern.) Vollm. f. albus (Makino) H. Hara;白花田野丝毛飞廉■☆

73341 Carduus cymosus Greene = Cirsium cymosum (Greene) J. T. Howell ■☆

73342 Carduus desertorum Fisch. ex Steud. = Cirsium alatum (S. G. Gmel.) Bobrov ■

73343 Carduus discolor (Muhl. ex Willd.) Nutt. = Cirsium discolor (Muhl. ex Willd.) Spreng. ■☆

73344 Carduus duriaei Boiss. et Reut. = Carduus meonanthus Hoffmanns. et Link ■☆

73345 Carduus echinatus Desf. = Cirsium echinatum (Desf.) DC. ■☆

73346 Carduus ellenbeckii R. E. Fr. = Carduus schimperi Sch. Bip. ■☆

73347 Carduus eremocephalus Chiov. = Carduus schimperi Sch. Bip. ■☆

73348 Carduus euosmus Forrest ex W. W. Sm. = Xanthopappus subacaulis C. Winkl. ■

73349 Carduus flodmanii Rydb. = Cirsium flodmanii (Rydb.) Arthur ■☆

73350 Carduus foliosus Hook. = Cirsium foliosum (Hook.) DC. ■☆

73351 Carduus giganteus Desf. = Cirsium scabrum (Poir.) Bonnet ■☆

73352 Carduus gigas Ucria;巨飞廉■☆

73353 Carduus glaber Nutt. = Cirsium nuttallii DC. ■☆

73354 Carduus glaucus Baumg.;灰绿飞廉■☆

73355 Carduus granatensis Willk. var. maurus (Emb. et Maire) Pau = Carduus platypus Lange subsp. maurus (Emb. et Maire) Maire ■☆

73356 Carduus hajastanicus Tamamsch.;哈贾斯坦飞廉■☆

73357 Carduus hamulosus Ehrh.;钩飞廉■☆

73358 Carduus helenioides L. = Cirsium helenioides (L.) Hill ■

73359 Carduus helleri Small = Cirsium undulatum (Nutt.) Spreng. ■☆

73360 Carduus hookerianus Nutt. var. eriocephalus (A. Gray) A. Nelson = Cirsium eatonii (A. Gray) B. L. Rob. var. eriocephalum (A. Gray) D. J. Keil ■☆
73361 Carduus horridus B. Fedtsch. = Schmalhausenia nidulans (Regel) Petr. ■
73362 Carduus hydrophilus Greene = Cirsium hydrophilum (Greene) Jeps. ■☆
73363 Carduus hystrix C. A. Mey.;豪猪飞廉■☆
73364 Carduus inamoenus Greene = Cirsium inamoenum (Greene) D. J. Keil ■☆
73365 Carduus japonicus (DC.) Franch. = Cirsium japonicum Fisch. ex DC. ■
73366 Carduus japonicus (Fisch. ex DC.) Franch. = Cirsium japonicum Fisch. ex DC. ■
73367 Carduus japonicus Franch. = Cirsium japonicum Fisch. ex DC. ■
73368 Carduus kahenae Pomel = Carduus macrocephalus Desf. ■☆
73369 Carduus karelinii B. Fedtsch. = Alfredia nivea Kar. et Kir. ■
73370 Carduus keniensis R. E. Fr.;肯尼飞廉■☆
73371 Carduus keniensis R. E. Fr. subsp. elgonensis (R. E. Fr.) Kazmi = Carduus keniensis R. E. Fr. ■☆
73372 Carduus keniensis R. E. Fr. subsp. hedbergii Kazmi = Carduus keniensis R. E. Fr. ■☆
73373 Carduus keniensis R. E. Fr. subsp. kilimandscharicus (R. E. Fr.) Kazmi = Carduus keniensis R. E. Fr. ■☆
73374 Carduus keniensis R. E. Fr. var. aberdaricus? = Carduus keniensis R. E. Fr. ■☆
73375 Carduus keniensis R. E. Fr. var. elgonensis? = Carduus keniensis R. E. Fr. ■☆
73376 Carduus keniensis R. E. Fr. var. kilimandscharicus (R. E. Fr.) Hedberg = Carduus keniensis R. E. Fr. ■☆
73377 Carduus kerneri Simonk.;克纳飞廉;Kerner Bristle Thistle ■☆
73378 Carduus kikuyorum R. E. Fr. = Carduus nyassanus (S. Moore) R. E. Fr. subsp. kikuyorum (R. E. Fr.) C. Jeffrey ■☆
73379 Carduus kikuyorum R. E. Fr. var. goetzei? = Carduus nyassanus (S. Moore) R. E. Fr. subsp. kikuyorum (R. E. Fr.) C. Jeffrey ■☆
73380 Carduus kilimandscharicus R. E. Fr. = Carduus keniensis R. E. Fr. ■☆
73381 Carduus lanatus Roxb. ex Willd. = Breea arvensis (L.) Less. ■
73382 Carduus lanatus Roxb. ex Willd. = Cirsium lanatum (Roxb. ex Willd.) Spreng. ■
73383 Carduus lanceolatus L. = Cirsium vulgare (Savi) Ten. ■
73384 Carduus lanipes C. Winkl. = Olgaea lanipes (C. Winkl.) Iljin ■
73385 Carduus lanuriensis De Wild. = Carduus ruwenzoriensis S. Moore ■☆
73386 Carduus lecontei (Torr. et A. Gray) Pollard = Cirsium lecontei Torr. et A. Gray ■☆
73387 Carduus leptacanthus Fresen.;细刺飞廉■☆
73388 Carduus leptacanthus Fresen. var. nyassanus S. Moore = Carduus nyassanus (S. Moore) R. E. Fr. ■☆
73389 Carduus leptacanthus Fresen. var. steudneri Engl. = Carduus nyassanus (S. Moore) R. E. Fr. subsp. kikuyorum (R. E. Fr.) C. Jeffrey ■☆
73390 Carduus leptocladus Durieu;细枝飞廉■☆
73391 Carduus leucographus L. = Tyrimnus leucographus (L.) Cass. ■☆
73392 Carduus leucophyllus Turcz. = Olgaea leucophylla (Turcz.) Iljin ■
73393 Carduus linearis Thunb. = Cirsium lineare (Thunb.) Sch. Bip. ■
73394 Carduus linearis Thunb. ex A. Murray = Cirsium lineare (Thunb.) Sch. Bip. ■
73395 Carduus lomonosowii Trautv. = Olgaea lomonosowii (Trautv.) Iljin ■
73396 Carduus macounii Greene = Cirsium edule Nutt. var. macounii (Greene) D. J. Keil ■☆
73397 Carduus macracanthus Sch. Bip. ex Kazmi;大花飞廉■☆
73398 Carduus macrocephalus Desf.;大头飞廉;Giant Thistle ■☆
73399 Carduus macrocephalus Desf. = Carduus nutans L. ■
73400 Carduus macrocephalus Desf. subsp. inconstrictus (O. Schwarz) Kazmi;缢缩大头飞廉■☆
73401 Carduus macrocephalus Desf. subsp. scabrisquamus (Arènes) Kazmi;糙鳞飞廉■☆
73402 Carduus macrocephalus Desf. var. atlanticus (Pomel) Maire = Carduus macrocephalus Desf. ■☆
73403 Carduus macrocephalus Desf. var. kabylicus Maire = Carduus macrocephalus Desf. ■☆
73404 Carduus macrocephalus Desf. var. kahenae (Pomel) Batt. = Carduus macrocephalus Desf. ■☆
73405 Carduus macrocephalus Desf. var. narcissii Sennen = Carduus ballii Hook. f. ■☆
73406 Carduus macrolepis Peterm. = Carduus nutans L. ■
73407 Carduus mariae Crantz = Silybum marianum (L.) Gaertn. ■
73408 Carduus marianus L. = Silybum marianum (L.) Gaertn. ■
73409 Carduus maroccanus (Arènes) Kazmi;摩洛哥飞廉■☆
73410 Carduus martinezii Pau;马丁内斯飞廉■☆
73411 Carduus megacephalus (A. Gray) Nutt. = Cirsium undulatum (Nutt.) Spreng. ■☆
73412 Carduus megalatlanticus Emb. = Carduus spachianus Durieu subsp. megalatlanticus (Emb.) Greuter ■☆
73413 Carduus meonanthus Hoffmanns. et Link;细花飞廉■☆
73414 Carduus meonanthus Hoffmanns. et Link subsp. spachianus (Durieu) Maire = Carduus spachianus Durieu ■☆
73415 Carduus meonanthus Hoffmanns. et Link var. duriaei (Boiss. et Reut.) Maire = Carduus meonanthus Hoffmanns. et Link ■☆
73416 Carduus millefolius R. E. Fr.;粟草叶飞廉■☆
73417 Carduus mohavensis Greene = Cirsium mohavense (Greene) Petr. ■☆
73418 Carduus multijugus K. Koch;多对飞廉■☆
73419 Carduus muticus (Michx.) Pers. = Cirsium muticum Michx. ■☆
73420 Carduus muticus Pers.;秃飞廉;Swamp Thistle Thistle ■☆
73421 Carduus myriacanthus DC.;多刺飞廉■☆
73422 Carduus myriacanthus DC. var. submyriacanthus Maire = Carduus myriacanthus DC. ■☆
73423 Carduus nanus R. E. Fr. = Carduus schimperi Sch. Bip. subsp. nanus (R. E. Fr.) C. Jeffrey ■☆
73424 Carduus narcissii Sennen = Carduus ballii Hook. f. ■☆
73425 Carduus navaschinii Bordz.;那瓦飞廉■☆
73426 Carduus nervosus K. Koch;多脉飞廉■☆
73427 Carduus nikitinii Tamamsch.;尼氏飞廉■☆
73428 Carduus numidicus Coss. et Durieu var. propinquus Pomel = Carduus numidicus Durieu ■☆
73429 Carduus numidicus Durieu;努米底亚飞廉■☆
73430 Carduus nutans L.;飞廉(垂花飞廉,垂头飞廉,俯垂飞廉,具色飞廉,麝香飞廉);Bank Thistle, Buck Thistle, Common Bristlethistle, Father Bigface, Musk Bristle Thistle, Musk Bristlethistle, Musk Bristle-thistle, Musk Thistle, Nodding Plumeless Thistle, Nodding Thistle, Queen Anne's Thistle, Scotch Thistle, Teaser ■

73431 Carduus nutans L. subsp. leiophyllus (Petrovic) Stoj. et Stefani = Carduus nutans L. ■

73432 Carduus nutans L. subsp. macrocephalus (Desf.) K. Richt. = Carduus macrocephalus Desf. ■☆

73433 Carduus nutans L. subsp. macrocephalus (Desf.) Nyman = Carduus nutans L. ■

73434 Carduus nutans L. subsp. macrolepis (Peterm.) Kazmi = Carduus nutans L. ■

73435 Carduus nutans L. subsp. maroccanus Arènes = Carduus maroccanus (Arènes) Kazmi ■☆

73436 Carduus nutans L. subsp. maurus (Emb. et Maire) Greuter = Carduus maroccanus (Arènes) Kazmi ■☆

73437 Carduus nutans L. subsp. numidicus (Coss. et Durieu) Arènes = Carduus numidicus Durieu ■☆

73438 Carduus nutans L. subsp. scabrisquamus Arènes = Carduus macrocephalus Desf. subsp. scabrisquamus (Arènes) Kazmi ■☆

73439 Carduus nutans L. var. leiophyllus (Petrovic) Arènes = Carduus nutans L. ■

73440 Carduus nutans L. var. macrocephalus (Desf.) B. Boivin = Carduus nutans L. ■

73441 Carduus nutans L. var. songaricus C. Winkl. = Carduus nutans L. ■

73442 Carduus nutans L. var. vestitus (Hallier) B. Boivin = Carduus nutans L. ■

73443 Carduus nuttallii (DC.) Pollard = Cirsium nuttallii DC. ■☆

73444 Carduus nyassanus (S. Moore) R. E. Fr.;尼萨飞廉■☆

73445 Carduus nyassanus (S. Moore) R. E. Fr. subsp. kikuyorum (R. E. Fr.) C. Jeffrey;菊阳飞廉■☆

73446 Carduus nyassanus (S. Moore) R. E. Fr. var. ruandensis R. E. Fr. = Carduus nyassanus (S. Moore) R. E. Fr. subsp. kikuyorum (R. E. Fr.) C. Jeffrey ■☆

73447 Carduus oblanceolatus Rydb. = Cirsium flodmanii (Rydb.) Arthur ■☆

73448 Carduus occidentalis Nutt. = Cirsium occidentale (Nutt.) Jeps. ■☆

73449 Carduus onopordioides Fisch. ex M. Bieb.;大翅蓟飞廉■☆

73450 Carduus osterhoutii Rydb. = Cirsium clavatum (M. E. Jones) Petr. var. osterhoutii (Rydb.) D. J. Keil ■☆

73451 Carduus palustris L. = Cirsium palustre (L.) Scop. ■☆

73452 Carduus pectinatus Popov = Olgaea pectinata Iljin ■

73453 Carduus perplexans Rydb. = Cirsium perplexans (Rydb.) Petr. ■☆

73454 Carduus personatus (L.) Jacq.;沼地飞廉(人面飞廉);Great Marsh Thistle ■☆

73455 Carduus platyphyllus R. E. Fr. = Carduus schimperi Sch. Bip. subsp. platyphyllus (R. E. Fr.) C. Jeffrey ■☆

73456 Carduus platypus Lange subsp. adpressus Andr. = Carduus maroccanus (Arènes) Kazmi ■☆

73457 Carduus platypus Lange subsp. maurus (Emb. et Maire) Maire;晚熟飞廉■☆

73458 Carduus poliochrus Trautv.;灰色飞廉■☆

73459 Carduus pseudocollinus (Schmalh.) Klokov;假山飞廉■☆

73460 Carduus pteracanthus Durieu = Carduus spachianus Durieu ■☆

73461 Carduus pteracanthus Durieu var. erythrolepis? = Carduus spachianus Durieu ■☆

73462 Carduus pteracanthus Durieu var. leptocladus (Durieu) Batt. = Carduus spachianus Durieu ■☆

73463 Carduus pteracanthus Durieu var. submyriacanthus Maire = Carduus spachianus Durieu ■☆

73464 Carduus pteracanthus Durieu var. tunetanus Murb. = Carduus spachianus Durieu ■☆

73465 Carduus pulcherrimus Rydb. = Cirsium pulcherrimum (Rydb.) K. Schum. ■☆

73466 Carduus pumilus Nutt. = Cirsium pumilum Spreng. ■☆

73467 Carduus pycnocephalus L.;密头飞廉(小头飞廉);Italian Plumeless Thistle, Italian Thistle, Plymouth Thistle ■☆

73468 Carduus pycnocephalus L. subsp. arabicus (Murray) Nyman = Carduus arabicus Murray ■☆

73469 Carduus pycnocephalus L. subsp. tenuiflorus (Curtis) Batt. = Carduus tenuiflorus Curtis ■☆

73470 Carduus pycnocephalus L. var. bourgeanus (Boiss. et Reut.) Maire = Carduus meonanthus Hoffmanns. et Link ■☆

73471 Carduus pycnocephalus L. var. corbariensis (Timb.-Lagr. et Thévenau) Rouy = Carduus tenuiflorus Curtis ■☆

73472 Carduus pycnocephalus L. var. spinulosus Pau = Carduus tenuiflorus Curtis ■☆

73473 Carduus pycnocephalus L. var. tenuiflorus (Curtis) Ball = Carduus tenuiflorus Curtis ■☆

73474 Carduus pycnocephalus L. var. tenuiflorus (Curtis) Fiori = Carduus tenuiflorus Curtis ■☆

73475 Carduus remotifolius Hook. = Cirsium remotifolium (Hook.) DC. ■☆

73476 Carduus repandus (Michx.) Pers. = Cirsium repandum Michx. ■☆

73477 Carduus ruwenzoriensis S. Moore;鲁文佐里飞廉■☆

73478 Carduus ruwenzoriensis S. Moore var. lanuriensis (De Wild.) Hedberg = Carduus ruwenzoriensis S. Moore ■☆

73479 Carduus scaber Poir. = Cirsium scabrum (Poir.) Bonnet ■☆

73480 Carduus schimperi Sch. Bip.;欣珀飞廉■☆

73481 Carduus schimperi Sch. Bip. subsp. nanus (R. E. Fr.) C. Jeffrey;矮欣珀飞廉■☆

73482 Carduus schimperi Sch. Bip. subsp. platyphyllus (R. E. Fr.) C. Jeffrey;宽叶欣珀飞廉■☆

73483 Carduus schischkinii Tamamsch. = Carduus nutans L. ■

73484 Carduus segetum (Bunge) Franch. = Cirsium setosum (Willd.) M. Bieb. ■

73485 Carduus semiensis Pic. Serm. = Carduus leptacanthus Fresen. ■☆

73486 Carduus seminudus M. Bieb.;半裸飞廉■☆

73487 Carduus serratuloides L. = Cirsium serratuloides (L.) Hill ■

73488 Carduus sinensis S. Moore = Olgaea lomonosowii (Trautv.) Iljin ■

73489 Carduus smallii (Britton) H. E. Ahles = Cirsium horridulum Michx. var. vittatum (Small) R. W. Long ■☆

73490 Carduus songaricus (C. Winkl.) Tamamsch. = Carduus nutans L. ■

73491 Carduus songaricus C. Winkl. = Carduus nutans L. ■

73492 Carduus songoricus Tamamsch.;准噶尔飞廉■☆

73493 Carduus spachianus Durieu;榉飞廉■☆

73494 Carduus spachianus Durieu subsp. atlantis (Humbert et Maire) Greuter;亚特兰大飞廉■☆

73495 Carduus spachianus Durieu subsp. megalatlanticus (Emb.) Greuter;大柱飞廉■☆

73496 Carduus spinosissimus Walter = Cirsium horridulum Michx. ■☆

73497 Carduus squarrosus (DC.) Lowe;粗鳞飞廉■☆

73498 Carduus stenocephalus Tamamsch.;窄头飞廉■☆

73499 Carduus steudneri (Engl.) R. E. Fr. = Carduus nyassanus (S. Moore) R. E. Fr. subsp. kikuyorum (R. E. Fr.) C. Jeffrey ■☆

73500 Carduus steudneri (Engl.) R. E. Fr. subsp. buchingeri Kazmi = Carduus nyassanus (S. Moore) R. E. Fr. subsp. kikuyorum (R. E.

Fr.) C. Jeffrey ■☆
73501 Carduus stolzii R. E. Fr. = Carduus leptacanthus Fresen. ■☆
73502 Carduus subalpinus R. E. Fr. = Carduus keniensis R. E. Fr. ■☆
73503 Carduus syriacus L. = Notobasis syriaca (L.) Cass. ■☆
73504 Carduus tenuiflorus Curtis;纤花飞廉;Seaside Thistle, Slender Bristlethistle, Slender-flowered Thistle, Winged Plumeless Thistle ■☆
73505 Carduus theodori R. E. Fr. = Carduus schimperi Sch. Bip. ■☆
73506 Carduus theodori R. E. Fr. var. serratus? = Carduus schimperi Sch. Bip. ■☆
73507 Carduus theodori R. E. Fr. var. serrulatus? = Carduus schimperi Sch. Bip. ■☆
73508 Carduus thoermeri Weinm.;帖氏飞廉■☆
73509 Carduus thoermeri Weinm. = Carduus nutans L. ■
73510 Carduus thoermeri Weinm. subsp. numidicus (Coss. et Durieu) Kazmi = Carduus numidicus Durieu ■☆
73511 Carduus thomsonii Hook. f. = Olgaea thomsonii (Hook. f.) Iljin ■
73512 Carduus tianschanicum B. Fedtsch. = Alfredia acantholepis Kar. et Kir. ■
73513 Carduus tracyi Rydb. = Cirsium tracyi (Rydb.) Petr. ■☆
73514 Carduus transcaspicus Gand.;里海飞廉■☆
73515 Carduus tsianschanicus B. Fedtsch. = Alfredia acantholepis Kar. et Kir. ■
73516 Carduus uncinatus M. Bieb.;钩髯飞廉■☆
73517 Carduus undulatus Nutt. = Cirsium undulatum (Nutt.) Spreng. ■☆
73518 Carduus undulatus Nutt. var. canescens (Nutt.) Porter = Cirsium canescens Nutt. ■☆
73519 Carduus undulatus Nutt. var. nevadensis Greene = Cirsium inamoenum (Greene) D. J. Keil ■☆
73520 Carduus uniflorus Turcz. ex DC. = Serratula marginata Tausch ■
73521 Carduus validus Greene = Cirsium scariosum Nutt. var. citrinum (Petr.) D. J. Keil ■☆
73522 Carduus venustus Greene = Cirsium occidentale (Nutt.) Jeps. var. venustum (Greene) Jeps. ■☆
73523 Carduus versonatus (L.) Jacq.;大沼生飞廉;Great Marsh Bristlethistle ■☆
73524 Carduus vinaceus Wooton et Standl. = Cirsium vinaceum (Wooton et Standl.) Wooton et Standl. ■☆
73525 Carduus virginianus L. = Cirsium virginianum (L.) Michx. ■☆
73526 Carduus vittatum (Small) Small = Cirsium horridulum Michx. var. vittatum (Small) R. W. Long ■☆
73527 Carduus vittatus Small = Cirsium horridulum Michx. var. vittatum (Small) R. W. Long ■☆
73528 Carduus vulgaris Sav. = Cirsium vulgare (Savi) Ten. ■
73529 Cardwellia F. Muell. (1865);澳洲银桦树属(昆士兰山龙眼属);Australian Silky-oak ●☆
73530 Cardwellia sublimis F. Muell.;澳洲银桦树;Australian Silky-oak, Silk Oak, Silky Oak ●☆
73531 Carelia Adans. = Ageratum L. ■●
73532 Carelia Cav. = Mikania Willd. (保留属名)■
73533 Carelia Fabr. = Ageratum L. ■●
73534 Carelia Juss. ex Cav. = Mikania Willd. (保留属名)■
73535 Carelia Less. = Radlkoferotoma Kuntze ●☆
73536 Carelia Moehring = Ageratum L. ■●
73537 Carelia Moehring = Carelia Adans. ■●
73538 Carelia Ponted. ex Fabr. = Ageratum L. ■●
73539 Carenidium Baptista = Oncidium Sw. (保留属名)■☆
73540 Carenophila Ridl. = Geostachys (Baker) Ridl. ■☆
73541 Cares Dombey ex Lam. = Embothrium J. R. Forst. et G. Forst. ●☆
73542 Careum Adans. = Carum L. ■
73543 Carex L. (1753);苔草属(苔属);New Zealand Sedge, Sedge ■
73544 Carex × akitaensis Fujiw.;秋田苔草■☆
73545 Carex × arakanei T. Koyama;荒金苔草■☆
73546 Carex × caudatifrons Akiyama;尾叶苔草■☆
73547 Carex × ciliatifructa Ohwi;缘毛果苔草■☆
73548 Carex × dandoensis T. Koyama;段户苔草■☆
73549 Carex × enshuensis T. Koyama;远州苔草■☆
73550 Carex × fulleri H. E. Ahles = Carex heterostachya Bunge ■
73551 Carex × furusei T. Koyama;古施苔草■☆
73552 Carex × goroi T. Koyama;后吕苔草■☆
73553 Carex × goyozanensis M. Kikuchi et Mochizuki;五叶山苔草■☆
73554 Carex × hanasakensis T. Koyama;花咲苔草■☆
73555 Carex × hokariensis T. Koyama;小山苔草■☆
73556 Carex × hosoii T. Koyama;细井苔草■☆
73557 Carex × inaensis T. Koyama;伊那苔草■☆
73558 Carex × kurilensis Ohwi;千岛苔草■☆
73559 Carex × leiogona Franch.;光蕊苔草■☆
73560 Carex × macilenta Fr.;细瘦苔草■☆
73561 Carex × mihashii T. Koyama;三阶苔草■☆
73562 Carex × moriyoshiensis Fujiw. et Y. Matsuda;森吉苔草■☆
73563 Carex × musashiensis Ohwi;武蔵苔草■☆
73564 Carex × nikaii T. Koyama;二阶苔草■☆
73565 Carex × okushirensis Akiyama;奥尻苔草■☆
73566 Carex × paludicola Ohwi;沼生苔草■☆
73567 Carex × pseudoaphanolepis Ohwi;假隐鳞苔草■☆
73568 Carex × pseudosadoensis Akiyama;假佐渡苔草■☆
73569 Carex × rikuchiuensis Akiyama;陆中苔草■☆
73570 Carex × sakaguchii Ohwi;坂口苔草■☆
73571 Carex × shakushizawensis Akiyama;沙库苔草■☆
73572 Carex × subnigra Lepage = Carex vacillans Drejer ■☆
73573 Carex × sumikawaensis Fujiw. et Y. Matsuda;澄川苔草■☆
73574 Carex × takoensis Y. Endo et Yashiro;田子苔草■☆
73575 Carex × uzenensis Koidz.;羽前苔草■☆
73576 Carex × xenostachya T. Koyama;外穗苔草■☆
73577 Carex aa Kom.;阿氏苔草■☆
73578 Carex abacta L. H. Bailey = Carex michauxiana Boeck. ■☆
73579 Carex abbreviata J. D. Prescott = Carex torreyi Tuck. ■☆
73580 Carex abdita E. B. Bicknell = Carex umbellata Schkuhr ex Willd. ■☆
73581 Carex abramsii Mack. = Carex lemmonii W. Boott ■☆
73582 Carex abscondita Mack.;模糊苔草;Sedge ■☆
73583 Carex abyssinica Chiov. = Carex cognata Kunth var. abyssinica (Chiov.) Lye ■☆
73584 Carex accedens T. Holm = Carex aperta Boott ■
73585 Carex accrescens Ohwi = Carex pallida C. A. Mey. ■
73586 Carex acocksii C. Archer;阿科苔草■☆
73587 Carex acrifolia V. I. Krecz.;顶叶苔草■☆
73588 Carex acrolepis Ledeb. = Carex gmelinii Hook. et Arn. ■
73589 Carex acuta L.;急尖苔草(褐鞘苔草,尖苔);Acute Sedge, Shaped Sedge, Slender Tufted-sedge ■
73590 Carex acuta L. var. appendiculata Trautv. = Carex appendiculata (Trautv.) Kük. ■
73591 Carex acuta L. var. appendiculata Trautv. et C. A. Mey. = Carex appendiculata (Trautv. et C. A. Mey.) Kük. ■
73592 Carex acuta L. var. erecta Dewey = Carex haydenii Dewey ■☆

73593 Carex acuta L. var. nigra L. = Carex nigra (L.) Reichard ■☆
73594 Carex acuta L. var. sparsiflora Dewey = Carex torta Boott ■☆
73595 Carex acuta Poir. = Carex hispida Willd. ■☆
73596 Carex acutatiformis H. E. Hess;锐苔草■☆
73597 Carex acutiformis Ehrh.; 刺叶苔草(尖形苔); Lesser Pond Sedge, Lesser Pond-sedge, Marsh Sedge, Spiny-leaved Sedge ■
73598 Carex acutiformis Ehrh. = Carex paludosa Gooden. ■
73599 Carex acutina L. H. Bailey = Carex nudata W. Boott ■☆
73600 Carex acutina L. H. Bailey var. tenuior L. H. Bailey = Carex aperta Boott ■
73601 Carex acutinella Mack. = Carex aquatilis Wahlenb. ■☆
73602 Carex adelostoma Krecz.;暗苔草(暗苔)■☆
73603 Carex adrienii E. G. Camus; 广东苔草; Guangdong Sedge, Kwangtung Sedge ■
73604 Carex adusta Boott; 褐卵叶苔草; Brown Oval Sedge, Burnt Sedge, Lesser Brown Sedge ■☆
73605 Carex aenea Fernald = Carex foenea Willd. ■☆
73606 Carex aequa C. B. Clarke = Carex serratodens W. Boott ■☆
73607 Carex aequialta Kük. = Carex aequialta Kük. ex Boott ■
73608 Carex aequialta Kük. ex Boott;等高苔草■
73609 Carex aequivoca V. I. Krecz. = Carex aequialta Kük. ex Boott ■
73610 Carex aestivalis M. A. Curtis ex A. Gray;夏苔草;Summer Sedge ■☆
73611 Carex aethiopica Schkuhr;埃塞俄比亚苔草■☆
73612 Carex aethiopica Schkuhr var. stolonifera Boeck. = Carex simensis Hochst. ex A. Rich. var. stolonifera (Boeck.) Kük. ■☆
73613 Carex affinis R. Br. = Kobresia myosuroides (Vill.) Fiori et Paol. ■
73614 Carex agglomerata C. B. Clarke; 团穗苔草(团序苔草); Congested Sedge ■
73615 Carex agglomerata C. B. Clarke var. rhizomata Y. C. Yang;疏丛苔草;Rhizomate Congested Sedge ■
73616 Carex agglomerata Mack. = Carex aggregata Mack. ■☆
73617 Carex aggregata Mack.;集生苔草;Glomerate Sedge, Sedge ■☆
73618 Carex akanensis Franch. = Carex drymophila Turcz. ex Steud. var. abbreviata (Kük.) Ohwi ■
73619 Carex akiensis Okamoto;秋苔草■☆
73620 Carex alajica Litv.;葱岭苔草;Congling Sedge ■
73621 Carex alata Torr.;翅苔草;Sedge ■☆
73622 Carex alba Scop.; 白鳞苔草(白苔,白苔草); White-scaled Sedge ■
73623 Carex alba Scop. subsp. ussuriensis (Kom.) Kük. = Carex ussuriensis Kom. ■
73624 Carex albata Boott ex Franch. = Carex nubigena D. Don ex Okamoto et Taylor subsp. albata (Boott ex Franch. et Sav.) T. Koyama ■
73625 Carex albata Boott ex Franch. et Sav. = Carex nubigena D. Don ex Okamoto et Taylor subsp. albata (Boott ex Franch. et Sav.) T. Koyama ■
73626 Carex albicans Willd. ex Spreng.;西方灰白苔草;Blunt-scaled Oak Sedge, Dry Woods Sedge, Oak Sedge, Sedge, White-tinge Sedge ■☆
73627 Carex albicans Willd. ex Spreng. var. emmonsii (Dewey ex Torr.) Rettig;爱蒙灰白苔草;Emmons' Sedge, Oak Sedge, Sharp-scaled Oak Sedge ■☆
73628 Carex albida L. H. Bailey;白苔草;White Sedge ■☆
73629 Carex albidibasis T. Koyama = Carex japonica Thunb. ■
73630 Carex albolutescens Schwein.;黄白苔草;Sedge ■☆
73631 Carex albomas C. B. Clarke = Carex pisiformis Boott ■
73632 Carex albula Allan;澳洲灰白苔草;Frosted Curls, Sedge ■☆
73633 Carex albursina E. Sheld.;北美苔草;Blunt-scaled Wood Sedge, Sedge, White Bear Sedge, White-bear Sedge ■☆
73634 Carex alexeenkoana Litv.;刺苞苔草;Spiny-bracted Sedge ■
73635 Carex allegheniensis Mack. = Carex debilis Michx. var. rudgei L. H. Bailey ■☆
73636 Carex alliiformis C. B. Clarke; 葱状苔草(林下苔); Alliform Sedge ■
73637 Carex alliiformis C. B. Clarke = Carex metallica H. Lév. et Vaniot ■
73638 Carex allivescens V. I. Krecz.;祁连苔草;Qilianshan Sedge ■
73639 Carex alopecoidea Tuck.;褐头苔草;Brown-headed Fox Sedge, Foxtail Sedge ■☆
73640 Carex alopecuroides D. Don = Carex alopecuroides D. Don ex Okamoto et Taylor ■
73641 Carex alopecuroides D. Don ex Okamoto et Taylor;禾状苔草(川上氏苔,大穗日本苔,高山日本苔,高山穗序苔);Brown-headed Fox Sedge, Foxtail Sedge ■
73642 Carex alopecuroides D. Don ex Okamoto et Taylor subsp. subtransvera (C. B. Clarke) T. Koyama = Carex subtransversa C. B. Clarke ■
73643 Carex alopecuroides D. Don ex Okamoto et Taylor subsp. subtransversa (C. B. Clarke) T. Koyama;高山日本苔■
73644 Carex alopecuroides D. Don ex Okamoto et Taylor subsp. subtransversa (C. B. Clarke) T. Koyama = Carex alopecuroides D. Don ex Okamoto et Taylor ■
73645 Carex alopecuroides D. Don ex Okamoto et Taylor var. chlorostachya C. B. Clarke = Carex doniana Spreng. ■
73646 Carex alopecuroides D. Don ex Okamoto et Taylor var. chlorostachys C. B. Clarke;绿穗禾状苔草■☆
73647 Carex alpina Lilj. = Carex norvegica Retz. ■☆
73648 Carex alpina Lilj. var. holostoma (Drejer) L. H. Bailey = Carex holostoma Drejer ■☆
73649 Carex alpina Lilj. var. inferalpina Wahlenb. = Carex media R. Br. ex Richardson ■☆
73650 Carex alpina Lilj. var. stevenii T. Holm = Carex stevenii (T. Holm) Kalela ■☆
73651 Carex alpina Sw. = Carex angarae Steud. ■
73652 Carex alpina Sw. ex Wahlenb.;高山苔草;Alpine Sedge ■☆
73653 Carex alpina Sw. ex Wahlenb. = Carex angarae Steud. ■
73654 Carex alpina Sw. ex Wahlenb. subsp. infuscata (Nees) Boott var. gracilenta (Boott ex Strachey) Kük. = Carex infuscata Nees var. gracilenta (Boott ex Strachey) P. C. Li ■
73655 Carex alpina Sw. var. gracilenta (Boott ex Strachey) C. B. Clarke = Carex infuscata Nees var. gracilenta (Boott ex Strachey) P. C. Li ■
73656 Carex alpina Sw. var. infuscata (Nees) Boott = Carex infuscata Nees var. gracilenta (Boott ex Strachey) P. C. Li ■
73657 Carex alpine Sw. subsp. infuscate Kük. var. gracilenta (Boott ex Strachey) Kük. = Carex infuscata Nees var. gracilenta (Boott ex Strachey) P. C. Li ■
73658 Carex alta Boott;高秆苔草;Highstem Sedge ■
73659 Carex alta Boott var. latialata Kük. = Carex alta Boott var. latialata Kük. ex Hand. -Mazz. ■
73660 Carex alta Boott var. latialata Kük. ex Hand. -Mazz.;有翅高秆苔草;Winged Highstem Sedge ■
73661 Carex alta Boott var. latialata Kük. ex Hand. -Mazz. = Carex alta Boott ■

73662 Carex altaica Gorodkov;阿尔泰苔草;Altai Sedge ■
73663 Carex alterniflora Franch.;宜兰宿柱苔草(线叶宿柱苔)■
73664 Carex alterniflora Franch. = Carex pisiformis Boott ■
73665 Carex alterniflora Franch. var. elongatula Ohwi = Carex sachalinensis Eastw. var. elongatula (Ohwi) Ohwi ■☆
73666 Carex alterniflora Franch. var. fulva Ohwi;褐色宜兰宿柱苔草■
73667 Carex alticola Popl. ex Sukaczev;高原苔草■☆
73668 Carex amblyorhyncha V. I. Krecz.;钝喙苔草■☆
73669 Carex amblyorhyncha V. I. Krecz. = Carex marina Dewey ■☆
73670 Carex ambusta Boott = Carex saxatilis L. ■☆
73671 Carex amgunensis Eastw. = Carex amgunensis F. Schmidt ■
73672 Carex amgunensis Eastw. var. chloroleuca (Meinsh.) Kük. = Carex amgunensis Eastw. ■
73673 Carex amgunensis F. Schmidt;阿姆贡苔草(阿姆贡苔,膨柱苔草,球穗苔草);Inflatedstyle Sedge ■
73674 Carex amgunensis F. Schmidt var. chloroleuca (Meinsh.) Kük. = Carex amgunensis F. Schmidt ■
73675 Carex amphibola Steud.;东部窄叶苔草;Eastern Narrow-leaved Sedge, Sedge ■☆
73676 Carex amphibola Steud. = Carex corrugata Fernald ■☆
73677 Carex amphibola Steud. var. globosa (L. H. Bailey) L. H. Bailey = Carex bulbostylis Mack. ■☆
73678 Carex amphibola Steud. var. rigida (L. H. Bailey) Fernald = Carex planispicata Naczi ■☆
73679 Carex amphibola Steud. var. turgida Fernald = Carex grisea Wahlenb. ■☆
73680 Carex amphigena (Fernald) Mack. = Carex glareosa Wahlenb. ■☆
73681 Carex amplifolia Boott subsp. dispalata (Boott) T. Koyama et Calder = Carex dispalata Boott ex A. Gray ■
73682 Carex amplisquama F. J. Herm. = Carex communis L. H. Bailey var. amplisquama (F. J. Herm.) Rettig ■☆
73683 Carex ampullacea Good. = Carex rostrata Stokes ex With. ■
73684 Carex amurensis Kük. = Carex drymophila Turcz. ex Steud. var. abbreviata (Kük.) Ohwi ■
73685 Carex amurensis Kük. var. abbreviata Kük. = Carex drymophila Turcz. ex Steud. var. abbreviata (Kük.) Ohwi ■
73686 Carex anagare Steud. var. stevenii (T. Holm) A. E. Porsild = Carex stevenii (T. Holm) Kalela ■☆
73687 Carex anceps Muhl. = Carex laxiflora Lam. ■
73688 Carex aneurocarpa V. I. Krecz.;歪嘴苔草;Aneurocarp Sedge ■
73689 Carex aneurocarpa V. I. Krecz. = Carex pediformis C. A. Mey. ■
73690 Carex angarae Steud.;圆穗苔草(安加拉苔,紫鳞苔草);Roundspike Sedge ■
73691 Carex angarae Steud. = Carex media R. Br. ex Richardson ■☆
73692 Carex angolensis Nelmes;安哥拉苔草■☆
73693 Carex anguillata Drejer = Carex bigelowii Torr. et Schwein. ex Boott ■
73694 Carex angustata Boott;尖果苔草■☆
73695 Carex angustifructa (Kük.) V. I. Krecz.;窄果苔草;Narrowfruit Sedge ■
73696 Carex angustior Mack.;小星穗苔草;Angustior Sedge, Narrow Sedge ■
73697 Carex angustior Mack. = Carex echinata Murray ■☆
73698 Carex angustior Mack. var. gracilenta R. T. Clausen et Wahl = Carex echinata Murray ■☆
73699 Carex angustisquama Franch.;窄鳞苔草■☆
73700 Carex angustiutricula F. T. Wang et Ts. Tang ex L. K. Dai;狭果囊苔草;Narrow-fruited Sedge ■
73701 Carex anhuiensis S. W. Su et S. W. Xu = Carex chungii C. P. Wang ■
73702 Carex anisoneura V. I. Krecz.;异脉苔草■☆
73703 Carex annectans (E. P. Bicknell) E. P. Bicknell;短舌苔草;Sedge, Yellow-headed Fox Sedge ■☆
73704 Carex annectans (E. P. Bicknell) E. P. Bicknell var. xanthocarpa (Kük.) Wiegand;黄果短舌苔草;Yellow-headed Fox Sedge ■☆
73705 Carex annectens (E. P. Bicknell) E. P. Bicknell var. ambigua (Barratt ex Boott) Gleason = Carex annectans (E. P. Bicknell) E. P. Bicknell ■☆
73706 Carex annectens (E. P. Bicknell) E. P. Bicknell var. xanthcarpa (Kük.) Wiegand = Carex annectans (E. P. Bicknell) E. P. Bicknell ■☆
73707 Carex anningensis F. T. Wang et Ts. Tang ex P. C. Li;安宁苔草;Anning Sedge ■
73708 Carex anomala Steud. = Carex petitiana A. Rich. ■
73709 Carex anomocarya Nelmes = Carex harlandii Boott ■
73710 Carex anomoea Hand.-Mazz.;无名苔草■
73711 Carex anthericoides J. Presl et C. Presl = Carex macrocephala Willd. ex Kunth ■☆
73712 Carex antoniensis A. Chev.;安东尼亚苔草■☆
73713 Carex aomorensis Franch. = Carex capillacea Boott var. sachalinensis (Eastw.) Ohwi ■☆
73714 Carex aperta Boott;亚美苔草■
73715 Carex aperta Boott var. umbrosa Kük. = Carex aperta Boott ■
73716 Carex aperta Boott var. viridans Kük. = Carex aperta Boott ■
73717 Carex aphanandra Franch. et Sav. = Carex leucochlora Bunge var. aphanandra (Franch. et Sav.) T. Koyama ■☆
73718 Carex aphanolepis Franch. et Sav.;匿鳞苔草■
73719 Carex aphyllopus Kük.;无叶柄苔草■☆
73720 Carex aphyllopus Kük. var. impura (Ohwi) T. Koyama = Carex impura Ohwi ■☆
73721 Carex apoda Clokey = Carex atrosquama Mack. ■☆
73722 Carex apodostachya Ohwi = Carex atrata L. subsp. apodostachys (Ohwi) T. Koyama ■
73723 Carex apodostachys Ohwi = Carex atrata L. ■
73724 Carex apoiensis Akiyama;阿伯伊苔草■☆
73725 Carex appalachica J. M. Webber et P. W. Ball;阿山苔草■☆
73726 Carex appendiculata (Trautv. et C. A. Mey.) Kük. = Carex thunbergii Steud. var. appendiculata (Trautv.) Ohwi ■
73727 Carex appendiculata (Trautv. et C. A. Mey.) Kük. var. sacculiformis Y. L. Chang et Y. L. Yang;小囊灰脉苔草■
73728 Carex appendiculata (Trautv.) Kük. = Carex thunbergii Steud. var. appendiculata (Trautv.) Ohwi ■
73729 Carex appropinquata K. Schum.;近缘苔草;Fibrous Tussock-sedge ■☆
73730 Carex approximata Bellardi;近苔草(近苔)■☆
73731 Carex aquanigra B. Boivin = Carex nigra (L.) Reichard ■☆
73732 Carex aquatilis Wahlenb.;水苔草(水苔);Aquatic Sedge, Long-bracted Tussock Sedge, Sedge, Water Sedge ■☆
73733 Carex aquatilis Wahlenb. subsp. altior (Rydb.) Hultén = Carex aquatilis Wahlenb. ■☆
73734 Carex aquatilis Wahlenb. subsp. stans (Drejer) Hultén = Carex aquatilis Wahlenb. var. stans (Drejer) Boott ■☆
73735 Carex aquatilis Wahlenb. var. altior (Rydb.) Fernald = Carex aquatilis Wahlenb. ■☆

73736 Carex aquatilis Wahlenb. var. dives (T. Holm) Kük.;丰富苔草■☆

73737 Carex aquatilis Wahlenb. var. minor Boott;小水苔草■☆

73738 Carex aquatilis Wahlenb. var. stans (Drejer) Boott;直立水苔草■☆

73739 Carex aquatilis Wahlenb. var. substricta Kük.;近直立水苔草;Long-bracted Tussock Sedge, Water Sedge ■☆

73740 Carex arcatica Meinsh.;北疆苔草■

73741 Carex arcatica Meinsh. = Carex orbicularis Boott ■

73742 Carex arcta Boott;北熊苔草;Bear Sedge, Northern Clustered Sedge ■☆

73743 Carex arctata Boott;林地俯卧苔草(扁苔草);Bear Sedge, Compressed Sedge, Drooping Woodland Sedge, Northern Clustered Sedge ■☆

73744 Carex arctata Boott var. faxonii L. H. Bailey = Carex arctata Boott ■☆

73745 Carex arctica Dewey = Carex parryana Dewey ■☆

73746 Carex arctogena Harry Sm. = Carex capitata L. ■☆

73747 Carex arenaria L.;沙苔(沙生苔草);Sand Sedge ■☆

73748 Carex arenicola Eastw.;沙生苔草■☆

73749 Carex arenicola F. Schmidt subsp. pansa (L. H. Bailey) T. Koyama et Calder = Carex pansa L. H. Bailey ■☆

73750 Carex argunensis Turcz. ex Ledeb.;额尔古纳苔草;Erguna Sedge ■

73751 Carex argyi H. Lév. et Vaniot;阿齐苔草(红穗苔草);Argy Sedge ■

73752 Carex argyrantha Tuck. ex Dewey;银花苔草;Silvery-flowered Sedge ■☆

73753 Carex argyroglochin Hornem.;银鳞苔草■☆

73754 Carex aridula V. I. Krecz.;干生苔草;Arid Sedge, Dryland Sedge ■

73755 Carex arisanensis Hayata;阿里山苔草(阿里山疏花苔);Alishan Sedge, Arishan Sedge ■

73756 Carex aristata D. Don ex Okamoto et Taylor subsp. orthostachys (C. A. Mey.) Kük. = Carex orthostachys C. A. Mey. ■

73757 Carex aristata D. Don ex Okamoto et Taylor subsp. raddei (Kük.) Kük. var. eriophylla Kük. = Carex eriophylla (Kük.) Kom. ■

73758 Carex aristata D. Don ex Okamoto et Taylor subsp. raddei (Kük.) Kük. = Carex raddei Kük. ■

73759 Carex aristata D. Don ex Okamoto et Taylor var. lanceisquama Hand. -Mazz.;芸苞苔草■

73760 Carex aristata R. Br. subsp. orthostachys (C. A. Mey.) Kük. = Carex orthostachys C. A. Mey. ■

73761 Carex aristata R. Br. subsp. raddei (Kük.) Kük. = Carex raddei Kük. ■

73762 Carex aristata R. Br. subsp. raddei Kük. var. eriophylla Kük. = Carex eriophylla (Kük.) Kom. ■

73763 Carex aristatisquamata Ts. Tang et F. T. Wang ex L. K. Dai;芒鳞苔草(长芒鳞苔草)■

73764 Carex aristulifera P. C. Li;具芒苔草(长芒苔草)■

73765 Carex arkansana (L. H. Bailey) L. H. Bailey;阿肯色苔草;Sedge ■☆

73766 Carex arnelli H. Christ = Carex arnelli H. Christ ex Scheutz ■

73767 Carex arnelli H. Christ ex Scheutz;麻根苔草(阿尔奈苔草,阿氏苔);Arnell Sedge ■

73768 Carex arnellii H. Christ ex Scheutz subsp. hondoensis (Ohwi) T. Koyama;本州苔草■☆

73769 Carex arnellii H. Christ ex Scheutz var. hondoensis (Ohwi) T. Koyama = Carex arnelli H. Christ subsp. hondoensis (Ohwi) T. Koyama ■☆

73770 Carex arrhyncha Franck;无喙苔草■☆

73771 Carex artitecta Mack. = Carex albicans Willd. ex Spreng. ■☆

73772 Carex artitecta Mack. var. subtilirostris F. J. Herm. = Carex albicans Willd. ex Spreng. ■☆

73773 Carex asa-grayi L. H. Bailey = Carex grayi J. Carey ■☆

73774 Carex ascotreta C. B. Clarke ex Franch.;宜昌苔草(宝岛宿柱苔);Yichang Sedge ■

73775 Carex aspartilis V. I. Krecz.;糙苔草■☆

73776 Carex asperifructus Kük.;粗糙囊苔草■

73777 Carex asperinervis T. Koyama = Carex bilateralis Hayata ■

73778 Carex asperner T. Koyama = Carex bilateralis Hayata ■

73779 Carex astrachanica Willd. ex Kunth;阿斯特拉哈苔草■☆

73780 Carex aterrima Hoppe = Carex atrata L. subsp. aterrima (Hoppe) S. Yun Liang, L. K. Dai et S. Yun Liang ■

73781 Carex athabascensis F. J. Herm. = Carex scirpoidea Michx. ■☆

73782 Carex atherodes Spreng.;直穗苔;Atlantic Star Sedge, Bird's-foot Sedge, Hairy-leaved Lake Sedge, Orthotropic-spike Sedge, Prickly Bog Sedge, Sedge, Slough Sedge ■

73783 Carex atherodes Spreng. = Carex orthostachys C. A. Mey. ■

73784 Carex atlantica L. H. Bailey;大西洋苔草;Atlantic Sedge, Atlantic Star Sedge, Prickly Bog Sedge ■☆

73785 Carex atlantica L. H. Bailey = Carex atherodes Spreng. ■

73786 Carex atlantica L. H. Bailey subsp. capillacea (L. H. Bailey) Reznicek;线状大西洋苔草■☆

73787 Carex atlasica (H. Lindb.) Inb Tattou = Carex ovalis Gooden. ■☆

73788 Carex atrata L.;黑穗苔草(黑苔,南湖扁果苔);Black Alpine-sedge, Black Sedge, Black-spiked Sedge ■

73789 Carex atrata L. subsp. apodostachys (Ohwi) T. Koyama;南湖苔草(南湖扁果苔)■

73790 Carex atrata L. subsp. apodostachys (Ohwi) T. Koyama = Carex atrata L. ■

73791 Carex atrata L. subsp. aterrima (Hoppe) S. Yun Liang = Carex atrata L. subsp. aterrima (Hoppe) S. Yun Liang, L. K. Dai et S. Yun Liang ■

73792 Carex atrata L. subsp. aterrima (Hoppe) S. Yun Liang, L. K. Dai et S. Yun Liang;大桥苔草■

73793 Carex atrata L. subsp. atratiformis (Britton) Kük. = Carex atratiformis Britton ■☆

73794 Carex atrata L. subsp. atrosquama (Mack.) Hultén = Carex atrosquama Mack. ■☆

73795 Carex atrata L. subsp. caucasica var. longistolonifera Kük. = Carex atrata L. subsp. longistolonifera (Kük.) S. Yun Liang ■

73796 Carex atrata L. subsp. japonalpina (T. Koyama) T. Koyama = Carex perfusca V. I. Krecz. var. japonalpina (T. Koyama) Kitag. ■☆

73797 Carex atrata L. subsp. longistolonifera (Kük.) S. Yun Liang;长匍匐茎苔草■

73798 Carex atrata L. subsp. pullata (Boott) Kük.;尖鳞黑穗苔草(黑被苔草,尖鳞苔草)■

73799 Carex atrata L. subsp. pullata (Boott) Kük. var. sinensis Kük. = Carex atrata L. subsp. pullata (Boott) Kük. ■

73800 Carex atrata L. var. atrosquama (Mack.) Cronquist = Carex atrosquama Mack. ■☆

73801 Carex atrata L. var. chalciolepis (T. Holm) Kük. = Carex chalciolepis T. Holm ■☆

73802 Carex atrata L. var. discolor (L. H. Bailey) Kük. = Carex bella L. H. Bailey ■☆

73803 Carex atrata L. var. discolor L. H. Bailey = Carex bella L. H. Bailey ■☆

73804 Carex atrata L. var. erecta W. Boott = Carex heteroneura W. Boott ■☆

73805 Carex atrata L. var. glacialis Boott = Carex atrata L. subsp. pullata (Boott) Kük. ■

73806 Carex atrata L. var. japonalpina T. Koyama = Carex atrata L. ■

73807 Carex atrata L. var. japonalpina T. Koyama = Carex perfusca V. I. Krecz. var. japonalpina (T. Koyama) Kitag. ■☆

73808 Carex atrata L. var. larimerana (Kelso) Kelso = Carex chalciolepis T. Holm ■☆

73809 Carex atrata L. var. nigra W. Boott = Carex henryi C. B. Clarke ex Franch. ■

73810 Carex atrata L. var. pullata Boott = Carex atrata L. subsp. pullata (Boott) Kük. ■

73811 Carex atrata L. var. subgracilenta Kük.;近黑苔草■

73812 Carex atratiformis Britton;拟黑穗苔草;Black Sedge ■☆

73813 Carex atratiformis Britton subsp. raymondii (Calder) A. E. Porsild = Carex atratiformis Britton ■☆

73814 Carex atrofusca Schkuhr;暗褐苔草(白尖苔草,黑棕苔);Deepbrown Sedge,Scorched Alpine-sedge ■

73815 Carex atrofusca Schkuhr subsp. minor (Boott) T. Koyama;黑褐穗苔草(黑褐苔草)■

73816 Carex atrofusca Schkuhr var. angustifructus (Boott) Kük. = Carex atrofusca Schkuhr subsp. minor (Boott) T. Koyama ■

73817 Carex atrofusca Schkuhr var. coriophora (Fisch.) Kük. = Carex coriophora Fisch. et C. A. Mey. ex Kunth ■

73818 Carex atrofusca Schkuhr var. minor (Boott) Kük. = Carex atrofusca Schkuhr subsp. minor (Boott) T. Koyama ■

73819 Carex atrofuscoides K. T. Ku;类黑褐穗苔草;Deepbrown-like Sedge ■

73820 Carex atronucula Hayata = Carex sociata Boott ■

73821 Carex atrosquama Mack.;暗鳞苔草■☆

73822 Carex atrovirens Boeck. = Carex aureolensis Steud. ■☆

73823 Carex atroviridis Ohwi = Carex multifolia Ohwi ■☆

73824 Carex augustini Tuyama;奥古斯丁苔草■☆

73825 Carex augustinowiczii Meinsh. = Carex augustinowiczii Meinsh. ex Korsh. ■

73826 Carex augustinowiczii Meinsh. ex Korsh.;短鳞苔草(奥古苔草,钝鳞苔草);Obtusescale Sedge ■

73827 Carex augustinowiczii Meinsh. ex Korsh. subsp. soyaeensis (Kük.) T. V. Egorova = Carex augustinowiczii Meinsh. ex Korsh. ■

73828 Carex augustinowiczii Meinsh. ex Korsh. var. sharensis (Franch.) Ohwi;斜里苔草■☆

73829 Carex aurea Nutt.;金黄苔草;Elk Sedge,Golden Sedge,Golden-fruited Sedge ■☆

73830 Carex aurea Nutt. var. androgyna Olney = Carex garberi Fernald ■☆

73831 Carex aureolensis Steud.;暗脉苔草■☆

73832 Carex auriculata Franch. = Carex pilosa Scop. var. auriculata Kük. ■

73833 Carex austrina Mack.;澳洲苔草■☆

73834 Carex austro-africana (Kük.) Raymond;南非苔草■☆

73835 Carex austrocaroliniana L. H. Bailey;卡罗来纳苔草■☆

73836 Carex austro-occidentalis F. T. Wang et Ts. Tang ex Y. C. Tang;西南苔草;SW. China Sedge ■

73837 Carex austrosinensis Ts. Tang et F. T. Wang ex S. Yun Liang;华南苔草;S. China Sedge ■

73838 Carex austrozhejiangsis C. Z. Zheng et X. F. Jin;浙南苔草;S. Zhejiang Sedge ■

73839 Carex autumnalis Ohwi;秋生苔草;Autumn Sedge ■

73840 Carex ayouensis X. Y. Mao et Y. C. Yang = Carex pediformis C. A. Mey. ■

73841 Carex baccans Nees = Carex baccans Nees ex Wight ■

73842 Carex baccans Nees ex Wight;浆果苔草(芭茅草,旱稗,红稗,红果莎,红果苔,红米,山稗子,山高粱,山红稗,水高粱,稊,土稗子,乌禾,野高粱,野红米草,野鸡稗);Bacca Sedge ■

73843 Carex baccans Nees var. pallida Satake = Carex baccans Nees ex Wight ■

73844 Carex backii Boott var. saximontana (Mack.) B. Boivin = Carex saximontana Mack. ■☆

73845 Carex backii var. subrostrata (Bates) Dorn = Carex saximontana Mack. ■☆

73846 Carex backii W. Boott;巴克苔草;Back's Sedge,Rocky Mountain Sedge ■☆

73847 Carex baimaensis S. W. Su;白马苔草;Baima Sedge ■

73848 Carex baiposhanensis P. C. Li;百坡山苔草;Baiposhan Sedge ■

73849 Carex baohuashanica Ts. Tang et F. T. Wang ex L. K. Dai;宝华山苔草;Baohuashan Sedge ■

73850 Carex baranovii Chou ex Liou et al. = Carex eleusinoides Turcz. ex Kunth ■

73851 Carex barbarae Dewey;巴尔巴拉苔草■☆

73852 Carex basiantha Steud.;基花苔草■☆

73853 Carex basiflora C. B. Clarke = Carex brevicuspis C. B. Clarke var. basiflora (C. B. Clarke) Kük. ■

73854 Carex basilaris Jord. = Carex depressa Link ■☆

73855 Carex basilata Ohwi = Carex angustior Mack. ■

73856 Carex basilata Ohwi = Carex echinata Murray ■☆

73857 Carex bauhwaensis Z. P. Wang = Carex pisiformis Boott ■

73858 Carex bayardii Fernald = Carex crus-corvi Shuttlew. ex Kunze ■☆

73859 Carex bebbii (L. H. Bailey) Olney ex Fernald;拜勃苔草;Bebb's Oval Sedge,Bebb's Sedge ■☆

73860 Carex behringensis C. B. Clarke = Carex podocarpa R. Br. ex Richardson ■☆

73861 Carex bella L. H. Bailey;雅致苔草■☆

73862 Carex bellardii All. = Kobresia myosuroides (Vill.) Fiori et Paol. ■

73863 Carex bengalensis Roxb. = Carex cruciata Wahlenb. ■

73864 Carex bengalensis Roxb. var. scaberrina Boeck. = Carex rafflesiana Boott ■

73865 Carex bequaertii De Wild.;贝卡尔苔草■☆

73866 Carex bequaertii De Wild. var. maxima Lye;大贝卡尔苔草■☆

73867 Carex beringiana Cham. ex Steud. = Carex stylosa C. A. Mey. ■☆

73868 Carex bicknelii Camus = Carex annectans (E. P. Bicknell) E. P. Bicknell ■☆

73869 Carex bicknellii Britton;毕克奈尔苔草;Bicknell's Oval Sedge,Bicknell's Sedge,Sedge ■☆

73870 Carex bicknellii Britton var. opaca F. J. Herm. = Carex opaca (F. J. Herm.) P. Rothr. et Reznicek ■☆

73871 Carex bicknellii Camus = Carex annectans (E. P. Bicknell) E. P. Bicknell ■☆

73872 Carex bicolor All. = Carex bicolor Bell. ex All. ■☆

73873 Carex bicolor Bell. ex All.;双色花苔草■☆

73874 Carex bifida W. Boott = Carex serratodens W. Boott ■☆

73875 Carex bigelowii Torr. et Schwein. ex Boott;箭叶苔草(北方苔);Arrowleaf Sedge,Stiff Sedge,Swordleaf Sedge ■

73876 Carex bijiangensis S. Yun Liang et S. R. Zhang;碧江苔草■

73877 Carex bilateralis Hayata;台湾苔草(短叶二柱苔);Taiwan Sedge ■
73878 Carex binervis Sm.;双脉苔草;Green-ribbed Sedge ■☆
73879 Carex binervis Sm. var. tingitana Maire = Carex paulo-vargasii Luceno et Marín ■☆
73880 Carex bipartida All. = Carex lachenalii Schkuhr ■
73881 Carex bipartita All. var. austromontana F. J. Herm. = Carex lachenalii Schkuhr ■
73882 Carex bisexualis C. B. Clarke = Schoenoxiphium ecklonii Nees var. unisexuale Kük. ■☆
73883 Carex bishallii C. B. Clarke = Carex nudata W. Boott ■☆
73884 Carex bitchuensis T. Hoshino et H. Ikeda;备中苔草■☆
73885 Carex biwensis Franch.;比温苔草■
73886 Carex biwensis Franch. = Carex rara Boott ■
73887 Carex blanda Dewey;东部林苔草;Common Wood Sedge, Eastern Woodland Sedge, Sedge, Smooth Sedge, Wood Sedge ■☆
73888 Carex blepharicarpa Franch.;睫毛果苔草■☆
73889 Carex blepharicarpa Franch. var. dueensis (Meinsh.) Akiyama = Carex blepharicarpa Franch. ■☆
73890 Carex blepharicarpa Franch. var. hirtifructus (Kük.) Ohwi = Carex blepharicarpa Franch. var. stenocarpa Ohwi ■☆
73891 Carex blepharicarpa Franch. var. stenocarpa Ohwi;窄睫毛果苔草■☆
73892 Carex blinii H. Lév. et Vaniot;石林苔草(白里苔草,白尼苔草);Blin Sedge ■
73893 Carex bodinieri Franch.;滨海苔草(锈点苔草);Bodinier Sedge ■
73894 Carex bohemica Schreb.;莎苔草(莎状苔草);Cyperuslike Sedge ■
73895 Carex bolanderi Olney;鲍兰苔草■☆
73896 Carex bolusii C. B. Clarke = Schoenoxiphium sparteum (Wahlenb.) C. B. Clarke ■☆
73897 Carex bonanzensis Britton = Carex bonanzensis Britton ex Britton et Rydb. ■☆
73898 Carex bonanzensis Britton ex Britton et Rydb.;博楠查苔草■☆
73899 Carex bongardii Boott = Carex wahuensis C. A. Mey. subsp. robusta (Franch. et Sav.) T. Koyama ■
73900 Carex bongardii Boott var. robusta Franch. et Sav. = Carex wahuensis C. A. Mey. subsp. robusta (Franch. et Sav.) T. Koyama ■
73901 Carex boottiana Hook. et Arn. = Carex wahuensis C. A. Mey. subsp. robusta (Franch. et Sav.) T. Koyama ■
73902 Carex boottiana Hook. et Arn. = Carex wahuensis C. A. Mey. var. bongardii (Boott) Franch. et Sav. ■
73903 Carex bordzilowskii Krecz.;保德氏苔草;Bordzilowski Sedge ■☆
73904 Carex borealihinganica Y. L. Chang;北兴安苔草;N. Xing'an Sedge ■
73905 Carex borealihinganica Y. L. Chang et Y. L. Yang = Carex borealihinganica Y. L. Chang ■
73906 Carex bostrychostigma Maxim.;卷柱头苔草■
73907 Carex brachyanthera Ohwi;短穗苔草(垂穗苔,垂穗苔草);Shortspike Sedge ■
73908 Carex brachyglossa Mack. = Carex annectans (E. P. Bicknell) E. P. Bicknell ■☆
73909 Carex brachypoda T. Holm = Carex scopulorum T. Holm var. bracteosa (L. H. Bailey) F. J. Herm. ■☆
73910 Carex brassii Nelmes;布拉斯苔草■☆
73911 Carex breviaristata K. T. Fu;短芒苔草;Shortawn Sedge ■
73912 Carex brevicollis DC. ex Lam. et DC.;短颈苔草(短颈苔)■☆
73913 Carex breviculmis R. Br.;短茎宿柱苔(青菅,青绿苔草,硬短茎宿柱苔)■
73914 Carex breviculmis R. Br. f. aphanandra (Franch. et Sav.) Kük. = Carex leucochlora Bunge var. aphanandra (Franch. et Sav.) T. Koyama ■☆
73915 Carex breviculmis R. Br. f. fibrillosa (Franch. et Sav.) Kük. = Carex breviculmis R. Br. var. fibrillosa (Franch. et Sav.) Kük. ex Matsum. et Hayata ■
73916 Carex breviculmis R. Br. f. filiculmis (Franch. et Sav.) Kük. = Carex breviculmis R. Br. ■
73917 Carex breviculmis R. Br. f. longearistata Kük. = Carex breviculmis R. Br. ■
73918 Carex breviculmis R. Br. subsp. fibrillosa (Franch. et Sav.) T. Koyama = Carex breviculmis R. Br. ■
73919 Carex breviculmis R. Br. subsp. fibrillosa (Franch. et Sav.) T. Koyama = Carex breviculmis R. Br. var. fibrillosa (Franch. et Sav.) Kük. ex Matsum. et Hayata ■
73920 Carex breviculmis R. Br. subsp. royleana (Nees) Kük. = Carex breviculmis R. Br. ■
73921 Carex breviculmis R. Br. subsp. royleana (Nees) Kük. f. longearistata Kük. = Carex breviculmis R. Br. ■
73922 Carex breviculmis R. Br. subsp. royleana (Nees) Kük. var. pluricostata Kük. = Carex fibrillosa Franch. et Sav. ■
73923 Carex breviculmis R. Br. var. cupulifera (Hayata) Y. C. Tang et S. Yun Liang;直蕊苔草(直蕊宿柱苔)■
73924 Carex breviculmis R. Br. var. discoidea (Boott) Boott = Carex leucochlora Bunge var. discoidea (Boott) T. Koyama ■☆
73925 Carex breviculmis R. Br. var. fibrillosa (Franch. et Sav.) Kük. ex Matsum. et Hayata;纤维青菅(硬短茎宿柱苔)■
73926 Carex breviculmis R. Br. var. fibrillosa (Franch. et Sav.) Kük. ex Matsum. et Hayata = Carex fibrillosa Franch. et Sav. ■
73927 Carex breviculmis R. Br. var. leucochlora (Bunge) Makino = Carex leucochlora Bunge ■
73928 Carex breviculmis R. Br. var. lonchophora (Ohwi) T. Koyama;矛梗苔草■☆
73929 Carex breviculmis R. Br. var. lonchophora (Ohwi) T. Koyama = Carex lonchophora Ohwi ■☆
73930 Carex brevicuspis C. B. Clarke;短尖苔草(大山宿柱苔);Shortcuspidate Sedge ■
73931 Carex brevicuspis C. B. Clarke var. basiflora (C. B. Clarke) Kük.;基花短尖苔草;Baseflower Shortcuspidate Sedge ■
73932 Carex breviliguata Mack. = Carex densa (L. H. Bailey) L. H. Bailey ■☆
73933 Carex brevior (Dewey) Mack. ex Lunell;短牛毛苔草;Fescue Sedge, Plains Oval Sedge, Sedge, Short-headed Sedge ■☆
73934 Carex brevior (Dewey) Mack. ex Lunell var. pseudofestucacea Farw. = Carex merritt-fernaldii Mack. ■☆
73935 Carex brevior (Dewey) Mack. var. pseudofestucacea Farw. = Carex merritt-fernaldii Mack. ■☆
73936 Carex brevipes W. Boott ex Mack. = Carex deflexa Hornem. var. boottii L. H. Bailey ■☆
73937 Carex breviscapa C. B. Clarke;短葶苔草(宽果宿柱苔,疏林苔草);Shortscape Sedge ■
73938 Carex breweri Boott var. paddoensis (Suksd.) Cronquist = Carex engelmannii L. H. Bailey ■☆
73939 Carex brittoniana L. H. Bailey = Carex tetrastachya Scheele ■☆
73940 Carex brizoides Jusl. ex L.;风凌草苔草■☆
73941 Carex brizoides Poir. = Carex divisa Huds. ■

73942 Carex bromoides Schkuhr ex Willd.;波罗氏苔草;Brome-like Sedge,Sedge ■☆

73943 Carex brongniartii Kunth var. densa L. H. Bailey = Carex densa (L. H. Bailey) L. H. Bailey ■☆

73944 Carex brotherorum H. Christ = Carex bromoides Schkuhr ex Willd. ■☆

73945 Carex brownii Tuck.;亚澳苔草(白郎苔,布朗苔,三方草,亚大苔草);Brown Sedge ■

73946 Carex brownii Tuck. subsp. dissosciata (Franch. et Sav.) T. Koyama = Carex transversa Boott ■

73947 Carex brownii Tuck. subsp. transversa (Boott) Kern. = Carex transversa Boott ■

73948 Carex brownii Tuck. var. dissosciata (Franch. et Sav.) T. Koyama = Carex transversa Boott ■

73949 Carex brownii Tuck. var. transversa (Boott) Kük. = Carex transversa Boott ■

73950 Carex brownii Tuck. var. transversa (Boott) Kük. ex Matsum. = Carex transversa Boott ■

73951 Carex brunnea Merr. = Carex hattoriana Nakai ex Tuyama ■

73952 Carex brunnea Thunb.;褐果苔草(布朗苔,栗褐苔草,囊草,莎草,束草);Chestnut-colored Sedge,Greater Brown Sedge ■

73953 Carex brunnea Thunb. subsp. occidentalis Lye;西方褐果苔草■☆

73954 Carex brunnea Thunb. var. abscondita T. Koyama = Carex brunnea Thunb. ■

73955 Carex brunnea Thunb. var. nakiri Ohwi = Carex brunnea Thunb. ■

73956 Carex brunnea Thunb. var. nakiri Ohwi = Carex lenta D. Don ■☆

73957 Carex brunnea Thunb. var. sendaica Kük. = Carex sendaica Franch. ■

73958 Carex brunnea Thunb. var. stipitinux (C. B. Clarke) Kük. = Carex stipitinux C. B. Clarke ex Franch. ■

73959 Carex brunnescens (Pers.) Poir.;棕苔(布尔津苔草,褐鳞苔草);Brown Sedge,Brownish Sedge,Green Bog Sedge ■

73960 Carex brunnescens (Pers.) Poir. subsp. alaskana Kalela = Carex brunnescens (Pers.) Poir. ■

73961 Carex brunnescens (Pers.) Poir. subsp. pacifica Kalela;太平洋苔草■☆

73962 Carex brunnescens (Pers.) Poir. subsp. pacifica Kalela = Carex brunnescens (Pers.) Poir. ■

73963 Carex brunnescens (Pers.) Poir. subsp. sphaerostachys (Tuck.) Kalela;球穗棕苔草;Brownish Sedge,Green Bog Sedge ■☆

73964 Carex brunnescens (Pers.) Poir. subsp. vitilis (Fr.) Kalela = Carex brunnescens (Pers.) Poir. ■

73965 Carex brunnescens (Pers.) Poir. var. gracilior Britton = Carex brunnescens (Pers.) Poir. subsp. sphaerostachys (Tuck.) Kalela ■☆

73966 Carex brunnescens (Pers.) Poir. var. sphaerostachya (Tuck.) Kük. = Carex brunnescens (Pers.) Poir. subsp. sphaerostachys (Tuck.) Kalela ■☆

73967 Carex buchananii Berggr.;革叶苔草;Leatherleaf Sedge ■☆

73968 Carex buchananii C. B. Clarke = Schoenoxiphium rufum Nees ■☆

73969 Carex bucharica Kük.;布哈尔苔草■☆

73970 Carex buckleyi Dewey = Carex brunnescens (Pers.) Poir. subsp. sphaerostachys (Tuck.) Kalela ■☆

73971 Carex buekii Wimm.;毕克苔草■☆

73972 Carex bulbostylis Mack.;美洲球柱苔草■☆

73973 Carex bullata Schkuhr var. laevirostris Blytt ex Fries = Carex rhynchophysa C. A. Mey. ■

73974 Carex burchelliana Boeck.;伯切尔苔草■☆

73975 Carex burjatorum Rorotky;布里雅特苔草■☆

73976 Carex bushii Mack.;布什苔草;Bush's Sedge,Long-scaled Green Sedge,Sedge ■☆

73977 Carex buxbaumii Wahlenb.;巴克斯保姆苔草;Buxbaum's Sedge,Sedge ■☆

73978 Carex buxbaumii Wahlenb. = Carex tarumensis Franch. ■

73979 Carex buxbaumii Wahlenb. f. dilutior Kük. = Carex buxbaumii Wahlenb. ■☆

73980 Carex buxbaumii Wahlenb. f. heterostachya Andersson = Carex buxbaumii Wahlenb. ■☆

73981 Carex buxbaumii Wahlenb. subsp. mutica (Hartm.) Isov. = Carex adelostoma Krecz. ■☆

73982 Carex buxbaumii Wahlenb. var. alpicola Hartm. = Carex adelostoma Krecz. ■☆

73983 Carex buxbaumii Wahlenb. var. alpina Hartm. = Carex adelostoma Krecz. ■☆

73984 Carex buxbaumii Wahlenb. var. anticostensis Raymond = Carex buxbaumii Wahlenb. ■☆

73985 Carex buxbaumii Wahlenb. var. mutica Hartm. = Carex adelostoma Krecz. ■☆

73986 Carex caespititia Nees;丛生苔草;Clustered Sedge ■

73987 Carex caespitosa L.;丛苔草(丛生苔草,丛苔);Tufted Sedge ■

73988 Carex caespitosa L. var. ramosa Dewey = Carex torta Boott ■☆

73989 Carex cajanderi Kük. = Carex bonanzensis Britton ex Britton et Rydb. ■☆

73990 Carex calcicola Ts. Tang et F. T. Wang;灰岩生苔草;Limestone Sedge ■

73991 Carex calderae Hansen = Carex paniculata L. subsp. calderae (Hansen) Lewej. et Lobin ■☆

73992 Carex callitrichos V. I. Krecz.;羊胡子苔草(羊须草)■

73993 Carex callitrichos V. I. Krecz. = Carex humilis Leyss. var. callitrichos (V. I. Krecz.) Ohwi ■☆

73994 Carex callitrichos V. I. Krecz. var. austrohinganica Y. L. Chang et Y. L. Yang;兴安羊胡子苔草■

73995 Carex callitrichos V. I. Krecz. var. austrohinganica Y. L. Chang et Y. L. Yang = Carex humilis Leyss. ■

73996 Carex callitrichos V. I. Krecz. var. nana (H. Lév. et Vaniot) Ohwi;矮羊须草(矮丛苔草)■

73997 Carex camporum Mack. = Carex praegracilis W. Boott ■☆

73998 Carex campylocarpa subsp. affinis Maguire et A. H. Holmgren = Carex scopulorum T. Holm var. bracteosa (L. H. Bailey) F. J. Herm. ■☆

73999 Carex campylocarpa T. Holm = Carex scopulorum T. Holm var. bracteosa (L. H. Bailey) F. J. Herm. ■☆

74000 Carex campylorhina V. I. Krecz. = Carex pilosa Scop. ■

74001 Carex canaliculata P. C. Li;沟囊苔草;Canaliculate Sedge ■

74002 Carex canariensis Kük.;加那利苔草■☆

74003 Carex candolleana H. Lév. et Vaniot = Carex leucochlora Bunge var. aphanandra (Franch. et Sav.) T. Koyama ■☆

74004 Carex canescens L.;灰白苔草(白山苔草,灰白苔,灰色苔草,青河苔草);Gray Bog Sedge,Grey Sedge,Silvery Sedge,Whitish Sedge ■

74005 Carex canescens L. = Carex curta Gooden. ■

74006 Carex canescens L. subsp. brunnescens (Pers.) Asch. et Graebn. = Carex brunnescens (Pers.) Poir. ■

74007 Carex canescens L. subsp. disjuncta (Fernald) Toivonen;叉灰白苔草;Gray Bog Sedge,Silvery Sedge ■☆

74008 Carex canescens L. var. alpicola Wahlenb. = Carex brunnescens (Pers.) Poir. ■

74009 Carex canescens L. var. brunnescens (Pers.) W. D. J. Koch = Carex brunnescens (Pers.) Poir. ■

74010 Carex canescens L. var. disjuncta Fernald = Carex canescens L. subsp. disjuncta (Fernald) Toivonen ■☆

74011 Carex canescens L. var. dubia L. H. Bailey = Carex praeceptorum Mack. ■☆

74012 Carex canescens L. var. oregana L. H. Bailey = Carex arctata Boott ■☆

74013 Carex canescens L. var. persoonii (Sieber) H. Christ = Carex brunnescens (Pers.) Poir. ■

74014 Carex canescens L. var. polystachya Boott = Carex arcta Boott ■☆

74015 Carex canescens L. var. robustina Macoun = Carex canescens L. ■

74016 Carex canescens L. var. robustior Blytt ex Andersson = Carex canescens L. ■

74017 Carex canescens L. var. sphaerostachya Tuck. = Carex brunnescens (Pers.) Poir. subsp. sphaerostachys (Tuck.) Kalela ■☆

74018 Carex canescens L. var. subloliacea Laest. = Carex lapponica O. Lang ■☆

74019 Carex canescens L. var. subloliacea sensu Fernald = Carex canescens L. ■

74020 Carex canescens L. var. vitilis (Fr.) J. Carey = Carex brunnescens (Pers.) Poir. subsp. sphaerostachys (Tuck.) Kalela ■☆

74021 Carex canescens L. var. vulgaris L. H. Bailey = Carex brunnescens (Pers.) Poir. subsp. sphaerostachys (Tuck.) Kalela ■☆

74022 Carex capensis Thunb. = Schoenoxiphium rufum Nees ■☆

74023 Carex capillacea Boott;发秆苔草(单穗苔);Capillaceous Sedge, Capillaryculm Sedge ■

74024 Carex capillacea Boott subsp. aomorensis (Franch.) T. V. Egorova = Carex capillacea Boott var. sachalinensis (Eastw.) Ohwi ■☆

74025 Carex capillacea Boott var. aomorensis (Franch.) Ohwi = Carex capillacea Boott var. sachalinensis (Eastw.) Ohwi ■☆

74026 Carex capillacea Boott var. linzensis Y. C. Yang = Carex capillacea Boott ■

74027 Carex capillacea Boott var. linziensis Y. C. Yang; 林芝苔草; Linzhi Sedge ■

74028 Carex capillacea Boott var. linziensis Y. C. Yang = Carex capillacea Boott ■

74029 Carex capillacea Boott var. sachalinensis (Eastw.) Ohwi;库页苔草(线叶宿柱苔);Sachalin Sedge ■☆

74030 Carex capillacea Boott var. yunnanensis Franch.;苍山发秆苔草; Cangshan Sedge ■

74031 Carex capillacea Boott var. yunnanensis Franch. = Carex capillacea Boott ■

74032 Carex capillaris L.;细秆苔草(绿穗苔,绿穗苔草,丝柄苔草,细毛苔,纤弱苔草); Capillary Sedge, Capillarystem Sedge, Greenspike Sedge, Hair Sedge, Hairlike Sedge, Hair-like Sedge ■

74033 Carex capillaris L. = Carex chlorostachys Steven ■

74034 Carex capillaris L. subsp. chlorostachys (Steven) Á. Löve, D. Löve et Raymond = Carex chlorostachys Steven ■

74035 Carex capillaris L. subsp. karoi Freyn. = Carex karoi (Freyn) Freyn ■

74036 Carex capillaris L. subsp. krausei (Böcher) Böcher = Carex krausei Boeck. ■☆

74037 Carex capillaris L. var. elongata Olney ex Fernald = Carex capillaris L. ■

74038 Carex capillaris L. var. ledebouriana Kük. = Carex ledeboudana C. A. Mey. ex Trevis. ■

74039 Carex capillaris L. var. ledebourii E. Schmidt = Carex ledeboudana C. A. Mey. ex Trevis. ■

74040 Carex capillaris L. var. major Blytt = Carex capillaris L. ■

74041 Carex capillaris L. var. parvirostris Kük. = Carex karoi (Freyn) Freyn ■

74042 Carex capillaris L. var. pohuanshanensis Y. Yabe = Carex chlorostachys Steven ■

74043 Carex capilliculmis S. R. Zhang;丝秆苔草;Filamentose Sedge ■

74044 Carex capilliformis Franch.;丝叶苔草(毛状苔草);Capillaryleaf Sedge ■

74045 Carex capilliformis Franch. var. major Kük. = Carex capilliformis Franch. ■

74046 Carex capitata L.;头状苔草;Capitate Sedge ■☆

74047 Carex capitata L. subsp. arctogena (Harry Sm.) Böcher = Carex capitata L. ■☆

74048 Carex capitata L. var. arctogena (Harry Sm.) Hultén = Carex capitata L. ■☆

74049 Carex capitellata Boiss. et Balansa ex Boiss.;小头状苔草■☆

74050 Carex capituliformis Meinsh. ex Maxim. = Carex onoei Franch. et Sav. ■

74051 Carex capricornis Meinsh. ex Maxim.;弓喙苔草(弓嘴苔草,羊角苔草);Arcuatestigma Sedge, Sheephorn Sedge ■

74052 Carex cardiolepis Nees;藏东苔草;E. Xizang Sedge ■

74053 Carex careyana Torr. ex Dewey;卡氏苔草;Carey's Sedge, Carey's Wood Sedge, Sedge ■☆

74054 Carex caricina (D. Don) Ghildyal et U. C. Bhattach. = Carex filicina Nees ex Wight ■

74055 Carex caricinus D. Don = Carex filicina Nees ex Wight ■

74056 Carex carltonia Dewey = Carex heleonastes Ehrh. ex L. f. ■☆

74057 Carex caroliniana Buckley = Carex austrocaroliniana L. H. Bailey ■☆

74058 Carex caroliniana Schwein. var. cuspidata (Dewey) Shinners = Carex bushii Mack. ■☆

74059 Carex caryophylla Latourr.; 石竹苔草; Spring Sedge, Vernal Sedge ■☆

74060 Carex caryophyllea Latourr. subsp. microtricha (Franch.) T. Koyama = Carex microtricha Franch. ■☆

74061 Carex caryophyllea Latourr. subsp. nervata (Franch. et Sav.) Kük. = Carex nervata Franch. et Sav. ■

74062 Carex caryophyllea Latourr. subsp. nervata (Franch. et Sav.) Kük. ex Matsum. = Carex nervata Franch. et Sav. ■

74063 Carex caryophyllea Latourr. var. microtricha (Franch.) Kük. = Carex microtricha Franch. ■☆

74064 Carex caryophyllea Latourr. var. vidalii (Franch. et Sav.) T. Koyama = Carex nervata Franch. et Sav. ■

74065 Carex castanea Elliott = Carex elliottii Schwein. et Torr. ■☆

74066 Carex castanea Wahlenb.;栗色苔草;Chestnut Sedge, Chestnut Woodland Sedge ■☆

74067 Carex castanostachya K. Schum.;栗穗苔草■☆

74068 Carex caucasica Steven;高加索苔草(扁果苔草,大井氏扁果苔,高加索苔);Caucasia Sedge ■

74069 Carex caucasica Steven subsp. jisaburo-ohwiana (T. Koyama) T. Koyama;大井氏扁果苔(扁果苔草)■

74070 Carex caucasica Steven subsp. jisaburo-ohwiana (T. Koyama) T.

Koyama =Carex caucasica Steven ■
74071 Carex caucasica Steven var. jisaburo-ohwiana (T. Koyama) T. Koyama =Carex caucasica Steven ■
74072 Carex caudispicata F. T. Wang et Ts. Tang ex P. C. Li;尾穗苔草;Caudate-spiked Sedge ■
74073 Carex cavaleriei H. Lév. =Carex laticeps C. B. Clarke ex Franch. ■
74074 Carex cavaleriei H. Lév. et Vaniot =Carex laticeps C. B. Clarke ex Franch. ■
74075 Carex cephalantha (L. H. Bailey) E. P. Bicknell =Carex echinata Murray ■☆
74076 Carex cephalantha E. P. Bicknell;头花苔草;Large-headed Sedge ■☆
74077 Carex cephaloidea (Dewey) Dewey;头花丛生苔草;Clustered Bracted Sedge ■☆
74078 Carex cephalophora Muhl. ex Willd.;卵头苔草;Oval-headed Sedge, Sedge, Short-headed Bracted Sedge, Wood-bank Sedge ■☆
74079 Carex cephalophora Muhl. ex Willd. var. angustifolia Boott =Carex leavenworthii Dewey ■☆
74080 Carex cephalophora Muhl. ex Willd. var. mesochorea (Mack.) Gleason =Carex mesochorea Mack. ■☆
74081 Carex cercostachys Franch. =Kobresia cercostachya (Franch.) C. B. Clarke ■
74082 Carex cernua Boott =Carex dimorpholepis Steud. ■
74083 Carex cernua Boott var. austro-africana Kük. =Carex austro-africana (Kük.) Raymond ■☆
74084 Carex cernuum L. var. nepalense (Less.) C. B. Clarke =Carpesium nepalense Less. ■
74085 Carex cespitosa L.;天蓝苔草■☆
74086 Carex cespitosa L. var. minuta (Franch.) Kük.;微小天蓝苔草■☆
74087 Carex chalciolepis T. Holm;鱼鳞苔草■☆
74088 Carex chalciolepis T. Holm var. larimerana Kelso =Carex chalciolepis T. Holm ■☆
74089 Carex chamiesonis Meinsh.;哈米氏苔草■☆
74090 Carex changmuensis Ts. Tang et F. T. Wang ex Y. C. Yang;樟木苔草;Changmu Sedge ■
74091 Carex chaofangii C. Z. Zheng et X. F. Jin;朝芳苔草;Chaofang's Sedge ■
74092 Carex chapmanii Steud.;查普曼苔草■☆
74093 Carex chelungkiangnica A. I. Baranov et Skvortsov =Carex cinerascens Kük. ■
74094 Carex cheniana Ts. Tang et F. T. Wang ex S. Yun Liang;陈氏苔草;Chen Sedge ■
74095 Carex cherokeensis Schwein.;切罗基苔草;Cherokee Sedge, Wolftail Sedge ■☆
74096 Carex chienii F. T. Wang et Ts. Tang =Carex adrienii E. G. Camus ■
74097 Carex chimaphila T. Holm =Carex scopulorum T. Holm ■☆
74098 Carex chinensis Retz.;中华苔草;China Sedge, Chinese Sedge ■
74099 Carex chinensis Retz. =Carex sociata Boott ■
74100 Carex chinensis Retz. var. longkiensis (Franch.) Kük.;龙溪苔草(龙奇苔草);Longqi Sedge ■
74101 Carex chinganensis Litv.;兴安苔草;Xing'an Sedge ■
74102 Carex chinoi Ohwi et T. Koyama;千布苔草■☆
74103 Carex chiwuana F. T. Wang et Ts. Tang ex P. C. Li;启无苔草;Chiwu Sedge ■
74104 Carex chlorocarpa Mack. =Carex viridula Michx. ■☆
74105 Carex chlorocephalula F. T. Wang et Ts. Tang ex P. C. Li;绿头苔草;Greenhead Sedge ■
74106 Carex chlorochystis Boeck. =Carex harlandii Boott ■
74107 Carex chloroleuca Meinsh. =Carex amgunensis Eastw. ■
74108 Carex chlorophila Mack. =Carex viridula Michx. ■☆
74109 Carex chlorosaccua C. B. Clarke;绿苔草■
74110 Carex chlorostachys D. Don =Carex doniana Spreng. ■
74111 Carex chlorostachys Steven;绿穗苔草(绿穗细秆苔草);Greenspike Sedge ■
74112 Carex chlorostachys Steven =Carex capillaris L. subsp. chlorostachys (Steven) Á. Löve, D. Löve et Raymond ■
74113 Carex chlorostachys Steven =Carex capillaris L. ■
74114 Carex chlorostachys Steven var. conferta F. T. Wang et Ts. Tang;无喙绿穗苔草■
74115 Carex chlorostachys Steven var. conferta Ts. Tang et F. T. Wang;青海绿穗苔草■
74116 Carex cholorleuca Meinsh. =Carex amgunensis F. Schmidt ■
74117 Carex chorda H. Lév. et Vaniot;柯达苔草■
74118 Carex chordorrhiza Ehrh. ex L. f.;索根苔草;Cord-root Sedge, Cord-rooted Sedge, Creeping Sedge, String Sedge ■☆
74119 Carex chordorrhiza Ehrh. ex L. f. var. pseudocuraica (Schmidt) Trautv. =Carex pseudocuraica Eastw. ■
74120 Carex chordorrhiza Ehrh. var. pseudocuraica (F. Schmidt) Trautv. =Carex pseudocuraica F. Schmidt ■
74121 Carex chosenica Ohwi;朝鲜苔草■☆
74122 Carex chrysolepis Franch. et Sav.;黄花苔草(黄花苔);Yellow-spiked Sedge ■
74123 Carex chrysolepis Franch. et Sav. subsp. glabrior (Ohwi) T. Koyama =Carex chrysolepis Franch. et Sav. var. glabrior (Ohwi) Ohwi ■☆
74124 Carex chrysolepis Franch. et Sav. var. glabrior (Ohwi) Ohwi;光黄花苔草■☆
74125 Carex chrysolepis Franch. et Sav. var. odontostoma (Kük.) Ohwi;齿孔黄花苔草■
74126 Carex chrysolepis Franch. et Sav. var. odontostoma (Kük.) Ohwi =Carex chrysolepis Franch. et Sav. ■
74127 Carex chrysoleuca T. Holm =Carex densa (L. H. Bailey) L. H. Bailey ■☆
74128 Carex chuiana F. T. Wang et Ts. Tang ex P. C. Li;桂龄苔草;Guiling Sedge ■
74129 Carex chuii Nelmes;曲氏苔草;Chu Sedge ■
74130 Carex chungii C. P. Wang;仲氏苔草(安徽苔草,皱苞苔草);Anhui Sedge, Chung Sedge ■
74131 Carex chungii C. P. Wang var. rigida Y. C. Tang et S. Yun Liang;坚硬苔草;Rigid Chung Sedge ■
74132 Carex chuniana F. T. Wang et Ts. Tang =Carex commixta Steud. ■
74133 Carex ciliatomarginata Nakai;睫毛苔草■☆
74134 Carex ciliatomarginata Nakai =Carex siderosticta Hance var. pilosa H. Lév. et Nakai ■
74135 Carex cinerascens Kük.;灰化苔草(匍枝苔草);Ashgrey Sedge ■
74136 Carex cinerascens Kük. =Carex micrantha Kük. ■
74137 Carex cinerea Pollich =Carex canescens L. ■
74138 Carex circinata C. A. Mey.;卷须苔草■☆
74139 Carex clavata Thunb.;棍棒苔草■☆
74140 Carex clivicola Fernald et Weath. =Carex peckii Howe ■☆
74141 Carex clivorum Ohwi;山岗苔草■☆
74142 Carex coacta Boott =Carex divisa Huds. ■
74143 Carex cobresiiformis A. I. Baranov et Skvortsov =Carex

gynocrates Wormsk. ex Drejer ■
74144 Carex cognata Kunth;共生苔草■☆
74145 Carex cognata Kunth var. abyssinica (Chiov.) Lye;阿比西尼亚苔草■☆
74146 Carex cognata Kunth var. congolensis (Turrill) Lye;刚果苔草■☆
74147 Carex cognata Kunth var. drakensbergensis (C. B. Clarke) Kük. = Carex cognata Kunth ■☆
74148 Carex colchica J. Gay;克里其苔草(克里其苔)■☆
74149 Carex collifera Ohwi;琉球苔草■☆
74150 Carex collinsii Nutt.;克林斯苔草■☆
74151 Carex colorata Mack. = Carex woodii Dewey ■☆
74152 Carex columbiana Dewey = Carex mertensii J. D. Prescott ex Bong. ■☆
74153 Carex comaroviana A. I. Baranov et Skvortsov = Carex angarae Steud. ■
74154 Carex commixta Steud.;细长喙苔草■
74155 Carex communis L. H. Bailey;普通苔草;Colonial Oak Sedge, Fibrous-root Sedge, Sedge, Common Sedge ■☆
74156 Carex communis L. H. Bailey var. amplisquama (F. J. Herm.) Rettig;富鳞苔草■☆
74157 Carex comosa Boott;刚毛苔草;Bristly Sedge, Sedge ■☆
74158 Carex compacta Lam.;密苔草■☆
74159 Carex compacta R. Br. ex Dewey = Carex saxatilis L. ■☆
74160 Carex complanata Torr. et Hook.;扁平苔草;Sedge ■☆
74161 Carex complanata Torr. et Hook. var. hirsuta (L. H. Bailey) Gleason = Carex hirsutella Mack. ■☆
74162 Carex composita Boott;复序苔草;Composite Sedge ■
74163 Carex composita Boott var. emineus (Nees) Boeck. = Carex emineus Nees ■
74164 Carex concinna R. Br.;北方雅致苔草;Beautiful Sedge, Elegant Sedge, Low Northern Sedge, Northern Elegant Sedge ■☆
74165 Carex concinnoides Mack.;拟北方雅致苔草;Northwest Sedge ■☆
74166 Carex concolor R. Br. = Carex bigelowii Torr. et Schwein. ex Boott ■
74167 Carex condensata C. B. Clarke = Carex zuluensis C. B. Clarke ■☆
74168 Carex condensata Nees = Carex zuluensis C. B. Clarke ■☆
74169 Carex conferta Hochst. ex A. Rich.;群集苔草■☆
74170 Carex conferta Hochst. ex A. Rich. var. leptosaccus (C. B. Clarke) Kük.;细囊苔草■☆
74171 Carex confertiflora Boott;密花苔草;Denseflower Sedge ■
74172 Carex congesta C. A. Mey.;聚集苔草■☆
74173 Carex congolensis Turrill = Carex cognata Kunth var. congolensis (Turrill) Lye ■☆
74174 Carex conica Boott;圆锥苔草;Hime Kan Suga ■☆
74175 Carex conica Boott var. densa Kük. = Carex kiangsuensis Kük. ■
74176 Carex conica Boott var. latifolia Hatus.;宽叶圆锥苔草■☆
74177 Carex conica Boott var. scabrocaudata (T. Koyama) Hatus.;糙尾苔草■☆
74178 Carex conicoides Honda = Carex pisiformis Boott ■
74179 Carex conjuncta Boott;接合苔草;Sedge, Soft Fox Sedge ■☆
74180 Carex conoidea Willd.;旷野苔草;Open-field Sedge, Prairie Gray Sedge, Sedge ■☆
74181 Carex consanguinea Kunth = Carex divisa Huds. ■
74182 Carex conspissata V. I. Krecz.;团集苔草■☆
74183 Carex continua C. B. Clarke;连续苔草;Continuous Sedge ■
74184 Carex convoluta Mack. = Carex rosea Schkuhr ex Willd. ■☆
74185 Carex copulata (L. H. Bailey) Mack. = Carex laxiculmis Schwein. ■☆
74186 Carex coreana Bailey = Carex dickinsii Franch. et Sav. ■
74187 Carex coriophora Fisch. et C. A. Mey. ex Kunth;扁囊苔草(贝加尔苔草);Flatutricle Sedge ■
74188 Carex coriophora Fisch. et C. A. Mey. ex Kunth subsp. langtaodianensis S. Yun Liang;浪淘殿苔草;Langtaodian Flatutricle Sedge ■
74189 Carex corrugata Fernald;皱折苔草■☆
74190 Carex courtallensis Nees ex Boott;隐穗柄苔草(庭园苔草)■
74191 Carex cranaocarpa Nelmes;鹤果苔草;Oblongfruit Sedge ■
74192 Carex crandallii Gand. = Carex micropoda C. A. Mey. ■☆
74193 Carex craspedotricha Nelmes;缘毛苔草;Ciliate Sedge ■
74194 Carex crassinervia Franch. = Carex meyeriana Kunth ■
74195 Carex crawei Dewey;克劳氏苔草;Crawe's Sedge, Early Fen Sedge, Sedge ■☆
74196 Carex crawfordii Fernald;克劳福德苔草;Crawford's Oval Sedge, Crawford's Sedge, Sedge ■☆
74197 Carex crawfordii Fernald var. vigens Fernald = Carex crawfordii Fernald ■☆
74198 Carex crebra V. I. Krecz.;密生苔草;Dense Sedge ■
74199 Carex cremostachys Franch.;燕子苔草(垂穗苔草)■
74200 Carex crinita Lam.;北美流苏苔草;Fringed Sedge ■☆
74201 Carex crinita Lam. var. brevicrinis Fernald = Carex crinita Lam. ■☆
74202 Carex crinita Lam. var. gynandra (Schwein.) Schwein. et Torr. = Carex gynandra Schwein. ■☆
74203 Carex crinita Lam. var. minor Boott = Carex crinita Lam. ■☆
74204 Carex crinita Lam. var. mitchelliana (M. A. Curtis) Gleason = Carex mitchelliana M. A. Curtis ■☆
74205 Carex crinita Lam. var. morbida J. Carey = Carex crinita Lam. ■☆
74206 Carex crinita Lam. var. paleacea (Schreb. ex Wahlenb.) Dewey = Carex paleacea Schreb. ex Wahlenb. ■☆
74207 Carex crinita Lam. var. simulans Fernald = Carex gynandra Schwein. ■☆
74208 Carex cristata Schwein. = Carex cristatella Britton ■☆
74209 Carex cristatella Britton;冠毛苔草;Crested Oval Sedge, Crested Sedge, Sedge ■☆
74210 Carex cristatella Britton f. catelliformis (Farw.) Fernald = Carex cristatella Britton ■☆
74211 Carex cruciata Wahlenb.;十字苔草(三角草,三棱草,烟火苔,油草);Cruciate Sedge ■
74212 Carex cruciata Wahlenb. subsp. rubro-brunnea (Ohwi) T. Koyama;红色烟火苔■
74213 Carex cruciata Wahlenb. subsp. rubro-brunnea (Ohwi) T. Koyama = Carex cruciata Wahlenb. ■
74214 Carex cruciata Wahlenb. subsp. rubro-brunnea (Ohwi) T. Koyama = Carex subfilicinoides Kük. ■
74215 Carex cruciata Wahlenb. var. rafflesiana (Boott) Kern et Noot. = Carex rafflesiana Boott ■
74216 Carex cruciata Wahlenb. var. rubro-brunea Ohwi = Carex cruciata Wahlenb. ■
74217 Carex cruenta Nees;红鳞苔草(狭囊苔草)■
74218 Carex crus-corvi Shuttlew. ex Kuntze var. virginiana Fernald = Carex crus-corvi Shuttlew. ex Kunze ■☆
74219 Carex crus-corvi Shuttlew. ex Kunze;鸟足苔草;Crow-foot Fox Sedge, Crow-spur Sedge, Raven's-foot Sedge, Sedge ■☆
74220 Carex cryptantha T. Holm = Carex glareosa Wahlenb. ■☆
74221 Carex cryptocarpa C. A. Mey.;隐果苔草■
74222 Carex cryptocarpa C. A. Mey. = Carex lyngbyei Hornem. ■

74223 Carex cryptolepis Mack.;隐苞苔草;Northeastern Sedge, Small Yellow Sedge ■☆

74224 Carex cryptostachys Brongn.;隐穗苔草(多序宿柱苔,隐柱苔);Cryptostachys Sedge ■

74225 Carex cumulata (L. H. Bailey) Mack.;簇生苔草;Clustered Sedge,Crowded Oval Sedge,Dense Sedge ■☆

74226 Carex cuneata Ohwi = Carex stenostachys Franch. et Sav. var. cuneata (Ohwi) Ohwi et T. Koyama ■☆

74227 Carex cuprea (Kük.) Nelmes = Carex petitiana A. Rich. ■

74228 Carex cuprina (Heuff.) A. Kern.;铜苔草■☆

74229 Carex cuprina (Sand. ex Heuff.) Nendtv. ex A. Kern. = Carex otrubae Podp. ■

74230 Carex curaica Kunth;库地苔草■

74231 Carex curaica Kunth = Carex pseudocuraica Eastw. ■

74232 Carex curaica Kunth = Carex pycnostachys Kar. et Kir. ■

74233 Carex curaica Kunth = Carex vulpinaris Nees ■

74234 Carex curaica Kunth var. pycnostachya (Kar. et Kir.) Kük. = Carex pycnostachys Kar. et Kir. ■

74235 Carex curta Gooden.;白山苔草;White Sedge ■

74236 Carex curta Gooden. = Carex canescens L. ■

74237 Carex curta Gooden. var. brunnescens Pers. = Carex brunnescens (Pers.) Poir. ■

74238 Carex curticeps C. B. Clarke = Kobresia curticeps (C. B. Clarke) Kük. ■

74239 Carex curvata Boott = Kobresia curvata (Boott) Kük. ■

74240 Carex curvata Boott = Kobresia fragilis C. B. Clarke ■

74241 Carex curvicollis Franch. et Sav.;弯颈苔草■☆

74242 Carex curvula All.;曲苔草■☆

74243 Carex cusickii Mack. ex Piper et Beattie;库西克苔草■☆

74244 Carex cuspidata Host;硬尖苔草■☆

74245 Carex cylindrostachys Franch.;柱穗苔草■

74246 Carex cyouensis X. Y. Mao et Y. C. Yang;阿右苔草■

74247 Carex cyperoides Murray = Carex bohemica Schreb. ■

74248 Carex cyperoides Murray ex L. = Carex bohemica Schreb. ■

74249 Carex cyrtosaccus C. B. Clarke;弯苔草■☆

74250 Carex czarwakensis Litv.;契尔瓦克苔草■☆

74251 Carex czarwakensis Litv. = Carex diluta M. Bieb. ■

74252 Carex dabieensis S. W. Su;大别苔草;Dabieshan Sedge ■

74253 Carex dahurica Kük.;针苔草;Dahurian Sedge ■

74254 Carex daibuensis Hayata = Carex sociata Boott ■

74255 Carex daibuensis Hayata = Carex transalpine Hayata ■

74256 Carex dailingensis Y. L. Chou;带岭苔草;Dailing Sedge ■

74257 Carex daisenensis Nakai;大山苔草■☆

74258 Carex davalliana Sm.;达氏苔草;Davall Sedge,Davall's Sedge ■☆

74259 Carex davidii Franch.;无喙囊苔草(长芒苔草);David Sedge ■

74260 Carex davidii Franch. var. ascocetra (C. B. Clarke) Kük. = Carex ascotreta C. B. Clarke ex Franch. ■

74261 Carex davisii Schwein. et Torr.;戴维斯苔草;Awned Graceful Sedge,Davis' Sedge ■☆

74262 Carex dayunshanensis L. K. Ling et Y. Z. Huang = Carex glossostigma Hand.-Mazz. ■

74263 Carex dayuongensis Z. P. Wang;大庸苔草;Dayong Sedge ■

74264 Carex dayuongensis Z. P. Wang = Carex glossostigma Hand.-Mazz. ■

74265 Carex debilis Michx.;柔弱苔草;Sedge,Weak Sedge,White-edge Sedge ■☆

74266 Carex debilis Michx. var. intercursa Fernald = Carex debilis Michx. ■☆

74267 Carex debilis Michx. var. intercursa Fernald = Carex debilis Michx. var. rudgei L. H. Bailey ■☆

74268 Carex debilis Michx. var. interjecta L. H. Bailey = Carex debilis Michx. var. rudgei L. H. Bailey ■☆

74269 Carex debilis Michx. var. pubera A. Gray = Carex debilis Michx. var. rudgei L. H. Bailey ■☆

74270 Carex debilis Michx. var. puesa A. Gray = Carex debilis Michx. ■☆

74271 Carex debilis Michx. var. rudgei L. H. Bailey;卢德柔弱苔草;Northern Weak Sedge,Weak Sedge,White-edge Sedge ■☆

74272 Carex debilis Michx. var. strictior L. H. Bailey = Carex debilis Michx. var. rudgei L. H. Bailey ■☆

74273 Carex deciduisquama F. T. Wang et Ts. Tang ex P. C. Li;落鳞苔草■

74274 Carex decomposita Muhl.;多裂苔草;Sedge ■☆

74275 Carex decora Boott;多穗苔草■

74276 Carex decurticaulis Ohwi = Carex reptabunda (Trautv.) V. I. Krecz. ■

74277 Carex deflexa Hornem.;外折苔草;Depressed Sedge, Northern Oak Sedge,Northern Sedge ■☆

74278 Carex deflexa Hornem. var. boottii L. H. Bailey;布特苔草■☆

74279 Carex deflexa Hornem. var. deanii L. H. Bailey = Carex deflexa Hornem. ■☆

74280 Carex deflexa Hornem. var. farwellii Britton = Carex rossii Boott ■☆

74281 Carex deflexa Hornem. var. media L. H. Bailey = Carex rossii Boott ■☆

74282 Carex deflexa Hornem. var. rossii (Boott) L. H. Bailey = Carex rossii Boott ■☆

74283 Carex delavayi Franch.;年佳苔草(虾蟆地苔草);Delavay Sedge ■

74284 Carex demissa Hornem. = Carex viridula Michx. subsp. oedocarpa (Andersson) B. Schmid ■☆

74285 Carex densa (L. H. Bailey) L. H. Bailey;密集苔草■☆

74286 Carex densefimbriata F. T. Wang et Ts. Tang ex S. Yun Liang;流苏苔草(兰花七,密缘毛苔草,棕树七);Densefimbriate Sedge ■

74287 Carex densefimbriata F. T. Wang et Ts. Tang ex S. Yun Liang var. hirsuta P. C. Li;粗毛流苏苔草;Hirsute Densefimbriate Sedge ■

74288 Carex densenervosa Chiov.;密脉苔草■☆

74289 Carex densicaespitosa L. K. Dai;密丛苔草;Dense Sedge ■

74290 Carex densipilosa C. Z. Zheng et X. F. Jin;密毛苔草;Densehair Sedge ■

74291 Carex depauperata Curtis;荒苔草;Starved Wood-sedge ■☆

74292 Carex depressa Link;凹陷苔草■☆

74293 Carex depressa Link subsp. basilaris (Jord.) Kerguélen;基生苔草■☆

74294 Carex depressa Link var. basilaris (Jord.) Asch. et Graebn. = Carex depressa Link subsp. basilaris (Jord.) Kerguélen ■☆

74295 Carex deqinensis L. K. Dai;德钦苔草;Deqin Sedge ■

74296 Carex descendens Kük.;俯垂苔草■☆

74297 Carex deweyana Schwein.;杜威苔草;Dewey's Sedge ■☆

74298 Carex deweyana Schwein. subsp. leptopoda (Mack.) Calder et R. L. Taylor = Carex leptopoda Mack. ■☆

74299 Carex deweyana Schwein. subsp. senanensis (Ohwi) T. Koyama;信浓苔草■☆

74300 Carex deweyana Schwein. var. bolanderi (Olney) W. Boott = Carex bolanderi Olney ■☆

74301 Carex deweyana Schwein. var. leptopoda (Mack.) B. Boivin = Carex leptopoda Mack. ■☆

74302 Carex deweyana Schwein. var. senanensis (Ohwi) T. Koyama = Carex deweyana Schwein. subsp. senanensis (Ohwi) T. Koyama ■☆

74303 Carex deweyana Schwein. var. sparsiflora L. H. Bailey = Carex laeviculmis Meinsh. ■☆

74304 Carex diandra Schrank;二蕊苔草(圆锥苔草);Bog Panicled Sedge, Diandrous Sedge, Lesser Panicled Sedge, Lesser Tussock-sedge ■

74305 Carex diandra Schrank var. ramosa (Boott) Fernald = Carex prairea Dewey ■☆

74306 Carex dichroa Franch. = Carex thibetica Franch. ■

74307 Carex dichroa Freyn;小穗苔草(二色苔草,双花苔草)■

74308 Carex dichroandra V. I. Krecz.;二色蕊苔草■☆

74309 Carex dickinsii Franch. et Sav.;迪金斯苔草(朝鲜苔草);Dickins Sedge ■

74310 Carex dielsiana Kük.;丽江苔草;Lijiang Sedge ■

74311 Carex digitalis Kük. var. asymmetrica Fernald = Carex digitalis Willd. var. floridana (L. H. Bailey) Naczi et Bryson ■☆

74312 Carex digitalis Willd.;北美窄叶苔草;Narrow-leaved Wood Sedge, Sedge, Slender Woodland Sedge ■☆

74313 Carex digitalis Willd. var. asymmetrica Fernald = Carex digitalis Willd. ■☆

74314 Carex digitalis Willd. var. copulata L. H. Bailey = Carex laxiculmis Schwein. var. copulata (L. H. Bailey) Fernald ■☆

74315 Carex digitalis Willd. var. copulata L. H. Bailey = Carex laxiculmis Schwein. ■☆

74316 Carex digitalis Willd. var. floridana (L. H. Bailey) Naczi et Bryson;佛罗里达北美窄叶苔草■☆

74317 Carex digitalis Willd. var. macropoda Fernald = Carex digitalis Willd. ■☆

74318 Carex digitata L. = Carex quadriflora (Kük.) Ohwi ■

74319 Carex digitata L. subsp. quadriflora Kük. = Carex quadriflora (Kük.) Ohwi ■

74320 Carex digitata L. var. pallida Meinsh. = Carex quadriflora (Kük.) Ohwi ■

74321 Carex digyne (Kük.) Ts. Tang et F. T. Wang = Carex przewalskii T. V. Egorova ■

74322 Carex diluta (Ball) Maire et Weiller subsp. fissirostris = Carex fissirostris Ball ■☆

74323 Carex diluta M. Bieb.;碱苔(八脉苔草,卡氏苔)■

74324 Carex diluta M. Bieb. var. bottae C. B. Clarke ex Blatt. = Carex distans L. f. sinaica (Nees ex Steud.) Kük. ■☆

74325 Carex diluta M. Bieb. var. fissirostris (Ball) Kük. = Carex fissirostris Ball ■☆

74326 Carex dimorpha Brot. = Carex infuscata Nees var. gracilenta (Boott ex Strachey) P. C. Li ■

74327 Carex dimorpholepis Steud.;二形鳞苔草(垂穗苔草);Dimorphicscale Sedge ■

74328 Carex dimorphotheca Stschegl. = Carex stenophylloides V. I. Krecz. ■☆

74329 Carex dineuros C. B. Clarke = Carex jaluensis Kom. ■

74330 Carex dioica L.;欧洲异株苔草;Dioecious Sedge ■☆

74331 Carex dioica L. subsp. gynocrates (Wormsk. ex Drejer) Hultén = Carex gynocrates Wormsk. ex Drejer ■

74332 Carex dioica L. var. gynocrates (Wormsk. ex Drejer) Ostenf. = Carex gynocrates Wormsk. ex Drejer ■

74333 Carex diplasiocarpa V. I. Krecz.;狭囊苔草;Narrow-utricle Sedge ■

74334 Carex diplasiocarpa V. I. Krecz. = Carex yamatsutana Ohwi ■

74335 Carex diplodon Nelmes;秦岭苔草;Qinling Sedge ■

74336 Carex discoidea Boott = Carex leucochlora Bunge var. discoidea (Boott) T. Koyama ■☆

74337 Carex discolor F. Nyl.;异色苔草■☆

74338 Carex disjuncta (Fernald) E. P. Bicknell = Carex canescens L. subsp. disjuncta (Fernald) Toivonen ■☆

74339 Carex dispalata Boott ex A. Gray;皱果苔草(苔,苔草,弯囊苔草,弯嘴苔草);Curvedutricle Sedge ■

74340 Carex dispalata Boott ex A. Gray var. costata Kük. = Carex dispalata Boott ex A. Gray ■

74341 Carex dispalata Boott var. costata Kük. = Carex dispalata Boott ex A. Gray ■

74342 Carex disperma Dewey;二籽苔草(二籽苔,少囊苔草);Biseed Sedge, Soft-leaf Sedge, Two-seeded Bog Sedge, Two-seeded Sedge ■

74343 Carex dissitiflora Franch.;台湾疏花苔■

74344 Carex dissitiflora Franch. subsp. taiwanensis (Ohwi) T. Koyama = Carex taiwanensis (Ohwi) Akiyama ■

74345 Carex dissitiflora Franch. var. taiwanensis Ohwi = Carex taiwanensis (Ohwi) Akiyama ■

74346 Carex dissitispicula Ohwi = Carex kujuzana Ohwi ■☆

74347 Carex distachya Desf.;双穗苔草■☆

74348 Carex distans L.;远布苔草;Distant Sedge ■☆

74349 Carex distans L. f. sinaica (Nees ex Steud.) Kük.;支那苔草■☆

74350 Carex distans L. subsp. oranensis (Trab.) Jahand. et Maire;奥兰苔草■☆

74351 Carex distans L. var. binervis (Sm.) Coss. et Durieu = Carex distans L. ■☆

74352 Carex distans L. var. oranensis Trab. = Carex distans L. subsp. oranensis (Trab.) Jahand. et Maire ■☆

74353 Carex distans L. var. sinaica (Nees ex Steud.) Boeck. = Carex distans L. f. sinaica (Nees ex Steud.) Kük. ■☆

74354 Carex disticha Huds.;间苔(二列苔草);Brown Sedge ■☆

74355 Carex disticha Huds. subsp. lithophila (Turcz.) Hämet-Ahti = Carex lithophila Turcz. ■

74356 Carex distichoidea H. Lév. et Vaniot = Carex lithophila Turcz. ■

74357 Carex divaricata Kük. = Carex mollissima H. Christ ex Meinsh. ■

74358 Carex diversicolor Crantz;杂花苔草■☆

74359 Carex dives T. Holm = Carex aquatilis Wahlenb. var. dives (T. Holm) Kük. ■☆

74360 Carex divisa Huds.;单行苔草(分苔,全裂苔草);Divided Sedge, Separated Sedge ■

74361 Carex divisa Huds. var. ammophila (Willd.) Kük. = Carex divisa Huds. ■

74362 Carex divisa Huds. var. chaetophylla (Steud.) Daveau = Carex divisa Huds. ■

74363 Carex divisa Huds. var. platyphylla Braun-Blanq. et Trab. = Carex divisa Huds. ■

74364 Carex divisa Huds. var. setifolia (Godr.) Trab. = Carex divisa Huds. ■

74365 Carex divulsa Stokes = Carex divulsa Stokes ex With. ■

74366 Carex divulsa Stokes ex With.;疏穗苔草(断苔);Grassland Sedge, Grey Sedge ■

74367 Carex divulsa Stokes subsp. leersii (Kneuck.) W. Koch;利尔斯苔草■☆

74368 Carex doenitzii Boeck.;德尼茨苔草■☆

74369 Carex doenitzii Boeck. subsp. okuboi (Franch.) T. Koyama;大久

苔草■☆
74370 Carex doenitzii Boeck. var. okuboi (Franch.) Kük. ex Matsum. = Carex doenitzii Boeck. subsp. okuboi (Franch.) T. Koyama ■☆
74371 Carex doistepensis T. Koyama;景洪苔草(北秦苔草)■
74372 Carex dolichocarpa C. A. Mey. ex Kom.;长果苔草■☆
74373 Carex dolichocarpa C. A. Mey. ex V. I. Krecz. = Carex michauxiana Boeck. subsp. asiatica Hultén ■☆
74374 Carex dolichostachys Hayata;长穗苔草(长穗宿柱苔);Longspike Sedge ■
74375 Carex dolichostachys Hayata = Carex transalpine Hayata ■
74376 Carex dolichostachys Hayata = Diplocarex matsudai Hayata ■☆
74377 Carex dolichostachys Hayata f. connatisquama T. Koyama;合鳞长穗苔草■☆
74378 Carex dolichostachys Hayata f. pallidisquama (Ohwi) Ohwi = Carex multifolia Ohwi var. pallidisquama Ohwi ■☆
74379 Carex dolichostachys Hayata subsp. multifolia (Ohwi) T. Koyama = Carex multifolia Ohwi ■☆
74380 Carex dolichostachys Hayata subsp. trichosperma (Ohwi) T. Koyama;阿里山宿柱苔■
74381 Carex dolichostachys Hayata subsp. trichosperma (Ohwi) T. Koyama = Carex transalpine Hayata ■
74382 Carex dolichostachys Hayata var. glaberrima (Ohwi) T. Koyama = Carex multifolia Ohwi ■☆
74383 Carex doniana Spreng.;签草(大穗日本苔,绿穗禾状苔草,芒尖苔,芒尖苔草);Mucronate Sedge ■
74384 Carex doniana Spreng. = Carex alopecuroides D. Don ex Okamoto et Taylor var. chlorostachys C. B. Clarke ■☆
74385 Carex douglasii Boott;道格拉斯苔草;Sedge ■☆
74386 Carex drakensbergensis C. B. Clarke = Carex cognata Kunth ■☆
74387 Carex dregeana Kunth = Schoenoxiphium sparteum (Wahlenb.) C. B. Clarke ■☆
74388 Carex dregeana Kunth var. major C. B. Clarke = Schoenoxiphium sparteum (Wahlenb.) C. B. Clarke ■☆
74389 Carex drepanorhyncha Franch.;镰喙苔草(垂嘴苔草)■
74390 Carex drummondiana Dewey = Carex rupestris Turcz. ex Ledeb. ■☆
74391 Carex drymophila Turcz. = Carex drymophila Turcz. ex Steud. ■
74392 Carex drymophila Turcz. ex Steud.;野笠苔草■
74393 Carex drymophila Turcz. ex Steud. subsp. abbreviata (Kük.) T. Koyama = Carex drymophila Turcz. ex Steud. var. abbreviata (Kük.) Ohwi ■
74394 Carex drymophila Turcz. ex Steud. var. abbreviata (Kük.) Ohwi;黑水苔草(阿穆尔苔草)■
74395 Carex drymophila Turcz. var. abbreviata (Kük.) Ohwi = Carex drymophila Turcz. ex Steud. var. abbreviata (Kük.) Ohwi ■
74396 Carex drymophila Turcz. var. akanensis (Franch.) Kük. = Carex drymophila Turcz. ex Steud. var. abbreviata (Kük.) Ohwi ■
74397 Carex drymophila Turcz. var. glabrescens (Kük.) Kitag. = Carex glabrescens (Kük.) Ohwi ■
74398 Carex drymophila Turcz. var. pilifera Kük. = Carex glabrescens (Kük.) Ohwi ■
74399 Carex drymophilia Turcz. var. abbreviata (Kük.) Ohwi;短序野笠苔草■
74400 Carex dudleyi Mack. = Carex densa (L. H. Bailey) L. H. Bailey ■☆
74401 Carex dunnii Hayata = Carex perakensis C. B. Clarke ■
74402 Carex dunnii Hayata = Carex tatewakiana Ohwi ■
74403 Carex durifolia L. H. Bailey;硬叶苔草;Tough-leaved Sedge ■☆
74404 Carex durifolia L. H. Bailey = Carex backii W. Boott ■☆
74405 Carex durifolia L. H. Bailey var. subrostrata Bates = Carex saximontana Mack. ■☆
74406 Carex duriuscula C. A. Mey.;寸草(寸草苔,卵穗苔草,羊胡子草);Eggspike Sedge, Needleleaf Sedge ■
74407 Carex duriuscula C. A. Mey. subsp. rigescens (Franch.) S. Yun Liang et Y. C. Tang;白颖寸草(白颖苔草);Rigescent Sedge ■
74408 Carex duriuscula C. A. Mey. subsp. stenophylloides (V. I. Krecz.) S. Yun Liang et Y. C. Tang;细叶寸草(细叶苔草,针叶苔草,中亚苔草);Mid-Asia Sedge ■
74409 Carex duriuscula C. A. Mey. var. rigescens (Franch.) S. Yun Liang et Y. C. Tang = Carex duriuscula C. A. Mey. subsp. rigescens (Franch.) S. Yun Liang et Y. C. Tang ■
74410 Carex duriuscula C. A. Mey. var. stenophylloides (V. I. Krecz.) S. Yun Liang et Y. C. Tang = Carex duriuscula C. A. Mey. subsp. stenophylloides (V. I. Krecz.) S. Yun Liang et Y. C. Tang ■
74411 Carex duriuscula C. A. Mey. var. tenuispica X. Y. Yuan;细穗寸草■
74412 Carex duriusculiformis V. I. Krecz. = Carex duriuscula C. A. Mey. subsp. stenophylloides (V. I. Krecz.) S. Yun Liang et Y. C. Tang ■
74413 Carex duriusculiformis V. I. Krecz. = Carex stenophylloides V. I. Krecz. ■☆
74414 Carex duthiei sensu Ohwi = Carex caucasica Steven var. jisaburo-ohwiana (T. Koyama) T. Koyama ■
74415 Carex duvaliana Franch. et Sav.;三阳苔草■
74416 Carex duvaliana Franch. et Sav. var. alterniflora (Franch.) Kük. = Carex pisiformis Boott ■
74417 Carex duvaliana Franch. et Sav. var. alterniflora (Franch.) Kük. ex Matsum. = Carex pisiformis Boott ■
74418 Carex earistata F. T. Wang et Y. L. Chang ex S. Yun Liang;无芒苔草;Awnless Sedge ■
74419 Carex eastwoodiana Stacey = Carex tahoensis Smiley ■☆
74420 Carex ebracteata Trautv. = Carex subebracteata (Kük.) Ohwi ■
74421 Carex eburnea Boott;象牙苔草;Bristle-leaf Sedge, Sedge ■☆
74422 Carex echinata Desf. = Carex hispida Willd. ■☆
74423 Carex echinata Murray;刺苔草;Large-fruited Star Sedge, Prickly Sedge, Star Sedge ■☆
74424 Carex echinata Murray subsp. phyllomanica (W. Boott) Reznicek;绿叶苔草■☆
74425 Carex echinata Murray var. angustata (J. Carey) L. H. Bailey = Carex echinata Murray ■☆
74426 Carex echinochloe Kuntze subsp. nyasensis (C. B. Clarke) Lye;尼亚斯苔草■☆
74427 Carex echinochloe Kuntze var. chlorosaccua (C. B. Clarke) Kük. = Carex chlorosaccua C. B. Clarke ■
74428 Carex echinochloe Kuntze var. nyasensis (C. B. Clarke) Kük. = Carex echinochloe Kuntze subsp. nyasensis (C. B. Clarke) Lye ■☆
74429 Carex echinochloiformis Y. I. Chang et Y. L. Yang;类稗苔草;Echinochloaeform Sedge ■
74430 Carex ecklonii Nees;埃氏苔草■☆
74431 Carex egena H. Lév. et Vaniot;少囊苔草(少穗苔草);Fewspike Sedge ■
74432 Carex egena H. Lév. et Vaniot = Carex filipes Franch. et Sav. var. oligostachys Kük. ■
74433 Carex egregia Mack. = Carex angustata Boott ■☆
74434 Carex elachycarpa Fernald = Carex sterilis Willd. ■☆

74435 Carex elata All.;草丛苔草;Common Tussock Sedge,Hummock Sedge,Stiff Sedge, Tufted Grass, Tufted Sedge, Tussock Sedge, Uptight Sedge ■☆

74436 Carex elata All. 'Aurea';金叶丛生苔草;Bowles' Golden Sedge ■☆

74437 Carex eleocharis L. H. Bailey = Carex duriuscula C. A. Mey. ■

74438 Carex eleusinoides Turcz. ex Kunth;蟋蟀苔草■

74439 Carex eleusinoides Turcz. ex Kunth var. subalpina Y. L. Chou;高山蟋蟀苔草■

74440 Carex elgonensis Nelmes;埃尔贡苔草■☆

74441 Carex elliottii Schwein. et Torr.;埃里奥特苔草■☆

74442 Carex elongata L.;长苔草;Elongated Sedge ■☆

74443 Carex elrodii M. E. Jones = Carex hallii Olney ■☆

74444 Carex eluta M. E. Jones;高苔草■☆

74445 Carex elyniformis A. E. Porsild = Carex nardina Fr. ■☆

74446 Carex elynoides T. Holm;贫弱苔草■☆

74447 Carex emasculata V. I. Krecz. = Carex hartmanii Cajander ■☆

74448 Carex emineus Nees;显异苔草■

74449 Carex emmonsii Dewey ex Torr. = Carex albicans Willd. ex Spreng. var. emmonsii (Dewey ex Torr.) Rettig ■☆

74450 Carex emmonsii Dewey ex Torr. var. australis (L. H. Bailey) Rettig = Carex albicans Willd. ex Spreng. ■☆

74451 Carex emmonsii Dewey ex Torr. var. muhlenbergii (A. Gray) Rettig = Carex albicans Willd. ex Spreng. ■☆

74452 Carex emoryi Dewey;埃默里苔草;Emory's Sedge,Sedge ■☆

74453 Carex enervis C. A. Mey.;无脉苔草;Emory Sedge,Emory's Sedge,Enerv Sedge,Veinless Sedge ■

74454 Carex enervis C. A. Mey. subsp. chuanxibeiensis S. Yun Liang et Y. C. Tang;川西北苔草;NW. Sichuan Sedge ■

74455 Carex enervis C. A. Mey. subsp. chuanxibeiensis S. Yun Liang et Y. C. Tang = Carex enervis C. A. Mey. ■

74456 Carex engelmannii L. H. Bailey;恩格尔曼苔草■☆

74457 Carex engelmannii L. H. Bailey var. paddoensis (Suksd.) Kneuck. = Carex engelmannii L. H. Bailey ■☆

74458 Carex ensifolia Turcz. ex Ledeb. = Carex bigelowii Torr. et Schwein. ex Boott ■

74459 Carex ereica Ts. Tang et F. T. Wang ex L. K. Dai;二峨苔草;Ere Sedge ■

74460 Carex eremopyroides V. I. Krecz.;离穗苔草■

74461 Carex ericetorum Pollard;泥炭苔草;Rare Spring-sedge ■☆

74462 Carex eriophora Fisch. ex Steud.;毛梗苔草■☆

74463 Carex eriophylla (Kük.) Kom.;毛叶苔草;Hairy-leaved Sedge ■

74464 Carex erxlebeniana Kelso = Carex inops L. H. Bailey subsp. heliophila (Mack.) Crins ■☆

74465 Carex erythrobasis H. Lév. et Vaniot;红鞘苔草■

74466 Carex erythrorrhiza Boeck.;淡红苔草■☆

74467 Carex erythrorrhiza Boeck. var. curva Chiov.;内折苔草■☆

74468 Carex erythrorrhiza Boeck. var. scabrida Kük. = Carex conferta Hochst. ex A. Rich. var. leptosaccus (C. B. Clarke) Kük. ■☆

74469 Carex esanbeckii Kunth = Kobresia esenbeckii (Kunth) F. T. Wang et Ts. Tang ex P. C. Li ■

74470 Carex esenbeckiana Boeck. = Schoenoxiphium lehmannii (Nees) Steud. ■☆

74471 Carex esenbeckiana Boeck. var. elongata? = Schoenoxiphium sparteum (Wahlenb.) C. B. Clarke ■☆

74472 Carex esenbeckii Kunth = Kobresia esenbeckii (Kunth) Noltie ■

74473 Carex esquirolii H. Lév. et Vaniot = Kyllinga brevifolia Rottb. ■

74474 Carex estesiana Kelso = Carex nelsonii Mack. ■☆

74475 Carex eurycarpa T. Holm = Carex angustata Boott ■☆

74476 Carex euxina (Woronow et Morcow.) F. P. Metcalf ex Kneuck.;黑海苔草■☆

74477 Carex exerta Chu = Carex glossostigma Hand. -Mazz. ■

74478 Carex exilis Dewey;沿海苔草;Coast Sedge,Coastal Sedge,Coastal Star Sedge,Starved Sedge ■☆

74479 Carex exserta Mack. = Carex filifolia Nutt. var. erostrata Kük. ■☆

74480 Carex exsiccata L. H. Bailey;干苔草■☆

74481 Carex extensa Good;广布苔草;Longbract Sedge,Long-bracted Sedge ■☆

74482 Carex extensa Gooden.;伸展苔草■☆

74483 Carex falcata Turcz.;镰苔草;Falcate Sedge ■

74484 Carex falcata Turcz. = Carex sparsiflora Steud. var. petersi (C. A. Mey.) Kük. ■

74485 Carex falcata Turcz. = Carex vaginata Tausch ■☆

74486 Carex fallax var. pseudoarenicola (Hayata) Ohwi = Carex nubigena D. Don ex Okamoto et Taylor subsp. pseudoarenicola (Hayata) T. Koyama ■

74487 Carex fargesii Franch.;川东苔草(亮鞘苔草);Farges Sedge ■

74488 Carex farwellii (Britton) Mack. = Carex rossii Boott ■☆

74489 Carex fascicularis Sol. ex Hook. f.;澳洲丛苔草;Tassel Sedge ■☆

74490 Carex fastigiata Franch.;簇穗苔草(帚状苔草);Broom Sedge,Fastigiate Sedge ■

74491 Carex fedia Nees ex Wight;鸡山苔草;Jishan Sedge ■

74492 Carex fedia Nees ex Wight subsp. miyabei (Franch.) T. Koyama = Carex miyabei Franch. ■

74493 Carex fedia Nees ex Wight var. miyabei (Franch.) T. Koyama = Carex miyabei Franch. ■

74494 Carex fedia Nees ex Wight var. pilifera (Kük.) T. Koyama = Carex glabrescens (Kük.) Ohwi ■

74495 Carex fedtscheakoana Kük.;范氏苔草■☆

74496 Carex fenghuangshanica F. T. Wang et Ts. Tang ex P. C. Li;凤凰山苔草;Fenghuangshan Sedge ■

74497 Carex fernaldiana H. Lév. et Vaniot;线苔草(线叶宿柱苔)■

74498 Carex fernaldiana H. Lév. et Vaniot = Carex pisiformis Boott ■

74499 Carex ferruginea Scop.;锈色苔草■

74500 Carex ferruginea Scop. var. tatsiensis Franch. = Carex tatsiensis (Franch.) Kük. ■

74501 Carex ferruspicalata Chu = Carex glossostigma Hand. -Mazz. ■

74502 Carex festivella Mack. = Carex microptera Mack. ■☆

74503 Carex festucacea Schkuhr ex Willd.;牛毛苔草;Fescue Oval Sedge,Fescue Sedge,Sedge ■☆

74504 Carex festucacea Schkuhr ex Willd. var. brevior (Dewey) Fernald = Carex brevior (Dewey) Mack. ex Lunell ■☆

74505 Carex feta L. H. Bailey;混杂苔草■☆

74506 Carex fexilis Rudge = Carex castanea Wahlenb. ■☆

74507 Carex fibrillosa Franch. et Sav. = Carex breviculmis R. Br. var. fibrillosa (Franch. et Sav.) Kük. ex Matsum. et Hayata ■

74508 Carex fidia Nees = Carex fidia Nees ex Wight ■

74509 Carex fidia Nees ex Wight;南亚苔草■

74510 Carex fidia Nees ex Wight var. miyabei (Franch.) T. Koyama = Carex miyabei Franch. ■

74511 Carex filamentosa K. T. Fu = Carex capilliculmis S. R. Zhang ■

74512 Carex filicina Nees = Carex filicina Nees ex Wight ■

74513 Carex filicina Nees ex Wight;蕨状苔草(红鞘苔);Fernlike Sedge ■

74514 Carex filicina Nees ex Wight subsp. pseudofilicina (Hayata) T. Koyama;假蕨状苔草■

74515 Carex filicina Nees ex Wight subsp. pseudofilicina (Hayata) T. Koyama = Carex filicina Nees ex Wight ■

74516 Carex filicina Nees ex Wight var. meiogyna (Nees) Strachey = Carex filicina Nees ex Wight ■

74517 Carex filicina Nees ex Wight var. pseudofilicina (Hayata) T. Koyama = Carex filicina Nees ex Wight subsp. pseudofilicina (Hayata) T. Koyama ■

74518 Carex filicina Nees ex Wight var. subdensa F. T. Wang et Ts. Tang = Carex filicina Nees ex Wight ■

74519 Carex filiculmis Franch. et Sav. = Carex breviculmis R. Br. ■

74520 Carex filifolia Nutt. var. erostrata Kük.;无喙线叶苔草■☆

74521 Carex filifolia Nutt. var. miser L. H. Bailey = Carex elynoides T. Holm ■☆

74522 Carex filiformis Good = Carex lasiocarpa Ehrh. ■

74523 Carex filipedunculata S. W. Su;丝梗苔草;Filipedunculate Sedge ■

74524 Carex filipes Franch. et Sav.;线柄苔草(羊胡子草);Sheep's Beard-like Sedge ■

74525 Carex filipes Franch. et Sav. subsp. arisanensis (Hayata) T. Koyama = Carex arisanensis Hayata ■

74526 Carex filipes Franch. et Sav. subsp. kuzakaiensis M. Kikuchi = Carex filipes Franch. et Sav. var. kuzakaiensis (M. Kikuchi) T. Koyama ■☆

74527 Carex filipes Franch. et Sav. subsp. oligostachys (Kük.) T. Koyama = Carex filipes Franch. et Sav. var. oligostachys Kük. ■

74528 Carex filipes Franch. et Sav. subsp. rouyana (Franch.) T. Koyama = Carex filipes Franch. et Sav. var. rouyana (Franch.) Kük. ■☆

74529 Carex filipes Franch. et Sav. var. arakiana (Ohwi) Ohwi;荒木苔草■☆

74530 Carex filipes Franch. et Sav. var. arisanensis (Hayata) T. Koyama = Carex arisanensis Hayata ■

74531 Carex filipes Franch. et Sav. var. kuzakaiensis (M. Kikuchi) T. Koyama;岩手苔草■☆

74532 Carex filipes Franch. et Sav. var. oligostachys (Meinsh. ex Maxim.) Kük. = Carex egena H. Lév. et Vaniot ■

74533 Carex filipes Franch. et Sav. var. oligostachys Kük. = Carex egena H. Lév. et Vaniot ■

74534 Carex filipes Franch. et Sav. var. rouyana (Franch.) Kük.;卢伊苔草■☆

74535 Carex filipes Franch. et Sav. var. rouyana (Franch.) Kük. = Carex filipes Franch. et Sav. ■

74536 Carex filipes Franch. et Sav. var. rouyana (Franch.) Kük. = Carex filipes Franch. et Sav. var. sparsinux (C. B. Clarke ex Franch.) Kük. ■

74537 Carex filipes Franch. et Sav. var. sparsinux (C. B. Clarke ex Franch.) Kük.;丝柄苔草■

74538 Carex filipes Franch. et Sav. var. sparsinux (C. B. Clarke) Kük. = Carex filipes Franch. et Sav. var. sparsinux (C. B. Clarke ex Franch.) Kük. ■

74539 Carex filipes Franch. et Sav. var. tremula (Ohwi) Ohwi;颤线柄苔草■☆

74540 Carex finitima Boott;亮绿苔草(长柱苔);Lightgreen Sedge, Shininggreenutricle Sedge ■

74541 Carex finitima Boott var. attenuata C. B. Clarke;短叶亮绿苔草■

74542 Carex firma Host;灰绿苔草;Blue Sedge, Carnation Grass, Glaucous Sedge, Heath Sedge ■☆

74543 Carex fischeri K. Schum. = Carex petitiana A. Rich. ■

74544 Carex fissa Mack.;半裂苔草;Sedge ■☆

74545 Carex fissirostris Ball;裂喙苔草■☆

74546 Carex flabellata H. Lév. et Vaniot;扇状苔草■☆

74547 Carex flacca Schreb.;康乃馨苔草;Blue Sedge, Carnation Grass, Glaucous Sedge, Heath Sedge ■☆

74548 Carex flacca Schreb. subsp. erythrostachys (Hoppe) Holub = Carex flacca Schreb. subsp. serrulata (Biv.) Greuter ■☆

74549 Carex flacca Schreb. subsp. serrulata (Biv.) Greuter;细齿康乃馨苔草■☆

74550 Carex flacca Schreb. var. arrecta (Drejer) Emb. et Maire = Carex flacca Schreb. subsp. serrulata (Biv.) Greuter ■☆

74551 Carex flacca Schreb. var. attenuata (Ball) Emb. et Maire = Carex flacca Schreb. ■☆

74552 Carex flacca Schreb. var. glauca Asch. et Graebn. = Carex flava L. ■☆

74553 Carex flaccidula Steud. = Carex rosea Schkuhr ex Willd. ■☆

74554 Carex flaccifolia Mack. = Carex whitneyi Olney ■☆

74555 Carex flaccosperma Dewey;细籽苔草;Blue Wood Sedge, Thinfruit Sedge ■☆

74556 Carex flaccosperma Dewey var. glaucodea (Tuck. ex Olney) Kük. = Carex glaucodea Tuck. ex Olney ■☆

74557 Carex flava L.;黄苔草;Large Yellow Sedge, Yellow Sedge ■☆

74558 Carex flava L. subsp. brachyrrhycha Celak. = Carex viridula Michx. subsp. brachyrrhyncha (Celak.) B. Schmid ■☆

74559 Carex flava L. var. elatior Schltdl. = Carex viridula Michx. var. elatior (Schltdl.) Crins ■☆

74560 Carex flava L. var. fertilis Peck = Carex flava L. ■☆

74561 Carex flava L. var. gaspensis Fernald = Carex flava L. ■☆

74562 Carex flava L. var. graminis L. H. Bailey = Carex flava L. ■☆

74563 Carex flava L. var. laxior (Kük.) Gleason = Carex flava L. ■☆

74564 Carex flava L. var. lepidocarpa (Tausch) Godr. = Carex lepidocarpa Tausch ■☆

74565 Carex flava L. var. nevadensis (Boiss. et Reut.) Briq. = Carex viridula Michx. subsp. nevadensis (Boiss. et Reut.) B. Schmid ■☆

74566 Carex flava L. var. rectirostra Gaudin = Carex flava L. ■☆

74567 Carex flava L. var. vulgaris Döll = Carex flava L. ■☆

74568 Carex flavella Krecz.;淡黄苔草■☆

74569 Carex flavocuspis Franch. et Sav.;黄尖苔草■☆

74570 Carex flavocuspis Franch. et Sav. var. platycarpa (Kük.) Akiyama;宽果黄尖苔草■☆

74571 Carex flexilis Rudge = Carex castanea Wahlenb. ■☆

74572 Carex flexuosa Muhl. ex Willd. = Carex debilis Michx. var. rudgei L. H. Bailey ■☆

74573 Carex floribunda (Korsh.) Meinsh. = Carex lanceolata Boott ■

74574 Carex floribunda Boeck. = Carex emineus Nees ■

74575 Carex floridana Schwein.;佛罗里达苔草■☆

74576 Carex floridana Schwein. = Carex nigromarginata Schwein. ■☆

74577 Carex fluviatilis Boott;溪生苔草■

74578 Carex fluviatilis Boott var. unisexualis (C. B. Clarke) Kük. = Carex unisexualis C. B. Clarke ■

74579 Carex foenea Willd.;铜色头苔草;Bronze-headed Oval Sedge, Hay Sedge ■☆

74580 Carex foenea Willd. = Carex siccata Dewey ■☆

74581 Carex foetida All.;烈味苔草■☆

74582 Carex foetida All. var. vernacula (L. H. Bailey) Kük. = Carex

vernacula L. H. Bailey ■☆
74583 Carex foliosa D. Don;密叶苔草;Manyleves Sedge ■
74584 Carex foliosissima Eastw.;繁叶苔草■☆
74585 Carex foliosissima Eastw. var. latissima (Ohwi) Akiyama = Carex foliosissima Eastw. ■☆
74586 Carex foliosissima Eastw. var. pallidivaginata J. Oda et Nagam.;苍白繁叶苔草■☆
74587 Carex folliculata L.;北方长苔草;Long Sedge, Northern Long Sedge ■☆
74588 Carex folliculata L. var. australis L. H. Bailey = Carex lonchocarpa Willd. ex Spreng. ■☆
74589 Carex fontinalis Y. L. Chang = Carex enervis C. A. Mey. ■
74590 Carex foraminata C. B. Clarke;穿孔苔草;Foramenbearing Sedge ■
74591 Carex foraminatiformis Y. C. Tang et S. Yun Liang;拟穿孔苔草;Foraminatiform Sedge ■
74592 Carex forficula Franch. et Sav.;溪水苔草■
74593 Carex forficula Franch. et Sav. var. melinacra (Franch.) Kük. = Carex melinacra Franch. ■
74594 Carex forficula Franch. et Sav. var. scabrida Kük.;粗糙溪水苔草■☆
74595 Carex formosa Dewey;美丽无芒苔草;Awnless Graceful Sedge, Handsome Sedge ■☆
74596 Carex formosensis H. Lév. et Vaniot;宝岛宿柱苔草(宝岛宿柱苔);Taiwan Sedge ■
74597 Carex formosensis H. Lév. et Vaniot = Carex ascotreta C. B. Clarke ex Franch. ■
74598 Carex formosensis H. Lév. et Vaniot var. vigens (Kük.) Ohwi = Carex genkaiensis Ohwi ■
74599 Carex forrestii Kük.;刺喙苔草(绿穗苔草);Forrest Sedge ■
74600 Carex forsicula Kük. var. scabrida Kük.;糙囊苔草;Scabrous-utricle Sedge ■☆
74601 Carex frankii Kunth;弗兰克苔草■☆
74602 Carex franklinii Boott = Carex petricosa Dewey ■☆
74603 Carex fraseri Andréws = Cymophyllus fraserianus (Ker Gawl.) Kartesz et Gandhi ■☆
74604 Carex fraserianus Ker Gawl. = Cymophyllus fraserianus (Ker Gawl.) Kartesz et Gandhi ■☆
74605 Carex fujimakii M. Kikuchi = Carex kujuzana Ohwi ■☆
74606 Carex fujitae Kudo;滕田苔草■☆
74607 Carex fuliniginosa Schkuhr;黑色苔草■
74608 Carex fulta Franch.;支撑苔草■☆
74609 Carex fulvescens Mack. = Carex hostiana DC. ■☆
74610 Carex fulvorubescens Hayata;茶色苔草(茶色扁果苔,长梗扁果苔)■
74611 Carex fulvorubescens Hayata subsp. longistipes (Hayata) T. Koyama;长梗茶色苔草(长梗扁果苔,长梗苔草)■
74612 Carex fulvorubescens Hayata subsp. longistipes (Hayata) T. Koyama = Carex longistipes Hayata ■
74613 Carex funhuangshanica F. T. Wang et Ts. Tang ex P. C. Li = Carex fenghuangshanica F. T. Wang et Ts. Tang ex P. C. Li ■
74614 Carex funingensis Ts. Tang et F. T. Wang ex S. Yun Liang;富宁苔草;Funing Sedge ■
74615 Carex fusanensis Ohwi = Carex gifuensis Franch. var. koreana Nakai ■☆
74616 Carex fusca All. = Carex nigra (L.) Reichard ■☆
74617 Carex fusca All. subsp. goodenowii (Gay) Maire et Weiller = Carex nigra (L.) Reichard ■☆
74618 Carex fusca All. subsp. intricata (Tineo) Maire et Weiller = Carex nigra (L.) Reichard subsp. intricata (Tineo) Rivas Mart. ■☆
74619 Carex fusca All. var. atlantica Litard. et Maire = Carex nigra (L.) Reichard subsp. intricata (Tineo) Rivas Mart. ■☆
74620 Carex fuscidula V. I. Krecz. ex T. V. Egorova = Carex capillaris L. ■
74621 Carex fuscocuprea (Kük.) V. I. Krecz.;铜色苔草■☆
74622 Carex fusco-vaginata Kük.;棕鞘苔草■☆
74623 Carex fusiformis Chapm. ex Boott var. moijishanica Y. C. Yang;麦碛山苔草;Maiqishan Sedge ■
74624 Carex fusiformis Chapm. ex Dewey = Carex chapmanii Steud. ■☆
74625 Carex gaoligongshanensis P. C. Li;高黎贡山苔草;Gaoligongshan Sedge ■
74626 Carex garberi Fernald;麋鹿苔草;Elk Sedge, False Golden Sedge ■☆
74627 Carex garberi Fernald subsp. bifaria (Fernald) Hultén = Carex garberi Fernald ■☆
74628 Carex garberi Fernald var. bifaria Fernald = Carex garberi Fernald ■☆
74629 Carex gardneri Lepage = Carex paleacea Schreb. ex Wahlenb. ■☆
74630 Carex gaudichaudiana Kunth subsp. appendiculata (Trautv.) Á. Löve et D. Löve = Carex thunbergii Steud. var. appendiculata (Trautv.) Ohwi ■
74631 Carex gaudichaudiana Kunth var. appendiculata (Trautv.) T. Koyama = Carex thunbergii Steud. var. appendiculata (Trautv.) Ohwi ■
74632 Carex gaudichaudiana Kunth var. thunbergii (Steud.) Kük. = Carex thunbergii Steud. ■
74633 Carex gaudichaudiana Kunth var. thunbergii (Steud.) Kük. ex Matsum. = Carex thunbergii Steud. ■
74634 Carex gauthieri Lepage = Carex paleacea Schreb. ex Wahlenb. ■☆
74635 Carex gebhardii Hoppe = Carex brunnescens (Pers.) Poir. ■
74636 Carex geihokuensis Okamoto;艺北苔草■☆
74637 Carex genkaiensis Ohwi = Carex formosensis H. Lév. et Vaniot var. vigens (Kük.) Ohwi ■
74638 Carex gentilis Franch.;亲族苔草(莎草,束草,小鳞苔草,中原氏二柱苔);Smalscale Sedge ■
74639 Carex gentilis Franch. subsp. nakaharae (Hayata) T. Koyama = Carex gentilis Franch. var. nakaharae (Hayata) T. Koyama ■
74640 Carex gentilis Franch. subsp. nakaharai (Hayata) T. Koyama = Carex brunnea Thunb. ■
74641 Carex gentilis Franch. var. intermedia F. T. Wang et Ts. Tang ex L. K. Dai;宽叶亲族苔草(中间苔草);Broad-leaved Smalscale Sedge ■
74642 Carex gentilis Franch. var. macrocarpa Ts. Tang et F. T. Wang ex L. K. Dai;大果亲族苔草;Bigfruit Smalscale Sedge, Largefruit Smalscale Sedge ■
74643 Carex gentilis Franch. var. nakaharae (Hayata) T. Koyama;短喙亲族苔草;Short-beaked Smalscale Sedge ■
74644 Carex gibba Wahlenb.;穹隆苔草(基膨苔);Hunchy Sedge ■
74645 Carex gifuensis Franch.;歧阜苔草■☆
74646 Carex gifuensis Franch. var. koreana Nakai;朝鲜歧阜苔草■☆
74647 Carex gigas (T. Holm) Mack. = Carex scabrirostris Kük. ■
74648 Carex giraldiana Kük.;涝峪苔草;Laoyu Sedge ■
74649 Carex giraudiasi H. Lév.;大海苔草■
74650 Carex glabrescens (Kük.) Ohwi;辽东苔草;Liaodong Sedge ■
74651 Carex glabrescens Kük. = Carex glabrescens (Kük.) Ohwi ■

74652 Carex glacialis Mack.;冰河苔草■☆
74653 Carex glandulifolia Kük. = Carex lancifolia C. B. Clarke ■
74654 Carex glareosa Wahlenb.;碎石苔草(碎石苔);Gravel Sedge ■☆
74655 Carex glareosa Wahlenb. var. amphigena Fernald = Carex glareosa Wahlenb. ■☆
74656 Carex glauca Bosc ex Boott;灰苔草;Blue Sedge, Carnation Grass, Glaucous Sedge, Heath Sedge ■☆
74657 Carex glauca Scop. = Carex flacca Schreb. ■☆
74658 Carex glauca Scop. var. acuminata Steud. = Carex flacca Schreb. ■☆
74659 Carex glauca Scop. var. algerica Trab. = Carex flacca Schreb. ■☆
74660 Carex glauca Scop. var. arrecta Drejer = Carex flacca Schreb. ■☆
74661 Carex glauca Scop. var. attenuata Ball = Carex flacca Schreb. ■☆
74662 Carex glauca Scop. var. erythrostachys Anders. = Carex flacca Schreb. ■☆
74663 Carex glauca Scop. var. serrulata Coss. et Durieu = Carex flacca Schreb. ■☆
74664 Carex glaucaeformis Meinsh.;米柱苔草;Glaucous Sedge ■
74665 Carex glaucescens Elliott;海绿苔草■☆
74666 Carex glaucescens Elliott var. androgyna M. A. Curtis = Carex verrucosa Muhl. ■☆
74667 Carex glaucodea Tuck. ex Olney;灰齿苔草■☆
74668 Carex glehnii Eastw.;格莱氏苔草■☆
74669 Carex globistylosa P. C. Li;球柱苔草;Globularstyle Sedge ■
74670 Carex globosa Boott;球形苔草■☆
74671 Carex globosa Boott var. brevipes W. Boott ex Mack. = Carex deflexa Hornem. var. boottii L. H. Bailey ■☆
74672 Carex globularis L.;玉簪苔草(球穗苔草,球苔);Globularspike Sedge ■
74673 Carex glomerabilis Krecz.;群生苔草■☆
74674 Carex glomerata Thunb. = Carex glomerabilis Krecz. ■☆
74675 Carex glossostigma Hand.-Mazz.;长梗苔草(戴云山苔草);Longpedicel Sedge ■
74676 Carex gmelinii Hook. et Arn.;长芒苔草;Gmelin Sedge ■
74677 Carex gokwanensis Hayata = Carex chrysolepis Franch. et Sav. ■
74678 Carex goligongshanensis P. C. Li;高黎贡苔草;Gaoligongshan Sedge ■
74679 Carex gonggaensis P. C. Li;贡嘎苔草;Gongga Sedge ■
74680 Carex gongshanensis Ts. Tang et F. T. Wang ex Y. C. Yang;贡山苔草;Gongshan Sedge ■
74681 Carex goodenoughii Asch. et Graebn.;欧苔草■☆
74682 Carex goodenovii Gay var. limnophila (T. Holm) M. E. Jones = Carex lenticularis Michx. var. limnophila (T. Holm) Cronquist ■☆
74683 Carex goodenowii J. Gay = Carex nigra (L.) Reichard ■☆
74684 Carex gorodkovii V. I. Krecz.;高劳德苔草■☆
74685 Carex gotoi Ohwi;叉齿苔草(红穗苔草,苏卡苔草);Twotoothedbeak Sedge ■
74686 Carex gracilenta Boott ex Strachey = Carex infuscata Nees var. gracilenta (Boott ex Strachey) P. C. Li ■
74687 Carex gracilescens Steud.;细疏花苔草;Slender Loose-flower Sedge, Slender Sedge, Slender Wood Sedge ■☆
74688 Carex graciliculmis Ohwi = Carex kirganica Kom. ■
74689 Carex gracilis Curtis = Carex acuta L. ■
74690 Carex gracilis Curtis var. mucronata Maire = Carex acuta L. ■
74691 Carex gracilis Ehrh. = Carex brunnescens (Pers.) Poir. ■
74692 Carex gracilispica Hayata;细穗宿柱苔(细穗柱苔)■
74693 Carex gracilispica Hayata = Carex truncatigluma C. B. Clarke ■
74694 Carex gracillima Schwein.;紫鞘线苔草;Filiform Sedge, Graceful Sedge, Purple-sheathed Graceful Sedge ■☆
74695 Carex gracillima Schwein. var. macerrima Fernald et Wiegand = Carex gracillima Schwein. ■☆
74696 Carex grahamii Boott;格拉氏苔草(窄鳞苔);Graham Sedge, Mountain Bladder-sedge ■☆
74697 Carex grallatoria Maxim.;菱果苔草(水鸟苔,异型菱果苔)■
74698 Carex grallatoria Maxim. subsp. heteroclita (Franch.) T. Koyama = Carex grallatoria Maxim. ■
74699 Carex grallatoria Maxim. subsp. heteroclita (Franch.) T. Koyama = Carex grallatoria Maxim. var. heteroclita (Franch.) Kük. ■
74700 Carex grallatoria Maxim. var. heteroclita (Franch.) Kük.;异型菱果苔(水鸟苔)■
74701 Carex grallatoria Maxim. var. heteroclita (Franch.) Kük. ex Matsum. = Carex grallatoria Maxim. var. heteroclita (Franch.) Kük. ■
74702 Carex grallatoria Maxim. var. heteroclita (Franch.) Kük. ex Matsum. = Carex grallatoria Maxim. subsp. heteroclita (Franch.) T. Koyama ■
74703 Carex grallatoria Maxim. var. heteroclita (Franch.) T. Koyama = Carex grallatoria Maxim. var. heteroclita (Franch.) Kük. ■
74704 Carex graminiculmis T. Koyama;禾秆苔草■
74705 Carex grandiligulata Kük.;大舌苔草;Grandligulate Sedge ■
74706 Carex granularis Muhl. ex Willd.;石灰岩苔草;Limestone Meadow Sedge ■☆
74707 Carex granularis Muhl. ex Willd. var. haleana (Olney) Porter;哈勒石灰岩苔草■☆
74708 Carex granularis Muhl. ex Willd. var. haleana (Olney) Porter = Carex granularis Muhl. ex Willd. ■☆
74709 Carex gravida L. H. Bailey;长芒苞叶苔草;Heavy Sedge, Long-awned Bracted Sedge ■☆
74710 Carex gravida L. H. Bailey var. lunelliana (Mack.) F. J. Herm. = Carex gravida L. H. Bailey ■☆
74711 Carex grayi J. Carey;格雷苔草;Bur Sedge, Gray's Bur Sedge, Gray's Sedge, Mace Sedge ■☆
74712 Carex grayi J. Carey var. hispidula A. Gray = Carex grayi J. Carey ■☆
74713 Carex griffithii Boott;山苔草;Griffith Sedge ■
74714 Carex grioletii Roem. ex Schkuhr;格里奥苔草■☆
74715 Carex grisea Wahlenb.;格雷斯苔草;Eastern Narrow-leaved Sedge, Gray Sedge ■☆
74716 Carex grisea Wahlenb. var. amphibola (Steud.) Kük. = Carex amphibola Steud. ■☆
74717 Carex grisea Wahlenb. var. globosa L. H. Bailey = Carex bulbostylis Mack. ■☆
74718 Carex grossheimii V. I. Krecz.;格罗苔草■☆
74719 Carex gymnoclada T. Holm = Carex scopulorum T. Holm var. bracteosa (L. H. Bailey) F. J. Herm. ■☆
74720 Carex gynandra Schwein.;低垂苔草;Nodding Sedge ■☆
74721 Carex gynandra Schwein. var. simulans (Fernald) Roll.-Germ. = Carex gynandra Schwein. ■☆
74722 Carex gynocrates Wormsk. ex Drejer;异株苔草;Northern Bog Sedge, Ridged Sedge ■
74723 Carex gynodynama Olney;奥尔尼苔草;Olney's Hairy Sedge ■☆
74724 Carex hachijoensis Akiyama;八丈岛苔草■☆
74725 Carex hachijoensis Akiyama 'Evergold';黄纹八丈岛苔草■☆
74726 Carex haematostachys H. Lév. et Vaniot;血穗苔草■☆
74727 Carex haematostoma Nees;红嘴苔草■

74728 Carex haematostoma Nees var. digyna Kük. = Carex przewalskii T. V. Egorova ■

74729 Carex haematostoma Nees var. hirtellolides Kük. = Carex hirtelloides (Kük.) F. T. Wang et Ts. Tang ex P. C. Li ■

74730 Carex hainanensis Merr. = Carex commixta Steud. ■

74731 Carex hainanensis Merr. ex Chun et F. C. How;海南苔草;Hainan Sedge ■

74732 Carex hainanensis Merr. ex Chun et F. C. How = Carex commixta Steud. ■

74733 Carex hakkodensis Franch.;八甲田苔草■☆

74734 Carex hakkuensis Hayata = Carex cruciata Wahlenb. ■

74735 Carex hakodatensis H. Lév. et Vaniot = Carex pilosa Scop. ■

74736 Carex hakonemontana Katsuy.;箱根山苔草■☆

74737 Carex hakonensis Franch. et Sav.;箱根苔草■☆

74738 Carex hakonensis Franch. et Sav. var. capituliformis (Meinsh. ex Maxim.) Ohwi = Carex onoei Franch. et Sav. ■

74739 Carex hakonensis Franch. et Sav. var. macrocarpa Akiyama = Carex hakonensis Franch. et Sav. ■☆

74740 Carex hakonensis Franch. et Sav. var. onoei (Franch. et Sav.) Ohwi = Carex onoei Franch. et Sav. ■

74741 Carex haleana Olney = Carex granularis Muhl. ex Willd. var. haleana (Olney) Porter ■☆

74742 Carex haleana Olney = Carex granularis Muhl. ex Willd. ■☆

74743 Carex halei J. Carey ex Chapm. = Carex louisianica L. H. Bailey ■☆

74744 Carex hallaisanensis H. Lév. et Vaniot = Carex erythrobasis H. Lév. et Vaniot ■

74745 Carex hallerana Asso;哈氏苔草(哈氏苔)■☆

74746 Carex halleri Gunnerus;哈勒苔草(哈勒氏苔);Haller Sedge ■☆

74747 Carex halleriana Asso var. lerinensis H. Christ = Carex halleriana Asso ■☆

74748 Carex halliana L. H. Bailey;霍氏苔草■☆

74749 Carex hallii L. H. Bailey = Carex nudata W. Boott ■☆

74750 Carex hallii Olney;霍尔苔草■☆

74751 Carex hancei C. B. Clarke = Carex laticeps C. B. Clarke ex Franch. ■

74752 Carex hancockiana Maxim.;点叶苔草(华北苔草,尖叶苔草);Hancock Sedge ■

74753 Carex hancockiana Maxim. var. peiktusani (Kom.) Kük. = Carex peiktusani Kom. ■

74754 Carex handelii Kük.;双脉囊苔草(其宗苔草)■

74755 Carex hangdongensis H. Lév. et Vaniot;杭东苔草■

74756 Carex hankaensis Kitag. = Carex pediformis C. A. Mey. ■

74757 Carex harfordii Mack.;哈尔苔草■☆

74758 Carex harlandii Boott;长囊苔草;Harland Sedge, Longutricle Sedge ■

74759 Carex harlandii Boott var. liuguensis S. Yun Liang et D. Y. Liu = Carex harlandii Boott ■

74760 Carex harlandii Boott var. liuquensis S. Yun Liang et D. Y. Liu;六股长囊苔草;Liugu ■

74761 Carex harlandii Boott var. xiuningensis S. W. Su = Carex harlandii Boott ■

74762 Carex harrysmithii Kük.;哈利苔草;Harrysmith Sedge ■

74763 Carex hartmanii Cajander;柱苔草(柱苔);Hartmann Sedge ■☆

74764 Carex hashimotoi Ohwi;端元苔草■☆

74765 Carex hassei L. H. Bailey;哈瑟苔草■☆

74766 Carex hastata Kük.;戟叶苔草;Hastateleaf Sedge ■

74767 Carex hattoriana Nakai = Carex hattoriana Nakai ex Tuyama ■

74768 Carex hattoriana Nakai ex Tuyama;长叶苔草(服部氏二柱苔);Hattor Sedge ■

74769 Carex hattoriana Nakai ex Tuyama = Carex brunnea Thunb. ■

74770 Carex hatusimana Ohwi = Carex truncatigluma C. B. Clarke subsp. rhynchachaenium (C. B. Clarke ex Merr.) Y. C. Tang et S. Yun Liang ■

74771 Carex hawaiiensis H. St. John = Carex echinata Murray ■☆

74772 Carex hayatana Honda = Carex subtransversa C. B. Clarke ■

74773 Carex haydeniana Olney;海氏苔草■☆

74774 Carex haydenii Dewey;海登苔草;Hayden's Sedge, Long-scaled Tussock Sedge, Sedge ■☆

74775 Carex hebecarpa C. A. Mey.;疏果苔草(毛果苔草,密叶苔草)■

74776 Carex hebecarpa C. A. Mey. var. ligulata (Nees ex Wight) Kük. = Carex ligulata Nees ex Wight ■

74777 Carex hebecarpa C. A. Mey. var. ligulata (Nees) Kük. = Carex ligulata Nees ex Wight ■

74778 Carex hebecarpa C. A. Mey. var. maubertiana (Boott) Franch. = Carex maubertiana Boott ■

74779 Carex hebecarpa C. A. Mey. var. maubertiana (Boott) Franch. f. latifolia Makino = Carex phyllocephala T. Koyama ■

74780 Carex heilongjiangensis Y. L. Chou = Carex reptabunda (Trautv.) V. I. Krecz. ■

74781 Carex heleonastes Ehrh. ex L. f.;沼泽苔草■☆

74782 Carex heleonastes L. f. = Carex heleonastes Ehrh. ex L. f. ■☆

74783 Carex heleonastes L. f. var. dubia (L. H. Bailey) B. Boivin = Carex praeceptorum Mack. ■☆

74784 Carex heleonastes L. f. var. scabriuscula Kük. = Carex arctata Boott ■☆

74785 Carex heleonastes subsp. neurochlaena (T. Holm) Böcher = Carex heleonastes Ehrh. ex L. f. ■☆

74786 Carex heliophila Mack. = Carex inops L. H. Bailey subsp. heliophila (Mack.) Crins ■☆

74787 Carex heliophila Mack. = Carex inops L. H. Bailey ■☆

74788 Carex helleri Mack. = Carex henryi C. B. Clarke ex Franch. ■

74789 Carex helodes Link var. maurusia Font Quer et Maire = Carex halleriana Asso ■☆

74790 Carex henryi C. B. Clarke ex Franch.;长腿苔草(亨氏苔草,湖北苔草);Henry Sedge ■

74791 Carex hepburnii Boott;海普苔草■☆

74792 Carex hepburnii Boott = Carex nardina Fr. ■☆

74793 Carex heshuonensis S. Yun Liang = Carex heshuonensis S. Yun Liang, L. K. Dai et S. Yun Liang ■

74794 Carex heshuonensis S. Yun Liang, L. K. Dai et S. Yun Liang;和硕苔草;Heshuo Sedge ■

74795 Carex heteroclita Franch. = Carex grallatoria Maxim. var. heteroclita (Franch.) Kük. ■

74796 Carex heteroclita Franch. = Carex grallatoria Maxim. var. heteroclita (Franch.) T. Koyama ■

74797 Carex heterolepis Bunge;异鳞苔草;Heterolepidote Sedge, Heteroscale Sedge ■

74798 Carex heterolepis Bunge var. abtegens Kük.;陕西溪水苔草■

74799 Carex heteroneura W. Boott;互脉苔草■☆

74800 Carex heteroneura W. Boott var. chalciolepis (T. Holm) F. J. Herm. = Carex chalciolepis T. Holm ■☆

74801 Carex heterostachya Bunge;异穗苔草(黑穗草,异穗苔);Different-spike Sedge, Heterostachys Sedge ■

74802 Carex heudesii H. Lév. et Vaniot;长安苔草;Heudes Sedge ■

74803 Carex hindsii C. B. Clarke = Carex lenticularis Michx. var. limnophila (T. Holm) Cronquist ■☆
74804 Carex hirsuta Willd. = Carex hirsutella Mack. ■☆
74805 Carex hirsutella Mack.;绒毛苔草;Fuzzy Wuzzy Sedge ■☆
74806 Carex hirta L.;微硬毛苔草(微硬毛苔);Hairy Sedge,Hammer Sedge,Sharp-toothed Woolly Sedge ■☆
74807 Carex hirtella Drejer;硬毛苔草■
74808 Carex hirtelloides (Kük.) F. T. Wang et Ts. Tang ex P. C. Li;流石苔草■
74809 Carex hirticaulis P. C. Li;毛茎苔草(密毛苔草)■
74810 Carex hirtifolia Mack.;西方毛叶苔草;Hairy Sedge,Hairy Wood Sedge,Pubescent Sedge,Sedge ■☆
74811 Carex hirtifructus Kük. = Carex blepharicarpa Franch. var. stenocarpa Ohwi ■☆
74812 Carex hirtissima W. Boott;多毛齿苔草;Fuzzy Sedge,Hairy Sierra Sedge ■☆
74813 Carex hirtiutriculata L. K. Dai;糙毛囊苔草■
74814 Carex hispida Willd.;粗毛苔草■☆
74815 Carex hispida Willd. var. soleirolii (Duby) Trab. = Carex hispida Willd. ■☆
74816 Carex hitchcockiana Dewey;希氏苔草;Hairy Wood Sedge,Hitchcock's Sedge,Sedge ■☆
74817 Carex holmiana Mack. = Carex buxbaumii Wahlenb. ■☆
74818 Carex holostoma Drejer;全孔苔草(全孔苔)■☆
74819 Carex holotricha Ohwi = Carex lasiolepis Franch. var. lata Ohwi ■☆
74820 Carex hondae Akiyama = Carex deweyana Schwein. subsp. senanensis (Ohwi) T. Koyama ■☆
74821 Carex hondoensis Ohwi = Carex arnellii H. Christ ex Scheutz subsp. hondoensis (Ohwi) T. Koyama ■☆
74822 Carex hongkongensis Franch. = Carex ligata Boott ex Benth. ■
74823 Carex hongyuanensis Y. C. Tang et S. Yun Liang;红原苔草;Hongyuan Sedge ■
74824 Carex hoozanensis Hayata;凤凰宿柱苔草■
74825 Carex hoppneri Boott = Carex subspathacea Wormsk. ■☆
74826 Carex hordeistichus Vill.;麦穗苔草(麦穗苔)■☆
74827 Carex horikawae Okamoto = Carex leucochlora Bunge var. horikawae (Okamoto) Katsuy. ■☆
74828 Carex hornschuchiana Hoppe = Carex hostiana DC. ■☆
74829 Carex hosokawae Kitam. = Carpesium minum Hemsl. ■
74830 Carex hostiana DC.;豪思特苔草(何氏苔,何氏苔草);Host's Sedge,Tawny Sedge ■☆
74831 Carex hostiana DC. var. laurentiana (Fernald et Wiegand) Fernald et Wiegand = Carex hostiana DC. ■☆
74832 Carex houghtoniana Torr. ex Dewey;霍顿苔草;Houghton's Sedge,Houghton's Woolly Sedge ■☆
74833 Carex howei Mack. = Carex atlantica L. H. Bailey subsp. capillacea (L. H. Bailey) Reznicek ■☆
74834 Carex howellii L. H. Bailey = Carex aquatilis Wahlenb. var. dives (T. Holm) Kük. ■☆
74835 Carex hriveri Britton = Carex granularis Muhl. ex Willd. var. haleana (Olney) Porter ■☆
74836 Carex huashanica Ts. Tang et F. T. Wang ex L. K. Dai;华山苔草;Huashan Sedge ■
74837 Carex hudsonii A. Benn. = Carex elata All. ■☆
74838 Carex huetiana Boiss.;休氏苔草■☆
74839 Carex humbertiana Ohwi;亨伯特苔草■☆
74840 Carex humbertii F. T. Wang et Ts. Tang = Carex commixta Steud. ■
74841 Carex humida Y. L. Chang et Y. L. Yang;湿苔草■
74842 Carex humilis Leyss.;低矮苔草(矮苔);Dwarf Sedge ■
74843 Carex humilis Leyss. f. callitrichos (V. I. Krecz.) T. Koyama = Carex callitrichos V. I. Krecz. var. nana (H. Lév. et Vaniot) Ohwi ■
74844 Carex humilis Leyss. f. callitrichos (V. I. Krecz.) T. Koyama = Carex humilis Leyss. var. callitrichos (V. I. Krecz.) Ohwi ■☆
74845 Carex humilis Leyss. subsp. lanceolata (Boott) T. Koyama = Carex lanceolata Boott ■
74846 Carex humilis Leyss. subsp. nana (H. Lév. et Vaniot) T. Koyama = Carex humilis Leyss. var. nana (H. Lév. et Vaniot) Ohwi ■
74847 Carex humilis Leyss. var. callitrichos (V. I. Krecz.) Ohwi;丽毛低矮苔草■☆
74848 Carex humilis Leyss. var. callitrichos (V. I. Krecz.) Ohwi = Carex callitrichos V. I. Krecz. ■
74849 Carex humilis Leyss. var. nana (H. Lév. et Vaniot) Ohwi;矮丛苔草(矮苔草,羊胡子草);Dwarf Sedge,Low Sedge ■
74850 Carex humilis Leyss. var. nana (H. Lév. et Vaniot) Ohwi = Carex humilis Leyss. ■
74851 Carex humilis Leyss. var. scirrobasis (Kitag.) Y. L. Chang et Y. L. Yang;雏田苔草■
74852 Carex humilis Leyss. var. subpediformis (Kük.) T. Koyama = Carex lanceolata Boott ■
74853 Carex humpatensis H. E. Hess;洪帕塔苔草■☆
74854 Carex huolushanensis P. C. Li;火炉山苔草;Huolushan Sedge ■
74855 Carex hyalinolepis Steud.;晶鳞苔草;Sedge ■☆
74856 Carex hylaea Krecz.;林苔草■☆
74857 Carex hymenodon Ohwi;膜齿苔草■☆
74858 Carex hymenolepis Nees;膜鳞苔草■☆
74859 Carex hyperborea Drejer = Carex bigelowii Torr. et Schwein. ex Boott ■
74860 Carex hypochlora Freyn = Carex sabynensis Less. ex Kunth ■
74861 Carex hypolytrifolia T. Koyama = Carex commixta Steud. ■
74862 Carex hystericina Muhl. ex Willd.;子宫苔草;Bottlebrush Sedge,Porcupine Sedge,Sedge ■☆
74863 Carex hystricina Muhl. ex Willd. = Carex hystericina Muhl. ex Willd. ■☆
74864 Carex ichanqensis C. B. Clarke = Carex ascotreta C. B. Clarke ex Franch. ■
74865 Carex idahoa L. H. Bailey;爱达荷苔草■☆
74866 Carex idzuroei Franch. et Sav.;马菅;Stable-broom Sedge ■
74867 Carex iljinii V. I. Krecz.;伊尔金苔草■☆
74868 Carex immanis C. B. Clarke = Carex fargesii Franch. ■
74869 Carex impura Ohwi;不洁苔草■☆
74870 Carex inanis Kunth;毛囊苔草;Hispidutricle Sedge ■
74871 Carex inanoviae T. V. Egorova;西部苔草(青海苔草);Qinghai Sedge ■
74872 Carex incisa Boott;锐裂苔草■☆
74873 Carex incomperta E. P. Bicknell = Carex atherodes Spreng. ■
74874 Carex incurva Lightf. = Carex maritima Gunnerus ■☆
74875 Carex indica L.;印度苔草;India Sedge ■
74876 Carex indica L. = Carex indiciformis F. T. Wang et Ts. Tang ex P. C. Li ■
74877 Carex indiciformis F. T. Wang et Ts. Tang ex P. C. Li;印度型苔草;India-form Sedge ■
74878 Carex inflata Huds.;膨苔草■☆
74879 Carex inflata Huds. = Carex vesicaria L. ■
74880 Carex inflata Huds. var. utriculata (Boott) Druce = Carex

utriculata Boott ■☆

74881 Carex inflata Suter. = Carex rostrata Stokes ex With. ■

74882 Carex infossa C. P. Wang;隐匿苔草;Sunken Sedge ■

74883 Carex infossa C. P. Wang var. extensa S. W. Su;显穗苔草;Extented Sunken Sedge ■

74884 Carex infuscata Nees;淡色苔草(棕色苔草);Brownish Sedge ■

74885 Carex infuscata Nees var. gracilenta (Boott ex Strachey) P. C. Li;高山淡色苔草(高山苔草);Alpine Brownish Sedge ■

74886 Carex inops L. H. Bailey;长匍匐枝苔草;Long-stolon Sedge, Sedge ■☆

74887 Carex inops L. H. Bailey subsp. heliophila (Mack.) Crins;喜阳长匍匐枝苔草;Long-stolon Sedge, Round-bodied Oak Sedge, Sun Sedge ■☆

74888 Carex insaniae Koidz.;错乱苔草■☆

74889 Carex insaniae Koidz. subsp. subdita (Ohwi) T. Koyama = Carex insaniae Koidz. var. subdita (Ohwi) Ohwi ■☆

74890 Carex insaniae Koidz. var. papillaticulmis (Ohwi) Ohwi;乳突错乱苔草■☆

74891 Carex insaniae Koidz. var. subdita (Ohwi) Ohwi;大井错乱苔草■☆

74892 Carex insignis Boott;秆叶苔草■

74893 Carex interimus Maguire = Carex aquatilis Wahlenb. ■☆

74894 Carex interior L. H. Bailey;内地苔草;Inland Sedge, Inland Star Sedge, Sedge ■☆

74895 Carex interior L. H. Bailey f. keweenawensis (F. J. Herm.) Fernald = Carex interior L. H. Bailey ■☆

74896 Carex interior L. H. Bailey subsp. charlestonensis Clokey = Carex interior L. H. Bailey ■☆

74897 Carex interior L. H. Bailey var. capillacea L. H. Bailey = Carex atlantica L. H. Bailey subsp. capillacea (L. H. Bailey) Reznicek ■☆

74898 Carex interior L. H. Bailey var. keweenawensis F. J. Herm. = Carex interior L. H. Bailey ■☆

74899 Carex intermedia Good.;间型苔草;Intermediate Sedge ■☆

74900 Carex interrupta Boeck.;间断苔草■☆

74901 Carex interrupta Boeck. var. distenta Kük. = Carex interrupta Boeck. ■☆

74902 Carex interrupta Boeck. var. impressa L. H. Bailey = Carex lenticularis Michx. var. impressa (L. H. Bailey) L. A. Standl. ■☆

74903 Carex intricata Tineo = Carex nigra (L.) Reichard subsp. intricata (Tineo) Rivas Mart. ■☆

74904 Carex intumescens Rudge;拟肿胀苔草;Bladder Sedge, Greater Bladder Sedge, Shining Bur Sedge, Swollen Sedge ■☆

74905 Carex intumescens Rudge var. fernaldii L. H. Bailey = Carex intumescens Rudge ■☆

74906 Carex intumescens Rudge var. globularis A. Gray = Carex grayi J. Carey ■☆

74907 Carex inumbrata Krecz.;阴苔草■☆

74908 Carex inversa R. Br.;瘤苔草;Knob Sedge ■☆

74909 Carex invisa L. H. Bailey = Carex spectabilis Dewey ■☆

74910 Carex involucrata Boeck. = Carex aureolensis Steud. ■☆

74911 Carex iridifolia Kunth = Carex aethiopica Schkuhr ■☆

74912 Carex irregularis Schwein. = Carex viridula Michx. ■☆

74913 Carex irrigua Sm.;浸苔草■☆

74914 Carex ischnostachya Steud. var. subtimida Kük. = Carex subtumida (Kük.) Ohwi ■

74915 Carex ischnostachys Steud.;狭穗苔草(珠穗苔草);Thinspiculate Sedge ■

74916 Carex ischnostachys Steud. var. fastigiata T. Koyama;帚状狭穗苔草■☆

74917 Carex ischnostachys Steud. var. subtumida Kük. = Carex subtumida (Kük.) Ohwi ■

74918 Carex ivanoviae T. V. Egorova;无穗柄苔草■

74919 Carex jacens C. B. Clarke;横卧苔草■☆

74920 Carex jacens C. B. Clarke f. pubescens (Akiyama) M. Mizush.;短柔毛横卧苔草■☆

74921 Carex jackiana Boott;雅克苔草■☆

74922 Carex jackiana Boott f. oxyphylla (Franch.) Kük.;尖叶雅克苔草;Sharpleaf Sedge ■

74923 Carex jackiana Boott f. oxyphylla (Franch.) Kük. = Carex oxyphylla Franch. ■☆

74924 Carex jackiana Boott subsp. macroglossa (Franch. et Sav.) T. Koyama = Carex parciflora Boott var. macroglossa (Franch. et Sav.) Ohwi ■☆

74925 Carex jackiana Boott subsp. parciflora (Boott) Kük. = Carex parciflora Boott ■☆

74926 Carex jackiana Boott subsp. parciflora (Boott) Kük. var. macroglossa (Franch. et Sav.) Kük. ex Matsum. f. subsessilis (Ohwi) T. Koyama = Carex parciflora Boott var. macroglossa (Franch. et Sav.) Ohwi f. subsessilis Ohwi ■☆

74927 Carex jackiana Boott subsp. parciflora (Boott) Kük. var. macroglossa (Franch. et Sav.) Kük. ex Matsum. = Carex parciflora Boott var. macroglossa (Franch. et Sav.) Ohwi ■☆

74928 Carex jackiana Boott subsp. parciflora (Boott) Kük. var. tsukudensis (T. Koyama) T. Koyama = Carex parciflora Boott var. tsukudensis T. Koyama ■☆

74929 Carex jackiana Boott subsp. parciflora (Boott) Kük. var. vaniotii (H. Lév.) T. Koyama = Carex parciflora Boott var. vaniotii (H. Lév.) Ohwi ■☆

74930 Carex jackiana Boott subsp. vaniotii (H. Lév.) T. Koyama = Carex parciflora Boott var. vaniotii (H. Lév.) Ohwi ■☆

74931 Carex jacobi-peteri Hultén = Carex micropoda C. A. Mey. ■☆

74932 Carex jacutica V. I. Krecz.;雅库特苔草■☆

74933 Carex jakiana V. I. Krecz. f. oxyphylla (Franch.) Kük. = Carex oxyphylla Franch. ■☆

74934 Carex jaluensis Kom.;鸭绿苔草;Yalu Sedge ■

74935 Carex jamesii Schwein.;詹姆斯苔草;Grass Sedge, James' Sedge, Sedge ■☆

74936 Carex jamesii Torr. = Carex nebrascensis Dewey ■☆

74937 Carex jankowskii Gorodkov;雅恩苔草■☆

74938 Carex japonalpina (T. Koyama) T. Koyama = Carex atrata L. ■

74939 Carex japonica Boott = Carex doniana Spreng. ■

74940 Carex japonica Boott subsp. subtransversa (C. B. Clarke) T. Koyama = Carex subtransversa C. B. Clarke ■

74941 Carex japonica Thunb.;日本苔草(软苔草);Japan Sedge, Japanese Sedge ■

74942 Carex japonica Thunb. subsp. chlorostachys (C. B. Clarke) T. Koyama;芒尖苔草(签草)■

74943 Carex japonica Thunb. subsp. chlorostachys (Kük. ex Matsum.) T. Koyama = Carex alopecuroides D. Don ex Okamoto et Taylor var. chlorostachys C. B. Clarke ■☆

74944 Carex japonica Thunb. subsp. subtransversa (C. B. Clarke) T. Koyama = Carex subtransversa C. B. Clarke ■

74945 Carex japonica Thunb. var. alopecuroides (D. Don) C. B. Clarke = Carex alopecuroides D. Don ex Okamoto et Taylor ■

74946 Carex japonica Thunb. var. alopecuroides (D. Don) C. B. Clarke = Carex japonica Thunb. subsp. chlorostachys (C. B. Clarke) T. Koyama ■
74947 Carex japonica Thunb. var. aphanolepis (Franch. et Sav.) Kük. ex Matsum. = Carex aphanolepis Franch. et Sav. ■
74948 Carex japonica Thunb. var. chlorostachys (D. Don) Kük. = Carex doniana Spreng. ■
74949 Carex japonica Thunb. var. chlorostachys Kük. ex Matsum. = Carex alopecuroides D. Don ex Okamoto et Taylor var. chlorostachys C. B. Clarke ■☆
74950 Carex japonica Thunb. var. chlorostachys Ts. Tang = Carex planiculmis Kom. ■
74951 Carex japonica Thunb. var. humilis Franch. = Carex aphanolepis Franch. et Sav. ■
74952 Carex japonicai Thunb. var. alopecuroides (D. Don) C. B. Clarke = Carex alopecuroides D. Don ex Okamoto et Taylor ■
74953 Carex jepsonii J. T. Howell = Carex whitneyi Olney ■☆
74954 Carex jiaodongensis Y. M. Zhang et X. D. Chen;胶东苔草;Jiaodong Sedge ■
74955 Carex jimcalderi B. Boivin = Carex leptalea Wahlenb. ■☆
74956 Carex jinfoshanensis Ts. Tang et F. T. Wang ex S. Yun Ling;金佛山苔草;Jinfoshan Sedge ■
74957 Carex jisaburo-ohwiana T. Koyama = Carex caucasica Steven var. jisaburo-ohwiana (T. Koyama) T. Koyama ■
74958 Carex jisaburo-ohwiana T. Koyama = Carex caucasica Steven ■
74959 Carex jiuhaensis S. W. Su = Carex manca Boott ex Benth. subsp. jiuhuaensis (S. W. Su) S. Yun Liang ■
74960 Carex jiuhuaensis S. W. Su = Carex manca Boott ex Benth. subsp. jiuhuaensis (S. W. Su) S. Yun Liang ■
74961 Carex jiuxianshanensis L. K. Dai et Y. Z. Huang;九仙山苔草;Jiuxianshan Sedge ■
74962 Carex jizhuangensis S. Yun Liang;季庄苔草;Jizhuang Sedge ■
74963 Carex johnstonii Boeck.;约翰斯顿苔草■☆
74964 Carex jonesii L. H. Bailey;琼斯苔草■☆
74965 Carex josselynii (Fernald) Mack. ex Pease = Carex echinata Murray ■☆
74966 Carex jubozanensis J. Oda et A. Tanaka;鹫峰山苔草■☆
74967 Carex juncella Th. Fr.;假莞草型苔草■☆
74968 Carex juniperorum Catling, Reznicek et Crins;刺柏状苔草;Juniper Sedge ■☆
74969 Carex kabanovii V. I. Krecz.;卡巴诺夫苔草■☆
74970 Carex kamagariensis Okamoto;蒲刈苔草■☆
74971 Carex kamtschatica (Gorodkov) Gorodkov ex V. I. Krecz. = Carex bigelowii Torr. et Schwein. ex Boott ■
74972 Carex kamtschatica Gorodkov;勘察加苔草■☆
74973 Carex kansuensis Nelmes;甘肃苔草;Gansu Sedge, Kansu Sedge ■
74974 Carex kaoi Ts. Tang et F. T. Wang ex S. Yun Liang;高氏苔草■
74975 Carex karafutoana Ohwi = Carex lanceolata Boott var. laxa Ohwi ■
74976 Carex karashidaniensis Akiyama;秋山苔草■☆
74977 Carex karelinii Meinsh.;卡莱林苔草■☆
74978 Carex karelinii Meinsh. = Carex diluta M. Bieb. ■
74979 Carex karisimbiensis Cherm.;卡里辛比苔草■☆
74980 Carex karlongensis Kük.;卡郎苔草;Karlong Sedge ■
74981 Carex karlongensis Kük. subsp. handelii (Kük.) P. C. Li = Carex handelii Kük. ■
74982 Carex karoi (Freyn) Freyn;小粒苔草(多花苔草)■
74983 Carex karoi (Freyn) Freyn var. parvirostris (Kük.) Loeve et Raymond = Carex karoi (Freyn) Freyn ■
74984 Carex katahdinensis Fernald;卡塔丁山粒苔草;Mount Katahdin Sedge ■☆
74985 Carex katahdinensis Fernald = Carex conoidea Willd. ■☆
74986 Carex kattegatensis Fr.;卡特加特苔草■☆
74987 Carex kattegatensis Fr. ex Lindm. = Carex recta Boott ■☆
74988 Carex kawakamii Hayata = Carex alopecuroides D. Don ex Okamoto et Taylor ■
74989 Carex kawakamii Hayata = Carex subtransversa C. B. Clarke ■
74990 Carex kengiana Z. P. Wang = Carex davidii Franch. ■
74991 Carex kengii Kük. = Carex paxii Kük. ■
74992 Carex kenkolensis Litv.;肯科尔苔草■☆
74993 Carex kiangsuensis Kük.;江苏苔草;Jiangsu Sedge ■
74994 Carex kilinowii Turcz. = Carex pediformis C. A. Mey. ■
74995 Carex killickii Nelmes;基利克苔草■☆
74996 Carex kimurae Ohwi et T. Koyama = Carex dolichostachys Hayata ■
74997 Carex kiotensis Franch. et Sav.;褐柄苔(斑囊果苔,沙地苔草)■
74998 Carex kirganica Kom.;显脉苔草(长秆苔草)■
74999 Carex kirganica Kom. var. mukdenensis (Kitag.) Kitag. = Carex kirganica Kom. ■
75000 Carex kirilowii Turcz.;吉里苔草■☆
75001 Carex kirinensis F. T. Wang et Y. L. Chang;吉林苔草;Jilin Sedge ■
75002 Carex knorringiae Kük.;克诺氏苔草■☆
75003 Carex kobomugi Ohwi;筛草(海米,砂贡子,砂砧苔草,砂钻苔草,筛草实,筛实,禹余粮,自然谷);Japanese Sedge, Sea Isle Japanese Sedge, Sieve Sedge ■
75004 Carex kobomugi Ohwi f. longibracteata (Oliv.) Ohwi = Carex kobomugi Ohwi ■
75005 Carex koidzumiana Ohwi = Carex thunbergii Steud. var. appendiculata (Trautv.) Ohwi ■
75006 Carex koidzumii Honda = Carex lasiocarpa Ehrh. subsp. occultans (Franch.) Hultén ■
75007 Carex koidzumii Honda var. fuscata (Ohwi) Ohwi = Carex lasiocarpa Ehrh. ■
75008 Carex kokrinensis A. E. Porsild = Carex eleusinoides Turcz. ex Kunth ■
75009 Carex komaroviana A. I. Baranov et Skvortsov = Carex angarae Steud. ■
75010 Carex koraginensis Meinsh.;科拉金苔草■☆
75011 Carex koreana Kom. = Carex tenuiformis H. Lév. et Vaniot ■
75012 Carex korshinskyi Kom.;黄囊苔草;Korshinsky Sedge ■
75013 Carex koshewnikowii Litv.;密刺苞苔草;Koshewnikov Sedge ■
75014 Carex kotoensis Hayata = Carex bilateralis Hayata ■
75015 Carex kotschyana Boiss. et Hohen.;考奇苔草■☆
75016 Carex kowakamii Hayata = Carex subtransversa C. B. Clarke ■
75017 Carex krausei Boeck.;克劳斯苔草■☆
75018 Carex kuchunensis Ts. Tang et F. T. Wang ex S. Yun Liang;古城苔草;Gucheng Sedge ■
75019 Carex kucyniakii Raymond;棕叶苔草■
75020 Carex kuekenthalii K. Schum. ex C. B. Clarke = Carex johnstonii Boeck. ■☆
75021 Carex kujuzana Ohwi;久住苔草■☆
75022 Carex kujuzana Ohwi var. dissitispicula (Ohwi) T. Koyama = Carex kujuzana Ohwi ■☆
75023 Carex kunioi T. Koyama = Carex kujuzana Ohwi ■☆
75024 Carex kunlunsanensis N. R. Cui;昆仑苔草;Kunlunshan Sedge ■

75025 Carex kunzei Olney = Carex arctata Boott ■☆
75026 Carex kwangsiensis F. T. Wang et Ts. Tang ex P. C. Li;广西苔草;Guangxi Sedge ■
75027 Carex kwangtoushanica K. T. Fu;光头山苔草;Guangtoushan Sedge ■
75028 Carex kwangtungensis F. T. Wang et Ts. Tang = Carex adrienii E. G. Camus ■
75029 Carex kweichowense Chang = Carpesium minum Hemsl. ■
75030 Carex lachenalii Schkuhr;二裂苔草;Hare's-foot Sedge ■
75031 Carex lachnosperma Wall. = Carex hebecarpa C. A. Mey. ■
75032 Carex laciniata Boott = Carex barbarae Dewey ■☆
75033 Carex lacunarum T. Holm = Carex barbarae Dewey ■☆
75034 Carex lacustris Willd.;湖畔苔草;Common Lake Sedge, Lake Sedge, Rip-gut Sedge, Sedge ■☆
75035 Carex lacustris Willd. var. laxiflora Dewey = Carex hyalinolepis Steud. ■☆
75036 Carex laeta Boott;明亮苔草;Brightcolored Sedge ■
75037 Carex laeviconica Dewey;长齿湖苔草;Long-toothed Lake Sedge, Sedge, Smooth-cone Sedge ■☆
75038 Carex laeviculmis Meinsh.;稀花苔草■☆
75039 Carex laevigata Sm.;光滑苔草;Smooth-stalked Sedge ■☆
75040 Carex laevirostris (Blytt ex Fries) Fries = Carex rhynchophysa C. A. Mey. ■
75041 Carex laevirostris (Blytt) Blytt ex Blytt et Fries = Carex rhynchophysa Fisch., Mey. et Avé-Lall. ■
75042 Carex laevissima Nakai;假尖嘴苔草■
75043 Carex laevivaginata (Kük.) Mack.;光鞘苔草;Sedge, Smooth-sheathed Fox Sedge, Smooth-sheathed Sedge ■☆
75044 Carex lagopina Wahlenb. = Carex lachenalii Schkuhr ■
75045 Carex lamprosandra Franch. = Carex tapintzensis Franch. ■
75046 Carex lancangensis S. Yun Liang;澜沧苔草;Lancang Sedge ■
75047 Carex lanceata Dewey = Carex salina Wahlenb. ■☆
75048 Carex lanceolata Boott;大披针苔草(披叶苔,披针苔草,凸脉苔草,羊胡髭草,羊胡子草,羊毛胡子);Lanceolate Sedge ■
75049 Carex lanceolata Boott var. alashanica T. V. Egorova;阿拉善凸脉苔草;Alashan Lanceolate Sedge ■
75050 Carex lanceolata Boott var. alashanica T. V. Egorova = Carex lanceolata Boott ■
75051 Carex lanceolata Boott var. albomediana Makino;白条大披针苔草■☆
75052 Carex lanceolata Boott var. laxa Ohwi;少花大披针苔草(少花凸脉苔草)■
75053 Carex lanceolata Boott var. macrosandra Franch. = Carex macrosandra (Franch.) V. I. Krecz. ■
75054 Carex lanceolata Boott var. nana H. Lév. et Vaniot = Carex callitrichos V. I. Krecz. var. nana (H. Lév. et Vaniot) Ohwi ■
75055 Carex lanceolata Boott var. subpediformis Kük.;亚柄苔草(早春苔草);Subpetiolar Sedge ■
75056 Carex lanceus (Thunb.) Baill. = Schoenoxiphium lanceum (Thunb.) Kük. ■☆
75057 Carex lancifolia C. B. Clarke;披针叶苔草(披针苔草)■
75058 Carex lancisquamata L. K. Dai;披针鳞苔草;Lanci-scaled Sedge ■
75059 Carex lanuginosa Michx.;绢毛苔草;Woolly Sedge ■☆
75060 Carex lanuginosa Michx. = Carex lasiocarpa Ehrh. ■
75061 Carex lanuginosa Michx. var. americana (Fernald) B. Boivin = Carex lasiocarpa Ehrh. ■
75062 Carex lapponica O. Lang;拉普兰苔草;Lapland Sedge ■☆
75063 Carex laricetorum Y. L. Chou;落叶松林苔草■
75064 Carex laricina Mack. ex Bright;落叶松苔草■
75065 Carex lasiocarpa Ehrh.;毛果苔草(毛果苔,毛苔草,绒毛苔草);American Woolly-fruit Sedge, Downy Sedge, Downy-fruited Sedge, Narrow-leaved Woolly Sedge, Slender Sedge, Wiregrass, Woolly-fruit Sedge ■
75066 Carex lasiocarpa Ehrh. subsp. americana (Fernald) D. Loe et Bernard;美洲毛苔草;American Woolly-fruit Sedge, Narrow-leaved Woolly Sedge, Woolly-fruit Sedge ■☆
75067 Carex lasiocarpa Ehrh. subsp. occultans (Franch.) Hultén = Carex lasiocarpa Ehrh. ■
75068 Carex lasiocarpa Ehrh. var. americana Fernald = Carex lasiocarpa Ehrh. ■
75069 Carex lasiocarpa Ehrh. var. fuscata Ohwi = Carex lasiocarpa Ehrh. ■
75070 Carex lasiocarpa Ehrh. var. latifolia (Boeck.) Gilly = Carex pellita Willd. ■☆
75071 Carex lasiocarpa Ehrh. var. occultans (Franch.) Kük. = Carex lasiocarpa Ehrh. ■
75072 Carex lasiocarpa Ehrh. var. occultans (Franch.) Kük. = Carex lasiocarpa Ehrh. subsp. occultans (Franch.) Hultén ■
75073 Carex lasiolepis Franch.;毛鳞苔草■☆
75074 Carex lasiolepis Franch. var. lata Ohwi;宽毛鳞苔草■☆
75075 Carex lateralis Nees;侧生苔草■☆
75076 Carex laticeps C. B. Clarke ex Franch.;弯喙苔草;Bentbeaked Sedge ■
75077 Carex laticuspis Franch. = Carex gmelinii Hook. et Arn. ■
75078 Carex latifrons Xl. Krecz.;宽苔草■☆
75079 Carex latisquamea Kom.;宽鳞苔草■
75080 Carex laxa Wahlenb.;松散苔草(疏苔草,松苔)■
75081 Carex laxiculmis Schwein.;铺散苔草;Sedge, Spreading Sedge, Weak-stemmed Wood Sedge ■☆
75082 Carex laxiculmis Schwein. var. copulata (L. H. Bailey) Fernald = Carex laxiculmis Schwein. ■☆
75083 Carex laxiculmis Schwein. var. floridana L. H. Bailey = Carex digitalis Willd. var. floridana (L. H. Bailey) Naczi et Bryson ■☆
75084 Carex laxiflora Lam.;疏花苔草;Beach Wood Sedge, Broad Loose-flower Sedge, Wood Sedge ■
75085 Carex laxiflora Lam. var. blanda (Dewey) Boott = Carex blanda Dewey ■☆
75086 Carex laxiflora Lam. var. gracillima Boott = Carex gracilescens Steud. ■☆
75087 Carex laxiflora Lam. var. latifolia Boott = Carex albursina E. Sheld. ■☆
75088 Carex laxiflora Lam. var. leptonervia Fernald = Carex leptonervia (Fernald) Fernald ■☆
75089 Carex laxiflora Lam. var. ormostachya (Wiegand) Gleason = Carex ormostachya Wiegand ■☆
75090 Carex laxiflora Lam. var. patulifolia (Dewey) J. Carey = Carex laxiflora Lam. ■
75091 Carex laxiflora Lam. var. serrulata F. J. Herm. = Carex laxiflora Lam. ■
75092 Carex laxiflora Lam. var. varians L. H. Bailey = Carex leptonervia (Fernald) Fernald ■☆
75093 Carex laxior (Kük.) Mack. = Carex flava L. ■☆
75094 Carex laxisquamata Ts. Tang et F. T. Wang = Carex breviscapa C. B. Clarke ■

75095 Carex leavenworthii Dewey; 莱温苔草; Leavenworth's Bracted Sedge, Leavenworth's Sedge, Sedge ■☆

75096 Carex ledeboudana C. A. Mey. ex Trevis.;棒穗苔草■

75097 Carex ledebouriana C. A. Mey. ex Trevir. subsp. tenuiformis (H. Lév. et Vaniot) T. V. Egorova = Carex tenuiformis H. Lév. et Vaniot ■

75098 Carex leersii Willd. = Carex echinata Murray ■☆

75099 Carex lehmannii Drejer;膨囊苔草;Lehman Sedge ■

75100 Carex leiophylla Mack. = Carex sabulosa Turcz. ex Kunth ■

75101 Carex leiorhyncha C. A. Mey.;尖嘴苔草;Sharpbeak Sedge ■

75102 Carex lemmonii W. Boott;莱蒙苔草;Lemmon's Sedge ■☆

75103 Carex lenta D. Don;树荫苔草■☆

75104 Carex lenta D. Don f. simplex (Kük.) T. Koyama;单枝树荫苔草■☆

75105 Carex lenta D. Don subsp. sendaica (Franch.) T. Koyama = Carex sendaica Franch. ■

75106 Carex lenticularis Michx.;凸镜苔草(湖畔苔草);Lakeshore Sedge, Lenticular Sedge, Shore Sedge, Tufted Sedge ■☆

75107 Carex lenticularis Michx. var. albimontana Dewey = Carex lenticularis Michx. ■☆

75108 Carex lenticularis Michx. var. blakei Dewey = Carex lenticularis Michx. ■☆

75109 Carex lenticularis Michx. var. eucycla Fernald et Wiegand = Carex lenticularis Michx. ■☆

75110 Carex lenticularis Michx. var. impressa (L. H. Bailey) L. A. Standl.;凹陷凸镜苔草(凹陷湖畔苔草)■☆

75111 Carex lenticularis Michx. var. limnophila (T. Holm) Cronquist;沼泽凸镜苔草■☆

75112 Carex lenticularis Michx. var. merens Howe = Carex lenticularis Michx. ■☆

75113 Carex leonura Wahlenb. = Carex crinita Lam. ■☆

75114 Carex lepageana Raymond = Carex petricosa Dewey ■☆

75115 Carex lepidocarpa Tausch;鳞果苔草(鳞果苔)■☆

75116 Carex leporina L.;兔耳苔草(卵形苔草);Hare's-foot Sedge ■

75117 Carex leporina L. = Carex ovalis Gooden. ■☆

75118 Carex leporina L. subsp. ovalis (Gooden.) Maire = Carex ovalis Gooden. ■☆

75119 Carex leporina Poir. = Carex divulsa Stokes ex With. ■

75120 Carex leptalea Wahlenb.;细毛茎苔草;Bristle-stalked Sedge, Sedge, Slender Sedge ■☆

75121 Carex leptalea Wahlenb. subsp. harperi (Fernald) W. Stone = Carex leptalea Wahlenb. ■☆

75122 Carex leptalea Wahlenb. subsp. pacifica Calder et R. L. Taylor = Carex leptalea Wahlenb. ■☆

75123 Carex leptalea Wahlenb. var. harperi (Fernald) Weath. et Griscom = Carex leptalea Wahlenb. ■☆

75124 Carex leptalea Wahlenb. var. tayloris B. Boivin = Carex leptalea Wahlenb. ■☆

75125 Carex leptocladus C. B. Clarke;细枝苔草■☆

75126 Carex leptonervia (Fernald) Fernald;细脉苔草(寡脉苔草);Few-nerved Wood Sedge, Nerveless Woodland Sedge ■☆

75127 Carex leptophylla Heuff. var. linearibracteata F. H. Chen et C. M. Hu = Carpesium longifolium F. H. Chen et C. M. Hu ■

75128 Carex leptopoda Mack.;细梗苔草■☆

75129 Carex leptosaccus C. B. Clarke = Carex conferta Hochst. ex A. Rich. var. leptosaccus (C. B. Clarke) Kük. ■☆

75130 Carex leribensis Nelmes = Carex glomerabilis Krecz. ■☆

75131 Carex leucochlora Bunge;青绿苔草(等穗苔草,青菅,青苔草,哮喘草);Whitegreen Sedge ■

75132 Carex leucochlora Bunge = Carex breviculmis R. Br. ■

75133 Carex leucochlora Bunge f. fibrillosa (Franch. et Sav.) K. T. Fu = Carex breviculmis R. Br. var. fibrillosa (Franch. et Sav.) Kük. ex Matsum. et Hayata ■

75134 Carex leucochlora Bunge f. longearistata (Kük.) Kitag. = Carex breviculmis R. Br. ■

75135 Carex leucochlora Bunge var. aphanandra (Franch. et Sav.) T. Koyama;毛蕊青绿苔草■☆

75136 Carex leucochlora Bunge var. discoidea (Boott) T. Koyama;盘状苔草■☆

75137 Carex leucochlora Bunge var. fibrillosa (Franch. et Sav.) T. Koyama = Carex fibrillosa Franch. et Sav. ■

75138 Carex leucochlora Bunge var. filiculmis (Franch. et Sav.) Kitag. = Carex breviculmis R. Br. ■

75139 Carex leucochlora Bunge var. horikawae (Okamoto) Katsuy.;堀川苔草■☆

75140 Carex leucochlora Bunge var. lonchophora (Ohwi) T. Koyama;矛梗青绿苔草■☆

75141 Carex leucochlora Bunge var. lonchophora (Ohwi) T. Koyama = Carex lonchophora Ohwi ■☆

75142 Carex leucochlora Bunge var. longiaristata (Kük.) Kitag. = Carex breviculmis R. Br. ■

75143 Carex leucochlora Bunge var. petrogena Kitag. = Carex breviculmis R. Br. ■

75144 Carex leucochlora Bunge var. setouchiensis Okamoto;濑户内苔草■☆

75145 Carex leucodonta T. Holm = Carex turbinata Liebm. ■☆

75146 Carex licentii Nemer = Carex karoi (Freyn) Freyn ■

75147 Carex liddonii Boott = Carex petasata Dewey ■☆

75148 Carex lienchengensis S. Yun Liang et Y. Z. Huang;连城苔草;Liancheng Sedge ■

75149 Carex ligata Boott ex Benth.;香港苔草(宝岛宿柱苔)■

75150 Carex ligata Boott ex Benth. subsp. formosensis (H. Lév. et Vaniot) T. Koyama;宝岛宿柱苔■

75151 Carex ligata Boott ex Benth. subsp. formosensis (H. Lév. et Vaniot) T. Koyama = Carex formosensis H. Lév. et Vaniot ■

75152 Carex ligata Boott ex Benth. subsp. formosensis (H. Lév. et Vaniot) T. Koyama = Carex ascotreta C. B. Clarke ex Franch. ■

75153 Carex ligata Boott ex Benth. subsp. vigens (Kük.) T. Koyama = Carex genkaiensis Ohwi ■

75154 Carex ligata Boott subsp. formosensis (H. Lév. et Vaniot) T. Koyama = Carex ascotreta C. B. Clarke ex Franch. ■

75155 Carex ligata Boott var. formosensis (H. Lév. et Vaniot) Kük. = Carex ascotreta C. B. Clarke ex Franch. ■

75156 Carex ligata Boott var. nexa (Boott) Kük. = Carex ligata Boott ex Benth. ■

75157 Carex ligata Boott var. strictior (Kük.) Kük. = Carex sociata Boott ■

75158 Carex ligulata Nees = Carex ligulata Nees ex Wight ■

75159 Carex ligulata Nees ex Wigh = Carex hebecarpa C. A. Mey. var. ligulata (Nees) Kük. ■

75160 Carex ligulata Nees ex Wigh var. glabriutriculata Q. S. Wang;光囊苔草■

75161 Carex ligulata Nees ex Wight;舌叶苔草(贝舌苔,具舌苔,舌叶苔);Ligulate Sedge ■

75162 Carex ligulata Nees ex Wight var. austrokoreensis (Ohwi) Ohwi;

南朝鲜苔草■☆

75163 Carex limnaea T. Holm = Carex lenticularis Michx. var. impressa (L. H. Bailey) L. A. Standl. ■☆

75164 Carex limnophila F. J. Herm. = Carex microptera Mack. ■☆

75165 Carex limosa L.;湿生苔草(拟地苔,沼生苔草,沼苔草);Bog-sedge,Marshy Sedge,Muck Sedge,Mud Sedge ■

75166 Carex limosa L. var. irrigua Wahlenb. = Carex limosa L. ■

75167 Carex limosa L. var. irrigua Wahlenb. = Carex magellanica Lam. subsp. irrigua (Wahlenb.) Hiitonen ■☆

75168 Carex limosa L. var. livida Wahlenb. = Carex livida Willd. ■☆

75169 Carex limosa L. var. rariflora Wahlenb. = Carex rariflora Wahlenb. ■☆

75170 Carex limprichtiana Kük.;小果囊苔草■

75171 Carex linearis Boott = Kobresia nepalensis (Nees) Kük. ■

75172 Carex linearis Boott var. elachista C. B. Clarke = Kobresia nepalensis (Nees) Kük. ■

75173 Carex lingii F. T. Wang et Ts. Tang;林氏苔草;Ling Sedge ■

75174 Carex linkii Schkuhr = Carex distachya Desf. ■☆

75175 Carex liouana F. T. Wang et Ts. Tang;刘氏苔草;Liou Sedge ■

75176 Carex liparocarpos Gaudin;闪光苔(草原苔草,亮苔)■

75177 Carex liqingii Ts. Tang et F. T. Wang ex S. Yun Liang;立卿苔草;Liqing Sedge ■

75178 Carex lithophila Turcz.;二柱苔草(卵囊苔草,岩地苔草);Eggbag Sedge ■

75179 Carex litorhyncha Franch.;坚喙苔草(小嘴苔草)■

75180 Carex litwinowii Kük.;利特氏苔草■☆

75181 Carex liuii T. Koyama et J. M. Chuang;台中苔草(刘氏苔)■

75182 Carex livida (Wahlenb.) Willd. var. grayana (Dewey) Fernald = Carex livida Willd. ■☆

75183 Carex livida (Wahlenb.) Willd. var. radicaulis Paine = Carex livida Willd. ■☆

75184 Carex livida Willd.;铅色苔草;Livid Sedge,Pale Stiff Sedge ■☆

75185 Carex loliacea L.;间穗苔草(黑麦草型苔);Rye-grass Sedge ■

75186 Carex lonchocarpa Willd. ex Spreng.;矛果苔草■☆

75187 Carex lonchophora Ohwi = Carex breviculmis R. Br. var. lonchophora (Ohwi) T. Koyama ■☆

75188 Carex longepedicellata Boeck. = Carex duriuscula C. A. Mey. var. stenophylloides (V. I. Krecz.) S. Yun Liang et Y. C. Tang ■

75189 Carex longerostrata C. A. Mey.;长嘴苔草;Longbeaked Sedge ■

75190 Carex longerostrata C. A. Mey. var. exaristata X. F. Jin et C. Z. Zheng;无芒长嘴苔草;Awnless Longbeaked Sedge ■

75191 Carex longerostrata C. A. Mey. var. hoi Chü ex S. Yun Liang;城弯苔草;Ho's Longbeaked Sedge ■

75192 Carex longerostrata C. A. Mey. var. pallida (Kitag.) Ohwi;细穗长嘴苔草(细穗苔草)■

75193 Carex longerostrata C. A. Mey. var. tsinlingensis K. T. Fu = Carex longerostrata C. A. Mey. ■

75194 Carex longicruris Diels = Carex henryi C. B. Clarke ex Franch. ■

75195 Carex longicruris Diels var. henryi C. B. Clarke = Carex henryi C. B. Clarke ex Franch. ■

75196 Carex longicuspis Boeck.;长尖苔草■☆

75197 Carex longii Mack.;朗氏苔草;Greenish-white Sedge,Long's Sedge ■☆

75198 Carex longipedunculata K. Schum. = Carex petitiana A. Rich. ■

75199 Carex longipedunculata K. Schum. subsp. cuprea Kük. = Carex petitiana A. Rich. ■

75200 Carex longipedunculata K. Schum. var. ninagongensis Kük. = Carex petitiana A. Rich. ■

75201 Carex longipes D. Don = Carex longipes D. Don ex Okamoto et Taylor ■

75202 Carex longipes D. Don ex Okamoto et Taylor;长穗柄苔草(长柄苔草)■

75203 Carex longipes D. Don ex Okamoto et Taylor var. sessilis Ts. Tang et F. T. Wang ex L. K. Dai;短穗柄苔草■

75204 Carex longiquamata Mainsh. ex Komarv = Carex lanceolata Boott ■

75205 Carex longirostrata C. A. Mey. = Carex longerostrata C. A. Mey. ■

75206 Carex longirostrata C. A. Mey. subsp. pallida (Kitag.) T. Koyama = Carex longerostrata C. A. Mey. var. pallida (Kitag.) Ohwi ■

75207 Carex longirostrata C. A. Mey. subsp. pallida (Kitag.) T. Koyama = Carex tenuistachys Nakai ■

75208 Carex longirostrata C. A. Mey. var. pallida (Kitag.) Ohwi = Carex longerostrata C. A. Mey. var. pallida (Kitag.) Ohwi ■

75209 Carex longirostrata C. A. Mey. var. pallida (Kitag.) Ohwi = Carex tenuistachys Nakai ■

75210 Carex longirostris Torr. = Carex sprengelii Dewey ex Spreng. ■☆

75211 Carex longiseta Brot. = Carex distachya Desf. ■☆

75212 Carex longispica Boeck. = Carex speciosa Kunth ■

75213 Carex longispiculata Y. C. Yang;长密花穗苔草(长穗苔草);Long-spiked Sedge ■

75214 Carex longisquamata Meinsh. ex Kom. = Carex lanceolata Boott ■

75215 Carex longistipes Hayata = Carex fulvorubescens Hayata subsp. longistipes (Hayata) T. Koyama ■

75216 Carex longistolon C. B. Clarke = Carex sendaica Franch. ■

75217 Carex longistolon C. B. Clarke ex Franch. = Carex sendaica Franch. ■

75218 Carex longistolonifera Kük. = Carex atrata L. subsp. longistolonifera (Kük.) S. Yun Liang ■

75219 Carex longkiensis Franch. = Carex chinensis Retz. var. longkiensis (Franch.) Kük. ■

75220 Carex longpanlaensis S. Yun Liang;龙盘拉苔草;Longpanla Sedge ■

75221 Carex longshengensis Y. C. Tang et S. Yun Liang;龙胜苔草;Longsheng Sedge ■

75222 Carex longxishanensis S. Yun Liang;龙栖山苔草(陇栖山苔草);Longxishan Sedge ■

75223 Carex louisianica L. H. Bailey;路易斯安娜苔草;Louisiana Sedge ■☆

75224 Carex lowei Bech.;洛氏苔草■☆

75225 Carex lucidula Franch. = Carex sabynensis Less. ex Kunth var. rostrata (Maxim.) Ohwi ■☆

75226 Carex lucorum Willd. ex Link;蓝脊苔草;Blue Ridge Sedge,Forest Sedge,Long-beaked Oak Sedge ■☆

75227 Carex luctuosa Franch.;城口苔草;Chengkou Sedge ■

75228 Carex luctuosa Franch. f. brevisquama K. T. Fu = Carex luctuosa Franch. ■

75229 Carex luctuosa Franch. f. mucronata K. T. Fu = Carex luctuosa Franch. ■

75230 Carex lumnitzeri Rouy;卢氏苔草■☆

75231 Carex lunelliana Mack. = Carex gravida L. H. Bailey ■☆

75232 Carex lupuliformis Sartwell ex Dewey;蛇麻草状苔草;False Hop Sedge,Hoplike Sedge,Hop-like Sedge,Knobbed Hop Sedge ■☆

75233 Carex lupulina Muhl. ex Willd.;蛇麻草苔草;Common Hop Sedge,Hop Sedge,Shallow Sedge,Shining Sedge ■☆

75234 Carex lupulina Muhl. ex Willd. var. pedunculata A. Gray = Carex lupulina Muhl. ex Willd. ■☆

75235 Carex lupulina Willd. var. pedunculata A. Gray = Carex lupulina Muhl. ex Willd. ■☆
75236 Carex lurida C. B. Clarke = Carex obscuriceps Kük. ■
75237 Carex lurida L. H. Bailey = Carex lupulina Muhl. ex Willd. ■☆
75238 Carex lurida Wahlenb.;浅黄苔草;Sedge,Shallow Sedge ■☆
75239 Carex lurida Wahlenb. = Carex lupulina Muhl. ex Willd. ■☆
75240 Carex lushanensis Kük.;庐山苔草;Lushan Sedge ■
75241 Carex lutchuensis Ohwi = Carex breviscapa C. B. Clarke ■
75242 Carex lutensis Kunth = Carex clavata Thunb. ■☆
75243 Carex luzulifolia W. Boott;地杨梅苔草■☆
75244 Carex lyallii Boott = Carex raynoldsii Dewey ■☆
75245 Carex lyngbyei Hornem. = Carex cryptocarpa C. A. Mey. ■
75246 Carex lyngbyei Hornem. subsp. cryptocarpa (C. A. Mey.) Hultén = Carex cryptocarpa C. A. Mey. ■
75247 Carex lyngbyei Hornem. subsp. cryptocarpa (C. A. Mey.) Hultén = Carex lyngbyei Hornem. ■
75248 Carex lyngbyei Hornem. subsp. prionocarpa (Franch.) Kitag. = Carex cryptocarpa C. A. Mey. ■
75249 Carex lyngbyei Hornem. var. cryptocarpa (C. A. Mey.) Hultén = Carex lyngbyei Hornem. ■
75250 Carex lyngbyei Hornem. var. prionocarpa (Franch.) Kük. = Carex cryptocarpa C. A. Mey. ■
75251 Carex lyngbyei Hornem. var. robusta (L. H. Bailey) Cronquist = Carex lyngbyei Hornem. ■
75252 Carex maackii Maxim.;卵果苔草(翅囊苔草);Maack Sedge ■
75253 Carex mackenziei V. I. Krecz.;马氏苔草■☆
75254 Carex macloviana d'Urv. subsp. festivella (Mack.) Á. Löve et D. Löve = Carex microptera Mack. ■☆
75255 Carex macloviana d'Urv. subsp. haydeniana (Olney) R. L. Taylor et MacBryde = Carex haydeniana Olney ■☆
75256 Carex macloviana d'Urv. subsp. pachystachya (Cham. ex Steud.) Hultén = Carex pachystachya Cham. ex Steud. ■☆
75257 Carex macounii Dewey = Carex lupulina Muhl. ex Willd. ■☆
75258 Carex macrandrolepis H. Lév. et Vaniot;和平菱果苔草(和平菱果苔,菱果苔)■
75259 Carex macrocephala Willd. ex Kunth;大头苔草(莔草)■☆
75260 Carex macrocephala Willd. ex Kunth var. kobomugi Miyabe et Kudo = Carex kobomugi Ohwi ■
75261 Carex macrocephala Willd. ex Kunth var. longibracteata Oliv. = Carex kobomugi Ohwi ■
75262 Carex macrocephala Willd. ex Spreng. var. kobomugi (Ohwi) Miyabe et Kudo = Carex kobomugi Ohwi ■
75263 Carex macrocephala Willd. var. kobomugi Miyabe et Kudo = Carex kobomugi Ohwi ■
75264 Carex macrocephala Willd. var. longibracteata Oliv. = Carex kobomugi Ohwi ■
75265 Carex macrochaeta C. A. Mey.;大毛苔草■☆
75266 Carex macrocystis Boeck. = Carex clavata Thunb. ■☆
75267 Carex macroglossa Franch. et Sav. = Carex parciflora Boott var. macroglossa (Franch. et Sav.) Ohwi ■☆
75268 Carex macrogyna Turcz. ex Besser;大柱苔草■☆
75269 Carex macrogyna Turcz. ex Besser var. nana?;矮大柱苔草■☆
75270 Carex macrokolea Steud. = Carex verrucosa Muhl. ■☆
75271 Carex macrolepis D. Don = Carex longipes D. Don ex Okamoto et Taylor ■
75272 Carex macrophyllidion Nelmes;拟大叶苔草■☆
75273 Carex macrosandra (Franch.) V. I. Krecz.;大雄苔草■
75274 Carex macrostachys Bertol.;大穗苔草;Largespike Sedge ■☆
75275 Carex macrostigmatica Kit.;大柱头■☆
75276 Carex macroura (Meinsh.) Kük.;梨囊苔草■
75277 Carex macroura (Meinsh.) Kük. = Carex pediformis C. A. Mey. ■
75278 Carex macroura Meinsh. = Carex pediformis C. A. Mey. ■
75279 Carex maculata Boott;斑点果苔草(斑点苔,斑点苔草,宽囊果苔);Maculate Sedge ■
75280 Carex maculata Boott var. tetsuoi (Ohwi) T. Koyama;彻翁苔草■☆
75281 Carex magellanica Lam.;北沼苔草;Boreal Bog Sedge,Stunted Sedge,Tall Bog-sedge ■☆
75282 Carex magellanica Lam. subsp. irrigua (Wahlenb.) Hiitonen;北阴苔草;Boreal Bog Sedge ■☆
75283 Carex magellanica Lam. var. irrigua (Wahlenb.) Britton,Sterns et Poggenb. = Carex magellanica Lam. subsp. irrigua (Wahlenb.) Hiitonen ■☆
75284 Carex magnoutriculata Ts. Tang et F. T. Wang ex L. K. Dai;大果囊苔草;Bigfruit Sedge,Largefruit Sedge ■
75285 Carex magnursina Raymond = Carex petricosa Dewey ■☆
75286 Carex mainensis Porter;缅因苔草;Maine Sedge ■☆
75287 Carex makinoensis Franch.;牧野氏苔草(牧野氏苔);Makino Sedge ■
75288 Carex makuensis P. C. Li;马库苔草;Maku Sedge ■
75289 Carex manca Boott ex Benth.;弯柄苔草(梦佳宿柱苔,缺如苔草)■
75290 Carex manca Boott ex Benth. subsp. jiuhuaensis (S. W. Su) S. Yun Liang;九华苔草;Jiuhuashan Sedge ■
75291 Carex manca Boott ex Benth. subsp. takasagoana (Akiyama) T. Koyama;梦佳弯柄苔草(梦佳宿柱苔,梦佳苔草)■
75292 Carex manca Boott ex Benth. subsp. takasagoana (Akiyama) T. Koyama = Carex takasagoana Akiyama ■
75293 Carex manca Boott ex Benth. subsp. wichurae (Boeck.) S. Yun Liang;短叶弯柄苔草(短叶苔草)■
75294 Carex manca Boott ex Benth. subsp. wichurai (Kük.) S. Yun Liang = Carex manca Boott ex Benth. subsp. wichurae (Boeck.) S. Yun Liang ■
75295 Carex manca Boott ex Benth. var. wichurai Kük. = Carex manca Boott ex Benth. subsp. wichurai (Kük.) S. Yun Liang ■
75296 Carex mancaeformis C. B. Clarke ex Franch.;鄂西苔草;W. Hubei Sedge ■
75297 Carex mandarinorum Raymond = Carex glossostigma Hand.-Mazz. ■
75298 Carex mandshurica Meinsh.;东北苔草;Mongolia Sedge ■☆
75299 Carex mannii E. A. Bruce;曼氏苔草■☆
75300 Carex maorshanica Y. L. Chou;帽儿山苔草;Maorshan Sedge ■
75301 Carex maquensis Y. C. Yang;玛曲苔草;Maqu Sedge ■
75302 Carex marginata Willd. = Carex pensylvanica Lam. ■☆
75303 Carex marina Dewey;海苔草;Marine Sedge ■☆
75304 Carex maritima Gunnerus;海滨苔草;Curved Sedge ■☆
75305 Carex maritima Gunnerus subsp. yukonensis A. E. Porsild = Carex maritima Gunn ■☆
75306 Carex maritima Gunnerus var. setina (H. Christ ex Scheutz) Fernald = Carex maritima Gunnerus ■☆
75307 Carex martinii H. Lév. et Vaniot;马特苔草■
75308 Carex matsudae (Hayata) Hayata ex Makino et Nemoto = Carex dolichostachys Hayata ■
75309 Carex matsudai (Hayata) Hayata ex Makino et Nemoto = Diplocarex matsudai Hayata ■☆

75310 Carex matsumurae Franch.;松村氏苔草■☆

75311 Carex maubertiana Boott;套鞘苔草(毛囊苔草,密叶苔草,三角草,山马鞭,山马鞭草);Denseleaf Sedge ■

75312 Carex mauritanica Boiss. et Reut. = Carex acuta L. ■

75313 Carex maxima Scop. = Carex pendula Huds. ■☆

75314 Carex maximowiczii Miq.;乳突苔草(麦苔草);Maximowicz Sedge ■

75315 Carex maximowiczii Miq. subsp. suifunensis (Kom.) Vorosch. = Carex maximowiczii Miq. var. suifunensis (Kom.) Nakai ex Kitag. ■☆

75316 Carex maximowiczii Miq. var. levisaccus Ohwi;光囊乳突苔草■☆

75317 Carex maximowiczii Miq. var. suifunensis (Kom.) Nakai ex Kitag.;水府苔草■☆

75318 Carex mayebarana Ohwi;马屋原苔草■☆

75319 Carex meadii Dewey;米德苔草;Mead's Sedge, Mead's Stiff Sedge, Sedge ■☆

75320 Carex media R. Br. ex Richardson;全叶苔草;Intermediate Sedge ■☆

75321 Carex media R. Br. var. stevenii (T. Holm) Fernald = Carex stevenii (T. Holm) Kalela ■☆

75322 Carex mediterranea Mack. = Carex mesochorea Mack. ■☆

75323 Carex medwedewii Leskov;麦德苔草■☆

75324 Carex meihsienica K. T. Fu;眉县苔草;Meixian Sedge ■

75325 Carex meinshauseniana Krecx.;梅恩氏苔草■☆

75326 Carex melanantha C. A. Mey.;黑花苔草(黑花苔);Blackflowered Sedge ■

75327 Carex melanantha C. A. Mey. var. moorcroflii (Boott) Kük. = Carex moorcroftii Falc. ex Boott ■

75328 Carex melananthiformis Litv.;尤尔都斯苔草■

75329 Carex melancephala Turcz. = Carex melanocephala Turcz. ex Kunth ■

75330 Carex melanocarpa Cham. ex Trautv.;黑果苔草■☆

75331 Carex melanocarpa Cham. ex Trautv. var. rigidior T. Koyama = Carex melanocarpa Cham. ex Trautv. ■☆

75332 Carex melanocephala Turcz. ex Kunth;黑鳞苔草■

75333 Carex melanostachys M. Bieb. ex Willd.;凹脉苔草(黑穗苔)■

75334 Carex melanostoma Fisch.;黑嘴苔草■☆

75335 Carex melinacra Franch.;扭喙苔草(小剪苔草)■

75336 Carex melinacra Franch. var. changning S. Yun Liang;昌宁苔草;Changning Sedge ■

75337 Carex melozitnensis A. E. Porsild = Carex rotundata Wahlenb. ■☆

75338 Carex membranacea Hook.;膜质苔草■☆

75339 Carex meridiana (Akiyama) Akiyama;南方苔草■☆

75340 Carex merritt-fernaldii Mack.;弗纳尔德苔草;Fernald's Oval Sedge, Fernald's Sedge ■☆

75341 Carex mertensii J. D. Prescott = Carex mertensii J. D. Prescott ex Bong. ■☆

75342 Carex mertensii J. D. Prescott ex Bong.;默顿苔草■☆

75343 Carex mertensii J. D. Prescott ex Bong. var. urostachys (Franch.) Kük.;尾穗默顿苔草■☆

75344 Carex mertensii J. D. Prescott var. urostachys (Franch.) Kük. = Carex mertensii J. D. Prescott ex Bong. var. urostachys (Franch.) Kük. ■☆

75345 Carex merxmuelleri Podlech = Carex zuluensis C. B. Clarke ■☆

75346 Carex mesochorea Mack.;中间苔草;Synonyms ■☆

75347 Carex metallica H. Lév. et Vaniot;锈果苔草(宽穗苔)■

75348 Carex meyeriana Kunth;乌拉草(靰鞡草);Meyer Sedge ■

75349 Carex michauxiana Boeck.;米氏苔草;Michaux's Sedge ■☆

75350 Carex michauxiana Boeck. subsp. asiatica Hultén;亚洲米氏苔草■☆

75351 Carex michauxiana Boeck. var. asiatica (Hultén) Ohwi = Carex michauxiana Boeck. subsp. asiatica Hultén ■☆

75352 Carex michauxiana C. B. Clarke = Carex idzuroei Franch. et Sav. ■

75353 Carex michauxii Dewey = Carex collinsii Nutt. ■☆

75354 Carex michauxii Schwein. = Carex scirpoidea Michx. ■☆

75355 Carex michelii Host;米奇里苔草(米氏苔,米氏苔草);Michel Sedge ■☆

75356 Carex michiganensis Dewey = Carex lucorum Willd. ex Link ■☆

75357 Carex micrantha Kük.;滑茎苔草(灰化苔草,匍枝苔草);Ashgrey Sedge ■

75358 Carex microchaeta T. Holm;小刚毛苔草■☆

75359 Carex microchaeta T. Holm subsp. nesophila (T. Holm) D. F. Murray;岛屿苔草■☆

75360 Carex microdonta Torr. et Hook.;小齿苔草;Sedge ■☆

75361 Carex microglochin Wahlenb.;尖苞苔草(小刺苔);Bristle Sedge, Racheoled Sedge ■

75362 Carex micropoda C. A. Mey.;小梗苔草(小足苔草)■☆

75363 Carex micropodioides V. I. Krecz.;假小足苔草■☆

75364 Carex microptera Mack.;小翅苔草■☆

75365 Carex microptera Mack. var. crassinervia F. J. Herm. = Carex microptera Mack. ■☆

75366 Carex microrhyncha Mack. = Carex umbellata Schkuhr ex Willd. ■☆

75367 Carex microsperma Wahlenb. = Carex vulpinoidea Michx. ■☆

75368 Carex microstoma Franch.;小口苔草■

75369 Carex microtricha Franch.;小毛苔草■☆

75370 Carex middendorffii Eastw.;高鞘苔草■

75371 Carex middendorffii Eastw. var. kirigaminensis (Ohwi) Ohwi = Carex × leiogona Franch. ■☆

75372 Carex mildbraediana K. Schum. var. friesiorum Kük. = Carex elgonensis Nelmes ■☆

75373 Carex mildbraediana Kük.;米尔德苔草■☆

75374 Carex miliaris Michx. = Carex saxatilis L. ■☆

75375 Carex millegrana T. Holm = Carex emoryi Dewey ■☆

75376 Carex minganinsularum Raymond = Carex sterilis Willd. ■☆

75377 Carex minquinensis Z. P. Wang;闽清苔草;Minqin Sedge ■

75378 Carex minuta Franch.;褐鞘苔草;Minute Sedge ■☆

75379 Carex minuta Franch. = Carex caespitosa L. ■

75380 Carex minuta Franch. = Carex cespitosa L. var. minuta (Franch.) Kük. ■☆

75381 Carex minuticulmis S. W. Su et S. M. Xu;矮秆苔草■

75382 Carex minutiscabra Kük. ex B. Fedtsch. et V. I. Krecz.;粗糙苔草■

75383 Carex minxianensis S. Yun Liang;岷县苔草;Minxian Sedge ■

75384 Carex mira Kük.;奇异苔草■☆

75385 Carex mirabilis Dewey = Carex normalis Mack. ■☆

75386 Carex mirabilis Dewey var. tinca Fernald = Carex tincta (Fernald) Fernald ■☆

75387 Carex misandra R. Br.;冷苔草■☆

75388 Carex misandra R. Br. = Carex fuliniginosa Schkuhr ■

75389 Carex misera Buckley;脏苔草;Wretched Sedge ■☆

75390 Carex mitchelliana M. A. Curtis;米切尔苔草■☆

75391 Carex mitrata Franch.;灰帽苔草(具芒宿柱苔)■

75392 Carex mitrata Franch. subsp. aristata (Ohwi) T. Koyama = Carex mitrata Franch. var. aristata Ohwi ■

75393 Carex mitrata Franch. var. aristata Ohwi;具芒灰帽苔草(具芒宿

柱苔)■

75394 Carex mitrata Franch. var. bingoensis Okamoto;备后苔草■☆

75395 Carex mitrata subsp. aristata (Ohwi) T. Koyama = Carex mitrata Franch. var. aristata Ohwi ■

75396 Carex miyabei Franch.;柔叶苔草;Miyabe Sedge ■

75397 Carex miyabei Franch. var. maopengensis S. W. Su;毛果柔叶苔草(毛果苔草)■

75398 Carex mohriana Mack. = Carex atlantica L. H. Bailey subsp. capillacea (L. H. Bailey) Reznicek ■☆

75399 Carex molesta Mack. ex Bright;田间苔草;Field Oval Sedge, Sedge, Troublesome Sedge ■☆

75400 Carex mollicula Boott;柔果苔草■

75401 Carex mollicula Boott subsp. planiculmis (Kom.) T. Koyama = Carex planiculmis Kom. ■

75402 Carex mollicula Boott var. naipiangensis (H. Lév. et Vaniot) T. Koyama = Carex planiculmis Kom. ■

75403 Carex mollissima H. Christ ex Meinsh.;柄苔草(柔苔)■

75404 Carex mongolica A. I. Baranov et Skvortsov ex Liou et al. = Carex lithophila Turcz. ■

75405 Carex monile Tuck. = Carex vesicaria L. var. monile (Tuck.) Boeck. ■☆

75406 Carex monile Tuck. = Carex vesicaria L. ■

75407 Carex monostachya A. Rich.;单穗苔草■☆

75408 Carex monotropa Nelmes;单棱苔草■☆

75409 Carex montana L.;山生苔草;Mountain Sedge, Soft-leaved Sedge ■☆

75410 Carex montana Ohwi = Carex ulobasis V. I. Krecz. ■

75411 Carex montana Ohwi var. manshuriensis Kom. = Carex ulobasis V. I. Krecz. ■

75412 Carex montanensis L. H. Bailey = Carex podocarpa R. Br. ex Richardson ■☆

75413 Carex montereyensis Mack. = Carex harfordii Mack. ■☆

75414 Carex montis-everestii Kük.;窄叶苔草■

75415 Carex montis-wutaii T. Koyama;五台山苔草;Wutaishan Sedge ■

75416 Carex moorcroftii Falc. ex Boott;青藏苔草;Moorcroft Sedge ■

75417 Carex morii Hayata;森氏苔草(森氏苔);Mori Sedge ■

75418 Carex morrisonicola Hayata;玉山苔草(阿里山苔草);Alishan Sedge ■

75419 Carex morrisonicola Hayata = Carex breviculmis R. Br. ■

75420 Carex morrisseyi A. E. Porsild = Carex adelostoma Krecz. ■☆

75421 Carex morrowii Boott;毛氏苔草;Japanese Sedge, Morrow Sedge, Morrow's Sedge, Variegated Japanese Sedge ■☆

75422 Carex morrowii Boott f. expallida (Ohwi) T. Koyama = Carex morrowii Boott var. albomarginata Makino ■☆

75423 Carex morrowii Boott var. albomarginata Makino;白边毛氏苔草■☆

75424 Carex morrowii Boott var. expallida Ohwi = Carex morrowii Boott var. albomarginata Makino ■☆

75425 Carex morrowii Boott var. laxa Ohwi;疏松毛氏苔草■☆

75426 Carex morrowii Boott var. temnolepis (Franch.) Ohwi ex Araki = Carex temnolepis Franch. ■☆

75427 Carex mosoynensis Franch.;滇西苔草;W. Yunnan Sedge ■

75428 Carex mossii Nelmes;莫西苔草■☆

75429 Carex motuoensis Y. C. Yang;墨脱苔草;Motuo Sedge ■

75430 Carex moupinensis Franch.;宝兴苔草;Baoxing Sedge, Mouping Sedge ■

75431 Carex mucronata All.;钝尖苔草■☆

75432 Carex mucronatiformis Ts. Tang et F. T. Wang ex S. Yun Liang;类短尖苔草■

75433 Carex muehlenbergii Schkuhr ex Willd. var. australis Olney = Carex austrina Mack. ■☆

75434 Carex muehlenbergii Schkuhr ex Willd. var. australis Olney ex L. H. Bailey = Carex austrina Mack. ■☆

75435 Carex muehlenbergii Schkuhr ex Willd. var. enervis Boott;无脉米伦苔草;muehlenberg's Sedge, Muhlenberg's Bracted Sedge, Sand Bracted Sedge, Sand Sedge ■☆

75436 Carex muhlenbergii Schkuhr ex Willd.;米伦苔草;Muhlenberg's Bracted Sedge, Muhlenberg's Sedge, Sand Bracted Sedge, Sand Sedge, Sedge ■☆

75437 Carex mukdenensis Kitag. = Carex kirganica Kom. ■

75438 Carex muliensis Hand.-Mazz.;木里苔草;Muli Sedge ■

75439 Carex multicaulis L. H. Bailey;多茎苔草■☆

75440 Carex multicostata Mack.;多脉苔草■☆

75441 Carex multiflora Willd. = Carex vulpinoidea Michx. ■☆

75442 Carex multiflora Willd. var. microsperma (Wahlenb.) Dewey = Carex vulpinoidea Michx. ■☆

75443 Carex multifolia Ohwi;西方多叶苔草■☆

75444 Carex multifolia Ohwi var. glaberrima Ohwi;光秃多叶苔草■☆

75445 Carex multifolia Ohwi var. imbecillis Ohwi;衰弱苔草■☆

75446 Carex multifolia Ohwi var. pallidisquama Ohwi;苍白多叶苔草■☆

75447 Carex multifolia Ohwi var. stolonifera Ohwi;匍匐多叶苔草■☆

75448 Carex multifolia Ohwi var. toriiana T. Koyama;鸟居苔草■☆

75449 Carex munda Boott;秀丽苔草■

75450 Carex munda Boott var. mundaeformis T. Koyama = Carex munda Boott ■

75451 Carex muricata L.;糙刺苔草(糙刺苔);Prickly Sedge, Rough Sedge ■☆

75452 Carex muricata L. subsp. divulsa (Stokes) Syme = Carex divulsa Stokes ex With. ■

75453 Carex muricata L. subsp. lamprocarpa Celak.;亮果糙刺苔草■☆

75454 Carex muricata L. subsp. pairaei (Schultz) Celak. = Carex muricata L. subsp. lamprocarpa Celak. ■☆

75455 Carex muricata L. var. angustata (J. Carey) J. Carey ex Gleason = Carex echinata Murray ■☆

75456 Carex muricata L. var. basilata (Ohwi) Y. L. Chou = Carex angustior Mack. ■

75457 Carex muricata L. var. cephalantha (L. H. Bailey) Wiegand et Eames = Carex echinata Murray ■☆

75458 Carex muricata L. var. cephaloidea Dewey = Carex cephaloidea (Dewey) Dewey ■☆

75459 Carex muricata L. var. chaberti (Schultz) Maire et Weiller = Carex divulsa Stokes subsp. leersi (Kneuck.) W. Koch ■☆

75460 Carex muricata L. var. divulsa Kunth = Carex divulsa Stokes ex With. ■

75461 Carex muricata L. var. foliosa? = Carex foliosa D. Don ■

75462 Carex muricata L. var. laricina (Mack. ex Bright) Gleason = Carex laricina Mack. ex Bright ■

75463 Carex muricata L. var. leersi Schinz = Carex divulsa Stokes subsp. leersi (Kneuck.) W. Koch ■☆

75464 Carex muricata L. var. ruthii (Mack.) Gleason = Carex ruthii Mack. ■☆

75465 Carex muricata L. var. sterilis (Willd.) Gleason = Carex sterilis Willd. ■☆

75466 Carex muskingumensis Schwein.;掌苔草;Muskingum Sedge, Palm Sedge, Swamp Oval Sedge ■☆

75467 Carex mutans Boott ex C. B. Clarke = Kobresia esenbeckii (Kunth) Noltie ■

75468 Carex myosuroides Villari = Kobresia myosuroides (Vill.) Fiori et Paol. ■

75469 Carex myosurus Nees;鼠尾苔草■

75470 Carex myosurus Nees subsp. spiculata (Boott) Kük. = Carex myosurus Nees ■

75471 Carex nachiana Ohwi;日南苔草■

75472 Carex nakaharai Hayata = Carex gentilis Franch. var. nakaharae (Hayata) T. Koyama ■

75473 Carex nakaoana T. Koyama;钝鳞苔草■

75474 Carex nanchuanensis K. L. Chü ex S. Yun Liang;南川苔草;Nanchuan Sedge ■

75475 Carex nanella Ohwi = Carex callitrichos V. I. Krecz. var. nana (H. Lév. et Vaniot) Ohwi ■

75476 Carex nanella Ohwi = Carex humilis Leyss. var. nana (H. Lév. et Vaniot) Ohwi ■

75477 Carex nardina Fr.;甘松苔草■☆

75478 Carex nardina Fr. subsp. hepburnii (Boott) Á. Löve, D. Löve et B. M. Kapoor = Carex nardina Fr. ■☆

75479 Carex nardina Fr. var. atriceps Kük. = Carex nardina Fr. ■☆

75480 Carex nardina Fr. var. hepburnii (Boott) Kük. = Carex nardina Fr. ■☆

75481 Carex nebrascensis Dewey;内布拉斯加苔草;Nebraska Sedge ■☆

75482 Carex nebrascensis Dewey var. eruciformis Suksd. = Carex nebrascensis Dewey ■☆

75483 Carex nebrascensis Dewey var. praevia L. H. Bailey = Carex nebrascensis Dewey ■☆

75484 Carex nebrascensis Dewey var. ultiformis L. H. Bailey = Carex nebrascensis Dewey ■☆

75485 Carex negrii Chiov.;内格里苔草■☆

75486 Carex nelsonii Mack.;纳尔逊苔草■☆

75487 Carex nemorosa Rebent. = Carex otrubae Podp. ■

75488 Carex nemostachys Steud.;条穗苔草(毛囊果苔,线穗苔草);Linearspike Sedge, Nebraska Sedge ■

75489 Carex nemostachys Steud. var. crassispiculosa Kük.;粗条穗苔草;Thickspike Sedge ■

75490 Carex nemurensis Franch.;根室苔草■☆

75491 Carex neochevalieri Kük.;舍瓦利耶苔草■☆

75492 Carex neodigyna P. C. Li;双柱苔草■

75493 Carex neopaleacea Lepage = Carex paleacea Schreb. ex Wahlenb. ■☆

75494 Carex neopolycephala Ts. Tang et F. T. Wang ex L. K. Dai;新多穗苔草■

75495 Carex neopolycephala Ts. Tang et F. T. Wang ex L. K. Dai var. simplex F. T. Wang et Ts. Tang ex L. K. Dai;简序苔草(单穗苔草)■

75496 Carex nervata Franch. et Sav.;截嘴苔草■

75497 Carex nervina L. H. Bailey var. jonesii (L. H. Bailey) Kük. = Carex jonesii L. H. Bailey ■☆

75498 Carex nervosa Desf. = Carex extensa Gooden. ■☆

75499 Carex nesophila T. Holm = Carex microchaeta T. Holm subsp. nesophila (T. Holm) D. F. Murray ■☆

75500 Carex neurocarpa Maxim.;翼果苔草;Wingfruit Sedge ■

75501 Carex neurochlaena T. Holm = Carex heleonastes Ehrh. ex L. f. ■☆

75502 Carex neurophora Mack.;脉梗苔草■☆

75503 Carex nevadensis Boiss. et Reut. = Carex viridula Michx. subsp. nevadensis (Boiss. et Reut.) B. Schmid ■☆

75504 Carex nevadensis Boiss. et Reut. subsp. flavella (Krecz.) Janch. = Carex flava L. ■☆

75505 Carex nexa Boott = Carex ligata Boott ex Benth. ■

75506 Carex nexa Boott var. strictior Kük. = Carex sociata Boott ■

75507 Carex nigella Boott = Carex spectabilis Dewey ■☆

75508 Carex nigra (L.) Reichard;黑苔草;Common Sedge, Smooth Black Sedge ■☆

75509 Carex nigra (L.) Reichard subsp. intricata (Tineo) Rivas Mart.;缠结黑苔草■☆

75510 Carex nigra (L.) Reichard var. strictiformis (L. H. Bailey) Fernald = Carex nigra (L.) Reichard ■☆

75511 Carex nigricans C. A. Mey.;变黑苔草■☆

75512 Carex nigritella Drejer = Carex stylosa C. A. Mey. ■☆

75513 Carex nigromarginata Schwein.;黑边苔草;Black-edge Sedge, Sedge ■☆

75514 Carex nigromarginata Schwein. var. elliptica (Boott) Gleason = Carex peckii Howe ■☆

75515 Carex nigromarginata Schwein. var. floridana (Schwein.) Kük. = Carex floridana Schwein. ■☆

75516 Carex nigromarginata Schwein. var. minor (Boott) Gleason = Carex peckii Howe ■☆

75517 Carex nigromarginata Schwein. var. muhlenbergii (A. Gray) Gleason = Carex albicans Willd. ex Spreng. ■☆

75518 Carex nikkoensis Franch. et Sav. = Carex satsumensis Franch. et Sav. ■

75519 Carex ninagongensis (Kük.) Nelmes = Carex petitiana A. Rich. ■

75520 Carex nipposinica Ohwi = Carex brownii Tuck. ■

75521 Carex nipposinica Ohwi = Carex recurvisaccus T. Koyama ■

75522 Carex nitida Host = Carex liparocarpos Gaudin ■

75523 Carex nitida Host var. aspera (Boeck.) Kük. = Carex turkestanica Regel ■

75524 Carex nitidiutriculata L. K. Dai;亮果苔草■

75525 Carex nivalis Boott;喜马拉雅苔草(黑穗苔草,雪苔草)■

75526 Carex nodaeana A. I. Baranov et Skvortsov ex Liou et al. = Carex pseudolongirostrata Y. L. Chang et Y. L. Yang ■

75527 Carex normalis Mack.;苔草;Greater Straw Sedge, Intermediate Sedge, Normal Sedge, Sedge, Spreading Oval Sedge ■☆

75528 Carex normalis Mack. var. perlonga Fernald = Carex normalis Mack. ■☆

75529 Carex norvegica Retz.;密头苔草;Close-headed Alpine-sedge, Norway Sedge ■☆

75530 Carex norvegica Retz. subsp. conicorostrata Kalela = Carex norvegica Retz. ■☆

75531 Carex norvegica Retz. subsp. inferalpina (Wahlenb.) Hultén = Carex media R. Br. ex Richardson ■☆

75532 Carex norvegica Retz. subsp. inserrulata Kalela = Carex norvegica Retz. ■☆

75533 Carex norvegica Retz. subsp. stevenii (T. Holm) D. F. Murray = Carex stevenii (T. Holm) Kalela ■☆

75534 Carex norvegica Retz. var. inferalpina (Wahlenb.) B. Boivin = Carex media R. Br. ex Richardson ■☆

75535 Carex norvegica Retz. var. inserrulata (Kalela) Raymond = Carex norvegica Retz. ■☆

75536 Carex norvegica Retz. var. stevenii (T. Holm) Dorn = Carex stevenii (T. Holm) Kalela ■☆

75537 Carex norvegica Willd. ex Schkuhr = Carex mackenziei V. I. Krecz. ■☆

75538 Carex nova L. H. Bailey;新苔草■☆

75539 Carex novae-angliae Schwein.;新英格兰苔草;New England Oak Sedge, New England Sedge ■☆

75540 Carex novae-angliae Schwein. var. rossii (Boott) L. H. Bailey = Carex rossii Boott ■☆

75541 Carex novograblenovii Kom.;诺沃氏苔草■☆

75542 Carex nubicola Mack. = Carex haydeniana Olney ■☆

75543 Carex nubigena D. Don = Carex fluviatilis Boott ■

75544 Carex nubigena D. Don ex Okamoto et Taylor;云雾苔草(聚生穗序苔,聚生穗序苔草);Cloud Sedge ■

75545 Carex nubigena D. Don ex Okamoto et Taylor f. viridens Kük.;绿云雾苔草■

75546 Carex nubigena D. Don ex Okamoto et Taylor f. viridens Kük. = Carex nubigena D. Don ex Okamoto et Taylor ■

75547 Carex nubigena D. Don ex Okamoto et Taylor subsp. albata (Boott ex Franch. et Sav.) T. Koyama;褐红脉苔草(白云雾苔草)■

75548 Carex nubigena D. Don ex Okamoto et Taylor subsp. pseudoarenicola (Hayata) T. Koyama = Carex nubigena D. Don ex Okamoto et Taylor ■

75549 Carex nubigena D. Don ex Okamoto et Taylor var. albata (Boott ex Franch. ex Sav.) Kük. ex Matsum. = Carex nubigena D. Don ex Okamoto et Taylor subsp. albata (Boott ex Franch. et Sav.) T. Koyama ■

75550 Carex nubigena D. Don ex Okamoto et Taylor var. fallax (Steud.) C. B. Clarke;假云雾苔草■

75551 Carex nubigena D. Don var. albata (Boott) Kük. ex Matsum. = Carex nubigena D. Don ex Okamoto et Taylor subsp. albata (Boott ex Franch. et Sav.) T. Koyama ■

75552 Carex nudata W. Boott;裸露苔草■☆

75553 Carex nudata W. Boott var. anomala L. H. Bailey = Carex nudata W. Boott ■☆

75554 Carex nugata Ohwi;横纹苔草■

75555 Carex nutans Boott = Kobresia esenbeckii (Kunth) F. T. Wang et Ts. Tang ex P. C. Li ■

75556 Carex nutans Host.;悬垂苔草;Drooping Sedge ■☆

75557 Carex nutans Host. = Carex heterostachys Bunge ■

75558 Carex nutans Host. = Carex melanostachys M. Bieb. ex Willd. ■

75559 Carex nyasensis C. B. Clarke = Carex echinochloe Kuntze subsp. nyasensis (C. B. Clarke) Lye ■☆

75560 Carex obesa Meinsh. var. aspera Beler = Carex turkestanica Regel ■

75561 Carex obesa Schleich. ex Kunth var. aspera Boeck. = Carex turkestanica Regel ■

75562 Carex obliquetruncata Y. C. Tang et S. Yun Liang;斜口苔草■

75563 Carex oblita Steud. = Carex venusta Dewey ■☆

75564 Carex obovatosquamata F. T. Wang et Y. L. Chang ex P. C. Li;倒卵鳞苔草■

75565 Carex obovoidea Cronquist = Carex cusickii Mack. ex Piper et Beattie ■☆

75566 Carex obscura Nees;暗昧苔草■

75567 Carex obscura Nees var. brachycarpa C. B. Clarke;刺囊苔草(短果苔草)■

75568 Carex obscuriceps Kük.;褐紫鳞苔草■

75569 Carex obscuriceps Kük. var. pamirica (O. Fedtsch.) Kük. = Carex pamirensis C. B. Clarke ■

75570 Carex obtusata Lilj.;北苔草(圆苔)■

75571 Carex obtuso-bracteata Hayata = Carex breviscapa C. B. Clarke ■

75572 Carex occultans (Franch.) V. I. Krecz.;隐蔽苔草■☆

75573 Carex odontostoma Kük. = Carex chrysolepis Franch. et Sav. ■

75574 Carex oederi Ehrh. = Carex viridula Michx. ■☆

75575 Carex oederi Retz.;埃德尔苔草■☆

75576 Carex oederi Retz. = Carex serotina Merat ■

75577 Carex oederi Retz. = Carex viridula Michx. subsp. oedocarpa (Andersson) B. Schmid ■☆

75578 Carex oederi Retz. subsp. viridula (Michx.) Hultén = Carex viridula Michx. ■☆

75579 Carex oederi Retz. var. pumila (Coss. et Germ.) Fernald = Carex viridula Michx. ■☆

75580 Carex oederi Retz. var. recterostrata (Vict.) Dorn = Carex viridula Michx. ■☆

75581 Carex oederi Retz. var. subglosa Fernald = Carex viridula Michx. var. saxilittoralis (Robertson) Crins ■☆

75582 Carex oederi Retz. var. viridula (Michx.) Kük. = Carex viridula Michx. ■☆

75583 Carex oedorrhampha Nelmes;肿喙苔草(膨果苔草)■

75584 Carex okamotoi Ohwi = Carex glossostigma Hand. -Mazz. ■

75585 Carex oklahomensis Mack.;俄克拉何马苔草■☆

75586 Carex okuboi Franch. = Carex doenitzii Boeck. subsp. okuboi (Franch.) T. Koyama ■☆

75587 Carex olbiensis Jord.;奥尔比亚苔草■☆

75588 Carex oligantha Stead;贫花苔草■☆

75589 Carex oligocarpa Willd.;少果苔草;Few-fruited Gray Sedge, Rich-woods Sedge, Sedge ■☆

75590 Carex oligocarpa Willd. var. hitchcockiana (Dewey) Kük. = Carex hitchcockiana Dewey ■☆

75591 Carex oligosperma Michx.;少籽苔草;Few-seeded Hop Sedge, Few-seeded Sedge ■☆

75592 Carex oligosperma Michx. subsp. tsuishikarensis (Koidz. et Ohwi) T. Koyama et Calder;对雁苔草■☆

75593 Carex oligosperma Michx. var. tsuishikarensis (Koidz. et Ohwi) B. Boivin = Carex oligosperma Michx. subsp. tsuishikarensis (Koidz. et Ohwi) T. Koyama et Calder ■☆

75594 Carex oligostachys Meinsh. et Maxim. = Carex egena H. Lév. et Vaniot ■

75595 Carex oligostachys Nees;少穗苔草■

75596 Carex olivacea Boott;榄绿果苔草(橄色苔草)■

75597 Carex olivacea Boott subsp. confertiflora (Boott) T. Koyama = Carex confertiflora Boott ■

75598 Carex olivacea Boott var. angustior Kük. = Carex confertiflora Boott ■

75599 Carex olivacea Boott var. angustior Kük. = Carex olivacea Boott subsp. confertiflora (Boott) T. Koyama ■

75600 Carex oliveri Boeck. = Carex griffithii Boott ■

75601 Carex oliveri W. Becker;奥里氏苔草■☆

75602 Carex omeiensis Ts. Tang et F. T. Wang;峨眉苔草(多囊苔草);Emei Sedge ■

75603 Carex omeiensis Ts. Tang et F. T. Wang var. multifascula Q. S. Wang = Carex omeiensis Ts. Tang et F. T. Wang ■

75604 Carex omiana Franch. et Sav.;星穗苔草■

75605 Carex omiana Franch. et Sav. subsp. monticola (Ohwi) T. Koyama = Carex omiana Franch. et Sav. var. monticola Ohwi ■☆

75606 Carex omiana Franch. et Sav. var. monticola Ohwi;山地星穗苔草■☆

75607 Carex omiana Franch. et Sav. var. yakushimana Ohwi;屋久岛星穗苔草■☆

75608 Carex omskiana Meinsh.;鄂木斯克苔草■☆

75609 Carex omurae T. Koyama;小村苔草■☆

75610 Carex onoei Franch. et Sav.;针叶苔草(阴地针苔草);Needleleaf Sedge ■

75611 Carex onoei Franch. et Sav. subsp. capituliformis (Meinsh. ex Maxim.) T. V. Egorova = Carex onoei Franch. et Sav. ■

75612 Carex onoei Franch. et Sav. subsp. krameri (Franch. et Sav.) T. Koyama = Carex hakonensis Franch. et Sav. ■☆

75613 Carex onoei Franch. et Sav. var. krameri (Franch. et Sav.) Kük. ex Matsum. = Carex hakonensis Franch. et Sav. ■☆

75614 Carex onoei Franch. et Sav. var. macrogyna Ts. Tang = Carex onoei Franch. et Sav. ■

75615 Carex onusta Mack. = Carex muehlenbergii Schkuhr ex Willd. var. enervis Boott ■☆

75616 Carex opaca (F. J. Herm.) P. Rothr. et Reznicek;暗色苔草■

75617 Carex orbicularinucis L. K. Dai;圆坚果苔草■

75618 Carex orbicularis Boott;圆囊苔草(红苔草);Beikiang Sedge, N. Xingjiang Sedge, Orbicular Sedge ■

75619 Carex orbicularis Boott var. taldycola (Meinsh.) Kük. = Carex taldycola Meinsh. ■

75620 Carex oregonensis Olney ex L. H. Bailey = Carex halliana L. H. Bailey ■☆

75621 Carex oreocharis T. Holm;山栖苔草■☆

75622 Carex oreocharis T. Holm var. enantiomorpha Kelso = Carex oreocharis T. Holm ■☆

75623 Carex oreophila C. A. Mey.;喜山苔草■☆

75624 Carex ormantha (Fernald) Mack. = Carex echinata Murray ■☆

75625 Carex ormostachya Wiegand;串穗苔草;Bead-like Sedge, Necklace-spike Sedge, Necklace-spike Wood Sedge ■☆

75626 Carex oronensis Fernald;奥罗诺苔草;Orono Sedge ■☆

75627 Carex orthostachys C. A. Mey.;直穗苔草■

75628 Carex orthostachys C. A. Mey. = Carex atherodes Spreng. ■

75629 Carex orthostachys C. A. Mey. var. drymophila (Turcz.) Maxim. = Carex drymophila Turcz. ex Steud. ■

75630 Carex orthostachys C. A. Mey. var. spuria Y. L. Chang et Y. L. Yang;疑直穗苔草■

75631 Carex orthostachys Trevir. var. drymophilia (Turcz.) Maxim. = Carex drymophila Turcz. ex Steud. ■

75632 Carex orthostemon Hayata;直蕊宿柱苔草■

75633 Carex orthostemon Hayata = Carex breviculmis R. Br. var. cupulifera (Hayata) Y. C. Tang et S. Yun Liang ■

75634 Carex orthostemon Hayata var. cupulifera Hayata = Carex breviculmis R. Br. var. cupulifera (Hayata) Y. C. Tang et S. Yun Liang ■

75635 Carex oshimensis Nakai;奄美苔草■☆

75636 Carex oshimensis Nakai 'Evergold';金奄美苔草;Evergold Striped Weeping Sedge, Oshima Kan Suge ■☆

75637 Carex oshimensis Nakai f. variegata Hid. Takah.;杂色奄美苔草■☆

75638 Carex otaruensis Franch.;鹞落苔草■

75639 Carex otaruensis Franch. var. kadzuoana T. Koyama;小鹞落苔草(小山苔草)■☆

75640 Carex otayae Ohwi;御旅屋苔草■☆

75641 Carex otrubae Podp.;捷克苔草(铜苔);False Fox-sedge ■

75642 Carex otrubae Podp. = Carex cuprina (Heuff.) A. Kern. ■☆

75643 Carex ouensanensis Ohwi = Carex micrantha Kük. ■

75644 Carex ovalis Gooden.;卵苔草;Oval Sedge ■☆

75645 Carex ovata Burm. f. = Abildgaardia ovata (Burm. f.) Král ■

75646 Carex ovata Burm. f. = Fimbristylis ovata (Burm. f.) J. Kern ■

75647 Carex ovata Rudge = Carex atratiformis Britton ■☆

75648 Carex ovatispiculata F. T. Wang et Y. L. Chang ex S. Yun Liang;卵穗苔草;Ovate-spiked Sedge ■

75649 Carex oxyandra (Franch. et Sav.) Kudo;球穗苔草(南投苔,球穗苔)■

75650 Carex oxyandra (Franch. et Sav.) Kudo var. lanceata (Kük.) Ohwi;披针状球穗苔草■☆

75651 Carex oxycarpa T. Holm = Carex angustata Boott ■☆

75652 Carex oxylepis Torr. et Hook.;尖鳞苔草;Sedge, Sharpscale Sedge ■☆

75653 Carex oxylepis Torr. et Hook. var. pubescens J. K. Underw. = Carex oxylepis Torr. et Hook. ■☆

75654 Carex oxyleuca V. I. Krecz. = Carex atrofusca Schkuhr subsp. minor (Boott) T. Koyama ■

75655 Carex oxyleuca V. I. Krecz. = Carex atrofusca Schkuhr ■

75656 Carex oxyphylla Franch.;尖叶苔草■☆

75657 Carex pachinensis Hayata = Carex metallica H. Lév. et Vaniot ■

75658 Carex pachygyna Franch. et Sav.;粗蕊苔草■☆

75659 Carex pachyneura Kitag.;肋脉苔草■

75660 Carex pachyrrhiza Franch.;粗根苔草;Thickroot Sedge ■☆

75661 Carex pachyrrhiza Franch. = Carex setosa Boott ■

75662 Carex pachystachya Cham. ex Steud.;粗穗苔草■☆

75663 Carex pachystoma T. Holm = Carex aquatilis Wahlenb. ■☆

75664 Carex pachystylis J. Gay;粗柱苔草(粉柱苔)■

75665 Carex paddoensis Suksd. = Carex engelmannii L. H. Bailey ■☆

75666 Carex pairaei F. W. Schultz = Carex muricata L. subsp. lamprocarpa Celak. ■☆

75667 Carex paishanensis Nakai = Carex atrata L. ■

75668 Carex paleacea Schreb. ex Wahlenb.;膜片苔草;Scaly Sedge ■☆

75669 Carex paleacea Schreb. ex Wahlenb. var. transatlantica Fernald = Carex paleacea Schreb. ex Wahlenb. ■☆

75670 Carex paleaceoides Lepage = Carex paleacea Schreb. ex Wahlenb. ■☆

75671 Carex pallens Z. P. Wang = Carex breviculmis R. Br. var. fibrillosa (Franch. et Sav.) Kük. ex Matsum. et Hayata ■

75672 Carex pallescens L.;苍苔草;Pale Green Sedge, Pale Sedge ■☆

75673 Carex pallescens L. var. leucantha (Schur) Asch. et Graebn. = Carex pallescens L. ■☆

75674 Carex pallescens L. var. neogaea Fernald = Carex pallescens L. ■☆

75675 Carex pallida C. A. Mey.;疣囊苔草■

75676 Carex pallida C. A. Mey. var. angustifolia Y. L. Chang;狭叶疣囊苔草■

75677 Carex paludosa Gooden. = Carex acutiformis Ehrh. ■

75678 Carex pamirensis C. B. Clarke;帕米尔苔草;Pamir Sedge ■

75679 Carex pamirensis C. B. Clarke ex B. Fedtsch. = Carex pamirensis C. B. Clarke ■

75680 Carex pamirensis C. B. Clarke var. angustispicata Y. C. Yang;窄穗帕米尔苔草(窄穗苔草)■

75681 Carex pamirica (O. Fedtsch.) B. Fedtseh. = Carex pamirensis C. B. Clarke ■

75682 Carex panda C. B. Clarke = Carex aquatilis Wahlenb. var. dives (T. Holm) Kük. ■☆

75683 Carex pandanophylla E. G. Camus = Carex scaposa C. B. Clarke ■

75684 Carex panicea L.;黍苔(黍状苔草);Carnation Sedge, Grasslike

Sedge ■

75685 Carex panicea L. var. sparsiflora Wahlenb. = Carex sparsiflora (Wahlenb.) Steud. ■

75686 Carex paniculata L.;圆锥花苔草;Greater Tussock-sedge, Panicled Sedge ■☆

75687 Carex paniculata L. subsp. calderae (Hansen) Lewej. et Lobin; 卡尔德拉苔草■☆

75688 Carex paniculata L. subsp. hansenii Lewej. et Lobin;汉森苔草■☆

75689 Carex paniculata L. subsp. lusitanica (Schkuhr) Maire;葡萄牙圆锥花苔草■☆

75690 Carex pansa L. H. Bailey;铺展苔草■☆

75691 Carex papillosissima Nelmes;大乳突苔草■☆

75692 Carex papulosa Boott;多乳突苔草■☆

75693 Carex paracuraica F. T. Wang et Y. L. Chang = Carex paracuraica F. T. Wang et Y. L. Chang ex S. Yun Liang ■

75694 Carex paracuraica F. T. Wang et Y. L. Chang ex S. Yun Liang;陇县苔草■

75695 Carex parakensis C. B. Clarke;黄穗苔■

75696 Carex paralia Krecz.;白海苔草■☆

75697 Carex parallela Laest.;平行脉苔草■☆

75698 Carex parciflora Boott;小花苔草■☆

75699 Carex parciflora Boott var. macroglossa (Franch. et Sav.) Ohwi; 大舌小花苔草■☆

75700 Carex parciflora Boott var. macroglossa (Franch. et Sav.) Ohwi f. subsessilis Ohwi;无柄大舌小花苔草■☆

75701 Carex parciflora Boott var. tsukudensis T. Koyama;津久田苔草■☆

75702 Carex parciflora Boott var. vaniotii (H. Lév.) Ohwi;瓦尼苔草■☆

75703 Carex parryana Dewey;帕里苔草■☆

75704 Carex parryana Dewey subsp. hallii (Olney) D. F. Murray = Carex hallii Olney ■☆

75705 Carex parryana Dewey subsp. idahoa (L. H. Bailey) D. F. Murray = Carex idahoa L. H. Bailey ■☆

75706 Carex parryana Dewey var. hallii (Olney) Kük. = Carex hallii Olney ■☆

75707 Carex parryana Dewey var. statonii M. E. Jones = Carex idahoa L. H. Bailey ■☆

75708 Carex parryana Dewey var. unica L. H. Bailey = Carex hallii Olney ■☆

75709 Carex parva Nees;小苔草■

75710 Carex paucicostata Mack. = Carex lenticularis Michx. var. impressa (L. H. Bailey) L. A. Standl. ■☆

75711 Carex pauciflora Lightf.;寡花苔草;Few-flowered Bog Sedge, Few-flowered Sedge, Pauciflorous Sedge, Star Sedge ■☆

75712 Carex pauciflora Lightf. var. microglochin (Wahlenb.) Poir. ex Lam. = Carex microglochin Wahlenb. ■

75713 Carex paulovargasii Luceno et Marín;瓦尔加斯苔草■☆

75714 Carex paupercula Michx. = Carex magellanica Lam. subsp. irrigua (Wahlenb.) Hiitonen ■☆

75715 Carex paupercula Michx. = Carex magellanica Lam. ■☆

75716 Carex paupercula Michx. var. brevisquama Fernald = Carex magellanica Lam. subsp. irrigua (Wahlenb.) Hiitonen ■☆

75717 Carex paupercula Michx. var. irrigua (Wahlenb.) Fernald = Carex magellanica Lam. subsp. irrigua (Wahlenb.) Hiitonen ■☆

75718 Carex paupercula Michx. var. pallens Fernald = Carex magellanica Lam. subsp. irrigua (Wahlenb.) Hiitonen ■☆

75719 Carex pauxilla V. I. Krecz.;极小苔草■☆

75720 Carex paxii Kük.;短苞苔草(短芒苔草);Pax Sedge ■

75721 Carex peckii Howe;佩克苔草;Peck's Oak Sedge, Peck's Sedge ■☆

75722 Carex pedata L.;欧洲鸟足苔草■☆

75723 Carex pedata Wahlenburg = Carex glacialis Mack. ■☆

75724 Carex pediformis C. A. Mey.;柄状苔草(短柄苔草,根苔,脚苔,脚苔草,毛根苔草,日荫菅,硬叶苔草);Hardleaf Sedge, Reddishbrown Sedge, Rigidleaf Sedge ■

75725 Carex pediformis C. A. Mey. var. floribunda Korsh. = Carex pediformis C. A. Mey. ■

75726 Carex pediformis C. A. Mey. var. genuina Maxim. = Carex pediformis C. A. Mey. ■

75727 Carex pediformis C. A. Mey. var. macrosandra (Franch.) C. B. Clarke = Carex macrosandra (Franch.) V. I. Krecz. ■

75728 Carex pediformis C. A. Mey. var. macroura (Meinsh.) Kük. = Carex pediformis C. A. Mey. ■

75729 Carex pediformis C. A. Mey. var. pedunculata Maxim.;柞苔草■

75730 Carex pediformis C. A. Mey. var. rhizina Kük. = Carex pediformis C. A. Mey. var. pedunculata Maxim. ■

75731 Carex pedunculata Muchl. ex Willd. var. erythrobasis (H. Lév. et Vaniot) T. Koyama = Carex erythrobasis H. Lév. et Vaniot ■

75732 Carex pedunculata Muhl. ex Schkuhr var. erythrobasis (H. Lév. et Vaniot) T. Koyama = Carex erythrobasis H. Lév. et Vaniot ■

75733 Carex pedunculata Muhl. ex Willd.;具梗苔草;Long-stalk Sedge, long-stalked Sedge, Peduncled Sedge, Pedunculate Sedge ■☆

75734 Carex pedunculifera Kom.;梗花苔草■☆

75735 Carex pedunculosum DC. = Carpesium cernuum L. ■

75736 Carex peiana F. T. Wang et Ts. Tang = Carex emineus Nees ■

75737 Carex peiktusani Kom.;白头山苔草(长白苔草);Baitoushan Sedge, Peiktushan Sedge ■

75738 Carex peliosanthifolia F. T. Wang et Ts. Tang ex P. C. Li;扇叶苔草■

75739 Carex pellita Muhl. = Carex pellita Willd. ex Willd. ■☆

75740 Carex pellita Muhl. ex Willd.;阔叶毛苔草;Broad-leaved Woolly Sedge, Sedge ■☆

75741 Carex pellita Willd. = Carex pellita Muhl. ex Willd. ■☆

75742 Carex pendula Huds.;下垂苔草(下垂苔);Hanging Sedge, Pendulous Sedge ■☆

75743 Carex pendula Huds. var. myosuroides (Lowe) Boott = Carex pendula Huds. ■☆

75744 Carex pensylvanica Lam.;宾州苔草;Common Oak Sedge, Early Sedge, Penn Sedge, Pennsylvania Sedge, Sedge, Yellow Sedge ■☆

75745 Carex pensylvanica Lam. subsp. heliophila (Mack.) W. A. Weber = Carex inops L. H. Bailey subsp. heliophila (Mack.) Crins ■☆

75746 Carex pensylvanica Lam. var. amblyolepis (Trautv. et C. A. Mey.) Kük. = Carex vanheurckii Müll. Arg. ■

75747 Carex pensylvanica Lam. var. digyna Boeck. = Carex inops L. H. Bailey subsp. heliophila (Mack.) Crins ■☆

75748 Carex pensylvanica Lam. var. digyna Boeck. = Carex inops L. H. Bailey ■☆

75749 Carex pensylvanica Lam. var. distans Peck = Carex lucorum Willd. ex Link ■☆

75750 Carex pensylvanica Lam. var. glumabunda Peck = Carex pensylvanica Lam. ■☆

75751 Carex pensylvanica Lam. var. lucorum (Willd. ex Link) Fernald = Carex lucorum Willd. ex Link ■☆

75752 Carex pensylvanica Lam. var. marginata (Willd.) Dewey = Carex pensylvanica Lam. ■☆

75753 Carex pensylvanica Lam. var. separans Peck = Carex lucorum Willd. ex Link ■☆
75754 Carex pensylvanica Lam. var. verspertina L. H. Bailey = Carex inops L. H. Bailey ■☆
75755 Carex pentastachys Fisch. ex Steud.;五穗苔草■☆
75756 Carex perakensis C. B. Clarke;霹雳苔草(黄穗苔)■
75757 Carex peregrina Link;外来苔草■
75758 Carex perfusca V. I. Krecz. = Carex aterrima Hoppe ■
75759 Carex perfusca V. I. Krecz. var. japonalpina (T. Koyama) Kitag.;日山苔草■☆
75760 Carex pergracilis Nelmes;纤细苔草■
75761 Carex persistens Ohwi;宿存苔草■☆
75762 Carex persistens Ohwi var. watanabei Ohwi;渡边苔草■☆
75763 Carex persoonii Sieber = Carex brunnescens (Pers.) Poir. ■
75764 Carex petasata Dewey;帽苔草■☆
75765 Carex petersii C. A. Mey. = Carex sparsiflora Steud. var. petersi (C. A. Mey.) Kük. ■
75766 Carex petersii C. A. Mey. ex F. Schmidt = Carex sparsiflora Steud. var. petersi (C. A. Mey.) Kük. ■
75767 Carex petitiana A. Rich.;佩蒂蒂苔草■
75768 Carex petricosa Dewey;岩地苔草■☆
75769 Carex phacota Spreng.;镜子苔草(大三方草,镜子苔,七星斑囊果苔,三棱草,三棱马尾,仙鹤草,有喙红苞苔);Convexutricle Sedge,Mirror Sedge ■
75770 Carex phaenocarpa Franch.;显果苔草(色果苔草)■
75771 Carex phaeodon T. Koyama;褐齿苔草■☆
75772 Carex phaeopoda Ohwi = Carex kiotensis Franch. et Sav. ■
75773 Carex philocrena V. I. Krecz. = Carex serotina Merat ■
75774 Carex phragmitoides Kük.;芦苇苔草■☆
75775 Carex phyllocephala T. Koyama;密苞叶苔草;Palm Sedge, Sparkler Sedge,Tenjiku Sedge ■
75776 Carex phyllomanica W. Boott = Carex echinata Murray subsp. phyllomanica (W. Boott) Reznicek ■☆
75777 Carex phyllomanica W. Boott var. angustata (J. Carey) B. Boivin = Carex echinata Murray ■☆
75778 Carex phyllomanica W. Boott var. ormantha (Fernald) B. Boivin = Carex echinata Murray ■☆
75779 Carex phyllostachys C. A. Mey.;叶穗苔草■☆
75780 Carex physocarpa C. Presl = Carex saxatilis L. ■☆
75781 Carex physochlaena T. Holm = Carex membranacea Hook. ■☆
75782 Carex physodes M. Bieb.;囊果苔草(囊泡苔)■
75783 Carex physorhyncha Liebm. ex Steud. = Carex albicans Willd. ex Spreng. ■☆
75784 Carex pierotii Miq. = Carex scabrifolia Steud. ■
75785 Carex pilosa Scop.;毛缘苔草;Hairy Sedge ■
75786 Carex pilosa Scop. var. auriculata Kük.;刺毛缘苔草■
75787 Carex pilosa Scop. var. auriculata Kük. = Carex pilosa Scop. ■
75788 Carex piluifera L.;球腺苔草;Pill Sedge,Pill-headed Sedge ■☆
75789 Carex piperi Mack. = Carex praticola Rydb. ■☆
75790 Carex pisanensis T. Koyama = Carex laeta Boott ■
75791 Carex pisformis Boott var. major (Kük.) T. Koyama = Carex capilliformis Franch. ■
75792 Carex pisiformis Boott;豌豆形苔草(白鳞苔草,白穗苔草,白雄穗苔草,类圆锥苔草,线苔草,线叶宿柱苔,宜兰宿柱苔);Whitescale Sedge ■
75793 Carex pisiformis Boott f. polyschoena (H. Lév. et Vaniot) Kük. = Carex pisiformis Boott ■
75794 Carex pisiformis Boott f. polyschoena (H. Lév. et Vaniot) Kük. = Carex polyschoena H. Lév. et Vaniot ■
75795 Carex pisiformis Boott subsp. alterniflora (Franch.) T. Koyama = Carex alterniflora Franch. ■
75796 Carex pisiformis Boott subsp. aureobrunnea (Ohwi) T. Koyama = Carex sachalinensis Eastw. var. aureobrunnea (Ohwi) Ohwi ■☆
75797 Carex pisiformis Boott subsp. cuneata (Ohwi) T. Koyama = Carex stenostachys Franch. et Sav. var. cuneata (Ohwi) Ohwi et T. Koyama ■☆
75798 Carex pisiformis Boott subsp. duvaliana (Franch. et Sav.) T. Koyama = Carex duvaliana Franch. et Sav. ■
75799 Carex pisiformis Boott subsp. fernaldiana (H. Lév. et Vaniot) T. Koyama = Carex fernaldiana H. Lév. et Vaniot ■
75800 Carex pisiformis Boott subsp. fulva (Ohwi) T. Koyama = Carex sachalinensis Eastw. var. fulva (Ohwi) Ohwi ■☆
75801 Carex pisiformis Boott subsp. mayebarana (Ohwi) T. Koyama = Carex mayebarana Ohwi ■☆
75802 Carex pisiformis Boott subsp. polyschoena (H. Lév. et Vaniot) T. Koyama = Carex polyschoena H. Lév. et Vaniot ■
75803 Carex pisiformis Boott subsp. polyschoena (H. Lév. et Vaniot) T. Koyama = Carex pisiformis Boott ■
75804 Carex pisiformis Boott subsp. sikokiana (Franch. et Sav.) T. Koyama = Carex sachalinensis Eastw. var. sikokiana (Franch. et Sav.) Ohwi ■☆
75805 Carex pisiformis Boott subsp. stenostachys (Franch. et Sav.) T. Koyama = Carex stenostachys Franch. et Sav. ■☆
75806 Carex pisiformis Boott var. alterniflora (Franch.) T. Koyama = Carex alterniflora Franch. ■
75807 Carex pisiformis Boott var. alterniflora (Franch.) T. Koyama = Carex pisiformis Boott ■
75808 Carex pisiformis Boott var. alterniflora (Franch.) T. Koyama f. musashiensis (Hiyama) T. Koyama = Carex sachalinensis Eastw. var. musashiensis Hiyama ■☆
75809 Carex pisiformis Boott var. aureobrunnea (Ohwi) T. Koyama = Carex sachalinensis Eastw. var. aureobrunnea (Ohwi) Ohwi ■☆
75810 Carex pisiformis Boott var. coreana (Nakai) T. Koyama = Carex polyschoena H. Lév. et Vaniot ■
75811 Carex pisiformis Boott var. cuneata (Ohwi) T. Koyama = Carex stenostachys Franch. et Sav. var. cuneata (Ohwi) Ohwi et T. Koyama ■☆
75812 Carex pisiformis Boott var. duvaliana (Franch. et Sav.) T. Koyama = Carex duvaliana Franch. et Sav. ■
75813 Carex pisiformis Boott var. elongatula (Ohwi) T. Koyama = Carex sachalinensis Eastw. var. elongatula (Ohwi) Ohwi ■☆
75814 Carex pisiformis Boott var. fernaldiana (H. Lév. et Vaniot) T. Koyama = Carex fernaldiana H. Lév. et Vaniot ■
75815 Carex pisiformis Boott var. fernaldiana (H. Lév. et Vaniot) T. Koyama = Carex pisiformis Boott ■
75816 Carex pisiformis Boott var. koreana (Nakai) T. Koyama = Carex pisiformis Boott ■
75817 Carex pisiformis Boott var. major (Kük.) T. Koyama = Carex capilliformis Franch. ■
75818 Carex pisiformis Boott var. major (Kük.) T. Koyama = Carex pisiformis Boott ■
75819 Carex pisiformis Boott var. mayebarana (Ohwi) T. Koyama = Carex mayebarana Ohwi ■☆
75820 Carex pisiformis Boott var. pineticola (Ohwi) T. Koyama = Carex

sachalinensis Eastw. var. pineticola (Ohwi) Ohwi ■☆
75821 Carex pisiformis Boott var. sachalinensis (F. Schmidt) Kük. ex Matsum. = Carex pisiformis Boott ■
75822 Carex pisiformis Boott var. sikokiana (Franch. et Sav.) T. Koyama = Carex sachalinensis Eastw. var. sikokiana (Franch. et Sav.) Ohwi ■☆
75823 Carex pisiformis Boott var. subebracteata Kük. = Carex subebracteata (Kük.) Ohwi ■
75824 Carex plana Mack. = Carex muehlenbergii Schkuhr ex Willd. var. enervis Boott ■☆
75825 Carex plana Mack. = Carex muhlenbergii Schkuhr ex Willd. ■☆
75826 Carex planata Franch. et Sav.;平展苔草■☆
75827 Carex planiculmis Kom.;扁秆苔草;Flatculm Sedge ■
75828 Carex planiculmis Kom. var. urasawae Ohwi;浦泽苔草■☆
75829 Carex planiscapa Chun et F. C. How;扁茎苔草■
75830 Carex planispicata Naczi;扁穗苔草;Sedge ■☆
75831 Carex plantaginea Lam.;蕉叶苔草;Plantain Leaf Sedge, Plantain Leaved Sedge, Plantain-leaved Sedge, Plantain-leaved Wood Sedge ■☆
75832 Carex platylepis Mack. = Carex praticola Rydb. ■☆
75833 Carex platyphylla J. Carey;阔叶苔草;Broad-leaf Sedge, Broad-leaved Sedge, Broad-leaved Wood Sedge, Thicket Sedge ■☆
75834 Carex platyrhyncha Franch. et Sav.;宽喙苔草■☆
75835 Carex platysperma Y. L. Chang et Y. L. Yang;双辽苔草■
75836 Carex platysperma Y. L. Chang et Y. L. Yang var. sungareensis Y. L. Chang et Y. L. Yang;松花江苔草■
75837 Carex pleistogyna V. I. Krecz.;无翅苔草;Wingless Sedge ■
75838 Carex pleistogyna V. I. Krecz. = Carex nubigena D. Don ex Okamoto et Taylor ■
75839 Carex plumbea Vahl;普拉姆苔草■☆
75840 Carex pocilliformis Boott = Carex tristachys Thunb. var. pocilliformis (Boott) Kük. ■
75841 Carex poculisquama Kük.;杯鳞苔草■
75842 Carex podocarpa R. Br. ex Richardson;柄果苔草■☆
75843 Carex podogyna Franch. et Sav.;柄蕊苔草■☆
75844 Carex pollens C. B. Clarke = Carex dispalata Boott ex A. Gray ■
75845 Carex pollens C. B. Clarke var. angustioir C. B. Clarke = Carex dispalata Boott ex A. Gray ■
75846 Carex polycephala Boott var. simplex Kük.;简单多头苔草;Simplex Sedge ■
75847 Carex polygama Schkuhr = Carex buxbaumii Wahlenb. ■☆
75848 Carex polymascula P. C. Li;多雄苔草(多头苔草);Manymale Sedge ■
75849 Carex polyphylla Kar. et Kir.;多叶苔草(多叶苔);Manyleaf Sedge ■
75850 Carex polyschoena H. Lév. et Vaniot = Carex pisiformis Boott ■
75851 Carex polyschoena H. Lév. et Vaniot ex H. Lév. = Carex pisiformis Boott ■
75852 Carex polyschoenoides K. T. Fu;类白鳞苔草(类白穗苔草,类白苔草);Whitescale-like Sedge ■
75853 Carex polyschoenoides K. T. Fu var. fractiflexa Y. C. Yang;曲轴苔草■
75854 Carex pomiensis Y. C. Yang;波密苔草;Bomi Sedge ■
75855 Carex pontica Albov;蓬特苔草■☆
75856 Carex praeceptorum Mack.;可疑苔草■☆
75857 Carex praeclara Nelmes;沙漠苔草;Sandliving Sedge ■
75858 Carex praecox Jacq. = Carex caryophylla Latourr. ■☆
75859 Carex praecox Schreb.;早熟苔(早发苔草);Vernal Sedge ■
75860 Carex praegracilis W. Boott;簇生野苔草;Clustered Field Sedge, Freeway Sedge ■☆
75861 Carex praelonga C. B. Clarke;帚状苔草■
75862 Carex prainii C. B. Clarke;大序苔草;Prain Sedge ■
75863 Carex prainii C. B. Clarke = Carex perakensis C. B. Clarke ■
75864 Carex prairea Dewey;沼泽圆锥苔草;Fen Panicled Sedge, Prairie Sedge ■☆
75865 Carex prasina Wahlenb.;韭绿苔草;Drooping Sedge, Leek-green Sedge, Sedge ■☆
75866 Carex pratensis Drejer = Carex praticola Rydb. ■☆
75867 Carex praticola Rydb.;草原苔草■☆
75868 Carex prescottiana Boott subsp. flabellata (H. Lév. et Vaniot) T. Koyama = Carex flabellata H. Lév. et Vaniot ■☆
75869 Carex prescottiana Boott subsp. kiotensis (Franch. et Sav.) T. Koyama = Carex kiotensis Franch. et Sav. ■
75870 Carex prescottiana Boott var. cremostachys (Franch.) Kük. = Carex cremostachys Franch. ■
75871 Carex prescottiana Boott var. fargesii (Franch.) Kük. = Carex fargesii Franch. ■
75872 Carex prescottiana Boott var. flabellata (H. Lév. et Vaniot) Ohwi = Carex flabellata H. Lév. et Vaniot ■☆
75873 Carex prescottiana Boott var. kiotensis (Franch. et Sav.) Kük. ex Matsum. = Carex kiotensis Franch. et Sav. ■
75874 Carex preussii K. Schum.;普罗伊斯苔草■☆
75875 Carex prevernalis Kitag. = Carex lanceolata Boott var. subpediformis Kük. ■
75876 Carex prionocarpa Franch. = Carex cryptocarpa C. A. Mey. ■
75877 Carex prionocarpa Franch. = Carex lyngbyei Hornem. var. prionocarpa (Franch.) Kük. ■
75878 Carex procerula V. I. Krecz.;高大苔草■☆
75879 Carex projecta Mack.;疏头苔草;Loose-headed Oval Sedge, Necklace Sedge, Spreading Sedge ■☆
75880 Carex prolixa Fr.;悬苔草■☆
75881 Carex prolongata Kük.;延长苔草■
75882 Carex pruinosa Boott;粉被苔草;Pruinose Sedge ■
75883 Carex pruinosa Boott subsp. levisaccus (Ohwi) T. Koyama = Carex maximowiczii Miq. var. levisaccus Ohwi ■☆
75884 Carex pruinosa Boott subsp. maximowiczii (Miq.) Kük. = Carex maximowiczii Miq. ■
75885 Carex pruinosa Boott subsp. maximowiczii (Miq.) Kük. var. levisaccus (Ohwi) Makino et Nemoto = Carex maximowiczii Miq. var. levisaccus Ohwi ■☆
75886 Carex pruinosa Boott subsp. maximowiczii (Miq.) Kük. var. suifunensis (Kom.) Kük. = Carex maximowiczii Miq. var. suifunensis (Kom.) Nakai ex Kitag. ■☆
75887 Carex przewalskii T. V. Egorova;红棕苔草;Przewalsk Sedge, Red-brown Sedge ■
75888 Carex przewalskii T. V. Egorova var. ramosa Y. C. Yang;分枝苔草;Raceme Przewalsk Sedge ■
75889 Carex przewalskii T. V. Egorova var. ramosa Y. C. Yang = Carex przewalskii T. V. Egorova ■
75890 Carex pseudoarenicola Hayata = Carex nubigena D. Don ex Okamoto et Taylor subsp. pseudoarenicola (Hayata) T. Koyama ■
75891 Carex pseudoarenicola Hayata = Carex nubigena D. Don ex Okamoto et Taylor ■
75892 Carex pseudobiwensis Kitag. = Carex rara Boott ■
75893 Carex pseudocuraica Eastw. = Carex pseudocuraica F. Schmidt ■

75894 Carex pseudocuraica F. Schmidt;漂筏苔草■
75895 Carex pseudocyperus L.;似莎苔草(假莎草,假莎草苔);Cyperus Sedge,Cypress-like Sedge,False Bristly Sedge ■
75896 Carex pseudodispalata K. T. Fu;似皱果苔草■
75897 Carex pseudoesicaria H. Lév. et Vaniot = Carex idzuroei Franch. et Sav. ■
75898 Carex pseudofilicina Hayata = Carex filicina Nees ex Wight ■
75899 Carex pseudofoetida Kük.;无味苔草(褐鳞苔草)■
75900 Carex pseudofoetida Kük. subsp. afghanica;阿富汗苔草■☆
75901 Carex pseudohumilis F. T. Wang et Y. L. Chang ex P. C. Li;假矮苔草■
75902 Carex pseudohypochlora Y. L. Chang et Y. L. Yang;喙果苔草■
75903 Carex pseudohypochlora Y. L. Chang et Y. L. Yang f. denticulata Y. L. Chang et Y. L. Yang;小齿喙果苔草■
75904 Carex pseudojaponica C. B. Clarke = Carex luzulifolia W. Boott ■☆
75905 Carex pseudojaponica Hayata = Carex subtransversa C. B. Clarke ■
75906 Carex pseudolanceolata V. I. Krecz.;假剑叶苔草■☆
75907 Carex pseudolanceolata V. I. Krecz. = Carex lanceolata Boott var. subpediformis Kük. ■
75908 Carex pseudolaticeps Ts. Tang et F. T. Wang ex S. Yun Liang;弥勒山苔草■
75909 Carex pseudoligulata L. K. Dai;锈点苔草■
75910 Carex pseudololiacea Eastw.;假间穗苔草■☆
75911 Carex pseudolongirostrata Y. L. Chang et Y. L. Yang;假长嘴苔草■
75912 Carex pseudophyllocephala L. K. Dai;假头序苔草■
75913 Carex pseudosabynensis (T. V. Egorova) A. E. Kozhevn. = Carex sabynensis Less. ex Kunth ■
75914 Carex pseudosphaerogyna Nelmes = Carex cognata Kunth var. congolensis (Turrill) Lye ■☆
75915 Carex pseudosupina Y. C. Tang ex L. K. Dai;山地苔草(高山苔草)■
75916 Carex pseudotracheliifolium Ling = Carpesium triste Maxim. ■
75917 Carex pseudotristachia X. F. Jin et C. Z. Zheng;拟三穗苔草■
75918 Carex pseudovesicaria H. Lév. et Vaniot = Carex idzuroei Franch. et Sav. ■
75919 Carex psychrophila Nees;黄绿苔草■
75920 Carex pterocaulos Nelmes;翅茎苔草■
75921 Carex pterolepta Franch. = Carex fluviatilis Boott ■
75922 Carex ptychocarpa Steud. = Carex abscondita Mack. ■☆
75923 Carex pubescens DC. = Carpesium cernuum L. ■
75924 Carex pubescens Muhl. ex Willd. = Carex hirtifolia Mack. ■☆
75925 Carex pubescens Poir. = Fuirena pubescens (Poir.) Kunth ■
75926 Carex pudica Honda;羞涩苔草■☆
75927 Carex pulchella (E. J. Lönnr.) Lindm. = Carex viridula Michx. ■☆
75928 Carex pulchella E. J. Lönnr.;姬苔草■☆
75929 Carex pulicaris L.;蚤苔草;Flea Sedge ■☆
75930 Carex pumila Thunb.;矮生苔草(栓皮苔草,小海米);Dwarf Sedge ■
75931 Carex punctata Gaudin;斑点苔草;Dotted Sedge ■☆
75932 Carex punctata Gaudin var. laevicaulis (Hochst.) Boott = Carex punctata Gaudin ■☆
75933 Carex purplevaginalis Q. S. Wang;紫鞘苔草;Purple-sheathed Sedge ■
75934 Carex purpurascens Kük. = Carex alliiformis C. B. Clarke ■
75935 Carex purpurascens Kük. ex Matsum. = Carex alliiformis C. B. Clarke ■
75936 Carex purpureosquamata L. K. Dai;紫鳞苔草;Purple-scaled Sedge ■
75937 Carex purpureotincta Ohwi;太鲁阁苔草(太鲁阁苔)■
75938 Carex purpureovagina F. T. Wang et Y. C. Chang = Carex purpureovagina F. T. Wang et Y. L. Chang ex S. Yun Liang ■
75939 Carex purpureovagina F. T. Wang et Y. L. Chang ex S. Yun Liang;紫红鞘苔草■
75940 Carex putuoensis S. Yun Liang;普陀苔草;Putuo Sedge ■
75941 Carex pycnostachys Kar. et Kir.;密穗苔草;Densespike Sedge ■
75942 Carex pyrenaica Wahlenb.;核苔草■☆
75943 Carex pyrenaica Wahlenb. subsp. micropoda (C. A. Mey.) Hultén = Carex micropoda C. A. Mey. ■☆
75944 Carex pyrenaica Wahlenb. var. altior Kük. = Carex pyrenaica Wahlenb. ■☆
75945 Carex pyrenaica Wahlenb. var. mondsii Kelso = Carex micropoda C. A. Mey. ■☆
75946 Carex qimenensis S. W. Su et S. M. Xu = Carex dolichostachys Hayata ■
75947 Carex qingdaoensis F. Z. Li et S. J. Fan;青岛苔草;Qingdao Sedge ■
75948 Carex qinghaiensis Y. C. Yang;青海苔草;Qinghai Sedge ■
75949 Carex qingyangensis S. W. Su et S. M. Xu;青阳苔草;Qingyang Sedge ■
75950 Carex qiyunensis S. W. Su et S. M. Xu;齐云苔草;Qiyun Sedge ■
75951 Carex quadrifida L. H. Bailey = Carex heteroneura W. Boott ■☆
75952 Carex quadrifida L. H. Bailey var. caeca L. H. Bailey = Carex heteroneura W. Boott ■☆
75953 Carex quadrifida L. H. Bailey var. lenis L. H. Bailey = Carex heteroneura W. Boott ■☆
75954 Carex quadriflora (Kük.) Ohwi;四花苔草(指苔);Fingered Sedge,Fourflower Sedge ■
75955 Carex quanigra B. Boivin = Carex nigra (L.) Reichard ■☆
75956 Carex rachillis Maguire = Carex subnigricans Stacey ■☆
75957 Carex raddei Kük.;锥囊苔草(河沙苔草);Radde Sedge ■
75958 Carex raddei Liou et al. = Carex eriophylla (Kük.) Kom. ■
75959 Carex radiata (Wahlenb.) Small;东方星苔草;Eastern Star Sedge,Radiating Sedge,Sedge,Straight-styled Wood Sedge ■☆
75960 Carex radicalis Boott;根穗苔草■
75961 Carex radicalis Boott = Carex chuiana F. T. Wang et Ts. Tang ex P. C. Li ■
75962 Carex radiciflora Dunn;根花苔草■
75963 Carex radicina Z. P. Wang;细根茎苔草■
75964 Carex rafflesiana Boott;红头苔草(红头苔)■
75965 Carex rafflesiana Boott subsp. scaberrima (Boeck.) T. Koyama = Carex rafflesiana Boott ■
75966 Carex rafflesiana Boott var. continua (C. B. Clarke) Kük. = Carex continua C. B. Clarke ■
75967 Carex rafflesiana Boott var. continua Kük. = Carex continua C. B. Clarke ■
75968 Carex rafflesiana Boott var. scaberrina (Boeck.) Kük. = Carex rafflesiana Boott ■
75969 Carex ramenksii Kom. var. caudata Hultén = Carex ramenskii Kom. ■☆
75970 Carex ramenskii Kom.;拉门氏苔草■☆
75971 Carex ramosa Nees = Schoenoxiphium lanceum (Thunb.) Kük. ■☆
75972 Carex ramosipes Cherm.;枝梗苔草■☆
75973 Carex rankanensis Hayata = Carex dolichostachys Hayata ■
75974 Carex rara Boott;松叶苔草(独穗苔草,假松叶苔草)■

75975 Carex rara Boott f. yunnanensis (Franch.) Kük. = Carex capillacea Boott ■

75976 Carex rara Boott subsp. biwensis (Franch.) T. Koyama = Carex biwensis Franch. ■

75977 Carex rara Boott var. biwensis (Franch.) Kük. ex Matsum. = Carex biwensis Franch. ■

75978 Carex rara Boott var. biwensis (Franch.) Kük. ex Matsum. = Carex rara Boott ■

75979 Carex rara Boott var. capillacea (Boott) Kük. = Carex capillacea Boott ■

75980 Carex rariflora Wahlenb.;单花苔草(单花苔);Few-flowered Sedge,Mountain Bog-sedge ■☆

75981 Carex rariflora Wahlenb. var. pluriflora (Hultén) T. V. Egorova = Carex rariflora Wahlenb. ■☆

75982 Carex raymondii Calder = Carex atratiformis Britton ■☆

75983 Carex raynoldsii Dewey;雷诺兹苔草■☆

75984 Carex recta Boott;江口苔草;Erect Sedge,Estuarine Sedge ■☆

75985 Carex recticulmis Franch. = Carex sabynensis Less. ex Kunth ■

75986 Carex rectior Mack. = Carex granularis Muhl. ex Willd. var. haleana (Olney) Porter ■☆

75987 Carex rectior Mack. = Carex granularis Muhl. ex Willd. ■☆

75988 Carex recurvisaccus T. Koyama;垂果苔草■

75989 Carex redowskiana C. A. Mey.;瑞氏苔草■☆

75990 Carex reflexistyla Hayata = Carex wahuensis C. A. Mey. subsp. robusta (Franch. et Sav.) T. Koyama ■

75991 Carex regeliana Kük. ex Litv.;瘦果苔草(莱格氏苔草)■

75992 Carex reinii Franch. et Sav.;雷恩苔草■☆

75993 Carex remota L.;远穗苔草(散苔,书带草,薮苔草,秀墩草);Remote Sedge ■

75994 Carex remota L. subsp. alta (Boott) Kük. = Carex alta Boott ■

75995 Carex remota L. subsp. rochebruni (Franch. et Sav.) Kük. = Carex rochebrunii Franch. et Sav. ■

75996 Carex remota L. subsp. rochebruni (Franch. et Sav.) Kük. var. enervulosa Kük. = Carex rochebrunii Franch. et Sav. subsp. reptans (Franch.) S. Yun Liang et Y. C. Tang ■

75997 Carex remota L. subsp. rochebruni (Franch. et Sav.) Kük. var. reptans (Franch.) Kük. = Carex rochebrunii Franch. et Sav. subsp. reptans (Franch.) S. Yun Liang et Y. C. Tang ■

75998 Carex remota L. subsp. rochebrunii (Franch. et Sav.) Kük. var. remotaeformis (Kom.) Kük. = Carex remotiuscula Wahlenb. ■

75999 Carex remota L. subsp. rochebrunii Franch. et Sav. var. remotaeformis (Kom.) Kük. = Carex remotiuscula Wahlenb. ■

76000 Carex remota L. var. enervulosa Kük. = Carex rochebrunii Franch. et Sav. subsp. reptans (Franch.) S. Yun Liang et Y. C. Tang ■

76001 Carex remota L. var. reptans Franch. = Carex rochebrunii Franch. et Sav. subsp. reptans (Franch.) S. Yun Liang et Y. C. Tang ■

76002 Carex remotaeformis Kom. = Carex remotiuscula Wahlenb. ■

76003 Carex remotispicula Hayata = Carex rochebrunii Franch. et Sav. subsp. remotispicula (Hayata) T. Koyama ■

76004 Carex remotispicula Hayata = Carex rochebrunii Franch. et Sav. ■

76005 Carex remotiuscula Wahlenb.;丝引苔草(疏穗苔草);Remotespike Sedge ■

76006 Carex reniformis (L. H. Bailey) Small;肾形苔草;Sedge ■☆

76007 Carex repanda C. B. Clarke;浅波苔草■

76008 Carex reptabunda (Trautv.) V. I. Krecz.;走茎苔草(牛毛苔草);Heilongjiang Sedge,Reptant Sedge ■

76009 Carex reticulata? = Carex elata All. ■☆

76010 Carex retorta (Fr.) Krecz.;回转苔草■☆

76011 Carex retroflexa Muhl. ex Willd.;反折苔草;Sedge ■☆

76012 Carex retroflexa Muhl. ex Willd. var. texensis (Torr.) Fernald = Carex texensis (Torr. ex L. H. Bailey) L. H. Bailey ■☆

76013 Carex retrofracta Kük.;反折果苔草■

76014 Carex retrorsa Schwein.;下弯苔草;Deflexed Bottlebrush Sedge,Knot-sheath Sedge,Retrorse Sedge ■☆

76015 Carex reventa V. I. Krecz.;楔囊苔草(柞苔草);Obovateutricle Sedge ■

76016 Carex reventa V. I. Krecz. = Carex pediformis C. A. Mey. var. pedunculata Maxim. ■

76017 Carex rhizina Blytt ex Boott = Carex pediformis C. A. Mey. ■

76018 Carex rhizina Blytt ex Lindblom = Carex pediformis C. A. Mey. ■

76019 Carex rhizodes Blytt ex Boott var. obbreviata Meinsh. = Carex pediformis C. A. Mey. ■

76020 Carex rhizodes Blytt var. abbrevias Meinsh. = Carex pediformis C. A. Mey. ■

76021 Carex rhizomata Steud. = Carex oligostachys Nees ■

76022 Carex rhizopoda Maxim.;根足苔草■

76023 Carex rhodesiaca Nelmes;罗得西亚苔草■☆

76024 Carex rhomalea (Fernald) Mack. = Carex saxatilis L. ■☆

76025 Carex rhynchachaenium C. B. Clarke ex Merr.;初岛氏宿柱苔(初岛氏柱苔)■

76026 Carex rhynchachaenium C. B. Clarke ex Merr. = Carex truncatigluma C. B. Clarke subsp. rhynchachaenium (C. B. Clarke ex Merr.) Y. C. Tang et S. Yun Liang ■

76027 Carex rhynchophora Franch.;长颈苔草(长颈坚果苔)■

76028 Carex rhynchophysa C. A. Mey. = Carex utriculata Boott ■☆

76029 Carex rhynchophysa Fisch.,Mey. et Avé-Lall.;大穗喙苔草(喙苔);Largespike Sedge ■

76030 Carex rhynchophysa Fisch.,Mey. et Avé-Lall. = Carex utriculata Boott ■☆

76031 Carex riabushinskii Kom.;里亚氏苔草■☆

76032 Carex richardii Thuilling = Carex canescens L. ■

76033 Carex richardsonii R. Br.;理查森苔草;Prairie Hummock Sedge,Richardson's Sedge ■☆

76034 Carex richii (Fernald) Mack. = Carex straminea Willd. ex Schkuhr ■☆

76035 Carex ridongensis P. C. Li;日东苔草;Ridong Sedge ■

76036 Carex rigescens (Franch.) V. I. Krecz. = Carex duriuscula C. A. Mey. ■

76037 Carex rigescens (Franch.) V. I. Krecz. = Carex duriuscula C. A. Mey. subsp. rigescens (Franch.) S. Yun Liang et Y. C. Tang ■

76038 Carex rigida Gooden. = Carex bigelowii Torr. et Schwein. ex Boott ■

76039 Carex rigida Gooden. subsp. altaica Gorodkov = Carex altaica Gorodkov ■

76040 Carex rigida Gooden. var. bigelowii (Torr. ex Schwein.) Tuck. = Carex bigelowii Torr. et Schwein. ex Boott ■

76041 Carex rigida Gooden. var. concolor (R. Br.) Kük. = Carex bigelowii Torr. et Schwein. ex Boott ■

76042 Carex rigidioides Gorodkov;坚挺苔草■☆

76043 Carex riishirensis Franch. = Carex scita Maxim. var. riishirensis (Franch.) Kük. ■☆

76044 Carex riparia Curtis;泽生苔草(河岸苔,河岸苔草,水滨苔草);Greater Pond Sedge,Greater Pond-sedge,Stream-bank Sedge ■

76045 Carex riparia Curtis 'Variegata';杂色河岸苔草■☆

76046 Carex riparia Curtis var. rugulosa (Kük.) Kük. = Carex rugulosa

Kük. ■
76047 Carex ripariiformis Litv.;拟泽生苔草■☆
76048 Carex roanensis F. J. Herm.;罗恩山苔草;Roan Mountain Sedge ■☆
76049 Carex robinsonii Podlech;鲁滨逊苔草■☆
76050 Carex roborowskii V. I. Krecz. = Carex pseudofoetida Kük. ■
76051 Carex rochebrunii Franch. et Sav.;书带苔草(高山穗序苔);Rochebrun Sedge ■
76052 Carex rochebrunii Franch. et Sav. subsp. remotispicula (Hayata) T. Koyama;高山穗序苔草(高山日东苔)■
76053 Carex rochebrunii Franch. et Sav. subsp. remotispicula (Hayata) T. Koyama = Carex rochebrunii Franch. et Sav. ■
76054 Carex rochebrunii Franch. et Sav. subsp. reptans (Franch.) S. Yun Liang et Y. C. Tang;匍匐书带苔草■
76055 Carex rochebrunii Franch. et Sav. var. remotispicula (Hayata) Ohwi = Carex rochebrunii Franch. et Sav. subsp. remotispicula (Hayata) T. Koyama ■
76056 Carex rosea Schkuhr ex Willd.;粉苔草;Curly-styled Wood Sedge,Hwolute Sedge,Rosy Sedge,Sedge,Stellate Sedge ■☆
76057 Carex rosea Schkuhr ex Willd. = Carex radiata (Wahlenb.) Small ■☆
76058 Carex rosea Schkuhr ex Willd. var. arkansana L. H. Bailey = Carex arkansana (L. H. Bailey) L. H. Bailey ■☆
76059 Carex rosea Schkuhr ex Willd. var. pusilla Peck et Howe = Carex rosea Schkuhr ex Willd. ■☆
76060 Carex rosea Schkuhr ex Willd. var. radiata (Wahlenb.) Dewey = Carex radiata (Wahlenb.) Small ■☆
76061 Carex rosea Schkuhr ex Willd. var. texensis Torr. ex L. H. Bailey = Carex texensis (Torr. ex L. H. Bailey) L. H. Bailey ■☆
76062 Carex rossii Boott;罗斯苔草■☆
76063 Carex rossii Boott var. brevipes (W. Boott ex Mack.) Kük. = Carex deflexa Hornem. var. boottii L. H. Bailey ■☆
76064 Carex rostellifera Y. L. Chang et Y. L. Yang;轴苔草■
76065 Carex rostrata Boeck. = Carex obscuriceps Kük. ■
76066 Carex rostrata Michx. = Carex michauxiana Boeck. ■☆
76067 Carex rostrata Stokes = Carex rostrata Stokes ex With. ■
76068 Carex rostrata Stokes ex With.;灰株苔草(尖嘴苔草);Beaked Sedge,Bottle Sedge,Northern Yellow Lake Sedge ■
76069 Carex rostrata Stokes ex With. subsp. rotundata (Wahlenb.) Kük. = Carex rotundata Wahlenb. ■☆
76070 Carex rostrata Stokes ex With. var. borealis (Hartm.) Kük.;北方灰株苔草■☆
76071 Carex rostrata Stokes var. ambigens Fernald = Carex rostrata Stokes ex With. ■
76072 Carex rostrata Stokes var. utriculata (Boott) L. H. Bailey = Carex utriculata Boott ■☆
76073 Carex rotundata Wahlenb.;近圆苔(圆苔草)■☆
76074 Carex rousseaui Raymond = Carex haydenii Dewey ■☆
76075 Carex rouyana Franch. = Carex filipes Franch. et Sav. var. rouyana (Franch.) Kük. ■☆
76076 Carex rouyana Franch. = Carex filipes Franch. et Sav. var. sparsinux (C. B. Clarke ex Franch.) Kük. ■
76077 Carex rouyana Franch. = Carex filipes Franch. et Sav. ■
76078 Carex royleana Nees = Carex breviculmis R. Br. ■
76079 Carex rubra H. Lév. et Vaniot = Carex caespitosa L. ■
76080 Carex rubro-brunnea C. B. Clarke;点囊苔草(红褐苔草);Red-brown Sedge ■
76081 Carex rubro-brunnea C. B. Clarke var. brevibracteata T. Koyama;短苞点囊苔草(短苞大理苔草,短苞苔草)■
76082 Carex rubro-brunnea C. B. Clarke var. elineolata Merr. = Carex rubro-brunnea C. B. Clarke var. taliensis (Franch.) Kük. ■
76083 Carex rubro-brunnea C. B. Clarke var. taliensis (Franch.) Kük.;大理苔草(羊胡子草);Dali Sedge ■
76084 Carex rufina Drejer;浅红苔草■☆
76085 Carex rugata Fernald = Carex corrugata Fernald ■☆
76086 Carex rugata Ohwi;皱纹苔草■☆
76087 Carex rugosperma Mack.;糙籽苔草;Rough-fruited Sedge ■☆
76088 Carex rugosperma Mack. = Carex tonsa (Fernald) E. P. Bicknell var. rugosperma (Mack.) Crins ■☆
76089 Carex rugosperma Mack. var. tonsa (Fernald) E. G. Voss = Carex tonsa (Fernald) E. P. Bicknell ■☆
76090 Carex rugulosa Kük.;粗脉苔草;Thicknerve Sedge ■
76091 Carex rugulosa Kük. var. graciliculmis (Ohwi) Kitag. = Carex kirganica Kom. ■
76092 Carex runssoroensis K. Schum.;伦索罗苔草■☆
76093 Carex rupestris All. = Carex rupestris Turcz. ex Ledeb. ■☆
76094 Carex rupestris Turcz. ex Ledeb.;岩苔草;Rock Sedge ■☆
76095 Carex rusbyi Mack. = Carex vallicola Dewey ■☆
76096 Carex ruthenica V. I. Krecz.;俄罗斯苔草(俄罗斯苔);Russia Sedge ■☆
76097 Carex ruthii Mack.;鲁斯苔草■☆
76098 Carex sabulosa Turcz. ex Kunth;沙地苔草;Sandy Sedge ■
76099 Carex sabulosa Turcz. ex Kunth subsp. leiophylla (Mack.) A. E. Porsild = Carex sabulosa Turcz. ex Kunth ■
76100 Carex sabynensis Less. ex Kunth;萨滨苔草(绿囊苔草)■
76101 Carex sabynensis Less. ex Kunth = Carex hypochlora Freyn ■
76102 Carex sabynensis Less. ex Kunth = Carex praecox Schreb. ■
76103 Carex sabynensis Less. ex Kunth var. rostrata (Maxim.) Ohwi;喙状苔草■☆
76104 Carex sabynensis Less. ex Kunth var. rostrata (Maxim.) Ohwi = Carex subebracteata (Kük.) Ohwi ■
76105 Carex sachalinensis Eastw. = Carex capillacea Boott var. sachalinensis (Eastw.) Ohwi ■☆
76106 Carex sachalinensis Eastw. = Carex pisiformis Boott ■
76107 Carex sachalinensis Eastw. subsp. alterniflora (Franch.) T. Koyama = Carex pisiformis Boott ■
76108 Carex sachalinensis Eastw. subsp. alterniflora (Franch.) T. Koyama = Carex alterniflora Franch. ■
76109 Carex sachalinensis Eastw. subsp. fernaldiana (H. Lév. et Vaniot) T. Koyama = Carex fernaldiana H. Lév. et Vaniot ■
76110 Carex sachalinensis Eastw. subsp. fernaldiana (H. Lév. et Vaniot) T. Koyama = Carex pisiformis Boott ■
76111 Carex sachalinensis Eastw. var. alterniflora (Franch.) Ohwi = Carex alterniflora Franch. ■
76112 Carex sachalinensis Eastw. var. alterniflora (Franch.) Ohwi = Carex pisiformis Boott ■
76113 Carex sachalinensis Eastw. var. arimaensis (Ohwi) Ohwi = Carex sachalinensis Eastw. var. sikokiana (Franch. et Sav.) Ohwi ■☆
76114 Carex sachalinensis Eastw. var. aureobrunnea (Ohwi) Ohwi;黄褐库页苔草■☆
76115 Carex sachalinensis Eastw. var. conicoides (Honda) Ohwi = Carex pisiformis Boott ■
76116 Carex sachalinensis Eastw. var. duvaliana (Franch. et Sav.) T. Koyama = Carex duvaliana Franch. et Sav. ■

76117 Carex sachalinensis Eastw. var. elongatula (Ohwi) Ohwi;伸长库页苔草■☆

76118 Carex sachalinensis Eastw. var. fernaldiana (H. Lév. et Vaniot) T. Koyama = Carex pisiformis Boott ■

76119 Carex sachalinensis Eastw. var. fulva (Ohwi) Ohwi;黄库页苔草■☆

76120 Carex sachalinensis Eastw. var. iwakiana Ohwi;石城苔草■☆

76121 Carex sachalinensis Eastw. var. longiuscula Ohwi;长库页苔草■☆

76122 Carex sachalinensis Eastw. var. musashiensis Hiyama;武藏库页苔草■☆

76123 Carex sachalinensis Eastw. var. pineticola (Ohwi) Ohwi;松林苔草■☆

76124 Carex sachalinensis Eastw. var. pineticola (Ohwi) Ohwi f. calvescens Hiyama = Carex sachalinensis Eastw. var. pineticola (Ohwi) Ohwi ■☆

76125 Carex sachalinensis Eastw. var. sachalinensis sensu Akiyama = Carex sachalinensis Eastw. var. iwakiana Ohwi ■☆

76126 Carex sachalinensis Eastw. var. sikokiana (Franch. et Sav.) Ohwi;四国苔草■☆

76127 Carex sachalinensis Eastw. var. tenuinervis (Ohwi) T. Koyama = Carex sachalinensis Eastw. var. aureobrunnea (Ohwi) Ohwi ■☆

76128 Carex sachalinensis F. Schmidt = Carex pisiformis Boott ■

76129 Carex sachalinensis F. Schmidt var. duvaliana (Franch. et Sav.) T. Koyama = Carex duvaliana Franch. et Sav. ■

76130 Carex sacrosancta Honda;长叶二柱苔草(长叶二柱苔)■

76131 Carex sacrosancta Honda var. tamakii (T. Koyama) T. Koyama;钏苔草■☆

76132 Carex sadoensis Franch.;美丽苔草■

76133 Carex sagaensis Y. C. Yang;萨嘎苔草■

76134 Carex sajanensis V. I. Krecz.;萨因苔草■☆

76135 Carex salina Wahlenb.;盐沼苔草;Saltmarsh Sedge ■☆

76136 Carex salina Wahlenb. subsp. ramenskii (Kom.) T. V. Egorova = Carex ramenskii Kom. ■☆

76137 Carex salina Wahlenb. var. kattegatensis (Fr. ex Lindm.) Almq. = Carex recta Boott ■☆

76138 Carex salina Wahlenb. var. lanceata (Dewey) Kük. = Carex salina Wahlenb. ■☆

76139 Carex salina Wahlenb. var. robusta L. H. Bailey = Carex lyngbyei Hornem. ■

76140 Carex salina Wahlenb. var. subspathacea (Wormsk.) Tuck. = Carex subspathacea Wormsk. ■☆

76141 Carex salina Wahlenb. var. tristigmatica Kük. = Carex ramenskii Kom. ■☆

76142 Carex saliniformis Mack. = Carex hassei L. H. Bailey ■☆

76143 Carex saltuensis L. H. Bailey = Carex vaginata Tausch ■☆

76144 Carex sanguinea Boott;血红苔草■☆

76145 Carex sareptana Krecz.;萨里旦苔草(萨里旦苔)■☆

76146 Carex sartwelliana Olney;萨氏苔草■☆

76147 Carex sartwellii Dewey;萨尔特苔草;Running Marsh Sedge, Sartwell's Sedge, Sedge ■☆

76148 Carex sasakii Hayata = Carex doniana Spreng. ■

76149 Carex satakeana T. Koyama;藏北苔草;Xizang Sedge ■

76150 Carex satsumensis Franch. et Sav.;砂地苔草(油苔);Sandland Sedge ■

76151 Carex satsumensis Franch. et Sav. var. longiculma Hayata = Carex satsumensis Franch. et Sav. ■

76152 Carex satsumensis Franch. et Sav. var. nakaii Hayata = Carex satsumensis Franch. et Sav. ■

76153 Carex saxatilis L.;石苔草;Russet Sedge ■☆

76154 Carex saxenii Raymond var. ferruginea Lepage = Carex recta Boott ■☆

76155 Carex saxicola Ts. Tang et F. T. Wang;岩生苔草;Rocky Sedge ■

76156 Carex saxilittoralis Roberts. = Carex viridula Michx. var. saxilittoralis (Robertson) Crins ■☆

76157 Carex saximontana Mack.;落基山苔草;Rocky Mountain Sedge ■☆

76158 Carex scaberrina (Boeck.) C. B. Clarke = Carex rafflesiana Boott ■

76159 Carex scabior Dewey = Carex vulpinoidea Michx. ■☆

76160 Carex scabrata Schwein.;东部糙苔草;Eastern Rough Sedge, Rough Sedge, Sedge ■☆

76161 Carex scabriculmis (Kük.) Ohwi = Carex teinogyna Boott ■

76162 Carex scabricuspis V. I. Krecz.;糙尖苔草■☆

76163 Carex scabrifolia Steud.;糙叶苔草(碱苔);Scabrousleaf Sedge ■

76164 Carex scabrinervia Franch. = Carex scita Maxim. var. scabrinervia (Franch.) Kük. ■☆

76165 Carex scabrinervia Franck;糙脉苔草■☆

76166 Carex scabrior Dewey = Carex vulpinoidea Michx. ■☆

76167 Carex scabrirostris Kük.;糙喙苔草;Scabrousbeak Sedge ■

76168 Carex scandinavica E. W. Davies = Carex viridula Michx. ■☆

76169 Carex scaposa C. B. Clarke;花葶苔草(翻天红,花茎苔草,落地蜈蚣);Scapose Sedge ■

76170 Carex scaposa C. B. Clarke var. baviensis Franch. = Carex adrienii E. G. Camus ■

76171 Carex scaposa C. B. Clarke var. dolicostachys F. T. Wang et Ts. Tang;长雄苔草(长穗侧茎苔草)■

76172 Carex scaposa C. B. Clarke var. hirsuta P. C. Li;糙叶花葶苔草;Haired Scapose Sedge ■

76173 Carex scaposa C. B. Clarke var. marantacea Raymod = Carex scaposa C. B. Clarke ■

76174 Carex schimperiana Boeck. = Schoenoxiphium sparteum (Wahlenb.) C. B. Clarke ■☆

76175 Carex schlagintweitiana Boeck. = Carex setigera D. Don var. schlagintweitiana (Boeck.) Kük. ■

76176 Carex schlechteri Nelmes = Carex glomerabilis Krecz. ■☆

76177 Carex schliebenii Podlech;施利本苔草■☆

76178 Carex schmidtii Meinsh.;瘤囊苔草(瘤囊苔草);Schmidt Sedge ■

76179 Carex schneideri Nelmes;川滇苔草;Schneider Sedge ■

76180 Carex schoenoides Desf. = Carex divisa Huds. ■

76181 Carex schreberi Schrank = Carex praecox Schreb. ■

76182 Carex schweinitzii Dewey ex Schwein.;施韦尼茨苔草;Rough Sand Sedge, Schweinitz's Sedge, Sedge ■☆

76183 Carex scirpiformis Mack. = Carex scirpoidea Michx. ■☆

76184 Carex scirpina Tuck. = Carex scirpoidea Michx. ■☆

76185 Carex scirpoidea Michx.;藨草状苔草;Scirpoid Sedge ■☆

76186 Carex scirpoidea Michx. subsp. convoluta (Kük.) D. A. Dunlop;旋扭苔草■☆

76187 Carex scirpoidea Michx. subsp. stenochlaena (T. Holm) Á. Löve et D. Löve;狭被苔草■☆

76188 Carex scirpoidea Michx. var. europaea Kük. = Carex scirpoidea Michx. ■☆

76189 Carex scirpoidea Michx. var. scirpiformis (Mack.) O'Neill et Duman = Carex scirpoidea Michx. ■☆

76190 Carex scirpoidea Michx. var. stenochlaena T. Holm = Carex scirpoidea Michx. subsp. stenochlaena (T. Holm) Á. Löve et D.

Löve ■☆

76191 Carex scirpoides Michx. var. convoluta Kük. = Carex scirpoidea Michx. subsp. convoluta (Kük.) D. A. Dunlop ■☆

76192 Carex scirpoides Michx. var. gigas T. Holm = Carex scabrirostris Kük. ■

76193 Carex scirpoides Schkuhr ex Willd. = Carex interior L. H. Bailey ■☆

76194 Carex scirrobasis Kitag. = Carex humilis Leyss. var. scirrobasis (Kitag.) Y. L. Chang et Y. L. Yang ■

76195 Carex scita Maxim.;绮丽苔草■☆

76196 Carex scita Maxim. subsp. brevisquama (Koidz.) T. Koyama = Carex scita Maxim. var. tenuiseta (Franch.) Kük. ■☆

76197 Carex scita Maxim. subsp. parvisquama (T. Koyama) T. Koyama = Carex scita Maxim. var. parvisquama T. Koyama ■☆

76198 Carex scita Maxim. subsp. scabrinervia (Franch.) T. Koyama = Carex scita Maxim. var. scabrinervia (Franch.) Kük. ■☆

76199 Carex scita Maxim. var. brevisquama (Koidz.) Ohwi = Carex scita Maxim. var. tenuiseta (Franch.) Kük. ■☆

76200 Carex scita Maxim. var. parvisquama T. Koyama;短鳞绮丽苔草■☆

76201 Carex scita Maxim. var. riishirensis (Franch.) Kük.;利尻苔草■☆

76202 Carex scita Maxim. var. scabrinervia (Franch.) Kük.;糙脉绮丽苔草■☆

76203 Carex scita Maxim. var. tenuiseta (Franch.) Kük.;细刚毛绮丽苔草■☆

76204 Carex scitiformis Kük.;悦目苔草■☆

76205 Carex sclerocarpa Franch.;硬果苔草(太平山苔,硬果苔);Hardfruit Sedge ■

76206 Carex scolopendriformis F. T. Wang et Ts. Tang ex P. C. Li;蜈蚣苔草;Scalopendriform Sedge ■

76207 Carex scoparia Schkuhr ex Willd.;洋帚状苔草;Broom Sedge, Broom-sedge, Lance-fruited Oval Sedge, Sedge ■☆

76208 Carex scoparia Schkuhr ex Willd. f. moniliformis (Tuck.) Kük. = Carex scoparia Schkuhr ex Willd. ■☆

76209 Carex scoparia Schkuhr ex Willd. var. condensa Fernald = Carex scoparia Schkuhr ex Willd. ■☆

76210 Carex scoparia Schkuhr ex Willd. var. moniliformis Tuck. = Carex scoparia Schkuhr ex Willd. ■☆

76211 Carex scoparia Schkuhr ex Willd. var. tessellata Fernald et Wiegand;方格帚状苔草■☆

76212 Carex scopulorum T. Holm;岩栖苔草■☆

76213 Carex scopulorum T. Holm var. bracteosa (L. H. Bailey) F. J. Herm.;多苞片苔草■☆

76214 Carex scopulorum T. Holm var. subconcoloroides Kelso = Carex scopulorum T. Holm ■☆

76215 Carex secalina Wahlenb.;黑麦苔草■☆

76216 Carex secalina Wahlenb. = Carex eremopyroides V. I. Krecz. ■

76217 Carex secalina Wahlenb. = Carex hordeistichos Vill. ■☆

76218 Carex sedakovii C. A. Mey. ex Meinsh. = Carex sedakowii C. A. Mey. ex Meinsh. ■

76219 Carex sedakowii C. A. Mey. ex Meinsh.;沟叶苔草(细毛苔草)■

76220 Carex selengensis K. V. Ivanova = Carex karoi (Freyn) Freyn ■

76221 Carex semihyalofructa Tak. Shimizu;半透明果苔草■☆

76222 Carex semiplena Kük.;半重瓣苔草■☆

76223 Carex semperirens Schwein. subsp. tristis (M. Bieb.) Kük. = Carex stenocarpa Turcz. ex V. I. Krecz. ■

76224 Carex sempervirens Schwein. = Carex glaucescens Elliott ■☆

76225 Carex senanensis Ohwi = Carex deweyana Schwein. subsp. senanensis (Ohwi) T. Koyama ■☆

76226 Carex sendaica Franch.;仙台苔草(锈鳞苔草);Ferrousscale Sedge ■

76227 Carex sendaica Franch. var. nakiri (Ohwi) T. Koyama = Carex brunnea Thunb. ■

76228 Carex sendaica Franch. var. pseudosendaica T. Koyama;多穗仙台苔草■

76229 Carex seorsa Howe;微弱星苔草;Weak Stellate Sedge ■☆

76230 Carex serotina Mérat;晚苔草(奥氏苔)■

76231 Carex serotina Mérat = Carex viridula Michx. ■☆

76232 Carex serratodens W. Boott;具齿苔草■☆

76233 Carex serreana Hand.-Mazz.;紫喙苔草;Purplebeak Sedge ■

76234 Carex setacea Dewey;毛穗苔草;Bristly-spiked Sedge ■☆

76235 Carex setacea Dewey = Carex vulpinoidea Michx. ■☆

76236 Carex setacea Dewey var. ambigua (Barratt ex Boott) Fernald = Carex annectans (E. P. Bicknell) E. P. Bicknell ■☆

76237 Carex setigera D. Don;长茎苔草■

76238 Carex setigera D. Don var. schlagintweitiana (Boeck.) Kük.;小长茎苔草■

76239 Carex setina V. I. Krecz.;多刚毛苔草■☆

76240 Carex setosa Boott;刺毛苔草;Setose Sedge ■

76241 Carex setosa Boott var. mianxianica S. Yun Liang;沔县苔草;Mianxian Sedge ■

76242 Carex setosa Boott var. punctata S. Yun Liang;锈点刺毛苔草(锈点苔草);Punctate Setose Sedge ■

76243 Carex setulifolia Nelmes = Carex perakensis C. B. Clarke ■

76244 Carex shaanxiensis F. T. Wang et Ts. Tang ex P. C. Li;陕西苔草;Shaanxi Sedge ■

76245 Carex shandanica Y. C. Yang;山丹苔草;Shandan Sedge ■

76246 Carex shangchengensis S. Yun Liang;商城苔草;Shangcheng Sedge ■

76247 Carex shanghaiensis S. X. Qian et Y. Q. Liu;上海苔草;Shanghai Sedge ■

76248 Carex shanghangensis S. Yun Liang;上杭苔草;Shanghang Sedge ■

76249 Carex sharensis Franch. = Carex augustinowiczii Meinsh. ex Korsh. var. sharensis (Franch.) Ohwi ■☆

76250 Carex sharyotoensis Hayata = Carex macrandrolepis H. Lév. et Vaniot ■

76251 Carex shichiseitensis Hayata = Carex phacota Spreng. ■

76252 Carex shimadai Hayata = Carex makinoensis Franch. ■

76253 Carex shimadai Hayata var. longibracteata Hayata = Carex makinoensis Franch. ■

76254 Carex shimidzensis Franch.;清水峠苔草■☆

76255 Carex shimotsukensis Honda = Carex ascotreta C. B. Clarke ex Franch. ■

76256 Carex shortiana Dewey;肖特苔草;Sedge, Short's Sedge ■☆

76257 Carex shortii Steud. = Carex frankii Kunth ■☆

76258 Carex shriveri Britton = Carex granularis Muhl. ex Willd. ■☆

76259 Carex shuangbaiensis L. K. Dai;双柏苔草;Shuangbai Sedge ■

76260 Carex shuchengensis S. W. Su et Q. Zhang;舒城苔草;Shucheng Sedge ■

76261 Carex siccata Dewey;干穗苔草;Dry-spiked Sedge, Hay Sedge, Hillside Sedge, Running Savanna Sedge ■☆

76262 Carex sichouensis P. C. Li;西畴苔草;Xichou Sedge ■

76263 Carex siderosticata Hance var. ciliatomarginata Nakai;毛边西畴苔草■☆

76264 Carex siderosticta Hance;宽叶苔草(崖棕);Broadleaf Sedge,

Creeping Broad Leafed Sedge, Creeping Broad-leaf Sedge, Japanese Sedge ■

76265 Carex siderosticta Hance f. stenophylla Kitag. = Carex siderosticta Hance var. stenophylla (Kitag.) Kitag. ■☆

76266 Carex siderosticta Hance f. variegata (Akiyama) T. Koyama;斑点宽叶苔草■

76267 Carex siderosticta Hance f. variegata (Akiyama) T. Koyama = Carex siderosticta Hance ■

76268 Carex siderosticta Hance var. ciliatomarginata Nakai;毛缘宽叶苔草(毛崖棕);Pilose Broadleaf Sedge ■

76269 Carex siderosticta Hance var. pilosa H. Lév. ex Nakai = Carex siderosticata Hance var. ciliatomarginata Nakai ■☆

76270 Carex siderosticta Hance var. pilosa H. Lév. ex T. Koyama = Carex ciliatomarginata Nakai ■

76271 Carex siderosticta Hance var. stenophylla (Kitag.) Kitag.;窄叶崖棕■☆

76272 Carex siderosticta Hance var. variegata Akiyama = Carex siderosticta Hance ■

76273 Carex siegertiana Uechtr.;习氏苔草■☆

76274 Carex sikokiana Franch. et Sav. = Carex sachalinensis Eastw. var. sikokiana (Franch. et Sav.) Ohwi ■☆

76275 Carex silicea Olney;砂生苔草;Seabeach Sedge ■☆

76276 Carex silvatica Huds.;森林苔草■☆

76277 Carex simensis Hochst. ex A. Rich.;锡米苔草■☆

76278 Carex simensis Hochst. ex A. Rich. var. nemorum Chiov.;丝状苔草■☆

76279 Carex simensis Hochst. ex A. Rich. var. ninagongensis (Kük.) Kük. = Carex petitiana A. Rich. ■

76280 Carex simensis Hochst. ex A. Rich. var. stolonifera (Boeck.) Kük.;匍匐苔草■☆

76281 Carex similigena V. I. Krecz. = Carex enervis C. A. Mey. ■

76282 Carex simpliciuscula Wahlenb. = Kobresia simpliciuscula (Wahlenb.) Mack. ■

76283 Carex simulans C. B. Clarke;相仿苔草;Resembling Sedge ■

76284 Carex simulans C. B. Clarke var. densiflora F. T. Wang et Ts. Tang ex S. Yun Liang;密花相仿苔草;Denseflower Resembling Sedge ■

76285 Carex sinaica Nees ex Steud. = Carex distans L. f. sinaica (Nees ex Steud.) Kük. ■☆

76286 Carex sino-aristata Ts. Tang et F. T. Wang ex L. K. Dai;华芒鳞苔草(芒鳞苔草)■

76287 Carex sinocriapa Raymond = Carex lingii F. T. Wang et Ts. Tang ■

76288 Carex sinodisitiflora Ts. Tang et F. T. Wang ex L. K. Dai;华疏花苔草(远穗苔草)■

76289 Carex siroumensis Koidz.;冻原苔草■

76290 Carex sitbpediformis (Kük.) Suto et Suzuki = Carex lanceolata Boott var. subpediformis Kük. ■

76291 Carex sitchensis J. D. Prescott ex Bong. = Carex aquatilis Wahlenb. var. dives (T. Holm) Kük. ■☆

76292 Carex slobodovii V. I. Krecz.;思劳苔草■☆

76293 Carex smalliana Mack. = Carex lonchocarpa Willd. ex Spreng. ■☆

76294 Carex smirnovii V. I. Krecz. = Carex rugulosa Kük. ■

76295 Carex socialis Mohlenbr. et Schwegman;团聚苔草;Sedge ■☆

76296 Carex sociata Boott;伴生苔草(中国宿柱苔)■

76297 Carex sociata Boott var. tsushimensis (Ohwi) T. Koyama = Carex tsushimensis (Ohwi) Ohwi ■☆

76298 Carex soczavaena Gorodkov;索恰苔草■☆

76299 Carex songarica Kar. et Kir.;准噶尔苔草;Dzungar Sedge, Songaria Sedge ■

76300 Carex sonomensis Stacey = Carex albida L. H. Bailey ■☆

76301 Carex sordida Cham. ex Van Heurck et Müll. Arg.;污苔草■☆

76302 Carex sordida Van Heurck et Müll. Arg. = Carex drymophila Turcz. ex Steud. var. abbreviata (Kük.) Ohwi ■

76303 Carex sordida Van Heurck et Müll. Arg. = Carex sordida Cham. ex Van Heurck et Müll. Arg. ■☆

76304 Carex souliei Franch. = Carex obscura Nees var. brachycarpa C. B. Clarke ■

76305 Carex spachiana Boott;澳门苔草■

76306 Carex spaniocarpa Steud.;寡果苔草■☆

76307 Carex sparganioides Muhl. ex Willd.;寡头苔草;Bur-reed Sedge, Loose-headed Bracted Sedge, Sedge ■☆

76308 Carex sparganioides Muhl. ex Willd. var. aggregata (Mack.) Gleason = Carex aggregata Mack. ■☆

76309 Carex sparganioides Muhl. ex Willd. var. cephaloidea (Dewey) J. Carey ex A. Gray = Carex cephaloidea (Dewey) Dewey ■☆

76310 Carex sparganioides Muhl. ex Willd. var. cephaloidea (Dewey) J. Carey = Carex cephaloidea (Dewey) Dewey ■☆

76311 Carex sparsiflora (Wahlenb.) Steud.;少花苔草■

76312 Carex sparsiflora Steud. var. petersi (C. A. Mey.) Kük.;大少花苔草■

76313 Carex sparsinux C. B. Clarke ex Franch. = Carex filipes Franch. et Sav. var. sparsinux (C. B. Clarke ex Franch.) Kük. ■

76314 Carex sparsinux C. B. Clarke ex Franch. = Carex filipes Franch. et Sav. ■

76315 Carex spartea Wahlenb. = Schoenoxiphium sparteum (Wahlenb.) C. B. Clarke ■☆

76316 Carex spatiosa Boott;广场苔草(石柏苔草)■

76317 Carex spatiosa Boott = Carex commixta Steud. ■

76318 Carex speciosa Kunth;翠丽苔草(美丽苔草)■

76319 Carex speciosa Kunth var. abscondita Kük. = Carex speciosa Kunth ■

76320 Carex speciosa Kunth var. angustifolia Boott = Carex speciosa Kunth ■

76321 Carex speciosa Kunth var. courtallensis (Nees) Kük. = Carex courtallensis Nees ex Boott ■

76322 Carex speciosa Kunth var. monor Boeck. = Carex speciosa Kunth ■

76323 Carex spectabilis Dewey;美花苔草■☆

76324 Carex spectabilis Dewey subsp. flavocuspis (Franch. et Sav.) T. Koyama = Carex flavocuspis Franch. et Sav. ■☆

76325 Carex spectabilis Dewey subsp. flavocuspis (Franch. et Sav.) T. Koyama f. platycarpa (Kük.) T. Koyama = Carex flavocuspis Franch. et Sav. var. platycarpa (Kük.) Akiyama ■☆

76326 Carex spectabilis Dewey var. superba T. Holm = Carex spectabilis Dewey ■☆

76327 Carex sphaerostachya (Tuck.) Dewey = Carex brunnescens (Pers.) Poir. subsp. sphaerostachys (Tuck.) Kalela ■☆

76328 Carex spicata Huds.;穗苔草;Prickly Sedge, Spiked Bracted Sedge ■☆

76329 Carex spicato-paniculata C. B. Clarke;圆锥穗苔草■☆

76330 Carex spissa L. H. Bailey;厚密苔草■☆

76331 Carex spissa L. H. Bailey var. ultra (L. H. Bailey) Kük. = Carex spissa L. H. Bailey ■☆

76332 Carex sprengelii Boeck. = Schoenoxiphium sparteum (Wahlenb.) C. B. Clarke ■☆

76333 Carex sprengelii Dewey ex Spreng.;斯普苔草;Long-beaked Sedge, Sprengel's Sedge ■☆

76334 Carex squamata V. I. Krecz. = Carex craspedotricha Nelmes ■

76335 Carex squarrosa L.;糙鳞苔草;Sedge, Squarrose Sedge ■☆

76336 Carex squarrosa L. var. typhina (Michx.) Nutt. = Carex typhina Michx. ■☆

76337 Carex stachydesma Franch. = Carex rubro-brunnea C. B. Clarke var. taliensis (Franch.) Kük. ■

76338 Carex stans Drejer;直茎苔(直苔)■☆

76339 Carex stans Drejer = Carex aquatilis Wahlenb. var. minor Boott ■☆

76340 Carex stans Drejer = Carex aquatilis Wahlenb. var. stans (Drejer) Boott ■☆

76341 Carex stantonensis M. E. Jones = Carex nardina Fr. ■☆

76342 Carex stellulata Gooden.;小星苔草(星苔);Smallstar Sedge, Star-headed Sedge ■

76343 Carex stellulata Gooden. = Carex echinata Murray ■☆

76344 Carex stellulata Gooden. var. conferta Chapm. = Carex atlantica L. H. Bailey ■☆

76345 Carex stellulata Gooden. var. omiana (Franch. et Sav.) Kük. = Carex omiana Franch. et Sav. ■

76346 Carex stellulata Gooden. var. radiata Wahlenb. = Carex radiata (Wahlenb.) Small ■☆

76347 Carex stenantha C. B. Clarke = Carex bostrychostigma Maxim. ■

76348 Carex stenantha Franch. et Sav.;狭花苔草■☆

76349 Carex stenantha Franch. et Sav. var. taisetsuensis Akiyama;大雪山苔草■☆

76350 Carex stenocarpa Turcz. ex V. I. Krecz.;细果苔草■

76351 Carex stenochlaena (T. Holm) Mack. = Carex scirpoidea Michx. subsp. stenochlaena (T. Holm) Á. Löve et D. Löve ■☆

76352 Carex stenolepis Torr. = Carex frankii Kunth ■☆

76353 Carex stenolepis Torr. = Carex grahamii Boott ■☆

76354 Carex stenophylla Wahlenb.;细叶苔(柄囊苔草,狭叶苔草);Slenderleaf Sedge ■

76355 Carex stenophylla Wahlenb. subsp. eleocharis (L. H. Bailey) Hultén = Carex duriuscula C. A. Mey. ■

76356 Carex stenophylla Wahlenb. subsp. stenophylloides (V. I. Krecz.) T. V. Egorova;假细叶苔■☆

76357 Carex stenophylla Wahlenb. var. duriuscula (C. A. Mey.) Trautv. = Carex duriuscula C. A. Mey. ■

76358 Carex stenophylla Wahlenb. var. eleocharis (L. H. Bailey) Breitung = Carex duriuscula C. A. Mey. ■

76359 Carex stenophylla Wahlenb. var. enervis (C. A. Mey.) Kük. = Carex enervis C. A. Mey. ■

76360 Carex stenophylla Wahlenb. var. longepedicellata (Boeck.) Kük. = Carex duriuscula C. A. Mey. var. stenophylloides (V. I. Krecz.) S. Yun Liang et Y. C. Tang ■

76361 Carex stenophylla Wahlenb. var. reptabunda Trautv. = Carex reptabunda (Trautv.) V. I. Krecz. ■

76362 Carex stenophylla Wahlenb. var. rigescens Franch. = Carex duriuscula C. A. Mey. subsp. rigescens (Franch.) S. Yun Liang et Y. C. Tang ■

76363 Carex stenophylloides V. I. Krecz.;假狭叶苔草■☆

76364 Carex stenophylloides V. I. Krecz. = Carex duriuscula C. A. Mey. subsp. stenophylloides (V. I. Krecz.) S. Yun Liang et Y. C. Tang ■

76365 Carex stenostachys Franch. et Sav.;窄穗苔草■☆

76366 Carex stenostachys Franch. et Sav. var. cuneata (Ohwi) Ohwi et T. Koyama;楔形狭穗苔草■☆

76367 Carex stenostachys Franch. et Sav. var. ikegamiana T. Koyama;池上苔草■☆

76368 Carex sterilis Willd.;不育苔草;Dioecious Sedge, Fen Star Sedge, Sedge, Sterile Sedge ■☆

76369 Carex steudelii Kunth = Carex jamesii Schwein. ■☆

76370 Carex steudneri Boeck.;斯托德苔草■☆

76371 Carex stevenii (T. Holm) Kalela;斯蒂文苔草■☆

76372 Carex stilbophaea V. I. Krecz.;槌状苔草■☆

76373 Carex stipata Muhl. = Carex stipata Muhl. ex Willd. ■

76374 Carex stipata Muhl. ex Schkuhr = Carex stipata Muhl. ex Willd. ■

76375 Carex stipata Muhl. ex Willd.;海绵基苔草(白绿苔草);Common Fox Sedge, Owl-fruit Sedge, Sedge, Stipule Sedge ■

76376 Carex stipata Muhl. ex Willd. var. crassicurta Peck = Carex stipata Muhl. ex Willd. ■

76377 Carex stipata Muhl. ex Willd. var. laevivaginata Kük. = Carex laevivaginata (Kük.) Mack. ■☆

76378 Carex stipata Muhl. ex Willd. var. maxima Chapm. ex Boott;大海绵基苔草■☆

76379 Carex stipata Muhl. ex Willd. var. maxima Chapm. ex Boott = Carex stipata Muhl. ex Willd. ■

76380 Carex stipata Muhl. ex Willd. var. oklahomensis (Mack.) Gleason = Carex oklahomensis Mack. ■☆

76381 Carex stipata Muhl. ex Willd. var. subsecuta Peck = Carex stipata Muhl. ex Willd. ■

76382 Carex stipata Muhl. ex Willd. var. uberior C. Mohr = Carex stipata Muhl. ex Willd. var. maxima Chapm. ex Boott ■☆

76383 Carex stipitinux C. B. Clarke = Carex stipitinux C. B. Clarke ex Franch. ■

76384 Carex stipitinux C. B. Clarke ex Franch.;褐绿苔草(柄果苔草);Browngreen Sedge, Browngreenutricle Sedge Sedge ■

76385 Carex stipitiutriculata P. C. Li;柄囊苔草■

76386 Carex stolonifera Schwein. = Carex pensylvanica Lam. ■☆

76387 Carex stramentitia Boott ex Boeck.;草黄苔草■

76388 Carex straminea Willd. ex Schkuhr;芒卵苔草;Awned Oval Sedge, Eastern Straw Sedge, Sedge, Straw Sedge ■☆

76389 Carex straminea Willd. ex Schkuhr var. brevior Dewey = Carex brevior (Dewey) Mack. ex Lunell ■☆

76390 Carex straminea Willd. ex Schkuhr var. crawei Boott = Carex bicknellii Britton ■☆

76391 Carex straminea Willd. ex Schkuhr var. cumulata L. H. Bailey = Carex cumulata (L. H. Bailey) Mack. ■☆

76392 Carex straminea Willd. ex Schkuhr var. echinodes Fernald = Carex tenera Dewey var. echinodes (Fernald) Wiegand ■☆

76393 Carex straminea Willd. ex Schkuhr var. maxima L. H. Bailey = Carex tetrastachya Scheele ■☆

76394 Carex straminea Willd. ex Schkuhr var. reniformis L. H. Bailey = Carex reniformis (L. H. Bailey) Small ■☆

76395 Carex straminea Willd. ex Schkuhr var. tenera (Dewey) Barratt = Carex tenera Dewey ■☆

76396 Carex striata Michx. = Carex stricta Lam. ■☆

76397 Carex stricta Gooden. = Carex elata All. ■☆

76398 Carex stricta L. var. haydenii (Dewey) Kük. = Carex haydenii Dewey ■☆

76399 Carex stricta Lam.;笔直苔草;Common Tussock Sedge, Hummock Sedge, Tussock Sedge, Uptight Sedge ■☆

76400 Carex stricta Lam. 'Aurea' = Carex elata All. 'Aurea' ■☆

76401 Carex stricta Lam. = Carex elata All. ■☆

76402 Carex stricta Lam. var. curtissima Peck = Carex stricta Lam. ■☆
76403 Carex stricta Lam. var. decora L. H. Bailey = Carex haydenii Dewey ■☆
76404 Carex stricta Lam. var. elongata (Boeck.) Gleason = Carex emoryi Dewey ■☆
76405 Carex stricta Lam. var. emoryi (Dewey) L. H. Bailey = Carex emoryi Dewey ■☆
76406 Carex stricta Lam. var. strictior (Dewey) J. Carey = Carex stricta Lam. ■☆
76407 Carex stricta Lam. var. xerocarpa (S. H. Wright) Britton = Carex stricta Lam. ■☆
76408 Carex strictior Dewey;刚直苔草;Narrower Sedge ■☆
76409 Carex strictior Dewey = Carex stricta Lam. ■☆
76410 Carex strigosa Huds.;糙伏毛苔草;Thin-spiked Wood-sedge ■☆
76411 Carex stupenda H. Lév. et Vaniot = Carex wahuensis C. A. Mey. var. bongardii (Boott) Franch. et Sav. ■
76412 Carex stygia Fr.;希腊苔草■☆
76413 Carex stylosa C. A. Mey.;富柱苔草■☆
76414 Carex stylosa C. A. Mey. var. nigritella (Drejer) Fernald = Carex stylosa C. A. Mey. ■☆
76415 Carex stylosa C. A. Mey. var. virens L. H. Bailey = Carex aperta Boott ■
76416 Carex subcernua Ohwi;武义苔草■
76417 Carex subconcolor Kitag. = Carex arnelli H. Christ ex Scheutz ■
76418 Carex subebracteata (Kük.) Ohwi;小苞叶苔草■
76419 Carex subebracteata (Kük.) Ohwi = Carex sabynensis Less. ex Kunth var. rostrata (Maxim.) Ohwi ■☆
76420 Carex suberecta (Olney) Britton;牧场苔草;Prairie Straw Sedge, Sedge, Wedge-fruited Oval Sedge ■☆
76421 Carex subfilicinoides Kük.;近蕨苔草(红色烟火苔,亚红鞘苔);Fern-like Sedge ■
76422 Carex subfusca W. Boott;暗棕苔草■☆
76423 Carex subglobosa Miel. = Carex viridula Michx. ■☆
76424 Carex subinflata Nelmes;膨胀苔草■☆
76425 Carex sublimosa Lepage = Carex paleacea Schreb. ex Wahlenb. ■☆
76426 Carex sublivida Norrl.;肝色苔草■☆
76427 Carex subloliacea (Fernald) E. P. Bicknell = Carex canescens L. ■
76428 Carex submollicula Ts. Tang et F. T. Wang ex L. K. Dai;拟柔果苔草(似柔果苔草)■
76429 Carex subnigricans Stacey;浅黑苔草■☆
76430 Carex suborbiculata Mack. = Carex nudata W. Boott ■☆
76431 Carex subpediformis (Kük.) Suto et Suzuki;早春苔草■
76432 Carex subpediformis (Kük.) Suto et Suzuki = Carex lanceolata Boott var. subpediformis Kük. ■
76433 Carex subpediformis (Kük.) Suto et Suzuki = Carex lanceolata Boott ■
76434 Carex subperakensis L. K. Ling et Y. Z. Huang;类霹雳苔草■
76435 Carex subphysodes Popov ex V. I. Krecz.;小囊果苔草■
76436 Carex subpumila Ts. Tang et F. T. Wang ex L. K. Dai;似矮生苔草■
76437 Carex subspathacea Wormsk.;苞苔草■☆
76438 Carex subspathacea Wormsk. subsp. ramenskii (Kom.) T. V. Egorova = Carex ramenskii Kom. ■☆
76439 Carex substricta (Kük.) Mack. = Carex aquatilis Wahlenb. var. substricta Kük. ■☆
76440 Carex substricta (Kük.) Mack. = Carex aquatilis Wahlenb. ■☆
76441 Carex subteinogyna Ohwi = Carex bilateralis Hayata ■
76442 Carex subtransversa C. B. Clarke;似横果苔草■
76443 Carex subtransversa C. B. Clarke = Carex alopecuroides D. Don ex Okamoto et Taylor ■
76444 Carex subtumida (Kük.) Ohwi;肿胀果苔草■
76445 Carex subulata Michx. = Carex collinsii Nutt. ■☆
76446 Carex subumbellata Meinsh.;亚伞状苔草■☆
76447 Carex subumbellata Meinsh. var. verecunda Ohwi;羞涩亚伞状苔草■☆
76448 Carex suifunensis Kom. = Carex maximowiczii Miq. var. suifunensis (Kom.) Nakai ex Kitag. ■☆
76449 Carex sukaczovii V. I. Krecz. = Carex gotoi Ohwi ■
76450 Carex suksdorfii Kük. = Carex aquatilis Wahlenb. ■☆
76451 Carex superata Naczi, Reznicek et B. A. Ford;超级苔草■☆
76452 Carex super-goodenoughii (Kük.) Lepage = Carex vacillans Drejer ■☆
76453 Carex supermascula V. I. Krecz. = Carex pediformis C. A. Mey. ■
76454 Carex supina Wahlenb. var. costata Meinsh. = Carex korshinskyi Kom. ■
76455 Carex supina Wahlenb. var. korshinskyi (Kom.) Kük. = Carex korshinskyi Kom. ■
76456 Carex supina Willd. ex Wahlenb.;矮苔草(矮苔,仰卧苔草)■
76457 Carex supina Willd. ex Wahlenb. var. costata Meinsh. = Carex korshinskyi Kom. ■
76458 Carex supina Willd. ex Wahlenb. var. korshinskii (Kom.) Kük. = Carex korshinskyi Kom. ■
76459 Carex surculosa Raymond = Carex tsiangii F. T. Wang et Ts. Tang ■
76460 Carex sutschanensis Kom. = Carex pediformis C. A. Mey. ■
76461 Carex sutschuensis Franch.;四川苔草;Sichuan Sedge ■
76462 Carex svensonis Skottsb. = Carex echinata Murray ■☆
76463 Carex swanii (Fernald) Mack.;斯万苔草;Downy Green Sedge, Sedge, Swan's Sedge ■☆
76464 Carex sychnocephala J. Carey;北美密头苔草;Compact Sedge, Many-headed Oval Sedge, Many-headed Sedge, Sedge ■☆
76465 Carex sylvatica Huds.;欧洲林苔草;European Woodland Sedge, Wood Sedge, Wood-sedge ■☆
76466 Carex sylvatica Huds. subsp. paui (Sennen) A. Bolòs et O. Bolòs;波氏苔草■☆
76467 Carex sylvatica Huds. var. algeriensis (Nelmes) Maire et Weiller = Carex sylvatica Huds. subsp. paui (Sennen) A. Bolòs et O. Bolòs ■☆
76468 Carex sylvatica Maxim. = Carex arnelli H. Christ ex Scheutz ■
76469 Carex szovitsii V. I. Krecz.;邵氏苔草(邵氏苔)■☆
76470 Carex tagawana T. Koyama = Carex fulvorubescens Hayata subsp. longistipes (Hayata) T. Koyama ■
76471 Carex tagawana T. Koyama = Carex longistipes Hayata ■
76472 Carex tahoensis Smiley;塔哈苔草■☆
76473 Carex taihokuensis Hayata;台北苔草(锐果苔,台北苔)■
76474 Carex taihokuensis Hayata = Carex tatsutakensis Hayata ■
76475 Carex taihuensis S. W. Su et S. M. Xu;太湖苔草;Taihu Sedge ■
76476 Carex taipaishanica K. T. Fu;太白山苔草;Taibaishan Sedge ■
76477 Carex taiwanensis (Ohwi) Akiyama;台湾疏花苔草;Taiwan Sedge ■
76478 Carex taiwanensis (Ohwi) Akiyama = Carex dissitiflora Franch. subsp. taiwanensis (Ohwi) T. Koyama ■
76479 Carex takasagoana Akiyama;梦佳苔草(梦佳宿柱苔)■
76480 Carex takasagoana Akiyama = Carex manca Boott ex Benth.

subsp. takasagoana (Akiyama) T. Koyama ■
76481 Carex takenakai Nakai = Carex karoi (Freyn) Freyn ■
76482 Carex taldycola Meinsh.;南疆苔草■
76483 Carex taliensis Franch. = Carex rubro-brunnea C. B. Clarke var. taliensis (Franch.) Kük. ■
76484 Carex tangiana Ohwi;唐进苔草(东陵苔草);Tang Sedge ■
76485 Carex tangii Kük.;河北苔草;Hebei Sedge ■
76486 Carex tangulashanensis Y. C. Yang;唐古拉苔草;Tangula Sedge ■
76487 Carex tapintzensis Franch.;大坪子苔草;Dapingzi Sedge, Tapintze Sedge ■
76488 Carex tapintzensis Franch. var. lamprosandra (Franch.) Kük.;光药大坪子苔草■
76489 Carex tapintzensis Franch. var. lamprosandra (Franch.) Kük. = Carex tapintzensis Franch. ■
76490 Carex tarumensis Franch.;长鳞苔草;Buxbaum's Sedge, Club Sedge ■
76491 Carex tashiroana Ohwi;田代苔草■☆
76492 Carex tatewakiana Ohwi;北越苔草;Tonkin Sedge ■
76493 Carex tatewakiana Ohwi = Carex perakensis C. B. Clarke ■
76494 Carex tato Y. L. Chang = Carex appendiculata (Trautv. et C. A. Mey.) Kük. ■
76495 Carex tatsatakensis Hayata = Carex macrandrolepis H. Lév. et Vaniot ■
76496 Carex tatsiensis (Franch.) Kük.;打箭苔草;Dajianlu Sedge ■
76497 Carex tatsutakensis Hayata;锐果苔草(锐果苔)■
76498 Carex taylori Nelmes = Carex phragmitoides Kük. ■☆
76499 Carex tchenkeouensis E. G. Camus = Carex thibetica Franch. ■
76500 Carex teansalpina Hayata = Carex dolichostachys Hayata subsp. trichosperma (Ohwi) T. Koyama ■
76501 Carex teinogyna Boott;长柱头苔草(细梗苔草);Slenderpedisel Sedge ■
76502 Carex teinogyna Boott var. scabriculmis Kük. = Carex teinogyna Boott ■
76503 Carex temnolepis Franch.;割鳞苔草;Hosoba Kan Suge ■☆
76504 Carex tenacissima Suksd. = Carex nudata W. Boott ■☆
76505 Carex tenebrosa Boott;芒尖鳞苔草■
76506 Carex tenella Schkuhr = Carex disperma Dewey ■
76507 Carex tenella Thuill. = Carex disperma Dewey ■
76508 Carex tenera Dewey;沼地窄叶苔草;Marsh Straw Sedge, Narrow-leaved Oval Sedge, Quill Sedge, Sedge, Weak Sedge ■☆
76509 Carex tenera Dewey var. echinodes (Fernald) Wiegand;刺沼地窄叶苔草;Marsh Straw Sedge, Quill Sedge ■☆
76510 Carex tenera Dewey var. suberecta Olney = Carex suberecta (Olney) Britton ■☆
76511 Carex teneriformis Mack. = Carex subfusca W. Boott ■☆
76512 Carex tentaculata Muhl. var. gracilis Boott = Carex backii W. Boott ■☆
76513 Carex tenuiflora Wahlenb.;细花苔草(瘦花苔);Small-headed Bog Sedge, Sparse-flowered Sedge, Thin-flowered Sedge ■
76514 Carex tenuifolia Poir. = Carex halleriana Asso ■☆
76515 Carex tenuiformis H. Lév. et Vaniot;细形苔草(朝鲜苔草)■
76516 Carex tenuiformis H. Lév. et Vaniot f. puberula Akiyama;毛细形苔草■☆
76517 Carex tenuinervis Ohwi;纤脉苔草■☆
76518 Carex tenuior T. Koyama et J. M. Chuang;瘦苔草■☆
76519 Carex tenuipaniculata P. C. Li;细序苔草;Thin-stachysed Sedge ■
76520 Carex tenuis Rudge = Carex debilis Michx. var. rudgei L. H. Bailey ■☆
76521 Carex tenuiseta Franch. = Carex scita Maxim. var. tenuiseta (Franch.) Kük. ■☆
76522 Carex tenuispicula Ts. Tang ex S. Yun Liang;细穗苔草;Thin-spiked Sedge ■
76523 Carex tenuissima Boott var. duvaliana (Franch. et Sav.) Kük. = Carex duvaliana Franch. et Sav. ■
76524 Carex tenuistachys Nakai = Carex longerostrata C. A. Mey. var. pallida (Kitag.) Ohwi ■
76525 Carex tenuistachys Nakai f. pallida (Kitag.) Kitag. = Carex longerostrata C. A. Mey. var. pallida (Kitag.) Ohwi ■
76526 Carex tenuistachys Nakai var. pallida Kitag. = Carex longerostrata C. A. Mey. var. pallida (Kitag.) Ohwi ■
76527 Carex teramotoi T. Koyama = Carex multifolia Ohwi var. stolonifera Ohwi ■☆
76528 Carex teres Boott;糙芒苔草■
76529 Carex teretiuscula Gooden. var. ampla L. H. Bailey = Carex cusickii Mack. ex Piper et Beattie ■☆
76530 Carex ternaria Meinsh. = Carex tuminensis Kom. ■
76531 Carex terrae-novae Fernald = Carex glacialis Mack. ■☆
76532 Carex tetanica Schkuhr;普通硬苔草;Common Stiff Sedge, Rigid Sedge, Sedge ■☆
76533 Carex tetanica Schkuhr var. meadii (Dewey) L. H. Bailey = Carex meadii Dewey ■☆
76534 Carex tetanica Schkuhr var. woodii (Dewey) A. W. Wood = Carex woodii Dewey ■☆
76535 Carex tetrastachya Scheele;四穗苔草■☆
76536 Carex tetsuoi Ohwi = Carex maculata Boott var. tetsuoi (Ohwi) T. Koyama ■☆
76537 Carex texensis (Torr. ex L. H. Bailey) L. H. Bailey;得州苔草■☆
76538 Carex thibetica Franch.;藏苔草(川藏苔草);Tibet Sedge, Xizang Sedge ■
76539 Carex thibetica Franch. var. minor Kük.;小藏苔草;Small Xizang Sedge ■
76540 Carex thibetica Franch. var. pauciflora F. T. Wang et Ts. Tang;少花藏苔草;Few-flowered Sedge ■
76541 Carex thomasii Nelmes;托马斯苔草■☆
76542 Carex thompsonii Franch.;球结苔草■
76543 Carex thomsonii Boott;高节苔草(块茎苔草);Tuber Sedge ■
76544 Carex thomsonii Boott = Carex thompsonii Franch. ■
76545 Carex thunbergii Steud.;陌上菅苔草;Thunberg Sedge ■
76546 Carex thunbergii Steud. f. appendiculata (Trautv.) T. Koyama = Carex thunbergii Steud. var. appendiculata (Trautv.) Ohwi ■
76547 Carex thunbergii Steud. var. appendiculata (Trautv.) Ohwi;灰脉苔草;Greynerve Sedge ■
76548 Carex thymicola Sennen = Carex halleriana Asso ■☆
76549 Carex tianmushanica C. Z. Zheng et X. F. Jin;天目山苔草;Tianmushan Sedge ■
76550 Carex tianschanica T. V. Egorova;天山苔草;Tianshan Sedge ■
76551 Carex tincta (Fernald) Fernald;淡色卵叶苔草;Coloured Sedge, Tinged Oval Sedge, Tinged Sedge ■☆
76552 Carex titovii V. I. Krecz.;山羊苔草■
76553 Carex tokarensis T. Koyama;陶卡尔苔草■☆
76554 Carex tolmiei Boott = Carex spectabilis Dewey ■☆
76555 Carex tolmiei Boott var. invisa (L. H. Bailey) Kük. = Carex spectabilis Dewey ■☆
76556 Carex tolmiei Boott var. subsessilis L. H. Bailey = Carex

scopulorum T. Holm ■☆
76557 Carex tomentosa L. = Carex filiformis Good ■☆
76558 Carex tonkinensis Boott = Carex perakensis C. B. Clarke ■
76559 Carex tonkinensis Franch. = Carex perakensis C. B. Clarke ■
76560 Carex tonkinensis Franch. = Carex tatewakiana Ohwi ■
76561 Carex tonsa (Fernald) E. P. Bicknell;光果苔草;Glabrous-fruited Sedge, Sedge, Shaved Sedge, Smooth-fruited Oak Sedge ■☆
76562 Carex tonsa (Fernald) E. P. Bicknell var. rugosperma (Mack.) Crins;糙籽光果苔草;Parachute Sedge, Wrinkled-seeded Oak Sedge ■☆
76563 Carex torreyi Tuck.;托里苔草;Red-sheathed Green Sedge, Torrey's Sedge ■☆
76564 Carex torta Boott;扭曲苔草;Sedge, Twisted Sedge ■☆
76565 Carex torta Boott var. composita Porter = Carex torta Boott ■☆
76566 Carex torta Boott var. staminata Peck = Carex torta Boott ■☆
76567 Carex tosaensis Akiyama = Carex polyschoenoides K. T. Fu ■
76568 Carex toyoshimae Tuyama;丰岛苔草■☆
76569 Carex tracyi Mack. = Carex ovalis Gooden. ■☆
76570 Carex traiziscana Eastw.;来知志苔草■☆
76571 Carex transalpina Hayata;大武宿柱苔(高山苔)■
76572 Carex transalpina Hayata = Carex sociata Boott ■
76573 Carex transversa Boott;横果苔草(柔菅);Villi-transverse Sedge ■
76574 Carex trautvetteriana Kom.;特拉苔草■☆
76575 Carex tremula (Ohwi) Ohwi = Carex filipes Franch. et Sav. var. tremula (Ohwi) Ohwi ■☆
76576 Carex triangularis Boeck.;三角苔草;Sedge ■☆
76577 Carex tribuloides Wahlenb.;锥果卵叶苔草;Awl-fruited Oval Sedge, Blunt Broom-sedge, Blunt-broom Sedge, Sedge ■☆
76578 Carex tribuloides Wahlenb. var. bebbii L. H. Bailey = Carex bebbii (L. H. Bailey) Olney ex Fernald ■☆
76579 Carex tribuloides Wahlenb. var. moniliformis (Tuck.) Britton = Carex scoparia Schkuhr ex Willd. ■☆
76580 Carex tribuloides Wahlenb. var. sangamonensis Clokey = Carex tribuloides Wahlenb. ■☆
76581 Carex tricephala Boeck.;三头苔草;Threehead Sedge ■
76582 Carex trichocarpa Muhl. ex Willd.;美国毛果苔草;Hairy-fruit Lake Sedge, Hairy-fruit Sedge, Sedge ■☆
76583 Carex trichosperma Ohwi = Carex dolichostachys Hayata subsp. trichosperma (Ohwi) T. Koyama ■
76584 Carex trinervis Nees = Kobresia esenbeckii (Kunth) Noltie ■
76585 Carex tripartita All.;三棱苔草(三裂苔)■
76586 Carex tripartita All. = Carex curvula All. ■☆
76587 Carex tripartita All. = Carex lachenalii Schkuhr ■
76588 Carex triquetrifolia Boeck. = Carex monostachya A. Rich. ■☆
76589 Carex trisperma Dewey;三籽苔草;Three-fruited Sedge, Three-seeded Bog Sedge, Three-seeded Sedge ■☆
76590 Carex trisperma Dewey var. billingsii O. W. Knight = Carex trisperma Dewey ■☆
76591 Carex tristachys Thunb.;三穗苔草(抱鳞宿柱苔);Threespike Sedge, Trispike Sedge ■
76592 Carex tristachys Thunb. subsp. pocilliformis (Boott) T. Koyama = Carex tristachys Thunb. var. pocilliformis (Boott) Kük. ■
76593 Carex tristachys Thunb. var. pocilliformis (Boott) Kük.;合鳞苔草(抱鳞宿柱苔,杯鳞苔草)■
76594 Carex tristiforme Hand.-Mazz. = Carpesium triste Maxim. ■
76595 Carex tristis M. Bieb.;苦苔草■☆
76596 Carex tristis M. Bieb. var. manshuricum Kitam. = Carpesium triste Maxim. ■
76597 Carex tristis M. Bieb. var. sinense Diels = Carpesium triste Maxim. ■
76598 Carex trucatirostris S. W. Su et S. M. Xu = Carex chungii C. P. Wang ■
76599 Carex trucatirostris S. W. Su et S. M. Xu f. erostris S. W. Su et S. M. Xu = Carex chungii C. P. Wang ■
76600 Carex truchananii;羽叶苔草;Leatherleaf Sedge ■☆
76601 Carex truncatigluma C. B. Clarke;截鳞苔草;Truncateglume Sedge ■
76602 Carex truncatigluma C. B. Clarke subsp. rhynchachaenium (C. B. Clarke ex Merr.) Y. C. Tang et S. Yun Liang;初岛氏苔草(喙果苔草)■
76603 Carex truncatirostris S. W. Su et S. M. Xu;截喙苔草;Truncatirostris Sedge ■
76604 Carex truncatirostris S. W. Su et S. M. Xu = Carex chungii C. P. Wang ■
76605 Carex truncatirostris S. W. Su et S. M. Xu f. erostris S. W. Su et S. M. Xu;平喙苔草(无喙苔草)■
76606 Carex truncatirostris S. W. Su et S. M. Xu var. erostris S. W. Su et S. M. Xu = Carex chungii C. P. Wang ■
76607 Carex tsaiana F. T. Wang et Ts. Tang ex P. C. Li;希陶苔草;Tsai Sedge ■
76608 Carex tsangensis Franch. = Carex setigera D. Don var. schlagintweitiana (Boeck.) Kük. ■
76609 Carex tsiangii F. T. Wang et Ts. Tang;三念苔草(三念草);Tsiang Sedge ■
76610 Carex tsoi Merr. et Chun;线茎苔草;Tso Sedge ■
76611 Carex tsuishikarensis Koidz. et Ohwi = Carex oligosperma Michx. subsp. tsuishikarensis (Koidz. et Ohwi) T. Koyama et Calder ■☆
76612 Carex tsushimensis (Ohwi) Ohwi;对马苔草■☆
76613 Carex tuckermanii Dewey;塔克曼苔草;Bent-seeded Hop Sedge, Tuckerman's Sedge ■☆
76614 Carex tumida Boott = Carex oedorrhampha Nelmes ■
76615 Carex tumidicarpa Andersson;胀果苔草■☆
76616 Carex tumidicarpa Andersson = Carex viridula Michx. subsp. oedocarpa (Andersson) B. Schmid ■☆
76617 Carex tumidula Ohwi;肿胀苔草■☆
76618 Carex tuminensis Kom.;图门苔草;Tumen Sedge ■
76619 Carex tungfangensis L. K. Dai et S. M. Hwang;东方苔草;Dongfang Sedge ■
76620 Carex turbinata Liebm.;白齿苔草■☆
76621 Carex turgidula L. H. Bailey = Carex aperta Boott ■
76622 Carex turkestanica Regel;新疆苔草(短柱苔草);Turkestan Sedge, Xinjiang Sedge ■
76623 Carex typhina Michx.;猫尾苔草;Cat-tail Sedge, Sedge ■☆
76624 Carex typhinoides Schwein. = Carex typhina Michx. ■☆
76625 Carex uberior (C. Mohr) Mack. = Carex stipata Muhl. ex Willd. var. maxima Chapm. ex Boott ■☆
76626 Carex uda Maxim.;大针苔草■
76627 Carex uhligii C. B. Clarke = Schoenoxiphium lehmannii (Nees) Steud. ■☆
76628 Carex ulobasis V. I. Krecz.;卷叶苔草■
76629 Carex ultra L. H. Bailey = Carex spissa L. H. Bailey ■☆
76630 Carex umbellata Schkuhr ex Willd.;伞状苔草;Early Oak Sedge, Hidden Sedge, Parasol Sedge, Sedge, Umbel-like Sedge ■☆
76631 Carex umbellata Schkuhr ex Willd. f. vicina (Dewey) Wiegand

= Carex umbellata Schkuhr ex Willd. ■☆
76632 Carex umbellata Schkuhr ex Willd. var. globosa (Boott) Kük. = Carex globosa Boott ■☆
76633 Carex umbellata Schkuhr ex Willd. var. tonsa Fernald = Carex tonsa (Fernald) E. P. Bicknell ■☆
76634 Carex umbellata sensu Fernald = Carex tonsa (Fernald) E. P. Bicknell var. rugosperma (Mack.) Crins ■☆
76635 Carex umbrosa Host;阴地苔草■☆
76636 Carex umbrosa Host subsp. pseudosabynensis T. V. Egorova = Carex sabynensis Less. ex Kunth ■
76637 Carex umbrosa Host subsp. sabynensis (Less. ex Kunth) Kük. = Carex sabynensis Less. ex Kunth ■
76638 Carex umbrosa Host var. koreana Nakai = Carex pisiformis Boott ■
76639 Carex uncinoides Boott = Kobresia uncinoides (Boott) C. B. Clarke ex Hook. f. ■
76640 Carex uncompahgre Kelso = Carex bella L. H. Bailey ■☆
76641 Carex ungurensis Litv.;绿囊苔草(安古苔草)■
76642 Carex unifoliata Kük. et Hand. -Mazz.;单叶苔草;Singleleaf Sedge ■
76643 Carex unifoliata Kük. et Hand. -Mazz. = Carex parva Nees ■
76644 Carex unisexualis C. B. Clarke;单性苔草;Unisexual Sedge ■
76645 Carex uraiensis C. B. Clarke = Carex stenophylla Wahlenb. ■
76646 Carex uraiensis Hayata = Carex sociata Boott ■
76647 Carex uralensis C. B. Clarke;乌拉尔苔草■☆
76648 Carex urelytra Ohwi;扁果苔草(扁果苔)■
76649 Carex urostachys Franch. = Carex mertensii J. D. Prescott var. urostachys (Franch.) Kük. ■☆
76650 Carex ursina Dewey;熊苔草■☆
76651 Carex ussuriensis Kom.;乌苏里苔草;Ussuri Sedge ■
76652 Carex ustulata Wahlenb. var. minor Boott = Carex atrofusca Schkuhr subsp. minor (Boott) T. Koyama ■
76653 Carex utriculata Boott;袋苔草;Common Yellow Lake Sedge, Northwest Territory Sedge ■☆
76654 Carex uzoni Kom.;乌祖苔草■☆
76655 Carex vacillans Drejer;颤苔草■☆
76656 Carex vaginata Tausch;鞘苔草;Sheathed Sedge, Wood Sedge ■☆
76657 Carex vaginata Tausch = Carex sparsiflora (Wahlenb.) Steud. ■
76658 Carex vaginata Tausch var. petersii (C. A. Mey. ex Eastw.) Akiyama;彼得鞘苔草■☆
76659 Carex vahlii Schkuhr = Carex norvegica Retz. ■☆
76660 Carex vahlii Schkuhr var. inferalpina Wahlenb. = Carex media R. Br. ex Richardson ■☆
76661 Carex vahlii Schkuhr var. stevenii (T. Holm) Fernald = Carex stevenii (T. Holm) Kalela ■☆
76662 Carex valida Nees = Carex cruciata Wahlenb. ■
76663 Carex vallicola Dewey;山谷苔草■☆
76664 Carex vallicola Dewey var. rusbyi (Mack.) F. J. Herm. = Carex vallicola Dewey ■☆
76665 Carex vanheurckii Müll. Arg.;鳞苞苔草■
76666 Carex varia Muhl. = Carex albicans Willd. ex Spreng. ■☆
76667 Carex variabilis L. H. Bailey = Carex aquatilis Wahlenb. ■☆
76668 Carex variabilis L. H. Bailey var. elatior L. H. Bailey = Carex emoryi Dewey ■☆
76669 Carex venusta Dewey;黑绿苔草;Dark-green Sedge ■☆
76670 Carex venusta Dewey var. minor Boeck. = Carex venusta Dewey ■☆
76671 Carex venustula T. Holm = Carex podocarpa R. Br. ex Richardson ■☆
76672 Carex verecunda T. Holm = Carex inops L. H. Bailey ■☆
76673 Carex verna Chaix et Vill.;春苔草■☆
76674 Carex vernacula L. H. Bailey;土著苔草■☆
76675 Carex vernacula L. H. Bailey var. hobsonii Maguire = Carex neurophora Mack. ■☆
76676 Carex verrucosa Muhl.;瘤突苔草■☆
76677 Carex vesicaria L.;胀囊苔草(膜囊苔草,乌拉草);Bladder Sedge, Blister Sedge, Inflated Sedge, Tufted Lake Sedge ■
76678 Carex vesicaria L. subsp. vesicata (Meinsh.) T. V. Egorova = Carex vesicaria L. ■
76679 Carex vesicaria L. var. alpigena Kük. = Carex dichroa Freyn ■
76680 Carex vesicaria L. var. distenta Fr. = Carex vesicaria L. ■
76681 Carex vesicaria L. var. jejuna Fernald = Carex vesicaria L. ■
76682 Carex vesicaria L. var. major Boott = Carex exsiccata L. H. Bailey ■☆
76683 Carex vesicaria L. var. monile (Tuck.) Boeck.;串珠胀囊苔草■☆
76684 Carex vesicaria L. var. monile (Tuck.) Fernald = Carex vesicaria L. ■
76685 Carex vesicaria L. var. pamirica O. Fedtsch. = Carex pamirensis C. B. Clarke ■
76686 Carex vesicaria L. var. tenuistachys Kük. = Carex vesicaria L. ■
76687 Carex vesicata Meinsh.;褐黄鳞苔草(尖苔草);Vesica Sedge ■
76688 Carex vesicata Meinsh. = Carex vesicaria L. ■
76689 Carex vespertina (L. H. Bailey) Howell = Carex inops L. H. Bailey ■☆
76690 Carex vicaria L. H. Bailey = Carex densa (L. H. Bailey) L. H. Bailey ■☆
76691 Carex vicaria L. H. Bailey var. costata L. H. Bailey = Carex densa (L. H. Bailey) L. H. Bailey ■☆
76692 Carex vidua Boott ex C. B. Clarke = Kobresia vidua (Boott ex C. B. Clarke) Kük. ■
76693 Carex villosa Boott = Carex latisquamea Kom. ■
76694 Carex villosa Boott var. latisquamea (Kom.) Kük. = Carex latisquamea Kom. ■
76695 Carex violacea C. B. Clarke = Carex nova L. H. Bailey ■☆
76696 Carex virens Lam. = Carex divulsa Stokes ex With. ■
76697 Carex virescens Muhl. ex Willd.;变绿苔草;Sedge ■☆
76698 Carex virescens Muhl. ex Willd. var. swanii Fernald = Carex swanii (Fernald) Mack. ■☆
76699 Carex virginiana Woods var. elongata Boeck. = Carex emoryi Dewey ■☆
76700 Carex viridimarginata Kük.;绿边苔草■
76701 Carex viridula Michx.;小绿苔草;Green Yellow Sedge, Greenish Sedge, Little Green Sedge, Yellow Sedge ■☆
76702 Carex viridula Michx. subsp. brachyrrhyncha (Celak.) B. Schmid;短喙小绿苔草;Long-stalked Yellow Sedge ■☆
76703 Carex viridula Michx. subsp. brachyrrhyncha (Celak.) B. Schmid = Carex lepidocarpa Tausch ■☆
76704 Carex viridula Michx. subsp. nevadensis (Boiss. et Reut.) B. Schmid;内华达苔草■☆
76705 Carex viridula Michx. subsp. oedocarpa (Andersson) B. Schmid;胀果小绿苔草■☆
76706 Carex viridula Michx. var. elatior (Schltdl.) Crins;较高苔草;Long-stalked Yellow Sedge ■☆
76707 Carex viridula Michx. var. lepidocarpa (Tausch) B. Schmid = Carex lepidocarpa Tausch ■☆
76708 Carex viridula Michx. var. pumila (Coss. et Germ.) Fernald =

Carex viridula Michx. ■☆
76709 Carex viridula Michx. var. saxilittoralis (Robertson) Crins;滨海小绿苔草■☆
76710 Carex vitilis Fr.;卷褶苔草■☆
76711 Carex vitilis Fr. = Carex brunnescens (Pers.) Poir. ■
76712 Carex volkensii K. Schum. = Carex johnstonii Boeck. ■☆
76713 Carex vulgaris Fr. var. strictiformis L. H. Bailey = Carex nigra (L.) Reichard ■☆
76714 Carex vulgaris L. H. Bailey var. bracteosa L. H. Bailey = Carex scopulorum T. Holm var. bracteosa (L. H. Bailey) F. J. Herm. ■☆
76715 Carex vulgaris L. H. Bailey var. limnophila T. Holm = Carex lenticularis Michx. var. limnophila (T. Holm) Cronquist ■☆
76716 Carex vulpina L.;狐狸苔草(狐苔,虎尾苔草);Fox Sedge,Fox-sedge,True Fox-sedge ■
76717 Carex vulpinaris Nees var. pseudofoetida (Kük.) Y. C. Yang = Carex pseudofoetida Kük. ■
76718 Carex vulpinoidea Michx.;肖狐狸苔草;Brown Fox Sedge,Fox Sedge ■☆
76719 Carex vulpinoidea Michx. var. ambigua Barratt ex Boott = Carex annectans (E. P. Bicknell) E. P. Bicknell ■☆
76720 Carex vulpinoidea Michx. var. ambigua Boott = Carex annectans (E. P. Bicknell) E. P. Bicknell ■☆
76721 Carex vulpinoidea Michx. var. annectens (E. P. Bicknell) Farw. = Carex annectans (E. P. Bicknell) E. P. Bicknell ■☆
76722 Carex vulpinoidea Michx. var. drummondiana Boeck. = Carex triangularis Boeck. ■☆
76723 Carex vulpinoidea Michx. var. microsperma (Wahlenb.) Dewey = Carex vulpinoidea Michx. ■☆
76724 Carex vulpinoidea Michx. var. pycnocephala F. J. Herm. = Carex vulpinoidea Michx. ■☆
76725 Carex vulpinoidea Michx. var. scabrior (Dewey) A. W. Wood = Carex vulpinoidea Michx. ■☆
76726 Carex vulpinoidea Michx. var. segregata Farw. = Carex vulpinoidea Michx. ■☆
76727 Carex vulpinoidea Michx. var. vicaria (L. H. Bailey) Kük. = Carex densa (L. H. Bailey) L. H. Bailey ■☆
76728 Carex vulpinoidea Michx. var. xanthocarpa (E. P. Bicknell) Kük. = Carex annectans (E. P. Bicknell) E. P. Bicknell ■☆
76729 Carex vulpinoidea Nees var. microsperma (Wahlenb.) Dewey = Carex vulpinoidea Michx. ■☆
76730 Carex vulpinoidea Nees var. pycnocephala F. J. Herm. = Carex vulpinoidea Michx. ■☆
76731 Carex vulpinoidea Nees var. scabrior (Dewey) A. W. Wood = Carex vulpinoidea Michx. ■☆
76732 Carex vulpinoidea Nees var. segregata Farw. = Carex vulpinoidea Michx. ■☆
76733 Carex vulpinoidea Nees var. setacea (Dewey) Kük. = Carex vulpinoidea Michx. ■☆
76734 Carex vulpinoidea Nees var. triangularis (Boeck.) Kük. = Carex triangularis Boeck. ■☆
76735 Carex vulpinoides Michx.;假狐狸苔草;American Fox-sedge, Brown Fox Sedge,Fox Sedge ■☆
76736 Carex wahlenbergiana C. B. Clarke = Carex zuluensis C. B. Clarke ■☆
76737 Carex wahuensis C. A. Mey.;瓦屋宿柱苔(布氏宿柱苔,瓦屋苔草);Wawu Sedge ■
76738 Carex wahuensis C. A. Mey. subsp. robusta (Franch. et Sav.) T. Koyama;健壮苔草■
76739 Carex wahuensis C. A. Mey. subsp. robusta (Franch. et Sav.) T. Koyama = Carex wahuensis C. A. Mey. var. bongardii (Boott) Franch. et Sav. ■
76740 Carex wahuensis C. A. Mey. var. bongardii (Boott) Franch. et Sav.;布氏宿柱苔(滨海苔草)■
76741 Carex wahuensis C. A. Mey. var. robusta (Franch. et Sav.) Franch. et Sav. = Carex wahuensis C. A. Mey. subsp. robusta (Franch. et Sav.) T. Koyama ■
76742 Carex wallichiana J. D. Prescott ex Nees = Carex fidia Nees ex Wight ■
76743 Carex wallichiana J. D. Prescott ex Nees var. miyabei (Franch.) Kük. = Carex miyabei Franch. ■
76744 Carex wallichiana J. D. Prescott ex Nees var. miyabei (Franch.) Kük. f. glabrescens Kük. = Carex glabrescens (Kük.) Ohwi ■
76745 Carex wallichiana J. D. Prescott var. miyabei (Franch.) Kük. = Carex miyabei Franch. ■
76746 Carex walteriana L. H. Bailey = Carex striata Michx. ■☆
76747 Carex warburgiana Kük. = Carex makinoensis Franch. ■
76748 Carex washingtoniana Dewey = Carex bigelowii Torr. et Schwein. ex Boott ■
76749 Carex wawuensis K. L. Chu et S. Yun Liang;瓦屋苔草■
76750 Carex wenchenii F. T. Wang et Ts. Tang = Carex serreana Hand.-Mazz. ■
76751 Carex wenshanensis L. K. Dai;文山苔草;Wenshan Sedge ■
76752 Carex whitneyi Olney;惠特尼苔草■☆
76753 Carex wichurae Boeck. = Carex manca Boott ex Benth. subsp. wichurae (Boeck.) S. Yun Liang ■
76754 Carex wichurai Kük. = Carex manca Boott ex Benth. subsp. wichurai (Kük.) S. Yun Liang ■
76755 Carex wiegandii Mack.;威甘德苔草■☆
76756 Carex willdenowii Schkuhr ex Willd.;威尔苔草;Willdenow's sedge ■☆
76757 Carex willdenowii Schkuhr ex Willd. var. megarrhyncha F. J. Herm. = Carex superata Naczi,Reznicek et B. A. Ford ■☆
76758 Carex willdenowii Schkuhr ex Willd. var. pauciflora Olney ex L. H. Bailey = Carex basiantha Steud. ■☆
76759 Carex williamsii Britton;威廉斯苔草■☆
76760 Carex wiluica Meinsh. ex Maack;苇陆苔草(威吕斯克苔)■
76761 Carex woodii Dewey;伍得苔草;Pretty Sedge,Wood's Stiff Sedge ■☆
76762 Carex wormskioldiana Hornem. = Carex scirpoidea Michx. ■☆
76763 Carex wrightii Olney = Carex tetrastachya Scheele ■☆
76764 Carex wui C. M. Chi ex L. K. Dai;沙坪苔草;Wu Sedge ■
76765 Carex wushanensis S. Yun Liang;武山苔草;Wushan Sedge ■
76766 Carex wutuensis K. T. Fu;武都苔草;Wudu Sedge ■
76767 Carex wuyishanensis Y. C. Tang ex S. Yun Liang;武夷山苔草;Wuyishan Sedge ■
76768 Carex xanchengensis S. W. Su et S. M. Xu = Carex chungii C. P. Wang ■
76769 Carex xanthocarpa E. P. Bicknell = Carex annectans (E. P. Bicknell) E. P. Bicknell ■☆
76770 Carex xanthocarpa E. P. Bicknell var. annectens E. P. Bicknell = Carex annectans (E. P. Bicknell) E. P. Bicknell ■☆
76771 Carex xerocarpa S. H. Wright = Carex stricta Lam. ■☆
76772 Carex xiangxiensis Z. P. Wang;湘西苔草;Xiangxi Sedge ■
76773 Carex xiphium Kom.;稗苔草■

76774 Carex xuanchengensis S. W. Su et S. M. Xu;宣城苔草;Xuancheng Sedge ■
76775 Carex xuanchengensis S. W. Su et S. M. Xu = Carex chungii C. P. Wang ■
76776 Carex yajiangensis Ts. Tang et F. T. Wang;雅江苔草;Yajiang Sedge ■
76777 Carex yakusimensis Masam.;屋久岛苔草■☆
76778 Carex yamatsutana Ohwi;山林苔草■
76779 Carex yangshuoensis Ts. Tang et F. T. Wang ex S. Yun Liang;阳朔苔草;Yangshuo Sedge ■
76780 Carex yingkiliensis A. I. Baranov et Skvortsov;英吉利苔草;England Sedge ■☆
76781 Carex yingkiliensis A. I. Baranov et Skvortsov = Carex mollissima H. Christ ex Meinsh. ■
76782 Carex yosemitana L. H. Bailey = Carex sartwelliana Olney ■☆
76783 Carex ypsilandrifolia F. T. Wang et Ts. Tang;丫蕊苔草■
76784 Carex yuexiensis S. W. Su et S. M. Xu;岳西苔草(秃喙苔草);Yuexi Sedge ■
76785 Carex yulungshanensis P. C. Li;玉龙苔草(红稗子);Yulong Sedge ■
76786 Carex yungningensis Hand.-Mazz. et Kük. ex Hand.-Mazz.;永宁苔草;Yongning Sedge ■
76787 Carex yungningensis Hand.-Mazz. et Kük. ex Hand.-Mazz. = Carex fluviatilis Boott ■
76788 Carex yunlingensis P. C. Li;云岭苔草;Yunling Sedge ■
76789 Carex yunnanensis Franch.;云南苔草(滇苔草);Yunnan Sedge ■
76790 Carex yushuensis Y. C. Yang;玉树苔草;Yushu Sedge ■
76791 Carex zekogensis Y. C. Yang;泽库苔草;Zeku Sedge ■
76792 Carex zeyheri C. B. Clarke = Schoenoxiphium ecklonii Nees ■☆
76793 Carex zhenkangensis F. T. Wang et Ts. Tang;镇康苔草;Zhenkang Sedge ■
76794 Carex zhonghaiensis S. Yun Liang;中海苔草;Zhonghai Sedge ■
76795 Carex zizaniifolia Raymond;菰叶苔草■
76796 Carex zuluensis C. B. Clarke;祖卢苔草■☆
76797 Carex zunyiensis Ts. Tang et F. T. Wang;遵义苔草;Zunyi Sedge ■
76798 Careya Roxb.(1811)(保留属名);印度玉蕊属(卡里玉蕊属)●☆
76799 Careya arborea Roxb.;印度玉蕊(卡里玉蕊);Ceylon Oak, Patana Oak, Wild Guava ●☆
76800 Cargila Raf. = Melampodium L. ■●
76801 Cargilia Hassk. = Cargillia R. Br. ●
76802 Cargilla Adans. = Chrysogonum A. Juss. ●■
76803 Cargilla Adans. = Leontice L. ●■
76804 Cargillia R. Br. = Diospyros L. ●
76805 Cargyllia Steud. = Cargillia R. Br. ●
76806 Caribea Alain(1960);抱茎茉莉属●☆
76807 Caribea litoralis Alain;抱茎茉莉■☆
76808 Carica L.(1753);番木瓜属;Papaya, Pawpaw ●
76809 Carica × heilborni V. M. Badillo;杂种番木瓜;Babaco, Mountain Papaya ●☆
76810 Carica candamarcensis Hook. f. = Carica cundinamarcensis Linden ex Hook. f. ●☆
76811 Carica cundinamarcensis Linden ex Hook. f.;坎大番木瓜(坎木番木瓜);Mountain Papaw, Mountain Pawpaw ●☆
76812 Carica goudotiana (Triana et Planch.) Solms;五棱番木瓜●☆
76813 Carica papaya L.;番木瓜(冬瓜树,番瓜,番瓜树,广西木瓜,麻菖蒲,马菖坡,满山抛,缅芭蕉,缅瓜,木冬瓜,木瓜,木瓜瓜,奶匏,蓬生果,乳瓜,乳果,石瓜,树冬瓜,树冬瓜树,土木瓜,万寿果,万寿匏);Melon Tree, Mexican Papaya, Papain, Papaw, Papaya, Pawpaw, Paw-paw, Pawpaw-tree, Tree Melon ●
76814 Carica pentagona Heilborn;五角番木瓜●☆
76815 Carica pubescens Lenné et K. Koch;高山番木瓜;Chamburo, Mountain Pawpaw ●☆
76816 Carica quercifolia Solms;栎叶番木瓜●☆
76817 Carica stipulata V. M. Badillo;斯克拉龙番木瓜;Siglaton ●☆
76818 Caricaceae Bercht. et J. Presl = Caricaceae Dumort.(保留科名)●
76819 Caricaceae Burnett = Cyperaceae Juss.(保留科名)■
76820 Caricaceae Dumort.(1829)(保留科名);番木瓜科(番瓜树科,万寿果科);Carica Family, Papaw Family, Papaya Family, Pawpaw Family ●
76821 Caricella Ehrh. = Carex L. ■
76822 Caricina St.-Lag. = Carex L. ■
76823 Caricinella St.-Lag. = Carex L. ■
76824 Caricteria Scop. = Corchorus L. ■●
76825 Caridochloa Endl. = Alloteropsis J. Presl ex C. Presl ■
76826 Caridochloa Endl. = Coridochloa Nees ex Graham ■
76827 Carigola Raf. = Monochoria C. Presl ■
76828 Carima Raf. = Adhatoda Mill. ●
76829 Carima sulcata (Vahl) Raf. = Justicia flava (Vahl) Vahl ■☆
76830 Carinavalva Ising(1955);澳洲灰绿芥属■☆
76831 Carinavalva glauca Ising;澳洲灰绿芥■☆
76832 Cariniana Casar.(1842);龙头木属(卡林玉蕊属,龙木属);Abarco, Abarco Wood, Albarco, Bacu, Jequitiba, Jiquitiba, Monkey Pot ●☆
76833 Cariniana brasiliensis Casar.;巴西龙头木(巴西卡林玉蕊)●☆
76834 Cariniana domestica Miers;家龙木●☆
76835 Cariniana estrellensis (Raddi) Kuntze;龙头木(卡林玉蕊)●☆
76836 Cariniana excelsa Casar.;白龙头木(白卡林玉蕊)●☆
76837 Cariniana integrifolia Ducke;全缘叶龙头木(全缘叶卡林玉蕊)●☆
76838 Cariniana legalis Kuntze;合法龙头木(合法卡林玉蕊)●☆
76839 Cariniana micrantha Ducke;小花龙头木(小花卡林玉蕊)●☆
76840 Cariniana pyriformis Miers;梨状龙头木(梨状卡林玉蕊);Abarco, Abarco Wood, Colombian Mahogany ●☆
76841 Cariniana uahupensis Miers;亚马逊龙头木(亚马逊卡林玉蕊)●☆
76842 Carinivalva Airy Shaw = Carinavalva Ising ■☆
76843 Carinta W. Wight = Geophila D. Don(保留属名)■
76844 Carinta cordiformis (A. Chev. ex Hutch. et Dalziel) G. Taylor = Geophila afzelii Hiern ■☆
76845 Carinta herbacea (Jacq.) W. Wight = Geophila repens (L.) I. M. Johnst. ■☆
76846 Carinta ioides Pic. Serm. = Geophila obvallata (Schumach.) Didr. subsp. ioides (K. Schum.) Verdc. ■☆
76847 Carinta obvallata (Schumach.) G. Taylor = Geophila obvallata (Schumach.) Didr. ■☆
76848 Carinta repens (L.) L. B. Sm. et Downs = Geophila repens (L.) I. M. Johnst. ■☆
76849 Carinta uniflora (Hiern) G. Taylor = Geophila repens (L.) I. M. Johnst. ■☆
76850 Carionia Naudin = Medinilla Gaudich. ex DC. ●
76851 Carionia Naudin(1851);菲律宾酸脚杆属●☆
76852 Carionia elegans Naudin;菲律宾酸脚杆(卡香木)●☆
76853 Carionia elegans Naudin = Medinilla coronata Regalado ●☆

76854 Carissa L.(1767)(保留属名);假虎刺属(刺黄果属);Carissa,Congaberry,Conkerberry ●

76855 Carissa abyssinica R. Br. = Acokanthera schimperi (A. DC.) Schweinf. ●☆

76856 Carissa abyssinica R. Br. = Carissa spinarum L. ●

76857 Carissa acokanthera Pichon = Acokanthera oppositifolia (Lam.) Codd ●

76858 Carissa acuminata (E. Mey.) A. DC. = Carissa bispinosa (L.) Desf. ex Brenan ●☆

76859 Carissa africana A. DC. = Carissa macrocarpa (Eckl.) A. DC. ●

76860 Carissa africana A. DC. = Carissa spinarum L. ●

76861 Carissa arduina Lam.;南非假虎刺;Hedge-thorn ●☆

76862 Carissa arduina Lam. = Carissa bispinosa (L.) Desf. ex Brenan ●☆

76863 Carissa axillaris Roxb. = Carissa spinarum L. ●

76864 Carissa bispinosa (L.) Desf. ex Brenan;二叉假虎刺(果李);Amatnngulu,Amatungula,Hedge Thorn,Num-num ●☆

76865 Carissa bispinosa (L.) Desf. ex Brenan subsp. zambesiensis Kupicha = Carissa bispinosa (L.) Desf. ex Brenan ●☆

76866 Carissa bispinosa (L.) Desf. ex Brenan var. acuminata (E. Mey.) Codd = Carissa bispinosa (L.) Desf. ex Brenan ●☆

76867 Carissa bispinosa (L.) Merxm. = Carissa bispinosa (L.) Desf. ex Brenan ●☆

76868 Carissa bispinosa Desf. ex Steud. = Carissa bispinosa (L.) Desf. ex Brenan ●☆

76869 Carissa boivinianum Leeuwenb.;博伊文假虎刺●☆

76870 Carissa brownii F. Muell. = Carissa spinarum L. ●

76871 Carissa campenonii (Drake) Palacky = Carissa spinarum L. ●

76872 Carissa candolleana Jaub. et Spach = Carissa spinarum L. ●

76873 Carissa carandas L.;刺黄果(瓜子金,假虎刺,纳达尔梅子);Christ's Thorn,Christ's-thorn,Karanda,Karanda Carissa ●

76874 Carissa carandas L. var. congesta (Wight) Bedd. = Carissa spinarum L. ●

76875 Carissa carandas L. var. paucinervia (A. DC.) Bedd. = Carissa spinarum L. ●

76876 Carissa cochinchinensis Pierre ex Pit. = Carissa spinarum L. ●

76877 Carissa comorensis (Pichon) Markgr. = Carissa spinarum L. ●

76878 Carissa congesta Wight;密假虎刺●☆

76879 Carissa congesta Wight = Carissa carandas L. ●

76880 Carissa congesta Wight = Carissa spinarum L. ●

76881 Carissa cordata (Mill.) Fourc. = Carissa bispinosa (L.) Desf. ex Brenan ●☆

76882 Carissa cordata Dinter = Carissa bispinosa (L.) Desf. ex Brenan ●☆

76883 Carissa coriacea Wall. = Carissa spinarum L. ●

76884 Carissa cornifolia Jaub. et Spach = Carissa spinarum L. ●

76885 Carissa cryptophlebia Baker;隐脉假虎刺●☆

76886 Carissa cryptophlebia Baker = Petchia cryptophlebia (Baker) Leeuwenb. ●☆

76887 Carissa curandas L.;卡兰假虎刺●☆

76888 Carissa dalzellii Bedd. = Carissa spinarum L. ●

76889 Carissa deflersii (Schweinf. ex Lewin) Pichon = Acokanthera schimperi (A. DC.) Schweinf. ●☆

76890 Carissa densiflora Baker var. microphylla Danguy ex Lecomte = Carissa spinarum L. ●

76891 Carissa diffusa Roxb. = Carissa spinarum L. ●

76892 Carissa dinteri Markgr. = Carissa bispinosa (L.) Desf. ex Brenan ●☆

76893 Carissa dulcis Schumach. et Thonn. = Carissa spinarum L. ●

76894 Carissa edulis (Forssk.) Vahl;甜假虎刺(埃及假虎刺,果李);Egyp Carissa,Egyptian Carissa,Small Num-num ●

76895 Carissa edulis (Forssk.) Vahl = Carissa spinarum L. ●

76896 Carissa edulis (Forssk.) Vahl f. continentalis = Carissa spinarum L. ●

76897 Carissa edulis (Forssk.) Vahl f. pubescens (A. DC.) Pichon = Carissa spinarum L. ●

76898 Carissa edulis (Forssk.) Vahl f. revoluta? = Carissa spinarum L. ●

76899 Carissa edulis (Forssk.) Vahl f. typica Pichon = Carissa spinarum L. ●

76900 Carissa edulis (Forssk.) Vahl subsp. continentalis Pichon = Carissa spinarum L. ●

76901 Carissa edulis (Forssk.) Vahl var. ambungana Pichon = Carissa spinarum L. ●

76902 Carissa edulis (Forssk.) Vahl var. densiflora (Baker) Pichon = Carissa spinarum L. ●

76903 Carissa edulis (Forssk.) Vahl var. horrida (Pichon) Markgr. = Carissa spinarum L. ●

76904 Carissa edulis (Forssk.) Vahl var. major Stapf = Carissa spinarum L. ●

76905 Carissa edulis (Forssk.) Vahl var. microphylla Pichon = Carissa spinarum L. ●

76906 Carissa edulis (Forssk.) Vahl var. tomentosa (A. Rich.) Stapf = Carissa spinarum L. ●

76907 Carissa edulis Vahl = Carissa edulis (Forssk.) Vahl ●

76908 Carissa erythrocarpa (Eckl.) A. DC. = Carissa bispinosa (L.) Desf. ex Brenan ●☆

76909 Carissa ferox (E. Mey.) A. DC. = Carissa bispinosa (L.) Desf. ex Brenan ●☆

76910 Carissa friesiorum (Markgr.) Cufod. = Acokanthera schimperi (A. DC.) Schweinf. ●☆

76911 Carissa grandiflora (E. Mey.) A. DC. = Carissa macrocarpa (Eckl.) A. DC. ●

76912 Carissa grandiflora A. DC. 'Horizontalis';平枝刺李●

76913 Carissa grandiflora A. DC. 'Minima';小叶刺李●

76914 Carissa grandiflora A. DC. 'Nana';矮刺李●

76915 Carissa grandiflora A. DC. 'Prostrana';卧刺李●

76916 Carissa grandiflora A. DC. = Carissa macrocarpa (Eckl.) A. DC. ●

76917 Carissa grandis Berteroex A. DC. = Fagraea berterianaA. Gray ex Benth ●☆

76918 Carissa haematocarpa (Eckl.) A. DC. = Carissa bispinosa (L.) Desf. ex Brenan ●☆

76919 Carissa hirsuta Roth = Carissa spinarum L. ●

76920 Carissa horrida Pichon = Carissa spinarum L. ●

76921 Carissa ineptaPerrot et Vogt = Acokanthera schimperi (A. DC.) Schweinf. ●☆

76922 Carissa inermis Vahl = Carissa spinarum L. ●

76923 Carissa lanceolata R. Br. = Carissa spinarum L. ●

76924 Carissa laxiflora Benth. = Carissa spinarum L. ●

76925 Carissa longiflora (Stapf) Lawr. = Acokanthera oppositifolia (Lam.) Codd ●

76926 Carissa macrocarpa (Eckl.) A. DC.;大果假虎刺(大花刺李,大花假虎刺,巨花假虎刺);Amatungula,Bigflower Carissa,Bigfruit Carissa,Big-fruited Carissa,Largeflower Carissa,Natal Plum,Natal-plum ●

76927 Carissa macrocarpa (Eckl.) A. DC. 'Boxwood Beauty';黄杨美

大果假虎刺●☆

76928 Carissa macrocarpa (Eckl.) A. DC. 'Boxwood Variegata';黄杨花叶大果假虎刺●☆

76929 Carissa macrocarpa (Eckl.) A. DC. 'Fancy';幻想大果假虎刺●☆

76930 Carissa macrocarpa (Eckl.) A. DC. 'Horizontalis';水平大果假虎刺●☆

76931 Carissa macrocarpa (Eckl.) A. DC. 'Nana';紧凑大果假虎刺●☆

76932 Carissa macrocarpa (Eckl.) A. DC. 'Prostrata';平卧大果假虎刺●☆

76933 Carissa macrocarpa (Eckl.) A. DC. = Carissa grandiflora A. DC. ●

76934 Carissa macrophylla Wall. = Carissa spinarum L. ●

76935 Carissa madagascariensis Thouars ex Poir. = Carissa spinarum L. ●

76936 Carissa megaphylla Gand. = Carissa bispinosa (L.) Desf. ex Brenan ●☆

76937 Carissa myrtoides Desf. = Carissa bispinosa (L.) Desf. ex Brenan ●☆

76938 Carissa oblongifolia Hochst. = Acokanthera oblongifolia (Hochst.) Codd ●☆

76939 Carissa obovata Markgr. = Carissa spinarum L. ●

76940 Carissa opaca Stapf = Carissa opaca Stapf ex Haines ●☆

76941 Carissa opaca Stapf ex Haines;阴生假虎刺●☆

76942 Carissa oppositifolia (Lam.) Pichon = Acokanthera oppositifolia (Lam.) Codd ●

76943 Carissa ovata R. Br.;卵叶假虎刺●☆

76944 Carissa ovata R. Br. = Carissa spinarum L. ●

76945 Carissa paucinervia A. DC. = Carissa spinarum L. ●

76946 Carissa pichoniana Leeuwenb.;皮雄假虎刺●☆

76947 Carissa pilosa Schinz = Carissa spinarum L. ●

76948 Carissa praetermissa Kupicha = Carissa macrocarpa (Eckl.) A. DC. ●

76949 Carissa pubescens A. DC. = Carissa spinarum L. ●

76950 Carissa revoluta Scott-Elliot = Carissa spinarum L. ●

76951 Carissa richardiana Jaub. et Spach = Carissa spinarum L. ●

76952 Carissa scabra R. Br. = Carissa spinarum L. ●

76953 Carissa schimperi A. DC.;箭毒假虎刺●☆

76954 Carissa schimperi A. DC. = Acokanthera schimperi (A. DC.) Schweinf. ●☆

76955 Carissa sechellensis Baker = Carissa spinarum L. ●

76956 Carissa sessiliflora Brongn. ex Pichon = Carissa boivinianum Leeuwenb. ●☆

76957 Carissa sessiliflora Brongn. ex Pichon var. grandiflora Markgr. = Carissa boivinianum Leeuwenb. ●☆

76958 Carissa sessiliflora Brongn. ex Pichon var. meridionalis Pichon = Carissa boivinianum Leeuwenb. ●☆

76959 Carissa sessiliflora Brongn. ex Pichon var. orientalis Pichon = Carissa boivinianum Leeuwenb. ●☆

76960 Carissa sessiliflora Brongn. ex Pichon var. septentrionalis Pichon = Carissa boivinianum Leeuwenb. ●☆

76961 Carissa spectabilis (Sond.) Pichon = Acokanthera oblongifolia (Hochst.) Codd ●☆

76962 Carissa spinarum L.;假虎刺(刺椰果,刺檀香,黑奶奶果,老虎刺,三颗针,绣花针,云南假虎刺);Spiny Carissa ●

76963 Carissa suavissima Bedd. ex Hook. f. = Carissa spinarum L. ●

76964 Carissa tetramera (Sacleux) Stapf;四数假虎刺●☆

76965 Carissa tomentosa A. Rich. = Carissa spinarum L. ●

76966 Carissa verticillata Pichon = Carissa pichoniana Leeuwenb. ●☆

76967 Carissa villosa Roxb. = Carissa spinarum L. ●

76968 Carissa wyliei N. E. Br. = Carissa bispinosa (L.) Desf. ex Brenan ●☆

76969 Carissa xylopicron Thouars = Carissa spinarum L. ●

76970 Carissa yunnanensis Tsiang = Carissa spinarum L. ●

76971 Carissa yunnanensis Tsiang et P. T. Li = Carissa spinarum L. ●

76972 Carissaceae Bertol. = Apocynaceae Juss. (保留科名)●■

76973 Carissophyllum Pichon = Tachiadenus Griseb. ●■☆

76974 Carissophyllum longiflorum Pichon = Tachiadenus tubiflorus (Thouars ex Roem. et Schult.) Griseb. ●☆

76975 Carlea C. Presl = Symplocos Jacq. ●

76976 Carlemannia Benth. (1853);香茜属;Carlemannia ■

76977 Carlemannia henryi H. Lév. = Carlemannia tetragona Hook. f. ■

76978 Carlemannia tetragona Hook. f.;香茜;Carlemannia ■

76979 Carlemanniaceae Airy Shaw = Caprifoliaceae Juss. (保留科名)●■

76980 Carlemanniaceae Airy Shaw(1965);香茜科■●

76981 Carlephyton Jum. (1919);沼石南星属■☆

76982 Carlephyton Jum. et Buchet = Carlephyton Jum. ■☆

76983 Carlephyton diegoense Bogner;沼石南星■☆

76984 Carlephyton glaucophyllum Bogner;灰绿沼石南星■☆

76985 Carlephyton madagascariense Jum.;马岛沼石南星■☆

76986 Carlesia Dunn(1902);山茴香属;Carlesia ●★

76987 Carlesia sinensis Dunn;山茴香(拉拉柏,山茴芹,岩茴香);China Carlesia ■

76988 Carlina L. (1753);刺苞菊属(刺苞木属,刺苞术属,刺菊属);Carlina, Carline Thistle, Carline-thistle, Thistle ■●

76989 Carlina acanthifolia All.;刺叶刺苞菊(刺叶蓟);Acanthus-leaved Carline Thistle ■☆

76990 Carlina acaulis L.;无茎刺苞菊(朝鲜蓟,无茎刺苞木);Alpine Thistle, Dwarf Thistle, Fair Weather Thistle, Silver Thistle, Smooth Carlina, Stemless Carline Thistle, Weather Fair Thistle ■☆

76991 Carlina aculeata Burm. f. = Heterorhachis aculeata (Burm. f.) Rössler ●☆

76992 Carlina atlantica Pomel;大西洋刺苞菊■☆

76993 Carlina atlantica Pomel var. claryi Batt. = Carlina atlantica Pomel ■☆

76994 Carlina biebersteinii Bernh. ex Hornem.;刺苞菊(卡林菊,新疆刺苞菊,窄叶刺菊);Bieberstein's Carlina, Xinjiang Carlina ■

76995 Carlina brachylepis (Batt.) Meusel et Kastner;短鳞刺苞菊■☆

76996 Carlina brachylepis (Batt.) Meusel et Kastner var. lanigera (Faure et Maire) D. P. Petit = Carlina brachylepis (Batt.) Meusel et Kastner ■☆

76997 Carlina canariensis Pit.;加那利刺苞菊■☆

76998 Carlina caulescens Lam.;有茎刺苞菊■☆

76999 Carlina cirsioides Klokov;蓟刺苞菊■☆

77000 Carlina cynara Pourr. ex DC.;菜蓟刺菊■☆

77001 Carlina falcata Svent. = Carlina salicifolia (L. f.) Cav. ■☆

77002 Carlina gummifera (L.) Less.;产胶刺苞菊■☆

77003 Carlina hispanica Lam.;西班牙刺苞菊■☆

77004 Carlina hispanica Lam. subsp. major (Lange) Meusel et Kastner;大西班牙刺苞菊■☆

77005 Carlina involucrata Poir.;总苞刺苞菊■☆

77006 Carlina involucrata Poir. subsp. corymbosa Quézel et Santa = Carlina hispanica Lam. ■☆

77007 Carlina involucrata Poir. var. lanigera Faure et Maire = Carlina brachylepis (Batt.) Meusel et Kastner ■☆

77008 Carlina lanata L.;绵毛刺苞菊■☆

77009 Carlina longifolia Rchb. = Carlina biebersteinii Bernh. ex Hornem. ■

77010 Carlina longifolia Rchb. var. pontica Boiss. = Carlina biebersteinii Bernh. ex Hornem. ■

77011 Carlina onopordifolia Besser ex Szafer, Kulcz. et Pawl.；大翅蓟刺苞菊■☆

77012 Carlina racemosa L.；总状刺苞菊■

77013 Carlina reboudiana Pomel = Carlina racemosa L. ■

77014 Carlina salicifolia（L. f.）Cav.；柳叶刺苞菊■☆

77015 Carlina salicifolia（L. f.）Cav. var. inermis Lowe = Carlina salicifolia（L. f.）Cav. ■☆

77016 Carlina sicula Ten.；西西里刺苞菊■☆

77017 Carlina sicula Ten. subsp. mareotica（Asch. et Schweinf.）Greuter；马雷奥特刺苞菊■☆

77018 Carlina sicula Ten. var. libyca Pamp. = Carlina sicula Ten. subsp. mareotica（Asch. et Schweinf.）Greuter ■☆

77019 Carlina sicula Ten. var. longibracteata（Cavara）Meusel et Karstner = Carlina sicula Ten. ■☆

77020 Carlina sicula Ten. var. mareotica（Asch. et Schweinf.）Meusel et Karstner = Carlina sicula Ten. ■☆

77021 Carlina sulfurea Desf. = Carlina racemosa L. ■

77022 Carlina vulgaris L.；欧洲刺苞菊；Carline Thistle, Common Carline, Earline Thistle, Weather Thistle ■☆

77023 Carlina vulgaris L. = Carlina biebersteinii Bernh. ex Hornem. ■

77024 Carlina vulgaris L. subsp. longifolia（C. Rchb.）Nyman = Carlina vulgaris L. ■☆

77025 Carlina vulgaris L. var. longifolia Kir. = Carlina biebersteinii Bernh. ex Hornem. ■

77026 Carlina xeranthemoides L. f.；干花菊状刺苞菊■

77027 Carlinaceae Bercht. et J. Presl = Asteraceae Bercht. et J. Presl（保留科名）●■

77028 Carlinaceae Bercht. et J. Presl = Compositae Giseke（保留科名）●■

77029 Carlinodes Kuntze = Berkheya Ehrh.（保留属名）●■☆

77030 Carlomohria Greene = Halesia J. Ellis ex L.（保留属名）●

77031 Carlostephania Bubani = Circaea L. ■

77032 Carlowitzia Moench = Carlina L. ■●

77033 Carlowizia Moench = Carlina L. ■●

77034 Carlowrightia A. Gray（1878）（保留属名）；卡洛爵床属☆

77035 Carlowrightia arizonica A. Gray；亚马逊卡洛爵床；Carlowrightia☆

77036 Carlquistia B. G. Baldwin（1999）；星盘菊属■☆

77037 Carlquistia muirii（A. Gray）B. G. Baldwin；星盘菊■☆

77038 Carludovica Ruiz et Pav.（1794）；巴拿马草属；Carludovica ●■

77039 Carludovica atrovirens H. Wendl.；暗绿巴拿马草；Darkgreen Carludovica ●☆

77040 Carludovica insignis Duchass. ex Griseb.；显著巴拿马草；Panama-hat Palm ●☆

77041 Carludovica jamaicensis Lodd. ex Fawc. et Harris；牙买加巴拿马草；Jamaica Carludovica ●☆

77042 Carludovica latifolia Ruiz et Pav.；宽叶巴拿马草●☆

77043 Carludovica palmata Ruiz et Pav.；巴拿马草；Palmate Carludovica, Panama Hat Palm, Panama Hat Plant, Panama Screw Pine, Panama-hat Palm, Toquilla ●■

77044 Carludovica plumerii Kunth；布美氏巴拿马草；Plumeri Carludovica ●☆

77045 Carludovicaceae A. Kern. = Cyclanthaceae Poit. ex A. Rich.（保留科名）●■

77046 Carmelita Gay = Chaetanthera Ruiz et Pav. ■☆

77047 Carmenocania Wernham = Pogonopus Klotzsch ■☆

77048 Carmenta Noronha = Viburnum L. ●

77049 Carmichaela Rchb. = Carmichaelia R. Br. ●☆

77050 Carmichaelia R. Br.（1825）；假金雀花属（扁枝豆属）；New Zealand Broom ●☆

77051 Carmichaelia arborea Druce；树假金雀花（大扁枝豆）●☆

77052 Carmichaelia australis R. Br.；澳大利亚假金雀花●☆

77053 Carmichaelia australis R. Br. = Carmichaelia arborea Druce ●☆

77054 Carmichaelia enysii Kirk；小扁枝豆●☆

77055 Carmichaelia flagelliformis Colenso = Carmichaelia australis R. Br. ●☆

77056 Carmichaelia glabrescens（Petrie）Heenan；无毛假金雀花；Pink Tree Broom ●☆

77057 Carmichaelia grandiflora Hook. f.；大花假金雀花（大假金雀花）●☆

77058 Carmichaelia odorata Colenso ex Hook. f.；香假金雀花；New Zealand Scented Broom, Scented Broom ●☆

77059 Carmichaelia stevensonii（Cheeseman）Heenan；美假金雀花●☆

77060 Carmichaelia williamsi Kirk；斑假金雀花；Giant-flowered Broom ●☆

77061 Carminatia Moc. ex DC.（1838）；羽冠肋泽兰属■☆

77062 Carminatia tenuiflora DC.；羽冠肋泽兰■☆

77063 Carmona Cav.（1799）；基及树属（满福木属）；Carmona ●

77064 Carmona Cav. = Ehretia P. Browne ●

77065 Carmona heterophylla Cav. = Carmona microphylla（Lam.）G. Don ●

77066 Carmona heterophylla Cav. = Carmona retusa（Vahl）Masam. ●

77067 Carmona microphylla（Lam.）G. Don；基及树（凹基及树，福建茶，满福木，小叶厚壳树）；Falsetea Ehretia, Fukien Tea, Scorpionbush, Smallleaf Carmona, Small-leaved Carmona ●

77068 Carmona microphylla（Lam.）G. Don = Carmona retusa（Vahl）Masam. ●

77069 Carmona retusa（Vahl）Masam. = Carmona microphylla（Lam.）G. Don ●

77070 Carmona viminea（Wall.）G. Don = Rotula aquatica Lour. ●

77071 Carmonea Pers. = Carmona Cav. ●

77072 Carmorea Steud. = Carmona Cav. ●

77073 Carnarvonia F. Muell.（1867）；卡尔山龙眼属●☆

77074 Carnarvonia araliifolia F. Muell.；卡尔山龙眼●☆

77075 Carnegiea Britton et Rose（1908）；巨人柱属；Saguaro ●☆

77076 Carnegiea Perkins = Carnegieodoxa Perkins ●☆

77077 Carnegiea gigantea（Engelm.）Britton et Rose；巨人柱（弁庆柱，巨大肉仙掌，巨仙人柱）；Arizona-giant, Giant Cactus, Giant Cactus of Arizona, Giant Cactus of Califoruia, Pitahaya, Saguared Cactus, Saguaro ●☆

77078 Carnegieodoxa Perkins（1914）；卡香木属●☆

77079 Carnegieodoxa eximia（Perkins）Perkins；卡香木●☆

77080 Carolifritschia Post et Kuntze = Carolofritschia Engl. ■☆

77081 Caroli-Gmelina P. Gaertn., B. Mey. et Scherb. = Radicula Hill ■

77082 Caroli-Gmelina P. Gaertn., B. Mey. et Scherb. = Rorippa Scop. ■

77083 Carolinea L. f. = Pachira Aubl. ●

77084 Carolinea insignis Sw. = Pachira insignis（Sw.）Sw. ex Sav. ●☆

77085 Carolinea macrocarpa Schltdl. et Cham. = Pachira aquatica Aubl. ●

77086 Carolinea princeps L. f. = Pachira aquatica Aubl. ●

77087 Carolinella Hemsl. = Primula L. ■

77088 Carolinella cordifolia Hemsl. = Primula partschiana Pax ■

77089 Carolinella henryi Hemsl. = Primula henryi (Hemsl.) Pax ■
77090 Carolinella obovata Hemsl. = Primula rugosa N. P. Balakr. ■
77091 Carolofritschia Engl. = Acanthonema Hook. f.(保留属名)■☆
77092 Carolofritschia diandra Engl. = Acanthonema diandrum (Engl.) B. L. Burtt ■☆
77093 Carolus W. R. Anderson = Hiraea Jacq. ●☆
77094 Carolus W. R. Anderson(2006);巴西藤翅果属●☆
77095 Caromba Steud. = Cabomba Aubl. ■
77096 Caropodium Stapf et Wettst. (1886);头足草属■☆
77097 Caropodium Stapf et Wettst. = Grammosciadium DC. ■☆
77098 Caropodium Stapf et Wettst. ex Stapf = Caropodium Stapf et Wettst. ■☆
77099 Caropodium armenum (Bordz.) Schischk.;头足草■☆
77100 Caropodium platycarpum (Boiss. et Hausskn.) Schischk.;宽果头足草■☆
77101 Caropsis (Rouy et Camus) Rauschert(1982);头状草属■☆
77102 Caropsis verticillatoinundata (Thore) Rauschert;头状草■☆
77103 Caropyxis Benth. et Hook. f. = Calopyxis Tul. ●☆
77104 Caropyxis Benth. et Hook. f. = Combretum Loefl.(保留属名)●
77105 Caroselinum Griseb. = Peucedanum L. ■
77106 Carota Rupr. = Daucus L. ■
77107 Caroxylon Thunb. = Salsola L. ●■
77108 Caroxylon articulatum Moq. = Haloxylon articulatum (Moq.) Bunge ●☆
77109 Caroxylon bottae (Jaub. et Spach) Moq. = Halothamnus bottae Jaub. et Spach ●☆
77110 Caroxylon brevifolium St. -Lag. = Salsola aphylla L. f. ●■☆
77111 Caroxylon divaricatum Moq. = Salsola capensis Botsch. ■☆
77112 Caroxylon glaucum (M. Bieb.) Moq. = Halothamnus glaucus (M. Bieb.) Botsch. ●
77113 Caroxylon imbricatum (Forssk.) Moq. = Salsola imbricata Forssk. ■
77114 Caroxylon orientale (S. G. Gmel.) Tzvelev = Salsola orientalis S. G. Gmel. ●
77115 Caroxylon salicornicum Moq. = Hammada salicornica (Moq.) Iljin ●☆
77116 Caroxylon salsola Thunb. = Salsola aphylla L. f. ●■☆
77117 Caroxylon tuberculatum Fenzl ex Moq. = Salsola tuberculata (Fenzl ex Moq.) Schinz ■☆
77118 Caroxylon tuberculatum Fenzl ex Moq. var. flavo-virens Moq. = Salsola gemmifera Botsch. ■☆
77119 Caroxylon zeyheri Moq. = Salsola zeyheri (Moq.) Bunge ■☆
77120 Carpacoce Sond. (1865);尖果茜属■●☆
77121 Carpacoce burchellii Puff;伯切尔尖果茜●☆
77122 Carpacoce curvifolia Puff;折叶尖果茜●☆
77123 Carpacoce gigantea Puff;巨大尖果茜●☆
77124 Carpacoce heteromorpha (H. Buek) L. Bolus;异形尖果茜●☆
77125 Carpacoce scabra (Thunb.) Sond.;粗糙尖果茜●☆
77126 Carpacoce scabra (Thunb.) Sond. subsp. rupestris Puff;岩生尖果茜■☆
77127 Carpacoce spermacocea (Rchb. f.) Sond.;鸭舌癀舅尖果茜●☆
77128 Carpacoce spermacocea (Rchb. f.) Sond. subsp. orientalis Puff;东方鸭舌癀舅尖果茜●☆
77129 Carpacoce vaginellata T. M. Salter;具鞘尖果茜■☆
77130 Carpangis Thouars = Angraecum Bory ■
77131 Carpanthea N. E. Br. (1925);隐子玉属■☆
77132 Carpanthea calendulacea (Haw.) L. Bolus = Carpanthea pomeridiana (L.) N. E. Br. ■☆
77133 Carpanthea pilosa (Haw.) L. Bolus = Carpanthea pomeridiana (L.) N. E. Br. ■☆
77134 Carpanthea pomeridiana (L.) N. E. Br.;隐子玉■☆
77135 Carparomorchis M. A. Clem. et D. L. Jones = Bulbophyllum Thouars(保留属名)■
77136 Carparomorchis M. A. Clem. et D. L. Jones(2002);澳洲石豆兰属■☆
77137 Carpentaria Becc. (1885);东澳棕属(北澳椰属,北澳棕属,木匠椰属)●☆
77138 Carpentaria acuminata (H. Wendl. et Drude) Becc.;东澳棕●☆
77139 Carpentaria acuminata Becc. = Carpentaria acuminata (H. Wendl. et Drude) Becc. ●☆
77140 Carpenteria Torr. (1851);树银莲花属(茶花常山属);Anemone ●☆
77141 Carpenteria californica Torr.;树银莲花(茶花常山);Bush Anemone,Tree Anemone ●☆
77142 Carpenteria californica Torr. 'Elizabeth';伊丽莎白树银莲花●☆
77143 Carpentia Ewart = Cressa L. ■☆
77144 Carpentiera Steud. = Charpentiera Gaudich. ●☆
77145 Carpesium L. (1753);天名精属(金挖耳属);Carpesium ■
77146 Carpesium abrotanoides L.;天名精(拔子盖子,北鹤虱,蟾蜍兰,臭草,地葱,地菘,埊松,癫蟖草,杜牛膝,鹄虱,鬼虱,蚵蚾草,鹤虱,鹤虱草,鸡踝子草,觐,葵松,癞格宝草,癞蛤蟆草,癞蟖草,癞头草,菵蘸,鹿活草,麦句姜,麦句名姜,母猪草,母猪芥,山烟,豕首,天蔓菁,天蔓精,天蔓青,天门精,天明精,天芜菁,土牛膝,挖耳草,虾蟆蓝,烟袋草,烟管头草,野烟,野叶子烟,玉门精,彘颅,皱面草,皱面地菘草);Carpesium,Common Carpesium ■
77147 Carpesium acutum Hayata = Carpesium nepalense Less. ■
77148 Carpesium atkinsonianum Hemsl. = Carpesium divaricatum Siebold et Zucc. ■
77149 Carpesium cernuum L.;烟管头草(杓儿菜,蛋黄草,倒提壶,金挖耳,六氏草,毛叶草,毛叶云香草,挖耳草,烟袋草,野朝阳柄,野葵花,野思草,野烟,野烟叶,云香草);Drooping Carpesium ■
77150 Carpesium cernuum L. var. glandulosum? = Carpesium nepalense Less. ■
77151 Carpesium cernuum L. var. lanatum Hook. f. et Thomson ex C. B. Clarke = Carpesium nepalense Less. var. lanatum (Hook. f. et Thomson ex C. B. Clarke) Kitam. ■
77152 Carpesium cernuum L. var. nepalense? = Carpesium nepalense Less. ■
77153 Carpesium cernuum L. var. tracheliifolium (Less.) C. B. Clarke = Carpesium tracheliifolium Less. ■
77154 Carpesium ciliatum DC. = Carpesium cernuum L. ■
77155 Carpesium cordatum F. H. Chen et C. M. Hu;心叶天名精(杓儿菜);Cordate Carpesium,Heartleaf Carpesium ■
77156 Carpesium divaricatum Siebold et Zucc.;金挖耳(除州鹤虱,大挖耳草,倒盖菊,耳瓢草,翻天印,金空耳,劳伤草,朴地菊,山烟筒头,蛇王菊,铁骨消,铁抓子草,挖耳草,烟管草,野葵花,野向日葵,野烟,野烟头);Divaricate Carpesium ■
77157 Carpesium divaricatum Siebold et Zucc. var. abrotanoides (Matsum. et Koidz.) H. Koyama;青蒿金挖耳■☆
77158 Carpesium divaricatum Siebold et Zucc. var. matsuei (Tatew. et Kitam.) Kitam.;松江天名精■☆
77159 Carpesium eximum C. Winkl. = Carpesium macrocephalum Franch. et Sav. ■
77160 Carpesium faberi C. Winkl.;贵州天名精(天名精,银挖耳子

草);Faber's Carpesium,Guizhou Carpesium ■
77161 Carpesium faberi C. Winkl. = Carpesium minum Hemsl. ■
77162 Carpesium glossophyllum Maxim.;舌叶天名精■☆
77163 Carpesium hosocawae Kitam. = Carpesium minum Hemsl. ■
77164 Carpesium hosokawae Kitam.;细川氏天名精■
77165 Carpesium hosokawae Kitam. = Carpesium faberi C. Winkl. ■
77166 Carpesium humile Winkl.;矮天名精;Dwarf Carpesium ■
77167 Carpesium koidzumii Makino var. matsuei (Tatew. et Kitam.) H. Hara = Carpesium divaricatum Siebold et Zucc. var. matsuei (Tatew. et Kitam.) Kitam. ■☆
77168 Carpesium kweichowense C. C. Chang = Carpesium faberi C. Winkl. ■
77169 Carpesium leptophyllum F. H. Chen et C. M. Hu;薄叶天名精;Thinleaf Carpesium ■
77170 Carpesium leptophyllum F. H. Chen et C. M. Hu = Carpesium longifolium F. H. Chen et C. M. Hu ■
77171 Carpesium leptophyllum F. H. Chen et C. M. Hu var. linearibracteatum F. H. Chen et C. M. Hu;狭苞薄叶天名精;Linearbract Thinleaf Carpesium ■
77172 Carpesium lipskyi Winkl.;高原天名精(高山金挖耳,高山天名精,高原金挖耳,金挖耳,挖耳子草);Lipsky's Carpesium,Plateum Carpesium ■
77173 Carpesium lipskyi Winkl. var. hotonense Winkl. = Carpesium lipskyi Winkl. ■
77174 Carpesium lipskyi Winkl. var. potaninii Winkl. = Carpesium lipskyi Winkl. ■
77175 Carpesium lipskyi Winkl. var. przewalskyi Winkl. = Carpesium lipskyi Winkl. ■
77176 Carpesium lipskyi Winkl. var. przewalskyi Winkl. = Carpesium velutinum C. Winkl. ■
77177 Carpesium longifolium F. H. Chen et C. M. Hu;长叶天名精(乌金野烟,烟管草,野烟);Longleaf Carpesium ■
77178 Carpesium macrocephalum Franch. et Sav.;大花金挖耳(大烟锅草,千日草,神灵草,仙草,香油罐);Bighead Carpesium ■
77179 Carpesium manshuricum Kitam. = Carpesium triste Maxim. ■
77180 Carpesium matsuei Tatew. et Kitam. = Carpesium divaricatum Siebold et Zucc. var. matsuei (Tatew. et Kitam.) Kitam. ■☆
77181 Carpesium minum Hemsl.;小花金挖耳(冬葵花,茄叶细辛,散血草,细川氏天名精,小金挖耳,止血药);Small Carpesium ■
77182 Carpesium nepalense Less.;尼泊尔天名精(黄金珠,挖耳草);Nepal Carpesium ■
77183 Carpesium nepalense Less. var. lanatum (Hook. f. et Thomson ex C. B. Clarke) Kitam.;绵毛尼泊尔天名精(倒提壶,地朝阳,绵毛天名精,绵毛烟管头草,绵毛云香草,挖耳子草,野葵花,野向阳花,野烟,野叶子烟);Hairy Nepal Carpesium ■
77184 Carpesium pseudotracheliifolium Y. Ling = Carpesium triste Maxim. var. sinense Diels ■
77185 Carpesium rosulatum Miq.;莲座天名精■☆
77186 Carpesium scapiforme F. H. Chen et C. M. Hu;葶茎天名精;Scapose Carpesium ■
77187 Carpesium szechuanense F. H. Chen et C. M. Hu;四川天名精;Sichuan Carpesium ■
77188 Carpesium thunbergianum Siebold et Zucc. = Carpesium abrotanoides L. ■
77189 Carpesium tracheliifolium Less.;粗齿天名精;Grosseserrate Carpesium ■
77190 Carpesium triste Maxim.;暗花金挖耳(东北金挖耳);Darkcoloured Carpesium,Dullcolored Carpesium ■
77191 Carpesium triste Maxim. var. manshuricum (Kitam.) Kitam.;东北金挖耳■
77192 Carpesium triste Maxim. var. manshuricum (Kitam.) Kitam. = Carpesium triste Maxim. ■
77193 Carpesium triste Maxim. var. sinense Diels;毛暗花金挖耳;China Dullcolored Carpesium ■
77194 Carpesium tristeforme Hand.-Mazz. = Carpesium triste Maxim. var. sinense Diels ■
77195 Carpesium velutinum C. Winkl.;绒毛天名精;Tomentose Carpesium,Velutinous Carpesium ■
77196 Carpesium zhouquense J. Q. Fu;舟曲天名精;Zhouqu Carpesium ■
77197 Carpesium zhouquense J. Q. Fu = Carpesium velutinum C. Winkl. ■
77198 Carpezium Gouan = Carpesium L. ■
77199 Carpha Banks et Sol. ex R. Br. (1810);壳莎属■☆
77200 Carpha angustissima Cherm.;狭壳莎■☆
77201 Carpha aristata Kük.;具芒壳莎■☆
77202 Carpha bracteosa C. B. Clarke;多苞片壳莎■☆
77203 Carpha capensis (Steud.) Pfeiff. = Trianoptiles capensis (Steud.) Harv. ■☆
77204 Carpha capitellata (Nees) Boeck.;小头壳莎■☆
77205 Carpha capitellata (Nees) Boeck. var. bracteosa (C. B. Clarke) Kük. = Carpha bracteosa C. B. Clarke ■☆
77206 Carpha eminii (K. Schum.) C. B. Clarke;埃明壳莎■☆
77207 Carpha eminii (K. Schum.) C. B. Clarke var. angustissima (Cherm.) Kük. = Carpha angustissima Cherm. ■☆
77208 Carpha filifolia Reid et T. H. Arnold;线叶壳莎■☆
77209 Carpha glomerata (Thunb.) Nees;团集壳莎■☆
77210 Carpha hexandra Nees = Cyathocoma hexandra (Nees) Browning ■☆
77211 Carpha schlechteri C. B. Clarke;施莱壳莎■☆
77212 Carpha schweinfurthiana Boeck. = Coleochloa schweinfurthiana (Boeck.) Nelmes ■☆
77213 Carphalea Juss. (1789);卡尔茜属■☆
77214 Carphalea angulata Baill.;棱角卡尔茜■☆
77215 Carphalea cloiselii Homolle;克卢塞尔卡尔茜■☆
77216 Carphalea glaucescens (Hiern) Verdc.;灰绿卡尔茜■☆
77217 Carphalea glaucescens (Hiern) Verdc. subsp. angustifolia Verdc.;窄叶卡尔茜■☆
77218 Carphalea linearifolia Homolle;线叶卡尔茜■☆
77219 Carphalea madagascariensis Lam.;马岛卡尔茜■☆
77220 Carphalea pervilleana Baill.;佩尔卡尔茜■☆
77221 Carphalea pubescens (Klotzsch) Verdc.;短柔毛卡尔茜■☆
77222 Carphalea somaliensis Puff;索马里卡尔茜☆
77223 Carphephorus Cass. (1816);托鞭菊属■☆
77224 Carphephorus baicalensis (Adams) DC. = Saussurea baicalensis (Adams) B. L. Rob. ■
77225 Carphephorus baicalensis DC. = Saussurea baicalensis (Adams) B. L. Rob. ■
77226 Carphephorus bellidifolius (Michx.) Torr. et A. Gray;沙托鞭菊;Sandy-woods Chaffhead ■☆
77227 Carphephorus carnosus (Small) C. W. James;松林托鞭菊;Pineland Chaffhead ■☆
77228 Carphephorus corymbosus (Nutt.) Torr. et A. Gray;伞序托鞭菊(佛罗里达托鞭菊);Florida Paintbrush ■☆
77229 Carphephorus junceus Benth. = Bebbia juncea (Benth.) Greene ●☆

77230 Carphephorus odoratissimus (J. F. Gmel.) H. J. -C. Hebert;香托鞭菊;Vanillaleaf ■☆

77231 Carphephorus odoratissimus (J. F. Gmel.) H. J. -C. Hebert var. subtropicanus (DeLaney N. Bissett et Weidenh.) Wunderlin et B. F. Hansen;亚热带托鞭菊■☆

77232 Carphephorus paniculatus (J. F. Gmel.) H. J. -C. Hebert;圆锥托鞭菊;Hairy Chaffhead ■☆

77233 Carphephorus pseudoliatris Cass.;刚毛托鞭菊(托鞭菊);Bristleleaf Chaffhead ■☆

77234 Carphephorus subtropicanus DeLaney, N. Bissett et Weidenh. = Carphephorus odoratissimus (J. F. Gmel.) H. J. -C. Hebert var. subtropicanus (DeLaney N. Bissett et Weidenh.) Wunderlin et B. F. Hansen ■☆

77235 Carphephorus tomentosus (Michx.) Torr. et A. Gray;毛托鞭菊;Woolly Chaffhead ■☆

77236 Carphephorus tomentosus var. walteri (Elliott) Fernald = Carphephorus tomentosus (Michx.) Torr. et A. Gray ■☆

77237 Carphobolus Schott = Piptocarpha Hook. et Arn. ●☆

77238 Carphochaete A. Gray = Cronquistia R. M. King ■☆

77239 Carphochaete A. Gray(1849);肖长芒菊属■☆

77240 Carphochaete bigelovii A. Gray;肖长芒菊;Bristlehead ■☆

77241 Carpholoma D. Don = Lachnospermum Willd. ●☆

77242 Carphopappus Sch. Bip. = Iphiona Cass.(保留属名)●■☆

77243 Carphopappus baccharidifolius (Less.) Sch. Bip. = Pegolettia baccharidifolia Less. ■☆

77244 Carphophorus Post et Kuntze = Carphephorus Cass. ■☆

77245 Carphostephium Cass. = Tridax L. ■●

77246 Carpidopterix H. Karst. = Thouinia Poit.(保留属名)●☆

77247 Carpinaceae Kuprian.;鹅耳枥科●

77248 Carpinaceae Kuprian. = Betulaceae Gray(保留科名)●

77249 Carpinaceae Kuprian. = Corylaceae Mirb.(保留科名)●

77250 Carpinaceae Vest = Betulaceae Gray(保留科名)●

77251 Carpinum Raf. = Carpinus L. ●

77252 Carpinus L.(1753);鹅耳枥属(千金榆属);Hornbeam, Ironwood ●

77253 Carpinus americana Michx. = Carpinus caroliniana Walter ●☆

77254 Carpinus austrosinensis Hu = Carpinus pubescens Burkill ●

77255 Carpinus austroyunnanensis Hu = Carpinus kweichowensis Hu ●

77256 Carpinus betulus L.;欧洲鹅耳枥(角木,欧洲角木,欧洲千金榆,七叶树,七夜树,西方鹅耳枥);Buckeye, Common Hornbeam, European Hornbeam, Harber, Harbur, Hay Beech, Horn Beech, Hornbeam, Horned Beech, Horse Beech, Horse Chestnut, Horst Beech, Husbeech, Ironwood, Lanterns, Musclewood, White Beech, Witch Hazel, Wych Hazel, Yoke Elm ●

77257 Carpinus betulus L. 'Fastigiata';倾斜欧洲鹅耳枥(塔形欧洲鹅耳枥);Columnar European Hornbeam ●☆

77258 Carpinus betulus L. 'Pendula';垂枝欧洲鹅耳枥●☆

77259 Carpinus betulus L. 'Purpurea';紫叶欧洲鹅耳枥●☆

77260 Carpinus betulus L. 'Pyramidalis' = Carpinus betulus L. 'Fastigiata' ●☆

77261 Carpinus betulus L. 'Variegata';斑叶欧洲鹅耳枥●☆

77262 Carpinus betulus L. var. virginiana Marshall = Carpinus caroliniana Walter subsp. virginiana (Marshall) Furlow ●☆

77263 Carpinus caroliniana Walter;美洲鹅耳枥(北美鹅耳枥,卡州鹅耳枥,美国榛树);American Hornbeam, Blue Beech, Blue-beech, Hornbeam, Ironwood, Muscle Tree, Musclewood, Muscle-wood, Water Beech, Water-beech ●☆

77264 Carpinus caroliniana Walter subsp. virginiana (Marshall) Furlow;弗州鹅耳枥;American Hornbeam, Blue Beech, Blue-beech, Hornbeam, Ironwood, Musclewood ●☆

77265 Carpinus caroliniana Walter var. virginiana (Marshall) Fernald = Carpinus caroliniana Walter ●☆

77266 Carpinus caroliniana Walter var. virginiana (Marshall) Fernald = Carpinus caroliniana Walter subsp. virginiana (Marshall) Furlow ●☆

77267 Carpinus carpinoides Makino = Carpinus japonica Blume ●☆

77268 Carpinus carpinoides Makino var. caudata Makino et Nemoto;长尾日本鹅耳枥●☆

77269 Carpinus carpinoides Makino var. cordifolia Winkl. = Carpinus japonica Blume var. cordifolia (Winkl.) Makino ●☆

77270 Carpinus carpinus Sarg. = Carpinus japonica Blume ●☆

77271 Carpinus caucasica Grossh.;高加索鹅耳枥;Caucasian Hornbeam ●☆

77272 Carpinus chinensis (Franch.) C. P'ei = Carpinus cordata Blume var. chinensis Franch. ●

77273 Carpinus chowii Hu = Carpinus turczaninowii Hance ●

77274 Carpinus chuniana Hu;粤北鹅耳枥(陈氏鹅耳枥,滇粤鹅耳枥);Chun Hornbeam ●

77275 Carpinus chuniana Hu var. sichourensis (Hu) T. Hong;西畴鹅耳枥;Xichou Hornbeam ●

77276 Carpinus cordata Blume;千金榆(半拉子,鹅耳枥,见风干,金丝榆,千金鹅耳枥,穗子榆,心叶鹅耳枥);Caucasian Hornbeam, Heartleaf Hornbeam, Heart-leaved Hornbeam ●

77277 Carpinus cordata Blume var. brevistachya S. L. Tung;直穗千金榆;Shortstachys Heartleaf Hornbeam ●

77278 Carpinus cordata Blume var. chinensis Franch.;南方千金榆(大叶马料,华鹅耳枥,华千金榆,千金榆,小果千金榆,野梅树,中华鹅耳枥);China Heartleaf Hornbeam, Chinese Heartleaf Hornbeam ●

77279 Carpinus cordata Blume var. microcarpa (Hayashi) Hayashi = Carpinus cordata Blume ●

77280 Carpinus cordata Blume var. mollis (Rehder) W. C. Cheng ex F. H. Chen;毛叶千金榆(软毛鹅耳枥);Hairyleaf Hornbeam ●

77281 Carpinus cordata Blume var. pseudojaponica H. J. P. Winkl.;假日本鹅耳枥(日本鹅耳枥)●☆

77282 Carpinus cordata Blume var. velutina (Hayashi) Honda = Carpinus cordata Blume var. chinensis Franch. ●

77283 Carpinus coreana Nakai;朝鲜千金榆;Korea Hornbeam ●☆

77284 Carpinus daginensis Hu = Carpinus fargesiana H. Winkl. ●

77285 Carpinus dayongina K. W. Liu et Q. Z. Lin;大庸鹅耳枥;Dayong Hornbeam ●

77286 Carpinus densispica Hu = Carpinus monbeigiana Hand. -Mazz. ●

77287 Carpinus erosa Blume = Carpinus cordata Blume ●

77288 Carpinus eximia Nakai;优异鹅耳枥●☆

77289 Carpinus falcatibrateata Hu = Carpinus tschonoskii Maxim. ●

77290 Carpinus fangiana Hu;川黔千金榆(长穗鹅耳枥,川黔鹅耳枥,方氏鹅耳枥);Fang Hornbeam ●

77291 Carpinus fargesiana H. Winkl.;川陕鹅耳枥(大金鹅耳枥,华纪氏鹅耳枥,千筋树);Chinese Hornbeam, Farges Hornbeam ●

77292 Carpinus fargesiana H. Winkl. var. hwai (Hu et W. C. Cheng) P. C. Li;狭叶鹅耳枥;Narrow-leaf Farges Hornbeam ●

77293 Carpinus fargesii Franch. = Carpinus viminea Wall. ●

77294 Carpinus fargesii Franch. var. latifolia S. Y. Wang et C. L. Chang;宽叶鹅耳枥●

77295 Carpinus fauriei Nakai;法氏鹅耳枥●☆

77296 Carpinus firmifolia (H. Winkl.) Hu;硬叶鹅耳枥(厚叶鹅耳枥);Firmleaf Hornbeam, Thick-leaf Pubescent Hornbeam ●

77297 Carpinus firmifolia Hu = Carpinus firmifolia (H. Winkl.) Hu ●

77298 Carpinus firmifolia Hu = Carpinus pubescens Burkill ●

77299 Carpinus funiushanensis P. C. Kuo = Carpinus hupeana Hu ●

77300 Carpinus glandulosopunctata (C. J. Qi) C. J. Qi;蜜腺鹅耳枥;Dense-glandule Hornbeam ●

77301 Carpinus handelii Rehder = Carpinus polyneura Franch. ●

77302 Carpinus hebestroma Gogelein;新城鹅耳枥(单齿鹅耳枥,太鲁阁鹅耳枥,太鲁阁千金榆);Hwalien Hornbeam, Tailuko Hornbeam, Velvetymouth Hornbeam, Xincheng Hornbeam ●

77303 Carpinus henryana H. Winkl.;川鄂鹅耳枥(亨利鹅耳枥);Chinese Hornbeam, Henry Hornbeam ●

77304 Carpinus hogoensis Hayata = Carpinus kawakamii Hayata ●

77305 Carpinus huana W. C. Cheng = Carpinus hupeana Hu ●

77306 Carpinus hupeana Hu;湖北鹅耳枥(长柄鹅耳枥,鄂鹅耳枥,河南鹅耳枥,华氏鹅耳枥,崖刷子);Hubei Hornbeam, Hupeh Hornbeam ●

77307 Carpinus hupeana Hu var. henryana (H. Winkl.) P. C. Li = Carpinus henryana H. Winkl. ●

77308 Carpinus hupeana Hu var. simplicidentata (Hu) P. C. Li = Carpinus simplicidentata Hu ●

77309 Carpinus hupeana Hu var. simplicidentata (Hu) P. C. Li = Carpinus stipulata H. Winkl. ●

77310 Carpinus hwai Hu et W. C. Cheng = Carpinus fargesiana H. Winkl. var. hwai (Hu et W. C. Cheng) P. C. Li ●

77311 Carpinus hwai Hu et W. C. Cheng = Carpinus hupeana Hu ●

77312 Carpinus japonica Blume;日本鹅耳枥(日本千金榆);Japanese Hornbeam ●☆

77313 Carpinus japonica Blume var. cordifolia (Winkl.) Makino;心叶日本鹅耳枥●☆

77314 Carpinus kawakamii Hayata;阿里山鹅耳枥(阿里山千金榆,川上鹅耳枥,鸡油舅);Alishan Hornbeam, Arishan Hornbeam, Kawakami Hornbeam, Minute-serrated Hornbeam ●

77315 Carpinus kweichowensis Hu;贵州鹅耳枥(黔鹅耳枥);Guizhou Hornbeam, Kweichow Hornbeam ●

77316 Carpinus kweitingensis Hu;贵定鹅耳枥;Guiding Hornbeam ●

77317 Carpinus kweitingensis Hu = Carpinus pubescens Burkill ●

77318 Carpinus kweiyangensis Hu;贵阳鹅耳枥;Guiyang Hornbeam ●

77319 Carpinus kweiyangensis Hu = Carpinus pubescens Burkill ●

77320 Carpinus lanceolata Hand.-Mazz. = Carpinus londoniana H. Winkl. var. lanceolata (Hand.-Mazz.) P. C. Li ●

77321 Carpinus lancilimba Hu = Carpinus pubescens Burkill ●

77322 Carpinus laponica Blume = Carpinus carpinoides Makino ●

77323 Carpinus laxiflora (Siebold et Zucc.) Blume;疏花鹅耳枥(见风干)●☆

77324 Carpinus laxiflora (Siebold et Zucc.) Blume = Carpinus hupeana Hu ●

77325 Carpinus laxiflora (Siebold et Zucc.) Blume f. pendula (Miyoshi) Sugim.;垂枝疏花鹅耳枥●☆

77326 Carpinus laxiflora (Siebold et Zucc.) Blume var. davidii Franch. = Carpinus viminea Wall. ●

77327 Carpinus laxiflora (Siebold et Zucc.) Blume var. macrosmchya Oliv. = Carpinus londoniana H. Winkl. var. lanceolata (Hand.-Mazz.) P. C. Li ●

77328 Carpinus laxiflora (Siebold et Zucc.) Blume var. macrostachya Oliv. = Carpinus viminea Wall. ●

77329 Carpinus laxiflora (Siebold et Zucc.) Blume var. macrostachya Oliv. ex Hu = Carpinus viminea Wall. ●

77330 Carpinus laxiflora (Siebold et Zucc.) Blume var. pendula Miyoshi = Carpinus laxiflora (Siebold et Zucc.) Blume f. pendula (Miyoshi) Sugim. ●☆

77331 Carpinus laxiflora (Siebold et Zucc.) Blume var. tientaiensis Hu = Carpinus tientaiensis W. C. Cheng ●◇

77332 Carpinus laxiflora Siebold et Zucc. var. tientaiensis (W. C. Cheng) Hu. = Carpinus tientaiensis W. C. Cheng ●◇

77333 Carpinus likiangensis Hu = Carpinus londoniana H. Winkl. var. lanceolata (Hand.-Mazz.) P. C. Li ●

77334 Carpinus likiangensis Hu = Carpinus monbeigiana Hand.-Mazz. ●

77335 Carpinus londoniana H. Winkl.;白皮鹅耳枥(短穗鹅耳枥,短尾鹅耳枥,见风干,岷江鹅耳枥);Shorttail Hornbeam ●

77336 Carpinus londoniana H. Winkl. var. lanceolata (Hand.-Mazz.) P. C. Li;海南鹅耳枥(披针叶鹅耳枥);Hainan Hornbeam, Lanceolateleaf Hornbeam ●

77337 Carpinus londoniana H. Winkl. var. latifolius P. C. Li;宽叶白皮鹅耳枥(宽叶鹅耳枥);Broadleaf Shorttail Hornbeam ●

77338 Carpinus londoniana H. Winkl. var. xiphobracteata P. C. Li;剑苞鹅耳枥;Xiphobracteate Shorttail Hornbeam ●

77339 Carpinus longipes Hu = Carpinus hupeana Hu ●

77340 Carpinus luochengensis J. Y. Liang;罗城鹅耳枥;Luocheng Hornbeam ●

77341 Carpinus marlipoensis Hu = Carpinus pubescens Burkill ●

77342 Carpinus mengshanensis S. B. Liang et F. Z. Zhao;蒙山鹅耳枥;Mengshan Hornbeam ●

77343 Carpinus mianningensis T. P. Yi;冕宁鹅耳枥;Mianning Hornbeam ●

77344 Carpinus mianningensis T. P. Yi = Carpinus tschonoskii Maxim. ●

77345 Carpinus microphylla Z. C. Chen ex Y. S. Wang et J. P. Huang;小叶鹅耳枥(田阳鹅耳枥);Smallleaf Hornbeam, Small-leaved Hornbeam ●

77346 Carpinus minutiserrata Hayata;细齿鹅耳枥(细齿千金榆);Minutely-serrate Hornbeam, Minutserrate Hornbeam, Serrulate Hornbeam ●

77347 Carpinus minutiserrata Hayata = Carpinus kawakamii Hayata ●

77348 Carpinus mollicoma Hu;柔毛鹅耳枥(软毛鹅耳枥);Puberulous Hornbeam, Softhair Hornbeam, Yellowpubescent Hornbeam, Yellow-pubescent Hornbeam ●

77349 Carpinus mollis Rehder = Carpinus cordata Blume var. mollis (Rehder) W. C. Cheng ex F. H. Chen ●

77350 Carpinus monbeigiana Hand.-Mazz.;云南鹅耳枥(滇鹅耳枥,丽江鹅耳枥,满氏鹅耳枥,密穗鹅耳枥,维西鹅耳枥);Monbeig Hornbeam, Yunnan Hornbeam ●

77351 Carpinus monbeigiana Hand.-Mazz. var. weisiensis Hu = Carpinus londoniana H. Winkl. var. lanceolata (Hand.-Mazz.) P. C. Li ●

77352 Carpinus monbeigiana Hand.-Mazz. var. weisiensis Hu = Carpinus monbeigiana Hand.-Mazz. ●

77353 Carpinus oblongifolia (Hu) Hu et W. C. Cheng;宝华鹅耳枥(长椭圆鹅耳枥,矩圆叶鹅耳枥);Baohuashan Hornbeam, Oblongleaf Hornbeam ●

77354 Carpinus oblongifolia (Hu) Hu et W. C. Cheng = Carpinus hupeana Hu ●

77355 Carpinus oblongifolia Hu et W. C. Cheng = Carpinus oblongifolia (Hu) Hu et W. C. Cheng ●

77356 Carpinus obovatifolia Hu;倒卵叶鹅耳枥;Obovateleaf Hornbeam ●
77357 Carpinus obovatifolia Hu = Carpinus tschonoskii Maxim. ●
77358 Carpinus omeiensis Hu;峨眉鹅耳枥(白木,千金榆);Emei Hornbeam,Omei Mountain Hornbeam ●
77359 Carpinus orientalis Mill.;南欧鹅耳枥(东方鹅耳枥);Easrern Hornbeam,Oriental Hornbeam,Turkish Hornbeam ●☆
77360 Carpinus paohsingensis Hu = Carpinus tschonoskii Maxim. ●
77361 Carpinus paoshingensis W. Y. Hsai = Carpinus tschonoskii Maxim. ●
77362 Carpinus parva Hu = Carpinus pubescens Burkill ●
77363 Carpinus paxii Kük. = Carpinus turczaninowii Hance ●
77364 Carpinus pilosinucula Hu = Carpinus pubescens Burkill ●
77365 Carpinus pinfaensis H. Lév. et Vaniot = Carpinus pubescens Burkill ●
77366 Carpinus pinfaensis Hu = Carpinus pubescens Burkill ●
77367 Carpinus pingpienensis Hu = Carpinus pubescens Burkill ●
77368 Carpinus poilanei A. Camus = Carpinus londoniana H. Winkl. ●
77369 Carpinus polyneura Franch.;多脉鹅耳枥(角枥木,岩刷子);Manynerve Hornbeam,Multinerved Hornbeam ●
77370 Carpinus polyneura Franch. var. glandulosopunctata C. J. Qi = Carpinus glanduloso-punctata (C. J. Qi) C. J. Qi ●
77371 Carpinus polyneura Franch. var. sungpanensis (W. Y. Hsai) P. C. Li = Carpinus sungpanensis W. Y. Hsai ●
77372 Carpinus polyneura Franch. var. tsunyihensis (Hu) P. C. Li = Carpinus tsunyihensis Hu ●
77373 Carpinus pubescens Burkill;云贵鹅耳枥(毕节鹅耳枥,毛鹅耳枥,小鹅耳枥);Pubescent Hornbeam ●
77374 Carpinus pubescens Burkill var. bigiehensis Hu = Carpinus firmifolia (H. Winkl.) Hu ●
77375 Carpinus pubescens Burkill var. bigiehensis Hu = Carpinus pubescens Burkill var. firmifolia (H. Winkl.) Hu ex P. C. Li ●
77376 Carpinus pubescens Burkill var. bigiehensis Hu = Carpinus pubescens Burkill ●
77377 Carpinus pubescens Burkill var. firmifolia (H. Winkl.) Hu ex P. C. Li = Carpinus firmifolia (H. Winkl.) Hu ●
77378 Carpinus pubescens Burkill var. firmifolia (H. Winkl.) Hu ex P. C. Li = Carpinus pubescens Burkill ●
77379 Carpinus purpurinervis Hu;紫脉鹅耳枥(鹅耳枥,肥喉);Purplenerved Hornbeam ●
77380 Carpinus putoensis W. C. Cheng;普陀鹅耳枥;Puto Hornbeam,Putuo Hornbeam ●◇
77381 Carpinus rankanensis Hayata;兰邯千金榆(兰嵌鹅耳枥,兰嵌千金榆,台湾鹅耳枥);Lanhan Hornbeam,Rankan Hornbeam ●
77382 Carpinus rankanensis Hayata var. matsudae Gogelein;窄苞千金榆(矩果鹅耳枥,兰邯千金榆,细叶兰邯千金榆,窄苞鹅耳枥);Matsuda Hornbeam ●
77383 Carpinus rupestris A. Camus;岩生鹅耳枥(岩鹅耳枥);Rockliving Hornbeam,Rupestrine Hornbeam,Saxicolous Hornbeam ●
77384 Carpinus seemeniana Diels = Carpinus pubescens Burkill ●
77385 Carpinus seemeniana Hu = Carpinus stipulata H. Winkl. ●
77386 Carpinus sekii Gogelein;关氏鹅耳枥;Taiwan Hornbeam ●
77387 Carpinus sekii Gogelein = Carpinus kawakamii Hayata ●
77388 Carpinus shensiensis Hu;陕西鹅耳枥(陕鹅耳枥);Shaanxi Hornbeam,Shensi Hornbeam ●◇
77389 Carpinus shensiensis Hu var. paucineura S. C. Qu et K. Y. Wang;少脉鹅耳枥;Few-veined Shaanxi Hornbeam ●
77390 Carpinus shimenensis C. J. Qi;石门鹅耳枥;Shimen Hornbeam ●
77391 Carpinus sichourensis Hu = Carpinus tsaiana Hu ●
77392 Carpinus simplicidentata Hu = Carpinus stipulata H. Winkl. ●
77393 Carpinus stipulata H. Winkl.;单齿鹅耳枥(小叶鹅耳枥);Simpledental Hornbeam,Simpletoothed Hornbeam ●
77394 Carpinus stunyihensis Hu = Carpinus tsunyihensis Hu ●
77395 Carpinus sungpanensis W. Y. Hsai;松潘鹅耳枥;Songpan Hornbeam ●
77396 Carpinus tanakaeana Makino = Carpinus turczaninowii Hance ●
77397 Carpinus tehchingensis Hu;德钦鹅耳枥;Deqin Hornbeam ●
77398 Carpinus tehchingensis Hu = Carpinus viminea Wall. ●
77399 Carpinus tientaiensis W. C. Cheng;天台鹅耳枥;Tiantai Hornbeam,Tiantaishan Hornbeam ●◇
77400 Carpinus tsaiana Hu;宽苞鹅耳枥(蔡氏鹅耳枥,大扫把栗);H. T. Tsai Hornbeam,Tsai Hornbeam ●
77401 Carpinus tschonoskii Maxim.;昌化鹅耳枥(宝兴鹅耳枥,昌化枥,镰苞鹅耳枥);Changhua Hornbeam,Tschonosk Hornbeam,Yeddo Hornbeam ●
77402 Carpinus tschonoskii Maxim. f. pendula Hayashi;垂枝昌化鹅耳枥●☆
77403 Carpinus tschonoskii Maxim. var. eximia (Nakai) Hatus. = Carpinus eximia Nakai ●☆
77404 Carpinus tschonoskii Maxim. var. falcatibracteata (Hu) P. C. Li = Carpinus tschonoskii Maxim. ●
77405 Carpinus tschonoskii Maxim. var. henryana H. Winkl. = Carpinus henryana H. Winkl. ●
77406 Carpinus tsiangiana Hu = Carpinus pubescens Burkill ●
77407 Carpinus tsoongiana Hu = Carpinus pubescens Burkill ●
77408 Carpinus tsunyihensis Hu;遵义鹅耳枥;Zunyi Hornbeam ●
77409 Carpinus tuczaninowii Hance var. chungnanensis P. C. Kuo = Carpinus turczaninowii Hance ●
77410 Carpinus tungtzeensis Hu = Carpinus pubescens Burkill ●
77411 Carpinus turczaninowii Hance;鹅耳枥(北鹅耳枥,杜氏鹅耳枥,见风干,千金榆,穗子榆,土姜树,小叶鹅耳枥,周氏鹅耳枥);Chinese Hornbeam,Turczaninow Hornbeam ●
77412 Carpinus turczaninowii Hance var. chungnanensis P. C. Kuo = Carpinus turczaninowii Hance ●
77413 Carpinus turczaninowii Hance var. firmifolia H. Winkl. = Carpinus firmifolia (H. Winkl.) Hu ●
77414 Carpinus turczaninowii Hance var. firmifolia H. Winkl. = Carpinus pubescens Burkill var. firmifolia (H. Winkl.) Hu ex P. C. Li ●
77415 Carpinus turczaninowii Hance var. oblongifolia Hu = Carpinus hupeana Hu ●
77416 Carpinus turczaninowii Hance var. oblongifolia Hu = Carpinus oblongifolia (Hu) Hu et W. C. Cheng ●
77417 Carpinus turczaninowii Hance var. polyneura (Franch.) H. Winkl. = Carpinus polyneura Franch. ●
77418 Carpinus turczaninowii Hance var. stipulata (H. Winkl.) H. Winkl. = Carpinus turczaninowii Hance ●
77419 Carpinus turczaninowii Hance var. stipulata (H. Winkl.) H. Winkl. = Carpinus stipulata H. Winkl. ●
77420 Carpinus viminea Wall.;大穗鹅耳枥(贡山鹅耳枥,牯岭鹅耳枥,雷公鹅耳枥,雷公枥,雷公椖,疏果鹅耳枥,细丝枥,岩刷子);Big-spiked Hornbeam,Chinese Hornbeam,Himalayan Hornbeam,Largespike Hornbeam,Thunder God Hornbeam,Vimineous Hornbeam ●
77421 Carpinus viminea Wall. var. chiukiangensis Hu;贡山鹅耳枥;Gongshan Hornbeam,Himalayan Hornbeam,Kiukiang Hornbeam ●

77422 Carpinus viminea Wall. var. chungnanensis P. C. Kuo;终南鹅耳枥;Zhongnan Hornbeam ●
77423 Carpinus virginiana (Marshall) Sudw. = Carpinus caroliniana Walter subsp. virginiana (Marshall) Furlow ●☆
77424 Carpinus virginiana Mill. = Ostrya virginiana (Mill.) K. Koch ●☆
77425 Carpinus wangii Hu et W. C. Cheng = Carpinus pubescens Burkill ●
77426 Carpinus wilsoniana Hu = Carpinus fangiana Hu ●
77427 Carpinus yedoensis Maxim. = Carpinus tschonoskii Maxim. ●
77428 Carpiphea Raf. = Cordia L.(保留属名)●
77429 Carpobrotus N. E. Br. (1925);佛手掌属(果食草属,食用昼花属,松叶菊属);Ice Plant, Hottentot-fig, Fig-marigold ●☆
77430 Carpobrotus acinaciformis (L.) L. Bolus;长刀佛手掌(长刀日中花);Gooseberry Fig, Sally-my-handsome, Sour-fig ●☆
77431 Carpobrotus chilensis (Molina) N. E. Br.;智利佛手掌(智利松叶菊);Hottentot Fig, Sea Fig, Sea-fig ●☆
77432 Carpobrotus concavus L. Bolus = Carpobrotus acinaciformis (L.) L. Bolus ●☆
77433 Carpobrotus deliciosus (L. Bolus) L. Bolus;姣美佛手掌●☆
77434 Carpobrotus dimidiatus (Haw.) L. Bolus;半片佛手掌●☆
77435 Carpobrotus dulcis L. Bolus = Carpobrotus deliciosus (L. Bolus) L. Bolus ●☆
77436 Carpobrotus edulis (L.) L. Bolus = Carpobrotus edulis (L.) N. E. Br. ●☆
77437 Carpobrotus edulis (L.) L. Bolus subsp. parviflorus Wisura et Glen;小花佛手掌(小花松叶菊)●☆
77438 Carpobrotus edulis (L.) L. Bolus var. chrysophthalmus C. D. Preston et P. D. Sell = Carpobrotus edulis (L.) N. E. Br. ●☆
77439 Carpobrotus edulis (L.) L. Bolus var. rubescens Druce;淡红佛手掌(淡红松叶菊);Hottentot Fig, Hottentot-fig ●☆
77440 Carpobrotus edulis (L.) L. Bolus var. rubescens Druce = Carpobrotus edulis (L.) N. E. Br. ●☆
77441 Carpobrotus edulis (L.) N. E. Br.;佛手掌(短剑,果食草,莫邪菊,食用昼花,松叶菊);Hottentot Fig, Hottentot-fig, Kaffir Fig, Pigface ●☆
77442 Carpobrotus edulis (L.) N. E. Br. = Mesembryanthemum edule L. ■
77443 Carpobrotus fourcadei L. Bolus;富尔卡德佛手掌(富尔卡德松叶菊)●☆
77444 Carpobrotus fourcadei L. Bolus = Carpobrotus deliciosus (L. Bolus) L. Bolus ●☆
77445 Carpobrotus fourcadei L. Bolus var. alba? = Carpobrotus deliciosus (L. Bolus) L. Bolus ●☆
77446 Carpobrotus glaucescens (Haw.) Schwantes;灰佛手掌(灰松叶菊);Angular Sea-fig, Pig Face ●☆
77447 Carpobrotus juritzii (L. Bolus) L. Bolus = Carpobrotus dimidiatus (Haw.) L. Bolus ●☆
77448 Carpobrotus laevigatus (Haw.) N. E. Br. = Carpobrotus acinaciformis (L.) L. Bolus ●☆
77449 Carpobrotus laevigatus (Haw.) Schwantes = Carpobrotus acinaciformis (L.) L. Bolus ●☆
77450 Carpobrotus mellei (L. Bolus) L. Bolus;梅勒佛手掌(梅勒松叶菊)●☆
77451 Carpobrotus muirii (L. Bolus) L. Bolus;缪里佛手掌(缪里松叶菊)●☆
77452 Carpobrotus pageae L. Bolus = Carpobrotus mellei (L. Bolus) L. Bolus ●☆
77453 Carpobrotus praecox (F. Muell.) G. D. Rowley;早佛手掌(早松叶菊)●☆
77454 Carpobrotus pulleinei J. M. Black;普氏佛手掌(普氏松叶菊)●☆
77455 Carpobrotus quadrifidus L. Bolus;四裂佛手掌●☆
77456 Carpobrotus quadrifidus L. Bolus f. rosea (L. Bolus) G. D. Rowley = Carpobrotus quadrifidus L. Bolus ●☆
77457 Carpobrotus rubrocinctus (Haw.) N. E. Br. = Carpobrotus acinaciformis (L.) L. Bolus ●☆
77458 Carpobrotus sauerae Schwantes = Carpobrotus quadrifidus L. Bolus ●☆
77459 Carpobrotus subalatus (Haw.) N. E. Br. = Carpobrotus acinaciformis (L.) L. Bolus ●☆
77460 Carpobrotus vanzijliae L. Bolus = Carpobrotus acinaciformis (L.) L. Bolus ●☆
77461 Carpocalymna Zipp. = Epithema Blume ■
77462 Carpoceras (DC.) Link = Thlaspi L. ■
77463 Carpoceras (DC.) Link(1831);角果菥蓂属■☆
77464 Carpoceras A. Rich. = Martynia L. ■
77465 Carpoceras Boiss. = Carpoceras (DC.) Link ■☆
77466 Carpoceras Boiss. = Thlaspi L. ■
77467 Carpoceras Link = Thlaspi L. ■
77468 Carpoceras brevistylum N. Busch;短柱肖菥蓂■☆
77469 Carpoceras ceratocarpum (Pall.) Busch;角果肖菥蓂■☆
77470 Carpoceras ceratocarpum (Pall.) Busch = Lepidium ceratocarpum Pall. ■☆
77471 Carpoceras griffithianum Boiss. = Thlaspi griffithianum (Boiss.) Boiss. ■☆
77472 Carpoceras hastulatum (Steven) Boiss.;戟形肖菥蓂■☆
77473 Carpoceras longiflora A. Rich. = Martynia diandra Gloxin ■
77474 Carpoceras stenocarpum Boiss.;窄果肖菥蓂■☆
77475 Carpodetaceae Fenzl = Brexiaceae Lindl. ●☆
77476 Carpodetaceae Fenzl = Escalloniaceae R. Br. ex Dumort.(保留科名)●
77477 Carpodetaceae Fenzl = Grossulariaceae DC.(保留科名)●
77478 Carpodetaceae Fenzl = Rousseaceae DC. ●☆
77479 Carpodetaceae Fenzl(1841);腕带花科●☆
77480 Carpodetes Herb. = Stenomesson Herb. ■☆
77481 Carpodetes recurvata Herb.;腕带花■☆
77482 Carpodetus J. R. Forst. et G. Forst. (1775);腕带花属(卡尔珀图属);Marble Leaf ●☆
77483 Carpodetus serratus J. R. Forst. et G. Forst.;齿叶腕带花(齿卡尔珀图,齿叶卡尔珀图);Putaputaweta ●☆
77484 Carpodinopsis Pichon = Pleiocarpa Benth. ●☆
77485 Carpodinopsis picralimoides Pichon = Pleiocarpa picralimoides (Pichon) Omino ●☆
77486 Carpodinopsis rostrata (Benth.) Pichon = Pleiocarpa rostrata Benth. ●☆
77487 Carpodinopsis talbotii (Wernham) Pichon = Pleiocarpa rostrata Benth. ●☆
77488 Carpodinopsis uniflora Pichon = Pleiocarpa rostrata Benth. ●☆
77489 Carpodinus R. Br. ex G. Don = Landolphia P. Beauv.(保留属名)●☆
77490 Carpodinus R. Br. ex Sabine = Landolphia P. Beauv.(保留属名)●☆
77491 Carpodinus acida Sabine = Landolphia dulcis (Sabine) Pichon ●☆
77492 Carpodinus alnifolia A. Chev. = Landolphia jumellei (Pierre ex Jum.) Pichon ●☆

77493 Carpodinus barteri Stapf = Landolphia dulcis (Sabine) Pichon ●☆
77494 Carpodinus baumannii Hutch. et Dalziel = Landolphia dulcis (Sabine) Pichon ●☆
77495 Carpodinus bequaertii De Wild. = Landolphia violacea (K. Schum. ex Hallier f.) Pichon ●☆
77496 Carpodinus bruneelii De Wild. = Landolphia bruneelii (De Wild.) Pichon ●☆
77497 Carpodinus camptoloba K. Schum. = Landolphia camptoloba (K. Schum.) Pichon ●☆
77498 Carpodinus chylorrhiza K. Schum. ex Stapf = Landolphia thollonii Dewèvre ●☆
77499 Carpodinus cirilus Guymer = Orthopichonia cirrhosa (Radlk.) H. Huber ●☆
77500 Carpodinus cirrhosa (Radlk.) Radlk. ex K. Schum. = Orthopichonia cirrhosa (Radlk.) H. Huber ●☆
77501 Carpodinus cirrosa Radlk. ex A. Chev. = Orthopichonia cirrhosa (Radlk.) H. Huber ●☆
77502 Carpodinus congolensis Stapf = Landolphia congolensis (Stapf) Pichon ●☆
77503 Carpodinus decipiens Pierre = Landolphia uniflora (Stapf) Pichon ●☆
77504 Carpodinus dulcis Sabine = Landolphia dulcis (Sabine) Pichon ●☆
77505 Carpodinus eetveldeana De Wild. et Gentil = Landolphia foretiana (Pierre ex Jum.) Pichon ●☆
77506 Carpodinus exserens K. Schum. = Landolphia incerta (K. Schum.) Pers. ●☆
77507 Carpodinus flava Pierre = Landolphia dulcis (Sabine) Pichon ●☆
77508 Carpodinus flavidiflora K. Schum. = Landolphia flavidiflora (K. Schum.) Pers. ●☆
77509 Carpodinus foretiana Pierre ex Jum. = Landolphia foretiana (Pierre ex Jum.) Pichon ●☆
77510 Carpodinus friabilis Pierre = Landolphia subrepanda (K. Schum.) Pichon ●☆
77511 Carpodinus fulva Pierre ex Hallier f. = Landolphia jumellei (Pierre ex Jum.) Pichon ●☆
77512 Carpodinus gentilii De Wild. = Landolphia villosa Pers. ●☆
77513 Carpodinus glabra Pierre ex Stapf = Landolphia glabra (Pierre ex Stapf) Pichon ●☆
77514 Carpodinus glandulosa Pellegr. = Landolphia glandulosa (Pellegr.) Pichon ●☆
77515 Carpodinus globulifera K. Schum. = Landolphia jumellei (Pierre ex Jum.) Pichon ●☆
77516 Carpodinus goosweileri A. Chev. = Landolphia gossweileri (Stapf) Pichon ●☆
77517 Carpodinus gossweileri Stapf = Landolphia gossweileri (Stapf) Pichon ●☆
77518 Carpodinus gracilis Stapf = Landolphia camptoloba (K. Schum.) Pichon ●☆
77519 Carpodinus hirsuta Hua = Landolphia hirsuta (Hua) Pichon ●☆
77520 Carpodinus incerta K. Schum. = Landolphia incerta (K. Schum.) Pers. ●☆
77521 Carpodinus jesperseni De Wild. = Landolphia ligustrifolia (Stapf) Pichon ●☆
77522 Carpodinus jumellei Pierre ex Jum. = Landolphia jumellei (Pierre ex Jum.) Pichon ●☆
77523 Carpodinus klaineana Pierre = Landolphia foretiana (Pierre ex Jum.) Pichon ●☆
77524 Carpodinus klainei Pierre ex Stapf = Landolphia foretiana (Pierre ex Jum.) Pichon ●☆
77525 Carpodinus lanceolata K. Schum. = Landolphia lanceolata (K. Schum.) Pichon ●☆
77526 Carpodinus lanceolata K. Schum. var. angustifolia A. Chev. = Landolphia lanceolata (K. Schum.) Pichon ●☆
77527 Carpodinus lanceolata K. Schum. var. latifolia A. Chev. = Landolphia lanceolata (K. Schum.) Pichon ●☆
77528 Carpodinus landolphioides (Hallier f.) Stapf = Landolphia landolphioides (Hallier f.) A. Chev. ●☆
77529 Carpodinus laxiflora K. Schum. = Landolphia incerta (K. Schum.) Pers. ●☆
77530 Carpodinus leptantha K. Schum. = Landolphia leptantha (K. Schum.) Pers. ●☆
77531 Carpodinus leptantha Stapf = Landolphia glabra (Pierre ex Stapf) Pichon ●☆
77532 Carpodinus leucantha K. Schum. = Landolphia camptoloba (K. Schum.) Pichon ●☆
77533 Carpodinus ligustifolia Stapf ex Pellegr. = Landolphia ligustrifolia (Stapf) Pichon ●☆
77534 Carpodinus ligustrifolia Stapf = Landolphia ligustrifolia (Stapf) Pichon ●☆
77535 Carpodinus ligustrifolia Stapf var. angusta De Wild. = Landolphia ligustrifolia (Stapf) Pichon ●☆
77536 Carpodinus littoralis A. Chev. = Landolphia dulcis (Sabine) Pichon ●☆
77537 Carpodinus macrantha K. Schum. = Landolphia macrantha (K. Schum.) Pichon ●☆
77538 Carpodinus macrophylla A. Chev. = Dictyophleba leonensis (Stapf) Pichon ●☆
77539 Carpodinus maximus K. Schum. ex Hallier f. = Landolphia maxima (K. Schum. ex Hallier f.) Pichon ●☆
77540 Carpodinus myriantha K. Schum. = Landolphia robustior (K. Schum.) Pers. ●☆
77541 Carpodinus nigerina A. Chev. = Dictyophleba leonensis (Stapf) Pichon ●☆
77542 Carpodinus oocarpa Stapf = Landolphia dulcis (Sabine) Pichon ●☆
77543 Carpodinus oxyanthoides Wernham = Landolphia congolensis (Stapf) Pichon ●☆
77544 Carpodinus parviflora Stapf = Landolphia dulcis (Sabine) Pichon ●☆
77545 Carpodinus parvoflurus Guymer = Landolphia dulcis (Sabine) Pichon ●☆
77546 Carpodinus pauciflora K. Schum. = Landolphia dulcis (Sabine) Pichon ●☆
77547 Carpodinus rufescens De Wild. = Landolphia rufescens (De Wild.) Pichon ●☆
77548 Carpodinus rufinervis Pierre ex Stapf = Landolphia foretiana (Pierre ex Jum.) Pichon ●☆
77549 Carpodinus sassandrae A. Chev. = Landolphia foretiana (Pierre ex Jum.) Pichon ●☆
77550 Carpodinus schlechteri K. Schum. ex Stapf = Landolphia congolensis (Stapf) Pichon ●☆
77551 Carpodinus schlecteri K. Schum. ex Pichon = Landolphia congolensis (Stapf) Pichon ●☆
77552 Carpodinus subrepanda K. Schum. = Landolphia subrepanda (K. Schum.) Pichon ●☆

77553 Carpodinus talbotii Wernham = Landolphia stenogyna Pichon ●☆
77554 Carpodinus tenuifolia Pierre ex Stapf = Landolphia dulcis (Sabine) Pichon ●☆
77555 Carpodinus trichanthera Pierre ex Stapf = Landolphia subrepanda (K. Schum.) Pichon ●☆
77556 Carpodinus turbinata Stapf = Landolphia congolensis (Stapf) Pichon ●☆
77557 Carpodinus umbellata K. Schum. = Hunteria umbellata (K. Schum.) Hallier f. ●☆
77558 Carpodinus uniflora Stapf = Landolphia uniflora (Stapf) Pichon ●☆
77559 Carpodinus utilis A. Chev. = Landolphia utilis (A. Chev.) Pichon ●☆
77560 Carpodinus verticillata De Wild. = Landolphia foretiana (Pierre ex Jum.) Pichon ●☆
77561 Carpodinus violaceus K. Schum. ex Hallier f. = Landolphia violacea (K. Schum. ex Hallier f.) Pichon ●☆
77562 Carpodinus watsoniana (Roxb.) Vogtherr = Landolphia watsoniana Roxb. ●☆
77563 Carpodiptera Griseb. (1860);双翅果属●☆
77564 Carpodiptera Griseb. = Berrya Roxb. (保留属名)●
77565 Carpodiptera africana Mast.;非洲双翅果●☆
77566 Carpodiptera sansibarensis Burret = Carpodiptera africana Mast. ●☆
77567 Carpodiptera schomburgkii Baill. = Christiana africana DC. ●☆
77568 Carpodontos Labill. = Eucryphia Cav. ●☆
77569 Carpolepis (J. W. Dawson) J. W. Dawson = Metrosideros Banks ex Gaertn. (保留属名)●☆
77570 Carpoliza Steud. = Carpolyza Salisb. ■☆
77571 Carpolobia G. Don(1831);片果远志属●☆
77572 Carpolobia afzeliana Oliv. = Atroxima afzeliana (Oliv.) Stapf ●☆
77573 Carpolobia alba G. Don;白片果远志●☆
77574 Carpolobia caudata Burtt Davy = Carpolobia lutea G. Don ●☆
77575 Carpolobia conradsiana Engl. = Carpolobia goetzei Gürke ●☆
77576 Carpolobia delvauxii E. M. Petit = Carpolobia alba G. Don ●☆
77577 Carpolobia glabrescens Hutch. et Dalziel = Carpolobia alba G. Don ●☆
77578 Carpolobia goetzei Gürke;高氏片果远志●☆
77579 Carpolobia gossweileri (Exell) E. M. Petit;高斯片果远志●☆
77580 Carpolobia leandriana (Desc.) Breteler = Carpolobia goetzei Gürke ●☆
77581 Carpolobia lutea G. Don;黄片果远志●☆
77582 Carpolobia suaveolens Meikle = Carpolobia goetzei Gürke ●☆
77583 Carpolobium Post et Kuntze = Carpolobia G. Don ●☆
77584 Carpolyza Salisb. (1807);口果石蒜属(口果属)■☆
77585 Carpolyza spiralis (L'Hér.) Salisb.;口果石蒜■☆
77586 Carpolyza spiralis (L'Hér.) Salisb. = Strumaria spiralis L'Hér. ■☆
77587 Carponema (DC.) Eckl. et Zeyh. = Heliophila Burm. f. ex L. ●■☆
77588 Carponema Eckl. et Zeyh. = Heliophila Burm. f. ex L. ●■☆
77589 Carponema aggregata Eckl. et Zeyh. = Heliophila digitata L. f. ■☆
77590 Carponema filiforme (L. f.) Eckl. et Zeyh. = Heliophila coronopifolia L. ■☆
77591 Carpophillus Neck. = Pereskia Mill. ●
77592 Carpophora Klotzsch = Silene L. (保留属名)■
77593 Carpophyllum Miq. = Sterculia L. ●
77594 Carpophyllum Neck. = Carpophillus Neck. ●
77595 Carpophyllum Neck. = Pereskia Mill. ●
77596 Carpopodium (DC.) Eckl. et Zeyh. = Heliophila Burm. f. ex L. ●■☆
77597 Carpopodium Eckl. et Zeyh. = Heliophila Burm. f. ex L. ●■☆
77598 Carpopodium carnosum Eckl. et Zeyh. = Heliophila brachycarpa Meisn. ■☆
77599 Carpopogon Roxb. = Mucuna Adans. (保留属名)●■
77600 Carpopogon bracteatum Roxb. = Mucuna bracteata DC. ex Kurz ●
77601 Carpopogon capitatum Roxb. = Mucuna pruriens (L.) DC. var. utilis (Wall. ex Wight) Baker ex Burck ■
77602 Carpopogon niveum Roxb. = Mucuna pruriens (L.) DC. var. utilis (Wall. ex Wight) Baker ex Burck ■
77603 Carpothalis E. Mey. = Kraussia Harv. ●☆
77604 Carpothalis E. Mey. = Tricalysia A. Rich. ex DC. ●
77605 Carpotheca Tamamshyan = Echinophora L. ■☆
77606 Carpotriche Rchb. = Carpotroche Endl. ●☆
77607 Carpotroche Endl. (1839);轮果大风子属●☆
77608 Carpotroche amazonica Mart. ex Eichler;亚马逊轮果大风子●☆
77609 Carpotroche angustifolia Pittier;窄叶轮果大风子●☆
77610 Carpotroche brasiliensis Endl.;巴西轮果大风子●☆
77611 Carpotroche grandiflora Spruce ex Eichler;大花轮果大风子●☆
77612 Carpotroche integrifolia Kuhlm.;全缘叶轮果大风子●☆
77613 Carpotroche laxiflora Benth.;疏花轮果大风子●☆
77614 Carpotroche mollis J. F. Macbr.;软轮果大风子●☆
77615 Carpotroche parvifolia J. F. Macbr.;小叶轮果大风子●☆
77616 Carpotroche platyptera Pittier;宽翅轮果大风子●☆
77617 Carpoxis Raf. = Forestiera Poir. (保留属名)●☆
77618 Carpoxylon H. Wendl. et Drude(1875);木果椰属(硬果椰属)●☆
77619 Carpoxylon macrospermum H. Wendl. et Drude;木果椰●☆
77620 Carprifolium elisae Kuntze = Lonicera elisae Franch. ●
77621 Carprifolium ligustrinum Kuntze = Lonicera ligustrina Wall. ●
77622 Carptotepala Moldenke = Syngonanthus Ruhland ■☆
77623 Carpunya C. Presl = Piper L. ●■
77624 Carpupica Raf. = Piper L. ●■
77625 Carradoria A. DC. = Globularia L. ●☆
77626 Carramboa Cuatrec. (1976);巨叶菊属●☆
77627 Carramboa Cuatrec. = Espeletia Mutis ex Humb. et Bonpl. ●☆
77628 Carramboa pittieri (Cuatrec.) Cuatrec.;巨叶菊●☆
77629 Carregnoa Boiss. = Braxireon Raf. ■
77630 Carregnoa Boiss. = Tapeinanthus Herb. (废弃属名)■
77631 Carregnoa humilis J. Gay = Narcissus cavanillesii Barra et G. López ■☆
77632 Carria Gardner = Gordonia J. Ellis(保留属名)●
77633 Carria V. P. Castro et K. G. Lacerda = Oncidium Sw. (保留属名)■☆
77634 Carrichtera Adans. (废弃属名) = Carrichtera DC. (保留属名)■☆
77635 Carrichtera Adans. (废弃属名) = Vella L. ●☆
77636 Carrichtera DC. (1821)(保留属名);杓喙芥属■☆
77637 Carrichtera Post et Kuntze = Caricteria Scop. ■●
77638 Carrichtera Post et Kuntze = Corchorus L. ■●
77639 Carrichtera annua (L.) DC.;杓喙芥■☆
77640 Carrichtera vellae DC. = Carrichtera annua (L.) DC. ■☆
77641 Carrichteria Wittst. = Carrichtera DC. (保留属名)■☆
77642 Carriella V. P. Castro et K. G. Lacerda = Oncidium Sw. (保留属名)■☆
77643 Carrierea Franch. (1896);山羊角树属(山羊角属);Carrierea, Goathorntree ●
77644 Carrièrea calycina Franch.;山羊角树(红木,嘉丽树,嘉利树,山丁木,山羊果);Calyxshaped Carrierea, Calyx-shaped Carrierea, Calyxshaped Goathorntree ●

77645 Carrièrea dunniana H. Lév. ;云贵山羊角树(贵州嘉丽树,贵州山羊角树);Dunn Goathorntree,Dunn Hornbeam ●
77646 Carrierea rehderiana Sleumer = Carrierea calycina Franch. ●
77647 Carrierea rehderiana Sleumer = Carrierea dunniana H. Lév. ●
77648 Carrietea vieillardii Gagnep. = Itoa orientalis Hemsl. ●
77649 Carrissoa Baker f. (1933);安哥拉雀脷珠属■☆
77650 Carrissoa angolensis Baker f. ;安哥拉雀脷珠■☆
77651 Carroa C. Presl = Dalea L. (保留属名)●■☆
77652 Carroa C. Presl = Marina Liebm. ■☆
77653 Carroa C. Presl = Trichopodium C. Presl ■☆
77654 Carronia F. Muell. (1875);卡罗藤属●☆
77655 Carronia multisepalea F. Muell. ;卡罗藤●☆
77656 Carruanthus (Schwantes) Schwantes(1927);菊波属■☆
77657 Carruanthus Schwantes = Carruanthus (Schwantes) Schwantes ■☆
77658 Carruanthus Schwantes ex N. E. Br. = Carruanthus (Schwantes) Schwantes ■☆
77659 Carruanthus caninus (Haw.) Schwantes = Carruanthus ringens (L.) Boom ■☆
77660 Carruanthus ringens (L.) Boom;菊波■☆
77661 Carruanthus vulpinus N. E. Br. ex Graebn. = Carruanthus ringens (L.) Boom ■☆
77662 Carruthersia Seem. (1866);卡竹桃属●☆
77663 Carruthersia axilliflora Merr. ;腋花卡竹桃●☆
77664 Carruthersia glabra D. J. Middleton;光卡竹桃●☆
77665 Carruthersia hirsuta Elmer;粗毛卡竹桃●☆
77666 Carruthersia laevis Elmer;平滑卡竹桃●☆
77667 Carruthersia latifolia Gillespie;宽叶卡竹桃●☆
77668 Carruthersia scandens Seem. ;卡竹桃●☆
77669 Carruthia Kuntze = Nymania Lindb. ●☆
77670 Carsonia Greene = Cleome L. ●■
77671 Cartalinia Szov. ex Kunth = Paris L. ■
77672 Carterella Terrell(1987);卡特茜属●☆
77673 Carterella alexanderae (A. Carter) Terrell;卡特茜●☆
77674 Carteretia A. Rich. = Acampe Lindl. (保留属名)■
77675 Carteretia A. Rich. = Cleisostoma Blume ■
77676 Carteretia A. Rich. = Sarcanthus Lindl. (废弃属名)■
77677 Carteria Small = Basiphyllaea Schltr. ■☆
77678 Carteria corallicola Small = Basiphyllaea corallicola (Small) Ames ■☆
77679 Carteruthamnus R. M. King = Hofmeisteria Walp. ■●☆
77680 Cartesia Cass. = Stokesia L'Hér. ■☆
77681 Carthamnodes Kuntze = Carduncellus Adans. ■☆
77682 Carthamodes Kuntze = Carduncellus Adans. ■☆
77683 Carthamoides Wolf = Carduus L. + Carduncellus Adans. + Carthamus L. + Centaurea L. + Cnicus L. (保留属名)■●
77684 Carthamus L. (1753);红花属(红蓝花属);Distaff Thistle,False Saffron,Safflower ■
77685 Carthamus albus Desf. ;白红花■☆
77686 Carthamus ambiguus Heldr. ex Halácsy;欧洲红花■☆
77687 Carthamus arborescens L. ;树红花■☆
77688 Carthamus atractyloides (Pomel) Greuter;苍术红花■☆
77689 Carthamus baeticus Boiss. et Reut. = Carthamus creticus L. ■☆
77690 Carthamus brasiliensis Vell. ;巴西红花■☆
77691 Carthamus caeruleus L. ;天蓝红花■☆
77692 Carthamus caeruleus L. subsp. tingitanus (L.) Batt. ;丹吉尔红花■☆
77693 Carthamus caeruleus L. var. dentatus (DC.) Rouy = Carthamus caeruleus L. ■☆
77694 Carthamus caeruleus L. var. tingitanus (L.) Ball = Carthamus caeruleus L. ■☆
77695 Carthamus calvus (Boiss. et Reut.) Batt. ;光秃红花■☆
77696 Carthamus calvus (Boiss. et Reut.) Batt. subsp. carlinoides (Pomel) Batt. = Carthamus calvus (Boiss. et Reut.) Batt. ■☆
77697 Carthamus calvus (Boiss. et Reut.) Batt. subsp. depauperatus (Pomel) Batt. = Carthamus calvus (Boiss. et Reut.) Batt. ■☆
77698 Carthamus calvus (Boiss. et Reut.) Batt. var. depauperatus (Pomel) Quézel et Santa = Carthamus calvus (Boiss. et Reut.) Batt. ■☆
77699 Carthamus calvus (Boiss. et Reut.) Batt. var. glaucescens (Faure et Maire) Maire = Carthamus calvus (Boiss. et Reut.) Batt. ■☆
77700 Carthamus canescens Sol. ;灰红花■☆
77701 Carthamus carthamoides (Pomel) Batt. ;染色红花■☆
77702 Carthamus chouletteanus (Pomel) Greuter;舒莱红花■☆
77703 Carthamus creticus L. ;光滑红花;Smooth Distaff Thistle ■☆
77704 Carthamus dentatus Vahl;齿红花■☆
77705 Carthamus divaricatus Bég. et Vacc. = Carthamus lanatus L. ■
77706 Carthamus eriocephalus (Boiss.) Greuter;毛头红花■☆
77707 Carthamus fruticosus Maire;灌丛红花■☆
77708 Carthamus glaber Burm. f. ;光红花■☆
77709 Carthamus glaucescens Faure et Maire = Carthamus calvus (Boiss. et Reut.) Batt. ■☆
77710 Carthamus glaucus M. Bieb. ;灰绿红花■☆
77711 Carthamus glaucus M. Bieb. subsp. alexandrinus (Boiss. et Heldr.) Hanelt;亚历山大红花■☆
77712 Carthamus glaucus M. Bieb. var. alexandrinus (Boiss. et Heldr.) Boiss. = Carthamus glaucus M. Bieb. subsp. alexandrinus (Boiss. et Heldr.) Hanelt ■☆
77713 Carthamus helenoides Desf. ;拟海伦娜红花■☆
77714 Carthamus ilicifolius (Pomel) Greuter;冬青叶红花■☆
77715 Carthamus laevis Hill = Stokesia laevis (Hill) Greene ■☆
77716 Carthamus lanatus L. ;毛红花(绵毛红花);Distaff Thistle, Downy Safflower, Hair Safflower, Hairy Safflower, Saffron Thistle, Woolly Distaff Thistle ■
77717 Carthamus lanatus L. subsp. baeticus (Boiss. et Reut.) Nyman = Carthamus creticus L. ■☆
77718 Carthamus lanatus L. subsp. creticus (L.) Holmboe = Carthamus creticus L. ■☆
77719 Carthamus lanatus L. subsp. montanus (Pomel) Batt. ;山地毛红花■☆
77720 Carthamus lanatus L. var. abyssinicus (A. Rich.) Sch. Bip. ex Schweinf. = Carthamus lanatus L. ■
77721 Carthamus lanatus L. var. algeriensis Batt. = Carthamus lanatus L. ■
77722 Carthamus lanatus L. var. divaricatus (Bég. et Vacc.) Pamp. = Carthamus lanatus L. ■
77723 Carthamus lanatus L. var. gracilis Schweinf. = Carthamus nitidus Boiss. ■☆
77724 Carthamus lanatus L. var. longifolius Pamp. = Carthamus lanatus L. ■
77725 Carthamus leucocauloides Schweinf. = Carthamus nitidus Boiss. ■☆
77726 Carthamus leucocaulos Sibth. et Sm. ;白茎红花;Whitestem Distaff Thistle,White-stem Distaff Thistle ■☆
77727 Carthamus linearifolius Molina;线叶红花■☆
77728 Carthamus lucens (Ball) Greuter = Carthamus pinnatus Desf. ■☆

77729 Carthamus maculatus (Scop.) Lam. = Silybum marianum (L.) Gaertn. ■
77730 Carthamus mareoticus Delile;马雷奥特红花■☆
77731 Carthamus multifidus Desf.;多裂红花■☆
77732 Carthamus multifidus Desf. var. simplex Maire = Carthamus multifidus Desf. ■☆
77733 Carthamus nitidus Boiss.;亮红花■☆
77734 Carthamus oxyacanthus M. Bieb.;尖红花;Jeweled Distaff Thistle, Wild Safflower ■☆
77735 Carthamus pectinatus Desf.;篦状红花■☆
77736 Carthamus pectinatus Desf. var. maroccanus Faure et Maire = Carthamus pectinatus Desf. ■☆
77737 Carthamus persicus Willd.;波斯红花■☆
77738 Carthamus pinnatus Desf.;羽裂红花■☆
77739 Carthamus pinnatus Desf. subsp. lucens (Ball) Dobignard;光亮羽裂红花■☆
77740 Carthamus plumosus (Pomel) Greuter;羽状红花■☆
77741 Carthamus pomelianus (Batt.) Batt.;波梅尔红花■☆
77742 Carthamus reboudianus Batt.;雷博红花■☆
77743 Carthamus rhaponticoides (Pomel) Greuter;缘膜菊红花■☆
77744 Carthamus rhiphaeus Font Quer et Pau;山红花■☆
77745 Carthamus strictus (Pomel) Batt.;刚直红花■☆
77746 Carthamus tenuis (Boiss. et Blanche) Bornm.;细红花■☆
77747 Carthamus tenuis (Boiss. et Blanche) Bornm. subsp. foliosus Hanelt;多叶细红花■☆
77748 Carthamus tinctorius L.;红花(草红花,刺红花,大红花,丹花,杜红花,红花菜,红花草,红花毛,红兰,红兰花,红蓝,红蓝花,怀红花,淮红花,黄蓝,黄蓝花,药花,摘花);Bastard Saffron, Carthamine, Catalonia Saffron, Distaff Thistle, Dyeing Carthamus, Dyer's Saffron, False Saffron, Kurdee, Mock Saffron, Parrot-seed, Rouge Plant, Safflor, Safflower, Saffron Thistle, Spanish Saffron, Stokes' Aster, Wild Saffron ■
77749 Carthamus tinctorius L. var. spinosus Kitam.;刺红花■☆
77750 Carthamus tingitanus L. = Carthamus caeruleus L. subsp. tingitanus (L.) Batt. ■☆
77751 Cartiera Greene = Streptanthus Nutt. ■☆
77752 Cartiera Greene(1906);卡蒂芥属■☆
77753 Cartiera cordata (Nutt.) Greene;卡蒂芥■☆
77754 Cartodium Sol. ex R. Br. = Craspedia G. Forst. ■☆
77755 Cartonema R. Br. (1810);黄剑草属(彩花草属)■☆
77756 Cartonema brachyantherum Benth.;短花黄剑草■☆
77757 Cartonema parviflorum Hassk.;小花黄剑草■☆
77758 Cartonema tenue Caruel;细黄剑草■☆
77759 Cartonemataceae Pichon(1946)(保留科名);黄剑草科(彩花草科)■☆
77760 Cartonemataceae Pichon(保留科名) = Commelinaceae Mirb.(保留科名)●■
77761 Cartrema Raf. = Osmanthus Lour. ●
77762 Caruelia Parl. = Melomphis Raf. ■☆
77763 Caruelia Parl. = Ornithogalum L. ■
77764 Caruelina Kuntze = Chomelia Jacq.(保留属名)●☆
77765 Carui Mill. = Carum L. ■
77766 Carum L. (1753);葛缕子属(黄蒿属);Caraway ■
77767 Carum ajowan Benth.;印度藏茴香■☆
77768 Carum alpinum (M. Bieb.) Benth.;高山葛缕子■☆
77769 Carum ammoides L. = Ammoides pusilla (Brot.) Breistr. ■☆
77770 Carum angolense C. Norman = Aframmi angolense (C. Norman) C. Norman ■☆
77771 Carum angustissimum Kitag. = Carum buriaticum Turcz. ■
77772 Carum anisum (L.) Baill. = Pimpinella anisum L. ■
77773 Carum atlanticum Litard. et Maire;大西洋葛缕子■☆
77774 Carum atrosanguineum Kar. et Kir.;暗红葛缕子(血色葛缕子);Darkred Caraway ■
77775 Carum bretschneideri H. Wolff;河北葛缕子(旱芹菜);Hebei Caraway ■
77776 Carum bupleuroides Schrenk ex Fisch. et C. A. Mey. = Hyalolaena bupleuroides (Schrenk ex Fisch. et C. A. Mey.) Pimenov et Kljuykov ■
77777 Carum buriaticum Turcz.;田葛缕子(抽麻苔,田贡蒿,田黄蒿);Field Caraway ■
77778 Carum buriaticum Turcz. f. angustissimum (Kitag.) H. Wolff = Carum buriaticum Turcz. ■
77779 Carum buriaticum Turcz. f. angustissimum (Kitag.) R. H. Shan et F. T. Pu;丝叶葛缕子;Filiformleaf Carum ■
77780 Carum candolleanum (Wight et Arn.) Franch. = Pimpinella candolleana Wight et Arn. ■
77781 Carum capense Thunb. = Chamarea capensis (Thunb.) Eckl. et Zeyh. ■☆
77782 Carum capensis (Thunb.) Sond. = Chamarea capensis (Thunb.) Eckl. et Zeyh. ■☆
77783 Carum cardiocarpum Franch. = Pternopetalum cardiocarpum (Franch.) Hand.-Mazz. ■
77784 Carum carvi L.;葛缕子(藏茴香,防风,贡蒿,贡牛,郭乌,黄蒿,马婴子,小防风,野胡萝卜);Caraway, Caraway Seed, Caruwaie, Carver, Carvie, Carvy, Common Caraway, Seedcake ■
77785 Carum carvi L. f. gracile (Lindl.) H. Wolff;细葛缕子;Thin Caraway ■
77786 Carum carvi L. f. rubriflorum H. Wolff. = Carum carvi L. ■
77787 Carum carvi L. f. var. gracile (Lindl.) H. Wolff = Carum carvi L. ■
77788 Carum caucasicum (M. Bieb.) Boiss.;高加索葛缕子;Caucasia Caraway ■☆
77789 Carum caudatum Franch. = Pimpinella caudata (Franch.) H. Wolff ■
77790 Carum chinense M. Hiroe = Sinocarum filicinum H. Wolff ■
77791 Carum coloratum Diels = Sinocarum coloratum (Diels) H. Wolff ex R. H. Shan et F. T. Pu ■
77792 Carum copticum (L.) Benth. et Hook. f. = Trachyspermum ammi (L.) Sprague ■
77793 Carum copticum (L.) C. B. Clarke = Trachyspermum ammi (L.) Sprague ■
77794 Carum coriaceum Franch. = Pimpinella coriacea (Franch.) H. Boissieu ■
77795 Carum crinitum (Pall.) Koso-Pol. = Schulzia crinita (Pall.) Spreng. ■
77796 Carum cruciatum Franch. = Sinocarum cruciatum (Franch.) H. Wolff ex R. H. Shan et F. T. Pu ■
77797 Carum cruciatum Franch. var. linearilobum Franch. = Sinocarum cruciatum (Franch.) H. Wolff ex R. H. Shan et F. T. Pu var. linearilobum (Franch.) R. H. Shan et F. T. Pu ■
77798 Carum curvatum C. B. Clarke ex H. Wolff = Carum buriaticum Turcz. ■
77799 Carum cylindricum Boiss. et Hausskn. = Bunium cylindricum (Boiss. et Hausskn.) Drude ■

77800 Carum delavayi Franch. = Pternopetalum delavayi (Franch.) Hand. -Mazz. ■
77801 Carum delicatulum H. Wolff = Pternopetalum delicatulum (H. Wolff) Hand. -Mazz. ■
77802 Carum dichotomum (L.) Benth. et Hook. f. = Stoibrax dichotomum (L.) Raf. ■☆
77803 Carum dolichopodum Diels = Sinocarum dolichopodum (Diels) H. Wolff ex R. H. Shan et F. T. Pu ■
77804 Carum filicinum Franch. = Pternopetalum filicinum (Franch.) Hand. -Mazz. ■
77805 Carum flaccidum (C. B. Clarke) Franch. = Pimpinella flaccida C. B. Clarke ■
77806 Carum foetidum (Batt.) Drude;臭葛缕子■☆
77807 Carum franchetii M. Hiroe = Harrysmithia franchetii (M. Hiroe) M. L. Sheh ■
77808 Carum furcatum H. Wolff = Carum buriaticum Turcz. ■
77809 Carum gairdneri A. Gray;嘎尔葛缕子;Ipo, Squawroot, Yamp, Yampa ■☆
77810 Carum gracile Lindl. = Carum carvi L. ■
77811 Carum graveolens (L.) Koso-Pol. = Apium graveolens L. ■
77812 Carum grossheimii Schischk.;格罗葛缕子■☆
77813 Carum hispidum (Thunb.) Koso-Pol. = Sonderina hispida (Thunb.) H. Wolff ■☆
77814 Carum imbricatum Schinz = Afrocarum imbricatum (Schinz) Rauschert ■☆
77815 Carum incrassatum Boiss. = Bunium pachypodum P. W. Ball ■☆
77816 Carum jahandiezii Litard. et Maire;贾汉葛缕子■☆
77817 Carum kelloggii A. Gray;凯洛格葛缕子(凯勒葛缕子);Wild Anise ■☆
77818 Carum komarovii Karjagin;科马罗夫葛缕子■☆
77819 Carum leptocladum Aitch. et Hemsl. = Aphanopleura leptoclada (Aitch. et Hemsl.) Lipsky ■
77820 Carum loloense Franch. = Tongoloa loloensis (H. Boissieu) H. Wolff ■
77821 Carum mauritanicum Boiss. et Reut. = Bunium fontanesii (Pers.) Maire ■☆
77822 Carum meisneri (Sond.) M. Hiroe = Sonderina humilis (Meisn.) H. Wolff ■☆
77823 Carum montanum (Batt.) Benth. et Hook. f.;山地葛缕子■☆
77824 Carum neurophyllum (Maxim.) Franch. et Sav. = Pterygopleurum neurophyllum (Maxim.) Kitag. ■
77825 Carum oreganum S. Watson;俄勒冈葛缕子;False Carraway ■☆
77826 Carum persicum Boiss. = Bunium persicum (Boiss.) B. Fedtsch. ■☆
77827 Carum petroselinum (L.) Benth. = Petroselinum crispum (Mill.) Nyman ex A. W. Hill ■
77828 Carum petroselinum Benth. et Hook. f.;黄蒿(石蛇床);Garden Parsley ■☆
77829 Carum pimpinelloides Balf. f. = Trachyspermum pimpinelloides (Balf. f.) H. Wolff ■☆
77830 Carum pityophilum Diels = Sinocarum pityophilum (Diels) H. Wolff ■
77831 Carum proliferum Maire;多育葛缕子■☆
77832 Carum pseudoburiaticum H. Wolff. = Carum buriaticum Turcz. ■
77833 Carum purpureum Franch. = Pimpinella purpurea (Franch.) H. Boissieu ■
77834 Carum ridolfia Benth. et Hook. f. = Ridolfia segetum (L.) Moris ■☆
77835 Carum roxburghianum (DC.) Kurz = Trachyspermum roxburghianum (DC.) H. Wolff ■
77836 Carum saxicola Albov;岩地葛缕子■☆
77837 Carum scaberulum Franch. = Trachyspermum scaberulum (Franch.) H. Wolff ex Hand. -Mazz. ■
77838 Carum scaberulum Franch. var. ambrosiifolium Franch. = Trachyspermum scaberulum (Franch.) H. Wolff ex Hand. -Mazz. var. ambrosiifolium (Franch.) R. H. Shan ■
77839 Carum schizopetalum Franch. = Sinocarum schizopetalum (Franch.) H. Wolff ex R. H. Shan et F. T. Pu ■
77840 Carum segetum Benth. et Hook. f.;田间葛缕子■☆
77841 Carum setaceum Schrenk;刚毛葛缕子■☆
77842 Carum sinense Franch. = Pternopetalum sinense (Franch.) Hand. -Mazz. ■
77843 Carum stictocarpum C. B. Clarke = Trachyspermum roxburghianum (DC.) H. Wolff ■
77844 Carum tanakae Franch. et Sav. = Pternopetalum tanakae (Franch. et Sav.) Hand. -Mazz. ■
77845 Carum tenuifolium (Coss.) Benth. et Hook. f.;细叶葛缕子■☆
77846 Carum trichomanifolium Franch. = Pternopetalum trichomanifolium (Franch.) Hand. -Mazz. ■
77847 Carum trichophyllum Schrenk = Hyalolaena trichophylla (Schrenk) Pimenov et Kljuykov ■
77848 Carum vaginatum (H. Wolff) M. Hiroe = Sinocarum vaginatum H. Wolff ■
77849 Carum velenovskyi Rohlena = Carum carvi L. ■
77850 Carum verticillatum (L.) Koch;轮生葛缕子■☆
77851 Carum verticillatum Koch;轮状葛缕子;Whorled Carraway ■☆
77852 Carum yunnanense Franch. = Pimpinella yunnanensis (Franch.) H. Wolff ■
77853 Carumbium Kurz = Sapium Jacq. (保留属名)●
77854 Carumbium Reinw. = Homalanthus A. Juss. (保留属名)●
77855 Caruncularia Haw. = Stapelia L. (保留属名)■
77856 Caruncularia aperta (Masson) Sweet = Tromotriche aperta (Masson) Bruyns ■☆
77857 Caruncularia jacquinii Sweet = Tromotriche pedunculata (Masson) Bruyns ■☆
77858 Caruncularia massonii Sweet = Tromotriche pedunculata (Masson) Bruyns ■☆
77859 Caruncularia pedunculata (Masson) Haw. = Tromotriche pedunculata (Masson) Bruyns ■☆
77860 Caruncularia penduliflora Sweet = Tromotriche pedunculata (Masson) Bruyns ■☆
77861 Caruncularia serrulata (Jacq.) G. Don = Piaranthus geminatus (Masson) N. E. Br. subsp. decorus (Masson) Bruyns ■☆
77862 Caruncularia simsii Sweet = Tromotriche pedunculata (Masson) Bruyns ■☆
77863 Carusia Mart. ex Nied. = Burdachia Juss. ex Endl. ●☆
77864 Carvalhoa K. Schum. (1895);小钟夹竹桃属●☆
77865 Carvalhoa campanulata K. Schum.;小钟夹竹桃●☆
77866 Carvalhoa macrophylla K. Schum. = Carvalhoa campanulata K. Schum. ●☆
77867 Carvalhoa petiolata K. Schum. = Carvalhoa campanulata K. Schum. ●☆
77868 Carvi Bernh. = Selinum L. (保留属名)■
77869 Carvi Bubani = Carui Mill. ■

77870 Carvi Bubani = Carum L. ■
77871 Carvia Bremek. = Strobilanthes Blume ●■
77872 Carvifolia C. Bauh. ex Vill. = Selinum L.(保留属名)■
77873 Carvifolia Vill. = Selinum L.(保留属名)■
77874 Carya Nutt.(1818)(保留属名);山核桃属;Caryer, Hickory, Hicorier, Pecan-tree, Pican ●
77875 Carya × laneyi Sarg.;拉尼山核桃;Laney's Hickory ●☆
77876 Carya × lecontei Little;里康山核桃;Bitter Pecan ●☆
77877 Carya alba Britton, Sterns et Poggenb. = Carya tomentosa (Poir.) Nutt. ●
77878 Carya alba K. Koch = Carya tomentosa (Poir.) Nutt. ●
77879 Carya alba L. = Carya tomentosa (Poir.) Nutt. ●
77880 Carya amara (F. Michx.) Nutt. ex Elliott = Carya cordiformis (Wangenh.) K. Koch ●
77881 Carya aquatica (F. Michx.) Nutt. ex Elliott;苦山核桃;Bitter Peach, Bitter Pecan, Swamp Hickory, Water Hickory ●
77882 Carya arkansana Sarg. = Carya texana C. DC. ●
77883 Carya australis Ashe = Carya ovata (Mill.) K. Koch var. australis (Ashe) Little ●☆
77884 Carya buckleyi Durand = Carya texana C. DC. ●
77885 Carya carolinae-septentrionalis (Ashe) Engl. et Graebn.;北方山核桃;Carolina Hickory ●☆
77886 Carya carolinae-septentrionalis (Ashe) Engl. et Graebn. = Carya ovata (Mill.) K. Koch var. australis (Ashe) Little ●☆
77887 Carya cathayensis Sarg.;山核桃(核桃,山核,山蟹,小核桃,野漆树);Cathay Hickory, Chinese Hickory ●
77888 Carya cordiformis (Wangenh.) C. Koch = Carya cordiformis (Wangenh.) K. Koch ●
77889 Carya cordiformis (Wangenh.) K. Koch;心果山核桃(苦山核桃);Bitter Nut, Bitternut, Bitternut Hickory, Bitter-nut Hickory, Butternut, Pignut, Pig-nut, Pignut Hickory, Swamp Hickory, Yellow-bud Hickory ●
77890 Carya cordiformis (Wangenh.) K. Koch var. latifolia Sarg. = Carya cordiformis (Wangenh.) K. Koch ●
77891 Carya cordiformis K. Koch = Carya cordiformis (Wangenh.) K. Koch ●
77892 Carya dabieshanensis M. C. Liu et Z. J. Li;大别山山核桃;Dabieshan Hickory ●
77893 Carya floridana Sarg.;佛罗里达山核桃;Scrub Hickory ●
77894 Carya glabra (Mill.) Sweet;光叶山核桃(光滑山核木);Black Hickory, False Shagbark Hickory, Hog Nut, Hognut, Hognut Broom Hickory, Pig Nut, Pignut, Pignut Hickory, Pig-nut Hickory, Red Hickory, Smoothbark Hickory, Smooth-bark Hickory, Sweet Pignut ●☆
77895 Carya glabra (Mill.) Sweet var. megacarpa (Sarg.) Sarg.;大果山核桃;Coast Pignut Hickory ●
77896 Carya glabra (Mill.) Sweet var. megacarpa Sarg. = Carya glabra (Mill.) Sweet ●☆
77897 Carya glabra (Mill.) Sweet var. odorata (Marshall) Little;红山核桃;Red Hickory ●
77898 Carya glabra (Mill.) Sweet var. odorata (Marshall) Little = Carya glabra (Mill.) Sweet ●☆
77899 Carya glabra (Mill.) Sweet var. villosa (Sarg.) B. L. Rob. = Carya texana C. DC. ●
77900 Carya hunanensis W. C. Cheng et R. H. Chang ex R. H. Chang et A. M. Lu;湖南山核桃;Hunan Hickory ●
77901 Carya illinoinensis (Wangenh.) C. Koch = Carya illinoinensis (Wangenh.) K. Koch ●
77902 Carya illinoinensis (Wangenh.) K. Koch;薄壳山核桃(长山核桃,美国山核桃,美洲核桃,美洲胡桃,美洲山核桃,培甘,山核桃,甜山核桃);Peach, Pecan, Pecan Hickory, Pecan Nut, Pecan-tree, Sweet Pecan ●
77903 Carya integrifoliolata (Kuang) Hjelmq. = Annamocarya sinensis (Dode) Leroy ●◇
77904 Carya kweichowensis Kuang et A. M. Lu ex R. H. Chang et A. M. Lu;贵州山核桃;Guizhou Hickory, Guizhou Pecan ●◇
77905 Carya laciniosa (F. Michx.) Loudon;条纹山核桃(美国大山核桃,条裂皮山核桃,条裂山核桃);Big Shellbark, Big Shellbark Hickory, Bottom Shellbark, Kingnut, King-nut, Shellbark Hickory ●☆
77906 Carya laneyi Sarg.;莱尼山核桃;Laney's Hickory ●☆
77907 Carya leiodermis Sarg.;沼泽山核桃;Swamp Hickory ●☆
77908 Carya leiodermis Sarg. = Carya glabra (Mill.) Sweet ●☆
77909 Carya magnifloridana Murrill = Carya glabra (Mill.) Sweet ●☆
77910 Carya myristicaeformis (F. Michx.) Nutt.;肉豆蔻山核桃;Bitter Water Hickory, Nutmeg Hickory, Swamp Hickory ●
77911 Carya oliviformis (Michx.) Nutt. = Carya illinoinensis (Wangenh.) K. Koch ●
77912 Carya ovalis (Wangenh.) Sarg.;甜山核桃;False Shagbark, Pignum Hickory, Red Hickory, Sweet Pignum ●☆
77913 Carya ovalis (Wangenh.) Sarg. = Carya glabra (Mill.) Sweet ●☆
77914 Carya ovalis (Wangenh.) Sarg. var. hirsuta (Ashe) Sarg. = Carya glabra (Mill.) Sweet ●☆
77915 Carya ovalis (Wangenh.) Sarg. var. obcordata (Muhl. ex Willd.) Sarg. = Carya glabra (Mill.) Sweet ●☆
77916 Carya ovalis (Wangenh.) Sarg. var. obcordata (Muhl. ex Willd.) Sarg. f. vestita Sarg. = Carya glabra (Mill.) Sweet ●☆
77917 Carya ovalis (Wangenh.) Sarg. var. obovalis Sarg. = Carya glabra (Mill.) Sweet ●☆
77918 Carya ovalis (Wangenh.) Sarg. var. odorata (Marshall) Sarg. = Carya glabra (Mill.) Sweet ●☆
77919 Carya ovata (Mill.) C. Koch = Carya ovata (Mill.) K. Koch ●
77920 Carya ovata (Mill.) K. Koch;粗皮山核桃(卵形山核桃,小糙皮山核桃);Hickory, Little Shellbark Hickory, Scalybark Hickory, Shagbark Hickory, Shellbark Hickory, Thick Shellbark Hickory, Upland Hickory ●
77921 Carya ovata (Mill.) K. Koch = Carya tomentosa (Poir.) Nutt. ●
77922 Carya ovata (Mill.) K. Koch var. australis (Ashe) Little;加州粗皮山核桃;Carolina Hickory ●☆
77923 Carya ovata (Mill.) K. Koch var. fraxinifolia Sarg. = Carya ovata (Mill.) K. Koch ●
77924 Carya ovata (Mill.) K. Koch var. nuttallii Sarg. = Carya ovata (Mill.) K. Koch ●
77925 Carya ovata (Mill.) K. Koch var. pubescens Sarg. = Carya ovata (Mill.) K. Koch ●
77926 Carya pallida (Ashe) Engl. et Graebn.;沙地山核桃;Pale Hickory, Pignut Hickory, Sand Hickory ●
77927 Carya palmeri Manning;墨西哥山核桃;Mexican Hickory, Mexico Hickory ●
77928 Carya pecan Engl. et Graebn. = Carya illinoinensis (Wangenh.) K. Koch ●
77929 Carya porcina Nutt. = Carya glabra (Mill.) Sweet ●☆
77930 Carya sinensis Dode = Annamocarya sinensis (Dode) Leroy ●◇
77931 Carya texana Buckley var. villosa (Sarg.) Little = Carya texana C. DC. ●
77932 Carya texana C. DC.;黑山核桃;Black Hickory, Buckley

Hickory, Ozark Pignut Hickory, Pignut Hickory ●

77933 Carya texana C. DC. var. villosa (Sarg.) Little = Carya texana C. DC. ●

77934 Carya tomentosa (Poir.) Nutt.;毛山核桃(绒毛山核桃,柔毛山核桃,毡毛山核桃);Big-bud Hickory, Bullnut, Hognut, Mocker Nut, Mockernut, Mockernut Hickory, Mocknut Hickory, Scatybark Hickory, Shagbark Hickory, Shellbark Hickory, White Hickory, White Mockernut, Whitehart Mockernut, White-heart Hickory ●

77935 Carya tonkinensis Lecomte;越南山核桃(安南山核桃,老鼠核桃,云南山核桃);Tonkin Hickory, Vietnam Hickory ●

77936 Carya tsiangiana Chun ex Lee = Annamocarya sinensis (Dode) Leroy ●◇

77937 Carya tsiangii Chun = Annamocarya sinensis (Dode) Leroy ●◇

77938 Caryella Bourn. ex Parm.;马岛柿属●☆

77939 Carynephyllum Rose = Sedum L. ●■

77940 Caryocar F. Allam. ex L. (1771);多柱树属(油桃木属)●☆

77941 Caryocar L. = Caryocar F. Allam. ex L. ●☆

77942 Caryocar amygdaliferum Mutis;膀胱多柱树;Suari, Swarri Nut ●☆

77943 Caryocar brasiliense A. St. -Hil.;巴西油桃木;Pequi, Piqui ●☆

77944 Caryocar glabrum Pers.;无毛油桃木;Soapwood ●☆

77945 Caryocar microcarpum Ducke;小果油桃木●☆

77946 Caryocar nuciferum L.;多柱树;Butternut, Butter-nut, Butter-nut Tree, Pekea Nut, Peruvian Almond, Sawara Nut, Sawari Nut, Souari-nut Tree, Souvari Nut ●☆

77947 Caryocar tomentosum Willd.;毛多柱树;Butter Nut, Butternut, Souari-nut ●☆

77948 Caryocar villosum Pers.;柔毛油桃木;Piqui ●☆

77949 Caryocaraceae Szyszyl. = Caryocaraceae Voigt(保留科名)●☆

77950 Caryocaraceae Voigt(1845)(保留科名);多柱树科(油桃木科)●☆

77951 Caryochloa Spreng. = Oryzopsis Michx. ■

77952 Caryochloa Spreng. = Piptochaetium J. Presl(保留属名)■☆

77953 Caryochloa Trin. = Luziola Juss. ■☆

77954 Caryococca Willd. ex Roem. et Schult. = Gonzalagunia Ruiz et Pav. ●☆

77955 Caryodaphne Blume ex Nees = Cryptocarya R. Br. (保留属名)●

77956 Caryodaphnopsis Airy Shaw (1940);檬果樟属(桂果樟属);Caryodaphnopsis ●

77957 Caryodaphnopsis baviensis (Lecomte) Airy Shaw;巴围檬果樟(巴围假檬果);Bavi Caryodaphnopsis ●

77958 Caryodaphnopsis burgeri N. Zamora et Poveda;哥斯达黎加檬果樟●☆

77959 Caryodaphnopsis cogolloi van der Werff et H. G. Richt.;哥伦比亚檬果樟●☆

77960 Caryodaphnopsis fosteri van der Werff;福氏檬果樟;Foster's Caryodaphnopsis ●☆

77961 Caryodaphnopsis henryi Airy Shaw;小花檬果樟(亨利假檬果,亨利檬果樟);Henry Caryodaphnopsis ●

77962 Caryodaphnopsis inaequialis (A. C. Sm.) van der Werff et H. G. Richt.;异被檬果樟●☆

77963 Caryodaphnopsis laotica Airy Shaw;老挝檬果樟;Laos Caryodaphnopsis ●

77964 Caryodaphnopsis latifolia W. T. Wang;宽叶檬果樟(宽叶假檬果,宽叶檬果);Broadleaf Caryodaphnopsis, Broad-leaved Caryodaphnopsis ●◇

77965 Caryodaphnopsis latifolia W. T. Wang = Caryodaphnopsis tonkinensis (Lecomte) Airy Shaw ●

77966 Caryodaphnopsis metallica Kosterm.;缘毛檬果樟●☆

77967 Caryodaphnopsis poilanei Kosterm.;赤毛檬果樟;Redhairs Caryodaphnopsis ●☆

77968 Caryodaphnopsis theabromifolia (A. H. Gentry) van der Werff et H. G. Richt.;可可叶檬果樟●☆

77969 Caryodaphnopsis tonkinensis (Lecomte) Airy Shaw;檬果樟(假檬果,檬果);Caryodaphnopsis, Tonkin Caryodaphnopsis ●

77970 Caryodendron H. Karst. (1860);核果大戟属●☆

77971 Caryodendron orinocense H. Karst.;核果大戟;Lacy Nut, Tacay ●☆

77972 Caryolobis Gaertn. (废弃属名) = Doona Thwaites(保留属名)●☆

77973 Caryolobis Gaertn. (废弃属名) = Shorea Roxb. ex C. F. Gaertn. ●

77974 Caryolobium Steven = Astragalus L. ●■

77975 Caryolopha Fisch. ex Trautv. = Anchusa L. ■

77976 Caryolopha Fisch. ex Trautv. = Pentaglottis Tausch ■☆

77977 Caryomene Barneby et Krukoff(1971);月实藤属●☆

77978 Caryomene grandifolia Barneby et Krukoff;大叶月实藤●☆

77979 Caryomene prumnoides Barneby et Krukoff;月实藤●☆

77980 Caryophyllaceae Juss. (1789)(保留科名);石竹科;Pink Family ■●

77981 Caryophyllata Mill. = Geum L. ■

77982 Caryophyllata Tourn. ex Scop. = Geum L. ■

77983 Caryophyllea Opiz = Aira L. (保留属名)■

77984 Caryophyllus L. (废弃属名) = Syzygium R. Br. ex Gaertn. (保留属名)●

77985 Caryophyllus Mill. = Dianthus L. ■

77986 Caryophyllus Tourn. ex Moench = Dianthus L. ■

77987 Caryophyllus aromaticus L. = Syzygium aromaticum (L.) Merr. et L. M. Perry ●

77988 Caryophyllus racemosus Mill. = Pimenta racemosa (Mill.) J. W. Moore ●☆

77989 Caryopitys Small = Pinus L. ●

77990 Caryopitys edulis (Engelm.) Small = Pinus edulis Engelm. ●☆

77991 Caryopitys monophylla (Torr. et Frém.) Rydb. = Pinus monophylla Torr. et Frém. ●

77992 Caryopteris Bunge(1835);莸属(莸草属);Blue Mist Shrub, Bluebeard, Blue-beard ●

77993 Caryopteris × clandonensis Hort.;蓝莸;Blue Mist, Blue Mist Shrub, Blue Spiraea, Bluebeard, Blue-mist Shrub ●

77994 Caryopteris × clandonensis Hort. 'Arthur Simmonds';阿苏尔蒙兹蓝莸●

77995 Caryopteris × clandonensis Hort. 'Blue Mist';蓝雾蓝莸;Blue Mist Shrub ●

77996 Caryopteris × clandonensis Hort. 'Dark Knight';暗骑士蓝莸;Blue Mist Shrub ●

77997 Caryopteris × clandonensis Hort. 'Heavenly Blue';天蓝蓝莸●

77998 Caryopteris × clandonensis Hort. 'Kew Blue';邱园蓝莸●

77999 Caryopteris aureoglandulosa (Vaniot) C. Y. Wu;金腺莸(跌打接骨菜,八瓜金);Golden-glandular Bluebeard, Golden-glandular Blue-beard ●

78000 Caryopteris bicolor (Roxb. ex Hardw.) Mabb.;香莸;Fragrant Bluebeard, Fragrant Blue-beard ●

78001 Caryopteris cordifolia C. Y. Wu = Caryopteris siccanea W. W. Sm. ■

78002 Caryopteris divaricata (Siebold et Zucc.) Maxim.;莸(叉枝莸);Divaricate Bluebeard ■

78003 Caryopteris divaricata Maxim. = Caryopteris divaricata (Siebold et Zucc.) Maxim. ■

78004 Caryopteris divaricata Siebold et Zucc. = Caryopteris divaricata (Siebold et Zucc.) Maxim. ■

78005 Caryopteris esquirolii H. Lév. = Pogostemon esquirolii (H. Lév.) C. Y. Wu et Y. C. Huang ●■

78006 Caryopteris fluminis H. Lév. = Colquhounia seguinii Vaniot ●

78007 Caryopteris foetida Thell.;臭莸●☆

78008 Caryopteris forrestii Diels;灰毛莸(白叶莸,兰香草);Forrest Bluebeard,Forrest Blue-beard,Greyhair Bluebeard ●

78009 Caryopteris forrestii Diels var. minor C. P'ei et S. L. Chen ex C. Y. Wu;小叶灰毛莸;Small-leaf Bluebeard ●

78010 Caryopteris glutinosa Rehder;黏叶莸;Glutinous Bluebeard,Glutinous Blue-beard,Nepal Bluebeard ●

78011 Caryopteris grata Benth. = Caryopteris foetida Thell. ●☆

78012 Caryopteris incana (Houtt.) Miq.;兰香草(白花山薄荷,白鸡婆捎,宝塔草,避蛇虫,齿瓣兰香草,地罗珠,独脚求,独脚球,短菊,段菊,对对花,黄鸦柴,灰叶莸,假仙草,节节花,金石香,九层楼,九层塔,酒饼草,酒药草,兰花草,蓝花草,卵叶莸,马蒿,婆绒草,婆绒花,七盆香,山薄荷,石黄精,石将军,石兰香,石母草,石上香,石仙草,狭叶兰香草,小六月寒,血汗草,岩薄荷,野薄荷,野金花,野仙草,茵陈草,莸,子附莲,紫罗球,紫罗毬,走马风);Blue Spiraea,Bluebeard,Common Bluebeard,Common Blue-beard,Moustache Plant,Silver Blue-beard ■

78013 Caryopteris incana (Houtt.) Miq. f. candida (C. K. Schneid.) H. Hara;白兰香草●☆

78014 Caryopteris incana (Houtt.) Miq. f. rosea Sugim.;粉兰香草●☆

78015 Caryopteris incana (Houtt.) Miq. var. angustifolia S. L. Chen et Y. L. Kuo;狭叶兰香草;Narrowleaf Bluebeard,Narrowleaf Common Bluebeard ●

78016 Caryopteris incana (Houtt.) Miq. var. brachyodonta (Hand.-Mazz.) Moldenke = Caryopteris trichosphaera W. W. Sm. ●

78017 Caryopteris incana (Thunb. ex Houtt.) Miq. var. brachyodonta (Hand.-Mazz.) Moldenke = Caryopteris trichosphaera W. W. Sm. ●

78018 Caryopteris incana (Thunb.) Miq. = Caryopteris incana (Houtt.) Miq. ■

78019 Caryopteris jinshajiangensis Y. K. Yang et X. D. Cong;金沙江莸;Jinshajiang Bluebeard ●

78020 Caryopteris mairei H. Lév. = Rubiteucris palmata (Benth. ex Hook. f.) Kudo ■

78021 Caryopteris martinii H. Lév. = Caryopteris paniculata C. B. Clarke ●

78022 Caryopteris mastacantnus Schaner = Caryopteris incana (Houtt.) Miq. ■

78023 Caryopteris mongholica Bunge;蒙古莸(白蒿,白沙蒿,兰花茶,蓝花菜,蓝花茶,蒙莸,山狼毒);Mongol Bluebeard,Mongolian Bluebeard,Mongolian Blue-beard ●◇

78024 Caryopteris mongholica Bunge var. serrata Maxim. = Caryopteris mongholica Bunge ●◇

78025 Caryopteris nepetifolia (Benth.) Maxim.;单花莸(半支莲,边兰,倒挂金钟,方梗金钱草,野苋菜,野苋草,莸);Oneflowered Bluebeard ■

78026 Caryopteris nepetifolia (Benth.) Maxim. f. brevipes C. Y. Wu et H. Li;短柄单花莸●

78027 Caryopteris ningpoensis Hemsl. = Comanthosphace ningpoensis (Hemsl.) Hand.-Mazz. ■

78028 Caryopteris odorata (D. Don) B. L. Rob. = Caryopteris bicolor (Roxb. ex Hardw.) Mabb. ●

78029 Caryopteris odorata (Ham. ex Roxb.) Rob. = Caryopteris bicolor (Roxb. ex Hardw.) Mabb. ●

78030 Caryopteris odorata (Ham.) B. L. Rob. = Caryopteris bicolor (Roxb. ex Hardw.) Mabb. ●

78031 Caryopteris ovata Miq. = Caryopteris incana (Houtt.) Miq. ■

78032 Caryopteris paniculata C. B. Clarke;锥花莸(密花莸,紫红鞭);Paniculate Bluebeard,Paniculate Blue-beard ●

78033 Caryopteris parvifolia Batalin = Isodon parvifolius (Batalin) H. Hara ●

78034 Caryopteris parvifolia Batalin = Rabdosia parvifolia (Batalin) H. Hara ●

78035 Caryopteris siccanea W. W. Sm.;腺毛莸;Glandularhair Bluebeard ■

78036 Caryopteris sinensis (Lour.) Dippel = Caryopteris incana (Houtt.) Miq. ■

78037 Caryopteris tangutica Maxim.;光果莸(白鸡婆梢,老鼠精,山薄荷,唐古特莸,西北莸,小六月寒);Tangut Bluebeard,Tangut Blue-beard ●

78038 Caryopteris tangutica Maxim. var. brachyodonta Hand.-Mazz. = Caryopteris trichosphaera W. W. Sm. ●

78039 Caryopteris terniflora Maxim.;三花莸(大风寒草,风寒草,蜂子草,红花野芝麻,化骨丹,黄刺泡,金谷草,金钱风,金线草,兰香草,六月寒,路边梢,毛老虎,气草,山卷莲,血汗草,野荆芥,野芝麻);Threeflowered Bluebeard,Triflorous Blue-beard ●

78040 Caryopteris terniflora Maxim. f. brevipedunculata C. P'ei et S. L. Chen;短梗三花莸;Shortpeduncle Bluebeard ●

78041 Caryopteris terniflora Maxim. f. brevipedunculata C. P'ei et S. L. Chen = Caryopteris terniflora Maxim. ●

78042 Caryopteris trichosphaera W. W. Sm.;毛球莸(毛莸,香薷);Hairysphere Bluebeard,Hairyspherical Blue-beard ●

78043 Caryopteris wallichiana Schauer = Caryopteris bicolor (Roxb. ex Hardw.) Mabb. ●

78044 Caryospermum Blume = Perrottetia Kunth ●

78045 Caryota L. (1753);鱼尾葵属(假桄榔属,假椰属,孔雀椰子属,鱼尾椰属);Bastard Sago Palm,Fishtail Palm,Fish-tail Palm,Fish-tail Palms,Fishtailpalm,Wine Palm ●

78046 Caryota cumingii Lodd. ex Mart.;菲岛鱼尾葵●☆

78047 Caryota mitis Lour.;短穗鱼尾葵(丛立孔雀椰子,短序鱼尾葵,桄榔,酒椰子);Burmese Fishtail Palm,Clustered Fishtail Palm,Clustered Fishtail-palm,Fishtail Palm,Shortspike Fishtailpalm,Tufted Fishtail Palm,Tufted Fishtailpalm ●

78048 Caryota monostachya Becc.;单穗鱼尾葵(单序鱼尾葵);Singlespike Fishtailpalm,Single-spiked Fishtail Palm ●

78049 Caryota no Becc. = Caryota no Becc. ex J. Dransf. ●

78050 Caryota no Becc. ex J. Dransf.;孔雀椰;Ciant Fishtail Palm ●

78051 Caryota obtusa Griff. var. aequatorialis Becc.;巨大山地鱼尾葵;Giant Mountain Fishtailpalm ●

78052 Caryota ochlandra Hance;鱼尾葵(桄根,桄榔,假桄榔,青棕);Chinese Fishtail Palm,Common Fishtail Palm,Common Fishtailpalm,Fish Tail Palm,Fishtailpalm ●

78053 Caryota rumphiana Mart.;紫果鱼尾葵;Fishtail Palm ●☆

78054 Caryota rumphiana Mart. var. borneensis Becc. = Caryota no Becc. ex J. Dransf. ●

78055 Caryota sobolifera Wall. = Caryota mitis Lour. ●

78056 Caryota urens L.;酒鱼尾葵(董棕,光榔,酒假桄榔,孔雀椰,孔雀椰子);Bastard Sago,Ceylon Piassava,East Indian Wine Palm,Fish Tail Palm,Fishtail Palm,Fish-tail Palm,Hill Palm,Indian Sago Palm,Jaggery Palm,Kittool,Kittool Palm,Kittool-palm,Kitul,Kitul Palm,Sago Palm,Solitary Fishtail Palm,Sting Fishtailpalm,Toddy

Fishtail Palm, Toddy Fishtail-palm, Toddy Palm, Wine Palm ●

78057 Caryotaceae O. F. Cook = Arecaceae Bercht. et J. Presl(保留科名)●

78058 Caryotaceae O. F. Cook = Palmae Juss.(保留科名)●

78059 Caryotaceae O. F. Cook;鱼尾葵科●

78060 Caryotaxus Zucc. ex Endl. = Torreya Arn. ●

78061 Caryotaxus Zucc. ex Henk. et Hochst. = Torreya Arn. ●

78062 Caryotaxus Zucc. ex Henk. et Hochst. = Tumion Raf. ●

78063 Caryotaxus grandis (Fortune ex Lindl.) Henkel et W. Hochst. = Torreya grandis Fortune ex Lindl. ●◇

78064 Caryotaxus grandis Henkel et Hochst. = Torreya grandis Fortune ex Lindl. ●◇

78065 Caryotophora Leistner(1958);长瓣玉属■☆

78066 Caryotophora skiatophytoides Leistner;长瓣玉■☆

78067 Casabitoa Alain(1980);海地大戟属☆

78068 Casabitoa perfae Alain;海地大戟☆

78069 Casalea A. St. -Hil. = Ranunculus L. ■

78070 Casanophorum Neck. = Castanea Mill. ●

78071 Casarettoa Walp. = Vitex L. ●

78072 Casasia A. Rich.(1850);卡萨茜属●☆

78073 Casasia calophylla A. Rich.;美叶卡萨茜☆

78074 Casasia nigrescens Wright ex B. L. Rob.;黑卡萨茜☆

78075 Casasia parvifolia Britton;小叶卡萨茜☆

78076 Cascabela Raf. = Thevetia L.(保留属名)●

78077 Cascabela peruviana (Pers.) Raf. = Cascabela thevetia (L.) Lippold ●

78078 Cascabela thevetia (L.) Lippold = Thevetia peruviana (Pers.) K. Schum. ●

78079 Cascadia A. M. Johnson = Saxifraga L. ■

78080 Cascadia A. M. Johnson(1927);肖虎耳草属■☆

78081 Cascadia nuttallii (Small) A. M. Johnson;肖虎耳草■☆

78082 Cascarilla (Endl.) Wedd. = Ladenbergia Klotzsch ●☆

78083 Cascarilla Adans. = Croton L. ●

78084 Cascarilla Ruiz ex Steud. = Cinchona L. ■●

78085 Cascarilla Wedd. = Ladenbergia Klotzsch ●☆

78086 Cascarilla carua Wedd. = Ladenbergia carua (Wedd.) Standl. ●☆

78087 Cascarilla magnifolia Wedd. var. caduciflora (Bonpl.) Wedd. = Ladenbergia oblongifolia (Humb. ex Mutis) L. Andersson ●☆

78088 Cascarilla magnifolia Wedd. var. rostrata (Wedd.) Wedd. = Ladenbergia oblongifolia (Humb. ex Mutis) L. Andersson ●☆

78089 Cascarilla magnifolia Wedd. var. vulgaris Wedd. = Ladenbergia oblongifolia (Humb. ex Mutis) L. Andersson ●☆

78090 Cascaronia Griseb.(1879);紫云英豆属■☆

78091 Cascoelytrum P. Beauv. = Briza L. ■

78092 Cascoelytrum P. Beauv. = Chascolytrum Desv. ■

78093 Casearia Griseb. = Casearia Jacq. ●

78094 Casearia Griseb. = Gossypiospermum (Griseb.) Urb. ●☆

78095 Casearia Jacq.(1760);脚骨脆属(嘉赐木属,嘉赐树属);Casearia ●

78096 Casearia aculeata Jacq.;刺脚骨脆●☆

78097 Casearia aequilateralis Merr.;海南脚骨脆(海南嘉赐木,海南嘉赐树);Equeal-sided Casearia, Hainan Casearia ●

78098 Casearia aequilateralis Merr. = Casearia membranacea Hance ●

78099 Casearia balansae Gagnep.;脚骨脆(嘉赐树,毛嘉赐树,毛叶嘉赐树);Balansa Casearia ●

78100 Casearia balansae Gagnep. = Casearia velutina Blume ●

78101 Casearia balansae Gagnep. var. cuneifolia Gagnep. = Casearia balansae Gagnep. ●

78102 Casearia balansae Gagnep. var. cuneifolia Gagnep. = Casearia velutina Blume ●

78103 Casearia balansae Gagnep. var. subglabra S. Y. Bao;景东脚骨脆(景东嘉赐树);Jingdong Casearia ●

78104 Casearia balansae Gagnep. var. subglabra S. Y. Bao = Casearia velutina Blume ●

78105 Casearia barteri Mast.;巴特嘉赐木●☆

78106 Casearia battiscombei R. E. Fr.;巴提嘉赐木●☆

78107 Casearia bridelioides Gilg ex Engl. = Keayodendron bridelioides Léandri ●☆

78108 Casearia bridelioides Mildbr. ex Hutch. et Dalziel = Keayodendron bridelioides Léandri ●☆

78109 Casearia bule Gilg = Casearia barteri Mast. ●☆

78110 Casearia calciphila C. Y. Wu et Y. C. Huang ex S. Y. Bao = Casearia tardieuae Lescot et Sleumer ●

78111 Casearia calodendron Gilg;美丽脚骨脆●☆

78112 Casearia chirindensis Engl. = Casearia battiscombei R. E. Fr. ●☆

78113 Casearia congensis Gilg;康格脚骨脆●☆

78114 Casearia dinklagei Gilg = Casearia barteri Mast. ●☆

78115 Casearia elliptica Willd. = Casearia tomentosa Roxb. ●☆

78116 Casearia engleri Gilg;恩格嘉赐木●☆

78117 Casearia esculenta Roxb.;可食脚骨脆;Edible Casearia ●☆

78118 Casearia flexuosa Craib;曲枝脚骨脆(蜿枝嘉赐树,云南嘉赐树);Flexuose Casearia, Sinuate Casearia ●

78119 Casearia gladiiformis Mast.;刀形嘉赐木●☆

78120 Casearia glomerata Roxb.;球花脚骨脆(扁鱼腩,嘉赐木,嘉赐树,脚骨脆,熊胆树皮);Clustered Casearia ●

78121 Casearia glomerata Roxb. f. pubinervis F. C. How et W. C. Ko;毛脉脚骨脆(毛嘉赐树,毛脉嘉赐树);Hairy-nerve Clustered Casearia ●

78122 Casearia glomerata Roxb. f. pubinervis F. C. How et W. C. Ko = Casearia glomerata Roxb. ●

78123 Casearia graveolens Dalzell;烈味脚骨脆(香味嘉赐木,香味嘉赐树);Strong-smelling Casearia ●

78124 Casearia graveolens Dalzell var. lingtsangensis S. Y. Bao;临沧脚骨脆(临沧嘉赐树);Lincang Casearia ●

78125 Casearia graveolens Dalzell var. lintsangensis S. Y. Bao = Casearia graveolens Dalzell ●

78126 Casearia guineensis G. Don;几内亚脚骨脆●☆

78127 Casearia harmandiana Pierre ex Gagnep. = Casearia flexuosa Craib ●

78128 Casearia hexagona Pierre ex A. Chev. = Casearia barteri Mast. ●☆

78129 Casearia holtzii Gilg = Casearia gladiiformis Mast. ●☆

78130 Casearia inaequalis Hutch. et Dalziel = Casearia calodendron Gilg ●☆

78131 Casearia junodii Schinz = Casearia gladiiformis Mast. ●☆

78132 Casearia klaineana Pierre ex A. Chev. = Casearia barteri Mast. ●☆

78133 Casearia kurzii C. B. Clarke;印度脚骨脆(滇南脚骨脆,印度嘉赐树);Kurz Casearia ●

78134 Casearia kurzii C. B. Clarke var. gracilis S. Y. Bao;细柄脚骨脆(细柄嘉赐木,细柄嘉赐树);Slender Kurz Casearia ●

78135 Casearia macrodendron Gilg = Casearia gladiiformis Mast. ●☆

78136 Casearia mannii Mast.;曼氏脚骨脆●☆

78137 Casearia membranacea Hance;薄叶脚骨脆(薄叶嘉赐木,薄叶嘉赐树,麦氏嘉赐树,膜叶嘉赐木,膜叶嘉赐树,膜叶脚骨脆,台湾嘉赐树,望山棵);Membranaceous Casearia, Thinleaf Casearia ●

78138 Casearia membranacea Hance var. nigrescens S. S. Lai;黑叶脚骨

脆;Blackleaf Casearia ●
78139 Casearia merrillii Hayata = Casearia glomerata Roxb. ●
78140 Casearia merrillii Hayata = Casearia membranacea Hance ●
78141 Casearia nitida Jacq. ;光亮脚骨脆;Smooth Honeytree ●☆
78142 Casearia noldei A. Fern. et Diniz = Casearia calodendron Gilg ●☆
78143 Casearia petelotii Merr. = Casearia balansae Gagnep. ●
78144 Casearia petelotii Merr. = Casearia velutina Blume ●
78145 Casearia praecox Griseb. ; 早生脚骨脆; Colombian Box, Colombian Boxwood, Maracaibo Box, Maracaibo Boxwood, Venezuelan Box, West Indian Box, West Indian Boxwood, Zapeteru ●☆
78146 Casearia prismatocarpa Mast. = Casearia barteri Mast. ●☆
78147 Casearia runssorica Gilg;伦索脚骨脆●☆
78148 Casearia schlechteri Gilg = Casearia congensis Gilg ●☆
78149 Casearia spruceana Benth. ex Eichler;云杉脚骨脆●☆
78150 Casearia stipitata Mast. ;具柄脚骨脆●☆
78151 Casearia subrhombea Hance = Xylosma congesta (Lour.) Merr. ●
78152 Casearia sylvestris Sw. ;野生脚骨脆●☆
78153 Casearia tardieuae Lescot et Sleumer;石生脚骨脆(钙生嘉赐树,石生嘉赐木);Calciphile Casearia, Saxicolous Casearia ●
78154 Casearia thonneri De Wild. = Casearia barteri Mast. ●☆
78155 Casearia tomentosa Roxb. ;椭圆脚骨脆;Elliptic Casearia ●☆
78156 Casearia velutina Blume;毛叶脚骨脆(毛叶嘉赐木,毛叶嘉赐树,箐黄果,爪哇嘉赐树,爪哇脚骨脆); Velvet-like Casearia, Velvety Casearia ●
78157 Casearia villilimba Merr. ;毛脚骨脆(脚骨脆,毛叶嘉赐树); Villous Casearia ●
78158 Casearia villilimba Merr. = Casearia balansae Gagnep. ●
78159 Casearia villilimba Merr. = Casearia velutina Blume ●
78160 Casearia virescens Pierre ex Gagnep. ;中越脚骨脆●
78161 Casearia yunnansnsis F. C. How et W. C. Ko;云南脚骨脆(云南嘉赐树);Yunnan Casearia ●
78162 Casearia yunnansnsis F. C. How et W. C. Ko = Casearia flexuosa Craib ●
78163 Casearia zenkeri Gilg = Casearia stipitata Mast. ●☆
78164 Caseola Noronha = Sonneratia L. f. (保留属名)●
78165 Cashalia Standl. = Dussia Krug et Urb. ex Taub. ■☆
78166 Casia Duhamel = Osyris L. ●
78167 Casia Gagnebin = Osyris L. ●
78168 Casimira Scop. = Melicoccus L. ●
78169 Casimirella Hassl. (1913);卡氏茶茱萸属●☆
78170 Casimirella guaranitica Hassl. ;卡氏茶茱萸●☆
78171 Casimiroa Dombey ex Baill. = Cervantesia Ruiz et Pav. ●☆
78172 Casimiroa La Lave et Lex. (1825);香肉果属(加锡弥罗果属);Casimiroa, Savoryfruit ●
78173 Casimiroa La Llave = Casimiroa La Lave et Lex. ●
78174 Casimiroa edulis La Llave = Casimiroa edulis La Llave et Lex. ●
78175 Casimiroa edulis La Llave et Lex. ;香肉果(掌叶瓜橘);Cochil Sapote, Edible Casimiroa, Mexican Apple, Mexican-apple, Savoryfruit, White Casimiroa, White Sapote ●
78176 Casimiroa sapota Oerst. ;掌叶香肉果(掌叶瓜橘)●☆
78177 Casimiroa tetrameria Millsp. ;四基香肉果●☆
78178 Casinga Griseb. = Laetia Loefl. ex L. (保留属名)●☆
78179 Casiostega Galeotti = Opizia J. Presl et C. Presl ■☆
78180 Casiostega Rupr. ex Galeotti = Opizia J. Presl et C. Presl ■☆
78181 Casiostega dactyloides (Nutt.) E. Fourn. = Buchloe dactyloides (Nutt.) Engelm. ■
78182 Casparea Kunth = Bauhinia L. ●
78183 Caspareopsis Britton et Rose = Bauhinia L. ●
78184 Caspareopsis monandra (Kurz) Britton et Rose = Bauhinia monandra Kurz ●☆
78185 Casparia Kunth = Bauhinia L. ●
78186 Casparya Klotzsch = Begonia L. ●■
78187 Casparya silletensis A. DC. = Begonia silletensis (A. DC.) C. B. Clarke ■
78188 Caspia Galushko = Salsola L. ●■
78189 Caspia Scop. = Vismia Vand. (保留属名)●☆
78190 Caspia foliosa (L.) Galushko = Salsola foliosa (L.) Schrad. ex Schult. ■
78191 Cassandra D. Don = Chamaedaphne Moench(保留属名)●
78192 Cassandra Spach = Eubotrys Nutt. ●☆
78193 Cassandra Spach = Leucothoe D. Don ●
78194 Cassandra calyculata (L.) D. Don = Cassandra calyculata (L.) Moench ●
78195 Cassandra calyculata (L.) D. Don = Chamaedaphne calyculata (L.) Moench ●
78196 Cassandra calyculata (L.) D. Don var. angustifolia (Aiton) Seym. = Chamaedaphne calyculata (L.) Moench var. angustifolia (Aiton) Rehder ●☆
78197 Cassandra calyculata (L.) Moench = Lyonia ovalifolia (Wall.) Drude var. formosana (Komatsu) T. Yamaz. ●
78198 Cassebeeria Dennst. = Codigi Augier ●■
78199 Cassebeeria Dennst. = Sonerila Roxb. (保留属名)●■
78200 Casselia Dumort. (废弃属名) = Casselia Nees et Mart. (保留属名)■●☆
78201 Casselia Dumort. (废弃属名) = Mertensia Roth(保留属名)■
78202 Casselia Nees et Mart. (1823)(保留属名);卡斯尔草属■●☆
78203 Casselia serrata Nees et Mart. ;卡斯尔草■●☆
78204 Cassia L. (1753)(保留属名);决明属(假含羞草属,山扁豆属,铁刀木属,铁刀苏木属);Cassia, Senna, Shower Tree ●■
78205 Cassia abbreviata Oliv. ;缩短决明●☆
78206 Cassia abbreviata Oliv. subsp. beareana (Holmes) Brenan;比尔决明●☆
78207 Cassia abbreviata Oliv. subsp. kassneri (Baker f.) Brenan;卡斯纳决明●☆
78208 Cassia abbreviata Oliv. var. glabrifructifera Steyaert = Cassia abbreviata Oliv. subsp. beareana (Holmes) Brenan ●☆
78209 Cassia abbreviata Oliv. var. granitica (Baker f.) Baker f. = Cassia abbreviata Oliv. subsp. beareana (Holmes) Brenan ●☆
78210 Cassia absus L. ;阿苏决明;Four-leaved Senna, Tropical Sensitive Pea ●☆
78211 Cassia absus L. = Chamaecrista absus (L.) H. S. Irwin et Barneby ■☆
78212 Cassia acutifolia Delile;旃那;Alexandrian Senna ●
78213 Cassia acutifolia Delile = Cassia angustifolia Wahlenb. ●
78214 Cassia acutifolia Delile = Cassia senna L. ●
78215 Cassia acutifolia Delile = Senna alexandrina Mill. ●
78216 Cassia adenensis Benth. ;亚丁决明●☆
78217 Cassia adenensis Benth. var. corneliana (Vatke) Chiov. = Cassia adenensis Benth. ●☆
78218 Cassia africana (Steyaert) Mendonça et Torre;非洲决明●☆
78219 Cassia afrofistula Brenan;肯尼亚决明(非洲腊肠树);African Laburnum, Kenyan Shower ●☆
78220 Cassia agnes (de Wit) Brenan = Cassia javanica L. subsp. agnes (de Wit) K. Larsen ●

78221 Cassia agnes (de Wit) Brenan = Cassia javanica L. subsp. nodosa (Buch. -Ham. ex Roxb.) K. Larsen et S. S. Larsen ●

78222 Cassia alata L. = Senna alata (L.) Roxb. ●

78223 Cassia angolensis Hiern = Cassia angolensis Welw. ex Hiern ●☆

78224 Cassia angolensis Welw. ex Hiern;安哥拉决明●☆

78225 Cassia angustifolia Vahl = Cassia senna L. ●

78226 Cassia angustifolia Vahl = Senna alexandrina Mill. ●

78227 Cassia angustifolia Wahlenb.;尖叶番泻(地熏叶,弟兄叶,番泻叶,尖叶番泻树,尖叶番泻叶,泡竹叶,通幽,狭叶番泻,狭叶番泻树,泻叶,泻叶茶,辛拿叶,亚历山大番泻,亚历山大番泻叶,印度番泻叶,旃那,旃那叶);Aden Senna, Alexandrian Senna, Arabian Senna, Congo Senna, Nubian Senna, Senna, Sharp-leaf Senna, Tinnevelly Senna, True Senna ●

78228 Cassia anthoxantha Capuron = Senna anthoxantha (Capuron) Du Puy ●☆

78229 Cassia arachoides Burch. = Cassia italica (Mill.) Spreng. subsp. arachoides (Burch.) Brenan ●☆

78230 Cassia armata S. Watson;沙地决明;Desert Cassia ●☆

78231 Cassia artemisioides DC. = Senna artemisioides (Gaudich. ex DC.) Randell ●☆

78232 Cassia artemisioides Gaudich. ex DC. = Senna artemisioides (Gaudich. ex DC.) Randell ●☆

78233 Cassia ashrek Forssk. = Senna italica Mill. ●☆

78234 Cassia atroreticulata Chiov. = Cassia abbreviata Oliv. subsp. beareana (Holmes) Brenan ●☆

78235 Cassia aubrevillei Pellegr.;奥布决明●☆

78236 Cassia auriculata L.;耳叶决明(耳叶番泻,麻榍勒茶,麻脱勒茶);Auriculate Senna, Auriculate-leaf Senna, Avaram Bark, Avaram Senna, Avaram Senna Bark, Matarab Tea, Matura Tea, Tanner's Cassia, Tanner's Cassia Bark, Turwad Bark ●

78237 Cassia australis Sims;澳大利亚决明●☆

78238 Cassia bacillaris L. = Cassia fruticosa Mill. ●

78239 Cassia beareana Holmes;大果山扁豆(贝尔决明)●☆

78240 Cassia beareana Holmes = Cassia abbreviata Oliv. subsp. beareana (Holmes) Brenan ●☆

78241 Cassia beccarinii Chiov.;贝卡林决明●☆

78242 Cassia bequaertii De Wild. = Cassia abbreviata Oliv. subsp. beareana (Holmes) Brenan ●☆

78243 Cassia bicapsularis L. = Senna bicapsularis (L.) Roxb. ●

78244 Cassia biciliaris L. f. = Cassia fruticosa Mill. ●

78245 Cassia biflora Mill.;双花决明;Twin Flowered Cassia ●☆

78246 Cassia brachiata (Pollard) J. F. Macbr.;对枝决明●☆

78247 Cassia brachiata (Pollard) J. F. Macbr. = Cassia fasciculata Michx. ●☆

78248 Cassia brevifolia Lam. = Chamaecrista brevifolia (Lam.) Greene ●☆

78249 Cassia brewsteri (F. Muell.) Benth. = Cassia brewsteri (F. Muell.) F. Muell. ex Benth. ●☆

78250 Cassia brewsteri (F. Muell.) F. Muell. ex Benth.;澳洲决明;Leichhardt Bean ●☆

78251 Cassia burmannii Wall. = Senna italica Mill. ●☆

78252 Cassia candenatensis Dennst. = Dalbergia candenatensis (Dennst.) Prain ●

78253 Cassia capensis Thunb. = Chamaecrista capensis (Thunb.) E. Mey. ●☆

78254 Cassia capensis Thunb. var. keiensis Steyaert = Chamaecrista capensis (Thunb.) E. Mey. var. flavescens E. Mey. ●☆

78255 Cassia caryophyllata L. = Dicypellium caryophyllaceum (C. Mart.) Nees et C. Mart. ●☆

78256 Cassia chamaecrista L. = Cassia fasciculata Michx. ●☆

78257 Cassia chamaecrista L. = Chamaecrista fasciculata (Michx.) Greene ●☆

78258 Cassia chamaecrista L. = Chamaecytisus chamaecrista Britton ●☆

78259 Cassia coluteoides Collad. = Cassia pendula Humb. et Bonpl. ex Willd. ●☆

78260 Cassia coluteoides Collad. = Senna pendula (Humb. et Bonpl. ex Willd.) H. S. Irwin et Barneby ●☆

78261 Cassia comosa (E. Mey.) Vogel = Chamaecrista comosa E. Mey. ●☆

78262 Cassia corymbosa Lam. = Senna corymbosa (Lam.) H. S. Irwin et Barneby ●☆

78263 Cassia coryophyllata Fresen.;巴西丁香树●☆

78264 Cassia delagoensis Harv. = Cassia petersiana Bolle ●☆

78265 Cassia densistipulata Taub. = Cassia auriculata L. ●

78266 Cassia dentata Vogel;齿状决明●☆

78267 Cassia didymobotrya Fresen. = Senna didymobotrya (Fresen.) H. S. Irwin et Barneby ●

78268 Cassia dimidiata Roxb.;二半决明(西南山决明)●☆

78269 Cassia dimidiata Roxb. = Cassia wallichiana DC. ●

78270 Cassia dimidiata Roxb. = Chamaecrista stricta E. Mey. ●☆

78271 Cassia droogmansiana De Wild. = Cassia abbreviata Oliv. subsp. beareana (Holmes) Brenan ●☆

78272 Cassia duboisii Steyaert;杜氏决明●☆

78273 Cassia dumaziana Brenan = Chamaecrista dumaziana (Brenan) Du Puy ●☆

78274 Cassia emarginata L.;微缺决明●☆

78275 Cassia eremophila Vogel;沙漠决明●☆

78276 Cassia excelsa Schrad.;高大决明;Crown of Gold Tree ●☆

78277 Cassia exilis Vatke;瘦小决明●☆

78278 Cassia falcinella Oliv.;小花决明●☆

78279 Cassia falcinella Oliv. var. intermedia Brenan;间型小花决明●☆

78280 Cassia falcinella Oliv. var. longifolia Ghesq. = Cassia parva Steyaert ●☆

78281 Cassia fallacina Chiov.;假山扁豆●☆

78282 Cassia fallacina Chiov. var. gracilior Ghesq.;纤细假山扁豆●☆

78283 Cassia fallacina Chiov. var. katangensis Ghesq. = Cassia katangensis (Ghesq.) Steyaert ●☆

78284 Cassia fasciculata Michx.;簇生决明;Golden Cassia, Golden Cassia Bark, Locust-weed, Partridge Pea, Partridge-pea, Prairie Senna, Senna, Showy Partridge Pea, Sleeping-plant ●☆

78285 Cassia fasciculata Michx. = Chamaecrista fasciculata (Michx.) Greene ●☆

78286 Cassia fasciculata Michx. var. brachiata (Pollard) Pullen ex Isely = Cassia fasciculata Michx. ●☆

78287 Cassia fasciculata Michx. var. depressa (Pollard) J. F. Macbr. = Cassia fasciculata Michx. ●☆

78288 Cassia fasciculata Michx. var. depressa (Pollard) J. F. Macbr. = Chamaecrista fasciculata (Michx.) Greene ●☆

78289 Cassia fasciculata Michx. var. fasciculata f. jensenii E. J. Palmer et Steyerm. = Chamaecrista fasciculata (Michx.) Greene ●☆

78290 Cassia fasciculata Michx. var. ferrisiae (Britton ex Britton et Rose) B. L. Turner = Cassia fasciculata Michx. ●☆

78291 Cassia fasciculata Michx. var. macrosperma Fernald = Cassia fasciculata Michx. ●☆

78292 Cassia fasciculata Michx. var. puberula (Greene) J. F. Macbr. = Cassia fasciculata Michx. ●☆

78293 Cassia fasciculata Michx. var. robusta (Pollard) J. F. Macbr. = Cassia fasciculata Michx. ●☆

78294 Cassia fasciculata Michx. var. robusta (Pollard) J. F. Macbr. = Chamaecrista fasciculata (Michx.) Greene ●☆

78295 Cassia fasciculata Michx. var. rostrata (Wooton et Standl.) B. L. Turner = Cassia fasciculata Michx. ●☆

78296 Cassia fasciculata Michx. var. tracyi (Pollard) J. F. Macbr. = Cassia fasciculata Michx. ●☆

78297 Cassia fenarolii Mendonça et Torre;费纳罗利决明●☆

78298 Cassia ferruginea Schrad. ex DC.;锈色铁刀苏木●☆

78299 Cassia filipendula Bojer ex Bouton = Chamaecrista stricta E. Mey. ●☆

78300 Cassia fistula L.;腊肠树(阿勃参,阿勃勒,阿勃簕,阿勒勃,阿黎,波斯皂荚,长果子树,黄槐花树,腊肠苏木,牛角树,婆罗门皂荚,清泻山扁豆);Casse Fistuleuse, Cassia Bean, Cassiasistre, Drumstick Tree, Drumstick-tree, Golden Shower, Golden Shower Tree, Golden-shower, Goldenshower Senna, Golden-shower Senna, Golden-shower Tree, Indian Laburnum, Indian Senna, Pipe-tree, Pudding Pipe Tree, Pudding-pine Tree, Purging Cassia, Purging Cassia Bark, Purging Fistula ●

78301 Cassia floribunda Cav.;光叶决明(大花黄槐,光决明,怀花米,平滑决明);Flowery Senna, Hedionda-macho, Profuse-flowering Senna ●

78302 Cassia fruticosa Mill. = Senna fruticosa (Mill.) H. S. Irwin et Barneby ●

78303 Cassia garambiensis Hosok.;鹅銮鼻决明;Eluanbi Senna ●

78304 Cassia garambiensis Hosok. = Chamaecrista garambiensis (Hosok.) H. Ohashi, Tateishi et T. Nemoto ●

78305 Cassia garrettiana Craib;加雷决明●☆

78306 Cassia ghesquiereana Brenan;盖斯基埃决明●☆

78307 Cassia glandulosa L. var. swartzii J. F. Macbr.;具腺决明;Wild Tamarind ●☆

78308 Cassia glauca Lam.;粉绿决明(粉叶决明,黄槐);Glaucous Senna, Glaucous-leaved Senna, Pale Senna ●

78309 Cassia glauca Lam. = Cassia surattensis Burm. f. subsp. glauca (Lam.) K. Larsen et S. S. Larsen ●

78310 Cassia glauca Lam. = Cassia surattensis Burm. f. ●

78311 Cassia glauca Lam. = Senna sulfurea (Collad.) H. S. Irwin et Barneby ●

78312 Cassia glauca Lam. var. suffruticosa (Koen. ex Roth) Baker = Cassia surattensis Burm. f. ●

78313 Cassia goratensis Fresen. = Cassia singueana Delile ●☆

78314 Cassia gossweileri Baker f.;戈斯决明●☆

78315 Cassia gracilior (Ghesq.) Steyaert;纤细决明●☆

78316 Cassia grandis L. f.;红花铁刀木(大果铁刀木,大决明,大铁刀苏木,巨决明);Appleblossom Cassia, Horse Cassia, Horse Cassia Bark, Pink Shower, Pink-shower, Pinkshower Senna ●

78317 Cassia granitica Baker f. = Cassia abbreviata Oliv. subsp. beareana (Holmes) Brenan ●☆

78318 Cassia grantii Oliv.;格兰特决明●☆

78319 Cassia grantii Oliv. var. pilosula? = Cassia grantii Oliv. ●☆

78320 Cassia hebecarpa Fernald = Senna hebecarpa (Fernald) H. S. Irwin et Barneby ●☆

78321 Cassia hebecarpa Fernald var. longipila E. L. Braun = Senna hebecarpa (Fernald) H. S. Irwin et Barneby ●☆

78322 Cassia hildebrandtii Vatke;希尔德决明●☆

78323 Cassia hildebrandtii Vatke var. crispata Serrato;皱波决明●☆

78324 Cassia hirsuta L. = Senna hirsuta (L.) H. S. Irwin et Barneby ●

78325 Cassia hochstetteri Ghesq.;霍赫决明●☆

78326 Cassia holosericea Fresen.;全绢毛决明●☆

78327 Cassia huillensis Mendonça et Torre;威拉决明●☆

78328 Cassia humifusa Brenan;平伏决明●☆

78329 Cassia indecora Kunth var. glabrata Vogel = Cassia pendula Humb. et Bonpl. ex Willd. ●☆

78330 Cassia italica (Mill.) F. W. Andréws = Cassia italica (Mill.) Spreng. ●☆

78331 Cassia italica (Mill.) Lam. ex F. W. Andréws = Cassia italica (Mill.) Spreng. ●☆

78332 Cassia italica (Mill.) Spreng. = Senna italica Mill. ●☆

78333 Cassia italica (Mill.) Spreng. subsp. arachoides (Burch.) Brenan = Senna italica Mill. subsp. micrantha (Brenan) Lock ●☆

78334 Cassia italica (Mill.) Spreng. subsp. micrantha Brenan = Senna italica Mill. subsp. micrantha (Brenan) Lock ●☆

78335 Cassia jaegeri Keay;耶格决明●☆

78336 Cassia javanica L.;爪哇决明(缤纷决明,腊肠豆,排钱豆,雄黄豆,爪哇旃那);Apple Blossom, Apple-blossom Cassia, Pink Shower, Rainbow Shower ●☆

78337 Cassia javanica L. subsp nodosa (Buch.-Ham. ex Roxb.) K. Larsen et S. S. Larsen;节荚决明●

78338 Cassia javanica L. subsp. agnes (de Wit) K. Larsen;神黄豆(回回豆,腊肠豆,排钱豆,雄黄豆,越南决明);Apple Blossom, God Senna, Indo-China Senna ●

78339 Cassia javanica L. subsp. nodosa (Buch.-Ham. ex Roxb.) K. Larsen et S. S. Larsen = Cassia agnes (de Wit) Brenan ●

78340 Cassia javanica L. var. agnes de Wit = Cassia agnes (de Wit) Brenan ●

78341 Cassia javanica L. var. agnes de Wit = Cassia javanica L. subsp. agnes (de Wit) K. Larsen ●

78342 Cassia javanica L. var. agnes de Wit = Cassia javanica L. subsp. nodosa (Buch.-Ham. ex Roxb.) K. Larsen et S. S. Larsen ●

78343 Cassia javanica L. var. indochinensis Gagnep. = Cassia agnes (de Wit) Brenan ●

78344 Cassia javanica L. var. indochinensis Gagnep. = Cassia javanica L. subsp. nodosa (Buch.-Ham. ex Roxb.) K. Larsen et S. S. Larsen ●

78345 Cassia kalulensis Steyaert;卡卢莱决明●☆

78346 Cassia kassneri Baker f. = Cassia abbreviata Oliv. subsp. kassneri (Baker f.) Brenan ●☆

78347 Cassia katangensis (Ghesq.) Steyaert;加丹加决明●☆

78348 Cassia katangensis (Ghesq.) Steyaert = Cassia fallacina Chiov. var. katangensis Ghesq. ●☆

78349 Cassia katangensis (Ghesq.) Steyaert var. nuda Steyaert;裸露决明●☆

78350 Cassia kethulleana De Wild. = Cassia singueana Delile ●☆

78351 Cassia kirkii Oliv.;柯克决明●☆

78352 Cassia kirkii Oliv. var. glabra Steyaert;光滑决明●☆

78353 Cassia kirkii Oliv. var. guineensis Steyaert;几内亚决明●☆

78354 Cassia kirkii Oliv. var. microphylla Dewèvre = Cassia robynsiana Ghesq. ●☆

78355 Cassia kirkii Oliv. var. quarrei Ghesq. = Cassia quarrei (Ghesq.) Steyaert ●☆

78356 Cassia kirkii Oliv. var. quarrei Ghesq. = Chamaecrista stricta E. Mey. ●☆

78357 Cassia kituiensis Vatke = Cassia grantii Oliv. ●☆
78358 Cassia kotschyana Oliv. = Cassia sieberiana DC. ●☆
78359 Cassia lactea Vatke = Senna lactea (Vatke) Du Puy ●☆
78360 Cassia laevigata Willd. = Cassia floribunda Cav. ●
78361 Cassia laevigata Willd. = Senna septemtrionalis (Viv.) H. S. Irwin et Barneby ●
78362 Cassia lancangensis Y. Y. Qian;澜沧决明;Lancang Senna ●
78363 Cassia lanceolata Forssk. = Cassia senna L. ●
78364 Cassia leandrii Ghesq. = Senna leandrii (Ghesq.) Du Puy ●☆
78365 Cassia leandrii Ghesq. var. maesta Ghesq. = Senna leandrii (Ghesq.) Du Puy ●☆
78366 Cassia lechenaultiana DC.;短叶决明(篦子草,大叶假含羞草,大叶山扁豆,地油甘,牛旧藤,铁箭矮陀,铁皂角,野皂角,夜合草);Shortleaf Senna ●
78367 Cassia lechenaultiana DC. = Chamaecrista lechenaultiana (DC.) O. Deg. ■
78368 Cassia leptophylla Vogel;细叶决明(狭叶决明);Gold Medallion Tree ●☆
78369 Cassia leschenaultiana DC.;大叶假含羞草(大叶山扁豆)●
78370 Cassia leschenaultiana DC. = Chamaecrista lechenaultiana (DC.) O. Deg. ■
78371 Cassia lignea?;木决明●☆
78372 Cassia ligustrina L.;女贞决明●☆
78373 Cassia longiracemosa Vatke;长序决明●☆
78374 Cassia macranthera DC. ex Collad.;大花药决明●
78375 Cassia mannii Oliv.;曼氏决明●☆
78376 Cassia mannii Oliv. var. van-houttei De Wild. = Cassia angolensis Hiern ●☆
78377 Cassia marginata Roxb.;印度决明(边缘决明,缘生铁刀苏木)●☆
78378 Cassia marginata Roxb. = Cassia roxburghii DC. ●☆
78379 Cassia marilandica L. = Senna marilandica (L.) Link ●☆
78380 Cassia marksiana (F. M. Bailey) Domin;亮黄决明●☆
78381 Cassia medsgeri Shafer = Cassia marilandica L. ●☆
78382 Cassia medsgeri Shafer = Senna marilandica (L.) Link ●☆
78383 Cassia meelii Steyaert;米尔决明●☆
78384 Cassia meridionalis R. Vig. = Senna meridionalis (R. Vig.) Du Puy ●☆
78385 Cassia mimosoides L. = Chamaecrista mimosoides (L.) Greene ●
78386 Cassia mimosoides L. subsp. lechenaultiana (DC.) H. Ohashi = Chamaecrista lechenaultiana (DC.) O. Deg. ■
78387 Cassia mimosoides L. subsp. leschenaultiana (DC.) H. Ohashi = Cassia lechenaultiana DC. ●
78388 Cassia mimosoides L. subsp. nomame (Siebold) H. Ohashi = Cassia nomame (Siebold) Kitag. ●
78389 Cassia mimosoides L. subsp. nomame (Siebold) H. Ohashi = Chamaecrista nomame (Siebold) H. Ohashi ●
78390 Cassia mimosoides L. var. africana Steyaert = Cassia africana (Steyaert) Mendonça et Torre ●☆
78391 Cassia mimosoides L. var. ankaratrensis R. Vig. = Chamaecrista ankaratrensis (R. Vig.) Du Puy ●☆
78392 Cassia mimosoides L. var. arenicola R. Vig. = Chamaecrista arenicola (R. Vig.) Du Puy ●☆
78393 Cassia mimosoides L. var. capensis (Thunb.) Harv. = Chamaecrista capensis (Thunb.) E. Mey. ●☆
78394 Cassia mimosoides L. var. comosa (E. Mey.) Harv. = Chamaecrista comosa E. Mey. ●☆
78395 Cassia mimosoides L. var. dimidiata? = Cassia mimosoides L. subsp. lechenaultiana (DC.) H. Ohashi ●
78396 Cassia mimosoides L. var. garambiensis (Hosok.) S. S. Ying = Cassia garambiensis Hosok. ●
78397 Cassia mimosoides L. var. garambiensis (Hosok.) S. S. Ying = Chamaecrista garambiensis (Hosok.) H. Ohashi, Tateishi et T. Nemoto ●
78398 Cassia mimosoides L. var. hygrophila R. Vig. = Chamaecrista stricta E. Mey. ●☆
78399 Cassia mimosoides L. var. lateriticola R. Vig. = Chamaecrista lateriticola (R. Vig.) Du Puy ●☆
78400 Cassia mimosoides L. var. myriophylla? = Cassia mimosoides L. ●
78401 Cassia mimosoides L. var. nomame (Siebold) Makino = Chamaecrista nomame (Siebold) H. Ohashi ●
78402 Cassia mimosoides L. var. pacifica (Ohwi) Tawada = Chamaecrista garambiensis (Hosok.) H. Ohashi, Tateishi et T. Nemoto ●
78403 Cassia mimosoides L. var. pratensis R. Vig. = Chamaecrista pratensis (R. Vig.) Du Puy ●☆
78404 Cassia mimosoides L. var. stricta (E. Mey.) Harv. = Cassia quarrei (Ghesq.) Steyaert ●☆
78405 Cassia mimosoides L. var. telfairiana Hook. f. = Cassia mimosoides L. ●
78406 Cassia mimosoides L. var. telfairiana Hook. f. = Cassia telfairiana (Hook. f.) Polhill ●☆
78407 Cassia mimosoides L. var. telfairiana Hook. f. = Chamaecrista telfairiana (Hook. f.) Lock ●☆
78408 Cassia mimosoides L. var. wallichiana (DC.) Baker;短叶山扁豆(大叶山扁豆,短叶决明)●
78409 Cassia mimosoides L. var. wallichiana (DC.) Baker = Cassia leschenaultiana DC. ●
78410 Cassia mimosoides L. var. wallichiana (DC.) Baker = Cassia mimosoides L. subsp. lechenaultiana (DC.) H. Ohashi ●
78411 Cassia mimosoides L. var. wallichiana (DC.) Baker = Chamaecrista lechenaultiana (DC.) O. Deg. ■
78412 Cassia mimosoides L. var. wallichiana DC. = Cassia mimosoides L. var. wallichiana (DC.) Baker ●
78413 Cassia mississippiensis Pollard = Cassia fasciculata Michx. ●☆
78414 Cassia moschata Kunth;麝香决明●☆
78415 Cassia multiglandulosa Jacq. = Senna multiglandulosa (Jacq.) H. S. Irwin et Barneby ●☆
78416 Cassia multijuga Rich. = Senna multijuga (Rich.) H. S. Irwin et Barneby ●
78417 Cassia mututu De Wild. = Cassia singueana Delile ●☆
78418 Cassia nairobensis L. H. Bailey = Cassia didymobotrya Fresen. ●☆
78419 Cassia nairobensis L. H. Bailey = Senna didymobotrya (Fresen.) H. S. Irwin et Barneby ●
78420 Cassia newtonii Mendonça et Torre;纽敦决明■☆
78421 Cassia nictitans L.;闪烁决明;Partridge Pea, Sensitive Partridge Pea, Sensitive Pea, Wild Sensitive Plant, Wild Sensitive-plant ●☆
78422 Cassia nictitans L. = Cassia lechenaultiana DC. ●
78423 Cassia nictitans L. = Chamaecrista nictitans (L.) Moench ●☆
78424 Cassia nigricans Vahl;黑决明●☆
78425 Cassia nodosa Buch.-Ham. ex Roxb.;节果决明(粉花决明,粉花山扁豆,回回豆,神黄豆);Joint Wood, Jointwood, Jointwood Senna, Joinwood, Knotty Senna, Pink Cassia ●
78426 Cassia nodosa Buch.-Ham. ex Roxb. = Cassia javanica L. subsp.

nodosa (Buch. -Ham. ex Roxb.) K. Larsen et S. S. Larsen ●
78427 Cassia nodosa Buch. -Ham. ex Roxb. = Cassia javanica L. var. indochinensis Gagnep. ●
78428 Cassia nomame (Siebold) Honda = Chamaecrista nomame (Siebold) H. Ohashi ●
78429 Cassia nomame (Siebold) Kitag.;豆茶决明(关门草,江芒决明,金豆子,莲子草,山扁豆,山茶叶,山梅豆,山野扁豆,水通,水皂角,夜关草);Nomame Senna ●
78430 Cassia nomame (Siebold) Kitag. = Senna nomame (Siebold) T. Chen ●
78431 Cassia obovata Collad. = Cassia italica (Mill.) F. W. Andréws ●☆
78432 Cassia obovata Collad. = Senna italica Mill. ●☆
78433 Cassia obovata Collad. var. mucronata Burtt Davy = Senna italica Mill. subsp. arachoides (Burch.) Lock ●☆
78434 Cassia obovata Collad. var. pallidiflora Dinter = Senna italica Mill. subsp. micrantha (Brenan) Lock ●☆
78435 Cassia obovata Collad. var. pilosa Burtt Davy = Senna italica Mill. subsp. arachoides (Burch.) Lock ●☆
78436 Cassia obovata Hayne;倒卵叶番泻树(倒卵叶番泻,卵叶番泻);Dog Senna, Italian Senna, Jamaica Senna, Obovate-leaved Senna, Port Royal Senna, Senegal Senna, Spanish Senna, Tripoli Senna ●☆
78437 Cassia obovata Hayne = Senna italica Mill. ●☆
78438 Cassia obtusifolia L. = Senna obtusifolia (L.) H. S. Irwin et Barneby ●☆
78439 Cassia occidentalis L. = Senna occidentalis (L.) Link ●
78440 Cassia odorata R. Morris = Senna odorata (Morris) Randell ●☆
78441 Cassia pachycarpa de Wit = Cassia grandis L. f. ●
78442 Cassia parahyba Vell. = Schizolobium parahybum (Vell.) Blake ●☆
78443 Cassia paralias Brenan;蓝决明●☆
78444 Cassia parva Steyaert;小决明●☆
78445 Cassia pendula Humb. et Bonpl. ex Willd.;下垂决明●☆
78446 Cassia perrieri R. Vig. ex Ghesq. = Senna perrieri (R. Vig.) Du Puy ●☆
78447 Cassia petersiana Bolle = Senna petersiana (Bolle) Lock ●☆
78448 Cassia phyllodinea R. Br. = Senna phyllodinea (R. Br.) Symon ●☆
78449 Cassia planitiicola Domin;茳芒决明(茶花儿,槐叶决明,山扁豆)●
78450 Cassia plumosa (E. Mey.) Vogel;羽状决明●☆
78451 Cassia plumosa (E. Mey.) Vogel var. diffusa? = Chamaecrista plumosa E. Mey. ●☆
78452 Cassia plumosa (E. Mey.) Vogel var. erecta Schorn et Gordon-Gray = Chamaecrista plumosa E. Mey. var. erecta (Schorn et Gordon-Gray) Lock ●☆
78453 Cassia podocarpa Guillaumin et Perr.;柄果决明●☆
78454 Cassia polyphylla Jacq.;多叶决明●☆
78455 Cassia polytricha Brenan;多毛决明●☆
78456 Cassia polytricha Brenan var. pauciflora?;少花美丽多毛决明●☆
78457 Cassia polytricha Brenan var. pulchella?;美丽多毛决明●☆
78458 Cassia psilocarpa Welw.;毛果决明●☆
78459 Cassia puccioniana Chiov.;普乔尼决明●☆
78460 Cassia pulchella Bojer = Cassia telfairiana (Hook. f.) Polhill ●☆
78461 Cassia pumila Lam.;柄腺山扁豆(柄脉山扁豆);Stipitategland Senna ●
78462 Cassia pumila Lam. = Chamaecrista pumila (Lam.) K. Larsen ●■
78463 Cassia purpurea Roxb.;紫决明●☆
78464 Cassia quarrei (Ghesq.) Steyaert = Chamaecrista stricta E. Mey. ●☆
78465 Cassia queenslandica C. T. White;昆士兰决明●☆
78466 Cassia quinquangulata Rich.;五角决明●☆
78467 Cassia reducta Brenan = Chamaecrista reducta (Brenan) Du Puy ●☆
78468 Cassia reticulata Willd.;网状决明●☆
78469 Cassia retusa Vogel;微凹决明●☆
78470 Cassia robusta (Pollard) Pollard = Cassia fasciculata Michx. ●☆
78471 Cassia robynsiana Ghesq.;罗宾斯决明●☆
78472 Cassia rogeonii Ghesq.;罗容决明●☆
78473 Cassia rostrata (Wooton et Standl.) Tidestr. ex Tidestr. = Cassia fasciculata Michx. ●☆
78474 Cassia rotundifolia Pers.;钝叶决明●☆
78475 Cassia roxburghii DC.;罗氏决明●☆
78476 Cassia ruspolii Chiov.;鲁斯波利决明●☆
78477 Cassia sabak Delile = Cassia singueana Delile ●☆
78478 Cassia schmitzii Steyaert;施密茨决明●☆
78479 Cassia scleroxylon Ducke;坚硬铁刀苏木●☆
78480 Cassia senna L. = Senna alexandrina Mill. ●
78481 Cassia senna L. var. obtusata Brenan;钝决明●☆
78482 Cassia septemtrionalis Viv. = Senna septemtrionalis (Viv.) H. S. Irwin et Barneby ●
78483 Cassia siamea Lam. = Senna siamea (Lam.) H. S. Irwin et Barneby ●
78484 Cassia sieberiana DC.;西博决明;African Laburnum ●☆
78485 Cassia singueana Delile;东非决明(东非山扁豆)●☆
78486 Cassia somalensis Serrato = Cassia adenensis Benth. ●☆
78487 Cassia sophera L. = Senna sophera (L.) Roxb. ●
78488 Cassia sophera L. var. purpurea? = Cassia purpurea Roxb. ●☆
78489 Cassia sophora L. var. penghuana Y. C. Liu et F. Y. Lu;澎湖决明(苦参类决明)●
78490 Cassia sparsa Steyaert = Cassia quarrei (Ghesq.) Steyaert ●☆
78491 Cassia spectabilis DC. = Senna spectabilis (DC.) H. S. Irwin et Barneby ●
78492 Cassia splendida Vogel;闪光决明●☆
78493 Cassia stricta (E. Mey.) Steud. = Chamaecrista stricta E. Mey. ●☆
78494 Cassia stuhlmannii Taub. = Cassia zambesica Oliv. ●☆
78495 Cassia suarezensis Capuron = Senna suarezensis (Capuron) Du Puy ●☆
78496 Cassia suffruticosa König ex Roth = Cassia surattensis Burm. f. ●
78497 Cassia sulfurea Collad. = Senna sulfurea (Collad.) H. S. Irwin et Barneby ●
78498 Cassia surattensis Burm. f. = Senna surattensis (Burm. f.) H. S. Irwin et Barneby ●
78499 Cassia surattensis Burm. f. subsp. glauca (Lam.) K. Larsen et S. S. Larsen = Cassia glauca Lam. ●
78500 Cassia surattensis Burm. f. subsp. glauca (Lam.) K. Larsen et S. S. Larsen = Senna sulfurea (Collad.) H. S. Irwin et Barneby ●
78501 Cassia surattensis Burm. f. subsp. glauca K. Klarsen, S. S. Larsen et Vidal = Cassia glauca Lam. ●
78502 Cassia telfairiana (Hook. f.) Polhill;泰尔决明●☆
78503 Cassia telfairiana (Hook. f.) Polhill = Chamaecrista telfairiana (Hook. f.) Lock ●☆
78504 Cassia tettensis Bolle = Cassia singueana Delile ●☆
78505 Cassia thyrsoidea Brenan;聚伞决明●☆

78506 Cassia timorensis DC.;帝汶决明●☆
78507 Cassia tomentosa L. f.;绵毛决明;Woolly Senna ●☆
78508 Cassia tomentosa L. f. = Cassia multiglandulosa Jacq. ●☆
78509 Cassia tora L. = Senna obtusifolia (L.) H. S. Irwin et Barneby ●☆
78510 Cassia tora L. = Senna tora (L.) Roxb. ●
78511 Cassia tora L. var. obtusifolia (L.) Haines = Senna obtusifolia (L.) H. S. Irwin et Barneby ●☆
78512 Cassia torosa Cav.;草决明(大本羊角豆,羊角豆,珠节决明)●☆
78513 Cassia transversali-seminata De Wild. = Senna bicapsularis (L.) Roxb. ●
78514 Cassia truncata Brenan;截叶决明●☆
78515 Cassia undulata Benth.;波状决明●☆
78516 Cassia usambarensis Taub.;乌桑巴拉决明●☆
78517 Cassia verdickii De Wild. = Cassia didymobotrya Fresen. ●☆
78518 Cassia viguierella Ghesq. = Senna viguierella (Ghesq.) Du Puy ●☆
78519 Cassia viguierella Ghesq. var. meridionalis Ghesq. = Senna meridionalis (R. Vig.) Du Puy ●☆
78520 Cassia villosa Mill.;柔毛决明●☆
78521 Cassia wallichiana DC. = Cassia leschenaultiana DC. ●
78522 Cassia wallichiana DC. = Chamaecrista lechenaultiana (DC.) O. Deg. ■
78523 Cassia wildemaniana Ghesq. = Cassia kirkii Oliv. ●☆
78524 Cassia wittei Ghesq.;维特决明●☆
78525 Cassia zambesica Oliv.;赞比西决明●☆
78526 Cassiaceae Link = Fabaceae Lindl.(保留科名)●■
78527 Cassiaceae Link = Leguminosae Juss.(保留科名)●■
78528 Cassiaceae Vest = Fabaceae Lindl.(保留科名)●■
78529 Cassiaceae Vest = Leguminosae Juss.(保留科名)●■
78530 Cassiana Raf. = Cassia L.(保留属名)●■
78531 Cassida Hill = Scutellaria L. ●■
78532 Cassida Ség. = Scutellaria L. ●■
78533 Cassida Tourn. ex Adans. = Scutellaria L. ●■
78534 Cassidispermum Hemsl. = Burckella Pierre ●☆
78535 Cassidocarpus C. Presl ex DC. = Asteriscium Cham. et Schltdl. ■☆
78536 Cassidospermum Post et Kuntze = Burckella Pierre ●☆
78537 Cassidospermum Post et Kuntze = Cassidispermum Hemsl. ●☆
78538 Cassiera Raeusch. = Cansjera Juss.(保留属名)●
78539 Cassine Kuntze = Otherodendron Makino ●
78540 Cassine L.(1753)(保留属名);藏红卫矛属;Caxsine ●☆
78541 Cassine L. = Elaeodendron L. ●☆
78542 Cassine Loes. = Elaeodendron J. Jacq. ●☆
78543 Cassine aethiopica Thunb. = Mystroxylon aethiopicum (Thunb.) Loes. ●☆
78544 Cassine affinis Sond. = Cassine peragua L. subsp. affinis (Sond.) R. H. Archer ●☆
78545 Cassine albanensis Sond. = Lauridia tetragona (L. f.) R. H. Archer ●☆
78546 Cassine albivenosa (Chiov.) Cufod. = Elaeodendron schweinfurthianum (Loes.) Loes. ●☆
78547 Cassine aquifolia Fiori = Elaeodendron aquifolium (Fiori) Chiov. ●☆
78548 Cassine barbara L. = Cassine peragua L. subsp. barbara (L.) R. H. Archer ●☆
78549 Cassine buchananii Loes. = Elaeodendron buchananii (Loes.) Loes. ●☆
78550 Cassine burchellii Loes. = Cassine parvifolia Sond. ●☆
78551 Cassine capensis L. = Cassine peragua L. ●☆
78552 Cassine capensis L. var. colpoon (L.) DC. = Cassine peragua L. ●☆
78553 Cassine colpoon (L.) Thunb. = Cassine peragua L. ●☆
78554 Cassine confertiflora (Tul.) Loes. = Mystroxylon aethiopicum (Thunb.) Loes. ●☆
78555 Cassine crocea (Thunb.) Kuntze;藏红卫矛木;Saffron-wood ●☆
78556 Cassine crocea (Thunb.) Kuntze = Elaeodendron croceum (Thunb.) DC. ●☆
78557 Cassine discolor Wall. = Microtropis discolor Wall. ●
78558 Cassine engleriana Loes. = Mystroxylon aethiopicum (Thunb.) Loes. ●☆
78559 Cassine eucleiformis (Eckl. et Zeyh.) Kuntze = Robsonodendron eucleiforme (Eckl. et Zeyh.) R. H. Archer ●☆
78560 Cassine excelsa Wall. = Ilex excelsa (Wall.) Hook. f. ●
78561 Cassine glauca (Pers.) Kuntze = Cassine glauca (Rottb.) Kuntze ●☆
78562 Cassine glauca (Pers.) Kuntze var. kamerunensis (Loes.) R. Wilczek = Elaeodendron kamerunense (Loes.) Villiers ●☆
78563 Cassine glauca (Rottb.) Kuntze;灰蓝藏红卫矛●☆
78564 Cassine holstii Loes. = Mystroxylon aethiopicum (Thunb.) Loes. ●☆
78565 Cassine illicifolia Hayata = Microtropis fokienensis Dunn ●
78566 Cassine japonica (Franch. et Sav.) Kuntze = Microtropis japonica (Franch. et Sav.) Hallier f. ●
78567 Cassine kotoensis Hayata = Microtropis japonica (Franch. et Sav.) Hallier f. ●
78568 Cassine kraussiana Bernh. = Cassine peragua L. ●☆
78569 Cassine lacinulata Loes. = Elaeodendron schlechterianum (Loes.) Loes. ●☆
78570 Cassine latifolia Eckl. et Zeyh. = Lauridia tetragona (L. f.) R. H. Archer ●☆
78571 Cassine latifolia Eckl. et Zeyh. var. heterophylla Sond. = Lauridia tetragona (L. f.) R. H. Archer ●☆
78572 Cassine laurifolia (Harv.) Davison = Allocassine laurifolia (Harv.) N. Robson ●☆
78573 Cassine macrocarpa (Sond.) Kuntze;大果藏红卫矛●☆
78574 Cassine maritima (Bolus) F. Bolus et L. Bolus = Robsonodendron maritimum (Bolus) R. H. Archer ●☆
78575 Cassine matsudai Hayata = Microtropis fokienensis Dunn ●
78576 Cassine maurocenia L. = Maurocenia frangula Mill. ●☆
78577 Cassine micrantha Hayata = Microtropis micrantha (Hayata) Koidz. ●
78578 Cassine papillosa (Hochst.) Kuntze = Elaeodendron croceum (Thunb.) DC. ●☆
78579 Cassine parvifolia E. Mey. = Elaeodendron zeyheri Spreng. ex Turcz. ●☆
78580 Cassine parvifolia Sond.;小叶藏红卫矛●☆
78581 Cassine peragua L.;美国藏红卫矛●☆
78582 Cassine peragua L. subsp. affinis (Sond.) R. H. Archer;近缘藏红卫矛●☆
78583 Cassine peragua L. subsp. barbara (L.) R. H. Archer;外来藏红卫矛●☆
78584 Cassine pubescens (Eckl. et Zeyh.) Kuntze = Mystroxylon aethiopicum (Thunb.) Loes. ●☆
78585 Cassine reticulata (Eckl. et Zeyh.) Codd = Lauridia reticulata Eckl. et Zeyh. ●☆

78586 Cassine scandens Eckl. et Zeyh. = Lauridia tetragona (L. f.) R. H. Archer ●☆
78587 Cassine scandens Eckl. et Zeyh. var. latifolia Sond. = Lauridia tetragona (L. f.) R. H. Archer ●☆
78588 Cassine scandens Eckl. et Zeyh. var. laxa Loes. = Lauridia tetragona (L. f.) R. H. Archer ●☆
78589 Cassine schinoides (Spreng.) R. H. Archer;好望角藏红卫矛●☆
78590 Cassine schinziana (Loes.) Loes. = Cassine parvifolia Sond. ●☆
78591 Cassine schlechteri (Loes.) Davison = Mystroxylon aethiopicum (Thunb.) Loes. ●☆
78592 Cassine schlechteriana Loes. = Elaeodendron schlechterianum (Loes.) Loes. ●☆
78593 Cassine schweinfurthiana Loes. = Elaeodendron schweinfurthianum (Loes.) Loes. ●☆
78594 Cassine sphaerophylla (Eckl. et Zeyh.) Kuntze = Mystroxylon aethiopicum (Thunb.) Loes. ●☆
78595 Cassine stuhlmannii (Loes.) Blakelock = Elaeodendron schlechterianum (Loes.) Loes. ●☆
78596 Cassine tetragona (L. f.) Druce = Lauridia tetragona (L. f.) R. H. Archer ●☆
78597 Cassine tetragona (L. f.) Loes. = Lauridia tetragona (L. f.) R. H. Archer ●☆
78598 Cassine tetragona (Thunb.) Loes. = Lauridia tetragona (L. f.) R. H. Archer ●☆
78599 Cassine tetragona Thunb. var. laxa (Loes.) Loes. = Lauridia tetragona (L. f.) R. H. Archer ●☆
78600 Cassine transvaalensis (Burtt Davy) Codd = Elaeodendron transvaalense (Burtt Davy) R. H. Archer ●☆
78601 Cassine velutinum (Harv.) Loes. = Mystroxylon aethiopicum (Thunb.) Loes. ●☆
78602 Cassinia R. Br. (1813)(保留属名);滨篱菊属(比迪木属)●☆
78603 Cassinia R. Br. ex Aiton = Angianthus J. C. Wendl. (保留属名) ■●☆
78604 Cassinia R. Br. ex Aiton = Cassinia R. Br. (保留属名)●☆
78605 Cassinia aculeata (Labill.) R. Br.;皮刺滨篱菊;Cassinia ●☆
78606 Cassinia aculeata R. Br. = Cassinia aculeata (Labill.) R. Br. ●☆
78607 Cassinia alba O. Hoffm. = Helichrysum platypterum DC. ●☆
78608 Cassinia arcuata R. Br.;比迪木;Biddy Bush, Chinese Shrub, Sifton Bush ●☆
78609 Cassinia aureonitens N. A. Wakef.;金花比迪木●☆
78610 Cassinia fulvida Hook. f.;金叶比迪木(黄枝滨篱菊);Golden Cotton Wood, Golden Cottonwood, Golden Tauhium ●☆
78611 Cassinia leptophylla R. Br.;狭叶比迪木●☆
78612 Cassinia phylicifolia (DC.) J. M. Wood = Tenrhynea phylicifolia (DC.) Hilliard et B. L. Burtt ■☆
78613 Cassinia vauvilliersii Hook. f.;绒背叶滨篱菊●☆
78614 Cassiniaceae Sch. Bip.;滨篱菊科●
78615 Cassiniaceae Sch. Bip. = Asteraceae Bercht. et J. Presl(保留科名)●■
78616 Cassiniaceae Sch. Bip. = Compositae Giseke(保留科名)●■
78617 Cassiniola F. Muell. = Helipterum DC. ex Lindl. ■☆
78618 Cassinopsis Sond. (1860);拟滨篱菊属●☆
78619 Cassinopsis capensis Sond. = Cassinopsis ilicifolia (Hochst.) Kuntze ●☆
78620 Cassinopsis ilicifolia (Hochst.) Kuntze;拟滨篱菊●☆
78621 Cassinopsis tinifolia Harv.;细叶拟滨篱菊●☆
78622 Cassiope D. Don(1834);锦绦花属(岩须属);Cassiope ●
78623 Cassiope abbreviata Hand. -Mazz.;短梗锦绦花(短梗岩须,短叶岩须,水灵芝);Shortleaves Cassiope, Short-pediceled Cassiope, Shortstipe Cassiope ●
78624 Cassiope argyrotricha T. Z. Hsu;银毛锦绦花(银毛岩须);Silver-hair Cassiope, Silver-haired Cassiope ●
78625 Cassiope dendrotricha Hand. -Mazz.;睫毛锦绦花(睫毛岩须);Ciliate Cassiope, Eyebrow Leaves Cassiope ●
78626 Cassiope dendrotricha Hand. -Mazz. = Cassiope pectinata Stapf ●
78627 Cassiope ericoides (Pall.) D. Don;石南状锦绦花●☆
78628 Cassiope fastigiata (Wall.) D. Don;扫帚锦绦花(扫帚岩须,雪灵芝,血地红);Broom-like Cassiope, Himalaya Cassiope, Himalayan Cassiope ●
78629 Cassiope fujianensis L. K. Ling et G. S. Hoo;福建锦绦花(福建岩须);Fujian Cassiope ●
78630 Cassiope hypnoides (L.) D. Don;灰藓锦绦花(灰藓状岩须,羽藓岩须);Arctic Cassiope, Like Hypnum Cassiope, Mossplant ●
78631 Cassiope lycopodioides (Pall.) D. Don;石松状锦绦花(石松岩须,石松状岩须,岩须);Like Lycopodum Cassiope ●
78632 Cassiope lycopodioides (Pall.) D. Don f. globularis H. Hara;小球锦绦花●☆
78633 Cassiope lycopodioides (Pall.) D. Don var. laxa Nakai = Cassiope lycopodioides (Pall.) D. Don ●
78634 Cassiope macratha Hand. -Mazz. = Cassiope pectinata Stapf ●
78635 Cassiope mariei H. Lév. = Cassiope selaginoides Hook. f. et Thomson ●
78636 Cassiope membranifolia R. C. Fang;膜叶锦绦花(膜叶岩须)●
78637 Cassiope mertensiana (Bong.) D. Don;白石南锦绦花(白石南岩须,毛梗岩须);White Heather, White Mountain-heather ●
78638 Cassiope myosuroides W. W. Sm.;鼠尾锦绦花(鼠尾岩须,勿忘草状岩须);Mouse-tail Cassiope, Mousetaillike Cassiope, Myosotis-like Cassiope ●
78639 Cassiope nana T. Z. Hsu;矮小锦绦花(矮小岩须);Dwarf Cassiope ●
78640 Cassiope palpebrata W. W. Sm.;朝天锦绦花(朝天岩须);Erect Flower Cassiope, Erectflower Cassiope ●
78641 Cassiope pectinata Stapf;篦叶锦绦花(篦叶岩须);Combshapeleaf Cassiope, Pectinate Cassiope ●
78642 Cassiope pulvinalis T. Z. Hsu;垫状锦绦花(垫状岩须);Cashionshape Cassiope, Cushion Cassiope, Cushion-shaped Cassiope ●
78643 Cassiope pulvinalis T. Z. Hsu = Cassiope palpebrata W. W. Sm. ●
78644 Cassiope redowskii G. Don;雷氏锦绦花(雷氏岩须)●☆
78645 Cassiope selaginoides Hook. f. et Thomson;锦绦花(八股绳,草灵芝,长梗岩须,水灵芝,水麻黄,铁刷把,万年青,雪灵芝,岩须);Common Cassiope, Dwarf-dubmoss Cassiope, Like Selago Cassiope ●
78646 Cassiope stelleriana (Pall.) DC.;狼毒状锦绦花(狼毒状岩须);Like Stellera Cassiope ●
78647 Cassiope stelleriana (Pall.) DC. = Harrimanella stelleriana (Pall.) Coville ●
78648 Cassiope tetragona (L.) D. Don;四棱锦绦花(四角岩须,四棱岩须);Fire Moss, Fourangular Cassiope, Tetragon Cassiope ●
78649 Cassiope wardii C. Marquand et Airy Shaw;长毛锦绦花(长毛岩须);Long-hairs Cassiope, Ward Cassiope ●
78650 Cassiphone Rchb. = Leucothoe D. Don ●
78651 Cassipourea Aubl. (1775);红柱树属●☆
78652 Cassipourea abyssinica (Engl.) Alston = Cassipourea malosana (Baker) Alston ●☆

78653 Cassipourea acuminata Liben;渐尖红柱树●☆
78654 Cassipourea adamii Jacq. -Fél. ;阿达姆红柱树●☆
78655 Cassipourea africana Benth. = Cassipourea congoensis DC. ●☆
78656 Cassipourea afzelii (Oliv.) Alston;阿芙泽尔红柱树●☆
78657 Cassipourea alba Griseb. ;蒜红柱树;Garlic Wood ●☆
78658 Cassipourea annobonensis Mildbr. ex Alston;安诺本红柱树●☆
78659 Cassipourea barteri (Hook. f. ex Oliv.) N. E. Br. ;巴特红柱树●☆
78660 Cassipourea caesia Stapf = Cassipourea afzelii (Oliv.) Alston ●☆
78661 Cassipourea celastroides Alston;南蛇藤红柱树●☆
78662 Cassipourea congoensis DC. ;刚果红柱树●☆
78663 Cassipourea dinklagei (Engl.) Alston;丁克红柱树●☆
78664 Cassipourea eickii (Engl.) Alston = Cassipourea malosana (Baker) Alston ●☆
78665 Cassipourea eketensis Baker f. ;埃凯特红柱树●☆
78666 Cassipourea elliottii (Engl.) Alston = Cassipourea malosana (Baker) Alston ●☆
78667 Cassipourea euryoides Alston;宽红柱树●☆
78668 Cassipourea evrardii Floret;埃夫拉尔红柱树●☆
78669 Cassipourea fanshawei Torre et A. E. Gonc. ;范肖红柱树●☆
78670 Cassipourea flanaganii (Schinz) Alston;弗拉纳根红柱树●☆
78671 Cassipourea gerrardii (Schinz) Alston = Cassipourea malosana (Baker) Alston ●☆
78672 Cassipourea glomerata Alston;团集红柱树●☆
78673 Cassipourea gossweileri Exell;戈斯红柱树●☆
78674 Cassipourea gummiflua Tul. ;产胶红柱树●☆
78675 Cassipourea gummiflua Tul. subsp. ugandensis (Stapf) Lye = Cassipourea gummiflua Tul. var. ugandensis (Stapf) J. Lewis ●☆
78676 Cassipourea gummiflua Tul. var. mannii (Hook. f. ex Oliv.) J. Lewis;曼氏红柱树●☆
78677 Cassipourea gummiflua Tul. var. ugandensis (Stapf) J. Lewis;乌干达红柱树●☆
78678 Cassipourea gummiflua Tul. var. verticillata (N. E. Br.) J. Lewis;轮生红柱树●☆
78679 Cassipourea honeyi Alston = Cassipourea euryoides Alston ●☆
78680 Cassipourea huillensis (Engl.) Alston;威拉红柱树●☆
78681 Cassipourea kamerunensis (Engl.) Alston;喀麦隆红柱树●☆
78682 Cassipourea korupensis Kenfack et Sainge;科鲁普红柱树●☆
78683 Cassipourea leptoneura Floret;细脉红柱树●☆
78684 Cassipourea letestui Pellegr. ;莱泰斯图红柱树●☆
78685 Cassipourea louisii Liben;路易斯红柱树●☆
78686 Cassipourea malosana (Baker) Alston;红柱树;Pillarwood ●☆
78687 Cassipourea malosana Alston = Cassipourea malosana (Baker) Alston ●☆
78688 Cassipourea mannii (Hook. f. ex Oliv.) Engl. = Cassipourea gummiflua Tul. var. mannii (Hook. f. ex Oliv.) J. Lewis ●☆
78689 Cassipourea mildbraedii (Engl.) Alston = Cassipourea ruwensorensis (Engl.) Alston ●☆
78690 Cassipourea mollis (R. E. Fr.) Alston;柔软红柱树●☆
78691 Cassipourea mossambicensis (Brehmer) Alston;莫桑比克红柱树●☆
78692 Cassipourea nodosa Alston = Cassipourea barteri (Hook. f. ex Oliv.) N. E. Br. ●☆
78693 Cassipourea obovata Alston;倒卵红柱树●☆
78694 Cassipourea paludosa Hutch. et Dalziel ex Jacq. -Fél. ;沼泽红柱树●☆
78695 Cassipourea parvifolia (Scott-Elliot) Stapf = Cassipourea afzelii (Oliv.) Alston ●☆
78696 Cassipourea plumosa (Oliv.) Alston;羽状红柱树●☆
78697 Cassipourea pumila Floret;偃伏红柱树●☆
78698 Cassipourea redslobii Engl. = Cassipourea gummiflua Tul. var. verticillata (N. E. Br.) J. Lewis ●☆
78699 Cassipourea rotundifolia (Engl.) Alston;圆叶红柱树●☆
78700 Cassipourea ruwensorensis (Engl.) Alston;鲁文红柱树●☆
78701 Cassipourea salvago-raggei (Chiov.) Alston = Cassipourea malosana (Baker) Alston ●☆
78702 Cassipourea schizocalyx C. H. Wright;裂萼红柱树●☆
78703 Cassipourea sericea (Engl.) Alston;绢毛红柱树●☆
78704 Cassipourea swaziensis Compton;斯威士红柱树●☆
78705 Cassipourea trichosticha Alston;毛红柱树●☆
78706 Cassipourea ugandensis (Stapf) Engl. = Cassipourea gummiflua Tul. var. ugandensis (Stapf) J. Lewis ●☆
78707 Cassipourea verticillata N. E. Br. = Cassipourea gummiflua Tul. var. verticillata (N. E. Br.) J. Lewis ●☆
78708 Cassipourea vilhenae Cavaco;维列纳红柱树●☆
78709 Cassipourea zenkeri (Engl.) Alston;岑克尔红柱树●☆
78710 Cassipoureaceae J. Agardh = Rhizophoraceae Pers. (保留科名)●
78711 Cassipureaceae J. Agardh = Rhizophoraceae Pers. (保留科名)●
78712 Cassitha Hill = Cassytha L. ■●
78713 Cassumbium Benth. et Hook. f. = Cussambium Buch. -Ham. ●☆
78714 Cassumbium Benth. et Hook. f. = Schleichera Willd. (保留属名)●☆
78715 Cassumunar Colla = Zingiber Mill. (保留属名)■
78716 Cassupa Bonpl. = Isertia Schreb. ●☆
78717 Cassupa Humb. et Bonpl. = Isertia Schreb. ●☆
78718 Cassutha Des Moul. = Cuscuta L. ■
78719 Cassuviaceae Juss. ex R. Br. = Anacardiaceae R. Br. (保留科名)●
78720 Cassuviaceae R. Br. = Anacardiaceae R. Br. (保留科名)●
78721 Cassuvium Kuntze = Semecarpus L. f. ●
78722 Cassuvium Lam. = Anacardium L. ●
78723 Cassyta J. M. Mill. = Rhipsalis Gaertn. (保留属名)●
78724 Cassyta L. = Cassytha L. ■●
78725 Cassytha Gray = Cuscuta L. ■
78726 Cassytha L. (1753);无根藤属(无根草属);Dodder-laurel, Rootless Vine ■●
78727 Cassytha Mill. = Rhipsalis Gaertn. (保留属名)●
78728 Cassytha baccifera J. S. Muell. = Rhipsalis baccifera (J. S. Muell.) Stearn ●☆
78729 Cassytha capensis Meisn. = Cassytha ciliolata Nees ●☆
78730 Cassytha ciliolata Nees;好望角无根藤●☆
78731 Cassytha filiformis L. ;无根藤(半天雪,半天云,飞天藤,飞扬藤,过天藤,黄鱼藤,金丝藤,流离网,罗网藤,马尾丝,青丝藤,无地生根,无根草,无娘藤,无头草,无头藤,无爷藤,无叶草,蜈蚣藤,雾水藤);Filiform Rootless Vine, Love Vine, Love-vine ●■
78732 Cassytha filiformis L. var. duripraticola Hatus. = Cassytha pubescens R. Br. ■☆
78733 Cassytha glabella R. Br. ;光无根藤■☆
78734 Cassytha guineensis Schumach. et Thonn. = Cassytha filiformis L. ●■
78735 Cassytha pergracilis (Hatus.) Hatus. = Cassytha glabella R. Br. ■☆
78736 Cassytha pondoensis Engl. ;庞多无根藤●☆
78737 Cassytha pondoensis Engl. var. schliebenii (Robyns et R. Wilczek) Diniz;施利本无根藤●☆

78738 Cassytha pubescens R. Br.;毛无根藤;Devil's Twine ■☆
78739 Cassytha rubiginosa E. Mey. = Cassytha pondoensis Engl. ●☆
78740 Cassytha schliebenii Robyns et R. Wilczek = Cassytha pondoensis Engl. var. schliebenii (Robyns et R. Wilczek) Diniz ●☆
78741 Cassytha senegalensis A. Chev. = Cassytha filiformis L. ●■
78742 Cassythaceae Bartl. ex Lindl. (1833)(保留科名);无根藤科●■
78743 Cassythaceae Bartl. ex Lindl.(保留科名)= Lauraceae Juss.(保留科名)●■
78744 Castalia Salisb. = Nymphaea L.(保留属名)■
78745 Castalia ampla Salisb. = Nymphaea ampla (Salisb.) DC. ■☆
78746 Castalia crassifolia Hand. -Mazz. = Nymphaea tetragona Georgi ■
78747 Castalia elegans (Hook.) Greene = Nymphaea elegans Hook. ■☆
78748 Castalia flava (Leitn.) Greene = Nymphaea mexicana Zucc. ■
78749 Castalia leibergii Morong = Nymphaea leibergii Morong ■☆
78750 Castalia lekophylla Small = Nymphaea odorata Aiton ■
78751 Castalia minor (Sims) DC. = Nymphaea odorata Aiton ■
78752 Castalia mystica Salisb. = Nymphaea lotus L. ■
78753 Castalia odorata (Aiton) Woodv. et A. W. Wood = Nymphaea odorata Aiton ■
78754 Castalia reniformis (DC.) Trel. ex Branner et Coville = Nymphaea odorata Aiton ■
78755 Castalia scutifolia Salisb. = Nymphaea nouchalii Burm. f. var. caerulea (Savigny) Verdc. ■☆
78756 Castalia tuberosa (Paine) Greene = Nymphaea tuberosa Paine ■☆
78757 Castalis Cass. (1824);洁菊属■☆
78758 Castalis Cass. = Dimorphotheca Vaill.(保留属名)■●☆
78759 Castalis nudicaulis (L.) Norl.;裸茎洁菊■☆
78760 Castalis nudicaulis (L.) Norl. = Dimorphotheca nudicaulis (L.) DC. ■☆
78761 Castalis nudicaulis (L.) Norl. var. graminifolia? = Dimorphotheca nudicaulis (L.) DC. var. graminifolia (L.) Harv. ■☆
78762 Castalis spectabilis (Schltr.) Norl. = Dimorphotheca spectabilis Schltr. ■☆
78763 Castalis tragus (Aiton) Norl. = Dimorphotheca tragus (Aiton) B. Nord. ■☆
78764 Castalis tragus (Aiton) Norl. var. pinnatifida Norl. = Dimorphotheca tragus (Aiton) B. Nord. ■☆
78765 Castalis ventenati Cass. = Dimorphotheca tragus (Aiton) B. Nord. ■☆
78766 Castanea Mill. (1754);栗属(板栗属);Chestnut, Chinkapin, Chinquapin, Spanish Chestnut, Sweet Chestnut ●
78767 Castanea alnifolia Nutt. = Castanea pumila Mill. ●☆
78768 Castanea alnifolia Nutt. et Ashe = Castanea pumila Mill. ●☆
78769 Castanea alnifolia Nutt. var. floridana Sarg. = Castanea pumila Mill. ●☆
78770 Castanea americana (Michx.) Raf. = Castanea dentata (Marshall) Borkh. ●☆
78771 Castanea arkansana Ashe = Castanea ozarkensis Ashe ●☆
78772 Castanea bodinieri H. Lév. et Vaniot = Castanopsis hystrix Hook. f. et Thomson ex A. DC. ●
78773 Castanea bungeana Blume = Castanea mollissima Blume ●
78774 Castanea chinensis Spreng. = Castanopsis chinensis (Spreng.) Hance ●
78775 Castanea chrysophylla Douglas ex Hook. var. minor Benth. = Chrysolepis chrysophylla var. minor (Benth.) Munz ●☆
78776 Castanea concinna Champ. ex Benth. = Castanopsis concinna (Champ. ex Benth.) A. DC. ●◇
78777 Castanea crenata Siebold et Zucc.;日本栗(栗子,茅栗,日本板栗);Japan Chestnut, Japanese Chestnut ●
78778 Castanea crenata Siebold et Zucc. f. foemina (Makino) Sugim.;牧野日本栗●☆
78779 Castanea crenata Siebold et Zucc. f. gigantea Makino;大日本栗●☆
78780 Castanea crenata Siebold et Zucc. f. imperfecta (Makino) Sugim.;不全日本栗■☆
78781 Castanea crenata Siebold et Zucc. f. pendula (Miyoshi) Sugim.;垂枝日本栗;Pendulous Japanese Chestnut ●
78782 Castanea crenata Siebold et Zucc. f. pleiocarpa (Makino) Sugim.;光果日本栗●☆
78783 Castanea crenata Siebold et Zucc. f. pulchella (Honda) Sugim.;美丽日本栗●☆
78784 Castanea crenata Siebold et Zucc. f. sakyacephala (Makino) Sugim.;佛头日本栗●☆
78785 Castanea crenata Siebold et Zucc. var. kusakuri (Blume) Nakai = Castanea crenata Siebold et Zucc. ●
78786 Castanea crenata Siebold et Zucc. var. pleiocarpa Makino = Castanea crenata Siebold et Zucc. f. pleiocarpa (Makino) Sugim. ●☆
78787 Castanea crenata Siebold et Zucc. var. pulchella Honda = Castanea crenata Siebold et Zucc. f. pulchella (Honda) Sugim. ●☆
78788 Castanea crenata Siebold et Zucc. var. sakyacephala Makino = Castanea crenata Siebold et Zucc. f. sakyacephala (Makino) Sugim. ●☆
78789 Castanea davidii Dode = Castanea seguinii Dode ●
78790 Castanea dentata (Marshall) Borkh.;美洲栗(齿栗,美国栗,美栗,甜栗);American Chestnut, American Sweet Chestnut, Chestnut, Sweet Chestnut ●☆
78791 Castanea duclouxii Dode = Castanea mollissima Blume ●
78792 Castanea fargesii Dode = Castanea mollissima Blume ●
78793 Castanea floridana (Sarg.) Ashe = Castanea pumila Mill. ●☆
78794 Castanea formosans (Hayata) Hayata = Castanea mollissima Blume ●
78795 Castanea henryi (V. Naray.) Rehder et E. H. Wilson;锥栗(尖栗,箭栗,旋栗,珍珠栗,真栗);Chinese Timber Chinquapin, Henry Chestnut, Henry Chinkapin ●
78796 Castanea henryi (V. Naray.) Rehder et E. H. Wilson var. omeiensis W. P. Fang;峨眉锥栗;Emei Chestnut, Omei Chestnut ●
78797 Castanea hupehensis Dode = Castanea mollissima Blume ●
78798 Castanea indica Roxb. = Castanopsis indica (Roxb.) A. DC. ●
78799 Castanea indica Roxb. ex Lindl. = Castanopsis indica (Roxb. ex Lindl.) A. DC. ●
78800 Castanea japonica Blume = Castanea crenata Siebold et Zucc. ●
78801 Castanea javanica Blume;爪哇栗●☆
78802 Castanea mollissima Blume;板栗(大栗,风栗,厚六,家栗,魁栗,栗,栗果,栗子,毛板栗,毛栗,毛栗子,瓦栗子树);Chestnut, Chinese Chestnut, Hairy Chestnut ●
78803 Castanea mollissima Blume var. pendula X. Y. Zhou et Z. D. Zhou = Castanea mollissima Blume ●
78804 Castanea ozarkensis Ashe;北美栗;Ozark Chestnut, Ozark Chinkapin ●☆
78805 Castanea ozarkensis Ashe = Castanea pumila Mill. ●☆
78806 Castanea pubinervis C. K. Schneid. = Castanea crenata Siebold et Zucc. ●
78807 Castanea pumila Mill.;毛枝栗(矮栗,矮小栗);Allegany Chinkapin, Allegheny Chinkapin, Chinkapin, Chinquapin, Common Chinquapin, Dwarf Chestnut, Glorida Chinkapin, Ozark Chestnut, Ozark Chinquapin, Trailing Chinkapin, Tree Chinquapin, Virginian

Chestnut ●☆

78808 Castanea pumila Mill. var. ashei Sudw. = Castanea pumila Mill. ●☆

78809 Castanea pumila Mill. var. ozarkensis (Ashe) G. E. Tucker = Castanea ozarkensis Ashe ●☆

78810 Castanea satica V. Naray. = Castanea mollissima Blume ●

78811 Castanea sativa Mill.;欧洲栗(欧栗,欧洲板栗,甜栗,西班牙栗,锥栗子);Chaste Nut,Chastey,Chestnut,Common Chestnut,Congo Stick,Edible Chesmut,Eurasian Chestnut,European Chestnut,French Chestnut,French Nut,Husked Nut,Italian Chestnut,Lady Nut,Meat Nut,Polly Nut,Rough Nut,Sardian Nut,Spanish Chestnut,Stover Nut,Stover-nut,Sweet Chestnut ●☆

78812 Castanea sativa Mill. 'Albomarginata';白边欧洲栗●☆

78813 Castanea sativa Mill. var. acuminatissima Seemen = Castanea henryi (V. Naray.) Rehder et E. H. Wilson ●

78814 Castanea sativa Mill. var. bungeana Pamp. = Castanea seguinii Dode ●

78815 Castanea sativa Mill. var. formosana Hayata = Castanea mollissima Blume ●

78816 Castanea sativa Mill. var. japonica Seem. = Castanea seguinii Dode ●

78817 Castanea sativa Mill. var. mollissima (Blume) Pamp. = Castanea mollissima Blume ●

78818 Castanea sativa Mill. var. mollissima Pamp. = Castanea mollissima Blume ●

78819 Castanea sativa Mill. var. typica Seem. = Castanea mollissima Blume ●

78820 Castanea sativa V. Naray. = Castanea henryi (V. Naray.) Rehder et E. H. Wilson ●

78821 Castanea sativa V. Naray. = Castanea mollissima Blume ●

78822 Castanea sativa V. Naray. = Castanea vesca Gaertn. ●

78823 Castanea seguinii Dode;茅栗(楠栗,尖栗,金栗,栵栗,毛凹栗子,毛板栗,毛栗,野栗子,野茅栗);Chinese Dwarf Chinquapin,Seguin Chinkapin ●

78824 Castanea sempervirens Kellogg = Chrysolepis sempervirens (Kellogg) Hjelmq. ●☆

78825 Castanea stricta Siebold et Zucc. = Castanea crenata Siebold et Zucc. ●

78826 Castanea tribuloides (Sm.) Lindl. = Castanopsis tribuloides (Sm.) A. DC. ●

78827 Castanea vesca Bunge = Castanea mollissima Blume ●

78828 Castanea vesca Gaertn. = Castanea sativa Mill. ●☆

78829 Castanea vilmoriniana Dode = Castanea henryi (V. Naray.) Rehder et E. H. Wilson ●

78830 Castanea vulgaris Hance = Castanea mollissima Blume ●

78831 Castanea vulgaris Hance var. japonica Hance = Castanea seguinii Dode ●

78832 Castanea vulgaris Hance var. yunnanensis Franch. = Castanea mollissima Blume ●

78833 Castanea vulgaris Lam. = Castanea sativa Mill. ●☆

78834 Castanea vulgaris Lam. var. yunnanensis Franch. = Castanea mollissima Blume ●

78835 Castaneaceae Baill. = Fagaceae Dumort.(保留科名)●

78836 Castaneaceae Link = Fagaceae Dumort.(保留科名)●

78837 Castaneaceae Link = Hippocastanaceae A. Rich.(保留科名)●

78838 Castanella Spruce ex Benth. et Hook. f. = Paullinia L. ●☆

78839 Castanella Spruce ex Hook. f. = Paullinia L. ●☆

78840 Castanocarpus Sweet = Castanospermum A. Cunn. ex Hook. ●☆

78841 Castanola Llanos = Agelaea Sol. ex Planch. ●

78842 Castanola glabrifolia Schellenb. = Agelaea trinervis (Llanos) Merr. ●

78843 Castanola obliqua Schellenb. = Agelaea trinervis (Llanos) Merr. ●

78844 Castanola paradoxa (Gilg) G. Schellenb. ex Hutch. et Dalziel = Agelaea paradoxa Gilg var. microcarpa Jongkind ●☆

78845 Castanola trinervis Llanos = Agelaea trinervis (Llanos) Merr. ●

78846 Castanophorum Pfeiff. = Casanophorum Neck. ●

78847 Castanophorum Pfeiff. = Castanea Mill. ●

78848 Castanopsis (D. Don) Spach(1841)(保留属名);锥栗属(栲属,苦槠属,椎属);Chinkapin,Chinquapin,Evergreen Chinkapin,Evergreen Chinkapiu,Evergreenchinkapin,Oat Chestnut,Oatchestnut ●

78849 Castanopsis D. Don = Castanopsis (D. Don) Spach(保留属名)●

78850 Castanopsis × kuchugouzhui C. C. Huang et Y. T. Chang;苦槠钩锥;Kuchugouzhui Evergreen Chinkapin ●

78851 Castanopsis amabilis W. C. Cheng et C. S. Chao var. brevispinosa W. C. Cheng et C. S. Chao = Castanopsis amabilis W. C. Cheng et H. C. Chao ●

78852 Castanopsis amabilis W. C. Cheng et H. C. Chao;南宁锥(南宁栲);Lovely Evergreen Chinkapin,Nanning Evergreenchinkapin ●

78853 Castanopsis amabilis W. C. Cheng et H. C. Chao var. brevispinosa W. C. Cheng et H. C. Chao = Castanopsis amabilis W. C. Cheng et H. C. Chao ●

78854 Castanopsis angustifolia C. C. Huang et Y. T. Chang = Castanopsis choboensis Hickel et A. Camus ●

78855 Castanopsis annamensis Hickel et A. Camus;越南栲;Indochinese Chinquapin,Indochinese Evergreenchinkapin ●

78856 Castanopsis annamensis Hickel et A. Camus = Castanopsis boisii Hickel et A. Camus ●

78857 Castanopsis argentea Blume;印尼栲(银叶锥)●☆

78858 Castanopsis argyracantha A. Camus = Castanopsis fargesii Franch. ●

78859 Castanopsis argyrophylla King ex Hook. f.;银叶锥(银叶栲);Silverleaf Evergreenchinkapin,Silver-leaved Evergreen Chinkapin ●

78860 Castanopsis armata (Roxb.) Spach. = Castanopsis lamontii Hance ●

78861 Castanopsis asymetrica H. Lév. = Castanopsis eyrei (Champ. ex Benth.) Tutcher ●

78862 Castanopsis balansae (Drake) Schottky = Lithocarpus balansae (Drake) A. Camus ●

78863 Castanopsis bodinieri (H. Lév. et Vaniot) Koidz. = Castanopsis hystrix Hook. f. et Thomson ex A. DC. ●

78864 Castanopsis boisii Hickel et A. Camus;榄壳锥(越南栲);Bois Evergreen Chinkapin,Bois Evergreenchinkapin ●

78865 Castanopsis borneensis King;赤栲(赤校)●

78866 Castanopsis borneensis King = Castanopsis kawakamii Hayata ●◇

78867 Castanopsis brachyacantha Hayata = Castanopsis eyrei (Champ. ex Benth.) Tutcher ●

78868 Castanopsis brachyacantha Hayata ex Koidz.;反刺槠(反刺椎栗,淋漓柯);Reflexed Spines Chinquapin ●

78869 Castanopsis brachyacantha Hayata ex Koidz. = Castanopsis eyrei (Champ.) Tutcher ●

78870 Castanopsis brevispina Hayata = Castanopsis fabri Hance ●

78871 Castanopsis brevispina Hayata et Kaneh. ex A. Camus = Castanopsis fabri Hance ●

78872 Castanopsis brevistella Hayata et Kaneh. ex A. Camus =

Castanopsis fabri Hance ●

78873 Castanopsis brunnea (H. Lév.) A. Camus = Castanopsis hystrix Hook. f. et Thomson ex A. DC. ●

78874 Castanopsis calathiformis (V. Naray.) Rehder et P. Wilson;枹丝锥(杯状栲,枹丝栲,黄栗,山枇杷,丝锥);Calyciform Clemati, Calyciform Evergreen Chinkapin,Cup-shaped Evergreen Chinkapin ●

78875 Castanopsis camelliifolia H. Lév. = Gordonia axillaris (Roxb. ex Ker Gawl.) Endl. ●

78876 Castanopsis camelliifolia H. Lév. = Polyspora axillaris (Roxb. ex Ker Gawl.) Sweet ex G. Don ●

78877 Castanopsis carlesii (Hemsl.) Hayata;小红栲(白栲,白校欑,白橼,长尾尖叶槠,长尾尖槠,长尾栲,长尾柯,单刺苦槠,锯叶长尾栲,卡氏槠,柯子,米槠,米锥,米子子槠,石槠,细米橼,小叶槠,朱槠); Carlese Chinquapin, Carlese Evergreen Chinkapin, Carlese Evergreenchinkapin, Caudate-leaved Chinquapin, Long-leaf Evergreenchinkapin ●

78878 Castanopsis carlesii (Hemsl.) Hayata var. sessilis Nakai;锯叶长尾栲(白校欑)●

78879 Castanopsis carlesii (Hemsl.) Hayata var. sessilis Nakai = Castanopsis carlesii (Hemsl.) Hayata ●

78880 Castanopsis carlesii (Hemsl.) Hayata var. spinulosa W. C. Cheng et C. S. Chao;小叶栲(短刺米槠,西南米槠);Shortspine Carles Evergreenchinkapin ●

78881 Castanopsis caudata Franch. = Castanopsis eyrei (Champ. ex Benth.) Tutcher ●

78882 Castanopsis cavaleriei H. Lév. = Sloanea sinensis (Hance) Hemsl. ●

78883 Castanopsis ceradacantha Rehder et E. H. Wilson var. semiserrata (Hickel et A. Camus) A. Camus = Castanopsis fabri Hance ●

78884 Castanopsis ceratacantha Rehder et E. H. Wilson;瓦山栲(瓦山锥);Washan Evergreen Chinkapin,Washan Evergreenchinkapin ●

78885 Castanopsis cerebrina (Hickel et A. Camus) Barnett;毛叶杯锥(毛叶杯状栲); Hairyleaf Evergreenchinkapin, Hairy-leaved Evergreen Chinkapin ●

78886 Castanopsis chaysophylla (Hook.) DC.;金叶锥●☆

78887 Castanopsis chengfengensis Hu = Castanopsis tibetana Hance ●

78888 Castanopsis chevalieri Hickel et A. Camus = Castanopsis rockii A. Camus ●

78889 Castanopsis chinensis (Spreng.) Hance;桂林栲(钩栗,狗栗,桂林锥,栲栗,勒翠,栗,米锥,山锥,小板栗,中华栲,锥,锥栗,锥子树);Chinese Evergreen Chinkapin,Chinese Evergreenchinkapin, Oatchestnut ●

78890 Castanopsis chinensis Hance = Castanopsis chinensis (Spreng.) Hance ●

78891 Castanopsis chinensis Hance = Castanopsis orthacantha Franch. ●

78892 Castanopsis chingii A. Camus = Castanopsis eyrei (Champ. ex Benth.) Tutcher ●

78893 Castanopsis choboensis Hickel et A. Camus;窄叶锥;Narrowleaf Evergreenchinkapin,Narrow-leaved Evergreen Chinkapin ●

78894 Castanopsis chrysophylla (Douglas) A. DC. = Chrysolepis chrysophylla (Douglas ex Hook.) Hjelmq. ●

78895 Castanopsis chuniana W. P. Fang = Castanopsis ceratacantha Rehder et E. H. Wilson ●

78896 Castanopsis chunii W. C. Cheng;厚皮栲(厚皮丝栗,厚皮锥,锥树);Chun Evergreen Chinkapin,Chun Evergreenchinkapin ●

78897 Castanopsis clarkei King ex Hook. f.;棱刺锥(棱果锥);C. B. Clarke's Evergreenchinkapin, Ribspine Angulate-fruited Evergreen Chinkapin ●

78898 Castanopsis clarkei King ex Hook. f. = Castanopsis indica (Roxb.) A. DC. ●

78899 Castanopsis concinna (Champ. ex Benth.) A. DC.;华南栲(华南锥);S. China Evergreen Chinkapin, South China Chinquapin, South China Evergreenchinkapin ●◇

78900 Castanopsis concolor Rehder et E. H. Wilson = Castanopsis orthacantha Franch. ●

78901 Castanopsis crasifolia Hickel et A. Camus;厚叶锥;Thickleaf Evergreenchinkapin,Thick-leaved Evergreen Chinkapin ●

78902 Castanopsis cryptoneuron (H. Lév.) A. Camus ex Rehder = Castanopsis fargesii Franch. ●

78903 Castanopsis cuspidata (Thunb. ex A. Murray) Schottky;甜栲(尖叶栲,米槠);Copper False Chestnut,Japanese Chinquapin ●☆

78904 Castanopsis cuspidata (Thunb. ex A. Murray) Schottky var. carlesii (Hemsl.) T. Yamaz.;单刺苦槠(白校欑)●

78905 Castanopsis cuspidata (Thunb. ex A. Murray) Schottky var. carlesii (Hemsl.) T. Yamaz. = Castanopsis carlesii (Hemsl.) Hayata ●

78906 Castanopsis cuspidata (Thunb.) Schottky f. angustifolia Nakai = Castanopsis cuspidata (Thunb.) Schottky ●☆

78907 Castanopsis cuspidata (Thunb.) Schottky f. awanoi (Yanagita) Nakai;淡野栲●☆

78908 Castanopsis cuspidata (Thunb.) Schottky f. lanceolata Sugim. = Castanopsis sieboldii (Makino) Hatus. ex T. Yamaz. et Mashiba ●☆

78909 Castanopsis cuspidata (Thunb.) Schottky subsp. sieboldii (Makino) Sugim. = Castanopsis sieboldii (Makino) Hatus. ex T. Yamaz. et Mashiba ●☆

78910 Castanopsis cuspidata (Thunb.) Schottky var. carlesii (Hemsl.) T. Yamaz. = Castanopsis carlesii (Hemsl.) Hayata ●

78911 Castanopsis cuspidata (Thunb.) Schottky var. longicaudata (Hayata) S. S. Ying = Castanopsis carlesii (Hemsl.) Hayata ●

78912 Castanopsis cuspidata (Thunb.) Schottky var. longicaudata (Hayata) S. S. Ying = Castanopsis cuspidata (Thunb. ex A. Murray) Schottky ●☆

78913 Castanopsis cuspidata (Thunb.) Schottky var. lutchuensis Masam.;琉球米槠●☆

78914 Castanopsis cuspidata (Thunb.) Schottky var. palustris L.;沼生槠●☆

78915 Castanopsis cuspidata (Thunb.) Schottky var. sieboldii (Makino) Nakai = Castanopsis sieboldii (Makino) Hatus. ex T. Yamaz. et Mashiba ●☆

78916 Castanopsis cuspidata (Thunb.) Schottky var. sieboldii Nakai;大果米槠(柯,米槠,甜槠,席氏米槠,椎)●☆

78917 Castanopsis cuspidata (Thunb.) Schottky var. thunbergii Nakai;通贝里栲(柯木,通氏栲);Thunberg Evergreenchinkapin ●☆

78918 Castanopsis damingshanensis S. L. Mo ex C. C. Huang et Y. T. Chang;大明山锥(卷叶米锥); Damingshan Chinkapinga, Damingshan Evergreenchinkapin ●

78919 Castanopsis daxinensis C. C. Huang et Y. T. Chang = Castanopsis crasifolia Hickel et A. Camus ●

78920 Castanopsis delavayi Franch.;高山栲(白栎,白栗,白猪栗,刺栗,滇锥,滇锥栗,高山栲树,高山栗,高山锥,毛栗,丝栗,椎栗); Delavay Chinquapin, Delavay Evergreen Chinkapin, Delavay Evergreenchinkapin ●

78921 Castanopsis densispinosa Y. C. Hsu et H. Wei Jen;密刺锥(密刺栲); Densethorn Evergreenchinkapin, Densispined Evergreen Chinkapin,Densithorned Evergreen Chinkapin ●

78922 Castanopsis diversifolia (Kurz) King ex Hook. f.;滇栲;Diversityleaf Evergreenchinkapin ●

78923 Castanopsis diversifolia (Kurz) King ex Hook. f. = Castanopsis mekongensis A. Camus ●

78924 Castanopsis echidnocarpa Miq.;短刺锥(长刺锥,短刺栲,红椆栗,锥栗);Shortspined Evergreen Chinkapin, Shortspine Chinquapin, Shortspine Evergreenchinkapin ●

78925 Castanopsis echidnocarpa Miq. var. semiduda W. C. Cheng et C. S. Chao;裸果栲;Naked-fruit Shortspine Evergreenchinkapin ●

78926 Castanopsis echinocarpa Miq. var. seminuda W. C. Cheng et C. S. Chao = Castanopsis echidnocarpa Miq. ●

78927 Castanopsis eyrei (Champ. ex Benth.) Tutcher;甜槠(曹槠,反刺苦槠,反刺槠,茅丝栗,丝栗,酸橡槠,甜槠栲,甜锥,小黄橡,槠柴,锥子);Eyre Chinquapin, Eyre Evergreen Chinkapin, Eyre Evergreenchinkapin, Reflexive-spined Evergreen Chinkapin ●

78928 Castanopsis eyrei (Champ. ex Benth.) Tutcher var. brachyacantha (Hayata) C. F. Shen = Castanopsis eyrei (Champ. ex Benth.) Tutcher ●

78929 Castanopsis eyrei (Champ.) Tutcher = Castanopsis eyrei (Champ. ex Benth.) Tutcher ●

78930 Castanopsis eyrei (Champ.) Tutcher var. brachyacantha (Hayata) C. F. Shen ex S. S. Ying = Castanopsis eyrei (Champ. ex Benth.) Tutcher ●

78931 Castanopsis eyrei (Champ.) Tutcher var. caudata (Franch.) W. C. Cheng = Castanopsis eyrei (Champ. ex Benth.) Tutcher ●

78932 Castanopsis fabri Hance;罗浮栲(白橡,白锥,草野氏锥栗,赤黎,短刺锥栗,狗牙锥,红黎,红缘栲,红缘木,红锥,酒枹,黎木,罗浮锥,毛氏栲,三检锥,星刺栲);Faber Chinquapin, Faber Evergreen Chinkapin, Faber Evergreenchinkapin, Hichel Evergreen-chinkapin●

78933 Castanopsis fargesii Franch.;丝栗栲(赤校欑,刺栲,钩栗,红背槠,红栲,红叶栲,火烧柯,栲,栲树,丝栗树,台栲,锥);Farges Chinquapin, Farges Evergreen Chinkapin, Farges Evergreenchinkapin, Taiwan Evergreenchinkapin ●

78934 Castanopsis ferox Spach;思茅栲(思茅锥);Simao Evergreen Chinkapin, Simao Evergreenchinkapin ●

78935 Castanopsis ferox Spach var. longispina (King ex Hook. f.) A. Camus = Castanopsis longispina (King ex Hook. f.) C. C. Huang et Y. T. Chang ●

78936 Castanopsis fissa (Champ. ex Benth.) Rehder et E. H. Wilson;黧蒴锥(大叶枹,大叶栎,大叶槠栗,大叶锥,黧蒴栲,裂壳锥);Breakingfruit Chinquapin, Breakingfruit Evergreenchinkapin, Breaking-fruited Evergreen Chinkapin ●

78937 Castanopsis fissoides Chun et C. C. Huang ex Luong = Castanopsis fissa (Champ. ex Benth.) Rehder et E. H. Wilson ●

78938 Castanopsis fissus (Champ. ex Benth.) A. Camus = Castanopsis fissa (Champ. ex Benth.) Rehder et E. H. Wilson ●

78939 Castanopsis fleuryi Hickel et A. Camus;小果锥(小果栲);Fleury Evergreen Chinkapin, Fleury Evergreenchinkapin ●

78940 Castanopsis fohaiensis Hu = Castanopsis mekongensis A. Camus ●

78941 Castanopsis fordii Hance;南岭栲(毛栲,毛锥);Ford Chinquapin, Ford Evergreen Chinkapin, Ford Evergreenchinkapin ●

78942 Castanopsis formosana (V. Naray.) Hayata;台湾锥(海南栲树,黄栲,黄椆锥,黄锥,台湾栲,台湾苦槠,台湾槠,台湾椎栗,台湾锥栗,校力);Taiwan Chinkapin, Taiwan Evergreen Chinkapin, Taiwan Evergreenchinkapin ●

78943 Castanopsis formosana (V. Naray.) Hayata = Castanopsis jucunda Hance ●

78944 Castanopsis globigemmata Chun et C. C. Huang;圆芽锥(大刺麻栗,小枹丝栗);Globigemmate Evergreenchinkapin, Round-bud Evergreen Chinkapin ●

78945 Castanopsis goniacantha A. Camus = Castanopsis lamontii Hance ●

78946 Castanopsis greenii Chun = Castanopsis kawakamii Hayata ●◇

78947 Castanopsis hainanensis Merr.;海南栲(刺锥,海南锥,坡锥);Hainan Evergreen Chinkapin, Hainan Evergreenchinkapin ●

78948 Castanopsis hamata Duanmu = Castanopsis boisii Hickel et A. Camus ●

78949 Castanopsis henryi V. Naray. = Castanea henryi (V. Naray.) Rehder et E. H. Wilson ●

78950 Castanopsis hichelii A. Camus = Castanopsis fabri Hance ●

78951 Castanopsis humata Duanmu = Castanopsis boisii Hickel et A. Camus ●

78952 Castanopsis hupehensis C. S. Chao;湖北锥(川鄂丝栗,湖北栲,锥栗果);Hubei Evergreenchinkapin, Hupeh Evergreen Chinkapin ●

78953 Castanopsis hystrix A. Camus = Castanopsis fargesii Franch. ●

78954 Castanopsis hystrix A. DC. = Castanopsis hystrix Hook. f. et Thomson ex A. DC. ●

78955 Castanopsis hystrix Hook. f. et Thomson ex A. DC.;刺栲(赤校,椆栗,刺锥,刺锥栗,钩栗,红背槠,红黎,红锥,红锥栗,火烧柯,栲树,锥栗,锥丝栗);Katus, Red Chinkapin, Red Evergreen Chinkapin, Spinybract Chinquapin, Spinybract Evergreenchinkapin, Spiny-bracted Evergreen Chinkapin ●

78956 Castanopsis hystrix Miq. = Castanopsis hystrix Hook. f. et Thomson ex A. DC. ●

78957 Castanopsis hystrix Miq. subsp. rufescens (Hook. f. et Thunb.) A. Camus = Castanopsis wattii (King) A. Camus ●

78958 Castanopsis incana A. Camus = Castanopsis eyrei (Champ. ex Benth.) Tutcher ●

78959 Castanopsis indica (Roxb. ex Lindl.) A. DC.;印度锥(黄椆栲,坡锥,山针锥,丝丝锥,印度栲,印度苦槠,印度锥栗);Indian Chinquapin, Indian Evergreen Chinkapin, Indian Evergreenchinkapin ●

78960 Castanopsis indica (Roxb.) A. DC. = Castanopsis indica (Roxb. ex Lindl.) A. DC. ●

78961 Castanopsis indica (Roxb.) Miq. = Castanopsis indica (Roxb. ex Lindl.) A. DC. ●

78962 Castanopsis jianfenglingensis Duanmu;尖峰岭锥;Jianfeng Evergreenchinkapin, Jianfengling Evergreen Chinkapin ●

78963 Castanopsis jucunda Hance;乌椆栲(东南栲,东南锥,秀丽栲,秀丽锥);Elegant Evergreen Chinkapin, Elegant Evergreenchinkapin ●

78964 Castanopsis jucunda Hance var. annularis Hickel et A. Camus;石山秀丽栲;Annular Elegant Evergreenchinkapin ●

78965 Castanopsis jucunda Hance var. versicilor C. C. Huang et Y. T. Chang = Castanopsis jucunda Hance ●

78966 Castanopsis kawakamii Hayata;青钩栲(赤栲,赤校,川上氏槠,大叶苦槠,大叶校力,吊皮栲,吊皮锥,格林锥,格氏栲,青钩锥栗,蓑衣栲);Kawakami Chinquapin, Kawakami Evergreen Chinkapin, Kawakami Evergreenchinkapin ●◇

78967 Castanopsis kusanoi Hayata;草野氏槠(草野锥果,细刺苦槠,星状短刺槠);Kusano Chinkapin ●

78968 Castanopsis kusanoi Hayata = Castanopsis fabri Hance ●

78969 Castanopsis kweichowensis Hu;贵州锥(贵州栲);Guizhou Evergreenchinkapin, Kweichow Evergreen Chinkapin ●

78970 Castanopsis lamontii Hance;鹿角锥(臭栲,黑炭木,红勾栲,鹿角栲,箐板栗);Lamont Chinquapin, Lamont Evergreen Chinkapin,

Lamont Evergreenchinkapin ●

78971 Castanopsis lamontii Hance var. shanghanensis Q. F. Zheng;上杭锥;Shanghang Evergreen Chinkapin,Shanghang Evergreenchinkapin ●

78972 Castanopsis lamontii Hance var. shanghanensis Q. F. Zheng = Castanopsis lamontii Hance ●

78973 Castanopsis lantsangensis Hu = Castanopsis mekongensis A. Camus ●

78974 Castanopsis ledongensis C. C. Huang et Y. T. Chang;乐东锥;Ledong Evergreen Chinkapin ●

78975 Castanopsis lohfauensis Hu = Castanopsis hystrix Hook. f. et Thomson ex A. DC. ●

78976 Castanopsis longicauclata (Hayata) Nakai = Castanopsis carlesii (Hemsl.) Hayata ●

78977 Castanopsis longicaudata (Hayata) Nakai;长尾栲●☆

78978 Castanopsis longicaudata (Hayata) Nakai = Castanopsis carlesii (Hemsl.) Hayata ●

78979 Castanopsis longispicata Hu = Castanopsis tribuloides (Lindl.) A. DC. var. echidnocarpa King ex Hook. f. ●

78980 Castanopsis longispicata Hu = Castanopsis echidnocarpa Miq. ●

78981 Castanopsis longispina (King ex Hook. f.) C. C. Huang et Y. T. Chang;长刺锥●

78982 Castanopsis longzhouica C. C. Huang et Y. T. Chang;龙州锥;Longzhou Evergreen Chinkapin,Longzhou Evergreenchinkapin ●

78983 Castanopsis lunglingensis Hu = Castanopsis rockii A. Camus ●

78984 Castanopsis macrostachya Hu = Castanopsis indica (Roxb.) A. DC. ●

78985 Castanopsis macrostachys A. Chev. = Castanopsis indica (Roxb.) A. DC. ●

78986 Castanopsis matsudai Hayata ex A. Camus = Castanopsis fabri Hance ●

78987 Castanopsis megaphylla Hu;大叶锥(大叶栲);Bigleaf Evergreenchinkapin, Big-leaved Evergreen Chinkapin, Largeleaf Evergreenchinkapin ●

78988 Castanopsis megaphylla Hu = Castanopsis boisii Hickel et A. Camus ●

78989 Castanopsis mekongensis A. Camus;湄公锥(澜沧栲,湄公栲);Mekong Chinquapin, Mekong Evergreen Chinkapin, Mekong Evergreenchinkapin ●

78990 Castanopsis mianningensis Hu = Castanopsis orthacantha Franch. ●

78991 Castanopsis microcarpa Hu = Castanopsis fleuryi Hickel et A. Camus ●

78992 Castanopsis neocaraleriei A. Camus;红背甜槠(红背栲);Redback Evergreenchinkapin,Red-backed Evergreen Chinkapin ●

78993 Castanopsis neocavaleriei A. Camus = Castanopsis eyrei (Champ. ex Benth.) Tutcher ●

78994 Castanopsis nigrescens Chun et C. C. Huang;黑叶锥(黑叶栲,墨叶锥,岩槠);Blackleaf Evergreenchinkapin,Black-leaved Evergreen Chinkapin ●

78995 Castanopsis ninbienensis Hickel et A. Camus = Castanopsis fabri Hance ●

78996 Castanopsis oblonga Y. C. Hsu et H. Wei Jen;矩叶锥(矩叶栲);Oblong Evergreen Chinkapin,Oblong Evergreenchinkapin ●

78997 Castanopsis oblongifolia W. C. Cheng et C. S. Chao = Castanopsis concinna (Champ. ex Benth.) A. DC. ●◇

78998 Castanopsis oerstedii Hickel et A. Camus = Castanopsis kawakamii Hayata ●◇

78999 Castanopsis orthacantha Franch.;毛果栲(扁栗,毛锥栗,元江栲,元江锥,猪栗,锥栗);Straightspine Evergreenchinkapin, Straight-spined Evergreen Chinkapin ●

79000 Castanopsis ouonbiensis Hickel et A. Camus;屏边锥(屏边栲);Pingbian Evergreen Chinkapin,Pingbian Evergreenchinkapin ●

79001 Castanopsis pachyrachis Hickel et A. Camus = Castanopsis lamontii Hance ●

79002 Castanopsis pinfaensis H. Lév. = Castanopsis fargesii Franch. ●

79003 Castanopsis pingchangensis G. A. Fu et S. X. Feng;平昌锥;Pingchang Evergreen Chinkapin ●

79004 Castanopsis platyacantha Rehder et E. H. Wilson;扁刺栲(白丝栗,扁刺锥,峨眉栲,黑铲栗,猴栗,石栗,丝栗);Flatspine Evergreenchinkapin,Flat-spined Evergreen Chinkapin ●

79005 Castanopsis poilanei Hickel et A. Camus;棕毛锥(野板栗,棕毛栲);Chequer-shaped Evergreen Chinkapin, Poilane Evergreen Chinkapin,Poilane Evergreenchinkapin ●

79006 Castanopsis poilanei Hickel et A. Camus = Castanopsis mekongensis A. Camus ●

79007 Castanopsis poilanei Hu = Castanopsis mekongensis A. Camus ●

79008 Castanopsis pseudoconcinna W. C. Cheng et C. S. Chao = Castanopsis hystrix Hook. f. et Thomson ex A. DC. ●

79009 Castanopsis quangtriensis Hickel et A. Camus = Castanopsis fabri Hance ●

79010 Castanopsis remotidenticulata Hu;疏齿锥(黑锥栗,细齿栲,细齿锥);Remote-denticulate Evergreen Chinkapin, Remotidenticulate Evergreenchinkapin,Serrulate Evergreenchinkapin ●

79011 Castanopsis remotiserrata Hu = Castanopsis chinensis (Spreng.) Hance ●

79012 Castanopsis robustispina Hu = Castanopsis lamontii Hance ●

79013 Castanopsis rockii A. Camus;龙岭锥(龙岭栲);Rock Evergreen Chinkapin,Rock Evergreenchinkapin ●

79014 Castanopsis rufescens (Hook. f. et Thomson) C. C. Huang et Y. T. Chang;变色锥;Reddish Evergreen Chinkapin, Rufescent Evergreenchinkapin ●

79015 Castanopsis rufescens (Hook. f. et Thomson) C. C. Huang et Y. T. Chang = Castanopsis wattii (King) A. Camus ●

79016 Castanopsis rufotomentosa Hu;红壳锥(千叶子刺栗,红毛栲,红毛锥);Red-haired Evergreen Chinkapin, Redtomentose Evergreenchinkapin ●

79017 Castanopsis sclerophylla (Lindl.) Schottky;苦槠栲(结节锥栗,苦栲,苦槠,苦槠栗,苦槠锥,苦槠子,血槠,株子,槠,槠栎,槠栗,槠子);Bitter Chinquapin, Bitter Evergreen Chinkapin, Bitter Evergreenchinkapin ●

79018 Castanopsis semiserrata Hickel et A. Camus = Castanopsis fabri Hance ●

79019 Castanopsis sempervirens (Kellogg) Dudley;常绿锥栗;Bush Chinkapin, Bush Chinquapin, Sempervirent Evergreenchinkapin, Sierra Chinkapin ●☆

79020 Castanopsis sempervirens (Kellogg) Dudley = Chrysolepis sempervirens (Kellogg) Hjelmq. ●☆

79021 Castanopsis sichouensis C. C. Huang et Y. T. Chang = Castanopsis xichouensis C. C. Huang et Y. T. Chang ●

79022 Castanopsis sieboldii (Makino) Hatus. ex T. Yamaz. et Mashiba;西氏栲(席氏栲)●☆

79023 Castanopsis sieboldii (Makino) Hatus. ex T. Yamaz. et Mashiba subsp. lutchuensis (Koidz.) H. Ohba;琉球栲●☆

79024 Castanopsis sieboldii (Makino) Hatus. ex T. Yamaz. et Mashiba var. lutchuensis (Koidz.) T. Yamaz. et Mashiba = Castanopsis sieboldii (Makino) Hatus. ex T. Yamaz. et Mashiba subsp. lutchuensis (Koidz.) H. Ohba ●☆

79025 Castanopsis sinensis A. Chev. = Castanopsis indica (Roxb. ex Lindl.) A. DC. ●

79026 Castanopsis sinsuiensis Kaneh. = Castanopsis fabri Hance ●

79027 Castanopsis stellatospina Hayata;星刺锥(短刺槠);Short Spine Chinkapin, Star-spined Evergreen Chinkapin, Stellato-spiny Evergreenchinkapin ●

79028 Castanopsis stellatospina Hayata = Castanopsis fabri Hance ●

79029 Castanopsis stipitata (Hayata ex Koidz.) Nakai = Castanopsis carlesii (Hemsl.) Hayata ●

79030 Castanopsis stipitata (Hayata) Nakai;单刺槠(单刺锥栗,水柯仔);Sharp Spines Chinkapin ●

79031 Castanopsis stipitata (Hayata) Nakai = Castanopsis carlesii (Hemsl.) Hayata ●

79032 Castanopsis subacuminata Hayata;微尖栲(恒春椎栗,恒春锥栗,渐尖叶槠,亚尖叶锥);Acuminate Leaved Chinkapin, Subacuminate Evergreen Chinkapin, Subacuminate Evergreenchinkapin ●

79033 Castanopsis subacuminata Hayata = Castanopsis indica (Roxb. ex Lindl.) A. DC. ●

79034 Castanopsis subuliformis Chun et C. C. Huang;钻刺锥(锥刺锥);Awl-shaped Evergreen Chinkapin, Subulate Evergreenchinkapin ●

79035 Castanopsis taiwaniana Hayata;台栲(赤校,钩栗,火烧柯);Taiwan Evergreenchinkapin ●

79036 Castanopsis taiwaniana Hayata = Castanopsis fargesii Franch. ●

79037 Castanopsis tapuensis Hu = Castanopsis hystrix Hook. f. et Thomson ex A. DC. ●

79038 Castanopsis tcheponensis Hickel et A. Camus;薄叶锥(薄叶栲);Thinleaf Evergreenchinkapin, Thin-leaved Evergreen Chinkapin ●

79039 Castanopsis tenuinervis A. Camus = Castanopsis orthacantha Franch. ●

79040 Castanopsis tenuispinula Hickel et A. Camus = Castanopsis fabri Hance ●

79041 Castanopsis tesselata Hickel et A. Camus = Castanopsis poilanei Hickel et A. Camus ●

79042 Castanopsis thunkinensis (Drake) Barnett = Castanopsis fissa (Champ. ex Benth.) Rehder et E. H. Wilson ●

79043 Castanopsis tibetana Hance;钩栲(巴栗,大叶钩栗,大叶槠,大叶锥栗,钩栲栗,钩栗,钩锥,猴板栗,猴栗,厚果栗,厚栗,葫芦树,假板栗,木栗,青柴栗,青叶槠,野板栗,槠栗,锥栗);Tibet Chinquapin, Tibet Evergreen Chinkapin, Tibet Evergreenchinkapin ●

79044 Castanopsis tonkinensis Seem. var. laocaiensis Luong;云南公孙锥●

79045 Castanopsis tonkinensis Seemen;公孙锥(斧柄锥,细刺栲);Tonkin Evergreen Chinkapin, Tonkin Evergreenchinkapin ●

79046 Castanopsis traninhensis Hickel et A. Camus = Castanopsis fabri Hance ●

79047 Castanopsis tribuloides (Lindl.) A. DC. = Castanopsis tribuloides (Sm.) A. DC. ●

79048 Castanopsis tribuloides (Lindl.) A. DC. var. echidnocarpa King ex Hook. f. = Castanopsis echidnocarpa Miq. ●

79049 Castanopsis tribuloides (Lindl.) A. DC. var. ferox King ex Hook. f. = Castanopsis ferox Spach ●

79050 Castanopsis tribuloides (Lindl.) A. DC. var. formosana V. Naray. = Castanopsis jucunda Hance ●

79051 Castanopsis tribuloides (Lindl.) A. DC. var. formosana V. Naray. = Castanopsis formosana (V. Naray.) Hayata ●

79052 Castanopsis tribuloides (Lindl.) A. DC. var. longispina King ex Hook. f. = Castanopsis longispina (King ex Hook. f.) C. C. Huang et Y. T. Chang ●

79053 Castanopsis tribuloides (Lindl.) A. DC. var. wattii King ex Hook. f. = Castanopsis wattii (King) A. Camus ●

79054 Castanopsis tribuloides (Sm.) A. DC.;蒺藜栲(蒺藜锥);Caltroplike Evergreen Chinkapin, Caltroplike Evergreenchinkapin ●

79055 Castanopsis tribuloides (Sm.) A. DC. var. echinocarpa (Hook. f. et Thomson ex Miq.) King ex Hook. f. = Castanopsis echidnocarpa Miq. ●

79056 Castanopsis tribuloides (Sm.) A. DC. var. ferox King ex Hook. f. = Castanopsis ferox Spach ●

79057 Castanopsis tribuloides (Sm.) A. DC. var. formosana V. Naray.;美丽蒺藜栲●☆

79058 Castanopsis tribuloides (Sm.) A. DC. var. longispina King ex Hook. f. = Castanopsis longispina (King ex Hook. f.) C. C. Huang et Y. T. Chang ●

79059 Castanopsis tribuloides (Sm.) A. DC. var. wattii King ex Hook. f. = Castanopsis wattii (King) A. Camus ●

79060 Castanopsis tribuloides A. Camus = Castanopsis ouonbiensis Hickel et A. Camus ●

79061 Castanopsis tsaii Hu = Castanopsis delavayi Franch. ●

79062 Castanopsis tunkinensis (Drake) Barnett = Castanopsis fissa (Champ. ex Benth.) Rehder et E. H. Wilson ●

79063 Castanopsis undulatifolia G. A. Fu;波叶锥;Undulateleaf Evergreen Chinkapin ●

79064 Castanopsis uraiana (Hayata) Kaneh. et Hatus.;吡漓锥(红肉杜,淋漓,淋漓柯,淋漓石栎,淋漓锥,鳞苞栲,鳞苞锥,菱果石栎,峦大山石栎,思仔,槵,乌来栲,乌来柯);Randaishan Tanoak, Urai Evergreen Chinkapin, Urai Evergreenchinkapin, Urai Tanoak ●

79065 Castanopsis wangii Hu = Castanopsis mekongensis A. Camus ●

79066 Castanopsis wangii Hu et W. C. Cheng = Castanopsis mekongensis A. Camus ●

79067 Castanopsis wattii (King) A. Camus;腾冲栲(变色锥);Tengchong Evergreenchinkapin ●

79068 Castanopsis wenchangensis G. A. Fu et C. C. Huang;文昌锥(杠锥,油锥);Wenchang Evergreen Chinkapin, Wenchang Evergreenchinkapin ●

79069 Castanopsis xichouensis C. C. Huang et Y. T. Chang;西畴锥;Xichou Evergreen Chinkapin, Xichou Evergreenchinkapin ●

79070 Castanopsis yanshanensis Hu = Castanopsis echidnocarpa Miq. ●

79071 Castanopsis yanshanensis Hu = Castanopsis orthacantha Franch. ●

79072 Castanospermum A. Cunn. = Castanospermum A. Cunn. ex Hook. ●☆

79073 Castanospermum A. Cunn. ex Hook. (1830);栗豆木属(昆士兰黑豆属,栗豆树属,栗果豆属,栗籽豆属);Moreton Bay Chestnut ●☆

79074 Castanospermum A. Cunn. ex Mudie = Castanospermum A. Cunn. ex Hook. ●☆

79075 Castanospermum australe A. Cunn. et C. Fraser;栗豆木(澳洲栗,澳洲栗籽豆,黑豆木,黑豆树,昆士兰黑豆,栗豆树);Australian Chestnut, Bean Tree, Beantree, Black Bean, Black Bean Tree, Black Bean-tree, Moreton Bay Bean, Moreton Bay Chestnut ●☆

79076 Castanospermum australe A. Cunn. et C. Fraser ex Hook. = Castanospermum australe A. Cunn. et C. Fraser ●☆

79077 Castanospermum australe A. Cunn. ex Mudie = Castanospermum australe A. Cunn. et C. Fraser ●☆

79078 Castanospora F. Muell. (1875);栗果无患子属●☆

79079 Castantmpora alphandii (F. Muell.) F. Muell.;栗果无患子●☆
79080 Castela Turpin(1806)(保留属名);堡树属(卡斯得拉属)●
79081 Castela alaternifolia Planch.;堡树●☆
79082 Castela erecta Turpin;直立堡树●☆
79083 Castela macrophylla Urb.;大叶堡树●☆
79084 Castela nicholsonii Hook.;尼氏堡树●☆
79085 Castela polyandra Moran et Felger;多蕊堡树●☆
79086 Castela texana (Torr. et Gray) Rose;得州堡树●☆
79087 Castela tortuosa Liebm.;扭旋堡树●☆
79088 Castelaceae J. Agardh = Simaroubaceae DC.(保留科名)●
79089 Castelaria Small = Castela Turpin(保留属名)●
79090 Castelia Cav.(废弃属名) = Castela Turpin(保留属名)●
79091 Castelia Cav.(废弃属名) = Pitraea Turcz. ■☆
79092 Castelia Liebm. = Castela Turpin(保留属名)●
79093 Castellanoa Traub(1953);卡斯石蒜属■☆
79094 Castellanoa marginata (R. E. Fr.) Traub;卡斯石蒜■☆
79095 Castellanosia Cárdenas = Browningia Britton et Rose ●☆
79096 Castellanosia Cárdenas(1951);钟花柱属■☆
79097 Castellanosia caineana Cárdenas;钟花柱■☆
79098 Castellia Tineo(1846);堡垒草属■☆
79099 Castellia tuberculata Tineo = Castellia tuberculosa (Moris) Bor ■☆
79100 Castellia tuberculosa (Moris) Bor;堡垒草■☆
79101 Castelnavia Tul. et Wedd.(1849);巴西苔草属■☆
79102 Castelnavia fruticulosa Tul. et Wedd.;巴西苔草■☆
79103 Castenedia R. M. King et H. Rob.(1978);细柱亮泽兰属■☆
79104 Castenedia R. M. King et H. Rob. = Eupatorium L. ■●
79105 Castenedia santamartensis R. M. King et H. Rob.;细柱亮泽兰■☆
79106 Castiglionia Ruiz et Pav. = Curcas Adans. ●■
79107 Castiglionia Ruiz et Pav. = Jatropha L.(保留属名)●■
79108 Castilla Cerv.(1794);橡胶桑属(美胶木属,美胶属,美洲胶属,美洲橡胶树属);Castilla, Gum Tree, Gum-tree ●☆
79109 Castilla Sessé = Castilla Cerv. ●☆
79110 Castilla elastica Cerv.;巴拿马橡胶桑(巴拿马橡胶树,弹性卡斯桑,弹性美胶树,卡斯提橡胶树,美国橡皮树);C. American Rubber, Castilloa Rubber, Central American Rubber, Central American Rubber Tree, Mexican Rubber Tree, Panama Gum Tree, Panama Gum-tree, Panama Rubber, Panama Rubber Tree, Panama Rubbertree, Ulé Rubber ●☆
79111 Castilla fallax O. F. Cook;橡胶桑(假美胶树)●☆
79112 Castilla ulei Warb.;亚马逊橡胶桑(卡斯桑);Uli ●☆
79113 Castilleja Mutis ex L. f.(1782);火焰草属(卡斯蒂属);Flamegrass, Indian Paint-Brosh, Indian Paintbrush, Painted Cup, Paintedcup ■
79114 Castilleja angustifolia A. Gray = Castilleja integra A. Gray ■☆
79115 Castilleja angustifolia G. Don = Castilleja parviflora Bong. ■☆
79116 Castilleja angustifolia M. Martens et Galeotti;狭叶火焰草;Desert Paintbrush, Indian Paintbrush ☆
79117 Castilleja arctica Krylov et Serg.;北极火焰草■☆
79118 Castilleja arvensis Cham. et Schltdl.;田野火焰草;Field Indian Paintbrush ■☆
79119 Castilleja chromosa A. Nelson;杂色火焰草;Desert Paintbrush ■☆
79120 Castilleja coccinea (L.) Spreng.;红火焰草;Indian Blanket, Indian Paintbrush, Painted Cup, Scarlet Painted-cup ■☆
79121 Castilleja coccinea (L.) Spreng. f. alba Farw. = Castilleja coccinea (L.) Spreng. ■☆
79122 Castilleja coccinea (L.) Spreng. f. lutescens Farw. = Castilleja coccinea (L.) Spreng. ■☆
79123 Castilleja coccinea (L.) Spreng. f. pallens (Michx.) Pennell = Castilleja coccinea (L.) Spreng. ■☆
79124 Castilleja exserta Eastw.;凸火焰草;Exserted Indian Paintbrush, Purple Owl's Clover ■☆
79125 Castilleja ferox;多刺火焰草■☆
79126 Castilleja foliolosa Hook. et Arn.;毛火焰草;Woolly Paintbrush ■☆
79127 Castilleja integra A. Gray;印度火焰草;Indian Paintbrush ■☆
79128 Castilleja linariifolia Benth.;长叶火焰草;Long-leaved Paintbrush, Painted Cup ☆
79129 Castilleja miniata Douglas ex Hook.;朱红火焰草;Giant Red Paintbrush ■☆
79130 Castilleja pallida (L.) Kunth;苍白火焰草;Flamegrass, Painted Cup, Paintedcup, Pale Painted-cup ■
79131 Castilleja parviflora Bong.;小花火焰草■☆
79132 Castilleja plagiotoma A. Gray;黄火焰草;Yellow Paintbrush ■☆
79133 Castilleja purpurea (Nutt.) G. Don;紫火焰草;Purple Paintbrush ■☆
79134 Castilleja septentrionalis Lindl.;北方火焰草;Pale Painted-cup ■☆
79135 Castilleja sessiliflora Pursh;无梗火焰草;Downy Paintbrush, Downy Painted Cup, Downy Painted-cup, Downy Yellow Painted-cup, Sessile Paintbrush ■☆
79136 Castilleja sulphurea Rydb.;硫黄火焰草;Sulfur Paintbrush ■☆
79137 Castillejoa Post et Kuntze = Castilleja Mutis ex L. f. ■
79138 Castilloa Cervant. = Castilla Cerv. ●☆
79139 Castilloa Endl. = Castilla Cerv. ●☆
79140 Castilloa elastica Cerv. = Castilla elastica Cerv. ●☆
79141 Castilloa fallax O. F. Cook = Castilla fallax O. F. Cook ●☆
79142 Castorea Mill. = Duranta L. ●
79143 Castra Vell. = Trixis P. Browne ■●☆
79144 Castratella Naudin(1850);雄黄牡丹属☆
79145 Castratella piloselloides Naudin;雄黄牡丹☆
79146 Castrea A. St. -Hil. = Phoradendron Nutt. ●☆
79147 Castrilanthemum Vogt et Oberpr.(1996);丁毛菊属■☆
79148 Castrilanthemum debeauxii (Degen, Hervier et E. Rev.) Vogt et Oberpr.;丁毛菊■☆
79149 Castroa Guiard = Oncidium Sw.(保留属名)■☆
79150 Castronia Noronha = Helicia Lour. ●
79151 Castroviejoa Galbany, L. Sáez et Benedí = Xeranthemum L. ■☆
79152 Castroviejoa Galbany, L. Sáez et Benedí(2004);岛蜡菊属■☆
79153 Castroviejoa frigida (Labill.) Galbany, L. Sáez et Benedí;岛蜡菊■☆
79154 Casuarina Adans. = Casuarina L. ●
79155 Casuarina L.(1759);木麻黄属;Australian-pine, Austrian Pine, Beef Wood, Beefwood, Casuarina, Ironwood, She Oak, She-oak, Swamp Oak ●
79156 Casuarina cristata Miq.;白木麻黄(南方木麻黄);Belah ●☆
79157 Casuarina cunninghamiana Miq.;细枝木麻黄(滨海木麻黄,肯氏木麻黄,银线木麻黄);Australian Beefwood, Australian Pine, Australian River Oak, Beefwood, Cunningham Beefwood, Cunningham Casuarina, Cunningham She-oak, Cunningham's Beefwood, Fire She Oak, Fire She-oak, Ri-oak Casuarina, River Oak, River She Oak, River Sheoak, River She-oak ●
79158 Casuarina decaisneana F. Muell.;沙地木麻黄;Desert Oak ●☆
79159 Casuarina deplancheana Miq. = Gymnostoma deplancheanum (Miq.) L. A. S. Johnson ●☆
79160 Casuarina equisetifolia J. R. Forst. et G. Forst. = Casuarina equisetifolia L. ●

79161 Casuarina equisetifolia L.;木麻黄(驳骨树,驳骨松,短枝木麻黄,马尾树,木贼叶木麻黄);Australian Pine,Beach Sheoak,Beach She-oak,Beef Wood,Beefsteak,Beefwood,Bull Oak,Coast Oak,Forest Oak,Horsetail Beefwood,Horsetail Tree,Iron Wood,Jau,Mile Tree,Polynesia Ironwood,Polynesian Iron Wood,Shingle Oak,South Sea Ironwood,Swamp She Oak,Swamp She-oak,Whistling Pine,Yar ●

79162 Casuarina equisetifolia L. subsp. incana (Benth.) L. A. S. Johnson;灰毛木麻黄;Horsetail She-oak ●☆

79163 Casuarina equisetifolia L. var. incana Benth. = Casuarina equisetifolia L. subsp. incana (Benth.) L. A. S. Johnson ●☆

79164 Casuarina fraseriana Miq.;澳洲木麻黄(佛勒塞木麻黄,佛氏木麻黄);Fraser Beefwood,Fraser's She Oak,Fraser's She-oak,She Oak,She-oak ●

79165 Casuarina glauca Siebold ex Spreng.;粗枝木麻黄(长叶木麻黄,坚木麻黄,蓝枝木麻黄,银木麻黄);Brazilian Beefwood,Cassowary Tree,Desert Oak,Gray Sheoak,Gray She-oak,Grey Buloke,Longleaf Beefwood,Longleaved Beefwood,River She Oak,Scaly-bark Beefwood,Suckering Australian Pine,Suckering Australian-pine,Swamp Oak,Swamp She-oak ●

79166 Casuarina huegeliana Miq.;虎氏木麻黄(休氏木麻黄);Huegel's She-oak,Murchion River Oak ●

79167 Casuarina lepidophloia F. Muell. = Casuarina cristata Miq. ●☆

79168 Casuarina leptoclada Miq. = Casuarina littoralis Salisb. ●☆

79169 Casuarina litorea L. ex Fosberg et Sachet = Casuarina equisetifolia L. ●

79170 Casuarina littoralis Salisb.;滨海木麻黄;Forest Oak ●☆

79171 Casuarina montana Lesch. = Casuarina montana Lesch. ex Miq. ●

79172 Casuarina montana Lesch. ex Miq.;山木麻黄;Mountain She Oak,Mountain She-oak ●

79173 Casuarina muricata Roxb. = Casuarina equisetifolia L. ●

79174 Casuarina muricata Roxb. ex Hornem. = Casuarina equisetifolia L. ●

79175 Casuarina nana Sieber ex Spreng.;千头木麻黄●☆

79176 Casuarina obesa Miq.;肥胖木麻黄●☆

79177 Casuarina pauper F. Muell. ex L. A. S. Johnson;矮生木麻黄;Belah,Black Oak ●☆

79178 Casuarina quadrivalvis Labill.;方苞木麻黄(小木麻黄);Coast She Oak,Square-valved She-oak ●

79179 Casuarina quadrivalvis Labill. = Casuarina stricta Aiton ●☆

79180 Casuarina quadrivalvis Labill. = Casuarina verticillata Lam. ●☆

79181 Casuarina rotundifolia;圆叶木麻黄;Bull Oak ●☆

79182 Casuarina stricta Aiton;小木麻黄;Coast Casuarina ●☆

79183 Casuarina stricta Aiton = Allocasuarina verticillata (Lam.) L. A. S. Johnson ●

79184 Casuarina stricta Aiton = Casuarina verticillata Lam. ●☆

79185 Casuarina suberosa Otto et F. Dietr.;栓皮木麻黄(直立木麻黄);Beef Wood,Erect Beefwood,Forest Oak,River Black Oak ●☆

79186 Casuarina suberosa Otto et F. Dietr. = Casuarina littoralis Salisb. ●☆

79187 Casuarina sumatrana Jungh.;苏门答腊木麻黄●☆

79188 Casuarina tortuosa A. Henry;链节木麻黄●☆

79189 Casuarina torulosa Aiton;林木麻黄;Forest She Oak,Forest She-oak ●☆

79190 Casuarina torulosa Vent. = Casuarina suberosa Otto et F. Dietr. ●☆

79191 Casuarina trichodon Miq.;毛齿木麻黄;Hairy Toothed She Oak,Hairy-bracted She-oak,Hairy-toothed She-oak ●

79192 Casuarina verticillata Lam.;轮生木麻黄●☆

79193 Casuarinaceae R. Br. (1814)(保留科名);木麻黄科;Beefwood Family,Casuarina Family,She-oak Family ●

79194 Catabrosa P. Beauv. (1812);沿沟草属;Brookgrass,Whorl-grass ■

79195 Catabrosa altaica (Trin.) Boiss. = Colpodium altaicum Trin. ■

79196 Catabrosa altaica (Trin.) Boiss. = Paracolpodium altaicum (Trin.) Tzvelev ■

79197 Catabrosa angusta (Stapf) L. Liou = Catabrosa aquatica (L.) P. Beauv. var. angusta Stapf ■

79198 Catabrosa aquatica (L.) P. Beauv.;沿沟草(水沿沟草);Brook Grass,Brookgrass,Brook-grass,Water Hair,Water Whorl-grass,Whorl-grass ■

79199 Catabrosa aquatica (L.) P. Beauv. subsp. capusii (Franch.) Tzvelev = Catabrosa capusii Franch. ■

79200 Catabrosa aquatica (L.) P. Beauv. var. angusta Stapf;窄沿沟草■

79201 Catabrosa aquatica (L.) P. Beauv. var. angusta Stapf = Catabrosa aquatica (L.) P. Beauv. ■

79202 Catabrosa aquatica P. Beauv. = Catabrosa aquatica (L.) P. Beauv. ■

79203 Catabrosa capusii Franch.;长颖沿沟草(紧穗沿沟草)■

79204 Catabrosa himalaica (Hook. f.) Stapf = Catabrosella himalaica (Hook. f.) Tzvelev ■☆

79205 Catabrosa humilis (M. Bieb.) Trin. = Catabrosella humilis (M. Bieb.) Tzvelev ■

79206 Catabrosa humilis (M. Bieb.) Trin. = Colpodium humile (M. Bieb.) Griseb. ■

79207 Catabrosa nutans Stapf;点头沿沟草■☆

79208 Catabrosa sikkimensis Stapf = Catabrosa aquatica (L.) P. Beauv. ■

79209 Catabrosa wallichii Hook. f. = Catabrosa capusii Franch. ■

79210 Catabrosa wallichii Stapf = Colpodium wallichii (Stapf) Bor ■

79211 Catabrosella (Tzvelev) Tzvelev = Colpodium Trin. ■

79212 Catabrosella (Tzvelev) Tzvelev(1965);小沿沟草属■

79213 Catabrosella himalaica (Hook. f.) Tzvelev;喜马拉雅小沿沟草■☆

79214 Catabrosella humilis (M. Bieb.) Tzvelev;小沿沟草(矮小沿沟草)■

79215 Catabrosella humilis (M. Bieb.) Tzvelev = Colpodium humile (M. Bieb.) Griseb. ■

79216 Catabrosella humilis (M. Bieb.) Tzvelev subsp. songorica Tzvelev = Colpodium humile (M. Bieb.) Griseb. ■

79217 Catabrosella songarica (Schrenk) Czerep. = Poa diaphora Trin. ■

79218 Catabrosella songorica (Tzvelev) Czerep. = Colpodium humile (M. Bieb.) Griseb. ■

79219 Catabrosia Roem. et Schult. = Catabrosella (Tsveiev) Tsveiev ■

79220 Catachaenia Griseb. = Miconia Ruiz et Pav. (保留属名)●☆

79221 Catachaetum Hoffmanns. = Catasetum Rich. ex Kunth ■☆

79222 Catachaetum Hoffmanns. ex Rchb. = Catasetum Rich. ex Kunth ■☆

79223 Catachenia Griseb. = Miconia Ruiz et Pav. (保留属名)●☆

79224 Catacline Edgew. = Tephrosia Pers. (保留属名)●■

79225 Catacolea B. G. Briggs et L. A. S. Johnson(1998);扁秆帚灯草属■☆

79226 Catacolea enodis B. G. Briggs et L. A. S. Johnson;扁秆帚灯草■☆

79227 Catacoma Walp. = Bredemeyera Willd. ●☆

79228 Catacoma Walp. = Catocoma Benth. ●☆

79229 Catadysia O. E. Schulz (1929);秘鲁莲座芥属■☆

79230 Catadysia rosulans O. E. Schulz;秘鲁莲座芥■☆

79231 Catagyna Beauv. = Coleochloa Gilly ■☆

79232 Catagyna Hutch. et Dalzell = Coleochloa Gilly ■☆

79233 Catagyna Hutch. et Dalziel = Scleria P. J. Bergius ■

79234 Catagyna pilosa (Boeck.) Hutch. = Afrotrilepis pilosa (Boeck.) J. Raynal ■☆

79235 Catakidozamia W. Hill = Macrozamia Miq. ●☆

79236 Catalepidia P. H. Weston = Helicia Lour. ●

79237 Catalepis Stapf et Stent(1929);实心草属■☆

79238 Catalepis gracilis Stapf et Stent;实心草■☆

79239 Cataleuca Hort. ex K. Koch = Onoseris Willd. ●■☆

79240 Catalium Buch. -Ham. ex Wall. = Carallia Roxb. (保留属名)●

79241 Catalpa Juss. = Catalpa Scop. ●

79242 Catalpa Scop. (1777);梓属(楸属,梓树属);Beantree,Catalpa,Catawba,Indian Bean,Indian Catalpa ●

79243 Catalpa bignonioides Walter;美国梓(美国木豆树,紫葳楸,紫葳梓);American Catalpa, Bean Tree, Bean-tree, Catalpa, Catawba, Caterpillar Tree,Cigar Tree,Cigartree,Common Catalpa,Indian Bean, Indian Bean Tree, Indian Bean-tree, Indian-bean, Locust Bean, Red Indian Bean-tree,Southern Catalpa ●☆

79244 Catalpa bignonioides Walter 'Aurea';金叶美国梓(黄叶美国梓,黄叶紫葳楸);Golden Indian Bean-tree,Golden-leaved Catalpa ●☆

79245 Catalpa bignonioides Walter var. speciosa Ward. ex Barney = Catalpa speciosa (Warder ex Barney) Warder ex Engelm. ●

79246 Catalpa bingonioides Walter 'Nana';矮美国梓;Dwarf Catalpa ●☆

79247 Catalpa bungei C. A. Mey.;楸(榎椅,角楸,金丝楸,木王,楸木,楸树,水桐,梓,梓桐);Beijing Catalpa,Indian Bean,Locust Bean,Manchurian Catalpa,Manchurian Catawba ●

79248 Catalpa cordifolia J. St. -Hil. = Catalpa speciosa (Warder ex Barney) Warder ex Engelm. ●

79249 Catalpa duclouxii Dode = Catalpa fargesii Bureau ●

79250 Catalpa fargesii Bureau;灰楸(白楸,川楸,滇楸,法氏楸,光灰楸,泡桐木,楸木,紫花楸,紫楸);Farges Catalpa ●

79251 Catalpa fargesii Bureau f. alba Q. Q. Liu et H. Y. Ye;白花灰楸;White-flower Farges Catalpa ●

79252 Catalpa fargesii Bureau f. duclouxii (Dode) Gilmour = Catalpa fargesii Bureau ●

79253 Catalpa fargesii Wilson = Catalpa fargesii Bureau ●

79254 Catalpa henryi Dode = Catalpa ovata G. Don ●

79255 Catalpa kaempferi Siebold et Zucc. = Catalpa ovata G. Don ●

79256 Catalpa longissima Sims;牙买加楸(极长梓);Bois-chene,Haitian Catalpa,Jamaican Oak ●☆

79257 Catalpa ovata G. Don;梓(臭梧桐,河楸,花楸,黄花楸,黄金树,豇豆树,筷子树,雷电木,木角豆,木王,楸,水桐,水桐楸,椅,梓树);Catalpa, Chinese Catalpa, Japanese Catalpa, Ovate Catalpa, Yellow Catalpa ●

79258 Catalpa punctata Griseb.;紫斑楸;Robillo,Roble de Olor ●☆

79259 Catalpa speciosa (Warder ex Barney) Engelm. = Catalpa speciosa (Warder ex Barney) Warder ex Engelm. ●

79260 Catalpa speciosa (Warder ex Barney) Warder ex Engelm.;黄金树(白花梓树,北方梓木,北美楸树,卡托巴木,西部梓木,雪茄树,梓木);Beantree, Catalpa, Catawba, Catawba Tree, Cigar Tree, Cigartree,Cigar-tree,Gold Tree,Hardy Catalpa,Indian Bean,Indian-bean, Northern Catalpa, Shawnee Wood, Western Catalpa, Western Catawba ●

79261 Catalpa speciosa Engelm. = Catalpa speciosa (Warder ex Barney) Warder ex Engelm. ●

79262 Catalpa speciosa Ward. ex Barney = Catalpa speciosa (Warder ex Barney) Warder ex Engelm. ●

79263 Catalpa sutchuenensis Dode = Catalpa fargesii Bureau ●

79264 Catalpa syringifolia Bunge = Catalpa bungei C. A. Mey. ●

79265 Catalpa tibetica Forrest;藏楸;Tibet Catalpa,Xizang Catalpa ●

79266 Catalpa vestita Diels = Catalpa fargesii Bureau ●

79267 Catalpium Raf. = Catalpa Scop. ●

79268 Catamixis Thomson(1867);簇黄菊属●☆

79269 Catamixis baccharoides Thomson;簇黄菊■☆

79270 Catanance St. -Lag. = Catananche L. ■☆

79271 Catananche L. (1753);蓝箭菊属(蓝苣属,琉璃菊属);Blue Cupidone, Cupidone, Cupid's Dart, Cupids-dart, Cupid's-dart, Cupid's-darts ■☆

79272 Catananche arenaria Coss. et Durieu;沙地琉璃菊■☆

79273 Catananche arenaria Coss. et Durieu var. atricha Murb. = Catananche arenaria Coss. et Durieu ■☆

79274 Catananche arenaria Coss. et Durieu var. aurea Maire = Catananche arenaria Coss. et Durieu ■☆

79275 Catananche arenaria Coss. et Durieu var. dimorpha Dobignard = Catananche arenaria Coss. et Durieu ■☆

79276 Catananche caerulea L.;蓝箭菊(爱神箭,蓝苣);Blue Cupidone,Blue Cupids-dart,Blue Cupid's-dart,Buck's Horn Weld, Buck's-horn Weld, Cupidone, Cupid's Dart, Cupid's Darts, Cupid's-dart ■☆

79277 Catananche caerulea L. 'Major';大花蓝箭菊(大花蓝苣)■☆

79278 Catananche caerulea L. var. alba Hort;白花蓝箭菊■☆

79279 Catananche caerulea L. var. bicolor Hort;二色蓝箭菊■☆

79280 Catananche caerulea L. var. ochroleuca Maire = Catananche caerulea L. ■☆

79281 Catananche caerulea L. var. propinqua (Pomel) Hochr. = Catananche caerulea L. ■☆

79282 Catananche caerulea L. var. tenuis Ball = Catananche caerulea L. var. propinqua (Pomel) Hochr. ■☆

79283 Catananche caerulea L. var. tlemcenensis Faure = Catananche caerulea L. ■☆

79284 Catananche caespitosa Desf.;丛生琉璃菊■☆

79285 Catananche coerulea L. = Catananche caerulea L. ■☆

79286 Catananche lutea L.;黄蓝箭菊(黄琉璃菊)■☆

79287 Catananche montana Coss. et Durieu;山地琉璃菊■☆

79288 Catananche propinqua Pomel = Catananche caerulea L. ■☆

79289 Catanga Steud. = Cananga (DC.) Hook. f. et Thomson(保留属名)●

79290 Catanga Steud. = Guatteria Ruiz et Pav. (保留属名)●☆

79291 Catanthera F. Muell. (1886);垂药野牡丹属●☆

79292 Catanthera brassii (Nayar) Nayar;巴西垂药野牡丹●☆

79293 Catanthera lysipetala F. Muell.;垂药野牡丹●☆

79294 Catanthera multiflora (Stapf) Nayar;多花垂药野牡丹●☆

79295 Catapodium Link (1827) ('Catopodium');绳柄草属;Catapodium,Fern-grass ■☆

79296 Catapodium demnatense (Murb.) Maire et Weiller = Wangenheimia demnatensis (Murb.) Stace ■☆

79297 Catapodium filiforme Nees ex Duthie = Tripogon filiformis Nees ex Steud. ■

79298 Catapodium hemipoa (Spreng.) Lainz;半草绳柄草■☆

79299 Catapodium hemipoa (Spreng.) Lainz subsp. occidentale (Paunero) H. Scholz et S. Scholz;西方绳柄草■☆

79300 Catapodium loliaceum (Huds.) Link = Catapodium marinum (L.) C. E. Hubb. ■☆

79301 Catapodium loliaceum (Huds.) Link subsp. loliaceum Maire et Weiller = Catapodium marinum (L.) C. E. Hubb. ■☆

79302 Catapodium loliaceum (Huds.) Link subsp. syrticum Barratte et Murb. = Catapodium marinum (L.) C. E. Hubb. subsp. syrticum (Murb.) H. Scholz ■☆

79303 Catapodium lolium (Balansa) Trab. = Agropyropsis lolium (Balansa) A. Camus ■☆

79304 Catapodium mamoraeum (Maire) Maire et Weiller = Micropyrum mamoraeum (Maire) Stace ■☆

79305 Catapodium marinum (L.) C. E. Hubb.;滨海绳柄草;Sea Fern-grass, Seashore Catapodium, Stiff Sand-grass ■☆

79306 Catapodium marinum (L.) C. E. Hubb. subsp. syrticum (Murb.) H. Scholz;瑟尔特绳柄草■☆

79307 Catapodium pungens Boiss. = Aeluropus macrostachyus Hack. ■☆

79308 Catapodium rigidum (L.) C. E. Hubb. = Catapodium rigidum (L.) C. E. Hubb. ex Dony ■☆

79309 Catapodium rigidum (L.) C. E. Hubb. ex Dony = Desmazeria rigida (L.) Tutin ■☆

79310 Catapodium rigidum (L.) C. E. Hubb. ex Dony = Scleropoa rigida (L.) Griseb. ■☆

79311 Catapodium rigidum (L.) C. E. Hubb. subsp. hemipoa (Spreng.) Kerguélen = Catapodium hemipoa (Spreng.) Lainz ■☆

79312 Catapodium salzmanni (Boiss.) Coss. = Narduroides salzmannii (Boiss.) Rouy ■☆

79313 Catapodium tenellum (L.) Trab. = Micropyrum tenellum (L.) Link ■☆

79314 Catapodium tenellum (L.) Trab. var. aristatum (Tausch) Trab. = Micropyrum tenellum (L.) Link ■☆

79315 Catapodium tenellum (L.) Trab. var. muticum (Tausch) Maire = Micropyrum tenellum (L.) Link ■☆

79316 Catapodium tuberculosum (Moris) Bor = Castellia tuberculosa (Moris) Bor ■☆

79317 Catapodium tuberculosum Moris = Castellia tuberculosa (Moris) Bor ■☆

79318 Catappa Gaertn. = Terminalia L.(保留属名)●

79319 Catapuntia Müll. Arg. = Cataputia Boehm. ●■

79320 Cataputia Boehm. = Ricinus L. ●■

79321 Cataputia Ludw. = Ricinus L. ●■

79322 Cataria Adans. = Nepeta L. ■●

79323 Cataria Mill. = Nepeta L. ■●

79324 Catarsis Post et Kuntze = Gypsophila L. ■●

79325 Catarsis Post et Kuntze = Katarsis Medik. ■●

79326 Catas Domb. ex Lam. = Embothrium J. R. Forst. et G. Forst. ●☆

79327 Catasetum L. = Catasetum Rich. ex Kunth ■☆

79328 Catasetum Rich. = Catasetum Rich. ex Kunth ■☆

79329 Catasetum Rich. ex Kunth (1822); 龙须兰属; Catasetum, Monkflower ■☆

79330 Catasetum adnatum Steud. = Catasetum atratum Lindl. ■☆

79331 Catasetum atratum Lindl.; 黑斑唇龙须兰; Black-spotted Catasetum ■☆

79332 Catasetum bicolor Klotzsch; 二色龙须兰; Two-coloured Catasetum ■☆

79333 Catasetum bungerothii N. E. Br. = Catasetum pileatum Rchb. f. ■☆

79334 Catasetum callosum Lindl.;硬皮龙须兰■☆

79335 Catasetum cernuum Rchb. f.;垂花龙须兰;Nodding Catasetum ■☆

79336 Catasetum cliftonii Hort.;克氏龙须兰■☆

79337 Catasetum darwiniana Rolfe;达尔文龙须兰■☆

79338 Catasetum discolor (Lindl.) Lindl.; 异色龙须兰; Discolor Catasetum ■☆

79339 Catasetum fimbriatum Rchb. f.; 流苏龙须兰; Fimbriate Catasetum ■☆

79340 Catasetum globiflorum Hook.; 球花龙须兰; Globularflower Catasetum ■☆

79341 Catasetum longifolium Lindl.;长叶龙须兰;Longleaf Catasetum ■☆

79342 Catasetum macrocarpum Rich. ex Kunth; 大果龙须兰; Bigfruit Catasetum ■☆

79343 Catasetum maculatum Kunth;斑点龙须兰■☆

79344 Catasetum pileatum Rchb. f.;具帽龙须兰■☆

79345 Catasetum russellianum Hook.;拉氏龙须兰■☆

79346 Catasetum viridiflavum Hook.;绿黄三星兰■

79347 Cataterophora Steud. = Catatherophora Steud. ■

79348 Catatherophora Steud. = Pennisetum Rich. ■

79349 Catatia Humbert(1923);尖柱鼠麴木属●☆

79350 Catatia attenuata Humbert;尖柱鼠麴木●☆

79351 Catatia cordata Humbert;马岛尖柱鼠麴木●☆

79352 Catcareoboea C. Y. Wu = Platyadenia B. L. Burtt ■☆

79353 Catdamine sachalinensis Miyabe et Miyake = Cardamine macrophylla Willd. ■

79354 Catenaria Benth. = Desmodium Desv.(保留属名)●■

79355 Catenaria Benth. = Ohwia H. Ohashi ●

79356 Catenaria caudata (Thunb.) Schindl. = Desmodium caudatum (Thunb.) DC. ●

79357 Catenaria caudata (Thunb.) Schindl. = Ohwia caudata (Thunb.) H. Ohashi ●

79358 Catenaria laburnifolia (Poir.) Benth. = Desmodium caudatum (Thunb.) DC. ●

79359 Catenaria laburnifolia (Poir.) Benth. = Ohwia caudata (Thunb.) H. Ohashi ●

79360 Catenaria laburnifolia Benth. = Desmodium laburnifolium (Poir.) DC. ●

79361 Catenularia Botsch. = Catenulina Soják ■☆

79362 Catenulina Soják(1980);塔吉克芥属■☆

79363 Catenulina hedysaroides (Botsch.) Soják;塔吉克芥■☆

79364 Catesbaea L. (1753);卡德藤属(卡德斯巴牙藤属)■☆

79365 Catesbaea elliptica Spreng. ex DC.;椭圆卡德藤■☆

79366 Catesbaea erecta Moc. et Sessé ex DC.;直立卡德藤■☆

79367 Catesbaea fasciculata Northr.;簇生卡德藤■☆

79368 Catesbaea flaviflora Urb.;黄花卡德藤■☆

79369 Catesbaea glabra Urb.;光卡德藤■☆

79370 Catesbaea longiflora Sw.;长花卡德藤■☆

79371 Catesbaea melanocarpa Krug et Urb.;黑果卡德藤■☆

79372 Catesbaea microcarpa Urb.;小果卡德藤■☆

79373 Catesbaea nana Greenm.;矮卡德藤■☆

79374 Catesbaea parviflora Sw.;小花卡德藤■☆

79375 Catesbaea phyllacantha Griseb. = Phyllacanthus grisebachianus Hook. f. ■☆

79376 Catesbaea spinosa L.; 卡德斯巴牙藤; Lily Thorn, Spanish Guava, Thorn Lily ■☆

79377 Catesbaea triacantha Spreng.;三刺卡德藤■☆

79378 Catesbaeaceae Martinov = Rubiaceae Juss.(保留科名)●■

79379 Catesbya Cothen. = Catesbaea L. ■☆

79380 Catevala Medik.(废弃属名) = Aloe L. + Haworthia Duval(保留属名)■☆

79381 Catevala Medik.(废弃属名) = Haworthia Duval(保留属名)■☆

79382 Catevala arborescens (Mill.) Medik. = Aloe arborescens Mill. ●☆

79383 Catevala atroviridis Medik. = Haworthia herbacea (Mill.)

Stearn ■☆
79384 Catevala humilis (L.) Medik. = Aloe humilis (L.) Mill. ●☆
79385 Catha Forssk.(废弃属名) = Catha Forssk. ex Scop.(废弃属名)●
79386 Catha Forssk.(废弃属名) = Maytenus Molina ●
79387 Catha Forssk. ex Schreb.(1777)(废弃属名);巧茶属(阿拉伯茶属,卡茶属);Arebian-tea, Artfulttea, Cafta, Chat, Khat, Khate Tree ●
79388 Catha Forssk. ex Scop.(废弃属名) = Gymnosporia (Wight et Arn.) Benth. et Hook. f.(保留属名)●
79389 Catha G. Don(废弃属名) = Celastrus L.(保留属名)●
79390 Catha abbottii A. E. van Wyk et M. Prins = Lydenburgia abbottii (A. E. van Wyk et M. Prins) Steenkamp, A. E. van Wyk et M. Prins ●☆
79391 Catha acuminata (L. f.) C. Presl = Gymnosporia acuminata (L. f.) Szyszyl. ●☆
79392 Catha campestris (Eckl. et Zeyh.) C. Presl = Putterlickia pyracantha (L.) Szyszyl. ●☆
79393 Catha cassinoides (L'Hér.) Webb et Berthault = Maytenus canariensis (Loes.) G. Kunkel et Sunding ●☆
79394 Catha cassinoides (N. Robson) Codd = Lydenburgia cassinoides N. Robson ●☆
79395 Catha edulis (Vahl) Forssk. ex Endl.;巧茶(阿拉伯茶,埃塞俄比亚茶,也门茶);Abyssinian Tea, Arabian Tea, Arabian-tea, Artfulttea, Bushraan's Tea, Cafta, Chat, Kat, Khat, Miraa, Qat, Somali Tea ●
79396 Catha edulis Forssk. = Catha edulis (Vahl) Forssk. ex Endl. ●
79397 Catha emarginata (Willd.) G. Don = Gymnosporia emarginata (Willd.) Thwaites ●
79398 Catha fasciculata Tul. = Maytenus undata (Thunb.) Blakelock ●☆
79399 Catha forskalii A. Rich. = Catha edulis (Vahl) Forssk. ex Endl. ●
79400 Catha inermis J. F. Gmel. = Catha edulis (Vahl) Forssk. ex Endl. ●
79401 Catha integrifolia (L. f.) G. Don = Gloveria integrifolia (L. f.) Jordaan ●☆
79402 Catha monosperma Benth. = Celastrus hindsii Benth. ●
79403 Catha montana (Roth) G. Don = Gymnosporia senegalensis (Lam.) Loes. ●☆
79404 Catha montana (Roth) G. Don = Maytenus senegalensis (Lam.) Exell ●☆
79405 Catha spinosa Forssk. = Gymnosporia parviflora (Vahl) Chiov. ●☆
79406 Catha transvaalensis Codd = Lydenburgia cassinoides N. Robson ●☆
79407 Catha wallichii G. Don = Maytenus wallichiana (Spreng.) D. C. S. Raju et Babu ●☆
79408 Cathanthes Rich. = Tetroncium Willd. ■☆
79409 Catharanthus G. Don = Vinca L. ■
79410 Catharanthus G. Don(1837);长春花属;Periwinkle ●■
79411 Catharanthus coriaceus Markgr.;革质长春花■☆
79412 Catharanthus lanceus (Bojer ex A. DC.) Pichon;剑形长春花■☆
79413 Catharanthus longifolius (Pichon) Pichon;长叶长春花;Longleaf Periwinkle ■☆
79414 Catharanthus ovalis Markgr.;卵圆长春花;Ovate Periwinkle ■☆
79415 Catharanthus ovalis Markgr. subsp. grandiflorus Markgr. = Catharanthus ovalis Markgr. ■☆
79416 Catharanthus pusillus (Murray) G. Don;小长春花■☆
79417 Catharanthus pusillus G. Don = Catharanthus pusillus (Murray) G. Don ■☆
79418 Catharanthus roseus (L.) G. Don;长春花(长春,花海棠,日日草,日日春,日日新,三万花,四时春,四时花,雁来红);Cayenne Jasmine, Madagascar Periwinkle, Old Maid, Rose Periwinkle, South African Periwinkle, Vinca ■
79419 Catharanthus roseus (L.) G. Don 'Albus';白长春花(长春花);Whiteflower Periwinkle ●
79420 Catharanthus roseus (L.) G. Don 'Flavus';黄长春花;Yellow Periwinkle ●
79421 Catharanthus roseus (L.) G. Don var. albus G. Don = Catharanthus roseus (L.) G. Don ■
79422 Catharanthus roseus (L.) G. Don var. albus G. Don = Catharanthus roseus (L.) G. Don 'Albus' ●
79423 Catharanthus roseus (L.) G. Don var. flavus F. P. Metcalf = Catharanthus roseus (L.) G. Don 'Flavus' ●
79424 Catharanthus scitulus (Pichon) Pichon;绮丽长春花■☆
79425 Catharanthus trichophyllus (Baker) Pichon;毛叶长春花■☆
79426 Cathariostachys S. Dransf.(1998);洁穗禾属●☆
79427 Cathariostachys capitata (Kunth) S. Dransf.;头状洁穗禾●☆
79428 Cathariostachys madagascariensis (A. Camus) S. Dransf.;洁穗禾●☆
79429 Cathartocarpus Pers. = Cassia L.(保留属名)●■
79430 Cathartocarpus brewsteri F. Muell. = Cassia brewsteri (F. Muell.) F. Muell. ex Benth. ●☆
79431 Cathartolinum Rchb. = Linum L. ●■
79432 Cathartolinum Rchb. = Mesynium Raf. ●■
79433 Cathartolinum curtissii (Small) Small = Linum medium (Planch.) Britton var. texanum (Planch.) Fernald ■☆
79434 Cathartolinum sulcatum (Riddell) Small = Linum sulcatum Riddell ■☆
79435 Cathastrum Turcz. = Pleurostylia Wight et Arn. ●
79436 Cathastrum capense Turcz. = Pleurostylia capensis (Turcz.) Loes. ●☆
79437 Cathaya Chun et Kuang(1962);银杉属;Cathay Silver Fir ●★
79438 Cathaya argyrophylla Chun et Kuang;银杉(杉公子);Cathay Silver Fir ●◇
79439 Cathaya nanchuanensis Chun et Kuang = Cathaya argyrophylla Chun et Kuang ●◇
79440 Cathayambar (Harms) Nakai = Liquidambar L. ●
79441 Cathayanthe Chun(1946);扁蒴苣苔属;Cathayanthe ●★
79442 Cathayanthe biflora Chun;扁蒴苣苔草;Twoflowered Cathayanthe ■
79443 Cathayeia Ohwi = Idesia Maxim.(保留属名)●
79444 Cathayeia polycarpa (Maxim.) Ohwi = Idesia polycarpa Maxim. ●
79445 Cathcartia Hook. f. = Meconopsis R. Vig. ■
79446 Cathcartia betonicifolia (Franch.) Prain = Meconopsis betonicifolia Franch. ■
79447 Cathcartia delavayi Franch. = Meconopsis delavayi (Franch.) Franch. et Prain ■
79448 Cathcartia integrifolia Maxim. = Meconopsis integrifolia (Maxim.) Franch. ■
79449 Cathcartia lancifolia Franch. = Meconopsis lancifolia (Franch.) Franch. ex Prain ■
79450 Cathcartia lyrata Cummins et Prain ex Praln = Meconopsis lyrata (Cummins et Prain) Fedde ■
79451 Cathcartia polygonoides Prain = Meconopsis lyrata (Cummins et Prain) Fedde ■
79452 Cathcartia smithiana Hand. -Mazz. = Cremanthodium smithianum (Hand. -Mazz.) Hand. -Mazz. ■

79453　Cathea Salisb. = Calopogon R. Br.(保留属名)■☆
79454　Cathea pulchella Salisb. = Calopogon tuberosus (L.) Britton, Sterns et Poggenb. ■☆
79455　Cathea tuberosa (L.) Morong = Calopogon tuberosus (L.) Britton, Sterns et Poggenb. ■☆
79456　Cathedra Miers(1852);椅树属●☆
79457　Cathedra acuminata Miers;渐尖椅树●☆
79458　Cathedra crassifolia Miers;厚叶椅树●☆
79459　Cathedra oblonga Sleumer;矩圆椅树●☆
79460　Cathedraceae Tiegh. = Olacaceae R. Br.(保留科名)●
79461　Cathestecum C. Presl(1830);假格兰马草属■☆
79462　Cathestecum annuum Swallen;一年假格兰马草■☆
79463　Cathestecum brevifolium Swallen;短叶假格兰马草■☆
79464　Cathetostema Blume = Hoya R. Br. ●
79465　Cathetostemma Blume = Hoya R. Br. ●
79466　Cathetus Lour. = Phyllanthus L. ●■
79467　Cathetus cochinchinensis Lour. = Phyllanthus cochinchinensis (Lour.) Spreng. ●
79468　Cathissa Salisb. (1866);短梗风信子属■☆
79469　Cathissa Salisb. = Ornithogalum L. ■
79470　Cathissa broteroi (Lainz) Speta;布洛短梗风信子■☆
79471　Cathissa reverchonii (Lange) Speta;短梗风信子■☆
79472　Cathormion (Benth.) Hassk. (1855);链合欢属●☆
79473　Cathormion (Benth.) Hassk. = Albizia Durazz. ●
79474　Cathormion Hassk. = Albizia Durazz. ●
79475　Cathormion altissimum (Hook. f.) Hutch. et Dandy;高大链合欢●☆
79476　Cathormion dinklagei (Harms) Hutch. et Dandy = Albizia dinklagei (Harms) Harms ●☆
79477　Cathormion eriorhachis (Harms) Dandy;毛兰链合欢●☆
79478　Cathormion leptophyllum (Harms) Keay = Samanea leptophylla (Harms) Brenan et Brummitt ●☆
79479　Cathormion leptophyllum (Harms) Keay subsp. guineense (G. C. C. Gilbert et Boutique) Cavaco = Samanea leptophylla (Harms) Brenan et Brummitt ●☆
79480　Cathormion leptophyllum (Harms) Keay var. guineense (G. C. C. Gilbert et Boutique) G. C. C. Gilbert et Boutique = Samanea leptophylla (Harms) Brenan et Brummitt ●☆
79481　Cathormion obliquifoliolatum (De Wild.) G. C. C. Gilbert et Boutique;斜叶链合欢●☆
79482　Cathormion rhombifolium (Benth.) Hutch. et Dandy;菱叶链合欢●☆
79483　Catila Ravenna = Calydorea Herb. ■☆
79484　Catimbium Holtt. = Alpinia Roxb.(保留属名)■
79485　Catimbium Juss. = Alpinia Roxb.(保留属名)■
79486　Catimbium Juss. = Renealmia L. f.(保留属名)■☆
79487　Catinga Aubl. = Calycorectes O. Berg ●☆
79488　Catinga Aubl. = Eugenia L. ●
79489　Catis O. F. Cook = Euterpe Mart.(保留属名)●☆
79490　Catoblastus H. Wendl. (1860);巴帕椰属●☆
79491　Catoblastus pubescens H. Wendl.;毛巴帕椰●☆
79492　Catocoma Benth. = Bredemeyera Willd. ●☆
79493　Catocoryne Hook. f. (1867);蔓牡丹属■☆
79494　Catocoryne linneoides Hook. f.;蔓牡丹■☆
79495　Catodiacrum Dulac = Orobanche L. ■
79496　Catoferia (Benth.) Benth. (1867);疏蕊无梗花属●■☆
79497　Catolesia D. J. N. Hind(2000);落苞柄泽兰属■☆
79498　Catolesia mentiens D. J. N. Hind;落苞柄泽兰■☆
79499　Catolobus (C. A. Mey.) Al-Shehbaz(2005);垂片芥属■☆
79500　Catolobus pendula (L.) Al-Shehbaz;垂片芥■☆
79501　Catonia Moench = Crepis L. ■
79502　Catonia P. Browne = Miconia Ruiz et Pav.(保留属名)●☆
79503　Catonia Raf. = Cordia L.(保留属名)●
79504　Catonia Vahl = Erycibe Roxb. ●
79505　Catonia Vell. = Symplocos Jacq. ●
79506　Catopheria Benth. = Catoferia (Benth.) Benth. ●■☆
79507　Catophractes D. Don(1839);南非刺葳属●☆
79508　Catophractes alexandri D. Don;南非刺葳●☆
79509　Catophractes kolbeana Harv. = Catophractes alexandri D. Don ●☆
79510　Catophractes welwitschii Seem. = Catophractes alexandri D. Don ●☆
79511　Catophyllum Poht ex Baker = Mikania Willd.(保留属名)■
79512　Catopodium Link = Catapodium Link ■☆
79513　Catopsis Griseb. (1864);卡凤梨属(拟卡铁属)■☆
79514　Catopsis berteroniana (Schult. f.) Mez;贝尔卡凤梨■☆
79515　Catopsis nutans (Sw.) Griseb.;卡凤梨■☆
79516　Catosperma Benth. = Goodenia Sm. ●■☆
79517　Catospermum Benth. = Goodenia Sm. ●■☆
79518　Catostemma Benth. (1843);垂冠木棉属■●☆
79519　Catostemma altsoni Sandwith;奥氏垂冠木棉●☆
79520　Catostemma commune Sandwith;垂冠木棉●☆
79521　Catostemma fragrans Benth.;芳香垂冠木棉●☆
79522　Catostigma O. F. Cook et Doyle = Catoblastus H. Wendl. ●☆
79523　Catsjopiri Rumph. = Gardenia Ellis(保留属名)●
79524　Cattimarus Kuntze = Kleinhovia L. ●
79525　Cattimarus Rumph. = Kleinhovia L. ●
79526　Cattleya Lindl. (1821);卡特兰属(布袋兰属,嘉德利亚兰属,卡特丽亚兰属);Cattleya ■
79527　Cattleya aclandiae Lindl.;阿柯兰德卡特兰(阿卡兰德卡特兰);Acland Cattleya ■☆
79528　Cattleya aurandiaca (Bateman ex Lindl.) P. N. Don;橙黄卡特兰(橙红布袋兰);Yellow Cattleya ■☆
79529　Cattleya bicolor Lindl.;二色卡特兰(双色布袋兰);Two-colour Cattleya ■☆
79530　Cattleya bowringiana Veitch;波氏卡特兰(博氏卡特兰,多花布袋兰,洪都拉斯卡特兰);Bowring Cattleya ■☆
79531　Cattleya bulbosa Lindl. = Cattleya walkeriana Gardner ■☆
79532　Cattleya citrina (La Llave et Lex.) Lindl.;柠檬黄卡特兰;Lemon-yellow Cattleya, Tulip Orchid ■☆
79533　Cattleya citrina Lindl. = Cattleya citrina (La Llave et Lex.) Lindl. ■☆
79534　Cattleya dowiana Bateman;秀丽卡特兰(道卫卡特兰,女王布袋兰)■☆
79535　Cattleya dowiana Bateman var. aurea (Linden) B. S. Williams et T. Moore;金色秀丽卡特兰■☆
79536　Cattleya dowiana Bateman var. aurea T. Moore = Cattleya dowiana Bateman var. aurea (Linden) B. S. Williams et T. Moore ■☆
79537　Cattleya elatior Lindl. = Cattleya guttata Lindl. ■☆
79538　Cattleya forbesii Lindl.;佛毕斯卡特兰;Forbes Cattleya ■☆
79539　Cattleya gaskelliana Rchb. f.;加斯克尔卡特兰■☆
79540　Cattleya granulosa Lindl.;粒斑卡特兰;Grain-spotted Cattleya ■☆
79541　Cattleya guttata Lindl.;褐红斑卡特兰(斑点布袋兰);Brown-red-spotted Cattleya ■☆
79542　Cattleya harrisoniana Rchb. f.;哈里森卡特兰■☆

79543 Cattleya hybrida H. J. Veitch;嘉德利亚兰(卡特兰);Cattleya, Hybrid Cattleya ■☆

79544 Cattleya intermedia Graham = Cattleya intermedia Graham ex Hook. ■☆

79545 Cattleya intermedia Graham ex Hook.;双叶布袋兰(早花卡特兰)■☆

79546 Cattleya labiata Lindl.;卡特兰(布袋兰,大唇卡特兰);Autumn Cattleya, Common Cattleya ■☆

79547 Cattleya labiata Lindl. var. aurea (Lindl.) H. J. Veitch;黄花卡特兰;Yellowflower Cattleya ■☆

79548 Cattleya labiata Lindl. var. dowiana (Bateman) H. J. Veitch = Cattleya dowiana Bateman ■☆

79549 Cattleya labiata Lindl. var. eldorado (Lindl.) H. J. Veitch;艾多里多卡特兰(多里多卡特兰);Eldorad Cattleya ■☆

79550 Cattleya labiata Lindl. var. gaskeliana (Sander) H. J. Veitch;夏布袋兰■☆

79551 Cattleya labiata Lindl. var. mendelii (Backh.) Rchb. f. ex Veitch;紫唇布袋兰■☆

79552 Cattleya labiata Lindl. var. mossiae (C. Parker ex Hook.) Lindl. = Cattleya mossiae C. Parker ex Hook. ■☆

79553 Cattleya labiata Lindl. var. quadricolor (Lindl.) A. D. Hawkes;四色卡特兰;Fourcolor Cattleya ■☆

79554 Cattleya labiata Lindl. var. rex (O'Brien) Schltr.;王卡特兰;Prince Cattleya ■☆

79555 Cattleya labiata Lindl. var. trianae (Linden et Rchb. f.) Duch. ex H. Karst.;迎春布袋兰(哥伦比亚卡特兰);Winter Cattleya ■☆

79556 Cattleya labiata Lindl. var. warneri (T. Moore ex Warner) H. J. Veitch;送春布袋兰(瓦尔纳卡特兰)■☆

79557 Cattleya labiata Lindl. var. warscewiczii (Rchb. f.) Rchb. f. ex H. J. Veitch;大花布袋兰(瓦氏卡特兰)■☆

79558 Cattleya lawrenceana Rchb. f.;劳伦氏卡特兰(高地布袋兰,劳伦卡特兰);Lawrence Cattleya ■☆

79559 Cattleya lemoniana Lindl. = Cattleya labiata Lindl. ■☆

79560 Cattleya loddigesii Lindl.;劳德基氏卡特兰(罗氏卡特兰);Loddiges Cattleya ■☆

79561 Cattleya loddigesii Lindl. var. harrisoniae H. J. Veitch = Cattleya harrisoniana Rchb. f. ■☆

79562 Cattleya lueddemanniana Rchb. f.;吕德曼氏卡特兰■☆

79563 Cattleya luteola Lindl.;淡黄卡特兰;Yellowish Cattleya ■☆

79564 Cattleya maxima Lindl.;巨大卡特兰;Maximum Cattleya ■☆

79565 Cattleya mendelii L. Linden et Rodigas;门氏卡特兰■☆

79566 Cattleya mossiae C. Parker ex Hook.;花叶卡特兰(粉红布袋兰,小卡特兰);Small Cattleya ■☆

79567 Cattleya mossiae C. Parker ex Hook. var. wageneri (Rchb. f.) Braem;瓦氏花叶卡特兰■☆

79568 Cattleya ovata Lindl. = Cattleya loddigesii Lindl. ■☆

79569 Cattleya papeiansiana Morren = Cattleya harrisoniana Rchb. f. ■☆

79570 Cattleya percivaliana Rchb. f.;珀氏卡特兰■☆

79571 Cattleya regnellii Warner = Cattleya schilleriana Rchb. f. ■☆

79572 Cattleya schilleriana Rchb. f.;席勒氏卡特兰■☆

79573 Cattleya schroederae Sander;施罗德卡特兰■☆

79574 Cattleya skinneri Bateman;危地马拉卡特兰;Flower of St. Sebastian ■☆

79575 Cattleya trianae Linden et Rchb. f. = Cattleya labiata Lindl. var. trianae (Linden et Rchb. f.) Duch. ex H. Karst. ■☆

79576 Cattleya violacea (Kunth) Rolfe;堇色卡特兰;Violet Cattleya ■☆

79577 Cattleya walkeriana Gardner;侧花卡特兰(侧花布袋兰)■☆

79578 Cattleya walneri T. Moore = Cattleya labiata Lindl. var. warneri (T. Moore ex Warner) H. J. Veitch ■☆

79579 Cattleya walneri T. Moore ex Warner = Cattleya labiata Lindl. var. warneri (T. Moore ex Warner) H. J. Veitch ■☆

79580 Cattleya warscewiczii Rchb. f. = Cattleya labiata Lindl. var. warscewiczii (Rchb. f.) Rchb. f. ex H. J. Veitch ■☆

79581 Cattleyella Van den Berg et M. W. Chase = Cattleya Lindl. ■

79582 Cattleyella Van den Berg et M. W. Chase(2003);小卡特兰属■☆

79583 Cattleyopsis Lem. (1853);拟卡特兰属■☆

79584 Cattleyopsis Lem. = Broughtonia R. Br. ■

79585 Cattleyopsis delicatula Lem.;拟卡特兰■☆

79586 Cattutella Rchb. = Katoutheka Adans. (废弃属名)●

79587 Cattutella Rchb. = Wendlandia Bartl. ex DC. (保留属名)●

79588 Catu-Adamboe Adans. = Lagerstroemia L. ●

79589 Catunaregam Adans. ex Wolf = Randia L. ●

79590 Catunaregam Wolf(1776);山石榴属;Wild Pomegranate ●

79591 Catunaregam nilotica (Stapf) Tirveng.;尼罗河山石榴●☆

79592 Catunaregam obovata (Hochst.) A. E. Gonc.;卵山石榴●☆

79593 Catunaregam pentandra (Gürke) Bridson;五蕊山石榴●☆

79594 Catunaregam pygmaea Vollesen;小山石榴●☆

79595 Catunaregam spinosa (Thunb.) Tirveng.;山石榴(鼻血刺,刺榴,刺仔,刺子,对面花,对面芳,假石榴,簕牯树,簕泡木,牛头簕,山黄皮,山葡萄,山蒲桃,猪肚簕);Malabar Randia, Spine Randia, Spine Wild Pomegranate, Spiny Randia ●

79596 Catunaregam spinosa (Thunb.) Tirveng. = Randia spinosa (Thunb.) Poir. ●

79597 Catunaregam spinosa (Thunb.) Tirveng. subsp. taylorii (S. Moore) Verdc. = Catunaregam obovata (Hochst.) A. E. Gonc. ●☆

79598 Catunaregam stenocarpa Bridson;细果山石榴●☆

79599 Catunaregam swynnertonii (S. Moore) Bridson;斯温纳顿山石榴●☆

79600 Catunaregam taylorii (S. Moore) Bridson;泰勒山石榴●☆

79601 Caturus L. = Acalypha L. ●■

79602 Caturus Lour. = Malaisia Blanco ●

79603 Caturus scandens Lour. = Malaisia scandens (Lour.) Planch. ●

79604 Catutsjeron Kuntze = Holigarna Buch. -Ham. ex Roxb. (保留属名)●☆

79605 Catyona Lindl. = Crepis L. ■

79606 Catyona Lindl. = Gatyona Cass. ■

79607 Caucaea Schltr. (1920);考卡兰属(高加兰属)■☆

79608 Caucaea Schltr. et Mansf. = Caucaea Schltr. ■☆

79609 Caucalidaceae Bercht. et J. Presl = Apiaceae Lindl. (保留科名)●■

79610 Caucalidaceae Bercht. et J. Presl = Umbelliferae Juss. (保留科名)■●

79611 Caucaliopsis H. Wolff = Agrocharis Hochst. ■☆

79612 Caucaliopsis H. Wolff. (1921);拟小窃衣属■☆

79613 Caucaliopsis stolzii H. Wolff = Agrocharis pedunculata (Baker f.) Heywood et Jury ■☆

79614 Caucalis L. (1753);小窃衣属(高加利属,高卡利属);False Carrot ■☆

79615 Caucalis L. = Agrocharis Hochst. ■☆

79616 Caucalis africana (L.) Crantz = Capnophyllum africanum (L.) W. D. J. Koch ■☆

79617 Caucalis africana Thunb. = Torilis arvensis (Huds.) Link var. purpurea (Ten.) Thell. ■☆

79618 Caucalis anthriscus (L.) C. B. Clarke = Torilis japonica

(Houtt.) DC. ■
79619 Caucalis anthriscus (L.) Desf. = Torilis japonica (Houtt.) DC. ■
79620 Caucalis anthriscus (L.) Huds. = Torilis japonica (Houtt.) DC. ■
79621 Caucalis anthriscus Huds. = Torilis japonica (Houtt.) DC. ■
79622 Caucalis arvensis Huds. = Torilis arvensis (Huds.) Link ■☆
79623 Caucalis bifrons (Pomel) Maire = Torilis elongata (Hoffmanns. et Link) Samp. ■☆
79624 Caucalis bifrons (Pomel) Maire var. cordisepala (Murb.) Emb. et Maire = Caucalis bifrons (Pomel) Maire ■☆
79625 Caucalis bifrons (Pomel) Maire var. heterocarpa (Ball) Maire = Torilis elongata (Hoffmanns. et Link) Samp. ■☆
79626 Caucalis bifrons (Pomel) Maire var. subtuberculata Emb. et Maire = Torilis elongata (Hoffmanns. et Link) Samp. ■☆
79627 Caucalis bischoffii Koso-Pol.;毕氏小窃衣(毕氏高卡利)■☆
79628 Caucalis capensis Lam. = Stoibrax capense (Lam.) B. L. Burtt ■☆
79629 Caucalis coerulescens Boiss. = Torilis elongata (Hoffmanns. et Link) Samp. ■☆
79630 Caucalis coniifolia Wall. ex DC. = Torilis japonica (Houtt.) DC. ■
79631 Caucalis cordisepala Murb. = Torilis arvensis (Huds.) Link ■☆
79632 Caucalis daucoides L. = Caucalis platycarpos L. ■☆
79633 Caucalis elata D. Don = Torilis japonica (Houtt.) DC. ■
79634 Caucalis gracilis (Engl.) Engl. = Agrocharis incognita (C. Norman) Heywood et Jury ■☆
79635 Caucalis heterophylla Guss. = Torilis arvensis (Huds.) Link subsp. purpurea (Ten.) Hayek ■☆
79636 Caucalis homoephila Coincy = Torilis elongata (Hoffmanns. et Link) Samp. ■☆
79637 Caucalis humilis Jacq. = Torilis leptophylla (L.) Rchb. f. ■☆
79638 Caucalis incognita C. Norman = Agrocharis incognita (C. Norman) Heywood et Jury ■☆
79639 Caucalis infesta (L.) Curtis = Torilis arvensis (Huds.) Link subsp. recta Jury ■☆
79640 Caucalis infesta (L.) Curtis var. elatior Gaudin = Torilis arvensis (Huds.) Link var. elatior (Gaudin) Thell. ■☆
79641 Caucalis infesta (L.) Curtis var. neglecta (Schult.) Ball = Torilis arvensis (Huds.) Link subsp. neglecta (Spreng.) Thell. ■☆
79642 Caucalis japonica Houtt. = Torilis japonica (Houtt.) DC. ■
79643 Caucalis lappula Grande = Caucalis platycarpos L. ■☆
79644 Caucalis latifolia (L.) L. = Turgenia latifolia (L.) Hoffm. ■
79645 Caucalis latifolia L. = Turgenia latifolia (L.) Hoffm. ■
79646 Caucalis latifolia L. var. megalocarpa Jahand. et Maire = Turgenia latifolia (L.) Hoffm. ■
79647 Caucalis latifolia L. var. multiflora (DC.) Thell. = Turgenia latifolia (L.) Hoffm. ■
79648 Caucalis latifolia L. var. tuberculata Boiss. = Turgenia latifolia (L.) Hoffm. ■
79649 Caucalis latifolia Lam. = Caucalis daucoides L. ■
79650 Caucalis latifolia Lam. = Turgenia latifolia (L.) Hoffm. ■
79651 Caucalis leptophylla L.;薄叶小窃衣(薄叶高卡利)■☆
79652 Caucalis leptophylla L. = Torilis leptophylla (L.) Rchb. f. ■☆
79653 Caucalis leptophylla L. var. bifrons (Pomel) Coss. et Durieu = Torilis elongata (Hoffmanns. et Link) Samp. ■☆
79654 Caucalis leptophylla L. var. heterocarpa Ball = Torilis leptophylla (L.) Rchb. f. ■☆
79655 Caucalis longisepala Engl. ex H. Wolff = Agrocharis pedunculata (Baker f.) Heywood et Jury ■☆
79656 Caucalis maritima Lam. = Daucus carota L. subsp. maritimus (Lam.) Batt. ■☆
79657 Caucalis mauritanica L. = Torilis elongata (Hoffmanns. et Link) Samp. ■☆
79658 Caucalis melanantha (Hochst.) Hiern = Agrocharis melanantha Hochst. ■☆
79659 Caucalis mossamedensis Welw. ex Hiern = Angoseseli mossamedensis (Welw. ex Hiern) C. Norman ■☆
79660 Caucalis nodosa (L.) Huds. = Torilis nodosa (L.) Gaertn. ■☆
79661 Caucalis nodosa (L.) Huds. var. heterocarpa Ball = Torilis nodosa (L.) Gaertn. ■☆
79662 Caucalis nodosa Crantz = Torilis nodosa (L.) Gaertn. ■☆
79663 Caucalis pedunculata Baker f. = Agrocharis pedunculata (Baker f.) Heywood et Jury ■☆
79664 Caucalis platycarpos L.;扁果小窃衣(扁果高卡利);Bastard Parsley, Bur Parsley, Carrot Burr Parsley, Hedgehog Parsley, Hen's Foot, Small Bur-parsley ■☆
79665 Caucalis praetermissa (Hance) Franch. = Torilis japonica (Houtt.) DC. ■
79666 Caucalis purpurea Ten. = Torilis arvensis (Huds.) Link var. purpurea (Ten.) Thell. ■☆
79667 Caucalis scabra (Thunb.) Makino = Torilis scabra (Thunb.) DC. ■
79668 Caucalis scabra Makino = Torilis scabra (Thunb.) DC. ■
79669 Caucalis scandix Scop. = Anthriscus scandicina (Weber) Mansf. ■☆
79670 Caucalis tenella Delile = Torilis tenella (Delile) Rchb. f. ■☆
79671 Caucalis virgata Poir. = Daucus virgatus (Poir.) Maire ■☆
79672 Caucaloides Fabr. = ? Caucalis L. sp. ■☆
79673 Caucaloides Heist. ex Fabr. = ? Caucalis L. sp. ■☆
79674 Caucanthus Forssk. (1775);考卡花属●☆
79675 Caucanthus Raf. = Sterculia L. ●
79676 Caucanthus albidus (Nied.) Nied.;白考卡花●☆
79677 Caucanthus argenteus Chiov. = Caucanthus albidus (Nied.) Nied. ●☆
79678 Caucanthus argenteus Nied. = Caucanthus auriculatus (Radlk.) Nied. ●☆
79679 Caucanthus auriculatus (Radlk.) Nied.;耳状考卡花●☆
79680 Caucanthus edulis Forssk.;可食考卡花●☆
79681 Caucanthus edulis Forssk. var. benadirensis (Fiori) Chiov.;贝纳迪尔考卡花●☆
79682 Caucanthus squarrosus (Radlk.) Nied. var. benadirensis Fiori = Caucanthus edulis Forssk. var. benadirensis (Fiori) Chiov. ●☆
79683 Caucasalia B. Nord. (1997);高加索菊属■☆
79684 Caucasalia macrophylla (M. Bieb.) B. Nord.;大叶高加索菊■☆
79685 Caucasalia parviflora (M. Bieb.) B. Nord.;小叶高加索菊■☆
79686 Caudanthera Plowes(1995);尾药萝藦属■☆
79687 Caudanthera edulis (Edgew.) Meve et Liede;可食高加索菊■☆
79688 Caudicia Ham. ex Wight = Parsonsia R. Br. (保留属名)●
79689 Caudoleucaena Britton et Rose = Leucaena Benth. (保留属名)●
79690 Caudoxalis Small = Oxalis L. ■●
79691 Caulangis Thouars = Angraecum Bory ■
79692 Caulanthus S. Watson(1871);甘蓝花属■☆
79693 Caulanthus amplexicaulis Watson;抱茎甘蓝花■☆
79694 Caulanthus annuus M. E. Jones;一年甘蓝花■☆
79695 Caulanthus cooperi Payson;库珀甘蓝花■☆
79696 Caulanthus crassicaulis S. Watson;甘蓝花;Wild Cabbage ■☆

79697 Caulanthus glaucus Watson;灰绿甘蓝花■☆

79698 Caulanthus inflatus S. Watson;沙地甘蓝花;Desert Candle ■☆

79699 Caulanthus sulfureus Payson;硫色甘蓝花■☆

79700 Caularthron Raf.(1837);双角兰属■☆

79701 Caularthron amazonicum (Schltr.) H. G. Jones;亚马逊双角兰■☆

79702 Caularthron umbellatum Raf.;双角兰■☆

79703 Caulinia DC. = Posidonia K. D. König(保留属名)■

79704 Caulinia Moench = Kennedia Vent. ●☆

79705 Caulinia Willd. = Najas L. ■

79706 Caulinia amurensis (Tzvelev) Tzvelev = Najas gracillima (A. Braun ex Engelm.) Magnus ■

79707 Caulinia flexilis Willd. = Najas flexilis (Willd.) Rostk. et Schmidt ■☆

79708 Caulinia fragilis Willd. = Najas minor All. ■

79709 Caulinia graminea (Delile) Batt. = Najas graminea Delile ■

79710 Caulinia graminea (Delile) Tzvelev = Najas graminea Delile ■

79711 Caulinia guadalupensis Spreng. = Najas guadalupensis (Spreng.) Magnus ■☆

79712 Caulinia japonica (Nakai) Nakai = Najas gracillima (A. Braun ex Engelm.) Magnus ■

79713 Caulinia minor (All.) Coss. et Germ. = Najas minor All. ■

79714 Caulinia muricata (Delile) Spreng. = Najas marina L. subsp. armata (H. Lindb.) Horn ■☆

79715 Caulinia orientalis (Triest et Uotila) Tzvelev = Najas chinensis N. Z. Wang ■

79716 Caulinia ovalis R. Br. = Halophila ovalis (R. Br.) Hook. f. ■

79717 Caulinia pectinata Parl. = Najas pectinata (Parl.) Magnus ■☆

79718 Caulinia serrulata R. Br. = Cymodocea serrulata (R. Br.) Asch. et Magnus ■☆

79719 Caulinia tenuissima (A. Braun) Tzvelev subsp. amurensis Tzvelev = Najas gracillima (A. Braun ex Engelm.) Magnus ■

79720 Caulipsolon Klak = Mesembryanthemum L.(保留属名)■●

79721 Caullinia Raf. = Hippuris L. ■

79722 Caulobryon Klotzsch ex C. DC. = Piper L. ●■

79723 Caulocarpus Baker f.(1926);茎果豆属●☆

79724 Caulocarpus Baker f. = Tephrosia Pers.(保留属名)●■

79725 Caulocarpus gossweileri Baker f.;茎果豆●☆

79726 Caulokaempferia K. Larsen(1964);大苞姜属;Bigbractginger ■

79727 Caulokaempferia coenobialis (Hance) K. Larsen;黄花大苞姜(石竹花,水马鞭,土山奈,岩白姜,岩竹叶);Yellow Bigbractginger,Yellowfloer Bigbractginger ■

79728 Caulokaempferia yunnanensis (Gagnep.) R. M. Sm. = Pyrgophyllum yunnanense (Gagnep.) T. L. Wu et Z. Y. Chen ■

79729 Cauloma Raf. = Verbesina L.(保留属名)●■☆

79730 Caulophyllum Michx.(1803);红毛七属(红三七属,类叶杜鹃属,类叶牡丹属,威岩仙属,葳岩仙属,岩威仙属);Blue Cohosh,Cohosh,Papoose Root,Squaw-root ●

79731 Caulophyllum giganteum (Farw.) Loconte et W. H. Blackw.;大红毛七●☆

79732 Caulophyllum robustum Maxim.;红毛七(棒槌幌子,背阳草,藏严仙,赤芍幌子,灯笼草,海椒七,黑汗腿,红毛漆,红毛三七,红毛细辛,火焰叉,鸡骨升麻,金丝七,类叶杜鹃,类叶牡丹,搜山猫,威岩仙,葳严仙,细毛细辛,竹参七);Blue Cohosh,Robust Blue Cohosh ●

79733 Caulophyllum thalictroides (L.) Michx.;蓝籽红毛七(唐松草状红毛七);Blue Cohosh,Blue Cohush,Blue Ginseng,Blueberry Root,Papoose Root,Papoose-root,Squaw Root,Squawroot,Squaw-root,Yellow Ginseng Papoose Root ●☆

79734 Caulophyllum thalictroides (L.) Michx. subsp. robustum (Maxim.) Kitam. = Caulophyllum robustum Maxim. ●

79735 Caulophyllum thalictroides (L.) Michx. var. giganteum Farw. = Caulophyllum giganteum (Farw.) Loconte et W. H. Blackw. ●☆

79736 Caulopsis Fourr. = Arabis L. ●■

79737 Caulostramina Rollins(1973);石缝铁线芥属■☆

79738 Caulostramina jaegeri (Rollins) Rollins;石缝铁线芥■☆

79739 Caulotretus Rich. ex Spreng. = Bauhinia L. ●

79740 Caulotulis Raf. = Ipomoea L.(保留属名)●■

79741 Causea Scop. = Hirtella L. ●☆

79742 Causonia Raf. = Cayratia Juss.(保留属名)●

79743 Causonia japonica (Thunb.) Raf. = Cayratia japonica (Thunb.) Gagnep. ●

79744 Causonia japonica Raf. = Cayratia japonica (Thunb.) Gagnep. ●

79745 Caustis R. Br.(1810);枝莎属■☆

79746 Caustis flexuosa R. Br.;曲枝莎;Curly Sedge,Curly Wig,Old Man's Beard ■☆

79747 Caustis pentandra R. Br.;五蕊枝莎;Tall Sedge ■☆

79748 Cautlea Royle = Cautleya Royle ■

79749 Cautleya (Benth.) Hook. f.(1888);距药姜属;Cautleya ■

79750 Cautleya Hook. f. = Cautleya (Benth.) Hook. f. ■

79751 Cautleya Royle = Cautleya (Benth.) Hook. f. ■

79752 Cautleya cathcarti Baker;多花距药姜;Manyflower Cautleya ■

79753 Cautleya gracilis (Sm.) Dandy;距药姜;Slender Cautleya ■

79754 Cautleya lutea (Royle) Hook. f. = Cautleya gracilis (Sm.) Dandy ■

79755 Cautleya lutea (Royle) Hook. f. var. robusta? = Cautleya gracilis (Sm.) Dandy ■

79756 Cautleya spicata (Sm.) Baker;红苞距药姜;Redbract Cautleya ■

79757 Cavacoa J. Léonard(1955);卡瓦大戟属☆

79758 Cavacoa aurea (Cavaco) J. Léonard;黄卡瓦大戟☆

79759 Cavacoa baldwinii (Keay et Cavaco) J. Léonard;巴氏卡瓦大戟☆

79760 Cavacoa quintasii (Pax et K. Hoffm.) J. Léonard;昆塔斯卡瓦大戟☆

79761 Cavalam Adans. = Sterculia L. ●

79762 Cavalcantia R. M. King et H. Rob.(1980);宽片菊属■☆

79763 Cavalcantia R. M. King et H. Rob. = Eupatorium L. ■●

79764 Cavaleriea H. Lév. = Ribes L. ●

79765 Cavaleriea enkianthoidea H. Lév. = Ribes laurifolium Jancz. ●

79766 Cavaleviella H. Lév. = Aspidopterys A. Juss. ex Endl. + Dipelta Maxim. ●★

79767 Cavallium Schott = Sterculia L. ●

79768 Cavanalia Griseb. = Canavalia Adans.(保留属名)●■

79769 Cavanilla J. F. Gmel. = Dombeya Cav.(保留属名)●☆

79770 Cavanilla Salisb. = Malachodendron Mitch. ●

79771 Cavanilla Salisb. = Stewartia L. ●

79772 Cavanilla Thunb. = Pyrenacantha Wight(保留属名)●

79773 Cavanilla Vell. = Caperonia A. St. -Hil. ■☆

79774 Cavanillea Desr. = Diospyros L. ●

79775 Cavanillea Desr. = Mabola Raf. ●

79776 Cavanillea Medik. = Anoda Cav. ■●☆

79777 Cavanillea philippensis Desr. = Diospyros philippensis (Desr.) Gurke ●

79778 Cavanillesia Ruiz et Pav.(1794);卡夫木棉属●☆

79779 Cavanillesia arborea K. Schum.;乔状卡夫木棉●☆

79780 Cavanillesia platanifolia Kunth;悬铃木叶卡夫木棉;Cuipo ●☆

79781 Cavaraea Speg. = Tamarindus L. ●
79782 Cavaria Steud. = Tovaria Ruiz et Pav.(保留属名)●■
79783 Cavea W. W. Sm. = Cavea W. W. Sm. et J. Small ■
79784 Cavea W. W. Sm. et J. Small(1917);葶菊属(莛菊属);Cavea ■
79785 Cavea tanguensis (Drumm.) W. W. Sm. = Cavea tanguensis (Drumm.) W. W. Sm. et J. Small ■☆
79786 Cavea tanguensis (Drumm.) W. W. Sm. et J. Small;葶菊;Tangut Cavea ■
79787 Cavendishia Gray(废弃属名) = Cavendishia Lindl.(保留属名)●☆
79788 Cavendishia Lindl.(1835)(保留属名);艳苞莓属(类越橘属)●☆
79789 Cavendishia acuminata Benth. ex Hemsl.;尖叶艳苞莓(尖叶类越橘)●☆
79790 Cavendishia bracteata (Ruiz et Pav. ex J. St. -Hil.) Hoerold;具苞艳苞莓(具苞类越橘)●☆
79791 Cavendishia grandifolia Hoerold;大叶艳苞莓(大叶类越橘)●☆
79792 Cavendishia martii (Meisn.) A. C. Sm.;马尔特艳苞莓(马尔特类越橘)●☆
79793 Cavendishia nobilis Lindl.;名贵艳苞莓(名贵类越橘)●☆
79794 Cavendishia orthosepala A. C. Sm.;直萼艳苞莓(直萼类越橘)●☆
79795 Cavendishia pubescens (Kunth) Hemsl.;毛艳苞莓(毛类越橘)●☆
79796 Cavendishia pubescens A. C. Sm. var. boliviensis Hoerold = Cavendishia pubescens (Kunth) Hemsl. ●☆
79797 Cavendishia spectabilis Bull;卓越艳苞莓●☆
79798 Cavinium Thouars = Vaccinium L. ●
79799 Cavinium madagascariense Thouars ex Poir. = Vaccinium madagascariense (Thouars ex Poir.) Sleumer ●☆
79800 Cavolinia Raf. = Caulinia Willd. ■
79801 Cavolinia Raf. = Najas L. ■
79802 Cavollana Raf. = Cavolinia Raf. ■
79803 Caxamarca Dillon et Sagást.(1999);臭根菊属■☆
79804 Caxamarca sanchezii Dillon et Sagást.;臭根菊■☆
79805 Cayaponia Silva Manso(1836)(保留属名);泻瓜属■☆
79806 Cayaponia africana (Hook. f.) Exell;泻瓜■☆
79807 Cayaponia africana (Hook. f.) Exell var. madagascariensis?;马岛泻瓜●☆
79808 Cayaponia alata Cogn.;具翅泻瓜■☆
79809 Cayaponia americana Cogn.;美洲泻瓜■☆
79810 Cayaponia biflora Cogn. ex Harms;双花泻瓜■☆
79811 Cayaponia glandulosa (Poepp. et Endl.) Cogn.;哥伦比亚臭根菊■☆
79812 Cayaponia globosa Silva Manso;圆果泻瓜■☆
79813 Cayaponia grandiflora Cogn.;大花泻瓜■☆
79814 Cayaponia micrantha Cogn.;小花泻瓜■☆
79815 Cayaponia multiglandulosa R. Fern.;多腺泻瓜■☆
79816 Cayaponia ophthalmica R. E. Schult.;眼泻瓜■☆
79817 Cayaponia pendulina?;下垂泻瓜■☆
79818 Cayaponia pilosa Cogn.;柔毛泻瓜■☆
79819 Cayaponia sessiliflora Wunderlin;无梗泻瓜■☆
79820 Cayaponia tayuya Cogn.;塔尤泻瓜■☆
79821 Cayatia pseudotrifolia W. T. Wang = Cayratia japonica (Thunb.) Gagnep. var. pseudotrifolia (W. T. Wang) C. L. Li ●
79822 Caylusea A. St. -Hil.(1837)(保留属名);凯吕斯草属■☆
79823 Caylusea abyssinica (Fresen.) Fisch. et C. A. Mey.;阿比西尼亚凯吕斯草■☆
79824 Caylusea canescens (L.) A. St. -Hil. = Caylusea hexagyna (Forssk.) M. L. Green ■☆
79825 Caylusea hexagyna (Forssk.) M. L. Green;凯吕斯草■☆
79826 Caylusea hexagyna (Forssk.) M. L. Green var. glabra Maire = Caylusea hexagyna (Forssk.) M. L. Green ■☆
79827 Caylusea hexagyna (Forssk.) M. L. Green var. glabrescens Maire = Caylusea hexagyna (Forssk.) M. L. Green ■☆
79828 Caylusea hexagyna (Forssk.) M. L. Green var. papillosa Maire = Caylusea hexagyna (Forssk.) M. L. Green ■☆
79829 Caylusea hexagyna (Forssk.) M. L. Green var. rigida Maire = Caylusea hexagyna (Forssk.) M. L. Green ■☆
79830 Caylusea latifolia P. Taylor;宽叶凯吕斯草■☆
79831 Cayratia Juss.(1818)(保留属名);乌蔹莓属(虎葛属);Cayratia ●
79832 Cayratia Juss. ex Guill. = Cayratia Juss.(保留属名)●
79833 Cayratia albifolia C. L. Li;白毛乌蔹莓●
79834 Cayratia albifolia C. L. Li var. glabra (Gagnep.) C. L. Li;脱毛乌蔹莓●
79835 Cayratia albifolia C. L. Li var. glabra (Gagnep.) C. L. Li = Cayratia albifolia C. L. Li ●
79836 Cayratia cannabina Gagnep. = Cayratia ciliifera (Merr.) Chun ●
79837 Cayratia cardiospermoides (Planch.) Gagnep.;短柄乌蔹莓(大九节铃);Short-stalked Cayratia ●
79838 Cayratia carnosa Gagnep. = Cissus carnosa (L.) Lam. ●
79839 Cayratia ciliifera (Merr.) Chun;节毛乌蔹莓;Ciliate Cayratia, Heart-leaf Cayratia ●
79840 Cayratia cordifolia C. Y. Wu ex C. L. Li;心叶乌蔹莓;Cordateleaf Cayratia ●
79841 Cayratia corniculata (Benth.) Gagnep.;角花乌蔹莓(九龙根,九牛根,九牛薯,九牛子,菱茎野葡萄,野葡萄,钻地羊);Corniculate Cayratia ●
79842 Cayratia daliensis C. L. Li;大理乌蔹莓●
79843 Cayratia debilis (Baker) Suess.;弱小乌蔹莓●☆
79844 Cayratia delicatula (Willems) Desc.;姣美乌蔹莓●☆
79845 Cayratia elongata (Roxb.) Suess. = Cissus elongata Roxb. ●
79846 Cayratia fugongensis C. L. Li;福贡乌蔹莓●
79847 Cayratia geniculata (Blume) Gagnep.;膝曲乌蔹莓(大麻藤果);Geniculate Cayratia ●
79848 Cayratia gracilis (Guillaumin et Perr.) Suess.;纤细乌蔹莓●☆
79849 Cayratia imerinensis (Baker) Desc.;伊梅里纳乌蔹莓●☆
79850 Cayratia japonica (Thunb.) Gagnep.;乌蔹莓(拔,白果葡萄,赤葛,赤泼藤,地老鼠,地五加,过江龙,过山龙,红母猪藤,虎葛,虎莓,黄眼藤,鸡丝藤,鲫鱼藤,绞股兰,老鸦藤,老鸦眼睛藤,蔹,龙草,龙尾,茏葛,笼草,母猪藤,木竹藤,酸甲藤,铁秤砣,铁散仙,乌蔹草,五甲藤,五将草,五龙草,五叶莓,五叶茑,五叶藤,五月五,五爪金龙,五爪龙,五爪龙草,五爪藤,细叶乌蔹莓,小母猪藤,血五甲,野葡萄藤,止血藤,猪婆藤,紫葛);Bushkiller, Japan Cayratia, Japanese Cayratia, Whitefruited Grape ●
79851 Cayratia japonica (Thunb.) Gagnep. var. canescens W. T. Wang;灰毛乌蔹莓;Grey-haired Japanese Cayratia ●
79852 Cayratia japonica (Thunb.) Gagnep. var. canescens W. T. Wang = Cayratia japonica (Thunb.) Gagnep. var. mollis (Wall. ex M. A. Lawson) Momiy. ●
79853 Cayratia japonica (Thunb.) Gagnep. var. canescens W. T. Wang = Cayratia japonica (Thunb.) Gagnep. var. mollis (Wall.) C. L. Li ●
79854 Cayratia japonica (Thunb.) Gagnep. var. dentata (Makino) Honda = Cayratia tenuifolia (Heyne ex Planch.) Gagnep. ●☆

79855 Cayratia japonica (Thunb.) Gagnep. var. ferruginea W. T. Wang;锈毛乌蔹莓;Rusty-haired Japanese Cayratia ●

79856 Cayratia japonica (Thunb.) Gagnep. var. ferruginea W. T. Wang = Cayratia japonica (Thunb.) Gagnep. var. mollis (Wall. ex M. A. Lawson) Momiy. ●

79857 Cayratia japonica (Thunb.) Gagnep. var. ferruginea W. T. Wang = Cayratia japonica (Thunb.) Gagnep. var. mollis (Wall.) C. L. Li ●

79858 Cayratia japonica (Thunb.) Gagnep. var. mollis (Wall. ex M. A. Lawson) Momiy.;毛乌蔹莓(小麻藤果)●

79859 Cayratia japonica (Thunb.) Gagnep. var. mollis (Wall.) C. L. Li = Cayratia japonica (Thunb.) Gagnep. var. mollis (Wall. ex M. A. Lawson) Momiy. ●

79860 Cayratia japonica (Thunb.) Gagnep. var. pseudotrifolia (W. T. Wang) C. L. Li;尖叶乌蔹莓(过路边,母猪藤,蜈蚣藤);Sharpleaf Cayratia,Sharp-leaved Cayratia ●

79861 Cayratia japonica (Thunb.) Gagnep. var. pseudotrifolia (W. T. Wang) C. L. Li = Cayratia pseudotrifolia W. T. Wang ●

79862 Cayratia japonica (Thunb.) Gagnep. var. pubifolia Merr. et Chun;毛叶乌蔹莓(车索藤,红母猪藤);Hairyleaf Japanese Cayratia ●

79863 Cayratia japonica (Thunb.) Gagnep. var. pubifolia Merr. et Chun = Cayratia japonica (Thunb.) Gagnep. var. mollis (Wall.) C. L. Li ●

79864 Cayratia japonica (Thunb.) Gagnep. var. pubifolia Merr. et Chun = Cayratia japonica (Thunb.) Gagnep. var. mollis (Wall. ex M. A. Lawson) Momiy. ●

79865 Cayratia javanica (Thunb.) Gagnep.;爪哇乌蔹莓●☆

79866 Cayratia kiujiangense C. Y. Wu ex W. T. Wang;俅江乌蔹莓;Qiujiang Cayratia ●

79867 Cayratia kiujiangense C. Y. Wu ex W. T. Wang = Tetrastigma rumicispermum (Lawson) Planch. ●

79868 Cayratia lanceolata (C. L. Li) J. Wen et Z. D. Chen;狭叶乌蔹莓;Narrowleaf Cayratia ●

79869 Cayratia longiflora Desc.;长花乌蔹莓●☆

79870 Cayratia longzhouensis W. T. Wang = Cayratia pedata (Lam.) Juss. ex Gagnep. ●

79871 Cayratia maritima Jackes;海岸乌蔹莓●

79872 Cayratia medoensis C. L. Li;墨脱乌蔹莓;Motuo Cayratia ●

79873 Cayratia mekongensis C. Y. Wu ex W. T. Wang = Cayratia timoriensis (DC.) C. L. Li var. mekongensis (C. Y. Wu ex W. T. Wang) C. L. Li ●

79874 Cayratia menglaensis C. L. Li;勐腊乌蔹莓;Mengla Cayratia ●

79875 Cayratia mollis (Wall. ex M. A. Lawson) C. Y. Wu = Cayratia japonica (Thunb.) Gagnep. var. mollis (Wall. ex M. A. Lawson) Momiy. ●

79876 Cayratia mollis (Wall.) C. Y. Wu = Cayratia japonica (Thunb.) Gagnep. var. mollis (Wall. ex M. A. Lawson) Momiy. ●

79877 Cayratia mollissima Gagnep. var. lanceolata C. L. Li = Cayratia lanceolata (C. L. Li) J. Wen et Z. D. Chen ●

79878 Cayratia oligocarpa (H. Lév. et Vaniot) Gagnep.;华中乌蔹莓(大母猪藤,大叶乌蔹莓,绿叶扁担藤,喜果野葡萄,野葡萄);Bigleaf Cayratia,Big-leaved Cayratia ●

79879 Cayratia oligocarpa (H. Lév. et Vaniot) Gagnep. f. glabra Gagnep. = Cayratia albifolia C. L. Li ●

79880 Cayratia oligocarpa (H. Lév. et Vaniot) Gagnep. var. glabra (Gagnep.) Rehder;樱叶乌蔹莓;Glabrous Bigleaf Cayratia ●

79881 Cayratia oligocarpa (H. Lév. et Vaniot) Gagnep. var. glabra (Gagnep.) Rehder = Cayratia albifolia C. L. Li ●

79882 Cayratia oligocarpa Gagnep. = Cayratia oligocarpa (H. Lév. et Vaniot) Gagnep. ●

79883 Cayratia papillata (Hance) Merr. et Chun = Tetrastigma papillatum (Hance) C. Y. Wu ●

79884 Cayratia pedata (Lam.) Juss. ex Gagnep.;乌足乌蔹莓(龙州乌蔹莓,七叶大麻藤)●

79885 Cayratia pseudotrifolia W. T. Wang = Cayratia japonica (Thunb.) Gagnep. var. pseudotrifolia (W. T. Wang) C. L. Li ●

79886 Cayratia rhombiformis W. T. Wang;菱叶乌蔹莓;Rhombic-leaved Cayratia ●

79887 Cayratia ruspolii (Gilg) Suess. = Cissus ruspolii Gilg ●☆

79888 Cayratia tenuifolia (Heyne ex Planch.) Gagnep.;细叶乌蔹莓●☆

79889 Cayratia tenuifolia (Heyne ex Planch.) Gagnep. var. cinema Gagnep. = Cayratia japonica (Thunb.) Gagnep. var. mollis (Wall.) C. L. Li ●

79890 Cayratia tenuifolia (Wight et Arn.) Gagnep. = Cayratia japonica (Thunb.) Gagnep. ●

79891 Cayratia tenuifolia (Wight et Arn.) Gagnep. var. cinerea Gagnep. = Cayratia japonica (Thunb.) Gagnep. var. mollis (Wall. ex M. A. Lawson) Momiy. ●

79892 Cayratia thalictrifolia (Planch.) Suess. = Cayratia triternata (Baker) Desc. ●☆

79893 Cayratia thomsonii (M. A. Lawson) Suess. = Yua thomsonii (M. A. Lawson) C. L. Li ●

79894 Cayratia timoriensis (DC.) C. L. Li;南亚乌蔹莓;Timor Ryssopterys ●

79895 Cayratia timoriensis (DC.) C. L. Li var. mekongensis (C. Y. Wu ex W. T. Wang) C. L. Li;澜沧乌蔹莓;Lancangjiang Cayratia,Mekong Cayratia ●

79896 Cayratia trifolia (L.) Domin;三叶乌蔹莓(狗脚迹,过路边,母猪藤,三叶野葡萄,三爪龙,蜈蚣藤,小拉蛇);Treeleaf Cayratia,Tree-leaved Cayratia ●

79897 Cayratia trifolia (L.) Domin var. quinquefoliola W. T. Wang;圆叶乌蔹莓;Round-tree-leaved Cayratia ●

79898 Cayratia trifolia (L.) Domin var. quinquefoliola W. T. Wang = Cayratia japonica (Thunb.) Gagnep. ●

79899 Cayratia trifolia (L.) Domin. = Cissus carnosa (L.) Lam. ●

79900 Cayratia triternata (Baker) Desc.;三出乌蔹莓●☆

79901 Cayratia yoshimurae (Makino) Honda;义村乌蔹莓●☆

79902 Ceanothus L. (1753);美洲茶属(蓟木属,曲萼茶属,野丁香属);Blue Ceanothus,California Lilac,Californian Lilac,Ceanothus,Cogwood,Mountain Lilac,New Jersey Tea,Wild Lilac ●☆

79903 Ceanothus Wall. = Rhamnus L. ●

79904 Ceanothus africanus L. = Noltea africana (L.) Rchb. f. ●☆

79905 Ceanothus americanus L.;美洲茶(红根鼠李,蓟木);Blue Tea Bush,Jepson Ceanothus,Mountain Snowbell,Mountain-aweet,New Jersey Tea,New-jersey-tea,Redroot,Red-root,Redshanks,Redwood,Wild Snowball ●☆

79906 Ceanothus americanus L. var. intermedius (Pursh) Torr. et A. Gray;乔治亚美洲茶;Georgia Ceanothus ●☆

79907 Ceanothus americanus L. var. intermedius (Pursh) Torr. et A. Gray = Ceanothus americanus L. ●☆

79908 Ceanothus americanus L. var. pitcheri Torr. et A. Gray;瓶状叶美洲茶;Pitcher New-jersey-tea ●☆

79909 Ceanothus americanus L. var. pitcheri Torr. et A. Gray = Ceanothus americanus L. ●☆

79910 Ceanothus arborescens Mill. = Colubrina arborescens (Mill.)

Sarg. ●

79911 Ceanothus arboreus Greene;树状美洲茶;Catalina Ceanothus, Catalina Mountain Lilav, Feltleaf Ceanothus, Tree Ceanothus ●☆

79912 Ceanothus asiaticus L. = Colubrina asiatica (L.) Brongn. ●

79913 Ceanothus azureus Desf. = Ceanothus caeruleus Lag. ●☆

79914 Ceanothus azureus Desf. ex Paxton = Ceanothus caeruleus Lag. ●☆

79915 Ceanothus bicolor Willd. = Ceanothus caeruleus Lag. ●☆

79916 Ceanothus caeruleus Lag.;天蓝美洲茶;Azure Ceanothus ●☆

79917 Ceanothus circumscissa L. f. = Scutia myrtina (Burm. f.) Kurz ●

79918 Ceanothus circumscissus (L. f.) Gaertn. = Scutia myrtina (Burm. f.) Kurz ●

79919 Ceanothus cordulatus Kellogg;山白刺美洲茶;Ceanothus Snowbush, Mountain Whitethorn ●☆

79920 Ceanothus crassifolius Torr.;厚叶美洲茶;Hoaryleaf Ceanothus ●☆

79921 Ceanothus cuneatus Nutt.;楔状美洲茶;Buckbrush, Buckbrush Ceanothus ●☆

79922 Ceanothus cyaneus Eastw.;蓝色美洲茶(圣地亚哥美洲茶);San Diego Ceanothus ●☆

79923 Ceanothus delilianus Spach;德利尔美洲茶;Delisle Ceanothus, French Hybrid Ceanothus ●☆

79924 Ceanothus dentatus Torr. et A. Gray;齿叶美洲茶;Cropleaf Ceanothus, Santa Barbara Ceanothus ●☆

79925 Ceanothus divergens Parry;略叉开美洲茶;Mount St. Helena Ceanothus ●☆

79926 Ceanothus fendleri A. Gray;芬德勒美洲茶;Fendler Ceanothus ●☆

79927 Ceanothus fendleri A. Gray var. venosus Trel.;显脉芬德勒美洲茶;Venosus Fendler Ceanothus ●☆

79928 Ceanothus foliosus Parry;波叶美洲茶;Wavyleaf Ceanothus ●☆

79929 Ceanothus gloriosus J. T. Howell;华丽美洲茶;Point Reyes Creeper, Pointreyes Ceanothus ●☆

79930 Ceanothus gloriosus J. T. Howell 'Anchor Bay';紧凑华丽美洲茶●☆

79931 Ceanothus gloriosus J. T. Howell var. exaltatus J. T. Howell;高大华丽美洲茶●☆

79932 Ceanothus greggii A. Gray;沙漠美洲茶;Desert Ceanothus ●☆

79933 Ceanothus greggii A. Gray subsp. perplexans (Trel.) R. M. Beauch.;杯叶沙漠美洲茶;Cup-leaf Desert Ceanothus ●☆

79934 Ceanothus griseus (Trel. ex B. L. Rob.) McMinn;灰叶美洲茶;Carmel Ceanothus, Carmel Creeper ●☆

79935 Ceanothus griseus (Trel. ex B. L. Rob.) McMinn 'Diamond Heights';钻石灰叶美洲茶;Variegated Ceanothus ●☆

79936 Ceanothus griseus (Trel. ex B. L. Rob.) McMinn 'Santa Ana';圣安娜灰叶美洲茶●☆

79937 Ceanothus griseus (Trel. ex B. L. Rob.) McMinn 'Yankee Point';美国点灰叶美洲茶●☆

79938 Ceanothus griseus (Trel. ex B. L. Rob.) McMinn var. horizontalis McMinn;平枝灰叶美洲茶●☆

79939 Ceanothus guineensis DC. = Dichapetalum madagascariense Poir. ●☆

79940 Ceanothus herbaceus Raf.;草绿色美洲茶;Inland Ceanothus, Inland New Jersey Tea, Jersey Tea, Prairie Red-root, Redroot ●☆

79941 Ceanothus herbaceus Raf. var. pubescens (Torr. et A. Gray ex S. Watson) Shinners;柔毛草绿色美洲茶;Hairy Inland Ceanothus ●☆

79942 Ceanothus herbaceus Raf. var. pubescens (Torr. et A. Gray ex S. Watson) Shinners = Ceanothus herbaceus Raf. ●☆

79943 Ceanothus impressus Trel.;耐寒美洲茶(陷脉美洲茶);Santa Barbara Ceanothus ●☆

79944 Ceanothus incanus Torr. et A. Gray;海岸白刺美洲茶(白背美洲茶);Coast Whitethorn, Coast Whitethorn Ceanothus ●☆

79945 Ceanothus integerrimus Hook. et Arn.;全缘叶美洲茶;California Lilac, Deer Brush, Deerbrush, Mogollon Ceanothus ●☆

79946 Ceanothus integerrimus Hook. et Arn. var. californicus G. T. Benson;加州全缘叶美洲茶;California Mogollon Ceanothus ●☆

79947 Ceanothus integerrimus Hook. et Arn. var. peduncularis Jeps.;窄花序全缘叶美洲茶;Narrow-inflorescence Ceanothus ●☆

79948 Ceanothus integerrimus Hook. et Arn. var. puberulus Abrams;宽花序全缘叶美洲茶;Ceanothus ●☆

79949 Ceanothus intermedius Pursh = Ceanothus americanus L. ●☆

79950 Ceanothus jepsonii Greene;杰普森美洲茶;Jepson Ceanothus ●☆

79951 Ceanothus lemmoni Parry;莱蒙美洲茶;Lemmons Ceanothus ●☆

79952 Ceanothus leucodermis Greene;白皮美洲茶;Chaparral Ceanothus ●☆

79953 Ceanothus lobbianus Hook.;洛布氏美洲茶(罗宾美洲茶);Lobb Ceanothus ●☆

79954 Ceanothus megacarpus Nutt.;大果美洲茶;Bigpod Ceanothus ●☆

79955 Ceanothus nepalensis Wall. = Rhamnus nepalensis (Wall.) M. A. Lawson ●

79956 Ceanothus oliganthus Nutt.;毛美洲茶;Hairy Ceanothus ●☆

79957 Ceanothus ovatus Desf.;卵叶美洲茶;Inland Ceanothus, Inland Jersey Tea, Oval-leaved Red-root, Redroot, Small Redroot ●☆

79958 Ceanothus ovatus Desf. = Ceanothus herbaceus Raf. ●☆

79959 Ceanothus ovatus Desf. f. pubescens (Torr. et A. Gray ex S. Watson) Soper = Ceanothus herbaceus Raf. ●☆

79960 Ceanothus ovatus Desf. var. pubescens Torr. et A. Gray = Ceanothus herbaceus Raf. ●☆

79961 Ceanothus ovatus Desf. var. pubescens Torr. et A. Gray ex S. Watson = Ceanothus herbaceus Raf. ●☆

79962 Ceanothus pallidus Lindl.;苍白美洲茶;Ceanothus, Pallidus Ceanothus ●☆

79963 Ceanothus pallidus Lindl. 'Marie Simon';粉白美洲茶;Pink ●☆

79964 Ceanothus palmeri Trel.;帕麦尔美洲茶;Palmer Ceanothus ●☆

79965 Ceanothus papillosus Torr. et A. Gray;乳突叶美洲茶●☆

79966 Ceanothus prostratus Benth.;平卧美洲茶;Manala Mats, Squam Carpet, Squaw's Carpet ●☆

79967 Ceanothus pubescens (Torr. et A. Gray ex S. Watson) Rydb. ex Small = Ceanothus herbaceus Raf. ●☆

79968 Ceanothus purpureus Jeps.;紫花美洲茶;Holly-leaf Ceanothus ●☆

79969 Ceanothus reclinatus Bosc ex Steud.;卷叶美洲茶●☆

79970 Ceanothus rigidus Nutt.;蒙特里美洲茶(硬叶美洲茶);Montery Ceanothus ●☆

79971 Ceanothus sanguineus Pursh;红茎美洲茶;Redstem Ceanothus ●☆

79972 Ceanothus spinosus Nutt.;刺美洲茶;Greenbark Ceanothus, Redheart Ceanothus ●☆

79973 Ceanothus thyrsiflorus Eschw.;蓝花美洲茶(聚花美洲茶,兰花美洲茶);Blue Blossom, Blue Blossom Ceanothus, Blue Shrub, Blueblossom, Blue-blossom, Californian Lilac ●☆

79974 Ceanothus thyrsiflorus Eschw. var. repens McMinn;匍匐蓝花美洲茶;Creeping Blue Blossom ●☆

79975 Ceanothus trinervus Moench = Ceanothus americanus L. ●☆

79976 Ceanothus triquetrus Wall. = Rhamnus triquetra (Wall.) Brandis ●☆

79977 Ceanothus veitchianus Hook.;维奇氏美洲茶(楔叶美洲茶);Veitch Ceanothus ●☆

79978 Ceanothus velutinus Douglas;毡毛美洲茶;Snowbrush

Ceanothus, Tobacco Brush, Tobacco-brush ●☆
79979 Ceanothus velutinus Douglas var. laevigatus (Hook.) Torr. et A. Gray;平滑毡毛美洲茶●☆
79980 Ceanothus vestitus Greene;包被美洲茶;Mohave Buck Brush, Mohave Buckbrush ●☆
79981 Cearanthes Ravenna(2000);巴西堇石蒜属■☆
79982 Cearia Dumort. = Eurycles Salisb. ■☆
79983 Cearia Dumort. = Proiphys Herb. ■☆
79984 Ceballosia G. Kunkel = Messerschmidia L. ex Hebenstr. ●■
79985 Ceballosia G. Kunkel = Tournefortia L. ●■
79986 Ceballosia G. Kunkel ex Förther = Messerschmidia L. ex Hebenstr. ●■
79987 Ceballosia G. Kunkel ex Förther = Tournefortia L. ●■
79988 Ceballosia G. Kunkel ex Förther(1980);墨西哥紫丹属●☆
79989 Ceballosia fruticosa (L. f.) G. Kunkel ex Förther;墨西哥紫丹●☆
79990 Ceballosia fruticosa (L. f.) G. Kunkel ex Förther = Messerschmidia fruticosa L. f. ●☆
79991 Ceballosia fruticosa (L. f.) G. Kunkel ex Förther = Tournefortia fruticosa (L. f.) Ortega ●☆
79992 Ceballosia fruticosa (L. f.) G. Kunkel ex Förther var. angustifolia (Lam.) G. Kunkel ex Förther = Ceballosia fruticosa (L. f.) G. Kunkel ex Förther ●☆
79993 Cebatha Forssk.(废弃属名) = Cocculus DC.(保留属名)●
79994 Cebatha hirsuta (L.) Kuntze = Cocculus hirsutus (L.) Diels ●☆
79995 Cebatha pendula (J. R. Forst. et G. Forst.) Kuntze = Cocculus pendulus (J. R. Forst. et G. Forst.) Diels ●☆
79996 Cebipira Juss. ex Kuntze = Bowdichia Kunth ●☆
79997 Cecarria Barlow(1973);切卡寄生属●☆
79998 Cecarria obtusifolia (Merr) Barlow;切卡寄生●☆
79999 Cecarria obtusifolia Barlow = Cecarria obtusifolia (Merr) Barlow ●☆
80000 Cecchia Chiov. = Oldfieldia Benth. et Hook. f. ●☆
80001 Cecchia somalensis Chiov. = Oldfieldia somalensis (Chiov.) Milne-Redh. ●☆
80002 Cecidodaphne Nees = Cinnamomum Schaeff.(保留属名)●
80003 Cecropia Loefl.(1758)(保留属名);蚁栖树属(号角树属,南美伞树属,轻桑属,伞树属,砂纸桑属);Pumpwood, Snakewood-tree, Snakewood Tree, Trumpet Tree, Trumpet-tree ●☆
80004 Cecropia adenopus Mart. ex Miq.;腺柄蚁栖树(腺柄号角树,腺柄砂纸桑);Ambay Pumpwood, Sandpaper Tree ●☆
80005 Cecropia leucocoma Miq.;白毛蚁栖树(白毛砂纸桑)●☆
80006 Cecropia obtusa Trécul;钝蚁栖树树(钝号角树)●☆
80007 Cecropia obtusifolia Bertol.;倒卵叶蚁栖树(钝叶号角树);Trumpet Tree ●☆
80008 Cecropia palmata Willd.;蚁栖树(掌状砂纸桑);Snakewood Tree, Trumpet Tree ●☆
80009 Cecropia peltata L.;盾叶蚁栖树(盾状砂纸桑,号角树,伞树);Cuaromo, Trumpet Tree ●☆
80010 Cecropia sciadophylla Mart.;伞叶蚁栖树(伞叶砂纸桑)●☆
80011 Cecropiaceae C. C. Berg = Urticaceae Juss.(保留科名)●■
80012 Cecropiaceae C. C. Berg(1978);蚁牺树科(号角树科,南美伞科,南美伞树科,伞树科,锥头麻科)●
80013 Cedraceae C. C. Berg = Pinaceae Spreng. ex F. Rudolphi(保留科名)●
80014 Cedraceae Vest = Pinaceae Spreng. ex F. Rudolphi(保留科名)●
80015 Cedrela P. Browne(1756);洋椿属(椿属);Bastard Cedar, Cedrela, Chinese Cedar, Foreigntoona ●
80016 Cedrela australis F. Muell. = Toona australis Harms ●
80017 Cedrela balansae C. DC.;巴氏洋椿●☆
80018 Cedrela calantas (Merr. et Rolfe) Burkill;热带南美洲椿●☆
80019 Cedrela chinensis Franch. = Toona sinensis (A. Juss.) M. Roem. ●◇
80020 Cedrela fissilis Vell.;裂果椿(劈裂洋椿);South Ameriean Cedar ●☆
80021 Cedrela glaziovii C. DC.;洋椿;Glaziov Cedrela, Glaziov Foreigntoona ●
80022 Cedrela kotschyi Schweinf. = Pseudocedrela kotschyi (Schweinf.) Harms ●☆
80023 Cedrela lilloi C. DC.;阿根廷洋椿●☆
80024 Cedrela mahagoni L. = Swietenia mahagoni (L.) Jacq. ●
80025 Cedrela mexicana M. Roem.;墨西哥洋椿(墨西哥椿,墨西哥雪松);Central American Cedar, Cigar-box Cedar, Mexican Cedar, Mexican Cedrela ●☆
80026 Cedrela mexicana M. Roem. = Cedrela odorata L. ●
80027 Cedrela microcarpa C. DC. = Toona ciliata M. Roem. ●◇
80028 Cedrela microcarpa C. DC. = Toona microcarpa (C. DC.) Harms ●
80029 Cedrela mollis Hand. -Mazz. = Toona ciliata M. Roem. var. pubescens (Franch.) Hand. -Mazz. ●
80030 Cedrela odorata L.;南美香椿(巴西洋椿,芳香洋椿,美国香椿,美洲香椿,香洋椿,烟洋椿,中美香椿);Brazilian Cedar, Cedar, Cigarbox Cedar, Cigar-box Cedar, Cigarbox Cedrela, Cigar-box Cedrella, Jamaican Red Cedar, South American Cedar, Spanish Cedar, West Indian Cedar ●
80031 Cedrela rosmarinus Lour. = Baeckea frutescens L. ●
80032 Cedrela salvadorensis Standl.;萨尔瓦多洋椿●☆
80033 Cedrela serrata Royle;马来洋椿●☆
80034 Cedrela serrata Royle = Toona sinensis (A. Juss.) M. Roem. ●◇
80035 Cedrela sinensis A. Juss. = Toona sinensis (A. Juss.) M. Roem. ●◇
80036 Cedrela sinensis A. Juss. var. hupehana C. DC. = Toona sinensis (A. Juss.) M. Roem. var. hupehana (C. DC.) P. Y. Chen ●
80037 Cedrela sinensis A. Juss. var. schensiana C. DC. = Toona sinensis (A. Juss.) M. Roem. var. schensiana (C. DC.) P. Y. Chen ●
80038 Cedrela toona Rottler et Willd. = Toona ciliata M. Roem. ●◇
80039 Cedrela toona Roxb. ex Rottler et Willd. = Toona ciliata M. Roem. ●◇
80040 Cedrela toona Roxb. ex Rottler et Willd. var. henryi C. DC. = Toona ciliata M. Roem. var. henryi (C. DC.) C. Y. Wu ●
80041 Cedrela toona Roxb. ex Rottler et Willd. var. pubescens Franch. = Toona ciliata M. Roem. var. pubescens (Franch.) Hand. -Mazz. ●
80042 Cedrela toona Roxb. ex Rottler et Willd. var. sublaxiflora C. DC. = Toona ciliata M. Roem. var. sublaxiflora (C. DC.) C. Y. Wu ●
80043 Cedrela toona Roxb. ex Willd. = Toona ciliata M. Roem. ●◇
80044 Cedrela tubiflora Bertero;管花椿(管花洋椿)●☆
80045 Cedrela whitfordii S. F. Blake;怀氏洋椿●☆
80046 Cedrela yunnanensis C. DC. = Toona ciliata M. Roem. var. yunnanensis (C. DC.) C. Y. Wu ●
80047 Cedrelaceae R. Br. = Meliaceae Juss.(保留科名)●
80048 Cedrelinga Ducke(1922);椿豆属(亚马逊豆属)●☆
80049 Cedrelinga cateniformis Ducke;椿豆(链椿豆)●☆
80050 Cedrella Scop. = Cedrela P. Browne ●
80051 Cedrelopsis Baill.(1893);拟洋椿属●☆
80052 Cedrelopsis gracilis J. -F. Leroy;纤细拟洋椿●☆
80053 Cedrelopsis grevei Baill.;拟洋椿●☆

80054 Cedrelopsis longibracteata J. -F. Leroy;长苞拟洋椿●☆

80055 Cedrelopsis microfoliolata J. -F. Leroy;微小叶拟洋椿●☆

80056 Cedrelopsis procera J. -F. Leroy;高大拟洋椿●☆

80057 Cedrelopsis rakotozafyi Cheek et Lescot;拉库图扎非拟洋椿●☆

80058 Cedrelopsis trivalvis J. -F. Leroy;三分果片拟洋椿●☆

80059 Cedro Loefl. = Cedrela P. Browne ●

80060 Cedronella Moench(1794);柠檬草属●☆

80061 Cedronella Riv. ex Rupp. = Cedronella Moench ●☆

80062 Cedronella cana Hook. = Agastache cana (Hook.) Wooton et Standl. ■☆

80063 Cedronella canariensis (L.) Webb et Berthel.;柠檬草(加那利柠檬草,三叶柠檬草);Canary Balm,Herb of Gilead ●☆

80064 Cedronella triphylla Moench;三叶柠檬草;Balm of Gilead ●☆

80065 Cedronella triphylla Moench = Cedronella canariensis (L.) Webb et Berthel. ●☆

80066 Cedronella urticifolia (Miq.) Maxim. = Meehania urticifolia (Miq.) Makino ■

80067 Cedronella urticifolia Maxim. = Meehania urticifolia (Miq.) Makino ■

80068 Cedronia Cuatrec. = Picrolemma Hook. f. ●☆

80069 Cedrostis Post et Kuntze = Kedrostis Medik. ■☆

80070 Cedrota Schreb. = Aniba Aubl. ●☆

80071 Cedrus Duhamel(废弃属名)= Juniperus L. ●

80072 Cedrus Loud. = Pinus L. ●

80073 Cedrus Mill. = Cedrela P. Browne ●

80074 Cedrus Trew(1757)(保留属名);雪松属;Cedar,True Cedar ●

80075 Cedrus africana Gordon = Cedrus atlantica (Endl.) Manetti ex Carrière ●

80076 Cedrus africana Gordon ex Knight = Cedrus atlantica (Endl.) Manetti ex Carrière ●

80077 Cedrus argentea Carrière = Cedrus atlantica (Endl.) Carrière ●

80078 Cedrus atlantica (Endl.) Carrière = Cedrus atlantica (Endl.) Manetti ex Carrière ●

80079 Cedrus atlantica (Endl.) Manetti = Cedrus atlantica (Endl.) Manetti ex Carrière ●

80080 Cedrus atlantica (Endl.) Manetti ex Carrière;北非雪松(大西洋雪松);African Cedar, Algerian Cedar, Atlantic Cedar, Atlas Cedar, Atlas Mountain Cedar, Mount Atlas Cedar, Mountain Atlas Cedar, Satinwood ●

80081 Cedrus atlantica (Endl.) Manetti ex Carrière 'Aurea';金叶北非雪松●☆

80082 Cedrus atlantica (Endl.) Manetti ex Carrière 'Glauca Pendula';垂枝蓝北非雪松;Weeping Blue Atlas Cedar ●☆

80083 Cedrus atlantica (Endl.) Manetti ex Carrière 'Glauca';蓝叶北非雪松;Blue Atlantic Cedar,Blue Atlas Cedar,Blue Cedar ●☆

80084 Cedrus atlantica (Endl.) Manetti ex Carrière 'Pendula';垂枝北非雪松●☆

80085 Cedrus brevifolia (Hook. f.) Elwes et Henry;短叶雪松(塞浦路斯雪松,小叶雪松);Cyprian Cedar,Cyprus Cedar,Shortleaf Cedar ●☆

80086 Cedrus brevifolia (Hook. f.) Henry = Cedrus brevifolia (Hook. f.) Elwes et Henry ●☆

80087 Cedrus deodara (Roxb.) G. Don;雪松(喜马拉雅香柏,香柏);California Christmas-tree, Deodar, Deodar Cedar, Himalayan Cedar, India Cedar, Indian Cedar ●

80088 Cedrus deodara (Roxb.) G. Don 'Aurea';金叶雪松(黄叶雪松);Golden Deodar ●☆

80089 Cedrus deodara (Roxb.) G. Don 'Pendula';垂枝雪松;Weeping Himalayan Cedar ●☆

80090 Cedrus deodara (Roxb.) G. Don 'Snow Sprite';精灵雪松;Snow Sprite Deodara Cedar ●☆

80091 Cedrus deodara (Roxb.) G. Don 'Umbraculiformis';伞形雪松;Umbrellate Deodar Cedar ●

80092 Cedrus deodara Loudon = Cedrus deodara (Roxb.) G. Don ●

80093 Cedrus indica Chambray = Cedrus deodara (Roxb.) G. Don ●

80094 Cedrus libani A. Rich. = Cedrus libani Lawson ●☆

80095 Cedrus libani A. Rich. subsp. atlantica (Endl.) Batt. et Trab. = Cedrus atlantica (Endl.) Manetti ex Carrière ●

80096 Cedrus libani A. Rich. subsp. brevifolia (Hook. f.) Meikle = Cedrus brevifolia (Hook. f.) Elwes et Henry ●☆

80097 Cedrus libani A. Rich. subsp. deodara (Roxb.) P. D. Sell = Cedrus deodara (Roxb.) G. Don ●

80098 Cedrus libani A. Rich. var. atlantica (Endl.) Hook. f. = Cedrus atlantica (Endl.) Manetti ex Carrière ●

80099 Cedrus libani A. Rich. var. deodara (Roxb.) Hook. f. = Cedrus deodara (Roxb.) G. Don ●

80100 Cedrus libani Lawson;黎巴嫩雪松;Cedar of Lebanon, Cedar-of-Lebanon, Lebanon Cedar, Lebanonzeder ●☆

80101 Cedrus libani Lawson 'Aurea-Prostrata';平卧金黎巴嫩雪松●☆

80102 Cedrus libani Lawson 'Comte de Dijon';矮生黎巴嫩雪松●☆

80103 Cedrus libani Lawson 'Sargentii';圆冠黎巴嫩雪松●☆

80104 Cedrus libani Lawson subsp. atlantica (Endl.) Bart. et Trab. = Cedrus atlantica (Endl.) Manetti ex Carrière ●

80105 Cedrus libani Lawson subsp. deodara (Roxb.) P. D. Sell = Cedrus deodara (Roxb.) G. Don ●

80106 Cedrus libani Lawson subsp. stenocoma (O. Schwarz) Greuter et Burdet;狭冠黎巴嫩雪松●☆

80107 Cedrus libani Lawson subsp. stenocoma (O. Schwarz) Greuter et Burdet = Cedrus libani Lawson ●☆

80108 Cedrus libani Lawson var. atlaniica (Endl.) Hook. f. = Cedrus atlantica (Endl.) Manetti ex Carrière ●

80109 Cedrus libani Lawson var. deodara (Roxb.) Hook. f. = Cedrus deodara (Roxb.) G. Don ●

80110 Cedrus libanotica Link = Cedrus libani Lawson ●☆

80111 Cedrus libanotica Link subsp. atlantica (Endl.) Jahand. et Maire = Cedrus atlantica (Endl.) Carrière ●

80112 Cedrus libanotica Link var. atrovirens Maire et Weiller = Cedrus atlantica (Endl.) Carrière ●

80113 Cedrus libanotica Link var. glauca Carrière = Cedrus atlantica (Endl.) Carrière ●

80114 Ceiba Mill. (1754);吉贝属(爪哇木棉属);Ceiba, Kapok, Silk-Cotton ●

80115 Ceiba acuminata Rose;尖叶吉贝;Kapok ●☆

80116 Ceiba aesculifolia (Kunth) Britten et Baker f.;七叶树叶吉贝;Pochote ●☆

80117 Ceiba casearia Medik. = Ceiba pentandra (L.) Gaertn. ●

80118 Ceiba insignis (Kunth) P. E. Gibbs et Semir;南美吉贝;Floss-silk Tree, South American Bottle Tree, Yachan ●☆

80119 Ceiba occidentalis (Spreng.) Burkill;洪都拉斯吉贝木●☆

80120 Ceiba pentandra (L.) Gaertn.;吉贝(吉贝棉,美洲木棉,娑罗木,娑罗树,五雄吉贝,爪哇木棉);Cotton Tree, Kapok, Kapok Ceiba, Kapok Tree, Kapok-tree, Silk Cotton, Silk Cotton Tree, Silk Cotton-tree, Silk-cotton Tree, Sillcotton Tree, White Silk Cotton Tree ●

80121 Ceiba speciosa (A. St. -Hil., A. Juss. et Cambess.) Ravenna;美丽吉贝;Pink Floss-silk Tree ●☆

80122 Ceiba thonningii A. Chev. = Ceiba pentandra (L.) Gaertn. ●

80123 Celaena Wedd. = Oligandra Less. ■☆

80124 Celaenodendron Standl. (1927);黑大戟属●☆

80125 Celaenodendron mexicanum Standl.;墨西哥黑大戟●☆

80126 Celasine Pritz. = Gelasine Herb. ■☆

80127 Celastraceae R. Br. (1814)(保留科名);卫矛科;Spindle Family, Stafftree Family, Staff-Tree Family ●

80128 Celastrus Baill. = Denhamia Meisn.(保留属名)●☆

80129 Celastrus L. (1753)(保留属名);南蛇藤属;Climbing Bittersweet, Staff Vine, Bittersweet, Staff Tree, Staff-tree ●

80130 Celastrus aculeatus Merr.;过山枫(穿山龙,落霜红);Aculeate Bittersweet, Aculeate Staff Tree ●

80131 Celastrus aculeatus Merr. var. oblanceifolius (F. T. Wang et P. C. Tsoong) Y. C. Hsu = Celastrus oblanceifolius F. T. Wang et P. C. Tsoong ●

80132 Celastrus acuminatus L. f. = Gymnosporia acuminata (L. f.) Szyszyl. ●☆

80133 Celastrus acuminatus L. f. var. microphyllus Sond. = Gymnosporia acuminata (L. f.) Szyszyl. ●☆

80134 Celastrus adenophyllus Miq. = Ilex crenata Thunb. ●

80135 Celastrus alatus Thunb. = Celastrus scandens L. ●

80136 Celastrus alatus Thunb. = Euonymus alatus (Thunb.) Siebold ●

80137 Celastrus albatus N. E. Br. = Maytenus albata (N. E. Br.) Ernst Schmidt et Jordaan ●☆

80138 Celastrus alnifolius D. Don = Celastrus paniculatus Willd. ●

80139 Celastrus andongensis Oliv. = Gymnosporia heterophylla (Eckl. et Zeyh.) Loes. ●☆

80140 Celastrus angularis Sond. = Gymnosporia heterophylla (Eckl. et Zeyh.) Loes. ●☆

80141 Celastrus angulatus Maxim.;苦皮藤(菜虫药,菜药,大钓鱼竿,大马桑,吊干麻,吊杆麻,苦皮树,苦树皮,苦通,老虎麻,老虎麻藤,老麻藤,棱枝南蛇藤,萝卜药,马断肠,南山叶,南蛇根,酸枣子藤);Angle Bittersweet, Angled Bittersweet, Anglestem, Angular Staff-tree, Angustem Bittersweet, Bittersweet ●

80142 Celastrus angulatus Maxim. var. latifolius Hort. = Celastrus angulatus Maxim. ●

80143 Celastrus arbutifolius Hochst. ex A. Rich. = Gymnosporia arbutifolia (Hochst. ex A. Rich.) Loes. ●☆

80144 Celastrus arbutifolius Hochst. ex A. Rich. var. major A. Rich. = Gymnosporia arbutifolia (Hochst. ex A. Rich.) Loes. ●☆

80145 Celastrus articulatus Thunb. = Celastrus orbiculatus Thunb. ex A. Murray ●

80146 Celastrus articulatus Thunb. var. cuneatus Rehder et E. H. Wilson = Celastrus cuneatus (Rehder et E. H. Wilson) C. Y. Cheng et T. C. Kao ●

80147 Celastrus articulatus Thunb. var. pubescens Makino = Celastrus orbiculatus Thunb. ex A. Murray ●

80148 Celastrus articulatus Thunb. var. pubescens Makino = Celastrus stephanotifolius (Makino) Makino ●☆

80149 Celastrus articulatus Thunb. var. punctatus (Thunb.) Makino = Celastrus punctatus Thunb. ●

80150 Celastrus articulatus Thunb. var. stephanotifolius? = Celastrus stephanotifolius (Makino) Makino ●☆

80151 Celastrus atkaio A. Rich. = Gymnosporia arbutifolia (Hochst. ex A. Rich.) Loes. ●☆

80152 Celastrus benthamii Rehder et E. H. Wilson = Celastrus monospermus Roxb. ●

80153 Celastrus bodinieri L'Hér. = Ilex chinensis Sims ●

80154 Celastrus buxifolius L. = Gymnosporia buxifolia (L.) Szyszyl. ●☆

80155 Celastrus campestris Eckl. et Zeyh. = Putterlickia pyracantha (L.) Szyszyl. ●☆

80156 Celastrus cantoniensis Hance = Celastrus hindsii Benth. ●

80157 Celastrus capitatus E. Mey. ex Sond. = Gymnosporia capitata (E. Mey. ex Sond.) Loes. ●☆

80158 Celastrus cavaleriei H. Lév. = Myrsine semiserrata Wall. ●

80159 Celastrus championi Benth. = Celastrus monospermus Roxb. ●

80160 Celastrus ciliidens Miq. = Celastrus flagellaris Rupr. ●

80161 Celastrus concinnus N. E. Br. = Gymnosporia harveyana Loes. ●☆

80162 Celastrus contonensis Hance = Celastrus hindsii Benth. ●

80163 Celastrus cordatus E. Mey. ex Sond. = Maytenus cordata (E. Mey. ex Sond.) Loes. ●☆

80164 Celastrus coriaceus Guillaumin et Perr. = Gymnosporia senegalensis (Lam.) Loes. ●☆

80165 Celastrus crassifolia C. H. Wang = Celastrus stylosus Wall. ●

80166 Celastrus crispulus Regel = Celastrus orbiculatus Thunb. ex A. Murray ●

80167 Celastrus crispus Thunb. = Euclea crispa (Thunb.) Gürke ●☆

80168 Celastrus cuneatus (Rehder et E. H. Wilson) C. Y. Cheng et T. C. Kao;楔叶南蛇藤(小南蛇藤);Cuneate Staff-tree, Dwarf Staff-tree, Small Bittersweet ●

80169 Celastrus cymosus Sol. = Gymnosporia buxifolia (L.) Szyszyl. ●☆

80170 Celastrus dependens Wall. = Celastrus paniculatus Willd. ●

80171 Celastrus diffusus G. Don = Ventilago diffusa (G. Don) Exell ●☆

80172 Celastrus dilatatus Thunb. = Orixa japonica Thunb. ●

80173 Celastrus discolor H. Lév. = Sabia parviflora Wall. ex Roxb. ●

80174 Celastrus diversifolius (Maxim.) Hemsl. = Gymnosporia diversifolia Maxim. ●

80175 Celastrus diversifolius Hemsl. = Maytenus diversifolia (Maxim.) Ding Hou ●

80176 Celastrus edulis Vahl = Catha edulis (Vahl) Forssk. ex Endl. ●

80177 Celastrus elevativenius Hayata = Celastrus punctatus Thunb. ●

80178 Celastrus ellipticus Thunb. = Gymnosporia elliptica (Thunb.) Schönland ●☆

80179 Celastrus emarginatus Willd. = Gymnosporia emarginata (Willd.) Thwaites ●

80180 Celastrus emarginatus Willd. = Maytenus emarginata (Willd.) Ding Hou ●

80181 Celastrus esquirolianus H. Lév. = Rhamnus crenata Siebold et Zucc. ●

80182 Celastrus esquirolii H. Lév. = Sabia parviflora Wall. ex Roxb. ●

80183 Celastrus euonymoidea H. Lév. = Grewia biloba Wall. ex G. Don ●

80184 Celastrus euonymoides Welw. ex Oliv. = Gymnosporia putterlickioides Loes. subsp. euonymoides (Welw. ex Oliv.) Jordaan ●☆

80185 Celastrus euphlebiphyllus (Hayata) Makino et Nemoto = Celastrus paniculatus Willd. ●

80186 Celastrus europaeus Boiss. = Maytenus senegalensis (Lam.) Exell subsp. europaea (Boiss.) Güemes et M. B. Crespo ●☆

80187 Celastrus excisus Thunb. = Scutia myrtina (Burm. f.) Kurz ●

80188 Celastrus filiformis L. f. = Secamone filiformis (L. f.) J. H. Ross ●☆

80189 Celastrus flagellaris Rupr.;刺苞南蛇藤(刺南蛇藤,刺叶南蛇藤,爬山虎);Flagellate Bittersweet, Hookedspine Bittersweet, Korean Bittersweet, Whip-like Staff-tree ●

80190 Celastrus flexuosus Thunb. = Dovyalis rhamnoides (Burch. ex DC.) Burch. et Harv. ●☆
80191 Celastrus franchetianus Loes.;洱源南蛇藤;Franchet Staff-tree ●☆
80192 Celastrus geminiflorus Hayata = Celastrus punctatus Thunb. ●
80193 Celastrus gemmatus Loes.;大芽南蛇藤(白花藤,穿山龙,哥兰叶,米汤叶,绵条子,三叶泡,霜红藤,钻地风);Budded Bittersweet,Gemmate Staff-tree ●
80194 Celastrus glaber (Hatus.) T. Yamaz. = Celastrus kusanoi Hayata ●
80195 Celastrus glaucophyllus Rehder et E. H. Wilson;灰叶南蛇藤(过山枫藤,麻麻藤,霜叶,藤木);Glaucous Bittersweet,Glaucous-leaved Staff-tree,Greyleaf Bittersweet ●
80196 Celastrus glaucophyllus Rehder et E. H. Wilson var. angustus Q. H. Chen = Celastrus glaucophyllus Rehder et E. H. Wilson ●
80197 Celastrus glaucophyllus Rehder et E. H. Wilson var. puberulus Y. C. Hsu = Celastrus stylosus Wall. var. puberulus (Y. C. Hsu) C. Y. Cheng et T. C. Kao ●
80198 Celastrus glaucophyllus Rehder et E. H. Wilson var. rugosus (Rehder et E. H. Wilson) C. Y. Wu = Celastrus rugosus Rehder et E. H. Wilson ●
80199 Celastrus glaucus R. Br. = Cassine glauca (Pers.) Kuntze ●☆
80200 Celastrus gracilipes Welw. ex Oliv. = Gymnosporia gracilipes (Welw. ex Oliv.) Loes. ●☆
80201 Celastrus gracillimus Hayata = Celastrus punctatus Thunb. ●
80202 Celastrus heterophyllus Eckl. et Zeyh. = Gymnosporia heterophylla (Eckl. et Zeyh.) Loes. ●☆
80203 Celastrus hindsii Benth.;青江藤(横脉南蛇藤,黄果藤,麻藤,面藤,南华南蛇藤,青杠藤,野茶藤);Chinese Bitter Sweet,Chinese Bittersweet,Hinds Bittersweet,Hinds Staff-tree,Tibetan Staff Tree,Xizang Staff-tree ●
80204 Celastrus hindsii Benth. var. henry Loes.;短梗青江藤;Henry Staff-tree ●
80205 Celastrus hirsutus Comber;硬毛南蛇藤;Hardhair Bittersweet,Hirsute Staff-tree ●
80206 Celastrus homaliifolius P. S. Hsu;小果南蛇藤(多花南蛇藤);Littlefruit Bittersweet,Smallfruit Bittersweet ●
80207 Celastrus hookeri Prain;滇边南蛇藤(尖药南蛇藤,毛枝南蛇藤,藤麻);Hooker Bittersweet,Hooker Staff-tree ●
80208 Celastrus huillensis Welw. ex Oliv. = Maytenus undata (Thunb.) Blakelock ●☆
80209 Celastrus hypoglaucus Hemsl. = Celastrus hypoleucus (Oliv.) Warb. ex Loes. ●
80210 Celastrus hypoleucoides P. L. Chiu;薄叶南蛇藤(拟粉背南蛇藤);Thinleaf Bittersweet ●
80211 Celastrus hypoleucus (Oliv.) Warb. ex Loes.;粉背南蛇藤(博根藤,落霜红,麻妹藤,麻妹条,绵藤);Pale Bittersweet,Pale Staff-tree,Paleback Bittersweet ●
80212 Celastrus hypoleucus (Oliv.) Warb. ex O. Loes. argutior Loes. = Celastrus hypoleucus (Oliv.) Warb. ex Loes. ●
80213 Celastrus hypoleucus (Oliv.) Warb. ex O. Loes. puberula Loes. = Celastrus stylosus Wall. ●
80214 Celastrus ilicinus Burch. = Maytenus ilicina (Burch.) Loes. ●☆
80215 Celastrus insularis Koidz. = Celastrus orbiculatus Thunb. ex A. Murray ●
80216 Celastrus integrifolius L. f. = Gloveria integrifolia (L. f.) Jordaan ●☆
80217 Celastrus japonicus K. Koch = Orixa japonica Thunb. ●
80218 Celastrus jeholensis Nakai = Celastrus orbiculatus Thunb. ex A. Murray ●
80219 Celastrus kiusianus Franch. et Sav. = Celastrus punctatus Thunb. ●
80220 Celastrus kouytchensis H. Lév. = Rhamnus crenata Siebold et Zucc. ●
80221 Celastrus kusanoi Hayata;圆叶南蛇藤(草野南蛇藤,秤星蛇,大叶南蛇藤,过山枫,双虎排牙,铁包金);Kusano Bitter Sweet,Kusano Staff-tree,Roundleaf Bittersweet ●
80222 Celastrus kusanoi Hayata var. glaber Hatus. = Celastrus kusanoi Hayata ●
80223 Celastrus lanceolatus E. Mey. ex Sond. = Gymnosporia linearis (L. f.) Loes. subsp. lanceolata (E. Mey. ex Sond.) Jordaan ●☆
80224 Celastrus lancifolius Thonn. = Maytenus undata (Thunb.) Blakelock ●☆
80225 Celastrus laoticus Pit.;老挝南蛇藤;Laos Staff-tree ●☆
80226 Celastrus latifolius Hemsl. = Celastrus angulatus Maxim. ●
80227 Celastrus laurifolius A. Rich. = Maytenus undata (Thunb.) Blakelock ●☆
80228 Celastrus laurinus Thunb. = Maytenus oleoides (Lam.) Loes. ●☆
80229 Celastrus leiocarpus Hayata = Celastrus punctatus Thunb. ●
80230 Celastrus linearis L. f. = Gymnosporia linearis (L. f.) Loes. ●☆
80231 Celastrus littoralis A. Chev. = Gymnosporia buchananii Loes. ●☆
80232 Celastrus longe-racemosus Hayata = Celastrus hindsii Benth. ●
80233 Celastrus longe-racemosus Hayata = Celastrus punctatus Thunb. ●
80234 Celastrus loseneri Rehder et E. H. Wilson = Celastrus rosthornianus Loes. var. loeseneri (Rehder et E. H. Wilson) C. Y. Wu ex Y. C. Ho ●
80235 Celastrus lucidus L. = Maytenus lucida (L.) Loes. ●☆
80236 Celastrus luteolus Delile = Maytenus undata (Thunb.) Blakelock ●☆
80237 Celastrus lyi H. Lév. = Rhamnus esquirolii H. Lév. ●
80238 Celastrus mairei H. Lév. = Sabia yunnanensis Franch. ●
80239 Celastrus maritimus Bolus = Robsonodendron maritimum (Bolus) R. H. Archer ●☆
80240 Celastrus monospermus Roxb. = Celastrus hindsii Benth. ●
80241 Celastrus monospermus Roxb. = Monocelastrus monosperma (Roxb.) F. T. Wang et Ts. Tang ●
80242 Celastrus montanus Roth = Gymnosporia senegalensis (Lam.) Loes. ●☆
80243 Celastrus montanus Roth, Roem. et Schult. = Maytenus senegalensis (Lam.) Exell ●☆
80244 Celastrus mossambicensis Klotzsch = Gymnosporia mossambicensis (Klotzsch) Loes. ●☆
80245 Celastrus mucronatus Eckl. et Zeyh. = Gymnosporia acuminata (L. f.) Szyszyl. ●☆
80246 Celastrus multiflorus Lam. = Gymnosporia heterophylla (Eckl. et Zeyh.) Loes. ●☆
80247 Celastrus multiflorus Roxb. = Celastrus paniculatus Willd. subsp. multiflorus (Roxb.) Ding Hou ●
80248 Celastrus multiflorus Roxb. = Celastrus paniculatus Willd. ●
80249 Celastrus ndelleensis A. Chev. = Gymnosporia buchananii Loes. ●☆
80250 Celastrus nemorosus Eckl. et Zeyh. = Gymnosporia nemorosa (Eckl. et Zeyh.) Szyszyl. ●☆
80251 Celastrus nutans Roxb. = Celastrus paniculatus Willd. ●
80252 Celastrus oblanceifolius F. T. Wang et P. C. Tsoong;窄叶南蛇藤(倒披针叶南蛇藤);Narrowleaf Bittersweet,Oblanceolate-leaved Staff-tree ●
80253 Celastrus oblongifolius Hayata = Celastrus hindsii Benth. ●

80254 Celastrus oblongifolius Hayata = Celastrus punctatus Thunb. ●

80255 Celastrus obscurus A. Rich. = Gymnosporia obscura (A. Rich.) Loes. ●☆

80256 Celastrus obtusus Thunb. = Putterlickia pyracantha (L.) Szyszyl. ●☆

80257 Celastrus oleoides Lam. = Maytenus oleoides (Lam.) Loes. ●☆

80258 Celastrus oppositus Wall. = Pleurostylia opposita (Wall.) Alston ●

80259 Celastrus orbiculata Thunb. = Celastrus orbiculatus Thunb. ex A. Murray ●

80260 Celastrus orbiculatus Thunb. ex A. Murray;南蛇藤(白龙,臭花椒,穿山龙,大伦藤,大南蛇,地南蛇,挂廊边,挂廊鞭,果山藤,过山风,过山枫,过山龙,合欢花,黄豆瓣,黄果藤,黄藤,降龙草,金红树,金银柳,苦树皮,老龙皮,老牛筋,老石棵子,蔓性落霜红,明开夜合,南蛇风,牛老筋,七寸麻,穷搅藤,香龙草,药狗旦子);Asian Bittersweet, Asiatic Bittersweet, Bittersweet, Chinese Bittersweet, Oriental Bittersweet, Round-leaved Staff Tree, Staff Vine, Staff-tree ●

80261 Celastrus orbiculatus Thunb. ex A. Murray f. aureoarillatus (Honda) Ohwi;黄假种皮南蛇藤●☆

80262 Celastrus orbiculatus Thunb. ex A. Murray f. ellipticus (Honda) H. Hara;椭圆果南蛇藤●☆

80263 Celastrus orbiculatus Thunb. ex A. Murray f. papillosus (Nakai ex H. Hara) H. Hara;毛脉南蛇藤●☆

80264 Celastrus orbiculatus Thunb. ex A. Murray var. cancatus (Rehder et E. H. Wilson) C. Y. Wu = Celastrus cuneatus (Rehder et E. H. Wilson) C. Y. Cheng et T. C. Kao ●

80265 Celastrus orbiculatus Thunb. ex A. Murray var. humilis Maxim.;小南蛇藤;Dwarf Staff-tree ●

80266 Celastrus orbiculatus Thunb. ex A. Murray var. papillosus (Nakai ex H. Hara) Ohwi = Celastrus orbiculatus Thunb. ex A. Murray f. papillosus (Nakai ex H. Hara) H. Hara ●☆

80267 Celastrus orbiculatus Thunb. ex A. Murray var. pubescens Makino;柔毛南蛇藤;Pubescent-veined Staff-tree ●

80268 Celastrus orbiculatus Thunb. ex A. Murray var. pubescens Makino = Celastrus orbiculatus Thunb. ex A. Murray ●

80269 Celastrus orbiculatus Thunb. ex A. Murray var. punctatus (Thunb.) Rehder = Celastrus punctatus Thunb. ●

80270 Celastrus orbiculatus Thunb. ex A. Murray var. strigillosus (Nakai) H. Hara = Celastrus strigillosus Nakai ●☆

80271 Celastrus orbiculatus Thunb. ex A. Murray var. strigillosus (Nakai) H. Hara f. lancifolius (Nakai) Sugim. = Celastrus strigillosus Nakai ●☆

80272 Celastrus orixa Siebold et Zucc. = Orixa japonica Thunb. ●

80273 Celastrus paniculatus Willd.;灯油藤(打油果,滇南蛇藤,多花滇南蛇藤,多花毛叶南蛇藤,红果藤,小黄果,悬垂南蛇藤,圆锥花南蛇藤,圆锥南蛇藤,锥序南蛇藤);Dependent Staff-tree, Lampoil Bittersweet, Panicled Bittersweet, Paniculate Staff-tree ●

80274 Celastrus paniculatus Willd. subsp. multiflorus (Roxb.) Ding Hou;多花南蛇藤(多花滇南蛇藤,多花毛叶灯油藤,厚叶南蛇藤,花灯油藤);Many-flowered Bitter Sweet, Manyflowered Bittersweet, Manyflowered Paniculate Staff-tree ●

80275 Celastrus paniculatus Willd. subsp. multiflorus (Roxb.) Ding Hou = Celastrus paniculatus Willd. ●

80276 Celastrus paniculatus Willd. subsp. serratus (Blanco) Ding Hou;具齿灯油藤;Serrate Paniculate Staff-tree ●☆

80277 Celastrus parviflorus Vahl = Gymnosporia parviflora (Vahl) Chiov. ●☆

80278 Celastrus patens Eckl. et Zeyh. = Gymnosporia heterophylla (Eckl. et Zeyh.) Loes. ●☆

80279 Celastrus patentiflorus Hayata;展花南蛇藤(疏花南蛇藤,疏花蛇藤);Spreading-flowered Staff-tree ●

80280 Celastrus peduncularis Sond. = Maytenus peduncularis (Sond.) Loes. ●☆

80281 Celastrus polyacanthus Sond. = Gymnosporia polyacanthus (Sond.) Szyszyl. ●☆

80282 Celastrus polyanthemos Eckl. et Zeyh. = Gymnosporia heterophylla (Eckl. et Zeyh.) Loes. ●☆

80283 Celastrus populifolius Lam. = Gymnosporia acuminata (L. f.) Szyszyl. ●☆

80284 Celastrus procumbens L. f. = Maytenus procumbens (L. f.) Loes. ●☆

80285 Celastrus punctatus Thunb.;东南南蛇藤(斑叶南蛇藤,光果南蛇藤,细点南蛇藤,皱脉灰叶南蛇藤,皱叶南蛇藤);American Bittersweet, Christmas Oriental Bitter Sweet, Punctate Staff-tree, Rugose Glaucous-leaved Staff-tree ●

80286 Celastrus punctatus Thunb. var. microphyllus H. L. Li et Ding Hou ex Ding Hou = Celastrus punctatus Thunb. ●

80287 Celastrus punctatus Thunb. var. microphyllus H. L. Li et Ding Hou. = Celastrus punctatus Thunb. ●

80288 Celastrus pyracanthus L. = Putterlickia pyracantha (L.) Szyszyl. ●☆

80289 Celastrus racemulosa Franch. = Celastrus franchetianus Loes. ●☆

80290 Celastrus reticulatus Chun H. Wang = Celastrus rosthornianus Loes. ●

80291 Celastrus rhombifolius Eckl. et Zeyh. = Gymnosporia heterophylla (Eckl. et Zeyh.) Loes. ●☆

80292 Celastrus rigida Wall. ex Spreng. = Maytenus wallichiana (Spreng.) D. C. S. Raju et Babu ●☆

80293 Celastrus rigidus Thunb. = Rhigozum obovatum Burch. ●☆

80294 Celastrus robustus Roxb. = Bhesa robusta (Roxb.) Ding Hou ●

80295 Celastrus rosthornianus Loes.;短梗南蛇藤(白花藤,丛花南蛇藤,大藤菜,褐柄南蛇藤,黄绳儿,山货榔,少果南蛇藤);Rosthorn Bittersweet, Rosthorn Staff-tree, Self-fruitful Bittrswt, Shortstalk Bittersweet ●

80296 Celastrus rosthornianus Loes. var. loeseneri (Rehder et E. H. Wilson) C. Y. Wu ex Y. C. Ho;宽叶短梗南蛇藤(丛花南蛇藤);Loesener Bittersweet, Loesener Staff-tree ●

80297 Celastrus rostratus Thunb. = Pterocelastrus rostratus (Thunb.) Walp. ●☆

80298 Celastrus rothianus Roem. et E. H. Wilson = Celastrus paniculatus Willd. ●

80299 Celastrus rotundifolius Thunb. = Dovyalis rotundifolia (Thunb.) Thunb. et Harv. ●☆

80300 Celastrus royleanus Wall. = Maytenus royleana (Wall.) Cufod. ●

80301 Celastrus ruber Harv. = Gymnosporia rubra (Harv.) Loes. ●☆

80302 Celastrus rufus Wall. = Gymnosporia rufa (Wall.) M. A. Lawson ●

80303 Celastrus rufus Wall. = Maytenus rufa (Wall. ex Roxb.) D. C. S. Raju et Babu ●

80304 Celastrus rugosus Rehder et E. H. Wilson;皱叶南蛇藤(南蛇藤,皱脉灰叶南蛇藤);Wrinkledleaf Bittersweet ●

80305 Celastrus rupestris Eckl. et Zeyh. = Gymnosporia acuminata (L. f.) Szyszyl. ●☆

80306 Celastrus saharae Batt. = Gymnosporia senegalensis (Lam.) Loes. ●☆

80307 Celastrus salicifolia H. Lév. = Ilex macrocarpa Oliv. ●

80308 Celastrus saxatilis Burch. = Putterlickia saxatilis (Burch.) Jordaan ●☆

80309 Celastrus scandens L.;美洲南蛇藤(北美南蛇藤,美南蛇藤,攀缘南蛇藤);America Bittersweet, American Bittersweet, Bittersweet, Climbing Bittersweet, David's Root, False Bittersweet, Fever-twig, Staff Vine, Stafftree, Staff-tree, Waxwort, Wax-wort, Waxworts ●

80310 Celastrus scandens L. = Celastrus punctatus Thunb. ●

80311 Celastrus schimperi Hochst. ex A. Rich. = Gymnosporia serrata (Hochst. ex A. Rich.) Loes. ●☆

80312 Celastrus seguinii H. Lév. = Myrsine semiserrata Wall. ●

80313 Celastrus senegalensis Lam. = Gymnosporia senegalensis (Lam.) Loes. ●☆

80314 Celastrus senegalensis Lam. = Maytenus senegalensis (Lam.) Exell ●☆

80315 Celastrus senegalensis Lam. var. europeaus (Boiss.) Ball = Gymnosporia senegalensis (Lam.) Loes. ●☆

80316 Celastrus senegalensis Lam. var. inermis A. Rich. = Gymnosporia senegalensis (Lam.) Loes. ●☆

80317 Celastrus serratus Hochst. ex A. Rich. = Gymnosporia serrata (Hochst. ex A. Rich.) Loes. ●☆

80318 Celastrus serratus Hochst. ex A. Rich. var. steudneri Engl. = Maytenus gracilipes (Welw. ex Oliv.) Exell subsp. arguta (Loes.) Sebsebe ●☆

80319 Celastrus spiciformis Rehder et E. H. Wilson = Celastrus vaniotii (H. Lév.) Rehder ●

80320 Celastrus spiciformis Rehder et E. H. Wilson var. laevis Rehder et E. H. Wilson = Celastrus vaniotii (H. Lév.) Rehder var. laevis (Rehder et E. H. Wilson) Rehder ●

80321 Celastrus spiciformis Rehder et E. H. Wilson var. laevis Rehder et E. H. Wilson = Celastrus vaniotii (H. Lév.) Rehder ●

80322 Celastrus spinosus Royle = Maytenus royleanus (Wall. ex Lawson) Cufod. ●

80323 Celastrus stenophyllus Eckl. et Zeyh. = Gymnosporia linearis (L. f.) Loes. ●☆

80324 Celastrus stephanotifolius (Makino) Makino;海岛南蛇藤●☆

80325 Celastrus striatus Thunb. = Euonymus alatus (Thunb.) Siebold ●

80326 Celastrus striatus Thunb. ex A. Murray = Euonymus alatus (Thunb.) Siebold f. striatus (Thunb.) Makino ●☆

80327 Celastrus strigillosus Nakai;粗毛南蛇藤●☆

80328 Celastrus stylosus Wall.;显柱南蛇藤(茎花南蛇藤,山货榔);Styled Bittersweet, Stylose Staff-tree ●

80329 Celastrus stylosus Wall. subsp. glaber Ding Hou;无毛南蛇藤(茎花南蛇藤,山货榔,无毛明柱南蛇藤);Smooth Stylose Staff-tree ●

80330 Celastrus stylosus Wall. var. angustifolius C. Y. Cheng et T. C. Kao;窄叶显柱南蛇藤(狭叶显柱南蛇藤);Narrow-leaved Stylose Staff-tree ●

80331 Celastrus stylosus Wall. var. puberulus (Y. C. Hsu) C. Y. Cheng et T. C. Kao;毛脉显柱南蛇藤;Puberulous Stylos Staff-tree ●

80332 Celastrus suaveolens H. Lév. = Ilex suaveolens (H. Lév.) Loes. ●

80333 Celastrus tatarinowii Rupr. = Celastrus orbiculatus Thunb. ex A. Murray ●

80334 Celastrus tenuispinus Sond. = Gymnosporia tenuispina (Sond.) Szyszyl. ●☆

80335 Celastrus tetragonus Thunb. = Lauridia tetragona (L. f.) R. H. Archer ●☆

80336 Celastrus tonkinensis Pit.;皱果南蛇藤(东京南蛇藤);Tonkin Staff-tree, Wrinklefruit Bittersweet ●☆

80337 Celastrus tricuspidatus Lam. = Pterocelastrus tricuspidatus (Lam.) Walp. ●☆

80338 Celastrus tristis H. Lév. = Rhamnus nepalensis (Wall.) M. A. Lawson ●

80339 Celastrus umbellatus R. Br. = Maytenus umbellata (R. Br.) Mabb. ●☆

80340 Celastrus undatus Thunb. = Maytenus undata (Thunb.) Blakelock ●☆

80341 Celastrus vaniotii (H. Lév.) Rehder;长序南蛇藤(棉花藤,穗花南蛇藤);Vaniot Staff-tree, Vaniot's Bittersweet ●

80342 Celastrus vaniotii (H. Lév.) Rehder var. laevis (Rehder et E. H. Wilson) Rehder;平滑南蛇藤;Smooth Vaniot Staff-tree ●

80343 Celastrus variabilis Hemsl. = Gymnosporia variabilis (Hemsl.) Loes. ●

80344 Celastrus variabilis Hemsl. = Maytenus variabilis (Hemsl.) C. Y. Cheng ●

80345 Celastrus verrucosus E. Mey. ex Sond. = Putterlickia verrucosa (E. Mey. ex Sond.) Szyszyl. ●☆

80346 Celastrus versicolor Nakai = Celastrus strigillosus Nakai ●☆

80347 Celastrus verticillatus Roxb. = Pittosporum napaulense (DC.) Rehder et E. H. Wilson ●

80348 Celastrus virens (F. T. Wang et Ts. Tang) C. Y. Cheng et T. C. Kao;绿独子藤;Green Monocelastrus, Green One-seed Staff-tree, Green Singleseed Bittersweet ●

80349 Celastrus virens (F. T. Wang et Ts. Tang) C. Y. Cheng et T. C. Kao = Monocelastrus virens F. T. Wang et Ts. Tang ●

80350 Celastrus wallichiana Spreng. = Maytenus wallichiana (Spreng.) D. C. S. Raju et Babu ●☆

80351 Celastrus xizangensis Y. R. Li = Celastrus hindsii Benth. ●

80352 Celastrus yunnanensis H. Lév. = Premna parvilimba C. P'ei ●

80353 Celastrus zeyheri Sond. = Maytenus undata (Thunb.) Blakelock ●☆

80354 Celebnia Noronha = Saraca L. ●

80355 Celeri Adans. = Apium L. ■

80356 Celerina Benoist(1964);马爵床属☆

80357 Celerina seyrigii Benoist;马爵床☆

80358 Celestina Raf. = Ageratum L. ■●

80359 Celestina Raf. = Coelestina Cass. ■●

80360 Celianella Jabl. (1965);山酒珠属☆

80361 Celianella montana Jabl.;山酒珠☆

80362 Celiantha Maguire(1981);瘤花龙胆属■☆

80363 Celiantha imthurniana (Oliv.) Maguire;瘤花龙胆■☆

80364 Celmisia Cass. (1817)(废弃属名) = Capelio B. Nord. ■☆

80365 Celmisia Cass. (1817)(废弃属名) = Celmisia Cass. (1825)(保留属名)■☆

80366 Celmisia Cass. (1825)(保留属名);寒菀属■☆

80367 Celmisia asteliifolia Hook. f.;长叶寒菀;Snow Daisy ■☆

80368 Celmisia bellidioides Hook. f.;雏菊状寒菀■☆

80369 Celmisia longifolia Cass. = Celmisia asteliifolia Hook. f. ■☆

80370 Celmisia ramulosa Hook. f.;多枝寒菀■☆

80371 Celmisia rotundifolia Cass. = Capelio tabularis (Thunb.) B. Nord. ■☆

80372 Celmisia semicordata Petrie;银叶寒菀■☆

80373 Celmisia spathulata A. Cunn. ex DC.;匙状寒菀■☆

80374 Celmisia spectabilis Hook. f.;美丽寒菀;Cotton Daisy ■☆

80375 Celmisia traversii Hook. f.;剑叶寒菀■☆

80376 Celmisia walkeri Kirk;黏叶寒菀■☆
80377 Celome Greene = Cleome L. ●■
80378 Celosia L.（1753）;青葙属（鸡冠花属）;Cock's comb, Cockscomb, Woolflower ■
80379 Celosia acroprosodes Hochst. = Celosia anthelminthica Asch. ■☆
80380 Celosia anthelminthica Asch.;驱虫青葙■☆
80381 Celosia argentea L.;青葙（百日红，草蒿，草蒿，草决明，狗尾巴子，狗尾草，狗尾花，狗尾鸡冠苋，狗尾苋，狐狸尾，鸡公花，鸡公花草，鸡冠菜，鸡冠花，昆仑草，狼尾巴果，牛母窝，牛尾巴花，牛尾花子，萋蒿，青葙子，青箱，犬尾鸡冠花，陶朱术，天灵草，土鸡冠，野鸡冠，野鸡冠花，指天笔）;Cockscomb, Feather Cockscomb, Mfungu, Pink Candle, Quail Grass, Silver Cockscomb, Silver Cock's-comb ■
80382 Celosia argentea L.'Fairy Fountains';仙境喷泉青葙■
80383 Celosia argentea L. f. cristata（L.）Schinz = Celosia cristata L. ■
80384 Celosia argentea L. var. cristata（L.）Benth. = Celosia cristata L. ■
80385 Celosia argenteiformis（Schinz）Schinz = Hermbstaedtia argenteiformis Schinz ■☆
80386 Celosia baccata Retz. = Deeringia amaranthoides（Lam.）Merr. ●
80387 Celosia bakeri C. C. Towns.;贝克青葙■☆
80388 Celosia baronii Cavaco = Celosia spicata（Thouars）Spreng. ■☆
80389 Celosia benguellensis C. C. Towns.;本格拉青葙■☆
80390 Celosia boivinii Benth. et Hook. f. = Lagrezia boivinii（Benth. et Hook. f.）Schinz ■☆
80391 Celosia brevispicata C. C. Towns.;短穗青葙■☆
80392 Celosia cernua Roxb. = Celosia cristata L. ■
80393 Celosia chenopodiifolia Baker;藜叶青葙■☆
80394 Celosia coccinea L. = Celosia cristata L. ■
80395 Celosia cristata L.;鸡冠花（红鸡冠，红鸡冠花，鸡公花，鸡冠，鸡冠苗，鸡冠头，鸡冠苋，鸡冠子，鸡髻花，鸡角枪，鸡脚枪，老来少，青葙）;Celosia, Cock's Comb, Cockscomb, Common Cockscomb, Common Feather Cockscomb, Feather Amaranth ■
80396 Celosia cristata L.'Childsii';黄鸡冠花（矮性赤玉鸡冠花）;Hutton Cockscomb ■☆
80397 Celosia cristata L. = Celosia argentea L. f. cristata（L.）Schinz ■
80398 Celosia cristata L. var. childsii Hort. = Celosia cristata L. 'Childsii' ■☆
80399 Celosia cristata L. var. nana Hort.;矮小鸡冠花■☆
80400 Celosia cristata L. var. plumosa Hort.;羽状鸡冠花;Feather Cockscomb ■☆
80401 Celosia cristata L. var. splendens Moq. = Celosia cristata L. ■
80402 Celosia cuneifolia Baker = Celosia pandurata Baker ■☆
80403 Celosia debilis S. Moore = Celosia argentea L. ■
80404 Celosia dewevreana Schinz = Celosia loandensis Baker ■☆
80405 Celosia digyna Suess. = Celosia trigyna L. ■☆
80406 Celosia digyna Suess. var. cordata? = Celosia anthelminthica Asch. ■☆
80407 Celosia digyna Suess. var. fusca = Celosia loandensis Baker ■☆
80408 Celosia echinulata Hauman = Celosia loandensis Baker ■☆
80409 Celosia elegantissima Hauman;雅致鸡冠花■☆
80410 Celosia exellii Suess. = Hermbstaedtia exellii（Suess.）C. C. Towns. ■☆
80411 Celosia expansifila C. C. Towns.;扩展鸡冠花■☆
80412 Celosia falcata Lopr. = Hermbstaedtia linearis Schinz ■☆
80413 Celosia fleckii Schinz = Hermbstaedtia fleckii（Schinz）Baker et C. B. Clarke ■☆
80414 Celosia glauca J. C. Wendl. = Hermbstaedtia glauca（J. C. Wendl.）Rchb. ex Steud. ■☆
80415 Celosia globosa Schinz;球形青葙■☆
80416 Celosia globosa Schinz var. porphyrostachya C. C. Towns.;紫穗球形青葙■☆
80417 Celosia globosa Schinz var. spicata C. C. Towns.;穗状球形青葙■☆
80418 Celosia gracilenta Suess. et Overkott = Celosia vanderystii Schinz ■☆
80419 Celosia hastata Lopr.;戟叶青葙■☆
80420 Celosia humbertiana Cavaco;亨伯特青葙■☆
80421 Celosia humilis Suess. = Hermbstaedtia fleckii（Schinz）Baker et C. B. Clarke ■☆
80422 Celosia huttonii Mast.;赫顿氏青葙;Hutton Cockscomb, Taiwan Cockscomb ■
80423 Celosia intermedia Hochst. = Celosia anthelminthica Asch. ■☆
80424 Celosia intermedia Schinz = Hermbstaedtia spathulifolia（Engl.）Baker ■☆
80425 Celosia lanata L. = Aerva javanica（Burm. f.）Juss. ex Schult. ■☆
80426 Celosia lanata L. var. latifolia Vahl = Aerva javanica（Burm. f.）Juss. ex Schult. ■☆
80427 Celosia laxa Schumach. et Thonn. = Celosia trigyna L. ■☆
80428 Celosia leptostachya Benth.;细穗青葙■☆
80429 Celosia linearis（Schinz）Schinz = Hermbstaedtia linearis Schinz ■☆
80430 Celosia loandensis Baker;罗安达青葙■☆
80431 Celosia loandensis Baker var. angustifolia Suess. et Beyerle = Celosia chenopodiifolia Baker ■☆
80432 Celosia longistyla（C. B. Clarke）Suess. = Hermbstaedtia argenteiformis Schinz ■☆
80433 Celosia macrocarpa Lopr. = Celosia schweinfurthiana Schinz ■☆
80434 Celosia madagascariensis Poir. = Lagrezia madagascariensis（Poir.）Moq. ■☆
80435 Celosia margaritacea L. = Celosia argentea L. ■
80436 Celosia melanocarpos Poir. = Celosia trigyna L. ■☆
80437 Celosia minutiflora Baker = Celosia trigyna L. ■☆
80438 Celosia monsoniae（L. f.）Retz. = Trichuriella monsoniae（L. f.）Bennet ■
80439 Celosia monsoniae Retz. = Trichurus monsoniae（L. f.）C. C. Towns. ■
80440 Celosia namaensis Schinz = Hermbstaedtia fleckii（Schinz）Baker et C. B. Clarke ■☆
80441 Celosia nana Baker = Celosia bakeri C. C. Towns. ■☆
80442 Celosia nervosa C. C. Towns.;多脉青葙■☆
80443 Celosia nitida Vahl;西印度青葙;West Indian cockscomb ■☆
80444 Celosia nodiflora L. = Allmania nodiflora（L.）R. Br. ■
80445 Celosia oblongocarpa Schinz = Celosia schweinfurthiana Schinz ■☆
80446 Celosia odorata Burch. = Hermbstaedtia odorata（Burch. ex Moq.）T. Cooke ■☆
80447 Celosia odorata Burch. ex Moq. = Hermbstaedtia odorata（Burch. ex Moq.）T. Cooke ■☆
80448 Celosia palmeri S. Watson;帕默青葙;Palmer's Cockscomb ■☆
80449 Celosia pandurata Baker;琴形青葙■☆
80450 Celosia pandurata Baker f. trigyna Suess. = Celosia hastata Lopr. ■☆
80451 Celosia pandurata Baker var. elobata Suess. = Celosia globosa Schinz ■☆

80452 Celosia patentiloba C. C. Towns.;浅裂青葙■☆
80453 Celosia plumosa Burv.;羽状青葙;Plume Cockscomb ■☆
80454 Celosia polysperma Roxb. = Cladostachys polysperma (Roxb.) K. C. Kuan ■
80455 Celosia polysperma Roxb. = Deeringia polysperma (Roxb.) Moq. ■
80456 Celosia polystachia (Forssk.) C. C. Towns.;多穗青葙■☆
80457 Celosia populifolia Moq. = Celosia polystachia (Forssk.) C. C. Towns. ■☆
80458 Celosia populifolia Moq. var. pluriovulata Suess. = Celosia polystachia (Forssk.) C. C. Towns. ■☆
80459 Celosia pseudovirgata Schinz;假枝青葙■☆
80460 Celosia pyramidalis Burm. f. var. plumosa Hort. = Celosia cristata L. var. plumosa Hort. ■☆
80461 Celosia recurva Burch. = Hermbstaedtia odorata (Burch. ex Moq.) T. Cooke ■☆
80462 Celosia richardsiae C. C. Towns.;理查兹青葙■☆
80463 Celosia scabra (Schinz) Schinz = Hermbstaedtia scabra Schinz ■☆
80464 Celosia schinzii (C. B. Clarke) Suess. = Hermbstaedtia linearis Schinz ■☆
80465 Celosia schweinfurthiana Schinz;施韦青葙■☆
80466 Celosia schweinfurthiana Schinz var. sansibarensis? = Celosia schweinfurthiana Schinz ■☆
80467 Celosia semperflorens Baker = Celosia trigyna L. ■☆
80468 Celosia spathulifolia Engl. = Hermbstaedtia spathulifolia (Engl.) Baker ■☆
80469 Celosia spicata (Thouars) Spreng.;线叶青葙■☆
80470 Celosia stuhlmanniana Schinz;斯图尔曼青葙■☆
80471 Celosia swinhoei Hemsl.;海南青葙■
80472 Celosia swinhoei Hemsl. = Celosia argentea L. ■
80473 Celosia taitoensis Hayata;台湾青葙(台东青葙);Taiwan Cockscomb ■
80474 Celosia texana Scheele = Celosia nitida Vahl ■☆
80475 Celosia tonjesii Schinz = Hermbstaedtia argenteiformis Schinz ■☆
80476 Celosia trigyna L.;非洲青葙;Woolflower ■☆
80477 Celosia trigyna L. f. leptostachya (Benth.) Cavaco = Celosia leptostachya Benth. ■☆
80478 Celosia trigyna L. subvar. brevifilamentosa Suess. = Celosia trigyna L. ■☆
80479 Celosia trigyna L. subvar. convexa Suess. = Celosia trigyna L. ■☆
80480 Celosia trigyna L. var. adoensis Moq. = Celosia trigyna L. ■☆
80481 Celosia trigyna L. var. brevifilamentosa Suess. = Celosia trigyna L. ■☆
80482 Celosia trigyna L. var. convexa Suess. = Celosia trigyna L. ■☆
80483 Celosia trigyna L. var. fasciculiflora Moq. = Celosia trigyna L. ■☆
80484 Celosia trigyna L. var. longistyla Suess. = Celosia trigyna L. ■☆
80485 Celosia trigyna L. var. pauciflora Moq. = Celosia trigyna L. ■☆
80486 Celosia triloba E. Mey. ex Meisn. = Celosia trigyna L. ■☆
80487 Celosia vanderystii Schinz;范德青葙■☆
80488 Celosia welwitschii Schinz = Hermbstaedtia angolensis C. B. Clarke ■☆
80489 Celosiaceae Martinov = Amaranthaceae Juss.(保留科名)●■
80490 Celsa Vell. = ? Casearia Jacq. ●
80491 Celsia Boehm. = Bulbocodium L. ■☆
80492 Celsia Fabr. = Ornithogalum L. + Gagea Salisb. ■
80493 Celsia Heist. ex Fabr. = Ornithogalum L. + Gagea Salisb. ■
80494 Celsia Heist. ex Fabr. = Ornithogalum L. ■
80495 Celsia L. (1753);肖毛蕊花属;Celsia ■☆
80496 Celsia L. = Verbascum L. ■●
80497 Celsia acanthifolia Pau = Verbascum faurei (Murb.) Hub. -Mor. ■☆
80498 Celsia affinis A. Rich. = Celsia floccosa Benth. ■☆
80499 Celsia arbuscula A. Rich.;树状肖毛蕊花■☆
80500 Celsia arcturus (L.) Jacq. = Verbascum arcturus L. ■☆
80501 Celsia arcturus Jacq. = Verbascum arcturus L. ■☆
80502 Celsia ballii Batt. = Lappula patula (Lehm.) Gürke ■☆
80503 Celsia ballii Batt. var. brevipes? = Lappula patula (Lehm.) Gürke ■☆
80504 Celsia battandieri Murb. = Verbascum battandieri (Murb.) Hub. -Mor. ■☆
80505 Celsia betonicifolia Desf. = Verbascum betonicifolium (Desf.) Kuntze ■☆
80506 Celsia betonicifolia Desf. f. glabra Bég. = Verbascum capitis-viridis Hub. -Mor. ■☆
80507 Celsia brevipedicellata Engl. = Rhabdotosperma brevipedicellata (Engl.) Hartl ●☆
80508 Celsia brevipedicellata Engl. var. hararensis Murb. = Rhabdotosperma brevipedicellata (Engl.) Hartl ●☆
80509 Celsia brevipedicellata Engl. var. heterostemon Murb. = Rhabdotosperma brevipedicellata (Engl.) Hartl ●☆
80510 Celsia commixta Murb. = Verbascum commixtum (Murb.) Hub. -Mor. ■☆
80511 Celsia coromandeliana Vahl = Verbascum chinense (L.) Santapau ■
80512 Celsia coromandeliana Vahl = Verbascum coromandelianum (Vahl) Kuntze ■
80513 Celsia cretica L. = Verbascum creticum (L.) Cav. ■☆
80514 Celsia densifolia Hook. f. = Rhabdotosperma densifolia (Hook. f.) Hartl ●☆
80515 Celsia ellenbeckii Engl. = Rhabdotosperma scrophulariifolia (Hochst. ex A. Rich.) Hartl subsp. foliosa (Chiov.) Hartl ●☆
80516 Celsia faurei Murb. = Verbascum faurei (Murb.) Hub. -Mor. ■☆
80517 Celsia faurei Murb. var. acanthifolia (Pau) Maire = Verbascum faurei (Murb.) Hub. -Mor. subsp. acanthifolium (Pau) Benedi et J. M. Monts. ■☆
80518 Celsia floccosa Benth.;丛毛肖毛蕊花■☆
80519 Celsia floccosa Benth. = Verbascum benthamianum Hepper ■☆
80520 Celsia foliosa Chiov. = Rhabdotosperma scrophulariifolia (Hochst. ex A. Rich.) Hartl subsp. foliosa (Chiov.) Hartl ●☆
80521 Celsia heterophylla Desf.;互叶肖毛蕊花■☆
80522 Celsia insularis Murb. = Verbascum capitis-viridis Hub. -Mor. ■☆
80523 Celsia interrupta Engl. = Celsia scabrida V. Naray. ■☆
80524 Celsia interrupta Engl. var. pedunculosa = Celsia scabrida V. Naray. ■☆
80525 Celsia interrupta Fresen.;间断肖毛蕊花■☆
80526 Celsia interrupta Fresen. var. pedunculosa (Steud. et Hochst. ex Benth.) Vatke ex Engl. = Celsia pedunculosa Steud. et Hochst. ex Benth. ■☆
80527 Celsia interrupta Fresen. var. pubescens V. Naray. = Celsia pubescens (V. Naray.) Murb. ■☆
80528 Celsia interrupta Fresen. var. sudanica Murb. = Celsia sudanica (Murb.) Wickens ■☆
80529 Celsia keniensis Murb. = Rhabdotosperma keniensis (Murb.) Hartl ●☆

80530 Celsia laciniata Poir. = Verbascum erosum Cav. ■☆

80531 Celsia ledermannii Murb. = Rhabdotosperma ledermannii (Murb.) Hartl ●☆

80532 Celsia longirostris Murb. = Verbascum longirostre (Murb.) Hub. -Mor. ■☆

80533 Celsia longirostris Murb. var. antiatlantica Emb. = Verbascum longirostre (Murb.) Hub. -Mor. ■☆

80534 Celsia longirostris Murb. var. atlantica Maire = Verbascum longirostre (Murb.) Hub. -Mor. ■☆

80535 Celsia mairei Murb. = Verbascum mairei (Murb.) Hub. -Mor. ■☆

80536 Celsia maroccana Ball = Verbascum maroccanum (Ball) Hub. -Mor. ■☆

80537 Celsia masguindalii Pau = Verbascum masguindalii (Pau) Benedi et J. M. Monts. ■☆

80538 Celsia mauretanica (Pau) Sennen et Mauricio = Verbascum battandieri (Murb.) Hub. -Mor. ■☆

80539 Celsia micrantha Chiov.;小花肖毛蕊花■☆

80540 Celsia nudicaulis (Wydler) B. Fedtsch.;裸茎肖毛蕊花■☆

80541 Celsia oranensis Murb. = Verbascum oranense (Murb.) Dobignard ■☆

80542 Celsia orientalis L.;东方肖毛蕊花■☆

80543 Celsia parvifolia Engl. = Alectra orobanchoides Benth. ■☆

80544 Celsia pedunculosa Steud. et Hochst. ex Benth.;梗花肖毛蕊花■☆

80545 Celsia pedunculosa Steud. et Hochst. ex Benth. var. emarginata Murb.;微缺肖毛蕊花■☆

80546 Celsia pinnatisecta Batt. = Verbascum battandieri (Murb.) Hub. -Mor. ■☆

80547 Celsia pubescens (V. Naray.) Murb.;短柔毛肖毛蕊花■☆

80548 Celsia rhiphaea Murb. = Verbascum rhiphaeum (Murb.) Hub. -Mor. ■☆

80549 Celsia scabrida V. Naray.;微糙肖毛蕊花■☆

80550 Celsia scrophulariifolia Hochst. ex A. Rich. = Rhabdotosperma scrophulariifolia (Hochst. ex A. Rich.) Hartl ●☆

80551 Celsia scrophulariifolia Hochst. ex A. Rich. var. ellenbeckii Murb. = Rhabdotosperma scrophulariifolia (Hochst. ex A. Rich.) Hartl subsp. foliosa (Chiov.) Hartl ●☆

80552 Celsia sinuata Cav. = Verbascum erosum Cav. ■☆

80553 Celsia sinuata Cav. var. demnatensis Maire et Murb. = Verbascum erosum Cav. ■☆

80554 Celsia sudanica (Murb.) Wickens;苏丹肖毛蕊花■☆

80555 Celsia suworowiana K. Koch;苏沃罗夫肖毛蕊花■☆

80556 Celsia tibestica Quézel;提贝斯提肖毛蕊花■☆

80557 Celsia tomentosa Hochst. ex Benth. = Celsia floccosa Benth. ■☆

80558 Celsia valerianiifolia A. Rich.;缬草肖毛蕊花■☆

80559 Celsia zaianensis Murb. = Verbascum zaianense (Murb.) Hub. -Mor. ■☆

80560 Celtica F. M. Vázquez et Barkworth = Stipa L. ■

80561 Celtica gigantea (Link) F. M. Vàzquez et Barkworth = Macrochloa arenaria (Brot.) Kunth ■☆

80562 Celtica gigantea (Link) F. M. Vàzquez et Barkworth subsp. donyanae (F. M. Vàzquez et Devesa) F. M. Vàzquez = Macrochloa arenaria (Brot.) Kunth ■☆

80563 Celtica gigantea (Link) F. M. Vàzquez et Barkworth subsp. maroccana (Font Quer) F. M. Vàzquez et Barkworth = Macrochloa arenaria (Brot.) Kunth ■☆

80564 Celtidaceae Endl. = Ulmaceae Mirb.(保留科名)●

80565 Celtidaceae Link = Cannabaceae Martinov(保留科名)■

80566 Celtidaceae Link = Ulmaceae Mirb.(保留科名)●

80567 Celtidaceae Link;朴科●

80568 Celtidopsis Priemer = Celtis L. ●

80569 Celtis L.(1753);朴树属(朴属);Hackberries,Hackberry,Nettle Tree,Nettletree,Nettle-tree,Sugar Berry,Sugarberry ●

80570 Celtis adolfi-friderici Engl.;阿多朴;African Celtis ●☆

80571 Celtis africana Burm. f.;白朴;White Stinkwood ●☆

80572 Celtis amboinensis Willd. = Trema cannabina Lour. ●

80573 Celtis amphibola C. K. Schneid. = Celtis bungeana Blume ●

80574 Celtis angustifolia Lindl. = Trema angustifolia (Planch.) Blume ●

80575 Celtis aspera Lodd. ex G. Don;糙朴●☆

80576 Celtis aurantiaca Nakai = Celtis koraiensis Nakai ●

80577 Celtis australis L.;欧洲朴(地中海朴,南欧朴);European Hackberry,European Nettle Tree,European Nettle-tree,Lore-tree,Lotusberry,Nettle Tree,Nettle-tree,Oriental Hackberry,South European Hackberry,Southern Nettle-tree ●

80578 Celtis australis L. var. eriocarpa (Decne.) Hook. f. = Celtis eriocarpa Decne. ●☆

80579 Celtis bifida J. -F. Leroy;二裂朴●☆

80580 Celtis biondii Pamp.;紫弹朴(粗壳榔,黄果朴,毛果朴,牛筋树,沙楠子树,沙糖果,紫弹树);Biond Hackberry,Biond Nettletree,Hackberry ●

80581 Celtis biondii Pamp. f. holophylla (Nakai) Koji Ito = Celtis biondii Pamp. var. holophylla (Nakai) E. W. Ma ●

80582 Celtis biondii Pamp. var. cavaleriei (H. Lév.) C. K. Schneid.;阔叶紫弹朴(阔叶紫弹树);Cavalerie Biond Hackberry ●

80583 Celtis biondii Pamp. var. cavaleriei (H. Lév.) C. K. Schneid. = Celtis biondii Pamp. ●

80584 Celtis biondii Pamp. var. heterophylla (H. Lév.) C. K. Schneid.;异叶紫弹朴(异叶紫弹树);Diverseleaf Biond Hackberry ●

80585 Celtis biondii Pamp. var. heterophylla (H. Lév.) C. K. Schneid. = Celtis biondii Pamp. ●

80586 Celtis biondii Pamp. var. holophylla (Nakai) E. W. Ma;全缘叶紫弹朴(全缘叶紫弹树);Entire-leaved Hackberry ●

80587 Celtis biondii Pamp. var. holophylla (Nakai) E. W. Ma = Celtis biondii Pamp. ●

80588 Celtis bodinieri H. Lév. = Celtis sinensis Pers. ●

80589 Celtis boliviensis Planch. = Celtis brasiliensis (Gardner) Planch. ●☆

80590 Celtis boninensis Koidz.;小笠原朴●☆

80591 Celtis brasiliensis (Gardner) Planch.;巴西朴●☆

80592 Celtis brevipes S. Watson = Celtis reticulata Torr. ●

80593 Celtis brieyi De Wild.;刚果朴●☆

80594 Celtis brieyi De Wild. = Celtis tessmannii Rendle ●☆

80595 Celtis brownii Rendle = Celtis philippensis Blanco ●

80596 Celtis bungeana Blume;黑弹朴(白麻树,白麻子,棒棒木,棒棒树,棒子木,棒子树,黑弹木,黑弹树,木黄瓜树,朴树,土黄瓜树,小叶朴);Bunge Hackberry ●

80597 Celtis bungeana Blume var. deqinensis Xiang W. Li et G. Sh. Fan;德钦黑弹朴;Deqin Hackberry ●

80598 Celtis bungeana Blume var. deqinensis Xiang W. Li et G. Sh. Fan = Celtis bungeana Blume ●

80599 Celtis bungeana Blume var. heterophylla H. Lév. = Celtis biondii Pamp. ●

80600 Celtis bungeana Blume var. jessoensis Kudo;虾夷朴(狭叶朴);Yesso Hackberry ●☆

80601 Celtis bungeana Blume var. lanceolata E. W. Ma;披针叶黑弹朴

(披针叶小叶朴);Lanceleaf Hackberry ●

80602 Celtis bungeana Blume var. lanceolata E. W. Ma = Celtis bungeana Blume ●

80603 Celtis bungeana Blume var. pubipedicella G. H. Wang;毛梗小叶朴●

80604 Celtis bungeana Blume var. pubipedicella G. H. Wang = Celtis sinensis Pers. ●

80605 Celtis burmannii Planch. = Celtis africana Burm. f. ●☆

80606 Celtis canina Raf. = Celtis occidentalis L. ●☆

80607 Celtis caucasica Willd.;高加索朴;Caucasian Hackberry, Caucasian Nettle Tree ●☆

80608 Celtis caudata Hance = Cerasus pogonostyla (Maxim.) Te T. Yu et C. L. Li ●

80609 Celtis cavaleriei H. Lév. = Celtis biondii Pamp. ●

80610 Celtis cerasifera C. K. Schneid.;樱果朴(小果朴,樱桃朴);Cherryfruit Hackberry,Cherry-fruited Hackberry,Smallfruit Nettletree ●

80611 Celtis cercidifolia C. K. Schneid. = Celtis sinensis Pers. ●

80612 Celtis chekiangensis W. C. Cheng;天目朴;Chekiang Hackberry, Tianmushan Nettletree,Zhejiang Hackberry ●◇

80613 Celtis chichape (Wedd.) Miq.;玻利维亚朴●☆

80614 Celtis chinensis Bunge = Celtis bungeana Blume ●

80615 Celtis chuanchowensis F. P. Metcalf;泉州朴;Quanzhou Hackberry ●

80616 Celtis chuanchowensis F. P. Metcalf = Celtis biondii Pamp. ●

80617 Celtis cinnamonea Lindl. ex Planch. = Celtis timorensis Span. ●

80618 Celtis cinnamonifolia Nakai;樟叶朴;Cinnamonleaf Hackberry ●☆

80619 Celtis collinsae Craib = Celtis philippensis Blanco var. consimilis (Blume) Leroy ●

80620 Celtis collinsae Craib = Celtis philippensis Blanco var. wightii (Planch.) Soepadmo ●

80621 Celtis crassifolia Lam. = Celtis occidentalis L. ●☆

80622 Celtis crenata (Wedd.) Miq. = Celtis brasiliensis (Gardner) Planch. ●☆

80623 Celtis davidiana Carrière = Celtis bungeana Blume ●

80624 Celtis dioica S. Moore = Celtis gomphophylla Baker ●☆

80625 Celtis discolor Brongn. = Trema orientalis (L.) Blume ●

80626 Celtis douglasii Planch.;美西朴;Douglas Hackberry ●

80627 Celtis douglasii Planch. = Celtis reticulata Torr. ●

80628 Celtis durandii Engl.;杜氏朴;African Celtis ●☆

80629 Celtis durandii Engl. = Celtis gomphophylla Baker ●☆

80630 Celtis durandii Engl. var. ugandensis (Rendle) Rendle = Celtis gomphophylla Baker ●☆

80631 Celtis ehrenbergiana (Klotzsch) Liebm.;爱伦堡朴●☆

80632 Celtis emuyaca F. P. Metcalf = Celtis biondii Pamp. ●

80633 Celtis emuyaca F. P. Metcalf var. cuspidatophylla (F. P. Metcalf) C. P'ei = Celtis biondii Pamp. ●

80634 Celtis eriantha E. Mey. ex Planch. = Celtis africana Burm. f. ●☆

80635 Celtis eriocarpa Decne.;喜马拉雅朴●☆

80636 Celtis fengqingensis Hu ex E. W. Ma;凤庆朴;Fengqing Hackberry ●

80637 Celtis fengqingensis Hu ex E. W. Ma = Celtis tetrandra Roxb. ●

80638 Celtis fooningensis W. C. Cheng;富宁朴;Funing Hackberry ●

80639 Celtis formosana Hayata;石朴●

80640 Celtis formosana Hayata = Celtis tetrandra Roxb. ●

80641 Celtis franksiae N. E. Br. = Celtis mildbraedii Engl. ●☆

80642 Celtis georgiana Small = Celtis tenuifolia Nutt. ●☆

80643 Celtis glabrata Steven ex Planch.;滑朴(光叶朴,无毛朴)●☆

80644 Celtis gomphophylla Baker;棍叶朴●☆

80645 Celtis gongshanensis Xiang W. Li et G. Sh. Fan;贡山朴;Gongshan Hackberry ●

80646 Celtis gongshanensis Xiang W. Li et G. Sh. Fan = Celtis bungeana Blume ●

80647 Celtis guangxiensis Chun = Celtis biondii Pamp. ●

80648 Celtis guineensis Schumach. et Thonn. = Trema orientalis (L.) Blume ●

80649 Celtis guineensis Schumach. et Thonn. var. parvifolia? = Trema orientalis (L.) Blume ●

80650 Celtis helleri Small = Celtis lindheimeri Engelm. ex K. Koch ●☆

80651 Celtis holtzii Engl. = Celtis africana Burm. f. ●☆

80652 Celtis hunanensis Hand. -Mazz. = Celtis sinensis Pers. ●

80653 Celtis iguanaea (Jacq.) Sarg.;刺朴●☆

80654 Celtis ilicifolia Engl. = Populus ilicifolia (Engl.) Rouleau ●☆

80655 Celtis integrifolia Lam.;全缘朴;African False Elm, African Falseelm,African Nettle Tree,African Nettletree ●☆

80656 Celtis integrifolia Lam. = Celtis toka (Forssk.) Hepper et J. R. I. Wood ●☆

80657 Celtis jessoensis Koidz. = Celtis bungeana Blume var. jessoensis Kudo ●☆

80658 Celtis jessoensis Koidz. = Celtis bungeana Blume ●

80659 Celtis jessoensis Koidz. f. angustifolia (Nakai) Hayashi;狭叶虾夷朴●☆

80660 Celtis jessoensis Koidz. f. hashimotoi (Koidz.) Hayashi;端元朴●☆

80661 Celtis julianae C. K. Schneid.;珊瑚朴(沙棠子,棠壳子树);Coral Nettletree,Hackberry,Julian's Hackberry ●

80662 Celtis julianae C. K. Schneid. var. calvescens C. K. Schneid. = Celtis julianae C. K. Schneid. ●

80663 Celtis koidzumii Nakai;琉球朴●☆

80664 Celtis koraiensis Nakai;大叶朴(朝鲜朴,橙黄朴,大青榆,大叶白麻,山灰枣,石榆子);Bigleaf Nettletree, Korean Hackberry, Orangecoloured Hackberry ●

80665 Celtis koraiensis Nakai var. aurantiaca (Nakai) Kitag. = Celtis koraiensis Nakai ●

80666 Celtis kraussiana Bernh. = Celtis africana Burm. f. ●☆

80667 Celtis kraussiana Bernh. var. stolzii Peter = Celtis africana Burm. f. ●☆

80668 Celtis kunmingensis W. C. Cheng et T. Hong;昆明朴;Kunming Hackberry ●

80669 Celtis kunmingensis W. C. Cheng et T. Hong = Celtis tetrandra Roxb. ●

80670 Celtis labilis C. K. Schneid.;黄果朴;Yellowfruit Hackberry ●

80671 Celtis labilis C. K. Schneid. = Celtis sinensis Pers. ●

80672 Celtis lactea Sim = Morus mesozygia Stapf ●☆

80673 Celtis laevigata Willd.;密西西比朴(糖朴);American Sugarberry, Hackberry, Mississippi Hackberry, Southern Hackberry, Sugar Hackberry,Sugarberry ●☆

80674 Celtis laevigata Willd. var. anomala Sarg. = Celtis laevigata Willd. ●☆

80675 Celtis laevigata Willd. var. brachyphylla Sarg. = Celtis laevigata Willd. ●☆

80676 Celtis laevigata Willd. var. reticulata L. D. Benson = Celtis reticulata Torr. ●

80677 Celtis laevigata Willd. var. smallii (Beadle) Sarg. = Celtis laevigata Willd. ●☆

80678 Celtis laevigata Willd. var. texana Sarg. = Celtis laevigata Willd. ●☆

80679 Celtis lamarckiana Roem. et Schult. = Trema lamarckiana (Roem. et Schult.) Blume ●☆

80680 Celtis leveillei Nakai = Celtis biondii Pamp. ●

80681 Celtis leveillei Nakai f. holophylla Nakai = Celtis biondii Pamp. f. holophylla (Nakai) Koji Ito ●

80682 Celtis leveillei Nakai var. cuspidatophylla F. P. Metcalf = Celtis biondii Pamp. ●

80683 Celtis leveillei Nakai var. heterophylla (H. Lév.) Nakai = Celtis biondii Pamp. ●

80684 Celtis leveillei Nakai var. hirtifolia Hand. -Mazz. = Celtis biondii Pamp. ●

80685 Celtis leveillei Nakai var. holophylla (Nakai) E. W. Ma = Celtis biondii Pamp. ●

80686 Celtis leveillei Nakai var. holophylla Nakai = Celtis biondii Pamp. ●

80687 Celtis lindheimeri Engelm. ex K. Koch; 林氏朴; Lindheimer Hackberry, Palo Blanco ●☆

80688 Celtis luzonica Warb.; 吕宋朴; Luzon Hackberry ●☆

80689 Celtis madagascariensis Sattarian; 马岛朴●☆

80690 Celtis mairei H. Lév. = Celtis bungeana Blume ●

80691 Celtis micranthus (L.) Sw. = Trema micrantha (L.) Blume ●☆

80692 Celtis mildbraedii Engl.; 米氏朴●☆

80693 Celtis mississippiensis Bosc = Celtis laevigata Willd. ●☆

80694 Celtis morifolia Planch. = Celtis iguanaea (Jacq.) Sarg. ●☆

80695 Celtis nervosa Hemsl.; 石涩朴(朴树,石涩,小叶朴); Small-leaf Hackberry ●

80696 Celtis nervosa Hemsl. = Celtis sinensis Pers. ●

80697 Celtis oblongifolia Hu et W. C. Cheng; 长圆叶朴; Oblongleaf Hackberry ●

80698 Celtis occidentalis L.; 美洲朴(北美朴,环岑树,美国朴树,密西西比朴,荨麻树,西方朴,西朴,杂交榆); American Hackberry, Bastard Elm, Beaver Wood, Beaverwood, Common Hackberry, False Elm, Hackberry, Hoop Ash, Micocoulier occidental, Netde-tree, Nettle Tree, Nettletree, North American Hackberry, Northern Hackberry, Occidental Hackberry, Rim Ash, Sugarberry ●☆

80699 Celtis occidentalis L. var. canina (Raf.) Sarg. = Celtis occidentalis L. ●☆

80700 Celtis occidentalis L. var. crassifolia (Lam.) A. Gray; 大叶美洲朴; Bigleaf American Hackberry ●

80701 Celtis occidentalis L. var. crassifolia (Lam.) A. Gray = Celtis occidentalis L. ●☆

80702 Celtis occidentalis L. var. georgiana (Small) Ahles = Celtis tenuifolia Nutt. ●☆

80703 Celtis occidentalis L. var. pumila (Pursh) A. Gray = Celtis occidentalis L. ●☆

80704 Celtis occidentalis L. var. reticulata (Torr.) Sarg. = Celtis reticulata Torr. ●

80705 Celtis opegrapha Planch. = Celtis africana Burm. f. ●☆

80706 Celtis orientalis L. = Trema orientalis (L.) Blume ●

80707 Celtis pallida Torr.; 沙地朴; Desert Hackberry ●☆

80708 Celtis paniculata Planch.; 澳洲朴; Australian Celtis, Desert Hackberry, Granjeno, Spiny Hackberry ●☆

80709 Celtis philippensis Blanco; 菲律宾朴(大果油朴,大叶朴树,菲律宾朴树,假玉桂,香胶木,油朴); Lanyu Hackberry, Philippine Hackberry, Philippine Nettletree ●

80710 Celtis philippensis Blanco = Celtis timorensis Span. ●

80711 Celtis philippensis Blanco var. consimilis (Blume) Leroy = Celtis philippensis Blanco var. wightii (Planch.) Soepadmo ●

80712 Celtis philippensis Blanco var. wightii (Planch.) Soepadmo; 铁灵花; Collins Hackberry, Collins Nettletree ●

80713 Celtis philippensis Blanco var. wightii (Planch.) Soepadmo = Celtis philippensis Blanco var. consimilis (Blume) Leroy ●

80714 Celtis politoria Wall. = Trema politoria (Planch.) Blume ●☆

80715 Celtis polycarpa H. Lév. = Bischofia polycarpa (H. Lév.) Airy Shaw ●

80716 Celtis prantlii Priemer ex Engl. f. parviflora Hauman = Celtis philippensis Blanco ●

80717 Celtis pruniputaminea E. W. Ma; 杏核朴; Pruniputamen Hackberry ●

80718 Celtis pruniputaminea E. W. Ma = Celtis vandervoetiana C. K. Schneid. ●

80719 Celtis pubescens S. Y. Wang et C. L. Chang; 毛叶朴; Pubescent Hackberry ●

80720 Celtis pubescens S. Y. Wang et C. L. Chang = Celtis sinensis Pers. ●

80721 Celtis pubescens Spreng. = Celtis iguanaea (Jacq.) Sarg. ●☆

80722 Celtis pubescens Spreng. var. chichape (Wedd.) Baehni = Celtis chichape (Wedd.) Miq. ●☆

80723 Celtis pumila Pursh; 矮朴; Hackberry, Low Hackberry ●

80724 Celtis pumila Pursh = Celtis occidentalis L. ●☆

80725 Celtis pumila Pursh var. deamii Sarg. = Celtis occidentalis L. ●☆

80726 Celtis pumila Pursh var. georgiana (Small) Sarg. = Celtis tenuifolia Nutt. ●☆

80727 Celtis rendleana G. Taylor = Celtis philippensis Blanco ●

80728 Celtis reticulata Torr.; 网脉朴(网脉); Bastard Elm, Beaverwood, False Elm, Netleaf Hackberry, Net-leaf Hackberry, Nettle Tree, One-berry, Sugarberry, Western Hackberry ●

80729 Celtis reticulata Torr. var. vestita Sarg. = Celtis reticulata Torr. ●

80730 Celtis rhamnifolia C. Presl = Rhamnus prinoides L'Hér. ●☆

80731 Celtis rigida Blume = Trema orientalis (L.) Blume ●

80732 Celtis rockii Rehder = Celtis biondii Pamp. ●

80733 Celtis salvatiana C. K. Schneid.; 曼版朴; Salvat Hackberry ●

80734 Celtis salvatiana C. K. Schneid. = Celtis tetrandra Roxb. ●

80735 Celtis schippii Standl.; 施氏朴●☆

80736 Celtis scottelioides A. Chev. = Celtis philippensis Blanco ●

80737 Celtis shunningensis W. C. Cheng; 顺宁朴; Shunning Hackberry ●

80738 Celtis sinensis Dunn et Tutcher = Aphananthe aspera (Thunb. ex A. Murray) Planch. ●

80739 Celtis sinensis Pers.; 朴树(榎,毛叶朴,木树果子,粕仔,魄,朴,朴仔树,朴子树,青檀,沙朴,榽樜,湘朴,小叶朴,崖枣树,云贵朴,紫荆朴); Bodinier Hackberry, China Nettletree, Chinese Hackberry, Chinese Nettle-tree, Hunan Hackberry, Hunan Nettletree ●

80740 Celtis sinensis Pers. f. pendula Honda; 垂枝朴; Pendulous China Nettletree ●☆

80741 Celtis sinensis Pers. var. japonica (Planch.) Nakai; 日本朴●

80742 Celtis sinensis Pers. var. japonica (Planch.) Nakai = Celtis sinensis Pers. ●

80743 Celtis sinensis Pers. var. japonica (Planch.) Nakai f. longifolia Uyeki = Celtis sinensis Pers. var. japonica (Planch.) Nakai ●

80744 Celtis sinensis Pers. var. japonica (Planch.) Nakai f. pendula (Miyoshi) Honda = Celtis sinensis Pers. f. pendula Honda ●☆

80745 Celtis sinensis Pers. var. japonica (Planch.) Nakai f. rotundata (Nakai) Nakai = Celtis sinensis Pers. var. japonica (Planch.) Nakai ●

80746 Celtis sinensis Pers. var. japonica Nakai = Celtis sinensis Pers. ●

80747 Celtis sinensis Pers. var. pendula Miyoshi = Celtis sinensis Pers. f. pendula Honda ●☆

80748 Celtis smallii Beadle = Celtis laevigata Willd. ●☆

80749 Celtis soyauxii Engl. = Celtis mildbraedii Engl. ●☆

80750 Celtis soyauxii Pers.;帕松朴;Africa Nettletree, African Celtis, African Hackberry, Ohia ●☆

80751 Celtis spinosa Spreng. = Celtis iguanaea (Jacq.) Sarg. ●☆

80752 Celtis spinosa Spreng. var. pallida (Torr.) M. C. Johnst. = Celtis pallida Torr. ●☆

80753 Celtis spinosa Spreng. var. weddelliana (Planch.) Baehni = Celtis ehrenbergiana (Klotzsch) Liebm. ●☆

80754 Celtis taiyuanensis E. W. Ma;太原朴;Taiyuan Hackberry ●

80755 Celtis taiyuanensis E. W. Ma = Celtis cerasifera C. K. Schneid. ●

80756 Celtis tala Gillies ex Planch. = Celtis ehrenbergiana (Klotzsch) Liebm. ●☆

80757 Celtis tala Gillies ex Planch. f. velutina Herzog = Celtis chichape (Wedd.) Miq. ●☆

80758 Celtis tala Gillies ex Planch. var. chichape (Wedd.) Planch. = Celtis chichape (Wedd.) Miq. ●☆

80759 Celtis tala Gillies ex Planch. var. weddelliana Planch. = Celtis ehrenbergiana (Klotzsch) Liebm. ●☆

80760 Celtis tenuifolia Nutt.; 乔治亚朴; Dwarf Hackberry, Georgia Hackberry ●☆

80761 Celtis tenuifolia Nutt. var. georgiana (Small) Fernald et B. G. Schub. = Celtis tenuifolia Nutt. ●☆

80762 Celtis tenuifolia Nutt. var. soperi B. Boivin = Celtis tenuifolia Nutt. ●☆

80763 Celtis tessmannii Rendle;泰斯曼朴●☆

80764 Celtis tetrandra Roxb.;四蕊朴(滇朴,石博,石朴,台湾朴,台湾朴树); Fourpistil Nettletree, Fourstamen Hackberry, Taiwan Hackberry, Taiwan Nettletree, Tetradrous Hackberry ●

80765 Celtis tetrandra Roxb. f. pendula Y. Q. Zhu;垂枝四蕊朴(垂枝朴);Pendulous Fourstamen Hackberry ●

80766 Celtis tetrandra Roxb. subsp. sinensis (Pers.) Y. C. Tang = Celtis sinensis Pers. ●

80767 Celtis timorensis Span.;假玉桂(菲岛朴,桂叶朴,相思树,香粉木,玉桂朴,樟叶朴); Cinnamomum Hackberry, Cinnamonleaf Hackberry, Cinnamonleaf Nettletree, Cinnamon-leaved Hackberry, Stink Celtis, Timor Hackberry, Timor Nettletree ●

80768 Celtis toka (Forssk.) Hepper et J. R. I. Wood;托卡朴●☆

80769 Celtis tomentosa Roxb. = Trema tomentosa (Roxb.) H. Hara ●

80770 Celtis tournefortii Lam. = Celtis australis L. ●

80771 Celtis trichocarpa W. C. Cheng et E. W. Ma;毛果朴;Hackberry, Hairyfruit Hackberry ●

80772 Celtis trichocarpa W. C. Cheng et E. W. Ma = Celtis biondii Pamp. ●

80773 Celtis triflora (Ruiz ex Klotzsch) Miq. = Celtis iguanaea (Jacq.) Sarg. ●☆

80774 Celtis trinervia Roxb.;三脉朴●☆

80775 Celtis ugandensis Rendle = Celtis gomphophylla Baker ●☆

80776 Celtis usambarensis Engl. = Celtis mildbraedii Engl. ●☆

80777 Celtis vandervoetiana C. K. Schneid.; 西川朴; Vandervoet Hackberry, W. Sichuan Nettletree ●

80778 Celtis virgata Roxb. ex Wall. = Trema cannabina Lour. ●

80779 Celtis wangii Hu et W. C. Cheng;王氏朴;Wang Hackberry ●

80780 Celtis wightii Planch.;油朴;Wight Hackberry ●◇

80781 Celtis wightii Planch. = Celtis philippensis Blanco var. wightii (Planch.) Soepadmo ●

80782 Celtis wightii Planch. = Celtis philippensis Blanco ●

80783 Celtis wightii Planch. var. consimilis (Blume) Gagnep. = Celtis philippensis Blanco var. consimilis (Blume) Leroy ●

80784 Celtis wightii Planch. var. consimilis (Blume) Gagnep. = Celtis philippensis Blanco var. wightii (Planch.) Soepadmo ●

80785 Celtis williamsii Rusby = Celtis iguanaea (Jacq.) Sarg. ●☆

80786 Celtis xizangensis E. W. Ma;西藏朴;Xizang Hackberry ●

80787 Celtis xizangensis E. W. Ma = Celtis tetrandra Roxb. ●

80788 Celtis yangquanensis E. W. Ma;阳泉朴;Yangquan Hackberry ●

80789 Celtis yangquanensis E. W. Ma = Celtis bungeana Blume ●

80790 Celtis yui W. C. Cheng;俞氏朴;Yu Hackberry ●

80791 Celtis yunconensis W. C. Cheng; 银坑朴; Yinkeng Hackberry, Yuncon Hackberry ●

80792 Celtis yunnanensis C. K. Schneid. = Celtis tetrandra Roxb. ●

80793 Celtis yunnanensis E. W. Ma = Celtis bungeana Blume ●

80794 Celtis zenkeri Engl.;热非朴;Esa African Celtis ●☆

80795 Cembra (Spach) Opiz = Pinus L. ●

80796 Cembra Opiz = Pinus L. ●

80797 Cenarium L. = Canarium L. ●

80798 Cenarrhenes Labill. (1805);空雄龙眼属●☆

80799 Cenarrhenes nitida Labill.;空雄龙眼●☆

80800 Cenchrinaceae Link = Gramineae Juss. (保留科名)■●

80801 Cenchrinaceae Link = Poaceae Barnhart(保留科名)■●

80802 Cenchropsis Nash = Cenchrus L. ■

80803 Cenchrus L. (1753);蒺藜草属; Bur Grass, Hedgehog Grass, Sandbur ■

80804 Cenchrus aequiglumis Chiov. = Pennisetum pennisetiforme (Hochst. et Steud. ex Steud.) Wipff ■☆

80805 Cenchrus annularis Andersson = Cenchrus biflorus Roxb. ■☆

80806 Cenchrus australis R. Br.;南方蒺藜草■☆

80807 Cenchrus bambusoides Caro et E. A. Sanchez;竹叶蒺藜草■☆

80808 Cenchrus barbatus Schumach. = Cenchrus biflorus Roxb. ■☆

80809 Cenchrus biflorus Hook. f. = Cenchrus setigerus Vahl ■

80810 Cenchrus biflorus Roxb.;双花蒺藜草;Indian Sandburr ■☆

80811 Cenchrus brownii Roem. et Schult.;布氏蒺藜草■☆

80812 Cenchrus bulbifer Hochst. ex Boiss. = Cenchrus setigerus Vahl ■

80813 Cenchrus calyculatus Cav.;光梗蒺藜草(蒺藜草);Sandbur ■

80814 Cenchrus catharticus Delile = Cenchrus biflorus Roxb. ■☆

80815 Cenchrus ciliaris L.; 睫毛蒺藜草(水牛草); Buffel Grass, Buffelgrass ■

80816 Cenchrus ciliaris L. var. anachoreticum Chiov. = Cenchrus ciliaris L. ■

80817 Cenchrus ciliaris L. var. leptostachys (Leeke) Maire et Weiller = Cenchrus ciliaris L. ■

80818 Cenchrus ciliaris L. var. nubicus Fig. et De Not. = Cenchrus ciliaris L. ■

80819 Cenchrus ciliaris L. var. pallens (Leeke) Maire et Weiller = Cenchrus ciliaris L. ■

80820 Cenchrus ciliaris L. var. setigerus (Vahl) Maire et Weiller = Pennisetum setigerum (Vahl) Wipff ■☆

80821 Cenchrus echinatus L.;蒺藜草(刺蒺藜草);Common Sandbur, Southern Sandbur, Spiny Sandbur ■

80822 Cenchrus echinoides Wight ex Steud. = Cenchrus pennisetiformis Hochst. et Steud. ex Steud. ■☆

80823 Cenchrus geniculatus Thunb. = Pennisetum thunbergii Kunth ■☆

80824 Cenchrus granularis L. = Hackelochloa granularis (L.) Kuntze ■

80825 Cenchrus hystrix Fig. et De Not. = Cenchrus prieurii (Kunth) Maire ■☆

80826 Cenchrus incertus M. A. Curtis;可疑蒺藜草(光梗蒺藜草)■
80827 Cenchrus lappaceus L. =Centotheca lappacea (L.) Desv. ■
80828 Cenchrus leptacanthus A. Camus =Cenchrus biflorus Roxb. ■☆
80829 Cenchrus longifolius Hochst. ex Steud. =Cenchrus ciliaris L. ■
80830 Cenchrus longispinus (Hack.) Fernald;田野蒺藜草;Field Sandbur,Innocent-weed,Mat Sandbur,Sandbur ■☆
80831 Cenchrus macrostachyus Hochst. ex Steud. =Cenchrus prieurii (Kunth) Maire ■☆
80832 Cenchrus mitis Andersson;柔软蒺藜草■☆
80833 Cenchrus montanus Nees =Cenchrus setigerus Vahl ■
80834 Cenchrus montanus Nees ex Royle =Cenchrus setigerus Vahl ■
80835 Cenchrus ovatus Lam. ex Poir.;卵形蒺藜草■☆
80836 Cenchrus parviflorus Poir. =Setaria parviflora (Poir.) Kerguélen ■
80837 Cenchrus pauciflorus Benth.;少花蒺藜草;Mat Sandbur ■
80838 Cenchrus pauciflorus Benth. =Cenchrus incertus M. A. Curtis ■
80839 Cenchrus pennisetiformis Hochst. et Steud. =Pennisetum pennisetiforme (Hochst. et Steud. ex Steud.) Wipff ■☆
80840 Cenchrus pennisetiformis Hochst. et Steud. ex Steud.;羽毛蒺藜草■☆
80841 Cenchrus pennisetiformis Hochst. et Steud. ex Steud. subsp. glabrata Chrtek et Osb.-Kos.;光滑羽毛蒺藜草■☆
80842 Cenchrus perinvolucratus Stapf et C. E. Hubb. =Cenchrus biflorus Roxb. ■☆
80843 Cenchrus prieurii (Kunth) Maire;普氏蒺藜草■☆
80844 Cenchrus pubescens Steud. =Anthephora pubescens Nees ■☆
80845 Cenchrus purpurascens Thunb. =Pennisetum alopecuroides (L.) Spreng. ■
80846 Cenchrus pycnostachyus Steud. =Pennisetum glaucum (L.) R. Br. ■
80847 Cenchrus racemosus L. =Tragus racemosus (L.) All. ■
80848 Cenchrus racemosus Saut. =Tragus berteronianus Schult. et Schult. f. ■
80849 Cenchrus rigidifolius Fig. et De Not. =Pennisetum pennisetiforme (Hochst. et Steud. ex Steud.) Wipff ■☆
80850 Cenchrus setigerus Vahl;倒刺蒺藜草;Birdwood Grass ■
80851 Cenchrus setosus Sw. =Pennisetum polystachion (L.) Schult. ■
80852 Cenchrus setosus Sw. =Pennisetum setosum (Sw.) Rich. ■
80853 Cenchrus somalensis Clayton =Pennisetum somalensis (Clayton) Wipff ■☆
80854 Cenchrus tomentosus Poir.;绒毛蒺藜草■☆
80855 Cenchrus tribuloides L.;沙丘蒺藜草;Bur-grass,Dune Sandbur, Sandbur ■
80856 Cenchrus uniflorus Ehrenb. ex Boiss. =Cenchrus setigerus Vahl ■
80857 Cenekia Opiz =Campanula L. ■●
80858 Cenesmon Gagnep. =Cnesmone Blume ●
80859 Cenesmon hainanense Merr. et Chun =Cnesmone hainanensis (Merr. et Chun) Croizat ●
80860 Cenesmon tonkinensis Gagnep. =Cnesmone tonkinensis (Gagnep.) Croizat ●
80861 Cenia Comm. ex Juss. =Cotula L. ■
80862 Cenia Comm. ex Juss. =Lancisia Fabr. ■
80863 Cenia albovillosa S. Moore =Cotula microglossa (DC.) O. Hoffm. et Kuntze ex Kuntze ■☆
80864 Cenia discoidea Less. =Cotula nudicaulis Thunb. ■☆
80865 Cenia duckittiae L. Bolus =Cotula duckittiae (L. Bolus) K. Bremer et Humphries ■☆
80866 Cenia expansa Compton =Cotula duckittiae (L. Bolus) K. Bremer et Humphries ■☆
80867 Cenia microglossa DC. =Cotula microglossa (DC.) O. Hoffm. et Kuntze ex Kuntze ■☆
80868 Cenia sericea (L. f.) DC. =Cotula sericea L. f. ■☆
80869 Cenia turbinata (L.) Pers. =Cotula turbinata L. ■☆
80870 Cennarrhenes Steud. =Cenarrhenes Labill. ●☆
80871 Cenocentrum Gagnep. (1909);大萼葵属;Cenocentrum ■●
80872 Cenocentrum tonkinense Gagnep.;大萼葵;Tonkin Cenocentrum ■
80873 Cenocline C. Koch =Cotula L. ■
80874 Cenocline K. Koch =Matricaria L. ■
80875 Cenolophium W. D. J. Koch =Cenolophium W. D. J. Koch ex DC. ■
80876 Cenolophium W. D. J. Koch ex DC. =Cenolophium W. D. J. Koch ■
80877 Cenolophium W. D. J. Koch (1824-1825);空棱芹属;Cenolophium ■
80878 Cenolophium chinense M. Hiroe =Cyclorhiza peucedanifolia (Franch.) Constance ■
80879 Cenolophium denudatum (Hornem.) Tutin;空棱芹;Fischer Cenolophium,Toothshape Cenolophium ■
80880 Cenolophium fischeri (Spreng.) W. D. J. Koch =Cenolophium denudatum (Hornem.) Tutin ■
80881 Cenolophium fischeri (Spreng.) W. D. J. Koch ex DC. =Cenolophium denudatum (Hornem.) Tutin ■
80882 Cenolophon Blume =Alpinia Roxb.(保留属名)■
80883 Cenopleurum Post et Kuntze =Ferula L. ■
80884 Cenopleurum Post et Kuntze =Kenopleurum P. Candargy ■
80885 Cenostigma Tul. (1843);空柱豆属(星毛苏木属)■☆
80886 Cenostigma gardnerianum Tul.;空柱豆■☆
80887 Cenothus Raf. =Ceanothus L. ●☆
80888 Cenotis Raf. =Cenotus Raf. ■●
80889 Cenotus Raf. =Caenotus Raf. ■●
80890 Cenotus Raf. =Erigeron L. ■●
80891 Censcus Gaertn. =Catunaregam Wolf ●
80892 Centaurea L. (1753)(保留属名);矢车菊属;Bachelor's Button, Blue Bottle, Bluebottle, Bluet, Centaurea, Centaury Knapweed,Cornflower,Knapweed,Star Thistle ●■
80893 Centaurea × moncktonii C. E. Britton;草地矢车菊;Hybrid Knapweed,Meadow Knapweed,Protean Knapweed ■☆
80894 Centaurea abbreviata (K. Koch) Hand.-Mazz.;缩短矢车菊■☆
80895 Centaurea abchasica (Albov) Sosn.;阿伯哈斯矢车菊■☆
80896 Centaurea abdelkaderi Sennen et Mauricio =Centaurea pubescens Willd. ■☆
80897 Centaurea abnormis Czerep.;异常矢车菊■☆
80898 Centaurea abyssinica (Boiss.) Sch. Bip. =Plectocephalus varians (A. Rich.) C. Jeffrey ex Cufod. ■☆
80899 Centaurea acaulis L. subsp. balansae (Boiss. et Reut.) Murb. =Centaurea balansae Boiss. et Reut. ■☆
80900 Centaurea acaulis L. subsp. boissieri Maire =Centaurea oranensis Greuter et M. V. Agab. ■☆
80901 Centaurea acaulis L. var. balansae (Boiss. et Reut.) Batt. =Centaurea balansae Boiss. et Reut. ■☆
80902 Centaurea acmophylla Boiss.;砧叶矢车菊■☆
80903 Centaurea acutangula Boiss. et Reut.;棱角矢车菊■☆
80904 Centaurea adamii Willd.;亚当斯矢车菊■☆
80905 Centaurea adpressa L. var. angustata Ledeb. =Centaurea adpressa Ledeb. ex Steud. ■
80906 Centaurea adpressa Ledeb. =Centaurea adpressa Ledeb. ex

Steud. ■

80907 Centaurea adpressa Ledeb. ex Steud.;糙叶矢车菊(压密矢车菊);Ruggeed Centaurea,Scabrous Centaurea ■

80908 Centaurea aegyptiaca L.;埃及矢车菊■☆

80909 Centaurea aemulans Klokov;匹敌矢车菊■☆

80910 Centaurea africana Lam. = Rhaponticoides africana (Lam.) M. V. Agab. et Greuter ■☆

80911 Centaurea africana Lam. var. tagana (Brot.) Maire = Rhaponticoides africana (Lam.) M. V. Agab. et Greuter ■☆

80912 Centaurea aggregata Fisch. et C. A. Mey.;密集矢车菊■☆

80913 Centaurea alaica Iljin;阿赖矢车菊■☆

80914 Centaurea alba L. var. mauritanica Batt. = Centaurea djebel-amouri Greuter ■☆

80915 Centaurea albida K. Koch;白矢车菊■☆

80916 Centaurea albispina (Bunge) B. Fedtsch. = Schischkinia albispina (Bunge) Iljin ■

80917 Centaurea albovii Sosn.;阿尔保夫矢车菊■☆

80918 Centaurea alexandri Bordz.;阿莱矢车菊■☆

80919 Centaurea alexandrina Delile;亚历山大矢车菊■☆

80920 Centaurea algeriensis Coss. = Centaurea diluta Aiton subsp. algeriensis (Coss. et Durieu) Maire ■☆

80921 Centaurea alpina L.;高山矢车菊■☆

80922 Centaurea amara L.;苦矢车菊■☆

80923 Centaurea amara L. subsp. angustifolia Gremli = Centaurea vinyalsii Sennen ■☆

80924 Centaurea amara L. subsp. ropalon (Pomel) Arènes = Centaurea ropalon Pomel ■☆

80925 Centaurea amara L. var. mairei Arènes = Centaurea vinyalsii Sennen ■☆

80926 Centaurea amblyolepis Ledeb.;钝鳞矢车菊■☆

80927 Centaurea americana Nutt. = Plectocephalus americanus (Nutt.) D. Don ■☆

80928 Centaurea ammocyanus Boiss.;蓝色矢车菊■☆

80929 Centaurea amourensis Pomel = Centaurea pubescens Willd. subsp. amourensis (Pomel) Batt. ■☆

80930 Centaurea androssovii Iljin;安德罗矢车菊■☆

80931 Centaurea angulosa Pomel = Centaurea resupinata Coss. subsp. spachii (Willk.) Fern. Casas et Susanna ■☆

80932 Centaurea apiculata Ledeb.;细尖矢车菊■☆

80933 Centaurea appendicata Klokov;附属物矢车菊■☆

80934 Centaurea arborea Webb et Berthel. = Cheirolophus arboreus (Webb et Berthel.) Holub ●☆

80935 Centaurea arenaria M. Bieb. ex Willd.;沙生矢车菊■☆

80936 Centaurea argentea L. = Centaurea gymnocarpa Moris et De Not. ■☆

80937 Centaurea arguta Nees = Cheirolophus argutus (Nees) Holub ●☆

80938 Centaurea armena Boiss.;亚美尼亚矢车菊■☆

80939 Centaurea aspera L.;粗糙矢车菊;Rough Star Thistle,Rough Star-thistle ■☆

80940 Centaurea aspera L. subsp. gentilii (Braun-Blanq.) Dobignard;让蒂矢车菊■☆

80941 Centaurea aspera L. var. parcespinosa Sennen = Centaurea aspera L. ■☆

80942 Centaurea aspera L. var. praetermissa DC. = Centaurea aspera L. ■☆

80943 Centaurea aspera L. var. subinermis DC. = Centaurea aspera L. ■☆

80944 Centaurea atakorensis A. Chev. = Centaurea praecox Oliv. et Hiern ■☆

80945 Centaurea atlantica (Font Quer) Figuerola, Peris et Stübing = Centaurea boissieri DC. subsp. atlantica (Font Quer) Blanca ■☆

80946 Centaurea atlantica (Font Quer) Figuerola, Peris et Stübing subsp. calvescens (Maire) Figuerola, Peris et Stübing = Centaurea boissieri DC. ■☆

80947 Centaurea atlantis Maire et Weiller;亚特兰大矢车菊■☆

80948 Centaurea atriplicifolia (Torr.) Matsum. = Synurus deltoides (Aiton) Nakai ■

80949 Centaurea atriplicifolia (Trevissan) Matsum. = Synurus deltoides (Aiton) Nakai ■

80950 Centaurea atropurpurea Waldst. et Kit.;深紫矢车菊■☆

80951 Centaurea austriaca Willd. = Centaurea phrygia L. ■☆

80952 Centaurea austromaroccana (Förther et Podlech) Gómiz = Centaurea pungens Pomel subsp. austromaroccana Förther et Podlech ■☆

80953 Centaurea aylmeri Baker = Volutaria abyssinica (A. Rich.) C. Jeffrey subsp. aylmeri (Baker) Wagenitz ■☆

80954 Centaurea babylonica (L.) L.;叙利亚矢车菊;Syrian Knapweed ■☆

80955 Centaurea babylonica M. Bieb. = Centaurea babylonica (L.) L. ■☆

80956 Centaurea balansae Boiss. et Reut.;巴兰萨矢车菊■☆

80957 Centaurea balsamita Lam.;香膏矢车菊(刺冠菊)■☆

80958 Centaurea barbara Pomel = Centaurea nicaeensis All. ■☆

80959 Centaurea belangeriana (DC.) Stapf;贝朗热矢车菊■☆

80960 Centaurea bella Trautv.;美丽矢车菊■☆

80961 Centaurea benedicta (L.) L. = Cnicus benedictus L. ■

80962 Centaurea benedicta L. = Cnicus benedictus L. ■

80963 Centaurea benoistii Humbert = Cheirolophus benoistii (Humbert) Holub ■☆

80964 Centaurea berbeyi (Albov) Sosn.;贝氏矢车菊■☆

80965 Centaurea besseriana DC.;白氏矢车菊;Besser Centaurea ■☆

80966 Centaurea beyrichii?;山地矢车菊;Mountain Pink ■☆

80967 Centaurea bicolor K. Koch;二色矢车菊■☆

80968 Centaurea biebersteinii DC.;毕氏矢车菊;Spotted Knapweed ■☆

80969 Centaurea biebersteinii DC. = Centaurea stoebe L. ■☆

80970 Centaurea boissieri DC.;布瓦西耶矢车菊■☆

80971 Centaurea boissieri DC. subsp. atlantica (Font Quer) Blanca;大西洋矢车菊■☆

80972 Centaurea boissieri DC. subsp. calvescens (Maire) Greuter = Centaurea boissieri DC. subsp. atlantica (Font Quer) Blanca ■☆

80973 Centaurea boissieri DC. subsp. transmalvana (Emb. et Maire) Breitw. et Podlech;外马尔文矢车菊■☆

80974 Centaurea boissieri DC. var. atlantica Font Quer = Centaurea boissieri DC. subsp. atlantica (Font Quer) Blanca ■☆

80975 Centaurea boissieri DC. var. calvescens (Maire) Blanca = Centaurea boissieri DC. ■☆

80976 Centaurea boissieri DC. var. rifana Emb. et Maire = Centaurea boissieri DC. ■☆

80977 Centaurea boissieri DC. var. transmalvana Emb. et Maire = Centaurea boissieri DC. subsp. transmalvana (Emb. et Maire) Breitw. et Podlech ■☆

80978 Centaurea borysthenica Gruner;第聂伯矢车菊■☆

80979 Centaurea bovina?;草原矢车菊;Pasture Knapweed ■☆

80980 Centaurea breviceps Iljin;短梗矢车菊■☆

80981 Centaurea calcitrapa L.;星苞矢车菊(卡西矢车菊);Caltra,

Caltrap, Caltrops, Chausse-trappe, Jersey Thistle, Purple Star-thistle, Red Star Thistle, Red Star-thistle, Star Bur, Star Thistle, Star-bur, Star-thistle ■☆

80982 Centaurea calcitrapoides DC.;拟星苞矢车菊;Smallhead Star-thistle ■☆

80983 Centaurea calocephala Willd.;美头矢车菊■☆

80984 Centaurea calycosa;多萼矢车菊;Buckley Centaury, Mountain Pink ■☆

80985 Centaurea canariensis Huber ex Regel;加那利矢车菊(加拿列矢车菊)■☆

80986 Centaurea canariensis Willd. = Cheirolophus canariensis (Willd.) Holub ●☆

80987 Centaurea canariensis Willd. var. subexpinnata Burch. = Cheirolophus canariensis (Willd.) Holub ●☆

80988 Centaurea candidissima Lam. = Centaurea cineraria L. ■☆

80989 Centaurea capillata L.;头状矢车菊;Dumpy Centaury ■☆

80990 Centaurea caprina Steven;山羊矢车菊■☆

80991 Centaurea carbonata Klokov;煤色矢车菊■☆

80992 Centaurea carbonellii Sennen et Mauricio = Centaurea pubescens Willd. ■☆

80993 Centaurea carduiformis DC.;飞廉矢车菊■☆

80994 Centaurea cardunculus Pall. = Serratula carduncula (Pall.) Schischk. ■

80995 Centaurea carpatica (Porcius) Porcius;卡尔帕特矢车菊■☆

80996 Centaurea carthamoides (Willd.) Benth. = Rhaponticum carthamoides (Willd.) Iljin ■

80997 Centaurea carthamoides Benth. = Stemmacantha carthamoides (Willd.) Dittrich ■

80998 Centaurea caspica Grossh.;里海矢车菊■☆

80999 Centaurea chamaerhaponticum Ball = Rhaponticum acaule (L.) DC. ■☆

81000 Centaurea chamaerhaponticum Ball var. ochreuleuca Emb. = Rhaponticum acaule (L.) DC. ■☆

81001 Centaurea chamaerhaponticum Ball var. purpurea Emb. et Maire = Rhaponticum acaule (L.) DC. ■☆

81002 Centaurea cheiranthifolia Willd.;桂竹香叶矢车菊■☆

81003 Centaurea choulettiana Pomel = Centaurea balansae Boiss. et Reut. ■☆

81004 Centaurea cineraria L.;灰叶矢车菊;Dusty Miller ■☆

81005 Centaurea circassica (Albov) Sosn.;切尔卡西亚矢车菊■☆

81006 Centaurea ciscaucasica Sosn.;北高加索矢车菊■☆

81007 Centaurea claryi Debeaux = Centaurea pullata L. ■☆

81008 Centaurea clementei Boiss.;克莱门特矢车菊■☆

81009 Centaurea colchica (Sosn.) Sosn.;黑海矢车菊■☆

81010 Centaurea comperiana Steven;康氏矢车菊;Comper Centaurea ■☆

81011 Centaurea conocephala Bolle;束头矢车菊■☆

81012 Centaurea consolida L.;疏花矢车菊■☆

81013 Centaurea contracta Viv. = Centaurea glomerata Vahl ■☆

81014 Centaurea coriacea Waldst. et Kit.;革矢车菊■☆

81015 Centaurea cossoniana Ball = Rhaponticum cossonianum (Ball) Greuter ■☆

81016 Centaurea cossoniana Batt. = Centaurea granatensis Boiss. subsp. battandieri (Hochr.) Maire ■☆

81017 Centaurea crupina L. = Crupina vulgaris Pers. ex Cass. ■

81018 Centaurea crupinoides Desf. = Volutaria crupinoides (Desf.) Cass. ex Maire ■☆

81019 Centaurea cyanocephala Velen. = Centaurea cyanus L. ■

81020 Centaurea cyanus L.;矢车菊(车轮花,苦蒿,蓝芙蓉);Bachelor's Buttons, Bachelor's Button, Bachelor's-button, Barbeau, Blaver, Blaverole, Blawort, Bleuet, Blow-ball, Blow-flower, Blue Blow, Blue Bonnets, Blue Bottle, Blue Bow, Blue Buttons, Blue Cap, Blue Jack, Blue Poppy, Blueblaw, Bluebonnets, Bluebottle, Blue-bottle, Blue-poppy, Bluet, Blunt-sickle, Bobby's Butions, Bobby's Buttons, Bottle-of-all-sorts, Brush, Brushes, Centaurea, Centaury, Corn Blinks, Corn Bluebottle, Corn Bottle, Corn Centaury, Corn Flower, Corn Pinks, Corn-bottle, Cornflower, Cornflower Bachelor's Button, Cornflower Centaurea, Cuckoo Hood, Cuckoo-hood, Devil-in-a-bush, Devil-in-thebush, Flake Flower, Garden Cornflower, Gold Knop, Horse Knop, Hurt Sickle, Hurtsickle, Hurt-sickle, Knobweed, Knopweed, Knotweed, Ladder Love, Miller's Delight, Pincushion, Thimbles, Thumble, Witch Bells, Witch's Bells, Witches' Thimbles ■

81021 Centaurea cyrenaica Bég. et Vacc.;昔兰尼矢车菊■☆

81022 Centaurea cyrenaica Bég. et Vacc. var. cavarae Grande = Centaurea cyrenaica Bég. et Vacc. ■☆

81023 Centaurea daghestanica (Lipsky) Czerep.;达赫斯坦矢车菊■☆

81024 Centaurea dealbata Willd.;白毛矢车菊(羽裂矢车菊);Persian Centaurea, Persian Cornflower ■☆

81025 Centaurea dealbata Willd. 'Steenbergii';斯腾伯格羽裂矢车菊■☆

81026 Centaurea debeauxii Gren. et Godr.;德氏矢车菊;Meadow Knapweed ■☆

81027 Centaurea debeauxii Gren. et Godr. subsp. thuillieri Dostál = Centaurea × moncktonii C. E. Britton ■☆

81028 Centaurea decipiens Thuill.;伪矢车菊■☆

81029 Centaurea declinata M. Bieb.;外折矢车菊■☆

81030 Centaurea degenii Sennen = Centaurea resupinata Coss. subsp. degenii (Sennen) Fern. Casas et Susanna ■☆

81031 Centaurea delicatula Breitw. et Podlech;姣美矢车菊■☆

81032 Centaurea delilei Godr. = Centaurea furfuracea Coss. et Durieu ■☆

81033 Centaurea depressa M. Bieb.;扁矢车菊(凹陷矢车菊);Iranian Knapweed, Low Cornflower ■☆

81034 Centaurea diffusa Lam.;铺散矢车菊(披散矢车菊);Diffuse Knapweed, Prostrate Centaurea, Spreading Knapweed, Tumble Knapweed, White Knapweed ■

81035 Centaurea diluta Aiton;北飞矢车菊;Knapweed, Lesser Star-thisde, North African Knapweed, Star Thistle ■☆

81036 Centaurea diluta Aiton subsp. algeriensis (Coss. et Durieu) Maire;阿尔及利亚矢车菊■☆

81037 Centaurea diluta Aiton var. micracantha Maire = Centaurea diluta Aiton ■☆

81038 Centaurea dimitriewiae Sosn.;迪米矢车菊■☆

81039 Centaurea dimorpha Viv.;二型矢车菊■☆

81040 Centaurea dimorpha Viv. var. kralickii Batt. et Trab. = Centaurea dimorpha Viv. ■☆

81041 Centaurea dimorpha Viv. var. laevibracteata Hochr. = Centaurea dimorpha Viv. ■☆

81042 Centaurea dimorpha Viv. var. major Pamp. = Centaurea dimorpha Viv. ■☆

81043 Centaurea dissecta Ten. var. orthoacantha (Pau et Font Quer) Maire = Centaurea monticola DC. ■☆

81044 Centaurea dissecta Ten. var. parlatoris (Heldr.) Quézel et Santa = Centaurea resupinata Coss. subsp. spachii (Willk.) Fern. Casas et Susanna ■☆

81045 Centaurea dissecta Ten. var. perplexans Emb. et Maire =

Centaurea resupinata Coss. subsp. simulans (Emb. et Maire) Breitw. et Podlech ■☆
81046 Centaurea dissecta Ten. var. vesceritensis (Boiss. et Reut.) Quézel et Santa = Centaurea resupinata Coss. subsp. spachii (Willk.) Fern. Casas et Susanna ■☆
81047 Centaurea djebel-amouri Greuter;德阿矢车菊■☆
81048 Centaurea drucei C. E. Britton = Centaurea × moncktonii C. E. Britton ■☆
81049 Centaurea dschungarica C. Shih;准噶尔矢车菊(多裂矢车菊);Dschungar Centaurea,Dzungar Centaurea ■
81050 Centaurea dubia Suter subsp. nigrescens (Willd.) Hayek = Centaurea nigrescens Willd. ■
81051 Centaurea dubia Suter subsp. vochinensis (Bernh. ex Rchb.) Hayek = Centaurea nigrescens Willd. ■
81052 Centaurea dubjanskyi Iljin;杜氏矢车菊■☆
81053 Centaurea ducellieri Batt. et Trab.;迪塞利耶矢车菊■☆
81054 Centaurea dufourii (Dostál) Blanca = Centaurea resupinata Coss. subsp. dufourii (Dostál) Greuter ■☆
81055 Centaurea dufourii (Dostál) Blanca = Centaurea resupinata Coss. subsp. spachii (Willk.) Fern. Casas et Susanna ■☆
81056 Centaurea dufourii (Dostál) Blanca subsp. lagascae (Nyman) Blanca = Centaurea resupinata Coss. subsp. spachii (Willk.) Fern. Casas et Susanna ■☆
81057 Centaurea dufourii (Dostál) Blanca subsp. rifana (Emb. et Maire) Blanca = Centaurea resupinata Coss. subsp. degenii (Sennen) Fern. Casas et Susanna ■☆
81058 Centaurea elongata Schousb. = Centaurea diluta Aiton ■☆
81059 Centaurea ericeticola Font Quer = Rhaponticum longifolium (Hoffmanns. et Link) Dittrich subsp. ericeticola (Font Quer) Greuter ■☆
81060 Centaurea eriocephala Boiss. = Centaurea dimorpha Viv. ■☆
81061 Centaurea eriophora L.;绵毛矢车菊;Wild Sandheath ■☆
81062 Centaurea eriosiphon Emb. et Maire = Rhaponticoides eriosiphon (Emb. et Maire) M. V. Agab. et Greuter ■☆
81063 Centaurea erivanensis (Lipsky) Bordz.;埃里温矢车菊■☆
81064 Centaurea eryngioides Lam.;刺芹矢车菊■☆
81065 Centaurea erythraea Raf.;普红矢车菊;Ball-of-the-earth, Bitter Herb, Centaury, Centaury Gentian, Centre-of-the-sun, Christ's Ladder, Cristaldre, Earthgall, Fellwort, Feverfew, Feverfuge, Feverwort, Fieldwort, Gall-of-the Earth, Gall-of-the-earth, Gentian, Hurdreve, Lesser Centaury, Lesser Churmel, Lesser Curmel, Little Centaury, Mountain Flax, Red Gentian, Sanctuary, Senna-pods, Spikenard ■☆
81066 Centaurea fatoui Gómiz = Centaurea pubescens Willd. subsp. amourensis (Pomel) Batt. ■☆
81067 Centaurea faurei Sennen et Mauricio = Centaurea pubescens Willd. ■☆
81068 Centaurea ferox Desf.;多刺矢车菊■☆
81069 Centaurea fischeri Willd.;费氏矢车菊■☆
81070 Centaurea fontanesii (Spach) Durieu = Centaurea sphaerocephala L. ■☆
81071 Centaurea fragilis Durieu;脆矢车菊■☆
81072 Centaurea fragilis Durieu var. integrifolia Ball = Centaurea gentilii Braun-Blanq. et Maire ■☆
81073 Centaurea fragilis Durieu var. subinermis Pau = Centaurea fragilis Durieu ■☆
81074 Centaurea furfuracea Coss. et Durieu;糠皮矢车菊■☆
81075 Centaurea fuscescens Pomel = Centaurea nicaeensis All. ■☆
81076 Centaurea fuscimarginata (K. Koch) Juz.;棕边矢车菊■☆
81077 Centaurea galactites L. = Galactites elegans (All.) Soldano ■☆
81078 Centaurea gattefossei Maire;加特福塞矢车菊■☆
81079 Centaurea gentilii Braun-Blanq. et Maire = Centaurea aspera L. subsp. gentilii (Braun-Blanq.) Dobignard ■☆
81080 Centaurea gentilii Braun-Blanq. et Maire var. ecillata Maire = Centaurea gentilii Braun-Blanq. et Maire ■☆
81081 Centaurea gentilii Braun-Blanq. et Maire var. integrifolia (Ball) Emb. et Maire = Centaurea gentilii Braun-Blanq. et Maire ■☆
81082 Centaurea georgica Klokov;乔治矢车菊■☆
81083 Centaurea gerberi Steven;格氏矢车菊■☆
81084 Centaurea glastifolia L. = Chartolepis glastifolia (L.) Cass. ■
81085 Centaurea glauca Willd. = Amberboa glauca (Willd.) Grossh. ■
81086 Centaurea glehnii Trautv.;格莱矢车菊■☆
81087 Centaurea glomerata Moris var. papposa Coss. = Centaurea papposa (Coss.) Greuter ■☆
81088 Centaurea glomerata Vahl;团集矢车菊■☆
81089 Centaurea glomerata Vahl var. glabriceps Asch. et Schweinf. = Centaurea glomerata Vahl ■☆
81090 Centaurea goetzeana O. Hoffm. = Centaurea praecox Oliv. et Hiern ■☆
81091 Centaurea gontscharovii Iljin;高恩恰洛夫矢车菊■☆
81092 Centaurea granatensis Boiss. subsp. battandieri (Hochr.) Maire = Centaurea malinvaldiana Batt. ■☆
81093 Centaurea granatensis Boiss. subsp. malinvaldiana (Batt.) Maire = Centaurea malinvaldiana Batt. ■☆
81094 Centaurea granatensis Boiss. var. citrina Maire = Centaurea malinvaldiana Batt. ■☆
81095 Centaurea granatensis Boiss. var. purpurea Maire = Centaurea malinvaldiana Batt. ■☆
81096 Centaurea grandiflora Pall. = Rhaponticum uniflorum (L.) DC. ■
81097 Centaurea grandiflora Pall. = Stemmacantha uniflora (L.) Dittrich ■
81098 Centaurea grossheimii Sosn.;格罗矢车菊■☆
81099 Centaurea gueryi Maire = Centaurea nigra L. subsp. gueryi (Maire) Maire ■☆
81100 Centaurea gueryi Maire var. bekritensis? = Centaurea nigra L. subsp. gueryi (Maire) Maire ■☆
81101 Centaurea guilhelmi (Pau et Sennen) Maire;吉列尔姆矢车菊■☆
81102 Centaurea gulissashvilii Dumbadze;古里矢车菊■☆
81103 Centaurea gymnocarpa Moris et De Not.;裸果矢车菊;Dusty Miller ■☆
81104 Centaurea hajastana Tzvelev;哈贾斯坦矢车菊■☆
81105 Centaurea hochstetteri Oliv. et Hiern = Volutaria abyssinica (A. Rich.) C. Jeffrey ■☆
81106 Centaurea hochstetteri Oliv. et Hiern subsp. boranensis Cufod. = Volutaria boranensis (Cufod.) Wagenitz ■☆
81107 Centaurea holophylla Soczava et Lipatova;全叶矢车菊■☆
81108 Centaurea hyalolepis Boiss.;无色鳞矢车菊■☆
81109 Centaurea hymenolepis Trautv.;膜鳞矢车菊■☆
81110 Centaurea hyrcanica Bornm.;西加矢车菊■☆
81111 Centaurea hyssopifolia Vahl;神香草叶矢车菊(线叶矢车菊)■☆
81112 Centaurea iberica Trevir. ex Spreng.;针叶矢车菊(针刺矢车菊);Georgia Centaurea, Iberian Knapweed, Iberian Star Knapweed, Iberian Star Thistle, Russian Centaurea, Spanish Thistle, Thistle ■
81113 Centaurea iljinii Czerniak.;伊尔金矢车菊■☆

81114 Centaurea imatongensis Philipson = Ochrocephala imatongensis (Philipson) Dittrich ■☆

81115 Centaurea imperialis Hausskn. = Centaurea imperialis Hausskn. ex Bornm. ■☆

81116 Centaurea imperialis Hausskn. ex Bornm.;帝王矢车菊;Giant Sweet Sultan,Sultan Flower,Sweet Sultan ■☆

81117 Centaurea incana Desf. = Centaurea pubescens Willd. ■☆

81118 Centaurea incana Desf. subsp. amourensis (Pomel) Batt. = Centaurea pubescens Willd. subsp. amourensis (Pomel) Batt. ■☆

81119 Centaurea incana Desf. subsp. omphalotricha (Coss. et Durieu) Batt. = Centaurea pubescens Willd. subsp. omphalotricha Batt. ■☆

81120 Centaurea incana Desf. subsp. ornata (Willd.) Maire = Centaurea pubescens Willd. ■☆

81121 Centaurea incana Desf. subsp. pubescens (Willd.) Maire = Centaurea pubescens Willd. ■☆

81122 Centaurea incana Desf. var. abdel-kaderi Sennen et Mauricio = Centaurea pubescens Willd. ■☆

81123 Centaurea incana Desf. var. amourensis (Pomel) Sennen et Pau = Centaurea pubescens Willd. ■☆

81124 Centaurea incana Desf. var. antiatlantica Emb. et Maire = Centaurea pubescens Willd. ■☆

81125 Centaurea incana Desf. var. battandieri (Sennen) Maire = Centaurea pubescens Willd. ■☆

81126 Centaurea incana Desf. var. carbonelli Sennen et Mauricio = Centaurea pubescens Willd. ■☆

81127 Centaurea incana Desf. var. discolor Maire = Centaurea pubescens Willd. ■☆

81128 Centaurea incana Desf. var. fulgida Maire = Centaurea pubescens Willd. ■☆

81129 Centaurea incana Desf. var. hookeriana Ball = Centaurea pubescens Willd. ■☆

81130 Centaurea incana Desf. var. ignea Emb. et Maire = Centaurea pubescens Willd. ■☆

81131 Centaurea incana Desf. var. leucophylla Alleiz. = Centaurea pubescens Willd. ■☆

81132 Centaurea incana Desf. var. litoralis Batt. = Centaurea pubescens Willd. ■☆

81133 Centaurea incana Desf. var. monocephala Hochr. = Centaurea pubescens Willd. ■☆

81134 Centaurea incana Desf. var. pauana Maire = Centaurea pubescens Willd. ■☆

81135 Centaurea incana Desf. var. pseudacaulis Maire = Centaurea pubescens Willd. ■☆

81136 Centaurea incana Desf. var. purpurea Maire = Centaurea malinvaldiana Batt. ■☆

81137 Centaurea incana Desf. var. rupicola (Pomel) Maire = Centaurea pubescens Willd. ■☆

81138 Centaurea incana Desf. var. saharae (Pomel) Hochr. = Centaurea pubescens Willd. subsp. saharae (Pomel) Dobignard ■☆

81139 Centaurea incana Desf. var. virens Pau = Centaurea pubescens Willd. ■☆

81140 Centaurea incana Lag. = Centaurea resupinata Coss. ■☆

81141 Centaurea incana Lag. var. angulosa (Pomel) Batt. = Centaurea resupinata Coss. subsp. spachii (Willk.) Fern. Casas et Susanna ■☆

81142 Centaurea incana Lag. var. polyphylla (Pomel) Batt. = Centaurea resupinata Coss. subsp. spachii (Willk.) Fern. Casas et Susanna ■☆

81143 Centaurea infestans Coss. et Durieu var. longispina Faure et Maire = Centaurea infestans Durieu ■☆

81144 Centaurea infestans Durieu;有害矢车菊■☆

81145 Centaurea integrifolia Tausch;全缘矢车菊■☆

81146 Centaurea involucrata Desf.;总苞矢车菊■☆

81147 Centaurea involucrata Desf. var. discolor Maire = Centaurea pullata L. ■☆

81148 Centaurea involucrata Desf. var. paulini Maire et Sennen = Centaurea involucrata Desf. ■☆

81149 Centaurea ixodes Pomel = Centaurea sphaerocephala L. ■☆

81150 Centaurea jacea L.;棕矢车菊(棕鳞矢车菊);Brown Knapweed,Brown-headed Knapweed,Brown-ray Knapweed,Brown-rayed Knapweed,Brownscale Centaurea,Brown-scale Centaurea,French Hardhead,Hardheads,Knapweed,Knapweed Harshweed,Knapwort Harshweed ■☆

81151 Centaurea jacea L. subsp. nigra (L.) Bonnier et Layens = Centaurea nigra L. ■☆

81152 Centaurea jacea L. subsp. nigrescens (Willd.) Celak. = Centaurea nigrescens Willd. ■

81153 Centaurea jacea L. subsp. ropalon (Pomel) Maire = Centaurea ropalon Pomel ■☆

81154 Centaurea jacea L. subsp. x pratensis (W. D. J. Koch) Celak. = Centaurea × moncktonii C. E. Britton ■☆

81155 Centaurea jacea L. var. illudens Maire = Centaurea ropalon Pomel ■☆

81156 Centaurea jacea L. var. pratensis W. D. J. Koch = Centaurea × moncktonii C. E. Britton ■☆

81157 Centaurea jeffreyana Mesfin;杰弗里矢车菊■☆

81158 Centaurea josiae Humbert;约西亚矢车菊■☆

81159 Centaurea josiae Humbert var. mgounica Quézel = Centaurea josiae Humbert ■☆

81160 Centaurea karabaghensie (Sosn.) Sosn.;卡拉巴赫矢车菊■☆

81161 Centaurea kasakorum Iljin;天山矢车菊(哈萨克矢车菊);Tianshan Centaurea ■

81162 Centaurea koktebelica Klokov;科克矢车菊■☆

81163 Centaurea kolakovskyi Sosn.;科拉科夫斯基矢车菊■☆

81164 Centaurea konkae Klokov;孔卡矢车菊■☆

81165 Centaurea kopetdashensis Iljin;科佩特矢车菊■☆

81166 Centaurea kotschyana Heuff.;考奇矢车菊■☆

81167 Centaurea kubanica Klokov;古巴矢车菊■☆

81168 Centaurea kultiassovii Iljin;库尔蒂矢车菊■☆

81169 Centaurea lasiopoda Popov et Kult.;毛梗矢车菊■☆

81170 Centaurea latiloba Klokov;宽裂矢车菊■☆

81171 Centaurea lavrenkoana Klokov;拉氏矢车菊■☆

81172 Centaurea leucolepis Ledeb. ex Nyman;白膜鳞矢车菊■☆

81173 Centaurea leucophylla M. Bieb.;白叶矢车菊■☆

81174 Centaurea leuzeoides Walp.;刘子菊状矢车菊■☆

81175 Centaurea linifolia L.;狭叶矢车菊■☆

81176 Centaurea lippii L. = Volutaria lippii (L.) Cass. ■☆

81177 Centaurea litardierei Jahand. et Maire;利塔矢车菊■☆

81178 Centaurea litardierei Jahand. et Maire var. villosa Maire = Centaurea litardierei Jahand. et Maire ■☆

81179 Centaurea littorea Gand.;滨海矢车菊;Narrow-leaved Centaury ■☆

81180 Centaurea longifolia (Hoffmanns. et Link) Cout. = Rhaponticum longifolium (Hoffmanns. et Link) Dittrich ■☆

81181 Centaurea longifolia (Hoffmanns. et Link) Cout. var. ericeticola Font Quer = Rhaponticum longifolium (Hoffmanns. et Link) Dittrich subsp. ericeticola (Font Quer) Greuter ■☆

81182 Centaurea macrocephala Muss. Puschk. = Centaurea macrocephala Muss. Puschk. ex Willd. ■☆

81183 Centaurea macrocephala Muss. Puschk. ex Willd.;长头矢车菊(大花矢车菊);Bighead Knapweed, Big-head Knapweed, Globe Centaurea, Great Knapweed, Knapweed, Yellow Bachelor's Button, Yellow Bachelor's Cornflower ■☆

81184 Centaurea maculosa Lam.;斑矢车菊;Spotted Centaurea, Spotted Knapweed ■☆

81185 Centaurea maculosa Lam. = Centaurea stoebe L. ■☆

81186 Centaurea maculosa Lam. subsp. micranthos S. G. Gmel. ex Gugler = Centaurea stoebe L. ■☆

81187 Centaurea maireana Emb.;迈雷矢车菊■☆

81188 Centaurea majorovii Dumbadze;马约罗夫矢车菊■☆

81189 Centaurea malinvaldiana Batt.;马林矢车菊■☆

81190 Centaurea malinvaldiana Batt. = Centaurea granatensis Boiss. subsp. malinvaldiana (Batt.) Maire ■☆

81191 Centaurea malinvaldiana Batt. subsp. battandieri (Hochr.) Maire = Centaurea granatensis Boiss. subsp. battandieri (Hochr.) Maire ■☆

81192 Centaurea malinvaldiana Batt. subsp. cossoniana (Batt.) Batt. = Centaurea granatensis Boiss. subsp. malinvaldiana (Batt.) Maire ■☆

81193 Centaurea margarita-alba Klokov;白边矢车菊■☆

81194 Centaurea margaritacea Ten.;珍珠矢车菊■☆

81195 Centaurea maroccana Ball;摩洛哥矢车菊■☆

81196 Centaurea marschalliana Spreng.;马氏矢车菊■☆

81197 Centaurea mauritanica (Font Quer) Pau = Cheirolophus mauritanicus (Font Quer) Susanna ●☆

81198 Centaurea melitensis L.;马耳他矢车菊;Centaury, Cockspur, Malta Thistle, Maltese Cockspur, Maltese Star, Maltese Star Thistle, Maltese Star-thistle, Maltese Thistle, Napa Thistle, Saucy Jack, Tocalote ■☆

81199 Centaurea melitensis L. var. apula Rouy = Centaurea melitensis L. ■☆

81200 Centaurea membranacea Lam. = Rhaponticum uniflorum (L.) DC. ■

81201 Centaurea membranacea Lam. = Stemmacantha uniflora (L.) Dittrich ■

81202 Centaurea meskhetica Sosn.;迈斯亥特矢车菊■☆

81203 Centaurea mexicana DC. = Centaurea americana Nutt. ■☆

81204 Centaurea meyeriana Tzvelev;迈氏矢车菊■☆

81205 Centaurea micranthos I. F. Gugler = Centaurea stoebe L. ■☆

81206 Centaurea microcarpa Batt.;小果矢车菊■☆

81207 Centaurea minima (Boiss.) B. Fedtsch. = Oligochaeta minima (Boiss.) Briq. ■

81208 Centaurea missionis H. Lév. = Serratula chinensis S. Moore ■

81209 Centaurea modestii Fed.;适度矢车菊■☆

81210 Centaurea mollis Waldst. et Kit.;柔软矢车菊■☆

81211 Centaurea monantha Georgi = Rhaponticum uniflorum (L.) DC. ■

81212 Centaurea monantha Georgi = Stemmacantha uniflora (L.) Dittrich ■

81213 Centaurea monodii Arènes;莫诺矢车菊■☆

81214 Centaurea montana L.;山矢车菊;Loggerheads, Mountain Bluet, Mountain Brush, Mountain Cornflower, Mountain Knapweed, Mountain-bluet, Perennial Cornflower ■☆

81215 Centaurea monticola DC.;山生矢车菊■☆

81216 Centaurea monticola DC. subsp. orthoacantha (Pau et Font Quer) Ibn Tattou = Centaurea monticola DC. ■☆

81217 Centaurea monticola DC. subsp. perplexans (Emb. et Maire) Ibn Tattou = Centaurea resupinata Coss. subsp. simulans (Emb. et Maire) Breitw. et Podlech ■☆

81218 Centaurea monticola DC. var. orthoacantha Pau et Font Quer = Centaurea monticola DC. ■☆

81219 Centaurea moschata L.;香矢车菊(香芙蓉);Sweet Sultan, Sweet-sultan ■

81220 Centaurea moschata L. = Amberboa moschata (L.) DC. ■

81221 Centaurea moschata L. var. alba Hort.;白花香芙蓉■☆

81222 Centaurea moschata L. var. coerulea Hort.;蓝花香芙蓉■☆

81223 Centaurea moschata L. var. rosea Hort.;粉红花香芙蓉■☆

81224 Centaurea moschata L. var. rubra Hort.;红花香芙蓉■☆

81225 Centaurea murbeckiana Maire = Volutaria maroccana (Barratte et Murb.) Maire ■☆

81226 Centaurea muricata L. = Volutaria muricata (L.) Maire ■☆

81227 Centaurea nana Desf.;矮小矢车菊■☆

81228 Centaurea napifolia L.;芜菁叶矢车菊■☆

81229 Centaurea nathadzeae Sosn.;纳氏矢车菊■☆

81230 Centaurea nemoralis Jord.;纤细矢车菊;Slender Hardhead ■☆

81231 Centaurea nemoralis Jord. = Centaurea nigra L. ■☆

81232 Centaurea nervosa Willd.;显脉矢车菊;Plume Knapweed ■☆

81233 Centaurea nicaeensis All.;尼西亚矢车菊■☆

81234 Centaurea nicaeensis All. var. barbara (Pomel) Batt. = Centaurea nicaeensis All. ■☆

81235 Centaurea nidulans Pomel;巢状矢车菊■☆

81236 Centaurea nigerica Hutch.;尼日利亚矢车菊■☆

81237 Centaurea nigra L.;黑矢车菊;Bachelor's Buttons, Ballweed, Bell-weed, Belweed, Black Centaurea, Black Eentaurea, Black Knapweed, Black Matfellon, Black Soap, Blackhead, Blue Jack, Blue Top, Bluebottle, Bobby's Butions, Bobby's Buttons, Boleweed, Bollwood, Bow-weed, Bow-wood, Brush, Bull Head, Bull Thistle, Bullweed, Bully Head, Bund, Bundweed, Buttonweed, Chimney Sweep, Churl's Head, Churrs Head, Clobweed, Clover-knob, Clubweed, Codweed, Common Knapweed, Cropweed, Darbottle, Devil's Bit, Devil's Spit, Dromedary, Drummer Boy, Drummer Boys, Drummer Head, Drumsticks, Gnat Flower, Gold Knop, Hackymore, Hairy Head, Hard Heads, Hardhack, Hardhead, Hardhead Knapweed, Hardheads, Hardine, Harebottle, Hickymore, Horse Hardhead, Horse Knob, Horse Knop, Horse Knot, Horse Nop, Horse Snap, Hurt Sickle, Hurt-sickle, Iron Head, Iron Knob, Iron-hard, Iron-head, Iron-knob, Iron-weed, Knapweed, Knobweed, Knopweed, Knotgrass, Knotweed, Lady's Ball, Lady's Cushion, Lesser Knapweed, Loggerheads, Loggerum, Matfellon, Paintbrush, Shaving Brush, Slender Knapweed, Spanish Buttons, Sweep, Tarbottle, Tassel, Thumble, Top Knot, Top-knot ■☆

81238 Centaurea nigra L. subsp. gueryi (Maire) Maire;盖里矢车菊■☆

81239 Centaurea nigra L. var. bekritensis (Jahand. et Maire) Maire = Centaurea nigra L. ■☆

81240 Centaurea nigra L. var. mairei Arènes = Centaurea nigra L. ■☆

81241 Centaurea nigra L. var. radiata DC. = Centaurea × moncktonii C. E. Britton ■☆

81242 Centaurea nigra L. var. radiata DC. = Centaurea nigra L. ■☆

81243 Centaurea nigrescens Willd.;穗裂矢车菊;Blackish Centaurea, Knapweed, Short-fringed Knapweed, Tyrol Knapweed, Vochin Knapweed ■

81244 Centaurea nigriceps Dobrocz.;黑头矢车菊■☆

81245 Centaurea nigrifimbria (K. Koch) Sosn.;黑流苏矢车菊■☆

81246 Centaurea novorossica Klokov;新罗西矢车菊■☆

81247 Centaurea obtusiloba Batt.;钝裂矢车菊■☆
81248 Centaurea ochrolopha Costa subsp. guilhelmi Pau et Sennen = Centaurea guilhelmi (Pau et Sennen) Maire ■☆
81249 Centaurea ochrolopha Costa var. mauritii Pau et Sennen = Centaurea guilhelmi (Pau et Sennen) Maire ■☆
81250 Centaurea odorata (DC.) Sch. Bip. = Centaurea moschata L. ■
81251 Centaurea odorata Burm. f. = Centaurea moschata L. ■
81252 Centaurea olivieri Pomel = Centaurea resupinata Coss. subsp. spachii (Willk.) Fern. Casas et Susanna ■☆
81253 Centaurea oltensis Sosn.;奥尔特矢车菊■☆
81254 Centaurea omphalodes Coss. et Durieu = Stephanochilus omphalodes (Benth. et Hook. f.) Maire ■☆
81255 Centaurea omphalotricha (Coss. et Durieu) Batt. = Centaurea pubescens Willd. subsp. omphalotricha Batt. ■☆
81256 Centaurea oranensis Greuter et M. V. Agab.;奥兰矢车菊■☆
81257 Centaurea orientalis L.;东方矢车菊;Caucasian Centaurea ■☆
81258 Centaurea ossethica Sosn.;骨质矢车菊■☆
81259 Centaurea ovina Pall. ex Steud.;绵羊矢车菊■☆
81260 Centaurea paczoskii Kotov ex Klokov;帕氏矢车菊■☆
81261 Centaurea pallescens Delile;苍白矢车菊■☆
81262 Centaurea pallescens Delile var. hyalolepis (Boiss.) Maire = Centaurea pallescens Delile ■☆
81263 Centaurea paniculata L.;圆锥矢车菊;Jersey Knapweed, Panicled Knapweed ■☆
81264 Centaurea paniculata L. subsp. guilhelmi (Pau et Sennen) Arènes = Centaurea resupinata Coss. subsp. spachii (Willk.) Fern. Casas et Susanna ■☆
81265 Centaurea pannonica (Heuff.) Hayek;帕地矢车菊■☆
81266 Centaurea papposa (Coss.) Greuter;冠毛矢车菊■☆
81267 Centaurea parviflora Besser = Centaurea diffusa Lam. ■
81268 Centaurea parviflora Desf.;小花矢车菊■☆
81269 Centaurea pauciloba Trautv.;少裂矢车菊■☆
81270 Centaurea pecho Albov;佩霍矢车菊■☆
81271 Centaurea pectinata L.;疏齿矢车菊■☆
81272 Centaurea pelia DC.;比立山矢车菊■☆
81273 Centaurea perraldierana Maire = Volutaria sinaica (DC.) Wagenitz ■☆
81274 Centaurea perrottettii DC.;佩氏矢车菊■☆
81275 Centaurea phaeolepis Coss.;暗鳞矢车菊■☆
81276 Centaurea phaeopappoides Bordz.;暗冠毛矢车菊■☆
81277 Centaurea phrygia L.;夫立基矢车菊;Wig Knapweed ■☆
81278 Centaurea phyllopoda Iljin;叶梗矢车菊■☆
81279 Centaurea picris Pall. = Acroptilon repens (L.) DC. ■
81280 Centaurea picris Pall. ex Willd. = Acroptilon repens (L.) DC. ■
81281 Centaurea picris Pall. ex Willd. = Centaurea repens L. ■
81282 Centaurea pineticola Iljin;松林矢车菊■☆
81283 Centaurea polyacantha Willd. var. adpressa Maire = Centaurea polyacantha Willd. ■☆
81284 Centaurea polyphylla Pomel = Centaurea resupinata Coss. subsp. spachii (Willk.) Fern. Casas et Susanna ■☆
81285 Centaurea polypodiifolia Boiss.;多足叶矢车菊■☆
81286 Centaurea pomeliana Batt.;波梅尔矢车菊■☆
81287 Centaurea pomeliana Batt. subsp. rouxiana (Maire) Breitw. et Podlech;鲁矢车菊■☆
81288 Centaurea pomeliana Batt. var. rouxiana Maire = Centaurea pomeliana Batt. subsp. rouxiana (Maire) Breitw. et Podlech ■☆
81289 Centaurea praecox Oliv. et Hiern;早矢车菊■☆
81290 Centaurea pratensis Thuill.;草甸矢车菊■☆
81291 Centaurea pratensis Thuill. subsp. gueryi (Maire) Arènes = Centaurea nigra L. subsp. gueryi (Maire) Maire ■☆
81292 Centaurea pseudoleucolepis Kleopow;假白鳞矢车菊■☆
81293 Centaurea pseudomaculosa Dobrocz.;假斑矢车菊■☆
81294 Centaurea pseudophrygia C. A. Mey.;假眉兰车菊■☆
81295 Centaurea pseudoscabiosa Boiss. et Buhse;假糙矢车菊■☆
81296 Centaurea pterocaulis Trautv.;翅茎矢车菊■☆
81297 Centaurea pterocaulos Pomel = Centaurea sphaerocephala L. ■☆
81298 Centaurea pterodonta Pomel = Centaurea maroccana Ball ■☆
81299 Centaurea pubescens Willd.;短柔毛矢车菊■☆
81300 Centaurea pubescens Willd. subsp. amourensis (Pomel) Batt.;阿穆尔矢车菊■☆
81301 Centaurea pubescens Willd. subsp. omphalotricha Batt.;脐毛矢车菊■☆
81302 Centaurea pubescens Willd. subsp. saharae (Pomel) Dobignard;左原矢车菊■☆
81303 Centaurea pubescens Willd. var. integrifolia Alleiz. = Centaurea pubescens Willd. ■☆
81304 Centaurea pubescens Willd. var. leucophylla Alleiz. = Centaurea pubescens Willd. ■☆
81305 Centaurea pulchella Ledeb. = Hyalea pulchella (Ledeb.) K. Koch ■
81306 Centaurea pulchella Ledeb. var. viminea (Less.) DC. = Hyalea pulchella (Ledeb.) K. Koch ■
81307 Centaurea pulcherrima Wight ex DC.;美艳矢车菊■☆
81308 Centaurea pulchra DC. = Centaurea cyanus L. ■
81309 Centaurea pullata L.;黑色矢车菊■☆
81310 Centaurea pullata L. subsp. claryi (Debeaux) Batt. = Centaurea pullata L. var. claryi (Debeaux) Batt. ■☆
81311 Centaurea pullata L. subsp. discolor (Maire) Mathez = Centaurea pullata L. ■☆
81312 Centaurea pullata L. subsp. involucrata (Desf.) Talavera = Centaurea involucrata Desf. ■☆
81313 Centaurea pullata L. var. albiflora (Font Quer) Dobignard = Centaurea pullata L. ■☆
81314 Centaurea pullata L. var. claryi (Debeaux) Batt. = Centaurea pullata L. ■☆
81315 Centaurea pullata L. var. discolor (Maire) Dobignard = Centaurea pullata L. ■☆
81316 Centaurea pullata L. var. minor Ball = Centaurea pullata L. ■☆
81317 Centaurea pullata L. var. zatii Maubert = Centaurea pullata L. ■☆
81318 Centaurea pumilio L.;弱小矢车菊■☆
81319 Centaurea pungens Pomel;刺矢车菊■☆
81320 Centaurea pungens Pomel subsp. austromaroccana Förther et Podlech;南摩洛哥矢车菊■☆
81321 Centaurea pungens Pomel var. conillii Sennen et Mauricio = Centaurea pungens Pomel ■☆
81322 Centaurea pygmaea Hoffm. = Centaurea depressa M. Bieb. ■☆
81323 Centaurea razdorskyi Karjagin;拉兹多尔斯基矢车菊■☆
81324 Centaurea reflexa Lam.;反折矢车菊■☆
81325 Centaurea reflexa Lam. subsp. pubescens (Willd.) Ball = Centaurea pubescens Willd. ■☆
81326 Centaurea reflexa Lam. var. hookeriana Ball = Centaurea pubescens Willd. ■☆
81327 Centaurea repens L. = Acroptilon repens (L.) DC. ■
81328 Centaurea resupinata Coss.;倒置矢车菊■☆

81329 Centaurea resupinata Coss. subsp. degenii (Sennen) Fern. Casas et Susanna;德根矢车菊■☆

81330 Centaurea resupinata Coss. subsp. dufourii (Dostál) Greuter;迪富尔矢车菊■☆

81331 Centaurea resupinata Coss. subsp. lagascae (Nyman) Fern. Casas et Susanna = Centaurea resupinata Coss. subsp. dufourii (Dostál) Greuter ■☆

81332 Centaurea resupinata Coss. subsp. rifana (Emb. et Maire) Breitw. et Podlech = Centaurea resupinata Coss. subsp. degenii (Sennen) Fern. Casas et Susanna ■☆

81333 Centaurea resupinata Coss. subsp. simulans (Emb. et Maire) Breitw. et Podlech;相似倒置矢车菊■☆

81334 Centaurea resupinata Coss. subsp. spachii (Willk.) Fern. Casas et Susanna;斯帕赫矢车菊■☆

81335 Centaurea rhizantha C. A. Mey.;根花矢车菊■☆

81336 Centaurea rhizanthoides Tzvelev;假根花矢车菊■☆

81337 Centaurea rhizocephala Oliv. et Hiern = Centaurea praecox Oliv. et Hiern ■☆

81338 Centaurea ropalon Pomel;棍棒矢车菊■☆

81339 Centaurea ropalon Pomel var. tunizensis Maire = Centaurea ropalon Pomel ■☆

81340 Centaurea rothrockii Greenm. = Plectocephalus rothrockii (Greenm.) D. J. N. Hind ■☆

81341 Centaurea rubescens Besser ex DC.;红矢车菊■☆

81342 Centaurea rubriflora N. B. Illar.;红花矢车菊■☆

81343 Centaurea rupicola Pomel = Centaurea pubescens Willd. ■☆

81344 Centaurea ruprechtii (Boiss.) Czerep.;卢氏矢车菊■☆

81345 Centaurea ruthenica Lam.;欧亚矢车菊(俄罗斯矢车菊,黄花漏芦,漏芦);Eurasia Centaurea, Euro-asiatic Centaurea, Ruthenian Centaurea ■

81346 Centaurea rutifolia Sibth. et Sm. = Centaurea cineraria L. ■☆

81347 Centaurea saharae Pomel = Centaurea pubescens Willd. subsp. saharae (Pomel) Dobignard ■☆

81348 Centaurea salicifolia M. Bieb. ex Willd.;柳叶矢车菊■☆

81349 Centaurea salmantica L. = Mantisalca salmantica (L.) Briq. et Cavill. ■☆

81350 Centaurea salmantica L. var. clusii (Spach) Ball = Mantisalca salmantica (L.) Briq. et Cavill. ■☆

81351 Centaurea salmantica L. var. leptoloncha (Spach) Ball = Mantisalca salmantica (L.) Briq. et Cavill. ■☆

81352 Centaurea salonitana Vis.;厅矢车菊;Yellow Knapweed ■☆

81353 Centaurea saltii Philipson = Volutaria abyssinica (A. Rich.) C. Jeffrey ■☆

81354 Centaurea salvlifolia (Boiss.) Sosn.;鼠尾草矢车菊■☆

81355 Centaurea sarandinakiae N. B. Illar.;萨拉恩矢车菊■☆

81356 Centaurea scabiosa L.;大矢车菊(粗糙矢车菊,山萝矢车菊);Bachelor's Buttons, Black Top, Bow-wood, Churl Head, Cornflower, Cow-weed, Dromedary, Drumsticks, Field Scabious, Gold Knop, Great Horse Knob, Great Knapweed, Greater Centaury, Greater Knapweed, Hardhead, Hardheads, Hard-heads, Horse Knop, Horse Knot, Knobweed, Knopweed, Knotweed, Loggerheads, Ltarshweed, Matfellon, Mop, Scabriosa Centaurea ■☆

81357 Centaurea scabiosa L. var. adpressa (Ledeb.) DC. = Centaurea adpressa Ledeb. ex Steud. ■

81358 Centaurea scabiosa L. var. angustata Ledeb. = Centaurea adpressa Ledeb. ex Steud. ■

81359 Centaurea schelkovnikovii Sosn.;谢氏矢车菊■☆

81360 Centaurea scopalia L.;帚枝矢车菊■☆

81361 Centaurea scoparia Spreng.;帚状矢车菊■☆

81362 Centaurea segetalis Salisb. = Centaurea cyanus L. ■

81363 Centaurea sempervirens L. = Cheirolophus mauritanicus (Font Quer) Susanna ●☆

81364 Centaurea sempervirens L. subsp. mauritanica (Font Quer) Jahand. et Maire = Cheirolophus mauritanicus (Font Quer) Susanna ●☆

81365 Centaurea sempervirens L. var. algerica Maire = Cheirolophus mauritanicus (Font Quer) Susanna ●☆

81366 Centaurea sempervirens L. var. mauritanica Font Quer = Cheirolophus mauritanicus (Font Quer) Susanna ●☆

81367 Centaurea senegalensis DC.;塞内加尔矢车菊■☆

81368 Centaurea sergii Klokov;赛尔格矢车菊■☆

81369 Centaurea seridis L.;苣状矢车菊■☆

81370 Centaurea seridis L. var. auriculata (Balb.) Ball = Centaurea seridis L. ■☆

81371 Centaurea seridis L. var. calva Maire et Sauvage = Centaurea seridis L. ■☆

81372 Centaurea seridis L. var. epapposa Caball. = Centaurea seridis L. ■☆

81373 Centaurea seridis L. var. maritima Lange;海滨矢车菊(西班牙矢车菊)■☆

81374 Centaurea seridis L. var. pterocaulos (Pomel) Maire = Centaurea seridis L. ■☆

81375 Centaurea seridis L. var. sonchifolia (L.) Briq. = Centaurea seridis L. ■☆

81376 Centaurea seridis L. var. subferox Pau et Font Quer = Centaurea seridis L. ■☆

81377 Centaurea setosa Cav.;多刚毛矢车菊■☆

81378 Centaurea sibirica L.;西伯利亚小矢车菊;Dwarf Centaurea, Siberian Centaurea ■

81379 Centaurea simplex Cav. = Rhaponticoides africana (Lam.) M. V. Agab. et Greuter ■☆

81380 Centaurea simplicicaulis Boiss. et Huet;单茎矢车菊■☆

81381 Centaurea sintenisiana Gand.;西恩矢车菊■☆

81382 Centaurea solstitialis L.;夏至矢车菊;Barnaby Star-thistle, Barnaby's Thistle, Midsummer Thistle, Saint Barnaby's Thistle, St. Barnaby's Thistle, Yellow Centaurea, Yellow Star Thistle, Yellow Star-thistle ■☆

81383 Centaurea somalensis Oliv. et Hiern = Volutaria abyssinica (A. Rich.) C. Jeffrey ■☆

81384 Centaurea sonchifolia L. var. dimorpha (Viv.) DC. = Centaurea dimorpha Viv. ■☆

81385 Centaurea sophiae Klokov;索非矢车菊■☆

81386 Centaurea sosnovskyi Grossh.;索斯矢车菊■☆

81387 Centaurea spachii Willk. = Centaurea resupinata Coss. subsp. spachii (Willk.) Fern. Casas et Susanna ■☆

81388 Centaurea spachii Willk. = Centaurea resupinata Coss. ■☆

81389 Centaurea spachii Willk. subsp. lagascae (Nyman) Figuerola et Peris et Stübing = Centaurea resupinata Coss. subsp. spachii (Willk.) Fern. Casas et Susanna ■☆

81390 Centaurea spachii Willk. subsp. resupinata (Coss.) Figuerola et Peris et Stübing = Centaurea resupinata Coss. ■☆

81391 Centaurea spachii Willk. subsp. rifana (Emb. et Maire) Figuerola et Peris et Stübing = Centaurea resupinata Coss. subsp. degenii (Sennen) Fern. Casas et Susanna ■☆

81392 Centaurea sparmannii DC. = Centaurea perrottettii DC. ■☆
81393 Centaurea sphaerocephala L.;球头矢车菊■☆
81394 Centaurea sphaerocephala L. subsp. lusitanica (Boiss. et Reut.) Nyman;葡萄牙矢车菊■☆
81395 Centaurea sphaerocephala L. var. fontanesii (Durieu) Batt. = Centaurea sphaerocephala L. ■☆
81396 Centaurea sphaerocephala L. var. oligocentra Maire = Centaurea sphaerocephala L. ■☆
81397 Centaurea sphaerocephala L. var. transiens Faure et Maire = Centaurea sphaerocephala L. ■☆
81398 Centaurea squarrosa Willd.;广展矢车菊(小花矢车菊);Smallflower Centaurea ■
81399 Centaurea squarrosa Willd. = Centaurea virgata Lam. subsp. squarrosa (Boiss.) Gugler ■☆
81400 Centaurea stenolepis J. Kern.;窄苞矢车菊■☆
81401 Centaurea stereophylla Besser;硬叶矢车菊■☆
81402 Centaurea sterilis Steven;不实矢车菊■☆
81403 Centaurea steveniana Klokov;斯氏矢车菊■☆
81404 Centaurea stevenii M. Bieb.;斯迪温矢车菊■☆
81405 Centaurea stoebe L.;斑纹矢车菊;Spotted Knapweed ■☆
81406 Centaurea stoebe L. subsp. micranthos?;小花斑纹矢车菊;Spotted Knapweed ■☆
81407 Centaurea stricta Waldst. et Kit.;刚直矢车菊■☆
81408 Centaurea suaveolens Willd.;甜矢车菊;Sweet Sultan ■
81409 Centaurea suaveolens Willd. = Centaurea moschata L. ■
81410 Centaurea sulphurea Willd.;硫黄矢车菊;Sicilian Star-thistle, Sulphur Knapweed, Sulphur-colored Sicilian Thistle ■☆
81411 Centaurea sumensis Kalen.;苏马矢车菊■☆
81412 Centaurea szovitsiana Boiss.;瑟维茨矢车菊■☆
81413 Centaurea tagana Brot. = Rhaponticoides africana (Lam.) M. V. Agab. et Greuter ■☆
81414 Centaurea tagana Brot. var. africana (Lam.) Batt. = Rhaponticoides africana (Lam.) M. V. Agab. et Greuter ■☆
81415 Centaurea talievii Kleopow;塔氏矢车菊■☆
81416 Centaurea tanaitica Klokov;塔奈特矢车菊■☆
81417 Centaurea tananica Maire = Cheirolophus tananicus (Maire) Holub ●☆
81418 Centaurea tenuiflora DC.;细花矢车菊;Channel Centaury ■☆
81419 Centaurea tenuifolia Dufour = Centaurea resupinata Coss. subsp. dufourii (Dostál) Greuter ■☆
81420 Centaurea tenuifolia Dufour subsp. boissieri (DC.) Emb. et Maire = Centaurea boissieri DC. ■☆
81421 Centaurea tenuifolia Dufour subsp. spachii (Willk.) Emb. et Maire = Centaurea resupinata Coss. subsp. spachii (Willk.) Fern. Casas et Susanna ■☆
81422 Centaurea tenuifolia Dufour var. atlantica (Font Quer) Emb. et Maire = Centaurea boissieri DC. subsp. atlantica (Font Quer) Blanca ■☆
81423 Centaurea tenuifolia Dufour var. calvescens (Maire) Emb. et Maire = Centaurea boissieri DC. subsp. atlantica (Font Quer) Blanca ■☆
81424 Centaurea tenuifolia Dufour var. rifana (Emb. et Maire) Maire = Centaurea resupinata Coss. subsp. degenii (Sennen) Fern. Casas et Susanna ■☆
81425 Centaurea tenuifolia Dufour var. simulans (Emb. et Maire) Emb. et Maire = Centaurea resupinata Coss. subsp. simulans (Emb. et Maire) Breitw. et Podlech ■☆
81426 Centaurea tenuifolia Dufour var. transmalvana (Emb. et Maire) Emb. et Maire = Centaurea boissieri DC. subsp. transmalvana (Emb. et Maire) Breitw. et Podlech ■☆
81427 Centaurea ternopoliensis Dobrocz.;三出矢车菊■☆
81428 Centaurea thuillieri (Dostál) J. Duvign. et Lambinon = Centaurea × moncktonii C. E. Britton ■☆
81429 Centaurea tisserantii Philipson = Centaurea praecox Oliv. et Hiern ■☆
81430 Centaurea tougourensis Boiss. et Reut.;图古尔矢车菊■☆
81431 Centaurea tougourensis Boiss. et Reut. var. brevimucronata Maire = Centaurea tougourensis Boiss. et Reut. ■☆
81432 Centaurea tougourensis Boiss. et Reut. var. medians Maire = Centaurea tougourensis Boiss. et Reut. ■☆
81433 Centaurea tougourensis Boiss. et Reut. var. transiens Maire = Centaurea tougourensis Boiss. et Reut. ■☆
81434 Centaurea transalpina Schleich. ex DC. = Centaurea nigrescens Willd. ■
81435 Centaurea transcaucasica Sosn. ex Grossh.;外高加索矢车菊■☆
81436 Centaurea trichocephala M. Bieb. ex Willd.;毛头矢车菊;Feather-head Knapweed ■☆
81437 Centaurea trifurcata Pomel = Stephanochilus omphalodes (Benth. et Hook. f.) Maire ■☆
81438 Centaurea trinervia Steph.;三脉矢车菊■☆
81439 Centaurea troitzkyi (Sosn.) Sosn.;特罗伊斯基矢车菊■☆
81440 Centaurea turgaica Klokov;图尔嘎矢车菊■☆
81441 Centaurea turkestanica Franch.;土耳其斯坦矢车菊■☆
81442 Centaurea umbrosa Huet et Reut. = Centaurea cyanus L. ■
81443 Centaurea uniflora L.;单花矢车菊;Plume Knapweed, Singleflower Knapweed ■☆
81444 Centaurea uniflora L. subsp. nervosa?;多脉单花矢车菊;Singleflower Knapweed ■☆
81445 Centaurea vankovii Klokov;瓦恩矢车菊■☆
81446 Centaurea varians A. Rich. = Plectocephalus varians (A. Rich.) C. Jeffrey ex Cufod. ■☆
81447 Centaurea varians A. Rich. var. macrocephala Vatke = Plectocephalus varians (A. Rich.) C. Jeffrey ex Cufod. ■☆
81448 Centaurea variegata Lam.;杂色矢车菊■☆
81449 Centaurea vedenskyi Popov = Oligochaeta minima (Boiss.) Briq. ■
81450 Centaurea vicina Lipsky;邻近矢车菊■☆
81451 Centaurea viminea Less. = Hyalea pulchella (Ledeb.) K. Koch ■
81452 Centaurea vinyalsii Sennen;比尼矢车菊■☆
81453 Centaurea virgata Lam.;葡萄蔓矢车菊;Squarrose Knapweed ■☆
81454 Centaurea virgata Lam. subsp. squarrosa (Boiss.) Gugler;糠秕葡萄蔓矢车菊;Squarrose knapweed ■☆
81455 Centaurea virgata Lam. var. squarrosa Boiss. = Centaurea virgata Lam. subsp. squarrosa (Boiss.) Gugler ■☆
81456 Centaurea vochinensis Bernh. ex Rchb. = Centaurea nigrescens Willd. ■
81457 Centaurea vvedenskyi Popov = Oligochaeta minima (Boiss.) Briq. ■
81458 Centaurea webbiana Sch. Bip.;韦布矢车菊■☆
81459 Centaurea webbiana Sch. Bip. = Cheirolophus webbianus (Sch. Bip.) Holub ●☆
81460 Centaurea willdenowii Czerep.;威尔矢车菊■☆
81461 Centaurea woronowii Bornm. ex Sosn.;沃氏矢车菊■☆
81462 Centaurea xanthocephala (DC.) Sosn.;黄头矢车菊■☆

81463 Centaurea xanthocephaloides Tzvelev;假黄头矢车菊■☆
81464 Centaurea xaveri N. Garcia et Susanna;克萨维尔矢车菊■☆
81465 Centaurea zansezuri (Sosn.) Sosn.;扎氏矢车菊■☆
81466 Centaurea zuvandica (Sosn.) Sosn.;祖万德矢车菊■☆
81467 Centaureaceae Bercht. et J. Presl = Asteraceae Bercht. et J. Presl (保留科名)●■
81468 Centaureaceae Bercht. et J. Presl = Compositae Giseke(保留科名)●■
81469 Centaureaceae Martinov = Asteraceae Bercht. et J. Presl(保留科名)●■
81470 Centaureaceae Martinov = Compositae Giseke(保留科名)●■
81471 Centaureaceae Martinov;矢车菊科■
81472 Centaurella Delarbre = Centaurium Hill ■
81473 Centaurella Michx. = Bartonia Muhl. ex Willd.(保留属名)■☆
81474 Centaureum Rupp. = Erythraea Borkh. ■
81475 Centauria L. = Centaurea L.(保留属名)●■
81476 Centauridium Torr. et A. Gray = Xanthisma DC. ●■☆
81477 Centauridium drummondii Torr. et A. Gray = Xanthisma texanum DC. var. drummondii (Torr. et A. Gray) A. Gray ■☆
81478 Centaurion Adans. = Centaurium Hill ■
81479 Centaurium Borkh. = Canscora Lam. ■
81480 Centaurium Borkh. = Heteroclita Raf. ■
81481 Centaurium Cass. = Centaurea L.(保留属名)●■
81482 Centaurium Cass. = Centaurium Hill ■
81483 Centaurium Haller = Centaurea L.(保留属名)●■
81484 Centaurium Haller f. = Rhaponticum Ludw. ■
81485 Centaurium Hill(1756);百金花属(埃蕾属,白金花属,百金属);Centaurium,Centaury ■
81486 Centaurium Pers. = Bartonia Muhl. ex Willd.(保留属名)■☆
81487 Centaurium Pers. = Centaurella Michx. ■☆
81488 Centaurium alpinum Pomel;阿尔及利亚百金花■☆
81489 Centaurium barrelieroides Pau;拟巴雷百金花■☆
81490 Centaurium calycosum (Buckley) Fernald;北美百金花;Centaury,Rosita ■☆
81491 Centaurium calycosum Fernald = Centaurium calycosum (Buckley) Fernald ■☆
81492 Centaurium canchalagum?;药用百金花■☆
81493 Centaurium candelabrum H. Lindb.;烛台百金花■☆
81494 Centaurium capitatum (Willd.) Borbás = Centaurium erythraea Raf. ■☆
81495 Centaurium erythraea Raf.;红百金花(百金花,遍生百金花,粉红埃蕾,浅红百金花,头状百金花);Century,Common Centaurium,Common Centaury,European Centaury ■☆
81496 Centaurium erythraea Raf. subsp. bifrons (Pau) Greuter;双叶红百金花■☆
81497 Centaurium erythraea Raf. subsp. grandiflorum (Pers.) Melderis = Centaurium erythraea Raf. ■☆
81498 Centaurium erythraea Raf. subsp. rhodense (Boiss. et Reut.) Melderis;罗得西亚百金花●☆
81499 Centaurium erythraea Raf. subsp. suffruticosum (Griseb.) Greuter;亚灌木百金花●☆
81500 Centaurium flexuosum (Maire) Lebrun et Marais;曲折百金花■☆
81501 Centaurium floribundum (Benth.) B. L. Rob.;繁花百金花■☆
81502 Centaurium japonicum (Maxim.) Druce;日本百金花(百金,百金花);Japan Centaurium,Japanese Centaurium ■
81503 Centaurium japonicum (Maxim.) Druce = Schenkia japonica (Maxim.) G. Mans. ■
81504 Centaurium latifolium Druce = Centaurium erythraea Raf. ■☆
81505 Centaurium laxiflorum H. Lindb. = Centaurium tenuiflorum (Hoffmanns. et Link) Fritsch ■☆
81506 Centaurium littorale (Turner) Gilmour;小百金花;Seaside Centaury,Small Centaurium ■☆
81507 Centaurium ludovici Sennen = Centaurium tenuiflorum (Hoffmanns. et Link) Fritsch ■☆
81508 Centaurium mairei Zeltner;迈雷百金花■☆
81509 Centaurium maritimum (L.) Fritsch;滨海百金花■☆
81510 Centaurium massonii Sw. = Centaurium scilloides (L. f.) Samp. ■☆
81511 Centaurium meyeri (Bunge) Druce = Centaurium pulchellum (Sw.) Druce subsp. meyeri (Bunge) Tzvelev ■☆
81512 Centaurium meyeri (Bunge) Druce = Centaurium pulchellum (Sw.) Druce var. altaicum (Griseb.) Kitag. et H. Hara ■
81513 Centaurium minus Moench = Centaurium erythraea Raf. ■☆
81514 Centaurium minus Moench var. grandiflorum (Biv.) Pau et Font Quer = Centaurium erythraea Raf. ■☆
81515 Centaurium minutissimum Maire;微小百金花■☆
81516 Centaurium portense Butcher = Centaurium scilloides (L. f.) Samp. ■☆
81517 Centaurium pulchellum (Sw.) Druce;美丽百金花(埃蕾百金花);Beautiful Centaurium, Branching Centaury, Lesser Centaury, Showy Centaury ●■
81518 Centaurium pulchellum (Sw.) Druce subsp. grandiflorum (Batt.) Maire = Centaurium candelabrum H. Lindb. ■☆
81519 Centaurium pulchellum (Sw.) Druce subsp. laxiflorum (H. Lindb.) Maire = Centaurium tenuiflorum (Hoffmanns. et Link) Fritsch ■☆
81520 Centaurium pulchellum (Sw.) Druce subsp. meyeri (Bunge) Tzvelev;迈尔美丽百金花■☆
81521 Centaurium pulchellum (Sw.) Druce subsp. tenuiflorum (Hoffmanns. et Link) Maire = Centaurium tenuiflorum (Hoffmanns. et Link) Fritsch ■☆
81522 Centaurium pulchellum (Sw.) Druce var. affine (Rouy) Emb. et Maire = Centaurium tenuiflorum (Hoffmanns. et Link) Fritsch ■☆
81523 Centaurium pulchellum (Sw.) Druce var. altaicum (Griseb.) Kitag. et H. Hara;百金花(埃蕾,东北,麦氏埃蕾);Altai Centaurium,Altai Mountain Centaurium ■
81524 Centaurium pulchellum (Sw.) Druce var. altaicum (Griseb.) Kitag. et H. Hara = Centaurium meyeri (Bunge) Druce ■
81525 Centaurium pulchellum (Sw.) Druce var. altaicum (Griseb.) Kitag. et H. Hara = Centaurium japonicum (Maxim.) Druce ■
81526 Centaurium pulchellum (Sw.) Druce var. intermedium (Rouy) Sauvage et Vindt = Centaurium pulchellum (Sw.) Druce ●■
81527 Centaurium pulchellum (Sw.) Druce var. ramosissimum (Vill.) Sauvage et Vindt = Centaurium pulchellum (Sw.) Druce ●■
81528 Centaurium pulchellum (Sw.) Hayek var. lauriolii Maire = Centaurium mairei Zeltner ■☆
81529 Centaurium ruthenicum (Lam.) K. Koch = Centaurea ruthenica Lam. ■
81530 Centaurium scilloides (L. f.) Samp.;绵枣状百金花(矮甸百金花,簇百金花);Centaury,Perennial Centaury ■☆
81531 Centaurium scilloides Druce = Centaurium scilloides (L. f.) Samp. ■☆
81532 Centaurium spicatum (L.) Fritsch = Centaurium japonicum (Maxim.) Druce ■

81533 Centaurium spicatum (L.) Fritsch =Schenkia spicata (L.) G. Mans. ●☆

81534 Centaurium spicatum (L.) Fritsch subsp. japonicum (Maxim.) Toyok. =Schenkia japonica (Maxim.) G. Mans. ■

81535 Centaurium spicatum (L.) Fritsch var. japonicum Maxim. = Schenkia japonica (Maxim.) G. Mans. ■

81536 Centaurium spicatum Druce;穗状百金花;Spicate Centaurium, Spiked Centaury ■

81537 Centaurium spicatum Fritsch =Centaurium japonicum (Maxim.) Druce ■

81538 Centaurium tenuiflorum (Hoffmanns. et Link) Fritsch =Centaurium tenuiflorum (Hoffmanns. et Link) Fritsch ex E. Jansen ■☆

81539 Centaurium tenuiflorum (Hoffmanns. et Link) Fritsch ex E. Jansen;细花百金花;Slender Centaury ■☆

81540 Centaurium tenuiflorum (Hoffmanns. et Link) Fritsch subsp. acutiflorum (Schott) Zeltner;尖细花百金花■☆

81541 Centaurium tenuiflorum (Hoffmanns. et Link) Fritsch subsp. viridense (Bolle) A. Hansen et Sunding;绿齿细花百金花■☆

81542 Centaurium terneri (Wheldon et Salman) Butcher;沙丘百金花■☆

81543 Centaurium texense (Griseb.) Fernald;得州百金花;Texas Centaury ■☆

81544 Centaurium umbellatum Gilib.;伞状百金花;Centaury, Drug Centaurium, Umbellate Centaurium ■☆

81545 Centaurium umbellatum Gilib. = Erythraea centaurium (L.) Pers. ■☆

81546 Centaurium vulgare Raf.;欧百金花;Common Centaurium ■☆

81547 Centaurodendron Johow(1896);矢车木属●☆

81548 Centaurodendron dracaenoides Johow;矢车木●☆

81549 Centaurodes Möhring ex Kuntze =Centaurium Hill ■

81550 Centauropsis Bojer ex DC. (1836);矢车鸡菊花属(拟矢车菊属)●☆

81551 Centauropsis antanossi (Scott-Elliot) Humbert;安塔诺斯矢车鸡菊花●☆

81552 Centauropsis boivinii Drake = Centauropsis antanossi (Scott-Elliot) Humbert ●☆

81553 Centauropsis cuspidata Humbert;骤尖矢车鸡菊花●☆

81554 Centauropsis decaryi Humbert;德卡里矢车鸡菊花●☆

81555 Centauropsis fruticosa Bojer ex DC.;灌木矢车鸡菊花●☆

81556 Centauropsis lanuginosa Bojer ex DC. = Oliganthes lanuginosa (Bojer ex DC.) Humbert ●☆

81557 Centauropsis laurifolia Humbert;月桂叶矢车鸡菊花●☆

81558 Centauropsis perrieri Humbert;佩里耶矢车鸡菊花●☆

81559 Centauropsis rhaponticoides (Baker) Drake;缘膜矢车鸡菊花●☆

81560 Centauropsis rutenbergiana Vatke =Oliganthes lanuginosa (Bojer ex DC.) Humbert ●☆

81561 Centauropsis vilersii Humbert;维莱尔斯矢车鸡菊花●☆

81562 Centaurothamnus Wagenitz et Dittrich(1982);小矢车木属■☆

81563 Centaurothamnus maximus (Forssk.) Wagenitz et Dittrich;小矢车木■☆

81564 Centella L. (1763);积雪草属(雷公根属);Pennywort ■

81565 Centella abbreviata (A. Rich.) Nannf.;缩短积雪草■☆

81566 Centella affinis (Eckl. et Zeyh.) Adamson;近缘积雪草■☆

81567 Centella affinis (Eckl. et Zeyh.) Adamson var. oblonga Adamson;矩圆近缘积雪草■☆

81568 Centella annua M. T. R. Schub. et B. -E. van Wyk;一年积雪草■☆

81569 Centella arbuscula (Schltr.) Domin =Centella rupestris (Eckl. et Zeyh.) Adamson ■☆

81570 Centella asiatica (L.) Urb.;积雪草(半边钱,半边月,蚌壳草,崩大碗,崩口碗,遍地金钱草,遍地香,草如意,大金钱草,大马蹄草,大叶金钱草,大叶伤筋草,灯盏菜,地浮萍,地钱草,地棠草,地细辛,鼎盖草,复老碗草,蚶壳草,鸳圭草,葫瓜草,节节连,九杯菜,酒杯菜,葵蓬菜,老公根,老鸦碗,雷公根,连钱草,落得打,落地梅花,马脚迹,马蹄草,牛浴菜,盘龙草,破铜钱,钱齿草,钱凿口,四棱草,四芝麻稞,铁灯盏,铜钱草,土细辛,野冬苋菜,野荠菜);Asiatic Centella,Asiatic Pennywort,Spadeleaf ■

81571 Centella asiatica (L.) Urb. f. crispata (Maxim.) H. Hara = Centella asiatica (L.) Urb. ■

81572 Centella asiatica (L.) Urb. var. boninensis (Nakai ex Tuyama) H. Hara f. crispata? =Centella asiatica (L.) Urb. ■

81573 Centella asiatica (L.) Urb. var. boninensis (Nakai ex Tuyama) H. Hara =Centella asiatica (L.) Urb. ■

81574 Centella boninensis Nakai ex Tuyama =Centella asiatica (L.) Urb. ■

81575 Centella brachycarpa M. T. R. Schub. et B. -E. van Wyk;短果积雪草■☆

81576 Centella bupleurifolia (A. Rich.) Adamson =Centella glabrata L. ■☆

81577 Centella caespitosa Adamson;丛生积雪草■☆

81578 Centella calcaria M. T. R. Schub. et B. -E. van Wyk;距积雪草■☆

81579 Centella capensis (L.) Domin;好望角积雪草■☆

81580 Centella capensis (L.) Domin var. micrantha Adamson = Centella annua M. T. R. Schub. et B. -E. van Wyk ■☆

81581 Centella cochlearia (Domin) Adamson;螺状积雪草■☆

81582 Centella comptonii Adamson;康普顿积雪草■☆

81583 Centella coriacea Nannf. =Centella asiatica (L.) Urb. ■

81584 Centella cryptocarpa M. T. R. Schub. et B. -E. van Wyk;隐果积雪草■☆

81585 Centella debilis (Eckl. et Zeyh.) Drude;弱小积雪草■☆

81586 Centella dentata Adamson;具齿积雪草■☆

81587 Centella didymocarpa Adamson;双果积雪草■☆

81588 Centella difformis (Eckl. et Zeyh.) Adamson;不齐积雪草■☆

81589 Centella dolichocarpa M. T. R. Schub. et B. -E. van Wyk;长果积雪草■☆

81590 Centella dregeana (Sond.) Domin =Centella tridentata (L. f.) Drude ex Domin var. dregeana (Sond.) M. T. R. Schub. et B. -E. van Wyk ■☆

81591 Centella eriantha (Rich.) Drude;毛花积雪草■☆

81592 Centella eriantha (Rich.) Drude var. orientalis Adamson;东方毛花积雪草■☆

81593 Centella eriantha (Rich.) Drude var. rotundifolia Adamson;圆叶毛花积雪草■☆

81594 Centella filicaulis (Baker) Domin = Centella tussilaginifolia (Baker) Domin ■☆

81595 Centella flexuosa (Eckl. et Zeyh.) Drude;曲折积雪草■☆

81596 Centella fourcadei Adamson;富尔卡德积雪草■☆

81597 Centella fusca (Eckl. et Zeyh.) Adamson;棕色积雪草■☆

81598 Centella glabrata L.;光滑积雪草■☆

81599 Centella glabrata L. var. bracteata Adamson;具苞积雪草■☆

81600 Centella glabrata L. var. cochlearia Domin = Centella cochlearia (Domin) Adamson ■☆

81601 Centella glabrata L. var. natalensis Adamson;纳塔尔光滑积雪草■☆

81602 Centella glauca M. T. R. Schub. et B. -E. van Wyk;灰绿积雪草■☆

81603 Centella graminifolia Adamson;禾叶积雪草■☆

81604 Centella gymnocarpa M. T. R. Schub. et B. -E. van Wyk;裸果积雪草■☆
81605 Centella hederifolia (Burch.) Drude = Centella macrodus (Spreng.) B. L. Burtt ■☆
81606 Centella hermanniifolia (Eckl. et Zeyh.) Domin = Centella tridentata (L. f.) Drude ex Domin var. hermanniifolia (Eckl. et Zeyh.) M. T. R. Schub. et B. -E. van Wyk ■☆
81607 Centella hermanniifolia (Eckl. et Zeyh.) Domin var. littoralis? = Centella tridentata (L. f.) Drude ex Domin var. litoralis (Eckl. et Zeyh.) M. T. R. Schub. et B. -E. van Wyk ■☆
81608 Centella laevis Adamson;平滑积雪草■☆
81609 Centella lanata Compton;绵毛积雪草■☆
81610 Centella linifolia (L. f.) Drude;亚麻叶积雪草■☆
81611 Centella linifolia (L. f.) Drude var. depressa Adamson;凹陷积雪草■☆
81612 Centella longifolia (Adamson) M. T. R. Schub. et B. -E. van Wyk;长叶积雪草■☆
81613 Centella macrocarpa (Rich.) Adamson;大果积雪草■☆
81614 Centella macrocarpa (Rich.) Adamson var. saxatilis Adamson;岩生大果积雪草■☆
81615 Centella macrodus (Spreng.) B. L. Burtt;大齿积雪草■☆
81616 Centella montana (Cham. et Schltdl.) Domin;山地积雪草■☆
81617 Centella montana (Cham. et Schltdl.) Domin var. longifolia Adamson = Centella longifolia (Adamson) M. T. R. Schub. et B. -E. van Wyk ■☆
81618 Centella obtriangularis Cannon;倒三角积雪草■☆
81619 Centella pilosa M. T. R. Schub. et B. -E. van Wyk;疏毛积雪草■☆
81620 Centella pottebergensis Adamson;波太伯格积雪草■☆
81621 Centella recticarpa Adamson;直果积雪草■☆
81622 Centella restioides Adamson;绳草积雪草■☆
81623 Centella rupestris (Eckl. et Zeyh.) Adamson;岩石积雪草■☆
81624 Centella scabra Adamson;粗糙积雪草■☆
81625 Centella sessilis Adamson;无柄积雪草■☆
81626 Centella solandra Drude = Centella capensis (L.) Domin ■☆
81627 Centella stenophylla Adamson;窄叶积雪草■☆
81628 Centella stipitata Adamson;具柄积雪草■☆
81629 Centella ternata M. T. R. Schub. et B. -E. van Wyk;三出积雪草■☆
81630 Centella thesioides M. T. R. Schub. et B. -E. van Wyk;百蕊草积雪草■☆
81631 Centella tridentata (L. f.) Drude ex Domin;三齿积雪草■☆
81632 Centella tridentata (L. f.) Drude ex Domin var. dregeana (Sond.) M. T. R. Schub. et B. -E. van Wyk;德雷积雪草■☆
81633 Centella tridentata (L. f.) Drude ex Domin var. hermanniifolia (Eckl. et Zeyh.) M. T. R. Schub. et B. -E. van Wyk;密钟木叶积雪草■☆
81634 Centella tridentata (L. f.) Drude ex Domin var. litoralis (Eckl. et Zeyh.) M. T. R. Schub. et B. -E. van Wyk;滨海三齿积雪草■☆
81635 Centella triloba (Thunb.) Drude;三裂积雪草■☆
81636 Centella tussilaginifolia (Baker) Domin;款冬叶积雪草■☆
81637 Centella ulugurensis (Engl.) Domin;乌卢古尔积雪草■☆
81638 Centella umbellata M. T. R. Schub. et B. -E. van Wyk;小伞积雪草■☆
81639 Centella verticillata (Thunb.) Fourc. = Hydrocotyle verticillata Thunb. ■☆
81640 Centella villosa L.;毛叶积雪草■☆
81641 Centella villosa L. var. latifolia (Eckl. et Zeyh.) Adamson;宽毛叶积雪草■☆
81642 Centella villosa L. var. major Sond. = Centella villosa L. var. latifolia (Eckl. et Zeyh.) Adamson ■☆
81643 Centella virgata (L. f.) Drude;条纹积雪草■☆
81644 Centella virgata (L. f.) Drude var. congesta Adamson;密集积雪草■☆
81645 Centella virgata (L. f.) Drude var. gracilescens Domin;纤细条纹积雪草■☆
81646 Centema Hook. f. (1880);花刺苋属■●☆
81647 Centema alternifolia Schinz = Neocentema alternifolia (Schinz) Schinz ■☆
81648 Centema angolensis Hook. f.;安哥拉花刺苋■☆
81649 Centema biflora Schinz = Centemopsis biflora (Schinz) Schinz ■☆
81650 Centema cruciata Schinz = Kyphocarpa cruciata (Schinz) Schinz ■☆
81651 Centema gracilenta Hiern = Centemopsis gracilenta (Hiern) Schinz ■☆
81652 Centema kirkii Hook. f. = Centemopsis kirkii (Hook. f.) Schinz ■☆
81653 Centema polygonoides Lopr. = Centemopsis biflora (Schinz) Schinz ■☆
81654 Centema rubra Lopr. = Centemopsis kirkii (Hook. f.) Schinz ■☆
81655 Centema stefaninii Chiov. = Eriostylos stefaninii (Chiov.) C. C. Towns. ■☆
81656 Centema subfusca (Moq.) T. Cooke;花刺苋■☆
81657 Centemopsis Schinz(1911);类花刺苋属■☆
81658 Centemopsis biflora (Schinz) Schinz;双花类花刺苋■☆
81659 Centemopsis clausii Schinz = Centemopsis kirkii (Hook. f.) Schinz ■☆
81660 Centemopsis conferta (Schinz) Suess.;密集类花刺苋■☆
81661 Centemopsis fastigiata (Suess.) C. C. Towns.;帚状类花刺苋■☆
81662 Centemopsis filiformis (E. A. Bruce) C. C. Towns.;线形类花刺苋■☆
81663 Centemopsis glomerata (Lopr.) Schinz;紧密类花刺苋■☆
81664 Centemopsis gracilenta (Hiern) Schinz;细黏类花刺苋■☆
81665 Centemopsis graminea (Suess. et Overkott) C. C. Towns.;禾叶类花刺苋■☆
81666 Centemopsis kirkii (Hook. f.) Schinz;柯克类花刺苋■☆
81667 Centemopsis kirkii (Hook. f.) Schinz f. intermedia Suess. = Centemopsis kirkii (Hook. f.) Schinz ■☆
81668 Centemopsis longipedunculata (Peter) C. C. Towns.;长梗类花刺苋■☆
81669 Centemopsis micrantha Chiov.;小花类花刺苋■☆
81670 Centemopsis myurus Suess. = Centemopsis gracilenta (Hiern) Schinz ■☆
81671 Centemopsis polygonoides (Lopr.) Suess. = Centemopsis biflora (Schinz) Schinz ■☆
81672 Centemopsis rubra (Lopr.) Schinz = Centemopsis kirkii (Hook. f.) Schinz ■☆
81673 Centemopsis sordida C. C. Towns.;污浊类花刺苋■☆
81674 Centemopsis trichotoma Suess. = Centemopsis fastigiata (Suess.) C. C. Towns. ■☆
81675 Centemopsis trinervis Hauman;三脉类花刺苋■☆
81676 Centhriscus Spreng. ex Steud. = Anthriscus Pers. (保留属名)■
81677 Centinodia (Rchb.) Rchb. = Polygonum L. (保留属名)■●
81678 Centinodia Rchb. = Polygonum L. (保留属名)■●
81679 Centinodium (Rchb.) Montandon = Centinodia (Rchb.) Rchb. ■●

81680 Centinodium (Rchb.) Montandon = Polygonum L. (保留属名) ■●

81681 Centipeda Lour. (1790);石胡荽属;Centipeda,Sneezeweed ■●

81682 Centipeda capensis Less. = Dichrocephala integrifolia (L. f.) Kuntze ■

81683 Centipeda cunninghamii (DC.) A. Braun et Asch.;杉石胡荽;Cunningham Centipeda ■☆

81684 Centipeda minima (L.) A. Braun et Asch. = Centipeda orbicularis Lour. ■

81685 Centipeda minuta (Less.) C. B. Clarke = Bidens pilosa L. ■

81686 Centipeda minuta (Less.) C. B. Clarke = Centipeda minima (L.) A. Braun et Asch. ■

81687 Centipeda orbicularis Lour.;石胡荽(白地茜,白顶顶,百珠子草,不食草,大救驾,地胡椒,地芫荽,地杨梅,杜网草,鹅不食,鹅不食草,鹅仔不食草,二郎草,二郎戟,二郎箭,杠网草,鸡肠草,连地稗,满天星,猫沙,疟疾草,球子草,三节剑,三牙戟,三牙钻,散星草,沙飞草,砂药草,山胡椒,食胡荽,铁拳头,通天窍,吐金草,蚊子草,雾水沙,小救驾,小拳头,小石胡荽,野园荽,珠子草,猪尿草,猪屎草,猪屎潺);Small Centipeda,Spreading Sneezeweed ■

81688 Centipeda orbicularis Lour. = Centipeda minima (L.) A. Braun et Asch. ■

81689 Centipeda thespidioides F. Muell.;澳石胡荽■☆

81690 Centopodium Burch. = Emex Campd. (保留属名) ■☆

81691 Centosteca Desv. = Centotheca Desv. (保留属名) ■

81692 Centotheca Desv. (1810)(保留属名)('Centosteca');假淡竹叶属(牛蒡芒属,酸模芒属);Centotheca ■

81693 Centotheca P. Beauv. = Centotheca Desv. (保留属名) ■

81694 Centotheca lappacea (L.) Desv.;假淡竹叶(山鸡谷,酸模芒);Common Centotheca ■

81695 Centotheca lappacea L. subsp. inermis (Rendle) T. Koyama = Centotheca lappacea (L.) Desv. ■

81696 Centotheca lappacea L. var. inermis Rendle = Centotheca lappacea (L.) Desv. ■

81697 Centotheca lappacea L. var. longilamina (Ohwi) Bor = Centotheca lappacea (L.) Desv. ■

81698 Centotheca latifolia Trin. = Centotheca lappacea (L.) Desv. ■

81699 Centotheca latifolia Trin. = Megastachya mucronata (Poir.) P. Beauv. ■☆

81700 Centotheca longilamina Ohwi = Centotheca lappacea (L.) Desv. ■

81701 Centotheca maxima Peter = Megastachya mucronata (Poir.) P. Beauv. ■☆

81702 Centotheca mucronata (Poir.) Kuntze = Megastachya mucronata (Poir.) P. Beauv. ■☆

81703 Centotheca owariensis Hack. = Megastachya mucronata (Poir.) P. Beauv. ■☆

81704 Centotheca parviflora Peters = Centotheca lappacea (L.) Desv. ■

81705 Centrachaena Less. = Centrachena Schott ■●

81706 Centrachena Schott = Chrysanthemum L. (保留属名) ■●

81707 Centrachena Schott ex Rchb. = Chrysanthemum L. (保留属名) ■●

81708 Centradenia G. Don(1832);距药花属●■☆

81709 Centradenia floribunda Planch.;圆锥距药花;Spanish Shawl Flower ●☆

81710 Centradenia grandifolia Endl. ex Walp.;大花距药花;Sea of Flowers ●☆

81711 Centradeniastrum Cogn. (1908);小距药花属■☆

81712 Centradeniastrum roseum Cogn.;小距药花■☆

81713 Centrandra H. Karst. = Julocroton Mart. (保留属名) ●■☆

81714 Centranthera R. Br. (1810);胡麻草属;Centranthera ■

81715 Centranthera Scheidw. = Pleurothallis R. Br. ■☆

81716 Centranthera brunoniana Wall. = Centranthera cochinchinensis (Lour.) Merr. ■

81717 Centranthera chevalieri Bonati = Centranthera cochinchinensis (Lour.) Merr. subsp. lutea (H. Hara) T. Yamaz. ■

81718 Centranthera cochinchinensis (Lour.) Merr.;胡麻草(金锁匙,兰胡麻草);Cochichina Centranthera ■

81719 Centranthera cochinchinensis (Lour.) Merr. subsp. lutea (H. Hara) T. Yamaz.;中南胡麻草■

81720 Centranthera cochinchinensis (Lour.) Merr. subsp. lutea (H. Hara) T. Yamaz. f. alba Hiyama;白花中南胡麻草■☆

81721 Centranthera cochinchinensis (Lour.) Merr. subsp. lutea (H. Hara) T. Yamaz. = Dicentra peregrina (Rudolphi) Makino ■

81722 Centranthera cochinchinensis (Lour.) Merr. var. longiflora (Merr.) P. C. Tsoong;长花胡麻草;Longflower Cochichina Centranthera ■

81723 Centranthera cochinchinensis (Lour.) Merr. var. longiflora (Merr.) P. C. Tsoong = Centranthera cochinchinensis (Lour.) Merr. ■

81724 Centranthera cochinchinensis (Lour.) Merr. var. lutea (H. Hara) H. Hara = Dicentra peregrina (Rudolphi) Makino ■

81725 Centranthera cochinchinensis (Lour.) Merr. var. lutea (H. Hara) H. Hara = Centranthera cochinchinensis (Lour.) Merr. subsp. lutea (H. Hara) T. Yamaz. ■

81726 Centranthera cochinchinensis (Lour.) Merr. var. nepalensis (D. Don) Merr.;西南胡麻草;Nepal Centranthera ■

81727 Centranthera grandiflora Benth.;大花胡麻草(滑野蚕豆,化血丹,金猫头,灵芝草,小红药,野蚕豆);Bigflower Centranthera ■

81728 Centranthera hispida R. Br. = Centranthera cochinchinensis (Lour.) Merr. ■

81729 Centranthera hispida R. Br. = Centranthera cochinchinensis (Lour.) Merr. var. nepalensis (D. Don) Merr. ■

81730 Centranthera humifusa Wall. = Centranthera tranquebarica (Spreng.) Merr. ■

81731 Centranthera longiflora (Merr.) Merr. = Centranthera cochinchinensis (Lour.) Merr. ■

81732 Centranthera nepalensis D. Don = Centranthera cochinchinensis (Lour.) Merr. var. nepalensis (D. Don) Merr. ■

81733 Centranthera rubra H. L. Li. = Centranthera cochinchinensis (Lour.) Merr. subsp. lutea (H. Hara) T. Yamaz. ■

81734 Centranthera tonkinensis Bonati;细瘦胡麻草;Slender Centranthera ■

81735 Centranthera tonkinensis Bonati = Centranthera tranquebarica (Spreng.) Merr. ■

81736 Centranthera tranquebarica (Spreng.) Merr.;矮胡麻草;Dwarf Centranthera ■

81737 Centranthera tranquebarica (Spreng.) T. Yamaz. = Centranthera tranquebarica (Spreng.) Merr. ■

81738 Centrantheropsis Bonati = Phtheirospermum Bunge ex Fisch. et C. A. Mey. ■

81739 Centrantheropsis Bonati(1914);假胡麻草属;Centrantheropsis ●★

81740 Centrantheropsis rigida Bonati;假胡麻草;Centrantheropsis ■

81741 Centranthus DC. = Centranthus Lam. et DC. ■

81742 Centranthus Lam. et DC. = Centranthus Neck. ex Lam. et DC. ■

81743 Centranthus Neck. ex Lam. et DC. (1805);距药草属(距花属,中花属);Centranth,Jupiter's Beard,Red Valerian,Valerian ■

81744 Centranthus albus?;白距药草(白距花);Delicate Bess ■☆

81745 Centranthus angustifolius DC. subsp. battandieri (Maire) Maire = Centranthus nevadensis Boiss. ■☆

81746 Centranthus angustifolius DC. subsp. maroccanus (Rouy) Maire = Centranthus nevadensis Boiss. subsp. marrocanus (Rouy) Dobignard ■☆

81747 Centranthus angustifolius DC. subsp. nevadensis (Boiss.) Maire = Centranthus nevadensis Boiss. ■☆

81748 Centranthus angustifolius DC. subsp. rifanus (Emb. et Maire) Maire = Centranthus nevadensis Boiss. ■☆

81749 Centranthus angustifolius DC. var. macrocentron Maire = Centranthus nevadensis Boiss. ■☆

81750 Centranthus battandieri Maire = Centranthus nevadensis Boiss. subsp. battandieri (Maire) Fern. Casas et Molero ■☆

81751 Centranthus calcitrapae (L.) Dufr.;矢车菊距药草;Annual Valerian ■☆

81752 Centranthus calcitrapae (L.) Dufr. subsp. trichocarpus I. Richardson = Centranthus calcitrapae (L.) Dufr. ■☆

81753 Centranthus calcitrapae (L.) Dufr. var. clausonis (Pomel) Batt. = Centranthus calcitrapae (L.) Dufr. ■☆

81754 Centranthus calcitrapae (L.) Dufr. var. orbiculatus (Sibth. et Sm.) DC. = Centranthus calcitrapae (L.) Dufr. ■☆

81755 Centranthus calcitrapae (L.) Dufr. var. trichocarpus (I. Richardson) O. Bolòs et Vigo = Centranthus calcitrapae (L.) Dufr. ■☆

81756 Centranthus clausonis Pomel = Centranthus calcitrapae (L.) Dufr. ■☆

81757 Centranthus lecoqii Jord. subsp. maroccanus (Rouy) I. Richardson = Centranthus nevadensis Boiss. subsp. marrocanus (Rouy) Dobignard ■☆

81758 Centranthus longiflorus Steven;长花距花■☆

81759 Centranthus longiflorus Steven subsp. atlanticus I. Richardson = Centranthus nevadensis Boiss. subsp. marrocanus (Rouy) Dobignard ■☆

81760 Centranthus macrosiphon Boiss.;大管距花;Spur-Valarian ■☆

81761 Centranthus macrosiphon Boiss. var. micranthus Willk. = Centranthus calcitrapae (L.) Dufr. ■☆

81762 Centranthus maroccanus Rouy = Centranthus nevadensis Boiss. subsp. marrocanus (Rouy) Dobignard ■☆

81763 Centranthus maroccanus Rouy var. macrocentron Maire = Centranthus nevadensis Boiss. subsp. marrocanus (Rouy) Dobignard ■☆

81764 Centranthus nevadensis Boiss.;内华达距花■☆

81765 Centranthus nevadensis Boiss. subsp. battandieri (Maire) Fern. Casas et Molero;巴坦距药草■☆

81766 Centranthus nevadensis Boiss. subsp. marrocanus (Rouy) Dobignard;摩洛哥距花■☆

81767 Centranthus nevadensis Boiss. var. maroccanus Pau = Centranthus nevadensis Boiss. ■☆

81768 Centranthus ruber (L.) DC.;红距花(红穿心排草,红中花,距药草);American Lilac, Bloody Butcher, Bouncing Bess, Bouncing Betty, Bovisand Sailor, Bovisand Sailors, Capon's Tail, Capon's Tails, Cat Bed, Cat's Love, Convict Grass, Devon Pride, Drunkards, Drunkard's Nose, Drunken Sailor, Drunken Willy, Drunkits, Fox's Brush, German Lilac, Good Neighbourhood, Good Neighbours, Ground Lilac, Gypsy Maids, Jupiter's Beard, Jupiter's-beard, Jupiter's-beard Centranthus, Kiss-behind-the-pantry-door, Kissing Kind, Kiss-me, Kiss-me-love, Kiss-me-quick, Kiss-tile-garden-door, Lady Betty, Lady's Needlework, Midsummer Men, Neighbours, Old Woman's Needlework, Pinheads, Pretty Baby, Pretty Betsy, Prince of Wales' Feathers, Queen Anne's Needlework, Quiet Neighbours, Red Cow Basil, Red Money, Red Valerian, Red-spurred Valerian, Roguery, Saucy Bet, Scarlet Lightning, Soldier Boys, Soldier's Pride, Spur Valerian, Sweet Betsy, Sweet Betty, Sweet Mary, Wall Lilac ■☆

81769 Centrapalus Cass.(1817);糙毛菊属■☆

81770 Centrapalus Cass. = Vernonia Schreb.(保留属名)●■

81771 Centrapalus acrocephalus (Klatt) H. Rob. = Vernonia acrocephala Klatt ■☆

81772 Centrapalus africanus (Sond.) H. Rob.;非洲糙毛菊■☆

81773 Centrapalus chthonocephalus (O. Hoffm.) H. Rob. = Vernonia chthonocephala O. Hoffm. ●☆

81774 Centrapalus denudatus (Hutch. et B. L. Burtt) H. Rob. = Vernonia denudata Hutch. et B. L. Burtt ■☆

81775 Centrapalus galamensis Cass. = Vernonia galamensis (Cass.) Less. ■☆

81776 Centrapalus kirkii (Oliv. et Hiern) H. Rob. = Vernonia kirkii Oliv. et Hiern ●☆

81777 Centrapalus pauciflorus (Willd.) H. Rob. = Vernonia galamensis (Cass.) Less. ■☆

81778 Centrapalus praemorsus (Muschl.) H. Rob. = Vernonia praemorsa Muschl. ■●☆

81779 Centrapalus purpureus (Sch. Bip. ex Walp.) H. Rob. = Vernonia purpurea Sch. Bip. ex Walp. ■☆

81780 Centrapalus subaphyllus (Baker) H. Rob. = Vernonia subaphylla Baker ■☆

81781 Centratherum Cass.(1817);中芒菊属(蓝冠菊属);Larkdaisy ■☆

81782 Centratherum angustifolium (Benth.) C. D. Adams = Kinghamia angustifolia (Benth.) C. Jeffrey ■☆

81783 Centratherum anthelminticum (Willd.) Kuntze;驱虫中芒菊■☆

81784 Centratherum englerianum Muschl. = Kinghamia engleriana (Muschl.) C. Jeffrey ■☆

81785 Centratherum intermedium Less.;间型中芒菊;Brazilian Bachelor Button ■☆

81786 Centratherum muticum (Kunth) Less.;无尖中芒菊■☆

81787 Centratherum punctatum Cass.;斑点中芒菊;Larkdaisy ■☆

81788 Centridobotryon Klotzsch ex Pfeiff. = Centridobryon Klotzsch ex Pfeiff. ●■

81789 Centridobryon Klotzsch ex Pfeiff. = Piper L. ●■

81790 Centrilla Lindau = Justicia L. ●■

81791 Centrocarpha D. Don = Rudbeckia L. ■

81792 Centrocarpha grandiflora Sweet = Rudbeckia grandiflora (Sweet) C. C. Gmel. ex DC. ■☆

81793 Centrochilus Schauer = Habenaria Willd. ■

81794 Centrochilus Schauer = Platanthera Rich.(保留属名)■

81795 Centrochilus gracilis Schauer = Habenaria linguella Lindl. ■

81796 Centrochloa Swallen(1935);巴西雀稗属■☆

81797 Centrochloa singularis Swallen;巴西雀稗■☆

81798 Centrochrosia Post et Kuntze = Kentrochrosia Lauterb. et K. Schum. ●

81799 Centroclinium D. Don = Onoseris Willd. ●■☆

81800 Centrodiscus Müll. Arg. = Caryodendron H. Karst. ●☆

81801 Centrogenium Schltr. = Eltroplectris Raf. ■☆

81802 Centrogenium Schltr. = Stenorrhynchos Rich. ex Spreng. ■☆

81803 Centrogenium setaceum (Lindl.) Schltr. = Eltroplectris calcarata

(Sw.) Garay et H. R. Sweet ■☆
81804 Centrogenium setaceum Lindl. = Eltroplectris calcarata (Sw.) Garay et H. R. Sweet ■☆
81805 Centroglossa Barb. Rodr. (1882); 距舌兰属 ■☆
81806 Centroglossa greeniana Cogn.; 格林距舌兰 ■☆
81807 Centroglossa macroceras Barb. Rodr.; 距舌兰 ■☆
81808 Centrogonium Willis = Centrogenium Schltr. ■☆
81809 Centrogyne Welw. ex Benth. et Hook. f. = Bosqueia Thouars ex Baill. ●☆
81810 Centrolepidaceae Desv. = Centrolepidaceae Endl. (保留科名) ■
81811 Centrolepidaceae Endl. (1836) (保留科名); 刺鳞草科; Centrolepis Family ■
81812 Centrolepis Labill. (1804); 刺鳞草属; Centrolepis ■
81813 Centrolepis asiatica Merr. ex Gagnep. = Centrolepis banksii (R. Br.) Roem. et Schult. ■
81814 Centrolepis banksii (R. Br.) Roem. et Schult.; 刺鳞草; Banks Centrolepis ■
81815 Centrolepis hainanensis Merr. et Metcalf = Centrolepis banksii (R. Br.) Roem. et Schult. ■
81816 Centrolepis miboroides Gagnep. = Centrolepis banksii (R. Br.) Roem. et Schult. ■
81817 Centrolobium Mart. ex Benth. (1837); 刺片豆属; Porcupine Pod Tree, Porcupine-pod Tree ●☆
81818 Centrolobium minus C. Presl; 小刺片豆 ●☆
81819 Centrolobium ochroxylum Rose ex Rudd; 刺片豆 ●☆
81820 Centrolobium paraense Tul.; 帕拉州刺片豆 ●☆
81821 Centrolobium robustum Mart. ex Benth.; 粗壮刺片豆; Zebrawood ●☆
81822 Centrolobium sclerophyllum H. C. Lima; 硬叶刺片豆 ●☆
81823 Centrolobium tomentosum Guillaumin ex Benth.; 毛刺片豆 ●☆
81824 Centromadia Greene = Hemizonia DC. ■☆
81825 Centromadia Greene (1894); 星刺菊属; Spikeweed ■☆
81826 Centromadia fitchii (A. Gray) Greene; 菲奇星刺菊; Fitch's spikeweed ■☆
81827 Centromadia maritima Greene = Centromadia pungens (Hook. et Arn.) Greene ■☆
81828 Centromadia parryi (Greene) Greene; 帕里星刺菊 ■☆
81829 Centromadia parryi (Greene) Greene subsp. australis (D. D. Keck) B. G. Baldwin; 南方星刺菊 ■☆
81830 Centromadia parryi (Greene) Greene subsp. congdonii (B. L. Rob. et Greenm.) B. G. Baldwin; 康登星刺菊 ■☆
81831 Centromadia parryi (Greene) Greene subsp. rudis (Greene) B. G. Baldwin; 粗糙星刺菊 ■☆
81832 Centromadia pungens (Hook. et Arn.) Greene; 星刺菊; Common Spikeweed ■☆
81833 Centromadia pungens (Hook. et Arn.) Greene subsp. laevis (D. D. Keck) B. G. Baldwin; 平滑星刺菊 ■☆
81834 Centromadia pungens (Hook. et Arn.) Greene subsp. maritima (Greene) B. G. Baldwin = Centromadia pungens (Hook. et Arn.) Greene ■☆
81835 Centromadia pungens (Hook. et Arn.) Greene subsp. septentrionalis (D. D. Keck) B. G. Baldwin = Centromadia pungens (Hook. et Arn.) Greene ■☆
81836 Centromadia rudis Greene = Centromadia parryi (Greene) Greene subsp. rudis (Greene) B. G. Baldwin ■☆
81837 Centronia Blume = Aeginetia L. ■
81838 Centronia Blume = Centronota A. DC. ■
81839 Centronia D. Don (1823); 刺萼野牡丹属 ●☆
81840 Centronia laurifolia D. Don; 刺萼野牡丹 ●☆
81841 Centronia pulchra Cogn.; 美丽刺萼野牡丹 ●☆
81842 Centronia reticulata Triana; 网脉刺萼野牡丹 ●☆
81843 Centronia sessilifolia Cogn.; 无柄刺萼野牡丹 ●☆
81844 Centronia tomentosa Cogn.; 毛刺萼野牡丹 ●☆
81845 Centronota A. DC. = Aeginetia L. ■
81846 Centropappus Hook. f. (1847); 泌液菊属 ■☆
81847 Centropappus Hook. f. = Senecio L. ■●
81848 Centropappus brunonis Hook. f.; 泌液菊 ■☆
81849 Centropetalum Lindl. = Fernandezia Ruiz et Pav. ■☆
81850 Centrophorum Trin. (废弃属名) = Chrysopogon Trin. (保留属名) ■
81851 Centrophorum chinense Trin. = Chrysopogon aciculatus (Retz. ex Roem. et Schult.) Trin. ■
81852 Centrophyllum Dumort. = Carthamus L. ■
81853 Centrophyta Rchb. = Astragalus L. ●■
81854 Centrophyta Rchb. = Kentrophyta Nutt. ●■
81855 Centroplacaceae Doweld et Reveal = Pandaceae Engl. et Gilg (保留科名) ●
81856 Centroplacaceae Doweld et Reveal (2005); 裂药树科 ●☆
81857 Centroplacus Pierre (1899); 裂药树属 (小花木属) ●☆
81858 Centroplacus glaucinus Pierre; 裂药树 (小花木) ●☆
81859 Centropodia (R. Br.) Rchb. (1829); 白霜草属 ■☆
81860 Centropodia (R. Br.) Rchb. = Danthonia DC. (保留属名) ■
81861 Centropodia Rchb. = Centropodia (R. Br.) Rchb. ■☆
81862 Centropodia forskalii (Vahl) Cope; 福斯科尔白霜草 ■☆
81863 Centropodia fragilis (Guinet et Sauvage) Cope; 脆白霜草 ■☆
81864 Centropodia glauca (Nees) Cope; 灰绿白霜草 ■☆
81865 Centropodia mossamedensis (Rendle) Cope; 莫萨梅迪白霜草 ■☆
81866 Centropodium Lindl. = Centopodium Burch. ■☆
81867 Centropodium Lindl. = Emex Campd. (保留属名) ■☆
81868 Centropogon C. Presl (1836); 须距桔梗属 ●■☆
81869 Centropogon alatus Gleason; 翅须距桔梗 ■☆
81870 Centropogon angustus Gleason; 窄须距桔梗 ■☆
81871 Centropsis Endl. = Kentropsis Moq. ●☆
81872 Centrosema (DC.) Benth. (1837) (保留属名); 距瓣豆属 (山珠豆属); Butter Pea, Butterflypea, Butterfly-pea, Centrosema, Conchita, Spurstandard ●■☆
81873 Centrosema Benth. = Centrosema (DC.) Benth. (保留属名) ●■☆
81874 Centrosema DC. = Centrosema (DC.) Benth. (保留属名) ●■☆
81875 Centrosema plumieri (Pers.) Benth.; 普吕米距瓣豆 ●☆
81876 Centrosema pubescens Benth.; 距瓣豆 (山珠豆); Centro, Pubescent Butterflypea, Pubescent Centrosema, Spurstandard ●
81877 Centrosema sagittatum (Humb. et Bonpl. ex Willd.) Brandegee; 箭头距瓣豆; Arrowleaf Butterfly Pea ●☆
81878 Centrosema virginianum (L.) Benth.; 北美距瓣豆; Piedmont Butterfly-pea, Spurred Butterfly Pea ●☆
81879 Centrosema virginianum Benth. = Centrosema virginianum (L.) Benth. ●☆
81880 Centrosepis R. Hedw. = Centrolepis Labill. ■
81881 Centrosia A. Rich. = Calanthe R. Br. (保留属名) ■
81882 Centrosia A. Rich. = Centrosis Sw. ■☆
81883 Centrosia auberti A. Rich. = Calanthe sylvatica (Thouars) Lindl. ■
81884 Centrosis Sw. = Limodorum Boehm. (保留属名) ■☆
81885 Centrosis Sw. ex Thouars = Alismorkis Thouars (废弃属名) ■
81886 Centrosis Thouars = Calanthe R. Br. (保留属名) ■

81887 Centrosis abortiva Sw. = Limodorum abortivum (L.) Sw. ■☆
81888 Centrosis sylvatica Thouars = Calanthe sylvatica (Thouars) Lindl. ■
81889 Centrosolenia Benth. (废弃属名) = Nautilocalyx Linden ex Hanst. (保留属名) ■☆
81890 Centrospermum Kunth (废弃属名) = Acanthospermum Schrank (保留属名) ■
81891 Centrospermum Spreng. = Chrysanthemum L. (保留属名) ■●
81892 Centrospermum xanthoides Kunth = Acanthospermum australe (Loefl.) Kuntze ■
81893 Centrosphaera Post et Kuntze = Kentrosphaera Volkens ■☆
81894 Centrosphaera Post et Kuntze = Volkensinia Schinz ■☆
81895 Centrostachys Wall. (1824); 湿生苋属 ■☆
81896 Centrostachys aquatica (R. Br.) Wall.; 湿生苋 ■☆
81897 Centrostachys conferta (Schinz) Standl. = Centemopsis conferta (Schinz) Suess. ■☆
81898 Centrostachys schinzii Standl. = Pandiaka lanuginosa (Schinz) Schinz ■☆
81899 Centrostegia A. Gray ex Benth. = Centrostegia A. Gray ■☆
81900 Centrostegia A. Gray (1856); 刺苞蓼属; Red Triangles ■☆
81901 Centrostegia insignis (Curran) A. Heller = Aristocapsa insignis (Curran) Reveal et Hardham ■☆
81902 Centrostegia leptoceras A. Gray = Dodecahema leptoceras (A. Gray) Reveal et Hardham ■☆
81903 Centrostegia thurberi (A. Gray) S. Watson var. macrotheca J. T. Howell = Centrostegia thurberi A. Gray ■☆
81904 Centrostegia thurberi A. Gray = Centrostegia thurberi A. Gray ex Benth. ■☆
81905 Centrostegia thurberi A. Gray ex Benth.; 刺苞蓼 ■☆
81906 Centrostegia thurberi A. Gray ex Benth. var. macrotheca (J. T. Howell) Goodman = Centrostegia thurberi A. Gray ex Benth. ■☆
81907 Centrostegia vortriedei (Brandegee) Goodman = Systenotheca vortriedei (Brandegee) Reveal et Hardham ■☆
81908 Centrostemma Baill. = Ceratostema Juss. ●☆
81909 Centrostemma Decne. (1838); 蜂出巢属(飞凤花属); Centrostemma ●
81910 Centrostemma Decne. = Hoya R. Br. ●
81911 Centrostemma multiflorum (Blume) Decne.; 蜂出巢(飞凤花); Hoya Shooting Star, Manyflower Centrostemma, Multiflorous Waxplant, Multiflower Waxplant ●
81912 Centrostemma multiflorum (Blume) Decne. = Hoya multiflora Blume ●
81913 Centrostemma platypetalum Merr. = Hoya multiflora Blume ●
81914 Centrostemma yunnanense P. T. Li; 云南蜂出巢; Yunnan Centrostemma ●
81915 Centrostemma yunnanense P. T. Li = Hoya lii C. M. Burton ●
81916 Centrostigma Schltr. (1915); 距柱兰属 ■☆
81917 Centrostigma clavatum Summerh.; 棒状距柱兰 ■☆
81918 Centrostigma nyassanum Schltr. = Centrostigma occultans (Welw. ex Rchb. f.) Schltr. ■☆
81919 Centrostigma occultans (Welw. ex Rchb. f.) Schltr.; 距柱兰 ■☆
81920 Centrostigma papillosum Summerh.; 乳突距柱兰 ■☆
81921 Centrostigma schlechteri (Kraenzl.) Schltr. = Centrostigma occultans (Welw. ex Rchb. f.) Schltr. ■☆
81922 Centrostylis Baill. (1858); 距柱大戟属 ■☆
81923 Centrostylis Baill. = Adenochlaena Boiss. ex Baill. ■☆
81924 Centrostylis zeylanica Baill.; 距柱大戟 ■☆
81925 Centunculus Adans. = Cerastium L. ■
81926 Centunculus L. = Anagallis L. ■
81927 Centunculus minimus L. = Anagallis minima (L.) E. H. L. Krause ■☆
81928 Centunculus tenellus Duby = Anagallis pumila Sw. ■☆
81929 Ceodes J. R. Forst. et G. Forst. (1775); 胶果木属(肖腺果藤属) ●
81930 Ceodes J. R. Forst. et G. Forst. = Pisonia L. ●
81931 Ceodes grandis (R. Br.) D. Q. Lu = Pisonia grandis R. Br. ●
81932 Ceodes longirostris (Teijsm. et Binn.) Merr. et L. M. Perry; 长喙胶果木 ●
81933 Ceodes umbellifera J. R. Forst. et G. Forst.; 胶果木(大叶避霜花,皮孙木,皮孙树,伞花腺果藤,伞形花腺果藤,水冬瓜); Bird-card Catchbird Tree, Bird-catcher, Bird-catcher Tree, Birdlime Tree, Malay Catchbird Tree, Para Para, Para-para ●
81934 Ceodes umbellifera J. R. Forst. et G. Forst. = Pisonia umbellifera (J. R. Forst. et G. Forst.) Seem. ●
81935 Cepa Kuntze = Eurycles Salisb. ■☆
81936 Cepa Kuntze = Proiphys Herb. ■☆
81937 Cepa Mill. = Allium L. ■
81938 Cepa prolifera Moench. = Allium cepa L. 'Proliferum' ■
81939 Cepa prolifera Moench. = Allium cepa L. var. proliferum Regel ■
81940 Cepaceae Salisb. = Alliaceae Borkh. (保留科名) ■
81941 Cepaea Caesalp. ex Fourr. = Sedum L. ●■
81942 Cepaea Fabr. = Sedum L. ●■
81943 Cepaeaceae Salisb. = Alliaceae Borkh. (保留科名) ■
81944 Cepalaria Raf. = Cephalaria Schrad. (保留属名) ■
81945 Cephaelis Sw. (1788) (保留属名); 头九节属(吐根属,头花属); Cephaelis, Ninenode ●
81946 Cephaelis Sw. (保留属名) = Psychotria L. (保留属名) ●
81947 Cephaelis abouabouensis Schnell = Psychotria abouabouensis (Schnell) Verdc. ■☆
81948 Cephaelis acuminata H. Karst.; 尖叶头九节(巴西吐根,尖叶吐根); Acuminate Cephaelis ●
81949 Cephaelis baillehachei Aké Assi = Psychotria peduncularis (Salisb.) Steyerm. var. tabouensis (Schnell) Verdc. ●☆
81950 Cephaelis biaurita (Hutch. et Dalziel) Hepper = Psychotria biaurita (Hutch. et Dalziel) Verdc. ●☆
81951 Cephaelis bidentata Benth.; 二齿头九节 ●☆
81952 Cephaelis bidentata Thunb. ex Roem. et Schult. = Psychotria bidentata (Thunb. ex Roem. et Schult.) Hiern ●☆
81953 Cephaelis bieleri (De Wild.) Bremek. = Gaertnera bieleri (De Wild.) E. M. Petit ●☆
81954 Cephaelis camponutans Dwyer et Hayden; 巴拿马头九节(巴拿马吐根) ●☆
81955 Cephaelis castaneo-pilosa Aké Assi; 栗毛头九节 ●☆
81956 Cephaelis condensata A. Chev. = Psychotria biaurita (Hutch. et Dalziel) Verdc. ●☆
81957 Cephaelis congensis Hiern = Psychotria psychotrioides (DC.) Roberty ●☆
81958 Cephaelis coriacea G. Don = Psychotria peduncularis (Salisb.) Steyerm. var. guineensis (Schnell) Verdc. ●☆
81959 Cephaelis cornuta Hiern = Psychotria vogeliana Benth. ●☆
81960 Cephaelis debauxii (De Wild. ex Schnell) Schnell = Psychotria peduncularis (Salisb.) Steyerm. ●☆
81961 Cephaelis densinervia (K. Krause) Hepper = Psychotria densinervia (K. Krause) Verdc. ●☆

81962 Cephaelis elata Sw.;高头九节(高吐根)●☆

81963 Cephaelis emetica (L. f.) Pers. = Psychotria yapoensis (Schnell) Verdc. ●☆

81964 Cephaelis ferruginea G. Don = Sabicea ferruginea (G. Don) Benth. ●☆

81965 Cephaelis fuscescens Hiern;浅棕色头九节●☆

81966 Cephaelis goetzei (K. Schum.) Hepper = Psychotria megalopus Verdc. ●☆

81967 Cephaelis gossweileri Cavaco;戈斯头九节●☆

81968 Cephaelis guerzeensis (Schnell) Schnell = Psychotria peduncularis (Salisb.) Steyerm. ●☆

81969 Cephaelis guineensis (Schnell) Schnell = Psychotria peduncularis (Salisb.) Steyerm. var. guineensis (Schnell) Verdc. ●☆

81970 Cephaelis ipecacuatha (Brot.) A. Rich.;吐根;Ipecac, Ipecacuanha ●

81971 Cephaelis ituriensis (De Wild.) Cavaco = Psychotria ituriensis De Wild. ex E. M. Petit ●☆

81972 Cephaelis ivorensis (Schnell) Schnell = Psychotria peduncularis (Salisb.) Steyerm. var. ivorensis (Schnell) Verdc. ●☆

81973 Cephaelis konkourensis Schnell = Psychotria rufipilis De Wild. ●☆

81974 Cephaelis laui (Merr. et F. P. Metcalf) Chun et F. C. How = Cephaelis laui (Merr. et F. P. Metcalf) F. C. How et W. C. Ko ●

81975 Cephaelis laui (Merr. et F. P. Metcalf) F. C. How et W. C. Ko;头九节;Lau Cephaelis, Lau Ninenode ●

81976 Cephaelis mangenotii Aké Assi = Psychotria mangenotii (Aké Assi) Verdc. ●☆

81977 Cephaelis mannii (Hook. f.) Hiern = Psychotria camptopus Verdc. ●☆

81978 Cephaelis micheliae J. -G. Adam;米歇尔头九节●☆

81979 Cephaelis nimbana (Schnell) Schnell = Psychotria peduncularis (Salisb.) Steyerm. var. suaveolens (Hiern) Verdc. ●☆

81980 Cephaelis ombrophila (Schnell) Schnell = Psychotria ombrophila (Schnell) Verdc. ●☆

81981 Cephaelis peduncularis Salisb. = Psychotria peduncularis (Salisb.) Steyerm. ●☆

81982 Cephaelis peduncularis Salisb. var. guineensis (Schnell) Hepper = Psychotria peduncularis (Salisb.) Steyerm. var. guineensis (Schnell) Verdc. ●☆

81983 Cephaelis peduncularis Salisb. var. ivorensis (Schnell) Hepper = Psychotria peduncularis (Salisb.) Steyerm. var. ivorensis (Schnell) Verdc. ●☆

81984 Cephaelis peduncularis Salisb. var. suaveolens (Hiern) Hepper = Psychotria peduncularis (Salisb.) Steyerm. var. suaveolens (Hiern) Verdc. ●☆

81985 Cephaelis peduncularis Salisb. var. tabouensis (Schnell) Hepper = Psychotria peduncularis (Salisb.) Steyerm. var. tabouensis (Schnell) Verdc. ●☆

81986 Cephaelis pedunculata G. Don = Psychotria peduncularis (Salisb.) Steyerm. ●☆

81987 Cephaelis rubescens Hiern;变红九节●☆

81988 Cephaelis sangalkamensis (Schnell) Schnell = Psychotria bidentata (Thunb. ex Roem. et Schult.) Hiern ●☆

81989 Cephaelis schnellii Aké Assi = Psychotria schnellii (Aké Assi) Verdc. ●☆

81990 Cephaelis siamica Craib = Psychotria prainii H. Lév. ●

81991 Cephaelis spathacea Hiern = Psychotria spathacea (Hiern) Verdc. ●☆

81992 Cephaelis suaveolens Hiern = Psychotria peduncularis (Salisb.) Steyerm. var. suaveolens (Hiern) Verdc. ●☆

81993 Cephaelis tabouensis Schnell = Psychotria peduncularis (Salisb.) Steyerm. var. tabouensis (Schnell) Verdc. ●☆

81994 Cephaelis talbotii Wernham = Psychotria abouabouensis (Schnell) Verdc. ■☆

81995 Cephaelis tomentosa (Aubl.) Vahl;绒毛头九节(绒毛吐根)●☆

81996 Cephaelis tomentosa Vahl = Cephaelis tomentosa (Aubl.) Vahl ●☆

81997 Cephaelis yapoensis (Schnell) Schnell = Psychotria yapoensis (Schnell) Verdc. ●☆

81998 Cephalacanthus Lindau(1905);头刺爵床属■☆

81999 Cephalacanthus maculatus Lindau;头刺爵床■☆

82000 Cephalandra Eckl. et Zeyh. = Coccinia Wight et Arn. ■

82001 Cephalandra Schrad. = Coccinia Wight et Arn. ■

82002 Cephalandra Schrad. ex Eckl. et Zeyh. = Coccinia Wight et Arn. ■

82003 Cephalandra decipiens Hook. f. = Diplocyclos decipiens (Hook. f.) C. Jeffrey ■☆

82004 Cephalandra indica (Wight. et Arn.) Naudin = Coccinia grandis (L.) Voigt ■

82005 Cephalandra indica Naudin = Coccinia grandis (L.) Voigt ■

82006 Cephalandra ivorensis A. Chev. = Ruthalicia eglandulosa (Hook. f.) C. Jeffrey ■☆

82007 Cephalandra mackennii Naudin = Coccinia palmata (Sond.) Cogn. ■☆

82008 Cephalandra palmata Sond. = Coccinia palmata (Sond.) Cogn. ■☆

82009 Cephalandra pubescens Sond. = Coccinia adoensis (A. Rich.) Cogn. ■☆

82010 Cephalandra quinqueloba (Thunb.) Schrad. = Coccinia quinqueloba (Thunb.) Cogn. ■☆

82011 Cephalandra senensis Klotzsch = Coccinia senensis (Klotzsch) Cogn. ■☆

82012 Cephalandra sessilifolia Sond. = Coccinia sessilifolia (Sond.) Cogn. ■☆

82013 Cephalandra sylvatica A. Chev. = Ruthalicia eglandulosa (Hook. f.) C. Jeffrey ■☆

82014 Cephalangraecum Schltr. = Ancistrorhynchus Finet ■☆

82015 Cephalangraecum braunii (T. Durand et Schinz) Summerh. = Ancistrorhynchus metteniae (Kraenzl.) Summerh. ■☆

82016 Cephalangraecum glomeratum (Ridl.) Schltr. = Ancistrorhynchus cephalotes (Rchb. f.) Summerh. ■☆

82017 Cephalanophlos Fourr. (1869);刺儿菜属;Cephalanoplos ■

82018 Cephalanophlos Fourr. = Cirsium Mill. ■

82019 Cephalanophlos Neck. = Cephalanophlos Fourr. (1869) ■

82020 Cephalanophlos Neck. = Cirsium Mill. + Serratula L. ■

82021 Cephalanophlos Neck. = Cnicus L. (保留属名) ■●

82022 Cephalanophlos segetum (Bunge) Kitam.;小刺儿菜(白鸡角刺,刺菜,刺菜芽,刺刺牙,刺儿菜,刺儿草,刺儿蓟,刺杆菜,刺蓟菜,刺尖头草,刺角菜,刺萝卜,刺杀草,刻叶刺儿菜,猫蓟,木刺艾,牛戳刺,萋萋菜,萋萋芽,荠荠毛,千针草,枪刀菜,青刺蓟,青青菜,曲曲菜,小刺盖,小刺头,小恶鸡婆,小鸡角刺,小蓟,小蓟草,小蓟姆,小牛扎口,野红花);Common Cephalanoplos ■

82023 Cephalanophlos segetum (Bunge) Kitam. = Cirsium segetum Bunge ■

82024 Cephalanophlos setosum (Willd.) Kitam.;大刺儿菜(刺儿菜,大蓟,刻叶刺儿菜);Setose Cephalanoplos ■

82025 Cephalanophlos setosum (Willd.) Kitam. = Cirsium setosum

(Willd.) M. Bieb. ■

82026 Cephalanthaceae Dumort. = Rubiaceae Juss.(保留科名)●■

82027 Cephalanthaceae Raf. = Rubiaceae Juss.(保留科名)●■

82028 Cephalanthera Rich. (1817); 头蕊兰属(金兰属); Cephalanthera, Helleborine, Phantom Orchid, Skull Orchid ■

82029 Cephalanthera acuminata Lindl. = Cephalanthera longifolia (L.) Fritsch ■

82030 Cephalanthera acuminata Lindl. ex Wall. = Cephalanthera longifolia (L.) Fritsch ■

82031 Cephalanthera alpicola Fukuy.;高山头蕊兰;Alp Cephalanthera ■

82032 Cephalanthera austiniae (A. Gray) A. Heller; 雪白头蕊兰; Phantom Orchid, Snow Orchid ■☆

82033 Cephalanthera bijiangensis S. C. Chen; 碧江头蕊兰; Bijiang Cephalanthera ■

82034 Cephalanthera bijiangensis S. C. Chen = Cephalanthera falcata (Thunb. ex A. Murray) Blume ■

82035 Cephalanthera calcalata S. C. Chen et K. Y. Lang;硕距头蕊兰; Spur Cephalanthera ■

82036 Cephalanthera caucasica Kraenzl.;高加索头蕊兰■☆

82037 Cephalanthera cucullata Boiss. et Heldr.; 僧帽头蕊兰; Hooded Helleborine ■☆

82038 Cephalanthera damasonium (Mill.) Druce;大花头蕊兰(苍白头蕊兰,大马逊头蕊兰); Bigflower Cephalanthera, Broad Helleborine, Egg Orchid, Large White Helleborine, Largeflower Cephalanthera, Poached-egg Plant, White Helleborine, Wood Lily ■

82039 Cephalanthera elegans Schltr. = Cephalanthera falcata (Thunb. ex A. Murray) Blume ■

82040 Cephalanthera ensifolia (Murray) Rich. = Cephalanthera longifolia (L.) Fritsch ■

82041 Cephalanthera ensifolia (Murray) Rich. var. acuminata (Lindl.) Ts. Tang et F. T. Wang = Cephalanthera longifolia (L.) Fritsch ■

82042 Cephalanthera ensifolia (Sw.) Rich. var. acuminata (Lindl.) Ts. Tang et F. T. Wang = Cephalanthera longifolia (L.) Fritsch ■

82043 Cephalanthera epipactoides Fisch.;东方头蕊兰;Eastern Hooded Helleborine ■☆

82044 Cephalanthera erecta (Thunb. ex A. Murray) Blume;银兰(白花草,小毛钩儿花兰,鱼头兰花); Erect Cephalanthera, Silver Cephalanthera ■

82045 Cephalanthera erecta (Thunb. ex A. Murray) Blume = Cephalanthera longifolia (L.) Fritsch ■

82046 Cephalanthera erecta (Thunb.) Blume var. elegans (Schltr.) Masam.;雅致银兰■☆

82047 Cephalanthera erecta (Thunb.) Blume var. shizuoi (F. Maek.) Ohwi;静冈银兰■☆

82048 Cephalanthera erecta (Thunb.) Blume var. subaphylla (Miyabe et Kudo) Ohwi;无叶银兰■☆

82049 Cephalanthera erecta Blume = Cephalanthera erecta (Thunb. ex A. Murray) Blume ■

82050 Cephalanthera erecta Blume var. szechuanica Schltr. = Cephalanthera erecta (Thunb. ex A. Murray) Blume ■

82051 Cephalanthera falcata (Thunb. ex A. Murray) Blume;金兰(黄花兰,头蕊兰,桠雀兰);Falcate Cephalanthera, Gold Cephalanthera ■

82052 Cephalanthera falcata (Thunb. ex A. Murray) Blume f. albescens S. Kobay.;白金兰■☆

82053 Cephalanthera gracilis S. C. Chen et G. H. Zhu;细头蕊兰(纤细头蕊兰)■

82054 Cephalanthera grandiflora (L.) Babal. = Cephalanthera damasonium (Mill.) Druce ■

82055 Cephalanthera grandiflora Babal. = Cephalanthera damasonium (Mill.) Druce ■

82056 Cephalanthera japonica A. Gray = Cephalanthera falcata (Thunb. ex A. Murray) Blume ■

82057 Cephalanthera latifolia Druce = Cephalanthera damasonium (Mill.) Druce ■

82058 Cephalanthera longibracteata Blume; 长苞头蕊兰; Longbract Cephalanthera ■

82059 Cephalanthera longibracteata Blume f. lurida Hayashi;光亮长苞头蕊兰■☆

82060 Cephalanthera longifolia (L.) Fritsch;头蕊兰(长叶头蕊兰); Longleaf Cephalanthera, Long-leaved Helleborine, Narrow Helleborine, Narrow-leaved Helleborine, Sword-leaved Helleborine ■

82061 Cephalanthera longifolia (L.) Fritsch var. pilosa Harz = Cephalanthera longifolia (L.) Fritsch ■

82062 Cephalanthera mairei Schltr. = Cephalanthera longifolia (L.) Fritsch ■

82063 Cephalanthera nanlingensis A. Q. Hu et F. W. Xing;南岭头蕊兰■

82064 Cephalanthera pallens (Willd.) Fritsch = Cephalanthera damasonium (Mill.) Druce ■

82065 Cephalanthera pallens Loisel. = Cephalanthera damasonium (Mill.) Druce ■

82066 Cephalanthera platycheila Rchb. f. = Cephalanthera falcata (Thunb. ex A. Murray) Blume ■

82067 Cephalanthera raymondiae Schltr. = Cephalanthera falcata (Thunb. ex A. Murray) Blume ■

82068 Cephalanthera royleana (Lindl.) Regel = Epipactis royleana Lindl. ■

82069 Cephalanthera royleana (Regel) Boiss. = Epipactis gigantea Douglas ex Hook. ■☆

82070 Cephalanthera rubra (L.) Rich.;红头蕊兰(红花头蕊兰);Red Helleborine, Redflower Cephalanthera ■☆

82071 Cephalanthera rubra Rich. = Cephalanthera rubra (L.) Rich. ■☆

82072 Cephalanthera shizuoi F. Maek. = Cephalanthera erecta (Thunb.) Blume var. shizuoi (F. Maek.) Ohwi ■☆

82073 Cephalanthera subaphylla Miyabe et Kudo = Cephalanthera erecta (Thunb.) Blume var. subaphylla (Miyabe et Kudo) Ohwi ■☆

82074 Cephalanthera szechuanica (Schltr.) Schltr. = Cephalanthera erecta (Thunb. ex A. Murray) Blume ■

82075 Cephalanthera taiwaniana S. S. Ying; 台湾头蕊兰; Taiwan Cephalanthera ■

82076 Cephalanthera thomsonii Rchb. f.;云南头蕊兰■

82077 Cephalanthera thomsonii Rchb. f. = Cephalanthera longifolia (L.) Fritsch ■

82078 Cephalanthera xylophyllum L. f.;剑叶头蕊兰■☆

82079 Cephalanthera xyphophyllum (L. f.) Rchb. = Cephalanthera longifolia (L.) Fritsch ■

82080 Cephalanthera xyphophyllum (L. f.) Rchb. var. latifolia Maire = Cephalanthera longifolia (L.) Fritsch ■

82081 Cephalanthera yunnanensis Hand. -Mazz. = Cephalanthera damasonium (Mill.) Druce ■

82082 Cephalantheropsis Guill. (1960);肖头蕊兰属(黄兰属)■

82083 Cephalantheropsis calanthoides (Ames) Tang S. Liu et H. J. Su = Cephalantheropsis halconensis (Ames) S. S. Ying ■

82084 Cephalantheropsis gracilis (Lindl.) S. Y. Hu = Cephalantheropsis obcordata (Lindl.) Ormerod ■

82085 Cephalantheropsis gracilis (Lindl.) S. Y. Hu var. calanthoides (Ames) T. P. Lin = Cephalantheropsis halconensis (Ames) S. S. Ying ■

82086 Cephalantheropsis halconensis (Ames) S. S. Ying;铃花黄兰(白花肖头蕊兰)■

82087 Cephalantheropsis longipes (Hook. f.) Ormerod;白花黄兰■

82088 Cephalantheropsis obcordata (Lindl.) Ormerod;黄花肖头蕊兰(长柄鹤顶兰,长茎虾脊兰,长轴鹤顶兰,黄兰,铃花肖头蕊兰,绿花肖头蕊兰,细茎鹤顶兰,细葶虾脊兰);Longstalk Phaius,Slender Calanthe ■

82089 Cephalantheropsis venusta (Schltr.) S. Y. Hu = Cephalantheropsis gracilis (Lindl.) S. Y. Hu ■

82090 Cephalantheropsis venusta (Schltr.) S. Y. Hu = Cephalantheropsis obcordata (Lindl.) Ormerod ■

82091 Cephalanthus L. (1753);风箱树属(风箱属);Button Bush, Buttonbush ●

82092 Cephalanthus africanus Rchb. ex DC. = Mitragyna inermis (Willd.) K. Schum. ●☆

82093 Cephalanthus chinensis Lam. = Breonia chinensis (Lam.) Capuron ●☆

82094 Cephalanthus coriaceus K. Schum. = Breonadia salicina (Vahl) Hepper et J. R. I. Wood ●☆

82095 Cephalanthus glabrifolius Hayata = Cephalanthus tetrandrus (Roxb.) Ridsdale et Bakh. f. ●

82096 Cephalanthus natalensis Oliv.;纳塔尔风箱;Tree Strwberry ●☆

82097 Cephalanthus naucleoides DC. = Cephalanthus tetrandrus (Roxb.) Ridsdale et Bakh. f. ●

82098 Cephalanthus occidentalis L.;西方风箱树(风箱,风箱树,鸡仔木,假番桃,马烟木,水抱木,水杨梅,水杨梅蕴,獭狗耳,杨梅树,珠花树);Bachelor's Buttons,Butter Bush,Button Bush,Buttonbush,Common Buttonbush,Globe Flower,Globe-flowers,Honeyball,Honey-balls,Pond Dogwood,Wild Liquorice ●☆

82099 Cephalanthus occidentalis L. = Cephalanthus tetrandrus (Roxb.) Ridsdale et Bakh. f. ●

82100 Cephalanthus occidentalis L. f. lanceolatus Fernald = Cephalanthus occidentalis L. ●☆

82101 Cephalanthus occidentalis L. var. californicus Benth. = Cephalanthus occidentalis L. ●☆

82102 Cephalanthus occidentalis L. var. pubescens Raf. = Cephalanthus occidentalis L. ●☆

82103 Cephalanthus pilulifera Lam. = Adina pilulifera (Lam.) Franch. ex Drake ●

82104 Cephalanthus ratoensis Hayata = Cephalanthus tetrandrus (Roxb.) Ridsdale et Bakh. f. ●

82105 Cephalanthus spathelliferus Baker = Breonadia salicina (Vahl) Hepper et J. R. I. Wood ●☆

82106 Cephalanthus tetrandrus (Roxb.) Ridsdale = Cephalanthus tetrandrus (Roxb.) Ridsdale et Bakh. f. ●

82107 Cephalanthus tetrandrus (Roxb.) Ridsdale et Bakh. f.;风箱树(大叶柳,大叶水杨梅,红扎树,假杨梅,马烟树,山杨梅,水泡木,水杨梅,珠花树);Asiatic Button-bush,Buttonbush,Common Buttonbush ●

82108 Cephalantus piluliferus Lam. = Adina pilulifera (Lam.) Franch. ex Drake ●

82109 Cephalaralia Harms(1897);头楤木属●☆

82110 Cephalaralia cephalobotrys (F. Muell.) Harms;头楤木●☆

82111 Cephalaralia cephalobotrys Harms = Cephalaralia cephalobotrys (F. Muell.) Harms ●☆

82112 Cephalaria Roem. et Schult. = Cephalaria Schrad.(保留属名)■

82113 Cephalaria Schrad.(1818)(保留属名);头花草属(刺头草属,头刺草属,头序花属,蝇毒草属);Cephalanthera,Cephalaria,Giant Scabious,Skull Orchid ■

82114 Cephalaria Schrad. ex Roem. et Schult. = Cephalaria Schrad.(保留属名)■

82115 Cephalaria acaulis Steud. ex A. Rich. = Dipsacus pinnatifidus Steud. ex A. Rich. ■☆

82116 Cephalaria alpina Schrad.;黄花头花草;Yellow Cephalaria ■☆

82117 Cephalaria aristata K. Koch;具芒头花草■☆

82118 Cephalaria armeniaca Bordz.;亚美尼亚头花草■☆

82119 Cephalaria armerioides Szabó;海石竹头花草■☆

82120 Cephalaria atlantica Coss. et Durieu = Cephalaria mauritanica Pomel ■☆

82121 Cephalaria attenuata (L. f.) Roem. et Schult.;渐狭头花草■☆

82122 Cephalaria attenuata (L. f.) Roem. et Schult. var. oblongifolia Kuntze = Cephalaria oblongifolia (Kuntze) Szabó ■☆

82123 Cephalaria attenuata (Thunb.) Roem. et Schult. var. longifolia De Wild. = Cephalaria katangensis Napper ■☆

82124 Cephalaria beijiangensis Y. K. Yang, J. K. Wu et Sayit = Dipsacus azureus Schrenk ■

82125 Cephalaria boisseri Reut. = Cephalaria syriaca (L.) Roem. et Schult. ■☆

82126 Cephalaria brevipalea Litv.;短瓣头花草■☆

82127 Cephalaria cachemirica Decne. = Dipsacus inermis Wall. var. mitis (D. Don) Y. J. Nasir ■

82128 Cephalaria cachemirica Decne. = Dipsacus inermis Wall. ■

82129 Cephalaria calcarea Alb.;石灰头花草■☆

82130 Cephalaria coriacea Steud.;革质头花草■☆

82131 Cephalaria cretacea Roem. et Schult.;垩白蝇毒草■☆

82132 Cephalaria decurrens (Thunb.) Roem. et Schult.;下延头花草■☆

82133 Cephalaria demetrii Bobrov;戴氏头花草■☆

82134 Cephalaria dipsacoides Kar. et Kir. = Dipsacus azureus Schrenk ■

82135 Cephalaria foliosa Compton;多叶头花草■☆

82136 Cephalaria fragosoana Pau = Cephalaria leucantha (L.) Roem. et Schult. ■☆

82137 Cephalaria galpiniana Szabó;盖尔头花草■☆

82138 Cephalaria galpiniana Szabó subsp. simplicior B. L. Burtt;单一盖尔头花草■☆

82139 Cephalaria gigantea (Ledeb.) Bobrov;头花草(大刺头草,大聚首花);Cephalaria, Giant Scabious, Tartarian Cephalaria, Yellow Scabious ■

82140 Cephalaria goetzei Engl.;格兹头花草■☆

82141 Cephalaria grossheimii Bobrov;格氏头花草■☆

82142 Cephalaria humilis (Thunb.) Roem. et Schult.;低矮头花草■☆

82143 Cephalaria integrifolia Napper;全缘叶头花草■☆

82144 Cephalaria katangensis Napper;加丹加头花草■☆

82145 Cephalaria kotschyi Boiss. et Hohen.;考氏头花草■☆

82146 Cephalaria lavandulacea Sond. = Cephalaria attenuata (L. f.) Roem. et Schult. ■☆

82147 Cephalaria leucantha (L.) Roem. et Schult.;头刺草■☆

82148 Cephalaria leucantha (L.) Roem. et Schult. var. fragosoana (Pau) Maire = Cephalaria leucantha (L.) Roem. et Schult. ■☆

82149 Cephalaria leucantha Schrad. = Cephalaria leucantha (L.) Roem. et Schult. ■☆

82150 Cephalaria linearifolia Lange;线叶头花草■☆

82151 Cephalaria litvinovii Bobrov;鞑靼蝇毒草;Tatarian Cephalaria ■☆

82152 Cephalaria maroccana (Coss.) Batt. var. glabrescens Emb. et Maire = Cephalaria mauritanica Pomel ■☆

82153 Cephalaria maroccana (Coss.) Batt. var. mesatlantica Maire = Cephalaria mauritanica Pomel ■☆

82154 Cephalaria maroccana Batt. = Cephalaria mauritanica Pomel subsp. maroccana (Batt.) Maire ■☆

82155 Cephalaria mauritanica Pomel;毛里塔尼亚头花草■☆

82156 Cephalaria mauritanica Pomel subsp. atlantica (Coss. et Durieu) Quézel et Santa = Cephalaria mauritanica Pomel ■☆

82157 Cephalaria mauritanica Pomel subsp. maroccana (Batt.) Maire;摩洛哥头花草■☆

82158 Cephalaria mauritanica Pomel subsp. rifana Maire;里夫头花草■☆

82159 Cephalaria mauritanica Pomel var. atlantica Coss. et Durieu = Cephalaria mauritanica Pomel ■☆

82160 Cephalaria media Litv.;间型头花草■☆

82161 Cephalaria microdonta Bobrov;小齿头花草■☆

82162 Cephalaria natalensis Kuntze;纳塔尔头花草■☆

82163 Cephalaria oblongifolia (Kuntze) Szabó;矩圆叶头花草■☆

82164 Cephalaria petiolata Compton;柄叶头花草■☆

82165 Cephalaria pilosa Boiss. et Huet;疏毛头花草■☆

82166 Cephalaria procera Fisch.,C. A. Mey. et Avé-Lall.;高头花草■☆

82167 Cephalaria pungens Szabó;刚毛头花草■☆

82168 Cephalaria retrosetosa Engl. et Gilg;倒刚毛头花草■☆

82169 Cephalaria rigida (L.) Roem. et Schult.;硬头花草■☆

82170 Cephalaria scabra (L. f.) Roem. et Schult.;粗糙头花草■☆

82171 Cephalaria sparsipilosa Matthews;散毛头刺草■☆

82172 Cephalaria sublanata Szabo;近无毛头花草■☆

82173 Cephalaria syriaca (L.) Roem. et Schult.;叙利亚头花草(叙利亚蝇毒草);Syrian Cephalaria,Syrian Scabious ■☆

82174 Cephalaria syriaca (L.) Roem. et Schult. subsp. phoeniciaca Bobrov;凤凰头花草■☆

82175 Cephalaria syriaca (L.) Roem. et Schult. var. boissieri (Reut.) Boiss. = Cephalaria syriaca (L.) Roem. et Schult. ■☆

82176 Cephalaria syriaca Schrad. = Cephalaria syriaca (L.) Roem. et Schult. ■☆

82177 Cephalaria syriaca Schrad. var. boissieri (Reut.) Boiss. = Cephalaria syriaca (L.) Roem. et Schult. ■☆

82178 Cephalaria tatarica (L.) Roem. et Schult. = Cephalaria gigantea (Ledeb.) Bobrov ■

82179 Cephalaria tatarica Schrad. = Cephalaria gigantea (Ledeb.) Bobrov ■

82180 Cephalaria tatarica Schrad. = Cephalaria litvinovii Bobrov ■☆

82181 Cephalaria tchihatchewi Boiss.;齐氏头花草■☆

82182 Cephalaria transylvanica (L.) Schrad.;特兰蝇毒草■☆

82183 Cephalaria uralensis (Murray) Schrad.;乌拉尔蝇毒草;Ural Cephalaria ■☆

82184 Cephalaria uralensis Roem. et Schult. = Cephalaria uralensis (Murray) Schrad. ■☆

82185 Cephalaria ustulata (Thunb.) Roem. et Schult. = Cephalaria decurrens (Thunb.) Roem. et Schult. ■☆

82186 Cephalaria velutina Bobrov;短绒毛蝇毒草■☆

82187 Cephalaria wilmsiana Szabó;威尔头花草■☆

82188 Cephalaria zeyheriana Szabó;泽赫头花草■☆

82189 Cephaleis Vahl = Cephaelis Sw.(保留属名)●

82190 Cephaleis Vahl = Psychotria L.(保留属名)●

82191 Cephalidium A. Rich. = Anthocephalus A. Rich. ●☆

82192 Cephalidium A. Rich. ex DC. = Breonia A. Rich. ex DC. ●☆

82193 Cephalidium citrifolium (Poir.) A. Rich. = Breonia chinensis (Lam.) Capuron ●☆

82194 Cephalina Thonn. = Sarcocephalus Afzel. ex Sabine ●☆

82195 Cephalina esculenta (Afzel. ex Sabine) Schumach. et Thonn. = Sarcocephalus latifolius (Sm.) E. A. Bruce ●☆

82196 Cephalina richardii (Drake) Palacky = Breonia chinensis (Lam.) Capuron ●☆

82197 Cephalipterum A. Gray(1852);顶羽鼠麴草属■☆

82198 Cephalipterum druramondii A. Gray;顶羽鼠麴草■☆

82199 Cephalobembix Rydb. = Schkuhria Roth(保留属名)■☆

82200 Cephalocarpus Nees(1842);头果莎属■☆

82201 Cephalocarpus clarkei H. Pfeiff.;克氏头果莎■☆

82202 Cephalocarpus dracaenula Nees;头果莎■☆

82203 Cephalocarpus rigidus Gilly;硬头果莎■☆

82204 Cephalocereus Pfeiff.(1838);翁柱属;Cephalocereus ●

82205 Cephalocereus deeringii Small = Pilosocereus robinii (Lem.) Byles et G. D. Rowley ●☆

82206 Cephalocereus dybowskii (Rol. -Goss.) Britton et Rose;丽翁柱;Dybowsk Cephalocereus ●

82207 Cephalocereus keyensis Britton et Rose = Pilosocereus robinii (Lem.) Byles et G. D. Rowley ●☆

82208 Cephalocereus militaris (Audot) H. E. Moore;金毛翁柱;Goldenhair Cephalocereus ●

82209 Cephalocereus polylophus Britton et Rose;多冠翁柱●

82210 Cephalocereus scoparius (Poselg.) Britton et Rose;舞翁柱;Broomshiped Cephalocereus ●

82211 Cephalocereus senilis (Haw.) Pfeiff.;翁柱(白头翁,翁头仙人柱);Cabeza De Viejo,Old Man Cactus,Old Man of Mexico,Old-man Cactus ●

82212 Cephalochloa Coss. et Durieu = Ammochloa Boiss. ■☆

82213 Cephalocleistocactus F. Ritter = Cleistocactus Lem. ●☆

82214 Cephalocroton Hochst.(1841);肖巴豆属●☆

82215 Cephalocroton cordifolius Baker = Adenochlaena leucocephala Baill ●☆

82216 Cephalocroton cordofanum Hochst.;肖巴豆●☆

82217 Cephalocroton depauperatum Pax et K. Hoffm. = Cephalocroton molle Klotzsch ●☆

82218 Cephalocroton incanum M. G. Gilbert;灰毛肖巴豆●☆

82219 Cephalocroton molle Klotzsch;柔软肖巴豆●☆

82220 Cephalocroton nudum Pax et K. Hoffm. = Cephalocroton cordofanum Hochst. ●☆

82221 Cephalocroton polygynum Pax et K. Hoffm. = Cephalocroton cordofanum Hochst. ●☆

82222 Cephalocroton pueschelii Pax = Cephalocroton molle Klotzsch ●☆

82223 Cephalocroton scabridums Pax et K. Hoffm. = Cephalocroton cordofanum Hochst. ●☆

82224 Cephalocroton velutinum Pax et K. Hoffm. = Cephalocroton cordofanum Hochst. ●☆

82225 Cephalocrotonopsis Pax = Cephalocroton Hochst. ●☆

82226 Cephalocrotonopsis Pax(1910);类巴豆属●☆

82227 Cephalocrotonopsis socotrana Pax;类巴豆●☆

82228 Cephalodendron Steyerm.(1972);头木茜属●☆

82229 Cephalodendron aracamuniense Steyerm;头木茜●☆

82230 Cephalodendron globosum Steyerm;球头木茜●☆

82231 Cephalodes St. -Lag. = Cephalaria Schrad.(保留属名)■

82232 Cephalohibiscus Ulbr.(1935);头木槿属●☆

82233 Cephalohibiscus Ulbr. = Thespesia Sol. ex Corrêa(保留属名)●

82234 Cephalohibiscus peekelii Ulbr.;头木槿●☆

82235 Cephaloma Neck. = Dracocephalum L.(保留属名)■●

82236 Cephalomamillaria Fric. = Cephalomammillaria Frič ●

82237 Cephalomammillaria Frič = Epithelantha F. A. C. Weber ex Britton et Rose ●

82238 Cephalomappa Baill. (1874);肥牛木属(肥牛树属);Cephalomappa ●

82239 Cephalomappa baccariana Baill.;刺果肥牛树●

82240 Cephalomappa sinensis (Chun et F. C. How) Kosterm.;肥牛木(肥牛树);China Cephalomappa, Chinese Cephalomappa ●◇

82241 Cephalomedinilla Merr. = Medinilla Gaudich. ex DC. ●

82242 Cephalonema K. Schum. = Clappertonia Meisn. ●☆

82243 Cephalonema polyandrum K. Schum. ex Sprague = Clappertonia polyandra (K. Schum. ex Sprague) Bech. ●☆

82244 Cephalonoplos (Neck. ex DC.) Fourr. = Breea Less. ■

82245 Cephalonoplos Fourr. = Cirsium Mill. ■

82246 Cephalonoplos Neck. = Cirsium Mill. ■

82247 Cephalonoplos arvense (L.) Fourr. = Cirsium arvense (L.) Scop. ■

82248 Cephalonoplos arvense (L.) Fourr. var. alpestre (Naegeli) Kitam. = Cirsium lanatum (Roxb. ex Willd.) Spreng. ■

82249 Cephalonoplos segetum (Bunge) Kitam. = Cirsium setosum (Willd.) M. Bieb. ■

82250 Cephalonoplos setosum (M. Bieb.) Kitam. = Cirsium setosum (Willd.) M. Bieb. ■

82251 Cephalopanax Baill. = Acanthopanax Miq. ●

82252 Cephalopappus Nees et Mart. (1824);毛头钝柱菊属■☆

82253 Cephalopappus sonchifolius Nees et Mart.;毛头钝柱菊■☆

82254 Cephalopentandra Chiov. (1929);五头蕊属■☆

82255 Cephalopentandra ecirrhosa (Cogn.) C. Jeffrey;无须五头蕊■☆

82256 Cephalopentandra obbiadensis Chiov. = Cephalopentandra ecirrhosa (Cogn.) C. Jeffrey ■☆

82257 Cephalophilon (Meisn.) Spach = Persicaria (L.) Mill. ■

82258 Cephalophilon capitatum (Buch. -Ham. ex D. Don) Tzvelev = Persicaria capitata (Buch. -Ham. ex D. Don) H. Gross ■

82259 Cephalophilon capitatum (Buch. -Ham. ex D. Don) Tzvelev = Polygonum capitatum Buch. -Ham. ex D. Don ■

82260 Cephalophilon malaicum (Danser) Borodina = Polygonum chinense L. var. ovalifolium Meisn. ■

82261 Cephalophilon nepalense (Meisn.) Tzvelev = Persicaria nepalensis (Meisn.) H. Gross ■

82262 Cephalophilon nepalense (Meisn.) Tzvelev = Polygonum nepalense Meisn. ■

82263 Cephalophilon palmatum (Dunn) Borodina = Polygonum palmatum Dunn ■

82264 Cephalophilon runcinatum (Buch. -Ham. ex D. Don) Tzvelev = Polygonum runcinatum Buch. -Ham. ex D. Don ■

82265 Cephalophilum (Meisn.) Börner = Polygonum L.(保留属名)■●

82266 Cephalophilum (Meisn.) Börner = Tasoba Raf. ■●

82267 Cephalophilum Börner = Polygonum L.(保留属名)■●

82268 Cephalophilum Börner = Tasoba Raf. ■●

82269 Cephalophilum Meisn. ex Börner = Echinocaulon (Meisn.) Spach ■

82270 Cephalophilum Meisn. ex Börner = Tasoba Raf. ■●

82271 Cephalophora Cav. = Helenium L. ■

82272 Cephalophora scaposa DC. = Tetraneuris scaposa (DC.) Greene ■☆

82273 Cephalophorus Lem. = Cephalocereus Pfeiff. ●

82274 Cephalophorus Lem. = Cereus Mill. ●

82275 Cephalophorus Lem. ex Boom = Cephalocereus Pfeiff. ●

82276 Cephalophyllum (Haw.) N. E. Br. (1925);帝王花属(绘岛属);Red Spike Ice Plant ■☆

82277 Cephalophyllum Haw. = Cephalophyllum (Haw.) N. E. Br. ■☆

82278 Cephalophyllum N. E. Br. = Cephalophyllum (Haw.) N. E. Br. ■☆

82279 Cephalophyllum 'Red Spike';红穗帝王花;Red Spike Ice Plant ■☆

82280 Cephalophyllum acutum L. Bolus = Cephalophyllum subulatoides (Haw.) N. E. Br. ■☆

82281 Cephalophyllum albertiniense (L. Bolus) Schwantes = Jordaaniella dubia (Haw.) H. E. K. Hartmann ■☆

82282 Cephalophyllum alstonii Marloth ex L. Bolus;旭峰花(旭峰);Red Spike ■☆

82283 Cephalophyllum anemoniflorum (L. Bolus) N. E. Br. = Jordaaniella dubia (Haw.) H. E. K. Hartmann ■☆

82284 Cephalophyllum apiculatum L. Bolus = Cephalophyllum loreum (L.) Schwantes ■☆

82285 Cephalophyllum artum L. Bolus = Cephalophyllum curtophyllum (L. Bolus) Schwantes ■☆

82286 Cephalophyllum aurantiacum L. Bolus = Cephalophyllum purpureo-album (Haw.) Schwantes ■☆

82287 Cephalophyllum aureorubrum L. Bolus;赫丽花(赫丽)■☆

82288 Cephalophyllum aureorubrum L. Bolus = Cephalophyllum rigidum L. Bolus ■☆

82289 Cephalophyllum ausense L. Bolus = Cephalophyllum ebracteatum (Pax ex Schltr. et Diels) Dinter et Schwantes ■☆

82290 Cephalophyllum baylissii L. Bolus = Cephalophyllum diversiphyllum (Haw.) H. E. K. Hartmann ■☆

82291 Cephalophyllum bredasdorpense L. Bolus = Cephalophyllum diversiphyllum (Haw.) H. E. K. Hartmann ■☆

82292 Cephalophyllum breviflorum L. Bolus = Cephalophyllum pulchellum L. Bolus ■☆

82293 Cephalophyllum brevifolium L. Bolus ex Jacobsen = Cephalophyllum pulchellum L. Bolus ■☆

82294 Cephalophyllum caespitosum H. E. K. Hartmann;丛生帝王花■☆

82295 Cephalophyllum caledonicum L. Bolus;黎明花(黎明)■☆

82296 Cephalophyllum caledonicum L. Bolus = Cephalophyllum diversiphyllum (Haw.) H. E. K. Hartmann ■☆

82297 Cephalophyllum calvinianum L. Bolus = Leipoldtia rosea L. Bolus ●☆

82298 Cephalophyllum cauliculatum (Haw.) N. E. Br. = Cephalophyllum diversiphyllum (Haw.) H. E. K. Hartmann ■☆

82299 Cephalophyllum cedrimontanum L. Bolus = Cephalophyllum loreum (L.) Schwantes ■☆

82300 Cephalophyllum ceresianum L. Bolus;启丽花(启丽)■☆

82301 Cephalophyllum ceresianum L. Bolus = Cephalophyllum corniculatum (L.) Schwantes ■☆

82302 Cephalophyllum clavifolium (L. Bolus) L. Bolus = Jordaaniella clavifolia (L. Bolus) H. E. K. Hartmann ■☆

82303 Cephalophyllum compactum L. Bolus = Cephalophyllum loreum (L.) Schwantes ■☆

82304 Cephalophyllum compressum L. Bolus;扁帝王花■☆

82305 Cephalophyllum comptonii N. E. Br. = Cephalophyllum tricolorum (Haw.) Schwantes ■☆

82306 Cephalophyllum concinnum L. Bolus = Cephalophyllum curtophyllum (L. Bolus) Schwantes ■☆

82307 Cephalophyllum confusum (Dinter) Dinter et Schwantes;妆炎花(妆炎)■☆

82308 Cephalophyllum confusum Dinter et Schwantes = Cephalophyllum confusum (Dinter) Dinter et Schwantes ■☆

82309 Cephalophyllum conicum L. Bolus ex H. Jacobsen = Cephalophyllum curtophyllum (L. Bolus) Schwantes ■☆

82310 Cephalophyllum corniculatum (L.) Schwantes;圆锥帝王花■☆

82311 Cephalophyllum crassum L. Bolus = Cephalophyllum tricolorum (Haw.) Schwantes ■☆

82312 Cephalophyllum cupreum L. Bolus;新丽■☆

82313 Cephalophyllum cupreum L. Bolus = Jordaaniella cuprea (L. Bolus) H. E. K. Hartmann ■☆

82314 Cephalophyllum curtophyllum (L. Bolus) Schwantes;短叶帝王花■☆

82315 Cephalophyllum decipiens (Haw.) L. Bolus = Cephalophyllum loreum (L.) Schwantes ■☆

82316 Cephalophyllum densum N. E. Br. = Cephalophyllum framesii L. Bolus ■☆

82317 Cephalophyllum diminutum (Haw.) L. Bolus = Cephalophyllum subulatoides (Haw.) N. E. Br. ■☆

82318 Cephalophyllum dissimile (N. E. Br.) Schwantes = Cephalophyllum corniculatum (L.) Schwantes ■☆

82319 Cephalophyllum diversifolium (Haw.) Schwantes = Cephalophyllum diversiphyllum (Haw.) H. E. K. Hartmann ■☆

82320 Cephalophyllum diversifolium Schwantes;帝王花■☆

82321 Cephalophyllum diversiphyllum (Haw.) H. E. K. Hartmann;异叶帝王花■☆

82322 Cephalophyllum dubium (Haw.) N. E. Br. = Jordaaniella dubia (Haw.) H. E. K. Hartmann ■☆

82323 Cephalophyllum ebracteatum (Pax ex Schltr. et Diels) Dinter et Schwantes;无苞片帝王花■☆

82324 Cephalophyllum ernii L. Bolus = Cephalophyllum ebracteatum (Pax ex Schltr. et Diels) Dinter et Schwantes ■☆

82325 Cephalophyllum framesii L. Bolus;弗雷斯帝王花■☆

82326 Cephalophyllum franciscii L. Bolus = Cephalophyllum alstonii Marloth ex L. Bolus ■☆

82327 Cephalophyllum frutescens L. Bolus;摇炎花(摇炎)■☆

82328 Cephalophyllum frutescens L. Bolus = Leipoldtia frutescens (L. Bolus) H. E. K. Hartmann ●☆

82329 Cephalophyllum frutescens L. Bolus var. decumbens? = Leipoldtia frutescens (L. Bolus) H. E. K. Hartmann ●☆

82330 Cephalophyllum fulleri L. Bolus;富勒帝王花■☆

82331 Cephalophyllum goodii L. Bolus;古德帝王花■☆

82332 Cephalophyllum gracile L. Bolus;秀炎花(秀炎)■☆

82333 Cephalophyllum gracile L. Bolus = Cephalophyllum purpureo-album (Haw.) Schwantes ■☆

82334 Cephalophyllum gracile L. Bolus var. longisepalum? = Cephalophyllum purpureo-album (Haw.) Schwantes ■☆

82335 Cephalophyllum hallii L. Bolus;霍尔帝王花■☆

82336 Cephalophyllum herrei L. Bolus;赫勒帝王花■☆

82337 Cephalophyllum herrei L. Bolus f. decumbens? = Cephalophyllum numeesense H. E. K. Hartmann ■☆

82338 Cephalophyllum herrei L. Bolus var. decumbens? = Cephalophyllum numeesense H. E. K. Hartmann ■☆

82339 Cephalophyllum inaequale L. Bolus;不等帝王花■☆

82340 Cephalophyllum insigne L. Bolus = Cephalophyllum rigidum L. Bolus ■☆

82341 Cephalophyllum kliprandense L. Bolus = Cephalophyllum parvibracteatum (L. Bolus) H. E. K. Hartmann ■☆

82342 Cephalophyllum laetulum L. Bolus = Cephalophyllum ebracteatum (Pax ex Schltr. et Diels) Dinter et Schwantes ■☆

82343 Cephalophyllum latipetalum L. Bolus = Cephalophyllum curtophyllum (L. Bolus) Schwantes ■☆

82344 Cephalophyllum littlewoodii L. Bolus = Cephalophyllum purpureo-album (Haw.) Schwantes ■☆

82345 Cephalophyllum loreum (L.) Schwantes;楼炎花(楼炎)■☆

82346 Cephalophyllum loreum Schwantes = Cephalophyllum loreum (L.) Schwantes ■☆

82347 Cephalophyllum luxurians Dinter = Jordaaniella cuprea (L. Bolus) H. E. K. Hartmann ■☆

82348 Cephalophyllum maritimum (L. Bolus) Schwantes = Jordaaniella dubia (Haw.) H. E. K. Hartmann ■☆

82349 Cephalophyllum middlemostii L. Bolus = Cephalophyllum purpureo-album (Haw.) Schwantes ■☆

82350 Cephalophyllum namaquanum L. Bolus = Cephalophyllum ebracteatum (Pax ex Schltr. et Diels) Dinter et Schwantes ■☆

82351 Cephalophyllum niveum L. Bolus;雪白帝王花■☆

82352 Cephalophyllum numeesense H. E. K. Hartmann;努米斯帝王花■☆

82353 Cephalophyllum pallens L. Bolus = Cephalophyllum herrei L. Bolus ■☆

82354 Cephalophyllum parvibracteatum (L. Bolus) H. E. K. Hartmann;小苞帝王花■☆

82355 Cephalophyllum parviflorum L. Bolus;彩炎花(彩炎)■☆

82356 Cephalophyllum parviflorum L. Bolus var. proliferum? = Cephalophyllum parviflorum L. Bolus ■☆

82357 Cephalophyllum parvulum (Schltr.) H. E. K. Hartmann;较小帝王花■☆

82358 Cephalophyllum paucifolium L. Bolus = Cephalophyllum purpureo-album (Haw.) Schwantes ■☆

82359 Cephalophyllum pillansii L. Bolus;皮氏帝王花■☆

82360 Cephalophyllum pillansii L. Bolus var. grandiflorum? = Cephalophyllum pillansii L. Bolus ■☆

82361 Cephalophyllum pittenii L. Bolus = Jordaaniella dubia (Haw.) H. E. K. Hartmann ■☆

82362 Cephalophyllum platycalyx (L. Bolus) Schwantes = Cephalophyllum curtophyllum (L. Bolus) Schwantes ■☆

82363 Cephalophyllum primulinum (L. Bolus) Schwantes = Cephalophyllum loreum (L.) Schwantes ■☆

82364 Cephalophyllum procumbens (Haw.) Schwantes = Jordaaniella dubia (Haw.) H. E. K. Hartmann ■☆

82365 Cephalophyllum procumbens Schwantes;翠炎花(翠炎)■☆

82366 Cephalophyllum pulchellum L. Bolus;丛丽花(丛丽)■☆

82367 Cephalophyllum pulchrum L. Bolus;美丽帝王花■☆

82368 Cephalophyllum punctatum (Haw.) N. E. Br. = Cephalophyllum subulatoides (Haw.) N. E. Br. ■☆

82369 Cephalophyllum purpureo-album (Haw.) Schwantes;紫白帝王花■☆

82370 Cephalophyllum ramosum N. E. Br. = Cephalophyllum framesii L. Bolus ■☆

82371 Cephalophyllum rangei (Engl.) L. Bolus ex H. Jacobsen = Cephalophyllum ebracteatum (Pax ex Schltr. et Diels) Dinter et Schwantes ■☆

82372 Cephalophyllum regale L. Bolus;装炎花(装炎)■☆

82373 Cephalophyllum rhodandrum L. Bolus = Cephalophyllum curtophyllum (L. Bolus) Schwantes ■☆

82374 Cephalophyllum rigidum L. Bolus;硬帝王花■☆

82375 Cephalophyllum roseum (L. Bolus) L. Bolus = Leipoldtia rosea L. Bolus ●☆

82376 Cephalophyllum rostellum (L. Bolus) H. E. K. Hartmann;喙帝王花■☆

82377 Cephalophyllum serrulatum L. Bolus = Cephalophyllum purpureo-album (Haw.) Schwantes ■☆

82378 Cephalophyllum spissum H. E. K. Hartmann;密集帝王花■☆

82379 Cephalophyllum spongiosum (L. Bolus) L. Bolus = Jordaaniella spongiosa (L. Bolus) H. E. K. Hartmann ■☆

82380 Cephalophyllum spongiosum L. Bolus;龙炎花(龙炎)■☆

82381 Cephalophyllum stayneri L. Bolus;斯泰纳帝王花;Red Spike Ice Plant ■☆

82382 Cephalophyllum stayneri L. Bolus = Cephalophyllum framesii L. Bolus ■☆

82383 Cephalophyllum stayneri L. Bolus var. latipetalum? = Cephalophyllum framesii L. Bolus ■☆

82384 Cephalophyllum stayneri L. Bolus var. palladium? = Cephalophyllum framesii L. Bolus ■☆

82385 Cephalophyllum subulatoides (Haw.) N. E. Br.;彰炎花(彰炎)■☆

82386 Cephalophyllum subulatoides N. E. Br. = Cephalophyllum subulatoides (Haw.) N. E. Br. ■☆

82387 Cephalophyllum tenuifolium L. Bolus;典丽花(典丽)■☆

82388 Cephalophyllum tenuifolium L. Bolus = Cephalophyllum tricolorum (Haw.) Schwantes ■☆

82389 Cephalophyllum tetrastichum H. E. K. Hartmann;四列帝王花■☆

82390 Cephalophyllum tricolorum (Haw.) Schwantes;止利巧花(止利巧)■☆

82391 Cephalophyllum tricolorum N. E. Br. = Cephalophyllum tricolorum (Haw.) Schwantes ■☆

82392 Cephalophyllum truncatum L. Bolus = Cephalophyllum niveum L. Bolus ■☆

82393 Cephalophyllum uniflorum L. Bolus = Jordaaniella uniflora (L. Bolus) H. E. K. Hartmann ■☆

82394 Cephalophyllum vandermerwei L. Bolus = Cephalophyllum diversiphyllum (Haw.) H. E. K. Hartmann ■☆

82395 Cephalophyllum vanheerdei L. Bolus = Cephalophyllum regale L. Bolus ■☆

82396 Cephalophyllum vanputtenii L. Bolus = Jordaaniella dubia (Haw.) H. E. K. Hartmann ■☆

82397 Cephalophyllum watermeyeri L. Bolus = Jordaaniella dubia (Haw.) H. E. K. Hartmann ■☆

82398 Cephalophyllum weigangianum (Dinter) Dinter et Schwantes = Leipoldtia weigangiana (Dinter) Dinter et Schwantes ●☆

82399 Cephalophyllum worcesterense L. Bolus;耀丽花(耀丽)■☆

82400 Cephalophyllum worcesterense L. Bolus = Cephalophyllum purpureo-album (Haw.) Schwantes ■☆

82401 Cephalophyton Hook. f. ex Baker = Thonningia Vahl ■☆

82402 Cephalophyton parkeri Hook. f. ex Baker = Langsdorffia malagasica (Fawc.) B. Hansen ■☆

82403 Cephalopodum Korovin(1973);头梗芹属■☆

82404 Cephalopodum afghanicum (Rech. f. et Riedl) Pimenov et Kljuykov;阿富汗头梗芹■☆

82405 Cephalopodum badachshanicum Korovin;头梗芹■☆

82406 Cephalopterum T. Post et Kuntze = Cephalipterum A. Gray ■☆

82407 Cephalorhizum Popov et Korovin(1923);粗根补血草属●☆

82408 Cephalorhizum oopodum Popov et Korovin;粗根补血草■☆

82409 Cephalorhizum turcomanicum Popov;土库曼粗根补血草■☆

82410 Cephalorhyncus Boiss. = Cephalorrhynchus Boiss. ■

82411 Cephalorrhynchus Boiss. (1844);头嘴菊属(头喙苣属,头咀菊属,头嘴苣属);Cephalorrhynchus ■

82412 Cephalorrhynchus albiflorus C. Shih;白花头嘴菊;Whiteflower Cephalorrhynchu ■

82413 Cephalorrhynchus candolleanus Boiss.;康多勒头嘴菊■☆

82414 Cephalorrhynchus kirpicznikovii Grossh.;吉氏头嘴菊■☆

82415 Cephalorrhynchus kossinskyi (Krasch.) Kirp.;考氏头嘴菊■☆

82416 Cephalorrhynchus macrorhizus (Royle) Tuisl = Cephalorrhynchus macrorrhizus (Royle) Tuisl ■

82417 Cephalorrhynchus macrorhizus (Royle) Tuisl = Cicerbita macrorrhiza (Willd.) Wallr. ■

82418 Cephalorrhynchus macrorrhizus (Royle) Tuisl;头嘴菊(蓝岩参菊,岩参,扎赤)■

82419 Cephalorrhynchus microcephalus (DC.) Schchian;小头头嘴菊■☆

82420 Cephalorrhynchus polycladus (Boiss.) Kirp.;多枝头嘴菊■☆

82421 Cephalorrhynchus saxatilis (Edgew.) C. Shih;岩生头嘴菊;Rocky Cephalorrhynchus ■

82422 Cephalorrhynchus soongoricus (Regel) Kovalevsk.;准噶尔头嘴菊■☆

82423 Cephalorrhynchus subplumosus Kovalevsk.;羽状头嘴菊■☆

82424 Cephalorrhynchus takhtadzhianii (Sosn.) Kirp.;塔氏头嘴菊■☆

82425 Cephalorrhynchus talyschensis Kirp.;塔里什头嘴菊■☆

82426 Cephalorrhynchus tuberosus (Steven) Schchian;块状头嘴菊■☆

82427 Cephaloschefflera (Harms) Merr. = Schefflera J. R. Forst. et G. Forst. (保留属名)●

82428 Cephaloschefflera Merr. = Schefflera J. R. Forst. et G. Forst. (保留属名)●

82429 Cephaloschoenus Nees = Rhynchospora Vahl(保留属名)■

82430 Cephaloscirpus Kurz = Mapania Aubl. ■

82431 Cephaloseris Poepp. ex Rchb. = Polyachyrus Lag. ●■☆

82432 Cephalosiachyum Munro = Schizostachyum Nees ●

82433 Cephalosorus A. Gray = Angianthus J. C. Wendl. (保留属名)■●☆

82434 Cephalosorus A. Gray(1851);鳞冠鼠麴草属■☆

82435 Cephalosorus carpesioides (Turcz.) Short;鳞冠鼠麴草■☆

82436 Cephalosphaera Warb. (1903);球花肉豆蔻属(球花蔻属,头花楠属)●☆

82437 Cephalosphaera usambarensis (Warb.) Warb.;球花肉豆蔻(球花蔻);Mtambara ●☆

82438 Cephalosphaera usambarensis Warb. = Cephalosphaera usambarensis (Warb.) Warb. ●☆

82439 Cephalostachyum Munro = Schizostachyum Nees ●

82440 Cephalostachyum Munro(1868);空竹属(头穗竹属,香竹属);Hollow Bamboo,Hollowbamboo,Hollow-bamboo ●

82441 Cephalostachyum burmanicum R. Parker et C. E. Parkinson;缅甸空竹;Burma Bamboo Shrub ●☆

82442 Cephalostachyum capitatum (Wall. et Griff.) Munro;头状空竹;Bamboo Shrub,Hollow Bamboo,Tufted Bamboo ●☆

82443 Cephalostachyum capitatum (Wall. et Griff.) Munro = Cephalostachyum pallidum Munro ●

82444 Cephalostachyum chapelieri Munro;沙普空竹●☆

82445 Cephalostachyum fuchsianum Gamble;空竹;Hollow Bamboo,

Hollowbamboo, Hollow-bamboo ●

82446 Cephalostachyum latifolium Munro; 大叶空竹; Large-leaved Bamboo Shrub, Large-leaved Hollow Bamboo ●☆

82447 Cephalostachyum madagascariense A. Camus = Cathariostachys madagascariensis (A. Camus) S. Dransf. ●☆

82448 Cephalostachyum mannii (Gamble) Stapleton et D. Z. Li;独龙江空竹;Mann's Hollowbamboo ●

82449 Cephalostachyum mindorensis Gamble;菲律宾空竹;Philippines Scrambling Bamboo ●☆

82450 Cephalostachyum pallidum Munro; 小空竹(空竹,山空竹); Small Hollowbamboo, Small Hollow-bamboo ●

82451 Cephalostachyum peclardii A. Camus = Cathariostachys capitata (Kunth) S. Dransf. ●☆

82452 Cephalostachyum pergracile Munro;糯竹(糯米饭竹,糯米香竹,纤细头穗竹,香竹);Burmese Bamboo, Rrice-cooking Bamboo, Thin Holow Bamboo, Thinest Hollow Bamboo, Thinest Hollowbamboo, Thinest Hollow-bamboo, Thin-headed Hollow Bamboo, Tinwa Bamboo ●

82453 Cephalostachyum perrieri A. Camus;佩里耶空竹●☆

82454 Cephalostachyum scandens Bor; 真麻竹(贡麻竹); Bending Hollow-bamboo ●

82455 Cephalostachyum scandens J. R. Xue et C. M. Hui = Clematoclethra scandens (Franch.) Maxim. ●

82456 Cephalostachyum viguieri A. Camus;马岛空竹●☆

82457 Cephalostachyum virgatum (Munro) Kurz;金毛空竹;Goldenhair Hollowbamboo, Golden-haired Bamboo ●

82458 Cephalostemon R. H. Schomb. (1845);头蕊偏穗草属■☆

82459 Cephalostemon Rob. = Cephalostemon R. H. Schomb. ■☆

82460 Cephalostemon affinis Körn.;近缘头蕊偏穗草■☆

82461 Cephalostemon angustatus Malme;狭头蕊偏穗草■☆

82462 Cephalostemon flavus (Link) Govaerts;黄头蕊偏穗草■☆

82463 Cephalostemon gracilis Rob.;纤细头蕊偏穗草■☆

82464 Cephalostigma A. DC. (1830);星花草属;Cephalostigma ●

82465 Cephalostigma A. DC. = Wahlenbergia Schrad. ex Roth(保留属名)■●

82466 Cephalostigma candolleanum Hiern = Wahlenbergia candolleana (Hiern) Thulin ■☆

82467 Cephalostigma diaguisse A. Chev. = Wahlenbergia perrottetii (A. DC.) Thulin ■☆

82468 Cephalostigma erectum (Roth ex Roem. et Schult.) Vatke = Wahlenbergia erecta (Roth ex Roem. et Schult.) Tuyn ■☆

82469 Cephalostigma erectum (Roth ex Roem. et Schult.) Vatke var. coeruleum Chiov. = Wahlenbergia hirsuta (Edgew.) Tuyn ■☆

82470 Cephalostigma erectum (Roth ex Roem. et Schult.) Vatke var. luteum Chiov. = Wahlenbergia flexuosa (Hook. f. et Thomson) Thulin ■☆

82471 Cephalostigma flexuosum Hook. f. et Thomson = Wahlenbergia flexuosa (Hook. f. et Thomson) Thulin ■☆

82472 Cephalostigma hirsutum Edgew. = Wahlenbergia hirsuta (Edgew.) Tuyn ■☆

82473 Cephalostigma hookeri C. B. Clarke = Wahlenbergia hookeri (C. B. Clarke) Tuyn ■

82474 Cephalostigma nanellum R. E. Fr. = Monopsis zeyheri (Sond.) Thulin ■☆

82475 Cephalostigma paniculatum A. DC.; 圆锥星花草; Paniculate Cephalostigma ●☆

82476 Cephalostigma paniculatum A. DC. = Cephalostigma hookeri C. B. Clarke ■

82477 Cephalostigma perotifolium (Wight et Arn.) Hutch. et Dalziel = Wahlenbergia erecta (Roth ex Roem. et Schult.) Tuyn ■☆

82478 Cephalostigma perrottetii A. DC. = Wahlenbergia perrottetii (A. DC.) Thulin ■☆

82479 Cephalostigma prieuri A. DC. = Wahlenbergia perrottetii (A. DC.) Thulin ■☆

82480 Cephalostigma pyramidale Schinz = Wahlenbergia ramosissima (Hemsl.) Thulin subsp. lateralis (Brehmer) Thulin ■☆

82481 Cephalostigma ramossisimum Hemsl. = Wahlenbergia ramosissima (Hemsl.) Thulin ■☆

82482 Cephalostigma schimperi Hochst. ex A. Rich. = Wahlenbergia erecta (Roth ex Roem. et Schult.) Tuyn ■☆

82483 Cephalostigma spathulatum Thwaites = Campanula dimorphantha Schweinf. ■

82484 Cephalostigmaton (Yakovlev) Yakovlev = Sophora L. ●■

82485 Cephalostigmaton Yakovlev = Sophora L. ●■

82486 Cephalostigmaton Yakovlev(1967);东京槐属●

82487 Cephalostigmaton tonkinensis (Gagnep.) Yakovlev = Sophora tonkinensis Gagnep. ●

82488 Cephalosurus C. Muell. = Angianthus J. C. Wendl. (保留属名)■●☆

82489 Cephalosurus C. Muell. = Cephalosorus A. Gray ■☆

82490 Cephalotaceae Dumort. (1829)(保留科名);土瓶草科(捕蝇草科,囊叶草科)■☆

82491 Cephalotaceae Dumort. (保留科名) = Rubiaceae Juss. (保留科名)●■

82492 Cephalotaceae Neger = Cephalotaceae Dumort. (保留科名)■☆

82493 Cephalotaxaceae Neger(1907)(保留科名);三尖杉科(粗榧科);Cowtail Pine Family, Plumyew Family, Plum-Yew Family ●

82494 Cephalotaxaceae Neger(保留科名) = Taxaceae Gray(保留科名)●

82495 Cephalotaxus Siebold et Zucc. = Cephalotaxus Siebold et Zucc. ex Endl. ●

82496 Cephalotaxus Siebold et Zucc. ex Endl. (1842);三尖杉属(粗榧属);Chinese Cow's Tall Pine, Plum Yew, Plumyew, Plum-Yew ●

82497 Cephalotaxus alpina (H. L. Li) L. K. Fu = Cephalotaxus fortunei Hook. f. var. alpina H. L. Li ●

82498 Cephalotaxus argotaenia (Hance) Pilg. = Amentotaxus argotaenia (Hance) Pilg. ●◇

82499 Cephalotaxus drupacea Siebold et Zucc.;核果粗榧(粗榘,粗榧,日本粗榧,杉);Japanese Plum-yew, Japenese Plumyew ●☆

82500 Cephalotaxus drupacea Siebold et Zucc. = Cephalotaxus harringtonia (Knight ex J. Forbes) K. Koch ●☆

82501 Cephalotaxus drupacea Siebold et Zucc. = Cephalotaxus latifolia W. C. Cheng et L. K. Fu ex L. K. Fu et al. ●

82502 Cephalotaxus drupacea Siebold et Zucc. = Cephalotaxus sinensis (Rehder et E. H. Wilson) H. L. Li ●

82503 Cephalotaxus drupacea Siebold et Zucc. var. sinensis Rehder et E. H. Wilson = Cephalotaxus sinensis (Rehder et E. H. Wilson) H. L. Li ●

82504 Cephalotaxus drupacea Siebold et Zucc. var. sinensis Rehder et E. H. Wilson f. globosa Rehder et E. H. Wilson = Cephalotaxus sinensis (Rehder et E. H. Wilson) H. L. Li ●

82505 Cephalotaxus drupacea Siebold et Zucc. var. sinensis Siebold et Zucc. = Cephalotaxus mannii Hook. f. ●◇

82506 Cephalotaxus filiformis Knight ex Gordon = Cephalotaxus fortunei Hook. f. ●

82507 Cephalotaxus fortunei Hook. f.;三尖杉(白头杉,藏杉,臭杉,榧子,狗尾松,红心杉,尖松,三尖松,山榧树,杉孔刺树,石榧,水柏子,桃松,头形杉,血榧,崖头杉,崖头杉树,岩杉,岩杉木);Chinese Cow's Tail Pine, Chinese Cow's-tail Pine, Chinese Cowtail Pine, Chinese Plum Yew, Chinese Plumyew, Chinese Plum-yew, Fortune Plumyew, Fortune Plum-yew, Fortune's Plum Yew, Plum Yew ●

82508 Cephalotaxus fortunei Hook. f. 'Brevifolia' = Cephalotaxus fortunei Hook. f. ●

82509 Cephalotaxus fortunei Hook. f. 'Longifolia' = Cephalotaxus fortunei Hook. f. ●

82510 Cephalotaxus fortunei Hook. f. f. globosa S. Y. Hu = Cephalotaxus fortunei Hook. f. ●

82511 Cephalotaxus fortunei Hook. f. var. alpina H. L. Li;高山三尖杉(密油果);Alpine Fortune Plumyew ●

82512 Cephalotaxus fortunei Hook. f. var. concolor Franch.;绿背三尖杉(小叶三尖杉);Greenback Fortune Plumyew ●

82513 Cephalotaxus fortunei Hook. f. var. concolor Franch. = Cephalotaxus fortunei Hook. f. ●

82514 Cephalotaxus fortunei Hook. f. var. globosa S. Y. Hu = Cephalotaxus fortunei Hook. f. ●

82515 Cephalotaxus fortunei Hook. f. var. longifolia Dallim. = Cephalotaxus fortunei Hook. f. ●

82516 Cephalotaxus fortunei Hook. f. var. longifolia Dallim. et Jacks. = Cephalotaxus fortunei Hook. f. ●

82517 Cephalotaxus fortunei Hook. var. lanceolata (K. M. Feng) Silba = Cephalotaxus lanceolata K. M. Feng ex W. C. Cheng et L. K. Fu ●◇

82518 Cephalotaxus griffithii Hook. f. = Cephalotaxus harringtonia (Knight ex J. Forbes) K. Koch ●☆

82519 Cephalotaxus griffithii Hook. f. = Cephalotaxus mannii Hook. f. ●◇

82520 Cephalotaxus griffithii Hook. f. = Cephalotaxus oliveri Mast. ●◇

82521 Cephalotaxus hainanensis H. L. Li;海南粗榧(薄叶篦子杉,红壳松);Cow's Tail Pine, Hainan Plumyew, Hainan Plum-yew ●

82522 Cephalotaxus hainanensis H. L. Li = Cephalotaxus mannii Hook. f. ●◇

82523 Cephalotaxus harringtonia (Knight ex J. Forbes) K. Koch;日本粗榧(长梗粗榧,粗榧,柱冠粗榧,柱冠日本粗榧);Assam Plum Yew, Cow's-tail Pine, Harrington Plum Yew, Harrington Plumyew, Japan Plumyew, Japanese Plum Yew, Japanese Plumyew, Japanese Plum-yew, Plum Yew ●☆

82524 Cephalotaxus harringtonia (Knight ex J. Forbes) K. Koch 'Drupacea';硬核柱冠日本粗榧(长梗粗榧,粗榧);Cow's Tall Pine, Cow's-tail Pine, Japanese Plum Yew ●☆

82525 Cephalotaxus harringtonia (Knight ex J. Forbes) K. Koch 'Fastigiana';柱冠日本粗榧(帚枝粗榧,帚状柱冠粗榧,柱冠粗榧); Columnar Plum-yew, Fastigiate Japan Plumyew, Fastigiate Japanese Plumyew, Upright Japanese Plum-yew ●

82526 Cephalotaxus harringtonia (Knight ex J. Forbes) K. Koch 'Gnome';侏儒日本粗榧●☆

82527 Cephalotaxus harringtonia (Knight ex J. Forbes) K. Koch 'Prostrata';俯卧日本粗榧;Prostrate Japanese Plum Yew ●☆

82528 Cephalotaxus harringtonia (Knight ex J. Forbes) K. Koch f. angustifolia Makino;狭叶日本粗榧●☆

82529 Cephalotaxus harringtonia (Knight ex J. Forbes) K. Koch f. fastigiana Rehder = Cephalotaxus harringtonia (Knight ex J. Forbes) K. Koch 'Fastigiana' ●

82530 Cephalotaxus harringtonia (Knight ex J. Forbes) K. Koch var. drupacea Koidz. = Cephalotaxus harringtonia (Knight ex J. Forbes) K. Koch 'Drupacea' ●☆

82531 Cephalotaxus harringtonia (Knight ex J. Forbes) K. Koch var. fastigiana Rehder = Cephalotaxus harringtonia (Knight ex J. Forbes) K. Koch 'Fastigiana' ●

82532 Cephalotaxus harringtonia (Knight ex J. Forbes) K. Koch var. nana (Nakai) Rehder;矮小日本粗榧●☆

82533 Cephalotaxus harringtonia (Knight ex J. Forbes) K. Koch var. sinensis (Rehder et E. H. Wilson) Rehder = Cephalotaxus sinensis (Rehder et E. H. Wilson) H. L. Li ●

82534 Cephalotaxus harringtonia (Knight ex J. Forbes) K. Koch var. wilsoniana (Hayata) Kitam. = Cephalotaxus wilsoniana Hayata ●

82535 Cephalotaxus harringtonia K. Koch = Cephalotaxus harringtonia (Knight ex J. Forbes) K. Koch ●☆

82536 Cephalotaxus kaempferi Anon = Cephalotaxus fortunei Hook. f. ●

82537 Cephalotaxus koreana Nakai = Cephalotaxus harringtonia (Knight ex J. Forbes) K. Koch 'Drupacea' ●☆

82538 Cephalotaxus lanceolata K. M. Feng = Cephalotaxus lanceolata K. M. Feng ex W. C. Cheng et L. K. Fu ●◇

82539 Cephalotaxus lanceolata K. M. Feng ex W. C. Cheng et L. K. Fu;贡山三尖杉(三尖杉);Gongshan Plumyew, Kungshan Plum-yew ●◇

82540 Cephalotaxus latifolia W. C. Cheng et L. K. Fu ex L. K. Fu et al.;宽叶粗榧;Broadleaf China Plumyew, Broadleaf Chinese Plumyew ●

82541 Cephalotaxus mannii Hook. f.;西双版纳粗榧(藏杉,刺油杉树,海南粗榧,印度粗榧,印度三尖杉); Griffith Plumyew, India Plumyew, Indian Plumyew, Mann Plumyew, Mann Plum-yew ●◇

82542 Cephalotaxus oliveri Mast.;篦子三尖杉(阿里杉,篦子粗榧,花枝杉,杉,梳叶圆头杉);Oliver Plumyew, Oliver Plum-yew ●◇

82543 Cephalotaxus pedunculata Siebold et Zucc. = Cephalotaxus harringtonia (Knight ex J. Forbes) K. Koch ●☆

82544 Cephalotaxus pedunculata Siebold et Zucc. var. fastigiana Carrière = Cephalotaxus harringtonia (Knight ex J. Forbes) K. Koch 'Fastigiana' ●

82545 Cephalotaxus sinensis (Rehder et E. H. Wilson) H. L. Li;粗榧(白头杉,藏杉,粗榧杉,粗榧子,鄂西粗榧,榧子,狗尾松,红壳松,尖松,木榧,三尖杉,山榧树,山榧子,石榧,水柏子,水松,跳松,土香榧,血榧,崖头杉,岩杉,野榧,中国粗榧,中华粗榧杉,竹叶粗榧);China Plumyew, Chinese Plumyew, Chinese Plum-yew ●

82546 Cephalotaxus sinensis (Rehder et E. H. Wilson) H. L. Li f. globosa (Rehder et E. H. Wilson) H. L. Li = Cephalotaxus sinensis (Rehder et E. H. Wilson) H. L. Li ●

82547 Cephalotaxus sinensis (Rehder et E. H. Wilson) H. L. Li f. globosa Rehder et E. H. Wilson = Cephalotaxus sinensis (Rehder et E. H. Wilson) H. L. Li ●

82548 Cephalotaxus sinensis (Rehder et E. H. Wilson) H. L. Li var. latifolia W. C. Cheng et L. K. Fu = Cephalotaxus latifolia W. C. Cheng et L. K. Fu ex L. K. Fu et al. ●

82549 Cephalotaxus sinensis (Rehder et E. H. Wilson) H. L. Li var. wilsoniana (Hayata) L. K. Fu et Nan Li = Cephalotaxus wilsoniana Hayata ●

82550 Cephalotaxus wilsoniana Hayata;台湾粗榧(台湾三尖杉);E. H. Wilson Plumyew, Taiwan Plumyew ●

82551 Cephalotaxus wilsoniana Hayata = Cephalotaxus sinensis (Rehder et E. H. Wilson) H. L. Li var. wilsoniana (Hayata) L. K. Fu et Nan Li ●

82552 Cephalotes Lehm. = Cephalotus Labill. (保留属名) ■☆

82553 Cephalotomandra H. karst. et Triana(1855);木果茉莉属■☆

82554 Cephalotomandra Triana = Cephalotomandra H. Karst. et Triana ■☆

82555 Cephalotomandra fragrans Triana;木果茉莉■☆

82556 Cephalotos Adans.(废弃属名)= Cephalotus Labill.(保留属名)■☆

82557 Cephalotos Adans.(废弃属名)= Thymus L. ●

82558 Cephalotrophis Blume = Malania Chun et S. K. Lee ●★

82559 Cephalotrophis Blume = Trophis P. Browne(保留属名)●☆

82560 Cephalotus Labill.(1806)(保留属名);土瓶草属(捕蝇草属,囊叶草属);Albany Pitcher Plant, Australian Pitcher Plant, Cephalotus ■☆

82561 Cephalotus follicularis Labill.;土瓶草;Albany Pitcher Plant, Australian Pitcher Plant ■☆

82562 Cephaloxis Desv. = Cephaloxys Desv. ■

82563 Cephaloxys Desv. = Juncus L. ■

82564 Ceradia Lindl. = Othonna L. ●■☆

82565 Ceradia furcata Lindl. = Othonna furcata (Lindl.) Druce ●☆

82566 Ceraia Lour.(废弃属名)= Dendrobium Sw.(保留属名)■

82567 Ceraia batanensis (Ames et Quisumb.) M. A. Clem. = Dendrobium equitans Kraenzl. ■

82568 Ceraia equitans (Kraenzl.) M. A. Clem. = Dendrobium equitans Kraenzl. ■

82569 Ceraia exilis (Schltr.) M. A. Clem. = Dendrobium exile Schltr. ■

82570 Ceraia parviflora (Ames et C. Schweinf.) M. A. Clem. = Dendrobium crumenatum Sw. ■

82571 Ceraia pseudotenella (Guillaumin) M. A. Clem. = Dendrobium pseudotenellum Guillaumin ■

82572 Ceramanthe (Rchb. f.) Dumort. = Scrophularia L. ■●

82573 Ceramanthe (Rchb.) Dumort. = Scrophularia L. ■●

82574 Ceramanthe Dumort. = Scrophularia L. ■●

82575 Ceramanthus (Kunze) Malme = Funastrum E. Fourn. ■

82576 Ceramanthus (Kunze) Malme = Sarcostemma R. Br. ■

82577 Ceramanthus Hassk. = Phyllanthus L. ●■

82578 Ceramanthus Malme = Ceramanthus (Kunze) Malme ■

82579 Ceramanthus Post et Kuntze = Adenia Forssk. ●

82580 Ceramanthus Post et Kuntze = Keramanthus Hook. f. ●

82581 Ceramia D. Don = Erica L. ●☆

82582 Ceramicalyx Blume = Osbeckia L. ●■

82583 Ceramiocephalum Sch. Bip. = Crepis L. ■

82584 Ceramium Blume = Apama Lam. ●

82585 Ceramium Blume = Munnickia Rchb. ●

82586 Ceramium Blume = Thottea Rottb. ●

82587 Ceramocalyx Post et Kuntze = Ceramicalyx Blume ●■

82588 Ceramocalyx Post et Kuntze = Osbeckia L. ●■

82589 Ceramocarpium Nees ex Meisn. = Ocotea Aubl. ●☆

82590 Ceramocarpus Wittst. = Coriandrum L. ■

82591 Ceramocarpus Wittst. = Keramocarpus Fenzl ■

82592 Ceramophora Nees ex Meisn. = Ocotea Aubl. ●☆

82593 Ceranthe (Rchb.) Opiz = Cerinthe L. ■☆

82594 Ceranthe Opiz = Cerinthe L. ■☆

82595 Ceranthera Elliott = Dicerandra Benth. ●■☆

82596 Ceranthera Endl. = Ceratanthera Hornem. ■

82597 Ceranthera Endl. = Colebrookia Donn ex T. Lestib. ■

82598 Ceranthera Endl. = Globba L. ■

82599 Ceranthera P. Beauv. = Rinorea Aubl.(保留属名)●

82600 Ceranthera Raf. = Androcera Nutt. ●■

82601 Ceranthera Raf. = Solanum L. ●■

82602 Ceranthera dentata P. Beauv. = Rinorea dentata (P. Beauv.) Kuntze ●☆

82603 Ceranthera subintegrifolia P. Beauv. = Rinorea subintegrifolia (P. Beauv.) Kuntze ●☆

82604 Cerantheraceae Dulac = Ericaceae Juss.(保留科名)●

82605 Ceranthus Schreb.(废弃属名)= Chionanthus L. ●

82606 Ceranthus Schreb.(废弃属名)= Linociera Sw. ex Schreb.(保留属名)●

82607 Cerapadus Buia = Prunus L. ●

82608 Ceraria H. Pearson et Stephens(1912);长寿城属(单性树马齿苋属)●☆

82609 Ceraria carrissoana Exell et Mendonça;卡里索长寿城●☆

82610 Ceraria fruticulosa H. Pearson et Stephens;灌木状长寿城●☆

82611 Ceraria gariepina H. Pearson et Stephens = Ceraria namaquensis (Sond.) H. Pearson et Stephens ●☆

82612 Ceraria longipedunculata Merxm. et Podlech;长梗长寿城●☆

82613 Ceraria namaquensis (Sond.) H. Pearson et Stephens;纳马夸长寿城●☆

82614 Ceraria schaeferi Engl. et Schltr. = Ceraria fruticulosa H. Pearson et Stephens ●☆

82615 Ceraseidos Siebold et Zucc. = Prunus L. ●

82616 Ceraselma Wittst. = Euphorbia L. ●■

82617 Ceraselma Wittst. = Keraselma Neck. ●■

82618 Cerasiocarpum Hook. f.(1867);角果葫芦属■☆

82619 Cerasiocarpum Hook. f. = Kedrostis Medik. ■☆

82620 Cerasiocarpum zeylanicum Hook. f.;角果葫芦■☆

82621 Cerasiocarpus Post et Kuntze = Cerasiocarpum Hook. f. ■☆

82622 Cerasites Steud. = Cerastites Gray ■

82623 Cerasites Steud. = Papaver L. ■

82624 Cerasophora Neck. = Cerasus Mill. ●

82625 Cerasophora Neck. = Prunus L. ●

82626 Cerastiaceae Vest = Caryophyllaceae Juss.(保留科名)■●

82627 Cerastites Gray = Meconopsis R. Vig. ■

82628 Cerastites Gray = Papaver L. ■

82629 Cerastium L.(1753);卷耳属(寄奴花属);Cerastium, Chickweed, Hornkraut, Mouse Ear, Mouseear Chickweed, Mouse-ear Chickweed, Snow-in-summer ■

82630 Cerastium aberdaricum T. C. E. Fr. et Weim. = Cerastium afromontanum T. C. E. Fr. et Weim. ■☆

82631 Cerastium acutatum Suksd. = Cerastium glomeratum Thuill. ■

82632 Cerastium adnivale Chiov. = Cerastium octandrum Hochst. ex A. Rich. var. adnivale (Chiov.) Möschl ■☆

82633 Cerastium adsurgens Greene = Cerastium brachypodum (Engelm. ex A. Gray) B. L. Rob. ■☆

82634 Cerastium adsurgens Greene = Cerastium fontanum Baumg. subsp. vulgare (Hartm.) Greuter et Burdet ■

82635 Cerastium africanum (Hook. f.) Oliv. = Cerastium indicum Wight et Arn. ■☆

82636 Cerastium africanum (Hook. f.) Oliv. var. ruwenzoriense F. N. Williams = Cerastium indicum Wight et Arn. ■☆

82637 Cerastium africanum Oliv. var. jaegeri Engl. = Cerastium afromontanum T. C. E. Fr. et Weim. ■☆

82638 Cerastium africanum Oliv. var. kilimanjarensis Williams = Cerastium afromontanum T. C. E. Fr. et Weim. ■☆

82639 Cerastium afromontanum T. C. E. Fr. et Weim.;非洲山生卷耳■☆

82640 Cerastium afromontanum T. C. E. Fr. et Weim. f. granvikii (T. C. E. Fr. et Weim.) Möschl;格兰维克卷耳■☆

82641 Cerastium afromontanum T. C. E. Fr. et Weim. var. granvikii? = Cerastium afromontanum T. C. E. Fr. et Weim. f. granvikii (T. C. E.

Fr. et Weim.) Möschl ■☆
82642 Cerastium aleuticum Hultén;阿留申卷耳;Aleutian Mouse-ear Chickweed ■☆
82643 Cerastium alexandrei Emb. = Cerastium arvense L. subsp. strictum (W. D. J. Koch) Schinz et Keller ■☆
82644 Cerastium alexeenkoanum Schischk.;阿氏卷耳■☆
82645 Cerastium algericum (Batt.) Batt. = Cerastium diffusum Pers. ■☆
82646 Cerastium alpinum L.;高山卷耳;Alpine Cerastium, Alpine Chickweed, Alpine Mouse Ear, Alpine Mouseear, Alpine Mouse-ear, Alpine Mouse-ear Chickweed, Chickweed ■☆
82647 Cerastium alpinum L. subsp. lanatum (Lam.) Cesati;绵毛高山卷耳■☆
82648 Cerastium alpinum L. subsp. squalidum (Raymond) Hultén = Cerastium alpinum L. subsp. lanatum (Lam.) Cesati ■☆
82649 Cerastium alpinum L. var. beeringianum Regel = Cerastium beeringianum Cham. et Schltdl. ■☆
82650 Cerastium alpinum L. var. caespitosum Malmgren = Cerastium regelii Ostenf. ■☆
82651 Cerastium alpinum L. var. capillare (Fernald et Wiegand) B. Boivin = Cerastium beeringianum Cham. et Schltdl. ■☆
82652 Cerastium alpinum L. var. fischerianum (Ser.) Torr. et A. Gray = Cerastium fischerianum Ser. ■
82653 Cerastium alpinum L. var. glanduliferum Trautv. = Cerastium pusillum Ser. ■
82654 Cerastium alpinum L. var. lanatum (Lam.) Hegetschw. = Cerastium alpinum L. subsp. lanatum (Lam.) Cesati ■☆
82655 Cerastium alpinum L. var. procerum Lange = Cerastium arcticum Lange ■☆
82656 Cerastium alpinum L. var. uniflorum Durand = Cerastium arcticum Lange ■☆
82657 Cerastium alsophilum Greene = Cerastium arvense L. subsp. strictum Gaudin ■☆
82658 Cerastium alsophilum Greene = Cerastium arvense L. ■
82659 Cerastium amplexicaule Sims = Cerastium dahuricum Fisch. ■
82660 Cerastium amurense Ohwi = Cerastium furcatum Cham. et Schltdl. ■
82661 Cerastium angustatum Greene = Cerastium arvense L. subsp. strictum Gaudin ■☆
82662 Cerastium angustatum Greene = Cerastium arvense L. ■
82663 Cerastium anomalum Waldst. et Kit. = Cerastium dubium (Bastard) Guépin ■☆
82664 Cerastium anomalum Waldst. et Kit. ex Willd.;反常卷耳;Anomal Mouseear ■☆
82665 Cerastium anomalum Willd. = Cerastium dubium (Bastard) Guépin ■☆
82666 Cerastium aquaticum L. = Myosoton aquaticum (L.) Moench ■
82667 Cerastium aquaticum L. = Stellaria aquatica (L.) Scop. ■☆
82668 Cerastium arabidis E. Mey. ex Fenzl;阿拉伯卷耳■☆
82669 Cerastium arcticum L. subsp. hyperboreum (Tolm.) Böcher = Cerastium arcticum Lange ■☆
82670 Cerastium arcticum L. subsp. procerum (Lange) Böcher = Cerastium arcticum Lange ■☆
82671 Cerastium arcticum L. var. procerum (Lange) Hultén = Cerastium arcticum Lange ■☆
82672 Cerastium arcticum L. var. vestitum Hultén = Cerastium arcticum Lange ■☆
82673 Cerastium arcticum Lange;北极卷耳;Arctic Mouse-ear, Shetland Mouse-ear ■☆
82674 Cerastium arcticum Lange var. sordidum Hultén = Cerastium bialynickii Tolm. ■☆
82675 Cerastium arenarioides Crantz = Arenaria cerastioides Poir. ■☆
82676 Cerastium argenteum M. Bieb.;银白卷耳■☆
82677 Cerastium argenteum M. Bieb. = Cerastium fontanum Baumg. subsp. grandiflorum (Buch.-Ham. ex D. Don) H. Hara ■
82678 Cerastium arisanense Hayata = Stellaria arisanensis (Hayata) Hayata ■
82679 Cerastium armeniacum Gren.;亚美尼亚卷耳■☆
82680 Cerastium arvense L.;卷耳(田卷耳,田野卷耳,野卷耳);Field Chickweed, Field Mouse Ear, Field Mouse-ear, Field Mouse-ear Chickweed, Meadow Chickweed, Mouseear, Powder-horn, Prairie Mouse-ear Chickweed, Starry Cerastium, Starry Grasswort ■
82681 Cerastium arvense L. f. oblongifolium (Torr.) Pennell = Cerastium arvense L. ■
82682 Cerastium arvense L. subsp. maximum (Hollick et Britton) Ugbor. = Cerastium viride A. Heller ■☆
82683 Cerastium arvense L. subsp. molle (Vill.) Arcang.;绢毛卷耳■☆
82684 Cerastium arvense L. subsp. strictum (L.) Gaudin = Cerastium strictum L. ■
82685 Cerastium arvense L. subsp. strictum (L.) Ugbor. = Cerastium arvense L. ■
82686 Cerastium arvense L. subsp. strictum (W. D. J. Koch) Schinz et Keller;直立卷耳■☆
82687 Cerastium arvense L. subsp. strictum Gaudin = Cerastium strictum L. ■
82688 Cerastium arvense L. subsp. velutinum (Raf.) Ugbor.;星卷耳;Field Chickweed, Starry Grasswort ■☆
82689 Cerastium arvense L. subsp. velutinum (Raf.) Ugbor. = Cerastium velutinum Raf. ■☆
82690 Cerastium arvense L. subsp. velutinum (Raf.) Ugbor. var. villosum (Muhl. ex Darl.) Ugbor. = Cerastium arvense L. ■
82691 Cerastium arvense L. subsp. velutinum (Raf.) Ugbor. var. villosum (Muhl. ex Darl.) Ugbor. f. oblongifolium (Torr.) Pennell = Cerastium arvense L. ■
82692 Cerastium arvense L. var. angustifolium Fenzl;狭叶卷耳(细叶卷耳);Narrowleaf Mouseear ■
82693 Cerastium arvense L. var. angustifolium Fenzl = Cerastium arvense L. ■
82694 Cerastium arvense L. var. angustifolium Fenzl = Cerastium arvense L. subsp. strictum (L.) Gaudin ■
82695 Cerastium arvense L. var. bracteatum (Raf.) MacMill. = Cerastium velutinum Raf. ■☆
82696 Cerastium arvense L. var. fuegianum Hook. f. = Cerastium arvense L. subsp. strictum Gaudin ■☆
82697 Cerastium arvense L. var. glabrllum (Turcz.) Fenzl;无毛卷耳■
82698 Cerastium arvense L. var. japonicum H. Hara = Cerastium arvense L. var. ovatum (Miyabe) E. Miki ■☆
82699 Cerastium arvense L. var. latifolium Fenzl = Cerastium arvense L. subsp. strictum Gaudin ■☆
82700 Cerastium arvense L. var. latifolium Fenzl = Cerastium arvense L. ■
82701 Cerastium arvense L. var. maximum Hollick et Britton = Cerastium viride A. Heller ■☆
82702 Cerastium arvense L. var. oblongifolium (Torr.) Hollick et Britton = Cerastium arvense L. subsp. velutinum (Raf.) Ugbor. ■☆
82703 Cerastium arvense L. var. oblongifolium (Torr.) Hollick et

Britton = Cerastium velutinum Raf. ■☆
82704 Cerastium arvense L. var. ophiticola Raymond = Cerastium arvense L. subsp. strictum Gaudin ■☆
82705 Cerastium arvense L. var. ovatum (Miyabe) E. Miki;卵状卷耳■☆
82706 Cerastium arvense L. var. purpurascens B. Boivin = Cerastium arvense L. ■
82707 Cerastium arvense L. var. purpurascens B. Boivin = Cerastium arvense L. subsp. strictum Gaudin ■☆
82708 Cerastium arvense L. var. sonnei (Greene) Smiley = Cerastium arvense L. subsp. strictum Gaudin ■☆
82709 Cerastium arvense L. var. strictum (Gaudin) W. D. J. Koch = Cerastium arvense L. subsp. strictum Gaudin ■☆
82710 Cerastium arvense L. var. strictum W. D. J. Koch = Cerastium arvense L. subsp. strictum (W. D. J. Koch) Schinz et Keller ■☆
82711 Cerastium arvense L. var. strictum W. D. J. Koch = Cerastium arvense L. subsp. strictum (L.) Gaudin ■
82712 Cerastium arvense L. var. velutinum (Raf.) Britton = Cerastium velutinum Raf. ■☆
82713 Cerastium arvense L. var. villosissimum Pennell = Cerastium velutinum Raf. var. villossissimum (Pennell) J. K. Morton ■☆
82714 Cerastium arvense L. var. villosum (Muhl. ex Darl.) Hollick et Britton = Cerastium arvense L. ■
82715 Cerastium arvense L. var. villosum (Muhl. ex Darl.) Hollick et Britton = Cerastium velutinum Raf. ■☆
82716 Cerastium arvense L. var. villosum Hollick et Britton = Cerastium arvense L. ■
82717 Cerastium arvense L. var. villosum Hollick et Britton = Cerastium velutinum Raf. ■☆
82718 Cerastium arvense L. var. viscidulum Gremli = Cerastium arvense L. ■
82719 Cerastium arvense L. var. viscidulum Gremli = Cerastium arvense L. subsp. strictum Gaudin ■☆
82720 Cerastium arvense L. var. webbii Jenn. = Cerastium velutinum Raf. ■☆
82721 Cerastium atlanticum Durieu;大西洋卷耳■☆
82722 Cerastium atlanticum Durieu subsp. longipes (Batt.) Möschl = Cerastium atlanticum Durieu ■☆
82723 Cerastium atlanticum Durieu var. brachypetalum Batt. et Trab. = Cerastium atlanticum Durieu ■☆
82724 Cerastium atlanticum Durieu var. longipes (Batt.) Maire = Cerastium atlanticum Durieu ■☆
82725 Cerastium atrovirens Bab. = Cerastium diffusum Pers. ■☆
82726 Cerastium atrovirens Bab. = Cerastium dubium (Bastard) O. Schwarz ■☆
82727 Cerastium axillare Correll;腋生卷耳;Trans Pecos Mouse-ear Chickweed ■☆
82728 Cerastium baischanense Y. C. Chu;长白卷耳;Changbaishan Mouseear ■
82729 Cerastium balearicum F. Herm. = Cerastium semidecandrum L. ■☆
82730 Cerastium ballsii Maire = Cerastium arvense L. ■
82731 Cerastium bambuseti (T. C. E. Fr. et Weim.) Weim. = Cerastium afromontanum T. C. E. Fr. et Weim. ■☆
82732 Cerastium beeringianum Cham. et Schltdl.;毕氏卷耳;Beering Mouseear, Bering Mouse-ear Chickweed, Mouseear ■☆
82733 Cerastium beeringianum Cham. et Schltdl. subsp. terrae-novae (Fernald et Wiegand) Hultén = Cerastium terrae-novae Fernald et Wiegand ■☆
82734 Cerastium beeringianum Cham. et Schltdl. var. aleuticum (Hultén) S. L. Welsh = Cerastium aleuticum Hultén ■☆
82735 Cerastium beeringianum Cham. et Schltdl. var. capillare Fernald et Wiegand = Cerastium beeringianum Cham. et Schltdl. ■☆
82736 Cerastium beeringianum Cham. et Schltdl. var. glabratum Hultén = Cerastium beeringianum Cham. et Schltdl. ■☆
82737 Cerastium beeringianum Cham. et Schltdl. var. grandiflorum Hultén = Cerastium beeringianum Cham. et Schltdl. ■☆
82738 Cerastium bialynickii Tolm.;比亚卷耳;Bialynick's Mouse-ear Chickweed ■☆
82739 Cerastium biebersteinii DC.;毕伯史坦氏卷耳(毕伯氏卷耳);Boreal Chickweed, Snow-in-summer, Taurus Cerastium ■☆
82740 Cerastium boissieri Gren. = Cerastium gibraltaricum Boiss. ■☆
82741 Cerastium boissierianum Greuter et Burdet = Cerastium gibraltaricum Boiss. ■☆
82742 Cerastium brachypetalum Pers.;短瓣卷耳;Chickweed, Gray Chickweed, Gray Mouse-ear Chickweed, Grey Mouse-ear ■☆
82743 Cerastium brachypetalum Pers. f. glandulosum W. D. J. Koch = Cerastium brachypetalum Pers. ■☆
82744 Cerastium brachypetalum Pers. subsp. luridum (Guss.) Nyman = Cerastium brachypetalum Pers. subsp. roeseri (Boiss. et Heldr.) Nyman ■☆
82745 Cerastium brachypetalum Pers. subsp. roeseri (Boiss. et Heldr.) Nyman;光亮短瓣卷耳■☆
82746 Cerastium brachypetalum Pers. subsp. tauricum (Spreng.) Murb. = Cerastium brachypetalum Pers. ■☆
82747 Cerastium brachypetalum Pers. subsp. tenoreanum (Ser.) Soó;泰诺雷卷耳■☆
82748 Cerastium brachypetalum Pers. var. glandulosum Koch = Cerastium brachypetalum Pers. subsp. roeseri (Boiss. et Heldr.) Nyman ■☆
82749 Cerastium brachypetalum Pers. var. viscosum Guss. = Cerastium brachypetalum Pers. subsp. roeseri (Boiss. et Heldr.) Nyman ■☆
82750 Cerastium brachypodum (Engelm. ex A. Gray) B. L. Rob.;短柄卷耳(短梗卷耳);Short-stalk Chickweed, Short-stalked Mouse-ear Chickweed ■☆
82751 Cerastium brachypodum (Engelm. ex A. Gray) B. L. Rob. var. compactum B. L. Rob. = Cerastium brachypodum (Engelm. ex A. Gray) B. L. Rob. ■☆
82752 Cerastium brachypodum var. compactum B. L. Rob. = Cerastium brachypodum (Engelm. ex A. Gray) B. L. Rob. ■☆
82753 Cerastium bracteatum Raf. = Cerastium velutinum Raf. ■☆
82754 Cerastium buffumiae A. Nelson = Cerastium beeringianum Cham. et Schltdl. ■☆
82755 Cerastium bungeanum Vved.;镰状卷耳■
82756 Cerastium bungeanum Vved. = Cerastium falcatum Bunge ex Fenzl ■
82757 Cerastium caespitosum Gilbert var. kilimandscharicum Engl. = Cerastium afromontanum T. C. E. Fr. et Weim. ■☆
82758 Cerastium caespitosum Gilib.;簇生卷耳(高脚鼠耳草,卷耳草,婆婆指甲草,破花絮草);Common Mouse Ear Chickweed, Fellon-herb, Field Cerastlum, Mouse Ear, Robin-under-the-hedge, Tufted Cerastium ■
82759 Cerastium caespitosum Gilib. = Cerastium fontanum Baumg. subsp. vulgare (Hartm.) Greuter et Burdet ■
82760 Cerastium caespitosum Gilib. = Cerastium fontanum Baumg. ■
82761 Cerastium caespitosum Gilib. subsp. alpestre (H. Lindb.) Murb.

= Cerastium fontanum Baumg. ■

82762 Cerastium caespitosum Gilib. subsp. fontanum (Baumg.) Schinz et Thell. = Cerastium fontanum Baumg. ■

82763 Cerastium caespitosum Gilib. subsp. triviale (Link) Hiitonen = Cerastium fontanum Baumg. subsp. triviale (Link) Jalas ■

82764 Cerastium caespitosum Gilib. var. glandulosum Wirtg.;丛生腺毛卷耳■

82765 Cerastium caespitosum Gilib. var. ianthes (F. N. Williams) H. Hara = Cerastium fontanum Baumg. subsp. vulgare (Hartm.) Greuter et Burdet var. angustifolium (Franch.) H. Hara ■☆

82766 Cerastium caespitosum Gilib. var. mauretanicum Maire = Cerastium fontanum Baumg. ■

82767 Cerastium caespitosum Gilib. var. scandens Engl. = Cerastium octandrum Hochst. ex A. Rich. var. scandens (Engl.) Cufod. ■☆

82768 Cerastium campestre Greene = Cerastium arvense L. subsp. strictum Gaudin ■☆

82769 Cerastium campestre Greene = Cerastium arvense L. ■

82770 Cerastium capense Sond.;好望角卷耳■☆

82771 Cerastium cerastioides (L.) Brown = Minuartia macrocarpa (Pursh) Ostenf. var. koreana (Nakai) H. Hara ■

82772 Cerastium cerastoides (L.) Britton;六齿卷耳(鼠卷耳);Sixtooth Mouseear, Starwert Chickweed, Starwort Mouse Ear, Starwort Mouse-ear, Starwort Mouse-ear Chickweed ■

82773 Cerastium cerastoides (L.) Britton var. eglandulosum Maire = Cerastium cerastoides (L.) Britton ■

82774 Cerastium cerastoides (L.) Britton var. foliosum Kozhevn. = Cerastium cerastoides (L.) Britton ■

82775 Cerastium cerastoides (L.) Britton var. subglaberrimum Norman = Cerastium cerastoides (L.) Britton ■

82776 Cerastium chlorifolium Fisch. et C. A. Mey.;色叶卷耳■☆

82777 Cerastium ciliarum Ohwi = Cerastium furcatum Cham. et Schltdl. ■

82778 Cerastium ciliatum Turcz. var. acutifolium (Fr.) Hand.-Mazz.;尖叶卷耳■

82779 Cerastium ciliatum Turcz. var. acutifolium (Franch.) Hand.-Mazz. = Cerastium furcatum Cham. et Schltdl. ■

82780 Cerastium ciliatum Turcz. var. brevifolium (Franch.) Hand.-Mazz. = Cerastium furcatum Cham. et Schltdl. ■

82781 Cerastium ciliatum Waldst. et Kit. = Cerastium furcatum Cham. et Schltdl. ■

82782 Cerastium comatum Desv.;束毛卷耳■☆

82783 Cerastium confertum Greene = Cerastium arvense L. subsp. strictum Gaudin ■☆

82784 Cerastium confertum Greene = Cerastium arvense L. ■

82785 Cerastium cordifolium Roxb. = Stellaria aquatica (L.) Scop. ■☆

82786 Cerastium dahuricum Fisch.;达乌里卷耳;Dahur Mouseear ■

82787 Cerastium dentatum Möschl;锯齿卷耳■☆

82788 Cerastium dentatum Möschl = Cerastium semidecandrum L. ■☆

82789 Cerastium dichotomum L.;二歧卷耳;Forked Chickweed, Forked Mouse-ear Chickweed ■

82790 Cerastium dichotomum L. subsp. inflatum (Link) Cullen;膨萼卷耳■

82791 Cerastium dichotomum L. subsp. inflatum (Link) Cullen = Cerastium inflatum Link ■☆

82792 Cerastium diehlii M. E. Jones = Cerastium nutans Raf. var. obtectum Kearney et Peebles ■☆

82793 Cerastium diffusum Pers.;深灰卷耳;Dark Grey Mouse Ear, Dark-green Mouse-ear Chickweed, Fourstamen Chickweed, Mouse-eared Chickweed, Sea Mouse-ear ■☆

82794 Cerastium diffusum Pers. subsp. gussonei (Lojac.) P. D. Sell et Whitehead;古索内卷耳■☆

82795 Cerastium dregeanum Fenzl = Cerastium arabidis E. Mey. ex Fenzl ■☆

82796 Cerastium dubium (Bastard) Guépin;可疑卷耳;Anomalous Mouse-ear Chickweed, Doubtful Chickweed ■☆

82797 Cerastium dubium (Bastard) O. Schwarz = Cerastium dubium (Bastard) Guépin ■☆

82798 Cerastium earlei Rydb. = Cerastium beeringianum Cham. et Schltdl. ■☆

82799 Cerastium echinulatum Batt. = Cerastium gracile L. Dufour ■☆

82800 Cerastium edmonstonii Murb. et Ostenf. = Cerastium arcticum Lange ■☆

82801 Cerastium effusum Greene = Cerastium arvense L. subsp. strictum Gaudin ■☆

82802 Cerastium effusum Greene = Cerastium arvense L. ■

82803 Cerastium elongatum Pursh = Cerastium arvense L. subsp. strictum Gaudin ■☆

82804 Cerastium elongatum Pursh = Cerastium arvense L. ■

82805 Cerastium erectum (L.) Coss. et Germ. = Moenchia erecta (L.) P. Gaertn., B. Mey. et Scherb. ■☆

82806 Cerastium erectum (L.) Coss. et Germ. var. linneanum Maire et Weiller = Moenchia erecta (L.) P. Gaertn., B. Mey. et Scherb. ■☆

82807 Cerastium erectum (L.) Coss. et Germ. var. octandrum (Ziz) Gren. = Moenchia erecta (L.) P. Gaertn., B. Mey. et Scherb. ■☆

82808 Cerastium falcatum Bunge ex Fenzl;镰刀叶卷耳(披针叶卷耳);Sickleleaf Mouseear ■

82809 Cerastium fallax Guss. = Cerastium pentandrum L. ■☆

82810 Cerastium fastigiatum Greene;帚状卷耳;Fastigiate Mouse-ear Chickweed ■☆

82811 Cerastium fimbriatum E. Pritz. = Arenaria fimbriata (E. Pritz.) Mattf. ■

82812 Cerastium fimbriatum Ledeb. = Stellaria radians L. ■

82813 Cerastium fischerianum Ser.;长蒴卷耳;Fischer's Mouse-ear Chickweed ■

82814 Cerastium fischerianum Ser. = Cerastium furcatum Cham. et Schltdl. ■

82815 Cerastium fischerianum Ser. ex DC. var. beeringianum (Cham. et Schltdl.) Hultén = Cerastium beeringianum Cham. et Schltdl. ■☆

82816 Cerastium fischerianum Ser. f. molle (Ohwi) M. Mizush. = Cerastium fischerianum Ser. var. molle Ohwi ■☆

82817 Cerastium fischerianum Ser. var. molle Ohwi;毛长蒴卷耳■☆

82818 Cerastium fontanum Baumg.;喜泉卷耳(簇生卷耳);Common Chickweed, Common Mouse-ear, Common Mouse-ear Chickweed, Common Mouse-eared Chickweed, Mouse-ear Chickweed, Spring Mouseear ■

82819 Cerastium fontanum Baumg. subsp. grandiflorum (Buch.-Ham. ex D. Don) H. Hara;大花喜泉卷耳(大花卷耳);Bigflower Spring Mouseear, Largeflower Mouseear Chickweed ■

82820 Cerastium fontanum Baumg. subsp. holosteoides (Fr.) Salman et al. = Cerastium fontanum Baumg. subsp. vulgare (Hartm.) Greuter et Burdet ■

82821 Cerastium fontanum Baumg. subsp. scandicum Gartner = Cerastium fontanum Baumg. ■

82822 Cerastium fontanum Baumg. subsp. triviale (Link) Jalas = Cerastium fontanum Baumg. subsp. vulgare (Hartm.) Greuter et

Burdet ■
82823 Cerastium fontanum Baumg. subsp. triviale (Spenn.) Jalas = Cerastium fontanum Baumg. subsp. vulgare (Hartm.) Greuter et Burdet ■
82824 Cerastium fontanum Baumg. subsp. vulgare (Hartm.) Greuter et Burdet var. angustifolium (Franch.) H. Hara;狭叶簇生喜泉卷耳■☆
82825 Cerastium fontanum Baumg. subsp. vulgare (Hartm.) Greuter et Burdet var. angustifolium (Franch.) H. Hara = Cerastium holosteoides Fr. var. angustifolium (Franch.) M. Mizush. ■☆
82826 Cerastium fontanum Baumg. subsp. vulgare (Hartm.) Greuter et Burdet;簇生喜泉卷耳(簇生卷耳);Big Chickweed, Big Mouse-ear Chickweed, Common Chickweed, Common Mouse Ear, Common Mouse-ear, Mouse-ear Chickweed ■
82827 Cerastium fontanum Baumg. var. angustifolium (Franch.) H. Hara = Cerastium fontanum Baumg. subsp. vulgare (Hartm.) Greuter et Burdet ■
82828 Cerastium fontanum Baumg. var. tibeticum (Edgew. et Hook. f.) C. Y. Wu et L. H. Zhou = Cerastium fontanum Baumg. subsp. vulgare (Hartm.) Greuter et Burdet ■
82829 Cerastium formosanum (Ohwi) Ohwi;台湾卷耳■
82830 Cerastium formosanum (Ohwi) Ohwi = Cerastium morrisonense Hayata ■
82831 Cerastium formosanum Ohwi = Cerastium formosanum (Ohwi) Ohwi ■
82832 Cerastium fuegianum (Hook. f.) A. Nelson = Cerastium arvense L. subsp. strictum Gaudin ■☆
82833 Cerastium fulvum Raf. = Cerastium glomeratum Thuill. ■
82834 Cerastium furcatum Cham. et Schltdl.;缘毛卷耳(高山卷耳);Fork Mouseear, Forked Cerastium ■
82835 Cerastium furcatum Cham. et Schltdl. var. ibukiense Ohwi;伊吹山卷耳■☆
82836 Cerastium furcatum Cham. et Schltdl. var. tetraschistum Ohwi = Cerastium rubescens Mattf. var. koreanum (Nakai) E. Miki ■☆
82837 Cerastium geniculatum Braun-Blanq. = Cerastium pentandrum L. ■☆
82838 Cerastium gibraltaricum Boiss.;直布罗陀卷耳■☆
82839 Cerastium gibraltaricum Boiss. var. boissieri (Gren.) Pau = Cerastium gibraltaricum Boiss. ■☆
82840 Cerastium gibraltaricum Boiss. var. glabrifolium Pau = Cerastium gibraltaricum Boiss. ■☆
82841 Cerastium gibraltaricum Boiss. var. lanuginosum (Gren.) F. N. Williams = Cerastium gibraltaricum Boiss. ■☆
82842 Cerastium gibraltaricum Boiss. var. viridulum (Pau) Font Quer = Cerastium gibraltaricum Boiss. ■☆
82843 Cerastium gibraltaricum Boiss. var. vulgare Willk. = Cerastium gibraltaricum Boiss. ■☆
82844 Cerastium glabellum Turcz. = Cerastium arvense L. var. glabrllum (Turcz.) Fenzl ■
82845 Cerastium glabratum (Wahlb.) Hartm.;光卷耳■☆
82846 Cerastium glaucum Gren. = Moenchia erecta (L.) P. Gaertn., B. Mey. et Scherb. ■☆
82847 Cerastium glaucum Gren. var. octandrum? = Moenchia erecta (L.) Gaertn. et B. Mey. et Scherb. subsp. octandra (Moris) Cout. ■☆
82848 Cerastium glaucum Gren. var. quaternellum Gren. et Godr. = Moenchia erecta (L.) P. Gaertn., B. Mey. et Scherb. ■☆
82849 Cerastium glomeratum Thuill.;球序卷耳(高脚鼠耳草,瓜子草,集球卷耳,卷耳,婆婆指甲菜,婆婆指甲草,圆序卷耳,粘毛卷耳);Broad-leaved Mouse Ear Chickweed, Clammy Chickweed, Clustered Mouse Ear Chickweed, Glomerate Mouseear, Mouse-ear Chickweed, Sticky Chickweed, Sticky Mouse Ear, Sticky Mouse-ear, Sticky Mouse-ear Chickweed ■
82850 Cerastium glomeratum Thuill. var. apetalum (Dumort.) Fenzl = Cerastium glomeratum Thuill. ■
82851 Cerastium glomeratum Thuill. var. apetalum (Dumort.) Mert. et Koch = Cerastium glomeratum Thuill. ■
82852 Cerastium glomeratum Thuill. var. brachycarpum L. H. Zhou et Q. Zh. Han;短果卷耳;Shortfruit Glomerate Mouseear ■
82853 Cerastium glomeratum Thuill. var. corollinum Fenzl = Cerastium glomeratum Thuill. ■
82854 Cerastium glomeratum Thuill. var. eglandulosum Mert. et Koch = Cerastium glomeratum Thuill. ■
82855 Cerastium glomeratum Thuill. var. kotulae Zapal. = Cerastium glomeratum Thuill. ■
82856 Cerastium glutinosum Fr.;黏卷耳■☆
82857 Cerastium glutinosum Fr. = Cerastium pumilum W. M. Curfis ■☆
82858 Cerastium glutinosum Fr. var. cedretorum Sennen = Cerastium glutinosum Fr. ■☆
82859 Cerastium gorodkovianum Schischk.;葛罗氏卷耳;Gorodkov Mouseear ■☆
82860 Cerastium gorodkovianum Schischk. = Cerastium regelii Ostenf. ■☆
82861 Cerastium gracile L. Dufour;纤细卷耳;Slender Chickweed ■☆
82862 Cerastium gracile L. Dufour subsp. ramosissimum (Boiss.) Font Quer = Cerastium ramosissimum Boiss. ■☆
82863 Cerastium gracile L. Dufour var. kebdanense Font Quer = Cerastium gracile L. Dufour ■☆
82864 Cerastium gracile Wallr. = Cerastium gracile L. Dufour ■☆
82865 Cerastium graminifolium Rydb. = Cerastium arvense L. subsp. strictum Gaudin ■☆
82866 Cerastium graminifolium Rydb. = Cerastium arvense L. ■
82867 Cerastium grandiflorum Buch. -Ham. ex D. Don = Cerastium fontanum Baumg. subsp. grandiflorum (Buch. -Ham. ex D. Don) H. Hara ■
82868 Cerastium grandiflorum D. Don = Cerastium fontanum Baumg. subsp. grandiflorum (Buch. -Ham. ex D. Don) H. Hara ■
82869 Cerastium grandiflorum Gilib. = Cerastium fontanum Baumg. subsp. grandiflorum (Buch. -Ham. ex D. Don) H. Hara ■
82870 Cerastium grandiflorum Pourr. ex Willk. et Lange = Cerastium fontanum Baumg. subsp. grandiflorum (Buch. -Ham. ex D. Don) H. Hara ■
82871 Cerastium grandiflorum Waldst. et Kit. = Cerastium fontanum Baumg. subsp. grandiflorum (Buch. -Ham. ex D. Don) H. Hara ■
82872 Cerastium gussonei Lojac. = Cerastium diffusum Pers. subsp. gussonei (Lojac.) P. D. Sell et Whitehead ■☆
82873 Cerastium hemschinicum Schischk.;海姆什卷耳■☆
82874 Cerastium hirtellum Pomel = Cerastium gracile L. Dufour ■☆
82875 Cerastium hirtellum Pomel subsp. echinulatum (Coss. et Durieu) Maire = Cerastium gracile L. Dufour ■☆
82876 Cerastium hirtellum Pomel subsp. subechinulatum Maire = Cerastium gracile L. Dufour ■☆
82877 Cerastium holosteoides Fr. = Cerastium caespitosum Gilib. ■
82878 Cerastium holosteoides Fr. = Cerastium fontanum Baumg. subsp. triviale (Link) Jalas ■
82879 Cerastium holosteoides Fr. = Cerastium fontanum Baumg. subsp. vulgare (Hartm.) Greuter et Burdet ■

82880 Cerastium holosteoides Fr. subsp. triviale (Link) Möschl = Cerastium fontanum Baumg. subsp. vulgare (Hartm.) Greuter et Burdet ■

82881 Cerastium holosteoides Fr. subsp. triviale (Link) Möschl f. glandulosum (Boenn.) Möschl = Cerastium fontanum Baumg. ■

82882 Cerastium holosteoides Fr. subsp. triviale (Link) Möschl var. grandiflorum Majumdar = Cerastium fontanum Baumg. subsp. grandiflorum (Buch. -Ham. ex D. Don) H. Hara ■

82883 Cerastium holosteoides Fr. subsp. triviale (Murb.) Möschl = Cerastium fontanum Baumg. subsp. triviale (Link) Jalas ■

82884 Cerastium holosteoides Fr. var. angustifolium (Franch.) M. Mizush. = Cerastium fontanum Baumg. subsp. vulgare (Hartm.) Greuter et Burdet var. angustifolium (Franch.) H. Hara ■☆

82885 Cerastium holosteoides Fr. var. hallaisanense (Nakai) M. Mizush. = Cerastium fontanum Baumg. subsp. vulgare (Hartm.) Greuter et Burdet var. angustifolium (Franch.) H. Hara ■☆

82886 Cerastium holosteoides Fr. var. hallaisanense (Nakai) Mizush. = Cerastium fontanum Baumg. subsp. vulgare (Hartm.) Greuter et Burdet ■

82887 Cerastium holosteoides Fr. var. vulgare (Hartm.) Hyl. = Cerastium fontanum Baumg. subsp. vulgare (Hartm.) Greuter et Burdet ■

82888 Cerastium holosteum Fisch. ex Hornem.;勘察加卷耳■☆

82889 Cerastium holotrichum Sennen et Mauricio;全毛卷耳■☆

82890 Cerastium hyperboreum Tolm. = Cerastium arcticum Lange ■☆

82891 Cerastium ianthes F. N. Williams = Cerastium fontanum Baumg. subsp. vulgare (Hartm.) Greuter et Burdet ■

82892 Cerastium ianthes F. N. Williams = Cerastium holosteoides Fr. var. hallaisanense (Nakai) Mizush. ■

82893 Cerastium illyricum Ard. subsp. comatum (Desv.) P. D. Sell et Whitehead = Cerastium comatum Desv. ■☆

82894 Cerastium incanum Ledeb. = Cerastium fontanum Baumg. subsp. grandiflorum (Buch. -Ham. ex D. Don) H. Hara ■

82895 Cerastium indicum Wight et Arn.;印度卷耳■☆

82896 Cerastium indicum Wight et Arn. var. ruwenzoriense (F. N. Williams) Möschl = Cerastium indicum Wight et Arn. ■☆

82897 Cerastium inflatum Link = Cerastium dichotomum L. subsp. inflatum (Link) Cullen ■

82898 Cerastium inflatum Link = Cerastium inflatum Link ex Sweet ■☆

82899 Cerastium inflatum Link ex Sweet;膨胀卷耳■☆

82900 Cerastium inflatum Link ex Sweet = Cerastium dichotomum L. ■

82901 Cerastium jenisejense Hultén = Cerastium regelii Ostenf. ■☆

82902 Cerastium kasbek Parrot;卡斯拜克卷耳■☆

82903 Cerastium keniense T. C. E. Fr. et Weim. = Cerastium octandrum Hochst. ex A. Rich. var. adnivale (Chiov.) Möschl ■☆

82904 Cerastium kilimandscharicum (Engl.) T. C. E. Fr. et Weim. = Cerastium afromontanum T. C. E. Fr. et Weim. ■☆

82905 Cerastium lanatum Lam. = Cerastium alpinum L. subsp. lanatum (Lam.) Cesati ■☆

82906 Cerastium lapponicum Crantz = Cerastium cerastoides (L.) Britton ■

82907 Cerastium latifolium Fenzl = Cerastium arvense L. ■

82908 Cerastium ledebourianum Ser. = Cerastium pauciflorum Steven ex Ser. ■

82909 Cerastium leibergii Rydb. = Cerastium arvense L. subsp. strictum Gaudin ■☆

82910 Cerastium limprichtii Pax et K. Hoffm.;华北卷耳(椭圆叶卷耳);N. China Mouseear ■

82911 Cerastium lithospermifolium Fisch.;紫草叶卷耳;Gromwellleaf Mouseear ■

82912 Cerastium longepedunculatum Muhl. ex Britton var. sordidum (B. L. Rob.) Briq. = Cerastium texanum Britton ■☆

82913 Cerastium longifolium Willd.;长叶卷耳■☆

82914 Cerastium longipedunculatum Muhl. ex Britton = Cerastium nutans Raf. ■☆

82915 Cerastium luridum Guss. = Cerastium brachypetalum Pers. subsp. roeseri (Boiss. et Heldr.) Nyman ■☆

82916 Cerastium mairei H. Lév. = Arenaria iochanensis C. Y. Wu ■

82917 Cerastium mauritanicum Pomel = Cerastium dubium (Bastard) Guépin ■☆

82918 Cerastium maximum L.;大卷耳;Big Mouseear, Great Mouse-ear Chickweed ■

82919 Cerastium maximum L. var. falcatum Gren. = Cerastium falcatum Bunge ex Fenzl ■

82920 Cerastium melanandrum Maxim. = Arenaria melanandra (Maxim.) Mattf. ex Hand. -Mazz. ■

82921 Cerastium meyerianum Rupr.;迈氏卷耳■☆

82922 Cerastium microspermum C. A. Mey.;小籽卷耳■☆

82923 Cerastium molle Vill. = Cerastium arvense L. subsp. molle (Vill.) Arcang. ■☆

82924 Cerastium morrisonense Hayata;玉山卷耳(合欢卷耳,台湾卷耳);Morrison Mouseear, Taiwan Mouseear, Yushan Mouseear ■

82925 Cerastium morrisonense Hayata var. formosanum Ohwi = Cerastium morrisonense Hayata ■

82926 Cerastium multiflorum C. A. Mey.;繁花卷耳■☆

82927 Cerastium murbeckii Maire = Cerastium pumilum Curtis ■☆

82928 Cerastium nemorale M. Bieb.;林生卷耳■☆

82929 Cerastium nigrescens (H. C. Watson) Edmondston ex H. C. Watson subsp. arcticum (Lange) P. S. Lusby = Cerastium arcticum Lange ■☆

82930 Cerastium nigrescens (H. C. Watson) H. C. Watson = Cerastium arcticum Lange ■☆

82931 Cerastium nigrescens (H. C. Watson) H. C. Watson subsp. arcticum (Lange) P. S. Lusby = Cerastium arcticum Lange ■☆

82932 Cerastium nipaulense Wall. ex G. Don = Cerastium fontanum Baumg. subsp. grandiflorum (Buch. -Ham. ex D. Don) H. Hara ■

82933 Cerastium nitidum Greene = Cerastium arvense L. subsp. strictum Gaudin ■☆

82934 Cerastium nitidum Greene = Cerastium arvense L. ■

82935 Cerastium nutans Raf.;悬垂卷耳;Nodding Chickweed, Nodding Mouse-ear Chickweed ■☆

82936 Cerastium nutans Raf. var. brachypodum Engelm. ex A. Gray = Cerastium brachypetalum Pers. ■☆

82937 Cerastium nutans Raf. var. brachypodum Engelm. ex A. Gray = Cerastium brachypodum (Engelm. ex A. Gray) B. L. Rob. ■☆

82938 Cerastium nutans Raf. var. obtectum Kearney et Peebles;覆被悬垂卷耳■☆

82939 Cerastium nutans Raf. var. occidentale B. Boivin = Cerastium nutans Raf. ■☆

82940 Cerastium oblongifolium Torr. = Cerastium arvense L. ■

82941 Cerastium oblongifolium Torr. = Cerastium velutinum Raf. ■☆

82942 Cerastium obscurum Chaub. = Cerastium pentandrum L. ■☆

82943 Cerastium occidentale Greene = Cerastium arvense L. subsp. strictum Gaudin ■☆

82944 Cerastium occidentale Greene = Cerastium arvense L. ■

82945 Cerastium octandrum Hochst. ex A. Rich.;八雄蕊卷耳■☆

82946 Cerastium octandrum Hochst. ex A. Rich. var. adnivale (Chiov.) Möschl;续卷耳■☆

82947 Cerastium octandrum Hochst. ex A. Rich. var. scandens (Engl.) Cufod.;攀缘八雄蕊卷耳■☆

82948 Cerastium oreades Schischk.;山地卷耳■☆

82949 Cerastium oreophilum Greene = Cerastium arvense L. subsp. strictum Gaudin ■☆

82950 Cerastium oreophilum Greene = Cerastium arvense L. ■

82951 Cerastium oxalidiflorum Makino = Cerastium pauciflorum Steven ex Ser. var. oxalidiflorum (Makino) Ohwi ■

82952 Cerastium parvipetalum Hosok.;小瓣卷耳;Small-petaled Mouseear ■

82953 Cerastium patulum Greene = Cerastium arvense L. subsp. strictum Gaudin ■☆

82954 Cerastium patulum Greene = Cerastium arvense L. ■

82955 Cerastium pauciflorum Steven ex Ser.;疏花卷耳(寄奴花,少花卷耳);Fewflower Mouseear ■

82956 Cerastium pauciflorum Steven ex Ser. var. amurense (Regel) M. Mizush.;阿穆尔疏花卷耳■☆

82957 Cerastium pauciflorum Steven ex Ser. var. amurense (Regel) M. Mizush. = Cerastium pauciflorum Steven ex Ser. var. oxalidiflorum (Makino) Ohwi ■

82958 Cerastium pauciflorum Steven ex Ser. var. oxalidiflorum (Makino) Ohwi;毛蕊卷耳■

82959 Cerastium pauciflorum Steven ex Ser. var. triviale (Link) Jalas;簇生疏花卷耳■

82960 Cerastium pensylvanicum Hook. = Cerastium arvense L. subsp. strictum Gaudin ■☆

82961 Cerastium pentandrum L.;五蕊卷耳■☆

82962 Cerastium pentandrum L. subsp. fallax (Guss.) Maire et Weiller = Cerastium pentandrum L. ■☆

82963 Cerastium pentandrum L. subsp. gracile (Dufour) Maire et Weiller = Cerastium gracile L. Dufour ■☆

82964 Cerastium pentandrum L. subsp. gussonei (Lojac.) Maire et Weiller = Cerastium diffusum Pers. subsp. gussonei (Lojac.) P. D. Sell et Whitehead ■☆

82965 Cerastium pentandrum L. subsp. obscurum (Chaub.) Maire et Weiller = Cerastium pentandrum L. ■☆

82966 Cerastium pentandrum L. subsp. tetrandrum (Curtis) Maire et Weiller = Cerastium diffusum Pers. ■☆

82967 Cerastium pentandrum L. var. herbaceum (Gren.) Maire = Cerastium pentandrum L. ■☆

82968 Cerastium perfoliatum L.;抱茎卷耳(抱茎叶卷耳,穿叶卷耳);Perfoliate Mouseear ■

82969 Cerastium pilosum Greene = Cerastium beeringianum Cham. et Schltdl. ■☆

82970 Cerastium pilosum Ledeb. = Cerastium pauciflorum Steven ex Ser. ■

82971 Cerastium pilosum Sibth. et Sm. = Cerastium pauciflorum Steven ex Ser. var. oxalidiflorum (Makino) Ohwi ■

82972 Cerastium pilosum Sibth. et Sm. var. amurense Regel = Cerastium pauciflorum Steven ex Ser. var. oxalidiflorum (Makino) Ohwi ■

82973 Cerastium polymorphum Rupr.;多形卷耳■☆

82974 Cerastium ponticum Albov;桥卷耳■☆

82975 Cerastium pubescens Goldie = Cerastium arvense L. subsp. strictum Gaudin ■☆

82976 Cerastium pulchellum Rydb. = Cerastium beeringianum Cham. et Schltdl. ■☆

82977 Cerastium pumilum Curtis;矮小卷耳■☆

82978 Cerastium pumilum Curtis subsp. algericum Batt. = Cerastium diffusum Pers. ■☆

82979 Cerastium pumilum Curtis subsp. fallax (Guss.) Maire = Cerastium pentandrum L. ■☆

82980 Cerastium pumilum Curtis subsp. glutinosum (Fr.) Corb. = Cerastium glutinosum Fr. ■☆

82981 Cerastium pumilum Curtis subsp. gussonei (Lojac.) Maire = Cerastium diffusum Pers. subsp. gussonei (Lojac.) P. D. Sell et Whitehead ■☆

82982 Cerastium pumilum Curtis subsp. murbeckii (Maire) Maire = Cerastium pumilum Curtis ■☆

82983 Cerastium pumilum Curtis subsp. siculum (Guss.) Maire = Cerastium siculum Guss. ■☆

82984 Cerastium pumilum Curtis subsp. subechinulatum (Maire et Wilczek) Maire = Cerastium gracile L. Dufour ■☆

82985 Cerastium pumilum Curtis subsp. tetrandrum (Curtis) Maire = Cerastium diffusum Pers. ■☆

82986 Cerastium pumilum Curtis var. algeriense Batt. = Cerastium diffusum Pers. ■☆

82987 Cerastium pumilum Curtis var. gussonei (Lojac.) Batt. = Cerastium diffusum Pers. subsp. gussonei (Lojac.) P. D. Sell et Whitehead ■☆

82988 Cerastium pumilum Curtis var. tetrandrum (Curtis) Batt. = Cerastium diffusum Pers. ■☆

82989 Cerastium pumilum Griseb. = Cerastium glutinosum Fr. ■☆

82990 Cerastium pumilum Raf. = Cerastium glomeratum Thuill. ■

82991 Cerastium pumilum W. M. Curfis;柯氏卷耳;Chickweed, Curtis' Mouse-ear, Dwarf Mouse-ear, Sticky Mouse-ear Chickweed ■☆

82992 Cerastium pumilum W. M. Curfis subsp. glutinosum (Fr.) Jalas = Cerastium pumilum W. M. Curfis ■☆

82993 Cerastium purpurascens Adams;紫卷耳■☆

82994 Cerastium pusillum Ser.;山卷耳(卷耳,细小卷耳,小卷耳);Dwarf Mouse-ear, European Chickweed, Wild Mouseear ■

82995 Cerastium pycnophyllum Peter = Cerastium afromontanum T. C. E. Fr. et Weim. ■☆

82996 Cerastium ramosissimum Boiss.;多枝卷耳■☆

82997 Cerastium regelii Ostenf.;瑞氏卷耳;Regel Mouseear, Regel's Mouse-ear Chickweed ■☆

82998 Cerastium regelii Ostenf. subsp. caespitosum (Malmgren) Tolm. = Cerastium regelii Ostenf. ■☆

82999 Cerastium repens L. = Cerastium tomentosum L. ■☆

83000 Cerastium riaei Des Moul. = Cerastium ramosissimum Boiss. ■☆

83001 Cerastium riaei Des Moul. subsp. brevicorollinum Maire et Weiller = Cerastium gracile L. Dufour ■☆

83002 Cerastium riaei Des Moul. subsp. echinulatum (Coss. et Durieu) Maire = Cerastium gracile L. Dufour ■☆

83003 Cerastium riaei Des Moul. subsp. subechinulatum Maire = Cerastium gracile L. Dufour ■☆

83004 Cerastium rigidum Ledeb. = Cerastium furcatum Cham. et Schltdl. ■

83005 Cerastium roeseri Boiss. et Heldr. = Cerastium brachypetalum Pers. subsp. roeseri (Boiss. et Heldr.) Nyman ■☆

83006 Cerastium rubescens Mattf. = Cerastium furcatum Cham. et

Schltdl. ■

83007 Cerastium rubescens Mattf. var. koreanum (Nakai) E. Miki;朝鲜卷耳■☆

83008 Cerastium rubescens Mattf. var. ovatum (Miyabe) M. Mizush. = Cerastium arvense L. var. ovatum (Miyabe) E. Miki ■☆

83009 Cerastium rubescens Mattf. var. ovatum (Miyabe) M. Mizush. f. tetraschistum M. Mizush.;四层卷耳■☆

83010 Cerastium ruderale M. Bieb.;荒地卷耳■☆

83011 Cerastium scammamiae Polunin = Cerastium beeringianum Cham. et Schltdl. ■☆

83012 Cerastium schimperi (Engl.) De Wild. = Cerastium indicum Wight et Arn. ■☆

83013 Cerastium schizopetalum Maxim.;裂瓣卷耳■☆

83014 Cerastium schizopetalum Maxim. var. bifidum Takeda ex M. Mizush.;双裂瓣卷耳■☆

83015 Cerastium schizopetalum Maxim. var. rupicola (Ohwi) Akiyama = Cerastium schizopetalum Maxim. var. bifidum Takeda ex M. Mizush. ■☆

83016 Cerastium schmalhausenii Pacz.;施麻氏卷耳■☆

83017 Cerastium schmidtianum Takeda = Cerastium fischerianum Ser. ■

83018 Cerastium scopulorum Greene = Cerastium arvense L. subsp. strictum Gaudin ■☆

83019 Cerastium scopulorum Greene = Cerastium arvense L. ■

83020 Cerastium semidecandrum L.;五雄卷耳;Chickweed, Fivestamen Chickweed, Five-stamen Chickweed, Five-stamen Mouse-ear Chickweed, Least Mouse Ear Chickweed, Little Mouse Ear, Little Mouse-ear, Small Mouse-ear Chickweed, Wayside Cerastium ■☆

83021 Cerastium semidecandrum L. subsp. dentatum (Möschl) Maire et Weiller = Cerastium semidecandrum L. ■☆

83022 Cerastium semidecandrum L. subsp. glutinosum (Fr.) Maire et Weiller = Cerastium glutinosum Fr. ■☆

83023 Cerastium semidecandrum L. subsp. linneanum Maire et Weiller = Cerastium semidecandrum L. ■☆

83024 Cerastium semidecandrum L. var. senneni Font Quer = Cerastium semidecandrum L. ■☆

83025 Cerastium sericeum S. Watson = Cerastium nutans Raf. var. obtectum Kearney et Peebles ■☆

83026 Cerastium siculum Guss.;西西里卷耳■☆

83027 Cerastium sonnei Greene = Cerastium arvense L. subsp. strictum Gaudin ■☆

83028 Cerastium sonnei Greene = Cerastium arvense L. ■

83029 Cerastium sordidum B. L. Rob. = Cerastium texanum Britton ■☆

83030 Cerastium sosnowskyi Schischk.;索氏卷耳■☆

83031 Cerastium squalidum Raymond = Cerastium alpinum L. subsp. lanatum (Lam.) Cesati ■☆

83032 Cerastium stevenii Schischk.;司梯氏卷耳;Steven Mouseear ■☆

83033 Cerastium strictum Haenke = Cerastium arvense L. subsp. strictum (L.) Gaudin ■

83034 Cerastium strictum L. = Cerastium arvense L. ■

83035 Cerastium subpilosa Hayata var. takasagomontana (Masam.) S. S. Ying = Arenaria takasagomontana (Masam.) S. S. Ying ■

83036 Cerastium subpilosum Hayata;毛卷耳(细叶卷耳,亚毛无心菜);Thinleaf Mouseear ■

83037 Cerastium subpilosum Hayata = Arenaria subpilosa (Hayata) Ohwi ■

83038 Cerastium subpilosum Hayata var. takasagomontanum (Masam.) S. S. Ying = Arenaria takasagomontana (Masam.) S. S. Ying ■

83039 Cerastium subulatum Greene = Cerastium arvense L. subsp. strictum Gaudin ■☆

83040 Cerastium subulatum Greene = Cerastium arvense L. ■

83041 Cerastium sventenii Jalas;斯文顿卷耳■☆

83042 Cerastium szechuense Williams;四川卷耳;Sichuan Mouseear ■

83043 Cerastium szowitsii Boiss.;绍氏卷耳■☆

83044 Cerastium taiwanense Tang S. Liu = Arenaria subpilosa (Hayata) Ohwi ■

83045 Cerastium takasagomontanum Masam.;山生卷耳;Alp Mouseear ■

83046 Cerastium takasagomontanum Masam. = Arenaria takasagomontana (Masam.) S. S. Ying ■

83047 Cerastium tauricum Spreng. = Cerastium brachypetalum Pers. subsp. tauricum (Spreng.) Murb. ■☆

83048 Cerastium tauricum Spreng. = Cerastium brachypetalum Pers. ■☆

83049 Cerastium tauricum Spreng. ex Ser.;克里木卷耳;Klimu Mouseear ■☆

83050 Cerastium tenoreanum Ser. = Cerastium brachypetalum Pers. subsp. tenoreanum (Ser.) Soó ■☆

83051 Cerastium tenuifolium Pursh = Cerastium arvense L. subsp. strictum Gaudin ■☆

83052 Cerastium tenuifolium Pursh = Cerastium arvense L. ■

83053 Cerastium terrae-novae Fernald et Wiegand;纽芬兰卷耳;Newfoundland Mouse-ear Chickweed ■☆

83054 Cerastium tetrandrum Curtis = Cerastium diffusum Pers. ■☆

83055 Cerastium tetrandrum Curtis = Cerastium dubium (Bastard) O. Schwarz ■☆

83056 Cerastium texanum Britton;奇瓦瓦卷耳;Chihuahuan Mouse-ear Chickweed ■☆

83057 Cerastium thermale Rydb. = Cerastium arvense L. subsp. strictum Gaudin ■☆

83058 Cerastium thomsonii Hook. f.;藏南卷耳;Xizang Mouseear ■

83059 Cerastium tianschanicum Schischk.;天山卷耳;Tianshan Mouseear ■

83060 Cerastium tomentosum L.;绒毛卷耳(银角);Asmanian Snow Gum, Dusty Husband, Dusty Miller, Jerusalem Star, Silver Moss, Snow in Summer, Snow-in-harvest, Snow-in-summer, Snow-on-the-Mountain, Snow-plant ■☆

83061 Cerastium trigynum Vill. = Cerastium cerastoides (L.) Britton ■

83062 Cerastium trigynum Vill. var. morrisonense (Hayata) Hayata = Cerastium morrisonense Hayata ■

83063 Cerastium trigynum Vill. var. taiwanianum S. S. Ying = Cerastium morrisonense Hayata ■

83064 Cerastium triviale Link;路旁卷耳;Chickenweed, Narrow-leaved Mouse Ear, Wayside Mouse Ear Chickweed, Wayside Mouse-ear Chickweed ■☆

83065 Cerastium triviale Link = Cerastium fontanum Baumg. subsp. triviale (Link) Jalas ■

83066 Cerastium triviale Link = Cerastium fontanum Baumg. subsp. vulgare (Hartm.) Greuter et Burdet ■

83067 Cerastium triviale Link = Cerastium pauciflorum Steven ex Ser. var. triviale (Link) Jalas ■

83068 Cerastium triviale Link var. glandulosum (Schur) Koch = Cerastium fontanum Baumg. subsp. vulgare (Hartm.) Greuter et Burdet ■

83069 Cerastium triviale Link var. nipaulense F. N. Williams = Cerastium fontanum Baumg. subsp. grandiflorum (Buch. -Ham. ex D. Don) H. Hara ■

83070 Cerastium unalaschkense Takeda = Cerastium fischerianum Ser. ■
83071 Cerastium undulatifolium Sommier et H. Lév. ;波叶卷耳■☆
83072 Cerastium vagans Lowe;漫游卷耳■☆
83073 Cerastium valgatum L. var. hallaisanense Nakai = Cerastium holosteoides Fr. var. hallaisanense (Nakai) Mizush. ■
83074 Cerastium variabile Goodd. = Cerastium beeringianum Cham. et Schltdl. ■☆
83075 Cerastium velutinum Raf. ;短毛卷耳;Barren Chickweed, Large field Mouse-ear Chickweed ■☆
83076 Cerastium velutinum Raf. var. villossissimum (Pennell) J. K. Morton;密长毛卷耳;Octoraro Creek Chickweed ■☆
83077 Cerastium verticifolium R. L. Dang et X. M. Pi;轮叶卷耳;Verticileaf Mouseear ■
83078 Cerastium vestitum Greene = Cerastium arvense L. subsp. strictum Gaudin ■☆
83079 Cerastium vestitum Greene = Cerastium arvense L. ■
83080 Cerastium villosum Muhl. ex Darl. = Cerastium arvense L. ■
83081 Cerastium viride A. Heller;绿卷耳;Western field Mouse-ear Chickweed ■☆
83082 Cerastium viscosum L. = Cerastium fontanum Baumg. subsp. vulgare (Hartm.) Greuter et Burdet ■
83083 Cerastium viscosum L. = Cerastium glomeratum Thuill. ■
83084 Cerastium viscosum L. f. apetalum (Dumort.) Mert. et W. D. J. Koch = Cerastium glomeratum Thuill. ■
83085 Cerastium vulgare Hartm. = Cerastium fontanum Baumg. subsp. vulgare (Hartm.) Greuter et Burdet ■
83086 Cerastium vulgare Hartm. subsp. triviale (Link) Murb. = Cerastium fontanum Baumg. subsp. triviale (Link) Jalas ■
83087 Cerastium vulgatum L. = Cerastium fontanum Baumg. subsp. vulgare (Hartm.) Greuter et Burdet ■
83088 Cerastium vulgatum L. = Cerastium fontanum Baumg. ■
83089 Cerastium vulgatum L. = Cerastium glomeratum Thuill. ■
83090 Cerastium vulgatum L. f. glandulosum (Boenn.) Druce = Cerastium fontanum Baumg. ■
83091 Cerastium vulgatum L. subsp. atlanticum (Durieu) Batt. = Cerastium atlanticum Durieu ■☆
83092 Cerastium vulgatum L. subsp. caespitosum Dostal = Cerastium fontanum Baumg. subsp. vulgare (Hartm.) Greuter et Burdet ■
83093 Cerastium vulgatum L. var. acutifolium Franch. = Cerastium furcatum Cham. et Schltdl. ■
83094 Cerastium vulgatum L. var. angustifolium Franch. = Cerastium fontanum Baumg. subsp. vulgare (Hartm.) Greuter et Burdet ■
83095 Cerastium vulgatum L. var. beeringianum (Cham. et Schltdl.) Fenzl = Cerastium beeringianum Cham. et Schltdl. ■☆
83096 Cerastium vulgatum L. var. brevifolium Franch. = Cerastium furcatum Cham. et Schltdl. ■
83097 Cerastium vulgatum L. var. glomeratum (Thuill.) DC. = Cerastium glomeratum Thuill. ■
83098 Cerastium vulgatum L. var. glomeratum (Thuill.) Edgew. et Hook. f. = Cerastium glomeratum Thuill. ■
83099 Cerastium vulgatum L. var. grandiflorum Edgew. et Hook. f. = Cerastium fontanum Baumg. subsp. grandiflorum (Buch. -Ham. ex D. Don) H. Hara ■
83100 Cerastium vulgatum L. var. hallaisanense Nakai = Cerastium fontanum Baumg. subsp. vulgare (Hartm.) Greuter et Burdet ■
83101 Cerastium vulgatum L. var. hirsutum Fr. = Cerastium fontanum Baumg. subsp. vulgare (Hartm.) Greuter et Burdet ■
83102 Cerastium vulgatum L. var. hirsutum Fr. f. glandulosum (Boenn.) Druce = Cerastium fontanum Baumg. subsp. vulgare (Hartm.) Greuter et Burdet ■
83103 Cerastium vulgatum L. var. holosteoides (Fr.) Wahlenb. = Cerastium fontanum Baumg. subsp. vulgare (Hartm.) Greuter et Burdet ■
83104 Cerastium vulgatum L. var. leiopetalum Fenzl = Cerastium pusillum Ser. ■
83105 Cerastium vulgatum L. var. longipes Batt. = Cerastium atlanticum Durieu ■☆
83106 Cerastium vulgatum L. var. tianschanicum (Schischk.) Kozhevn. = Cerastium tianschanicum Schischk. ■
83107 Cerastium vulgatum L. var. tibeticum Edgew. et Hook. f. = Cerastium fontanum Baumg. subsp. vulgare (Hartm.) Greuter et Burdet ■
83108 Cerastium wilsonii Takeda;卵叶卷耳(鄂西卷耳,威氏卷耳);W. Hubei Mouseear ■
83109 Cerastium winkleri Briq. = Stellaria winkleri (Briq.) Schischk. ■
83110 Cerasus Mill. (1754);樱属(樱桃属,郁李属);Cherry ●
83111 Cerasus Mill. = Prunus L. ●
83112 Cerasus 'Autumnalis';秋早樱(秋季大叶早樱,十月彼岸樱);Autumn Blooming Cherry, Autumn Cherry, Winter Cherry, Winter-flowering Cherry ●☆
83113 Cerasus 'Autumnalis' = Prunus subhirtella Miq. 'Autumnalis' ●☆
83114 Cerasus 'Morioka-pendula';盛冈垂枝樱●☆
83115 Cerasus × chichibuensis (Kubota et Moriya) H. Ohba;秩父樱●☆
83116 Cerasus × chichibuensis (Kubota et Moriya) H. Ohba nothovar. aizuensis (Kawas.) Yonek. ;会津樱●☆
83117 Cerasus × chichibuensis (Kubota et Moriya) H. Ohba var. uyekii (Kubota) H. Ohba;植木樱●☆
83118 Cerasus × compta (Koidz.) H. Ohba;装饰樱●☆
83119 Cerasus × furuseana (Ohwi) H. Ohba;古施樱●☆
83120 Cerasus × furuseana (Ohwi) H. Ohba nothovar. pseudaffinis (Kawas.) H. Ohba;假近缘樱●☆
83121 Cerasus × gondouini Poit. et Turpin;公爵樱桃;Duke Cherry ●☆
83122 Cerasus × hisauchiana (Koidz. ex Hisauti) H. Ohba;久内樱●☆
83123 Cerasus × juddii (E. S. Anderson) H. Ohba;贾德樱●☆
83124 Cerasus × kubotana (Kawas.) H. Ohba;久保樱●☆
83125 Cerasus × mitsuminensis (Moriya) H. Ohba;三峰山樱●☆
83126 Cerasus × miyasakana (H. Kubota) H. Ohba;宫坂樱●☆
83127 Cerasus × miyoshii (Ohwi) H. Ohba;三好学樱●☆
83128 Cerasus × mochizukiana (Nakai) H. Ohba;望月樱●☆
83129 Cerasus × moniwana (Kawas.) Yonek. ;茂庭樱●☆
83130 Cerasus × oneyamensis (Hayashi) H. Ohba;大根山樱●☆
83131 Cerasus × oneyamensis (Hayashi) H. Ohba nothovar. takasawana (Kubota et Funatsu) H. Ohba;高泽樱●☆
83132 Cerasus × sacra (Miyoshi) H. Ohba;神樱●☆
83133 Cerasus × syodoi (Nakai) H. Ohba;赛多樱●☆
83134 Cerasus × takasawana (Kubota et Funatsu) H. Ohba = Cerasus × oneyamensis (Hayashi) H. Ohba nothovar. takasawana (Kubota et Funatsu) H. Ohba ●☆
83135 Cerasus × takinoensis (Kawas.) Yonek. ;泷野樱●☆
83136 Cerasus × tschoniskii (Maxim.) H. Ohba nothovar. pseudoverecunda (H. Kubota et Moriya) H. Ohba;假羞涩樱●☆
83137 Cerasus × tschonoskii (Koehne) H. Ohba;须川樱桃(柴氏樱桃)●☆
83138 Cerasus × yanashimana (H. Kubota et Moriya) H. Ohba;梁岛

樱●☆
83139 Cerasus × yedoensis (Matsum.) A. V. Vassil.;江户樱桃●☆
83140 Cerasus × yuyamae (Sugim.) H. Ohba;汤山樱●☆
83141 Cerasus × yuyamae (Sugim.) H. Ohba nothovar. bukosanensis (Moriya) H. Ohba;武甲山樱●☆
83142 Cerasus acuminata Wall. = Laurocerasus undulata (Buch. -Ham. ex D. Don) M. Roem. ●
83143 Cerasus alaica Pojark.;阿赖樱●☆
83144 Cerasus amygdaliflora Nevski;膀胱花樱●☆
83145 Cerasus apetala (Siebold et Zucc.) Ohle ex H. Ohba;日本樱桃(丁字樱,无瓣樱桃)●☆
83146 Cerasus apetala (Siebold et Zucc.) Ohle ex H. Ohba = Prunus apetala (Siebold et Zucc.) Franch. et Sav. ●☆
83147 Cerasus apetala (Siebold et Zucc.) Ohle ex H. Ohba var. montica (Kawas. et H. Koyama) H. Ohba;山地日本樱桃●☆
83148 Cerasus apetala (Siebold et Zucc.) Ohle ex H. Ohba var. pilosa (Koidz.) H. Ohba;毛日本樱桃●☆
83149 Cerasus apetala (Siebold et Zucc.) Ohle ex H. Ohba var. pilosa (Koidz.) H. Ohba f. multipetala (Kawas.) H. Ohba;多瓣日本樱桃●☆
83150 Cerasus apetala (Siebold et Zucc.) Ohle ex H. Ohba var. pilosa (Koidz.) H. Ohba = Prunus apetala (Siebold et Zucc.) Franch. et Sav. subsp. pilosa (Koidz.) H. Ohba ●☆
83151 Cerasus apetala (Siebold et Zucc.) Ohle ex H. Ohba var. pilosa (Koidz.) H. Ohba = Prunus apetala (Siebold et Zucc.) Franch. et Sav. var. pilosa (Koidz.) E. H. Wilson ●☆
83152 Cerasus apetala (Siebold et Zucc.) Ohle ex H. Ohba var. pilosa (Koidz.) H. Ohba f. multipetala (Kawas.) H. Ohba = Prunus apetala (Siebold et Zucc.) Franch. et Sav. var. pilosa (Koidz.) E. H. Wilson f. multipetala Kawas. ●☆
83153 Cerasus arasoides (D. Don) Sokoloff = Cerasus cerasoides (D. Don) Sokoloff ●
83154 Cerasus arasoides (D. Don) Sokoloff = Prunus cerasoides D. Don ●
83155 Cerasus avium (L.) Moench;欧洲甜樱桃(尖尾樱桃,欧洲樱桃,甜樱桃,野樱桃);Bird Cherry, Black Merry, Brandy Mazzard, Cherry, Crab Cherry, Europe Cherry, European Cherry, Gaskins, Gean, Geen, Guind, Hagberry, Hawkberry, Hegberry, Kerroon, Kerroon Coroon, Mazzard, Mazzard Cherry, Mazzud, Merry-tree, Mezard, Sweet Cherry, Wild Cherry, Wild Heart Cherry ●
83156 Cerasus avium (L.) Moench = Prunus avium (L.) L. ●
83157 Cerasus besseyi (L. H. Bailey) Smyth;西部沙地樱桃(贝西樱桃,沙樱桃);Bessey Cherry, Bessey's Cherry, Sand Cherry, Western Sand Cherry ●☆
83158 Cerasus campanulata (Maxim.) A. V. Vassil.;钟花樱桃(绯樱,福建山樱花,福建山樱桃,福建樱桃,山樱花,山樱桃,台湾山樱花,钟花樱);Bell-flower Cherry, Bellflowered Cherry, Bell-flowered Cherry, Taiwan Cherry, Taiwan Flowering Cherry ●
83159 Cerasus campanulata (Maxim.) Te T. Yu et C. L. Li = Cerasus campanulata (Maxim.) A. V. Vassil. ●
83160 Cerasus campanulata (Maxim.) Te T. Yu et C. L. Li = Prunus campanulata Maxim. ●
83161 Cerasus canescens (Bois) S. Y. Sokolov;灰樱桃;Hoary Cherry ●
83162 Cerasus caudata (Franch.) Te T. Yu et C. L. Li;尖尾樱桃(尖尾樱);Caudate Cherry ●
83163 Cerasus caudata (Franch.) Te T. Yu et C. L. Li = Prunus caudata Franch. ●
83164 Cerasus cerasoides (Buch. -Ham. ex D. Don) Sokoloff = Cerasus cerasoides (D. Don) Sokoloff ●
83165 Cerasus cerasoides (D. Don) Sokoloff;高盆樱桃(滇樱桃,高盆李,高盆樱,箐樱桃,山樱桃,云南欧李);Cerasoid Cherry, Salver-shaped Cherry ●
83166 Cerasus cerasoides (D. Don) Sokoloff = Prunus cerasoides D. Don ●
83167 Cerasus cerasoides (D. Don) Sokoloff var. rubea (Ingram) Te T. Yu et C. L. Li;红花高盆樱桃(西府海棠);Redflower Salver-shaped Cherry ●
83168 Cerasus cerasoides (D. Don) Sokoloff var. rubea (Ingram) Te T. Yu et C. L. Li = Prunus cerasoides Ehrh. var. rubea Ingram ●
83169 Cerasus cerasoides (D. Don) Sokoloff var. rubea (Ingram) Te T. Yu et C. L. Li = Cerasus cerasoides (D. Don) Sokoloff ●
83170 Cerasus cerasus Eaton et Wright = Cerasus vulgaris Mill. ●
83171 Cerasus clarofolia (C. K. Schneid.) Te T. Yu et C. L. Li;微毛樱桃(微毛野樱桃,西南樱桃);Fewhairy Cherry ●
83172 Cerasus clarofolia (C. K. Schneid.) Te T. Yu et C. L. Li = Prunus clarofolia C. K. Schneid. ●
83173 Cerasus claviculata Te T. Yu et C. L. Li = Cerasus dolichadenia (Cardot) S. Y. Jiang et C. L. Li ●
83174 Cerasus collina Lej. et Court;山丘樱桃●☆
83175 Cerasus conadenia (Koehne) Te T. Yu et C. L. Li;锥腺樱桃(锥腺樱);Subulate Cherry, Subulate Gland Cherry ●
83176 Cerasus conadenia (Koehne) Te T. Yu et C. L. Li = Prunus conadenia Koehne ●
83177 Cerasus conradinae (Koehne) Te T. Yu et C. L. Li;华中樱桃(单齿樱花,华中樱,康拉樱);Conradina Cherry ●
83178 Cerasus cornuta Wall. = Padus cornuta (Wall. ex Royle) Carrière ●
83179 Cerasus cornuta Wall. = Prunus cornuta (Wall. ex Royle) Steud. ●
83180 Cerasus cornuta Wall. ex Royle = Padus cornuta (Wall. ex Royle) Carrière ●
83181 Cerasus cornuta Wall. ex Royle = Prunus cornuta (Wall. ex Royle) Steud. ●
83182 Cerasus crataegifolia (Hand. -Mazz.) Te T. Yu et C. L. Li;山楂叶樱桃(山楂叶樱);Hawthorn Cherry, Hawthornleaf Cherry, Hawthorn-leaved Cherry ●
83183 Cerasus crataegifolia (Hand. -Mazz.) Te T. Yu et C. L. Li = Prunus crataegifolia Hand. -Mazz. ●
83184 Cerasus cyclamina (Koehne) Te T. Yu et C. L. Li;襄阳樱桃(双花山樱桃,襄阳山樱,襄阳山樱桃,襄阳樱);Cyclamen Cherry, Hsiangyang Cherry, Xiangyang Cherry ●
83185 Cerasus cyclamina (Koehne) Te T. Yu et C. L. Li = Prunus cyclamina Koehne ●
83186 Cerasus cyclamina (Koehne) Te T. Yu et C. L. Li var. biflora (Koehne) Te T. Yu et C. L. Li;双花山樱桃;Twoflower Cherry ●
83187 Cerasus cyclamina (Koehne) Te T. Yu et C. L. Li var. biflora (Koehne) Te T. Yu et C. L. Li = Prunus cyclamina Koehne var. biflora Koehne ●
83188 Cerasus demissa Nutt.;垂樱;Western Choke Cherry ●☆
83189 Cerasus dictyoneura (Diels) Te T. Yu;毛叶欧李(脉欧李,网脉欧李,显脉欧李);Hairyleaf Cherry, Hairy-leaved Cherry ●
83190 Cerasus dictyoneura (Diels) Te T. Yu = Prunus dictyoneura Diels ●
83191 Cerasus dielsiana (C. K. Schneid.) Te T. Yu et C. L. Li;尾叶樱桃(尾叶欧李,尾叶樱);Diels Cherry ●
83192 Cerasus dielsiana (C. K. Schneid.) Te T. Yu et C. L. Li = Prunus dielsiana C. K. Schneid. ●

83193 Cerasus dielsiana (C. K. Schneid.) Te T. Yu et C. L. Li var. abbreviana (Cardot) Te T. Yu et C. L. Li;短梗尾叶樱桃(短柄尾叶樱,短梗尾叶欧李);Shortstalk Cherry,Shortstalk Diels Cherry ●

83194 Cerasus dielsiana (C. K. Schneid.) Te T. Yu et C. L. Li var. abbreviana (Cardot) Te T. Yu et C. L. Li = Prunus dielsiana C. K. Schneid. var. abbreviata Cardot ●

83195 Cerasus discadenia (Koehne) S. Y. Jiang et C. L. Li;盘腺樱桃●

83196 Cerasus discoidea Te T. Yu et C. L. Li;迎春樱桃;Discoid Cherry,Disc-shaped Cherry ●

83197 Cerasus dolichadenia (Cardot) S. Y. Jiang et C. L. Li;长腺樱桃;Clubshaped Cherry,Longgland Cherry ●

83198 Cerasus duclouxii (Koehne) Te T. Yu et C. L. Li;西南樱桃(西南樱);Ducloux Cherry ●

83199 Cerasus duclouxii (Koehne) Te T. Yu et C. L. Li = Cerasus yunnanensis (Franch.) Te T. Yu et C. L. Li ●

83200 Cerasus duclouxii (Koehne) Te T. Yu et C. L. Li = Prunus duclouxii Koehne ●

83201 Cerasus erythrocarpa Nevski;红果樱桃●

83202 Cerasus fruticosa (Pall.) Woronow;灌木樱桃(草原樱桃,灌木樱,金老梅,欧洲草原酸樱桃);Dwarf Cherry, European Ground Cherry,Grassland Cherry, Ground Cherry, Ground-cherry, Mongolian Cherry ●

83203 Cerasus fruticosa (Pall.) Woronow = Prunus fruticosa Pall. ●

83204 Cerasus glabra (Pamp.) Te T. Yu et C. L. Li;光叶樱桃(光叶樱);Glabrousleaf Cherry,Glabrous-leaved Cherry ●

83205 Cerasus glabra (Pamp.) Te T. Yu et C. L. Li = Cerasus conradinae (Koehne) Te T. Yu et C. L. Li ●

83206 Cerasus glabra (Pamp.) Te T. Yu et C. L. Li = Prunus glabra (Pamp.) Koehne ●

83207 Cerasus glandulifolia (Rupr. et Maxim.) Kom.;腺叶樱桃●☆

83208 Cerasus glandulosa (Thunb.) Loisel.;麦李;Almont Cherry, Chinese Bush Cherry, Dwarf Flowering Almond, Dwarf Flowering Cherry,Flowering Almond ●

83209 Cerasus glandulosa (Thunb.) Loisel. = Prunus glandulosa Thunb. ●

83210 Cerasus glandulosa (Thunb.) Loisel. f. alboplena Koehne;白花重瓣麦李●

83211 Cerasus glandulosa (Thunb.) Loisel. f. rosea Koehne;粉花麦李●

83212 Cerasus glandulosa (Thunb.) Loisel. f. sinensis (Pers.) Koehne;粉花重瓣麦李●

83213 Cerasus glandulosa (Thunb.) Loisel. var. trichostyla (Koehne) J. X. Yang = Prunus glandulosa Thunb. var. trichostyla Koehne ●☆

83214 Cerasus glandulosa (Thunb.) Loisel. var. trichostyla (Koehne) J. X. Yang;毛柱麦李●☆

83215 Cerasus hainanensis G. A. Fu et Y. S. Lin;海南樱桃;Hainan Cherry ●

83216 Cerasus henryi (C. K. Schneid.) Te T. Yu et C. L. Li;蒙自樱桃(蒙自樱);Henry Plum ●

83217 Cerasus henryi (C. K. Schneid.) Te T. Yu et C. L. Li = Prunus henryi (C. K. Schneid.) Koehne ●

83218 Cerasus hortensis Mill. = Cerasus vulgaris Mill. ●

83219 Cerasus humilis (Bunge) A. I. Baranov et Liou = Cerasus humilis (Bunge) Sokoloff ●

83220 Cerasus humilis (Bunge) Sokoloff;欧李(欧梨,欧李儿,酸丁,乌拉纳,乌拉奈,乌喇奈,郁李);Bunge Cherry, China Dwarf Cherry,Chinese Dwarf Cherry ●

83221 Cerasus humilis (Bunge) Sokoloff = Prunus humilis Bunge ●

83222 Cerasus incana (Pall.) Spach = Prunus prostrata Labill. ●☆

83223 Cerasus incana Boiss. = Prunus incisa Thunb. ●

83224 Cerasus incisa (Thunb.) Loisel.;豆樱(锯叶豆樱,日本樱桃);Fuji Cherry,Madagascar Plum ●

83225 Cerasus incisa (Thunb.) Loisel. f. chrysantha H. Ohba;金花豆樱●☆

83226 Cerasus incisa (Thunb.) Loisel. f. globosa (Kawas.) H. Ohba;球形樱桃●☆

83227 Cerasus incisa (Thunb.) Loisel. f. urceolata (Koidz.) H. Ohba;坛状樱桃●☆

83228 Cerasus incisa (Thunb.) Loisel. f. yamadae (Makino) H. Ohba;山田氏豆樱●☆

83229 Cerasus incisa (Thunb.) Loisel. var. bukosanensis (Honda) H. Ohba;武甲山豆樱●☆

83230 Cerasus incisa (Thunb.) Loisel. var. incisa f. globosa (Kawas.) H. Ohba;球豆樱●☆

83231 Cerasus incisa (Thunb.) Loisel. var. kinkiensis (Koidz.) H. Ohba;近畿樱●☆

83232 Cerasus incisa (Thunb.) Loisel. var. kinkiensis (Koidz.) H. Ohba f. plena (Satomi) H. Ohba;重瓣近畿樱●☆

83233 Cerasus jacquemontii (Hook. f.) Buser;雅克樱桃●☆

83234 Cerasus jamasakura (Siebold ex Koidz.) H. Ohba;日本山樱桃(日本山樱,山樱,山樱花,野生福岛樱);Cherry, Hill Cherry, Japanese Hill Cherry,Underbrown Japanese Cherry ●☆

83235 Cerasus jamasakura (Siebold ex Koidz.) H. Ohba 'Humilis';矮日本山樱桃●☆

83236 Cerasus jamasakura (Siebold ex Koidz.) H. Ohba f. pubescens (Makino) H. Ohba;毛日本山樱桃●☆

83237 Cerasus jamasakura (Siebold ex Koidz.) H. Ohba var. chikusiensis (Koidz.) H. Ohba;筑紫樱●☆

83238 Cerasus japonica (Thunb.) Loisel.;郁李(奥李,常棣,车下李,赤李子,棣,多叶郁李,爵李,爵梅,苦李子,林生梅,马鞭花,麦李,雀李,雀梅,山里黄,时,寿李,唐棣,棠棣,秧李,野李子,野樱桃树,英梅,栯木,御园李,郁);China Cherry, Chinese Bush Cherry,Chinese Bushcherry,Dwarf Flowering Cherry ●

83239 Cerasus japonica (Thunb.) Loisel. = Prunus japonica Thunb. ●

83240 Cerasus japonica (Thunb.) Loisel. var. glandulosa Kom. et Aliss. = Cerasus glandulosa (Thunb.) Loisel. ●

83241 Cerasus japonica (Thunb.) Loisel. var. nakaii (H. Lév.) Te T. Yu et C. L. Li;长梗郁李(中井郁李);Longpedicel Chinese Bushcherry ●

83242 Cerasus japonica (Thunb.) Loisel. var. nakaii (H. Lév.) Te T. Yu et C. L. Li = Cerasus nakai (H. Lév.) A. I. Baranov et Liou ●

83243 Cerasus japonica (Thunb.) Loisel. var. nakaii (H. Lév.) Te T. Yu et C. L. Li = Prunus japonica Thunb. var. nakaii (H. Lév.) Rehder ●

83244 Cerasus kurilensis (Miyabe) De Moor = Cerasus nipponica (Matsum.) Ohle ex H. Ohba var. kurilensis (Miyabe) H. Ohba ●☆

83245 Cerasus lannesiana Carrière = Cerasus serrulata (Lindl.) G. Don ex Loudon var. lannesiana (Carrière) Makino ●

83246 Cerasus laxiflora (Koehne) C. L. Li et S. Y. Jiang;疏花樱;Loose-flower Cherry ●

83247 Cerasus laxiflora (Koehne) C. L. Li et S. Y. Jiang = Padus laxiflora (Koehne) T. C. Ku ●

83248 Cerasus laxiflora (Koehne) C. L. Li et S. Y. Jiang = Prunus laxiflora Koehne ●

83249 Cerasus leveilleana (Koehne) H. Ohba = Cerasus serrulata

(Lindl.) G. Don ex Loudon var. pubescens (Makino) Te T. Yu et C. L. Li ●

83250 Cerasus leveilleana (Koehne) H. Ohba f. pendula (H. Hara) H. Ohba;垂枝山樱花●☆

83251 Cerasus maackii (Rupr.) Eremin et Simagin = Padus maackii (Rupr.) Kom. ●

83252 Cerasus maackii (Rupr.) Eremin et Simagin = Prunus maackii Rupr. ●

83253 Cerasus mahaleb (L.) Mill.;圆叶樱桃(麻哈勒布樱桃,马哈勒布樱桃,马哈雷樱,马哈利酸樱桃,马哈利樱桃,爪瓣樱);Mahaleb, Mahaleb Cherry, Perfumed Cherry, Roundleaf Cherry, Saint Lucie Cherry, St. Lucie Cherry, St. Lucie's Cherry, St. Lucy Cherry ●

83254 Cerasus mahaleb (L.) Mill. = Prunus mahaleb L. ●

83255 Cerasus maximowiczii (Rupr.) Kom.;黑樱桃(毛樱桃,深山樱,野樱桃);Maximowicz Cherry, Miyama Cherry ●

83256 Cerasus maximowiczii (Rupr.) Kom. = Prunus maximowiczii Rupr. ●

83257 Cerasus microcarpa (C. A. Mey.) Boiss.;小果樱桃●☆

83258 Cerasus mugus (Hand.-Mazz.) Te T. Yu et C. L. Li;偃樱桃(偃樱);Decline Cherry, Dwarf Cherry ●

83259 Cerasus mugus (Hand.-Mazz.) Te T. Yu et C. L. Li = Prunus mugus Hand.-Mazz. ●

83260 Cerasus nakai (H. Lév.) A. I. Baranov et Liou = Cerasus japonica (Thunb.) Loisel. var. nakaii (H. Lév.) Te T. Yu et C. L. Li ●

83261 Cerasus nakai (H. Lév.) A. I. Baranov et Liou = Prunus japonica Thunb. var. nakaii (H. Lév.) Rehder ●

83262 Cerasus napaulensis Ser. = Padus napaulensis (Ser.) C. K. Schneid. ●

83263 Cerasus napaulensis Ser. = Prunus napaulensis (Ser.) Steud. ●

83264 Cerasus nigra Mill. = Cerasus avium (L.) Moench ●

83265 Cerasus nigra Mill. = Prunus avium L. ●

83266 Cerasus nikaii (Honda) H. Ohba;二阶氏樱桃●☆

83267 Cerasus nipponica (Matsum.) Ohle ex H. Ohba;本州樱桃(高岭樱,岭樱,日本樱);Japanese Alpine Cherry ●☆

83268 Cerasus nipponica (Matsum.) Ohle ex H. Ohba var. alpina (Koidz.) H. Ohba;山地本州樱桃●☆

83269 Cerasus nipponica (Matsum.) Ohle ex H. Ohba var. kurilensis (Miyabe) H. Ohba;千岛樱●☆

83270 Cerasus padus (L.) Delarbre = Padus avium Mill. ●

83271 Cerasus parvifolia (Matsum.) H. Ohba;小叶樱桃(小叶茶碗樱)●☆

83272 Cerasus patentipila (Hand.-Mazz.) Te T. Yu et C. L. Li;散毛樱桃(散毛樱);Scatterhair Cherry, Spread-haired Cherry ●

83273 Cerasus pausilliflora (Cardot) Te T. Yu et C. L. Li = Prunus pseudocerasus Lindl. ●

83274 Cerasus pensylvanica (L. f.) Loisel.;野生红樱桃(北美樱,宾州樱);Bird Cherry, Fire Cherry, Fire-cherry, Pin Cherry, Wild Red Cherry ●

83275 Cerasus pleiocerasus (Koehne) Te T. Yu et C. L. Li;雕核樱桃(雕核樱);Furrowedstone Cherry, Furrowed-stone Cherry ●

83276 Cerasus pleiocerasus (Koehne) Te T. Yu et C. L. Li = Prunus pleiocerasus Koehne ●

83277 Cerasus pogonostyla (Maxim.) Te T. Yu et C. L. Li;毛柱樱桃(高岭梅花,毛柱樱,毛柱郁李,庭梅);Hairystyle Cherry, Hairy-styled Cherry ●

83278 Cerasus pogonostyla (Maxim.) Te T. Yu et C. L. Li = Prunus pogonostyla Maxim. ●

83279 Cerasus pogonostyla (Maxim.) Te T. Yu et C. L. Li var. obovata (Koehne) Te T. Yu et C. L. Li;长尾毛柱樱桃(长尾毛樱桃,长尾毛柱郁李,毛柱樱,毛柱郁李);Obovate Hairystyle Cherry ●

83280 Cerasus pogonostyla (Maxim.) Te T. Yu et C. L. Li var. obovata (Koehne) Te T. Yu et C. L. Li = Prunus pogonostyla Maxim. var. obovata Koehne ●

83281 Cerasus polytricha (Koehne) Te T. Yu et C. L. Li;多毛樱桃(多毛野樱桃);Densepubescent Cherry, Dense-pubescent Cherry, Polytrichous Cherry ●

83282 Cerasus polytricha (Koehne) Te T. Yu et C. L. Li = Prunus polytricha Koehne ●

83283 Cerasus prostrata (Labill.) Ser. var. concolor Boiss. = Cerasus tianschanica Pojark. ●

83284 Cerasus pseudocerasus (Lindl.) G. Don;樱桃(含桃,家樱桃,荆桃,蜡樱,牛桃,楔,楔桃,莺桃,樱珠,支那樱珠,朱果,朱樱,紫樱);Cherry, Chinese Cherry, Falsesour Cherry ●

83285 Cerasus pseudoprostrata Pojark.;平卧樱桃●☆

83286 Cerasus pseudoserasus (Lindl.) G. Don = Prunus pseudocerasus Lindl. ●

83287 Cerasus puddum Roxb. ex Ser. = Cerasus cerasoides (D. Don) Sokoloff ●

83288 Cerasus puddum Ser. = Cerasus cerasoides (D. Don) Sokoloff ●

83289 Cerasus pumila (L.) Michx.;矮樱桃●☆

83290 Cerasus pumila Pall. = Cerasus fruticosa (Pall.) Woronow ●

83291 Cerasus pumila Pall. = Prunus fruticosa Pall. ●

83292 Cerasus pusilliflora (Cardot) Te T. Yu et C. L. Li;细花樱桃(细花樱);Slenderflower Cherry, Slender-flowered Cherry, Thinflower Cherry, Thin-flower Cherry ●

83293 Cerasus pusilliflora (Cardot) Te T. Yu et C. L. Li = Prunus pusilliflora Cardot ●

83294 Cerasus rufa (Hook. f.) Te T. Yu et C. L. Li var. trichantha (Koehne) Te T. Yu et C. L. Li = Cerasus trichantha (Koehne) S. Y. Jiang et C. L. Li ●

83295 Cerasus rufa Wall.;红毛樱桃(红毛樱);Faxy-red Cherry, Himalayan Cherry, Himalayas Cherry ●

83296 Cerasus rufa Wall. = Prunus rufa (Wall.) Hook. f. ●

83297 Cerasus rufa Wall. var. trichantha (Koehne) Te T. Yu et C. L. Li;毛花红毛樱(毛萼红毛樱桃);Hairycalyx Himalayan Cherry, Hairy-flower Cherry ●

83298 Cerasus rufa Wall. var. trichantha (Koehne) Te T. Yu et C. L. Li = Prunus rufa (Wall.) Hook. f. var. trichantha (Koehne) Hara ●

83299 Cerasus sachalinensis (Eastw.) Kom. = Cerasus serrulata (Lindl.) G. Don ex Loudon ●

83300 Cerasus sachalinensis (Eastw.) Kom. et Aliss.;库页樱桃;Sachalin Cherry ●☆

83301 Cerasus sargentii (Rehder) H. Ohba;山樱桃(大山樱,大山樱花,萨金特氏樱,萨金特樱桃);N. Japanese Hill Cherry, Sargent Cherry ●☆

83302 Cerasus sargentii (Rehder) H. Ohba f. albida (Miyoshi) H. Ohba;白山樱桃●☆

83303 Cerasus sargentii (Rehder) H. Ohba f. pendula (Honda) Yonek.;垂枝山樱桃●☆

83304 Cerasus sargentii (Rehder) H. Ohba f. pubescens (Tatew.) H. Ohba;毛山樱桃●☆

83305 Cerasus sargentii (Rehder) H. Ohba var. akimotoi H. Ohba et Mas. Saito;秋元樱●☆

83306 Cerasus schneideriana (Koehne) Te T. Yu et C. L. Li;浙闽樱桃;Scheneider Cherry ●

83307 Cerasus schneideriana (Koehne) Te T. Yu et C. L. Li = Prunus schneideriana Koehne ●

83308 Cerasus scopulorum (Koehne) Te T. Yu et C. L. Li;崖樱桃(岩生樱桃);Cliff Cherry ●

83309 Cerasus scopulorum (Koehne) Te T. Yu et C. L. Li = Prunus scopulorum Koehne ●

83310 Cerasus scopulorum (Koehne) Te T. Yu et L. C. Li = Cerasus pseudocerasus (Lindl.) G. Don ●

83311 Cerasus serotina (Ehrh.) Loisel.;野黑樱(迟熟李,迟樱桃,黑果樱桃,黑野樱,黑樱桃,兰姆樱桃,柳叶野黑樱,美国黑果稠李,晚樱,新英格兰桃心木,野核桃,野黑樱桃);American Cherry, Black Cherry, Cabinet Cherry, Capulin, New England Mahogany, Rum Cherry, Rum-cherry, Wild Black Cherry, Wild Cherry, Wild Rum-cherry ●

83312 Cerasus serrula (Franch.) Te T. Yu et C. L. Li;细齿樱桃(西藏野樱桃,细齿樱,云南樱花,云南樱桃);Birch-bark Cherry, Oriental Cherry, Serrulate Cherry, Slendertoothed Cherry, Tibetan Cherry ●

83313 Cerasus serrula (Franch.) Te T. Yu et C. L. Li = Prunus serrula Franch. ●

83314 Cerasus serrulata (Lindl.) G. Don ex Loudon;樱花(锯缘樱,山樱,山樱花,山樱桃,小樱桃,野生福岛樱,野樱花,樱桃);Flowering Cherry, Japanese Cherry, Japanese Flowering Cherry, Oriental Cherry, Underbrown Cherry ●

83315 Cerasus serrulata (Lindl.) G. Don ex Loudon = Prunus serrulata Lindl. ●

83316 Cerasus serrulata (Lindl.) G. Don ex Loudon var. lannesiana (Carrière) Makino;日本晚樱;Japanese Late Cherry ●

83317 Cerasus serrulata (Lindl.) G. Don ex Loudon var. lannesiana (Carrière) Makino = Prunus serrulata Lindl. var. lannesiana (Carrière) Makino ●

83318 Cerasus serrulata (Lindl.) G. Don ex Loudon var. lannesiana (Carrière) Makino = Cerasus lannesiana Carrière ●

83319 Cerasus serrulata (Lindl.) G. Don ex Loudon var. pubescens (Makino) Te T. Yu et C. L. Li;毛叶山樱花(毛叶福岛樱,毛叶福樱,毛叶山樱桃,毛叶樱花,毛叶樱桃);Hairyleaf Japanese Cherry ●

83320 Cerasus serrulata (Lindl.) G. Don ex Loudon var. pubescens (Makino) Te T. Yu et C. L. Li = Prunus serrulata Lindl. var. pubescens (Makino) E. H. Wilson ●

83321 Cerasus serrulata (Lindl.) G. Don ex Loudon var. taishanensis Y. Zhang et C. D. Shi;泰山野樱花●

83322 Cerasus setulosa (Batalin) Te T. Yu et C. L. Li;刺毛樱桃(刺毛山樱花);Bristle Cherry, Setulose Cherry ●

83323 Cerasus setulosa (Batalin) Te T. Yu et C. L. Li = Prunus setulosa Batalin ●

83324 Cerasus shikokuensis (Moriya) H. Ohba;四国山地樱●☆

83325 Cerasus sibirica Hort. ex K. Koch;西伯利亚樱桃●☆

83326 Cerasus sieboldii Carrière;西氏樱桃(席氏樱桃)●☆

83327 Cerasus spachiana Lavalée ex H. Otto;早樱●☆

83328 Cerasus spachiana Lavalée ex H. Otto f. ascendens (Makino) H. Ohba;大叶早樱●

83329 Cerasus spachiana Lavalée ex H. Otto f. ascendens (Makino) H. Ohba = Prunus subhirtella Miq. var. ascendens (Makino) E. H. Wilson ●

83330 Cerasus spachiana Lavalée ex H. Otto var. koshiensis (Koidz.) H. Ohba;高志樱●☆

83331 Cerasus speciosa (Koidz.) H. Ohba;伊豆樱(奄美樱桃);Blackthorn, Oshima Cherry, Sloe ●☆

83332 Cerasus stipulacea (Maxim.) Te T. Yu et C. L. Li;托叶樱桃(托叶樱);Stipulaceous Cherry, Stipulate Cherry ●

83333 Cerasus stipulacea (Maxim.) Te T. Yu et C. L. Li = Prunus stipulacea Maxim. ●

83334 Cerasus subhirtella (Miq.) Sokoloff;日本早樱(彼岸樱,大叶早樱,早樱);Higan Cherry, Rosebud Cherry, Spring Cherry, Winter-flowering Cherry ●

83335 Cerasus subhirtella (Miq.) Sokoloff = Prunus subhirtella Miq. ●

83336 Cerasus subhirtella (Miq.) Sokoloff f. autumnalis (Makino) H. Ohba = Cerasus 'Autumnalis' ●☆

83337 Cerasus subhirtella (Miq.) Sokoloff var. pendula (Tanaka) Te T. Yu et C. L. Li;垂枝大叶早樱(垂枝大叶樱桃,垂枝樱花);Pendulous Higan Cherry ●

83338 Cerasus subhirtella (Miq.) Sokoloff var. pendula (Tanaka) Te T. Yu et C. L. Li = Prunus subhirtella Miq. var. pendula Tanaka ●

83339 Cerasus subhirtella (Miq.) Sokoloff var. pendula (Tanaka) Te T. Yu et C. L. Li = Prunus taiwaniana Hayata ●

83340 Cerasus szechuanica (Batalin) Te T. Yu et C. L. Li;四川樱桃(缠条子,盘腺野樱桃,盘腺樱桃,四川樱,条子,野樱桃);Sichuan Cherry, Szechuan Cherry, Szechwan Cherry ●

83341 Cerasus szechuanica (Batalin) Te T. Yu et C. L. Li = Prunus szechuanica Batalin ●

83342 Cerasus tama-clivorum (Oohara, Seriz. et Wakab.) Yonek.;山岗樱桃●☆

83343 Cerasus tatsieenensis (Batalin) Te T. Yu et C. L. Li;康定樱桃;Kangding Cherry ●

83344 Cerasus tatsieenensis (Batalin) Te T. Yu et C. L. Li = Prunus tatsienensis Batalin ●

83345 Cerasus tianschanica Pojark.;天山樱桃(天山樱);Tianshan Cherry ●

83346 Cerasus tianschanica Pojark. = Prunus tianschanica (Pojark.) Te T. Yu et C. L. Li ●

83347 Cerasus tomentosa (Thunb.) Wall.;毛樱桃(华皜树,李桃,麦樱,梅李,梅桃,奈桃,牛桃,山豆子,山婴桃,山樱桃,野樱桃,英豆,英桃,婴桃,樱桃,朱桃);Chinese Bush Fruit, Downy Cherry, Hansen's Bush Fruit, Manchu Cherry, Nanjing Cherry, Nanking Cherry ●

83348 Cerasus tomentosa (Thunb.) Wall. = Prunus tomentosa Thunb. ●

83349 Cerasus trichantha (Koehne) S. Y. Jiang et C. L. Li;毛瓣藏樱●

83350 Cerasus trichostoma (Koehne) Te T. Yu et C. L. Li;川西樱桃(川西樱);Trichostomous Cherry, West Sichuan Cherry, West Szechuan Cherry, West Szechwan Cherry ●

83351 Cerasus trichostoma (Koehne) Te T. Yu et C. L. Li = Prunus trichostoma Koehne ●

83352 Cerasus triloba (Lindl.) A. I. Baranov et Liou var. truncata Kom. = Prunus triloba Lindl. var. truncata Kom. ●

83353 Cerasus turcomanica Pojark.;土库曼樱桃●

83354 Cerasus undulata (Buch.-Ham. ex D. Don) Ser. = Laurocerasus undulata (Buch.-Ham. ex D. Don) M. Roem. ●

83355 Cerasus undulata (D. Don) Ser. = Laurocerasus undulata (D. Don) M. Roem. ●

83356 Cerasus undulata (D. Don) Ser. = Prunus undulata Buch.-Ham. ●

83357 Cerasus verecunda (Koidz.) H. Ohba = Cerasus serrulata (Lindl.) G. Don ex Loudon var. pubescens (Makino) Te T. Yu et

C. L. Li ●

83358 Cerasus verecunda (Koidz.) H. Ohba f. pendula (H. Hara) H. Ohba = Cerasus leveilleana (Koehne) H. Ohba f. pendula (H. Hara) H. Ohba ●☆

83359 Cerasus verecunda (Koidz.) H. Ohba f. pubipes (H. Hara) H. Ohba = Cerasus serrulata (Lindl.) G. Don ex Loudon var. pubescens (Makino) Te T. Yu et C. L. Li ●

83360 Cerasus verrucosa (Franch.) Nevski;多疣樱桃●☆

83361 Cerasus vulgaris Mill.;欧洲酸樱桃;Amareile Cherry, Broomdashers, Cherry, Cherry Gum, Cherry Tree, Common Cherry, Dwarf Cherry, Morello, Morello Cherry, Pie Cherry, Pie-cherry, Sour Cherry, Tutties ●

83362 Cerasus vulgaris Mill. = Prunus cerasus L. ●

83363 Cerasus wallichii (Steud.) M. Roem. = Laurocerasus undulata (Buch. -Ham. ex D. Don) M. Roem. ●

83364 Cerasus yaoiana W. L. Cheng;姚氏樱桃;Yao's Cherry ●

83365 Cerasus yedoensis (Matsum.) A. V. Vassil.;东京樱花(江户樱,日本樱花,樱花);Flowering Cherry, Japanese Flowering Cherry, Someiyoshine, Tokyo Cherry, Yashino-zakura, Yoshino Cherry ●

83366 Cerasus yedoensis (Matsum.) Te T. Yu et C. L. Li = Cerasus yedoensis (Matsum.) A. V. Vassil. ●

83367 Cerasus yedoensis (Matsum.) Te T. Yu et C. L. Li = Prunus yedoensis Matsum. ●

83368 Cerasus yunnanensis (Franch.) Te T. Yu et C. L. Li;云南樱桃(云南樱);Yunnan Cherry ●

83369 Cerasus yunnanensis (Franch.) Te T. Yu et C. L. Li = Prunus yunnanensis Franch. ●

83370 Cerasus yunnanensis (Franch.) Te T. Yu et C. L. Li var. polybotrys (Koehne) Te T. Yu et C. L. Li;多花云南樱桃(多花樱桃);Manyflower Yunnan Cherry ●

83371 Cerasus yunnanensis (Franch.) Te T. Yu et C. L. Li var. polybotrys (Koehne) Te T. Yu et C. L. Li = Prunus yunnanensis Franch. var. polybotrys Koehne ●

83372 Ceratandra Eckl. ex F. A. Bauer(1837);角雄兰属■☆

83373 Ceratandra Lindl. = Ceratandra Eckl. ex F. A. Bauer ■☆

83374 Ceratandra affinis Sond. = Ceratandra harveyana Lindl. ■☆

83375 Ceratandra atrata (L.) T. Durand et Schinz;变黑角雄兰■☆

83376 Ceratandra auriculata Lindl. = Ceratandra atrata (L.) T. Durand et Schinz ■☆

83377 Ceratandra bicolor Sond. ex Bolus;二色角雄兰■☆

83378 Ceratandra chloroleuca Eckl. ex F. A. Bauer = Ceratandra atrata (L.) T. Durand et Schinz ■☆

83379 Ceratandra globosa Lindl.;球形角雄兰■☆

83380 Ceratandra grandiflora Lindl.;大花角雄兰■☆

83381 Ceratandra harveyana Lindl.;哈维角雄兰■☆

83382 Ceratandra harveyana Lindl. ex Sond. = Ceratandra bicolor Sond. ex Bolus ■☆

83383 Ceratandra parviflora Lindl. = Ceratandra globosa Lindl. ■☆

83384 Ceratandra venosa (Lindl.) Schltr.;多脉角雄兰■☆

83385 Ceratandropais Rolfe = Ceratandra Eckl. ex F. A. Bauer ■☆

83386 Ceratandropsis globosa (Lindl.) Rolfe = Ceratandra globosa Lindl. ■☆

83387 Ceratandropsis grandiflora (Lindl.) Rolfe = Ceratandra grandiflora Lindl. ■☆

83388 Ceratanthera Hornem. = Globba L. ■

83389 Ceratanthus F. Muell. = Ceratanthus F. Muell. ex G. Taylor ■

83390 Ceratanthus F. Muell. = Platostoma P. Beauv. ■☆

83391 Ceratanthus F. Muell. ex G. Taylor = Platostoma P. Beauv. ■☆

83392 Ceratanthus F. Muell. ex G. Taylor(1936);角花属;Hornflower ■

83393 Ceratanthus calcaratus (Hemsl.) G. Taylor;角花(癞子药);Calcarate Hornflower ■

83394 Ceratella Hook. f. = Abrotanella Cass. ■☆

83395 Ceratephorus de Vriese = Payena A. DC. ●☆

83396 Ceratia Adans. = Ceratonia L. ●

83397 Ceratiola Michx.(1803);岩角兰属(角石南属,沙石南属);Sand Heath ●☆

83398 Ceratiola ericoides Michx.;岩角兰(沙石南);Florida Rosemary, Sandhill Rosemary ■☆

83399 Ceratiosicyos Nees(1836);落冠藤属●☆

83400 Ceratiosicyos ecklonii Nees = Ceratiosicyos laevis (Thunb.) A. Meeuse ●☆

83401 Ceratiosicyos laevis (Thunb.) A. Meeuse;落冠藤●☆

83402 Ceratistes Hort. = Eriosyce Phil. ●☆

83403 Ceratistes Labour. = Eriosyce Phil. ●☆

83404 Ceratites Miers = Rudgea Salisb. ■☆

83405 Ceratites Sol. ex Miers = Rudgea Salisb. ■☆

83406 Ceratium Blume = Cylindrolobus Blume ■

83407 Ceratium Blume = Eria Lindl.(保留属名)■

83408 Ceratobium (Lindl.) M. A. Clem. et D. L. Jones = Dendrobium Sw.(保留属名)■

83409 Ceratocalyx Coss. = Boulardia F. Schultz ■

83410 Ceratocalyx Coss. = Orobanche L. ■

83411 Ceratocalyx macrolepis Coss. = Orobanche latisquama (F. W. Schultz) Batt. ■☆

83412 Ceratocapnos Durieu(1844);藤堇属;Climbing Corydalis ■☆

83413 Ceratocapnos claviculata (L.) Lidén;卷须藤堇■☆

83414 Ceratocapnos heterocarpa Durieu;藤堇■☆

83415 Ceratocapnos umbrosus Durieu = Ceratocapnos heterocarpa Durieu ■☆

83416 Ceratocarpus Buxb. ex L. = Ceratocarpus L. ■

83417 Ceratocarpus Durieu = Ceratocapnos Durieu ■☆

83418 Ceratocarpus L.(1753);角果藜属;Ceratocarpus ■

83419 Ceratocarpus arenarius L.;角果藜(沙生角果藜);Sandloving Ceratocarpus ■

83420 Ceratocarpus caputmedusae Bluket = Ceratocarpus areanrius L. ■

83421 Ceratocarpus turkestanicus Sav. -Rycz. = Ceratocarpus areanrius L. ■

83422 Ceratocarpus turkestanicus Sav. -Rycz. ex Iljin = Ceratocarpus areanrius L. ■

83423 Ceratocarpus utriculosus Bluket = Ceratocarpus areanrius L. ■

83424 Ceratocaryum Nees(1836);角果帚灯草属■☆

83425 Ceratocaryum argenteum Kunth;阿根廷角果帚灯草■☆

83426 Ceratocaryum caespitosum H. P. Linder;丛生角果帚灯草■☆

83427 Ceratocaryum decipiens (N. E. Br.) H. P. Linder;迷惑角果帚灯草■☆

83428 Ceratocaryum fimbriatum (Kunth) H. P. Linder;流苏角果帚灯草■☆

83429 Ceratocaryum fistulosum Mast.;管角果帚灯草■☆

83430 Ceratocaryum persistens H. P. Linder;宿存角果帚灯草■☆

83431 Ceratocaryum pulchrum H. P. Linder;美丽角果帚灯草■☆

83432 Ceratocaryum xerophilum (Pillans) H. P. Linder;沙地角果帚灯草■☆

83433 Ceratocaulos (Bernh.) Rchb. = Datura L. ●■

83434 Ceratocaulos Rchb. = Apemon Raf. ●■

83435 Ceratocaulos Rchb. = Datura L. ●■
83436 Ceratocentron Senghas(1989);弓距兰属■☆
83437 Ceratocentron fesselii Senghas;弓距兰■☆
83438 Ceratocephala Moench(1794);角果毛茛属(角茛属)■
83439 Ceratocephala falcata (L.) Pers.;弯喙角果毛茛(镰状角果毛茛);Falcate Ceratocephalus ■
83440 Ceratocephala falcata (L.) Pers. orthoceras (DC.) Aitch. et Hemsl. = Ceratocephala falcata (L.) Pers. ■
83441 Ceratocephala orthoceras DC. = Ceratocephala falcata (L.) Pers. ■
83442 Ceratocephala orthoceras DC. = Ceratocephala testiculatus (Crantz) Roth ■
83443 Ceratocephala testiculata (Crantz) Roth;角果毛茛;Bur Buttercup, Ceratocephalus, Common Ceratocephalus, Curveseed Butterwort, Curve-seed Butterwort, Hornseed ■
83444 Ceratocephala testiculata (Crantz) Roth = Ranunculus testiculatus Crantz ■
83445 Ceratocephala testiculata Hook. et Arn. = Ranunculus testiculatus Crantz ■
83446 Ceratocephalus Burm. ex Kuntze = Spilanthes Jacq. ■
83447 Ceratocephalus Cass. = Bidens L. ■●
83448 Ceratocephalus Kuntze = Acmella Rich. ex Pers. ■
83449 Ceratocephalus Kuntze = Spilanthes Jacq. ■
83450 Ceratocephalus Pers. = Ceratocephala Moench ■
83451 Ceratocephalus Vaill. ex Cass. = Bidens L. ■●
83452 Ceratocephalus exasperatus (Jacq.) Kuntze = Acmella radicans (Jacq.) R. K. Jansen ■☆
83453 Ceratocephalus falcatus (L.) Pers. = Ceratocephala falcata (L.) Pers. ■
83454 Ceratocephalus falcatus (L.) Pers. var. orthoceras (DC.) Batt. et Trab. = Ceratocephalus testiculatus (Crantz) Roth ■
83455 Ceratocephalus glaber (Beck.) Janisch. = Ceratocephala testiculata (Crantz) Roth ■
83456 Ceratocephalus incurvus Steven = Ceratocephalus falcatus (L.) Pers. subsp. incurvus (Steven) Chrtek et Chrtková ■☆
83457 Ceratocephalus orthoceras DC. = Ceratocephalus testiculatus (Crantz) Roth ■
83458 Ceratocephalus orthoceras DC. = Ranunculus testiculatus Crantz ■
83459 Ceratocephalus testiculatus (Crantz) Roth = Ceratocephala testiculata (Crantz) Roth ■
83460 Ceratocephalus testiculatus (Crantz) Roth = Ranunculus testiculatus Crantz ■
83461 Ceratochaete Lunell = Zizania L. ■
83462 Ceratochilus Blume(1825);角唇兰属■☆
83463 Ceratochilus Lindl. = Stanhopea J. Frost ex Hook. ■☆
83464 Ceratochilus biglandulosus Blume;角唇兰■☆
83465 Ceratochloa DC. et P. Beauv. = Bromus L. (保留属名)■
83466 Ceratochloa P. Beauv. (1812);角雀麦属;Brome, Brome Grass ■☆
83467 Ceratochloa P. Beauv. = Bromus L. (保留属名)■
83468 Ceratochloa brevis (Nees ex Steud.) B. D. Jacks.;短角雀麦;Patagoniau Brome ■☆
83469 Ceratochloa carinata (Hook. et Arn.) Tutin. = Bromus carinatus Hook. et Arn. ■
83470 Ceratochloa cathartica (Vahl) Herter;角雀麦;Rescue Grass ■☆
83471 Ceratochloa cathartica (Vahl) Herter = Bromus catharticus Vahl ■
83472 Ceratochloa marginata (Nees ex Steud.) W. A. Weber = Bromus marginatus Nees ex Steud. ■
83473 Ceratochloa sitchensis (Trin.) Cope et Ryves = Bromus sitchensis Trin. ■
83474 Ceratochloa staminea (E. Desv.) Stace = Bromus stamineus E. Desv. ■
83475 Ceratochloa unioloides (Willd.) P. Beauv. = Bromus catharticus Vahl ■
83476 Ceratocnemum Coss. et Balansa(1873);摩洛哥野蔓菁属■☆
83477 Ceratocnemum rapistroides Coss. et Balansa;摩洛哥野蔓菁■☆
83478 Ceratocnemum rapistroides Coss. et Balansa var. glabrescens Maire = Ceratocnemum rapistroides Coss. et Balansa ■☆
83479 Ceratococca Wllld. ex Roem. et Schult. = Microtea Sw. ■☆
83480 Ceratococcus Meisn. = Pterococcus Hassk. (保留属名)●☆
83481 Ceratodes Kuntze = Ceratocarpus L. ■
83482 Ceratodiscus T. Durand et Jacks. = Corallodiscus Batalin ■
83483 Ceratoealyx Coss. = Orobanche L. ■
83484 Ceratogonon Meisn. = Oxygonum Burch. ex Campd. ●■☆
83485 Ceratogonon atriplicifolium Meisn. = Oxygonum atriplicifolium (Meisn.) Martelli ●☆
83486 Ceratogonum C. A. Mey. = Ceratogonon Meisn. ●■☆
83487 Ceratogonum cordofanum Meisn. = Oxygonum sinuatum (Hochst. et Steud. ex Meisn.) Dammer ●☆
83488 Ceratogonum sinuatum Hochst. et Steud. ex Meisn. = Oxygonum sinuatum (Hochst. et Steud. ex Meisn.) Dammer ●☆
83489 Ceratogyna T. Post et Kuntze = Ceratogyne Turcz. ■☆
83490 Ceratogyne Turcz. (1851);角果菊属■☆
83491 Ceratogyne obionoides Turcz.;角果菊■☆
83492 Ceratogynum Wight = Sauropus Blume ●■
83493 Ceratoides (Tourn.) Gagnebin = Axyris L. ■
83494 Ceratoides (Tourn.) Gagnebin = Ceratocarpus L. ■
83495 Ceratoides Gagnebin = Axyris L. ■
83496 Ceratoides Gagnebin = Ceratocarpus L. ■
83497 Ceratoides arborescens (Losinsk.) C. P. Tsien et C. G. Ma = Krascheninnikovia arborescens (Losinsk.) Czerep. ●
83498 Ceratoides compacta (Losinsk.) C. P. Tsien et C. G. Ma = Krascheninnikovia compacta (Losinsk.) Grubov ●
83499 Ceratoides compacta (Losinsk.) C. P. Tsien et C. G. Ma var. longipilosa C. P. Tsien et C. G. Ma = Krascheninnikovia compacta (Losinsk.) Grubov var. longipilosa (C. P. Tsien et C. G. Ma) Mosyakin ●
83500 Ceratoides ewersmanniana (Stschegl. ex Losinsk.) Botsch. et Ikonn. = Krascheninnikovia ewersmannia (Stschegl. ex Losinsk.) Grubov ●
83501 Ceratoides intramongolica H. C. Fu, J. Y. Yang et S. Y. Zhao;内蒙驼绒藜(内蒙古驼绒藜);Innermongolian Ceratoides ●
83502 Ceratoides lanata (Pursh) J. T. Howell = Krascheninnikovia lanata (Pursh) A. Meeuse et A. Smit ●☆
83503 Ceratoides latens (J. F. Gmel.) Reveal et N. H. Holmgren = Krascheninnikovia ceratoides (L.) Gueldenst. ●
83504 Ceratoides papposa (Pers.) Botsch. et Ikonn. = Ceratoides latens (J. F. Gmel.) Reveal et N. H. Holmgren ●
83505 Ceratoides papposa (Pers.) Botsch. et Ikonn. = Krascheninnikovia ceratoides (L.) Gueldenst. ●
83506 Ceratolacis (Tul.) Wedd. (1873);空角川苔草属■☆
83507 Ceratolacis Wedd. = Ceratolacis (Tul.) Wedd. ■☆
83508 Ceratolacis erythrolichen (Tul. et Wedd.) Wedd.;空角川苔草■☆
83509 Ceratolacis erythrolichen Wedd. = Ceratolacis erythrolichen

(Tul. et Wedd.) Wedd. ■☆

83510 Ceratolepis Cass. (1819);角鳞菊属■☆

83511 Ceratolepis Cass. = Pamphalea DC. ■☆

83512 Ceratolepis Cass. = Panphalea Lag. ■☆

83513 Ceratolimon M. B. Crespo et M. D. Lledó(2000);角匙丹属■☆

83514 Ceratolimon feei (Girard) Crespo et Lledò;菲角匙丹■☆

83515 Ceratolimon feei (Girard) Crespo et Lledò var. grandiflorum (Maire et Wilczek) Crespo et Lledò = Ceratolimon feei (Girard) Crespo et Lledò ■☆

83516 Ceratolimon migiurtinum (Chiov.) M. B. Crespo et Lledò;索马里角匙丹■☆

83517 Ceratolimon rechingeri (J. R. Edm.) M. B. Crespo et Lledò = Ceratolimon migiurtinum (Chiov.) M. B. Crespo et Lledò ■☆

83518 Ceratolimon weygandiorum (Maire et Wilczek) M. B. Crespo et Lledò;魏刚角匙丹■☆

83519 Ceratolobus Blume(1830);角裂棕属(角裂藤属,距裂藤属,孔苞藤属);Ceratolobus ●☆

83520 Ceratolobus kingianus Becc. et Hook. f.;庆氏角裂棕;King Ceratolobus ●☆

83521 Ceratolobus laevigatus Becc. et Hook. f.;平滑角裂棕;Smooth Ceratolobus ●☆

83522 Ceratominthe Briq. = Satureja L. ●■

83523 Ceratonia L. (1753);长角豆属(角豆树属,角豆苏木属);Carob, Locust-tree ●

83524 Ceratonia chilensis Molina = Prosopis chilensis (Molina) Stuntz ●☆

83525 Ceratonia oreothauma Hillc. et G. P. Lewis et Verdc. subsp. somalensis Hillc., G. P. Lewis et Verdc.;索马里长角豆●☆

83526 Ceratonia oreothauma Hillc., G. P. Lewis et Verdc.;非洲长角豆●☆

83527 Ceratonia siliqua L.;长角豆(长角豆苏木,角豆树);Algaroba, Bean Tree, Carob, Carob Bean, Carob Seed Gum, Carob Tree, Carob-tree, Chilean Mesquite, Karoub, Locust, Locust Bean, Locust Been Gum, Locust Gum, Locust Tree, Locust-tree, Old World Locust, St. John's Bread, St. John's-bread, St. John's Sweet Bread, Sweet Bean, Tragasol ●

83528 Ceratoniaceae Link = Fabaceae Lindl. (保留科名)●■

83529 Ceratoniaceae Link = Leguminosae Juss. (保留科名)●■

83530 Ceratonychia Edgew. = Cometes L. ■☆

83531 Ceratonychia nidus Edgew. = Cometes surattensis L. ■☆

83532 Ceratopetalorchis Szlach., Górniak et Tukallo = Habenaria Willd. ■

83533 Ceratopetalum Sm. (1793);角瓣木属;Christmas Bush ●☆

83534 Ceratopetalum apetalum D. Don;角瓣木(香缎木);Coachwood, Lightwood Scented Satinwood ●☆

83535 Ceratopetalum gummiferum Sm.;新南威尔士角瓣木;Christmas Bush, New South Wales Christmas Bush, Sydney Christmas Bush ●☆

83536 Ceratophorus Hassk. = Payena A. DC. ●☆

83537 Ceratophorus Sond. = Suregada Roxb. ex Rottler ●

83538 Ceratophorus africanus Sond. = Suregada africana (Sond.) Kuntze ●☆

83539 Ceratophyllaceae Gray(1822)(保留科名);金鱼藻科;Hornwort Family ■

83540 Ceratophyllum L. (1753);金鱼藻属;Coontail, Cornifle, Horn Wort, Hornwort ■

83541 Ceratophyllum apiculatum Cham. = Ceratophyllum demersum L. var. apiculatum (Cham.) Asch. ■

83542 Ceratophyllum apiculatum Cham. = Ceratophyllum demersum L. ■

83543 Ceratophyllum australe Griseb. = Ceratophyllum muricatum Cham. subsp. australe (Griseb.) Les. ■☆

83544 Ceratophyllum cristatum Guillaumin et Perr. = Ceratophyllum muricatum Cham. ■☆

83545 Ceratophyllum demersum L.;金鱼藻(灯笼丝,聚藻,软草,松藻,细草,虾须草,鱼草,扎毛);Common Hornwort, Coon's-tail, Coontail, Cornifle Nageante, Falling Star, Falling Stars, Flower of Heaven, Horn Wort, Hornwort, Hornwort Pondweed, Rigid Hornwort ■

83546 Ceratophyllum demersum L. f. missionis (Wight et Arn.) Wilmot-Dear = Ceratophyllum demersum L. ■

83547 Ceratophyllum demersum L. f. quadrispinum (Makino) Kitag. = Ceratophyllum platyacanthum Cham. subsp. oryzetorum (Kom.) Les ■

83548 Ceratophyllum demersum L. f. quadrispinum (Makino) Kitag. = Ceratophyllum oryzetorum Kom. ■

83549 Ceratophyllum demersum L. var. apiculatum (Cham.) Asch. = Ceratophyllum demersum L. ■

83550 Ceratophyllum demersum L. var. apiculatum (Cham.) Garcke = Ceratophyllum demersum L. ■

83551 Ceratophyllum demersum L. var. cristatum K. Schum. = Ceratophyllum muricatum Cham. subsp. australe (Griseb.) Les ■☆

83552 Ceratophyllum demersum L. var. echinatum (A. Gray) A. Gray = Ceratophyllum echinatum A. Gray ■☆

83553 Ceratophyllum demersum L. var. inerme Radcl. -Sm.;无刺金鱼藻■☆

83554 Ceratophyllum demersum L. var. inflatum R. E. Fr. = Ceratophyllum demersum L. ■

83555 Ceratophyllum demersum L. var. pentacorne Kitag. = Ceratophyllum platyacanthum Cham. subsp. oryzetorum (Kom.) Les ■

83556 Ceratophyllum demersum L. var. quadrispinum Makino = Ceratophyllum oryzetorum Kom. ■

83557 Ceratophyllum demersum L. var. quadrispinum Makino = Ceratophyllum platyacanthum Cham. subsp. oryzetorum (Kom.) Les ■

83558 Ceratophyllum echinatum A. Gray;刺金鱼藻;Coontail, Hornwort, Prickly Hornwort, Spiny Hornwort ■☆

83559 Ceratophyllum echinatum A. Gray = Ceratophyllum muricatum Cham. ■☆

83560 Ceratophyllum floridanum Fassett = Ceratophyllum muricatum Cham. subsp. australe (Griseb.) Les ■☆

83561 Ceratophyllum inflatum C. C. Jao ex K. C. Kuan = Ceratophyllum muricatum Cham. subsp. kossinskyi (Kuzen.) Les ■

83562 Ceratophyllum inflatum C. C. Jao ex Kuanin;宽叶金鱼藻;Inflated Hornwort ■

83563 Ceratophyllum inflatum C. C. Jao ex Kuanin = Ceratophyllum maricatum Cham. subsp. kossinskyi (Kuzen.) Les ■

83564 Ceratophyllum komarovii Kuzen.;科马罗夫金鱼藻■☆

83565 Ceratophyllum kossinskyi Kuzen. = Ceratophyllum muricatum Cham. subsp. kossinskyi (Kuzen.) Les ■

83566 Ceratophyllum llerenae Fassett = Ceratophyllum muricatum Cham. subsp. australe (Griseb.) Les ■☆

83567 Ceratophyllum manschuricum (Miki) Kitag. = Ceratophyllum muricatum Cham. subsp. kossinskyi (Kuzen.) Les ■

83568 Ceratophyllum manshuricum (Miki) Kitag.;东北金鱼藻;Manchurian Hornwort, NE. China Hornwort ■

83569 Ceratophyllum maricatum Cham. subsp. kossinskyi (Kuzen.) Les;粗糙金鱼藻■

83570 Ceratophyllum missionis Wight et Arn. = Ceratophyllum

demersum L. ■

83571 Ceratophyllum muricatum Cham. = Ceratophyllum echinatum A. Gray ■☆

83572 Ceratophyllum muricatum Cham. subsp. australe (Griseb.) Les;南方粗糙金鱼藻■☆

83573 Ceratophyllum muricatum Cham. subsp. kossinskyi (Kuzen.) Les;科氏粗糙金鱼藻(贫角金鱼藻,细金鱼藻);Kossinsky Hornwort ■

83574 Ceratophyllum oryzetorum Kom. = Ceratophyllum platyacanthum Cham. subsp. oryzetorum (Kom.) Les ■

83575 Ceratophyllum pentacanthum Hayata;日本金鱼藻;Japan Hornwort ■

83576 Ceratophyllum pentacanthum Hayata = Ceratophyllum platyacanthum Cham. subsp. oryzetorum (Kom.) Les ■

83577 Ceratophyllum platyacanthum Cham. subsp. oryzetorum (Kom.) Les;五刺金鱼藻(十叶金鱼藻,五角金鱼藻,五叶金鱼藻,五针金鱼藻);Fivespined Hornwort ■

83578 Ceratophyllum submersum L.;细金鱼藻(沉水金鱼藻,黄角金鱼藻);Coomail, Hornwort, Slender Hornwort, Soft Hornwort, Spineless Hornwort, Unarmed Hornwort ■

83579 Ceratophyllum submersum L. subsp. muricatum (Cham.) Wilmot-Dear = Ceratophyllum muricatum Cham. ■☆

83580 Ceratophyllum submersum L. var. echinatum (A. Gray) Wilmot-Dear = Ceratophyllum echinatum A. Gray ■☆

83581 Ceratophyllum submersum L. var. manschuricum Miki = Ceratophyllum maricatum Cham. subsp. kossinskyi (Kuzen.) Les ■

83582 Ceratophyllum submersum L. var. squamosum Wilmot-Dear = Ceratophyllum maricatum Cham. subsp. kossinskyi (Kuzen.) Les ■

83583 Ceratophyllum tanaiticum Sapjegin;塔奈特金鱼藻■☆

83584 Ceratophytum Pittier(1928);角紫葳属●☆

83585 Ceratophytum brachycarpum Pittier;角紫葳●☆

83586 Ceratopsis Lindl. = Epipogium J. G. Gmel. ex Borkh. ■

83587 Ceratopsis rosea (D. Don) Lindl. = Epipogium roseum (D. Don) Lindl. ■

83588 Ceratopyxis Hook. f. (1872);古巴角果茜属☆

83589 Ceratopyxis verbenacea Hook. f.;古巴角果茜☆

83590 Ceratosanthes Adans. = Ceratosanthes Burm. ex Adans. ■☆

83591 Ceratosanthes Burm. ex Adans. (1763);角花葫芦属■☆

83592 Ceratosanthes angustiloba Ridl.;窄裂角花葫芦■☆

83593 Ceratosanthes parviflora Cogn.;小花角花葫芦■☆

83594 Ceratosanthes tomentosa Cogn.;毛角花葫芦■☆

83595 Ceratosanthus Schur = Consolida Gray ■

83596 Ceratosanthus Schur = Delphinium L. ■

83597 Ceratoschoenus Nees = Rhynchospora Vahl(保留属名)■

83598 Ceratoschoenus Nees(1834);角莎属■☆

83599 Ceratoschoenus capitatus Chapm. = Rhynchospora tracyi Britton ■☆

83600 Ceratoschoenus corniculatus (Lam.) Nees = Rhynchospora corniculata (Lam.) A. Gray ■☆

83601 Ceratoschoenus corniculatus Nees;角莎■☆

83602 Ceratoschoenus longirostris (Michx.) A. Gray = Rhynchospora corniculata (Lam.) A. Gray ■☆

83603 Ceratoschoenus macrostachys (Torr. ex A. Gray) A. Gray var. patulus Chapm. = Rhynchospora careyana Fernald ■☆

83604 Ceratoschoenus macrostachyus (Torr. ex A. Gray) A. Gray = Rhynchospora macrostachya Torr. ex A. Gray ■☆

83605 Ceratoschoenus macrostachyus (Torr. ex A. Gray) A. Gray var. inundatus Oakes = Rhynchospora inundata (Oakes) Fernald ■☆

83606 Ceratoscyphus Chun = Chirita Buch. -Ham. ex D. Don ●■

83607 Ceratoscyphus Chun = Ornithoboea Parish ex C. B. Clarke ■

83608 Ceratoscyphus caeruleus Chun = Chirita ceratoscyphus B. L. Burtt ■

83609 Ceratoscyphus caeruleus Chun = Chirita corniculata Pellegr. ■

83610 Ceratosepalum Oerst. (1894);角萼西番莲属●☆

83611 Ceratosepalum Oerst. = Passiflora L. ●■

83612 Ceratosepalum Oliv. = Triumfetta Plum. ex L. ●■

83613 Ceratosepalum Oliv. = Triumfettoides Rauschert ●■

83614 Ceratosepalum digitatum Oliv. = Triumfetta digitata (Oliv.) Sprague et Hutch. ■☆

83615 Ceratosepalum micranthum Oerst.;角萼西番莲●☆

83616 Ceratosicyus Post et Kuntze = Ceratiosicyos Nees ●☆

83617 Ceratospermum Pers. (1807);角籽藜属●☆

83618 Ceratospermum Pers. = Axyris L. ■

83619 Ceratospermum Pers. = Ceratoides (Tourn.) Gagnebin ■

83620 Ceratospermum Pers. = Eurotia Adans. ■☆

83621 Ceratospermum papposum Pers.;角籽藜●☆

83622 Ceratostachys Blume = Nyssa L. ●

83623 Ceratostachys arborea Blume = Nyssa javanica (Blume) Wangerin ●

83624 Ceratostanthus B. D. Jacks. = Ceratosanthus Schur ■

83625 Ceratostanthus B. D. Jacks. = Delphinium L. ■

83626 Ceratostanthus Schur = Delphinium L. ■

83627 Ceratostema G. Don = Pellegrinia Sleumer ●☆

83628 Ceratostema Juss. (1789);角蕊莓属(囊冠莓属)●☆

83629 Ceratostema alatum (Hoerold) Sleumer;翅角蕊莓●☆

83630 Ceratostema angulatum Griff. = Agapetes angulata (Griff.) Benth. et Hook. f. ●

83631 Ceratostema biflorum Poepp. et Endl.;双花角蕊莓●☆

83632 Ceratostema ellipticum Benth. et Hook. f.;椭圆角蕊莓●☆

83633 Ceratostema longiflorum Lindl. ex Lem.;长花角蕊莓●☆

83634 Ceratostema microphyllum Hoerold;小叶角蕊莓●☆

83635 Ceratostema miniature Griff. = Agapetes miniata (Griff.) Benth. et Hook. f. ●

83636 Ceratostema nanum Griff.;矮角蕊莓●☆

83637 Ceratostema rigidum Benth.;硬角蕊莓●☆

83638 Ceratostema vacciniacea Roxb. = Vaccinium vacciniaceum (Roxb.) Sleumer ●

83639 Ceratostema vacciniaceum Roxb. = Vaccinium vacciniaceum (Roxb.) Sleumer ●

83640 Ceratostemma Spreng. = Ceratostema Juss. ●☆

83641 Ceratostigma Bunge(1833);蓝雪花属(角柱花属,蓝雪属,蓝血花属,蓝血属);Bluesnow, Ceratostigma, Hardy Plumbago, Leadwort, Plumbago ●■

83642 Ceratostigma abyssinicum (Hochst.) Schweinf. et Asch.;阿比西尼亚蓝雪花●☆

83643 Ceratostigma griffithii C. B. Clarke;毛蓝雪花(星毛角柱花);Burmese Plumbago, Griffith Bluesnow, Griffith Ceratostigma, Griffith Leadwort, Plumbago ●

83644 Ceratostigma minus Stapf ex Prain;小蓝雪花(东南菊,对节兰,对叶兰,风湿草,红花紫金标,九节莲,蓝花岩陀,小角柱花,小蓝雪,岩五姜,紫金标);Creeping Bluesnow, Creeping Ceratostigma, Creeping Leadwort ●

83645 Ceratostigma minus Stapf ex Prain f. lasaense Z. X. Peng;拉萨小蓝雪花;Lasa Creeping Leadwort ●

83646 Ceratostigma plumbaginoides Bunge;蓝雪花(假靛,角柱花,山灰柴);Blue Bluesnow, Blue Ceratostigma, Blue Leadwood,

Ceratostigma, Dwarf Blue Plumbago, Dwarf Plumbago, Indigo Flower, Larpente Plumbago, Leadwort, Lumbago, Plumbago ●■

83647 Ceratostigma polhilli Bulley = Ceratostigma minus Stapf ex Prain ●

83648 Ceratostigma speciosum Prain = Ceratostigma abyssinicum (Hochst.) Schweinf. et Asch. ●☆

83649 Ceratostigma ulicinum Prain;刺鳞蓝雪花(荆豆角桂花);Ulex Ceratostigma ●

83650 Ceratostigma willmottianum Stapf;岷江蓝雪花(搬倒甑,拌倒甑,角柱花,九节莲,蓝花丹,蓝雪花,攀倒甑,七星箭,铁丝岩陀,叶叶兰,转子莲,紫金标,紫金莲);Chinese Plumbago, Willmott Bluesnow, Willmott Ceratostigma, Willmott Leadwort ●

83651 Ceratostylis Blume(1825);牛角兰属;Ceratostylis, Hornstyle ■

83652 Ceratostylis caespitosa (Rolfe) Ts. Tang et F. T. Wang = Ceratostylis hainanensis Z. H. Tsi ■

83653 Ceratostylis hainanensis Z. H. Tsi;牛角兰(集束牛角兰);Hainan Ceratostylis, Hainan Hornstyle ■

83654 Ceratostylis himalaica Hook. f.;叉枝牛角兰;Himalayan Ceratostylis, Himalayas Hornstyle ■

83655 Ceratostylis pendula Hook. f.;垂茎牛角兰;Pendulous Ceratostylis ■☆

83656 Ceratostylis rubra Ames;红花牛角兰;Redflower Ceratostylis ■☆

83657 Ceratostylis subulata Blume;管叶牛角兰;Tubeleaf Ceratostylis, Tubeleaf Hornstyle ■

83658 Ceratostylis teres (Griff.) Rchb. f. = Ceratostylis subulata Blume ■

83659 Ceratotheca Endl. (1832);角囊胡麻属■●☆

83660 Ceratotheca elliptica Schinz = Ceratotheca integribracteata Engl. subsp. elliptica (Schinz) Ihlenf. ■☆

83661 Ceratotheca integribracteata Engl.;全苞角囊胡麻■☆

83662 Ceratotheca integribracteata Engl. subsp. elliptica (Schinz) Ihlenf.;椭圆全苞角囊胡麻■☆

83663 Ceratotheca lamiifolia (Engl.) Engl. = Ceratotheca triloba (Bernh.) Hook. f. ■☆

83664 Ceratotheca melanosperma Hochst. ex Bernh. = Ceratotheca sesamoides Endl. ■☆

83665 Ceratotheca reniformis Abels;肾形角囊胡麻■☆

83666 Ceratotheca saxicola E. A. Bruce;岩石角囊胡麻■☆

83667 Ceratotheca sesamoides Endl.;拟芝麻;Bungu ■☆

83668 Ceratotheca sesamoides Endl. f. latifolia Engl. = Ceratotheca sesamoides Endl. ■☆

83669 Ceratotheca sesamoides Endl. var. baoulensis A. Chev. = Ceratotheca sesamoides Endl. ■☆

83670 Ceratotheca sesamoides Endl. var. grandiflora Berhaut = Ceratotheca sesamoides Endl. ■☆

83671 Ceratotheca sesamoides Endl. var. melanoptera A. DC. = Ceratotheca sesamoides Endl. ■☆

83672 Ceratotheca triloba (Bernh.) Hook. f.;三裂角囊胡麻;Wild Foxglove ■☆

83673 Ceratotheca vanderystii De Wild. = Ceratotheca integribracteata Engl. ■☆

83674 Ceratoxalis (Dumort.) Lunell = Oxalis L. ■●

83675 Ceratoxalis (Dumort.) Lunell = Xanthoxalis Small ■●

83676 Ceratoxalis Lunell = Oxalis L. ■●

83677 Ceratoxalis Lunell = Xanthoxalis Small ■●

83678 Ceratoxalis coloradensis (Rydb.) Lunell = Oxalis stricta L. ■

83679 Ceratoxalis cymosa (Small) Lunell = Oxalis stricta L. ■

83680 Ceratozamia Brongn. (1846);角果铁属(角果泽米属,角铁属,有角坚果凤尾蕉属);Ceratozamia, Horncone ●☆

83681 Ceratozamia hildae G. P. Landry et M. C. Wilson;海氏角果铁(海氏角果泽米);Bamboo Cycad ●☆

83682 Ceratozamia kuesteriana Regel;库氏角果铁(库氏角铁);Horncone ●☆

83683 Ceratozamia latifolia Miq.;宽叶角果铁(宽叶角铁);Broadleaf Ceratozamia ●☆

83684 Ceratozamia longifolia Miq.;长叶角果铁(长叶角铁);Longleaf Ceratozamia ●☆

83685 Ceratozamia mexicana Brongn.;墨西哥角果铁(刺苏铁,角果铁,墨西哥角果泽米,墨西哥角铁);Mexican Ceratozamia, Mexican Horncone ●☆

83686 Ceratozamia microstrobila Vovides et J. D. Rees;小果角果铁(小果角果泽米)●☆

83687 Ceratozamia norstogii D. W. Steven;诺氏角果铁(诺氏角果泽米)●☆

83688 Ceratozamia robusta Miq.;粗壮角果铁(巨型角果泽米,壮角铁);Robust Ceratozamia ●☆

83689 Ceraunia Noronha = Aegiceras Gaertn. ●

83690 Cerbera L. (1753);海杧果属(海檬果属);Cerberustree, Cerberus-tree, Cerberus Tree ●

83691 Cerbera Lour. = Cerbera L. ●

83692 Cerbera Lour. = Scaevola L. (保留属名)●■

83693 Cerbera ahouai L. = Thevetia ahouai (L.) A. DC. ●

83694 Cerbera chinensis Spreng. = Rauvolfia verticillata (Lour.) Baill. ●

83695 Cerbera dilatata S. T. Blake;大海杧果●☆

83696 Cerbera floribunda K. Schum.;多花海杧果;Manyflowered Cerberustree ●☆

83697 Cerbera fruticosa Ker Gawl. = Kopsia fruticosa (Ker Gawl.) A. DC. ●

83698 Cerbera manghas L.;海杧果(海芒果,海檬果,猴欢喜,黄金调,黄金茄,牛金茄,牛心荔,牛心茄子,山杭果,山样子,山样子,香军树);Cerberus Tree, Common Cerberustree, Common Cerberus-tree, Odollam Cerberus-tree ●

83699 Cerbera manghas L. f. sakuyana K. Nakaj.;作屋海杧果●☆

83700 Cerbera odollam Gaertn.;奥道拉姆海杧果(海南海杧果,猴欢喜,山样仔);Odollam Cerberustree ●☆

83701 Cerbera odollam Gaertn. = Cerbera manghas L. ●

83702 Cerbera peruviana Pers. = Cascabela thevetia (L.) Lippold ●

83703 Cerbera peruviana Pers. = Thevetia peruviana (Pers.) K. Schum. ●

83704 Cerbera thevetia L. = Cascabela thevetia (L.) Lippold ●

83705 Cerbera thevetia L. = Thevetia peruviana (Pers.) K. Schum. ●

83706 Cerbera venenifera Steud.;马达加斯加海杧果;Madagascar Ordeal Tree, Tanghin ●☆

83707 Cerberaceae Martinov = Apocynaceae Juss. (保留科名)●■

83708 Cerberiopsis Vieill. ex Pancher et Sebert(1874);拟海杧果属●☆

83709 Cerberiopsis candelabra Vieill. ex Pancheret Sebert;拟海杧果●☆

83710 Cercaceae Dulac = Ceratophyllaceae Gray(保留科名)■

83711 Cercanthemum Tiegh. = Ouratea Aubl. (保留属名)●

83712 Cercanthemum sacleuxii Tiegh. = Campylospermum sacleuxii (Tiegh.) Farron ●☆

83713 Cercestis Schott(1857);网纹芋属;Rhektophyllum ■☆

83714 Cercestis afzelii Schott;阿芙泽尔网纹芋■☆

83715 Cercestis alepensis A. Chev. = Cercestis dinklagei Engl. ■☆

83716 Cercestis camerunensis (Ntepe-Nyame) Bogner;喀麦隆网纹芋■☆

83717 Cercestis congensis Engl.;康格网纹芋■☆

83718 Cercestis dinklagei Engl.;丁克网纹芋■☆

83719 Cercestis elliotii Engl. = Cercestis dinklagei Engl. ■☆
83720 Cercestis gabunensis Engl. = Cercestis kamerunianus (Engl.) N. E. Br. ■☆
83721 Cercestis hastifolia A. Chev. = Cercestis dinklagei Engl. ■☆
83722 Cercestis ivorensis A. Chev.;伊沃里网纹芋■☆
83723 Cercestis kamerunianus (Engl.) N. E. Br.;卡氏网纹芋■☆
83724 Cercestis lanceolata Engl. = Cercestis ivorensis A. Chev. ■☆
83725 Cercestis ledermannii Engl. = Cercestis dinklagei Engl. ■☆
83726 Cercestis mirabilis (N. E. Br.) Bogner = Rhektophyllum mirabile N. E. Br. ■☆
83727 Cercestis sagittata Engl. = Cercestis dinklagei Engl. ■☆
83728 Cercestis scaber A. Chev. = Cercestis afzelii Schott ■☆
83729 Cercestis stigmatica N. E. Br. = Cercestis dinklagei Engl. ■☆
83730 Cercidiopsis Britton et Rose = Cercidium Tul. ●☆
83731 Cercidiopsis Britton et Rose = Parkinsonia L. ●
83732 Cercidiphyllaceae Engl. (1907)(保留科名);连香树科;Cercidiphyllum Family, Katsura Tree Family, Katsuratree Family, Katsura-tree Family ●
83733 Cercidiphyllaceae Tiegh. = Cercidiphyllaceae Engl.(保留科名)●
83734 Cercidiphyllum Siebold et Zucc. (1846);连香树属;Cercidiphyllum, Katsura Tree, Katsuratree, Katsura-tree ●
83735 Cercidiphyllum japonicum Siebold et Zucc. = Cercidiphyllum japonicum Siebold et Zucc. ex Hoffm. et Schult. ●
83736 Cercidiphyllum japonicum Siebold et Zucc. ex Hoffm. et Schult.;连香树(芭蕉香清,白果,饼木,连香木,日本连香树,山白果,王君树,五君树,心木,圆槿);Cake Tree, China Katsuratree, Chinese Katsuratree, Heart Tree, Japanese Katsura Tree, Japanese Katsura-tree, Katsura, Katsura Tree, Katsuratree, Katsura-tree ●
83737 Cercidiphyllum japonicum Siebold et Zucc. ex Hoffm. et Schult. f. pendulum (Miyoshi ex Makino et Nemoto) Ohwi;垂枝连香树●☆
83738 Cercidiphyllum japonicum Siebold et Zucc. ex Hoffm. et Schult. f. pendulum (Miyoshi ex Makino et Nemoto) Ohwi = Cercidiphyllum magnificum Nakai f. pendulum (Miyoshi ex Makino et Nemoto) Spongberg ●☆
83739 Cercidiphyllum japonicum Siebold et Zucc. ex Hoffm. et Schult. var. magnificum Nakai = Cercidiphyllum magnificum (Nakai) Nakai ●☆
83740 Cercidiphyllum japonicum Siebold et Zucc. ex Hoffm. et Schult. var. sinense Rehder et E. H. Wilson;毛叶连香树(芭蕉香清,连香树,山白果,五君树,银叶连香树,圆檀,中国连香树);Chinese Katsura Tree, Chinese Katsuratree, Katsura Tree ●
83741 Cercidiphyllum japonicum Siebold et Zucc. var. sinense Rehder et E. H. Wilson = Cercidiphyllum japonicum Siebold et Zucc. ex Hoffm. et Schult. ●
83742 Cercidiphyllum japonicum Siebold et Zucc. var. sinense Rehder et E. H. Wilson = Cercidiphyllum japonicum Siebold et Zucc. ex Hoffm. et Schult. var. sinense Rehder et E. H. Wilson ●
83743 Cercidiphyllum magnificum (Nakai) Nakai;大叶连香树;Bigleaf Katsura Tree, Bigleaf Katsuratree ●☆
83744 Cercidiphyllum magnificum Nakai f. pendulum (Miyoshi ex Makino et Nemoto) Spongberg;垂枝日本连香树;Pendulous Katsuratree ●☆
83745 Cercidium Tul. (1844);假紫荆属●☆
83746 Cercidium Tul. = Parkinsonia L. ●
83747 Cercidium floridum Benth. ex A. Gray;多花假紫荆;Blue Palo Verde, Blue Paloverde, Border Palo Verde, Palo Verde ●☆
83748 Cercidium microphyllum Rose et I. M. Johnst.;小叶假紫荆;Foothill Yellow Palo Verde, Foothills Paloverde, Little-leaved Horse Bean, Palo Verde ●☆
83749 Cercidium praecox (Ruiz et Pav. ex Hook. et Arn.) Harms;早生假紫荆;Palo Brea, Sonoran Palo Verde ●☆
83750 Cercidophyllum Post et Kuntze = Cercidiphyllum Siebold et Zucc. ●
83751 Cercinia Tiegh. = Ouratea Aubl.(保留属名)●
83752 Cercis L. (1753);紫荆属;Cercis, Judas-tree, Red Bud, Redbud ●
83753 Cercis californica Torr. ex Benth.;加州紫荆;California Redbud, Western Redbud ●☆
83754 Cercis canadensis (Rehder) Bean f. alba Rehder = Cercis canadensis L. 'Alba' ●☆
83755 Cercis canadensis (Rose) M. Hopkins var. orbiculata (Greene) Barneby = Cercis orbiculata Greene ●☆
83756 Cercis canadensis (S. Watson) A. E. Murray var. texensis (S. Watson) M. Hopkins = Cercis canadensis L. subsp. texensis (S. Watson) A. E. Murray ●☆
83757 Cercis canadensis L.;加拿大紫荆(美国紫荆);American Judas Tree, Canada Redbud, Eastern Redbud, Judas Tree, Judas-tree, Red Bud, Redbud ●
83758 Cercis canadensis L. 'Alba';白花加拿大紫荆●☆
83759 Cercis canadensis L. 'Convey';弯曲加拿大紫荆;Convey Redbud ●☆
83760 Cercis canadensis L. 'Forest Pansy';森林紫加拿大紫荆(森林三色堇加拿大紫荆);Forest Pansy Redbud ●☆
83761 Cercis canadensis L. f. glabrifolia Fernald;光叶加拿大紫荆●☆
83762 Cercis canadensis L. subsp. mexicana (Rose) A. E. Murray = Cercis mexicana Rose ●☆
83763 Cercis canadensis L. subsp. retisus?;俄克拉何马紫荆;Oklahoma Redbud ●☆
83764 Cercis canadensis L. subsp. texensis (S. Watson) A. E. Murray;得州紫荆;Texas Redbud ●☆
83765 Cercis canadensis L. var. alba (Rehder) Bean = Cercis canadensis L. 'Alba' ●☆
83766 Cercis canadensis L. var. mexicana (Rose) M. Hopkins = Cercis mexicana Rose ●☆
83767 Cercis canadensis L. var. plena Sudw.;重瓣加拿大紫荆●☆
83768 Cercis canadensis L. var. pubescens Pursh;毛加拿大紫荆●☆
83769 Cercis chinensis Bunge;紫荆(扁头翁,川紫荆,箩筐桑,箩筐树,裸枝树,满条红,内消,清明花,肉红,乌桑,紫花树,紫今皮,紫金盘,紫金皮,紫荆木,紫珠);Avondale Redbud, China Red Bud, China Redbud, Chinese Judas Tree, Chinese Red Bud, Chinese Redbud ●
83770 Cercis chinensis Bunge 'Alba' = Cercis chinensis Bunge f. alba S. C. Hsu ●
83771 Cercis chinensis Bunge f. alba S. C. Hsu;白花紫荆;White-flowered Redbud ●
83772 Cercis chinensis Bunge f. alba S. C. Hsu = Cercis chinensis Bunge ●
83773 Cercis chinensis Bunge f. leucantha Sugim. = Cercis chinensis Bunge f. alba S. C. Hsu ●
83774 Cercis chinensis Bunge f. pubescens C. F. Wei;短毛紫荆;Pubescent Chinese Redbud ●
83775 Cercis chinensis Bunge f. pubescens C. F. Wei = Cercis chinensis Bunge ●
83776 Cercis chinensis Bunge f. rosea P. S. Hsu;粉红紫荆;Rose China Redbud ●

83777 Cercis chinensis Bunge f. rosea P. S. Hsu = Cercis chinensis Bunge ●

83778 Cercis chingii Chun;黄山紫荆(秦氏紫荆,浙皖紫荆,紫荆);Ching Redbud,Huangshan Redbud,Redbud ●

83779 Cercis chuniana F. P. Metcalf;岭南紫荆(陈氏紫荆,广西紫荆);Chun Redbud,Guangxi Redbud ●

83780 Cercis ellipsoidea Greene;俄州紫荆●☆

83781 Cercis florida Salisb. = Cercis siliquastrum L. ●

83782 Cercis funiushanensis S. Y. Wang et T. B. Chao;伏牛紫荆;Funiushan Redbud ●

83783 Cercis funiushanensis S. Y. Wang et T. B. Chao = Cercis glabra Pamp. ●

83784 Cercis georgiana Greene;乔治亚紫荆●☆

83785 Cercis gigantea W. C. Cheng et P. C. Keng;巨紫荆(罗钱树,天目紫荆,乌桑);Giant Redbud ●

83786 Cercis gigantea W. C. Cheng et P. C. Keng = Cercis glabra Pamp. ●

83787 Cercis glabra Pamp.;湖北紫荆(箩筐树,乌桑树,云南紫荆);Glabrous Redbud,Hubei Redbud ●

83788 Cercis griffithii Boiss.;格氏紫荆●☆

83789 Cercis japonica Siebold ex Planch. = Cercis chinensis Bunge ●

83790 Cercis latissima Greene;宽紫荆●☆

83791 Cercis liangkwangensis Chun = Cercis chuniana F. P. Metcalf ●

83792 Cercis likiangensis Chun = Cercis chuniana F. P. Metcalf ●

83793 Cercis likiangensis Chun ex Y. Chen = Cercis chuniana F. P. Metcalf ●

83794 Cercis mexicana Rose;墨西哥紫荆;Mexican Redbud ●☆

83795 Cercis nephrophylla Greene;肾叶紫荆;Texan Redbud ●☆

83796 Cercis nigricans (Vahl) Greene;变黑紫荆●☆

83797 Cercis nitida Greene;光亮紫荆●☆

83798 Cercis occidentalis Torr.;西方紫荆(加州紫荆);California Redbud,Judas-tree,Redbud,Western Redbud ●☆

83799 Cercis occidentalis Torr. var. orbiculata (Greene) Tidestr. = Cercis orbiculata Greene ●☆

83800 Cercis occidentalis Torr. var. texensis S. Watson = Cercis orbiculata Greene ●☆

83801 Cercis orbiculata Greene;圆叶紫荆●☆

83802 Cercis pauciflora H. L. Li;少花紫荆;Fewflower Redbud ●

83803 Cercis pauciflora H. L. Li = Cercis chinensis Bunge ●

83804 Cercis pubescens S. Y. Wang;毛紫荆;Pubescent Redbud ●

83805 Cercis pubescens S. Y. Wang = Cercis glabra Pamp. ●

83806 Cercis pumila W. Young;矮小紫荆●☆

83807 Cercis racemosa Oliv.;垂丝紫荆(垂枝紫荆);Chain Flowered Redbud,Raceme Redbud ●

83808 Cercis reniformis Engelm. ex A. Gray = Cercis texensis Sarg. ●☆

83809 Cercis siliquastrum L.;南欧紫荆(地中海紫荆,西亚紫荆);Bean Tree,Carobe-tree,Judaes Tree,Judas Tree,Judas-tree,Love Tree,Love-tree,Mediterranean Redbud ●

83810 Cercis siliquosa St. -Lag. = Cercis siliquastrum L. ●

83811 Cercis texensis Sarg. = Cercis occidentalis Torr. ●☆

83812 Cercis yunnanensis Hu et W. C. Cheng;云南紫荆;Yunnan Redbud ●

83813 Cercis yunnanensis Hu et W. C. Cheng = Cercis glabra Pamp. ●

83814 Cercocarpaceae J. Agardh = Rosaceae Juss.(保留科名)●■

83815 Cercocarpaceae J. Agardh;山桃花心木科●

83816 Cercocarpus Kunth(1824);山桃花心木属;Mountain-mahogany ●☆

83817 Cercocarpus betuloides;桦叶山桃花心木;Birch-leafed Mountain Mahogany,Mountain Mahogany ●☆

83818 Cercocarpus intricatus S. Watson;小叶山桃花心木;Curl-leaf Mountain-mahogany,Little-leaved Mountain Mahogany ●☆

83819 Cercocarpus ledifolius Nutt.;杜香叶山桃花心木;Birchleaf Mountain-mahogany,Curl-leaf Cercocarpus,Curl-leaf Mountain Mahogany,Curl-leaf Mountain-mahogany,Mountain Mahogany,Mountain-mahogany ●☆

83820 Cercocarpus montanus Raf.;山桃花心木;Alder-leaf Mountain Mahogany,Hard-tack,Mountain Mahogany ●☆

83821 Cercocodia Post et Kuntze = Cercodia Murr. ■●

83822 Cercocodia Post et Kuntze = Haloragis J. R. Forst. et G. Forst. ■●

83823 Cercocoma Miq. = Rhynchodia Benth. ●■

83824 Cercocoma Wall. = Strophanthus DC. ●

83825 Cercocoma Wall. ex G. Don = Strophanthus DC. ●

83826 Cercodea Sol. ex Lam. = Cercodia Murr. ■●

83827 Cercodia Murr. = Haloragis J. R. Forst. et G. Forst. ■●

83828 Cercodiaceae Juss. = Haloragaceae R. Br.(保留科名)●■

83829 Cercopetalum Gilg = Pentadiplandra Baill. ●☆

83830 Cercopetalum dasyanthum Gilg = Pentadiplandra brazzeana Baill. ●☆

83831 Cercophora Miers = Lecythis Loefl. ●☆

83832 Cercophora Miers = Strailia T. Durand ●☆

83833 Cercostylos Less. = Gaillardia Foug. ■

83834 Cercouratea Tiegh. = Ouratea Aubl.(保留属名)●

83835 Cerdana Ruiz et Pav. = Cordia L.(保留属名)●

83836 Cerdana alliodora Ruiz et Pav. = Cordia alliodora (Ruiz et Pav.) Oken ●☆

83837 Cerdia DC. = Cerdia Moc. et Sessé ex DC. ■☆

83838 Cerdia Moc. et Sessé = Cerdia Moc. et Sessé ex DC. ■☆

83839 Cerdia Moc. et Sessé ex DC. (1828);单蕊莲豆草属■☆

83840 Cerdia glauca Hemsl.;灰单蕊莲豆草■☆

83841 Cerdia purpurascens Moc. et Sessé ex DC.;紫单蕊莲豆草■☆

83842 Cerdia virescens Moc. et Sessé;绿单蕊莲豆草■☆

83843 Cerdosurus Ehth. = Alopecurus L. ■

83844 Cerea Schltdl. = Paspalum L. ■

83845 Cerea Thouars = Elaeocarpus L. ●

83846 Cereaceae DC. et Spreng. = Cactaceae Juss.(保留科名)●■

83847 Cereaceae Spreng. ex DC. et Spreng. = Cactaceae Juss.(保留科名)●■

83848 Cereaceae Spreng. ex Jameson = Cactaceae Juss.(保留科名)●■

83849 Cerefolium Fabr.(废弃属名) = Anthriscus Pers.(保留属名)■

83850 Cereopsis Blanco = Coreopsis L. ●■

83851 Cereopsis Raf. = Coreopsis L. ●■

83852 Ceresia Pers. = Paspalum L. ■

83853 Cereus Haw. = Cereus Mill. ●

83854 Cereus L. = Cereus Mill. ●

83855 Cereus Mill. (1754);仙影掌属(天轮柱属,仙人拳属,仙人柱属);Cereus ●

83856 Cereus Raf. = Cereus Mill. ●

83857 Cereus aboriginum (Small ex Britton et Rose) Little = Harrisia aboriginum Small ex Britton et Rose ●☆

83858 Cereus assurgens Wright ex Griseb. = Leptocereus assurgens (Wright ex Griseb.) Britton et Rose ●☆

83859 Cereus aureus Meyen;金黄仙影掌●☆

83860 Cereus californicus Torr. et A. Gray = Cylindropuntia californica (Torr. et A. Gray) F. M. Knuth ■☆

83861 Cereus candelaris Meyen = Browningia candelaris (Meyen)

Britton et Rose ●☆
83862 Cereus dayamii Speg.;冲天柱;Dayam Cereus ●
83863 Cereus diguetii F. A. C. Weber = Peniocereus striatus (Brandegee) Buxb. ●☆
83864 Cereus divaricatus (Lam.) DC.;分叉仙影拳●☆
83865 Cereus emoryi Engelm. = Bergerocactus emoryi (Engelm.) Britton et Rose ■☆
83866 Cereus eriophorus Sweet var. fragrans (Small ex Britton et Rose) L. D. Benson = Harrisia fragrans Small ex Britton et Rose ●☆
83867 Cereus fendleri Engelm. = Echinocereus fendleri (Engelm.) Sencke ex J. N. Haage ●☆
83868 Cereus flagelliformis Mill. = Aporocactus flagelliformis (L.) Lem. ●
83869 Cereus fragrans (Small ex Britton et Rose) Little = Harrisia fragrans Small ex Britton et Rose ●☆
83870 Cereus geometrizans Mart. ex Pfeiff.;浆果仙人掌●☆
83871 Cereus giganteus Engelm. = Carnegiea gigantea (Engelm.) Britton et Rose ●☆
83872 Cereus gracilis Mill. var. aboriginus (Small ex Britton et Rose) L. D. Benson = Harrisia aboriginum Small ex Britton et Rose ●☆
83873 Cereus gracilis Mill. var. simpsonii (Small ex Britton et Rose) L. D. Benson = Harrisia simpsonii Small ex Britton et Rose ●☆
83874 Cereus grandiflorus (L.) Mill.;大花仙人掌(大花蛇鞭柱,大花月光掌,大轮柱,西施仙人柱,夜花仙人掌);Largeflower Queen-of-night, Night-blooming Cereus, Night-flowering Cereus, Organillo, Queen of the Night, Queen-of-night, Queen-of-the-night ●
83875 Cereus grandiflorus (L.) Mill. var. armatus (K. Schum.) L. D. Benson = Cereus grandiflorus (L.) Mill. ●
83876 Cereus grandiflorus Mill. = Cereus grandiflorus (L.) Mill. ●
83877 Cereus grandiflorus Mill. = Selenicereus grandiflorus (L.) Britton et Rose ●☆
83878 Cereus greggii Engelm. = Peniocereus greggii (Engelm.) Britton et Rose ●
83879 Cereus greggii Engelm. var. cismontanus Engelm. = Peniocereus greggii (Engelm.) Britton et Rose ●
83880 Cereus greggii Engelm. var. roseiflorus Kunze = Peniocereus greggii (Engelm.) Britton et Rose var. transmontanus (Engelm.) Backeb. ●☆
83881 Cereus greggii Engelm. var. transmontanus Engelm. = Peniocereus greggii (Engelm.) Britton et Rose var. transmontanus (Engelm.) Backeb. ●☆
83882 Cereus hexagonus Mill.;六角天轮柱(六棱柱);Lady of the Night Cactus ●
83883 Cereus imbricatus Haw. = Cylindropuntia imbricata (Haw.) F. M. Knuth ■☆
83884 Cereus imbricatus Haw. = Opuntia imbricata (Haw.) DC. ●■
83885 Cereus jamacaru DC.;牙买加冲天柱;Jamacar Cereus, Mandacaru, Pleated Cereus ●
83886 Cereus martinii Labour. = Harrisia martinii (Labour.) Britton ●☆
83887 Cereus oxypetalus DC. = Epiphyllum oxypetalum (DC.) Haw. ■
83888 Cereus pectinatus Engelm. var. rigidissimus Engelm. = Echinocereus rigidissimus (Engelm.) F. Haage ●
83889 Cereus pentalophus DC. = Echinocereus pentalophus (DC.) H. P. Kelsey et Dayton ●
83890 Cereus peruvianus (L.) Mill.;秘鲁冲天柱(鬼面角,六角柱,六棱仙人鞭,天轮柱);Night Blooming Cereus, Peruvian Apple Cactus, Peruvian Cereus ●
83891 Cereus peruvianus Mill. var. monstrosus DC.;岩狮子;Curiosity Plant, Peruvian Torch ●
83892 Cereus phyllanthus (L.) DC. = Epiphyllum phyllanthus (L.) Haw. ■☆
83893 Cereus pitahaya (L.) DC.;仙影掌(山影拳,仙人柱);Cereus ●
83894 Cereus pottsii Salm-Dyck = Peniocereus greggii (Engelm.) Britton et Rose ●
83895 Cereus pteranthus Link ex A. Dietr. = Selenicereus pteranthus (Link ex A. Dietr.) Britton et Rose ●
83896 Cereus repandus Mill.;波状仙影掌;Hedge Cactus, Peruvian Apple Cactus ●☆
83897 Cereus robinii (Lem.) L. D. Benson = Pilosocereus robinii (Lem.) Byles et G. D. Rowley ●☆
83898 Cereus robinii (Lem.) L. D. Benson var. deeringii (Small) L. D. Benson = Pilosocereus robinii (Lem.) Byles et G. D. Rowley ●☆
83899 Cereus robinii (Lem.) L. D. Benson var. keyensis (Britton et Rose) L. D. Benson ex R. W. Long et Lakela = Pilosocereus robinii (Lem.) Byles et G. D. Rowley ●☆
83900 Cereus schottii Engelm. = Pachycereus schottii (Engelm.) D. R. Hunt ●☆
83901 Cereus scopa Salm-Dyck ex DC. = Echinocactus scopa Link et Otto ■
83902 Cereus scopa Salm-Dyck ex DC. = Notocactus scopa (Link et Otto) A. Berger ■
83903 Cereus spachianus Lem. = Echinopsis spachiana (Lem.) H. Friedrich et G. D. Rowley ■☆
83904 Cereus spinulosus DC. = Selenicereus spinulosus (DC.) Britton et Rose ●☆
83905 Cereus squarrosus Vaupel = Erdisia squarrosa (Vaupel) Britton et Rose ●☆
83906 Cereus stramineus Engelm. = Echinocereus stramineus (Engelm.) Engelm. ex Rümpler ●
83907 Cereus striatus Brandegee = Peniocereus striatus (Brandegee) Buxb. ●☆
83908 Cereus tetragonus Haw.;连城角●☆
83909 Cereus thurberi Engelm. = Stenocereus thurberi (Engelm.) Buxb. ●☆
83910 Cereus undatus Haw. = Hylocereus undatus (Haw.) Britton et Rose ●
83911 Cereus uruguayanus F. Ritter ex R. Kiesling;绿萼仙人柱(仙人柱);Torch Cactus ●☆
83912 Cereus validus Haw.;刚柱(黄刺神代,金狮子,有力柱)●☆
83913 Cereus variabilis Pfeiff.;神代柱;Variable Cereus ●
83914 Cerinozoma Post et Kuntze = Kerinozoma Steud. ex Zoll. ■☆
83915 Cerinthaceae Bercht. et J. Presl = Boraginaceae Juss.(保留科名)■●
83916 Cerinthaceae Martinov = Boraginaceae Juss.(保留科名)■●
83917 Cerinthaceae Martinov;琉璃紫草科■
83918 Cerinthe L. (1753);琉璃紫草属(琉璃苣属);Honeywort ■☆
83919 Cerinthe alpina Kit.;高山琉璃紫草■☆
83920 Cerinthe aspera Roth = Cerinthe major L. ■☆
83921 Cerinthe gymnandra Gasp.;裸蕊琉璃紫草■☆
83922 Cerinthe gymnandra Gasp. var. dubia Maire = Cerinthe gymnandra Gasp. ■☆
83923 Cerinthe major L.;大琉璃紫草(琉璃苣,琉璃紫草);Honey Wort, Honeywort, Lesser Honeywort ■☆
83924 Cerinthe major L. subsp. gymnandra (Gasp.) Rouy = Cerinthe

gymnandra Gasp. ■☆
83925 Cerinthe major L. var. aspera Rchb. = Cerinthe major L. ■☆
83926 Cerinthe major L. var. gymnandra? = Cerinthe gymnandra Gasp. ■☆
83927 Cerinthe major L. var. macrosiphonia Murb. = Cerinthe gymnandra Gasp. ■☆
83928 Cerinthe major L. var. oranensis (Batt.) Murb. = Cerinthe gymnandra Gasp. ■☆
83929 Cerinthe minor L.;小琉璃紫草;Lesser Honeywort ■☆
83930 Cerinthe minor L. 'Purpurascens';紫晕琉璃苣■☆
83931 Cerinthe oranensis Batt. = Cerinthe gymnandra Gasp. ■☆
83932 Cerinthe strigosa Rchb. = Cerinthe major L. ■☆
83933 Cerinthodes Kuntze = Mertensia Roth(保留属名)■
83934 Cerinthodes Ludwig = Mertensia Roth(保留属名)■
83935 Cerinthodes Ludwig ex Kuntze = Mertensia Roth(保留属名)■
83936 Cerinthopsis Kotschy ex Benth. et Hook. f. = Solenanthus Ledeb. ■
83937 Cerinthopsis Kotschy ex Paine = Lindelofia Lehm. ■
83938 Cerionanthus Schott ex Roem. et Schult. = Cephalaria Schrad. (保留属名)■
83939 Ceriops Arn. (1838);角果木属(细蕊红树属);Ceriops ●
83940 Ceriops boiviniana Tul. = Ceriops tagal (Perr.) C. B. Rob. ●
83941 Ceriops candolleana Arn. = Ceriops tagal (Perr.) C. B. Rob. ●
83942 Ceriops candolleana Arn. var. sassakii Hayata = Ceriops tagal (Perr.) C. B. Rob. ●
83943 Ceriops mossambicensis Klotzsch = Ceriops tagal (Perr.) C. B. Rob. ●
83944 Ceriops roxburghiana Arn.;罗氏角果木;Prong, Roxburgh Ceriops ●
83945 Ceriops somalensis Chiov. = Ceriops tagal (Perr.) C. B. Rob. ●
83946 Ceriops tagal (Perr.) C. B. Rob.;角果木(大卡红树,海淀仔,海枷子,剪子树,细蕊红树);Ceriops, Common Ceriops, Prong Ceriops, Robinson Ceriops ●
83947 Ceriops tagal (Perr.) C. B. Rob. var. australis C. T. White = Ceriops tagal (Perr.) C. B. Rob. ●
83948 Ceriops timoriensis (DC.) Domin = Ceriops tagal (Perr.) C. B. Rob. ●
83949 Ceriosperma (O. E. Schulz) Greuter et Burdet = Rorippa Scop. ■
83950 Ceriosperma (O. E. Schulz) Greuter et Burdet(1983);叙利亚豆瓣菜属■☆
83951 Ceriosperma macrocarpum (Boiss.) Greuter et Burdet;叙利亚豆瓣菜■☆
83952 Ceriscoides (Benth. et Hook. f.) Tirveng. = Ceriscoides (Hook. f.) Tirveng. ●
83953 Ceriscoides (Hook. f.) Tirveng. (1978);木瓜榄属;Ceriscoides ●
83954 Ceriscoides howii H. C. Lo;木瓜榄;Ceriscoides ●
83955 Ceriscus Gaertn. = Catunaregam Wolf ●
83956 Ceriscus Gaertn. ex Nees = Catunaregam Wolf ●
83957 Ceriseus Gaertn. = Randia L. ●
83958 Ceriseus Nees = Tarenna Gaertn. ●
83959 Cerium Lour. = Lysimachia L. ●■
83960 Cernohorskya Á. Löve et D. Löve = Arenaria L. ■
83961 Cerocarpus Colebr. ex Hassk. = Syzygium R. Br. ex Gaertn. (保留属名)●
83962 Cerocarpus Hassk. = Malidra Raf. ●
83963 Cerocarpus Hassk. = Syzygium R. Br. ex Gaertn. (保留属名)●
83964 Cerochilus Lindl. = Hetaeria Blume(保留属名)■
83965 Cerochilus rubens Lindl. = Hetaeria affinis (Griff.) Seidenf. et Ormerod ■
83966 Cerochilus rubens Lindl. = Hetaeria rubens (Lindl.) Benth. et Hook. f. ■
83967 Cerochlamys N. E. Br. (1928);玉细鳞属(蜡波属)■☆
83968 Cerochlamys gemina (L. Bolus) H. E. K. Hartmann;玉细鳞■☆
83969 Cerochlamys pachyphylla (L. Bolus) L. Bolus;毛叶玉细鳞■☆
83970 Cerochlamys pachyphylla (L. Bolus) L. Bolus var. albiflora H. Jacobsen = Cerochlamys pachyphylla (L. Bolus) L. Bolus ■☆
83971 Cerochlamys purpureostyla (L. Bolus) H. E. K. Hartmann = Acrodon purpureostylus (L. Bolus) Burgoyne ■☆
83972 Cerochlamys trigona N. E. Br.;三角玉细鳞■☆
83973 Cerolepis Pierre = Camptostylus Gilg ●☆
83974 Cerolepis petiolaris Pierre = Oncoba mannii Oliv. ●☆
83975 Ceropegia L. (1753);吊灯花属(吊金钱属,金雀马尾参属);Ceropegia, Pendentlamp ■
83976 Ceropegia aberrans Schltr. = Ceropegia stenoloba Hochst. ex Chiov. ■☆
83977 Ceropegia abyssinica Decne.;阿比西尼亚吊灯花■☆
83978 Ceropegia acacietorum Schltr. = Ceropegia pachystelma Schltr. ■☆
83979 Ceropegia achtenii De Wild. subsp. togoensis H. Huber;多哥吊灯花■☆
83980 Ceropegia acuminata Roxb. = Ceropegia bulbosa Roxb. ■☆
83981 Ceropegia affinis Vatke;近缘吊灯花■☆
83982 Ceropegia africana R. Br.;非洲吊灯花■☆
83983 Ceropegia africana R. Br. subsp. barklyi Bruyns;巴克利吊灯花■☆
83984 Ceropegia albertina S. Moore = Ceropegia aristolochioides Decne. subsp. albertina (S. Moore) H. Huber ■☆
83985 Ceropegia albisepta Jum. et H. Perrier;白萼吊灯花■☆
83986 Ceropegia albisepta Jum. et H. Perrier var. robynsiana (Werderm.) H. Huber;罗宾斯吊灯花■☆
83987 Ceropegia ampliata E. Mey.;膨大吊灯花■☆
83988 Ceropegia ampliata E. Mey. var. oxyloba H. Huber;尖裂吊灯花■☆
83989 Ceropegia angusta N. E. Br. = Ceropegia racemosa N. E. Br. ■☆
83990 Ceropegia angustifolia Vahl ex Decne. = Ceropegia longifolia Wall. ■
83991 Ceropegia angustilimba Merr. = Ceropegia trichantha Hemsl. ■
83992 Ceropegia angustiloba De Wild. = Ceropegia stenantha K. Schum. ■☆
83993 Ceropegia apiculata Schltr. = Ceropegia lugardae N. E. Br. ■☆
83994 Ceropegia arenaria R. A. Dyer;沙地吊灯花■☆
83995 Ceropegia aridicola W. W. Sm.;丽江吊灯花(旱地马尾参);Lijiang Ceropegia, Parched Pendentlamp ■
83996 Ceropegia aristolochioides Decne.;马兜铃吊灯花■☆
83997 Ceropegia aristolochioides Decne. subsp. albertina (S. Moore) H. Huber;阿尔伯特马兜铃吊灯花■☆
83998 Ceropegia aristolochioides Decne. var. wittei Werderm. = Ceropegia aristolochioides Decne. subsp. albertina (S. Moore) H. Huber ■☆
83999 Ceropegia aristolochioides Hutch. et Dalziel = Ceropegia sankuruensis Schltr. ■☆
84000 Ceropegia assimilis N. E. Br. = Ceropegia cancellata Rchb. ■☆
84001 Ceropegia atacorensis A. Chev. = Ceropegia racemosa N. E. Br. ■☆
84002 Ceropegia bajana Schltr. ex Bullock = Ceropegia racemosa N. E. Br. ■☆
84003 Ceropegia balfouriana Schltr. = Ceropegia mairei (H. Lév.) H. Huber ■
84004 Ceropegia ballyana Bullock;博利吊灯花■☆
84005 Ceropegia barbata R. A. Dyer;髯毛吊灯花■☆

84006 Ceropegia barbertonensis N. E. Br. = Ceropegia linearis E. Mey. subsp. woodii (Schltr.) H. Huber ■☆
84007 Ceropegia barbigera Bruyns;缘毛吊灯花■☆
84008 Ceropegia barklyi Hook. f. = Ceropegia africana R. Br. subsp. barklyi Bruyns ■☆
84009 Ceropegia barklyi Hook. f. var. tugelensis N. E. Br. = Ceropegia africana R. Br. subsp. barklyi Bruyns ■☆
84010 Ceropegia batesii S. Moore = Ceropegia sankuruensis Schltr. ■☆
84011 Ceropegia beccariana Martelli = Ceropegia aristolochioides Decne. ■☆
84012 Ceropegia bequaertii De Wild. = Ceropegia abyssinica Decne. ■☆
84013 Ceropegia biddumana K. Schum. = Ceropegia affinis Vatke ■☆
84014 Ceropegia boerhaaviifolia Schinz = Ceropegia pachystelma Schltr. ■☆
84015 Ceropegia bonafouxii K. Schum.;博纳吊灯花■☆
84016 Ceropegia botrys K. Schum. = Ceropegia subaphylla K. Schum. ■☆
84017 Ceropegia boussingaultiflora Dinter = Ceropegia nilotica Kotschy ■☆
84018 Ceropegia bowkeri Harv.;鲍克吊灯花■☆
84019 Ceropegia bowkeri Harv. subsp. sororia (Harv. ex Hook. f.) R. A. Dyer;堆积吊灯花■☆
84020 Ceropegia brachyceras Schltr.;短角吊灯花■☆
84021 Ceropegia brachysiphon H. Huber = Riocreuxia aberrans R. A. Dyer ■☆
84022 Ceropegia brevirostris P. R. O. Bally et D. V. Field;短喙吊灯花■☆
84023 Ceropegia brownii Ledger = Ceropegia nilotica Kotschy ■☆
84024 Ceropegia bulbosa Roxb.;印巴吊灯花■☆
84025 Ceropegia bulbosa Roxb. var. esculenta (Edgew.) Hook. f. = Ceropegia bulbosa Roxb. ■☆
84026 Ceropegia bulbosa Roxb. var. lushii (Grab.) Hook. f. = Ceropegia bulbosa Roxb. ■☆
84027 Ceropegia burchellii (K. Schum.) H. Huber subsp. profusa (N. E. Br.) H. Huber = Riocreuxia polyantha Schltr. ■☆
84028 Ceropegia burgeri M. G. Gilbert;伯格吊灯花■☆
84029 Ceropegia butaguensis De Wild. = Ceropegia racemosa N. E. Br. ■☆
84030 Ceropegia butayei De Wild. = Orthanthera butayei (De Wild.) Werderm. ■☆
84031 Ceropegia caffrorum Schltr. = Ceropegia linearis E. Mey. ■☆
84032 Ceropegia caffrorum Schltr. var. dubia N. E. Br. = Ceropegia linearis E. Mey. ■☆
84033 Ceropegia calcarata N. E. Br. = Ceropegia meyeri-johannis Engl. var. verdickii (De Wild.) Werderm. ■☆
84034 Ceropegia campanulata G. Don;风铃草状吊灯花■☆
84035 Ceropegia campanulata G. Don var. porphyrotricha (W. W. Sm.) H. Huber;紫毛吊灯花■☆
84036 Ceropegia campanulata G. Don var. pulchella H. Huber = Ceropegia insignis R. A. Dyer ■☆
84037 Ceropegia cancellata Rchb.;格纹吊灯花■☆
84038 Ceropegia carnosa E. Mey.;肉质吊灯花■☆
84039 Ceropegia ceratophora Svent. = Ceropegia dichotoma Haw. ■☆
84040 Ceropegia christenseniana Hand. -Mazz.;短序吊灯花(吊灯花,小鹅儿肠);Christensen Ceropegia,Christensen Pendentlamp ■
84041 Ceropegia chrysantha Svent. = Ceropegia dichotoma Haw. ■☆
84042 Ceropegia chrysochroma H. Huber = Riocreuxia chrysochroma (H. Huber) A. R. Sm. ■☆
84043 Ceropegia claviloba Werderm.;棒吊灯花■☆
84044 Ceropegia collaricorona Werderm. = Ceropegia linearis E. Mey. subsp. woodii (Schltr.) H. Huber ■☆
84045 Ceropegia connivens R. A. Dyer = Ceropegia fimbriata E. Mey. subsp. connivens (R. A. Dyer) Bruyns ■☆
84046 Ceropegia connivens R. A. Dyer f. angustata R. A. Dyer = Ceropegia fimbriata E. Mey. subsp. connivens (R. A. Dyer) Bruyns ■☆
84047 Ceropegia constricta N. E. Br. = Ceropegia nilotica Kotschy ■☆
84048 Ceropegia convolvuloides A. Rich.;旋花吊灯花■☆
84049 Ceropegia convolvulus Hochst. ex Werderm. = Ceropegia ringens A. Rich. ■☆
84050 Ceropegia cordiloba Werderm. = Ceropegia papillata N. E. Br. var. cordiloba (Werderm.) H. Huber ■
84051 Ceropegia crassifolia Schltr.;厚叶吊灯花■☆
84052 Ceropegia crassula Schltr. = Ceropegia aristolochioides Decne. ■☆
84053 Ceropegia criniticaulis Werderm. = Ceropegia meyeri-johannis Engl. var. verdickii (De Wild.) Werderm. ■☆
84054 Ceropegia crispata N. E. Br. = Ceropegia crassifolia Schltr. ■☆
84055 Ceropegia cufodontii Chiov.;卡佛吊灯花■☆
84056 Ceropegia cynanchoides Schltr. = Ceropegia racemosa N. E. Br. subsp. secamonoides (S. Moore) H. Huber ■☆
84057 Ceropegia cyrtoidea Werderm. = Ceropegia distincta N. E. Br. ■☆
84058 Ceropegia dalzielii N. E. Br. = Ceropegia campanulata G. Don ■☆
84059 Ceropegia debilis N. E. Br. = Ceropegia linearis E. Mey. subsp. debilis (N. E. Br.) H. Huber ■☆
84060 Ceropegia decidua E. A. Bruce;脱落吊灯花■☆
84061 Ceropegia decidua E. A. Bruce subsp. pretoriensis R. A. Dyer;比勒陀利亚吊灯花■☆
84062 Ceropegia deightonii Hutch. et Dalziel;戴顿吊灯花■☆
84063 Ceropegia deightonii Hutch. et Dalziel subsp. conjuncta H. Huber;接合吊灯花■☆
84064 Ceropegia deightonii Hutch. et Dalziel subsp. tisserantii H. Huber;蒂斯朗特吊灯花■☆
84065 Ceropegia denticulata K. Schum. = Ceropegia nilotica Kotschy var. simplex H. Huber ■☆
84066 Ceropegia denticulata K. Schum. var. brownii (Ledger) P. R. O. Bally = Ceropegia nilotica Kotschy ■☆
84067 Ceropegia de-vecchii Chiov. = Ceropegia variegata (Forssk.) Decne. ■☆
84068 Ceropegia de-vecchii Chiov. var. adelaidae P. R. O. Bally = Ceropegia variegata (Forssk.) Decne. ■☆
84069 Ceropegia dewevrei De Wild. = Ceropegia volubilis N. E. Br. ■☆
84070 Ceropegia dichotoma Haw.;双叉吊金钱■☆
84071 Ceropegia dichroantha K. Schum. = Ceropegia filipendula K. Schum. ■☆
84072 Ceropegia dimorpha Humbert;二型吊灯花■☆
84073 Ceropegia dinteri Schltr.;丁特吊灯花■☆
84074 Ceropegia distincta N. E. Br.;离生吊灯花■☆
84075 Ceropegia distincta N. E. Br. f. pubescens H. Huber = Ceropegia lugardae N. E. Br. ■☆
84076 Ceropegia distincta N. E. Br. subsp. haygarthii (Schltr.) H. Huber = Ceropegia haygarthii Schltr. ■☆
84077 Ceropegia distincta N. E. Br. subsp. lugardae (N. E. Br.) H. Huber = Ceropegia lugardae N. E. Br. ■☆
84078 Ceropegia distincta N. E. Br. subsp. verruculosa R. A. Dyer = Ceropegia lugardae N. E. Br. ■☆
84079 Ceropegia dolichophylla Schltr.;剑叶吊灯花(长叶吊灯花,蕵参,双剪菜,双剪草);Lanceo-leaf Ceropegia,Longleaf Pendentlamp ■
84080 Ceropegia dolichophylla Schltr. var. brachyloba Hand. -Mazz. =

Ceropegia dolichophylla Schltr. ■

84081 Ceropegia dolichophylla Schltr. var. purpureo-barbata W. W. Sm. = Ceropegia dolichophylla Schltr. ■

84082 Ceropegia driophila C. K. Schneid.; 巴东吊灯花; Badong Ceropegia, Badong Pendentlamp ■

84083 Ceropegia dubia R. A. Dyer; 可疑吊灯花■☆

84084 Ceropegia effusa H. Huber = Riocreuxia torulosa (E. Mey.) Decne. ■☆

84085 Ceropegia ellenbeckii K. Schum.; 埃伦吊灯花■☆

84086 Ceropegia esculenta Edgew. = Ceropegia bulbosa Roxb. ■☆

84087 Ceropegia estelleana R. A. Dyer = Ceropegia fimbriata E. Mey. ■☆

84088 Ceropegia euryacme Schltr. = Ceropegia linearis E. Mey. subsp. woodii (Schltr.) H. Huber ■☆

84089 Ceropegia exigua (H. Huber) M. G. Gilbert et P. T. Li; 四川吊灯花; Sichuan Ceropegia ■

84090 Ceropegia filicalyx Bullock = Ceropegia abyssinica Decne. ■☆

84091 Ceropegia filiformis (Burch.) Schltr.; 丝形吊灯花■☆

84092 Ceropegia filipendula K. Schum.; 悬丝吊灯花■☆

84093 Ceropegia fimbriata E. Mey.; 流苏吊灯花■☆

84094 Ceropegia fimbriata E. Mey. subsp. connivens (R. A. Dyer) Bruyns; 靠合吊灯花■☆

84095 Ceropegia fimbriata E. Mey. subsp. geniculata (R. A. Dyer) Bruyns; 膝曲吊灯花■☆

84096 Ceropegia fimbriata Schltr. = Ceropegia sandersonii Decne. ex Hook. ■☆

84097 Ceropegia flanaganii (Schltr.) H. Huber; 弗拉纳根吊灯花■☆

84098 Ceropegia flanaganii (Schltr.) H. Huber = Riocreuxia flanaganii Schltr. ■☆

84099 Ceropegia flanaganii (Schltr.) H. Huber var. alexandrina H. Huber = Riocreuxia flanaganii Schltr. var. alexandrina (H. Huber) Masinde ■☆

84100 Ceropegia flanaganii (Schltr.) H. Huber var. fallax H. Huber = Riocreuxia woodii N. E. Br. ■☆

84101 Ceropegia floribunda N. E. Br.; 繁花吊灯花■☆

84102 Ceropegia furcata Werderm.; 叉分吊灯花■☆

84103 Ceropegia fusca Bolle; 棕色吊灯花■☆

84104 Ceropegia fusiformis N. E. Br.; 纺锤形吊灯花■☆

84105 Ceropegia galeata H. Huber; 盔形吊灯花■☆

84106 Ceropegia gemmifera K. Schum. = Ceropegia nilotica Kotschy var. simplex H. Huber ■☆

84107 Ceropegia geniculata R. A. Dyer = Ceropegia fimbriata E. Mey. subsp. geniculata (R. A. Dyer) Bruyns ■☆

84108 Ceropegia gilgiana Werderm.; 吉尔格吊灯花■☆

84109 Ceropegia gilletii De Wild. et T. Durand = Ceropegia abyssinica Decne. ■☆

84110 Ceropegia glabripedicellata De Wild. = Ceropegia racemosa N. E. Br. ■☆

84111 Ceropegia gossweileri S. Moore = Ceropegia nilotica Kotschy var. simplex H. Huber ■☆

84112 Ceropegia gourmacea A. Chev. = Ceropegia racemosa N. E. Br. ■☆

84113 Ceropegia grandis E. A. Bruce = Ceropegia nilotica Kotschy ■☆

84114 Ceropegia gymnopoda Schltr. = Brachystelma gymnopodum (Schltr.) Bruyns ■☆

84115 Ceropegia hastata N. E. Br. = Ceropegia linearis E. Mey. subsp. woodii (Schltr.) H. Huber ■☆

84116 Ceropegia haygarthii Schltr.; 哈氏吊灯花■☆

84117 Ceropegia helicoides E. A. Bruce et P. R. O. Bally = Ceropegia ballyana Bullock ■☆

84118 Ceropegia hians Svent. = Ceropegia dichotoma Haw. ■☆

84119 Ceropegia hians Svent. var. hians? = Ceropegia dichotoma Haw. ■☆

84120 Ceropegia hians Svent. var. striata? = Ceropegia dichotoma Haw. ■☆

84121 Ceropegia hirsuta Hochst. ex Decne. = Ceropegia abyssinica Decne. ■☆

84122 Ceropegia hispidipes S. Moore = Ceropegia abyssinica Decne. ■☆

84123 Ceropegia hochstetteri Chiov. = Ceropegia racemosa N. E. Br. ■☆

84124 Ceropegia hookeri C. B. Clarke ex Hook. f.; 匙冠吊灯花; Hooker's Ceropegia ■

84125 Ceropegia imbricata E. A. Bruce et P. R. O. Bally; 覆瓦吊灯花■☆

84126 Ceropegia infausta N. E. Br. = Ceropegia stenantha K. Schum. ■☆

84127 Ceropegia inflata Hochst. ex Chiov.; 膨胀吊灯花■☆

84128 Ceropegia infundibuliformis E. Mey. = Ceropegia filiformis (Burch.) Schltr. ■☆

84129 Ceropegia inornata P. R. O. Bally ex Masinde; 无饰吊灯花■☆

84130 Ceropegia insignis R. A. Dyer; 显著吊灯花■☆

84131 Ceropegia intracolor L. E. Newton = Ceropegia imbricata E. A. Bruce et P. R. O. Bally ■☆

84132 Ceropegia johnsonii N. E. Br.; 约翰斯顿吊灯花■☆

84133 Ceropegia jucunda Kerr = Ceropegia trichantha Hemsl. ■

84134 Ceropegia juncea Roxb.; 灯芯草吊灯花(吊灯花)■☆

84135 Ceropegia kamerunensis Schltr. = Ceropegia affinis Vatke ■☆

84136 Ceropegia kassneri S. Moore = Ceropegia purpurascens K. Schum. ■

84137 Ceropegia keniensis Masinde; 肯尼亚吊灯花■☆

84138 Ceropegia kerstingii K. Schum. = Ceropegia campanulata G. Don ■☆

84139 Ceropegia kroboensis N. E. Br. = Ceropegia nigra N. E. Br. ■☆

84140 Ceropegia kundelunguensis Malaisse; 昆德龙吊灯花■☆

84141 Ceropegia kwebensis N. E. Br. = Ceropegia purpurascens K. Schum. ■

84142 Ceropegia laikipiensis Masinde; 莱基皮吊灯花■☆

84143 Ceropegia lanceolata Wight et Arn. = Ceropegia longifolia Wall. ■

84144 Ceropegia ledermannii Schltr.; 莱德吊灯花■☆

84145 Ceropegia leptocarpa Schltr. = Ceropegia linearis E. Mey. subsp. woodii (Schltr.) H. Huber ■☆

84146 Ceropegia leptophylla Bruyns; 细叶吊灯花■☆

84147 Ceropegia leucotaenia K. Schum. = Ceropegia abyssinica Decne. ■☆

84148 Ceropegia lindenii Lavranos; 林登吊灯花■☆

84149 Ceropegia linearis E. Mey.; 线形吊灯花■☆

84150 Ceropegia linearis E. Mey. subsp. debilis (N. E. Br.) H. Huber; 弱小线形吊灯花■☆

84151 Ceropegia linearis E. Mey. subsp. tenuis (N. E. Br.) Bruyns; 细线形吊灯花■☆

84152 Ceropegia linearis E. Mey. subsp. woodii (Schltr.) H. Huber; 伍得吊灯花■☆

84153 Ceropegia linophyllum H. Huber; 塞内加尔吊灯花■☆

84154 Ceropegia longifolia Wall.; 长叶吊灯花; Longleaf Ceropegia ■

84155 Ceropegia longifolia Wall. subsp. exigua H. Huber = Ceropegia exigua (H. Huber) M. G. Gilbert et P. T. Li ■

84156 Ceropegia longifolia Wall. subsp. sinensis H. Huber = Ceropegia dolichophylla Schltr. ■

84157 Ceropegia loranthiflora K. Schum.; 革质吊灯花■☆

84158 Ceropegia lucida Wall. subsp. driophila (C. K. Schneid.) H. Huber = Ceropegia driophila C. K. Schneid. ■

84159 Ceropegia lugardae N. E. Br.; 卢格德吊灯花■☆

84160 Ceropegia lujai De Wild. = Ceropegia johnsonii N. E. Br. ■☆
84161 Ceropegia macrantha Wight;大花吊灯花■☆
84162 Ceropegia mafekingensis (N. E. Br.) R. A. Dyer.;马地吊灯花■☆
84163 Ceropegia mairei (H. Lév.) H. Huber;普吉藤(金雀马尾参,太子参);Maire Ceropegia, Maire Pendentlamp ■
84164 Ceropegia mairei (H. Lév.) H. Huber var. tenella H. Huber = Ceropegia mairei (H. Lév.) H. Huber ■
84165 Ceropegia manderensis Masinde;曼德拉吊灯花■☆
84166 Ceropegia mazoensis S. Moore = Ceropegia stenantha K. Schum. ■☆
84167 Ceropegia medoensis N. E. Br. = Ceropegia filipendula K. Schum. ■☆
84168 Ceropegia mendesii Stopp;门代斯吊灯花■☆
84169 Ceropegia meyeri Decne.;迈尔吊灯花■☆
84170 Ceropegia meyeri-johannis Engl.;迈尔约翰吊灯花■☆
84171 Ceropegia meyeri-johannis Engl. var. verdickii (De Wild.) Werderm.;韦尔吊灯花■☆
84172 Ceropegia micrantha Merr. = Ceropegia driophila C. K. Schneid. ■
84173 Ceropegia microgaster M. G. Gilbert;小囊吊灯花■☆
84174 Ceropegia mirabilis H. Huber;奇异吊灯花■☆
84175 Ceropegia monteiroae Hook. f. = Ceropegia sandersonii Decne. ex Hook. ■☆
84176 Ceropegia monticola W. W. Sm.; 白马吊灯花; Montane Ceropegia, Montane Pendentlamp ■
84177 Ceropegia moyalensis (H. Huber) M. G. Gilbert;莫亚莱吊灯花■☆
84178 Ceropegia mozambicensis Schltr. = Ceropegia nilotica Kotschy ■☆
84179 Ceropegia mozambicensis Schltr. var. ulugurensis Werderm. = Ceropegia nilotica Kotschy ■☆
84180 Ceropegia muliensis W. W. Sm.;木里吊灯花;Muli Ceropegia, Muli Pendentlamp ■
84181 Ceropegia multiflora Baker;多花吊灯花■☆
84182 Ceropegia multiflora Baker f. puberula (Hiern) H. Huber;微毛吊灯花■☆
84183 Ceropegia multiflora Baker f. tentaculata?;触角吊灯花■☆
84184 Ceropegia multiflora Baker subsp. tentaculata (N. E. Br.) H. Huber = Ceropegia multiflora Baker f. puberula (Hiern) H. Huber ■☆
84185 Ceropegia multiflora Baker var. latifolia N. E. Br. = Ceropegia multiflora Baker ■☆
84186 Ceropegia namaquensis Bruyns;纳马夸吊灯花■☆
84187 Ceropegia nigra N. E. Br.;黑吊灯花■☆
84188 Ceropegia nilotica Kotschy;尼罗河吊灯花■☆
84189 Ceropegia nilotica Kotschy var. plicata (E. A. Bruce) H. Huber;折扇吊灯花■☆
84190 Ceropegia nilotica Kotschy var. simplex H. Huber;简单吊灯花(多哥吊灯花);Togo Tangle ■☆
84191 Ceropegia nuda Hutch. et E. A. Bruce = Ceropegia subaphylla K. Schum. ■☆
84192 Ceropegia obscura N. E. Br. = Ceropegia pachystelma Schltr. ■☆
84193 Ceropegia occidentalis R. A. Dyer;西方吊灯花■☆
84194 Ceropegia occulta R. A. Dyer;隐蔽吊灯花■☆
84195 Ceropegia pachystelma Schltr.;粗冠吊灯花■☆
84196 Ceropegia paoshingensis Tsiang et P. T. Li;宝兴吊灯花;Baoxing Pendentlamp, Paoxing Ceropegia ■
84197 Ceropegia papillata N. E. Br.;乳突吊灯花■☆
84198 Ceropegia papillata N. E. Br. var. cordiloba (Werderm.) H. Huber;心裂吊灯花■
84199 Ceropegia patersoniae N. E. Br. = Ceropegia zeyheri Schltr. ■☆
84200 Ceropegia patriciae Rauh et Buchloh = Ceropegia mafekingensis (N. E. Br.) R. A. Dyer ■☆
84201 Ceropegia pedunculata Turrill = Ceropegia racemosa N. E. Br. ■☆
84202 Ceropegia perrottetii N. E. Br. = Ceropegia aristolochioides Decne. ■☆
84203 Ceropegia peteri Stopp ex Werderm.;彼得吊灯花■☆
84204 Ceropegia peulhorum A. Chev. var. breviloba H. Huber;浅裂吊灯花■
84205 Ceropegia picta (Schltr.) H. Huber = Riocreuxia picta Schltr. ■☆
84206 Ceropegia plicata E. A. Bruce = Ceropegia nilotica Kotschy ■☆
84207 Ceropegia porphyrotricha W. W. Sm. = Ceropegia campanulata G. Don var. porphyrotricha (W. W. Sm.) H. Huber ■☆
84208 Ceropegia praetermissa J. Raynal;疏忽吊灯花■☆
84209 Ceropegia profundorum Hand.-Mazz. = Ceropegia dolichophylla Schltr. ■
84210 Ceropegia pubescens E. Mey. = Ceropegia meyeri Decne. ■☆
84211 Ceropegia pubescens Wall.;西藏吊灯花(底线参,对叶林,柔毛吊灯花,莪参);Pubescent Ceropegia, Pubescent Pendentlamp ■
84212 Ceropegia purpurascens K. Schum.;浅紫裂吊灯花■
84213 Ceropegia purpurascens K. Schum. subsp. thysanotos (Werderm.) H. Huber;流苏浅紫裂吊灯花■
84214 Ceropegia quarrei De Wild. = Ceropegia stenantha K. Schum. ■☆
84215 Ceropegia racemosa N. E. Br.;多枝吊灯花■☆
84216 Ceropegia racemosa N. E. Br. subsp. secamonoides (S. Moore) H. Huber;鲫鱼藤吊灯花■☆
84217 Ceropegia racemosa N. E. Br. subsp. setifera (Schltr.) H. Huber = Ceropegia carnosa E. Mey. ■☆
84218 Ceropegia radicans Schltr.;具根吊灯花■☆
84219 Ceropegia radicans Schltr. subsp. smithii (M. R. Hend.) R. A. Dyer;史密斯具根吊灯花■☆
84220 Ceropegia radicans Schltr. var. smithii (M. R. Hend.) H. Huber = Ceropegia radicans Schltr. subsp. smithii (M. R. Hend.) R. A. Dyer ■☆
84221 Ceropegia rara S. Moore = Ceropegia vanderystii De Wild. ■☆
84222 Ceropegia recurvata M. G. Gilbert;反折吊灯花■☆
84223 Ceropegia rendallii N. E. Br.;伦德尔吊灯花■☆
84224 Ceropegia renzii Stopp;伦兹吊灯花■☆
84225 Ceropegia rhynchantha Schltr.;喙状吊灯花■☆
84226 Ceropegia ringens A. Rich.;张开吊灯花■☆
84227 Ceropegia ringens Vatke = Ceropegia convolvuloides A. Rich. ■☆
84228 Ceropegia ringoetii De Wild.;林戈吊灯花■☆
84229 Ceropegia robynsiana Werderm. = Ceropegia albisepta Jum. et H. Perrier var. robynsiana (Werderm.) H. Huber ■☆
84230 Ceropegia rostrata E. A. Bruce = Ceropegia umbraticola K. Schum. ■☆
84231 Ceropegia rudatisii Schltr.;鲁达蒂斯吊灯花■☆
84232 Ceropegia ruspoliana K. Schum. = Ceropegia affinis Vatke ■☆
84233 Ceropegia sagittata L. = Microloma sagittatum (L.) R. Br. ■☆
84234 Ceropegia salicifolia H. Huber; 柳叶吊灯花; Willowleaf Ceropegia, Willowleaved Pendentlamp ■
84235 Ceropegia sandersonii Decne. ex Hook.;醉龙吊灯花(醉龙);Fountain Flower, Parachute Plant ■☆
84236 Ceropegia sankuruensis Schltr.;桑库鲁吊灯花■☆
84237 Ceropegia saxatilis S. Moore = Ceropegia bonafouxii K. Schum. ■☆
84238 Ceropegia scabriflora N. E. Br.;糙叶吊灯花■☆
84239 Ceropegia scandens N. E. Br. = Ceropegia volubilis N. E. Br. ■☆
84240 Ceropegia schinziana Bullock = Ceropegia pachystelma Schltr. ■☆
84241 Ceropegia schlechteriana Werderm. = Ceropegia ringoetii De

Wild. ■☆
84242 Ceropegia schliebenii Markgr. = Ceropegia stenoloba Hochst. ex Chiov. ■☆
84243 Ceropegia schoenlandii N. E. Br. = Ceropegia linearis E. Mey. subsp. woodii (Schltr.) H. Huber ■☆
84244 Ceropegia secamonoides S. Moore = Ceropegia racemosa N. E. Br. subsp. secamonoides (S. Moore) H. Huber ■☆
84245 Ceropegia senegalensis H. Huber = Ceropegia linophyllum H. Huber ■☆
84246 Ceropegia serpentina E. A. Bruce = Ceropegia stapeliiformis Haw. subsp. serpentina (E. A. Bruce) R. A. Dyer ■☆
84247 Ceropegia seticorona E. A. Bruce;毛冠吊灯花■☆
84248 Ceropegia seticorona E. A. Bruce = Ceropegia aristolochioides Decne. ■☆
84249 Ceropegia seticorona E. A. Bruce var. dilatiloba P. R. O. Bally = Ceropegia aristolochioides Decne. ■☆
84250 Ceropegia setifera Schltr. = Ceropegia carnosa E. Mey. ■☆
84251 Ceropegia setifera Schltr. var. natalensis N. E. Br. = Ceropegia affinis Vatke ■☆
84252 Ceropegia siamensis Kerr. = Ceropegia driophila C. K. Schneid. ■
84253 Ceropegia siamensis Kerr. = Ceropegia monticola W. W. Sm. ■
84254 Ceropegia sinoerecta M. G. Gilbert et P. T. Li;鹤庆吊灯花;Heqing Ceropegia ■
84255 Ceropegia sinuata Decne. ex A. Rich. = Ceropegia ringens A. Rich. ■☆
84256 Ceropegia smithii M. R. Hend. = Ceropegia radicans Schltr. subsp. smithii (M. R. Hend.) R. A. Dyer ■☆
84257 Ceropegia somalensis Chiov.;索马里吊灯花■☆
84258 Ceropegia somalensis Chiov. f. erostrata H. Huber = Ceropegia somalensis Chiov. ■☆
84259 Ceropegia sootepensis Craib;河坝吊灯花(河坝口吊灯花);Hebakou Ceropegia,Soótep Pendentlamp ■
84260 Ceropegia sororia Harv. ex Hook. f. = Ceropegia bowkeri Harv. subsp. sororia (Harv. ex Hook. f.) R. A. Dyer ■☆
84261 Ceropegia speciosa H. Huber;美丽吊灯花■☆
84262 Ceropegia splendida (K. Schum.) H. Huber = Riocreuxia splendida K. Schum. ■☆
84263 Ceropegia stapeliiformis Haw.;拟犀角吊金钱(薄云);Carrion-flower Form Ceropegia ■
84264 Ceropegia stapeliiformis Haw. subsp. serpentina (E. A. Bruce) R. A. Dyer;蛇形吊灯花■☆
84265 Ceropegia stenantha K. Schum.;窄花吊灯花■☆
84266 Ceropegia stenantha K. Schum. var. parviflora N. E. Br. = Ceropegia stenantha K. Schum. ■☆
84267 Ceropegia stenoloba Hochst. ex Chiov.;窄裂吊灯花■☆
84268 Ceropegia stenoloba Hochst. ex Chiov. var. australis H. Huber = Ceropegia stenoloba Hochst. ex Chiov. ■☆
84269 Ceropegia stenophylla C. K. Schneid.;狭叶吊灯花;Narrowleaf Ceropegia,Narrowleaved Pendentlamp ■
84270 Ceropegia steudneri Vatke = Ceropegia abyssinica Decne. ■☆
84271 Ceropegia steudneriana K. Schum. = Ceropegia abyssinica Decne. ■☆
84272 Ceropegia subaphylla K. Schum.;近无叶吊灯花■☆
84273 Ceropegia succulenta E. A. Bruce = Ceropegia albisepta Jum. et H. Perrier ■☆
84274 Ceropegia swaziorum D. V. Field;斯威士吊灯花■☆
84275 Ceropegia talbotii S. Moore;塔尔博特吊灯花■☆
84276 Ceropegia teniana Hand.-Mazz.;马鞍山吊灯花(盐丰吊灯花);Maanshan Ceropegia,Ten Pendentlamp ■
84277 Ceropegia tentaculata N. E. Br. = Ceropegia multiflora Baker f. tentaculata? ■☆
84278 Ceropegia tentaculata N. E. Br. var. puberula Hiern = Ceropegia multiflora Baker f. puberula (Hiern) H. Huber ■☆
84279 Ceropegia tenuifolia L. = Microloma tenuifolium (L.) K. Schum. ■☆
84280 Ceropegia tenuis N. E. Br. = Ceropegia linearis E. Mey. subsp. tenuis (N. E. Br.) Bruyns ■☆
84281 Ceropegia tenuissima S. Moore = Ceropegia stenantha K. Schum. ■☆
84282 Ceropegia thorncroftii N. E. Br. = Ceropegia crassifolia Schltr. ■☆
84283 Ceropegia thysanotos Werderm. = Ceropegia purpurascens K. Schum. subsp. thysanotos (Werderm.) H. Huber ■
84284 Ceropegia tomentosa Schltr.;绒毛吊灯花■☆
84285 Ceropegia torulosa E. Mey. = Riocreuxia torulosa (E. Mey.) Decne. ■☆
84286 Ceropegia tourana A. Chev.;土伦吊灯花■☆
84287 Ceropegia trichantha Hemsl.;毛花吊灯花(吊灯花,狭瓣吊灯花);Common Ceropegia,Common Pendentlamp ■
84288 Ceropegia tristis Hutch. = Ceropegia haygarthii Schltr. ■☆
84289 Ceropegia tsaiana Tsiang = Ceropegia pubescens Wall. ■
84290 Ceropegia tsaiana Tsiang = Ceropegia trichantha Hemsl. ■
84291 Ceropegia tuberculata Dinter = Ceropegia crassifolia Schltr. ■☆
84292 Ceropegia tuberosa Dalzell et Gibson = Ceropegia bulbosa Roxb. ■☆
84293 Ceropegia umbraticola K. Schum.;荫蔽吊灯花■☆
84294 Ceropegia undulata N. E. Br. = Ceropegia pachystelma Schltr. ■☆
84295 Ceropegia vaduliae Lavranos;瓦杜尔吊灯花■☆
84296 Ceropegia vanderystii De Wild.;范德吊灯花■☆
84297 Ceropegia variegata (Forssk.) Decne.;斑叶吊灯花■☆
84298 Ceropegia variegata (Forssk.) Decne. var. cornigera H. Huber = Ceropegia variegata (Forssk.) Decne. ■☆
84299 Ceropegia verdickii De Wild. = Ceropegia meyeri-johannis Engl. var. verdickii (De Wild.) Werderm. ■☆
84300 Ceropegia verruculosa (R. A. Dyer) D. V. Field = Ceropegia lugardae N. E. Br. ■☆
84301 Ceropegia verticillata Masinde;轮生吊灯花■☆
84302 Ceropegia volubilis N. E. Br.;缠绕吊灯花■☆
84303 Ceropegia volubilis N. E. Br. var. crassicaulis Huber = Ceropegia aristolochioides Decne. ■☆
84304 Ceropegia wellmannii N. E. Br. = Ceropegia umbraticola K. Schum. ■☆
84305 Ceropegia woodii Schltr.;吊金钱(吊灯花,可爱藤,天邪鬼);Heart Vine,Hearts-entangled,Rosary Vine,String of Hearts,String-of-hearts,Wood Ceropegia,Woods Ceropegia ■
84306 Ceropegia woodii Schltr. = Ceropegia linearis E. Mey. subsp. woodii (Schltr.) H. Huber ■☆
84307 Ceropegia yorubana Schltr.;约鲁巴吊灯花■☆
84308 Ceropegia yunnanensis Schltr. et Hand.-Mazz. = Ceropegia monticola W. W. Sm. ■
84309 Ceropegia zambesiaca Masinde et Meve;赞比西吊灯花■☆
84310 Ceropegia zeyheri Schltr.;泽赫吊灯花■☆
84311 Cerophora Raf. = Myrica L. ●
84312 Cerophora lanceolata Raf. = Myrica cerifera L. ●☆
84313 Cerophyllum Spach = Ribes L. ●

84314 Cerothamnus Tidestr. = Myrica L. ●
84315 Cerothamnus arborescens (Castigl.) Tidestr. = Myrica cerifera L. ●☆
84316 Cerothamnus carolinensis (Mill.) Tidestr. = Myrica heterophylla Raf. ●☆
84317 Cerothamnus ceriferus (L.) Small = Myrica cerifera L. ●☆
84318 Cerothamnus inodorus (W. Bartram) Small = Myrica inodora W. Bartram ●☆
84319 Cerothamnus pensylvanica (Mirb.) Moldenke = Myrica pensylvanica Mirb. ●☆
84320 Cerothamnus pumilus (Michx.) Small = Myrica cerifera L. ●☆
84321 Ceroxylaceae O. F. Cook = Arecaceae Bercht. et J. Presl(保留科名)●
84322 Ceroxylaceae O. F. Cook = Palmae Juss.(保留科名)●
84323 Ceroxylaceae Vines = Arecaceae Bercht. et J. Presl(保留科名)●
84324 Ceroxylaceae Vines = Palmae Juss.(保留科名)●
84325 Ceroxylon Bonpl. = Ceroxylon Bonpl. ex DC. ●☆
84326 Ceroxylon Bonpl. ex DC.(1804);蜡棕属(安地斯蜡椰子属,腊棕属,蜡材榈属,蜡椰属,蜡椰子属);Andean Wax Palm,Wax Palm ●☆
84327 Ceroxylon Humb. et Bonpl. = Ceroxylon Bonpl. ex DC. ●☆
84328 Ceroxylon alpinum Steud.;山蜡棕(高山腊椰子,腊棕);Colombian Wax Palm,Wax Palm ●☆
84329 Ceroxylon andicola Humb. et Bonpl.;蜡棕;Wax Palm ●☆
84330 Ceroxylon andicola Humb. et Bonpl. = Ceroxylon alpinum Steud. ●☆
84331 Ceroxylon quindinsnse (H. Karst.) H. Wendl.;哥伦比亚蜡棕●☆
84332 Ceroxylon utile (H. Karst.) H. Wendl.;厄瓜多尔蜡棕●☆
84333 Cerqueiria O. Berg = Gomidesia O. Berg ●☆
84334 Cerqueiria O. Berg = Myrcia DC. ex Guill. ●☆
84335 Cerquieria Benth. et Hook. f. = Cerqueiria O. Berg ●☆
84336 Cerraria Tausch = Cervaria L. ■☆
84337 Cerraria Tausch = Peucedanum L. ■
84338 Cerris Raf. = Quercus L. ●
84339 Cerseidos Siebold et Zucc. = Prunus L. ●
84340 Certoides arborescens (Losinsk.) C. P. Tsien et C. G. Ma = Krascheninnikovia arborescens (Losinsk.) Czerep. ●
84341 Certoides compacta (Losinsk.) C. P. Tsien et C. G. Ma = Krascheninnikovia compacta (Losinsk.) Grubov ●
84342 Ceruana Forssk.(1775);草基黄属■☆
84343 Ceruana pratensis Forssk.;草基黄■☆
84344 Ceruana rotundifolia Cass. = Ceruana pratensis Forssk. ■☆
84345 Ceruana schimperi Boiss. = Asteriscus graveolens (Forssk.) Less. ■☆
84346 Ceruana senegalensis DC. = Ceruana pratensis Forssk. ■☆
84347 Ceruchis Gaertn. ex Schreb. = Spilanthes Jacq. ■
84348 Cervantesia Ruiz et Pav.(1794);塞檀香属●☆
84349 Cervantesia bicolor Cav.;二色塞檀香●☆
84350 Cervantesia glabrata Stapf;无毛塞檀香●☆
84351 Cervantesia macrocarpa Cuatrec.;大果塞檀香●☆
84352 Cervantesia tomentosa Ruiz et Pav.;毛塞檀香●☆
84353 Cervantesiaceae Nickrent et Der;塞檀香科●☆
84354 Cervaria L. = Ortegia L. ■☆
84355 Cervaria Wolf = Libanotis Haller ex Zinn(保留属名)■
84356 Cervaria Wolf = Peucedanum L. ■
84357 Cervaria Wolf(1781);鹿芹属■☆
84358 Cervaria angustifolia Andrz. ex Trautv.;窄鹿芹■☆
84359 Cervaria glauca Gaudin;灰鹿芹■☆
84360 Cervaria laevis Gaudin;平滑鹿芹■☆
84361 Cervaria latifolia Andrz. ex Trautv.;宽叶鹿芹■☆
84362 Cervia Rodr. ex Lag. = Rochelia Rchb.(保留属名)■
84363 Cervia disperma (L. f.) Hayek = Rochelia disperma (L. f.) Hochr. ■
84364 Cervicina Delile(废弃属名) = Wahlenbergia Schrad. ex Roth(保留属名)■●
84365 Cervicina campanuloides Delile = Wahlenbergia campanuloides (Delile) Vatke ■☆
84366 Cervispina Ludw. = Rhamnus L. ●
84367 Cervispina Moench = Rhamnus L. ●
84368 Cerynella DC. = Poitea Vent. ●☆
84369 Cesatia Endl. = Trachymene Rudge ■☆
84370 Cesdelia DC. ex Raf. = Ammannia L. ■
84371 Cespa Hill = Eriocaulon L. ■
84372 Cespa aquatica Hill = Eriocaulon aquaticum (Hill) Druce ■☆
84373 Cespedesia Goudot(1844);同萼树属●☆
84374 Cespedesia bonplandi Goudot;同萼树●☆
84375 Cespedesia discolor Bull.;异色同萼树●☆
84376 Cespedesia spathulata (Ruiz et Pav.) Planch.;匙形同萼树●☆
84377 Cestichis Pfitzer = Liparis Rich.(保留属名)■
84378 Cestichis Thouars = Liparis Rich.(保留属名)■
84379 Cestichis Thouars ex Pfitzer = Liparis Rich.(保留属名)■
84380 Cestichis caespitosa (Thouars) Ames = Liparis caespitosa (Thouars) Lindl. ■
84381 Cestichis cespitosa (Thouars) Ames = Liparis cespitosa (Thouars) Lindl. ■
84382 Cestichis condylobulbon (Rchb. f.) M. A. Clem. et D. L. Jones = Liparis condylobulbon Rchb. f. ■
84383 Cestichis dolichopoda Hayata = Liparis condylobulbon Rchb. f. ■
84384 Cestichis hensoaensis (Kudo) Maek. = Liparis hensoaensis Kudo ■
84385 Cestichis hensoaensis Kudo = Liparis hensoaensis Kudo ■
84386 Cestichis kawakamii (Hayata) Maek. = Liparis nakaharai Hayata ■
84387 Cestichis latifolia (Lindl.) Pfitzer = Liparis latifolia (Blume) Lindl. ■
84388 Cestichis laurisilvatica (Fukuy.) Maek. = Liparis laurisilvatica Fukuy. ■
84389 Cestichis longipes (Lindl.) Ames = Liparis viridiflora (Blume) Lindl. ■
84390 Cestichis nakaharai (Hayata) Kudo = Liparis distans C. B. Clarke ■
84391 Cestichis nakaharai (Hayata) Kudo = Liparis nakaharai Hayata ■
84392 Cestichis nokoensis (Fukuy.) Maek. = Liparis nakaharai Hayata ■
84393 Cestichis platybulba (Hayata) Kudo = Liparis elliptica Wight ■
84394 Cestichis plicata (Franch. et Sav.) Maek. = Liparis bootanensis Griff. ■
84395 Cestichis taiwaniana (Hayata) Nakai = Liparis distans C. B. Clarke ■
84396 Cestichis taiwaniana (Hayata) Nakai = Liparis nakaharai Hayata ■
84397 Cestraceae Schltdl. = Solanaceae Juss.(保留科名)●■
84398 Cestrinus Cass. = Centaurea L.(保留属名)●■
84399 Cestron St. -Lag. = Cestrum L. ●
84400 Cestrum L.(1753);夜香树属(夜香花属,夜香木属);Cestrum,Jessamine,Red Cestrum ●
84401 Cestrum alternifolium (Jacq.) O. E. Schulz;互叶夜香树●☆
84402 Cestrum atrovirens Dunal;墨绿夜香树●☆

84403 Cestrum aurantiacum F. K. Mayer = Freylinia lanceolata (L. f.) G. Don ●☆

84404 Cestrum aurantiacum Lindl.;黄花夜香树(黄丁子,黄花洋素馨,黄瓶子花,黄瓶子树);Grange Cestrum,Orange Cestrum,Orange Jessamine ●

84405 Cestrum corymbosum Schltdl. = Cestrum endlicheri Miers ●☆

84406 Cestrum cultum Francey;紫花夜香树;Purple Cestrum ●☆

84407 Cestrum depauperatum Dunal = Cestrum alternifolium (Jacq.) O. E. Schulz ●☆

84408 Cestrum diurnum L.;墨水夜香树(昼花夜香树);Day Cestrum,Day Jessamine,Day-blooming Jessamine,Inkberry,Jessamine ●☆

84409 Cestrum elegans (Brongn.) Schltdl.;毛茎夜香树(美丽夜香树,瓶儿花);Bastard Jasmine,Red Cestrum ●

84410 Cestrum elegans (Brongn.) Schltdl. 'Exotica';美丽夜来香●☆

84411 Cestrum endlicheri Miers;恩氏夜香树●☆

84412 Cestrum fasciculatum Miers;早花夜香树;Early Jessamine,Newell Cestrum,Red Night-blooming Jasmine ●☆

84413 Cestrum fasciculatum Miers var. newellii Bailey;瓶子花;Newell Fascicled Cestrum ●

84414 Cestrum jamaicense Lam. = Cestrum alternifolium (Jacq.) O. E. Schulz ●☆

84415 Cestrum laevigatum Schltdl.;光滑夜香树●☆

84416 Cestrum lycioides Roem. et Schult. = Acokanthera lycioides (Roem. et Schult.) G. Don ●☆

84417 Cestrum nocturnum L.;夜香树(素清花,洋素馨,夜丁香,夜来香,夜香花);Lady of the Night,Lady-of-the-night,Night Cestrum,Night Jassamine,Nightblooming Cestrum,Night-blooming Cestrum,Nightblooming Jassamine,Night-blooming Jassamine,Night-scented Jassamine ●

84418 Cestrum oppositifolium Lam. = Acokanthera oppositifolia (Lam.) Codd ●

84419 Cestrum parqui L'Hér.;帕克夜香树(柳叶夜香木,智利夜来香,智利夜香树);Chilean Jessamine,Green Cestrum,Green Jessamine,Parque Cestrum,Willow-leaved Jessamine,Willowleaved-jassamine,Yellow Syringa ●☆

84420 Cestrum parviflorum Dunal;小花夜香树●☆

84421 Cestrum pauciflorum Willd. ex Roem. et Schult.;疏花夜香树●☆

84422 Cestrum pedunculatum Sessé et Moc.;垂枝夜香树●☆

84423 Cestrum peruvianum Willd. ex Roem. et Schult.;秘鲁夜香树●☆

84424 Cestrum pubescens Roem. et Schult. = Acokanthera pubescens (Roem. et Schult.) G. Don ●☆

84425 Cestrum purpureum (Lindl.) Standl.;紫色夜香树(瓶子花,紫瓶子花);Purple Cestrum ●

84426 Cestrum purpureum (Lindl.) Standl. = Cestrum elegans (Brongn.) Schltdl. ●

84427 Cestrum vespertinum L. = Cestrum alternifolium (Jacq.) O. E. Schulz ●☆

84428 Cetra Noronha = Syzygium R. Br. ex Gaertn. (保留属名)●

84429 Ceuthocarpus Aiello(1979);古巴隐果茜属●☆

84430 Ceuthocarpus involucratus (Wernham) Aiello;古巴隐果茜●☆

84431 Ceuthostoma L. A. S. Johnson(1988);隐口木麻黄属(新几内亚木麻黄属)●☆

84432 Ceuthostoma palawanense L. A. S. Johnson;隐口木麻黄●☆

84433 Cevallia Lag. (1805);墨西哥刺莲花属●☆

84434 Cevallia sinuata Lag.;墨西哥刺莲花(深波);Stinging Cevallia ●☆

84435 Cevalliaceae Griseb. = Loasaceae Juss. (保留科名)●■☆

84436 Ceytosis Munro = Crypsis Aiton(保留属名)■

84437 Chaboissaea Benth. et Hook f. = Muhlenbergia Schreb. ■

84438 Chaboissaea E. Fourn. = Muhlenbergia Schreb. ■

84439 Chaboissaea E. Fourn. ex Benth. et Hook f. = Muhlenbergia Schreb. ■

84440 Chabraea Adans. = Peplis L. ■

84441 Chabraea Bubani = Lythrum L. ●■

84442 Chabraea DC. = Leucheria Lag. ■☆

84443 Chabrea Raf. (1840);沙布尔芹属■☆

84444 Chabrea Raf. = Peucedanum L. ■

84445 Chabrea carvifolia (Crantz ex Jacq.) Raf.;沙布尔芹■☆

84446 Chacaya Escal. = Discaria Hook. ●☆

84447 Chacaya Escal. = Ochetophila Poepp. ex Reissek ●☆

84448 Chacoa R. M. King et H. Rob. (1975);腺瓣亮泽兰属●☆

84449 Chacoa pseudoprasiifolia (Hassl.) R. M. King et H. Rob.;腺瓣亮泽兰■☆

84450 Chadara Forssk. = Grewia L. ●

84451 Chadara arborea Forssk. = Grewia arborea (Forssk.) Lam. ●☆

84452 Chadara betulifolia Juss. = Grewia tenax (Forssk.) Fiori ●☆

84453 Chadara erythraea Schweinf. = Grewia tenax (Forssk.) Fiori ●☆

84454 Chadara tenax Forssk. = Grewia tenax (Forssk.) Fiori ●☆

84455 Chadara velutina Forssk. = Grewia velutina (Forssk.) Lam. ●☆

84456 Chadra T. Anderson = Chadara Forssk. ●

84457 Chadsia Bojer(1842);灌木查豆属●☆

84458 Chadsia coluteifolia Baill.;膀胱豆灌木查豆●☆

84459 Chadsia flammea Bojer;焰红灌木查豆●☆

84460 Chadsia flammea Bojer subsp. acutidentata Du Puy et Labat;尖齿灌木查豆●☆

84461 Chadsia flammea Bojer subsp. parviflora Du Puy et Labat;小花灌木查豆●☆

84462 Chadsia grandidieri Baill. = Sylvichadsia grandidieri (Baill.) Du Puy et Labat ●☆

84463 Chadsia grandifolia R. Vig. = Sylvichadsia grandifolia (R. Vig.) Du Puy et Labat ●☆

84464 Chadsia granitica Baill. = Chadsia versicolor Bojer ●☆

84465 Chadsia grevei Drake;格雷弗灌木查豆●☆

84466 Chadsia irodoensis Du Puy et Labat;伊鲁杜灌木查豆●☆

84467 Chadsia jullyana Dubard et Dop = Chadsia flammea Bojer ●☆

84468 Chadsia lantziana Baill. = Strongylodon madagascariensis Baker ●☆

84469 Chadsia longidentata R. Vig.;长齿灌木查豆●☆

84470 Chadsia magnifica R. Vig.;华丽灌木查豆●☆

84471 Chadsia majungensis Drake = Chadsia flammea Bojer ●☆

84472 Chadsia perrieri Dubard et Dop = Chadsia flammea Bojer ●☆

84473 Chadsia racemosa Drake;总花灌木查豆●☆

84474 Chadsia salicina Baill.;柳叶灌木查豆●☆

84475 Chadsia versicolor Bojer;异色灌木查豆●☆

84476 Chaelanthus Poir. = Chaetanthus R. Br. ■☆

84477 Chaelothilus Beck. = Gentiana L. ■

84478 Chaenactis DC. (1836);针垫菊属;Pincushion ■●☆

84479 Chaenactis alpigena Sharsm.;南方山地针垫菊;Mountain Pincushion,Sharsmith Pincushion,Southern Pincushion Sierra ■☆

84480 Chaenactis alpina (A. Gray) M. E. Jones = Chaenactis douglasii (Hook.) Hook. et Arn. var. alpina A. Gray ■☆

84481 Chaenactis alpina (A. Gray) M. E. Jones var. leucopsis (Greene) Stockw. = Chaenactis douglasii (Hook.) Hook. et Arn. var. alpina A. Gray ■☆

84482 Chaenactis alpina (A. Gray) M. E. Jones var. rubella (Greene) Stockw. = Chaenactis douglasii (Hook.) Hook. et Arn. var. alpina A. Gray ■☆

84483 Chaenactis angustifolia Greene = Chaenactis douglasii (Hook.) Hook. et Arn. ■☆

84484 Chaenactis artemisiifolia (Harv. et A. Gray) A. Gray;白针垫菊;White Pincushion ■☆

84485 Chaenactis carphoclinia A. Gray var. attenuata (A. Gray) M. E. Jones = Chaenactis carphoclinia A. Gray ■☆

84486 Chaenactis carphoclinia A. Gray var. peirsonii (Jeps.) Munz;皮尔逊针垫菊;Peirson Pincushion ■☆

84487 Chaenactis cusickii A. Gray;库西克针垫菊;Cusick's Pincushion, Morning Brides ■☆

84488 Chaenactis douglasii (Hook.) Hook. et Arn.;道格拉斯针垫菊;Douglas' Dustymaiden, False Yarrow, Hoary Pincushion ■☆

84489 Chaenactis douglasii (Hook.) Hook. et Arn. var. achilleifolia (Hook. et Arn.) A. Gray = Chaenactis douglasii (Hook.) Hook. et Arn. ■☆

84490 Chaenactis douglasii (Hook.) Hook. et Arn. var. alpina A. Gray;高山针垫菊;Alpine Dustymaidens, Alpine Pincushion ■☆

84491 Chaenactis douglasii (Hook.) Hook. et Arn. var. glandulosa Cronquist = Chaenactis douglasii (Hook.) Hook. et Arn. ■☆

84492 Chaenactis douglasii (Hook.) Hook. et Arn. var. montana M. E. Jones = Chaenactis douglasii (Hook.) Hook. et Arn. ■☆

84493 Chaenactis douglasii (Hook.) Hook. et Arn. var. rubricaulis (Rydb.) Ferris = Chaenactis douglasii (Hook.) Hook. et Arn. ■☆

84494 Chaenactis evermannii Greene;埃弗曼针垫菊;Evermann's Pincushion ■☆

84495 Chaenactis fremontii A. Gray;沙漠针垫菊;Desert Pincushion, Fremont Pincushion, Pincushion Flower ■☆

84496 Chaenactis furcata Stockw. = Chaenactis stevioides Hook. et Arn. ■☆

84497 Chaenactis gillespiei Stockw. = Chaenactis stevioides Hook. et Arn. ■☆

84498 Chaenactis glabriuscula DC.;黄针垫菊;Yellow Pincushion ■☆

84499 Chaenactis glabriuscula DC. var. curta (A. Gray) Jeps. = Chaenactis glabriuscula DC. ■☆

84500 Chaenactis glabriuscula DC. var. denudata (Nutt.) Munz = Chaenactis glabriuscula DC. var. lanosa (DC.) H. M. Hall ■☆

84501 Chaenactis glabriuscula DC. var. gracilenta (Greene) D. D. Keck = Chaenactis glabriuscula DC. var. heterocarpha (Torr. et A. Gray) H. M. Hall ■☆

84502 Chaenactis glabriuscula DC. var. heterocarpha (Torr. et A. Gray) H. M. Hall;异果针垫菊;Inner Coast Range Chaenactis ■☆

84503 Chaenactis glabriuscula DC. var. lanosa (DC.) H. M. Hall;南海岸针垫菊;Sand buttons, South Coast Ranges pincushion ■☆

84504 Chaenactis glabriuscula DC. var. megacephala A. Gray;大头针垫菊;Yellow Pincushion ■☆

84505 Chaenactis glabriuscula DC. var. orcuttiana (Greene) H. M. Hall;奥克特针垫菊;Orcutt's Pincushion ■☆

84506 Chaenactis glabriuscula DC. var. tenuifolia (Nutt.) H. M. Hall = Chaenactis glabriuscula DC. ■☆

84507 Chaenactis heterocarpha Torr. et A. Gray = Chaenactis glabriuscula DC. var. heterocarpha (Torr. et A. Gray) H. M. Hall ■☆

84508 Chaenactis heterocarpha Torr. et A. Gray var. tanacetifolia (A. Gray) A. Gray = Chaenactis glabriuscula DC. var. heterocarpha (Torr. et A. Gray) H. M. Hall ■☆

84509 Chaenactis lanosa DC. = Chaenactis glabriuscula DC. var. lanosa (DC.) H. M. Hall ■☆

84510 Chaenactis latifolia Stockw. = Chaenactis stevioides Hook. et Arn. ■☆

84511 Chaenactis macrantha D. C. Eaton;光亮针垫菊;Bighead Dustymaidens, Mojave Pincushion, Showy Dustymaidens ■☆

84512 Chaenactis mexicana Stockw. = Chaenactis stevioides Hook. et Arn. ■☆

84513 Chaenactis nevadensis (Kellogg) A. Gray;内华达针垫菊;Nevada Dustymaidens, Sierra Pincushion ■☆

84514 Chaenactis nevadensis (Kellogg) A. Gray var. mainsiana (A. Nelson et J. F. Macbr.) Stockw. = Chaenactis evermannii Greene ■☆

84515 Chaenactis nevii A. Gray;内维针垫菊;John Day Ppincushion ■☆

84516 Chaenactis panamintensis Stockw. = Chaenactis douglasii (Hook.) Hook. et Arn. var. alpina A. Gray ■☆

84517 Chaenactis parishii A. Gray;帕里什针垫菊;Parish Chaenactis ●☆

84518 Chaenactis pedicularia Greene = Chaenactis douglasii (Hook.) Hook. et Arn. ■☆

84519 Chaenactis peirsonii Jeps. = Chaenactis carphoclinia A. Gray var. peirsonii (Jeps.) Munz ■☆

84520 Chaenactis pumila Greene = Chaenactis douglasii (Hook.) Hook. et Arn. ■☆

84521 Chaenactis ramosa Stockw. = Chaenactis douglasii (Hook.) Hook. et Arn. ■☆

84522 Chaenactis santolinoides Greene;圣麻针垫菊;Santolina Pincushion ■☆

84523 Chaenactis scaposa Eastw. = Chamaechaenactis scaposa (Eastw.) Rydb. ■☆

84524 Chaenactis stevioides Hook. et Arn.;针垫菊;Broad-flower Pincushion, Desert Pincushion, Esteve Pincushion, Esteve's Pincushion, False Yarrow ■☆

84525 Chaenactis stevioides Hook. et Arn. var. brachypappa (A. Gray) H. M. Hall = Chaenactis stevioides Hook. et Arn. ■☆

84526 Chaenactis stevioides Hook. et Arn. var. thornberi Stockw. = Chaenactis stevioides Hook. et Arn. ■☆

84527 Chaenactis suffrutescens A. Gray;灌木针垫菊;Shasta Pincushion ●■☆

84528 Chaenactis suffrutescens A. Gray var. incana Stockw. = Chaenactis suffrutescens A. Gray ●■☆

84529 Chaenactis tanacetifolia A. Gray = Chaenactis glabriuscula DC. var. heterocarpha (Torr. et A. Gray) H. M. Hall ■☆

84530 Chaenactis tanacetifolia A. Gray var. gracilenta (Greene) Stockw. = Chaenactis glabriuscula DC. var. heterocarpha (Torr. et A. Gray) H. M. Hall ■☆

84531 Chaenactis tenuifolia Nutt. = Chaenactis glabriuscula DC. ■☆

84532 Chaenactis tenuifolia Nutt. var. orcuttiana Greene = Chaenactis glabriuscula DC. var. orcuttiana (Greene) H. M. Hall ■☆

84533 Chaenactis thompsonii Cronquist;汤普森针垫菊;Thompson's Pincushion ■☆

84534 Chaenactis thysanocarpha A. Gray = Orochaenactis thysanocarpha (A. Gray) Coville ■☆

84535 Chaenactis xantiana A. Gray;肉色针垫菊;Fleshcolor Pincushion, Fleshy Pincushion, Xantus Pincushion ■☆

84536 Chaenanthe Lindl. (1838);裂花兰属■☆

84537 Chaenanthe Lindl. = Diadenium Poepp. et Endl. ■☆

84538 Chaenanthe barkeri Lindl.;裂花兰■☆

84539 Chaenanthe micrantha Kuntze;小花裂花兰■☆

84540 Chaenanthera Rich. ex DC. = Charianthus D. Don ●☆

84541 Chaenarrhinum Rchb. = Chaenorhinum (DC.) Rchb. ■☆

84542 Chaenesthes Miers = Diplusion Raf. (废弃属名) ●

84543 Chaenesthes Miers = Iochroma Benth. (保留属名) ●☆

84544 Chaenocarpus Juss. = Spermacoce L. ●■

84545 Chaenocephalus Griseb. = Verbesina L. (保留属名) ●■☆

84546 Chaenolobium Miq. = Ormosia Jacks. (保留属名) ●

84547 Chaenolobus Small = Pterocaulon Elliott ■

84548 Chaenomeles Bartl. = Chaenomeles Lindl. (保留属名) ●

84549 Chaenomeles Lindl. (1821) (保留属名) ('Choenomeles'); 贴梗海棠属(木瓜属); Cydonia, Flowering Quince, Floweringquince, Flowering-quince, Japanese Quince, Japonica ●

84550 Chaenomeles × californica W. B. Clarke ex C. Weber; 加州木瓜海棠●☆

84551 Chaenomeles × vilmoriniana C. Weber; 法国贴梗海棠●☆

84552 Chaenomeles cardinalis Carrière; 绯红贴梗海棠●☆

84553 Chaenomeles cathayensis (Hemsl.) C. K. Schneid.; 毛叶木瓜(和圆子, 木瓜, 木瓜海棠, 木桃, 樝子, 西南木瓜); Hairyleaf Flowering Quince, Hairyleaf Floweringquince, Hairy-leaved Flowering-quince ●

84554 Chaenomeles eburnea Nakai; 象牙色贴梗海棠●☆

84555 Chaenomeles hybrida Hort.; 杂种木瓜; Chaenomelea, Red Chief, Toyo Nishiki ●☆

84556 Chaenomeles japonica (Thunb.) Lindl. ex Spach; 日本木瓜(白海棠, 和木瓜, 和圆子, 栌子, 木瓜花, 木桃, 日本贴梗海棠, 倭海棠, 樝子); Dwarf Flowering-quince, Dwarf Japanese Floweringquince, Dwarf Japanese Flowering-quince, Japan Floweringquince, Japan Quince, Japanese Flowering Quince, Japanese Floweringquince, Japanese Quince, Japonica, Lesser Flowering Quince, Maule's Quince ●

84557 Chaenomeles japonica (Thunb.) Lindl. ex Spach = Cydonia japonica (Thunb.) Pers. ●

84558 Chaenomeles japonica (Thunb.) Lindl. ex Spach f. alba (Nakai) Ohwi = Chaenomeles japonica (Thunb.) Lindl. ex Spach var. alba (Nakai) Ohwi ●☆

84559 Chaenomeles japonica (Thunb.) Lindl. ex Spach var. alba (Nakai) Ohwi; 白花日本木瓜; White Japan Floweringquince ●☆

84560 Chaenomeles lagenaria (Loisel.) Koidz. = Chaenomeles speciosa (Sweet) Nakai ●

84561 Chaenomeles lagenaria (Loisel.) Koidz. var. cathayensis (Hemsl.) Rehder = Chaenomeles cathayensis (Hemsl.) C. K. Schneid. ●

84562 Chaenomeles lagenaria (Loisel.) Koidz. var. wilsonii Rehder = Chaenomeles cathayensis (Hemsl.) C. K. Schneid. ●

84563 Chaenomeles lagenaria Koidz. var. cathayensis (Hemsl.) Rehder = Chaenomeles cathayensis (Hemsl.) C. K. Schneid. ●

84564 Chaenomeles lagenaria Koidz. var. wilsonii Rehder = Chaenomeles cathayensis (Hemsl.) C. K. Schneid. ●

84565 Chaenomeles lagenaria Loisel. = Chaenomeles speciosa (Sweet) Nakai ●

84566 Chaenomeles maulei (Mast.) C. K. Schneid. = Chaenomeles japonica (Thunb.) Lindl. ex Spach ●

84567 Chaenomeles maulei C. K. Schneid. = Chaenomeles japonica (Thunb.) Lindl. ex Spach ●

84568 Chaenomeles maulei C. K. Schneid. var. alba Nakai = Chaenomeles japonica (Thunb.) Lindl. ex Spach var. alba (Nakai) Ohwi ●☆

84569 Chaenomeles maulei C. K. Schneid. var. tortuosa Nakai; 红花木桃●☆

84570 Chaenomeles maulei H. Lév. = Chaenomeles japonica (Thunb.) Lindl. ex Spach ●

84571 Chaenomeles sinensis (Thouin) Koehne = Pseudocydonia sinensis C. K. Schneid. ●

84572 Chaenomeles sinensis Koehne = Chaenomeles sinensis (Thouin) Koehne ●

84573 Chaenomeles speciosa (Sweet) Nakai; 皱皮木瓜(陈木瓜, 川木瓜, 蛮楂, 楙, 木瓜, 秋木瓜, 山木瓜, 酸木瓜, 贴梗海棠, 贴梗木瓜, 铁脚梨, 香木瓜, 宣木瓜, 云木瓜, 资木瓜); Beautiful Floweringquince, Beautiful Flowering-quince, Chinese Flowering Quince, Chinese Quince, Common Flowering Quince, Common Floweringquince, Fairy's Fire, Flowering Quince, Flowering-quince, Japanese Flowering Quince, Japanese Quince, Japonica, Wrinkle Floweringquince ●

84574 Chaenomeles speciosa (Sweet) Nakai 'Moerloosei'; 莫尔罗斯皱皮木瓜●☆

84575 Chaenomeles speciosa (Sweet) Nakai 'Nivalis'; 雪白贴梗海棠(赛雪皱皮木瓜)●☆

84576 Chaenomeles speciosa (Sweet) Nakai 'Simonii'; 西蒙皱皮木瓜●☆

84577 Chaenomeles speciosa (Sweet) Nakai 'Toyo Nishiki'; 多彩贴梗海棠●☆

84578 Chaenomeles speciosa (Sweet) Nakai var. cathayensis (Hemsl.) Hara = Chaenomeles cathayensis (Hemsl.) C. K. Schneid. ●

84579 Chaenomeles speciosa (Sweet) Nakai var. wilsonii (Rehder) Hara = Chaenomeles cathayensis (Hemsl.) C. K. Schneid. ●

84580 Chaenomeles speciosa Nakai var. cathayensis (Hemsl.) H. Hara = Chaenomeles cathayensis (Hemsl.) C. K. Schneid. ●

84581 Chaenomeles speciosa Nakai var. wilsonii (Rehder) H. Hara = Chaenomeles cathayensis (Hemsl.) C. K. Schneid. ●

84582 Chaenomeles superba Rehder; 傲大贴梗海棠●☆

84583 Chaenomeles superba Rehder 'Cameo'; 浮雕傲大贴梗海棠; Common Flowering Quince ●☆

84584 Chaenomeles superba Rehder 'Crimson and Gold'; 猩红与金黄傲大贴梗海棠(华丽木瓜)●☆

84585 Chaenomeles superba Rehder 'Crimson Beauty'; 红美人傲大贴梗海棠●☆

84586 Chaenomeles superba Rehder 'Etna'; 埃特纳华丽木瓜●☆

84587 Chaenomeles superba Rehder 'Glowing Embers'; 余热傲大贴梗海棠●☆

84588 Chaenomeles superba Rehder 'Knap Hill Scarlet'; 猩红华丽木瓜●☆

84589 Chaenomeles superba Rehder 'Nicoline'; 尼考林傲大贴梗海棠(尼科林华丽木瓜)●☆

84590 Chaenomeles superba Rehder 'Rowallane'; 洛娃拉内傲大贴梗海棠(洛瓦兰华丽木瓜)●☆

84591 Chaenomeles thibelica Te T. Yu; 西藏木瓜; Tibet Floweringquince, Tibet Flowering-quince, Xizang Floweringquince ●

84592 Chaenophora Rich. ex Crueger = Miconia Ruiz et Pav. (保留属名)●☆

84593 Chaenopleura Rich. ex DC. = Miconia Ruiz et Pav. (保留属名)●☆

84594 Chaenorhinum (DC.) Rchb. (1829); 云兰参属; Toadflax ■☆

84595 Chaenorhinum klokovii Kotov.; 克氏云兰参■☆

84596 Chaenorhinum minus (L.) Lange; 小云兰参; Dwarf Snapdragon, Jack-by-the-hedge, Lesser Toadflax, Small Snapdragon,

Small Toadflax ■☆

84597 Chaenorrhinum (DC.) Rchb. = Chaenorhinum (DC.) Rchb. ■☆

84598 Chaenorrhinum Lange = Chaenorhinum (DC.) Rchb. ■☆

84599 Chaenorrhinum flexuosum (Desf.) Lange;之字云兰参■☆

84600 Chaenorrhinum grandiflorum (Coss.) Willk.;大花云兰参■☆

84601 Chaenorrhinum hians Murb. = Chaenorrhinum grandiflorum (Coss.) Willk. ■☆

84602 Chaenorrhinum klokovii Kotov. = Chaenorhinum klokovii Kotov. ■☆

84603 Chaenorrhinum minus (L.) Lange = Chaenorhinum minus (L.) Lange ■☆

84604 Chaenorrhinum origanifolium (L.) Fourr.;橘叶云兰参;Dwarf Snapdragon, Mailing Toadflax ■☆

84605 Chaenorrhinum origanifolium (L.) Fourr. subsp. flexuosum (Desf.) Romo = Chaenorrhinum flexuosum (Desf.) Lange ■☆

84606 Chaenorrhinum origanifolium (L.) Fourr. subsp. maroccanum (Pau) Dobignard;摩洛哥橘叶云兰参■☆

84607 Chaenorrhinum origanifolium (L.) Fourr. var. maroccanum Pau = Chaenorrhinum origanifolium (L.) Fourr. ■☆

84608 Chaenorrhinum rubrifolium (DC.) Fourr.;红叶云兰参■☆

84609 Chaenorrhinum rubrifolium (DC.) Fourr. subsp. imintalense (Murb.) Font Quer = Chaenorrhinum rubrifolium (DC.) Fourr. ■☆

84610 Chaenorrhinum rubrifolium (DC.) Fourr. var. bianorii Knoche = Chaenorrhinum rubrifolium (DC.) Fourr. ■☆

84611 Chaenorrhinum rubrifolium (DC.) Fourr. var. grandiflorum (Coss.) Lange = Chaenorrhinum grandiflorum (Coss.) Willk. ■☆

84612 Chaenorrhinum rubrifolium (DC.) Fourr. var. imintalensis Murb. = Chaenorrhinum rubrifolium (DC.) Fourr. ■☆

84613 Chaenorrhinum rubrifolium (DC.) Fourr. var. maroccanum Font Quer = Chaenorrhinum rubrifolium (DC.) Fourr. ■☆

84614 Chaenorrhinum rupestre (Guss.) Maire;岩生云兰参■☆

84615 Chaenorrhinum spicatum Korovin;穗状云兰参■☆

84616 Chaenorrhinum suttonii Benedi et J. M. Monts.;萨顿云兰参■☆

84617 Chaenorrhinum villosum (L.) Lange;长柔毛云兰参■☆

84618 Chaenorrhinum villosum (L.) Lange subsp. granatensis (Willk.) Valdés;格拉云兰参■☆

84619 Chaenorrhinum viscidum (Moench) Simonk.;黏云兰参■☆

84620 Chaenostoma Benth. (1836)(保留属名);裂口玄参属■☆

84621 Chaenostoma Benth. (保留属名) = Sutera Roth ■●☆

84622 Chaenostoma acutilobum (Pilg.) Thell. = Jamesbrittenia acutiloba (Pilg.) Hilliard ■☆

84623 Chaenostoma aethiopicum (L.) Benth. = Sutera aethiopica (L.) Kuntze ■☆

84624 Chaenostoma affine Bernh. = Sutera affinis (Bernh.) Kuntze ■☆

84625 Chaenostoma ambleophyllum Thell. = Jamesbrittenia pallida (Pilg.) Hilliard ■☆

84626 Chaenostoma annuum Schltr. ex Hiern = Manulea annua (Hiern) Hilliard ■☆

84627 Chaenostoma annuum Schltr. ex Hiern var. laxum = Manulea paucibarbata Hilliard ■☆

84628 Chaenostoma aspalathoides (Benth.) Wettst. ex Diels = Jamesbrittenia aspalathoides (Benth.) Hilliard ■☆

84629 Chaenostoma aurantiacum (Burch.) Thell. = Jamesbrittenia aurantiaca (Burch.) Hilliard ■☆

84630 Chaenostoma breviflorum (Schltr.) Diels = Jamesbrittenia breviflora (Schltr.) Hilliard ■☆

84631 Chaenostoma burkeanum (Benth.) Wettst. ex Diels = Jamesbrittenia burkeana (Benth.) Hilliard ■☆

84632 Chaenostoma calycinum Benth. = Sutera calycina (Benth.) Kuntze ■☆

84633 Chaenostoma calycinum Benth. var. laxiflorum = Sutera calycina (Benth.) Kuntze ■☆

84634 Chaenostoma campanulatum Benth. = Sutera campanulata (Benth.) Kuntze ■☆

84635 Chaenostoma canescens (Benth.) Wettst. ex Diels = Jamesbrittenia canescens (Benth.) Hilliard ■☆

84636 Chaenostoma cooperi (Hiern) Thell. = Sutera cooperi Hiern ■☆

84637 Chaenostoma cordatum (Thunb.) Benth. = Sutera cordata (Thunb.) Kuntze ■☆

84638 Chaenostoma cordatum (Thunb.) Benth. var. hirsutior Benth. = Sutera cooperi Hiern ■☆

84639 Chaenostoma corymbosum Marloth et Engl. = Camptoloma rotundifolia Benth. ■☆

84640 Chaenostoma corymbosum Marloth et Engl. var. huillanum Diels = Camptoloma rotundifolia Benth. ■☆

84641 Chaenostoma crassicaule (Benth.) Thell. = Jamesbrittenia crassicaulis (Benth.) Hilliard ■☆

84642 Chaenostoma croceum (Eckl. ex Benth.) Wettst. ex Diels = Jamesbrittenia atropurpurea (Benth.) Hilliard ■☆

84643 Chaenostoma cuneatum (Benth.) Wettst. ex Diels = Jamesbrittenia argentea (L. f.) Hilliard ■☆

84644 Chaenostoma cuneatum Benth. = Sutera hispida (Thunb.) Druce ■☆

84645 Chaenostoma cymbalarifolium (Chiov.) Cufod. = Stemodiopsis buchananii V. Naray. ■☆

84646 Chaenostoma denudatum Benth. = Sutera denudata (Benth.) Kuntze ■☆

84647 Chaenostoma dielsianum (Hiern) Thell. = Jamesbrittenia racemosa (Benth.) Hilliard ■☆

84648 Chaenostoma dissectum (Delile) Thell. = Jamesbrittenia dissecta (Delile) Kuntze ■☆

84649 Chaenostoma divaricatum Diels = Manulea annua (Hiern) Hilliard ■☆

84650 Chaenostoma fasciculatum Hort. = Sutera uncinata (Desr.) Hilliard ■☆

84651 Chaenostoma fastigiatum Benth. = Sutera aethiopica (L.) Kuntze ■☆

84652 Chaenostoma fastigiatum Benth. var. glabratum? = Sutera aethiopica (L.) Kuntze ■☆

84653 Chaenostoma fleckii Thell. = Jamesbrittenia fleckii (Thell.) Hilliard ■☆

84654 Chaenostoma floribundum Benth. = Sutera floribunda (Benth.) Kuntze ■☆

84655 Chaenostoma foetidum (Andréws) Benth. = Sutera foetida Roth ■☆

84656 Chaenostoma fraternum (Hiern) Thell. = Jamesbrittenia thunbergii (G. Don) Hilliard ■☆

84657 Chaenostoma fruticosum (Benth.) Wettst. = Jamesbrittenia fruticosa (Benth.) Hilliard ■☆

84658 Chaenostoma glabratum Benth. = Sutera glabrata (Benth.) Kuntze ■☆

84659 Chaenostoma gracile Diels = Jamesbrittenia thunbergii (G. Don) Hilliard ■☆

84660 Chaenostoma halimifolium Benth. = Sutera halimifolia (Benth.) Kuntze ■☆

84661 Chaenostoma hereroense Engl. = Jamesbrittenia hereroensis

(Engl.) Hilliard ■☆

84662 Chaenostoma heucherifolium Diels = Jamesbrittenia heucherifolia (Diels) Hilliard ■☆

84663 Chaenostoma hispidum (Thunb.) Benth. = Sutera hispida (Thunb.) Druce ■☆

84664 Chaenostoma huillanum Diels = Jamesbrittenia huillana (Diels) Hilliard ■☆

84665 Chaenostoma integrifolium (L. f.) Benth. = Sutera integrifolia (L. f.) Kuntze ■☆

84666 Chaenostoma integrifolium (L. f.) Benth. var. parvifolium Benth. = Sutera hispida (Thunb.) Druce ■☆

84667 Chaenostoma kraussianum Bernh. = Jamesbrittenia kraussiana (Bernh.) Hilliard ■☆

84668 Chaenostoma laxiflorum Benth. = Sutera halimifolia (Benth.) Kuntze ■☆

84669 Chaenostoma linifolium (Thunb.) Benth. = Sutera uncinata (Desr.) Hilliard ■☆

84670 Chaenostoma linifolium (Thunb.) Benth. var. hispidum Bernh. = Sutera uncinata (Desr.) Hilliard ■☆

84671 Chaenostoma litorale (Schinz) Wettst. ex Diels = Jamesbrittenia fruticosa (Benth.) Hilliard ■☆

84672 Chaenostoma lyperiiflorum (Vatke) Wettst. = Camptoloma lyperiiflorum (Vatke) Hilliard ■●☆

84673 Chaenostoma lyperioides Engl. = Jamesbrittenia lyperioides (Engl.) Hilliard ■☆

84674 Chaenostoma macrosiphon Schltr. = Sutera macrosiphon (Schltr.) Hiern ■☆

84675 Chaenostoma marifolium Benth. = Sutera marifolia (Benth.) Kuntze ■☆

84676 Chaenostoma micranthum (Klotzsch) Engl. = Jamesbrittenia micrantha (Klotzsch) Hilliard ■☆

84677 Chaenostoma microphyllum (L. f.) Wettst. = Jamesbrittenia microphylla (L. f.) Hilliard ■☆

84678 Chaenostoma molle (Benth.) Wettst. ex Diels = Jamesbrittenia pinnatifida (L. f.) Hilliard ■☆

84679 Chaenostoma montanum Diels = Jamesbrittenia montana (Diels) Hilliard ■☆

84680 Chaenostoma natalense Bernh. = Sutera floribunda (Benth.) Kuntze ■☆

84681 Chaenostoma neglectum J. M. Wood et M. S. Evans = Sutera neglecta (J. M. Wood et M. S. Evans) Hiern ■☆

84682 Chaenostoma pauciflorum Benth. = Sutera pauciflora (Benth.) Kuntze ■☆

84683 Chaenostoma pedicellatum (Klotzsch) Engl. = Jamesbrittenia micrantha (Klotzsch) Hilliard ■☆

84684 Chaenostoma pedunculosum Benth. = Jamesbrittenia pedunculosa (Benth.) Hilliard ■☆

84685 Chaenostoma phlogiflorum (Benth.) Wettst. ex Diels = Jamesbrittenia phlogiflora (Benth.) Hilliard ■☆

84686 Chaenostoma pinnatifidum (L. f.) Wettst. ex Diels = Jamesbrittenia pinnatifida (L. f.) Hilliard ■☆

84687 Chaenostoma polyanthum Benth. = Sutera polyantha (Benth.) Kuntze ■☆

84688 Chaenostoma primuliflorum Thell. = Jamesbrittenia primuliflora (Thell.) Hilliard ■☆

84689 Chaenostoma procumbens Benth. = Sutera polyantha (Benth.) Kuntze ■☆

84690 Chaenostoma pumilum Benth. = Sutera halimifolia (Benth.) Kuntze ■☆

84691 Chaenostoma racemosum (Benth.) Wettst. ex Diels = Jamesbrittenia racemosa (Benth.) Hilliard ■☆

84692 Chaenostoma racemosum Benth. = Sutera racemosa (Benth.) Kuntze ■☆

84693 Chaenostoma revolutum (Thunb.) Benth. = Sutera revoluta (Thunb.) Kuntze ■☆

84694 Chaenostoma revolutum (Thunb.) Benth. var. pubescens Benth. = Sutera paniculata Hilliard ■☆

84695 Chaenostoma rotundifolium Benth. = Sutera rotundifolia (Benth.) Kuntze ■☆

84696 Chaenostoma schinzianum Thell. = Jamesbrittenia pallida (Pilg.) Hilliard ■☆

84697 Chaenostoma sessilifolium Diels = Jamesbrittenia sessilifolia (Diels) Hilliard ■☆

84698 Chaenostoma stenopetalum Diels = Jamesbrittenia incisa (Thunb.) Hilliard ●☆

84699 Chaenostoma subnudum N. E. Br. = Sutera subnuda (N. E. Br.) Hiern ■☆

84700 Chaenostoma subspicatum Benth. = Sutera subspicata (Benth.) Kuntze ■☆

84701 Chaenostoma tomentosum Thell. = Jamesbrittenia thunbergii (G. Don) Hilliard ■☆

84702 Chaenostoma triste (L. f.) Wettst. = Lyperia tristis (L. f.) Benth. ■☆

84703 Chaenostoma triste (L. f.) Wettst. var. montanum Diels = Lyperia tristis (L. f.) Benth. ■☆

84704 Chaenostoma violaceum Schltr. = Sutera violacea (Schltr.) Hiern ■☆

84705 Chaenostoma woodianum Diels = Jamesbrittenia breviflora (Schltr.) Hilliard ■☆

84706 Chaenotheca Urb. = Chascotheca Urb. ●☆

84707 Chaenotheca Urb. = Securinega Comm. ex Juss. (保留属名)●☆

84708 Chaeradoplectron Benth. et Hook. f. = Choeradoplectron Schauer ■

84709 Chaeradoplectron Benth. et Hook. f. = Habenaria Willd. ■

84710 Chaerefolium Haller = Anthriscus Pers. (保留属名)■

84711 Chaerefolium Hoffm. = Anthriscus Pers. (保留属名)■

84712 Chaerefolium anthriscus (L.) Schinz et Thell. = Anthriscus caucalis M. Bieb. ■☆

84713 Chaerefolium cerefolium (L.) Schinz et Thell. = Anthriscus cerefolius (L.) Hoffm. ■☆

84714 Chaerefolium sylvestre (L.) Schinz et Thell. = Anthriscus sylvestris (L.) Hoffm. ■

84715 Chaerefolium sylvestris (L.) Schinz et Thell. subsp. mollis (Boiss. et Reut.) Maire = Anthriscus sylvestris (L.) Hoffm. ■

84716 Chaerefolium sylvestris (L.) Schinz et Thell. var. glabricaulis Maire = Anthriscus sylvestris (L.) Hoffm. ■

84717 Chaerefolium sylvestris (L.) Schinz et Thell. var. villicaulis Maire = Anthriscus sylvestris (L.) Hoffm. ■

84718 Chaerophyllastrum Fabr. = Chaerophyllastrum Heist. ex Fabr. ■☆

84719 Chaerophyllastrum Fabr. = Myrrhis Mill. ■☆

84720 Chaerophyllastrum Heist. ex Fabr. = Myrrhis Mill. ■☆

84721 Chaerophyllopsis H. Boissieu (1909);滇藏细叶芹属(滇细叶芹属,假香叶芹属);Falsechervil ●★

84722 Chaerophyllopsis huai H. Boissieu;滇藏细叶芹(假香叶芹);Falsechervil ■

84723 Chaerophyllum L.(1753);细叶芹属(香叶芹属);Chervil ■

84724 Chaerophyllum aristatum Thunb. = Osmorhiza aristata (Thunb.) Makino et Y. Yabe ■

84725 Chaerophyllum aristatum Thunb. ex A. Murray = Osmorhiza aristata (Thunb.) Makino et Y. Yabe ■

84726 Chaerophyllum aromaticum L.;芳香细叶芹;Fragrant Chervil ■

84727 Chaerophyllum atlanticum Batt.;大西洋细叶芹■☆

84728 Chaerophyllum aureum L.;黄细叶芹;Golden Chervil ■☆

84729 Chaerophyllum bobrovii Schischk.;鲍勃细叶芹■☆

84730 Chaerophyllum borodinii Albov;保罗细叶芹■☆

84731 Chaerophyllum bulbosum L.;鳞茎细叶芹;Bulbose Chervil, Tuberous Chervil, Turnip-rooted Chervil ■

84732 Chaerophyllum bulbosum L. = Chaerophyllum prescottii DC. ■

84733 Chaerophyllum bulbosum L. subsp. prescottii (DC.) Nyman. = Chaerophyllum prescottii DC. ■

84734 Chaerophyllum capense Thunb. = Annesorhiza thunbergii B. L. Burtt ■☆

84735 Chaerophyllum capnoides (Decne.) Benth.;延胡索细叶芹■☆

84736 Chaerophyllum caucasicum (Fisch.) Schischk.;高加索细叶芹■☆

84737 Chaerophyllum claytonii (Michx.) Pers. = Osmorhiza aristata (Thunb.) Makino et Y. Yabe ■

84738 Chaerophyllum confusum Woronow;混乱细叶芹■☆

84739 Chaerophyllum crinitum Boiss.;长软毛细叶芹■☆

84740 Chaerophyllum cyminum Fisch. = Sphallerocarpus gracilis (Trevir.) Koso-Pol. ■

84741 Chaerophyllum gracile Besser ex Trevir. = Sphallerocarpus gracilis (Trevir.) Koso-Pol. ■

84742 Chaerophyllum gracillum Klotzsch = Vicatia coniifolia Wall. ex DC. ■

84743 Chaerophyllum hirsutum L.;毛细叶芹;Hairy Chervil ■☆

84744 Chaerophyllum humile Stev.;小细叶芹■☆

84745 Chaerophyllum khorossanicum Czerniak.;呼罗珊细叶芹■☆

84746 Chaerophyllum kiapazi Woronow;基亚帕兹细叶芹■☆

84747 Chaerophyllum longilobum (Kar. et Kir.) O. Fedtsch. et B. Fedtsch. = Krasnovia longiloba (Kar. et Kir.) Popov ex Schischk. ■

84748 Chaerophyllum macrospermum (Willd.) Fisch. et C. A. Mey. ex Hohen.;大籽细叶芹■☆

84749 Chaerophyllum maculatum Willd.;斑点细叶芹■☆

84750 Chaerophyllum meyeri Boiss. et Buhse;迈氏细叶芹■☆

84751 Chaerophyllum millefolium Klotzsch = Vicatia coniifolia Wall. ex DC. ■

84752 Chaerophyllum nemorosum Hoffm. = Anthriscus nemorosa (M. Bieb.) Spreng. ■

84753 Chaerophyllum nemorosum Lag. ex DC. = Anthriscus cerefolius Hoffm. ■☆

84754 Chaerophyllum nemorosum M. Bieb. = Anthriscus nemorosa (M. Bieb.) Spreng. ■

84755 Chaerophyllum nemorosum M. Bieb. = Anthriscus sylvestris (L.) Hoffm. ■

84756 Chaerophyllum nemorosum M. Bieb. = Anthriscus sylvestris (L.) Hoffm. var. nemorosa (M. Bieb.) Trautv. ■

84757 Chaerophyllum nodosum Lam. = Myrrhoides nodosa (L.) Cannon ■☆

84758 Chaerophyllum prescottii DC.;新疆细叶芹;Prescot Chervil, Xinjiang Chervil ■

84759 Chaerophyllum procumbens (L.) Crantz;匍匐细叶芹;Spreading Chervil, Wild Chervil ■☆

84760 Chaerophyllum reflexum Lindl. = Chaerophyllum villosum Wall. ex DC. ■

84761 Chaerophyllum roseum M. Bieb.;粉红细叶芹■☆

84762 Chaerophyllum rubellum Albov;红细叶芹■☆

84763 Chaerophyllum scabrum Thunb. = Torilis scabra (Thunb.) DC. ■

84764 Chaerophyllum sphallerocarpus Kar. et Kir. = Krasnovia longiloba (Kar. et Kir.) Popov ex Schischk. ■

84765 Chaerophyllum sylvestre L. = Anthriscus sylvestris (L.) Hoffm. ■

84766 Chaerophyllum tainturieri Hook.;谈蒂里耶细叶芹;Wild Chervil ■☆

84767 Chaerophyllum tainturieri Hook. var. floridanum J. M. Coult. et Rose = Chaerophyllum tainturieri Hook. ■☆

84768 Chaerophyllum temulentum L.;细叶芹(毒细叶芹,绵毛细叶芹);Chervil, Cow Mumble, Cow Parsley, Rough Chervil, Sheep's Parsley, Turnip-rooted Chervil ■

84769 Chaerophyllum temuloides Boiss.;假毒细叶芹■☆

84770 Chaerophyllum temulum L. = Chaerophyllum temulentum L. ■

84771 Chaerophyllum texanum J. M. Coult. et Rose = Chaerophyllum tainturieri Hook. ■☆

84772 Chaerophyllum villosum DC. = Chaerophyllum villosum Wall. ex DC. ■

84773 Chaerophyllum villosum Wall. ex DC.;香叶芹(反卷细叶芹,细叶芹);Villous Chervil ■

84774 Chaetacanthus Nees(1836);刺毛爵床属■☆

84775 Chaetacanthus burchellii Lindau = Dyschoriste radicans Nees ■☆

84776 Chaetacanthus burchellii Nees;布尔刺毛爵床■☆

84777 Chaetacanthus costatus Nees;单脉刺毛爵床■☆

84778 Chaetacanthus glandulosus Nees = Chaetacanthus setiger (Pers.) Lindl. ■☆

84779 Chaetacanthus persoonii Nees = Chaetacanthus setiger (Pers.) Lindl. ■☆

84780 Chaetacanthus setiger (Pers.) Lindl.;刚毛刺毛爵床■☆

84781 Chaetachlaena D. Don = Onoseris Willd. ●■☆

84782 Chaetachme Planch.(1848);非洲朴属●☆

84783 Chaetachme aristata E. Mey. ex Planch. var. kamerunensis Engl. = Chaetachme aristata Planch. ●☆

84784 Chaetachme aristata E. Mey. ex Planch. var. longifolia De Wild. et T. Durand = Chaetachme aristata Planch. ●☆

84785 Chaetachme aristata E. Mey. ex Planch. var. nitida (Planch. et Harv.) Engl. = Chaetachme aristata Planch. ●☆

84786 Chaetachme aristata Planch.;非洲朴●☆

84787 Chaetachme madagascariensis Baker = Chaetachme aristata Planch. ●☆

84788 Chaetachme microcarpa Rendle = Chaetachme aristata Planch. ●☆

84789 Chaetachme microcarpa Rendle var. crenata Hutch. et Dalziel = Chaetachme aristata Planch. ●☆

84790 Chaetachme nitida Planch. et Harv. = Chaetachme aristata Planch. ●☆

84791 Chaetachme serrata Engl. = Chaetachme aristata Planch. ●☆

84792 Chaetacme Planch. = Chaetachme Planch. ●☆

84793 Chaetadelpha A. Gray = Chaetadelpha A. Gray ex S. Watson ■☆

84794 Chaetadelpha A. Gray ex S. Watson (1873);骨苣属;Skeletonweed ■☆

84795 Chaetadelpha wheeleri A. Gray;骨苣■☆

84796 Chaetaea Jacq. = Byttneria Loefl.(保留属名)●

84797 Chaetaea Post et Kuntze = Chaitaea Sol. ex Seem. ■

84798 Chaetaea Post et Kuntze = Tacca J. R. Forst. et G. Forst.（保留属名）■

84799 Chaetagastra Crueg. = Chaetogastra DC. ●■☆

84800 Chaetagastra Crueg. = Tibouchina Aubl. ●■☆

84801 Chaetantera Less. = Chaetanthera Ruiz et Pav. ■☆

84802 Chaetanthera Nutt. = Chaetopappa DC. ■☆

84803 Chaetanthera Ruiz et Pav.（1794）；毛药菊属（寒绒菊属，毛花属）■☆

84804 Chaetanthera asteroides Nutt. = Chaetopappa asteroides（Nutt.）DC. ■☆

84805 Chaetanthera leptocephala Cabrera；细头毛药菊（细头毛花）■☆

84806 Chaetanthera linearis Poepp. ex Less.；线叶毛药菊（线叶毛花）■☆

84807 Chaetanthera montana Phil.；山地毛药菊（山地毛花）■☆

84808 Chaetanthera multicaulis DC.；多茎毛药菊（多茎毛花）■☆

84809 Chaetanthera nana Phil.；小毛药菊（小毛花）■☆

84810 Chaetanthera ramosissima D. Don ex Taylor et Phillips；智利毛药菊■☆

84811 Chaetanthera spathulata Poepp. ex Less.；匙叶毛药菊（匙叶毛花）■☆

84812 Chaetanthera sphaeroidalis Hicken；圆球毛药菊（圆球毛花）■☆

84813 Chaetanthera tenuifolia D. Don；细叶毛药菊（细叶毛花）■☆

84814 Chaetanthera villosa D. Don；柔毛毛药菊（毛毛花）■☆

84815 Chaetanthus R. Br.（1810）；齿瓣帚灯草属■☆

84816 Chaetanthus leptocarpoides R. Br.；齿瓣帚灯草■☆

84817 Chaetaphora Nutt. = Chaetanthera Nutt. ■☆

84818 Chaetaria P. Beauv. = Aristida L. ■

84819 Chaetaria adscensionis（L.）P. Beauv. = Aristida adscensionis L. ■

84820 Chaetaria bipartita Nees = Aristida bipartita（Nees）Trin. et Rupr. ■☆

84821 Chaetaria curvata Nees = Aristida adscensionis L. ■

84822 Chaetaria curvata Nees var. minor = Aristida adscensionis L. ■

84823 Chaetaria depressa（Retz.）P. Beauv. = Aristida depressa Retz. ■

84824 Chaetaria vestita（Thunb.）P. Beauv. = Aristida vestita Thunb. ■☆

84825 Chaethymenia Hook. et Arn.（1841）；毛棱菊属■☆

84826 Chaethymenia Hook. et Arn. = Jaumea Pers. ■●☆

84827 Chaethymenia peduncularis Hook. et Arn.；毛棱菊■☆

84828 Chaetium Nees（1829）；刚毛禾属■☆

84829 Chaetium festucoides Nees；刚毛禾■☆

84830 Chaetobromus Nees（1836）；南非雀麦属■☆

84831 Chaetobromus dregeanus Barker = Chaetobromus involucratus（Schrad.）Nees ■☆

84832 Chaetobromus dregeanus Nees = Chaetobromus involucratus（Schrad.）Nees subsp. dregeanus（Nees）Verboom ■☆

84833 Chaetobromus fascicularis Nees = Merxmuellera stricta（Schrad.）Conert ■☆

84834 Chaetobromus interceptus Nees = Chaetobromus involucratus（Schrad.）Nees subsp. dregeanus（Nees）Verboom ■☆

84835 Chaetobromus involucratus（Schrad.）Nees；南非雀麦■☆

84836 Chaetobromus involucratus（Schrad.）Nees subsp. dregeanus（Nees）Verboom；德雷南非雀麦■☆

84837 Chaetobromus involucratus（Schrad.）Nees subsp. sericeus（Nees）Verboom；绢毛南非雀麦■☆

84838 Chaetobromus involucratus（Schrad.）Nees var. sericeus Nees = Chaetobromus involucratus（Schrad.）Nees subsp. sericeus（Nees）Verboom ■☆

84839 Chaetobromus schlechteri Pilg. = Chaetobromus involucratus（Schrad.）Nees subsp. dregeanus（Nees）Verboom ■☆

84840 Chaetobromus schraderi Stapf = Chaetobromus involucratus（Schrad.）Nees ■☆

84841 Chaetocalyx DC.（1825）；鬃萼豆属（毛萼豆属）■☆

84842 Chaetocalyx longiflorus Benth. ex A. Gray；长花鬃萼豆■☆

84843 Chaetocalyx magniflorus Pittier；大花鬃萼豆■☆

84844 Chaetocalyx nigrescens Pittier；黑鬃萼豆■☆

84845 Chaetocalyx polyphylla Benth.；多叶鬃萼豆■☆

84846 Chaetocalyx pubescens DC.；鬃萼豆■☆

84847 Chaetocalyx tomentosa（Gardner）Rudd；毛鬃萼豆■☆

84848 Chaetocapnia Sweet = Coetocapnia Link et Otto ■

84849 Chaetocapnia Sweet = Polianthes L. ■

84850 Chaetocarpus Schreb.（废弃属名）= Chaetocarpus Thwaites（保留属名）●

84851 Chaetocarpus Schreb.（废弃属名）= Pouteria Aubl. ●

84852 Chaetocarpus Thwaites（1854）（保留属名）；刺果树属（白大凤属，毛果大戟属）；Chestnutfruit，Chestnut-fruit，Setafruit ●

84853 Chaetocarpus africanus Pax；非洲刺果树●

84854 Chaetocarpus castanocarpus（Roxb.）Thwaites；刺果树（白大凤，刺果大戟，毛果大戟）；Chestnut Fruit，Chestnut Setafruit，Chestnutfruit，Chestnut-fruit，Japan Quince ●

84855 Chaetocarpus gabonensis Breteler；加蓬刺果树●

84856 Chaetocephala Barb. Rodr. = Myoxanthus Poepp. et Endl. ■☆

84857 Chaetochilus Vahl = Schwenckia L. ■●☆

84858 Chaetochlaena Post et Kuntze = Chaetachlaena D. Don ●■☆

84859 Chaetochlaena Post et Kuntze = Onoseris Willd. ●■☆

84860 Chaetochlamys Lindau = Justicia L. ●■

84861 Chaetochloa Scribn. = Setaria P. Beauv.（保留属名）■

84862 Chaetochloa brevispica Scribn. et Merr. = Setaria verticillata（L.）P. Beauv. ■

84863 Chaetochloa chondrachne（Steud.）Honda = Setaria chondrachne（Steud.）Honda ■

84864 Chaetochloa forbesiana（Nees ex Steud.）Scribn. et Merr. = Setaria forbesiana（Nees）Hook. f. ■

84865 Chaetochloa forbesiana（Nees）Scribn. et Merr. = Setaria forbesiana（Nees）Hook. f. ■

84866 Chaetochloa geniculata（Lam.）Millsp. et Chase = Setaria geniculata（Lam.）P. Beauv. ■

84867 Chaetochloa geniculata（Poir.）Millsp. et Chase = Setaria parviflora（Poir.）Kerguélen ■

84868 Chaetochloa germanica（Mill.）Smyth = Setaria italica（L.）P. Beauv. ■

84869 Chaetochloa germanica（Mill.）Smyth = Setaria italica P. Beauv. 'Major' ■

84870 Chaetochloa germanica（Mill.）Smyth = Setaria italica P. Beauv. var. germanica（Mill.）Schrad. ■

84871 Chaetochloa gigantea（Franch. et Sav.）Honda = Setaria viridis（L.）P. Beauv. subsp. pycnocoma（Steud.）Tzvelev ■

84872 Chaetochloa glauca（L.）Scribn. = Setaria glauca（L.）P. Beauv. ■

84873 Chaetochloa glauca（L.）Scribn. var. aures Wight = Setaria pallidifusca（Schumach.）Stapf et C. E. Hubb. ■

84874 Chaetochloa italica（L.）Scribn. = Setaria italica（L.）P. Beauv. ■

84875 Chaetochloa italica（L.）Scribn. var. germanica（Mill.）Scribn. = Setaria italica（L.）P. Beauv. ■

84876 Chaetochloa italica Scribn. = Setaria italica (L.) P. Beauv. ■

84877 Chaetochloa italica Scribn. var. germanica Scribn. = Setaria italica (L.) P. Beauv. var. germanica (Mill.) Schrad. ■☆

84878 Chaetochloa lutescens Stuntze = Setaria glauca (L.) P. Beauv. ■

84879 Chaetochloa matsumurae (Hack.) Keng = Setaria chondrachne (Steud.) Honda ■

84880 Chaetochloa matsumurae (Matsum.) Keng = Setaria chondrachne (Steud.) Honda ■

84881 Chaetochloa palmifolia (Willd.) Hitchc. et Chase = Setaria palmifolia (J. König) Stapf ■

84882 Chaetochloa palmifolia Hitchc. et Chase = Setaria palmifolia (J. König) Stapf ■

84883 Chaetochloa verticillata (L.) Scribn. = Setaria verticillata (L.) P. Beauv. ■

84884 Chaetochloa verticillata Scribn. = Setaria verticillata (L.) P. Beauv. ■

84885 Chaetochloa verticillata? var. breviseta (Godr.) Farw. = Setaria verticillata (L.) P. Beauv. ■

84886 Chaetochloa viridis (L.) Scribn. = Setaria viridis (L.) P. Beauv. ■

84887 Chaetochloa viridis (L.) Scribn. var. pachystachys (Franch. et Sav.) Honda = Setaria viridis (L.) P. Beauv. subsp. pachystachys (Franch. et Sav.) Masam. et Yanagita ■

84888 Chaetochloa virids (L.) Scribn. var. pachystachys (Franch. et Sav.) Honda = Setaria viridis (L.) P. Beauv. subsp. pachystachys (Franch. et Sav.) Masam. et Yanagita ■

84889 Chaetocladus J. Nelson = Ephedra Tourn. ex L. ●■

84890 Chaetocrater Ruiz et Pav. = Casearia Jacq. ●

84891 Chaetocyperus Nees = Eleocharis R. Br. ■

84892 Chaetocyperus baldwinii Torr. = Eleocharis baldwinii (Torr.) Chapm. ■☆

84893 Chaetocyperus jamesonii Steud. = Eleocharis minima Kunth ■☆

84894 Chaetocyperus limnocharis Nees = Eleocharis chaetaria Roem. et Schult. ■

84895 Chaetocyperus limnocharis Nees = Heleocharis chaetaria Roem. et Schult. ■

84896 Chaetocyperus niveus Liebm. = Eleocharis retroflexa (Poir.) Urb. ■☆

84897 Chaetocyperus polymorphus Lindl. et Nees = Eleocharis minima Kunth ■☆

84898 Chaetocyperus polymorphus Lindl. et Nees var. depauperatus Nees = Eleocharis retroflexa (Poir.) Urb. ■☆

84899 Chaetocyperus punctatus Nees = Eleocharis nana Kunth ■☆

84900 Chaetocyperus rugulosus Nees = Eleocharis retroflexa (Poir.) Urb. ■☆

84901 Chaetocyperus setaceus Nees = Eleocharis chaetaria Roem. et Schult. ■

84902 Chaetocyperus setaceus Nees = Heleocharis chaetaria Roem. et Schult. ■

84903 Chaetocyperus viviparus Liebm. = Eleocharis retroflexa (Poir.) Urb. ■☆

84904 Chaetocyperus viviparus Nees = Eleocharis minima Kunth ■☆

84905 Chaetodiscus Steud. = Eriocaulon L. ■

84906 Chaetogastra DC. = Tibouchina Aubl. ●■☆

84907 Chaetolepis (DC.) Miq. (1840);毛鳞野牡丹属●☆

84908 Chaetolepis Miq. = Chaetolepis (DC.) Miq. ●☆

84909 Chaetolepis alpina Naudin;高山毛鳞野牡丹●☆

84910 Chaetolepis gentianoides (Naudin) Jacq. -Fél.;毛鳞野牡丹●☆

84911 Chaetolepis nana Standl.;矮毛鳞野牡丹●☆

84912 Chaetolimon (Bunge) Lincz. (1940);刚毛彩花属■☆

84913 Chaetolimon Lincz. = Chaetolimon (Bunge) Lincz. ■☆

84914 Chaetolimon limbatum Lincz.;具边刚毛彩花■☆

84915 Chaetolimon setiferum (Bunge) Lincz.;刚毛彩花■☆

84916 Chaetonychia (DC.) Sweet = Paronychia Mill. ■

84917 Chaetonychia (DC.) Sweet(1839);异萼醉人花属■☆

84918 Chaetonychia Sweet = Chaetonychia (DC.) Sweet ■☆

84919 Chaetonychia Sweet = Paronychia Mill. ■

84920 Chaetonychia cymosa (L.) Sweet;异萼醉人花■☆

84921 Chaetopappa DC. (1836);毛冠雏菊属;Lazy Daisy ■☆

84922 Chaetopappa alsinoides (Greene) D. D. Keck = Pentachaeta alsinoides Greene ■☆

84923 Chaetopappa asteroides (Nutt.) DC.;小毛冠雏菊;Tiny Lazy Daisy ■☆

84924 Chaetopappa asteroides (Nutt.) DC. var. grandis Shinners;大毛冠雏菊■☆

84925 Chaetopappa asteroides (Nutt.) DC. var. imberbis A. Gray = Chaetopappa imberbis (A. Gray) G. L. Nesom ■☆

84926 Chaetopappa aurea (Nutt.) D. D. Keck = Pentachaeta aurea Nutt. ■☆

84927 Chaetopappa bellidiflora (Greene) D. D. Keck = Pentachaeta bellidiflora Greene ■☆

84928 Chaetopappa bellidifolia (A. Gray et Engelm.) Shinners;爱氏毛冠雏菊;Edwards Lazy Daisy ■☆

84929 Chaetopappa bellioides (A. Gray) Shinners;拟爱氏毛冠雏菊■☆

84930 Chaetopappa bellioides (A. Gray) Shinners var. hirticaulis Shinners = Chaetopappa bellioides (A. Gray) Shinners ■☆

84931 Chaetopappa effusa (A. Gray) Shinners;铺散毛冠雏菊;Spreading Lazy Daisy ■☆

84932 Chaetopappa elegans Soreng et Spellenb. = Ionactis elegans (Soreng et Spellenb.) G. L. Nesom ■☆

84933 Chaetopappa ericoides (Torr.) G. L. Nesom;粉毛冠雏菊;Rose Heath, Rose-heath ■☆

84934 Chaetopappa exilis (A. Gray) D. D. Keck = Pentachaeta exilis (A. Gray) A. Gray ■☆

84935 Chaetopappa fragilis (Brandegee) D. D. Keck = Pentachaeta fragilis Brandegee ■☆

84936 Chaetopappa hersheyi S. F. Blake;赫尔希毛冠雏菊;Guadalupe Lazy Daisy ■☆

84937 Chaetopappa imberbis (A. Gray) G. L. Nesom;无须希毛冠雏菊;Awnless Lazy Daisy ■☆

84938 Chaetopappa lyonii (A. Gray) D. D. Keck = Pentachaeta lyonii A. Gray ■☆

84939 Chaetopappa modesta (DC.) A. Gray = Chaetopappa asteroides (Nutt.) DC. var. grandis Shinners ■☆

84940 Chaetopappa parryi A. Gray;帕里毛冠雏菊;Parry's Lazy Daisy ■☆

84941 Chaetophora Nutt. ex DC. = Chaetopappa DC. ■☆

84942 Chaetopoa C. E. Hubb. (1967);东非早熟禾属■☆

84943 Chaetopoa pilosa Clayton;疏毛东非早熟禾■☆

84944 Chaetopoa taylorii C. E. Hubb.;东非早熟禾■☆

84945 Chaetopogon Janch. (1913);刚须草属■☆

84946 Chaetopogon Janch. = Chaeturus Link ■☆

84947 Chaetopogon fasciculatus (Link) Hayek;刚须草■☆

84948 Chaetoptelea Liebm. (1850);墨西哥榆属●☆

84949 Chaetoptelea mexicana Liebm.;墨西哥榆●☆

84950 Chaetosciadium Boiss.(1872);刚毛伞芹属■☆
84951 Chaetosciadium trichospermunl(L.)Boiss.;刚毛伞芹■☆
84952 Chaetoseris C. Shih(1991);毛鳞菊属■
84953 Chaetoseris beesiana(Diels)C. Shih;羽裂毛鳞菊(线叶毛鳞菊,紫毛毛鳞菊)■
84954 Chaetoseris bonatii((P. Beauv.)C. Shih;东川毛鳞菊■
84955 Chaetoseris ciliata C. Shih;景东毛鳞菊■
84956 Chaetoseris cyanea(D. Don)C. Shih;蓝花毛鳞菊(大桦口草,戟叶锡莎菊,苦参,蓝花岩参,蓝锡莎菊,蓝岩参,蓝岩参菊,犁头尖,雪兰山菊)■
84957 Chaetoseris dolichophylla C. Shih;长叶毛鳞菊■
84958 Chaetoseris grandiflora(Franch.)C. Shih;大花毛鳞菊(大花蓝岩参,大花苺苣)■
84959 Chaetoseris hastata(Wall. ex DC.)C. Shih;滇藏毛鳞菊(密毛毛鳞菊)■
84960 Chaetoseris hirsuta(Franch.)C. Shih;鹤庆毛鳞菊(毛苺苣,硬毛毛鳞菊);Hairy Lettuce ■
84961 Chaetoseris hirsuta(Franch.)C. Shih = Lactuca hirsuta Franch. ■
84962 Chaetoseris hispida C. Shih;粗毛毛鳞菊■
84963 Chaetoseris hispida C. Shih = Chaetoseris cyanea(D. Don)C. Shih ■
84964 Chaetoseris leiolepis C. Shih;光苞毛鳞菊■
84965 Chaetoseris likiangensis(Franch.)C. Shih;丽江毛鳞菊(丽江蓝岩参菊)■
84966 Chaetoseris lutea(Hand. -Mazz.)C. Shih;黄花毛鳞菊(黄花蓝岩参)■
84967 Chaetoseris lyriformis C. Shih;毛鳞菊■
84968 Chaetoseris macrantha(C. B. Clarke)C. Shih;缘毛毛鳞菊(大花岩参)■
84969 Chaetoseris macrocephala C. Shih;大头毛鳞菊■
84970 Chaetoseris pectiniformis C. Shih;栉齿毛鳞菊■
84971 Chaetoseris qiliangshanensis S. W. Liu et T. N. Ho;祁连山毛鳞菊■
84972 Chaetoseris rhombiformis C. Shih;菱裂毛鳞菊■
84973 Chaetoseris roborowskii(Maxim.)C. Shih;川甘毛鳞菊(青甘岩参)■
84974 Chaetoseris sichuanensis C. Shih;四川毛鳞菊■
84975 Chaetoseris taliensis C. Shih;戟裂毛鳞菊(戟叶毛鳞菊)■
84976 Chaetoseris teniana(P. Beauv.)C. Shih;盐丰毛鳞菊(紫苞毛鳞菊)■
84977 Chaetoseris yunnanensis C. Shih;云南毛鳞菊■
84978 Chaetospermum(M. Roem.)Swingle = Limonia L. ●☆
84979 Chaetospermum(M. Roem.)Swingle = Swinglea Merr. ●☆
84980 Chaetospermum Swingle = Limonia L. ●☆
84981 Chaetospermum Swingle = Swinglea Merr. ●☆
84982 Chaetospermum Swingle(1913);毛籽枳属(毛籽橘属);Tabog ●☆
84983 Chaetospermum glutinosum Swingle;毛籽橘;Common Tabog ●☆
84984 Chaetospira S. F. Blake = Pseudelephantopus Rohr(保留属名)■
84985 Chaetospira S. F. Blake = Spirochaeta Turcz. ■
84986 Chaetospora Faurel et Schotter(1965);毛子莎属■☆
84987 Chaetospora Kunth = Rhynchospora Vahl(保留属名)■
84988 Chaetospora R. Br. = Schoenus L. ■
84989 Chaetospora alpina(R. Br.)F. Muell.;毛子莎■☆
84990 Chaetospora burmannii(Vahl)Schrad. = Tetraria burmannii(Vahl)C. B. Clarke ■☆
84991 Chaetospora calostachya R. Br. = Schoenus calostachyus(R. Br.)Poir. ■
84992 Chaetospora capillacea(Thunb.)Nees = Tetraria capillacea(Thunb.)C. B. Clarke ■☆
84993 Chaetospora distachya Nees = Abildgaardia hispidula(Vahl)Lye ■☆
84994 Chaetospora fimbriolata Nees = Tetraria fimbriolata(Nees)C. B. Clarke ■☆
84995 Chaetospora hexandra Boeck. = Tetraria triangularis(Boeck.)C. B. Clarke ■☆
84996 Chaetospora robusta Kunth = Tetraria robusta(Kunth)C. B. Clarke ■☆
84997 Chaetostachydium Airy Shaw(1965);小毛穗茜属●■☆
84998 Chaetostachydium versteegii(Valeton)Airy Shaw;小毛穗茜●■☆
84999 Chaetostachys Benth. = Lavandula L. ●■
85000 Chaetostachys Valeton = Chaetostachydium Airy Shaw ●■☆
85001 Chaetostemma Rchb. = Chaetostoma DC. ●☆
85002 Chaetostichium C. E. Hubb. = Oropetium Trin. ■☆
85003 Chaetostichium majusculum C. E. Hubb. = Oropetium minimum(Hochst.)Pilg. ■☆
85004 Chaetostichium minimum(Hochst.)C. E. Hubb. = Oropetium minimum(Hochst.)Pilg. ■☆
85005 Chaetostichium minimum(Hochst.)C. E. Hubb. var. microchaetum Chiov. = Oropetium minimum(Hochst.)Pilg. ■☆
85006 Chaetostoma DC.(1828);毛口野牡丹属●☆
85007 Chaetostoma albiflorum(Naudin)Koschnitzke et A. B. Martins;白花毛口野牡丹●☆
85008 Chaetostoma aureum Glaz.;黄毛口野牡丹●☆
85009 Chaetostoma gracile Glaz.;纤细毛口野牡丹●☆
85010 Chaetostoma tetrastichum DC.;毛口野牡丹●☆
85011 Chaetosus Benth. = Parsonsia R. Br.(保留属名)●
85012 Chaetothylax Nees = Justicia L. ●■
85013 Chaetothylopsis Oerst. = Chaetothylax Nees ●■
85014 Chaetothylopsis Oerst. = Justicia L. ●■
85015 Chaetotropis Kunth = Polypogon Desf. ■
85016 Chaetotropis Kunth(1829);智利刺毛禾属;Chaetotropis ■☆
85017 Chaetotropis chilensis Kunth;智利刺毛禾;Chile Chaetotropis ■☆
85018 Chaetotropsis D. Dietr. = Chaetotropis Kunth ■☆
85019 Chaeturus Host ex St. -Lag. = Chaiturus Ehrh. ex Willd. ■
85020 Chaeturus Host ex St. -Lag. = Leonurus L. ■
85021 Chaeturus Link = Chaetopogon Janch. ■☆
85022 Chaeturus Rchb. = Chaiturus Ehrh. ex Willd. ■
85023 Chaeturus fasciculatus Link = Chaetopogon fasciculatus(Link)Hayek ■☆
85024 Chaetymenia Hook. et Arn. = Chaethymenia Hook. et Arn. ■☆
85025 Chaetymenia peduncularis Hook. et Arn. = Chaethymenia peduncularis Hook. et Arn. ■☆
85026 Chaffeyopuntia Frič et Schelle = Opuntia Mill. ●
85027 Chailletia DC. = Dichapetalum Thouars ●
85028 Chailletia affinis Planch. ex Benth. = Dichapetalum affine(Planch. ex Benth.)Breteler ●☆
85029 Chailletia bangii Didr. = Dichapetalum bangii(Didr.)Engl. ●☆
85030 Chailletia cymosa Hook. = Dichapetalum cymosum(Hook.)Engl. ●☆
85031 Chailletia deflexa Klotzsch = Dichapetalum deflexum(Klotzsch)Engl. ●☆
85032 Chailletia dichapetalum DC. f. macrophylla Tul. = Dichapetalum madagascariense Poir. ●☆
85033 Chailletia dichapetalum DC. var. pubescens Desc. =

Dichapetalum madagascariense Poir. ●☆

85034 Chailletia floribunda Planch. = Dichapetalum floribundum (Planch.) Engl. ●☆

85035 Chailletia gelonioides Bedd. = Dichapetalum gelonioides (Roxb.) Engl. ●

85036 Chailletia hainanensis Hance = Dichapetalum longipetalum (Turcz.) Engl. ●

85037 Chailletia heudelotii Planch. ex Oliv. = Dichapetalum heudelotii (Planch. ex Oliv.) Baill. ●☆

85038 Chailletia hispida Oliv. = Dichapetalum heudelotii (Planch. ex Oliv.) Baill. var. hispidum (Oliv.) Breteler ●☆

85039 Chailletia longipetala Turcz. = Dichapetalum longipetalum (Turcz.) Engl. ●

85040 Chailletia macrophylla Oliv. = Dichapetalum heudelotii (Planch. ex Oliv.) Baill. var. hispidum (Oliv.) Breteler ●☆

85041 Chailletia mossambicensis Klotzsch = Dichapetalum mossambicense (Klotzsch) Engl. ●☆

85042 Chailletia oblonga Hook. f. ex Benth. = Dichapetalum oblongum (Hook. f. ex Benth.) Engl. ●☆

85043 Chailletia pallida Oliv. = Dichapetalum pallidum (Oliv.) Engl. ●☆

85044 Chailletia rufipilis Turcz. = Dichapetalum bangii (Didr.) Engl. ●☆

85045 Chailletia subauriculata Oliv. = Dichapetalum heudelotii (Planch. ex Oliv.) Baill. ●☆

85046 Chailletia subcordata Hook. f. ex Benth. = Dichapetalum madagascariense Poir. ●☆

85047 Chailletia thomsonii Oliv. = Dichapetalum madagascariense Poir. ●☆

85048 Chailletia toxicaria G. Don = Dichapetalum toxicarium (G. Don) Baill. ●☆

85049 Chailletia toxicaria G. Don var. elliptica Oliv. = Dichapetalum toxicarium (G. Don) Baill. ●☆

85050 Chailletia whytei Stapf = Dichapetalum pallidum (Oliv.) Engl. ●☆

85051 Chailletiaceae R. Br. = Dichapetalaceae Baill. (保留科名)●

85052 Chaitaea Sol. ex Seem. = Tacca J. R. Forst. et G. Forst. (保留属名)■

85053 Chaitea S. Parkinson = Chaitaea Sol. ex Seem. ■

85054 Chaiturus Ehrh. ex Willd. (1787);鬃尾草属;Chaiturus ■

85055 Chaiturus Willd. = Chaiturus Ehrh. ex Willd. ■

85056 Chaiturus leonuroides Willd. = Chaiturus marrubiastrum (L.) Spenn. ■

85057 Chaiturus marrubiastrum (L.) Rchb. = Chaiturus marrubiastrum (L.) Spenn. ■

85058 Chaiturus marrubiastrum (L.) Spenn.;鬃尾草(夏毛益母草);Biennial Motherwort, Hoarhound-like Chaiturus, Horehound Motherwort, Lion's-tail ■

85059 Chaixia Lapeyr. = Ramonda Pers. ■☆

85060 Chaixia Lapeyr. = Ramonda Rich. (保留属名)■☆

85061 Chaklatella DC. = Chatiakella Cass. ■☆

85062 Chaklatella DC. = Wulffia Neck. ex Cass. ■☆

85063 Chalarium DC. = Desmodium Desv. (保留属名)●■

85064 Chalarium DC. = Edusaron Medik. ●■

85065 Chalarium Poit. ex DC. = Eleutheranthera Poit. ex Bosc ■☆

85066 Chalarium Poit. ex DC. = Ogiera Cass. ■☆

85067 Chalarothyrsus Lindau(1904);柔茎爵床属☆

85068 Chalarothyrsus amplexicaulis Lindau;柔茎爵床☆

85069 Chalazocarpus Hiern = Schumanniophyton Harms ●☆

85070 Chalazocarpus hirsutus Hiern = Schumanniophyton hirsutum (Hiern) R. D. Good ●☆

85071 Chalcanthus Boiss. (1867);中亚铜花芥属■☆

85072 Chalcanthus renifolius Boiss.;中亚铜花芥■☆

85073 Chalcanthus tuberosus (Kom.) Kom. = Chalcanthus renifolius Boiss. ■☆

85074 Chalcas L. = Camunium Adans. ●

85075 Chalcas L. = Murraya J. König ex L. (保留属名)●

85076 Chalcas cammuneng Burm. f. = Murraya paniculata (L.) Jack ●

85077 Chalcas crenulata Tanaka = Murraya crenulata (Turcz.) Oliv. ●

85078 Chalcas euchrestifolia Tanaka = Murraya euchrestifolia Hayata ●

85079 Chalcas exotica Millsp. = Murraya paniculata (L.) Jack ●

85080 Chalcas japonensis Lour. = Murraya paniculata (L.) Jack ●

85081 Chalcas koenigii (L.) Kurz = Murraya koenigii (L.) Spreng. ●

85082 Chalcas koenigii Kurz. ex Swingle = Murraya koenigii (L.) Spreng. ●

85083 Chalcas paniculata L. = Murraya paniculata (L.) Jack ●

85084 Chalcitis Post et Kuntze = ? Aster L. ●■

85085 Chalcitis Post et Kuntze = Xalkitis Raf. ●■

85086 Chalcoelytrum Lunell = Chrysopogon Trin. (保留属名)■

85087 Chalcoelytrum Lunell = Sorghastrum Nash ■☆

85088 Chaldia Bojer = Chadsia Bojer ●☆

85089 Chaleas N. T. Burb. = Chalcas L. ●

85090 Chaleas N. T. Burb. = Murraya J. König ex L. (保留属名)●

85091 Chalebus Raf. = Salix L. (保留属名)●

85092 Chalema Dieterle(1980);聚药瓜属■☆

85093 Chalema synanthera Dieterle;聚药瓜■☆

85094 Chalepoa Hook. f. = Tribeles Phil. ●☆

85095 Chalepophyllum Hook. f. (1873);三齿叶茜属(亮叶茜属)☆

85096 Chalepophyllum guyanense Hook. f.;圭亚那三齿叶茜☆

85097 Chalepophyllum latifolium Standl.;宽三齿叶茜☆

85098 Chalinanthus Briq. = Lagochilus Bunge ex Benth. ●■

85099 Chalmersia F. Muell. ex S. Moore = Dichrotrichum Reinw. ■☆

85100 Chalmysporum Salisb. = Thysanotus R. Br. (保留属名)■

85101 Chalybea Naudin = Pachyanthus A. Rich. ●☆

85102 Chalybea Naudin(1851);钢灰野牡丹属●☆

85103 Chalybea corymbifera Naudin;钢灰野牡丹☆

85104 Chalynochlamys Franch. = Arundinella Raddi ■

85105 Chalynochlamys anomala (Steud.) Franch. = Arundinella anomala Steud. ■

85106 Chamabainia Wight(1853);微柱麻属(虫蚁麻属,张麻属);Chamabainia, Ministylenettle ■

85107 Chamabainia cuspidata Wight;微柱麻(虫蚁菜,虫蚁麻,地水麻,红四楞麻,水水苏麻,台湾虫蚁麻,小米麻草,张麻,止血草,虫蚁菜);Cuspid Chamabainia, Ministylenettle ■

85108 Chamabainia cuspidata Wight var. denticulosa W. T. Wang et C. J. Chen;多齿微柱麻;Denticulate Chamabainia, Manytooth Ministylenettle ■

85109 Chamabainia cuspidata Wight var. denticulosa W. T. Wang et C. J. Chen = Chamabainia cuspidata Wight ■

85110 Chamabainia cuspidata Wight var. morri (Hayata) W. T. Wang;小叶微柱麻;Sallleaf Chamabainia, Sallleaf Ministylenettle ■

85111 Chamabainia cuspidata Wight var. morri (Hayata) W. T. Wang = Chamabainia cuspidata Wight ■

85112 Chamabainia morri Hayata = Chamabainia cuspidata Wight ■

85113 Chamabainia squamigera Wedd. = Chamabainia cuspidata Wight ■

85114 Chamaeacanthus Chiov. = Campylanthus Roth ●☆

85115 Chamaeacanthus pumilus Chiov. = Campylanthus spinosus Balf.

f. ●☆

85116 Chamaealoe A. Berger = Aloe L. ●■

85117 Chamaealoe A. Berger = Bowiea Haw. (废弃属名)●■

85118 Chamaealoe africana (Haw.) A. Berger = Aloe bowiea Schult. et Schult. f. ■☆

85119 Chamaeangis Schltr. (1918);矮船兰属■☆

85120 Chamaeangis dewevrei (De Wild.) Schltr. = Chamaeangis odoratissima (Rchb. f.) Schltr. ■☆

85121 Chamaeangis gabonensis Summerh.;加蓬矮船兰■☆

85122 Chamaeangis gracilis Schltr.;纤细矮船兰■☆

85123 Chamaeangis ichneumonea (Lindl.) Schltr.;蜂矮船兰■☆

85124 Chamaeangis lanceolata Summerh.;披针矮船兰■☆

85125 Chamaeangis lecomtei (Finet) Schltr.;勒孔特矮船兰■☆

85126 Chamaeangis letouzeyi Szlach. et Olszewski;勒图矮船兰■☆

85127 Chamaeangis odoratissima (Rchb. f.) Schltr.;极香矮船兰■☆

85128 Chamaeangis orientalis Summerh. = Chamaeangis sarcophylla Schltr. ■☆

85129 Chamaeangis pauciflora Perez-Vera;少花矮船兰■☆

85130 Chamaeangis sarcophylla Schltr.;肉叶矮船兰■☆

85131 Chamaeangis schliebenii Mansf. = Rhipidoglossum rutilum (Rchb. f.) Schltr. ■☆

85132 Chamaeangis thomensis (Rolfe) Schltr.;汤姆矮船兰■☆

85133 Chamaeangis vagans (Lindl.) Schltr.;漫游矮船兰■☆

85134 Chamaeangis vesicata (Lindl.) Schltr.;瘦弱矮船兰■☆

85135 Chamaeanthus Schltr. = Chamaeanthus Schltr. ex J. J. Sm. ■

85136 Chamaeanthus Schltr. ex J. J. Sm. (1905);低药兰属(微花兰属);Lowanther-orchis ■

85137 Chamaeanthus Ule = Geogenanthus Ule ■☆

85138 Chamaeanthus Ule = Uleopsis Fedde ■

85139 Chamaeanthus wenzelii Ames;低药兰(威氏细花兰);Lowanther-orchis ■

85140 Chamaebatia Benth. (1849);矮灌蔷薇属●☆

85141 Chamaebatia foliosa Benth.;矮灌蔷薇;Mountain Misery ●☆

85142 Chamaebatiaria (Porter ex W. H. Brewer et S. Watson) Maxim. (1879);蕨木属(蓍叶木属);Fernbush ●☆

85143 Chamaebatiaria (Porter) Maxim. = Chamaebatiaria (Porter ex W. H. Brewer et S. Watson) Maxim. ●☆

85144 Chamaebatiaria (W. H. Brewer et S. Watson) Maxim. = Chamaebatiaria (Porter ex W. H. Brewer et S. Watson) Maxim. ●☆

85145 Chamaebatiaria Maxim. = Chamaebatiaria (Porter ex W. H. Brewer et S. Watson) Maxim. ●☆

85146 Chamaebatiaria millefolium (Torr.) Maxim.;蕨木(蓍叶木);Fernbush ●☆

85147 Chamaebatiaria millefolium Maxim. = Chamaebatiaria millefolium (Torr.) Maxim. ●☆

85148 Chamaebetula Opiz = Betula L. ●

85149 Chamaebuxus (DC.) Spach = Polygaloides Haller ●☆

85150 Chamaebuxus (Tourn.) Spach = Polygala L. ●■

85151 Chamaebuxus Spach = Polygala L. ●■

85152 Chamaebuxus arillata Hassk. = Polygala arillata Buch.-Ham. ex D. Don ●

85153 Chamaebuxus arillata Hassk. var. brachybotrya Hassk. = Polygala arillata Buch.-Ham. ex D. Don ●

85154 Chamaebuxus arillata Hassk. var. robusta Hassk. = Polygala arillata Buch.-Ham. ex D. Don ●

85155 Chamaebuxus paniculata Hassk. = Polygala tricholopha Chodat ●

85156 Chamaecalamus Meyen = Calamagrostis Adans. ■

85157 Chamaecalamus Meyen = Deyeuxia Clarion ■

85158 Chamaecassia Link = Cassia L. (保留属名)●■

85159 Chamaecerasus Duhamel = Lonicera L. ●■

85160 Chamaecerasus alberti (Regel) Carrière = Lonicera albertii Regel ●

85161 Chamaecerasus alberti Carrière = Lonicera albertii Regel ●

85162 Chamaecereus Britton et Rose = Echinopsis Zucc. ●

85163 Chamaecereus Britton et Rose (1922);白檀属(白檀柱属,仙人柱属);Peanut Cactus ●☆

85164 Chamaecereus silvestrii Britton et Rose = Chamaecereus sylvestri (Speg.) Britton et Rose ●☆

85165 Chamaecereus sylvestri (Speg.) Britton et Rose;白檀(葫芦拳白檀,金牛掌,山吹,小仙人鞭);Peanut Cactus ●☆

85166 Chamaechaenactis Rydb. (1906);矮针垫菊属;Fullstem ■☆

85167 Chamaechaenactis scaposa (Eastw.) Rydb.;矮针垫菊;Fullstem ■☆

85168 Chamaechaenactis scaposa (Eastw.) Rydb. var. parva Preece et B. L. Turner = Chamaechaenactis scaposa (Eastw.) Rydb. ■☆

85169 Chamaecissos Lunell = Glechoma L. (保留属名)■

85170 Chamaecistus (G. Don) Regel = Rhododendron L. ●

85171 Chamaecistus (G. Don) Regel = Rhodothamnus Rchb. (保留属名)●☆

85172 Chamaecistus Fabr. = Helianthemum Mill. ●■

85173 Chamaecistus Gray = Loiseleuria Desv. (保留属名)●☆

85174 Chamaecistus Oeder = Loiseleuria Desv. (保留属名)●☆

85175 Chamaecistus Regel = Rhododendron L. ●

85176 Chamaecistus Regel = Rhodothamnus Rchb. (保留属名)●☆

85177 Chamaecistus procumbens (L.) Kuntze = Loiseleuria procumbens (L.) Desv. ex Loisel. ●☆

85178 Chamaecladon Miq. = Homalomena Schott ■

85179 Chamaeclema Boehm. = Glechoma L. (保留属名)■

85180 Chamaeclema Moench = Nepeta L. ■●

85181 Chamaeclitandra (Stapf) Pichon (1953);非洲斜蕊夹竹桃属●☆

85182 Chamaeclitandra henriquesiana (Hallier f.) Pichon;非洲斜蕊夹竹桃●☆

85183 Chamaecnide Nees et Mart. ex Miq. = Pilea Lindl. (保留属名)■

85184 Chamaecostus C. D. Specht et D. W. Stev. = Globba L. ■

85185 Chamaecrinum Diels = Hensmania W. Fitzg. ■☆

85186 Chamaecrinum Diels ex Diels et Pritz. = Hensmania W. Fitzg. ■☆

85187 Chamaecrista (L.) Moench (1794);茶豆属(山扁豆属,鹧鸪豆属)■●

85188 Chamaecrista Moench = Cassia L. (保留属名)●■

85189 Chamaecrista Moench = Chamaecrista (L.) Moench ■●

85190 Chamaecrista absus (L.) H. S. Irwin et Barneby = Cassia absus L. ●☆

85191 Chamaecrista africana (Steyaert) Lock = Cassia africana (Steyaert) Mendonça et Torre ●☆

85192 Chamaecrista ankaratrensis (R. Vig.) Du Puy;安卡拉特拉茶豆●☆

85193 Chamaecrista arenicola (R. Vig.) Du Puy;沙生茶豆●☆

85194 Chamaecrista brachiata Pollard = Cassia fasciculata Michx. ●☆

85195 Chamaecrista brevifolia (Lam.) Greene;短叶茶豆●☆

85196 Chamaecrista capensis (Thunb.) E. Mey.;好望角茶豆●☆

85197 Chamaecrista capensis (Thunb.) E. Mey. var. flavescens E. Mey.;黄好望角茶豆●☆

85198 Chamaecrista capensis (Thunb.) E. Mey. var. flavescens E. Mey. = Cassia capensis Thunb. var. flavescens (E. Mey.) Vogel ●☆

85199 Chamaecrista comosa E. Mey.;簇毛茶豆●☆

85200 Chamaecrista comosa E. Mey. = Cassia comosa (E. Mey.) Vogel ●☆

85201 Chamaecrista depressa (Pollard) Greene = Cassia fasciculata Michx. ●☆

85202 Chamaecrista dimidiata (Roxb.) Lock = Cassia hochstetteri Ghesq. ●☆

85203 Chamaecrista duboisii (Steyaert) Lock = Cassia duboisii Steyaert ●☆

85204 Chamaecrista dumaziana (Brenan) Du Puy;杜马茶豆●☆

85205 Chamaecrista dunensis Thulin;砂丘茶豆●☆

85206 Chamaecrista exilis (Vatke) Lock = Cassia exilis Vatke ●☆

85207 Chamaecrista falcinella (Oliv.) Lock = Cassia falcinella Oliv. ●☆

85208 Chamaecrista falcinella (Oliv.) Lock var. intermedia (Brenan) Lock = Cassia falcinella Oliv. var. intermedia Brenan ●☆

85209 Chamaecrista falcinella (Oliv.) Lock var. parviflora (Steyaert) Lock;小花茶豆●☆

85210 Chamaecrista fallacina (Chiov.) Lock = Cassia fallacina Chiov. ●☆

85211 Chamaecrista fasciculata (Michx.) Greene = Cassia fasciculata Michx. ●☆

85212 Chamaecrista fasciculata (Michx.) Greene var. macrosperma (Fernald) C. F. Reed = Cassia fasciculata Michx. ●☆

85213 Chamaecrista fenarolii (Mendonça et Torre) Lock;费纳罗利茶豆●☆

85214 Chamaecrista garambiensis (Hosok.) H. Ohashi, Tateishi et T. Nemoto = Cassia garambiensis Hosok. ●

85215 Chamaecrista ghesquiereana (Brenan) Lock = Cassia ghesquiereana Brenan ●☆

85216 Chamaecrista gracilior (Ghesq.) Lock = Cassia gracilior (Ghesq.) Steyaert ●☆

85217 Chamaecrista grantii (Oliv.) Standl. = Cassia grantii Oliv. ●☆

85218 Chamaecrista hildebrandtii (Vatke) Lock = Cassia hildebrandtii Vatke ●☆

85219 Chamaecrista huillensis (Mendonça et Torre) Lock;威拉茶豆●☆

85220 Chamaecrista jaegeri (Keay) Lock = Cassia jaegeri Keay ●☆

85221 Chamaecrista kalulensis (Steyaert) Lock = Cassia kalulensis Steyaert ●☆

85222 Chamaecrista katangensis (Ghesq.) Lock = Cassia katangensis (Ghesq.) Steyaert ●☆

85223 Chamaecrista kirkii (Oliv.) Standl.;柯克茶豆●☆

85224 Chamaecrista kirkii (Oliv.) Standl. = Cassia kirkii Oliv. ●☆

85225 Chamaecrista kirkii (Oliv.) Standl. var. glabra (Steyaert) Lock = Cassia kirkii Oliv. var. glabra Steyaert ●☆

85226 Chamaecrista kirkii (Oliv.) Standl. var. guineensis (Steyaert) Lock = Cassia kirkii Oliv. var. guineensis Steyaert ●☆

85227 Chamaecrista lateriticola (R. Vig.) Du Puy;砖红茶豆●☆

85228 Chamaecrista lechenaultiana (DC.) O. Deg.;大叶山扁豆■

85229 Chamaecrista lechenaultiana (DC.) O. Deg. = Cassia lechenaultiana DC. ●

85230 Chamaecrista littoralis Pollard = Cassia fasciculata Michx. ●☆

85231 Chamaecrista meelii (Steyaert) Lock = Cassia meelii Steyaert ●☆

85232 Chamaecrista mimosoides (L.) Greene;山扁豆(篦子草,地柏草,疳草,含羞草决明,红杠木,红霜石,还瞳子,黄瓜香,鸡毛箭,假含羞草,假牛柑,金蜈蚣,梦草,砂子草,水皂角,挞地沙,挞地砂,铁箭矮陀,望江南,细杠木,下通草,野皂角,鱼骨折);Sensitiveplant-like Senna, Tea Senna ●

85233 Chamaecrista mimosoides (L.) Greene = Cassia mimosoides L. ●

85234 Chamaecrista mississippiensis (Pollard) Pollard ex A. Heller = Cassia fasciculata Michx. ●☆

85235 Chamaecrista newtonii (Mendonça et Torre) Lock = Cassia newtonii Mendonça et Torre ■☆

85236 Chamaecrista nicticans (L.) Moench subsp. patellaria (DC. ex Collad.) H. S. Irwin et Barneby var. glabrata (Vogel) H. S. Irwin et Barneby = Chamaecrista lechenaultiana (DC.) O. Deg. ■

85237 Chamaecrista nictitans (L.) Moench = Cassia lechenaultiana DC. ●

85238 Chamaecrista nictitans (L.) Moench = Cassia nictitans L. ●☆

85239 Chamaecrista nigricans (Vahl) Greene = Cassia nigricans Vahl ●☆

85240 Chamaecrista nomame (Siebold) H. Ohashi = Cassia nomame (Siebold) Kitag. ●

85241 Chamaecrista paralias (Brenan) Lock = Cassia paralias Brenan ●☆

85242 Chamaecrista parva (Steyaert) Lock;较小茶豆●☆

85243 Chamaecrista plumosa E. Mey.;羽状茶豆●☆

85244 Chamaecrista plumosa E. Mey. var. diffusa = Chamaecrista plumosa E. Mey. ●☆

85245 Chamaecrista plumosa E. Mey. var. erecta (Schorn et Gordon-Gray) Lock;直立茶豆●☆

85246 Chamaecrista polytricha (Brenan) Lock;多毛茶豆●☆

85247 Chamaecrista polytricha Lock = Chamaecrista polytricha (Brenan) Lock ●☆

85248 Chamaecrista pratensis (R. Vig.) Du Puy;草原茶豆●☆

85249 Chamaecrista puccioniana (Chiov.) Lock = Cassia puccioniana Chiov. ●☆

85250 Chamaecrista pumila (Lam.) K. Larsen;小柄腺山扁豆●■

85251 Chamaecrista reducta (Brenan) Du Puy;退缩茶豆●☆

85252 Chamaecrista robusta (Pollard) Pollard ex A. Heller = Cassia fasciculata Michx. ●☆

85253 Chamaecrista robynsiana (Ghesq.) Lock = Cassia robynsiana Ghesq. ●☆

85254 Chamaecrista rostrata Wooton et Standl. = Cassia fasciculata Michx. ●☆

85255 Chamaecrista schmitzii (Steyaert) Lock = Cassia schmitzii Steyaert ●☆

85256 Chamaecrista stricta E. Mey.;刚直茶豆●☆

85257 Chamaecrista telfairiana (Hook. f.) Lock = Cassia telfairiana (Hook. f.) Polhill ●☆

85258 Chamaecrista tracyi Pollard = Cassia fasciculata Michx. ●☆

85259 Chamaecrista usambarensis (Taub.) Standl. = Cassia usambarensis Taub. ●☆

85260 Chamaecrista wittei (Ghesq.) Lock = Cassia wittei Ghesq. ●☆

85261 Chamaecrista zambesica (Oliv.) Lock = Cassia zambesica Oliv. ●☆

85262 Chamaecrypta Schltr. et Diels = Diascia Link et Otto ■☆

85263 Chamaecyparis Spach(1841);扁柏属(花柏属);Cypress, False Cypress, Falsecypress, False-cypress, Lawson Cypress, White-cedar, Yellow Cedar ●

85264 Chamaecyparis breviramea Maxim. = Chamaecyparis obtusa (Siebold et Zucc.) Siebold et Zucc. ex Endl. ●

85265 Chamaecyparis breviramea Maxim. = Chamaecyparis obtusa (Siebold et Zucc.) Siebold et Zucc. ex Endl. 'Breviramea' ●

85266 Chamaecyparis cyano-virde;蓝绿扁柏;False Cypress ●☆

85267 Chamaecyparis filifera Veitch ex Sénükl. = Chamaecyparis pisifera (Siebold et Zucc.) Siebold et Zucc. ex Endl. 'Filifera' ●

85268 Chamaecyparis formosensis Matsum.;红桧(薄皮,薄皮松萝,松

梧，台湾扁柏）；Formosan Cypress，Formosan False Cypress，Formosan Red Cypress，Taiwan Cypress，Taiwan False Cypress，Taiwan Red Cypress，Taiwan Red False Cypress，Tree of God ●◇

85269 Chamaecyparis funebris（Endl.）Franco = Cupressus funebris Endl. ●

85270 Chamaecyparis henryae L. H. Li = Chamaecyparis thyoides（L.）Britton，Sterns et Poggenb. ●

85271 Chamaecyparis lawsoniana（A. Murray）Parl.；美国扁柏（美国红桧，美国花柏，美国尖叶扁柏）；Cedar，Ginger-pine，Lawson Cypress，Lawson False Cypress，Lawson's Cypress，Lawsons-cypress，Match Wood，Oregon Cedar，Oxford Cedar，Port Orford Cedar，Port Orford White Cedar，Port-orford-cedar，White Cedar ●

85272 Chamaecyparis lawsoniana（A. Murray）Parl. 'Allumi'；金龟子叶花柏；Blue Lawson Cypress，Scaraba Cypress ●

85273 Chamaecyparis lawsoniana（A. Murray）Parl. 'Argentea'；银白美国扁柏（银白花柏，银色美国扁柏）；Silver Lawson Cypress ●

85274 Chamaecyparis lawsoniana（A. Murray）Parl. 'Aurea'；金色美国扁柏●

85275 Chamaecyparis lawsoniana（A. Murray）Parl. 'Chilworth Silver'；蓝灰美国扁柏●

85276 Chamaecyparis lawsoniana（A. Murray）Parl. 'Columnaris'；柱状美国扁柏（柱冠美国花柏）●☆

85277 Chamaecyparis lawsoniana（A. Murray）Parl. 'Ellwoodii'；爱乌德花柏（埃尔伍德美国花柏）；Ellwood Cypress ●

85278 Chamaecyparis lawsoniana（A. Murray）Parl. 'Erecta Viridis'；绿柱美国扁柏（绿柱美国花柏）；Erect Green Cedar ●

85279 Chamaecyparis lawsoniana（A. Murray）Parl. 'Filiformis'；线叶美国扁柏●☆

85280 Chamaecyparis lawsoniana（A. Murray）Parl. 'Fletcheri'；弗雷彻花柏（弗莱彻美国花柏）；Fletcher Cypress ●

85281 Chamaecyparis lawsoniana（A. Murray）Parl. 'Forsteckensis'；弗斯特克花柏；Forsteck Cypress ●

85282 Chamaecyparis lawsoniana（A. Murray）Parl. 'Fraseri'；蓝叶美国扁柏（蓝叶美国花柏）；Fraser Cedar ●

85283 Chamaecyparis lawsoniana（A. Murray）Parl. 'Glauca'；灰叶美国扁柏（灰叶美国花柏）；Gray-leaf Cedar ●

85284 Chamaecyparis lawsoniana（A. Murray）Parl. 'Gnome'；侏儒美国花柏●☆

85285 Chamaecyparis lawsoniana（A. Murray）Parl. 'Green Pillar'；绿柱美国花柏●☆

85286 Chamaecyparis lawsoniana（A. Murray）Parl. 'Intertexta'；交织美国扁柏（垂枝美国花柏）●☆

85287 Chamaecyparis lawsoniana（A. Murray）Parl. 'Kilmacurragh'；柱冠美国花柏●☆

85288 Chamaecyparis lawsoniana（A. Murray）Parl. 'Lanei'；拉尼美国花柏●☆

85289 Chamaecyparis lawsoniana（A. Murray）Parl. 'Lutea'；金花柏（黄叶美国扁柏，金黄凤羽柏）；Golden Lawson Cypress ●

85290 Chamaecyparis lawsoniana（A. Murray）Parl. 'Minima Glauca'；兰花柏（极小美国扁柏，袖珍美国花柏）；Little Blue Cypress ●

85291 Chamaecyparis lawsoniana（A. Murray）Parl. 'Nana'；矮生美国扁柏●☆

85292 Chamaecyparis lawsoniana（A. Murray）Parl. 'Nidiformis'；鸟巢花柏；Bird Nest Cypress ●

85293 Chamaecyparis lawsoniana（A. Murray）Parl. 'Pempury Blue'；银蓝美国扁柏（孔雀美国花柏）；Blue Lawson Cypress ●☆

85294 Chamaecyparis lawsoniana（A. Murray）Parl. 'Pendula'；垂枝美国扁柏●☆

85295 Chamaecyparis lawsoniana（A. Murray）Parl. 'Silver Queen'；银后花柏；Lawson's Cypress，Silver Lawson Cypress ●

85296 Chamaecyparis lawsoniana（A. Murray）Parl. 'Stardust'；安乐美国扁柏●☆

85297 Chamaecyparis lawsoniana（A. Murray）Parl. 'Stewartii'；紫茎美国扁柏（黄绿花柏）；Stewart Lawson Cypress ●

85298 Chamaecyparis lawsoniana（A. Murray）Parl. 'Tamariscifolia'；柽柳叶美国花柏●☆

85299 Chamaecyparis lawsoniana（A. Murray）Parl. 'Triump of Boskkvop'；凯旋美国扁柏●☆

85300 Chamaecyparis lawsoniana（A. Murray）Parl. 'Wisselii'；卫塞尔花柏（维西尔美国花柏，卫斯里美国扁柏）；Wissel Cypress ●

85301 Chamaecyparis nootkatensis（D. Don）Spach；黄扁柏（阿拉斯加扁柏，努特卡扁柏）；Alaska Cedar，Alaska Cypress，Alaska Yellow Cedar，Alaska-cedar，Nootka Cedar，Nootka Cypress，Nootka False Cypress，Nootka Sound Cypress，Sitka Cypress，Slinking Cedar，Yellow Cedar，Yellow Cypress，Yellow-cypress ●

85302 Chamaecyparis nootkatensis（D. Don）Spach 'Aurea'；金叶黄扁柏●☆

85303 Chamaecyparis nootkatensis（D. Don）Spach 'Compacta'；柱黄扁柏（紧密黄扁柏）；Pyramidal Yellow Cedar ●

85304 Chamaecyparis nootkatensis（D. Don）Spach 'Glauca'；灰叶黄扁柏●☆

85305 Chamaecyparis nootkatensis（D. Don）Spach 'Green Arrow'；绿箭黄扁柏；Columnar Alaska Cedar ●☆

85306 Chamaecyparis nootkatensis（D. Don）Spach 'Pendula'；垂枝黄扁柏（垂叶黄扁柏，垂枝努特卡扁柏）；Weeping Alaska Cedar，Weeping Yellow Cedar ●

85307 Chamaecyparis nootkatensis（D. Don）Sudw. = Chamaecyparis nootkatensis（D. Don）Spach ●

85308 Chamaecyparis nootkatensis Spach = Chamaecyparis nootkatensis（D. Don）Spach ●

85309 Chamaecyparis obtusa（Siebold et Zucc.）Endl. = Chamaecyparis obtusa（Siebold et Zucc.）Siebold et Zucc. ex Endl. ●

85310 Chamaecyparis obtusa（Siebold et Zucc.）Siebold et Zucc. ex Endl.；日本扁柏（白柏，白松，扁柏，扁松，钝叶扁柏，钝叶扁松，桧木，日本柏木）；Fire Cypress，Hinoki Cedar，Hinoki Cypress，Hinoki False Cypress，Hinoki Falsecypress，Japan Cedar，Japanese Cedar，Japanese Cedar Cypress，Japanese Cypress，Japanese False Cypress，Japanese Hinoki Chamaecyparis，Obtuse Ground Cypress ●

85311 Chamaecyparis obtusa（Siebold et Zucc.）Siebold et Zucc. ex Endl. 'Aurea'；金黄日本扁柏；Hinoki Cypress ●

85312 Chamaecyparis obtusa（Siebold et Zucc.）Siebold et Zucc. ex Endl. 'Breviramea'；云片柏；Cloud Nootka Cypress ●

85313 Chamaecyparis obtusa（Siebold et Zucc.）Siebold et Zucc. ex Endl. 'Breviramea Aurea'；金叶云片柏●

85314 Chamaecyparis obtusa（Siebold et Zucc.）Siebold et Zucc. ex Endl. 'Crippsii'；塔形日本扁柏（黄叶日本扁柏）；Crippsi Golden Cypress，Golden Hinoke Cypress ●

85315 Chamaecyparis obtusa（Siebold et Zucc.）Siebold et Zucc. ex Endl. 'Coralliformis'；珊瑚日本扁柏；Coralliformis Hinoke Cypress ●☆

85316 Chamaecyparis obtusa（Siebold et Zucc.）Siebold et Zucc. ex Endl. 'Filicoides'；凤尾柏（凤尾日本扁柏，蕨枝柏）；Fernspray Cypress ●

85317 Chamaecyparis obtusa（Siebold et Zucc.）Siebold et Zucc. ex

Endl.‘Filicoides-aurea’;黄凤尾柏●☆

85318 Chamaecyparis obtusa (Siebold et Zucc.) Siebold et Zucc. ex Endl.‘Gracillis’;纤细日本扁柏;Gracillis Hinoke Cypress, Slender Hinoki Cypress ●

85319 Chamaecyparis obtusa (Siebold et Zucc.) Siebold et Zucc. ex Endl.‘Intermedia’;亮绿日本扁柏●☆

85320 Chamaecyparis obtusa (Siebold et Zucc.) Siebold et Zucc. ex Endl.‘Kosteri’;蔓生日本扁柏;Koster's Hinoke Falsecypress ●☆

85321 Chamaecyparis obtusa (Siebold et Zucc.) Siebold et Zucc. ex Endl.‘Lycopodioides’;石松日本扁柏(水龙骨叶扁柏);Club-moss Cypress ●

85322 Chamaecyparis obtusa (Siebold et Zucc.) Siebold et Zucc. ex Endl.‘Nana Aurea’;矮黄日本扁柏(矮金日本扁柏,日光柏);Golden Dwarf Hinoki Cypress, Nana Lutea Chamaecyparis ●

85323 Chamaecyparis obtusa (Siebold et Zucc.) Siebold et Zucc. ex Endl.‘Nana’;矮生日本扁柏;Dwarf Hinoki Cypress ●

85324 Chamaecyparis obtusa (Siebold et Zucc.) Siebold et Zucc. ex Endl.‘Nana Gracilis’;矮绿日本扁柏(寿星日本扁柏);Nana Gracilis Falsecypress ●

85325 Chamaecyparis obtusa (Siebold et Zucc.) Siebold et Zucc. ex Endl.‘Nana Pyramidalis’;矮锥日本扁柏●☆

85326 Chamaecyparis obtusa (Siebold et Zucc.) Siebold et Zucc. ex Endl.‘Pygmeae’;低矮日本扁柏●

85327 Chamaecyparis obtusa (Siebold et Zucc.) Siebold et Zucc. ex Endl.‘Pygmeae Aurescens’;低矮绿日本扁柏●

85328 Chamaecyparis obtusa (Siebold et Zucc.) Siebold et Zucc. ex Endl.‘Reis Dwarf’;里斯矮日本扁柏;Hinoke Cypress, Reis Dwarf ●☆

85329 Chamaecyparis obtusa (Siebold et Zucc.) Siebold et Zucc. ex Endl.‘Tetragona Aurea’;金黄四方柏;Cypress, Golden Fernleaf ●☆

85330 Chamaecyparis obtusa (Siebold et Zucc.) Siebold et Zucc. ex Endl.‘Tetragona’;孔雀柏(洒金孔雀柏,四方柏,四棱日本扁柏);Tetragonous Hinoki False Cypress ●

85331 Chamaecyparis obtusa (Siebold et Zucc.) Siebold et Zucc. ex Endl. f. hasegawana Hayashi;长谷河柏●☆

85332 Chamaecyparis obtusa (Siebold et Zucc.) Siebold et Zucc. ex Endl. f. barronii Rehder = Chamaecyparis obtusa (Siebold et Zucc.) Siebold et Zucc. ex Endl.‘Tetragona’●

85333 Chamaecyparis obtusa (Siebold et Zucc.) Siebold et Zucc. ex Endl. f. breviramea (Maxim.) Rehder = Chamaecyparis obtusa (Siebold et Zucc.) Siebold et Zucc. ex Endl.‘Breviramea’●

85334 Chamaecyparis obtusa (Siebold et Zucc.) Siebold et Zucc. ex Endl. f. formosana Hayata = Chamaecyparis obtusa (Siebold et Zucc.) Siebold et Zucc. ex Endl. var. formosana (Hayata) Rehder ●

85335 Chamaecyparis obtusa (Siebold et Zucc.) Siebold et Zucc. ex Endl. f. formosana Hayata = Chamaecyparis obtusa (Siebold et Zucc.) Siebold et Zucc. ex Endl. ●

85336 Chamaecyparis obtusa (Siebold et Zucc.) Siebold et Zucc. ex Endl. subsp. formosana (Hayata) H. L. Li = Chamaecyparis obtusa (Siebold et Zucc.) Siebold et Zucc. ex Endl. var. formosana (Hayata) Rehder ●

85337 Chamaecyparis obtusa (Siebold et Zucc.) Siebold et Zucc. ex Endl. subsp. formosana (Hayata) H. L. Li = Chamaecyparis obtusa (Siebold et Zucc.) Siebold et Zucc. ex Endl. ●

85338 Chamaecyparis obtusa (Siebold et Zucc.) Siebold et Zucc. ex Endl. var. formosana (Hayata) Rehder;台湾扁柏(扁柏,厚壳,黄桧,松萝);Taiwan False Cypress, Taiwan Hinoki Falsecypress ●

85339 Chamaecyparis obtusa (Siebold et Zucc.) Siebold et Zucc. ex Endl. var. takeuchii Hayashi;竹内扁柏●☆

85340 Chamaecyparis obtusa (Siebold et Zucc.) Siebold et Zucc. ex Endl. var. filicoides Hartwig et Rümpler = Chamaecyparis obtusa (Siebold et Zucc.) Siebold et Zucc. ex Endl.‘Filicoides’●

85341 Chamaecyparis obtusa (Siebold et Zucc.) Siebold et Zucc. ex Endl. var. plumosa Carrière = Chamaecyparis pisifera (Siebold et Zucc.) Siebold et Zucc. ex Endl.‘Plumosa’●

85342 Chamaecyparis obtusa (Siebold et Zucc.) Siebold et Zucc. ex Endl. var. tetragona Hornibr. = Chamaecyparis obtusa (Siebold et Zucc.) Siebold et Zucc. ex Endl.‘Tetragona’●

85343 Chamaecyparis obtusa (Siebold et Zucc.) Siebold et Zucc. ex Endl. var. filicoides Hartwig et Rome = Chamaecyparis obtusa (Siebold et Zucc.) Siebold et Zucc. ex Endl.‘Filicoides’●

85344 Chamaecyparis pendula Maxim. = Chamaecyparis obtusa (Siebold et Zucc.) Siebold et Zucc. ex Endl. ●

85345 Chamaecyparis pisifera (Siebold et Zucc.) Siebold et Zucc. ex Endl.;日本花柏(五彩松,羽花柏);Boulevard Cypress, Japanese Falsecypress, Pea-fruited Cypress, Retinospora, Sawara Chamaecyparis, Sawara Cypress, Sawara False Cypress, Sawara Falsecypress ●

85346 Chamaecyparis pisifera (Siebold et Zucc.) Siebold et Zucc. ex Endl.‘Aurea’;金叶日本花柏(金叶花柏);Golden Sawara False Cypress ●

85347 Chamaecyparis pisifera (Siebold et Zucc.) Siebold et Zucc. ex Endl.‘Boulevard’;波尔瓦日本花柏(灰柏);Blue Moss Cypress, Boulevard Cypress, Boulevard Falsecypress ●☆

85348 Chamaecyparis pisifera (Siebold et Zucc.) Siebold et Zucc. ex Endl.‘Filifera’;线柏;Thred Cypress, Thredbranch Cypress ●

85349 Chamaecyparis pisifera (Siebold et Zucc.) Siebold et Zucc. ex Endl.‘Plumosa Aurea’;黄孔雀柏(金叶凤尾柏,金羽叶日本花柏);Golden Plumose Sawara False Cypress ●

85350 Chamaecyparis pisifera (Siebold et Zucc.) Siebold et Zucc. ex Endl.‘Plumosa’;羽叶花柏(凤尾柏,孔雀柏);Plumose False Cypress ●

85351 Chamaecyparis pisifera (Siebold et Zucc.) Siebold et Zucc. ex Endl.‘Snow’;白尖日本花柏;Snow Falsecypress, White False Cypress ●

85352 Chamaecyparis pisifera (Siebold et Zucc.) Siebold et Zucc. ex Endl.‘Squarrosa Sulphurea’;黄叶绒柏;Golden Mos Cypress ●

85353 Chamaecyparis pisifera (Siebold et Zucc.) Siebold et Zucc. ex Endl.‘Squarrosa’;绒柏(绒柏日本花柏);Mos Cypress ●

85354 Chamaecyparis pisifera (Siebold et Zucc.) Siebold et Zucc. ex Endl.‘Squarrosa Lutea’;粗鳞日本花柏;Squarrosa Falsecypress ●☆

85355 Chamaecyparis pisifera (Siebold et Zucc.) Siebold et Zucc. ex Endl.‘Filifera Aurea’;金线日本花柏(金线柏);Golden Threadleaf Sawara Cypress ●☆

85356 Chamaecyparis pisifera (Siebold et Zucc.) Siebold et Zucc. ex Endl.‘Filifera Nana’;矮线柏;Dwarf Thread Sawara Falsecypress ●☆

85357 Chamaecyparis pisifera (Siebold et Zucc.) Siebold et Zucc. ex Endl.‘Pendula’;垂枝日本花柏;Weeping Sawara False Cypress ●

85358 Chamaecyparis pisifera (Siebold et Zucc.) Siebold et Zucc. ex Endl.‘Filifera Aurea Nana’;金线矮日本花柏●☆

85359 Chamaecyparis pisifera (Siebold et Zucc.) Siebold et Zucc. ex Endl.‘Nana’;矮花柏●☆

85360 Chamaecyparis pisifera (Siebold et Zucc.) Siebold et Zucc. ex Endl.‘Plumosa Compressa’;羽叶矮日本花柏●☆

85361 Chamaecyparis pisifera (Siebold et Zucc.) Siebold et Zucc. ex Endl. 'Plumosa Argentea';银叶凤尾柏(羽叶日本花柏)●☆
85362 Chamaecyparis pisifera (Siebold et Zucc.) Siebold et Zucc. ex Endl. 'Plumosa Rogersii';金羽柏(金凤柏)●☆
85363 Chamaecyparis pisifera (Siebold et Zucc.) Siebold et Zucc. ex Endl. 'Squarrosa Veitchii' = Chamaecyparis pisifera (Siebold et Zucc.) Siebold et Zucc. ex Endl. 'Squarrosa'●
85364 Chamaecyparis pisifera (Siebold et Zucc.) Siebold et Zucc. ex Endl. f. crassa Hayashi;粗日本花柏●☆
85365 Chamaecyparis pisifera (Siebold et Zucc.) Siebold et Zucc. ex Endl. f. plumosa (Carrière) Beissn. = Chamaecyparis pisifera (Siebold et Zucc.) Siebold et Zucc. ex Endl. 'Plumosa'●
85366 Chamaecyparis pisifera (Siebold et Zucc.) Siebold et Zucc. ex Endl. var. filifera (Veitch ex Sénükl.) Hartwiss et Rümpler = Chamaecyparis pisifera (Siebold et Zucc.) Siebold et Zucc. ex Endl. 'Filifera Aurea Nana'●☆
85367 Chamaecyparis pisifera (Siebold et Zucc.) Siebold et Zucc. ex Endl. var. filifera (Veitch) Hartwig et Rümpler = Chamaecyparis pisifera (Siebold et Zucc.) Siebold et Zucc. ex Endl. 'Filifera'●
85368 Chamaecyparis pisifera (Siebold et Zucc.) Siebold et Zucc. ex Endl. var. filifera (Veitch ex Sénükl.) Hartwiss et Rümpler = Chamaecyparis pisifera (Siebold et Zucc.) Siebold et Zucc. ex Endl. 'Filifera'●
85369 Chamaecyparis pisifera (Siebold et Zucc.) Siebold et Zucc. ex Endl. var. plumosa (Carrière) Otto = Chamaecyparis pisifera (Siebold et Zucc.) Siebold et Zucc. ex Endl. 'Plumosa'●
85370 Chamaecyparis pisifera (Siebold et Zucc.) Siebold et Zucc. ex Endl. var. plumosa Otto = Chamaecyparis pisifera (Siebold et Zucc.) Siebold et Zucc. ex Endl. 'Plumosa'●
85371 Chamaecyparis pisifera (Siebold et Zucc.) Siebold et Zucc. ex Endl. var. squarrosa (Siebold et Zucc.) Beissn. et Hochst. = Chamaecyparis pisifera (Siebold et Zucc.) Siebold et Zucc. ex Endl. 'Squarrosa'●☆
85372 Chamaecyparis squarrosa (Zucc.) Siebold et Zucc. ex Endl. = Chamaecyparis pisifera (Siebold et Zucc.) Siebold et Zucc. ex Endl. 'Squarrosa'●
85373 Chamaecyparis taiwanensis Masam. et S. Suzuki = Chamaecyparis obtusa (Siebold et Zucc.) Siebold et Zucc. ex Endl.●
85374 Chamaecyparis taiwanensis Masam. et S. Suzuki = Chamaecyparis obtusa (Siebold et Zucc.) Siebold et Zucc. ex Endl. var. formosana (Hayata) Rehder●
85375 Chamaecyparis thyoides (L.) Britton, Sterns et Poggenb.;美国尖叶扁柏(猴掌柏,美国扁柏,沼柏);Atlantic White Cedar, Atlantic White-cedar, Coast West Cedar, False Cypress, Southern White Cedar, Southern White-cedar, Swamp Cedar, White Cedar, White Cedar False Cypress, White Chamaecyparis, White Cypress●
85376 Chamaecyparis thyoides (L.) Britton, Sterns et Poggenb. 'Andelyensis';安德列美国尖叶扁柏(扇枝美国扁柏,扇枝美国尖叶扁柏,直立美尖柏);Andley White Cedar●
85377 Chamaecyparis thyoides (L.) Britton, Sterns et Poggenb. 'Ericoides';欧石南美国尖叶扁柏(丛生美尖柏);Heath White Cedar●
85378 Chamaecyparis thyoides (L.) Britton, Sterns et Poggenb. 'Heatherbum';石南美国尖叶扁柏;Heatherbun Chamaecyparis●☆
85379 Chamaecyparis thyoides (L.) Britton, Sterns et Poggenb. 'Red Star';红星美国尖叶扁柏;Rubicon White Cedar●☆
85380 Chamaecyparis thyoides (L.) Britton, Sterns et Poggenb. 'Rubicon' = Chamaecyparis thyoides (L.) Britton, Sterns et Poggenb. 'Red Star'●☆
85381 Chamaecyparis thyoides (L.) Britton, Sterns et Poggenb. f. andelyensis C. K. Schneid. = Chamaecyparis thyoides (L.) Britton, Sterns et Poggenb. 'Andelyensis'●
85382 Chamaecyparis thyoides (L.) Britton, Sterns et Poggenb. f. ericoides (Carrière) Sudw. = Chamaecyparis thyoides (L.) Britton, Sterns et Poggenb. 'Ericoides'●
85383 Chamaecyparis thyoides (L.) Britton, Sterns et Poggenb. f. glauca (Endl.) Sudw.;蓝叶美尖扁柏;Blue White Cedar●
85384 Chamaecyparis thyoides (L.) Britton, Sterns et Poggenb. f. variegata Sudw.;黄斑美尖扁柏;Variegated White Cedar●
85385 Chamaecyparis thyoides (L.) Britton, Sterns et Poggenb. subsp. henryae (H. L. Li) E. Murray = Chamaecyparis thyoides (L.) Britton, Sterns et Poggenb.●
85386 Chamaecyparis thyoides (L.) Britton, Sterns et Poggenb. var. henryae (H. L. Li) Little = Chamaecyparis thyoides (L.) Britton, Sterns et Poggenb.●
85387 Chamaecytisus Link(1831);假金雀儿属(山雀花属,小金雀属)●☆
85388 Chamaecytisus Vis. = Argyrolobium Eckl. et Zeyh.(保留属名)●☆
85389 Chamaecytisus albidus (DC.) Rothm. = Chamaecytisus mollis (Cav.) Greuter et Burdet●☆
85390 Chamaecytisus albus (Jacq.) Rothm. = Cytisus albus Jacq.●☆
85391 Chamaecytisus chamaecrista Britton = Cassia fasciculata Michx.●☆
85392 Chamaecytisus hirsutus Link;柔毛假金雀儿●☆
85393 Chamaecytisus mollis (Cav.) Greuter et Burdet;柔软假金雀儿●☆
85394 Chamaecytisus palmensis (Christ) F. A. Bisby et K. W. Nicholls;加那利假金雀儿;Tagasaste, Tucerne●☆
85395 Chamaecytisus proliferus (L. f.) Link;多育假金雀儿;Escabon●☆
85396 Chamaecytisus proliferus (L. f.) Link var. canariae (Christ) G. Kunkel = Chamaecytisus proliferus (L. f.) Link●☆
85397 Chamaecytisus proliferus (L. f.) Link var. palmensis (Christ) A. Hansen et Sunding = Chamaecytisus proliferus (L. f.) Link●☆
85398 Chamaecytisus proliferus (L. f.) Link var. palmensis (Christ) A. Hansen et Sunding = Chamaecytisus palmensis (Christ) F. A. Bisby et K. W. Nicholls●☆
85399 Chamaecytisus proliferus (L. f.) Link var. perezii (Hutch.) G. Kunkel = Chamaecytisus proliferus (L. f.) Link●☆
85400 Chamaecytisus pulvinatus (Quézel) Raynaud;叶枕假金雀儿●☆
85401 Chamaecytisus purpureus Link;暗斑假金雀儿(红花金雀花,紫金雀花,紫雀花);Purple Broom●☆
85402 Chamaecytisus purpureus Link = Cytisus purpureus Scop.●☆
85403 Chamaecytisus supinus (L.) Link;平卧假金雀儿(黄矮雀花);Big-flowered Broom●☆
85404 Chamaecytisus supinus (L.) Link = Cytisus supinus L.●☆
85405 Chamaedactylis T. Nees = Aeluropus Trin.■
85406 Chamaedadon Miq. = Homalomena Schott■
85407 Chamaedaphne Catesby = Kalmia L.●
85408 Chamaedaphne Catesby ex Kuntze = Kalmia L.●
85409 Chamaedaphne Kuntze = Kalmia L.●
85410 Chamaedaphne Mitch.(废弃属名) = Chamaedaphne Moench(保留属名)●
85411 Chamaedaphne Mitch.(废弃属名) = Mitchella L.■
85412 Chamaedaphne Moench(1794)(保留属名);地桂属(矮绿属,矮踯躅属,甸杜属,湿地踯躅属);Cassandra, Chamaedaphne,

Leatherleaf ●
85413 Chamaedaphne Moench(保留属名) = Cassandra D. Don ●
85414 Chamaedaphne calyculata (L.) Moench;地桂(矮踯躅,甸杜,副萼金叶子,湿地踯躅,湿原踯躅);Calycular Cassandra, Calycular Chamaedaphne, Chamaedaphne, Leatherleaf, Leather-leaf ●
85415 Chamaedaphne calyculata (L.) Moench = Cassandra calyculata (L.) Moench ●
85416 Chamaedaphne calyculata (L.) Moench var. angustifolia (Aiton) Rehder;窄叶地桂(狭叶地桂);Leather-leaf ●☆
85417 Chamaedaphne glauca (Aiton) Kuntze = Kalmia polifolia Wangenh. ●☆
85418 Chamaedoraceae O. F. Cook = Arecaceae Bercht. et J. Presl(保留科名)●
85419 Chamaedoraceae O. F. Cook = Palmae Juss. (保留科名)●
85420 Chamaedorea Willd. (1806)(保留属名);袖珍椰子属(矮椰子属,茶马椰子属,凯美多利属,坎棕属,客室葵属,客室棕属,客厅棕属,玲珑椰子属,墨西哥棕属,欧洲矮棕属,唐棕榈属,袖珍椰属,竹节椰属,竹棕属);Bamboo Palm, Chamaedorea, Dorea Palm, Moreno Palm, Pacaya, Parlor Palm ●☆
85421 Chamaedorea arenbergiana H. Wendl.;美洲袖珍椰子(美洲榈)●☆
85422 Chamaedorea cataractarum Mart.;缨络椰子(璎珞椰)●☆
85423 Chamaedorea costaricana Oerst.;叶舌袖珍椰子(哥斯达黎加坎棕);Bamboo Palm ●☆
85424 Chamaedorea donnell-smithii Dammer = Chamaedorea seifrizii Burret ●☆
85425 Chamaedorea elatior Mart.;袖珍椰子;Good-luck Palm, Parlour Palm ●☆
85426 Chamaedorea elegans Mart.;雅致袖珍椰子(客室棕,美丽坎棕,袖珍椰子,雅致茶马椰子);Dwarf Mountain Palm, Parlor Palm, Parlour Palm ●☆
85427 Chamaedorea ernesti-augusti H. Wendl.;薄瓣袖珍椰子●☆
85428 Chamaedorea erumpens H. E. Moore;竹茎袖珍椰子(裂坎棕,竹茎玲珑椰子,竹榈);Bamboo Palm ●☆
85429 Chamaedorea erumpens H. E. Moore = Chamaedorea seifrizii Burret ●☆
85430 Chamaedorea falcifera H. E. Moore;镰叶袖珍椰子(镰叶坎棕)●☆
85431 Chamaedorea latifolia W. Watson;宽叶袖珍椰子●☆
85432 Chamaedorea latisecta (H. E. Moore) A. H. Gentry;魔力棕;Moreno Palm ●☆
85433 Chamaedorea metallica O. F. Cook ex H. E. Moore;银玲珑椰子(玲珑椰子)●☆
85434 Chamaedorea microspadix Burret;红果袖珍椰子(小穗坎棕)●☆
85435 Chamaedorea oblongata Mart.;长叶袖珍椰子(长叶坎棕)●☆
85436 Chamaedorea parvifolia Burret;小叶袖珍椰子●☆
85437 Chamaedorea seifrizii Burret;攀缘袖珍椰子(雪佛里椰子);Bamboo Palm, Reed Palm, Seifriz's Chamaedorea ●☆
85438 Chamaedorea tepejilote Liebm.;巴卡亚袖珍椰子(巴卡亚椰子,胀节坎棕);Pacaya Palm, Tepejilote Palm ●☆
85439 Chamaedorea vistae Hodel et N. W. Uhl;维斯塔袖珍椰子●☆
85440 Chamaedoreaceae O. F. Cook = Arecaceae Bercht. et J. Presl(保留科名)●
85441 Chamaedoreaceae O. F. Cook = Palmae Juss. (保留科名)●
85442 Chamaedryfolia Kuntze = Forsskaolea L. ■☆
85443 Chamaedryfolium Post et Kuntze = Chamaedryfolia Kuntze ■☆
85444 Chamaedrys Mill. = Teucrium L. ●■
85445 Chamaedrys Moench = Teucrium L. ●■
85446 Chamaefistula (DC.) G. Don = Cassia L. (保留属名)●■
85447 Chamaefistula G. Don = Cassia L. (保留属名)●■
85448 Chamaegastrodia Makino et F. Maek. (1935);叠鞘兰属(迭鞘兰属);Dualsheathorchis ■
85449 Chamaegastrodia exigua (Rolfe) F. Maek. = Chamaegastrodia vaginata (Hook. f.) Seidenf. ■
85450 Chamaegastrodia exigua (Rolfe) F. Maek. ex Ormerod = Chamaegastrodia vaginata (Hook. f.) Seidenf. ■
85451 Chamaegastrodia inverta (W. W. Sm.) Seidenf.;川滇叠鞘兰(西南翻唇兰);SW. China Dualsheathorchis ■
85452 Chamaegastrodia nanlingensis H. Z. Tian et F. W. Xing = Odontochilus guangdongensis S. C. Chen ■
85453 Chamaegastrodia poilanei (Gagnep.) Seidenf. = Odontochilus poilanei (Gagnep.) Ormerod ■
85454 Chamaegastrodia poilanei (Gagnep.) Seidenf. et A. N. Rao = Odontochilus poilanei (Gagnep.) Ormerod ■
85455 Chamaegastrodia shikokiana Makino et F. Maek.;叠鞘兰(小喙翻唇兰);Japan Dualsheathorchis ■
85456 Chamaegastrodia vaginata (Hook. f.) Seidenf.;戟唇叠鞘兰;Halbert Dualsheathorchis ■
85457 Chamaegeron Schrenk(1845);矮蓬属■☆
85458 Chamaegeron asterellus (Bornm.) Botsch.;紫菀矮蓬■☆
85459 Chamaegeron bungei (Boiss.) Botsch.;邦奇矮蓬■☆
85460 Chamaegeron oligocephalum Schrenk.;寡头矮蓬■☆
85461 Chamaegigas Dinter ex Heil = Chamaegigas Dinter ■☆
85462 Chamaegigas Dinter ex Heil = Lindernia All. ■
85463 Chamaegigas Dinter(1924);南非母草属■☆
85464 Chamaegigas intrepidus Dinter ex Heil;南非母草■☆
85465 Chamaegigas intrepidus Dinter ex Heil = Lindernia intrepidus (Dinter ex Heil) Oberm. ■☆
85466 Chamaegyne Suess. = Eleocharis R. Br. ■
85467 Chamaeiasma Gmel. = Cymbaria L. ■
85468 Chamaeiris Medik. = Iris L. ■
85469 Chamaejasme Amm. = Stellera L. ■●
85470 Chamaejasme Amm. ex Kuntze = Stellera L. ■●
85471 Chamaejasme Kuntze = Stellera L. ■●
85472 Chamaejasme stelleriana Kuntze = Stellera chamaejasme L. ■
85473 Chamaejesme formosana Hayata = Stellera formosana Hayata ex H. L. Li ●
85474 Chamaelauciaceae Lindl. = Myrtaceae Juss. (保留科名)●
85475 Chamaelaucium DC. = Chamelaucium Desf. ●☆
85476 Chamaelaucium Desf. = Chamelaucium Desf. ●☆
85477 Chamaele Miq. (1867);俯卧叠鞘兰属■☆
85478 Chamaele decumbens (Thunb.) Makino;俯卧叠鞘兰■☆
85479 Chamaele decumbens (Thunb.) Makino f. dilatata Satake et Okuyama;膨大俯卧叠鞘兰■☆
85480 Chamaele decumbens (Thunb.) Makino f. flabellifoliata Y. Kimura;扇叶俯卧叠鞘兰■☆
85481 Chamaele decumbens (Thunb.) Makino f. gracillima (H. Wolff) Sugim. = Chamaele decumbens (Thunb.) Makino var. gracillima H. Wolff ■☆
85482 Chamaele decumbens (Thunb.) Makino f. japonica (Y. Yabe) Ohwi;日本俯卧叠鞘兰■☆
85483 Chamaele decumbens (Thunb.) Makino var. gracillima H. Wolff;纤细日本俯卧叠鞘兰■☆
85484 Chamaele decumbens (Thunb.) Makino var. japonica (Y. Yabe) Makino = Chamaele decumbens (Thunb.) Makino f. japonica (Y.

Yabe) Ohwi ■☆
85485 Chamaele decumbens (Thunb.) Makino var. micrantha Masam.;小花俯卧叠鞘兰■☆
85486 Chamaelea Adans. = Cneorum L. ●☆
85487 Chamaelea Duhamel = Cneorum L. ●☆
85488 Chamaelea Gagnebin = Cneorum L. ●☆
85489 Chamaelea Tiegh. = Cneorum L. ●☆
85490 Chamaelea Tiegh. = Neochamaelea (Engl.) Erdtman ●☆
85491 Chamaeleaceae Bertol. = Cneoraceae Vest(保留科名)●☆
85492 Chamaeleaceae Bertol. = Rutaceae Juss.(保留科名)●■
85493 Chamaeledon Link = Loiseleuria Desv.(保留属名)●☆
85494 Chamaeleon Cass.(1827);小狮菊属■☆
85495 Chamaeleon Cass. = Atractylis L. ■☆
85496 Chamaeleon Cass. = Carlina L. ■●
85497 Chamaeleon Tausch = Picnomon Adans. ■☆
85498 Chamaeleon gummifer (L.) Cass. = Carlina gummifera (L.) Less. ■☆
85499 Chamaeleon macrophyllus (Desf.) Sch. Bip. = Atractylis macrophylla Desf. ■☆
85500 Chamaeleorchis Senghas et Lückel = Miltonia Lindl.(保留属名)■☆
85501 Chamaelinum Guett. = Radiola Hill ■☆
85502 Chamaelinum Host = Camelina Crantz ■
85503 Chamaelinum Host = Neslia Desv.(保留属名)■
85504 Chamaelirium Willd.(1808);矮百合属■☆
85505 Chamaelirium luteum A. Gray;黄矮百合;Blazing Star, Blazing-star, Devil's-bit, Devil's Bit, Devil's Bit Fairy-wand, Fairy Wand, Fairy-wand, Rattlesnake-root, Unicom Root ■☆
85506 Chamaelobivia Y. Ito = Echinopsis Zucc. ●
85507 Chamaelum Baker = Chamelum Phil. ■☆
85508 Chamaemeles Lindl.(1821);矮果蔷薇属●☆
85509 Chamaemeles coriacea Lindl.;矮果蔷薇●☆
85510 Chamaemelum Mill.(1754);果香菊属(甘菊属,黄金菊属);Chamaemelum, Chamomile, Spicedaisy ■
85511 Chamaemelum Tourn. ex Adans. = Anthemis L. ■
85512 Chamaemelum Vis. = Matricaria L. ■
85513 Chamaemelum Vis. = Tripleurospermum Sch. Bip. ■
85514 Chamaemelum ambiguum (Ledeb.) Boiss. = Tripleurospermum ambiguum (Ledeb.) Franch. et Sav. ■
85515 Chamaemelum arvense (L.) Hoffmanns. et Link = Anthemis arvensis L. ■
85516 Chamaemelum caucasicum Boiss. = Chrysanthemum caucasicum Pers. ■☆
85517 Chamaemelum cotula (L.) All. = Anthemis cotula L. ■
85518 Chamaemelum eriolepis (Maire) Benedi = Cladanthus eriolepis (Maire) Oberpr. et Vogt ●☆
85519 Chamaemelum flahaultii (Emb.) Benedi = Cladanthus flahaultii (Emb.) Oberpr. et Vogt ●☆
85520 Chamaemelum fuscatum (Brot.) Vasc.;深棕果香菊;Chamomile ■☆
85521 Chamaemelum incrassatum Hoffmanns. et Link = Anacyclus clavatus (Desf.) Pers. ■☆
85522 Chamaemelum limosum Maxim. = Tripleurospermum limosum (Maxim.) Pobed. ■
85523 Chamaemelum mixtum (L.) All. = Cladanthus mixtus (L.) Oberpr. et Vogt ●☆
85524 Chamaemelum mixtum (L.) All. var. aureum (Durieu) Benedi = Cladanthus mixtus (L.) Oberpr. et Vogt ●☆
85525 Chamaemelum mixtum (L.) All. var. glabrescens (Maire) Benedi = Cladanthus mixtus (L.) Oberpr. et Vogt ●☆
85526 Chamaemelum nobile (L.) All.;果香菊(白花春黄菊,春黄菊,黄金菊,罗马措暮米辣,银黄菊);Camomill, Camomine, Camovyne, Camowyne, Chamomile, Chamomile Tea, Chamomiletea, Common Chamomile, Cuntblows, Double Camomile, Dusky Dogfennel, English Camomile, Noble Chamaemelum, Roman Camomile, Roman Chamomile, Scotch Camomile, Spicedaisy, Sweet Chamomile, Weedy Dogfennel, White Camomile, Wild Chamomile ■
85527 Chamaemelum nobile (L.) All. 'Treneague';特纳盖黄金菊■☆
85528 Chamaemelum nobile (L.) All. var. discoideum (Boiss.) P. Silva = Chamaemelum nobile (L.) All. ■
85529 Chamaemelum scariosum (Ball) Benedi = Cladanthus scariosus (Ball) Oberpr. et Vogt ●☆
85530 Chamaemelum tetragomospermum F. Schmidt = Tripleurospermum tetragonospermum (F. Schmidt) Pobed. ■
85531 Chamaemespilus Medik. = Sorbus L. ●
85532 Chamaemoraceae Lilja = Rosaceae Juss.(保留科名)●■
85533 Chamaemorus Ehrh. = Rubus L. ●■
85534 Chamaemorus Greene = Rubus L. ●■
85535 Chamaemorus Hill. = Rubus L. ●■
85536 Chamaemyrrhis Endl. ex Heynh. = Oreomyrrhis Endl. ■
85537 Chamaenerion Adans. = Epilobium L. ■
85538 Chamaenerion Hill = Epilobium L. ■
85539 Chamaenerion Ség. = Epilobium L. ■
85540 Chamaenerion Ség. emend. Gray = Epilobium L. ■
85541 Chamaenerion Spach = Chamaenerion Ség. emend. Gray ■
85542 Chamaenerion Spach = Epilobium L. ■
85543 Chamaenerion angustifolioum (L.) Scop. = Epilobium angustifolium L. ■☆
85544 Chamaenerion angustifolium (L.) Holub subsp. circumvagum (Mosquin) Moldenke = Chamerion angustifolium (L.) Holub subsp. circumvagum (Mosquin) Hoch ■
85545 Chamaenerion angustifolium (L.) Holub var. platyphyllum Daniels = Chamerion angustifolium (L.) Holub subsp. circumvagum (Mosquin) Hoch ■
85546 Chamaenerion angustifolium (L.) Scop. = Chamerion angustifolium (L.) Holub ■
85547 Chamaenerion angustifolium (L.) Scop. = Epilobium angustifolium L. ■☆
85548 Chamaenerion angustifolium (L.) Scop. subsp. circumvagum (Mosquin) Moldenke = Epilobium angustifolium L. subsp. circumvagum Mosquin ■☆
85549 Chamaenerion angustifolium (L.) Scop. subsp. circumvagum (Mosquin) Moldenke = Chamerion angustifolium (L.) Holub subsp. circumvagum (Mosquin) Hoch ■
85550 Chamaenerion angustifolium (L.) Scop. var. albium Yue Zhang et J. Y. Ma;白花柳兰;White Rosebay ■
85551 Chamaenerion angustifolium (L.) Scop. var. album Yue Zhang et J. Y. Ma = Chamerion angustifolium (L.) Holub ■
85552 Chamaenerion angustifolium (L.) Scop. var. platyphyllum Daniels = Epilobium angustifolium L. subsp. circumvagum Mosquin ■☆
85553 Chamaenerion angustissimum (Weber) Grossh.;狭柳兰■☆
85554 Chamaenerion caucasicum (Hausskn.) Sosn. ex Grossh.;高加索柳兰■☆
85555 Chamaenerion colchicum (Albov) Steinb.;黑海柳兰■☆

85556 Chamaenerion conspersum (Hausskn.) H. Li = Chamerion conspersum (Hausskn.) Holub ■
85557 Chamaenerion conspersum (Hausskn.) Kitam. = Chamerion conspersum (Hausskn.) Holub ■
85558 Chamaenerion conspersum (Hausskn.) Kitam. = Epilobium conspersum Hausskn. ■
85559 Chamaenerion halimifolium Salisb.;哈利木叶柳兰■☆
85560 Chamaenerion hirsutum (L.) Scop. = Epilobium hirsutum L. ■
85561 Chamaenerion latifolium (L.) Franch. et Lange = Chamerion latifolium (L.) Holub ■
85562 Chamaenerion latifolium (L.) Franch. et Lange = Epilobium latifolium L. ■
85563 Chamaenerion latifolium (L.) Holub = Epilobium latifolium L. ■
85564 Chamaenerion latifolium (L.) Sweet = Chamerion latifolium (L.) Holub ■
85565 Chamaenerion latifolium (L.) Sweet = Epilobium latifolium L. ■
85566 Chamaenerion nobile (L.) All. = Chamaemelum nobile (L.) All. ■
85567 Chamaenerion reticulatum (C. B. Clarke) Kitam. = Chamerion conspersum (Hausskn.) Holub ■
85568 Chamaenerion reticulatum (C. B. Clarke) Kitam. = Epilobium conspersum Hausskn. ■
85569 Chamaenerion speciosum (Decne.) Holub = Epilobium speciosum Decne. ■
85570 Chamaenerium Spach = Chamaenerion Spach ■
85571 Chamaeorchis Rich. = Chamorchis Rich. ■☆
85572 Chamaeorchis W. D. J. Koch = Chamorchis Rich. ■☆
85573 Chamaeorchis W. D. J. Koch = Herminium L. ■
85574 Chamaeorchis alpinus (L.) Rich. = Chamorchis alpina (L.) Rich. ■☆
85575 Chamaepentas Bremek. (1952);矮五星花属☆
85576 Chamaepentas greenwayi Bremek.;矮五星花☆
85577 Chamaepentas greenwayi Bremek. var. glabra? 光滑矮五星花☆
85578 Chamaepericlimenum Asch. et Graebn. = Chamaepericlymenum Asch. et Graebn. ■
85579 Chamaepericlymenum Asch. et Graebn. (1898);草茱萸属(御膳橘属);Grasscoal ■
85580 Chamaepericlymenum Hill = Cornus L. ●
85581 Chamaepericlymenum canadense (L.) Asch. et Graebn.;草茱萸(红串果,红瑞木,加拿大草茱萸,加拿大山茱萸,御膳橘);Bunch Berry, Bunchberry, Bunchberry Dogwood, Bunch-of-keys, Canada Grasscoal, Crackerberry, Creeping Dogwood, Dwarf Cornel, Dwarf Dogwood, Low Cornel, Pudding-berry ■
85582 Chamaepericlymenum canadense (L.) Asch. et Graebn. = Cornus canadensis L. ■
85583 Chamaepericlymenum suecicum (L.) Asch. et Graebn. = Cornus suecica L. ●☆
85584 Chamaepeuce DC. = Ptilostemon Cass. ■☆
85585 Chamaepeuce Zucc. = Chamaecyparis Spach ●
85586 Chamaepeuce abylensis Pau et Font Quer = Ptilostemon abylensis (Pau et Font Quer) Greuter ■☆
85587 Chamaepeuce leptophylla Pau et Font Quer = Ptilostemon leptophyllus (Pau et Font Quer) Greuter ■☆
85588 Chamaepeuce macrantha Schrenk var. bracteata Rupr. = Cirsium semenovii Regel et Schmalh. ■
85589 Chamaepeuce macrantha Schrnek = Cirsium lamyroides Tamamsch. ■
85590 Chamaepeuce rhipaea Pau et Font Quer = Ptilostemon rhiphaeus (Pau et Font Quer) Greuter ■☆
85591 Chamaephoenix Curtiss = Pseudophoenix H. Wendl. ex Sarg.(废弃属名)●☆
85592 Chamaephoenix H. Wendl. ex Curtiss = Pseudophoenix H. Wendl. ex Sarg.(废弃属名)●☆
85593 Chamaephyton Fourr. = Potentilla L. ■●
85594 Chamaepitys Hill = Ajuga L. ■●
85595 Chamaepitys Tourn. ex Rupp. = Ajuga L. ■●
85596 Chamaeplium Wallr. = Kibera Adans. ■
85597 Chamaeplium Wallr. = Sisymbrium L. ■
85598 Chamaepus Spreng. = Herminium L. ■
85599 Chamaepus Wagenitz(1980);骨苞紫绒草属■☆
85600 Chamaepus afghanicus Wagenitz;骨苞紫绒草■☆
85601 Chamaeranthemum Nees = Chameranthemum Nees ■☆
85602 Chamaeraphis Kuntze = Setaria P. Beauv.(保留属名)■
85603 Chamaeraphis R. Br. (1810);短针狗尾草属■☆
85604 Chamaeraphis brunoniana (Wall. et Griff.) A. Camus = Pseudoraphis brunoniana (Wall. et Griff.) Pilg. ■
85605 Chamaeraphis depauperata Nees = Pseudoraphis spinescens (R. Br.) Vickery var. depauperata (Nees) Bor ■
85606 Chamaeraphis hordeacea R. Br.;短针狗尾草■☆
85607 Chamaeraphis italica (L.) Kuntze = Setaria italica (L.) P. Beauv. ■
85608 Chamaeraphis italica (L.) Kuntze var. germanica (Mill.) Kuntze = Setaria italica (L.) P. Beauv. ■
85609 Chamaeraphis italica (L.) Kuntze var. germanica (Mill.) Kuntze = Setaria italica P. Beauv. 'Major' ■
85610 Chamaeraphis italica (L.) Kuntze var. germanica (Mill.) Kuntze = Setaria italica P. Beauv. var. germanica (Mill.) Schrad. ■
85611 Chamaeraphis italica (L.) Kuntze var. verticillata (L.) Kuntze = Setaria verticillata (L.) P. Beauv. ■
85612 Chamaeraphis italica Kuntze = Setaria italica (L.) P. Beauv. ■
85613 Chamaeraphis palmifolia Kuntze = Setaria palmifolia (J. König) Stapf ■
85614 Chamaeraphis spinescens (R. Br.) Poir. var. brunoniana (Wall. et Griff.) Hook. f. = Pseudoraphis brunoniana (Wall. et Griff.) Pilg. ■
85615 Chamaeraphis spinescens (R. Br.) Poir. var. depauperata Nees ex Hook. f. = Pseudoraphis sordida (Thwaites) S. M. Phillips et S. L. Chen ■
85616 Chamaeraphis spinescens var. depauperata (Nees) Hook. f. = Pseudoraphis spinescens (R. Br.) Vickery var. depauperata (Nees) Bor ■
85617 Chamaeraphis verticillata (L.) Porter = Setaria verticillata (L.) P. Beauv. ■
85618 Chamaeraphis verticillata Porter = Setaria verticillata (L.) P. Beauv. ■
85619 Chamaerepes Spreng. = Chamorchis Rich. ■☆
85620 Chamaerepes Spreng. = Herminium L. ■
85621 Chamaerhodendron Bubani = Chamaerhododendron Mill. ●
85622 Chamaerhodendron Bubani = Rhododendron L. ●
85623 Chamaerhodiola Nakai = Rhodiola L. ■
85624 Chamaerhodiola Nakai = Sedum L. ●■
85625 Chamaerhodiola atuntsuensis (Praeger) Nakai = Rhodiola atuntsuensis (Praeger) S. H. Fu ■
85626 Chamaerhodiola crassipes (Hook. f. et Thomson) Nakai =

Rhodiola wallichiana (Hook.) S. H. Fu ■

85627 Chamaerhodiola crassipes (Wall. ex Hook. f. et Thomson) Nakai = Rhodiola wallichiana (Hook.) S. H. Fu ■

85628 Chamaerhodiola cretinii (Raym. -Hamet) Nakai = Rhodiola cretinii (Raym. -Hamet) H. Ohba ■

85629 Chamaerhodiola dumulosa (Franch.) Nakai = Rhodiola dumulosa (Franch.) S. H. Fu ■

85630 Chamaerhodiola eurycarpa (Fröd.) Nakai = Rhodiola macrocarpa (Praeger) S. H. Fu ■

85631 Chamaerhodiola fastigiata (Hook. f. et Thomson) Fröd. = Rhodiola fastigiata (Hook. f. et Thomson) S. H. Fu ■

85632 Chamaerhodiola gelida (Schrenk) Nakai = Rhodiola gelida Schrenk ■

85633 Chamaerhodiola himalensis (D. Don) Nakai = Rhodiola himalensis (D. Don) S. H. Fu ■

85634 Chamaerhodiola horrida (Praeger) Nakai = Rhodiola nobilis (Franch.) S. H. Fu ■

85635 Chamaerhodiola humilis (Hook. f. et Thomson) Nakai = Rhodiola humilis (Hook. f. et Thomson) S. H. Fu ■

85636 Chamaerhodiola nobilis (Franch.) Nakai = Rhodiola nobilis (Franch.) S. H. Fu ■

85637 Chamaerhodiola quadridida (Pall.) Nakai = Rhodiola quadrifida (Pall.) Fisch. et C. A. Mey. ■

85638 Chamaerhodiola scabrida (Franch.) Nakai = Rhodiola coccinea (Royle) Boriss. subsp. scabrida (Franch.) H. Ohba ■

85639 Chamaerhodiola stephanii (Cham.) Nakai = Rhodiola stephanii (Cham.) Trautv. et C. A. Mey. ■

85640 Chamaerhodiola stracheyi (Hook. f. et Thomson) Nakai = Rhodiola tibetica (Hook. f. et Thomson) S. H. Fu ■

85641 Chamaerhodiola tibetica (Hook. f. et Thomson) Nakai = Rhodiola tibetica (Hook. f. et Thomson) S. H. Fu ■

85642 Chamaerhodiola wulingensis Nakai = Rhodiola dumulosa (Franch.) S. H. Fu ■

85643 Chamaerhododendron Bubani = Rhododendron L. ●

85644 Chamaerhododendron Mill. = Rhododendron L. ●

85645 Chamaerhododendros Duhamel = Chamaerhododendron Mill. ●

85646 Chamaerhodos Bunge (1829); 地蔷薇属; Chamaerhodos, Minorrose ■●

85647 Chamaerhodos altaica (Laxm.) Bunge; 阿尔泰地蔷薇; Altai Chamaerhodos, Altai Minorrose ●

85648 Chamaerhodos canescens J. Krause; 灰毛地蔷薇(毛地蔷薇); Hoary Chamaerhodos, Hoaryhair Minorrose ●

85649 Chamaerhodos corymbosa Murav. = Chamaerhodos canescens J. Krause ●

85650 Chamaerhodos corymbosa Murav. var. brevifolia Murav. = Chamaerhodos canescens J. Krause ●

85651 Chamaerhodos erecta (L.) Bunge; 地蔷薇(茵陈狼牙,直立地蔷薇,追风蒿); Erect Chamaerhodos, Erect Minorrose ●

85652 Chamaerhodos grandiflora (Pall.) Ledeb.; 大花地蔷薇●☆

85653 Chamaerhodos grandiflora (Pall.) Ledeb. = Chamaerhodos canescens J. Krause ●

85654 Chamaerhodos klementzii Murav. = Chamaerhodos trifida Ledeb. ●

85655 Chamaerhodos micrantha J. Krause = Chamaerhodos erecta (L.) Bunge ●

85656 Chamaerhodos mongolica Bunge = Chamaerhodos trifida Ledeb. ●

85657 Chamaerhodos sabulosa Bunge; 砂生地蔷薇; Sandy Chamaerhodos, Sandy Minorrose ●

85658 Chamaerhodos songorica Juz.; 准噶尔地蔷薇●

85659 Chamaerhodos songorica Juz. = Chamaerhodos erecta (L.) Bunge ●

85660 Chamaerhodos trifida Ledeb.; 三裂地蔷薇(矮地蔷薇); Trifid Chamaerhodos, Trifid Minorrose ●

85661 Chamaeriphe Steck = Chamaerops L. ●☆

85662 Chamaeriphes Dill. ex Kuntze = Hyphaene Gaertn. ●☆

85663 Chamaeriphes Kuntze = Hyphaene Gaertn. ●☆

85664 Chamaeriphes Ponted. ex Gaertn. = Chamaerops L. ●☆

85665 Chamaerops L. (1753); 欧洲矮棕属(矮棡属,矮棕属,丛棡属,低丛棕榈属,发棕榈属,欧矮棕属,欧洲榈属,扇葵属,扇棕属); Dwarf Fan Palm, European Fan Palm, Fan Palm, Hair Palm, Mediterranean Palm ●☆

85666 Chamaerops acaulis Michx. = Sabal minor (Jacq.) Pers. ●

85667 Chamaerops excelsa Thunb. = Rhapis exselsa (Thunb.) Henry ex Rehder ●

85668 Chamaerops excelsa Thunb. = Trachycarpus fortunei (Hook.) H. Wendl. ●

85669 Chamaerops excelsa Thunb. var. humilior Thunb. = Rhapis humilis (Thunb.) Blume ●

85670 Chamaerops fortunei Hook. = Trachycarpus fortunei (Hook.) H. Wendl. ●

85671 Chamaerops humilis L.; 欧洲矮棕(矮棕,丛棡,欧洲扇棕,欧洲棕,意大利扇棕); Algerian Fibre, Dwarf Fan Palm, Dwarf Fan-palm, European Fan Palm, European Fan-palm, Hair Palm, Lady Palm, Mediterranean Fan Palm, Mediterranean Palm ●☆

85672 Chamaerops humilis L. var. argentea André = Chamaerops humilis L. ●☆

85673 Chamaerops khasyana Griff. = Trachycarpus martianus (Wall.) H. Wendl. ●◇

85674 Chamaerops louisiana Darby = Sabal minor (Jacq.) Pers. ●

85675 Chamaerops martiana Wall. = Trachycarpus martianus (Wall.) H. Wendl. ●◇

85676 Chamaerops palmetto (Walter) Michx. = Sabal palmetto (Walter) Lodd. ex Roem. et Schult. f. ●

85677 Chamaerops palmetto Michx. = Sabal palmetto (Walter) Lodd. ex Roem. et Schult. f. ●

85678 Chamaerops ritchiana Griff. = Nannorrhops ritchiana (Griff.) Aiton ●☆

85679 Chamaerops serrulata Michx. = Serenoa repens (Bartram) Small ●☆

85680 Chamaesaracha (A. Gray) Benth. = Chamaesaracha (A. Gray) Benth. et Hook. f. ■☆

85681 Chamaesaracha (A. Gray) Benth. et Hook. f. (1876); 刺酸浆属 ■☆

85682 Chamaesaracha A. Gray ex Franch. et Sav. = Chamaesaracha (A. Gray) Benth. et Hook. f. ■☆

85683 Chamaesaracha A. Gray ex Franch. et Sav. = Physaliastrum Makino ■

85684 Chamaesaracha echinata Yatabe = Physaliastrum echinatum (Yatabe) Makino ■

85685 Chamaesaracha echinata Yatabe = Physaliastrum japoncum (Franch. et Sav.) Honda ■

85686 Chamaesaracha grandiflora (Hook.) Fernald = Leucophysalis grandiflora (Hook.) Rydb. ●☆

85687 Chamaesaracha heterophylla Hemsl. = Physaliastrum heterophyllum (Hemsl.) Migo ■

85688 Chamaesaracha japonica Franch. et Sav. = Physaliastrum echinatum (Yatabe) Makino ■

85689 Chamaesaracha japonica Franch. et Sav. = Physaliastrum japonicum (Franch. et Sav.) Honda ■

85690 Chamaesaracha savatieri Makino = Physaliastrum savatieri (Makino) Makino ■

85691 Chamaesaracha sinensis Hemsl. = Physaliastrum sinense (Hemsl.) D'Arcy et Zhi Y. Zhang ■

85692 Chamaesaracha sordida A. Gray; 暗色刺酸浆; Dingy Chamaesaracha ■☆

85693 Chamaesaracha watanabei Yatabe = Physaliastrum japonicum (Franch. et Sav.) Honda ■

85694 Chamaesarachia Franch. et Sav. = Chamaesaracha A. Gray ex Franch. et Sav. ■☆

85695 Chamaeschoenus Ehrh. = Scirpus L. (保留属名) ■

85696 Chamaesciadium C. A. Mey. (1831); 矮伞芹属(矮泽芹属); Chamaesciadium ■

85697 Chamaesciadium acaule (M. Bieb.) Boiss.; 矮伞芹■☆

85698 Chamaesciadium acaule C. A. Mey. = Chamaesciadium acaule (M. Bieb.) Boiss. ■☆

85699 Chamaesciadium acaule C. A. Mey. var. simplex R. H. Shan = Chamaesciadium acaule C. A. Mey. var. simplex R. H. Shan et F. T. Pu ■

85700 Chamaesciadium acaule C. A. Mey. var. simplex R. H. Shan = Dimorphosciadium shenii Pimenov et Kljuykov ■

85701 Chamaesciadium acaule C. A. Mey. var. simplex R. H. Shan et F. T. Pu; 单羽矮伞芹; Simplepinnata Chamaesciadium ■

85702 Chamaesciadium albiflorum Kar. et Kir. = Schultzia albiflora (Kar. et Kir.) Popov ■

85703 Chamaesciadium flavescens C. A. Mey. = Chamaesciadium acaule C. A. Mey. ■☆

85704 Chamaesciadium subnudum (C. B. Clarke ex H. Wolff) C. Norman = Trachydium subnudum C. B. Clarke ex H. Wolff ■

85705 Chamaescilla F. Muell. = Chamaescilla F. Muell. ex Benth. ■☆

85706 Chamaescilla F. Muell. ex Benth. (1878); 绵枣兰属■☆

85707 Chamaesenna (DC.) Raf. ex Pittier = Senna Mill. ●■

85708 Chamaesenna Pittier = Cassia L. (保留属名) ●■

85709 Chamaesenna Raf. ex Pittier = Senna Mill. ●■

85710 Chamaesenna laevigata (Willd.) Pittier; 平滑绵枣兰■☆

85711 Chamaesium H. Wolff (1925); 矮泽芹属(矮芹属, 地芹属); Chamaesium ●★

85712 Chamaesium delavayi (Franch.) R. H. Shan et S. L. Liou; 鹤庆矮泽芹; Delavay Chamaesium ■

85713 Chamaesium frigidum (Hand.-Mazz.) R. H. Shan ex F. T. Pu. = Sium frigidum Hand.-Mazz. ■

85714 Chamaesium mallaeanum Farille et S. B. Malla; 聂拉木矮泽芹■

85715 Chamaesium markgrafianum (Fedde ex H. Wolff) C. Norman = Chamaesium viridiflorum (Franch.) H. Wolff ex R. H. Shan ■

85716 Chamaesium novem-jugum (C. B. Clarke) C. Norman; 粗棱矮泽芹(九对叶矮泽芹) ■

85717 Chamaesium novem-jugum (C. B. Clarke) C. Norman var. delavayi (Franch.) C. Norman = Chamaesium delavayi (Franch.) R. H. Shan et S. L. Liou ■

85718 Chamaesium paradoxum H. Wolff; 矮泽芹; Low Chamaesium ■

85719 Chamaesium spatuliferum (W. W. Sm.) C. Norman; 大苞矮泽芹(小矮泽芹); Bigbract Chamaesium, Dwarf Bigbract Chamaesium ■

85720 Chamaesium spatuliferum (W. W. Sm.) C. Norman = Chamaesium novem-jugum (C. B. Clarke) C. Norman ■

85721 Chamaesium spatuliferum (W. W. Sm.) C. Norman var. minor R. H. Shan et S. L. Liou; 小矮泽芹; Small Chamaesium ■

85722 Chamaesium spatuliferum (W. W. Sm.) C. Norman var. minus R. H. Shan et S. L. Liou = Chamaesium novem-jugum (C. B. Clarke) C. Norman ■

85723 Chamaesium thalictrifolium H. Wolff; 松潘矮泽芹; Songpan Chamaesium ■

85724 Chamaesium viridiflorum (Franch.) H. Wolff ex R. H. Shan; 细叶矮泽芹(绿花矮泽芹); Greenflower Chamaesium ■

85725 Chamaespartium Adans. = Genista L. ●

85726 Chamaespartium sagittale (L.) P. E. Gibbs = Genista sagittalis L. ●☆

85727 Chamaespartium tridentatum (L.) P. E. Gibbs = Pterospartum tridentatum (L.) Willk. ●☆

85728 Chamaesparton Fourr. = Chamaespartium Adans. ●

85729 Chamaesphacos Schrenk = Chamaesphacos Schrenk ex Fisch. et C. A. Mey. ■

85730 Chamaesphacos Schrenk ex Fisch. et C. A. Mey. (1841); 矮刺苏属; Chamaesphacos ■

85731 Chamaesphacos ilicifolius Schrenk; 矮刺苏; Holleyleaf Chamaesphacos ■

85732 Chamaesphacos longiflorus Bornm. et Sint. = Chamaesphacos ilicifolius Schrenk ■

85733 Chamaesphaerion A. Gray = Chthonocephalus Steetz ■☆

85734 Chamaespilus Fourr. = Chamaemespilus Medik. ●

85735 Chamaespilus Fourr. = Sorbus L. ●

85736 Chamaestephanum Willd. = Schkuhria Roth (保留属名) ■☆

85737 Chamaesyce Gray = Euphorbia L. ●■

85738 Chamaesyce Gray (1821); 地锦苗属●■

85739 Chamaesyce albicaulis (Rydb.) Rydb. = Chamaesyce serpyllifolia (Pers.) Small ■☆

85740 Chamaesyce arabica (Hochst. et Steud. ex T. Anderson) Soják = Euphorbia arabica Hochst. et Steud. ex T. Anderson ●■☆

85741 Chamaesyce atoto (G. Forst.) Croizat = Euphorbia atoto G. Forst. ■

85742 Chamaesyce bifida (Hook. et Arn.) T. Kuros. = Euphorbia bifida Hook. et Arn. ■

85743 Chamaesyce canescens (L.) Prokh.; 灰白地锦苗●☆

85744 Chamaesyce canescens (L.) Prokh. subsp. massiliensis (DC.) Soják; 马西利地锦苗■☆

85745 Chamaesyce chamaesycoides (B. Nord.) Koutnik = Euphorbia chamaesycoides B. Nord. ■☆

85746 Chamaesyce eylesii (Rendle) Koutnik = Euphorbia eylesii Rendle ■☆

85747 Chamaesyce forsskalii (J. Gay) Soják; 福斯科尔地锦苗●☆

85748 Chamaesyce garanbiensis (Hayata) Hara = Euphorbia garanbiensis Hayata ■

85749 Chamaesyce geyeri (Engelm.) Small; 盖氏地锦苗; Geyer's Sand-mat, Geyer's Spurge ■☆

85750 Chamaesyce glanduligera (Pax) Koutnik = Euphorbia glanduligera Pax ■☆

85751 Chamaesyce glaucophylla (Poir.) Croizat = Euphorbia glaucophylla Poir. ■☆

85752 Chamaesyce glomerifera Millsp. = Euphorbia glomerifera (Millsp.) L. C. Wheeler ■☆

85753 Chamaesyce glyptosperma (Engelm.) Small; 沙生地锦苗; Ribseed Sand-mat, Rib-seeded Sand Mat, Ridge-seeded Spurge, Sand Mat ■☆

85754 Chamaesyce glyptosperma (Engelm.) Small = Euphorbia glyptosperma Engelm. ■☆

85755 Chamaesyce granulata (Forssk.) Soják;颗粒地锦苗■☆

85756 Chamaesyce hirta (L.) Mill. = Euphorbia hirta L. ■

85757 Chamaesyce hirta (L.) Mill. f. glaberrima (Koidz.) Hurus. = Chamaesyce hirta (L.) Mill. var. glaberrima (Koidz.) H. Hara ■

85758 Chamaesyce hirta (L.) Mill. var. glaberrima (Koidz.) H. Hara = Euphorbia hirta L. var. glaberrima Koidz. ■

85759 Chamaesyce hsinchuensis S. C. Lin et Chaw = Euphorbia hsinchuensis (S. C. Lin et Chaw) C. Y. Wu et J. S. Ma ■

85760 Chamaesyce humifusa (Willd. ex Schltdl.) Prokh. = Euphorbia humifusa Willd. ex Schltdl. ■

85761 Chamaesyce humifusa (Willd. ex Schltdl.) Prokh. var. glabra (C. A. Mey.) H. Hara = Chamaesyce humifusa (Willd. ex Schltdl.) Prokh. ■

85762 Chamaesyce humifusa (Willd. ex Schltdl.) Prokh. var. pseudochamaesyce (Fisch., C. A. Mey. et Avé-Lall.) Hurus. = Chamaesyce humifusa (Willd. ex Schltdl.) Prokh. ■

85763 Chamaesyce humifusa (Willd.) Prokh. = Euphorbia humifusa Willd. ex Schltdl. ■

85764 Chamaesyce humistrata (Engelm.) Small; 铺散地锦苗; Spreading Sand-mat, Spurge ■

85765 Chamaesyce hyssopifolia (L.) Small = Euphorbia hyssopifolia L. ■

85766 Chamaesyce inaequilatera (Sond.) Soják;偏基地锦苗■☆

85767 Chamaesyce indica (Lam.) Croizat = Euphorbia indica Lam. ■☆

85768 Chamaesyce liukiuensis (Hayata) H. Hara;琉球地锦苗■☆

85769 Chamaesyce livida (E. Mey. ex Boiss.) Koutnik = Euphorbia livida E. Mey. ex Boiss. ■☆

85770 Chamaesyce maculata (L.) Small;斑点地锦苗;Milk Purslane, Milk-purslane, Prostrate Spurge, Spotted Sand-mat, Wart-weed ■☆

85771 Chamaesyce maculata (L.) Small = Chamaesyce nutans (Lag.) Small ■☆

85772 Chamaesyce maculata (L.) Small = Euphorbia maculata L. ■☆

85773 Chamaesyce makinoi (Hayata) H. Hara = Euphorbia makinoi Hayata ■

85774 Chamaesyce mathewsii Small = Chamaesyce maculata (L.) Small ■☆

85775 Chamaesyce missurica (Raf.) Shinners; 密苏里地锦苗; Missouri Spurge, Prairie Spurge ■☆

85776 Chamaesyce mossambicensis (Klotzsch et Garcke) Koutnik = Euphorbia mossambicensis (Klotzsch et Garcke) Boiss. ■●☆

85777 Chamaesyce neomexicana (Greene) Standl. = Chamaesyce serpyllifolia (Pers.) Small ■☆

85778 Chamaesyce neopolycnemoides (Pax et K. Hoffm.) Koutnik = Euphorbia neopolycnemoides Pax et K. Hoffm. ■☆

85779 Chamaesyce nutans (Lag.) Small; 俯垂地锦苗; Eye-bane, Nodding Spurge ■☆

85780 Chamaesyce peplis (L.) Prokh.;孛艾地锦苗■☆

85781 Chamaesyce pilulifera (L.) Small var. glaberrima (Koidz.) Tuyama = Chamaesyce hirta (L.) Mill. var. glaberrima (Koidz.) H. Hara ■

85782 Chamaesyce polycnemoides (Hochst. ex Boiss.) Soják = Euphorbia polycnemoides Hochst. ex Boiss. ☆

85783 Chamaesyce polygonifolia (L.) Small; 蓼叶地锦苗; Seaside Spurge ■☆

85784 Chamaesyce preslii (Guss.) Arthur = Chamaesyce nutans (Lag.) Small ■☆

85785 Chamaesyce prieuriana (Baill.) Soják = Euphorbia convolvuloides Hochst. ex Benth. ■☆

85786 Chamaesyce prostrata (Aiton) Small = Euphorbia prostrata Aiton ■

85787 Chamaesyce pseudochamaesyce (Fisch., C. A. Mey. et Avé-Lall.) Kom. = Chamaesyce humifusa (Willd. ex Schltdl.) Prokh. ■

85788 Chamaesyce rafinesquii (Greene) Arthur = Chamaesyce vermiculata (Raf.) House ■☆

85789 Chamaesyce schlechteri (Pax) Koutnik = Euphorbia schlechteri Pax ■☆

85790 Chamaesyce scordifolia (Jacq.) Croizat = Euphorbia scordifolia Jacq. ■☆

85791 Chamaesyce serpens (Kunth) Small = Euphorbia serpens Kunth ■

85792 Chamaesyce serpyllifolia (Pers.) Small; 百里香叶地锦苗; Thyme-leaved Spurge ■☆

85793 Chamaesyce serpyllifolia (Pers.) Small = Euphorbia serpyllifolia Pers. ■☆

85794 Chamaesyce sparrmannii (Boiss.) Hurus. = Euphorbia sparrmannii Boiss. ●■

85795 Chamaesyce stictospora (Engelm.) Small; 席子地锦苗; Mat Spurge ■☆

85796 Chamaesyce supina (Raf.) Moldenke = Chamaesyce maculata (L.) Small ■☆

85797 Chamaesyce supina (Raf.) Moldenke = Euphorbia maculata L. ■

85798 Chamaesyce taihsiensis Chaw et Koutnik = Euphorbia taihsiensis (Chaw et Koutnik) Oudejans ■

85799 Chamaesyce tashiroi (Hayata) Hara = Euphorbia tashiroi Hayata ●

85800 Chamaesyce tashiroi Hara = Euphorbia humifusa Willd. ex Schltdl. ■

85801 Chamaesyce tettensis (Klotzsch) Koutnik = Euphorbia tettensis Klotzsch ☆

85802 Chamaesyce thymifolia (L.) Millsp. = Euphorbia thymifolia L. ●■

85803 Chamaesyce tracyi Small = Chamaesyce maculata (L.) Small ■☆

85804 Chamaesyce vachelli (Hook. et Arn.) Hurus. = Chamaesyce bifida (Hook. et Arn.) T. Kuros. ■

85805 Chamaesyce vachellii (Hook. et Arn.) H. Hara = Euphorbia bifida Hook. et Arn. ■

85806 Chamaesyce vermiculata (Raf.) House = Euphorbia vermiculata Raf. ■☆

85807 Chamaesyce vulgaris Prokh. = Chamaesyce canescens (L.) Prokh. ●☆

85808 Chamaesyce vulgaris Prokh. subsp. massiliensis (DC.) Benedi et Orell = Chamaesyce canescens (L.) Prokh. subsp. massiliensis (DC.) Soják ■☆

85809 Chamaesyce vulgaris Prokh. subsp. vulgaris = Chamaesyce canescens (L.) Prokh. ●☆

85810 Chamaesyce wigthiana V. S. Raju et P. N. Rao = Euphorbia agowensis Hochst. ex Boiss. ■☆

85811 Chamaesyce zambesiana (Benth.) Koutnik;赞比西地锦苗●☆

85812 Chamaesyce zambesiana (Benth.) Koutnik = Euphorbia zambesiana Benth. ■☆

85813 Chamaetaxus Bubani = Empetrum L. ●

85814 Chamaetaxus Rupr. = Empetrum L. ●

85815 Chamaethrinax H. Wendl. ex R. Pfister = Trithrinax Mart. ●☆

85816 Chamaexeros Benth. (1878);矮点柱花属■☆

85817 Chamaexeros fimbriata Benth.;矮点柱花■☆
85818 Chamaexiphion Hochst. ex Steud. = Ficinia Schrad.(保留属名)■☆
85819 Chamaexiphium Hochst. = Ficinia Schrad.(保留属名)■☆
85820 Chamaexyphium Pfeiff. = Chamaexiphium Hochst. ■☆
85821 Chamaexyphium clandestinum (Steud.) Hochst. = Ficinia clandestina (Steud.) Boeck. ■☆
85822 Chamaexyphium dregeanum Steud. = Ficinia pygmaea Boeck. ■☆
85823 Chamaezelum Link = Antennaria Gaertn.(保留属名)■●
85824 Chamagrostidaceae Link = Gramineae Juss.(保留科名)■●
85825 Chamagrostidaceae Link = Poaceae Barnhart(保留科名)■●
85826 Chamagrostis Borkh. = Mibora Adans. ■☆
85827 Chamagrostis Borkh. ex Wibel = Mibora Adans. ■☆
85828 Chamalirium Raf. = Chamaelirium Willd. ■☆
85829 Chamalium Cass. = Atractylis L. ■☆
85830 Chamalium Cass. = Chamaeleon Cass. ■☆
85831 Chamalium Juss. = Cardopatium Juss. ■☆
85832 Chamamelum tetragonospermum Eastw. = Tripleurospermum tetragonospermum (F. Schmidt) Pobed. ■
85833 Chamarea Eckl. et Zeyh.(1837);矮缕子属■☆
85834 Chamarea caffra Eckl. et Zeyh. = Chamarea capensis (Thunb.) Eckl. et Zeyh. ■☆
85835 Chamarea capensis (Thunb.) Eckl. et Zeyh.;矮缕子■☆
85836 Chamarea esterhuyseniae B. L. Burtt;埃斯特矮缕子■☆
85837 Chamarea gracillima (H. Wolff) B. L. Burtt;细长矮缕子■☆
85838 Chamarea longipedicellata B. L. Burtt;长梗矮缕子■☆
85839 Chamartemisia Rydb. = Artemisia L. ●■
85840 Chamartemisia Rydb. = Sphaeromeria Nutt. ■☆
85841 Chamartemisia Rydb. = Tanacetum L. ■●
85842 Chambeyronia Vieill.(1873);红心椰属(茶梅椰属,禅比罗棕属,肖肯棕属);Chambeyronia Palm ●☆
85843 Chambeyronia macrocarpa Vieill. ex Becc.;大果红心椰(大果禅比罗棕,大果肖肯棕)●☆
85844 Chambeyronia morieri Vieill.;红心椰●☆
85845 Chamedrys Raf.(1836) = Spiraea L. ●
85846 Chamedrys Raf.(1837) = Chamaedrys Moench ●■
85847 Chamedrys Raf.(1837) = Teucrium L. ●■
85848 Chamelaea Post et Kuntze = Chamaelea Duhamel ●☆
85849 Chamelaea Post et Kuntze = Cneorum L. ●☆
85850 Chamelauciaceae DC. ex F. Rudolphi = Myrtaceae Juss.(保留科名)●
85851 Chamelauciaceae F. Rudolphi = Myrtaceae Juss.(保留科名)●
85852 Chamelaucium Desf.(1819);澳蜡花属(玉梅属);Esperance Waxflower, Wax Flower, Geraldton Wax Flower ●☆
85853 Chamelaucium gracile F. Muell.;纤细澳蜡花●☆
85854 Chamelaucium hallii Ewart;哈尔澳蜡花●☆
85855 Chamelaucium heterandrum Benth.;异蕊澳蜡花●☆
85856 Chamelaucium micropetalum (F. Muell.) F. Muell.;小瓣澳蜡花●☆
85857 Chamelaucium pauciflorum Benth.;少花澳蜡花●☆
85858 Chamelaucium uncinatum Schauer = Darwinia uncinata (Schauer) F. Muell. ●☆
85859 Chamelophyton Garay(1974);枝变兰属■☆
85860 Chamelophyton kegelii (Rchb. f.) Garay;枝变兰■☆
85861 Chamelum Phil. = Olsynium Raf. ■☆
85862 Chamepeuce Raf. = Chamaepeuce DC. ■☆
85863 Chamepeuce Raf. = Cirsium Mill. ■
85864 Chameranthemum Nees;小可爱花属■☆
85865 Chameranthemum gaudichaudii Nees;巴西小可爱花■☆
85866 Chamerasia Raf. = Lonicera L. ●■
85867 Chamerion (Raf.) Raf. = Epilobium L. ■
85868 Chamerion (Raf.) Raf. ex Holub = Chamaenerion Ség. emend. Gray ■
85869 Chamerion (Raf.) Raf. ex Holub = Epilobium L. ■
85870 Chamerion (Raf.) Raf. ex Holub(1972);柳兰属;Willowweed ■
85871 Chamerion Raf. = Epilobium L. ■
85872 Chamerion Raf. = Lonicera L. ●■
85873 Chamerion Raf. ex Holub = Chamerion (Raf.) Raf. ex Holub ■
85874 Chamerion Raf. ex Holub = Epilobium L. ■
85875 Chamerion angustifolium (L.) Holub;柳兰(遍山红,大救驾,红筷子,火烧兰,糯芋,山麻条,铁筷子,土秦艽);Apple Pie, Bay Willow, Blood Vine, Blooming Sally, Blooming Willow, Bullock's Eyes, Cat's Eyes, Cherry Pie, Cherry-pie, Eyebright, Feeneh Saugh, Fireweed, Fire-weed, Flowering Withy, French Bay, French Saugh, French Willow, Great Willow Herb, Great Willowherb, Great Willow-herb, Great Willowherb Willowweed, Narrow-leaf Fireweed, Narrow-leaf Willowherb, Persian Willow, Purple Rocket, Ranting Widow, Rosebay, Rose-bay, Rosebay Willowherb, Spiked Willow, Spiked Willowherb, Tame Withy, Wickup, Wicopy, Wild Snapdragon, Willow Herb, Willowweed ■
85876 Chamerion angustifolium (L.) Holub = Chamaenerion angustifolium (L.) Scop. ■
85877 Chamerion angustifolium (L.) Holub = Epilobium angustifolium L. ■☆
85878 Chamerion angustifolium (L.) Holub f. pleniflorum (Nakai) Yonek. = Epilobium angustifolium L. f. pleniflorum (Nakai) H. Hara ●☆
85879 Chamerion angustifolium (L.) Holub f. pleniflorum (Nakai) Yonek. = Rhododendron kaempferi Planch. f. komatsui (Nakai) H. Hara ●☆
85880 Chamerion angustifolium (L.) Holub subsp. circumvagum (Mosquin) Hoch;毛脉柳兰;Fireweed, Great Willow-herb, Hairyvein ■
85881 Chamerion angustifolium (L.) Holub subsp. circumvagum (Mosquin) Hoch = Epilobium angustifolium L. subsp. circumvagum Mosquin ■☆
85882 Chamerion conspersum (Hausskn.) Holub;网脉柳兰(网叶柳兰)■
85883 Chamerion conspersum (Hausskn.) Hulub = Epilobium conspersum Hausskn. ■
85884 Chamerion latifolium (L.) Holub;宽叶柳兰(阔叶柳叶菜);Broadleaf Willowweed, Red Willow Weed ■
85885 Chamerion latifolium (L.) Holub = Epilobium latifolium L. ■
85886 Chamerion speciosum (Decne.) Holub;喜马拉雅柳兰;Himalayas Willowweed ■
85887 Chamerion speciosum (Decne.) Holub = Epilobium latifolium L. subsp. speciosum (Decne.) P. H. Raven ■
85888 Chamerion speciosum (Decne.) Holub = Epilobium speciosum Decne. ■
85889 Chamerops Raf. = Chamaerops L. ●☆
85890 Chamguava Landrum(1991);美樱木属●☆
85891 Chamguava gentlei (Lundell) Landrum;美樱木●☆
85892 Chamira Thunb.(1782);南非角状芥属■☆
85893 Chamira circaeoides (L. f.) Zahlbr.;南非角状芥■☆
85894 Chamira cornuta Thunb. = Chamira circaeoides (L. f.) Zahlbr. ■☆

85895 Chamisme Nieuwl. = Houstonia L. ■☆
85896 Chamisme Raf. = Houstonia L. ■☆
85897 Chamisme Raf. ex Steud. = Houstonia L. ■☆
85898 Chamissoa Kunth(1818)(保留属名);弓枝苋属■●☆
85899 Chamissoa altissima (Jacq.) Nees et Mart.;弓枝苋■●☆
85900 Chamissomneia Kuntze = Schlechtendalia Less.(保留属名)■☆
85901 Chamissonia Endl. = Camissonia Link ■☆
85902 Chamissonia Endl. = Oenothera L. ●■
85903 Chamissonia Raim. = Camissonia Link ■☆
85904 Chamissoniophila Brand(1929);喜查花属■☆
85905 Chamissoniophila cruciata Brand;喜查花■☆
85906 Chamitea (Dumort.) A. Kern. = Salix L.(保留属名)●
85907 Chamitea A. Kern. = Nectusion Raf. ●
85908 Chamitea A. Kern. = Salix L.(保留属名)●
85909 Chamitis Banks ex Gaertn. = Azorella Lam. ■☆
85910 Chamoletta Adans. = Iris L. ■
85911 Chamoletta Adans. = Xiphion Mill. ■
85912 Chamomilla Godr. = Anthemis L. ■
85913 Chamomilla Godr. = Chamaemelum Mill. ■
85914 Chamomilla Gray = Matricaria L. ■
85915 Chamomilla aurea (Loefl.) Coss. et Kralik = Matricaria aurea (Loefl.) Sch. Bip. ■☆
85916 Chamomilla aurea (Loefl.) Coss. et Kralik var. coronata Coss. et Kralik = Matricaria aurea (Loefl.) Sch. Bip. ■☆
85917 Chamomilla chamomilla (L.) Rydb. = Matricaria recutita L. ■
85918 Chamomilla discoidea (DC.) J. Gay ex A. Braun = Matricaria matricarioides (Less.) Ced. Porter ex Britton ■
85919 Chamomilla inodora (L.) Gilib. = Matricaria maritima L. ■☆
85920 Chamomilla inodora (L.) Gilib. = Tripleurospermum maritimum (L.) W. D. J. Koch ■☆
85921 Chamomilla inodora (L.) K. Koch = Tripleurospermum inodorum (L.) Sch. Bip. ■
85922 Chamomilla maritima (L.) Rydb. = Tripleurospermum maritimum (L.) W. D. J. Koch ■☆
85923 Chamomilla occidentalis (Greene) Rydb. = Matricaria occidentalis Greene ■☆
85924 Chamomilla pubescens (Desf.) Alavi = Aaronsohnia pubescens (Desf.) K. Bremer et Humphries ■☆
85925 Chamomilla recutita (L.) Rauschert = Matricaria recutita L. ■
85926 Chamomilla suaveolens (Pursh) Rydb. = Matricaria discoidea DC. ■
85927 Chamomilla suaveolens (Pursh) Rydb. = Santolina suaveolens Pursh ●☆
85928 Chamorchis Rich.(1817);偃伏兰属■☆
85929 Chamorchis Rich. = Herminium L. ■
85930 Chamorchis alpina (L.) Rich.;偃伏兰(矮兰)■☆
85931 Champaca Adans. = Michelia L. ●
85932 Champereia Griff.(1843);台湾山柚属(拟常山属,詹柏木属);Champereia ●
85933 Champereia griffithiana Planch. ex Kurz = Champereia manillana (Blume) Merr. ●
85934 Champereia longistaminea (W. Z. Li) D. D. Tao;四数台湾山柚(长蕊甜菜树,茎花山柚);Longstamen Champereia ●
85935 Champereia manillana (Blume) Merr.;台湾山柚(马尼拉詹柏木,拟常山,山柑,山柑仔,山柚,山柚仔,詹柏木);Manila Champereia,Taiwan Champereia ●
85936 Champereia manillana (Blume) Merr. var. longistaminea (W. Z. Li) H. S. Kiu = Champereia longistaminea (W. Z. Li) D. D. Tao ●
85937 Championella Bremek.(1944);黄猄草属(黄琼草属,棱果马兰属,梭果爵床属);Championella ●■
85938 Championella Bremek. = Strobilanthes Blume ●■
85939 Championella dalzellii (W. W. Sm.) Bremek. var. glaber R. Ben.? = Pteroptychia dalziellii (Sm.) H. S. Lo ●■
85940 Championella debilis (Hemsl.) Bremek. = Championella tetrasperma (Champ. ex Benth.) Bremek. ●
85941 Championella debilis (Hemsl.) Bremek. = Strobilanthes tetraspermus (Champ. ex Benth.) Druce ●
85942 Championella fauriei (Benoist) C. Y. Wu et C. C. Hu;台湾黄猄草;Faurie Championella ●
85943 Championella fulvihispida (D. Fang et H. S. Lo) C. Y. Wu et C. C. Hu;锈毛黄猄草(锈毛马蓝);Rusty-haired Championella ●
85944 Championella japonica (Thunb.) Bremek.;日本黄猄草(长苞蓝,长苞马蓝,垂序马蓝,红泽兰,马兰,拟马蓝,日本马蓝,山泽兰,泽兰);Japan Championella,Japan Conehead,Japanese Conehead ●■
85945 Championella japonica (Thunb.) Bremek. = Strobilanthes japonica (Thunb.) Miq. ■
85946 Championella labordei (H. Lév.) E. Hossain;贵阳黄猄草(薄叶马蓝);Guiyang Championella ●■
85947 Championella longiflora (Benoist) C. Y. Wu et C. C. Hu;长花黄猄草(长花紫云英);Longflower Conehead ●
85948 Championella maclurei (Merr.) C. Y. Wu et H. S. Lo;海南黄猄草(汗斑草);Hainan Championella,Maclure Championella ●
85949 Championella oligantha (Miq.) Bremek.;少花黄猄草(少花马蓝,紫云菜,紫云英马蓝);Fewflower Conehead ●
85950 Championella oligantha (Miq.) Bremek. = Strobilanthes oligantha Miq. ■
85951 Championella sarcorrhiza C. Ling;肉根马蓝(菜头肾,土太子参)●
85952 Championella tetrasperma (Champ. ex Benth.) Bremek.;黄琼草(狗肝菜,黄猄草,九头狮子草,四籽马蓝,四子马蓝,岩冬菜);Championella,Common Championella,Fourseed Conehead ●
85953 Championella xanthantha (Diels) Bremek.;黄花黄猄草(黄花梭果爵床);Yellowflower Championella ●
85954 Championia C. B. Clarke = Leptobaea Benth. ●
85955 Championia Gardner(1846);斯里兰卡苣苔属●☆
85956 Championia multiflora C. B. Clarke = Leptobaea multiflora (C. B. Clarke) C. B. Clarke ●
85957 Championia reticulata Gardner;斯里兰卡苣苔●☆
85958 Chamula Noronha = Lobelia L. ●■
85959 Chamysyke Raf. = Chamaesyce Gray ●■
85960 Chandrasekharania V. J. Nair, V. S. Ramach. et Sreek.(1982);喀拉草属(喀拉拉草属)■☆
85961 Chandrasekharania keralensis V. J. Nair, V. S. Ramach. et Sreek.;喀拉草■☆
85962 Chanekia Lundell = Licaria Aubl. ●☆
85963 Changiodendron R. H. Miao = Sabia Colebr. ●
85964 Changiodendron R. H. Miao(1995);岐花鼠刺属●
85965 Changiodendron guangxiense R. H. Miao;广西岐花鼠刺(广西鼠刺)●
85966 Changiodendron guangxiense R. H. Miao = Sabia parviflora Wall. ex Roxb. ●
85967 Changiostyrax Tao Chen = Sinojackia Hu ●★
85968 Changiostyrax Tao Chen(1995);长果安息香属●
85969 Changiostyrax dolichocarpus (C. J. Qi) Tao Chen;长果安息香(长果秤锤树);Longfruit Sinojackia, Longfruit Weigttree, Long-

fruited Sinojackia ●

85970 Changiostyrax dolichocarpus (C. J. Qi) Tao Chen = Sinojackia dolichocarpa C. J. Qi ●

85971 Changium H. Wolff(1924);明党参属;Changium ●★

85972 Changium smyrnioides H. Wolff;明党参(百丈光,粉沙参,红党参,金鸡爪,明参,明沙参,山花,山萝卜,天瓠,土人参);Medicinal Changium ■

85973 Changnienia S. S. Chien (1935);独花兰属(长年兰属);Uniflower Orchid,Uniflowerorchid ●★

85974 Changnienia amoena S. S. Chien;独花兰(半边锣,长年兰,带血独叶一枝枪,山慈姑);Uniflower Orchid,Uniflowerorchid ■

85975 Changruicaoia Z. Y. Zhu = Heterolamium C. Y. Wu ■★

85976 Changruicaoia Z. Y. Zhu(2001);长蕊草属;Changruicaoia ■

85977 Changruicaoia flaviflora Z. Y. Zhu;黄花长蕊草;Golden Cassia,Yellowflower Changruicaoia ■

85978 Changruicaoia flaviflora Z. Y. Zhu = Heterolamium debile (Hemsl.) C. Y. Wu var. tochauense (Kudo) C. Y. Wu ■

85979 Chapeliera Meisn. = Chapelieria A. Rich. ex DC. ●☆

85980 Chapelieria A. Rich. = Chapelieria A. Rich. ex DC. ●☆

85981 Chapelieria A. Rich. ex DC. (1830);沙普茜属●☆

85982 Chapelieria madagascariensis A. Rich.;沙普茜●☆

85983 Chapelliera Nees = Cladium P. Browne ■

85984 Chapmannia Torr. et A. Gray(1838);佛罗里达豆属■☆

85985 Chapmannia somalensis (Hillc. et J. B. Gillett) Thulin;佛罗里达豆■☆

85986 Chapmanolirion Dinter = Pancratium L. ■

85987 Chapmanolirion juttae Dinter = Pancratium tenuifolium Hochst. ex A. Rich. ■☆

85988 Chaptalia Royle = Gerbera L. (保留属名)■

85989 Chaptalia Vent. (1802)(保留属名);阳帽菊属(沙普塔菊属);Sunbonnet ■☆

85990 Chaptalia albicans (Sw.) Vent. ex B. D. Jacks.;白阳帽菊(白沙普塔菊);White Sunbonnet ■☆

85991 Chaptalia alsophila Greene = Leibnitzia lyrata (Sch. Bip.) G. L. Nesom ■☆

85992 Chaptalia leiocarpa (DC.) Urb. = Chaptalia albicans (Sw.) Vent. ex B. D. Jacks. ■☆

85993 Chaptalia leucocephala Greene = Leibnitzia lyrata (Sch. Bip.) G. L. Nesom ■☆

85994 Chaptalia lyrata D. Don = Leibnitzia lyrata (Sch. Bip.) G. L. Nesom ■☆

85995 Chaptalia maxima D. Don = Gerbera maxima (D. Don) Beauverd ■

85996 Chaptalia nutans (L.) Pol.;下垂阳帽菊(下垂沙普塔菊)■☆

85997 Chaptalia nutans (L.) Pol. var. texana (Greene) Burkart = Chaptalia texana Greene ■☆

85998 Chaptalia texana Greene;银色阳帽菊(银色沙普塔菊);Silverpuff ■☆

85999 Chaptalia tomentosa Vent.;毛阳帽菊(毛沙普塔菊);Woolly Sunbonnet ■☆

86000 Chaquepiria Endl. = Caquepiria J. F. Gmel. ●

86001 Chaquepiria Endl. = Gardenia Ellis(保留属名)●

86002 Charachera Forssk. = Lantana L. (保留属名)●

86003 Charachera tetragona Forssk. = Lantana viburnoides (Forssk.) Vahl ●☆

86004 Charachera viburnoides Forssk. = Lantana viburnoides (Forssk.) Vahl ●☆

86005 Characias Gray = Euphorbia L. ●■

86006 Charadra Scop. = Chadara Forssk. ●

86007 Charadra Scop. = Grewia L. ●

86008 Charadranaetes Janovec et H. Rob. (1997);裸托千里光属●☆

86009 Charadranaetes durandii (Klatt) Janovec et H. Rob.;裸托千里光■☆

86010 Charadrophila Marloth(1899);喜沟玄参属■☆

86011 Charadrophila capensis Marloth;喜沟玄参■☆

86012 Chardinia Desf. (1817);外翅菊属■☆

86013 Chardinia macrocarpa K. Koch;大果外翅菊■☆

86014 Chardinia orientalis (L.) Kuntze;东方外翅菊■☆

86015 Chareis N. T. Burb. = Charieis Cass. ■☆

86016 Charesia E. A. Busch = Silene L. (保留属名)■

86017 Charia C. DC. = Ekebergia Sparrm. ●☆

86018 Charia C. E. C. Fisch. = Ekebergia Sparrm. ●☆

86019 Charia chevalieri C. DC. = Ekebergia capensis Sparrm. ●☆

86020 Charia indeniensis A. Chev. = Ekebergia capensis Sparrm. ●☆

86021 Charianthus D. Don(1823);雅花野牡丹属●☆

86022 Charianthus alpinus (Sw.) R. A. Howard;雅花野牡丹●☆

86023 Charidia Baill. = Savia Willd. ●☆

86024 Charidion Bong. = Luxemburgia A. St. -Hil. ●☆

86025 Charieis Cass. (1817);佳丽菊属(小非洲菊属)■☆

86026 Charieis caerulea Cass. = Felicia heterophylla (Cass.) Grau ■☆

86027 Charieis heterophylla Cass.;佳丽菊;Blue Daisy,Kaulfussia ■☆

86028 Charieis heterophylla Cass. = Felicia heterophylla (Cass.) Grau ■☆

86029 Charieis heterophylla Cass. var. atroviolacea Hort.;紫堇佳丽菊■☆

86030 Charieis neesii Cass. = Felicia heterophylla (Cass.) Grau ■☆

86031 Chariessa Miq. = Citronella D. Don ●☆

86032 Chariomma Miers = Echites P. Browne ●☆

86033 Charisma D. Don = Chorisis DC. ■

86034 Charisma repens (L.) D. Don = Chorisis repens (L.) DC. ■

86035 Charistemma Janka = Scilla L. ■

86036 Charlwoodia Sweet = Cordyline Comm. ex R. Br. (保留属名)●

86037 Charpentiera Gaudich. (1826);穗苋树属●☆

86038 Charpentiera Vieill. = Ixora L. ●

86039 Charpentiera australis Sohmer;澳洲穗苋树●☆

86040 Charpentiera elliptica A. Heller;椭圆穗苋树●☆

86041 Charpentiera obovata Gaudich.;穗苋树●☆

86042 Charpentiera tomentosa Sohmer;毛穗苋树●☆

86043 Chartacalyx Maingay ex Mast. = Schoutenia Korth. ●☆

86044 Chartocalyx Regel = Harmsiella Briq. ●☆

86045 Chartocalyx Regel = Otostegia Benth. ●☆

86046 Chartolepis Cass. (1826);薄鳞菊属;Chartolepis ■

86047 Chartolepis Cass. = Centaurea L. (保留属名)●■

86048 Chartolepis biebersteinii Jaub. et Spach.;毕氏薄鳞菊■☆

86049 Chartolepis glastifolia (L.) Cass.;薄鳞菊;Intermediate Chartolepis ■

86050 Chartolepis intermedia Boiss. = Chartolepis glastifolia (L.) Cass. ■

86051 Chartolepis pterocaulis (Trautv.) Czerep.;翼茎薄鳞菊■☆

86052 Chartoloma Bunge(1844);薄缘芥属■☆

86053 Chartoloma platycarpum Bunge;薄缘芥■☆

86054 Charybdis Speta(1998);西西里风信子属■☆

86055 Charybdis anthericoides (Poir.) Dobignard et Vela;花篱风信子■☆

86056 Charybdis maritima (L.) Speta;西西里风信子■☆

86057 Charybdis maura (Maire) Speta;晚熟西西里风信子■☆

86058 Charybdis numidica (Jord. et Fourr.) Speta;努米底亚风信子■☆
86059 Charybdis simensis (Hochst. ex A. Rich.) Speta;锡米风信子■☆
86060 Charybdis tazensis (Maire) Speta;塔兹西西里风信子■☆
86061 Charybdis undulata (Desf.) Speta;波状西西里风信子■☆
86062 Chasalia Comm. ex DC. = Chasallia Comm. ex Poir. ■
86063 Chasalia DC. = Chasallia Comm. ex Poir. ■
86064 Chasallia Comm. ex Poir. = Chassalia Comm. ex Poir. ■
86065 Chascanum E. Mey. (1838)(保留属名);胀萼马鞭草属●☆
86066 Chascanum adenostachyum (Schauer) Moldenke;腺穗胀萼马鞭草●☆
86067 Chascanum africanum Moldenke = Chascanum hildebrandtii (Vatke) J. B. Gillett ●☆
86068 Chascanum angolense Moldenke;安哥拉胀萼马鞭草●☆
86069 Chascanum angolense Moldenke subsp. zambesiacum (R. Fern.) R. Fern.;赞比西胀萼马鞭草●☆
86070 Chascanum arabicum Moldenke = Chascanum laetum Walp. ●☆
86071 Chascanum caespitosum (H. Pearson) Moldenke;丛生胀萼马鞭草●☆
86072 Chascanum cernuum (L.) E. Mey.;俯垂胀萼马鞭草●☆
86073 Chascanum cuneifolium (L. f.) E. Mey.;楔叶胀萼马鞭草●☆
86074 Chascanum dehiscens (L. f.) Moldenke = Chascanum cuneifolium (L. f.) E. Mey. ●☆
86075 Chascanum garipense E. Mey.;加里普胀萼马鞭草●☆
86076 Chascanum gillettii Moldenke;吉莱特胀萼马鞭草●☆
86077 Chascanum glandulosum Thulin;具腺胀萼马鞭草●☆
86078 Chascanum hanningtonii (Oliv.) Moldenke;汉宁顿胀萼马鞭草●☆
86079 Chascanum hederaceum (Sond.) Moldenke;常春藤胀萼马鞭草●☆
86080 Chascanum hederaceum (Sond.) Moldenke var. natalense (H. Pearson) Moldenke;纳塔尔胀萼马鞭草●☆
86081 Chascanum hildebrandtii (Vatke) J. B. Gillett;希尔德胀萼马鞭草●☆
86082 Chascanum incisum (H. Pearson) Moldenke;锐裂胀萼马鞭草●☆
86083 Chascanum incisum (H. Pearson) Moldenke var. canescens Moldenke = Chascanum pumilum E. Mey. ●☆
86084 Chascanum integrifolium (H. Pearson) Moldenke;全叶胀萼马鞭草●☆
86085 Chascanum krookii (Gürke ex Zahlbr.) Moldenke;克鲁科胀萼马鞭草●☆
86086 Chascanum laetum Walp.;阿拉伯胀萼马鞭草●☆
86087 Chascanum latifolium (Harv.) Moldenke;宽叶胀萼马鞭草●☆
86088 Chascanum latifolium (Harv.) Moldenke var. glabrescens (H. Pearson) Moldenke;光滑宽叶胀萼马鞭草●☆
86089 Chascanum latifolium (Harv.) Moldenke var. transvaalense Moldenke;德兰士瓦胀萼马鞭草●☆
86090 Chascanum lignosum Dinter ex Moldenke = Chascanum pumilum E. Mey. ●☆
86091 Chascanum marrubiifolium Fenzl ex Walp.;夏至草叶胀萼马鞭草●☆
86092 Chascanum mixtum Thulin;混杂胀萼马鞭草●☆
86093 Chascanum moldenkei (J. B. Gillett) Sebsebe et Verdc.;莫尔登克胀萼马鞭草●☆
86094 Chascanum namaquanum (Bolus ex H. Pearson) Moldenke;纳马夸胀萼马鞭草●☆
86095 Chascanum obovatum Sebsebe;倒卵胀萼马鞭草●☆
86096 Chascanum obovatum Sebsebe subsp. glaucum? = Chascanum gillettii Moldenke ●☆
86097 Chascanum pinnatifidum (L. f.) E. Mey.;羽裂胀萼马鞭草●☆
86098 Chascanum pinnatifidum (L. f.) E. Mey. var. racemosum Schinz ex Moldenke;总花胀萼马鞭草●☆
86099 Chascanum pumilum E. Mey.;矮胀萼马鞭草●☆
86100 Chascanum rariflorum (A. Terracc.) Moldenke;稀花胀萼马鞭草●☆
86101 Chascanum schlechteri (Gürke) Moldenke;施莱胀萼马鞭草●☆
86102 Chascanum schlechteri (Gürke) Moldenke f. torrei R. Fern. = Chascanum schlechteri (Gürke) Moldenke var. torrei Moldenke ●☆
86103 Chascanum schlechteri (Gürke) Moldenke var. torrei Moldenke;托雷马鞭草●☆
86104 Chascanum sessilifolium (Vatke) Moldenke;无柄叶胀萼马鞭草●☆
86105 Chascanum sulcatum Sebsebe;纵沟胀萼马鞭草●☆
86106 Chascolytrum Desv. = Briza L. ■
86107 Chascolytrum subaristatum (Lam.) Desv. = Briza subaristata Lam. ■☆
86108 Chascotheca Urb. (1904);裂果大戟属●☆
86109 Chascotheca Urb. = Securinega Comm. ex Juss. (保留属名)●☆
86110 Chascotheca neopeltandra Urb.;裂果大戟●☆
86111 Chasea Nieuwl. = Panicum L. ■
86112 Chasea virgata (L.) Nieuwl. = Panicum virgatum L. ■
86113 Chasechloa A. Camus = Echinolaena Desv. ■☆
86114 Chasechloa A. Camus(1949);肖刺衣黍属■☆
86115 Chasechloa egregia (Mez) A. Camus;优秀肖刺衣黍■☆
86116 Chasechloa humbertiana A. Camus;亨伯特肖刺衣黍■☆
86117 Chasechloa madagascariensis (Baker) A. Camus;马岛肖刺衣黍■☆
86118 Chaseella Summerh. (1961);沙塞兰属■☆
86119 Chaseella pseudohydra Summerh.;沙塞兰■☆
86120 Chasmanthe N. E. Br. (1932);裂冠花属(豁裂花属);Chasmanthe, Pennants ■☆
86121 Chasmanthe aethiopica (L.) N. E. Br.;埃塞俄比亚裂冠花(豁裂花);Ethiopia Chasmanthe ■☆
86122 Chasmanthe bicolor (Gasp. ex Ten.) N. E. Br.;二色裂冠花;Chasmanthe ■☆
86123 Chasmanthe bicolor (Gasp.) N. E. Br. = Chasmanthe bicolor (Gasp. ex Ten.) N. E. Br. ■☆
86124 Chasmanthe caffra (Ker Gawl. ex Baker) N. E. Br. = Tritoniopsis caffra (Ker Gawl. ex Baker) Goldblatt ■☆
86125 Chasmanthe floribunda (Salisb.) N. E. Br.;繁锦豁裂花;African Cornflag ■☆
86126 Chasmanthe floribunda (Salisb.) N. E. Br. var. duckittii G. J. Lewis ex L. Bolus;杜克繁锦豁裂花■☆
86127 Chasmanthe fucata (Herb.) N. E. Br. = Crocosmia fucata (Herb.) M. P. de Vos ■☆
86128 Chasmanthe intermedia (Baker) N. E. Br. = Tritoniopsis intermedia (Baker) Goldblatt ■☆
86129 Chasmanthe peglerae N. E. Br. = Chasmanthe aethiopica (L.) N. E. Br. ■☆
86130 Chasmanthe spectabilis (Schinz) N. E. Br. = Gladiolus magnificus (Harms) Goldblatt ■☆
86131 Chasmanthe vittigera (Salisb.) N. E. Br. = Chasmanthe aethiopica (L.) N. E. Br. ■☆
86132 Chasmanthera Hochst. (1844);裂药防己属(张口藤属)■☆
86133 Chasmanthera Hochst. = Tinospora Miers(保留属名)●■

86134 Chasmanthera dependens Hochst. ;裂药防己■☆
86135 Chasmanthera nervosa Miers = Rhigiocarya racemifera Miers ■☆
86136 Chasmanthera strigosa Welw. ex Hiern = Chasmanthera welwitschii Troupin ■☆
86137 Chasmanthera uviformis Baill. = Tinospora uviforme (Baill.) Troupin ■☆
86138 Chasmanthera welwitschii Troupin;韦氏裂药防己■☆
86139 Chasmanthium Link(1827);裂口草属(海竹属);Sea Oats ■☆
86140 Chasmanthium latifolium (Michx.) H. O. Yates;宽叶裂口草(宽叶海竹);Bamboo Grass, Broad-leaf Chasmanthium, Broadleaf Spike Grass, Indian Wood Oats, Indian Woodoats, Inland Sea Oats, Northern Sea Oats, River Oats, Wild Oats ■☆
86141 Chasmanthium laxum (L.) H. O. Yates;疏松裂口草;Spike Grass ■☆
86142 Chasmanthium sessiliflorum (Poir.) H. O. Yates = Chasmanthium laxum (L.) H. O. Yates ■☆
86143 Chasmatocallis R. C. Foster = Lapeirousia Pourr. ■☆
86144 Chasmatophyllum (Schwantes) Dinter et Schwantes(1927);裂叶番杏属(开叶玉属)●■☆
86145 Chasmatophyllum Dinter et Schwantes = Chasmatophyllum (Schwantes) Dinter et Schwantes ●■☆
86146 Chasmatophyllum braunsii Schwantes;布氏裂叶番杏●☆
86147 Chasmatophyllum braunsii Schwantes var. majus L. Bolus = Chasmatophyllum braunsii Schwantes ●☆
86148 Chasmatophyllum musculinum (Haw.) Dinter et Schwantes;老鼠裂叶番杏●☆
86149 Chasmatophyllum nelii Schwantes;尼尔裂叶番杏●☆
86150 Chasmatophyllum rouxii L. Bolus;鲁裂叶番杏●☆
86151 Chasmatophyllum stanleyi (L. Bolus) H. E. K. Hartmann;斯坦利裂叶番杏●☆
86152 Chasmatophyllum verdoorniae (N. E. Br.) L. Bolus;韦尔裂叶番杏●☆
86153 Chasmatophyllum willowmorense (L. Bolus) L. Bolus;维洛莫尔裂叶番杏●☆
86154 Chasme Salisb. = Leucadendron R. Br. (保留属名)●
86155 Chasme spiralis Knight = Leucadendron spirale (Salisb. ex Knight) I. Williams ●☆
86156 Chasme spiralis Salisb. ex Knight = Leucadendron spirale (Salisb. ex Knight) I. Williams ●☆
86157 Chasmia Schott ex Spreng. = Arrabidaea DC. + Tynnanthus Miers ●☆
86158 Chasmia Schott ex Spreng. = Arrabidaea DC. ●☆
86159 Chasmone E. Mey. = Argyrolobium Eckl. et Zeyh. (保留属名)●☆
86160 Chasmone andrewsiana E. Mey. = Argyrolobium tomentosum (Andréws) Druce ●☆
86161 Chasmone angustissima E. Mey. = Argyrolobium angustissimum (E. Mey.) T. J. Edwards ●☆
86162 Chasmone apiculata E. Mey. = Argyrolobium molle Eckl. et Zeyh. ●☆
86163 Chasmone ascendens E. Mey. = Argyrolobium ascendens (E. Mey.) Walp. ●☆
86164 Chasmone baptisioides E. Mey. = Argyrolobium baptisioides (E. Mey.) Walp. ●☆
86165 Chasmone barbata Meisn. = Argyrolobium barbatum (Meisn.) Walp. ●☆
86166 Chasmone crassifolia (E. Mey.) E. Mey. = Argyrolobium crassifolium (E. Mey.) Eckl. et Zeyh. ●☆
86167 Chasmone crinita E. Mey. = Argyrolobium crinitum (E. Mey.) Walp. ●☆
86168 Chasmone cuneifolia E. Mey. = Argyrolobium polyphyllum Eckl. et Zeyh. ●☆
86169 Chasmone diversifolia E. Mey. = Argyrolobium speciosum Eckl. et Zeyh. ●☆
86170 Chasmone goodioides Meisn. = Argyrolobium crassifolium (E. Mey.) Eckl. et Zeyh. ●☆
86171 Chasmone holosericea E. Mey. = Argyrolobium trifoliatum (Thunb.) Druce ●☆
86172 Chasmone holosericea E. Mey. var. incana Meisn. = Argyrolobium incanum Eckl. et Zeyh. ●☆
86173 Chasmone lanceolata (E. Mey.) E. Mey. = Argyrolobium lunare (L.) Druce subsp. sericeum (Thunb.) T. J. Edwards ●☆
86174 Chasmone longifolia Meisn. = Argyrolobium longifolium (Meisn.) Walp. ●☆
86175 Chasmone obcordata E. Mey. = Argyrolobium trifoliatum (Thunb.) Druce ●☆
86176 Chasmone petiolaris E. Mey. = Argyrolobium petiolare (E. Mey.) Steud. ●☆
86177 Chasmone pumila (Eckl. et Zeyh.) Meisn. = Argyrolobium pumilum Eckl. et Zeyh. ●☆
86178 Chasmone rupestris E. Mey. = Argyrolobium rupestre (E. Mey.) Walp. ●☆
86179 Chasmone sessiliflora E. Mey. = Argyrolobium candicans Eckl. et Zeyh. ●☆
86180 Chasmone splendens Meisn. = Argyrolobium splendens (Meisn.) Walp. ●☆
86181 Chasmone stricta E. Mey. = Argyrolobium pauciflorum Eckl. et Zeyh. ●☆
86182 Chasmone tenuis E. Mey. = Argyrolobium tenue (E. Mey.) Walp. ●☆
86183 Chasmone tuberosa (Eckl. et Zeyh.) Meisn. = Argyrolobium tuberosum Eckl. et Zeyh. ●☆
86184 Chasmone venosa E. Mey. = Argyrolobium molle Eckl. et Zeyh. ●☆
86185 Chasmone venosa E. Mey. var. obscura = Argyrolobium molle Eckl. et Zeyh. ●☆
86186 Chasmone verticillata E. Mey. = Argyrolobium stipulaceum Eckl. et Zeyh. ●☆
86187 Chasmonia C. Presl = Moluccella L. ■☆
86188 Chasmopodium Stapf(1917);假叶柄草属■☆
86189 Chasmopodium afzelii (Hack.) Stapf;阿芙泽尔假叶柄草■☆
86190 Chasmopodium caudatum (Hack.) Stapf;尾假叶柄草■☆
86191 Chasmopodium purpurascens (Robyns) Clayton;紫假叶柄草■☆
86192 Chassalia Comm. ex Poir. (1812);弯管花属(柴沙榈属);Chasalia ■
86193 Chassalia acutiflora Bremek. ;尖花弯管花■☆
86194 Chassalia afzelii (Hiern) K. Schum. ;阿芙泽尔弯管花■☆
86195 Chassalia albiflora K. Krause;白花弯管花■☆
86196 Chassalia androrangensis Bremek. ;安德鲁兰加弯管花■☆
86197 Chassalia assimilis Bremek. ;相似弯管花■☆
86198 Chassalia betsilensis Bremek. ;贝齐尔弯管花■☆
86199 Chassalia bipindensis Sonké et Nguembou et A. P. Davis;比平迪弯管花■☆
86200 Chassalia bojeri Bremek. ;博耶尔弯管花■☆
86201 Chassalia buchwaldii K. Schum. ;布赫弯管花■☆
86202 Chassalia campyloneura Mildbr. ;弯脉弯管花■☆

86203 Chassalia caudifolia Bremek.;尾叶弯管花■☆
86204 Chassalia coursii Bremek.;库尔斯弯管花■☆
86205 Chassalia cristata (Hiern) Bremek.;冠状弯管花■☆
86206 Chassalia cupularis Hutch. et Dalziel;杯状弯管花■☆
86207 Chassalia curviflora Thwaites;弯管花(柴桫椤,假九节,假蓝枕,山椒,水松萝,银锦);Curvedflower Chasalia, Curved-flowered Chasalia ●
86208 Chassalia curviflora Thwaites var. longifolia Hook. f.;长叶弯管花;Longleaf Chasalia ●
86209 Chassalia densiflora Bremek.;密花弯管花■☆
86210 Chassalia discolor K. Schum.;异色弯管花■☆
86211 Chassalia discolor K. Schum. subsp. grandifolia Verdc.;大花异色弯管花■☆
86212 Chassalia discolor K. Schum. subsp. taitensis Verdc.;泰特弯管花■☆
86213 Chassalia doniana (Benth.) G. Taylor;唐弯管花■☆
86214 Chassalia elongata Hutch. et Dalziel;伸长弯管花■☆
86215 Chassalia euchlora (K. Schum.) Figueiredo;良芽弯管花■☆
86216 Chassalia eurybotrya Bremek.;宽序弯管花■☆
86217 Chassalia grandistipula Bremek.;大弯管花■☆
86218 Chassalia hiernii (Kuntze) G. Taylor;希尔恩弯管花■☆
86219 Chassalia hiernii (Kuntze) G. Taylor = Uragoga hiernii Kuntze ●☆
86220 Chassalia hiernii (Kuntze) G. Taylor var. glandulosa G. Taylor;具腺弯管花■☆
86221 Chassalia humbertii Bremek.;亨伯特弯管花■☆
86222 Chassalia ischnophylla (K. Schum.) Hepper;细长叶弯管花■☆
86223 Chassalia kenyensis Verdc.;肯尼亚弯管花■☆
86224 Chassalia kolly (Schumach.) Hepper;科利弯管花■☆
86225 Chassalia laikomensis Cheek;莱基皮弯管花■☆
86226 Chassalia laxiflora Benth.;疏花弯管花■☆
86227 Chassalia leandrii Bremek.;利安弯管花■☆
86228 Chassalia leptothyrsa Bremek.;细序弯管花■☆
86229 Chassalia longiloba Borhidi et Verdc.;长裂片弯管花■☆
86230 Chassalia lukwangulensis Thulin;卢夸古尔弯管花■☆
86231 Chassalia macrodiscus K. Schum.;大盘弯管花■☆
86232 Chassalia magnifolia Bremek.;大叶弯管花■☆
86233 Chassalia moramangensis Bremek.;莫拉芒弯管花■☆
86234 Chassalia parva Bremek.;较小弯管花■☆
86235 Chassalia parviflora Benth. = Chassalia kolly (Schumach.) Hepper ■☆
86236 Chassalia parvifolia K. Schum.;小叶弯管花■☆
86237 Chassalia pentachotoma Bremek.;五叉弯管花■☆
86238 Chassalia perrieri Bremek.;佩里耶弯管花■☆
86239 Chassalia petitiana Piessch.;佩蒂蒂弯管花■☆
86240 Chassalia pteropetala (K. Schum.) Cheek;翼瓣弯管花■☆
86241 Chassalia simplex K. Krause;简单弯管花■☆
86242 Chassalia subcordatifolia (De Wild.) Piessch.;亚心叶弯管花■☆
86243 Chassalia subherbacea (Hiern) Hepper;草本弯管花■☆
86244 Chassalia subnuda (Hiern) Hepper;亚裸弯管花■☆
86245 Chassalia subochreata (De Wild.) Robyns;亚鞘状托叶弯管花■☆
86246 Chassalia subspicata K. Schum.;穗状弯管花■☆
86247 Chassalia tchibangensis Pellegr.;奇班加弯管花■☆
86248 Chassalia ternifolia (Baker) Bremek.;三叶弯管花■☆
86249 Chassalia ugandensis Verdc.;乌干达弯管花■☆
86250 Chassalia umbraticola Vatke;荫蔽弯管花■☆
86251 Chassalia vanderystii (De Wild.) Verdc.;范德弯管花■☆
86252 Chassalia violacea K. Schum.;堇色弯管花■☆
86253 Chassalia violacea K. Schum. var. parviflora Verdc.;小花弯管花■☆
86254 Chassalia yorubensis K. Schum. = Chassalia kolly (Schumach.) Hepper ■☆
86255 Chassalia zenkeri K. Schum. et K. Krause;岑克尔弯管花■☆
86256 Chassalia zimmermannii Verdc.;齐默尔曼弯管花■☆
86257 Chasseloupia Vieill. = Symplocos Jacq. ●
86258 Chastenaea DC. = Axinaea Ruiz et Pav. ●☆
86259 Chastoloma Lindl. = Chartoloma Bunge ■☆
86260 Chataea Sol. = Chaitaea Sol. ex Seem. ■
86261 Chataea Sol. = Tacca J. R. Forst. et G. Forst. (保留属名)■
86262 Chataea Sol. ex Seem. = Chaitaea Sol. ex Seem. ■
86263 Chataea Sol. ex Seem. = Tacca J. R. Forst. et G. Forst. (保留属名)■
86264 Chatelania Neck. = Tolpis Adans. ●■☆
86265 Chatiakella Cass. = Wulffia Neck. ex Cass. ■☆
86266 Chatinia Tiegh. = Psittacanthus Mart. ●☆
86267 Chaubardia Rchb. f. (1852);肖巴尔兰属(乔巴兰属)■☆
86268 Chaubardia surinamensis Rchb. f.;肖巴尔兰■☆
86269 Chaubardiella Garay(1969);拟乔巴兰属■☆
86270 Chaubardiella calceolaris Garay;拟乔巴兰■☆
86271 Chauliodon Summerh. (1943);突齿兰属■☆
86272 Chauliodon buntingii Summerh. = Chauliodon deflexicalcaratum (De Wild.) L. Jonss. ■☆
86273 Chauliodon deflexicalcaratum (De Wild.) L. Jonss.;弯距突齿兰■☆
86274 Chaulmoogra Roxb. = Gynocardia R. Br. ●
86275 Chaulmoogra odorata Roxb. = Gynocardia odorata R. Br. ●◇
86276 Chaunanthus O. E. Schulz = Iodanthus (Torr. et A. Gray) Steud. ■☆
86277 Chaunanthus O. E. Schulz(1924);口花芥属■☆
86278 Chaunanthus petiolatus (Hemsl.) O. E. Schulz;口花芥■☆
86279 Chaunanthus petiolatus O. E. Schulz = Chaunanthus petiolatus (Hemsl.) O. E. Schulz ■☆
86280 Chaunochiton Benth. (1867);张口木属●☆
86281 Chaunochiton angustifolium Sleumer;窄叶张口木●☆
86282 Chaunochiton breviflorum Ducke;短花张口木●☆
86283 Chaunochiton loranthoides Benth.;张口木●☆
86284 Chaunochiton purpurascens Rizzini;紫张口木●☆
86285 Chaunochitonaceae Tiegh. = Olacaceae R. Br. (保留科名)●
86286 Chaunostoma Donn. Sm. (1895);开口草属●☆
86287 Chaunostoma mecistandrum Donn. Sm.;开口草●☆
86288 Chautemsia A. O. Araujo et V. C. Souza(2010);巴西苣苔属☆
86289 Chauvinia Steud. = Spartina Schreb. ex J. F. Gmel. ■
86290 Chavannesia A. DC. = Urceola Roxb. (保留属名)●
86291 Chavica Miq. = Piper L. ●■
86292 Chavica boehmeriifolia Miq. = Piper boehmeriifolium (Miq.) C. DC. ●
86293 Chavica hainana C. DC. = Piper sarmentosum Roxb. ■
86294 Chavica leptostachya Hance = Piper hancei Maxim. ●■
86295 Chavica mullesua (Buch.-Ham. ex D. Don) Miq. = Piper mullesua D. Don ●
86296 Chavica mullesua (D. Don) Miq. = Piper mullesua D. Don ●
86297 Chavica officinarum Miq. = Piper retrofractum Vahl ■
86298 Chavica peepuloides sunsu Wight = Piper retrofractum Vahl ■
86299 Chavica puberula Benth. = Piper hongkongense C. DC. ■

86300 Chavica roxburghii Miq. = Piper longum L. ■
86301 Chavica sarmentosa (Roxb.) Miq. = Piper sarmentosum Roxb. ■
86302 Chavica sarmentosa Miq. = Piper sarmentosum Roxb. ■
86303 Chavica sinensis (Champ.) Benth. = Piper cathayanum M. G. Gilbert et N. H. Xia ■
86304 Chavica sinensis Champ. = Piper cathayanum M. G. Gilbert et N. H. Xia ■
86305 Chavica sphaerostachya Miq. = Piper mullesua D. Don ●
86306 Chavica sphaerostachya Wall. ex Miq. = Piper mullesua D. Don ●
86307 Chavica suipigua (Buch. -Ham. ex D. Don) Miq. = Piper suipigua Buch. -Ham. ex D. Don ■
86308 Chavica sylvatica (Roxb.) Miq. = Piper sylvaticum Roxb. ■
86309 Chavica sylvatica Miq. = Piper sylvaticum Roxb. ■
86310 Chavica thomsonii C. DC. = Piper thomsonii (C. DC.) Hook. f. ■
86311 Chavica wallichii Miq. = Piper wallichii (Miq.) Hand. -Mazz. ■
86312 Chavinia Gand. = Rosa L. ●
86313 Chaydaia Pit. (1912);苞叶木属;Chaydaia ●
86314 Chaydaia Pit. = Rhamnella Miq. ●
86315 Chaydaia berchemiifolia Koidz. = Berchemiella berchemiifolia (Makino) Nakai ●
86316 Chaydaia crenulata Hand. -Mazz. = Chaydaia rubrinervis (H. Lév.) C. Y. Wu ●
86317 Chaydaia crenulata Miq. = Chaydaia rubrinervis (H. Lév.) C. Y. Wu ●
86318 Chaydaia rubrinervis (H. Lév.) C. Y. Wu;苞叶木(红脉麦果,十两叶);Rednerved Chaydaia,Red-nerved Chaydaia ●
86319 Chaydaia rubrinervis (H. Lév.) C. Y. Wu ex Y. L. Chen et P. K. Chou = Rhamnella rubrinervis (H. Lév.) Rehder ●
86320 Chaydaia tonkinensis Pit.;越南苞叶木;Tonkin Chaydaia,Vietnam Chaydaia ●
86321 Chaydaia wilsonii C. K. Schneid. = Berchemiella berchemiifolia (Makino) Nakai ●
86322 Chaydaia wilsonii C. K. Schneid. = Berchemiella wilsonii (C. K. Schneid.) Nakai ●◇
86323 Chayota Jacq. = Sechium P. Browne(保留属名)■
86324 Chayota edulis (Jacq.) Jacq. = Sechium edule (Jacq.) Sw. ■
86325 Chazaliella E. Petit et Verdc. (1975);沙扎尔茜属●☆
86326 Chazaliella abrupta (Hiern) E. M. Petit et Verdc.;平截沙扎尔茜●☆
86327 Chazaliella abrupta (Hiern) E. M. Petit et Verdc. var. parvifolia Verdc.;小叶平截沙扎尔茜●☆
86328 Chazaliella coffeosperma (K. Schum.) Verdc.;咖啡籽沙扎尔茜●☆
86329 Chazaliella coffeosperma (K. Schum.) Verdc. subsp. longipedicellata Verdc.;长梗咖啡籽沙扎尔茜●☆
86330 Chazaliella cupulicalyx Verdc.;杯萼沙扎尔茜●☆
86331 Chazaliella gossweileri (Cavaco) E. M. Petit et Verdc.;戈斯沙扎尔茜●☆
86332 Chazaliella insidens (Hiern) E. M. Petit et Verdc. subsp. liberica Verdc.;离生沙扎尔茜●☆
86333 Chazaliella letouzeyi Robbr.;勒图沙扎尔茜●☆
86334 Chazaliella longistylis (Hiern) E. M. Petit et Verdc.;长柱沙扎尔茜●☆
86335 Chazaliella lophoclada (Hiern) E. M. Petit et Verdc.;冠枝沙扎尔茜●☆
86336 Chazaliella macrocarpa Verdc.;大果沙扎尔茜●☆
86337 Chazaliella obanensis (Wernham) E. M. Petit et Verdc.;奥班沙扎尔茜●☆
86338 Chazaliella obovoidea Verdc.;倒卵沙扎尔茜●☆
86339 Chazaliella obovoidea Verdc. subsp. longipedunculata Verdc.;长梗倒卵沙扎尔茜●☆
86340 Chazaliella obovoidea Verdc. subsp. rhytidophloea Verdc.;皱皮倒卵沙扎尔茜●☆
86341 Chazaliella obovoidea Verdc. subsp. villosistipula Verdc.;毛托叶沙扎尔茜●☆
86342 Chazaliella oddonii (De Wild.) E. M. Petit et Verdc.;奥顿沙扎尔茜●☆
86343 Chazaliella oddonii (De Wild.) E. M. Petit et Verdc. var. cameroonensis Verdc.;喀麦隆沙扎尔茜●☆
86344 Chazaliella oddonii (De Wild.) E. M. Petit et Verdc. var. grandifolia Verdc.;大叶沙扎尔茜●☆
86345 Chazaliella parviflora (R. D. Good) Verdc.;小花沙扎尔茜●☆
86346 Chazaliella pilosula (De Wild.) E. M. Petit et Verdc. = Chazaliella oddonii (De Wild.) E. M. Petit et Verdc. ●☆
86347 Chazaliella poggei (K. Schum.) E. M. Petit et Verdc.;波格沙扎尔茜●☆
86348 Chazaliella ramisulca Verdc.;枝生沙扎尔茜●☆
86349 Chazaliella rotundifolia (R. D. Good) E. M. Petit et Verdc.;圆叶沙扎尔茜●☆
86350 Chazaliella sciadephora (Hiern) E. M. Petit et Verdc.;伞形沙扎尔茜●☆
86351 Chazaliella sciadephora (Hiern) E. M. Petit et Verdc. var. condensata Verdc.;密集沙扎尔茜●☆
86352 Chazaliella subcordatifolia (De Wild.) E. M. Petit et Verdc. = Chassalia subcordatifolia (De Wild.) Piessch. ■☆
86353 Chazaliella viridicalyx (R. D. Good) Verdc.;绿萼沙扎尔茜●☆
86354 Chazaliella wildemaniana (T. Durand ex De Wild.) E. M. Petit et Verdc.;怀尔扎尔茜●☆
86355 Cheesemania O. E. Schulz(1929);澳洲芥属■☆
86356 Cheesemania fastigiata O. E. Schulz;澳洲芥■☆
86357 Cheiloclinium Miers(1872);斜唇卫矛属●☆
86358 Cheiloclinium lineolatum (A. C. Sm.) A. C. Sm.;线形斜唇卫矛●☆
86359 Cheiloclinium lucidum A. C. Sm.;光亮斜唇卫矛●☆
86360 Cheiloclinium obtusum A. C. Sm.;钝斜唇卫矛●☆
86361 Cheiloclinium parviflorum (Miers) A. C. Sm.;小花斜唇卫矛●☆
86362 Cheiloclinium serratum (Cambess. ex St. Hil.) A. C. Sm.;齿斜唇卫矛●☆
86363 Cheilococca Salisb. = Platylobium Sm. ●■☆
86364 Cheilococca Salisb. ex Sm. = Platylobium Sm. ●■☆
86365 Cheilocostus C. D. Specht = Banksea J. König ■
86366 Cheilocostus C. D. Specht = Costus L. ■
86367 Cheilodiscus Triana = Pectis L. ■☆
86368 Cheilophyllum Pennell = Cheilophyllum Pennell. ex Britton ■☆
86369 Cheilophyllum Pennell. ex Britton(1920);唇叶玄参属■☆
86370 Cheilophyllum dentatum Urb.;齿唇叶玄参■☆
86371 Cheilophyllum jamaicense Pennell;唇叶玄参■☆
86372 Cheilophyllum micranthum Urb.;小花唇叶玄参■☆
86373 Cheilophyllum microphyllum Pennell;小叶唇叶玄参■☆
86374 Cheilophyllum radicans Pennell;辐射唇叶玄参■☆
86375 Cheilopsis Moq. = Acanthus L. ●■
86376 Cheilopsis montana Nees = Acanthus montanus (Nees) T. Anderson ●☆
86377 Cheilopsis polystachya (Delile) Moq. = Acanthus polystachyus

Delile ●☆

86378 Cheilopsis steudneri Schweinf. = Acanthus sennii Chiov. ●☆

86379 Cheilosa Blume(1826);山唇木属●☆

86380 Cheilosa montana Blume;山唇木●☆

86381 Cheilosaceae Doweld;山唇木科●☆

86382 Cheilosandra Griff. ex Lindl. = Rhynchotechum Blume ●

86383 Cheilotheca Hook. f. (1876);假水晶兰属(拟水晶兰属,水晶兰属);Cheilotheca ■

86384 Cheilotheca hkasiana Hook. f.;假水晶兰■☆

86385 Cheilotheca humilis (D. Don) H. Keng = Monotropastrum humile (D. Don) H. Hara ■

86386 Cheilotheca humilis (D. Don) H. Keng var. baranovii (Y. L. Chang et Y. L. Chou) Y. L. Chou;矮假水晶兰■

86387 Cheilotheca humilis (D. Don) H. Keng var. glaberrima (H. Hara) H. Keng et C. F. Hsieh = Monotropastrum humile (D. Don) H. Hara ■

86388 Cheilotheca humilis (D. Don) H. Keng var. glaberrima (H. Hara) H. Keng et C. F. Hsieh = Cheilotheca macrocarpa (Andréws) Y. L. Chou ■

86389 Cheilotheca humilis (D. Don) H. Keng var. pubescens (K. F. Wu) C. Ling = Monotropastrum humile (D. Don) H. Hara ■

86390 Cheilotheca macrocarpa (Andres) Y. L. Chou;大果假水晶兰(阿里山水晶兰,大果拟水晶兰,假水晶兰,拟水晶兰,台湾假水晶兰,台湾拟水晶兰,浙江假水晶兰);Bigfruit Cheilotheca ■

86391 Cheilotheca macrocarpa (Andres) Y. L. Chou = Monotropastrum humile (D. Don) H. Hara ■

86392 Cheilotheca pubescens (K. F. Wu) Y. L. Chou;毛花假水晶兰;Hairflower Cheilotheca ■

86393 Cheilotheca pubescens (K. F. Wu) Y. L. Chou = Monotropastrum humile (D. Don) H. Hara ■

86394 Cheilyctis (Raf.) Spach = Monarda L. ■

86395 Cheilyctis Benth. = Monarda L. ■

86396 Cheiradenia Lindl. (1853);手腺兰属■☆

86397 Cheiradenia cuspidata Lindl.;手腺兰■☆

86398 Cheiranthera A. Cunn. ex Brongn. (1834);毛花海桐属●☆

86399 Cheiranthera A. Cunn. ex Lindl. = Cheiranthera A. Cunn. ex Brongn. ●☆

86400 Cheiranthera Brongn. = Cheiranthera A. Cunn. ex Brongn. ●☆

86401 Cheiranthera Endl. = Chiranthodendron Sessé ex Larreat. ●☆

86402 Cheiranthera alternifolia E. M. Benn.;互叶毛花海桐●☆

86403 Cheiranthera borealis (E. M. Benn.) L. W. Cayzer et Crisp;北方毛花海桐●☆

86404 Cheiranthera brevifolia F. Muell.;短叶毛花海桐●☆

86405 Cheiranthera filifolia Turcz.;线叶毛花海桐●☆

86406 Cheiranthodendraceae A. Gray = Sterculiaceae Vent. (保留科名)●■

86407 Cheiranthodendron Benth. et Hook. f. = Chiranthodendron Sessé ex Larreat. ●☆

86408 Cheiranthodendrum Steud. = Chiranthodendron Sessé ex Larreat. ●☆

86409 Cheiranthus L. (1753);桂竹香属(紫罗兰花属);Wall Flower, Wallflower ●■

86410 Cheiranthus L. = Erysimum L. ■●

86411 Cheiranthus acaulis Hand. -Mazz. = Cheiranthus forrestii (W. W. Sm.) Hand. -Mazz. var. acaulis (Hand. -Mazz.) K. C. Kuan ●

86412 Cheiranthus acaulis Hand. -Mazz. = Erysimum handel-mazzettii Polatschek ■

86413 Cheiranthus aculis Hand. -Mazz. = Erysimum handel-mazzettii Polatschek ■

86414 Cheiranthus africanus L. = Heliophila africana (L.) Marais ■☆

86415 Cheiranthus albiflorus T. Anderson = Phaeonychium albiflorum (T. Anderson) Jafri ■

86416 Cheiranthus alpinus L.;高山桂竹香■

86417 Cheiranthus annuus L. = Matthiola incana (L.) R. Br. ■

86418 Cheiranthus apricus L. var. trichosepalus (Turcz.) Franch. = Clausia trichosepala (Turcz.) Dvorák ■

86419 Cheiranthus apricus Steph. = Clausia aprica (Steph.) Korn.-Trotzky ■

86420 Cheiranthus apricus Steph. var. trichosepalus (Turcz.) Franch. = Clausia trichosepala (Turcz.) Dvorák ■

86421 Cheiranthus arbuscula Lowe = Erysimum arbuscula (Lowe) Snogerup ■☆

86422 Cheiranthus aurantiacus Bunge = Erysimum amurense Kitag. ■

86423 Cheiranthus bicolor Hornem. = Erysimum bicolor (Hornem.) DC. ■☆

86424 Cheiranthus bicornis Sibth. et Sm. = Matthiola bicornis (Sibth. et Sm.) DC. ■☆

86425 Cheiranthus carnosus Thunb. = Heliophila carnosa (Thunb.) Steud. ■☆

86426 Cheiranthus caspicus Lam. = Sterigmostemum caspicum (Lam.) Rupr. ■

86427 Cheiranthus cheiri L.;桂竹香(黄紫罗兰花,五彩糖芥);Aegean Wallflower, Bee Flower, Bellflower, Bleeding Heart, Bleeding Warrior, Bleeding Wire, Bleedy Warrior, Bleedy War-rior, Blood Wall, Bloody Wall, Bloody Wallier, Bloody Walls, Bloody Warrior, Bloody Wires, Bloody-warrior, Chare, Cheiry, Cherisaunce, Chevisaunce, Churl, Common Wallflower, Cross Flower, Cross-flower, Crucifix-flower, Devil's Gillofer, English Wallflower, Gellalfred, Geraflour, Gilliflower, Gilliver, Gilly, Gillyfer, Gilly-flower, Gilver, Handflower, Harcher, Heartsease, Jelly Stock, Jelly-flower, Jelly-stock, Jeroffleris, Jilaffer, Jilliver, Jilloffer, Jilly Offers, July-flower, Keiry, King's Cross, Old Maid, Soldier's Cross, Sweet William, Wall Flower, Wall Gilliflower, Wall July Flower, Wall July-flower, Wallflower, Wallwort, Wild Chier, Winter Gilliflower, Yellow Stock Gilliflower, Yellow Violet, Yellow Wallflower ●■

86428 Cheiranthus cheiri L. = Erysimum cheiri (L.) Crantz ●■

86429 Cheiranthus corinthius Boiss. = Cheiranthus cheiri L. ●■

86430 Cheiranthus coronopifolius Sibth. et Sm. = Matthiola fruticulosa (L.) Maire ●☆

86431 Cheiranthus elongatus Thunb. = Heliophila elongata (Thunb.) DC. ■☆

86432 Cheiranthus farsetia L. = Farsetia aegyptia Turra ■☆

86433 Cheiranthus fenestralis L. = Matthiola incana (L.) R. Br. ■

86434 Cheiranthus forrestii (W. W. Sm.) Hand. -Mazz. = Erysimum forrestii (W. W. Sm.) Polatschek ■

86435 Cheiranthus forrestii (W. W. Sm.) Hand. -Mazz. var. acaulis (Hand. -Mazz.) K. C. Kuan = Erysimum handel-mazzettii Polatschek ■

86436 Cheiranthus fruticulosus L. = Matthiola fruticulosa (L.) Maire ●☆

86437 Cheiranthus gramineus Thunb. = Heliophila carnosa (Thunb.) Steud. ■☆

86438 Cheiranthus himalaicus Hook. f. et Thomson = Desideria himalayensis (Cambess.) Al-Shehbaz ■

86439 Cheiranthus himalayensis Cambess. = Christolea himalayensis

(Cambess.) Jafri ■
86440 Cheiranthus himalayensis Cambess. = Desideria himalayensis (Cambess.) Al-Shehbaz ■
86441 Cheiranthus hirnalaicus Hook. f. et Thomson = Christolea himalayensis (Cambess.) Jafri ■
86442 Cheiranthus incanus L. = Erysimum incanum Kuntze ■☆
86443 Cheiranthus incanus L. = Matthiola incana (L.) W. T. Aiton ■
86444 Cheiranthus kewensis Hort.;邱园桂竹香●☆
86445 Cheiranthus lacerus L. = Malcolmia triloba (L.) Spreng. ■☆
86446 Cheiranthus linearis Thunb. = Heliophila linearis (Thunb.) DC. ■☆
86447 Cheiranthus linifolius Pers.;线叶桂竹香●☆
86448 Cheiranthus linifolius Pers. = Erysimum linifolium J. Gay ■☆
86449 Cheiranthus littoreus L. = Malcolmia littorea (L.) R. Br. ■☆
86450 Cheiranthus lividus Delile = Matthiola longipetala (Vent.) DC. subsp. livida (Delile) Maire ■☆
86451 Cheiranthus longipetalus Vent. = Matthiola longipetala (Vent.) DC. ■☆
86452 Cheiranthus maritimus L. = Malcolmia maritima (L.) R. Br. ■☆
86453 Cheiranthus muricatus Weinm. = Dontostemon integrifolius (L.) Ledeb. ■
86454 Cheiranthus parryoides Kurz ex Hook. f. et T. Anderson = Phaeonychium parryoides (Kurz ex Hook. f. et T. Anderson) O. E. Schulz ■
86455 Cheiranthus parviflorus Schousb. = Matthiola parviflora (Schousb.) R. Br. ■☆
86456 Cheiranthus pinnatifidus Willd. = Dontostemon pinnatifidus (Willd.) Al-Shehbaz et H. Ohba ■
86457 Cheiranthus pumilio Sibth. et Sm. = Matthiola longipetala (Vent.) DC. subsp. hirta (Conti) Greuter et Burdet ■☆
86458 Cheiranthus roseus Maxim. = Erysimum roseum (Maxim.) Polatschek ■
86459 Cheiranthus roseus Maxim. var. glabrescens Danguy = Erysimum roseum (Maxim.) Polatschek ■
86460 Cheiranthus scapiger Adams = Parrya nudicaulis (L.) Regel ■
86461 Cheiranthus scoparius Brouss. ex Willd.;帚状桂竹香;Broom Wallflower ●
86462 Cheiranthus scoparius Brouss. ex Willd. = Erysimum scoparium (Willd.) Wettst. ■☆
86463 Cheiranthus semperflorens Schousb. = Erysimum semperflorens (Schousb.) Wettst. ■☆
86464 Cheiranthus siliculosus M. Bieb. = Erysimum siliculosum (M. Bieb.) DC. ■
86465 Cheiranthus sinuatus L. = Matthiola sinuata (L.) R. Br. ■☆
86466 Cheiranthus stewartii T. Anderson = Christolea stewartii (T. Anderson) Jafri ■
86467 Cheiranthus stewartii T. Anderson = Desideria stewartii (T. Anderson) Al-Shehbaz ■
86468 Cheiranthus strictus L. f. = Heliophila scoparia Burch. ex DC. ■☆
86469 Cheiranthus taraxacifolius Balb. = Malcolmia africana (L.) R. Br. ■
86470 Cheiranthus tomentosus Willd. = Sterigmostemum caspicum (Lam.) Rupr. ■
86471 Cheiranthus torulosus M. Bieb. = Sterigmostemum incanum M. Bieb. ■
86472 Cheiranthus torulosus Thunb. = Matthiola torulosa (Thunb.) DC. ■☆
86473 Cheiranthus tricuspidatus L. = Matthiola tricuspidata (L.) R. Br. ■☆
86474 Cheiranthus trilobus L. = Malcolmia triloba (L.) Spreng. ■☆
86475 Cheiranthus tristis L. = Matthiola fruticulosa (L.) Maire ●☆
86476 Cheiranthus tristis L. = Matthiola tristis (L.) R. Br. ■☆
86477 Cheiranthus younghusbandii Prain;拉萨桂竹香;Lasa Wallflower ●
86478 Cheiri Adans. = Cheiranthus L. ●■
86479 Cheiri Ludw. = Cheiranthus L. ●■
86480 Cheiridopsis N. E. Br. (1925);虾疳花属(鞘袖属);Cheiridopsis ■☆
86481 Cheiridopsis acuminata L. Bolus;京鱼虾疳花(京鱼)■☆
86482 Cheiridopsis alata L. Bolus = Cheiridopsis robusta (Haw.) N. E. Br. ■☆
86483 Cheiridopsis albiflora L. Bolus;白花虾疳花■☆
86484 Cheiridopsis albiflora L. Bolus = Cheiridopsis derenbergiana Schwantes ■☆
86485 Cheiridopsis albirosea L. Bolus = Ihlenfeldtia excavata (L. Bolus) H. E. K. Hartmann ■☆
86486 Cheiridopsis altitecta Schwantes = Cheiridopsis derenbergiana Schwantes ■☆
86487 Cheiridopsis amabilis S. A. Hammer;秀丽虾疳花■☆
86488 Cheiridopsis ampliata L. Bolus = Cheiridopsis derenbergiana Schwantes ■☆
86489 Cheiridopsis angustipetala L. Bolus = Cheiridopsis derenbergiana Schwantes ■☆
86490 Cheiridopsis aspera L. Bolus;朱蟹玉■☆
86491 Cheiridopsis aurea L. Bolus;金黄虾疳花■☆
86492 Cheiridopsis aurea L. Bolus = Cheiridopsis robusta (Haw.) N. E. Br. ■☆
86493 Cheiridopsis aurea L. Bolus var. lutea? = Cheiridopsis robusta (Haw.) N. E. Br. ■☆
86494 Cheiridopsis ausensis L. Bolus = Cheiridopsis caroli-schmidtii (Dinter et A. Berger) N. E. Br. ■☆
86495 Cheiridopsis bibracteata (Haw.) N. E. Br. = Cheiridopsis rostrata (L.) N. E. Br. ■☆
86496 Cheiridopsis bifida (Haw.) N. E. Br.;虾疳花;Cheiridopsis ■☆
86497 Cheiridopsis bifida (Haw.) N. E. Br. = Cheiridopsis rostrata (L.) N. E. Br. ■☆
86498 Cheiridopsis borealis L. Bolus = Cheiridopsis caroli-schmidtii (Dinter et A. Berger) N. E. Br. ■☆
86499 Cheiridopsis brachystigma L. Bolus = Cheiridopsis pillansii L. Bolus ■☆
86500 Cheiridopsis braunsii Schwantes = Argyroderma fissum (Haw.) L. Bolus ●☆
86501 Cheiridopsis breachiae L. Bolus = Cheiridopsis namaquensis (Sond.) H. E. K. Hartmann ■☆
86502 Cheiridopsis brevipes L. Bolus = Cheiridopsis robusta (Haw.) N. E. Br. ■☆
86503 Cheiridopsis brevis L. Bolus = Cheiridopsis namaquensis (Sond.) H. E. K. Hartmann ■☆
86504 Cheiridopsis brownii Schick et Tischer;布朗虾疳花■☆
86505 Cheiridopsis campanulata G. Will.;钟状虾疳花■☆
86506 Cheiridopsis candidissima (Haw.) N. E. Br.;慈晃锦■☆
86507 Cheiridopsis candidissima (Haw.) N. E. Br. = Cheiridopsis denticulata (Haw.) N. E. Br. ■☆
86508 Cheiridopsis carinata L. Bolus = Cheiridopsis acuminata L. Bolus ■☆

86509 Cheiridopsis carnea N. E. Br. = Cheiridopsis rostrata (L.) N. E. Br. ■☆
86510 Cheiridopsis caroli-schmidtii (Dinter et A. Berger) N. E. Br.;大双剑■☆
86511 Cheiridopsis cigarettifera (A. Berger) N. E. Br.;逆锋■☆
86512 Cheiridopsis cigarettifera (A. Berger) N. E. Br. = Cheiridopsis namaquensis (Sond.) H. E. K. Hartmann ■☆
86513 Cheiridopsis citrina L. Bolus = Cheiridopsis derenbergiana Schwantes ■☆
86514 Cheiridopsis compressa L. Bolus = Cheiridopsis derenbergiana Schwantes ■☆
86515 Cheiridopsis comptonii L. Bolus ex H. Jacobsen = Cheiridopsis speciosa L. Bolus ■☆
86516 Cheiridopsis crassa L. Bolus = Cheiridopsis pillansii L. Bolus ■☆
86517 Cheiridopsis cuprea (L. Bolus) N. E. Br.;晚光■☆
86518 Cheiridopsis cuprea (L. Bolus) N. E. Br. = Cephalophyllum caespitosum H. E. K. Hartmann ■☆
86519 Cheiridopsis curta L. Bolus = Cheiridopsis namaquensis (Sond.) H. E. K. Hartmann ■☆
86520 Cheiridopsis delphinoides S. A. Hammer;翠雀虾疳花■☆
86521 Cheiridopsis denticulata (Haw.) N. E. Br.;冰岭(齿鞘袖)■☆
86522 Cheiridopsis denticulata (Haw.) N. E. Br. var. glauca? = Cheiridopsis denticulata (Haw.) N. E. Br. ■☆
86523 Cheiridopsis derenbergiana Schwantes;戴伦虾疳花■☆
86524 Cheiridopsis difforme N. E. Br. = Cheiridopsis namaquensis (Sond.) H. E. K. Hartmann ■☆
86525 Cheiridopsis dilatata L. Bolus = Ihlenfeldtia excavata (L. Bolus) H. E. K. Hartmann ■☆
86526 Cheiridopsis duplessii L. Bolus = Cheiridopsis namaquensis (Sond.) H. E. K. Hartmann ■☆
86527 Cheiridopsis eburnea L. Bolus = Cheiridopsis derenbergiana Schwantes ■☆
86528 Cheiridopsis excavata L. Bolus = Ihlenfeldtia excavata (L. Bolus) H. E. K. Hartmann ■☆
86529 Cheiridopsis framesii L. Bolus = Cheiridopsis namaquensis (Sond.) H. E. K. Hartmann ■☆
86530 Cheiridopsis gibbosa Schick et Tischer;拟宝珠■☆
86531 Cheiridopsis gibbosa Schick et Tischer = Cheiridopsis pillansii L. Bolus ■☆
86532 Cheiridopsis glabra L. Bolus = Cheiridopsis pearsonii N. E. Br. ■☆
86533 Cheiridopsis glomerata S. A. Hammer;团集虾疳花■☆
86534 Cheiridopsis graessneri Schick et Tischer = Cheiridopsis brownii Schick et Tischer ■☆
86535 Cheiridopsis grandiflora L. Bolus = Cheiridopsis robusta (Haw.) N. E. Br. ■☆
86536 Cheiridopsis hallii L. Bolus = Cheiridopsis brownii Schick et Tischer ■☆
86537 Cheiridopsis herrei L. Bolus;弥生■☆
86538 Cheiridopsis hilmarii L. Bolus = Deilanthe hilmarii (L. Bolus) H. E. K. Hartmann ■☆
86539 Cheiridopsis hutchinsonii L. Bolus = Cheiridopsis namaquensis (Sond.) H. E. K. Hartmann ■☆
86540 Cheiridopsis inaequalis L. Bolus = Cheiridopsis derenbergiana Schwantes ■☆
86541 Cheiridopsis inconspicua N. E. Br. = Cheiridopsis denticulata (Haw.) N. E. Br. ■☆
86542 Cheiridopsis insignis Schwantes = Cheiridopsis brownii Schick et Tischer ■☆
86543 Cheiridopsis inspersa (N. E. Br.) N. E. Br. = Cheiridopsis rostrata (L.) N. E. Br. ■☆
86544 Cheiridopsis inspersa N. E. Br.;虹鱼■☆
86545 Cheiridopsis intrusa L. Bolus = Cheiridopsis namaquensis (Sond.) H. E. K. Hartmann ■☆
86546 Cheiridopsis johannis-winkleri Schwantes = Cheiridopsis schlechteri Tischer ■☆
86547 Cheiridopsis latifolia L. Bolus = Cheiridopsis namaquensis (Sond.) H. E. K. Hartmann ■☆
86548 Cheiridopsis lecta (N. E. Br.) N. E. Br. = Cheiridopsis robusta (Haw.) N. E. Br. ■☆
86549 Cheiridopsis leptopetala L. Bolus = Cheiridopsis robusta (Haw.) N. E. Br. ■☆
86550 Cheiridopsis littlewoodii L. Bolus = Cheiridopsis denticulata (Haw.) N. E. Br. ■☆
86551 Cheiridopsis longipes L. Bolus;朗月■☆
86552 Cheiridopsis longipes L. Bolus = Cheiridopsis namaquensis (Sond.) H. E. K. Hartmann ■☆
86553 Cheiridopsis luckhoffii L. Bolus = Cheiridopsis namaquensis (Sond.) H. E. K. Hartmann ■☆
86554 Cheiridopsis macrocalyx L. Bolus = Cheiridopsis derenbergiana Schwantes ■☆
86555 Cheiridopsis macrophylla L. Bolus = Cheiridopsis denticulata (Haw.) N. E. Br. ■☆
86556 Cheiridopsis marlothii N. E. Br. = Cheiridopsis namaquensis (Sond.) H. E. K. Hartmann ■☆
86557 Cheiridopsis meyeri N. E. Br.;迈尔虾疳花■☆
86558 Cheiridopsis meyeri N. E. Br. var. minor L. Bolus = Cheiridopsis minor (L. Bolus) H. E. K. Hartmann ■☆
86559 Cheiridopsis minima Tischer = Antimima minima (Tischer) H. E. K. Hartmann ■☆
86560 Cheiridopsis minor (L. Bolus) H. E. K. Hartmann;小虾疳花■☆
86561 Cheiridopsis mirabilis Schwantes = Cheiridopsis verrucosa L. Bolus ■☆
86562 Cheiridopsis multiseriata L. Bolus = Cheiridopsis robusta (Haw.) N. E. Br. ■☆
86563 Cheiridopsis namaquensis (Sond.) H. E. K. Hartmann;纳马夸虾疳花■☆
86564 Cheiridopsis nelii Schwantes;尼尔虾疳花■☆
86565 Cheiridopsis olivacea Schwantes = Cheiridopsis robusta (Haw.) N. E. Br. ■☆
86566 Cheiridopsis pachyphylla Schwantes = Cheiridopsis brownii Schick et Tischer ■☆
86567 Cheiridopsis papillata L. Bolus = Ihlenfeldtia vanzylii (L. Bolus) H. E. K. Hartmann ■☆
86568 Cheiridopsis parvibracteata L. Bolus = Cephalophyllum parvibracteatum (L. Bolus) H. E. K. Hartmann ■☆
86569 Cheiridopsis parvula (Schltr.) N. E. Br. = Cephalophyllum parvulum (Schltr.) H. E. K. Hartmann ■☆
86570 Cheiridopsis paucifolia L. Bolus = Cheiridopsis schlechteri Tischer ■☆
86571 Cheiridopsis pearsonii N. E. Br.;皮尔逊虾疳花■☆
86572 Cheiridopsis peculiaris N. E. Br.;翔凤■☆
86573 Cheiridopsis peersii L. Bolus = Cheiridopsis turbinata L. Bolus ■☆
86574 Cheiridopsis perdecora N. E. Br. = Cheiridopsis robusta (Haw.) N. E. Br. ■☆

86575 Cheiridopsis pillansii L. Bolus;神风玉■☆
86576 Cheiridopsis pillansii L. Bolus var. crassa (L. Bolus) G. D. Rowley = Cheiridopsis pillansii L. Bolus ■☆
86577 Cheiridopsis pilosula L. Bolus;疏毛虾疳花■☆
86578 Cheiridopsis ponderosa S. A. Hammer;笨重虾疳花■☆
86579 Cheiridopsis pressa (N. E. Br.) N. E. Br. = Cheiridopsis robusta (Haw.) N. E. Br. ■☆
86580 Cheiridopsis puberula Dinter = Cheiridopsis caroli-schmidtii (Dinter et A. Berger) N. E. Br. ■☆
86581 Cheiridopsis pulverulenta L. Bolus = Cheiridopsis schlechteri Tischer ■☆
86582 Cheiridopsis purpurascens (Salm-Dyck) N. E. Br.;变紫虾疳花■☆
86583 Cheiridopsis purpurascens (Salm-Dyck) N. E. Br. = Cheiridopsis rostrata (L.) N. E. Br. ■☆
86584 Cheiridopsis purpurascens (Salm-Dyck) N. E. Br. var. leipoldtii L. Bolus = Cheiridopsis rostrata (L.) N. E. Br. ■☆
86585 Cheiridopsis purpurascens N. E. Br.;凌云■☆
86586 Cheiridopsis purpurata L. Bolus = Cheiridopsis purpurea L. Bolus ■☆
86587 Cheiridopsis purpurea L. Bolus;春意玉(红紫鞘袖)■☆
86588 Cheiridopsis quadrifida (Haw.) L. Bolus ex Schwantes = Cheiridopsis rostrata (L.) N. E. Br. ■☆
86589 Cheiridopsis quadrifolia (Haw.) L. Bolus;四叶玉■☆
86590 Cheiridopsis quadrifolia (Haw.) L. Bolus = Cheiridopsis derenbergiana Schwantes ■☆
86591 Cheiridopsis quaternifolia L. Bolus = Cheiridopsis namaquensis (Sond.) H. E. K. Hartmann ■☆
86592 Cheiridopsis resurgens L. Bolus = Cheiridopsis pearsonii N. E. Br. ■☆
86593 Cheiridopsis richardiana L. Bolus = Ihlenfeldtia vanzylii (L. Bolus) H. E. K. Hartmann ■☆
86594 Cheiridopsis robusta (Haw.) N. E. Br.;粗壮虾疳花■☆
86595 Cheiridopsis roodiae N. E. Br. = Cheiridopsis robusta (Haw.) N. E. Br. ■☆
86596 Cheiridopsis rostrata (L.) N. E. Br.;喙状虾疳花■☆
86597 Cheiridopsis rostratoides (Haw.) N. E. Br. = Cheiridopsis rostrata (L.) N. E. Br. ■☆
86598 Cheiridopsis rudis L. Bolus;粗糙虾疳花■☆
86599 Cheiridopsis scabra L. Bolus;糙虾疳花■☆
86600 Cheiridopsis scabra L. Bolus = Cheiridopsis namaquensis (Sond.) H. E. K. Hartmann ■☆
86601 Cheiridopsis schickiana Tischer = Cheiridopsis robusta (Haw.) N. E. Br. ■☆
86602 Cheiridopsis schinziana Dinter = Cheiridopsis robusta (Haw.) N. E. Br. ■☆
86603 Cheiridopsis schlechteri Schwantes = Cheiridopsis brownii Schick et Tischer ■☆
86604 Cheiridopsis schlechteri Tischer;施莱虾疳花■☆
86605 Cheiridopsis serrulata L. Bolus = Cheiridopsis namaquensis (Sond.) H. E. K. Hartmann ■☆
86606 Cheiridopsis splendens L. Bolus = Cheiridopsis purpurea L. Bolus ■☆
86607 Cheiridopsis subaequalis L. Bolus = Cheiridopsis derenbergiana Schwantes ■☆
86608 Cheiridopsis subalba L. Bolus = Cheiridopsis namaquensis (Sond.) H. E. K. Hartmann ■☆
86609 Cheiridopsis tenuifolia L. Bolus = Cheiridopsis namaquensis (Sond.) H. E. K. Hartmann ■☆
86610 Cheiridopsis truncata L. Bolus;彩帆■☆
86611 Cheiridopsis truncata L. Bolus = Cheiridopsis robusta (Haw.) N. E. Br. ■☆
86612 Cheiridopsis tuberculata (Mill.) N. E. Br.;雄飞玉■☆
86613 Cheiridopsis tuberculata (Mill.) N. E. Br. = Cheiridopsis rostrata (L.) N. E. Br. ■☆
86614 Cheiridopsis turbinata L. Bolus;陀螺虾疳花■☆
86615 Cheiridopsis turbinata L. Bolus var. minor? = Cheiridopsis turbinata L. Bolus ■☆
86616 Cheiridopsis turgida L. Bolus = Cheiridopsis brownii Schick et Tischer ■☆
86617 Cheiridopsis umbrosa S. A. Hammer et Desmet;耐荫虾疳花■☆
86618 Cheiridopsis vanbredai L. Bolus = Ihlenfeldtia excavata (L. Bolus) H. E. K. Hartmann ■☆
86619 Cheiridopsis vanheerdei L. Bolus = Cheiridopsis denticulata (Haw.) N. E. Br. ■☆
86620 Cheiridopsis vanzylii L. Bolus;丽玉■☆
86621 Cheiridopsis vanzylii L. Bolus = Ihlenfeldtia vanzylii (L. Bolus) H. E. K. Hartmann ■☆
86622 Cheiridopsis velutina L. Bolus = Cheiridopsis rostrata (L.) N. E. Br. ■☆
86623 Cheiridopsis ventricosa (L. Bolus) N. E. Br. = Antimima ventricosa (L. Bolus) H. E. K. Hartmann ■☆
86624 Cheiridopsis verrucosa L. Bolus;神风玉■☆
86625 Cheiridopsis verrucosa L. Bolus var. minor? = Cheiridopsis verrucosa L. Bolus ■☆
86626 Cheirinia Link = Erysimum L. + Sisymbrium L. ■
86627 Cheirinia Link = Erysimum L. ■●
86628 Cheirinia cheiranthoides (L.) Link = Erysimum cheiranthoides L. ■
86629 Cheirinia inconspicua (S. Watson) Britton = Erysimum inconspicuum (S. Watson) MacMill. ■☆
86630 Cheirisanthera Hort. ex Lindl. et Paxt. = Heppiella Regel ■☆
86631 Cheirodendron Nutt. ex Seem. (1867);手参木属●☆
86632 Cheirodendron dominii Krajina;多明手参木●☆
86633 Cheirodendron laetevirens Seem.;手参木●☆
86634 Cheirodendron platyphylla Seem.;宽叶手参木●☆
86635 Cheirolaena Benth. (1862);手苞梧桐属■●☆
86636 Cheirolaena linearis Benth.;手苞梧桐■●☆
86637 Cheirolepis Boiss. (1849);手鳞菊属●■☆
86638 Cheirolepis Boiss. = Centaurea L. (保留属名)●■
86639 Cheirolepis persica Boiss.;手鳞菊☆
86640 Cheiroloma F. Muell. = Calotis R. Br. ■
86641 Cheirolophus Cass. (1827);齿菊木属●■☆
86642 Cheirolophus Cass. = Centaurea L. (保留属名)●■
86643 Cheirolophus arboreus (Webb et Berthel.) Holub;乔木齿菊木●☆
86644 Cheirolophus arbutifolius (Svent.) G. Kunkel;浆果鹃叶齿菊木●☆
86645 Cheirolophus argutus (Nees) Holub;亮齿菊木●☆
86646 Cheirolophus canariensis (Willd.) Holub;加那利齿菊木●☆
86647 Cheirolophus canariensis (Willd.) Holub var. subexpinnatus (Burch.) A. Hansen et Sunding = Cheirolophus canariensis (Willd.) Holub ●☆
86648 Cheirolophus ghomerytus (Svent.) Holub var. integrifolius (Svent.) G. Kunkel = Cheirolophus ghomerytus (Svent.) Holub ●☆

86649 Cheirolophus junonianus (Svent.) G. Kunkel var. isoplexiphyllus (Svent.) G. Kunkel = Cheirolophus junonianus (Svent.) G. Kunkel ●☆

86650 Cheirolophus junonianus (Svent.) G. Kunkel var. junonianus (Svent.) G. Kunkel = Cheirolophus junonianus (Svent.) G. Kunkel ●☆

86651 Cheirolophus massonianus (Lowe) A. Hansen et Sunding;马森齿菊木●☆

86652 Cheirolophus mauritanicus (Font Quer) Susanna;毛里塔尼亚齿菊木●☆

86653 Cheirolophus metlesicsii Montel.;梅特齿菊木●☆

86654 Cheirolophus sventenii (Santos) G. Kunkel;斯文顿齿菊木●☆

86655 Cheirolophus tananicus (Maire) Holub;泰南齿菊木●☆

86656 Cheirolophus webbianus (Sch. Bip.) Holub;韦布齿菊木●☆

86657 Cheiropetalum E. Fries = Silene L.(保留属名)■

86658 Cheiropetalum E. Fries ex Schltdl. = Silene L.(保留属名)■

86659 Cheiropsis (DC.) Bercht. et J. Presl = Clematis L. ●■

86660 Cheiropsis (DC.) Bercht. et J. Presl = Muralta Adans.(废弃属名)●■

86661 Cheiropsis Bercht. et J. Presl = Clematis L. ●■

86662 Cheiropsis Bercht. et J. Presl = Muralta Adans.(废弃属名)●■

86663 Cheiropterocephalus Barb. Rodr. = Malaxis Sol. ex Sw. ■

86664 Cheirorchis Carr = Cordiglottis J. J. Sm. ■☆

86665 Cheirostemon Bonpl. = Chiranthodendron Sessé ex Larreat. ●☆

86666 Cheirostemon Humb. et Bonpl. = Chiranthodendron Sessé ex Larreat. ●☆

86667 Cheirostylis Blume(1825);叉柱兰属(指柱兰属);Cheirostylis ■

86668 Cheirostylis anomala Ohwi = Cheirostylis takeoi (Hayata) Schltr. ■

86669 Cheirostylis calcarata X. H. Jin et S. C. Chen;短距叉柱兰■

86670 Cheirostylis chinensis Rolfe;中华叉柱兰(台湾指柱兰,指柱兰,中国指柱兰);China Cheirostylis ■

86671 Cheirostylis chinensis Rolfe var. tortilacinia (C. S. Leou) S. S. Ying = Cheirostylis thailandica Seidenf. ■

86672 Cheirostylis clibborndyeri S. Y. Hu et Barretto;叉柱兰■

86673 Cheirostylis clibborndyeri S. Y. Hu et Barretto = Cheirostylis takeoi (Hayata) Schltr. ■

86674 Cheirostylis clibborndyeri S. Y. Hu et Barretto = Cheirostylis tatewakii Masam. ■

86675 Cheirostylis cochinchinensis Blume;雉尾指柱兰;Cochichina Cheirostylis ■

86676 Cheirostylis cochinchinensis Blume = Cheirostylis taichungensis S. S. Ying ■

86677 Cheirostylis demhiensis S. S. Ying;德基叉柱兰(德基指柱兰);Deji Cheirostylis ■

86678 Cheirostylis derchiensis S. S. Ying = Cheirostylis clibborndyeri S. Y. Hu et Barretto ■

86679 Cheirostylis divina (Guinea) Summerh.;神叉柱兰■☆

86680 Cheirostylis eglandulosa Aver. = Cheirostylis takeoi (Hayata) Schltr. ■

86681 Cheirostylis flabellata Wight = Cheirostylis inabai Hayata ■

86682 Cheirostylis franchetiana King et Pantl. = Myrmechis pumila (Hook. f.) Ts. Tang et F. T. Wang ■

86683 Cheirostylis griffithii Lindl.;大花叉柱兰;Bigflower Cheirostylis ■

86684 Cheirostylis gymnochiloides (Ridl.) Rchb. f.;裸唇叉柱兰■☆

86685 Cheirostylis gymnochiloides (Ridl.) Rchb. f. = Cheirostylis nuda (Thouars) Ormerod ■☆

86686 Cheirostylis heterosepala Rchb. f. = Zeuxine heterosepala (Rchb. f.) Geerinck ■☆

86687 Cheirostylis humblotii Rchb. f. = Cheirostylis nuda (Thouars) Ormerod ■☆

86688 Cheirostylis hungyehensis T. P. Lin = Cheirostylis clibborndyeri S. Y. Hu et Barretto ■

86689 Cheirostylis hungyrhensis P. T. Lin = Cheirostylis tatewakii Masam. ■

86690 Cheirostylis hungyrhensis T. P. Lin;斑叶指柱兰;Punctate-leaved Cheirostylis ■

86691 Cheirostylis inabae Hayata = Cheirostylis octodactyla Ames ■

86692 Cheirostylis jamesleungii S. Y. Hu et Barretto;粉红叉柱兰;Pink Cheirostylis ■

86693 Cheirostylis josephii Schltr. = Cheirostylis chinensis Rolfe ■

86694 Cheirostylis josephii Schltr. = Cheirostylis yunnanensis Rolfe ■

86695 Cheirostylis kanarensis Blatter et McCann = Chamaegastrodia vaginata (Hook. f.) Seidenf. ■

86696 Cheirostylis kanashiroi Ohwi = Cheirostylis liukiuensis Masam. ■

86697 Cheirostylis lepida (Rchb. f.) Rolfe;鳞片叉柱兰■☆

86698 Cheirostylis liukiuensis Masam.;琉球叉柱兰(琉球指柱兰,墨绿指柱兰);Likiu Cheirostylis ■

86699 Cheirostylis macrantha Schltr. = Cheirostylis griffithii Lindl. ■

86700 Cheirostylis malipoensis X. H. Jin et S. C. Chen;麻栗坡叉柱兰■

86701 Cheirostylis micrantha Schltr. = Cheirostylis nuda (Thouars) Ormerod ■☆

86702 Cheirostylis monteiroi S. Y. Hu et Barretto;箭药叉柱兰;Arrowanther Cheirostylis ■

86703 Cheirostylis munnacampensis A. N. Rao = Cheirostylis yunnanensis Rolfe ■

86704 Cheirostylis nemorosa Fukuy. = Cheirostylis tabiyahanensis (Hayata) Pearce et Cribb ■

86705 Cheirostylis nemorosa Fukuy. = Zeuxine nemorosa (Fukuy.) T. P. Lin ■

86706 Cheirostylis nuda (Thouars) Ormerod;裸指柱兰■☆

86707 Cheirostylis octodactyla Ames;羽唇叉柱兰(扇叶指柱兰,雾社指柱兰,羽唇指柱兰);Inaba Cheirostylis ■

86708 Cheirostylis okabeana Tuyama = Cheirostylis liukiuensis Masam. ■

86709 Cheirostylis oligantha Masam. et Fukuy. = Cheirostylis inabai Hayata ■

86710 Cheirostylis oligantha Masam. et Fukuy. = Cheirostylis octodactyla Ames ■

86711 Cheirostylis pabongensis Lucksom = Cheirostylis yunnanensis Rolfe ■

86712 Cheirostylis philippinensis Ames = Cheirostylis chinensis Rolfe ■

86713 Cheirostylis phillipinensis Ames;雾社指柱兰■

86714 Cheirostylis phillipinensis Ames = Cheirostylis chinensis Rolfe ■

86715 Cheirostylis pingbianensis K. Y. Lang;屏边叉柱兰;Pingbian Cheirostylis ■

86716 Cheirostylis pusilla Lindl.;细小叉柱兰■

86717 Cheirostylis sarcopus Schltr. = Cheirostylis gymnochiloides (Ridl.) Rchb. f. ■☆

86718 Cheirostylis sarcopus Schltr. = Cheirostylis nuda (Thouars) Ormerod ■☆

86719 Cheirostylis tabiyahanensis (Hayata) Pearce et Cribb;东部叉柱兰(东部线柱兰)■

86720 Cheirostylis taichungensis S. S. Ying;台中叉柱兰(台中指柱兰,雉尾指柱兰);Taizhong Cheirostylis ■

86721 Cheirostylis taichungensis S. S. Ying = Cheirostylis

cochinchinensis Blume ■

86722 Cheirostylis tairae (Fukuy.) Masam. = Cheirostylis takeoi (Hayata) Schltr. ■

86723 Cheirostylis taiwanensis Gogelein;台湾指柱兰■

86724 Cheirostylis taiwanensis Gogelein = Cheirostylis chinensis Rolfe ■

86725 Cheirostylis takeoi (Hayata) Schltr.;全唇叉柱兰(阿里山指柱兰,斑叶指柱兰,卵唇指柱兰,全唇指柱兰,太鲁阁叉柱兰,太鲁阁指柱兰,无叶指柱兰);Entirelip Cheirostylis, Tatewaki Cheirostylis ■

86726 Cheirostylis tatewakii Masam.;卵唇指柱兰■

86727 Cheirostylis tatewakii Masam. = Cheirostylis takeoi (Hayata) Schltr. ■

86728 Cheirostylis thailandica Seidenf.;反瓣叉柱兰■

86729 Cheirostylis tortilacinia C. S. Leou;和社叉柱兰(和社指柱兰);Twistysplit Cheirostylis ■

86730 Cheirostylis yunnanensis Rolfe;云南叉柱兰(石头虾);Yunnan Cheirostylis ■

86731 Chelidoniaceae Martinov = Papaveraceae Juss.(保留科名)●■

86732 Chelidoniaceae Nakai = Papaveraceae Juss.(保留科名)●■

86733 Chelidonium L. (1753);白屈菜属;Celandine, Celandine Poppy, Greater Celandine ■

86734 Chelidonium Tourn. ex L. = Chelidonium L. ■

86735 Chelidonium asiaticum (H. Hara) Krahulc. = Chelidonium majus L. subsp. asiaticum H. Hara ■

86736 Chelidonium asiaticum (H. Hara) Krahulc. = Chelidonium majus L. ■

86737 Chelidonium corniculatum L. = Glaucium corniculatum (L.) Rudolph ■☆

86738 Chelidonium dicranostigma Prain = Dicranostigma lactucoides Hook. f. et Thomson ■

86739 Chelidonium dicranostigma Prain = Dicranostigma platycarpum C. Y. Wu et H. Chuang ■

86740 Chelidonium diphyllum Michx. = Stylophorum diphyllum (Michx.) Nutt. ■☆

86741 Chelidonium dodecandrum Forssk. = Roemeria hybrida (L.) DC. subsp. dodecandra (Forssk.) Maire ■☆

86742 Chelidonium franchetianum Prain = Dicranostigma leptopodum (Maxim.) Fedde ■

86743 Chelidonium glaucium L. = Glaucium flavum Crantz ■☆

86744 Chelidonium grandiflorum DC. = Chelidonium majus L. ■

86745 Chelidonium hybridum L. = Roemeria hybrida (L.) DC. ■

86746 Chelidonium hylomeconoides (Nakai) Ohwi = Coreanomecon hylomeconoides Nakai ■☆

86747 Chelidonium japonicum Thunb. = Hylomecon japonica (Thunb.) Prantl et Kündig ■

86748 Chelidonium japonicum Thunb. f. dissectum (Franch. et Sav.) Ohwi = Hylomecon japonica (Thunb.) Prantl et Kündig f. dissecta (Franch. et Sav.) Okuyama ■

86749 Chelidonium japonicum Thunb. f. lanceolatum (Yatabe) Ohwi = Hylomecon japonica (Thunb.) Prantl et Kündig f. subintegra (Fedde) Okuyama ■

86750 Chelidonium japonicum Thunb. f. palliflavidum Moriya = Hylomecon japonica (Thunb.) Prantl et Kündig f. palliflavens Honda ■☆

86751 Chelidonium lactucoides (Hook. f. et Thomson) Prain = Dicranostigma lactucoides Hook. f. et Thomson ■

86752 Chelidonium lasiocarpum Oliv. = Stylophorum lasiocarpum (Oliv.) Fedde ■

86753 Chelidonium leptopodium Prain = Dicranostigma leptopodum (Maxim.) Fedde ■

86754 Chelidonium leptopodum (Maxim.) Prain = Dicranostigma leptopodum (Maxim.) Fedde ■

86755 Chelidonium majus L.;白屈菜(八步紧,地黄连,断肠草,观音草,黄连,假黄连,见肿消,牛金花,山黄连,山西瓜,山野人血草,水黄草,水黄连,土黄连,小黄连,小人血草,小人血七,小野人血草,雄黄草);Celandine, Celandine Poppy, Cock Foot, Cockfoot, Common Celandine, Devil's Milk, Fellon-wort, Fenugreek, Greater Celandine, Grecian Hay, Jacob's Ladder, Kenning-wort, Kill Wart, Kill-wart, Major Celandine, May-flower, Rock-poppy, Saladine, Solandine, Sollandine, St. John's Wort, Swallow Wort, Swallow-grass, Swallow-herb, Swallow-wort, Tetter, Tetterwort, Wart-curer, Wart-flower, Wart-plant, Wartweed, Wartwort, Witches' Flower, Wretweed, Yellow Spit ■

86756 Chelidonium majus L. 'Flore Pleno';重瓣白屈菜■☆

86757 Chelidonium majus L. subsp. asiaticum H. Hara;亚洲白屈菜■

86758 Chelidonium majus L. subsp. asiaticum H. Hara = Chelidonium majus L. ■

86759 Chelidonium majus L. var. asiaticum (H. Hara) Ohwi ex W. T. Lee = Chelidonium majus L. subsp. asiaticum H. Hara ■

86760 Chelidonium majus L. var. asiaticum (H. Hara) Ohwi ex W. T. Lee = Chelidonium majus L. ■

86761 Chelidonium majus L. var. grandiflorum DC.;大花白屈菜;Bigflower Celandine ■

86762 Chelidonium majus L. var. grandiflorum DC. = Chelidonium majus L. ■

86763 Chelidonium majus L. var. hirsutum Trautv. et C. A. Mey. = Chelidonium majus L. subsp. asiaticum H. Hara ■

86764 Chelidonium sutchuense Franch. = Stylophorum sutchuense (Franch.) Fedde ■

86765 Chelidospermum Zipp. ex Blume = Pittosporum Banks ex Gaertn. (保留属名)●

86766 Cheliusia Sch. Bip. = Vernonia Schreb.(保留属名)●■

86767 Chelona Post et Kuntze = Chelone L. ■☆

86768 Chelonaceae D. Don = Plantaginaceae Juss.(保留科名)■

86769 Chelonaceae D. Don = Scrophulariaceae Juss.(保留科名)●■

86770 Chelonaceae Martinov = Scrophulariaceae Juss.(保留科名)●■

86771 Chelonanthera Blume = Pholidota Lindl. ex Hook. ■

86772 Chelonanthus (Griseb.) Gilg = Irlbachia Mart. ■☆

86773 Chelonanthus (Griseb.) Gilg(1895);龟花龙胆属■☆

86774 Chelonanthus Gilg = Irlbachia Mart. ■☆

86775 Chelonanthus Raf. = Chelone L. ■☆

86776 Chelonanthus alatus Pulle;翅龟花龙胆■☆

86777 Chelonanthus albus (Spruce ex Progel) V. M. Badillo;白龟花龙胆■☆

86778 Chelonanthus angustifolius Gilg;窄叶龟花龙胆■☆

86779 Chelonanthus bifidus (Kunth) Gilg;双裂龟花龙胆■☆

86780 Chelonanthus viridiflorus (Mart.) Gilg;绿花龟花龙胆■☆

86781 Chelone L. (1753);龟头花属;Chelone, Shellflower, Turtlehead, Turtle-head ■☆

86782 Chelone chlorantha Pennell et Wherry = Chelone glabra L. ●■☆

86783 Chelone glabra L.;龟头花(窄叶蛇头草);Balmony, Bitter Herb, Chelone, Hummingbird Tree, Salt Rheum Weed, Salt Rheumweed, Shell Flower, Snakehead, Snake-head, Snakeshead, Turtlehead, Turtle-head, White Turtlehead, White Turtle-head ●■☆

86784 Chelone glabra L. f. tomentosa (Raf.) Pennell = Chelone glabra L. ●■☆

86785 Chelone glabra L. var. chlorantha (Pennell et Wherry) Cooperr. = Chelone glabra L. ●■☆

86786 Chelone glabra L. var. dilatata Fernald et Wiegand = Chelone glabra L. ●■☆

86787 Chelone glabra L. var. elatior Raf. = Chelone glabra L. ●■☆

86788 Chelone glabra L. var. elongata Pennell et Wherry = Chelone glabra L. ●■☆

86789 Chelone glabra L. var. linifolia N. Coleman = Chelone glabra L. ●■☆

86790 Chelone glabra L. var. linifolia N. Coleman f. velutina Pennell et Wherry = Chelone glabra L. ●■☆

86791 Chelone glabra L. var. ochroleuca Pennell et Wherry = Chelone glabra L. ●■☆

86792 Chelone glabra L. var. typica Pennell = Chelone glabra L. ●■☆

86793 Chelone lyonii Pursh;粉色龟头花;Lyon's Turtlehead, Pink Turtlehead, Pink Turtle-head, Turtlehead ●☆

86794 Chelone montana (Raf.) Pennell et Wherry = Chelone glabra L. ●■☆

86795 Chelone obliqua L.;斜叶龟头花;Pink Turtlehead, Purple Turtlehead, Red Turtlehead, Red Turtle-head, Rose Turtlehead, Rose Turtle-head, Turtle-head ■☆

86796 Chelone obliqua L. var. speciosa Pennell et Wherry;美丽斜叶龟头花;Purple Turtlehead, Red Turtlehead ■☆

86797 Chelonecarya Pierre = Rhaphiostylis Planch. ex Benth. ●☆

86798 Chelonecarya fusca Pierre = Rhaphiostylis fusca (Pierre) Pierre ●☆

86799 Chelonespermum Hemsl. = Burckella Pierre ●☆

86800 Chelonistele Pfitzer = Panisea (Lindl.) Lindl.(保留属名)■

86801 Chelonistele Pfitzer et Carr = Chelonistele Pfitzer. ■☆

86802 Chelonistele Pfitzer. (1907);龟柱兰属(角柱兰属)■☆

86803 Chelonistele biflora (E. C. Parish et Rchb. f.) Pfitzer = Panisea uniflora (Lindl.) Lindl. ■

86804 Chelonistele biflora Pfitzer;双花龟柱兰■☆

86805 Chelonopsis Miq. (1865);铃子香属(麝香草属);Chelonopsis ●■

86806 Chelonopsis abbreviata C. Y. Wu et H. W. Li;缩序铃子香;Shortened Chelonopsis, Shortenspike Chelonopsis ●

86807 Chelonopsis albiflora Pax et K. Hoffm. ex H. Limpr.;白花铃子香(铃子香);White Chelonopsis, Whiteflower Chelonopsis, White-flowered Chelonopsis ●

86808 Chelonopsis benthamiana Hemsl. = Bostrychanthera deflexa Benth. ■

86809 Chelonopsis bracteata W. W. Sm.;具苞铃子香;Bracteate Chelonopsis ●

86810 Chelonopsis chekiangensis C. Y. Wu;浙江铃子香(铃子三七);Chekiang Chelonopsis, Zhejiang Chelonopsis ■

86811 Chelonopsis chekiangensis C. Y. Wu var. brevipes C. Y. Wu et H. W. Li;短梗浙江铃子香(短枝浙江铃子香)■

86812 Chelonopsis deflexa (Benth.) Diels;华麝香草●

86813 Chelonopsis deflexa (Benth.) Diels = Bostrychanthera deflexa Benth. ■

86814 Chelonopsis deflexa (Benth.) Diels var. matsudae Kudo = Bostrychanthera deflexa Benth. ■

86815 Chelonopsis forrestii Anthony;大萼铃子香;Bigcalyx Chelonopsis, Forrest Chelonopsis ●

86816 Chelonopsis giraldii Diels;小叶铃子香;Girald Chelonopsis, Littleflower Chelonopsis ●

86817 Chelonopsis lichiangensis W. W. Sm.;丽江铃子香;Lijiang Chelonopsis ●

86818 Chelonopsis longipes Makino;长梗铃子香■☆

86819 Chelonopsis longipes Makino f. albiflora Honda;白花长梗铃子香■☆

86820 Chelonopsis mollissima C. Y. Wu;多毛铃子香;Hairy Chelonopsis ●

86821 Chelonopsis moschata Miq.;麝香铃子香■☆

86822 Chelonopsis moschata Miq. var. jesoensis (Koidz.) Miyabe et Tatew. = Chelonopsis moschata Miq. ■☆

86823 Chelonopsis moschata Miq. var. lasiocalyx Hayata = Chelonopsis yagiharana Hisauti et Matsuno ●☆

86824 Chelonopsis moschata Miq. var. longipes? = Chelonopsis longipes Makino ■☆

86825 Chelonopsis odontochila Diels;齿唇铃子香;Odontochilous Chelonopsis ●

86826 Chelonopsis odontochila Diels subsp. bracteata (W. W. Sm.) Kudo = Chelonopsis bracteata W. W. Sm. ●

86827 Chelonopsis odontochila Diels subsp. forrestii (Anthony) Kudo = Chelonopsis forrestii Anthony ●

86828 Chelonopsis odontochila Diels subsp. lichiangensis (W. W. Sm.) Kudo = Chelonopsis lichiangensis W. W. Sm. ●

86829 Chelonopsis odontochila Diels subsp. rosea (W. Sm.) Kudo = Chelonopsis rosea W. W. Sm. ●

86830 Chelonopsis odontochila Diels subsp. smithii Kudo = Chelonopsis odontochila Diels var. smithii (Kudo) C. Y. Wu ●

86831 Chelonopsis odontochila Diels var. smithii (Kudo) C. Y. Wu;钝齿唇铃子香(钝齿铃子香)●

86832 Chelonopsis pseudobracteata C. Y. Wu et H. W. Li;假具苞铃子香;False Bracteate Chelonopsis, Falsebracteate Chelonopsis ●

86833 Chelonopsis pseudobracteata C. Y. Wu et H. W. Li var. rubra C. Y. Wu et H. W. Li;红花假具苞铃子香;Red-flowered False Bracteate Chelonopsis ●

86834 Chelonopsis pseudobracteata C. Y. Wu et H. W. Li var. rubra C. Y. Wu et H. W. Li = Chelonopsis pseudobracteata C. Y. Wu et H. W. Li ●

86835 Chelonopsis rosea W. W. Sm.;玫红铃子香;Roseate Chelonopsis, Rosered Chelonopsis, Rose-red Chelonopsis ●

86836 Chelonopsis siccanea W. W. Sm.;干生铃子香;Dryland Chelonopsis, Dryliving Chelonopsis, Siccocolous Chelonopsis ●

86837 Chelonopsis souliei (Bonati) Merr.;轮叶铃子香;Soulie Chelonopsis ●

86838 Chelonopsis yagiharana Hisauti et Matsuno;八木氏铃子香●☆

86839 Chelonopsis yagiharana Hisauti et Matsuno var. jesoensis? = Chelonopsis moschata Miq. ■☆

86840 Chelrostylis Pritz. = Cheirostylis Blume ■

86841 Chelyocarpus Dammer(1920);龟果榈属(龟果棕属,契里榈属)●☆

86842 Chelyocarpus ulei Dammer;龟果榈●☆

86843 Chelyorchis Dressler et N. H. Williams = Oncidium Sw.(保留属名)■☆

86844 Chelyorchis Dressler et N. H. Williams(2000);中美瘤瓣兰属■☆

86845 Chemnicia Scop. = Rouhamon Aubl. ●

86846 Chemnicia Scop. = Strychnos L. ●

86847 Chemnitzia Post et Kuntze = Chemnicia Scop. ●

86848 Chemnizia Fabr. = Lagoecia L. ●☆

86849 Chemnizia Heist. ex Fabr. = Lagoecia L. ●☆

86850 Chemnizia Steud. = Chemnicia Scop. ●

86851 Chemnizia Steud. = Strychnos L. ●

86852 Chemria Steud. = Chetastrum Neck. ●■

86853 Chengiopanax C. B. Shang et J. Y. Huang(1993);人参木属;Chengiopanax ●

86854 Chengiopanax fargesii (Franch.) C. B. Shang et J. Y. Huang;华人参木(人参木,中华五加);Chinese Chengiopanax, Farges Chengiopanax ●

86855 Chengiopanax sciadophylloides (Franch. et Sav.) C. B. Shang et J. Y. Huang;日本人参木(人参木);Japanese Chengiopanax ●☆

86856 Chennapyrum Á. Löve = Aegilops L. (保留属名)■

86857 Chenocarpus Neck. = Chaenocarpus Juss. ●■

86858 Chenocarpus Neck. = Spermacoce L. ●■

86859 Chenolea Thunb. (1781);膜被雾冰藜属●☆

86860 Chenolea Thunb. = Bassia All. ■●

86861 Chenolea arabica Boiss. = Bassia arabica (Boiss.) Maire et Weiller ■☆

86862 Chenolea canariensis Moq. = Bassia tomentosa (Lowe) Maire et Weiller ■☆

86863 Chenolea diffusa (Thunb.) Thunb.;松散膜被雾冰藜■☆

86864 Chenolea diffusa (Thunb.) Thunb. = Bassia diffusa (Thunb.) Kuntze ■☆

86865 Chenolea dinteri Botsch. = Bassia dinteri (Botsch.) A. J. Scott ■☆

86866 Chenolea divaricata (Kar. et Kir.) Hook. f. = Bassia dasyphylla (Fisch. et C. A. Mey.) Kuntze ■

86867 Chenolea divaricata Hook. f. = Bassia dasyphylla (Fisch. et C. A. Mey.) Kuntze ■

86868 Chenolea sedoides (Schrad.) Hook. f. = Bassia sedoides (Schrad.) Asch. ■

86869 Chenolea tomentosa (Lowe) Maire = Bassia tomentosa (Lowe) Maire et Weiller ■☆

86870 Chenoleoides Botsch. = Bassia All. ■●

86871 Chenoleoides dinteri (Botsch.) Botsch. = Bassia dinteri (Botsch.) A. J. Scott ■☆

86872 Chenoleoides tomentosa (Lowe) Botsch. = Bassia tomentosa (Lowe) Maire et Weiller ■☆

86873 Chenopodiaceae Vent. (1799)(保留科名);藜科;Goosefoot Family ●■

86874 Chenopodiaceae Vent. (保留科名) = Amaranthaceae Juss. (保留科名)●■

86875 Chenopodiaceae Vent. (保留科名) = Chionographidaceae Takht. ■

86876 Chenopodina (Moq.) Moq. = Suaeda Forssk. ex J. F. Gmel. (保留属名)●■

86877 Chenopodina Moq. = Suaeda Forssk. ex J. F. Gmel. (保留属名)●■

86878 Chenopodina dendroides (C. A. Mey.) Moq. = Suaeda dendroides (C. A. Mey.) Moq. ●

86879 Chenopodina dendroides Moq. = Suaeda dendroides (C. A. Mey.) Moq. ●

86880 Chenopodina glauca (Bunge) Moq. = Suaeda glauca (Bunge) Bunge ■

86881 Chenopodina glauca Moq. = Suaeda glauca (Bunge) Bunge ■

86882 Chenopodina maritima (L.) Moq. var. vulgaris (Moq.) Moq. = Suaeda prostrata Pall. ■

86883 Chenopodina microphylla (Pall.) Moq. = Suaeda microphylla (C. A. Mey.) Pall. ●

86884 Chenopodina microphylla Moq. = Suaeda microphylla (C. A. Mey.) Pall. ●

86885 Chenopodina salsa (L.) Moq. = Suaeda salsa (L.) Pall. ■

86886 Chenopodiopsis Hilliard et B. L. Burtt = Chenopodiopsis Hilliard ■☆

86887 Chenopodiopsis Hilliard(1990);假藜属■☆

86888 Chenopodiopsis chenopodioides (Diels) Hilliard;拟藜■☆

86889 Chenopodiopsis hirta (L. f.) Hilliard;毛假藜■☆

86890 Chenopodiopsis retrorsa Hilliard;假藜■☆

86891 Chenopodium L. (1753);藜属(灰菜属);Blite, Goosefoot, Pigweed ■●

86892 Chenopodium acuminatum Willd.;尖头叶藜(变叶藜,渐尖藜,绿珠藜,也西风古,油杓杓);Acuminate Goosefoot ■

86893 Chenopodium acuminatum Willd. subsp. virgatum (Thunb.) Kitam.;狭叶尖头叶藜(变叶藜,圆头藜);Narrowleaf Acuminate Goosefoot, Narrowleaf Goosefoot, Round-leaved Goosefoot ■

86894 Chenopodium acuminatum Willd. subsp. virgatum (Thunb.) Kitam. = Chenopodium acuminatum Willd. ■

86895 Chenopodium acuminatum Willd. var. ovatum Fenzl = Chenopodium acuminatum Willd. ■

86896 Chenopodium acuminatum Willd. var. vachelii (Hook. et Arn.) Moq. = Chenopodium acuminatum Willd. ■

86897 Chenopodium acuminatum Willd. var. virgatum (Thunb.) Moq. = Chenopodium acuminatum Willd. ■

86898 Chenopodium acuminatum Willd. var. virgatum Moq. = Chenopodium acuminatum Willd. subsp. virgatum (Thunb.) Kitam. ■

86899 Chenopodium acutifolium Sm. = Chenopodium polyspermum L. var. acutifolium (Sm.) Gaudin ■☆

86900 Chenopodium aegyptiacum Hasselq. = Suaeda aegyptiaca (Hasselq.) Zohary ■☆

86901 Chenopodium album L.;藜(白藜,菜,赤藜,飞扬草,粉仔菜,鹤顶草,红落藜,红心灰藋,灰菜,灰藋,灰藋苋,灰堆头菜,灰灰菜,灰藜,灰蓼头草,灰天苋,灰条,灰条菜,灰苋菜,灰胭草,金锁天,厘,落藜,蔓华,蒙华,蓬子菜,舜芒谷,胭脂菜,野灰菜,银粉菜,猪灰头菜);Bacon-weed, Beaconweed, Common Lamb's-quarters, Confetti, Dash Bagger, Dirtweed, Dirty Dick, Dirty Jack, Dirty John, Dock Flower, Dumbweed, Fat Hen, Fat-hen, Goose Foot, Goosefoot, Johnny O'neele, Lamb's Quarters, Lamb's Quarters Goosefoot, Lamb's Tongue, Lambsquarters, Lambs-quarters, Lamb's-quarters, Malls, Meldweed, Melgs, Midden Mylies, Muckhill Weed, Muckweed, Mutton Chops, Mutton Tops, Myles, Pigweed, Rag Jack, Rag-Jack, Sea Spinach, White Goosefoot, White Pigweed, Wild Orache, Wild Spinach ■

86902 Chenopodium album L. 'Blite';清淡藜;Frost-blite ■☆

86903 Chenopodium album L. f. heterophyllum Wang-Wei et P. Y. Fu;异叶藜;Diverseleaf Lambsquarters ■

86904 Chenopodium album L. f. lanceolatum (Muhl. ex Willd.) Schinz et Thell. = Chenopodium album L. ■

86905 Chenopodium album L. subsp. amaranthicolor H. J. Coste et Reyn. = Chenopodium album L. var. centrorubrum Makino ■

86906 Chenopodium album L. subsp. amaranthicolor H. J. Coste et Reyn. = Chenopodium giganteum D. Don ■

86907 Chenopodium album L. subsp. karoi Murr = Chenopodium karoi (Murr) Aellen ■

86908 Chenopodium album L. subsp. karoi Murr = Chenopodium prostratum Bunge ex Herder ■

86909 Chenopodium album L. subsp. opulifolium (W. D. J. Koch et Ziz)

Batt. = Chenopodium opulifolium Schrad. ex W. D. J. Koch et Ziz ■☆

86910 Chenopodium album L. subsp. striatum (Krasan) Murr = Chenopodium strictum Roth ■

86911 Chenopodium album L. var. acuminatum (Willd.) Kuntze = Chenopodium acuminatum Willd. ■

86912 Chenopodium album L. var. berlandieri (Moq.) Mack. et Bush = Chenopodium berlandieri Moq. ■☆

86913 Chenopodium album L. var. centrorubrum Makino;日本藜■

86914 Chenopodium album L. var. centrorubrum Makino = Chenopodium album L. ■

86915 Chenopodium album L. var. concatenatum (Thuill.) Gaudin = Chenopodium album L. ■

86916 Chenopodium album L. var. glomerulosum (Rchb.) Peterm. = Chenopodium album L. ■

86917 Chenopodium album L. var. lanceolatum (Muhl. ex Willd.) Coss. et Germ. = Chenopodium album L. ■

86918 Chenopodium album L. var. lanceolatum Coss. et Germ. = Chenopodium album L. ■

86919 Chenopodium album L. var. laxiflorum Wang-Wei et P. Y. Fu;散花藜;Laxflower Lambsquarters ■

86920 Chenopodium album L. var. leptophyllum Moq. = Chenopodium leptophyllum (Nutt. ex Moq.) Nutt. ex S. Watson ■☆

86921 Chenopodium album L. var. microphyllum Boenn.;小叶藜;Lambsquarters ■☆

86922 Chenopodium album L. var. microphyllum Boenn. = Chenopodium strictum Roth ■

86923 Chenopodium album L. var. missouriense (Aellen) Bassett et Crompton = Chenopodium album L. ■

86924 Chenopodium album L. var. missouriense Aellen = Chenopodium missouriense Aellen ■☆

86925 Chenopodium album L. var. pseudoborbasii (Murr) Beck = Chenopodium strictum Roth ■

86926 Chenopodium album L. var. pseudoborbasii (Murr) Hayek = Chenopodium strictum Roth var. pseudoborbasii (Murr) Cufod. ■☆

86927 Chenopodium album L. var. stenophyllum Makino;狭叶藜■☆

86928 Chenopodium album L. var. stevensii Aellen = Chenopodium album L. ■

86929 Chenopodium album L. var. striatum Krasan = Chenopodium album L. subsp. striatum (Krasan) Murr ■

86930 Chenopodium album L. var. striatum Krasan = Chenopodium strictum Roth ■

86931 Chenopodium album L. var. viride (L.) Wahlenb. = Chenopodium suecicum Murr ■☆

86932 Chenopodium altissimum L. = Suaeda altissima (L.) Pall. ■

86933 Chenopodium amaranticolor H. J. Coste et Reyn. = Chenopodium giganteum D. Don ■

86934 Chenopodium ambrosioides L.;美洲藜(墨西哥茶);American Wormseed, Epazote, Mexican Tea, Wormseed ■☆

86935 Chenopodium ambrosioides L. = Dysphania ambrosioides (L.) Mosyakin et Clemants ■

86936 Chenopodium ambrosioides L. subsp. eu-ambrosioides Aellen = Dysphania ambrosioides (L.) Mosyakin et Clemants ■

86937 Chenopodium ambrosioides L. subsp. eu-ambrosioides Aellen var. suffruticosum (Willd.) Aellen = Dysphania ambrosioides (L.) Mosyakin et Clemants ■

86938 Chenopodium ambrosioides L. subsp. eu-ambrosioides Aellen var. typicum (Speg.) Aellen f. integrifolium (Fenzl) Aellen = Dysphania ambrosioides (L.) Mosyakin et Clemants ■

86939 Chenopodium ambrosioides L. subsp. eu-ambrosioides Aellen var. typicum (Speg.) Aellen = Dysphania ambrosioides (L.) Mosyakin et Clemants ■

86940 Chenopodium ambrosioides L. var. anthelminticum (L.) A. Gray = Dysphania anthelmintica (L.) Mosyakin et Clemants ■☆

86941 Chenopodium ambrosioides L. var. anthelminticum A. Gray;驱虫土荆芥■

86942 Chenopodium ambrosioides L. var. chilense (Schrad.) Speg. = Dysphania chilensis (Schrad.) Mosyakin et Clemants ■☆

86943 Chenopodium ambrosioides L. var. comosum Willk. = Chenopodium ambrosioides L. ■☆

86944 Chenopodium ambrosioides L. var. dentatum Fenzl = Chenopodium ambrosioides L. ■☆

86945 Chenopodium ambrosioides L. var. integrifolium Fenzl = Dysphania ambrosioides (L.) Mosyakin et Clemants ■

86946 Chenopodium ambrosioides L. var. obovata Speg.;墨西哥土荆芥(墨西哥茶);Mexican Tea ■☆

86947 Chenopodium ambrosioides L. var. pubescens (Makino) Makino = Chenopodium ambrosioides L. ■☆

86948 Chenopodium ambrosioides L. var. suffruticosom (Willd.) Asch. et Graebn. = Dysphania ambrosioides (L.) Mosyakin et Clemants ■

86949 Chenopodium ambrosioides L. var. typicum Speg. = Dysphania ambrosioides (L.) Mosyakin et Clemants ■

86950 Chenopodium ambrosioides L. var. vagans (Standl.) J. T. Howell = Dysphania chilensis (Schrad.) Mosyakin et Clemants ■☆

86951 Chenopodium anthelminticum L. = Dysphania anthelmintica (L.) Mosyakin et Clemants ■☆

86952 Chenopodium aristatum L. = Dysphania aristata (L.) Mosyakin et Clemants ■

86953 Chenopodium aristatum L. var. inerme W. Z. Di;无刺藜(无刺刺藜)■

86954 Chenopodium atripliciforme Murr = Chenopodium bryoniifolium Bunge ■

86955 Chenopodium atrovirens Rydb.;暗绿藜■☆

86956 Chenopodium australe R. Br. = Suaeda australis (R. Br.) Moq. ●

86957 Chenopodium baryosmum Schult. ex Roem. et Schult. = Salsola imbricata Forssk. ■

86958 Chenopodium berlandieri Moq.;伯兰藜;Berlandier's Goosefoot, Pitseed Goosefoot, Pit-seed Goosefoot ■☆

86959 Chenopodium berlandieri Moq. subsp. platyphyllum (Issler) Ludw. = Chenopodium berlandieri Moq. var. zschackei (Murr) Murr ex Asch. ■☆

86960 Chenopodium berlandieri Moq. subsp. zschackei (Murr) Zobel = Chenopodium berlandieri Moq. var. zschackei (Murr) Murr ex Asch. ■☆

86961 Chenopodium berlandieri Moq. var. boscianum (Moq.) Wahl;鸭状伯兰藜■☆

86962 Chenopodium berlandieri Moq. var. bushianum (Aellen) Cronquist;布什藜;Bush's Goosefoot, Pigweed, Pit-seed Goosefoot, Soya-bean Goosefoot ■☆

86963 Chenopodium berlandieri Moq. var. farinosum Ludw. ex Aellen = Chenopodium berlandieri Moq. var. zschackei (Murr) Murr ex Asch. ■☆

86964 Chenopodium berlandieri Moq. var. macrocalycium (Aellen) Cronquist;大萼伯兰藜■☆

86965 Chenopodium berlandieri Moq. var. sinuatum (Murr) Wahl;深波

伯兰藜■☆

86966 Chenopodium berlandieri Moq. var. zschackei (Murr) Murr ex Asch.;恰克藜;Pit-seed Goosefoot,Zschack's Goosefoot ■☆

86967 Chenopodium betaceum Andrz. = Chenopodium strictum Roth ■

86968 Chenopodium blitum F. Muell. = Chenopodium foliosum Asch. ■

86969 Chenopodium blitum Hook. f. = Chenopodium foliosum (Moench) Asch. ■

86970 Chenopodium bonus-henricus L.;王藜(英国贡草);All Good, Allgood, Bleets, Blite, English Marquery, English Mercury, False Mercury, Fat Goose, Fat Hen, Flowery Docken, Good Henry, Good King Harry, Good King Henry, Good Neighbourhood, Good-king-henry, Good-king-henry Goosefoot, Goosefoot Mercury, Johnny O'neele, Lincolnshire Asparagus, Lincolnshire Spinach, Mar Querry, Marcaram, Margery, Markery, Mercury, Mercury Docken, Mercury Goosefoot, Mercury-docken, Midden Mylies, Milder, Mutton Dock, Perennial Goosefoot, Roman Plant, Sea Spinach, Shoemaker's Heel, Smear Docken, Smearwort, Wild Asparagus, Wild Spinach ■☆

86971 Chenopodium borbasii Murr;博尔巴什藜■☆

86972 Chenopodium boscianum Moq. = Chenopodium berlandieri Moq. var. boscianum (Moq.) Wahl ■☆

86973 Chenopodium boscianum Moq. = Chenopodium standleyanum Aellen ■☆

86974 Chenopodium botryodes Sm. = Chenopodium chenopodioides (L.) Aellen ■

86975 Chenopodium botrys L. = Chenopodium foetidum Schrad. ■

86976 Chenopodium botrys L. = Dysphania botrys (L.) Mosyakin et Clemants ■

86977 Chenopodium bryoniifolium Bunge; 菱叶藜; Rhombicleaf Goosefoot ■

86978 Chenopodium bryoniifolium Bunge var. kapelleriae Aellen ex Iljin = Chenopodium iljinii Golosk. ■

86979 Chenopodium bushianum Aellen = Chenopodium berlandieri Moq. var. bushianum (Aellen) Cronquist ■☆

86980 Chenopodium bushianum Aellen var. acutidentatum Aellen = Chenopodium berlandieri Moq. var. bushianum (Aellen) Cronquist ■☆

86981 Chenopodium calceoliforme Hook. = Suaeda calceoliformis (Hook.) Moq. ■☆

86982 Chenopodium californicum (S. Watson) S. Watson; 加州藜; California Goosefoot, Soap Plant ■☆

86983 Chenopodium capitatum (L.) Ambrosi var. parvicapitatum S. L. Welsh;小头状藜■☆

86984 Chenopodium capitatum Asch.;头状藜;Blite Goosefoot, Capitate Goosefoot, Indian Ink, Indian Paint, Indian-paint, Strawberry Blite, Strawberry Spinach, Strawberry-blight, Strawberry-blite ■☆

86985 Chenopodium carinatum R. Br. = Dysphania carinata (R. Br.) Mosyakin et Clemants ■☆

86986 Chenopodium carnosulum Moq.;肉藜;Ridged Goosefoot ■☆

86987 Chenopodium caudatum Jacq. = Amaranthus viridis L. ■

86988 Chenopodium caudatum Jacq. = Chenopodium viridis L. ■☆

86989 Chenopodium centrorubrum (Makino) Nakai = Chenopodium album L. var. centrorubrum Makino ■

86990 Chenopodium chenopodioides (L.) Aellen;合被藜(厚皮藜); Gamotepalous Goosefoot, Saltmarsh Goosefoot, Small Red Goosefoot ■

86991 Chenopodium chenopodioides (L.) Aellen var. degenianum (Aellen) Aellen = Chenopodium chenopodioides (L.) Aellen ■

86992 Chenopodium chilense Schrad. = Dysphania chilensis (Schrad.) Mosyakin et Clemants ■☆

86993 Chenopodium concatenatum Thuill. = Chenopodium album L. var. concatenatum (Thuill.) Gaudin ■

86994 Chenopodium congolanum (Hauman) Brenan;刚果藜■☆

86995 Chenopodium coronopus Moq.;臭荠藜■☆

86996 Chenopodium crassifolium Hornem. = Chenopodium chenopodioides (L.) Aellen ■

86997 Chenopodium cristatum (F. Muell.) F. Muell.;鸡冠藜;Crested Goosefoot ■☆

86998 Chenopodium cristatum F. Muell. = Chenopodium cristatum (F. Muell.) F. Muell. ■☆

86999 Chenopodium deltoideum Lam. = Chenopodium urbicum L. var. deltoideum (Lam.) Neilr. ■

87000 Chenopodium desiccatum A. Nelson;干藜;Pigweed ■☆

87001 Chenopodium desiccatum A. Nelson var. leptophylloides (Murr) Wahl = Chenopodium pratericola Rydb. ■☆

87002 Chenopodium elegantissimum Koidz. = Chenopodium purpurascens Jacq. ■☆

87003 Chenopodium farinosum Standl. = Chenopodium macrospermum Hook. f. ■☆

87004 Chenopodium fasciculosum Aellen;簇藜■☆

87005 Chenopodium fasciculosum Aellen var. muraliforme? = Chenopodium fasciculosum Aellen var. schimperi (Asch.) M. G. Gilbert ■☆

87006 Chenopodium fasciculosum Aellen var. schimperi (Asch.) M. G. Gilbert;欣珀藜■☆

87007 Chenopodium ficifolium Sm.;榕叶藜(小藜);Figleaf Goosefoot, Fig-leaved Goosefoot, Pigweed ■☆

87008 Chenopodium ficifolium Sm. = Chenopodium bryoniifolium Bunge ■

87009 Chenopodium ficifolium Sm. = Chenopodium serotinum L. ■

87010 Chenopodium foetidum Schrad. = Chenopodium schraderianum Roem. et Schult. ■

87011 Chenopodium foetidum Schrad. = Dysphania schraderiana (Roem. et Schult.) Mosyakin et Clemants ■

87012 Chenopodium foetidum Schrad. subsp. gracile Aellen = Chenopodium schraderianum Roem. et Schult. ■

87013 Chenopodium foetidum Schrad. subsp. pseudomultiflorum Murr = Chenopodium schraderianum Roem. et Schult. ■

87014 Chenopodium foetidum Schrad. subsp. tibetanum Murr = Chenopodium foetidum Schrad. ■

87015 Chenopodium foetidum Schrad. subsp. tibetanum Murr = Dysphania schraderiana (Roem. et Schult.) Mosyakin et Clemants ■

87016 Chenopodium foliosum (Moench) Asch.;球花藜(多叶藜); Foliose Goosefoot, Leafy Goosefoot, Strawberry Blite, Strawberry Goosefoot, Strawberry-blite ■

87017 Chenopodium foliosum Asch. = Chenopodium foliosum (Moench) Asch. ■

87018 Chenopodium foliosum Asch. var. virgatum? = Blitum virgatum L. ■☆

87019 Chenopodium formosanum Koidz.;台湾藜;Taiwan Goosefoot ■

87020 Chenopodium fremontii S. Watson; 沙漠藜; Desert Lamb's Quarters ■☆

87021 Chenopodium fremontii S. Watson var. atrovirens (Rydb.) Fosberg = Chenopodium atrovirens Rydb. ■☆

87022 Chenopodium fremontii S. Watson var. incanum S. Watson = Chenopodium incanum (S. Watson) A. Heller ■☆

87023 Chenopodium frutescens C. A. Mey.;灌状藜■☆

87024 Chenopodium fruticosum L. = Suaeda fruticosa (L.) Forssk. ●☆

87025 Chenopodium giganteum D. Don; 杖藜（红盐菜）; Gigant Goosefoot, Tree Spinach ■

87026 Chenopodium gigantospermum Aellen = Chenopodium simplex (Torr.) Raf. ■☆

87027 Chenopodium gigantospermum Aellen var. standleyanum (Aellen) Aellen = Chenopodium standleyanum Aellen ■☆

87028 Chenopodium glaucophyllum Aellen = Chenopodium strictum Roth ■

87029 Chenopodium glaucum L.; 灰绿藜（灰藜，水灰菜）; Glaucous Goosefoot, Oakleaf Goosefoot, Oak-leaved Goosefoot, Pigweed ■

87030 Chenopodium glaucum L. subsp. congolanum Hauman = Chenopodium congolanum (Hauman) Brenan ■☆

87031 Chenopodium glaucum L. subsp. euglaucum Aellen = Chenopodium glaucum L. ■

87032 Chenopodium glaucum L. subsp. marlothianum (Murr) Thell. et Aellen = Chenopodium glaucum L. ■

87033 Chenopodium glaucum L. var. salinum (Standl.) B. Boivin; 碱地灰绿藜■☆

87034 Chenopodium glaucum Standl. = Chenopodium glaucum L. ■

87035 Chenopodium glaucum Standl. subsp. salinum (Standl.) Aellen = Chenopodium glaucum L. var. salinum (Standl.) B. Boivin ■☆

87036 Chenopodium glomerulosum Rchb. = Chenopodium album L. var. glomerulosum (Rchb.) Peterm. ■

87037 Chenopodium gracilispicum H. W. Kung; 细穗藜（小叶野灰菜）; Slenderspicate Goosefoot ■

87038 Chenopodium gracilispicum H. W. Kung var. longifolium C. S. Zhu et X. D. Li; 长叶藜■

87039 Chenopodium graveolens Willd.; 气味藜■☆

87040 Chenopodium guineense Jacq. = Chenopodium murale L. ■☆

87041 Chenopodium hederiforme (Murr) Aellen; 常春藤藜■☆

87042 Chenopodium hederiforme (Murr) Aellen var. dentatum Aellen; 尖齿常春藤藜■☆

87043 Chenopodium hederiforme (Murr) Aellen var. undulatum Aellen; 波状常春藤藜■☆

87044 Chenopodium hians Standl.; 开裂藜■☆

87045 Chenopodium hircinum Schrad.; 山羊藜; Avian Goosefoot, Foetid Goosefoot ■☆

87046 Chenopodium hirsutum L. = Bassia hirsuta (L.) Asch. ■☆

87047 Chenopodium humile Hook. = Chenopodium rubrum L. var. humile (Hook.) S. Watson ■☆

87048 Chenopodium hybridum L.; 杂配藜（八角灰菜，大叶灰菜，大叶藜，黄靛花，灰菜，血见愁，野角尖草，杂灰藜）; Mapleleaf Goosefoot, Maple-leaved Goosefoot, Sowbane, Thorn-apple-leaved Goosefoot ■

87049 Chenopodium hybridum L. subsp. gigantospermum (Aellen) Hultén = Chenopodium simplex (Torr.) Raf. ■☆

87050 Chenopodium hybridum L. var. gigantospermum (Aellen) Rouleau = Chenopodium simplex (Torr.) Raf. ■☆

87051 Chenopodium hybridum L. var. simplex Torr. = Chenopodium simplex (Torr.) Raf. ■☆

87052 Chenopodium hybridum L. var. standleyanum (Aellen) Fernald = Chenopodium standleyanum Aellen ■☆

87053 Chenopodium iljinii Golosk.; 小白藜; Iljin Goosefoot ■

87054 Chenopodium incanum (S. Watson) A. Heller; 灰白毛藜; Pigweed ■☆

87055 Chenopodium incognitum Wahl = Chenopodium hians Standl. ■☆

87056 Chenopodium intermedium Mert. et Koch = Chenopodium urbicum L. var. intermedium (Mert. et Koch) Koch ■☆

87057 Chenopodium jenissejense Aellen et Iljin; 热尼斯藜■☆

87058 Chenopodium karoi (Murr) Aellen; 平卧藜; Prostrate Goosefoot ■

87059 Chenopodium karoi Aellen = Chenopodium karoi (Murr) Aellen ■

87060 Chenopodium karoi Aellen = Chenopodium prostratum Bunge ex Herder ■

87061 Chenopodium khnggraeffii Aellen; 克林氏藜■☆

87062 Chenopodium koraiense Nakai = Chenopodium bryoniifolium Bunge ■

87063 Chenopodium koraiense Nakai = Chenopodium gracilispicum H. W. Kung ■

87064 Chenopodium korshinskyi Litv.; 考尔藜■☆

87065 Chenopodium korshinskyi Litv. = Chenopodium foliosum (Moench) Asch. ■

87066 Chenopodium lanceolatum Muhl. ex Willd. = Chenopodium album L. ■

87067 Chenopodium leptophyllum (Moq.) Nutt. ex S. Watson = Chenopodium leptophyllum (Nutt. ex Moq.) Nutt. ex S. Watson ■☆

87068 Chenopodium leptophyllum (Moq.) Nutt. ex S. Watson var. oblongifolium S. Watson = Chenopodium desiccatum A. Nelson ■☆

87069 Chenopodium leptophyllum (Moq.) Nutt. ex S. Watson var. subglabrum S. Watson = Chenopodium subglabrum (S. Watson) A. Nelson ■☆

87070 Chenopodium leptophyllum (Nutt. ex Moq.) Nutt. ex S. Watson; 细叶藜; Fine-leaf Goosefoot, Narrow-leaved Lamb's Quarters ■☆

87071 Chenopodium leptophyllum (Nutt. ex Moq.) S. Watson = Chenopodium leptophyllum (Nutt. ex Moq.) Nutt. ex S. Watson ■☆

87072 Chenopodium leptophyllum (Nutt. ex Moq.) S. Watson var. oblongifolium S. Watson = Chenopodium pratericola Rydb. ■☆

87073 Chenopodium leptophyllum Nutt. ex Moq. var. subglabrum S. Watson; 光细叶藜■☆

87074 Chenopodium leptophyllum Nutt. ex S. Watson = Chenopodium pratericola Rydb. ■☆

87075 Chenopodium leptophyllum Nutt. ex S. Watson sensu Swink et Wilh. = Chenopodium pratericola Rydb. ■☆

87076 Chenopodium linifolium (Pall.) Roem. et Schult. = Suaeda linifolia Pall. ■

87077 Chenopodium linifolium Schult. = Suaeda linifolia Pall. ■

87078 Chenopodium longidjawense Peter = Chenopodium murale L. ■☆

87079 Chenopodium macrocalycium Aellen = Chenopodium berlandieri Moq. var. macrocalycium (Aellen) Cronquist ■☆

87080 Chenopodium macrospermum Hook. f.; 大籽藜; Largeseed Goosefoot ■☆

87081 Chenopodium macrospermum Hook. f. var. halophilum (Phil.) Aellen = Chenopodium macrospermum Hook. f. ■☆

87082 Chenopodium macrospermum Hook. f. var. halophilum Standl.; 喜盐大籽藜; Saltloving Goosefoot ■☆

87083 Chenopodium mairei H. Lév. = Chenopodium giganteum D. Don ■

87084 Chenopodium maritimum L. = Suaeda maritima (L.) Dumort. ■

87085 Chenopodium marlothianum Murr = Chenopodium glaucum L. ■

87086 Chenopodium microphyllum Thunb. = Exomis microphylla (Thunb.) Aellen ■☆

87087 Chenopodium minimum Wang-Wei et P. Y. Fu; 矮藜; Little Goosefoot ■

87088 Chenopodium minimum Wang-Wei et P. Y. Fu = Chenopodium aristatum L. ■

87089 Chenopodium minimum Wang-Wei et P. Y. Fu = Dysphania aristata (L.) Mosyakin et Clemants ■
87090 Chenopodium missouriense Aellen;密苏里藜;Missouri Pigweed ■☆
87091 Chenopodium missouriense Aellen = Chenopodium album L. ■
87092 Chenopodium mucronatum Thunb.;短尖藜■☆
87093 Chenopodium mucronatum Thunb. subsp. olukondae (Murr) Murr = Chenopodium olukondae (Murr) Murr ■☆
87094 Chenopodium mucronatum Thunb. var. subintegrum Aellen = Chenopodium opulifolium Schrad. ex W. D. J. Koch et Ziz ■☆
87095 Chenopodium multifidum L.;多裂藜;Cutleaf Goosefoot, Scented Goosefoot ■☆
87096 Chenopodium murale L.;荨麻叶藜(壁生藜);Lamb's Quarters, Nettleleaf Goosefoot, Nettle-leaf Goosefoot, Nettle-leaved Goosefoot, Sowbane ■☆
87097 Chenopodium murale L. var. acutidentatum Aellen = Chenopodium murale L. ■☆
87098 Chenopodium murale L. var. albescens Moq. = Chenopodium murale L. ■☆
87099 Chenopodium murale L. var. farinosum S. Watson = Chenopodium macrospermum Hook. f. ■☆
87100 Chenopodium murale L. var. microphyllum Boiss. = Chenopodium murale L. ■☆
87101 Chenopodium murale L. var. paucidentatum Beck = Chenopodium murale L. ■☆
87102 Chenopodium murale L. var. schimperi Asch. = Chenopodium fasciculosum Aellen var. schimperi (Asch.) M. G. Gilbert ■☆
87103 Chenopodium murale L. var. spissidentatum Murr = Chenopodium murale L. ■☆
87104 Chenopodium nigrum Raf. = Suaeda nigra (Raf.) J. F. Macbr. ■☆
87105 Chenopodium nitrariaceum (F. Muell.) Benth.;白刺藜;Nitre Goosefoot ■☆
87106 Chenopodium nitrariaceum F. Muell. ex Benth. = Chenopodium nitrariaceum (F. Muell.) Benth. ■☆
87107 Chenopodium nuttalliae Saff.;纳托尔藜;Pigweed ■☆
87108 Chenopodium olidum Curtis = Chenopodium vulvaria L. ■☆
87109 Chenopodium olidum S. Watson = Chenopodium watsonii A. Nelson ■☆
87110 Chenopodium olukondae (Murr) Murr;奥卢孔达藜■☆
87111 Chenopodium opulifolium Schrad. ex DC. = Chenopodium opulifolium Schrad. ex Koch et Ziz. ■☆
87112 Chenopodium opulifolium Schrad. ex Koch et Ziz.;荚迷叶藜;Grey Goosefoot, Pigweed, Seaport Goosefoot ■☆
87113 Chenopodium opulifolium Schrad. ex W. D. J. Koch et Ziz subsp. amboanum Murr = Chenopodium amboanum (Murr) Aellen ■☆
87114 Chenopodium opulifolium Schrad. ex W. D. J. Koch et Ziz subsp. hederiforme Murr = Chenopodium hederiforme (Murr) Aellen var. dentatum Aellen ■☆
87115 Chenopodium opulifolium Schrad. ex W. D. J. Koch et Ziz subsp. olukondae Murr = Chenopodium olukondae (Murr) Murr ■☆
87116 Chenopodium opulifolium Schrad. ex W. D. J. Koch et Ziz subsp. petiolariforme Aellen = Chenopodium petiolariforme (Aellen) Aellen ■☆
87117 Chenopodium opulifolium Schrad. ex W. D. J. Koch et Ziz subsp. ugandae Aellen = Chenopodium ugandae (Aellen) Aellen ■☆
87118 Chenopodium opulifolium Schrad. ex W. D. J. Koch et Ziz var. typicum Beck = Chenopodium opulifolium Schrad. ex W. D. J. Koch et Ziz ■☆
87119 Chenopodium overi Aellen = Chenopodium capitatum (L.) Ambrosi var. parvicapitatum S. L. Welsh ■☆
87120 Chenopodium pallescens Standl.;苍白藜;Pigweed ■☆
87121 Chenopodium pallidicaule Aellen;白茎藜;Canihua ■☆
87122 Chenopodium pamiricum Iljin;帕米尔藜■☆
87123 Chenopodium petiolare Kunth var. leptophylloides Murr = Chenopodium pratericola Rydb. ■☆
87124 Chenopodium petiolare Kunth var. sinuatum Murr = Chenopodium berlandieri Moq. var. sinuatum (Murr) Wahl ■☆
87125 Chenopodium petiolariforme (Aellen) Aellen;柄叶藜■☆
87126 Chenopodium phillipsianum Aellen;菲利藜■☆
87127 Chenopodium physophora Moq. = Suaeda physophora Pall. ●
87128 Chenopodium physophorum (Pall.) Moq. = Suaeda physophora Pall. ●
87129 Chenopodium platyphyllum Issler = Chenopodium berlandieri Moq. var. zschackei (Murr) Murr ex Asch. ■☆
87130 Chenopodium polyspermum L.;多籽藜;Allseed Goosefoot, Manyseed Goosefoot, Many-seed Goosefoot, Many-seeded Goosefoot, Round-leaved Goosefoot ■☆
87131 Chenopodium polyspermum L. var. acutifolium (Sm.) Gaudin;尖叶多籽藜;Manyseed Goosefoot, Many-seed Goosefoot ■☆
87132 Chenopodium polyspermum L. var. obtusifolium?;钝叶多籽藜;Manyseed Goosefoot ■☆
87133 Chenopodium pratericola Rydb.;沙藜;Desert Goosefoot, Narrowleaf Goosefoot, Narrow-leaf Goosefoot, Pigweed ■☆
87134 Chenopodium pratericola Rydb. subsp. eupratericola Aellen = Chenopodium pratericola Rydb. ■☆
87135 Chenopodium pratericola Rydb. var. leptophylloides (Murr) Aellen = Chenopodium pratericola Rydb. ■☆
87136 Chenopodium preissmannii Murr;普赖斯曼藜■☆
87137 Chenopodium probstii Aellen;普罗藜;Probst's Goosefoot ■☆
87138 Chenopodium procerum Hochst. ex Moq.;高藜■☆
87139 Chenopodium prostratum Bunge ex Herder = Chenopodium karoi (Murr) Aellen ■
87140 Chenopodium prostratum Bunge ex Herder subsp. karoi (Murr) Lomon. = Chenopodium karoi (Murr) Aellen ■
87141 Chenopodium pseudoborbasii Murr = Chenopodium album L. var. pseudoborbasii (Murr) Beck ■☆
87142 Chenopodium pumilio R. Br.;黏藜;Clammy Goosefoot, Pigweed, Small Crumbweed ■☆
87143 Chenopodium pumilio R. Br. = Dysphania pumilio (R. Br.) Mosyakin et Clemants ■☆
87144 Chenopodium purpurascens Jacq.;紫藜■☆
87145 Chenopodium quinoa Willd.;昆诺阿藜;Pigweed, Quinoa, Quinua ■☆
87146 Chenopodium rhadinostachyum F. Muell.;穗藜■☆
87147 Chenopodium rubrum L.;红叶藜(红藜);Alkali-blite, Coast Blite, Coast-blite, Pigweed, Red Goosefoot, Red Pigweed, Sowbane ■
87148 Chenopodium rubrum L. = Chenopodium botryodes Sm. ■
87149 Chenopodium rubrum L. subsp. crassifolium (Hornem.) Maire = Chenopodium chenopodioides (L.) Aellen ■
87150 Chenopodium rubrum L. subsp. humile (Hook.) Aellen = Chenopodium rubrum L. var. humile (Hook.) S. Watson ■☆
87151 Chenopodium rubrum L. var. humile (Hook.) S. Watson;小红叶藜;Marshland Goosefoot ■☆
87152 Chenopodium salinum Standl. = Chenopodium glaucum L. var. salinum (Standl.) B. Boivin ■☆

87153 Chenopodium salsum L. = Suaeda salsa (L.) Pall. ■

87154 Chenopodium schraderianum Roem. et Schult. = Dysphania schraderiana (Roem. et Schult.) Mosyakin et Clemants ■

87155 Chenopodium schraderianum Schult. = Chenopodium foetidum Schrad. ■

87156 Chenopodium scoparium L. = Kochia scoparia (L.) Schrad. ■

87157 Chenopodium sericeum (Aiton) Spreng. = Bassia diffusa (Thunb.) Kuntze ■☆

87158 Chenopodium serotinum L.;小藜(粉子菜,灰藋,灰灰菜,灰藜,灰蓼,灰蒴,灰条,灰条菜,灰苋,灰苋菜,金锁天,球花藜,水落藜,小叶灰藋,小叶藜);Small Goosefoot ■

87159 Chenopodium simplex (Torr.) Raf.;槭叶藜;Maple-leaved Goosefoot ■☆

87160 Chenopodium sinense Moq. = Chenopodium aristatum L. ■

87161 Chenopodium sinense Moq. = Dysphania aristata (L.) Mosyakin et Clemants ■

87162 Chenopodium sosnowskyi Kapeller;索斯诺夫藜■☆

87163 Chenopodium spinosum Hook. = Grayia spinosa (Hook.) Moq. ■☆

87164 Chenopodium standleyanum Aellen;斯氏藜;Pigweed, Standley's Goosefoot, Woodland Goosefoot ■☆

87165 Chenopodium stellulatum (Benth.) Aellen;星藜■☆

87166 Chenopodium stenophyllum (Makino) Koidz. = Chenopodium album L. var. stenophyllum Makino ■☆

87167 Chenopodium striatiforme Murr = Chenopodium strictum Roth ■

87168 Chenopodium striatum (Krasan) Murr = Chenopodium strictum Roth ■

87169 Chenopodium striatum Murr var. pseudoborbasii (Murr) Graebn. = Chenopodium strictum Roth var. pseudoborbasii (Murr) Cufod. ■☆

87170 Chenopodium strictum (Krasan) Murr subsp. striatiforme (Murr) Uotila = Chenopodium strictum Roth ■

87171 Chenopodium strictum Roth;圆头藜(直立藜);Late-flowering Goosefoot, Pigweed, Strict Goosefoot, Striped Goosefoot ■

87172 Chenopodium strictum Roth subsp. glaucophyllum (Aellen) Aellen et K. Just = Chenopodium strictum Roth ■

87173 Chenopodium strictum Roth var. glaucophyllum (Aellen) Wahl = Chenopodium strictum Roth ■

87174 Chenopodium strictum Roth var. pseudoborbasii (Murr) Cufod.;假博尔巴什藜■☆

87175 Chenopodium strictum Roth var. striatum (Krasan) Aellen et Iljin;条纹藜■☆

87176 Chenopodium suberifolium Murr = Chenopodium procerum Hochst. ex Moq. ■☆

87177 Chenopodium subglabrum (S. Watson) A. Nelson;光滑藜■☆

87178 Chenopodium subglabrum (S. Watson) A. Nelson = Chenopodium pratericola Rydb. ■☆

87179 Chenopodium suecicum Murr;瑞典藜;Green Goosefoot, Swedish Goosefoot ■☆

87180 Chenopodium suffruticosom Willd. = Dysphania ambrosioides (L.) Mosyakin et Clemants ■

87181 Chenopodium suffruticosum Pall. = Flueggea suffruticosa (Pall.) Baill. ●

87182 Chenopodium tibeticum A. J. Li;西藏藜;Xizang Goosefoot ■

87183 Chenopodium tibeticum A. J. Li = Dysphania aristata (L.) Mosyakin et Clemants ■

87184 Chenopodium triangulare R. Br. var. stellulatum Benth. = Chenopodium stellulatum (Benth.) Aellen ■☆

87185 Chenopodium ugandae (Aellen) Aellen;乌干达藜■☆

87186 Chenopodium urbicum L.;市藜;City Goosefoot, Lamb Tongue, Lamb-tongue, Upright Goosefoot ■

87187 Chenopodium urbicum L. subsp. sinicum H. W. Kung et G. L. Chu;东亚市藜;Dongya Goosefoot, East Asia Goosefoot ■

87188 Chenopodium urbicum L. var. deltoideum (Lam.) Neilr. = Chenopodium urbicum L. ■

87189 Chenopodium urbicum L. var. intermedium (Mert. et W. D. J. Koch) W. D. J. Koch = Chenopodium urbicum L. ■

87190 Chenopodium urbicum L. var. intermedium Koch = Chenopodium urbicum L. subsp. sinicum H. W. Kung et G. L. Chu ■

87191 Chenopodium vachelii Hook. et Arn.;瓦氏藜■

87192 Chenopodium vachelii Hook. et Arn. = Chenopodium acuminatum Willd. subsp. virgatum (Thunb.) Kitam. ■

87193 Chenopodium virgatum (L.) Ambrosi = Blitum virgatum L. ■☆

87194 Chenopodium virgatum (L.) Ambrosi = Chenopodium foliosum (Moench) Asch. ■

87195 Chenopodium virgatum Thunb. = Chenopodium acuminatum Willd. subsp. virgatum (Thunb.) Kitam. ■

87196 Chenopodium virgatum Thunb. = Chenopodium acuminatum Willd. var. vachelii (Hook. et Arn.) Moq. ■

87197 Chenopodium virgatum Thunb. = Chenopodium foliosum Asch. ■

87198 Chenopodium viride L. = Chenopodium album L. var. viride (L.) Wahlenb. ●☆

87199 Chenopodium viridis L.;绿藜■☆

87200 Chenopodium vulvaria L.;臭藜;Dirty John, Dog's Airach, Dog's Arrach, Dog's Orach, Dog's Orache, Fat Hen, Goat's Airach, Goat's Arrach, Notchweed, Stinking Airach, Stinking Arag, Stinking Arrach, Stinking Blite, Stinking Goosefoot, Stinking Motherwort, Stinking Orach, Stinking Orache, Wild Arrach ■☆

87201 Chenopodium vulvaria L. var. incisum Maire;锐裂藜■☆

87202 Chenopodium vulvaria L. var. microphyllum Moq. = Chenopodium vulvaria L. ■☆

87203 Chenopodium watsonii A. Nelson;瓦特森藜;Pigweed ■☆

87204 Chenopodium zahnii J. Murray;察恩假藜■☆

87205 Chenopodium zschackei Murr = Chenopodium berlandieri Moq. ■☆

87206 Chenopodium zshcackei Murr = Chenopodium berlandieri Moq. var. zschackei (Murr) Murr ex Asch. ■☆

87207 Chenorchis Z. J. Liu, K. W. Liu et L. J. Chen = Penkimia Phukan et Odyuo ■

87208 Chenorchis singchii Z. J. Liu, K. W. Liu et L. J. Chen = Penkimia nagalandensis Phukan et Odyuo ■

87209 Cheobula Vell. = Cleobula Vell. ☆

87210 Cheramela Rumph. = Cicca L. ●

87211 Cherimolia Raf. = Annona L. ●

87212 Cherina Cass. = Chaetanthera Ruiz et Pav. ■☆

87213 Cherleria Haller = Arenaria L. ■

87214 Cherleria Haller ex L. = Arenaria L. ■

87215 Cherleria L. = Minuartia L. ■

87216 Cherleria dicranoides Cham. et Schltdl. = Stellaria dicranoides (Cham. et Schltdl.) Fenzl ■☆

87217 Cherleria juniperina D. Don = Arenaria densissima Wall. ex Edgew. et Hook. f. ■

87218 Cherleria sedoides L. = Minuartia sedoides (L.) Hiern ■☆

87219 Cherophilum Nocca = Cherophylum Raf. ■

87220 Cherophylum Raf. = Chaerophyllum L. ■

87221 Chersodoma Phil. (1891);山绒菊属■●☆

87222 Chersodoma candida Phil.;山绒菊■●☆

87223 Chersydrtum Schott = Dracontium L. ■☆

87224 Chesmone Bubani = Argyrolobium Eckl. et Zeyh.(保留属名)●☆

87225 Chesmone Bubani = Chasmone E. Mey. ●☆

87226 Chesnea Scop. = Carapichea Aubl.(废弃属名)●

87227 Chesnea Scop. = Cephaelis Sw.(保留属名)●

87228 Chesnea Scop. = Psychotria L.(保留属名)●

87229 Chesneya Bertol. = Gaytania Münter ■

87230 Chesneya Bertol. = Pimpinella L. ■

87231 Chesneya Lindl. = Chesneya Lindl. ex Endl. ●

87232 Chesneya Lindl. ex Endl.(1840);雀儿豆属;Birdlingbran, Chesneya ●

87233 Chesneya acaulis (Baker) Popov;无茎雀儿豆;Stemless Birdlingbran, Stemless Chesneya ●

87234 Chesneya crassipes Boriss.;长梗雀儿豆;Long Pedicel Chesneya, Longstalk Birdlingbran ●

87235 Chesneya cuneata (Benth.) Ali;截叶雀儿豆;Cunealleaf Birdlingbran, Cunealleaf Chesneya ●

87236 Chesneya depressa (Oliv.) Popov;凹陷雀儿豆●☆

87237 Chesneya elegans Fomin;雅致雀儿豆●☆

87238 Chesneya ferganensis Korsh. = Chesniella ferganensis (Korsh.) Boriss. ■

87239 Chesneya gansuensis Y. X. Liou;甘肃雀儿豆(戈壁雀儿豆);Chesneya, Gansu Birdlingbran, Gansu Kansu ●

87240 Chesneya gansuensis Y. X. Liou = Chesniella ferganensis (Korsh.) Boriss. ■

87241 Chesneya grubovii Yakovlev = Chesneya gansuensis Y. X. Liou ●

87242 Chesneya grubovii Yakovlev = Chesniella ferganensis (Korsh.) Boriss. ■

87243 Chesneya hissarica Boriss.;希萨尔雀儿豆●☆

87244 Chesneya intermedia (Yakovlev et R. E. Ulziykh. ex Yakovlev) Z. G. Qian;察隅雀儿豆;Chayu Birdlingbran, Chayu Chesneya ●

87245 Chesneya intermedia (Yakovlev) Z. G. Qian = Chesneya nubigena (D. Don) Ali ●☆

87246 Chesneya kopetdaghensis Boriss.;科佩特雀儿豆●☆

87247 Chesneya linczevskii Boriss.;林氏雀儿豆●☆

87248 Chesneya macrantha H. S. Cheng ex H. C. Fu;大花雀儿豆(红花海绵豆,红花雀儿豆);Bigflower Birdlingbean, Redflower Chesneya ●■

87249 Chesneya macrantha H. S. Cheng ex P. C. Li = Chesneya macrantha H. S. Cheng ex H. C. Fu ●■

87250 Chesneya mongolica Maxim. = Chesniella mongolica (Maxim.) P. C. Li ●

87251 Chesneya nubigena (D. Don) Ali;云雾雀儿豆;Cloud Chesneya, Mist Birdlingbran ●☆

87252 Chesneya nubigena (D. Don) Ali subsp. purpurea (P. C. Li) X. Y. Zhu = Chesneya purpurea P. C. Li ●

87253 Chesneya parviflora Jaub. et Spach;小花雀儿豆■☆

87254 Chesneya paucifoliolata (Yakovlev et R. E. Ulziykh. ex Yakovlev) Z. G. Qian;疏叶雀儿豆;Laxleaf Birdlingbran, Laxleaf Chesneya ●

87255 Chesneya paucifoliolata (Yakovlev et R. E. Ulziykh.) Z. G. Qian = Chesneya paucifoliolata (Yakovlev et R. E. Ulziykh. ex Yakovlev) Z. G. Qian ●

87256 Chesneya paucifoliolata (Yakovlev et R. E. Ulziykh.) Z. G. Qian = Chesneya nubigena (D. Don) Ali ●☆

87257 Chesneya polystichoides (Hand. -Mazz.) Ali;川滇雀儿豆(滇康丽豆);Cuan-Dian Birdlingbran, Sichuan-Yunnan Chesneya ●

87258 Chesneya polystichoides (Hand. -Mazz.) H. S. Cheng = Chesneya polystichoides (Hand. -Mazz.) Ali ●

87259 Chesneya potaninii (N. Ulziykh.) Govaerts = Chesneya macrantha S. H. Cheng ex H. C. Fu ●

87260 Chesneya purpurea P. C. Li;紫花雀儿豆;Purple Birdlingbran, Purpleflower Chesneya ●

87261 Chesneya purpurea P. C. Li = Chesneya nubigena (D. Don) Ali subsp. purpurea (P. C. Li) X. Y. Zhu ●

87262 Chesneya rysidosperma Jaub. et Spach;皱籽雀儿豆●

87263 Chesneya spinosa P. C. Li;刺柄雀儿豆;Spine Birdlingbran, Spiny Chesneya ●

87264 Chesneya tribuloides Nevski;三尖雀儿豆●☆

87265 Chesneya turkestanica Franch.;土耳其斯坦雀儿豆●☆

87266 Chesneya yunnanensis (Yakovlev et R. E. Ulziykh. ex Yakovlev et Sviaz.) Z. G. Qian;云南雀儿豆;Yunnan Chesneya ●

87267 Chesneya yunnanensis (Yakovlev et R. E. Ulziykh.) Z. G. Qian = Chesneya nubigena (D. Don) Ali ●☆

87268 Chesniella Boriss.(1964);旱雀豆属;Chesniella, Drybirdbean ■

87269 Chesniella Boriss. = Chesneya Lindl. ex Endl. ●

87270 Chesniella depressa (Oliv.) Boiss. = Chesneya depressa (Oliv.) Popov ●☆

87271 Chesniella ferganensis (Korsh.) Boriss.;甘肃旱雀豆(费尔干雀儿豆,甘肃雀儿豆,戈壁雀儿豆);Chesneya, Gansu Birdlingbran, Gansu Chesniella, Gansu Drybirdbean, Gansu Kansu ■

87272 Chesniella gansuensis (Y. X. Liou) P. C. Li = Chesniella ferganensis (Korsh.) Boriss. ■

87273 Chesniella mongolica (Maxim.) Boriss.;蒙古旱雀豆(蒙古切思豆,蒙古雀儿豆);Mongol Drybirdbean, Mongolia Chesneya, Mongolian Chesniella ■

87274 Chesniella mongolica (Maxim.) P. C. Li = Chesniella mongolica (Maxim.) Boriss. ■

87275 Chesnya Rchb. = Chesneya Lindl. ex Endl. ●

87276 Chetanthera Raf. = Chaetanthera Ruiz et Pav. ■☆

87277 Chetanthera Raf. = Chaetopappa DC. ■☆

87278 Chetaria Steud. = Scabiosa L. ●■

87279 Chetastrum Neck. = Scabiosa L. ●■

87280 Chetocrater Raf. = Casearia Jacq. ●

87281 Chetocrater Raf. = Chaetocrater Ruiz et Pav. ●

87282 Chetopappua Raf. = Chaetopappa DC. ■☆

87283 Chetropis Raf. = Arenaria L. ■

87284 Chetyson Raf. = Sedum L. ●■

87285 Chevaliera magdalenae André = Aechmea magdalenae André ex Baker ■

87286 Chevalierella A. Camus(1933);隐节草属■☆

87287 Chevalierella congoensis A. Camus = Chevalierella dewildemanii (Vanderyst) Van der Veken ex Compère ■☆

87288 Chevalierella dewildemanii (Vanderyst) Van der Veken ex Compère;隐节草■☆

87289 Chevalieria Gaudich. = Aechmea Ruiz et Pav.(保留属名)■☆

87290 Chevalieria Gaudich. = Chevalieria Gaudich. ex Beer ■☆

87291 Chevalieria Gaudich. ex Beer = Aechmea Ruiz et Pav.(保留属名)■☆

87292 Chevalieria Gaudich. ex Beer(1852);雪佛凤梨属(雪佛兰属)■☆

87293 Chevalieria ornata Gaudich.;雪佛凤梨■☆

87294 Chevalierodendron J. -F. Leroy = Streblus Lour. ●

87295 Chevalliera Carrière = Chevalieria Gaudich. ex Beer ■☆

87296 Chevreulia Cass.(1817);钝柱紫绒草属■☆

87297 Chevreulia acuminata Less.;尖钝柱紫绒草■☆
87298 Chevreulia elegans Rusby;雅致钝柱紫绒草■☆
87299 Chevreulia filiformis Hook. et Arn.;线形钝柱紫绒草■☆
87300 Chevreulia nivea Phil.;雪白钝柱紫绒草■☆
87301 Cheynia Harv. = Balaustion Hook. ●☆
87302 Cheynia J. Drumm. ex Harv. = Balaustion Hook. ●☆
87303 Cheyniana Rye = Balaustion Hook. ●☆
87304 Cheyniana Rye(2009);澳洲石榴花属●☆
87305 Chiangiodendron T. Wendt(1988);墨西哥大风子属●☆
87306 Chiangiodendron mexicanum Wendt;墨西哥大风子●☆
87307 Chianthemum Kuntze = Galanthus L. ■☆
87308 Chiapasia Britton et Rose = Disocactus Lindl. ●☆
87309 Chiapasia Britton et Rose(1923);恰帕斯掌属(恰帕西亚属)●☆
87310 Chiapasia nelsonii (Britton et Rose) Britton et Rose;恰帕斯掌(奈尔孙)●☆
87311 Chiapasophyllum Doweld = Epiphyllum Haw. ●
87312 Chiapasophyllum Doweld(2001);墨西哥昙花属●☆
87313 Chiapasophyllum chrysocardium (Alexander) Doweld;墨西哥昙花●☆
87314 Chiarinia Chiov. = Lecaniodiscus Planch. ex Benth. ●☆
87315 Chiastophyllum (Ledeb.) A. Berger (1930);对叶景天属;Chiastophyllum, Cotyledon ■☆
87316 Chiastophyllum (Ledeb.) Stapf = Chiastophyllum (Ledeb.) A. Berger ■☆
87317 Chiastophyllum Stapf = Chiastophyllum (Ledeb.) A. Berger ■☆
87318 Chiastophyllum oppositifolium (Ledeb.) A. Berger;对叶景天;Chiastophyllum, Lamb's-tail ■☆
87319 Chiastophyllum oppositifolium A. Berger = Chiastophyllum oppositifolium (Ledeb.) A. Berger ■☆
87320 Chiazospermum Bernh. (1833);节角茴属■
87321 Chiazospermum Bernh. = Hypecoum L. ■
87322 Chiazospermum erectum (L.) Bernh. = Hypecoum erectum L. ■
87323 Chiazospermum lactiflorum Kar. et Kir.;节角茴■☆
87324 Chibaca Bertol. (废弃属名) = Warburgia Engl. (保留属名)●☆
87325 Chibaca G. Bertol. (废弃属名) = Warburgia Engl. (保留属名)●☆
87326 Chibaca salutaris G. Bertol. = Warburgia salutaris (G. Bertol.) Chiov. ●☆
87327 Chichaea C. Presl = Brachychiton Schott et Endl. ●☆
87328 Chicharronia A. Rich. = Terminalia L. (保留属名)●
87329 Chichicaste Weigend(1997);长叶刺莲花属■☆
87330 Chichicaste grandis (Standl.) Weigend;长叶刺莲花■☆
87331 Chichipia Backeb. = Polaskia Backeb. ●☆
87332 Chichipia Marn. -Lap. = Polaskia Backeb. ●☆
87333 Chickassia Wight et Arn. = Chukrasia A. Juss. ●
87334 Chiclea Lundell = Manilkara Adans. (保留属名)●
87335 Chicoca Augier = Chiococca P. Browne ex L. ●☆
87336 Chicoinaea Comm. ex DC. = Psathura Comm. ex Juss. ■☆
87337 Chidlowia Hoyle(1932);奇罗维豆属●☆
87338 Chidlowia sanguinea Hoyle;奇罗维豆●☆
87339 Chienia W. T. Wang = Delphinium L. ■
87340 Chienia honanensis W. T. Wang = Delphinium grandiflorum L. ■
87341 Chieniodendron Tsiang et P. T. Li = Meiogyne Miq. ●
87342 Chieniodendron Tsiang et P. T. Li = Oncodostigma Diels ●
87343 Chieniodendron Tsiang et P. T. Li(1964);蕉木属(钱木属)●
87344 Chieniodendron hainanense (Merr.) Tsiang et P. T. Li;蕉木(海南山指甲,鹿茸木,山蕉);Hainan Meiogyne, Hainan Oncodostigma ●◇
87345 Chieniodendron hainanense (Merr.) Tsiang et P. T. Li = Oncodostigma hainanense (Merr.) Tsiang et P. T. Li ●◇
87346 Chienodoxa Y. Z. Sun = Schnabelia Hand. -Mazz. ■●★
87347 Chienodoxa tetrodonta Y. Z. Sun = Schnabelia tetrodonta (Y. Z. Sun) C. Y. Wu et C. Chen ■
87348 Chifolium Hamm. = Anthriscus Pers. (保留属名)■
87349 Chigua D. W. Stev. (1990);哥伦比亚苏铁属●☆
87350 Chihuahuana Urbatsch et R. P. Roberts(2004);簇黄花属●☆
87351 Chihuahuana purpusii (Brandegee) Urbatsch et R. P. Roberts;簇黄花■☆
87352 Chikusichloa Koidz. (1925);山涧草属;Chikusichloa, Gullygrass ●★
87353 Chikusichloa aqutica Koidz.;山涧草;Aquatic Chikusichloa, Aquatic Gullygrass ■
87354 Chikusichloa brachyanthera Ohwi;小花山涧草■☆
87355 Chikusichloa mutica Keng;无芒山涧草(无芒涧草);Awnless Chikusichloa, Awnless Gullygrass ■
87356 Childsia Childs = Hidalgoa La Llave ■☆
87357 Chilechium Pfeiff. = Chilochium Raf. ■☆
87358 Chilechium Pfeiff. = Echiochilon Desf. ■☆
87359 Chilenia Backeb. (1935) = Neoporteria Britton et Rose ●■
87360 Chilenia Backeb. (1938) = Neoporteria Britton et Rose ●■
87361 Chilenia Backeb. (1939) = Neoporteria Britton et Rose ●■
87362 Chilenia Backeb. (1939) = Nichelia Bullock ●■
87363 Chileniopsis Backeb. = Neoporteria Britton et Rose ●■
87364 Chileocactus Frič = Horridocactus Backeb. ■☆
87365 Chileocactus Frič = Neoporteria Britton et Rose ●■
87366 Chileorchis Szlach. = Chloraea Lindl. ■☆
87367 Chileorebutia F. Ritter = Pyrrhocactus (A. Berger) Backeb. et F. M. Knuth ●■
87368 Chileorebutia Frič = Neoporteria Britton et Rose ●■
87369 Chileranthemum Oerst. (1854);智利喜花草属■☆
87370 Chileranthemum trifidum Oerst.;智利喜花草■☆
87371 Chiliadenus Cass. (1825);千腺菊属(干腺菊属)■●☆
87372 Chiliadenus Cass. = Jasonia (Cass.) Cass. ■☆
87373 Chiliadenus antiatlanticus (Emb. et Maire) Gómiz = Jasonia glutinosa (L.) DC. subsp. antiatlantica (Emb. et Maire) Dobignard ■☆
87374 Chiliadenus candicans (Delile) Brullo;纯白千腺菊■☆
87375 Chiliadenus glutinosus (L.) Fourr. = Jasonia glutinosa (L.) DC. ■☆
87376 Chiliadenus hesperius (Maire et Wilczek) Brullo = Varthemia hesperia (Maire et Wilczek) Dobignard ■☆
87377 Chiliadenus montanus (Vahl) Brullo = Jasonia montana (Delile) Botsch. ■☆
87378 Chiliadenus rupestris (Pomel) Brullo;岩地千腺菊■☆
87379 Chiliadenus saxatilis (Lam.) Brullo;岩生千腺菊■☆
87380 Chiliadenus sericeus (Batt. et Trab.) Brullo;绢毛千腺菊■☆
87381 Chiliandra Griff. = Rhynchotechum Blume ●
87382 Chiliandra obovata Griff. = Rhynchotechum ellipticum (Wall. ex D. Dietr.) A. DC. ●
87383 Chiliandra obovata Griff. = Rhynchotechum obovatum (Griff.) B. L. Burtt ●
87384 Chilianthus Burch. (1822);千花醉鱼草属●☆
87385 Chilianthus Burch. = Buddleja L. ●■
87386 Chilianthus arboreus (L. f.) A. DC. = Buddleja saligna Willd. ●☆

87387 Chilianthus arboreus (L. f.) A. DC. var. rosmarinaceus Kuntze = Buddleja saligna Willd. ●☆
87388 Chilianthus corrugatus (Benth.) A. DC. = Buddleja loricata Leeuwenb. ●☆
87389 Chilianthus dysophyllus (Benth.) A. DC. = Buddleja dysophylla (Benth.) Radlk. ●☆
87390 Chilianthus dysophyllus (Benth.) A. DC. var. rufescens Sond. = Buddleja dysophylla (Benth.) Radlk. ●☆
87391 Chilianthus lobulatus (Benth.) A. DC. = Buddleja glomerata H. L. Wendl. ●☆
87392 Chilianthus oleaceus Burch. = Buddleja saligna Willd. ●☆
87393 Chiliocephalum Benth.(1873);千头草属(光果金绒草属,千头属)■☆
87394 Chiliocephalum schimperi Benth.;光果金绒草■☆
87395 Chiliocephalum tegetum Mesfin;千头草■☆
87396 Chiliophyllum DC.(废弃属名)= Chiliophyllum Phil.(保留属名)●☆
87397 Chiliophyllum DC.(废弃属名)= Hybridella Cass. ●☆
87398 Chiliophyllum DC.(废弃属名)= Zaluzania Pers. ■☆
87399 Chiliophyllum Phil.(1864)(保留属名);黄帚菀属●☆
87400 Chiliophyllum densifolium Phil.;黄帚菀●☆
87401 Chiliorebutia Frič = Neoporteria Britton et Rose ●■
87402 Chiliotrichiopsis Cabrera(1937);胶帚菀属●☆
87403 Chiliotrichiopsis keideli Cabrera;胶帚菀●☆
87404 Chiliotrichum Cass.(1817);绒帚菀属●☆
87405 Chiliotrichum diffusurn (G. Forst.) Kuntze;铺散绒帚菀●☆
87406 Chilita Orcutt = Mammillaria Haw.(保留属名)●
87407 Chillania Roiv. = Eleocharis R. Br. ■
87408 Chilmoria Buch. -Ham. = Gynocardia R. Br. ●
87409 Chilmoria dodecandra Buch. -Ham. = Gynocardia odorata R. Br. ●◇
87410 Chilmoria odorata Buch. -Ham. = Gynocardia odorata R. Br. ●◇
87411 Chilocalyx Hook. f. = Chylocalyx Hassk. ex Miq. ●■
87412 Chilocalyx Hook. f. = Echinocaulos (Meisn. ex Endl.) Hassk. ●■
87413 Chilocalyx Klotzsch = Cleome L. ●■
87414 Chilocalyx Turcz. = Atalantia Corrêa(保留属名)●
87415 Chilocalyx Turcz. = Pamburus Swingle ●☆
87416 Chilocalyx macrophyllus Klotzsch = Cleome macrophylla (Klotzsch) Briq. ■☆
87417 Chilocalyx maculatus (Sond.) Gilg et Gilg-Ben. = Cleome maculata (Sond.) Szyszyl. ■☆
87418 Chilocalyx tenuifolius Klotzsch = Cleome macrophylla (Klotzsch) Briq. ■☆
87419 Chilocardamum O. E. Schulz = Sisymbrium L. ■
87420 Chilocardamum O. E. Schulz(1924);千碎荠属(巴塔哥尼亚芥属)■☆
87421 Chilocardamum longistylum (Romanczuk) Al-Shehbaz;长柱千碎荠■☆
87422 Chilocardamum patagonicum O. E. Schulz;千碎荠■☆
87423 Chilocarpus Blume(1823);唇果夹竹桃属●☆
87424 Chilocarpus atroviridis Blume;墨绿唇果夹竹桃●☆
87425 Chilocarpus australis F. Muell.;南方唇果夹竹桃●☆
87426 Chilocarpus cuneifolius Kerr;楔叶唇果夹竹桃●☆
87427 Chilocarpus globosus Elmer;球唇果夹竹桃●☆
87428 Chilocarpus gracilis Markgr.;纤细唇果夹竹桃●☆
87429 Chilocarpus hirtus D. J. Middleton;硬毛唇果夹竹桃●☆
87430 Chilocarpus minutiflorus King et Gamble;小花唇果夹竹桃●☆
87431 Chilochium Raf. = Echiochilon Desf. ■☆
87432 Chilochloa P. Beauv. = Phleum L. ■
87433 Chilochloa paniculata (Huds.) P. Beauv. = Phleum paniculatum Huds. ■
87434 Chilococca Post et Kuntze = Cheilococca Salisb. ●■☆
87435 Chilococca Post et Kuntze = Platylobium Sm. ●■☆
87436 Chilodia R. Br. = Prostanthera Labill. ●☆
87437 Chilodiscus Post et Kuntze = Cheilodiscus Triana ■☆
87438 Chilodiscus Post et Kuntze = Pectis L. ■☆
87439 Chilogloasa Oerst. = Justicia L. ●■
87440 Chiloglossa Oerst. = Dianthera L. ■☆
87441 Chiloglottis R. Br.(1810);喉唇兰属■☆
87442 Chiloglottis bifolia (Hook. f.) Schltr.;双叶喉唇兰■☆
87443 Chiloglottis platyptera D. L. Jones;宽翅喉唇兰■☆
87444 Chilopogon Schltr. = Appendicula Blume ■
87445 Chiloporus Naudin = Miconia Ruiz et Pav.(保留属名)●☆
87446 Chilopsis D. Don(1823);沙漠紫葳属(沙漠柳属,沙漠葳属);Desert Willow ●☆
87447 Chilopsis Post et Kuntze = Acanthus L. ●■
87448 Chilopsis Post et Kuntze = Cheilopsis Moq. ●■
87449 Chilopsis linearis (Cav.) Sweet;沙漠紫葳(沙漠柳);Desert Catalpa, Desert Willow, Flowering Willow ●☆
87450 Chilopsis linearis Sweet = Chilopsis linearis (Cav.) Sweet ●☆
87451 Chilosa Blume(1826);爪哇大戟属●☆
87452 Chilosa Post et Kuntze = Cheilosa Blume ●☆
87453 Chilosa montana Blume;爪哇大戟●☆
87454 Chilosandra Post et Kuntze = Cheilosandra Griff. ex Lindl. ●
87455 Chilosandra Post et Kuntze = Rhynchotechum Blume ●
87456 Chiloschista Lindl.(1832);异唇兰属(大蜘蛛兰属,异型兰属);Chilosehista ■
87457 Chiloschista guangdongensis Z. H. Tsi;广东异型兰;Guangdong Chilosehista ■
87458 Chiloschista hoi S. S. Yin;何氏梅兰■
87459 Chiloschista hoi S. S. Ying = Chiloschista segawai (Masam.) Masam. et Fukuy. ■
87460 Chiloschista lunifera (Rchb. f.) J. J. Sm.;短蕊柱异唇兰;Moon Chilosehista ■
87461 Chiloschista lunifera (Rchb. f.) J. J. Sm. = Chiloschista yunnanensis Schltr. ■
87462 Chiloschista pusilla (Willd.) Schltr. = Taeniophyllum pusillum (Willd.) Seidenf. et Ormerod ■
87463 Chiloschista segawae (Masam.) Masam. et Fukuy.;台湾异型兰(大蜘蛛兰,梅兰,小花台湾异型兰);Taiwan Chilosehista ■
87464 Chiloschista segawae (Masam.) Masam. et Fukuy. f. taiwaniana (S. S. Ying) S. S. Ying = Chiloschista segawae (Masam.) Masam. et Fukuy. ■
87465 Chiloschista segawae (Masam.) Masam. et Fukuy. var. taiwaniana S. S. Ying = Chiloschista segawae (Masam.) Masam. et Fukuy. ■
87466 Chiloschista segawae (Masam.) S. S. Ying = Chiloschista segawae (Masam.) Masam. et Fukuy. ■
87467 Chiloschista segawae (Masam.) S. S. Ying var. taiwaniana S. S. Ying = Chiloschista segawae (Masam.) Masam. et Fukuy. ■
87468 Chiloschista usneoides (D. Don) Lindl. = Chiloschista yunnanensis Schltr. ■
87469 Chiloschista yunnanensis Schltr.;异型兰;Yunnan Chilosehista ■
87470 Chilostigma Hochst. = Aptosimum Burch. ex Benth.(保留属名)■●☆

87471 Chilostigma pumilum Hochst. = Aptosimum pumilum (Hochst.) Benth. ■☆

87472 Chiloterus D. L. Jones et M. A. Clem. = Prasophyllum R. Br. ■☆

87473 Chilotheca Post et Kuntze = Cheilotheca Hook. f. ■

87474 Chilyathum Post et Kuntze = Oncidium Sw.(保留属名)■☆

87475 Chilyathum Post et Kuntze = Xeilyathum Raf. ■☆

87476 Chilyctis Post et Kuntze = Cheilyctis (Raf.) Spach ■

87477 Chilyctis Post et Kuntze = Monarda L. ■

87478 Chimantaea Maguire, Steyerm. et Wurdack(1957);直瓣菊属●☆

87479 Chimanthus Raf. = Lauro-Cerasus Duhamel ●

87480 Chimaphila Pursh(1814);喜冬草属(爱冬叶属,梅笠草属);Ground Holly, Pipsissewa, Prince's Pine, Waxflower, Wintergreen ●■

87481 Chimaphila astyla Maxim. = Chimaphila japonica Miq. ■

87482 Chimaphila corymbosa Pursh = Chimaphila umbellata (L.) W. P. C. Barton subsp. cisatlantica (S. F. Blake) Hultén ■☆

87483 Chimaphila corymbosa Pursh = Chimaphila umbellata (L.) W. P. C. Barton ■

87484 Chimaphila fukuyamai Masam. = Chimaphila japonica Miq. ■

87485 Chimaphila japonica Miq.;喜冬草(罗汉草,梅笠草,日本爱冬叶,喜冬树,细叶鹿蹄草);Japan Pipsissewa, Japanese Pipsissewa ■

87486 Chimaphila japonica Miq. var. taiwaniana (Masam.) C. F. Hsieh = Chimaphila monticola Andres subsp. taiwaniana (Masam.) Hid. Takah. ■

87487 Chimaphila japonica Miq. var. taiwaniana (Masam.) C. F. Hsieh = Chimaphila monticola Andres ■

87488 Chimaphila maculata (L.) Pursh;斑点喜冬草(斑点梅笠草);Spotted Wintergreen, Striped Pipsissewa, Striped Wintergreen, Twinflower, William-and-Mary ■

87489 Chimaphila menziesii R. Br. et D. Don;孟席斯梅笠草(墨叙梅笠草,墨叙喜冬草);Little Pipsissewa, Menzies Pipsissewa ■

87490 Chimaphila monticola Andres;川西喜冬草(川西梅笠草,台湾爱冬叶);Montane Pipsissewa ■

87491 Chimaphila monticola Andres subsp. taiwaniana (Masam.) Hid. Takah.;台湾喜冬草(台湾爱冬叶)■

87492 Chimaphila monticola Andres subsp. taiwaniana (Masam.) Hid. Takah. = Chimaphila monticola Andres ■

87493 Chimaphila monticola Andres var. taiwaniana (Masam.) S. S. Ying = Chimaphila monticola Andres ■

87494 Chimaphila rhombifolia Hayata = Moneses uniflora (L.) A. Gray ■

87495 Chimaphila taiwaniana Masam. = Chimaphila monticola Andres ■

87496 Chimaphila umbellata (L.) DC. ex Hegi = Chimaphila umbellata (L.) W. P. C. Barton ■

87497 Chimaphila umbellata (L.) Nutt. = Chimaphila umbellata (L.) W. P. C. Barton ■

87498 Chimaphila umbellata (L.) W. P. C. Barton;伞形喜冬草(伞形梅笠草);Butter Winter, Common Pipsissewa, Love-in-winter, Pipsissewa, Prince's Pine, Princespine, Prince's-pine, Rheumatism Weed, Umbellate Wintergreen ■

87499 Chimarhis Raf. = Chimarrhis Jacq. ■☆

87500 Chimarrhis Jacq.(1763);急流茜属■☆

87501 Chimarrhis barbata (Ducke) Bremek.;髯毛急流茜■☆

87502 Chimarrhis ferruginea Standl.;锈色急流茜■☆

87503 Chimarrhis latifolia Standl.;宽叶急流茜■☆

87504 Chimarrhis microcarpa Standl.;小果急流茜■☆

87505 Chimaza R. Br. ex DC. = Chimaphila Pursh ●■

87506 Chimborazoa H. T. Beck = Paullinia L. ●☆

87507 Chimocarpus Baill. = Chymocarpus D. Don ■

87508 Chimocarpus Baill. = Tropaeolum L. ■

87509 Chimonanthaceae Perleb = Calycanthaceae Lindl.(保留科名)●

87510 Chimonanthus Lindl.(1819)(保留属名);蜡梅属(腊梅属);Wintersweet ●★

87511 Chimonanthus baokanensis D. M. Chen et Z. I. Dai = Chimonanthus praecox (L.) Link ●

87512 Chimonanthus baokanensis D. M. Chen et Z. I. Dai var. yupiensis D. M. Chen et Z. I. Dai = Chimonanthus praecox (L.) Link ●

87513 Chimonanthus caespitosus T. B. Chao, Z. X. Chen et Z. Q. Li;簇花蜡梅;Fascicleflower Wintersweet ●

87514 Chimonanthus caespitosus T. B. Chao, Z. X. Chen et Z. Q. Li = Chimonanthus praecox (L.) Link ●

87515 Chimonanthus campanulatus R. H. Chang et C. S. Ding;西南蜡梅;Bellshaped Wintersweet, Bell-shaped Wintersweet ●

87516 Chimonanthus campanulatus R. H. Chang et C. S. Ding var. guizhouensis R. H. Chang = Chimonanthus campanulatus R. H. Chang et C. S. Ding ●

87517 Chimonanthus fragrans Lindl. = Chimonanthus praecox (L.) Link ●

87518 Chimonanthus fragrans Lindl. var. grandiflorus Lindl. = Chimonanthus praecox (L.) Link ●

87519 Chimonanthus fragrans Lindl. var. lutens Bean = Chimonanthus praecox (L.) Link ●

87520 Chimonanthus grammatus M. C. Liu;突托蜡梅;Bare-receptacled Wintersweet ●

87521 Chimonanthus nitens Oliv.;山蜡梅(白花蜡梅,臭蜡梅,亮叶蜡梅,毛山茶,秋蜡梅,山腊梅,香风茶,雪里花,野蜡梅,野蜡梅花);Shining Wintersweet ●

87522 Chimonanthus nitens Oliv. var. ovatus T. B. Chao et Z. Q. Li = Chimonanthus nitens Oliv. ●

87523 Chimonanthus nitens Oliv. var. salicifolius (S. Y. Hu) H. D. Zhang = Chimonanthus salicifolius S. Y. Hu ●

87524 Chimonanthus parviflorus Raf. = Chimonanthus praecox (L.) Link ●

87525 Chimonanthus praecox (L.) Link;蜡梅(臭腊梅,大叶蜡梅,狗蝇梅,荷花蜡梅,黄腊梅,黄梅花,金黄茶,九英梅,腊梅,腊梅花,腊木,蜡花,蜡梅花,蜡木,磬口梅,石凉茶,素心蜡梅,檀香梅,铁钢叉,铁筷子,瓦乌柴,雪里花,岩马桑,钻石风);Chinese Snowdrop Tree, Chinese Snowdrop-tree, Fragrant Wintersweet, Japanese Allspice, Winter Sweet, Wintersweet, Winter-sweet ●

87526 Chimonanthus praecox (L.) Link 'Concolor';素心蜡梅(素心腊梅,同色蜡梅);Concolor Wintersweet ●

87527 Chimonanthus praecox (L.) Link 'Grandiflora-concolor';磬口素心蜡梅;Grandiflower-concolor Wintersweet ●

87528 Chimonanthus praecox (L.) Link 'Grandiflorus';磬口蜡梅(磬口腊梅);Grandiflower Wintersweet ●

87529 Chimonanthus praecox (L.) Link 'Luteus' = Chimonanthus praecox (L.) Link 'Concolor' ●

87530 Chimonanthus praecox (L.) Link 'Parviflorus';小花蜡梅●☆

87531 Chimonanthus praecox (L.) Link f. concolor (Makino) Makino = Chimonanthus praecox (L.) Link 'Concolor' ●

87532 Chimonanthus praecox (L.) Link f. intermedius (Makino) Okuyama;狗蝇蜡梅(狗牙腊梅,狗牙蜡梅,狗英梅);Intermediate Wintersweet ●

87533 Chimonanthus praecox (L.) Link f. luteus (Makino) Okuyama = Chimonanthus praecox (L.) Link 'Concolor' ●

87534 Chimonanthus praecox (L.) Link var. concolor Makino =

Chimonanthus praecox (L.) Link ●

87535 Chimonanthus praecox (L.) Link var. concolor Makino = Chimonanthus praecox (L.) Link 'Concolor' ●

87536 Chimonanthus praecox (L.) Link var. grandiflorus (Lindl.) Makino = Chimonanthus praecox (L.) Link ●

87537 Chimonanthus praecox (L.) Link var. grandiflorus (Lindl.) Makino = Chimonanthus praecox (L.) Link 'Grandiflorus' ●

87538 Chimonanthus praecox (L.) Link var. intermedius Makino = Chimonanthus praecox (L.) Link ●

87539 Chimonanthus praecox (L.) Link var. intermedius Makino = Chimonanthus praecox (L.) Link f. intermedius (Makino) Okuyama ●

87540 Chimonanthus praecox (L.) Link var. pilosus L. Q. Chen. = Chimonanthus salicifolius S. Y. Hu ●

87541 Chimonanthus praecox (L.) Link var. reflexus B. Zhao = Chimonanthus praecox (L.) Link ●

87542 Chimonanthus salicifolius S. Y. Hu;柳叶蜡梅(柳叶腊梅,毛山茶,秋蜡梅,山蜡梅,香风茶);Willowleaf Wintersweet, Willow-leaved Wintersweet ●

87543 Chimonanthus yunnanensis W. W. Sm.;云南蜡梅●

87544 Chimonanthus yunnanensis W. W. Sm. = Chimonanthus praecox (L.) Link ●

87545 Chimonanthus zhejiangensis M. C. Liu;浙江蜡梅(石凉茶);Zhejiang Wintersweet ●

87546 Chimoncalamus bicorniculatus S. F. Li et Z. P. Wang = Chimonocalamus pallens J. R. Xue et T. P. Yi ●

87547 Chimonobambusa Makino(1914);方竹属(寒竹属,四方竹属);Bamboo, Square Bamboo, Squarebamboo, Square-bamboo, Square-stemmed Bamboo ●

87548 Chimonobambusa angulata (Munro) Nakai = Bambusa tuldoides Munro ●

87549 Chimonobambusa angulata Nakai = Chimonobambusa quadrangularis (Franch.) Makino ex Nakai ●

87550 Chimonobambusa angustifolia C. D. Chu et C. S. Chao;狭叶方竹;Narrowleaf Squarebamboo, Narrow-leaved Square-bamboo ●

87551 Chimonobambusa armata (Gamble) J. R. Xue et T. P. Yi;缅甸方竹;Burma Square Bamboo, Burma Square-bamboo ●

87552 Chimonobambusa armata (Gamble) J. R. Xue et T. P. Yi f. tuberculata (J. R. Xue et L. Z. Gao) T. H. Wen ex Ohrnb. = Chimonobambusa tuberculata J. R. Xue et L. Z. Gao ●

87553 Chimonobambusa brevinoda J. R. Xue et W. P. Zhang;短节方竹;Shortnode Square Bamboo, Short-noded Square-bamboo ●

87554 Chimonobambusa communis (J. R. Xue et T. P. Yi) T. H. Wen et Ohrnb. = Chimonobambusa communis (J. R. Xue et T. P. Yi) T. H. Wen et Ohrnb. ex Ohrnb. ●

87555 Chimonobambusa communis (J. R. Xue et T. P. Yi) T. H. Wen et Ohrnb. ex Ohrnb.;平竹;Common Qiongzhu, Common Swollennoded Cane ●

87556 Chimonobambusa communis (J. R. Xue et T. P. Yi) T. H. Wen et Ohrnb. ex Ohrnb. = Qiongzhuea communis J. R. Xue et T. P. Yi ●

87557 Chimonobambusa convoluta Q. H. Dai et X. L. Tao;小方竹;Convolule Square-bamboo, Little Square Bamboo, Small Squarebamboo ●

87558 Chimonobambusa damingshanensis J. R. Xue et W. P. Zhang;大明山方竹;Damingshan Square Bamboo, Damingshan Square-bamboo ●

87559 Chimonobambusa falcata (Nees) Nakai = Arundinaria falcata Nees ●☆

87560 Chimonobambusa grandifolia J. R. Xue et W. P. Zhang;大叶方竹;Bigleaf Square Bamboo, India Aromatic Bamboo, Large-leaved Square-bamboo ●

87561 Chimonobambusa griffithiana (Munro) Nakai = Chimonocalamus griffithianus (Munro) J. R. Xue et T. P. Yi ●

87562 Chimonobambusa hejiangensis C. D. Chu et C. S. Chao;合江方竹;Hejiang Square Bamboo, Hejiang Square-bamboo ●

87563 Chimonobambusa hirtinoda C. S. Chao et K. M. Lan;毛环方竹;Hairnode Squarebamboo, Hispid-node Square Bamboo, Hispid-noded Square-bamboo ●

87564 Chimonobambusa hsuehiana D. Z. Li et H. Q. Yang;细秆筇竹(冷水竹,冷竹,细竿筇竹);Intermediate Swollennoded Cane, Thinculm Qiongzhu ●

87565 Chimonobambusa lactistriata W. D. Li et Q. X. Wu;乳环方竹(乳纹方竹);Lactistriate Square Bamboo, Milkline Squarebamboo, White-striped Square-bamboo ●

87566 Chimonobambusa leishanensis T. P. Yi;雷山方竹;Leishan Fragrant-bamboo, Leishan Square-bamboo ●

87567 Chimonobambusa linearifolia W. D. Li et Q. X. Wu;线叶方竹(狭叶方竹);Linearleaf Squarebamboo ●

87568 Chimonobambusa linearifolia W. D. Li et Q. X. Wu = Chimonobambusa angustifolia C. D. Chu et C. S. Chao ●

87569 Chimonobambusa luzhiensis (J. R. Xue et T. P. Yi) T. H. Wen et Ohrnb. = Chimonobambusa luzhiensis (J. R. Xue et T. P. Yi) T. H. Wen et Ohrnb. ex Ohrnb. ●

87570 Chimonobambusa luzhiensis (J. R. Xue et T. P. Yi) T. H. Wen et Ohrnb. ex Ohrnb.;光竹;Luzhi Swollennoded Cane, Smooth Qiongzhu ●

87571 Chimonobambusa luzhiensis (J. R. Xue et T. P. Yi) T. H. Wen et Ohrnb. ex Ohrnb. = Qiongzhuea luzhiensis J. R. Xue et T. P. Yi ●

87572 Chimonobambusa macrophylla (J. R. Xue et T. P. Yi) T. H. Wen et Ohrnb. = Chimonobambusa macrophylla (J. R. Xue et T. P. Yi) T. H. Wen et Ohrnb. ex Ohrnb. ●

87573 Chimonobambusa macrophylla (J. R. Xue et T. P. Yi) T. H. Wen et Ohrnb. = Chimonobambusa macrophylla (J. R. Xue et T. P. Yi) T. H. Wen et Ohrnb. ex Ohrnb. f. leiboensis (J. R. Xue et T. P. Yi) T. H. Wen et Ohrnb. ex Ohrnb. ●

87574 Chimonobambusa macrophylla (J. R. Xue et T. P. Yi) T. H. Wen et Ohrnb. ex Ohrnb.;大叶筇竹;Bigleaf Qiongzhu, Largeleaf Swollennoded Cane, Large-leaved Swollennoded Cane ●

87575 Chimonobambusa macrophylla (J. R. Xue et T. P. Yi) T. H. Wen et Ohrnb. ex Ohrnb. f. leiboensis (J. R. Xue et T. P. Yi) T. H. Wen et Ohrnb. ex Ohrnb.;雷波大叶筇竹;Leibo Qiongzhu, Swollennoded Cane ●

87576 Chimonobambusa macrophylla (J. R. Xue et T. P. Yi) T. H. Wen et Ohrnb. ex Ohrnb. = Qiongzhuea macrophylla J. R. Xue et T. P. Yi ●

87577 Chimonobambusa macrophylla (J. R. Xue et T. P. Yi) T. H. Wen et Ohrnb. ex Ohrnb. f. leiboensis (J. R. Xue et T. P. Yi) T. H. Wen et Ohrnb. ex Ohrnb. = Qiongzhuea macrophylla J. R. Xue et T. P. Yi f. leiboensis J. R. Xue et D. Z. Li ●

87578 Chimonobambusa macrophylla (J. R. Xue et T. P. Yi) T. H. Wen et Ohrnb. f. intermedia T. H. Wen et Ohrnb. = Chimonobambusa hsuehiana D. Z. Li et H. Q. Yang ●

87579 Chimonobambusa maculata (T. H. Wen) T. H. Wen = Chimonobambusa opienensis (J. R. Xue et T. P. Yi) T. H. Wen et Ohrnb. ex Ohrnb. ●

87580 Chimonobambusa marmorea (Mitford) Makino = Chimonobambusa marmorea (Mitford) Makino ex Nakai ●◇

87581 Chimonobambusa marmorea (Mitford) Makino ex Nakai;寒竹

(观音竹,黑刺竹);Coldbamboo,Kan Chiku,Marble Bamboo,Marbled Bamboo,Mottled Bamboo ●◇

87582 Chimonobambusa marmorea (Mitford) Makino ex Nakai 'Variegata';白条寒竹●☆

87583 Chimonobambusa marmorea (Mitford) Makino f. variegata Ohwi = Chimonobambusa marmorea (Mitford) Makino 'Variegata' ●

87584 Chimonobambusa marmorea (Mitford) Makino var. variegata Makino = Chimonobambusa marmorea (Mitford) Makino f. variegata Ohwi ●☆

87585 Chimonobambusa marmorea Makino = Chimonobambusa marmorea (Mitford) Makino ex Nakai ●◇

87586 Chimonobambusa metuoensis J. R. Xue et T. P. Yi;墨脱方竹;Medog Square-bamboo,Motuo Square Bamboo ●

87587 Chimonobambusa microfloscula McClure;小花方竹;Little-flower Square Bamboo,Smallflower Squarebamboo,Small-flowered Square-bamboo,Small-spiked Square-bamboo ●

87588 Chimonobambusa montigena (T. P. Yi) Ohrnb.;荆竹;Swollennoded Cane Mountain ●

87589 Chimonobambusa naibunense (Hayata) McClure et W. C. Lin = Drepanostachyum naibunense (Hayata) P. C. Keng ●

87590 Chimonobambusa naibunensis (Hayata) McClure et W. C. Lin = Ampelocalamus naibunensis (Hayata) T. H. Wen ●

87591 Chimonobambusa neopurpurea T. P. Yi = Chimonobambusa purpurea J. R. Xue et T. P. Yi ●

87592 Chimonobambusa ningnanica J. R. Xue et L. Z. Gao;宁南方竹;Ningnan Square-bamboo ●

87593 Chimonobambusa ningnanica J. R. Xue et L. Z. Gao = Chimonobambusa tuberculata J. R. Xue et L. Z. Gao ●

87594 Chimonobambusa opienensis (J. R. Xue et T. P. Yi) T. H. Wen et Ohrnb. = Chimonobambusa opienensis (J. R. Xue et T. P. Yi) T. H. Wen et Ohrnb. ex Ohrnb. ●

87595 Chimonobambusa opienensis (J. R. Xue et T. P. Yi) T. H. Wen et Ohrnb. ex Ohrnb.;三月竹(湖南冷竹);March Qiongzhu,Opien Swollennoded Cane,Hunan Qiongzhu,Hunan Swollennoded Cane ●

87596 Chimonobambusa opienensis (J. R. Xue et T. P. Yi) T. H. Wen et Ohrnb. ex Ohrnb. = Qiongzhuea opienensis J. R. Xue et T. P. Yi ●

87597 Chimonobambusa pachystachys J. R. Xue et T. P. Yi;刺竹子;Spine Square-bamboo,Thick-spike Square Bamboo ●

87598 Chimonobambusa paucispinosa T. P. Yi;少刺方竹;Few-spiny Fragrant-bamboo ●

87599 Chimonobambusa puberula (J. R. Xue et T. P. Yi) T. H. Wen et Ohrnb. = Chimonobambusa puberula (J. R. Xue et T. P. Yi) T. H. Wen et Ohrnb. ex Ohrnb. ●

87600 Chimonobambusa puberula (J. R. Xue et T. P. Yi) T. H. Wen et Ohrnb. ex Ohrnb.;柔毛筇竹;Hairy Swollennoded Cane,Puberulent Swollennoded Cane,Soft Hair Qiongzhu ●

87601 Chimonobambusa puberula (J. R. Xue et T. P. Yi) T. H. Wen et Ohrnb. ex Ohrnb. = Qiongzhuea puberula J. R. Xue et T. P. Yi ●

87602 Chimonobambusa pubescens T. H. Wen;十月寒竹;Pubescent Squarebamboo ●

87603 Chimonobambusa purpurea J. R. Xue et T. P. Yi;刺黑竹(刺刺竹,刺竹子,牛尾竹);Deeppurple Square Bamboo,Purple Square-bamboo,Spine Squarebamboo ●

87604 Chimonobambusa purpurea J. R. Xue et T. P. Yi = Chimonobambusa marmorea (Mitford) Makino ●

87605 Chimonobambusa purpurea J. R. Xue et T. P. Yi = Chimonobambusa neopurpurea T. P. Yi ●

87606 Chimonobambusa quadrangularis (Franceschi) Makino = Chimonobambusa quadrangularis (Franceschi) Makino ex Nakai ●

87607 Chimonobambusa quadrangularis (Franceschi) Makino ex Nakai;方竹(标竹,方苦竹,角竹,箬竹,四方竹,四季竹,四角竹,疣竹,棕竹);Square Bamboo,Square-bamboo,Square-culmed Bamboo,Square-stem Bamboo,Square-stemmed Bamboo ●

87608 Chimonobambusa rigidula (J. R. Xue et T. P. Yi) T. H. Wen et Ohrnb. = Chimonobambusa rigidula (J. R. Xue et T. P. Yi) T. H. Wen et Ohrnb. ex Ohrnb. ●

87609 Chimonobambusa rigidula (J. R. Xue et T. P. Yi) T. H. Wen et Ohrnb. ex Ohrnb.;实竹子;Firm Qiongzhu,Rigid Swollennoded Cane ●

87610 Chimonobambusa rigidula (J. R. Xue et T. P. Yi) T. H. Wen et Ohrnb. ex Ohrnb. = Qiongzhuea rigidula J. R. Xue et T. P. Yi ●

87611 Chimonobambusa rivularis T. P. Yi = Chimonobambusa lactistriata W. D. Li et Q. X. Wu ●

87612 Chimonobambusa setiformis T. H. Wen;武夷山方竹;Setiform Square-bamboo,Wuyishan Square Bamboo ●

87613 Chimonobambusa setiformis T. H. Wen = Chimonobambusa marmorea (Mitford) Makino ex Nakai ●◇

87614 Chimonobambusa sichuanensis (T. P. Yi) T. H. Wen = Monstruocalamus sichuanensis (T. P. Yi) T. P. Yi ●

87615 Chimonobambusa solida B. M. Yang et C. Y. Zhang = Chimonobambusa pubescens T. H. Wen ●

87616 Chimonobambusa szechuanensis (Rendle) P. C. Keng;八月竹(川方竹,瓦山方竹);August Squarebamboo,Sichuan Square Bamboo,Sichuan Square-bamboo,Szechwan Square Bamboo ●◇

87617 Chimonobambusa szechuanensis (Rendle) P. C. Keng f. flexuosa (J. R. Xue et C. Li) T. H. Wen et Ohrnb. = Chimonobambusa szechuanensis (Rendle) P. C. Keng var. flexuosa J. R. Xue et C. Li ●

87618 Chimonobambusa szechuanensis (Rendle) P. C. Keng var. flexuosa J. R. Xue et C. Li;龙拐竹;Curved Sichuan Square Bamboo ●

87619 Chimonobambusa tortuosa J. R. Xue et T. P. Yi = Chimonocalamus griffithianus (Munro) J. R. Xue et T. P. Yi ●

87620 Chimonobambusa tuberculata J. R. Xue et L. Z. Gao;永善方竹;Tuberculate Square Bamboo,Yongshan Squarebamboo,Yongshan Square-bamboo ●

87621 Chimonobambusa tumidinoda J. R. Xue et T. P. Yi ex Ohrnb. = Qiongzhuea tumidinoda (Ohrnb.) J. R. Xue et T. P. Yi ●

87622 Chimonobambusa tumidissinoda J. R. Xue et T. P. Yi ex Ohrnb.;筇竹(罗汉竹);Qiongzhu,Swollennoded Cane ●◇

87623 Chimonobambusa tumidissinoda J. R. Xue et T. P. Yi ex Ohrnb. = Qiongzhuea tumidinoda (Ohrnb.) J. R. Xue et T. P. Yi ●

87624 Chimonobambusa unifolia (T. P. Yi) T. H. Wen et Ohrnb.;半边罗汉竹;Unifoliate Swollennoded Cane ●

87625 Chimonobambusa utilis (Keng) P. C. Keng;金佛山方竹;Chinfu Mountain Square Bamboo,Chinfushan Square-bamboo,Jinfoshan Square Bamboo ●

87626 Chimonobambusa verruculosa (T. P. Yi) T. H. Wen et Ohrnb.;瘤箨筇竹;Verruculose Swollennoded Cane ●

87627 Chimonobambusa yunnanensis J. R. Xue et W. P. Zhang;云南方竹;Yunnan Square Bamboo,Yunnan Square-bamboo ●

87628 Chimonobambusa yunnanensis J. R. Xue et W. P. Zhang = Chimonobambusa ningnanica J. R. Xue et L. Z. Gao ●

87629 Chimonocalamus J. R. Xue et T. P. Yi = Sinarundinaria Nakai ●

87630 Chimonocalamus J. R. Xue et T. P. Yi(1979);香竹属;Fragrant Bamboo,Fragrantbamboo,Fragrant-bamboo ●

87631 Chimonocalamus bicornicalamus S. Y. Li et Z. P. Wang;角香竹●

87632 Chimonocalamus burmaensis (C. S. Chao et Renv.) D. Z. Li;缅甸香竹;Burmese Bamboo ●☆

87633 Chimonocalamus delicatus J. R. Xue et T. P. Yi;香竹;Aromatic Bamboo, Chinese Fragrant Bamboo, Delicate Bamboo, Delicate Fragrant-bamboo, Fragrant Bamboo, Fragrantbamboo ●

87634 Chimonocalamus dumosus J. R. Xue et T. P. Yi;小香竹(刺竹,香竹);Little Fragrant Bamboo, Shrubby Fragrant-bamboo, Small Aromatic Bamboo, Small Fragrantbamboo, Small Fragrant-bamboo ●

87635 Chimonocalamus dumosus J. R. Xue et T. P. Yi var. pygmaeus J. R. Xue et T. P. Yi;耿马小香竹(耿马香竹,小刺竹);Pygny Little Fragrant Bamboo ●

87636 Chimonocalamus fimbriatus J. R. Xue et T. P. Yi;流苏香竹(流苏小香竹,香竹);Fimbriate Fragrant Bamboo, Fimbriate Fragrant-bamboo, Tassel Fragrantbamboo ●

87637 Chimonocalamus griffithianus (Munro) J. R. Xue et T. P. Yi;西藏青篱竹●

87638 Chimonocalamus leishanensis T. P. Yi = Chimonobambusa leishanensis T. P. Yi ●

87639 Chimonocalamus longiligulatus J. R. Xue et T. P. Yi;长舌香竹;Longtongue Fragrant Bamboo, Long-tongue Fragrant-bamboo ●

87640 Chimonocalamus longiusculus J. R. Xue et T. P. Yi;长节香竹(香竹);Long-node Fragrant Bamboo, Long-noded Fragrant-bamboo ●

87641 Chimonocalamus makuanensis J. R. Xue et T. P. Yi;马关香竹(香竹);Maguan Fragrant Bamboo, Maguan Fragrant-bamboo ●

87642 Chimonocalamus montanus J. R. Xue et T. P. Yi;山香竹;Montane Fragrantbamboo, Mountain Fragrant Bamboo, Mountain Fragrant-bamboo ●

87643 Chimonocalamus pallens J. R. Xue et T. P. Yi;灰香竹(灰竹);Chinese Grey Bamboo, Grey Fragrant Bamboo, Pale Fragrant-bamboo ●

87644 Chimonocalamus quadranligulis (Fenzl) Makino f. purpureiculma T. H. Wen;紫秆方竹●

87645 Chimonocalamus recurva T. P. Yi;弯刺方竹●

87646 Chimonocalamus tortuosus J. R. Xue et T. P. Yi;西藏香竹;Tibet Aromatic Bamboo, Tibet Fragrant Bamboo, Xizang Fragrant Bamboo, Xizang Fragrant-bamboo ●

87647 Chimophila Radius = Chimaphila Pursh ●■

87648 Chincharronia A. Rich. = Chicharronia A. Rich. ●

87649 Chincharronia A. Rich. = Terminalia L. (保留属名)●

87650 Chinchona Howard = Cinchona L. ■●

87651 Chingiacanthus Hand. -Mazz. = Isoglossa Oerst. (保留属名)●★

87652 Chingiacanthus glaber Hand. -Mazz. = Isoglossa glaber (Hand. -Mazz.) B. Hansen ■

87653 Chingiacanthus patulus Hand. -Mazz. = Isoglossa collina (T. Anderson) B. Hansen ■

87654 Chingithamnaceae Hand. -Mazz. = Celastraceae R. Br. (保留科名)●

87655 Chingithamnus Hand. -Mazz. = Microtropis Wall. ex Meisn. (保留属名)●

87656 Chingithamnus osmanthoides Hand. -Mazz. = Microtropis osmanthoides (Hand. -Mazz.) Hand. -Mazz. ●

87657 Chingyungia T. M. Ai = Melampyrum L. ■

87658 Chingyungia scutellarioidea T. M. Ai = Melampyrum klebelsbergianum Soó ■

87659 Chiococca L. = Chiococca P. Browne ex L. ●☆

87660 Chiococca P. Browne = Chiococca P. Browne ex L. ●☆

87661 Chiococca P. Browne ex L. (1756);雪果木属;Milkberry ●☆

87662 Chiococca alba (L.) Hitchc.;白雪果木●☆

87663 Chiococca alba L. = Chiococca alba (L.) Hitchc. ●☆

87664 Chiococca axillaris Moc. et Sessé;墨西哥雪果木●☆

87665 Chiococca belizensis Lundell;伯明兹雪果木●☆

87666 Chiococca bermudiana S. Br.;百慕大雪果木●☆

87667 Chiococca cubensis Urb.;古巴雪果木●☆

87668 Chiogenes Salisb. (1843);伏地杜鹃属(伏地杜属);Chiogenes ●

87669 Chiogenes Salisb. = Gaultheria L. ●

87670 Chiogenes Salisb. et Torr. = Gaultheria L. ●

87671 Chiogenes Salisb. et Torr. = Glyciphylla Raf. ●

87672 Chiogenes Salisb. ex Torr. = Gaultheria L. ●

87673 Chiogenes hispidula (L.) Torr. et A. Gray;硬毛伏地杜鹃(葡根伏地杜);Hispid Chiogenes, Snowberry ●

87674 Chiogenes hispidula (L.) Torr. et A. Gray = Gaultheria hispidula (L.) Muhl. ex Bigelow ●☆

87675 Chiogenes hispidula (L.) Torr. et A. Gray subsp. japonica (A. Gray) T. Shimizu = Chiogenes japonica A. Gray ●☆

87676 Chiogenes hispidula (L.) Torr. et A. Gray var. japonica (A. Gray) Makino = Chiogenes japonica A. Gray ●☆

87677 Chiogenes japonica A. Gray = Gaultheria japonica (A. Gray) Sleumer ●☆

87678 Chiogenes serpyllifolia (Pursh) Salisb.;北美伏地杜鹃(百里香叶伏地杜);Maidenhair Barry ●☆

87679 Chiogenes suborbicularis (W. W. Sm.) Ching ex T. Z. Hsu;伏地杜鹃(伏地杜);Suborbicular Chiogenes ●

87680 Chiogenes suborbicularis (W. W. Sm.) Ching ex T. Z. Hsu = Gaultheria suborbicularis W. W. Sm. ●

87681 Chiogenes suborbicularis (W. W. Sm.) Ching ex T. Z. Hsu var. albiflorus T. Z. Hsu;白花伏地杜鹃;Whiteflower Suborbicular Chiogenes ●

87682 Chiogenes suborbicularis (W. W. Sm.) Ching ex T. Z. Hsu var. albiflorus T. Z. Hsu = Gaultheria suborbicularis W. W. Sm. ●

87683 Chionachne R. Br. (1838);葫芦草属■

87684 Chionachne barbata (Roxb.) R. Br. = Chionachne koenigii (Spreng.) Thwaites ■☆

87685 Chionachne barbata R. Br. = Chionachne koenigii (Spreng.) Thwaites ■☆

87686 Chionachne barbata R. Br. = Polytoca bracteata R. Br. ■

87687 Chionachne cyathopoda F. Muell.;饲料葫芦草■☆

87688 Chionachne koenigii (Spreng.) Thwaites;柯氏葫芦草■☆

87689 Chionachne koenigii Thwaites = Chionachne koenigii (Spreng.) Thwaites ■☆

87690 Chionachne massiei Balansa;葫芦草;Gourd grass, Mass Polytoca ■

87691 Chionachne massiei Balansa = Polytoca massiei (Balansa) Schenk ■

87692 Chionanthula Börner = Carex L. ■

87693 Chionanthus Gaertn. = Linociera Sw. ex Schreb. (保留属名)●

87694 Chionanthus L. (1753);流苏树属(流苏木属,牛金子属);Fringe-tree, Fringe Flower, Fringe Tree, Fringe-flower, Tasseltree, Fringetree ●

87695 Chionanthus Royen ex L. = Chionanthus L. ●

87696 Chionanthus africanus (Welw. ex Knobl.) Stearn;非洲流苏树●☆

87697 Chionanthus battiscombei (Hutch.) Stearn;巴蒂葫芦草■☆

87698 Chionanthus brachythyrsus (Merr.) P. S. Green;白枝流苏树●

87699 Chionanthus camptoneurus (Gilg et G. Schellenb.) Stearn;曲脉流苏树●☆

87700 Chionanthus chinensis Maxim. = Chionanthus retusus Lindl. et Paxton ●

87701 Chionanthus coreanus H. Lév. = Chionanthus retusus Lindl. et Paxton ●

87702 Chionanthus coriaceus (S. Vidal) Yuen P. Yang et S. Y. Lu;厚叶李榄●

87703 Chionanthus coriaceus (S. Vidal) Yuen P. Yang et S. Y. Lu = Chionanthus retusus Lindl. et Paxton ●

87704 Chionanthus duclouxii Hickel = Chionanthus retusus Lindl. et Paxton ●

87705 Chionanthus foveolatus (E. Mey.) Stearn;蜂窝流苏树●☆

87706 Chionanthus foveolatus (E. Mey.) Stearn subsp. major (I. Verd.) Stearn;大蜂窝流苏树●☆

87707 Chionanthus foveolatus (E. Mey.) Stearn subsp. tomentellus (I. Verd.) Stearn;绒毛蜂窝流苏树●☆

87708 Chionanthus guangxiensis B. M. Miao;广西流苏树(广西插柚紫,广西李榄);Guangxi Linociera, Guangxi Liolive ●

87709 Chionanthus guangxiensis B. M. Miao = Linociera guangxiensis (B. M. Miao) B. M. Miao ●

87710 Chionanthus hainanensis (Merr. et Chun) B. M. Miao;海南流苏树(海南李榄,尾叶李榄);Caudate-leaf Blackbark Linociera, Caudate-leaf Liolive, Hainan Linociera, Hainan Liolive ●

87711 Chionanthus hainanensis (Merr. et Chun) B. M. Miao = Linociera hainanensis Merr. et Chun ●

87712 Chionanthus henryanus P. S. Green = Linociera insignis C. B. Clarke ●

87713 Chionanthus leucocladus (Merr. et Chun) B. M. Miao;白枝李榄;White-branch Linociera, Whitebranch Liolive, White-branched Linociera ●

87714 Chionanthus leucocladus (Merr. et Chun) B. M. Miao = Chionanthus brachythyrsus (Merr.) P. S. Green ●

87715 Chionanthus leucocladus (Merr. et Chun) B. M. Miao = Linociera leucoclada Merr. et Chun ●

87716 Chionanthus longiflorus (H. L. Li) B. M. Miao;长花流苏树(长花李榄);Long-flower Linociera, Longflower Liolive, Long-flowered Linociera ●

87717 Chionanthus longiflorus (H. L. Li) B. M. Miao = Linociera longiflora H. L. Li ●

87718 Chionanthus luzoniacus Blume = Chionanthus ramiflorus Roxb. ●

87719 Chionanthus macrophyllus Blume = Chionanthus ramiflorus Roxb. ●

87720 Chionanthus mannii (Soler.) Stearn;曼氏流苏树●☆

87721 Chionanthus mannii (Soler.) Stearn subsp. congesta (Baker) Stearn;密集曼氏流苏树●☆

87722 Chionanthus mildbraedii (Gilg et G. Schellenb.) Stearn;米尔德流苏树●☆

87723 Chionanthus montanus Blume = Chionanthus henryanus P. S. Green ●

87724 Chionanthus niloticus (Oliv.) Stearn;尼罗河流苏树●☆

87725 Chionanthus peglerae (C. H. Wright) Stearn;佩格拉流苏树●☆

87726 Chionanthus ramiflorus Roxb.;枝花流苏树(黑皮插柚柴,红头李榄,毛萼李榄,枝花李榄);Blackbark Linociera, Blackbark Liolive, Black-barked Linociera, Hairy-sepal Blackbark Linociera, Hairy-sepal Liolive ●

87727 Chionanthus ramiflorus Roxb. = Linociera ramiflora (Roxb.) Wall. ex G. Don ●

87728 Chionanthus ramiflorus Roxb. var. grandiflora B. M. Miao;大花流苏树(大花李榄);Bigflower Linociera, Bigflower Liolive ●

87729 Chionanthus retusus Lindl. et Paxton;流苏树(流苏,炭栗树,铁树);Chinese Fringe Tree, Chinese Fringetree, Chinese Fringe-tree Tasseltree ●

87730 Chionanthus retusus Lindl. et Paxton var. coreanus (H. Lév.) Nakai = Chionanthus retusus Lindl. et Paxton ●

87731 Chionanthus retusus Lindl. et Paxton var. fauriei H. Lév. = Chionanthus retusus Lindl. et Paxton ●

87732 Chionanthus retusus Lindl. et Paxton var. serrulatus (Hayata) Koidz. = Chionanthus retusus Lindl. et Paxton ●

87733 Chionanthus retusus Lindl. et Paxton var. serrulatus Koidz. = Chionanthus retusus Lindl. et Paxton ●

87734 Chionanthus richardsiae Stearn;理查兹流苏树●☆

87735 Chionanthus serrulatus Hayata = Chionanthus retusus Lindl. et Paxton ●

87736 Chionanthus virginicus L.;北美流苏树(美国流苏树,维州流苏树);American Fringetree, Fringe Tree, Fringetree, Fringe-tree, Grandsir-greybeard, North American Fringe Tree, Old Man's Beard, Oldman's Beard, Poison Ash, Snowdrop-tree, Snowflower, White Fringe Tree, White Fringetree ●☆

87737 Chione DC. (1830);雪茜属■☆

87738 Chione Salisb. = Narcissus L. ■

87739 Chione allenii L. O. Williams;阿伦雪茜■☆

87740 Chione buxifolia Dwyer et M. V. Hayden;黄杨叶雪茜■☆

87741 Chione glabra DC.;无毛雪茜■☆

87742 Chione lucida Griseb.;光亮雪茜■☆

87743 Chione mexicana Standl.;墨西哥雪茜■☆

87744 Chionice Bunge ex Ledeb. = Potentilla L. ■●

87745 Chionocarpium Brand = Adesmia DC. (保留属名)■☆

87746 Chionocharis I. M. Johnst. (1924);垫紫草属;Chionocharis ■

87747 Chionocharis hookeri (C. B. Clarke) I. M. Johnst.;垫紫草;Hooker Chionocharis ■

87748 Chionochlaena Post et Kuntze = Chionolaena DC. ●☆

87749 Chionochloa Zotov (1963);白穗茅属■☆

87750 Chionochloa conspicua (G. Forst.) Zotov;显著白穗茅;Hunangemoho-grass ■☆

87751 Chionochloa crassiuscula (Kirk) Zotov;白穗茅;Hunangemoho Grass ■☆

87752 Chionochloa flavescens Zotov;黄白穗茅;Dwarf Toitoi ■☆

87753 Chionochloa flavicans Zotov;小白穗茅;Dwarf Toitoi ■☆

87754 Chionochloa pallida (R. Br.) S. W. L. Jacobs;灰白穗茅;Dwarf Toitoi ■☆

87755 Chionochloa rigida (Raoul) Zotov;硬白穗茅■☆

87756 Chionochloa rubra Zotov;红白穗茅;Dwarf Toitoi ■☆

87757 Chionodoxa Boiss. (1844);雪光花属(雪百合属,雪宝花属,雪花百合属);Chionodoxa, Glory-of-the-snow ■☆

87758 Chionodoxa Boiss. = Scilla L. ■

87759 Chionodoxa forbesii Baker;雪光花(雪百合,雪宝花,雪花百合,耀斑雪百合);Glory of the Snow, Glory-of-the-snow ■☆

87760 Chionodoxa forbesii Baker 'Pink Giant';粉巨人耀斑雪百合■☆

87761 Chionodoxa gigantea Hort. = Chionodoxa luciliae Boiss. ■☆

87762 Chionodoxa gigantea L. H. Bailey;蓝光花;Glory-of-the-blue ■☆

87763 Chionodoxa gigantea L. H. Bailey var. alba Hort.;白雪光花;Glory-of-the-white-snow ■☆

87764 Chionodoxa gigantea L. H. Bailey var. pink Hort.;粉光花;Glory-of-the-pink ■☆

87765 Chionodoxa luciliae Boiss.;卢氏雪光花■☆

87766 Chionodoxa luciliae Boiss. = Scilla luciliae (Boiss.) Speta ■☆

87767 Chionodoxa luciliae Boiss. var. alba Hort.;白花卢氏雪光花■☆

87768 Chionodoxa luciliae Boiss. var. gigantea Hort.;大卢氏雪光花■☆

87769 Chionodoxa luciliae Boiss. var. rosea Hort.;粉红卢氏雪光花■☆

87770 Chionodoxa sardensis Barr et Sugden;深蓝雪百合(撒丁光花);Lesser Glory-of-the-snow,Sardinia Glory-of-the-blue ■☆

87771 Chionodoxa siehei Stapf = Chionodoxa forbesii Baker ■☆

87772 Chionoglochin Gand. = Carex L. ■

87773 Chionographidaceae Takht.(1994);白丝草科■

87774 Chionographidaceae Takht. = Melanthiaceae Batsch ex Borkh.(保留科名)■

87775 Chionographis Maxim.(1867)(保留属名);白丝草属;Chionographis,Whitesilkgrass ■

87776 Chionographis chinensis K. Krause;中国白丝草(白花莱,白丝草);China Whitesilkgrass,Chinese Chionographis ■

87777 Chionographis japonica Maxim.;日本白丝草■☆

87778 Chionographis merrilliana Hara = Chionographis chinensis K. Krause ■

87779 Chionohebe B. G. Briggs et Ehrend.(1976);雪婆婆纳属●☆

87780 Chionohebe ciliolata(Hook. f.)B. G. Briggs et Ehrend.;雪婆婆纳●☆

87781 Chionohebe densifolia(F. Muell.)B. G. Briggs et Ehrend.;密叶雪婆婆纳●☆

87782 Chionohebe glabra(Cheeseman)Heads;无毛雪婆婆纳●☆

87783 Chionolaena DC.(1836);雪衣鼠麴木属●☆

87784 Chionolaena capitata(Baker)C. V. Freire;头状雪衣鼠麴木●☆

87785 Chionolaena latifolia Baker;宽叶雪衣鼠麴木●☆

87786 Chionolaena longifolia Baker;长叶雪衣鼠麴木●☆

87787 Chionolaena mexicana S. E. Freire;墨西哥雪衣鼠麴木●☆

87788 Chionolaena salicifolia(Bertol.)G. L. Nesom;柳叶雪衣鼠麴木●☆

87789 Chionopappus Benth.(1873);羽冠黄安菊属●☆

87790 Chionopappus benthamii S. F. Blake;羽冠黄安菊●☆

87791 Chionophila Benth.(1846);喜寒婆婆纳属(喜寒玄参属)■☆

87792 Chionophila Miers ex Lindl. = Boopis Juss. ■☆

87793 Chionophila jamesii Benth.;喜寒婆婆纳■☆

87794 Chionoptera DC. = Pachylaena D. Don ex Hook. et Arn. ●☆

87795 Chionothrix Hook. f.(1880);白苋木属●☆

87796 Chionothrix hyposericea Chiov. = Dasysphaera hyposericea(Chiov.)C. C. Towns. ■☆

87797 Chionothrix latifolia Rendle;宽叶白苋木●☆

87798 Chionothrix somalensis(S. Moore)Hook. f.;索马里白苋木●☆

87799 Chionotria Jack = Glycosmis Corrêa(保留属名)●

87800 Chiophila Raf. = Gentiana L. ■

87801 Chiovendaea Speg. = Coursetia DC. ●☆

87802 Chiradenia Post et Kuntze = Cheiradenia Lindl. ■☆

87803 Chiradenia Post et Kuntze = Zygopetalum Hook. ■☆

87804 Chiranthera Post et Kuntze = Cheiranthera A. Cunn. ex Brongn. ●☆

87805 Chiranthodendraceae A. Gray = Malvaceae Juss.(保留科名)●■

87806 Chiranthodendron Cerv. ex Cav. = Chiranthodendron Sessé ex Larreat. ●☆

87807 Chiranthodendron Larreat. = Chiranthodendron Sessé ex Larreat. ●☆

87808 Chiranthodendron Sessé ex Larreat.(1795);手药木属●☆

87809 Chiranthodendron pentadactylon Larreat.;手药木(五指手药木);Hand Plant Tree,Monkey's-hand ●☆

87810 Chirata G. Don = Chirita Buch. -Ham. ex D. Don ●■

87811 Chiratia Montrouz. = Sonneratia L. f.(保留属名)●

87812 Chiratia leucantha Montrouz. = Sonneratia alba Sm. ●

87813 Chiridium Tiegh. = Helixanthera Lour. ●

87814 Chirita Buch. -Ham. = Chirita Buch. -Ham. ex D. Don ●■

87815 Chirita Buch. -Ham. ex D. Don(1822);唇柱苣苔属(蚂蝗七属,双心皮草属);Chirita ●■

87816 Chirita acaulis Merr. = Opithandra acaulis(Merr.)B. L. Burtt ■

87817 Chirita acuminata R. Br. = Chirita oblongifolia(Roxb.)J. Sinclair ■

87818 Chirita acuminata Wall. ex R. Br. = Chirita oblongifolia(Roxb.)J. Sinclair ■

87819 Chirita adenocalyx Chatterjee;腺萼唇柱苣苔■

87820 Chirita anachoreta Hance;光萼唇柱苣苔(长蒴苣苔,双心皮草);Smoothcalyx Chirita ■

87821 Chirita atroglandulosa W. T. Wang;黑腺唇柱苣苔■

87822 Chirita atropurpurea W. T. Wang;紫萼唇柱苣苔草;Purplecalyx Chirita ■

87823 Chirita baishouensis Y. G. Wei,H. Q. Wen et S. H. Zhong;百寿唇柱苣苔草;Baishou Chirita ■

87824 Chirita balansae Drake = Chirita swinglei(Merr.)W. T. Wang ■

87825 Chirita barbata Sprague;髯毛唇柱苣苔草;Bearded Chirita ■

87826 Chirita bicolor W. T. Wang;二色唇柱苣苔草;Bicolor Chirita ■

87827 Chirita bicornuta Hayata = Hemiboea bicornuta(Hayata)Ohwi ■

87828 Chirita botusidentata W. T. Wang;钝齿唇柱苣苔■

87829 Chirita brachystigma W. T. Wang;短头唇柱苣苔(红花根);Shortstigma Chirita ■

87830 Chirita brachytricha W. T. Wang et D. Y. Chen;短毛唇柱苣苔草;Shorthair Chirita ■

87831 Chirita brachytricha W. T. Wang et D. Y. Chen var. magnibracteata W. T. Wang et D. Y. Chen;大苞短毛唇柱苣苔草;Bigbract Chirita ■

87832 Chirita brassicoides W. T. Wang;芥状唇柱苣苔草;Brassicalike Chirita ■

87833 Chirita brevipes C. B. Clarke = Chirita speciosa Kurz ■

87834 Chirita brevipes Clarke = Chirita speciosa Kurz ■

87835 Chirita briggsioides W. T. Wang;鹤峰唇柱苣苔草;Hefeng Chirita ■

87836 Chirita carnosifolia C. Y. Wu ex H. W. Li;肉叶唇柱苣苔草;Succulentleaf Chirita ■

87837 Chirita ceratoscyphus B. L. Burtt = Chirita corniculata Pellegr. ■

87838 Chirita chlamydata W. W. Sm. = Didissandra begoniifolia H. Lév. ■

87839 Chirita cicatricosa W. T. Wang;多痕唇柱苣苔■

87840 Chirita cicatricosa W. T. Wang = Chirita minutihamata D. Wood ■

87841 Chirita clarkei Hook. f. = Chirita lachenensis C. B. Clarke ■

87842 Chirita cordifolia D. Fang et W. T. Wang;心叶唇柱苣苔草;Heartleaf Chirita ■

87843 Chirita corniculata Pellegr.;角萼唇柱苣苔草;Hornedcalyx Chirita ■

87844 Chirita cortusifolia Hance = Didymocarpus cortusifolius(Hance)W. T. Wang ■

87845 Chirita crassituba W. T. Wang;粗筒唇柱苣苔草;Crassitube Chirita ■

87846 Chirita cruciformis(Chun)W. T. Wang;十字唇柱苣苔草;Crisscross Chirita ■

87847 Chirita cyrtocarpa D. Fang et L. Zeng;弯果唇柱苣苔草;Bentfruit Chirita ■

87848 Chirita dalzielii W. W. Sm. = Opithandra dalzielii(W. W. Sm.)B. L. Burtt ■

87849 Chirita demissa(Hance)W. T. Wang;巨柱唇柱苣苔(绒毛长

蒴苣苔);Nappy Didymocarpus ■
87850 Chirita depressa Hook. f.;短序唇柱苣苔草;Shortcyme Chirita ■
87851 Chirita dielsii (Borza) B. L. Burtt;圆叶唇柱苣苔草;Roundleaf Chirita ■
87852 Chirita dimidiata R. Br. ex C. B. Clarke = Chirita anachoreta Hance ■
87853 Chirita dimidiata Wall. ex C. B. Clarke;墨脱唇柱苣苔(薄叶唇柱苣苔);Motuo Chirita,Thinleaf Chirita ■
87854 Chirita dryas Dunn = Chirita sinensis Lindl. ■
87855 Chirita eburnea Hance;牛耳朵(光白菜,猫耳朵,牛耳岩白菜,爬面虎,山金兜菜,山石兰,石虎耳,石三七,岩白菜,岩青菜);Ivory-white Chirita ■
87856 Chirita fangii W. T. Wang;方氏唇柱苣苔草;Fang Chirita ■
87857 Chirita fasciculiflora W. T. Wang;簇花唇柱苣苔草;Fascicleflower Chirita ■
87858 Chirita fauriei Franch. = Chirita eburnea Hance ■
87859 Chirita fimbrisepala Hand. -Mazz.;蚂蝗七(飞天蜈蚣,红蚂蝗七,睫萼长蒴苣苔,蚂蟥七,石棉,石螃蟹,石蜈蚣,岩白菜,岩蚂蝗);Fimbriatesepal Chirita,Fringed Chirita ■
87860 Chirita fimbrisepala Hand. -Mazz. var. mollis W. T. Wang;密毛蚂蝗七(石蚂蝗);Hair Fimbriatesepal Chirita ■
87861 Chirita flava R. Br. = Chirita pumila D. Don ■
87862 Chirita flavimaculata W. T. Wang;黄斑唇柱苣苔草;Yellowspotted Chirita ■
87863 Chirita floribunda W. T. Wang;多花唇柱苣苔草;Flowery Chirita ■
87864 Chirita fordii (Hemsl.) D. Wood;桂粤唇柱苣苔草;Ford Chirita ■
87865 Chirita fordii (Hemsl.) D. Wood var. dolichotricha (W. T. Wang) W. T. Wang;鼎湖唇柱苣苔草;Dinghu Chirita ■
87866 Chirita forrestii Anthony;滇川唇柱苣苔草;Forrest Chirita ■
87867 Chirita forrestii Anthony var. acutidentata W. T. Wang;锐齿滇川唇柱苣苔草;Sharptooth Forrest Chirita ■
87868 Chirita forrestii Anthony var. acutidentata W. T. Wang = Chirita forrestii Anthony ■
87869 Chirita fruticola H. W. Li;灌丛唇柱苣苔草;Shrublover Chirita ■
87870 Chirita glabrescens W. T. Wang et D. Y. Chen;少毛唇柱苣苔草;Fewhir Chirita ■
87871 Chirita grandidentata W. T. Wang = Didymocarpus grandidentatus (W. T. Wang) W. T. Wang ■
87872 Chirita grandiflora Wall. = Chirita urticifolia Buch. -Ham. ex D. Don ■
87873 Chirita guangxiensis S. Z. Huang = Pseudochirita guangxiensis (S. Z. Huang) W. T. Wang ■
87874 Chirita gueilinensis W. T. Wang var. brachycarpa W. T. Wang = Chirita juliae Hance ■
87875 Chirita gueilinensis W. T. Wang var. dolichotricha W. T. Wang = Chirita fordii (Hemsl.) D. Wood var. dolichotricha (W. T. Wang) W. T. Wang ■
87876 Chirita guilinensis W. T. Wang;桂林唇柱苣苔(蚂蝗七,山蚂蝗);Guilin Chirita ■
87877 Chirita guilinensis W. T. Wang var. brachycalpa W. T. Wang;宁化唇柱苣苔草;Ninghua Chirita ■
87878 Chirita guilinensis W. T. Wang var. brachycalpa W. T. Wang = Chirita juliae Hance ■
87879 Chirita guilinensis W. T. Wang var. dolichotricha W. T. Wang = Chirita fordii (Hemsl.) D. Wood var. dolichotricha (W. T. Wang) W. T. Wang ■
87880 Chirita hamosa R. Br.;钩序唇柱苣苔(叶序唇柱苣苔);Hookcyme Chirita ■
87881 Chirita hedyotidea (Chun) W. T. Wang;肥牛草(矮脚甘松,耳草长蒴苣苔,红接骨草,石上莲);Fatox Chirita ■
87882 Chirita heterotricha Merr.;烟叶唇柱苣苔(异毛唇柱苣苔);Tobaccoleaf Chirita ■
87883 Chirita heucherifolia (Hand. -Mazz.) D. Wood = Didymocarpus heucherifolius Hand. -Mazz. ■
87884 Chirita heucherifolia (Hand. -Mazz.) Wood. = Didymocarpus heucherifolius Hand. -Mazz. ■
87885 Chirita hochiensis C. C. Huang et X. X. Chen;河池唇柱苣苔草;Hechi Chirita ■
87886 Chirita infundibuliformis W. T. Wang;合苞唇柱苣苔草;Funnel Chirita ■
87887 Chirita jiuwanshanica W. T. Wang;九万山唇柱苣苔草;Jiuwanshan Chirita ■
87888 Chirita juliae Hance;大齿唇柱苣苔草;Largetooth Chirita ■
87889 Chirita kurzii (C. B. Clarke) C. B. Clarke = Briggsia kurzii (C. B. Clarke) W. E. Evans ■
87890 Chirita lachenensis C. B. Clarke;卧茎唇柱苣苔草;Creeping Chirita ■
87891 Chirita laifengensis W. T. Wang;来凤唇柱苣苔草;Laifeng Chirita ■
87892 Chirita laifengensis W. T. Wang = Chirita obtusidentata W. T. Wang ■
87893 Chirita langshanica W. T. Wang;崀山唇柱苣苔草;Langshan Chirita ■
87894 Chirita latinervis W. T. Wang;宽脉唇柱苣苔草;Broad-veined Chirita ■
87895 Chirita lavandulacea Stapf;蓝檐唇柱苣苔草■☆
87896 Chirita laxiflora W. T. Wang;疏花唇柱苣苔草;Sparseflower Chirita ■
87897 Chirita leiophylla W. T. Wang;光叶唇柱苣苔草;Smoothleaf Chirita ■
87898 Chirita liboensis W. T. Wang et D. Y. Chen;荔波唇柱苣苔草;Libo Chirita ■
87899 Chirita lienxienensis W. T. Wang;连县唇柱苣苔草;Lianxian Chirita ■
87900 Chirita liguliformis W. T. Wang;舌柱唇柱苣苔草;Tongue Chirita ■
87901 Chirita linearifolia W. T. Wang;线叶唇柱苣苔(红接骨,石上莲);Linearleaf Chirita ■
87902 Chirita linglingensis W. T. Wang;零陵唇柱苣苔草;Lingling Chirita ■
87903 Chirita liujiangensis D. Fang et D. H. Qin;柳江唇柱苣苔草;Liujiang Chirita ■
87904 Chirita longgangensis W. T. Wang;弄岗唇柱苣苔草;Longgang Chirita ■
87905 Chirita longgangensis W. T. Wang var. hongyao S. Z. Huang;红药;Hongyao Chirita ■
87906 Chirita longipedunculata W. T. Wang = Lysionotus longipedunculatus (W. T. Wang) W. T. Wang ●
87907 Chirita longistyla W. T. Wang;长柱唇柱苣苔草;Longstyle Chirita ■
87908 Chirita longistyla W. T. Wang = Chirita fordii (Hemsl.) D. Wood ■
87909 Chirita lunglinensis W. T. Wang;隆林唇柱苣苔草;Longlin Chirita ■

87910 Chirita lunglinensis W. T. Wang var. amblyosepala W. T. Wang;钝萼唇柱苣苔草;Obtusesepal Chirita ■

87911 Chirita lungzhouensis W. T. Wang;龙州唇柱苣苔草;Longzhou Chirita ■

87912 Chirita lutea Yan Liu et Y. G. Wei;黄花牛耳朵;Yellow flower Chirita ■

87913 Chirita macrodonta D. Fang et D. H. Qin;粗齿唇柱苣苔草;Bigtooth Chirita ■

87914 Chirita macrophylla Wall.;大叶唇柱苣苔草;Bigleaf Chirita ■

87915 Chirita macrorhiza D. Fang et D. H. Qin;大根唇柱苣苔草;Bigroot Chirita ■

87916 Chirita macrosiphon Hance = Didissandra macrosiphon (Hance) W. T. Wang ■

87917 Chirita mangshanensis W. T. Wang;莽山唇柱苣苔草;Mangshan Chirita ■

87918 Chirita mangshanensis W. T. Wang = Chirita juliae Hance ■

87919 Chirita mangshanensis W. T. Wang var. lasiandra W. T. Wang;广丰唇柱苣苔草;Guangfeng Chirita ■

87920 Chirita mangshanensis W. T. Wang var. lasiandra W. T. Wang = Chirita juliae Hance ■

87921 Chirita martinii H. Lév. et Vaniot = Paraboea martinii (H. Lév. et Vaniot) B. L. Burtt ■

87922 Chirita medica D. Fang ex W. T. Wang;药用唇柱苣苔(牛耳朵);Medicinal Chirita ■

87923 Chirita minuteserrulata Hayata;双心皮草■

87924 Chirita minuteserrulata Hayata = Chirita anachoreta Hance ■

87925 Chirita minutihamata D. Wood;多痕唇柱苣苔草;Minihook Chirita ■

87926 Chirita minutimaculata D. Fang et W. T. Wang;微斑唇柱苣苔(上石蚂蝗);Minispotted Chirita ■

87927 Chirita minutiserrulata Hayata = Chirita anachoreta Hance ■

87928 Chirita mollifolia D. Fang, Y. G. Wei et Eggl.;软叶唇柱苣苔草;Softleaf Chirita ■

87929 Chirita monantha W. T. Wang;单花唇柱苣苔草;Singleflower Chirita ■

87930 Chirita naponensis Z. Y. Li;那坡唇柱苣苔草;Napo Chirita ■

87931 Chirita oblongifolia (Roxb.) J. Sinclair;长圆叶唇柱苣苔草;Oblongleaf Chirita ■

87932 Chirita obtusa C. B. Clarke = Didymostigma obtusum (C. B. Clarke) W. T. Wang ■

87933 Chirita obtusidentata W. T. Wang;钝齿唇柱苣苔草;Blunttooth Chirita ■

87934 Chirita obtusidentata W. T. Wang var. mollipes W. T. Wang;毛序唇柱苣苔■

87935 Chirita ophiopogoides D. Fang et W. T. Wang;条叶唇柱苣苔(耳羊);Beltleaf Chirita ■

87936 Chirita orbicularis W. W. Sm. = Chirita dielsii (Borza) B. L. Burtt ■

87937 Chirita orthandra W. T. Wang;直蕊唇柱苣苔草;Erectthrum Chirita ■

87938 Chirita parvifolia W. T. Wang;小叶唇柱苣苔(胡连);Smallleaf Chirita ■

87939 Chirita pellegriniata B. L. Burtt = Chirita swinglei (Merr.) W. T. Wang ■

87940 Chirita pinnata W. T. Wang;复叶唇柱苣苔草;Pinnate Chirita ■

87941 Chirita pinnatifida (Hand.-Mazz.) Burtt;羽裂唇柱苣苔(石岩菜);Pinnatifid Chirita ■

87942 Chirita polycephala (Chun) W. T. Wang;多葶唇柱苣苔草;Manyscape Chirita ■

87943 Chirita polyneura (Chun) W. T. Wang var. amabilis C. B. Clarke = Chirita anachoreta Hance ■

87944 Chirita polyneura (Chun) W. T. Wang var. amabilis C. B. Clarke = Chirita dimidiata R. Br. ex C. B. Clarke ■

87945 Chirita primuloides (Miq.) Ohwi = Opithandra primuloides (Miq.) B. L. Burtt ■☆

87946 Chirita primuloides (Miq.) Ohwi f. albiflora (Makino) H. Hara = Campanula lasiocarpa Cham. f. albiflora Tatew. ■☆

87947 Chirita pseudoeburnea D. Fang et W. T. Wang;紫纹唇柱苣苔(假牛耳朵,岩白菜);Purplestriped Chirita ■

87948 Chirita pteropoda W. T. Wang;翅柄唇柱苣苔草;Wingstipe Chirita ■

87949 Chirita puerensis Y. Y. Qian;普洱唇柱苣苔草;Puer Chirita ■

87950 Chirita pumila D. Don;斑叶唇柱苣苔(虎须草);Spottedleaf Chirita, Variegated-leaf Chirita ■

87951 Chirita pungentisepala W. T. Wang;尖萼唇柱苣苔■

87952 Chirita pycnantha W. T. Wang;密花唇柱苣苔■

87953 Chirita quercifolia D. Wood = Chirita pinnatifida (Hand.-Mazz.) Burtt ■

87954 Chirita ronganensis D. Fang et Y. G. Wei;融安唇柱苣苔草;Rongan Chirita ■

87955 Chirita roseo-alba W. T. Wang;粉花唇柱苣苔草;Pink Chirita ■

87956 Chirita rotundifolia (Hemsl.) D. Wood;卵圆唇柱苣苔草;Roundleaf Chirita ■

87957 Chirita rupestris Ridl.;岩石唇柱苣苔草;Rockliving Chirita ■

87958 Chirita sclerophylla W. T. Wang;硬叶唇柱苣苔草;Hardleaf Chirita ■

87959 Chirita secundiflora (Chun) W. T. Wang;青镇唇柱苣苔草;Secundflower Chirita ■

87960 Chirita sericea (H. Lév.) H. Lév. et Vaniot = Oreocharis auricula (S. Moore) C. B. Clarke ■

87961 Chirita sericea H. Lév. et Vaniot = Oreocharis auricula (S. Moore) C. B. Clarke ■

87962 Chirita shennungjiaensis W. T. Wang = Chirita tenuituba (W. T. Wang) W. T. Wang ■

87963 Chirita shouchengensis Z. Y. Li;寿城唇柱苣苔草;Shoucheng Chirita ■

87964 Chirita shuii Z. Y. Li;税氏唇柱苣苔草;Shui's Chirita ■

87965 Chirita sichuanensis W. T. Wang;四川唇柱苣苔草;Sichuan Chirita ■

87966 Chirita sinensis Lindl.;唇柱苣苔(两广唇柱苣苔);China Chirita, Chinese Chirita ■

87967 Chirita sinensis Lindl. var. angustifolia Dunn = Chirita sinensis Lindl. ■

87968 Chirita sinensis Lindl. var. bodinieri H. Lév. = Chirita sinensis Lindl. ■

87969 Chirita skogiana Z. Y. Li;斯氏唇柱苣苔■

87970 Chirita spadiciformis W. T. Wang;焰苞唇柱苣苔草;Spathose Chirita ■

87971 Chirita speciosa Kurz;美丽唇柱苣苔草;Lovely Chirita ■

87972 Chirita speluncae (Hand.-Mazz.) D. Wood;小唇柱苣苔(小长蒴苣苔);Small Chirita, Small Didissandra, Small Didymocarpus ■

87973 Chirita sphagnicola H. Lév. et Vaniot = Chirita pumila D. Don ■

87974 Chirita spinulosa D. Fang et W. T. Wang;刺齿唇柱苣苔(山芭蕉);Spinetooth Chirita ■

87975 Chirita subacaulis (Hand. -Mazz.) B. L. Burtt = Hemiboea subacaulis Hand. -Mazz. ■
87976 Chirita subrhomboides W. T. Wang;菱叶唇柱苣苔草;Subrhombicleaf Chirita ■
87977 Chirita subulatisepala W. T. Wang;钻萼唇柱苣苔(飞蛾树);Drilsepal Chirita ■
87978 Chirita swinglei (Merr.) W. T. Wang;钟冠唇柱苣苔(白芨药,肥知母);Swingle Chirita ■
87979 Chirita tenuifolia W. T. Wang;薄叶唇柱苣苔草;Thinleaf Chirita ■
87980 Chirita tenuituba (W. T. Wang) W. T. Wang;神农架唇柱苣苔(舌唇苣苔,小岩白菜);Shennongjia Chirita ■
87981 Chirita tibetica (Franch.) B. L. Burtt;康定唇柱苣苔草;Kangding Chirita ■
87982 Chirita trailliana Forrest et W. W. Sm. = Chirita speciosa Kurz ■
87983 Chirita tribracteata W. T. Wang;三苞唇柱苣苔草;Threebract Chirita ■
87984 Chirita umbrophila C. Y. Wu ex H. W. Li;喜荫唇柱苣苔(喜阴唇柱苣苔);Shady Chirita ■
87985 Chirita urticifolia Buch. -Ham. ex D. Don;麻叶唇柱苣苔草;Nettleleaf Chirita ■
87986 Chirita varicolor D. Fang et D. H. Qin;变色唇柱苣苔草;Varicolor Chirita ■
87987 Chirita verecunda (Chun) W. T. Wang;齿萼唇柱苣苔草;Toothsepal Chirita ■
87988 Chirita vestita D. Wood;细筒唇柱苣苔草;Thintube Chirita ■
87989 Chirita villosissima W. T. Wang;长毛唇柱苣苔草;Longvelvety Chirita ■
87990 Chirita wangiana Z. Y. Li;王氏唇柱苣苔草;Wang's Chirita ■
87991 Chirita wentsaii D. Fang et L. Zeng;文采唇柱苣苔草;Wentsai Chirita ■
87992 Chirita xinningensis W. T. Wang;新宁唇柱苣苔草;Xinning Chirita ■
87993 Chirita yungfuensis W. T. Wang;永福唇柱苣苔草;Yungfu Chirita ■
87994 Chiritopsis W. T. Wang(1981);小花苣苔属;Chiritopsis ■★
87995 Chiritopsis bipinnatifida W. T. Wang;羽裂小花苣苔(白疔芋);Bipinnatifid Chiritopsis ■
87996 Chiritopsis confertiflora W. T. Wang;密小花苣苔草;Denseflower Chiritopsis ■
87997 Chiritopsis cordifolia D. Fang et W. T. Wang;心叶小花苣苔草;Heartleaf Chiritopsis ■
87998 Chiritopsis glandulosa D. Fang, L. Zeng et D. H. Qin;紫腺小花苣苔(紫脉小花苣苔);Glandule Chiritopsis ■
87999 Chiritopsis lingchuanensis Yan Liu et Y. G. Wei;灵川小花苣苔■
88000 Chiritopsis lobulata W. T. Wang;浅裂小花苣苔草;Lobulate Chiritopsis ■
88001 Chiritopsis molifolia D. Fang et W. T. Wang;密毛小花苣苔草;Densehairy Chiritopsis ■
88002 Chiritopsis repanda W. T. Wang;小花苣苔(水滴,阴山泽);Chiritopsis ■
88003 Chiritopsis repanda W. T. Wang var. guilinensis W. T. Wang;桂林小花苣苔草;Guilin Chiritopsis ■
88004 Chiritopsis subulata W. T. Wang;钻丝小花苣苔草;Drilthrum Chiritopsis ■
88005 Chiritopsis subulata W. T. Wang var. yangchunensis W. T. Wang;阳春小花苣苔草;Yangchun Chiritopsis ■
88006 Chiritopsis xiuningensis X. L. Liu et X. H. Guo;休宁小花苣苔草;Xiuning Chiritopsis ■
88007 Chirocalyx Meisn. = Erythrina L. ●■
88008 Chirocalyx abyssinicus (Lam. ex DC.) Hochst. = Erythrina abyssinica Lam. ex DC. ●☆
88009 Chirocalyx mollissimus Meisn. = Erythrina latissima E. Mey. ●☆
88010 Chirocalyx tomentosus Hochst. = Erythrina abyssinica Lam. ex DC. ●☆
88011 Chirocarpus A. Braun ex Pfeiff. = Caylusea A. St. -Hil.(保留属名)■☆
88012 Chirochlaena Post et Kuntze = Cheirolaena Benth. ■●☆
88013 Chirodendrum Post et Kuntze = Cheirodendron Nutt. ex Seem. ●☆
88014 Chirolepis T. Post et Kuntze = Centaurea L.(保留属名)●■
88015 Chirolepis T. Post et Kuntze = Cheirolepis Boiss. ●■☆
88016 Chiroloma T. Post et Kuntze = Calotis R. Br. ■
88017 Chiroloma T. Post et Kuntze = Cheiroloma F. Muell. ■
88018 Chirolophus Cass. = Centaurea L.(保留属名)●■
88019 Chirolophus Cass. = Cheirolophus Cass. ●■☆
88020 Chironea Raf. = Chironia L. ●■☆
88021 Chironia F. W. Schmidt = Gentiana L. ■
88022 Chironia L.(1753);圣诞果属(蚩龙属);Star Pink, Star-pink ●■☆
88023 Chironia P. Gaertn., B. Mey. et Scherb. = Centaurium Hill ■
88024 Chironia albiflora Hilliard;白花圣诞果■☆
88025 Chironia angolensis Gilg;安哥拉圣诞果■☆
88026 Chironia angustifolia Sims = Orphium frutescens (L.) E. Mey. ●☆
88027 Chironia arenaria E. Mey.;沙地圣诞果■☆
88028 Chironia arenaria E. Mey. var. mediocris (Schoch) Prain = Chironia arenaria E. Mey. ■☆
88029 Chironia baccifera L.;圣诞果;Chritmas Berry, Wild Gentian ●☆
88030 Chironia bachmannii Gilg = Chironia purpurascens (E. Mey.) Benth. et Hook. f. ■☆
88031 Chironia barclayana Berol. ex Griseb. = Chironia peduncularis Lindl. ■☆
88032 Chironia baumiana Gilg;鲍姆圣诞果■☆
88033 Chironia caryophylloides L. = Orphium frutescens (L.) E. Mey. ●☆
88034 Chironia decumbens Levyns;外倾圣诞果■☆
88035 Chironia decussata Vent. = Orphium frutescens (L.) E. Mey. ●☆
88036 Chironia densiflora Scott-Elliot = Chironia krebsii Griseb. ■☆
88037 Chironia ecklonii Schoch = Chironia linoides L. ■☆
88038 Chironia elgonensis Bullock;埃尔贡圣诞果■☆
88039 Chironia emarginata Jaroscz = Chironia linoides L. subsp. emarginata (Jaroscz) I. Verd. ■☆
88040 Chironia erythraea (Raf.) Schousb. = Centaurium erythraea Raf. ■☆
88041 Chironia erythraeoides Hiern;浅红圣诞果■☆
88042 Chironia exigua Oliv. = Sebaea exigua (Oliv.) Schinz ■☆
88043 Chironia flexuosa Baker;之字圣诞果■☆
88044 Chironia floribunda Paxton;多花圣诞果■☆
88045 Chironia frutescens L. = Orphium frutescens (L.) E. Mey. ●☆
88046 Chironia gracilis Salisb. ex Prain = Chironia linoides L. subsp. nana I. Verd. ■☆
88047 Chironia gracilis Salisb. ex Prain var. macrocalyx Prain = Chironia linoides L. subsp. macrocalyx (Prain) I. Verd. ■☆
88048 Chironia gratissima S. Moore;可爱圣诞果■☆
88049 Chironia humilis Gilg = Chironia purpurascens (E. Mey.) Benth. et Hook. f. subsp. humilis (Gilg) I. Verd. ■☆

88050 Chironia humilis Gilg var. wilmsii (Gilg) Prain = Chironia purpurascens (E. Mey.) Benth. et Hook. f. subsp. humilis (Gilg) I. Verd. ■☆

88051 Chironia humilis Gilg var. zuluensis Prain = Chironia purpurascens (E. Mey.) Benth. et Hook. f. subsp. humilis (Gilg) I. Verd. ■☆

88052 Chironia ixifera Garden = Chironia linoides L. ■☆

88053 Chironia jasminoides L.;茉莉圣诞果■☆

88054 Chironia katangensis De Wild.;加丹加圣诞果■☆

88055 Chironia katangensis De Wild. subsp. verdickii (De Wild.) Boutique;韦尔圣诞果■☆

88056 Chironia krebsii Griseb.;克雷布斯圣诞果■☆

88057 Chironia lancifolia Baker = Ornichia lancifolia (Baker) Klack. ●☆

88058 Chironia latifolia Donn = Orphium frutescens (L.) E. Mey. ●☆

88059 Chironia latifolia E. Mey. = Chironia peduncularis Lindl. ■☆

88060 Chironia laxa Gilg;疏松圣诞果■☆

88061 Chironia laxiflora Baker;疏花圣诞果■☆

88062 Chironia linoides L.;亚麻圣诞果■☆

88063 Chironia linoides L. subsp. emarginata (Jaroscz) I. Verd.;微缺亚麻圣诞果■☆

88064 Chironia linoides L. subsp. macrocalyx (Prain) I. Verd.;大萼圣亚麻诞果■☆

88065 Chironia linoides L. subsp. nana I. Verd.;低矮亚麻圣诞果■☆

88066 Chironia lychnoides Berg = Chironia linoides L. ■☆

88067 Chironia madagascariensis Baker = Ornichia madagascariensis (Baker) Klack. ●☆

88068 Chironia maritima (L.) Schousb. = Centaurium maritimum (L.) Fritsch ■☆

88069 Chironia maritima Eckl. = Chironia decumbens Levyns ■☆

88070 Chironia maxima Schoch = Chironia palustris Burch. subsp. rosacea (Gilg) I. Verd. ■☆

88071 Chironia mediocris Schoch = Chironia arenaria E. Mey. ■☆

88072 Chironia melampyrifolia Lam.;山罗花叶圣诞果■☆

88073 Chironia nudicaulis L. f. = Chironia jasminoides L. ■☆

88074 Chironia ovata Spreng. ex Griseb. = Chironia serpyllifolia Lehm. ■☆

88075 Chironia palustris Burch.;沼泽圣诞果■☆

88076 Chironia palustris Burch. subsp. rosacea (Gilg) I. Verd.;粉红圣诞果■☆

88077 Chironia palustris Burch. subsp. transvaalensis (Gilg) I. Verd.;德兰士瓦圣诞果■☆

88078 Chironia parvifolia E. Mey. = Chironia serpyllifolia Lehm. ■☆

88079 Chironia peduncularis Lindl.;梗花圣诞果■☆

88080 Chironia peglerae Prain;佩格拉圣诞果■☆

88081 Chironia purpurascens (E. Mey.) Benth. et Hook. f.;浅紫圣诞果■☆

88082 Chironia purpurascens (E. Mey.) Benth. et Hook. f. subsp. humilis (Gilg) I. Verd.;矮小浅紫圣诞果■☆

88083 Chironia rosacea Gilg = Chironia palustris Burch. subsp. rosacea (Gilg) I. Verd. ■☆

88084 Chironia rubrocoerulea Gilg = Chironia laxiflora Baker ■☆

88085 Chironia scabrida Griseb. = Chironia tetragona L. f. ■☆

88086 Chironia schlechteri Schoch = Chironia laxa Gilg ■☆

88087 Chironia serpyllifolia Lehm.;百里香叶圣诞果■☆

88088 Chironia speciosa E. Mey. = Chironia melampyrifolia Lam. ■☆

88089 Chironia spicata (L.) Schousb. = Centaurium spicatum (L.) Fritsch ■

88090 Chironia stokoei I. Verd.;斯托克圣诞果■☆

88091 Chironia tabularis Page = Chironia tetragona L. f. ■☆

88092 Chironia tetragona L. f.;四角圣诞果■☆

88093 Chironia transvaalensis Gilg = Chironia palustris Burch. subsp. transvaalensis (Gilg) I. Verd. ■☆

88094 Chironia tysonii Gilg = Chironia purpurascens (E. Mey.) Benth. et Hook. f. ■☆

88095 Chironia uniflora A. W. Hill = Chironia katangensis De Wild. subsp. verdickii (De Wild.) Boutique ■☆

88096 Chironia uniflora Lam. = Chironia tetragona L. f. ■☆

88097 Chironia verdickii De Wild. = Chironia katangensis De Wild. subsp. verdickii (De Wild.) Boutique ■☆

88098 Chironia viscosa Zeyh. ex Griseb. = Chironia tetragona L. f. ■☆

88099 Chironia wilmsii Gilg = Chironia purpurascens (E. Mey.) Benth. et Hook. f. subsp. humilis (Gilg) I. Verd. ■☆

88100 Chironia zeyheri Prain = Chironia linoides L. ■☆

88101 Chironiaceae Bercht. et J. Presl = Gentianaceae Juss.(保留科名)●■

88102 Chironiaceae Horan.;圣诞果科■

88103 Chironiaceae Horan. = Gentianaceae Juss.(保留科名)●■

88104 Chironiella Braem = Cattleya Lindl. ■

88105 Chiropetalum A. Juss. = Argythamnia P. Browne ●☆

88106 Chirostemon Cerv. = Chiranthodendron Larreat. ●☆

88107 Chirostemum Cerv. = Chiranthodendron Sessé ex Larreat. ●☆

88108 Chirostylis Post et Kuntze = Cheirostylis Blume ■

88109 Chirripoa Suess. = Guzmania Ruiz et Pav. ■☆

88110 Chirvnia Raf. = Chironia L. ●■☆

88111 Chisocheton Blume(1825);溪杪属(拟樫木属);Chisocheton ●

88112 Chisocheton chinensis Merr. = Chisocheton paniculatus (Roxb.) Hiern ●

88113 Chisocheton erythrocarpus Hayata et Kaneh. ex Hayata = Chisocheton patens Blume ●

88114 Chisocheton erythrocarpus Hayata et Kaneh. ex Hayata = Dysoxylum kanehirai (Sasaki) Kaneh. et Hatus. ●

88115 Chisocheton erythrocarpus Hiern = Chisocheton patens Blume ●

88116 Chisocheton erythrocarpus Hiern = Dysoxylum kanehirai (Sasaki) Kaneh. et Hatus. ●

88117 Chisocheton hongkengensis Tutcher = Dysoxylum hongkongense (Tutcher) Merr. ●

88118 Chisocheton kanehirai Sasaki = Chisocheton patens Blume ●

88119 Chisocheton kanehirai Sasaki = Dysoxylum kanehirai (Sasaki) Kaneh. et Hatus. ●

88120 Chisocheton kusukusuense Hayata = Dysoxylum kusukusens (Hayata) Kaneh. et Hatus. ●

88121 Chisocheton paniculatus (Roxb.) Hiern;溪杪(滇南溪杪);Paniculate Chisocheton ●

88122 Chisocheton patens Blume;兰屿拟樫木;Lanyu Chisocheton ●

88123 Chisocheton pentandrus Merr.;五雄蕊溪杪●☆

88124 Chisocheton siamensis Craib = Chisocheton paniculatus (Roxb.) Hiern ●

88125 Chisocheton tetrapetalus DC. = Dysoxylum kanehirai (Sasaki) Kaneh. et Hatus. ●

88126 Chisocheton tetrapetalus Turcz.;兰屿楝树●

88127 Chithonanthus Lehm. = Acacia Mill.(保留属名)●■

88128 Chitonanthera Schltr. = Octarrhena Thwaites ■☆

88129 Chitonia D. Don = Miconia Ruiz et Pav.(保留属名)●☆

88130 Chitonia DC. = Morkillia Rose et J. H. Painter ●☆

88131 Chitonia Moc. et Sessé = Morkillia Rose et J. H. Painter ●☆

88132 Chitonia Moc. et Sessé ex DC. = Morkillia Rose et J. H. Painter ●☆
88133 Chitonia Salisb. = Zigadenus Michx. ■
88134 Chitonochilus Schltr. (1905);隐唇兰属■☆
88135 Chitonochilus Schltr. = Agrostophyllum Blume ■
88136 Chitonochilus papuanum Schltr.;隐唇兰■☆
88137 Chizocheton A. Juss. = Chisocheton Blume ●
88138 Chlaenaceae Thouars = Sarcolaenaceae Caruel(保留科名)●☆
88139 Chlaenandra Miq. (1868);被蕊藤属●☆
88140 Chlaenandra ovata Miq.;被蕊藤●☆
88141 Chlaenanthus Post et Kuntze = Chlainanthus Briq. ●■
88142 Chlaenanthus Post et Kuntze = Lagochilus Bunge ex Benth. ●■
88143 Chlaenobolus Cass. = Pterocaulon Elliott ■
88144 Chlaenosciadium C. Norman(1938);篷伞芹属■☆
88145 Chlaenosciadium gardneri C. Norman;篷伞芹■☆
88146 Chlainanthus Briq. = Lagochilus Bunge ex Benth. ●■
88147 Chlamidacanthus Lindau = Chlamydacanthus Lindau ■☆
88148 Chlamydacanthus Lindau = Theileamea Baill. ■☆
88149 Chlamydacanthus Lindau(1893);刺被爵床属■☆
88150 Chlamydacanthus dichrostachyus Mildbr.;二色穗刺被爵床☆
88151 Chlamydacanthus euphorbioides Lindau;大戟刺被爵床☆
88152 Chlamydacanthus lindavianus H. Winkl.;林达维刺被爵床☆
88153 Chlamydanthus C. A. Mey. = Tartonia Raf. ●■
88154 Chlamydanthus C. A. Mey. = Thymelaea Mill.(保留属名)●■
88155 Chlamydia Banks ex Gaertn. = Phormium J. R. Forst. et G. Forst. ■☆
88156 Chlamydia Gaertn. = Phormium J. R. Forst. et G. Forst. ■☆
88157 Chlamydioboea Stapf = Paraboea (C. B. Clarke) Ridl. ■
88158 Chlamydites J. R. Drumm. (1907);厚毛紫菀属;Chitalpa, Chlamydites ●
88159 Chlamydites J. R. Drumm. = Aster L. ●■
88160 Chlamydites prainii J. R. Drumm.;厚毛紫菀(厚棉紫菀,绵毛紫菀);Prain Aster, Prain's Chlamydites, Thickly Woolly Aster ●■
88161 Chlamydites prainii J. R. Drumm. = Aster prainii (Drumm.) Y. L. Chen ■
88162 Chlamydobalanus (Endl.) Koidz. = Castanopsis (D. Don) Spach(保留属名)●
88163 Chlamydoboea Stapf = Paraboea (C. B. Clarke) Ridl. ■
88164 Chlamydoboea Stapf (1913);宽萼苣苔属(被萼苣苔属);Chlamydoboea ■
88165 Chlamydoboea connata Craib;合生宽萼苣苔■
88166 Chlamydoboea sinensis (Oliv.) Stapf;宽萼苣苔(厚脸皮,华被萼苣苔,门听,石青菜,石头菜,石头草,中华被萼苣苔);China Chlamydoboea, Chinese Chlamydoboea ■
88167 Chlamydoboea sinensis (Oliv.) Stapf = Paraboea sinensis (Oliv.) B. L. Burtt ●
88168 Chlamydoboea sinensis (Oliv.) Stapf f. macra Stapf = Paraboea sinensis (Oliv.) B. L. Burtt ●
88169 Chlamydoboea sinensis (Oliv.) Stapf f. macrophylla Stapf = Paraboea sinensis (Oliv.) B. L. Burtt ●
88170 Chlamydocardia Lindau(1894);心被爵床属☆
88171 Chlamydocardia buettneri Lindau;比特纳心被爵床☆
88172 Chlamydocardia subrhomboidea Lindau;亚菱形心被爵床☆
88173 Chlamydocarya Baill. (1872);篷果茱萸属☆
88174 Chlamydocarya capitata Baill. = Polycephalium capitatum (Baill.) Keay ●☆
88175 Chlamydocarya glabrescens Engl. = Pyrenacantha glabrescens (Engl.) Engl. ●☆
88176 Chlamydocarya gossweileri Exell;戈斯篷果茱萸●☆
88177 Chlamydocarya klaineana Pierre = Chlamydocarya soyauxii Engl. ●☆
88178 Chlamydocarya lobata Pierre = Polycephalium lobatum (Pierre) Pierre ex Engl. ●☆
88179 Chlamydocarya macrocarpa A. Chev. ex Hutch. et Dalziel;大果篷果茱萸●☆
88180 Chlamydocarya rostrata Bullock = Chlamydocarya thomsoniana Baill. ●☆
88181 Chlamydocarya soyauxii Engl.;索亚篷果茱萸●☆
88182 Chlamydocarya staudtii Engl. = Pyrenacantha staudtii (Engl.) Engl. ●☆
88183 Chlamydocarya tenuis Engl. = Pyrenacantha acuminata Engl. ●☆
88184 Chlamydocarya tessmannii Engl. = Pyrenacantha cordicula Villiers ●☆
88185 Chlamydocarya thomsoniana Baill.;托马森篷果茱萸●☆
88186 Chlamydocola (K. Schum.) Bodard = Cola Schott et Endl.(保留属名)●☆
88187 Chlamydocola (K. Schum.) Bodard(1954);斗篷木属●☆
88188 Chlamydocola chlamydantha (K. Schum.) M. Bodard;斗篷木●☆
88189 Chlamydocola lastoursvillensis (M. Bodard et Pellegr.) N. Hallé;加蓬斗篷木●☆
88190 Chlamydojatropha Pax et K. Hoffm. (1912);斗篷麻疯树属●☆
88191 Chlamydojatropha kamerunica Pax et K. Hoffm.;斗篷麻疯树●☆
88192 Chlamydophora Ehrenb. = Cotula L. ■
88193 Chlamydophora Ehrenb. ex Less. (1832);齿芫荽属■☆
88194 Chlamydophora pubescens (Desf.) Coss. et Durieu = Chamomilla pubescens (Desf.) Alavi ■☆
88195 Chlamydophora tridentata (Delile) Ehrenb. ex Less.;齿芫荽■☆
88196 Chlamydophora tridentata (Delile) Less. = Chlamydophora tridentata (Delile) Ehrenb. ex Less. ■☆
88197 Chlamydophytum Mildbr. (1925);斗篷菰属■☆
88198 Chlamydophytum aphyllum Mildbr.;斗篷菰■☆
88199 Chlamydosperma A. Rich. = Stegnosperma Benth. ●☆
88200 Chlamydostachya Mildbr. (1934);篷穗爵床属☆
88201 Chlamydostachya spectabilis Mildbr.;篷穗爵床☆
88202 Chlamydostylus Baker = Nemastylis Nutt. ■☆
88203 Chlamyphorus Klatt = Gomphrena L. ●■
88204 Chlamysperma Less. = Villanova Lag.(保留属名)■☆
88205 Chlamyspermum F. Muell. = Chlamysporum Salisb.(废弃属名)■
88206 Chlamyspermum F. Muell. = Thysanotus R. Br.(保留属名)■
88207 Chlamysporum Salisb.(废弃属名) = Thysanotus R. Br.(保留属名)■
88208 Chlanis Klotzsch = Xylotheca Hochst. ●☆
88209 Chlanis macrophylla Klotzsch = Oncoba tettensis (Klotzsch) Harv. var. macrophylla (Klotzsch) Hul et Breteler ●☆
88210 Chlanis tettensis Klotzsch = Oncoba tettensis (Klotzsch) Harv. ●☆
88211 Chlaotrachelus Hook. f. = Claotrachelus Zoll. ●■
88212 Chlaotrachelus Hook. f. = Vernonia Schreb.(保留属名)●■
88213 Chleterus Raf. = Boea Comm. ex Lam. ■
88214 Chlevax Cesati ex Boiss. = Ferula L. ■
88215 Chlidanthus Herb. (1821);黛玉花属(千花属)■☆
88216 Chlidanthus fragrans Herb.;黛玉花(秘鲁千花,秘鲁水仙);Delicate Lily, Perfumed Fairy Lily ■☆
88217 Chlidanthus luteus D. Dietr. = Chlidanthus fragrans Herb. ■☆
88218 Chloachne Stapf = Poecilostachys Hack. ■☆
88219 Chloachne oplismenoides (Hack.) Stapf ex Robyns =

Poecilostachys oplismenoides (Hack.) Clayton ■☆
88220 Chloammia Raf. = Festuca L. ■
88221 Chloammia Raf. = Vulpia C. C. Gmel. ■
88222 Chloamnia Raf. = Festuca L. ■
88223 Chloamnia Schltdl. = Chloammia Raf. ■
88224 Chloanthaceae Hutch. (1959);连药灌科●■☆
88225 Chloanthaceae Hutch. = Dicrastylidaceae J. Drumm. ex Harv. ●■☆
88226 Chloanthaceae Hutch. = Labiatae Juss. (保留科名)●■
88227 Chloanthaceae Hutch. = Lamiaceae Martinov(保留科名)●■
88228 Chloanthes R. Br. (1810);连药灌属(连药属)●☆
88229 Chloanthes stoechadis R. Br.;连药灌●☆
88230 Chloerum Willd. ex Link = Abolboda Humb. ■☆
88231 Chloerum Willd. ex Spreng. = Abolboda Humb. ■☆
88232 Chloidia Lindl. = Corymborkis Thouars + Tropidia Lindl. ■
88233 Chloidia Lindl. = Tropidia Lindl. ■
88234 Chloidia polystachya (Sw.) Rchb. f. = Tropidia polystachya (Sw.) Ames ■☆
88235 Chloidia vernalis Lindl. = Tropidia polystachya (Sw.) Ames ■☆
88236 Chlonanthes Raf. = Chelonanthus Raf. ■☆
88237 Chlonanthus Raf. = Chelone L. ■☆
88238 Chlonanthus montana Raf. = Chelone glabra L. ●■☆
88239 Chlonanthus tomentosa Raf. = Chelone glabra L. ●■☆
88240 Chloopsis Blume = Ophiopogon Ker Gawl. (保留属名)■
88241 Chloothamnus Büse = Nastus Juss. ●☆
88242 Chlora Adans. (1763);克劳拉草属;Yellowwort ■☆
88243 Chlora Adans. = Blackstonia Huds. ■☆
88244 Chlora Ren. ex Adans. = Blackstonia Huds. ■☆
88245 Chlora Ren. ex Adans. = Chlora Adans. ■☆
88246 Chlora grandiflora Viv. = Blackstonia grandiflora (Viv.) Pau ■☆
88247 Chlora imperfoliata L. f. = Blackstonia imperfoliata (L. f.) Samp. ■☆
88248 Chlora perfoliata Gorter ex Steud.;贯叶克劳拉草;Common Yellowwort ■☆
88249 Chlora perfoliata L. = Blackstonia perfoliata (L.) Huds. ■☆
88250 Chlora perfoliata L. var. longidens H. Lindb. = Blackstonia perfoliata (L.) Huds. ■☆
88251 Chlora perfoliata L. var. mascariensis Desf. = Blackstonia grandiflora (Viv.) Pau ■☆
88252 Chlora serotina Koch = Blackstonia acuminata (Koch et Ziz) Domin ■☆
88253 Chloracantha G. L. Nesom, Y. B. Suh, D. R. Morgan, S. D. Sundb. et B. B. Simpson(1991);刺菀属;Mexican Devilweed, Spiny Aster ●■☆
88254 Chloracantha spinosa (Benth.) G. L. Nesom;刺菀;Mexican Devilweed, Spiny Aster ●■☆
88255 Chloradenia Baill. = Adenogynum Rchb. f. et Zoll. ●
88256 Chloradenia Baill. = Cladogynos Zipp. ex Span. ●
88257 Chloraea Lindl. (1827);绿丝兰属(科劳里亚兰属);Chloraea ■☆
88258 Chloraea austiniae A. Gray = Cephalanthera austiniae (A. Gray) A. Heller ■☆
88259 Chloraea penicillata Rchb. f.;绿丝兰(科劳里亚兰);Common Chloraea ■☆
88260 Chlorantha Nesom et al. = Boltonia L'Hér. ■☆
88261 Chloranthaceae R. Br. = Chloranthaceae R. Br. ex Sims(保留科名)●■
88262 Chloranthaceae R. Br. ex Lindl. = Chloranthaceae R. Br. ex Sims (保留科名)●■
88263 Chloranthaceae R. Br. ex Sims(1820)(保留科名);金粟兰科;Chloranth Family, Chloranthus Family ●■
88264 Chloranthus Sw. (1787);金粟兰属;Chloranth, Chloranthus ■●
88265 Chloranthus angustifolius Oliv.;狭叶金粟兰(四叶细辛,小四块瓦);Narrowleaf Chloranthus ■
88266 Chloranthus anhuiensis K. F. Wu;安徽金粟兰(黑细辛);Anhui Chloranthus ■
88267 Chloranthus brachystachys Blume = Sarcandra glabra (Thunb.) Nakai subsp. brachystachys (Blume) Verdc. ●
88268 Chloranthus brachystachys Blume = Sarcandra glabra Nakai subsp. brachystachys (Blume) Verdc. ●
88269 Chloranthus brachystachys sensu Merr. = Sarcandra glabra Nakai subsp. brachystachys (Blume) Verdc. ●
88270 Chloranthus dentialatus Cordem. = Sarcandra glabra (Thunb.) Nakai ●
88271 Chloranthus elatior R. Br. = Chloranthus officinalis Blume ■
88272 Chloranthus erectus (Buch.-Ham.) Verdc.;鱼子兰(节节茶,九节风,石风节);Roe Chloranthus, Tall Chloranthus ■
88273 Chloranthus erectus (Buch.-Ham.) Verdc. = Chloranthus officinalis Blume ■
88274 Chloranthus erectus Sweet = Chloranthus officinalis Blume ■
88275 Chloranthus esquirolii H. Lév. = Sarcandra glabra (Thunb.) Nakai ●
88276 Chloranthus fortunei (A. Gray) Solms;丝穗金粟兰(白开口箭,豆皮香,黑细辛,平头细辛,水晶花,四大金刚,四大天王,四对草,四块瓦,四叶对,四子莲,土细辛,银线草);Fortune Chloranthus, Silkspike Chloranthus ■
88277 Chloranthus fortunei (A. Gray) Solms var. holostegius Hand.-Mazz. = Chloranthus holostegius (Hand.-Mazz.) C. P'ei et R. H. Shan ■
88278 Chloranthus glaber (Thunb.) Makino = Sarcandra glabra (Thunb.) Nakai ●
88279 Chloranthus glaber (Thunb.) Makino var. flavus (Makino) Makino = Sarcandra glabra (Thunb.) Nakai f. flava (Makino) Okuyama ●☆
88280 Chloranthus glaber Makino = Sarcandra glabra (Thunb.) Nakai ●
88281 Chloranthus hainanensis C. P'ei = Sarcandra glabra Nakai subsp. brachystachys (Blume) Verdc. ●
88282 Chloranthus henryi Hemsl.;宽叶金粟兰(大四块瓦,大叶及已,四大金刚,四大天王,四儿风,四块瓦,四匹瓦,四叶对,四叶细辛,狭叶排草,重楼排草);Broadleaf Chloranthus, Henry Chloranthus ■
88283 Chloranthus henryi Hemsl. var. hupehensis (Pamp.) K. F. Wu;湖北金粟兰(四叶七);Hubei Chloranthus, Hupeh Chloranthus ■
88284 Chloranthus holostegius (Hand.-Mazz.) C. P'ei et R. H. Shan;全缘金粟兰(对叶四块瓦,黑细辛,平头细辛,四大金刚,四大天王,四块瓦,四叶金,土细辛);Entire Chloranthus, Integrifolious Chloranthus ■
88285 Chloranthus holostegius (Hand.-Mazz.) C. P'ei et R. H. Shan var. shimianensis K. F. Wu;石棉金粟兰;Shimian Chloranthus ■
88286 Chloranthus holostegius (Hand.-Mazz.) C. P'ei et R. H. Shan var. trichoneurus K. F. Wu;毛脉金粟兰;Hairvein Chloranthus ■
88287 Chloranthus hupehensis Pamp. = Chloranthus henryi var. hupehensis (Pamp.) K. F. Wu ■
88288 Chloranthus inconspicuus Sw. = Chloranthus spicatus (Thunb.) Makino ●
88289 Chloranthus japonicus Siebold;银钱草(白毛七,灯笼花,独摇草,分叶芹,拐拐细辛,鬼都邮,鬼督邮,鬼独摇草,胡苋眼,及已,假细辛,山油菜,四大金刚,四大天王,四代草,四季香,四块瓦,四匹瓦,四叶草,四叶对,四叶金,四叶七,四叶细辛,苏叶蒿,天王七,土细辛,万根丹,杨梅草,一黄珊,银线草);Japan Chloranthus,Japanese Chloranthus ■

88290 Chloranthus kiangsiensis F. P. Metcalf = Ardisia gigantifolia Stapf ●

88291 Chloranthus mandshuricus Rupr. = Chloranthus japonicus Siebold ■

88292 Chloranthus multistachys C. P'ei;多穗金粟兰(白毛七,大四块瓦,四大天王,四块瓦,四眼牛夕,四叶对,四叶细辛);Manyspike Chloranthus ■

88293 Chloranthus officinalis Blume = Chloranthus erectus (Buch.-Ham.) Verdc. ■

88294 Chloranthus oldhami sensu C. P'ei = Chloranthus sessifolius K. F. Wu var. austro-sinensis K. F. Wu ■

88295 Chloranthus oldhamii Solms;台湾金粟兰(东南金粟兰,台湾及已);Oldham Chloranthus,Taiwan Chloranthus ■

88296 Chloranthus pernyanus Solms;贵州金粟兰■

88297 Chloranthus serratus (Thunb.) Roem. et Schult.;及已(对叶四块瓦,金薄荷,老君须,牛细辛,四大金刚,四大天王,四大王,四儿风,四角金,四块瓦,四皮风,四叶对,四叶箭,四叶金,四叶莲,四叶麻,四叶细辛,獐耳细辛);Serrate Chloranthus ■

88298 Chloranthus serratus (Thunb.) Roem. et Schult. var. taiwanensis K. F. Wu;台湾及已;Taiwan Serrate Chloranthus ■

88299 Chloranthus sessifolius K. F. Wu;四川金粟兰(红毛七,四大天王,四块瓦);Sichuan Chloranthus ■

88300 Chloranthus sessifolius K. F. Wu var. austro-sinensis K. F. Wu;华南金粟兰;S. China Chloranthus,South China Chloranthus ■

88301 Chloranthus spicatus (Thunb.) Makino;金粟兰(茶兰,鸡脚兰,鸡爪兰,鱼子兰,珍珠菊,珍珠兰,真珠兰,珠兰);Chu-lan Tree ●

88302 Chloranthus tianmushanensis K. F. Wu;天目金粟兰;Tianmushan Chloranthus ■

88303 Chloraster Haw. = Narcissus L. ■

88304 Chlorea Nyl. = Rhytidocaulon Nyl. ex Elenkin(废弃属名)●☆

88305 Chloridaceae (Rchb.) Herter = Gramineae Juss.(保留科名)■●

88306 Chloridaceae (Rchb.) Herter = Poaceae Barnhart(保留科名)■●

88307 Chloridaceae Bercht. et J. Presl = Gramineae Juss.(保留科名)■●

88308 Chloridaceae Bercht. et J. Presl = Poaceae Barnhart(保留科名)■●

88309 Chloridaceae Herter = Gramineae Juss.(保留科名)■●

88310 Chloridaceae Herter = Poaceae Barnhart(保留科名)■●

88311 Chloridion Stapf = Stereochlaena Hack. ■☆

88312 Chloridion cameronii Stapf = Stereochlaena cameronii (Stapf) Pilg. ■☆

88313 Chloridiopsis J. Gay ex Scribn. = Trichloris E. Fourn. ex Benth. ■☆

88314 Chloridopsis Hack. = Trichloris E. Fourn. ex Benth. ■☆

88315 Chloridopsis Hort. ex Hack. = Chloridiopsis J. Gay ex Scribn. ■☆

88316 Chloris Sw. (1788);虎尾草属(棒槌草属,棒锤草属);Chloris, Finger Grass, Fingergrass, Finger-grass, Green Grass, Windmill Grass,Windmillgrass,Windmill-grass ●■

88317 Chloris abyssinica Hochst. ex A. Rich. = Chloris gayana Kunth ■

88318 Chloris alba J. Presl et C. Presl = Chloris virgata Sw. ■

88319 Chloris anomala B. S. Sun et Z. H. Hu = Chloris pycnothrix Trin. ■

88320 Chloris barbata Sw.;孟仁草;Barbat Fingergrass,Peacock-plume Grass,Swellen Chloris ■

88321 Chloris barbata Sw. var. formosana Honda = Chloris formosana (Honda) Keng ■

88322 Chloris boivinii A. Camus = Daknopholis boivinii (A. Camus) Clayton ■☆

88323 Chloris brachystachys Andersson = Chloris virgata Sw. ■

88324 Chloris breviseta Benth. = Chloris pilosa Schumach. ■☆

88325 Chloris canterai Arechav.;巴拉圭虎尾草;Paraguayan Windmill Grass ■☆

88326 Chloris canterai Arechav. var. grandiflora (Roseng. et Izag.) D. E. Anderson;大花巴拉圭虎尾草;Paraguayan Windmill Grass ■☆

88327 Chloris capensis (Houtt.) Thell. = Eustachys paspaloides (Vahl) Lanza et Mattei ■☆

88328 Chloris caudata Trin. ex Bunge = Chloris virgata Sw. ■

88329 Chloris cheesemanii Hack. ex Cheeseman = Enteropogon unispiceus (F. Muell.) Clayton ■

88330 Chloris compressa DC. = Chloris virgata Sw. ■

88331 Chloris cruciata (L.) Sw.;十字虎尾草;Cross-shaped Chloris ■

88332 Chloris cryptostachya J. A. Schmidt = Enteropogon prieurii (Kunth) Clayton ■☆

88333 Chloris curtipendula Michx. = Bouteloua curtipendula (Michx.) Torr. ■

88334 Chloris decora Nees ex Steud. = Chloris virgata Sw. ■

88335 Chloris digitata (Roxb.) Steud. = Chloris dolichostachya Lag. ■

88336 Chloris diluta Renvoize;稀薄虎尾草■☆

88337 Chloris distichophylla Lag. var. acuminata (Trin.) Hack. = Eustachys distichophylla (Lag.) Nees ■☆

88338 Chloris distichophylla Lag. var. argentina Hack. = Eustachys retusa (Lag.) Kunth ■☆

88339 Chloris divaricata R. Br.;铺散虎尾草;Australian Rhodes-grass, Spreading Windmill Grass ■☆

88340 Chloris divaricata R. Br. var. cynodontoides (Balansa) Lazarides;齿状铺散虎尾草■☆

88341 Chloris dolichostachya Lag. = Enteropogon dolichostachyus (Lag.) Keng ■

88342 Chloris elata Desv.;高虎尾草;Tall Windmill Grass ■☆

88343 Chloris elegans (Kunth) Roberty = Ctenium elegans Kunth ■☆

88344 Chloris elegans Kunth;雅致虎尾草;Feather Finger-grass, Feather-fingergrass ■☆

88345 Chloris elegans Kunth = Chloris virgata Sw. ■

88346 Chloris equitans Trin. = Eustachys paspaloides (Vahl) Lanza et Mattei ■☆

88347 Chloris falcata (L. f.) Sw. = Harpochloa falx (L. f.) Kuntze ■☆

88348 Chloris ferruginea Renvoize = Tetrapogon ferrugineus (Renvoize) S. M. Phillips ■☆

88349 Chloris filiformis Poir.;丝状虎尾草■☆

88350 Chloris flabellata (Hack.) Launert;扇状虎尾草■☆

88351 Chloris formosana (Honda) Keng;台湾虎尾草;Taiwan Chloris, Taiwan Fingergrass ■

88352 Chloris gayana Kunth;非洲虎尾草(盖氏虎尾草,无芒虎尾草);Africa Fingergrass,Rhodes Grass ■

88353 Chloris gayana Kunth subsp. oligostachys Barratte et Murb. = Chloris gayana Kunth ■

88354 Chloris glabrata Andersson = Chloris gayana Kunth ■

88355 Chloris guineensis Schumach. = Dactyloctenium aegyptium (L.) Willd. ■☆

88356 Chloris incompleta Roth = Chloris dolichostachya Lag. ■

88357 Chloris incompleta Roth = Enteropogon dolichostachyus (Lag.) Keng ■

88358 Chloris inflata Link = Chloris barbata Sw. ■

88359 Chloris intermedia A. Rich. = Chloris pycnothrix Trin. ■

88360 Chloris jubaensis Cope;朱巴虎尾草■☆

88361 Chloris lamproparia Stapf;亮虎尾草■☆
88362 Chloris leptostachya Hochst. ex A. Rich. = Chloris pycnothrix Trin. ■
88363 Chloris longiaristata Napper = Enteropogon longiaristata (Napper) Clayton ■☆
88364 Chloris macrantha Jaub. et Spach = Tetrapogon tenellus (J. König ex Roxb.) Chiov. ■☆
88365 Chloris macrostachya Hochst. ex A. Rich. = Enteropogon macrostachyus (Hochst. ex A. Rich.) Munro ex Benth. ■☆
88366 Chloris meccana Hochst. = Chloris barbata Sw. ■
88367 Chloris meccana Hochst. et Steud. ex Schltdl. = Chloris virgata Sw. ■
88368 Chloris meccana Schltdl. = Chloris virgata Sw. ■
88369 Chloris mensensis (Schweinf.) Cufod.;芒斯疏毛虎尾草■☆
88370 Chloris mossambicensis K. Schum.;莫桑比克虎尾草■☆
88371 Chloris mucronata Michx. = Dactyloctenium aegyptium (L.) P. Beauv. ■
88372 Chloris mucronata Michx. = Dactyloctenium aegyptium (L.) Willd. ■
88373 Chloris multiradiata Hochst. = Chloris virgata Sw. ■
88374 Chloris multiradiata Hochst. var. ragazzii Pirotta = Chloris virgata Sw. ■
88375 Chloris myosuroides Hook. f. = Schoenefeldia gracilis Kunth ■☆
88376 Chloris myriostachya Hochst. = Chloris roxburghiana Schult. ■☆
88377 Chloris myriostachya Hochst. var. minor Chiov. = Chloris roxburghiana Schult. ■☆
88378 Chloris nigra Hack. = Chloris pilosa Schumach. var. nigra (Hack.) Vanden Berghen ■☆
88379 Chloris notocoma Hochst. = Chloris virgata Sw. ■
88380 Chloris pallida (Edgew.) Hook. f. = Schoenefeldia gracilis Kunth ■☆
88381 Chloris parva Mimeur = Enteropogon prieurii (Kunth) Clayton ■☆
88382 Chloris paspaloides Hochst. = Eustachys paspaloides (Vahl) Lanza et Mattei ■☆
88383 Chloris pectinata Benth.;梳状虎尾草;Comb Windmill Grass ■☆
88384 Chloris penicellata Willd. ex Steud. = Chloris virgata Sw. ■
88385 Chloris perrieri A. Camus = Daknopholis boivinii (A. Camus) Clayton ■☆
88386 Chloris perrieri A. Camus var. aristata A. Camus = Daknopholis boivinii (A. Camus) Clayton ■☆
88387 Chloris petraea Thunb. = Eustachys paspaloides (Vahl) Lanza et Mattei ■☆
88388 Chloris pilosa Schumach.;疏毛虎尾草■☆
88389 Chloris pilosa Schumach. var. nigra (Hack.) Vanden Berghen;黑疏毛虎尾草■☆
88390 Chloris prieurii Kunth = Enteropogon prieurii (Kunth) Clayton ■☆
88391 Chloris pubescens Lag. = Chloris virgata Sw. ■
88392 Chloris pulchra Schumach. = Ctenium canescens Benth. ■☆
88393 Chloris punctulata Hochst. ex Steud. = Enteropogon prieurii (Kunth) Clayton ■☆
88394 Chloris pycnothrix Trin.;异序虎尾草;Differentspike Fingergrass ■
88395 Chloris quinquesetica Bhide;五毛虎尾草■☆
88396 Chloris radiata (L.) Sw.;辐射虎尾草■☆
88397 Chloris ramosissima A. Camus = Daknopholis boivinii (A. Camus) Clayton ■☆
88398 Chloris refuscens Steud. = Chloris barbata Sw. ■
88399 Chloris repens Hochst. = Chloris gayana Kunth ■
88400 Chloris robusta Stapf;粗壮虎尾草■☆
88401 Chloris rogeonii A. Chev. = Chloris virgata Sw. ■
88402 Chloris roxburghiana Schult.;罗克斯伯勒虎尾草■☆
88403 Chloris ruahensis Renvoize;鲁阿哈虎尾草■☆
88404 Chloris simplex Schumach. et Thonn. = Enteropogon macrostachyus (Hochst. ex A. Rich.) Munro ex Benth. ■☆
88405 Chloris somalensis Rendle = Chloris mensensis (Schweinf.) Cufod. ■☆
88406 Chloris spathacea Hochst. ex Steud. = Tetrapogon cenchriformis (A. Rich.) Clayton ■☆
88407 Chloris subaequigluma Rendle = Chrysochloa subaequigluma (Rendle) Swallen ■☆
88408 Chloris subtriflora Steud. = Chloris pilosa Schumach. ■☆
88409 Chloris tenella J. König ex Roxb. = Tetrapogon tenellus (J. König ex Roxb.) Chiov. ■☆
88410 Chloris tenera (J. Presl) Scribn. = Eustachys tener (J. Presl) A. Camus ■
88411 Chloris tibestica Quézel = Chloris virgata Sw. ■
88412 Chloris transiens Pilg. = Schoenefeldia transiens (Pilg.) Chiov. ■☆
88413 Chloris triangulata Hochst. ex A. Rich. = Tetrapogon tenellus (J. König ex Roxb.) Chiov. ■☆
88414 Chloris truncata R. Br.;截形虎尾草;Australian Fingergrass, Windmill-grass ■☆
88415 Chloris unispicea F. Muell. = Enteropogon unispiceus (F. Muell.) Clayton ■
88416 Chloris ventricosa R. Br.;澳洲虎尾草;Australian Windmill Grass ■☆
88417 Chloris verticillata Nutt.;轮生虎尾草;Showy Windmill Grass ■☆
88418 Chloris villosa (Desf.) Pers. = Tetrapogon villosus Desf. ■☆
88419 Chloris villosa (Desf.) Pers. var. sinaicus Decne. = Tetrapogon villosus Desf. ■☆
88420 Chloris villosa Pers. = Tetrapogon villosus Desf. ■☆
88421 Chloris virgata Sw.;虎尾草(棒锤草,盘草,刷头草,刷子头);Feather Finger Grass, Fingergrass, Showy Chloris ■
88422 Chloris virgata Sw. var. elegans (Kunth) Stapf = Chloris virgata Sw. ■
88423 Chloris woodii Renvoize;伍得虎尾草■☆
88424 Chlorita Raf. = Blackstonia Huds. ■☆
88425 Chlorita Raf. = Chlora Adans. ■☆
88426 Chloriza Salisb. = Lachenalia J. Jacq. ex Murray ■☆
88427 Chlorocalymma Clayton(1970);东非绿苞草属■☆
88428 Chlorocalymma cryptacanthum Clayton;东非绿苞草■☆
88429 Chlorocardium Rohwer, H. G. Richt. et van der Werff(1991);毒樟属(绿心樟属)●☆
88430 Chlorocardium rodiaei (R. H. Schomb.) Rohwer;罗迪毒樟(罗迪绿心樟);Greenheart ●☆
88431 Chlorocardium venenosum (Kosterm. et Pinkley) Rohwer, H. G. Richt. et van der Werff;毒樟●☆
88432 Chlorocarpa Alston(1931);绿果木属●☆
88433 Chlorocarpa pentaschista Alston;绿果木●☆
88434 Chlorocaulon Klotzsch = Chiropetalum A. Juss. ●☆
88435 Chlorocharis Rildi = Eleocharis R. Br. ■
88436 Chlorocharis vivipara (Link) Rikli = Eleocharis vivipara Link ■☆
88437 Chlorochlamys Miq. = Marsdenia R. Br. (保留属名)●
88438 Chlorochorion Puff et Robbr. (1989);绿膜茜属■☆
88439 Chlorochorion Puff et Robbr. = Pentanisia Harv. ■☆
88440 Chlorochorion foetidum (Verdc.) Puff et Robbr. = Pentanisia foetida Verdc. ■☆

88441 Chlorochorion monticola (K. Krause) Puff et Robbr. = Pentanisia monticola (K. Krause) Verdc. ■☆
88442 Chlorocodon (DC.) Fourr. = Erica L. ●☆
88443 Chlorocodon Fourr. = Erica L. ●☆
88444 Chlorocodon Hook. f. (1871);绿钟草属●■☆
88445 Chlorocodon Hook. f. = Mondia Skeels ●☆
88446 Chlorocodon ecornutus N. E. Br. = Mondia ecornuta (N. E. Br.) Bullock ●☆
88447 Chlorocodon whitei Hook. f.;非洲绿钟草■☆
88448 Chlorocodon whitei Hook. f. = Mondia whitei (Hook. f.) Skeels ●☆
88449 Chlorocrambe Rydb. (1907);绿色两节荠属■☆
88450 Chlorocrambe hastata (S. Watson) Rydb.;绿色两节荠■☆
88451 Chlorocrambe hastata Rydb. = Chlorocrambe hastata (S. Watson) Rydb. ■☆
88452 Chlorocrepis Griseb. = Hieracium L. ■
88453 Chlorocrepis albiflora (Hook.) W. A. Weber = Hieracium albiflorum Hook. ■☆
88454 Chlorocrepis fendleri (Sch. Bip.) W. A. Weber = Hieracium fendleri Sch. Bip. ■☆
88455 Chlorocrepis fendleri (Sch. Bip.) W. A. Weber var. discolor A. Gray = Hieracium fendleri Sch. Bip. ■☆
88456 Chlorocrepis tristis (Willd. ex Spreng.) Á. Löve et D. Löve = Hieracium triste Willd. ex Spreng. ■☆
88457 Chlorocyathus Oliv. (1887);绿杯萝藦属■☆
88458 Chlorocyathus lobulata (Venter et R. L. Verh.) Venter;绿杯萝藦■☆
88459 Chlorocyathus monteiroae Oliv. = Raphionacme monteiroae (Oliv.) N. E. Br. ■☆
88460 Chlorocyperus Rikli = Cyperus L. ■
88461 Chlorocyperus Rikli = Pycreus P. Beauv. ■
88462 Chlorocyperus inflexus (Muhl.) Palla = Mariscus aristatus (Rottb.) Ts. Tang et F. T. Wang ■
88463 Chlorocyperus laevigatus (L.) Palla = Cyperus laevigatus L. ■☆
88464 Chlorocyperus phymatodes (Muhl.) Palla = Cyperus esculentus L. var. leptostachyus Boeck. ■☆
88465 Chlorocyperus serotinus (Rottb.) Palla = Cyperus serotinus Rottb. ■
88466 Chlorodes Post et Kuntze = Chloris Sw. ●■
88467 Chlorodes Post et Kuntze = Chloroides Fisch. ●■
88468 Chlorogalaceae Doweld et Reveal = Agavaceae Dumort. (保留科名)●■
88469 Chlorogalum (Lindl.) Kunth (1843) (保留属名);皂百合属;Soap Plant ■☆
88470 Chlorogalum Kunth = Chlorogalum (Lindl.) Kunth (保留属名) ■☆
88471 Chlorogalum angustifolium Kellogg;窄叶皂百合■☆
88472 Chlorogalum divaricatum (Lindl.) Kunth = Chlorogalum pomeridianum Kunth var. divaricatum (Lindl.) Hoover ■☆
88473 Chlorogalum grandiflorum Hoover;大花皂百合;Red Hills Soap Plant ■☆
88474 Chlorogalum leichtlinii Baker = Camassia leichtlinii (Baker) S. Watson ■☆
88475 Chlorogalum parviflorum S. Watson;小花皂百合■☆
88476 Chlorogalum pomeridianum (DC.) Kunth;皂百合;Soap-plant, Soaproot, Wild Potato ■☆
88477 Chlorogalum pomeridianum (DC.) Kunth var. divaricatum (Lindl.) Hoover;宽叉皂百合■☆
88478 Chlorogalum pomeridianum (DC.) Kunth var. minus Hoover;小皂百合;Small Soap Plant ■☆
88479 Chlorogalum pomeridianum Kunth = Chlorogalum pomeridianum (DC.) Kunth ■☆
88480 Chlorogalum pomeridianum Kunth var. divaricatum (Lindl.) Hoover = Chlorogalum pomeridianum (DC.) Kunth var. divaricatum (Lindl.) Hoover ■☆
88481 Chlorogalum purpureum Brandegee;紫色皂百合;Purple Amole ■☆
88482 Chlorogalum purpureum Brandegee var. reductum Hoover;褪色紫色皂百合■☆
88483 Chloroides Fisch. = Chloris Sw. ●■
88484 Chloroides Fisch. ex Regel = Chloris Sw. ●■
88485 Chloroides Regel = Eustachys Desv. ■
88486 Chlorolepis Nutt. = Maschalanthus Nutt. ●☆
88487 Chloroleucon (Benth.) Britton et Rose = Pithecellobium Mart. (保留属名)●
88488 Chloroleucon (Benth.) Record = Albizia Durazz. ●
88489 Chloroleucon Britton et Rose ex Record = Pithecellobium Mart. (保留属名)●
88490 Chloroleucum (Benth.) Record = Chloroleucon (Benth.) Record ●
88491 Chloroluma Baill. = Chrysophyllum L. ●
88492 Chloromeles (Decne.) Decne. = Malus Mill. ●
88493 Chloromyron Pers. = Rheedia L. ●☆
88494 Chloromyron Pers. = Verticillaria Ruiz et Pav. ●☆
88495 Chloromyrtus Pierre = Eugenia L. ●
88496 Chloromyrtus klaineana Pierre = Eugenia klaineana (Pierre) Engl. ●☆
88497 Chloropatane Engl. = Erythrococca Benth. ●☆
88498 Chloropatane africana Engl. = Erythrococca rivularis (Müll. Arg.) Prain ●☆
88499 Chloropatane batesii C. H. Wright = Erythrococca welwitschiana (Müll. Arg.) Prain ●☆
88500 Chlorophora Gaudich. (1830);绿柄桑属(黄颜木属);Fustic Tree, Fustic-tree ●☆
88501 Chlorophora Gaudich. = Broussonetia L'Hér. ex Vent. (保留属名)●
88502 Chlorophora Gaudich. = Maclura Nutt. (保留属名)●
88503 Chlorophora alba A. Chev. = Milicia excelsa (Welw.) C. C. Berg ●☆
88504 Chlorophora brasiliensis (C. Mart.) Standl. ex J. F. Macbr. = Maclura brasiliensis Endl. ●☆
88505 Chlorophora excelsa (Welw.) Benth. et Hook. f. = Milicia excelsa (Welw.) C. C. Berg ●☆
88506 Chlorophora excelsa Benth. et Hook. f.;大绿柄桑;African Oak, African Teak, Bush Oak, Counter-wood, Iroko, Iroko Fustic Wood, Iroko Fustic-tree, Odum, Rock Elm, Swamp Mahogany, West African Mulberry ●☆
88507 Chlorophora mollis Fernald = Maclura tinctoria (L.) D. Don ex Steud. ●☆
88508 Chlorophora regia A. Chev.;高贵绿柄桑●☆
88509 Chlorophora regia A. Chev. = Milicia regia (A. Chev.) C. C. Berg ●☆
88510 Chlorophora reticulata Herzog = Maclura tinctoria (L.) D. Don ex Steud. ●☆
88511 Chlorophora scandens Standl. et L. O. Williams = Maclura

brasiliensis Endl. ●☆

88512 Chlorophora tenuifolia Engl. = Milicia excelsa (Welw.) C. C. Berg ●☆

88513 Chlorophora tinctoria (L.) Gaudich. = Maclura tinctoria (L.) D. Don ex Steud. ●☆

88514 Chlorophora tinctoria (L.) Gaudich. ex Benth. et Hook. f.;染料绿柄桑(黄颜木);Drug Fustic Tree, Drug Fustic-tree, Dyer's Mulberry, Fustic, Old Fustic ●☆

88515 Chlorophora tinctoria (L.) Gaudich. f. glabrescens Huber = Maclura tinctoria (L.) D. Don ex Steud. ●☆

88516 Chlorophora tinctoria (L.) Gaudich. f. miqueliana Hassl. = Maclura tinctoria (L.) D. Don ex Steud. ●☆

88517 Chlorophora tinctoria (L.) Gaudich. f. polyneura (Miq.) Hassl. = Maclura tinctoria (L.) D. Don ex Steud. ●☆

88518 Chlorophora tinctoria (L.) Gaudich. f. tataiiba Hassl. = Maclura tinctoria (L.) D. Don ex Steud. ●☆

88519 Chlorophora tinctoria (L.) Gaudich. subsp. mora (Griseb.) Hassl. = Maclura tinctoria (L.) D. Don ex Steud. ●☆

88520 Chlorophora tinctoria (L.) Gaudich. subsp. zanthoxylon (L.) Hassl. = Maclura tinctoria (L.) D. Don ex Steud. ●☆

88521 Chlorophora tinctoria (L.) Gaudich. var. acuminatissima Huber = Maclura tinctoria (L.) D. Don ex Steud. ●☆

88522 Chlorophora tinctoria (L.) Gaudich. var. affinis (Miq.) Hassl. = Maclura tinctoria (L.) D. Don ex Steud. ●☆

88523 Chlorophora tinctoria (L.) Gaudich. var. mora (Griseb.) Lillo = Maclura tinctoria (L.) D. Don ex Steud. ●☆

88524 Chlorophora tinctoria (L.) Gaudich. var. ovata (Bureau) Chodat = Maclura tinctoria (L.) D. Don ex Steud. ●☆

88525 Chlorophora tinctoria (L.) Gaudich. var. zanthoxylon (L.) Chodat = Maclura tinctoria (L.) D. Don ex Steud. ●☆

88526 Chlorophyllum Liais = Chrysophyllum L. ●

88527 Chlorophytum Ker Gawl. (1807);吊兰属(青百合属);Bernard's Lily, Bracketplant, Chlorophytum ■

88528 Chlorophytum Pohl ex DC. = Spermacoce L. ●■

88529 Chlorophytum acutum (C. H. Wright) Nordal;尖吊兰■☆

88530 Chlorophytum affine Baker;近缘吊兰■☆

88531 Chlorophytum affine Baker var. curviscapum (Poelln.) Hanid;弯茎近缘吊兰■☆

88532 Chlorophytum africanum (Baker) Engl.;非洲吊兰■☆

88533 Chlorophytum anceps (Baker) Kativu;二棱吊兰■☆

88534 Chlorophytum andongense Baker;安东吊兰■☆

88535 Chlorophytum angulicaule (Baker) Kativu;角茎吊兰■☆

88536 Chlorophytum angustissimum (Poelln.) Nordal;极细吊兰■☆

88537 Chlorophytum bichetii Baker;条纹叶吊兰(白纹草,银边吊兰);Dwarf Bernard's Lily ■☆

88538 Chlorophytum bifolium Dammer;双叶吊兰■☆

88539 Chlorophytum blepharophyllum Schweinf. ex Baker;百簕花叶吊兰■☆

88540 Chlorophytum bowkeri Baker;鲍克吊兰■☆

88541 Chlorophytum brachystachyum Baker;短穗吊兰;Shortspoke Bracketplant ■

88542 Chlorophytum calyptrocarpum (Baker) Kativu;盖果吊兰■☆

88543 Chlorophytum cameronii (Baker) Kativu;喀麦隆吊兰■☆

88544 Chlorophytum cameronii (Baker) Kativu var. grantii (Baker) Nordal;格兰特吊兰■☆

88545 Chlorophytum cameronii (Baker) Kativu var. pterocaulon (Welw. ex Baker) Nordal;翅茎吊兰■☆

88546 Chlorophytum capense (L.) Kuntze;高吊兰(八叶兰,大吊兰,大叶吊兰,吊兰,钓兰,金边贝母,金边草,金边吊兰,宽叶吊兰,阔叶吊兰,兰草,银边吊兰,银边兰,硬叶吊兰);Bracketplant, Bracket-plant, Spider Plant ■

88547 Chlorophytum capense (L.) Kuntze var. variegatum Hort.;银边吊兰(金边草,金边吊兰,银边兰)■

88548 Chlorophytum capense (L.) Voss = Chlorophytum capense (L.) Kuntze ■

88549 Chlorophytum caulescens (Baker) Marais et Reilly;无茎吊兰■☆

88550 Chlorophytum chinense Bureau et Franch.;狭叶吊兰;Chinese Bracketplant ■

88551 Chlorophytum collinum (Poelln.) Nordal;山丘吊兰■☆

88552 Chlorophytum colubrinum (Welw. ex Baker) Engl.;长穗吊兰■☆

88553 Chlorophytum comosum (Thunb.) Baker;吊兰(丛毛吊兰,倒挂兰,钓兰,挂兰,匍匐兰,树蕉瓜,洋吊兰,折鹤兰);Airplane Plant, Bracket Plant, Ribbon Plant, Spider Plant, Tufted Bracketplant ●■

88554 Chlorophytum comosum (Thunb.) Baker 'Mediopictum';多心吊兰■

88555 Chlorophytum comosum (Thunb.) Baker 'Milky Way';乳白吊兰■

88556 Chlorophytum comosum (Thunb.) Baker 'Picturatum';彩叶吊兰(斑心吊兰)■

88557 Chlorophytum comosum (Thunb.) Baker 'Variegatum';镶边吊兰(金边吊兰,银边吊兰);Spider Plant ■

88558 Chlorophytum comosum (Thunb.) Baker 'Vittatum';中斑吊兰(金心吊兰,条纹吊兰)■

88559 Chlorophytum comosum (Thunb.) Jacques = Chlorophytum comosum (Thunb.) Baker ●■

88560 Chlorophytum comosum (Thunb.) Jacques var. bipindense (Engl. et K. Krause) A. D. Poulsen et Nordal;比平迪吊兰(双齿吊兰)■☆

88561 Chlorophytum comosum (Thunb.) Jacques var. sparsiflorum (Baker) A. D. Poulsen et Nordal;疏花吊兰■☆

88562 Chlorophytum crassinerve (Baker) Oberm.;粗脉吊兰■☆

88563 Chlorophytum crispum (Thunb.) Baker;皱波吊兰■☆

88564 Chlorophytum dalzielii (Hutch. ex Hepper) Nordal;戴尔吊兰■☆

88565 Chlorophytum elatum R. Br.;大吊兰■☆

88566 Chlorophytum elatum R. Br. var. mediapictum Hort.;金心吊兰■☆

88567 Chlorophytum elatum R. Br. var. variegatum Voss;斑纹大吊兰■☆

88568 Chlorophytum fasciculatum (Baker) Kativu;簇生吊兰■☆

88569 Chlorophytum fischeri (Baker) Baker;菲舍尔吊兰■☆

88570 Chlorophytum flaccidum W. W. Sm. = Chlorophytum nepalense (Lindl.) Baker ■

88571 Chlorophytum floribundum Baker;繁花吊兰■☆

88572 Chlorophytum gallabatense Schweinf. ex Baker;加拉吊兰■☆

88573 Chlorophytum galpinii (Baker) Kativu;盖尔吊兰■☆

88574 Chlorophytum galpinii (Baker) Kativu var. matabalense (Baker) Kativu;马塔贝莱吊兰■☆

88575 Chlorophytum galpinii (Baker) Kativu var. norlindhii (Weim.) Kativu;诺尔吊兰■☆

88576 Chlorophytum geophilum Peter ex Poelln.;地花吊兰■☆

88577 Chlorophytum goetzei Engl.;戈策吊兰■☆

88578 Chlorophytum haygarthii J. M. Wood et M. S. Evans;海加斯吊兰■☆

88579 Chlorophytum holstii Engl.;霍尔吊兰■☆

88580 Chlorophytum immaculatum (Hepper) Nordal;无斑吊兰■☆

88581 Chlorophytum inconspicuum (Baker) Nordal;显著吊兰■☆

88582 Chlorophytum inornatum Ker Gawl.;无饰吊兰■☆
88583 Chlorophytum khasianum Hook. f. = Chlorophytum nepalense (Lindl.) Baker ■
88584 Chlorophytum krauseanum (Dinter) Kativu;克鲁斯吊兰■☆
88585 Chlorophytum krookianum Zahlbr.;克鲁科吊兰■☆
88586 Chlorophytum lancifolium Welw. ex Baker;披针叶吊兰■☆
88587 Chlorophytum lancifolium Welw. ex Baker subsp. cordatum (Engl.) A. D. Poulsen et Nordal;心叶克鲁科吊兰■☆
88588 Chlorophytum lancifolium Welw. ex Baker subsp. togoense (Engl.) A. D. Poulsen et Nordal;多哥吊兰■☆
88589 Chlorophytum laxum R. Br.;小花吊兰(三角草,山韭菜,疏花吊兰,水三棱草,丝毛草,土冬青,土麦冬,银边草);Smallflower Bracketplant ■
88590 Chlorophytum leptoneurum (C. H. Wright) Poelln.;细脉吊兰■☆
88591 Chlorophytum limosum (Baker) Nordal;沼生吊兰■☆
88592 Chlorophytum linearifolium Marais et Reilly;线叶吊兰●☆
88593 Chlorophytum longifolium Schweinf. ex Baker;长叶吊兰■☆
88594 Chlorophytum macrophyllum (A. Rich.) Asch.;大叶吊兰■☆
88595 Chlorophytum macrosporum Baker;大籽吊兰■☆
88596 Chlorophytum madagscariensis Baker;马达加斯加吊兰■☆
88597 Chlorophytum majus (Hemsl.) Marais et Reilly = Diuranthera major Hemsl. ■
88598 Chlorophytum malayense Ridl.;马拉吊兰;Bigleaf Bracketplant ■
88599 Chlorophytum mekongense W. W. Sm. = Chlorophytum nepalense (Lindl.) Baker ■
88600 Chlorophytum modestum Baker;适度吊兰■☆
88601 Chlorophytum namaquense Poelln.;纳马夸吊兰■☆
88602 Chlorophytum nepalense (Lindl.) Baker;西南吊兰;Nepal Bracketplant ■
88603 Chlorophytum nidulans (Baker) Brenan;鸟巢状吊兰■☆
88604 Chlorophytum nubicum (Baker) Kativu;鸟巢吊兰■☆
88605 Chlorophytum nyassae (Rendle) Kativu;尼萨吊兰■☆
88606 Chlorophytum orchidastrum Lindl. = Chlorophytum malayense Ridl. ■
88607 Chlorophytum oreogenes W. W. Sm. = Chlorophytum nepalense (Lindl.) Baker ■
88608 Chlorophytum parviflorum (Wight) Dalzell = Chlorophytum laxum R. Br. ■
88609 Chlorophytum paucinervatum (Poelln.) Nordal;寡脉吊兰■☆
88610 Chlorophytum plarystemon Diels = Chlorophytum chinense Bureau et Franch. ■
88611 Chlorophytum polystachys Baker;多穗吊兰■☆
88612 Chlorophytum pubiflorum Baker;毛花吊兰■☆
88613 Chlorophytum pusillum Schweinf. ex Baker;微小吊兰■☆
88614 Chlorophytum pygmaeum (Weim.) Kativu;矮小吊兰■☆
88615 Chlorophytum pygmaeum (Weim.) Kativu subsp. rhodesianum (Rendle) Kativu;罗得西亚矮小吊兰■☆
88616 Chlorophytum radula (Baker) Nordal;刮刀吊兰■☆
88617 Chlorophytum rangei (Engl. et K. Krause) Nordal;朗热吊兰■☆
88618 Chlorophytum recurvifolium (Baker) C. Archer et Kativu;卷叶吊兰■☆
88619 Chlorophytum rigidum Kunth;硬吊兰■☆
88620 Chlorophytum ruahense Engl.;鲁阿哈吊兰■☆
88621 Chlorophytum rubribracteatum (De Wild.) Kativu;红苞吊兰■☆
88622 Chlorophytum saundersiae (Baker) Nordal;桑德斯吊兰■☆
88623 Chlorophytum scabrum Baker;粗糙吊兰■☆
88624 Chlorophytum senegalense (Baker) Hepper;塞内加尔吊兰●☆
88625 Chlorophytum silvaticum Dammer;林吊兰■☆
88626 Chlorophytum somaliense Baker;索马里吊兰■☆
88627 Chlorophytum sphacelatum (Baker) Kativu;毒吊兰■☆
88628 Chlorophytum sphacelatum (Baker) Kativu var. hockii (De Wild.) Nordal;霍克吊兰■☆
88629 Chlorophytum sphacelatum (Baker) Kativu var. milanjianum (Rendle) Nordal;米拉吊兰■☆
88630 Chlorophytum stenopetalum Baker;窄瓣吊兰■●☆
88631 Chlorophytum stolzii (K. Krause) Kativu;斯托尔兹吊兰■☆
88632 Chlorophytum subpetiolatum (Baker) Kativu;微梗吊兰■☆
88633 Chlorophytum suffruticosum Baker;亚灌木吊兰●☆
88634 Chlorophytum superpositum (Baker) Marais et Reilly;后吊兰■☆
88635 Chlorophytum tetraphyllum (L. f.) Baker;四叶吊兰■☆
88636 Chlorophytum transvaalense (Baker) Kativu;德兰士瓦吊兰■☆
88637 Chlorophytum trichophlebium (Baker) Nordal;毛脉吊兰■☆
88638 Chlorophytum triflorum (Aiton) Kunth;三花吊兰■☆
88639 Chlorophytum tuberosum (Roxb.) Baker;块根吊兰■☆
88640 Chlorophytum undulatum (Jacq.) Oberm.;波状吊兰■☆
88641 Chlorophytum vestitum Baker;被吊兰■☆
88642 Chlorophytum viridescens Engl.;绿吊兰■☆
88643 Chlorophytum viscosum Kunth;黏吊兰■☆
88644 Chlorophytum warneckei (Engl.) Marais et Reilly;沃内克吊兰■☆
88645 Chlorophytum zavattari (Cufod.) Nordal;扎瓦吊兰■☆
88646 Chloropsis Hack. ex Kuntze = Trichloris E. Fourn. ex Benth. ■☆
88647 Chloropsis Kuntze = Trichloris E. Fourn. ex Benth. ■☆
88648 Chloropyron Behr = Cordylanthus Nutt. ex Benth. (保留属名)■☆
88649 Chlorosa Blume = Cryptostylis R. Br. ■
88650 Chlorosa latifolia Blume = Cryptostylis arachnites (Blume) Blume ■
88651 Chlorospatha Engl. (1878);绿苞南星属■☆
88652 Chlorospatha atropurpurea (Madison) Madison;深紫绿苞南星■☆
88653 Chlorospatha longipoda (K. Krause) Madison;长足绿苞南星■☆
88654 Chlorostelma Welw. ex Rendle = Asclepias L. ■
88655 Chlorostelma fritillarioides Welw. ex Rendle = Glossostelma lisianthoides (Decne.) Bullock ■☆
88656 Chlorostemma (Lange) Fourr. = Asperula L. (保留属名)■
88657 Chlorostemma Fourr. = Galium L. ■●
88658 Chlorostis Raf. = Chloris Sw. ●■
88659 Chloroxylon DC. (1824)(保留属名);绿木树属●☆
88660 Chloroxylon Raf. = Diospyros L. ●
88661 Chloroxylon Rumph. ex Scop. = Chloroxylon DC. (保留属名)●☆
88662 Chloroxylon Scop. = Chloroxylon DC. (保留属名)●☆
88663 Chloroxylon faho Capuron;绿木树(东印度缎,东印沙木,沙丁木,锡兰缎);Ceylon Satinwood, East Indian Satinwood, Satin Wood, Satinwood ●☆
88664 Chloroxylon falcatum Capuron;镰形绿木树●☆
88665 Chloroxylon swietenia Capuron = Chloroxylon faho Capuron ●☆
88666 Chloroxylon swietenia DC. = Chloroxylon faho Capuron ●☆
88667 Chloroxylum P. Browne(废弃属名) = Chloroxylon DC. (保留属名)●☆
88668 Chloroxylum P. Browne(废弃属名) = Ziziphus Mill. ●
88669 Chloroxylum Post et Kuntze = Chloroxylon DC. (保留属名)●☆
88670 Chloryllis E. Mey. = Dolichos L. (保留属名)■
88671 Chloryllis pratensis E. Mey. = Dolichos pratensis (E. Mey.) Taub. ●☆
88672 Chloryta Raf. = Blackstonia Huds. ■☆
88673 Chloryta Raf. = Chlora Adans. ■☆

88674 Chnoanthus Phil. = Gomphrena L. ●■

88675 Choananthus Rendle = Scadoxus Raf. ■

88676 Chocho Adans. = Sechium P. Browne(保留属名)■

88677 Chodanthus Hassl. = Mansoa DC. ●☆

88678 Chodaphyton Minod = Stemodia L.(保留属名)■☆

88679 Chodondendron Bosc = Chondrodendron Ruiz et Pav. ●☆

88680 Chodsha-Kasiana Rauschert = Catenularia Botsch. ■☆

88681 Chodsha-Kasiana Rauschert = Catenulina Soják ■☆

88682 Choenomeles Lindl. = Chaenomeles Lindl.(保留属名)●

88683 Choeradodia Herb. = Strumaria Jacq. ■☆

88684 Choeradoplectron Schauer = Habenaria Willd. ■

88685 Choeradoplectron Schauer = Peristylus Blume(保留属名)■

88686 Choeradoplectron spiranthes Schauer = Peristylus lacertifer (Lindl.) J. J. Sm. ■

88687 Choerophillum Neck. = Choerophyllum Brongn. ■

88688 Choerophyllum Brongn. = Chaerophyllum L. ■

88689 Choeroseris Link = Picris L. ■

88690 Choerospondias B. L. Burtt et A. W. Hill(1937);南酸枣属;Choerospondias, Southern Wildjujube ●

88691 Choerospondias axillaris (Roxb.) B. L. Burtt et A. W. Hill;南酸枣(鼻涕果,鼻子果,醋酸果,广枣,花心木,黄酸枣,货郎果,啃不死,连麻树,棉麻树,人面子,山桉果,山枣,山枣木,山枣树,山枣子,四眼果,酸醋树,酸枣,酸枣树,五根果,五眼果,五眼睛果,枣,枣子);Axillary Choerospondias, Caja, Choerospondias, Cirnela, Golden Apple, Hog-plum, Jamaica Plum, Jobo, Mombin, Southern Choerospondias, Spanish Plum, Yellow Mombin ●

88692 Choerospondias axillaris (Roxb.) B. L. Burtt et A. W. Hill var. japonica (Ohwi) Ohwi = Choerospondias axillaris (Roxb.) B. L. Burtt et A. W. Hill ●

88693 Choerospondias axillaris (Roxb.) B. L. Burtt et A. W. Hill var. pubinervis (Rehder et E. H. Wilson) Burtt et Hill;毛脉南酸枣;Hairynerve Choerospondias, Southern ●

88694 Choetophora Franch. et Sav. = Chaetospora R. Br. ■

88695 Choetophora Franch. et Sav. = Schoenus L. ■

88696 Choisya Kunth(1823);墨西哥橘属;Mexican Orange, Mexican Orange Blossom ●☆

88697 Choisya arizonica Standl.;矮墨西哥橘●☆

88698 Choisya dumosa A. Gray;灌丛橘;Mexican Orange ●☆

88699 Choisya mollis Standl.;墨西哥毛橘●☆

88700 Choisya ternata Kunth;墨西哥橘;American Orange-flower, Mexican Orange, Mexican Orange Blossom, Mexican Orange Flower ●☆

88701 Choisya ternata Kunth 'Sundance';金叶墨西哥橘;Golden-leaved Mexican Orange ●☆

88702 Cholisma Greene = Lyonia Nutt.(保留属名)●

88703 Cholisma Greene = Xolisma Raf. ●

88704 Chomelia Jacq.(1760)(保留属名);肖乌口树属●☆

88705 Chomelia Jacq. = Anisomeris C. Presl ●☆

88706 Chomelia L.(废弃属名) = Chomelia Jacq.(保留属名)●☆

88707 Chomelia L.(废弃属名) = Tarenna Gaertn. ●

88708 Chomelia Vell. = Ilex L. ●

88709 Chomelia affinis K. Schum. = Tarenna pavettoides (Harv.) Sim subsp. affinis (K. Schum.) Bridson ●☆

88710 Chomelia alleizettei Dubard et Dop = Tarenna alleizettei (Dubard et Dop) De Block ●☆

88711 Chomelia angolensis (Hiern) Kuntze = Crossopteryx febrifuga (Afzel. ex G. Don) Benth. ■☆

88712 Chomelia apiculata De Wild.;细尖乌口树●☆

88713 Chomelia bipindensis K. Schum. = Tarenna bipindensis (K. Schum.) Bremek. ●☆

88714 Chomelia buchananii K. Schum. = Crossopteryx febrifuga (Afzel. ex G. Don) Benth. ■☆

88715 Chomelia conferta (Hiern) Kuntze = Tarenna conferta (Benth.) Hiern ●☆

88716 Chomelia congensis (Hiern) Kuntze = Tarenna congensis Hiern ●☆

88717 Chomelia fusco-flava K. Schum. = Tarenna fusco-flava (K. Schum.) S. Moore ●☆

88718 Chomelia gilletii De Wild. et T. Durand = Tarenna gilletii (De Wild. et T. Durand) N. Hallé ex Gereau ●☆

88719 Chomelia gracilipes Hayata = Tarenna gracilipes (Hayata) Ohwi ●

88720 Chomelia grandiflora (Benth.) Kuntze = Tarenna grandiflora (Benth.) Hiern ●☆

88721 Chomelia junodii Schinz = Tarenna junodii (Schinz) Bremek. ●☆

88722 Chomelia kotocensis Hayata = Tarenna zeylanica Gaertn. ●

88723 Chomelia lancifolia Hayata = Tarenna gracilipes (Hayata) Ohwi ●

88724 Chomelia lasioclada K. Krause = Pavetta lasioclada (K. Krause) Mildbr. ex Bremek. ●☆

88725 Chomelia laurentii De Wild. = Tarenna laurentii (De Wild.) J. G. Garcia ●☆

88726 Chomelia laxissima K. Schum. = Tarenna fusco-flava (K. Schum.) S. Moore ●☆

88727 Chomelia leucodermis K. Krause = Vangueriella spinosa (Schumach. et Thonn.) Verdc. ●☆

88728 Chomelia longifolia De Wild. = Pavetta wildemannii Bremek. ●☆

88729 Chomelia mechowiana K. Schum.;梅休乌口树●☆

88730 Chomelia mossambicensis (Hiern) Kuntze = Crossopteryx febrifuga (Afzel. ex G. Don) Benth. ■☆

88731 Chomelia neurocarpa K. Schum. = Tarenna grandiflora (Benth.) Hiern ●☆

88732 Chomelia nigrescens (Hook. f.) Kuntze = Coptosperma nigrescens Hook. f. ●☆

88733 Chomelia nilotica (Hiern) Kuntze = Tarenna nilotica Hiern ●☆

88734 Chomelia nitidula (Hiern) Kuntze = Tarenna nitidula (Benth.) Hiern ●☆

88735 Chomelia odora K. Krause = Coptosperma neurophyllum (S. Moore) Degreef ●☆

88736 Chomelia oligantha K. Schum. et K. Krause = Coptosperma graveolens (S. Moore) Degreef ●☆

88737 Chomelia oligoneura K. Schum. = Tarenna pallidula Hiern var. oligoneura (K. Schum.) N. Hallé ●☆

88738 Chomelia pallidula (Hiern) Kuntze = Tarenna pallidula Hiern ●☆

88739 Chomelia subcapitata K. Schum. et K. Krause = Cladoceras subcapitatum (K. Schum. et K. Krause) Bremek. ●☆

88740 Chomelia tetramera (Hiern) Kuntze = Pavetta tetramera (Hiern) Bremek. ●☆

88741 Chomelia ulugurensis K. Schum. = Tarenna pavettoides (Harv.) Sim subsp. affinis (K. Schum.) Bridson ●☆

88742 Chomutowia B. Fedtsch. = Acantholimon Boiss.(保留属名)●

88743 Chona D. Don = Erica L. ●☆

88744 Chonais Salisb. = Hippeastrum Herb.(保留属名)■

88745 Chondilophyllum Panch. ex Guillaumin = Meryta J. R. Forst. et G. Forst. ●☆

88746 Chondodendron Benth. et Hook. f. = Odontocarya Miers ●☆

88747 Chondodendron Ruiz et Pav. = Chondrodendron Ruiz et Pav. ●☆

88748 Chondodendron macrophyllum Hiern = Triclisia sacleuxii (Pierre) Diels ●☆
88749 Chondodendron tomentosum Ruiz et Pav.;绒毛乌口树;Pareira ●☆
88750 Chondrachna T. Post et Kuntze = Chondrachne R. Br. ■
88751 Chondrachne R. Br. = Lepironia Pers. ■
88752 Chondrachyrum Nees = Briza L. ■
88753 Chondrachyrum Nees = Melica L. ■
88754 Chondradenia Maxim. ex Maekawa = Orchis L. ■
88755 Chondradenia Maxim. ex Makino = Orchis L. ■
88756 Chondradenia doyonensis (Hand.-Mazz.) Verm. = Platanthera roseotincta (W. W. Sm.) Ts. Tang et F. T. Wang ■
88757 Chondradenia fauriei (Finet) Sawada ex F. Maek. = Orchis fauriei Finet ■☆
88758 Chondraphylla A. Nelson = Gentiana L. ■
88759 Chondrilla L. (1753);粉苞菊属(苞粉菊属,粉苞苣属);Gum Succory,Skeletonweed,Spanish Succory ■
88760 Chondrilla acantholepis Boiss.;刚毛粉苞苣■☆
88761 Chondrilla ambigua Fisch. ex Kar. et Kir.;沙地粉苞菊(无喙粉苞苣,疑粉苞苣);Sandy Skeletonweed ■
88762 Chondrilla articulata Rodin = Chondrilla lejosperma Kar. et Kir. ■
88763 Chondrilla aspera (Schrad. ex Willd.) Poir.;硬叶粉苞菊(卵叶粉苞苣)■
88764 Chondrilla baicalensis (Ledeb.) Sch. Bip. = Youngia tenuifolia (Willd.) Babc. et Stebbins ■
88765 Chondrilla bosseana Iljin;鲍氏粉苞菊■☆
88766 Chondrilla brevirostris Fisch. et C. A. Mey.;短喙粉苞菊(短喙粉苞苣);Short-beaked Skeletonweed ■
88767 Chondrilla canescens Kar. et Kir.;灰白粉苞菊■
88768 Chondrilla chinensis (Thunb.) Poir. = Ixeridium chinense (Thunb.) Tzvelev ■
88769 Chondrilla debilis (Thunb.) Poir. = Ixeris japonica (Burm. f.) Nakai ■
88770 Chondrilla debilis Poir. = Ixeris japonica (Burm. f.) Nakai ■
88771 Chondrilla filifolia Iljin;线叶粉苞苣;Linearleaf Skeletonweed ■☆
88772 Chondrilla graminea M. Bieb.;禾叶粉苞苣■☆
88773 Chondrilla gummifera Iljin;胶粉苞苣■☆
88774 Chondrilla hastata Wall. = Chaetoseris hastata (Wall. ex DC.) C. Shih ■
88775 Chondrilla japonica (L.) Lam. = Youngia japonica (L.) DC. ■
88776 Chondrilla juncea L.;灯芯草粉苞苣(狭叶粉苞苣);Chondrilla,Gum Succory,Rush Skeletonweed,Skeleton Weed ■☆
88777 Chondrilla juncea L. var. canescens (Kar. et Kir.) Steud. = Chondrilla canescens Kar. et Kir. ■
88778 Chondrilla juncea L. var. canescens (Kar. et Kir.) Trautv. = Chondrilla canescens Kar. et Kir. ■
88779 Chondrilla kossinskyi Iljin;柯辛氏粉苞苣■☆
88780 Chondrilla kusnezovii Iljin;库兹涅佐夫粉苞苣■☆
88781 Chondrilla lanceolata (Houtt.) Poir. = Crepidiastrum lanceolatum (Houtt.) Nakai ■
88782 Chondrilla laticoronata Leonova;宽冠粉苞苣■
88783 Chondrilla latifolia M. Bieb.;宽叶粉苞苣■☆
88784 Chondrilla lejosperma Kar. et Kir.;刺苞粉苞苣(北疆粉苞菊);N. Xingjiang Skeletonweed ■
88785 Chondrilla longifolia Wall. = Lactuca dolichophylla Kitam. ■
88786 Chondrilla lyrata Poir. = Youngia japonica (L.) DC. ■
88787 Chondrilla macra Iljin;大粉苞苣■☆
88788 Chondrilla macrocarpa Leonova;大果粉苞苣;Bigfruit Skeletonweed ■
88789 Chondrilla maracandica Bunge;上节粉苞苣■
88790 Chondrilla nudicaulis L. = Launaea nudicaulis (L.) Hook. f. ■☆
88791 Chondrilla ornata Iljin;中亚粉苞苣■
88792 Chondrilla pauciflora D. Don = Pyrrhopappus pauciflorus (D. Don) DC. ■☆
88793 Chondrilla pauciflora Ledeb.;少花粉苞菊(少花粉苞苣);Fewflower Skeletonweed ■
88794 Chondrilla phaeocephala Rupr.;暗苞粉苞菊(暗苞粉苞苣)■
88795 Chondrilla piptocoma Fisch. et Mey. = Chondrilla piptocoma Fisch. Mey. et Avé-Lall. ■
88796 Chondrilla piptocoma Fisch. Mey. et Avé-Lall.;粉苞菊(粉苞苣);Common Skeletonweed ■
88797 Chondrilla polydichotoma Ostenf. = Hexinia polydichotoma (Ostenf.) H. L. Yang ■
88798 Chondrilla polydichotoma Ostenf. = Zollikoferia polydichotoma (Ostenf.) Iljin ■
88799 Chondrilla rouillieri Kar. et Kir.;基叶粉苞菊(基节粉苞苣)■
88800 Chondrilla sagittata Wall. = Ixeridium sagittaroides (C. B. Clarke) C. Shih ■
88801 Chondrilla stricta Ledeb. = Chondrilla aspera (Schrad. ex Willd.) Poir. ■
88802 Chondrocarpus Nutt. = Hydrocotyle L. ■
88803 Chondrocarpus Steven = Astragalus L. ●■
88804 Chondrochilus Phil. = Chaetanthera Ruiz et Pav. ■☆
88805 Chondrochlaena Kuntze = Prionanthium Desv. ■☆
88806 Chondrochlaena Post et Kuntze = Chondrolaena Nees ■☆
88807 Chondrochlaena Post et Kuntze = Prionanthium Desv. ■☆
88808 Chondrococcus Steyerm. = Coccochondra Rauschert☆
88809 Chondrococcus laevis Steyerm. = Coccochondra laevis (Steyerm.) Rauschert☆
88810 Chondrodendron Ruiz et Pav. (1794);粉毒藤属(甘蜜树属,谷树属,南美防己属)●☆
88811 Chondrodendron Spreng. = Chondrodendron Ruiz et Pav. ●☆
88812 Chondrodendron platyphyllum (A. St.-Hil.) Miers;阔叶粉毒藤(阔叶南美防己,南美防己,南美木防己)●☆
88813 Chondrodendron polyanthum Diels;多花粉毒藤(多花谷树)●☆
88814 Chondrodendron tomentocarpum (Rusby) Moldenke;毛果粉毒藤(阔叶南美防己,毛果南美防己)●☆
88815 Chondrodendron tomentosum Ruiz et Pav.;粉毒藤(南美防己);Pareirs Brava,Pareirs Root ●☆
88816 Chondrodendron toxicoferum (Wedd.) Krukoff et Moldenke;毒箭藤●☆
88817 Chondrolaena Nees = Prionachne Nees ■☆
88818 Chondrolaena Nees = Prionanthium Desv. ■☆
88819 Chondrolomia Nees = Scleria P. J. Bergius ■
88820 Chondropetalon Raf. = Chondropetalum Rottb. ■☆
88821 Chondropetalum Rottb. (1772);软骨瓣属■☆
88822 Chondropetalum acockii Pillans;阿科科软骨瓣■☆
88823 Chondropetalum aggregatum (Mast.) Pillans;聚集软骨瓣■☆
88824 Chondropetalum albo-aristatum Pillans = Askidiosperma albo-aristatum (Pillans) H. P. Linder ■☆
88825 Chondropetalum capitatum (Steud.) Pillans = Askidiosperma capitatum Steud. ■☆
88826 Chondropetalum chartaceum (Pillans) Pillans = Askidiosperma chartaceum (Pillans) H. P. Linder ■☆
88827 Chondropetalum decipiens Esterh.;迷惑软骨瓣■☆

88828 Chondropetalum deustum Rottb.;焦色软骨瓣■☆
88829 Chondropetalum ebracteatum (Kunth) Pillans;无苞软骨瓣■☆
88830 Chondropetalum esterhuyseniae Pillans = Askidiosperma esterhuyseniae (Pillans) H. P. Linder ■☆
88831 Chondropetalum hookerianum (Mast.) Pillans;胡克软骨瓣■☆
88832 Chondropetalum longiflorum Pillans = Askidiosperma longiflorum (Pillans) H. P. Linder ■☆
88833 Chondropetalum macrocarpum (Kunth) Pillans = Dovea macrocarpa Kunth ■☆
88834 Chondropetalum marlothii (Pillans) Pillans;马洛斯软骨瓣■☆
88835 Chondropetalum microcarpum (Kunth) Pillans;小果软骨瓣■☆
88836 Chondropetalum mucronatum (Nees) Pillans;短尖软骨瓣■☆
88837 Chondropetalum nitidum (Mast.) Pillans = Askidiosperma nitidum (Mast.) H. P. Linder ■☆
88838 Chondropetalum nudum Rottb.;裸露软骨瓣■☆
88839 Chondropetalum paniculatum (Mast.) Pillans = Askidiosperma paniculatum (Mast.) H. P. Linder ■☆
88840 Chondropetalum rectum (Mast.) Pillans;直立软骨瓣■☆
88841 Chondropetalum tectorum (L. f.) Pillans = Chondropetalum tectorum (L. f.) Raf. ■☆
88842 Chondropetalum tectorum (L. f.) Raf.;软骨瓣;Cape Rush ■☆
88843 Chondrophora Raf. = Bigelowia DC.(保留属名)●☆
88844 Chondrophora Raf. ex Porter et Britton = Bigelowia DC.(保留属名)●☆
88845 Chondrophylla (Bunge) A. Nelson = Gentiana L. ■
88846 Chondrophylla (Bunge) A. Nelson(1904);脆叶龙胆属■☆
88847 Chondrophylla A. Nelson = Chondrophylla (Bunge) A. Nelson ■☆
88848 Chondrophylla A. Nelson = Gentiana L. ■
88849 Chondrophylla aquatica (L.) W. A. Weber;脆叶龙胆■☆
88850 Chondropsis Raf. = Exacum L. ●■
88851 Chondropyxis D. A. Cooke(1986);长果鼠麹草属■☆
88852 Chondropyxis halophila D. A. Cooke;长果鼠麹草■☆
88853 Chondrorhyncha (Rchb. f.) Garay = Stenia Lindl. ■☆
88854 Chondrorhyncha Lindl. (1846);喙柱兰属(康多兰属);Chondrorhyncha ■☆
88855 Chondrorhyncha amazonica (Rchb. f. et Warsz.) A. D. Hawkes;亚马逊喙柱兰(亚马逊康多兰);Amazon Chondrorhyncha ■☆
88856 Chondrorhyncha aromatica (Rchb. f.) P. H. Allen;芳香喙柱兰(芳香康多兰);Aromatic Chondrorhyncha ■☆
88857 Chondrorhyncha chestertonii Rchb. f.;切氏喙柱兰(切氏康多兰)■☆
88858 Chondrorhyncha discolor (Lindl.) P. H. Allen = Warszewiczella discolor Rchb. f. ■☆
88859 Chondrorhyncha fimbriata (Lindl. et Rchb. f.) Rchb. f.;流苏喙柱兰(流苏康多兰);Fimbriate Chondrorhyncha ■☆
88860 Chondrorhyncha flabelliformis (Sw.) Alain;鞭形喙柱兰(鞭形康多兰)■☆
88861 Chondrosaceae Link = Gramineae Juss.(保留科名)■●
88862 Chondrosaceae Link = Poaceae Barnhart(保留科名)■●
88863 Chondroscaphe (Dressler) Senghas et G. Gerlach = Chondrorhyncha Lindl. ■☆
88864 Chondroscaphe (Dressler) Senghas et G. Gerlach = Zygopetalum Hook. ■☆
88865 Chondrosea Haw. = Saxifraga L. ■
88866 Chondrosea likiangensis (Franch.) Losinsk. = Saxifraga likiangensis Franch. ■
88867 Chondrosea pulchra (Engl. et Irmsch.) Losinsk. = Saxifraga pulchra Engl. et Irmsch. ■
88868 Chondrosea saxatilis (Harry Sm.) Losinsk. = Saxifraga kansuensis Mattf. ■
88869 Chondrosea saxatilis (Harry Sm.) Losinsk. = Saxifraga saxatilis Harry Sm. ■
88870 Chondrosea saxatilis (Harry Sm.) Losinsk. = Saxifraga unguipetala Engl. et Irmsch. ■
88871 Chondrosea unguipetala (Engl. et Irmsch.) Losinsk. = Saxifraga unguipetala Engl. et Irmsch. ■
88872 Chondrosia Benth. = Chondrosium Desv. ■☆
88873 Chondrosium Desv. = Chondrosum Desv. ■☆
88874 Chondrosium gracile Kunth = Bouteloua gracilis (Kunth) Lag. ex Steud. ■
88875 Chondrospermum Wall. = Myxopyrum Blume ●
88876 Chondrospermum Wall. ex G. Don = Myxopyrum Blume ●
88877 Chondrostylis Boerl. (1897);骨柱大戟属●☆
88878 Chondrostylis bancana Boerl.;骨柱大戟●☆
88879 Chondrosum Desv. (1810);砂垂穗草属■☆
88880 Chondrosum Desv. = Botelua Lag. ■
88881 Chondrosum gracile Kunth = Bouteloua gracilis (Kunth) Lag. ex Steud. ■
88882 Chondrosum gracile Willd. ex Kunth;纤细砂垂穗草;Blue Grama ■☆
88883 Chondrosum gracile Willd. ex Kunth = Bouteloua gracilis (Kunth) Lag. ex Steud. ■
88884 Chondrosum hirsutum (Lag.) Kunth = Bouteloua hirsuta Lag. ■
88885 Chondrosum oligostachyum (Nutt.) Torr. = Bouteloua gracilis (Kunth) Lag. ex Steud. ■
88886 Chondylophyllum Panch. ex R. Vig. = Meryta J. R. Forst. et G. Forst. ●☆
88887 Chone Dulac = Eupatorium L. ■●
88888 Chonemorpha G. Don (1837)(保留属名);鹿角藤属;Antlevine, Chonemorpha ●
88889 Chonemorpha antidysenterica (L.) G. Don = Holarrhena pubescens (Buch. -Ham.) Wall. ex G. Don ●
88890 Chonemorpha antidysenterica (Roth) G. Don = Holarrhena pubescens (Buch. -Ham.) Wall. ex G. Don ●
88891 Chonemorpha eriostylis Pit.;鹿角藤(黄藤,毛柱鹿角藤);Antlevine, Common Chonemorpha ●
88892 Chonemorpha floccosa Tsiang et P. T. Li;丛毛鹿角藤;Fascicularhair Antlevine, Floccose Chonemorpha ●
88893 Chonemorpha fragrans (Moon) Alston;大叶鹿角藤;Bigleaf Antlevine, Bigleaf Chonemorpha, Big-leaved Chonemorpha ●
88894 Chonemorpha grandiflora G. Don = Chonemorpha fragrans (Moon) Alston ●
88895 Chonemorpha griffithii Hook. f.;漾濞鹿角藤;Griffith Antlevine, Griffith Chonemorpha ●
88896 Chonemorpha macrophylla G. Don = Chonemorpha fragrans (Moon) Alston ●
88897 Chonemorpha megacalyx Pierre;长萼鹿角藤(大萼鹿角菜,大萼鹿角藤,金丝杜仲,藤仲,土杜仲);Longcalyx Antlevine, Longcalyx Chonemorpha, Long-calyxed Chonemorpha ●
88898 Chonemorpha mollis Miq. = Chonemorpha fragrans (Moon) Alston ●
88899 Chonemorpha parviflora Tsiang et P. T. Li;小花鹿角藤;Smallflower Antlevine, Smallflower Chonemorpha, Small-flowered Chonemorpha ●

88900 Chonemorpha rheedei Ridl. = Chonemorpha fragrans (Moon) Alston ●

88901 Chonemorpha splendens Chun et Tsiang;海南鹿角藤;Hainan Antlevine,Hainan Chonemorpha ●

88902 Chonemorpha valvata Chatterjee;毛叶藤仲(瓣裂鹿角藤,大叶鹿角藤,枪花药,藤仲,土杜仲);Hairyleaf Antlevine,Hairyleaf Chonemorpha,Hairy-leaved Chonemorpha ●

88903 Chonemorpha valvata Chatterjee = Chonemorpha griffithii Hook. f. ●

88904 Chonemorpha verrucosa (Blume) D. J. Middleton = Rhynchodia rhynchosperma (Wall.) K. Schum. ●

88905 Chonocentrum Pierre ex Pax et K. Hoffm. (1922);管距大戟属●☆

88906 Chonocentrum cyathophorum (Müll. Arg.) Pierre ex Pax et K. Hoffm.;管距大戟●☆

88907 Chonolea Kuntze = Chenolea Thunb. ●☆

88908 Chonopetalum Radlk. (1920);管瓣无患子属●☆

88909 Chonopetalum stenodictyum Radlk.;管瓣无患子●☆

88910 Chontalesia Lundell = Ardisia Sw. (保留属名)●■

88911 Chordifex B. G. Briggs et L. A. S. Johnson(1998);乳突帚灯草属■☆

88912 Chordifex abortivus (Nees) B. G. Briggs et L. A. S. Johnson;乳突帚灯草■☆

88913 Chordifex laxus (R. Br.) B. G. Briggs et L. A. S. Johnson;松散乳突帚灯草■☆

88914 Chordifex stenandrus B. G. Briggs et L. A. S. Johnson;窄蕊乳突帚灯草■☆

88915 Chordorrhiza Ehrh. = Carex L. ■

88916 Chordospartium Cheeseman(1910);裸枝豆属(新西兰裸枝豆属)●☆

88917 Chordospartium stevensoni Cheeseman;裸枝豆;Weeping Broom ●☆

88918 Choretis Herb. = Hymenocallis Salisb. ■

88919 Choretis galvestonensis Herb. = Hymenocallis liriosme (Raf.) Shinners ■☆

88920 Choretrum R. Br. (1810);垂酸木属●☆

88921 Choretrum lateriflorum R. Br.;垂酸木;Dwarf Sour Bush ●☆

88922 Chorianthа Riedl(1961);分花紫草属☆

88923 Chorianthа popoviana Riedl;分花紫草☆

88924 Choribaena Steud. = Chorilaena Endl. ●☆

88925 Choribena Endl. ex Steud. = Chorilaena Endl. ●☆

88926 Choribena Steud. = Chorilaena Endl. ●☆

88927 Choricarpha Boeck. = Lepironia Pers. ■

88928 Choricarpia Domin(1928);分果桃金娘属●☆

88929 Choricarpia leptopetala (F. Muell.) Domin;分果桃金娘●☆

88930 Choricarpia subargentea (C. T. White) L. A. S. Johnson;亚银色分果桃金娘;Ironwood ●☆

88931 Choriceras Baill. (1873);分角大戟属☆

88932 Choriceras australiana Baill.;分角大戟☆

88933 Choriceras majus Airy Shaw;大分角大戟☆

88934 Choriceras tricorne (Benth.) Airy Shaw;三分角大戟☆

88935 Chorichlaena T. Post et Kuntze = Chorilaena Endl. ●☆

88936 Chorigyne R. Erikss. (1989);分蕊草属■☆

88937 Chorigyne cylindrica R. Erikss.;圆柱分蕊草■☆

88938 Chorigyne densiflora R. Erikss.;密花分蕊草■☆

88939 Chorigyne paucinervis R. Erikss.;寡脉分蕊草■☆

88940 Chorigyne pterophylla R. Erikss.;翅叶分蕊草■☆

88941 Chorilaena Endl. (1837);分被芸香属●☆

88942 Chorilaena quercifolia Endl.;分被芸香●☆

88943 Chorilepidella Tiegh. = Lepidaria Tiegh. ●☆

88944 Chorilepis Tiegh. = Lepidaria Tiegh. ●☆

88945 Chorioluma Baill. = Pycnandra Benth. ●☆

88946 Choriophyllum Benth. = Austrobuxus Miq. ●☆

88947 Choriophyllum Benth. = Longetia Baill. ●☆

88948 Choriosphaera Melch. = ? Pseudocalymma A. Samp. et Kuhlm. ●☆

88949 Choriozandra Steud. = Chorizandra R. Br. ■☆

88950 Choripetalum A. DC. = Embelia Burm. f. (保留属名)●■

88951 Choripetalum benthamii Hance = Embelia laeta (L.) Mez ●

88952 Choripetalum obovatum Benth. = Embelia laeta (L.) Mez ●

88953 Choripetalum undulatum (Wall.) A. DC. = Embelia undulata (Wall.) Mez ●

88954 Choripetalum undulatum A. DC. = Embelia undulata (Wall.) Mez ●

88955 Choriptera Botsch. (1967);离翅蓬属●☆

88956 Choriptera semhahensis (Vierh.) Botsch.;离翅蓬●☆

88957 Choriptera semhahensis (Vierh.) Botsch. = Lagenantha cycloptera (Stapf) M. G. Gilbert et Friis ■☆

88958 Chorisandra Benth. = Chorizandra R. Br. ■☆

88959 Chorisandra Benth. et Hook. f. = Chorizandra R. Br. ■☆

88960 Chorisandra Wight = Phyllanthus L. ●■

88961 Chorisandra pinnata Wight = Phyllanthus pinnatus (Wight) G. L. Webster ●☆

88962 Chorisandrachne Airy Shaw(1969);分蕊鞘属☆

88963 Chorisandrachne diplosperma Airy Shaw;分蕊鞘☆

88964 Chorisanthera Oerst. = Pentarhaphia Lindl. ■☆

88965 Chorisema Fisch. = Chorizema Labill. ●■☆

88966 Chorisepalum Gleason et Wodehouse(1931);分萼龙胆属■☆

88967 Chorisepalum breweri Steyerm. et Maguire;分萼龙胆■☆

88968 Chorisepalum ovatum Gleason;卵形分萼龙胆■☆

88969 Chorisia Kunth = Ceiba Mill. ●

88970 Chorisia Kunth(1822);美人树属(郝瑞棉属,郝瑞希阿属,南美木棉属)●☆

88971 Chorisia insignis Kunth;显著美人树(显著郝瑞木棉);White Floss Silk Tree,White Silk Floss Tree ●☆

88972 Chorisia monstruosa?;奇形美人树;Floss-silk Tree ●☆

88973 Chorisia speciosa A. St. -Hil.;美人树(美丽郝瑞木棉,南美木棉,丝木棉);Brazilian Floss Silk Tree,Floss Silk Tree,Floss-silk Tree,Silk Floss Tree,Silk-floss Tree ●☆

88974 Chorisia speciosa A. St. -Hil. 'Arcadia';艳丽美人树(美丽木棉,美人树)●☆

88975 Chorisis DC. (1838);沙苦荬属(厚肋苦荬菜属);Chorisis ■

88976 Chorisis DC. = Lactuca L. ■

88977 Chorisis repens (L.) DC.;沙苦荬(滨剪刀股,滨苦荬,厚肋苦荬菜,匍匐苦荬菜,沙苦荬菜);Creeping Chorisis,Creeping Ixeris ■

88978 Chorisis repens (L.) DC. = Ixeris repens (L.) A. Gray ■

88979 Chorisiva (A. Gray) Rydb. (1922);内华达伊瓦菊属(肖伊瓦菊属)■☆

88980 Chorisiva (A. Gray) Rydb. = Iva L. ■☆

88981 Chorisiva Rydb. = Euphrosyne DC. ■☆

88982 Chorisiva nevadensis (M. E. Jones) Rydb. = Iva nevadensis M. E. Jones ■☆

88983 Chorisma D. Don = Chorisis DC. ■

88984 Chorisma D. Don = Lactuca L. ■

88985 Chorisma Lindl. = Pelargonium L'Hér. ex Aiton ●■

88986 Chorisma repens (L.) D. Don = Chorisis repens (L.) DC. ■

88987 Chorisma repens (L.) D. Don = Ixeris repens (L.) A. Gray ■
88988 Chorisochora Vollesen(1994);赭爵床属■☆
88989 Chorisochora minor (Balf. f.) Vollesen;小赭爵床■☆
88990 Chorisochora striata (Balf. f.) Vollesen;斑纹赭爵床■☆
88991 Chorisochora transvaalensis (A. Meeuse) Vollesen;德兰士瓦赭爵床■☆
88992 Chorispermum R. Br. = Chorispora R. Br. ex DC. (保留属名)■
88993 Chorispermum W. T. Aiton(废弃属名) = Chorispora R. Br. ex DC. (保留属名)■
88994 Chorispermum tenellus R. Br. = Chorispora tenella (Pall.) DC. ■
88995 Chorispora DC. = Chorispora R. Br. ex DC. (保留属名)■
88996 Chorispora R. Br. ex DC. (1821)(保留属名);离子芥属(离子草属);Chorispora ■
88997 Chorispora bungeana Fisch. et C. A. Mey.;高山离子芥;Alpine Chorispora ■
88998 Chorispora elegans Cambess.;优雅离子芥■☆
88999 Chorispora elegans Cambess. = Chorispora sabulosa Cambess. ■
89000 Chorispora elegans Cambess. var. integrifolia O. E. Schulz = Chorispora sabulosa Cambess. ■
89001 Chorispora elegans Cambess. var. sabulosa (Cambess.) O. E. Schulz = Chorispora sabulosa Cambess. ■
89002 Chorispora elegans Cambess. var. stenophylla O. E. Schulz = Chorispora sabulosa Cambess. ■
89003 Chorispora exscapa Bunge ex Ledeb. = Chorispora bungeana Fisch. et C. A. Mey. ■
89004 Chorispora gracilis Alfons. = Chorispora sibirica (L.) DC. ■
89005 Chorispora gracilis Ernst = Chorispora sibirica (L.) DC. ■
89006 Chorispora greigii Regel;具葶离子芥;Scape Chorispora ■
89007 Chorispora iberica (M. Bieb.) DC.;高加索离子草;Caucasia Chorispora ■☆
89008 Chorispora macropoda Trautv.;小花离子芥(大果离子芥,大花离子芥);Smallflower Chorispora ■
89009 Chorispora pamirica Pakhom. = Chorispora songarica Schrenk ■
89010 Chorispora pectinata Hadac. = Chorispora macropoda Trautv. ■
89011 Chorispora sabulosa Cambess.;砂生离子芥■
89012 Chorispora sabulosa Cambess. var. elganclulosa V. Naray. ex Naithani et Uniyal = Chorispora sabulosa Cambess. ■
89013 Chorispora sibirica (L.) DC.;西伯利亚离子芥;Siberia Chorispora, Siberian Chorispora ■
89014 Chorispora sibirica (L.) DC. var. songarica (Schrenk) O. Fedtsch. = Chorispora songarica Schrenk ■
89015 Chorispora sibirica (L.) DC. var. songorica O. Fedtsch. = Chorispora sibirica (L.) DC. ■
89016 Chorispora songarica Schrenk;准噶尔离子芥;Soóngar Chorispora ■
89017 Chorispora stenopetala Regel et Schmalh. = Diptychocarpus strictus (Fisch. ex M. Bieb.) Trautv. ■
89018 Chorispora stricta (Fisch. ex M. Bieb.) DC. = Diptychocarpus strictus (Fisch. ex M. Bieb.) Trautv. ■
89019 Chorispora tashkorganica Al-Shehbaz et al.;塔什离子芥■
89020 Chorispora tenella (Pall.) DC.;离子芥(红花荠菜,离子草,荠儿菜);Blue Mustard, Crossflower, Tender Chorispora ■
89021 Chorispora tianschanica C. H. An;腺毛离子芥;Tianshan Chorispora ■
89022 Chorispora tianschanica C. H. An = Chorispora bungeana Fisch. et C. A. Mey. ■
89023 Choristanthus K. Schum. = Eleutheranthus K. Schum. ■☆
89024 Choristanthus K. Schum. = Eleuthranthes F. Muell. ■☆
89025 Choristea Thunb. = Didelta L'Hér. (保留属名)■☆
89026 Choristega Tiegh. = Lepeostegeres Blume ●☆
89027 Choristegeres Tiegh. = Choristega Tiegh ●☆
89028 Choristegeres Tiegh. = Lepeostegeres Blume ●☆
89029 Choristegia Tiegh. = Lepeostegeres Blume ●☆
89030 Choristemon H. B. Will. (1924);分蕊尖苞木属●☆
89031 Choristemon humilis H. B. Will.;分蕊尖苞木●☆
89032 Choristes Benth. = Deppea Cham. et Schltdl. ●☆
89033 Choristigma (Baill.) Baill. = Tetrastylidium Engl. ●☆
89034 Choristigma Baill. = Tetrastylidium Engl. ●☆
89035 Choristigma Kurtz = Stuckertia Kuntze ●☆
89036 Choristigma Kurtz ex Heger = Stuckertia Kuntze ●☆
89037 Choristigma stuckertianum Kurtz ex Heger = Stuckertia stuckertianum Kuntze ☆
89038 Choristylis Harv. (1842);分柱鼠刺属●☆
89039 Choristylis rhamnoides Harv.;分柱鼠刺●☆
89040 Choristylis shirensis Baker f. = Choristylis rhamnoides Harv. ●☆
89041 Choristylis ulugurensis Mildbr. = Choristylis rhamnoides Harv. ●☆
89042 Choritaenia Benth. (1867);分带芹属●☆
89043 Choritaenia capensis (Sond. et Harv.) Benth.;分带芹●☆
89044 Chorizandra Benth. et Hook. f. = Chorisandra Wight ●■
89045 Chorizandra Benth. et Hook. f. = Phyllanthus L. ●■
89046 Chorizandra Griff. ex C. B. Clarke = Boeica T. Anderson ex C. B. Clarke ●■
89047 Chorizandra R. Br. (1810);分蕊莎草属■☆
89048 Chorizandra cymbaria R. Br.;分蕊莎草■☆
89049 Chorizandra orientalis Craib;东方分蕊莎草■☆
89050 Chorizanthe R. Br. = Chorizanthe R. Br. ex Benth. ■●☆
89051 Chorizanthe R. Br. ex Benth. (1836);刺花蓼属;Spineflower ■●☆
89052 Chorizanthe andersonii Parry = Chorizanthe diffusa Benth. ■☆
89053 Chorizanthe angustifolia Nutt.;窄叶刺花蓼;Narrow-leaf Spineflower■☆
89054 Chorizanthe angustifolia Nutt. var. eastwoodiae Goodman = Chorizanthe angustifolia Nutt. ■☆
89055 Chorizanthe biloba Goodman;双裂刺花蓼;Two-lobe Spineflower ■☆
89056 Chorizanthe biloba Goodman var. immemora Reveal et Hardham;埃尔刺花蓼;Hernandez's Spineflower ■☆
89057 Chorizanthe blakleyi Hardham;布氏刺花蓼;Blakley's Spineflower■☆
89058 Chorizanthe brevicornu Torr.;脆弱刺花蓼;Brittle Spineflower ■☆
89059 Chorizanthe brevicornu Torr. subsp. spathulata (Small ex Rydb.) Munz = Chorizanthe brevicornu Torr. var. spathulata (Small ex Rydb.) C. L. Hitchc. ■☆
89060 Chorizanthe brevicornu Torr. var. spathulata (Small ex Rydb.) C. L. Hitchc.;匙状脆弱刺花蓼;Great Basin Brittle Spineflower ■☆
89061 Chorizanthe breweri S. Watson;布鲁尔刺花蓼;Brewer's Spineflower■☆
89062 Chorizanthe californica (Benth.) A. Gray = Mucronea californica Benth. ■☆
89063 Chorizanthe californica (Benth.) A. Gray var. suksdorfii J. F. Macbr. = Mucronea californica Benth. ■☆
89064 Chorizanthe chaetophora Goodman = Chorizanthe procumbens Nutt. ■☆
89065 Chorizanthe chrysacantha Goodman = Chorizanthe staticoides Benth. ■☆

89066 Chorizanthe chrysacantha Goodman var. compacta Goodman = Chorizanthe staticoides Benth. ■☆

89067 Chorizanthe clevelandii Parry; 克莱刺花蓼; Cleveland's Spineflower ■☆

89068 Chorizanthe coriacea Goodman = Lastarriaea coriacea (Goodman) Hoover ■☆

89069 Chorizanthe corrugata (Torr.) Torr. et A. Gray; 皱折刺花蓼; Wrinkled Spineflower ■☆

89070 Chorizanthe cuspidata S. Watson; 骤尖刺花蓼; San Francisco Spineflower ■☆

89071 Chorizanthe cuspidata S. Watson var. villosa (Eastw.) Munz; 绒毛骤尖刺花蓼; Villose Spineflower ■☆

89072 Chorizanthe diffusa Benth.; 铺散刺花蓼; Diffuse Spineflower ■☆

89073 Chorizanthe discolor Nutt. = Chorizanthe staticoides Benth. ■☆

89074 Chorizanthe douglasii Benth.; 道格拉斯刺花蓼; Douglas' Spineflower ■☆

89075 Chorizanthe douglasii Benth. var. hartwegii Benth. = Chorizanthe robusta Parry var. hartwegii (Benth.) Reveal et Rand. Morgan ■☆

89076 Chorizanthe fernandina S. Watson = Chorizanthe parryi S. Watson var. fernandina (S. Watson) Jeps. ■☆

89077 Chorizanthe fimbriata Nutt.; 流苏刺花蓼; Fringed Spineflower ■☆

89078 Chorizanthe fimbriata Nutt. var. laciniata (Torr.) Jeps.; 撕裂流苏刺花蓼; Lacinate Spineflower ■☆

89079 Chorizanthe howellii Goodman; 豪厄尔刺花蓼; Howell's Spineflower ■☆

89080 Chorizanthe insignis Curran = Aristocapsa insignis (Curran) Reveal et Hardham ■☆

89081 Chorizanthe insularis R. Hoffm. = Chorizanthe wheeleri S. Watson ■☆

89082 Chorizanthe jonesiana Goodman = Chorizanthe procumbens Nutt. ■☆

89083 Chorizanthe laciniata Torr. = Chorizanthe fimbriata Nutt. var. laciniata (Torr.) Jeps. ■☆

89084 Chorizanthe leptoceras (A. Gray) S. Watson = Dodecahema leptoceras (A. Gray) Reveal et Hardham ■☆

89085 Chorizanthe leptotheca Goodman = Dodecahema leptoceras (A. Gray) Reveal et Hardham ■☆

89086 Chorizanthe membranacea Benth.; 粉刺花蓼; Pink Spineflower ■☆

89087 Chorizanthe nortonii Greene = Chorizanthe douglasii Benth. ■☆

89088 Chorizanthe nudicaulis Nutt. = Chorizanthe staticoides Benth. ■☆

89089 Chorizanthe obovata Goodman; 匙瓣刺花蓼; Spoon-sepal Spineflower ■☆

89090 Chorizanthe orcuttiana Parry; 奥克特刺花蓼; Orcutt Spineflower ■☆

89091 Chorizanthe palmeri S. Watson; 帕默刺花蓼; Palmer's Spineflower ■☆

89092 Chorizanthe palmeri S. Watson var. biloba (Goodman) Munz = Chorizanthe biloba Goodman ■☆

89093 Chorizanthe palmeri S. Watson var. ventricosa (Goodman) Munz = Chorizanthe ventricosa Goodman ■☆

89094 Chorizanthe parryi S. Watson; 帕里刺花蓼; Parry's Spineflower ■☆

89095 Chorizanthe parryi S. Watson var. fernandina (S. Watson) Jeps.; 费尔南德刺花蓼; San Fernando Spineflower ■☆

89096 Chorizanthe perfoliata A. Gray = Mucronea perfoliata (A. Gray) A. Heller ■☆

89097 Chorizanthe polygonoides Torr. et A. Gray; 紫菀刺花蓼; Knotweed Spineflower ■☆

89098 Chorizanthe polygonoides Torr. et A. Gray subsp. longispina (Goodman) Munz = Chorizanthe polygonoides Torr. et A. Gray var. longispina (Goodman) Munz ■☆

89099 Chorizanthe polygonoides Torr. et A. Gray var. longispina (Goodman) Munz; 长芒紫菀刺花蓼; Long-awned Spineflower ■☆

89100 Chorizanthe procumbens Nutt.; 伏卧刺花蓼; Prostrate Spineflower ■☆

89101 Chorizanthe procumbens Nutt. var. albiflora Goodman = Chorizanthe procumbens Nutt. ■☆

89102 Chorizanthe procumbens Nutt. var. mexicana Goodman = Chorizanthe procumbens Nutt. ■☆

89103 Chorizanthe pungens Benth.; 蒙特里刺花蓼; Monterey Spineflower ■☆

89104 Chorizanthe pungens Benth. var. cuspidata (S. Watson) Parry = Chorizanthe cuspidata S. Watson ■☆

89105 Chorizanthe pungens Benth. var. diffusa (Benth.) Parry = Chorizanthe diffusa Benth. ■☆

89106 Chorizanthe pungens Benth. var. hartwegiana Reveal et Hardham; 哈特刺花蓼; Ben Lomand Spineflower ■☆

89107 Chorizanthe pungens Benth. var. hartwegii (Benth.) Goodman = Chorizanthe robusta Parry var. hartwegii (Benth.) Reveal et Rand. Morgan ■☆

89108 Chorizanthe pungens Benth. var. nivea Curran = Chorizanthe diffusa Benth. ■☆

89109 Chorizanthe pungens Benth. var. robusta (Parry) Jeps. = Chorizanthe robusta Parry ■☆

89110 Chorizanthe rectispina Goodman; 直芒刺花蓼; Prickly Spineflower ■☆

89111 Chorizanthe rigida (Torr.) Torr. et A. Gray; 硬刺花蓼; Devil's Spineflower ■☆

89112 Chorizanthe robusta Parry; 刺花蓼; Robust Spineflower ■☆

89113 Chorizanthe robusta Parry var. hartwegii (Benth.) Reveal et Rand. Morgan; 斯科特刺花蓼; Scott Valley Spineflower ■☆

89114 Chorizanthe spathulata Small ex Rydb. = Chorizanthe brevicornu Torr. var. spathulata (Small ex Rydb.) C. L. Hitchc. ■☆

89115 Chorizanthe spinosa S. Watson; 莫哈韦刺花蓼; Mojave Spineflower ■☆

89116 Chorizanthe staticoides Benth.; 土耳其刺花蓼; Turkish Rugging ■☆

89117 Chorizanthe staticoides Benth. subsp. chrysacantha (Goodman) Munz = Chorizanthe staticoides Benth. ■☆

89118 Chorizanthe staticoides Benth. var. brevispina Goodman = Chorizanthe staticoides Benth. ■☆

89119 Chorizanthe staticoides Benth. var. elata Goodman = Chorizanthe staticoides Benth. ■☆

89120 Chorizanthe staticoides Benth. var. latiloba Goodman = Chorizanthe staticoides Benth. ■☆

89121 Chorizanthe staticoides Benth. var. nudicaulis (Nutt.) Jeps. = Chorizanthe staticoides Benth. ■☆

89122 Chorizanthe stellulata Benth.; 莲座刺花蓼; Starlite Spineflower ■☆

89123 Chorizanthe thurberi (A. Gray) S. Watson = Centrostegia thurberi A. Gray ■☆

89124 Chorizanthe uncinata Nutt. = Chorizanthe procumbens Nutt. ■☆

89125 Chorizanthe uniaristata Torr. et A. Gray; 单芒刺花蓼; One-awn Spineflower ■☆

89126 Chorizanthe valida S. Watson; 强壮刺花蓼; Sonoma Spineflower ■☆

89127 Chorizanthe ventricosa Goodman; 单臌刺花蓼; Priest Valley

Spineflower ■☆
89128 Chorizanthe villosa Eastw. = Chorizanthe cuspidata S. Watson var. villosa (Eastw.) Munz ■☆
89129 Chorizanthe vortriedei Brandegee = Systenotheca vortriedei (Brandegee) Reveal et Hardham ■☆
89130 Chorizanthe watsonii Torr. et A. Gray;瓦氏刺花蓼;Watson's Spineflower ■☆
89131 Chorizanthe wheeleri S. Watson;惠勒刺花蓼;Wheeler's Spineflower ■☆
89132 Chorizanthe xanti S. Watson;松林刺花蓼;Pinyon Spineflower ■☆
89133 Chorizanthe xanti S. Watson var. leucotheca Goodman;白松林刺花蓼;Whitewater Spineflower ■☆
89134 Chorizema Labill. (1800);甘泉豆属(橙花豆属,火豌豆属);Flame Pea,Tango-plant ●■☆
89135 Chorizema cordatum Lindl.;心叶甘泉豆(心叶火豌豆);Flame Pea,Heart-leafed Flame Pea,Heart-leafed Flame-pea,Tango-plant ●☆
89136 Chorizema ilicifolium Labill.;甘泉豆(橙花豆,冬青叶火豌豆);Flame Pea,Holly Flame Pea,Holly Flame-pea ●☆
89137 Chorizema manglesii Hort. = Chorizema cordatum Lindl. ●☆
89138 Chorizema superbum Lem. = Chorizema cordatum Lindl. ●☆
89139 Chorizema varium Benth.;变色甘泉豆(变色火豌豆);Variable Tango-plant ●☆
89140 Chorizonema Jean F. Brunel = Chorisandra Wight ●■
89141 Chorizonema Jean F. Brunel = Phyllanthus L. ●■
89142 Chorizospermum Post et Kuntze = Casearia Jacq. ●
89143 Chorizospermum Post et Kuntze = Corizospermum Zipp. ex Blume ●
89144 Chorizotheca Müll. Arg. = Pseudanthus Sieber ex A. Spreng. ■☆
89145 Chorobanche B. D. Jacks. = Orobanche L. ■
89146 Chorobane C. Presl = Orobanche L. ■
89147 Chorosema Brongn. = Chorizema Labill. ●■☆
89148 Chorozema Sm. = Chorizema Labill. ●■☆
89149 Chortolirion A. Berger = Haworthia Duval(保留属名)■☆
89150 Chortolirion A. Berger(1908);园白花属■☆
89151 Chortolirion angolense (Baker) A. Berger = Haworthia angolensis Baker ■☆
89152 Chortolirion bergerianum Dinter = Chortolirion angolense (Baker) A. Berger ■☆
89153 Chortolirion stenophyllum (Baker) A. Berger = Chortolirion angolense (Baker) A. Berger ■☆
89154 Chortolirion subspicatum (Baker) A. Berger = Chortolirion angolense (Baker) A. Berger ■☆
89155 Chortolirion tenuifolium (Engl.) A. Berger = Chortolirion angolense (Baker) A. Berger ■☆
89156 Choryzema Bosc = Chorizema Labill. ●■☆
89157 Choryzemum Bosc = Chorizema Labill. ●■☆
89158 Chosenia Nakai = Salix L. (保留属名)●
89159 Chosenia Nakai(1920);钻天柳属(朝鲜柳属);Chosenia ●
89160 Chosenia arbutifolia (Pall.) A. K. Skvortsov;钻天柳(朝鲜柳,上天柳,顺河柳);Arbuteleaf Chosenia, Arbute-leaved Chosenia, Awlleaf Chosenia, Bracteole Chosenia, Largescale Chosenia, Large-scaled Chosenia ●◇
89161 Chosenia arbutifolia (Pall.) A. K. Skvortsov = Salix arbutifolia Pall. ●◇
89162 Chosenia bracteosa (Trautv.) Nakai = Chosenia arbutifolia (Pall.) A. K. Skvortsov ●◇
89163 Chosenia bracteosa (Turcz. ex Trautv.) Nakai = Chosenia arbutifolia (Pall.) A. K. Skvortsov ●◇
89164 Chosenia eucalyptoides (Mey. ex C. K. Schneid.) Nakai = Chosenia arbutifolia (Pall.) A. K. Skvortsov ●◇
89165 Chosenia eucalyptoides (Schneid.) Nakai = Chosenia arbutifolia (Pall.) A. K. Skvortsov ●◇
89166 Chosenia macrolepis (Turcz.) Kom. = Chosenia arbutifolia (Pall.) A. K. Skvortsov ●◇
89167 Chosenia splendida (Nakai) Nakai = Chosenia arbutifolia (Pall.) A. K. Skvortsov ●◇
89168 Chotchia Benth. = Chotekia Opiz et Corda ●■
89169 Chotchia Benth. = Pogostemon Desf. ●■
89170 Choteckia Steud. = Chotekia Opiz et Corda ●■
89171 Chotekia Opiz et Corda = Pogostemon Desf. ●■
89172 Chotekia Steud. = Pogostemon Desf. ●■
89173 Chotellia Hook. f. = Pogostemon Desf. ●■
89174 Chouardia Speta(1998);舒氏风信子属■☆
89175 Chouardia litardierei (Breistr.) Speta;舒氏风信子■☆
89176 Choulettia Pomel = Gaillonia A. Rich. ex DC. ■☆
89177 Choulettia Pomel = Jaubertia Guill. ■☆
89178 Choulettia reboudiana (Coss. et Durieu) Pomel = Jaubertia reboudiana (Coss. et Durieu) Ehrend. et Schönb. -Tem. ☆
89179 Chouxia Capuron(1969);干序木属●☆
89180 Chouxia borealis G. E. Schatz, Gereau et Lowry;北方干序木●☆
89181 Chouxia macrophylla G. E. Schatz, Gereau et Lowry;大叶干序木●☆
89182 Chouxia mollis G. E. Schatz, Gereau et Lowry;绢毛干序木●☆
89183 Chouxia saboureaui Capuron ex G. E. Schatz, Gereau et Lowry;萨布罗干序木●☆
89184 Chouxia sorindeioides Capuron;索林漆干序木●☆
89185 Chresta Vell. = Eremanthus Less. ●☆
89186 Chresta Vell. ex DC. (1836);长管菊属■●☆
89187 Chresta alpestris Gardner;高山长管菊●☆
89188 Chresta angustifolia Gardner;狭叶长管菊●☆
89189 Chresta cordata Vell.;心形长管菊●☆
89190 Chresta lanceolata Vell.;剑形长管菊●☆
89191 Chresta pinnatifida (Philipson) H. Rob.;羽裂长管菊●☆
89192 Chrestienia Montrouz. = Pseuderanthemum Radlk. ●■
89193 Chretomeris Nutt. ex J. G. Sm. = Sitanion Raf. ■☆
89194 Chrisanthemum Neck. = Chrysanthemum L. (保留属名)■●
89195 Chrisosplenium Neck. = Chrysosplenium L. ■
89196 Christannia C. Presl = Pineda Ruiz et Pav. ●☆
89197 Christannia Walp. = Christiana DC. ●☆
89198 Christensonella Szlach., Mytnik, Górniak et Ś miszek = Maxillaria Ruiz et Pav. ■☆
89199 Christensonella Szlach., Mytnik, Górniak et Ś miszek(2006);拟越南兰属■☆
89200 Christensonia Haager(1993);越南兰属■☆
89201 Christia Moench (1802);蝙蝠草属(萝藟草属);Christia, Batweed ■●
89202 Christia campanulata (Benth.) Thoth.;台湾蝙蝠草(蝙蝠草);Taiwan Batweed, Taiwan Christia ●
89203 Christia campanulata (Wall.) Thoth. = Christia campanulata (Benth.) Thoth. ●
89204 Christia constricta (C. K. Schneid.) T. C. Chen;长管蝙蝠草;Longtube Batweed, Long-tube Christia ■●
89205 Christia hainanensis Yen C. Yang et P. H. Huang;海南蝙蝠草;Hainan Batweed, Hainan Christia ■
89206 Christia lunata Moench;月见罗瑞草■

89207 Christia obcordata (Poir.) Bakh. f. ex Meeuwen;铺地蝙蝠草(半边钱,蝴蝶叶,罗藟草,罗瑞草,萝藟草,马蹄金,马蹄香,钱凿草,三脚虎,纱帽草,土豆草);Obcordate Christia,Spread Batweed ■

89208 Christia vespertilionis (L. f.) Bakh. f. = Christia vespertilionis (L. f.) Bakh. f. ex Meeuwen ■

89209 Christia vespertilionis (L. f.) Bakh. f. ex Meeuwen;蝙蝠草(飞锡草,蝴蝶草,蝴蝶风,鸡子草,雷州蝴蝶草,双飞蝴蝶,鹞子草,月见萝藟草);Batweed,Common Christia,East Indian Island Pea,Evening Christia ■

89210 Christiana DC.(1824);翅果片椴属(非洲椴属)●☆

89211 Christiana africana DC.;翅果片椴(非洲椴)●☆

89212 Christiana cordifolia Hook. f. = Christiana africana DC. ●☆

89213 Christiana madagascariensis Baill. = Christiana africana DC. ●☆

89214 Christianella W. R. Anderson = Mascagnia (Bertero ex DC.) Colla ●☆

89215 Christianella W. R. Anderson(2006);小克利木属●☆

89216 Christiania Rchb. = Christiana DC. ●☆

89217 Christisonia Gardner (1847);假野菰属(彩花菰属);Christisonia ■

89218 Christisonia calcarata Wight;印巴假野菰■☆

89219 Christisonia hookeri C. B. Clarke ex Hook. f.;假野菰(川蔗寄生,花菰,竹花,竹子花);Common Christisonia,Hooker Christisonia ■

89220 Christisonia sinensis Beck = Christisonia hookeri C. B. Clarke ex Hook. f. ■

89221 Christisonia stocksii Hook. f. = Christisonia calcarata Wight ■☆

89222 Christmannia Dennst. = Salacia L.(保留属名)●

89223 Christolea Cambess.(1839);高原芥属(北疆芥属,新疆芥属);Plateaucress Christolea ■

89224 Christolea Cambess. ex Jacquem. = Christolea Cambess. ■

89225 Christolea afghanica (Rech. f.) Rech. f. = Christolea crassifolia Cambess. ■

89226 Christolea albiflora (T. Anderson) Jafri = Phaeonychium albiflorum (T. Anderson) Jafri ■

89227 Christolea baiogoinensis K. C. Kuan et C. H. An;藏北高原芥(藏北扇叶芥);N. Xizang Plateaucress,North Tibet Christolea ■

89228 Christolea baiogoinensis K. C. Kuan et C. H. An = Desideria baiogoinensis (K. C. Kuan et C. H. An) Al-Shehbaz ■

89229 Christolea crassifolia Cambess.;高原芥;Common Christolea,Plateaucress ■

89230 Christolea crassifolia Cambess. var. pamirica (Korsh.) Korsh. = Christolea crassifolia Cambess. ■

89231 Christolea flabellata (Regel) N. Busch;长毛高原芥(长毛扇叶芥);Longhair Christolea,Longhair Plateaucress ■

89232 Christolea flabellata (Regel) N. Busch = Desideria flabellata (Regel) Al-Shehbaz ■

89233 Christolea himalayensis (Cambess.) Jafri;喜马拉雅高原芥(须弥扇叶芥);Himalayan Christolea,Himalayas Plateaucress ■

89234 Christolea himalayensis (Cambess.) Jafri = Desideria himalayensis (Cambess.) Al-Shehbaz ■

89235 Christolea incisa O. E. Schulz = Christolea crassifolia Cambess. ■

89236 Christolea karakorumensis Y. H. Wu et C. H. An;喀拉昆仑高原芥;Kalakunlun Christolea,Kalakunlun Plateaucress ■

89237 Christolea karakorumensis Y. H. Wu et C. H. An = Desideria mirabilis Pamp. ■

89238 Christolea kashgarica (Botsch.) C. H. An;喀什高原芥;Kashgar Christolea,Kashgar Plateaucress ■

89239 Christolea kashgarica (Botsch.) C. H. An = Phaeonychium kashgaricum (Botsch.) Al-Shehbaz ■

89240 Christolea lanuginosa (Hook. f. et Thomson) Ovcz. = Eurycarpus lanuginosus (Hook. f. et Thomson) Botsch. ■

89241 Christolea linearis N. Busch;线形高原芥■☆

89242 Christolea linearis N. Busch = Desideria himalayensis (Cambess.) Al-Shehbaz ■

89243 Christolea linearis N. Busch = Desideria linearis (N. Busch) Al-Shehbaz ■

89244 Christolea longmucoensis Y. H. Wu et C. H. An;龙木错高原芥;Longmucuo Christolea,Longmucuo Plateaucress ■

89245 Christolea longmucoensis Y. H. Wu et C. H. An = Eurycarpus marinellii (Pamp.) Al-Shehbaz et G. Yan ■

89246 Christolea maidantalica (Popov et Baran.) N. Busch;迈丹塔尔高原芥■☆

89247 Christolea mirabilis (Pamp.) Jafri = Desideria mirabilis Pamp. ■

89248 Christolea niyaica C. H. An;尼亚高原芥(尼雅高原芥)■

89249 Christolea pamirica Korsh.;帕米尔高原芥■☆

89250 Christolea pamirica Korsh. = Christolea crassifolia Cambess. ■

89251 Christolea parkeri (O. E. Schulz) Jafri;线果高原芥;Parker Christolea,Parker Plateaucress ■

89252 Christolea parkeri (O. E. Schulz) Jafri = Desideria linearis (N. Busch) Al-Shehbaz ■

89253 Christolea parryoides (Cham.) N. Busch;条果高原芥■☆

89254 Christolea pinnatifida R. F. Huang;羽裂高原芥;Pinnatifid Christolea,Pinnatifid Plateaucress ■

89255 Christolea pinnatifida R. F. Huang = Desideria flabellata (Regel) Al-Shehbaz ■

89256 Christolea prolifera (Maxim.) Jafri;丛生高原芥(丛生扇叶芥);Tufted Christolea,Tufted Plateaucress ■

89257 Christolea prolifera (Maxim.) Jafri = Desideria prolifera (Maxim.) Al-Shehbaz ■

89258 Christolea prolifera (Maxim.) Ovcz. = Desideria prolifera (Maxim.) Al-Shehbaz ■

89259 Christolea pumila (Kurz) Jafri;矮高原芥(矮扇叶芥);Dwarf Christolea,Dwarf Plateaucress ■

89260 Christolea pumila (Kurz) Jafri = Desideria pumila (Kurz) Al-Shehbaz ■

89261 Christolea rosularia K. C. Kuan et C. H. An;莲座高原芥;Rosette Christolea,Rosette Plateaucress ■

89262 Christolea rosularis K. C. Kuan et C. H. An. = Lepidostemon rosularis (K. C. Kuan et C. H. An) Al-Shehbaz ■

89263 Christolea rosularis K. C. Kuan et C. H. An. = Torularia rosulifolia K. C. Kuan et C. H. An ■

89264 Christolea scaposa Jafri = Desideria mirabilis Pamp. ■

89265 Christolea stewartii (T. Anderson) Jafri;少花高原芥(少花扇叶芥);Fewflower Christolea,Fewflower Plateaucress ■

89266 Christolea stewartii (T. Anderson) Jafri = Desideria stewartii (T. Anderson) Al-Shehbaz ■

89267 Christolea suslovaeana Jafri = Desideria mirabilis Pamp. ■

89268 Christolea villosa (Maxim.) Jafri;柔毛高原芥;Villose Plateaucress,Villous Christolea ■

89269 Christolea villosa (Maxim.) Jafri = Phaeonychium villosum (Maxim.) Al-Shehbaz ■

89270 Christolea villosa (Maxim.) Jafri var. platifilamenta K. C. Kuan et C. H. An;宽丝高原芥;Broadfilament Christolea,Platifilament Plateaucress ■

89271 Christolea villosa (Maxim.) Jafri var. platyfilamenta K. C. Kuan

et C. H. An = Phaeonychium villosum (Maxim.) Al-Shehbaz ■
89272 Christolia Post et Kuntze = Chrystolia Montrouz. ■
89273 Christolia Post et Kuntze = Glycine Willd. (保留属名) ■
89274 Christophoriana Burm. = Knowltonia Salisb. ■☆
89275 Christophoriana Burm. ex Kuntze = Knowltonia Salisb. ■☆
89276 Christophoriana Kuntze = Knowltonia Salisb. ■☆
89277 Christophoriana Mill. = Actaea L. ■
89278 Christophoriana Tourn. ex Rupp. = Actaea L. ■
89279 Christya Ward et Harv. = Strophanthus DC. ●
89280 Christya speciosa Ward et Harv. = Strophanthus speciosus (Ward et Harv.) Reber ●☆
89281 Chritmum Brot. = Crithmum L. ■☆
89282 Chroesthes Benoist(1927);色萼花属(色萼木属);Chroesthes, Colorcalyx ●
89283 Chroesthes lanceolata (T. Anderson) B. Hansen;色萼花(林生色萼花);Forest Chroesthes, Forest Colorcalyx, Hairyflower Chroesthes, Hairyflower Colorcalyx, Hairy-flowered Chroesthes ●
89284 Chroesthes pubiflora Benoist = Chroesthes lanceolata (T. Anderson) B. Hansen ●
89285 Chroesthes racemiflora Bremek. = Chroesthes lanceolata (T. Anderson) B. Hansen ●
89286 Chroesthes silvicola (W. W. Sm.) E. Hassain = Chroesthes lanceolata (T. Anderson) B. Hansen ●
89287 Chroesthes silvicola (W. W. Sm.) E. Hossain;林生色萼花;Wooded Chroesthes ●
89288 Chroilema Bernh. = Haplopappus Cass. (保留属名) ■●☆
89289 Chromanthus Phil. = Talinum Adans. (保留属名) ■●
89290 Chromatolepis Dulac = Carlina L. ■●
89291 Chromatopogon F. W. Schmidt = Scorzonera L. ■
89292 Chromatotriccum M. A. Clem. et D. L. Jones = Dendrobium Sw. (保留属名) ■
89293 Chromoehiton Cass. = Cassinia R. Br. (保留属名) ●☆
89294 Chromolaena DC. (1836);香泽兰属(飞机草属,色衣菊属) ●■
89295 Chromolaena bigelovii (A. Gray) R. M. King et H. Rob.;毕氏香泽兰;Bigelow's False Thoroughwort ■☆
89296 Chromolaena frustrata (B. L. Rob.) R. M. King et H. Rob.;迷惑香泽兰;Cape Sable False Thoroughwort ■☆
89297 Chromolaena ivifolia (L.) R. M. King et H. Rob.;常春藤叶香泽兰;Ivy-leaf False Thoroughwort ■☆
89298 Chromolaena moritziana (Sch. Bip. ex Hieron.) R. M. King et H. Rob.;莫里色衣菊■☆
89299 Chromolaena odorata (L.) R. M. King et H. Rob.;飞机草(香色衣菊,香泽兰);Aircraft Grass, Christmas Bush, Fragrant Bogorchid, Fragrant Eupatorium, Siam Weed, Siamweed, Triffid Weed ■
89300 Chromolepis Benth. (1840);彩鳞菊属■☆
89301 Chromolepis heterophylla Benth.;彩鳞菊■☆
89302 Chromolucuma Ducke(1925);大托叶山榄属●☆
89303 Chromolucuma rubriflora Ducke;大托叶山榄●☆
89304 Chromophora Post et Kuntze = Chrozophora A. Juss. (保留属名) ●
89305 Chronanthos (DC.) K. Koch = Cytisus Desf. (保留属名) ●
89306 Chronanthos (DC.) K. Koch = Genista L. ●
89307 Chronanthos K. Koch = Genista L. ●
89308 Chronanthus K. Koch = Cytisus Desf. (保留属名) ●
89309 Chronanthus biflorus (Desf.) Frodin et Heywood = Cytisus fontanesii Spach ●☆
89310 Chrone Dulac = Eupatorium L. ■●
89311 Chroniochilus J. J. Sm. (1918);迟花兰属■☆
89312 Chroniochilus minimus J. J. Sm.;迟花兰■☆
89313 Chronobasis DC. ex Benth. et Hook. f. = Ursinia Gaertn. (保留属名) ●■☆
89314 Chronopappus DC. (1836);泡叶巴西菊属●☆
89315 Chronopappus bifrons (DC. ex Pers.) DC.;泡叶巴西菊■☆
89316 Chrosothamnus Post et Kuntze = Aster L. ●■
89317 Chrosothamnus Post et Kuntze = Chrysothamnus Nutt. (保留属名) ■●☆
89318 Chrosperma Raf. (废弃属名) = Amianthium A. Gray (保留属名) ■☆
89319 Chrosperma muscitoxicum (Walter) Kuntze = Amianthium muscaetoxicum (Walter) A. Gray ■☆
89320 Chrozophora A. Juss. (1824)(保留属名)('Crozophora');沙戟属(苏染草属,星毛戟属);Turnsole, Turnsole Crozophore ●
89321 Chrozophora Neck. = Chrozophora A. Juss. (保留属名) ●
89322 Chrozophora Neck. ex A. Juss. = Chrozophora A. Juss. (保留属名) ●
89323 Chrozophora gracilis Fisch. et C. A. Mey.;纤细沙戟●☆
89324 Chrozophora obliqua (Vahl) A. Juss. ex Spreng.;斜沙戟●☆
89325 Chrozophora oblongifolia (Delile) A. Juss. ex Spreng.;矩圆叶沙戟●☆
89326 Chrozophora plicata (Vahl) A. Juss. ex Spreng.;折扇沙戟●☆
89327 Chrozophora plicata (Vahl) A. Juss. ex Spreng. var. obliquifolia (Vis.) Prain = Chrozophora plicata (Vahl) A. Juss. ex Spreng. ●☆
89328 Chrozophora sabulosa Kar. et Kir.;沙戟(星毛戟);Turnsole ●
89329 Chrozophora senegalensis (Lam.) A. Juss. ex Spreng.;塞内加尔沙戟●☆
89330 Chrozophora tinctoria (L.) A. Juss.;染料沙戟;Dyer's Croton, Giradol, Tournesol ●☆
89331 Chrozophora verbascifolia (Willd.) Spreng. = Chrozophora obliqua (Vahl) A. Juss. ex Spreng. ●☆
89332 Chrozorrhiza Ehrh. = Galium L. ■●
89333 Chryostoma Lilja = Mentzelia L. ●■☆
89334 Chrysa Raf. = Chryza Raf. ■
89335 Chrysa Raf. = Coptis Salisb. ■
89336 Chrysactinia A. Gray(1849);墨西哥金星菊属(金线菊属)●☆
89337 Chrysactinia mexicana A. Gray;墨西哥金星菊;Damianita, Damianita Daisy ■☆
89338 Chrysactinium (Kunth) Wedd. (1857);白冠黑药菊属■☆
89339 Chrysactinium (Kunth) Wedd. = Liabum Adans. ■●☆
89340 Chrysactinium Wedd. = Liabum Adans. ■●☆
89341 Chrysactinium acaule (Kunth) Wedd.;无茎白冠黑药菊■☆
89342 Chrysactinium bicolor (S. F. Blake) H. Rob. et Brettell;二色白冠黑药菊■☆
89343 Chrysactinium tenuius (S. F. Blake) H. Rob. et Brettell;细白冠黑药菊■☆
89344 Chrysaea Nieuwl. et Lunell = Impatiens L. ■
89345 Chrysalidocarpus H. Wendl. (1878);散尾葵属(黄椰属,黄椰子属);Butterfly Palm, Chrysalidocarpus, Golden Cane Palm, Madagascar Palm, Madagascarpalm ●
89346 Chrysalidocarpus H. Wendl. = Dypsis Noronha ex Mart. ●☆
89347 Chrysalidocarpus acuminum Jum. = Dypsis acuminum (Jum.) Beentje et J. Dransf. ●☆
89348 Chrysalidocarpus ankaizinensis Jum. = Dypsis ankaizinensis (Jum.) Beentje et J. Dransf. ●☆
89349 Chrysalidocarpus arenarum Jum. = Dypsis arenarum (Jum.) Beentje et J. Dransf. ●☆

89350 Chrysalidocarpus auriculatus Jum. = Dypsis perrieri (Jum.) Beentje et J. Dransf. ●☆
89351 Chrysalidocarpus baronii Becc. = Dypsis baronii (Becc.) Beentje et J. Dransf. ●☆
89352 Chrysalidocarpus baronii Becc. var. littoralis Jum. et H. Perrier = Dypsis lutescens (H. Wendl.) Beentje et J. Dransf. ●☆
89353 Chrysalidocarpus brevinodis H. Perrier = Dypsis onilahensis (Jum. et H. Perrier) Beentje et J. Dransf. ●☆
89354 Chrysalidocarpus cabadae H. E. Moore;马达加斯加散尾葵●☆
89355 Chrysalidocarpus canescens Jum. et H. Perrier = Dypsis canescens (Jum. et H. Perrier) Beentje et J. Dransf. ●☆
89356 Chrysalidocarpus decipiens Becc. = Dypsis decipiens (Becc.) Beentje et J. Dransf. ●☆
89357 Chrysalidocarpus fibrosus Jum. = Dypsis mananjarensis (Jum. et H. Perrier) Beentje et J. Dransf. ●☆
89358 Chrysalidocarpus glaucescens Waby = Dypsis lutescens (H. Wendl.) Beentje et J. Dransf. ●☆
89359 Chrysalidocarpus lucubensis Becc. = Dypsis madagascariensis (Becc.) Beentje et J. Dransf. ●☆
89360 Chrysalidocarpus lutescens H. Wendl.;散尾葵(黄椰子);Areca Palm, Bamboo Palm, Butterfly Palm, Cane Palm, Golden Butterfly Palm, Golden Cane Palm, Golden Cane-palm, Golden-feather Palm, Madagascar Palm, Madagascarpalm, Yellow Areca Palm, Yellow Butterfly Palm, Yellow Palm, Yellowish Chrysalidocarpus ●
89361 Chrysalidocarpus lutescens H. Wendl. = Dypsis lutescens (H. Wendl.) Beentje et J. Dransf. ●☆
89362 Chrysalidocarpus madagascariensis Becc. = Dypsis madagascariensis (Becc.) Beentje et J. Dransf. ●☆
89363 Chrysalidocarpus madagascariensis Becc. f. oleraceus (Jum. et H. Perrier) Jum. = Dypsis madagascariensis (Becc.) Beentje et J. Dransf. ●☆
89364 Chrysalidocarpus madagascariensis Becc. var. lucubensis (Becc.) Jum. = Dypsis madagascariensis (Becc.) Beentje et J. Dransf. ●☆
89365 Chrysalidocarpus mananjarensis Jum. et H. Perrier = Dypsis mananjarensis (Jum. et H. Perrier) Beentje et J. Dransf. ●☆
89366 Chrysalidocarpus midongensis Jum. = Dypsis onilahensis (Jum. et H. Perrier) Beentje et J. Dransf. ●☆
89367 Chrysalidocarpus nossibensis Becc. = Dypsis nossibensis (Becc.) Beentje et J. Dransf. ●☆
89368 Chrysalidocarpus oleraceus Jum. et H. Perrier = Dypsis madagascariensis (Becc.) Beentje et J. Dransf. ●☆
89369 Chrysalidocarpus onilahensis Jum. et H. Perrier = Dypsis onilahensis (Jum. et H. Perrier) Beentje et J. Dransf. ●☆
89370 Chrysalidocarpus paucifolius Jum. = Dypsis pilulifera (Becc.) Beentje et J. Dransf. ●☆
89371 Chrysalidocarpus pembanus H. E. Moore;彭贝散尾葵●☆
89372 Chrysalidocarpus pilulifera Becc. = Dypsis pilulifera (Becc.) Beentje et J. Dransf. ●☆
89373 Chrysalidocarpus propinquus Jum. = Dypsis baronii (Becc.) Beentje et J. Dransf. ●☆
89374 Chrysalidocarpus rivularis Jum. et H. Perrier = Dypsis rivularis (Jum. et H. Perrier) Beentje et J. Dransf. ●☆
89375 Chrysalidocarpus ruber Jum. = Dypsis perrieri (Jum.) Beentje et J. Dransf. ●☆
89376 Chrysalidocarpus sahanofensis (Jum. et H. Perrier) Jum. = Dypsis sahonofensis (Jum. et H. Perrier) Beentje et J. Dransf. ●☆
89377 Chrysalidocarpus sambiranensis (Jum. et H. Perrier) Jum. = Dypsis pinnatifrons Mart. ●☆
89378 Chrysallidosperma H. E. Moore = Syagrus Mart. ●
89379 Chrysamphora Greene = Darlingtonia Torr. (保留属名)■☆
89380 Chrysangia Link = Musschia Dumort. ●☆
89381 Chrysanthellina Cass. = Chrysanthellum Rich. ex Pers. ■☆
89382 Chrysanthellum Pers. = Chrysanthellum Rich. ex Pers. ■☆
89383 Chrysanthellum Rich. = Chrysanthellum Rich. ex Pers. ■☆
89384 Chrysanthellum Rich. ex Pers. (1807);苏头菊属■☆
89385 Chrysanthellum abyssinicum Sch. Bip. = Bidens setigera (Sch. Bip. ex Walp.) Sherff ■☆
89386 Chrysanthellum indicum DC.;印度苏头菊■☆
89387 Chrysanthellum indicum DC. var. afroamericanum B. L. Turner;非美苏头菊■☆
89388 Chrysanthemodes Post et Kuntze = Chrysanthemoides Medik. ●☆
89389 Chrysanthemoides Fabr. (1759);核果菊属(菊状木属)●☆
89390 Chrysanthemoides Fabr. = Osteospermum L. ●■☆
89391 Chrysanthemoides Medik. = Chrysanthemoides Fabr. ●☆
89392 Chrysanthemoides Thourn. ex Medik. = Chrysanthemoides Fabr. ●☆
89393 Chrysanthemoides incana (Burm. f.) Norl.;灰毛核果菊●☆
89394 Chrysanthemoides monilifera (L.) Norl.;核果菊(菊状木);Boneseed ●☆
89395 Chrysanthemoides monilifera (L.) Norl. subsp. canescens (DC.) Norl.;灰白核果菊●☆
89396 Chrysanthemoides monilifera (L.) Norl. subsp. rotundata (DC.) Norl.;圆叶核果菊(圆叶菊状木)●☆
89397 Chrysanthemoides monilifera (L.) Norl. subsp. septentrionalis Norl.;北方核果菊●☆
89398 Chrysanthemoides monilifera (L.) Norl. subsp. subcanescens (DC.) Norl.;浅灰核果菊●☆
89399 Chrysanthemoides pisiformis Medik.;念珠核果菊●☆
89400 Chrysanthemopsis Rech. f. = Smelowskia C. A. Mey. ex Ledebour (保留属名)■
89401 Chrysanthemopsis koelzii Rech. f. = Smelowskia calycina (Stephan ex Willd.) C. A. Mey. ■
89402 Chrysanthemum L. (1753)(保留属名);茼蒿属;Chrysanthemum, Crown Daisy, Matricaria, Mum, Oxeyedaisy, Shasta Daisy, Tansy ■●
89403 Chrysanthemum × aphrodite Kitam.;爱神茼蒿■☆
89404 Chrysanthemum × cuneifolium Kitam.;楔叶茼蒿■☆
89405 Chrysanthemum × konoanum Makino;小野茼蒿■☆
89406 Chrysanthemum × leucanthum (Makino) Makino;白花茼蒿■☆
89407 Chrysanthemum × marginatum (Miq.) Matsum.;具边茼蒿■☆
89408 Chrysanthemum × miyatojimense Kitam.;宫户岛茼蒿■☆
89409 Chrysanthemum × ogawae Kitam.;小川茼蒿■☆
89410 Chrysanthemum × shimotomaii Makino;下斗米蒿■☆
89411 Chrysanthemum × superbum Bergmans ex J. W. Ingram = Leucanthemum maximum (Ramond) DC. ■
89412 Chrysanthemum abrotanifolium (Bunge ex Ledeb.) Krylov = Pyrethrum abrotanifolium Bunge ex Ledeb. ■
89413 Chrysanthemum achilloides (Turcz.) Hand.-Mazz. = Ajania achilloides (Turcz.) Poljakov ex Grubov ■
89414 Chrysanthemum adenanthum (Diels) Hand.-Mazz. = Ajania adenantha (Diels) Y. Ling et C. Shih ■
89415 Chrysanthemum alashanense (Y. Ling) Y. Ling = Hippolytia alashanensis (Y. Ling) C. Shih ●◇
89416 Chrysanthemum alashanense (Y. Ling) Y. Ling = Hippolytia kaschgarica (Krasch.) Poljakov ■

89417　Chrysanthemum alatavicum (Herder) B. Fedtsch. = Pyrethrum alatavicum (Herder) O. Fedtsch. et B. Fedtsch. ■

89418　Chrysanthemum alatavicum (Herder) Sch. Bip. = Pyrethrum alatavicum (Herder) O. Fedtsch. et B. Fedtsch. ■

89419　Chrysanthemum alpinum L.;高山茼蒿;Alpine Chrysanthemum ■☆

89420　Chrysanthemum alpinum L. = Leucanthemopsis alpina (L.) Heywood ■☆

89421　Chrysanthemum anethifolium Brouss. ex Willd.;玛格丽菊;Mediterranean Chrysanthemum ■☆

89422　Chrysanthemum anethifolium Willd. = Argyranthemum foeniculaceum (Willd.) Webb ex Sch. Bip. ●☆

89423　Chrysanthemum arassanicum C. Winkl. = Pyrethrum pyrethroides (Kar. et Kir.) B. Fedtsch. ex Krasch. ■

89424　Chrysanthemum arcticum L.;北极茼蒿;Arctic Chrysanthemum ■☆

89425　Chrysanthemum arcticum L. = Arctanthemum arcticum (L.) Tzvelev ■☆

89426　Chrysanthemum arcticum L. = Dendranthema zawadskii (Herb.) Tzvelev ■

89427　Chrysanthemum arcticum L. subsp. maekawanum Kitam. = Chrysanthemum arcticum L. var. maekawanum Kitam. ■☆

89428　Chrysanthemum arcticum L. subsp. yezoense (Maek.) H. Ohashi et Yonek.;北海道茼蒿■☆

89429　Chrysanthemum arcticum L. var. maekawanum Kitam.;前川氏北极茼蒿■☆

89430　Chrysanthemum arcticum L. var. yezoense Maek. = Chrysanthemum arcticum L. subsp. yezoense (Maek.) H. Ohashi et Yonek. ■☆

89431　Chrysanthemum argyrophyllum Y. Ling = Dendranthema argyrophyllum (Y. Ling) Y. Ling et C. Shih ■

89432　Chrysanthemum arisanense Hayata = Dendranthema arisanense (Hayata) Y. Ling et C. Shih ■

89433　Chrysanthemum arrasanicum C. Winkl. = Pyrethrum arrasanicum (C. Winkl.) O. Fedtsch. et B. Fedtsch. ■

89434　Chrysanthemum artemisioides (Less.) Kitam. = Crossostephium chinense (L.) Makino ●

89435　Chrysanthemum assakae Caball. = Ismelia carinata (Schousb.) Sch. Bip. ■

89436　Chrysanthemum atkinsonii C. B. Clarke = Pyrethrum atkinsonii (C. B. Clarke) Y. Ling et C. Shih ■

89437　Chrysanthemum atlanticum Ball = Rhodanthemum atlanticum (Ball) B. H. Wilcox, K. Bremer et Humphries ■☆

89438　Chrysanthemum aureoglobosum (W. W. Sm. et Farrer) Hand.-Mazz. = Ajania fruticulosa (Ledeb.) Poljakov ●

89439　Chrysanthemum balsamita (L.) Baill. = Balsamita major Desf. ■☆

89440　Chrysanthemum balsamita (L.) Baill. = Tanacetum balsamita L. ■☆

89441　Chrysanthemum balsamita L. = Tanacetum balsamita L. ■☆

89442　Chrysanthemum balsamita L. f. tanacetoides (Boiss.) B. Boivin = Balsamita major Desf. ■☆

89443　Chrysanthemum balsamita L. var. tanacetoides Boiss. = Balsamita major Desf. ■☆

89444　Chrysanthemum balsamita L. var. tanacetoides Boiss. = Tanacetum balsamita L. ■☆

89445　Chrysanthemum bellum Grüning = Dendranthema lavandulifolium (Fisch. ex Trautv.) Kitam. ■

89446　Chrysanthemum bellum Grüning = Dendranthema lavandulifolium (Fisch. ex Trautv.) Y. Ling et C. Shih ■

89447　Chrysanthemum bellum Grüning var. glabriusculum Y. Ling = Dendranthema lavandulifolium (Fisch. ex Trautv.) Kitam. ■

89448　Chrysanthemum bellum Grüning var. jucundum (Nakai et Kitag.) Hand.-Mazz. = Dendranthema lavandulifolium (Fisch. ex Trautv.) Y. Ling et C. Shih ■

89449　Chrysanthemum bellum Grüning var. jucundum (Nakai et Kitag.) Kitam. = Dendranthema lavandulifolium (Fisch. ex Trautv.) Kitam. ■

89450　Chrysanthemum bipinnatum L. = Tanacetum bipinnatum (L.) Sch. Bip. ■☆

89451　Chrysanthemum bipinnatum L. subsp. huronense (Nutt.) Hultén = Tanacetum huronense Nutt. ■☆

89452　Chrysanthemum bipinnatum L. subsp. huronense (Nutt.) Hultén = Tanacetum bipinnatum (L.) Sch. Bip. ■☆

89453　Chrysanthemum boreale (Makino) Makino = Chrysanthemum seticuspe (Maxim.) Hand.-Mazz. f. boreale (Makino) H. Ohashi et Yonek. ■☆

89454　Chrysanthemum boreale (Makino) Makino = Dendranthema lavandulifolium (Fisch. ex Trautv.) Kitam. ■

89455　Chrysanthemum boreale (Makino) Makino = Dendranthema lavandulifolium (Fisch. ex Trautv.) Y. Ling et C. Shih var. seticuspe (Maxim.) C. Shih ■

89456　Chrysanthemum boreale (Makino) Makino var. tomentellum (Hand.-Mazz.) Kitam. = Dendranthema lavandulifolium (Fisch. ex Trautv.) Y. Ling et C. Shih var. tomentellum (Hand.-Mazz.) Y. Ling et C. Shih ■

89457　Chrysanthemum boreale Makino var. tomentellum (Hand.-Mazz.) Kitam. = Dendranthema lavandulifolium (Fisch. ex Trautv.) Y. Ling et C. Shih var. tomentellum (Hand.-Mazz.) Y. Ling et C. Shih ■

89458　Chrysanthemum brachyglossum Y. Ling = Dendranthema glabriusculum (W. W. Sm.) C. Shih ■

89459　Chrysanthemum brevilobum (Franch. ex Diels) Hand.-Mazz. = Ajania breviloba (Franch. ex Hand.-Mazz.) Y. Ling et C. Shih ■

89460　Chrysanthemum brevilobum (Franch.) Hand.-Mazz. = Ajania breviloba (Franch. ex Hand.-Mazz.) Y. Ling et C. Shih ■

89461　Chrysanthemum brousonetii Pers. = Argyranthemum broussonetii (Pers.) Humphries ●☆

89462　Chrysanthemum bulbosum (Hand.-Mazz.) Hand.-Mazz. = Hippolytia delavayi (Franch. ex W. W. Sm.) C. Shih ●■

89463　Chrysanthemum burbankii Makino = Leucanthemum maximum (Ramond) DC. ■

89464　Chrysanthemum canariense (Sch. Bip.) Christ = Argyranthemum adauctum (Link) Humphries subsp. canariense (Sch. Bip.) Humphries ●☆

89465　Chrysanthemum canariense (Sch. Bip.) Christ var. tenuisecta Christ = Argyranthemum adauctum (Link) Humphries subsp. canariense (Sch. Bip.) Humphries ●☆

89466　Chrysanthemum carinatum Schousb.;蒿子秆(粉环菊,花环菊,三色菊);Annual Chrysanthemum, Annual Oxeyedaisy, Tricolor Chrysanthemum, Tricolor Daisy ■

89467　Chrysanthemum carinatum Schousb. = Glebionis carinata (Schousb.) Tzvelev ■

89468　Chrysanthemum carinatum Schousb. = Ismelia carinata (Schousb.) Sch. Bip. ■

89469　Chrysanthemum carinatum Schousb. var. chrysoporphyreum Maire et al. = Ismelia carinata (Schousb.) Sch. Bip. ■

89470 Chrysanthemum carnosulum DC. var. filifolium Harv. = Cymbopappus adenosolen (Harv.) B. Nord. ■☆
89471 Chrysanthemum caroliniana Walter;卡罗来纳茼蒿■☆
89472 Chrysanthemum caucasicum Pers.;高加索茼蒿■☆
89473 Chrysanthemum chanetii H. Lév. = Chrysanthemum zawadskii Herb. var. latilobum (Maxim.) Kitam. ■
89474 Chrysanthemum chanetii H. Lév. = Dendranthema chanetii (H. Lév.) C. Shih ■
89475 Chrysanthemum cinerariifolium (Trevir.) Vis. = Pyrethrum cinerariifolium Trevis. ■
89476 Chrysanthemum cinerariifolium (Trevir.) Vis. = Tanacetum cinerariifolium (Trevir.) Sch. Bip. ■
89477 Chrysanthemum clausonis (Pomel) Batt. = Coleostephus paludosus (Durieu) Alavi ■☆
89478 Chrysanthemum coccineum Willd. = Pyrethrum coccineum (Willd.) Vorosch. ■
89479 Chrysanthemum coccineum Willd. = Tanacetum coccineum (Willd.) Grierson ■
89480 Chrysanthemum coreanum Nakai = Chrysanthemum sibiricum Fisch. ex Turcz. ■
89481 Chrysanthemum coronarium L.;茼蒿(艾菜,春菊,蒿菜,花环菊,菊花菜,蓬蒿,蓬蒿菜,同蒿,同蒿菜,茼蒿菜,茼蒿菊,铜蒿);Annual Chrysanthemum, Chrysanihemum Greens, Corn Marigold of Candy, Corn of Candy Marigold, Crown Daisy, Crown Daisy Chrysanthemum, Crowndaisy, Crowndaisy Chrysanthemum, Crowndaisy Chrysanthemum, Crowndaisy Oxeyedaisy, Garden Chrysanthemum, Garland Chrysanthemum, Shungiku, Summer Chrysanthemum ■
89482 Chrysanthemum coronarium L. = Glebionis coronaria (L.) Cass. ex Spach ■
89483 Chrysanthemum coronarium L. f. spatiosum (L. H. Bailey) Kitam. = Chrysanthemum segetum L. ■
89484 Chrysanthemum coronarium L. f. spatiosum (L. H. Bailey) Kitam. = Glebionis segetum (L.) Fourr. ■
89485 Chrysanthemum coronarium L. var. concolor Batt. = Chrysanthemum coronarium L. ■
89486 Chrysanthemum coronarium L. var. discolor d'Urv. = Glebionis coronaria (L.) Tzvelev ■
89487 Chrysanthemum coronarium L. var. spatiosum L. H. Bailey = Chrysanthemum segetum L. ■
89488 Chrysanthemum coronarium L. var. spatiosum L. H. Bailey = Glebionis segetum (L.) Fourr. ■
89489 Chrysanthemum coronarium L. var. subdiscolor Maire = Glebionis coronaria (L.) Tzvelev ■
89490 Chrysanthemum coronopifolium (Willd.) Christ = Argyranthemum coronopifolium (Willd.) Humphries ●☆
89491 Chrysanthemum coronopifolium (Willd.) Christ var. angusta Christ = Argyranthemum coronopifolium (Willd.) Humphries ●☆
89492 Chrysanthemum corymbosum L. = Chrysanthemum anethifolium Brouss. ex Willd. ■☆
89493 Chrysanthemum corymbosum L. = Pyrethrum corymbosum Link ■☆
89494 Chrysanthemum corymbosum L. = Tanacetum corymbosum (L.) Sch. Bip. ■☆
89495 Chrysanthemum corymbosum L. subsp. achilleae (L.) Murb. = Tanacetum corymbosum (L.) Sch. Bip. ■☆
89496 Chrysanthemum corymbosum L. var. pumila Batt. et Jahand. = Aaronsohnia pubescens (Desf.) K. Bremer et Humphries ■☆
89497 Chrysanthemum cossonianum Batt. = Aaronsohnia pubescens (Desf.) K. Bremer et Humphries ■☆
89498 Chrysanthemum crassipes (Stschegl.) B. Fedtseh. = Tanacetum crassipes (Stschegl.) Tzvelev ■
89499 Chrysanthemum crassum (Kitam.) Kitam.;粗茼蒿■☆
89500 Chrysanthemum crithmifolium (Link) Christ = Argyranthemum frutescens (L.) Sch. Bip. ●
89501 Chrysanthemum cuneifolium Kitam. = Dendranthema cuneifolium (Kitam.) H. Koyama ■☆
89502 Chrysanthemum decurrens Hutch. = Adenoglossa decurrens (Hutch.) B. Nord. ■☆
89503 Chrysanthemum delavayi (Franch. et W. W. Sm.) Hand.-Mazz. = Hippolytia delavayi (Franch. ex W. W. Sm.) C. Shih ●■
89504 Chrysanthemum deserticola Batt. et Trab. = Plagius maghrebinus Vogt et Greuter ■☆
89505 Chrysanthemum deserticola Murb. = Chrysanthoglossum deserticola (Murb.) B. H. Wilcox, K. Bremer et Humphries ■☆
89506 Chrysanthemum djilgense Franch. = Pyrethrum djilgense (Franch.) Tzvelev ■
89507 Chrysanthemum dugourii (Bolle) Christ = Argyranthemum adauctum (Link) Humphries subsp. dugourii (Bolle) Humphries ●☆
89508 Chrysanthemum elegantulum (W. W. Sm.) S. Y. Hu = Ajania elegantyla (W. W. Sm.) C. Shih ■
89509 Chrysanthemum erubescens Stapf = Dendranthema chanetii (H. Lév.) C. Shih ■
89510 Chrysanthemum erubescens Stapf = Dendranthema zawadskii (Herb.) Tzvelev var. latilobum (Maxim.) Kitam. ■
89511 Chrysanthemum falcatolobatum Krasch. = Cancrinia maximowiczii C. Winkl. ●
89512 Chrysanthemum foeniculaceum (Willd.) Steud. = Argyranthemum foeniculaceum (Willd.) Webb ex Sch. Bip. ●☆
89513 Chrysanthemum frutescens L. = Argyranthemum frutescens (L.) Sch. Bip. ●
89514 Chrysanthemum frutescens L. var. canariae Christ = Argyranthemum frutescens (L.) Sch. Bip. subsp. canariae (Christ) Humphries ●☆
89515 Chrysanthemum frutescens L. var. gracilescens Christ = Argyranthemum frutescens (L.) Sch. Bip. subsp. gracilescens (Christ) Humphries ●☆
89516 Chrysanthemum fruticulosum Ledeb. = Brachanthemum fruticulosum (Ledeb.) DC. ●
89517 Chrysanthemum fuscatum Desf. = Heteromera fuscata (Desf.) Pomel ■☆
89518 Chrysanthemum fuscatum Desf. var. tripolitanum Pamp. = Heteromera philaenorum Maire et Weiller ■☆
89519 Chrysanthemum gayanum (Coss. et Durieu) Batt. = Rhodanthemum gayanum (Coss. et Durieu) B. H. Wilcox, K. Bremer et Humphries ■☆
89520 Chrysanthemum glabratum Thunb. = Oncosiphon africanum (P. J. Bergius) Källersjö ■☆
89521 Chrysanthemum glabriusculum (W. W. Sm.) Hand.-Mazz. = Dendranthema glabriusculum (W. W. Sm.) C. Shih ■
89522 Chrysanthemum glabrum Poir. = Mauranthemum paludosum (Poir.) Vogt et Oberpr. ■☆
89523 Chrysanthemum gmelinii Ledeb. ex Turcz. = Dendranthema zawadskii (Herb.) Tzvelev ■
89524 Chrysanthemum grandiflorum (Desf.) Batt. = Plagius grandis

(L.) Alavi et Heywood ■☆

89525 Chrysanthemum hirtum Thunb. = Oncosiphon africanum (P. J. Bergius) Källersjö ■☆

89526 Chrysanthemum holophyllum Pau = Glebionis segetum (L.) Fourr. ■

89527 Chrysanthemum horaimontanum Masam. = Dendranthema horaimontana (Masam.) S. S. Ying ■

89528 Chrysanthemum hortorum W. Mill.;园圃茼蒿■☆

89529 Chrysanthemum huronense (Nutt.) Hultén = Tanacetum huronense Nutt. ■☆

89530 Chrysanthemum hwangshanense Y. Ling = Dendranthema zawadskii (Herb.) Tzvelev ■

89531 Chrysanthemum hybridum Guss.;杂种茼蒿;Chrysanthemum ■☆

89532 Chrysanthemum hypargyreum Diels = Dendranthema hypargyreum (Diels) Y. Ling et C. Shih ■

89533 Chrysanthemum incanum Thunb. = Pentzia incana (Thunb.) Kuntze ■☆

89534 Chrysanthemum indicum L. = Dendranthema indicum (L.) Des Moul. ■

89535 Chrysanthemum indicum L. f. albescens (Makino) T. B. Lee = Chrysanthemum indicum L. var. albescens Makino ■☆

89536 Chrysanthemum indicum L. f. lactiflorum Hiyama;乳白花野菊(白花野菊)■☆

89537 Chrysanthemum indicum L. subsp. aphrodite (Kitam.) Kitam. = Chrysanthemum × aphrodite Kitam. ■☆

89538 Chrysanthemum indicum L. var. acutum Uyeki = Dendranthema indicum (L.) Des Moul. var. acutum Uyeki ■

89539 Chrysanthemum indicum L. var. acutum Uyeki = Dendranthema indicum (L.) Des Moul. ■

89540 Chrysanthemum indicum L. var. acutum Uyeki = Dendranthema lavandulifolium (Fisch. ex Trautv.) Y. Ling et C. Shih ■

89541 Chrysanthemum indicum L. var. albescens Makino;白花野菊■☆

89542 Chrysanthemum indicum L. var. aphrodite (Kitam.) Kitam. = Chrysanthemum × aphrodite Kitam. ■☆

89543 Chrysanthemum indicum L. var. boreale Makino = Dendranthema lavandulifolium (Fisch. ex Trautv.) Y. Ling et C. Shih var. seticuspe (Maxim.) C. Shih ■

89544 Chrysanthemum indicum L. var. coreanum H. Lév. = Dendranthema indicum (L.) Des Moul. ■

89545 Chrysanthemum indicum L. var. edule Kitam. = Dendranthema indicum (L.) Des Moul. ■

89546 Chrysanthemum indicum L. var. iyoense Kitam.;伊予山菊■☆

89547 Chrysanthemum indicum L. var. iyoense Kitam. f. album T. Yamanaka;白伊予山菊■☆

89548 Chrysanthemum indicum L. var. litorale Y. Ling = Dendranthema indicum (L.) Des Moul. ■

89549 Chrysanthemum indicum L. var. lushanense (Kitam.) Hand.-Mazz. = Dendranthema indicum (L.) Des Moul. ■

89550 Chrysanthemum indicum L. var. maruyamanum Kitam.;丸山菊■☆

89551 Chrysanthemum indicum L. var. procumbens (Lour.) Nakai;伸展野菊■

89552 Chrysanthemum indicum L. var. tsurugisanense Kitam.;剑山菊■☆

89553 Chrysanthemum inodorum L. = Matricaria inodora L. ■☆

89554 Chrysanthemum inodorum L. = Tripleurospermum inodorum (L.) Sch. Bip. ■

89555 Chrysanthemum integrifolium Richardson = Hulténiella integrifolia (Richardson) Tzvelev ■☆

89556 Chrysanthemum japonense (Makino) Nakai;野路菊■☆

89557 Chrysanthemum japonense (Makino) Nakai var. ashizuriense Kitam.;足折山菊■☆

89558 Chrysanthemum japonense (Makino) Nakai var. crassum Kitam. = Chrysanthemum crassum (Kitam.) Kitam. ■☆

89559 Chrysanthemum japonense (Makino) Nakai var. debile Kitam. = Chrysanthemum japonense (Makino) Nakai ■☆

89560 Chrysanthemum japonicola Makino = Chrysanthemum kinokuniense (Shimot. et Kitam.) H. Ohashi et Yonek. ■☆

89561 Chrysanthemum japonicum Thunb. = Artemisia japonica Thunb. ■

89562 Chrysanthemum jucundum Nakai et Kitag. = Dendranthema lavandulifolium (Fisch. ex Trautv.) Kitam. ■

89563 Chrysanthemum jugorum (W. W. Sm.) var. tanacetopsis W. W. Sm. = Pyrethrum tatsienense (Bureau et Franch.) Y. Ling et C. Shih var. tanacetopsis (W. W. Sm.) Y. Ling et C. Shih ■

89564 Chrysanthemum jugorum W. W. Sm. = Pyrethrum tatsienense (Bureau et Franch.) Y. Ling et C. Shih ■

89565 Chrysanthemum jugorum W. W. Sm. var. tanacetopsis W. W. Sm. = Pyrethrum tatsienense (Bureau et Franch.) Y. Ling et C. Shih var. tanacetopsis (W. W. Sm.) Y. Ling et C. Shih ■

89566 Chrysanthemum kennedyi (Dunn) Kitam. = Hippolytia kennedyi (Dunn) Y. Ling ■

89567 Chrysanthemum kinokuniense (Shimot. et Kitam.) H. Ohashi et Yonek.;木国菊■☆

89568 Chrysanthemum lacustre Brot. = Leucanthemum lacustre (Brot.) Samp. ■☆

89569 Chrysanthemum lasiopodum Hutch. = Cymbopappus piliferus (Thell.) B. Nord. ■☆

89570 Chrysanthemum lavandulifolium (Fisch. ex Trautv.) Makino = Dendranthema lavandulifolium (Fisch. ex Trautv.) Kitam. ■

89571 Chrysanthemum lavandulifolium (Fisch. ex Trautv.) Makino = Dendranthema lavandulifolium (Fisch. ex Trautv.) Y. Ling et C. Shih ■

89572 Chrysanthemum lavandulifolium (Fisch. ex Trautv.) Makino var. discoideum Hand.-Mazz. = Dendranthema lavandulifolium (Fisch. ex Trautv.) Kitam. var. discoideum (Hand.-Mazz.) Y. Ling ■

89573 Chrysanthemum lavandulifolium (Fisch. ex Trautv.) Makino var. discoideum Hand.-Mazz. = Dendranthema lavandulifolium (Fisch. ex Trautv.) Y. Ling et C. Shih var. discoideum (Maxim.) C. Shih ■

89574 Chrysanthemum lavandulifolium (Fisch. ex Trautv.) Makino var. glabriusculum (Y. Ling) Kitam. = Dendranthema lavandulifolium (Fisch. ex Trautv.) Y. Ling et C. Shih ■

89575 Chrysanthemum lavandulifolium (Fisch. ex Trautv.) Makino var. jucundum (Nakai et Kitag.) Kitam. = Dendranthema lavandulifolium (Fisch. ex Trautv.) Kitam. ■

89576 Chrysanthemum lavandulifolium (Fisch. ex Trautv.) Makino var. sianense Kitam. = Dendranthema lavandulifolium (Fisch. ex Trautv.) Kitam. ■

89577 Chrysanthemum lavandulifolium (Fisch. ex Trautv.) Makino var. sianense Kitam. = Dendranthema lavandulifolium (Fisch. ex Trautv.) Y. Ling et C. Shih ■

89578 Chrysanthemum lavandulifolium (Fisch. ex Trautv.) Makino var. tomentellum Hand.-Mazz. = Dendranthema lavandulifolium (Fisch. ex Trautv.) Kitam. var. tomentellum (Hand.-Mazz.) Y. Ling et C. Shih ■

89579 Chrysanthemum lavandulifolium (Fisch. ex Trautv.) Makino var. tomentellum Hand.-Mazz. = Dendranthema lavandulifolium (Fisch.

ex Trautv.) Y. Ling et C. Shih var. tomentellum (Hand.-Mazz.) Y. Ling et C. Shih ■
89580 Chrysanthemum ledebourianum Y. Ling = Cancrinia discoides (Ledeb.) Poljakov ex Tzvelev ■
89581 Chrysanthemum leicentianum W. C. Wu = Dendranthema hypargyrum (Diels) Y. Ling et C. Shih ■
89582 Chrysanthemum leptophyllum DC. = Leucoptera nodosa (Thunb.) B. Nord. ●☆
89583 Chrysanthemum leptophyllum DC. var. indivisum? = Leucoptera subcarnosa B. Nord. ●☆
89584 Chrysanthemum leucanthemum L. = Leucanthemum vulgare Lam. ■
89585 Chrysanthemum leucanthemum L. var. boecheri B. Boivin = Leucanthemum vulgare Lam. ■
89586 Chrysanthemum leucanthemum L. var. pinnatifidum Lecoq et Lamotte = Leucanthemum vulgare Lam. ■
89587 Chrysanthemum licentianum W. C. Wu = Dendranthema hypargyreum (Diels) Y. Ling et C. Shih ■
89588 Chrysanthemum lidbeckioides Less. = Osteospermum pinnatum (Thunb.) Norl. ■☆
89589 Chrysanthemum lineare Matsum. = Dendranthema maximowiczii (Kom.) Tzvelev ■
89590 Chrysanthemum lineare Matsum. = Leucanthemella linearis (Matsum.) Tzvelev ■
89591 Chrysanthemum lineare Matsum. var. manshuricum Kom. = Leucanthemella linearis (Matsum.) Tzvelev ■
89592 Chrysanthemum linearifolium C. C. Chang = Ajania salicifolia (Mattf. ex Rehder et Kobuski) Poljakov ■
89593 Chrysanthemum lushanense Kitam. = Dendranthema indicum (L.) Des Moul. ■
89594 Chrysanthemum macrocarpum (Sch. Bip.) Coss. et Kralik ex Batt. = Endopappus macrocarpus Sch. Bip. ■☆
89595 Chrysanthemum macrocarpum (Sch. Bip.) Coss. et Kralik ex Batt. subsp. maroccanum Jahand. et Weiller = Endopappus macrocarpus Sch. Bip. subsp. maroccanus (Jahand. et Weiller) Ibn Tattou ■☆
89596 Chrysanthemum macrocarpum (Sch. Bip.) Coss. et Kralik ex Batt. var. aureum L. Chevall. = Endopappus macrocarpus Sch. Bip. ■☆
89597 Chrysanthemum macrocephalum Viv. = Chrysanthoglossum trifurcatum (Desf.) B. H. Wilcox, K. Bremer et Humphries ■☆
89598 Chrysanthemum macrophyllum King et Pantl. = Diglyphosa latifolia Blume ■
89599 Chrysanthemum macrophyllum Waldst. et Kit.;大叶茼蒿;Tansy Chrysanthemum ■☆
89600 Chrysanthemum macrotum (Durieu) Ball = Glossopappus macrotus (Durieu) Briq. ■☆
89601 Chrysanthemum mairei (H. Lév.) Hand.-Mazz. = Ajania myriantha (Franch.) Y. Ling ex C. Shih ■
89602 Chrysanthemum majus Asch.;巨茼蒿■☆
89603 Chrysanthemum makinoi Matsum. et Nakai;牧野茼蒿(龙脑菊)■☆
89604 Chrysanthemum makinoi Matsum. et Nakai var. wakasaense (Shimot. ex Kitam.) Kitam. = Chrysanthemum wakasaense Shimot. ex Kitam. ■☆
89605 Chrysanthemum maresii (Coss.) Ball = Rhodanthemum maresii (Coss.) B. H. Wilcox, K. Bremer et Humphries ■☆
89606 Chrysanthemum maresii Batt. = Rhodanthemum gayanum (Coss. et Durieu) B. H. Wilcox, K. Bremer et Humphries ■☆
89607 Chrysanthemum maresii Batt. var. hosmariense Ball = Rhodanthemum maresii (Coss.) B. H. Wilcox, K. Bremer et Humphries ■☆
89608 Chrysanthemum maroccanum Batt. = Rhodanthemum gayanum (Coss. et Durieu) B. H. Wilcox, K. Bremer et Humphries ■☆
89609 Chrysanthemum marschallii Asch. = Pyrethrum coccineum (Willd.) Vorosch. ■
89610 Chrysanthemum marschallii Asch. ex O. Hoffm.;马氏茼蒿(马氏菊)■☆
89611 Chrysanthemum mawii Hook. f. = Rhodanthemum gayanum (Coss. et Durieu) B. H. Wilcox, K. Bremer et Humphries ■☆
89612 Chrysanthemum maximoviczianum Y. Ling var. aristato-mucronatum Y. Ling = Dendranthema chanetii (H. Lév.) C. Shih ■
89613 Chrysanthemum maximoviczianum Y. Ling var. dissectum Y. Ling = Dendranthema zawadskii (Herb.) Tzvelev ■
89614 Chrysanthemum maximowiczianum Y. Ling = Dendranthema chanetii (H. Lév.) C. Shih ■
89615 Chrysanthemum maximowiczii Kom. = Dendranthema maximowiczii (Kom.) Tzvelev ■
89616 Chrysanthemum maximum Ramond = Leucanthemum maximum (Ramond) DC. ■
89617 Chrysanthemum merzabacheri B. Fedtsch. ex Merzb. = Pyrethrum richterioides (C. Winkl.) Krasn. ■
89618 Chrysanthemum mongolicum Y. Ling = Dendranthema mongolicum (Y. Ling) Tzvelev ■
89619 Chrysanthemum morifolium Ramat.;桑叶茼蒿;Florist's Daisy, Garden Chrysanthemum, Garden Mum ■☆
89620 Chrysanthemum morifolium Ramat. = Dendranthema grandiflorum (Ramat.) Kitam. ■
89621 Chrysanthemum morifolium Ramat. = Dendranthema morifolium (Ramat.) Tzvelev ■
89622 Chrysanthemum morifolium Ramat. f. japonense Makino = Chrysanthemum japonense (Makino) Nakai ■☆
89623 Chrysanthemum morifolium Ramat. var. gracile Hemsl. = Dendranthema grandiflorum (Ramat.) Kitam. ■
89624 Chrysanthemum morifolium Ramat. var. sinense (Sabine) Makino = Dendranthema morifolium (Ramat.) Tzvelev ■
89625 Chrysanthemum morifolium Ramat. var. spontaneum (Makino) Makino = Chrysanthemum japonense (Makino) Nakai ■☆
89626 Chrysanthemum morii Hayata = Dendranthema morii (Hayata) Kitam. ■
89627 Chrysanthemum muirei (H. Lév.) Hand.-Mazz. = Ajania myriantha (Franch.) Y. Ling ex C. Shih ■
89628 Chrysanthemum multicaule Desf.;多茎鞘冠菊(春俏菊);Nippon Chrysanthemum, Nippon Daisy ■☆
89629 Chrysanthemum multicaule Desf. = Coleostephus multicaulis (Desf.) Durieu ■☆
89630 Chrysanthemum mutellina (Hand.-Mazz.) Hand.-Mazz. = Dendranthema mutellina (Hand.-Mazz.) Kitam. ■☆
89631 Chrysanthemum mutellinum (Hand.-Mazz.) Hand.-Mazz. = Ajania khartensis (Dunn) C. Shih ■
89632 Chrysanthemum myconis L. = Coleostephus myconis (L.) Cass. ■
89633 Chrysanthemum myrianthum (Franch.) Y. Ling = Ajania myriantha (Franch.) Y. Ling ex C. Shih ■
89634 Chrysanthemum myrianthum (Franch.) Y. Ling var. sericocephalum Hand.-Mazz. = Ajania myriantha (Franch.) Y. Ling ex C. Shih ■
89635 Chrysanthemum myrianthum (Franch.) Y. Ling var. wardii

(Marquart et Shaw) Hand. -Mazz. = Ajania myriantha (Franch.) Y. Ling ex C. Shih ■

89636 Chrysanthemum nactongense Nakai var. dissectum (Y. Ling) Hand. -Mazz. = Dendranthema zawadskii (Herb.) Tzvelev ■

89637 Chrysanthemum naktongense Nakai = Chrysanthemum zawadskii Herb. var. latilobum (Maxim.) Kitam. ■

89638 Chrysanthemum nalaongense Nakai = Dendranthema naktongense (Nakai) Tzvelev ■

89639 Chrysanthemum namikawanum Kitam. = Dendranthema lavandulifolium (Fisch. ex Trautv.) Kitam. ■

89640 Chrysanthemum nankingense Hand. -Mazz. = Dendranthema indicum (L.) Des Moul. ■

89641 Chrysanthemum nanum Hook. = Blennosperma nanum (Hook.) S. F. Blake ■☆

89642 Chrysanthemum nematolobum Hand. -Mazz. = Ajania nematoloba (Hand. -Mazz.) Y. Ling et C. Shih ■

89643 Chrysanthemum neofruticulosum (Ledeb.) Y. Ling = Ajania fruticulosa (Ledeb.) Poljakov ●

89644 Chrysanthemum neofruticulosum Y. Ling = Ajania fruticulosa (Ledeb.) Poljakov ●

89645 Chrysanthemum neooreastrum C. C. Chang = Dendranthema hypargyrum (Diels) Y. Ling et C. Shih ■

89646 Chrysanthemum nipponicum (Franch. ex Maxim.) Matsum.;浜茼蒿(浜菊);Nippon Chrysanthemum,Nippon Daisy ■☆

89647 Chrysanthemum nipponicum (Franch. ex Maxim.) Matsum. = Nipponanthemum nipponicum (Franch. ex Maxim.) Kitam. ■☆

89648 Chrysanthemum nivellei Braun-Blanq. et Maire = Nivellea nivellei (Braun-Blanq. et Maire) X. Wilcox, K. Bremer et Humphries ■☆

89649 Chrysanthemum nodosum (Thunb.) DC. = Leucoptera nodosa (Thunb.) B. Nord. ●☆

89650 Chrysanthemum oreastrum Hance = Dendranthema oreastrum (Hance) Y. Ling ■

89651 Chrysanthemum oresbium (W. W. Sm.) Hand. -Mazz. = Ajania myriantha (Franch.) Y. Ling ex C. Shih ■

89652 Chrysanthemum ornatum Hemsl.;装饰茼蒿(装饰菊)■☆

89653 Chrysanthemum ornatum Hemsl. var. ashizuriense (Kitam.) Kitam. = Chrysanthemum japonense (Makino) Nakai var. ashizuriense Kitam. ■☆

89654 Chrysanthemum ornatum Hemsl. var. crassum (Kitam.) Kitam. = Chrysanthemum crassum (Kitam.) Kitam. ■☆

89655 Chrysanthemum ornatum Hemsl. var. spontaneum (Makino) Kitam. = Chrysanthemum japonense (Makino) Nakai ■☆

89656 Chrysanthemum ornatum Hemsl. var. spontaneum (Makino) Kitam. f. debile (Kitam.) Kitam. = Chrysanthemum japonense (Makino) Nakai ■☆

89657 Chrysanthemum ornatum Hemsl. var. tokarense (M. Hotta et Hirai) H. Ohashi et Yonek.;陶卡尔茼蒿(陶卡尔菊)■☆

89658 Chrysanthemum osmitoides Harv. = Adenanthellum osmitoides (Harv.) B. Nord. ■☆

89659 Chrysanthemum pacificum Nakai;太平洋茼蒿(太平洋菊);Gold And Silver Chrysanthemum ■☆

89660 Chrysanthemum pacificum Nakai = Dendranthema pacificum (Nakai) Kitam. ■☆

89661 Chrysanthemum pallasianum (Fisch. ex Besser) Kom. = Ajania pallasiana (Fisch. ex Besser) Poljakov ■

89662 Chrysanthemum pallasianum (Fisch. ex Besser) Kom. var. brevilobum (Franch.) Hand. -Mazz. = Ajania breviloba (Franch. ex Hand. -Mazz.) Y. Ling et C. Shih ■

89663 Chrysanthemum pallasianum (Fisch. ex Besser) Kom. var. japonicum (Franch. et Sav.) Matsum. = Chrysanthemum rupestre Matsum. et Koidz. ■☆

89664 Chrysanthemum pallasianum (Fisch.) Kom. var. brevilobum (Franch. ex Diels) Hand. -Mazz. = Ajania breviloba (Franch. ex Hand. -Mazz.) Y. Ling et C. Shih ■

89665 Chrysanthemum paludosum Poir. = Leucoglossum paludosum (Poir.) B. H. Wilcox, K. Bremer et Humphries ■☆

89666 Chrysanthemum paludosum Poir. = Mauranthemum paludosum (Poir.) Vogt et Oberpr. ■☆

89667 Chrysanthemum paludosum Poir. subsp. decipiens (Pomel) Quézel et Santa = Mauranthemum decipiens (Pomel) Vogt et Oberpr. ■☆

89668 Chrysanthemum paludosum Poir. subsp. glabrum (Maire) Quézel et Santa = Mauranthemum paludosum (Poir.) Vogt et Oberpr. ■☆

89669 Chrysanthemum parthenifolium (Willd.) Pers. = Pyrethrum parthenifolium Willd. ■

89670 Chrysanthemum parthenium (L.) Benth. = Pyrethrum parthenium (L.) Sm. ■

89671 Chrysanthemum parthenium (L.) Pers. = Pyrethrum parthenium (L.) Sm. ■

89672 Chrysanthemum parthenium (L.) Pers. ex Benth. = Pyrethrum parthenium (L.) Sm. ■

89673 Chrysanthemum parthenium L. = Pyrethrum parthenium (L.) Sm. ■

89674 Chrysanthemum parthenium Pers. = Pyrethrum parthenium (L.) Sm. ■

89675 Chrysanthemum parviflorum Grüning = Ajania parviflora (Grüning) Y. Ling ■

89676 Chrysanthemum parvifolium C. C. Chang = Dendranthema parvifolium (C. C. Chang) C. Shih ■

89677 Chrysanthemum potaninii (Krasch.) Hand. -Mazz. = Ajania potaninii (Krasch.) Poljakov ■

89678 Chrysanthemum potaninii (Krasch.) Hand. -Mazz. var. amphiseriaceum Hand. -Mazz. = Ajania amphiseriacea (Hand. -Mazz.) C. Shih ■

89679 Chrysanthemum potaninii (Krasch.) Hand. -Mazz. var. amphiseriaceum Hand. -Mazz. = Ajania potaninii (Krasch.) Poljakov ■

89680 Chrysanthemum potentilloides Hand. -Mazz. = Dendranthema potentilloides (Hand. -Mazz.) C. Shih ■

89681 Chrysanthemum procumbens Lour. = Dendranthema indicum (L.) Des Moul. ■

89682 Chrysanthemum pulchrum (Ledeb.) C. Winkl. = Pyrethrum pulchrum Ledeb. ■

89683 Chrysanthemum pulchrum (Ledeb.) Y. Ling = Pyrethrum pulchrum Ledeb. ■

89684 Chrysanthemum pullum Hand. -Mazz. = Pyrethrum tatsienense (Bureau et Franch.) Y. Ling et C. Shih var. tanacetopsis (W. W. Sm.) Y. Ling et C. Shih ■

89685 Chrysanthemum pulvinatum Hand. -Mazz. = Brachanthemum pulvinatum (Hand. -Mazz.) C. Shih ■

89686 Chrysanthemum pyrethroides (Kar. et Kir.) B. Fedtsch. = Pyrethrum pyrethroides (Kar. et Kir.) B. Fedtsch. ex Krasch. ■

89687 Chrysanthemum quercifolium (W. W. Sm.) Hand. -Mazz. = Ajania quercifolia (W. W. Sm.) Y. Ling et C. Shih ■

89688 Chrysanthemum reboudianum (Pomel) Quézel et Santa = Mauranthemum reboudianum (Pomel) Vogt et Oberpr. ■☆

89689 Chrysanthemum remotipinnum Hand. -Mazz. = Ajania remotipinna (Hand. -Mazz.) Y. Ling et C. Shih ■

89690 Chrysanthemum richteria Benth. = Pyrethrum pyrethroides (Kar. et Kir.) B. Fedtsch. ex Krasch. ■

89691 Chrysanthemum richterioides C. Winkl. = Pyrethrum richterioides (C. Winkl.) Krasn. ■

89692 Chrysanthemum rockii (Mattf.) Y. Ling = Ajania potaninii (Krasch.) Poljakov ■

89693 Chrysanthemum roseum Adam = Chrysanthemum coccineum Willd. ■

89694 Chrysanthemum roseum Adam = Pyrethrum coccineum (Willd.) Vorosch. ■

89695 Chrysanthemum rotundifolium Waldst. et Kit.;匈牙利茼蒿;Hungarian Chrysanthemum ■☆

89696 Chrysanthemum roxburghii Desf.;罗氏茼蒿■☆

89697 Chrysanthemum rubellum Sealy = Dendranthema zawadskii (Herb.) Tzvelev var. latilobum (Maxim.) Kitam. ■

89698 Chrysanthemum rupestre Matsum. et Koidz.;岩生茼蒿■☆

89699 Chrysanthemum sabinii Lindl. = Dendranthema indicum (L.) Des Moul. ■

89700 Chrysanthemum salicifolium (Mattf.) Hand. -Mazz. = Ajania salicifolia (Mattf. ex Rehder et Kobuski) Poljakov ■

89701 Chrysanthemum santolina (C. Winkl.) B. Fedtsch. = Tanacetum santolina C. Winkl. ■

89702 Chrysanthemum satsumense (Yatabe) Makino = Dendranthema ornatum (Hemsl.) Kitam. ■☆

89703 Chrysanthemum scharnhorstii (Regel et Schmalh.) B. Fedtsch. = Ajania scharnhorstii (Regel et Schmalh.) Tzvelev ■

89704 Chrysanthemum segetum L.;南茼蒿(蒿菜,金银菊,菊花菜,孔雀菊,蓬蒿,蓬蒿菜,田地菊,同蒿,同蒿菜,茼蒿,珍珠菊);Basthag-bwee, Bigold, Boddle, Boodle, Bossell, Boswell, Botham, Bothen, Botherum, Bottle, Bozen, Bozzel, Bozzom, Corn Chrysanthemum, Corn Gold, Corn Marigold, Corndaisy, Corn-gold, Corn-marigold, Fat Hen, Field Marigold, Gale Gowan, Geal Gowan, Geal-gowan, Geal-seed, Gill Gowan, Gold, Golden Cornflower, Golden Daisy, Goldings, Gole, Golland, Gools, Goulan, Gouls, Gowan, Gowlan, Gowland, Gowles, Guild, Guild Weed, Guile, Guills, Gules, Gules-gowan, Gull, Harvest Flower, Ling Gowland, Manelet, Marigold, Marigold Goldins, Marigold-goldins, Mary Gowlan, Mogue Tobin, Ox Eyes, Ox-eye, Southern Chrysanthemum, Southern Oxeyedaisy, St. John's Bloom, Sunflower, Tansy, Wild Marigold, Yellow Bottle, Yellow Bozzom, Yellow Bussell, Yellow Corn Flower, Yellow Gold, Yellow Gowan, Yellow Horse Daisy, Yellow Ox Eyes, Yellow Ox-eye, Yellowby ■

89705 Chrysanthemum segetum L. = Glebionis segetum (L.) Fourr. ■

89706 Chrysanthemum serotinum L. = Leucanthemella serotina (L.) Tzvelev ■☆

89707 Chrysanthemum seticuspe (Maxim.) Hand. -Mazz. = Dendranthema lavandulifolium (Fisch. ex Trautv.) Kitam. ■

89708 Chrysanthemum seticuspe (Maxim.) Hand. -Mazz. = Dendranthema lavandulifolium (Fisch. ex Trautv.) Y. Ling et C. Shih var. seticuspe (Maxim.) C. Shih ■

89709 Chrysanthemum seticuspe (Maxim.) Hand. -Mazz. f. boreale (Makino) H. Ohashi et Yonek.;北方茼蒿■☆

89710 Chrysanthemum seticuspe (Maxim.) Hand. -Mazz. f. boreale (Makino) H. Ohashi et Yonek. = Dendranthema boreale (Makino) Y. Ling ex Kitam. ■☆

89711 Chrysanthemum seticuspe (Maxim.) Hand. -Mazz. var. boreale (Makino) Hand. -Mazz. = Chrysanthemum seticuspe (Maxim.) Hand. -Mazz. f. boreale (Makino) H. Ohashi et Yonek. ■☆

89712 Chrysanthemum seticuspe Maxim. = Dendranthema lavandulifolium (Fisch. ex Trautv.) Y. Ling et C. Shih var. seticuspe (Maxim.) C. Shih ■

89713 Chrysanthemum seticuspe Maxim. var. boreale (Makano) Hand. -Mazz. = Dendranthema lavandulifolium (Fisch. ex Trautv.) Y. Ling et C. Shih var. seticuspe (Maxim.) C. Shih ■

89714 Chrysanthemum seticuspe Maxim. var. boreale (Makino) Hand. -Mazz. = Dendranthema lavandulifolium (Fisch. ex Trautv.) Kitam. ■

89715 Chrysanthemum shiwogiku Kitam.;盐茼蒿■☆

89716 Chrysanthemum shiwogiku Kitam. = Dendranthema shiwogiku (Kitam.) Kitam. ■☆

89717 Chrysanthemum shiwogiku Kitam. var. kinokuniense Shimot. et Kitam. = Chrysanthemum kinokuniense (Shimot. et Kitam.) H. Ohashi et Yonek. ■☆

89718 Chrysanthemum sibiricum (DC.) Fisch. et Turcz. var. alpinum Nakai = Dendranthema oreastrum (Hance) Y. Ling ■

89719 Chrysanthemum sibiricum (DC.) Fisch. et Turcz. var. latilobum (Maxim.) Kom. = Dendranthema naktongense (Nakai) Tzvelev ■

89720 Chrysanthemum sibiricum Fisch. ex Turcz. = Dendranthema zawadskii (Herb.) Tzvelev ■

89721 Chrysanthemum sibiricum Fisch. ex Turcz. var. alpinum Nakai = Chrysanthemum zawadskii Herb. var. alpinum (Nakai) Kitam. ■

89722 Chrysanthemum sibiricum Fisch. ex Turcz. var. latilobum (Maxim.) Kom. = Chrysanthemum zawadskii Herb. var. latilobum (Maxim.) Kitam. ■

89723 Chrysanthemum sibiricum Turcz. ex DC.;西伯利亚茼蒿(朝鲜茼蒿);Korean Chrysanthemum ■☆

89724 Chrysanthemum sibiricum Turcz. ex DC. = Dendranthema zawadskii (Herb.) Tzvelev ■

89725 Chrysanthemum sibiricum Turcz. ex DC. var. alpinum Nakai = Dendranthema oreastrum (Hance) Y. Ling ■

89726 Chrysanthemum sibiricum Turcz. ex DC. var. aristato-mucronatum Y. Ling = Dendranthema zawadskii (Herb.) Tzvelev ■

89727 Chrysanthemum sibiricum Turcz. ex DC. var. latilobum (Maxim.) Kom. = Dendranthema naktongense (Nakai) Tzvelev ■

89728 Chrysanthemum sibiricum Turcz. ex DC. var. sinoalpinum Nakai = Dendranthema chanetii (H. Lév.) C. Shih ■

89729 Chrysanthemum sichotense (Tzvelev) Vorosch. = Chrysanthemum zawadskii Herb. var. alpinum (Nakai) Kitam. ■

89730 Chrysanthemum sinense Sabine = Dendranthema morifolium (Ramat.) Tzvelev ■

89731 Chrysanthemum sinense Sabine var. hortense Makino ex Matsum. = Dendranthema morifolium (Ramat.) Tzvelev ■

89732 Chrysanthemum sinense Sabine var. vestitum Hemsl. = Dendranthema vestitum (Hemsl.) Y. Ling ■

89733 Chrysanthemum spatiosum Bailey = Chrysanthemum coronarium L. ■

89734 Chrysanthemum squamiferum Y. Ling = Brachanthemum pulvinatum (Hand. -Mazz.) C. Shih ■

89735 Chrysanthemum stenolobum Hand. -Mazz. = Ajania tenuifolia (Jacq.) Tzvelev ■

89736 Chrysanthemum taihangense Y. Ling = Opisthopappus

taihangensis (Y. Ling) C. Shih ■
89737 Chrysanthemum tanacetoides (DC.) B. Fedtsch. = Tanacetum tanacetoides (DC.) Tzvelev ■
89738 Chrysanthemum tanacetum Vis. = Tanacetum vulgare L. ■
89739 Chrysanthemum tatsienense Bureau et Franch. = Pyrethrum tatsienense (Bureau et Franch.) Y. Ling et C. Shih ■
89740 Chrysanthemum tatsienense Bureau et Franch. var. tanacetopsis (W. W. Sm.) Marquart et Shaw = Pyrethrum tatsienense (Bureau et Franch.) Y. Ling et C. Shih var. tanacetopsis (W. W. Sm.) Y. Ling et C. Shih ■
89741 Chrysanthemum tatsienense Bureau et Franch. var. tanacetopsis (W. W. Sm.) Marquand = Pyrethrum tatsienense (Bureau et Franch.) Y. Ling et C. Shih var. tanacetopsis (W. W. Sm.) Y. Ling et C. Shih ■
89742 Chrysanthemum tibeticum (Hook. f. et Thomson ex C. B. Clarke) Hoffman = Ajania tibetica (Hook. f. et Thomson ex C. B. Clarke) Tzvelev ■
89743 Chrysanthemum tibeticum (Hook. f. et Thomson ex C. B. Clarke) S. Y. Hu = Ajania tibetica (Hook. f. et Thomson ex C. B. Clarke) Tzvelev ■
89744 Chrysanthemum togakushiense Kitag. et Nagami = Chrysanthemum pallasianum (Fisch. ex Besser) Kom. ■
89745 Chrysanthemum tricolor Andr. = Chrysanthemum carinatum Schousb. ■
89746 Chrysanthemum trifidum (DC.) Krasch. = Hippolytia trifida (Turcz.) Poljakov ●
89747 Chrysanthemum trifidum (Turcz.) Krasch. = Hippolytia trifida (Turcz.) Poljakov ■
89748 Chrysanthemum trifurcatum (Desf.) Batt. et Trab. var. macrocephalum (Viv.) Bég. = Chrysanthoglossum trifurcatum (Desf.) B. H. Wilcox, K. Bremer et Humphries ■☆
89749 Chrysanthemum trifurcatum (Desf.) Batt. et Trab. var. microcephalum Bég. et Vacc. = Chrysanthoglossum trifurcatum (Desf.) B. H. Wilcox, K. Bremer et Humphries ■☆
89750 Chrysanthemum trifurcatum Desf. = Chrysanthoglossum trifurcatum (Desf.) B. H. Wilcox, K. Bremer et Humphries ■☆
89751 Chrysanthemum trinioides Hand. -Mazz. = Filifolium sibiricum (L.) Kitam. ■
89752 Chrysanthemum tripolitanum (Pamp.) Le Houér. = Heteromera philaenorum Maire et Weiller ■☆
89753 Chrysanthemum truncatum Hand. -Mazz. = Ajania potaninii (Krasch.) Poljakov ■
89754 Chrysanthemum truncatum Hand. -Mazz. = Ajania truncata (Hand. -Mazz.) Y. Ling ex C. Shih ■
89755 Chrysanthemum uliginosum (Waldst. et Kit. Pers. ex Willd.) Pers. = Leucanthemella serotina (L.) Tzvelev ■☆
89756 Chrysanthemum uliginosum Pers. = Leucanthemella serotina (L.) Tzvelev ■☆
89757 Chrysanthemum variifolium C. C. Chang = Ajania variifolia (C. C. Chang) Tzvelev ■
89758 Chrysanthemum variifolium C. C. Chang var. ramosum C. C. Chang = Ajania ramosa (C. C. Chang) C. Shih ■
89759 Chrysanthemum vestitum (Hemsl.) Stapf = Dendranthema vestitum (Hemsl.) Y. Ling ■
89760 Chrysanthemum viscidehirtum (Schott) Thell. = Heteranthemis viscidehirta Schott ■☆
89761 Chrysanthemum viscosum Desf. = Heteranthemis viscidehirta Schott ■☆
89762 Chrysanthemum vulgare (L.) Bernh. = Tanacetum vulgare L. ■
89763 Chrysanthemum vulgare (L.) Bernh. var. boreale (Fisch. ex DC.) Makino = Tanacetum vulgare L. var. boreale (Fisch. ex DC.) Trautv. et C. A. Mey. ■☆
89764 Chrysanthemum vulgare (L.) Bernh. var. boreale (Fisch. ex DC.) Makino = Tanacetum vulgare L. ■
89765 Chrysanthemum vulgatum (L.) Bernh. var. boreale (Fisch. ex DC.) Makino et Nemoto = Tanacetum vulgare L. ■
89766 Chrysanthemum wakasaense Shimot. ex Kitam.;若樱茼蒿■☆
89767 Chrysanthemum webbianum Ball = Tanacetum corymbosum (L.) Sch. Bip. ■☆
89768 Chrysanthemum weyrichii (Maxim.) Miyabe et T. Miyake;魏里希茼蒿■☆
89769 Chrysanthemum weyrichii (Maxim.) Miyabe et T. Miyake var. littorale (Maek.) Kudo = Chrysanthemum weyrichii (Maxim.) Miyabe et T. Miyake ■☆
89770 Chrysanthemum wilsonianum Hand. -Mazz. = Dendranthema lavandulifolium (Fisch. ex Trautv.) Kitam. ■
89771 Chrysanthemum wilsonianum Hand. -Mazz. = Dendranthema lavandulifolium (Fisch. ex Trautv.) Y. Ling et C. Shih var. seticuspe (Maxim.) C. Shih ■
89772 Chrysanthemum yezoense Maek. = Chrysanthemum arcticum L. subsp. yezoense (Maek.) H. Ohashi et Yonek. ■☆
89773 Chrysanthemum yoshinaganthum Makino ex Kitam.;吉永茼蒿■☆
89774 Chrysanthemum yunnanense (Jeffrey) Hand. -Mazz. = Hippolytia yunnanensis (Jeffrey) C. Shih ■
89775 Chrysanthemum zawadskii Herb. = Dendranthema zawadskii (Herb.) Tzvelev ■
89776 Chrysanthemum zawadskii Herb. subsp. acutilobum (DC.) Kitag. = Dendranthema maximowiczii (Kom.) Tzvelev ■
89777 Chrysanthemum zawadskii Herb. subsp. acutilobum Kitag. = Dendranthema zawadskii (Herb.) Tzvelev ■
89778 Chrysanthemum zawadskii Herb. subsp. acutilobum Kitag. var. alpinum (Nakai) Kitag. = Dendranthema oreastrum (Hance) Y. Ling ■
89779 Chrysanthemum zawadskii Herb. subsp. latilobum (Maxim.) Kitag. = Chrysanthemum zawadskii Herb. var. latilobum (Maxim.) Kitam. ■
89780 Chrysanthemum zawadskii Herb. subsp. latilobum (Maxim.) Kitag. = Dendranthema naktongense (Nakai) Tzvelev ■
89781 Chrysanthemum zawadskii Herb. subsp. latilobum (Maxim.) Kitag. var. dissectum (Y. Ling) Kitag. = Chrysanthemum zawadskii Herb. ■
89782 Chrysanthemum zawadskii Herb. subsp. latilobum Kitag. = Dendranthema chanetii (H. Lév.) C. Shih ■
89783 Chrysanthemum zawadskii Herb. var. acutilobum (DC.) Sealy f. alpinum (Nakai) Kitag. = Chrysanthemum zawadskii Herb. var. alpinum (Nakai) Kitam. ■
89784 Chrysanthemum zawadskii Herb. var. alpinum (Nakai) Kitam. = Dendranthema chanetii (H. Lév.) C. Shih ■
89785 Chrysanthemum zawadskii Herb. var. alpinum (Nakai) Kitam. = Dendranthema oreastrum (Hance) Y. Ling ■
89786 Chrysanthemum zawadskii Herb. var. campanulatum (Makino) Kitam.;风铃草状茼蒿■☆
89787 Chrysanthemum zawadskii Herb. var. latilobum (Maxim.) Kitam. = Dendranthema zawadskii (Herb.) Tzvelev var. latilobum

(Maxim.) Kitam. ■
89788 Chrysanthemum zawadskii Herb. var. latilobum (Maxim.) Kitam. = Dendranthema naktongense (Nakai) Tzvelev ■
89789 Chrysanthoglossum B. H. Wilcox, K. Bremer et Humphries (1993);黄舌菊属■☆
89790 Chrysanthoglossum deserticola (Murb.) B. H. Wilcox, K. Bremer et Humphries;沙地黄舌菊■☆
89791 Chrysanthoglossum trifurcatum (Desf.) B. H. Wilcox, K. Bremer et Humphries;黄舌菊■☆
89792 Chrysapsis Pascher = Trifolium L. ■
89793 Chrysaspis Desv. = Trifolium L. ■
89794 Chrysastrum Willd. ex Wedd. = Liabum Adans. ■●☆
89795 Chryseis Cass.(废弃属名) = Amberboa (Pers.) Less. ■
89796 Chryseis Cass.(废弃属名) = Centaurea L.(保留属名)●■
89797 Chryseis Lindl. = Eschscholtzia Cham. ■
89798 Chryseis glauca (Willd.) Cass. = Amberboa glauca (Willd.) Grossh. ■
89799 Chrysion Spach = Viola L. ■●
89800 Chrysiphiala Ker Gawl. = Stenomesson Herb. ■☆
89801 Chrysis DC. = Helianthus L. ■
89802 Chrysis Renealm. ex DC. = Helianthus L. ■
89803 Chrysithrix L. (1771);金黄莎草属■☆
89804 Chrysitrix L. = Chrysithrix L. ■☆
89805 Chrysitrix capensis L.;好望角金黄莎草■☆
89806 Chrysitrix capensis L. var. subteres C. B. Clarke;圆柱金黄莎草■☆
89807 Chrysitrix dodii C. B. Clarke;多德金黄莎草■☆
89808 Chrysitrix junciformis Nees;灯芯草黄莎草■☆
89809 Chrysitrix subteres (C. B. Clarke) Pfeiff. = Chrysitrix capensis L. var. subteres C. B. Clarke ■☆
89810 Chrysobactron Hook. f. = Bulbinella Kunth ■☆
89811 Chrysobalanaceae R. Br. (1818)(保留科名);金壳果科(金棒科,金橡实科,可可李科)●☆
89812 Chrysobalanus L. (1753);金壳果属(金棒属,可可李属,可口梅属);Coco Plum ●☆
89813 Chrysobalanus atacorensis A. Chev. = Chrysobalanus icaco L. subsp. atacorensis (A. Chev.) F. White ●☆
89814 Chrysobalanus chariensis A. Chev. = Chrysobalanus icaco L. subsp. atacorensis (A. Chev.) F. White ●☆
89815 Chrysobalanus ellipticus Sol. ex Sabine = Chrysobalanus icaco L. ●☆
89816 Chrysobalanus icaco L.;金壳果(金棒,金果梅,可口梅);Cocoplum, Coco-plum, Cocos Palm, Icaco, Icaco Coco Plum, Icaco Coco-plum, Icacos, West Indian Cocoa Palm ●☆
89817 Chrysobalanus icaco L. subsp. atacorensis (A. Chev.) F. White;阿塔金壳果●☆
89818 Chrysobalanus icaco L. subsp. ellipticus (Sol. ex Sabine) Souza = Chrysobalanus icaco L. ●☆
89819 Chrysobalanus icaco L. var. chariensis (A. Chev.) Souza = Chrysobalanus icaco L. subsp. atacorensis (A. Chev.) F. White ●☆
89820 Chrysobalanus icaco L. var. luteus (Sabine) Souza = Chrysobalanus icaco L. ●☆
89821 Chrysobalanus icaco L. var. macrocarpus Souza = Chrysobalanus icaco L. ●☆
89822 Chrysobalanus icaco L. var. orbicularis (Schumach.) Souza = Chrysobalanus icaco L. ●☆
89823 Chrysobalanus icaco L. var. pellocarpus (G. Mey.) Souza = Chrysobalanus icaco L. ●☆
89824 Chrysobalanus icaco L. var. roseus Souza = Chrysobalanus icaco L. ●☆
89825 Chrysobalanus orbicularis Schumach. = Chrysobalanus icaco L. ●☆
89826 Chrysobaphus Wall. = Anoectochilus Blume(保留属名)■
89827 Chrysobaphus roxburghii Wall. = Anoectochilus roxburghii (Wall.) Lindl. ●■
89828 Chrysobotrya Spach = Ribes L. ●
89829 Chrysobotrya odorata (H. L. Wendl.) Rydb. = Ribes odoratum H. L. Wendl. ●
89830 Chrysobotrya odorata Rydb. = Ribes odoratum H. L. Wendl. ●
89831 Chrysobotrya revoluta Spach = Ribes odoratum H. L. Wendl. ●
89832 Chrysobraya H. Hara = Lepidostemon Hook. f. et Thomson(保留属名)■
89833 Chrysobraya H. Hara(1974);金色肉叶芥属(金肉叶芥属)■☆
89834 Chrysobraya glaricola H. Hara;金色肉叶芥■☆
89835 Chrysocactus Y. Ito = Notocactus (K. Schum.) A. Berger et Backeb. ■
89836 Chrysocactus Y. Ito = Parodia Speg.(保留属名)●
89837 Chrysocalyx Guill. et Perr. (1831);金萼豆属■☆
89838 Chrysocalyx Guill. et Perr. = Crotalaria L. ●■
89839 Chrysocalyx ebenoides Guillaumin et Perr. = Crotalaria ebenoides (Guillaumin et Perr.) Walp. ●☆
89840 Chrysocalyx gracilis Guillaumin et Perr. = Crotalaria perrottetii DC. ●☆
89841 Chrysocalyx perrottetii (DC.) Guillaumin et Perr. = Crotalaria perrottetii DC. ●☆
89842 Chrysocalyx petitiana A. Rich. = Crotalaria petitiana (A. Rich.) Walp. ■☆
89843 Chrysocephalum Walp. (1841);金头菊属■☆
89844 Chrysocephalum Walp. = Helichrysum Mill.(保留属名)●■
89845 Chrysocephalum brevicilium (DC.) Walp.;金头菊■☆
89846 Chrysocephalum brevicilium Walp. = Chrysocephalum brevicilium (DC.) Walp. ■☆
89847 Chrysocephalum canescens Turcz.;灰金头菊■☆
89848 Chrysochamela (Fenzl) Boiss. (1867);金角芥属(金角状芥属)■☆
89849 Chrysochamela Boiss. = Chrysochamela (Fenzl) Boiss. ■☆
89850 Chrysochamela draboides Woronow;金角芥■☆
89851 Chrysochlamys Poepp. (1840);金被藤黄属●☆
89852 Chrysochlamys Poepp. et Endl. = Chrysochlamys Poepp. ●☆
89853 Chrysochlamys Poepp. et Endl. = Tovomitopsis Planch. et Triana ●☆
89854 Chrysochlamys micrantha Engl.;小花金被藤黄●☆
89855 Chrysochlamys multiflora Poepp. et Endl.;多花金被藤黄●☆
89856 Chrysochlamys pachypoda Planch. et Triana;粗梗金被藤黄●☆
89857 Chrysochlamys tenuifolia Cuatrec.;细叶金被藤黄●☆
89858 Chrysochloa Swallen(1941);金草属■☆
89859 Chrysochloa hubbardiana R. Germ. et Risop.;哈伯德金草■☆
89860 Chrysochloa lucida (Swallen) Swallen = Chrysochloa subaequigluma (Rendle) Swallen ■☆
89861 Chrysochloa orientalis (C. E. Hubb.) Swallen;东方金草■☆
89862 Chrysochloa subaequigluma (Rendle) Swallen;近等颖金草■☆
89863 Chrysocoma L. (1753);金毛菀属;Shrub Goldilocks ●☆
89864 Chrysocoma acaulis Walter = Vernonia acaulis (Walter) Gleason ■☆
89865 Chrysocoma acicularis Ehr. Bayer;针形金毛菀■☆
89866 Chrysocoma aurea Salisb. = Chrysocoma coma-aurea L. ●☆

89867 Chrysocoma candelabrum Ehr. Bayer;烛台金毛菀■☆
89868 Chrysocoma cernua L.;俯垂金毛菀■☆
89869 Chrysocoma ciliata E. Mey. = Chrysocoma oblongifolia DC. ■☆
89870 Chrysocoma ciliata L.;睫毛金毛菀;Fine-leaved Goldilocks ■☆
89871 Chrysocoma coma-aurea L.;灌木金毛菀;Shrub Goldilocks ●☆
89872 Chrysocoma coma-aurea L. var. cernua (L.) DC. = Chrysocoma cernua L. ■☆
89873 Chrysocoma coma-aurea L. var. patula (P. J. Bergius) DC. = Chrysocoma coma-aurea L. ●☆
89874 Chrysocoma decurrens DC. = Heteromma decurrens (DC.) O. Hoffm. ■☆
89875 Chrysocoma decurrens DC. var. pterocaulis (DC.) Harv. = Heteromma decurrens (DC.) O. Hoffm. ■☆
89876 Chrysocoma dinteri Muschl. = Nolletia arenosa O. Hoffm. ■☆
89877 Chrysocoma esterhuyseniae Ehr. Bayer;埃斯特金毛菀■☆
89878 Chrysocoma flava Ehr. Bayer;黄金毛菀■☆
89879 Chrysocoma gigantea Walter = Vernonia gigantea (Walter) Trel. ex Branner et Coville ■☆
89880 Chrysocoma graminifolia L. = Euthamia graminifolia (L.) Nutt. ■☆
89881 Chrysocoma graveolens Nutt. = Ericameria nauseosa (Pall. ex Pursh) G. L. Nesom et G. I. Baird var. graveolens (Nutt.) Reveal et Schuyler ●☆
89882 Chrysocoma hantamensis J. C. Manning et Goldblatt;汉塔姆金毛菀■☆
89883 Chrysocoma incana Burm. = Pteronia incana (Burm.) DC. ●☆
89884 Chrysocoma linosyris L. = Aster linosyris (L.) Bernh. ■☆
89885 Chrysocoma longifolia DC.;长叶金毛菀■☆
89886 Chrysocoma longifolia DC. var. patula? = Chrysocoma longifolia DC. ■☆
89887 Chrysocoma microcephala DC. = Chrysocoma ciliata L. ■☆
89888 Chrysocoma microphylla Thunb.;小叶金毛菀■☆
89889 Chrysocoma mozambicensis Ehr. Bayer;莫桑比克金毛菀■☆
89890 Chrysocoma nauseosa Pall. ex Pursh = Ericameria nauseosa (Pall. ex Pursh) G. L. Nesom et G. I. Baird ●☆
89891 Chrysocoma oblongifolia DC.;矩圆叶金毛菀■☆
89892 Chrysocoma obtusata (Thunb.) Ehr. Bayer;钝金毛菀■☆
89893 Chrysocoma odoratissima J. F. Gmel. = Carphephorus odoratissimus (J. F. Gmel.) H. J. -C. Hebert ■☆
89894 Chrysocoma paniculata J. F. Gmel. = Carphephorus paniculatus (J. F. Gmel.) H. J. -C. Hebert ■☆
89895 Chrysocoma patula P. J. Bergius = Chrysocoma coma-aurea L. ●☆
89896 Chrysocoma peduncularis DC. = Chrysocoma microphylla Thunb. ■☆
89897 Chrysocoma pinnatifida DC. = Chrysocoma tridentata DC. ■☆
89898 Chrysocoma polygalifolia S. Moore = Chrysocoma obtusata (Thunb.) Ehr. Bayer ■☆
89899 Chrysocoma pterocaulis DC. = Heteromma decurrens (DC.) O. Hoffm. ■☆
89900 Chrysocoma puberula Merxm.;微毛金毛菀■☆
89901 Chrysocoma puberula Schltr. ex Hutch. = Chrysocoma schlechteri Ehr. Bayer ■☆
89902 Chrysocoma rigidula (DC.) Ehr. Bayer;硬金毛菀■☆
89903 Chrysocoma scabra Thunb. = Polyarrhena reflexa (L.) Cass. ●☆
89904 Chrysocoma schlechteri Ehr. Bayer;施莱金毛菀■☆
89905 Chrysocoma sparsifolia Hutch.;稀叶金毛菀■☆
89906 Chrysocoma spatulata Forssk. = Vernonia spatulata (Forssk.) Sch. Bip. ■☆
89907 Chrysocoma spicata Forssk. = Ifloga spicata (Forssk.) Sch. Bip. ■☆
89908 Chrysocoma strigosa Ehr. Bayer;糙伏毛金毛菀■☆
89909 Chrysocoma subumbellata Thell. = Chrysocoma longifolia DC. ■☆
89910 Chrysocoma tatarica Less. = Linosyris tatarica (Less.) C. A. Mey. ■
89911 Chrysocoma tenuifolia P. J. Bergius = Chrysocoma ciliata L. ■☆
89912 Chrysocoma tomentosa L.;绒毛金毛菀■☆
89913 Chrysocoma tridentata DC.;三齿金毛菀■☆
89914 Chrysocoma undulata Thunb. = Nidorella undulata (Thunb.) Sond. ex Harv. ■☆
89915 Chrysocoma valida Ehr. Bayer;刚直金毛菀■☆
89916 Chrysocoma villosa L. = Linosyris villosa (L.) DC. ■
89917 Chrysocome St. -Lag. = Chrysocoma L. ●☆
89918 Chrysocoptis Nutt. = Coptis Salisb. ■
89919 Chrysocoptis occidentalis Nutt. = Coptis occidentalis (Nutt.) Torr. et A. Gray ■☆
89920 Chrysocoryne Endl. = Angianthus J. C. Wendl. (保留属名)■●☆
89921 Chrysocoryne Endl. = Crossolepis Benth. ■☆
89922 Chrysocoryne Endl. = Gnephosis Cass. ■☆
89923 Chrysocoryne Zoellner = Leucocoryne Lindl. ■☆
89924 Chrysocoryne Zoellner = Pabellonia Quezada et Martic. ■☆
89925 Chrysocyathus Falc. = Adonis L. (保留属名)■
89926 Chrysocyathus Falc. = Calathodes Hook. f. et Thomson ●★
89927 Chrysocyathus nepalensis (Simonov.) Chrtek et Slavíková = Adonis nepalensis Simonov. ■☆
89928 Chrysocycnis Linden et Rchb. f. (1854);金鹅兰属■☆
89929 Chrysocycnis schlimii Linden et Rchb. f.;金鹅兰■☆
89930 Chrysodendron Meisn. = Protea L. (保留属名)●☆
89931 Chrysodendron Teran et Beriand. = Mahonia Nutt. (保留属名)●
89932 Chrysodendron Vaill. ex Meisn. = Protea L. (保留属名)●☆
89933 Chrysodiscus Steetz = Athrixia Ker Gawl. ●■☆
89934 Chrysoglossella Hatus. (1967);小金唇兰属■☆
89935 Chrysoglossella Hatus. = Hancockia Rolfe ■
89936 Chrysoglossella japonica Hatus.;小金唇兰■☆
89937 Chrysoglossella japonica Hatus. = Hancockia uniflora Rolfe ■
89938 Chrysoglossum Blume (1825);金唇兰属(黄唇兰属);Chrysoglossum,Goldlip-orchis ■
89939 Chrysoglossum assamicum Hook. f.;锚钩金唇兰(锚钩吻兰,中国吻兰);Anchor Collabium ■
89940 Chrysoglossum assamicum Hook. f. = Collabium assamicum (Hook. f.) Seidenf. ■
89941 Chrysoglossum chapaense (Gagnep.) Ts. Tang et F. T. Wang = Collabium formosanum Hayata ■
89942 Chrysoglossum delavayi (Gagnep.) Ts. Tang et F. T. Wang = Collabium delavayi (Gagnep.) Seidenf. ■
89943 Chrysoglossum delavayi (Gagnep.) Ts. Tang et F. T. Wang = Collabium formosanum Hayata ■
89944 Chrysoglossum erraticum Hook. f. = Chrysoglossum ornatum Blume ■
89945 Chrysoglossum formosanum Hayata = Chrysoglossum ornatum Blume ■
89946 Chrysoglossum latifolium (Blume) Benth. et Hook. f. = Diglyphosa latifolia Blume ■
89947 Chrysoglossum maculatum (Thwaites) Hook. f. = Chrysoglossum ornatum Blume ■
89948 Chrysoglossum ornatum Blume;金唇兰(黄唇兰,金蝉兰,美丽

金唇兰,台湾黄唇兰);Goldlip-orchis,Pretty Chrysoglossum ■
89949 Chrysoglossum robinsonii Ridl. = Collabium chinense (Rolfe) Ts. Tang et F. T. Wang ■
89950 Chrysoglossum sinense Mansf. = Collabium assamicum (Hook. f.) Seidenf. ■
89951 Chrysogonum A. Juss. = Leontice L. ●■
89952 Chrysogonum L. (1753);金星菊属;Golden Star,Green and Gold ●■☆
89953 Chrysogonum australe Alexander ex Small = Chrysogonum virginianum L. var. australe (Alexander ex Small) H. E. Ahles ■☆
89954 Chrysogonum leandrii Humbert;利安金星菊■☆
89955 Chrysogonum madagascariense Humbert;马岛金星菊■☆
89956 Chrysogonum perrieri (Humbert) Humbert;佩里耶金星菊■☆
89957 Chrysogonum peruvianum L. = Zinnia peruviana (L.) L. ■
89958 Chrysogonum stenocephalum Humbert;细头金星菊●☆
89959 Chrysogonum virginianum L.;金星菊;Delosperma Nubigenum, Golden Knee, Golden Star, Goldenstar, Gold-in-green, Green and Gold,Green-and-Gold,Hardy Ice,Trailing Ice Plant ■☆
89960 Chrysogonum virginianum L. var. australe (Alexander ex Small) H. E. Ahles;南方金星菊■☆
89961 Chrysogonum virginianum L. var. dentatum A. Gray = Chrysogonum virginianum L. ■☆
89962 Chrysolaena H. Rob. (1988);黄毛斑鸠菊属■☆
89963 Chrysolaena H. Rob. = Vernonia Schreb.(保留属名)●■
89964 Chrysolaena flexuosa (Sims) H. Rob.;黄毛斑鸠菊■☆
89965 Chrysolarix H. E. Moore = Pseudolarix Gordon ●★
89966 Chrysolarix amabilis (J. Nelson) H. E. Moore = Pseudolarix amabilis (J. Nelson) Rehder ●◇
89967 Chrysolarix amabilis (Nelson) Moore = Pseudolarix kaempferi (Lindl.) Gordon ●◇
89968 Chrysolepis Hjelmq. (1948);金栗属(黄鳞栗属);Chinquapin, Western Chinkapin ●☆
89969 Chrysolepis chrysophylla (Douglas ex Hook.) Hjelmq.;金栗(黄叶锥栲,金叶锥栗,美国栲树);Chinquapin,Giant Chinkapin,Giant Chinquapin, Giant Golden Chinkapin, Golden Chestnut, Golden Chinkapin,Golden Evergreenchinkapin,Goldenleaf Chinkapin ●
89970 Chrysolepis chrysophylla (Douglas ex Hook.) Hjelmq. = Castanopsis chrysophylla (Douglas) A. DC. ●
89971 Chrysolepis chrysophylla (Douglas ex Hook.) Hjelmq. var. minor (Benth.) Munz;小金栗●☆
89972 Chrysolepis megacarpa?;大果金栗;Greater Malayan Chestnut ●☆
89973 Chrysolepis sempervirens (Kellogg) Hjelmq.;齿金栗;Bush Golden Chinquapin,Sierra Chinkapin ●☆
89974 Chrysolepis sempervirens (Kellogg) Hjelmq. = Castanea sempervirens Kellogg ●☆
89975 Chrysoliga Willd. ex DC. = Nesaea Comm. ex Kunth(保留属名) ■●☆
89976 Chrysolinum Fourr. = Linum L. ●■
89977 Chrysolyga Willd. ex Steud. = Nesaea Comm. ex Kunth(保留属名)■●☆
89978 Chrysoma Nutt. (1834);木黄花属;Woody Goldenrod ●☆
89979 Chrysoma Nutt. = Solidago L. ■
89980 Chrysoma fasciculata Eastw. = Ericameria fasciculata (Eastw.) J. F. Macbr. ●☆
89981 Chrysoma pauciflosculosa (Michx.) Greene;木黄花;Woody Goldenrod ●☆
89982 Chrysoma pumila Nutt. = Petradoria pumila (Nutt.) Greene ■☆
89983 Chrysoma uniligulata (DC.) Nutt. = Solidago uliginosa Nutt. ■☆
89984 Chrysomallum Thouars = Vitex L. ●
89985 Chrysomallum integrifolium Bojer ex Schauer = Vitex bojeri Schau ●☆
89986 Chrysomallum lanuginosum Bojer ex Schauer = Vitex lanigera Schauer ●☆
89987 Chrysomallum madagascariense Thouars ex Steud. = Vitex chrysomallum Steud. ●☆
89988 Chrysomelea Tausch = Coreopsis L. ●■
89989 Chrysomelea Tausch = Coreopsoides Moench ●■
89990 Chrysomelon J. R. Forst. et G. Forst. ex A. Gray = Spondias L. ●
89991 Chrysomu biflora L. = Galatella biflora (L.) Nees ■
89992 Chrysonias Benth. ex Steud. = Chrysoscias E. Mey. ■☆
89993 Chrysonias Benth. ex Steud. = Rhynchosia Lour.(保留属名)●■
89994 Chrysopappus Takht. = Centaurea L.(保留属名)●■
89995 Chrysopelta Tausch = Achillea L. ■
89996 Chrysophae Koso-Pol. = Chaerophyllum L. ■
89997 Chrysophania Kunth ex Less. = Zaluzania Pers. ■☆
89998 Chrysophiala Post et Kuntze = Chrysiphiala Ker Gawl. ■☆
89999 Chrysophiala Post et Kuntze = Stenomesson Herb. ■☆
90000 Chrysophora Cham. ex Triana = Leandra Raddi ●■☆
90001 Chrysophora Cham. ex Triana = Oxymeris DC. ●■☆
90002 Chrysophthalmum Phil. = Grindelia Willd. ●■☆
90003 Chrysophthalmum Sch. Bip. = Chrysophthalmum Sch. Bip. ex Walp. ■☆
90004 Chrysophthalmum Sch. Bip. ex Walp. (1843);金眼菊属■☆
90005 Chrysophthalmum leptocladum Rech. f.;细枝金眼菊■☆
90006 Chrysophthalmum sternutatorium Sch. Bip.;金眼菊■☆
90007 Chrysophyllum L. (1753);金叶树属(金叶山榄属,星苹果属);Golden Leaf, Goldenleaf, Goldleaftree, Star Apple, Starapple, Star-apple ●
90008 Chrysophyllum africanum A. DC.;非洲金叶树;African Apple, African Star-apple,Star Apple ●☆
90009 Chrysophyllum africanum A. DC. var. aubrevillei (Pellegr.) Aubrév. = Chrysophyllum africanum A. DC. ●☆
90010 Chrysophyllum africanum A. DC. var. casteelsii De Wild. = Chrysophyllum africanum A. DC. ●☆
90011 Chrysophyllum africanum A. DC. var. likimensis De Wild. = Chrysophyllum africanum A. DC. ●☆
90012 Chrysophyllum africanum A. DC. var. multinervatum De Wild.;多脉非洲金叶树●☆
90013 Chrysophyllum africanum A. DC. var. orientale Engl.;东部非洲金叶树●☆
90014 Chrysophyllum akuase A. Chev. = Chrysophyllum ubangiense (De Wild.) D. J. Harris ●☆
90015 Chrysophyllum albidum G. Don;白金叶树;Star Apple, White Apple,White Star-apple ●☆
90016 Chrysophyllum ambrense (Aubrév.) G. E. Schatz et L. Gaut.;昂布尔金叶树●☆
90017 Chrysophyllum analalavense (Aubrév.) G. E. Schatz et L. Gaut.;阿纳拉拉瓦金叶树●☆
90018 Chrysophyllum antunesii Engl. = Englerophytum magalismontanum (Sond.) T. D. Penn. ●☆
90019 Chrysophyllum argenteum Jacq.;安的列斯金叶树;Antilles Fear,Antilles Pear,Damoceen ●☆
90020 Chrysophyllum argyrophyllum Hiern = Englerophytum magalismontanum (Sond.) T. D. Penn. ●☆

90021 Chrysophyllum autranianum A. Chev. = Chrysophyllum lacourtianum De Wild. ●☆

90022 Chrysophyllum azaguieanum J. Miège;阿扎金叶树●☆

90023 Chrysophyllum bangweolense R. E. Fr.;班韦金叶树●☆

90024 Chrysophyllum batangense C. H. Wright = Synsepalum brevipes (Baker) T. D. Penn. ●☆

90025 Chrysophyllum beguei Aubrév. et Pellegr.;贝格金叶树●☆

90026 Chrysophyllum belemba De Wild. = Chrysophyllum ubangiense (De Wild.) D. J. Harris ●☆

90027 Chrysophyllum boivinianum (Pierre) Baehni;博伊文金叶树●☆

90028 Chrysophyllum boukokoense (Aubrév. et Pellegr.) L. Gaut.;布科科金叶树●☆

90029 Chrysophyllum brieyi De Wild. = Chrysophyllum subnudum Baker ●☆

90030 Chrysophyllum buchholzii Engl. = Chrysophyllum pruniforme Pierre ex Engl. ●☆

90031 Chrysophyllum cacondense Greves = Chrysophyllum bangweolense R. E. Fr. ●☆

90032 Chrysophyllum cainito L.;星苹果(牛奶果,星苹果);Bubby Water, Cainito, Cainito Star Apple, Cainito Star-apple, Star Apple, Star Plum, Starapple, Star-apple ●

90033 Chrysophyllum calophyllum Exell;美叶金叶树●☆

90034 Chrysophyllum capuronii G. E. Schatz et L. Gaut.;凯普伦金叶树●☆

90035 Chrysophyllum carvalhoi Engl. = Englerophytum magalismontanum (Sond.) T. D. Penn. ●☆

90036 Chrysophyllum cerasiferum (Welw.) Hiern = Synsepalum cerasiferum (Welw.) T. D. Penn. ●☆

90037 Chrysophyllum cinereum Engl. = Synsepalum brevipes (Baker) T. D. Penn. ●☆

90038 Chrysophyllum claessensii De Wild. = Chrysophyllum ubangiense (De Wild.) D. J. Harris ●☆

90039 Chrysophyllum congoense Pierre ex A. Chev. = Tridesmostemon congoense (Pierre ex A. Chev.) Aubrév. et Pellegr. ●☆

90040 Chrysophyllum delphinense (Aubrév.) G. E. Schatz et L. Gaut.;德尔芬金叶树●☆

90041 Chrysophyllum disaco Hiern = Synsepalum cerasiferum (Welw.) T. D. Penn. ●☆

90042 Chrysophyllum ealaense De Wild. = Chrysophyllum welwitschii Engl. ●☆

90043 Chrysophyllum edule Hoyle = Chrysophyllum africanum A. DC. ●☆

90044 Chrysophyllum ellipticum A. Chev. = Chrysophyllum welwitschii Engl. ●☆

90045 Chrysophyllum farannense A. Chev. = Englerophytum magalismontanum (Sond.) T. D. Penn. ●☆

90046 Chrysophyllum fenerivense (Aubrév.) G. E. Schatz et L. Gaut.;费内里沃金叶树●☆

90047 Chrysophyllum ferrugineo-tomentosum Engl. = Pouteria alnifolia (Baker) Roberty ●☆

90048 Chrysophyllum giganteum A. Chev.;巨大金叶树●☆

90049 Chrysophyllum glomeruliferum Hutch. et Dalziel = Englerophytum oblanceolatum (S. Moore) T. D. Penn. ●☆

90050 Chrysophyllum gossweileri De Wild. = Neoboivinella gossweileri (De Wild.) Liben ●☆

90051 Chrysophyllum gracile A. Chev. = Chrysophyllum welwitschii Engl. ●☆

90052 Chrysophyllum guyanense Klotzsch ex Miq.;圭亚那金叶木●☆

90053 Chrysophyllum henriquesii Engl.;亨利克斯金叶木●☆

90054 Chrysophyllum holtzii Engl.;霍尔茨金叶木●☆

90055 Chrysophyllum imperiale (Linden) Benth. et Hook. f.;壮丽金叶树●☆

90056 Chrysophyllum iturense Engl. = Englerophytum iturense (Engl.) L. Gaut. ●☆

90057 Chrysophyllum kayei S. Moore = Chrysophyllum albidum G. Don ●☆

90058 Chrysophyllum kilimandscharicum G. M. Schulze = Englerophytum natalense (Sond.) T. D. Penn. ●☆

90059 Chrysophyllum klainei Engl. = Chrysophyllum welwitschii Engl. ●☆

90060 Chrysophyllum lacourtianum De Wild.;拉库尔特金叶木●☆

90061 Chrysophyllum lanceolatum (Blume) A. DC.;多花金叶树;Manyflower Starapple, Stellate-frited Star-apple ●

90062 Chrysophyllum lanceolatum (Blume) A. DC. var. stellatocarpon P. Royen;金叶树(大横纹);Goldleaftree, Stellatefruit Star Apple, Stellatefruit Starapple ●

90063 Chrysophyllum leptospermum (Baehni) Roberty = Breviea sericea Aubrév. et Pellegr. ●☆

90064 Chrysophyllum letestuanum A. Chev. = Chrysophyllum ubangiense (De Wild.) D. J. Harris ●☆

90065 Chrysophyllum longepedicellatum De Wild. = Zeyherella longepedicellata (De Wild.) Aubrév. et Pellegr. ●☆

90066 Chrysophyllum longifolium De Wild. = Zeyherella longepedicellata (De Wild.) Aubrév. et Pellegr. ●☆

90067 Chrysophyllum longipes Engl.;长梗金叶树●☆

90068 Chrysophyllum lujai De Wild. = Englerophytum magalismontanum (Sond.) T. D. Penn. ●☆

90069 Chrysophyllum lungii De Wild.;伦格金叶树●☆

90070 Chrysophyllum macrophyllum Sabine = Chrysophyllum africanum A. DC. ●☆

90071 Chrysophyllum magalismontanum Sond. = Englerophytum magalismontanum (Sond.) T. D. Penn. ●☆

90072 Chrysophyllum masoalense (Aubrév.) G. E. Schatz et L. Gaut.;马苏阿拉金叶树●☆

90073 Chrysophyllum metallicum Hutch. et Dalziel = Chrysophyllum subnudum Baker ●☆

90074 Chrysophyllum millenianum Engl. = Chrysophyllum albidum G. Don ●☆

90075 Chrysophyllum mohorense Engl. = Englerophytum magalismontanum (Sond.) T. D. Penn. ●☆

90076 Chrysophyllum monopyrenum Sw.;石星苹果(金枣李)●☆

90077 Chrysophyllum mortehanii De Wild. = Chrysophyllum pruniforme Pierre ex Engl. ●☆

90078 Chrysophyllum natalense Sond. = Englerophytum natalense (Sond.) T. D. Penn. ●☆

90079 Chrysophyllum normandii A. Chev. = Chrysophyllum subnudum Baker ●☆

90080 Chrysophyllum obovatum Sabine et G. Don = Manilkara obovata (Sabine et G. Don) J. H. Hemsl. ●☆

90081 Chrysophyllum ogoouense A. Chev.;奥果韦金叶树●☆

90082 Chrysophyllum oliviforme L.;卵形金叶树;Damson-plum, Satin Leaf ●☆

90083 Chrysophyllum omumu J. D. Kenn. = Chrysophyllum africanum A. DC. ●☆

90084 Chrysophyllum pentagonocarpum Engl. et K. Krause = Chrysophyllum ubangiense (De Wild.) D. J. Harris ●☆

90085 Chrysophyllum perpulchrum Mildbr. ex Hutch. et Dalziel;猴子金叶树;Monkey Apple,Monkey Star-apple,Star Apple ●☆

90086 Chrysophyllum perrieri (Lecomte) G. E. Schatz et L. Gaut.;佩里耶金叶树●☆

90087 Chrysophyllum prunifolium Baker;李叶金叶树●☆

90088 Chrysophyllum pruniforme Pierre ex Engl.;李形金叶树●☆

90089 Chrysophyllum renieri De Wild. = Chrysophyllum subnudum Baker ●☆

90090 Chrysophyllum roxburghii G. Don;罗氏星苹果(金叶树);Roxburth Starapple ●☆

90091 Chrysophyllum roxburghii G. Don = Chrysophyllum lanceolatum (Blume) A. DC. var. stellatocarpon Royen ●

90092 Chrysophyllum rwandense Troupin = Afrosersalisia rwandensis (Troupin) Liben ●☆

90093 Chrysophyllum sapinii De Wild. = Zeyherella longepedicellata (De Wild.) Aubrév. et Pellegr. ●☆

90094 Chrysophyllum sericeum A. Chev. = Breviea sericea Aubrév. et Pellegr. ●☆

90095 Chrysophyllum stuhlmannii Engl. = Synsepalum brevipes (Baker) T. D. Penn. ●☆

90096 Chrysophyllum subnudum Baker;近裸星苹果●☆

90097 Chrysophyllum taiense Aubrév. et Pellegr.;塔亚星苹果●☆

90098 Chrysophyllum tessmannii Engl. et K. Krause = Englerophytum magalismontanum (Sond.) T. D. Penn. ●☆

90099 Chrysophyllum ubangiense (De Wild.) D. J. Harris;乌班吉星苹果●☆

90100 Chrysophyllum ulugurense Engl. = Synsepalum ulugurense (Engl.) Engl. ●☆

90101 Chrysophyllum vermoesenii De Wild. = Englerophytum iturense (Engl.) L. Gaut. ●☆

90102 Chrysophyllum viridifolium J. M. Wood et Franks;绿叶星苹果●☆

90103 Chrysophyllum welwitschii Engl.;韦氏星苹果●☆

90104 Chrysophyllum wilmsii Engl. = Englerophytum magalismontanum (Sond.) T. D. Penn. ●☆

90105 Chrysophyllum zimmermannii Engl.;齐默尔曼星苹果●☆

90106 Chrysopia Noronha ex Thouars = Symphonia L. f. ●☆

90107 Chrysopogon Trin. (1820)(保留属名);金须茅属(金丝草属,竹节草属);Chrysopogon,False Beardgrass,Scented Grass,Scented-grass,Sugar Grass ■

90108 Chrysopogon aciculatus (Retz. ex Roem. et Schult.) Trin.;竹节草(草谷子,草子花,鬼谷草,过路蜈蚣草,鸡谷草,鸡骨草,铺地蜈蚣,埔虾头,蜈蚣草,粘人草,粘身草,紫穗茅香);Aciculate Chrysopogon,Golden False Beardgrass,Love Grass ■

90109 Chrysopogon aciculatus (Retz.) Trin. = Chrysopogon aciculatus (Retz. ex Roem. et Schult.) Trin. ■

90110 Chrysopogon argutus Trin. ex Steud.;亮金须茅■☆

90111 Chrysopogon aucheri (Boiss.) Stapf;奥切尔金须茅■☆

90112 Chrysopogon aucheri (Boiss.) Stapf var. chrysopus (Coss.) Maire et Weiller = Chrysopogon aucheri (Boiss.) Stapf ■☆

90113 Chrysopogon aucheri (Boiss.) Stapf var. pulvinatus Stapf = Chrysopogon plumulosus Hochst. ■☆

90114 Chrysopogon aucheri (Boiss.) Stapf var. quinqueplumis (A. Rich.) Stapf = Chrysopogon plumulosus Hochst. ■☆

90115 Chrysopogon aucheri (Boiss.) Stapf var. quinqueplumis (Hack.) Stapf = Chrysopogon aucheri (Boiss.) Stapf ■☆

90116 Chrysopogon ciliolatus (Nees ex Steud.) Boiss. = Chrysopogon serrulatus Trin. ■☆

90117 Chrysopogon ciliolatus (Nees ex Steud.) Boiss. var. aucheri (Boiss.) Boiss. = Chrysopogon aucheri (Boiss.) Stapf ■☆

90118 Chrysopogon echinulatus (Nees ex Steud.) W. Watson = Chrysopogon echinulatus (Steud.) W. Watson ■

90119 Chrysopogon echinulatus (Nees) W. Watson = Chrysopogon gryllus (L.) Trin. subsp. echinulatus (Nees) Cope ■☆

90120 Chrysopogon echinulatus (Steud.) W. Watson;刺金须茅;Spine Chrysopogon ■

90121 Chrysopogon fulvus (Spreng.) Chiov. var. migiurtinus (Chiov.) Chiov. = Chrysopogon aucheri (Boiss.) Stapf ■☆

90122 Chrysopogon fulvus (Spreng.) Chiov. var. tremulus (Hack.) Chiov. = Chrysopogon serrulatus Trin. ■☆

90123 Chrysopogon fulvus Chiov.;红金须茅;Red False Beardgrass ■☆

90124 Chrysopogon fulvus Chiov. var. serrulatus (Trin.) R. R. Stewart = Chrysopogon serrulatus Trin. ■☆

90125 Chrysopogon gryllus (L.) Trin.;欧洲金须茅(蝉金须茅,刺金须茅);Cricket Rhaphis,European Chrysopogon,French Whisk ■☆

90126 Chrysopogon gryllus (L.) Trin. subsp. echinulatus (Nees) Cope;小刺欧洲金须茅■☆

90127 Chrysopogon gryllus Trin. = Chrysopogon gryllus (L.) Trin. ■☆

90128 Chrysopogon gryllus Trin. subsp. echinulatus (Nees ex Steud.) Cope = Chrysopogon echinulatus (Steud.) W. Watson ■

90129 Chrysopogon gryllus Trin. subsp. echinulatus (Nees) Cope = Chrysopogon echinulatus (Steud.) W. Watson ■

90130 Chrysopogon montanus Trin. var. migiurtinus Chiov. = Chrysopogon aucheri (Boiss.) Stapf ■☆

90131 Chrysopogon montanus Trin. var. tremulus (Hack.) Stapf = Chrysopogon serrulatus Trin. ■☆

90132 Chrysopogon nigritanus (Benth.) Veldkamp;尼格里塔金须茅■☆

90133 Chrysopogon orientalis (Desv.) A. Camus;金须茅(金须草);Oriental Chrysopogon ■

90134 Chrysopogon parviflorus (R. Br.) Benth. var. spicigerus Benth. = Capillipedium spicigerum S. T. Blake ■

90135 Chrysopogon parviflorus (R. Br.) Nees = Capillipedium parviflorum (R. Br.) Stapf ■

90136 Chrysopogon parviflorus (R. Br.) Nees var. spicigerus Benth. = Capillipedium parviflorum (R. Br.) Stapf var. spicigerum (Benth.) C. C. Hsu ■

90137 Chrysopogon parvispicus (R. Br.) Nees = Capillipedium parviflorum (R. Br.) Stapf ■

90138 Chrysopogon parvispicus (Steud.) Watson = Capillipedium parviflorum (R. Br.) Stapf ■

90139 Chrysopogon pictus Hance = Capillipedium parviflorum (R. Br.) Stapf ■

90140 Chrysopogon plumulosus Hochst.;羽状金须茅■☆

90141 Chrysopogon quinqueplumis A. Rich. = Chrysopogon plumulosus Hochst. ■☆

90142 Chrysopogon serrulatus Trin.;小齿金须茅■☆

90143 Chrysopogon sinensis Rendle = Chrysopogon orientalis (Desv.) A. Camus ■

90144 Chrysopogon villosulus (Nees ex Steud.) W. Watson = Capillipedium parviflorum (R. Br.) Stapf ■

90145 Chrysopogon wightianus (Nees ex Steud.) W. Watson var. leucanthus Thwaites = Chrysopogon serrulatus Trin. ■☆

90146 Chrysopogon zizanioides (L.) Roberty = Vetiveria zizanioides (L.) Nash ■

90147 Chrysoprenanthes (Sch. Bip.) Bramwell = Prenanthes L. ■

90148 Chrysoprenanthes (Sch. Bip.) Bramwell(2003);金盘果菊属■☆

90149 Chrysoprenanthes pendula (Sch. Bip.) Bramwell = Sonchus pendulus (Sch. Bip.) Sennikov ■☆

90150 Chrysopsis (Nutt.) Elliott = Pityopsis Nutt. ■☆

90151 Chrysopsis (Nutt.) Elliott(1823)(保留属名);金菊属(金菀属);Goldaster,Golden Aster ■☆

90152 Chrysopsis Elliott = Pityopsis Nutt. ■☆

90153 Chrysopsis acaulis Nutt. = Stenotus acaulis (Nutt.) Nutt. ■☆

90154 Chrysopsis adenolepis Fernald = Pityopsis aspera (A. Gray) Small var. adenolepis (Fernald) Semple et F. D. Bowers ■☆

90155 Chrysopsis alpicola Rydb. = Heterotheca pumila (Greene) Semple ■☆

90156 Chrysopsis alpicola Rydb. var. glomerata A. Nelson = Heterotheca pumila (Greene) Semple ■☆

90157 Chrysopsis alpina Nutt. = Ionactis alpina (Nutt.) Greene ■☆

90158 Chrysopsis amplifolia Rydb. = Heterotheca fulcrata (Greene) Shinners var. amplifolia (Rydb.) Semple ■☆

90159 Chrysopsis amygdalina (Lam.) Nutt. ex Elliott = Doellingeria umbellata (Mill.) Nees ■☆

90160 Chrysopsis angustifolia Rydb.;狭叶金菊■☆

90161 Chrysopsis arenaria Elmer = Heterotheca sessiliflora (Nutt.) Shinners subsp. bolanderi (A. Gray) Semple ■☆

90162 Chrysopsis arenicola Alexander ex Small = Chrysopsis gossypina (Michx.) Elliott ■☆

90163 Chrysopsis argentea (Pers.) Elliott = Pityopsis graminifolia (Michx.) Nutt. var. latifolia (Fernald) Semple et F. D. Bowers ■☆

90164 Chrysopsis arida A. Nelson = Heterotheca villosa (Pursh) Shinners var. minor (Hook.) Semple ■☆

90165 Chrysopsis aspera (A. Gray) Shuttlew. ex Small = Pityopsis aspera (A. Gray) Small ■☆

90166 Chrysopsis asprella Greene = Heterotheca villosa (Pursh) Shinners var. minor (Hook.) Semple ■☆

90167 Chrysopsis bakeri Greene = Heterotheca villosa (Pursh) Shinners var. minor (Hook.) Semple ■☆

90168 Chrysopsis ballardii Rydb. = Heterotheca villosa (Pursh) Shinners var. ballardii (Rydb.) Semple ■☆

90169 Chrysopsis berlandieri Greene = Heterotheca canescens (DC.) Shinners ■☆

90170 Chrysopsis bolanderi A. Gray = Heterotheca sessiliflora (Nutt.) Shinners subsp. bolanderi (A. Gray) Semple ■☆

90171 Chrysopsis breweri A. Gray = Eucephalus breweri (A. Gray) G. L. Nesom ■☆

90172 Chrysopsis butleri Rydb. = Heterotheca villosa (Pursh) Shinners var. foliosa (Nutt.) V. L. Harms ■☆

90173 Chrysopsis californica Elmer = Heterotheca sessiliflora (Nutt.) Shinners subsp. echioides (Benth.) Semple ■☆

90174 Chrysopsis camphorata Eastw. = Heterotheca sessiliflora (Nutt.) Shinners subsp. echioides (Benth.) Semple ■☆

90175 Chrysopsis camporum Greene;弯金菊;Hairy Golden Aster, Prairie Golden Aster ■☆

90176 Chrysopsis camporum Greene = Heterotheca camporum (Greene) Shinners ■☆

90177 Chrysopsis camporum Greene var. glandulissima (Semple) Cronquist = Heterotheca camporum (Greene) Shinners var. glandulissima Semple ■☆

90178 Chrysopsis canescens (DC.) Torr. et A. Gray = Heterotheca canescens (DC.) Shinners ■☆

90179 Chrysopsis canescens (DC.) Torr. et A. Gray var. nana A. Gray;矮小金菊■☆

90180 Chrysopsis chrysophylla?;黄叶金菊;Golden Chestnut ■☆

90181 Chrysopsis columbiana Greene = Heterotheca villosa (Pursh) Shinners var. minor (Hook.) Semple ■☆

90182 Chrysopsis compacta Greene = Heterotheca villosa (Pursh) Shinners var. minor (Hook.) Semple ■☆

90183 Chrysopsis cooperi A. Nelson = Heterotheca pumila (Greene) Semple ■☆

90184 Chrysopsis correllii Fernald = Pityopsis graminifolia (Michx.) Nutt. var. latifolia (Fernald) Semple et F. D. Bowers ■☆

90185 Chrysopsis cruiseana Dress = Chrysopsis gossypina (Michx.) Elliott subsp. cruiseana (Dress) Semple ■☆

90186 Chrysopsis cryptocephala Wooton et Standl. = Heterotheca fulcrata (Greene) Shinners ■☆

90187 Chrysopsis decumbens Chapm. = Chrysopsis gossypina (Michx.) Elliott ■☆

90188 Chrysopsis delaneyi Wunderlin et Semple;德莱尼金菊;DeLaney's Goldenaster ■☆

90189 Chrysopsis dentata Elliott = Chrysopsis gossypina (Michx.) Elliott ■☆

90190 Chrysopsis depressa Rydb. = Heterotheca villosa (Pursh) Shinners var. depressa (Rydb.) Semple ■☆

90191 Chrysopsis echioides Benth. = Heterotheca sessiliflora (Nutt.) Shinners subsp. echioides (Benth.) Semple ■☆

90192 Chrysopsis elata Osterh. = Heterotheca fulcrata (Greene) Shinners ■☆

90193 Chrysopsis falcata (Nutt.) Elliott;镰叶金菊;Sickleleaf Goldaster ●☆

90194 Chrysopsis falcata (Pursh) Elliott = Pityopsis falcata (Pursh) Nutt. ■☆

90195 Chrysopsis fastigiata Greene = Heterotheca sessiliflora (Nutt.) Shinners subsp. fastigiata (Greene) Semple ■☆

90196 Chrysopsis flexuosa Nash = Pityopsis flexuosa (Nash) Small ■☆

90197 Chrysopsis floribunda Greene = Heterotheca villosa (Pursh) Shinners var. minor (Hook.) Semple ■☆

90198 Chrysopsis floridana Small;佛罗里达金菊;Florida Goldenaster ●■☆

90199 Chrysopsis foliosa Nutt. = Heterotheca villosa (Pursh) Shinners var. foliosa (Nutt.) V. L. Harms ■☆

90200 Chrysopsis foliosa Nutt. var. amplifolia (Rydb.) A. Nelson = Heterotheca fulcrata (Greene) Shinners var. amplifolia (Rydb.) Semple ■☆

90201 Chrysopsis foliosa Nutt. var. imbricata (A. Nelson) A. Nelson = Heterotheca villosa (Pursh) Shinners var. foliosa (Nutt.) V. L. Harms ■☆

90202 Chrysopsis foliosa Nutt. var. sericeovillosissima A. Gray = Heterotheca rutteri (Rothr.) Shinners ■☆

90203 Chrysopsis fulcrata Greene = Heterotheca fulcrata (Greene) Shinners ■☆

90204 Chrysopsis gigantea Small = Chrysopsis gossypina (Michx.) Elliott subsp. hyssopifolia (Nutt.) Semple ■☆

90205 Chrysopsis godfreyi Semple;戈氏金菊;Godfrey's Goldenaster ■☆

90206 Chrysopsis gossypina (Michx.) Elliott;棉金菊;Cottony Goldenaster ■☆

90207 Chrysopsis gossypina (Michx.) Elliott subsp. cruiseana (Dress) Semple;金菊;Cruise's Goldenaster ■☆

90208 Chrysopsis gossypina (Michx.) Elliott subsp. hyssopifolia (Nutt.) Semple;神香草叶金菊■☆

90209 Chrysopsis gossypina (Michx.) Elliott var. dentata (Elliott) Torr. et A. Gray = Chrysopsis gossypina (Michx.) Elliott ■☆

90210 Chrysopsis graminifolia (Michx.) Elliott = Pityopsis graminifolia (Michx.) Nutt. ■☆

90211 Chrysopsis graminifolia (Michx.) Elliott var. latifolia Fernald = Pityopsis graminifolia (Michx.) Nutt. var. latifolia (Fernald) Semple et F. D. Bowers ■☆

90212 Chrysopsis graminifolia (Michx.) Elliott var. microcephala (Small) Cronquist = Pityopsis graminifolia (Michx.) Nutt. var. tenuifolia (Torr.) Semple et F. D. Bowers ■☆

90213 Chrysopsis graminifolia (Pursh) Elliott var. aspera A. Gray = Pityopsis aspera (A. Gray) Small ■☆

90214 Chrysopsis graminifolia Elliott = Pityopsis graminifolia (Michx.) Nutt. ■☆

90215 Chrysopsis grandis Rydb. = Heterotheca villosa (Pursh) Shinners var. minor (Hook.) Semple ■☆

90216 Chrysopsis highlandsensis Delaney et Wunderlin;高地金菊;Highlands Goldenaster ■☆

90217 Chrysopsis hirsuta Greene = Heterotheca villosa (Pursh) Shinners var. minor (Hook.) Semple ■☆

90218 Chrysopsis hirsutissima Greene = Heterotheca villosa (Pursh) Shinners var. minor (Hook.) Semple ■☆

90219 Chrysopsis hirtella DC. = Erigeron chrysopsidis A. Gray ■☆

90220 Chrysopsis hispida (Hook.) DC.;硬毛金菊;Golden Aster, Hispid Aster ●☆

90221 Chrysopsis hispida (Hook.) DC. = Heterotheca villosa (Pursh) Shinners var. minor (Hook.) Semple ■☆

90222 Chrysopsis hispida (Hook.) DC. var. stenophylla A. Gray = Heterotheca stenophylla (A. Gray) Shinners ■☆

90223 Chrysopsis horrida Rydb. = Heterotheca villosa (Pursh) Shinners var. nana (A. Gray) Semple ■☆

90224 Chrysopsis hyssopifolia Nutt. = Chrysopsis gossypina (Michx.) Elliott subsp. hyssopifolia (Nutt.) Semple ■☆

90225 Chrysopsis imbricata A. Nelson = Heterotheca villosa (Pursh) Shinners var. foliosa (Nutt.) V. L. Harms ■☆

90226 Chrysopsis jonesii S. F. Blake = Heterotheca jonesii (S. F. Blake) S. L. Welsh et N. D. Atwood ■☆

90227 Chrysopsis lanuginosa Small;绵毛金菊;Lynn Haven Goldenaster ■☆

90228 Chrysopsis latifolia (Fernald) Small = Pityopsis graminifolia (Michx.) Nutt. var. latifolia (Fernald) Semple et F. D. Bowers ■☆

90229 Chrysopsis latisquamea Pollard;松林金菊;Pineland Goldenaster ■☆

90230 Chrysopsis linearifolia Semple;窄叶金菊;Narrowleaf Goldenaster ■☆

90231 Chrysopsis linearifolia Semple subsp. dressii Semple;德雷斯金菊;Dress' Goldenaster ■☆

90232 Chrysopsis longii Fernald = Chrysopsis gossypina (Michx.) Elliott ■☆

90233 Chrysopsis mariana (L.) Elliott;马里兰金菊;Broad-leaved Golden Aster, Golden Aster, Maryland Aster, Maryland Golden Aster, Maryland Goldenaster ■☆

90234 Chrysopsis mariana (L.) Elliott var. floridana (Small) Fernald = Chrysopsis floridana Small ●■☆

90235 Chrysopsis mariana (Small) Fernald = Chrysopsis mariana (L.) Elliott ■☆

90236 Chrysopsis mariana (Small) Fernald var. macradenia Fernald = Chrysopsis mariana (L.) Elliott ■☆

90237 Chrysopsis mixta Dress = Chrysopsis gossypina (Michx.) Elliott subsp. hyssopifolia (Nutt.) Semple ■☆

90238 Chrysopsis mollis Nutt. = Heterotheca villosa (Pursh) Shinners ■☆

90239 Chrysopsis nervosa (Willd.) Fernald = Pityopsis graminifolia (Michx.) Nutt. var. latifolia (Fernald) Semple et F. D. Bowers ■☆

90240 Chrysopsis nervosa (Willd.) Fernald var. stenolepis Fernald = Pityopsis graminifolia (Michx.) Nutt. var. latifolia (Fernald) Semple et F. D. Bowers ■☆

90241 Chrysopsis nervosa (Willd.) Fernald var. virgata Fernald = Pityopsis graminifolia (Michx.) Nutt. var. latifolia (Fernald) Semple et F. D. Bowers ■☆

90242 Chrysopsis nitidula Wooton et Standl. = Heterotheca fulcrata (Greene) Shinners var. amplifolia (Rydb.) Semple ■☆

90243 Chrysopsis nuttallianum Britton = Bradburia pilosa (Nutt.) Semple ■☆

90244 Chrysopsis obovata Nutt. = Oclemena reticulata (Pursh) G. L. Nesom ■☆

90245 Chrysopsis oligantha Chapm. ex Torr. et A. Gray = Pityopsis oligantha (Chapm. ex Torr. et A. Gray) Small ■☆

90246 Chrysopsis oregona (Nutt.) A. Gray = Heterotheca oregona (Nutt.) Shinners ■☆

90247 Chrysopsis oregona (Nutt.) A. Gray var. compacta D. D. Keck = Heterotheca oregona (Nutt.) Shinners var. compacta (D. D. Keck) Semple ■☆

90248 Chrysopsis oregona (Nutt.) A. Gray var. rudis (Greene) Jeps. = Heterotheca oregona (Nutt.) Shinners var. rudis (Greene) Semple ■☆

90249 Chrysopsis oregona (Nutt.) A. Gray var. scaberrima A. Gray = Heterotheca oregona (Nutt.) Shinners var. scaberrima (A. Gray) Semple ■☆

90250 Chrysopsis pedunculata Greene = Heterotheca villosa (Pursh) Shinners var. pedunculata (Greene) V. L. Harms ex Semple ■☆

90251 Chrysopsis pilosa (Walter) Britton = Chrysopsis gossypina (Michx.) Elliott ■☆

90252 Chrysopsis pilosa Nutt. = Bradburia pilosa (Nutt.) Semple ■☆

90253 Chrysopsis pinifolia Elliott = Pityopsis pinifolia (Elliott) Nutt. ■☆

90254 Chrysopsis pumila Greene = Heterotheca pumila (Greene) Semple ■☆

90255 Chrysopsis resinolens A. Nelson = Heterotheca fulcrata (Greene) Shinners ■☆

90256 Chrysopsis resinolens A. Nelson var. ciliata A. Nelson = Heterotheca fulcrata (Greene) Shinners ■☆

90257 Chrysopsis rudis Greene = Heterotheca oregona (Nutt.) Shinners var. rudis (Greene) Semple ■☆

90258 Chrysopsis ruthii Small = Pityopsis ruthii (Small) Small ■☆

90259 Chrysopsis rutteri (Rothr.) Greene = Heterotheca rutteri (Rothr.) Shinners ■☆

90260 Chrysopsis scabra Elliott = Heterotheca subaxillaris (Lam.) Britton et Rusby ■☆

90261 Chrysopsis scabrella Torr. et A. Gray;海岸金菊;Coastalplain Goldenaster ■☆

90262 Chrysopsis scabrifolia A. Nelson = Heterotheca stenophylla (A. Gray) Shinners ■☆

90263 Chrysopsis senilis Wooton et Standl. = Heterotheca fulcrata

(Greene) Shinners var. senilis (Wooton et Standl.) Semple ■☆

90264 Chrysopsis sessiliflora Nutt. = Heterotheca sessiliflora (Nutt.) Shinners ■☆

90265 Chrysopsis shastensis Jeps. = Arnica viscosa A. Gray ■☆

90266 Chrysopsis stenophylla (A. Gray) Greene = Heterotheca stenophylla (A. Gray) Shinners ■☆

90267 Chrysopsis subulata Small;灌丛金菊;Scrubland Goldenaster ■☆

90268 Chrysopsis texana G. L. Nesom = Bradburia hirtella Torr. et A. Gray ■☆

90269 Chrysopsis tracyi Small = Pityopsis graminifolia (Michx.) Nutt. var. tracyi (Small) Semple ■☆

90270 Chrysopsis trichophylla (Nutt.) Elliott = Chrysopsis gossypina (Michx.) Elliott ■☆

90271 Chrysopsis trichophylla (Nutt.) Elliott var. hyssopifolia (Nutt.) Torr. et A. Gray = Chrysopsis gossypina (Michx.) Elliott subsp. hyssopifolia (Nutt.) Semple ■☆

90272 Chrysopsis vestita Greene = Heterotheca sessiliflora (Nutt.) Shinners subsp. echioides (Benth.) Semple ■☆

90273 Chrysopsis villosa (Pursh) Nutt. = Heterotheca villosa (Pursh) Shinners ■☆

90274 Chrysopsis villosa (Pursh) Nutt. ex DC. = Heterotheca villosa (Pursh) Shinners ■☆

90275 Chrysopsis villosa (Pursh) Nutt. ex DC. var. angustifolia (Rydb.) Cronquist = Heterotheca stenophylla (A. Gray) Shinners ■☆

90276 Chrysopsis villosa (Pursh) Nutt. ex DC. var. angustifolia (Rydb.) Cronquist = Heterotheca stenophylla (A. Gray) Shinners var. angustifolia (Rydb.) Semple ■☆

90277 Chrysopsis villosa (Pursh) Nutt. ex DC. var. bolanderi (A. Gray) A. Gray = Heterotheca sessiliflora (Nutt.) Shinners subsp. bolanderi (A. Gray) Semple ■☆

90278 Chrysopsis villosa (Pursh) Nutt. ex DC. var. camphorata (Eastw.) Jeps. = Heterotheca sessiliflora (Nutt.) Shinners subsp. echioides (Benth.) Semple ■☆

90279 Chrysopsis villosa (Pursh) Nutt. ex DC. var. camporum (Greene) Cronquist = Heterotheca camporum (Greene) Shinners ■☆

90280 Chrysopsis villosa (Pursh) Nutt. ex DC. var. canescens (DC.) A. Gray = Heterotheca canescens (DC.) Shinners ■☆

90281 Chrysopsis villosa (Pursh) Nutt. ex DC. var. discoidea A. Gray = Heterotheca villosa (Pursh) Shinners var. minor (Hook.) Semple ■☆

90282 Chrysopsis villosa (Pursh) Nutt. ex DC. var. echioides (Benth.) A. Gray = Heterotheca sessiliflora (Nutt.) Shinners subsp. echioides (Benth.) Semple ■☆

90283 Chrysopsis villosa (Pursh) Nutt. ex DC. var. fastigiata (Greene) H. M. Hall = Heterotheca sessiliflora (Nutt.) Shinners subsp. fastigiata (Greene) Semple ■☆

90284 Chrysopsis villosa (Pursh) Nutt. ex DC. var. foliosa (Nutt.) Cronquist = Heterotheca villosa (Pursh) Shinners var. foliosa (Nutt.) V. L. Harms ■☆

90285 Chrysopsis villosa (Pursh) Nutt. ex DC. var. hispida (Hook.) A. Gray = Heterotheca villosa (Pursh) Shinners var. minor (Hook.) Semple ■☆

90286 Chrysopsis villosa (Pursh) Nutt. ex DC. var. minor Hook. = Heterotheca villosa (Pursh) Shinners var. minor (Hook.) Semple ■☆

90287 Chrysopsis villosa (Pursh) Nutt. ex DC. var. rutteri Rothr. = Heterotheca rutteri (Rothr.) Shinners ■☆

90288 Chrysopsis villosa (Pursh) Nutt. ex DC. var. scabra Eastw. = Heterotheca villosa (Pursh) Shinners var. scabra (Eastw.) Semple ■☆

90289 Chrysopsis villosa (Pursh) Nutt. ex DC. var. sessiliflora (Nutt.) A. Gray = Heterotheca sessiliflora (Nutt.) Shinners ■☆

90290 Chrysopsis villosa (Pursh) Nutt. ex DC. var. stenophylla (A. Gray) A. Gray = Heterotheca stenophylla (A. Gray) Shinners ■☆

90291 Chrysopsis villosa (Pursh) Nutt. ex DC. var. viscida A. Gray = Heterotheca viscida (A. Gray) V. L. Harms ■☆

90292 Chrysopsis villosa (Pursh) Nutt. var. hispida (Hook.) A. Gray = Heterotheca villosa (Pursh) Shinners var. minor (Hook.) Semple ■☆

90293 Chrysopsis villosa (Pursh) Nutt. var. minor Hook. = Heterotheca villosa (Pursh) Shinners var. minor (Hook.) Semple ■☆

90294 Chrysopsis villosa DC.;毛金菊;Golden Aster,Hairy Goldaster ●☆

90295 Chrysopsis viscida (A. Gray) Greene = Heterotheca viscida (A. Gray) V. L. Harms ■☆

90296 Chrysopsis viscida (A. Gray) Greene subsp. ciliata (A. Nelson) S. F. Blake = Heterotheca fulcrata (Greene) Shinners ■☆

90297 Chrysopsis viscida (A. Gray) Greene subsp. cinerascens S. F. Blake = Heterotheca villosa (Pursh) Shinners var. scabra (Eastw.) Semple ■☆

90298 Chrysopsis wisconsinensis Shinners = Heterotheca villosa (Pursh) Shinners var. minor (Hook.) Semple ■☆

90299 Chrysorhoe Lindl. = Verticordia DC. (保留属名)●☆

90300 Chrysosciadium Tamamsch. = Echinophora L. ■☆

90301 Chrysoscias E. Mey. (1836);金影鹿藿属■☆

90302 Chrysoscias E. Mey. = Rhynchosia Lour. (保留属名)●■

90303 Chrysoscias angustifolia (Jacq.) C. A. Sm. = Rhynchosia angustifolia (Jacq.) DC. ■☆

90304 Chrysoscias argentea (Thunb.) C. A. Sm. = Rhynchosia argentea (Thunb.) Harv. ■☆

90305 Chrysoscias calycina E. Mey. = Rhynchosia leucoscias Benth. ex Harv. ■☆

90306 Chrysoscias erecta (Thunb.) C. A. Sm. = Rhynchosia chrysoscias Benth. ex Harv. ■☆

90307 Chrysoscias floribunda Lem.;繁花金影鹿藿■☆

90308 Chrysoscias grandiflora E. Mey. = Rhynchosia chrysoscias Benth. ex Harv. ■☆

90309 Chrysoscias parviflora E. Mey. = Rhynchosia microscias Benth. ex Harv. ■☆

90310 Chrysoscias pauciflora (Bolus) C. A. Sm. = Rhynchosia pauciflora Bolus ■☆

90311 Chrysosperma T. Durand et Jacks. = Amianthium A. Gray(保留属名)■☆

90312 Chrysosperma T. Durand et Jacks. = Chrosperma Raf. (废弃属名)■☆

90313 Chrysosperma T. Durand et Jacks. = Zigadenus Michx. ■

90314 Chrysospermum Rchb. = Anthospermum L. ●☆

90315 Chrysosphaerium Willd. ex DC. = Calea L. ●■☆

90316 Chrysospleniaceae Bercht. et J. Presl = Saxifragaceae Juss. (保留科名)●■

90317 Chrysosplenium L. (1753);金腰属(金腰子属,猫儿眼睛草属,猫眼草属);Golden Saxifrage, Golden-saxifrage, Goldsaxifrage, Goldwaist ■

90318 Chrysosplenium Tourn. ex L. = Chrysosplenium L. ■

90319 Chrysosplenium absconditicapsulum J. T. Pan;蔽果金腰;Hidefruit Goldwaist ■

90320 Chrysosplenium adoxoides Hook. f. et Thomson ex Maxim. =

Chrysosplenium lanuginosum Hook. f. et Thomson ■
90321 Chrysosplenium album Maxim.;白花金腰■☆
90322 Chrysosplenium album Maxim. var. flavum H. Hara;黄花金腰■☆
90323 Chrysosplenium album Maxim. var. nachiense H. Hara;那智金腰■☆
90324 Chrysosplenium album Maxim. var. stamineum (Franch.) H. Hara;长蕊金腰■☆
90325 Chrysosplenium alternifolium L.;金腰(互叶金腰,金腰子,猫儿眼睛草);Altelnateleaf Goldsaxifrage, Altelnateleaf Goldwaist, Alternate-leaf Golden Saxifrage, Alternate-leaved Golden-saxifrage ■
90326 Chrysosplenium alternifolium L. = Chrysosplenium serreanum Hand. -Mazz. ■
90327 Chrysosplenium alternifolium L. subsp. sibiricum (Ser.) Hultén = Chrysosplenium serreanum Hand. -Mazz. ■
90328 Chrysosplenium alternifolium L. var. chinense H. Hara = Chrysosplenium chinense (H. Hara) J. T. Pan ■
90329 Chrysosplenium alternifolium L. var. japonicum Maxim. = Chrysosplenium japonicum (Maxim.) Makino ■
90330 Chrysosplenium alternifolium L. var. sibiricum Ser. = Chrysosplenium serreanum Hand. -Mazz. ■
90331 Chrysosplenium alternifolium L. var. sibiricum Ser. ex DC.;互叶金腰(互叶金腰子,西伯利亚互叶金腰);Siberian Altelnateleaf Goldsaxifrage ■
90332 Chrysosplenium alternifolium L. var. sibiricum Ser. ex DC. = Chrysosplenium griffithii Hook. f. et Thomson var. intermedium (H. Hara) J. T. Pan ■
90333 Chrysosplenium alternifolium L. var. sibiricum Ser. ex DC. = Chrysosplenium alternifolium L. ■
90334 Chrysosplenium amabile Kitag. = Chrysosplenium lectus-cochleae Kitag. ■
90335 Chrysosplenium americanum Schwein. ex Hook.;美洲金腰;American Golden Saxifrage, Golden-saxifrage, Water Carpet ■☆
90336 Chrysosplenium axillare Maxim.;长梗金腰(腋花金腰子);Longpedicelled Goldsaxifrage, Longpedicelled Goldwaist ■
90337 Chrysosplenium baicalense Maxim.;拜卡尔金腰■☆
90338 Chrysosplenium baicalense Maxim. var. lectus-cochleae (Kitag.) A. I. Baranov et Skvortsov = Chrysosplenium lectus-cochleae Kitag. ■
90339 Chrysosplenium barbeyi Terracc. = Chrysosplenium macrophyllum Oliv. ■
90340 Chrysosplenium biondianum Engl.;秦岭金腰(红筋草,秦岭金腰子);Biond Goldsaxifrage, Biond Goldwaist ■
90341 Chrysosplenium briquetii A. Terracc. = Chrysosplenium davidianum Decne. ex Maxim. ■
90342 Chrysosplenium carnosulum Maxim. = Chrysosplenium carnosum Hook. f. et Thomson ■
90343 Chrysosplenium carnosum Hook. f. et Thomson;肉质金腰;Carnose Goldsaxifrage, Carnose Goldwaist ■
90344 Chrysosplenium cavaleriei H. Lév. et Vaniot;滇黔金腰;Cavalerie Goldsaxifrage, Cavalerie Goldwaist ■
90345 Chrysosplenium chamaedryoides Engl. ex Diels = Chrysosplenium sinicum Maxim. ■
90346 Chrysosplenium chinense (H. Hara) J. T. Pan;乳突金腰;China Goldwaist, Chinese Altelnateleaf Goldsaxifrage ■
90347 Chrysosplenium chingii H. Hara ex E. Walker = Chrysosplenium sinicum Maxim. ■
90348 Chrysosplenium ciliatum Franch. = Chrysosplenium lanuginosum Hook. f. et Thomson var. ciliatum (Franch.) J. T. Pan ■
90349 Chrysosplenium davidianum Decne. ex Maxim.;锈毛金腰(红小虎耳草);David Goldsaxifrage, David Goldwaist ■
90350 Chrysosplenium davidianum Decne. ex Maxim. var. alpinum H. Hara = Chrysosplenium davidianum Decne. ex Maxim. ■
90351 Chrysosplenium delavayi Franch.;肾萼金腰(大虎耳草,丽江猫眼草,青猫儿眼睛草);Delavay Goldsaxifrage, Delavay Goldwaist ■
90352 Chrysosplenium dubium J. Gay;可疑金腰■☆
90353 Chrysosplenium dunnianum H. Lév. = Chrysosplenium lanuginosum Hook. f. et Thomson ■
90354 Chrysosplenium dunnianum H. Lév. et Vaniot = Chrysosplenium lanuginosum Hook. f. et Thomson ■
90355 Chrysosplenium duplocrenatum Hand. -Mazz. = Chrysosplenium biondianum Engl. ■
90356 Chrysosplenium echinus Maxim.;刺金腰■☆
90357 Chrysosplenium esquirolii H. Lév. = Chrysosplenium hydrocotylifolium H. Lév. et Vaniot ■
90358 Chrysosplenium fauriei Franch.;法氏金腰■☆
90359 Chrysosplenium fauriei Franch. f. ferruginiflorum Wakab. et H. Ohba;锈花法氏金腰■☆
90360 Chrysosplenium fauriei Franch. var. kiotoense (Ohwi) Ohwi = Chrysosplenium kiotoense Ohwi ■☆
90361 Chrysosplenium filipes Kom.;线梗金腰■☆
90362 Chrysosplenium flagelliferum Eastw.;蔓金腰(蔓金腰子);Flagellate Goldwaist ■
90363 Chrysosplenium formosanum Hayata = Chrysosplenium lanuginosum Hook. f. et Thomson var. formosanum (Hayata) H. Hara ■
90364 Chrysosplenium forrestii Diels;贡山金腰(滇西猫眼草);Forrest Goldwaist ■
90365 Chrysosplenium fuscopuncticulosum Z. P. Jien;褐点金腰(褐点猫眼草);Brownspot Goldwaist ■
90366 Chrysosplenium giraldianum Engl.;纤细金腰;Slender Goldwaist ■
90367 Chrysosplenium glaberrimum W. T. Wang;无毛金腰;Hairless Goldwaist ■
90368 Chrysosplenium glossophyllum H. Hara;舌叶金腰;Liguleleaf Goldwaist ■
90369 Chrysosplenium gracile Franch. = Chrysosplenium lanuginosum Hook. f. et Thomson var. gracile (Franch.) H. Hara ■
90370 Chrysosplenium grayarum Maxim.;格雷氏金腰■☆
90371 Chrysosplenium griffithii Hook. f. et Thomson;肾叶金腰(高山金腰子);Griffith Goldsaxifrage, Griffith Goldwaist ■
90372 Chrysosplenium griffithii Hook. f. et Thomson var. intermedium (H. Hara) J. T. Pan;居间金腰;Intermediate Goldwaist ■
90373 Chrysosplenium guangxiense H. G. Ye et G. C. Zhang = Chrysosplenium glossophyllum H. Hara ■
90374 Chrysosplenium guebriantianum Hand. -Mazz. = Chrysosplenium nepalense D. Don ■
90375 Chrysosplenium hebetatum Ohwi;大武金腰(大武猫儿眼睛草);Hebetate Goldwaist ■
90376 Chrysosplenium henryi Franch.;蜕叶金腰(蛇叶金腰,野包耳菜);Henry Goldsaxifrage, Henry Goldwaist ■
90377 Chrysosplenium henryi Franch. = Chrysosplenium lanuginosum Hook. f. et Thomson ■
90378 Chrysosplenium holochlorum Ohwi;青猫目草■
90379 Chrysosplenium holochlorum Ohwi = Chrysosplenium delavayi Franch. ■
90380 Chrysosplenium hydrocotylifolium H. Lév. et Vaniot;天胡荽金腰(大虎耳草);Pennywortleaf Goldsaxifrage, Pennywortleaf Goldwaist ■

90381 Chrysosplenium hydrocotylifolium H. Lév. et Vaniot var. emeiense J. T. Pan;峨眉金腰;Emei Goldsaxifrage,Emei Goldwaist ■

90382 Chrysosplenium hydrocotylifolium H. Lév. et Vaniot var. guangdongense X. J. Xu et Z. X. Li;广东金腰;Guangdong Goldsaxifrage,Guangdong Goldwaist ■

90383 Chrysosplenium japonicum (Maxim.) Makino;日本金腰(珠芽金腰);Japanese Goldsaxifrage,Japanese Goldwaist ■

90384 Chrysosplenium japonicum (Maxim.) Makino f. tetrandrum H. Hara;四蕊日本金腰■☆

90385 Chrysosplenium japonicum (Maxim.) Makino var. cuneifolium X. H. Guo et X. P. Zhang;楔叶金腰;Cuneate-leaf Goldsaxifrage ■

90386 Chrysosplenium jienningense W. T. Wang;建宁金腰;Jianning Goldsaxifrage,Jianning Goldwaist ■

90387 Chrysosplenium kamtschaticum Fisch. ex Ser.;勘察加金腰■☆

90388 Chrysosplenium kamtschaticum Fisch. ex Ser. f. tobishimense Wakab.;飞岛金腰■☆

90389 Chrysosplenium kamtschaticum Fisch. ex Ser. var. aomorense (Franch.) H. Hara;青森金腰■☆

90390 Chrysosplenium kiotense Ohwi f. xanthandrum (Araki) Wakab. et H. Ohba;黄蕊金腰■☆

90391 Chrysosplenium kiotoense Ohwi;京都金腰■☆

90392 Chrysosplenium komarovii Losinsk.;科马罗夫金腰■☆

90393 Chrysosplenium lanuginosum Hook. f. et Thomson;绵毛金腰(红地棉);Shortwoolled Goldsaxifrage,Shortwoolled Goldwaist ■

90394 Chrysosplenium lanuginosum Hook. f. et Thomson var. ciliatum (Franch.) J. T. Pan;睫毛金腰;Ciliate Goldsaxifrage,Ciliate Goldwaist ■

90395 Chrysosplenium lanuginosum Hook. f. et Thomson var. dunnianum (H. Lév. et Vaniot) H. Hara = Chrysosplenium lanuginosum Hook. f. et Thomson ■

90396 Chrysosplenium lanuginosum Hook. f. et Thomson var. formosanum (Hayata) H. Hara;台湾金腰(台湾猫儿眼睛草);Taiwan Goldwaist ■

90397 Chrysosplenium lanuginosum Hook. f. et Thomson var. gracile (Franch.) H. Hara;细弱金腰;Weak Goldsaxifrage ■

90398 Chrysosplenium lanuginosum Hook. f. et Thomson var. pilosomarginatum (H. Hara) J. T. Pan;毛边金腰;Pilosemarginate Shortwoolled Goldwaist ■

90399 Chrysosplenium lanuginosum Hook. f. et Thomson var. yunnanense H. Hara = Chrysosplenium lanuginosum Hook. f. et Thomson ■

90400 Chrysosplenium lectus-cochleae Kitag.;林金腰(林金腰子);Forest Goldsaxifrage,Forest Goldwaist ■

90401 Chrysosplenium lixianense Z. P. Jien ex J. T. Pan;理县金腰;Lixian Goldsaxifrage,Lixian Goldwaist ■

90402 Chrysosplenium ludlowii H. Hara = Chrysosplenium oxygraphoides Hand. -Mazz. ■

90403 Chrysosplenium lushanense W. T. Wang;庐山金腰;Lushan Goldsaxifrage,Lushan Goldwaist ■

90404 Chrysosplenium lushanense W. T. Wang = Chrysosplenium sinicum Maxim. ■

90405 Chrysosplenium macrophyllum Oliv.;大叶金腰(大虎耳草,肺心草,虎皮草,龙舌草,龙香草,马耳朵草,牛耳朵,岩窝鸡,岩乌金莱,猪耳朵);Bigleaf Goldsaxifrage,Bigleaf Goldwaist ■

90406 Chrysosplenium macrostemon Maxim.;大蕊金腰■☆

90407 Chrysosplenium macrostemon Maxim. var. atrandrum H. Hara;黑大蕊金腰■☆

90408 Chrysosplenium macrostemon Maxim. var. calicitrapa (Franch.) H. Hara;杯萎金腰■☆

90409 Chrysosplenium macrostemon Maxim. var. shiobarense (Franch.) H. Hara;盐原金腰■☆

90410 Chrysosplenium macrostemon Maxim. var. viridescens (Suto) H. Hara;浅绿大蕊金腰■☆

90411 Chrysosplenium maximowiczii Franch. et Sav.;马氏金腰■☆

90412 Chrysosplenium microspermum Franch.;微子金腰;Smallseed Goldsaxifrage,Smallseed Goldwaist ■

90413 Chrysosplenium nagasei Wakab. et H. Ohba;长亩金腰■☆

90414 Chrysosplenium nagasei Wakab. et H. Ohba var. luteoflorum Wakab. et H. Ohba;黄花长亩金腰■☆

90415 Chrysosplenium nagasei Wakab. et H. Ohba var. porphyranthes Wakab. et H. Ohba;紫花长亩金腰■☆

90416 Chrysosplenium nepalense D. Don;山溪金腰(尼泊尔猫眼草);Nepal Goldsaxifrage,Nepal Goldwaist ■

90417 Chrysosplenium nepalense D. Don var. cavaleriei (H. Lév. et Vaniot) H. Hara = Chrysosplenium cavaleriei H. Lév. et Vaniot ■

90418 Chrysosplenium nepalense D. Don var. vegetum H. Hara = Chrysosplenium cavaleriei H. Lév. et Vaniot ■

90419 Chrysosplenium nepalense D. Don var. yunnanense (Franch.) Franch. = Chrysosplenium nepalense D. Don ■

90420 Chrysosplenium niitakayamense Masam.;玉山猫儿眼睛草■

90421 Chrysosplenium nudicaule Bunge;裸茎金腰(金腰草,裸茎金腰子,猫眼草,亚吉玛);Nakedcaule Goldsaxifrage,Nakedcaule Goldwaist ■

90422 Chrysosplenium nudicaule Bunge var. intermedium H. Hara = Chrysosplenium griffithii Hook. f. et Thomson var. intermedium (H. Hara) J. T. Pan ■

90423 Chrysosplenium oppositifolium L.;对叶金腰子;Buttered Eggs,Creeping Jenny,Golden Spleenwort,Lady's Cushion,Opposite-leaved Golden Saxifrage ■☆

90424 Chrysosplenium ovalifolium M. Bieb. ex Bunge;卵形金腰■☆

90425 Chrysosplenium oxygraphoides Hand. -Mazz.;鸦跖花金腰(米林金腰);Oxygraphislike Goldsaxifrage,Oxygraphislike Goldwaist ■

90426 Chrysosplenium peltatum Turcz.;盾状金腰■☆

90427 Chrysosplenium pilosomarginatum H. Hara = Chrysosplenium lanuginosum Hook. f. et Thomson var. pilosomarginatum (H. Hara) J. T. Pan ■

90428 Chrysosplenium pilosopetiolatum Z. P. Jien = Chrysosplenium pilosum Maxim. var. pilosopetiolatum (Z. P. Jien) J. T. Pan ■

90429 Chrysosplenium pilosum Maxim.;毛金腰(毛金腰子);Hairy Goldwaist,Pilose Goldsaxifrage ■

90430 Chrysosplenium pilosum Maxim. var. fulvum (Terracc.) H. Hara;黄毛金腰■☆

90431 Chrysosplenium pilosum Maxim. var. pilosopetiolatum (Z. P. Jien) J. T. Pan;毛柄金腰(毛柄猫眼草)■

90432 Chrysosplenium pilosum Maxim. var. sphaerospermum (Maxim.) H. Hara;圆籽毛金腰■☆

90433 Chrysosplenium pilosum Maxim. var. valdepilosum Ohwi;柔毛金腰(柔毛金腰子);Hairy Goldsaxifrage ■

90434 Chrysosplenium pseudofauriei H. Lév.;异叶金腰;Diversifolious Goldsaxifrage ■

90435 Chrysosplenium pseudofauriei H. Lév. = Chrysosplenium sinicum Maxim. ■

90436 Chrysosplenium pseudofauriei H. Lév. var. nipponense Wakab.;日本异叶金腰■☆

90437 Chrysosplenium pseudopilosum Wakab. et Hir. Takah.;假毛金

腰■☆
90438 Chrysosplenium pseudopilosum Wakab. et Hir. Takah. var. divaricatistylosum Wakab. et Hir. Takah.;叉柱金腰■☆
90439 Chrysosplenium pumilum Franch.;矮小猫眼草;Dwarf Goldsaxifrage ■
90440 Chrysosplenium pumilum Franch. = Chrysosplenium delavayi Franch. ■
90441 Chrysosplenium qinlingense Z. P. Jien et J. T. Pan;陕甘金腰;Qinling Goldsaxifrage,Qinling Goldwaist ■
90442 Chrysosplenium ramosum Maxim.;多枝金腰(多枝金腰子);Manybranched Goldsaxifrage,Manybranched Goldwaist ■
90443 Chrysosplenium ramosum Maxim. f. macrophyllum H. Hara = Chrysosplenium ramosum Maxim. ■
90444 Chrysosplenium ramosum Maxim. f. microphyllum (Tatew. et Suto) H. Hara = Chrysosplenium ramosum Maxim. ■
90445 Chrysosplenium rhabdospermum Maxim.;棒籽金腰■☆
90446 Chrysosplenium rhabdospermum Maxim. var. shikokianum Wakab.;四国金腰■☆
90447 Chrysosplenium sedakowii Turcz.;塞达金腰■☆
90448 Chrysosplenium serreanum Hand. -Mazz.;五台金腰;Wutai Goldsaxifrage,Wutai Goldwaist ■
90449 Chrysosplenium serreanum Hand. -Mazz. = Chrysosplenium alternifolium L. var. sibiricum Ser. ■
90450 Chrysosplenium sibiricum (Ser.) Kharkev. = Chrysosplenium serreanum Hand. -Mazz. ■
90451 Chrysosplenium sikangense H. Hara;西康金腰;Xikang Goldsaxifrage,Xikang Goldwaist ■
90452 Chrysosplenium sinicum Maxim.;中华金腰(华金腰子,金钱苦叶草,金腰子,猫眼草,异叶金腰,中华金腰子);Chinese Goldsaxifrage,Chinese Goldwaist ■
90453 Chrysosplenium sphaerospermum Maxim. = Chrysosplenium lectus-cochleae Kitag. ■
90454 Chrysosplenium sphaerospermum Maxim. = Chrysosplenium pilosum Maxim. var. sphaerospermum (Maxim.) H. Hara ■☆
90455 Chrysosplenium sphaerospermum Maxim. var. amabile (Kitag.) A. I. Baranov et Skvortsov = Chrysosplenium lectus-cochleae Kitag. ■
90456 Chrysosplenium sphaerospermum Maxim. var. fulvum (Terracc.) H. Hara = Chrysosplenium pilosum Maxim. var. fulvum (Terracc.) H. Hara ■☆
90457 Chrysosplenium subargenteum H. Lév. et Vaniot = Chrysosplenium delavayi Franch. ■
90458 Chrysosplenium sulcatum Maxim. = Chrysosplenium nepalense D. Don ■
90459 Chrysosplenium taibaishanense J. T. Pan;太白金腰;Taibaishan Goldsaxifrage,Taibaishan Goldwaist ■
90460 Chrysosplenium tetrandrum Fr.;四雄金腰子■☆
90461 Chrysosplenium tianschanicum Krasn.;天山金腰子;Tianshan Goldsaxifrage ■
90462 Chrysosplenium tianschanicum Krasn. = Chrysosplenium axillare Maxim. ■
90463 Chrysosplenium tibeticum Limpr. = Chrysosplenium carnosum Hook. f. et Thomson ■
90464 Chrysosplenium tosaense (Makino) Makino ex Matsum.;土佐金腰■☆
90465 Chrysosplenium trachyspermum Maxim.;糙籽金腰■☆
90466 Chrysosplenium trachyspermum Maxim. = Chrysosplenium sinicum Maxim. ■
90467 Chrysosplenium umbellatum Kitag. = Chrysosplenium pilosum Maxim. ■
90468 Chrysosplenium uniflorum Maxim.;单花金腰;Uniflowered Goldsaxifrage,Uniflowered Goldwaist ■
90469 Chrysosplenium villosum Franch. = Chrysosplenium pilosum Maxim. var. valdepilosum Ohwi ■
90470 Chrysosplenium viridescens (Suto) Suto ex H. Hara = Chrysosplenium macrostemon Maxim. var. viridescens (Suto) H. Hara ■☆
90471 Chrysosplenium wrightii Franch. et Sav.;赖特金腰■☆
90472 Chrysosplenium wuwenchenii Z. P. Jien;韫珍金腰(韫珍猫眼草);Wuyuzhen Goldsaxifrage,Wuyuzhen Goldwaist ■
90473 Chrysosplenium yunnanense Franch. = Chrysosplenium nepalense D. Don ■
90474 Chrysostachys Poepp. ex Baill. = Sclerolobium Vogel ■☆
90475 Chrysostachys Pohl = Combretum Loefl.(保留属名)●
90476 Chrysostachys Pohl(1830);金花使君子属●☆
90477 Chrysostachys ovatifolia Pohl;金花使君子●☆
90478 Chrysostemma E. Mey. ex Spach = Gorteria L. ■☆
90479 Chrysostemma Less.(1832);金冠菊属■
90480 Chrysostemma Less. = Coreopsis L. ●■
90481 Chrysostemma tripteris (L.) Less. = Coreopsis tripteris L. ■
90482 Chrysostemon Klotzsch = Pseudanthus Sieber ex A. Spreng. ■☆
90483 Chrysostoma Lilja = Mentzelia L. ●■☆
90484 Chrysothamnus Nutt.(1840)(保留属名);金灌菊属(兔黄花属);Rabbit Brush,Rabbitbrush ■●☆
90485 Chrysothamnus affinis A. Nelson = Ericameria parryi (A. Gray) G. L. Nesom et G. I. Baird var. affinis (A. Nelson) G. L. Nesom et G. I. Baird ●☆
90486 Chrysothamnus albidus (M. E. Jones ex A. Gray) Greene = Ericameria albida (M. E. Jones ex A. Gray) L. C. Anderson ●☆
90487 Chrysothamnus asper Greene = Ericameria parryi (A. Gray) G. L. Nesom et G. I. Baird var. aspera (Greene) G. L. Nesom et G. I. Baird ●☆
90488 Chrysothamnus axillaris D. D. Keck = Chrysothamnus viscidiflorus (Hook.) Nutt. subsp. axillaris (D. D. Keck) L. C. Anderson ●☆
90489 Chrysothamnus baileyi Wooton et Standl. = Lorandersonia baileyi (Wooton et Standl.) Urbatsch,R. P. Roberts et Neubig ●☆
90490 Chrysothamnus bakeri Greene;巴氏金灌菊●☆
90491 Chrysothamnus bolanderi (A. Gray) Greene;波兰金灌菊●☆
90492 Chrysothamnus confinis Greene;道氏金灌菊;Douglas Rabbit Brush,Douglas Rabbitbrush ☆
90493 Chrysothamnus depressus Nutt.;长花金灌菊;Long-flowered rabbitbrush ●☆
90494 Chrysothamnus eremobius L. C. Anderson;沙地金灌菊;Pintwater Rabbitbrush ●☆
90495 Chrysothamnus gramineus H. M. Hall = Cuniculotinus gramineus (H. M. Hall) Urbatsch,R. P. Roberts et Neubig ●☆
90496 Chrysothamnus greenei (A. Gray) Greene;格林金灌菊;Greene's Rabbitbrush ■☆
90497 Chrysothamnus humilis Greene;矮金灌菊;Truckee Rabbitbrush ■☆
90498 Chrysothamnus lanceolatus Nutt. = Chrysothamnus viscidiflorus (Hook.) Nutt. subsp. lanceolatus (Nutt.) H. M. Hall et Clem. ●☆
90499 Chrysothamnus linifolius Greene = Lorandersonia linifolia (Greene) Urbatsch ●☆

90500 Chrysothamnus molestus (S. F. Blake) L. C. Anderson;亚利桑那金灌菊;Arizona Rabbitbrush ■☆

90501 Chrysothamnus monocephalus A. Nelson et P. B. Kenn. = Ericameria parryi (A. Gray) G. L. Nesom et G. I. Baird var. monocephala (A. Nelson et P. B. Kenn.) G. L. Nesom et G. I. Baird ●☆

90502 Chrysothamnus nauseosus (Pall. ex Pursh) Britton;臭金灌菊;Rubber Rabbit Brush, Rubber Rabbitbrush, Stinking Rabbitbrush ●☆

90503 Chrysothamnus nauseosus (Pall. ex Pursh) Britton = Ericameria nauseosa (Pall. ex Pursh) G. L. Nesom et G. I. Baird ●☆

90504 Chrysothamnus nauseosus (Pall. ex Pursh) Britton subsp. albicaulis (Nutt.) H. M. Hall et Clem. = Ericameria nauseosa (Pall. ex Pursh) G. L. Nesom et G. I. Baird var. speciosa (Nutt.) G. L. Nesom et G. I. Baird ●☆

90505 Chrysothamnus nauseosus (Pall. ex Pursh) Britton subsp. arenarius L. C. Anderson = Ericameria nauseosa (Pall. ex Pursh) G. L. Nesom et G. I. Baird var. arenaria (L. C. Anderson) G. L. Nesom et G. I. Baird ●☆

90506 Chrysothamnus nauseosus (Pall. ex Pursh) Britton subsp. bigelovii (A. Gray) H. M. Hall et Clem. = Ericameria nauseosa (Pall. ex Pursh) G. L. Nesom et G. I. Baird var. bigelovii (A. Gray) G. L. Nesom et G. I. Baird ●☆

90507 Chrysothamnus nauseosus (Pall. ex Pursh) Britton subsp. ceruminosus (Durand et Hilg.) H. M. Hall et Clem. = Ericameria nauseosa (Pall. ex Pursh) G. L. Nesom et G. I. Baird var. ceruminosa (Durand et Hilg.) G. L. Nesom et G. I. Baird ●☆

90508 Chrysothamnus nauseosus (Pall. ex Pursh) Britton subsp. consimilis (Greene) H. M. Hall et Clem. = Ericameria nauseosa (Pall. ex Pursh) G. L. Nesom et G. I. Baird var. oreophila (A. Nelson) G. L. Nesom et G. I. Baird ●☆

90509 Chrysothamnus nauseosus (Pall. ex Pursh) Britton subsp. graveolens (Nutt.) H. M. Hall et Clem. = Ericameria nauseosa (Pall. ex Pursh) G. L. Nesom et G. I. Baird var. graveolens (Nutt.) Reveal et Schuyler ●☆

90510 Chrysothamnus nauseosus (Pall. ex Pursh) Britton subsp. iridis L. C. Anderson = Ericameria nauseosa (Pall. ex Pursh) G. L. Nesom et G. I. Baird var. iridis (L. C. Anderson) G. L. Nesom et G. I. Baird ●☆

90511 Chrysothamnus nauseosus (Pall. ex Pursh) Britton subsp. latisquameus (A. Gray) H. M. Hall et Clem. = Ericameria nauseosa (Pall. ex Pursh) G. L. Nesom et G. I. Baird var. latisquamea (A. Gray) G. L. Nesom et G. I. Baird ●☆

90512 Chrysothamnus nauseosus (Pall. ex Pursh) Britton subsp. leiospermus (A. Gray) H. M. Hall et Clem. = Ericameria nauseosa (Pall. ex Pursh) G. L. Nesom et G. I. Baird var. leiosperma (A. Gray) G. L. Nesom et G. I. Baird ●☆

90513 Chrysothamnus nauseosus (Pall. ex Pursh) Britton subsp. nanus (Cronquist) D. D. Keck = Ericameria nauseosa (Pall. ex Pursh) G. L. Nesom et G. I. Baird var. nana (Cronquist) G. L. Nesom et G. I. Baird ●☆

90514 Chrysothamnus nauseosus (Pall. ex Pursh) Britton subsp. nitidus L. C. Anderson = Ericameria nauseosa (Pall. ex Pursh) G. L. Nesom et G. I. Baird var. nitida (L. C. Anderson) G. L. Nesom et G. I. Baird ●☆

90515 Chrysothamnus nauseosus (Pall. ex Pursh) Britton subsp. psilocarpus (S. F. Blake) L. C. Anderson = Ericameria nauseosa (Pall. ex Pursh) G. L. Nesom et G. I. Baird var. psilocarpa (S. F. Blake) G. L. Nesom et G. I. Baird ●☆

90516 Chrysothamnus nauseosus (Pall. ex Pursh) Britton subsp. salicifolius (Rydb.) H. M. Hall et Clem. = Ericameria nauseosa (Pall. ex Pursh) G. L. Nesom et G. I. Baird var. salicifolia (Rydb.) G. L. Nesom et G. I. Baird ●☆

90517 Chrysothamnus nauseosus (Pall. ex Pursh) Britton subsp. speciosus (Nutt.) H. M. Hall et Clem. = Ericameria nauseosa (Pall. ex Pursh) G. L. Nesom et G. I. Baird var. speciosa (Nutt.) G. L. Nesom et G. I. Baird ●☆

90518 Chrysothamnus nauseosus (Pall. ex Pursh) Britton subsp. texensis L. C. Anderson = Ericameria nauseosa (Pall. ex Pursh) G. L. Nesom et G. I. Baird var. texensis (L. C. Anderson) G. L. Nesom et G. I. Baird ●☆

90519 Chrysothamnus nauseosus (Pall. ex Pursh) Britton var. arenarius (L. C. Anderson) S. L. Welsh = Ericameria nauseosa (Pall. ex Pursh) G. L. Nesom et G. I. Baird var. arenaria (L. C. Anderson) G. L. Nesom et G. I. Baird ●☆

90520 Chrysothamnus nauseosus (Pall. ex Pursh) Britton var. artus (A. Nelson) Cronquist = Ericameria nauseosa (Pall. ex Pursh) G. L. Nesom et G. I. Baird var. oreophila (A. Nelson) G. L. Nesom et G. I. Baird ●☆

90521 Chrysothamnus nauseosus (Pall. ex Pursh) Britton var. bernardinus H. M. Hall = Ericameria nauseosa (Pall. ex Pursh) G. L. Nesom et G. I. Baird var. bernardina (H. M. Hall) G. L. Nesom et G. I. Baird ●☆

90522 Chrysothamnus nauseosus (Pall. ex Pursh) Britton var. bigelovii (A. Gray) H. M. Hall;毕氏臭金灌菊;Rabbit Brush, Rabbitbrush ●☆

90523 Chrysothamnus nauseosus (Pall. ex Pursh) Britton var. bigelovii H. M. Hall = Chrysothamnus nauseosus (Pall. ex Pursh) Britton var. bigelovii (A. Gray) H. M. Hall ●☆

90524 Chrysothamnus nauseosus (Pall. ex Pursh) Britton var. ceruminosus (Durand et Hilg.) H. M. Hall = Ericameria nauseosa (Pall. ex Pursh) G. L. Nesom et G. I. Baird var. ceruminosa (Durand et Hilg.) G. L. Nesom et G. I. Baird ●☆

90525 Chrysothamnus nauseosus (Pall. ex Pursh) Britton var. consimilis (Greene) H. M. Hall = Ericameria nauseosa (Pall. ex Pursh) G. L. Nesom et G. I. Baird var. oreophila (A. Nelson) G. L. Nesom et G. I. Baird ●☆

90526 Chrysothamnus nauseosus (Pall. ex Pursh) Britton var. graveolens (Nutt.) H. M. Hall = Ericameria nauseosa (Pall. ex Pursh) G. L. Nesom et G. I. Baird var. graveolens (Nutt.) Reveal et Schuyler ●☆

90527 Chrysothamnus nauseosus (Pall. ex Pursh) Britton var. hololeucus (A. Gray) H. M. Hall = Bigelowia graveolens Nutt. var. hololeuca A. Gray ●☆

90528 Chrysothamnus nauseosus (Pall. ex Pursh) Britton var. iridis (L. C. Anderson) S. L. Welsh = Ericameria nauseosa (Pall. ex Pursh) G. L. Nesom et G. I. Baird var. iridis (L. C. Anderson) G. L. Nesom et G. I. Baird ●☆

90529 Chrysothamnus nauseosus (Pall. ex Pursh) Britton var. junceus (Greene) H. M. Hall = Ericameria nauseosa (Pall. ex Pursh) G. L. Nesom et G. I. Baird var. juncea (Greene) G. L. Nesom et G. I. Baird ●☆

90530 Chrysothamnus nauseosus (Pall. ex Pursh) Britton var. mohavensis (Greene) H. M. Hall = Ericameria nauseosa (Pall. ex Pursh) G. L. Nesom et G. I. Baird var. mohavensis (Greene) G. L. Nesom et G. I. Baird ●☆

90531 Chrysothamnus nauseosus (Pall. ex Pursh) Britton var.

mohavensis H. M. Hall;莫哈维金灌菊;Mohave Rubberbrush ●☆

90532 Chrysothamnus nauseosus (Pall. ex Pursh) Britton var. nanus Cronquist = Ericameria nauseosa (Pall. ex Pursh) G. L. Nesom et G. I. Baird var. nana (Cronquist) G. L. Nesom et G. I. Baird ●☆

90533 Chrysothamnus nauseosus (Pall. ex Pursh) Britton var. nitidus (L. C. Anderson) S. L. Welsh = Ericameria nauseosa (Pall. ex Pursh) G. L. Nesom et G. I. Baird var. nitida (L. C. Anderson) G. L. Nesom et G. I. Baird ●☆

90534 Chrysothamnus nauseosus (Pall. ex Pursh) Britton var. psilocarpus S. F. Blake = Ericameria nauseosa (Pall. ex Pursh) G. L. Nesom et G. I. Baird var. psilocarpa (S. F. Blake) G. L. Nesom et G. I. Baird ●☆

90535 Chrysothamnus nauseosus (Pall. ex Pursh) Britton var. turbinatus (M. E. Jones) H. M. Hall = Ericameria nauseosa (Pall. ex Pursh) G. L. Nesom et G. I. Baird var. turbinata (M. E. Jones) G. L. Nesom et G. I. Baird ●☆

90536 Chrysothamnus nauseosus (Pall.) Britton = Chrysothamnus nauseosus (Pall. ex Pursh) Britton ●☆

90537 Chrysothamnus nauseosus J. T. Howell var. mohavensis H. M. Hall = Chrysothamnus nauseosus (Pall. ex Pursh) Britton var. mohavensis H. M. Hall ●☆

90538 Chrysothamnus oreophilus A. Nelson = Ericameria nauseosa (Pall. ex Pursh) G. L. Nesom et G. I. Baird var. oreophila (A. Nelson) G. L. Nesom et G. I. Baird ●☆

90539 Chrysothamnus oreophilus A. Nelson var. artus A. Nelson = Ericameria nauseosa (Pall. ex Pursh) G. L. Nesom et G. I. Baird var. oreophila (A. Nelson) G. L. Nesom et G. I. Baird ●☆

90540 Chrysothamnus paniculatus (A. Gray) H. M. Hall = Ericameria paniculata (A. Gray) Rydb. ●☆

90541 Chrysothamnus parryi (A. Gray) Greene subsp. affinis (A. Nelson) L. C. Anderson = Ericameria parryi (A. Gray) G. L. Nesom et G. I. Baird var. affinis (A. Nelson) G. L. Nesom et G. I. Baird ●☆

90542 Chrysothamnus parryi (A. Gray) Greene subsp. asper (Greene) H. M. Hall et Clem. = Ericameria parryi (A. Gray) G. L. Nesom et G. I. Baird var. aspera (Greene) G. L. Nesom et G. I. Baird ●☆

90543 Chrysothamnus parryi (A. Gray) Greene subsp. attenuatus (M. E. Jones) H. M. Hall et Clem. = Ericameria parryi (A. Gray) G. L. Nesom et G. I. Baird var. attenuata (M. E. Jones) G. L. Nesom et G. I. Baird ●☆

90544 Chrysothamnus parryi (A. Gray) Greene subsp. howardii (Parry ex A. Gray) H. M. Hall et Clem. = Ericameria parryi (A. Gray) G. L. Nesom et G. I. Baird var. howardii (Parry ex A. Gray) G. L. Nesom et G. I. Baird ●☆

90545 Chrysothamnus parryi (A. Gray) Greene subsp. imulus H. M. Hall et Clem. = Ericameria parryi (A. Gray) G. L. Nesom et G. I. Baird var. imula (H. M. Hall et Clem.) G. L. Nesom et G. I. Baird ●☆

90546 Chrysothamnus parryi (A. Gray) Greene subsp. latior H. M. Hall et Clem. = Ericameria parryi (A. Gray) G. L. Nesom et G. I. Baird var. latior (H. M. Hall et Clem.) G. L. Nesom et G. I. Baird ●☆

90547 Chrysothamnus parryi (A. Gray) Greene subsp. monocephalus (A. Nelson et P. B. Kenn.) H. M. Hall et Clem. = Ericameria parryi (A. Gray) G. L. Nesom et G. I. Baird var. monocephala (A. Nelson et P. B. Kenn.) G. L. Nesom et G. I. Baird ●☆

90548 Chrysothamnus parryi (A. Gray) Greene subsp. montanus L. C. Anderson = Ericameria parryi (A. Gray) G. L. Nesom et G. I. Baird var. montana (L. C. Anderson) G. L. Nesom et G. I. Baird ●☆

90549 Chrysothamnus parryi (A. Gray) Greene subsp. nevadensis (A. Gray) H. M. Hall et Clem. = Ericameria parryi (A. Gray) G. L. Nesom et G. I. Baird var. nevadensis (A. Gray) G. L. Nesom et G. I. Baird ●☆

90550 Chrysothamnus parryi (A. Gray) Greene subsp. salmonensis L. C. Anderson = Ericameria parryi (A. Gray) G. L. Nesom et G. I. Baird var. salmonensis (L. C. Anderson) G. L. Nesom et G. I. Baird ●☆

90551 Chrysothamnus parryi (A. Gray) Greene subsp. vulcanicus (Greene) H. M. Hall et Clem. = Ericameria parryi (A. Gray) G. L. Nesom et G. I. Baird var. vulcanica (Greene) G. L. Nesom et G. I. Baird ●☆

90552 Chrysothamnus parryi (A. Gray) Greene var. affinis (A. Nelson) Cronquist = Ericameria parryi (A. Gray) G. L. Nesom et G. I. Baird var. affinis (A. Nelson) G. L. Nesom et G. I. Baird ●☆

90553 Chrysothamnus pulchellus (A. Gray) Greene = Lorandersonia pulchella (A. Gray) Urbatsch, R. P. Roberts et Neubig ●☆

90554 Chrysothamnus pulchellus (A. Gray) Greene subsp. baileyi (Wooton et Standl.) H. M. Hall et Clem. = Lorandersonia baileyi (Wooton et Standl.) Urbatsch, R. P. Roberts et Neubig ●☆

90555 Chrysothamnus pulchellus (A. Gray) Greene subsp. elatior (Standl.) H. M. Hall et Clem. = Lorandersonia pulchella (A. Gray) Urbatsch, R. P. Roberts et Neubig ●☆

90556 Chrysothamnus pulchellus Greene;黑金灌菊;Black-banded Rabbit Brush, Blackbanded Rabbitbrush, Rubber Rabbitbush, Squarestem ●☆

90557 Chrysothamnus pulchellus Greene = Bigelowia pulchella A. Gray ●☆

90558 Chrysothamnus salicifolius Rydb. = Ericameria nauseosa (Pall. ex Pursh) G. L. Nesom et G. I. Baird var. salicifolia (Rydb.) G. L. Nesom et G. I. Baird ●☆

90559 Chrysothamnus scopulorum (M. E. Jones) Urbatsch, R. P. Roberts et Neubig;峡谷金灌菊;Grand Canyon Evening-daisy, Grand Canyon Glowweed ●☆

90560 Chrysothamnus spathulatus L. C. Anderson = Lorandersonia spathulata (L. C. Anderson) Urbatsch, R. P. Roberts et Neubig ●☆

90561 Chrysothamnus speciosus Nutt. = Ericameria nauseosa (Pall. ex Pursh) G. L. Nesom et G. I. Baird var. speciosa (Nutt.) G. L. Nesom et G. I. Baird ●☆

90562 Chrysothamnus stylosus (Eastw.) Urbatsch, R. P. Roberts et Neubig = Vanclevea stylosa (Eastw.) Greene ●☆

90563 Chrysothamnus teretifolius (Durand et Hilg.) H. M. Hall = Ericameria teretifolia (Durand et Hilg.) Jeps. ●☆

90564 Chrysothamnus vaseyi (A. Gray) Greene;瓦齐金灌菊;Vasey's Rabbitbrush ●☆

90565 Chrysothamnus viscidiflorus (Hook.) Nutt.;黏花金灌菊;Sticky-leaf Rabbitbrush, Yellow Rabbitbrush ●☆

90566 Chrysothamnus viscidiflorus (Hook.) Nutt. subsp. axillaris (D. D. Keck) L. C. Anderson;腋生黏花金灌菊;Inyo Rabbitbrush ●☆

90567 Chrysothamnus viscidiflorus (Hook.) Nutt. subsp. elegans (Greene) H. M. Hall et Clem. = Chrysothamnus viscidiflorus (Hook.) Nutt. subsp. lanceolatus (Nutt.) H. M. Hall et Clem. ●☆

90568 Chrysothamnus viscidiflorus (Hook.) Nutt. subsp. humilis H. M. Hall et Clem. = Chrysothamnus humilis Greene ■☆

90569 Chrysothamnus viscidiflorus (Hook.) Nutt. subsp. lanceolatus (Nutt.) H. M. Hall et Clem.;剑叶黏花金灌菊●☆

90570 Chrysothamnus viscidiflorus (Hook.) Nutt. subsp. linifolius (Greene) H. M. Hall et Clem. = Lorandersonia linifolia (Greene) Urbatsch ●☆

90571 Chrysothamnus viscidiflorus (Hook.) Nutt. subsp. puberulus (D. C. Eaton) H. M. Hall et Clem.;绒毛黏花金灌菊●☆

90572 Chrysothamnus viscidiflorus (Hook.) Nutt. subsp. pumilus (Nutt.) H. M. Hall et Clem. = Chrysothamnus viscidiflorus (Hook.) Nutt. ●☆

90573 Chrysothamnus viscidiflorus (Hook.) Nutt. var. lanceolatus (Nutt.) Greene = Chrysothamnus viscidiflorus (Hook.) Nutt. subsp. lanceolatus (Nutt.) H. M. Hall et Clem. ●☆

90574 Chrysothamnus viscidiflorus (Hook.) Nutt. var. latifolius (D. C. Eaton) Greene = Chrysothamnus viscidiflorus (Hook.) Nutt. ●☆

90575 Chrysothamnus viscidiflorus (Hook.) Nutt. var. molestus S. F. Blake = Chrysothamnus molestus (S. F. Blake) L. C. Anderson ■☆

90576 Chrysothamnus viscidiflorus (Hook.) Nutt. var. puberulus (D. C. Eaton) Jeps. = Chrysothamnus viscidiflorus (Hook.) Nutt. subsp. puberulus (D. C. Eaton) H. M. Hall et Clem. ●☆

90577 Chrysothamnus viscidiflorus (Hook.) Nutt. var. stenophyllus (A. Gray) H. M. Hall = Chrysothamnus viscidiflorus (Hook.) Nutt. ●☆

90578 Chrysothamnus vulcanicus Greene = Ericameria parryi (A. Gray) G. L. Nesom et G. I. Baird var. vulcanica (Greene) G. L. Nesom et G. I. Baird ●☆

90579 Chrysothemis Decne. (1849);金红花属■☆

90580 Chrysothemis friedrich-sthaliana (Hanst.) H. E. Moore;金红花■☆

90581 Chrysothesium (Jaub. et Spach) Hendrych = Thesium L. ■

90582 Chrysothesium (Jaub. et Spach) Hendrych(1994);金黄百蕊草属■☆

90583 Chrysothrix Roam. et Schult. = Chrysithrix L. ■☆

90584 Chrysotolia N. T. Burb. = Chrystolia Montrouz. ex Beauvis. ■

90585 Chrysoxylon Casar. = Plathymenia Benth. (保留属名)●★

90586 Chrysoxylon Wedd. = Howardia Wedd. ■☆

90587 Chrysoxylon Wedd. = Pogonopus Klotzsch ■☆

90588 Chrystolia Montrouz. = Glycine Willd. (保留属名)■

90589 Chrystolia Montrouz. ex Beauvis. = Glycine Willd. (保留属名)■

90590 Chrysurus Pers. = Lamarckia Moench(保留属名)■☆

90591 Chrytotheca G. Don = Ammannia L. ■

90592 Chrytotheca G. Don = Cryptotheca Blume ■

90593 Chryza Raf. = Coptis Salisb. ■

90594 Chthamalia Decne. = Lachnostoma Kunth ●☆

90595 Chthonia Cass. = Pectis L. ■☆

90596 Chthonia glaucescens Cass. = Pectis glaucescens (Cass.) D. J. Keil ■☆

90597 Chthonocephalus Steetz(1845);对叶鼠麹草属■☆

90598 Chthonocephalus multiceps J. H. Willis;多头对叶鼠麹草■☆

90599 Chthonocephalus pseudevax Steetz;对叶鼠麹草■☆

90600 Chuanminshen M. L. Sheh et R. H. Shan (1980);川明参属;Chuanminshen ●★

90601 Chuanminshen violaceum M. L. Sheh et R. H. Shan;川明参(川明党,明参,明沙参,沙参,土明参);Chuanminshen ■

90602 Chucoa Cabrera(1955);黄菊木属●☆

90603 Chucoa ilicifolia Cabrera;黄菊木●☆

90604 Chukrasia A. Juss. (1830);麻楝属;Chittagong Chickrassy ●

90605 Chukrasia tabularis A. Juss.;麻楝(白椿,茸毛麻楝);Bastard Cedar, Chickrassy, Chittagong, Chittagong Chickrassy, Chittagong Ehickrassy, Chittagong Wood, Indian Redwood, White Cedar, Yinma, Yomhin ●

90606 Chukrasia tabularis A. Juss. var. velutina (Wall.) King;毛麻楝;Hairy Chittagong Chickrassy ●

90607 Chulustum Raf. = Polygonum L. (保留属名)■●

90608 Chumsriella Bor = Germainia Balansa et Poitr. ■

90609 Chunchoa Pers. = Chuncoa Pav. ex Juss. ●

90610 Chuncoa Pav. ex Juss. = Terminalia L. (保留属名)●

90611 Chunechites Tsiang = Urceola Roxb. (保留属名)●

90612 Chunechites Tsiang(1937);乐东藤属;Chunechites, Lotungvine ●★

90613 Chunechites xylinabariopsoides Tsiang;乐东藤;Common Chunechites, Lotungvine ●

90614 Chunechites xylinabariopsoides Tsiang = Urceola xylinabariopsoides (Tsiang) D. J. Middleton ●

90615 Chunia Hung T. Chang(1948);山铜材属(陈木属,假马蹄荷属);Chunia ●★

90616 Chunia bucklandinides Hung T. Chang;山铜材(陈木,假马蹄荷);Bucklandia-like Chunia ●◇

90617 Chuniodendmn Hu = Aphanamixis Blume ●

90618 Chuniodendron yunnanense Hu = Aphanamixis polystachya (Wall.) R. Parker ●

90619 Chuniophoenix Burret (1937);琼棕属(掌叶海枣属);Chuniophoenix, Qiongpalm ●

90620 Chuniophoenix hainanensis Burret;琼棕;Hainan Chuniophoenix, Qiongpalm ●◇

90621 Chuniophoenix humilis C. Z. Tang et T. L. Wu = Chuniophoenix nana Burret ●◇

90622 Chuniophoenix nana Burret;矮琼棕(小琼棕);Dwarf Chuniophoenix, Dwarf Qiongpalm, Hanous Chuniophoenix ●◇

90623 Chupalon Adans. (废弃属名) = Cavendishia Lindl. (保留属名) ●☆

90624 Chupalones Nieremb. ex Steud. = Cavendishia Lindl. (保留属名)●☆

90625 Chupalones Nieremb. ex Steud. = Chupalon Adans. (废弃属名) ●☆

90626 Chuquiraga Juss. (1789);多枝刺菊木属(丘奎菊属)●☆

90627 Chuquiraga insignis Humb. et Bonpl.;显丘多枝刺菊木(显丘奎菊)●☆

90628 Churumaya Raf. = Piper L. ●■

90629 Chusquea Kunth(1822);楚氏竹属(楚氏库竹属,南美高原竹属,丘斯夸竹属,朱丝贵竹属,朱丝奎竹属);Chusquea ●☆

90630 Chusquea andina Phil.;安第斯楚氏竹;Andean Weeping Bamboo, Chilean Bamboo ●☆

90631 Chusquea coronalis Soderstr. et C. E. Calderón;哥斯达黎加楚氏竹(哥斯达黎加楚氏库竹);Costa Rican Weeping Bamboo ●☆

90632 Chusquea culeou E. Desv.;智利楚氏竹(智利楚氏库竹,朱丝贵竹);Chilean Bamboo ●☆

90633 Chusquea pittieri Hack.;皮特楚氏竹;Chusquea ●☆

90634 Chusquea quila Kunth;智利垂枝楚氏竹;Chilean Weeping Bamboo ●☆

90635 Chusua Nevski = Orchis L. ■

90636 Chusua Nevski = Ponerorchis Rchb. f. ■

90637 Chusua brevicalcarata (Finet) P. F. Hunt = Orchis brevicalcarata (Finet) Schltr. ■

90638 Chusua brevicalcarata (Finet) P. F. Hunt = Ponerorchis brevicalcarata (Finet) Soó ■

90639 Chusua chidori (Makino) P. F. Hunt = Ponerorchis chidori (Makino) Ohwi ■☆

90640 Chusua chrysea (W. W. Sm.) P. F. Hunt = Orchis chrysea (W. W. Sm.) Schltr. ■

90641 Chusua chrysea (W. W. Sm.) P. F. Hunt = Ponerorchis chrysea

(W. W. Sm.) Soó ■
90642 Chusua crenulata (Schltr.) P. F. Hunt = Orchis crenulata Schltr. ■
90643 Chusua crenulata (Soó) P. F. Hunt = Orchis crenulata Schltr. ■
90644 Chusua curtipes (Ohwi) P. F. Hunt = Ponerorchis chidori (Makino) Ohwi var. curtipes (Ohwi) F. Maek. ■☆
90645 Chusua donii Nevski = Orchis chusua D. Don ■
90646 Chusua donii Nevski = Ponerorchis chusua (D. Don) Soó ■
90647 Chusua hui (Ts. Tang et F. T. Wang) P. F. Hunt = Orchis limprichtii Schltr. ■
90648 Chusua hui (Ts. Tang et F. T. Wang) P. F. Hunt = Ponerorchis limprichtii (Schltr.) Soó ■
90649 Chusua joo-iokiana (Makino) P. F. Hunt = Ponerorchis joo-iokiana (Makino) Nakai ■☆
90650 Chusua kiraishiensis (Hayata) P. F. Hunt = Orchis kiraishiensis Hayata ■
90651 Chusua kiraishiensis (Hayata) P. F. Hunt = Ponerorchis kiraishiensis (Hayata) Ohwi ■
90652 Chusua kunihikoana (Masam. et Fukuy.) P. F. Hunt = Ponerorchis tominagai (Hayata) H. J. Su et J. J. Chen ■
90653 Chusua limprichtii (Schltr.) P. F. Hunt = Orchis limprichtii Schltr. ■
90654 Chusua limprichtii (Schltr.) P. F. Hunt = Ponerorchis limprichtii (Schltr.) Soó ■
90655 Chusua monophylla (Collett et Hemsl.) P. F. Hunt = Orchis monophylla (Collett et Hemsl.) Rolfe ■
90656 Chusua monophylla (Collett et Hemsl.) P. F. Hunt = Ponerorchis monophylla (Collett et Hemsl.) Soó ■
90657 Chusua nana (King et Pantl.) Pradhan = Ponerorchis chusua (D. Don) Soó ■
90658 Chusua pauciflora (Lindl.) P. F. Hunt = Orchis chusua D. Don ■
90659 Chusua pauciflora (Lindl.) P. F. Hunt = Ponerorchis chusua (D. Don) Soó ■
90660 Chusua pauciflora (Lindl.) P. F. Hunt = Ponerorchis pauciflora (Lindl.) Ohwi ■
90661 Chusua pulchella (Hand.-Mazz.) P. F. Hunt = Orchis chusua D. Don ■
90662 Chusua pulchella (Hand.-Mazz.) P. F. Hunt = Ponerorchis chusua (D. Don) Soó ■
90663 Chusua roborovskii (Maxim.) P. F. Hunt = Orchis roborowskii Maxim. ■
90664 Chusua roborovskii (Maxim.) P. F. Hunt var. delavayi (Schltr.) P. F. Hunt = Orchis chusua D. Don ■
90665 Chusua roborovskii (Maxim.) P. F. Hunt var. giraldiana (Kraenzl.) P. F. Hunt = Orchis chusua D. Don ■
90666 Chusua roborovskii (Maxim.) P. F. Hunt var. tenii (Schltr.) P. F. Hunt = Orchis chusua D. Don ■
90667 Chusua roborovskii (Maxim.) P. F. Hunt var. unifoliata (Schltr.) P. F. Hunt = Orchis chusua D. Don ■
90668 Chusua roborowskyi (Maxim.) P. F. Hunt = Galearis roborowskyi (Maxim.) S. C. Chen, P. J. Cribb et S. W. Gale ■
90669 Chusua roborowskyi (Maxim.) P. F. Hunt var. delavayi (Schltr.) P. F. Hunt = Ponerorchis chusua (D. Don) Soó ■
90670 Chusua roborowskyi (Maxim.) P. F. Hunt var. giraldiana (Kraenzl.) P. F. Hunt = Ponerorchis chusua (D. Don) Soó ■
90671 Chusua roborowskyi (Maxim.) P. F. Hunt var. nana (King et Pantl.) P. F. Hunt = Ponerorchis chusua (D. Don) Soó ■
90672 Chusua roborowskyi (Maxim.) P. F. Hunt var. tenii (Schltr.) P. F. Hunt = Ponerorchis chusua (D. Don) Soó ■
90673 Chusua roborowskyi (Maxim.) P. F. Hunt var. unifoliata (Schltr.) P. F. Hunt = Ponerorchis chusua (D. Don) Soó ■
90674 Chusua secunda Nevski = Orchis chusua D. Don ■
90675 Chusua secunda Nevski = Ponerorchis chusua (D. Don) Soó ■
90676 Chusua taiwanensis (Fukuy.) P. F. Hunt = Orchis taiwanensis Fukuy. ■
90677 Chusua taiwanensis (Fukuy.) P. F. Hunt = Platanthera tipuloides (L. f.) Lindl. ■
90678 Chusua taiwanensis (Fukuy.) P. F. Hunt = Ponerorchis taiwanensis (Fukuy.) Ohwi ■
90679 Chusua takasagomontana (Masam.) P. F. Hunt = Orchis takasagomontana Masam. ■
90680 Chusua takasagomontana (Masam.) P. F. Hunt = Ponerorchis takasagomontana (Masam.) Ohwi ■
90681 Chwenkfeldia Willd. = Sabicea Aubl. ●☆
90682 Chydenanthus Miers(1875);节毛玉蕊属●☆
90683 Chydenanthus excelsus Miers;节毛玉蕊●☆
90684 Chylaceae Dulac = Fumariaceae Marquis(保留科名)■☆
90685 Chylisma Nutt. ex Torr. et A. Gray = Camissonia Link ■☆
90686 Chylismia (Torr. et A. Gray) Nutt. ex Raim. = Camissonia Link ■☆
90687 Chylismia Nutt. = Camissonia Link ■☆
90688 Chylismia Nutt. ex Torr. et A. Gray = Camissonia Link ■☆
90689 Chylismiella (Munz) W. L. Wagner et Hoch = Oenothera L. ●■
90690 Chylismiella (Munz) W. L. Wagner et Hoch(2007);翅籽月见草属■☆
90691 Chylocalyx Hassk. = Echinocaulos (Meisn. ex Endl.) Hassk. ●■
90692 Chylocalyx Hassk. ex Miq. = Echinocaulos (Meisn. ex Endl.) Hassk. ●■
90693 Chylocalyx perfoliatus (L.) Hassk. ex Miq. = Persicaria perfoliata (L.) H. Gross ■
90694 Chylocalyx perfoliatus (L.) Hassk. ex Miq. = Polygonum perfoliatum L. ■
90695 Chylocalyx senticosus Meisn. = Polygonum senticosum (Meisn.) Franch. et Sav. ■
90696 Chylocalyx senticosus Meisn. ex Miq. = Persicaria senticosa (Meisn.) H. Gross ■
90697 Chylocalyx senticosus Meisn. ex Miq. = Polygonum senticosum (Meisn.) Franch. et Sav. ■
90698 Chylodia Rich. ex Cass. = Chatiakella Cass. ■☆
90699 Chylodia Rich. ex Cass. = Wulffia Neck. ex Cass. ■☆
90700 Chylogala Fourr. = Euphorbia L. ●■
90701 Chymaceae Dulac = Papaveraceae Juss.(保留科名)●■
90702 Chymocarpus D. Don = Tropaeolum L. ■
90703 Chymocarpus D. Don ex Brewster, R. Taylor et R. Phillips = Tropaeolum L. ■
90704 Chymococca Meisn. = Passerina L. ●☆
90705 Chymococca empetroides Meisn. = Passerina ericoides L. ●☆
90706 Chymocormus Harv. = Fockea Endl. ●☆
90707 Chymocormus edulis (Thunb.) Harv. = Fockea edulis (Thunb.) K. Schum. ●☆
90708 Chymsydia Albov = Agasyllis Spreng. ■☆
90709 Chymsydia Albov(1895);高加索芹属■☆
90710 Chymsydia agasylloides Alb.;高加索芹■☆
90711 Chysis Lindl.(1837);长足兰属(长脚兰属);Baby Orchid, Chysis ■☆

90712 Chysis aurea Lindl.;黄长足兰(黄长脚兰);Golden Chysis ■☆
90713 Chysis bractescens Lindl.;苞状长脚兰■☆
90714 Chysis laevis Lindl.;平滑长足兰(平滑长脚兰);Smooth Chysis ■☆
90715 Chytra C. F. Gaertn.(废弃属名)= Agalinis Raf.(保留属名)■☆
90716 Chytra C. F. Gaertn.(废弃属名)= Gerardia Benth. ■☆
90717 Chytraculia P. Browne(废弃属名)= Calyptranthes Sw.(保留属名)●☆
90718 Chytralia Adans. = Calyptranthes Sw.(保留属名)●☆
90719 Chytranthus Hook. f.(1862);壶花无患子属●☆
90720 Chytranthus angustifolius Exell;窄叶壶花无患子●☆
90721 Chytranthus atroviolaceus Baker f. ex Hutch. et Dalziel;暗堇壶花无患子●☆
90722 Chytranthus bracteosus Radlk. = Chytranthus angustifolius Exell ●☆
90723 Chytranthus brunneotomentosus Gilg ex Radlk. = Chytranthus atroviolaceus Baker f. ex Hutch. et Dalziel ●☆
90724 Chytranthus calophyllus Radlk.;美叶壶花无患子●☆
90725 Chytranthus carneus Radlk.;肉色壶花无患子●☆
90726 Chytranthus carneus Radlk. var. secundiflorus Hauman = Chytranthus carneus Radlk. ●☆
90727 Chytranthus cauliflorus (Hutch. et Dalziel) Wickens;茎花壶花无患子●☆
90728 Chytranthus dasystachys Gilg ex Radlk.;毛穗壶花无患子●☆
90729 Chytranthus dinklagei Gilg ex Engl.;丁克壶花无患子●☆
90730 Chytranthus edulis Pierre;可食壶花无患子●☆
90731 Chytranthus ellipticus Hutch. et Dalziel;椭圆壶花无患子●☆
90732 Chytranthus flavoviridis Radlk.;黄绿壶花无患子●☆
90733 Chytranthus fouilloyanus Pellegr. = Chytranthus mortehanii (De Wild.) de Voldere ex Hauman ●☆
90734 Chytranthus gerardii De Wild. = Chytranthus stenophyllus Gilg var. gerardii (De Wild.) Hauman ●☆
90735 Chytranthus gilletii De Wild.;吉勒特壶花无患子●☆
90736 Chytranthus imenoensis Pellegr.;伊梅诺壶花无患子●☆
90737 Chytranthus klaineanus Radlk.;克莱恩壶花无患子●☆
90738 Chytranthus laurentii De Wild. = Pancovia laurentii (De Wild.) Gilg ex De Wild. ●☆
90739 Chytranthus ledermannii Gilg ex Radlk.;莱德壶花无患子●☆
90740 Chytranthus longibracteatus F. G. Davies;长苞壶花无患子●☆
90741 Chytranthus longiracemosus Gilg ex Radlk. = Chytranthus carneus Radlk. ●☆
90742 Chytranthus macrobotrys (Gilg) Exell et Mendonça;大穗壶花无患子●☆
90743 Chytranthus macrophyllus Gilg;大叶壶花无患子●☆
90744 Chytranthus macrophyllus Gilg var. obanensis Baker f. = Chytranthus atroviolaceus Baker f. ex Hutch. et Dalziel ●☆
90745 Chytranthus mangenotii N. Hallé et Aké Assi = Chytranthus cauliflorus (Hutch. et Dalziel) Wickens ●☆
90746 Chytranthus mannii Hook. f.;曼氏壶花无患子●☆
90747 Chytranthus mayumbensis Exell = Chytranthus macrobotrys (Gilg) Exell et Mendonça ●☆
90748 Chytranthus micranthus Gilg ex Radlk.;小花壶花无患子●☆
90749 Chytranthus mortehanii (De Wild.) de Voldere ex Hauman;莫特汉壶花无患子●☆
90750 Chytranthus obliquinervis Radlk.;斜脉壶花无患子●☆
90751 Chytranthus pilgerianus (Gilg) Pellegr. = Chytranthus talbotii (Baker f.) Keay ●☆
90752 Chytranthus prieureanus Baill.;普里厄壶花无患子●☆
90753 Chytranthus prieurianus Baill. subsp. longiflorus (Verdc.) N. Hallé = Chytranthus prieureanus Baill. ●☆
90754 Chytranthus punctatus Radlk.;斑点壶花无患子●☆
90755 Chytranthus sacleuxii Pierre = Chytranthus prieureanus Baill. ●☆
90756 Chytranthus sacleuxii Pierre ex Sacleux subsp. longiflorus Verdc. = Chytranthus prieureanus Baill. ●☆
90757 Chytranthus setosus Radlk.;刚毛壶花无患子●☆
90758 Chytranthus sexlocularis Radlk. = Chytranthus talbotii (Baker f.) Keay ●☆
90759 Chytranthus stenophyllus Gilg;狭叶壶花无患子●☆
90760 Chytranthus stenophyllus Gilg var. gerardii (De Wild.) Hauman;杰勒德壶花无患子●☆
90761 Chytranthus strigosus Radlk.;糙伏毛壶花无患子●☆
90762 Chytranthus talbotii (Baker f.) Keay;塔尔博特壶花无患子●☆
90763 Chytranthus verecundus N. Hallé et Aké Assi;羞涩无患子●☆
90764 Chytranthus welwitschii Exell = Chytranthus carneus Radlk. ●☆
90765 Chytranthus xanthophyllus Radlk.;黄叶壶花无患子●☆
90766 Chytranthus zenkeri Gilg = Chytranthus macrophyllus Gilg ●☆
90767 Chytroglossa Rchb. f.(1863);壶舌兰属■☆
90768 Chytroglossa aurata Rchb. f.;壶舌兰■☆
90769 Chytroma Miers = Lecythis Loefl. ●☆
90770 Chytroma Miers(1874);热美玉蕊属●☆
90771 Chytroma jarana Huber;热美玉蕊●☆
90772 Chytropsia Bremek. = Psychotria L.(保留属名)●
90773 Cianitis Reinw. = Cyanitis Reinw. ●
90774 Cianitis Reinw. = Dichroa Lour. ●
90775 Cibirhiza Bruyns(1988);囊根萝藦属☆
90776 Cibirhiza albersiana Kuntze et Meve et Liede;囊根萝藦☆
90777 Cibotarium O. E. Schulz = Sphaerocardamum Nees et Schauer ■☆
90778 Cicca Adans.(废弃属名)= Julocroton Mart.(保留属名)●■☆
90779 Cicca L.(1767);醋栗属(核果叶下珠属)●
90780 Cicca L. = Phyllanthus L. ●■
90781 Cicca acida (L.) Merr.;西印度醋栗●
90782 Cicca discoidea Baill. = Margaritaria discoidea (Baill.) G. L. Webster ●☆
90783 Cicca disticha L. = Phyllanthus acidus Skeels ●☆
90784 Cicca flexuosa Siebold et Zucc. = Phyllanthus flexuosus (Siebold et Zucc.) Müll. Arg. ●
90785 Cicca leucopyra (Willd.) Kurz = Flueggea leucopyrus Willd. ●
90786 Cicca microcarpa Benth. = Phyllanthus reticulatus Poir. ●
90787 Cicca obovata Kurz = Flueggea virosa (Roxb. ex Willd.) Voigt ●
90788 Cicca reticulata Kurz = Phyllanthus reticulatus Poir. ●
90789 Cicca sinica Baill. = Margaritaria indica (Dalzell) Airy Shaw ●
90790 Cicendia Adans.(1763);百黄花属;Yellow Centaury ■☆
90791 Cicendia Griseb. = Exaculum Caruel ■☆
90792 Cicendia filiformis (L.) Delarbre;百黄花;Slender Cicendia, Yellow Centaury ■☆
90793 Cicendia microphylla Edgew. = Sebaea microphylla (Edgew.) Knobl. ■
90794 Cicendia pusilla Lam. = Exaculum pusillum (Lam.) Caruel ■☆
90795 Cicendiola Bubani = Cicendia Adans. ■☆
90796 Cicendiopsis Kuntze = Exaculum Caruel ■☆
90797 Cicer L.(1753);鹰嘴豆属(鸡豆属,鹰咀豆属);Chick Pea, Chikpea ■
90798 Cicer acanthophyllum Boriss.;刺叶鹰嘴豆■☆
90799 Cicer acanthophyllum Boriss. = Cicer macranthum Popov ■☆

90800 Cicer anatolicum Alef.;阿纳托里鹰嘴豆■☆
90801 Cicer arietinum L.;鹰嘴豆(胡豆,胡豆子,回鹘豆,回回豆,鸡豆,鸡儿豆,那合豆,桃豆,香豆子,香子豆,鹰咀豆);Bengal Gram,Chick Pea,Chick-pea,Chikpea,Cich,Garbanzo,Garbanzo Bean,Garbanzo Beans,Gram Chick Pea,Gram Chick-pea,Gram Chikpea,Spanish Pea ■
90802 Cicer arietinum L. f. album Gaudich.;白鹰嘴豆■☆
90803 Cicer atlanticum Maire;大西洋鹰嘴豆■☆
90804 Cicer baldshuanicum (Popov) Lincz.;巴地鹰嘴豆■☆
90805 Cicer bijugum Rech. f.;叙利亚鹰嘴豆■☆
90806 Cicer canariense A. Santos et G. P. Lewis;加那利鹰嘴豆■☆
90807 Cicer caucasicum Bornm.;高加索鹰嘴豆■☆
90808 Cicer cuneatum A. Rich.;楔形鹰嘴豆■☆
90809 Cicer echinospermum P. H. Davis;刺子鹰嘴豆■☆
90810 Cicer ervoides (Siebold) Fenzl;野豌豆状鹰嘴豆■☆
90811 Cicer ervoides Brign. = Lens ervoides (Brign.) Grande ■☆
90812 Cicer fedtschenkoi Lincz.;中亚鹰嘴豆■☆
90813 Cicer flexuosum Lipsky;曲折鹰嘴豆■☆
90814 Cicer floribundum Fenzl;多花鹰嘴豆■☆
90815 Cicer jacquemontii Jaub. et Spach;帕米尔鹰嘴豆■
90816 Cicer jacquemontii Jaub. et Spach = Cicer microphyllum Benth. ■
90817 Cicer kopetdaghense Lincz.;科佩特鹰嘴豆■☆
90818 Cicer korshinskyi Lincz.;科尔欣斯基鹰嘴豆■☆
90819 Cicer macracanthum Popov;大刺鹰嘴豆■☆
90820 Cicer macranthum Popov;大花鹰嘴豆■☆
90821 Cicer microphyllum Benth.;小叶鹰嘴豆;Littleleaf Chikpea ■
90822 Cicer minutum Boiss. et Hohen.;微小鹰嘴豆■☆
90823 Cicer mogoltavicum (Popov) A. S. Korol.;莫戈尔塔夫鹰嘴豆■☆
90824 Cicer nigrum Vindob. ex Zeyh.;黑鹰嘴豆■☆
90825 Cicer nuristanicum Kitam.;努里斯坦鹰嘴豆■☆
90826 Cicer oxyodou Boiss. et Hohen.;波斯鹰嘴豆■☆
90827 Cicer pungens Boiss.;刺鹰嘴豆■☆
90828 Cicer songaricum Jaub. et Spach;准噶尔鹰嘴豆■☆
90829 Cicer songaricum Jaub. et Spach var. spinosum Aitch. = Cicer macranthum Popov ■☆
90830 Cicer songaricum Steph. ex DC. = Cicer microphyllum Benth. ■
90831 Cicer soongaricum Steph. = Cicer microphyllum Benth. ■
90832 Cicer yamashitae Kitam.;阿富汗鹰嘴豆■☆
90833 Ciceraceae Steele = Fabaceae Lindl.(保留科名)●■
90834 Ciceraceae Steele = Leguminosae Juss.(保留科名)●■
90835 Cicerbita Wallr.(1822);岩参属(鸡豆菊属);Blue Sow-thistle, Sow-thistle ■
90836 Cicerbita alpina (L.) Wallr.;高山岩参(高山莴苣);Alpine Blue Sow-thistle,Alpine Lettuce,Alpine Sow Thistle,Alpine Sow-thistle,Blue Sow-thistle,Mountain Sow Thistle,Mountain Sow-thistle ■
90837 Cicerbita azurea (Ledeb.) Beauverd;岩参■
90838 Cicerbita bonatii Beauverd = Chaetoseris bonatii (Beauverd) C. Shih ■
90839 Cicerbita bourgaei Beauverd;紫蓝岩参;Pontic Blue Sow-thistle ■☆
90840 Cicerbita cyanea (D. Don) Beauverd = Chaetoseris cyanea (D. Don) C. Shih ■
90841 Cicerbita cyanea (D. Don) Beauverd var. glandulifera (Franch.) Beauverd = Chaetoseris cyanea (D. Don) C. Shih ■
90842 Cicerbita cyanea (D. Don) Beauverd var. hastata (Wall. ex DC.) Beauverd = Chaetoseris hastata (Wall. ex DC.) C. Shih ■
90843 Cicerbita cyanea (D. Don) Beauverd var. lutea Hand.-Mazz. = Chaetoseris lutea (Hand.-Mazz.) C. Shih ■
90844 Cicerbita cyanea (D. Don) Beauverd var. teniana P. Beauv. = Chaetoseris teniana (P. Beauv.) C. Shih ■
90845 Cicerbita deltoidea Beauverd;三角岩参■☆
90846 Cicerbita dutchieana Beauverd = Cephalorrhynchus albiflorus C. Shih ■
90847 Cicerbita grandiflora (Franch.) Beauverd = Chaetoseris grandiflora (Franch.) C. Shih ■
90848 Cicerbita kovalevskiana Kirp.;科瓦廖夫斯基岩参■☆
90849 Cicerbita likiangensis (Franch.) Beauverd = Chaetoseris likiangensis (Franch.) C. Shih ■
90850 Cicerbita macrantha (C. B. Clarke) Beauverd = Chaetoseris macrantha (C. B. Clarke) C. Shih ■
90851 Cicerbita macrophylla (Willd.) Wallr.;大叶岩参;Blue Sow Thistle,Blue Sow-thistle,Common Blue Sow-thistle ■☆
90852 Cicerbita macrorhiza (Royle) Beauverd var. saxatilis (Edgew.) Beauverd = Cephalorrhynchus saxatilis (Edgew.) C. Shih ■
90853 Cicerbita macrorrhiza (Royle) Beauverd = Cephalorrhynchus macrorrhizus (Royle) Tsuil ■
90854 Cicerbita macrorrhiza (Royle) Beauverd var. saxatilis (Edgew.) Beauverd = Cephalorrhynchus saxatilis (Edgew.) C. Shih ■
90855 Cicerbita macrorrhiza (Willd.) Wallr. = Cephalorrhynchus macrorrhizus (Royle) Tsuil ■
90856 Cicerbita oligolepis C. C. Chang ex C. Shih;大理岩参;Dali Sow-thistle ■
90857 Cicerbita plumieri Kirschl.;普吕米岩参;Hairless Blue Sow-thistle ■☆
90858 Cicerbita prenanthoides Beauverd;福王草岩参■☆
90859 Cicerbita racemosa Beauverd;总花岩参■☆
90860 Cicerbita roborowskii (Maxim.) Beauverd;青甘岩参(青甘莴苣)■
90861 Cicerbita roborowskii (Maxim.) Beauverd = Chaetoseris roborowskii (Maxim.) C. Shih ■
90862 Cicerbita rosea (Popov et Vved.) Krasch. ex Kovalevsk.;粉红岩参■☆
90863 Cicerbita scandens (C. C. Chang) C. Shih;攀缘岩参■
90864 Cicerbita sikkimensis (Hook. f.) C. Shih;西藏岩参;Xizang Sow-thistle ■
90865 Cicerbita taliensis (Franch.) Beauverd = Stenoseris taliensis (Franch.) C. Shih ■
90866 Cicerbita tenerrima (Pourr.) Beauverd = Lactuca tenerrima Pourr. ■☆
90867 Cicerbita tenerrima (Pourr.) Beauverd var. albiflora (Emb.) Maire = Lactuca tenerrima Pourr. ■☆
90868 Cicerbita tenerrima (Pourr.) Beauverd var. glabra Boiss. = Lactuca tenerrima Pourr. ■☆
90869 Cicerbita tenerrima (Pourr.) Beauverd var. micrantha Maire = Lactuca tenerrima Pourr. ■☆
90870 Cicerbita tenerrima (Pourr.) Beauverd var. scabra Boiss. = Lactuca tenerrima Pourr. ■☆
90871 Cicerbita thianschanica (Regel et Schmalh.) Beauverd;天山岩参;Tianshan Sow-thistle ■
90872 Cicerbita uralensis Beauverd;乌拉尔岩参■☆
90873 Cicerbita zeravschanica Popov;泽拉夫尚岩参■☆
90874 Cicercula Medik. = Lathyrus L. ■
90875 Cicerella DC. = Cicercula Medik. ■
90876 Ciceronia Urb.(1925);长冠亮泽兰属■☆

90877 Ciceronia chaptalioides Urb.;长冠亮泽兰■☆
90878 Cicheria Raf. = Cichorium L. ■
90879 Cichlanthus (Endl.) Tiegh. = Scurrula L.(废弃属名)●
90880 Cichlanthus Tiegh. = Scurrula L.(废弃属名)●
90881 Cichlanthus ferrugineus (Jack) Tiegh. = Scurrula ferruginea (Jack) Danser ●
90882 Cichlanthus philippensis (Cham. et Schltdl.) Tiegh. = Scurrula atropurpurea (Blume) Danser ●
90883 Cichlanthus philippinensis (Champ. et Schltdl.) Tiegh. = Scurrula philippensis (Cham. et Schltdl.) G. Don ●
90884 Cichlanthus pulverulentus (Wall.) Tiegh. = Scurrula pulverulenta (Wall.) G. Don ●
90885 Cichlanthus scurrula (L.) Tiegh. = Scurrula parasitica L. ●
90886 Cichoriaceae Juss.(1789)(保留科名);菊苣科;Cichorium Family ■
90887 Cichoriaceae Juss.(保留科名)= Asteraceae Bercht. et J. Presl (保留科名)●■
90888 Cichoriaceae Juss.(保留科名)= Compositae Giseke(保留科名)●■
90889 Cichorium L.(1753);菊苣属;Chicory,Succory,Wild Chicory ■
90890 Cichorium alatum Hochst. et Steud. = Geigeria alata (Hochst. et Steud.) Oliv. et Hiern ■☆
90891 Cichorium callosum Pomel = Cichorium endivia L. subsp. pumilum (Jacq.) Cout. ■☆
90892 Cichorium calvum Sch. Bip. ex Asch.;光秃菊苣■☆
90893 Cichorium divaricatum Schousb. = Cichorium endivia L. subsp. pumilum (Jacq.) Cout. ■☆
90894 Cichorium endivia L.;栽培菊苣(菊苣,苣荬菜,苦菜,苦苣,苦荬,荬菜,意大利菊苣);Broad-leaved Endive,Chicory,Common Endive,Cultivated Endive,Endive,Escarole,Garden Succory,Goat's Beard,White Chicory,White Endive ■
90895 Cichorium endivia L.'Crispa';皱波菊苣■☆
90896 Cichorium endivia L.'Latifolia';宽叶栽培菊苣■
90897 Cichorium endivia L. subsp. divaricatum (Schousb.) Ball = Cichorium endivia L. subsp. pumilum (Jacq.) Cout. ■☆
90898 Cichorium endivia L. subsp. pumilum (Jacq.) Cout.;矮菊苣■☆
90899 Cichorium endivia L. var. sativa DC. = Cichorium endivia L. ■
90900 Cichorium glandulosum Boiss. et Houtt.;腺毛菊苣(菊苣,卡斯尼,毛菊苣)■
90901 Cichorium intybus L.;菊苣(卡斯尼,苦苣,苦马草,荬菜,欧洲菊苣,苣菜,野生苦苣);Belgian Endive,Blue Dandelion,Blue Endive,Blue Sailor,Blue Sailors,Blue Succory,Blue-sailors,Bunks,Chicory,Coffee Root,Common Chicory,Endive,French Endive,Monk's Beard,Radiccio Radicchio,Strip-for-strip,Succory,Suckery,Turnsole,Wild Chicory,Wild Succory,Witloof,Witloof Chicory ■
90902 Cichorium intybus L. f. albiflorum Neuman = Cichorium intybus L. ■
90903 Cichorium intybus L. f. roseum Neuman = Cichorium intybus L. ■
90904 Cichorium intybus L. subsp. glabratum (C. Presl) Arcang.;光滑菊苣■☆
90905 Cichorium intybus L. subsp. pumilum (Jacq.) Ball = Cichorium endivia L. subsp. pumilum (Jacq.) Cout. ■☆
90906 Cichorium intybus L. var. callosum (Pomel) Batt. = Cichorium endivia L. subsp. pumilum (Jacq.) Cout. ■☆
90907 Cichorium intybus L. var. glabratum (C. Presl) Batt. = Cichorium intybus L. subsp. glabratum (C. Presl) Arcang. ■☆
90908 Cichorium intybus L. var. longipes Faure et Maire = Cichorium intybus L. ■
90909 Cichorium polystachyum Pomel = Cichorium endivia L. subsp. pumilum (Jacq.) Cout. ■☆
90910 Cichorium pumilum Jacq.;矮小菊苣■☆
90911 Cichorium pumilum Jacq. = Cichorium endivia L. subsp. pumilum (Jacq.) Cout. ■☆
90912 Cichorium pumilum Jacq. var. polystachyum (Pomel) Batt. = Cichorium endivia L. subsp. pumilum (Jacq.) Cout. ■☆
90913 Cichorium shansiense Petr.;牛口蓟;Shansi Chicory,Shanxi Chicory ■
90914 Cichorium spinosum L.;具刺菊苣■☆
90915 Ciclospermum Lag. = Cyclospermum Lag. ■
90916 Ciclospermum ammi Lag. = Apium leptophyllum (Pers.) F. Muell. ex Benth. ■
90917 Ciconium Sweet = Pelargonium L'Hér. ex Aiton ●■
90918 Cicuta L.(1753);毒芹属;Cowbane,Poisoncelery,Poisonhemlock,Water Hemlock,Waterhemlock ■
90919 Cicuta Mill. = Conium L. ■
90920 Cicuta africana (L.) Lam. = Capnophyllum africanum (L.) W. D. J. Koch ■☆
90921 Cicuta bulbifera L.;北美毒芹;Bulb-bearing Water Hemlock,Bulblet Water-hemlock ■☆
90922 Cicuta dahurica Fisch. ex Schultz = Sium suave Walter ■
90923 Cicuta douglasi Coult. et Rose;西部毒芹;Western Water Hemlock ■☆
90924 Cicuta maculata L.;鼠芹(水毒芹);American Cowbane,Beaver Poison,Common Water Hemlock,Common Water-hemlock,Cowbean,Death of Man,False Parsley,Fever Root,Mock-eel Root,Mushquash Root,Muskrat Weed,Musquash Root,Poison Hemlock,Snakeroot,Snakeweed,Spotted Cowbane,Spotted Hemlock,Spotted Parsley,Spotted Water Hemlock,Spotted Water-hemlock,Water Hemlock,Wild Carrot,Wild Parsnip ■☆
90925 Cicuta maculata L. var. angustifolia Hook.;窄叶鼠芹;Spotted Water-hemlock ■☆
90926 Cicuta maculata L. var. bolanderi (S. Watson) G. A. Mulligan;鲍氏鼠芹;Water-hemlock ■☆
90927 Cicuta maculata L. var. curtissii (J. M. Coult. et Rose) Fernald = Cicuta maculata L. ■☆
90928 Cicuta monnieri (L.) Crantz = Cnidium monnieri (L.) Cusson ■
90929 Cicuta nipponica Franch. = Cicuta virosa L. var. latisecta Celak. ■
90930 Cicuta officinalis Crantz = Conium maculatum L. ■
90931 Cicuta sinensis Zuccagni = Cnidium monnieri (L.) Cusson ■
90932 Cicuta virosa L.;毒芹(河毒,芹叶钩吻,野胡萝卜,野芹,野芹菜花,走马芹);Brook Tongue,Brook-tongue,Cow-bana,Cowbane,Deathin,Europe Waterhemlock,European Water Hemlock,Northern Water-hemlock,Poisoncelery,Water Hemlock ■
90933 Cicuta virosa L. f. longiinvolucellata Y. C. Chu;长苞毒芹■
90934 Cicuta virosa L. f. longiinvolucellata Y. C. Chu = Cicuta virosa L. ■
90935 Cicuta virosa L. var. angustifolia (Kit.) Schube;窄叶毒芹(细叶毒芹);Narrowleaf Poisoncelery ■
90936 Cicuta virosa L. var. latisecta Celak.;宽叶毒芹(毒芹);Broadleaf Poisoncelery ■
90937 Cicuta virosa L. var. nipponica (Franch.) Makino;日本毒芹;Japanese Poisoncelery ■
90938 Cicuta virosa L. var. nipponica (Franch.) Makino = Cicuta virosa L. ■
90939 Cicuta virosa L. var. stricta Schultz = Cicuta virosa L. ■

90940 Cicuta virosa L. var. tenuifolia (A. Froehl.) Koch;细叶毒芹;Thinleaf Poisoncelery ■
90941 Cicutaria Fabr. = Conium L. ■
90942 Cicutaria Heist. ex Fabr. = Conium L. ■
90943 Cicutaria Lam. = Cicuta L. ■
90944 Cicutaria Mill. = Molopospermum W. D. J. Koch ■☆
90945 Cicutastrum Fabr. = Thapsia L. ■☆
90946 Cieca Adans. = Julocroton Mart.(保留属名)●■☆
90947 Cieca Medik. = Passiflora L. ●■
90948 Cienfuegia Willd. = Cienfuegosia Cav. ■●☆
90949 Cienfuegosia Cav.(1786);美非棉属■●☆
90950 Cienfuegosia anomala (Wawra ex Wawra et Peyr.) Gürke = Gossypium anomalum Wawra ex Wawra et Peyr. ●☆
90951 Cienfuegosia bricchettii Ulbr. = Gossypium bricchettii (Ulbr.) Vollesen ■☆
90952 Cienfuegosia chiarugii Chiov. = Cienfuegosia welshii (T. Anderson) Garcke ■●☆
90953 Cienfuegosia digitata Cav.;指裂美非棉●☆
90954 Cienfuegosia digitata Cav. var. lineariloba Hochr. = Cienfuegosia digitata Cav. ●☆
90955 Cienfuegosia ellenbeckii Gürke = Gossypium somalense (Gürke) J. B. Hutch. ●☆
90956 Cienfuegosia gerrardii (Harv.) Hochr.;杰勒德美非棉■●☆
90957 Cienfuegosia hearnii Fryxell = Cienfuegosia welshii (T. Anderson) Garcke ■●☆
90958 Cienfuegosia heteroclada Sprague;互枝美非棉■●☆
90959 Cienfuegosia hildebrandtii Garcke;希尔德美非棉■●☆
90960 Cienfuegosia junciformis A. Chev. = Cienfuegosia digitata Cav. ●☆
90961 Cienfuegosia junciformis A. Chev. var. ruyssenii? = Cienfuegosia digitata Cav. ●☆
90962 Cienfuegosia pentaphylla K. Schum. = Gossypium anomalum Wawra ex Wawra et Peyr. ●☆
90963 Cienfuegosia somalensis Gürke = Gossypium somalense (Gürke) J. B. Hutch. ●☆
90964 Cienfuegosia somaliana Fryxell = Cienfuegosia welshii (T. Anderson) Garcke ■●☆
90965 Cienfuegosia triphylla (Harv.) Hochr. = Gossypium triphyllum (Harv.) Hochr. ■☆
90966 Cienfuegosia triphylla (Harv.) K. Schum. = Gossypium triphyllum (Harv.) Hochr. ■☆
90967 Cienfuegosia welshii (T. Anderson) Garcke;沃尔什美非棉■●☆
90968 Cienfugosia DC. = Cienfuegosia Cav. ■●☆
90969 Cienfugosia Giseke = Cienfuegosia Cav. ■●☆
90970 Cienkowskia Schweinf. = Cienkowskiella Y. K. Kam ■☆
90971 Cienkowskia Schweinf. = Siphonochilus J. M. Wood et Franks ■☆
90972 Cienkowskia aethiopica Schweinf. = Siphonochilus aethiopicus (Schweinf.) B. L. Burtt ■☆
90973 Cienkowskia kirkii Hook. f. = Siphonochilus kirkii (Hook. f.) B. L. Burtt ■☆
90974 Cienkowskiella Y. K. Kam = Siphonochilus J. M. Wood et Franks ■☆
90975 Cienkowskiella aethiopica (Schweinf.) Y. K. Kam = Siphonochilus aethiopicus (Schweinf.) B. L. Burtt ■☆
90976 Cienkowskiella brachystemon (K. Schum.) Y. K. Kam = Siphonochilus brachystemon (K. Schum.) B. L. Burtt ■☆
90977 Cienkowskiella evae (Briq.) Y. K. Kam = Siphonochilus evae (Briq.) B. L. Burtt ■☆
90978 Cienkowskiella kilimanensis (Gagnep.) Y. K. Kam = Siphonochilus kilimanensis (Gagnep.) B. L. Burtt ■☆
90979 Cienkowskiella kirkii (Hook. f.) Y. K. Kam = Siphonochilus kirkii (Hook. f.) B. L. Burtt ■☆
90980 Cienkowskiella nigerica (Hepper) Y. K. Kam = Siphonochilus nigericus (Hepper) B. L. Burtt ■☆
90981 Cienkowskya Regel et Rach = ? Ehretia L. ●
90982 Cienkowskya Schweinf. = Siphonochilus J. M. Wood et Franks ■☆
90983 Cienkowskya Solms = Kaempferia L. ■
90984 Cigarrilla Aiello = Nernstia Urb. ●☆
90985 Cigarrilla Aiello(1979);墨烟叶属■☆
90986 Cigarrilla mexicana (Zucc. et Mart. ex DC.) Aiello;墨烟叶■☆
90987 Ciliaria Haw. = Saxifraga L. ■
90988 Ciliaria bronchialis (L.) Haw. = Saxifraga bronchialis L. ■
90989 Ciliaria cinerascens (Engl. et Irmsch.) Losinsk. = Saxifraga cinerascens Engl. et Irmsch. ■
90990 Ciliosemina Antonelli = Cinchona L. ■●
90991 Ciliovallaceae Dulac = Campanulaceae Juss.(保留科名)■●
90992 Cimbaria Hill = Cymbaria L. ■
90993 Cimicifuga L. = Cimicifuga Wernisch. ●■
90994 Cimicifuga Wernisch. (1763);升麻属;Black Cohosh, Black Snakeroot, Bugbane, Cohosh, Snakeroot ●■
90995 Cimicifuga acerina (Siebold et Zucc.) Tanaka = Cimicifuga japonica (Thunb.) Spreng. ■
90996 Cimicifuga acerina (Siebold et Zucc.) Tanaka f. hispidula P. K. Hsiao = Cimicifuga japonica (Thunb.) Spreng. ■
90997 Cimicifuga acerina (Siebold et Zucc.) Tanaka f. purpurea P. K. Hsiao = Cimicifuga japonica (Thunb.) Spreng. ■
90998 Cimicifuga acerina (Siebold et Zucc.) Tanaka f. strigulosa P. K. Hsiao = Cimicifuga japonica (Thunb.) Spreng. ■
90999 Cimicifuga acerina Tanaka = Cimicifuga japonica (Thunb.) Spreng. ■
91000 Cimicifuga americana Michx.;美洲升麻;American Bugbane, Black Snakeroot ■☆
91001 Cimicifuga arizonica S. Watson;亚利桑那升麻;Arizona Bugbane ■☆
91002 Cimicifuga brachycarpa P. K. Hsiao;短果升麻(披针升麻);Lanceoleaf Bugbane, Shortfruit Bugbane ■
91003 Cimicifuga calthifolia Maxim. = Beesia calthifolia (Maxim.) Ulbr. ■
91004 Cimicifuga calthifolia Maxim. ex Oliv. = Beesia calthifolia (Maxim. ex Oliv.) Ulbr. ■
91005 Cimicifuga chinensis Koidz. = Cimicifuga japonica (Thunb.) Spreng. ■
91006 Cimicifuga cordifolia Pursh = Cimicifuga americana Michx. ■☆
91007 Cimicifuga cordifolia Pursh.;心叶升麻;Heart-leaved Snakeroot ■☆
91008 Cimicifuga dahurica (Turcz. ex Fisch. et C. A. Mey.) Maxim.;兴安升麻(北升麻,鬼脸升麻,鸡骨升麻,窟窿牙根,龙眼根,牤牛牙根,升麻,周麻,周升麻);Dahur Bugbane, Dahurian Bugbane ■
91009 Cimicifuga elata Nutt.;高升麻;Tall Bugbane ■☆
91010 Cimicifuga europaea Schipcz.;欧洲升麻;Europe Bugbane, European Bugbane ■☆
91011 Cimicifuga foetida L.;升麻(鬼脸升麻,黑升麻,鸡骨升麻,龙眼根,绿升麻,马氏升麻,西升麻,鸭脚七,周麻,周升麻);Bugbane, Maire Skunk Bugbane, Skunk Bugbane ■
91012 Cimicifuga foetida L. var. bifida W. T. Wang et P. K. Hsiao;两裂升麻■

91013 Cimicifuga foetida L. var. foliolosa P. K. Hsiao;多小叶升麻;Foliole Bugbane,Manysmallleaf Skunk Bugbane ■

91014 Cimicifuga foetida L. var. intermedia Huth = Cimicifuga simplex Wormsk. ex DC. ■

91015 Cimicifuga foetida L. var. intermedia Regel = Cimicifuga simplex Wormsk. ex DC. ■

91016 Cimicifuga foetida L. var. longibracteata P. K. Hsiao;长苞升麻;Longbract Bugbane,Longbract Skunk Bugbane ■

91017 Cimicifuga foetida L. var. mairei (H. Lév.) W. T. Wang et Zh. Wang = Cimicifuga foetida L. ■

91018 Cimicifuga foetida L. var. racemosa Regel = Cimicifuga simplex Wormsk. ex DC. ■

91019 Cimicifuga foetida L. var. racemosa Y. Yabe = Cimicifuga simplex Wormsk. ex DC. ■

91020 Cimicifuga foetida L. var. simplex Regel = Cimicifuga simplex Wormsk. ex DC. ■

91021 Cimicifuga foetida L. var. velutina Franch. ex Finet et Gagnep.;毛叶升麻;Hair Skunk Bugbane,Velvetyleaf Bugbane ■

91022 Cimicifuga frigida Royle = Cimicifuga foetida L. ■

91023 Cimicifuga heracleifolia Kom.;大三叶升麻(地龙菜,鬼脸升麻,鸡骨升麻,窟窿牙根,苦龙菜,龙眼根,周麻,周升麻);Cowparsnipleaf Bugbane,Largetrifoliolious Bugbane ■

91024 Cimicifuga japonica (Thunb.) Spreng.;小升麻(茶七,独叶八角草,伏毛紫花小升麻,拐三七,拐枣七,黑八角莲,回龙七,金龟草,金丝三七,开喉箭,绿升麻,帽瓣七,槭叶升麻,三面刀,三叶升麻,五角连,硬毛小升麻,竹根七,紫花小升麻);Hispid Small Bugbane,Japan Bugbane,Japanese Bugbane,Purple Small Bugbane,Small Bugbane,Strigose Small Bugbane ■

91025 Cimicifuga japonica (Thunb.) Spreng. var. acerina Huth = Cimicifuga japonica (Thunb.) Spreng. ■

91026 Cimicifuga laciniata S. Watson;裂叶升麻;Cut-leaved Bugbane,Mount Hood Bugbane ■☆

91027 Cimicifuga lancifoliolata X. F. Pu et M. R. Jia = Cimicifuga brachycarpa P. K. Hsiao ■

91028 Cimicifuga macrophylla Koidz. = Cimicifuga japonica (Thunb.) Spreng. ■

91029 Cimicifuga mairei H. Lév. = Cimicifuga foetida L. var. mairei (H. Lév.) W. T. Wang et Zh. Wang ■

91030 Cimicifuga mairei H. Lév. = Cimicifuga foetida L. ■

91031 Cimicifuga mairei H. Lév. var. foliolosa (P. K. Hsiao) J. Compton et Hedd. = Cimicifuga foetida L. var. foliolosa P. K. Hsiao ■

91032 Cimicifuga nanchuenensis P. K. Hsiao;南川升麻(绿豆升麻);Nanchun Bugbane ■

91033 Cimicifuga purpurea (P. K. Hsiao) C. W. Park et H. W. Lee = Cimicifuga japonica (Thunb.) Spreng. ■

91034 Cimicifuga racemosa (L.) Nutt.;总状升麻(美类叶升麻);Black Cohosh,Black Serpentaria,Black Snakeroot,Black Snake-root,Bugbane,Bugwort,False Bugbane,Racemose Bugbane,Rattle-root,Rattle-snakeroot,Rattlesnake-root,Rattleweed,Richweed,Serpentaria,Serpentary,Snakeroot,Squawroot ■☆

91035 Cimicifuga racemosa (L.) Nutt. var. cordifolia (Pursh) A. Gray = Cimicifuga cordifolia Pursh. ■☆

91036 Cimicifuga racemosa (L.) Nutt. var. cordifolia (Pursh) A. Gray = Cimicifuga americana Michx. ■☆

91037 Cimicifuga rubifolia Kearney;红叶升麻(心叶升麻);Appalachian Bugbane ■☆

91038 Cimicifuga simplex (DC.) Turcz.;单穗升麻(大叶毛狼,毛山七,升麻,野菜升麻,野升麻);Kamchatka Bugbane ■

91039 Cimicifuga simplex Wormsk. ex DC. = Cimicifuga simplex (DC.) Turcz. ■

91040 Cimicifuga simplex Wormsk. ex DC. var. ramosa Maxim.;多枝单穗升麻(单穗升麻)■

91041 Cimicifuga ussuriensis Oett. = Cimicifuga simplex Wormsk. ex DC. ■

91042 Cimicifuga yunnanensis P. K. Hsiao;云南升麻;Yunnan Bugbane ■

91043 Cimicifugaceae Arn. = Ranunculaceae Juss.(保留科名)●■

91044 Cimicifugaceae Bromhead = Ranunculaceae Juss.(保留科名)●■

91045 Ciminalis Adans. = Gentiana L. ■

91046 Ciminalis Raf. = Leiphaimos Cham. et Schltdl. ■☆

91047 Ciminalis linearis (Froel.) Bercht. et C. Presl = Gentiana linearis Froel. ■☆

91048 Ciminalis squarrosa (Ledeb.) Zuev = Gentiana squarrosa Ledeb. ■

91049 Cinara L. = Cynara L. ■

91050 Cinara Mill. = Cynara L. ■

91051 Cinarocephalaceae Juss. = Asteraceae Bercht. et J. Presl(保留科名)●■

91052 Cinarocephalaceae Juss. = Compositae Giseke(保留科名)●■

91053 Cinarocephalae Juss. = Cynaraceae Lindl. ■

91054 Cincetoxicum taihangense (Tsiang et H. D. Zhang) C. Y. Wu et D. Z. Li = Cynanchum taihangense Tsiang et H. D. Zhang ■

91055 Cinchona L.(1753);金鸡纳属(鸡纳树属,金鸡纳树属);Cinchon,Cinchona,Cinchona Tree,Druggists' Bark,Jesuit's Bark Tree,Peru Bark,Peruvian Bark,Quinine,Quinine Tree ■●

91056 Cinchona amygdalifolia Wedd. = Cinchona calisaya Wedd. ●☆

91057 Cinchona asperifolia Wedd.;糙叶金鸡纳●☆

91058 Cinchona australis Wedd. = Cinchona calisaya Wedd. ●☆

91059 Cinchona calisaya Wedd.;黄金鸡纳(白金鸡纳,白金鸡纳树,扁桃叶金鸡纳,金鸡纳树);Bolivian Bark,Calisaya,Calysasa,Ledger Bark,Yellow Bark,Yellow-bark ●☆

91060 Cinchona calisaya Wedd. var. josephiana Wedd. = Cinchona calisaya Wedd. ●☆

91061 Cinchona calisaya Wedd. var. josephiana Wedd. = Cinchona officinalis L. ●

91062 Cinchona calisaya Wedd. var. ledgeriana Howard = Cinchona ledgeriana Bern. Moens ●

91063 Cinchona calisaya Wedd. var. ledgeriana Howard = Cinchona officinalis L. ●

91064 Cinchona carua (Wedd.) Miq. = Ladenbergia carua (Wedd.) Standl. ●☆

91065 Cinchona chomeliana Wedd. = Cinchona pubescens Vahl ●

91066 Cinchona cordifolia Mutis ex Humb.;心叶金鸡纳(黄色金鸡纳)●☆

91067 Cinchona cordifolia Mutis ex Humb. var. macrocarpa Wedd. ex Howard = Cinchona pubescens Vahl ●

91068 Cinchona excelsa Roxb. = Hymenodictyon orixense (Roxb.) Mabb. ●

91069 Cinchona gratissima Wall. = Luculia gratissima (Wall.) Sweet ●

91070 Cinchona humboldtiana Lamb.;休氏金鸡纳●☆

91071 Cinchona hybrida Sasaki;杂种金鸡纳树(杂金鸡纳)●☆

91072 Cinchona josephiana (Wedd.) Wedd. = Cinchona calisaya Wedd. ●☆

91073 Cinchona josephiana (Wedd.) Wedd. = Cinchona officinalis L. ●

91074 Cinchona lancifolia Mutis;披针叶金鸡纳;Carthagena Bark,

Colombian Bark,Lanceoleaf Cinchona ●☆

91075 Cinchona ledgeriana (Howard) Bern. Moens ex Trimen;金鸡纳树(鸡纳树,金鸡勒,金鸡纳,金鸡纳木,奎宁树,莱氏金鸡纳,累哲氏金鸡纳,狭叶金鸡纳,小叶鸡纳树,杂种金鸡纳);Cinchona Tree,Ledger Bark,Ledger Cinchona,Ledgerbark Cinchona,Medicinal Cinchona ●

91076 Cinchona ledgeriana (Howard) Bern. Moens ex Trimen = Cinchona calisaya Wedd. ●☆

91077 Cinchona ledgeriana (Howard) Bern. Moens ex Trimen = Cinchona officinalis L. ●

91078 Cinchona ledgeriana Bern. Moens = Cinchona ledgeriana (Howard) Bern. Moens ex Trimen ●

91079 Cinchona micrantha Ruiz et Pav.;小花金鸡纳(小花金鸡纳树);Littleflower Cinchona ●☆

91080 Cinchona nitida Benth. = Ladenbergia oblongifolia (Humb. ex Mutis) L. Andersson ●☆

91081 Cinchona oblongifolia Lamb.;红色金鸡纳●☆

91082 Cinchona officinalis L.;正鸡纳树(褐皮金鸡纳,药用金鸡纳,棕金鸡纳,棕色金鸡纳);Crown Bark,Jesuits' Bark,Loxa Bark,Loxabark,Medicinal Cinchona,Peruvian Bark,Quinine Tree ●

91083 Cinchona orixensis Roxb. = Hymenodictyon orixense (Roxb.) Mabb. ●

91084 Cinchona ovata Ruiz et Pav. = Cinchona pubescens Vahl ●

91085 Cinchona pitayensis Wedd.;皮塔金鸡纳(棕金鸡纳);Pitay Cinchona ●☆

91086 Cinchona pubescens Vahl;大叶鸡纳树(鸡纳树,毛金鸡纳);Peruvian Bark Tree,Pubescent Cinchona,Quinine,Red Quina,Redbark Cinchona ●

91087 Cinchona purpurascens Wedd. = Cinchona pubescens Vahl ●

91088 Cinchona robusta Hort. = Cinchona officinalis L. ●

91089 Cinchona robusta Howard;大金鸡纳(大金鸡纳树)●☆

91090 Cinchona succirubra Pav. et Klotzsch;鸡纳树(莱氏金鸡纳树,大叶金鸡纳树,红金鸡纳,红皮金鸡纳,红色规那树,红色金鸡纳树,红色奎宁树,红汁金鸡纳,金鸡勒,金鸡纳,金鸡纳树,奎宁树);Peruviana Bark Tree,Red Bark,Red Bark Tree,Red Cinchona,Red Quina,Redbark Cinchona,Red-barked Cinchona ●

91091 Cinchona succirubra Pav. ex Klotzsch = Cinchona pubescens Vahl ●

91092 Cinchonaceae Batsch = Rubiaceae Juss.(保留科名)●■

91093 Cinchonaceae Juss.;金鸡纳科●

91094 Cinchonaceae Juss. = Rubiaceae Juss.(保留科名)●■

91095 Cinchonopsis L. Andersson(1995);拟金鸡纳属●☆

91096 Cincinnobotrys Gilg(1897);卷序牡丹属■☆

91097 Cincinnobotrys acaulis (Cogn.) Gilg;无茎卷序牡丹■☆

91098 Cincinnobotrys burttianus Pócs;伯特卷序牡丹■☆

91099 Cincinnobotrys felicis (A. Chev.) Jacq.-Fél.;多育卷序牡丹■☆

91100 Cincinnobotrys letouzeyi Jacq.-Fél.;勒图卷序牡丹■☆

91101 Cincinnobotrys oreophila Gilg;喜山卷序牡丹■☆

91102 Cincinnobotrys pulchella (Brenan) Jacq.-Fél.;美丽卷序牡丹■☆

91103 Cincinnobotrys seretii De Wild. = Cincinnobotrys acaulis (Cogn.) Gilg ■☆

91104 Cincinnobotrys speciosa (A. Fern. et R. Fern.) Jacq.-Fél.;艳丽卷序牡丹■☆

91105 Cinclia Hoffmanns. = Ceropegia L. ■

91106 Cinclidocarpus Zoll. et Moritzi = Caesalpinia L. ●

91107 Cineraria L.(1763);泽菊属(瓜叶菊属,黄花瓜叶菊属);Cineraria ■●☆

91108 Cineraria abrotanifolia (L.) P. J. Bergius = Euryops abrotanifolius (L.) DC. ●☆

91109 Cineraria abyssinica Sch. Bip. ex A. Rich.;阿比西尼亚泽菊■☆

91110 Cineraria abyssinica Sch. Bip. ex A. Rich. f. rothii Oliv. et Hiern = Cineraria abyssinica Sch. Bip. ex A. Rich. ■☆

91111 Cineraria aitoniana Spreng. = Cineraria canescens J. C. Wendl. ex Link ■☆

91112 Cineraria alabamensis Britton ex Small = Packera tomentosa (Michx.) C. Jeffrey ■☆

91113 Cineraria albicans N. E. Br.;微白泽菊■☆

91114 Cineraria albomontana Hilliard = Bolandia pedunculosa (DC.) Cron ■☆

91115 Cineraria alchemilloides DC.;羽衣草泽菊■☆

91116 Cineraria alchemilloides DC. subsp. namibiensis Cron;纳米布泽菊■☆

91117 Cineraria amelloides L. = Felicia amelloides (L.) Voss ■☆

91118 Cineraria anampoza (Baker) Baker f. hygrophila (Klatt) Cron;喜水泽菊■☆

91119 Cineraria angulosa Lam.;棱角泽菊■☆

91120 Cineraria arctotidea DC. = Cineraria mollis E. Mey. ex DC. ■☆

91121 Cineraria argillacea Cron = Bolandia argillacea (Cron) Cron ■☆

91122 Cineraria aspera Thunb.;粗糙泽菊■☆

91123 Cineraria atriplicifolia DC.;暗沟泽菊■☆

91124 Cineraria austrotransvaalensis Cron;南德兰士瓦泽菊■☆

91125 Cineraria bequaertii De Wild. = Cineraria deltoidea Sond. ■☆

91126 Cineraria bergeriana Spreng. = Felicia bergeriana (Spreng.) O. Hoffm. ■☆

91127 Cineraria bracteosa O. Hoffm. ex Engl. = Cineraria deltoidea Sond. ■☆

91128 Cineraria britteniae Hutch. et R. A. Dyer = Cineraria erodioides DC. ■☆

91129 Cineraria buchananii S. Moore = Cineraria deltoidea Sond. ■☆

91130 Cineraria burkei Burtt Davy et Hutch. = Cineraria aspera Thunb. ■☆

91131 Cineraria cacalioides L. f. = Othonna carnosa Less. ■☆

91132 Cineraria canescens J. C. Wendl. ex Link;灰白泽菊■☆

91133 Cineraria canescens J. C. Wendl. ex Link var. flabelliofolia Harv.;扇叶灰白泽菊■☆

91134 Cineraria capillacea L. f. = Steirodiscus capillaceus (L. f.) Less. ■☆

91135 Cineraria chamaedrifolia Lam. = Senecio cordifolius L. f. ■☆

91136 Cineraria chinensis Spreng. = Senecio scandens Buch.-Ham. ex D. Don ■

91137 Cineraria congesta R. Br. = Tephroseris palustris (L.) Fourr. ■

91138 Cineraria coronata Thunb. = Senecio coronatus (Thunb.) Harv. ■☆

91139 Cineraria crenata Spreng. = Senecio erosus L. f. ■☆

91140 Cineraria cruenta Masson ex L'Hér. = Pericallis hybrida B. Nord. ■

91141 Cineraria cymbalarifolia Thunb. = Senecio hastifolius (L. f.) Less. ■☆

91142 Cineraria decipiens Harv.;迷惑泽菊■☆

91143 Cineraria deltoidea Sond.;三角泽菊■☆

91144 Cineraria densiflora R. E. Fr. = Cineraria deltoidea Sond. ■☆

91145 Cineraria dieterlenii E. Phillips = Cineraria erodioides DC. ■☆

91146 Cineraria discolor Sw. = Cissus discolor Blume ●

91147 Cineraria discolor Sw. = Zemisia discolor (Sw.) B. Nord. ■☆

91148 Cineraria dregeana DC. = Senecio gariepiensis Cron ■☆

91149 Cineraria eenii S. Moore = Senecio eenii (S. Moore) Merxm. ■☆
91150 Cineraria erodioides DC. ;牻牛儿苗泽菊■☆
91151 Cineraria erodioides DC. var. tomentosa Cron;绒毛泽菊■☆
91152 Cineraria erosa (Thunb.) Harv. ;啮蚀状泽菊■☆
91153 Cineraria exilis DC. ;瘦小泽菊■☆
91154 Cineraria filifolia Thunb. = Senecio linifolius L. ■☆
91155 Cineraria fischeri Ledeb. = Ligularia fischeri (Ledeb.) Turcz. ■
91156 Cineraria foliosa O. Hoffm. ;富叶泽菊■☆
91157 Cineraria frigida Richardson = Tephroseris frigida (Richardson) Holub ■☆
91158 Cineraria frigida Richardson f. tomentosa Kjellm. = Tephroseris kjellmanii (A. E. Porsild) Holub ■☆
91159 Cineraria frigida Richardson var. robusta Herder = Tephroseris turczaninowii (DC.) Holub ■
91160 Cineraria glandulosa Cron;具腺泽菊■☆
91161 Cineraria gracilis O. Hoffm. ;纤细泽菊■☆
91162 Cineraria grandibracteata Hilliard;大苞泽菊■☆
91163 Cineraria grandiflora Vatke = Cineraria deltoidea Sond. ■☆
91164 Cineraria hamiltonii S. Moore = Cineraria aspera Thunb. ■☆
91165 Cineraria hastifolia L. f. = Senecio hastifolius (L. f.) Less. ■☆
91166 Cineraria hastifolia Thunb. = Senecio hastifolius (L. f.) Less. ■☆
91167 Cineraria hederifolia Cron = Senecio hederiformis Cron ■☆
91168 Cineraria hirsuta Vent. = Felicia cymbalariae (Aiton) Bolus et Wolley-Dod ex Adamson et T. M. Salter ■☆
91169 Cineraria humifusa L'Hér. ;平伏泽菊■☆
91170 Cineraria hybrida Willd. = Pericallis hybrida B. Nord. ■
91171 Cineraria hygrophila (Klatt) Klatt = Cineraria anampoza (Baker) Baker f. hygrophila (Klatt) Cron ■☆
91172 Cineraria hypoleuca Sieber = Senecio verbascifolius Burm. f. ■☆
91173 Cineraria integrifolia Jacq. ex Willd. var. minor Pursh = Packera tomentosa (Michx.) C. Jeffrey ■☆
91174 Cineraria kilimandscharica Engl. = Cineraria deltoidea Sond. ■☆
91175 Cineraria laricifolia Lam. = Senecio pinifolius (L.) Lam. ■☆
91176 Cineraria laxiflora R. E. Fr. = Cineraria deltoidea Sond. ■☆
91177 Cineraria lineata L. f. = Senecio lineatus (L. f.) DC. ■☆
91178 Cineraria linifolia (L.) L. = Euryops linifolius (L.) DC. ■☆
91179 Cineraria linifolia Zeyh. ex DC. = Felicia linifolia (Harv.) Grau ■☆
91180 Cineraria lobata L'Hér. ;浅裂泽菊■☆
91181 Cineraria lobata L'Hér. subsp. lasiocaulis Cron;毛茎浅裂泽菊■☆
91182 Cineraria lobata L'Hér. subsp. platyptera Cron;宽翅浅裂泽菊■☆
91183 Cineraria lobata L'Hér. subsp. soutpansbergensis Cron;索特潘泽菊■☆
91184 Cineraria longipes S. Moore;长梗泽菊■☆
91185 Cineraria lyrata DC. = Cineraria lyratiformis Cron ■☆
91186 Cineraria lyrata Ledeb. = Packera cymbalaria (Pursh) Á. Löve et D. Löve ■
91187 Cineraria lyratiformis Cron;大头羽裂泽菊■☆
91188 Cineraria lyratipartita (Sch. Bip. ex A. Rich.) Cufod. = Senecio lyratus Forssk. ■☆
91189 Cineraria macrophylla Ledeb. = Ligularia macrophylla (Ledeb.) DC. ■
91190 Cineraria magnicephala Cron;大头泽菊■☆
91191 Cineraria maritima (L.) L. = Senecio bicolor (Willd.) Tod. subsp. cineraria (DC.) Chater ■☆
91192 Cineraria mazoensis S. Moore;马索泽菊■☆
91193 Cineraria mazoensis S. Moore var. graniticola Cron;花岗岩泽菊■☆
91194 Cineraria microglossa DC. = Mesogramma apiifolium DC. ■☆
91195 Cineraria microphylla (Cass.) Vahl ex DC. = Felicia aethiopica (Burm. f.) Bolus et Wolley-Dod ex Adamson et T. M. Salter ■☆
91196 Cineraria mitellifolia L'Hér. = Senecio cordifolius L. f. ■☆
91197 Cineraria mollis E. Mey. ex DC. ;柔软泽菊■☆
91198 Cineraria mongolica Turcz. = Ligularia mongolica (Turcz.) DC. ■
91199 Cineraria montana Bolus = Senecio haygarthii Hilliard ■☆
91200 Cineraria monticola Hutch. = Cineraria deltoidea Sond. ■☆
91201 Cineraria ngwenyensis Cron;恩圭尼泽菊■☆
91202 Cineraria oppositifolia Moench = Felicia amelloides (L.) Voss ■☆
91203 Cineraria othonnoides Harv. ;厚敦菊状泽菊■☆
91204 Cineraria oxyodonta DC. = Cineraria erosa (Thunb.) Harv. ■☆
91205 Cineraria palmensis Nees = Bethencourtia palmensis (Nees) Link ■☆
91206 Cineraria palustris (L.) L. = Tephroseris palustris (L.) Fourr. ■
91207 Cineraria pandurata Thunb. = Senecio panduratus (Thunb.) Less. ■☆
91208 Cineraria parviflora Aiton = Cineraria canescens J. C. Wendl. ex Link ■☆
91209 Cineraria parvifolia Burtt Davy;小叶泽菊■☆
91210 Cineraria pedunculosa DC. = Bolandia pedunculosa (DC.) Cron ■☆
91211 Cineraria pinnata O. Hoffm. ex Schinz;羽状泽菊■☆
91212 Cineraria platycarpa DC. ;宽果泽菊■☆
91213 Cineraria polycephala DC. = Cineraria erodioides DC. ■☆
91214 Cineraria polyglossa DC. = Cineraria mollis E. Mey. ex DC. ■☆
91215 Cineraria pratensis Herder var. borealis Herder = Tephroseris subdentata (Bunge) Holub ■
91216 Cineraria prittwitzii O. Hoffm. ex Engl. = Cineraria deltoidea Sond. ■☆
91217 Cineraria pulchra Cron;美丽泽菊■☆
91218 Cineraria repanda Lour. = Senecio scandens Buch. -Ham. ex D. Don ■
91219 Cineraria salicifolia Kunth = Barkleyanthus salicifolius (Kunth) H. Rob. et Brettell ■☆
91220 Cineraria saxifraga DC. ;虎耳草泽菊■☆
91221 Cineraria saxifraga DC. var. axillipila? = Cineraria saxifraga DC. ■☆
91222 Cineraria schimperi Sch. Bip. ex Oliv. et Hiern = Senecio lyratus Forssk. ■☆
91223 Cineraria sebaldii Cufod. = Cineraria abyssinica Sch. Bip. ex A. Rich. ■☆
91224 Cineraria seminuda Klatt = Senecio cinerascens Aiton ■☆
91225 Cineraria sibirica L. = Ligularia sibirica (L.) Cass. ■
91226 Cineraria speciosa Schrad. ex Link. = Ligularia fischeri (Ledeb.) Turcz. ■
91227 Cineraria spinulosa Lam. = Othonna parviflora P. J. Bergius ●☆
91228 Cineraria subdentata Bunge = Tephroseris subdentata (Bunge) Holub ■
91229 Cineraria tenella (L.) Link = Felicia tenella (L.) Nees ●☆
91230 Cineraria thyrsoidea Ledeb. = Ligularia thyrsoidea (Ledeb.) DC. ■
91231 Cineraria tussilaginea Thunb. = Cineraria erodioides DC. ■☆
91232 Cineraria vagans Hilliard;漫游泽菊■☆
91233 Cinga Noronha = Cyrtandra J. R. Forst. et G. Forst. ●■
91234 Cinhona L. = Cinchona L. ■●
91235 Cinna L. (1753);单蕊草属;Wood Reed,Woodreed ■
91236 Cinna arundinacea L. ;苇状单蕊草;Common Wood-reed, Reed

Cinna, Stout Woodreed, Stout Wood-reed, Sweet Wood-reed, Wood Reed, Wood Reed Grass ■

91237 Cinna arundinacea L. var. inexpansa Fernald et Griscom = Cinna arundinacea L. ■

91238 Cinna filiformis Llanos = Pogonatherum crinitum (Thunb.) Kunth ■

91239 Cinna latifolia (Trevis.) Griseb.;单蕊草;Drooping Woodreed, Drooping Wood-reed, Wideleaved Cinna ■

91240 Cinna mexicana P. Beauv.;墨西哥单蕊草■☆

91241 Cinnabarinea F. Ritter = Echinopsis Zucc. ●

91242 Cinnabarinea Frič = Lobivia Britton et Rose ■

91243 Cinnabarinea Frič ex F. Ritter = Echinopsis Zucc. ●

91244 Cinnadenia Kosterm. (1973);圆锥樟属(喜马拉雅樟属)●☆

91245 Cinnadenia paniculata (Hook. f.) Kosterm.;圆锥樟●☆

91246 Cinnadenia paniculata (Hook. f.) Kosterm. = Litsea liyuyingi H. Liu ●

91247 Cinnagrostis Griseb. = Calamagrostis Adans. ■

91248 Cinnagrostis Griseb. = Cinna L. ■

91249 Cinnamodendron Endl. (1840);多蕊樟属(桂枝树属,辣树属,南美樟属)●

91250 Cinnamodendron corticosum Miers;多蕊樟(厚皮树桂枝,南美樟);False Winter's Bark, Mountain Cinnamon, Red Canella ●☆

91251 Cinnamomum Blume = Cinnamomum Schaeff. (保留属名)●

91252 Cinnamomum Schaeff. (1760)(保留属名);樟属;Camphor Tree, Cassia, Cinnamon ●

91253 Cinnamomum Spreng. = Cinnamomum Schaeff. (保留属名)●

91254 Cinnamomum Trew. = Cinnamomum Schaeff. (保留属名)●

91255 Cinnamomum × durifruticeticola Hatus.;灌丛樟●☆

91256 Cinnamomum × takushii Hatus.;卓志樟●☆

91257 Cinnamomum acuminatifolium Hayata = Cinnamomum japonicum Siebold ex Nees ●

91258 Cinnamomum acuminatissimum Hayata = Cinnamomum philippinense (Merr.) C. E. Chang ●

91259 Cinnamomum acuminatissimum Hayata = Machilus philippinense Merr. ●

91260 Cinnamomum albiflorum Hook. f. et Thomson ex Meisn. = Cinnamomum impressinervium Meisn. ●☆

91261 Cinnamomum albiflorum Nees = Cinnamomum tamala (Buch.-Ham.) T. Nees et C. H. Eberm. ●

91262 Cinnamomum albiflorum Nees var. kwangtungense H. Liu = Cinnamomum subavenium Miq. ●

91263 Cinnamomum albiflorum Nees var. tonkinensis Lecomte = Cinnamomum tonkinense (Lecomte) A. Chev. ●

91264 Cinnamomum albosericeum (Gamble) W. C. Cheng = Cinnamomum septentrionale Hand.-Mazz. ●

91265 Cinnamomum appelianum Schewe;毛桂(阿氏樟,柴桂,假桂皮,三条筋,山桂皮,山桂枝,山沾树,土肉桂,香桂子,香沾树);Hair Twig Cassia, Hairy Cinnamon, Hairy-twigged Cinnamon ●

91266 Cinnamomum appelianum Schewe var. tripartitum Yen C. Yang = Cinnamomum appelianum Schewe ●

91267 Cinnamomum argenteum Gamble = Cinnamomum mairei H. Lév. ●◇

91268 Cinnamomum argenteum H. Liu = Cinnamomum subavenium Miq. ●

91269 Cinnamomum aromaticum Nees = Cinnamomum cassia C. Presl ●

91270 Cinnamomum austrosinense Hung T. Chang;华南桂(大叶辣樟树,大叶樟,华南樟,牡丹叶桂皮,肉桂,野桂皮);S. China Cinnamon, South China Cinnamon ●

91271 Cinnamomum austroyunnanense H. W. Li;滇南桂(假桂枝,野肉桂);S. Yunnan Cinnamon, South-Yunnan Cassia ●

91272 Cinnamomum barbato-axillatum N. Chao;髯毛樟;Barbato-axillate Cinnamon ●

91273 Cinnamomum barbatoaxillatum N. Chao = Cinnamomum parthenoxylum (Jack) Meisn. ●

91274 Cinnamomum barbatoaxillatum N. Chao = Cinnamomum porrectum (Roxb.) Kosterm. ●

91275 Cinnamomum bartheifolium Hayata = Cinnamomum subavenium Miq. ●

91276 Cinnamomum bejolghota (Buch.-Ham.) Sweet;钝叶桂(大叶山桂,钝叶樟,奉楠,桂皮,假桂皮,老母楠,老母猪桂皮,泡木,青樟木,三条筋,山桂,山桂楠,山肉桂,山玉桂,土桂皮,土肉桂,香桂楠,小黏药,鸭母桂);Obtuseleaf Cassia, Obtuseleaf Cinnamon, Obtuse-leaved Cinnamon, Wild Cinnamon ●

91277 Cinnamomum bodinieri H. Lév.;猴樟(大胡椒树,猴夹木,猴楝木,楠木,牛筋条,牛荆树,香树,香樟,樟树);Bodinier Cinnamon, Monkey Cinnamon ●

91278 Cinnamomum bodinieri H. Lév. var. hupehanum (Gamble) G. F. Tao = Cinnamomum bodinieri H. Lév. ●

91279 Cinnamomum brachythyrsum J. Li;短序樟;Short-thyrse Cinnamon ●

91280 Cinnamomum brevifolium Miq.;小叶樟;Small-leaved Cinnamon ●☆

91281 Cinnamomum brevipedunculatum C. E. Chang = Cinnamomum rigidissimum Hung T. Chang ●◇

91282 Cinnamomum burmannii (Nees et T. Nees) Blume;阴香(阿尼茶,八角,炳继树,大叶樟,广东桂皮,桂树,桂秧,假桂树,假桂枝,坎香草,连粘树,山桂,山肉桂,山玉桂,土肉桂,土山肉桂,香柴,香桂,香胶叶,小桂皮,野桂树,野玉桂树,野樟树,阴草);Batavia Cinnamon, Burmann Cinnamon, Burmann's Cassia, Java Cinnamon, Padang Cassia, Padang Cinnamon ●

91283 Cinnamomum burmannii (Nees et T. Nees) Blume f. heyneanum (Nees) H. W. Li = Cinnamomum heyneanum Nees ●

91284 Cinnamomum burmannii (Nees et T. Nees) Blume var. angustifolium (Hemsl.) C. K. Allen = Cinnamomum burmannii (Nees et T. Nees) Blume f. heyneanum (Nees) H. W. Li ●

91285 Cinnamomum calcareum Y. K. Li;石山桂;Limy Cinnamon ●

91286 Cinnamomum calcareum Y. K. Li = Cinnamomum pauciflorum Nees ●

91287 Cinnamomum camphora (L.) J. Presl;樟树(臭樟,吹风散,大木姜子,东部樟,东南部樟,芳樟,甲沉香,梼樟,栳树,栳樟,山沉香,山臭樟,山乌樟,土沉香,乌樟,香蕊,香蕊木,香通,香樟,香樟木,小叶樟,瑶人柴,油樟,豫章,樟,樟公,樟木,樟脑树,走马胎);Camphor, Camphor Laurel, Camphor Tree, Camphor Wood, Camphora-tree, Camphortree, Camphor-tree, Fragrant Cinnamon, Ho Wood, Japanese Camphor, True Camphortree ●

91288 Cinnamomum camphora (L.) J. Presl = Cinnamomum camphora (L.) T. Nees et C. H. Eberm. ●

91289 Cinnamomum camphora (L.) J. Presl f. linaloolifera (Y. Fujita) Sugim.;滕田樟●☆

91290 Cinnamomum camphora (L.) J. Presl var. nominale Hayata subvar. hosyo Hatus. = Cinnamomum camphora (L.) J. Presl f. linaloolifera (Y. Fujita) Sugim. ●☆

91291 Cinnamomum camphora (L.) Siebold = Cinnamomum camphora (L.) T. Nees et C. H. Eberm. ●

91292 Cinnamomum camphora (L.) T. Nees et C. H. Eberm. = Cinnamomum camphora (L.) J. Presl ●

91293 Cinnamomum camphora (L.) T. Nees et C. H. Eberm. var. glaucescens (Hans Braun) Meisn. = Cinnamomum camphora (L.) T. Nees et C. H. Eberm. ●

91294 Cinnamomum camphora (L.) T. Nees et C. H. Eberm. var. glaucescens (Hans Braun) Meisn. = Cinnamomum camphora (L.) T. Nees et C. H. Eberm. var. nominale Hayata ex Matsum. et Hayata ●

91295 Cinnamomum camphora (L.) T. Nees et C. H. Eberm. var. glaucescens Nakai;芳樟●

91296 Cinnamomum camphora (L.) T. Nees et C. H. Eberm. var. lanata Nakai = Cinnamomum camphora (L.) T. Nees et C. H. Eberm. var. nominale Hayata ex Matsum. et Hayata ●

91297 Cinnamomum camphora (L.) T. Nees et C. H. Eberm. var. nominale Hayata ex Matsum. et Hayata;栳樟●

91298 Cinnamomum camphora (L.) T. Nees et C. H. Eberm. var. nominale Hayata ex Matsum. et Hayata = Cinnamomum camphora (L.) T. Nees et C. H. Eberm. ●

91299 Cinnamomum camphoroides Hayata = Cinnamomum camphora (L.) T. Nees et C. H. Eberm. ●

91300 Cinnamomum cassia (L.) D. Don subsp. pseudomelastoma J. C. Liao, Y. L. Kuo et C. C. Lin = Cinnamomum austrosinense Hung T. Chang ●

91301 Cinnamomum cassia (Nees) Blume = Cinnamomum tamala (Buch. -Ham.) T. Nees et C. H. Eberm. ●

91302 Cinnamomum cassia C. Presl;肉桂(大桂,桂,桂木,桂皮,桂枝,菌桂,糠桂,辣桂,柳桂,牡桂,木桂,梫,筒桂,玉桂,紫桂);Bastard Cinnamon, Canton Cassia, Cassia, Cassia Bark, Cassia Bark Tree, Cassia Buds, Cassia Cinnamon, Cassia-bark Tree, Cassiabarktree, Cassia-bark-tree, Chinese Cinnamon, Cinnamon-tree ●

91303 Cinnamomum cassia C. Presl subsp. pseudomelastoma J. C. Liao, Chin C. Kuo et C. C. Lin = Cinnamomum austrosinense Hung T. Chang ●

91304 Cinnamomum cassia Lour. = Cinnamomum cassia C. Presl ●

91305 Cinnamomum caudatifolium Hayata = Cinnamomum philippinense (Merr.) C. E. Chang ●

91306 Cinnamomum caudatum Nees = Neocinnamomum caudatum (Nees) Merr. ●

91307 Cinnamomum caudiferum Kosterm. = Cinnamomum foveolatum (Merr.) H. W. Li et J. Li ●

91308 Cinnamomum cavaleriei H. Lév. = Cinnamomum glanduliferum (Wall.) Nees ●

91309 Cinnamomum chartophyllum H. W. Li;坚叶樟(坚叶桂);Chartaceous-leaf Cinnamon, Chartaceous-leaved Cinnamon, Hardleaf Cinnamon ●

91310 Cinnamomum chekiangense Nakai = Cinnamomum japonicum Siebold ex Nees ●

91311 Cinnamomum chengkouense N. Chao = Cinnamomum platyphyllum (Diels) C. K. Allen ●◇

91312 Cinnamomum chenii Nakai = Cinnamomum japonicum Siebold ex Nees ●

91313 Cinnamomum chinense Blume = Cinnamomum burmannii (Nees et T. Nees) Blume ●

91314 Cinnamomum chingii F. P. Metcalf = Cinnamomum austrosinense Hung T. Chang ●

91315 Cinnamomum chingii F. P. Metcalf = Cinnamomum subavenium Miq. ●

91316 Cinnamomum contractum H. W. Li;聚花桂(柴桂,柴树,桂树);Contract Cinnamon, Contract Flower Cassia, Contract-flowered Cinnamon ●

91317 Cinnamomum culilawan Blume;柯樟;Culilawan ●

91318 Cinnamomum daphnoides Siebold et Zucc.;海岸樟●☆

91319 Cinnamomum delavayi Lecomte = Neocinnamomum delavayi (Lecomte) H. Liu ●

91320 Cinnamomum delavayi Lecomte var. aromaticum Lecomte ex S. K. Lee = Neocinnamomum mekongense (Hand. -Mazz.) Kosterm. ●

91321 Cinnamomum delavayi Lecomte var. mekongense Hand. -Mazz. = Neocinnamomum mekongense (Hand. -Mazz.) Kosterm. ●

91322 Cinnamomum doederleinii Engl.;德德莱因樟●☆

91323 Cinnamomum doederleinii Engl. var. pseudodaphnoides Hatus.;假海岸樟●☆

91324 Cinnamomum dulce (Roxb.) Sweet = Cinnamomum burmannii (Nees et T. Nees) Blume ●

91325 Cinnamomum esquirolii H. Lév. = Cocculus laurifolius DC. ●

91326 Cinnamomum fargesii Lecomte = Neocinnamomum fargesii (Lecomte) Kosterm. ●

91327 Cinnamomum foveolatum (Merr.) H. W. Li et J. Li;尾叶樟;Caudateleaf Cinnamon, Caudate-leaved Cinnamon ●

91328 Cinnamomum glanduliferum (Wall.) Meisn. = Cinnamomum porrectum (Roxb.) Kosterm. ●

91329 Cinnamomum glanduliferum (Wall.) Meisn. var. longipaniculatum Lecomte = Cinnamomum bodinieri H. Lév. ●

91330 Cinnamomum glanduliferum (Wall.) Nees;云南樟(白樟,臭樟,大黑叶樟,滇香樟,果东樟,红樟,青皮树,香叶树,香叶樟,香樟,樟木,樟脑树,樟叶树);Indian Camphor Tree, Indian Camphortree, Nepal Camphor Tree, Nepal Camphortree, Nepal Camphor-tree, Nepal Cinnamon, Nepal Sassafras ●

91331 Cinnamomum glanduliferum (Wall.) Nees var. longipaniculata Lecomte = Cinnamomum bodinieri H. Lév. ●

91332 Cinnamomum hainanense Nakai = Cinnamomum burmannii (Nees et T. Nees) Blume ●

91333 Cinnamomum heyneanum Nees;狭叶桂(大舒筋活血,三股筋,顺江木,狭叶天竺桂,狭叶阴香,狭叶樟);Korintji Cinnamon, Narrowleaf Batavia Cinnamon, Narrowleaf Burmann Cinnamon ●

91334 Cinnamomum heyneanum Nees = Cinnamomum burmannii (Nees et T. Nees) Blume f. heyneanum (Nees) H. W. Li ●

91335 Cinnamomum hiananense Nakai = Cinnamomum burmannii (Nees et T. Nees) Blume ●

91336 Cinnamomum hupehanum Gamble = Cinnamomum bodinieri H. Lév. ●

91337 Cinnamomum hurmannii (Nees) H. W. Li = Cinnamomum burmannii (Nees et T. Nees) Blume f. heyneanum (Nees) H. W. Li ●

91338 Cinnamomum ilicioides A. Chev.;八角樟;Anisetree-like Cinnamon ●

91339 Cinnamomum impressinervium Meisn.;陷脉樟●☆

91340 Cinnamomum iners Reinw. = Cinnamomum bejolghota (Buch. -Ham.) Sweet ●

91341 Cinnamomum iners Reinw. ex Blume;大叶桂(大叶樟,假桂皮);Bigleaf Cinnamon, Big-leaved Cinnamon ●

91342 Cinnamomum insularimontanum Hayata = Cinnamomum japonicum Siebold ex Nees ●

91343 Cinnamomum inunctum (Nees) Meisn. = Cinnamomum longepaniculatum (Gamble) N. Chao ex H. W. Li ●◇

91344 Cinnamomum inunctum (Nees) Meisn. var. albosericeum Gamble

= Cinnamomum septentrionale Hand. -Mazz. ●

91345 Cinnamomum inunctum (Nees) Meisn. var. fulvipilosum Yen C. Yang = Cinnamomum bodinieri H. Lév. ●

91346 Cinnamomum inunctum Meisn. var. fulvipilosa Yen C. Yang = Cinnamomum bodinieri H. Lév. ●

91347 Cinnamomum inunctum Meisn. var. longepaniculatum Gamble = Cinnamomum longepaniculatum (Gamble) N. Chao ex H. W. Li ●◇

91348 Cinnamomum japonicum Siebold ex Nakai = Cinnamomum tenuifolium (Makino) Sugim. ex H. Hara ●☆

91349 Cinnamomum japonicum Siebold ex Nakai f. pilosum Hatus.; 毛天竺桂●☆

91350 Cinnamomum japonicum Siebold ex Nakai f. tenuifolium (Makino) Sugim. = Malva sylvestris L. ■

91351 Cinnamomum japonicum Siebold ex Nees; 天竺桂(阿里美,大叶天竺桂,假桂,日本香桂,肉桂,山桂,山桂皮,山肉桂,山玉桂,台湾肉桂,土桂,土肉桂,月桂,浙桂,浙江桂,竺香); Japan Cinnamon, Japanese Cassia, Japanese Cinnamon, Mountain Cinnamon Tree, Taiwan Cinnamon, Yabunikki ●

91352 Cinnamomum japonicum Siebold ex Nees var. chekiangense (Nakai) M. B. Deng et G. Yao = Cinnamomum japonicum Siebold ex Nees ●

91353 Cinnamomum javanicum Blume; 爪哇肉桂; Java Cassia, Java Cinnamon ●

91354 Cinnamomum jensenianum Hand. -Mazz.; 野黄桂(桂皮树,三条筋树,山肉桂,稀花樟); Jensen Cinnamon ●

91355 Cinnamomum kanehirae Hayata = Cinnamomum micranthum (Hayata) Hayata ●

91356 Cinnamomum kiamis Nees = Cinnamomum burmannii (Nees et T. Nees) Blume ●

91357 Cinnamomum kotoense Kaneh. et Sasaki; 兰屿肉桂; Botel Tobago Cinnamon Tree, Lanyu Cinnamon ●

91358 Cinnamomum kwangtungensis Merr.; 红辣槁树(红叶辣汁树); Guangdong Cinnamon ●

91359 Cinnamomum liangii C. K. Allen; 软皮桂(软皮樟,向日樟); Liang Cinnamon ●

91360 Cinnamomum linearifolium Lecomte = Cinnamomum burmannii (Nees et T. Nees) Blume f. heyneanum (Nees) H. W. Li ●

91361 Cinnamomum linearifolium Lecomte = Cinnamomum heyneanum Nees ●

91362 Cinnamomum lioui C. K. Allen = Cinnamomum subavenium Miq. ●

91363 Cinnamomum longepaniculatum (Gamble) N. Chao ex H. W. Li; 油樟(黄葛树,香叶子树,香樟,雅樟,樟木); Longpaniculate Cinnamon, Long-paniculated Cinnamon, Oil Cinnamon, Polished Cinnamon ●◇

91364 Cinnamomum longicarpum Kaneh. = Cinnamomum subavenium Miq. ●

91365 Cinnamomum longipetiolatum H. W. Li; 长柄樟; Longpediole Cinnamon, Long-pedioled Cinnamon ●

91366 Cinnamomum loureirii Nees; 清化肉桂(桂,桂仔,留氏桂,牡桂,梫,肉桂,越南肉桂); Cassia-flower-tree, Saigon Cinnamon, Smgon Cinnamon, Vietnam Cassia ●

91367 Cinnamomum macranthum (Hayata) Hayata f. kanehirai (Hayata) S. S. Ying = Cinnamomum kanehirae Hayata ●

91368 Cinnamomum macranthum Hayata; 牛樟; Bigflowered Cinnamon ●

91369 Cinnamomum macrostemon Hayata; 胡氏肉桂●

91370 Cinnamomum macrostemon Hayata var. pseudoloureirii (Hayata) Gogelein = Cinnamomum japonicum Siebold ex Nees ●

91371 Cinnamomum macrostemon Hayata var. pseudoloureiroi (Hayata) Yamamoto = Cinnamomum japonicum Siebold ex Nees ●

91372 Cinnamomum mairei H. Lév.; 银叶桂(川桂,川桂皮,关桂,官桂皮,桂皮,桂皮树,山桂皮,银叶樟,樟桂); Silveryleaf Cassia, Silveryleaf Cinnamon, Silvery-leaved Cinnamon ●◇

91373 Cinnamomum marlipoense H. W. Li = Alseodaphne marlipoensis (H. W. Li) H. W. Li ●

91374 Cinnamomum massoia Schewe; 马索桂; Masso Cinnamon ●☆

91375 Cinnamomum merrillianum C. K. Allen = Cinnamomum tsangii Merr. ●

91376 Cinnamomum miaoshanense S. K. Lee et F. N. Wei; 苗山桂; Miaoshan Cassia ●

91377 Cinnamomum miaoshanense S. K. Lee et F. N. Wei = Cinnamomum burmannii (Nees et T. Nees) Blume ●

91378 Cinnamomum micranthum (Hayata) Hayata; 沉水樟(臭樟,黄樟树,冇樟,牛樟,水樟,樟牛); Smallflower Camphor Tree, Smallflower Cinnamon, Small-flowered Cinnamon, Stout Camphor Tree ●

91379 Cinnamomum micranthum (Hayata) Hayata f. kanehirae (Hayata) S. S. Ying = Cinnamomum micranthum (Hayata) Hayata ●

91380 Cinnamomum migao H. W. Li; 米槁(大果樟,麻槁); Migao Cinnamon ●

91381 Cinnamomum mollifolium H. W. Li; 毛叶樟(罗木来,罗木束,毛叶芳樟,平叶樟,香茅樟,中俄,中朗,中朗俄,中沙海); Hairyleaf Cinnamon, Hairy-leaved Cinnamon ●◇

91382 Cinnamomum mollissimum Hook. f.; 极软樟●☆

91383 Cinnamomum myrianthum Merr. = Cinnamomum kotoense Kaneh. et Sasaki ●

91384 Cinnamomum nominale (Hayata ex Matsum. et Hayata) Hayata = Cinnamomum camphora (L.) T. Nees et C. H. Eberm. var. nominale Hayata ex Matsum. et Hayata ●

91385 Cinnamomum nominale (Hayata) Hayata = Cinnamomum camphora (L.) T. Nees et C. H. Eberm. ●

91386 Cinnamomum obovatifolium Hayata = Machilus obovatifolia (Hayata) Kaneh. et Sasaki ●

91387 Cinnamomum obtusifolium (Roxb.) Nees = Cinnamomum bejolghota (Buch. -Ham.) Sweet ●

91388 Cinnamomum oliveri F. M. Bailey; 昆士兰樟(檫皮桂); Australian Cinnamon, Oliver Bark, Queensland Camphor ●☆

91389 Cinnamomum osmophloeum Kaneh.; 土肉桂(假肉桂,山肉桂); Indigenous Cinnamon Tree, Native Cassia Bark Tree, Odour Cassia-bark, Odour-bark Cinnamon ●

91390 Cinnamomum ovatum C. K. Allen = Cinnamomum rigidissimum Hung T. Chang ●◇

91391 Cinnamomum parthenoxylon Nees = Cinnamomum ilicioides A. Chev. ●

91392 Cinnamomum parthenoxylum (Jack) Meisn.; 黄樟(冰片树,臭樟,大叶樟,黄槁,假樟,梅崇,南安,蒲香树,山椒,香桂,香喉,香湖,香樟,油樟,樟木,樟木树,樟脑树,中折旺); Fragrant Cinnamon, Yellow Cinnamon ●

91393 Cinnamomum parthenoxylum (Jack) Meisn. = Cinnamomum porrectum (Roxb.) Kosterm. ●

91394 Cinnamomum parthenoxylum (Jack) Nees = Cinnamomum porrectum (Roxb.) Kosterm. ●

91395 Cinnamomum parvifolium Lecomte = Neocinnamomum delavayi (Lecomte) H. Liu ●

91396 Cinnamomum pauciflorum Hung T. Chang = Cinnamomum jensenianum Hand. -Mazz. ●

91397 Cinnamomum pauciflorum Nees;少花桂(臭乌桂,臭樟,三条筋,土桂皮,香桂,香叶子树,岩桂);Fewflower Cinnamon, Few-flowered Cinnamon, Rock-living Cinnamon ●

91398 Cinnamomum pedatinovium Meisn.;野桂●☆

91399 Cinnamomum pedunculatum Nees = Cinnamomum japonicum Siebold ex Nees ●

91400 Cinnamomum pedunculatum Nees Hayata = Cinnamomum japonicum Siebold ex Nees ●

91401 Cinnamomum pedunculatum Nees var. angustifolium Hemsl. = Cinnamomum heyneanum Nees ●

91402 Cinnamomum pedunculatum Nees var. angustifolium Hemsl. = Cinnamomum burmannii (Nees et T. Nees) Blume f. heyneanum (Nees) H. W. Li ●

91403 Cinnamomum pedunculatum Nees var. nervosum Meisn. = Cinnamomum tenuifolium (Makino) Sugim. ex H. Hara f. nervosum (Meisn.) H. Hara ●☆

91404 Cinnamomum petrophilum N. Chao = Cinnamomum pauciflorum Nees ●

91405 Cinnamomum philippinense (Merr.) C. E. Chang;菲律宾樟树(菲律宾楠);Philippine Cinnamon ●

91406 Cinnamomum philippinense (Merr.) C. E. Chang = Machilus philippinense Merr. ●

91407 Cinnamomum pingbienense H. W. Li;屏边桂;Pingbian Cinnamon ●

91408 Cinnamomum pittosporoides Hand.-Mazz.;刀把木(大果香樟,桂皮树);Pittosporum-like Cinnamon ●

91409 Cinnamomum platyphyllum (Diels) C. K. Allen;阔叶樟(城口樟);Broadleaf Cinnamon, Broad-leaved Cinnamon, Chengkou Cinnamon ●◇

91410 Cinnamomum porrectum (Roxb.) Kosterm. = Cinnamomum parthenoxylum (Jack) Meisn. ●

91411 Cinnamomum porrectum Roxb. = Cinnamomum parthenoxylum (Jack) Meisn. ●

91412 Cinnamomum pruciflorum Chun ex C. C. Chang;稀花樟树;Loose-flower Cinnamon ●

91413 Cinnamomum pseudoloureiroi Hayata = Cinnamomum japonicum Siebold ex Nees ●

91414 Cinnamomum pseudomelastoma (J. C. Liao, Y. L. Kuo et C. C. Lin) J. C. Liao = Cinnamomum austrosinense Hung T. Chang ●

91415 Cinnamomum pseudopedunculatum Hayata = Cinnamomum tenuifolium (Makino) Sugim. ex H. Hara f. nervosum (Meisn.) H. Hara ●☆

91416 Cinnamomum purpureum H. G. Ye et F. G. Wang = Cinnamomum parthenoxylum (Jack) Meisn. ●

91417 Cinnamomum randaiense Hayata = Cinnamomum subavenium Miq. ●

91418 Cinnamomum recurvatum (Roxb.) Wight = Cinnamomum pauciflorum Nees ●

91419 Cinnamomum recurvatum Roxb. = Cinnamomum pauciflorum Nees ●

91420 Cinnamomum reticulatum Hayata;网脉桂(土樟,香桂);Netvein Cinnamon, Reticular-veined Cinnamon Tree, Reticulate-veined Camphortree, Reticulate-veined Cinnamon Tree, Taiwan Camphor Tree ●

91421 Cinnamomum rigidissimum Hung T. Chang;卵叶桂(卵叶樟,小叶樟,硬叶桂,硬叶樟);Ovateleaf Cinnamon, Ovate-leaved Cinnamon, Small-leaf Cinnamon Tree ●◇

91422 Cinnamomum rufotomentosum K. M. Lan;绒毛樟;Redish-tomentose Cinnamon, Red-tomentose Cinnamon ●

91423 Cinnamomum saigonicum Farw.;西贡肉桂●☆

91424 Cinnamomum saxatile H. W. Li;岩樟(沟厚,楞蚬,米槁,米瓜,香楠);Rock Cinnamon, Rockdwelling Cinnamon, Rockd-welling Cinnamon ●

91425 Cinnamomum septentrionale Hand.-Mazz.;银木(土沉香,香棍子,香樟,樟树);North Cinnamon, Silverwood Cinnamon ●

91426 Cinnamomum sieboldii Meisn.;西氏肉桂(西博氏肉桂)●☆

91427 Cinnamomum simondii Lecomte = Cinnamomum camphora (L.) T. Nees et C. H. Eberm. ●

91428 Cinnamomum sintok Blume;辛脱克樟●☆

91429 Cinnamomum subavenium Miq.;香桂(长果桂,假桂皮,九芎舅,峦大桂,峦大山樟,三条筋,山肉桂,土肉桂,细叶香桂,细叶月桂,香槁树,香桂皮,香树皮,月桂);Fragrant Cinnamon, Luandashan Cinnamon Tree, Randa Cinnamon, Randaishan Cinnamon Tree ●

91430 Cinnamomum szechuanense Yen C. Yang = Cinnamomum appelianum Schewe ●

91431 Cinnamomum taimoshanicum Chun ex Hung T. Chang = Cinnamomum appelianum Schewe ●

91432 Cinnamomum tamala (Buch.-Ham.) T. Nees et C. H. Eberm.;柴桂(柴樟,桂,桂皮,辣皮树,肉桂,三股筋,三条筋,三条筋树,香叶子树,小华草);Firewood Cinnamon, Indian Cassia, Tamala Cassia ●

91433 Cinnamomum tamala (Buch.-Ham.) T. Nees et C. H. Eberm. = Cinnamomum wilsonii Gamble ●

91434 Cinnamomum tamala Nees = Cinnamomum jensenianum Hand.-Mazz. ●

91435 Cinnamomum taquetii H. Lév. = Cinnamomum camphora (L.) T. Nees et C. H. Eberm. ●

91436 Cinnamomum tavoyanum Meisn.;印度肉桂树;Indian Cinnamon ●☆

91437 Cinnamomum tenuifolium (Makino) Sugim. ex H. Hara;狭叶肉桂树●☆

91438 Cinnamomum tenuifolium (Makino) Sugim. ex H. Hara f. nervosum (Meisn.) H. Hara;台湾天竺桂(天竺桂)●☆

91439 Cinnamomum tenuipile Kosterm.;细毛樟;Thin Cinnamon, Thinhair Cinnamon, Thin-haired Cinnamon ●

91440 Cinnamomum tonkinense (Lecomte) A. Chev.;假桂皮树(东京樟,美高量);Tonkin Cinnamon ●

91441 Cinnamomum trinervatum Yen C. Yang = Cinnamomum appelianum Schewe ●

91442 Cinnamomum tsangii Merr.;辣汁树(海南樟,怀德樟,辣汁桂);Tsang Cinnamon ●

91443 Cinnamomum tsoi C. K. Allen;平托桂(景烈樟,乌身香槁);Tso Cassia, Tso Cinnamon ●

91444 Cinnamomum validinerve Hance;粗脉桂;Thicknerved Cinnamon, Thick-nerved Cinnamon ●

91445 Cinnamomum validinerve Hance var. poilanei H. Liu = Cinnamomum subavenium Miq. ●

91446 Cinnamomum verum J. Presl;锡兰肉桂(标准肉桂,斯里兰卡桂树,斯里兰卡肉桂,锡兰樟,正肉桂);Cannell, Ceylon Cinnamon, Ceylon Cinnamon Tree, Cinnamon, Cinnamon-tree, Quillings ●

91447 Cinnamomum villosulum S. K. Lee et E. N. Wei;锈毛桂;Villose Cinnamon ●

91448 Cinnamomum villosulum S. K. Lee et F. N. Wei = Cinnamomum

appelianum Schewe ●
91449 Cinnamomum wilsonii Gamble;川桂(柴桂,臭樟,臭樟木,大叶子树,官桂,桂皮树,三条筋,山肉桂);E. H. Wilson Cinnamon, Sichuan Cinnamon,Wilson Cinnamon ●
91450 Cinnamomum wilsonii Gamble var. multiflorum Gamble = Cinnamomum wilsonii Gamble ●
91451 Cinnamomum xanthoneurum Blume;黄脉樟;Yellow-veins Cinnamon ●☆
91452 Cinnamomum xanthophyllum H. W. Li = Cinnamomum micranthum (Hayata) Hayata ●
91453 Cinnamomum zeylanicum Blume = Cinnamomum verum J. Presl ●
91454 Cinnamosma Baill. (1867);合瓣樟属●☆
91455 Cinnamosma fragrans Baill.;合瓣樟;Taggar ●☆
91456 Cinnamosma fragrans Baill. var. bailloni Courchet = Cinnamosma fragrans Baill. ●☆
91457 Cinnamosma fragrans Baill. var. perrieri Courchet = Cinnamosma fragrans Baill. ●☆
91458 Cinnamosma macrocarpa H. Perrier;大果合瓣樟●☆
91459 Cinnamosma madagascariensis Danguy;马岛合瓣樟●☆
91460 Cinnastrum E. Fourn. = Cinna L. ■
91461 Cinnastrum E. Fourn. ex Benth. et Hook. f. = Cinna L. ■
91462 Cinogasum Neck. = Croton L. ●
91463 Cinsania Lavy. (1830);意大利杜鹃属●☆
91464 Cintia Kni? e et ? iha(1995);玻利维亚仙人掌属■☆
91465 Ciomena P. Beauv. = Muhlenbergia Schreb. ■
91466 Cionandra Griseb. = Cayaponia Silva Manso(保留属名)■☆
91467 Cionisaccus Breda = Goodyera R. Br. ■
91468 Cionomene Krukoff = Elephantomene Barneby et Krukoff ●☆
91469 Cionomene Krukoff(1979);月牙藤属●☆
91470 Cionomene javariensis Krukoff;月牙藤●☆
91471 Cionosicyos Griseb. (1860);柱葫芦属■☆
91472 Cionosicyos excisus (Griseb.) C. Jeffrey;柱葫芦■☆
91473 Cionosicyos macrantha (Pittier) C. Jeffrey;大花柱葫芦■☆
91474 Cionosicys Griseb. = Cionosicyos Griseb. ■☆
91475 Cionosicyus Post et Kuntze = Cionosicyos Griseb. ■☆
91476 Cionura Griseb. (1844);拟牛奶菜属■☆
91477 Cionura Griseb. = Marsdenia R. Br. (保留属名)●
91478 Cionura erecta Griseb.;拟牛奶菜■☆
91479 Cipadessa Blume(1825);浆果楝属;Baccamelia,Cipadessa ●
91480 Cipadessa baccifera (Roth) Miq.;浆果楝(苦亚罗椿,老鸦饭,老鸦树,埋皮纺,亚罗椿,秧勒);Baccamelia, Berry-bearing Cipadessa ●
91481 Cipadessa baccifera (Roth) Miq. var. sinensis Rehder et E. H. Wilson = Cipadessa cinerascens (Pell.) Hand.-Mazz. ●
91482 Cipadessa cinerascens (Pell.) Hand.-Mazz.;灰毛浆果楝(臭子,串黄皮,大苦木,假茶辣,假吴萸,罗汉香,毛浆果楝,软柏木,山黄皮,碎米青,亚洛轻,野茶辣,野桐椒,鱼胆木);Greyhair Baccamelia,Greyhair Cipadessa,Grey-haired Cipadessa ●
91483 Cipadessa cinerascens Hook. et Arn. = Cipadessa cinerascens (Pell.) Hand.-Mazz. ●
91484 Cipadessa fruticosa Blume = Cipadessa baccifera (Roth) Miq. ●
91485 Cipadessa fruticosa Blume var. cinerascens Pell. = Cipadessa cinerascens (Pell.) Hand.-Mazz. ●
91486 Cipadessa sinensis (Rehder et E. H. Wilson) Hand.-Mazz. = Cipadessa cinerascens Hook. et Arn. ●
91487 Cipocereus F. Ritter(1979);角棱柱属●☆
91488 Cipocereus minensis (Werderm.) F. Ritter;角棱柱●☆
91489 Cipoia C. T. Philbrick,Novelo et Irgang(2004);锡波川苔草属■☆
91490 Ciponima Aubl. = Symplocos Jacq. ●
91491 Ciposia Silveira(1918);巴西桃金娘属●☆
91492 Cipura Aubl. (1775);粗柱鸢尾属■☆
91493 Cipura Klotzsch ex Klatt = Alophia Herb. ■☆
91494 Cipura Klotzsch ex Klatt = Herbertia Sweet ■☆
91495 Cipura paludosa Aubl.;粗柱鸢尾■☆
91496 Cipuropsis Ule = Vriesea Lindl. (保留属名)■☆
91497 Cipuropsis Ule(1907);奇普凤梨属■☆
91498 Cipuropsis subandina Ule.;奇普凤梨■☆
91499 Circaea L. (1753);露珠草属(谷蓼属);Circaea,Dewdrograss, Enchanter's Nightshade,Enchanter's-nightshade ■
91500 Circaea × canadensis (L.) Hill var. rishiriensis H. Hara = Circaea × intermedia Ehrh. ■☆
91501 Circaea × decipiens Boufford;迷惑露珠草■☆
91502 Circaea × dubia H. Hara var. makinoi H. Hara = Circaea alpina L. subsp. caulescens (Kom.) Tatew. ■
91503 Circaea × intermedia Ehrh.;间型露珠草■☆
91504 Circaea × mentiens Boufford;髯毛露珠草■☆
91505 Circaea × ovata Boufford;卵叶露珠草■
91506 Circaea × skvortsovii Boufford;北方露珠草■
91507 Circaea × taronensis H. Li;贡山露珠草■
91508 Circaea alpina L.;高山露珠草;Alpine Circaea, Alpine Dewdrograss, Alpine Enchanter's-nightshade, Dwarf Enchanter's-nightshade, Northern Enchanter's-nightshade, Small Enchanter's-nightshade ■
91509 Circaea alpina L. f. pilosula (H. Hara) Kitag. = Circaea alpina L. subsp. caulescens (Kom.) Tatew. ■
91510 Circaea alpina L. subsp. angustifolia (Hand.-Mazz.) Boufford;狭叶露珠草■
91511 Circaea alpina L. subsp. caulescens (Kom.) Tatew.;深山露珠草■
91512 Circaea alpina L. subsp. imaicola (Asch. et Magnus) Kitam.;高原露珠草(高山露珠草,西南露珠草)■
91513 Circaea alpina L. subsp. imaicola Asch. et Magnus = Circaea alpina L. subsp. imaicola (Asch. et Magnus) Kitam. ■
91514 Circaea alpina L. subsp. micrantha (Skvortsov) Boufford;高寒露珠草■
91515 Circaea alpina L. var. caulescens Kom. = Circaea alpina L. subsp. caulescens (Kom.) Tatew. ■
91516 Circaea alpina L. var. himalaica C. B. Clarke = Circaea repens Wall. ex Asch. et Magnus ■
91517 Circaea alpina L. var. imaicola Asch. et Magnus = Circaea alpina L. subsp. imaicola (Asch. et Magnus) Kitam. ■
91518 Circaea alpina L. var. pilosula (H. Hara) H. Hara = Circaea alpina L. subsp. caulescens (Kom.) Tatew. ■
91519 Circaea alpina L. var. pilosula H. Hara = Circaea alpina L. subsp. caulescens (Kom.) Tatew. ■
91520 Circaea alpina L. var. robusta Kom.;粗壮高山露珠草■
91521 Circaea bodinieri H. Lév. = Circaea cordata Royle ■
91522 Circaea canadensis (L.) Hill;加拿大露珠草;Canadian Enchanter's-nightshade ■☆
91523 Circaea canadensis (L.) Hill = Circaea lutetiana L. subsp. canadensis (L.) Asch. et Magnus ■☆
91524 Circaea canadensis (L.) Hill subsp. quadrisulcata (Maxim.) Boufford;水珠草(谷蓼,华水珠草,露珠草);Enchanter's-nightshade,Four-angled Circaea ■

91525 Circaea canadensis (L.) Hill var. rishiriensis H. Hara = Circaea intermedia Ehrh. ■☆

91526 Circaea canadensis (L.) Hill var. virginiana Fernald = Circaea lutetiana L. subsp. canadensis (L.) Asch. et Magnus ■☆

91527 Circaea cardiophylla Makino = Circaea cordata Royle ■

91528 Circaea caucasica A. K. Skvortsov = Circaea alpina L. subsp. caulescens (Kom.) Tatew. ■

91529 Circaea caucasica Skvortsov = Circaea alpina L. subsp. caulescens (Kom.) Tatew. ■

91530 Circaea caulescens (Kom.) H. Hara = Circaea alpina L. subsp. caulescens (Kom.) Tatew. ■

91531 Circaea caulescens (Kom.) Nakai ex H. Hara = Circaea alpina L. subsp. caulescens (Kom.) Tatew. ■

91532 Circaea caulescens (Kom.) Nakai ex H. Hara f. ramosissima H. Hara = Circaea alpina L. ■

91533 Circaea caulescens (Kom.) Nakai ex H. Hara var. glabra H. Hara = Circaea alpina L. ■

91534 Circaea caulescens (Kom.) Nakai ex H. Hara var. pilosula H. Hara = Circaea alpina L. subsp. caulescens (Kom.) Tatew. ■

91535 Circaea caulescens (Kom.) Nakai ex H. Hara var. robusta Nakai ex H. Hara = Circaea alpina L. subsp. caulescens (Kom.) Tatew. ■

91536 Circaea caulescens (Kom.) Nakai ex H. Hara var. rosulata H. Hara = Circaea alpina L. ■

91537 Circaea cordata Royle;露珠草(牛泷草,曲毛露珠草,三角叶,心叶谷蓼,心叶露珠草,夜麻光,夜抹光);Cordate Circaea,Cordate Dewdrograss ■

91538 Circaea cordata Royle var. glabrescens Pamp. = Circaea glabrescens (Pamp.) Hand. -Mazz. ■

91539 Circaea coreana H. Lév. = Circaea mollis Siebold et Zucc. ■

91540 Circaea coreana H. Lév. var. sinensis H. Lév. = Circaea mollis Siebold et Zucc. ■

91541 Circaea delavayi H. Lév. = Circaea erubescens Franch. et Sav. ■

91542 Circaea dubia H. Hara;可疑露珠草■☆

91543 Circaea dubia H. Hara var. makinoi H. Hara;牧野氏露珠草■☆

91544 Circaea dubia H. Hara var. makinoi H. Hara = Circaea alpina L. subsp. caulescens (Kom.) Tatew. ■

91545 Circaea erubescens Franch. et Sav.;谷蓼(台湾露珠草);Reddening Circaea,Reddening Dewdrograss ■

91546 Circaea glabrescens (Pamp.) Hand. -Mazz.;秃梗露珠草(光梗露珠草)■

91547 Circaea hohuanensis S. S. Ying = Circaea alpina L. subsp. imaicola (Asch. et Magnus) Kitam. ■

91548 Circaea hybrida Hand. -Mazz. = Circaea cordata Royle ■

91549 Circaea imaicola (Asch. et Magnus) Hand. -Mazz. = Circaea alpina L. subsp. imaicola (Asch. et Magnus) Kitam. ■

91550 Circaea imaicola (Asch. et Magnus) Hand. -Mazz. = Circaea alpina L. subsp. angustifolia (Hand. -Mazz.) Boufford ■

91551 Circaea imaicola (Asch. et Magnus) Hand. -Mazz. var. angustifolia Hand. -Mazz. = Circaea alpina L. subsp. angustifolia (Hand. -Mazz.) Boufford ■

91552 Circaea imaicola (Asch. et Magnus) Hand. -Mazz. var. mairei (H. Lév.) Hand. -Mazz. = Circaea alpina L. subsp. angustifolia (Hand. -Mazz.) Boufford ■

91553 Circaea intermedia Ehrh.;全叶露珠草;Upland Enchanter's-nightshade ■☆

91554 Circaea kawakamii Hayata = Circaea erubescens Franch. et Sav. ■

91555 Circaea kitagawae H. Hara = Circaea cordata Royle ■

91556 Circaea latifolia Hill = Circaea lutetiana L. subsp. canadensis (L.) Asch. et Magnus ■☆

91557 Circaea lutetiana L.;牛泷草(巴黎露珠草,谷蓼,露珠草,水珠草);Bindweed Nightshade,Broad-leaf Enchanter's-nightshade,Common Enchanter's-nightshade,Dragon-root,Enchanter's Nightshade,Enchanter's-nightshade,Mandrake,Nightshade Bindweed,Paris Circaea,Philtrewort,Wild London Pride,Witchelower,Witchwort ■

91558 Circaea lutetiana L. = Circaea repens Wall. ex Asch. et Magnus ■

91559 Circaea lutetiana L. f. quadrisulcata Maxim. = Circaea canadensis (L.) Hill subsp. quadrisulcata (Maxim.) Boufford ■

91560 Circaea lutetiana L. f. quadrisulcata Maxim. = Circaea lutetiana L. subsp. quadrisulcata (Maxim.) Asch. et Magnus ■

91561 Circaea lutetiana L. subsp. alpina (L.) H. Lév. = Circaea alpina L. ■

91562 Circaea lutetiana L. subsp. canadensis (L.) Asch. et Magnus;加拿大牛泷草;Broad-leaf Enchanter's-nightshade ■☆

91563 Circaea lutetiana L. subsp. mediterranea Asch. et Magnus = Circaea lutetiana L. ■

91564 Circaea lutetiana L. subsp. quadrisulcata (Maxim.) Asch. et Magnus = Circaea canadensis (L.) Hill subsp. quadrisulcata (Maxim.) Boufford ■

91565 Circaea lutetiana L. var. canadensis L. = Circaea lutetiana L. subsp. canadensis (L.) Asch. et Magnus ■☆

91566 Circaea lutetiana L. var. mairei H. Lév. = Circaea alpina L. subsp. angustifolia (Hand. -Mazz.) Boufford ■

91567 Circaea lutetiana L. var. taquetii H. Lév. = Circaea mollis Siebold et Zucc. ■

91568 Circaea maximowiczii (H. Lév.) H. Hara = Circaea canadensis (L.) Hill subsp. quadrisulcata (Maxim.) Boufford ■

91569 Circaea maximowiczii (H. Lév.) H. Hara = Circaea lutetiana L. subsp. quadrisulcata (Maxim.) Asch. et Magnus ■

91570 Circaea maximowiczii (H. Lév.) H. Hara f. viridicalyx (H. Hara) Kitag. = Circaea lutetiana L. subsp. quadrisulcata (Maxim.) Asch. et Magnus ■

91571 Circaea maximowiczii (H. Lév.) H. Hara f. viridicalyx (H. Hara) Kitag. = Circaea canadensis (L.) Hill subsp. quadrisulcata (Maxim.) Boufford ■

91572 Circaea maximowiczii (H. Lév.) H. Hara var. viridicalyx H. Hara = Circaea canadensis (L.) Hill subsp. quadrisulcata (Maxim.) Boufford ■

91573 Circaea micrantha A. K. Skvortsov = Circaea alpina L. subsp. micrantha (Skvortsov) Boufford ■

91574 Circaea minutula Ohwi = Circaea alpina L. subsp. imaicola (Asch. et Magnus) Kitam. ■

91575 Circaea mollis Siebold et Zucc.;南方露珠草(白辣蓼草,白洋漆药,红节草,假蛇床子,辣椒七,三角叶,土灵仙,细毛谷蓼,野牛膝);South Circaea,Southern Dewdrograss ■

91576 Circaea mollis Siebold et Zucc. f. montana Hiyama;山地南方露珠草■☆

91577 Circaea mollis Siebold et Zucc. f. ovata (Honda) Okuyama = Circaea ovata (Honda) Boufford ■☆

91578 Circaea mollis Siebold et Zucc. var. maximowiczii H. Lév. = Circaea canadensis (L.) Hill subsp. quadrisulcata (Maxim.) Boufford ■

91579 Circaea mollis Siebold et Zucc. var. maximowiczii H. Lév. = Circaea lutetiana L. subsp. quadrisulcata (Maxim.) Asch. et Magnus ■

91580 Circaea ovata (Honda) Boufford;白花谷蓼■☆
91581 Circaea pacifica Asch. et Magnus;太平洋露珠草;Pacific Circaea ■☆
91582 Circaea pricei Hayata = Circaea alpina L. subsp. imaicola (Asch. et Magnus) Kitam. ■
91583 Circaea pricei Hayata var. mairei (H. Lév.) Hand. -Mazz. = Circaea alpina L. subsp. angustifolia (Hand. -Mazz.) Boufford ■
91584 Circaea quadrisulcata (Maxim.) Franch. et Sav. = Circaea × ovata Boufford ■
91585 Circaea quadrisulcata (Maxim.) Franch. et Sav. = Circaea canadensis (L.) Hill subsp. quadrisulcata (Maxim.) Boufford ■
91586 Circaea quadrisulcata (Maxim.) Franch. et Sav. = Circaea lutetiana L. subsp. quadrisulcata (Maxim.) Asch. et Magnus ■
91587 Circaea quadrisulcata (Maxim.) Franch. et Sav. = Circaea mollis Siebold et Zucc. ■
91588 Circaea quadrisulcata (Maxim.) Franch. et Sav. f. viridicalyx (H. Hara) Kitag.;绿萼水珠草■
91589 Circaea quadrisulcata (Maxim.) Franch. et Sav. var. canadensis (L.) H. Hara = Circaea lutetiana L. ■
91590 Circaea quadrisulcata (Maxim.) Franch. et Sav. var. canadensis (L.) H. Hara = Circaea lutetiana L. subsp. canadensis (L.) Asch. et Magnus ■☆
91591 Circaea repens Wall. = Circaea repens Wall. ex Asch. et Magnus ■
91592 Circaea repens Wall. ex Asch. et Magnus;匍匐露珠草(匍茎谷蓼)■
91593 Circaea taiwaniana S. S. Ying = Circaea alpina L. subsp. imaicola (Asch. et Magnus) Kitam. ■
91594 Circaeaceae Bercht. et J. Presl = Onagraceae Juss.(保留科名)■●
91595 Circaeaceae Lindl. = Onagraceae Juss.(保留科名)■●
91596 Circaeaster Maxim. (1882);星叶草属(星叶属);Circaester ■
91597 Circaeaster agrestis Maxim.;星叶草(星叶);Field Circaester ■
91598 Circaeasteraceae Hutch. (1926)(保留科名);星叶草科;Circaeaster Family ■
91599 Circaeasteraceae Kuntze ex Hutch. = Circaeasteraceae Hutch. (保留科名)■
91600 Circaeocarpus C. Y. Wu = Zippelia Blume ■
91601 Circaeocarpus saururoides C. Y. Wu = Zippelia begoniifolia Blume ex Schult. et Schult. f. ■
91602 Circandra N. E. Br. (1930);亮黄玉属●☆
91603 Circandra serrata (L.) N. E. Br.;亮黄玉■☆
91604 Circea Raf. = Circaea L. ■
91605 Circinnus Medik.(废弃属名) = Hymenocarpos Savi(保留属名)■☆
91606 Circinus Medik. = Circinnus Medik.(废弃属名)■☆
91607 Circinus Medik. = Hymenocarpos Savi(保留属名)■☆
91608 Circis Chapm. = Cercis L. ●
91609 Cirinosum Neck. = Cereus Mill. ●
91610 Ciripedium Zumagl. = Cypripedium L. ■
91611 Cirrhaea Lindl. (1832);须喙兰属(卷须兰属);Cirrhaea ■☆
91612 Cirrhaea dependens (Lodd.) Rchb. f.;悬垂须喙兰(悬垂卷须兰);Dependent Cirrhaea ■☆
91613 Cirrhaea fuscolutea Lindl. = Cirrhaea dependens (Lodd.) Rchb. f. ■☆
91614 Cirrhaea hoffmanseggii Heynh. ex Rchb. f. = Cirrhaea dependens (Lodd.) Rchb. f. ■☆
91615 Cirrhaea loddigesii Lindl. = Cirrhaea dependens (Lodd.) Rchb. f. ■☆
91616 Cirrhaea tristis Lindl. = Cirrhaea dependens (Lodd.) Rchb. f. ■☆
91617 Cirrhopetalum Lindl. (1830)(保留属名);卷瓣兰属■
91618 Cirrhopetalum Lindl.(保留属名) = Bulbophyllum Thouars(保留属名)■
91619 Cirrhopetalum aemulum W. W. Sm. = Bulbophyllum forrestii Seidenf. ■
91620 Cirrhopetalum africanum Schltr. = Bulbophyllum longiflorum Thouars ■☆
91621 Cirrhopetalum albociliatum Tang S. Liu et H. J. Su = Bulbophyllum albociliatum (Tang S. Liu et H. J. Su) Nackej. ■
91622 Cirrhopetalum amplifolium Rolfe = Bulbophyllum amplifolium (Rolfe) N. P. Balakr. et S. Chowdhury ■
91623 Cirrhopetalum andersonii Hook. f. = Bulbophyllum andersonii (Hook. f.) J. J. Sm. ■
91624 Cirrhopetalum aurantiacum W. W. Sm. = Bulbophyllum hirundinis (Gagnep.) Seidenf. ■
91625 Cirrhopetalum autumnale Fukuy. = Bulbophyllum macraei (Lindl.) Rchb. f. ■
91626 Cirrhopetalum bicolor (Lindl.) Rolfe = Bulbophyllum bicolor (Lindl.) Hook. f. ■
91627 Cirrhopetalum bicolor (Lindl.) Rolfe = Sunipia bicolor Lindl. ■
91628 Cirrhopetalum boninense Schltr. = Bulbophyllum boninense (Schltr.) J. J. Sm. ■☆
91629 Cirrhopetalum bootanense Griff. = Bulbophyllum spathulatum (Rolfe ex E. Cooper) Seidenf. ■
91630 Cirrhopetalum brevipes Hook. f. = Bulbophyllum emarginatum (Finet) J. J. Sm. ■
91631 Cirrhopetalum caudatum (Lindl.) King et Pantl. = Bulbophyllum caudatum Lindl. ■
91632 Cirrhopetalum chinense Lindl. = Bulbophyllum chinense (Lindl.) Rchb. f. ■
91633 Cirrhopetalum chondriophorum Gagnep. = Bulbophyllum chrondriophorum (Gagnep.) Seidenf. ■
91634 Cirrhopetalum clavigerum Fitzg. = Bulbophyllum longiflorum Thouars ■☆
91635 Cirrhopetalum delitescens (Hance) Rolfe = Bulbophyllum delitescens Hance ■
91636 Cirrhopetalum dyerianum King et Pantl. = Bulbophyllum rolfei (Kuntze) Seidenf. ■
91637 Cirrhopetalum elatum Hook. f. = Bulbophyllum elatum (Hook. f.) J. J. Sm. ■
91638 Cirrhopetalum emarginatum Finet = Bulbophyllum emarginatum (Finet) J. J. Sm. ■
91639 Cirrhopetalum farreri W. W. Sm. = Bulbophyllum farreri (W. W. Sm.) Seidenf. ■
91640 Cirrhopetalum flaviflorum (Tang S. Liu et H. J. Su) Seidenf. = Bulbophyllum pectenveneris (Gagnep.) Seidenf. ■
91641 Cirrhopetalum flaviflorum Tang S. Liu et H. J. Su = Bulbophyllum pectenveneris (Gagnep.) Seidenf. ■
91642 Cirrhopetalum flavisepalum (Hayata) Hayata = Bulbophyllum retusiusculum Rchb. f. ■
91643 Cirrhopetalum flavisepalum Hayata = Bulbophyllum retusiusculum Rchb. f. ■
91644 Cirrhopetalum fordii Rolfe = Bulbophyllum fordii (Rolfe) J. J. Sm. ■
91645 Cirrhopetalum formosanum Rolfe = Bulbophyllum formosanum (Rolfe) Seidenf. ■

91646 Cirrhopetalum gongshanense (Z. H. Tsi) Garay, Hamer et Siegerist = Bulbophyllum gongshanense Z. H. Tsi ■

91647 Cirrhopetalum guttulatum Hook. f. = Bulbophyllum guttulatum (Hook. f.) N. P. Balakr. ■

91648 Cirrhopetalum henryi Rolfe = Bulbophyllum andersonii (Hook. f.) J. J. Sm. ■

91649 Cirrhopetalum hirundinis Gagnep. = Bulbophyllum hirundinis (Gagnep.) Seidenf. ■

91650 Cirrhopetalum inabai Hayata = Bulbophyllum japonicum (Makino) Makino ■

91651 Cirrhopetalum insulsum Gagnep. = Bulbophyllum insulsum (Gagnep.) Seidenf. ■

91652 Cirrhopetalum insulsum Gagnep. = Bulbophyllum levinei Schltr. ■

91653 Cirrhopetalum japonicum Makino = Bulbophyllum japonicum (Makino) Makino ■

91654 Cirrhopetalum longibrachiatum (Z. H. Tsi) Garay, Hamer et Siegerist = Bulbophyllum longibrachiatum Z. H. Tsi ■

91655 Cirrhopetalum longiflorum (Thouars) Schltr. = Bulbophyllum longiflorum Thouars ■☆

91656 Cirrhopetalum macraei Lindl. = Bulbophyllum macraei (Lindl.) Rchb. f. ■

91657 Cirrhopetalum makinoanum Schltr. = Bulbophyllum macraei (Lindl.) Rchb. f. ■

91658 Cirrhopetalum melanoglossum (Hayata) Hayata = Bulbophyllum melanoglossum Hayata ■

91659 Cirrhopetalum melanoglossum Hayata = Bulbophyllum melanoglossum Hayata ■

91660 Cirrhopetalum melinanthum Schltr. = Bulbophyllum hirundinis (Gagnep.) Seidenf. ■

91661 Cirrhopetalum miniatum Rolfe = Bulbophyllum pectenveneris (Gagnep.) Seidenf. ■

91662 Cirrhopetalum omerandrum (Hayata) Hayata = Bulbophyllum omerandrum Hayata ■

91663 Cirrhopetalum omerandrum Hayata = Bulbophyllum omerandrum Hayata ■

91664 Cirrhopetalum oreogenes W. W. Sm. = Bulbophyllum retusiusculum Rchb. f. ■

91665 Cirrhopetalum parvulum Hook. f. = Bulbophyllum rolfei (Kuntze) Seidenf. ■

91666 Cirrhopetalum pecten-veneris Gagnep. = Bulbophyllum pectenveneris (Gagnep.) Seidenf. ■

91667 Cirrhopetalum picturatum Lodd. = Bulbophyllum picturatum (Lodd.) Rchb. f. ■

91668 Cirrhopetalum pingtungense S. S. Ying = Bulbophyllum wightii Rchb. f. ■

91669 Cirrhopetalum racemosum (Hayata) Hayata = Bulbophyllum insulsum (Gagnep.) Seidenf. ■

91670 Cirrhopetalum racemosum Hayata = Bulbophyllum insulsum (Gagnep.) Seidenf. ■

91671 Cirrhopetalum refractum Zoll. = Bulbophyllum refractum Rchb. f. ■☆

91672 Cirrhopetalum remotifolium Fukuy. = Bulbophyllum hirundinis (Gagnep.) Seidenf. ■

91673 Cirrhopetalum retusiusculum (Rchb. f.) Hemsl. = Bulbophyllum retusiusculum Rchb. f. ■

91674 Cirrhopetalum rothschildianum O ' Brien = Bulbophyllum rothschildianum (O'Brien) J. J. Sm. ■

91675 Cirrhopetalum saruwatarii (Hayata) Hayata = Bulbophyllum umbellatum Lindl. ■

91676 Cirrhopetalum spathulatum Rolfe ex Cooper = Bulbophyllum spathulatum (Rolfe ex E. Cooper) Seidenf. ■

91677 Cirrhopetalum stramineum Rolfe ex Cooper var. purpureum Gagnep. = Bulbophyllum obtusiangulum Z. H. Tsi ■

91678 Cirrhopetalum striatum Tang S. Liu et H. J. Su = Bulbophyllum melanoglossum Hayata ■

91679 Cirrhopetalum sutepense Rolfe ex Downie = Bulbophyllum sutepense (Rolfe ex Downie) Seidenf. ■

91680 Cirrhopetalum taeniophyllum (Parl. et Rchb. f.) Hook. f. = Bulbophyllum taeniophyllum Parl. et Rchb. f. ■

91681 Cirrhopetalum taichungianum S. S. Ying = Bulbophyllum albociliatum (Tang S. Liu et H. J. Su) Nackej. ■

91682 Cirrhopetalum taiwanense Fukuy. = Bulbophyllum taiwanense (Fukuy.) Nackej. ■

91683 Cirrhopetalum thomasii Otto et A. Dietr. = Bulbophyllum longiflorum Thouars ■☆

91684 Cirrhopetalum thouarsii Lindl. = Bulbophyllum lemurense Bosser et P. J. Cribb ■☆

91685 Cirrhopetalum thouarsii Lindl. = Bulbophyllum longiflorum Thouars ■☆

91686 Cirrhopetalum tigridum (Hance) Rolfe = Bulbophyllum retusiusculum Rchb. f. var. tigridum (Hance) Z. H. Tsi ■

91687 Cirrhopetalum tigridum (Hance) Rolfe = Bulbophyllum tigridum Hance ■

91688 Cirrhopetalum trichocephalum Schltr. = Bulbophyllum odoratissimum (Sm.) Lindl. ■

91689 Cirrhopetalum tseanum S. Y. Hu et Barretto = Bulbophyllum tsanum (S. Y. Hu et Barretto) Z. H. Tsi ■

91690 Cirrhopetalum umbellatum (G. Forst.) Hook. et Arn. = Bulbophyllum longiflorum Thouars ■☆

91691 Cirrhopetalum umbellatum (Sw.) A. Frapp. ex Cordem. = Bulbophyllum longiflorum Thouars ■☆

91692 Cirrhopetalum uraiense (Hayata) Hayata ex Schltr. = Bulbophyllum macraei (Lindl.) Rchb. f. ■

91693 Cirrhopetalum uraiense (Hayata) Hayata ex Schltr. var. tanegashimense (Masam.) F. Maek. = Euscaphis japonica (Thunb.) J. Buchholz f. lanata (Masam.) K. Iwats. et H. Ohba ●☆

91694 Cirrhopetalum wallichii Lindl. = Bulbophyllum wallichii Rchb. f. ■

91695 Cirselium Brot. = Cirsellium Gaertn. ■☆

91696 Cirsellium Gaertn. = Atractylis L. ■☆

91697 Cirsium Mill. (1754);蓟属;Plumed Thistle,Thistle ■

91698 Cirsium × celakovskyanum Knaf;塞拉考夫斯基蓟■☆

91699 Cirsium × glabratum Kitam.;光滑蓟■☆

91700 Cirsium × iburiense Kitam.;胆振蓟■☆

91701 Cirsium × mirabile Kitam.;奇异蓟■☆

91702 Cirsium × misawaense Nakai ex Kitam.;三泽蓟■☆

91703 Cirsium × patens Kitam.;铺展蓟■☆

91704 Cirsium × perplexissimum Kitam.;紊乱蓟■☆

91705 Cirsium × pilosum Kitam.;疏毛蓟■☆

91706 Cirsium × sugimotoi Kitam.;杉本蓟■☆

91707 Cirsium abkhasicum (Petr.) Grossh.;阿布哈兹蓟■☆

91708 Cirsium abukumense Kadota;阿武隈蓟■☆

91709 Cirsium abyssinicum Sch. Bip. ex A. Rich. = Cirsium vulgare (Savi) Ten. ■

91710 Cirsium acanthodontum S. F. Blake = Cirsium remotifolium

(Hook.) DC. var. rivulare Jeps. ■☆

91711 Cirsium acarnum (L.) Moench = Picnomon acarna (L.) Cass. ■☆

91712 Cirsium acarnum Moench = Cirsium acarnum (L.) Moench ■☆

91713 Cirsium acaule (L.) All.;无茎蓟;Chalk Thistle, Dwarf Carline Thistle, Dwarf Thistle, Ground Thistle, Picnic Thistle, Pod Thistle, Scotch Thistle, Stemless Thistle ■☆

91714 Cirsium acaule (L.) Scop. var. americanum A. Gray = Cirsium scariosum Nutt. var. americanum (A. Gray) D. J. Keil ■☆

91715 Cirsium acaule Ledeb. = Cirsium esculentum (Siev.) C. A. Mey. ■

91716 Cirsium acaule Ledeb. var. gmelini DC. = Cirsium esculentum (Siev.) C. A. Mey. ■

91717 Cirsium acaule Ledeb. var. sibiricum Ledeb. = Cirsium esculentum (Siev.) C. A. Mey. ■

91718 Cirsium acaulescens (A. Gray) K. Schum. = Cirsium scariosum Nutt. var. americanum (A. Gray) D. J. Keil ■☆

91719 Cirsium acuatum (Osterh.) Cockerell = Cirsium tracyi (Rydb.) Petr. ■☆

91720 Cirsium aduncum Fisch. et C. A. Mey. ex DC.;钩状蓟■☆

91721 Cirsium aggregatum Ledeb.;密蓟■☆

91722 Cirsium aidzuense Nakai ex Kitam.;会津蓟■☆

91723 Cirsium alatum (S. G. Gmel.) Bobrov;准噶尔蓟(刚毛蓟);Dzungar Thistle, Winged Thistle ■

91724 Cirsium albertii Regel et Schmalh.;天山蓟;Tianshan Thistle ■

91725 Cirsium albescens Kitam.;白毛蓟■

91726 Cirsium albescens Kitam. = Cirsium brevicaule A. Gray ■

91727 Cirsium albiflorum (Kitag.) Kitag. = Cirsium setosum (Willd.) M. Bieb. ■

91728 Cirsium albrechtii (Maxim.) Kudo ex Tatew. = Cirsium heiianum Koidz. ■☆

91729 Cirsium alpicola Nakai;日本山蓟■☆

91730 Cirsium altissimum (L.) Spreng.;高蓟;Roadside Thistle, Tall Thistle, Wood Thistle ■☆

91731 Cirsium altissimum (L.) Spreng. var. biltmoreanum Petr. = Cirsium altissimum (L.) Spreng. ■☆

91732 Cirsium altissimum Hill = Cirsium altissimum (L.) Spreng. ■☆

91733 Cirsium amblylepis Petr. = Cirsium remotifolium (Hook.) DC. var. odontolepis Petr. ■☆

91734 Cirsium americanum (A. Gray) K. Schum. = Cirsium scariosum Nutt. var. americanum (A. Gray) D. J. Keil ■☆

91735 Cirsium americanum (A. Gray) K. Schum. var. callilepis (Greene) Jeps. = Cirsium remotifolium (Hook.) DC. var. odontolepis Petr. ■☆

91736 Cirsium amplexifolium (Nakai) Kitam.;褶叶蓟■☆

91737 Cirsium amplexifolium (Nakai) Kitam. var. muraii (Kitam.) Kitam.;村井蓟■☆

91738 Cirsium anatolicum (Petr.) Grossh.;阿纳托里蓟■☆

91739 Cirsium andersonii (A. Gray) Petr.;安氏蓟;Anderson's Thistle, Rose Thistle ■☆

91740 Cirsium andrewsii (A. Gray) Jeps.;安德鲁斯蓟;Franciscan Thistle ■☆

91741 Cirsium aomorense Nakai;青森蓟■☆

91742 Cirsium aomorense Nakai f. albiflorum (Kitam.) Kitam.;白花安德鲁斯蓟■☆

91743 Cirsium aomorense Nakai f. echinoides Kitam.;具刺安德鲁斯蓟■☆

91744 Cirsium apiculatum DC.;细尖蓟■☆

91745 Cirsium apoense Nakai;阿泡蓟■☆

91746 Cirsium arachnoides (M. Bieb.) M. Bieb.;蛛毛蓟■☆

91747 Cirsium arachnoides M. Bieb. = Cirsium arachnoides (M. Bieb.) M. Bieb. ■☆

91748 Cirsium araneans Rydb. = Cirsium clavatum (M. E. Jones) Petr. var. osterhoutii (Rydb.) D. J. Keil ■☆

91749 Cirsium argillosum Petr. ex Charadze;白土蓟■☆

91750 Cirsium argunense DC. = Cirsium setosum (Willd.) M. Bieb. ■

91751 Cirsium argyracanthum DC.;南蓟;Southern Thistle ■

91752 Cirsium argyracanthum DC. = Cirsium verutum (D. Don) Spruner ■

91753 Cirsium arisanense Kitam.;阿里山蓟(污白冠毛蓟,紫花阿里山蓟);Alishan Thistle ■

91754 Cirsium arisanense Kitam. f. purpurescens Kitam.;紫花阿里山蓟■

91755 Cirsium arisanense Kitam. f. purpurescens Kitam. = Cirsium arisanense Kitam. ■

91756 Cirsium arizonicum (A. Gray) Petr.;亚利桑那蓟;Arizona Thistle ■☆

91757 Cirsium arizonicum (A. Gray) Petr. var. bipinnatum (Eastw.) D. J. Keil;十字蓟;Four Corners Thistle ■☆

91758 Cirsium arizonicum (A. Gray) Petr. var. chellyense (R. J. Moore et Frankton) D. J. Keil;纳瓦霍蓟;Navajo Thistle ■☆

91759 Cirsium arizonicum (A. Gray) Petr. var. nidulum (M. E. Jones) S. L. Welsh = Cirsium arizonicum (A. Gray) Petr. ■☆

91760 Cirsium arizonicum (A. Gray) Petr. var. rothrockii (A. Gray) D. J. Keil;罗氏亚利桑那蓟;Rothrock's Thistle ■☆

91761 Cirsium arizonicum (A. Gray) Petr. var. tenuisectum D. J. Keil;沙丘亚利桑那蓟;Desert Mountains Thistle ■☆

91762 Cirsium armenum DC.;亚美尼亚蓟■☆

91763 Cirsium arnese (L.) Scop. = Cirsium setosum (Willd.) M. Bieb. ■

91764 Cirsium arnese (L.) Scop. var. incanum Ledeb. = Cirsium setosum (Willd.) M. Bieb. ■

91765 Cirsium arvense (L.) Robson var. argenteum (Peyer ex Vest) Fiori = Cirsium arvense (L.) Scop. ■

91766 Cirsium arvense (L.) Robson var. horridum Wimm. et Grab. = Cirsium arvense (L.) Scop. ■

91767 Cirsium arvense (L.) Robson var. integrifolium Wimm. et Grab. = Cirsium arvense (L.) Scop. ■

91768 Cirsium arvense (L.) Robson var. mite Wimm. et Grab. = Cirsium arvense (L.) Scop. ■

91769 Cirsium arvense (L.) Robson var. vestitum Wimm. et Grab. = Cirsium arvense (L.) Scop. ■

91770 Cirsium arvense (L.) Scop.;丝路蓟(加拿大飞廉,野刺儿菜);Boar Thistle, Boar-thistle, Canada Thistle, Canadian Thistle, Corn Thistle, Cow's Thistle, Cramp Thistle, Creeping Thistle, Cursed Thistle, Dashel, Dassel, Dazzle, Dodger, Dog Thistle, Field Thistle, Gypsy Flower, Hard Thistle, Pricky Thistle, Sharp Thistle, Sheep's Thistle, Swamp Thistle, Way Thistle, Western Thistle ■

91771 Cirsium arvense (L.) Scop. f. albiflorum (E. L. Rand et Redfield) Ralph Hoffm.;白花丝路蓟■☆

91772 Cirsium arvense (L.) Scop. f. albiflorum (E. L. Rand et Redfield) Ralph Hoffm. = Cirsium arvense (L.) Scop. ■

91773 Cirsium arvense (L.) Scop. f. albiflorum (Redfield) Ralph Hoffm. = Cirsium arvense (L.) Scop. f. albiflorum (E. L. Rand et

Redfield) Ralph Hoffm. ■☆

91774 Cirsium arvense (L.) Scop. var. alpestre Naig. et S. Y. Hu = Cirsium lanatum (Roxb. ex Willd.) Spreng. ■

91775 Cirsium arvense (L.) Scop. var. argenteum (Vest) Fiori = Cirsium arvense (L.) Scop. ■

91776 Cirsium arvense (L.) Scop. var. horridum Wimm. et Grab. = Cirsium arvense (L.) Scop. ■

91777 Cirsium arvense (L.) Scop. var. incanum Ledeb. = Cirsium incanum (S. G. Gmel.) Fisch. ex M. Bieb. ■

91778 Cirsium arvense (L.) Scop. var. integrifolium Wimm. et Grab. = Cirsium arvense (L.) Scop. ■

91779 Cirsium arvense (L.) Scop. var. integrifolium Wimm. et Grab. = Cirsium setosum (Willd.) M. Bieb. ■

91780 Cirsium arvense (L.) Scop. var. mite Wimm. et Grab. = Cirsium arvense (L.) Scop. ■

91781 Cirsium arvense (L.) Scop. var. mite Wimm. et Grab. = Cirsium lanatum (Roxb. ex Willd.) Spreng. ■

91782 Cirsium arvense (L.) Scop. var. setosum (Willd.) Ledeb. = Cirsium setosum (Willd.) M. Bieb. ■

91783 Cirsium arvense (L.) Scop. var. setosum (Willd.) Ledeb. f. albbiflorum Kitag. = Cirsium setosum (Willd.) M. Bieb. ■

91784 Cirsium arvense (L.) Scop. var. vestitum Wimm. et Grab. = Cirsium arvense (L.) Scop. ■

91785 Cirsium ashinokuraense Kadota;芦仓蓟■☆

91786 Cirsium asiaticum Schischk. = Cirsium serratuloides (L.) Hill ■

91787 Cirsium asperum Nakai = Cirsium maackii Maxim. ■

91788 Cirsium austrinum (Small) E. D. Schulz = Cirsium texanum Buckley ■☆

91789 Cirsium austrokiushianum Kitam. = Cirsium sieboldii Miq. subsp. austrokiushianum (Kitam.) Kitam. ■☆

91790 Cirsium autumnale Kitam. = Cirsium oligophyllum (Franch. et Sav.) Matsum. ■☆

91791 Cirsium babanum Koidz.;马场蓟■☆

91792 Cirsium babanum Koidz. var. otayae (Kitam.) Kitam. = Cirsium otayae Kitam. ■☆

91793 Cirsium barnebyi S. L. Welsh et Neese;巴纳比蓟;Barneby's Thistle ■☆

91794 Cirsium belingshanicum Petr. ex Hand. -Mazz. = Cirsium japonicum Fisch. ex DC. ■

91795 Cirsium bernardinum (Greene) Petr. = Cirsium occidentale (Nutt.) Jeps. var. californicum (A. Gray) D. J. Keil et C. E. Turner ■☆

91796 Cirsium biebersteinii Charadze;毕氏蓟■☆

91797 Cirsium bigelovii DC. = Cirsium muticum Michx. ■☆

91798 Cirsium bipinnatum (Eastw.) Rydb. = Cirsium arizonicum (A. Gray) Petr. var. bipinnatum (Eastw.) D. J. Keil ■☆

91799 Cirsium bitchuense Nakai;备中蓟■☆

91800 Cirsium bitchuense Nakai var. manisanense Kitam.;摩尼山蓟■☆

91801 Cirsium blumeri Petr. = Cirsium wheeleri (A. Gray) Petr. ■☆

91802 Cirsium bodinieri (Vaniot) H. Lév. = Cirsium japonicum Fisch. ex DC. ■

91803 Cirsium bolocephalum Petr. ex Hand. -Mazz. = Cirsium eriophoroides (Hook. f.) Petr. ■

91804 Cirsium bolocephalum Petr. ex Hand. -Mazz. var. racemosum Petr. ex Hand. -Mazz. = Cirsium eriophoroides (Hook. f.) Petr. ■

91805 Cirsium bolocephalum Petr. ex Hand. -Mazz. var. setchwanicum Petr. ex Hand. -Mazz. = Cirsium eriophoroides (Hook. f.) Petr. ■

91806 Cirsium boninense Koidz.;小泉蓟■☆

91807 Cirsium boreale Kitam. = Cirsium kamtschaticum Ledeb. ex DC. var. boreale (Kitam.) Tatew. ■☆

91808 Cirsium borealinipponense Kitam.;北村蓟■☆

91809 Cirsium bornmfilleri Sint. ex Bornm.;鲍尔蓟■☆

91810 Cirsium botryodes Petr.;总状蓟■

91811 Cirsium botryodes Petr. = Cirsium griseum H. Lév. ■

91812 Cirsium botryodes Petr. ex Hand. -Mazz. = Cirsium griseum H. Lév. ■

91813 Cirsium botrys Petr. = Cirsium cymosum (Greene) J. T. Howell ■☆

91814 Cirsium bracteiferum C. Shih;刺蓟草(刺盖草,大刺盖);Bracteate Thistle ■

91815 Cirsium bracteosum DC.;多苞片蓟■☆

91816 Cirsium brevicaule A. Gray;大小蓟(岛蓟,鸡觞刺,统天草)■

91817 Cirsium brevicaule A. Gray f. albiflorum Kitam.;白花大小蓟■☆

91818 Cirsium brevicaule A. Gray var. irumtiense (Kitam.) Kitam. = Cirsium brevicaule A. Gray ■

91819 Cirsium brevicaule A. Gray var. oshimense Kitam. = Cirsium brevicaule A. Gray ■

91820 Cirsium brevifolium Nutt.;短叶蓟;Palouse Thistle ■☆

91821 Cirsium brevipapposum Tschern.;短柔毛蓟■☆

91822 Cirsium brevistylum Cronquist;短柱蓟;Clustered Thistle, Indian Thistle, Short-style Thistle ■☆

91823 Cirsium breweri (A. Gray) Jeps. = Cirsium douglasii DC. var. breweri (A. Gray) D. J. Keil et C. E. Turner ■☆

91824 Cirsium buchwaldii O. Hoffm.;布赫蓟■☆

91825 Cirsium buergeri Miq.;伯格蓟■☆

91826 Cirsium buergeri Miq. f. albiflorum Honda;白花伯格蓟■☆

91827 Cirsium buschianum Charadze;布什蓟■☆

91828 Cirsium butleri (Rydb.) Petr. = Cirsium scariosum Nutt. ■☆

91829 Cirsium calcareum (M. E. Jones) Wooton et Standl. = Cirsium arizonicum (A. Gray) Petr. var. bipinnatum (Eastw.) D. J. Keil ■☆

91830 Cirsium calcareum (M. E. Jones) Wooton et Standl. var. bipinnatum (Eastw.) S. L. Welsh = Cirsium arizonicum (A. Gray) Petr. var. bipinnatum (Eastw.) D. J. Keil ■☆

91831 Cirsium calcareum (M. E. Jones) Wooton et Standl. var. pulchellum (Greene ex Rydb.) S. L. Welsh = Cirsium arizonicum (A. Gray) Petr. var. bipinnatum (Eastw.) D. J. Keil ■☆

91832 Cirsium calcicola Nakai = Cirsium dipsacolepis (Maxim.) Matsum. var. calcicola (Nakai) Kitam. ■☆

91833 Cirsium californicum A. Gray = Cirsium occidentale (Nutt.) Jeps. var. californicum (A. Gray) D. J. Keil et C. E. Turner ■☆

91834 Cirsium californicum A. Gray subsp. pseudoreglense Petr. = Cirsium occidentale (Nutt.) Jeps. var. californicum (A. Gray) D. J. Keil et C. E. Turner ■☆

91835 Cirsium californicum A. Gray var. bernardinum (Greene) Petr. = Cirsium occidentale (Nutt.) Jeps. var. californicum (A. Gray) D. J. Keil et C. E. Turner ■☆

91836 Cirsium callilepis (Greene) Jeps. = Cirsium remotifolium (Hook.) DC. var. odontolepis Petr. ■☆

91837 Cirsium callilepis (Greene) Jeps. var. oregonense (Petr.) J. T. Howell = Cirsium remotifolium (Hook.) DC. var. odontolepis Petr. ■☆

91838 Cirsium callilepis (Greene) Jeps. var. pseudocarlinoides (Petr.) J. T. Howell = Cirsium remotifolium (Hook.) DC. var. odontolepis Petr. ■☆

91839 Cirsium campylon H. Sharsm. = Cirsium fontinale (Greene) Jeps. var. campylon (H. Sharsm.) Pilz ex D. J. Keil et C. E. Turner ■☆

91840 Cirsium canescens Nutt.;普拉特蓟;Platte Thistle,Prairie Thistle ■☆

91841 Cirsium canovirens (Rydb.) Petr. = Cirsium cymosum (Greene) J. T. Howell var. canovirens (Rydb.) D. J. Keil ■☆

91842 Cirsium canum M. Bieb.;灰蓝蓟;Queen Anne's Thistle ■☆

91843 Cirsium carolinianum (Walter) Fernald et B. G. Schub.;卡罗来纳蓟;Carolina Thistle, Purple Thistle, Smallhead Thistle, Soft Thistle, Thistle ■☆

91844 Cirsium carthamoides (Willd.) Link = Rhaponticum carthamoides (Willd.) Iljin ■

91845 Cirsium carthamoides Link = Stemmacantha carthamoides (Willd.) Dittrich ■

91846 Cirsium casabonae (L.) DC. = Ptilostemon casabonae (L.) Greuter ■☆

91847 Cirsium casabonae (L.) DC. subsp. abylense (Pau Maire = Ptilostemon abylensis (Pau et Font Quer) Greuter ■☆

91848 Cirsium casabonae (L.) DC. subsp. dyricola Maire = Ptilostemon dyricola (Maire) Greuter ■☆

91849 Cirsium casabonae (L.) DC. subsp. rhiphaeum (Pau et Font Quer) Maire = Ptilostemon rhiphaeus (Pau et Font Quer) Greuter ■☆

91850 Cirsium casabonae (L.) DC. subsp. trispinosum (Moench) Maire = Ptilostemon casabonae (L.) Greuter ■☆

91851 Cirsium casabonae (L.) DC. var. tetauense Font Quer = Ptilostemon rhiphaeus (Pau et Font Quer) Greuter ■☆

91852 Cirsium caucasicum (Adam) Petr.;高加索蓟■☆

91853 Cirsium cavaleriei (H. Lév.) H. Lév. = Cirsium monocephalum (Vaniot) H. Lév. ■

91854 Cirsium centaureae (Rydb.) K. Schum. = Cirsium clavatum (M. E. Jones) Petr. var. americanum (A. Gray) D. J. Keil ■☆

91855 Cirsium centauroides Willd. = Rhaponticum carthamoides (Willd.) Iljin ■

91856 Cirsium cephalotes Boiss.;头蓟■☆

91857 Cirsium cerberus (Vaniot) H. Lév. = Cirsium japonicum Fisch. ex DC. ■

91858 Cirsium chamaepeuce (L.) Ten. var. gnaphalioides (Cirillo) Pamp. = Ptilostemon gnaphaloides (Cirillo) Soják ■☆

91859 Cirsium charkeviczii Barkalov = Cirsium pectinellum A. Gray ■☆

91860 Cirsium chellyense R. J. Moore et Frankton = Cirsium arizonicum (A. Gray) Petr. var. chellyense (R. J. Moore et Frankton) D. J. Keil ■☆

91861 Cirsium chienii C. C. Chang = Cirsium leo Nakai et Kitag. ■

91862 Cirsium chikushiense Koidz.;筑紫蓟■☆

91863 Cirsium chikushiense Koidz. f. albiflorum Sakata;白花筑紫蓟■☆

91864 Cirsium chinense Gardner et Champ.;绿蓟(钩芙,华蓟,苦芙,苦芙子,苦板,苦藉,轮蓟,牛刺梨,狭叶蓟,小蓟,小样刺米草,中国蓟);Chinese Thistle,Green Thistle ■

91865 Cirsium chinense Gardner et Champ. = Cirsium lineare (Thunb.) Sch. Bip. ■

91866 Cirsium chinense Gardner et Champ. = Cirsium shansiense Petr. ■

91867 Cirsium chinense Gardner et Champ. ex Hook. = Cirsium chinense Gardner et Champ. ■

91868 Cirsium chinense Gardner et Champ. var. austale Diels = Cirsium shansiense Petr. ■

91869 Cirsium chinense Gardner et Champ. var. laushanense (Y. Yabe) Kitag. = Cirsium chinense Gardner et Champ. ■

91870 Cirsium chlorocomos Sommier et H. Lév.;绿簇毛蓟■☆

91871 Cirsium chlorolepis Petr. = Cirsium chlorolepis Petr. ex Hand.-Mazz. ■

91872 Cirsium chlorolepis Petr. ex Hand.-Mazz.;两面刺(白马刺,大蓟,滇大蓟,鸡脚刺,两面蓟,青刺蓟);Greenbract Thistle ■

91873 Cirsium chokaiense Kitam.;鸟海山蓟■☆

91874 Cirsium chrysacanthum (Ball) Jahand.;金花蓟■☆

91875 Cirsium chrysacanthum (Ball) Jahand. var. ornatum (Ball) Maire = Cirsium chrysacanthum (Ball) Jahand. ■☆

91876 Cirsium chrysacanthum (Ball) Jahand. var. subornatum Emb. et Maire = Cirsium chrysacanthum (Ball) Jahand. ■☆

91877 Cirsium chrysolepis C. Shih;黄苞蓟;Yellowbract Thistle ■

91878 Cirsium chuskaense R. J. Moore et Frankton = Cirsium arizonicum (A. Gray) Petr. var. chellyense (R. J. Moore et Frankton) D. J. Keil ■☆

91879 Cirsium ciliatum (Murray) M. Bieb.;纤毛蓟■☆

91880 Cirsium ciliolatum (L. F. Hend.) J. T. Howell;阿希兰蓟;Ashland Thistle ■☆

91881 Cirsium clavatum (M. E. Jones) Petr.;棍棒蓟;Fish Lake Thistle ■☆

91882 Cirsium clavatum (M. E. Jones) Petr. var. americanum (A. Gray) D. J. Keil;美洲棍棒蓟;Rocky Mountain Fringed Thistle ■☆

91883 Cirsium clavatum (M. E. Jones) Petr. var. markaguntense S. L. Welsh = Cirsium clavatum (M. E. Jones) Petr. ■☆

91884 Cirsium clavatum (M. E. Jones) Petr. var. osterhoutii (Rydb.) D. J. Keil;奥氏棍棒蓟;Osterhout's Thistle ■☆

91885 Cirsium clokeyi S. F. Blake = Cirsium eatonii (A. Gray) B. L. Rob. var. clokeyi (S. F. Blake) D. J. Keil ■☆

91886 Cirsium coccinatum Osterh. = Cirsium drummondii Torr. et A. Gray ■☆

91887 Cirsium coloradense (Rydb.) Cockerell ex Daniels = Cirsium scariosum Nutt. var. coloradense (Rydb.) D. J. Keil ■☆

91888 Cirsium coloradense (Rydb.) Cockerell ex Daniels subsp. acaulescens (A. Gray) Petr. = Cirsium scariosum Nutt. var. americanum (A. Gray) D. J. Keil ■☆

91889 Cirsium coloradense (Rydb.) Cockerell ex Daniels subsp. longissimum (A. Heller) Petr. = Cirsium scariosum Nutt. var. americanum (A. Gray) D. J. Keil ■☆

91890 Cirsium comosum (Franch. et Sav.) Matsum. = Cirsium nipponicum (Maxim.) Makino var. incomptum (Maxim.) Kitam. ■☆

91891 Cirsium comosum (Franch. et Sav.) Matsum. var. yatsugatakense (Nakai) Kitam. = Cirsium ovalifolium (Franch. et Sav.) Matsum. ■☆

91892 Cirsium confertissimum Nakai;极密蓟■☆

91893 Cirsium confertissimum Nakai var. herbicola Nakai = Cirsium confertissimum Nakai ■☆

91894 Cirsium congdonii R. J. Moore et Frankton = Cirsium scariosum Nutt. var. congdonii (R. J. Moore et Frankton) D. J. Keil ■☆

91895 Cirsium congestissimum Kitam.;密簇蓟■☆

91896 Cirsium congestum Fisch. et C. A. Mey. ex DC.;密集蓟■☆

91897 Cirsium coulteri Harv. et A. Gray = Cirsium occidentale (Nutt.) Jeps. var. coulteri (Harv. et A. Gray) Jeps. ■☆

91898 Cirsium crassicaule (Greene) Jeps.;粗茎蓟;Slough Thistle ■☆

91899 Cirsium creticum (Lam.) d'Urv.;克里特蓟■☆

91900 Cirsium cymosum (Greene) J. T. Howell;外来蓟;Peregrine Thistle ■☆

91901 Cirsium cymosum (Greene) J. T. Howell var. canovirens (Rydb.) D. J. Keil;灰绿蓟;Graygreen Thistle ■☆

91902 Cirsium daghestanicum Charadze;达赫斯坦蓟■☆

91903 Cirsium davisii Cronquist = Cirsium inamoenum (Greene) D. J. Keil var. davisii (Cronquist) D. J. Keil ■☆

91904 Cirsium dealbatum M. Bieb.;白蓟■☆

91905 Cirsium desertorum Fisch. ex Link = Cirsium alatum (S. G. Gmel.) Bobrov ■

91906 Cirsium desertorum Fisch. ex Link var. integerrimum Trautv. = Cirsium alatum (S. G. Gmel.) Bobrov ■

91907 Cirsium desertorum Fisch. ex Link var. sinuatolobatum Trautv. = Cirsium alatum (S. G. Gmel.) Bobrov ■

91908 Cirsium desertorum Fisch. ex Link var. subintegerrima Trautv. = Cirsium alatum (S. G. Gmel.) Bobrov ■

91909 Cirsium diabolicum Kitam. = Onopordum illyricum L. ■☆

91910 Cirsium diamentiacum (Nakai) Nakai = Cirsium schantarense Trautv. et C. A. Mey. ■

91911 Cirsium diffusum (Eastw.) Rydb. = Cirsium arizonicum (A. Gray) Petr. var. bipinnatum (Eastw.) D. J. Keil ■☆

91912 Cirsium dipsacolepis (Maxim.) Matsum. var. calcicola (Nakai) Kitam.;钙生蓟■☆

91913 Cirsium dipsacolepis Matsum.;续断状蓟(菊牛蒡)■☆

91914 Cirsium discolor (Muhl. ex Willd.) Spreng.;异色蓟;Field Thistle,Fieldthistle,Pasture Thistle,Prairie Thistle ■☆

91915 Cirsium discolor (Muhl. ex Willd.) Spreng. f. albiflorum (Britton) House = Cirsium discolor (Muhl. ex Willd.) Spreng. ■☆

91916 Cirsium dissectum (L.) Hill;草地蓟;Gentle Thistle, Marsh Plume Thistle,Meadow Thistle,Pig Leaves ■☆

91917 Cirsium douglasii DC.;道格拉斯蓟;California Swamp Thistle, Douglas' Thistle ■☆

91918 Cirsium douglasii DC. var. breweri (A. Gray) D. J. Keil et C. E. Turner;布鲁尔蓟;Brewer's Thistle ■☆

91919 Cirsium douglasii DC. var. canescens (Petr.) J. T. Howell = Cirsium douglasii DC. var. breweri (A. Gray) D. J. Keil et C. E. Turner ■☆

91920 Cirsium drummondii Torr. et A. Gray;德拉蒙德蓟;Drummond's Thistle,Dwarf Thistle Short-stemmed Thistle ■☆

91921 Cirsium drummondii Torr. et A. Gray subsp. latisquamum Petr. = Cirsium scariosum Nutt. var. americanum (A. Gray) D. J. Keil ■☆

91922 Cirsium drummondii Torr. et A. Gray subsp. latisquamum Petr. = Cirsium scariosum Nutt. var. citrinum (Petr.) D. J. Keil ■☆

91923 Cirsium drummondii Torr. et A. Gray subsp. vexans Petr. = Cirsium scariosum Nutt. var. americanum (A. Gray) D. J. Keil ■☆

91924 Cirsium drummondii Torr. et A. Gray var. acaulescens (A. Gray) J. F. Macbr. = Cirsium scariosum Nutt. var. americanum (A. Gray) D. J. Keil ■☆

91925 Cirsium drummondii Torr. et A. Gray var. oregonense Petr. = Cirsium scariosum Nutt. var. americanum (A. Gray) D. J. Keil ■☆

91926 Cirsium ducellieri Maire;迪塞利耶蓟■☆

91927 Cirsium dyris Jahand. et Maire;荒地蓟■☆

91928 Cirsium dyris Jahand. et Maire var. sidi-guinii (Pau et Font Quer) Maire = Cirsium dyris Jahand. et Maire ■☆

91929 Cirsium eatonii (A. Gray) B. L. Rob.;伊顿蓟;Eaton's Thistle, Mountaintop Thistle ■☆

91930 Cirsium eatonii (A. Gray) B. L. Rob. var. clokeyi (S. F. Blake) D. J. Keil;克洛基蓟;Clokey Thistle, Spring Mountains Thistle, White-spine Thistle ■☆

91931 Cirsium eatonii (A. Gray) B. L. Rob. var. eriocephalum (A. Gray) D. J. Keil;毛头蓟;Alpine Thistle,Mountain Thistle ■☆

91932 Cirsium eatonii (A. Gray) B. L. Rob. var. harrisonii S. L. Welsh = Cirsium eatonii (A. Gray) B. L. Rob. ■☆

91933 Cirsium eatonii (A. Gray) B. L. Rob. var. hesperium (Eastw.) D. J. Keil;高山蓟;Tall Mountain Thistle ■☆

91934 Cirsium eatonii (A. Gray) B. L. Rob. var. murdockii S. L. Welsh;北山蓟;Northern Mountain Thistle ■☆

91935 Cirsium eatonii (A. Gray) B. L. Rob. var. peckii (L. F. Hend.) D. J. Keil;佩克蓟;Ghost Thistle,Steens Mountain Thistle ■☆

91936 Cirsium eatonii (A. Gray) B. L. Rob. var. viperinum D. J. Keil;蝮蛇蓟;Snake Range Thistle ■☆

91937 Cirsium echinatum (Desf.) DC.;小刺蓟■☆

91938 Cirsium echinatum (Desf.) DC. var. willkomminianum (Porta et Rigo) Maire = Cirsium echinatum (Desf.) DC. ■☆

91939 Cirsium echinus (M. Bieb.) Hand.-Mazz.;刺蓟■☆

91940 Cirsium edule Nutt.;印度蓟;Edible Thistle, Hall's Thistle, Indian Thistle ■☆

91941 Cirsium edule Nutt. var. macounii (Greene) D. J. Keil;梅肯蓟;Macoun's Thistle ■☆

91942 Cirsium edule Nutt. var. wenatchense D. J. Keil;韦纳契蓟;Wenatchee Thistle ■☆

91943 Cirsium elbrusense Sommier et H. Lév.;厄尔布鲁士蓟■☆

91944 Cirsium elodes M. Bieb.;盐土蓟■☆

91945 Cirsium elodes M. Bieb. var. setigerum Krylov = Cirsium alatum (S. G. Gmel.) Bobrov ■

91946 Cirsium elodes M. Bieb. var. sinuatolobatum O. Fedtsch. = Cirsium alatum (S. G. Gmel.) Bobrov ■

91947 Cirsium elodes M. Bieb. var. subintegrerrima O. Fedtsch. = Cirsium alatum (S. G. Gmel.) Bobrov ■

91948 Cirsium elodes M. Bieb. var. subintegrerrima O. Fedtsch. et B. Fedtsch. = Cirsium alatum (S. G. Gmel.) Bobrov ■

91949 Cirsium engelmannii Rydb.;恩格尔曼蓟;Blackland Thistle, Engelmann's Thistle ■☆

91950 Cirsium englerianum O. Hoffm.;恩氏蓟■☆

91951 Cirsium eriocephalum A. Gray = Cirsium eatonii (A. Gray) B. L. Rob. var. eriocephalum (A. Gray) D. J. Keil ■☆

91952 Cirsium eriocephalum A. Gray var. leiocephalum D. C. Eaton = Cirsium eatonii (A. Gray) B. L. Rob. ■☆

91953 Cirsium eriophoroides (Hook. f.) Petr.;贡山蓟(藏大蓟,大刺儿菜,大蓟,毛头蓟);Gongshan Thistle,Kungshan Thistle ■

91954 Cirsium eriophorum (L.) Scop.;绵毛蓟;Cotton Thistle, Down Thistle, Fish Belly, Friar's Crown, Wool Thistle, Woolly Thistle, Woolly-headed Thistle ■☆

91955 Cirsium erisithales Scop.;糙黏蓟(黄花蓟);Yellow Melancholy Thistle,Yellow Thistle ■☆

91956 Cirsium erosum (Rydb.) K. Schum. = Cirsium scariosum Nutt. var. coloradense (Rydb.) D. J. Keil ■☆

91957 Cirsium erythrolepis K. Koch;红苞蓟■☆

91958 Cirsium esculentum (Siev.) C. A. Mey.;莲座蓟(食用蓟);Rosette Thistle ■

91959 Cirsium esculentum (Siev.) C. A. Mey. var. acaule Trautv. = Cirsium esculentum (Siev.) C. A. Mey. ■

91960 Cirsium euxinum Charadze;黑蓟■☆

91961 Cirsium falcatum Turcz. ex DC. = Cirsium pendulum Fisch. ex DC. ■

91962 Cirsium fangii Petr.;峨眉蓟;Emei Thistle,Fang's Thistle ■

91963 Cirsium fanjingshanense C. Shih;梵净蓟;Fanjingshan Thistle ■

91964 Cirsium fargesii (Franch.) Diels;等苞蓟(光苞蓟);Farges Thistle ■

91965 Cirsium fargesii (Franch.) Diels = Cirsium leo Nakai et Kitag. ■
91966 Cirsium fauriei Nakai;法氏蓟■☆
91967 Cirsium ferum Kitam.;鳞毛蓟(台湾野蓟,锥果蓟)■
91968 Cirsium ficifolium Fisch. = Synurus deltoides (Aiton) Nakai ■
91969 Cirsium flaccidum (Small) Petr. = Cirsium carolinianum (Walter) Fernald et B. G. Schub. ■☆
91970 Cirsium flavispina Boiss. = Cirsium pyrenaicum (Jacq.) All. ■☆
91971 Cirsium flavispina Boiss. subsp. perniveum H. Lindb. = Cirsium pyrenaicum (Jacq.) All. ■☆
91972 Cirsium flavispina Boiss. var. brachyacanthum Maire = Cirsium pyrenaicum (Jacq.) All. ■☆
91973 Cirsium flavispina Boiss. var. longespinosum Kunze = Cirsium pyrenaicum (Jacq.) All. ■☆
91974 Cirsium flavispina Boiss. var. perniveum (H. Lindb.) Maire = Cirsium pyrenaicum (Jacq.) All. ■☆
91975 Cirsium floccosum (Rydb.) Petr. = Cirsium tracyi (Rydb.) Petr. ■☆
91976 Cirsium flodmanii (Rydb.) Arthur;弗洛德曼蓟;Flodman's Thistle,Prairie Thistle ■☆
91977 Cirsium foliosum (Hook.) DC.;密叶蓟;Elk Thistle,Foliose Thistle,Leafy Thistle ■☆
91978 Cirsium fominii Petr.;福明蓟■☆
91979 Cirsium fontinale (Greene) Jeps.;泉蓟;Fountain Thistle ■☆
91980 Cirsium fontinale (Greene) Jeps. var. campylon (H. Sharsm.) Pilz ex D. J. Keil et C. E. Turner;哈密尔顿山蓟;Mt. Hamilton Thistle ■☆
91981 Cirsium fontinale (Greene) Jeps. var. obispoense J. T. Howell;袄维斯波蓟;Chorro Creek Bog Thistle ■☆
91982 Cirsium formosanum Sasaki;台湾蓟;Taiwan Thistle ■
91983 Cirsium forrestii (Diels) H. Lév. = Cirsium henryi (Franch.) Diels ■
91984 Cirsium frickii Fisch. et C. A. Mey.;弗里克蓟■☆
91985 Cirsium furusei Kitam.;古施蓟■☆
91986 Cirsium furusei Kitam. var. spinuliferum (Kitam.) Kitam.;小刺古施蓟■☆
91987 Cirsium fusco-trichum C. C. Chang;褐毛蓟;Brownhair Thistle ■
91988 Cirsium gagnidzei Charadze;嘎格蓟■☆
91989 Cirsium giganteum (Desf.) Batt. = Cirsium scabrum (Poir.) Bonnet ■☆
91990 Cirsium gilense (Wooton et Standl.) Wooton et Standl. = Cirsium parryi (A. Gray) Petr. ■☆
91991 Cirsium glabrifolium (C. Winkl.) O. Fedtsch. et B. Fedtsch.;无毛蓟;Glabrous Thistle ■
91992 Cirsium glabrifolium Petr. = Cirsium glabrifolium (C. Winkl.) O. Fedtsch. et B. Fedtsch. ■
91993 Cirsium gmelinii (Spreng.) Tausch = Cirsium esculentum (Siev.) C. A. Mey. ■
91994 Cirsium gmelinii Tausch. = Cirsium esculentum (Siev.) C. A. Mey. ■
91995 Cirsium grahamii A. Gray;格雷哈姆蓟■☆
91996 Cirsium grandirosuliferum Kadota;大莲座蓟■☆
91997 Cirsium gratiosum Kitam.;多姿蓟■☆
91998 Cirsium gratiosum Kitam. var. alpinum (Nakai) Kitam.;高山多姿蓟■☆
91999 Cirsium grayanum (Maxim.) Nakai;格雷蓟■☆
92000 Cirsium griseum (Rydb.) K. Schum. = Cirsium clavatum (M. E. Jones) Petr. var. americanum (A. Gray) D. J. Keil ■☆
92001 Cirsium griseum H. Lév.;灰蓟(总状蓟);Grey Thistle ■
92002 Cirsium hachijoense Nakai;八丈岛蓟■☆
92003 Cirsium hachimantaiense Kadota;八幡平蓟■☆
92004 Cirsium hadelii Petr. ex Hand.-Mazz. = Cirsium interpositum Petr. ■
92005 Cirsium hainanense Masam. = Cirsium japonicum Fisch. ex DC. ■
92006 Cirsium hallii (A. Gray) M. E. Jones = Cirsium edule Nutt. ■☆
92007 Cirsium hanamakiense Kitam.;花卷蓟■☆
92008 Cirsium handelii Petr. = Cirsium argyracanthum DC. ■
92009 Cirsium handelii Petr. ex Hand.-Mazz.;骆骑蓟(骆骑);Handel Thistle ■
92010 Cirsium happoense Kadota;八风蓟■☆
92011 Cirsium heiianum Koidz.;平氏蓟■☆
92012 Cirsium helenioides (L.) Hill;堆心蓟(忧郁蓟);Melancholy Thistle,Sneezeweedlike Thistle ■
92013 Cirsium helenioides (L.) Hill = Cirsium heterophyllum (L.) Hill ■
92014 Cirsium helenioides Willd. = Cirsium helenioides (L.) Hill ■
92015 Cirsium heleophilum Petr. ex Hand.-Mazz. = Cirsium griseum H. Lév. ■
92016 Cirsium helgendrofii (Franch. et Sav.) Makino = Cirsium pendulum Fisch. ex DC. ■
92017 Cirsium helleri (Small) Cory = Cirsium texanum Buckley ■☆
92018 Cirsium helleri (Small) Cory = Cirsium undulatum (Nutt.) Spreng. ■☆
92019 Cirsium henryi (Franch.) Diels;刺苞蓟(鄂西大蓟);Henry Thistle ■
92020 Cirsium hesperium (Eastw.) Petr. = Cirsium eatonii (A. Gray) B. L. Rob. var. hesperium (Eastw.) D. J. Keil ■☆
92021 Cirsium heterophylloides Pavlov = Cirsium helenioides (L.) Hill ■
92022 Cirsium heterophyllum (L.) Hill;异叶蓟;Carldoddie,Diverseleaf Thistle,Fish Belly,Melancholy Thistle,Tazzle ■
92023 Cirsium heterophyllum (L.) Hill = Cirsium helenioides (L.) Hill ■
92024 Cirsium heterophyllum (L.) Hill subsp. angarense Popov = Cirsium helenioides (L.) Hill ■
92025 Cirsium hidaense Kitam. = Cirsium tashiroi Kitam. var. hidaense (Kitam.) Kadota ■☆
92026 Cirsium hidakamontanum Kadota;日高山蓟■☆
92027 Cirsium hillii (Canby) Fernald;希尔蓟;Hill's Thistle,Hollow-rooted Thistle,Prairie Thistle ■☆
92028 Cirsium hillii (Canby) Fernald = Cirsium pumilum (Nutt.) Spreng. var. hillii (Canby) B. Boivin ■☆
92029 Cirsium homolepis Nakai;同鳞蓟■☆
92030 Cirsium hookerianum Nutt.;胡克蓟(虎克蓟);Hooker's Thistle,White Thistle ■☆
92031 Cirsium hookerianum Nutt. var. scariosum (Nutt.) B. Boivin = Cirsium scariosum Nutt. ■☆
92032 Cirsium horiianum Kadota;堀井蓟■☆
92033 Cirsium horridulum Michx.;毛黄蓟(厌蓟);Bristly Thistle,Bull Thistle,Horrid Thistle,Yellow Thistle ■☆
92034 Cirsium horridulum Michx. var. elliottii Torr. et A. Gray = Cirsium horridulum Michx. ■☆
92035 Cirsium horridulum Michx. var. megacanthum (Nutt.) D. J. Keil;大刺蓟;Bigspine Thistle ■☆
92036 Cirsium horridulum Michx. var. vittatum (Small) R. W. Long;佛罗里达蓟;Florida Thistle ■☆

92037 Cirsium hosokawae Kitam.;红花蓟(白冠毛蓟,细川氏蓟);Hosokawa Thistle ■
92038 Cirsium howellii Petr. = Cirsium ciliolatum (L. F. Hend.) J. T. Howell ■☆
92039 Cirsium hsiaowutaishanensis Chen = Olgaea lomonosowii (Trautv.) Iljin ■
92040 Cirsium hupehense Pamp.;湖北蓟;Hubei Thistle ■
92041 Cirsium hydrophiloides Charadze;拟喜湿蓟■☆
92042 Cirsium hydrophilum (Greene) Jeps.;喜湿蓟;Suisun Thistle ■☆
92043 Cirsium hydrophilum (Greene) Jeps. var. vaseyi (A. Gray) J. T. Howell;瓦赛蓟;Mount Tamalpais Thistle,Vasey's Thistle ■☆
92044 Cirsium hygrophilum Boiss.;喜水蓟■☆
92045 Cirsium hypoleucum DC.;里白蓟■☆
92046 Cirsium ibukiense Nakai = Cirsium japonicum Fisch. ex DC. ■
92047 Cirsium igniarium Spreng. = Ancathia igniaria (Spreng.) DC. ■
92048 Cirsium imereticum Boiss.;伊梅里特蓟■☆
92049 Cirsium inamoenum (Greene) D. J. Keil;格林蓟;Greene's Thistle ■☆
92050 Cirsium inamoenum (Greene) D. J. Keil var. davisii (Cronquist) D. J. Keil;戴维斯蓟;Davis' Thistle ■☆
92051 Cirsium incanum (S. G. Gmel.) Fisch. = Cirsium arvense (L.) Scop. ■
92052 Cirsium incanum (S. G. Gmel.) Fisch. ex M. Bieb.;阿尔泰蓟;Altai Mountain Thistle,Altai Thistle ■
92053 Cirsium incanum (S. G. Gmel.) Fisch. ex M. Bieb. = Cirsium arvense (L.) Scop. ■
92054 Cirsium incomptum (Maxim.) Nakai = Cirsium nipponicum (Maxim.) Makino var. incomptum (Maxim.) Kitam. ■☆
92055 Cirsium indefensum Kitam.;无护蓟■☆
92056 Cirsium inornatum (Wooton et Standl.) Wooton et Standl. = Cirsium parryi (A. Gray) Petr. ■☆
92057 Cirsium interpositum Petr.;披裂蓟;Lanceolatelobe Thistle ■
92058 Cirsium inundatum Makino;洪水地蓟■☆
92059 Cirsium inundatum Makino subsp. alpicola (Nakai) Kitam. = Cirsium alpicola Nakai ■☆
92060 Cirsium inundatum Makino var. alpicola (Nakai) Ohwi = Cirsium alpicola Nakai ■☆
92061 Cirsium involucratum DC. = Cirsium verutum (D. Don) Spruner ■
92062 Cirsium iowense (Pammel) Fernald = Cirsium altissimum (L.) Spreng. ■☆
92063 Cirsium iowense (Pammel) Fernald = Cirsium altissimum Hill ■☆
92064 Cirsium irumtiense Kitam. = Cirsium brevicaule A. Gray ■
92065 Cirsium ishizuchiense (Kitam.) Kadota;石锤山蓟■☆
92066 Cirsium japonicum Fisch. ex DC.;蓟(白花小蓟,草鞋刺,茨芥,刺盖草,刺蓟,刺蓟菜,刺秸子,刺橘子,刺萝卜,大刺刺菜,大刺儿菜,大刺盖,大刺角牙,大恶鸡,大恶鸡婆,大蓟,大蓟草,大居寒,大牛喳口,地丁草,地丁香,地疔,地萝卜,恶鸡婆,高雄蓟,鼓椎,虎蓟,鸡脚刺,鸡母刺,鸡姆刺,鸡项草,老虎刺,老虎脷,笋菜,六轮台,六月霜,驴扎嘴,罗平蓟,马刺草,马刺刺,马刺蓟,马刺口,马蓟,猫仔刺头,南国蓟,南国小蓟,南蓟,鸟不扑,牛不嗅,牛触嘴,牛刺笋菜,牛口参,牛口刺,牛口舌,牛溺刺,牛喳口,牛枝簕,山刺儿菜,山老鼠,山老鼠簕,山萝卜,山牛蒡,刷把头,台湾蓟,铁刺杆菜,土红花,土人参,土洋参,小蓟,野刺菜,野红花,蚁姆刺,猪妈菜,猪乸刺);Formosan Thistle,Japan Thistle,Japanese Thistle,Luoping Thistle,Taiwan Thistle ■
92067 Cirsium japonicum Fisch. ex DC. f. leucanthum Nakai;白花蓟■☆
92068 Cirsium japonicum Fisch. ex DC. subsp. maackii (Maxim.) Nakai = Cirsium maackii Maxim. ■
92069 Cirsium japonicum Fisch. ex DC. var. amurense Kitam. = Cirsium maackii Maxim. ■
92070 Cirsium japonicum Fisch. ex DC. var. australe Kitam.;南国小蓟■
92071 Cirsium japonicum Fisch. ex DC. var. australe Kitam. = Cirsium japonicum Fisch. ex DC. ■
92072 Cirsium japonicum Fisch. ex DC. var. diabolicum (Kitam.) Kitam. ex Ohwi;鬼蓟■☆
92073 Cirsium japonicum Fisch. ex DC. var. fukiense Kitam. = Cirsium japonicum Fisch. ex DC. ■
92074 Cirsium japonicum Fisch. ex DC. var. horridum Nakai;多刺蓟■☆
92075 Cirsium japonicum Fisch. ex DC. var. ibukiense (Nakai) Nakai ex Kitam.;伊吹山蓟■☆
92076 Cirsium japonicum Fisch. ex DC. var. intermeditan (Maxim.) Matsum. = Cirsium japonicum Fisch. ex DC. ■
92077 Cirsium japonicum Fisch. ex DC. var. maackii (Maxim.) Matsum. = Cirsium japonicum Fisch. ex DC. subsp. maackii (Maxim.) Nakai ■
92078 Cirsium japonicum Fisch. ex DC. var. okiense H. Koyama et Murata;大木蓟■☆
92079 Cirsium japonicum Fisch. ex DC. var. takaoense Kitam.;白花小蓟■
92080 Cirsium japonicum Fisch. ex DC. var. takaoense Kitam. = Cirsium japonicum Fisch. ex DC. ■
92081 Cirsium japonicum Fisch. ex DC. var. ussuriense (Regel) Kitam. = Cirsium maackii Maxim. ■
92082 Cirsium japonicum Fisch. ex DC. var. ussuriense (Regel) Kitam. = Cirsium japonicum Fisch. ex DC. subsp. maackii (Maxim.) Nakai ■
92083 Cirsium japonicum Fisch. ex DC. var. vestitum Kitam.;包被蓟■☆
92084 Cirsium japonicum Fisch. ex DC. var. vestitum Kitam. f. arakii (Kitam.) Kitam.;荒木蓟■☆
92085 Cirsium japonicum Fisch. ex DC. var. villosum Kadota;长柔毛蓟■☆
92086 Cirsium joannae S. L. Welsh,N. D. Atwood et L. C. Higgins;乔安娜蓟;Joanna's Thistle ■☆
92087 Cirsium kagamontanum Nakai;加贺山蓟■☆
92088 Cirsium kagamontanum Nakai f. albiflorum Ikegami;白花加贺山蓟■☆
92089 Cirsium kamtschaticum Ledch. ex DC.;勘察加蓟;Kamchatka Thistle ■☆
92090 Cirsium kamtschaticum Ledeb. ex DC. f. velutinum Kawano;短绒毛勘察加蓟■☆
92091 Cirsium kamtschaticum Ledeb. ex DC. subsp. apoense (Nakai) Kitam. = Cirsium apoense Nakai ■☆
92092 Cirsium kamtschaticum Ledeb. ex DC. subsp. boreale (Kitam.) Kitam. = Cirsium kamtschaticum Ledeb. ex DC. var. boreale (Kitam.) Tatew. ■☆
92093 Cirsium kamtschaticum Ledeb. ex DC. subsp. pectinellum (A. Gray) Kitam. = Cirsium pectinellum A. Gray ■☆
92094 Cirsium kamtschaticum Ledeb. ex DC. var. boreale (Kitam.) Tatew.;北方勘察加蓟■☆
92095 Cirsium kamtschaticum Ledeb. subsp. pectinellum (A. Gray) Kitam. var. alpinum (Koidz. ex Kitam.) Kitam. = Cirsium kamtschaticum Ledeb. ex DC. ■☆
92096 Cirsium kawakamii Hayata;玉山蓟(褐冠蓟,褐冠毛蓟);Kawakami Thistle ■

92097 Cirsium kelseyi (Rydb.) Petr. = Cirsium hookerianum Nutt. ■☆
92098 Cirsium kemulariae Charadze;凯姆蓟■☆
92099 Cirsium ketzkhovelii Charadze;凯兹蓟■☆
92100 Cirsium kitagoense Nakai = Cirsium japonicum Fisch. ex DC. ■
92101 Cirsium kiushianum Nakai = Cirsium suffultum (Maxim.) Matsum. et Koidz. ■☆
92102 Cirsium komarovii Schischk.;科马罗夫蓟■☆
92103 Cirsium kurobense Honda = Cirsium norikurense Nakai ■☆
92104 Cirsium kusnetzowianum Sommier et H. Lév.;库斯蓟■☆
92105 Cirsium lacerum (Rydb.) Petr. = Cirsium scariosum Nutt. ■☆
92106 Cirsium laciniatum Döll ex Nyman = Cirsium japonicum Fisch. ex DC. ■
92107 Cirsium lacinulatum Nakai = Cirsium japonicum Fisch. ex DC. ■
92108 Cirsium lactucinum Rydb. = Cirsium rydbergii Petr. ■☆
92109 Cirsium laevigatum Tausch. = Cirsium setosum (Willd.) M. Bieb. ■
92110 Cirsium lamyroides Tamamsch.;阿拉套蓟■
92111 Cirsium lanatum (Roxb. ex Willd.) Spreng.;藏蓟;Hairy Thistle,Lanose Thistle ■
92112 Cirsium lanceolatum (L.) Scop.;披针叶蓟;Bull Thistle,Lance-leaved Thistle ■
92113 Cirsium lanceolatum (L.) Scop. = Cirsium vulgare (Savi) Ten. ■
92114 Cirsium lanceolatum (L.) Scop. subsp. sylvaticum (Tausch) Arènes = Cirsium vulgare (Savi) Ten. ■
92115 Cirsium lanceolatum (L.) Scop. var. hypoleucum DC. = Cirsium vulgare (Savi) Ten. ■
92116 Cirsium lanceolatum (L.) Scop. var. rhiphaeum Pau et Font Quer = Cirsium vulgare (Savi) Ten. ■
92117 Cirsium lanceolatum Hill. = Cirsium vulgare (Savi) Ten. ■
92118 Cirsium lanceolatum Hill. var. nemorale Naegeli ex Koch = Cirsium vulgare (Savi) Ten. ■
92119 Cirsium lanceolatum Hill. var. vulagre Naegeli ex Koch = Cirsium vulgare (Savi) Ten. ■
92120 Cirsium laniflorum (M. Bieb.) M. Bieb.;绒花蓟■☆
92121 Cirsium lappaceum (M. Bieb.) M. Bieb.;钩毛蓟■☆
92122 Cirsium laterifolium (Osterh.) Petr. = Cirsium clavatum (M. E. Jones) Petr. var. americanum (A. Gray) D. J. Keil ■☆
92123 Cirsium latifolium Lowe;宽叶蓟■☆
92124 Cirsium laushanense Y. Yabe = Cirsium chinense Gardner et Champ. ■
92125 Cirsium lecontei Torr. et A. Gray;康特蓟;Black Thistle, Le Conte's Thistle ■☆
92126 Cirsium leducei (Franch.) H. Lév.;覆瓦蓟;Leduce's Thistle ■
92127 Cirsium leo Nakai et Kitag.;魁蓟;Giant Thistle ■
92128 Cirsium leo Nakai et Kitag. var. angustilobum Y. Ling = Cirsium leo Nakai et Kitag. ■
92129 Cirsium leptophyllum (Pau et Font Quer) Font Quer = Ptilostemon leptophyllus (Pau et Font Quer) Greuter ■☆
92130 Cirsium leucocephalum K. Koch;白头蓟■☆
92131 Cirsium lidjiangense Petr. et Hand.-Mazz.;丽江蓟;Lijiang Thistle ■
92132 Cirsium lineare (Thunb.) Sch. Bip.;条叶蓟(钩芙,华蓟,尖叶小蓟,苦芙,苦板,轮蓟,牛刺梨,山红花,狭叶蓟,线叶蓟,小样刺米草,野红花,中国蓟);Linearleaf Thistle ■
92133 Cirsium lineare (Thunb.) Sch. Bip. = Cirsium hupehense Pamp. ■
92134 Cirsium lineare (Thunb.) Sch. Bip. var. discolor Nakai = Cirsium lineare (Thunb.) Sch. Bip. ■
92135 Cirsium lineare (Thunb.) Sch. Bip. var. franchetii Kitam. = Cirsium hupehense Pamp. ■
92136 Cirsium lineare (Thunb.) Sch. Bip. var. franchetii Kitam. f. pallidum Kitam. = Cirsium lineare (Thunb.) Sch. Bip. ■
92137 Cirsium lineare (Thunb.) Sch. Bip. var. glabrescens Petr. = Cirsium chinense Gardner et Champ. ■
92138 Cirsium lineare (Thunb.) Sch. Bip. var. intermedium (Pamp.) Petr. = Cirsium shansiense Petr. ■
92139 Cirsium lineare (Thunb.) Sch. Bip. var. latifolium H. C. Fu;阴山条叶蓟;Bigleaf Linearleaf Thistle ■
92140 Cirsium lineare (Thunb.) Sch. Bip. var. laushanense (Y. Yabe) Kitam. = Cirsium chinense Gardner et Champ. ■
92141 Cirsium lineare (Thunb.) Sch. Bip. var. laushanense (Y. Yabe) Kitam. f. inciso-lobatum Kitam. = Cirsium shansiense Petr. ■
92142 Cirsium lineare (Thunb.) Sch. Bip. var. linearifolium f. tomentosum Petr. = Cirsium hupehense Pamp. ■
92143 Cirsium lineare (Thunb.) Sch. Bip. var. linearifolium f. viride Petr. = Cirsium lineare (Thunb.) Sch. Bip. ■
92144 Cirsium lineare (Thunb.) Sch. Bip. var. lushanense f. vestitum Kitam. = Cirsium shansiense Petr. ■
92145 Cirsium lineare (Thunb.) Sch. Bip. var. pallidum (Kitam.) Y. Ling;线叶蓟(刺儿菜,滇小蓟,轮蓟,条叶蓟,细叶蓟,小蓟)■
92146 Cirsium lineare (Thunb.) Sch. Bip. var. pallidum (Kitam.) Y. Ling = Cirsium lineare (Thunb.) Sch. Bip. ■
92147 Cirsium lineare (Thunb.) Sch. Bip. var. rigidum Petr. = Cirsium shansiense Petr. ■
92148 Cirsium lineare (Thunb.) Sch. Bip. var. rigidum Petr. f. subintegrifolium Petr. = Cirsium leducei (Franch.) H. Lév. ■
92149 Cirsium lineare (Thunb.) Sch. Bip. var. salicifolium Y. Ling = Cirsium hupehense Pamp. ■
92150 Cirsium lineare (Thunb.) Sch. Bip. var. spatulatum Petr. = Cirsium shansiense Petr. ■
92151 Cirsium lineare (Thunb.) Sch. Bip. var. tchefouense (Debeaux) Y. Ling = Cirsium chinense Gardner et Champ. ■
92152 Cirsium lineare (Thunb.) Sch. Bip. var. tenii Petr. = Cirsium shansiense Petr. ■
92153 Cirsium lineare (Thunb.) Sch. Bip. var. tsoongianum (Y. Ling) Y. Ling = Cirsium lineare (Thunb.) Sch. Bip. ■
92154 Cirsium lineare (Thunb.) Sch. Bip. var. typicum Nakai = Cirsium lineare (Thunb.) Sch. Bip. ■
92155 Cirsium lineare (Thunb.) Sch. Bip. var. yunnanense Petr. = Cirsium shansiense Petr. ■
92156 Cirsium littorale Maxim. = Cirsium schantarense Trautv. et C. A. Mey. ■
92157 Cirsium littorale Maxim. var. nudum Regel = Cirsium schantarense Trautv. et C. A. Mey. ■
92158 Cirsium littorale Maxim. var. ussuriense Regel = Cirsium maackii Maxim. ■
92159 Cirsium loncholepis Petr. = Cirsium scariosum Nutt. var. citrinum (Petr.) D. J. Keil ■☆
92160 Cirsium longiflorum Charadze;长花蓟■☆
92161 Cirsium longipedunculatum Kitam.;长花梗蓟■☆
92162 Cirsium longistylum R. J. Moore et Frankton;长柱蓟;Long-style Thistle ■☆
92163 Cirsium lucens Kitam.;光亮蓟■☆
92164 Cirsium lucens Kitam. var. bracteosum Imae et S. Watan.;多苞片光亮蓟■☆

92165 Cirsium lucens Kitam. var. opacum Kitam.;暗亮蓟■☆

92166 Cirsium lyratum Bunge = Hemistepta lyrata (Bunge) Bunge ■

92167 Cirsium maackii Maxim.;野蓟(刺蓟,大蓟,老牛锉,马氏蓟,牛戳口,千针草);Maack Thistle ■

92168 Cirsium maackii Maxim. f. albiflorum (Sakata) Sakata;白花野蓟■

92169 Cirsium maackii Maxim. f. albiflorum W. Wang et C. Y. Li = Cirsium maackii Maxim. f. albiflorum (Sakata) Sakata ■

92170 Cirsium maackii Maxim. f. koraiensis (Nakai) Nakai = Cirsium maackii Maxim. ■

92171 Cirsium maackii Maxim. var. horridum = Cirsium japonicum Fisch. ex DC. ■

92172 Cirsium maackii Maxim. var. intermedium (Maxim.) Nakai = Cirsium japonicum Fisch. ex DC. ■

92173 Cirsium maackii Maxim. var. kiusianum Nakai = Cirsium japonicum Fisch. ex DC. ■

92174 Cirsium maackii Maxim. var. koreiense (Nakai) Nakai = Cirsium maakii Maxim. ■

92175 Cirsium maackii Maxim. var. spiniferum Nakai = Cirsium schantarense Trautv. et C. A. Mey. ■

92176 Cirsium maackii Maxim. var. vulcani? = Cirsium japonicum Fisch. ex DC. ■

92177 Cirsium macounii (Greene) Petr. = Cirsium edule Nutt. var. macounii (Greene) D. J. Keil ■☆

92178 Cirsium macrobotrys (K. Koch) Boiss.;大穗蓟■☆

92179 Cirsium macrocephalum C. A. Mey.;大头蓟■☆

92180 Cirsium maculatum Scop. = Silybum marianum (L.) Gaertn. ■

92181 Cirsium magnificum (A. Nelson) Petr. = Cirsium scariosum Nutt. ■☆

92182 Cirsium magofukui Kitam.;孙福蓟■☆

92183 Cirsium mairei (H. Lév.) H. Lév. = Cirsium griseum H. Lév. ■

92184 Cirsium mairei (H. Lév.) H. Lév. = Cnicus mairei H. Lév. ■

92185 Cirsium manshuricum Kitag. = Cirsium chinense Gardner et Champ. ■

92186 Cirsium maritimum Makino;滨蓟■☆

92187 Cirsium maritimum Makino var. leucanthum Nakai ex Honda;白花滨蓟■☆

92188 Cirsium maroccanum Petr.;摩洛哥蓟■☆

92189 Cirsium maruyamanum Kitam.;丸山蓟■☆

92190 Cirsium matsumurae Nakai;松村氏蓟■☆

92191 Cirsium matsumurae Nakai var. dubium Kitam.;可疑松村氏蓟■☆

92192 Cirsium matsumurae Nakai var. pubescens Kitam. = Cirsium matsumurae Nakai ■☆

92193 Cirsium maximowiczii Nakai = Cirsium aomorense Nakai ■☆

92194 Cirsium megacanthum Nutt. = Cirsium horridulum Michx. var. megacanthum (Nutt.) D. J. Keil ■☆

92195 Cirsium megacephalum (A. Gray) Cockerell = Cirsium undulatum (Nutt.) Spreng. ■☆

92196 Cirsium megacephalum (A. Gray) Cockerell ex Daniels = Cirsium undulatum (Nutt.) Spreng. ■☆

92197 Cirsium melanolepis Petr. = Cirsium henryi (Franch.) Diels ■

92198 Cirsium microspicatum Nakai;小穗蓟■☆

92199 Cirsium microspicatum Nakai f. glutinosum Kitam.;黏性小穗蓟■☆

92200 Cirsium microspicatum Nakai var. kiotoense Kitam.;京都小穗蓟■☆

92201 Cirsium microspicatum Nakai var. yechizenense Kitam.;北村小穗蓟■☆

92202 Cirsium minganense Vict. = Cirsium scariosum Nutt. ■☆

92203 Cirsium modestum (Osterh.) Cockerell = Cirsium clavatum (M. E. Jones) Petr. var. americanum (A. Gray) D. J. Keil ■☆

92204 Cirsium mohavense (Greene) Petr.;莫哈韦蓟;Mojave Thistle ■☆

92205 Cirsium monocephalum (Vaniot) H. Lév.;马刺蓟;Singlehead Thistle ■

92206 Cirsium monspessulanum (L.) Hill;高卢蓟■☆

92207 Cirsium monspessulanum (L.) Hill subsp. ferox (Coss.) Talavera;多刺高卢蓟■☆

92208 Cirsium monspessulanum (L.) Hill var. laxum Rouy = Cirsium monspessulanum (L.) Hill ■☆

92209 Cirsium montigenum Petr. = Cirsium hydrophilum (Greene) Jeps. var. vaseyi (A. Gray) J. T. Howell ■☆

92210 Cirsium morii Hayata;羽冠蓟(森氏蓟,线冠蓟);Mori Thistle ■

92211 Cirsium muliense C. Shih;木里蓟;Muli Thistle ■

92212 Cirsium murdockii (S. L. Welsh) Cronquist = Cirsium eatonii (A. Gray) B. L. Rob. var. murdockii S. L. Welsh ■☆

92213 Cirsium muticum F. Michx.;沼生蓟;Dunce-nettle, Horsetops, Swamp Thistle ■☆

92214 Cirsium muticum F. Michx. lactiflorum Fernald = Cirsium muticum F. Michx. ■☆

92215 Cirsium muticum F. Michx. var. monticola Fernald = Cirsium muticum F. Michx. ■☆

92216 Cirsium muticum F. Michx. var. subpinnatifidum (Britton) Fernald = Cirsium muticum F. Michx. ■☆

92217 Cirsium nambuense Nakai;南部蓟■☆

92218 Cirsium navajoense R. J. Moore et Frankton = Cirsium arizonicum (A. Gray) Petr. var. chellyense (R. J. Moore et Frankton) D. J. Keil ■☆

92219 Cirsium nebraskense (Britton) Lunell = Cirsium canescens Nutt. ■☆

92220 Cirsium nebraskense (Britton) Lunell var. discissum Lunell = Cirsium flodmanii (Rydb.) Arthur ■☆

92221 Cirsium nelsonii (Pammel) Petr. = Cirsium canescens Nutt. ■☆

92222 Cirsium neomexicanum A. Gray;新墨西哥蓟;Desert Thistle, New Mexico Thistle, Yellow Thistle ■☆

92223 Cirsium neomexicanum A. Gray var. utahense (Petr.) S. L. Welsh = Cirsium neomexicanum A. Gray ■☆

92224 Cirsium nidulans Regel = Schmalhausenia nidulans (Regel) Petr. ■

92225 Cirsium nikkoense Nakai ex Matsum. et Koidz. = Cirsium oligophyllum (Franch. et Sav.) Matsum. var. nikkoense (Nakai ex Matsum. et Koidz.) Kitam. ■☆

92226 Cirsium nipponense (Nakai) Koidz.;日本蓟■☆

92227 Cirsium nipponense (Nakai) Koidz. = Cirsium borealinipponense Kitam. ■☆

92228 Cirsium nipponense (Nakai) Koidz. f. albiflorum Ikegami;白花日本蓟■☆

92229 Cirsium nipponense (Nakai) Koidz. var. spinulosum Kitam. = Cirsium borealinipponense Kitam. ■☆

92230 Cirsium nipponicum (Maxim.) Makino f. lanuginosum (Nakai) Kitam.;多绵毛日本蓟■☆

92231 Cirsium nipponicum (Maxim.) Makino var. alpestre Kitam. = Cirsium ovalifolium (Franch. et Sav.) Matsum. ■☆

92232 Cirsium nipponicum (Maxim.) Makino var. amplifolium (Kitam.) Kitam. = Cirsium nipponicum (Maxim.) Makino var. yoshinoi (Nakai) Kitam. ■☆

92233 Cirsium nipponicum (Maxim.) Makino var. comosum (Franch. et Sav.) Kitam. = Cirsium nipponicum (Maxim.) Makino var.

incomptum (Maxim.) Kitam. ■☆
92234 Cirsium nipponicum (Maxim.) Makino var. incomptum (Maxim.) Kitam.;装饰日本蓟■☆
92235 Cirsium nipponicum (Maxim.) Makino var. indefensum (Kitam.) Kitam. = Cirsium indefensum Kitam. ■☆
92236 Cirsium nipponicum (Maxim.) Makino var. sadoense T. Yamaz.;佐渡日本蓟■☆
92237 Cirsium nipponicum (Maxim.) Makino var. sawadae (Kitam.) Kitam. = Cirsium nipponicum (Maxim.) Makino var. incomptum (Maxim.) Kitam. ■☆
92238 Cirsium nipponicum (Maxim.) Makino var. shikokianum (Kitam.) Kitam.;四国蓟■☆
92239 Cirsium nipponicum (Maxim.) Makino var. shiroumense Kadota;白马岳蓟■☆
92240 Cirsium nipponicum (Maxim.) Makino var. yatsugatakense (Nakai) Kitam. = Cirsium ovalifolium (Franch. et Sav.) Matsum. ■☆
92241 Cirsium nipponicum (Maxim.) Makino var. yoshinoi (Nakai) Kitam.;吉野蓟■☆
92242 Cirsium norikurense Nakai;乘鞍蓟■☆
92243 Cirsium norikurense Nakai var. integrifolium Kitam. = Cirsium norikurense Nakai ■☆
92244 Cirsium norikurense Nakai var. kisoense T. Yamaz. et S. Asano = Cirsium furusei Kitam. ■☆
92245 Cirsium nuttallii DC.;纳托尔蓟;Nuttall's Thistle ■☆
92246 Cirsium oblanceolatum (Rydb.) K. Schum. = Cirsium flodmanii (Rydb.) Arthur ■☆
92247 Cirsium oblongifolium K. Koch;矩圆叶蓟■☆
92248 Cirsium obvallatum (M. Bieb.) DC.;密盖蓟■☆
92249 Cirsium occidentale (Nutt.) Jeps.;西方蓟;Cobwebby Thistle, Western Thistle ■☆
92250 Cirsium occidentale (Nutt.) Jeps. subsp. candidissimum (Greene) Petr. = Cirsium occidentale (Nutt.) Jeps. var. candidissimum (Greene) J. F. Macbr. ■☆
92251 Cirsium occidentale (Nutt.) Jeps. subsp. venustum (Greene) Petr. = Cirsium occidentale (Nutt.) Jeps. var. venustum (Greene) Jeps. ■☆
92252 Cirsium occidentale (Nutt.) Jeps. var. californicum (A. Gray) D. J. Keil et C. E. Turner;加州蓟;California Thistle ■☆
92253 Cirsium occidentale (Nutt.) Jeps. var. candidissimum (Greene) J. F. Macbr.;白西方蓟;Snowy Thistle ■☆
92254 Cirsium occidentale (Nutt.) Jeps. var. compactum Hoover;紧凑蓟;Compact cobwebby Thistle ■☆
92255 Cirsium occidentale (Nutt.) Jeps. var. coulteri (Harv. et A. Gray) Jeps.;库尔特蓟;Coulter's Thistle ■☆
92256 Cirsium occidentale (Nutt.) Jeps. var. lucianum D. J. Keil;西方硬蓟;Cuesta Ridge Thistle ■☆
92257 Cirsium occidentale (Nutt.) Jeps. var. venustum (Greene) Jeps.;雅致西方蓟;Venus Thistle ■☆
92258 Cirsium occidentalinipponense Kadota;西日本蓟■☆
92259 Cirsium ochrocentrum A. Gray;黄刺蓟;Yellowspine Thistle, Yellow-spirted Thistle ■☆
92260 Cirsium ochrocentrum A. Gray subsp. martinii Barlow-Irick = Cirsium ochrocentrum A. Gray var. martinii (Barlow-Irick) D. J. Keil ■☆
92261 Cirsium ochrocentrum A. Gray var. helleri (Small) Petr. = Cirsium undulatum (Nutt.) Spreng. ■☆
92262 Cirsium ochrocentrum A. Gray var. martinii (Barlow-Irick) D. J. Keil;马丁蓟;Martin's Thistle ■☆
92263 Cirsium ochrolepideum Juz. = Cirsium arvense (L.) Scop. ■
92264 Cirsium odontolepis Boiss.;齿鳞蓟■☆
92265 Cirsium odoratum (Muhl. ex W. P. C. Barton) Petr. = Cirsium pumilum Spreng. ■☆
92266 Cirsium okamotoi Kitam.;冈田蓟■☆
92267 Cirsium okamotoi Kitam. f. albiflorum Kadota;白花岡田蓟■☆
92268 Cirsium oleraceum Scop.;蔬菜蓟(菜蓟,圆白蓟);Cabbage Thistle, Meadow Cabbage, Siberian Thistle ■☆
92269 Cirsium oligophyllum (Franch. et Sav.) Matsum.;寡叶蓟■☆
92270 Cirsium oligophyllum (Franch. et Sav.) Matsum. f. albiflorum (Kitam.) Kitam.;白花寡叶蓟■☆
92271 Cirsium oligophyllum (Franch. et Sav.) Matsum. subsp. aomorense (Nakai) Kitam. f. leucanthum Kitam. = Cirsium aomorense Nakai f. albiflorum (Kitam.) Kitam. ■☆
92272 Cirsium oligophyllum (Franch. et Sav.) Matsum. subsp. aomorense (Nakai) Kitam. = Cirsium aomorense Nakai ■☆
92273 Cirsium oligophyllum (Franch. et Sav.) Matsum. subsp. nikkoense (Nakai ex Matsum. et Koidz.) Kitam. = Cirsium oligophyllum (Franch. et Sav.) Matsum. var. nikkoense (Nakai ex Matsum. et Koidz.) Kitam. ■☆
92274 Cirsium oligophyllum (Franch. et Sav.) Matsum. var. nikkoense (Nakai ex Matsum. et Koidz.) Kitam.;日光寡叶蓟■☆
92275 Cirsium olivescens (Rydb.) Petr. = Cirsium scariosum Nutt. var. coloradense (Rydb.) D. J. Keil ■☆
92276 Cirsium olivescens (Rydb.) Petr. = Cirsium wheeleri (A. Gray) Petr. ■☆
92277 Cirsium oreganum Piper = Cirsium remotifolium (Hook.) DC. var. rivulare Jeps. ■☆
92278 Cirsium osseticum (Adam) Petr.;骨质蓟■☆
92279 Cirsium osterhoutii (Rydb.) Petr. = Cirsium clavatum (M. E. Jones) Petr. var. osterhoutii (Rydb.) D. J. Keil ■☆
92280 Cirsium otayae Kitam.;御旅屋蓟■☆
92281 Cirsium ovalifolium (Franch. et Sav.) Matsum.;卵叶蓟■☆
92282 Cirsium ownbeyi S. L. Welsh;欧氏蓟;Ownbey's Thistle ■☆
92283 Cirsium pallidum (Wooton et Standl.) Wooton et Standl. = Cirsium parryi (A. Gray) Petr. ■☆
92284 Cirsium palousense (Piper) Piper = Cirsium brevifolium Nutt. ■☆
92285 Cirsium palustre (L.) Scop.;沼泽蓟;Black Thistle, Bog Thistle, European Swamp Thistle, Marsh Thistle, Moss Thistle, Red Thistle, Water Thistle ■☆
92286 Cirsium palustre Scop. = Cirsium palustre (L.) Scop. ■☆
92287 Cirsium pannonicum Gaudich.;潘城蓟■☆
92288 Cirsium parryi (A. Gray) Petr.;帕里蓟;Parry Thistle ■☆
92289 Cirsium parryi (A. Gray) Petr. subsp. mogollonicum Schaack et Goodwin = Cirsium parryi (A. Gray) Petr. ■☆
92290 Cirsium pastoris J. T. Howell;艳丽蓟;Showy Thisde ■☆
92291 Cirsium pastoris J. T. Howell = Cirsium occidentale (Nutt.) Jeps. var. candidissimum (Greene) J. F. Macbr. ■☆
92292 Cirsium peckii L. F. Hend. = Cirsium eatonii (A. Gray) B. L. Rob. var. peckii (L. F. Hend.) D. J. Keil ■☆
92293 Cirsium pectinellum A. Gray;篦齿蓟■☆
92294 Cirsium pectinellum A. Gray var. alpinum Koidz. ex Kitam. = Cirsium kamtschaticum Ledeb. ex DC. ■☆
92295 Cirsium pectinellum A. Gray var. apoense (Nakai) Okuyama = Cirsium apoense Nakai ■☆
92296 Cirsium pendulum Fisch. ex DC.;烟管蓟;Pendulate Thistle ■

92297 Cirsium pendulum Fisch. ex DC. f. albiflorum (Makino) Kitam.; 白花烟管蓟■☆

92298 Cirsium pendulum Fisch. ex DC. var. albiflorum Makino = Cirsium pendulum Fisch. ex DC. f. albiflorum (Makino) Kitam. ■☆

92299 Cirsium pendulum Fisch. ex DC. var. leucanthum? = Cirsium pendulum Fisch. ex DC. f. albiflorum (Makino) Kitam. ■☆

92300 Cirsium pendulum Fisch. ex DC. var. oligocephalum Regel et Till = Cirsium schantarense Trautv. et C. A. Mey. ■

92301 Cirsium penicillatum Fisch. ex DC. subsp. tenuilobum (K. Koch) Soldano;细裂烟管蓟■☆

92302 Cirsium penicillatum Fisch. ex DC. var. tomentosum (Boiss.) D. Heller;毛烟管蓟■☆

92303 Cirsium perennans (Greene) Wooton et Standl. = Cirsium wheeleri (A. Gray) Petr. ■☆

92304 Cirsium periacanthaceum C. Shih;川蓟;Sichuan Thistle ■

92305 Cirsium perplexans (Rydb.) Petr.;错乱蓟;Adobe Hills Thistle ■☆

92306 Cirsium pinnatibracteamm Y. Ling = Cirsium leo Nakai et Kitag. ■

92307 Cirsium pitcheri (Torr. ex Eaton) Torr. et A. Gray;沙丘蓟;Dune Thistle, Sand Dune Thistle, Sand-dune Thistle ■☆

92308 Cirsium plattense (Rydb.) Cockerell ex Daniels = Cirsium canescens Nutt. ■☆

92309 Cirsium polonicum (Petr.) Iljin;波兰蓟■☆

92310 Cirsium polyacanthum Hochst. ex A. Rich. = Cirsium straminispinum C. Jeffrey ex Cufod. ■☆

92311 Cirsium polyphyllum (Rydb.) Petr. = Cirsium eatonii (A. Gray) B. L. Rob. var. murdockii S. L. Welsh ■☆

92312 Cirsium praeteriens J. F. Macbr.;迷惑蓟;Lost Thistle, Palo Alto Thistle ■☆

92313 Cirsium pratense DC.;草原蓟(土人参)■☆

92314 Cirsium proteanum J. T. Howell = Cirsium occidentale (Nutt.) Jeps. var. venustum (Greene) Jeps. ■☆

92315 Cirsium provostii (Franch.) Petr. = Cirsium pendulum Fisch. ex DC. ■

92316 Cirsium provostii (Franch.) Petr. f. subulatum Petr. = Cirsium monocephalum (Vaniot) H. Lév. ■

92317 Cirsium provostii (Franch.) Petr. var. monocephalum (Vaniot) Petr. = Cirsium monocephalum (Vaniot) H. Lév. ■

92318 Cirsium provostii (Franch.) Petr. var. oleracioides Petr. = Cirsium monocephalum (Vaniot) H. Lév. ■

92319 Cirsium provostii (Franch.) Petr. var. racemosum Petr. = Cirsium monocephalum (Vaniot) H. Lév. ■

92320 Cirsium provostii (Franch.) Petr. var. spinosum Petr. = Cirsium monocephalum (Vaniot) H. Lév. ■

92321 Cirsium pseudolappaceum Charadze;假钩毛蓟■☆

92322 Cirsium pulchellum (Greene ex Rydb.) Wooton et Standl. = Cirsium arizonicum (A. Gray) Petr. var. bipinnatum (Eastw.) D. J. Keil ■☆

92323 Cirsium pulchellum (Greene ex Rydb.) Wooton et Standl. subsp. bipinnatum (Eastw.) Petr. = Cirsium arizonicum (A. Gray) Petr. var. bipinnatum (Eastw.) D. J. Keil ■☆

92324 Cirsium pulchellum (Greene ex Rydb.) Wooton et Standl. subsp. diffusum (Eastw.) Petr. = Cirsium arizonicum (A. Gray) Petr. var. bipinnatum (Eastw.) D. J. Keil ■☆

92325 Cirsium pulchellum (Greene ex Rydb.) Wooton et Standl. var. glabrescens Petr. = Cirsium arizonicum (A. Gray) Petr. var. bipinnatum (Eastw.) D. J. Keil ■☆

92326 Cirsium pulcherrimum (Rydb.) K. Schum.;怀俄明蓟;Wyoming Thistle ■☆

92327 Cirsium pumilum (Nutt.) Spreng. subsp. hillii (Canby) R. J. Moore et Frankton = Cirsium hillii (Canby) Fernald ■☆

92328 Cirsium pumilum (Nutt.) Spreng. var. hillii (Canby) B. Boivin; 希尔鱼蓟;Hill's Thistle ■☆

92329 Cirsium pumilum (Nutt.) Spreng. var. hillii (Canby) B. Boivin = Cirsium hillii (Canby) Fernald ■☆

92330 Cirsium pumilum Spreng.;鱼蓟;Fragrant Thistle, Pasture Thistle ■☆

92331 Cirsium pumilum Spreng. subsp. hillii (Canby) R. J. Moore et Frankton = Cirsium pumilum (Nutt.) Spreng. var. hillii (Canby) B. Boivin ■☆

92332 Cirsium purpuratum (Maxim.) Matsum.;紫蓟■☆

92333 Cirsium purpuratum (Maxim.) Matsum. f. albiflorum (Kitam.) Kitam.;白花紫蓟■☆

92334 Cirsium purpuratum (Maxim.) Matsum. var. albiflorum Kitam. = Cirsium purpuratum (Maxim.) Matsum. f. albiflorum (Kitam.) Kitam. ■☆

92335 Cirsium pyrenaicum (Jacq.) All.;核蓟;Pyrenean Thistle ■☆

92336 Cirsium pyrenaicum (Jacq.) All. var. longespinosum (Kuntze) Talavera et Valdés = Cirsium pyrenaicum (Jacq.) All. ■☆

92337 Cirsium quercetorum (A. Gray) Jeps.;阿拉米达蓟;Alameda County Thistle, Brownie Thistle ■☆

92338 Cirsium quercetorum (A. Gray) Jeps. var. citrinum Petr. = Cirsium scariosum Nutt. var. citrinum (Petr.) D. J. Keil ■☆

92339 Cirsium quercetorum (A. Gray) Jeps. var. walkerianum (Petr.) Jeps. = Cirsium quercetorum (A. Gray) Jeps. ■☆

92340 Cirsium quercetorum (A. Gray) Jeps. var. xerolepis Petr. = Cirsium quercetorum (A. Gray) Jeps. ■☆

92341 Cirsium racemiforme Y. Ling et C. Shih;总序蓟;Racemose Thistle ■

92342 Cirsium remotifolium (Hook.) DC.;少叶蓟;Fewleaf Thistle, Remote-leaved Thistle ■☆

92343 Cirsium remotifolium (Hook.) DC. subsp. oregonense Petr. = Cirsium remotifolium (Hook.) DC. var. odontolepis Petr. ■☆

92344 Cirsium remotifolium (Hook.) DC. subsp. pseudocarlinoides Petr. = Cirsium remotifolium (Hook.) DC. var. odontolepis Petr. ■☆

92345 Cirsium remotifolium (Hook.) DC. var. odontolepis Petr.;齿鳞寡叶蓟;Fringe-scaled Thistle, Pacific Fringed Thistle ■☆

92346 Cirsium remotifolium (Hook.) DC. var. rivulare Jeps.;可拉马斯蓟;Klamath Thistle ■☆

92347 Cirsium repandum Michx.;滨海平原蓟;Coastal-plain Thistle, Sand-hill Thistle ■☆

92348 Cirsium rhabdotolepis Petr.;棒鳞蓟■☆

92349 Cirsium rhizocephalum C. A. Mey.;根头蓟■☆

92350 Cirsium rivulare Link;溪蓟(河岸蓟)■☆

92351 Cirsium rivulare Link 'Atropurpureum';紫花河岸蓟■☆

92352 Cirsium rothrockii (A. Gray) Petr. = Cirsium arizonicum (A. Gray) Petr. var. rothrockii (A. Gray) D. J. Keil ■☆

92353 Cirsium rothrockii Petr.;深红蓟;Carmine Thistle ■☆

92354 Cirsium rusbyi (Greene) Petr. = Cirsium mohavense (Greene) Petr. ■☆

92355 Cirsium rydbergii Petr.;理德蓟;Alcove Thistle, Rydberg's Thistle ■☆

92356 Cirsium sairamense (C. Winkl.) O. Fedtsch. et B. Fedtsch.;赛里木蓟;Sairam Thistle, Sairim Thistle ■

92357 Cirsium salicifolium (Kitag.) C. Shih;块蓟(柳叶绒背蓟);Willowleaf Thistle, Willowleaf Vlassoviana Thistle ■

92358 Cirsium scabrum (Poir.) Bonnet;粗糙蓟;Rough Thistle ■☆

92359 Cirsium scapanolepis Petr. = Cirsium clavatum (M. E. Jones) Petr. var. americanum (A. Gray) D. J. Keil ■☆

92360 Cirsium scariosum Nutt.;牧场蓟;Elk Thistle, Meadow Thistle ■☆

92361 Cirsium scariosum Nutt. var. americanum (A. Gray) D. J. Keil;无梗蓟;Dinnerplate Thistle, Sessile Thistle, Stemless Thistle ■☆

92362 Cirsium scariosum Nutt. var. citrinum (Petr.) D. J. Keil;橘蓟;La Graciosa Thistle ■☆

92363 Cirsium scariosum Nutt. var. coloradense (Rydb.) D. J. Keil;科罗拉多蓟;Colorado Thistle ■☆

92364 Cirsium scariosum Nutt. var. congdonii (R. J. Moore et Frankton) D. J. Keil;玫瑰蓟;Rosette Thistle ■☆

92365 Cirsium scariosum Nutt. var. robustum D. J. Keil;粗壮牧场蓟;Shasta Valley Thistle ■☆

92366 Cirsium scariosum Nutt. var. thorneae S. L. Welsh;索恩蓟;Thorne's Thistle ■☆

92367 Cirsium schantarense Trautv. et C. A. Mey.;林蓟;Woodland Thistle ■

92368 Cirsium schimperi (Vatke) C. Jeffrey ex Cufod.;欣珀蓟■☆

92369 Cirsium schimperi (Vatke) C. Jeffrey ex Cufod. var. inerme (Oliv. et Hiern) Cufod. = Cirsium schimperi (Vatke) C. Jeffrey ex Cufod. ■☆

92370 Cirsium schischkinii Serg.;希施蓟■☆

92371 Cirsium scopulorum (Greene) Cockerell ex Daniels = Cirsium eatonii (A. Gray) B. L. Rob. var. eriocephalum (A. Gray) D. J. Keil ■☆

92372 Cirsium segetum Bunge = Breea segetum (Bunge) Kitam. ■

92373 Cirsium segetum Bunge = Cirsium setosum (Willd.) M. Bieb. ■

92374 Cirsium segetum Bunge f. albiflorum Nakai;白花谷地蓟■☆

92375 Cirsium semenovii Regel et Schmalh.;新疆蓟;Semenov Thistle ■

92376 Cirsium semenovii Regel et Schmalh. subsp. sairamense Petr. = Cirsium sairamense (C. Winkl.) O. Fedtsch. et B. Fedtsch. ■

92377 Cirsium senile Nakai = Cirsium japonicum Fisch. ex DC. ■

92378 Cirsium senjoense Kitam.;千丈蓟■☆

92379 Cirsium senjoense Kitam. var. kurosawae Kitam. = Cirsium senjoense Kitam. ■☆

92380 Cirsium serratuloides (L.) Hill;麻花头蓟;Sawwortlike Thistle ■

92381 Cirsium serrulatum (M. Bieb.) M. Bieb.;蓖齿形蓟■☆

92382 Cirsium setigerum Ledeb. = Cirsium alatum (S. G. Gmel.) Bobrov ■

92383 Cirsium setosum (Willd.) Besser ex M. Bieb. = Cirsium arvense (L.) Scop. ■

92384 Cirsium setosum (Willd.) M. Bieb.;刺儿菜(白鸡角刺,刺菜,刺刺菜,刺儿草,刺秆菜,刺蓟,刺蓟菜,刺尖头草,刺角菜,刺萝卜,刺杀草,大刺儿菜,大蓟,大小蓟,刻叶刺儿菜,猫蓟,木刺艾,牛戳刺,萋萋菜,荠荠菜,荠荠毛,千针草,枪刀菜,青刺蓟,青青菜,曲曲菜,细叶蓟,小恶鸡婆,小鸡角刺,小蓟,小蓟姆,小牛扎口,野红花);Setose Thistle, Spinegreens ■

92385 Cirsium setosum (Willd.) M. Bieb. f. albiflorum (Kitag.) Kitag.;白花刺儿菜■☆

92386 Cirsium shansiense Petr.;牛口刺(硬条叶蓟);Oxmouth Thistle ■

92387 Cirsium shidokimontanum Kadota;雫蓟■☆

92388 Cirsium shinanense T. Shimizu;信浓蓟■☆

92389 Cirsium sidi-guinii Pau et Font Quer = Cirsium dyris Jahand. et Maire ■☆

92390 Cirsium sieboldii Miq.;西氏蓟(席氏蓟)■☆

92391 Cirsium sieboldii Miq. f. leucanthum T. Shimizu;白花西氏蓟■☆

92392 Cirsium sieboldii Miq. f. pilosiusculum? = Cirsium sieboldii Miq. ■☆

92393 Cirsium sieboldii Miq. subsp. austrokiushianum (Kitam.) Kitam. = Cirsium sieboldii Miq. ■☆

92394 Cirsium sieboldii Miq. var. austrokiushianum (Kitam.) Kitam. = Cirsium sieboldii Miq. subsp. austrokiushianum (Kitam.) Kitam. ■☆

92395 Cirsium sieboldii Miq. var. austrokiushianum (Kitam.) Kitam. = Cirsium sieboldii Miq. ■☆

92396 Cirsium sieversii (Fisch. et C. A. Mey.) Sch. Bip.;附片蓟;Silvers Thistle ■

92397 Cirsium simithianum Petr. = Cirsium japonicum Fisch. ex DC. ■

92398 Cirsium simplex C. A. Mey.;简单蓟■☆

92399 Cirsium sinensis S. Moore = Olgaea lomonosowii (Trautv.) Iljin ■

92400 Cirsium sinuatum (Trautv.) Boiss.;深波蓟■☆

92401 Cirsium smallii Britton = Cirsium horridulum Michx. var. vittatum (Small) R. W. Long ■☆

92402 Cirsium smithianum Petr. = Cirsium japonicum Fisch. ex DC. ■

92403 Cirsium sorocephalum Fisch. et C. A. Mey.;堆头蓟■☆

92404 Cirsium sosnowskyi Charadze;索斯蓟■☆

92405 Cirsium souliei (Franch.) Mattf.;葵花大蓟(聚头蓟,犁头蓟);Soulie Thistle ■

92406 Cirsium spathulifolium Rydb. = Cirsium clavatum (M. E. Jones) Petr. var. americanum (A. Gray) D. J. Keil ■☆

92407 Cirsium spicatum Matsum.;虎蓟■☆

92408 Cirsium spinosissimum (L.) Scop.;密刺苞;Spinest Thistle ■☆

92409 Cirsium spinosissimum (L.) Scop. = Cirsium horridulum Michx. ■☆

92410 Cirsium spinosum Kitam.;显刺蓟■☆

92411 Cirsium straminispinum C. Jeffrey ex Cufod.;草黄蓟■☆

92412 Cirsium strigosum (M. Bieb.) M. Bieb.;糙伏毛蓟■☆

92413 Cirsium subinerme Fisch. et C. A. Mey.;近无刺蓟■☆

92414 Cirsium sublaniflorum Soják;亚毛蓟■☆

92415 Cirsium subuliforme C. Shih;钻苞蓟;Subulate Thistle ■

92416 Cirsium suffultum (Maxim.) Matsum. et Koidz.;支撑蓟■☆

92417 Cirsium suffultum (Maxim.) Matsum. et Koidz. f. albiflorum Kitam.;白花支撑■☆

92418 Cirsium suzukaense Kitam.;铃木蓟■☆

92419 Cirsium suzukii Kitam.;台湾大蓟(铃木蓟,铃木氏蓟,棕果蓟);Suzuki Thistle ■

92420 Cirsium sychnosanthum Petr.;密花蓟■☆

92421 Cirsium syriacum (L.) Gaertn. = Notobasis syriaca (L.) Cass. ■☆

92422 Cirsium szowitsii (K. Koch) Boiss.;瑟维茨蓟■☆

92423 Cirsium taliense (Jeffrey) H. Lév. = Cirsium henryi (Franch.) Diels ■

92424 Cirsium taliense Jeffrey = Cirsium henryi (Franch.) Diels ■

92425 Cirsium tanakae (Franch. et Sav.) Matsum.;田中氏蓟(田中蓟);Tanaka Thistle ■☆

92426 Cirsium tanakae (Franch. et Sav.) Matsum. subsp. aomorense (Nakai) Kitam. = Cirsium aomorense Nakai ■☆

92427 Cirsium tanakae (Franch. et Sav.) Matsum. subsp. nikkoense (Nakai ex Matsum. et Koidz.) Kitam. = Cirsium oligophyllum (Franch. et Sav.) Matsum. var. nikkoense (Nakai ex Matsum. et Koidz.) Kitam. ■☆

92428 Cirsium tanakae (Franch. et Sav.) Matsum. var. nikkoense (Nakai ex Matsum. et Koidz.) Kitam. = Cirsium oligophyllum

(Franch. et Sav.) Matsum. var. nikkoense (Nakai ex Matsum. et Koidz.) Kitam. ■☆

92429 Cirsium tanakae (Franch. et Sav.) Matsum. var. niveum (Kitam.) Kitam. = Cirsium oligophyllum (Franch. et Sav.) Matsum. var. nikkoense (Nakai ex Matsum. et Koidz.) Kitam. ■☆

92430 Cirsium tanegashimense Kitam. ex Kadota;种子岛蓟■☆

92431 Cirsium tashiroi Kitam.;田代蓟■☆

92432 Cirsium tashiroi Kitam. var. hidaense (Kitam.) Kadota;斐太蓟■☆

92433 Cirsium tauricum Soják;克里木蓟■☆

92434 Cirsium tchefouense Debeaux = Cirsium chinense Gardner et Champ. ■

92435 Cirsium tenue Kitam.;细蓟■☆

92436 Cirsium tenue Kitam. var. ishizuchiense Kitam. = Cirsium ishizuchiense (Kitam.) Kadota ■☆

92437 Cirsium tenuiflorus?;细花蓟;Seaside Thistle, Slender Flowered Thistle, Slender Thistle ■☆

92438 Cirsium tenuifolium C. Shih;薄叶蓟;Thinleaf Thistle ■

92439 Cirsium tenuipedunculatum Kadota;梗花蓟■☆

92440 Cirsium tenuisquamatum Kitam.;细鳞蓟■☆

92441 Cirsium terrae-nigrae Shinners = Cirsium engelmannii Rydb. ■☆

92442 Cirsium texanum Buckley;得州蓟;Southern Thistle, Texas Purple Thistle, Texas Thistle ■☆

92443 Cirsium texanum Buckley var. stenolepis Shinners = Cirsium texanum Buckley ■☆

92444 Cirsium tianmushanicum C. Shih;杭蓟;Tianmushan Thistle ■

92445 Cirsium tibeticum Kitam. = Cirsium argyracanthum DC. ■

92446 Cirsium tioganum (Congdon) Petr. = Cirsium scariosum Nutt. var. americanum (A. Gray) D. J. Keil ■☆

92447 Cirsium tioganum (Congdon) Petr. var. coloradense (Rydb.) Dorn = Cirsium scariosum Nutt. var. coloradense (Rydb.) D. J. Keil ■☆

92448 Cirsium tomentosum C. A. Mey.;绒毛蓟■☆

92449 Cirsium toyoshimae Koidz.;丰岛蓟■☆

92450 Cirsium tracyi (Rydb.) Petr.;特拉塞蓟;Tracy's Thistle ■☆

92451 Cirsium triacanthum Petr. = Cirsium cymosum (Greene) J. T. Howell ■☆

92452 Cirsium tricholoma Fisch. et C. A. Mey.;毛口蓟■☆

92453 Cirsium tsoongianum Y. Ling = Cirsium lineare (Thunb.) Sch. Bip. ■

92454 Cirsium tuberosum (L.) All.;块根蓟;Boyton Thistle, Tuberous Plume Thistle, Tuberous Thistle ■☆

92455 Cirsium turkestanicum Petr.;土耳其斯坦蓟■☆

92456 Cirsium turneri Warnock;悬崖蓟;Cliff Thistle ■☆

92457 Cirsium tweedyi (Rydb.) Petr. = Cirsium eatonii (A. Gray) B. L. Rob. var. murdockii S. L. Welsh ■☆

92458 Cirsium ugoense Nakai;羽后蓟■☆

92459 Cirsium ukranicum Besser;乌克兰蓟■☆

92460 Cirsium uliginosum DC.;泥潭蓟■☆

92461 Cirsium undulatum (Nutt.) Spreng.;波缘蓟;Gray Thistle, Pasture Thistle, Wavyleaf Thistle, Wavy-leaved Thistle ■☆

92462 Cirsium undulatum (Nutt.) Spreng. var. albescens D. C. Eaton = Cirsium neomexicanum A. Gray ■☆

92463 Cirsium undulatum (Nutt.) Spreng. var. ciliolatum L. F. Hend. = Cirsium ciliolatum (L. F. Hend.) J. T. Howell ■☆

92464 Cirsium undulatum (Nutt.) Spreng. var. megacephalum (A. Gray) Fernald = Cirsium undulatum (Nutt.) Spreng. ■☆

92465 Cirsium undulatus Nutt. var. tracyi (Rydb.) S. L. Welsh = Cirsium tracyi (Rydb.) Petr. ■☆

92466 Cirsium uniflorus Siev. = Rhaponticum carthamoides (Willd.) Iljin ■

92467 Cirsium utahense Petr. = Cirsium neomexicanum A. Gray ■☆

92468 Cirsium uzenense Kadota;羽前蓟■☆

92469 Cirsium vaseyi (A. Gray) Jeps. = Cirsium hydrophilum (Greene) Jeps. var. vaseyi (A. Gray) J. T. Howell ■☆

92470 Cirsium vaseyi (A. Gray) Jeps. var. hydrophilum (Greene) Petr. = Cirsium hydrophilum (Greene) Jeps. ■☆

92471 Cirsium vernale (Osterh.) Cockerell = Cirsium perplexans (Rydb.) Petr. ■☆

92472 Cirsium vernonioides C. Shih;斑鸠蓟;Ironweed Thistle ■

92473 Cirsium verutum (D. Don) Spruner;苞叶蓟;Bractleaf Thistle, Tibetian Thistle ■

92474 Cirsium vinaceum (Wooton et Standl.) Wooton et Standl.;萨山蓟;Sacramento Mountains Thistle ■☆

92475 Cirsium virginense S. L. Welsh = Cirsium mohavense (Greene) Petr. ■☆

92476 Cirsium virginianum (L.) Michx.;弗吉尼亚蓟(维吉尼亚蓟);Virginia Thistle ■☆

92477 Cirsium virginianum (L.) Michx. var. filipendulum A. Gray = Cirsium engelmannii Rydb. ■☆

92478 Cirsium viridifolium (Hand.-Mazz.) C. Shih = Cirsium salicifolium (Kitag.) C. Shih ■

92479 Cirsium vlassovianum Fisch. ex DC.;绒背蓟(猫腿姑,猫腿菇,斩龙草);Vlassoviana Thistle ■

92480 Cirsium vlassovianum Fisch. ex DC. var. bracteatum Ledeb. = Cirsium vlassovianum Fisch. ex DC. ■

92481 Cirsium vlassovianum Fisch. ex DC. var. genuinum Herder = Cirsium vlassovianum Fisch. ex DC. ■

92482 Cirsium vlassovianum Fisch. ex DC. var. salicifolium Kitag. = Cirsium salicifolium (Kitag.) C. Shih ■

92483 Cirsium vlassovianum Fisch. ex DC. var. salicifolium Kitag. = Cirsium vlassovianum Fisch. ex DC. ■

92484 Cirsium vlassovianum Fisch. ex DC. var. viridifolium Hand.-Mazz. = Cirsium salicifolium (Kitag.) C. Shih ■

92485 Cirsium vulgare (Savi) Ten.;翼蓟(欧洲蓟);Bank Thistle, Bell Thistle, Bird Thistle, Black Thistle, Blue Thistle, Bo Thistle, Boar Thistle, Boar-Thistle, Bow Thistle, Buck Brush, Buck Thistle, Bull Thistle, Bur, Bur thistle, Bur Thrissel, Bur Thrissil, Common Thistle, Counsellors, Cuckoo Buttons, Cuckoo-buttons, Dashel, Horse Dashel, Horse Thistle, Marian, Prickly Coat, Prickly Coats, Quat-vessel, Scotch Thistle, Scottish Thistle, Spear Plume Thistle, Spear Thistle, Spear-plume Thistle, Wing Thistle ■

92486 Cirsium wakasugianum Kadota;若杉蓟■☆

92487 Cirsium waldsteinii Rouy;瓦尔德蓟■☆

92488 Cirsium walkerianum Petr. = Cirsium quercetorum (A. Gray) Jeps. ■☆

92489 Cirsium wallichii DC. = Cirsium shansiense Petr. ■

92490 Cirsium wallichii DC. var. glabratum (Hook. f.) Raizada = Cirsium glabrifolium (C. Winkl.) O. Fedtsch. et B. Fedtsch. ■

92491 Cirsium wallichii DC. var. intermedium Pamp. = Cirsium shansiense Petr. ■

92492 Cirsium weyrichii Maxim.;韦里奇蓟■☆

92493 Cirsium wheeleri (A. Gray) Petr.;惠勒蓟;Wheeler's Thistle ■☆

92494 Cirsium wheeleri (A. Gray) Petr. var. salinense S. L. Welsh = Cirsium wheeleri (A. Gray) Petr. ■☆

92495 Cirsium willkomianum Porta et Rigo = Cirsium echinatum (Desf.) DC. ■☆

92496 Cirsium wrightii A. Gray;赖特蓟;Wright's Marsh Thistle ■☆

92497 Cirsium xanthacanthum Nakai = Cirsium japonicum Fisch. ex DC. ■

92498 Cirsium yakusimense Masam.;屋久岛蓟■☆

92499 Cirsium yatsualpicola Kadota et Y. Amano;八束蓟■☆

92500 Cirsium yatsugatakense Nakai = Cirsium ovalifolium (Franch. et Sav.) Matsum. ■☆

92501 Cirsium yezoense (Maxim.) Makino;北海道蓟■☆

92502 Cirsium yunnanense Petr. = Cirsium griseum H. Lév. ■

92503 Cirsium zawoense Kadota;藏奥蓟■☆

92504 Cischweinfia Dressler et N. H. Williams(1970);西宣兰属■☆

92505 Cischweinfia colombiana Garay;哥伦比亚西宣兰■☆

92506 Cischweinfia dasyandra (Rchb. f.) Dressler et N. H. Williams;西宣兰■☆

92507 Cischweinfia nana Dressler;矮西宣兰■☆

92508 Cissabryon Kuntze ex Poepp. = Viviania Cav. ■☆

92509 Cissabryon Meisn. = Cissarobryon Poepp. ■☆

92510 Cissabryon Meisn. = Viviania Cav. ■☆

92511 Cissaceae Drejer = Vitaceae Juss.(保留科名)●■

92512 Cissaceae Horan. = Vitaceae Juss.(保留科名)●■

92513 Cissampelopsis (DC.) Miq.(1856);藤菊属(大叶千里光属,菊藤属);Cissampelopsis ●■

92514 Cissampelopsis (DC.) Miq. = Senecio L. ■●

92515 Cissampelopsis Miq. = Cissampelopsis (DC.) Miq. ■●

92516 Cissampelopsis Miq. = Senecio L. ■●

92517 Cissampelopsis buimalia (Buch. -Ham. ex D. Don) C. Jeffrey et Y. L. Chen;尼泊尔藤菊(舌花藤菊);Ligulate Cissampelopsis ●

92518 Cissampelopsis corifolia C. Jeffrey et Y. L. Chen;革叶藤菊;Coriaceous Leaf Cissampelopsis ■

92519 Cissampelopsis erythrochaeta C. Jeffrey et Y. L. Chen;赤缨藤菊(红缨藤菊);Redpappus Cissampelopsis ●

92520 Cissampelopsis glandulosa C. Jeffrey et Y. L. Chen;腺毛藤菊;Glandulous Cissampelopsis ●■

92521 Cissampelopsis spelaeicola (Vaniot) C. Jeffrey et Y. L. Chen;岩穴藤菊(庐山藤,岩生藤菊,岩穴大叶千里光,岩穴千里光,岩叶千里光,岩页千里光);Rock Groundsel,Saxicolous Cissampelopsis,Shale Groundsel ●

92522 Cissampelopsis volubilis (Blume) Miq.;藤菊(大叶千里光,滇南大叶千里光,滇南千里光);Ho Groundsel,Voluble Cissampelopsis ●

92523 Cissampelos L.(1753);锡生藤属;Cissampelos,False Pareira Root ●

92524 Cissampelos angustifolia Burch. = Antizoma angustifolia (Burch.) Miers ex Harv. ●☆

92525 Cissampelos capensis L. f.;灌丛锡生藤●☆

92526 Cissampelos convolvulacea Willd. var. hirsuta (Buch. -Ham. ex DC.) Hassk. = Cissampelos pareira L. var. hirsuta (Buch. -Ham. ex DC.) Forman ●

92527 Cissampelos dinklagei Engl. = Stephania dinklagei (Engl.) Diels ●☆

92528 Cissampelos ellenbeckii Diels = Cissampelos pareira L. var. hirsuta (Buch. -Ham. ex DC.) Forman ●

92529 Cissampelos friesiorum Diels;弗里斯锡生藤●☆

92530 Cissampelos fruticosa L. f. = Cissampelos capensis L. f. ●☆

92531 Cissampelos glaberrima J. St. -Hil.;极光锡生藤●☆

92532 Cissampelos glabra Roxb. = Stephania glabra (Roxb.) Miers ■

92533 Cissampelos hernandifolia Willd. = Stephania hernandifolia (Willd.) Walp. ●■

92534 Cissampelos hernandiifolia Willd. = Stephania japonica (Thunb.) Miers var. discolor (Blume) Forman ●■

92535 Cissampelos hirsuta Buch. ex DC. = Cissampelos pareira L. var. hirsuta (Buch. -Ham. ex DC.) Forman ●

92536 Cissampelos hirsuta Buch. ex DC. = Cissampelos pareira L. ●

92537 Cissampelos hirsuta Buch. -Ham. ex DC. = Cissampelos pareira L. var. hirsuta (Buch. -Ham. ex DC.) Forman ●

92538 Cissampelos hirsuta DC. = Cissampelos pareira L. ●

92539 Cissampelos hirta Klotzsch;多毛锡生藤●☆

92540 Cissampelos humilis Poir. = Cissampelos capensis L. f. ●☆

92541 Cissampelos hypoglauca Schauer = Cyclea hypoglauca (Schauer) Diels ●

92542 Cissampelos insignis Alston;显著锡生藤●☆

92543 Cissampelos insolita Miers ex Oliv. = Cissampelos owariensis P. Beauv. ex DC. ●☆

92544 Cissampelos insularis Makino = Cyclea insularis (Makino) Hatus. ●■

92545 Cissampelos lycioides (Miers) T. Durand et Schinz;枸杞状锡生藤●☆

92546 Cissampelos macrosepala Diels;大萼锡生藤●☆

92547 Cissampelos macrostachya Klotzsch = Cissampelos mucronata A. Rich. ●☆

92548 Cissampelos mucronata A. Rich.;棘状锡生藤●☆

92549 Cissampelos nigrescens Diels;黑锡生藤●☆

92550 Cissampelos nigrescens Diels var. cardiophylla Troupin;心叶黑锡生藤●☆

92551 Cissampelos ochiaiana Gogelein = Cyclea ochiaiana (Yamam.) S. F. Huang et T. C. Huang ●

92552 Cissampelos orbiculata DC. = Cissampelos pareira L. var. orbiculata (DC.) Miq. ●☆

92553 Cissampelos ovalifolia DC.;卵叶锡生藤●☆

92554 Cissampelos owariensis P. Beauv. ex DC.;奥瓦锡生藤●☆

92555 Cissampelos pareira L.;锡生藤(老鼠耳朵草,美非锡生藤,美洲锡生藤,鼠耳草,亚红龙);Cissampelos,Common Cissampelos,False Pareira Root ●

92556 Cissampelos pareira L. subvar. crassifolia Engl. = Cissampelos pareira L. ●

92557 Cissampelos pareira L. subvar. hirta (Klotzsch) Engl. = Cissampelos hirta Klotzsch ●☆

92558 Cissampelos pareira L. subvar. rigidifolia Engl. = Cissampelos rigidifolia (Engl.) Diels ●☆

92559 Cissampelos pareira L. subvar. zairensis (Miers) Engl. = Cissampelos zairensis Miers ●☆

92560 Cissampelos pareira L. var. hirsuta (Buch. -Ham. ex DC.) Forman;毛锡生藤(锡生藤)●

92561 Cissampelos pareira L. var. klotzschii T. Durand et Schinz = Cissampelos hirta Klotzsch ●☆

92562 Cissampelos pareira L. var. mucronata (A. Rich.) Engl. = Cissampelos mucronata A. Rich. ●☆

92563 Cissampelos pareira L. var. orbiculata (DC.) Miq.;典型锡生藤●☆

92564 Cissampelos pareira L. var. owariensis (P. Beauv. ex DC.) Oliv. = Cissampelos owariensis P. Beauv. ex DC. ●☆

92565 Cissampelos pareira L. var. typica Diels = Cissampelos orbiculata

DC. ●☆

92566 Cissampelos pareira L. var. wildei P. Beauv.;维尔德锡生藤●☆

92567 Cissampelos pareira sensu Hook. f. et Thomson = Cissampelos pareira L. ●

92568 Cissampelos pauciflora Nutt. = Croomia pauciflora (Nutt.) Torr. ■☆

92569 Cissampelos rigidifolia (Engl.) Diels;硬叶锡生藤●☆

92570 Cissampelos rigidifolia (Engl.) Diels var. lanuginosa Troupin;多绵毛硬叶锡生藤●☆

92571 Cissampelos robertsonii Exell = Cissampelos owariensis P. Beauv. ex DC. ●☆

92572 Cissampelos senensis Klotzsch = Cissampelos mucronata A. Rich. ●☆

92573 Cissampelos sympodialis Eichler;巴西锡生藤●☆

92574 Cissampelos tamnifolia Miers = Cissampelos hirta Klotzsch ●☆

92575 Cissampelos tenuipes Engl.;细梗锡生藤●☆

92576 Cissampelos torulosa E. Mey. ex Harv.;结节锡生藤●☆

92577 Cissampelos truncata Engl.;平截锡生藤●☆

92578 Cissampelos umbellata E. Mey. ex Harv. = Stephania abyssinica (Quart.-Dill. et A. Rich.) Walp. var. tomentella (Oliv.) Diels ●☆

92579 Cissampelos wildemaniana Bossche ex De Wild. = Cissampelos torulosa E. Mey. ex Harv. ●☆

92580 Cissampelos zairensis Miers = Cissampelos mucronata A. Rich. ●☆

92581 Cissarobryon Kuntze = Viviania Cav. ■☆

92582 Cissarobryon Kuntze ex Poepp. = Viviania Cav. ■☆

92583 Cissarobryon Poepp. = Viviania Cav. ■☆

92584 Cissodendron F. Muell. = Kissodendron Seem. ●

92585 Cissodendrum T. Post et Kuntze = Cissodendron F. Muell ●

92586 Cissus L. (1753);白粉藤属(粉藤属,青紫葛属);Cissus, Grape Ivy, Ivy Treebine, Treebine ●

92587 Cissus acida L.;肉白粉藤;Acid Treebine ●☆

92588 Cissus adamii Dewit;阿达姆白粉藤●☆

92589 Cissus adenantha Fresen. = Cyphostemma adenanthum (Fresen.) Desc. ●☆

92590 Cissus adenocarpa Gilg et M. Brandt = Cyphostemma adenocarpum (Gilg et M. Brandt) Desc. ●☆

92591 Cissus adenocaulis Steud. ex A. Rich. = Cyphostemma adenocaule (Steud. ex A. Rich.) Desc. ex Wild et R. B. Drumm. ●☆

92592 Cissus adenocaulis Steud. ex A. Rich. var. eglandulosa Dewit = Cyphostemma adenocaule (Steud. ex A. Rich.) Desc. ex Wild et R. B. Drumm. ●☆

92593 Cissus adenocaulis Steud. ex A. Rich. var. eglanduloso-pubescens Dewit = Cyphostemma adenocaule (Steud. ex A. Rich.) Desc. ex Wild et R. B. Drumm. ●☆

92594 Cissus adenocaulis Steud. ex A. Rich. var. pubescens Dewit = Cyphostemma adenocaule (Steud. ex A. Rich.) Desc. ex Wild et R. B. Drumm. ●☆

92595 Cissus adenocephala Gilg et M. Brandt = Cyphostemma dembianense (Chiov.) Vollesen ●☆

92596 Cissus adenopoda Sprague;腺荚白粉藤;Adenopode Treebine ●☆

92597 Cissus adenopoda Sprague = Cyphostemma adenopodum (Sprague) Desc. ●☆

92598 Cissus adnata Roxb.;贴生白粉藤(红毛粉藤,锈毛白粉藤);Adnate Treebine ●

92599 Cissus aegirophylla Bunge. = Ampelopsis vitifolia (Boiss.) Planch. ●☆

92600 Cissus afzelii (Baker) Gilg et M. Brandt = Cissus diffusiflora (Baker) Planch. ●☆

92601 Cissus agnus-castus Planch. = Cyphostemma bororense (Klotzsch) Desc. ex Wild et R. B. Drumm. ●☆

92602 Cissus allophylloides Gilg et M. Brandt = Cyphostemma allophylloides (Gilg et M. Brandt) Desc. ●☆

92603 Cissus alnifolia Schweinf. ex Planch. = Cyphostemma alnifolium (Schweinf. ex Planch.) Desc. ●☆

92604 Cissus amoena Gilg et M. Brandt;秀丽白粉藤●☆

92605 Cissus ampelopsis Pers. = Ampelopsis cordata Michx. ●

92606 Cissus antandroyi Desc.;安坦德罗白粉藤●☆

92607 Cissus antarctica Vent.;南极白粉藤(澳大利亚白粉藤);Kangaroo, Kangaroo Treebine, Kangaroo Vine ●☆

92608 Cissus aphylla Chiov.;无叶白粉藤●☆

92609 Cissus aphyllantha Gilg et M. Brandt;无叶花白粉藤●☆

92610 Cissus aralioides (Baker) Planch.;楤木白粉藤●☆

92611 Cissus aralioides (Baker) Planch. subsp. orientalis Verdc.;东方白粉藤●☆

92612 Cissus arguta Hook. f.;亮白粉藤●☆

92613 Cissus arguta Hook. f. var. oliveri Engl. = Cissus oliveri (Engl.) Gilg ●☆

92614 Cissus aristata Blume;毛叶苦郎藤;Aristate Treebine ●

92615 Cissus aristolochiifolia Planch.;马兜铃叶白粉藤●☆

92616 Cissus assamica (Lawson) Craib;苦郎藤(粗壳藤,风叶藤,红背丝绸,葫芦叶,毛叶白粉藤,野葡萄,左边藤,左爬藤);Assam Treebine ●

92617 Cissus assamica (Lawson) Craib var. pilosissima Gagnep.;毛叶白粉藤;Hairyleaf Assam Treebine ●

92618 Cissus assamica (Lawson) Craib var. pilosissima Gagnep. = Cissus aristata Blume ●

92619 Cissus auricoma Desc.;金毛白粉藤●☆

92620 Cissus austroyunnanensis Y. H. Li et Y. Zhang;滇南青紫葛;South Yunnan Treebine ●

92621 Cissus bainesii Hook. f. = Cyphostemma bainesii (Hook. f.) Desc. ●☆

92622 Cissus bakeriana Planch. = Cyphostemma cymosum (Schumach. et Thonn.) Desc. ●☆

92623 Cissus bambuseti Gilg et M. Brandt = Cyphostemma bambuseti (Gilg et M. Brandt) Desc. ex Wild et R. B. Drumm. ●☆

92624 Cissus bambuseti Gilg et M. Brandt var. glandulosissima Dewit = Cyphostemma bambuseti (Gilg et M. Brandt) Desc. ex Wild et R. B. Drumm. var. glandulosissima (Dewit) Desc. ●☆

92625 Cissus barbeyana De Wild. et T. Durand;巴比白粉藤●☆

92626 Cissus barteri (Baker) Planch.;巴特白粉藤●☆

92627 Cissus baudiniana Brouss. = Cissus antarctica Vent. ●☆

92628 Cissus bequaertii Dewit;贝卡尔白粉藤●☆

92629 Cissus bignonioides Gilg et M. Brandt = Cissus petiolata Hook. f. ●☆

92630 Cissus biternata (Baker) Planch. = Cyphostemma microdiptera (Baker) Desc. ●☆

92631 Cissus bororensis Klotzsch = Cyphostemma bororense (Klotzsch) Desc. ex Wild et R. B. Drumm. ●☆

92632 Cissus bosseri Desc.;博瑟白粉藤●☆

92633 Cissus braunii Gilg et M. Brandt = Cyphostemma braunii (Gilg et M. Brandt) Desc. ●☆

92634 Cissus brevipedunculata Maxim. = Ampelopsis brevipedunculata (Maxim.) Trautv. ●

92635 Cissus brevipedunculata Maxim. = Ampelopsis glandulosa

(Wall.) Momiy. var. brevipedunculata (Maxim.) Momiy. ●
92636 Cissus brevipedunculata Maxim. = Ampelopsis heterophylla (Thunb.) Siebold et Zucc. var. brevipedunculata (Regel) C. L. Li ●
92637 Cissus brieyi De Wild. = Cyphostemma brieyi (De Wild.) Compère ●☆
92638 Cissus buchananii Planch. = Cyphostemma buchananii (Planch.) Desc. ex Wild et R. B. Drumm. ●☆
92639 Cissus bullata Gilg et M. Brandt = Cyphostemma bullatum (Gilg et M. Brandt) Desc. ●☆
92640 Cissus cactiformis Gilg; 仙素莲(翡翠葡萄,四角茎粉藤); Cactuform Cissus ●☆
92641 Cissus caesia Afzel.; 淡蓝白粉藤●☆
92642 Cissus caillei A. Chev. = Cissus aralioides (Baker) Planch. ●☆
92643 Cissus cantoniensis Hook. et Arn. = Ampelopsis cantoniensis (Hook. et Arn.) Planch. ●
92644 Cissus cantoniensis Hook. et Arn. = Ampelopsis hypoglauca (Hance) C. L. Li ●
92645 Cissus capensis Willd. = Rhoicissus tomentosa (Lam.) Wild et R. B. Drumm. ●☆
92646 Cissus capensis Willd. = Vitis capensis Thunb. ●☆
92647 Cissus capreolatus Royle = Tetrastigma serrulatum (Roxb.) Planch. ●■
92648 Cissus carnosa (L.) Lam. = Cayratia trifolia (L.) Domin ●
92649 Cissus carnosa Lam. = Cayratia trifolia (L.) Domin ●
92650 Cissus carrissoi Exell et Mendonça; 卡里索白粉藤●☆
92651 Cissus centrali-africana Gilg et R. E. Fr. = Cyphostemma mildbraedii (Gilg et M. Brandt) Desc. ex Wild et R. B. Drumm. ●☆
92652 Cissus chevalieri Gilg et M. Brandt = Cyphostemma flavicans (Baker) Desc. ●☆
92653 Cissus chrysadenia Gilg = Cyphostemma chrysadenium (Gilg) Desc. ●☆
92654 Cissus cinerea Lam. = Cissus carnosa (L.) Lam. ●
92655 Cissus cirrhosa (Thunb.) Willd. var. glabra Harv. = Cyphostemma quinatum (Dryand.) Desc. ex Wild et R. B. Drumm. ●☆
92656 Cissus cirrhosa (Thunb.) Willd. var. transvaalensis (Szyszyl.) Burtt Davy et R. Pott = Cyphostemma cirrhosum (Thunb.) Desc. ex Wild et R. B. Drumm. subsp. transvaalense (Szyszyl.) Wild et R. B. Drumm. ●☆
92657 Cissus comosa Desc.; 簇毛白粉藤●☆
92658 Cissus connivens Lam. = Cyphostemma natalitium (Szyszyl.) J. J. M. van der Merwe ●☆
92659 Cissus constricta (Baker) A. Chev. = Cissus aralioides (Baker) Planch. ●☆
92660 Cissus cordata Roxb. = Cissus repens (Wight et Arn.) Lam. ●
92661 Cissus cornifolia (Baker) Planch.; 角叶白粉藤●☆
92662 Cissus coursii Desc.; 库尔斯白粉藤●☆
92663 Cissus cramerana Schinz = Cyphostemma currorii (Hook. f.) Desc. ●☆
92664 Cissus crassifolia (Baker) Planch. = Cissus rotundifolia (Forssk.) Vahl ●☆
92665 Cissus crinita Planch. = Cyphostemma crinitum (Planch.) Desc. ●☆
92666 Cissus crithmifolia Chiov. = Cyphostemma crithmifolium (Chiov.) Desc. ●☆
92667 Cissus crotalarioides Planch. = Cyphostemma crotalarioides (Planch.) Desc. ex Wild et R. B. Drumm. ●☆
92668 Cissus cucumerifolia Planch.; 黄瓜叶白粉藤●☆
92669 Cissus cuneata Gilg et M. Brandt = Cyphostemma cuneatum (Gilg et M. Brandt) Desc. ●☆
92670 Cissus cuneifolia Eckl. et Zeyh. = Rhoicissus tridentata (L. f.) Wild et R. B. Drumm. subsp. cuneifolia (Eckl. et Zeyh.) Urton ●☆
92671 Cissus currorii Hook. f. = Cyphostemma currorii (Hook. f.) Desc. ●☆
92672 Cissus curvipoda (Baker) Planch. = Cyphostemma curvipodum (Baker) Desc. ●☆
92673 Cissus cuspidata Planch. = Cissus floribunda (Baker) Planch. ●☆
92674 Cissus cussonioides Schinz; 甘蓝白粉藤●☆
92675 Cissus cymosa Schumach. et Thonn. = Cyphostemma cymosum (Schumach. et Thonn.) Desc. ●☆
92676 Cissus cyphopetala Fresen. = Cyphostemma cyphopetalum (Fresen.) Desc. ex Wild et R. B. Drumm. ●☆
92677 Cissus dahomensis A. Chev. = Cyphostemma flavicans (Baker) Desc. ●☆
92678 Cissus dasyantha Gilg et M. Brandt; 毛花白粉藤●☆
92679 Cissus dasypleura C. A. Sm. = Cyphostemma dasypleurum (C. A. Sm.) J. J. M. van der Merwe ●☆
92680 Cissus davidiana Carrière = Ampelopsis heterophylla (Thunb.) Siebold et Zucc. ●
92681 Cissus davidiana Carrière = Ampelopsis humulifolia Bunge ●
92682 Cissus debilis (Baker) Planch. = Cayratia debilis (Baker) Suess. ●☆
92683 Cissus decurrens Gilg et M. Brandt = Cyphostemma chloroleucum (Welw. ex Baker) Desc. ex Wild et R. B. Drumm. ●☆
92684 Cissus delicatula Willems = Cayratia delicatula (Willems) Desc. ●☆
92685 Cissus dembianensis Chiov. = Cyphostemma dembianense (Chiov.) Vollesen ●☆
92686 Cissus dewevrei De Wild. et T. Durand; 德韦白粉藤●☆
92687 Cissus diffusiflora (Baker) Planch.; 散花白粉藤●☆
92688 Cissus diffusiflora (Baker) Planch. var. pilosocalyx Dewit; 毛萼白粉藤●☆
92689 Cissus dinklagei Gilg et M. Brandt; 丁克白粉藤●☆
92690 Cissus dinklagei Gilg et M. Brandt var. pilosa Desc.; 疏毛白粉藤●☆
92691 Cissus dinteri Schinz = Cissus nymphiifolia (Welw. ex Baker) Planch. ●☆
92692 Cissus discolor Blume; 锦叶葡萄; Begonia Treevine, Begonia Tree-vine, Rex-begonia Vine ●
92693 Cissus discolor Blume = Cissus javana DC. ●
92694 Cissus diversifolia Walp. = Ampelopsis cantoniensis (Hook. et Arn.) Planch. ●
92695 Cissus diversilobata C. A. Sm.; 异裂白粉藤●☆
92696 Cissus doeringii Gilg et M. Brandt; 多林白粉藤●☆
92697 Cissus dolichopus C. A. Sm. = Cyphostemma humile (N. E. Br.) Desc. ex Wild et R. B. Drumm. subsp. dolichopus (C. A. Sm.) Wild et R. B. Drumm. ●☆
92698 Cissus duboisii Dewit; 杜氏白粉藤●☆
92699 Cissus duparquetii Planch. = Cyphostemma duparquetii (Planch.) Desc. ●☆
92700 Cissus edulis Dalzell = Cissus quadrangularis L. ●☆
92701 Cissus egregia Gilg = Cyphostemma cyphopetalum (Fresen.) Desc. ex Wild et R. B. Drumm. ●☆
92702 Cissus ellenbeckii Gilg et M. Brandt; 埃伦白粉藤●☆
92703 Cissus elongata (Roxb.) Suess. = Cissus elongata Roxb. ●

92704 Cissus elongata Roxb.;五叶白粉藤(五叶粉藤)●

92705 Cissus eminii Gilg = Cyphostemma eminii (Gilg) Desc. ●☆

92706 Cissus engleri Gilg = Cyphostemma engleri (Gilg) Desc. ●☆

92707 Cissus erythreae Gilg et M. Brandt = Cyphostemma molle (Steud. ex Baker) Desc. ●☆

92708 Cissus erythrocephala Gilg et M. Brandt = Cyphostemma erythrocephalum (Gilg et M. Brandt) Desc. ●☆

92709 Cissus erythrochlora Gilg = Cyphostemma cyphopetalum (Fresen.) Desc. ex Wild et R. B. Drumm. ●☆

92710 Cissus fanshawei Wild et R. B. Drumm.;范肖白粉藤●☆

92711 Cissus farinosa Planch.;非洲白粉藤●☆

92712 Cissus fischeri Gilg = Cissus quadrangularis L. var. pubescens Dewit ●☆

92713 Cissus flavicans (Baker) Desc. = Cyphostemma flavicans (Baker) Desc. ●☆

92714 Cissus flaviflora Sprague = Cyphostemma flaviflorum (Sprague) Desc. ●☆

92715 Cissus fleckii Schinz = Cyphostemma congestum (Baker) Desc. ex Wild et R. B. Drumm. ●☆

92716 Cissus floribunda (Baker) Planch.;繁花白粉藤●☆

92717 Cissus fragilis E. Mey. ex Kunth;脆白粉藤●☆

92718 Cissus fugosioides Gilg = Cyphostemma fugosioides (Gilg) Desc. ex Wild et R. B. Drumm. ●☆

92719 Cissus furcifera Chiov.;叉白粉藤●☆

92720 Cissus gallaensis Gilg et M. Brandt = Cyphostemma cyphopetalum (Fresen.) Desc. ex Wild et R. B. Drumm. ●☆

92721 Cissus geniculata Blume = Cayratia geniculata (Blume) Gagnep. ●

92722 Cissus gilletii De Wild. et T. Durand = Cyphostemma gilletii (De Wild. et T. Durand) Desc. ●☆

92723 Cissus glaberrima (Wall.) Planch. = Cissus luzoniensis (Merr.) C. L. Li ●

92724 Cissus glaberrima Planch.;粉藤果(白葡萄,大绿藤,光叶白粉藤,尖光叶白粉藤,小花叶,野葡萄)●

92725 Cissus glandulosissima Gilg et M. Brandt = Cyphostemma glandulosissimum (Gilg et M. Brandt) Desc. ex Wild et R. B. Drumm. ●☆

92726 Cissus glauca Roxb. = Cissus repens (Wight et Arn.) Lam. ●

92727 Cissus glaucophylla Hook. f.;灰绿叶粉藤●☆

92728 Cissus glossopetala (Baker) Suess.;舌瓣粉藤●☆

92729 Cissus gongylodes Baker;瘤状粉藤;Marble Treebine ●☆

92730 Cissus gossweileri Exell et Mendonça;戈斯绿粉藤●☆

92731 Cissus gossweileri Exell et Mendonça var. hirsuto-pubescens Dewit;杂毛白粉藤●☆

92732 Cissus goudotii Planch. = Cyphostemma microdiptera (Baker) Desc. ●☆

92733 Cissus gracilis Guillaumin et Perr. = Cayratia gracilis (Guillaumin et Perr.) Suess. ●☆

92734 Cissus gracillima Werderm. = Cyphostemma gracillimum (Werderm.) Desc. ●☆

92735 Cissus gracillimoides Dewit = Cyphostemma fugosioides (Gilg) Desc. ex Wild et R. B. Drumm. ●☆

92736 Cissus grandistipulata Gilg et M. Brandt = Cyphostemma grandistipulatum (Gilg et M. Brandt) Desc. ●☆

92737 Cissus granitica Wild et R. B. Drumm. = Cyphostemma graniticum (Wild et R. B. Drumm.) Wild et R. B. Drumm. ●☆

92738 Cissus grisea (Baker) Planch.;灰粉藤●☆

92739 Cissus griseo-rubra Gilg et M. Brandt = Cyphostemma griseo-rubrum (Gilg et M. Brandt) Desc. ●☆

92740 Cissus guerkeana (Büttner) T. Durand et Schinz;盖尔克绿粉藤●☆

92741 Cissus hastata (Miq.) Planch.;戟叶白粉藤(春根藤,方藤,风藤,红宽筋藤,红四方藤,宽筋藤,蚂蝗藤,软筋藤,伸筋藤,舒筋藤,四方宽筋藤,四方藤,四方钻,万戈藤,翼枝白粉藤);Hastate Treebine,Hastateleaf Treebine ●

92742 Cissus haumanii Dewit = Cyphostemma haumanii (Dewit) Desc. ●☆

92743 Cissus hederacea Pers. = Parthenocissus quinquefolius (L.) Planch. ●

92744 Cissus helenae Buscal. et Muschl. = Cyphostemma hildebrandtii (Gilg) Desc. ex Wild et R. B. Drumm. ●☆

92745 Cissus hereroensis Schinz = Cyphostemma hereroense (Schinz) Desc. ex Wild et R. B. Drumm. ●☆

92746 Cissus heterotricha Gilg et R. E. Fr. = Cyphostemma heterotrichum (Gilg et R. E. Fr.) Desc. ex Wild et R. B. Drumm. ●☆

92747 Cissus hexangularis Thorel ex Planch.;翅茎白粉藤(春根藤,方茎宽筋,方茎宽筋藤,方藤,风藤,红宽筋藤,红四方藤,宽筋藤,拦河藤,六方藤,六棱粉藤,蚂蝗藤,软筋藤,散血龙,山坡瓜藤,伸筋藤,四方藤,五俭藤);Winged-stem Treebine ●

92748 Cissus hildebrandtii Gilg = Cyphostemma hildebrandtii (Gilg) Desc. ex Wild et R. B. Drumm. ●☆

92749 Cissus himalayana Walp. = Parthenocissus semicordata (Wall. ex Roxb.) Planch. var. roylei (King ex Parker) Nazim. et Qaiser ●☆

92750 Cissus hispidiflorus C. A. Sm. = Cyphostemma hispidiflorum (C. A. Sm.) J. J. M. van der Merwe ●☆

92751 Cissus hochstetteri (Miq.) Planch. = Cissus petiolata Hook. f. ●☆

92752 Cissus homblei De Wild. = Cyphostemma homblei (De Wild.) Desc. ●☆

92753 Cissus huillensis Exell et Mendonça = Cyphostemma huillense (Exell et Mendonça) Desc. ●☆

92754 Cissus humbertiana Desc.;亨氏白粉藤●☆

92755 Cissus humbertii Robyns et Lawalrée;亨伯特白粉藤●☆

92756 Cissus humilis (N. E. Br.) Planch. = Cyphostemma humile (N. E. Br.) Desc. ex Wild et R. B. Drumm. ●☆

92757 Cissus humulifolia (Bunge) Regel = Ampelopsis heterophylla (Thunb.) Siebold et Zucc. ●

92758 Cissus humulifolia (Bunge) Regel = Ampelopsis humulifolia Bunge ●

92759 Cissus humulifolia (Bunge) Regel var. brevipedunculata (Maxim.) Regel = Ampelopsis glandulosa (Wall.) Momiy. var. brevipedunculata (Maxim.) Momiy. ●

92760 Cissus humulifolia (Bunge) Regel var. brevipedunculata (Maxim.) Regel = Ampelopsis brevipedunculata (Maxim.) Trautv. ●

92761 Cissus humulifolia (Bunge) Regel var. brevpedunculata Regel = Ampelopsis brevipedunculata (Maxim.) Trautv. ●

92762 Cissus humulifolia (Bunge) Regel var. brevpedunculata Regel = Ampelopsis heterophylla (Thunb.) Siebold et Zucc. var. brevipedunculata (Regel) C. L. Li ●

92763 Cissus hypargyrea Gilg = Cyphostemma chloroleucum (Welw. ex Baker) Desc. ex Wild et R. B. Drumm. ●☆

92764 Cissus hypoglauca A. Gray;水白粉藤;Water Vine ●☆

92765 Cissus hypoglauca A. Gray = Vitis hypoglauca (A. Gray) F. Muell. ●☆

92766 Cissus hypoleuca Harv. = Cyphostemma hypoleucum (Harv.) Desc. ex Wild et R. B. Drumm. ●☆

92767 Cissus imerinensis (Baker) Suess. = Cayratia imerinensis (Baker) Desc. ●☆

92768 Cissus incisa (Torr. et Gray) Des Moul.;缺刻白粉藤(北美乌蔹莓,常春白粉藤);Marine Ivy,Marine Vine ●☆

92769 Cissus integrifolia (Baker) Planch.;全叶白粉藤●☆

92770 Cissus jaegeri Gilg et M. Brandt = Cyphostemma kilimandscharicum (Gilg) Desc. ex Wild et R. B. Drumm. var. jaegeri (Gilg et M. Brandt) Verdc. ●☆

92771 Cissus japonica (Thunb.) Willd. = Cayratia japonica (Thunb.) Gagnep. ●

92772 Cissus japonica (Thunb.) Willd. var. canescens W. T. Wang = Cayratia japonica (Thunb.) Gagnep. var. mollis (Wall.) C. L. Li ●

92773 Cissus japonica (Thunb.) Willd. var. mollis (Wall. ex M. A. Lawson) Planch. = Cayratia japonica (Thunb.) Gagnep. var. mollis (Wall. ex M. A. Lawson) Momiy. ●

92774 Cissus japonica (Thunb.) Willd. var. pubifolia Merr. et Chun = Cayratia japonica (Thunb.) Gagnep. var. mollis (Wall.) C. L. Li ●

92775 Cissus japonica Willd. = Cayratia japonica (Thunb.) Gagnep. ●

92776 Cissus jatrophoides (Baker) Planch. = Cyphostemma junceum (Webb) Wild et R. B. Drumm. subsp. jatrophoides (Baker) Verdc. ●☆

92777 Cissus javana DC.;青紫葛(白粉藤,变色白粉藤,粪虫叶,花斑叶,青紫藤);Begonia Treevine,Cissus,Java Treevine,Rex Begonia Vine,Trailing Begonia ●

92778 Cissus javana DC. var. pubescens C. L. Li = Cissus austroyunnanensis Y. H. Li et Y. Zhang ●

92779 Cissus javellensis Lanza = Cyphostemma cyphopetalum (Fresen.) Desc. ex Wild et R. B. Drumm. ●☆

92780 Cissus javellensis Lanza var. rotundata = Cyphostemma cyphopetalum (Fresen.) Desc. ex Wild et R. B. Drumm. ●☆

92781 Cissus johannis Exell et Mendonça = Cyphostemma johannis (Exell et Mendonça) Desc. ●☆

92782 Cissus juncea Webb = Cyphostemma junceum (Webb) Wild et R. B. Drumm. ●☆

92783 Cissus juttae Dinter et Gilg = Cyphostemma juttae (Dinter et Gilg) Desc. ●☆

92784 Cissus juttae Dinter et Gilg ex Gilg et M. Brandt = Cyphostemma juttae (Dinter et Gilg ex Gilg et M. Brandt) Desc. ●☆

92785 Cissus kaessneri Gilg et M. Brandt = Cyphostemma cyphopetalum (Fresen.) Desc. ex Wild et R. B. Drumm. ●☆

92786 Cissus kaniamae Dewit = Cyphostemma kaniamae (Dewit) Desc. ●☆

92787 Cissus kapiriensis Dewit = Cyphostemma kapiriense (Dewit) Desc. ●☆

92788 Cissus karaguensis Gilg = Cyphostemma cyphopetalum (Fresen.) Desc. ex Wild et R. B. Drumm. ●☆

92789 Cissus keilii Gilg et M. Brandt = Cyphostemma keilii (Gilg et M. Brandt) Desc. ●☆

92790 Cissus keniensis T. C. E. Fr. = Cyphostemma knittelii (Gilg) Desc. ●☆

92791 Cissus kerrii Craib;鸡心藤(白粉藤,白面水鸡,白薯藤,独脚乌桕,飞龙接骨,假葡萄,接骨藤,青龙跌打,山葫芦,山鸡蛋,万年薯,夜牵牛);Kerr Treebine ●

92792 Cissus kilimandscharica Gilg = Cyphostemma kilimandscharicum (Gilg) Desc. ex Wild et R. B. Drumm. ●☆

92793 Cissus kirkiana Planch. = Cyphostemma kirkianum (Planch.) Desc. ex Wild et R. B. Drumm. ●☆

92794 Cissus knittelii Gilg = Cyphostemma knittelii (Gilg) Desc. ●☆

92795 Cissus kouandeensis A. Chev.;宽代白粉藤●☆

92796 Cissus lageniflora Gilg et M. Brandt = Cyphostemma lageniflorum (Gilg et M. Brandt) Desc. ●☆

92797 Cissus lamprophylla Gilg et M. Brandt;亮叶白粉藤●☆

92798 Cissus landuk Hassk. = Parthenocissus dalzielii Gagnep. ●

92799 Cissus lanea Desc.;绵毛白粉藤●☆

92800 Cissus lanigera Harv. = Cyphostemma lanigerum (Harv.) Desc. ex Wild et R. B. Drumm. ●☆

92801 Cissus lanyuensis (C. C. Chang) F. Y. Lu;兰屿粉藤●

92802 Cissus lanyuensis (C. C. Chang) F. Y. Lu = Tetrastigma lanyuense C. C. Chang ●

92803 Cissus lebrunii Dewit;勒布伦白粉藤●☆

92804 Cissus ledermannii Gilg et M. Brandt = Cyphostemma ledermannii (Gilg et M. Brandt) Desc. ●☆

92805 Cissus leemansii Dewit;利曼斯白粉藤●☆

92806 Cissus lemurica Desc.;莱穆拉白粉藤●☆

92807 Cissus lenticellata (Baker) Suess. = Cissus floribunda (Baker) Planch. ●☆

92808 Cissus leonardii Dewit;莱奥白粉藤●☆

92809 Cissus leonensis Hook. f. = Ampelocissus leonensis (Hook. f.) Planch. ●☆

92810 Cissus leucadenia Suess. = Cyphostemma gigantophyllum (Gilg et M. Brandt) Desc. ex Wild et R. B. Drumm. ●☆

92811 Cissus leucocarpa Blume = Cayratia japonica (Thunb.) Gagnep. ●

92812 Cissus leucotricha Gilg et M. Brandt = Cyphostemma leucotrichum (Gilg et M. Brandt) Desc. ●☆

92813 Cissus libenii Dewit = Cyphostemma libenii (Dewit) Desc. ●☆

92814 Cissus loandensis Gilg et M. Brandt = Cyphostemma adenocaule (Steud. ex A. Rich.) Desc. ex Wild et R. B. Drumm. ●☆

92815 Cissus longzhouensis W. T. Wang = Cayratia pedata (Lam.) Juss. ex Gagnep. ●

92816 Cissus lonicerifolia C. A. Sm. = Cissus cornifolia (Baker) Planch. ●☆

92817 Cissus louisii Dewit;路易斯白粉藤●☆

92818 Cissus lutea Exell et Mendonça = Cyphostemma luteum (Exell et Mendonça) Desc. ●☆

92819 Cissus luzoniensis (Merr.) C. L. Li;粉果藤(兰屿粉藤)●

92820 Cissus lynesii Dewit = Cyphostemma lynesii (Dewit) Desc. ex Wild et R. B. Drumm. ●☆

92821 Cissus macrantha Werderm. = Cissus cucumerifolia Planch. ●☆

92822 Cissus macropus Welw.;大足白粉藤●☆

92823 Cissus macropus Welw. = Cyphostemma currorii (Hook. f.) Desc. ●☆

92824 Cissus macrothyrsa Gilg = Cyphostemma cyphopetalum (Fresen.) Desc. ex Wild et R. B. Drumm. ●☆

92825 Cissus madecassa Desc.;马德卡萨白粉藤●☆

92826 Cissus manikensis De Wild. = Cyphostemma manikense (De Wild.) Desc. ex Wild et R. B. Drumm. ●☆

92827 Cissus maranguensis Gilg = Cyphostemma maranguense (Gilg) Desc. ●☆

92828 Cissus marionae Exell et Mendonça = Cyphostemma subciliatum (Baker) Desc. ex Wild et R. B. Drumm. ●☆

92829 Cissus marlothii Dinter et Gilg = Cyphostemma congestum (Baker) Desc. ex Wild et R. B. Drumm. ●☆

92830 Cissus marunguensis Dewit = Cyphostemma marunguense (Dewit) Desc. ●☆

92831 Cissus mekongensis C. Y. Wu et W. T. Wang = Cayratia timoriensis (DC.) C. L. Li var. mekongensis (C. Y. Wu ex W. T. Wang) C. L. Li ●
92832 Cissus meyeri-johannis Gilg et M. Brandt = Cyphostemma meyeri-johannis (Gilg et M. Brandt) Verdc. ●☆
92833 Cissus michelii Dewit = Cyphostemma michelii (Dewit) Desc. ●☆
92834 Cissus micradenia Gilg et M. Brandt = Cyphostemma micradenium (Gilg et M. Brandt) Desc. ●☆
92835 Cissus microdonta (Baker) Planch.;小齿白粉藤●☆
92836 Cissus microphylla Turcz. = Rhoicissus microphylla (Turcz.) Gilg et M. Brandt ●☆
92837 Cissus migeodii Verdc.;米容德白粉藤●☆
92838 Cissus migiurtinorum Chiov. = Cyphostemma migiurtinorum (Chiov.) Desc. ●☆
92839 Cissus mildbraedii Gilg et M. Brandt = Cyphostemma mildbraedii (Gilg et M. Brandt) Desc. ex Wild et R. B. Drumm. ●☆
92840 Cissus milnei Verdc.;米尔恩白粉藤●☆
92841 Cissus modeccoides Planch. = Cissus triloba (Lour.) Merr. ●
92842 Cissus modeccoides Planch. var. kerrii Craib = Cissus kerrii Craib ●
92843 Cissus modeccoides Planch. var. subintegra Gagnep. = Cissus kerrii Craib ●
92844 Cissus mollis (Wall.) C. Y. Wu = Cayratia japonica (Thunb.) Gagnep. var. mollis (Wall.) C. L. Li ●
92845 Cissus mollis Steud. ex Baker = Cyphostemma molle (Steud. ex Baker) Desc. ●☆
92846 Cissus morifolia (Baker) Suess. = Cissus auricoma Desc. ●☆
92847 Cissus muhuluensis Mildbr. = Cyphostemma muhuluense (Mildbr.) Desc. ●☆
92848 Cissus myriantha Gilg et M. Brandt = Cissus tiliifolia Planch. ●☆
92849 Cissus nanella Gilg et R. E. Fr. = Cyphostemma nanellum (Gilg et R. E. Fr.) Desc. ex Wild et R. B. Drumm. ●☆
92850 Cissus napaulensis DC. = Tetrastigma napaulense (DC.) C. L. Li ●■
92851 Cissus napaulensis DC. = Tetrastigma serrulatum (Roxb.) Planch. ●■
92852 Cissus natalitia (Szyszyl.) Codd = Cyphostemma natalitium (Szyszyl.) J. J. M. van der Merwe ●☆
92853 Cissus neghelliensis Lanza = Cyphostemma cyphopetalum (Fresen.) Desc. ex Wild et R. B. Drumm. ●☆
92854 Cissus nieriensis T. C. E. Fr. = Cyphostemma cyphopetalum (Fresen.) Desc. ex Wild et R. B. Drumm. ●☆
92855 Cissus nigroglandulosa Gilg et M. Brandt = Cyphostemma nigroglandulosum (Gilg et M. Brandt) Desc. ●☆
92856 Cissus nigropilosa Dewit;黑毛白粉藤●☆
92857 Cissus nivea Hochst. ex Schweinf. = Cyphostemma niveum (Hochst. ex Schweinf.) Desc. ●☆
92858 Cissus nodiglandulosa T. C. E. Fr. = Cyphostemma cyphopetalum (Fresen.) Desc. ex Wild et R. B. Drumm. var. nodiglandulosum (T. C. E. Fr.) Verdc. ●☆
92859 Cissus nodosa Blume;多节白粉藤;Grape Ivy,Javanese Treebine ●☆
92860 Cissus nymphiifolia (Welw. ex Baker) Planch.;睡莲叶白粉藤●☆
92861 Cissus oblonga (Benth.) Planch.;矩圆白粉藤;Smooth Kangaroo-vine ●☆
92862 Cissus obovato-oblonga De Wild. = Cyphostemma obovato-oblongum (De Wild.) Desc. ex Wild et R. B. Drumm. ●☆
92863 Cissus odontadenia Gilg = Cyphostemma odontadenium (Gilg) Desc. ●☆
92864 Cissus okoutensis Berhaut;奥库特白粉藤●☆
92865 Cissus oleraceus Bolus = Cyphostemma oleraceum (Bolus) J. J. M. van der Merwe ●☆
92866 Cissus oligocarpa (H. Lév. et Vaniot) Bailey = Cayratia oligocarpa (H. Lév. et Vaniot) Gagnep. ●
92867 Cissus oligocarpa (H. Lév. et Vaniot) Gagnep. = Cayratia albifolia C. L. Li ●
92868 Cissus oligocarpa (H. Lév. et Vaniot) Gagnep. f. glabra Gagnep. = Cayratia albifolia C. L. Li var. glabra (Gagnep.) C. L. Li ●
92869 Cissus oligocarpa (H. Lév. et Vaniot) Gagnep. var. glabra (Gagnep.) Rehder = Cayratia albifolia C. L. Li var. glabra (Gagnep.) C. L. Li ●
92870 Cissus oliveri (Engl.) Gilg;奥里弗白粉藤●☆
92871 Cissus oliviformis Planch. = Cissus aralioides (Baker) Planch. ●☆
92872 Cissus omburensis Gilg et M. Brandt = Cyphostemma omburense (Gilg et M. Brandt) Desc. ●☆
92873 Cissus oreophila Gilg et M. Brandt;喜山白粉藤●☆
92874 Cissus ornata A. Chev. ex Hutch. et Dalziel = Cyphostemma ornatum (A. Chev. ex Hutch. et Dalziel) Desc. ●☆
92875 Cissus orondo Gilg et M. Brandt = Cyphostemma serpens (A. Rich.) Desc. ●☆
92876 Cissus overlaetii Dewit = Cyphostemma overlaetii (Dewit) Desc. ●☆
92877 Cissus oxyodonta (Baker) Desc.;尖齿白粉藤●☆
92878 Cissus oxyphylla (A. Rich.) Chiov. = Cyphostemma oxyphyllum (A. Rich.) Vollesen ●☆
92879 Cissus pachyantha Gilg et M. Brandt = Cyphostemma pachyanthum (Gilg et M. Brandt) Desc. ●☆
92880 Cissus pachyrrhachis Gilg et M. Brandt = Cyphostemma cyphopetalum (Fresen.) Desc. ex Wild et R. B. Drumm. ●☆
92881 Cissus palmatifida (Baker) Planch.;掌状半裂白粉藤●☆
92882 Cissus passargei Gilg et M. Brandt = Cyphostemma passargei (Gilg et M. Brandt) Desc. ●☆
92883 Cissus paucidentata Klotzsch = Cyphostemma paucidentatum (Klotzsch) Desc. ex Wild et R. B. Drumm. ●☆
92884 Cissus pauli-guilielmii Schweinf. = Ampelocissus abyssinica (Hochst. ex A. Rich.) Planch. ●☆
92885 Cissus pedata Lam. = Cayratia pedata (Lam.) Juss. ex Gagnep. ●
92886 Cissus penduloides Dewit = Cyphostemma hildebrandtii (Gilg) Desc. ex Wild et R. B. Drumm. ●☆
92887 Cissus perforata Louis ex Dewit = Cyphostemma perforatum (Louis ex Dewit) Desc. ●☆
92888 Cissus perrieri Desc.;佩里耶白粉藤●☆
92889 Cissus petiolata Hook. f.;柄叶白粉藤●☆
92890 Cissus petiolata Hook. f. var. pubescens Dewit = Cissus petiolata Hook. f. ●☆
92891 Cissus phyllomicron Chiov. = Cyphostemma phyllomicron (Chiov.) Desc. ●☆
92892 Cissus phymatocarpa Masinde et L. E. Newton;胀果白粉藤●☆
92893 Cissus pileata Desc.;帽状白粉藤●☆
92894 Cissus planchoniana Gilg;扁白粉藤●☆
92895 Cissus polyantha Gilg et M. Brandt;多花白粉藤●☆
92896 Cissus populnea Guillaumin et Perr.;杨叶白粉藤●☆
92897 Cissus producta Afzel.;伸展白粉藤●☆
92898 Cissus psammophila Gilg et M. Brandt = Cyphostemma setosum (Roxb.) Alston ●☆

92899 Cissus pseudocaesia Gilg et M. Brandt;假淡蓝白粉藤●☆

92900 Cissus pseudoguerkeana Verdc.;假盖尔克白粉藤●☆

92901 Cissus pseudomanikensis Dewit = Cyphostemma manikense (De Wild.) Desc. ex Wild et R. B. Drumm. ●☆

92902 Cissus pseudonivea Gilg et M. Brandt = Cyphostemma cyphopetalum (Fresen.) Desc. ex Wild et R. B. Drumm. ●☆

92903 Cissus pseudopolyantha Mildbr.;假多花白粉藤●☆

92904 Cissus pseudorhodesiae Dewit = Cyphostemma pseudorhodesiae (Dewit) Desc. ●☆

92905 Cissus pseudotrifolia W. T. Wang = Cayratia japonica (Thunb.) Gagnep. var. pseudotrifolia (W. T. Wang) C. L. Li ●

92906 Cissus pseudoupembaensis Dewit = Cyphostemma pseudoupembaense (Dewit) Desc. ●☆

92907 Cissus pteroclada Hayata;翼枝白粉藤(春根藤,方藤,风藤,红宽筋藤,红四方藤,宽筋藤,蚂蝗藤,软筋藤,伸筋藤,四方藤,翼茎粉藤,翼茎葡萄);Winged Branch Treebine ●

92908 Cissus puberula C. A. Sm. = Cyphostemma puberulum (C. A. Sm.) Wild et R. B. Drumm. ●☆

92909 Cissus pynaertii De Wild.;皮那白粉藤●☆

92910 Cissus quadrangula L.;方茎青紫葛;Veld Grape ●☆

92911 Cissus quadrangularis L.;四棱白粉藤●☆

92912 Cissus quadrangularis L. var. pubescens Dewit;短毛四棱白粉藤●☆

92913 Cissus quadrangularis Planch. = Vitis quadrangularis (L.) Wall. ●

92914 Cissus quarrei Dewit;卡雷白粉藤●☆

92915 Cissus quinata Dryand. = Cyphostemma quinatum (Dryand.) Desc. ex Wild et R. B. Drumm. ●☆

92916 Cissus quinquangularis Chiov.;五角白粉藤●☆

92917 Cissus reedii Dewit = Cyphostemma heterotrichum (Gilg et R. E. Fr.) Desc. ex Wild et R. B. Drumm. ●☆

92918 Cissus repanda Vahl;大叶白粉藤(粉藤)●

92919 Cissus repanda Vahl var. subferruginea (Merr. et Chun) C. L. Li;海南大叶白粉藤(琼南地锦);Hainan Treebine, South Hainan Creeper, Subrusty Creeper ●

92920 Cissus repens (Wight et Arn.) Lam.;白粉藤(白薯藤,独脚乌桕,粉藤,接骨藤,葡萄白粉藤,伸筋藤);Creeping Treebine ●

92921 Cissus repens (Wight et Arn.) Lam. var. luzoniensis Merr. = Cissus luzoniensis (Merr.) C. L. Li ●

92922 Cissus repens Lam. = Cissus assamica (Lawson) Craib ●

92923 Cissus repens Lam. var. luzoniensis Merr. = Cissus luzoniensis (Merr.) C. L. Li ●

92924 Cissus repens Lam. var. sinensis Hand. -Mazz. = Cissus repens (Wight et Arn.) Lam. ●

92925 Cissus rhodesiae Gilg et M. Brandt = Cyphostemma rhodesiae (Gilg et M. Brandt) Desc. ex Wild et R. B. Drumm. ●☆

92926 Cissus rhodotricha (Baker) Desc.;红毛白粉藤●☆

92927 Cissus rhombifolia Planch. 'Ellen Danica';栎叶白粉藤●☆

92928 Cissus rhombifolia Planch. = Vitis rhombifolia Khakhlov ●☆

92929 Cissus rhomboidea E. Mey. ex Harv. = Rhoicissus rhomboidea (E. Mey. ex Harv.) Planch. ●☆

92930 Cissus ringoetii De Wild. = Cyphostemma obovato-oblongum (De Wild.) Desc. ex Wild et R. B. Drumm. ●☆

92931 Cissus rivae Gilg = Cyphostemma rivae (Gilg) Desc. ●☆

92932 Cissus robynsii Dewit = Cyphostemma robynsii (Dewit) Desc. ●☆

92933 Cissus rocheana Planch. = Cissus incisa (Torr. et Gray) Des Moul. ●☆

92934 Cissus rotundifolia (Forssk.) Vahl;圆叶白粉藤;Roundleaf Treebine, Venezuelan Treebine ●☆

92935 Cissus rotundifolia (Forssk.) Vahl var. ferrugineo-pubescens Verdc.;锈毛圆叶白粉藤●☆

92936 Cissus rotundifolia Vahl = Cissus rotundifolia (Forssk.) Vahl ●☆

92937 Cissus rowlandii Gilg et M. Brandt = Cyphostemma adenopodum (Sprague) Desc. ●☆

92938 Cissus ruacanensis Exell et Mendonça = Cyphostemma ruacanense (Exell et Mendonça) Desc. ●☆

92939 Cissus rubiginosa (Welw. ex Baker) Planch.;锈红白粉藤●☆

92940 Cissus rubromarginata Gilg et M. Brandt = Cyphostemma rubromarginatum (Gilg et M. Brandt) Desc. ●☆

92941 Cissus rubrosetosa Gilg et M. Brandt = Cyphostemma rubrosetosum (Gilg et M. Brandt) Desc. ●☆

92942 Cissus rufescens Guillaumin et Perr.;浅红白粉藤●☆

92943 Cissus ruginosicarpa Desc.;皱果白粉藤●☆

92944 Cissus rupicola Gilg et M. Brandt = Cyphostemma rupicola (Gilg et M. Brandt) Desc. ●☆

92945 Cissus ruspolii Gilg;鲁斯波利白粉藤●☆

92946 Cissus sagittifera Desc.;箭状白粉藤●☆

92947 Cissus sandersonii Harv. = Cyphostemma cirrhosum (Thunb.) Desc. ex Wild et R. B. Drumm. subsp. transvaalense (Szyszyl.) Wild et R. B. Drumm. ●☆

92948 Cissus sandersonii Harv. var. transvaalensis (Szyszyl.) C. A. Sm. = Cyphostemma cirrhosum (Thunb.) Desc. ex Wild et R. B. Drumm. subsp. transvaalense (Szyszyl.) Wild et R. B. Drumm. ●☆

92949 Cissus sarcospathula Chiov. = Cyphostemma sarcospathulum (Chiov.) Desc. ●☆

92950 Cissus saxicola Gilg et R. E. Fr. = Cyphostemma saxicola (Gilg et R. E. Fr.) Desc. ex Wild et R. B. Drumm. ●☆

92951 Cissus scarlatina Gilg et M. Brandt = Cyphostemma scarlatinum (Gilg et M. Brandt) Desc. ●☆

92952 Cissus schimperi Hochst. ex Planch. = Cyphostemma oxyphyllum (A. Rich.) Vollesen ●☆

92953 Cissus schlechteri Gilg et M. Brandt = Cyphostemma schlechteri (Gilg et M. Brandt) Desc. ex Wild et R. B. Drumm. ●☆

92954 Cissus schliebenii Mildbr. = Cyphostemma schliebenii (Mildbr.) Desc. ●☆

92955 Cissus schmitzii Dewit;施密茨白粉藤●☆

92956 Cissus schweinfurthii Planch. = Cyphostemma serpens (A. Rich.) Desc. ●☆

92957 Cissus segmentata C. A. Sm. = Cyphostemma segmentatum (C. A. Sm.) J. J. M. van der Merwe ●☆

92958 Cissus seitziana Gilg et M. Brandt = Cyphostemma bainesii (Hook. f.) Desc. ●☆

92959 Cissus senegalensis Lavie;塞内加尔白粉藤●☆

92960 Cissus serianiifolia (Bunge) Walp. = Ampelopsis japonica (Thunb.) Makino ●

92961 Cissus serjanioides Planch. = Cyphostemma adenocaule (Steud. ex A. Rich.) Desc. ex Wild et R. B. Drumm. ●☆

92962 Cissus serpens A. Rich. = Cyphostemma serpens (A. Rich.) Desc. ●☆

92963 Cissus serrulata Roxb. = Tetrastigma serrulatum (Roxb.) Planch. ●■

92964 Cissus sesquipedalis Gilg = Cyphostemma serpens (A. Rich.) Desc. ●☆

92965 Cissus sessilifolia Dewit = Cyphostemma sessilifolium (Dewit) Desc. ●☆

92966 Cissus setosa Roxb. = Cyphostemma setosum (Roxb.) Alston ●☆
92967 Cissus sicyoides L.;四季藤(葫芦白粉藤,南美白粉藤);Blister Bush,Season Vine ●
92968 Cissus silvestris Tchoumé;森林白粉藤●☆
92969 Cissus simulans C. A. Sm. = Cyphostemma simulans (C. A. Sm.) Wild et R. B. Drumm. ●☆
92970 Cissus smithiana (Baker) Planch.;史密斯白粉藤●☆
92971 Cissus somaliensis Gilg = Cyphostemma ternatum (Forssk.) Desc. ●☆
92972 Cissus spinosopilosa Gilg et M. Brandt = Cyphostemma spinosopilosum (Gilg et M. Brandt) Desc. ●☆
92973 Cissus stans Pers. = Ampelopsis arborea (L.) Koehne ●☆
92974 Cissus stefaniniana Chiov. = Cyphostemma stefaninianum (Chiov.) Desc. ●☆
92975 Cissus stenopoda Gilg = Cyphostemma stenopodum (Gilg) Desc. ●☆
92976 Cissus stipulacea (Baker) Planch. var. hochstetteri Planch. = Cyphostemma dembianense (Chiov.) Vollesen ●☆
92977 Cissus striata Ruiz et Pav.;条纹白粉藤;Grape Ivy, Ivy of Uruguay, Ivy-of-uruguay, Miniature Grape Ivy, Striate Treebine ●☆
92978 Cissus strigosa Dewit = Cyphostemma strigosum (Dewit) Desc. ●☆
92979 Cissus subciliata (Baker) Planch. = Cyphostemma subciliatum (Baker) Desc. ex Wild et R. B. Drumm. ●☆
92980 Cissus subdiaphana Steud. ex A. Rich. = Cayratia gracilis (Guillaumin et Perr.) Suess. ●☆
92981 Cissus suberosa (Baker) Planch. = Cissus petiolata Hook. f. ●☆
92982 Cissus subglaucescens Planch. = Cyphostemma subciliatum (Baker) Desc. ex Wild et R. B. Drumm. ●☆
92983 Cissus subtetragona Planch.;近四棱白粉藤;Fourangle Treebine ●
92984 Cissus subtetragona Planch. = Tetrastigma subtetragonum C. L. Li ●
92985 Cissus succulenta (Galpin) Burtt Davy = Cissus cactiformis Gilg ●☆
92986 Cissus sulcatus C. A. Sm. = Cyphostemma sulcatum (C. A. Sm.) J. J. M. van der Merwe ●☆
92987 Cissus sulfurosa Desc.;硫色白粉藤●☆
92988 Cissus sylvicola Masinde et L. E. Newton;西尔维亚白粉藤●☆
92989 Cissus tenuifolia (Wight et Arn.) F. Heyne ex Planch.;薄叶白粉藤(薄叶野葡萄)●☆
92990 Cissus tenuifolia (Wight et Arn.) F. Heyne ex Planch. = Cayratia japonica (Thunb.) Gagnep. ●
92991 Cissus tenuifolia (Wight et Arn.) F. Heyne ex Planch. = Cayratia tenuifolia (Wight et Arn.) Gagnep. ●☆
92992 Cissus tenuipes Gilg et R. E. Fr. = Cyphostemma cyphopetalum (Fresen.) Desc. ex Wild et R. B. Drumm. ●☆
92993 Cissus tenuissima Gilg et R. E. Fr. = Cyphostemma tenuissimum (Gilg et R. E. Fr.) Desc. ex Wild et R. B. Drumm. ●☆
92994 Cissus termetophila De Wild. = Cyphostemma mildbraedii (Gilg et M. Brandt) Desc. ex Wild et R. B. Drumm. ●☆
92995 Cissus ternatomultifida Chiov. = Cyphostemma ternato-multifidum (Chiov.) Desc. ●☆
92996 Cissus tetragona Harv. = Cissus cactiformis Gilg ●☆
92997 Cissus thalictrifolia Planch. = Cayratia triternata (Baker) Desc. ●☆
92998 Cissus thomasii Gilg et M. Brandt = Cyphostemma thomasii (Gilg et M. Brandt) Desc. ●☆
92999 Cissus thomsonii (M. A. Lawson) Planch. = Yua thomsonii (M. A. Lawson) C. L. Li ●
93000 Cissus thunbergii Eckl. et Zeyh. = Rhoicissus digitata (L. f.) Gilg et M. Brandt ●☆
93001 Cissus thunbergii Siebold et Zucc. = Parthenocissus tricuspidatus (Siebold et Zucc.) Planch. ●
93002 Cissus tiliifolia Planch.;椴叶白粉藤●☆
93003 Cissus timoriensis DC. = Cayratia timoriensis (DC.) C. L. Li ●
93004 Cissus tomentosa Lam. = Rhoicissus tomentosa (Lam.) Wild et R. B. Drumm. ●☆
93005 Cissus trachyphylla Werderm. = Cyphostemma bororense (Klotzsch) Desc. ex Wild et R. B. Drumm. ●☆
93006 Cissus triangularis A. Chev. = Cissus palmatifida (Baker) Planch. ●☆
93007 Cissus trifoliata (L.) Lour.;三叶白粉藤;Arizona Grape Ivy, Cissus, Sorrelvine ●☆
93008 Cissus triloba (Lour.) Merr.;掌叶白粉藤●
93009 Cissus triternata (Baker) Suess. = Cayratia triternata (Baker) Desc. ●☆
93010 Cissus triumfettioides Gilg et M. Brandt = Cyphostemma rhodesiae (Gilg et M. Brandt) Desc. ex Wild et R. B. Drumm. ●☆
93011 Cissus trothae Gilg et M. Brandt;特罗塔白粉藤●☆
93012 Cissus ukerewensis Gilg = Cyphostemma ukerewense (Gilg) Desc. ●☆
93013 Cissus umbellata Lour. = Strychnos umbellata (Lour.) Merr. ●
93014 Cissus unguiformifolia C. A. Sm. = Cyphostemma schlechteri (Gilg et M. Brandt) Desc. ex Wild et R. B. Drumm. ●☆
93015 Cissus unifoliata Harv. = Rhoicissus microphylla (Turcz.) Gilg et M. Brandt ●☆
93016 Cissus upembaensis Dewit = Cyphostemma zombense (Baker) Desc. ex Wild et R. B. Drumm. ●☆
93017 Cissus urophylla Gilg et M. Brandt = Cyphostemma urophyllum (Gilg et M. Brandt) Desc. ●☆
93018 Cissus uter Exell et Mendonça = Cyphostemma uter (Exell et Mendonça) Desc. ●☆
93019 Cissus vandenbrandeana Dewit = Cyphostemma vandenbrandeanum (Dewit) Desc. ex Wild et R. B. Drumm. ●☆
93020 Cissus vanderbenii Dewit = Cyphostemma vanderbenii (Dewit) Desc. ●☆
93021 Cissus vanmeelii Lawalrée = Cyphostemma vanmeelii (Lawalrée) Wild et R. B. Drumm. ●☆
93022 Cissus veitchii? = Parthenocissus tricuspidatus (Siebold et Zucc.) Planch. ●
93023 Cissus villosiglandulosa Werderm. = Cyphostemma villosiglandulosum (Werderm.) Desc. ●☆
93024 Cissus viniferoides Mildbr. = Cissus cucumerifolia Planch. ●☆
93025 Cissus violaceoglandulosa Gilg = Cyphostemma violaceoglandulosum (Gilg) Desc. ●☆
93026 Cissus viscosa Gilg et R. E. Fr. = Cyphostemma viscosum (Gilg et R. E. Fr.) Desc. ex Wild et R. B. Drumm. ●☆
93027 Cissus viticifolia Salisb. = Cayratia japonica (Thunb.) Gagnep. ●
93028 Cissus viticilata (L.) Nicholson et Travis;轮生白粉藤●☆
93029 Cissus vitifolia Boiss. = Ampelopsis vitifolia (Boiss.) Planch. ●☆
93030 Cissus vitiginea L.;林地白粉藤(森林白粉藤,森林葡萄);Forest Grape ●☆
93031 Cissus vogelii Hook. f. = Cyphostemma vogelii (Hook. f.) Desc. ●☆
93032 Cissus voinieriana Ball. = Tetrastigma voinierianum (Ball.) Pierre ex Gagnep. ●
93033 Cissus vuilletii A. Chev. = Cissus palmatifida (Baker) Planch. ●☆

93034 Cissus wallacei Verdc.;瓦利亚塞白粉藤●☆
93035 Cissus waterlotii A. Chev. = Cyphostemma waterlotii (A. Chev.) Desc. ●☆
93036 Cissus wellmanii Gilg et M. Brandt;韦尔曼白粉藤●☆
93037 Cissus welwitschii (Baker) Planch.;韦尔白粉藤●☆
93038 Cissus wenshanensis C. L. Li;文山青紫葛;Wenshan Treebine ●
93039 Cissus wilmsii Gilg et M. Brandt = Cyphostemma wilmsii (Gilg et M. Brandt) Desc. ●☆
93040 Cissus wittei Staner = Cyphostemma wittei (Staner) Wild et R. B. Drumm. ●☆
93041 Cissus woodii Gilg et M. Brandt = Cyphostemma woodii (Gilg et M. Brandt) Desc. ●☆
93042 Cissus youngii Exell et Mendonça;扬氏白粉藤●☆
93043 Cissus zambesica Wild et R. B. Drumm. = Cyphostemma bororense (Klotzsch) Desc. ex Wild et R. B. Drumm. ●☆
93044 Cissus zechiana Gilg et M. Brandt = Cyphostemma zechianum (Gilg et M. Brandt) Desc. ●☆
93045 Cistaceae Adans. = Cistaceae Juss.(保留科名)●■
93046 Cistaceae Juss.(1789)(保留科名);半日花科(岩蔷薇科);Rock Rose Family, Rockrose Family, Rock-rose Family ●■
93047 Cistanche Hoffmanns. et Link(1813-1820);肉苁蓉属;Cistanche ■
93048 Cistanche ambigua (Bunge) Beck;迷肉苁蓉■☆
93049 Cistanche ambigua (Bunge) Beck = Cistanche salsa (C. A. Mey.) Beck ■
93050 Cistanche ambigua Beck = Cistanche deserticola Ma ■
93051 Cistanche brunneri (Webb) Beck = Cistanche phelypaea (L.) Cout. ■☆
93052 Cistanche carnosa Pax = Cistanche tubulosa (Schernk) Wight ■
93053 Cistanche compacta (Viv.) Bég. et Vacc.;紧密肉苁蓉■☆
93054 Cistanche deserticola Ma;肉苁蓉(苁蓉,寸蓉,寸芸,大芸,地精,金笋,迷肉苁蓉,肉松蓉,纵蓉);Desertliving Cistanche ■
93055 Cistanche feddeana K. S. Hao = Cistanche sinensis Beck ■
93056 Cistanche feddeana K. S. Hao = Orobanche cernua Loefl. var. hansii (A. Kern.) Beck ■
93057 Cistanche feddeana K. S. Hao = Orobanche cernua Loefl. ■
93058 Cistanche fissa (C. A. Mey.) Grossh.;半裂肉苁蓉■☆
93059 Cistanche flava (C. A. Mey.) Korsh.;黄花肉苁蓉■
93060 Cistanche hesperugo (Webb) Beck = Cistanche phelypaea (L.) Cout. ■☆
93061 Cistanche lanzhouensis Zhi Y. Zhang;兰州肉苁蓉;Lanzhou Cistanche ■
93062 Cistanche lutea (Desf.) Hoffmanns. et Link = Cistanche phelypaea (L.) Cout. ■☆
93063 Cistanche lutea (Desf.) Hoffmanns. et Link f. minor Bég. = Cistanche phelypaea (L.) Cout. ■☆
93064 Cistanche lutea (Desf.) Hoffmanns. et Link subsp. compacta (Viv.) Bég. et Vacc. = Cistanche phelypaea (L.) Cout. ■☆
93065 Cistanche lutea (Desf.) Hoffmanns. et Link subsp. compacta (Viv.) Pamp. = Cistanche compacta (Viv.) Bég. et Vacc. ■☆
93066 Cistanche lutea Hoffmanns. et Link = Cistanche tubulosa (Schernk) Wight ■
93067 Cistanche mauritanica (Coss. et Durieu) Beck;毛里塔尼亚肉苁蓉■☆
93068 Cistanche mongonica Beck;内蒙古肉苁蓉(管花肉苁蓉,内蒙肉苁蓉);Mongol Cistanche ■
93069 Cistanche ningxiaensis D. Z. Ma et J. A. Duan;宁夏肉苁蓉;Ningxia Cistanche ■
93070 Cistanche ningxiaensis D. Z. Ma et J. A. Duan = Cistanche lanzhouensis Zhi Y. Zhang ■
93071 Cistanche phelypaea (L.) Cout.;塞内加尔肉苁蓉■☆
93072 Cistanche phelypaea (L.) Cout. subsp. lutea (Desf.) Fern. Casas et Lainz = Cistanche phelypaea (L.) Cout. ■☆
93073 Cistanche phelypaea (L.) Cout. var. transiens Maire = Cistanche phelypaea (L.) Cout. ■☆
93074 Cistanche ridgewayana Aitch. et Hemsl.;里德肉苁蓉■☆
93075 Cistanche salsa (C. A. Mey.) Beck;盐生肉苁蓉(草苁蓉,苁蓉,肉苁蓉);Cistanche, Saline Cistanche ■
93076 Cistanche salsa (C. A. Mey.) Beck var. albiflora P. F. Tu et Z. C. Lou;白花盐苁蓉■
93077 Cistanche salsa (C. A. Mey.) Beck var. albiflora P. F. Tu et Z. C. Lou = Cistanche salsa (C. A. Mey.) Beck ■
93078 Cistanche senegalensis (Reut.) Beck = Cistanche phelypaea (L.) Cout. ■☆
93079 Cistanche sinensis Beck;沙苁蓉;China Cistanche, Chinese Cistanche ■
93080 Cistanche tinctoria (Forssk.) Beck = Cistanche phelypaea (L.) Cout. ■☆
93081 Cistanche tubulosa (Schernk) Hook. f. = Cistanche tubulosa (Schernk) Wight ■
93082 Cistanche tubulosa (Schernk) Wight;管花肉苁蓉;Tubeshaped Cistanche, Tubeshaped Flower Cistanche ■
93083 Cistanche tubulosa (Schernk) Wight var. tomentosa Hook. f. = Cistanche tubulosa (Schernk) Wight ■
93084 Cistanche tubulosa (Schrenk) Wight = Cistanche mongonica Beck ■
93085 Cistanche violacea (Desf.) Hoffmanns. et Link;堇色肉苁蓉■☆
93086 Cistanthe Spach = Calandrinia Kunth(保留属名)■☆
93087 Cistanthe Spach(1836);猫爪苋属;Pussypaws ■☆
93088 Cistanthe ambigua (S. Watson) Carolin ex Hershk.;可疑猫爪苋■☆
93089 Cistanthe grandiflora (Lindl.) Schltdl.;大花猫爪苋■☆
93090 Cistanthe maritima (Nutt.) Carolin ex Hershk.;沼泽猫爪苋■☆
93091 Cistanthe monandra (Nutt.) Hershk.;单蕊猫爪苋■☆
93092 Cistanthe monosperma (Greene) Hershk.;单籽猫爪苋■☆
93093 Cistanthe parryi (A. Gray) Hershk.;帕里猫爪苋■☆
93094 Cistanthe pulchella (Eastw.) Hershk.;美丽猫爪苋;Mariposa Pussypaws ■☆
93095 Cistanthe pygmaea (Parish ex Rydb.) Hershk.;矮小猫爪苋■☆
93096 Cistanthe quadripetala (S. Watson) Hershk.;四瓣猫爪苋;Four-petaled Pussypaws ■☆
93097 Cistanthe rosea (S. Watson) Hershk.;粉花猫爪苋■☆
93098 Cistanthe tweedyi (A. Gray) Hershk.;特威迪猫爪苋■☆
93099 Cistanthe umbellata (Torr.) Hershk.;伞形猫爪苋■☆
93100 Cistanthera K. Schum. = Nesogordonia Baill. et H. Perrier ●☆
93101 Cistanthera dewevrei De Wild. et T. Durand = Nesogordonia dewevrei (De Wild. et T. Durand) Capuron ex R. Germ. ●☆
93102 Cistanthera fouassieri A. Chev. = Nesogordonia leplaei (Vermoesen) Capuron ex R. Germ. ●☆
93103 Cistanthera holtzii Engl. = Nesogordonia holtzii (Engl.) Capuron ex L. C. Barnett et Dorr ●☆
93104 Cistanthera ituriensis De Wild. = Nesogordonia kabingaensis (K. Schum.) Capuron ex R. Germ. ●☆
93105 Cistanthera kabingaensis K. Schum. = Nesogordonia kabingaensis (K. Schum.) Capuron ex R. Germ. ●☆

93106 Cistanthera leplaei Vermoesen = Nesogordonia leplaei (Vermoesen) Capuron ex R. Germ. ●☆
93107 Cistanthera papaverifera A. Chev. = Nesogordonia papaverifera (A. Chev.) Capuron ex N. Hallé ●☆
93108 Cistanthera parvifolia M. B. Moss ex Milne-Redh. = Nesogordonia holtzii (Engl.) Capuron ex L. C. Barnett et Dorr ●☆
93109 Cistela Blume = Geodorum Jacks. ■
93110 Cistella Blume = Geodorum Jacks. ■
93111 Cistella cernua (Willd.) Blume = Geodorum densiflorum (Lam.) Schltr. ■
93112 Cisticapnos Adans. = Corydalis DC.(保留属名)■
93113 Cistocarpium Spach = Alyssoides Mill. ■●☆
93114 Cistocarpium Spach = Vesicaria Tourn. ex Adans. ■☆
93115 Cistocarpum Pfeiff. = Cistoearpus Kunth ●☆
93116 Cistoearpus Kunth = Balbisia Cav.(保留属名)●☆
93117 Cistomorpha Caley ex DC. = Hibbertia Andréws ●☆
93118 Cistrum Hill = Centaurea L.(保留属名)●■
93119 Cistula Noronha = Maesa Forssk. ●
93120 Cistus L.(1753);岩蔷薇属(爱花属,半日花属,午时葵属);Cistus, Rock Rose, Rockrose, Sunrose ●
93121 Cistus Medik. = Helianthemum Mill. ●■
93122 Cistus × dansereaui P. Silva;健壮岩蔷薇●☆
93123 Cistus aegyptiacus L. = Helianthemum aegyptiacum (L.) Mill. ●☆
93124 Cistus albidus L.;白叶岩蔷薇(白毛岩蔷薇,白毛叶岩蔷薇,白岩蔷薇,微白岩蔷薇);Grey-leaved Cistus, Rock Rose, White Rockrose, White-leaf Rock-rose, White-leaved Rockrose ●
93125 Cistus alpestre Crantz. = Helianthemum alpestre (Jacq.) DC. ●☆
93126 Cistus arabicus L. = Fumana arabica (L.) Spach ■☆
93127 Cistus asperifolius Pomel = Cistus heterophyllus Desf. ●☆
93128 Cistus berthelotianus Spach = Cistus symphytifolius Lam. ●☆
93129 Cistus berthelotianus Spach var. leucophyllus? = Cistus symphytifolius Lam. ●☆
93130 Cistus berthelotianus Spach var. pilosus Pit. = Cistus symphytifolius Lam. ●☆
93131 Cistus berthelotianus Spach var. symphytfolius (Lam.) Webb = Cistus symphytifolius Lam. ●☆
93132 Cistus canadensis L. = Helianthemum canadense (L.) Michx. ●☆
93133 Cistus canariensis Jacq. = Helianthemum canariense (Jacq.) Pers. ●☆
93134 Cistus canescens Sweet;红花杂种岩蔷薇●☆
93135 Cistus ciliatus Desf. = Helianthemum ciliatum (Desf.) Pers. ●☆
93136 Cistus clusii Dunal;克卢斯岩蔷薇●☆
93137 Cistus creticus L.;毛茎岩蔷薇(白色岩蔷薇,克里特灰白岩蔷薇,玫红岩蔷薇,岩蔷薇);Cretan Rockrose, Hairy Rock Rose, Myrrh, Rock Rose ●☆
93138 Cistus creticus L. subsp. corsicus (Loisel.) Greuter et Burdet;科西嘉岩蔷薇●☆
93139 Cistus creticus L. subsp. eriocephalus (Viv.) Greuter et Burdet;毛头岩蔷薇●☆
93140 Cistus creticus L. subsp. trabutii (Maire) Dobignard;特拉布特岩蔷薇●☆
93141 Cistus crispus L.;卷叶岩蔷薇;Curly-leaf Reckrose, Curly-leaved Rock Rose, Rock Rose, Wrinkle-leaf Rockrose ●☆
93142 Cistus croceus Desf. = Helianthemum croceum (Desf.) Pers. ●☆
93143 Cistus ellipticus Desf. = Helianthemum ellipticum (Desf.) Pers. ●☆
93144 Cistus eriocephalus Viv. = Cistus creticus L. subsp. eriocephalus (Viv.) Greuter et Burdet ●☆
93145 Cistus fastigiatus Guss. = Cistus clusii Dunal ●☆
93146 Cistus florentinus Lam.;佛罗伦萨岩蔷薇●☆
93147 Cistus glaucus Desf. = Helianthemum helianthemoides (Desf.) Grosser ●☆
93148 Cistus glutinosus L. = Fumana thymifolia (L.) Webb ■☆
93149 Cistus helianthemoides Desf. = Helianthemum helianthemoides (Desf.) Grosser ●☆
93150 Cistus heterophyllus Desf.;互叶岩蔷薇●☆
93151 Cistus heterophyllus Desf. var. asperifolius (Pomel) Batt. = Cistus heterophyllus Desf. ●☆
93152 Cistus hirsutus Lam.;疏毛岩蔷薇;Hairy Rockrose ●☆
93153 Cistus hybridus Vahl;杂种岩蔷薇;White Rockrose ●☆
93154 Cistus incanus L.;灰毛岩蔷薇;Hairy Rockrose, Hoary Cistus, Ladanum, Pink Rockrose, Rose Cistus, White-leaved Rock Rose ●☆
93155 Cistus incanus L. subsp. corsicus?;毛灰白岩蔷薇;Hairy Rockrose ●☆
93156 Cistus incanus L. subsp. creticus (L.) Heywood = Cistus creticus L. ●☆
93157 Cistus incanus L. var. creticus (L.) Boiss. = Cistus creticus L. ●☆
93158 Cistus incanus L. var. reichenbachii Hochr. = Cistus creticus L. ●☆
93159 Cistus incanus L. var. villosus (L.) Murb. = Cistus creticus L. ●☆
93160 Cistus incanus L. var. villosus (L.) Murb. = Cistus villosus L. ●☆
93161 Cistus ladanifer L.;岩蔷薇(胶蔷树,赖百当,石蔷薇,树脂半日花,树脂岩蔷薇);Common Gum Cistus, Cretan Rose, Crimson-spot Rockrose, Day Flower, Gum Cistus, Gum Rockrose, Ladanum, Ladon-shrub, Laudan, Laudanum, Rockrose, Sweet Cistus, Sweet Holly Rose ●
93162 Cistus ladanifer L. subsp. africanus Dans. = Cistus ladanifer L. subsp. mauritianus Pau et Sennen ●☆
93163 Cistus ladanifer L. subsp. mauritianus Pau et Sennen;毛里求斯岩蔷薇●☆
93164 Cistus ladanifer L. var. maculatus DC. = Cistus ladanifer L. ●
93165 Cistus ladanifer L. var. pedicellatus Maire = Cistus ladanifer L. ●
93166 Cistus ladanifer L. var. petiolatus Maire = Cistus ladanifer L. subsp. mauritianus Pau et Sennen ●☆
93167 Cistus ladanifer L. var. subangustifolius Sennen = Cistus ladanifer L. ●
93168 Cistus ladanifer L. var. tangerinus Pau = Cistus ladanifer L. ●
93169 Cistus ladaniferus L. var. sulcatus Demoly;矮岩蔷薇●☆
93170 Cistus laurifolius L.;月桂岩蔷薇(桂叶岩蔷薇);Laurel Rockrose, Laurel-leaved Cistus, Laurel-leaved Rock Rose ●☆
93171 Cistus laurifolius L. subsp. atlanticus (Pit.) Sennen;大西洋岩蔷薇●☆
93172 Cistus laurifolius L. var. atlanticus Pit. = Cistus laurifolius L. subsp. atlanticus (Pit.) Sennen ●☆
93173 Cistus laurifolius L. var. prostratus Sennen = Cistus laurifolius L. ●☆
93174 Cistus libanotis L.;红萼岩蔷薇●☆
93175 Cistus libanotis L. var. pruinosus Willk. = Cistus clusii Dunal ●☆
93176 Cistus libanotis L. var. sedjera (Pomel) Grosser = Cistus clusii Dunal ●☆
93177 Cistus libanotis L. var. viridis (Willk.) Maire = Cistus clusii Dunal ●☆
93178 Cistus lippii L. = Helianthemum lippii (L.) Pers. ●☆
93179 Cistus madagascariensis?;马岛岩蔷薇;Abyssinian Myrrh ●☆
93180 Cistus monspeliensis L.;蒙彼利埃岩蔷薇(蒙斯帕里岩蔷薇);

Montpelier Cistus, Montpelier Rock Rose ●☆

93181 Cistus monspeliensis L. var. feredjensis Batt. = Cistus monspeliensis L. ●☆

93182 Cistus monspeliensis Willk. = Cistus creticus L. ●☆

93183 Cistus monspeliensis Willk. var. corsicus (Loisel.) Ball = Cistus creticus L. subsp. corsicus (Loisel.) Greuter et Burdet ●☆

93184 Cistus monspeliensis Willk. var. creticus (L.) Ball = Cistus creticus L. ●☆

93185 Cistus munbyi Pomel;芒比岩蔷薇●☆

93186 Cistus niloticus L. = Helianthemum ledifolium (L.) Mill. ●☆

93187 Cistus nummularius L. = Helianthemum nummularium (L.) Mill. ●☆

93188 Cistus obtusifolius Sweet;钝叶岩蔷薇●☆

93189 Cistus ochreatus C. Sm.;鞘状托岩蔷薇●☆

93190 Cistus ocymoides Lam. = Halimium ocymoides (Lam.) Willk. ●☆

93191 Cistus parviflorus Gaterau = Cistus parviflorus Lam. ●☆

93192 Cistus parviflorus Lam.;小花岩蔷薇●☆

93193 Cistus polyanthus Desf. = Helianthemum polyanthum (Desf.) Pers. ●☆

93194 Cistus populifolius L.;心叶岩蔷薇(杨叶岩蔷薇);Poplar-leaved Rockrose ●☆

93195 Cistus populifolius L. subsp. major (Dunal) Heywood;大心叶岩蔷薇●☆

93196 Cistus populifolius L. var. lasiocalyx Willk. = Cistus populifolius L. subsp. major (Dunal) Heywood ●☆

93197 Cistus psilosepalus Sweet;毛萼岩蔷薇●☆

93198 Cistus pulverulentus Pourr.;矮生岩蔷薇;Poplar-leaved Rock Rose ●☆

93199 Cistus purpureus Lam.;紫花岩蔷薇(紫花杂种岩蔷薇);Orchid Rockrose, Purple Rockrose, Rockrose ●☆

93200 Cistus racemosuss Desf. = Helianthemum syriacum (Jacq.) Dum. Cours. ●☆

93201 Cistus salicifolius L. = Helianthemum salicifolium (L.) Mill. ●☆

93202 Cistus salviifolius L.;鼠尾草叶岩蔷薇;Bush Sage, Cistsage, Gallipoli Rose, Sage Rose, Sage-leaf Rockrose, Sage-leaved Cistus, Sage-leaved Rock Rose, Sage-leaved Rockrose, Salvia Cistus, Salvia Rockrose ●☆

93203 Cistus salviifolius L. 'Prostratus';平卧鼠尾草叶岩蔷薇●☆

93204 Cistus salviifolius L. var. biflorus Willk. = Cistus salviifolius L. ●☆

93205 Cistus salviifolius L. var. macrocalyx Willk. = Cistus salviifolius L. ●☆

93206 Cistus salviifolius L. var. pandoanus Font Quer = Cistus salviifolius L. ●☆

93207 Cistus sedjera Pomel = Cistus clusii Dunal ●☆

93208 Cistus sennenianus Pau = Cistus creticus L. ●☆

93209 Cistus sericeus Munby = Cistus munbyi Pomel ●☆

93210 Cistus sessiliflorus Desf. = Helianthemum lippii (L.) Dum. Cours. ●☆

93211 Cistus sessilours Desf. = Helianthemum lippii (L.) Pers. ●☆

93212 Cistus skanbergii Lojac.;希腊岩蔷薇(粉花岩蔷薇);Pink Rockrose, Rockrose ●☆

93213 Cistus stipulatus Forssk. = Helianthemum stipulatum (Forssk.) C. Chr. ●☆

93214 Cistus symphytifolius Lam.;聚叶岩蔷薇●☆

93215 Cistus symphytifolius Lam. var. leucophyllus (Spach) Dans. = Cistus symphytifolius Lam. ●☆

93216 Cistus tauricus J. Presl et C. Presl;克里木岩蔷薇●☆

93217 Cistus thymifolius L. = Fumana thymifolia (L.) Webb ■☆

93218 Cistus tuberaria L. = Tuberaria lignosa (Sweet) Samp. ■☆

93219 Cistus villosus L.;长柔毛岩蔷薇●☆

93220 Cistus villosus L. var. corsicus (Loisel.) Grosser = Cistus creticus L. subsp. corsicus (Loisel.) Greuter et Burdet ●☆

93221 Cistus villosus L. var. creticus (L.) Boiss. = Cistus creticus L. ●☆

93222 Cistus villosus L. var. eriocephalus (Viv.) Grosser = Cistus creticus L. subsp. eriocephalus (Viv.) Greuter et Burdet ●☆

93223 Cistus villosus L. var. genuinus Boiss. = Cistus creticus L. subsp. eriocephalus (Viv.) Greuter et Burdet ●☆

93224 Cistus villosus L. var. mauritanicus Grosser = Cistus creticus L. ●☆

93225 Cistus villosus L. var. tauricus (Dumort.) Grosser = Cistus creticus L. ●☆

93226 Cistus villosus L. var. trabutii Maire = Cistus creticus L. subsp. trabutii (Maire) Dobignard ●☆

93227 Cistus villosus L. var. undulatus (Spach) Grosser;波毛岩蔷薇●☆

93228 Cistus villosus L. var. undulatus (Spach) Grosser = Cistus creticus L. ●☆

93229 Cistus virgatus Desf. = Helianthemum virgatum (Desf.) Pers. ●☆

93230 Citharaexylon Adans. = Citharexylum L. ●☆

93231 Citharella Noronha = Eranthemum L. ●■

93232 Cithareloma Bunge(1845);对枝菜属(竖琴芥属)■

93233 Cithareloma gedrosiacum Rech. f. et Esfand. = Eremobium aegyptium (Spreng.) Boiss. ■☆

93234 Cithareloma lehmannii Bunge;赖氏对枝菜■

93235 Cithareloma vernum Bunge;对枝菜■

93236 Citharexylon Adans. = Citharexylum L. ●☆

93237 Citharexylum L. (1753);琴木属;Fiddlewood ●☆

93238 Citharexylum Mill. = Citharexylum L. ●☆

93239 Citharexylum caudatum L.;尾状琴木;Fiddlewood ●☆

93240 Citharexylum cinereum L.;灰琴木●☆

93241 Citharexylum fruticosum L.;灌木状琴木●☆

93242 Citharexylum fruticosum L. = Citharexylum spinosum L. ●☆

93243 Citharexylum laetum Hiern;琴木●☆

93244 Citharexylum macranthum Hayek;大花琴木●☆

93245 Citharexylum paniculatum Poir. = Premna corymbosa (Burm. f.) Rottl. et Willd. ●

93246 Citharexylum quadrangulare Jacq. = Citharexylum spinosum L. ●☆

93247 Citharexylum spinosum L.;多利琴木;Fiddlewood ●☆

93248 Citharexylum subserratum Sw. = Citharexylum spinosum L. ●☆

93249 Citinus All. = Cytinus L. (保留属名)■☆

93250 Citrabenis Thouars = Habenaria Willd. ■

93251 Citraceae Drude = Aurantiaceae Juss. ●■

93252 Citraceae Drude = Rutaceae Juss. (保留科名)●■

93253 Citraceae Roussel = Rutaceae Juss. (保留科名)●■

93254 Citrangis Thouars = Aerangis Rchb. f. ■☆

93255 Citrangis Thouars = Angraecum Bory ■

93256 Citreum Mill. = Citrus L. ●

93257 Citriobathus A. Juss. = Citriobatus A. Cunn. ex Putt. ●☆

93258 Citriobatus A. Cnnn. = Citriobatus A. Cunn. ex Putt. ●☆

93259 Citriobatus A. Cunn. et Putt. = Pittosporum Banks ex Gaertn. (保留属名)●

93260 Citriobatus A. Cunn. ex Loudon = Citriobatus A. Cunn. ex Putt. ●☆

93261 Citriobatus A. Cunn. ex Putt. (1839);橘海桐属●☆

93262 Citriobatus pauciflorus A. Cunn. ex Benth.;橘海桐;Orange Thorn ●☆

93263 Citriopsis Pierre ex A. Chev. (1961);类橘属●☆

93264 Citriosma Tul. = Citrosma Ruiz et Pav. ●☆
93265 Citriosma Tul. = Siparuna Aubl. ●☆
93266 Citronella D. Don(1832);橘茱萸属●☆
93267 Citronella moorei (Benth.) R. A. Howard;穆尔橘茱萸;Churnwood ●☆
93268 Citrophorum Neck. = Citrus L. ●
93269 Citropsis (Engl.) Swingle et M. Kellerm. (1914);樱桃橘属(非洲樱桃橘属);Cherry Orange ●☆
93270 Citropsis Swingle et M. Kellerm. = Citropsis (Engl.) Swingle et M. Kellerm. ●☆
93271 Citropsis angolensis Exell;安哥拉樱桃橘●☆
93272 Citropsis articulata (Spreng.) Swingle et M. Kellerm.;关节樱桃橘●☆
93273 Citropsis citrifolia Y. Tanaka;樱桃橘●☆
93274 Citropsis gabunensis (Engl.) Swingle et M. Kellerm.;加蓬樱桃橘●☆
93275 Citropsis gabunensis (Engl.) Swingle et M. Kellerm. var. lacourtiana (De Wild.) Swingle et M. Kellerm. = Citropsis gabunensis (Engl.) Swingle et M. Kellerm. ●☆
93276 Citropsis gilletiana Swingle et M. Kellerm.;吉莱特樱桃橘●☆
93277 Citropsis lanattts (Thunb.) Matsum. et Nakai;肉花非洲樱桃橘;Water Melon ●☆
93278 Citropsis letestui Pellegr.;莱泰斯图樱桃橘●☆
93279 Citropsis schweinfurthii (Engl.) Swingle et M. Kellerm.;施韦樱桃橘●☆
93280 Citropsis schweinfurthii Swingle et M. Kellerm.;施文樱桃橘;Schweinfurth Cherry Orange ●☆
93281 Citropsis tanakae Swingle et M. Kellerm. = Citropsis gabunensis (Engl.) Swingle et M. Kellerm. ●☆
93282 Citrosena Bose ex Steud. = Citrosma Ruiz et Pav. ●☆
93283 Citrosma Ruiz et Pav. = Siparuna Aubl. ●☆
93284 Citrullus Forssk. = Citrullus Schrad. ex Eckl. et Zeyh. (保留属名)■
93285 Citrullus Schrad. = Citrullus Schrad. ex Eckl. et Zeyh. (保留属名)■
93286 Citrullus Schrad. ex Eckl. et Zeyh. (1836)(保留属名);西瓜属;Citrullus, Watermelon ■
93287 Citrullus amarus Schrad. = Citrullus vulgaris Schrad. ex Eckl. et Zeyh. ■
93288 Citrullus aquosus Schur = Citrullus vulgaris Schrad. ex Eckl. et Zeyh. ■
93289 Citrullus battich Forssk.;台湾西瓜(西瓜);Taiwan Citrullus ■
93290 Citrullus battich Forssk. = Citrullus lanatus (Thunb.) Matsum. et Nakai ■
93291 Citrullus battich Forssk. var. lanatus? = Citrullus lanatus (Thunb.) Matsum. et Nakai ■
93292 Citrullus caffer Schrad.;卡菲尔西瓜;Spiked Cucumber ■☆
93293 Citrullus caffer Schrad. = Citrullus lanatus (Thunb.) Matsum. et Nakai ■
93294 Citrullus caffer Schrad. = Citrullus vulgaris Schrad. ex Eckl. et Zeyh. ■
93295 Citrullus caffrorum Schrad. = Citrullus vulgaris Schrad. ex Eckl. et Zeyh. ■
93296 Citrullus chodospermus Fal.? et Dun. = Citrullus vulgaris Schrad. ex Eckl. et Zeyh. ■
93297 Citrullus colocynthis (L.) Schrad.;药西瓜(西腰葫芦);Bitter Apple, Colocynth, Vine-of-sodom, Wild Gourd ■☆
93298 Citrullus colocynthis Schrad. var. lanatus (Thunb.) Matsum. et Nakai = Citrullus lanatus (Thunb.) Matsum. et Nakai ■
93299 Citrullus ecirrhosus Cogn.;无卷须西瓜■☆
93300 Citrullus edulis Spach = Citrullus lanatus (Thunb.) Matsum. et Nakai ■
93301 Citrullus edulis Spach = Citrullus vulgaris Schrad. ex Eckl. et Zeyh. ■
93302 Citrullus fistulosus Stocks = Citrullus vulgaris Schrad. ex Eckl. et Zeyh. ■
93303 Citrullus fistulosus Stocks = Praecitrullus fistulosus (Stocks) Pangalo ■☆
93304 Citrullus lanatus (Thunb.) Matsum. et Nakai;西瓜(寒瓜,青登瓜,水瓜,天然白虎汤,天生白虎汤,夏瓜);Bitter Apple, Bitter Cucumber, Bitter Gourd, Colocynth, Coloquintida, Tsama Melon, Water Melon, Watermelon ■
93305 Citrullus melo (L.) Ozenda = Cucumis melo L. ■
93306 Citrullus naudinianus (Sond.) Hook. f. = Acanthosicyos naudinianus (Sond.) C. Jeffrey ■☆
93307 Citrullus naudinianus Hook. f.;苦西瓜;Bitter Melon ■☆
93308 Citrullus pasteca Sageret = Citrullus lanatus (Thunb.) Matsum. et Nakai ■
93309 Citrullus rehmii De Winter;雷姆西瓜■☆
93310 Citrullus vulgaris Eckl. et Zeyh. = Citrullus lanatus (Thunb.) Matsum. et Nakai ■
93311 Citrullus vulgaris Schrad. = Citrullus lanatus (Thunb.) Matsum. et Nakai ■
93312 Citrullus vulgaris Schrad. ex Eckl. et Zeyh. = Citrullus lanatus (Thunb.) Matsum. et Nakai ■
93313 Citrullus vulgaris Schrad. var. fistulosus (Stocks) Stewart = Praecitrullus fistulosus (Stocks) Pangalo ■☆
93314 Citrus L. (1753);柑橘属;Citrus, Gold Fruit, Orange, Pomelo ●
93315 Citrus acida Pers. = Citrus aurantifolia (Christm.) Swingle ●
93316 Citrus acida Rottb. = Citrus aurantifolia (Christm.) Swingle ●
93317 Citrus alata (Y. Tanaka) Tanaka = Citrus medica L. ●
93318 Citrus alata Tanaka;翼枸橼;Winged Citron ●☆
93319 Citrus alata Tanaka = Citrus medica L. ●
93320 Citrus amara Link. = Citrus aurantium L. var. amara Engl. ●
93321 Citrus amara Link. = Citrus aurantium L. ●
93322 Citrus amblycarpa (Hassk.) Ochse;求罗克利檬;Djeroek Limoe ●☆
93323 Citrus ampullacea Y. Tanaka;日本瓢柑(瓢柑);Flask-shaped Orange, Hyokan ●☆
93324 Citrus anonyma Y. Tanaka;西莲宝●☆
93325 Citrus articulata Spreng. = Citropsis articulata (Spreng.) Swingle et M. Kellerm. ●☆
93326 Citrus asahikan Y. Tanaka;旭柑(赤夏);Asahikan ●☆
93327 Citrus assamensis S. Dutta et S. C. Bhattach.;阿萨姆橘(阿达·雅米);Ada-jamir ●☆
93328 Citrus aurantiaca Y. Tanaka;赤黄柑;Ogon-daidai ●☆
93329 Citrus aurantifolia (Christm.) Swingle;来檬(来母,赖母,绿檬,酸柠檬);Common Lime, Key Lime, Lime, Lime-fruit Tree, Lime-tree, Mexican Lime ●
93330 Citrus aurantifolia (Christm.) Swingle var. latifolia Y. Tanaka;宽叶来檬(塔西提来檬);Tahiti-lime ●☆
93331 Citrus aurantiifolia (Christm.) Swingle subsp. murgetana Garcia Lidón et al. = Citrus aurantifolia (Christm.) Swingle ●
93332 Citrus aurantium L.;酸橙(橙,来母,酸柑,香柑,枳壳,枳实);

Bigarade, Bitter Orange, Eau De Cologne, Neroli, Orange, Pelitgrain Oil, Seville Orange, Sour Orange ●

93333 Citrus aurantium L. 'Daidai';代代酸橙●

93334 Citrus aurantium L. 'Goutou Cheng';枸头橙;Goutou Orange ●

93335 Citrus aurantium L. 'Hongpi-Suanchen';红皮酸橙●

93336 Citrus aurantium L. 'Huangpi' = Citrus aurantium L. 'Huangpi-Suanchen' ●

93337 Citrus aurantium L. 'Huangpi-Suanchen';黄皮酸橙●

93338 Citrus aurantium L. 'Hutou Gan';虎头柑●

93339 Citrus aurantium L. 'Jiangjin-Suanchen';江津酸橙●

93340 Citrus aurantium L. 'Natsudaidai';日本夏橙(夏橙,夏蜜柑);Natsu-daidai, Summer Orange ●

93341 Citrus aurantium L. 'Taiwanica';南庄橙(南庄代代);Nansho-daidai, Taiwan Orange ●

93342 Citrus aurantium L. 'Tanchen';塘橙●

93343 Citrus aurantium L. 'Xiaohong Cheng';小红橙●

93344 Citrus aurantium L. 'Zhuluan';朱栾(小红橙)●

93345 Citrus aurantium L. = Citrus maxima (Burm. ex Rumph.) Merr. ●

93346 Citrus aurantium L. f. deliciosa (Tenore) Hiroe = Citrus reticulata Blanco ●

93347 Citrus aurantium L. f. grandis (L.) Hiroe = Citrus maxima (Burm.) Merr. ●

93348 Citrus aurantium L. f. natsudaidai (Tanaka) Hiroe;日本夏甜橙●☆

93349 Citrus aurantium L. subf. nobilis (Lour.) Hiroe = Citrus aurantium L. ●

93350 Citrus aurantium L. subf. sinensis (L.) Hiroe = Citrus aurantium L. ●

93351 Citrus aurantium L. subsp. amara Engl. = Citrus aurantium L. ●

93352 Citrus aurantium L. subsp. aurantiifolia (Christm.) Guillaumin = Citrus aurantifolia (Christm.) Swingle ●

93353 Citrus aurantium L. subsp. bergamia (Risso) Engl. = Citrus limon (L.) Burm. f. ●

93354 Citrus aurantium L. subsp. decumana (L.) Tanaka = Citrus maxima (Burm.) Merr. ●

93355 Citrus aurantium L. subsp. ichangensis (Swingle) Guillaumin = Citrus cavaleriei H. Lév. ex Cavaler ●

93356 Citrus aurantium L. subsp. japonica (Thunb.) Engl. = Citrus japonica Thunb. ●

93357 Citrus aurantium L. subsp. japonica (Thunb.) Engl. var. globifera subvar. margarita Engl. = Fortunella margarita (Lour.) Swingle ●

93358 Citrus aurantium L. subsp. junos (Siebold ex Tanaka) Makino = Citrus junos (Siebold) Siebold ex Tanaka ●☆

93359 Citrus aurantium L. subsp. junos Makino ex Y. Tanaka = Citrus junos (Siebold) Siebold ex Tanaka ●

93360 Citrus aurantium L. subsp. nobilis var. tachibana Makino = Citrus tachibana (Makino) C. Tanaka ●

93361 Citrus aurantium L. subsp. sinensis (L.) Engl. = Citrus aurantium L. ●

93362 Citrus aurantium L. subsp. suntra Engl. = Citrus reticulata Blanco ●

93363 Citrus aurantium L. subvar. amilbed Engl. = Citrus medica L. ●

93364 Citrus aurantium L. subvar. chakotra Engl. = Citrus medica L. ●

93365 Citrus aurantium L. subvar. madurensis (Lour.) Engl. = Citrus japonica Thunb. ●

93366 Citrus aurantium L. subvar. margarita (Lour.) Engl. = Citrus japonica Thunb. ●

93367 Citrus aurantium L. subvar. spinosa Siebold et Zuccarini ex Engl. = Citrus japonica Thunb. ●

93368 Citrus aurantium L. var. amara Engl.;代代花(代代,代代酸橙,玳玳花,玳玳橘,回青橘,苏枳壳,酸橙,酸橙花,香栾,枳壳花);Daidai, Daidaihua, Seville Orange, Sour Orange ●

93369 Citrus aurantium L. var. amara L. = Citrus aurantium L. ●

93370 Citrus aurantium L. var. aurantium Hook. f. = Citrus sinensis (L.) Osbeck ●

93371 Citrus aurantium L. var. bergamia (Risso) Brandis = Citrus bergamia Risso et Poit. ●

93372 Citrus aurantium L. var. bergamia (Risso) Brandis = Citrus limon (L.) Burm. f. ●

93373 Citrus aurantium L. var. bigaradia (Loisel.) Brandis = Citrus aurantium L. ●

93374 Citrus aurantium L. var. bigaradia Hook. f. = Citrus aurantium L. ●

93375 Citrus aurantium L. var. crassa Risso = Citrus aurantium L. ●

93376 Citrus aurantium L. var. crispa Y. Tanaka;花束酸橙;Bouquet ●☆

93377 Citrus aurantium L. var. daidai Makino = Citrus aurantium L. ●

93378 Citrus aurantium L. var. daidai Y. Tanaka = Citrus aurantium L. 'Daidai' ●

93379 Citrus aurantium L. var. decumana L. = Citrus maxima (Burm.) Merr. ●

93380 Citrus aurantium L. var. dulcis Hayne = Citrus aurantium L. ●

93381 Citrus aurantium L. var. fetifera Risso = Citrus aurantium L. ●

93382 Citrus aurantium L. var. globifera Engl. = Citrus japonica Thunb. ●

93383 Citrus aurantium L. var. grandis L. = Citrus maxima (Burm.) Merr. ●

93384 Citrus aurantium L. var. japonica (Thunb.) Hook. = Citrus japonica Thunb. ●

93385 Citrus aurantium L. var. japonica (Thunb.) Hook. = Fortunella japonica (Thunb.) Swingle ●

93386 Citrus aurantium L. var. japonica (Thunb.) Hook. = Fortunella margarita (Lour.) Swingle ●

93387 Citrus aurantium L. var. lusitanica Risso = Citrus aurantium L. ●

93388 Citrus aurantium L. var. mellarosa (Risso) Engl. = Citrus limon (L.) Burm. f. ●

93389 Citrus aurantium L. var. myrtifolia Ker Gawl.;印度支那代代花(印度支那代代,桃金娘叶橙);Chinotto Orange, Murtle-leaf Orange ●☆

93390 Citrus aurantium L. var. myrtifolia Ker Gawler = Citrus aurantium L. ●

93391 Citrus aurantium L. var. oliviformis Risso ex Loisel. = Citrus japonica Thunb. ●

93392 Citrus aurantium L. var. proper Guillaumin = Citrus aurantifolia (Christm.) Swingle ●

93393 Citrus aurantium L. var. sanguinea Engl. = Citrus aurantium L. ●

93394 Citrus aurantium L. var. sinensis L. = Citrus aurantium L. ●

93395 Citrus aurantium L. var. sinensis L. = Citrus maxima (Burm.) Merr. ●

93396 Citrus aurantium L. var. sinensis L. = Citrus sinensis (L.) Osbeck ●

93397 Citrus aurantium L. var. tachibana (Makino) Tanaka = Citrus tachibana (Makino) C. Tanaka ●

93398 Citrus aurantium L. var. tachibana Makino = Citrus reticulata Blanco ●

93399 Citrus aurantium L. var. tachibana Makino = Citrus tachibana (Makino) C. Tanaka ●

93400 Citrus aurantium L. var. vulgaris (Risso) Risso et Poiteau =

Citrus aurantium L. ●
93401 Citrus auraria Michel = Citrus hystrix DC. ●
93402 Citrus aurata Risso = Citrus aurantium L. ●
93403 Citrus aurata Risso.;爱尔康太拉橙(爱尔康太拉);Adam's Apple,El-kantara ●☆
93404 Citrus aurea Y. Tanaka;川田橙;Kawabata ●☆
93405 Citrus australasica F. Muell. = Microcitrus australasica (F. Muell.) Swingle ●☆
93406 Citrus balincolong Y. Tanaka;巴利古龙橙;Balincolong ●☆
93407 Citrus balotina Poit. et Turpin;田中香柠檬;Balotin Bergamot ●☆
93408 Citrus bergamia Risso et Poit.;贝加莫橙(巴柑檬,贝加毛橙,贝加蜜柑,佛手,香柠檬);Bergamot,Bergamot Orange,Bigarade,Bitter Orange,Lemmarosa,Seville Orange,Sour Orange ●
93409 Citrus bergamia Risso et Poit. = Citrus limon (L.) Burm. f. ●
93410 Citrus bergamia Risso subsp. mellarosa (Risso) Rivera et al. = Citrus limon (L.) Burm. f. ●
93411 Citrus bergamota Raf. = Citrus limon (L.) Burm. f. ●
93412 Citrus bigaradia Loisel. = Citrus aurantium L. ●
93413 Citrus bigaradia Risso et Poit. = Citrus aurantium L. ●
93414 Citrus bigaradia Risso. et Poit.;苦橙(臭橙,回青橙);Seville Orange,Sour Orange ●☆
93415 Citrus boholensis (Wester) Y. Tanaka;康西;Kansi ●☆
93416 Citrus buxifolia Poir. = Atalantia buxifolia (Poir.) Oliv. ●
93417 Citrus canaliculata Y. Tanaka;菊代代;Kikudaidai ●☆
93418 Citrus canaliculata Y. Tanaka = Citrus aurantium L. ●
93419 Citrus cavaleriei H. Lév. ex Cavaler = Citrus ichangensis Swingle ●
93420 Citrus cedra Link = Citrus medica L. ●
93421 Citrus cedrata Raf. = Citrus medica L. ●
93422 Citrus celebica Koord.;西里伯橙(塞里比橙);Celebes Citrus,Celebes Papeda ●☆
93423 Citrus celebica Koord. var. southwickii (Wester) Swingle;沙斯伟克大翼橙;Limao,Southwick's Papeda ●☆
93424 Citrus chachiensis Y. Tanaka = Citrus reticulata Blanco 'Chachiensis' ●
93425 Citrus changshan-huyou Y. B. Chang;常山胡柚;Changshan Orange ●
93426 Citrus changshan-huyou Y. B. Chang = Citrus aurantium L. ●
93427 Citrus chrysocarpa Lush. = Citrus reticulata Blanco ●
93428 Citrus chuana Y. Tanaka;土橘;Tuju Mandarin ●
93429 Citrus clementina Y. Tanaka;阿尔及利亚橘(克力迈丁红橘,克利檬橙);Algerian,Clementine Mandarin ●☆
93430 Citrus combara Raf.;安乃姆橙(安乃姆大翼橙);Annam Papeda,Sat-kara ●☆
93431 Citrus communis Poiteau et Turpin = Citrus aurantium L. ●
93432 Citrus compressa Y. Tanaka = Citrus reticulata Blanco 'Subcompressa' ●
93433 Citrus costata Raf. = Citrus maxima (Burm.) Merr. ●
93434 Citrus crassifolia Swingle;厚叶柑●
93435 Citrus crenatifolia Lush.;开翁拉(考拉);Kawla,Keonla ●☆
93436 Citrus daidai Siebold et Hayata = Citrus aurantium L. 'Daidai' ●
93437 Citrus daoxianensis S. W. He et G. F. Liu;道县野橘;Daoxian Citrus,Daoxian Wild Mandarin ●
93438 Citrus daoxianensis S. W. He et G. F. Liu = Citrus reticulata Blanco ●
93439 Citrus davaoensis Y. Tanaka;卡尔比;Kalpi ●☆
93440 Citrus decumana (L.) L. = Citrus maxima (Burm.) Merr. ●
93441 Citrus decumana (L.) L. var. paradisi (Macfad.) H. H. A. Nicholls = Citrus aurantium L. ●
93442 Citrus decumana L. = Citrus grandis (L.) Osbeck ●
93443 Citrus decumana L. = Citrus maxima (Burm.) Merr. ●
93444 Citrus decumana L. var. paradisi Nicholson = Citrus paradisi Macfad. ●
93445 Citrus decumana L. var. racemosa Roem. = Citrus paradisi Macfad. ●
93446 Citrus deliciosa Ten.;地中海柑(大红橘,地中海红橘,福橘,红橘,橘,宽皮橘);Loose Skinned Orange,Mandarin,Mandarine Orange,Tangerine,Tangerine Orange ●
93447 Citrus deliciosa Ten. = Citrus reticulata Blanco ●
93448 Citrus deprassa Hayata;台湾香檬(扁平橘);Depressed Orange,Shiikuwasha ●
93449 Citrus depressa Hayata = Citrus reticulata Blanco ●
93450 Citrus dulcis Persoon = Citrus aurantium L. ●
93451 Citrus duttae Y. Tanaka;哈许·苦利;Hash-khuli ●☆
93452 Citrus echinata Saint-Lager = Citrus hystrix DC. ●
93453 Citrus erythrocarpa Hayata = Glycosmis parviflora (Sims) Little ●
93454 Citrus erythrosa Y. Tanaka = Citrus reticulata Blanco 'Erythrosa' ●☆
93455 Citrus erythrosa Y. Tanaka = Citrus reticulata Blanco ●
93456 Citrus excelsa Wester.;利阿尔柠檬(菲律宾柠檬);Limon Real ●☆
93457 Citrus flavicarpa Y. Tanaka;前田柑;Maedakan ●☆
93458 Citrus florida Salisbury = Citrus aurantium L. ●
93459 Citrus fragrans Salisbury = Citrus medica L. ●
93460 Citrus fumida Y. Tanaka;富库来蜜柑;Fukulai-mikan ●☆
93461 Citrus funadoko Y. Tanaka;舟床;Funadoko ●☆
93462 Citrus fusca Lour.;四川橘;Sichuan Citrus ●
93463 Citrus genshokan Y. Tanaka;元宵橘(元宵柑);Genshokan,Genshokan Citrus ●
93464 Citrus glaberrima Y. Tanaka;神代橘(绢皮);Kinukawa ●☆
93465 Citrus glauca (Lindl.) Burkill;灰叶柠檬;Desert Lime ●☆
93466 Citrus grandis (L.) Osbeck 'Tengu';天狗;Tengu ●☆
93467 Citrus grandis (L.) Osbeck = Citrus maxima (Burm.) Merr. ●
93468 Citrus grandis (L.) Osbeck f. buntan Hayata;文旦柚(麻豆文旦,文旦);Buntan Shaddock ●
93469 Citrus grandis (L.) Osbeck f. hakunikuyu Hayata;白柚;White Shaddock ●
93470 Citrus grandis (L.) Osbeck var. anseikan Y. Tanaka;安政柑;Anseikan ●☆
93471 Citrus grandis (L.) Osbeck var. banokan Y. Tanaka;晚王柑;Banokan ●☆
93472 Citrus grandis (L.) Osbeck var. pinshanyu Hort. = Citrus maxima (Burm.) Merr. 'Pingshan Yu' ●
93473 Citrus grandis (L.) Osbeck var. pseudoanseikan Y. Tanaka;拟安政柑●☆
93474 Citrus grandis (L.) Osbeck var. pyriformis (Hassk.) Karaya = Citrus maxima (Burm.) Merr. ●
93475 Citrus grandis (L.) Osbeck var. sabon (Siebold ex Hayata) Hayata = Citrus maxima (Burm.) Merr. ●
93476 Citrus grandis (L.) Osbeck var. tanikawa Y. Tanaka;谷川文旦●☆
93477 Citrus grandis (L.) Osbeck var. tomentosa Hort.;化橘红(化州陈皮,化州橘红,化州柚,橘红,橘红,毛橘红);Tomentosa Pummelo ●
93478 Citrus grandis Osbeck = Citrus grandis (L.) Osbeck ●
93479 Citrus hainana Hort.;黄橘●☆

93480 Citrus hainanensis Y. Tanaka;海南野橘(海南野生橙);Hainan Wild Kat,Hainan Wild Orange ●☆

93481 Citrus hanayu Shirai ex Shirai = Citrus wilsonii Y. Tanaka ●☆

93482 Citrus hassaku Y. Tanaka;八朔蜜柑;Hassaku-mikan ●☆

93483 Citrus himekitsu Y. Tanaka;姬橘;Himekitsu ●☆

93484 Citrus hindsii (Champ. ex Benth.) Govaerts = Citrus japonica Thunb. ●

93485 Citrus hiroshimana Y. Tanaka;广岛夏柚;Hiroshima-natsuzabon ●☆

93486 Citrus hongheensis Y. Ye, X. Liu, S. Ding et M. Liang;红河橙(阿雷,阿蕾,树葫芦,枳壳);Honghe Orange ●

93487 Citrus hongheensis Y. Ye, X. Liu, S. Ding et M. Liang = Citrus cavaleriei H. Lév. ex Cavaler ●

93488 Citrus hsiangyuan Tanaka = Citrus junos (Siebold) Siebold ex Tanaka ●☆

93489 Citrus humilis (Mill.) Poiret = Citrus aurantium L. ●

93490 Citrus hyalopulpa Tanaka = Citrus hystrix DC. ●

93491 Citrus hyalopulpa Tanaka = Citrus macroptera Montrouz. var. kerrii Swingle ●

93492 Citrus hystrix DC.;箭叶橙(箭叶金橘,金钱吊葫芦,马蜂橙,马蜂柑,马蜂毛柑,毛里求斯苦橙);Arrowleaf Orange, Arrow-leaved Kumquat, Bitter Orange, Cabuyao, Caffir Lime, Caffre Lime, Leech Lime, Makrut Lime, Mauritius, Mauritius Papeda, Purrut, Rough-skinned Lime, Sagittate-leaf Kumquat, Swingle's Kumquat, Thai Bai Makrut ●

93493 Citrus hystrix DC. subsp. acida Engl. = Citrus aurantifolia (Christm.) Swingle ●

93494 Citrus hystrix H. Perrier = Citrus aurantium L. ●

93495 Citrus ichangensis Swingle;宜昌橙(罗汉柑,酸柑子,野柑子,宜昌柑);Ichang Bitter Orange, Ichang bitterorange, Ichang Lemon, Ichange Ihih, Ichant Papeda, Ketsa-shupfu, Yichang Bitterorange, Yichang Lemon, Yichang Orange ●

93496 Citrus ichangensis Swingle = Citrus cavaleriei H. Lév. ex Cavaler ●

93497 Citrus indica Y. Tanaka;印度野橘;Humutiatenga, Indian Wild Citrus, Indian wild Orange ●☆

93498 Citrus inermis Roxb. = Citrus japonica Thunb. ●

93499 Citrus inflata Y. Tanaka;饼柚;Mochi-yu ●☆

93500 Citrus inflato-rugosa Y. Tanaka;考夫客尔●☆

93501 Citrus intermedia Y. Tanaka;山蜜橘(山蜜柑);Intermediate Citrus, Intermediate Orange, Yamamikan ●☆

93502 Citrus iriomotensis Y. Tanaka;西表橘(富赛拉)●☆

93503 Citrus iwaikan Y. Tanaka;岩井橘;Iwaikan, Iwaikan Citrus ●☆

93504 Citrus iyo Y. Tanaka;伊予柑(伊予蜜柑);Iyo-mikan ●☆

93505 Citrus jambhiri Lush.;印度柠檬(粗柠檬);Bush Lemon, Jambhiri Orange, Rough Lemon ●☆

93506 Citrus japonica Thunb. = Fortunella japonica (Thunb.) Swingle ●

93507 Citrus japonica Thunb. subf. crassifolia (Swingle) Hiroe = Citrus japonica Thunb. ●

93508 Citrus japonica Thunb. subf. hindsii (Champ. ex Benth.) Hiroe = Citrus japonica Thunb. ●

93509 Citrus japonica Thunb. subf. margarita (Lour.) Hiroe = Citrus japonica Thunb. ●

93510 Citrus japonica Thunb. var. fructuelliptico Siebold et Zucc. = Fortunella margarita (Lour.) Swingle ●

93511 Citrus japonica Thunb. var. madurensis (Lour.) Guillaumin = Citrus japonica Thunb. ●

93512 Citrus japonica Thunb. var. margarita (Lour.) Guillaumin = Citrus japonica Thunb. ●

93513 Citrus javanica Blume;求罗克哈齐;Djeroek Hunje, Jeruk Hunje ●☆

93514 Citrus javanica Blume = Citrus aurantifolia (Christm.) Swingle ●

93515 Citrus junos (Siebold) Siebold ex Tanaka;香橙(橙,橙子,鹄壳,黄橙,金橙,金球,罗汉橙,蜜橙,糖橙,蟹橙,柚,柚子);Fragrant Citrus, Fragrant Orange, Hsiang Cheng, Yuzu ●☆

93516 Citrus junos Siebold ex Tanaka = Citrus junos (Siebold) Siebold ex Tanaka ●☆

93517 Citrus karna Raf.;卡塔檬;Karna Khatta, Khatta ●☆

93518 Citrus katokan Hayata;红头柑●

93519 Citrus keraji Y. Tanaka;花良治;Keraji ●☆

93520 Citrus keraji Y. Tanaka var. kabuchii Y. Tanaka;卡蒲橘;Kabuchii ●☆

93521 Citrus keraji Y. Tanaka var. unzoki Y. Tanaka;温蜀橘●☆

93522 Citrus kerrii (Swingle) Tanaka = Citrus hystrix DC. ●

93523 Citrus kerrii Tanaka;泰兰特大翼橙(大翼橙);Papeda, Thailand Papeda ●☆

93524 Citrus kinokuni Tanaka = Citrus japonica Thunb. ●

93525 Citrus kinokuni Tanaka = Citrus reticulata Blanco 'Kinokuni' ●

93526 Citrus kotokan Hayata;虎豆柑(虎头柑);Kotokan, Kotokan Citrus ●

93527 Citrus kwangsiensis Hu = Citrus maxima (Burm.) Merr. ●

93528 Citrus latifolia (Tanaka ex Y. Tanaka) Tanaka;塔希提柠檬(宽叶来母);Persian Lime, Tahiti Lime, Tahitian Lime ●☆

93529 Citrus latipes (Swingle) Tanaka;印缅橙(卡西大翼橙);Broad-stalked Orange, Khasi Papeda ●

93530 Citrus leiocarpa Y. Tanaka;日本土柑(柑子,光橘);Koji, Smooth-fruited Orange ●☆

93531 Citrus leiocarpa Y. Tanaka f. monoembryota Y. Tanaka;骏河柑子(骏河柚柑);Suruga-yuko ●☆

93532 Citrus lima Lunan;利马橙(柠檬)●☆

93533 Citrus lima Macfad. = Citrus aurantifolia (Christm.) Swingle ●

93534 Citrus limetta Risso;甜柠檬;Adam's Apple, Bitter Orange, Lime, Limetta, Limette, Mediterranean Sweet Lemon, Sweet Lemon, Sweet Lime, Sweetie ●☆

93535 Citrus limetta Risso = Citrus maxima (Burm.) Merr. ●

93536 Citrus limettioides Y. Tanaka;甜来母(蜜太柠蒲);Mita-nimbu, Sweetlime, Sweet-lime-of-India ●☆

93537 Citrus limettioides Y. Tanaka = Citrus limetta Risso ●☆

93538 Citrus limodulcis Rivera et al. = Citrus limon (L.) Burm. f. ●

93539 Citrus limon (L.) Burm. f.;柠檬(梨橡干,黎朦子,黎檬,黎檬子,黎子,里母子,梦子,柠果,西柠檬,香檬,洋柠檬,药果,宜濛子,宜母果,宜母子);Citron, Lemon, Rough Lemon, Wild Lemon ●

93540 Citrus limon (L.) Burm. f. 'Eureka';尤里卡柠檬●☆

93541 Citrus limon (L.) Burm. f. 'Lisbon';里斯本柠檬●☆

93542 Citrus limon (L.) Burm. f. 'Meyer';美亚柠檬●☆

93543 Citrus limon (L.) Osbeck = Citrus medica L. ●

93544 Citrus limon (L.) Osbeck var. digitata Risso = Citrus medica L. ●

93545 Citrus limonelloides Hayata = Citrus limonia Osbeck ●

93546 Citrus limonia (L.) Osbeck;黎檬(广东黎檬,广东柠檬,红柠檬,梨橡干,黎朦子,黎檬子,黎子,里母子,朦子,梦子,柠果,柠檬,药果,宜濛子,宜朦子,宜母,宜母果,宜母子);Canton Lemon, Hime Lemon, Lemon, Lemonlike Citrus, Lemon-like Citrus, Lemonlike Orange, Limonia, Rangpur Lime ●

93547 Citrus limonia (L.) Osbeck = Citrus medica L. ●

93548 Citrus limonia Osbeck = Citrus limonia (L.) Osbeck ●

93549 Citrus limonia Osbeck var. digitata Risso = Citrus medica L. var. sarcodactylis (Noot.) Swingle ●

93550 Citrus limonimedica Lush.;圆佛手柑(巴肖拉枸橼);Bajoura Citron,Maru-bushukan ●☆

93551 Citrus limonum Risso = Citrus limon (L.) Burm. f. ●

93552 Citrus longilimon Y. Tanaka;阿萨姆柠檬;Assam lemon ●☆

93553 Citrus longispina Wester;太拉蜜散;Talamisan ●☆

93554 Citrus lumia Risso. et Poit.;露蜜;Lumie ●☆

93555 Citrus luteo-turgida Y. Tanaka;台台富施●☆

93556 Citrus lycopersiciformis Y. Tanaka;科克尼;Kokni ●☆

93557 Citrus macrolimon Y. Tanaka;大加尔加尔;Galgal Large ●☆

93558 Citrus macrophylla Wester;大叶来檬(阿蕾檬,阿利莫来檬);Alemon,Alemow ●☆

93559 Citrus macroptera Montrouz.;大翼橙(马来西亚苦橙,美拉尼亚大翼橙);Cabuyao,Large-wing Citrus,Large-winged Orange,Malesia Bitterorange,Melanesian Papeda ●☆

93560 Citrus macroptera Montrouz. var. annamensis Y. Tanaka;安乃姆大翼橙;Annam Papeda ●☆

93561 Citrus macroptera Montrouz. var. kerrii Swingle;马蜂橙(盖尔泰国大翼橙,红河大翼橙,马蜂柑,石碌柑,树葫芦);Hornet Orange,Kerr Large-wing Citrus,Kerr's Thailand Papeda ●

93562 Citrus macroptera Montrouz. var. kerrii Swingle = Citrus hystrix DC. ●

93563 Citrus macrosperma T. C. Guo et Y. M. Ye;大种橙;Big-seed Citrus ●

93564 Citrus macrosperma T. C. Guo et Y. M. Ye = Citrus cavaleriei H. Lév. ex Cavaler ●

93565 Citrus maderaspatana Y. Tanaka;马都拉斯橙(气利橘);Kitchli,Vadlapudi Orange ●☆

93566 Citrus madurensis Lour.;月橘(金橘,四季橘);Calamondin,Shikikitsu,Tokinkan ●

93567 Citrus madurensis Lour. = Citrus japonica Thunb. ●

93568 Citrus madurensis Lour. = Citrus reticulata Blanco ●

93569 Citrus madurensis Lour. = Fortunella japonica (Thunb.) Swingle ●

93570 Citrus madurensis Lour. var. deliciosa (Tenore) Sagot = Citrus reticulata Blanco ●

93571 Citrus malaccensis Ridl. = Burkillanthus malaccensis (Ridl.) Swingle ●☆

93572 Citrus mangshanensis S. W. He et G. F. Liu;莽山野柑;Mangshan Citrus,Mangshan Wild Mandarin ●

93573 Citrus mangshanensis S. W. He et G. F. Liu = Citrus reticulata Blanco ●

93574 Citrus margarita Lour. = Citrus japonica Thunb. ●

93575 Citrus margarita Lour. = Fortunella margarita (Lour.) Swingle ●

93576 Citrus marginata Steud. = Fortunella japonica (Thunb.) Swingle ●

93577 Citrus matsudaidai Hayata;夏橙;Summer Citrus ●☆

93578 Citrus maxima (Burm. ex Rumph.) Merr. = Citrus maxima (Burm.) Merr. ●

93579 Citrus maxima (Burm.) Merr.;柚(苞,臭橙,臭柚,斗柚,胡柑,雷柚,栾,抛,抛栾,脬,气柑,条,文旦,香栾,柚子,朱栾);Buntan,Jabon,Narinjin,Pomelo,Pompelmous,Pummelo,Shaddock,Shaddock Pummelo,Sour Orange,Zabon ●

93580 Citrus maxima (Burm.) Merr. 'Anjiang Yu';安江香柚●

93581 Citrus maxima (Burm.) Merr. 'Jinlan Yu';金兰柚●

93582 Citrus maxima (Burm.) Merr. 'Jinxiang Yu';金香柚●

93583 Citrus maxima (Burm.) Merr. 'Liangping Yu';梁平柚●

93584 Citrus maxima (Burm.) Merr. 'Pingshan Yu';坪山柚●

93585 Citrus maxima (Burm.) Merr. 'Sangma Yu';桑麻柚●

93586 Citrus maxima (Burm.) Merr. 'Shatian Yu';沙田柚●

93587 Citrus maxima (Burm.) Merr. 'Szechipaw';四季抛●

93588 Citrus maxima (Burm.) Merr. 'Tomentosa';橘红(化州橘红,仙橘)●

93589 Citrus maxima (Burm.) Merr. 'Wanbei Yu';晚白柚●

93590 Citrus maxima (Burm.) Merr. 'Wentan';文旦;Buntan ●

93591 Citrus maxima (Burm.) Merr. f. banyu Hayata;大斗柚(晚柚)●

93592 Citrus maxima (Burm.) Merr. f. buntan Hayata = Citrus grandis (L.) Osbeck f. buntan Hayata ●

93593 Citrus maxima (Burm.) Merr. f. hakunikuyu Hayata;白肉柚●

93594 Citrus maxima (Burm.) Merr. f. hakuyu Hayata;麻豆白柚●

93595 Citrus maxima (Burm.) Merr. f. jiyu Hort.;时柚●

93596 Citrus maxima (Burm.) Merr. f. mitsuyu Hayata;蜜柚●

93597 Citrus maxima (Burm.) Merr. f. santiyu Hayata;石头柚(山东柚)●

93598 Citrus maxima (Burm.) Merr. f. soyu Hort.;早柚(白叶柚)●

93599 Citrus maxima (Burm.) Merr. f. uyoyu Hort.;乌叶柚(高墙柚,中柚)●

93600 Citrus maxima (Burm.) Merr. var. uvacarpa Merr. = Citrus aurantium L. ●

93601 Citrus maxima (Burm.) Merr. var. uvacarpa Merr. = Citrus maxima (Burm. ex Rumph.) Merr. ●

93602 Citrus maxima (Burm.) Merr. var. uvacarpa Merr. = Citrus paradisi Macfad. ●

93603 Citrus medica L.;香橼(陈香圆,佛手柑,钩橼,枸橼,枸橼子,丸佛手柑,香黄,香泡树,香圆);Buddha's Head Citron,Candied Peel,Cedrat,Citron,Median Apple,Medicinal Citron,Medicinal Citrus,Persian Apple ●

93604 Citrus medica L. 'Sarcodactylis';佛手柑(川佛手,飞穰,佛柑,佛手,佛手香柑,佛手香橼,福寿柑,广佛手,开佛手,蜜罗柑,蜜筩柑,拳佛手,十指柑,十指香圆,手柑,手桔,手橘,五指柑,五指香橼,香柚,香圆);Buddha's Hand,Finger Citron,Fingered Citron,Fingered Lemon,Fleshfingered Citron ●

93605 Citrus medica L. f. limon (L.) Hiroe = Citrus limon (L.) Burm. f. ●

93606 Citrus medica L. f. monstrosa Guillaumin = Citrus medica L. ●

93607 Citrus medica L. subf. aurantiifolia (Christm.) Hiroe = Citrus aurantifolia (Christm.) Swingle ●

93608 Citrus medica L. subf. junos (Siebold ex Tanaka) Hiroe = Citrus junos (Siebold) Siebold ex Tanaka ●☆

93609 Citrus medica L. subf. pyriformis (Hassk.) Hiroe = Citrus maxima (Burm.) Merr. ●

93610 Citrus medica L. subsp. limonia Hook. f. ex Engl. = Citrus limon (L.) Burm. f. ●

93611 Citrus medica L. subsp. limonum (Risso) J. D. Hooker = Citrus limon (L.) Burm. f. ●

93612 Citrus medica L. var. alata Y. Tanaka = Citrus medica L. ●

93613 Citrus medica L. var. digitata Risso = Citrus medica L. ●

93614 Citrus medica L. var. dulcis Risso et Poit.;山柠檬(甘实大枸橼);Madhkankur ●

93615 Citrus medica L. var. ethrog Engl. = Citrus medica L. ●

93616 Citrus medica L. var. gaoganensis (Hayata) Tanaka;高山柚●

93617 Citrus medica L. var. junos Siebold = Citrus junos (Siebold) Siebold ex Tanaka ●

93618 Citrus medica L. var. limetta Hook. f. = Citrus limetta Risso ●☆

93619 Citrus medica L. var. limon L. = Citrus limon (L.) Burm. f. ●

93620 Citrus medica L. var. limon L. = Citrus limonia Osbeck ●
93621 Citrus medica L. var. limon L. = Citrus medica L. ●
93622 Citrus medica L. var. limonum (Risso) Brandis = Citrus limon (L.) Burm. f. ●
93623 Citrus medica L. var. proper J. D. Hooker = Citrus medica L. ●
93624 Citrus medica L. var. sarcodactylis (Noot.) Swingle = Citrus medica L. ●
93625 Citrus medica L. var. sarcodactylis (Noot.) Swingle = Citrus medica L. 'Sarcodactylis' ●
93626 Citrus medica L. var. tarung Tanaka = Citrus medica var. acida Brandis ●
93627 Citrus medica L. var. yunnanensis S. Q. Ding ex C. C. Huang;云南香橼;Yunnan ●☆
93628 Citrus megaloxicarpa Lush.;亚米尔培特;Amilbed ●☆
93629 Citrus mellarosa Risso = Citrus limon (L.) Burm. f. ●
93630 Citrus mellarosa Risso.;米拉罗赛●☆
93631 Citrus meyeri Y. Tanaka;香柠檬(北京柠檬);Grant Lemon, Meyer Citrus, Meyer Lemon, Soh-long ●☆
93632 Citrus meyeri Y. Tanaka = Citrus limon (L.) Burm. f. ●
93633 Citrus miaray Wester;米阿雷;Miaray ●☆
93634 Citrus micrantha Wester;小花橙;Biasong, Littleflower Citrus, Littleflower Orange, Small-flowered Bitterorange ●☆
93635 Citrus micrantha Wester var. microcarpa Wester;小果苦橙;Samuyao, Smallflower Bitterorange ●☆
93636 Citrus microcarpa Bunge;四季橘(四季成金柑,唐金柑,月橘);Calamondin, Musk Lime ●
93637 Citrus microcarpa Bunge = Citrus madurensis Lour. ●
93638 Citrus microcarpa Bunge = Citrus mitis Blanco ●
93639 Citrus mitis Blanco;加拉蒙丁橘(蕃柑,公孙橘,加拉蒙地亚橘,角生橘,四季橘,唐金柑,唐金橘);Calamondin, Calamondin Orange, Musk Lime, Panama Orange ●
93640 Citrus mitsuharu Y. Tanaka;光春蜜柑(春光柑);Mitsuharu, Shunkokan ●☆
93641 Citrus montana Y. Tanaka;山来檬(皮罗罗);Bilolo ●☆
93642 Citrus myrtifolia (Ker Gawler) Raf. = Citrus aurantium L. ●
93643 Citrus myrtifolia Raf.;厚叶橙(桃金娘叶橙);Chinotto, Myrtle-leaf Orange, Thickleaf Citrus, Thickleaf Orange ●☆
93644 Citrus myrtifolia Raf. = Citrus aurantium L. ●
93645 Citrus nana Y. Tanaka;奈奈长尼;Nanachangney ●☆
93646 Citrus natsudaidai Hayata = Citrus aurantium L. 'Natsudaidai' ●
93647 Citrus neoaurantium Y. Tanaka;新酸橙(萨摩枳壳,小根占);Satsuma-kikoku ●☆
93648 Citrus nippokoreana Y. Tanaka;高雷立花橘;Kolitachibana ●☆
93649 Citrus nobilis Lour.;柑(本地广柑,川陈皮,川橘,芙蓉柑,广州沙柑,九年母,沙柑,王柑,香橙);Jeneru Tenga, King Mandarin, King Orange, King Tangor, Mandarin Orange, Tangor ●
93650 Citrus nobilis Lour. = Citrus reticulata Blanco ●
93651 Citrus nobilis Lour. subf. deliciosa (Tenore) Hiroe = Citrus reticulata Blanco ●
93652 Citrus nobilis Lour. subf. erythrosa (Y. Tanaka) Hiroe = Citrus reticulata Blanco ●
93653 Citrus nobilis Lour. subf. reticulata (Blanco) Hiroe = Citrus reticulata Blanco ●
93654 Citrus nobilis Lour. subf. succosa (Tanaka) Hiroe = Citrus reticulata Blanco ●
93655 Citrus nobilis Lour. subf. tachibana (Makino) Hiroe = Citrus reticulata Blanco ●
93656 Citrus nobilis Lour. subf. unshiu (Marcowicz) Hiroe = Citrus reticulata Blanco ●
93657 Citrus nobilis Lour. var. deliciosa (Tenore) Guillaumin = Citrus reticulata Blanco ●
93658 Citrus nobilis Lour. var. deliciosa Swingle = Citrus deliciosa Ten. ●
93659 Citrus nobilis Lour. var. inermis (Roxb.) Sagot = Citrus japonica Thunb. ●
93660 Citrus nobilis Lour. var. kunep Y. Tanaka;九年母;Kunenbo ●☆
93661 Citrus nobilis Lour. var. major Ker Gawler = Citrus reticulata Blanco ●
93662 Citrus nobilis Lour. var. ponki Hayata = Citrus reticulata Blanco ●
93663 Citrus nobilis Lour. var. spontanea Ito = Citrus reticulata Blanco ●
93664 Citrus nobilis Lour. var. subcompressa Hu = Citrus reticulata Blanco 'Subcompressa' ●
93665 Citrus nobilis Lour. var. sunki Hayata = Citrus reticulata Blanco var. sunki (Hayata) Hu ●
93666 Citrus nobilis Lour. var. sunki Hayata = Citrus reticulata Blanco ●
93667 Citrus nobilis Lour. var. tachibana (Makino) Ito = Citrus reticulata Blanco ●
93668 Citrus nobilis Lour. var. tanakan Hayata = Citrus reticulata Blanco 'Tankan' ●
93669 Citrus nobilis Lour. var. unshiu (Marcowicz) Tanaka ex Swingle = Citrus reticulata Blanco ●
93670 Citrus nobilis Lour. var. vangasy (Bojer) Guillaumin = Citrus reticulata Blanco ●
93671 Citrus oblonga Y. Tanaka;椭圆柑●☆
93672 Citrus obovoidea Y. Tanaka;金柑子(百寿柑,上柑);Kinkoji ●☆
93673 Citrus obovoidea Y. Tanaka = Citrus maxima (Burm.) Merr. ●
93674 Citrus obversa Hassk.;求罗克·巴利克;Djeroek Balik ●☆
93675 Citrus odorata Roussel = Citrus medica L. ●
93676 Citrus odorata Wester;太西太西;Tihi-tihi ●☆
93677 Citrus oleocarpa Tanaka;年橘(臭皮橘,太姆卡脱,油皮橘);Tim-kat ●☆
93678 Citrus omikanto Y. Tanaka;大身甘橙;Omikanto ●☆
93679 Citrus otachibana Y. Tanaka;大橘(寿柑);Otachibana ●☆
93680 Citrus oto Y. Tanaka;奥太橘;Oto ●☆
93681 Citrus oto Y. Tanaka var. crassiuscula Y. Tanaka;考亚橘●☆
93682 Citrus oto Y. Tanaka var. elegans Y. Tanaka;假太罗额柚●☆
93683 Citrus paniculata Schumach. = Afraegle paniculata (Schumach.) Engl. ●☆
93684 Citrus panuban Y. Tanaka;拍纽彭●☆
93685 Citrus papaya Hassk.;求罗克·派派耶;Djeroek Papaya ●☆
93686 Citrus papeda Miq. = Citrus hystrix DC. ●
93687 Citrus papillaris Blanco;贴松;Tizon ●☆
93688 Citrus paradisi Macfad. 'Changshanhuyou';胡柚;Changshan-huyou Tangelo ●☆
93689 Citrus paradisi Macfad. = Citrus aurantium L. ●
93690 Citrus paradisi Macfad.;葡萄柚;Grape Fruit, Grape Pomelo, Grapefruit, Grapefruit Citrus, Grape-fruit Citrus, Pomelo ●
93691 Citrus paradisi Macfad. = Citrus maxima (Burm. ex Rumph.) Merr. ●
93692 Citrus paratangerina Y. Tanaka;拉特橘;Beauty of Glen Retreat, Ladoo, Ladu ●☆
93693 Citrus pennivesiculata Y. Tanaka;盖加尼马柚;Carabao Lime, Gajanimma, Ganjanimma ●☆
93694 Citrus peretta Risso.;蓓雷太●☆
93695 Citrus platymamma Y. Tanaka;瓶橘(比乌西克尔橘);

Biuisikul ●☆

93696 Citrus polyandra Tanaka = Clymenia polyandra (Tanaka) Swingle ●☆

93697 Citrus pompelmos Risso = Citrus maxima (Burm.) Merr. ●

93698 Citrus ponki (Hayata) Y. Tanaka;椪橘(柑橘,桔仔,橘子,甜橘);Ponki Citrus,Ponki Mandarin ●

93699 Citrus ponki (Hayata) Y. Tanaka = Citrus reticulata Blanco ●

93700 Citrus ponki Y. Tanaka = Citrus reticulata Blanco ●

93701 Citrus poonensis Y. Tanaka = Citrus reticulata Blanco 'Ponkan' ●

93702 Citrus poonensis Y. Tanaka = Citrus reticulata Blanco ●

93703 Citrus pseudoaurantium Y. Tanaka;变化蜜柑●☆

93704 Citrus pseudograndis Shirai;假柚;Uzon-kunnebu ●☆

93705 Citrus pseudogulgul Shirai;大柚(狮子柚);Lion Citrus ●☆

93706 Citrus pseudolimon Y. Tanaka;加尔加尔柠檬;Galgal Lemon, Hill Lemon ●☆

93707 Citrus pseudolimonum Wester;假柠檬;Colo-colo,False Lemon ●☆

93708 Citrus pseudopapillaris Y. Tanaka;巴兰格橙;Balanga Orange ●☆

93709 Citrus pseudoparadisi Y. Tanaka;宇和橘;Uwa-pomelo ●☆

93710 Citrus pseudosunki Y. Tanaka;菲律宾酸橘●☆

93711 Citrus pseudoyulyui Shirai;狮子柚;Lion Citrus ●☆

93712 Citrus pyriformis Hassk.;梨形橙(胖大罗赛);Pearform Citrus, Ponderosa Lemon ●☆

93713 Citrus pyriformis Hassk. = Citrus maxima (Burm.) Merr. ●

93714 Citrus racemosa Marcow. = Citrus paradisi Macfad. ●

93715 Citrus racemosa Marcow. ex Tanaka = Citrus paradisi Macfad. ●

93716 Citrus reshni Y. Tanaka;雷须尼橘(印度酸橘);Cleopatra Mandarin ●☆

93717 Citrus reticulata Blanco;柑橘(甘子,柑,柑子,橘,宽皮橘,行柑,木奴,椪柑,青皮,青皮橘,松皮橘,酸橘,凸柑,西螺柑);Chinese Honey, Clementine, Japanese Orange, King Orange, Looseskinned Orange, Mandarin, Mandarine Orange, Orange, Ponkan Orange, Pummelo Orange, Satsuma, Satsuma Orange, Sour Mandarin, Sunki, Suntara, Tangerine ●

93718 Citrus reticulata Blanco 'Bian Gan';扁柑(肚脐柑,柿饼柑,州柑)●

93719 Citrus reticulata Blanco 'Chachiensis';茶枝柑(大红柑,柑子,金实,木奴,瑞金奴,新会柑)●

93720 Citrus reticulata Blanco 'Dahongpao';大红袍●

93721 Citrus reticulata Blanco 'Erythrosa';九月黄(小红橘,小红蜜柑,朱红橘,朱橘);Red Orange, Tangerine ●

93722 Citrus reticulata Blanco 'Hanggan';行柑(四会柑);Suhui Orange ●

93723 Citrus reticulata Blanco 'Jiangan';建柑●

93724 Citrus reticulata Blanco 'Kinokuni';南丰蜜橘(纪州蜜柑,金钱橘,蜜橘,乳橘,莳橘);Kinokuni, Kinokuni Citrus, Kinokuni Mandarin ●

93725 Citrus reticulata Blanco 'Manau Gan';玛瑙柑(皱皮柑)●

93726 Citrus reticulata Blanco 'Manau';黄柑●☆

93727 Citrus reticulata Blanco 'Nobilis';沙柑●

93728 Citrus reticulata Blanco 'Ponkan';椪柑(广东蜜柑,芦柑,蜜糖柑,汕头蜜柑,凸柑);Batangas Mandarin, Poonen Citrus ●

93729 Citrus reticulata Blanco 'Shiyue Ju';十月橘(冰糖橘)●

93730 Citrus reticulata Blanco 'Suavissima';瓯柑●

93731 Citrus reticulata Blanco 'Subcompressa';早橘(黄岩蜜橘);Sokitsu Mandarin, Subcompressed Orange, Zaoju Mandarin ●

93732 Citrus reticulata Blanco 'Succosa';本地早(本地早橘,地蜜柑,天台蜜橘,天台山蜜橘);Bendizao Mandarin, Pentitsao, Spicy Mandarin, Suppy Orange ●

93733 Citrus reticulata Blanco 'Tangerina';福橘(川橘,柑子,红柑,红橘,绿橘)●

93734 Citrus reticulata Blanco 'Tankan';蕉柑(潮州柑,海梨,橘柑,蜜桶柑,年柑,抬甘,桶柑,招柑);Sheo Kan, Tankan, Tankan Mandarine ●

93735 Citrus reticulata Blanco 'Tardiferax';慢橘;Huangyan Orange ●

93736 Citrus reticulata Blanco 'Unshiu';温州蜜柑(无核橘);Satsuma Mandarin, Unshiu Mandarin, Wase-satsuma Mandarin, Wase-unshiu, Wenzhou Migan ●

93737 Citrus reticulata Blanco 'Zaohong';早红(洞庭红,早橘子);Benikoji Mandarine ●

93738 Citrus reticulata Blanco 'Zhuhong';朱红(大红袍,朱橘,朱砂橘)●

93739 Citrus reticulata Blanco subsp. deliciosa (Tenore) Rivera et al. = Citrus reticulata Blanco ●

93740 Citrus reticulata Blanco subsp. tachibana (Tanaka) Rivera et al. = Citrus reticulata Blanco ●

93741 Citrus reticulata Blanco subsp. unshiu (Marcowicz) Rivera et al. = Citrus reticulata Blanco ●

93742 Citrus reticulata Blanco var. austera Swingle = Citrus reticulata Blanco ●

93743 Citrus reticulata Blanco var. chachiensis Hu = Citrus reticulata Blanco 'Chachiensis' ●☆

93744 Citrus reticulata Blanco var. erythrosa (Tanaka) Hu = Citrus reticulata Blanco 'Erythrosa' ●☆

93745 Citrus reticulata Blanco var. kinokumi (Tanaka) Hu;乳橘●

93746 Citrus reticulata Blanco var. poonensis (Hayata) Hu;冇柑●

93747 Citrus reticulata Blanco var. suhoiensis (Tanaka) Hu = Citrus reticulata Blanco 'Hanggan' ●

93748 Citrus reticulata Blanco var. sunki (Hayata) Hu;酸橘(本地早,本地早橘,酸桔,天台山蜜橘);Bendizao Mandarin, Honjiso Mandarin, Sour Orange ●

93749 Citrus reticulata Blanco var. szehuikan (Tanaka) Tseng = Citrus reticulata Blanco 'Hanggan' ●

93750 Citrus reticulata Blanco var. tankan (Hayata) Hu = Citrus reticulata Blanco 'Tankan' ●

93751 Citrus rissoi Risso.;利沙柚;Risso's pummelo ●☆

93752 Citrus rokugatsu Y. Tanaka;六月蜜柑(六月橙);Rokugatsu ●☆

93753 Citrus rugulosa Y. Tanaka;阿塔尼柚;Attani ●☆

93754 Citrus sabon Siebold;斗柚●

93755 Citrus sabon Siebold = Citrus maxima (Burm.) Merr. ●

93756 Citrus sabon Siebold ex Hayata = Citrus maxima (Burm.) Merr. ●

93757 Citrus sarbati Y. Tanaka;撒巴体;Sarbati ●☆

93758 Citrus sarcodactylis Noot. = Citrus medica L. 'Sarcodactylis' ●

93759 Citrus sarcodactylis Noot. = Citrus medica L. ●

93760 Citrus sauvissima Y. Tanaka = Citrus reticulata Blanco 'Suavissima' ●

93761 Citrus sccosa Y. Tanaka = Citrus reticulata Blanco 'Succosa' ●

93762 Citrus sechen subsp. sjanshen Kokaya = Citrus junos (Siebold) Siebold ex Tanaka ●☆

93763 Citrus semperflorans Lush.;赛达发耳;Sadaphal ●☆

93764 Citrus shunkokan Y. Tanaka;春光柑;Shunkokan ●☆

93765 Citrus sinensis (L.) Osbeck;橙(柑橙,广柑,广橘,黄果,甜橙,雪柑,印子柑);Blood Orange, China Orange, Chinois, Malta Orange, Orange, Orange Tree, Pelitgrain Oil, Portuguese Orange, Sweet Orange ●

93766 Citrus sinensis (L.) Osbeck 'Cadenera';卡特尼拉●

93767 Citrus sinensis (L.) Osbeck 'Dahong Cheng';大红橙●
93768 Citrus sinensis (L.) Osbeck 'Doublefine Amelioree';西班牙血橙●
93769 Citrus sinensis (L.) Osbeck 'Egyptian Blood';埃及血橙●
93770 Citrus sinensis (L.) Osbeck 'Gailiang Cheng';改良橙●
93771 Citrus sinensis (L.) Osbeck 'Hamlin';哈姆林●
93772 Citrus sinensis (L.) Osbeck 'Huangguo';黄果●
93773 Citrus sinensis (L.) Osbeck 'Huazhou Cheng';化州橙●
93774 Citrus sinensis (L.) Osbeck 'Jaffa';贾发●
93775 Citrus sinensis (L.) Osbeck 'Jin Cheng';锦橙●
93776 Citrus sinensis (L.) Osbeck 'Joppa';乔伯●
93777 Citrus sinensis (L.) Osbeck 'Liu Cheng';柳橙●
93778 Citrus sinensis (L.) Osbeck 'Lue Gim Gong';雷建刚●
93779 Citrus sinensis (L.) Osbeck 'Maltais';马尔台斯血橙●
93780 Citrus sinensis (L.) Osbeck 'Pushi Cheng';浦市橙●
93781 Citrus sinensis (L.) Osbeck 'Robertson';罗伯生脐橙●
93782 Citrus sinensis (L.) Osbeck 'Ruby';红玉血橙●
93783 Citrus sinensis (L.) Osbeck 'Taoye Cheng';桃叶橙●
93784 Citrus sinensis (L.) Osbeck 'Thomson';汤姆生脐橙●
93785 Citrus sinensis (L.) Osbeck 'Valencia';付令夏橙;Valencia Orange ●
93786 Citrus sinensis (L.) Osbeck 'Washington Navel';华盛顿脐橙●
93787 Citrus sinensis (L.) Osbeck 'Xinhui Cheng';新会橙●
93788 Citrus sinensis (L.) Osbeck 'Xue Cheng';雪橙●
93789 Citrus sinensis (L.) Osbeck = Citrus aurantium L. ●
93790 Citrus sinensis (L.) Osbeck = Citrus maxima (Burm. ex Rumph.) Merr. ●
93791 Citrus sinensis (L.) Osbeck f. kanton Y. Tanaka = Citrus sinensis (L.) Osbeck 'Xinhui Cheng' ●
93792 Citrus sinensis (L.) Osbeck f. sekkan Hayata = Citrus sinensis (L.) Osbeck 'Xue Cheng' ●
93793 Citrus sinensis (L.) Osbeck subsp. crassa (Risso) Rivera et al. = Citrus aurantium L. ●
93794 Citrus sinensis (L.) Osbeck subsp. fetifera (Risso) Rivera et al. = Citrus aurantium L. ●
93795 Citrus sinensis (L.) Osbeck subsp. lusitanica (Risso) Rivera et al. = Citrus aurantium L. ●
93796 Citrus sinensis (L.) Osbeck subsp. suntara (Engl.) Engl. = Citrus aurantium L. ●
93797 Citrus sinensis (L.) Osbeck var. brassiliensis Tanaka;美国脐橙(花旗蜜橘,华盛顿脐橙,美橘,脐橙);Navel Orange, Washington Navel Orange ●
93798 Citrus sinensis (L.) Osbeck var. brassiliensis Tanaka = Citrus aurantium L. ●
93799 Citrus sinensis (L.) Osbeck var. liucheng Hort. = Citrus sinensis (L.) Osbeck 'Liu Cheng' ●
93800 Citrus sinensis (L.) Osbeck var. sanguinea (Engl.) Engl. = Citrus aurantium L. ●
93801 Citrus sinensis (L.) Osbeck var. sekkan Hayata;雪柑●
93802 Citrus sinensis (L.) Osbeck var. sekkan Hayata = Citrus aurantium L. ●
93803 Citrus sinograndis Y. Tanaka;大唐柑(大唐蜜柑);Chinese Pummelo ●☆
93804 Citrus southwickii Wester = Citrus celebica Koord. var. southwickii (Wester) Swingle ●☆
93805 Citrus sphaerocarpa Y. Nakaj. ex H. Ohba;卡抱斯;Cabosu ●☆
93806 Citrus suavissima Y. Tanaka;瓯橘(长果瓯柑,柑,柑子,金实,木奴,瓯柑,瑞金奴,圆果瓯柑);Pleasant Orange ●☆
93807 Citrus suavissima Y. Tanaka = Citrus reticulata Blanco 'Suavissima' ●
93808 Citrus subcompressa (Tanaka) Tanaka = Citrus reticulata Blanco 'Subcompressa' ●
93809 Citrus subcompressa Tanaka = Citrus compressa Y. Tanaka ●
93810 Citrus subcompressa Tanaka = Citrus reticulata Blanco 'Subcompressa' ●
93811 Citrus sucata Tanaka = Citrus sulcata Takah. ●☆
93812 Citrus succosa Tanaka = Citrus reticulata Blanco 'Succosa' ●
93813 Citrus succosa Tanaka = Citrus reticulata Blanco ●
93814 Citrus sudachi Shirai;苏打其柑橘(酢橘,德岛酸橘,史达橘)●☆
93815 Citrus suhuiensis (Tanaka) Tseng;四会柑(四会十月橘);Suhui Orange, Szu-ui-kan ●
93816 Citrus suhuiensis Hayata = Citrus reticulata Blanco ●
93817 Citrus suhuiensis Tanaka = Citrus reticulata Blanco 'Hanggan' ●
93818 Citrus suizabon Y. Tanaka;素强抛●☆
93819 Citrus sukata Tanaka = Citrus sulcata Takah. ●☆
93820 Citrus sulcata Takah.;三宝柑;Grooved Orange, Sanbokan ●☆
93821 Citrus sunki Tanaka = Citrus reticulata Blanco var. sunki (Hayata) Hu ●
93822 Citrus sunki Tanaka = Citrus reticulata Blanco ●
93823 Citrus tachibana (Makino) C. Tanaka;立花橘(番橘,橘,橘柑,橘仔,日本橘,日本立花橘,山橘,台湾香檬);Japan Citrus, Japanese Tachibana, Tachibana, Taiwan Lemon ●
93824 Citrus tachibana (Makino) C. Tanaka = Citrus reticulata Blanco ●
93825 Citrus tachibana (Makino) C. Tanaka subf. depressa (Hayata) Hiroe = Citrus reticulata Blanco ●
93826 Citrus tachibana (Makino) C. Tanaka subf. ponki (Hayata) Hiroe = Citrus reticulata Blanco ●
93827 Citrus tachibana (Makino) C. Tanaka subf. suhuiensis (Hayata) Hiroe = Citrus reticulata Blanco ●
93828 Citrus tachibana (Makino) C. Tanaka subf. sunki (Hayata) Hiroe = Citrus reticulata Blanco ●
93829 Citrus tachibana (Makino) C. Tanaka var. attenuata C. Tanaka;太尼蒲太●☆
93830 Citrus tagerina Y. Tanaka = Citrus reticulata Blanco 'Tangerina' ●
93831 Citrus taiwanica Tanaka et Shimada = Citrus aurantium L. 'Taiwanica' ●
93832 Citrus taiwanica Tanaka et Shimada = Citrus aurantium L. ●
93833 Citrus takuma-sudachi Y. Tanaka;田熊史达橘;Naoshichi, Takuma-sudachi ●☆
93834 Citrus tamurana Takah.;小夏蜜柑(日向夏,日向夏蜜柑,新夏橙);Hyuganatsu ●☆
93835 Citrus tanakan Hayata = Citrus reticulata Blanco 'Tankan' ●
93836 Citrus tangelo Ingram et H. E. Moore = Citrus aurantium L. ●
93837 Citrus tangerina Tanaka;红橘(大红橘,丹西橘,福橘,福州蜜橘,红柑);Dancy Tangerine, Tangerine, Tangerine Orange ●
93838 Citrus tangerina Y. Tanaka = Citrus reticulata Blanco 'Tangerina' ●
93839 Citrus tangerina Y. Tanaka = Citrus reticulata Blanco ●
93840 Citrus tankan Hayata = Citrus reticulata Blanco 'Tankan' ●
93841 Citrus tankan Hayata = Citrus reticulata Blanco ●
93842 Citrus tankan Hayata f. hairi Hort.;海梨柑●
93843 Citrus tankan Hayata f. koshotankan Hort.;高墙桶柑●
93844 Citrus tardiferax Y. Tanaka;槾橘;Manju Mandarine, Mankich, Mankieh ●☆

93845 Citrus tardiferax Y. Tanaka = Citrus reticulata Blanco 'Tardiferax' ●
93846 Citrus tardiva Y. Tanaka;奇利蜜柑●☆
93847 Citrus tarogayo Y. Tanaka;太罗额柚;Tarogayo ●☆
93848 Citrus temple Y. Tanaka;寺院柑;Temple Orange ●☆
93849 Citrus tengu Tanaka;天狗橘(天狗);Tengu ●☆
93850 Citrus tosa-asahi Y. Tanaka;土佐旭;Tosa-asahi ●☆
93851 Citrus trifolia Thunb. = Citrus trifoliata L. ●
93852 Citrus trifolia Thunb. = Poncirus trifoliata (L.) Raf. ●
93853 Citrus trifoliata L. = Poncirus trifoliata (L.) Raf. ●
93854 Citrus trifoliata L. var. monstrosa T. Ito = Poncirus trifoliata (L.) Raf. ●
93855 Citrus triptera Desf. = Citrus trifoliata L. ●
93856 Citrus triptera Desf. = Poncirus trifoliata (L.) Raf. ●
93857 Citrus truncata Y. Tanaka;海红柑;Igorot lemon,Truncate Orange ●☆
93858 Citrus tuberosa Mill. = Citrus medica L. ●
93859 Citrus tumida Y. Tanaka;福克雷蜜柑;Fukure-mikan ●☆
93860 Citrus ujukitsu Y. Tanaka;宇树橘(宝来柑);Ujukitsu ●☆
93861 Citrus unshiu (Swingle) F. P. Metcalf = Citrus reticulata Blanco 'Unshiu' ●
93862 Citrus unshiu Marcow. = Citrus reticulata Blanco ●
93863 Citrus vangasy Bojer. = Citrus reticulata Blanco ●
93864 Citrus verrucosa Y. Tanaka = Citrus reticulata Blanco 'Manau Gan' ●
93865 Citrus vitiensis Y. Tanaka;维提橘(莫利古利古利);Molikurikur ●☆
93866 Citrus volkameriana Pasq.;沃尔卡默柠檬;Volkamer Lemon ●☆
93867 Citrus vulgaris Risso = Citrus aurantium L. ●
93868 Citrus webberi Wester;魏橘(山来檬,魏朴来母);Alsem, Mountain Lime ●☆
93869 Citrus wilsonii Y. Tanaka;香圆(陈香圆,枸橼,花柚,撒慕耶欧,香黄,宜昌柠檬);Hanayu,Ichang Lemon,Samuyao,Xiangyuan ●☆
93870 Citrus wilsonii Y. Tanaka = Citrus hanayu Shirai ●☆
93871 Citrus yamabukii Y. Tanaka;山吹蜜柑(山吹);Yamabuki Mandarin ●☆
93872 Citrus yanbaruensis Y. Tanaka;油克尼蒲●☆
93873 Citrus yatsushiro Y. Tanaka;八代蜜柑(八代);Yatsushiro ●☆
93874 Citrus yuge-hyokan Y. Tanaka;弓削瓢柑;Yuge-hyokan ●☆
93875 Citrus yuko Y. Tanaka;玉克柑橘(柚柑);Yuko,Yuko Citrus ●☆
93876 Citta Lour. = Mucuna Adans.(保留属名)●■
93877 Citta nigricans Lour. = Mucuna nigricans (Lour.) Steud. ●■
93878 Cittaronium Rchb. = Viola L. ■●
93879 Cittorhinchus Willd. ex Kunth = Ouratea Aubl.(保留属名)●
93880 Cittorhynchus Post et Kuntze = Cittorhinchus Willd. ex Kunth ●
93881 Cladandra O. F. Cook = Chamaedorea Willd.(保留属名)●☆
93882 Cladanthus (L.) Cass.(1816);羽叶香菊属(金凤菊属,枝花菊属);Cladanthus ●■☆
93883 Cladanthus Cass. = Cladanthus (L.) Cass. ●■☆
93884 Cladanthus arabicus (L.) Cass.;羽叶香菊(金凤菊,枝花菊);Palm Springs Daisy ●☆
93885 Cladanthus arabicus Cass. = Cladanthus arabicus (L.) Cass. ●☆
93886 Cladanthus eriolepis (Maire) Oberpr. et Vogt;毛鳞羽叶香菊●☆
93887 Cladanthus flahaultii (Emb.) Oberpr. et Vogt;弗拉奥羽叶香菊 ●☆
93888 Cladanthus geslinii Coss. = Mecomischus halimifolius (Munby) Hochr. ■☆
93889 Cladanthus ifniensis Caball. = Cladanthus arabicus (L.) Cass. ●☆
93890 Cladanthus mixtus (L.) Chevall. = Cladanthus mixtus (L.) Oberpr. et Vogt ●☆
93891 Cladanthus mixtus (L.) Oberpr. et Vogt;混杂羽叶香菊●☆
93892 Cladanthus pedunculatus Coss. et Durieu = Mecomischus pedunculatus (Coss. et Durieu) Oberpr. et Greuter ■☆
93893 Cladanthus proliferus DC. = Cladanthus arabicus (L.) Cass. ●☆
93894 Cladanthus scariosus (Ball) Oberpr. et Vogt;干膜质羽叶香菊●☆
93895 Cladapus Moeller(1899);爪哇川苔草属■☆
93896 Cladapus Tbis.-Dyer = Cladopus H. Möller ■
93897 Claderia Hook. f.(1890)(保留属名);绿花脆兰属■☆
93898 Claderia Raf.(废弃属名)= Claderia Hook. f.(保留属名)■☆
93899 Claderia Raf.(废弃属名)= Murraya J. König ex L.(保留属名)●
93900 Claderia viridiflora Hook. f.;绿花脆兰■☆
93901 Cladium P. Browne(1756);克拉莎属(一本芒属);Cladium, Great Fen-sedge, Marisque, Saw Grass, Sawgrass, Saw-grass, Twig Rush,Twigrush,Twig-rush ■
93902 Cladium boninsimae Nakai = Machaerina glomerata (Gaudich.) T. Koyama ■☆
93903 Cladium californicum (S. Watson) O'Neill;加州克拉莎■☆
93904 Cladium chinense Nees = Cladium jamaicense Crantz subsp. chinense (Nees) T. Koyama ■
93905 Cladium chinense Nees et Arn. = Cladium jamaicense Crantz ■
93906 Cladium chinense Nees ex Hook. et Arn. = Cladium jamacense Crantz subsp. chinense (Nees) T. Koyama ■
93907 Cladium durandoi Chabert = Cladium mariscus (L.) Pohl ■☆
93908 Cladium ensigerum Hance;剑叶克拉莎;Swordleaf Cladium, Swordleaf Twigrush ■
93909 Cladium ensigerum Hance = Machaerina ensigera (Hance) T. Koyama ■
93910 Cladium flexuosum (Boeck.) C. B. Clarke var. polyanthemum Kük. = Machaerina flexuosa (Boeck.) Kern subsp. polyanthemum (Kük.) Lye ■☆
93911 Cladium glomeratum C. B. Clarke = Cladium nipponense Ohwi ■
93912 Cladium jamaicense C. B. Clarke = Cladium chinense Nees ■
93913 Cladium jamaicense Crantz;克拉莎(华克拉莎,一本芝);China Twigrush,Chinese Cladium,Elk Sedge,Fen Sedge,Great Fen-sedge, Saw-grass,Sedge,Sticky Grass ■
93914 Cladium jamaicense Crantz = Cladium mariscus (L.) Pohl subsp. jamaicense (Crantz) Kük. ■
93915 Cladium jamaicense Crantz subsp. chinense (Nees) T. Koyama;华克拉莎(一本莎)■
93916 Cladium jamaicense Crantz subsp. chinense (Nees) T. Koyama = Cladium jamaicense Crantz ■
93917 Cladium jamaicense Crantz var. chinense (Nees) T. Koyama = Cladium jamaicense Crantz subsp. chinense (Nees) T. Koyama ■
93918 Cladium mariscoides (Muhl.) Torr.;光克拉莎;Smooth Sawgrass,Twig-rush ■☆
93919 Cladium mariscus (L.) Pohl;小沟克拉莎;Elk Sedge ■☆
93920 Cladium mariscus (L.) Pohl subsp. jamaicense (Crantz) Kük. = Cladium jamaicense Crantz ■
93921 Cladium mariscus (L.) Pohl var. californicum S. Watson = Cladium californicum (S. Watson) O'Neill ■☆
93922 Cladium mariscus (L.) R. Br. = Cladium mariscus (L.) Pohl ■☆
93923 Cladium mariscus Benth. = Cladium chinense Nees ■
93924 Cladium mariscus L. = Cladium jamaicense Crantz ■

93925 Cladium myrianthum Chun et F. C. How;多花克拉莎(多花一本芒);Manyflower Cladium,Manyflower Twigrush ■

93926 Cladium myrianthum Chun et F. C. How = Machaerina myriantha (Chun et F. C. How) Y. C. Tang ■

93927 Cladium nipponense Ohwi;毛喙克拉莎;Japan Twigrush, Japanese Cladium ■

93928 Cladium undulatum Henry = Schoenus falcatus R. Br. ■

93929 Cladium undulatum Thwaites = Tricostularia undulate (Thwaites) Kern ■

93930 Cladobium Lindl. = Scaphyglottis Poepp. et Endl.(保留属名)■☆

93931 Cladobium Schltr. = Lankesterella Ames ■☆

93932 Cladobium Schltr. = Stenorrhynchos Rich. ex Spreng. ■☆

93933 Cladocarpa (St. John) St. John = Sicyos L. ■

93934 Cladocaulon Gardn. = Paepalanthus Kunth(保留属名)■☆

93935 Cladoceras Bremek.(1940);弓枝茜属●☆

93936 Cladoceras subcapitatum (K. Schum. et K. Krause) Bremek.;弓枝茜●☆

93937 Cladochaeta DC.(1838);毛果棕鼠麹属■☆

93938 Cladochaeta candidissima DC.;白毛果棕鼠麹■☆

93939 Cladochaeta caspica Sosn.;毛果棕鼠麹■☆

93940 Cladochaeta velutina Anderb.;短毛毛果棕鼠麹■☆

93941 Cladocolea Tiegh.(1895);鞘枝寄生属●☆

93942 Cladocolea Tiegh. = Oryctanthus (Griseb.) Eichler ●☆

93943 Cladocolea andrieuxii Tiegh.;鞘枝寄生●☆

93944 Cladoda (Cladodea) Poir. = Cladodes Lour. ●

93945 Cladoda Poir = Cladodes Lour. ●

93946 Cladodea Poir. = Cladodes Lour. ●

93947 Cladodes Lour. = Alchornea Sw. ●

93948 Cladodes rugosa Lour. = Alchornea rugosa (Lour.) Müll. Arg. ●

93949 Cladogelonium Léandri(1939);枝白树属●☆

93950 Cladogelonium madagascariense Léandri;枝白树●☆

93951 Cladogynos Zipp. ex Span.(1841);枝实属(白大凤属);Cladogynos ●

93952 Cladogynos orientalis Zipp. ex Span.;枝实(白大凤);Oriental Cladogynos ●

93953 Cladogynos orientalis Zipp. ex Span. var. tonkinensis Gagnep.;越南枝实;Tonkin Cladogynos,Vietnam Cladogynos ●

93954 Cladolepis Moq. = Ofaiston Raf. ■☆

93955 Cladomisehus Klotzsch ex A. DC. = Begonia L. ●■

93956 Cladomyza Danser(1940);枝寄生属●☆

93957 Cladomyza acrosclera Danser;枝寄生●☆

93958 Cladomyza acutata (Pilg.) Danser;尖枝寄生●☆

93959 Cladomyza crassifolia (Gibbs) Danser;厚叶枝寄生●☆

93960 Cladomyza gracilis Danser;纤细枝寄生●☆

93961 Cladomyza laevis (Pilg.) Danser;平滑枝寄生●☆

93962 Cladomyza microphylla (Lauterb.) Danser;小叶枝寄生●☆

93963 Cladomyza multinervis Danser;多脉枝寄生●☆

93964 Cladomyza trinervia Stauffer;三脉枝寄生●☆

93965 Cladophyllaceae Dulac = Dioscoreaceae R. Br.(保留科名)●■

93966 Cladopogon Sch. Bip. = Senecio L. ■●

93967 Cladopogon Sch. Bip. ex Lehm. = Senecio L. ■●

93968 Cladopus H. Möller(1899);飞瀑草属(川苔草属,河苔草属);Cladopus ■

93969 Cladopus austro-osumiensis Kadono et N. Usui;南大隅飞瀑草■☆

93970 Cladopus austro-satsumensis (Koidz.) Ohwi;南萨摩飞瀑草■☆

93971 Cladopus chinensis (H. C. Chao) H. C. Chao;川苔草■

93972 Cladopus doianus (Koidz.) Koriba;道氏飞瀑草(道氏苔草)■☆

93973 Cladopus fukienensis (H. C. Chao) H. C. Chao = Cladopus nymani H. Moller ■

93974 Cladopus japonicus Imamura = Cladopus nymani H. Moller ■

93975 Cladopus nymani H. Moller;飞瀑草(川苔草,日本飞瀑草);Common Cladopus,Nyman Cladopus ■

93976 Cladoraphis Franch.(1887);木本画眉草属●☆

93977 Cladoraphis Franch. = Eragrostis Wolf ■

93978 Cladoraphis cyperoides (Thunb.) S. M. Phillips;木本画眉草;Bristly Lovegrass ●☆

93979 Cladoraphis spinosa (L. f.) S. M. Phillips;毛木本画眉草●☆

93980 Cladorhiza Raf. = Corallorhiza Gagnebin(保留属名)■

93981 Cladorhiza maculata Raf. = Corallorhiza maculata (Raf.) Raf. ■☆

93982 Cladoseris Spach = Onoseris Willd. ●■☆

93983 Cladosicyos Hook. f. = Cucumeropsis Naudin ■☆

93984 Cladosicyos edulis Hook. f. = Cucumeropsis mannii Naudin ■☆

93985 Cladosperma Griff.(1851);叉子棕属●☆

93986 Cladosperma Griff. = Pinanga Blume ●

93987 Cladostachys D. Don = Deeringia R. Br. ●■

93988 Cladostachys amaranthoides (Lam.) K. C. Kuan = Deeringia amaranthoides (Lam.) Merr. ●

93989 Cladostachys frutescens D. Don. = Deeringia amaranthoides (Lam.) Merr. ●

93990 Cladostachys polysperma (Roxb.) K. C. Kuan = Deeringia polysperma (Roxb.) Moq. ■

93991 Cladostemon A. Braun et Vatke(1877);枝蕊白花菜属●☆

93992 Cladostemon kirkii (Oliv.) Pax et Gilg;枝蕊白花菜●☆

93993 Cladostemon paradoxus A. Br. et Vatke = Cladostemon kirkii (Oliv.) Pax et Gilg ●☆

93994 Cladostemon paxianus Gilg = Cladostemon kirkii (Oliv.) Pax et Gilg ●☆

93995 Cladostigma Radlk.(1883);枝柱头旋花属■☆

93996 Cladostigma dioicum Radlk.;异株枝柱头旋花■☆

93997 Cladostigma hildebrandtioides Hallier f.;希尔德枝柱头旋花■☆

93998 Cladostyles Humb. et Bonpl. = Evolvulus L. ●■

93999 Cladothamnus Bong. = Elliottia Muhl. ex Elliott ●☆

94000 Cladothamnus bracteatus (Maxim.) T. Yamaz. = Tripetaleia bracteata Maxim. ●☆

94001 Cladotheea Steud. = Cryptangium Schrad. ex Nees ■☆

94002 Cladothrix (Moq.) Hook. f. = Tidestromia Standl. ■☆

94003 Cladothrix (Moq.) Nutt. ex Benth. et Hook. f. = Tidestromia Standl. ■☆

94004 Cladothrix (Nutt. ex Moq.) Benth. = Tidestromia Standl. ■☆

94005 Cladothrix Nutt. ex Hook. f. = Tidestromia Standl. ■☆

94006 Cladothrix Nutt. ex Moq. = Tidestromia Standl. ■☆

94007 Cladothrix Nutt. ex S. Watson = Tidestromia Standl. ■☆

94008 Cladothrix cryptantha S. Watson = Tidestromia suffruticosa (Torr.) Standl. var. oblongifolia (S. Watson) Sánch. Pino et Flores Olv. ●☆

94009 Cladothrix lanuginosa (Nutt.) Nutt. ex S. Watson = Tidestromia lanuginosa (Nutt.) Standl. ■☆

94010 Cladothrix lanuginosa (Nutt.) Nutt. ex S. Watson var. carnosa Steyerm. = Tidestromia carnosa (Steyerm.) I. M. Johnst. ■☆

94011 Cladothrix oblongifolia S. Watson = Tidestromia suffruticosa (Torr.) Standl. var. oblongifolia (S. Watson) Sánch. Pino et Flores Olv. ●☆

94012 Cladothrix suffruticosa (Torr.) Benth. et Hook. f. ex S. Watson = Tidestromia suffruticosa (Torr.) Standl. ●☆

94013 Cladotriehium Vogel = Caesalpinia L. ●

94014 Cladrastis Raf. (1824);香槐属;Yellow Wood, Yellowwood, Yellow-wood ●

94015 Cladrastis amurensis (Rupr. et Maxim.) Benth. ex Maxim. = Maackia amurensis Rupr. et Maxim. ●

94016 Cladrastis amurensis Benth. = Maackia amurensis Rupr. et Maxim. ●

94017 Cladrastis australis Dunn = Maackia australis (Dunn) Takeda ●

94018 Cladrastis chinensis?;支那香槐(中国香槐);Chinese Yellow-wood ●☆

94019 Cladrastis chingii Duley et Vincent;秦氏香槐●

94020 Cladrastis delavayi (Franch.) Prain;小花香槐(鸡足香槐,香槐);China Yellowwood, Chinese Yellowwood, Chinese Yellow-wood, Delavay Yellowwood ●

94021 Cladrastis fauriei H. Lév. = Maackia floribunda (Miq.) Takeda ●

94022 Cladrastis kentukea (Dum. Cours.) Rudd;美洲香槐(美国香槐);American Yellowwood, American Yellow-wood, Kentucky Yellowwood, Virgilia, Yellowwood ●☆

94023 Cladrastis kentukea (Dum. Cours.) Rudd = Virgilia lutea F. Michx. ●☆

94024 Cladrastis lichuanensis Q. W. Yao et G. G. Tang;利川香槐;Lichuan Yellowwood ●

94025 Cladrastis lichuanensis Q. W. Yao et G. G. Tung = Cladrastis wilsonii Takeda ●

94026 Cladrastis lutea (F. Michx.) K. Koch;黄香槐(黄维吉尔豆);American Yellowwood, American Yellow-wood, Kentucky, Kentucky Yellow Wood, Kentucky Yellow-wood, Yellow Ash, Yellow Wood, Yellowwood, Yellow-wood ●☆

94027 Cladrastis lutea (F. Michx.) K. Koch = Cladrastis kentukea (Dum. Cours.) Rudd ●☆

94028 Cladrastis lutea (F. Michx.) K. Koch = Virgilia lutea F. Michx. ●☆

94029 Cladrastis parvifolia C. Y. Ma;小叶香槐;Smallleaf Yellowwood, Small-leaved Yellowwood ●

94030 Cladrastis platycarpa (Maxim.) Makino;翅荚香槐;Broad-fruit Yellowwood, Flatfruited Yellowwood, Flat-fruited Yellowwood, Japanese Yellowwood, Japanese Yellow-wood ●

94031 Cladrastis scandens C. Y. Ma;藤香槐;Climbing Yellowwood, Vine Yellowwood ●

94032 Cladrastis shikokiana (Makino) Makino;四国香槐;Sikoku Yellowwood ●☆

94033 Cladrastis sinensis Hemsl. = Cladrastis delavayi (Franch.) Prain ●

94034 Cladrastis tinctoria Raf. = Cladrastis lutea (F. Michx.) K. Koch ●☆

94035 Cladrastis wilsonii Takeda;香槐(山荆,威氏槐,香近豆);Wilson Yellowwood ●

94036 Cladrastis yungchunii Xiang W. Li et G. S. Fan;永椿香槐;Yongchun Yellowwood ●

94037 Cladrastis yungchunii Xiang W. Li et G. S. Fan = Cladrastis platycarpa (Maxim.) Makino ●

94038 Clairisia Abat ex Benth. et Hook. f. = Anredera Juss. ●■

94039 Clairisia Benth. et Hook. f. = Anredera Juss. ●■

94040 Clairvillea DC. = Cacosmia Kunth ●☆

94041 Clambus Miers = Phyllanthus L. ●■

94042 Clamydanthus Fourr. = Chlamydanthus C. A. Mey. ●■

94043 Clamydanthus Fourr. = Thymelaea Mill. (保留属名)●■

94044 Clandestina Adans. = Lathraea L. ■

94045 Clandestina Hill = Lathraea L. ■

94046 Clandestina Tourn. ex Adans. = Lathraea L. ■

94047 Clandestina japonica Miq. = Lathraea japonica Miq. ■

94048 Clandestinaria Spach = Rorippa Scop. ■

94049 Clandestinaria indica (L.) Spach = Rorippa indica (L.) Hiern ■

94050 Claotrachelus Zoll. = Vernonia Schreb. (保留属名)●■

94051 Claoxylon A. Juss. (1824);白桐树属(假铁苋属,咸鱼头属);Claoxylon, Whitetung ●

94052 Claoxylon angolense Müll. Arg. = Erythrococca angolensis (Müll. Arg.) Prain ●☆

94053 Claoxylon atrovirens Pax = Erythrococca atrovirens (Pax) Prain ●☆

94054 Claoxylon barteri Hook. f. = Erythrococca africana (Baill.) Prain ●☆

94055 Claoxylon brachyandrum Pax et K. Hoffm.;台湾白桐树(假铁苋);Claoxylon, Taiwan Claoxylon, Taiwan Whitetung ●

94056 Claoxylon capense Baill. = Micrococca capensis (Baill.) Prain ●☆

94057 Claoxylon centinarium Koidz.;小泉白桐树●☆

94058 Claoxylon chevalieri Beille = Erythrococca chevalieri (Beille) Prain ●☆

94059 Claoxylon columnare = Erythrococca columnaris (Müll. Arg.) Prain ●☆

94060 Claoxylon deflersii Schweinf. ex Pax et K. Hoffm. = Erythrococca abyssinica Pax ●☆

94061 Claoxylon dewevrei Pax ex De Wild. = Erythrococca dewevrei (Pax ex De Wild.) Prain ●☆

94062 Claoxylon flaccidum Pax = Erythrococca atrovirens (Pax) Prain var. flaccida (Pax) Radcl. -Sm. ●☆

94063 Claoxylon hainanensis Pax et K. Hoffm.;海南白桐树;Hainan Claoxylon, Hainan Whitetung ●

94064 Claoxylon hexandrum Müll. Arg. = Discoclaoxylon hexandrum (Müll. Arg.) Pax et K. Hoffm. ●☆

94065 Claoxylon hispidum Pax = Erythrococca hispida (Pax) Prain ●☆

94066 Claoxylon holstii Pax = Micrococca holstii (Pax) Prain ●☆

94067 Claoxylon inaequilaterum Pax = Erythrococca atrovirens (Pax) Prain ●☆

94068 Claoxylon indicum (Reinw. ex Blume) Hassk.;白桐树(臭平桐,大叶大青,刁了棒,丢了棒,赶风柴,赶风债,泡平桐,咸鱼头,追风棍);Common Claoxylon, India Whitetung, Indian Claoxylon ●

94069 Claoxylon khasianum Hook. f.;喀西白桐树(喀西咸鱼头);Khas Whitetung, Khasia Claoxylon ●

94070 Claoxylon kirkii Müll. Arg. = Erythrococca kirkii (Müll. Arg.) Prain ●☆

94071 Claoxylon kotoense Hayata = Claoxylon brachyandrum Pax et K. Hoffm. ●

94072 Claoxylon lasiococcum Pax = Erythrococca trichogyne (Müll. Arg.) Prain ●☆

94073 Claoxylon longifolium (Blume) Endl. ex Hassk.;长叶白桐树(长叶咸鱼头);Longleaf Claoxylon, Longleaf Whitetung, Long-leaved Claoxylon ●

94074 Claoxylon macrophyllum Prain = Erythrococca macrophylla (Prain) Prain ●☆

94075 Claoxylon mannii Hook. f. = Erythrococca mannii (Hook. f.) Prain ●☆

94076 Claoxylon membranaceum Müll. Arg. = Erythrococca membranacea (Müll. Arg.) Prain ●☆

94077 Claoxylon menyharthii Pax = Erythrococca menyharthii (Pax) Prain ●☆

94078 Claoxylon mercurialis (L.) Thwaites = Micrococca mercurialis

(L.) Benth. ●☆
94079 Claoxylon mildbraedii Pax = Erythrococca trichogyne (Pax) Prain ●☆
94080 Claoxylon molleri Pax = Erythrococca molleri (Pax) Prain ●☆
94081 Claoxylon oleraceum Prain = Erythrococca atrovirens (Pax) Prain var. flaccida (Pax) Radcl. -Sm. ●☆
94082 Claoxylon oligandrum Müll. Arg. = Micrococca oligandra (Müll. Arg.) Prain ●☆
94083 Claoxylon parviflorum Hook. et Arn.;小花白桐树●
94084 Claoxylon parviflorum Hook. et Arn. = Claoxylon indicum (Reinw. ex Blume) Hassk. ●
94085 Claoxylon patulum Prain = Erythrococca patula (Prain) Prain ●☆
94086 Claoxylon pauciflorum Müll. Arg. = Erythrococca pauciflora (Müll. Arg.) Prain ●☆
94087 Claoxylon pedicellare Müll. Arg. = Discoclaoxylon pedicellare (Müll. Arg.) Pax et K. Hoffm. ●☆
94088 Claoxylon poggei Prain = Erythrococca poggei (Prain) Prain ●☆
94089 Claoxylon polot (Burm.) Merr. = Claoxylon indicum (Reinw. ex Blume) Hassk. ●
94090 Claoxylon polyandrum Pax et K. Hoffm. = Erythrococca polyandra (Pax et K. Hoffm.) Prain ●☆
94091 Claoxylon preussii Pax = Discoclaoxylon hexandrum (Pax) Pax et K. Hoffm. ●☆
94092 Claoxylon purpurascens Beille = Erythrococca molleri (Pax) Prain ●☆
94093 Claoxylon rivulare Müll. Arg. = Erythrococca rivularis (Müll. Arg.) Prain ●☆
94094 Claoxylon rubescens Miq. = Claoxylon brachyandrum Pax et K. Hoffm. ●
94095 Claoxylon schweinfurthii Pax = Erythrococca atrovirens (Pax) Prain ●☆
94096 Claoxylon sphaerocarpum Kuntze = Croton sylvaticum Hochst. ex Krauss ●☆
94097 Claoxylon trichogyne Müll. Arg. = Erythrococca trichogyne (Müll. Arg.) Prain ●☆
94098 Claoxylon triste Müll. Arg. = Erythrococca neglecta Pax et K. Hoffm. ●☆
94099 Claoxylon virens N. E. Br. = Erythrococca menyharthii (Pax) Prain ●☆
94100 Claoxylon volkensii Pax = Micrococca volkensii (Pax) Prain ●☆
94101 Claoxylon welwitschianum Müll. Arg. = Erythrococca welwitschiana (Müll. Arg.) Prain ●☆
94102 Claoxylopsis Léandri(1939);拟白桐树属●☆
94103 Claoxylopsis andapensis Radcl. -Sm.;安达帕拟白桐树●☆
94104 Claoxylopsis perrieri Léandri;佩里耶拟白桐树●☆
94105 Claoxylopsis purpurascens Radcl. -Sm.;紫拟白桐树●☆
94106 Clappertonia Meisn.(1837);合头椴属(克拉椴属)●☆
94107 Clappertonia ficifolia (Willd.) Decne.;合头椴;Bolo Bolo ●☆
94108 Clappertonia minor (Baill.) Bech.;小合头椴●☆
94109 Clappertonia polyandra (K. Schum. ex Sprague) Bech.;多蕊合头椴●☆
94110 Clappia A. Gray(1859);盐菊属●☆
94111 Clappia suaedifolia Wooton et Standl.;盐菊■☆
94112 Clara Kunth = Herreria Ruiz et Pav. ■☆
94113 Clarckia Pursh = Clarkia Pursh ■
94114 Clariona Spreng. = Clarionea Lag. ex DC. ■☆
94115 Clarionea Lag. = Perezia Lag. ■☆
94116 Clarionea Lag. ex DC. = Perezia Lag. ■☆
94117 Clarionella DC. ex Steud. = Clarionea Lag. ■☆
94118 Clarionema Phil. = Clarionea Lag. ex DC. ■☆
94119 Clarionia D. Don = Clarionea Lag. ex DC. ■☆
94120 Clarisia Abat(废弃属名) = Anredera Juss. ●■
94121 Clarisia Abat(废弃属名) = Clarisia Ruiz et Pav.(保留属名)●☆
94122 Clarisia Ruiz et Pav.(1794)(保留属名);无被桑属(克拉桑属)●☆
94123 Clarisia Ruiz, Pav. et Lanj. = Clarisia Ruiz et Pav.(保留属名)●☆
94124 Clarisia nitida J. F. Macbr.;光亮无被桑(光亮克拉桑)●☆
94125 Clarisia racemosa Ruiz et Pav.;总花无被桑(总花克拉桑)●☆
94126 Clarkeasia J. R. I. Wood = Echinacanthus Nees ●■
94127 Clarkeasia J. R. I. Wood(1994);喜马拉雅恋岩花属●☆
94128 Clarkeifedia Kuntze = Patrinia Juss.(保留属名)■
94129 Clarkella Hook. f.(1880);岩上珠属(矮独叶属);Rockpearl ■
94130 Clarkella nana (Edgew.) Hook. f.;岩上珠(矮独叶);Rockpearl ■
94131 Clarkella siamensis Craib = Clarkella nana (Edgew.) Hook. f. ■
94132 Clarkia Pursh(1814);克拉花属(春再来属,古代稀属,山字草属);Clarkia, Farewell-to-spring, Godetia ■
94133 Clarkia amoena (Lehm.) A. Nelson et J. F. Macbr.;可爱克拉花(古代稀,可爱春再来,愉悦山字草);Farewell to Spring, Farewell-to-spring, Godetia, Satin Flower, Summer's Darling ■☆
94134 Clarkia breweri Greene;仙女扇;Fairy Fans ■☆
94135 Clarkia concinna (Fisch. et C. A. Mey.) Greene;优雅克拉花(优雅春再来);Red Ribbons ■☆
94136 Clarkia concinna Greene = Clarkia concinna (Fisch. et C. A. Mey.) Greene ■☆
94137 Clarkia elegans Douglas = Clarkia elegans Douglas ex Lindl. ■☆
94138 Clarkia elegans Douglas ex Lindl.;山字草(粉妆花,粉粧花,绣衣花);Clarkia, Rose Clarkia ■☆
94139 Clarkia hybrida Hort.;杂种克拉花;Farewell-to-spring, Godetia, Rocky Mountain Garland ■☆
94140 Clarkia pulchella Pursh;克拉花(极美古代稀,美丽春再来,细叶山字草);Beautiful Clarkia ■
94141 Clarkia pulchella Pursh var. holopetala Voss;全萼春再来■☆
94142 Clarkia rhomboidea Douglas;菱形春再来■☆
94143 Clarkia superba A. Nelson et J. F. Macbr.;华美克拉花■☆
94144 Clarkia unguiculata Lindl.;有爪春再来;Clarkia, Elegant Clarkia ■☆
94145 Clarorivinia Pax et K. Hoffm. = Ptychopyxis Miq. ●☆
94146 Clasta Comm. ex Vent. = Casearia Jacq. ●
94147 Clastilix Raf. = Miconia Ruiz et Pav.(保留属名)●☆
94148 Clastopus Bunge ex Boiss.(1867);克拉芥属■☆
94149 Clastopus bicolor Stapf;二色克拉芥■☆
94150 Clastopus purpureus Bunge;克拉芥■☆
94151 Clathrospermum Planch. = Enneastemon Exell(保留属名)●☆
94152 Clathrospermum Planch. ex Benth. = Monanthotaxis Baill. ●☆
94153 Clathrospermum Planch. ex Benth. et Hook. f. = Enneastemon Exell(保留属名)●☆
94154 Clathrospermum Planch. ex Hook. f.(废弃属名) = Enneastemon Exell(保留属名)●☆
94155 Clathrospermum baillonii Scott-Elliot = Monanthotaxis mannii (Baill.) Verdc. ●☆
94156 Clathrospermum biovulatum S. Moore = Monanthotaxis fornicata (Baill.) Verdc. ●☆
94157 Clathrospermum heudelotii (Baill.) Scott-Elliot = Monanthotaxis

barteri (Baill.) Verdc. ●☆

94158 Clathrospermum mannii Oliv. = Monanthotaxis cauliflora (Chipp) Verdc. ●☆

94159 Clathrospermum vogelii (Hook. f.) Planch. = Monanthotaxis vogelii (Hook. f.) Verdc. ●☆

94160 Clathrotropis (Benth.) Harms(1901);龙骨豆属(格瓣豆属)●☆

94161 Clathrotropis Harms = Clathrotropis (Benth.) Harms ●☆

94162 Clathrotropis brachypetala (Tul.) Kleinh.;短瓣龙骨豆(短格瓣豆)●☆

94163 Clathrotropis grandiflora (Tul.) Harms;大叶亮龙骨豆●☆

94164 Clathrotropis nitida Harms;光亮龙骨豆●☆

94165 Claucena Burm. f. = Clausena Burm. f. ●

94166 Claudia Opiz = Beckeria Bernh. ■

94167 Claudia Opiz = Melica L. ■

94168 Claudia Opiz ex Panz. = Claudia Opiz ■

94169 Clausena Burm. f. (1768);黄皮属(黄皮果属);Wampee ●

94170 Clausena abyssinica Engl.;阿比西尼亚黄皮●☆

94171 Clausena anisata (Willd.) Hook. f. ex Benth.;异黄皮●☆

94172 Clausena anisata (Willd.) Hook. f. ex Benth. subsp. abyssinica (Engl.) Cufod. = Clausena anisata (Willd.) Hook. f. ex Benth. ●☆

94173 Clausena anisata Hook. f.;八角黄皮●☆

94174 Clausena anisumolens (Blanco) Merr.;细叶黄皮(短柱黄皮);Thin-leaf Wampee ●

94175 Clausena brevistyla Oliv.;短柱黄皮(满山香);Shortstyle Wampee ●☆

94176 Clausena brevistyla Oliv. = Clausena anisumolens (Blanco) Merr. ●

94177 Clausena cordata?;心形黄皮;Pickerelweed, Wampee ●☆

94178 Clausena dentata (Willd.) M. Roem. = Clausena dunniana H. Lév. ●

94179 Clausena dentata H. Lév. = Clausena dunniana H. Lév. ●

94180 Clausena dentata M. Roem. = Clausena dunniana H. Lév. ●

94181 Clausena dentata M. Roem. var. dunniana Swingle = Clausena dunniana H. Lév. ●

94182 Clausena dentata M. Roem. var. henryi Swingle = Clausena dunniana H. Lév. var. robusta (Tanaka) C. C. Huang ●

94183 Clausena dentata M. Roem. var. henryi Swingle = Clausena henryi (Swingle) C. C. Huang ●

94184 Clausena dentata M. Roem. var. robusta Y. Tanaka = Clausena dunniana H. Lév. var. robusta (Tanaka) C. C. Huang ●

94185 Clausena dunniana H. Lév.;齿叶黄皮(假黄皮,接骨木,山黄皮,山茴香,野黄皮);Dunn Wampee, Toothleaf Wampee ●

94186 Clausena dunniana H. Lév. var. henryi Swingle = Clausena dunniana H. Lév. var. robusta (Tanaka) C. C. Huang ●

94187 Clausena dunniana H. Lév. var. robusta (Tanaka) C. C. Huang;毛齿叶黄皮(大型黄皮);Robust Dunn Wampee ●

94188 Clausena emarginata C. C. Huang;小黄皮(凹叶黄皮,假鸡皮果,山黄皮,山鸡皮,十里香);Emarginate Wampee, Notched Wampee, Small Wampee ●

94189 Clausena esquirolii H. Lév. = Glycosmis esquirolii (H. Lév.) Tanaka ●

94190 Clausena euchrestifolia Kaneh. = Murraya euchrestifolia Hayata ●

94191 Clausena excavata Burm. f.;假黄皮(半边枫,臭黄皮,臭麻木,臭皮树,大棵,番仔香草,龟里㷄,过山香,鸡母黄,假黄皮树,山黄皮,山鸡皮,五暑叶,小叶臭黄皮,野黄皮);Curvedleaf Wampee, Hollowed Wampee, Pink Lime-berry, Sham Wampee ●

94192 Clausena excavata Burm. f. var. lunulata Tanaka = Clausena excavata Burm. f. ●

94193 Clausena excavata Burm. f. var. quadrangulata Z. J. Yu et C. Y. Wong;四棱黄皮;Fourangular Hollowed Wampee ●

94194 Clausena ferruginea C. C. Huang;锈毛黄皮;Rustyhair Wampee ●

94195 Clausena ferruginea C. C. Huang = Glycosmis esquirolii (H. Lév.) Tanaka ●

94196 Clausena hainanensis C. C. Huang et F. W. Xing;海南黄皮;Hainan Wampee ●

94197 Clausena henryi (Swingle) C. C. Huang;川鄂黄皮(亨利黄皮);Henry Wampee ●

94198 Clausena henryi (Swingle) C. C. Huang = Clausena dunniana H. Lév. var. robusta (Tanaka) C. C. Huang ●

94199 Clausena heptaphylla (DC.) Steud.;七叶黄皮●☆

94200 Clausena hildebrandtii Engl. = Fagaropsis hildebrandtii (Engl.) Milne-Redh. ●☆

94201 Clausena inaequalis (DC.) Benth. = Clausena anisata (Willd.) Hook. f. ex Benth. ●☆

94202 Clausena inaequalis (DC.) Benth. var. abyssinica Engl. = Clausena anisata (Willd.) Hook. f. ex Benth. ●☆

94203 Clausena indica (Dalzell) Oliv.;小叶黄皮(鸡皮果,假黄皮,细叶黄皮,印度黄皮);Indian Wampee ●

94204 Clausena indica (Dalzell) Oliv. = Clausena anisumolens (Blanco) Merr. ●

94205 Clausena inolida Z. J. Yu et C. Y. Wong;丽达黄皮●

94206 Clausena kwangsiensis C. C. Huang = Murraya kwangsiensis (C. C. Huang) C. C. Huang ●

94207 Clausena lansium (Lour.) Skeels;黄皮(斑点黄皮,果子黄,黄弹,黄弹子,黄淡子,黄柑,黄皮果,黄皮子,黄枇,黄檀子,金弹子,王坛子,王檀子,油梅);Chinese Wampee, Punctate Wampee, Wampee, Wampi ●

94208 Clausena lenis Drake;光滑黄皮(鸡皮);Smooth Wampee ●

94209 Clausena loheri Merr. = Clausena anisumolens (Blanco) Merr. ●

94210 Clausena lunulata Hayata = Clausena excavata Burm. f. ●

94211 Clausena melioides Hiern;楝状黄皮●☆

94212 Clausena microphylla Merr. et Chun = Murraya microphylla (Merr. et Chun) Swingle ●

94213 Clausena moningerae Merr. = Clausena excavata Burm. f. ●

94214 Clausena odorata C. C. Huang;香花黄皮;Fragrant Wampee, Fragrantflower Wampee ●

94215 Clausena pentaphylla DC.;五叶黄皮●☆

94216 Clausena punctata Wight et Arn. = Clausena lansium (Lour.) Skeels ●

94217 Clausena suffruticosa Wight et Arn. = Clausena dunniana H. Lév. var. robusta (Tanaka) C. C. Huang ●

94218 Clausena tetramera Hayata = Clausena excavata Burm. f. ●

94219 Clausena vestita D. D. Tao;毛叶黄皮;Clothed Wampee, Hairyleaf Wampee ●

94220 Clausena wampi (Blanco) Oliv. = Clausena lansium (Lour.) Skeels ●

94221 Clausena wampi Blanco = Clausena lansium (Lour.) Skeels ●

94222 Clausena wampi Oliv.;万氏黄皮●☆

94223 Clausena willdenovii Wight et Arn. = Clausena dentata H. Lév. ●

94224 Clausena willdienowii Wight et Arn. = Clausena dunniana H. Lév. ●

94225 Clausena yunnanensis C. C. Huang;云南黄皮;Yunnan Wampee ●

94226 Clausena yunnanensis C. C. Huang var. dolichocarpa C. F. Liang et A. M. Lu ex C. C. Huang = Clausena yunnanensis C. C. Huang

var. longgangensis C. F. Liang et Y. X. Lu ●
94227 Clausena yunnanensis C. C. Huang var. longgangensis C. F. Liang et Y. X. Lu;弄岗黄皮(毛云南黄皮);Nonggang Yunnan Wampee ●
94228 Clausenellia Á. Löve et D. Löve = Sedum L. ●■
94229 Clausenopsis (Engl.) Engl. = Fagaropsis Mildbr. ex Siebenl. ●☆
94230 Clausenopsis angolensis Engl. var. mollis Suess. = Fagaropsis angolensis (Engl.) Dale ●☆
94231 Clausia Korn.-Trotzky = Clausia Korn.-Trotzky ex Hayek ■
94232 Clausia Korn.-Trotzky ex Hayek (1911);香芥属(寇夕属);Aromcress,Clausia ■
94233 Clausia aprica (Steph.) Korn.-Trotzky;香芥;Rocket ■
94234 Clausia aprica (Steph.) Korn.-Trotzky var. trichosepala (Turcz.) Korn.-Trotzky = Clausia trichosepala (Turcz.) Dvorák ■
94235 Clausia aprica Steph. var. trichosepala (Turcz.) Korn.-Trotzky = Clausia trichosepala (Turcz.) Dvorák ■
94236 Clausia hispida (Regel) Lipsky;硬毛香芥■☆
94237 Clausia kasakorum Pavlov;哈萨克香芥■
94238 Clausia mollissima Lipsky;柔软香芥■☆
94239 Clausia olgae (Regel et Schmalh.) Lipsky;奥氏香芥■☆
94240 Clausia papillosa Vassilcz.;乳头香芥■☆
94241 Clausia trichosepala (Turcz.) Dvorák;香花芥(香花草);Hairy Sepal Rocket,Hairysepal Rocket ■
94242 Clausia turkestanica Lipsky = Pseudoclausia turkestanica (Lipsky) A. Vassilcz. ■
94243 Clausia turkestanica Lipsky var. glandulosissima Lipsky = Pseudoclausia turkestanica (Lipsky) A. Vassilcz. ■
94244 Clausia turkestanica Lipsky var. subintegrifolia Lipsky = Pseudoclausia turkestanica (Lipsky) A. Vassilcz. ■
94245 Clausia ussuriensis N. Busch = Dontostemon hispidus Maxim. ■
94246 Clausonia Pomel = Asphodelus L. ■☆
94247 Clausonia acaulis Pomel = Asphodelus acaulis Desf. ■☆
94248 Clausospicula Lazarides(1991);闭穗草属■☆
94249 Clausospicula extensa Lazarides;闭穗草■☆
94250 Clavapetalum Pulle = Dendrobangia Rusby ●☆
94251 Clavaria Steud. = Calvaria Comm. ex C. F. Gaertn. ●☆
94252 Clavarioidia Kreuz. = Opuntia Mill. ●
94253 Clavena DC. = Carduus L. ■
94254 Clavenna Neck. ex Standl. = Lucya DC.(保留属名)■☆
94255 Clavennaea Neck. ex Post et Kuntze = Lucya DC.(保留属名)■☆
94256 Claviga Regel = Clavija Ruiz et Pav. ●☆
94257 Clavigera DC. = Brickellia Elliott(保留属名)■●
94258 Clavija Ruiz et Pav. (1794);全缘轮叶属●☆
94259 Clavija allenii Lundell;阿伦全缘轮叶●☆
94260 Clavija boliviensis Mez;玻利维亚全缘轮叶●☆
94261 Clavija brachystachys Brongn. ex Guillaumin;短穗全缘轮叶●☆
94262 Clavija cauliflora Regel.;茎花全缘轮叶●☆
94263 Clavija clavata Decne.;棒状全缘轮叶●☆
94264 Clavija elliptica Mez;椭圆全缘轮叶●☆
94265 Clavija grandis Decne.;大全缘轮叶●☆
94266 Clavija integrifolia Mart. et Miq.;全缘轮叶●☆
94267 Clavija lancifolia Desf.;披针叶全缘轮叶●☆
94268 Clavija latifolia K. Koch;宽叶全缘轮叶●☆
94269 Clavija longifolia Mez;长叶全缘轮叶●☆
94270 Clavija macrocarpa Ruiz et Pav.;大果全缘轮叶●☆
94271 Clavija membranacea Mez;膜质全缘轮叶●☆
94272 Clavija nutans (Vell.) B. Stahl;俯垂全缘轮叶●☆
94273 Clavija parviflora Mez;小花全缘轮叶●☆
94274 Clavija reticulata Cuatrec.;网脉全缘轮叶●☆
94275 Clavimyrtus Blume = Syzygium R. Br. ex Gaertn.(保留属名)●
94276 Clavimyrtus latifolia Blume = Syzygium lineatum (DC.) Merr. et L. M. Perry ●
94277 Clavimyrtus lineata (DC.) Blume = Syzygium lineatum (DC.) Merr. et L. M. Perry ●
94278 Clavinodum T. H. Wen = Arundinaria Michx. ●
94279 Clavinodum T. H. Wen = Oligostachyum Z. P. Wang et G. H. Ye ●★
94280 Clavinodum globinodum (C. H. Hu) P. C. Keng = Oligostachyum gracilipes (McClure) G. H. Ye et Z. P. Wang ●
94281 Clavinodum globinodum (C. H. Hu) P. C. Keng et T. H. Wen = Pleioblastus globinodus C. H. Hu ●
94282 Clavinodum oedogonatum (Z. P. Wang et G. H. Ye) T. H. Wen = Oligostachyum oedogonatum (Z. P. Wang et G. H. Ye) Q. F. Zheng et K. F. Huang ●
94283 Clavipodium Desv. ex Grruening = Beyeria Miq. ☆
94284 Clavistylus J. J. Sm. = Megistostigma Hook. f. ■
94285 Clavophylis Thouars = Bulbophyllum Thouars(保留属名)■
94286 Clavula Dumort. = Eleocharis R. Br. ■
94287 Clavula Dumort. = Scirpus L.(保留属名)■
94288 Clavulium Desv. = Crotalaria L. ●■
94289 Clavulum G. Don = Clavulium Desv. ●■
94290 Clayomyza Whitmore = Cladomyza Danser ●☆
94291 Claytonia Gronov. ex L. = Claytonia L. ■☆
94292 Claytonia L. (1753);春美草属(春美苋属,克莱东苋属);Claytonia,Purslane,Spring Beauty,Spring-beauty ■☆
94293 Claytonia acutifolia Ledeb. = Claytonia arctica Adams ■☆
94294 Claytonia acutifolia Pall. ex Willd.;阿州春美草;Alaska Beauty,Siberian narrow-leaved Claytonia ■☆
94295 Claytonia acutifolia Pall. ex Willd. subsp. graminifolia Hultén = Claytonia acutifolia Pall. ex Willd. ■☆
94296 Claytonia acutifolia Pall. ex Willd. var. graminifolia (Hultén) B. Boivin = Claytonia acutifolia Pall. ex Willd. ■☆
94297 Claytonia alsinoides Sims;粉春美草;Pink Purslane ■☆
94298 Claytonia ambigua S. Watson = Cistanthe ambigua (S. Watson) Carolin ex Hershk. ■☆
94299 Claytonia arctica Adams;北极春美草;Arctic Claytonia ■☆
94300 Claytonia arctica Adams var. megarhiza A. Gray = Claytonia megarhiza (Hemsl.) Kuntze ■☆
94301 Claytonia arenicola L. F. Hend.;亨德森春美草;Henderson's claytonia ■☆
94302 Claytonia bostockii A. E. Porsild = Montia bostockii (A. E. Porsild) S. L. Welsh ■☆
94303 Claytonia calycina (Engelm.) Kuntze = Phemeranthus calycinus (Engelm.) Kiger ■
94304 Claytonia calycina (Engelm.) Kuntze = Talinum calycinum Engelm. ■☆
94305 Claytonia caroliniana Michx.;卡罗来纳春美草;Broad-leaved Spring Beauty,Carolina Spring Beauty,Carolina Spring-beauty,Spring Beauty,Spring-beauty,Wide-leaved Spring Beauty ■☆
94306 Claytonia caroliniana Michx. var. lewisii McNeill = Claytonia caroliniana Michx. ■☆
94307 Claytonia caroliniana Michx. var. peirsonii (Munz et I. M. Johnst.) B. Boivin = Claytonia lanceolata Pall. ex Pursh ■☆
94308 Claytonia caroliniana Michx. var. spatulifolia (Salisb.) W. H. Lewis = Claytonia caroliniana Michx. ■☆

94309 Claytonia caroliniana Michx. var. tuberosa (Pall. ex Willd.) B. Boivin = Claytonia tuberosa Pall. ex Willd. ■☆

94310 Claytonia chamissoi Ledeb. ex Spreng. = Montia chamissoi (Ledeb. ex Spreng.) Greene ■☆

94311 Claytonia cordifolia S. Watson; 心叶春美草; Cordate-leaved Spring Beauty ■☆

94312 Claytonia czukczorum Volkova = Claytonia multiscapa Rydb. ■☆

94313 Claytonia dichotoma Nutt. = Montia dichotoma (Nutt.) Howell ■☆

94314 Claytonia diffusa Nutt. = Montia diffusa (Nutt.) Greene ■☆

94315 Claytonia eschscholtzii Cham.; 埃绍春美草■☆

94316 Claytonia eschscholtzii Cham. = Claytonia acutifolia Pall. ex Willd. ■☆

94317 Claytonia exigua Douglas ex Torr. et A. Gray; 苍白春美草; Pale Claytonia ■☆

94318 Claytonia grayana Kuntze = Lewisia nevadensis (A. Gray) B. L. Rob. ■☆

94319 Claytonia gypsophiloides Fisch. et C. A. Mey.; 沿海春美草; Coast Range Claytonia ■☆

94320 Claytonia hallii A. Gray = Montia fontana L. ■☆

94321 Claytonia howellii (S. Watson) Piper = Montia howellii S. Watson ■☆

94322 Claytonia joanneana Roem. et Schult.; 伊奥春美草■☆

94323 Claytonia lanceolata Pall. ex Pursh; 披针春美草; Spring Beauty, Western Spring Beauty ■☆

94324 Claytonia lanceolata Pall. ex Pursh subsp. chrysantha (Greene) Ferris = Claytonia lanceolata Pall. ex Pursh ■☆

94325 Claytonia lanceolata Pall. ex Pursh var. chrysantha (Greene) C. L. Hitchc. = Claytonia lanceolata Pall. ex Pursh ■☆

94326 Claytonia lanceolata Pall. ex Pursh var. idahoensis R. J. Davis = Claytonia lanceolata Pall. ex Pursh ■☆

94327 Claytonia lanceolata Pall. ex Pursh var. peirsonii Munz et I. M. Johnst. = Claytonia lanceolata Pall. ex Pursh ■☆

94328 Claytonia lanceolata Pursh = Claytonia lanceolata Pall. ex Pursh ■☆

94329 Claytonia lanceolata Pursh var. flava (A. Nelson) C. L. Hitchc. = Claytonia multiscapa Rydb. ■☆

94330 Claytonia lanceolata Pursh var. multiscapa (Rydb.) C. L. Hitchc. = Claytonia multiscapa Rydb. ■☆

94331 Claytonia lanceolata Pursh var. pacifica McNeill = Claytonia multiscapa Rydb. ■☆

94332 Claytonia lanceolata Pursh var. rosea (Rydb.) R. J. Davis = Claytonia rosea Rydb. ■☆

94333 Claytonia linearis Douglas ex Hook. = Montia linearis (Douglas ex Hook.) Greene ■☆

94334 Claytonia megarhiza (Hemsl.) Kuntze; 高山春美草; Fell-fields Claytonia ■☆

94335 Claytonia megarhiza (Hemsl.) Kuntze var. bellidifolia (Rydb.) C. L. Hitchc. = Claytonia megarhiza (Hemsl.) Kuntze ■☆

94336 Claytonia megarhiza (Hemsl.) Kuntze var. nivalis (Engl.) C. L. Hitchc.; 雪地春美草■☆

94337 Claytonia megarhiza (Hemsl.) Kuntze var. nivalis (Engl.) C. L. Hitchc. = Claytonia megarhiza (Hemsl.) Kuntze ■☆

94338 Claytonia multiscapa Rydb.; 多葶春美草; Rydberg's Spring Beauty ■☆

94339 Claytonia nevadensis S. Watson; 内华达山春美草; Sierra Nevada Claytonia ■☆

94340 Claytonia nubigena Greene = Claytonia gypsophiloides Fisch. et C. A. Mey. ■☆

94341 Claytonia ogilviensis McNeill; 奥吉尔维春美草; Ogilvie Claytonia ■☆

94342 Claytonia palustris Swanson et Kelley; 湿地春美草; Marsh Claytonia ■☆

94343 Claytonia parviflora Douglas ex Hook.; 小花春美草; Indian-lettuce ■☆

94344 Claytonia parviflora Douglas ex Hook. subsp. grandiflora J. M. Mill. et K. L. Chambers; 大花春美草; Large-flowered Indian-lettuce ■☆

94345 Claytonia parviflora Douglas ex Hook. subsp. utahensis (Rydb.) J. M. Mill. et K. L. Chambers; 莫哈韦春美草; Mojave Indian-lettuce ■☆

94346 Claytonia parviflora Douglas ex Hook. subsp. viridis (Davidson) J. M. Mill. et K. L. Chambers; 戴维森春美草; Davidson's Indian-lettuce ■☆

94347 Claytonia parviflora Douglas ex Hook. var. depressa A. Gray = Claytonia rubra (Howell) Tidestr. subsp. depressa (A. Gray) J. M. Mill. et K. L. Chambers ■☆

94348 Claytonia parvifolia DC. = Montia parvifolia (DC.) Greene ■☆

94349 Claytonia perfoliata Donn ex Willd.; 野春美草; Buttonhole Plant, Miner's Lettuce, Spanish Lettuce, Spring Beauty, Wild Lettuce, Winter Purslane ■☆

94350 Claytonia perfoliata Donn ex Willd. subsp. intermontana J. M. Mill. et K. L. Chambers; 深山春美草; Intermountain Miner's-lettuce ■☆

94351 Claytonia perfoliata Donn ex Willd. subsp. mexicana (Rydb.) J. M. Mill. et K. L. Chambers; 墨西哥春美草; Madrean miner's-lettuce ■☆

94352 Claytonia perfoliata Donn ex Willd. var. depressa (A. Gray) Poelln. = Claytonia rubra (Howell) Tidestr. subsp. depressa (A. Gray) J. M. Mill. et K. L. Chambers ■☆

94353 Claytonia perfoliata Donn ex Willd. var. parviflora (Douglas ex Hook.) Torr. = Claytonia parviflora Douglas ex Hook. ■☆

94354 Claytonia perfoliata Donn ex Willd. var. utahensis (Rydb.) Poelln. = Claytonia parviflora Douglas ex Hook. subsp. utahensis (Rydb.) J. M. Mill. et K. L. Chambers ■☆

94355 Claytonia rosea Rydb.; 粉花春美草; Rocky Mountain Spring Bbeauty ■☆

94356 Claytonia rubra (Howell) Tidestr.; 红春美草; Erubescent Miner's-lettuce ■☆

94357 Claytonia rubra (Howell) Tidestr. subsp. depressa (A. Gray) J. M. Mill. et K. L. Chambers; 垫状春美草; Cushion Miner's-lettuce ■☆

94358 Claytonia sarmentosa C. A. Mey.; 匍匐春美草; Creeping spring beauty ■☆

94359 Claytonia saxosa Brandegee; 岩地春美草; Brandegee's claytonia ■☆

94360 Claytonia scammaniana Hultén; 斯卡曼春美草; Scamman's Claytonia ■☆

94361 Claytonia sessilifolia (Torr.) Henshaw = Claytonia lanceolata Pall. ex Pursh ■☆

94362 Claytonia sibirica L.; 西伯利亚春美草(西伯利亚蒙蒂苋); Candy Flower, Pink Purslane, Siberian Miners' Lettuce ■☆

94363 Claytonia sibirica L. = Montia sibirica (L.) Howell ■☆

94364 Claytonia sibirica L. var. bulbifera (A. Gray) B. L. Rob. = Claytonia sibirica L. ■☆

94365 Claytonia sibirica L. var. cordifolia (S. Watson) R. J. Davis = Claytonia cordifolia S. Watson ■☆

94366 Claytonia sibirica L. var. heterophylla (Torr. et A. Gray) B. L.

Rob. = Claytonia sibirica L. ■☆
94367 Claytonia teretifolia (Pursh) Kuntze = Phemeranthus teretifolius (Pursh) Raf. ■☆
94368 Claytonia triphylla S. Watson = Lewisia triphylla (S. Watson) B. L. Rob. ■☆
94369 Claytonia tuberosa Pall. ex Willd.;块状春美草;Beringian Spring Beauty ■☆
94370 Claytonia tuberosa Pall. ex Willd. var. czukczorum (Volkova) Hultén = Claytonia multiscapa Rydb. ■☆
94371 Claytonia umbellata S. Watson;大春美草;Great Basin Claytonia ■☆
94372 Claytonia vessilievii Kuzen. = Claytoniella vassilievii (Kuzen.) Jurtzev ■☆
94373 Claytonia virginica L.;弗吉尼亚春美草;Eastern Spring Beauty, Narrow-leaved Spring Beauty, Rose Elf, Spring Beauty, Spring-beauty, Virginia Spring Beauty, Virginia Spring-beauty ■☆
94374 Claytonia virginica L. f. robusta (Somes) E. J. Palmer et Steyerm. = Claytonia virginica L. ■☆
94375 Claytonia virginica L. var. hammondiae (Kalmb.) J. J. Doyle, W. H. Lewis et D. B. Snyder = Claytonia virginica L. ■☆
94376 Claytonia washingtoniana (Suksd.) Suksd.;华盛顿春美草;Lake Washington Claytonia ■☆
94377 Claytoniella Jurtsev = Claytoniella Jurtzev ■☆
94378 Claytoniella Jurtsev = Montia L. ■☆
94379 Claytoniella Jurtsev = Montiastrum (A. Gray) Rydb. ■☆
94380 Claytoniella Jurtzev(1972);小春美草属■☆
94381 Claytoniella bostockii (A. E. Porsild) Jurtzev;鲍氏小春美草■☆
94382 Claytoniella vassilievii (Kuzen.) Jurtzev;小春美草■☆
94383 Cleachne Adans. = Paspalum L. ■
94384 Cleachne Roland ex Steud. = Paspalum L. ■
94385 Cleanthe Salisb. = Aristea Sol. ex Aiton ■☆
94386 Cleanthe Salisb. ex Benth. = Aristea Sol. ex Aiton ■☆
94387 Cleanthe bicolor Salisb. = Aristea lugens (L. f.) Steud. ■☆
94388 Cleanthe lugens (L. f.) Asch. et Graebn. = Aristea lugens (L. f.) Steud. ■☆
94389 Cleanthe melaleuca (Thunb.) Salisb. = Aristea lugens (L. f.) Steud. ■☆
94390 Cleanthes D. Don = Trixis P. Browne ■●☆
94391 Cleghornia Wight(1848);金平藤属;Baissea ●
94392 Cleghornia acuminata Wight = Baissea acuminata (Wight) Benth. ex Hook. f. ●
94393 Cleghornia acuminata Wight = Cleghornia malaccensis (Hook. f.) King et Gamble ●
94394 Cleghornia chinensis (Merr.) P. T. Li = Sindechites chinensis (Merr.) Markgr. et Tsiang ●
94395 Cleghornia cymosa Wight = Cleghornia malaccensis (Hook. f.) King et Gamble ●
94396 Cleghornia henryi (Oliv.) P. T. Li = Sindechites henryi Oliv. ●
94397 Cleghornia malaccensis (Hook. f.) King et Gamble;金平藤;Acuminate Baissea ●
94398 Cleghornia malaccensis (Hook. f.) King et Gamble = Baissea acuminata (Wight) Benth. ex Hook. f. ●
94399 Cleghornia malaccensis (Hook. f.) Pichon = Cleghornia malaccensis (Hook. f.) King et Gamble ●
94400 Cleianthus Lour. ex Gomes = Clerodendrum L. ●■
94401 Cleidiocarpon Airy Shaw (1965);蝴蝶果属;Butterflyfruit, Cleidiocarpon ●
94402 Cleidiocarpon cavalerei (H. Lév.) Airy Shaw;蝴蝶果(山板栗);Butterflyfruit, Cavalerie Cleidiocarpon ●◇
94403 Cleidiocarpon laurinum Airy Shaw;缅甸蝴蝶果;Burma Butterflyfruit ●☆
94404 Cleidion Blume(1826);棒柄花属;Cleidion ●
94405 Cleidion bracteosum Gagnep.;灰岩棒柄花;Limestone Cleidion ●
94406 Cleidion brevipetiolatum Pax et K. Hoffm.;棒柄花(大树三台,三台花,三台树);Shortpetiole Cleidion, Short-petiole Cleidion ●
94407 Cleidion gabonicum Baill.;加蓬棒柄花●☆
94408 Cleidion javanicum Blume;长棒柄花;Java Cleidion ●
94409 Cleidion mannii Baker = Tetracarpidium conophorum (Müll. Arg.) Hutch. et Dalziel ●☆
94410 Cleidion preussii (Pax) Baker = Tetracarpidium conophorum (Müll. Arg.) Hutch. et Dalziel ●☆
94411 Cleidion ulmifolium Müll. Arg. = Discocleidion ulmifolium (Müll. Arg.) Pax et K. Hoffm. ●☆
94412 Cleidion xyphophylloidea Croizat = Trigonostemon xyphophylloides (Croizat) L. K. Dai et T. L. Wu ●
94413 Cleiemera Raf. = Ipomoea L.(保留属名)●■
94414 Cleiostoma Raf. = Ipomoea L.(保留属名)●■
94415 Cleisocentron Bruhl(1926);闭距兰属■☆
94416 Cleisocentron trichromum Bruhl;闭距兰■☆
94417 Cleisocratera Korth. = Saprosma Blume ●
94418 Cleisomeria Lindl. ex G. Don(1855);半闭兰属■☆
94419 Cleisomeria lanatum Lindl. ex G. Don;半闭兰■☆
94420 Cleisomeria pilosulum (Gagn.) Seidenf. et Garav;毛半闭兰■☆
94421 Cleisostoma B. D. Jacks. = Cleiostoma Raf. ●■
94422 Cleisostoma B. D. Jacks. = Ipomoea L.(保留属名)●■
94423 Cleisostoma Blume = Acampe Lindl.(保留属名)■
94424 Cleisostoma Blume = Sarcanthus Lindl.(废弃属名)■
94425 Cleisostoma Blume(1825);隔距兰属(闭口兰属,蜈蚣兰属);Cleisostoma, Closedspurorchis ■
94426 Cleisostoma Raf. = Convolvulus L. ■●
94427 Cleisostoma acuminatum Rolfe = Pomatocalpa acuminata (Rolfe) Schltr. ■
94428 Cleisostoma bifidus (Lindl.) Ames;二裂隔距兰■☆
94429 Cleisostoma birmanicum (Schltr.) Garay;美花隔距兰;Burma Cleisostoma, Burma Closedspurorchis ■
94430 Cleisostoma brachybotryum Hayata = Pomatocalpa acuminata (Rolfe) Schltr. ■
94431 Cleisostoma brachybotryum Hayata = Pomatocalpa undulatum (Lindl.) J. J. Sm. subsp. acuminatum (Rolfe) S. Watthana et S. W. Chung ■
94432 Cleisostoma brevipes Hook. f. = Cleisostoma striatum (Rchb. f.) Garay ■
94433 Cleisostoma breviracemum Hayata = Trichoglottis rosea (Lindl.) Ames var. breviracema (Hayata) Tang S. Liu et H. J. Su ■
94434 Cleisostoma breviracemum Hayata = Trichoglottis rosea (Lindl.) Ames ■
94435 Cleisostoma cerinum Hance = Cleisostoma paniculatum (Ker Gawl.) Garay ■
94436 Cleisostoma dawsonianum Rchb. f. = Staurochilus dawsonianus (Rchb. f.) Schltr. ■
94437 Cleisostoma elongatum (Rolfe) Garay = Cleisostoma williamsonii (Rchb. f.) Garay ■
94438 Cleisostoma filiforme (Lindl.) Garay;金塔隔距兰(丝状隔距

兰）；Filiform Cleisostoma，Filiform Closedspurorchis，Longleaf Closedspurorchis ■

94439 Cleisostoma flagellare（Schltr.）Garay = Cleisostoma fuerstenbergianum Kraenzl. ■

94440 Cleisostoma flagelliforme（Rolfe ex Downie）Garay = Cleisostoma filiforme（Lindl.）Garay ■

94441 Cleisostoma flagelliforme（Rolfe ex Downie）Garay = Cleisostoma fuerstenbergianum Kraenzl. ■

94442 Cleisostoma fordii Hance = Cleisostoma rostratum（Lodd.）Seidenf. ex Aver. ■

94443 Cleisostoma formosanum Hance = Cleisostoma paniculatum（Ker Gawl.）Garay ■

94444 Cleisostoma fuerstenbergianum Kraenzl.；长叶隔距兰（树葱）；Longleaf Cleisostoma，Longleaf Closedspurorchis ■

94445 Cleisostoma fuscomaculatum（Hayata）Garay = Cleisostoma paniculatum（Ker Gawl.）Garay ■

94446 Cleisostoma gemmatum（Lindl.）King et Pantl. = Schoenorchis gemmata（Lindl.）J. J. Sm. ■

94447 Cleisostoma hongkongense（Rolfe）Garay；香港花隔距兰（红花隔距兰）■

94448 Cleisostoma hongkongense（Rolfe）Garay = Cleisostoma williamsonii（Rchb. f.）Garay ■

94449 Cleisostoma ionosmum Lindl. f. lutschuense Makino = Staurochilus luchuensis（Rolfe）Fukuy. ■

94450 Cleisostoma ionosmum Lindl. f. lutschuense Makino = Trichoglottis lutchuensis（Rolfe）Garay et H. R. Sweet ■☆

94451 Cleisostoma longioperculatum Z. H. Tsi；长帽隔距兰；Longcap Cleisostoma，Longcap Closedspurorchis ■

94452 Cleisostoma medogense Z. H. Tsi；西藏隔距兰；Medog Cleisostoma，Medog Closedspurorchis ■

94453 Cleisostoma menghaiense Z. H. Tsi；勐海隔距兰；Menghai Cleisostoma，Menghai Closedspurorchis ■

94454 Cleisostoma micranthum（Lindl.）King et Pantl. = Smitinandia micrantha（Lindl.）Holttum ■

94455 Cleisostoma nangongense Z. H. Tsi；南贡隔距兰；Nangong Cleisostoma，Nangong Closedspurorchis ■

94456 Cleisostoma oblongisepala Hayata = Trichoglottis rosea（Lindl.）Ames var. breviracema（Hayata）Tang S. Liu et H. J. Su ■

94457 Cleisostoma oblongisepalum Hayata = Trichoglottis rosea（Lindl.）Ames ■

94458 Cleisostoma paniculatum（Ker Gawl.）Garay；大序隔距兰（虎皮隔距兰，虎纹兰，山吊兰）；Panicle Closedspurorchis，Paniculate Cleisostoma ■

94459 Cleisostoma parishii（Hook. f.）Garay；短茎隔距兰（巴氏隔距兰）；Parish Cleisostoma，Parish Closedspurorchis ■

94460 Cleisostoma poilanei Gagnep. = Smitinandia micrantha（Lindl.）Holttum ■

94461 Cleisostoma racemiferum（Lindl.）Garay；大叶隔距兰（总状序隔距兰）；Largeleaf Cleisostoma，Largeleaf Closedspurorchis ■

94462 Cleisostoma roseum Lindl. = Trichoglottis rosea（Lindl.）Ames ■

94463 Cleisostoma rostratum（Lindl.）Garay = Cleisostoma rostratum（Lodd.）Seidenf. ex Aver. ■

94464 Cleisostoma rostratum（Lodd.）Seidenf. ex Aver.；尖喙隔距兰（蜈蚣草）；Acute Closedspurorchis，Rosrate Cleisostoma ■

94465 Cleisostoma sagittiforme Garay；隔距兰；Cleisostoma，Common Closedspurorchis ■

94466 Cleisostoma scolopendrifolium（Makino）Garay = Pelatantheria scolopendrifolia（Makino）Aver. ■

94467 Cleisostoma simondii（Gagnep.）Seidenf.；毛柱隔距兰；Simond Cleisostoma，Simond Closedspurorchis ■

94468 Cleisostoma simondii（Gagnep.）Seidenf. var. guangdongense Z. H. Tsi；广东隔距兰；Guangdong Cleisostoma，Guangdong Closedspurorchis ■

94469 Cleisostoma spatulatum Blume = Robiquetia spatulata（Blume）J. J. Sm. ■

94470 Cleisostoma striatum（Rchb. f.）Garay；短序隔距兰■

94471 Cleisostoma striatum（Rchb. f.）N. E. Br. = Cleisostoma striatum（Rchb. f.）Garay ■

94472 Cleisostoma striolatum（Rchb. f.）Garay；条纹隔距兰；Striate Cleisostoma，Striate Closedspurorchis ■☆

94473 Cleisostoma taiwanianum（Hayata）Hayata = Sarcophyton taiwanianum（Hayata）Garay ■

94474 Cleisostoma taiwanianum Hayata = Sarcophyton taiwanianum（Hayata）Garay ■

94475 Cleisostoma teres Garay = Cleisostoma simondii（Gagnep.）Seidenf. ■

94476 Cleisostoma teres Garay = Vanda watsonii Rolfe ■☆

94477 Cleisostoma unciferum（Schltr.）Garay = Cleisostoma paniculatum（Ker Gawl.）Garay ■

94478 Cleisostoma uraiense（Hayata）Garay et H. R. Sweet；绿花隔距兰（乌来闭口兰，乌来隔距兰）；Greenflower Cleisostoma，Greenflower Closedspurorchis ■

94479 Cleisostoma uteriferum Hook. f. = Pomatocalpa spicatum Breda ■

94480 Cleisostoma virginale Hance = Robiquetia succisa（Lindl.）Seidenf. et Garay ■

94481 Cleisostoma viridescens（Fukuy.）Garay = Cleisostoma uraiense（Hayata）Garay et H. R. Sweet ■

94482 Cleisostoma wendlandorum Rchb. f. = Pomatocalpa spicatum Breda ■

94483 Cleisostoma williamsonii（Rchb. f.）Garay；红花隔距兰（滇缅隔距兰，光棍草，龙角草，马尾吊兰，木石草，树葱，圆叶吊兰）；Redflower Cleisostoma，Redflower Closedspurorchis ■

94484 Cleisostomopsis Seidenf.（1992）；拟隔距兰属■

94485 Cleisostomopsis eberhardtii（Finet）Seidenf. et Smitinand；拟隔距兰■

94486 Cleissocratera Miq. = Cleisocratera Korth. ●

94487 Cleissocratera Miq. = Saprosma Blume ●

94488 Cleistachne Benth.（1882）；闭壳草属■☆

94489 Cleistachne macrantha Stapf = Cleistachne sorghoides Benth. ■☆

94490 Cleistachne sorghoides Benth.；闭壳草■☆

94491 Cleistachne teretifolia Hack. = Miscanthus junceus（Stapf）Pilg. ■☆

94492 Cleistanthes Kuntze = Cleistanthus Hook. f. ex Planch. ●

94493 Cleistanthium Kuntze = Gerbera L.（保留属名）■

94494 Cleistanthium nepalense Kunze = Gerbera kunzeana A. Braun et Asch. ■

94495 Cleistanthium nepalense Kunze = Leibnitzia nepalensis（Kunze）Kitam. ■

94496 Cleistanthopsis Capuron = Allantospermum Forman ●☆

94497 Cleistanthopsis multicaulis Capuron = Allantospermum multicaule（Capuron）Noot. ●☆

94498 Cleistanthus Hook. f. ex Planch.（1848）；闭花木属（闭花属，尾叶木属）；Cleistanthus ●

94499 Cleistanthus amaniensis Jabl. = Cleistanthus polystachyus Hook.

f. ex Planch. ●☆

94500 Cleistanthus apetalus S. Moore = Cleistanthus polystachyus Hook. f. ex Planch. subsp. milleri (Dunkley) Radcl. -Sm. ●☆

94501 Cleistanthus beentjei Q. Luke;比特闭花木●☆

94502 Cleistanthus bipindensis Pax;比平迪闭花木●☆

94503 Cleistanthus camerunensis J. Léonard;喀麦隆闭花木●☆

94504 Cleistanthus caudatus Pax;尾状闭花木●☆

94505 Cleistanthus collinus (Roxb.) Benth. = Cleistanthus collinus (Roxb.) Hook. f. ●

94506 Cleistanthus collinus (Roxb.) Hook. f.;丘生闭花木(小丘闭花木)●

94507 Cleistanthus dongfangensis (P. T. Li) H. S. Kiu;东方闭花木;Dongfang Cleistanthus ●

94508 Cleistanthus eburneus Gagnep. = Cleistanthus tomentosus Hance ●

94509 Cleistanthus eburneus Gagnep. var. sordidus Gagnep. = Cleistanthus tomentosus Hance ●

94510 Cleistanthus evrardii J. Léonard;埃夫拉尔闭花木●☆

94511 Cleistanthus gabonensis Hutch.;加蓬闭花木●☆

94512 Cleistanthus glauca Hiern = Pseudolachnostylis maprouneifolia Pax var. glabra (Pax) Brenan ●☆

94513 Cleistanthus holtzii Pax = Cleistanthus schlechteri (Pax) Hutch. ●☆

94514 Cleistanthus integer C. B. Rob.;全缘闭花木●☆

94515 Cleistanthus inundatus J. Léonard;洪水地闭花木●☆

94516 Cleistanthus inundatus J. Léonard var. velutinus;短绒毛闭花木●☆

94517 Cleistanthus johnsonii Hutch. var. pubescens = Cleistanthus schlechteri (Pax) Hutch. var. pubescens (Hutch.) J. Léonard ●☆

94518 Cleistanthus kasaiensis J. Léonard;开赛闭花木●☆

94519 Cleistanthus kasaiensis J. Léonard var. paulopubescens;疏毛闭花木●☆

94520 Cleistanthus letouzeyi J. Léonard;勒图闭花木●☆

94521 Cleistanthus libericus N. E. Br.;离生闭花木●☆

94522 Cleistanthus macrophyllus Hook. f.;大叶闭花木;Big-leaf Cleistanthus, Big-leaved Cleistanthus ●

94523 Cleistanthus michelsonii J. Léonard;米歇尔松闭花木●☆

94524 Cleistanthus mildbraedii Jabl.;米氏闭花木●☆

94525 Cleistanthus mildbraedii Jabl. et J. Léonard = Cleistanthus mildbraedii Jabl ●☆

94526 Cleistanthus milleri Dunkley = Cleistanthus polystachyus Hook. f. ex Planch. subsp. milleri (Dunkley) Radcl. -Sm. ●☆

94527 Cleistanthus ngounyensis Pellegr.;恩戈尼亚闭花木●☆

94528 Cleistanthus nyasicus Dunkley = Cleistanthus polystachyus Hook. f. ex Planch. subsp. milleri (Dunkley) Radcl. -Sm. ●☆

94529 Cleistanthus pedicellatus Hook. f.;米咀闭花木(米咀);Long-pedicel Cleistanthus, Pedicellate Cleistanthus ●

94530 Cleistanthus petelotii Merr. ex Croizat;假肥牛树;Petelot Cleistanthus ●

94531 Cleistanthus pierlotii J. Léonard;皮氏闭花木●☆

94532 Cleistanthus polystachyus Hook. f. ex Planch.;多穗闭花木●☆

94533 Cleistanthus polystachyus Hook. f. ex Planch. subsp. milleri (Dunkley) Radcl. -Sm.;米勒多穗闭花木●☆

94534 Cleistanthus racemosus Pierre ex Hutch.;总花闭花木●☆

94535 Cleistanthus ripicola J. Léonard;岩石闭花木●☆

94536 Cleistanthus saichikii Merr. = Cleistanthus sumatranus (Miq.) Müll. Arg. ●

94537 Cleistanthus schlechteri (Pax) Hutch.;施莱闭花木●☆

94538 Cleistanthus schlechteri (Pax) Hutch. var. pubescens (Hutch.) J. Léonard;短柔毛施莱闭花木●☆

94539 Cleistanthus sumatranus (Miq.) Müll. Arg.;闭花木(闭花,火炭木,苏岛闭花木,尾叶木);Cleistanthus, Common Cleistanthus, Sumatra Cleistanthus ●

94540 Cleistanthus tomentosus Hance;锈毛闭花木(黄毛闭花木,毛闭花木,崖县闭花木);Tomentose Cleistanthus ●

94541 Cleistanthus tonkinensis Jabl.;馒头果(东京闭花木,河内闭花木,馒头闭花木,野茶叶,越南闭花木);Tonkin Cleistanthus, Vietnamese Cleistanthus ●

94542 Cleistanthus willmannianus J. Léonard;威尔闭花木●☆

94543 Cleistanthus zenkeri Jabl.;岑克尔闭花木●☆

94544 Cleistes Rich. = Cleistes Rich. ex Lindl. ■☆

94545 Cleistes Rich. ex Lindl. (1840);科雷兰属;Cleistes, Spreading Pogonia ■☆

94546 Cleistes bifaria (Fernald) Catling et Gregg;小科雷兰;Smaller Spreading Pogonia ■☆

94547 Cleistes divaricata (L.) Ames;叉唇科雷兰;Divaricate Cleistes, Rosebud Orchis, Spreading Pogonia ■☆

94548 Cleistes divaricata (L.) Ames var. bifaria Fernald = Cleistes bifaria (Fernald) Catling et Gregg ■☆

94549 Cleistes rosea Lindl.;红花科雷兰;Redflower Cleistes ■☆

94550 Cleistesiopsis Pansarin et F. Barros = Arethusa L. ■☆

94551 Cleistocactus Lem. (1861);管花柱属(吹雪柱属);Cleistocatus ●☆

94552 Cleistocactus anguinus Britton et Rose;蛇形柱(白闪)●☆

94553 Cleistocactus areolatus Riccob.;网地柱●☆

94554 Cleistocactus areolatus Riccob. var. herzogianus (Backeb.) Backeb.;光彩柱●☆

94555 Cleistocactus ayopayanus Cárdenas;乱吹雪●☆

94556 Cleistocactus baumanii Lem.;凌云柱(凌云阁);Firecracker Cactus, Scarlet Bugler, Scarlet-bugler ●☆

94557 Cleistocactus baumannii Lem. var. flavispinus Riccob.;蛇纹柱(黄刺柱)●☆

94558 Cleistocactus brookei Cárdenas;布路凯管花柱●☆

94559 Cleistocactus buchtienii Backeb.;山吹雪●☆

94560 Cleistocactus candelilla Cardenas;猩猩吹雪●☆

94561 Cleistocactus flavispinus (K. Schum.) Backeb.;真皇(里特里管花柱)●☆

94562 Cleistocactus jujuyensis (Backeb.) Backeb.;优吹雪柱●☆

94563 Cleistocactus parapetiensis Cárdenas;弥勒柱●☆

94564 Cleistocactus rojoi Cardenas;吉祥天●☆

94565 Cleistocactus smaragdiflorus (F. A. C. Weber) Britton et Rose;绿花柱(阿根廷雪柱,阿根廷银毛柱)●☆

94566 Cleistocactus strausii (Heese) Backeb.;吹雪柱(管花仙人柱,银毛柱);Silver Torch, Silver Torch Cactus, Silver-torch ●☆

94567 Cleistocactus tarijensis Cardenas;白闪柱(白闪,谷间灯)●☆

94568 Cleistocactus tupizensis (Vaupel) Backeb.;红吹雪●☆

94569 Cleistocactus winteri D. R. Hunt;黄金柱;Golden Rat Tail ●☆

94570 Cleistocalyx Blume = Syzygium R. Br. ex Gaertn. (保留属名)●

94571 Cleistocalyx Blume (1850);水翁属(水榕属,水蓊属);Cleistocalyx, Waterfig ●

94572 Cleistocalyx Steud. = Rhynchospora Vahl(保留属名)■

94573 Cleistocalyx cerasoides (Roxb.) I. M. Turner = Syzygium nervosum DC. ●

94574 Cleistocalyx conspersipunctatum Merr. et L. M. Perry;大果水翁(大果水榕);Bigfruit Cleistocalyx, Bigfruit Waterfig, Big-fruited Cleistocalyx ●

94575 Cleistocalyx conspersipunctatus Merr. et L. M. Perry = Syzygium conspersipunctatum (Merr. et L. M. Perry) Craven et Biffin ●

94576 Cleistocalyx operculatus (Roxb.) Merr. et L. M. Perry;水翁(大蛇药,酒翁,水榕,水榕树,水翁树,水蓊,水香,水雍花,土槿皮,有盖蒲桃);Lidded Cleistocalyx, Operculate Cleistocalyx, Operculate Waterfig, Water Banyan ●

94577 Cleistocalyx operculatus (Roxb.) Merr. et L. M. Perry = Syzygium nervosum DC. ●

94578 Cleistocereus Frič et Kreuz. = Cleistocactus Lem. ●☆

94579 Cleistochlamys Oliv. (1865);闭被木属●☆

94580 Cleistochlamys kirkii (Benth.) Oliv.;闭被木●☆

94581 Cleistochloa C. E. Hubb. (1933);澳隐黍属■☆

94582 Cleistochloa rigida (S. T. Blake) Clayton;挺澳隐黍■☆

94583 Cleistochloa sclerachne (F. M. Bailey) C. E. Hubb.;澳隐黍■☆

94584 Cleistogenes Keng = Kengia Packer ■

94585 Cleistogenes Keng (1934);隐子草属;Cleistogenes, Hideseedgrass ■

94586 Cleistogenes andropogonoides Honda = Cleistogenes squarrosa (Trin.) Keng ■

94587 Cleistogenes caespitosa Keng;丛生隐子草;Tufted Cleistogenes, Tufted Hideseedgrass ■

94588 Cleistogenes caespitosa Keng var. ramosa F. Z. Li et C. K. Ni;多枝隐子草;Manybranch Cleistogenes, Manybranch Hideseedgrass ■

94589 Cleistogenes caespitosa Keng var. ramosa F. Z. Li et C. K. Ni = Cleistogenes hackelii (Honda) Honda ■

94590 Cleistogenes chiennsis (Maxim.) Keng var. nakai Keng = Cleistogenes hackelii (Honda) Honda var. nakai (Keng) Ohwi ■

94591 Cleistogenes chinensis (Maxim.) Keng;中华隐子草(朝鲜青茅);China Hideseedgrass, Chinese Cleistogenes ■

94592 Cleistogenes chinensis (Maxim.) Keng = Cleistogenes hackelii (Honda) Honda ■

94593 Cleistogenes chinensis Keng = Cleistogenes kitagawai Honda ■

94594 Cleistogenes festucacea Honda;薄鞘隐子草■

94595 Cleistogenes foliosa Keng = Cleistogenes festucacea Honda ■

94596 Cleistogenes foliosa Keng = Cleistogenes kitagawai Honda var. foliosa (Keng) S. L. Chen et C. P. Wang ■

94597 Cleistogenes gatacrei (Stapf) Bor = Kengia gatacrei (Stapf) Cope ■☆

94598 Cleistogenes gracilis Keng ex P. C. Keng et L. Liou = Cleistogenes mucronata P. C. Keng ■

94599 Cleistogenes gracilis P. C. Keng;细弱隐子草;Slender Cleistogenes, Slender Hideseedgrass ■

94600 Cleistogenes hackeli (Honda) Honda var. chinensis (Maxim.) Ohwi = Cleistogenes chinensis (Maxim.) Keng ■

94601 Cleistogenes hackelii (Honda) Honda;朝阳隐子草(朝鲜莲座梅,朝阳青茅);Hackel Cleistogenes, Hackel Hideseedgrass ■

94602 Cleistogenes hackelii (Honda) Honda = Kengia hackelii (Honda) Packer ■

94603 Cleistogenes hackelii (Honda) Honda var. chinensis (Maxim.) Ohwi = Cleistogenes hackelii (Honda) Honda ■

94604 Cleistogenes hackelii (Honda) Honda var. nakaii (Keng) Ohwi;宽叶隐子草;Nakai Cleistogenes, Nakai Hideseedgrass ■

94605 Cleistogenes hancei Keng;北京隐子草;Beijing Hideseedgrass, Peking Cleistogenes ■

94606 Cleistogenes hancei Keng var. jeholensis (Honda) Kitag. = Cleistogenes polyphylla Keng ex P. C. Keng et L. Liou ■

94607 Cleistogenes hancei Keng var. jeholensis (Honda) Kitag. = Cleistogenes hancei Keng ■

94608 Cleistogenes kitagawae Honda;凌源隐子草;Kitagawa Hideseedgrass, Kitagawa Cleistogenes ■

94609 Cleistogenes kitagawae Honda var. foliosa (Keng) S. L. Chen et C. P. Wang;包鞘隐子草;Foliose Cleistogenes, Foliose Kitagawa Hideseedgrass ■

94610 Cleistogenes kitagawae Honda var. foliosa (Keng) S. L. Chen et C. P. Wang = Cleistogenes festucacea Honda ■

94611 Cleistogenes kokonorica K. S. Hao = Orinus kokonorica (K. S. Hao) Keng ex X. L. Yang ■

94612 Cleistogenes kokonorica K. S. Hao = Orinus kokonorica (K. S. Hao) Tzvelev ■

94613 Cleistogenes longiflora Keng ex P. C. Keng et L. Liou = Cleistogenes festucacea Honda ■

94614 Cleistogenes longiflora P. C. Keng ex P. C. Keng et L. Liou;长花隐子草;Longflower Cleistogenes, Longflower Hideseedgrass ■

94615 Cleistogenes mucronata P. C. Keng;小尖隐子草;Mucronate Hideseedgrass, Sharp-point Cleistogenes ■

94616 Cleistogenes mutica Keng = Cleistogenes songorica (Roshev.) Ohwi ■

94617 Cleistogenes nakaii (Keng) Honda = Cleistogenes hackelii (Honda) Honda var. nakaii (Keng) Ohwi ■

94618 Cleistogenes nakaii (Keng) Honda var. purpurascens Honda = Cleistogenes hancei Keng ■

94619 Cleistogenes polyphylla Keng ex P. C. Keng et L. Liou;多叶隐子草;Manyleaf Cleistogenes, Manyleaf Hideseedgrass ■

94620 Cleistogenes ramiflora Keng et C. P. Wang;枝花隐子草;Ramiflorus Cleistogenes, Ramiflorus Hideseedgrass ■

94621 Cleistogenes ramiflora Keng et C. P. Wang var. tianmushanensis F. Z. Li et C. K. Ni;天目山隐子草;Tianmushan Cleistogenes, Tianmushan Hideseedgrass ■

94622 Cleistogenes serotina (L.) Keng;秋隐子草■

94623 Cleistogenes serotina (L.) Keng = Cleistogenes kitagawai Honda ■

94624 Cleistogenes serotina (L.) Keng var. aristata (Hack.) Keng = Cleistogenes hackelii (Honda) Honda ■

94625 Cleistogenes serotina (L.) Keng var. chinensis (Maxim.) Hand. -Mazz. = Cleistogenes chinensis (Maxim.) Keng ■

94626 Cleistogenes serotina (L.) Keng var. jeholensis Honda = Cleistogenes hancei Keng ■

94627 Cleistogenes serotina (L.) Keng var. jeholensis Honda = Cleistogenes polyphylla Keng ex P. C. Keng et L. Liou ■

94628 Cleistogenes serotina (L.) Keng var. nakaii Keng = Cleistogenes hackelii (Honda) Honda var. nakai (Keng) Ohwi ■

94629 Cleistogenes serotina (L.) Keng var. sinensis (Hance) Keng = Cleistogenes hancei Keng ■

94630 Cleistogenes serotina (L.) Keng var. vivipara Honda = Cleistogenes hancei Keng ■

94631 Cleistogenes songorica (Roshev.) Ohwi;无芒隐子草(青海隐子草);Awnless Cleistogenes, Awnless Hideseedgrass, Songarian Cleistogenes ■

94632 Cleistogenes squarrosa (Trin. ex Ledeb.) Keng;糙隐子草;Scabrous Cleistogenes, Scabrous Hideseedgrass ■

94633 Cleistogenes squarrosa (Trin.) Keng = Cleistogenes squarrosa (Trin. ex Ledeb.) Keng ■

94634 Cleistogenes squarrosa (Trin.) Keng var. longe-aristata (Rendle) Keng = Cleistogenes squarrosa (Trin.) Keng ■

94635 Cleistogenes striata Honda;条纹隐子草;Striate Cleistogenes ■

94636 Cleistogenes striata Honda = Cleistogenes festucacea Honda ■
94637 Cleistogenes striata Honda = Cleistogenes kitagawai Honda ■
94638 Cleistogenes thoroldii (Stapf ex Hemsl.) Roshev. = Cleistogenes songorica (Roshev.) Ohwi ■
94639 Cleistogenes thoroldii (Stapf ex Hemsl.) Roshev. = Orinus thoroldii (Stapf ex Hemsl.) Bor ■
94640 Cleistoloranthus Merr. = Amyema Tiegh. ●☆
94641 Cleistopetalum H. Okada(1996);合瓣番荔枝属●☆
94642 Cleistopholis Pierre ex Engl. (1897);闭盔木属(闭鳞番荔枝属)●☆
94643 Cleistopholis Pierre. = Cleistopholis Pierre ex Engl. ●☆
94644 Cleistopholis albida (Engl.) Engl. et Diels = Friesodielsia gracilipes (Benth.) Steenis ●☆
94645 Cleistopholis albida (Engl.) Engl. et Diels var. longipedicellata Baker f. = Friesodielsia gracilipes (Benth.) Steenis ●☆
94646 Cleistopholis bequaerti De Wild. = Cleistopholis glauca Pierre ex Engl. et Diels ●☆
94647 Cleistopholis brevipetala Exell = Cleistopholis patens (Benth.) Engl. et Diels ●☆
94648 Cleistopholis discostigma Diels = Friesodielsia discostigma (Diels) Steenis ●☆
94649 Cleistopholis glauca Pierre ex Engl. et Diels;灰闭盔木(灰闭鳞番荔枝)●☆
94650 Cleistopholis gracilipes (Benth.) Engl. et Diels = Friesodielsia gracilipes (Benth.) Steenis ●☆
94651 Cleistopholis grandiflora De Wild. = Cleistopholis glauca Pierre ex Engl. et Diels ●☆
94652 Cleistopholis klaineana Pierre ex Engl. et Diels = Cleistopholis patens (Benth.) Engl. et Diels ●☆
94653 Cleistopholis lucens De Wild. = Cleistopholis patens (Benth.) Engl. et Diels ●☆
94654 Cleistopholis patens (Benth.) Engl. et Diels;闭盔木(闭鳞番荔枝);Otu ●☆
94655 Cleistopholis patens (Benth.) Engl. et Diels var. klaineana Pellegr. = Cleistopholis patens (Benth.) Engl. et Diels ●☆
94656 Cleistopholis patens Engl. et Diels = Cleistopholis patens (Benth.) Engl. et Diels ●☆
94657 Cleistopholis platypetala (Benth.) Engl. et Diels = Friesodielsia gracilis (Hook. f.) Steenis ●☆
94658 Cleistopholis pynaerti De Wild. = Cleistopholis patens (Benth.) Engl. et Diels ●☆
94659 Cleistopholis staudtii Engl. et Diels;施陶闭盔木(施陶番荔枝)●☆
94660 Cleistopholis verschuereni De Wild. = Cleistopholis patens (Benth.) Engl. et Diels ●☆
94661 Cleistoyucca Eastw. = Clistoyucca (Engelm.) Trel. ●☆
94662 Cleistoyucca arborescens (Torr.) Trel. = Yucca brevifolia Engelm. ●☆
94663 Cleistoyucca brevifolia (Engelm.) Rydb. = Yucca brevifolia Engelm. ●☆
94664 Cleithria Steud. = Cleitria Schrad. ■☆
94665 Cleitria Schrad. = Venidium Less. ■☆
94666 Cleitria Schrad. ex L. = Venidium Less. ■☆
94667 Clelandia J. M. Black = Hybanthus Jacq. (保留属名)●■
94668 Clelia Casar. = Calliandra Benth. (保留属名)●
94669 Clemanthus Klotzsch = Adenia Forssk. ●
94670 Clematepistephium N. Hallé(1977);歧冠兰属■☆
94671 Clematepistephium smilacifolium (Rchb. f.) N. Hallé;歧冠兰■☆
94672 Clematicissus Planch. (1887);多枝藤属●☆
94673 Clematicissus angustissima (F. Muell.) Planch.;多枝藤●☆
94674 Clematidaceae Martinov = Ranunculaceae Juss. (保留科名)●■
94675 Clematis Dill. ex L. = Clematis L. ●■
94676 Clematis L. (1753);铁线莲属;Clematis, Lether Flower, Old Man's Beard, Traveller's-joy, Virgin's Bower, Virgin's-bower ●■
94677 Clematis acerifolia Maxim.;槭叶铁线莲(岩花);Maple-leaf Clematis, Maple-leaved Clematis ●
94678 Clematis acerifolia Maxim. var. elobata S. X. Yan;无裂槭叶铁线莲●
94679 Clematis acuminata DC.;渐尖铁线莲■
94680 Clematis acuminata DC. subsp. leschenaultiana (DC.) Brühl = Clematis leschenaultiana DC. ●
94681 Clematis acuminata DC. subsp. leschenaultiana Brühl = Clematis leschenaultiana DC. ●
94682 Clematis acuminata DC. subsp. normalis Kuntze = Clematis siamensis Drumm. et Craib ●
94683 Clematis acuminata DC. subsp. sikkimensis (Hook. f. et Thomson) Brühl var. andersonii Brühl = Clematis siamensis Drumm. et Craib var. andersonii (Brühl) W. T. Wang ●☆
94684 Clematis acuminata DC. subsp. sikkimensis (Hook. f. et Thomson) Brühl var. hookeri Brühl = Clematis siamensis Drumm. et Craib ●
94685 Clematis acuminata DC. subsp. sikkimensis (Hook. f. et Thomson) Brühl var. klarkei (Kuntze) Brühl = Clematis siamensis Drumm. et Craib var. klarkei (Kuntze) W. T. Wang ●
94686 Clematis acuminata DC. subsp. sikkimensis (Hook. f. et Thomson) Brühl var. klarkei (Kuntze) W. T. Wang = Clematis siamensis Drumm. et Craib var. klarkei (Kuntze) W. T. Wang ●
94687 Clematis acuminata DC. subsp. sikkimensis (Hook. f. et Thomson) Brühl = Clematis siamensis Drumm. et Craib ●
94688 Clematis acuminata DC. subsp. sikkimensis Brühl = Clematis siamensis Drumm. et Craib ●
94689 Clematis acuminata DC. subsp. yunnanensis Brühl = Clematis yunnanensis Franch. ●
94690 Clematis acuminata DC. var. andersoni Brühl;滇缅铁线莲;Anderson Clematis ●
94691 Clematis acuminata DC. var. clarkei Kuntze = Clematis siamensis Drumm. et Craib var. klarkei (Kuntze) W. T. Wang ●
94692 Clematis acuminata DC. var. hirtella Hand.-Mazz. = Clematis siamensis Drumm. et Craib ●
94693 Clematis acuminata DC. var. leschenaultiana (DC.) Kuntze = Clematis leschenaultiana DC. ●
94694 Clematis acuminata DC. var. leschenaultiana Kuntze = Clematis leschenaultiana DC. ●
94695 Clematis acuminata DC. var. longicaudata W. T. Wang;长尾尖铁线莲(尾尖铁线莲);Longicaudate Clematis ●
94696 Clematis acuminata DC. var. multiflora Comber = Clematis multiflora (H. F. Comber) W. T. Wang ●
94697 Clematis acuminata DC. var. multiflora Comber = Clematis siamensis Drumm. et Craib ●
94698 Clematis acuminata DC. var. normalis = Clematis sikkimensis Drumm. ex Burkill ●
94699 Clematis acuminata DC. var. sikkimensis Hook. f. et Thomson = Clematis siamensis Drumm. et Craib ●
94700 Clematis acutangula Hook. f. et Thomson;锐角发汗藤(细麦藤);Acutangule Clematis ●

94701 Clematis acutangula Hook. f. et Thomson f. major W. T. Wang = Clematis ranunculoides Franch. ■

94702 Clematis acutangula Hook. f. et Thomson subsp. ranunculoides (Franch.) W. T. Wang = Clematis ranunculoides Franch. ■

94703 Clematis addisonii Britton; 阿迪森铁线莲; Addison Brown's Clematis, Addison Brown's Leather-flower, Addison's Leather-flower, Addison's Virgin's-bower ■☆

94704 Clematis aethusifolia Turcz.; 芹叶铁线莲(白拉拉秧,断肠草,驴断肠,铁线透骨草,透骨草,细叶铁线莲)■

94705 Clematis aethusifolia Turcz. var. latisecta Maxim.; 宽芹叶铁线莲(草地铁线莲,芹叶铁线莲); Broadleaf Clematis ■

94706 Clematis affinis A. St. -Hil. = Clematis smilacifolia Wall. ●

94707 Clematis akebioides (Maxim.) Veitch; 甘川铁线莲●

94708 Clematis akoensis Hayata; 屏东铁线莲(阿猴木通花藤,阿猴仙人草,大渡氏牡丹藤); Pingdong Clematis ●

94709 Clematis albicoma Wherry; 白毛铁线莲; Erect Mountain Clematis, White-haired Leather-flower ■☆

94710 Clematis albicoma Wherry var. coactilis Fernald = Clematis coactilis (Fernald) Keener ■☆

94711 Clematis alpina (L.) Mill. var. chinensis Maxim. = Clematis sibirica (L.) Mill. var. ochotensis (Pall.) S. H. Li et Y. Huei Huang ●

94712 Clematis alpina (L.) Mill. var. koreana (Kom.) Nakai = Clematis koreana Kom. ●

94713 Clematis alpina (L.) Mill. var. occidentalis (Hornem.) A. Gray subvar. tenuiloba A. Gray = Clematis columbiana (Nutt.) Torr. et A. Gray var. tenuiloba (A. Gray) J. S. Pringle ■☆

94714 Clematis alpina Mill.; 高山铁线莲; Alpine Clematis ●☆

94715 Clematis alpina Mill. subsp. macropetala var. albiflora Maxim. ex Kuntze = Clematis macropetala Ledeb. var. albiflora (Maxim.) Hand. -Mazz. ●

94716 Clematis alpina Mill. subsp. macropetala var. rupestris Turcz. ex Kuntze = Clematis macropetala Ledeb. ●

94717 Clematis alpina Mill. subsp. sibirica Kuntze = Clematis sibirica (L.) Mill. ●

94718 Clematis alpina Mill. var. chinensis Maxim. = Clematis sibirica (L.) Mill. var. ochotensis (Pall.) S. H. Li et Y. Huei Huang ●

94719 Clematis alpina Mill. var. koreana (Kom.) Nakai = Clematis koreana Kom. ●

94720 Clematis alpina Mill. var. macropetala Maxim. = Clematis macropetala Ledeb. ●

94721 Clematis alpina Mill. var. macropetala Maxim. subvar. albiflora Maxim. = Clematis macropetala Ledeb. var. albiflora (Maxim.) Hand. -Mazz. ●

94722 Clematis alpina Mill. var. macropetala Maxim. subvar. rupestris Maxim. = Clematis macropetala Ledeb. ●

94723 Clematis alpina Mill. var. ochotensis (Pall.) Kuntze = Clematis sibirica (L.) Mill. var. ochotensis (Pall.) S. H. Li et Y. Huei Huang ●

94724 Clematis alsomitrifolia Hayata; 台三叶铁线莲●

94725 Clematis alsomitrifolia Hayata = Clematis uncinata Champ. ex Benth. ●

94726 Clematis alternata Kitam. et Tamura = Archiclematis alternata (Kitam. et Tamura) Tamura ●

94727 Clematis amplexicaulis Edgew. = Clematis connata DC. ●

94728 Clematis anemoniflora D. Don = Clematis montana Buch. -Ham. ex DC. var. wilsonii Sprague ●

94729 Clematis angustifolia Hayata = Clematis leschenaultiana DC. ●

94730 Clematis angustifolia Jacq. = Clematis hexapetala Pall. ●

94731 Clematis angustifolia Jacq. f. dissecta (Y. Yabe) Kitag. = Clematis hexapetala Pall. ●

94732 Clematis angustifolia Jacq. var. breviloba Freyn = Clematis hexapetala Pall. ●

94733 Clematis angustifolia Jacq. var. dissecta Y. Yabe = Clematis hexapetala Pall. ●

94734 Clematis angustifolia Jacq. var. longiloba Freyn = Clematis hexapetala Pall. ●

94735 Clematis angustifolia Jacq. var. tchefouensis Debeaux = Clematis hexapetala Pall. var. tchefouensis (Debeaux) S. Y. Hu ●

94736 Clematis angustifoliola W. T. Wang = Clematis subfalcata C. P'ei ex M. Y. Fang var. stephylla (Hand. -Mazz.) W. T. Wang ●

94737 Clematis angustifoliola W. T. Wang = Clematis yunnanensis Franch. ●

94738 Clematis anhwuiensis M. C. Chang = Clematis chinensis Osbeck var. anhwuiensis (M. C. Chang) W. T. Wang ●

94739 Clematis anshunensis M. Y. Fang = Clematis clarkeana H. Lév. et Vaniot ●

94740 Clematis apiifolia DC.; 女萎(白棉纱,白木通,百根草,穿山藤,粗糠藤,方藤,风藤,花木通,蔓楚,牡丹蔓,木通草,千里光,山木通,苏木通,万年藤,小木通,小叶鸭脚力刚,一把抓,银匙藤,钥匙藤); October Clematis ●

94741 Clematis apiifolia DC. var. argentilucida (H. Lév. et Vaniot) W. T. Wang; 钝齿铁线莲(川木通,粗齿铁线莲,粗糠藤,黄藤通,毛木通); Bluntooth Clematis ●

94742 Clematis apiifolia DC. var. obtusidentata Rehder et E. H. Wilson = Clematis apiifolia DC. var. argentilucida (H. Lév. et Vaniot) W. T. Wang ●

94743 Clematis argentilucida (H. Lév. et Vaniot) H. Eichler = Clematis argentilucida (H. Lév. et Vaniot) W. T. Wang var. likiangensis (Rehder) W. T. Wang ●

94744 Clematis argentilucida (H. Lév. et Vaniot) W. T. Wang = Clematis grandidentata (Rehder et E. H. Wilson) W. T. Wang ●

94745 Clematis argentilucida (H. Lév. et Vaniot) W. T. Wang var. likiangensis (Rehder) W. T. Wang; 丽江大木通(丽江铁线莲); Lijiang Clematis ●

94746 Clematis argentilucida W. T. Wang var. likiangensis (Rehder) W. T. Wang = Clematis grandidentata (Rehder et E. H. Wilson) W. T. Wang var. likiangensis (Rehder) W. T. Wang ●

94747 Clematis aristata Ker Gawl.; 具芒铁线莲; Old Man's Beard ■☆

94748 Clematis armandii Franch.; 小木通(白花木通,白木通,川木通,大木通,大叶木通,海木通,海通,花木通,淮木通,黄防己,辣木通,老虎须,乱根草,木通,三叶木通,山木通,四朵梅,蓑衣藤,铁线莲,土木通,葳灵仙,油木通,紫木通); Armand Clematis, Evergreen Clematis ●

94749 Clematis armandii Franch. f. farquhariana Rehder et E. H. Wilson = Clematis armandii Franch. var. farquhariana (Rehder et E. H. Wilson) W. T. Wang ●

94750 Clematis armandii Franch. var. biondiana (Pavol.) Rehder = Clematis armandii Franch. ●

94751 Clematis armandii Franch. var. farquhariana (Rehder et E. H. Wilson) W. T. Wang; 大花小木通; Bigflower Armand Clematis ●

94752 Clematis armandii Franch. var. hefengensis (G. F. Tao) W. T. Wang; 鹤峰铁线莲; Heqing Clematis ●

94753 Clematis asplenifolia Schrenk = Clematis songarica Bunge var.

asplenifolia (Schrenk) Trautv. ●
94754 Clematis asplenifolia Schrenk = Clematis songarica Bunge ●
94755 Clematis asplenifolia Schrenk var. boissieriana (Korsh.) Krasch. = Clematis songarica Bunge var. asplenifolia (Schrenk) Trautv. ●
94756 Clematis azurea Turcz. = Clematis patens E. Morren et Decne. ●
94757 Clematis baldwinii Torr. et A. Gray;松林铁线莲;Pine-hyacinth, Pine-woods Clematis ■☆
94758 Clematis baldwinii Torr. et A. Gray var. latiuscula R. W. Long = Clematis baldwinii Torr. et A. Gray ■☆
94759 Clematis balearica Rich. = Clematis cirrhosa L. var. balearica (Rich.) Willk. ■☆
94760 Clematis balearica Rich. = Clematis cirrhosa L. ■☆
94761 Clematis baominiana W. T. Wang;多毛铁线莲(长毛铁线莲);Long-hair Clematis ●
94762 Clematis barbellata Edgew. var. obtusa Edgew. = Clematis barbellata Edgew. var. obtusa Kitam. et Tamura ●
94763 Clematis barbellata Edgew. var. obtusa Kitam. et Tamura;吉隆铁线莲;Jilong Clematis, Kilung Clematis, Obtuse Clematis ●
94764 Clematis bartlettii Gogelein = Clematis parviloba Gardner et Champ. ●
94765 Clematis bartlettii Gogelein = Clematis parviloba Gardner et Champ. var. bartlettii (Gogelein) W. T. Wang ●
94766 Clematis beadlei (Small) R. O. Erickson = Clematis viorna L. ■☆
94767 Clematis benthamiana Hemsl. = Clematis chinensis Osbeck ●
94768 Clematis benthamiana Hemsl. = Clematis pashanensis (M. C. Chang) W. T. Wang ●
94769 Clematis bigelovii Torr.;毕氏铁线莲;Bigelow's Clematis ■☆
94770 Clematis biondiana Pavol. = Clematis armandii Franch. ●
94771 Clematis biternata DC. = Clematis javana DC. ●
94772 Clematis boissieriana Korsh. = Clematis songarica Bunge var. asplenifolia (Schrenk) Trautv. ●
94773 Clematis brachiata Thunb.;臂形铁线莲;Old Man's Beard ■☆
94774 Clematis brachiata Thunb. var. burkei Burtt Davy = Clematis brachiata Thunb. ■☆
94775 Clematis brachyura Maxim.;朝鲜短尾铁线莲(朝鲜铁线莲)●
94776 Clematis bracteata (Roxb.) Kurz = Clematis cadmia Buch.-Ham. ex Hook. f. et Thomson ■
94777 Clematis bracteata (Roxb.) Kurz var. leptomera (Hance) Kuntze = Clematis florida Thunb. ●
94778 Clematis brevicaudata DC.;短尾铁线莲(短尾木通,红钉耙藤,连架拐,林地铁线莲,山木通,石通,铁脚灵仙,小木通);Shortplume Clematis ●
94779 Clematis brevicaudata DC. var. filipes Rehder et E. H. Wilson = Clematis puberula Hook. f. et Thomson var. tenuisepala (Maxim.) W. T. Wang ●
94780 Clematis brevicaudata DC. var. gangpiniana (H. Lév. et Vaniot) Hand.-Mazz. = Clematis puberula Hook. f. et Thomson var. ganpiniana (H. Lév. et Vaniot) W. T. Wang ●
94781 Clematis brevicaudata DC. var. leiophylla Hand.-Mazz. = Clematis puberula Hook. f. et Thomson var. ganpiniana (H. Lév. et Vaniot) W. T. Wang ●
94782 Clematis brevicaudata DC. var. lissocarpa Rehder et E. H. Wilson = Clematis puberula Hook. f. et Thomson var. ganpiniana (H. Lév. et Vaniot) W. T. Wang ●
94783 Clematis brevicaudata DC. var. malocotrocha W. T. Wang;密毛短尾铁线莲;Densehair Shortplume Clematis ●
94784 Clematis brevicaudata DC. var. subsericea Rehder et E. H. Wilson = Clematis puberula Hook. f. et Thomson var. tenuisepala (Maxim.) W. T. Wang ●
94785 Clematis brevicaudata DC. var. subsericea Rehder et E. H. Wilson = Clematis ganpiniana (H. Lév. et Vaniot) Tamura var. subsericea (Rehder et E. H. Wilson) C. T. Ting ●
94786 Clematis brevicaudata DC. var. subsericea Rehder et E. H. Wilson = Clematis puberula Hook. f. et Thomson var. subsericea (Rehder et E. H. Wilson) W. T. Wang ●
94787 Clematis brevicaudata DC. var. tenuisepala Maxim. = Clematis puberula Hook. f. et Thomson var. tenuisepala (Maxim.) W. T. Wang ●
94788 Clematis brevipes Rehder;短梗铁线莲;Short-stalk Clematis ●
94789 Clematis buchananiana DC.;毛木通;Buchanan Clematis ●
94790 Clematis buchananiana DC. subsp. conata (DC.) Kuntze = Clematis connata DC. ●
94791 Clematis buchananiana DC. subsp. connata Kuntze = Clematis connata DC. ●
94792 Clematis buchananiana DC. subsp. grewiiflora = Clematis grewiiflora DC. ●
94793 Clematis buchananiana DC. subsp. trullifera Franch. = Clematis connata DC. var. trullifera (Franch.) W. T. Wang ●
94794 Clematis buchananiana DC. var. rugosa Hook. f. et Thomson = Clematis buchananiana DC. ●
94795 Clematis buchananiana DC. var. trullifera Franch. = Clematis connata DC. var. trullifera (Franch.) W. T. Wang ●
94796 Clematis buchananiana DC. var. vitifolia Hook. f. et Thomson;膜叶毛木通;Membranousleaf Clematis ●
94797 Clematis burgensis Engl. = Clematis brachiata Thunb. ■☆
94798 Clematis burmanica Lace;缅甸铁线莲;Burma Clematis ●
94799 Clematis cadmia Buch.-Ham. ex Hook. f. et Thomson;短柱铁线莲■
94800 Clematis cadmia Buch.-Ham. ex Wall. = Clematis cadmia Buch.-Ham. ex Hook. f. et Thomson ■
94801 Clematis caerulea Lindl. = Clematis patens E. Morren et Decne. ●
94802 Clematis caesariata Hance = Clematis leschenaultiana DC. ●
94803 Clematis canadensis Mill. = Clematis virginiana L. ■☆
94804 Clematis canescens (Turcz.) W. T. Wang et M. C. Chang = Clematis fruticosa Turcz. var. canescens Turcz ●
94805 Clematis canescens (Turcz.) W. T. Wang et M. C. Chang subsp. viridis W. T. Wang et M. C. Chang = Clematis viridis (W. T. Wang et M. C. Chang) W. T. Wang ●
94806 Clematis canescens (Turcz.) W. T. Wang et M. C. Chang var. viridis (W. T. Wang et M. C. Chang) W. T. Wang = Clematis viridis (W. T. Wang et M. C. Chang) W. T. Wang ●
94807 Clematis catesbyana Pursh;北美西部铁线莲;Virgin's-bower ■☆
94808 Clematis caudigera W. T. Wang;尾尖铁线莲;Sharptail Clematis ●
94809 Clematis cavaleriei H. Lév. et Porter = Clematis chinensis Osbeck ●
94810 Clematis chanetii H. Lév. = Clematis kirilowii Maxim. var. chanetii (H. Lév.) Hand.-Mazz. ●
94811 Clematis chekiangensis C. P'ei;浙江铁线莲(浙江山木通);Chekiang Clematis, Zhejiang Clematis ●
94812 Clematis cheusanensis Plukenet = Veronicastrum stenostachyum (Hemsl.) T. Yamaz. subsp. plukenetii (T. Yamaz.) D. Y. Hong ■
94813 Clematis chinensis f. vestita Rehder et E. H. Wilson = Clematis chinensis Osbeck var. vestica (Rehder et E. H. Wilson) W. T. Wang ●
94814 Clematis chinensis Osbeck;威灵仙(白钱草,百条根,赤茎威灵仙,东北铁线莲,杜灵仙,粉灵仙,风车,黑骨头,黑脚威灵仙,黑

灵仙,黑木通,黑茜,黑薇,黑尾,黑须公,九草阶,九里火,九龙须,酒灵仙,辣椒藤,老虎须,灵仙,灵仙藤,毛柱铁线莲,绵团铁线莲,能消,牛杆草,牛闲草,七寸草,青风藤,青龙须,软灵仙,山蓼,寿祖,铁杆威灵仙,铁脚灵仙,铁脚威灵仙,铁灵仙,铁扫帚,铁掬帚,铁丝根,铁线根,土灵仙,葳灵仙,葳苓仙,乌骨胆草,乌头力刚,小木通,药王草,移星草,芝查藤根,中国铁线莲);Chinese Clematis,Weiling Xian ●

94815 Clematis chinensis Osbeck f. vestica Rehder et E. H. Wilson = Clematis chinensis Osbeck var. vestica (Rehder et E. H. Wilson) W. T. Wang ●

94816 Clematis chinensis Osbeck var. anhweiensis (M. C. Chang) W. T. Wang;安徽铁线莲;Anhui Clematis ●

94817 Clematis chinensis Osbeck var. dissecta (Y. Yabe) Kitag.;细叶威灵仙●

94818 Clematis chinensis Osbeck var. fujisanensis (Hisauti et Hara) W. T. Wang;富士山铁线莲;Fujisan Clematis ●☆

94819 Clematis chinensis Osbeck var. laeviachenium R. H. Miao;光果威灵仙●

94820 Clematis chinensis Osbeck var. uncinata (Champ. ex Benth.) Kuntze = Clematis uncinata Champ. ex Benth. ●

94821 Clematis chinensis Osbeck var. vestica (Rehder et E. H. Wilson) W. T. Wang;毛叶威灵仙;Hairy-leaf Chinese Clematis ●

94822 Clematis chinensis Retz. = Clematis chinensis Osbeck ●

94823 Clematis chingii W. T. Wang;两广铁线莲(康壳藤);Ching's Clematis ●

94824 Clematis chiupehensis M. Y. Fang;邱北铁线莲;Chiubei Clematis,Qiubei Clematis ●

94825 Clematis chlorantha Lindl. = Clematis grandiflora DC. ■☆

94826 Clematis chrysantha Ulbr. = Clematis tangutica (Maxim.) Korsh. ●

94827 Clematis chrysantha Ulbr. var. brevipes? = Clematis tibetana Kuntze ●

94828 Clematis chrysantha Ulbr. var. paucidentata = Clematis tibetana Kuntze ●

94829 Clematis chrysocarpa Welw. ex Oliv.;金果铁线莲■☆

94830 Clematis chrysocoma Franch.;金毛铁线莲(大风藤棵,大木通,风藤草,花木通,黄毛铁线莲,金毛木通,金丝木通,九头狮子草,马尾巴,钱丝木通,山棉花,铁脚威灵,五爪金龙,小木通,野棉花,野木通);Goldwool Clematis ●

94831 Clematis chrysocoma Franch. var. glabrescens H. F. Comber = Clematis montana Buch.-Ham. ex DC. var. glabrescens (H. F. Comber) W. T. Wang et M. C. Chang ●

94832 Clematis chrysocoma Franch. var. laxistrigosa W. T. Wang et M. C. Chang = Clematis laxistrigosa (W. T. Wang et M. C. Chang) W. T. Wang ●

94833 Clematis chrysocoma Franch. var. sericea (Franch. ex Finet et Gagnep.) C. K. Schneid. = Clematis chrysocoma Franch. ●

94834 Clematis chrysocoma Franch. var. sericea (Franch.) C. K. Schneid. = Clematis chrysocoma Franch. ●

94835 Clematis cirrhosa L.;卷须铁线莲;Fern-leaved Clematis,Vernal Clematis,Virgin's-bower ■☆

94836 Clematis cirrhosa L. var. balearica (Rich.) Willk. = Clematis cirrhosa L. ■☆

94837 Clematis cirrhosa L. var. dautezi Debeaux = Clematis cirrhosa L. ■☆

94838 Clematis cirrhosa L. var. napaulensis (DC.) Kuntze = Clematis napaulensis DC. ●

94839 Clematis cirrhosa L. var. semitriloba (Lag.) Batt. = Clematis cirrhosa L. ■☆

94840 Clematis clarkeana H. Lév. et Vaniot;平坝铁线莲(喉痛药,拦路虎);Clarke Clematis ●

94841 Clematis clarkeana H. Lév. et Vaniot var. stenophylla Hand.-Mazz.;川滇铁线莲;Narrowleaf Clark Clematis ●

94842 Clematis clarkeana H. Lév. et Vaniot var. stenophylla Hand.-Mazz. = Clematis yunnanensis Franch. ●

94843 Clematis clarkeana H. Lév. et Vaniot var. stenophylla Hand.-Mazz. = Clematis subfalcata C. P'ei ex M. Y. Fang var. stephylla (Hand.-Mazz.) W. T. Wang ●

94844 Clematis coactilis (Fernald) Keener;维吉尼亚铁线莲■☆

94845 Clematis coccinea Engelm.;绯红铁线莲;Scarlet Clematis ■☆

94846 Clematis coccinea Engelm. = Clematis texensis Buckley ●☆

94847 Clematis coerulea Lindl. = Clematis patens E. Morren et Decne. ●

94848 Clematis columbiana (Nutt.) Torr. et A. Gray;哥伦比亚铁线莲■☆

94849 Clematis columbiana (Nutt.) Torr. et A. Gray var. dissecta C. L. Hitchc. = Clematis occidentalis (Hornem.) DC. var. dissecta (C. L. Hitchc.) J. S. Pringle ●☆

94850 Clematis columbiana (Nutt.) Torr. et A. Gray var. tenuiloba (A. Gray) J. S. Pringle;细裂哥伦比亚铁线莲■☆

94851 Clematis commutata Kuntze;变异铁线莲■☆

94852 Clematis commutata Kuntze var. glabrisepala W. T. Wang;光萼铁线莲■☆

94853 Clematis connata DC.;合柄铁线莲(白木通,大发汗藤,接骨草);Connate Clematis ●

94854 Clematis connata DC. subsp. sublanata W. T. Wang = Clematis connata DC. var. trullifera (Franch.) W. T. Wang ●

94855 Clematis connata DC. var. bipinnata M. Y. Fang = Clematis connata DC. var. pseudoconata (Kuntze) W. T. Wang ●

94856 Clematis connata DC. var. lanceolata Biswas;印度铁线莲;Indian Clematis ●☆

94857 Clematis connata DC. var. pseudoconata (Kuntze) W. T. Wang;川藏铁线莲;Bipinnate Clematis ●

94858 Clematis connata DC. var. sublanata W. T. Wang = Clematis connata DC. var. trullifera (Franch.) W. T. Wang ●

94859 Clematis connata DC. var. trullifera (Franch.) W. T. Wang;杯柄铁线莲(大木通,大发汗藤);Chengtu Clematis,Cupshapedstalk Clematis,Cupshaped-stalk Clematis,Cupstalk Clematis ●

94860 Clematis cordata Pursh = Clematis catesbyana Pursh ■☆

94861 Clematis cordata Sims = Clematis simsii Sweet ●☆

94862 Clematis coriigera H. Lév. = Clematis connata DC. var. trullifera (Franch.) W. T. Wang ●

94863 Clematis corniculata W. T. Wang;角萼铁线莲;Corniculate Clematis ●

94864 Clematis corrigera H. Lév. = Clematis connata DC. var. trullifera (Franch.) W. T. Wang ●

94865 Clematis courtoisii Hand.-Mazz.;大花威灵仙(百木通,大花铁线莲,小脚威灵仙);Bigflower Clematis,Big-flowered Clematis ●

94866 Clematis crassifolia Benth.;厚叶铁线莲;September Clematis,Thickleaf Clematis ●

94867 Clematis crassipes Chun et F. C. How;粗柄铁线莲;Thickstipe Clematis,Thick-stiped Clematis ●

94868 Clematis crassipes Chun et F. C. How var. pubipes W. T. Wang;毛序粗柄铁线莲(毛梗铁线莲,三叶木通);Hairy-stalk Clematis ●

94869 Clematis crassipes Chun et F. C. How var. pubipes W. T. Wang =

Clematis pubipes (W. T. Wang) W. T. Wang ●

94870 Clematis crispa L.;沼泽铁线莲;Blue Jasmine, Blue-jasmine, Curly Clematis, Marsh Clematis, Swamp Leather Flower, Swamp Leather-flower ■☆

94871 Clematis dasyandra Maxim.;毛花铁线莲;Hairyflower Clematis ●

94872 Clematis dasyandra Maxim. var. polyantha Finet et Gagnep. = Clematis dasyandra Maxim. ●

94873 Clematis davidiana Decne. ex Verl. = Clematis heracleifolia DC. ●

94874 Clematis delavayi Franch.;银叶铁线莲(德氏铁线莲,银叶大蓼);Delavay Clematis ●

94875 Clematis delavayi Franch. var. calvescens C. K. Schneid.;疏毛银叶铁线莲(光秃银叶大蓼);Becomingnaked Delavay Clematis ●

94876 Clematis delavayi Franch. var. limprichtii (Ulbr.) M. C. Chang;裂银叶铁线莲;Limpricht Clematis ●

94877 Clematis delavayi Franch. var. spinescens Balf. f. ex Diels;刺铁线莲(刺叶铁线莲,刺银叶大蓼);Spiny Clematis ●

94878 Clematis dictyota Greene = Clematis pitcheri Torr. et Gray var. dictyota (Greene) W. M. Dennis ■☆

94879 Clematis dilatata C. P'ei;舟柄铁线莲;Boatshapedstalk Clematis, Dilatate Clematis, Extented Clematis ●

94880 Clematis dimorphophylla W. T. Wang;异叶铁线莲;Diversi-leaf Clematis ●☆

94881 Clematis dioscoreifolia H. Lév. et Vaniot;盘叶铁线莲■☆

94882 Clematis dioscoreifolia H. Lév. et Vaniot = Clematis terniflora DC. ●

94883 Clematis dioscoreifolia H. Lév. et Vaniot var. robusta (Carrière) Rehder = Clematis terniflora DC. ●

94884 Clematis dolichopoda Brenan = Clematis brachiata Thunb. ■☆

94885 Clematis dolichosepala Hayata = Clematis akoensis Hayata ●

94886 Clematis drakeana H. Lév. et Vaniot = Clematis uncinata Champ. ex Benth. ●

94887 Clematis drummondii Torr. et A. Gray;德拉蒙德铁线莲;Drummond's Clematis, Old Man Beard ■☆

94888 Clematis duclouxii H. Lév. = Clematis lancifolia Bureau et Franch. ●

94889 Clematis ethusifolia var. latisecta Maxim.;宽叶铁线莲●

94890 Clematis faberi Hemsl. et E. H. Wilson = Clematis pogonandra Maxim. ●

94891 Clematis fargesii Franch. = Clematis potaninii Maxim. ●

94892 Clematis fargesii Franch. var. souliei Finet et Gagnep. = Clematis potaninii Maxim. ●

94893 Clematis fargesii Franch. var. souliei Franch. ex Finet et Gagnep. = Clematis potaninii Maxim. ●

94894 Clematis fasciculiflora Franch.;滑叶藤(木通,三叶五香血藤,三爪金龙,山金银,小木通,小叶五香血藤,小粘药,小粘叶);Fascicled Clematis ●

94895 Clematis fasciculiflora Franch. var. angustifolia Comber;狭叶滑叶藤(狭叶山金银);Narrowleaf Fascicled Clematis ●

94896 Clematis fengii W. T. Wang;国楣铁线莲;Feng's Clematis ●

94897 Clematis filamentosa Dunn = Clematis loureiriana DC. ●

94898 Clematis filifera Benth. = Clematis pitscheri Torr. et Gray ■☆

94899 Clematis finetiana H. Lév. et Vaniot;山木通(冲倒山,大木通,大叶光板力刚,过山照,九里花,老虎毛,老虎须,千金拔,蓑衣藤,天仙菊,万年藤,威灵仙,西木通,雪球藤,硬骨灵仙);Finet Clematis ●

94900 Clematis finetii var. pedata W. T. Wang;鸟足叶铁线莲;Pendate Clematis ●

94901 Clematis flammula L.;羽果铁线莲;Fragrant Clematis, Plume Clematis, Sweet Virgin's Bower, Upright Virgin's Bower, Virgin's-bower ●☆

94902 Clematis flammula L. var. grandiflora Pomel = Clematis flammula L. ●☆

94903 Clematis flammula L. var. maritima (Lam.) DC. = Clematis flammula L. ●☆

94904 Clematis flammula L. var. parviflora Pomel = Clematis flammula L. ●☆

94905 Clematis flammula L. var. robusta Carrière = Clematis terniflora DC. ●

94906 Clematis flammula L. var. rotundifolia DC. = Clematis flammula L. ●☆

94907 Clematis flammula L. var. sancti-marini Pamp. = Clematis flammula L. ●☆

94908 Clematis floribunda (Hayata) Gogelein = Clematis uncinata Champ. ex Benth. ●

94909 Clematis floribunda Kurz = Clematis subumbellata Kurz ●

94910 Clematis florida Thunb.;铁线莲(大花威灵仙,番莲,龙须草,山木通,铜威灵仙,威灵仙);Asian Virginsbower, Cream Clematis ●

94911 Clematis florida Thunb. = Clematis courtoisii Hand. -Mazz. ●

94912 Clematis florida Thunb. var. hancockiana (Maxim.) Courtois = Clematis hancockiana Maxim. ●

94913 Clematis florida Thunb. var. lanuginosa (Lindl.) Kuntze = Clematis lanuginosa Lindl. et Paxton ●

94914 Clematis florida Thunb. var. plena D. Don;重瓣铁线莲;Doubleflower Clematis, Doublepetal Cream Clematis ●

94915 Clematis florida Thunb. var. simsii Honda;铁脚威灵仙;Sims Crean Clematis ●

94916 Clematis formosana Kuntze;宝岛铁线莲(台湾女萎,台湾铁线莲);Taiwan Clematis ●

94917 Clematis forrestii W. W. Sm. = Clematis napaulensis DC. ●

94918 Clematis fremontii S. Watson;涪雷蒙铁线莲;Fremont's Leather Flower ■☆

94919 Clematis fremontii S. Watson var. riehlii R. O. Erickson = Clematis fremontii S. Watson ■☆

94920 Clematis fruticosa Turcz.;灌木铁线莲;Shrubby Clematis ●

94921 Clematis fruticosa Turcz. = Clematis viridis (W. T. Wang et M. C. Chang) W. T. Wang ●

94922 Clematis fruticosa Turcz. f. atriplexifolia Koslovsky = Clematis fruticosa Turcz. var. lobata Maxim. ●

94923 Clematis fruticosa Turcz. f. chenopodiofolia Koslovsky = Clematis fruticosa Turcz. var. lobata Maxim. ●

94924 Clematis fruticosa Turcz. f. lanceifolia Koslovsky = Clematis tomentella (Maxim.) W. T. Wang et L. Q. Li ●

94925 Clematis fruticosa Turcz. var. canescens Turcz.;毛灌木铁线莲(灰叶铁线莲);Hoary Clematis ●

94926 Clematis fruticosa Turcz. var. lobata Maxim.;浅裂铁线莲●

94927 Clematis fruticosa Turcz. var. tomentella Maxim. = Clematis tomentella (Maxim.) W. T. Wang et L. Q. Li ●

94928 Clematis fruticosa Turcz. var. viridis Turcz. = Clematis fruticosa Turcz. ●

94929 Clematis fujisanensis Hisauti et H. Hara = Clematis chinensis Osbeck var. fujisanensis (Hisauti et Hara) W. T. Wang ●☆

94930 Clematis fulvicoma Rehder et E. H. Wilson;滇南铁线莲(黄毛木通,黄毛铁线莲);Mengzi Clematis, South Yunnan Clematis ●

94931 Clematis funebris H. Lév. = Clematis chinensis Osbeck ●

94932 Clematis funebris H. Lév. et Vaniot = Clematis chinensis Osbeck ●

94933 Clematis fusca Turcz.;褐毛铁线莲;Brownhair Clematis,Stanavoi Clematis ●

94934 Clematis fusca Turcz. f. obtusifoliola Kuntze = Clematis fusca Turcz. ●

94935 Clematis fusca Turcz. subsp. violacea (Maxim.) Kitag. = Clematis fusca Turcz. var. violacea Maxim. ●

94936 Clematis fusca Turcz. var. amurensis Kuntze = Clematis fusca Turcz. ●

94937 Clematis fusca Turcz. var. mandshurica Regel = Clematis fusca Turcz. ●

94938 Clematis fusca Turcz. var. violacea Maxim.;紫花铁线莲;Purpleflower Brownhair Clematis,Purpleflower Stanavoi Clematis ●

94939 Clematis gagnepainiana H. Lév. et Vaniot = Clematis uncinata Champ. ex Benth. ●

94940 Clematis ganpiniana (H. Lév. et Vaniot) Tamura = Clematis puberula Hook. f. et Thomson var. ganpiniana (H. Lév. et Vaniot) W. T. Wang ●

94941 Clematis ganpiniana (H. Lév. et Vaniot) Tamura var. subsericea (Rehder et E. H. Wilson) C. T. Ting = Clematis puberula Hook. f. et Thomson var. subsericea (Rehder et E. H. Wilson) W. T. Wang ●

94942 Clematis ganpiniana (H. Lév. et Vaniot) Tamura var. tenuisepala (Maxim.) C. T. Ting = Clematis puberula Hook. f. et Thomson var. tenuisepala (Maxim.) W. T. Wang ●

94943 Clematis garanbiensis Hayata = Clematis terniflora DC. var. garanbiensis DC. ●

94944 Clematis gebleHana Bong. = Clematis songarica Bunge ●

94945 Clematis glabrifolia K. Sun et M. S. Yan;光叶铁线莲;Glabrousleaf Clematis ●

94946 Clematis glauca Willd.;粉绿铁线莲(粉叶铁线莲,灰绿铁线莲,苒苒草,铁线莲,透骨草)●

94947 Clematis glauca Willd. var. akebioides (Maxim.) Rehder et E. H. Wilson = Clematis akebioides (Maxim.) Veitch ●

94948 Clematis glauca Willd. var. angustifolia Ledeb. = Clematis intricata Bunge ●

94949 Clematis glaucescens Fresen. = Clematis brachiata Thunb. ■☆

94950 Clematis glycinoides DC.;大豆铁线莲■☆

94951 Clematis gouriana Roxb. = Clematis javana DC. ●

94952 Clematis gouriana Roxb. = Clematis peterae Hand. -Mazz. var. trichocarpa W. T. Wang ●

94953 Clematis gouriana Roxb. ex DC.;小蓑衣藤(串鼻龙,大车藤,范氏木通,古氏铁线莲,茅衣藤,木通,小花木通);Gourian Clematis,Indian Traveller's Joy ●

94954 Clematis gouriana Roxb. ex DC. subsp. lishanensis T. Y. Yang et T. C. Huang = Clematis peterae Hand. -Mazz. var. lishanensis (T. Y. Yang et T. C. Huang) W. T. Wang ●

94955 Clematis gouriana Roxb. ex DC. var. finetii Rehder et E. H. Wilson = Clematis peterae Hand. -Mazz. ●

94956 Clematis gouriana Roxb. ex DC. var. malaiana Miq. = Clematis javana DC. ●

94957 Clematis gracilifolia Rehder et E. H. Wilson;薄叶铁线莲(细叶四喜牡丹);Graceful Clematis ●

94958 Clematis gracilifolia Rehder et E. H. Wilson var. dissectifolia W. T. Wang et M. C. Chang;狭裂薄叶铁线莲;Dissectedleaf Graceful Clematis ●

94959 Clematis gracilifolia Rehder et E. H. Wilson var. lasiocarpa W. T. Wang;毛果薄叶铁线莲;Hairfruit Graceful Clematis ●

94960 Clematis gracilifolia Rehder et E. H. Wilson var. macrantha W. T. Wang et M. C. Chang;大花薄叶铁线莲;Bigflower Graceful Clematis ●

94961 Clematis gracilifolia Rehder et E. H. Wilson var. pentaphylla (Maxim.) W. T. Wang = Clematis gracilifolia Rehder et E. H. Wilson ●

94962 Clematis gracilifolia Rehder et E. H. Wilson var. trifoliata M. Johnson = Clematis gracilifolia Rehder et E. H. Wilson ●

94963 Clematis gracilis Edgew. = Clematis connata DC. ●

94964 Clematis grandidentata (Rehder et E. H. Wilson) W. T. Wang;粗齿铁线莲(白头公公,川木通,大木通,大蓑衣藤,黄藤通,毛木通,山木通,线木通,小木通,银叶铁线莲);Grossedentate Clematis,Thick-tooth Clematis ●

94965 Clematis grandidentata (Rehder et E. H. Wilson) W. T. Wang var. likiangensis (Rehder) W. T. Wang;丽江铁线莲;Lijiang Thick-tooth Clematis ●

94966 Clematis grandiflora DC.;大花铁线莲■☆

94967 Clematis grandifolia (Staner et J. Léonard) M. Johnson = Clematis uhehensis Engl. ●☆

94968 Clematis granulata (Finet et Gagnep.) Ohwi = Clematis meyeniana Walp. var. granulata Finet et Gagnep. ●

94969 Clematis grata Wall.;秀丽铁线莲(串鼻龙,大齿铁线莲,大木通,毛牡丹藤,木通);Beautiful Clematis ●

94970 Clematis grata Wall. var. argentilucida (H. Lév. et Vaniot) Rehder = Clematis apiifolia DC. var. argentilucida (H. Lév. et Vaniot) W. T. Wang ●

94971 Clematis grata Wall. var. grandidentata Rehder et E. H. Wilson = Clematis grandidentata (Rehder et E. H. Wilson) W. T. Wang ●

94972 Clematis grata Wall. var. likiangensis Rehder = Clematis grandidentata (Rehder et E. H. Wilson) W. T. Wang var. likiangensis (Rehder) W. T. Wang ●

94973 Clematis grata Wall. var. lobulata Rehder et E. H. Wilson = Clematis gratopsis W. T. Wang ●

94974 Clematis grata Wall. var. lobulata Rehder et E. H. Wilson = Clematis javana DC. ●

94975 Clematis grata Wall. var. lobulata Rehder et E. H. Wilson = Clematis taiwaniana Hayata ●

94976 Clematis grata Wall. var. ryukiuensis Tamura = Clematis javana DC. ●

94977 Clematis gratopsis W. T. Wang;金佛铁线莲(绿木通);Jinfoshan Clematis,Kingfowshan Clematis ●

94978 Clematis gratopsis W. T. Wang var. integriloba W. T. Wang = Clematis wissmanniana Hand. -Mazz. ●

94979 Clematis graveclens Lindl. = Clematis tenuifolia Royle ●

94980 Clematis graveolens Lindl.;藏西铁线莲(西藏铁线莲)■☆

94981 Clematis graveolens Lindl. = Clematis zandaensis W. T. Wang ●

94982 Clematis grewiiflora DC.;黄毛铁线莲;Yellowhairy Clematis,Yellow-hairy Clematis ●

94983 Clematis hainanensis W. T. Wang;海南铁线莲;Hainan Clematis ●

94984 Clematis hakonensis Franch. et Sav. = Clematis cadmia Buch. -Ham. ex Hook. f. et Thomson ■

94985 Clematis hancockiana Maxim.;毛萼铁线莲;Hancock Clematis ●

94986 Clematis hastata Finet et Gagnep.;戟状铁线莲(戟叶小木通);Hastate Clematis ●

94987 Clematis hayatae Kudo et Masam. = Clematis henryi Oliv. ●

94988 Clematis hefengensis G. F. Tao = Clematis armandii Franch. var. hefengensis (G. F. Tao) W. T. Wang ●

94989 Clematis henryi Oliv.;单叶铁线莲(地雷,拐子药,亨利氏铁线

莲,蛇松子,雪里开,野灵仙);Henry Clematis ●

94990 Clematis henryi Oliv. var. leptophylla Hayata;薄单叶铁线莲;Thin Henry Clematis ●

94991 Clematis henryi Oliv. var. leptophylla Hayata = Clematis henryi Oliv. ●

94992 Clematis henryi Oliv. var. mollis W. T. Wang;毛单叶铁线莲;Hair Henry Clematis ●

94993 Clematis henryi Oliv. var. morii (Hayata) T. Y. Yang et T. C. Huang = Clematis morii Hayata ●

94994 Clematis henryi Oliv. var. ternata M. Y. Fang;陕南单叶铁线莲;Ternata Henry Clematis ●

94995 Clematis heracleifolia DC.;大叶铁线莲(草本女萎,草牡丹,大样十月泡,牡丹藤,木通,木通花,气死大夫);Tube Clematis ●

94996 Clematis heracleifolia DC. var. davidiana (Decne. ex Verl.) Kuntze;卷萼铁线莲●

94997 Clematis heracleifolia DC. var. davidiana (Decne. ex Verl.) Kuntze = Clematis heracleifolia DC. ●

94998 Clematis heracleifolia DC. var. ichangensis Rehder et E. H. Wilson = Clematis heracleifolia DC. ●

94999 Clematis heracleifolia DC. var. taiwanica S. Suzuki et Hosok. = Clematis psilandra Kitag. ●

95000 Clematis hexapetala Pall.;棉团铁线莲(山蓼,山棉花)●

95001 Clematis hexapetala Pall. f. breviloba (Freyn) Nakai = Clematis hexapetala Pall. ●

95002 Clematis hexapetala Pall. f. longiloba (Freyn) S. H. Li et Y. Huei Huang = Clematis hexapetala Pall. ●

95003 Clematis hexapetala Pall. var. elliptica S. Y. Hu = Clematis hexapetala Pall. var. tchefouensis (Debeaux) S. Y. Hu ●

95004 Clematis hexapetala Pall. var. insularis S. Y. Hu = Clematis hexapetala Pall. var. tchefouensis (Debeaux) S. Y. Hu ●

95005 Clematis hexapetala Pall. var. longiloba (Freyn) S. Y. Hu = Clematis hexapetala Pall. ●

95006 Clematis hexapetala Pall. var. smithiana S. Y. Hu = Clematis hexapetala Pall. ●

95007 Clematis hexapetala Pall. var. tchefouensis (Debeaux) S. Y. Hu;长冬草铁线莲(长冬草)●

95008 Clematis hexapetela Pall. f. dissecta (Y. Yabe) Kitag. = Clematis hexapetala Pall. ●

95009 Clematis hirsuta Guillaumin et Perr.;粗毛铁线莲■☆

95010 Clematis hirsuta Guillaumin et Perr. = Clematis brachiata Thunb. ■☆

95011 Clematis hirsuta Guillaumin et Perr. var. dolichopoda (Brenan) Staner et J. Léonard = Clematis brachiata Thunb. ■☆

95012 Clematis hirsuta Guillaumin et Perr. var. gallaensis (Engl. ex Mildbr.) Cufod. = Clematis brachiata Thunb. ■☆

95013 Clematis hirsuta Guillaumin et Perr. var. glabrescens A. Chev. = Clematis brachiata Thunb. ■☆

95014 Clematis hirsuta Guillaumin et Perr. var. incisodentata (A. Rich.) W. T. Wang;锐齿铁线莲■☆

95015 Clematis hirsuta Guillaumin et Perr. var. junodii (Burtt Davy) W. T. Wang;朱诺德铁线莲■☆

95016 Clematis hirsutissima Pursh;狮鬃铁线莲;Hairy Clematis, Lion's-beard, Vase Flower ■☆

95017 Clematis hirsutissima Pursh var. arizonica (A. Heller) R. O. Erickson = Clematis hirsutissima Pursh ■☆

95018 Clematis hirsutissima Pursh var. scottii (Porter) R. O. Erickson;欧洲铁线莲;Scott's Clematis ■☆

95019 Clematis holosericea Pursh = Clematis virginiana L. ■☆

95020 Clematis homblei De Wild. = Clematis uhehensis Engl. ●☆

95021 Clematis honanensis S. Y. Wang et C. L. Chang = Clematis pseudootophora M. Y. Fang ●

95022 Clematis huchouensis Tamura;吴兴铁线莲(大叶十月泡,河边威灵仙,湖州铁线莲,金剪刀,金剪刀草,铜脚威灵仙);Wuxing Clematis ■

95023 Clematis hupehensis Hemsl. et E. H. Wilson;湖北铁线莲;Hubei Clematis, Hupeh Clematis ●

95024 Clematis ianthina Koehne var. mandshurica (Regel) Nakai = Clematis fusca Turcz. ●

95025 Clematis ianthina Koehne var. violacea (Maxim.) Nakai = Clematis fusca Turcz. var. violacea Maxim. ●

95026 Clematis iliensis Y. S. Hou et W. H. Hou;伊犁铁线莲;Yili Clematis ●

95027 Clematis incisodentata A. Rich. = Clematis hirsuta Guillaumin et Perr. var. incisodentata (A. Rich.) W. T. Wang ■☆

95028 Clematis insularialpina Hayata = Clematis montana Buch. -Ham. ex DC. ●

95029 Clematis integrifolia L.;全缘铁线莲(全叶铁线莲,全缘叶铁线莲,铁线透骨草);Devil's Cut, Entire Clematis, Entireleaf Clematis, Hungarian Climber, Robin Hood's Feather, Robin Hood's Feathers, Robin Hood's Fetter, Solitary Clematis ●

95030 Clematis integrifolia L. var. normalis Kuntze = Clematis integrifolia L. ●

95031 Clematis intricata Bunge;黄花铁线莲(狗肠草,狗豆蔓,狗断肠,萝萝蔓,铁线透骨草,透骨草,狭叶灰绿铁线莲);Intricate Clematis ●

95032 Clematis intricata Bunge var. purpurea Y. Z. Zhao;紫萼铁线莲(变异黄花铁线莲)●

95033 Clematis intricata Bunge var. serram (Maxim.) Kom. = Clematis serratifolia Rehder ●

95034 Clematis iochanica Ulbr. = Clematis lancifolia Bureau et Franch. ●

95035 Clematis iringaensis Engl.;伊林加铁线莲●☆

95036 Clematis jackmanii Moore;雅克曼铁线莲;Jackman Clematis ●☆

95037 Clematis japonica Thunb.;日本铁线莲●☆

95038 Clematis japonica Thunb. var. simsii Makino;西姆斯铁线莲(铁线莲)●☆

95039 Clematis japonica Thunb. var. simsii Makino = Clematis florida Thunb. ●

95040 Clematis japonica Thunb. var. urophylla (Franch.) Kuntze = Clematis urophylla Franch. ●

95041 Clematis japonica Thunb. var. urophylla Kuntze = Clematis urophylla Franch. ●

95042 Clematis javana DC.;台湾铁线莲(川鼻龙,串鼻龙);Java Clematis, Taiwan Clematis ●

95043 Clematis jialasaensis W. T. Wang;加拉萨铁线莲(滇北铁线莲);Jialasa Clematis ●

95044 Clematis jialasaensis W. T. Wang var. macrantha W. T. Wang;滇北铁线莲;N. Yunnan Clematis, North Yunnan Clematis ●

95045 Clematis jingdongensis W. T. Wang;多花铁线莲;Jingdong Clematis ●

95046 Clematis jinzhaiensis Z. W. Xue et X. W. Wang;金寨铁线莲;Jinzhai Clematis ●

95047 Clematis karntscharica Bong. = Clematis fusca Turcz. ●

95048 Clematis keilii Engl. = Clematis commutata Kuntze ■☆

95049 Clematis kerriana Drumm. et Craib = Clematis subumbellata Kurz ●

95050 Clematis kilungensis W. T. Wang et M. Y. Fang = Clematis barbellata Edgew. var. obtusa Kitam. et Tamura ●

95051 Clematis kirilowii Maxim.;太行铁线莲(黑老筋,黑老婆秧,克氏威灵仙,老牛杆);Kirilow Clematis,Taihangshan Clematis ●

95052 Clematis kirilowii Maxim. var. chanetii (H. Lév.) Hand. -Mazz.;狭裂太行铁线莲(狭叶太行铁线莲);Narrowleaf Kirilow Clematis ●

95053 Clematis kirilowii Maxim. var. latisepala (M. C. Chang) W. T. Wang = Clematis pashanensis (M. C. Chang) W. T. Wang ●

95054 Clematis kirilowii Maxim. var. latisepala (M. C. Chang) W. T. Wang = Clematis pashanensis (M. C. Chang) W. T. Wang var. latisepala (M. C. Chang) W. T. Wang ●

95055 Clematis kirilowii Maxim. var. pashanensis M. C. Chang = Clematis pashanensis (M. C. Chang) W. T. Wang ●

95056 Clematis kirkii Oliv. = Clematis villosa DC. subsp. kirkii (Oliv.) Brummitt ■☆

95057 Clematis kirkii Oliv. var. octopetala Engl.;八瓣铁线莲■☆

95058 Clematis kockiana C. K. Schneid.;滇川铁线莲;Kock Clematis ●

95059 Clematis komaroviana Koidz. = Clematis koreana Kom. ●

95060 Clematis komarovii Koidz. = Clematis koreana Kom. ●

95061 Clematis koreana Kom.;朝鲜铁线莲;Korea Clematis,Korean Clematis ●

95062 Clematis kuntziana H. Lév. et Vaniot = Clematis montana Buch.-Ham. ex DC. ●

95063 Clematis kweichowensis C. P'ei;贵州铁线莲;Guizhou Clematis,Kweichow Clematis ●

95064 Clematis kyushuensis Tamura = Clematis chinensis Osbeck var. anhwuiensis (M. C. Chang) W. T. Wang ●

95065 Clematis lancifolia Bureau et Franch.;披针叶铁线莲(八瓜筋,长叶大蓼,披针铁线莲);Lanceolate Clematis ●

95066 Clematis lancifolia Bureau et Franch. var. ternata W. T. Wang et M. C. Chang;竹叶铁线莲;Ternate Clematis ●

95067 Clematis lanuginosa Lindl. et Paxton;毛叶铁线莲;Lanuginose Clematis, Ningpo Clematis, Woolly-leaf Clematis, Woolly-leaved Clematis ●

95068 Clematis lasiandra Maxim.;毛蕊铁线莲(过山龙,毛蕊发汗藤,丝瓜花,小木通)■

95069 Clematis lasiandra Maxim. var. nagasawai Hayata = Clematis lasiandra Maxim. ■

95070 Clematis lasiantha Nutt. = Clematis lasiantha Nutt. ex Torr. et Gray ■☆

95071 Clematis lasiantha Nutt. ex Torr. et Gray;北美毛花铁线莲;Pipestem,Pipe-stem ■☆

95072 Clematis laxipaniculata C. P'ei;串鼻龙●

95073 Clematis laxipaniculata C. P'ei = Clematis subumbellata Kurz ●

95074 Clematis laxistrigosa (W. T. Wang et M. C. Chang) W. T. Wang;糙毛铁线莲(疏金毛铁线莲);Lax Goldwool Clematis, Loose Goldwool Clematis ●

95075 Clematis leiocarpa Oliv. = Clematis uncinata Champ. ex Benth. var. coriacea Pamp. ●

95076 Clematis leptomera Hance = Clematis florida Thunb. ●

95077 Clematis leschenaultiana DC.;锈毛铁线莲(齿叶发汗藤,齿叶铁线莲,滇淮木通,毛木通,细叶女萎,野棉花);Leschenault Clematis,Narrowleaf Rustyhair Clematis,Rustyhair Clematis,Rusty-haired Clematis ●

95078 Clematis leschenaultiana DC. var. angustifolia Hayata = Clematis leschenaultiana DC. ●

95079 Clematis leschenaultiana DC. var. denticulata Merr.;细齿金盏藤●

95080 Clematis leschenaultiana DC. var. denticulata Merr. = Clematis leschenaultiana DC. ●

95081 Clematis leschenaultiana DC. var. rubifolia (C. H. Wright) W. T. Wang = Clematis rubifolia C. H. Wright ●

95082 Clematis liaotungensis Kitag. = Clematis mandshurica Rupr. ●

95083 Clematis liaotungensis Kitag. = Clematis terniflora DC. subsp. mandshurica (Rupr.) Ohwi ●

95084 Clematis liboensis Z. R. Xu;荔波铁线莲;Libo Clematis ●

95085 Clematis ligusticifolia Nutt.;女贞叶铁线莲;Old Man's Beard, Virgin's Bower,Virgin's-bower ■☆

95086 Clematis ligusticifolia Nutt. var. brevifolia Nutt. = Clematis ligusticifolia Nutt. ■☆

95087 Clematis ligusticifolia Nutt. var. californica S. Watson = Clematis ligusticifolia Nutt. ■☆

95088 Clematis limprichtii Ulbr. = Clematis delavayi Franch. var. limprichtii (Ulbr.) M. C. Chang ●

95089 Clematis lingyunensis W. T. Wang;凌云铁线莲;Lingyun Clematis ●

95090 Clematis liukiuensis Warb. = Clematis chinensis Osbeck ●

95091 Clematis loasifolia D. Don = Clematis grewiiflora DC. ●

95092 Clematis longicauda Steud. ex A. Rich. = Clematis grandiflora DC. ■☆

95093 Clematis longiloba DC. = Clematis chinensis Osbeck ●

95094 Clematis longisepala Hayata = Clematis tashiroi Maxim. ●

95095 Clematis longistyla Hand.-Mazz.;光柱铁线莲;Longstyle Clematis,Longstyled Clematis ■

95096 Clematis loureiriana DC.;丝铁线莲(菝葜叶铁线莲,盾叶铁线莲,甘木通,喉痹根,金丝木通,棉藤,紫木通);Filamentary Clematis,Loureiro Clematis ●

95097 Clematis loureiriana DC. subsp. subpeltata (Wall.) Hand.-Mazz. = Clematis loureiriana DC. ●

95098 Clematis loureiriana DC. var. peltata W. T. Wang = Clematis smilacifolia Wall. var. peltata (W. T. Wang) W. T. Wang ●

95099 Clematis lutchuensis Koidz. = Clematis meyeniana Walp. var. insularis Sprague ●

95100 Clematis macropetala Ledeb.;长瓣铁线莲(大瓣铁线莲,大萼铁线莲,石生长瓣铁线莲,四喜牡丹);Bigpetal Clematis, Big-petaled Clematis,Rockliving Clematis ●

95101 Clematis macropetala Ledeb. var. albiflora (Maxim.) Hand.-Mazz.;白花长瓣铁线莲;Whiteflower Clematis ●

95102 Clematis macropetala Ledeb. var. punicoflora Y. Z. Zhao;紫红花长瓣铁线莲●

95103 Clematis macropetala Ledeb. var. rupestris (Turcz.) Hand.-Mazz. = Clematis macropetala Ledeb. ●

95104 Clematis mandshurica Rupr. = Clematis terniflora DC. var. mandshurica (Rupr.) Ohwi ●

95105 Clematis martini H. Lév. = Clematis gouriana Roxb. ex DC. ●

95106 Clematis mashanensis W. T. Wang;马山铁线莲;Mashan Clematis ●

95107 Clematis matsumurana Y. Yabe = Clematis kirilowii Maxim. ●

95108 Clematis maximowicziana Franch. et Sav.;马氏铁线莲;Sweet Autumn Clematis ●☆

95109 Clematis maximowicziana Franch. et Sav. = Clematis terniflora DC. ●

95110 Clematis maximowicziana Franch. et Sav. var. robusta Nakai = Clematis terniflora DC. ●

95111 Clematis menglaensis M. C. Chang;勐腊铁线莲;Mengla Clematis ●

95112 Clematis metouensis M. Y. Fang;墨脱铁线莲;Medog Clematis,

Motuo Clematis ●

95113 Clematis meyeniana Walp.;毛柱铁线莲(吹风藤,老虎须藤,麦氏铁线莲,南铁线莲,山木通,威灵仙);Hairstyle Clematis,Meyer's Clematis ●

95114 Clematis meyeniana Walp. f. major Sprague = Clematis meyeniana Walp. ●

95115 Clematis meyeniana Walp. f. retusa Sprague = Clematis meyeniana Walp. ●

95116 Clematis meyeniana Walp. var. granulata Finet et Gagnep.;沙叶铁线莲(老虎须,软骨过山龙,沙叶山木通);Granulate Clematis ●

95117 Clematis meyeniana Walp. var. insularis Sprague;光梗毛柱铁线莲;Glabrous Meyer Clematis ●

95118 Clematis meyeniana Walp. var. pavoliniana (Pamp.) Sprague = Clematis finetiana H. Lév. et Vaniot ●

95119 Clematis meyeniana Walp. var. uniflora W. T. Wang;单花毛柱铁线莲;One-flower Meyer's Clematis ●

95120 Clematis micrantha Small = Clematis catesbyana Pursh ■☆

95121 Clematis microphylla DC.;澳洲铁线莲(小叶铁线莲)●☆

95122 Clematis minggangiana W. T. Wang = Clematis siamensis Drumm. et Craib ●

95123 Clematis minor Lour. = Clematis chinensis Osbeck ●

95124 Clematis missouriensis Rydb. = Clematis virginiana L. ■☆

95125 Clematis moisseenkoi (Serov) W. T. Wang;绒萼铁线莲●

95126 Clematis montana Buch. -Ham. = Clematis napaulensis DC. ●

95127 Clematis montana Buch. -Ham. ex DC.;绣球藤(白木通,柴木通,川木通,大淮通,海木通,海通,花木通,花叶木通,淮木通,三角枫,山铁线莲,四喜牡丹,油木通);Anemone Clematis, Himalayan Clematis,Indian's Virgin's Bower,Mountain Clematis ●

95128 Clematis montana Buch. -Ham. ex DC. var. brevifolia Kuntze;伏毛绣球藤;Strigose Anemone Clematis ●

95129 Clematis montana Buch. -Ham. ex DC. var. chumbica? = Clematis montana Buch. -Ham. ex DC. ●

95130 Clematis montana Buch. -Ham. ex DC. var. fasciculiflora (Franch.) Brühl = Clematis fasciculiflora Franch. ●

95131 Clematis montana Buch. -Ham. ex DC. var. glabrescens (H. F. Comber) W. T. Wang et M. C. Chang;毛果绣球藤(毛果四喜牡丹);Hairyfruit Anemone Clematis ●

95132 Clematis montana Buch. -Ham. ex DC. var. grandiflora Hook.;大花绣球藤(大花四喜牡丹,大绣球藤,红木通);Large Anemone Clematis ●

95133 Clematis montana Buch. -Ham. ex DC. var. grandiflora Hook. = Clematis montana Buch. -Ham. ex DC. ●

95134 Clematis montana Buch. -Ham. ex DC. var. incisa? = Clematis montana Buch. -Ham. ex DC. ●

95135 Clematis montana Buch. -Ham. ex DC. var. lilacina Lam.;淡紫花绣球藤●☆

95136 Clematis montana Buch. -Ham. ex DC. var. longipes W. T. Wang;长梗绣球藤(大花绣球藤);Long-stalk Anemone Clematis ●

95137 Clematis montana Buch. -Ham. ex DC. var. pentaphylla Maxim. = Clematis gracilifolia Rehder et E. H. Wilson ●

95138 Clematis montana Buch. -Ham. ex DC. var. potanini (Maxim.) Finet et Gagnep. = Clematis potaninii Maxim. ●

95139 Clematis montana Buch. -Ham. ex DC. var. praecox (Kuntze) Brühl;早花绣球藤●

95140 Clematis montana Buch. -Ham. ex DC. var. rubens E. H. Wilson;粉红绣球藤●

95141 Clematis montana Buch. -Ham. ex DC. var. rubens E. H. Wilson = Clematis montana Buch. -Ham. ex DC. var. grandiflora Hook. ●

95142 Clematis montana Buch. -Ham. ex DC. var. sericea Franch. = Clematis chrysocoma Franch. ●

95143 Clematis montana Buch. -Ham. ex DC. var. sterilis Hand. -Mazz.;小叶绣球藤(小叶四喜牡丹);Smallleaf Anemone Clematis ●

95144 Clematis montana Buch. -Ham. ex DC. var. trichogyna M. C. Chang = Clematis montana Buch. -Ham. ex DC. var. glabrescens (H. F. Comber) W. T. Wang et M. C. Chang ●

95145 Clematis montana Buch. -Ham. ex DC. var. wilsonii Sprague;晚花绣球藤(晚花四喜牡丹);E. H. Wilson Anemone Clematis ●

95146 Clematis montana Buch. -Ham. ex DC. var. wilsonii Sprague f. platysepala Rehder et E. H. Wilson = Clematis montana Buch. -Ham. ex DC. var. wilsonii Sprague ●

95147 Clematis morefieldii Král;莫里铁线莲;Morefield's Clematis, Morefield's leather-flower ■☆

95148 Clematis morii Hayata;森氏铁线莲(台湾丝瓜花);Mori Clematis,Taiwan Towelgourdflower ●

95149 Clematis multiflora (H. F. Comber) W. T. Wang = Clematis siamensis Drumm. et Craib ●

95150 Clematis munroiana Wight = Clematis smilacifolia Wall. ●

95151 Clematis nannophylla Maxim.;小叶铁线莲;Small-leaf Clematis, Small-leaved Clematis ●

95152 Clematis nannophylla Maxim. var. foliosa Maxim.;多叶铁线莲;Manyleaf Clematis ●

95153 Clematis nannophylla Maxim. var. pinnatisecta W. T. Wang et L. Q. Li;长小叶铁线莲●

95154 Clematis napaulensis DC.;合苞铁线莲(尼泊尔山金银,尼泊尔铁线莲);Nepal Clematis ●

95155 Clematis napoensis W. T. Wang;那坡铁线莲;Napo Clematis ●

95156 Clematis neomexicana Wooton et Standl. = Clematis ligusticifolia Nutt. ■☆

95157 Clematis nervata Benth. = Clematis drummondii Torr. et A. Gray ■☆

95158 Clematis ningjingshanica W. T. Wang;宁静山铁线莲;Ningjingshan Clematis ●

95159 Clematis nobilis Nakai = Clematis sibirica (L.) Mill. var. ochotensis (Pall.) S. H. Li et Y. Huei Huang ●

95160 Clematis nukiangensis M. Y. Fang;怒江铁线莲;Nujiang Clematis,Nukiang Clematis ●

95161 Clematis nutans Royle var. aethusifolia (Turcz.) Kuntze = Clematis aethusifolia Turcz. ■

95162 Clematis nutans Royle var. pseudoconnata Kuntze = Clematis connata DC. var. pseudoconata (Kuntze) W. T. Wang ●

95163 Clematis nutans Royle var. pseudoconnata Kuntze = Clematis connata DC. ●

95164 Clematis nutans Royle var. thyrsoidea Rehder et E. H. Wilson = Clematis rehderiana Craib ●

95165 Clematis obligocarpa H. Lév. = Clematis chinensis Osbeck ●

95166 Clematis obscura Maxim.;秦岭铁线莲;Qinling Clematis, Tsingling Mountain Clematis ●

95167 Clematis obtusidentata (Rehder et E. H. Wilson) H. Eichler = Clematis apiifolia DC. var. argentilucida (H. Lév. et Vaniot) W. T. Wang ●

95168 Clematis occidentalis (Hornem.) DC.;西方铁线莲;Purple Clematis,Western Blue Virgin's-bower ●☆

95169 Clematis occidentalis (Hornem.) DC. subsp. grosseserrata (Rydb.) R. L. Taylor et Macbryde = Clematis occidentalis

(Hornem.) DC. var. grosseserrata (Rydb.) J. S. Pringle ●☆

95170 Clematis occidentalis (Hornem.) DC. var. dissecta (C. L. Hitchc.) J. S. Pringle;深裂西方铁线莲●☆

95171 Clematis occidentalis (Hornem.) DC. var. grosseserrata (Rydb.) J. S. Pringle;粗齿西方铁线莲●☆

95172 Clematis ochotensis (Pall.) Poir. = Clematis sibirica (L.) Mill. var. ochotensis (Pall.) S. H. Li et Y. Huei Huang ●

95173 Clematis ochroleuca Aiton;西方直立铁线莲;Curly-heads, Erect Silky Leather-flower ●☆

95174 Clematis okinawensis Ohwi = Clematis uncinata Champ. ex Benth. var. okinawensis (Ohwi) Ohwi ●

95175 Clematis okinawensis Ohwi var. trichocarpa (Tamura) Tamura = Clematis uncinata Champ. ex Benth. var. okinawensis (Ohwi) Ohwi ●

95176 Clematis oligocarpa H. Lév. et Vaniot = Clematis chinensis Osbeck ●

95177 Clematis oreophila Hance = Clematis meyeniana Walp. ●

95178 Clematis oreophila Hance ex Walp. = Clematis meyeniana Walp. ●

95179 Clematis orientalis L.;东方铁线莲(铁线莲);Orange-peel Clematis, Oriental Clematis, Oriental Virginsbower ●

95180 Clematis orientalis L. subsp. graveolens = Clematis graveolens Lindl. ■☆

95181 Clematis orientalis L. var. akebioides Maxim. = Clematis akebioides (Maxim.) Veitch ●

95182 Clematis orientalis L. var. glauca Maxim. = Clematis glauca Willd. ●

95183 Clematis orientalis L. var. intricata (Bunge) Maxim. = Clematis intricata Bunge ●

95184 Clematis orientalis L. var. robusta W. T. Wang = Clematis orientalis L. var. sinorobusta W. T. Wang ●

95185 Clematis orientalis L. var. serrata Maxim. = Clematis serratifolia Rehder ●

95186 Clematis orientalis L. var. sinorobusta W. T. Wang;粗梗东方铁线莲;Thickstalk Oriental Clematis ●

95187 Clematis orientalis L. var. tangutica Maxim. = Clematis orientalis L. ●

95188 Clematis orientalis L. var. tangutica Maxim. = Clematis tangutica (Maxim.) Korsh. ●

95189 Clematis orientalis L. var. wilfordii Maxim. = Clematis serratifolia Rehder ●

95190 Clematis ornithopus Ulbr. = Clematis armandii Franch. ●

95191 Clematis otophora Franch. ex Finet et Gagnep.;宽柄铁线莲(抱茎铁线莲);Broadstipe Clematis ●

95192 Clematis otophora Franch. ex Finet et Gagnep. var. longnanensis K. Sun et M. S. Yan;陇南铁线莲;Longnan Clematis ●

95193 Clematis otophora Franch. ex Finet et Gagnep. var. nanensis K. Sun et M. S. Yan = Clematis otophora Franch. ex Finet et Gagnep. ●

95194 Clematis ovatifolia Ito;卵叶铁线莲;Ovateleaf Clematis ●☆

95195 Clematis owatarii Hayata;小渡铁线莲●

95196 Clematis owatarii Hayata = Clematis akoensis Hayata ●

95197 Clematis oweniae Harv.;欧文铁线莲●☆

95198 Clematis oweniae Harv. var. junodii Burtt Davy = Clematis hirsuta Guillaumin et Perr. var. junodii (Burtt Davy) W. T. Wang ■☆

95199 Clematis pamiralaica Grey-Wilson;帕米尔铁线莲;Pamir Clematis ●

95200 Clematis paniculata Thunb. = Clematis terniflora DC. ●

95201 Clematis parviloba Gardner et Champ.;裂叶铁线莲(大牡丹藤,岛屿小叶牡丹藤,裂叶威灵仙,皮氏铁线莲,小裂木通,小叶铁线莲);Pierott Clematis, Smalllobed Clematis ●

95202 Clematis parviloba Gardner et Champ. subsp. bartlettii (Gogelein) T. Y. Yang et T. C. Huang = Clematis parviloba Gardner et Champ. var. bartlettii (Gogelein) W. T. Wang ●

95203 Clematis parviloba Gardner et Champ. subsp. puberula (Hook. f. et Thomson) Kuntze = Clematis puberula Hook. f. et Thomson ●

95204 Clematis parviloba Gardner et Champ. var. bartlettii (Gogelein) W. T. Wang;巴氏裂叶铁线莲(巴氏铁线莲);Bartlett Smalllobed Clematis ●

95205 Clematis parviloba Gardner et Champ. var. ganpiniana (H. Lév. et Vaniot) Rehder = Clematis puberula Hook. f. et Thomson var. ganpiniana (H. Lév. et Vaniot) W. T. Wang ●

95206 Clematis parviloba Gardner et Champ. var. glabrescens Finet et Gagnep. = Clematis puberula Hook. f. et Thomson var. ganpiniana (H. Lév. et Vaniot) W. T. Wang ●

95207 Clematis parviloba Gardner et Champ. var. longianthera W. T. Wang;长药裂叶铁线莲;Longflower Smalllobed Clematis ●

95208 Clematis parviloba Gardner et Champ. var. puberula? = Clematis puberula Hook. f. et Thomson ●

95209 Clematis parviloba Gardner et Champ. var. rhombico-elliptica W. T. Wang;菱果裂叶铁线莲●

95210 Clematis parviloba Gardner et Champ. var. suboblonga W. T. Wang;长圆裂叶铁线莲;Suboblong Smalllobed Clematis ●

95211 Clematis parviloba Gardner et Champ. var. tenuipes (W. T. Wang) C. T. Ting;细柄木通;Thinstalk Smalllobed Clematis ●

95212 Clematis parviloba Gardner et Champ. var. tenuipes (W. T. Wang) C. T. Ting = Clematis tenuipes W. T. Wang ●

95213 Clematis parviloba Gardner et Champ. var. tenuipes (W. T. Wang) C. T. Ting = Clematis parviloba Gardner et Champ. var. longianthera W. T. Wang ●

95214 Clematis pashanensis (M. C. Chang) W. T. Wang;巴山铁线莲(尖药巴山铁线莲,老虎须,细尖太行铁线莲);Bashan Clematis, Latisepal Kirilow Clematis, Pashan Clematis ●

95215 Clematis pashanensis (M. C. Chang) W. T. Wang var. latisepala (M. C. Chang) W. T. Wang;尖药巴山铁线莲●

95216 Clematis patens E. Morren et Decne.;转子莲(大花铁线莲);Lilac Clematis ●

95217 Clematis patens E. Morren et Decne. 'Grandiflora';大花转子莲(大花铁线莲);Big-flower Clematis ●☆

95218 Clematis patens E. Morren et Decne. f. alba Makino;白花转子莲●☆

95219 Clematis patens E. Morren et Decne. f. coerulea Makino;紫花转子莲●☆

95220 Clematis patens E. Morren et Decne. subsp. tientaiensis M. Y. Fang = Clematis patens E. Morren et Decne. var. tientaiensis (M. Y. Fang) W. T. Wang ●

95221 Clematis patens E. Morren et Decne. var. monstrsa Planch.;奇形转子莲●☆

95222 Clematis patens E. Morren et Decne. var. tientaiensis (M. Y. Fang) W. T. Wang;天台铁线莲;Tiantai Clematis, Tientai Clematis ●

95223 Clematis pauciflora Nutt.;少花铁线莲;Ropevine ■☆

95224 Clematis pavoliniana Pamp. = Clematis finetiana H. Lév. et Vaniot ●

95225 Clematis petalotii Gagnep. = Clematis smilacifolia Wall. ●

95226 Clematis peterae Hand. -Mazz.;钝萼铁线莲(白血藤,柴木通,风藤草,花氏木通,木通,木通藤,山棉花,疏齿铁线莲,细木通,线木通,小果木通,小木通,小蓑衣藤);Obtusesepal Clematis ●

95227 Clematis peterae Hand. -Mazz. var. lishanensis (T. Y. Yang et T. C. Huang) W. T. Wang;梨山铁线莲(梨山小蓑衣藤);Lishan Obtusesepal Clematis ●

95228 Clematis peterae Hand. -Mazz. var. mollis W. T. Wang;毛叶小果木通(毛叶钝萼铁线莲);Hairyleaf Obtusesepal Clematis ●

95229 Clematis peterae Hand. -Mazz. var. mollis W. T. Wang = Clematis peterae Hand. -Mazz. ●

95230 Clematis peterae Hand. -Mazz. var. trichocarpa W. T. Wang;毛果钝萼铁线莲(毛果木通,毛果铁线莲);Hairyfruit Obtusesepal Clematis ●

95231 Clematis petersiana Klotzsch = Clematis brachiata Thunb. ■☆

95232 Clematis phaseolifolia W. T. Wang = Clematis vaniotii H. Lév. et Porter ●

95233 Clematis philippiana H. Lév. = Clematis ranunculoides Franch. ■

95234 Clematis pianmaensis W. T. Wang;片马铁线莲;Pianma Clematis ●

95235 Clematis pierotii Miq. = Clematis parviloba Gardner et Champ. ●

95236 Clematis pinchuanensis W. T. Wang et M. Y. Fang;宾川铁线莲;Binchuan Clematis, Pinchuan Clematis ●

95237 Clematis pinchuanensis W. T. Wang et M. Y. Fang var. tomentosa (Finet et Gagnep.) W. T. Wang;三出宾川铁线莲●

95238 Clematis pinnata Maxim.;羽叶铁线莲;Pinnate Clematis ●

95239 Clematis pinnata Maxim. var. tatarinowii (Maxim.) Kuntze = Clematis tatarinowii Maxim. ●

95240 Clematis pinnata Maxim. var. ternatifolia W. T. Wang;平谷铁线莲;Pinggu Pinnate Clematis ●

95241 Clematis pitcheri Torr. et Gray var. dictyota (Greene) W. M. Dennis;小网皮彻铁线莲■☆

95242 Clematis pitcheri Torr. et Gray var. filifera (Benth.) R. O. Erickson = Clematis pitscheri Torr. et Gray ■☆

95243 Clematis pitscheri Torr. et Gray;皮彻铁线莲;Bellflower Clematis, Bluebell, Leather Flower, Pitcher's Clematis ■☆

95244 Clematis platysepala (Trautv. et C. A. Mey.) Hand. -Mazz. = Clematis sibirica (L.) Mill. var. ochotensis (Pall.) S. H. Li et Y. Huei Huang ●

95245 Clematis pogonandra Maxim.;须蕊铁线莲(髯丝铁线莲);Tomentoseanther Clematis ●

95246 Clematis pogonandra Maxim. var. alata W. T. Wang et M. Y. Fang;雷波铁线莲;Winged Clematis ●

95247 Clematis pogonandra Maxim. var. pilosa Rehder et E. H. Wilson = Clematis pogonandra Maxim. var. pilosula Rehder et E. H. Wilson ●

95248 Clematis pogonandra Maxim. var. pilosula Rehder et E. H. Wilson;多毛须蕊铁线莲●

95249 Clematis potaninii Maxim.;美花铁线莲(藏四喜牡丹);Potanin Clematis ●

95250 Clematis potaninii Maxim. var. fargesii (Franch.) Hand. -Mazz. = Clematis potaninii Maxim. ●

95251 Clematis prattii Hemsl. = Clematis pogonandra Maxim. ●

95252 Clematis pseudoalpina (Kuntze) A. Nelson = Clematis columbiana (Nutt.) Torr. et A. Gray ■☆

95253 Clematis pseudoflammula Schmalh. ex Lipsky;假焰红铁线莲■☆

95254 Clematis pseudograndiflora Kuntze = Clematis grandiflora DC. ■☆

95255 Clematis pseudootophora M. Y. Fang;华中铁线莲;Central China Clematis ●

95256 Clematis pseudootophora M. Y. Fang var. integra W. T. Wang;全缘华中铁线莲;Entire Central Clematis ●

95257 Clematis pseudootophora M. Y. Fang var. integra W. T. Wang = Clematis pseudootophora M. Y. Fang ●

95258 Clematis pseudopogonandra Finet et Gagnep.;西南铁线莲(花木通,假须蕊铁线莲,少齿发汗藤,西藏花木通,须蕊发汗藤);Fewtooth SW. China Clematis, Southwestern China Clematis, SW. China Clematis ●

95259 Clematis pseudopogonandra Finet et Gagnep. var. paucidentata Finet et Gagnep. = Clematis pseudopogonandra Finet et Gagnep. ●

95260 Clematis psilandra Kitag.;光蕊铁线莲(台湾草牡丹,台湾牡丹藤);Smoothcalyx Clematis ●

95261 Clematis pterantha Dunn;思茅铁线莲;Simao Clematis ■

95262 Clematis pterantha Dunn = Clematis ranunculoides Franch. var. pterantha (Dunn) M. Y. Fang ●

95263 Clematis pterantha Dunn var. grossedentata Rehder et E. H. Wilson = Clematis ranunculoides Franch. ■

95264 Clematis puberula Hook. f. et Thomson;短毛铁线莲(淮通);Shorthair Clematis ●

95265 Clematis puberula Hook. f. et Thomson var. ganpiniana (H. Lév. et Vaniot) W. T. Wang;扬子铁线莲(小肥猪藤,镇平铁线莲);Ganpin Clematis, Ganpin Shorthair Clematis ●

95266 Clematis puberula Hook. f. et Thomson var. subsericea (Rehder et E. H. Wilson) W. T. Wang;毛叶扬子铁线莲;Hairyleaf Ganpin Clematis, Hairyleaf Shorthair Clematis ●

95267 Clematis puberula Hook. f. et Thomson var. tenuisepala (Maxim.) W. T. Wang;毛果扬子铁线莲;Hairy-fruit Short-hair Clematis, Thinsepal Clematis ●

95268 Clematis pubipes (W. T. Wang) W. T. Wang = Clematis crassipes Chun et F. C. How var. pubipes W. T. Wang ●

95269 Clematis pycnocoma W. T. Wang;密毛铁线莲;Densehair Clematis ●

95270 Clematis qingchengshanica W. T. Wang;青城铁线莲;Qingcheng Clematis ●

95271 Clematis quinquefoliolata Hutch.;五叶铁线莲(大舒筋活血,见血愁,辣药,柳叶见血飞);Fivefoliolate Clematis, Quinquefoliolate Clematis, Woodbine ●

95272 Clematis ranunculoides Franch.;毛茛铁线莲(白木通,山棉花,铁丝牡丹,细木通,小九狮子草,绣球藤);Buttercup-like Clematis ■

95273 Clematis ranunculoides Franch. var. cordata M. Y. Fang;心叶铁线莲■

95274 Clematis ranunculoides Franch. var. grossedentata (Rehder et E. H. Wilson) Hand. -Mazz. = Clematis ranunculoides Franch. ■

95275 Clematis ranunculoides Franch. var. pterantha (Dunn) M. Y. Fang = Clematis peterae Hand. -Mazz. ●

95276 Clematis ranunculoides Franch. var. tomentosa Finet et Gagnep. = Clematis pinchuanensis W. T. Wang et M. Y. Fang var. tomentosa (Finet et Gagnep.) W. T. Wang ●

95277 Clematis recta L.;直立威灵仙;Erect Clematis, Ground Clematis, Ground Virginsbower ●☆

95278 Clematis recta L. = Clematis mandshurica Rupr. ●

95279 Clematis recta L. = Clematis terniflora DC. subsp. mandshurica (Rupr.) Ohwi ●

95280 Clematis recta L. subsp. kirilowii (Maxim.) Kuntze = Clematis kirilowii Maxim. ●

95281 Clematis recta L. subsp. nannophylla (Maxim.) Kuntze = Clematis nannophylla Maxim. ●

95282 Clematis recta L. subsp. paniculata Kuntze = Clematis terniflora DC. ●

95283 Clematis recta L. subsp. songarica (Bunge) Kuntze = Clematis songarica Bunge ●

95284 Clematis recta L. var. mandshurica (Rupr.) Maxim. = Clematis mandshurica Rupr. ●

95285 Clematis recta L. var. mandshurica (Rupr.) Maxim. = Clematis terniflora DC. subsp. mandshurica (Rupr.) Ohwi ●

95286 Clematis rehderiana Craib;长花铁线莲(垂花发汗藤);Rehder Clematis ●

95287 Clematis repens Finet et Gagnep.;曲柄铁线莲●

95288 Clematis reticulata Walter;网状铁线莲■☆

95289 Clematis rubifolia C. H. Wright;莓叶铁线莲(薰叶发汗藤,红叶铁线莲);Raspberryleaf Clematis,Raspberry-leaved Clematis ●

95290 Clematis sapini De Wild. = Clematis villosa DC. ■☆

95291 Clematis sasakii Shimizu = Clematis formosana Kuntze ●

95292 Clematis scabiosifolia DC. = Clematis villosa DC. ■☆

95293 Clematis scottii Porter = Clematis hirsutissima Pursh var. scottii (Porter) R. O. Erickson ■☆

95294 Clematis serrata (Maxim.) Kom. = Clematis serratifolia Rehder ●

95295 Clematis serratifolia Rehder;齿叶铁线莲;Hermitgold Clematis ●

95296 Clematis shenlungchiaensis M. Y. Fang;神农架铁线莲;Shennongjia Clematis ●

95297 Clematis shensiensis W. T. Wang;陕西铁线莲(武当铁线莲);Shaanxi Clematis,Shensi Clematis ●

95298 Clematis siamensis Drumm. et Craib;锡金铁线莲(多花铁线莲,多花锡金铁线莲);Manyflower Sikkim Clematis,Many-flowered Sikkim Clematis,Sikkim Clematis ●

95299 Clematis siamensis Drumm. et Craib var. andersonii (Brühl) W. T. Wang;安德森铁线莲;Anderson's Clematis ●☆

95300 Clematis siamensis Drumm. et Craib var. klarkei (Kuntze) W. T. Wang;毛萼锡金铁线莲;Klarke's Sikkim Clematis ●

95301 Clematis siamensis Drumm. et Craib var. monantha (W. T. Wang et L. Q. Li) W. T. Wang et L. Q. Li;单花锡金铁线莲;One-flower Sikkim Clematis ●

95302 Clematis sibirica (L.) Mill.;西伯利亚铁线莲(花木通,天山木通,天祝铁线莲,新疆木通);Siberia Clematis,Siberian Clematis ●

95303 Clematis sibirica (L.) Mill. var. iliensis (Y. S. Hou et W. H. Hou) J. G. Liu = Clematis iliensis Y. S. Hou et W. H. Hou ●

95304 Clematis sibirica (L.) Mill. var. ochotensis (Pall.) S. H. Li et Y. Huei Huang;半钟铁线莲(高山铁线莲);Ochotsk Clematis ●

95305 Clematis sibirica (L.) Mill. var. ochotensis (Pall.) S. H. Li et Y. Huei Huang = Clematis ochotensis (Pall.) Poir. ●

95306 Clematis sibirica (L.) Mill. var. tianzhuensis M. S. Yan et K. Sun = Clematis sibirica (L.) Mill. ●

95307 Clematis sibirica Mill. var. iliensis (Y. S. Hou et W. H. Hou) J. G. Liu = Clematis iliensis Y. S. Hou et W. H. Hou ●

95308 Clematis sikkimensis Drumm. ex Burkill = Clematis siamensis Drumm. et Craib ●

95309 Clematis sikkimensis Drumm. ex Burkill var. clarkei (Kuntze) W. T. Wang = Clematis siamensis Drumm. et Craib var. klarkei (Kuntze) W. T. Wang ●

95310 Clematis sikkimensis Drumm. ex Burkill var. monantha W. T. Wang et L. Q. Li = Clematis siamensis Drumm. et Craib var. monantha (W. T. Wang et L. Q. Li) W. T. Wang et L. Q. Li ●

95311 Clematis simensis Fresen.;西门铁线莲■☆

95312 Clematis simsii Sweet;西氏铁线莲●☆

95313 Clematis sinensis Lour. = Clematis chinensis Osbeck ●

95314 Clematis sinii W. T. Wang;辛氏铁线莲;Sin's Clematis ●

95315 Clematis smilacifolia Wall.;菝葜叶铁线莲;Smilax-leaf Clematis ●

95316 Clematis smilacifolia Wall. var. peltata (W. T. Wang) W. T. Wang;盾叶铁线莲;Peltate Clematis,Peltate Smilax-leaf Clematis,Peltateleaf Clematis ●

95317 Clematis smilacifolia Wall. var. subpeltata (Wall.) Kuntze = Clematis loureiriana DC. ●

95318 Clematis socialis Král;阿拉巴马铁线莲;Alabama Leather-flower ■☆

95319 Clematis songarica Bunge;准噶尔铁线莲;Dzungar Clematis,Songar Clematis,Sungari Clematis ●

95320 Clematis songarica Bunge var. asplenifolia (Schrenk) Trautv.;蕨叶铁线莲;Fernleaf Clematis ●

95321 Clematis songarica Bunge var. asplenifolia (Schrenk) Trautv. = Clematis songarica Bunge ●

95322 Clematis songarica Bunge var. intermedia Truatv. = Clematis songarica Bunge var. asplenifolia (Schrenk) Trautv. ●

95323 Clematis souliei Franch. ex Finet et Gagnep. = Clematis potaninii Maxim. ●

95324 Clematis spathulifolia (Kuntze) Prantl;匙叶铁线莲■☆

95325 Clematis splendens H. Lév. et Vaniot = Clematis rubifolia C. H. Wright ●

95326 Clematis spooneri Rehder et E. H. Wilson = Clematis chrysocoma Franch. ●

95327 Clematis spooneri Rehder et E. H. Wilson var. subglabra S. Y. Hu = Clematis montana Buch. -Ham. ex DC. ●

95328 Clematis stanleyi Hook. = Clematis villosa DC. subsp. stanleyi (Hook.) Kuntze ■☆

95329 Clematis stans Siebold. et Zucc.;直立铁线莲(直铁线莲)■

95330 Clematis stewartiae Burtt Davy;斯图铁线莲■☆

95331 Clematis stronachii Hance = Clematis cadmia Buch. -Ham. ex Hook. f. et Thomson ■

95332 Clematis subfalcata C. P'ei ex M. Y. Fang;镰叶铁线莲;Falcate Clematis,Subfalcate Clematis,Subfalcateleaf Clematis,Subfalcate-leaved Clematis ●

95333 Clematis subfalcata C. P'ei ex M. Y. Fang var. pubipes W. T. Wang;毛梗镰叶铁线莲;Hairystalk Subfalcate Clematis ●

95334 Clematis subfalcata C. P'ei ex M. Y. Fang = Clematis yunnanensis Franch. ●

95335 Clematis subfalcata C. P'ei ex M. Y. Fang var. pubipes W. T. Wang = Clematis yunnanensis Franch. ●

95336 Clematis subfalcata C. P'ei ex M. Y. Fang var. stephylla (Hand. -Mazz.) W. T. Wang;序梗镰叶铁线莲●

95337 Clematis subfalcata C. P'ei ex M. Y. Fang var. stephylla (Hand. -Mazz.) W. T. Wang = Clematis yunnanensis Franch. ●

95338 Clematis subpeltata Wall. = Clematis loureiriana DC. ●

95339 Clematis subpeltata Wall. = Clematis smilacifolia Wall. ●

95340 Clematis subumbellata Kurz;细木通(基端铁线莲,小木通);Kerr Clematis ●

95341 Clematis suksdorfii B. L. Rob. = Clematis ligusticifolia Nutt. ■☆

95342 Clematis taiwaniana Hayata = Clematis javana DC. ●

95343 Clematis taiwanica Hayata = Clematis javana DC. ●

95344 Clematis tamurae T. Y. Yang et T. C. Huang;田村氏铁线莲(田村铁线莲);Tamura Clematis ■

95345 Clematis tangutica (Maxim.) Korsh.;甘青铁线莲(唐古特铁线莲,田村氏铁线莲);Golden Clematis,Orange-peel Clematis,Tangut Clematis ●

95346 Clematis tangutica (Maxim.) Korsh. subsp. mongolica Grey-Wilson = Clematis tangutica (Maxim.) Korsh. var. mongolica (Grey-Wilson) W. T. Wang ●☆

95347 Clematis tangutica (Maxim.) Korsh. var. mongolica (Grey-Wilson) W. T. Wang;蒙古铁线莲●☆

95348 Clematis tangutica (Maxim.) Korsh. var. obtusiuscula Rehder et E. H. Wilson;钝萼甘青铁线莲●

95349 Clematis tangutica (Maxim.) Korsh. var. pubescens M. C. Chang et P. P. Ling;毛萼甘青铁线莲●

95350 Clematis tashiroi Maxim.;长萼铁线莲(长叶铁线莲,琉球铁线莲,田代氏铁线莲);Longsepal Clematis, Long-sepaled Clematis, Tashiro Clematis ●

95351 Clematis tatarinowii Maxim.;细花铁线莲;Tatarinow Clematis ●

95352 Clematis tenuifolia Royle;西藏铁线莲●

95353 Clematis tenuifolia Royle = Clematis orientalis L. ●

95354 Clematis tenuiloba (A. Gray) C. L. Hitchc. = Clematis columbiana (Nutt.) Torr. et A. Gray var. tenuiloba (A. Gray) J. S. Pringle ■☆

95355 Clematis tenuipes W. T. Wang;细梗铁线莲;Thin-stalk Clematis ●

95356 Clematis tenuipes W. T. Wang = Clematis parviloba Gardner et Champ. var. tenuipes (W. T. Wang) C. T. Ting ●

95357 Clematis teretipes W. T. Wang;柱梗铁线莲●

95358 Clematis terniflora DC.;圆锥铁线莲(白花藤,黄药子,铁脚威灵仙,铜脚威灵仙,铜灵仙,铜威灵,小叶力刚,蟹珠眼草,芋叶铁线莲,锥花铁线莲);Bush Clematis, Japanese Clematis, Sweet Autumn Clematis, Sweet Autumn Virginsbower, Threeflower Clematis, Triflorous Clematis, Yam-leaved Clematis ●

95359 Clematis terniflora DC. subsp. garanbiensis (Hayata) M. C. Chang = Clematis terniflora DC. var. garanbiensis DC. ●

95360 Clematis terniflora DC. subsp. mandshurica (Rupr.) Ohwi = Clematis mandshurica Rupr. ●

95361 Clematis terniflora DC. var. garanbiensis DC.;鹅銮鼻铁线莲;Garanbi Clematis ●

95362 Clematis terniflora DC. var. latisepala M. C. Chang;宽萼圆锥铁线莲;Broadsepal Clematis, Broadsepal Threeflower Clematis ●

95363 Clematis terniflora DC. var. latisepala M. C. Chang = Clematis kirilowii Maxim. var. latisepala (M. C. Chang) W. T. Wang ●

95364 Clematis terniflora DC. var. latisepala M. C. Chang = Clematis pashanensis (M. C. Chang) W. T. Wang ●

95365 Clematis terniflora DC. var. latisepala M. C. Chang = Clematis pashanensis (M. C. Chang) W. T. Wang var. latisepala (M. C. Chang) W. T. Wang ●

95366 Clematis terniflora DC. var. mandshurica (Rupr.) Ohwi;辣蓼铁线莲(东北铁线莲);Manchurian Clematis ●

95367 Clematis terniflora DC. var. mandshurica (Rupr.) Ohwi = Clematis mandshurica Rupr. ●

95368 Clematis terniflora DC. var. robusta (Carrière) Tamura;粗壮铁线莲(鹅銮鼻铁线莲)●

95369 Clematis terniflora DC. var. robusta (Carrière) Tamura = Clematis terniflora DC. var. garanbiensis DC. ●

95370 Clematis terniflora DC. var. robusta (Carrière) Tamura = Clematis terniflora DC. ●

95371 Clematis texensis Buckley;红花半钟蔓;Scarlet Clematis ●☆

95372 Clematis thunbergii Steud. = Clematis triloba Thunb. ■☆

95373 Clematis tianschanica Pavlov;天山铁线莲●

95374 Clematis tianschanica Pavlov = Clematis sibirica (L.) Mill. ●

95375 Clematis tibestica Quézel = Clematis brachiata Thunb. ■☆

95376 Clematis tibetana Kuntze;中印铁线莲;Tibet Clematis ●

95377 Clematis tibetana Kuntze var. lineariloba W. T. Wang;狭裂中印铁线莲●

95378 Clematis tibetana Kuntze var. vernayi (C. E. C. Fisch.) Grey-Wilson = Clematis tibetana Kuntze var. vernayi (C. E. C. Fisch.) W. T. Wang ●

95379 Clematis tibetana Kuntze var. vernayi (C. E. C. Fisch.) W. T. Wang;厚萼中印铁线莲●

95380 Clematis tinghuensis C. T. Ting;鼎湖铁线莲;Dinghu Clematis, Tinghu Clematis ●

95381 Clematis tomentella (Maxim.) W. T. Wang et L. Q. Li;灰叶铁线莲●

95382 Clematis tongluensis C. T. Ting var. mollisepala W. T. Wang;软萼铁线莲(毛萼铁线莲);Hairy-sepal Clematis ●

95383 Clematis tozanensis Hayata = Clematis tashiroi Maxim. ●

95384 Clematis trichocarpa Tamura;台毛果铁线莲●

95385 Clematis trichocarpa Tamura = Clematis uncinata Champ. ex Benth. var. okinawensis (Ohwi) Ohwi ●

95386 Clematis trifollata Thunb. = Akebia trifoliata (Thunb.) Koidz. ●

95387 Clematis triloba Thunb.;三裂铁线莲■☆

95388 Clematis tripartita W. T. Wang; 深裂铁线莲; Deep-lobed Clematis ●

95389 Clematis trullifera (Franch.) Finet et Gagnep. = Clematis connata DC. var. trullifera (Franch.) W. T. Wang ●

95390 Clematis tsaii W. T. Wang;福贡铁线莲;Tsai's Clematis ●

95391 Clematis tsengiana Metcalf = Clematis hancockiana Maxim. ●

95392 Clematis tsugetorum Ohwi;台中铁线莲(高山铁线莲);Alpine Clematis, Taizhong Clematis ●

95393 Clematis tubulosa Turcz. = Clematis heracleifolia DC. ●

95394 Clematis tubulosa Turcz. var. davidiana (Decne. ex Veriot) Franch. = Clematis heracleifolia DC. ●

95395 Clematis uhehensis Engl.;乌赫铁线莲●☆

95396 Clematis uncinata Champ. ex Benth.;柱果铁线莲(白灵仙,钩铁线莲,黑木通,花木通,癞子藤,老虎师藤,三叶铁线莲,铁脚威灵仙,威灵仙,小叶光板力刚,岩木通一,一把扇,猪娘藤);Hooked Clematis, Stylefruit Clematis ●

95397 Clematis uncinata Champ. ex Benth. var. biternata W. T. Wang = Clematis uncinata Champ. ex Benth. ●

95398 Clematis uncinata Champ. ex Benth. var. coriacea Pamp.;皱叶铁线莲(革叶铁线莲);Wrinkleleaf Clematis ●

95399 Clematis uncinata Champ. ex Benth. var. floribunda Hayata = Clematis uncinata Champ. ex Benth. ●

95400 Clematis uncinata Champ. ex Benth. var. okinawensis (Ohwi) Ohwi;毛柱果铁线莲(毛果铁线莲)●

95401 Clematis urophylla Franch.;尾叶铁线莲(尾叶发汗藤,小齿发汗藤,小齿铁线莲);Obtuse-tooth Urophyllous Clematis, Smalltooth Clematis, Urophyllous Clematis ●

95402 Clematis urophylla Franch. var. heterophylla H. Lév. = Clematis ranunculoides Franch. ■

95403 Clematis urophylla Franch. var. obtusiuscula C. K. Schneid. = Clematis urophylla Franch. ●

95404 Clematis urophylla Franch. var. taitongensis Y. C. Liu et C. H. Ou;台东女萎■

95405 Clematis vaniotii H. Lév. et Porter;云贵铁线莲(粗糠藤,豆叶木通);Vaniot Clematis ●

95406 Clematis veitchiana Craib = Clematis rehderiana Craib ●

95407 Clematis velutina Edgew. = Clematis connata DC. ●

95408 Clematis venosa Royle = Clematis connata DC. ●

95409 Clematis venusta M. C. Chang; 丽叶铁线莲; Beautifulleaf Clematis, Pretty Clematis ●

95410 Clematis vernayi C. E. C. Fisch. = Clematis tibetana Kuntze var. vernayi (C. E. C. Fisch.) W. T. Wang ●

95411 Clematis vernayi C. E. C. Fisch. = Clematis tibetana Kuntze ●

95412 Clematis versicolor Small ex Rydb.;苍白铁线莲■☆

95413 Clematis verticillaris DC. = Clematis occidentalis (Hornem.) DC. ●☆

95414 Clematis verticillaris DC. var. cacuminis Fernald = Clematis occidentalis (Hornem.) DC. ●☆

95415 Clematis verticillaris DC. var. grandiflora B. Boivin = Clematis occidentalis (Hornem.) DC. ●☆

95416 Clematis villosa B. M. Yang = Clematis baominiana W. T. Wang ●

95417 Clematis villosa DC.;长毛铁线莲■☆

95418 Clematis villosa DC. subsp. kirkii (Oliv.) Brummitt;柯克长毛铁线莲■☆

95419 Clematis villosa DC. subsp. oliveri (Hutch.) Brummitt;奥里弗铁线莲■☆

95420 Clematis villosa DC. subsp. spathulifolia Kuntze = Clematis spathulifolia (Kuntze) Prantl ■☆

95421 Clematis villosa DC. subsp. stanleyi (Hook.) Kuntze;斯坦利铁线莲■☆

95422 Clematis villosa DC. var. scabiosifolia (DC.) Kuntze = Clematis villosa DC. ■☆

95423 Clematis viorna L.;革花铁线莲;Leather Flower, Leatherflower, Leatherflower Clematis ■☆

95424 Clematis viorna L. = Coriflora viorna (L.) W. A. Weber ■☆

95425 Clematis viorna L. var. flaccida (Small ex Rydb.) R. O. Erickson = Clematis viorna L. ■☆

95426 Clematis virginiana L.;北美铁线莲;Devil's Darning Needle, Devil's Darning-needle, Devil's-darning-needle, Virginia Virgin's-bower, Virgin's Bower, Virgin's Bower Clematis, Virgin's-bower, Wild Clematis, Woodbine ■☆

95427 Clematis virginiana L. f. missouriensis (Rydb.) Fernald = Clematis virginiana L. ■☆

95428 Clematis virginiana L. var. missouriensis (Rydb.) E. J. Palmer et Steyerm. = Clematis virginiana L. ■☆

95429 Clematis viridiflora Bertol.;绿花铁线莲■☆

95430 Clematis viridis (W. T. Wang et M. C. Chang) W. T. Wang;绿叶铁线莲(长梗灰叶铁线莲);Longpedicel Hoary Clematis ●

95431 Clematis vitalba L.;葡萄叶铁线莲(葡萄胞铁线莲);Bearbind, Bedvine, Bedwind, Bedwine, Beggar Brushes, Beggar's Brush, Beggar's Plant, Bellywind, Bethroot, Bethwine, Binder, Bithywind, Blind Man's Buff, Boy's Bacca, Brihywine, Bullbine, Burning Bush, Burning Climber, Bushy Beard, By-the-wind, Climbers, Consolation Flower, Crocodile, Daddy Man's Beard, Daddy's Beard, Daddy's Whiskers, Devil's Cut, Devil's Guts, Devil's Thread, Devil's Threads, Devil's Twine, Devil's Yam, Downivine, Evergreen Clematis, Father Time, Grandfather's Beard, Grandfather's Whiskers, Grandfy's Beard, Grey Beard, Grey-beard, Gypsy's Bacca, Gypsy's Tobacco, Hag Rope, Hag-rope, Halfwood, Hay Rope, Hedge Feathers, Hedge Vine, Honesty, Honeystick, Lady-in-the-bower, Lady's Bower, Lady's Bowler, Love-bind, Love-entangled, Loveman, Love-man, Love-me, Maiden's Hair, Maiden's Honesty, Old Man, Old Man's Beard, Old Man's Woozard, Pethwine, Pithywind, Poor Man's Friend, Robin Hood's Fetter, Shepherd's Delight, Silver Bush, Skipping Rope, Smokewood, Smoking Cane, Snow-in-harvest, Tasmanian Snow Gum, Tom Bacca, Traveler's Joy, Traveler's-joy, Tuzzy-muzzy, Virghl's-bower, Virgin's Bower, Vitis-cell Clematis, Wild Clematis, Wild Vine, Willow-wind, Withyvine, Withywind ●☆

95432 Clematis vitalba L. subsp. brevicaudata (DC.) Kuntze = Clematis brevicaudata DC. ●

95433 Clematis vitalba L. subsp. brevicaudata (DC.) Kuntze = Clematis javana DC. ●

95434 Clematis vitalba L. subsp. cumingii Kuntze = Clematis javana DC. ●

95435 Clematis vitalba L. subsp. gouriana (Roxb. ex DC.) Kuntze = Clematis gouriana Roxb. ex DC. ●

95436 Clematis vitalba L. subsp. gouriana (Roxb.) Kuntze = Clematis gouriana Roxb. ex DC. ●

95437 Clematis vitalba L. subsp. grata (Wall.) Kuntze = Clematis argentilucida (H. Lév. et Vaniot) W. T. Wang ●

95438 Clematis vitalba L. subsp. javana (DC.) Kuntze = Clematis javana DC. ●

95439 Clematis vitalba L. var. argentilucida H. Lév. = Clematis apiifolia DC. var. argentilucida (H. Lév. et Vaniot) W. T. Wang ●

95440 Clematis vitalba L. var. argentilucida H. Lév. et Vaniot = Clematis grandidentata (Rehder et E. H. Wilson) W. T. Wang ●

95441 Clematis vitalba L. var. ganpiniana H. Lév. et Vaniot = Clematis ganpiniana (H. Lév. et Vaniot) Tamura ●

95442 Clematis vitalba L. var. ganpiniana H. Lév. et Vaniot = Clematis puberula Hook. f. et Thomson var. ganpiniana (H. Lév. et Vaniot) W. T. Wang ●

95443 Clematis vitalba L. var. gouriana (Roxb. ex DC.) Finet et Gagnep. = Clematis gouriana Roxb. ex DC. ●

95444 Clematis vitalba L. var. integra DC. = Clematis vitalba L. ●☆

95445 Clematis vitalba L. var. micrantha H. Lév. et Vaniot = Clematis gouriana Roxb. ex DC. ●

95446 Clematis vitalba L. var. microcarpa Franch. = Clematis peterae Hand. -Mazz. ●

95447 Clematis vitalba L. var. syriaca Boiss. = Clematis vitalba L. ●☆

95448 Clematis viticaulis Steele;葡萄铁线莲;Grape Clematis, Grape Leather-flower, Millboro Leather-flower ■☆

95449 Clematis viticella L.;意大利铁线莲;Italian Clematis, Italian Leather Flower, Purple Clematis ●☆

95450 Clematis viticella L. 'Prince Charles';查尔斯王子铁线莲;Clematis 'Prince Charles' ●☆

95451 Clematis viticella L. var. gigantiflora Kuntze;大花意大利铁线莲●☆

95452 Clematis wallichii W. T. Wang;沃利克铁线莲;Wallich's Clematis ●☆

95453 Clematis welwitschii Hiern ex Kuntze;韦尔铁线莲■☆

95454 Clematis wenshanensis W. T. Wang;文山铁线莲;Wenshan Clematis ●

95455 Clematis wightiana Wall. var. gallaensis Engl. ex Mildbr. = Clematis brachiata Thunb. ■☆

95456 Clematis wilfordii (Maxim.) Kom. = Clematis serratifolia Rehder ●

95457 Clematis wissmanniana Hand. -Mazz.;厚萼铁线莲(元江木通);Thickcalyx Clematis, Wissmann Clematis ●

95458 Clematis wutangensis W. T. Wang = Clematis shensiensis W. T. Wang ●

95459 Clematis xinhuiensis R. J. Wang;新惠铁线莲●

95460 Clematis yingtzulinia S. S. Ying = Clematis tashiroi Maxim. ●

95461 Clematis yuanjiangensis W. T. Wang;元江铁线莲;Yuanjiang Clematis ●

95462 Clematis yui W. T. Wang;俞氏铁线莲;Yu's Clematis ●

95463 Clematis yunnanensis Franch.;云南铁线莲(大花木通,大叶木通,辣木通,牛角,三叶木通,云南发汗藤);Yunnan Clematis ●

95464 Clematis yunnanensis Franch. var. brevipedunculata W. T. Wang = Clematis kockiana C. K. Schneid. ●

95465 Clematis yunnanensis Franch. var. chingtungensis M. Y. Fang;景东铁线莲;Chingtung Clematis,Jingdong Clematis ●

95466 Clematis yunnanensis Franch. var. chingtungensis M. Y. Fang = Clematis kockiana C. K. Schneid. ●

95467 Clematis zandaensis W. T. Wang;扎达铁线莲;Zhada Clematis ●

95468 Clematis zygophylla Hand. -Mazz.;对叶铁线莲;Pairleaf Clematis ●

95469 Clematitaria Bureau = Pleonotoma Miers ●☆

95470 Clematitis Duhamel = Clematis L. ●■

95471 Clematoclethra (Franch.) Maxim. (1890);藤山柳属(铁线山柳属);Vineclethra ●★

95472 Clematoclethra Maxim. = Clematoclethra (Franch.) Maxim. ●★

95473 Clematoclethra actinidioides Maxim.;猕猴桃藤山柳(杨叶藤山柳);Actinialike Vineclethra, Actinia-like Vineclethra, Kiwifruitlike Vineclethra ●

95474 Clematoclethra actinidioides Maxim. = Clematoclethra scandens (Franch.) Maxim. subsp. actinidioides (Maxim.) Y. C. Tang et Q. Y. Xiang ●

95475 Clematoclethra actinidioides Maxim. var. integrifola (Maxim.) C. F. Liang et Y. C. Chen;全缘藤山柳;Entire Leaf Vineclethra ●

95476 Clematoclethra actinidioides Maxim. var. integrifola (Maxim.) C. F. Liang et Y. C. Chen = Clematoclethra scandens (Franch.) Maxim. subsp. actinidioides (Maxim.) Y. C. Tang et Q. Y. Xiang ●

95477 Clematoclethra actinidioides Maxim. var. populifolia C. F. Liang et Y. C. Chen = Clematoclethra populifolia C. F. Liang et Y. C. Chen ●

95478 Clematoclethra actinidioides Maxim. var. populifolia C. F. Liang et Y. C. Chen = Clematoclethra scandens (Franch.) Maxim. subsp. actinidioides (Maxim.) Y. C. Tang et Q. Y. Xiang ●

95479 Clematoclethra actinidioides var. integrifolia (Maxim.) C. F. Liang et Y. C. Chen = Clematoclethra scandens (Franch.) Maxim. subsp. actinidioides (Maxim.) Y. C. Tang et Q. Y. Xiang ●

95480 Clematoclethra actinidioides var. populifolia C. F. Liang et Y. C. Chen = Clematoclethra scandens (Franch.) Maxim. subsp. actinidioides (Maxim.) Y. C. Tang et Q. Y. Xiang ●

95481 Clematoclethra argentifolia C. F. Liang et Y. C. Chen;银叶藤山柳;Silver Leaf Vineclethra,Silver-leaved Vineclethra ●

95482 Clematoclethra argentifolia C. F. Liang et Y. C. Chen = Clematoclethra scandens (Franch.) Maxim. subsp. actinidioides (Maxim.) Y. C. Tang et Q. Y. Xiang ●

95483 Clematoclethra cordifolia Franch.;心叶藤山柳;Heart-leaf Vineclethra ●

95484 Clematoclethra cordifolia Franch. = Clematoclethra scandens (Franch.) Maxim. ●

95485 Clematoclethra cordifolia Franch. var. tiliacea (Kom.) C. Yu Chang = Clematoclethra scandens (Franch.) Maxim. subsp. tomentella (Franch.) Y. C. Tang et Q. Y. Xiang ●

95486 Clematoclethra disticha Hemsl. = Clematoclethra scandens (Franch.) Maxim. subsp. tomentella (Franch.) Y. C. Tang et Q. Y. Xiang ●

95487 Clematoclethra distincha Hemsl.;二裂藤山柳;Distinchous Vineclethra,Distingchus Vineclethra,Twolobed Vineclethra ●

95488 Clematoclethra distincha Hemsl. = Clematoclethra scandens (Franch.) Maxim. subsp. tomentella (Franch.) Y. C. Tang et Q. Y. Xiang ●

95489 Clematoclethra faberi Franch.;光萼藤山柳(尖叶藤山柳);Faber Vineclethra,Glabrous Vineclethra,Sharpleaf Vineclethra ●

95490 Clematoclethra faberi Franch. = Clematoclethra scandens (Franch.) Maxim. subsp. actinidioides (Maxim.) Y. C. Tang et Q. Y. Xiang ●

95491 Clematoclethra faberi Franch. var. emeiensis C. Yu Chang = Clematoclethra scandens (Franch.) Maxim. subsp. actinidioides (Maxim.) Y. C. Tang et Q. Y. Xiang ●

95492 Clematoclethra floribunda W. T. Wang ex C. F. Liang;多花藤山柳;Manyflower Vineclethra,Multiflorous Vineclethra ●

95493 Clematoclethra floribunda W. T. Wang ex C. F. Liang = Clematoclethra scandens (Franch.) Maxim. subsp. tomentella (Franch.) Y. C. Tang et Q. Y. Xiang ●

95494 Clematoclethra floribunda W. T. Wang ex C. F. Liang et Y. C. Chen = Clematoclethra scandens (Franch.) Maxim. subsp. tomentella (Franch.) Y. C. Tang et Q. Y. Xiang ●

95495 Clematoclethra franchetii Kom.;川西藤山柳(圆叶藤山柳);Franchet Vineclethra,Roundleaf Vineclethra ●

95496 Clematoclethra franchetii Kom. = Clematoclethra scandens (Franch.) Maxim. subsp. actinidioides (Maxim.) Y. C. Tang et Q. Y. Xiang ●

95497 Clematoclethra giraldii Diels = Actinidia tetramera Maxim. ●

95498 Clematoclethra grandis Hemsl. = Clematoclethra lasioclada Maxim. var. grandis (Hemsl.) Rehder ●

95499 Clematoclethra grandis Hemsl. = Clematoclethra scandens (Franch.) Maxim. subsp. actinidioides (Maxim.) Y. C. Tang et Q. Y. Xiang ●

95500 Clematoclethra guangxiensis C. F. Liang et Y. C. Chen;广西藤山柳;Guangxi Vineclethra ●

95501 Clematoclethra guangxiensis C. F. Liang et Y. C. Chen = Clematoclethra scandens (Franch.) Maxim. ●

95502 Clematoclethra guizhouensis C. F. Liang et Y. C. Chen;贵州藤山柳;Guizhou Vineclethra ●

95503 Clematoclethra guizhouensis C. F. Liang et Y. C. Chen = Clematoclethra scandens (Franch.) Maxim. ●

95504 Clematoclethra hemsleyana Baill. ex Kom. = Clematoclethra scandens (Franch.) Maxim. subsp. actinidioides (Maxim.) Y. C. Tang et Q. Y. Xiang ●

95505 Clematoclethra hemsleyi Baill.;繁花藤山柳;Hemsley Vineclethra,Hemsley Setose Vineclethra ●

95506 Clematoclethra hemsleyi Baill. = Clematoclethra scandens (Franch.) Maxim. subsp. hemsleyi (Baill.) Y. C. Tang et Q. Y. Xiang ●

95507 Clematoclethra hemsleyi Baill. ex Kom. = Clematoclethra scandens (Franch.) Maxim. subsp. actinidioides (Maxim.) Y. C. Tang et Q. Y. Xiang ●

95508 Clematoclethra henryi Franch. ex Kom. = Clematoclethra scandens (Franch.) Maxim. subsp. hemsleyi (Baill.) Y. C. Tang et Q. Y. Xiang ●

95509 Clematoclethra integrifola Maxim. = Clematoclethra actinidioides Maxim. var. integrifola (Maxim.) C. F. Liang et Y. C. Chen ●

95510 Clematoclethra integrifolia Maxim. = Clematoclethra scandens (Franch.) Maxim. subsp. actinidioides (Maxim.) Y. C. Tang et Q. Y. Xiang ●

95511 Clematoclethra lanceolata C. F. Liang et Y. C. Chen;披针叶藤山柳;Lanceleaf Vineclethra ●

95512 Clematoclethra lanosa Rehder;绵毛藤山柳;Woolly Vineclethra ●

95513 Clematoclethra lanosa Rehder = Clematoclethra scandens (Franch.) Maxim. subsp. hemsleyi (Baill.) Y. C. Tang et Q. Y. Xiang ●

95514 Clematoclethra lasioclada Maxim. = Clematoclethra scandens (Franch.) Maxim. subsp. actinidioides (Maxim.) Y. C. Tang et Q. Y. Xiang ●

95515 Clematoclethra lasioclada Maxim. var. grandis (Hemsl.) Rehder;大叶藤山柳;Largeleaf Vineclethra ●

95516 Clematoclethra lasioclada Maxim. var. grandis (Hemsl.) Rehder = Clematoclethra scandens (Franch.) Maxim. subsp. actinidioides (Maxim.) Y. C. Tang et Q. Y. Xiang ●

95517 Clematoclethra lasioclada Maxim. var. oblonga C. F. Liang et Y. C. Chen ex C. Y. Chang;短叶藤山柳;Oblongleaf Common Vineclethra ●

95518 Clematoclethra lasioclada Maxim. var. oblonga C. F. Liang et Y. C. Chen ex C. Y. Chang = Clematoclethra scandens (Franch.) Maxim. subsp. actinidioides (Maxim.) Y. C. Tang et Q. Y. Xiang ●

95519 Clematoclethra leiboensis C. F. Liang et Y. C. Chen ex C. Y. Chang;雷波藤山柳;Leibo Vineclethra ●

95520 Clematoclethra leiboensis C. F. Liang et Y. C. Chen ex C. Y. Chang = Clematoclethra scandens (Franch.) Maxim. ●

95521 Clematoclethra loniceroides C. F. Liang et Y. C. Chen;银花藤山柳;Honeysuchle-like Vineclethra,Silver Flower Vineclethra ●

95522 Clematoclethra loniceroides C. F. Liang et Y. C. Chen = Clematoclethra scandens (Franch.) Maxim. ●

95523 Clematoclethra maximowiczii Baill. = Clematoclethra scandens (Franch.) Maxim. ●

95524 Clematoclethra nanchuanensis W. T. Wang;南川藤山柳;Nanchuan Vineclethra ●

95525 Clematoclethra nanchuanensis W. T. Wang = Clematoclethra scandens (Franch.) Maxim. ●

95526 Clematoclethra nanchuanensis W. T. Wang ex C. F. Liang et Y. C. Chen = Clematoclethra scandens (Franch.) Maxim. ●

95527 Clematoclethra oliviformis C. F. Liang et Y. C. Chen;榄叶藤山柳;Oliveleaf Vineclethra,Olive-leaved Vineclethra ●

95528 Clematoclethra oliviformis C. F. Liang et Y. C. Chen = Clematoclethra scandens (Franch.) Maxim. ●

95529 Clematoclethra omeiensis W. P. Fang;峨眉藤山柳;Emei Vineclethra ●

95530 Clematoclethra pachyphylla C. F. Liang et Y. C. Chen;厚叶藤山柳;Thickleaf Vineclethra,Thick-leaved Vineclethra ●

95531 Clematoclethra pachyphylla C. F. Liang et Y. C. Chen = Clematoclethra scandens (Franch.) Maxim. ●

95532 Clematoclethra pauciseta C. Yu Chang;疏刚毛藤山柳;Paucisete Vineclethra ●

95533 Clematoclethra pauciseta C. Yu Chang = Clematoclethra scandens (Franch.) Maxim. ●

95534 Clematoclethra pingwuensis C. F. Liang et Y. C. Chen;平武藤山柳;Pingwu Vineclethra ●

95535 Clematoclethra pingwuensis C. Yu Chang et Y. C. Chen = Clematoclethra scandens (Franch.) Maxim. subsp. actinidioides (Maxim.) Y. C. Tang et Q. Y. Xiang ●

95536 Clematoclethra populifolia C. F. Liang et Y. C. Chen;杨叶藤山柳(柄叶藤山柳);Poplarleaf Vineclethra ●

95537 Clematoclethra prattii Kom. = Clematoclethra lasioclada Maxim. var. grandis (Hemsl.) Rehder ●

95538 Clematoclethra prattii Kom. = Clematoclethra scandens (Franch.) Maxim. subsp. actinidioides (Maxim.) Y. C. Tang et Q. Y. Xiang ●

95539 Clematoclethra pyrifolia C. Yu Chang;梨叶藤山柳;Pear-leaf Vineclethra ●

95540 Clematoclethra pyrifolia C. Yu Chang = Clematoclethra scandens (Franch.) Maxim. ●

95541 Clematoclethra racemosa H. Lév.;总状藤山柳●

95542 Clematoclethra racemosa H. Lév. = Gouania javanica Miq. ●

95543 Clematoclethra scandens (Franch.) Maxim.;藤山柳(刚毛藤山柳);Climbing Vineclethra, Common Vineclethra, Hispid Vineclethra,Setose Vineclethra ●

95544 Clematoclethra scandens (Franch.) Maxim. subsp. actinidioides (Maxim.) Y. C. Tang et Q. Y. Xiang = Clematoclethra actinidioides Maxim. ●

95545 Clematoclethra scandens (Franch.) Maxim. subsp. hemsleyi (Baill.) Y. C. Tang et Q. Y. Xiang = Clematoclethra hemsleyi Baill. ●

95546 Clematoclethra scandens (Franch.) Maxim. subsp. tomentella (Franch.) Y. C. Tang et Q. Y. Xiang = Clematoclethra tomentella Franch. ●

95547 Clematoclethra scandens Maxim. = Clematoclethra scandens (Franch.) Maxim. ●

95548 Clematoclethra sichuanensis C. Shih ex C. F. Liang et Y. C. Chen;四川藤山柳;Sichuan Vineclethra ●

95549 Clematoclethra sichuanensis C. Shih ex C. F. Liang et Y. C. Chen = Clematoclethra scandens (Franch.) Maxim. ●

95550 Clematoclethra strigillosa Franch.;粗毛藤山柳;Hirsute Vineclethra,Strigose Vineclethra ●

95551 Clematoclethra strigillosa Franch. = Clematoclethra scandens (Franch.) Maxim. ●

95552 Clematoclethra tiliacea Kom.;椴叶藤山柳●

95553 Clematoclethra tiliacea Kom. = Clematoclethra scandens (Franch.) Maxim. subsp. tomentella (Franch.) Y. C. Tang et Q. Y. Xiang ●

95554 Clematoclethra tomentella Franch.;绒毛藤山柳●

95555 Clematoclethra tomentella Franch. = Clematoclethra hemsleyi Baill. ●

95556 Clematoclethra tomentella Franch. = Clematoclethra scandens (Franch.) Maxim. subsp. tomentella (Franch.) Y. C. Tang et Q. Y. Xiang ●

95557 Clematoclethra variabilis C. F. Liang et Y. C. Chen;变异藤山柳;Variable Vineclethra,Various Vineclethra ●

95558 Clematoclethra variabilis C. F. Liang et Y. C. Chen = Clematoclethra scandens (Franch.) Maxim. ●

95559 Clematoclethra variabilis C. F. Liang et Y. C. Chen var. multinervis C. F. Liang et Y. C. Chen;多变藤山柳(多脉藤山柳);Sargent Vineclethra ●

95560 Clematoclethra variabilis C. F. Liang et Y. C. Chen var. multinervis C. F. Liang et Y. C. Chen = Clematoclethra scandens (Franch.) Maxim. ●

95561 Clematoclethra wilsonii Hemsl. = Clematoclethra scandens (Franch.) Maxim. ●

95562 Clematoclethra wilsonii Hemsl. = Clematoclethra strigillosa Franch. ●

95563 Clematopsis Boj. ex Hook. = Clematis L. ●■

95564 Clematopsis Bojer ex Hutch. = Clematis L. ●■

95565 Clematopsis chrysocarpa (Welw. ex Oliv.) Hutch. = Clematis

chrysocarpa Welw. ex Oliv. ■☆
95566 Clematopsis grandifolia Staner et J. Léonard = Clematis uhehensis Engl. ●☆
95567 Clematopsis homblei (De Wild.) Staner et J. Léonard = Clematis uhehensis Engl. ●☆
95568 Clematopsis kirkii (Oliv.) Hutch. = Clematis villosa DC. subsp. kirkii (Oliv.) Brummitt ■☆
95569 Clematopsis oliveri Hutch. = Clematis villosa DC. subsp. oliveri (Hutch.) Brummitt ■☆
95570 Clematopsis pulchra Weim. = Clematis villosa DC. subsp. kirkii (Oliv.) Brummitt ■☆
95571 Clematopsis sapinii (De Wild.) Staner et J. Léonard = Clematis villosa DC. ■☆
95572 Clematopsis scabiosifolia (DC.) Hutch. = Clematis villosa DC. ■☆
95573 Clematopsis scabiosifolia (DC.) Hutch. subsp. kirkii (Oliv.) Brummitt = Clematis villosa DC. subsp. kirkii (Oliv.) Brummitt ■☆
95574 Clematopsis scabiosifolia (DC.) Hutch. subsp. oliveri (Hutch.) Brummitt = Clematis villosa DC. subsp. oliveri (Hutch.) Brummitt ■☆
95575 Clematopsis scabiosifolia (DC.) Hutch. subsp. stanleyi (Hook.) Brummitt = Clematis villosa DC. ■☆
95576 Clematopsis scabiosifolia (DC.) Hutch. subsp. uhehensis (Engl.) Brummitt = Clematis uhehensis Engl. ●☆
95577 Clematopsis simplicifolia Hutch. et Summerh. = Clematis uhehensis Engl. ●☆
95578 Clematopsis spathulifolia (Kuntze) Staner et J. Léonard = Clematis spathulifolia (Kuntze) Prantl ■☆
95579 Clematopsis stanleyi (Hook.) Hutch. = Clematis villosa DC. subsp. stanleyi (Hook.) Kuntze ■☆
95580 Clematopsis uhehensis (Engl.) Staner et J. Léonard = Clematis uhehensis Engl. ●☆
95581 Clematopsis villosa (DC.) Hutch. = Clematis villosa DC. ■☆
95582 Clematopsis villosa (DC.) Hutch. subsp. kirkii (Oliv.) J. Raynal et Brummitt = Clematis villosa DC. subsp. kirkii (Oliv.) Brummitt ■☆
95583 Clematopsis villosa (DC.) Hutch. subsp. oliveri (Hutch.) J. Raynal et Brummitt = Clematis villosa DC. subsp. oliveri (Hutch.) Brummitt ■☆
95584 Clematopsis villosa (DC.) Hutch. subsp. stanleyi (Hook.) J. Raynal et Brummitt = Clematis villosa DC. subsp. stanleyi (Hook.) Kuntze ■☆
95585 Clematopsis villosa (DC.) Hutch. subsp. uhehensis (Engl.) J. Raynal et Brummitt = Clematis uhehensis Engl. ●☆
95586 Clemensia Merr. = Chisocheton Blume ●
95587 Clemensia Schltr. = Clemensiella Schltr. ■☆
95588 Clemensiella Schltr. (1915);克莱门斯兰萝藦属■☆
95589 Clemensiella mariae (Schltr.) Schltr.;克莱门斯兰萝藦■☆
95590 Clementea Cav. (1804) = Canavalia Adans. (保留属名)●■
95591 Clementsia Rose = Rhodiola L. ■
95592 Clementsia Rose ex Britton et Rose = Sedum L. ●■
95593 Clementsia semenovii (Regel et Herder) Boriss. = Rhodiola semenovii (Regel et Herder) Boriss. ■
95594 Cleobula Vell. (1829);巴西克豆属☆
95595 Cleobulia Mart. ex Benth. (1837);克利奥豆属■☆
95596 Cleobulia diocleoides Benth.;克利奥豆■☆
95597 Cleobulia multiflora Mart. ex Benth.;多花克利奥豆■☆
95598 Cleochroma Miers = Iochroma Benth. (保留属名)●☆
95599 Cleodora Klotzsch = Croton L. ●
95600 Cleomaceae Airy Shaw = Brassicaceae Burnett(保留科名)■●
95601 Cleomaceae Airy Shaw = Cruciferae Juss. (保留科名)■●
95602 Cleomaceae Bercht. et J. Presl = Capparaceae Juss. (保留科名)●■
95603 Cleomaceae Bercht. et J. Presl = Cleomaceae Airy Shaw ●■
95604 Cleomaceae Bercht. et J. Presl(1825);白花菜科(醉蝶花科)●■
95605 Cleomaceae Horan. = Capparaceae Juss. (保留科名)●■
95606 Cleome L. (1753);白花菜属(风蝶草属,紫龙须属,醉蝶花属);Cleome, Gleome, Spider Flower, Spider Herb, Spider Plant, Spiderflower ●■
95607 Cleome L. = Gynandropsis DC. (保留属名)■
95608 Cleome aculeata L.;皮刺白花菜■☆
95609 Cleome acuta Schumach. et Thonn. = Cleome gynandra L. ■
95610 Cleome africana Botsch. = Cleome amblyocarpa Barratte et Murb. ■☆
95611 Cleome albescens Franch.;微白白花菜■☆
95612 Cleome allamanii Chiov.;阿拉曼白花菜■☆
95613 Cleome amblyocarpa Barratte et Murb.;非洲白花菜■☆
95614 Cleome amblyocarpa Barratte et Murb. var. glandulosa (Forssk.) Botsch.;多腺非洲白花菜■☆
95615 Cleome angustifolia Forssk.;窄叶白花菜■☆
95616 Cleome angustifolia Forssk. subsp. petersiana (Klotzsch) Kers;彼得斯白花菜■☆
95617 Cleome angustifolia Forssk. var. damarensis Kers = Cleome angustifolia Forssk. var. diandra (Burch.) Kers ■☆
95618 Cleome angustifolia Forssk. var. diandra (Burch.) Kers;双蕊窄叶白花菜■☆
95619 Cleome angustifolia Forssk. var. namaquensis Kers = Cleome angustifolia Forssk. var. diandra (Burch.) Kers ■☆
95620 Cleome angustifolia Forssk. var. pteropoda (Welw. ex Oliv.) Kers;翼梗窄叶白花菜■☆
95621 Cleome angustifolia Forssk. var. pteropoda (Welw. ex Oliv.) Kers = Cleome angustifolia Forssk. var. diandra (Burch.) Kers ■☆
95622 Cleome aphylla Thunb. = Cadaba aphylla (Thunb.) Wild ●☆
95623 Cleome arabica L.;阿拉伯白花菜■☆
95624 Cleome arabica L. subsp. amblyocarpa (Barratte et Murb.) Ozenda = Cleome amblyocarpa Barratte et Murb. ■☆
95625 Cleome arabica L. var. stenocarpa Franch. = Cleome ramosissima Parl. ■☆
95626 Cleome armata Thunb. = Microloma armatum (Thunb.) Schltr. ■☆
95627 Cleome augustinensis (Hochr.) Briq.;奥古斯丁白花菜■☆
95628 Cleome augustinensis (Hochr.) Briq. var. goudotii (Briq.) Hadj-Moust. = Cleome augustinensis (Hochr.) Briq. ■☆
95629 Cleome bechuanensis Bremek. et Oberm. = Cleome hirta (Klotzsch) Oliv. ■☆
95630 Cleome benedictae Dinter = Cleome kalachariensis (Schinz) Gilg et Gilg-Ben. ■☆
95631 Cleome bicolor (Pax) Gilg = Cleome oxyphylla Burch. ■☆
95632 Cleome bororensis (Klotzsch) Oliv.;博罗雷白花菜■☆
95633 Cleome brachycarpa Vahl ex DC.;短果白花菜■☆
95634 Cleome brachycarpa Vahl ex DC. var. glauca? = Cleome brachycarpa Vahl ex DC. ■☆
95635 Cleome brachycarpa Vahl ex DC. var. longipetiolata? = Cleome brachycarpa Vahl ex DC. ■☆
95636 Cleome brachypoda Gilg et Gilg-Ben. = Cleome briquetii Polhill ■☆
95637 Cleome brachystyla Deflers ex Franch.;短柱白花菜■☆
95638 Cleome breyeri Burtt Davy = Cleome oxyphylla Burch. ■☆

95639 Cleome briquetii Polhill;短足白花菜■☆
95640 Cleome bungei Steud. = Cleome gynandra L. ■
95641 Cleome burmannii Wight et Arn. ;布尔曼白花菜■☆
95642 Cleome burttii R. A. Graham;伯特白花菜■☆
95643 Cleome candelabrum Sims = Cleome gynandra L. ■
95644 Cleome capensis L. = Heliophila subulata Burch. ex DC. ■☆
95645 Cleome carnosa (Pax) Gilg et Gilg-Ben. ;肉质白花菜■☆
95646 Cleome chevalieri Schinz = Cleome polyanthera Schweinf. et Gilg ■☆
95647 Cleome chilocalyx Oliv. = Cleome macrophylla (Klotzsch) Briq. ■☆
95648 Cleome chilocalyx Oliv. var. tenuifolia (Klotzsch) Oliv. = Cleome macrophylla (Klotzsch) Briq. ■☆
95649 Cleome chrysantha Decne. ;金花白花菜■☆
95650 Cleome ciliata Schumach. et Thonn. = Cleome rutidosperma DC. ■
95651 Cleome coeruleo-rosea Gilg et Gilg-Ben. ;蓝蔷薇白花菜■☆
95652 Cleome coluteoides Boiss. ;膀胱豆白花菜■☆
95653 Cleome confusa Dinter = Cleome elegantissima Briq. ■☆
95654 Cleome conrathii Burtt Davy;康拉特白花菜■☆
95655 Cleome cordata Burch. ex DC. = Cleome monophylla L. ■☆
95656 Cleome densifolia C. H. Wright;密花白花菜■☆
95657 Cleome denticulata (DC.) Schult. f. = Cleome gynandra L. ■
95658 Cleome diandra Burch. = Cleome angustifolia Forssk. var. diandra (Burch.) Kers ■☆
95659 Cleome diandra Burch. var. delagoensis Kuntze = Cleome angustifolia Forssk. subsp. petersiana (Klotzsch) Kers ■☆
95660 Cleome diandra Burch. var. pteropoda Welw. ex Oliv. = Cleome angustifolia Forssk. var. diandra (Burch.) Kers ■☆
95661 Cleome didynama Hochst. ex Oliv. = Cleome angustifolia Forssk. ■☆
95662 Cleome diffusa Banks ex DC. ; 松散白花菜; Spreading Spiderflower ■☆
95663 Cleome dodecandra L. = Polanisia dodecandra (L.) DC. ■☆
95664 Cleome dolichostyla Jafri;长柱白花菜■☆
95665 Cleome droserifolia (Forssk.) Delile = Cleome fimbriata Vicary ■☆
95666 Cleome eckloniana Schrad. = Cleome gynandra L. ■
95667 Cleome ehrenbergiana Schweinf. = Cleome scaposa DC. ■☆
95668 Cleome elegantissima Briq. ;雅致白花菜■☆
95669 Cleome elegantissima Chiov. = Cleome angustifolia Forssk. subsp. petersiana (Klotzsch) Kers ■☆
95670 Cleome epilobioides Baker = Cleome monophylla L. ■☆
95671 Cleome filifolia Vahl = Cleome angustifolia Forssk. ■☆
95672 Cleome fimbriata Vicary;流苏白花菜■☆
95673 Cleome fimbriata Vicary subsp. brachystyla Govaerts;短柱流苏白花菜■☆
95674 Cleome fischeri B. L. Rob. = Cleome usambarica Pax ●☆
95675 Cleome foliosa Hook. f. ;多叶白花菜■☆
95676 Cleome foliosa Hook. f. var. lutea (Sond.) Codd et Kers;黄多叶白花菜■☆
95677 Cleome foliosa Hook. f. var. namibensis (Kers) Codd;纳米布白花菜■☆
95678 Cleome fritzscheae Gilg et Gilg-Ben. = Cleome iberidella Welw. ex Oliv. ■☆
95679 Cleome gilletii De Wild. = Cleome aculeata L. ■☆
95680 Cleome giorgii De Wild. = Cleome hirta (Klotzsch) Oliv. ■☆
95681 Cleome glandulosissima Gilg = Cleome hirta (Klotzsch) Oliv. ■☆
95682 Cleome gordjagiaii Popov;高尔白花菜■☆
95683 Cleome gossweileri Exell;戈斯白花菜■☆
95684 Cleome goudotii Briq. = Cleome augustinensis (Hochr.) Briq. ■☆
95685 Cleome gracilis Edgew. = Cleome scaposa DC. ■☆
95686 Cleome grata Graham = Cleome allamanii Chiov. ■☆
95687 Cleome graveolens Raf. ;臭白花菜■☆
95688 Cleome griffithiana Rech. f. = Cleome fimbriata Vicary ■☆
95689 Cleome guineensis Hook. f. = Cleome rutidosperma DC. ■
95690 Cleome gynandra L. = Gynandropsis gynandra (L.) Briq. ■
95691 Cleome gynandrae L. = Gynandropsis gynandra (L.) Briq. ■
95692 Cleome hanburyana Penz. ;宽萼白花菜■☆
95693 Cleome hassleriana Chodat = Cleome spinosa Jacq. ■
95694 Cleome hassleriana Chodat = Tarenaya hassleriana (Chodat) Iltis ■
95695 Cleome heterochroma Briq. = Cleome oxyphylla Burch ■☆
95696 Cleome heterotricha Burch = Gynandropsis gynandra (L.) Briq. ■
95697 Cleome heterotricha Burch. = Cleome gynandra L. ■
95698 Cleome hildebrandtii Gilg et Gilg-Ben. = Cleome hirta (Klotzsch) Oliv. ■☆
95699 Cleome hirta (Klotzsch) Oliv. ;毛白花菜■☆
95700 Cleome hirta Oliv. = Cleome hirta (Klotzsch) Oliv. ■☆
95701 Cleome hochstetteri (Eichler) Cufod. = Cleome angustifolia Forssk. ■☆
95702 Cleome hotsonii Blatt. et Hallb. = Cleome oxypetala Boiss. ■☆
95703 Cleome hulletii King = Cleome aculeata L. ■☆
95704 Cleome iberica DC. ;伊比利亚白花菜;Iberian Spiderflower ■☆
95705 Cleome iberidella Welw. ex Oliv. ;小蜂室花白花菜■☆
95706 Cleome icosandra L. = Arivela viscosa (L.) Raf. ■
95707 Cleome icosandra L. = Cleome viscosa L. ■
95708 Cleome inconcinna Briq. = Cleome macrophylla (Klotzsch) Briq. var. maculatiflora (Merxm.) Wild ■☆
95709 Cleome isomeris Greene;球梗白花菜;Bladderpod Spiderflower, Burro Fat ■☆
95710 Cleome johnstonii Exell et Mendonça = Cleome foliosa Hook. f. ■☆
95711 Cleome juncea P. J. Bergius = Heliophila juncea (P. J. Bergius) Druce ■☆
95712 Cleome juncea Sparrm. = Cadaba aphylla (Thunb.) Wild ●☆
95713 Cleome kalachariensis (Schinz) Gilg et Gilg-Ben. ;卡拉恰白花菜■☆
95714 Cleome kalachariensis (Schinz) Gilg et Gilg-Ben. var. namibensis Kers = Cleome foliosa Hook. f. var. namibensis (Kers) Codd ■☆
95715 Cleome kelleriana (Schinz) Gilg et Gilg-Ben. ;凯勒白花菜■☆
95716 Cleome kermesina Kers ex Gilg et Gilg-Ben. ;克迈斯白花菜■☆
95717 Cleome kermesina Kers ex Gilg et Gilg-Ben. var. plebeia Kers;普通白花菜■☆
95718 Cleome laburnifolia Rössler;毒豆叶白花菜■☆
95719 Cleome latifolia Vahl;宽叶白花菜■☆
95720 Cleome laxa Thunb. = Brachycarpaea juncea (P. J. Bergius) Marais ■☆
95721 Cleome linearifolia (Stephens) Dinter = Cleome semitetranda Sond. ■☆
95722 Cleome linearis Stocks ex T. Anderson = Cleome scaposa DC. ■☆
95723 Cleome lipskyi Popov;利普斯基白花菜■☆
95724 Cleome luederitziana Schinz = Cleome foliosa Hook. f. ■☆
95725 Cleome lupinifolia Gilg et Gilg-Ben. ;羽扇豆白花菜■☆
95726 Cleome lutea (Sond.) Szyszyl. var. polyphylla Pax = Cleome foliosa Hook. f. ■☆
95727 Cleome lutea E. Mey. ex Szyszyl. ;黄醉蝶花;Yellow Bee Plant,

Yellow Beeplant, Yellow Spiderwort ■☆
95728 Cleome macrophylla (Klotzsch) Briq.;大叶白花菜■☆
95729 Cleome macrophylla (Klotzsch) Briq. var. maculatiflora (Merxm.) Wild;斑花大叶白花菜■☆
95730 Cleome maculata (Sond.) Szyszyl.;斑点大叶白花菜■☆
95731 Cleome massae Chiov.;马萨白花菜■☆
95732 Cleome microtatodonta Briq. = Cleome usambarica Pax ●☆
95733 Cleome minima Stephens = Heliophila minima (Stephens) Marais ■☆
95734 Cleome monophylla L.;单叶白花菜■☆
95735 Cleome monophylloides Wilczek;假单叶白花菜■☆
95736 Cleome montana A. Chev. ex Keay = Cleome iberidella Welw. ex Oliv. ■☆
95737 Cleome moschata Stocks ex T. Anderson = Cleome brachycarpa Vahl ex DC. ■☆
95738 Cleome mossamedensis Exell et Mendonça;莫萨梅迪白花菜■☆
95739 Cleome mullendersii Wilczek = Cleome iberidella Welw. ex Oliv. ■☆
95740 Cleome nationae Burtt Davy = Cleome macrophylla (Klotzsch) Briq. var. maculatiflora (Merxm.) Wild ■☆
95741 Cleome niamniamensis Schweinf. et Gilg;尼亚白花菜■☆
95742 Cleome noeana Boiss.;诺氏白花菜■☆
95743 Cleome noeana Boiss. = Cleome fimbriata Vicary ■☆
95744 Cleome noeana Boiss. subsp. brachystyla (Deflers) D. F. Chamb. et Lamond = Cleome fimbriata Vicary subsp. brachystyla Govaerts ■☆
95745 Cleome oligandra Kers;寡蕊醉蝶花■☆
95746 Cleome ornithopodioides L.;鸟足醉蝶花;Bird Spiderflower ■☆
95747 Cleome ovalifolia Franch.;卵叶醉蝶花■☆
95748 Cleome oxypetala Boiss.;尖瓣醉蝶花■☆
95749 Cleome oxyphylla Burch.;尖叶醉蝶花■☆
95750 Cleome oxyphylla Burch. var. robusta Kers;粗壮尖叶醉蝶花■☆
95751 Cleome pachycephala Gilg et Gilg-Ben. = Cleome schlechteri Briq. ■☆
95752 Cleome papillosa T. Anderson = Cleome scaposa DC. ■☆
95753 Cleome paradoxa R. Br. ex DC.;奇异醉蝶花■☆
95754 Cleome parvipetala R. A. Graham;小瓣醉蝶花■☆
95755 Cleome parvula R. A. Graham;较小醉蝶花■☆
95756 Cleome paxiana Gilg = Cleome polyanthera Schweinf. et Gilg ■☆
95757 Cleome paxii (Schinz) Gilg et Gilg-Ben.;宽果白花菜;Golden Spider Flower ■☆
95758 Cleome pentanervia? = Cleome fimbriata Vicary ■☆
95759 Cleome pentaphylla L.;五叶白花菜(白花菜,羊角菜);Cat's Whiskers ■☆
95760 Cleome pentaphylla L. = Cleome gynandra L. ■
95761 Cleome pentaphylla L. = Gynandropsis gynandra (L.) Briq. ■
95762 Cleome petersiana (Klotzsch) Briq. = Cleome angustifolia Forssk. subsp. petersiana (Klotzsch) Kers ■☆
95763 Cleome platycarpa Schinz = Cleome paxii (Schinz) Gilg et Gilg-Ben. ■☆
95764 Cleome platysepala Gilg et Gilg-Ben. = Cleome hanburyana Penz. ■☆
95765 Cleome polyanthera Schweinf. et Gilg;多药白花菜■☆
95766 Cleome polytricha Franch.;多毛白花菜■☆
95767 Cleome pulcherrima Buscal. et Muschl. = Cleome hirta (Klotzsch) Oliv. ■☆
95768 Cleome quinquenervia DC.;五脉白花菜■☆
95769 Cleome quinquenervia DC. = Cleome fimbriata Vicary ■☆
95770 Cleome quinquenervia DC. var. noeana? = Cleome fimbriata Vicary ■☆
95771 Cleome raddeana Trautv.;拉德白花菜■☆
95772 Cleome radula Fenzl = Cleome scaposa DC. ■☆
95773 Cleome ramosissima Parl.;多枝白花菜■☆
95774 Cleome roridula R. Br. = Cleome droserifolia (Forssk.) Delile ■☆
95775 Cleome rostrata Bobrov;喙状白花菜■☆
95776 Cleome rubella Burch.;微红白花菜■☆
95777 Cleome rubelloides Kers;拟微红白花菜■☆
95778 Cleome rupestris Sond. = Cleome rubella Burch. ■☆
95779 Cleome rupicola Vicary;岩生白花菜■☆
95780 Cleome ruta Cambess. = Cleome brachycarpa Vahl ex DC. ■☆
95781 Cleome rutidosperma DC.;皱子白花菜(平伏茎白花菜);Fringed Spiderflower, Wrinkle-seed Spiderflower ■
95782 Cleome rutidosperma DC. var. hainanensis J. L. Shan;海南皱子白花菜;Hainan Wrinkle-seed Spiderflower ■
95783 Cleome scaposa DC.;花茎白花菜■☆
95784 Cleome scheffleri Briq. = Cleome hirta (Klotzsch) Oliv. ■☆
95785 Cleome schimperi Pax;欣珀白花菜■☆
95786 Cleome schlechteri Briq.;施莱白花菜■☆
95787 Cleome schweinfurthii Gilg = Cleome ramosissima Parl. ■☆
95788 Cleome semitetranda Sond.;半四蕊白花菜■☆
95789 Cleome seretii De Wild. = Cleome polyanthera Schweinf. et Gilg ■☆
95790 Cleome serrulata Pax = Cleome usambarica Pax ●☆
95791 Cleome serrulata Pursh;齿白花菜;Blue Colorado Bee Plant, Clammy Weed, Rocky Mountain Bee Plant, Rocky Mountain Bee-plant, Rocky Mountain Beeweed, Skunkweed, Spider Plant, Stink Flower, Stinking Clover, Stinking-clover, Stinkweed, Toothed Spider-flower ■☆
95792 Cleome silvatica Gilg et Gilg-Ben.;森林白花菜●☆
95793 Cleome sonorae A. Gray;墨西哥臭白花菜;Skunkweed, Sonora Beeweed, Sonora Stink Flower, Stinking Clover ■☆
95794 Cleome speciosa Raf. = Cleoserrata speciosa (Raf.) Iltis ■
95795 Cleome speciosissima Deppe ex Lindl. = Cleoserrata speciosa (Raf.) Iltis ■
95796 Cleome spinosa Jacq.;西洋醉蝶花(西洋白花菜,西洋风蝶菜,西洋风蝶草,紫龙须,醉蝶花);Beautiful Spider Flower, Giant Spider Flower, Giant Spider Plant, Pink Queen, Pink-queen, Spider Flower, Spider Flower Cleome, Spider Plant, Spider-flower, Spiny Spiderflower, Volantines Preciosos ■
95797 Cleome stenopetala Gilg et Gilg-Ben.;窄瓣白花菜■☆
95798 Cleome stocksiana Boiss. = Cleome rupicola Vicary ■☆
95799 Cleome stricta (Klotzsch) R. A. Graham;刚直白花菜■☆
95800 Cleome strigosa (Bojer) Oliv.;糙伏毛白花菜■☆
95801 Cleome subcordata Steud. = Cleome monophylla L. ■☆
95802 Cleome suffruticosa Schinz;亚灌木白花菜●☆
95803 Cleome sulfurea Bremek. et Oberm. = Cleome kalachariensis (Schinz) Gilg et Gilg-Ben. ■☆
95804 Cleome tenella L.;柔弱白花菜●☆
95805 Cleome thyrsiflora De Wild. et T. Durand = Cleome rutidosperma DC. ■
95806 Cleome tomentella Popov;小毛白花菜■☆
95807 Cleome trachysperma (Torr. et A. Gray) Pax et K. Hoffm.;粗籽白花菜■☆
95808 Cleome trinervia Fresen. = Cleome arabica L. ■☆
95809 Cleome triphylla L. = Cleome gynandra L. ■
95810 Cleome turkmena Bobrov;土库曼白花菜■☆

95811 Cleome usambarica Pax;乌桑巴拉白花菜●☆
95812 Cleome vahliana Fresen. = Cleome brachycarpa Vahl ex DC. ■☆
95813 Cleome violacea L.;堇色白花菜■☆
95814 Cleome virgata Thunb. = Heliophila maraisiana Al-Shehbaz et Mummenhoff ■☆
95815 Cleome viscosa L. = Arivela viscosa (L.) Raf. ■
95816 Cleome viscosa L. f. deglabrata (Backer) Jacobs = Arivela viscosa (L.) Raf. var. deglabrata (Backer) M. L. Zhang et G. C. Tucker ■
95817 Cleome viscosa L. f. deglabrata (Backer) Jacobs = Cleome viscosa L. var. deglabrata (Backer) B. S. Sun ■
95818 Cleome viscosa L. var. deglabrata (Backer) B. S. Sun = Arivela viscosa (L.) Raf. var. deglabrata (Backer) M. L. Zhang et G. C. Tucker ■
95819 Cleome welwitschii Exell = Cleome elegantissima Briq. ■☆
95820 Cleome xanthopetala Briq. = Cleome foliosa Hook. f. var. lutea (Sond.) Codd et Kers ■☆
95821 Cleome yunnanensis W. W. Sm. = Cleoserrata speciosa (Raf.) Iltis ■
95822 Cleomella DC. (1824);小白花菜属■☆
95823 Cleomella mexicana DC.;墨西哥小白花菜■☆
95824 Cleomena Roem. et Schult. = Muhlenbergia Schreb. ■
95825 Cleomodendron Pax = Farsetia Turra ■☆
95826 Cleomodendron somalense Pax = Farsetia somalensis (Pax) Engl. ex Gilg et Gilg-Ben. ■☆
95827 Cleomopsideae Vill. = Stanleyaceae Nutt. ●■
95828 Cleonia L. (1763);克里昂草属(葹苞花属)■☆
95829 Cleonia lusitanica (L.) L.;克里昂草■☆
95830 Cleonia punica Beauverd;微红克里昂草■☆
95831 Cleopatra Pancher ex Baillon = Neoguillauminia Croizat☆
95832 Cleopatra Pancher ex Croizat = Neoguillauminia Croizat☆
95833 Cleophora Gaertn. = Latania Comm. ex Juss. ●☆
95834 Cleoserrata Iltis(2007);西洋白花菜属■
95835 Cleoserrata speciosa (Raf.) Iltis;西洋白花菜(滇白花菜,美丽白花菜);Spiffy Spiderflower, Yunnan Spiderflower ■
95836 Cleosma Urb. et Ekman ex Sandwith = Tynanthus Miers ●☆
95837 Clercia Vell. = Salacia L. (保留属名)●
95838 Cleretum N. E. Br. (1925);鸦嘴玉属■☆
95839 Cleretum bellidiforme (Burm. f.) G. D. Rowley = Dorotheanthus bellidiformis (Burm. f.) N. E. Br. ■☆
95840 Cleretum herrei (Schwantes) Ihlenf. et Struck;赫勒鸦嘴玉■☆
95841 Cleretum longipes L. Bolus = Cleretum papulosum (L. f.) L. Bolus subsp. schlechteri (Schwantes) Ihlenf. et Struck ■☆
95842 Cleretum lyratifolium Ihlenf. et Struck;大头羽裂鸦嘴玉■☆
95843 Cleretum papulosum (L. f.) L. Bolus;乳突鸦嘴玉■☆
95844 Cleretum papulosum (L. f.) L. Bolus subsp. schlechteri (Schwantes) Ihlenf. et Struck;施莱鸦嘴玉■☆
95845 Cleretum papulosum (L. f.) N. E. Br. = Cleretum papulosum (L. f.) L. Bolus ■☆
95846 Cleretum pinnatifidum (L. f.) L. Bolus = Aethephyllum pinnatifidum (L. f.) N. E. Br. ■☆
95847 Cleretum puberulum (Haw.) N. E. Br. = Mesembryanthemum aitonis Jacq. ■☆
95848 Cleretum schlechteri (Schwantes) N. E. Br. = Cleretum papulosum (L. f.) L. Bolus subsp. schlechteri (Schwantes) Ihlenf. et Struck ■☆
95849 Cleretum sessiliflorum (Aiton) N. E. Br. = Cleretum papulosum (L. f.) L. Bolus ■☆
95850 Clerkia Neck. = Tabernaemontana L. ●
95851 Clermontia Gaudich. (1829);克勒木属●☆
95852 Clermontia arborescens (H. Mann) Hillebr.;克勒木●☆
95853 Clermontia arborescens Hillebr. = Clermontia arborescens (H. Mann) Hillebr. ●☆
95854 Clermontia grandiflora Gaudich.;大花克勒木●☆
95855 Clermontia lindseyana Rock;林氏克勒草●☆
95856 Clermontia multiflora Hillebr.;多花克勒草●☆
95857 Clermontia oblongifolia Gaudich.;矩圆叶克勒草●☆
95858 Clermontia reticulata St. John;网状克勒草●☆
95859 Clerodendranthus Kudo = Orthosiphon Benth. ●■
95860 Clerodendranthus Kudo(1929);肾茶属;Clerodendranthus ●
95861 Clerodendranthus spicatus (Thunb.) C. Y. Wu = Clerodendranthus spicatus (Thunb.) C. Y. Wu ex H. W. Li ●
95862 Clerodendranthus spicatus (Thunb.) C. Y. Wu ex H. W. Li;肾茶(化石草,猫须草,猫须公,肾菜,肾草);Spicate Clerodendranthus ●
95863 Clerodendranthus stamineus (Benth.) Kudo = Clerodendranthus spicatus (Thunb.) C. Y. Wu ex H. W. Li ●
95864 Clerodendron Adans. = Clerodendrum L. ●■
95865 Clerodendron Burm. = Clerodendrum L. ●■
95866 Clerodendron L. = Clerodendrum L. ●■
95867 Clerodendron R. Br. = Clerodendrum L. ●■
95868 Clerodendron amplius Hance = Clerodendrum cyrtophyllum Turcz. ●
95869 Clerodendron bodinierii H. Lév. = Clerodendrum mandarinorum Diels ●
95870 Clerodendron calamitosum L. = Clerodendrum calamitosum L. ●
95871 Clerodendron castaneifolium Hook. et Arn. = Clerodendrum fortunatum L. ●
95872 Clerodendron cavaleriei H. Lév. = Clerodendrum mandarinorum Diels ●
95873 Clerodendron commersonii Spreng. = Clerodendrum inerme (L.) Gaertn. ●
95874 Clerodendron cyrtophyllum Turcz. = Clerodendrum cyrtophyllum Turcz. ●
95875 Clerodendron darrisii H. Lév. = Clerodendrum japonicum (Thunb.) Sweet ●
95876 Clerodendron disparifolium Blume = Clerodendrum disparifolium Blume ●☆
95877 Clerodendron divaricatum Jack = Clerodendrum serratum (L.) Moon var. wallichii C. B. Clarke ●
95878 Clerodendron divaricatum Siebold et Zucc. = Caryopteris divaricata (Siebold et Zucc.) Maxim. ■
95879 Clerodendron esquirolii H. Lév. = Clerodendrum japonicum (Thunb.) Sweet ●
95880 Clerodendron esquirolii H. Lév. = Tacca chantrieri André ■
95881 Clerodendron fargesii Dode = Clerodendrum trichotomum Thunb. ex A. Murray var. fargesii (Dode) Rehder ●
95882 Clerodendron foetidum D. Don = Clerodendrum bungei Steud. ●
95883 Clerodendron formosanum Maxim. = Clerodendrum cyrtophyllum Turcz. ●
95884 Clerodendron fortunatum L. = Clerodendrum fortunatum L. ●
95885 Clerodendron fragrans Vent. = Clerodendrum chinense (Osbeck) Mabb. ●
95886 Clerodendron fragrans Vent. = Clerodendrum chinense (Osbeck)

Mabb. var. simplex (Moldenke) S. L. Chen ●

95887 Clerodendron fragrans Vent. = Clerodendrum philippinum Schauer ●

95888 Clerodendron fragrans Vent. var. foetida (Bunge) Bakj. ? = Clerodendrum bungei Steud. ●

95889 Clerodendron fragrans Vent. var. multiplex Sweet = Clerodendrum chinense (Osbeck) Mabb. ●

95890 Clerodendron fragrans Vent. var. pleniflora Schauer = Clerodendrum chinense (Osbeck) Mabb. ●

95891 Clerodendron glandulosum Colebr. ex Wall. = Clerodendrum colebrokianum Walp. ●

95892 Clerodendron gratum Kurz = Caryopteris foetida Thell. ●☆

95893 Clerodendron gratum Kurz = Caryopteris paniculata C. B. Clarke ●

95894 Clerodendron haematocalyx Hance = Clerodendrum canescens Wall. ●

95895 Clerodendron herbacea (Roxb.) Wall. = Clerodendrum serratum (L.) Moon var. herbaceum (Roxb. ex Schauer) C. Y. Wu ●

95896 Clerodendron inerme (L.) Gaertn. = Clerodendrum inerme (L.) Gaertn. ●

95897 Clerodendron infortunatum L. = Clerodendrum japonicum (Thunb.) Sweet ●

95898 Clerodendron japonicum (Thunb.) Sweet = Clerodendrum japonicum (Thunb.) Sweet ●

95899 Clerodendron japonicum (Thunb.) Sweet var. album C. P'ei = Clerodendrum japonicum (Thunb.) Sweet ●

95900 Clerodendron kaempfer (Jacq.) Siebold var. album (C. P'ei) Moldenke = Clerodendrum japonicum (Thunb.) Sweet ●

95901 Clerodendron kaempferi (Jacq.) Siebold = Clerodendrum japonicum (Thunb.) Sweet ●

95902 Clerodendron koshunense Hayata = Clerodendrum trichotomum Thunb. ex A. Murray ●

95903 Clerodendron kwangtungense Hand.-Mazz. var. puberulum H. L. Li = Clerodendrum mandarinorum Diels ●

95904 Clerodendron leucosceptrum D. Don = Leucosceptrum canum Sm. ●

95905 Clerodendron leveillei Fedde ex H. Lév. = Clerodendrum japonicum (Thunb.) Sweet ●

95906 Clerodendron lividum Lindl. = Clerodendrum fortunatum L. ●

95907 Clerodendron longipetiolatum C. P'ei = Clerodendrum peii Moldenke ●

95908 Clerodendron molle Kunth = Clerodendrum villosum Blume ●

95909 Clerodendron moupinense Franch. = Microtoena moupinensis (Franch.) Prain ■

95910 Clerodendron nerifolium Wall. = Clerodendrum inerme (L.) Gaertn. ●

95911 Clerodendron nutans Jack = Clerodendrum wallichii Merr. ●

95912 Clerodendron odoratum D. Don = Caryopteris odorata (D. Don) B. L. Rob. ●

95913 Clerodendron oxysepalum Miq. = Clerodendrum fortunatum L. ●

95914 Clerodendron pentagonum Hance = Clerodendrum fortunatum L. ●

95915 Clerodendron pumilum (Lour.) Spreng. = Clerodendrum fortunatum L. ●

95916 Clerodendron pyramidale Andr. = Clerodendrum paniculatum L. ●

95917 Clerodendron sericeum Wall. = Hiptage sericea (Wall.) Hook. f. ●

95918 Clerodendron serotinum Carrière = Clerodendrum trichotomum Thunb. ex A. Murray ●

95919 Clerodendron siphonathus R. Br. = Clerodendrum indicum (L.) Kuntze ●

95920 Clerodendron spicatum Thunb. = Clerodendranthus spicatus (Thunb.) C. Y. Wu ex H. W. Li ●

95921 Clerodendron splendens G. Don = Clerodendrum splendens G. Don ●☆

95922 Clerodendron squamatum Vahl = Clerodendrum japonicum (Thunb.) Sweet ●

95923 Clerodendron thomsonae Balf. 'Variegatum';花叶龙吐珠●☆

95924 Clerodendron thomsonae Balf. 'Variegatum' = Clerodendrum thomsonae Balf. f. 'Variegatum'●☆

95925 Clerodendron thomsonae Balf. f. = Clerodendrum thomsonae Balf. f. ●

95926 Clerodendron trichotomum Thunb. ex A. Murray = Clerodendrum trichotomum Thunb. ex A. Murray ●

95927 Clerodendron trichotomum Thunb. ex A. Murray var. fargesii (Dode) Rehder = Clerodendrum trichotomum Thunb. ex A. Murray ●

95928 Clerodendron trichotomum Thunb. ex A. Murray var. villosum Hsu = Clerodendrum trichotomum Thunb. ex A. Murray ●

95929 Clerodendron trichotomum Thunb. ex A. Murray var. yakusimense = Clerodendrum trichotomum Thunb. ex A. Murray var. fargesii (Dode) Rehder ●

95930 Clerodendron tsaii H. L. Li = Clerodendrum mandarinorum Diels ●

95931 Clerodendron ugandense Prain = Clerodendrum ugandense Prain ●☆

95932 Clerodendron viscosum Vent. = Clerodendrum canescens Wall. ●

95933 Clerodendron yatschuense H. Winkl. = Clerodendrum bungei Steud. ●

95934 Clerodendrum L. (1753);赪桐属(臭牡丹属,大青属,海州常山属);Glory Bower, Glory Tree, Glorybower, Glory-bower, Glory-tree, Tuber Flower, Tuberflower ●■

95935 Clerodendrum acerbianum (Vis.) Benth.;阿切尔比赪桐●☆

95936 Clerodendrum aculeatum (L.) Griseb.;多刺赪桐;Coffee Fence, Prickly Wild Coffee ●☆

95937 Clerodendrum aculeatum (L.) Schltdl. = Clerodendrum aculeatum (L.) Griseb. ●☆

95938 Clerodendrum africanum Moldenke;非洲赪桐●☆

95939 Clerodendrum aggregatum Gürke;聚集赪桐●☆

95940 Clerodendrum alatum Gürke = Rotheca alata (Gürke) Verdc. ●☆

95941 Clerodendrum alatum Gürke var. adamauense B. Thomas = Rotheca alata (Gürke) Verdc. ●☆

95942 Clerodendrum alatum Gürke var. pubescens B. Thomas = Rotheca alata (Gürke) Verdc. ●☆

95943 Clerodendrum alboviolaceum Moldenke;浅堇色赪桐●☆

95944 Clerodendrum amplifolium S. Moore = Rotheca amplifolia (S. Moore) R. Fern. ●☆

95945 Clerodendrum amplius Hance = Clerodendrum cyrtophyllum Turcz. ●

95946 Clerodendrum angolense Gürke = Clerodendrum poggei Gürke ●☆

95947 Clerodendrum anomalum Letouzey;异常赪桐●☆

95948 Clerodendrum arenarium Baker;沙地赪桐●☆

95949 Clerodendrum asperatum B. Boivin = Clerodendrum boivinii Moldenke ●☆

95950 Clerodendrum assurgens (Hiern) K. Schum. = Rotheca sansibarensis (Gürke) Steane et Mabb. subsp. caesia (Gürke) Steane et Mabb. ●☆

95951 Clerodendrum attenuatum (De Wild.) De Wild. = Rotheca

sansibarensis (Gürke) Steane et Mabb. ●☆

95952 Clerodendrum aurantiacum Baker f. faulkneri R. Fern. = Rotheca aurantiaca (Baker) R. Fern. f. faulkneri (R. Fern.) R. Fern. ●☆

95953 Clerodendrum aurantium G. Don = Clerodendrum splendens G. Don ●☆

95954 Clerodendrum bakeri Gürke = Clerodendrum schweinfurthii Gürke ●☆

95955 Clerodendrum baronianum Oliv.;巴龙赪桐●☆

95956 Clerodendrum barteri Baker = Clerodendrum dusenii Gürke ●☆

95957 Clerodendrum baumii Gürke;鲍姆赪桐●☆

95958 Clerodendrum bellum Moldenke;雅致赪桐●☆

95959 Clerodendrum bequaertii De Wild. = Clerodendrum tanganyikense Baker ●☆

95960 Clerodendrum bequaertii De Wild. = Rotheca myricoides (Hochst.) Steane et Mabb. var. discolor (Klotzsch) Verdc. ●☆

95961 Clerodendrum bequaertii De Wild. var. debeerstii = Rotheca myricoides (Hochst.) Steane et Mabb. var. discolor (Klotzsch) Verdc. ●☆

95962 Clerodendrum bipindense Gürke;比平迪赪桐●☆

95963 Clerodendrum bodinierii H. Lév. = Clerodendrum mandarinorum Diels ●

95964 Clerodendrum bodinierii H. Lév. var. cavaleriei H. Lév. = Clerodendrum mandarinorum Diels ●

95965 Clerodendrum boivinii Moldenke;博伊文赪桐●☆

95966 Clerodendrum botryodes (Hiern) Baker = Clerodendrum silvanum Henriq. f. botryodes (Hiern) R. Fern. ●☆

95967 Clerodendrum brachystemon C. Y. Wu et R. C. Fang;短蕊大青(短蕊茉莉);Shortstamen Glorybower,Short-stamened Glorybower ●

95968 Clerodendrum bracteatum Wall.;苞花大青(苞花赪桐,苞片大青);Bracteous Glorybower ●

95969 Clerodendrum brazzavillense A. Chev. = Clerodendrum welwitschii Gürke ●☆

95970 Clerodendrum brunnescens Moldenke;浅褐赪桐●☆

95971 Clerodendrum brunsvigioides Baker = Boutonia cuspidata DC. ●☆

95972 Clerodendrum buchananii (Roxb.) Walp.;爪哇大青(布氏赪桐);Bogang,Kembang,Mata Ajam,Pagoda Flower ●☆

95973 Clerodendrum buchananii (Roxb.) Walp. var. fallax (Lindl.) Moldenke = Clerodendrum speciosissimum C. Morren ●☆

95974 Clerodendrum buchholzii Gürke = Clerodendrum silvanum Henriq. var. buchholzii (Gürke) Verdc. ●☆

95975 Clerodendrum buchneri Gürke;布赫纳赪桐●☆

95976 Clerodendrum buettneri Gürke;比特纳赪桐●☆

95977 Clerodendrum bukobense Gürke = Rotheca bukobensis (Gürke) Verdc. ●☆

95978 Clerodendrum bungei Steud.;臭牡丹(矮脚桐,矮桐,矮桐子,矮童子,臭八宝,臭草,臭灯桐,臭枫,臭枫草,臭枫根,臭芙蓉,臭树,臭梧桐,大红花,大红袍,丁香牡丹,逢仙草,鸡虱草,假龙船花,假真珠梧桐,野朱桐,野珠桐);Cashmere Bouquet, Clerodendrum, Glory Flower, Rose Glorybower, Strong-scented Glorybower ●

95979 Clerodendrum bungei Steud. var. megocalyx C. Y. Wu ex S. L. Chen;大萼臭牡丹;Bigcalyx Glorybower ●

95980 Clerodendrum cabrae De Wild.;卡布拉赪桐●☆

95981 Clerodendrum caeruleum N. E. Br. = Rotheca caerulea (N. E. Br.) Herman et Retief ●☆

95982 Clerodendrum caesium Gürke = Rotheca sansibarensis (Gürke) Steane et Mabb. subsp. caesia (Gürke) Steane et Mabb. ●☆

95983 Clerodendrum calamitosum L.;化石树●

95984 Clerodendrum canescens Wall.;灰毛大青(白花鬼灯笼,白毛臭牡丹,白蜻蜓,大叶白花鬼点火,灰毛臭茉莉,六灯笼,毛赪桐,人瘦木,山茉莉,狮子球,粘毛赪桐);Greyhair Glorybower, Grey-haired Glorybower ●

95985 Clerodendrum capense Eckl. et Zeyh. = Clerodendrum glabrum E. Mey. ●☆

95986 Clerodendrum capense G. Don = Clerodendrum glabrum E. Mey. ●☆

95987 Clerodendrum capitatum (Willd.) Schumach.;头状大青●☆

95988 Clerodendrum capitatum (Willd.) Schumach. var. butayei De Wild. = Clerodendrum poggei Gürke ●☆

95989 Clerodendrum capitatum (Willd.) Schumach. var. cephalanthum (Oliv.) Baker = Clerodendrum cephalanthum Oliv. ●☆

95990 Clerodendrum capitatum (Willd.) Schumach. var. conglobatum (Baker) B. Thomas = Clerodendrum capitatum (Willd.) Schumach. ●☆

95991 Clerodendrum capitatum (Willd.) Schumach. var. subcordatum De Wild. = Clerodendrum capitatum (Willd.) Schumach. ●☆

95992 Clerodendrum capitatum (Willd.) Schumach. var. talbotii (Wernham) B. Thomas = Clerodendrum capitatum (Willd.) Schumach. ●☆

95993 Clerodendrum capitatum Hook. = Clerodendrum whitfieldii Seem. ●☆

95994 Clerodendrum castaneifolium Hook. et Arn. = Clerodendrum fortunatum L. ●

95995 Clerodendrum cauliflorum De Wild. = Clerodendrum wildemanianum Exell ●☆

95996 Clerodendrum cauliflorum Vatke;茎花赪桐●☆

95997 Clerodendrum cavaleriei H. Lév. = Clerodendrum mandarinorum Diels ●

95998 Clerodendrum cavum De Wild. = Clerodendrum rotundifolium Oliv. ●☆

95999 Clerodendrum cecil-fischeri A. Rajendran et Daniel;菲舍尔大青●☆

96000 Clerodendrum cephalanthum Oliv.;头花大青●☆

96001 Clerodendrum cephalanthum Oliv. subsp. impensum (B. Thomas) Verdc.;宽头花大青●☆

96002 Clerodendrum cephalanthum Oliv. subsp. montanum (B. Thomas) Verdc.;山生头花大青●☆

96003 Clerodendrum cephalanthum Oliv. var. coriaceum B. Thomas = Clerodendrum cephalanthum Oliv. subsp. impensum (B. Thomas) Verdc. ●☆

96004 Clerodendrum cephalanthum Oliv. var. schliebenii (Mildbr.) Verdc.;施利本赪桐●☆

96005 Clerodendrum cephalanthum Oliv. var. swynnertonii (S. Moore) Verdc.;斯温纳顿赪桐●☆

96006 Clerodendrum cephalanthum Oliv. var. torrei R. Fern.;托雷赪桐●☆

96007 Clerodendrum chartaceum Moldenke;纸质赪桐●☆

96008 Clerodendrum chevalieri Moldenke = Clerodendrum velutinum A. Chev. ●☆

96009 Clerodendrum chinense (Osbeck) Mabb.;重瓣臭茉莉(臭茉莉,臭牡丹,臭矢茉莉,臭屎茉莉,臭梧桐,臭朱桐,大风草,大髻婆,冬地梅,过墙风,老虎草,龙船花,蜻蜓叶,山茉莉,小将军,走马风);Chinese Glory Bower, Chinese Glorybower, Fragrant Glorybower, Glory Bower, Glory Tree, Honolulu Rose, Philippine

Glorybower, Stickbush ●

96010 Clerodendrum chinense (Osbeck) Mabb. var. simplex (Moldenke) S. L. Chen;臭茉莉●

96011 Clerodendrum colebrokianum Walp.;腺茉莉(臭茉莉,臭牡丹,过墙风);Glandular Glorybower ●

96012 Clerodendrum commersonii H. H. Chung = Clerodendrum inerme (L.) Gaertn. ●

96013 Clerodendrum commiphoroides Verdc. = Rotheca commiphoroides (Verdc.) Steane et Mabb. ●☆

96014 Clerodendrum confine S. L. Chen et T. D. Zhuang;川黔大青;Chuan-Qian Glorybower, Sichuan-Guizhou Glorybower, Szechwan-Kweichow Glorybower ●

96015 Clerodendrum congense Baker = Clerodendrum schweinfurthii Gürke ●☆

96016 Clerodendrum congense Engl. = Clerodendrum umbellatum Poir. ●☆

96017 Clerodendrum congestum Gürke = Clerodendrum pleiosciadium Gürke ●☆

96018 Clerodendrum conglobatum Baker = Clerodendrum capitatum (Willd.) Schumach. ●☆

96019 Clerodendrum consors S. Moore = Clerodendrum tanganyikense Baker ●☆

96020 Clerodendrum cordifolium (Hochst.) A. Rich. = Clerodendrum umbellatum Poir. ●☆

96021 Clerodendrum costulatum (Hiern) K. Schum. = Clerodendrum silvanum Henriq. var. buchholzii (Gürke) Verdc. ●☆

96022 Clerodendrum crassifolium Rich. = Clerodendrum rubellum Baker ●☆

96023 Clerodendrum cubense Schauer;腋花大青;Axillaryflower Glorybower ●

96024 Clerodendrum cuneatum Gürke = Rotheca cuneiformis (Moldenke) Herman et Retief ●☆

96025 Clerodendrum cuneifolium Baker = Clerodendrum buchneri Gürke ●☆

96026 Clerodendrum cuneiforme Moldenke = Rotheca cuneiformis (Moldenke) Herman et Retief ●☆

96027 Clerodendrum cyaneum R. Fern. = Rotheca cyanea (R. Fern.) R. Fern. ●☆

96028 Clerodendrum cyrtophyllum Turcz.;大青(臭冲柴,臭根,臭屎青,臭婆根,臭腥公,臭叶树,大青木,大叶青,淡婆婆,观音灿,光花大青,鬼点灯,鸡屎青,蓝靛,路边青,木本大青,木大青,牛耳青,埔草欉,埔草样,青心草,山靛,山靛青,山漆,山尾花,土常山,土地骨皮,细叶臭牡丹,鸭公青,羊咪青,猪子菜,野地骨,野靛青,猪屎青);Axillaryflower Glorybower, Many Flower Glorybower, Manyflower Glorybower, Multiflorous Glorybower ●

96029 Clerodendrum cyrtophyllum Turcz. var. kwangsiensis S. L. Chen et W. D. Zhang;广西大青(地骨皮);Guangxi Glorybower, Kwangsi Glorybower ●

96030 Clerodendrum dalei Moldenke = Rotheca incisa (Klotzsch) Steane et Mabb. ●☆

96031 Clerodendrum darrisii H. Lév. = Clerodendrum japonicum (Thunb.) Sweet ●

96032 Clerodendrum dauphinense Moldenke;多芬大青●☆

96033 Clerodendrum decaryi Moldenke;德卡里大青●☆

96034 Clerodendrum deflexum Wall.;歪冠大青;Deflexed Glorybower ●☆

96035 Clerodendrum dekindtii Gürke var. dinteri B. Thomas = Rotheca myricoides (Hochst.) Steane et Mabb. ●☆

96036 Clerodendrum dependens Aug. DC.;悬垂大青●☆

96037 Clerodendrum dinklagei Gürke = Clerodendrum buettneri Gürke ●☆

96038 Clerodendrum discolor (Klotzsch) Vatke = Rotheca myricoides (Hochst.) Steane et Mabb. var. discolor (Klotzsch) Verdc. ●☆

96039 Clerodendrum discolor (Klotzsch) Vatke var. duemmeri B. Thomas = Rotheca myricoides (Hochst.) Steane et Mabb. var. discolor (Klotzsch) Verdc. ●☆

96040 Clerodendrum discolor (Klotzsch) Vatke var. kilimandscharense B. Thomas = Rotheca myricoides (Hochst.) Steane et Mabb. var. kilimandscharensis (Verdc.) Verdc. ●☆

96041 Clerodendrum discolor (Klotzsch) Vatke var. oppositifolium B. Thomas = Rotheca myricoides (Hochst.) Steane et Mabb. ●☆

96042 Clerodendrum discolor (Klotzsch) Vatke var. pluriflorum Gürke = Rotheca myricoides (Hochst.) Steane et Mabb. subsp. austromonticola (Verdc.) Verdc. ●☆

96043 Clerodendrum discolor (Klotzsch) Vatke var. verbascifolium Moldenke = Rotheca suffruticosa (Gürke) Verdc. ●☆

96044 Clerodendrum disparifolium Blume;马来臭梧桐●☆

96045 Clerodendrum divaricatum Jack. = Clerodendrum serratum (L.) Moon var. wallichii C. B. Clarke ●

96046 Clerodendrum divaricatum Jack. = Clerodendrum serratum (L.) Moon ●

96047 Clerodendrum divaricatum Siebold et Zucc. = Caryopteris divaricata (Siebold et Zucc.) Maxim. ■

96048 Clerodendrum dumale (Hiern) Baker = Rotheca myricoides (Hochst.) Steane et Mabb. var. dumalis (Hiern) R. Fern. ●☆

96049 Clerodendrum dusenii Gürke;杜森大青●☆

96050 Clerodendrum eburneum Chiov. ex Chiarugi = Clerodendrum robecchii Chiov. ●☆

96051 Clerodendrum eketense Wernham;埃凯特赪桐●☆

96052 Clerodendrum elachistanthum Merr. ex H. L. Li = Clerodendrum mandarinorum Diels ●

96053 Clerodendrum elachistanthum Merr. ex H. L. Li = Premna cavaleriei H. Lév. ●

96054 Clerodendrum elliotii Moldenke;埃利大青●☆

96055 Clerodendrum emirnense Bojer ex Hook.;埃米大青●☆

96056 Clerodendrum erectum De Wild. = Rotheca myricoides (Hochst.) Steane et Mabb. ●☆

96057 Clerodendrum eriophylloides Moldenke = Clerodendrum eriophyllum Gürke ●☆

96058 Clerodendrum eriophyllum Gürke;毛叶大青●☆

96059 Clerodendrum ervatamioides C. Y. Wu;狗牙大青(假狗牙花);Dogteeth Glorybower, Ervatamialike Glorybower, Ervatamia-like Glorybower ●

96060 Clerodendrum esquirolii H. Lév. = Clerodendrum japonicum (Thunb.) Sweet ●

96061 Clerodendrum esquirolii H. Lév. = Tacca chantrieri André ■

96062 Clerodendrum eucalycinum Oliv.;萼状大青●☆

96063 Clerodendrum eupatorioides Baker = Eremomastax speciosa (Hochst.) Cufod. ■☆

96064 Clerodendrum euryphyllum Mildbr. = Clerodendrum poggei Gürke ●☆

96065 Clerodendrum euryphyllum Mildbr. var. glabrum B. Thomas = Clerodendrum poggei Gürke ●☆

96066 Clerodendrum excavatum De Wild.;凹陷大青●☆

96067 Clerodendrum fallax Lindl. = Clerodendrum speciosissimum C. Morren ●☆

96068 Clerodendrum fargesii Dode = Clerodendrum trichotomum Thunb. ex A. Murray ●

96069 Clerodendrum fasciculatum B. Thomas;簇生大青●☆

96070 Clerodendrum faulkneri Moldenke;福克纳赪桐●☆

96071 Clerodendrum filipes Moldenke;线梗大青●☆

96072 Clerodendrum fischeri Gürke = Clerodendrum robustum Klotzsch var. fischeri (Gürke) Verdc. ●☆

96073 Clerodendrum fischeri Gürke var. robustum (Klotzsch) B. Thomas = Clerodendrum robustum Klotzsch ●☆

96074 Clerodendrum fischeri H. B. Naithani et Bennet = Clerodendrum cecil-fischeri A. Rajendran et Daniel ●☆

96075 Clerodendrum fleuryi A. Chev. = Rotheca alata (Gürke) Verdc. ●☆

96076 Clerodendrum floribundum R. Br.;丰花大青(多花赪桐);Lolly Bush ●☆

96077 Clerodendrum floribundum Schau = Clerodendrum emirnense Bojer ex Hook. ●☆

96078 Clerodendrum foetidum Bunge;臭大青(八仙花,臭枫根,臭牡丹,大红袍);Fetid Glorybower ●

96079 Clerodendrum foetidum Bunge = Clerodendrum bungei Steud. ●

96080 Clerodendrum formicarum Gürke;蚁赪桐●☆

96081 Clerodendrum formicarum Gürke var. sulcatum B. Thomas = Clerodendrum formicarum Gürke ●☆

96082 Clerodendrum formosanum Maxim. = Clerodendrum cyrtophyllum Turcz. ●

96083 Clerodendrum fortunatum L.;白花灯笼(白灯笼,白花鬼灯笼,灯笼草,岗灯笼,鬼把火,鬼灯笼,鬼点火,红灯笼,红萼灯笼草,红花大青,红花路边青,红羊米青,虎灯笼,尖萼常山,苦灯笼,苦丁茶,山顶企,土骨皮,土羚羊,夜鬼灯笼);Redcalyx Glorybower, White-flowered Glorybower ●

96084 Clerodendrum fragrans (Vent.) Willd. var. foetidum (Bunge) Bakhuizen = Clerodendrum bungei Steud. ●

96085 Clerodendrum fragrans (Vent.) Willd. var. multiplex Sweet = Clerodendrum chinense (Osbeck) Mabb. ●

96086 Clerodendrum fragrans (Vent.) Willd. var. pleniflorum Schauer = Clerodendrum chinense (Osbeck) Mabb. ●

96087 Clerodendrum fragrans Vent. = Clerodendrum chinense (Osbeck) Mabb. ●

96088 Clerodendrum fragrans Vent. = Clerodendrum philippinum Schauer ●

96089 Clerodendrum fragrans Vent. var. foetidum (Bunge) Baker = Clerodendrum bungei Steud. ●

96090 Clerodendrum fragrans Vent. var. multiplex Sweet = Clerodendrum chinense (Osbeck) Mabb. ●

96091 Clerodendrum fragrans Vent. var. pleniflorum Schauer = Clerodendrum chinense (Osbeck) Mabb. ●

96092 Clerodendrum fragrans Vent. var. pleniflorum Schauer = Clerodendrum philippinum Schauer ●

96093 Clerodendrum fragrans Willd. = Clerodendrum philippinum Schauer ●

96094 Clerodendrum francavilleanum Buchinger ex B. Thomas = Clerodendrum capitatum (Willd.) Schumach. ●☆

96095 Clerodendrum fuscum Gürke;棕色大青●☆

96096 Clerodendrum garrettianum Craib;泰国垂茉莉;Siam Glorybower ●

96097 Clerodendrum gibbosum Moldenke;浅囊大青●☆

96098 Clerodendrum gilletii De Wild. et T. Durand = Clerodendrum splendens G. Don ●☆

96099 Clerodendrum glabrum E. Mey.;光大青;Natal Glorybower ●☆

96100 Clerodendrum glabrum E. Mey. f. pubescens R. Fern.;毛大青●☆

96101 Clerodendrum glabrum E. Mey. var. angustifolium? = Clerodendrum glabrum E. Mey. ●☆

96102 Clerodendrum glabrum E. Mey. var. vagum (Hiern) Moldenke = Clerodendrum eriophyllum Gürke ●☆

96103 Clerodendrum globosum Moldenke;球大青●☆

96104 Clerodendrum globuliflorum B. Thomas;球花赪桐●☆

96105 Clerodendrum goossensii De Wild. = Clerodendrum silvanum Henriq. f. botryodes (Hiern) R. Fern. ●☆

96106 Clerodendrum gossweileri Exell = Clerodendrum schweinfurthii Gürke ●☆

96107 Clerodendrum grandicalyx Bruce = Clerodendrum fuscum Gürke ●☆

96108 Clerodendrum grandifolium Gürke = Clerodendrum excavatum De Wild. ●☆

96109 Clerodendrum grevei Moldenke;格雷弗赪桐●☆

96110 Clerodendrum greyi Baker = Clerodendrum speciosissimum C. Morren ●☆

96111 Clerodendrum griffithianum C. B. Clarke;西垂茉莉;Griffith Glorybower ●

96112 Clerodendrum guerkei Baker = Clerodendrum rotundifolium Oliv. ●☆

96113 Clerodendrum haematocalyx Hance = Clerodendrum canescens Wall. ●

96114 Clerodendrum hainanense Hand.-Mazz.;海南赪桐;Hainan Glorybower ●

96115 Clerodendrum harnierianum Schweinf. = Clerodendrum triflorum Vis. ●☆

96116 Clerodendrum henryi C. P'ei;南垂茉莉(都劳九,小刺毛树);Henry Glorybower ●

96117 Clerodendrum herbaceum Roxb. ex Schauer = Clerodendrum serratum (L.) Moon var. herbaceum (Roxb. ex Schauer) C. Y. Wu ●

96118 Clerodendrum hexagonum De Wild. = Rotheca alata (Gürke) Verdc. ●☆

96119 Clerodendrum hexangulatum B. Thomas;六角赪桐●☆

96120 Clerodendrum hildebrandtii Vatke;希尔德赪桐●☆

96121 Clerodendrum hildebrandtii Vatke var. puberula Verdc.;微毛赪桐●☆

96122 Clerodendrum hildebrandtii Vatke var. pubescens Moldenke = Clerodendrum rotundifolium Oliv. ●☆

96123 Clerodendrum hircinum Schauer;山羊赪桐●☆

96124 Clerodendrum hirsutum (Hochst.) H. Pearson = Rotheca hirsuta (Hochst.) R. Fern. ●☆

96125 Clerodendrum hirsutum G. Don = Clerodendrum umbellatum Poir. ●☆

96126 Clerodendrum hockii De Wild. = Clerodendrum buchneri Gürke ●☆

96127 Clerodendrum holstii Gürke ex Baker = Clerodendrum acerbianum (Vis.) Benth. ●☆

96128 Clerodendrum humbertii Moldenke;亨伯特赪桐●☆

96129 Clerodendrum humile Chiov. = Clerodendrum buchneri Gürke ●☆

96130 Clerodendrum hysteranthum Baker = Clerodendrum poggei Gürke ●☆

96131 Clerodendrum impensum B. Thomas = Clerodendrum cephalanthum Oliv. subsp. impensum (B. Thomas) Verdc. ●☆

96132 Clerodendrum impensum B. Thomas var. buchneroides = Clerodendrum buchneri Gürke ●☆

96133 Clerodendrum inaequipetiolatum R. D. Good;不等梗赪桐●☆

96134 Clerodendrum incisum Klotzsch = Rotheca incisa (Klotzsch) Steane et Mabb. ●☆

96135 Clerodendrum incisum Klotzsch var. longepedunculatum B. Thomas = Rotheca incisa (Klotzsch) Steane et Mabb. ●☆

96136 Clerodendrum incisum Klotzsch var. macrosiphon (Hook. f.) Baker = Rotheca incisa (Klotzsch) Steane et Mabb. ●☆

96137 Clerodendrum incisum Klotzsch var. vinosum Chiov. = Rotheca incisa (Klotzsch) Steane et Mabb. ●☆

96138 Clerodendrum indeniense A. Chev. = Rotheca violacea (Gürke) Verdc. ●☆

96139 Clerodendrum indicum (L.) Kuntze;长管大青(长管假茉莉,长管假牡丹,疟疾草,印度海州常山);India Glorybower,Indian Glorybower,Tube-flower,Turk's Turban,Turk's Turbin ●

96140 Clerodendrum inerme (L.) Gaertn.;苦郎树(白骨沿藤,白花苦林盘,草朗,臭苦蓢,臭苦蓢,臭矢茉莉,臭沿藤,海常山,虎狼草,黄藤,鸡公尾,假茉莉,见水生,九里苔,苦蓝盘,苦林盘,澎蜞盖,蟛蜞盖,水胡满,许树,猪屘怕);Embrert,Sea-side Clerodendrum,Unarmed Glorybower ●

96141 Clerodendrum infortunatum L.;欠愉大青;Unfortunate Glorybower ●

96142 Clerodendrum insolitum Moldenke = Vitex bracteata Scott-Elliot ●☆

96143 Clerodendrum intermedium B. Thomas = Clerodendrum dusenii Gürke ●☆

96144 Clerodendrum intermedium Champ.;垦丁苦林盘●

96145 Clerodendrum involucratum Vatke;总苞赪桐●☆

96146 Clerodendrum izuinsulare K. Inoue, Haseg. et S. Kobay.;伊豆岛赪桐●☆

96147 Clerodendrum japonicum (Thunb.) Sweet;赪桐(白菱,百日红,抽须红,雌雄树,大丹,大将军,大叶红花倒水莲,绯桐,合包花,荷包花,红池木,红虫木,红顶风,红花倒血莲,红苓蕴,红菱,红菱蕴,红龙船花,龙穿花,龙船花,唐桐,香带花,香斗花,香盏花,洋海棠,贞桐花,桢桐,真珠花,真珠梧桐,朱桐,状元红);Japan Glorybower,Japanese Glorybower,Japanese Tubeflower,Kaempfer's Glorybower ●

96148 Clerodendrum japonicum (Thunb.) Sweet var. album C. P'ei = Clerodendrum japonicum (Thunb.) Sweet ●

96149 Clerodendrum japonicum (Thunb.) Sweet var. pleniflorum (Schauer) Maheshw. = Clerodendrum chinense (Osbeck) Mabb. ●

96150 Clerodendrum jaundense Gürke = Clerodendrum formicarum Gürke ●☆

96151 Clerodendrum johnstonii Oliv.;约翰斯顿赪桐●☆

96152 Clerodendrum johnstonii Oliv. subsp. marsabitense Verdc.;马萨比特赪桐●☆

96153 Clerodendrum johnstonii Oliv. var. rubrum B. Thomas = Clerodendrum johnstonii Oliv. ●☆

96154 Clerodendrum kaempferi (Jacq.) Siebold = Clerodendrum japonicum (Thunb.) Sweet ●

96155 Clerodendrum kaempferi (Jacq.) Siebold ex Steud. = Clerodendrum japonicum (Thunb.) Sweet ●

96156 Clerodendrum kaempferi (Jacq.) Siebold var. album (C. P'ei) Moldenke = Clerodendrum japonicum (Thunb.) Sweet ●

96157 Clerodendrum kaichianum P. S. Hsu;浙江大青(黄山大青,凯基大青);Chekiang Glorybower,Zhejiang Glorybower ●

96158 Clerodendrum kalbreyeri Baker = Rotheca violacea (Gürke) Verdc. ●☆

96159 Clerodendrum katangense De Wild.;加丹加赪桐●☆

96160 Clerodendrum kentrocaule Baker = Clerodendrum silvanum Henriq. var. buchholzii (Gürke) Verdc. ●☆

96161 Clerodendrum kiangsiense Merr. ex H. L. Li;江西大青;Jiangxi Glorybower,Kiangsi Glorybower ●

96162 Clerodendrum kibwesense Moldenke = Premna oligotricha Baker ●☆

96163 Clerodendrum kirkii Baker;柯克大青●☆

96164 Clerodendrum kissakense Gürke = Rotheca kissakensis (Gürke) Verdc. ●☆

96165 Clerodendrum kissakense Gürke var. rovumense B. Thomas = Rotheca kissakensis (Gürke) Verdc. ●☆

96166 Clerodendrum koshunense Hayata = Clerodendrum trichotomum Thunb. ex A. Murray ●

96167 Clerodendrum kwangtungense Hand. -Mazz.;广东大青(广东赪桐,广东臭茉莉,广东臭牡丹,红花鬼灯笼,红花鬼点灯,野靛,野靛青);Guangdong Glorybower,Kwangtung Glorybower ●

96168 Clerodendrum kwangtungense Hand. -Mazz. var. puberulum H. L. Li = Clerodendrum mandarinorum Diels ●

96169 Clerodendrum laevifolium Blume;平滑叶赪桐●☆

96170 Clerodendrum lanceolatum F. Muell.;披针叶赪桐●☆

96171 Clerodendrum lanceolatum Gürke = Clerodendrum ternatum Schinz ●☆

96172 Clerodendrum lastellei Moldenke;拉斯泰勒赪桐●☆

96173 Clerodendrum laxicymosum De Wild. = Clerodendrum silvanum Henriq. var. nuxioides (S. Moore) Verdc. ●☆

96174 Clerodendrum laxiflorum Baker;疏花臭茉莉●☆

96175 Clerodendrum lelyi Hutch. = Rotheca alata (Gürke) Verdc. ●☆

96176 Clerodendrum leprieurii Moldenke;莱普里厄赪桐●☆

96177 Clerodendrum leucobotrys Breteler;白穗赪桐●☆

96178 Clerodendrum leucosceptrum D. Don = Leucosceptrum canum Sm. ●

96179 Clerodendrum lindiense Moldenke = Clerodendrum tricholobum Gürke ●☆

96180 Clerodendrum lindleyi Decne. ex Planch.;尖齿臭茉莉(臭黄根,臭茉莉,臭牡丹,大风叶,鬼点火,过墙风,尖齿大青);Lindley Glorybower,Lindley's Clerodendrum ●

96181 Clerodendrum lindleyi Decne. ex Planch. var. paniculatum Moldenke;圆锥尖齿臭茉莉;Lindley's Clerodendrum ●☆

96182 Clerodendrum lividum Lindl. = Clerodendrum fortunatum L. ●

96183 Clerodendrum longilimbum C. P'ei;长叶大青(长叶臭茉莉);Longleaf Glorybower,Long-leaved Glorybower ●

96184 Clerodendrum longitubum De Wild. et T. Durand = Clerodendrum schweinfurthii Gürke ●☆

96185 Clerodendrum luembense De Wild. = Rotheca luembensis (De Wild.) R. Fern. ●☆

96186 Clerodendrum luembense De Wild. f. herbaceum (Hiern) R. Fern. = Rotheca luembensis (De Wild.) R. Fern. ●☆

96187 Clerodendrum luembense De Wild. var. malawiensis R. Fern. = Rotheca luembensis (De Wild.) R. Fern. var. malawiensis (R. Fern.) R. Fern. ●☆

96188 Clerodendrum lujaei De Wild. = Clerodendrum formicarum Gürke ●☆

96189 Clerodendrum lupakense S. Moore = Clerodendrum tanganyikense Baker ●☆

96190 Clerodendrum lutambense Verdc.;卢塔波赪桐●☆

96191 Clerodendrum luteopunctatum C. P'ei et S. L. Chen;黄腺大青;Yellowdot Glorybower,Yellow-dotted Glorybower ●

96192 Clerodendrum macrocalycinum Baker;大萼赪桐●☆

96193 Clerodendrum macrocalyx De Wild. = Clerodendrum fuscum Gürke ●☆

96194 Clerodendrum macrosiphon (Baker) W. Piep. = Clerodendrum dusenii Gürke ●☆

96195 Clerodendrum macrosiphon Hook. f. = Rotheca incisa (Klotzsch) Steane et Mabb. ●☆

96196 Clerodendrum macrostegium Schauer; 大盖赪桐; Velvetleaf Glorybower ●☆

96197 Clerodendrum madagascariense Moldenke;马岛赪桐●☆

96198 Clerodendrum magnoliifolium Baker;木兰叶赪桐●☆

96199 Clerodendrum mananjariense Moldenke;马南扎里赪桐●☆

96200 Clerodendrum mandarinorum Diels;海通(白灯笼,臭梧桐,满大青,牡丹树,木常山,泡桐树,朴瓜树,水木通,铁枪桐,桐木树,土常山,线桐树,小花泡桐,鞋头树);Tomentose Glorybower ●

96201 Clerodendrum mandrarense Moldenke;曼德拉赪桐●☆

96202 Clerodendrum mannii Baker;曼氏赪桐●☆

96203 Clerodendrum megasepalum Baker = Clerodendrum poggei Gürke ●☆

96204 Clerodendrum melanocrater Gürke;黑杯赪桐●☆

96205 Clerodendrum melanophyllum (S. Moore) S. Moore = Clerodendrum melanocrater Gürke ●☆

96206 Clerodendrum mendesii R. Fern. = Rotheca mendesii (R. Fern.) R. Fern. ●☆

96207 Clerodendrum micans Gürke;弱光泽赪桐●☆

96208 Clerodendrum micranthum Gilli = Hoslundia opposita Vahl ●☆

96209 Clerodendrum microphyllum B. Thomas = Clerodendrum robecchii Chiov. ●☆

96210 Clerodendrum mildbraedii B. Thomas;米尔德赪桐●☆

96211 Clerodendrum milne-redheadii Moldenke = Rotheca prittwitzii (B. Thomas) Verdc. ●☆

96212 Clerodendrum mirabile Baker;奇异赪桐●☆

96213 Clerodendrum montanum B. Thomas = Clerodendrum cephalanthum Oliv. subsp. montanum (B. Thomas) Verdc. ●☆

96214 Clerodendrum moramangense Moldenke;莫拉芒赪桐●☆

96215 Clerodendrum mossambicense Klotzsch = Clerodendrum capitatum (Willd.) Schumach. ●☆

96216 Clerodendrum mossambicense Klotzsch var. glabrum B. Thomas = Clerodendrum cephalanthum Oliv. ●☆

96217 Clerodendrum moupinense Franch. = Microtoena moupinensis (Franch.) Prain ■

96218 Clerodendrum multiflorum G. Don = Clerodendrum volubile P. Beauv. ●☆

96219 Clerodendrum murigono Chiov. = Clerodendrum johnstonii Oliv. ●☆

96220 Clerodendrum myrianthum Mildbr.;多花赪桐●☆

96221 Clerodendrum myricoides (Hochst.) R. Br. ex Vatke = Rotheca myricoides (Hochst.) Steane et Mabb. ●☆

96222 Clerodendrum myricoides (Hochst.) Vatke = Rotheca myricoides (Hochst.) Steane et Mabb. ●☆

96223 Clerodendrum myricoides (Hochst.) Vatke f. angustilobatum R. Br. = Rotheca myricoides (Hochst.) Steane et Mabb. f. angustilobata (R. Fern.) R. Fern. ●☆

96224 Clerodendrum myricoides (Hochst.) Vatke f. brevilobatum R. Fern. = Rotheca myricoides (Hochst.) Steane et Mabb. f. brevilobata (R. Fern.) R. Fern. ●☆

96225 Clerodendrum myricoides (Hochst.) Vatke f. cubangense R. Fern. = Rotheca myricoides (Hochst.) Steane et Mabb. f. cubangensis (R. Fern.) R. Fern. ●☆

96226 Clerodendrum myricoides (Hochst.) Vatke f. lanceolatilobatum R. Fern. = Rotheca myricoides (Hochst.) Steane et Mabb. f. lanceolatilobata (R. Fern.) R. Fern. ●☆

96227 Clerodendrum myricoides (Hochst.) Vatke f. lobulatum R. Fern. = Rotheca myricoides (Hochst.) Steane et Mabb. f. lobulata (R. Fern.) R. Fern. ●☆

96228 Clerodendrum myricoides (Hochst.) Vatke f. reflexilobatum R. Fern. = Rotheca myricoides (Hochst.) Steane et Mabb. f. reflexilobata (R. Fern.) R. Fern. ●☆

96229 Clerodendrum myricoides (Hochst.) Vatke subsp. austromonticola Verdc. = Rotheca myricoides (Hochst.) Steane et Mabb. subsp. austromonticola (Verdc.) Verdc. ●☆

96230 Clerodendrum myricoides (Hochst.) Vatke subsp. mafiense Verdc. = Rotheca myricoides (Hochst.) Steane et Mabb. subsp. mafiensis (Verdc.) Verdc. ●☆

96231 Clerodendrum myricoides (Hochst.) Vatke subsp. namibiense R. Fern. = Rotheca myricoides (Hochst.) Steane et Mabb. ●☆

96232 Clerodendrum myricoides (Hochst.) Vatke var. attenuatum De Wild. = Rotheca sansibarensis (Gürke) Steane et Mabb. ●☆

96233 Clerodendrum myricoides (Hochst.) Vatke var. discolor (Klotzsch) Baker = Rotheca myricoides (Hochst.) Steane et Mabb. var. discolor (Klotzsch) Verdc. ●☆

96234 Clerodendrum myricoides (Hochst.) Vatke var. involutum B. Thomas = Rotheca myricoides (Hochst.) Steane et Mabb. ●☆

96235 Clerodendrum myricoides (Hochst.) Vatke var. kilimandscharense Verdc. = Rotheca myricoides (Hochst.) Steane et Mabb. var. kilimandscharensis (Verdc.) Verdc. ●☆

96236 Clerodendrum myricoides (Hochst.) Vatke var. niansanum B. Thomas = Rotheca myricoides (Hochst.) Steane et Mabb. ●☆

96237 Clerodendrum myricoides (Hochst.) Vatke var. viridiflorum Verdc. = Rotheca myricoides (Hochst.) Steane et Mabb. var. viridiflora (Verdc.) Verdc. ●☆

96238 Clerodendrum myrtifolium Moldenke;香桃木叶赪桐●☆

96239 Clerodendrum natalense Gürke = Rotheca hirsuta (Hochst.) R. Fern. ●☆

96240 Clerodendrum nereifolium (Roxb.) Schauer = Clerodendrum inerme (L.) Gaertn. ●

96241 Clerodendrum neumayeri Vatke = Rotheca myricoides (Hochst.) Steane et Mabb. ●☆

96242 Clerodendrum noiroti A. Chev. = Rotheca violacea (Gürke) Verdc. ●☆

96243 Clerodendrum nudiflorum Moldenke;裸花赪桐●☆

96244 Clerodendrum nutans D. Don = Clerodendrum wallichii Merr. ●

96245 Clerodendrum nutans Wall. ex D. Don = Clerodendrum wallichii Merr. ●

96246 Clerodendrum nuxioides (S. Moore) B. Thomas = Clerodendrum silvanum Henriq. var. nuxioides (S. Moore) Verdc. ●☆

96247 Clerodendrum obanense Wernham = Clerodendrum capitatum (Willd.) Schumach. ●☆

96248 Clerodendrum odoratum (Ham. ex Roxb.) D. Don. = Caryopteris odorata (Ham. ex Roxb.) Rob. ●

96249 Clerodendrum ohwii Kaneh. et Hatus.;花莲海州常山(大井氏海州常山)●

96250 Clerodendrum orbiculare Baker = Clerodendrum poggei Gürke ●☆

96251 Clerodendrum oreadum S. Moore = Clerodendrum formicarum Gürke ●☆

96252 Clerodendrum ornatum Wall. = Clerodendrum serratum (L.)

Moon ●

96253 Clerodendrum ovale Klotzsch = Clerodendrum glabrum E. Mey. ●☆

96254 Clerodendrum ovalifolium (Juss.) Bakh.;椭圆叶赪桐●☆

96255 Clerodendrum ovalifolium Engl. = Clerodendrum glabrum E. Mey. ●☆

96256 Clerodendrum oxysepalum Miq. = Clerodendrum fortunatum L. ●

96257 Clerodendrum paniculatum L.;圆锥大青(白童女,赪桐,癫婆花,龙船花,女贞草,贞桐花,朱桐花);Pagoda Flower,Paniculate Glorybower,Scarlet Glorybower ●

96258 Clerodendrum paniculatum L. var. albiflorum Hemsl.;白圆锥大青(白龙船花);White-flower Paniculate Glorybower ●

96259 Clerodendrum parvitubulatum B. Thomas;小管赪桐●☆

96260 Clerodendrum paucidentataum Moldenke;少齿大青●☆

96261 Clerodendrum pauciflorum Moldenke;少花大青●☆

96262 Clerodendrum pearsonii Moldenke = Rotheca hirsuta (Hochst.) R. Fern. ●☆

96263 Clerodendrum peii Moldenke;长梗大青(长柄臭牡丹);Longpetiole Glorybower,Pei Glorybower ●

96264 Clerodendrum penduliflorum Wall.;垂花大青;Pendulousflower Glorybower ●

96265 Clerodendrum pentagonum Hance = Clerodendrum fortunatum L. ●

96266 Clerodendrum peregrinum Moldenke;外来赪桐●☆

96267 Clerodendrum perrieri Moldenke;佩里耶大青●☆

96268 Clerodendrum petasites (Lour.) A. Meeuse;毛赪桐(白毛臭牡丹);Woolly Glorybower ●

96269 Clerodendrum petasites (Lour.) Moore = Clerodendrum canescens Wall. ●

96270 Clerodendrum petunioides Baker;碧冬茄赪桐●☆

96271 Clerodendrum philippinum Schauer = Clerodendrum chinense (Osbeck) Mabb. ●

96272 Clerodendrum philippinum Schauer var. simplex C. Y. Wu ex R. C. Fang = Clerodendrum chinense (Osbeck) Mabb. var. simplex (Moldenke) S. L. Chen ●

96273 Clerodendrum philippinum Schauer var. simplex Moldenke;单瓣臭茉莉(白花臭茉莉,白花臭牡丹,臭茉莉);Simple Glorybower ●

96274 Clerodendrum philippinum Schauer var. simplex Moldenke = Clerodendrum chinense (Osbeck) Mabb. var. simplex (Moldenke) S. L. Chen ●

96275 Clerodendrum phlebodes C. H. Wright = Rotheca myricoides (Hochst.) Steane et Mabb. var. discolor (Klotzsch) Verdc. ●☆

96276 Clerodendrum phlebodes C. H. Wright var. pilosocalyx B. Thomas = Rotheca myricoides (Hochst.) Steane et Mabb. var. discolor (Klotzsch) Verdc. ●☆

96277 Clerodendrum phlomidis L. f.;糙苏赪桐●☆

96278 Clerodendrum pilosum H. Pearson = Rotheca pilosa (H. Pearson) Herman et Retief ●☆

96279 Clerodendrum pleiosciadium Gürke;多伞赪桐●☆

96280 Clerodendrum pleiosciadium Gürke var. bussei B. Thomas = Clerodendrum pleiosciadium Gürke ●☆

96281 Clerodendrum pleiosciadium Gürke var. dentata B. Thomas = Clerodendrum pleiosciadium Gürke ●☆

96282 Clerodendrum poggei Gürke;波格赪桐●☆

96283 Clerodendrum polyanthum Gürke;繁花赪桐●☆

96284 Clerodendrum polycephalum Baker;多头赪桐●☆

96285 Clerodendrum premnoides A. Meeuse = Clerodendrum dusenii Gürke ●☆

96286 Clerodendrum premnoides Moldenke;豆腐柴赪桐●☆

96287 Clerodendrum preussii Gürke = Clerodendrum silvanum Henriq. var. buchholzii (Gürke) Verdc. ●☆

96288 Clerodendrum preussii Gürke var. silvanum (Henriq.) B. Thomas = Clerodendrum silvanum Henriq. ●☆

96289 Clerodendrum prittwitzii B. Thomas = Rotheca prittwitzii (B. Thomas) Verdc. ●☆

96290 Clerodendrum pulverulentum Engl.;粉粒赪桐●☆

96291 Clerodendrum pumilum (Lour.) Spreng. = Clerodendrum fortunatum L. ●

96292 Clerodendrum pusillum Gürke;微小赪桐●☆

96293 Clerodendrum pynaertii De Wild.;皮那赪桐●☆

96294 Clerodendrum pyramidale Andréws = Clerodendrum paniculatum L. ●

96295 Clerodendrum pyrifolium Baker;梨叶赪桐●☆

96296 Clerodendrum quadrangulatum B. Thomas = Rotheca sansibarensis (Gürke) Steane et Mabb. ●☆

96297 Clerodendrum ramosissimum Baker;多枝赪桐●☆

96298 Clerodendrum reflexum H. Pearson = Rotheca reflexa (H. Pearson) R. Fern. ●☆

96299 Clerodendrum rehmannii Gürke;拉赫曼赪桐●☆

96300 Clerodendrum rehmannii Gürke = Clerodendrum glabrum E. Mey. ●☆

96301 Clerodendrum revolutum Bosser;外卷赪桐●☆

96302 Clerodendrum ringoetii De Wild.;林戈赪桐●☆

96303 Clerodendrum robecchii Chiov.;罗贝克赪桐●☆

96304 Clerodendrum robecchii Chiov. var. macrophyllum = Clerodendrum robecchii Chiov. ●☆

96305 Clerodendrum robustum Klotzsch;粗壮赪桐●☆

96306 Clerodendrum robustum Klotzsch var. fischeri (Gürke) Verdc.;费氏粗壮赪桐●☆

96307 Clerodendrum robustum Klotzsch var. latilobum Verdc.;宽裂粗壮赪桐●☆

96308 Clerodendrum robustum Klotzsch var. macrocalyx R. Fern.;大萼粗壮赪桐●☆

96309 Clerodendrum roseiflorum Moldenke;粉花赪桐●☆

96310 Clerodendrum rotundifolium Oliv.;圆叶赪桐;Round-leaf Glorybower ●☆

96311 Clerodendrum rotundifolium Oliv. var. keniense T. C. E. Fr. = Clerodendrum rotundifolium Oliv. ●☆

96312 Clerodendrum rotundifolium Oliv. var. stuhlmannii (Gürke) Thomas = Clerodendrum rotundifolium Oliv. ●☆

96313 Clerodendrum rubellum Baker;微红赪桐●☆

96314 Clerodendrum rupicola Verdc. = Rotheca rupicola (Verdc.) Verdc. ●☆

96315 Clerodendrum sakaleonense Moldenke;萨卡莱乌纳赪桐●☆

96316 Clerodendrum sanguineum (Hiern) K. Schum. = Clerodendrum poggei Gürke ●☆

96317 Clerodendrum sansibarense Gürke;桑给巴尔赪桐●☆

96318 Clerodendrum sansibarense Gürke = Rotheca sansibarensis (Gürke) Steane et Mabb. ●☆

96319 Clerodendrum sansibarense Gürke f. tomentellum R. Br. = Rotheca sansibarensis (Gürke) Steane et Mabb. f. tomentella (R. Br.) R. Fern. ●☆

96320 Clerodendrum sansibarense Gürke subsp. caesium (Gürke) Verdc. = Rotheca sansibarensis (Gürke) Steane et Mabb. subsp. caesia (Gürke) Steane et Mabb. ●☆

96321 Clerodendrum sansibarense Gürke subsp. occidentale Verdc. =

Rotheca sansibarensis (Gürke) Steane et Mabb. subsp. occidentalis (Verdc.) Steane et Mabb. ●☆

96322 Clerodendrum sassandrense Jongkind;萨桑德拉赪桐●☆

96323 Clerodendrum savanorum De Wild. = Rotheca myricoides (Hochst.) Steane et Mabb. ●☆

96324 Clerodendrum scandens P. Beauv. = Clerodendrum umbellatum Poir. ●☆

96325 Clerodendrum scandens P. Beauv. var. asperifolium B. Thomas = Clerodendrum umbellatum Poir. ●☆

96326 Clerodendrum scandens P. Beauv. var. speciosum B. Thomas = Clerodendrum umbellatum Poir. ●☆

96327 Clerodendrum scheffleri Gürke = Rotheca sansibarensis (Gürke) Steane et Mabb. ●☆

96328 Clerodendrum scheffleri Gürke var. ellipticum Moldenke = Rotheca cuneiformis (Moldenke) Herman et Retief ●☆

96329 Clerodendrum scheffleri Gürke var. mahengianum B. Thomas = Rotheca sansibarensis (Gürke) Steane et Mabb. ●☆

96330 Clerodendrum schifferi A. Chev. = Clerodendrum silvanum Henriq. var. buchholzii (Gürke) Verdc. ●☆

96331 Clerodendrum schliebenii Mildbr. = Clerodendrum cephalanthum Oliv. var. schliebenii (Mildbr.) Verdc. ●☆

96332 Clerodendrum schultzei Mildbr. = Clerodendrum mannii Baker ●☆

96333 Clerodendrum schweinfurthii Gürke;施韦赪桐●☆

96334 Clerodendrum schweinfurthii Gürke var. bakeri (Gürke) Thomas = Clerodendrum schweinfurthii Gürke ●☆

96335 Clerodendrum schweinfurthii Gürke var. conradsii B. Thomas = Clerodendrum schweinfurthii Gürke ●☆

96336 Clerodendrum schweinfurthii Gürke var. longitubum (De Wild. et T. Durand) B. Thomas = Clerodendrum schweinfurthii Gürke ●☆

96337 Clerodendrum seretii De Wild. = Clerodendrum melanocrater Gürke ●☆

96338 Clerodendrum serotinum Carrière = Clerodendrum trichotomum Thunb. ex A. Murray ●

96339 Clerodendrum serratum (L.) Moon;三对节(齿叶赪桐,大常山,大罗伞,大叶土常山,对节生,火山麻,三百棒,三多,三台大药,三台红花,三台花,山利桐,山枇杷);Serrate Glorybower, Witches' Tongue, Witch's Tongue ●

96340 Clerodendrum serratum (L.) Moon var. amplexifolium Moldenke;三台花(抱茎三对节,三台红花);Amplexicaul Serrate Glorybower, Amplexifolious Glorybower ●

96341 Clerodendrum serratum (L.) Moon var. herbaceum (Roxb. ex Schauer) C. Y. Wu;草本三对节●

96342 Clerodendrum serratum (L.) Moon var. wallichii C. B. Clarke;大序三对节;Wallich Glorybower ●

96343 Clerodendrum sessilifolium Moldenke;无柄叶赪桐●☆

96344 Clerodendrum silvanum Henriq.;森林赪桐●☆

96345 Clerodendrum silvanum Henriq. f. botryodes (Hiern) R. Fern.;葡萄森林赪桐●☆

96346 Clerodendrum silvanum Henriq. var. brevitubum R. Fern.;短管森林赪桐●☆

96347 Clerodendrum silvanum Henriq. var. buchholzii (Gürke) Verdc.;布氏森林赪桐●☆

96348 Clerodendrum silvanum Henriq. var. nuxioides (S. Moore) Verdc.;坚果状赪桐●☆

96349 Clerodendrum silvestre B. Thomas;林赪桐●☆

96350 Clerodendrum simile H. Pearson = Clerodendrum transvaalense B. Thomas ●☆

96351 Clerodendrum simplex G. Don = Clerodendrum umbellatum Poir. ●☆

96352 Clerodendrum sinuatum Hook.;深波赪桐●☆

96353 Clerodendrum sinuatum Hook. var. aureum Berhaut = Clerodendrum sinuatum Hook. ●☆

96354 Clerodendrum siphonanthus R. Br. = Clerodendrum indicum (L.) Kuntze ●

96355 Clerodendrum somalense Chiov. = Clerodendrum glabrum E. Mey. ●☆

96356 Clerodendrum speciosissimum C. Morren;爪哇常山(唐桐,爪哇赪桐);Java Glorybower, Javanese Glorybower, Pagoda Flower ●☆

96357 Clerodendrum speciosissimum Drapiez = Clerodendrum speciosissimum C. Morren ●☆

96358 Clerodendrum speciosum Bullock;美丽赪桐●☆

96359 Clerodendrum speciosum Gürke = Clerodendrum poggei Gürke ●☆

96360 Clerodendrum spicatum Thunb. = Clerodendranthus spicatus (Thunb.) C. Y. Wu ex H. W. Li ●

96361 Clerodendrum spinescens (Oliv.) Gürke = Clerodendrum uncinatum Schinz ●☆

96362 Clerodendrum splendens A. Chev.;艳桢桐;Glorybower ●☆

96363 Clerodendrum splendens G. Don = Clerodendrum splendens A. Chev. ●☆

96364 Clerodendrum squamatum Vahl = Clerodendrum japonicum (Thunb.) Sweet ●

96365 Clerodendrum stenanthum Klotzsch = Clerodendrum capitatum (Willd.) Schumach. ●☆

96366 Clerodendrum streptocaulon Hutch. et Dalziel = Clerodendrum thyrsoideum Gürke ●☆

96367 Clerodendrum strictum Baker = Clerodendrum buchneri Gürke ●☆

96368 Clerodendrum stuhlmannii Gürke = Clerodendrum rotundifolium Oliv. ●☆

96369 Clerodendrum subpeltatum Wernham;亚盾状赪桐●☆

96370 Clerodendrum subscaposum Hemsl.;抽葶大青(抽葶赪桐);Subscapose Glorybower ●

96371 Clerodendrum subtruncatum Moldenke;平截赪桐●☆

96372 Clerodendrum suffruticosum Gürke = Rotheca suffruticosa (Gürke) Verdc. ●☆

96373 Clerodendrum swynnertonii S. Moore = Clerodendrum cephalanthum Oliv. var. swynnertonii (S. Moore) Verdc. ●☆

96374 Clerodendrum sylviae J. -G. Adam;西尔维亚赪桐●☆

96375 Clerodendrum syringiifolium Baker = Clerodendrum pleiosciadium Gürke ●☆

96376 Clerodendrum taborense Verdc. = Rotheca taborensis (Verdc.) Verdc. ●☆

96377 Clerodendrum taborense Verdc. var. latifolium Verdc. = Rotheca taborensis (Verdc.) Verdc. var. latifolia (Verdc.) Verdc. ●☆

96378 Clerodendrum talbotii Wernham = Clerodendrum capitatum (Willd.) Schumach. ●☆

96379 Clerodendrum tanganyikense Baker;坦噶尼卡赪桐●☆

96380 Clerodendrum tanneri Verdc. = Rotheca tanneri (Verdc.) Verdc. ●☆

96381 Clerodendrum teaguei Hutch. = Rotheca teaguei (Hutch.) R. Fern. ●☆

96382 Clerodendrum ternatum Schinz;三出赪桐●☆

96383 Clerodendrum ternatum Schinz f. glabricalyx Moldenke;光萼赪桐●☆

96384 Clerodendrum ternatum Schinz var. lanceolatum (Gürke)

Moldenke = Clerodendrum ternatum Schinz ●☆

96385 Clerodendrum ternifolium D. Don = Clerodendrum serratum (L.) Moon ●

96386 Clerodendrum thomasii Moldenke = Clerodendrum dusenii Gürke ●☆

96387 Clerodendrum thomsonae Balf. f.;龙吐珠(白萼赪桐,臭牡丹藤,九龙吐珠,汤氏赪桐);Bag Flower,Bleeding Heart Glorybower, Bleeding Heart Vine, Bleedingheart Glorybower, Bleeding-hearted Glorybower, Broken Heart, Broken Hearts, Glorybower, Thomson Glorybower ●

96388 Clerodendrum thomsoniae Balf.;托马森赪桐●☆

96389 Clerodendrum thonneri Gürke = Clerodendrum silvanum Henriq. var. buchholzii (Gürke) Verdc. ●☆

96390 Clerodendrum thyrsoideum Baker = Clerodendrum mannii Baker ●☆

96391 Clerodendrum thyrsoideum Gürke;聚伞赪桐●☆

96392 Clerodendrum tibetanum C. T. Wu et S. K. Wu;西藏大青(西藏赪桐);Xizang Glorybower ●

96393 Clerodendrum tomentellum Hutch. et Dalziel;绒毛赪桐●☆

96394 Clerodendrum toxicarium Baker;毒赪桐●☆

96395 Clerodendrum transvaalense B. Thomas;德兰士瓦赪桐●☆

96396 Clerodendrum trichanthum Bosser;毛花赪桐●☆

96397 Clerodendrum tricholobum Gürke;毛片赪桐●☆

96398 Clerodendrum trichotomum Thunb. ex A. Murray;海州常山(矮桐,八角梧桐,臭芙蓉,臭牡丹树,臭桐,臭桐柴,臭梧,臭梧桐,地梧桐,法氏海州常山,凤眼子,海桐,恒春海州常山,后庭花,龙船花,泡花桐,泡火桐,秋叶,楸茶叶,楸叶常山,绒毛海州常山,柔毛海州常山,山梧桐,山猪枷,香楸,小对叶草,锈毛海州常山,岩桐,追骨风);Farges Harlequin Glorybower, Glory Bower, Harlequin Glorybower, Villose Harlequin Glorybower ●

96399 Clerodendrum trichotomum Thunb. ex A. Murray f. albicarpum Satomi;白果海州常山●☆

96400 Clerodendrum trichotomum Thunb. ex A. Murray f. ferrugineum (Nakai) Ohwi;长叶海州常山(锈毛海州常山);Ferruginous Clerodendrum ●

96401 Clerodendrum trichotomum Thunb. ex A. Murray var. esculentum Makino;心叶海州常山●☆

96402 Clerodendrum trichotomum Thunb. ex A. Murray var. fargesii (Dode) Nakai = Clerodendrum trichotomum Thunb. ex A. Murray ●

96403 Clerodendrum trichotomum Thunb. ex A. Murray var. fargesii (Dode) Rehder;法氏海州常山(恒春海州常山)●

96404 Clerodendrum trichotomum Thunb. ex A. Murray var. ferrugineum Nakai = Clerodendrum trichotomum Thunb. ex A. Murray f. ferrugineum (Nakai) Ohwi ●

96405 Clerodendrum trichotomum Thunb. ex A. Murray var. ferrugineum Ohwi = Clerodendrum trichotomum Thunb. ex A. Murray ●

96406 Clerodendrum trichotomum Thunb. ex A. Murray var. izuinsulare (K. Inoue, Haseg. et S. Kobay.) H. Ohba et Akiyama = Clerodendrum izuinsulare K. Inoue, Haseg. et S. Kobay. ●☆

96407 Clerodendrum trichotomum Thunb. ex A. Murray var. villosum Y. C. Xu = Clerodendrum trichotomum Thunb. ex A. Murray ●

96408 Clerodendrum trichotomum Thunb. ex A. Murray var. yakusimense Ohwi;屋久岛海州常山●☆

96409 Clerodendrum triflorum Vis.;三花赪桐●☆

96410 Clerodendrum triphyllum (Harv.) H. Pearson;三叶赪桐●☆

96411 Clerodendrum triphyllum (Harv.) H. Pearson = Rotheca hirsuta (Hochst.) R. Fern. ●☆

96412 Clerodendrum triphyllum (Harv.) H. Pearson f. angustissimum Moldenke;狭三叶赪桐●☆

96413 Clerodendrum triplinerve Rolfe = Clerodendrum formicarum Gürke ●☆

96414 Clerodendrum tsaii H. L. Li = Clerodendrum mandarinorum Diels ●

96415 Clerodendrum tubulosum Moldenke;管状赪桐●☆

96416 Clerodendrum ubanghense A. Chev. = Clerodendrum poggei Gürke ●☆

96417 Clerodendrum ugandense Prain;蓝大青;Blue Butterfly Bush, Blue Glorybower, Blue Utterfly-bush ●☆

96418 Clerodendrum ugandense Prain = Rotheca myricoides (Hochst.) Steane et Mabb. ●☆

96419 Clerodendrum umbellatum Poir.;伞大青;Umbel Clerodendrum ●☆

96420 Clerodendrum umbellatum Poir. f. scandens (P. Beauv.) Moldenke = Clerodendrum umbellatum Poir. ●☆

96421 Clerodendrum uncinatum Schinz;钩大青●☆

96422 Clerodendrum uncinatum Schinz var. parviflorum? = Clerodendrum uncinatum Schinz ●☆

96423 Clerodendrum validipes S. Moore = Clerodendrum silvanum Henriq. var. nuxioides (S. Moore) Verdc. ●☆

96424 Clerodendrum variifolium De Wild. = Rotheca bukobensis (Gürke) Verdc. ●☆

96425 Clerodendrum variifolium De Wild. var. scandens? = Rotheca violacea (Gürke) Verdc. ●☆

96426 Clerodendrum varium B. Thomas = Rotheca sansibarensis (Gürke) Steane et Mabb. ●☆

96427 Clerodendrum velutinum A. Chev.;短绒毛大青●☆

96428 Clerodendrum velutinum B. Thomas = Clerodendrum villosum Blume ●

96429 Clerodendrum verdcourtii R. Fern. = Rotheca verdcourtii (R. Fern.) R. Fern. ●☆

96430 Clerodendrum verticillatum D. Don. = Clerodendrum indicum (L.) Kuntze ●

96431 Clerodendrum villosicalyx Moldenke = Vitex grandidiana W. Piep. ●☆

96432 Clerodendrum villosulum De Wild. = Rotheca myricoides (Hochst.) Steane et Mabb. var. discolor (Klotzsch) Verdc. ●☆

96433 Clerodendrum villosulum De Wild. var. debeerstii (De Wild.) De Wild. = Rotheca myricoides (Hochst.) Steane et Mabb. var. discolor (Klotzsch) Verdc. ●☆

96434 Clerodendrum villosum Blume;绢毛大青(长毛臭牡丹);Villous Glorybower ●

96435 Clerodendrum violaceum Gürke = Rotheca violacea (Gürke) Verdc. ●☆

96436 Clerodendrum viscosum Vent. = Clerodendrum petasites (Lour.) A. Meeuse ●

96437 Clerodendrum volubile P. Beauv.;扭赪桐●☆

96438 Clerodendrum volubile P. Beauv. var. grosseserratum Moldenke;粗齿赪桐●☆

96439 Clerodendrum wallichii Merr.;垂茉莉;Nodding Clerodendron, Wallich Glorybower, Wallich's Glorybower ●

96440 Clerodendrum wallii Moldenke;瓦尔赪桐●☆

96441 Clerodendrum welwitschii Gürke;韦尔赪桐●☆

96442 Clerodendrum whitfieldii Seem.;惠特赪桐●☆

96443 Clerodendrum wildemanianum Exell;怀尔赪桐●☆

96444 Clerodendrum wildemanianum Robyns = Rotheca myricoides (Hochst.) Steane et Mabb. ●☆

96445 Clerodendrum wildii Moldenke = Rotheca wildii (Moldenke) R. Fern. ●☆

96446 Clerodendrum wildii Moldenke f. glabrum R. Fern. = Rotheca wildii (Moldenke) R. Fern. f. glabra (R. Fern.) R. Fern. ●☆

96447 Clerodendrum wilmsii Gürke = Clerodendrum ternatum Schinz ●☆

96448 Clerodendrum yatschuense H. Winkl. = Clerodendrum bungei Steud. ●

96449 Clerodendrum yunnanense Hu ex Hand. -Mazz.;滇常山(矮赪桐,矮桐子,臭茉莉,臭牡丹,滇赪桐,乌药);Yunnan Glorybower ●

96450 Clerodendrum yunnanense Hu ex Hand. -Mazz. var. lineacilobum S. L. Chen et G. Y. Sheng;线齿滇常山;Linearlobe Yunnan Glorybower ●

96451 Clerodendrum zambesiacum Baker = Clerodendrum rotundifolium Oliv. ●☆

96452 Cleterus Raf. = Chleterus Raf. ■

96453 Clethra Bert. ex Steud. = Viviania Cav. ■☆

96454 Clethra Gronov. ex L. (1753);桤叶树属(山柳属);Clethra, Summer Sweet, Summersweet, Summer-sweet, White Alder, White-Alder ●

96455 Clethra L. = Clethra Gronov. ex L. ●

96456 Clethra acuminata Michx.;肉桂山柳;Cinnamon Bark Clethra, Cinnamon Clethra, Cinnamonbark Clethra, Mountain Pepperbush, White Alder ●☆

96457 Clethra alnifolia L.;桤叶树(桤叶山柳);Alder-leaf Clethra, Bush Pepper, Pepperbush, Pepperbush Clethra, Summer Sweet, Summersweet, Summersweet Clethra, Sweet Bush, Sweet Clethra, Sweet Pepper, Sweet Pepper Bush, Sweet Pepperbush, Sweet Pepperbush, Ummersweet, White Alder ●

96458 Clethra alnifolia L. 'Compacta';密丛桤叶树;Dwarf Summer-sweet ●☆

96459 Clethra alnifolia L. 'Hummingbird';蜂雀桤叶树;Summer-sweet ●☆

96460 Clethra alnifolia L. 'Pink Spires';粉尖桤叶树;Summer-sweet ●☆

96461 Clethra alnifolia L. 'Rosea';玫瑰桤叶树;Summer-sweet ●☆

96462 Clethra alnifolia L. 'Ruby Spice';红宝石桤叶树;Summer-sweet ●☆

96463 Clethra annamensis Dop = Clethra faberi Hance ●

96464 Clethra arborea Aiton;铃兰花山柳;Folhado, Lily of the Valley Tree, Lily-of-the-valley Tree ●☆

96465 Clethra arborea Vent. = Clethra arborea Aiton ●☆

96466 Clethra barbinervis Siebold et Zucc.;华东桤叶树(华东山柳,日本山柳,山柳,髭脉桤叶树,髭脉山柳);Japan Clethra, Japanese Clethra, Japanese Summersweet, Tree Clethra ●

96467 Clethra barbinervis Siebold et Zucc. var. kawadana (Yanagita) H. Hara = Clethra barbinervis Siebold et Zucc. ●

96468 Clethra barbinervis Siebold et Zucc. var. stolonifera (Nakai) Honda = Clethra barbinervis Siebold et Zucc. var. kawadana (Yanagita) H. Hara ●

96469 Clethra bodinieri H. Lév.;单毛桤叶树(单毛山柳,单柱山柳,小山柳);Bodinier Clethra ●

96470 Clethra bodinieri H. Lév. var. coriacea L. C. Hu;革叶桤叶树;Leather-leaf Bodinier Clethra ●

96471 Clethra bodinieri H. Lév. var. coriacea L. C. Hu = Clethra bodinieri H. Lév. ●

96472 Clethra bodinieri H. Lév. var. latifolia S. Y. Hu = Clethra bodinieri H. Lév. ●

96473 Clethra bodinieri H. Lév. var. parviflora W. P. Fang et L. C. Hu;小花桤叶树;Little-flower Bodinier Clethra ●

96474 Clethra bodinieri H. Lév. var. parviflora W. P. Fang et L. C. Hu = Clethra bodinieri H. Lév. ●

96475 Clethra bodinieri H. Lév. var. parvifolia S. Y. Hu = Clethra bodinieri H. Lév. ●

96476 Clethra brachypoda L. C. Hu;短柄桤叶树;Shortstalk Clethra, Short-stalked Clethra, Shortstipe Clethra ●

96477 Clethra brachypoda L. C. Hu = Clethra fargesii Franch. ●

96478 Clethra brachystachya W. P. Fang et L. C. Hu;短穗桤叶树;Short-spike Clethra, Short-spikeed Clethra ●

96479 Clethra brachystachya W. P. Fang et L. C. Hu = Clethra fargesii Franch. ●

96480 Clethra brammeriana Hand. -Mazz. = Clethra kaipoensis H. Lév. ●

96481 Clethra cavaleriei H. Lév.;贵定桤叶树(华中山柳,江南桤木树,江南桤叶树,江南山柳,南岭桤叶树);Cavalerie Clethra, Esquirol Clethra ●

96482 Clethra cavaleriei H. Lév. = Clethra delavayi Franch. ●

96483 Clethra cavaleriei H. Lév. var. leptophylla L. C. Hu;薄叶桤叶树;Thinleaf Clethra ●

96484 Clethra cavaleriei H. Lév. var. leptophylla L. C. Hu = Clethra delavayi Franch. ●

96485 Clethra cavaleriei H. Lév. var. subintegrifolia Ching ex L. C. Hu;全缘桤叶树;Subentire-leaf Thinleaf Clethra ●

96486 Clethra cavaleriei H. Lév. var. subintegrifolia Ching ex L. C. Hu = Clethra delavayi Franch. ●

96487 Clethra delavayi Franch.;云南桤叶树(滇西山柳,云南山柳);Delavay Clethra ●

96488 Clethra delavayi Franch. var. glabra S. Y. Hu = Clethra delavayi Franch. ●

96489 Clethra delavayi Franch. var. lanata S. Y. Hu;毛叶云南桤叶树;Hairy-leaf Delavay Clethra ●

96490 Clethra delavayi Franch. var. lanata S. Y. Hu = Clethra delavayi Franch. ●

96491 Clethra delavayi Franch. var. yuana (S. Y. Hu) C. Y. Wu et L. C. Hu = Clethra delavayi Franch. ●

96492 Clethra delavayi Franch. var. yuiana (S. Y. Hu) C. Y. Wu et L. C. Hu;大花云南桤叶树(大花桤叶树);Bigflower Delavay Clethra ●

96493 Clethra esquirolii H. Lév. = Clethra cavaleriei H. Lév. ●

96494 Clethra esquirolii H. Lév. = Clethra delavayi Franch. ●

96495 Clethra euosmoda Dop = Clethra delavayi Franch. ●

96496 Clethra faberi Hance;华南桤叶树(罗浮山柳,泡花,桤叶树,山柳,石河树);Faber Clethra, S. China Clethra, Tonkin Clethra, White Elder ●

96497 Clethra faberi Hance var. brevipes L. C. Hu;短梗华南桤叶树;Short-stalk Faber Clethra ●

96498 Clethra faberi Hance var. laxiflora W. P. Fang et L. C. Hu;疏花桤叶树;Loose-flower Faber Clethra ●

96499 Clethra fabri Hance var. brevipes L. C. Hu = Clethra faberi Hance ●

96500 Clethra fabri Hance var. laxiflora W. P. Fang et L. C. Hu = Clethra faberi Hance ●

96501 Clethra fargesii Franch.;城口桤叶树(城口山柳,构骨树,花培子,华中桤叶树,华中山柳);Farges Clethra ●

96502 Clethra glandulosa W. P. Fang et L. C. Hu;腺叶桤叶树;Glandleaf Clethra, Glandular Clethra ●

96503 Clethra glandulosa W. P. Fang et L. C. Hu. = Clethra bodinieri H. Lév. ●

96504 Clethra japonica Thunb. ex Steud. = Clethra barbinervis Siebold et Zucc. ●

96505 Clethra kaipoensis H. Lév.;贵州桤叶树(大叶桤木树,大叶桤叶树,大叶山柳,嘉宝山柳,毛叶桤叶树,黔桂桤叶树,脱壳树,正安山柳);Bigleaf Clethra,Big-leaved Clethra,Pinfa Clethra ●

96506 Clethra kaipoensis H. Lév. var. paucinervis L. C. Hu;稀脉桤叶树;Loose-vein Bigleaf Clethra ●

96507 Clethra kaipoensis H. Lév. var. paucinervis L. C. Hu = Clethra kaipoensis H. Lév. ●

96508 Clethra kaipoensis H. Lév. var. polyneara (H. L. Li) W. P. Fang et L. C. Hu;多肋桤叶树;Dense-vein Bigleaf Clethra, Manynerved Clethra ●

96509 Clethra kaipoensis H. Lév. var. polyneura (H. L. Li) W. P. Fang et L. C. Hu = Clethra kaipoensis H. Lév. ●

96510 Clethra kawadana Yanagita = Clethra barbinervis Siebold et Zucc. ●

96511 Clethra kwangsiensis S. Y. Hu = Clethra kaipoensis H. Lév. ●

96512 Clethra lanata M. Martens et Galeotti;绵毛山柳●☆

96513 Clethra lancilimba C. Y. Wu = Clethra delavayi Franch. ●

96514 Clethra lancilimba C. Y. Wu = Clethra monostachya Rehder et E. H. Wilson var. lancilimba (C. Y. Wu) C. Y. Wu et L. C. Hu ●

96515 Clethra liangii H. L. Li = Clethra faberi Hance ●

96516 Clethra lineata H. Lév. = Clethra cavaleriei H. Lév. ●

96517 Clethra lineata H. Lév. = Clethra delavayi Franch. ●

96518 Clethra longebracteata Sleumer = Clethra cavaleriei H. Lév. ●

96519 Clethra longebracteata Sleumer = Clethra delavayi Franch. ●

96520 Clethra magnifica W. P. Fang et L. C. Hu;壮丽桤叶树;Glorious Clethra, Magnificent Clethra, Magnified Clethra ●

96521 Clethra magnifica W. P. Fang et L. C. Hu = Clethra fargesii Franch. ●

96522 Clethra magnifica W. P. Fang et L. C. Hu var. trichocarpa L. C. Hu;毛果桤叶树;Hairyfruit Clethra ●

96523 Clethra magnifica W. P. Fang et L. C. Hu var. trichocarpa L. C. Hu. = Clethra fargesii Franch. ●

96524 Clethra minutistellata C. Y. Wu = Clethra delavayi Franch. ●

96525 Clethra minutistellata C. Y. Wu = Clethra monostachya Rehder et E. H. Wilson var. minutistellata (C. Y. Wu) C. Y. Wu et L. C. Hu ●

96526 Clethra monostachya Rehder et E. H. Wilson;单穗桤叶树(单穗山柳,西蜀山柳);Monospike Clethra, Monostachys Clethra ●

96527 Clethra monostachya Rehder et E. H. Wilson = Clethra delavayi Franch. ●

96528 Clethra monostachya Rehder et E. H. Wilson var. cuprescens W. P. Fang et L. C. Hu;铜色桤叶树;Cupri-coloured Clethra ●

96529 Clethra monostachya Rehder et E. H. Wilson var. cuprescens W. P. Fang et L. C. Hu = Clethra delavayi Franch. ●

96530 Clethra monostachya Rehder et E. H. Wilson var. lancilimba (C. Y. Wu) C. Y. Wu et L. C. Hu;披针桤叶树(批针叶山柳);Lanceoleaf Cupri-coloured Clethra ●

96531 Clethra monostachya Rehder et E. H. Wilson var. lancilimba (C. Y. Wu) C. Y. Wu et L. C. Hu = Clethra delavayi Franch. ●

96532 Clethra monostachya Rehder et E. H. Wilson var. minutistellata (C. Y. Wu) C. Y. Wu et L. C. Hu;细星毛桤叶树;Stellatehair Clethra ●

96533 Clethra monostachya Rehder et E. H. Wilson var. minutistellata (C. Y. Wu) C. Y. Wu et L. C. Hu = Clethra delavayi Franch. ●

96534 Clethra monostachya Rehder et E. H. Wilson var. trichopetala W. P. Fang et L. C. Hu;毛瓣桤叶树;Hairypetal Clethra ●

96535 Clethra monostachya Rehder et E. H. Wilson var. trichopetala W. P. Fang et L. C. Hu = Clethra delavayi Franch. ●

96536 Clethra nanchuanensis W. P. Fang et L. C. Hu;南川桤叶树;Nanchuan Clethra ●◇

96537 Clethra nanchuanensis W. P. Fang et L. C. Hu = Clethra delavayi Franch. ●

96538 Clethra nanchuanensis W. P. Fang et L. C. Hu var. albescens L. C. Hu;白毛桤叶树;Whitehairy Nanchuan Clethra ●

96539 Clethra nanchuanensis W. P. Fang et L. C. Hu var. albescens L. C. Hu = Clethra delavayi Franch. ●

96540 Clethra petelotii Dop et Troch. -Marquart;白背桤叶树(白背山柳);Petelot Clethra ●

96541 Clethra pinfaensis H. Lév. = Clethra kaipoensis H. Lév. ●

96542 Clethra polyneura H. L. Li = Clethra kaipoensis H. Lév. var. polyneara (H. L. Li) W. P. Fang et L. C. Hu ●

96543 Clethra polyneura H. L. Li. = Clethra kaipoensis H. Lév. ●

96544 Clethra purpurea W. P. Fang et L. C. Hu;紫花桤叶树;Purpleflower Clethra, Purple-flowered Clethra ●

96545 Clethra purpurea W. P. Fang et L. C. Hu = Clethra delavayi Franch. ●

96546 Clethra purpurea W. P. Fang et L. C. Hu var. microcarpa W. P. Fang et L. C. Hu;小果桤叶树;Littlefruit Clethra ●

96547 Clethra purpurea W. P. Fang et L. C. Hu var. microcarpa W. P. Fang et L. C. Hu = Clethra delavayi Franch. ●

96548 Clethra scandens Franch. = Clematoclethra scandens (Franch.) Maxim. ●

96549 Clethra sinica K. S. Hao = Clethra cavaleriei H. Lév. ●

96550 Clethra sinica K. S. Hao = Clethra delavayi Franch. ●

96551 Clethra sleumeriana K. S. Hao;湖南山柳(湖南桤叶树);Hunan Clethra ●

96552 Clethra smithiana W. P. Fang = Clethra faberi Hance ●

96553 Clethra smithiana W. P. Fang var. latifolia C. Y. Wu = Clethra faberi Hance ●

96554 Clethra tonkinensis Dop = Clethra faberi Hance ●

96555 Clethra wuyishanica Ching et L. C. Hu;武夷桤叶树;Wuyishan Clethra ●

96556 Clethra wuyishanica Ching et L. C. Hu var. erosa L. C. Hu;蚀瓣桤叶树;Erose Wuyishan Clethra ●

96557 Clethra wuyishanica Ching ex L. C. Hu = Clethra barbinervis Siebold et Zucc. ●

96558 Clethra wuyishanica Ching ex L. C. Hu var. erosa L. C. Hu. = Clethra barbinervis Siebold et Zucc. ●

96559 Clethra yuana S. Y. Hu. = Clethra delavayi Franch. ●

96560 Clethra yuiana S. Y. Hu = Clethra delavayi Franch. var. yuiana (S. Y. Hu) C. Y. Wu et L. C. Hu ●

96561 Clethraceae Klotzsch(1851)(保留科名);桤叶树科(山柳科);Pepperbush Family, White Alder Family, Clethra Family ●

96562 Clethraceae Klotzsch(保留科名) = Cyrillaceae Lindl.(保留科名)●☆

96563 Clethropsis Spach = Alnus Mill. ●

96564 Clethropsis nepalensis (D. Don) Spach. = Alnus nepalensis D. Don ●

96565 Clethropsis nepalensis Spach. = Alnus nepalensis D. Don ●

96566 Clethropsis nitida Spach = Alnus nitida (Spach) Endl. ●☆

96567 Clethrospermum Planch. = Clathrospermum Planch. ex Benth. ●☆

96568 Clethrospermum Planch. = Enneastemon Exell(保留属名)●☆

96569 Clethrosperum Planch. = Clethrospermum Planch. ●☆

96570 Clevelandia Greene ex Brandegee = ? Orthocarpus Nutt. ■☆

96571 Clevelandia Greene ex Brandegee = Clevelandia Greene ■☆
96572 Clevelandia Greene(1885);克利草属■☆
96573 Clevelandia beldingii (Greene) Greene;克利草■☆
96574 Cleyera Adans.(废弃属名)= Cleyera Thunb.(保留属名)●
96575 Cleyera Adans.(废弃属名)= Polypremum L. ■☆
96576 Cleyera Thunb.(1783)(保留属名);红淡比属(肖柃属,杨桐属);Cleyera,Eurya ●
96577 Cleyera Thunb. = Ternstroemia Mutis ex L. f.(保留属名)●
96578 Cleyera angustifolia S. H. Chun ex H. G. Ye = Adinandra angustifolia (S. H. Chun ex H. G. Ye) B. M. Barthol. et T. L. Ming ●
96579 Cleyera conocarpa Hung T. Chang = Cleyera obscurinervis (Merr. et Chun) Hung T. Chang ●
96580 Cleyera cuspidata Hung T. Chang et S. H. Shi;锐尖肖柃●
96581 Cleyera dubia Champ. = Ternstroemia japonica (Thunb.) Thunb. ●
96582 Cleyera fragrans Champ. = Ternstroemia japonica (Thunb.) Thunb. ●
96583 Cleyera grandiflora Wall. ex Choisy = Cleyera japonica Thunb. var. wallichiana (DC.) Sealy ●
96584 Cleyera gymnanthera Wight et Arn. = Ternstroemia gymnanthera (Wight et Arn.) Bedd. ●
96585 Cleyera hayatae Masam. et Gogelein = Cleyera japonica Thunb. var. hayatae (Masam. et Gogelein) Kobuski ●
96586 Cleyera incornuta Y. C. Wu;凹脉红淡比(肖柃);Impressed Cleyera ●
96587 Cleyera japonica Thunb.;红淡比(红淡,荣树,杨桐);Japan Cleyera,Japanese Cleyera,Sakaki ●
96588 Cleyera japonica Thunb. 'Tricolor';三色红淡比●☆
96589 Cleyera japonica Thunb. = Ternstroemia japonica (Thunb.) Thunb. ●
96590 Cleyera japonica Thunb. f. serrata (Hayashi) Sugim. = Cleyera japonica Thunb. ●
96591 Cleyera japonica Thunb. f. tricolor (G. Nicholson) Kobuski = Cleyera japonica Thunb. 'Tricolor' ●☆
96592 Cleyera japonica Thunb. var. grandiflora (Wall. ex Choisy) Kobuski = Cleyera japonica Thunb. var. wallichiana (DC.) Sealy ●
96593 Cleyera japonica Thunb. var. hayatae (Masam. et Gogelein) Kobuski;早田氏红淡比(森氏杨桐,松田氏红淡比,早田氏杨桐,早田野氏红淡比);Hayata Cleyera ●
96594 Cleyera japonica Thunb. var. hayatae (Masam. et Gogelein) Kobuski = Cleyera japonica Thunb. ●
96595 Cleyera japonica Thunb. var. lipingensis (Hand.-Mazz.) Kobuski. = Cleyera lipingensis (Hand.-Mazz.) T. L. Ming ●
96596 Cleyera japonica Thunb. var. longicarpa (Gogelein) L. K. Ling et C. F. Hsieh = Cleyera longicarpa (Gogelein) L. K. Ling ●
96597 Cleyera japonica Thunb. var. longicarpa (Yamam.) L. K. Ling et C. F. Hsieh = Cleyera longicarpa (Gogelein) L. K. Ling ●
96598 Cleyera japonica Thunb. var. morii (Gogelein) Masam.;森氏红淡比(大叶红淡比,森氏杨桐);Mori's Cleyera ●
96599 Cleyera japonica Thunb. var. morii (Gogelein) Masam. = Cleyera japonica Thunb. ●
96600 Cleyera japonica Thunb. var. morii (Yamam.) Masam. = Cleyera japonica Thunb. ●
96601 Cleyera japonica Thunb. var. parvifolia Kobuski = Cleyera parvifolia (Kobuski) Hu ex L. K. Ling ●
96602 Cleyera japonica Thunb. var. taipehensis H. Keng;台北红淡比(台北杨桐);Taibei Cleyera ●
96603 Cleyera japonica Thunb. var. taipingensis H. Keng = Cleyera lipingensis (Hand.-Mazz.) T. L. Ming var. taipinensis (H. Keng) T. L. Ming ●
96604 Cleyera japonica Thunb. var. wallichiana (DC.) Sealy;大花红淡比;Wallich Cleyera ●
96605 Cleyera lipingensis (Hand.-Mazz.) T. L. Ming;齿叶红淡比(长果红淡比,长果杨桐);Liping Cleyera,Long-fruit Cleyera ●
96606 Cleyera lipingensis (Hand.-Mazz.) T. L. Ming var. taipinensis (H. Keng) T. L. Ming;太平山红淡比(太平红淡比,太平山杨桐,太平杨桐);Tai-pin Cleyera,Taiping Cleyera ●
96607 Cleyera longicarpa (Gogelein) L. K. Ling;长果红淡比(长果杨桐);Longfruit Cleyera,Long-fruited Cleyera ●
96608 Cleyera longicarpa (Gogelein) L. K. Ling = Cleyera japonica Thunb. var. longicarpa (Gogelein) L. K. Ling et C. F. Hsieh ●
96609 Cleyera longicarpa (Yamam.) L. K. Ling = Cleyera japonica Thunb. var. longicarpa (Gogelein) L. K. Ling et C. F. Hsieh ●
96610 Cleyera lushia (Buch.-Ham. ex D. Don) G. Don = Cleyera japonica Thunb. var. wallichiana (DC.) Sealy ●
96611 Cleyera lushia (Ham. ex D. Don) G. Don = Cleyera japonica Thunb. var. wallichiana (DC.) Sealy ●
96612 Cleyera lushia (Ham. ex D. Don) G. Don var. wallichiana (DC.) G. Don = Cleyera japonica Thunb. var. wallichiana (DC.) Sealy ●
96613 Cleyera lushia Ham. ex G. Don = Cleyera japonica Thunb. ●
96614 Cleyera millettii Hook. et Arn. = Adinandra millettii (Hook. et Arn.) Benth. et Hook. f. ex Hance ●
96615 Cleyera obovata Hung T. Chang;倒卵叶红淡比(倒卵红淡比);Obovate Cleyera ●
96616 Cleyera obscurinervis (Merr. et Chun) Hung T. Chang;隐脉红淡比(红枧木);Obscurenerved Cleyera,Obscure-nerved Cleyera ●
96617 Cleyera ochnacea DC.;大果杨桐(荣树);Sakaki ●
96618 Cleyera ochnacea DC. = Cleyera japonica Thunb. ●
96619 Cleyera ochnacea DC. var. grandiflora (Wall. ex Choisy) Choisy = Cleyera japonica Thunb. var. wallichiana (DC.) Sealy ●
96620 Cleyera ochnacea DC. var. kaempferiana DC. = Cleyera japonica Thunb. var. wallichiana (DC.) Sealy ●
96621 Cleyera ochnacea DC. var. lushai (Ham. ex D. Don) Dyer = Cleyera japonica Thunb. var. wallichiana (DC.) Sealy ●
96622 Cleyera ochnacea DC. var. lushia (D. Don) Dyer = Cleyera japonica Thunb. ●
96623 Cleyera ochnacea DC. var. lushia (D. Don) Dyer = Cleyera japonica Thunb. var. wallichiana (DC.) Sealy ●
96624 Cleyera ochnacea DC. var. wallichiana DC. = Cleyera japonica Thunb. var. wallichiana (DC.) Sealy ●
96625 Cleyera pachyphylla Chun ex Hung T. Chang;厚叶红淡比(厚叶肖柃);Thikleaf Cleyera,Thik-leaved Cleyera ●
96626 Cleyera pachyphylla Chun ex Hung T. Chang var. epunctata Hung T. Chang = Cleyera pachyphylla Chun ex Hung T. Chang ●
96627 Cleyera parvifolia (Kobuski) Hu ex L. K. Ling;小叶红淡比(小叶杨桐);Small-leaf Cleyera,Small-leaved Cleyera ●
96628 Cleyera wallichiana (DC.) Siebold et Zucc. = Cleyera japonica Thunb. var. wallichiana (DC.) Sealy ●
96629 Cleyera wightii Choisy = Ternstroemia gymnanthera (Wight et Arn.) Bedd. var. wightii (Choisy) Hand.-Mazz. ●
96630 Cleyera yangchunensis L. K. Ling;阳春红淡比;Yangchun Cleyera ●
96631 Cleyria Neck. = Dialium L. ●☆
96632 Clianthus Banks et Sol. ex G. Don = Clianthus Sol. ex Lindl.(保

留属名)●

96633 Clianthus Sol. ex Lindl. (1835)(保留属名);鹦鹉嘴属(沙耀花豆属,所罗豆属,耀花豆属,原耀花豆属);Clianthus, Glory Pea, Glory Vine, Glorypea, Kakabeak, Parrot-beak, Parrotbill, Parrot-bill, Parrotheak, Parrothill, Sarcodum ●

96634 Clianthus binnedyckianum Kurz = Clianthus scandens (Lour.) Merr. ●■

96635 Clianthus binnedyckianum Kurz = Sarcodum scandens Lour. ●■

96636 Clianthus dampieri A. Cunn. = Clianthus formosus (G. Don) Ford et Vickery ●☆

96637 Clianthus formosus (G. Don) Ford et Vickery;澳洲鹦鹉嘴(美丽沙耀花豆,耀眼豆);Glory Pea, Glorypea, Patrot's Bill, Sturt's Desert Pea, Sturt's Desert-pea ●☆

96638 Clianthus formosus (G. Don) Ford et Vickery = Donia formosa G. Don ■☆

96639 Clianthus puniceus Lindl.;鹦鹉嘴(红耀花豆);Glory Pea, Kaka-beak, Lobster Claws, Lobster-claw, Lobster's Claw, New Zealand Glory-pea Kaka Beak, Parrot Beak, Parrot's Bill, Parrot's-bill, Red Kowhai, Red Parrot-beak ●

96640 Clianthus scandens (Lour.) Merr.;耀花豆(原耀花豆);Climbing Glorypea, Climbing Parrotbeak, Climbing Parrotbill, Climbing Sarcodum ●■

96641 Clianthus scandens (Lour.) Merr. = Sarcodum scandens Lour. ●■

96642 Clianthus speciosus (G. Don) Asch. et Graebn. = Clianthus dampieri A. Cunn. ●☆

96643 Clibadium F. Allam. ex L. (1771);白头菊属(克力巴菊属);Guado ●■☆

96644 Clibadium L. = Clibadium F. Allam. ex L. ●■☆

96645 Clibadium cujete L.;白头菊(克力巴菊)■☆

96646 Clibadium grandifolium S. F. Blake;大叶白头菊■☆

96647 Clibadium micranthum O. E. Schulz;小花白头菊(小花克力巴菊)●■☆

96648 Clibadium surinamense L.;苏里南白头菊(苏里南克力巴菊)■☆

96649 Clidanthera R. Br. = Glycyrrhiza L. ■

96650 Clidemia D. Don(1823);克利木属;Bush Currant ●☆

96651 Clidemia hirta (L.) D. Don;毛克利木(毛野牡丹);Koster's Curse ●

96652 Clidemia novemnervia (DC.) Triana;克利木●☆

96653 Clidemiastrum Nand. = Leandra Raddi ●■☆

96654 Clidemiastrum Nand. = Oxymeris DC. ●■☆

96655 Clidium Post et Kuntze = Cleidion Blume ●

96656 Clifordia Livera = Christisonia Gardner ■

96657 Cliffordiochloa B. K. Simon(1992);澳禾属■☆

96658 Cliffortia L. (1753);可利果属●☆

96659 Cliffortia acanthophylla C. Whitehouse;刺叶可利果●☆

96660 Cliffortia acockii Weim.;阿科科可利果●☆

96661 Cliffortia aculeata Weim.;皮刺可利果●☆

96662 Cliffortia acutifolia Weim.;尖叶可利果●☆

96663 Cliffortia aequatorialis R. E. Fr. et T. C. E. Fr. = Cliffortia nitidula (Engl.) R. E. Fr. et T. C. E. Fr. var. aequatorialis (R. E. Fr. et T. C. E. Fr.) Brenan ●☆

96664 Cliffortia alata N. E. Br.;具翅可利果●☆

96665 Cliffortia alnifolia Rchb. = Cliffortia odorata L. f. ●☆

96666 Cliffortia amplexistipula Schltr.;套折可利果●☆

96667 Cliffortia apiculata Weim.;细尖可利果●☆

96668 Cliffortia arachnoidea Lodd. = Cliffortia ruscifolia L. ●☆

96669 Cliffortia arborea Marloth;树状可利果●☆

96670 Cliffortia arcuata Weim.;拱可利果●☆

96671 Cliffortia atrata Weim.;黑可利果●☆

96672 Cliffortia berberidifolia Lam. = Cliffortia ferruginea L. f. ●☆

96673 Cliffortia bolusii Diels ex C. Whitehouse;博卢斯可利果●☆

96674 Cliffortia brevifolia Weim.;短叶可利果●☆

96675 Cliffortia browniana Burtt Davy;布朗可利果●☆

96676 Cliffortia burchellii Stapf;布尔可利果●☆

96677 Cliffortia burgersii E. G. H. Oliv. et Fellingham;伯吉斯可利果●☆

96678 Cliffortia carinata Weim.;龙骨可利果●☆

96679 Cliffortia castanea Weim.;栗色可利果●☆

96680 Cliffortia ceresana C. Whitehouse;塞里斯可利果●☆

96681 Cliffortia cinerea Thunb. = Nenax cinerea (Thunb.) Puff ●☆

96682 Cliffortia complanata E. Mey. ex Harv. et Sond.;扁平可利果●☆

96683 Cliffortia concavifolia Eckl. et Zeyh. = Aspalathus cytisoides Lam. ●☆

96684 Cliffortia concinna Weim.;整洁可利果●☆

96685 Cliffortia conoides Sparrm. = Cliffortia strobilifera L. ●☆

96686 Cliffortia cordifolia E. Mey. = Cliffortia virgata Weim. ●☆

96687 Cliffortia cordifolia Lam. = Cliffortia ilicifolia L. var. cordifolia (Lam.) Harv. ●☆

96688 Cliffortia crassinervis Weim.;粗脉可利果●☆

96689 Cliffortia crenata L. f.;圆齿可利果●☆

96690 Cliffortia crenulata Weim.;细圆齿可利果●☆

96691 Cliffortia cristata Weim.;冠状可利果●☆

96692 Cliffortia cruciata C. Whitehouse;十字形可利果●☆

96693 Cliffortia cuneata Dryand.;楔形可利果●☆

96694 Cliffortia cuneata Dryand. var. cylindrica C. Whitehouse;柱形可利果●☆

96695 Cliffortia curvifolia Weim.;折叶可利果●☆

96696 Cliffortia cymbifolia Weim.;舟叶可利果●☆

96697 Cliffortia densa Weim.;稠密可利果●☆

96698 Cliffortia dentata E. Mey. = Cliffortia gracilis Harv. ●☆

96699 Cliffortia dentata Eckl. et Zeyh. = Cliffortia triloba Harv. ●☆

96700 Cliffortia dentata Willd.;具齿可利果●☆

96701 Cliffortia dentata Willd. var. gracilis (Harv.) C. Whitehouse;纤细具齿可利果●☆

96702 Cliffortia denticulata (Weim.) C. Whitehouse;小齿可利果●☆

96703 Cliffortia dichotoma Fellingham;二歧可利果●☆

96704 Cliffortia discolor Weim. = Cliffortia odorata L. f. ●☆

96705 Cliffortia dispar Weim.;异型可利果●☆

96706 Cliffortia dodecandra Weim.;十二雄蕊可利果●☆

96707 Cliffortia dracomontana C. Whitehouse;德拉科可利果●☆

96708 Cliffortia dregeana C. Presl;德雷可利果●☆

96709 Cliffortia dregeana C. Presl var. meyeriana (C. Presl) Weim.;迈尔可利果●☆

96710 Cliffortia drepanoides Eckl. et Zeyh.;镰状可利果●☆

96711 Cliffortia erectisepala Weim.;直立萼可利果●☆

96712 Cliffortia ericifolia E. Mey. = Cliffortia eriocephalina Cham. ●☆

96713 Cliffortia ericifolia L. f. var. acutistipula Kuntze = Cliffortia linearifolia Eckl. et Zeyh. ●☆

96714 Cliffortia eriocephalina Cham.;红头可利果●☆

96715 Cliffortia esterhuyseniae Weim.;埃斯特可利果●☆

96716 Cliffortia exilifolia Weim.;瘦花可利果●☆

96717 Cliffortia falcata L. f.;镰形可利果●☆

96718 Cliffortia falcata Spreng. = Cliffortia drepanoides Eckl. et Zeyh. ●☆

96719 Cliffortia fasciculata Weim.;簇生可利果●☆

96720 Cliffortia ferruginea L. f.;锈色可利果●☆

96721 Cliffortia ferruginea L. f. var. flexuosa (E. Mey.) Harv. = Cliffortia ferruginea L. f. ●☆

96722 Cliffortia ferruginea L. f. var. latifolia Eckl. et Zeyh. = Cliffortia ferruginea L. f. ●☆

96723 Cliffortia ferruginea L. f. var. longifolia Eckl. et Zeyh. = Cliffortia ferruginea L. f. ●☆

96724 Cliffortia ferruginea L. f. var. villosa Harv. = Cliffortia ferruginea L. f. ●☆

96725 Cliffortia filicaulis E. Mey. = Cliffortia tricuspidata Harv. ●☆

96726 Cliffortia filicaulis Schltdl.;线茎可利果●☆

96727 Cliffortia filicaulis Schltdl. var. octandra (Cham.) Weim.;八蕊可利果●☆

96728 Cliffortia filicauloides Weim.;拟线茎可利果●☆

96729 Cliffortia filifolia L. f.;线叶可利果●☆

96730 Cliffortia filifolia L. f. var. subsetacea Eckl. et Zeyh. = Cliffortia subsetacea (Eckl. et Zeyh.) Diels ex Bolus et Wolley-Dod ●☆

96731 Cliffortia flabellifolia Sond. = Myrothamnus flabellifolia Welw. ●☆

96732 Cliffortia flexuosa E. Mey. = Cliffortia ferruginea L. f. ●☆

96733 Cliffortia galpinii N. E. Br. = Cliffortia paucistaminea Weim ●☆

96734 Cliffortia geniculata Weim.;膝曲可利果●☆

96735 Cliffortia glauca Weim.;灰绿可利果●☆

96736 Cliffortia gracilis Harv. = Cliffortia dentata Willd. var. gracilis (Harv.) C. Whitehouse ●☆

96737 Cliffortia gracillima C. Whitehouse;细长可利果●☆

96738 Cliffortia graminea L. f.;禾色可利果●☆

96739 Cliffortia graminea L. f. var. convoluta Weim.;旋扭可利果●☆

96740 Cliffortia graminea L. f. var. elegans Weim.;雅致可利果●☆

96741 Cliffortia grandifolia Eckl. et Zeyh.;大叶可利果●☆

96742 Cliffortia grandifolia Eckl. et Zeyh. var. denticulata Weim. = Cliffortia denticulata (Weim.) C. Whitehouse ●☆

96743 Cliffortia grandifolia Eckl. et Zeyh. var. recurvata Weim. = Cliffortia recurvata (Weim.) C. Whitehouse ●☆

96744 Cliffortia hantamensis Diels;汉塔姆可利果●☆

96745 Cliffortia hermaphroditica Weim. = Cliffortia juniperina L. f. ●☆

96746 Cliffortia heterophylla Weim.;互叶可利果●☆

96747 Cliffortia hexandra Weim.;六蕊可利果●☆

96748 Cliffortia hirsuta Eckl. et Zeyh.;粗毛可利果●☆

96749 Cliffortia hirta Burm. f.;多毛可利果●☆

96750 Cliffortia ilicifolia L.;冬青叶可利果●☆

96751 Cliffortia ilicifolia L. var. cordifolia (Lam.) Harv.;心叶可利果●☆

96752 Cliffortia ilicifolia L. var. incisa Harv.;锐裂可利果●☆

96753 Cliffortia ilicifolia L. var. reniformis Weim. = Cliffortia reniformis (Weim.) C. Whitehouse ●☆

96754 Cliffortia ilicifolia L. var. schlechteri Weim. = Cliffortia schlechteri (Weim.) C. Whitehouse ●☆

96755 Cliffortia incana Weim.;灰毛可利果●☆

96756 Cliffortia integerrima Weim.;全缘可利果●☆

96757 Cliffortia intermedia Eckl. et Zeyh.;间型可利果●☆

96758 Cliffortia juniperina L. f.;刺柏状可利果●☆

96759 Cliffortia juniperina L. f. var. brevifolia Harv. = Cliffortia paucistaminea Weim. ●☆

96760 Cliffortia juniperina L. f. var. foliis-brevioribus Eckl. et Zeyh. = Cliffortia paucistaminea Weim. ●☆

96761 Cliffortia juniperina L. f. var. muricata Harv. = Cliffortia tuberculata (Harv.) Weim. var. muricata ●☆

96762 Cliffortia juniperina L. f. var. pilosula Weim. = Cliffortia juniperina L. f. ●☆

96763 Cliffortia juniperina L. f. var. pterocarpa Harv. = Cliffortia pterocarpa (Harv.) Weim. ●☆

96764 Cliffortia juniperina L. f. var. serrulata Engl. = Cliffortia repens Schltr. ●☆

96765 Cliffortia juniperina L. f. var. tuberculata Harv. = Cliffortia tuberculata (Harv.) Weim. ●☆

96766 Cliffortia lanata Weim.;绵毛可利果●☆

96767 Cliffortia lanceolata Weim.;披针形可利果●☆

96768 Cliffortia laricina E. Mey. = Cliffortia juniperina L. f. ●☆

96769 Cliffortia laricina E. Mey. var. brevior Drège = Cliffortia tuberculata (Harv.) Weim. var. muricata ●☆

96770 Cliffortia laricina E. Mey. var. longifolia Meisn. = Cliffortia pterocarpa (Harv.) Weim. ●☆

96771 Cliffortia lepida Weim.;小鳞可利果●☆

96772 Cliffortia leptophylla Eckl. et Zeyh. = Cliffortia filifolia L. f. ●☆

96773 Cliffortia leptophylla Eckl. et Zeyh. var. longifolia = Cliffortia burchellii Stapf ●☆

96774 Cliffortia linearifolia Eckl. et Zeyh.;条叶可利果●☆

96775 Cliffortia linearifolia Eckl. et Zeyh. var. nitidula Engl. = Cliffortia nitidula (Engl.) R. E. Fr. et T. C. E. Fr. ●☆

96776 Cliffortia longifolia (Eckl. et Zeyh.) Weim.;长叶可利果●☆

96777 Cliffortia marginata Eckl. et Zeyh.;边缘可利果●☆

96778 Cliffortia meyeriana C. Presl = Cliffortia dregeana C. Presl var. meyeriana (C. Presl) Weim. ●☆

96779 Cliffortia micrantha Weim.;小花可利果●☆

96780 Cliffortia mirabilis Weim.;奇异可利果●☆

96781 Cliffortia monophylla Weim.;单叶可利果●☆

96782 Cliffortia montana Weim.;山地可利果●☆

96783 Cliffortia multiformis Weim.;多形可利果●☆

96784 Cliffortia natalensis J. M. Wood = Cliffortia repens Schltr. ●☆

96785 Cliffortia neglecta Schltr.;忽视可利果●☆

96786 Cliffortia nitidula (Engl.) R. E. Fr. et T. C. E. Fr.;光亮可利果●☆

96787 Cliffortia nitidula (Engl.) R. E. Fr. et T. C. E. Fr. subsp. angolensis Weim. = Cliffortia nitidula (Engl.) R. E. Fr. et T. C. E. Fr. var. angolensis (Weim.) Brenan ●☆

96788 Cliffortia nitidula (Engl.) R. E. Fr. et T. C. E. Fr. subsp. pilosa Weim.;疏毛可利果●☆

96789 Cliffortia nitidula (Engl.) R. E. Fr. et T. C. E. Fr. var. aequatorialis (R. E. Fr. et T. C. E. Fr.) Brenan;昼夜光亮可利果●☆

96790 Cliffortia nitidula (Engl.) R. E. Fr. et T. C. E. Fr. var. angolensis (Weim.) Brenan;安哥拉可利果●☆

96791 Cliffortia nivenioides Fellingham;类尼文可利果●☆

96792 Cliffortia obcordata L. f.;倒心形可利果●☆

96793 Cliffortia obliqua Spreng. = Cliffortia obcordata L. f. ●☆

96794 Cliffortia obovata E. Mey. ex Harv. et Sond.;倒卵可利果●☆

96795 Cliffortia octandra Cham. = Cliffortia filicaulis Schltdl. var. octandra (Cham.) Weim. ●☆

96796 Cliffortia odorata L. f.;锐齿可利果●☆

96797 Cliffortia odorata L. f. var. hypoleuca Harv. = Cliffortia hirsuta Eckl. et Zeyh. ●☆

96798 Cliffortia odorata L. f. var. reticulata Harv. = Cliffortia reticulata Eckl. et Zeyh. ●☆

96799 Cliffortia odorata L. f. var. vera Harv. = Cliffortia odorata L. f. ●☆

96800 Cliffortia oligodonta C. Whitehouse;寡齿可利果●☆

96801 Cliffortia ovalis Weim.;椭圆可利果●☆

96802 Cliffortia paucistaminea Weim.;少冠可利果●☆

96803 Cliffortia paucistaminea Weim. var. australis C. Whitehouse;南方少冠可利果●☆
96804 Cliffortia pedunculata Schltr.;梗花可利果●☆
96805 Cliffortia perpendicularis C. Whitehouse;垂直可利果●☆
96806 Cliffortia phillipsii Weim.;菲利可利果●☆
96807 Cliffortia phylicoides Eckl. et Zeyh. = Cliffortia eriocephalina Cham. ●☆
96808 Cliffortia phyllanthoides Schltr.;拟叶花可利果●☆
96809 Cliffortia pilifera Bolus;纤毛可利果●☆
96810 Cliffortia polita Weim.;亮可利果●☆
96811 Cliffortia polycephala E. Mey. = Cliffortia sericea Eckl. et Zeyh. ●☆
96812 Cliffortia polygonifolia L.;蓼叶可利果●☆
96813 Cliffortia polygonifolia L. var. membranifolia Weim.;膜质蓼叶可利果●☆
96814 Cliffortia polygonifolia L. var. pubescens Weim.;短柔毛可利果●☆
96815 Cliffortia polygonifolia L. var. ternata (L. f.) Harv. = Cliffortia polygonifolia L. ●☆
96816 Cliffortia polygonifolia L. var. trifoliata (L.) Harv.;三小叶蓼叶可利果●☆
96817 Cliffortia polyphylla Eckl. et Zeyh. = Cliffortia serpyllifolia Cham. et Schltdl. ●☆
96818 Cliffortia propinqua Eckl. et Zeyh.;邻近可利果●☆
96819 Cliffortia propinqua Eckl. et Zeyh. var. chamissonis (Harv.) Weim. = Cliffortia complanata E. Mey. ex Harv. et Sond. ●☆
96820 Cliffortia pterocarpa (Harv.) Weim.;翅果可利果●☆
96821 Cliffortia pulchella L. f.;美丽可利果●☆
96822 Cliffortia pungens C. Presl;刺可利果●☆
96823 Cliffortia ramosissima Schltr.;多分枝可利果●☆
96824 Cliffortia recurvata (Weim.) C. Whitehouse;反折可利果●☆
96825 Cliffortia reniformis (Weim.) C. Whitehouse;肾形可利果●☆
96826 Cliffortia repens Schltr.;匍匐可利果●☆
96827 Cliffortia reticulata Eckl. et Zeyh.;网状可利果●☆
96828 Cliffortia rigida Weim.;硬可利果●☆
96829 Cliffortia robusta Weim.;粗壮可利果●☆
96830 Cliffortia rogersii Burtt Davy = Cliffortia strobilifera L. ●☆
96831 Cliffortia rubricaulis C. Presl = Cliffortia ilicifolia L. ●☆
96832 Cliffortia ruscifolia L.;假叶树可利果●☆
96833 Cliffortia ruscifolia L. var. purpurea Weim.;紫可利果●☆
96834 Cliffortia sarmentosa L. = Cliffortia hirta Burm. f. ●☆
96835 Cliffortia scandens C. Whitehouse;攀缘可利果●☆
96836 Cliffortia schlechteri (Weim.) C. Whitehouse;施莱可利果●☆
96837 Cliffortia semiteres Weim.;半圆柱形可利果●☆
96838 Cliffortia sericea Eckl. et Zeyh.;绢毛可利果●☆
96839 Cliffortia serpyllifolia Cham. et Schltdl.;百里香叶可利果●☆
96840 Cliffortia serpyllifolia Cham. et Schltdl. var. angustifolia Drège = Cliffortia serpyllifolia Cham. et Schltdl. ●☆
96841 Cliffortia serpyllifolia Cham. et Schltdl. var. chamissonis Harv. = Cliffortia complanata E. Mey. ex Harv. et Sond. ●☆
96842 Cliffortia serpyllifolia Cham. et Schltdl. var. foliis-angustioribus Eckl. et Zeyh. = Cliffortia serpyllifolia Cham. et Schltdl. ●☆
96843 Cliffortia serpyllifolia Cham. et Schltdl. var. foliis-latioribus Eckl. et Zeyh. = Cliffortia propinqua Eckl. et Zeyh. ●☆
96844 Cliffortia serpyllifolia Cham. et Schltdl. var. obovata (E. Mey. ex Harv. et Sond.) Kuntze = Cliffortia obovata E. Mey. ex Harv. et Sond. ●☆
96845 Cliffortia serpyllifolia Cham. et Schltdl. var. penninervis Harv. = Cliffortia complanata E. Mey. ex Harv. et Sond. ●☆
96846 Cliffortia serpyllifolia Cham. et Schltdl. var. polyphylla (Eckl. et Zeyh.) Harv. = Cliffortia serpyllifolia Cham. et Schltdl. ●☆
96847 Cliffortia serpyllina Drège = Cliffortia serpyllifolia Cham. et Schltdl. ●☆
96848 Cliffortia serrata Thunb. = Cliffortia ferruginea L. f. ●☆
96849 Cliffortia setifolia Weim.;毛叶可利果●☆
96850 Cliffortia sparsa C. Whitehouse;散生可利果●☆
96851 Cliffortia spathulata Weim.;匙形可利果●☆
96852 Cliffortia stricta Weim.;刚直可利果●☆
96853 Cliffortia strigosa Weim.;糙伏毛可利果●☆
96854 Cliffortia strobilifera L.;球果可利果●☆
96855 Cliffortia strobilifera L. var. longifolia Eckl. et Zeyh. = Cliffortia longifolia (Eckl. et Zeyh.) Weim. ●☆
96856 Cliffortia subsetacea (Eckl. et Zeyh.) Diels ex Bolus et Wolley-Dod;亚刚毛可利果●☆
96857 Cliffortia tenuis Weim.;细可利果●☆
96858 Cliffortia teretifolia E. Mey. = Cliffortia pungens C. Presl ●☆
96859 Cliffortia teretifolia L. f.;柱叶可利果●☆
96860 Cliffortia teretifolia L. f. var. alata Compton = Cliffortia teretifolia L. f. ●☆
96861 Cliffortia teretifolia L. f. var. tenuior E. Mey. = Cliffortia teretifolia L. f. ●☆
96862 Cliffortia ternata L. f. = Cliffortia polygonifolia L. ●☆
96863 Cliffortia theodori-friesii Weim.;弗里斯可利果●☆
96864 Cliffortia theodori-friesii Weim. var. puberula?;微毛可利果●☆
96865 Cliffortia tricuspidata Harv.;三尖可利果●☆
96866 Cliffortia trifida Thunb. = Cliffortia dentata Willd. ●☆
96867 Cliffortia trifoliata L. = Cliffortia polygonifolia L. var. trifoliata (L.) Harv. ●☆
96868 Cliffortia triloba Harv.;三裂可利果●☆
96869 Cliffortia tuberculata (Harv.) Weim.;多疣可利果●☆
96870 Cliffortia tuberculata (Harv.) Weim. var. muricata?;粗糙多疣可利果●☆
96871 Cliffortia tychonis Weim. = Cliffortia nitidula (Engl.) R. E. Fr. et T. C. E. Fr. ●☆
96872 Cliffortia uncinata Weim.;具钩可利果●☆
96873 Cliffortia uncinata Weim. var. recta?;直立可利果●☆
96874 Cliffortia varians Weim.;杂色可利果●☆
96875 Cliffortia verrucosa Weim.;密疣可利果●☆
96876 Cliffortia virgata Weim.;条纹可利果●☆
96877 Cliffortia viridis Weim.;绿可利果●☆
96878 Cliffortiaceae Mart. = Rosaceae Juss.(保留科名)●■
96879 Cliffortioides Dryand. ex Hook. = Nothofagus Blume(保留属名)●☆
96880 Cliftonia Banks ex C. F. Gaertn.(1807);克利夫木属●☆
96881 Cliftonia monophylla (Lam.) Britton et Sarg.;克利夫木;Black Titi,Buckwheat Tree,Buckwheat-tree,Titi ●☆
96882 Climacandra Miq. = Ardisia Sw.(保留属名)●■
96883 Climacanthus Nees;梯花属☆
96884 Climacoptera Botsch.(1956);梯翅蓬属●☆
96885 Climacoptera Botsch. = Salsola L. ●■
96886 Climacoptera affinis (C. A. Mey. ex Schrenk) Botsch. = Salsola affinis C. A. Mey. ex Schrenk ■
96887 Climacoptera affinis (C. A. Mey.) Botsch. = Salsola affinis C. A. Mey. ex Schrenk ■
96888 Climacoptera brachiata (Pall.) Botsch. = Salsola brachiata (Pall.) Botsch. ■

96889 Climacoptera brachiata Pall. = Salsola brachiata (Pall.) Botsch. ■

96890 Climacoptera dampieri A. Cunn.;丹氏梯翅蓬;Glory Pea ●☆

96891 Climacoptera ferganica (Drobow) Botsch. = Salsola ferganica Drobow ■●

96892 Climacoptera korshinskyi (Drobow) Botsch. = Salsola korshinskyi Drobow ■

96893 Climacoptera lanata (Pall.) Botsch. = Salsola lanata Pall. ■

96894 Climacoptera obtusifolia (Schrenk) Botsch. = Salsola heptapotamica Iljin ●

96895 Climacoptera roborowskii (Iljin) Grubov = Salsola affinis C. A. Mey. ex Schrenk ■

96896 Climacoptera subcrassa (Popov ex Iljin) Botsch. = Salsola subcrassa Popov ex Iljin ■

96897 Climacoptera subcrassa (Popov) Botsch. = Salsola crassa Popov ■

96898 Climacoptera sukaczevii Botsch. = Salsola sukaczevii (Botsch.) A. J. Li ■

96899 Climacorachis Hemsl. et Rose = Aeschynomene L. ●■

96900 Climedia Raf. = Clidemia D. Don ●☆

96901 Clinacanthus Nees(1847);鳄嘴花属(鳄咀花属,扭序花属);Clinacanthus ■

96902 Clinacanthus angustus Nees = Isoglossa angusta (Nees) Baker ■☆

96903 Clinacanthus burmanni Nees = Clinacanthus nutans (Burm. f.) Lindau ■

96904 Clinacanthus nutans (Burm. f.) Lindau;鳄嘴花(扭序花,青箭,柔刺草,小接骨,竹节黄);Drooping Clinacanthus ■

96905 Clinacanthus nutans (Burm. f.) Lindau var. robinsonii Benoist;大花鳄嘴花;Robinson Drooping Clinacanthus ■

96906 Clinacanthus nutans Lindau et Groff. = Peristrophe lanceolaria (Roxb.) Nees ■

96907 Clinanthus Herb. = Stenomesson Herb. ■☆

96908 Clinelymus (Griseb.) Nevski = Elymus L. ■

96909 Clinelymus Nevski = Elymus L. ■

96910 Clinelymus atratus Nevski = Elymus atratus (Nevski) Hand. -Mazz. ■

96911 Clinelymus breviaristatus Keng = Elymus breviaristatus Keng ex P. C. Keng ■

96912 Clinelymus breviaristatus Keng = Elymus yilianus S. L. Chen ■

96913 Clinelymus canadensis (L.) Nevski = Elymus canadensis L. ■

96914 Clinelymus coreanus (Honda) Honda = Hystrix coreana (Honda) Ohwi ■

96915 Clinelymus cylindricus (Franch.) Honda = Elymus cylindricus (Franch.) Honda ■

96916 Clinelymus cylindricus (Franch.) Honda = Elymus dahuricus Turcz. ex Griseb. var. cyhndricus Franch. ■

96917 Clinelymus dahuricus (Turcz. ex Griseb.) Nevski = Elymus dahuricus Turcz. ex Griseb. ■

96918 Clinelymus dahuricus (Turcz.) Nevski = Elymus dahuricus Turcz. ex Griseb. ■

96919 Clinelymus excelsus (Turcz. ex Griseb.) Nevski = Elymus excelsus Turcz. ex Griseb. ■

96920 Clinelymus excelsus (Turcz.) Nevski = Elymus excelsus Turcz. ex Griseb. ■

96921 Clinelymus nutans (Griseb.) Nevski = Elymus nutans Griseb. ■

96922 Clinelymus sibiricus (L.) Nevski = Elymus sibiricus L. ■

96923 Clinelymus submuticus Keng = Elymus submuticus Keng ex P. C. Keng ■

96924 Clinelymus tangutorum Nevski = Elymus tangutorus (Nevski) Hand. -Mazz. ■

96925 Clinelymus yubaridakensis Honda = Elymus sibiricus L. ■

96926 Clinhymenia A. Rich. et Galeotti = Cryptarrhena R. Br. ■☆

96927 Clinhymenia A. Rich. et Galeotti = Orchidofunckia A. Rich. et Galeotti ■☆

96928 Clinogyne K. Schum. = Marantochloa Brongn. ex Gris ■☆

96929 Clinogyne Salisb. = Clinogyne Salisb. ex Benth. ■☆

96930 Clinogyne Salisb. ex Benth. = Donax Lour. + Schumannianthus Gagnep. + Marantochloa Brongn. ex Gris ■☆

96931 Clinogyne arcta Stapf = Marantochloa cuspidata (Roscoe) Milne-Redh. ■☆

96932 Clinogyne arillata (K. Schum.) K. Schum. = Marantochloa purpurea (Ridl.) Milne-Redh. ■☆

96933 Clinogyne baumannii K. Schum. = Marantochloa purpurea (Ridl.) Milne-Redh. ■☆

96934 Clinogyne chrysantha Gagnep. = Marantochloa cuspidata (Roscoe) Milne-Redh. ■☆

96935 Clinogyne congensis (K. Schum.) K. Schum. = Marantochloa congensis (K. Schum.) J. Léonard et Mullend. ■☆

96936 Clinogyne cordifolia K. Schum. = Marantochloa cordifolia (K. Schum.) Koechlin ■☆

96937 Clinogyne cuspidata K. Schum. ex A. Chev. = Marantochloa cuspidata (Roscoe) Milne-Redh. ■☆

96938 Clinogyne eburnea A. Chev. = Marantochloa filipes (Benth.) Hutch. ■☆

96939 Clinogyne filipes (Benth.) Benth. et Hook. = Marantochloa filipes (Benth.) Hutch. ■☆

96940 Clinogyne flexuosa (Benth.) K. Schum. = Marantochloa cuspidata (Roscoe) Milne-Redh. ■☆

96941 Clinogyne hensii (Baker) K. Schum. = Marantochloa mannii (Benth.) Milne-Redh. ■☆

96942 Clinogyne holostachya (Baker) K. Schum. = Marantochloa monophylla (K. Schum.) D'Orey ■☆

96943 Clinogyne inaequilatera (Baker) K. Schum. = Marantochloa congensis (K. Schum.) J. Léonard et Mullend. var. pubescens (Loes.) J. Léonard et Mullend. ■☆

96944 Clinogyne lasiocolea K. Schum. = Marantochloa leucantha (K. Schum.) Milne-Redh. var. lasiocolea (K. Schum.) Koechlin ■☆

96945 Clinogyne ledermannii Loes. = Marantochloa congensis (K. Schum.) J. Léonard et Mullend. var. pubescens (Loes.) J. Léonard et Mullend. ■☆

96946 Clinogyne ledermannii Loes. var. pubescens? = Marantochloa congensis (K. Schum.) J. Léonard et Mullend. var. pubescens (Loes.) J. Léonard et Mullend. ■☆

96947 Clinogyne leucantha (K. Schum.) K. Schum. = Marantochloa leucantha (K. Schum.) Milne-Redh. ■☆

96948 Clinogyne mildbraedii Loes. = Marantochloa mildbraedii Loes. ex Koechlin ■☆

96949 Clinogyne monophylla (K. Schum.) K. Schum. = Marantochloa monophylla (K. Schum.) D'Orey ■☆

96950 Clinogyne oligantha (K. Schum.) K. Schum. = Marantochloa filipes (Benth.) Hutch. ■☆

96951 Clinogyne pubescens Loes. = Marantochloa congensis (K. Schum.) J. Léonard et Mullend. var. pubescens (Loes.) J. Léonard et Mullend. ■☆

96952 Clinogyne purpurea Ridl. = Marantochloa purpurea (Ridl.) Milne-Redh. ■☆

96953 Clinogyne ramosissima (Benth.) K. Schum. = Marantochloa ramosissima (Benth.) Hutch. ■☆

96954 Clinogyne rubescens Gagnep. = Marantochloa leucantha (K. Schum.) Milne-Redh. ■☆

96955 Clinogyne schweinfurthiana K. Schum. = Marantochloa purpurea (Ridl.) Milne-Redh. ■☆

96956 Clinogyne similis Gagnep. = Marantochloa similis (Gagnep.) Pellegr. ■☆

96957 Clinogyne sulphurea (Baker) K. Schum. = Marantochloa sulphurea (Baker) Koechlin ■☆

96958 Clinogyne ubangiensis Gagnep. = Marantochloa congensis (K. Schum.) J. Léonard et Mullend. var. pubescens (Loes.) J. Léonard et Mullend. ■☆

96959 Clinogyne ugandensis (K. Schum.) K. Schum. = Marantochloa leucantha (K. Schum.) Milne-Redh. ■☆

96960 Clinopodium L. (1753); 风 轮 菜 属; Calamint, Clinopodium, Wildbasil ■●

96961 Clinopodium abyssinicum (Hochst. ex Benth.) Kuntze;阿比西尼亚风轮菜■☆

96962 Clinopodium abyssinicum (Hochst. ex Benth.) Kuntze var. condensata (Hedberg) Ryding;密集阿比西尼亚风轮菜■☆

96963 Clinopodium acinos (L.) Kuntze = Calamintha arvensis Lam. ■☆

96964 Clinopodium acinos Kuntze; 北美风轮菜; Basil Balm, Basil Thyme, Spring Savory ■☆

96965 Clinopodium arkansanum (Nutt.) House = Calamintha arkansana (Nutt.) Shinners ■☆

96966 Clinopodium ascendens Samp.;林地新风轮菜(上举林地新风轮菜,上举野生新风轮,野生新风轮); Ascending Wild Basil, Common Calamint, Mountain Balm, Mountain Mint, Wood Calamint, Woodland Calamint ■☆

96967 Clinopodium atlanticum (Ball) N. Galland;大西洋新风轮菜■☆

96968 Clinopodium biflorum (Buch.-Ham. ex D. Don) Kuntze = Micromeria imbricata (Forssk.) C. Chr. ■☆

96969 Clinopodium calamintha Kuntze;田野新风轮菜;Basil Thyme, Field Balm, Lesser Calamint ■☆

96970 Clinopodium chinense (Benth.) Kuntze;风轮菜(断血流,风轮草,蜂窝草,红九塔花,华风轮,节节草,九层塔,九塔草,苦胆草,苦刀草,苦地胆,龙胆草,落地梅花,山薄荷,山血胆,熊胆草,野薄荷,野凉粉草,野凉粉藤);Chinese Clinopodium, Chinese Savory, Chinese Wildbasil ■

96971 Clinopodium chinense (Benth.) Kuntze = Clinopodium urticifolium (Hance) C. Y. Wu et S. J. Hsuan ex H. W. Li ■

96972 Clinopodium chinense (Benth.) Kuntze subsp. grandiflorum (Maxim.) H. Hara;大花风轮菜(风车草,紫苏)■

96973 Clinopodium chinense (Benth.) Kuntze subsp. grandiflorum (Maxim.) H. Hara var. parviflorum (Kudo) H. Hara f. setilobum H. Hara;毛裂片风轮菜(毛裂片风车草)■☆

96974 Clinopodium chinense (Benth.) Kuntze subsp. grandiflorum (Maxim.) H. Hara var. parviflorum (Kudo) H. Hara f. albiflorum H. Hara;白花风轮菜(白花风车草)■☆

96975 Clinopodium chinense (Benth.) Kuntze subsp. grandiflorum (Maxim.) H. Hara var. parviflorum (Kudo) H. Hara = Clinopodium urticifolium (Hance) C. Y. Wu et S. J. Hsuan ex H. W. Li ■

96976 Clinopodium chinense (Benth.) Kuntze subsp. grandiflorum (Maxim.) H. Hara var. urticifolium (Hance) Koidz. = Clinopodium urticifolium (Hance) C. Y. Wu et S. J. Hsuan ex H. W. Li ■

96977 Clinopodium chinense (Benth.) Kuntze var. glabrescens (Nakai) Ohwi = Clinopodium chinense (Benth.) Kuntze var. shibetchense (H. Lév.) Koidz. ■☆

96978 Clinopodium chinense (Benth.) Kuntze var. macranthum Makino = Clinopodium macranthum (Makino) H. Hara ■☆

96979 Clinopodium chinense (Benth.) Kuntze var. shibetchense (H. Lév.) Koidz.;光风车草■☆

96980 Clinopodium chinense Kuntze subsp. grandiflorum H. Hara var. parviflorum (Kudo) H. Hara = Clinopodium polycephalum (Vaniot) C. Y. Wu et S. J. Hsuan ex H. W. Li ■

96981 Clinopodium confine (Hance) Kuntze;邻近风轮菜(泛红草,风轮菜,光风轮,光风轮菜,红薰草,廻文草,剪刀草,节节花,鲤鱼草,山薄荷,四季草,四季青,四季秋,塔花,铁籬散,土薄荷,小叶仙人草,野薄荷,野仙草,野仙人草,野香草,叶底红,玉如意); Adjoin Clinopodium, Adjoin Wildbasil, Pagoda Savory ■

96982 Clinopodium confine (Hance) Kuntze = Clinopodium gracile (Benth.) Matsum. ■

96983 Clinopodium confine (Hance) Kuntze var. globosum C. Y. Wu et S. J. Hsuan ex H. W. Li;球花邻近风轮菜;Clinopodium, Globose Adjoin Wildbasil ■

96984 Clinopodium confine (Hance) Kuntze var. globosum C. Y. Wu et S. J. Hsuan ex H. W. Li = Clinopodium confine (Hance) Kuntze ■

96985 Clinopodium coreanum (H. Lév.) Hara = Clinopodium urticifolium (Hance) C. Y. Wu et S. J. Hsuan ex H. W. Li ■

96986 Clinopodium cryptanthum (Vatke) Kuntze = Clinopodium simense (Benth.) Kuntze ■☆

96987 Clinopodium discolor (Diels) C. Y. Wu et S. J. Hsuan ex H. W. Li;异色风轮菜;Discolour Clinopodium, Discolour Wildbasil ■

96988 Clinopodium douglasii Kuntze = Satureja douglasii (Benth.) Briq. ■☆

96989 Clinopodium ellipticum Z. Y. Zhu; 椭圆风轮菜; Elliptical Wildbasil ■

96990 Clinopodium fruticosum Forssk. = Otostegia fruticosa (Forssk.) Schweinf. ex Penz. ●☆

96991 Clinopodium glabrescens Pomel = Clinopodium vulgare L. ■☆

96992 Clinopodium glabrum (Nutt.) Kuntze = Calamintha arkansana (Nutt.) Shinners ■☆

96993 Clinopodium gracile (Benth.) Kuntze 'Crispatum';皱波风轮菜■☆

96994 Clinopodium gracile (Benth.) Kuntze var. latifolium (H. Hara) Ohwi = Clinopodium multicaule (Maxim.) Kuntze var. latifolium H. Hara ■☆

96995 Clinopodium gracile (Benth.) Kuntze var. latifolium (H. Hara) Ohwi = Clinopodium latifolium (H. Hara) T. Yamaz. et Murata ■☆

96996 Clinopodium gracile (Benth.) Kuntze var. minimum (H. Hara) Ohwi = Clinopodium multicaule (Maxim.) Kuntze var. yakusimense (Masam.) Yahara ■☆

96997 Clinopodium gracile (Benth.) Kuntze var. multicaule (Maxim.) Ohwi = Clinopodium multicaule (Maxim.) Kuntze ■☆

96998 Clinopodium gracile (Benth.) Kuntze var. sachalinense (Eastw.) Ohwi = Enkianthus campanulatus (Miq.) G. Nicholson var. kikuchi-masaoi (Mochizuki) Sugim. ■☆

96999 Clinopodium gracile (Benth.) Matsum.;细风轮菜(臭草,光风轮,光风轮菜,花花王根草,假韩酸草,假仙菜,剪刀草,箭头草,苦草,瘦风轮,四季青,塔花,王如意,细密草,小叶仙人草,野薄荷,野凉粉草,野仙人草,野香草);Slender Clinopodium, Slender Wild Basil, Think Wildbasil ■

97000 Clinopodium grandiflorum Kuntze; 西方大花风轮菜; Greater

Calamint ■☆

97001 Clinopodium imbricatum (Forssk.) Kuntze = Micromeria imbricata (Forssk.) C. Chr. ■☆

97002 Clinopodium integerrimum Boriss.;全缘风轮菜■☆

97003 Clinopodium japonicum Makino = Clinopodium chinense (Benth.) Kuntze var. shibetchense (H. Lév.) Koidz. ■☆

97004 Clinopodium kilimandschari (Gürke) Ryding;基利风轮菜■☆

97005 Clinopodium kudoi (Hosok.) Mori = Clinopodium repens (D. Don) Vell. ■

97006 Clinopodium kudoi (Hosok.) Nemoto;工藤氏塔花■

97007 Clinopodium kudoi (Hosok.) Nemoto = Clinopodium repens (D. Don) Vell. ■

97008 Clinopodium latifolium (H. Hara) T. Yamaz. et Murata;宽叶风轮菜■☆

97009 Clinopodium latifolium (H. Hara) T. Yamaz. et Murata = Clinopodium multicaule (Maxim.) Kuntze var. latifolium H. Hara ■☆

97010 Clinopodium laxiflorum (Hayata) C. Y. Wu et S. J. Hsuan ex H. W. Li = Clinopodium laxiflorum (Hayata) Mori ■

97011 Clinopodium laxiflorum (Hayata) Mori;疏花风轮菜(疏花塔花);Laxflower Clinopodium, Sparseflower Wildbasil ■

97012 Clinopodium laxiflorum (Hayata) Mori var. taiwanianum T. H. Hsieh et T. C. Huang;台湾风轮菜■

97013 Clinopodium laxiflorum (Hayata) T. Shimizu var. parviflora T. Shimizu = Clinopodium laxiflorum (Hayata) Mori ■

97014 Clinopodium longipes C. Y. Wu et S. J. Hsuan ex H. W. Li;长梗风轮菜;Longstalk Clinopodium, Longstalk Wildbasil ■

97015 Clinopodium macranthum (Makino) H. Hara;日本大花风轮菜■☆

97016 Clinopodium martinicense Jacq. = Leucas martinicensis (Jacq.) R. Br. ■

97017 Clinopodium megalanthum (Diels) C. Y. Wu et S. J. Hsuan ex H. W. Li;寸金草(灯笼花,莲台夏枯草,麻布草,山夏枯草,蛇床子,土白芷,盐烟苏);Bigflower Clinopodium, Bigflower Wildbasil ■

97018 Clinopodium megalanthum (Diels) C. Y. Wu et S. J. Hsuan f. subglabrum C. Y. Wu et S. J. Hsuan ex H. W. Li;近无毛寸金草■

97019 Clinopodium megalanthum (Diels) C. Y. Wu et S. J. Hsuan var. intermedia C. Y. Wu et S. J. Hsuan ex H. W. Li;居间风车草(居间寸金草);Intermediate Clinopodium ■

97020 Clinopodium megalanthum (Diels) C. Y. Wu et S. J. Hsuan var. intermedia C. Y. Wu et S. J. Hsuan ex H. W. Li = Clinopodium megalanthum (Diels) C. Y. Wu et S. J. Hsuan ex H. W. Li ■

97021 Clinopodium megalanthum (Diels) C. Y. Wu et S. J. Hsuan var. lancifolium C. Y. Wu et S. J. Hsuan ex H. W. Li;披针叶风车草(披针叶寸金草);Lanveleaf Clinopodium ■

97022 Clinopodium megalanthum (Diels) C. Y. Wu et S. J. Hsuan var. lancifolium C. Y. Wu et S. J. Hsuan ex H. W. Li = Clinopodium megalanthum (Diels) C. Y. Wu et S. J. Hsuan ex H. W. Li ■

97023 Clinopodium megalanthum (Diels) C. Y. Wu et S. J. Hsuan var. robustum C. Y. Wu et S. J. Hsuan ex H. W. Li;粗壮风车草;Robust Clinopodium, Robust Wildbasil ■

97024 Clinopodium megalanthum (Diels) C. Y. Wu et S. J. Hsuan var. robustum C. Y. Wu et S. J. Hsuan ex H. W. Li = Clinopodium megalanthum (Diels) C. Y. Wu et S. J. Hsuan ex H. W. Li ■

97025 Clinopodium megalanthum (Diels) C. Y. Wu et S. J. Hsuan var. speciosum C. Y. Wu et S. J. Hsuan ex H. W. Li;美丽风车草;Pretty Clinopodium, Pretty Wildbasil ■

97026 Clinopodium megalanthum (Diels) C. Y. Wu et S. J. Hsuan var. speciosum C. Y. Wu et S. J. Hsuan ex H. W. Li = Clinopodium megalanthum (Diels) C. Y. Wu et S. J. Hsuan ex H. W. Li ■

97027 Clinopodium menthifolium (Host) Stace;薄荷叶风轮菜;Common Calamint, Wood Calamint ■☆

97028 Clinopodium mexicanum (Benth.) Govaerts;墨西哥风轮菜■☆

97029 Clinopodium micranthum (Regel) H. Hara;小花风轮菜■☆

97030 Clinopodium micranthum (Regel) H. Hara f. albiflorum (Honda) Honda;白小花风轮菜■☆

97031 Clinopodium micranthum (Regel) H. Hara var. sachalinense (Eastw.) T. Yamaz. et Murata;库页风轮菜;Sachalin Clinopodium ■☆

97032 Clinopodium micranthum (Regel) H. Hara var. yakusimense (Masam.) H. Hara = Clinopodium multicaule (Maxim.) Kuntze var. yakusimense (Masam.) Yahara ■☆

97033 Clinopodium minimum H. Hara = Clinopodium multicaule (Maxim.) Kuntze var. yakusimense (Masam.) Yahara ■☆

97034 Clinopodium multicaule (Maxim.) Kuntze;多茎剪刀草■☆

97035 Clinopodium multicaule (Maxim.) Kuntze var. latifolium H. Hara;宽叶多茎剪刀草■☆

97036 Clinopodium multicaule (Maxim.) Kuntze var. sachalinense (Eastw.) Ohwi = Enkianthus campanulatus (Miq.) G. Nicholson var. kikuchi-masaoi (Mochizuki) Sugim. ■☆

97037 Clinopodium multicaule (Maxim.) Kuntze var. yakusimense (Masam.) Yahara;屋久岛多茎剪刀草■☆

97038 Clinopodium multicaule Maxim.) Kuntze var. minimum (H. Hara) Ohwi = Clinopodium multicaule (Maxim.) Kuntze var. yakusimense (Masam.) Yahara ■☆

97039 Clinopodium myrianthum (Baker) Ryding;多花风轮菜■☆

97040 Clinopodium nocucenhacum (Vaniot) C. Y. Wu et S. J. Hsuan;多头风轮菜■

97041 Clinopodium omeiense C. Y. Wu et S. J. Hsuan ex H. W. Li;峨眉风轮菜;Emei Clinopodium, Emei Wildbasil, Omei Clinopodium ■

97042 Clinopodium paradoxum (Vatke) Ryding;奇异风轮菜■☆

97043 Clinopodium polycephalum (Vaniot) C. Y. Wu et S. J. Hsuan ex H. W. Li;灯笼草(大叶藿香,大叶香薷,第第菜,断血流,多头灯笼草,多头风轮菜,风轮菜,风轮草,蜂窝草,脚癣草,节节草,九层塔,楼台草,漫胆草,蒙锄草,山藿香,瘦风轮,田螺菜,土防风,土荆芥,夏枯草,小益母草,绣球草,野薄荷,野鱼腥草,阴风轮,荫风轮,走马灯笼草);Manyhead Clinopodium ■

97044 Clinopodium repens (D. Don) Vell.;匍匐风轮菜(工藤氏塔花,漫胆草);Cudo Clinopodium, Repent Wildbasil ■

97045 Clinopodium robustum (Hook. f.) Ryding;粗壮风轮菜■☆

97046 Clinopodium sachalinense (Eastw.) Koidz. = Clinopodium micranthum (Regel) H. Hara var. sachalinense (Eastw.) T. Yamaz. et Murata ■☆

97047 Clinopodium sachalinense (Eastw.) Koidz. = Enkianthus campanulatus (Miq.) G. Nicholson var. kikuchi-masaoi (Mochizuki) Sugim. ■☆

97048 Clinopodium simense (Benth.) Kuntze;西姆风轮菜■☆

97049 Clinopodium uhligii (Gürke) Ryding;乌里希风轮菜■☆

97050 Clinopodium uhligii (Gürke) Ryding var. obtusifolium (Avetta) Ryding;钝叶乌里希风轮菜■☆

97051 Clinopodium umbrosum (M. Bieb.) K. Koch;耐荫风轮菜■☆

97052 Clinopodium umbrosum (M. Bieb.) K. Koch = Clinopodium repens (D. Don) Vell. ■

97053 Clinopodium umbrosum Matsum. var. japonicum? = Clinopodium micranthum (Regel) H. Hara ■☆

97054 Clinopodium umbrosum Matsum. var. sachalinensis Eastw. =

Clinopodium micranthum (Regel) H. Hara var. sachalinense (Eastw.) T. Yamaz. et Murata ■☆

97055 Clinopodium umbrosum Matsum. var. shibetchense (H. Lév.) D. McKean = Clinopodium chinense (Benth.) Kuntze var. shibetchense (H. Lév.) Koidz. ■☆

97056 Clinopodium urticifolium (Hance) C. Y. Wu et S. J. Hsuan ex H. W. Li;荨麻叶风轮菜(风车草,麻叶风轮菜,紫苏);Nettleleaf Clinopodium, Nettleleaf Wildbasil ■

97057 Clinopodium urticifolium (Hance) C. Y. Wu et S. J. Hsuan ex H. W. Li = Clinopodium chinense (Benth.) Kuntze subsp. grandiflorum (Maxim.) H. Hara ■

97058 Clinopodium vernayanum (Brenan) Ryding;韦尔奈风轮菜■☆

97059 Clinopodium villosum Noë = Clinopodium vulgare L. ■☆

97060 Clinopodium vulgare L.;普通风轮菜(普通塔花,普通香草); Basil, Basil Thyme, Basilweed, Bedsfoot Flower, Bush Basil, Clinopodium Calamintha, Cushion Calamint, Field Basil, Hedge Basil, Hedge Calamint, Horse Thyme, Misteli Mistil, Poly-mountain, Stone Basil, Wild Basil, Wild Basil Savory, Wild Basil-savory ■☆

97061 Clinopodium vulgare L. subsp. arundanum (Boiss.) Nyman;芦苇风轮菜■☆

97062 Clinopodium vulgare L. subsp. villosum (Noë) Bothmer = Clinopodium vulgare L. subsp. arundanum (Boiss.) Nyman ■☆

97063 Clinopodium vulgare L. var. diminutum Simon = Clinopodium vulgare L. ■☆

97064 Clinopodium vulgare L. var. neogaea (Fernald) C. F. Reed = Clinopodium vulgare L. ■☆

97065 Clinopoidum repens Roxb. = Clinopodium repens (D. Don) Vell. ■

97066 Clinosperma Becc. (1921);斜籽榈属(克利诺椰属,细鳞椰属,斜子棕属)●☆

97067 Clinosperma bracteata (Brongn.) Becc.;斜籽榈●☆

97068 Clinostemon Kuhlm. et A. Samp. (1928);斜蕊樟属●☆

97069 Clinostemon Kuhlm. et A. Samp. = Licaria Aubl. ●☆

97070 Clinostemon Kuhlm. et A. Samp. = Mezilaurus Kuntze ex Taub. ●☆

97071 Clinostemon mahuba (A. Samp.) Kuhlm. et A. Samp.;斜蕊樟●☆

97072 Clinostigma H. Wendl. (1862);斜柱棕属(根柱椰属,曲嘴椰子属,萨摩亚棕属,西萨摩亚棕属,斜柱头榈属,斜柱椰属)●☆

97073 Clinostigma carolinense (Becc.) H. E. Moore et Fosberg;卡罗来纳斜柱棕●☆

97074 Clinostigma ponapensis (Becc.) H. E. Moore et Fosberg;波岛斜柱棕●☆

97075 Clinostigma samoense H. Wendl.;斜柱棕●☆

97076 Clinostigma savoryanum (Rehder et E. H. Wilson) H. E. Moore et Fosberg;小笠原斜柱棕●☆

97077 Clinostigmopsis Becc. (1934);拟斜柱棕属●☆

97078 Clinostigmopsis Becc. = Clinostigma H. Wendl. ●☆

97079 Clinostigmopsis harlandi (Becc.) Becc.;拟斜柱棕●☆

97080 Clinostylis Hochst. (1844);斜柱百合属■☆

97081 Clinostylis Hochst. = Gloriosa L. ■

97082 Clinostylis speciosa Hochst. = Gloriosa speciosa (Hochst.) Engl. ■☆

97083 Clinta Griff. = Chirita Buch.-Ham. ex D. Don ●■

97084 Clintonia Douglas = Downingia Torr. (保留属名)■☆

97085 Clintonia Douglas ex Lindl. = Downingia Torr. (保留属名)■☆

97086 Clintonia Raf. (1818);七筋姑属(七筋菇属);Bead Lily, Broadlily, Clintonia, Clintonis, Clinton's Lily, Wood Lily ■

97087 Clintonia allegheniensis Harned = Clintonia umbellulata (Michx.) Morong ■☆

97088 Clintonia alpina Kunth = Clintonia udensis Trautv. et C. A. Mey. ■

97089 Clintonia alpina Kunth var. udensis (Trautv. et C. A. Mey.) Macbr. = Clintonia udensis Trautv. et C. A. Mey. ■

97090 Clintonia andrewsiana Torr.;安德鲁斯七筋姑(安德鲁斯七筋菇,红七筋姑,红七筋菇);Adrews Broadlily, Red Climonia ■☆

97091 Clintonia borealis (Aiton) Raf.;黄花七筋姑(黄花七筋菇,黄七筋姑,黄七筋菇);Bead Lily, Blue Bead-lily, Bluebead, Bluebead Lily, Bluebead-lily, Blue-bead-lily, Boreal Broadlily, Clintonia, Clinton's Lily, Corn Lily, Corn-lily, Yellow Blue-bead-lily, Yellow Clintonia ■☆

97092 Clintonia borealis Raf. = Clintonia borealis (Aiton) Raf. ■☆

97093 Clintonia udensis Trautv. et C. A. Mey.;七筋姑(对口剪,螯韭茹,剪刀七,雷公七,七筋菇,搜山虎,竹叶七);Common Broadlily ■

97094 Clintonia udensis Trautv. et C. A. Mey. subsp. alpina (Kunth ex Baker) H. Hara = Clintonia udensis Trautv. et C. A. Mey. var. alpina (Kunth ex Baker) H. Hara ■

97095 Clintonia udensis Trautv. et C. A. Mey. var. alpina (Kunth ex Baker) H. Hara = Clintonia udensis Trautv. et C. A. Mey. ■

97096 Clintonia umbellulata (Michx.) Morong;白七筋姑(白七筋菇);Clinton's Lily, Speckled Wood Lily, Speckled Wood-lily, White Clintonia ■☆

97097 Clintonia umbellulata Shafer = Clintonia umbellulata (Michx.) Morong ■☆

97098 Clintonia uniflora (Menzies ex Schult.) Kunth;单花七筋姑(单花七筋菇);Bbride's-bonnet, Queen's-cup, Queen-cup ■☆

97099 Clintonia uniflora Kunth = Clintonia uniflora (Menzies ex Schult.) Kunth ■☆

97100 Cliocarpus Miers = Solanum L. ●■

97101 Cliococca Bab. (1841);穿果亚麻属■☆

97102 Cliococca tenuifolia Bab.;穿果亚麻■☆

97103 Cliococea Bab. = Linum L. ●■

97104 Cliomera Post et Kuntze = Cleiemera Raf. ●■

97105 Cliomera Post et Kuntze = Ipomoea L. (保留属名)●■

97106 Clipeola Hail. = Clypeola L. ■☆

97107 Clipteria Raf. = Eclipta L. (保留属名)■

97108 Clistanthium Post et Kuntze = Cleistanthium Kuntze ■

97109 Clistanthium Post et Kuntze = Gerbera L. (保留属名)■

97110 Clistanthocereus Backeb. = Borzicactus Riccob. ■☆

97111 Clistanthocereus Backeb. = Cleistocactus Lem. ●☆

97112 Clistanthus Müll. Arg. = Clistranthus Poit. ex Baill. ●☆

97113 Clistanthus Müll. Arg. = Pera Mutis ●☆

97114 Clistanthus Post et Kuntze = Cleistanthus Hook. f. ex Planch. ●

97115 Clistax Mart. (1829);坚冠爵床属(坚冠马兰属)☆

97116 Clistax brasiliensis Mart.;坚冠爵床☆

97117 Clistax speciosus Nees;美丽坚冠爵床☆

97118 Clistes Post et Kuntze = Cleistes Rich. ex Lindl. ■☆

97119 Clistoyucca (Engelm.) Trel. (1902);闭丝兰属(短叶丝兰属)●☆

97120 Clistoyucca (Engelm.) Trel. = Yucca L. ●■

97121 Clistoyucca Trel. = Clistoyucca (Engelm.) Trel. ●☆

97122 Clistoyucca arborescens (Torr.) Trel.;闭丝兰●☆

97123 Clistranthus Poit. ex Baill. = Pera Mutis ●☆

97124 Clitandra Benth. (1849);斜蕊夹竹桃属●☆

97125 Clitandra alba Stapf = Landolphia togolana (Hallier f.) Pichon ●☆

97126 Clitandra arnoldiana De Wild. = Clitandra cymulosa Benth. ●☆

97127 Clitandra arnoldiana De Wild. var. sereti = Clitandra cymulosa

Benth. ●☆

97128 Clitandra barteri Stapf = Orthopichonia barteri (Stapf) H. Huber ●☆

97129 Clitandra batesii Wernham = Orthopichonia cirrhosa (Radlk.) H. Huber ●☆

97130 Clitandra brazzavillensis A. Chev. = Landolphia dewevrei Stapf ●☆

97131 Clitandra buchananii Hallier f. = Landolphia buchananii (Hallier f.) Stapf ●☆

97132 Clitandra cirrhosa Radlk. = Orthopichonia cirrhosa (Radlk.) H. Huber ●☆

97133 Clitandra cymulosa Benth.;东方斜蕊夹竹桃●☆

97134 Clitandra cymulosa Stapf f. laxiflora A. Chev. = Landolphia incerta (K. Schum.) Pers. ●☆

97135 Clitandra elastica A. Chev. = Clitandra cymulosa Benth. ●☆

97136 Clitandra elastica A. Chev. var. micrantha (A. Chev.) A. Chev. ex Dalziel = Landolphia micrantha (A. Chev.) Pichon ●☆

97137 Clitandra eugenifolia A. Chev. = Orthopichonia indeniensis (A. Chev.) H. Huber ●☆

97138 Clitandra flavidiflora (K. Schum.) Hallier f. = Landolphia flavidiflora (K. Schum.) Pers. ●☆

97139 Clitandra gentilii De Wild. = Landolphia incerta (K. Schum.) Pers. ●☆

97140 Clitandra gilletii De Wild. = Clitandra cymulosa Benth. ●☆

97141 Clitandra gracilis (Stapf) Hallier f. = Landolphia camptoloba (K. Schum.) Pichon ●☆

97142 Clitandra henriquesiana K. Schum. ex Warb. = Chamaeclitandra henriquesiana (Hallier f.) Pichon ●☆

97143 Clitandra indeniensis A. Chev. = Orthopichonia indeniensis (A. Chev.) H. Huber ●☆

97144 Clitandra ivorensis A. Chev. ex Hutch. et Dalziel = Orthopichonia indeniensis (A. Chev.) H. Huber ●☆

97145 Clitandra kabulu De Wild. = Orthopichonia barteri (Stapf) H. Huber ●☆

97146 Clitandra kilimanjarica Warb. = Landolphia buchananii (Hallier f.) Stapf ●☆

97147 Clitandra lacourtiana De Wild. = Orthopichonia barteri (Stapf) H. Huber ●☆

97148 Clitandra landolphioides Hallier f. = Landolphia landolphioides (Hallier f.) A. Chev. ●☆

97149 Clitandra laurifolia A. Chev. = Landolphia utilis (A. Chev.) Pichon ●☆

97150 Clitandra laxiflora (K. Schum.) Hallier f. = Landolphia incerta (K. Schum.) Pers. ●☆

97151 Clitandra leptantha (K. Schum.) Hallier f. = Landolphia leptantha (K. Schum.) Pers. ●☆

97152 Clitandra letestui Pellegr. = Landolphia letestui (Pellegr.) Pichon ●☆

97153 Clitandra longituba Wernham = Orthopichonia indeniensis (A. Chev.) H. Huber ●☆

97154 Clitandra macrantha (K. Schum.) Hallier f. = Landolphia macrantha (K. Schum.) Pichon ●☆

97155 Clitandra mannii Stapf = Landolphia incerta (K. Schum.) Pers. ●☆

97156 Clitandra membranacea Stapf = Landolphia membranacea (Stapf) Pichon ●☆

97157 Clitandra micrantha A. Chev. = Landolphia micrantha (A. Chev.) Pichon ●☆

97158 Clitandra mildbraedii Markgr. = Orthopichonia schweinfurthii (Stapf) H. Huber ●☆

97159 Clitandra myriantha (K. Schum.) K. Schum. ex Pierre = Landolphia robustior (K. Schum.) Pers. ●☆

97160 Clitandra nitida Stapf = Landolphia nitidula Pers. ●☆

97161 Clitandra nzunde De Wild. = Clitandra cymulosa Benth. ●☆

97162 Clitandra oocarpa Stapf = Landolphia dulcis (Sabine) Pichon ●☆

97163 Clitandra orientalis K. Schum. = Clitandra cymulosa Benth. ●☆

97164 Clitandra parvifolia (Pierre) Stapf = Cylindropsis parvifolia Pierre ●☆

97165 Clitandra robustior K. Schum. = Landolphia robustior (K. Schum.) Pers. ●☆

97166 Clitandra schweinfurthii Stapf = Orthopichonia schweinfurthii (Stapf) H. Huber ●☆

97167 Clitandra schweinfurthii Stapf var. sclerophylla? = Orthopichonia visciflua (K. Schum. ex Hallier f.) Vonk ●☆

97168 Clitandra sclerophylla K. Schum. ex Stapf = Orthopichonia visciflua (K. Schum. ex Hallier f.) Vonk ●☆

97169 Clitandra semlikiensis Robyns et Boutique = Landolphia buchananii (Hallier f.) Stapf ●☆

97170 Clitandra seretii De Wild. = Orthopichonia seretii (De Wild.) Vonk ●☆

97171 Clitandra stapfiana A. Chev. = Landolphia kirkii R. A. Dyer ●☆

97172 Clitandra staudtii Stapf = Orthopichonia visciflua (K. Schum. ex Hallier f.) Vonk ●☆

97173 Clitandra togoensis Hallier f. ex A. Chev. = Landolphia togolana (Hallier f.) Pichon ●☆

97174 Clitandra togolana (Hallier f.) Stapf = Landolphia togolana (Hallier f.) Pichon ●☆

97175 Clitandra visciflua K. Schum. ex Hallier f. = Orthopichonia visciflua (K. Schum. ex Hallier f.) Vonk ●☆

97176 Clitandropsis S. Moore = Melodinus J. R. Forst. et G. Forst. ●

97177 Clitanthes Herb. = Chlidanthus Herb. ■☆

97178 Clitanthes Herb. = Stenomesson Herb. ■☆

97179 Clitanthum Benth. et Hook. f. = Chlidanthus Herb. ■☆

97180 Clithria Post et Kuntze = Cleitria Schrad. ■☆

97181 Clithria Post et Kuntze = Venidium Less. ■☆

97182 Clitocyamos St. -Lag. = Ipomoea L. (保留属名)●■

97183 Clitocyamos St. -Lag. = Quamoclit Moench ●■

97184 Clitoria L. (1753);蝶豆属(蝴蝶花豆属);Butterfly Pea, Pigeonwings, Pigeon-wings ●

97185 Clitoria alba G. Don = Vigna unguiculata (L.) Walp. subsp. alba (G. Don) Pasquet ■☆

97186 Clitoria albiflora Mattei = Clitoria ternatea L. f. albiflora (Mattei) Chiov. ●☆

97187 Clitoria cajanifolia (C. Presl) Benth. = Clitoria laurifolia Poir. ●

97188 Clitoria falcata Lam.;镰刀荚蝶豆;Falcate Pigeonwings ●

97189 Clitoria glycinoides DC. = Clitoria falcata Lam. ●

97190 Clitoria hanceana Hemsl.;广东蝶豆(白花蝶豆,韩氏蝶豆,假沙葛,立蝶豆,山岗荚,山葛薯,山萝卜);Guangdong Pigeonwings, Kwangtung Pigeonwings ●

97191 Clitoria heterophylla Lam.;异叶蝶豆●☆

97192 Clitoria kaessneri Harms;卡斯纳蝶豆●☆

97193 Clitoria laurifolia Poir.;棱荚蝶豆(棱叶蝶豆);Ribbed-pod Pigeonwings, Ribpod Butterfly Pea ●

97194 Clitoria macrophylla Wall. ex Benth. = Clitoria hanceana Hemsl. ●

97195 Clitoria mariana L.;三叶蝶豆(大山豆,三叶蝴蝶花豆,顺齐豆,顺气药,野黄豆);Atlantic Pigeonwings,Butterfly Pea,Pigeonwings,Trileaf Butterfly Pea ●
97196 Clitoria mearnsii De Wild. = Clitoria ternatea L. ●
97197 Clitoria pedunculata Bojer ex Benth. = Clitoria heterophylla Lam. ●☆
97198 Clitoria plumieri Pers. = Centrosema plumieri (Pers.) Benth. ●☆
97199 Clitoria racemosa G. Don = Vigna racemosa (G. Don) Hutch. et Dalziel ■☆
97200 Clitoria rubiginosa Pers.;锈红蝶豆●☆
97201 Clitoria rubiginosa Pers. var. glabrescens Verdc. = Clitoria falcata Lam. ●
97202 Clitoria tanganicensis Micheli = Clitoria ternatea L. ●
97203 Clitoria ternatea L.;蝶豆(蝴蝶花豆,兰蝴蝶,兰花豆,蓝蝴蝶,蓝花豆);Asian Pigeonwings,Blue Pea,Blue Vine,Butter Pea,Butterfly Pea,Mussel-shell Pea ●
97204 Clitoria ternatea L. f. albiflora (Mattei) Chiov.;白花蝶豆●☆
97205 Clitoria ternatea L. f. flaviflora Chiov.;黄花蝶豆●☆
97206 Clitoria ternatea L. var. angustifolia Hochst. ex Baker f. = Clitoria ternatea L. ●
97207 Clitoria virginiana L. = Centrosema virginianum (L.) Benth. ●☆
97208 Clitoria viridiflora Bouton ex Hook. = Macrotyloma axillare (E. Mey.) Verdc. var. glabrum? ■☆
97209 Clitoria zanzibarensis Vatke = Clitoria ternatea L. ●
97210 Clitoriastrum Heist. = Clitoria L. ●
97211 Clitoriopsis R. Wilczek(1954);苏丹豆属●☆
97212 Clitoriopsis mollis R. Wilczek;苏丹豆●☆
97213 Clivia Lindl. (1828);君子兰属;Flame Lily,Kaffir Lily,Kaffirlily,Kaffir-lily ■
97214 Clivia caulescens R. A. Dyer;具茎君子兰■☆
97215 Clivia gardenii Hook.;加登君子兰■☆
97216 Clivia gardenii Hook. var. citrina Swanevelder et A. E. van Wyk et Truter;橘黄加登君子兰■☆
97217 Clivia miniata (Hook.) Regel = Clivia miniata (Lindl.) Bosse ■
97218 Clivia miniata (Lindl.) Bosse;君子兰(达木兰,大花君子兰,大叶石蒜,红花君子兰,箭叶石蒜);Clivia,Scarlet Kaffir Lily,Scarlet Kaffirlily,Scarlet Kaffir-lily,September Lily,St. John's Lily ■
97219 Clivia miniata (Lindl.) Regel = Clivia miniata (Lindl.) Bosse ■
97220 Clivia miniata (Lindl.) Regel var. citrina Watson;橘黄君子兰■☆
97221 Clivia miniata (Lindl.) Regel var. flava E. Phillips = Clivia miniata (Lindl.) Regel var. citrina Watson ■☆
97222 Clivia mirabilis Rourke;奇异君子兰■☆
97223 Clivia nobilis Lindl.;垂笑君子兰(君子兰);Edle Clivie,Greentip Kaffir Lily,Greentip Kaffir-lily ■
97224 Clivia robusta B. G. Murray,Ran,de Lange,Hammett,Truter et Swanev.;粗壮君子兰■☆
97225 Clivia robusta B. G. Murray,Ran,de Lange,Hammett,Truter et Swanev. var. citrina Swanev.,Forb.-Hard.,Truter et A. E. van Wyk;橘黄粗壮君子兰■☆
97226 Cloanthe Nees = Chloanthes R. Br. ●☆
97227 Cloezia Brongn. et Gris(1864);新喀香桃属●☆
97228 Cloiselia S. Moore = Dicoma Cass. ●☆
97229 Cloiselia carbonaria S. Moore;煤色新喀香桃●☆
97230 Cloiselia humbertii S. Ortiz;亨伯特新喀香桃●☆
97231 Cloiselia madagascariensis S. Ortiz;马岛新喀香桃●☆
97232 Cloiselia oleifolia (Humbert) S. Ortiz;木犀榄叶新喀香桃●☆
97233 Clomena P. Beauv. = Muhlenbergia Schreb. ■
97234 Clomenocoma Cass. = Dyssodia Cav. ■☆
97235 Clomium Adans. = Carduus L. ■
97236 Clomopanus Steud. = Lonchocarpus Kunth(保留属名)●■☆
97237 Clompanus Aubl.(废弃属名) = Lonchocarpus Kunth(保留属名)●■☆
97238 Clompanus Raf. = Sterculia L. ●
97239 Clonium Post et Kuntze = Eryngium L. ■
97240 Clonium Post et Kuntze = Klonion Raf. ■
97241 Clonodia Griseb. (1858);八腺木属●☆
97242 Clonostachys Klotzsch = Sebastiania Spreng. ●
97243 Clonostylis S. Moore = Spathiostemon Blume ●☆
97244 Closaschima Korth. = Laplacea Kunth(保留属名)●☆
97245 Closia J. Rémy = Perityle Benth. ●■☆
97246 Closirospermum Neck. = Crepis L. ■
97247 Closirospermum Neck. ex Rupr. = Picris L. ■
97248 Closterandra Boiv. ex Bél. = Papaver L. ■
97249 Closteranthera Walp. = Closterandra Boiv. ex Bél. ■
97250 Clowesia Lindl. (1843);克洛斯兰属(克劳兰属)■☆
97251 Clowesia Lindl. = Catasetum Rich. ex Kunth ■☆
97252 Clowesia amazonica K. G. Lacerda et V. P. Castro;亚马逊克洛斯兰■☆
97253 Clowesia rosea Lindl.;克洛斯兰■☆
97254 Clozelia A. Chev. = Antrocaryon Pierre ●☆
97255 Clozella A. Chev. = Clozelia A. Chev. ●☆
97256 Clozella Courtet = Clozelia A. Chev. ●☆
97257 Cluacena Raf.(废弃属名) = Myrteola O. Berg(保留属名)●☆
97258 Clueria Raf. = Eremostachys Bunge ■
97259 Clugnia Comm. ex DC. = Dillenia L. ●
97260 Clusia L. (1753);猪胶树属(克鲁斯木属,克鲁希亚木属,书带木属);Clusia,Rock Balsam ●☆
97261 Clusia grandiflora Splitg.;大花猪胶树●☆
97262 Clusia major L.;红猪胶树(大克鲁斯木,书带木);Autograph Tree,Balsam Apple,Balsam Fig,Copey,Fat Pork Tree,Pitch Apple,Star of the Night ●☆
97263 Clusia nemorosa G. Mey.;丛林猪胶树●☆
97264 Clusia rosea Jacq. = Clusia major L. ●☆
97265 Clusiacaceae Lindl. = Guttiferae Juss.(保留科名)●■
97266 Clusiaceae Lindl. (1836)(保留科名);猪胶树科(金丝桃科,克鲁西科,山竹子科,藤黄科);Clusia Family,Garcinia Family,St. John's-wort Family ●■
97267 Clusiaceae Lindl.(保留科名) = Guttiferae Juss.(保留科名)●■
97268 Clusianthemum Vieill. = Garcinia L. ●
97269 Clusiella Planch. et Triana(1860);小猪胶树属●☆
97270 Clusiella elegans Planch. et Triana;小猪胶树●☆
97271 Clusiophyllea Baill. = Canthium Lam. ●
97272 Cluslophyllum Müll. Arg. = Cunuria Baill. ●☆
97273 Cluslophyllum Müll. Arg. = Micrandra Benth.(保留属名)●☆
97274 Clutia L. (1753);油芦子属■☆
97275 Clutia abyssinica Jaub. et Spach;阿比西尼亚油芦子■☆
97276 Clutia abyssinica Jaub. et Spach var. calvescens Pax = Clutia abyssinica Jaub. et Spach ■☆
97277 Clutia abyssinica Jaub. et Spach var. firma Pax = Clutia abyssinica Jaub. et Spach ■☆
97278 Clutia abyssinica Jaub. et Spach var. glabra Pax = Clutia abyssinica Jaub. et Spach ■☆
97279 Clutia abyssinica Jaub. et Spach var. ovalifolia Pax et K. Hoffm. = Clutia abyssinica Jaub. et Spach var. usambarica Pax et K. Hoffm. ■☆

97280 Clutia abyssinica Jaub. et Spach var. pedicellaris (Pax) Pax;梗花油芦子■☆

97281 Clutia abyssinica Jaub. et Spach var. usambarica Pax et K. Hoffm.;卵叶阿比西尼亚油芦子■☆

97282 Clutia acuminata E. Mey. = Clutia polifolia Jacq. ■☆

97283 Clutia acuminata Thunb. = Lachnostylis hirta (L. f.) Müll. Arg. ■●☆

97284 Clutia affinis Sond.;近缘油芦子■☆

97285 Clutia africana Poir.;非洲油芦子■☆

97286 Clutia alaternoides L.;互叶油芦子■☆

97287 Clutia alaternoides L. var. angustifolia E. Mey. ex Sond.;窄互叶油芦子■☆

97288 Clutia alaternoides L. var. brevifolia E. Mey. ex Sond.;短互叶油芦子■☆

97289 Clutia alpina Prain;高山油芦子■☆

97290 Clutia androgyna L. = Sauropus androgynus (L.) Merr. ●

97291 Clutia angustifolia Knauf;窄叶油芦子■☆

97292 Clutia anomala Pax et K. Hoffm. = Clutia abyssinica Jaub. et Spach ■☆

97293 Clutia benguelensis Müll. Arg.;本格拉油芦子■☆

97294 Clutia brachyadenia Volkens ex Pax = Clutia lanceolata Forssk. subsp. robusta (Pax) M. G. Gilbert ■☆

97295 Clutia brassii Brenan;布拉斯油芦子■☆

97296 Clutia brevifolia Sond. = Clutia polifolia Jacq. ■☆

97297 Clutia conferta Hutch.;密集油芦子■☆

97298 Clutia cordata Bernh.;心形油芦子■☆

97299 Clutia crassifolia Pax = Clutia thunbergii Sond. ■☆

97300 Clutia daphnoides Lam.;月桂油芦子■☆

97301 Clutia densifolia Gilli = Clutia whytei Hutch. ■☆

97302 Clutia dictyophlebodes A. R. Sm.;网脉油芦子■☆

97303 Clutia dregeana Scheele;德雷油芦子■☆

97304 Clutia eluteria L. = Croton eluteria (L.) Sw. ●☆

97305 Clutia ericoides Thunb.;石南状油芦子■☆

97306 Clutia ericoides Thunb. var. pachyphylla Prain;厚叶油芦子■☆

97307 Clutia ericoides Thunb. var. tenuis Sond.;薄叶油芦子■☆

97308 Clutia galpinii Pax = Clutia pulchella L. ■☆

97309 Clutia glabrata Pax = Clutia pubescens Thunb. ●☆

97310 Clutia glabrescens Knauf ex Pax = Clutia abyssinica Jaub. et Spach ■☆

97311 Clutia glauca Pax = Clutia rubricaulis Eckl. ex Sond. ■☆

97312 Clutia govaertsii Radcl. -Sm.;戈氏油芦子■☆

97313 Clutia gracilis Hutch. = Clutia paxii Knauf ex Pax ■☆

97314 Clutia heterophylla Thunb.;异叶油芦子■☆

97315 Clutia heterophylla Thunb. var. hirsuta Sond. = Clutia hirsuta Eckl. et Zeyh. ex Sond. ■☆

97316 Clutia hirsuta E. Mey. = Clutia hirsuta Eckl. et Zeyh. ex Sond. ■☆

97317 Clutia hirsuta Eckl. et Zeyh. ex Sond.;粗毛油芦子■☆

97318 Clutia hirsuta Eckl. et Zeyh. ex Sond. var. robusta Prain;粗壮粗毛油芦子■☆

97319 Clutia hybrida Pax et K. Hoffm. = Clutia hirsuta Eckl. et Zeyh. ex Sond. ■☆

97320 Clutia imbricata E. Mey. ex Sond.;覆瓦油芦子■☆

97321 Clutia impedita Prain;累赘油芦子■☆

97322 Clutia intertexta Pax et K. Hoffm. = Clutia pubescens Thunb. ●☆

97323 Clutia inyangensis Hutch. = Clutia hirsuta Eckl. et Zeyh. ex Sond. ■☆

97324 Clutia jaubertiana? = Clutia lanceolata Forssk. ■☆

97325 Clutia kamerunica Pax;喀麦隆油芦子■☆

97326 Clutia katharinae Pax;卡塔琳娜油芦子■☆

97327 Clutia kilimandscharica Engl. = Clutia lanceolata Forssk. subsp. robusta (Pax) M. G. Gilbert ■☆

97328 Clutia krookii Pax = Clutia hirsuta Eckl. et Zeyh. ex Sond. var. robusta Prain ■☆

97329 Clutia lanceolata Forssk.;披针形油芦子■☆

97330 Clutia lanceolata Forssk. subsp. robusta (Pax) M. G. Gilbert;粗壮披针形油芦子■☆

97331 Clutia lanceolata Forssk. var. angustifolia A. Rich. = Clutia lanceolata Forssk. ■☆

97332 Clutia lanceolata Forssk. var. glabra A. Rich. = Clutia abyssinica Jaub. et Spach ■☆

97333 Clutia lanceolata Forssk. var. pubescens A. Rich.;短毛披针形油芦子■☆

97334 Clutia laxa Eckl. ex Sond.;舒松油芦子■☆

97335 Clutia leuconeura Pax = Clutia abyssinica Jaub. et Spach var. usambarica Pax et K. Hoffm. ■☆

97336 Clutia marginata E. Mey. ex Sond.;具边油芦子■☆

97337 Clutia meyeriana? = Clutia polifolia Jacq. ■☆

97338 Clutia mollis Pax? = Clutia abyssinica Jaub. et Spach var. usambarica Pax et K. Hoffm. ■☆

97339 Clutia monticola S. Moore;山地油芦子■☆

97340 Clutia monticola S. Moore var. stelleroides (S. Moore) Radcl. - Sm.;星状油芦子■☆

97341 Clutia myricoides Jaub. et Spach = Clutia lanceolata Forssk. ■☆

97342 Clutia nana Prain;矮小油芦子■☆

97343 Clutia natalensis Bernh.;纳塔尔油芦子■☆

97344 Clutia ovalis (E. Mey. ex Sond.) Scheele = Andrachne ovalis (E. Mey. ex Sond.) Müll. Arg. ●☆

97345 Clutia paxii Knauf ex Pax;帕克斯油芦子■☆

97346 Clutia pedicellaris (Pax) Hutch. = Clutia abyssinica Jaub. et Spach var. pedicellaris (Pax) Pax ■☆

97347 Clutia phyllanthoides S. Moore = Clutia paxii Knauf ex Pax ■☆

97348 Clutia platyphylla Pax et K. Hoffm.;宽多叶油芦子■☆

97349 Clutia polifolia Jacq.;多叶油芦子■☆

97350 Clutia polyadenia Pax;多腺油芦子■☆

97351 Clutia polygonoides L.;多棱油芦子■☆

97352 Clutia polygonoides Thunb. = Clutia polygonoides L. ■☆

97353 Clutia polygonoides Willd. = Clutia rubricaulis Eckl. ex Sond. ■☆

97354 Clutia pterogona Müll. Arg.;翅角油芦子■☆

97355 Clutia pubescens Thunb.;短柔毛油芦子■☆

97356 Clutia pulchella L.;美丽油芦子■☆

97357 Clutia pulchella L. var. franksiae Prain;弗兰克斯油芦子■☆

97358 Clutia pulchella L. var. obtusata Sond.;钝油芦子■☆

97359 Clutia punctata Wild;斑点油芦子■☆

97360 Clutia retusa L. = Bridelia spinosa (Roxb.) Willd. ●

97361 Clutia richardiana Müll. Arg. = Clutia lanceolata Forssk. ■☆

97362 Clutia richardiana Müll. Arg. var. pedicellaris Pax = Clutia abyssinica Jaub. et Spach var. pedicellaris (Pax) Pax ■☆

97363 Clutia richardiana Müll. Arg. var. pubescens (A. Rich.) = Clutia lanceolata Forssk. var. pubescens A. Rich. ■☆

97364 Clutia richardiana Müll. Arg. var. trichophora? = Clutia lanceolata Forssk. ■☆

97365 Clutia robusta Pax = Clutia lanceolata Forssk. subsp. robusta (Pax) M. G. Gilbert ■☆

97366 Clutia robusta Pax var. acutifolia? = Clutia lanceolata Forssk.

subsp. robusta (Pax) M. G. Gilbert ■☆
97367 Clutia robusta Pax var. kilimandscharica (Engl.) Pax = Clutia lanceolata Forssk. subsp. robusta (Pax) M. G. Gilbert ■☆
97368 Clutia robusta Pax var. polyphylla? = Clutia lanceolata Forssk. subsp. robusta (Pax) M. G. Gilbert ■☆
97369 Clutia robusta Pax var. rhododendroides? = Clutia lanceolata Forssk. subsp. robusta (Pax) M. G. Gilbert ■☆
97370 Clutia robusta Pax var. salicifolia? = Clutia lanceolata Forssk. subsp. robusta (Pax) M. G. Gilbert ■☆
97371 Clutia rotundifolia Pax = Clutia abyssinica Jaub. et Spach var. usambarica Pax et K. Hoffm. ■☆
97372 Clutia rubricaulis Eckl. ex Sond.;红茎油芦子■☆
97373 Clutia rubricaulis Eckl. ex Sond. var. grandifolia Prain = Clutia rubricaulis Eckl. ex Sond. ■☆
97374 Clutia rubricaulis Eckl. ex Sond. var. microphylla Prain = Clutia rubricaulis Eckl. ex Sond. ■☆
97375 Clutia rubricaulis Eckl. ex Sond. var. tenuifolia Prain = Clutia rubricaulis Eckl. ex Sond. ■☆
97376 Clutia sericea Müll. Arg.;绢毛油芦子■☆
97377 Clutia sessilifolia Radcl. -Sm.;无柄叶油芦子■☆
97378 Clutia similis Müll. Arg. = Clutia heterophylla Thunb. ■☆
97379 Clutia sonderiana Müll. Arg. = Clutia dregeana Scheele ■☆
97380 Clutia steenkampianus Gerstner;斯滕油芦子■☆
97381 Clutia stellerioides S. Moore;小星状油芦子■☆
97382 Clutia stelleroides S. Moore = Clutia monticola S. Moore var. stelleroides (S. Moore) Radcl. -Sm. ■☆
97383 Clutia stenophylla Pax et K. Hoffm. = Clutia lanceolata Forssk. subsp. robusta (Pax) M. G. Gilbert ■☆
97384 Clutia stipularis L. = Bridelia stipularis (L.) Blume ●
97385 Clutia stuhlmannii Pax;斯图尔曼油芦子■☆
97386 Clutia swynnertonii S. Moore;斯温纳顿油芦子■☆
97387 Clutia tenuifolia Willd. = Clutia ericoides Thunb. var. tenuis Sond. ■☆
97388 Clutia thunbergii Sond.;通贝里油芦子■☆
97389 Clutia timpermaniana J. Léonard;廷珀曼油芦子■☆
97390 Clutia tomentosa L.;绒毛油芦子■☆
97391 Clutia usambarica Pax et K. Hoffm. = Clutia abyssinica Jaub. et Spach var. usambarica Pax et K. Hoffm. ■☆
97392 Clutia vaccinioides (Pax et K. Hoffm.) Prain = Clutia govaertsii Radcl. -Sm. ■☆
97393 Clutia virgata Pax et K. Hoffm.;条纹油芦子■☆
97394 Clutia volubilis Hutch. = Clutia hirsuta Eckl. et Zeyh. ex Sond. ■☆
97395 Clutia whytei Hutch.;怀特油芦子■☆
97396 Clutia whytei Hutch. var. monticoloides Radcl. -Sm.;拟山生油芦子■☆
97397 Cluytia Aiton = Clutia L. ■☆
97398 Cluytia Roxb. ex Steud. = Bridelia Willd. (保留属名)●
97399 Cluytia Steud. = Bridelia Willd. (保留属名)●
97400 Cluytia acuminata L. f. = Clutia acuminata E. Mey. ■☆
97401 Cluytia ambigua Pax et K. Hoffm. = Clutia ericoides Thunb. var. pachyphylla Prain ■☆
97402 Cluytia diosmoides Sond. = Clutia polygonoides L. ■☆
97403 Cluytia eckloniana? = Clutia pubescens Thunb. ●☆
97404 Cluytia fallacina Pax et K. Hoffm. = Clutia pubescens Thunb. ●☆
97405 Cluytia hirta L. f. = Lachnostylis hirta (L. f.) Müll. Arg. ■●☆
97406 Cluytia lasiococca Pax et K. Hoffm. = Clutia angustifolia Knauf ■☆
97407 Cluytia leuconeura Pax = Clutia abyssinica Jaub. et Spach var. usambarica Pax et K. Hoffm. ■☆
97408 Cluytia montana Roxb. = Bridelia montana (Roxb.) Willd. ●
97409 Cluytia scandens Roxb. = Bridelia stipularis (L.) Blume ●
97410 Cluytia schlechteri Pax = Clutia hirsuta Eckl. et Zeyh. ex Sond. ■☆
97411 Cluytia spinosa Roxb. = Bridelia spinosa (Roxb.) Willd. ●
97412 Cluytiandra Müll. Arg. = Meineckia Baill. ■☆
97413 Cluytiandra capillariformis (Vatke et Pax) Pax et K. Hoffm. = Meineckia phyllanthoides Baill. subsp. capillariformis (Vatke et Pax) G. L. Webster ■☆
97414 Cluytiandra engleri Pax = Meineckia fruticans (Pax) G. L. Webster var. engleri? ■☆
97415 Cluytiandra fruticans Pax = Meineckia fruticans (Pax) G. L. Webster ■☆
97416 Cluytiandra schinzii Pax = Phyllanthus pinnatus (Wight) G. L. Webster ●☆
97417 Cluytiandra somalensis Pax = Meineckia phyllanthoides Baill. subsp. somalensis (Pax) G. L. Webster ■☆
97418 Cluytiandra trichopoda Müll. Arg. = Meineckia phyllanthoides Baill. subsp. trichopoda (Müll. Arg.) G. L. Webster ■☆
97419 Clybatis Phil. = Leucheria Lag. ■☆
97420 Clymenia Swingle(1939);多蕊橘属●☆
97421 Clymenia polyandra (Tanaka) Swingle;多蕊橘●☆
97422 Clymenia polyandra (Tanaka) Swingle = Citrus polyandra Tanaka ●☆
97423 Clymenum Mill. = Lathyrus L. ■
97424 Clynhymenla A. Rich. et Galeotti = Clinhymenia A. Rich. et Galeotti ■☆
97425 Clynhymenla A. Rich. et Galeotti = Cryptarrhena R. Br. ■☆
97426 Clyostomanthus Pichon = Clytostoma Miers ex Bureau ●
97427 Clypea Blume = Stephania Lour. ●■
97428 Clypea abyssinica Quart. -Dill. et A. Rich. = Stephania abyssinica (Quart. -Dill. et A. Rich.) Walp. ●☆
97429 Clypea discolor Blume = Stephania japonica (Thunb.) Miers var. discolor (Blume) Forman ●■
97430 Clypeola Burm. ex DC. = Pterocarpus Jacq. (保留属名)●
97431 Clypeola Crantz = Biscutella L. + Alyssum L. ■●
97432 Clypeola L. (1753);盾果荠属■☆
97433 Clypeola Neck. = Adyseton Adans. ■
97434 Clypeola Neck. = Alyssum L. ■●
97435 Clypeola alyssoides (L.) L. = Alyssum alyssoides (L.) L. ■
97436 Clypeola alyssoides L. = Alyssum alyssoides (L.) L. ■
97437 Clypeola ambigua Jord. et Fourr. = Clypeola jonthlaspi L. subsp. microcarpa (Moris) Arcang. ■☆
97438 Clypeola aspera (Grauer) Turrill;粗糙盾果荠■☆
97439 Clypeola bruhnsii Gruner = Clypeola jonthlaspi L. ■☆
97440 Clypeola chaetocarpa Jaub. et Spach. = Clypeola aspera (Grauer) Turrill ■☆
97441 Clypeola cyclodontea Delile;环齿盾果荠■☆
97442 Clypeola dichotoma Boiss.;二歧盾果荠■☆
97443 Clypeola echinata DC. = Clypeola aspera (Grauer) Turrill ■☆
97444 Clypeola elegans Boiss. et Huet;雅致盾果荠■☆
97445 Clypeola glabra Boiss. = Clypeola jonthlaspi L. subsp. microcarpa (Moris) Arcang. ■☆
97446 Clypeola jonthlaspi L.;约斯盾果荠■☆
97447 Clypeola jonthlaspi L. subsp. macrocarpa Fiori = Clypeola jonthlaspi L. ■☆
97448 Clypeola jonthlaspi L. subsp. microcarpa (Moris) Arcang.;小果

约斯盾果荠■☆

97449 Clypeola jonthlaspi L. var. ambigua (Jord. et Fourr.) Fiori = Clypeola jonthlaspi L. subsp. microcarpa (Moris) Arcang. ■☆

97450 Clypeola jonthlaspi L. var. glabra (Boiss.) Halácsy = Clypeola jonthlaspi L. subsp. microcarpa (Moris) Arcang. ■☆

97451 Clypeola jonthlaspi L. var. glabriuscula Gruner = Clypeola jonthlaspi L. ■☆

97452 Clypeola jonthlaspi L. var. lasiocarpa Guss. = Clypeola jonthlaspi L. subsp. microcarpa (Moris) Arcang. ■☆

97453 Clypeola jonthlaspi L. var. minor Monnier = Clypeola jonthlaspi L. subsp. microcarpa (Moris) Arcang. ■☆

97454 Clypeola jonthlaspi L. var. petraea (Jord. et Fourr.) Rouy et Foucaud = Clypeola jonthlaspi L. ■☆

97455 Clypeola jonthlaspi L. var. pyrenaica (Bordère) Reyn. = Clypeola jonthlaspi L. subsp. microcarpa (Moris) Arcang. ■☆

97456 Clypeola maritima L. = Lobularia maritima (L.) Desv. ■

97457 Clypeola microcarpa Boiss.;小果盾果荠■☆

97458 Clypeola microcarpa Moris = Clypeola jonthlaspi L. subsp. microcarpa (Moris) Arcang. ■☆

97459 Clypeola minor L. = Alyssum simplex Rudolphi ■

97460 Clypeola petraea Jord. et Fourr. = Clypeola jonthlaspi L. ■☆

97461 Clypeola pyrenaica Bordère = Clypeola jonthlaspi L. subsp. microcarpa (Moris) Arcang. ■☆

97462 Clystomenon Müll. Arg. = Cyclostemon Blume ●

97463 Clystomenon Müll. Arg. = Drypetes Vahl ●

97464 Clytia Stokes = Clutia L. ■☆

97465 Clytoria J. Presl = Clitoria L. ●

97466 Clytostoma Bureau = Clytostoma Miers ex Bureau ●

97467 Clytostoma Miers = Clytostoma Miers ex Bureau ●

97468 Clytostoma Miers ex Bureau. (1868);连理藤属(美花藤属);Trumpet Vine, Trumpetvine, Trumpet-vine ●

97469 Clytostoma callistegioides (Cham.) Baill. = Clytostoma callistegioides (Cham.) Bureau ex Griseb. ●

97470 Clytostoma callistegioides (Cham.) Bureau et Schum. = Clytostoma callistegioides (Cham.) Bureau ex Griseb. ●

97471 Clytostoma callistegioides (Cham.) Bureau ex Griseb.;连理藤(刺实紫葳);Argentine Trumpet Vine, Trumpetvine, Trumpet-vine, Violet Trumpet Vine ●

97472 Clytostoma magnifica W. Bull = Bignonia magnifica W. Bull ●

97473 Clytostomanthus Pichon(1946);丽口紫葳属●☆

97474 Clytostomanthus decorus (S. Moore) Pichon;丽口紫葳●☆

97475 Cnema Post et Kuntze = Knema Lour. ●

97476 Cnemidia Lindl. = Tropidia Lindl. ■

97477 Cnemidia angulosa Lindl. = Tropidia angulosa (Lindl.) Blume ■

97478 Cnemidia semilibera Lindl. = Tropidia angulosa (Lindl.) Blume ■

97479 Cnemidiscus Pierre = Glenniea Hook. f. ●☆

97480 Cnemidophacos Rydb. = Astragalus L. ●■

97481 Cnemidostachys Mart. = Sebastiania Spreng. ●

97482 Cnemidostachys Mart. et Zucc. = Sebastiania Spreng. ●

97483 Cnenamum Tausch = Crenamum Adans. ■☆

97484 Cnenamum Tausch = Crepis L. ■

97485 Cneoraceae Link = Cneoraceae Vest(保留科名)●☆

97486 Cneoraceae Link = Rutaceae Juss.(保留科名)●■

97487 Cneoraceae Vest(1818)(保留科名);叶柄花科(拟荨麻科)●☆

97488 Cneoraceae Vest(保留科名) = Rutaceae Juss.(保留科名)●■

97489 Cneoridium Hook. f. (1862);小叶柄花属●☆

97490 Cneoridium dumosum (Nutt. ex Torr. et A. Gray) B. D. Jacks.;小叶柄花●☆

97491 Cneorum L. (1753);叶柄花属;Spurge Olive, Spurge-olive ●☆

97492 Cneorum pulverulentum Vent. = Neochamaelea pulverulenta (Vent.) Erdtman ●☆

97493 Cneorum tricoccon L.;叶柄花;Spurge-olive ●☆

97494 Cnesmocarpon Adema(1993);锉果藤属●☆

97495 Cnesmocarpon dasyantha (Radlk.) Adema;锉果藤●☆

97496 Cnesmocarpon dentata Adema;多齿锉果藤●☆

97497 Cnesmocarpon montana Adema;山地锉果藤●☆

97498 Cnesmocarpus Zipp. ex Blume = Pometia J. R. Forst. et G. Forst. ●

97499 Cnesmone Blume(1826);粗毛藤属;Cnesmone, Shagvine ●

97500 Cnesmone anisosepala (Merr. et Chun) Croizat = Cnesmone tonkinensis (Gagnep.) Croizat ●

97501 Cnesmone hainanensis (Merr. et Chun) Croizat;海南粗毛藤(痒藤);Hainan Cnesmone, Hainan Shagvine ●

97502 Cnesmone hainanensis Merr. et Chun = Cnesmone hainanensis (Merr. et Chun) Croizat ●

97503 Cnesmone javanica Blume;爪哇粗毛藤;Java Shagvine ●☆

97504 Cnesmone mairei (H. Lév.) Croizat;粗毛藤(刺痒藤);Maire Cnesmone, Maire Shagvine ●

97505 Cnesmone tonkinensis (Gagnep.) Croizat;灰岩粗毛藤(麻风藤,异萼粗毛藤);Diversisepal Shagvine, Limestone Shagvine, Saxatile Cnesmone, Tonkin Cnesmone ●

97506 Cnesmosa Blume = Cnesmone Blume ●

97507 Cnestidaceae Raf. = Connaraceae R. Br.(保留科名)●

97508 Cnestidium Planch. (1850);小螯毛果属●☆

97509 Cnestidium rufescens Planch.;小螯毛果●☆

97510 Cnestis Juss. (1789);螯毛果属;Cnestis ●

97511 Cnestis agelaeoides G. Schellenb. = Cnestis corniculata Lam. ●☆

97512 Cnestis angolensis G. Schellenb. = Cnestis corniculata Lam. ●☆

97513 Cnestis aurantiaca Gilg = Cnestis corniculata Lam. ●☆

97514 Cnestis boiviniana Baill. = Cnestis polyphylla Lam. ●☆

97515 Cnestis bomiensis Lemmens;博米螯毛果●☆

97516 Cnestis calantha G. Schellenb. = Cnestis corniculata Lam. ●☆

97517 Cnestis calocarpa Gilg = Cnestis corniculata Lam. ●☆

97518 Cnestis cinnabarina G. Schellenb. = Cnestis corniculata Lam. ●☆

97519 Cnestis claessensii De Wild. = Cnestis corniculata Lam. ●☆

97520 Cnestis confertiflora Gilg = Cnestis corniculata Lam. ●☆

97521 Cnestis congolana De Wild. = Cnestis corniculata Lam. ●☆

97522 Cnestis corniculata Lam.;小角螯毛果●☆

97523 Cnestis dinklagei G. Schellenb. = Cnestis corniculata Lam. ●☆

97524 Cnestis emarginata De Wild. et T. Durand = Cnestis corniculata Lam. ●☆

97525 Cnestis emarginata Jack = Roureopsis emarginata (Jack) Merr. ●

97526 Cnestis ferruginea Vahl ex DC.;锈色螯毛果●☆

97527 Cnestis fraterna Planch. = Cnestis ferruginea Vahl ex DC. ●☆

97528 Cnestis gabunensis G. Schellenb. = Cnestis corniculata Lam. ●☆

97529 Cnestis gimbiensis Troupin = Cnestis corniculata Lam. ●☆

97530 Cnestis glabra Lam. = Cnestis polyphylla Lam. ●☆

97531 Cnestis grandiflora Gilg = Cnestis corniculata Lam. ●☆

97532 Cnestis grandifoliolata De Wild. = Cnestis mannii (Baker) G. Schellenb. ●☆

97533 Cnestis grisea Baker = Cnestis corniculata Lam. ●☆

97534 Cnestis hirsuta Troupin = Cnestis corniculata Lam. ●☆

97535 Cnestis iomalla Gilg = Cnestis corniculata Lam. ●☆

97536 Cnestis iomalla Gilg var. grandifoliolata De Wild. = Cnestis corniculata Lam. ●☆

97537 Cnestis laurentii De Wild. = Cnestis urens Gilg ●☆
97538 Cnestis lescrauwaetii De Wild. = Cnestis corniculata Lam. ●☆
97539 Cnestis leucantha Gilg ex G. Schellenb. = Cnestis corniculata Lam. ●☆
97540 Cnestis leucanthoides Pellegr. = Cnestis corniculata Lam. ●☆
97541 Cnestis liberica G. Schellenb. = Cnestis racemosa Don ●☆
97542 Cnestis longiflora G. Schellenb. = Cnestis corniculata Lam. ●☆
97543 Cnestis lurida Baill. = Cnestis polyphylla Lam. ●☆
97544 Cnestis macrantha Baill.;大花螯毛果●☆
97545 Cnestis macrophylla Gilg ex G. Schellenb.;大叶螯毛果●☆
97546 Cnestis mannii (Baker) G. Schellenb.;曼氏螯毛果●☆
97547 Cnestis mildbraedii Gilg;米尔德螯毛果●☆
97548 Cnestis mullendersii Troupin = Cnestis corniculata Lam. ●☆
97549 Cnestis natalensis (Hochst.) Planch. et Sond. = Cnestis polyphylla Lam. ●☆
97550 Cnestis obliqua P. Beauv. = Agelaea pentagyna (Lam.) Baill. ●☆
97551 Cnestis oblongifolia Baker = Cnestis ferruginea Vahl ex DC. ●☆
97552 Cnestis palala (Lour.) Merr.;螯毛果;Common Cnestis, Stinghaired Cnestis, Stinginghair Cnestis ●
97553 Cnestis palatantha Griff.;广花螯毛果●☆
97554 Cnestis pinnata P. Beauv. = Rourea thomsonii (Baker) Jongkind ●☆
97555 Cnestis polyantha Gilg = Cnestis corniculata Lam. ●☆
97556 Cnestis polyphylla Lam.;多叶螯毛果●☆
97557 Cnestis prehensilis A. Chev. = Cnestis corniculata Lam. ●☆
97558 Cnestis pynaertii De Wild. = Cnestis corniculata Lam. ●☆
97559 Cnestis racemosa Don;总花螯毛果●☆
97560 Cnestis ramiflora Griff. = Cnestis palala (Lour.) Merr. ●
97561 Cnestis riparia Gilg = Cnestis corniculata Lam. ●☆
97562 Cnestis sapinii De Wild. = Cnestis corniculata Lam. ●☆
97563 Cnestis sapinii De Wild. var. claessensii (De Wild.) Troupin = Cnestis corniculata Lam. ●☆
97564 Cnestis setosa Gilg = Cnestis corniculata Lam. ●☆
97565 Cnestis togoensis Gilg = Cnestis ferruginea Vahl ex DC. ●☆
97566 Cnestis tomentosa Hepper = Cnestis mannii (Baker) G. Schellenb. ●☆
97567 Cnestis trichopoda Gilg ex Schellenb. = Cnestis corniculata Lam. ●☆
97568 Cnestis trifolia Lam. = Agelaea pentagyna (Lam.) Baill. ●☆
97569 Cnestis ugandensis G. Schellenb. = Cnestis mildbraedii Gilg ●☆
97570 Cnestis uncata Lemmens;具钩螯毛果●☆
97571 Cnestis urens Gilg;螫毛螯毛果●☆
97572 Cnestis vanderystii Troupin = Cnestis corniculata Lam. ●☆
97573 Cnestis yangambiensis Louis ex Troupin;扬甘比螯毛果●☆
97574 Cnestis zenkeri G. Schellenb. = Cnestis corniculata Lam. ●☆
97575 Cnicaceae Vest = Asteraceae Bercht. et J. Presl(保留科名)●■
97576 Cnicaceae Vest = Compositae Giseke(保留科名)●■
97577 Cnicothamnus Griseb. (1874);橙菊木属●☆
97578 Cnicothamnus lorentzii Griseb.;橙菊木●☆
97579 Cnicus D. Don = Theodorea (Cass.) Cass. ●■
97580 Cnicus Gaertn. = Carbenia Adans. ■●
97581 Cnicus L. (1753)(保留属名);藏掖花属(廉菊属);Blessed Thistle, Cnicus ■●
97582 Cnicus L. (保留属名) = Centaurea L. (保留属名)●■
97583 Cnicus acaulis Willd.;无茎藏掖花■☆
97584 Cnicus andersonii A. Gray = Cirsium andersonii (A. Gray) Petr. ■☆
97585 Cnicus andrewsii A. Gray = Cirsium andrewsii (A. Gray) Jeps. ■☆
97586 Cnicus argyracanthus (DC.) C. B. Clarke = Cirsium argyracanthum DC. ■
97587 Cnicus argyracanthus (DC.) C. B. Clarke = Cirsium verutum (D. Don) Spruner ■
97588 Cnicus argyracanthus (DC.) C. B. Clarke var. nepalensis? = Cirsium shansiense Petr. ■
97589 Cnicus argyracanthus (DC.) C. B. Clarke var. nepalensis? = Cirsium wallichii DC. ■
97590 Cnicus arizonicus A. Gray = Cirsium arizonicum (A. Gray) Petr. ■☆
97591 Cnicus arnensis O. Hoffm. = Cirsium setosum (Willd.) M. Bieb. ■
97592 Cnicus arthamoides Wall. = Hemistepta lyrata (Bunge) Bunge ■
97593 Cnicus arvensis (L.) Hoffm. = Cirsium arvense (L.) Scop. ■
97594 Cnicus arvensis (L.) Hoffm. var. setosus (Ledeb.) Maxim. = Cirsium setosum (Willd.) M. Bieb. ■
97595 Cnicus arvensis (L.) Roth = Cirsium arvense (L.) Scop. ■
97596 Cnicus arvensis (L.) Roth f. albiflorum E. L. Rand et Redfield = Cirsium arvense (L.) Scop. ■
97597 Cnicus arvensis O. Hoffm. = Cirsium arvense (L.) Scop. ■
97598 Cnicus auriculata Wall. = Saussurea auriculata (DC.) Sch. Bip. ■
97599 Cnicus benedictus L.;藏掖花(地中海蓟,地中海菊,蓟,廉菊);Blessed Thistle, Cnicus ■
97600 Cnicus benedictus L. = Centaurea benedicta (L.) L. ■
97601 Cnicus bodinieri Vaniot = Cirsium japonicum Fisch. ex DC. ■
97602 Cnicus breweri A. Gray = Cirsium douglasii DC. var. breweri (A. Gray) D. J. Keil et C. E. Turner ■☆
97603 Cnicus breweri A. Gray var. vaseyi A. Gray = Cirsium hydrophilum (Greene) Jeps. var. vaseyi (A. Gray) J. T. Howell ■☆
97604 Cnicus carlinoides Schrank var. americanus A. Gray = Cirsium clavatum (M. E. Jones) Petr. var. americanum (A. Gray) D. J. Keil ■☆
97605 Cnicus carthamoides Wall. = Hemistepta lyrata (Bunge) Bunge ■
97606 Cnicus carthamoides Willd. = Rhaponticum carthamoides (Willd.) Iljin ■
97607 Cnicus carthamoides Willd. = Stemmacantha carthamoides (Willd.) Dittrich ■
97608 Cnicus cavaleriei H. Lév. = Cirsium monocephalum (Vaniot) H. Lév. ■
97609 Cnicus cerberus Vaniot = Cirsium japonicum Fisch. ex DC. ■
97610 Cnicus cernuus L. = Alfredia cernua (L.) Cass. ■
97611 Cnicus chamaecephalus Vatke = Carduus schimperi Sch. Bip. ■☆
97612 Cnicus chinensis (Gardner et Champ.) Benth. = Cirsium shansiense Petr. ■
97613 Cnicus chinensis (Gardner et Champ.) C. B. Clarke = Cirsium shansiense Petr. ■
97614 Cnicus chinensis Benth. = Cirsium lineare (Thunb.) Sch. Bip. ■
97615 Cnicus chinensis Gardner et Champ. = Cirsium chinense Gardner et Champ. ■
97616 Cnicus chrysacanthus Ball = Cirsium chrysacanthum (Ball) Jahand. ■☆
97617 Cnicus clavatus M. E. Jones = Cirsium clavatum (M. E. Jones) Petr. ■☆
97618 Cnicus diacantha (Lab. ?) C. C. Gmel.;二刺藏掖花;Fishbone Thistle ■☆
97619 Cnicus diamentiacus Nakai = Cirsium schantarense Trautv. et C. A. Mey. ■

97620 Cnicus discolor Muhl. ex Willd. = Cirsium discolor (Muhl. ex Willd.) Spreng. ■☆

97621 Cnicus drummondii Torr. et A. Gray var. bipinnatus Eastw. = Cirsium arizonicum (A. Gray) Petr. var. bipinnatum (Eastw.) D. J. Keil ■☆

97622 Cnicus eatonii A. Gray = Cirsium eatonii (A. Gray) B. L. Rob. ■☆

97623 Cnicus echinatus (Desf.) Willd. = Cirsium echinatum (Desf.) DC. ■☆

97624 Cnicus eriocephalus A. Gray = Cirsium eatonii (A. Gray) B. L. Rob. var. eriocephalum (A. Gray) D. J. Keil ■☆

97625 Cnicus eriophoroides Hook. f. = Cirsium eriophoroides (Hook. f.) Petr. ■

97626 Cnicus esculentus Siev. = Cirsium esculentum (Siev.) C. A. Mey. ■

97627 Cnicus fargesii Franch. = Cirsium fargesii (Franch.) Diels ■

97628 Cnicus fontinalis Greene = Cirsium fontinale (Greene) Jeps. ■☆

97629 Cnicus forrestii Diels = Cirsium henryi (Franch.) Diels ■

97630 Cnicus giganteus (Desf.) Willd. = Cirsium scabrum (Poir.) Bonnet ■☆

97631 Cnicus glabrifolium C. Winkl. = Cirsium glabrifolium (C. Winkl.) O. Fedtsch. et B. Fedtsch. ■

97632 Cnicus gmelinii Spreng. = Cirsium esculentum (Siev.) C. A. Mey. ■

97633 Cnicus griffithii Hook. f. = Cirsium interpositum Petr. ■

97634 Cnicus helenioides (L.) Willd. = Cirsium helenioides (L.) Hill ■

97635 Cnicus helgendorfii Franch. et Sav. = Cirsium pendulum Fisch. ex DC. ■

97636 Cnicus henryi Franch. = Cirsium henryi (Franch.) Diels ■

97637 Cnicus hesperius Eastw. = Cirsium eatonii (A. Gray) B. L. Rob. var. hesperium (Eastw.) D. J. Keil ■☆

97638 Cnicus hillii Canby = Cirsium hillii (Canby) Fernald ■☆

97639 Cnicus hillii Canby = Cirsium pumilum (Nutt.) Spreng. var. hillii (Canby) B. Boivin ■☆

97640 Cnicus ignarius (Spreng.) Benth. = Ancathia igniaria (Spreng.) DC. ■

97641 Cnicus iowensis Pammel = Cirsium altissimum (L.) Spreng. ■☆

97642 Cnicus japonicus (DC.) Maxim. = Cirsium japonicum Fisch. ex DC. ■

97643 Cnicus japonicus (Fisch. ex DC.) Maxim. = Cirsium japonicum Fisch. ex DC. ■

97644 Cnicus japonicus (Fisch. ex DC.) Maxim. var. maackii Maxim. = Cirsium maackii Maxim. ■

97645 Cnicus japonicus (Fisch. ex DC.) Maxim. var. schantarensis (Trautv. et C. A. Mey.) Maxim. = Cirsium schantarense Trautv. et C. A. Mey. ■

97646 Cnicus japonicus Maxim. = Cirsium japonicum Fisch. ex DC. ■

97647 Cnicus japonicus Maxim. var. intermedius Maxim. = Cirsium japonicum Fisch. ex DC. ■

97648 Cnicus lanceolatus (L.) Willd. = Cirsium vulgare (Savi) Ten. ■

97649 Cnicus lanceolatus (L.) Willd. var. abyssinicus (Sch. Bip. ex A. Rich.) Vatke = Cirsium vulgare (Savi) Ten. ■

97650 Cnicus leducei Franch. = Cirsium leducei (Franch.) H. Lév. ■

97651 Cnicus linearis (Thunb.) Benth. = Cirsium lineare (Thunb.) Sch. Bip. ■

97652 Cnicus linearis Benth. et Hook. f. ex Franch. et Sav. = Cirsium lineare (Thunb.) Sch. Bip. ■

97653 Cnicus maackii (Maxim.) Nakai = Cirsium maakii Maxim. ■

97654 Cnicus maackii (Maxim.) Nakai var. koreiensis Nakai = Cirsium maakii Maxim. ■

97655 Cnicus mairei H. Lév. = Cirsium griseum H. Lév. ■

97656 Cnicus monocephalus Vaniot = Cirsium monocephalum (Vaniot) H. Lév. ■

97657 Cnicus multicaulis Wall. ex DC. = Hemistepta lyrata (Bunge) Bunge ■

97658 Cnicus niveus Wall. = Saussurea crispa Vaniot ■

97659 Cnicus ochrocentrus (A. Gray) A. Gray = Cirsium ochrocentrum A. Gray ■☆

97660 Cnicus ornatus Ball = Cirsium chrysacanthum (Ball) Jahand. ■☆

97661 Cnicus palustris (L.) Willd. = Cirsium palustre (L.) Scop. ■☆

97662 Cnicus parryi A. Gray = Cirsium parryi (A. Gray) Petr. ■☆

97663 Cnicus pendalus (Fisch. ex DC.) Maxim. = Cirsium pendulum Fisch. ex DC. ■

97664 Cnicus pexus Franch. et Sav. = Cirsium suffultum (Maxim.) Matsum. et Koidz. ■☆

97665 Cnicus pitcheri Torr. ex Eaton = Cirsium pitcheri (Torr. ex Eaton) Torr. et A. Gray ■☆

97666 Cnicus polyacanthus Vatke = Cirsium straminispinum C. Jeffrey ex Cufod. ■☆

97667 Cnicus provostii Franch. = Cirsium pendulum Fisch. ex DC. ■

97668 Cnicus quercetorum A. Gray = Cirsium quercetorum (A. Gray) Jeps. ■☆

97669 Cnicus rothrockii A. Gray = Cirsium arizonicum (A. Gray) Petr. var. rothrockii (A. Gray) D. J. Keil ■☆

97670 Cnicus sairamenisis C. Winkl. = Cirsium sairamense (C. Winkl.) O. Fedtsch. et B. Fedtsch. ■

97671 Cnicus schimperi Vatke = Cirsium schimperi (Vatke) C. Jeffrey ex Cufod. ■☆

97672 Cnicus schimperi Vatke var. inermis Oliv. et Hiern = Cirsium schimperi (Vatke) C. Jeffrey ex Cufod. ■☆

97673 Cnicus semenovii (Regel et Schmalh.) C. Winkl. = Cirsium semenovii Regel et Schmalh. ■

97674 Cnicus semenovii C. Winkl. = Cirsium semenovii Regel et Schmalh. ■

97675 Cnicus serratuloides (L.) Roth = Cirsium serratuloides (L.) Hill ■

97676 Cnicus setosus Besser = Cirsium setosum (Willd.) M. Bieb. ■

97677 Cnicus souliei Franch. = Cirsium souliei (Franch.) Mattf. ■

97678 Cnicus suffultus Maxim. = Cirsium suffultum (Maxim.) Matsum. et Koidz. ■☆

97679 Cnicus syriacus (L.) Willd. = Notobasis syriaca (L.) Cass. ■☆

97680 Cnicus taliensis Jeffrey = Cirsium henryi (Franch.) Diels ■

97681 Cnicus tchefouensis (Debeaux) Franch. = Cirsium chinense Gardner et Champ. ■

97682 Cnicus undulatus (Nutt.) A. Gray = Cirsium undulatum (Nutt.) Spreng. ■☆

97683 Cnicus undulatus (Nutt.) A. Gray var. megacephalus A. Gray = Cirsium undulatum (Nutt.) Spreng. ■☆

97684 Cnicus uniflorus L. = Rhaponticum uniflorum (L.) DC. ■

97685 Cnicus uniflorus L. = Stemmacantha uniflora (L.) Dittrich ■

97686 Cnicus uniflorus Siev. = Stemmacantha carthamoides (Willd.) Dittrich ■

97687 Cnicus verutus D. Don = Cirsium verutum (D. Don) Spruner ■

97688 Cnicus vlassovianum (Fisch. ex DC.) Maxim. = Cirsium vlassovianum Fisch. ex DC. ■

97689 Cnicus wallichii Hook. f. = Cirsium shansiense Petr. ■
97690 Cnicus wallichii Hook. f. = Cirsium wallichii DC. ■
97691 Cnicus wallichii Hook. f. var. nepalensis? = Cirsium shansiense Petr. ■
97692 Cnicus wallichii Hook. f. var. nepalensis? = Cirsium wallichii DC. ■
97693 Cnicus wheeleri A. Gray = Cirsium wheeleri (A. Gray) Petr. ■☆
97694 Cnidiocarpa Pimenov(1983);麻刺果属(荨麻刺果属)■☆
97695 Cnidiocarpa alaica Pimenov;麻刺果■☆
97696 Cnidium Cusson = Cnidium Cusson ex Juss. ■
97697 Cnidium Cusson ex Juss. (1787);蛇床属(芎穷属);Cnidium, Sakebed ■
97698 Cnidium Juss. = Cnidium Cusson ex Juss. ■
97699 Cnidium Juss. = Selinum L. (保留属名)■
97700 Cnidium ajanense (Regel et Tiling) Drude = Ligusticum ajanense (Regel et Tiling) Koso-Pol. ■
97701 Cnidium chinense (L.) Spreng. ex Steud. = Conioselinum chinense (L.) Britton, Sterns et Poggenb. ■
97702 Cnidium chinense Spreng. = Conioselinum chinense (L.) Britton, Sterns et Poggenb. ■
97703 Cnidium dahuricum (Jacq.) Turcz. ex Fisch. et C. A. Mey.;兴安蛇床(山胡萝卜);Dahur Sakebed ■
97704 Cnidium dichotomum Desf. = Krubera peregrinum (L.) Hoffm. ■☆
97705 Cnidium dubium (Schkuhr) Schmeil et Fitschen;疑蛇床■☆
97706 Cnidium filicinum Hara;蕨蛇床(蕨蛇床)■☆
97707 Cnidium filisectum Nakai et Kitag. = Ligusticum tachiroei (Franch. et Sav.) M. Hiroe et Constance ■
97708 Cnidium fischeri (Spreng.) Spreng. = Cenolophium denudatum (Hornem.) Tutin ■
97709 Cnidium formosanum Y. Yabe = Cnidium monnieri (L.) Cusson var. formosanum (Y. Yabe) Kitag. ■
97710 Cnidium grossheimii Manden.;格罗蛇床■☆
97711 Cnidium japonicum Miq.;滨蛇床;Japan Sakebed ■
97712 Cnidium jeholense Nakai et Kitag. = Ligusticum jeholense (Nakai et Kitag.) Nakai et Kitag. ■
97713 Cnidium kamelinii V. M. Vinogr. = Lithosciadium kamelinii (V. M. Vinogr.) Pimenov ex Gubanov ■
97714 Cnidium kraussianum (Meisn.) Sond. = Pimpinella caffra (Eckl. et Zeyh.) D. Dietr. ■☆
97715 Cnidium kraussianum (Meisn.) Sond. var. glabratum Sond. = Pimpinella caffra (Eckl. et Zeyh.) D. Dietr. ■☆
97716 Cnidium longiradiatum Y. Yabe = Angelica longiradiata (Maxim.) Kitag. ■☆
97717 Cnidium microcarpum Turcz. = Cnidium monnieri (L.) Cusson ■
97718 Cnidium mongolicum H. Wolff = Cnidium monnieri (L.) Cusson ■
97719 Cnidium monnieri (L.) Cusson;蛇床(赤木草,赤目草,鬼老子,虺床,建阳八座,癞头花子,连阳八座,马床,气果,蛇常,蛇米,蛇粟,蛇朱,蛇珠,绳毒,双肾子,思益,秃头花子,秃子花子,盱子,野胡萝卜,野茴香,枣棘);Monnier's Snowparsley, Sakebed ■
97720 Cnidium monnieri (L.) Cusson var. formosanum (Y. Yabe) Kitag.;台湾蛇床(流明草,台湾芎藭);Taiwan Sakebed ■
97721 Cnidium multicaule (Turcz.) Ledeb.;多茎蛇床■☆
97722 Cnidium nulivittatum K. T. Fu;无油管蛇床■
97723 Cnidium nullivittatum K. T. Fu = Ligusticum nullivittatum (K. T. Fu) F. T. Pu et M. F. Watson ■
97724 Cnidium officinale Makino = Ligusticum officinale (Makino) Kitag. ■
97725 Cnidium orientale Boiss.;东方蛇床■☆
97726 Cnidium pauciradiatum Sommier et H. Lév.;稀射线蛇床■☆
97727 Cnidium salinum Turcz.;碱蛇床;Saline Sakebed ■
97728 Cnidium salinum Turcz. = Kadenia salina (Turcz.) Lavrova et V. N. Tikhom. ■
97729 Cnidium salinum Turcz. = Peucedanum stepposum C. C. Huang ■
97730 Cnidium salinum Turcz. var. rhizomaticum Ma = Cnidium salinum Turcz. ■
97731 Cnidium sinchianum K. T. Fu;辛加山蛇床;Xinjiashan Sakebed ■
97732 Cnidium suffruticosum (P. J. Bergius) Cham. et Schltdl. = Dasispermum suffruticosum (P. J. Bergius) B. L. Burtt ■☆
97733 Cnidium tachiroei (Franch. et Sav.) Makino = Ligusticum tachiroei (Franch. et Sav.) M. Hiroe et Constance ■
97734 Cnidium tilingia (Maxim.) Takeda = Ligusticum ajanense (Regel et Tiling) Koso-Pol. ■
97735 Cnidium tilingia Takeda = Ligusticum ajanense (Regel et Tiling) Koso-Pol. ■
97736 Cnidium yakushimense Masam. et Ohwi = Angelica longiradiata (Maxim.) Kitag. var. yakushimensis (Masam. et Ohwi) Kitag. ■☆
97737 Cnidone E. Mey. ex Endl. = Kissenia R. Br. ex Endl. ●☆
97738 Cnidoscolus Pohl(1827);荨麻刺属;Chilte ●☆
97739 Cnidoscolus aconitifolius I. M. Johnst.;乌头叶荨麻刺;Chaya, Treadsoftly ●☆
97740 Cnidoscolus elashcus Lundell;墨西哥荨麻刺;Chilte, Chilte Rubber ●☆
97741 Cnidoscolus stimulosus Engelm. et A. Gray;具柄荨麻刺;Bull Nettle, Spurge Nettle ●☆
97742 Cnidoscolus tehuacanensis Breckon;特瓦坎荨麻刺●☆
97743 Cnidume E. Mey. ex Walp. = Cnidone E. Mey. ex Endl. ●☆
97744 Cnopos Raf. = Polygonum L. (保留属名)■●
97745 Coa Adans. = Hippocratea L. ●☆
97746 Coa Mill. = Hippocratea L. ●☆
97747 Coalisia Raf. = Coalisina Raf. ●■
97748 Coalisina Raf. = Cleome L. ●■
97749 Coatesia F. Muell. = Geijera Schott ●☆
97750 Coaxana J. M. Coult. et Rose(1895);紫美芹属■☆
97751 Coaxana purpurea J. M. Coult. et Rose;紫美芹■☆
97752 Cobaea Cav. (1791);电灯花属;Cobaea ●■
97753 Cobaea Neck. = Lonicera L. ●■
97754 Cobaea macrostema Pav.;大花电灯花●☆
97755 Cobaea minor M. Martens et Galeotti;小电灯花●☆
97756 Cobaea pringlei Standl.;普林格尔电灯花●☆
97757 Cobaea scandens Cav.;电灯花;Cathedral Bells, Climbing Cobaea, Coral Bells, Cup and Saucer Vine, Cup-and-saucer Flower, Cup-and-saucer Vine, Jack Beanstalk, Mexican Ivy, Monastery Bells, Purplebell Cobaea, Purple-bell Cobaea, Violet Ivy ●
97758 Cobaeaceae D. Don = Polemoniaceae Juss. (保留科名)●■
97759 Cobaeaceae D. Don;电灯花科●■
97760 Cobamba Blanco = Canscora Lam. ■
97761 Cobana Ravenna(1974);科沃鸢尾属■☆
97762 Cobana guatemalensis (Standl.) Ravenna;科沃鸢尾■☆
97763 Cobananthus Wiehler = Alloplectus Mart. (保留属名)●■☆
97764 Cobananthus Wiehler(1977);科沃苣苔属■☆
97765 Cobananthus calochlamys (Donn. Sm.) Wiehler;科沃苣苔■☆
97766 Cobresia Pars. = Kobresia Willd. ■
97767 Cobresia Willd. = Kobresia Willd. ■
97768 Cobresia bellardii (All.) Degl.;贝拉尔肖嵩草■☆
97769 Cobresia caricina?;卡里亚肖嵩草;Carex-like Kobresia ■☆

97770 Cobresia filifolia (Turcz.) Meinsh.;丝叶肖嵩草■☆
97771 Cobresia humilis (C. A. Mey.) Serg.;低矮肖嵩草■☆
97772 Cobresia paniculata Meinsh.;圆锥肖嵩草■☆
97773 Cobresia royleana (Nees) W. Becker;洛伊肖嵩草■☆
97774 Cobresia schoenoides (C. A. Mey.) Steud.;拟舍恩肖嵩草■☆
97775 Cobresia simpliciuscula (Wahlenb.) Maack;单一肖嵩草■☆
97776 Coburgia Herb. = Leopoldia Parl.(保留属名)■☆
97777 Coburgia Sweet = Stenomesson Herb. ■☆
97778 Coccanthera C. Koch et Hanst. = Codonanthe (Mart.) Hanst.(保留属名)●■☆
97779 Coccanthera K. Koch et Hanst. = Codonanthe (Mart.) Hanst.(保留属名)●■☆
97780 Coccineorchis Schltr. = Stenorrhynchos Rich. ex Spreng. ■☆
97781 Coccinia Wight et Arn.(1834);红瓜属(狸红瓜属);Ivygourd ■
97782 Coccinia abyssinica (Lam.) Cogn.;阿比西尼亚红瓜■☆
97783 Coccinia adoensis (A. Rich.) Cogn.;阿多红瓜■☆
97784 Coccinia aostae Buscal. et Muschl. = Coccinia adoensis (A. Rich.) Cogn. ■☆
97785 Coccinia barteri (Hook. f.) Keay;巴特红瓜■☆
97786 Coccinia buettneriana Cogn.;比特纳红瓜■☆
97787 Coccinia buikoensis A. Zimm. = Coccinia microphylla Gilg ■☆
97788 Coccinia calantha Gilg = Coccinia grandiflora Cogn. ■☆
97789 Coccinia calophylla Harms = Coccinia schliebenii Harms ■☆
97790 Coccinia cordifolia (L.) Cogn.;心叶红瓜■☆
97791 Coccinia cordifolia (L.) Cogn. = Coccinia grandis (L.) Voigt ■
97792 Coccinia cordifolia (L.) Cogn. = Mukia maderaspatana (L.) M. Roem. ■
97793 Coccinia cordifolia sensu Cogn. = Coccinia grandis (L.) Voigt ■
97794 Coccinia decipiens (Hook. f.) Cogn. = Diplocyclos decipiens (Hook. f.) C. Jeffrey ■☆
97795 Coccinia djurensis Gilg = Coccinia adoensis (A. Rich.) Cogn. ■☆
97796 Coccinia ecirrhosa Cogn. = Cephalopentandra ecirrhosa (Cogn.) C. Jeffrey ■☆
97797 Coccinia engleri Gilg = Coccinia grandiflora Cogn. ■☆
97798 Coccinia fernandesiana C. Jeffrey;费尔南红瓜■☆
97799 Coccinia gabonensis Rabenant.;加蓬红瓜■☆
97800 Coccinia grandiflora Cogn.;大花红瓜■☆
97801 Coccinia grandis (L.) Voigt;红瓜(狸红瓜,印度红瓜);India Ivygourd, Indian Ivygourd, Ivy Gourd, Ivygourd ■
97802 Coccinia helenae Buscal. et Muschl. = Coccinia grandis (L.) Voigt ■
97803 Coccinia hirtella Cogn.;多毛红瓜■☆
97804 Coccinia homblei Cogn. = Coccinia adoensis (A. Rich.) Cogn. ■☆
97805 Coccinia indica Wight et Arn. = Coccinia grandis (L.) Voigt ■
97806 Coccinia jatrophifolia (A. Rich.) Cogn. = Coccinia adoensis (A. Rich.) Cogn. ■☆
97807 Coccinia keayana R. Fern.;凯伊红瓜■☆
97808 Coccinia kilimandjarica A. Zimm. = Coccinia trilobata (Cogn.) C. Jeffrey ■☆
97809 Coccinia kilimandjarica A. Zimm. var. subintegrifolia Harms = Coccinia trilobata (Cogn.) C. Jeffrey ■☆
97810 Coccinia longicarpa Jongkind;长果红瓜■☆
97811 Coccinia longipetiolata Chiov.;长梗红瓜■☆
97812 Coccinia mackennii (Naudin) Cogn. = Coccinia palmata (Sond.) Cogn. ■☆
97813 Coccinia macrocarpa Cogn. = Coccinia barteri (Hook. f.) Keay ■☆
97814 Coccinia megarrhiza C. Jeffrey;大根红瓜■☆
97815 Coccinia microphylla Gilg;小叶红瓜■☆
97816 Coccinia mildbraedii Harms;米尔德红瓜■☆
97817 Coccinia obbiadensis (Chiov.) Cufod. = Cephalopentandra ecirrhosa (Cogn.) C. Jeffrey ■☆
97818 Coccinia ovifera Dinter et Gilg = Coccinia rehmannii Cogn. ■☆
97819 Coccinia palmata (Sond.) Cogn.;掌叶红瓜■☆
97820 Coccinia palmatisecta Kotschy = Coccinia grandis (L.) Voigt ■
97821 Coccinia parvifolia Cogn. = Coccinia adoensis (A. Rich.) Cogn. ■☆
97822 Coccinia petersii Gilg = Eureiandra fasciculata (Cogn.) C. Jeffrey ■☆
97823 Coccinia polyantha Gilg = Eureiandra fasciculata (Cogn.) C. Jeffrey ■☆
97824 Coccinia princeae Gilg = Coccinia adoensis (A. Rich.) Cogn. ■☆
97825 Coccinia pubescens (Sond.) Cogn. ex Harms = Coccinia adoensis (A. Rich.) Cogn. ■☆
97826 Coccinia quercifolia Hutch. et E. A. Bruce = Cephalopentandra ecirrhosa (Cogn.) C. Jeffrey ■☆
97827 Coccinia quinqueloba (Thunb.) Cogn.;五裂红瓜■☆
97828 Coccinia racemiflora Rabenant.;总花红瓜■☆
97829 Coccinia rehmannii Cogn.;拉赫曼红瓜■☆
97830 Coccinia rehmannii Cogn. var. littoralis A. Meeuse = Coccinia rehmannii Cogn. ■☆
97831 Coccinia rigida Gilg = Coccinia adoensis (A. Rich.) Cogn. ■☆
97832 Coccinia roseiflora Suess. = Coccinia adoensis (A. Rich.) Cogn. ■☆
97833 Coccinia schinzii Cogn. = Coccinia sessilifolia (Sond.) Cogn. ■☆
97834 Coccinia schliebenii Harms;施利本红瓜■☆
97835 Coccinia senensis (Klotzsch) Cogn.;塞纳红瓜■☆
97836 Coccinia sessilifolia (Sond.) Cogn.;无柄叶红瓜■☆
97837 Coccinia stefaninii Chiov. = Dactyliandra stefaninii (Chiov.) C. Jeffrey ■☆
97838 Coccinia stolzii Harms = Diplocyclos decipiens (Hook. f.) C. Jeffrey ■☆
97839 Coccinia subglabra C. Jeffrey;近光红瓜■☆
97840 Coccinia subhastata Rabenant.;亚戟形红瓜■☆
97841 Coccinia subsessiliflora Cogn.;近无柄红瓜■☆
97842 Coccinia subspicata Cogn. = Coccinia adoensis (A. Rich.) Cogn. ■☆
97843 Coccinia trilobata (Cogn.) C. Jeffrey;三裂红瓜■☆
97844 Coccinia ulugurensis Harms;乌卢古尔红瓜■☆
97845 Coccinia variifolia A. Meeuse;异叶红瓜■☆
97846 Coccobryon Klotzsch = Piper L. ●■
97847 Coccobryon capense (L. f.) Miq. = Piper capense L. f. ●☆
97848 Coccoceras Miq. = Mallotus Lour. ●
97849 Coccochondra Rauschert(1982);脆果茜属☆
97850 Coccochondra laevis (Steyerm.) Rauschert;脆果茜☆
97851 Coccocipsilum P. Browne = Coccocypselum P. Browne(保留属名)●☆
97852 Coccocypselum P. Browne(1756)(保留属名);蜂巢茜属●☆
97853 Coccocypselum Sw. = Coccocypselum P. Browne(保留属名)●☆
97854 Coccocypselum crassifolium Standl.;厚叶蜂巢茜●☆
97855 Coccocypselum montanum Mart.;山地蜂巢茜●☆
97856 Coccocypselum rotundifolium Glaz.;圆叶蜂巢茜●☆
97857 Coccocypselum uniflorum Willd.;单花蜂巢茜●☆
97858 Coccoglochidion K. Schum. = Glochidion J. R. Forst. et G. Forst.

(保留属名)●
97859 Coccoloba L. = Coccoloba P. Browne(保留属名)●
97860 Coccoloba P. Browne ex L. = Coccoloba P. Browne(保留属名)●
97861 Coccoloba P. Browne(1756)(保留属名)('Coccolobis');海葡萄属;Sea Grape, Seagrape, Sea-grape ●
97862 Coccoloba diversifolia Jacq.;异叶海葡萄;Dove-plum, Pigeon-plum, Tie-tongue, Uvilla ●☆
97863 Coccoloba sagittifolia Ortega = Muehlenbeckia sagittifolia (Ortega) Meisn. ●☆
97864 Coccoloba schiedeana Lindau;希特海葡萄●☆
97865 Coccoloba uvifera (L.) L.;海葡萄(树蓼);Common Seagrape, Jamaican King, Jamaican Kino, Platter Leaf, Sea Grae, Sea Grape, Sea Grape Tree, Sea-grape, Seaside Grape, Shore-grape ●
97866 Coccolobaceae Barkley = Polygonaceae Juss.(保留科名)●■
97867 Coccolobis P. Browne = Coccoloba P. Browne(保留属名)●
97868 Coccomelia Reinw. = Baccaurea Lour. ●
97869 Coccomelia Ridl. = Angelcsia Korth. ●☆
97870 Coccomelia Ridl. = Licania Aubl. ●☆
97871 Cocconerion Baill.(1873);桃果大戟属●☆
97872 Cocconerion balansae Baill.;桃果大戟●☆
97873 Cocconerion minus Baill.;小桃果大戟●☆
97874 Coccos Gaertn. = Cocos L. ●
97875 Coccosipsilum Sw. = Coccocypselum P. Browne(保留属名)●☆
97876 Coccosperma Klotzsch = Blaeria L. ●☆
97877 Coccosperma Klotzsch = Erica L. ●☆
97878 Coccosperma areolatum N. E. Br. = Erica areolata (N. E. Br.) E. G. H. Oliv. ●☆
97879 Coccosperma forbesianum Klotzsch = Erica subcapitata (N. E. Br.) E. G. H. Oliv. ●☆
97880 Coccosperma hexandrum (Klotzsch) Druce = Erica subcapitata (N. E. Br.) E. G. H. Oliv. ●☆
97881 Coccosperma rugosum Klotzsch = Erica rugata E. G. H. Oliv. ●☆
97882 Coccosperma subcapitatum N. E. Br. = Erica subcapitata (N. E. Br.) E. G. H. Oliv. ●☆
97883 Coccothrinax Sarg.(1899);银扇葵属(可可棕榈属,银榈属,银扇棕属,银叶葵属,银叶棕属,银棕属);Silver Palm, Thatch Palm ●☆
97884 Coccothrinax alta Becc.;波多黎哥银扇葵(波多黎哥银棕)●☆
97885 Coccothrinax argentata (Jacq.) L. H. Bailey;大银扇葵(银白可可棕榈);Florida Silver Palm, Guano Palm, Silver Palm ●☆
97886 Coccothrinax argentea K. Schum.;银扇葵(银扇棕);Guano Palm, Latarier Balai ●☆
97887 Coccothrinax crinita Becc.;老人棕;Old Man Palm, Thatch Palm ●☆
97888 Coccothrinax ekmanii Burret;依可玛银扇葵(银扇葵)●☆
97889 Coccothrinax garberi (Chapm.) Sarg. = Coccothrinax argentata (Jacq.) L. H. Bailey ●☆
97890 Coccothrinax jucunda Sarg. = Coccothrinax argentata (Jacq.) L. H. Bailey ●☆
97891 Coccothrinax miraguano Becc.;银尖棕●☆
97892 Coccothrinax radiata (Schult. et Schult. f.) Sarg. ex K. Schum. = Thrinax radiata Lodd. ex Schult. et Schult. f. ●☆
97893 Cocculidium Spach = Cocculus DC.(保留属名)●
97894 Cocculus DC.(1817)(保留属名);木防己属;Coral Beads, Coral-beads, Red Moons, Snailseed, Snail-seed ●
97895 Cocculus abuta Kostel. = Abuta rufescens Aubl. ●☆
97896 Cocculus acuminatus DC. = Tiliacora racemosa Colebr. ●☆
97897 Cocculus acutus (Thunb.) Makino = Sinomenium acutum (Thunb.) Rehder et E. H. Wilson ●
97898 Cocculus affinis Oliv. = Diploclisia affinis (Oliv.) Diels ●
97899 Cocculus bakis A. Rich. = Tinospora bakis (A. Rich.) Miers ●☆
97900 Cocculus carolinus (L.) DC.;美国青藤;Carolina Moonseed, Carolina Snailseed, Coral Beads, Coral Vine, Coral-beads, Fishberry, Margil, Red-berried Moonseed ●☆
97901 Cocculus carolinus DC. = Cocculus carolinus (L.) DC. ●☆
97902 Cocculus cordifolius DC.;心叶木防己(心叶青藤);Cordateleaf Snailseed ●☆
97903 Cocculus cuneatus Benth. = Cocculus orbiculatus (L.) DC. ●
97904 Cocculus diversifolius DC.;异叶木防己;Correhuela, Sarsaparilla ●☆
97905 Cocculus diversifolius DC. = Sinomenium acutum (Thunb.) Rehder et E. H. Wilson ●
97906 Cocculus diversifolius DC. var. cinereus Diels = Sinomenium acutum (Thunb.) Rehder et E. H. Wilson ●
97907 Cocculus diversifolius Miq. = Sinomenium acutum (Thunb.) Rehder et E. H. Wilson ●
97908 Cocculus diversifolius Miq. var. cinereus Diels = Sinomenium acutum (Thunb.) Rehder et E. H. Wilson ●
97909 Cocculus forsteri DC. = Stephania forsteri (DC.) A. Gray ●■
97910 Cocculus forsteri DC. = Stephania japonica (Thunb.) Miers var. timoriensis (DC.) Forman ●■
97911 Cocculus glaucescens Blume = Diploclisia glaucescens (Blume) Diels ●
97912 Cocculus gomphioides DC. = Orthogynium gomphioides (DC.) Baill. ●☆
97913 Cocculus heterophyllus Hemsl. et E. H. Wilson = Sinomenium acutum (Thunb.) Rehder et E. H. Wilson ●
97914 Cocculus hirsutus (L.) Diels;硬毛木防己;Hirsute Snailseed ●☆
97915 Cocculus hirsutus (L.) W. Theob. = Cocculus hirsutus (L.) Diels ●☆
97916 Cocculus incanus Colebr. = Pericampylus glaucus (Lam.) Merr. ●
97917 Cocculus japonicus DC. var. timoriensis DC. = Stephania japonica (Thunb.) Miers var. timoriensis (DC.) Forman ●■
97918 Cocculus kunstleri King = Diploclisia glaucescens (Blume) Diels ●
97919 Cocculus laurifolius DC.;樟叶木防己(矮脚樟,桂叶防己,衡州乌药,九皮英,木防己,十八症,土巴戟,土牛入石,托食茶,乌药,消食树);Hindu Laurel, Laurelleaf Snailseed, Laurel-leaf Snailseed, Laurel-leaved Snailseed ●
97920 Cocculus leaeba (Delile) DC. = Cocculus pendulus (J. R. Forst. et G. Forst.) Diels ●☆
97921 Cocculus lenissimus Gagnep. = Cocculus orbiculatus (L.) DC. var. mollis (Wall. ex Hook. f. et Thomson) Hara ●
97922 Cocculus lenissimus Gagnep. = Syagrus romanzoffiana (Cham.) Glassman ●
97923 Cocculus lucidus Teijsm. et Binn. = Pycnarrhena lucida (Teijsm. et Binn.) Miq. ●
97924 Cocculus macranthus Hook. f. = Jateorhiza macrantha (Hook. f.) Exell et Mendonça ●☆
97925 Cocculus macrocarpus Wight = Diploclisia glaucescens (Blume) Diels ●
97926 Cocculus macrocarpus Wight et Arn. = Diploclisia glaucescens (Blume) Diels ●
97927 Cocculus mokiangensis W. Y. Lien = Cocculus orbiculatus (L.) DC. var. mollis (Wall. ex Hook. f. et Thomson) Hara ●
97928 Cocculus mokiangensis W. Y. Lien = Syagrus romanzoffiana

(Cham.) Glassman ●

97929 Cocculus mollis Wall. = Cocculus orbiculatus (L.) DC. var. mollis (Wall. ex Hook. f. et Thomson) Hara ●

97930 Cocculus mollis Wall. = Cocculus orbiculatus (L.) DC. ●

97931 Cocculus mollis Wall. ex Hook. f. et Thomson = Cocculus orbiculatus (L.) DC. var. mollis (Wall. ex Hook. f. et Thomson) Hara ●

97932 Cocculus mollis Wall. ex Hook. f. et Thomson = Syagrus romanzoffiana (Cham.) Glassman ●

97933 Cocculus orbiculatus (L.) DC.;木防己(白木香,百解薯,百解薯,匐茎木防己,黑皮青木香,华南木防己,青风藤,青檀香,青藤,青藤根,青藤香,少花匐茎木防己,铁牛入石,土木香,土牛入石,细叶铁牛入石,小暗消,小青藤,钻骨龙);Fewflower Snailseed,Orbicular Snailseed,Snailseed ●

97934 Cocculus orbiculatus (L.) DC. var. mollis (Wall. ex Hook. f. et Thomson) H. Hara;毛木防己(臭藤子,穿心龙,防己,金石榄,金钥匙,密毛木防己,墨江木防己,追风散);Densehairs Snailseed,Hairy Orbicular Snailseed,Mojiang Snailseed,Ofbicular Snailseed ●

97935 Cocculus orbiculatus (L.) DC. var. mollis (Wall. ex Hook. f. et Thomson) H. Hara = Cocculus orbiculatus (L.) DC. ●

97936 Cocculus pendulus (Forssk.) Diels = Cocculus pendulus (J. R. Forst. et G. Forst.) Diels ●☆

97937 Cocculus pendulus (J. R. Forst. et G. Forst.) Diels;下垂木防己(下垂防己);Dropin Snailseed ●☆

97938 Cocculus sarmentosus (Lour.) Diels;匐茎木防己(华南木防己,毛木防己,铁牛入石,细叶铁牛入石)●

97939 Cocculus sarmentosus (Lour.) Diels = Cocculus orbiculatus (L.) DC. ●

97940 Cocculus sarmentosus (Lour.) Diels var. linearis Gogelein = Cocculus orbiculatus (L.) DC. ●

97941 Cocculus sarmentosus (Lour.) Diels var. pauciflorus Y. C. Wu = Cocculus orbiculatus (L.) DC. ●

97942 Cocculus sarmentosus (Lour.) Diels var. stenophyllus Merr. = Cocculus orbiculatus (L.) DC. ●

97943 Cocculus thunbergii DC. = Cocculus orbiculatus (L.) DC. ●

97944 Cocculus thunbergii Forman = Cocculus orbiculatus (L.) DC. ●

97945 Cocculus tomentosus Colebr. = Tinospora sinensis (Lour.) Merr. ■

97946 Cocculus trilobus (Thunb. ex A. Murray) DC. = Cocculus orbiculatus (L.) DC. ●

97947 Cocculus trilobus (Thunb.) DC.;日本木防己(白山番薯,百蛇基,防己,鼓儿藤,黑皮青木香,金锁匙,木防己,牛木香,青风藤,青木香,青檀香,青藤,青藤香,土防己,土木香,乌龙,小葛藤,小青藤,钻骨龙,钻龙骨);Japan Snailseed,Japanese Snailseed ●

97948 Cocculus trilobus (Thunb.) DC. = Cocculus orbiculatus (L.) DC. ●

97949 Cocculus villosus DC.;长毛木防己●☆

97950 Cocculus villosus DC. = Cocculus hirsutus (L.) Diels ●☆

97951 Coccus Mill. = Cocos L. ●

97952 Coccyganthe Rchb. = Lychnis L.(废弃属名)■

97953 Cochemiea (K. Brandegee) Walton = Mammillaria Haw.(保留属名)●

97954 Cochemiea Walton = Mammillaria Haw.(保留属名)●

97955 Cochiseia W. H. Earle = Coryphantha (Engelm.) Lem.(保留属名)●■

97956 Cochiseia W. H. Earle = Escobaria Britton et Rose ●☆

97957 Cochiseia robbinsorum W. H. Earle = Coryphantha robbinsorum (W. H. Earle) Zimmerman ●☆

97958 Cochlanthera Choisy = Clusia L. ●☆

97959 Cochlanthera Choisy(1851);螺药藤黄属(匙花藤黄属)●☆

97960 Cochlanthera lanceolata Choisy;螺药藤黄●☆

97961 Cochlanthus Balf. f. = Socotranthus Kuntze ●☆

97962 Cochlanthus socotranus Balf. f.;螺花藤●☆

97963 Cochleanthes Raf.(1838);匙花兰属(壳花兰属)■☆

97964 Cochleanthes amazonica (Rchb. f. et Warsz.) R. E. Schult. et Garay;亚马逊匙花兰■☆

97965 Cochleanthes bidentata (Rchb. f. ex Hemsl.) R. E. Schult. et Garay;双齿匙花兰■☆

97966 Cochleanthes fragrans Raf.;香匙花兰■☆

97967 Cochlearia L.(1753);岩荠属(假山葵属,辣根菜属,辣根属,岩芥属);Cragcress,Scurvy Weed,Scurvy-grass,Scurvyweed,Spoonwort ■

97968 Cochlearia acaulis Desf. = Jonopsidium acaule (Desf.) Rchb. ■☆

97969 Cochlearia acutangula O. E. Schulz;锐棱岩荠(锐棱阴山荠,阴山荠);Acuteangular Cragcress,Acuteangular Scurvyweed ■

97970 Cochlearia acutangula O. E. Schulz = Yinshania acutangula (O. E. Schulz) Y. H. Zhang ■

97971 Cochlearia alatipes Hand.-Mazz.;翅柄岩荠(阿拉泰阴山荠);Winged Cragcress,Winged Scurvyweed ■

97972 Cochlearia alatipes Hand.-Mazz. = Cardamine fragarifolia O. E. Schulz ■

97973 Cochlearia alatipes Hand.-Mazz. = Yinshania alatipes (Hand.-Mazz.) Y. Z. Zhao ■

97974 Cochlearia alpina Sweet = Cochlearia officinalis L. ■

97975 Cochlearia altaica (C. A. Mey.) Hook. f. et T. Anderson = Taphrospermum altaicum C. A. Mey. ■

97976 Cochlearia anglica L.;英国岩荠;English Scurvy Grass,English Scurvy-grass,Sea Scurvy-grass ■☆

97977 Cochlearia arctica Schltdl.;北极辣根;Arctic Scurvyweed ■☆

97978 Cochlearia armoracia L. = Armoracia rusticana (Lam.) Gaertn., B. Mey. et Scherb. ■

97979 Cochlearia changhuaensis (Y. H. Zhang) L. L. Lou = Yinshania lichuanensis (Y. H. Zhang) Al-Shehbaz, G. Yang, L. L. Lu et T. Y. Cheo ■

97980 Cochlearia coronopus L. = Coronopus squamatus (Forssk.) Asch. ■☆

97981 Cochlearia danica L.;丹麦岩荠;Danish Scurvy-grass,Ivy-leaved Scurvy-grass,Stalked Scurvy-grass ■☆

97982 Cochlearia draba (L.) L. = Cardaria draba (L.) Desv. subsp. chalepensis (L.) O. E. Schulz ■

97983 Cochlearia formosana Hayata;台湾岩荠(河岸阴山荠,台湾假山葵);Taiwan Cragcress,Taiwan Scurvyweed ■

97984 Cochlearia formosana Hayata = Hilliella rivulorum (Dunn) Y. H. Zhang et H. W. Li ■

97985 Cochlearia formosana Hayata = Yinshania rivulorum (Dunn) Al-Shehbaz et al. ■

97986 Cochlearia fumarioides Dunn;紫堇叶岩荠;Fumaria-like Cragcress,Fumaria-like Scurvyweed ■

97987 Cochlearia fumarioides Dunn = Hilliella fumarioides (Dunn) Y. H. Zhang et H. W. Li ■

97988 Cochlearia fumarioides Dunn = Yinshania fumarioides (Dunn) Y. Z. Zhao ■

97989 Cochlearia furcatopilosa K. C. Kuan;叉毛岩荠(叉毛阴山荠);Forkedpilose Cragcress,Forkedpilose Scurvyweed ■

97990 Cochlearia furcatopilosa K. C. Kuan = Yinshania furcatopilosa

(K. C. Kuan) Y. H. Zhang ■

97991 Cochlearia glastifolia L.;光叶岩荠■☆

97992 Cochlearia glastifolia L. var. echinosperma Maire = Cochlearia glastifolia L. ■☆

97993 Cochlearia glastifolia L. var. megalosperma Maire = Cochlearia megalosperma (Maire) Vogt ■☆

97994 Cochlearia globosa (Turcz.) Ledeb. = Rorippa globosa (Turcz. ex Fisch. et C. A. Mey.) Hayek ■

97995 Cochlearia globosa Ledeb. = Rorippa globosa (Turcz. ex Fisch. et C. A. Mey.) Hayek ■

97996 Cochlearia groenlandica L.;北部岩荠;Northern Scurvy-grass, Scottish Scurvy-grass ■☆

97997 Cochlearia groenlandica L. = Cochlearia officinalis L. ■

97998 Cochlearia henryi (Oliv.) O. E. Schulz = Yinshania henryi (Oliv.) Y. H. Zhang ■

97999 Cochlearia henryi (Oliv.) O. E. Schulz var. wilsonii O. E. Schulz = Yinshania acutangula (O. E. Schulz) Y. H. Zhang subsp. wilsonii (O. E. Schulz) Al-Shehbaz, G. Yang, L. L. Lu et T. Y. Cheo ■

98000 Cochlearia himalaica Hook. f. et Thomson = Taphrospermum himalaicum (Hook. f. et Thomson) Al-Shehbaz et G. Yang ■

98001 Cochlearia hobsonii H. Pearson = Lignariella hobsonii (H. Pearson) Baehni ■

98002 Cochlearia hui O. E. Schulz;武功山岩荠;Wugongshan Cragcress, Wugongshan Scurvyweed ■

98003 Cochlearia hui O. E. Schulz = Hilliella hui (O. E. Schulz) Y. H. Zhang et H. W. Li ■

98004 Cochlearia hui O. E. Schulz = Yinshania hui (O. E. Schulz) Y. Z. Zhao ■

98005 Cochlearia integrifolia DC. = Eutrema integrifolium (DC.) Bunge ■

98006 Cochlearia islandica Pobed. = Cochlearia officinalis L. ■

98007 Cochlearia lapathifolia Gilib. = Armoracia rusticana (Lam.) Gaertn., B. Mey. et Scherb. ■

98008 Cochlearia lenensis Adams;伦岩荠■☆

98009 Cochlearia lichuanensis (Y. H. Zhang) L. L. Lou = Hilliella changhuaensis Y. H. Zhang var. lichuanensis (Y. H. Zhang) Y. H. Zhang ■

98010 Cochlearia lichuanensis (Y. H. Zhang) L. L. Lou = Yinshania lichuanensis (Y. H. Zhang) Al-Shehbaz, G. Yang, L. L. Lu et T. Y. Cheo ■

98011 Cochlearia longistyla (Y. H. Zhang) L. L. Lou = Hilliella changhuaensis Y. H. Zhang var. longistyla (Y. H. Zhang) Y. H. Zhang ■

98012 Cochlearia longistyla (Y. H. Zhang) L. L. Lou = Yinshania lichuanensis (Y. H. Zhang) Al-Shehbaz, G. Yang, L. L. Lu et T. Y. Cheo ■

98013 Cochlearia megalosperma (Maire) Vogt;大籽岩荠■☆

98014 Cochlearia micacea E. S. Marshall;苏格兰岩荠;Mountain Scurvy-grass, Scottish Scurvy-grass ■☆

98015 Cochlearia microcarpa K. C. Kuan;小果岩荠(小果阴山荠);Smallfruit Cragcress, Smallfruit Scurvyweed ■

98016 Cochlearia microcarpa K. C. Kuan = Yinshania acutangula (O. E. Schulz) Y. H. Zhang subsp. microcarpa (K. C. Kuan) Al-Shehbaz, L. L. Lu et T. Y. Cheo ■

98017 Cochlearia microcarpa K. C. Kuan = Yinshania microcarpa (K. C. Kuan) Y. H. Zhang ■

98018 Cochlearia nilotica Delile = Coronopus niloticus (Delile) Spreng. ■☆

98019 Cochlearia oblongifolia DC. = Cochlearia officinalis L. subsp. oblongifolia (DC.) Hultén ■☆

98020 Cochlearia officinalis L.;岩荠(高山岩荠,辣根,辣根菜,岩芥);Alpine Scurvy-grass, Common Scurvy Weed, Common Scurvygrass, Common Scurvy-grass, Common Scurvyweed, Cragcress, Crawlers, Mountain Scurvy-grass, Scrooby Grass, Scrubby Grass, Scruby Grass, Scurvy Grass, Scurvy-grass, Scurvyweed, Spoonwort ■

98021 Cochlearia officinalis L. alpina = Cochlearia officinalis L. ■

98022 Cochlearia officinalis L. subsp. oblongifolia (DC.) Hultén;矩圆叶岩荠■☆

98023 Cochlearia paradoxa (Hance) O. E. Schulz;卵叶岩荠;Ovateleaf Cragcress, Ovateleaf Scurvyweed ■

98024 Cochlearia paradoxa (Hance) O. E. Schulz = Hilliella paradoxa (Hance) Y. H. Zhang et H. W. Li ■

98025 Cochlearia paradoxa (Hance) O. E. Schulz = Yinshania paradoxa (Hance) Y. Z. Zhao ■

98026 Cochlearia paucifolia Hand.-Mazz. = Cardamine trifoliolata Hook. f. et Thomson ■

98027 Cochlearia pyrenaica DC.;核岩荠;Pyrenean Scurvy-grass ■☆

98028 Cochlearia rivulorum (Dunn) O. E. Schulz;河岸岩荠;Rivebank Cragcress, Rivebank Scurvyweed ■

98029 Cochlearia rivulorum (Dunn) O. E. Schulz = Yinshania rivulorum (Dunn) Al-Shehbaz et al. ■

98030 Cochlearia rupicola D. C. Zhang et J. Z. Shao;石生岩荠;Rock Scurvyweed ■

98031 Cochlearia rupicola D. C. Zhang et J. Z. Shao = Yinshania rupicola (D. C. Zhang et J. Z. Shao) Al-Shehbaz, G. Yang, L. L. Lu et T. Y. Cheo ■

98032 Cochlearia rusticana Lam. = Armoracia rusticana (Lam.) Gaertn., B. Mey. et Scherb. ■

98033 Cochlearia scapiflora Hook. f. et Thomson = Pegaeophyton scapiflorum (Hook. f. et Thomson) C. Marquand et Airy Shaw ■

98034 Cochlearia scoriarum (W. W. Sm.) Hand.-Mazz. = Cardamine fragarifolia O. E. Schulz ■

98035 Cochlearia scoriarum (W. W. Sm.) Hand.-Mazz. = Cardamine trifoliolata Hook. f. et Thomson ■

98036 Cochlearia scotica Druce = Cochlearia officinalis L. ■

98037 Cochlearia serpens W. W. Sm. = Lignariella serpens (W. W. Sm.) Al-Shehbaz et al. ■

98038 Cochlearia sinuata K. C. Kuan;弯缺岩荠;Sinuate Cragcress, Sinuate Scurvyweed ■

98039 Cochlearia sinuata K. C. Kuan = Hilliella sinuata (K. C. Kuan) Y. H. Zhang et H. W. Li ■

98040 Cochlearia sinuata K. C. Kuan = Yinshania sinuata (K. C. Kuan) Al-Shehbaz, G. Yang, L. L. Lu et T. Y. Cheo ■

98041 Cochlearia warburgii O. E. Schulz;浙江岩荠;Zhejiang Cragcress, Zhejiang Scurvyweed ■

98042 Cochlearia warburgii O. E. Schulz = Cochleariopsis warburgii (O. E. Schulz) L. L. Lu ■

98043 Cochlearia warburgii O. E. Schulz = Hilliella warburgii (O. E. Schulz) Y. H. Zhang et H. W. Li ■

98044 Cochlearia warburgii O. E. Schulz = Yinshania fumarioides (Dunn) Y. Z. Zhao ■

98045 Cochlearia wasabi Siebold = Eutrema wasabii (Siebold) Maxim. ■

98046 Cochleariella Y. H. Zhang et Vogt = Yinshania Ma et Y. Z. Zhao ●★

98047 Cochleariella Y. H. Zhang et Voigt = Hilliella (O. E. Schulz) Y. H. Zhang et H. W. Li ●★

98048 Cochleariella Y. H. Zhang et Voigt(1989);棒毛荠属(棒毛芥属)●★

98049 Cochleariella zhejiangensis (Y. H. Zhang) Y. H. Zhang et Vogt. = Yinshania fumarioides (Dunn) Y. Z. Zhao ■

98050 Cochleariella zhejiangensis (Y. H. Zhang) Y. H. Zhang et Voigt ex Y. H. Zhang et S. X. Cai = Hilliella fumarioides (Dunn) Y. H. Zhang et H. W. Li ■

98051 Cochleariopsis Á. Löve et D. Löve = Cochlearia L. ■

98052 Cochleariopsis Y. H. Zhang = Cochleariella Y. H. Zhang et Voigt ●★

98053 Cochleariopsis Y. H. Zhang = Yinshania Ma et Y. Z. Zhao ●★

98054 Cochleariopsis Y. H. Zhang(1985);拟棒毛荠属■

98055 Cochleariopsis groenlandica (L.) Á. Löve et D. Löve subsp. oblongifolia (DC.) Á. Löve et D. Löve = Cochlearia officinalis L. subsp. oblongifolia (DC.) Hultén ■☆

98056 Cochleariopsis warburgii (O. E. Schulz) L. L. Lu;拟棒毛荠■

98057 Cochleariopsis warburgii (O. E. Schulz) L. L. Lu = Hilliella warburgii (O. E. Schulz) Y. H. Zhang et H. W. Li ■

98058 Cochleariopsis warburgii (O. E. Schulz) L. L. Lu = Yinshania fumarioides (Dunn) Y. Z. Zhao ■

98059 Cochleariopsis zhejiangensis Y. H. Zhang = Cochleariopsis warburgii (O. E. Schulz) L. L. Lu ■

98060 Cochleariopsis zhejiangensis Y. H. Zhang = Hilliella fumarioides (Dunn) Y. H. Zhang et H. W. Li ■

98061 Cochleariopsis zhejiangensis Y. H. Zhang = Yinshania fumarioides (Dunn) Y. Z. Zhao ■

98062 Cochleata Medik. = Medicago L.(保留属名)●■

98063 Cochlia Blume = Bulbophyllum Thouars(保留属名)■

98064 Cochlianthus Benth. (1852);旋花豆属;Cochlianthus, Turnflowerbean ■

98065 Cochlianthus gracilis Benth.;细茎旋花豆(旋花豆,野老鼠豆);Gracile Cochlianthus, Slenderstem Turnflowerbean ■

98066 Cochlianthus gracilis Benth. var. brevipes C. F. Wei;短柄旋花豆;Shortstalk Turnflowerbean ■

98067 Cochlianthus montanus (Diels) Harms;高山旋花豆;Alpine Turnflowerbean, Montane Cochlianthus ■

98068 Cochliasanthus Trew = Caracalla Tod. ■

98069 Cochliasanthus Trew = Phaseolus L. ■

98070 Cochlidiosperma (Rchb.) Rchb. (1837);旋子草属■

98071 Cochlidiosperma (Rchb.) Rchb. = Veronica L. ■

98072 Cochlidiosperma lycica (E. Lehm.) D. Y. Hong et S. Nilsson;小亚细亚旋子草■☆

98073 Cochlidiosperma panormitana (Tineo ex Guss.) D. Y. Hong et S. Nilsson;巴勒摩旋子草■☆

98074 Cochlidiosperma sibthorpioides (Debeaux ex Degen et Herv.) D. Y. Hong et S. Nilsson;西班牙旋子草■☆

98075 Cochlidiosperma stamatiadae (M. Fisch. et Greuter) D. Y. Hong et S. Nilsson;希腊旋子草■☆

98076 Cochlidiosperma stewartii (Pennell) D. Y. Hong et S. Nilsson;斯氏旋子草■☆

98077 Cochlidiosperma sublobata (M. Fisch.) D. Y. Hong et S. Nilsson;浅裂旋子草■☆

98078 Cochlidiosperma trichadena (Jord. et Fourr.) D. Y. Hong et S. Nilsson;毛腺旋子草■☆

98079 Cochlidiosperma triloba (Opiz) D. Y. Hong et S. Nilsson;三裂旋子草■☆

98080 Cochlidiospermum Opiz = Cochlidiosperma (Rchb.) Rchb. ■

98081 Cochlidiospermum Opiz = Veronica L. ■

98082 Cochlidiospermum Rchb. = Veronica L. ■

98083 Cochlidosperma (Rchb.) Rchb. = Veronica L. ■

98084 Cochlidosperma Rchb. = Veronica L. ■

98085 Cochlioda Lindl. (1853);蜗牛兰属;Cochlioda ■☆

98086 Cochlioda brasiliensis Rolfe;巴西蜗牛兰■☆

98087 Cochlioda chasei D. E. Benn. et Christenson;扎斯蜗牛兰■☆

98088 Cochlioda densiflora Lindl.;密花蜗牛兰■☆

98089 Cochlioda noezliana Rolfe;诺氏蜗牛兰■☆

98090 Cochlioda rosea (Lindl.) Benth. et Hook. f.;粉红蜗牛兰■☆

98091 Cochlioda rosea Benth. = Cochlioda rosea (Lindl.) Benth. et Hook. f. ■☆

98092 Cochlioda sanguinea (Rchb. f.) Benth. et Hook. f.;血红蜗牛兰■☆

98093 Cochlioda sanguinea Benth. et Hook. f. = Cochlioda sanguinea (Rchb. f.) Benth. et Hook. f. ■☆

98094 Cochliopetalum Beer = Pitcairnia L'Hér.(保留属名)■☆

98095 Cochliospermum Lag. = Suaeda Forssk. ex J. F. Gmel.(保留属名)●■

98096 Cochliostema Lem. (1859);旋蕊草属(鹤蕊花属)■☆

98097 Cochliostema odoratissimum Lem.;旋蕊草■☆

98098 Cochlospermaceae Planch. (1847)(保留科名);弯籽木科(卷胚科,弯胚树科,弯子木科)●■☆

98099 Cochlospermaceae Planch.(保留科名) = Bixaceae Kunth(保留科名)●■

98100 Cochlospermum Kunth(1822)(保留属名);弯籽木属(黄花木棉属,卷胚属,弯胚树属);Shellseed ●☆

98101 Cochlospermum Post et Kuntze = Cochliospermum Lag. ●■

98102 Cochlospermum Post et Kuntze = Suaeda Forssk. ex J. F. Gmel.(保留属名)●■

98103 Cochlospermum angolense Welw. = Cochlospermum angolense Welw. ex Oliv. ●☆

98104 Cochlospermum angolense Welw. ex Oliv.;弯籽木(安哥拉卷胚,黄花木棉)●☆

98105 Cochlospermum fraseri Planch.;弗氏弯籽木(黄花木棉);Western Kapok Bush, Yellow Kapok ●☆

98106 Cochlospermum gossypium DC. = Cochlospermum religiosum (L.) Alston ●☆

98107 Cochlospermum incanum Robyns = Cochlospermum wittei Robyns subsp. incanum (Robyns) Poppend. ●☆

98108 Cochlospermum intermedium Mildbr.;间型弯籽木(间型木棉)●☆

98109 Cochlospermum niloticum Oliv. = Cochlospermum tinctorium Perr. ex A. Rich. ●☆

98110 Cochlospermum noldei Poppend.;诺尔德弯籽木(诺尔德木棉)●☆

98111 Cochlospermum planchonii Hook. f. = Cochlospermum planchonii Hook. f. ex Planch. ●☆

98112 Cochlospermum planchonii Hook. f. ex Planch.;普氏弯籽木(普氏黄花木棉,普氏卷胚,普氏旋子)●☆

98113 Cochlospermum religiosum (L.) Alston;棉叶弯籽木(棉叶黄花木棉,棉叶卷胚,棉叶旋子);Buttercup Tree, Karaya, Karaya Gum, Kutira Gum, Silk Cotton Tree, Silk-cotton Tree, White Silk Cotton, White Silkcotton ●☆

98114 Cochlospermum tinctorium Perr. = Cochlospermum tinctorium

Perr. ex A. Rich. ●☆

98115 Cochlospermum tinctorium Perr. ex A. Rich.;染色弯籽木(染色黄花木棉,染色卷胚,染色旋子)●☆

98116 Cochlospermum vitifolium (Willd.) Spreng.;葡萄叶弯籽木(葡萄叶黄花木棉);Silk Cottontree ●☆

98117 Cochlospermum wittei Robyns;维特弯籽木(维特黄花木棉)●☆

98118 Cochlospermum wittei Robyns subsp. incanum (Robyns) Poppend.;灰毛弯籽木(灰毛黄花木棉)●☆

98119 Cochlostemon Post et Kuntze = Cochliostema Lem. ■☆

98120 Cochranea Miers = Heliotropium L. ●■

98121 Cockaynea Zntov = Stenostachys Turcz. ■☆

98122 Cockaynea Zotov = Hystrix Moench ■

98123 Cockburnia Balf. f. = Poskea Vatke ●☆

98124 Cockburnia socotrana Balf. f. = Poskea socotrana (Balf. f.) G. Taylor ●☆

98125 Cockburnia somalensis Chiov. = Poskea socotrana (Balf. f.) G. Taylor ●☆

98126 Cockerellia (R. T. Clausen et Uhl) Á. Löve et D. Löve = Sedum L. ●■

98127 Cockerellia Á. Löve et D. Löve = Sedum L. ●■

98128 Cocleorchis Szlach. (1994);巴拿马兰属■☆

98129 Cocoaceae Schultz Sch. = Arecaceae Bercht. et J. Presl(保留科名)●

98130 Cocoaceae Schultz Sch. = Palmae Juss. (保留科名)●

98131 Cococipsilum J. St. -Hil. = Coccocypselum P. Browne(保留属名)●☆

98132 Cocoloba Raf. = Coccoloba P. Browne(保留属名)●

98133 Cocops O. F. Cook = Calyptronoma Griseb. ●☆

98134 Cocos L. (1753);椰子属(可可椰子属);Coconut, Coco-nut, Coconut Palm ●

98135 Cocos capitata Mart. = Butia capitata (Mart.) Becc. ●☆

98136 Cocos nucifera L.;椰子(古古椰子,可可椰子,无叶,椰瓢);Coco Nut Palm, Coconut, Coconut Palm, Coconut Tree, Cokar-nut, Coker Nut, Jelly Palm, Pindo Palm, Porcupine Wood ●

98137 Cocos nucifera L. 'Malay Dwarf';马来矮椰子●☆

98138 Cocos nucifera L. 'Nino';尼瑙椰子●☆

98139 Cocos nypa Lour. = Nypa fructicans Wurmb. ●◇

98140 Cocos oleracea Mart. = Syagrus oleracea (Mart.) Becc. ●☆

98141 Cocos plumosa Lodd. ex Hook. = Syagrus romanzoffiana (Cham.) Glassman ●

98142 Cocos regia Liebm. = Scheelea liebmannii Becc. ●☆

98143 Cocos romanzoffiana Champ. = Syagrus romanzoffiana (Cham.) Glassman ●

98144 Cocos weddellianum H. Wendl. = Lytocaryum weddellianum (H. Wendl.) Toledo ●☆

98145 Cocosaceae Schultz Sch. = Arecaceae Bercht. et J. Presl(保留科名)●

98146 Cocosaceae Schultz Sch. = Palmae Juss. (保留科名)●

98147 Cocoucia Aubl. = Combretum Loefl. (保留属名)●

98148 Codanthera Raf. = Salvia L. ●■

98149 Codaria Kuntze = Lerchea L. (保留属名)●■

98150 Codaria L. ex Benn. = Lerchea L. (保留属名)●■

98151 Codaria L. ex Kuntze = Lerchea L. (保留属名)●■

98152 Codariocalyx Hassk. (1842);舞草属(钟萼豆属);Codariocalyx, Danceweed, Dancing Grass, Telegraph Plant ●

98153 Codariocalyx Hassk. = Desmodium Desv. (保留属名)●■

98154 Codariocalyx gyrans (L. f.) Hassk. = Codariocalyx motorius (Houtt.) H. Ohashi ●

98155 Codariocalyx gyrans (L. f.) Hassk. = Desmodium motorium (Houtt.) Merr. ●

98156 Codariocalyx gyroides (Roxb. ex Link) Hassk.;圆叶舞草(短萼豆,毛荚舞草,无风独摇草);False Tick Trefoil, Pubescentfruit Codariocalyx, Roundleaf Codariocalyx, Roundleaf Danceweed, Round-leaved Danceweed ●

98157 Codariocalyx microphyllus (Thunb.) H. Ohashi;小叶舞草●☆

98158 Codariocalyx microphyllus (Thunb.) H. Ohashi = Desmodium microphyllum (Thunb.) A. DC. ■

98159 Codariocalyx motorius (Houtt.) H. Ohashi;舞草(电信草,红毛母鸡,红母鸡药,接骨草,无风独摇草,舞莪,枝叶壮阳草,钟萼豆,壮阳草);Common Codariocalyx, Common Danceweed, Longleaf Codariocalyx, Semaphore Plant, Telegraph Plant, Telegraph Tick Clover ●

98160 Codariocalyx motorius (Houtt.) H. Ohashi = Desmodium motorium (Houtt.) Merr. ●

98161 Codariocalyx motorius (Houtt.) H. Ohashi var. roylei? = Codariocalyx motorius (Houtt.) H. Ohashi ●

98162 Codarium Sol. ex Vahl = Dialium L. ●☆

98163 Codazzia H. Karst. et Triana = Delostoma D. Don ●☆

98164 Coddampulli Adans. = Cambogia L. ●

98165 Coddampulli Adans. = Garcinia L. ●

98166 Codda-Pana Adans. = Corypha L. ●

98167 Coddia Verdc. (1981);科德茜属☆

98168 Coddia rudis (E. Mey. ex Harv.) Verdc.;科德茜●☆

98169 Coddingtonia Bowdich(1825);科丁茜属☆

98170 Codebo Raf. = Codiaeum A. Juss. (保留属名)●

98171 Codia Forst. = Codia J. R. Forst. et G. Forst. ●☆

98172 Codia J. R. Forst. et G. Forst. (1775);无瓣火把树属●☆

98173 Codia discolor (Brongn. et Gris) Guillaumin;异色无瓣火把树●☆

98174 Codia ferruginea Brongn. et Gris;锈色无瓣火把树●☆

98175 Codia montana Labill. ex D. Don;山地无瓣火把树●☆

98176 Codiaceae Tiegh. = Cunoniaceae R. Br. (保留科名)●☆

98177 Codiaeum A. Juss. (1824)(保留属名);变叶木属;Chengingleaf Tree, Croton, Leafcroton, Leaf-croton, Seaside-balsam ●

98178 Codiaeum Rumph. ex A. Juss. = Codiaeum A. Juss. (保留属名)●

98179 Codiaeum pentzii Müll. Arg. = Blachia pentzii (Müll. Arg.) Benth. ●

98180 Codiaeum pictum (Lodd.) Hook. = Codiaeum variegatum (L.) A. Juss. ●

98181 Codiaeum variegatum (L.) A. Juss. 'Interruptum';间断变叶木●☆

98182 Codiaeum variegatum (L.) A. Juss. = Codiaeum variegatum (L.) Blume ●

98183 Codiaeum variegatum (L.) A. Juss. f. ambiguum Pax = Codiaeum variegatum (L.) Blume var. pictum (Lodd.) Müll. Arg. f. ambiguum Pax. ex Bailey ●☆

98184 Codiaeum variegatum (L.) A. Juss. f. appendiculatum Celak. = Codiaeum variegatum (L.) Blume var. pictum (Lodd.) Müll. Arg. f. appediculatum Celak ex Bailey ●☆

98185 Codiaeum variegatum (L.) A. Juss. f. cornutum André = Codiaeum variegatum (L.) Blume var. pictum (Lodd.) Müll. Arg. f. cornutum Andr. ex Bailey ●☆

98186 Codiaeum variegatum (L.) A. Juss. f. crispum? = Codiaeum variegatum (L.) Blume var. pictum (Lodd.) Müll. Arg. f. crispum DC. ●☆

98187 Codiaeum variegatum (L.) A. Juss. f. lobatum Pax = Codiaeum variegatum (L.) Blume var. pictum (Lodd.) Müll. Arg. f. lobatum Pax. ex Bailey ●☆

98188 Codiaeum variegatum (L.) A. Juss. f. taeniosum? = Codiaeum variegatum (L.) Blume var. pictum (Lodd.) Müll. Arg. f. taeniosum Bailey ●☆

98189 Codiaeum variegatum (L.) A. Juss. var. pictum (Lodd.) Müll. Arg. = Codiaeum variegatum (L.) A. Juss. ●

98190 Codiaeum variegatum (L.) Blume;变叶木(斑纹变叶木,长叶变叶木,大叶变叶木,戟叶变叶木,角叶变叶木,阔叶变叶木,螺旋木变叶木,母子叶变叶木,洒金榕,细叶变叶木);Chengingleaf Tree,Croton,Garden Croton,Garden Spotted-leaf,South Sea Laurel,Variegaled Laurel,Variegaled Leaf Croton,Variegated Leafcroton,Variegated Leaf-croton ●

98191 Codiaeum variegatum (L.) Blume var. pictum (Lodd.) Müll. Arg.;宽叶变叶木●

98192 Codiaeum variegatum (L.) Blume var. pictum (Lodd.) Müll. Arg. = Codiaeum variegatum (L.) Blume ●

98193 Codiaeum variegatum (L.) Blume var. pictum (Lodd.) Müll. Arg. f. ambiguum Pax. ex Bailey;长叶变叶木(长叶洒金榕,可疑变叶木)●

98194 Codiaeum variegatum (L.) Blume var. pictum (Lodd.) Müll. Arg. f. appediculatum Celak ex Bailey;母子叶变叶木(蜂腰变叶木,纤叶洒金榕,附属物变叶木)●

98195 Codiaeum variegatum (L.) Blume var. pictum (Lodd.) Müll. Arg. f. appediculatum Celak = Codiaeum variegatum (L.) Blume var. pictum (Lodd.) Müll. Arg. f. appediculatum Celak ex Bailey ●

98196 Codiaeum variegatum (L.) Blume var. pictum (Lodd.) Müll. Arg. f. cornutum Andr. ex Bailey;角叶变叶木●

98197 Codiaeum variegatum (L.) Blume var. pictum (Lodd.) Müll. Arg. f. crispum DC.;螺旋木变叶木(扭叶变叶木,皱波变叶木)●

98198 Codiaeum variegatum (L.) Blume var. pictum (Lodd.) Müll. Arg. f. lobatum Pax. ex Bailey;戟叶变叶木(浅裂变叶木)●

98199 Codiaeum variegatum (L.) Blume var. pictum (Lodd.) Müll. Arg. f. lobatum Pax. = Codiaeum variegatum (L.) Blume var. pictum (Lodd.) Müll. Arg. f. lobatum Pax. ex Bailey ●

98200 Codiaeum variegatum (L.) Blume var. pictum (Lodd.) Müll. Arg. f. platyphyllum Pax. ex Bailey;阔叶变叶木(宽叶洒金榕)●

98201 Codiaeum variegatum (L.) Blume var. pictum (Lodd.) Müll. Arg. f. platyphyllum Pax. = Codiaeum variegatum (L.) Blume var. pictum (Lodd.) Müll. Arg. f. platyphyllum Pax. ex Bailey ●

98202 Codiaeum variegatum (L.) Blume var. pictum (Lodd.) Müll. Arg. f. taeniosum Bailey;细叶变叶木(细叶洒金溶)●

98203 Codiaminum Raf. = Narcissus L. ■

98204 Codieum Raf. = Codiaeum A. Juss.(保留属名)●

98205 Codigi Augier = Sonerila Roxb.(保留属名)●■

98206 Codiocarpus R. A. Howard(1943);钟果茱萸属●☆

98207 Codiocarpus andamanicus (Kurz) R. A. Howard;钟果茱萸●☆

98208 Codiphus Raf. = Prismatocarpus L'Hér.(保留属名)●■☆

98209 Codivalia Raf. = Pupalia Juss.(保留属名)■☆

98210 Codochisma Raf. = Convolvulus L. ■●

98211 Codochonia Dunal = Acnistus Schott ●☆

98212 Codocline A. DC. = Xylopia L.(保留属名)●

98213 Codomale Raf. = Polygonatum Mill. ■

98214 Codon L. = Codon Royen ex L. ☆

98215 Codon Royen ex L.(1767);钟基麻属☆

98216 Codon dregei E. Mey. = Codon schenckii Schinz☆

98217 Codon luteum Marloth et Engl. = Codon schenckii Schinz☆

98218 Codon royenii L.;澳非钟基麻☆

98219 Codon schenckii Schinz;申克钟基麻☆

98220 Codonacanthus Nees (1847);钟花草属(刺针草属);Codonacanthus ■

98221 Codonacanthus acuminatus Nees = Codonacanthus pauciflorus (Nees) Nees ■

98222 Codonacanthus albo-nervosa (Hosok.) Yuen P. Yang;银脉钟花草(银脉爵床)■

98223 Codonacanthus pauciflorus (Nees) Nees;钟花草(刺针草,青木香草,针刺草);Fewflower Codonacanthus ■

98224 Codonacanthus spicatus Hand.-Mazz. = Leptostachya wallichii Nees ■

98225 Codonachne Steud. = Tetrapogon Desf. ■☆

98226 Codonachne Wight et Arn. ex Steud. = Chloris Sw. ●■

98227 Codonandra H. Karst. = Calliandra Benth.(保留属名)●

98228 Codonanthe (Mart.) Hanst.(1854)(保留属名);钟花苣苔属;Codonanthe ●■☆

98229 Codonanthe Hanst. = Codonanthe (Mart.) Hanst.(保留属名)●■☆

98230 Codonanthe Mart. ex Steud. = Codonanthe (Mart.) Hanst.(保留属名)●■☆

98231 Codonanthe Mart. ex Steud. = Hypocyrta Mart. ●■☆

98232 Codonanthe crassifolia (Focke) C. V. Morton;厚叶钟花苣苔;Codonanthe ●☆

98233 Codonanthe elegans Wiehler;雅致钟花苣苔;Elegant Codonanthe ●☆

98234 Codonanthe gracilis Hanst.;纤细钟花苣苔;Slender Codonanthe ●☆

98235 Codonanthemum Klotzsch = Eremia D. Don ●☆

98236 Codonanthemum discolor (Klotzsch) Benth. = Erica anguliger (N. E. Br.) E. G. H. Oliv. ●☆

98237 Codonanthemum parviflorum (Klotzsch) Klotzsch = Erica anguliger (N. E. Br.) E. G. H. Oliv. ●☆

98238 Codonanthemum puberulum (Klotzsch) Klotzsch = Erica anguliger (N. E. Br.) E. G. H. Oliv. ●☆

98239 Codonanthemum tenue Benth. = Erica anguliger (N. E. Br.) E. G. H. Oliv. ●☆

98240 Codonanthes Raf. = Pitcairnia L'Hér.(保留属名)■☆

98241 Codonanthopsis Mansf.(1934);拟钟花苣苔属■●☆

98242 Codonanthopsis huebneri Mansf.;拟钟花苣苔■●☆

98243 Codonanthus G. Don(废弃属名) = Breweria R. Br. ●☆

98244 Codonanthus G. Don(废弃属名) = Codonanthe (Mart.) Hanst.(保留属名)●■☆

98245 Codonanthus Hassk. = Physostelma Wight ●☆

98246 Codonanthus africanus G. Don = Calycobolus africanus (G. Don) Heine ●☆

98247 Codonanthus alternifolius Planch. = Calycobolus africanus (G. Don) Heine ●☆

98248 Codonechites Markgr. = Odontadenia Benth. ●☆

98249 Codonemma Miers = Tabernaemontana L. ●

98250 Codonium Vahl = Schoepfia Schreb. ●

98251 Codonoboea Ridl. = Didymocarpus Wall.(保留属名)●■

98252 Codonocalyx Klotzsch ex Baill. = Croton L. ●

98253 Codonocalyx Miers = Psychotria L.(保留属名)●

98254 Codonocalyx Miers = Suteria DC. ●

98255 Codonocarpus A. Cunn. = Codonocarpus A. Cunn. ex Endl. ●☆

98256 Codonocarpus A. Cunn. ex Endl.(1837);钟果木属(铃果属);Bellfruit-tree ●☆

98257 Codonocarpus Endl. = Codonocarpus A. Cunn. ex Endl. ●☆

98258 Codonocarpus cotinfolius (Desf.) F. Muell.;黄栌叶钟果木(黄栌叶铃果);Horseradish-tree ●☆

98259 Codonocephalum Fenzl = Inula L. ●■

98260 Codonocephalum Fenzl(1843);西亚菊属■☆

98261 Codonocephalum grande (Schrenk) O. Fedtsch. et B. Fedtsch.;西亚菊■☆

98262 Codonocephalum grande (Schrenk.) O. Fedtsch. et B. Fedtsch. = Inula grandis Schrenk ex Fisch. et C. A. Mey. ■

98263 Codonocephalum peacockeanum Aitch. et Hemsl.;拍氏西亚菊■☆

98264 Codonocephalum serratuloides Gilli;锯齿西亚菊■☆

98265 Codonocephalus Fenzl = Inula L. ●■

98266 Codonochlamys Ulbr.(1915);钟被锦葵属☆

98267 Codonochlamys glaziovii Ulbr.;钟被锦葵☆

98268 Codonocrinum Willd. ex Schult. = Yucca L. ●■

98269 Codonocrinum Willd. ex Schult. f. = Yucca L. ●■

98270 Codonocroton E. Mey. ex Engl. et Diels = Combretum Loefl.(保留属名)●

98271 Codonophora Lindl. = Paliavana Vell. ex Vand. ●☆

98272 Codonoprasum Rchb. = Allium L. ■

98273 Codonopsis Wall.(1824);党参属(山奶草属,羊乳属);Asia Bell,Asiabell,Asian Bell,Bellwort,Bonnet Bellflower ■

98274 Codonopsis Wall. ex Roxb. = Codonopsis Wall. ■

98275 Codonopsis accrescenticalyx H. Lév. = Codonopsis tubulosa Kom. ■

98276 Codonopsis afiinis Hook. f. et Thomson;大叶党参(近缘党参);Bigleaf Asiabell ■

98277 Codonopsis albiflora Griff. = Campanumoea lancifolia (Roxb.) Merr. ■

98278 Codonopsis alpina Nannf.;高山党参;Alpine Asiabell ■

98279 Codonopsis argentea P. C. Tsoong;银背叶党参;Argent Asiabell,Siverback Asiabell ■

98280 Codonopsis benthamii Hook. f. et Thomson;大萼党参■

98281 Codonopsis bicolor Nannf.;二色党参;Twocolored Asiabell ■

98282 Codonopsis bicolor Nannf. = Codonopsis viridiflora Maxim. ■

98283 Codonopsis bodinieri H. Lév. = Codonopsis lanceolata (Siebold et Zucc.) Benth. et Hook. f. ■●

98284 Codonopsis bulleyana Forrest ex Diels;管钟党参;Bulley Asiabell ■

98285 Codonopsis canescens Nannf.;灰毛党参(北路蛇头党,甘孜党,灰白叶党参,蛇头党,紫党);Greyhair Asiabell ■

98286 Codonopsis cardiophylla Diels ex Kom.;光叶党参(臭参,大头党参,高山党参,小人参,心叶党参);Glabrousleaf Asiabell ■

98287 Codonopsis celabica (Blume) Miq. = Campanumoea celebica Blume ■

98288 Codonopsis chimillensis Anthony;滇缅党参(缅甸党参);N. Burma Asiabell,Yunnan-burma Asiabell ■

98289 Codonopsis chlorocodon C. Y. Wu;绿钟党参(臭绿钟参);Green Asiabell ■

98290 Codonopsis clematidea (Schrenk) C. B. Clarke;铁线莲状党参(新疆党参);Bonnet Bellflower,Clematis Asiabell,Climbing Bellflower ■

98291 Codonopsis clematidea (Schrenk) C. B. Clarke var. obtusa (Chipp) Kitam. = Codonopsis obtusa (Chipp) Nannf. ■☆

98292 Codonopsis convolvulacea Kurz;鸡蛋参(白地瓜,补血草,金线吊葫芦,金线壶卢,金线葫芦,牛尾参,山鸡蛋);Convolvulate Asiabell ■

98293 Codonopsis convolvulacea Kurz subsp. forrestii (Diels) D. Y. Hong et L. M. Ma;鸡腰参(白地瓜,大金线吊葫芦,珠儿参,珠子参);Forrest Asiabell ■

98294 Codonopsis convolvulacea Kurz subsp. vinciflora (Kom.) D. Y. Hong;薄叶鸡蛋参;Thinleaf Asiabell ■

98295 Codonopsis convolvulacea Kurz var. efilamentosa (W. W. Sm.) L. T. Shen;心叶鸡蛋参(缺花党参,心叶珠子参);Heartleaf Convolvulate Asiabell ■

98296 Codonopsis convolvulacea Kurz var. efilamentosa (W. W. Sm.) L. T. Shen = Codonopsis efilamentosa W. W. Sm. ■

98297 Codonopsis convolvulacea Kurz var. forrestii (Diels) Ballard = Codonopsis convolvulacea Kurz subsp. forrestii (Diels) D. Y. Hong et L. M. Ma ■

98298 Codonopsis convolvulacea Kurz var. heterophylla C. Y. Wu = Codonopsis convolvulacea Kurz subsp. vinciflora (Kom.) D. Y. Hong ■

98299 Codonopsis convolvulacea Kurz var. hirsuta (Hand.-Mazz.) Nannf.;毛叶鸡蛋参(兰花参,獭头参,土党参);Hairleaf Convolvulate Asiabell,Hirsuteleaf Asiabell ■

98300 Codonopsis convolvulacea Kurz var. hirsuta (Hand.-Mazz.) Nannf. = Codonopsis hirsuta (Hand.-Mazz.) D. Y. Hong et L. M. Ma ■

98301 Codonopsis convolvulacea Kurz var. limprichtii (Lingelsh. et Borza) Anthony;直立鸡蛋参;Erect Asiabell ■

98302 Codonopsis convolvulacea Kurz var. multifolia?;多叶鸡蛋参■☆

98303 Codonopsis convolvulacea Kurz var. pinifolia (Hand.-Mazz.) Nannf. = Codonopsis graminifolia H. Lév. ■

98304 Codonopsis convolvulacea Kurz var. typica Anthony;山鸡蛋■☆

98305 Codonopsis convolvulacea Kurz var. vinciflora (Kom.) L. T. Shen = Codonopsis convolvulacea Kurz subsp. vinciflora (Kom.) D. Y. Hong ■

98306 Codonopsis cordata Hassk. = Campanumoea javanica Blume ■

98307 Codonopsis cordifolia Kom. = Campanumoea javanica Blume ■

98308 Codonopsis cordifolloidea P. C. Tsoong;心叶党参(拟心叶党参);Cordateleaflet Asiabell,Heartleaf Asiabell ■

98309 Codonopsis deltoidea Chipp;三角叶党参(白党参,泡参,三角党参,土党参);Deltoid Asiabell ■

98310 Codonopsis dicentrifolia (C. B. Clarke) W. W. Sm.;珠峰党参;Bleedingheartleaf Asiabell,Jolmo Lungma Asiabell ■

98311 Codonopsis efilamentosa W. W. Sm. = Codonopsis convolvulacea Kurz var. efilamentosa (W. W. Sm.) L. T. Shen ■

98312 Codonopsis elematidea (Schrenk) C. B. Clarke;新疆党参;Xinjiang Asiabell ■

98313 Codonopsis farreri Anthony var. grandiflora S. H. Huang;大花秃叶党参;Bigflower Farrer Asiabell,Largeflower Farrer Asiabell ■

98314 Codonopsis farreri J. Anthony;秃叶党参;Farrer Asiabell ■

98315 Codonopsis farreri J. Anthony var. grandiflora S. H. Huang = Codonopsis farreri J. Anthony ■

98316 Codonopsis foetens Hook. f. et Thomson;臭党参;Fetid Asiabell,Stink Asiabell ■

98317 Codonopsis foetens Hook. f. et Thomson var. major Hand.-Mazz. = Codonopsis alpina Nannf. ■

98318 Codonopsis forrestii Diels = Codonopsis convolvulacea Kurz subsp. forrestii (Diels) D. Y. Hong et L. M. Ma ■

98319 Codonopsis forrestii Diels = Codonopsis convolvulacea Kurz var. forrestii (Diels) Ballard ■

98320 Codonopsis forrestii Diels var. heterophylla C. Y. Wu;异叶珠子参;Diverseleaf Asiabell ■

98321 Codonopsis forrestii Diels var. hirsuta P. C. Tsoong et L. T. Shen; 毛叶珠子参; Hairleaf Stink Asiabell ■

98322 Codonopsis forrestii Diels var. hirsuta P. C. Tsoong et L. T. Shen = Codonopsis convolvulacea Kurz var. forrestii (Diels) Ballard ■

98323 Codonopsis glaberrima Nannf. = Codonopsis pilosula (Franch.) Nannf. ■

98324 Codonopsis glaberrima Nannf. = Codonopsis pilosula (Franch.) Nannf. var. modesta (Nannf.) L. T. Shen ■

98325 Codonopsis gombalana C. Y. Wu; 贡山党参(贡山臭参); Gaoligong Asiabell, Gongshan Asiabell ■

98326 Codonopsis gracilis Hook. f. = Leptocodon gracilis (Hook. f. et Thomson) Lem. ■●

98327 Codonopsis graminifolia H. Lév.; 松叶鸡蛋参(松叶党参); Pineleaf Asiabell ■

98328 Codonopsis graminifolia H. Lév. = Codonopsis convolvulacea Kurz var. pinifolia (Hand. -Mazz.) Nannf. ■

98329 Codonopsis handeliana Nannf. = Codonopsis pilosula (Franch.) Nannf. subsp. handeliana (Nannf.) D. Y. Hong et L. M. Ma ■

98330 Codonopsis handeliana Nannf. = Codonopsis pilosula (Franch.) Nannf. var. handeliana (Nannf.) L. T. Shen ■

98331 Codonopsis henryi Oliv.; 川鄂党参; Henry Asiabell ■

98332 Codonopsis hirsuta (D. Y. Hong) K. E. Morris et Lammers = Leptocodon hirsutus D. Y. Hong ●

98333 Codonopsis hirsuta (Hand. -Mazz.) D. Y. Hong et L. M. Ma = Codonopsis convolvulacea Kurz var. hirsuta (Hand. -Mazz.) Nannf. ■

98334 Codonopsis hongii Lammers = Leptocodon hirsutus D. Y. Hong ●

98335 Codonopsis inflata Hook. f. = Campanumoea inflata (Hook. f.) C. B. Clarke ●

98336 Codonopsis japonica Maxim. = Campanumoea javanica Blume var. japonica Makino ■

98337 Codonopsis javanica (Blume) Hook. f. = Campanumoea javanica Blume ■

98338 Codonopsis javanica (Blume) Miq. subsp. japonica (Maxim. ex Makino) Lammers = Campanumoea javanica Blume var. japonica Makino ■

98339 Codonopsis javanica (Blume) Miq. var. japonica Makino = Campanumoea javanica Blume var. japonica Makino ■

98340 Codonopsis javanica (Blume) Miq. var. japonica Maxim. ex Makino = Codonopsis javanica (Blume) Miq. subsp. japonica (Maxim. ex Makino) Lammers ■

98341 Codonopsis kawakamii Hayata; 台湾党参(玉山山奶草); Kawakami Asiabell ■

98342 Codonopsis labordei H. Lév. = Campanumoea javanica Blume ■

98343 Codonopsis lanceolata (Siebold et Zucc.) Benth. et Hook. f.; 羊乳(白河车,白蟒肉,地黄,狗头参,轮叶党参,蔓参,奶参,奶奶头,奶薯,奶树,牛奶子,乳薯,乳树,乳头薯,山海螺,山胡萝卜,四叶参,天海螺,通乳草,土党参,土洋参,羊奶,羊奶参); Lance Asiabell ■●

98344 Codonopsis lanceolata (Siebold et Zucc.) Trautv. = Campanumoea javanica Blume ■

98345 Codonopsis lanceolata (Siebold et Zucc.) Trautv. f. emaculata (Honda) H. Hara; 无斑羊乳■☆

98346 Codonopsis lanceolata (Siebold et Zucc.) Trautv. var. omurae T. Koyama; 小村党参■☆

98347 Codonopsis lanceolata (Siebold et Zucc.) Trautv. var. ussuriensis Trautv. = Codonopsis ussuriensis (Rupr. et Maxim.) Hemsl. ■

98348 Codonopsis lancifolia (Roxb.) Moeliono = Campanumoea lancifolia (Roxb.) Merr. ■

98349 Codonopsis lancifolia (Roxb.) Moeliono = Cyclocodon lancifolius (Roxb.) Kurz ■

98350 Codonopsis lancifolia (Roxb.) Moeliono subsp. celabica Moeliono = Campanumoea celebica Blume ■

98351 Codonopsis lancifolia (Roxb.) Moeliono subsp. celebica (Blume) Moeliono = Cyclocodon celebicus (Blume) D. Y. Hong ■

98352 Codonopsis levicalyx L. T. Shen; 光萼党参(五花党参); Smoothcalyx Asiabell ■

98353 Codonopsis levicalyx L. T. Shen = Codonopsis henryi Oliv. ■

98354 Codonopsis levicalyx L. T. Shen var. hirsuticalyx L. T. Shen; 线党参; Hirsutecalyx Asiabell ■

98355 Codonopsis levicalyx L. T. Shen var. hirsuticalyx L. T. Shen = Codonopsis henryi Oliv. ■

98356 Codonopsis limprichtii Lingelsh. et Borza = Codonopsis convolvulacea Kurz var. limprichtii (Lingelsh. et Borza) Anthony ■

98357 Codonopsis limprichtii Lingelsh. et Borza var. hirsuta Hand. -Mazz. = Codonopsis convolvulacea Kurz var. hirsuta (Hand. -Mazz.) Nannf. ■

98358 Codonopsis limprichtii Lingelsh. et Borza var. hirsuta Hand. -Mazz. = Codonopsis hirsuta (Hand. -Mazz.) D. Y. Hong et L. M. Ma ■

98359 Codonopsis limprichtii Lingelsh. et Borza var. pinifolia Hand. -Mazz. = Codonopsis convolvulacea Kurz var. pinifolia (Hand. -Mazz.) Nannf. ■

98360 Codonopsis limprichtii Lingelsh. et Borza var. pinifolia Hand. -Mazz. = Codonopsis graminifolia H. Lév. ■

98361 Codonopsis lomonosovoi?; 劳氏党参■☆

98362 Codonopsis longifolia D. Y. Hong; 长叶党参; Longleaf Asiabell ■

98363 Codonopsis macrantha Nannf. = Codonopsis nervosa (Chipp) Nannf. subsp. macrantha (Nannf.) D. Y. Hong et L. M. Ma ■

98364 Codonopsis macrantha Nannf. = Codonopsis nervosa (Chipp) Nannf. var. macrantha (Nannf.) L. T. Shen ■

98365 Codonopsis macrocalyx Diels; 巨萼党参(大萼党参,党参,线党); Bigcalyx Asiabell ■

98366 Codonopsis macrocalyx Diels = Codonopsis benthamii Hook. f. et Thomson ■

98367 Codonopsis macrocalyx Diels var. coerulescens Hand. -Mazz. = Codonopsis tubulosa Kom. ■

98368 Codonopsis macrocalyx Diels var. parviloba J. Anthony = Codonopsis macrocalyx Diels ■

98369 Codonopsis mairei H. Lév. = Codonopsis convolvulacea Kurz subsp. forrestii (Diels) D. Y. Hong et L. M. Ma ■

98370 Codonopsis mairei H. Lév. = Codonopsis convolvulacea Kurz var. forrestii (Diels) Ballard ■

98371 Codonopsis maximowiczi Honda = Campanumoea javanica Blume var. japonica Makino ■

98372 Codonopsis meleagris Diels; 珠鸡斑党参; Guineafowl Asiabell ■

98373 Codonopsis micrantha Chipp; 小花党参(臭党参,党参,理党参,土党参,细条党参,野党参); Smallflower Asiabell ■

98374 Codonopsis microtubulosa Z. T. Wang et G. J. Xu; 小管花党参; Smalltube Asiabell ■

98375 Codonopsis microtubulosa Z. T. Wang et G. J. Xu = Codonopsis pilosula (Franch.) Nannf. ■

98376 Codonopsis modesta Nannf. = Codonopsis pilosula (Franch.) Nannf. var. modesta (Nannf.) L. T. Shen ■

98377 Codonopsis modesta Nannf. = Codonopsis pilosula (Franch.) Nannf. ■

98378 Codonopsis mollis Chipp = Codonopsis thalictrifolia Oliv. var. mollis (Chipp) L. T. Shen ■

98379 Codonopsis mollis Chipp = Codonopsis thalictrifolia Wall. ■

98380 Codonopsis nervosa (Chipp) Nannf.;脉花党参(柴党,柴党参,党参,绿花党参,紫党参);Nerved Asiabell ■

98381 Codonopsis nervosa (Chipp) Nannf. subsp. macrantha (Nannf.) D. Y. Hong et L. M. Ma;大花党参(柴党,大头党参,泡参,狮头党);Bigflower Asiabell ■

98382 Codonopsis nervosa (Chipp) Nannf. var. macrantha (Nannf.) L. T. Shen = Codonopsis nervosa (Chipp) Nannf. subsp. macrantha (Nannf.) D. Y. Hong et L. M. Ma ■

98383 Codonopsis nervosa (Chipp) Nannf. var. macrentha (Nannf.) L. T. Shen = Codonopsis nervosa (Chipp) Nannf. subsp. macrantha (Nannf.) D. Y. Hong et L. M. Ma ■

98384 Codonopsis obtusa (Chipp) Nannf.;粗壮党参■☆

98385 Codonopsis ovata Benth.;卵叶党参■☆

98386 Codonopsis ovata Benth. = Codonopsis clematidea (Schrenk) C. B. Clarke ■

98387 Codonopsis ovata Benth. var. cuspidata Chipp = Codonopsis clematidea (Schrenk) C. B. Clarke ■

98388 Codonopsis ovata Benth. var. nervosa Chipp = Codonopsis nervosa (Chipp) Nannf. ■

98389 Codonopsis ovata Benth. var. obtusa Chipp = Codonopsis clematidea (Schrenk) C. B. Clarke ■

98390 Codonopsis ovata Benth. var. obtusa Chipp = Codonopsis obtusa (Chipp) Nannf. ■☆

98391 Codonopsis parviflora Wall. ex A. DC. = Campanumoea parviflora (Wall. ex A. DC.) Benth. et Hook. f. ■

98392 Codonopsis parviflora Wall. ex A. DC. = Cyclocodon parviflorus (Wall. ex A. DC.) Hook. f. et Thomson ■

98393 Codonopsis pianmaensis S. H. Huang;片马党参;Pianma Asiabell ■

98394 Codonopsis pilosa Chipp = Codonopsis tubulosa Kom. ■

98395 Codonopsis pilosula (Franch.) Nannf.;党参(东党,凤党,黄参,晶党,口党,潞党,上党人参,狮头参,台参,台党防党,纹党,西党,仙草根,叶子菜,中灵草);Pilose Asiabell,Tangshen ■

98396 Codonopsis pilosula (Franch.) Nannf. ar. handeliana (Nannf.) L. T. Shen = Codonopsis pilosula (Franch.) Nannf. subsp. handeliana (Nannf.) D. Y. Hong et L. M. Ma ■

98397 Codonopsis pilosula (Franch.) Nannf. subsp. handeliana (Nannf.) D. Y. Hong et L. M. Ma;闪毛党参(小叶党参);Handel Asiabell ■

98398 Codonopsis pilosula (Franch.) Nannf. var. glaberrima (Nannf.) P. C. Tsoong = Codonopsis pilosula (Franch.) Nannf. ■

98399 Codonopsis pilosula (Franch.) Nannf. var. glaberrima (Nannf.) P. C. Tsoong = Codonopsis pilosula (Franch.) Nannf. var. modesta (Nannf.) L. T. Shen ■

98400 Codonopsis pilosula (Franch.) Nannf. var. handeliana (Nannf.) L. T. Shen = Codonopsis pilosula (Franch.) Nannf. subsp. handeliana (Nannf.) D. Y. Hong et L. M. Ma ■

98401 Codonopsis pilosula (Franch.) Nannf. var. modesta (Nannf.) L. T. Shen;素花党参(党参,晶党,南平党参,文元党,西党);Moderate Asiabell ■

98402 Codonopsis pilosula (Franch.) Nannf. var. modesta (Nannf.) L. T. Shen = Codonopsis pilosula (Franch.) Nannf. ■

98403 Codonopsis pilosula (Franch.) Nannf. var. volubilis (Nannf.) L. T. Shen;缠绕党参(白党,臭党,茂党,肉党,甜党);Twisting Asiabell ■

98404 Codonopsis pilosula (Franch.) Nannf. var. volubilis (Nannf.) L. T. Shen = Codonopsis pilosula (Franch.) Nannf. ■

98405 Codonopsis purpurea Wall.;紫花党参(岩人参);Purpleflower Asiabell ■

98406 Codonopsis retroserrata Z. T. Wang et G. J. Xu;倒齿党参;Retroserrate Asiabell ■

98407 Codonopsis retroserrata Z. T. Wang et G. J. Xu = Codonopsis convolvulacea Kurz subsp. forrestii (Diels) D. Y. Hong et L. M. Ma ■

98408 Codonopsis rosulata W. W. Sm.;莲座状党参;Rosette Asiabell, Rosulate Asiabell ■

98409 Codonopsis rotundifolia (Royle) Benth. = Codonopsis pilosula (Franch.) Nannf. var. handeliana (Nannf.) L. T. Shen ■

98410 Codonopsis rotundifolia Benth. = Codonopsis pilosula (Franch.) Nannf. var. handeliana (Nannf.) L. T. Shen ■

98411 Codonopsis silvestris Kom. = Codonopsis pilosula (Franch.) Nannf. var. handeliana (Nannf.) L. T. Shen ■

98412 Codonopsis silvestris Kom. = Codonopsis pilosula (Franch.) Nannf. ■

98413 Codonopsis subglobosa W. W. Sm.;球花党参(柴党,臭参,甘孜党,康南根,南路蛇头党,蛇头党);Subglobose Asiabell ■

98414 Codonopsis subscaposa Kom.;抽葶党参(党参,康南党,野党参);Subscapose Asiabell ■

98415 Codonopsis subsimplex Hook. f. et Thomson;藏南党参(近单一党参);South Tibet Asiabell ■

98416 Codonopsis sylvestris Kom. = Codonopsis pilosula (Franch.) Nannf. ■

98417 Codonopsis tangshen Oliv.;川党参(板党,单支党,党参,东党参,条党,巫山党,禹党);Sichuan Asiabell,Szechuan Asiabell ■

98418 Codonopsis thalictrifolia Oliv. var. mollis (Chipp) L. T. Shen = Codonopsis thalictrifolia Wall. ■

98419 Codonopsis thalictrifolia Wall.;唐松草党参;Meadowrueleaf Asiabell ■

98420 Codonopsis thalictrifolia Wall. var. mollis (Chipp) L. T. Shen;长花党参(藏党参);Longflower Asiabell ■

98421 Codonopsis truncata Wall. ex A. DC. = Cyclocodon lancifolius (Roxb.) Kurz ■

98422 Codonopsis tsinglingensis Pax et K. Hoffm.;秦岭党参(大头党参);Qinling Asiabell,Tsinling Asiabell ■

98423 Codonopsis tubulosa Kom.;管花党参(白党,党参,理党参,牛尾党参,甜党,西昌党参);Tubularflower Asiabell ■

98424 Codonopsis ussuriensis (Rupr. et Maxim.) Hemsl.;雀斑党参(奶树,山海螺);Ussuri Asiabell ■

98425 Codonopsis ussuriensis (Rupr. et Maxim.) Hemsl. f. viridiflora J. Ohara;绿花雀斑党参■☆

98426 Codonopsis vinciflora Kom. = Codonopsis convolvulacea Kurz subsp. vinciflora (Kom.) D. Y. Hong ■

98427 Codonopsis vinciflora Kom. = Codonopsis convolvulacea Kurz var. vinciflora (Kom.) L. T. Shen ■

98428 Codonopsis viridiflora Maxim.;绿花党参;Greenflower Asiabell ■

98429 Codonopsis volubilis Nannf. = Codonopsis pilosula (Franch.) Nannf. ■

98430 Codonopsis volubilis Nannf. = Codonopsis pilosula (Franch.) Nannf. var. volubilis (Nannf.) L. T. Shen ■

98431 Codonopsis xizangensis D. Y. Hong;西藏党参;Xizang Asiabell ■

98432 Codonoraphta Oerst. = Pentarhaphia Lindl. ■☆

98433 Codonorchis Lindl. (1840);毛唇钟兰属■☆

98434 Codonorchis lessonii Lindl.;毛唇钟兰■☆

98435 Codonosiphon Schltr. = Bulbophyllum Thouars(保留属名)■

98436 Codonostigma Klotzsch = Scyphogyne Decne. ●☆

98437 Codonostigma Klotzsch ex Benth. = Scyphogyne Decne. ●☆

98438 Codonostigma erinus Klotzsch ex Benth. = Erica erina (Klotzsch ex Benth.) E. G. H. Oliv. ●☆

98439 Codonura K. Schum. = Baissea A. DC. ●☆

98440 Codoriocalyx Hassk. = Codariocalyx Hassk. ●

98441 Codoriocalyx Hassk. = Desmodium Desv. (保留属名)●■

98442 Codornia Gand. = Helianthemum Mill. ●■

98443 Codosiphus Raf. = Convolvulus L. ■●

98444 Codylis Raf. = Solanum L. ●■

98445 Coelachna T. Post et Kuntze = Coelachne R. Br. ■

98446 Coelachne R. Br. (1810);小丽草属;Coelachne ■

98447 Coelachne R. Br. et C. E. Hubb. = Coelachne R. Br. ■

98448 Coelachne africana Pilg.;非洲小丽草■☆

98449 Coelachne angolensis (Rendle) Jacq. -Fél.;安哥拉小丽草■☆

98450 Coelachne friesiorum C. E. Hubb.;弗里斯小丽草■☆

98451 Coelachne japonica Hack.;日本小丽草;Japanese Coelachne ■☆

98452 Coelachne madagascariensis Baker = Coelachne simpliciuscula (Wight et Arn. ex Steud.) Munro ex Benth. ■

98453 Coelachne occidentalis Jacq. -Fél. = Coelachne angolensis (Rendle) Jacq. -Fél. ■☆

98454 Coelachne paludosa Peter = Coelachne africana Pilg. ■☆

98455 Coelachne pulchella R. Br.;肖丽草(美丽小丽草);Lovely Coelachne ■☆

98456 Coelachne pulchella R. Br. = Coelachne simpliciuscula (Wight et Arn. ex Steud.) Munro ex Benth. ■

98457 Coelachne simpliciuscula (Wight et Arn. ex Steud.) Munro ex Benth.;小丽草;Simple Coelachne ■

98458 Coelachne simpliciuscula (Wight et Arn.) Munro ex Benth. = Coelachne simpliciuscula (Wight et Arn. ex Steud.) Munro ex Benth. ■

98459 Coelachyropsis Bor = Coelachyrum Hochst. et Nees ■☆

98460 Coelachyrum Hochst. et Nees(1842);天壳草属■☆

98461 Coelachyrum Nees = Eragrostis Wolf ■

98462 Coelachyrum brevifolium Nees;短叶天壳草■☆

98463 Coelachyrum lagopoides (Burm. f.) Senaratna;兔天壳草■☆

98464 Coelachyrum longiglume Napper;长颖天壳草■☆

98465 Coelachyrum oligobrachiatum A. Camus = Coelachyrum brevifolium Nees ■☆

98466 Coelachyrum oligobrachiatum A. Camus var. villiglume Maire = Coelachyrum brevifolium Nees ■☆

98467 Coelachyrum piercei (Benth.) Bor;皮尔斯天壳草■☆

98468 Coelachyrum stoloniferum C. E. Hubb. = Coelachyrum piercei (Benth.) Bor ■☆

98469 Coelachyrum yemenicum (Schweinf.) S. M. Phillips;也门天壳草■☆

98470 Coeladena Post et Kuntze = Coiladena Raf. ●■

98471 Coeladena Post et Kuntze = Ipomoea L. (保留属名)●■

98472 Coelandria Fitzg. = Dendrobium Sw. (保留属名)■

98473 Coelanthe Griseb. = Coilantha Borkh. ■

98474 Coelanthe Griseb. = Gentiana L. ■

98475 Coelanthera Post et Kuntze = Coilanthera Raf. ●

98476 Coelanthera Post et Kuntze = Cordia L. (保留属名)●

98477 Coelanthum E. Mey. ex Fenzl(1840);连萼粟草属■☆

98478 Coelanthum grandiflorum E. Mey. ex Fenzl;大花连萼粟草■☆

98479 Coelanthum semiquinquefidum (Hook. f.) Druce;连萼粟草■☆

98480 Coelanthum verticillatum Adamson;轮生连萼粟草■☆

98481 Coelanthus Willd. ex Schult. f. = Lachenalia J. Jacq. ex Murray ■☆

98482 Coelarthron Hook. f. = Microstegium Nees ■

98483 Coelas Dulac = Sibbaldia L. ■

98484 Coelebogyne J. Sm. = Alchornea Sw. ●

98485 Coelestina Cass. = Ageratum L. ■●

98486 Coelestina Cass. = Caelestina Cass. ■●

98487 Coelestina Hill = Amellus L. (保留属名)■●☆

98488 Coelestinia Endl. = Ageratum L. ■●

98489 Coelestinia Endl. = Caelestina Cass. ■●

98490 Coelia Lindl. (1830);粉兰属■☆

98491 Coelia baueriana Lindl.;芳香粉兰■☆

98492 Coelia bella (Lem.) Rchb. f.;美丽粉兰■☆

98493 Coelia bella Rchb. f. = Coelia bella (Lem.) Rchb. f. ■☆

98494 Coelia macrostachya Lindl.;大穗粉兰■☆

98495 Coelidium Vogel ex Walp. (1840);天盛豆属■☆

98496 Coelidium amphithaleoides Dümmer = Amphithalea muraltioides (Benth.) A. L. Schutte ■☆

98497 Coelidium bowiei Benth. = Amphithalea bowiei (Benth.) A. L. Schutte ■☆

98498 Coelidium bullatum Benth. = Amphithalea bullata (Benth.) A. L. Schutte ■☆

98499 Coelidium cedarbergense Granby = Amphithalea cedarbergensis (Granby) A. L. Schutte ■☆

98500 Coelidium ciliare (Eckl. et Zeyh.) Walp. = Amphithalea ciliaris Eckl. et Zeyh. ●☆

98501 Coelidium ciliare (Eckl. et Zeyh.) Walp. var. brevifolium L. = Amphithalea flava (Granby) A. L. Schutte ■☆

98502 Coelidium cymbifolium C. A. Sm. = Amphithalea cymbifolia (C. A. Sm.) A. L. Schutte ■☆

98503 Coelidium dahlgrenii Granby = Amphithalea dahlgrenii (Granby) A. L. Schutte ■☆

98504 Coelidium esterhuyseniae Granby = Amphithalea esterhuyseniae (Granby) A. L. Schutte ■☆

98505 Coelidium euchaetioides Dümmer = Amphithalea villosa Schltr. ■☆

98506 Coelidium flavum Granby = Amphithalea flava (Granby) A. L. Schutte ■☆

98507 Coelidium fourcadei Compton = Amphithalea parvifolia (Thunb.) A. L. Schutte ■☆

98508 Coelidium humile Schltr. = Amphithalea monticola A. L. Schutte ■☆

98509 Coelidium minimum Granby = Amphithalea minima (Granby) A. L. Schutte ■☆

98510 Coelidium muirii Granby = Amphithalea muirii (Granby) A. L. Schutte ■☆

98511 Coelidium muraltioides Benth. = Amphithalea muraltioides (Benth.) A. L. Schutte ■☆

98512 Coelidium obtusilobum Granby = Amphithalea obtusiloba (Granby) A. L. Schutte ■☆

98513 Coelidium pageae L. Bolus = Amphithalea pageae (L. Bolus) A. L. Schutte ■☆

98514 Coelidium parviflorum (Thunb.) Druce = Amphithalea parvifolia (Thunb.) A. L. Schutte ■☆

98515 Coelidium perplexum (Eckl. et Zeyh.) Granby = Amphithalea perplexa Eckl. et Zeyh. ■☆

98516 Coelidium purpureum Granby = Amphithalea purpurea (Granby) A. L. Schutte ■☆

98517 Coelidium roseum (E. Mey.) Benth. = Amphithalea perplexa Eckl. et Zeyh. ■☆

98518 Coelidium spinosum Harv. = Amphithalea spinosa (Harv.) A. L. Schutte ■☆

98519 Coelidium thunbergii Harv. = Amphithalea parvifolia (Thunb.) A. L. Schutte ■☆

98520 Coelidium tortile (E. Mey.) Druce = Amphithalea tortilis (E. Mey.) Steud. ■☆

98521 Coelidium villosum (Schltr.) Granby = Amphithalea villosa Schltr. ■☆

98522 Coelidium vlokii A. L. Schutte et B. -E. van Wyk = Amphithalea vlokii (A. L. Schutte et B. -E. van Wyk) A. L. Schutte ■☆

98523 Coelidium vogelii Walp. = Amphithalea tortilis (E. Mey.) Steud. ■☆

98524 Coelina Noronha = Elaeocarpus L. ●

98525 Coeliopsis Rchb. f. (1872);拟粉兰属■☆

98526 Coeliopsis hyacinthosma Rchb. f.;拟粉兰■☆

98527 Coelobogyne J. Sm. = Alchornea Sw. ●

98528 Coelocarpum Balf. f. (1883);凹果马鞭草属●☆

98529 Coelocarpum africanum Moldenke;非洲凹果马鞭草●☆

98530 Coelocarpus Post et Kuntze = Coilocarpus F. Muell. ex Domin ●☆

98531 Coelocarpus Scott Elliot = Coelocarpum Balf. f. ●☆

98532 Coelocaryon Warb. (1895);凹果豆蔻属(天堂果属)●☆

98533 Coelocaryon botryoides Vermoesen;葡萄凹果豆蔻●☆

98534 Coelocaryon cuneatum Warb. = Coelocaryon preussii Warb. ●☆

98535 Coelocaryon klainei Pierre = Coelocaryon preussii Warb. ●☆

98536 Coelocaryon klainei Pierre ex Heckel = Coelocaryon preussii Warb. ●☆

98537 Coelocaryon multiflorum Warb. = Coelocaryon preussii Warb. ●☆

98538 Coelocaryon oxycarpum Stapf;尖凹果豆蔻●☆

98539 Coelocaryon preussii Warb.;凹果豆蔻●☆

98540 Coelocaryon sphaerocarpum Fouilloy;球凹果豆蔻●☆

98541 Coelocaryon staneri Ghesq. = Coelocaryon botryoides Vermoesen ●☆

98542 Coelochloa Hochst. = Coelachyrum Hochst. et Nees ■☆

98543 Coelochloa Hochst. ex Steud. = Coelachyrum Hochst. et Nees ■☆

98544 Coelochloa Steud. = Coelachyrum Hochst. et Nees ■☆

98545 Coelocline A. DC. = Xylopia L. (保留属名)●

98546 Coelocline acutiflora (Dunal) A. DC. = Xylopia acutiflora (Dunal) A. Rich. ●☆

98547 Coelocline oxypetala (DC.) A. DC. = Xylopia acutiflora (Dunal) A. Rich. ●☆

98548 Coelocline parviflora (A. Rich.) DC. = Xylopia parviflora (A. Rich.) Benth. ●☆

98549 Coelococcus H. Wendl. (1862);橡扣树属(波利西谷椰属);Ivory-nut Palm, Sago Palm ●☆

98550 Coelococcus H. Wendl. = Metroxylon Rottb. (保留属名)●

98551 Coelococcus amicarum W. Wight = Coelococcus carolinensis Dingley ●☆

98552 Coelococcus carolinensis Dingley;橡扣树(象牙椰子)●☆

98553 Coelodepas Hassk. = Koilodepas Hassk. ●

98554 Coelodepas hainanense (Merr.) Croizat = Koilodepas hainanense (Merr.) Airy Shaw ●

98555 Coelodiscus Baill. (1858);穴盘木属;Coelodiscus ●

98556 Coelodiscus Baill. = Mallotus Lour. ●

98557 Coelodiscus eriocarpoides Kurz = Mallotus decipiens Müll. Arg. ●

98558 Coelodiscus lappaceus Kurz;穴盘木;Stick Coelodiscus ●

98559 Coelodiscus speciosus Müll. Arg. = Sumbaviopsis albicans (Blume) J. J. Sm. ●

98560 Coeloglossum Hartm. (1820);凹舌兰属(凹唇兰属);Coeloglossum, Frog Orchid, Frog-orchis, Long-bracted Orchid ■

98561 Coeloglossum Hartm. = Dactylorhiza Neck. ex Nevski (保留属名)■

98562 Coeloglossum Hartm. = Satyrium L. (废弃属名)■

98563 Coeloglossum Hartm. = Satyrium Sw. (保留属名)■

98564 Coeloglossum Lindl. = Lindblomia Fr. ■

98565 Coeloglossum Lindl. = Peristylus Blume(保留属名)■

98566 Coeloglossum bracteatum (Muhl. ex Willd.) Parl. = Coeloglossum viride (L.) Hartm. ■

98567 Coeloglossum bracteatum (Muhl. ex Willd.) Parl. = Dactylorhiza viridis (L.) R. M. Bateman, Pridgeon et M. W. Chase ■

98568 Coeloglossum bracteatum (Muhl. ex Willd.) Parl. var. kashmiricum Soó = Coeloglossum viride (L.) Hartm. ■

98569 Coeloglossum bracteatum (Muhl. ex Willd.) Schltr. = Coeloglossum viride (L.) Hartm. ■

98570 Coeloglossum bracteatum (Muhl.) Parl. = Coeloglossum viride (L.) Hartm. var. virescens (Muhl.) Luer ■☆

98571 Coeloglossum bracteatum Parl. = Coeloglossum viride (L.) Hartm. ■

98572 Coeloglossum bracteatum Parl. var. kashmiricum? = Coeloglossum viride (L.) Hartm. ■

98573 Coeloglossum densum Lindl. = Peristylus densus (Lindl.) Santapau et Kapadia ■

98574 Coeloglossum flagelliferum (Makino) Maxim. ex Makino = Peristylus densus (Lindl.) Santapau et Kapadia ■

98575 Coeloglossum flagelliferum Maxim. ex Makino = Peristylus flagellifer (Makino) Ohwi ■

98576 Coeloglossum formosanum Makino et Hayata ex Hayata = Peristylus formosanus (Schltr.) T. P. Lin ■

98577 Coeloglossum kaschmirianum Schltr. = Coeloglossum viride (L.) Hartm. ■

98578 Coeloglossum lacertiferum Lindl. = Peristylus lacertifer (Lindl.) J. J. Sm. ■

98579 Coeloglossum mannii Rchb. f. = Peristylus mannii (Rchb. f.) Mukerjee ■

98580 Coeloglossum nankotaizanense (Masam.) Masam. = Coeloglossum viride (L.) Hartm. ■

98581 Coeloglossum nankotaizanense (Masam.) S. S. Ying;南湖大山凹舌兰■

98582 Coeloglossum nankotaizanense (Masam.) S. S. Ying = Coeloglossum viride (L.) Hartm. ■

98583 Coeloglossum nankotaizanense (Masam.) S. S. Ying = Dactylorhiza viridis (L.) R. M. Bateman, Pridgeon et M. W. Chase ■

98584 Coeloglossum taiwanianum S. S. Ying = Coeloglossum viride (L.) Hartm. ■

98585 Coeloglossum taiwanianum S. S. Ying = Dactylorhiza viridis (L.) R. M. Bateman, Pridgeon et M. W. Chase ■

98586 Coeloglossum viride (L.) Hartm.;凹舌兰(长苞凹舌兰,绿花凹舌兰,南湖大山凹舌兰,手参,手儿参,台湾裂唇兰);Bracteate Coeloglossum, Brarted Green Orchid, Coeloglossum, Frog Orchid, Long-bracted Green Orchid, Long-bracted Orchid, Rein Orchid, Rein Orchis ■

98587 Coeloglossum viride (L.) Hartm. = Dactylorhiza viridis (L.) R. M. Bateman, Pridgeon et M. W. Chase ■

98588 Coeloglossum viride (L.) Hartm. subsp. bracteatum (Muhl. ex Willd.) Hultén = Coeloglossum viride (L.) Hartm. var. bracteatum (Muhl. ex Willd.) Richt. ex Miyabe et T. Miyake ■

98589 Coeloglossum viride (L.) Hartm. subsp. bracteatum (Muhl.) Hultén = Coeloglossum viride (L.) Hartm. var. virescens (Muhl.) Luer ■☆

98590 Coeloglossum viride (L.) Hartm. subsp. bracteatum (Muhl.) Hultén = Coeloglossum viride (L.) Hartm. ■

98591 Coeloglossum viride (L.) Hartm. var. akaishimontanum Satomi;明石山凹舌兰■☆

98592 Coeloglossum viride (L.) Hartm. var. bracteatum (Muhl. ex Willd.) A. Gray = Dactylorhiza viridis (L.) R. M. Bateman, Pridgeon et M. W. Chase ■

98593 Coeloglossum viride (L.) Hartm. var. bracteatum (Muhl. ex Willd.) Richt. ex Miyabe et T. Miyake;长苞凹舌兰(凹舌兰);Longbract Coeloglossum, Long-bracted, Orchis ■

98594 Coeloglossum viride (L.) Hartm. var. interjectum (Fernald) Miyabe et Kudo = Coeloglossum viride (L.) Hartm. ■

98595 Coeloglossum viride (L.) Hartm. var. islandicum (Lindl.) M. Schulze = Coeloglossum viride (L.) Hartm. ■

98596 Coeloglossum viride (L.) Hartm. var. virescens (Muhl.) Luer;浅绿凹舌兰;Frog Orchid, Long-bracted Green Orchid ■☆

98597 Coeloglossum viride (L.) Hartm. var. virescens (Muhl.) Luer = Coeloglossum viride (L.) Hartm. ■

98598 Coeloglossum viride (L.) Richt. = Coeloglossum viride (L.) Hartm. ■

98599 Coeloglossum viride (L.) Richt. var. bracteatum (Muhl. ex Willd.) A. Gray = Coeloglossum viride (L.) Hartm. ■

98600 Coeloglossum viride (Willd.) Richt. = Coeloglossum viride (L.) Hartm. ■

98601 Coelogyne Lindl. (1821);贝母兰属;Coelogyne ■

98602 Coelogyne alba (Lindl.) Rchb. f. = Otochilus albus Lindl. ■

98603 Coelogyne alba Rchb. f. = Otochilus albus Lindl. ■

98604 Coelogyne annamensis Rolfe;越南贝母兰;Vietnam Coelogyne ■☆

98605 Coelogyne annamensis Rolfe = Coelogyne assamica Linden et Rchb. f. ■

98606 Coelogyne arthuriana Rchb. f. = Pleione maculata (Lindl.) Lindl. et Paxton ■

98607 Coelogyne articulata (Lindl.) Rchb. f. = Pholidota articulata Lindl. ■

98608 Coelogyne arunachalensis H. J. Chowdhery et G. D. Pal = Coelogyne fimbriata Lindl. ■

98609 Coelogyne asperata Lindl.;粗糙贝母兰■☆

98610 Coelogyne assamica Linden et Rchb. f.;云南贝母兰■

98611 Coelogyne assamica Linden et Rchb. f. = Coelogyne fuscescens Lindl. var. brunnea (Lindl.) Lindl. ■

98612 Coelogyne barbata Lindl. ex Griff.;髯毛贝母兰;Barbat Coelogyne ■

98613 Coelogyne biflora E. C. Parish et Rchb. f. = Panisea uniflora (Lindl.) Lindl. ■

98614 Coelogyne birmanica Rchb. f. = Pleione praecox (Sm.) D. Don ■

98615 Coelogyne borneensis Rolfe;婆罗州贝母兰;Borneo Coelogyne ■☆

98616 Coelogyne brevifolia Lindl. = Coelogyne punctulata Lindl. ■

98617 Coelogyne breviscapa Lindl.;短花茎贝母兰;Shortscape Coelogyne ■☆

98618 Coelogyne brunnea Lindl. = Coelogyne fuscescens Lindl. var. brunnea (Lindl.) Lindl. ■

98619 Coelogyne brunnea Lindl. = Coelogyne fuscescens Lindl. ■

98620 Coelogyne bulbocodioides Franch. = Pleione bulbocodioides (Franch.) Rolfe ■

98621 Coelogyne bulbodiscoides Franch.;球茎贝母兰(光菇,茅慈菰,茅菰,山茨菰,山慈姑,珍珠七)■☆

98622 Coelogyne calcicola Kerr;滇西贝母兰;W. Yunnan Coelogyne ■

98623 Coelogyne carinata Rolfe;龙骨唇贝母兰;Carinate Coelogyne ■☆

98624 Coelogyne carnea Hook. f.;肉红花贝母兰;Fleshred Coelogyne ■☆

98625 Coelogyne chinensis (Lindl.) Rchb. f. = Pholidota chinensis Lindl. ■

98626 Coelogyne chloptera Rchb. f. = Coelogyne nitida (Wall. ex D. Don) Lindl. ■

98627 Coelogyne chrysotropis Schltr. = Coelogyne fimbriata Lindl. ■

98628 Coelogyne cinnamomea Teijsm. et Binn.;褐色贝母兰;Brownlip Coelogyne ■☆

98629 Coelogyne convallariae E. C. Parish et Rchb. f. = Pholidota convallariae (Rchb. f.) Hook. f. ■

98630 Coelogyne convallariae Rchb. f. = Pholidota convallariae (Rchb. f.) Hook. f. ■

98631 Coelogyne coronaria Lindl. = Eria coronaria (Lindl.) Rchb. f. ■

98632 Coelogyne corrugata Wight;皱折贝母兰■☆

98633 Coelogyne corymbosa Lindl.;眼斑贝母兰(贝母兰,对叶果,果上叶,伞房贝母兰,石芭蕉,小绿芨,止血果);Corymblike Coelogyne, Corymbose Coelogyne ■

98634 Coelogyne cristata Lindl.;贝母兰(毛唇贝母兰);Cristate Coelogyne ■

98635 Coelogyne cristata Lindl. var. hololeuca Rchb.;全白贝母兰■☆

98636 Coelogyne cumingii Lindl.;卡明氏贝母兰■☆

98637 Coelogyne cycnoches E. C. Parish et Rchb. f. = Coelogyne fuscescens Lindl. ■

98638 Coelogyne dalatensis Gagnep. = Coelogyne assamica Linden et Rchb. f. ■

98639 Coelogyne darlacensis Gagnep. = Coelogyne sanderae Kraenzl. ex O'Brien ■

98640 Coelogyne dayana Rchb. f.;金璎珞■☆

98641 Coelogyne decora Wall. ex Voigt = Coelogyne ovalis Lindl. ■

98642 Coelogyne delavayi Rolfe = Pleione bulbocodioides (Franch.) Rolfe ■

98643 Coelogyne diphylla (Lindl. et Paxton) Lindl. = Pleione maculata (Lindl.) Lindl. et Paxton ■

98644 Coelogyne diphylla (Lindl.) Lindl. = Pleione maculata (Lindl.) Lindl. et Paxton ■

98645 Coelogyne elata Lindl.;长茎贝母兰;Tall Coelogyne ■☆

98646 Coelogyne elata Lindl. = Coelogyne stricta (D. Don) Schltr. ■

98647 Coelogyne elegantula Kraenzl. = Bletilla formosana (Hayata) Schltr. ■

98648 Coelogyne elegantula Kraenzl. = Bletilla striata (Thunb. ex A. Murray) Rchb. f. ■

98649 Coelogyne esquirolii Schltr. = Coelogyne flaccida Lindl. ■

98650 Coelogyne falcata T. Anderson ex Hook. f. = Panisea uniflora (Lindl.) Lindl. ■

98651 Coelogyne fimbriata Lindl.;流苏贝母兰(贝母兰,石仙桃);Fimbriate Coelogyne, Tassel Coelogyne ■

98652 Coelogyne fimbriata Lindl. var. annamica Finet ex Gagnep. = Coelogyne fimbriata Lindl. ■

98653 Coelogyne flaccida Lindl.;栗鳞贝母兰(大果上叶,鸡大腿);Flaccid Coelogyne ■

98654 Coelogyne flavida Hook. f. ex Lindl. = Coelogyne prolifera Lindl. ■
98655 Coelogyne flavida Wall. ex Lindl. = Coelogyne prolifera Lindl. ■
98656 Coelogyne flavida Wall. ex Lindl. = Coelogyne schultesii S. K. Jain et S. Das ■
98657 Coelogyne fuliginosa Lodd. ex Hook.;暗褐贝母兰■☆
98658 Coelogyne fuliginosa Lodd. ex Hook. = Coelogyne fimbriata Lindl. ■
98659 Coelogyne fusca (Lindl.) Rchb. f. = Otochilus fuscus Lindl. ■
98660 Coelogyne fuscescens Lindl.;褐唇贝母兰(褐色贝母兰);Brownlip Coelogyne ■
98661 Coelogyne fuscescens Lindl. var. assamica (Linden et Rchb. f.) Pfitzer et Kraenzl. = Coelogyne assamica Linden et Rchb. f. ■
98662 Coelogyne fuscescens Lindl. var. brunnea (Lindl.) Lindl.;斑唇贝母兰■
98663 Coelogyne gardneriana Lindl. = Neogyna gardneriana (Lindl.) Rchb. f. ■
98664 Coelogyne gongshanensis H. Li ex S. C. Chen;贡山贝母兰;Gongshan Coelogyne ■
98665 Coelogyne goweri Rchb. f. = Coelogyne punctulata Lindl. ■
98666 Coelogyne graminifolia C. S. Parish et Rchb. f. = Coelogyne viscosa Rchb. f. ■
98667 Coelogyne grandiflora Rolfe = Pleione grandiflora (Rolfe) Rolfe ■
98668 Coelogyne griffithii Hook. f.;格里菲斯贝母兰(格力贝母兰)■
98669 Coelogyne henryi Rolfe = Pleione bulbocodioides (Franch.) Rolfe ■
98670 Coelogyne hookeriana Lindl. = Pholidota imbricata Hook. ■
98671 Coelogyne hookeriana Lindl. = Pleione hookeriana (Lindl.) B. S. Williams ■
98672 Coelogyne hookeriana Lindl. var. brachyglossa Rchb. f. = Pholidota imbricata Hook. ■
98673 Coelogyne hookeriana Lindl. var. brachyglossa Rchb. f. = Pleione hookeriana (Lindl.) B. S. Williams ■
98674 Coelogyne huettneriana Rchb.;休氏贝母兰■☆
98675 Coelogyne humilis (Sm.) Lindl. = Pleione humilis (Sm.) D. Don ■
98676 Coelogyne humilis (Sm.) Lindl. var. albata Rchb. f. = Pleione humilis (Sm.) D. Don ■
98677 Coelogyne humilis (Sm.) Lindl. var. tricolor Rchb. f. = Pleione humilis (Sm.) D. Don ■
98678 Coelogyne humilis Lindl. = Pleione humilis (Sm.) D. Don ■
98679 Coelogyne imbricata (Hook.) Rchb. f. = Pholidota imbricata Hook. ■
98680 Coelogyne khaiana (Rchb. f.) Rchb. f. = Pholidota articulata Lindl. ■
98681 Coelogyne khasyana (Rchb. f.) Rchb. f. = Pholidota articulata Lindl. ■
98682 Coelogyne lactea Rchb. f. = Coelogyne flaccida Lindl. ■
98683 Coelogyne laotica Gagnep. = Coelogyne fimbriata Lindl. ■
98684 Coelogyne lawrenceana Rolfe;劳伦斯贝母兰■☆
98685 Coelogyne lentiginosa Lindl.;麝香贝母兰;Lentiginous Coelogyne ■☆
98686 Coelogyne leucantha W. W. Sm.;白花贝母兰(大果上叶,对叶果,果上叶,石芭蕉,石串莲);White Coelogyne ■
98687 Coelogyne leucantha W. W. Sm. var. heterophylla Ts. Tang et F. T. Wang = Coelogyne leucantha W. W. Sm. ■
98688 Coelogyne leungiana S. Y. Hu;单唇贝母兰;Singlelip Coelogyne ■
98689 Coelogyne longiciliata Teijsm. et Binn. = Coelogyne fimbriata Lindl. ■
98690 Coelogyne longipes Lindl.;长柄贝母兰;Longstalk Coelogyne ■
98691 Coelogyne longipes Lindl. var. verruculata S. C. Chen = Coelogyne schultesii S. K. Jain et S. Das ■
98692 Coelogyne lowii Paxton = Coelogyne asperata Lindl. ■☆
98693 Coelogyne maculata Lindl. = Pleione maculata (Lindl.) Lindl. et Paxton ■
98694 Coelogyne malipoensis Z. H. Tsi;麻栗坡贝母兰;Malipo Coelogyne ■
98695 Coelogyne mandarinorum Kraenzl. = Ischnogyne mandarinorum (Kraenzl.) Schltr. ■
98696 Coelogyne massangeana Rchb. f.;马氏贝母兰■☆
98697 Coelogyne miniata (Blume) Lindl.;朱红贝母兰;Red Coelogyne ■☆
98698 Coelogyne mooreana Sand.;穆尔贝母兰■☆
98699 Coelogyne nervosa A. Rich. = Coelogyne corrugata Wight ■☆
98700 Coelogyne nitida (Roxb.) Hook. f. = Coelogyne punctulata Lindl. ■
98701 Coelogyne nitida (Wall. ex D. Don) Lindl.;密茎贝母兰(亮花贝母兰,绿花贝母兰,赭黄贝母兰);Greenflower Coelogyne, Ochrecoloured Coelogyne, Shine Coelogyne ■
98702 Coelogyne nitida Lindl. = Coelogyne nitida (Wall. ex D. Don) Lindl. ■
98703 Coelogyne nitida Lindl. = Coelogyne punctulata Lindl. ■
98704 Coelogyne occultata Hook. f.;卵叶贝母兰(有瓜石斛);Ovateleaf Coelogyne ■
98705 Coelogyne ocellata Lindl. = Coelogyne punctulata Lindl. ■
98706 Coelogyne ocellata Lindl. var. boddaertiana Rchb. f. = Coelogyne punctulata Lindl. ■
98707 Coelogyne ocellata Lindl. var. maxima Rchb. f. = Coelogyne punctulata Lindl. ■
98708 Coelogyne ochracea Lindl. = Coelogyne nitida (Wall. ex D. Don) Lindl. ■
98709 Coelogyne ochracea Lindl. subsp. conferta C. P. Parish et Rchb. f. = Coelogyne nitida (Wall. ex D. Don) Lindl. ■
98710 Coelogyne odoratissima Lindl.;芳香贝母兰;Fragrant Coelogyne ■☆
98711 Coelogyne ovalis Lindl.;长鳞贝母兰(贝母兰,散生贝母兰);Common Coelogyne, Oval Coelogyne ■
98712 Coelogyne padangensis J. J. Sm. et Schltr. = Coelogyne fimbriata Lindl. ■
98713 Coelogyne pallens Ridl. = Coelogyne fimbriata Lindl. ■
98714 Coelogyne pallida (Lindl.) Rchb. f. = Pholidota pallida Lindl. ■
98715 Coelogyne pandurata Lindl.;小提琴状贝母兰■☆
98716 Coelogyne parviflora Lindl. = Panisea demissa (D. Don) Pfitzer ■
98717 Coelogyne pholas Rchb. f. = Pholidota chinensis Lindl. ■
98718 Coelogyne pilosissima Planch. = Coelogyne ovalis Lindl. ■
98719 Coelogyne pogonioides Rolfe = Pleione bulbocodioides (Franch.) Rolfe ■
98720 Coelogyne porrecta (Lindl.) Rchb. f. = Otochilus porrectus Lindl. ■
98721 Coelogyne praecox (Sm.) Lindl. = Pleione praecox (Sm.) D. Don ■
98722 Coelogyne prasina Ridl.;绿花贝母兰(绿色贝母兰);Greenflower Coelogyne ■☆
98723 Coelogyne primulina Barretto;报春贝母兰;Primroselike Coelogyne ■

98724 Coelogyne primulina Barretto = Coelogyne fimbriata Lindl. ■
98725 Coelogyne prolifera Lindl.;黄绿贝母兰;Yellow-green Coelogyne ■
98726 Coelogyne prolifera Lindl. = Coelogyne schultesii S. K. Jain et S. Das ■
98727 Coelogyne pulchella Rolfe;美丽贝母兰■
98728 Coelogyne punctulata Lindl.;狭瓣贝母兰(斑唇贝母兰,贝母兰,对叶果,果上叶,伞房贝母兰,石芭蕉,小绿芨,止血果);Stenopetal Coelogyne ■
98729 Coelogyne punctulata Lindl. = Coelogyne corymbosa Lindl. ■
98730 Coelogyne punctulata Lindl. var. conferta (E. C. Parish et Rchb. f.) Ts. Tang et F. T. Wang = Coelogyne nitida (Wall. ex D. Don) Lindl. ■
98731 Coelogyne punctulata Lindl. var. conferta (Pax. et Rchb. f.) Ts. Tang et F. T. Wang = Coelogyne nitida (Wall. ex D. Don) Lindl. ■
98732 Coelogyne punctulata Lindl. var. hysterantha Ts. Tang et F. T. Wang = Coelogyne punctulata Lindl. ■
98733 Coelogyne raizadae S. K. Jain et S. Das;三褶贝母兰■
98734 Coelogyne raizadae S. K. Jain et S. Das = Coelogyne longipes Lindl. ■
98735 Coelogyne reichenbachiana T. Moore et Veitch = Pleione praecox (Sm.) D. Don ■
98736 Coelogyne ridleyi Gagnep. = Coelogyne sanderae Kraenzl. ex O'Brien ■
98737 Coelogyne rigida Parl. et Rchb. f.;挺茎贝母兰;Rigid Coelogyne ■
98738 Coelogyne rossiana Rchb. f.;罗斯贝母兰■☆
98739 Coelogyne saigonensis Gagnep. = Coelogyne assamica Linden et Rchb. f. ■
98740 Coelogyne sanderae Kraenzl. ex O'Brien;撕裂贝母兰(褐唇贝母兰);Sandera Coelogyne ■
98741 Coelogyne schultesii S. K. Jain et S. Das;疣鞘贝母兰;Schultes Coelogyne ■
98742 Coelogyne siamensis Rolfe = Coelogyne assamica Linden et Rchb. f. ■
98743 Coelogyne speciosa (Blume) Lindl.;艳花贝母兰(美丽贝母兰);Beautiful Coelogyne ■☆
98744 Coelogyne stricta (D. Don) Schltr.;双褶贝母兰;Dualplicated Coelogyne ■
98745 Coelogyne suaveolens (Lindl.) Hook. f.;疏茎贝母兰;Poorstem Coelogyne ■
98746 Coelogyne taronensis Hand. -Mazz.;高山贝母兰■
98747 Coelogyne taronensis Hand. -Mazz. = Coelogyne corymbosa Lindl. ■
98748 Coelogyne thuniana Rchb. f. = Panisea uniflora (Lindl.) Lindl. ■
98749 Coelogyne tomentosa Lindl.;毛序贝母兰;Tomentose Coelogyne ■☆
98750 Coelogyne treutleri Hook. f. = Epigeneium treutleri (Hook. f.) Ormerod ■
98751 Coelogyne tsii X. H. Jin et H. Li;吉氏贝母兰■
98752 Coelogyne uniflora Lindl. = Panisea uniflora (Lindl.) Lindl. ■
98753 Coelogyne venusta Rolfe;多花贝母兰(云南贝母兰);Flowery Coelogyne, Yunnan Coelogyne ■
98754 Coelogyne viscosa Rchb. f.;禾叶贝母兰(石莲草,石仙桃,黏贝母兰);Grassleaf Coelogyne ■
98755 Coelogyne wallichi Hook. = Pleione praecox (Sm.) D. Don ■
98756 Coelogyne wallichiana Lindl. = Pleione praecox (Sm.) D. Don ■
98757 Coelogyne weixiensis X. H. Jin;维西贝母兰■
98758 Coelogyne xerophyta Hand. -Mazz. = Coelogyne fimbriata Lindl. ■
98759 Coelogyne xerophyta Hand. -Mazz. = Coelogyne ovalis Lindl. ■
98760 Coelogyne yunnanensis Rolfe = Pleione yunnanensis (Rolfe) Rolfe ■
98761 Coelogyne zhenkangensis S. C. Chen et K. Y. Lang;镇康贝母兰;Zhenkang Coelogyne ■
98762 Coelonema Maxim. (1880);穴丝荠属(穴丝芥属);Coelonema ●★
98763 Coelonema Maxim. = Draba L. ■
98764 Coelonema draboides Maxim.;穴丝荠(草原葶苈,穴丝芥);Drabalike Coelonema, Grassland Whitlowgrass, Steppe Draba ■
98765 Coelonema draboides Maxim. = Draba draboides (Maxim.) Al-Shehbaz ■
98766 Coeloneurum Radlk. (1889);凹脉茄属●☆
98767 Coeloneurum ferrugineum (Spreng.) Urb.;凹脉茄●☆
98768 Coelonox Post et Kuntze = Coilonox Raf. ■☆
98769 Coelonox Post et Kuntze = Ornithogalum L. ■
98770 Coelophragmus O. E. Schulz(1924);墨西哥大蒜芥属■☆
98771 Coelophragmus auriculatus O. E. Schulz;墨西哥大蒜芥■☆
98772 Coelopleurum Ledeb. (1844);高山芹属(高山芥属,空肋芥属);Alpparsley, Angelica ■
98773 Coelopleurum Ledeb. = Angelica L. ■
98774 Coelopleurum alpinum Kitag. = Coelopleurum saxatile (Turcz. ex Ledeb.) Drude ■
98775 Coelopleurum brevicaule (Rupr.) Drude = Archangelica brevicaulis (Rupr.) Rchb. f. ■
98776 Coelopleurum brevicaule Drude = Archangelica brevicaulis (Rupr.) Rchb. f. ■
98777 Coelopleurum gmelinii (DC.) Ledeb.;格氏高山芹(格氏当归,勘查加高山芹);Gmelin Alpparsley ■☆
98778 Coelopleurum lucidum (L.) Fernald;海岸高山芹;Sea-coast Angelica, Seaside Angelica ■☆
98779 Coelopleurum lucidum (L.) Fernald var. gmelinii (DC.) H. Hara = Coelopleurum gmelinii (DC.) Ledeb. ■☆
98780 Coelopleurum lucidum (L.) Fernald var. trichocarpum (H. Hara) H. Hara = Coelopleurum rupestre (Koidz.) T. Yamaz. ■☆
98781 Coelopleurum multisectum (Maxim.) Kitag.;多毛高山芹■☆
98782 Coelopleurum multisectum (Maxim.) Kitag. f. trichocarpum (H. Hara) Kitag. = Coelopleurum rupestre (Koidz.) T. Yamaz. ■☆
98783 Coelopleurum multisectum (Maxim.) Kitag. var. trichocarpum (H. Hara) H. Ohba = Coelopleurum rupestre (Koidz.) T. Yamaz. ■☆
98784 Coelopleurum nakaianum (Kitag.) Kitag.;长白高山芹(白山芹);Changbaishan Alpparsley ■
98785 Coelopleurum rupestre (Koidz.) T. Yamaz.;岩生高山芹■☆
98786 Coelopleurum saxatile (Turcz. ex Ledeb.) Drude;高山芹;Alpparsley ■
98787 Coelopleurum saxatile (Turcz.) Drude = Coelopleurum saxatile (Turcz. ex Ledeb.) Drude ■
98788 Coelopleurum trichocarpum (Hara) Kitag. = Coelopleurum multisectum (Maxim.) Kitag. var. trichocarpum (H. Hara) H. Ohba ■☆
98789 Coelopyrena Valeton(1909);柳叶茜属☆
98790 Coelopyrena salicifolia Valeton;柳叶茜☆
98791 Coelopyrum Jack(废弃属名) = Campnosperma Thwaites(保留属名)●☆
98792 Coelorachis Brongn. (1831);空轴茅属;Emptygrass ■
98793 Coelorachis Brongn. = Mnesithea Kunth ■
98794 Coelorachis afraurita (Stapf) Stapf;热非空轴茅■☆
98795 Coelorachis capensis Stapf;好望角空轴茅■☆

98796 Coelorachis cylindrica (Michx.) Nash;卡罗来纳空轴茅;Carolina Jointtail,Joint Grass ■☆
98797 Coelorachis khasiana (Hack.) Stapf ex Bor = Mnesithea khasiana (Hack.) de Koning et Sosef ■
98798 Coelorachis lepidura Stapf;雅致空轴茅■☆
98799 Coelorachis mollicoma (Hance) Bor = Mnesithea mollicoma (Hance) A. Camus ■
98800 Coelorachis striata (Nees ex Steud.) A. Camus;空轴茅(条花罗氏草);Striate Emptygrass,Striate Itchgrass,Striate Mnesithea ■
98801 Coelorachis striata (Nees ex Steud.) A. Camus = Mnesithea striata (Steud.) de Koning et Sosef ■
98802 Coelorachis striata (Nees ex Steud.) A. Camus var. pubescens (Hack.) Bor;毛空轴茅(毛罗氏草);Pubescent Emptygrass ■
98803 Coelorachis striata (Nees ex Steud.) A. Camus var. pubescens (Hack.) Bor = Mnesithea striata (Steud.) de Koning et Sosef var. pubescens (Hack.) S. M. Phillips et S. L. Chen ■
98804 Coelorhachis Endl. = Coelorachis Brongn. ■
98805 Coelosperma Post et Kuntze = Coilosperma Raf. ●■
98806 Coelosperma Post et Kuntze = Deeringia R. Br. ●■
98807 Coelospermum Blume = Caelospermum Blume ●
98808 Coelospermum Blume (1827);穴果木属;Caelospermum, Coelospermum ●
98809 Coelospermum decipiens Merr.;迷惑穴果木●☆
98810 Coelospermum kanehire Merr.;穴果木;Kanehira Caelospermum, Kanehira Coelospermum ●
98811 Coelospermum morindiforme Pierre ex Pit.;长叶穴果木;Longleaf Caelospermum,Longleaf Coelospermum ●
98812 Coelostegia Benth. (1862);凹顶木棉属(露冠树属)●☆
98813 Coelostegia griffithii Benth.;凹顶木棉(格氏露冠树)●☆
98814 Coelostelma E. Fourn. (1885);空柱萝藦属☆
98815 Coelostelma refractum E. Fourn.;空柱萝藦☆
98816 Coelostigma Post et Kuntze = Coilostigma Klotzsch ●☆
98817 Coelostigma Post et Kuntze = Salaxis Salisb. ●☆
98818 Coelostigmaceae Dulac = Berberidaceae Juss. (保留科名)●■
98819 Coelostylis (A. Juss.) Kuntze = Echinopterys A. Juss. ●☆
98820 Coelostylis Post et Kuntze = Coilostylis Raf. ■
98821 Coelostylis Post et Kuntze = Epidendrum L. (保留属名)■☆
98822 Coelostylis Torr. et A. Gray ex Endl. = Spigelia L. ■☆
98823 Coelostylis Torr. et A. Gray ex Endl. et Fenzl = Spigelia L. ■☆
98824 Coelotapalus Post et Kuntze = Cecropia Loefl. (保留属名)●☆
98825 Coelotapalus Post et Kuntze = Coilotapalus P. Browne(废弃属名)●☆
98826 Coemansia Marchal = Coudenbergia Marchal ●
98827 Coemansia Marchal = Pentapanax Seem. ●
98828 Coenadenium (Summerh.) Szlach. = Angraecopsis Kraenzl. ■☆
98829 Coenochlamys Post et Kuntze = Coinochlamys T. Anderson ex Benth. et Hook. f. ●☆
98830 Coenogyna Post et Kuntze = Coinogyne Less. ■●☆
98831 Coenogyna Post et Kuntze = Jaumea Pers. ■●☆
98832 Coenolophium Rchb. = Cenolophium W. D. J. Koch ■
98833 Coenotus Benth. et Hook. f. = Caenotus Raf. ■●
98834 Coenotus Benth. et Hook. f. = Erigeron L. ■●
98835 Coerulinia Fourr. = Veronica L. ■
98836 Coespeletia Cuatrec. = Espeletia Mutis ex Humb. et Bonpl. ●☆
98837 Coespiphylis Thouars = Bulbophyllum Thouars(保留属名)■
98838 Coestichis Thouars = Liparis Rich. (保留属名)■
98839 Coestichis Thouars = Malaxis Sol. ex Sw. ■
98840 Coetocapnia Link et Otto = Polianthes L. ■
98841 Cofeanthus A. Chev. = Coffea L. ●
98842 Cofeanthus A. Chev. = Psilanthus Hook. f. (保留属名)●☆
98843 Cofer Loefl. = Symplocos Jacq. ●
98844 Coffea L. (1753);咖啡属;Coffee ●
98845 Coffea abeocuta De Wild. = Coffea liberica Bull. ex Hiern ●
98846 Coffea abeokutae P. J. S. Cramer = Coffea liberica Bull. ex Hiern ●
98847 Coffea abeokutae P. J. S. Cramer var. camerunensis A. Chev. = Coffea liberica Bull. ex Hiern ●
98848 Coffea abeokutae P. J. S. Cramer var. indeniensis (Siebert) A. Chev. = Coffea liberica Bull. ex Hiern ●
98849 Coffea abeokutae P. J. S. Cramer var. longicarpa Portères = Coffea liberica Bull. ex Hiern ●
98850 Coffea abeokutae P. J. S. Cramer var. macrocarpa A. Chev. = Coffea liberica Bull. ex Hiern ●
98851 Coffea abeokutae P. J. S. Cramer var. microcarpa A. Chev. = Coffea liberica Bull. ex Hiern ●
98852 Coffea abeokutae P. J. S. Cramer var. sphaerocarpa Portères = Coffea liberica Bull. ex Hiern ●
98853 Coffea affinis De Wild. = Coffea stenophylla G. Don ●
98854 Coffea afzelii Hiern = Argocoffeopsis afzelii (Hiern) Robbr. ●☆
98855 Coffea angolensis R. D. Good = Gardenia brachythamnus (K. Schum.) Launert ●☆
98856 Coffea ankaranensis J. -F. Leroy ex A. P. Davis et Rakotonas.;安卡兰咖啡●☆
98857 Coffea arabica L.;咖啡(阿拉伯咖啡,咖啡树,小果咖啡,小粒咖啡);Abyssinian Coffee, Arab Coffee, Arabian Coffee, Arabian Coffee Plant, Coffee, Coffee Plant, Coffee Tree, Coffee-tree, Common Coffee ●
98858 Coffea arabica L. var. intermedia A. Froehner = Coffea eugenioides S. Moore ●☆
98859 Coffea arabica L. var. leucocarpa Hiern = Tricalysia reflexa Hutch. ●☆
98860 Coffea arabica L. var. mokka Cramer = Coffea arabica L. ●
98861 Coffea arabica L. var. stuhlmannii Warb. ex A. Froehner = Coffea canephora Pierre ex A. Froehner ●
98862 Coffea arabica L. var. typica P. J. S. Cramer = Coffea arabica L. ●
98863 Coffea arnoldiana De Wild. = Coffea liberica Bull. ex Hiern var. dewevrei (De Wild. et T. Durand) Lebrun ●
98864 Coffea aruwimiensis De Wild. = Coffea liberica Bull. ex Hiern var. dewevrei (De Wild. et T. Durand) Lebrun ●
98865 Coffea becquetii A. Chev. = Coffea eugenioides S. Moore ●☆
98866 Coffea benghalensis K. Heyne ex Roem. et Schult.;米什咖啡(米什米咖啡);Bengal Coffee ●
98867 Coffea boiviniana (Baill.) Drake;博伊文咖啡●☆
98868 Coffea brenanii Leroy = Argocoffeopsis rupestris (Hiern) Robbr. ●☆
98869 Coffea brevipes Hiern;短梗咖啡●☆
98870 Coffea brevipes Hiern var. heterocalyx A. Chev. = Coffea heterocalyx Stoffelen ●☆
98871 Coffea bukobensis Zimm. = Coffea canephora Pierre ex A. Froehner ●
98872 Coffea canephora Pierre ex A. Froehner;中果咖啡(中粒咖啡);Coffee Shrub, Congo Coffee, Medianfruit Coffee, Median-fruited Coffee,Robusta Coffee ●
98873 Coffea canephora Pierre ex A. Froehner f. sankuruensis De Wild. = Coffea canephora Pierre ex A. Froehner ●

98874 Coffea canephora Pierre ex A. Froehner subvar. robusta (Linden ex De Wild.) A. Chev. = Coffea canephora Pierre ex A. Froehner ●
98875 Coffea canephora Pierre ex A. Froehner var. crassifolia Laurent ex De Wild. = Coffea canephora Pierre ex A. Froehner ●
98876 Coffea canephora Pierre ex A. Froehner var. gossweileri A. Chev. = Coffea canephora Pierre ex A. Froehner ●
98877 Coffea canephora Pierre ex A. Froehner var. hiernii Pierre = Coffea canephora Pierre ex A. Froehner ●
98878 Coffea canephora Pierre ex A. Froehner var. hinaultii Pierre = Coffea canephora Pierre ex A. Froehner ●
98879 Coffea canephora Pierre ex A. Froehner var. kouilouensis Pierre = Coffea canephora Pierre ex A. Froehner ●
98880 Coffea canephora Pierre ex A. Froehner var. laurentii (De Wild.) A. Chev. = Coffea canephora Pierre ex A. Froehner ●
98881 Coffea canephora Pierre ex A. Froehner var. muniensis Pierre = Coffea canephora Pierre ex A. Froehner ●
98882 Coffea canephora Pierre ex A. Froehner var. nganda A. E. Haarer = Coffea canephora Pierre ex A. Froehner ●
98883 Coffea canephora Pierre ex A. Froehner var. oka A. Chev. = Coffea canephora Pierre ex A. Froehner ●
98884 Coffea canephora Pierre ex A. Froehner var. oligoneura Pierre = Coffea canephora Pierre ex A. Froehner ●
98885 Coffea canephora Pierre ex A. Froehner var. opaca Pierre = Coffea canephora Pierre ex A. Froehner ●
98886 Coffea canephora Pierre ex A. Froehner var. robusta (Linden ex De Wild.) Chev.;大叶咖啡(粗咖啡,刚果咖啡);Robust Coffee ●
98887 Coffea canephora Pierre ex A. Froehner var. robusta (Linden) A. Chev. = Coffea canephora Pierre ex A. Froehner ●
98888 Coffea canephora Pierre ex A. Froehner var. sankuruensis (De Wild.) De Wild. = Coffea canephora Pierre ex A. Froehner ●
98889 Coffea canephora Pierre ex A. Froehner var. stuhlmannii (A. Froehner) A. Chev. = Coffea canephora Pierre ex A. Froehner ●
98890 Coffea canephora Pierre ex A. Froehner var. trillesii Pierre = Coffea canephora Pierre ex A. Froehner ●
98891 Coffea canephora Pierre ex A. Froehner var. ugandae (P. J. S. Cramer) A. Chev. = Coffea canephora Pierre ex A. Froehner ●
98892 Coffea canephora Pierre ex A. Froehner var. welwitschii (De Wild.) A. Chev. = Coffea canephora Pierre ex A. Froehner ●
98893 Coffea canephora Pierre ex A. Froehner var. wildemanii Pierre = Coffea canephora Pierre ex A. Froehner ●
98894 Coffea carrissoi A. Chev.;卡里索咖啡●☆
98895 Coffea claessensii Lebrun = Argocoffeopsis subcordata (Hiern) Lebrun ●☆
98896 Coffea congensis A. Froehner;刚果咖啡;Congo Coffee ●
98897 Coffea costatifructa Bridson;肋果咖啡●☆
98898 Coffea dactylifera Robbr. et Stoffelen;指状咖啡●☆
98899 Coffea dewevrei De Wild. et T. Durand = Coffea liberica Bull. ex Hiern var. dewevrei (De Wild. et T. Durand) Lebrun ●
98900 Coffea dewevrei De Wild. et T. Durand var. aruwimiensis (De Wild.) A. Chev. = Coffea liberica Bull. ex Hiern var. dewevrei (De Wild. et T. Durand) Lebrun ●
98901 Coffea dewevrei De Wild. et T. Durand var. dybowskii (Pierre ex De Wild.) A. Chev. = Coffea liberica Bull. ex Hiern var. dewevrei (De Wild. et T. Durand) Lebrun ●
98902 Coffea dewevrei De Wild. et T. Durand var. excelsa (A. Chev.) A. Chev. = Coffea liberica Bull. ex Hiern var. dewevrei (De Wild. et T. Durand) Lebrun ●
98903 Coffea dewevrei De Wild. et T. Durand var. ituriensis A. Chev. = Coffea liberica Bull. ex Hiern var. dewevrei (De Wild. et T. Durand) Lebrun ●
98904 Coffea dewevrei De Wild. et T. Durand var. neoarnoldiana (A. Chev.) A. Chev. = Coffea liberica Bull. ex Hiern var. dewevrei (De Wild. et T. Durand) Lebrun ●
98905 Coffea dewevrei De Wild. et T. Durand var. sylvatica (A. Chev.) A. Chev. = Coffea liberica Bull. ex Hiern var. dewevrei (De Wild. et T. Durand) Lebrun ●
98906 Coffea dewevrei De Wild. et T. Durand var. zenkeri (De Wild.) A. Chev. = Coffea liberica Bull. ex Hiern var. dewevrei (De Wild. et T. Durand) Lebrun ●
98907 Coffea divaricata K. Schum. = Argocoffeopsis rupestris (Hiern) Robbr. ●☆
98908 Coffea dybowskii H. C. Hall = Coffea liberica Bull. ex Hiern var. dewevrei (De Wild. et T. Durand) Lebrun ●
98909 Coffea dybowskii Pierre ex De Wild. = Coffea liberica Bull. ex Hiern var. dewevrei (De Wild. et T. Durand) Lebrun ●
98910 Coffea ebracteolata (Hiern) Brenan;无苞片咖啡●☆
98911 Coffea eketensis Wernham = Argocoffeopsis eketensis (Wernham) Robbr. ●☆
98912 Coffea engleri K. Krause = Sericanthe andongensis (Hiern) Robbr. subsp. engleri (K. Krause) Bridson ●☆
98913 Coffea eugenioides S. Moore;番樱桃咖啡;Nandi Coffee ●☆
98914 Coffea excelsa A. Chev.;高咖啡(大咖啡,大叶咖啡);High Coffee ●
98915 Coffea excelsa A. Chev. = Coffea liberica Bull. ex Hiern var. dewevrei (De Wild. et T. Durand) Lebrun ●
98916 Coffea excelsioides Portères = Coffea liberica Bull. ex Hiern ●
98917 Coffea fadenii Bridson;法登咖啡●☆
98918 Coffea gilgiana A. Froehner = Psilanthus mannii Hook. f. ●☆
98919 Coffea heterocalyx Stoffelen;异萼咖啡●☆
98920 Coffea hirsuta G. Don = Cremaspora triflora (Thonn.) K. Schum. ●☆
98921 Coffea humilis A. Chev.;低矮咖啡●☆
98922 Coffea hypoglauca Welw. ex Hiern = Belonophora coffeoides Hook. f. subsp. hypoglauca (Welw. ex Hiern) S. E. Dawson et Cheek ■☆
98923 Coffea ibo A. Froehner = Coffea zanguebariae Lour. ●☆
98924 Coffea intermedia (A. Froehner) A. Chev. = Coffea eugenioides S. Moore ●☆
98925 Coffea ituriensis A. Chev. = Coffea liberica Bull. ex Hiern ●
98926 Coffea jasminoides Welw. ex Hiern = Argocoffeopsis eketensis (Wernham) Robbr. ●☆
98927 Coffea jenkinsii Hook. f.;藏咖啡;Jenkins Coffee ●
98928 Coffea kapakata (A. Chev.) Bridson;卡帕特咖啡●☆
98929 Coffea kivuensis Lebrun;基伍咖啡●☆
98930 Coffea klainei Pierre = Coffea liberica Bull. ex Hiern ●
98931 Coffea kraussiana Hochst. = Kraussia floribunda Harv. ●☆
98932 Coffea lasiodelphys K. Schum. et K. Krause = Tricalysia lasiodelphys (K. Schum. et K. Krause) A. Chev. ●☆
98933 Coffea laurentii De Wild. = Coffea canephora Pierre ex A. Froehner ●
98934 Coffea laurifolia Salisb. = Coffea arabica L. ●
98935 Coffea laurina Poir. = Craterispermum laurinum (Poir.) Benth. ●☆
98936 Coffea laurina Smeathman ex DC.;月桂咖啡●☆
98937 Coffea lebrunianus R. Germ. et Kesler = Psilanthus lebrunianus

(R. Germ. et Kesler) Leroy ex Bridson ●☆

98938 Coffea lemblinii (A. Chev.) Keay = Argocoffeopsis lemblinii (A. Chev.) Robbr. ●☆

98939 Coffea lemblinii A. Chev. = Argocoffeopsis lemblinii (A. Chev.) Robbr. ●☆

98940 Coffea leroyi A. P. Davis;勒罗伊咖啡●☆

98941 Coffea liberica Bull ex Hiern;大果咖啡(大咖啡,大粒咖啡,利比里亚咖啡树,利比亚咖啡,中粒咖啡);Abeokuta Coffee, Liberia Coffee, Liberian Coffee, Monrovia Coffee, Rio Nufiez Coffee, Rio Nunez Coffee ●

98942 Coffea liberica Bull. ex Hiern 'Brasillian Bush';巴西咖啡(巴西木)●

98943 Coffea liberica Bull. ex Hiern f. bwambensis Bridson = Coffea liberica Bull ex Hiern ●

98944 Coffea liberica Bull. ex Hiern var. aurantiaca A. Chev. = Coffea liberica Bull ex Hiern ●

98945 Coffea liberica Bull. ex Hiern var. dewevrei (De Wild. et T. Durand) Lebrun = Coffea liberica Bull ex Hiern ●

98946 Coffea liberica Bull. ex Hiern var. gossweileri A. Chev. = Coffea liberica Bull ex Hiern ●

98947 Coffea liberica Bull. ex Hiern var. grandifolia A. Chev. = Coffea liberica Bull ex Hiern ●

98948 Coffea liberica Bull. ex Hiern var. indeniensis Seibert = Coffea liberica Bull ex Hiern ●

98949 Coffea liberica Bull. ex Hiern var. ivorensis Siebert = Coffea liberica Bull ex Hiern ●

98950 Coffea liberica Bull. ex Hiern var. liberiensis Seibert = Coffea liberica Bull ex Hiern ●

98951 Coffea liberica Bull. ex Hiern var. pyriformis Fauchère = Coffea liberica Bull ex Hiern ●

98952 Coffea liberica Bull. ex K. Schum. = Coffea liberica Bull ex Hiern ●

98953 Coffea liberica Hiern = Coffea liberica Bull ex Hiern ●

98954 Coffea ligustrifolia Stapf = Argocoffeopsis afzelii (Hiern) Robbr. ●☆

98955 Coffea ligustroides S. Moore;女贞咖啡●☆

98956 Coffea littoralis A. P. Davis et Rakotonas.;滨海咖啡●☆

98957 Coffea lulandoensis Bridson;卢兰多咖啡●☆

98958 Coffea maclaudii A. Chev. = Coffea canephora Pierre ex A. Froehner ●

98959 Coffea macrochlamys K. Schum. = Calycosiphonia macrochlamys (K. Schum.) Robbr. ●☆

98960 Coffea madagascariensis Drake ex Dubard = Tricalysia madagascariensis (Drake ex Dubard) A. Chev. ●☆

98961 Coffea magnistipula Stoffelen et Robbr.;大托叶咖啡●☆

98962 Coffea mayombensis A. Chev.;马永贝咖啡●☆

98963 Coffea melanocarpa Welw. ex Hiern = Psilanthus melanocarpus (Welw. ex Hiern) J.-F. Leroy ●☆

98964 Coffea microcarpa DC. = Cremaspora triflora (Thonn.) K. Schum. ●☆

98965 Coffea millotii J.-F. Leroy = Coffea sambavensis ex A. P. Davis et Rakotonas. ●☆

98966 Coffea minutiflora A. P. Davis et Rakotonas.;微花咖啡●☆

98967 Coffea moka Heynh. = Coffea arabica L. ●

98968 Coffea moratii ex A. P. Davis et Rakotonas.;莫拉特咖啡●☆

98969 Coffea mozambicana DC. = Coffea racemosa Lour. ●☆

98970 Coffea mufindiensis Hutch. ex Bridson;穆芬迪咖啡●☆

98971 Coffea mufindiensis Hutch. ex Bridson subsp. australis Bridson;南方穆芬迪咖啡●☆

98972 Coffea mufindiensis Hutch. ex Bridson subsp. pawekiana (Bridson) Bridson;帕氏咖啡●☆

98973 Coffea myrtifolia Roxb. = Coffea arabica L. ●

98974 Coffea nandiensis Dowson = Coffea eugenioides S. Moore ●☆

98975 Coffea neoarnoldiana A. Chev. = Coffea liberica Bull. ex Hiern var. dewevrei (De Wild. et T. Durand) Lebrun ●

98976 Coffea nigerina A. Chev. = Coffea ebracteolata (Hiern) Brenan ●☆

98977 Coffea nudiflora Stapf = Argocoffeopsis rupestris (Hiern) Robbr. ●☆

98978 Coffea nufindiensis A. Chev. = Coffea mufindiensis Hutch. ex Bridson ●☆

98979 Coffea oyemensis A. Chev. = Coffea liberica Bull. ex Hiern ●

98980 Coffea paolia Bridson;保尔咖啡●☆

98981 Coffea pawekiana Bridson = Coffea mufindiensis Hutch. ex Bridson subsp. pawekiana (Bridson) Bridson ●☆

98982 Coffea perrottetii Steud. ex H. Buek;佩罗咖啡●☆

98983 Coffea pervilleana (Baill.) Drake;佩尔咖啡●☆

98984 Coffea pierrei Gentil;皮埃尔咖啡●☆

98985 Coffea pocsii Bridson;波奇咖啡●☆

98986 Coffea pseudozanguebariae Bridson;假赞古咖啡●☆

98987 Coffea pulchella K. Schum. = Argocoffeopsis pulchella (K. Schum.) Robbr. ●☆

98988 Coffea quillou Wester = Coffea canephora Pierre ex A. Froehner ●

98989 Coffea racemosa Lour.;总花咖啡●☆

98990 Coffea racemosa Lour. var. myrtoidea A. Chev. = Coffea racemosa Lour. ●☆

98991 Coffea ramosa Roem. et Schult. = Coffea racemosa Lour. ●☆

98992 Coffea resinosa (Hook. f.) Radlk.;树脂咖啡●☆

98993 Coffea rhamnifolia (Chiov.) Bridson;鼠李叶咖啡●☆

98994 Coffea rhamnifolia (Chiov.) Bridson = Coffea paolia Bridson ●☆

98995 Coffea robusta L. Linden = Coffea canephora Pierre ex A. Froehner ●

98996 Coffea robusta Linden = Coffea canephora Pierre ex A. Froehner var. robusta (Linden ex De Wild.) Chev. ●

98997 Coffea robusta Linden ex De Wild. = Coffea canephora Pierre ex A. Froehner var. robusta (Linden ex De Wild.) Chev. ●

98998 Coffea royauxii De Wild. = Coffea liberica Bull. ex Hiern var. dewevrei (De Wild. et T. Durand) Lebrun ●

98999 Coffea rupestris Hiern = Argocoffeopsis rupestris (Hiern) Robbr. ●☆

99000 Coffea rupestris Hiern var. thonneri (Lebrun) A. Chev. = Argocoffeopsis rupestris (Hiern) Robbr. subsp. thonneri (Lebrun) Robbr. ●☆

99001 Coffea sambavensis ex A. P. Davis et Rakotonas.;桑巴咖啡●☆

99002 Coffea scandens K. Schum. = Argocoffeopsis scandens (K. Schum.) Lebrun ●☆

99003 Coffea schliebenii Bridson;施利本咖啡●☆

99004 Coffea schumanniana Busse = Coffea zanguebariae Lour. ●☆

99005 Coffea sessiliflora Bridson;无梗花咖啡●☆

99006 Coffea spathicalyx K. Schum. = Calycosiphonia spathicalyx (K. Schum.) Robbr. ●☆

99007 Coffea staudtii A. Froehner = Coffea brevipes Hiern ●☆

99008 Coffea stenophylla G. Don;狭叶咖啡;Bush Coffee, Narrowleaf Coffee, Narrow-leaved Coffee, Rio Nufiez Coffee, Rio Nunez Coffee, Upland Coffee ●

99009 Coffea subcordata Hiern = Argocoffeopsis subcordata (Hiern)

Lebrun ●☆
99010 Coffea swynnertonii S. Moore = Coffea racemosa Lour. ●☆
99011 Coffea sylvatica A. Chev. = Coffea liberica Bull. ex Hiern var. dewevrei (De Wild. et T. Durand) Lebrun ●
99012 Coffea talbotii Wernham = Tricalysia wernhamiana (Hutch. et Dalziel) Keay ●☆
99013 Coffea tetrandra Roxb. = Prismatomeris tetrandra (Roxb.) K. Schum. ●
99014 Coffea thonneri Lebrun = Argocoffeopsis rupestris (Hiern) Robbr. subsp. thonneri (Lebrun) Robbr. ●☆
99015 Coffea togoensis A. Chev.;多哥咖啡●☆
99016 Coffea ugandae P. J. S. Cramer = Coffea canephora Pierre ex A. Froehner ●
99017 Coffea vanroechoudtii Lebrun ex Van Roech. = Tricalysia vanroechoudtii (Lebrun ex Van Roech.) Robbr. ●☆
99018 Coffea vohemarensis A. P. Davis et Rakotonas.;武海马尔咖啡●☆
99019 Coffea vulgaris Moench = Coffea arabica L. ●
99020 Coffea welwitschii Pierre = Coffea canephora Pierre ex A. Froehner ●
99021 Coffea zanguebariae Lour.;赞古咖啡●☆
99022 Coffea zanzibarensis R. M. Grey = Coffea zanguebariae Lour. ●☆
99023 Coffea zenkeri De Wild. = Coffea liberica Bull. ex Hiern var. dewevrei (De Wild. et T. Durand) Lebrun ●
99024 Coffeaceae Batsch = Rubiaceae Juss.(保留科名)●■
99025 Coffeaceae J. Agardh = Rubiaceae Juss.(保留科名)●■
99026 Coffeaceae J. Agardh;咖啡科●
99027 Cogniauxella Baill. = Cogniauxia Baill. ■☆
99028 Cogniauxia Baill.(1884);科葫芦属■☆
99029 Cogniauxia ampla Cogn. = Cogniauxia podolaena Baill. ■☆
99030 Cogniauxia auriculata Cogn. = Cogniauxia podolaena Baill. ■☆
99031 Cogniauxia brazzaei Cogn. = Cogniauxia podolaena Baill. ■☆
99032 Cogniauxia cordifolia Cogn. = Cogniauxia podolaena Baill. ■☆
99033 Cogniauxia podolaena Baill.;科葫芦■☆
99034 Cogniauxia trilobata Cogn.;三裂科葫芦■☆
99035 Cogniauxiella Baill. = Cogniauxia Baill. ■☆
99036 Cogniauxiocharis (Schltr.) Hoehne = Stenorrhynchos Rich. ex Spreng. ■☆
99037 Cogsvellia Raf. = Cogswellia Poem. et Schult. ■☆
99038 Cogswellia Poem. et Schult. = Lomatium Raf. ■☆
99039 Cogswellia Schult. = Peucedanum L. ■
99040 Cogswellia Spreng.(1820);肖北美前胡属■☆
99041 Cogswellia Spreng. = Lomatium Raf. ■☆
99042 Cogswellia daucifolia (Torr. et A. Gray) M. E. Jones;肖北美前胡;Love-seed ■☆
99043 Cogswellia flava J. M. Coult. et Rose;黄肖北美前胡■☆
99044 Cogswellia grayi J. M. Coult. et Rose;格雷肖北美前胡■☆
99045 Cogswellia latifolia (Nutt.) M. E. Jones;宽叶肖北美前胡■☆
99046 Cogswellia lucida M. E. Jones;光亮肖北美前胡■☆
99047 Cogswellia orientalis M. E. Jones;东方北美前胡;White-flowered Parsley ■☆
99048 Cogylia Molina = Lardizabala Ruiz et Pav. ●☆
99049 Cohautia Endl. = Kohautia Cham. et Schltdl. ■☆
99050 Cohautia Endl. = Oldenlandia L. ●■
99051 Cohiba Raf. = Wigandia Kunth(保留属名)●■☆
99052 Cohnia Kunth = Cordyline Comm. ex R. Br.(保留属名)●
99053 Cohnia Kunth(1850);空树属●☆
99054 Cohnia Rchb. f. = Cohniella Pfitzer ■☆
99055 Cohnia floribunda Kunth;多花空树●☆
99056 Cohnia macrophylla Kunth;大叶空树●☆
99057 Cohnia parviflora Kunth;小花空树●☆
99058 Cohniella Pfitzer(1889);科恩兰属■☆
99059 Cohniella quekettioides Pfitzer;科恩兰■☆
99060 Coiladena Raf. = Ipomoea L.(保留属名)●■
99061 Coilantha Borkh. = Gentiana L. ■
99062 Coilanthera Raf. = Cordia L.(保留属名)●
99063 Coilmeroa Endl. = Colmeiroa Reut. ●☆
99064 Coilmeroa Endl. = Securinega Comm. ex Juss.(保留属名)●☆
99065 Coilocarpus Domin = Sclerolaena R. Br. ●☆
99066 Coilocarpus F. Muell. ex Domin = Sclerolaena R. Br. ●☆
99067 Coilochilus Schltr.(1906);空唇兰属■☆
99068 Coilochilus neocaledonicum Schltr.;空唇兰■☆
99069 Coilomphis Raf. = Melaleuca L.(保留属名)●
99070 Coilonox Raf.(1837);钻丝风信子属■☆
99071 Coilonox Raf. = Ornithogalum L. ■
99072 Coilonox albucoides Raf.;钻丝风信子■☆
99073 Coilosperma Raf. = Deeringia R. Br. ●■
99074 Coilostigma Klotzsch = Erica L. ●☆
99075 Coilostigma Klotzsch = Salaxis Salisb. ●☆
99076 Coilostigma Klotzsch(1838);空柱杜鹃属●☆
99077 Coilostigma dregeanum Klotzsch = Erica zeyheriana (Klotzsch) E. G. H. Oliv. ●☆
99078 Coilostigma glabrum Benth. = Erica burchelliana E. G. H. Oliv. ●☆
99079 Coilostigma puberulum (Klotzsch) Benth. = Erica puberuliflora E. G. H. Oliv. ●☆
99080 Coilostigma tenuifolium Klotzsch = Erica zeyheriana (Klotzsch) E. G. H. Oliv. ●☆
99081 Coilostigma zeyherianum Klotzsch;泽赫空柱杜鹃●☆
99082 Coilostigma zeyherianum Klotzsch = Erica zeyheriana (Klotzsch) E. G. H. Oliv. ●☆
99083 Coilostigma zeyherianum Klotzsch var. tenuifolium (Klotzsch) E. G. H. Oliv. = Erica zeyheriana (Klotzsch) E. G. H. Oliv. ●☆
99084 Coilostylis Raf. = Epidendrum L.(保留属名)■☆
99085 Coilotapalus P. Browne(废弃属名) = Cecropia Loefl.(保留属名)●☆
99086 Coincya Rouy(1891);星芥属;Cabbage ■☆
99087 Coincya cheiranthos (Vill.) Greuter et Burdet = Coincya monensis (L.) Greuter et Burdet subsp. cheiranthos (Vill.) Aedo, Leadlay et Munoz Garm. ■☆
99088 Coincya concyoides (Humbert et Maire) Greuter et Burdet = Coincya monensis (L.) Greuter et Burdet ■☆
99089 Coincya monensis (L.) Greuter et Burdet;星芥;Isle-of-man Cabbage,Star-mustard ■☆
99090 Coincya monensis (L.) Greuter et Burdet subsp. cheiranthos (Vill.) Aedo,Leadlay et Munoz Garm.;反折星芥;Star-mustard, Wallflower Cabbage ■☆
99091 Coincya monensis (L.) Greuter et Burdet subsp. hispida (Cav.) Leadlay = Coincya monensis (L.) Greuter et Burdet subsp. orophila (Franco) Aedo et al. ■☆
99092 Coincya monensis (L.) Greuter et Burdet subsp. orophila (Franco) Aedo et al.;喜山星芥■☆
99093 Coincya monensis (L.) Greuter et Burdet subsp. recurvata (All.) Leadlay = Coincya monensis (L.) Greuter et Burdet subsp. cheiranthos (Vill.) Aedo,Leadlay et Munoz Garm. ■☆
99094 Coincya wrightii (O. E. Schulz) Stace;赖特星芥;Hutera

wrightii, Lundy Cabbage, Rhynchosinapis wrightii ■☆

99095 Coinochlamys T. Anderson ex Benth. et Hook. f. (1876);同被马钱属●☆

99096 Coinochlamys T. Anderson ex Benth. et Hook. f. = Mostuea Didr. ●☆

99097 Coinochlamys angolana S. Moore = Mostuea hirsuta (T. Anderson ex Benth. et Hook. f.) Baill. ex Baker ●☆

99098 Coinochlamys congolana Gilg = Mostuea hirsuta (T. Anderson ex Benth. et Hook. f.) Baill. ex Baker ●☆

99099 Coinochlamys congolana Gilg var. laurentii De Wild. = Mostuea hirsuta (T. Anderson ex Benth. et Hook. f.) Baill. ex Baker ●☆

99100 Coinochlamys gabonica (Baill.) Soler. ex Durand et Jacks. = Mostuea hirsuta (T. Anderson ex Benth. et Hook. f.) Baill. ex Baker ●☆

99101 Coinochlamys poggeana Gilg = Mostuea hirsuta (T. Anderson ex Benth. et Hook. f.) Baill. ex Baker ●☆

99102 Coinochlamys schweinfurthii Gilg = Mostuea hirsuta (T. Anderson ex Benth. et Hook. f.) Baill. ex Baker ●☆

99103 Coinogyne Less. = Jaumea Pers. ■●☆

99104 Coinogyne carnosa Less. = Jaumea carnosa (Less.) A. Gray ■☆

99105 Coix L. (1753);薏苡属;Jobstears ●■

99106 Coix agrestis Lour. = Coix lacryma-jobi L. ●■

99107 Coix aquatica Roxb.;水生薏苡;Water Jobstears ■

99108 Coix arundinacea Koen. ex Willd. = Chionachne koenigii (Spreng.) Thwaites ■☆

99109 Coix arundinacea Lam. = Coix lacryma-jobi L. ●■

99110 Coix barbata Roxb. = Chionachne koenigii (Spreng.) Thwaites ■☆

99111 Coix barbata Roxb. = Polytoca digitata (L. f.) Druce ■

99112 Coix chinensis Tod. = Coix lacryma-jobi L. var. mayuen (Rom. Caill.) Stapf ex Backer ■

99113 Coix chinensis Tod. ex Balansa = Coix lacryma-jobi L. var. mayuen (Rom. Caill.) Stapf ex Backer ■

99114 Coix chinensis Tod. ex Balansa var. formosana (Ohwi) L. Liou = Coix lacryma-jobi L. var. mayuen (Rom. Caill.) Stapf ex Backer ■

99115 Coix chinensis Tod. var. formosana (Ohwi) L. Liou;台湾薏苡;Taiwan Jobstears ■

99116 Coix exaltata Jacq. = Coix lacryma-jobi L. ●■

99117 Coix gigantea Roxb. = Coix aquatica Roxb. ■

99118 Coix gigantea Roxb. subsp. aquatica (Roxb.) Bhattacharya = Coix aquatica Roxb. ■

99119 Coix gigantea Roxb. var. aquatica (Roxb.) Watt = Coix aquatica Roxb. ■

99120 Coix heteroclita Roxb. = Polytoca digitata (L. f.) Druce ■

99121 Coix koenigii Spreng. = Chionachne koenigii (Spreng.) Thwaites ■☆

99122 Coix lacryma L. = Coix lacryma-jobi L. ●■

99123 Coix lacryma L. var. stenocarpa Oliv. = Coix lacryma-jobi L. var. stenocarpa (Oliv.) Stapf ■

99124 Coix lacryma-jobi L.;薏苡(必提珠,便婆菊,草菩提,草鱼目,草珠儿,草珠子,川谷,催生子,大碗子,竿珠,感米,赣米,赣珠,沟子米,鬼贮箭,桂珠黍,回回米,胶念珠,解蠡,老鸦珠,蓼茶子,六谷米,六谷子,米珠,尿端子,尿塘草,尿塘珠,尿珠子,菩提珠,菩提子,葡芦,蒲米,祁薏米,芑实,起目,起实,水玉米,素珠果,天谷,铁玉蜀黍,物莢,药玉米,野绿米,野薏米,苡米,苡仁米,益米,薏米,薏仁米,薏黍,薏苡仁,薏珠子,有乙梅,玉米,玉秫,玉珠,裕米,珠珠米,珠珠密,珠子,珠子米);Adlai, Adlay, Job's Tears, Jobstears, Job's-tears ●■

99125 Coix lacryma-jobi L. f. aquatica (Roxb.) Backer = Coix aquatica Roxb. ■

99126 Coix lacryma-jobi L. subsp. ma-yuen (Rom. Caill.) T. Koyama = Coix lacryma-jobi L. var. mayuen (Rom. Caill.) Stapf ex Backer ■

99127 Coix lacryma-jobi L. var. formosana Ohwi = Coix chinensis Tod. var. formosana (Ohwi) L. Liou ■

99128 Coix lacryma-jobi L. var. formosana Ohwi = Coix lacryma-jobi L. var. mayuen (Rom. Caill.) Stapf ex Backer ■

99129 Coix lacryma-jobi L. var. frumentacea Makino = Coix chinensis Tod. ■

99130 Coix lacryma-jobi L. var. frumentacea Makino = Coix lacryma-jobi L. var. mayuen (Rom. Caill.) Stapf ex Backer ■

99131 Coix lacryma-jobi L. var. maxima Makino;念珠薏苡■

99132 Coix lacryma-jobi L. var. maxima Makino = Coix lacryma-jobi L. ●■

99133 Coix lacryma-jobi L. var. mayuen (Rom. Caill.) Stapf = Coix lacryma-jobi L. var. mayuen (Rom. Caill.) Stapf ex Backer ■

99134 Coix lacryma-jobi L. var. mayuen (Rom. Caill.) Stapf ex Backer;薏米(芑实,川谷,感米,回回米,六谷米,绿谷,马圆薏苡,苡米,薏仁,薏苡仁);China Jobstears ■

99135 Coix lacryma-jobi L. var. mayuen (Rom. Caill.) Stapf ex Backer = Coix chinensis Tod. ■

99136 Coix lacryma-jobi L. var. puellarum (Balansa) A. Camus = Coix puellarum Balansa ■

99137 Coix lacryma-jobi L. var. stenocarpa (Oliv.) Stapf;窄果薏苡;Narrowfruit Jobstears ■

99138 Coix lacryma-jobi L. var. stenocarpa Stapf;狭果薏苡■☆

99139 Coix lacryma-jobi L. var. stenocarpa Stapf = Coix stenocarpa (Oliv.) Balansa ■

99140 Coix lacryma-jobi L. var. tubulosa K. Schum. et Lauterb. = Coix lacryma-jobi L. var. stenocarpa (Oliv.) Stapf ■

99141 Coix lacryoma L. = Coix lacryma-jobi L. ●■

99142 Coix lingulata Hack. = Coix aquatica Roxb. ■

99143 Coix ma-yuen Rom. Caill. = Coix chinensis Tod. ■

99144 Coix ma-yuen Rom. Caill. = Coix lacryma-jobi L. var. mayuen (Rom. Caill.) Stapf ex Backer ■

99145 Coix ovata Stokes = Coix lacryma-jobi L. ●■

99146 Coix puellarum Balansa;小珠薏苡;Pear Jobstears ■

99147 Coix puellarum Balansa = Coix lacryma-jobi L. var. puellarum (Balansa) A. Camus ■

99148 Coix stenocarpa (Oliv.) Balansa = Coix lacryma-jobi L. var. stenocarpa (Oliv.) Stapf ■

99149 Coix tubulosa Hack. = Coix lacryma-jobi L. var. stenocarpa (Oliv.) Stapf ■

99150 Coix tubutosa Hack. ex Warb. = Coix stenocarpa (Oliv.) Balansa ■

99151 Cojoba Britton et Rose = Pithecellobium Mart. (保留属名)●

99152 Cojoba Britton et Rose(1928);科若木属●☆

99153 Cojoba arborea (L.) Britton et Rose;乔木科若木●☆

99154 Cojoba glabra Britton et Rose;光科若木●☆

99155 Cojoba graciliflora (S. F. Blake) Britton et Rose;细花科若木;Guadeloupe Blackbead ●☆

99156 Cojoba micrantha (Benth.) Britton et Rose;小花科若木●☆

99157 Cola Schott et Endl. (1832)(保留属名);可拉木属(非洲梧桐属,可拉属,可乐果属,可乐树属,苏丹梧桐属);Cola, Cola Nut, Kola, Kola Nut ●☆

99158 Cola acuminata (Brenan) Schott et Endl.;可拉木(阿拉树,红可拉,尖叶可乐树,可拉,可拉豆,苏丹可乐果);Abata Cola, Bichy

Tree, Cola, Cola Nut, Kola, Sudan Cola Nut, Sudan Cola-nut ●☆
99159 Cola acuminata (P. Beauv.) Schott et Endl. = Cola acuminata (Brenan) Schott et Endl. ●☆
99160 Cola angustifolia K. Schum.;窄叶可拉木●☆
99161 Cola anomala K. Schum.;巴门达可拉木;Bamenda Cola ●☆
99162 Cola argentea Mast.;银白可拉木●☆
99163 Cola attiensis Aubrév. et Pellegr. var. bodardii (Pellegr.) N. Hallé;博达尔可拉木●☆
99164 Cola ballayi Cornu ex Heckel;博利可拉木●☆
99165 Cola bipindensis Engl. = Cola diversifolia Engl. ●☆
99166 Cola bodardii Pellegr. = Cola attiensis Aubrév. et Pellegr. var. bodardii (Pellegr.) N. Hallé ●☆
99167 Cola bracteata De Wild. = Cola congolana De Wild. et T. Durand ●☆
99168 Cola brevipes K. Schum.;短梗可拉木●☆
99169 Cola brevipes K. Schum. var. hirsuta (Pellegr.) N. Hallé;毛短梗可拉木●☆
99170 Cola bruneelii De Wild.;布吕内尔可拉木●☆
99171 Cola buesgenii Engl.;比斯根可拉木●☆
99172 Cola buntingii Baker f.;邦廷可拉木●☆
99173 Cola cabindensis Exell;卡宾达可拉木●☆
99174 Cola cauliflora Mast.;茎花可拉木●☆
99175 Cola chlamydantha K. Schum. = Chlamydocola chlamydantha (K. Schum.) M. Bodard ●☆
99176 Cola chlorantha F. White;绿花可拉木●☆
99177 Cola clavata Mast.;棍棒可拉木●☆
99178 Cola coccinea Engl. et K. Krause;绯红可拉木●☆
99179 Cola congolana De Wild. et T. Durand;刚果可拉木●☆
99180 Cola congolana De Wild. et T. Durand var. puberula R. Germ.;微毛可拉木●☆
99181 Cola cordifolia (Cav.) R. Br.;心叶可拉木(心叶非洲梧桐,心叶可拉);Monkey Kola ●☆
99182 Cola cordifolia (Cav.) R. Br. var. maclaudi A. Chev. = Cola lateritia K. Schum. var. maclaudi (A. Chev.) Brenan et Keay ●☆
99183 Cola cordifolia R. Br. = Cola cordifolia (Cav.) R. Br. ●☆
99184 Cola crispiflora K. Schum.;皱波花可拉木●☆
99185 Cola digitata Mast.;指裂可拉木●☆
99186 Cola diversifolia De Wild. et T. Durand;异叶可拉木●☆
99187 Cola diversifolia Engl. = Cola diversifolia De Wild. et T. Durand ●☆
99188 Cola duparquetiana Baill.;迪帕可拉木●☆
99189 Cola fibrillosa Engl. et K. Krause;须毛可拉木●☆
99190 Cola ficifolia Mast.;榕叶可拉木●☆
99191 Cola ficifolia Mast. var. macrantha (K. Schum.) N. Hallé;大花榕叶可拉木●☆
99192 Cola flavescens Engl. var. hirsuta Pellegr. = Cola brevipes K. Schum. var. hirsuta (Pellegr.) N. Hallé ●☆
99193 Cola flaviflora Engl. et K. Krause;黄花可拉木●☆
99194 Cola flavo-velutina K. Schum.;黄绒毛可拉木●☆
99195 Cola gabonensis Mast.;加蓬可拉木●☆
99196 Cola gigantea A. Chev.;巨大可拉木●☆
99197 Cola gigantea A. Chev. var. glabrescens Brenan et Keay;渐光可拉木●☆
99198 Cola gigas Baker f.;大可拉木●☆
99199 Cola gilletii De Wild.;吉勒特可拉木(吉勒特可拉)●☆
99200 Cola glabra Brenan et Keay;光滑可拉木●☆
99201 Cola glaucoviridis Pellegr.;灰绿可拉木●☆
99202 Cola greenwayi Brenan;小果可拉木●☆
99203 Cola greenwayi Brenan var. keniensis?;肯尼亚可拉木●☆
99204 Cola griseiflora De Wild.;灰花可拉木●☆
99205 Cola heterophylla (P. Beauv.) Schott et Endl.;互叶可拉木●☆
99206 Cola hispida Brenan et Keay;硬毛可拉木●☆
99207 Cola humilis A. Chev. = Cola heterophylla (P. Beauv.) Schott et Endl. ●☆
99208 Cola humilis Pierre ex Bodard;低矮可拉木●☆
99209 Cola hypochrysea K. Schum.;金背可拉木●☆
99210 Cola lasiantha Engl. et K. Krause;毛花可拉木●☆
99211 Cola lateritia K. Schum.;砖红可拉木●☆
99212 Cola lateritia K. Schum. var. maclaudi (A. Chev.) Brenan et Keay;马克洛可拉木●☆
99213 Cola laurifolia Mast.;月桂叶可拉木●☆
99214 Cola laurina Roberty = Cola laurifolia Mast. ●☆
99215 Cola leonensis Hutch. ex Lane-Poole = Cola lateritia K. Schum. var. maclaudi (A. Chev.) Brenan et Keay ●☆
99216 Cola letestui Pellegr.;莱泰斯图可拉木●☆
99217 Cola letouzeyana Nkongmeneck;勒图可拉木●☆
99218 Cola liberica Jongkind;离生可拉木●☆
99219 Cola limbengensis Pellegr. = Cola griseiflora De Wild. ●☆
99220 Cola lindensis Engl.;林德可拉木●☆
99221 Cola linearis Pierre ex Pellegr.;线状可拉木●☆
99222 Cola lizae N. Hallé;利扎可拉木●☆
99223 Cola louisii R. Germ.;路易斯可拉木●☆
99224 Cola maclaudii (A. Chev.) Aubrév. = Cola lateritia K. Schum. var. maclaudi (A. Chev.) Brenan et Keay ●☆
99225 Cola maclaudii A. Chev. = Cola lateritia K. Schum. var. maclaudi (A. Chev.) Brenan et Keay ●☆
99226 Cola macrantha K. Schum. = Cola ficifolia Mast. var. macrantha (K. Schum.) N. Hallé ●☆
99227 Cola mayumbensis Exell;马永巴可拉木●☆
99228 Cola megalophylla Brenan et Keay;大叶可拉木●☆
99229 Cola metallica Cheek;光泽可拉木●☆
99230 Cola microcarpa Brenan = Cola greenwayi Brenan ●☆
99231 Cola millenii K. Schum.;米伦可拉木●☆
99232 Cola minor Brenan;较小可拉木●☆
99233 Cola mirabilis A. Chev. = Chlamydocola chlamydantha (K. Schum.) M. Bodard ●☆
99234 Cola mossambicensis Wild;莫桑比克可拉木●☆
99235 Cola nana Engl. et K. Krause;矮小可拉木●☆
99236 Cola natalensis Oliv.;纳塔尔可拉木●☆
99237 Cola ndongensis Engl. et K. Krause;恩东加可拉木●☆
99238 Cola nigerica Brenan et Keay;尼日利亚可拉木●☆
99239 Cola nitida (Vent.) A. Chev. = Cola nitida (Vent.) Schott et Endl. ●☆
99240 Cola nitida (Vent.) Schott et Endl.;大叶可乐树(白可拉,光亮可乐果,可拉果,苏丹可乐果);Bissy Nut, Cola Nut, Gourou Nut, Karo Nut, Karoo Nut, Kola ●☆
99241 Cola nitida (Vent.) Schott et Endl. = Cola nitida (Vent.) A. Chev. ●☆
99242 Cola nitida (Vent.) Schott et Endl. subsp. pallida A. Chev. = Cola pallida (A. Chev.) M. Bodard ●☆
99243 Cola nitida Schott et Endl. = Cola nitida (Vent.) A. Chev. ●☆
99244 Cola noldeae Exell;诺尔德可拉木●☆
99245 Cola numbi Exell = Cola brevipes K. Schum. ●☆
99246 Cola obtusa Engl. et K. Krause;钝可拉木●☆
99247 Cola octoloboides Brenan;八裂可拉木●☆

99248 Cola pachycarpa K. Schum.;粗果可拉木●☆
99249 Cola pallida (A. Chev.) M. Bodard;苍白可拉木●☆
99250 Cola pierlotii R. Germ.;皮氏可拉木●☆
99251 Cola porphyrantha Brenan;紫花可拉木●☆
99252 Cola pugionifera K. Schum. = Cola duparquetiana Baill. ●☆
99253 Cola pulcherrima Engl.;美丽可拉木●☆
99254 Cola quinqueloba Garcke = Sterculia quinqueloba (Garcke) K. Schum. ●☆
99255 Cola reticulata A. Chev.;网状可拉木●☆
99256 Cola rhynchophylla K. Schum.;喙叶可拉木●☆
99257 Cola ricinifolia Engl. et K. Krause;蓖麻叶可拉木●☆
99258 Cola rolandi-principis A. Chev. = Octolobus spectabilis Welw. ●☆
99259 Cola rostrata K. Schum.;喙可拉木●☆
99260 Cola scheffleri K. Schum.;谢夫勒可拉木●☆
99261 Cola semecarpophylla K. Schum.;肉托果叶可拉木●☆
99262 Cola sphaerocarpa A. Chev.;球果可拉木●☆
99263 Cola subglaucescens Engl.;粉绿可拉木●☆
99264 Cola suboppositifolia Cheek;对叶可拉木●☆
99265 Cola sulcata Engl.;纵沟可拉木●☆
99266 Cola supfiana Busse;祖普夫可拉木●☆
99267 Cola talbotii Baker f. = Cola argentea Mast. ●☆
99268 Cola tessmannii Engl. et K. Krause;泰斯曼可拉木●☆
99269 Cola triloba (R. Br.) K. Schum.;三裂可拉木●☆
99270 Cola uloloma Brenan;卷边可拉木●☆
99271 Cola umbratilis Brenan et Keay;荫蔽可拉木●☆
99272 Cola urceolata K. Schum.;坛状可拉木●☆
99273 Cola usambarensis Engl.;乌桑巴拉可拉木●☆
99274 Cola vandersmisseniana R. Germ.;范德可拉木●☆
99275 Cola vera K. Schum. = Cola nitida (Vent.) Schott et Endl. ●☆
99276 Cola verticillata (Thonn.) A. Chev. = Cola verticillata (Thonn.) Stapf ex A. Chev. ●☆
99277 Cola verticillata (Thonn.) Stapf ex A. Chev.;轮叶可拉木●☆
99278 Cola winkleri Engl.;温克勒可拉木●☆
99279 Colania Gagnep. = Aspidistra Ker Gawl. ●■
99280 Colania tonkinensis Gagnep. = Aspidistra tonkinensis (Gagnep.) F. T. Wang et K. Y. Lang ■
99281 Colanthelia McClure et E. W. Sm. (1973);短序竹属●☆
99282 Colanthelia burchellii (Munro) McClure;短序竹●☆
99283 Colanthelia intermedia (McClure et L. B. Sm.) McClure;全缘短序竹●☆
99284 Colanthelia macrostachya (Nees) McClure;大穗短序竹●☆
99285 Colaria Raf. = Cola Schott et Endl. (保留属名)●☆
99286 Colax Lindl. = Colax Lindl. ex Spreng. ■☆
99287 Colax Lindl. = Pabstia Garay ■☆
99288 Colax Lindl. ex Spreng. (1843);巴西寄生兰属■☆
99289 Colax jugosus (Lindl.) Lindl.;巴西寄生兰■☆
99290 Colax jugosus Lindl. = Colax jugosus (Lindl.) Lindl. ■☆
99291 Colbertia Salisb. = Dillenia L. ●
99292 Colchicaceae DC. (1804)(保留科名);秋水仙科■
99293 Colchicum L. (1753);秋水仙属;Autumn Crocus, Autumn-Croctus, Colchicum, Meadow Saffron ■
99294 Colchicum agrippinum Baker;尖瓣秋水仙■☆
99295 Colchicum aitchisonii (Hook. f.) Nasir;爱奇秋水仙■☆
99296 Colchicum alpinum Lam.;高山秋水仙;Alpine Autumn-croctus ■☆
99297 Colchicum autumnale L.;秋水仙;Arsenic Plant, Autumn, Autumn Crocus, Autumnal Meadow Saffron, Autumn-crocus, Colchicum, Common Autumn Crocus, Crocus, Crocus Autumn, Dagger, Daughter-before-mother, Fall Crocus, Fog Crocus, Go-to-sleep-at-noon, Greet-wort, Kite's Legs, Meadow Crocus, Meadow Saffron, Michaelmas Crocus, Mysteria, Naked Boys, Naked Jack, Naked Ladies, Naked Maidens, Naked Men, Naked Nannies, Naked Virgin, Naked Virgins, Pop-up, Purple Crocus, Rams, Snake's Flower, Son-before-the-father, Star Naked Boys, Star Naked Ladies, Star-naked Boys, Strip-jack-naked, Tube-root, Upstart, Wild Saffron ■☆
99298 Colchicum autumnale L. 'Alboplenum';白重瓣秋水仙■☆
99299 Colchicum autumnale L. subsp. algeriense Batt. = Colchicum lusitanum Brot. ■☆
99300 Colchicum autumnale L. var. albiflorum Hort.;白花秋水仙;Whiteflower Autumn-crocus ■☆
99301 Colchicum autumnale L. var. album Hort. = Colchicum autumnale L. var. albiflorum Hort. ■☆
99302 Colchicum autumnale L. var. atropurpureum Hort.;紫花秋水仙■☆
99303 Colchicum autumnale L. var. florepleno Hort. = Colchicum autumnale L. var. plenum Hort. ■☆
99304 Colchicum autumnale L. var. majus Hort.;大秋水仙;Large Autumn-crocus ■☆
99305 Colchicum autumnale L. var. plenum Hort.;重瓣秋水仙;Doubleflower Autumn-crocus ■☆
99306 Colchicum bertolonii Stev. = Colchicum cupanii Guss. ■☆
99307 Colchicum biebersteinii Rouy;毕氏秋水仙■☆
99308 Colchicum bivonae Guss.;荷瓣秋水仙■☆
99309 Colchicum bivonae Willk. et Lange = Colchicum bivonae Guss. ■☆
99310 Colchicum bornmuelleri Freyn;早花秋水仙;Bornmueller Autumn-crocus ■☆
99311 Colchicum bowlesianum B. L. Burtt = Colchicum bivonae Willk. et Lange ■☆
99312 Colchicum bulbocodioides Kunth = Colchicum triphyllum Kuntze ■☆
99313 Colchicum byzantinum Ker Gawl.;东方秋水仙(拜占庭秋水仙,土荆芥);Oriental Meadow Saffron, Oriental Meadow-saffron ■☆
99314 Colchicum byzantinum Ker Gawl. var. cilicicum Boiss.;睫毛东方秋水仙■☆
99315 Colchicum cilicicum Dammer;土耳其秋水仙■☆
99316 Colchicum cornigerum (Sickenb.) Tackh. et Drar;角秋水仙■☆
99317 Colchicum crociflorum Anderson = Colchicum autumnale L. ■☆
99318 Colchicum cupanii Guss.;库潘秋水仙■☆
99319 Colchicum cupanii Guss. var. bertolonii (Steven) Maire et Weiller = Colchicum cupanii Guss. ■☆
99320 Colchicum cupanii Guss. var. pulverulentum Batt. = Colchicum cupanii Guss. ■☆
99321 Colchicum haussknechtii Boiss.;豪斯秋水仙■☆
99322 Colchicum hybridum Hort.;杂种秋水仙■☆
99323 Colchicum kesselringii Regel;雷格秋水仙■☆
99324 Colchicum laetum Steven;喜悦秋水仙■☆
99325 Colchicum latifolium Sibth. et Sm.;阔叶秋水仙■☆
99326 Colchicum longifolium Castagne = Colchicum neapolitanum (Ten.) Ten. ■☆
99327 Colchicum longifolium Castagne var. micranthum Emb. et Maire = Colchicum neapolitanum (Ten.) Ten. ■☆
99328 Colchicum lusitanum Brot.;路西特秋水仙■☆
99329 Colchicum luteum Baker;黄秋水仙(花叶秋水仙,黄花秋水仙) ■☆
99330 Colchicum macrophylla B. L. Burtt;大叶秋水仙■☆
99331 Colchicum multiflorum Brot.;多花秋水仙■☆

99332 Colchicum multiflorum Brot. = Colchicum autumnale L. ■☆
99333 Colchicum neapolitanum (Ten.) Ten.;长叶秋水仙■☆
99334 Colchicum neapolitanum (Ten.) Ten. var. castrense (De Laramb.) Debeaux = Colchicum neapolitanum (Ten.) Ten. ■☆
99335 Colchicum neapolitanum (Ten.) Ten. var. micranthum Emb. et Maire = Colchicum neapolitanum (Ten.) Ten. ■☆
99336 Colchicum nivale Boiss. et Huet ex Stefani;雪白秋水仙■☆
99337 Colchicum patens Schultz = Colchicum autumnale L. ■☆
99338 Colchicum persicum Hort. = Colchicum byzantinum Ker Gawl. ■☆
99339 Colchicum pusillum Sieber;微小秋水仙■☆
99340 Colchicum regelii Stefani = Colchicum kesselringii Regel ■☆
99341 Colchicum ritchii R. Br.;里奇秋水仙■☆
99342 Colchicum ritchii R. Br. var. pusillum Bég. et Vacc. = Colchicum ritchii R. Br. ■☆
99343 Colchicum serpentinum Miscz.;蛇形秋水仙■☆
99344 Colchicum sibthorpii Baker = Colchicum bivonae Willk. et Lange ■☆
99345 Colchicum speciosum Steven;美丽秋水仙(丽花秋水仙);Crocus, Pretty Autumn-crocus ■☆
99346 Colchicum speciosum Steven 'Album';白花丽花秋水仙■☆
99347 Colchicum speciosum Steven var. giganteum Hort.;粉花秋水仙;Pinkflower Autumn-crocus ■☆
99348 Colchicum szovitsii Fisch. et May.;绍氏秋水仙■☆
99349 Colchicum triphyllum Kuntze;三叶秋水仙■☆
99350 Colchicum umbrosum Steven;耐荫秋水仙■☆
99351 Colchicum variegatum L.;斑点秋水仙(缟花秋水仙);Variegated Meadow Saffron ■☆
99352 Colchicum vernum Kunth;春花秋水仙;Meadow Saffron, Meadow-saffron-of-the-spring, Spring Colchicum, Spring Meadow-saffron ■☆
99353 Colchicum vernum Kunth = Bulbocodium vernum L. ■☆
99354 Colchicum visianii Parl.;维氏秋水仙■☆
99355 Colchicum zangezurum Grossh.;赞格祖尔秋水仙■☆
99356 Coldenella Ellis = Coptis Salisb. ■
99357 Coldenella Ellis = Fibra Colden ■
99358 Coldenia L. (1753);双柱紫草属(刺锚草属,生果草属);Coldenia ■
99359 Coldenia angolensis Welw. = Coldenia procumbens L. ■
99360 Coldenia nuttallii Benth. ex Hook. = Tiquilia nuttallii (Benth. ex Hook.) A. T. Richardson ■☆
99361 Coldenia procumbens L.;双柱紫草(安哥拉双柱紫草,刺锚草,卧茎同篱生果草);Procumbent Coldenia ■
99362 Coldenia succulenta Peter = Heliotropium curassavicum L. ■☆
99363 Colea Bojer = Colea Bojer ex Meisn. (保留属名)●☆
99364 Colea Bojer ex Meisn. (1840)(保留属名);鞘葳属●☆
99365 Colea aberrans Baill. = Rhodocolea racemosa (Lam.) H. Perrier ●☆
99366 Colea alata H. Perrier;高鞘葳●☆
99367 Colea alba H. Perrier;白鞘葳●☆
99368 Colea asperrima H. Perrier;粗糙鞘葳●☆
99369 Colea barbatula H. Perrier;髯毛鞘葳●☆
99370 Colea bernieri Baill. ex H. Perrier;伯尼尔鞘葳●☆
99371 Colea boivinii Baill. = Rhodocolea boivinii (Baill.) H. Perrier ●☆
99372 Colea campenonii H. Perrier;康珀农鞘葳●☆
99373 Colea cauliflora A. DC. = Ophiocolea floribunda (Bojer ex Lindl.) H. Perrier ●☆
99374 Colea cava A. H. Gentry;凹地鞘葳●☆
99375 Colea chapelieri A. DC. = Rhodocolea racemosa (Lam.) H. Perrier ●☆
99376 Colea coccinea Scott-Elliot = Fernandoa coccinea (Scott-Elliot) A. H. Gentry ●☆
99377 Colea commersonii A. DC. = Ophiocolea floribunda (Bojer ex Lindl.) H. Perrier ●☆
99378 Colea concinna Baker;整洁鞘葳●☆
99379 Colea decora Seem. = Rhodocolea racemosa (Lam.) H. Perrier ●☆
99380 Colea floribunda Bojer ex Lindl. = Ophiocolea floribunda (Bojer ex Lindl.) H. Perrier ●☆
99381 Colea fusca H. Perrier;棕色鞘葳●☆
99382 Colea hirsuta Aug. DC.;粗毛鞘葳●☆
99383 Colea humblotiana Baill. = Rhodocolea racemosa (Lam.) H. Perrier ●☆
99384 Colea involucrata Bojer ex DC. = Rhodocolea involucrata (Bojer ex DC.) H. Perrier ●☆
99385 Colea lantziana Baill.;兰兹鞘葳●☆
99386 Colea longepetiolata Baker = Stereospermum euphorioides (Bojer) A. DC. ●☆
99387 Colea lutescens H. Perrier;黄鞘葳●☆
99388 Colea macrantha Baker = Fernandoa macrantha (Baker) A. H. Gentry ●☆
99389 Colea macrophylla Baker = Ophiocolea floribunda (Bojer ex Lindl.) H. Perrier ●☆
99390 Colea muricata H. Perrier;糙鞘葳●☆
99391 Colea myriaptera H. Perrier;多翅鞘葳●☆
99392 Colea nana H. Perrier;矮鞘葳●☆
99393 Colea nitida A. DC. = Rhodocolea racemosa (Lam.) H. Perrier ●☆
99394 Colea obtusifolia A. DC.;钝叶鞘葳●☆
99395 Colea parviflora Baker = Rhodocolea racemosa (Lam.) H. Perrier ●☆
99396 Colea poivrei Baill. = Rhodocolea racemosa (Lam.) H. Perrier ●☆
99397 Colea purpurascens Seem.;紫鞘葳●☆
99398 Colea racemosa Baker = Colea lutescens H. Perrier ●☆
99399 Colea rubra H. Perrier;红鞘葳●☆
99400 Colea telfairiae Bojer = Rhodocolea telfairiae (Bojer ex Hook.) H. Perrier ●☆
99401 Colea tetragona A. DC.;四角鞘葳●☆
99402 Colea tripinnata Seem. = Vitex tripinnata (Lour.) Merr. ●
99403 Colea undulata Regel = Ophiocolea floribunda (Bojer ex Lindl.) H. Perrier ●☆
99404 Coleachyron J. Gay ex Boiss. = Carex L. ■
99405 Coleactina N. Hallé (1970);射鞘茜属☆
99406 Coleactina papalis N. Hallé;射鞘茜☆
99407 Coleanthaceae Link = Gramineae Juss. (保留科名)■●
99408 Coleanthaceae Link = Poaceae Barnhart(保留科名)■●
99409 Coleanthera Stschegl. (1859);锥药石南属●☆
99410 Coleanthera myrtioides Stschegl.;锥药石南●☆
99411 Coleanthus Seidel (1817)(保留属名);莎禾属(沙禾属);Mudgrass, Mud Grass ■
99412 Coleanthus Seidl ex Roem. et Schult. = Coleanthus Seidel(保留属名)■
99413 Coleanthus subtilis (Tratt.) Seidl;莎禾;Mud Grass, Mudgrass, Mud-grass ■
99414 Coleataenia Griseb. = Panicum L. ■
99415 Colebrockia Steud. = Colebrookea Sm. ●
99416 Colebrockia Steud. = Colebrookia Donn ex T. Lestib. ■

99417　Colebrockia Steud. = Globba L. ■
99418　Colebrookea Sm. (1806);羽萼木属;Colebrookea,Colebrookia ●
99419　Colebrookea oppositifolia Lodd. = Elsholtzia fruticosa (D. Don) Rehder ●
99420　Colebrookea oppositifolia Sm.;羽萼木(黑羊巴巴,马垮皮,野山茶,羽萼);Oppositeleaf Colebrookea,Opposite-leaved Colebrookia ●
99421　Colebrookea ternifolia Roxb. = Colebrookea oppositifolia Sm. ●
99422　Colebrookia Donn = Globba L. ■
99423　Colebrookia Donn ex T. Lestib. = Globba L. ■
99424　Colebrookia Spreng. = Colebrookea Sm. ●
99425　Colema Raf. = Corema D. Don ●☆
99426　Colensoa Hook. f. = Pratia Gaudich. ■
99427　Coleobotrys Tiegh. = Helixanthera Lour. ●
99428　Coleocarya S. T. Blake(1943);鞘果灯草属■☆
99429　Coleocarya gracilis S. T. Blake;鞘果灯草■☆
99430　Coleocephalocereus Backeb. (1938);鞘头柱属(头天轮柱属)●☆
99431　Coleocephalocereus Backeb. = Austrocephalocereus (Backeb.) Backeb. ●☆
99432　Coleocephalocereus aureispinus Buining et Brederoo;黄刺鞘头柱(黄刺头柱)●☆
99433　Coleocephalocereus brevicylindricus (Buining) F. Ritter;短鞘头柱●☆
99434　Coleocephalocereus luetzelburgii (Vaupel) Buxb.;卢氏鞘头柱(卢氏头柱)●☆
99435　Coleocephalocereus paulensis F. Ritter;鞘头柱●☆
99436　Coleocephalocereus pluricostatus Buining et Brederoo;多肋鞘头柱(多肋头柱)●☆
99437　Coleochloa Gilly(1943);鞘芽莎草属■☆
99438　Coleochloa abyssinica (Hochst. ex A. Rich.) Gilly;阿比西尼亚鞘芽莎■☆
99439　Coleochloa abyssinica (Hochst. ex A. Rich.) Gilly var. castanea (C. B. Clarke) Pic. Serm.;栗色阿比西尼亚鞘芽莎■☆
99440　Coleochloa glabra Nelmes;光鞘芽莎■☆
99441　Coleochloa microcephala Nelmes;小头鞘芽莎■☆
99442　Coleochloa pallidior Nelmes;苍白鞘芽莎■☆
99443　Coleochloa rehmanniana (C. B. Clarke) Gilly = Coleochloa setifera (Ridl.) Gilly ■☆
99444　Coleochloa schweinfurthiana (Boeck.) Nelmes;施韦鞘芽莎■☆
99445　Coleochloa setifera (Ridl.) Gilly;刚毛鞘芽莎■☆
99446　Coleochloa virgata (K. Schum.) Nelmes;条纹鞘芽莎■☆
99447　Coleocoma F. Muell. (1857);鞘苞菊属(宽苞菊属)■☆
99448　Coleocoma centaurea F. Muell.;鞘苞菊(宽苞菊)■☆
99449　Coleogeton (Rchb.) Les et R. R. Haynes = Stuckenia Börner ■
99450　Coleogeton filiformis (Pers.) Les et R. R. Haynes = Stuckenia filiformis (Pers.) Börner ■
99451　Coleogeton filiformis (Pers.) Les et R. R. Haynes subsp. alpinus (Blytt) Les et R. R. Haynes = Stuckenia filiformis (Pers.) Börner subsp. alpina (Blytt) R. R. Haynes, Les et Král ■☆
99452　Coleogeton filiformis (Pers.) Les et R. R. Haynes subsp. occidentalis (J. W. Robbins) Les et R. R. Haynes = Stuckenia filiformis (Pers.) Börner subsp. occidentalis (J. W. Robbins) R. R. Haynes, Les et Král ■☆
99453　Coleogeton pectinatus (L.) Les et R. R. Haynes = Potamogeton pectinatus L. ■
99454　Coleogeton vaginatus (Turcz.) Les et R. R. Haynes = Stuckenia vaginata (Turcz.) Holub ■☆
99455　Coleogynaceae J. Agardh = Rosaceae Juss. (保留科名) ●■
99456　Coleogyne Torr. (1851);鞘蕊蔷薇属●☆
99457　Coleogyne ramosissima Torr.;鞘蕊蔷薇;Black Brush, Blackbrush ●☆
99458　Coleonema Bartl. et H. L. Wendl. (1824);糖果木属(石南芸木属,石楠芸木属);Diosma, White Breath of Heaven ●☆
99459　Coleonema album (Thunb.) Bartl. et H. L. Wendl.;白糖果木●☆
99460　Coleonema album (Thunb.) Bartl. et H. L. Wendl. var. virgatum Schltdl. = Coleonema virgatum (Schltdl.) Eckl. et Zeyh. ●☆
99461　Coleonema album Bartl. et H. L. Wendl. = Diosma ericoides L. ●☆
99462　Coleonema aspalathoides Juss. ex Don;芳香木糖果木●☆
99463　Coleonema calycinum (Steud.) I. Williams;萼状糖果木●☆
99464　Coleonema gracile Schltr. = Coleonema nubigenum Esterh. ●☆
99465　Coleonema juniperinum Sond.;刺柏状糖果木●☆
99466　Coleonema nubigenum Esterh.;云雾糖果木●☆
99467　Coleonema pulchellum I. Williams;糖果木(丽花石南芸木);Confetti Bush, Pink Breath of Heaven, Pink Diosma ●☆
99468　Coleonema pulchellum I. Williams 'Compactum';紧密糖果木●☆
99469　Coleonema pulchellum I. Williams 'Nanum';矮生糖果木●☆
99470　Coleonema pulchellum I. Williams 'Pinkie';粉红糖果木●☆
99471　Coleonema pulchellum I. Williams 'Rubrum';红花糖果木●☆
99472　Coleonema pulchellum I. Williams 'Sunset Gold';落日金糖果木●☆
99473　Coleonema pulchrum Hook.;美丽糖果木●☆
99474　Coleonema virgatum (Schltdl.) Eckl. et Zeyh.;条纹糖果木●☆
99475　Coleophora Miers = Daphnopsis Mart. ●☆
99476　Coleophyllum Klotzsch = Chlidanthus Herb. ■☆
99477　Coleophyllum Klotzsch(1840);鞘叶石蒜属■☆
99478　Coleophyllum ehrenbergii Klotzsch;鞘叶石蒜■☆
99479　Coleosanthus Cass. (废弃属名) = Brickellia Elliott(保留属名) ■●
99480　Coleosanthus chenopodinus Greene = Brickellia chenopodina (Greene) B. L. Rob. ■☆
99481　Coleosanthus cordifolius (Elliott) Kuntze = Brickellia cordifolia Elliott ■☆
99482　Coleosanthus lemmonii (A. Gray) Kuntze = Brickellia lemmonii A. Gray ■☆
99483　Coleosanthus venosus Wooton et Standl. = Brickellia venosa (Wooton et Standl.) B. L. Rob. ■☆
99484　Coleospadix Becc. = Drymophloeus Zipp. ●☆
99485　Coleostachys A. Juss. (1840);鞘穗花属●☆
99486　Coleostachys genipifolia A. Juss.;鞘穗花●☆
99487　Coleostephus Cass. (1826);鞘冠菊属■
99488　Coleostephus clausonis Pomel = Coleostephus paludosus (Durieu) Alavi ■☆
99489　Coleostephus macrotus Durieu = Glossopappus macrotus (Durieu) Briq. ■☆
99490　Coleostephus multicaulis (Desf.) Durieu = Chrysanthemum multicaule Desf. ■☆
99491　Coleostephus myconis (L.) Cass.;鞘冠菊;Common Coleostephus ■
99492　Coleostephus myconis (L.) Cass. = Chrysanthemum myconis L. ■
99493　Coleostephus myconis (L.) Rchb. f. = Chrysanthemum myconis L. ■
99494　Coleostephus paludosus (Durieu) Alavi;沼泽鞘冠菊■☆
99495　Coleostyles Benth. et Hook. f. = Coleostylis Sond. ■☆
99496　Coleostylis Sond. = Levenhookia R. Br. ■☆
99497　Coleotrype C. B. Clarke(1881);瓣鞘花属■☆

99498 Coleotrype baroni Baker;巴龙瓣鞘花■☆
99499 Coleotrype goudotii C. B. Clarke;古氏瓣鞘花■☆
99500 Coleotrype laurentii K. Schum.;洛朗瓣鞘花■☆
99501 Coleotrype lutea H. Perrier;黄瓣鞘花■☆
99502 Coleotrype madagascarica C. B. Clarke;马岛瓣鞘花■☆
99503 Coleotrype natalensis C. B. Clarke;纳塔尔瓣鞘花■☆
99504 Coleotrype synanthera H. Perrier;合药瓣鞘花■☆
99505 Coletia Vell. = Mayaca Aubl. ■☆
99506 Coleus Lour. (1790);鞘蕊花属(彩叶草属,锦紫苏属,鞘蕊属,小鞘蕊花属);Coleus, Flame Nettle, Flamenettle, Painted Leaves ●■
99507 Coleus Lour. = Plectranthus L'Hér. (保留属名)●■
99508 Coleus Lour. = Solenostemon Thonn. ■
99509 Coleus × hybridus Hort.;杂种鞘蕊花;Coleus, Painted Nettle ■☆
99510 Coleus acuminatus Benth. = Coleus scutellarioides (L.) Benth. ■
99511 Coleus adolfi-friderici Perkins;弗里德里西鞘蕊花■☆
99512 Coleus africanus (Baker ex Scott-Elliot) Roberty = Leocus africanus (Baker ex Scott-Elliot) J. K. Morton ■●☆
99513 Coleus albidus Vatke;白鞘蕊花■☆
99514 Coleus alpinus Vatke;高山鞘蕊花■☆
99515 Coleus amboinicus Lour.;安勃鞘蕊花(安倍那鞘蕊花,倒手香);Amboin Coleus, Indian Borage, Oregano ■
99516 Coleus amboinicus Lour. = Plectranthus amboinicus (Lour.) Spreng. ■☆
99517 Coleus amboinicus Lour. var. violaceus Gürke = Plectranthus amboinicus (Lour.) Spreng. ■☆
99518 Coleus aquaticus Gürke = Plectranthus edulis (Vatke) Agnew ■☆
99519 Coleus aromaticus Benth.;香鞘蕊花■☆
99520 Coleus aromaticus Benth. = Plectranthus amboinicus (Lour.) Spreng. ■☆
99521 Coleus ascendens Gilli;上升鞘蕊花■☆
99522 Coleus assurgens Baker = Plectranthus assurgens (Baker) J. K. Morton ■☆
99523 Coleus atropurpureus Benth.;暗紫鞘蕊花(黑紫彩叶草)■
99524 Coleus barbatus (Andréws) Benth. = Coleus forskohlii (Willd.) Briq. ■
99525 Coleus barbatus (Andréws) Benth. = Plectranthus barbatus Andréws ■☆
99526 Coleus baumii Gürke;鲍姆鞘蕊花■☆
99527 Coleus bernieri Briq. = Plectranthus bojeri (Benth.) Hedge ■☆
99528 Coleus betonicoides Baker ex Hiern;药水苏鞘蕊花■☆
99529 Coleus blumei Benth. = Coleus scutellarioides (L.) Benth. ■
99530 Coleus bracteatus Dunn; 光萼鞘蕊花; Bracteate Coleus, Smoothcalyx Flamenettle ■
99531 Coleus brazzavillensis A. Chev.;布拉柴维尔鞘蕊花■☆
99532 Coleus briquetii Baker = Solenostemon latifolius (Hochst. ex Benth.) J. K. Morton ■☆
99533 Coleus buchananii (Baker) Brenan = Plectranthus buchananii Baker ■☆
99534 Coleus bullulatus Briq.;泡状鞘蕊花■☆
99535 Coleus caillei A. Chev. ex Hutch. et Dalziel = Leocus caillei (A. Chev. ex Hutch. et Dalziel) J. K. Morton ■☆
99536 Coleus camporum Gürke;弯鞘蕊花■☆
99537 Coleus caninus (Roth) Vatke = Plectranthus caninus Roth ■☆
99538 Coleus caninus Vatke;犬锦紫苏;Scaredy Cat Plant ■☆
99539 Coleus carnosifolius (Hemsl.) Dunn;肉叶鞘蕊花(假回菜);Fleshyleaf Coleus, Fleshyleaf Flamenettle ■
99540 Coleus casamanicus A. Chev. ex Hutch. et Dalziel = Isodictyophorus reticulatus (A. Chev.) J. K. Morton ■☆
99541 Coleus chevalieri Briq.;舍瓦利耶鞘蕊花■☆
99542 Coleus claessensii De Wild.;克莱森斯鞘蕊花■☆
99543 Coleus clivicola S. Moore;山坡鞘蕊花■☆
99544 Coleus coerulescens Gürke = Plectranthus barbatus Andréws ■☆
99545 Coleus coeruleus Gürke = Plectranthus coeruleus (Gürke) Agnew ■☆
99546 Coleus collinus Lebrun et L. Touss. = Solenostemon collinus (Lebrun et L. Touss.) Troupin ■☆
99547 Coleus comosus Hochst. ex Gürke = Plectranthus ornatus Codd ■☆
99548 Coleus concinnus (Hiern) Baker;整洁鞘蕊花■☆
99549 Coleus conglomeratus (T. C. E. Fr.) Robyns et Lebrun = Plectranthus conglomeratus (T. C. E. Fr.) Hutch. et Dandy ■☆
99550 Coleus copiosiflorus Briq. = Solenostemon latifolius (Hochst. ex Benth.) J. K. Morton ■☆
99551 Coleus crassifolius Benth. = Plectranthus amboinicus (Lour.) Spreng. ■☆
99552 Coleus crispipilus (Merr.) Merr. = Coleus scutellarioides (L.) Benth. var. crispipilus (Merr.) H. Keng ■
99553 Coleus crispipilus Merr. = Coleus scutellarioides (L.) Benth. var. crispipilus (Merr.) H. Keng ■
99554 Coleus cuneatus Baker f. = Plectranthus cuneatus (Baker f.) Ryding ■☆
99555 Coleus cunenensis Baker;库内内鞘蕊花■☆
99556 Coleus darfurensis R. D. Good = Solenostemon autrani (Briq.) J. K. Morton ■☆
99557 Coleus daviesii E. A. Bruce = Plectranthus daviesii (E. A. Bruce) B. Mathew ■☆
99558 Coleus dazo A. Chev. = Plectranthus esculentus N. E. Br. ■☆
99559 Coleus decurrens Gürke = Plectranthus decurrens (Gürke) J. K. Morton ■☆
99560 Coleus delpierrei De Wild.;戴尔皮埃尔鞘蕊花■☆
99561 Coleus denudatus (A. Chev. ex Hutch. et Dalziel) Robyns = Englerastrum nigericum Alston ■☆
99562 Coleus dewevrei Briq.;德韦鞘蕊花■☆
99563 Coleus diffusus (Alston) Robyns et Lebrun = Plectranthus gracillimus (T. C. E. Fr.) Hutch. et Dandy ■☆
99564 Coleus dissitiflorus Gürke = Plectranthus dissitiflorus (Gürke) J. K. Morton ■☆
99565 Coleus djalonensis A. Chev. = Solenostemon latifolius (Hochst. ex Benth.) J. K. Morton ■☆
99566 Coleus edulis Vatke = Plectranthus edulis (Vatke) Agnew ■☆
99567 Coleus elatus Baker = Plectranthus decurrens (Gürke) J. K. Morton ■☆
99568 Coleus englerastrum Roberty = Englerastrum schweinfurthii Briq. ■☆
99569 Coleus entebbensis S. Moore = Plectranthus luteus Gürke ■☆
99570 Coleus equisetiformis E. A. Bruce = Plectranthus equisetiformis (E. A. Bruce) Launert ■☆
99571 Coleus esculentus (N. E. Br.) G. Taylor;开菲尔鞘蕊花;Kaffir Potato, Native Potato ■☆
99572 Coleus esculentus (N. E. Br.) G. Taylor = Plectranthus esculentus N. E. Br. ■☆
99573 Coleus esquirolii (H. Lév.) Dunn;毛萼鞘蕊花(白花紫苏,红靛,岩紫苏);Esquirol Coleus, Esquirol Flamenettle ■
99574 Coleus ferrugineus Robyns;锈色鞘蕊花■☆

99575 Coleus fimbriatus Lebrun et L. Touss. = Plectranthus fimbriatus (Lebrun et L. Touss.) Troupin et Ayob. ■☆
99576 Coleus flaccidus Vatke = Plectranthus flaccidus (Vatke) Gürke ■☆
99577 Coleus flavovirens Gürke = Plectranthus caninus Roth ■☆
99578 Coleus floribundus (N. E. Br.) Robyns et Lebrun = Plectranthus esculentus N. E. Br. ■☆
99579 Coleus floribundus (N. E. Br.) Robyns et Lebrun var. longipes = Plectranthus esculentus N. E. Br. ■☆
99580 Coleus floribundus Baker;繁花鞘蕊花■☆
99581 Coleus formosanus Hayata;台湾鞘蕊花(兰屿小鞘蕊花)■
99582 Coleus formosanus Hayata = Coleus scutellarioides (L.) Benth. var. crispipilus (Merr.) H. Keng ■
99583 Coleus forskohlii (Willd.) Briq.;毛喉鞘蕊花(髯毛鞘蕊花,束毛鞘蕊花);Forskohl Coleus, Forskohl Flamenettle, Kaffir Potato, Pained Nettle ■
99584 Coleus frederici G. Taylor = Plectranthus welwitschii (Briq.) Codd ■☆
99585 Coleus gallaensis Gürke = Plectranthus lanuginosus (Hochst. ex Benth.) Agnew ■☆
99586 Coleus gazensis S. Moore;加兹鞘蕊花■☆
99587 Coleus giorgii De Wild. = Solenostemon giorgii (De Wild.) Champl. ■☆
99588 Coleus glandulosus Hook. f. = Plectranthus punctatus (L. f.) L'Hér. ■☆
99589 Coleus gomphophyllus Baker = Plectranthus lanuginosus (Hochst. ex Benth.) Agnew ■☆
99590 Coleus goudotii Briq. = Plectranthus persoonii (Benth.) Hedge ■☆
99591 Coleus gracilentus S. Moore;细黏鞘蕊花■☆
99592 Coleus gracilifolius Briq. = Plectranthus bojeri (Benth.) Hedge ■☆
99593 Coleus gracilis Gürke = Plectranthus puberulentus J. K. Morton ■☆
99594 Coleus gracillimus (T. C. E. Fr.) Robyns et Lebrun = Plectranthus gracillimus (T. C. E. Fr.) Hutch. et Dandy ■☆
99595 Coleus grandicalyx E. A. Bruce = Plectranthus grandicalyx (E. A. Bruce) J. K. Morton ■☆
99596 Coleus grandis L. H. Cramer = Plectranthus barbatus Andréws var. grandis (L. H. Cramer) Lukhoba et A. J. Paton ■☆
99597 Coleus guidottii Chiov. = Plectranthus igniarius (Schweinf.) Agnew ■☆
99598 Coleus gymnostomus Gürke;裸口鞘蕊花■☆
99599 Coleus herbaceus (Briq.) G. Taylor;草本鞘蕊花■☆
99600 Coleus heterotrichus Briq.;异毛鞘蕊花■☆
99601 Coleus hjalmari (T. C. E. Fr.) Robyns et Lebrun = Plectranthus hjalmarii (T. C. E. Fr.) Hutch. et Dandy ■☆
99602 Coleus hockii De Wild.;霍克鞘蕊花■☆
99603 Coleus homblei De Wild.;洪布勒鞘蕊花■☆
99604 Coleus igniarius Schweinf. = Plectranthus igniarius (Schweinf.) Agnew ■☆
99605 Coleus inflatus Benth.;膨大鞘蕊花■☆
99606 Coleus insolitus (C. H. Wright) Robyns et Lebrun = Plectranthus insolitus C. H. Wright ■☆
99607 Coleus kapatensis R. E. Fr. = Plectranthus kapatensis (R. E. Fr.) J. K. Morton ■☆
99608 Coleus kasomenensis De Wild.;卡索门鞘蕊花■☆
99609 Coleus kassneri (T. C. E. Fr.) Robyns et Lebrun = Plectranthus kassneri (T. C. E. Fr.) Hutch. et Dandy ■☆
99610 Coleus keniensis Standl.;肯尼亚鞘蕊花■☆
99611 Coleus kilimandschari Gürke = Plectranthus kilimandschari (Gürke) C. A. Maass ■☆
99612 Coleus kisantuensis De Wild.;基桑图鞘蕊花■☆
99613 Coleus kivuensis Lebrun et L. Touss. = Plectranthus kivuensis (Lebrun et L. Touss.) R. H. Willemse ■☆
99614 Coleus koualensis A. Chev. ex Hutch. et Dalziel = Plectranthus koualensis (A. Chev. ex Hutch. et Dalziel) B. J. Pollard et A. J. Paton ■☆
99615 Coleus lactiflorus Vatke = Plectranthus lactiflorus (Vatke) Agnew ■☆
99616 Coleus lactiflorus Vatke var. velutinus Fiori = Plectranthus lactiflorus (Vatke) Agnew ■☆
99617 Coleus lageniocalyx Briq.;长颈瓶鞘蕊花■☆
99618 Coleus langouassiensis A. Chev. = Plectranthus esculentus N. E. Br. ■☆
99619 Coleus lanuginosus Hochst. ex Benth. = Plectranthus lanuginosus (Hochst. ex Benth.) Agnew ■☆
99620 Coleus lasianthus Gürke = Plectranthus lasianthus (Gürke) Vollesen ■☆
99621 Coleus latericola A. Chev. = Plectranthus monostachyus (P. Beauv.) B. J. Pollard subsp. latericola (A. Chev.) B. J. Pollard ■☆
99622 Coleus latidens S. Moore;宽齿大鞘蕊花■☆
99623 Coleus latifolius Hochst. ex Benth. = Plectranthus bojeri (Benth.) Hedge ■☆
99624 Coleus latifolius Hochst. ex Benth. = Solenostemon latifolius (Hochst. ex Benth.) J. K. Morton ■☆
99625 Coleus latifolius Hochst. ex Benth. var. madiensis Baker = Solenostemon latifolius (Hochst. ex Benth.) J. K. Morton ■☆
99626 Coleus laurentii De Wild.;洛朗鞘蕊花■☆
99627 Coleus laxiflorus (Benth.) Roberty = Plectranthus laxiflorus Benth. ■☆
99628 Coleus lebrunii Robyns;勒布伦鞘蕊花■☆
99629 Coleus leptophyllus Baker;细叶鞘蕊花■☆
99630 Coleus leucophyllus Baker = Plectranthus mirabilis (Briq.) Launert ■☆
99631 Coleus longipetiolatus Gürke;长梗鞘蕊花■☆
99632 Coleus luteus (Gürke) Staner = Plectranthus luteus Gürke ■☆
99633 Coleus lyratus (A. Chev.) Roberty = Leocus lyratus A. Chev. ■☆
99634 Coleus macranthus Merr. var. crispipilus Merr. = Coleus scutellarioides (L.) Benth. var. crispipilus (Merr.) H. Keng ■
99635 Coleus maculatus Gürke;斑点鞘蕊花■☆
99636 Coleus madagascariensis (Pers.) A. Chev. = Plectranthus madagascariensis (Pers.) Benth. ■☆
99637 Coleus mahonii Baker;马洪鞘蕊花■☆
99638 Coleus malabaricus Benth.;印度鞘蕊花(马拉巴锦)■☆
99639 Coleus malinvaldii (Briq.) Briq. = Plectranthus malinvaldii Briq. ■☆
99640 Coleus mannii Hook. f. = Plectranthus occidentalis B. J. Pollard ■☆
99641 Coleus maranguensis Gürke;马兰古鞘蕊花■☆
99642 Coleus matopensis S. Moore;马托鞘蕊花■☆
99643 Coleus mechowianus Briq.;梅休鞘蕊花■☆
99644 Coleus melanocarpus (Gürke) Robyns et Lebrun = Plectranthus tetragonus Gürke ■☆
99645 Coleus membranaceus Briq.;膜质鞘蕊花■☆
99646 Coleus menyarthii Briq.;梅尼鞘蕊花■☆
99647 Coleus microtrichus Chiov.;小毛鞘蕊花■☆
99648 Coleus mirabilis Briq. = Plectranthus mirabilis (Briq.) Launert ■☆

99649 Coleus mirabilis Briq. var. buchnerianus? = Plectranthus mirabilis (Briq.) Launert ■☆
99650 Coleus mirabilis Briq. var. hypisodontus? = Plectranthus mirabilis (Briq.) Launert ■☆
99651 Coleus mirabilis Briq. var. mechowianus? = Plectranthus mirabilis (Briq.) Launert ■☆
99652 Coleus mirabilis Briq. var. poggeanus? = Plectranthus mirabilis (Briq.) Launert ■☆
99653 Coleus modestus (Baker) Robyns et Lebrun = Plectranthus modestus Baker ■☆
99654 Coleus montanus Hochst. ex Ces.;山地鞘蕊花■☆
99655 Coleus monticola (Gürke) Gürke = Plectranthus monticola Gürke ■☆
99656 Coleus mucosus Hayata = Coleus esquirolii (H. Lév.) Dunn ■
99657 Coleus myrianthellus Briq.;多小花鞘蕊花■☆
99658 Coleus myrianthus (Briq.) Brenan = Plectranthus hereroensis Engl. ■☆
99659 Coleus nervosus Briq.;多脉鞘蕊花■☆
99660 Coleus newtonii Briq.;纽敦鞘蕊花■☆
99661 Coleus nigericus A. Chev. = Plectranthus chevalieri (Briq.) B. J. Pollard et A. J. Paton ■☆
99662 Coleus nyikensis Baker = Plectranthus malawiensis B. Mathew ■☆
99663 Coleus odoratus Gürke;芳香鞘蕊花■☆
99664 Coleus omahekense Dinter = Plectranthus caninus Roth ■☆
99665 Coleus orbicularis Baker;圆形鞘蕊花■☆
99666 Coleus ostinii Chiov. = Solenostemon autrani (Briq.) J. K. Morton ■☆
99667 Coleus pachyphyllus Gürke = Plectranthus caninus Roth ■☆
99668 Coleus pallidiflorus A. Chev. = Solenostemon rotundifolius (Poir.) J. K. Morton ■☆
99669 Coleus palustris Vatke = Plectranthus edulis (Vatke) Agnew ■☆
99670 Coleus parviflorus Benth.;小花鞘蕊花;Hausa Potato ■☆
99671 Coleus pentheri Gürke = Plectranthus pentheri (Gürke) Van Jaarsv. et T. J. Edwards ■☆
99672 Coleus penzigii Baker = Plectranthus barbatus Andréws ■☆
99673 Coleus persoonii Benth. = Plectranthus persoonii (Benth.) Hedge ■☆
99674 Coleus petersianus Vatke;彼得斯鞘蕊花■☆
99675 Coleus petiolatissimus Briq.;长柄鞘蕊花■☆
99676 Coleus petrophilus Gürke;喜岩鞘蕊花■☆
99677 Coleus peulhorum A. Chev. var. violacea? = Solenostemon latifolius (Hochst. ex Benth.) J. K. Morton ■☆
99678 Coleus phymatodes Briq. = Solenostemon latifolius (Hochst. ex Benth.) J. K. Morton ■☆
99679 Coleus platostomoides Robyns et Lebrun = Solenostemon platostomoides (Robyns et Lebrun) Troupin ■☆
99680 Coleus pobeguinii Hutch. et Dalziel = Leocus pobeguinii (Hutch. et Dalziel) J. K. Morton ■☆
99681 Coleus poggeanus Briq.;波格鞘蕊花■☆
99682 Coleus polyanthus S. Moore;多花鞘蕊花■☆
99683 Coleus praetermissus Bullock et Killick = Plectranthus punctatus (L. f.) L'Hér. ■☆
99684 Coleus preussii Gürke;普罗伊斯鞘蕊花■☆
99685 Coleus primulinus (Baker) A. Chev. = Plectranthus primulinus Baker ■☆
99686 Coleus pumilus Blanco;小叶彩纹草(小洋紫苏)■
99687 Coleus pumilus Blanco = Coleus scutellarioides (L.) Benth. var. crispipilus (Merr.) H. Keng ■
99688 Coleus pumilus Blanco = Coleus scutellarioides (L.) Benth. ■
99689 Coleus punctatus Baker = Solenostemon autrani (Briq.) J. K. Morton ■☆
99690 Coleus quarrei Robyns et Lebrun;卡雷鞘蕊花■☆
99691 Coleus ramosissimus (Hook. f.) Robyns = Isodon ramosissimus (Hook. f.) Codd ●☆
99692 Coleus rehmannii Briq. = Solenostemon latifolius (Hochst. ex Benth.) J. K. Morton ■☆
99693 Coleus rehneltianus A. Berger;小彩叶草(红斑洋紫苏)■
99694 Coleus reticulatus A. Chev. = Isodictyophorus reticulatus (A. Chev.) J. K. Morton ■☆
99695 Coleus ringoetii De Wild.;林戈鞘蕊花■☆
99696 Coleus rivularis Vatke = Plectranthus edulis (Vatke) Agnew ■☆
99697 Coleus rotundifolius (Poir.) A. Chev. et Perrot = Solenostemon rotundifolius (Poir.) J. K. Morton ■☆
99698 Coleus rotundifolius (Poir.) A. Chev. et Perrot var. nigra A. Chev. = Solenostemon rotundifolius (Poir.) J. K. Morton ■☆
99699 Coleus rotundifolius A. Chev. et Perrot;圆叶鞘蕊花;Daso, Fabirama, Fra-fra Potato, Fura-fura Potato, Hapotato, Hausa Potato ■☆
99700 Coleus rupestris (Vatke ex Baker) A. Chev. = Plectranthus rupestris Vatke ex Baker ■☆
99701 Coleus rupestris Hochst. = Plectranthus rupestris Vatke ex Baker ■☆
99702 Coleus ruwenzoriensis Baker;鲁文佐里鞘蕊花■☆
99703 Coleus salagensis Gürke = Solenostemon rotundifolius (Poir.) J. K. Morton ■☆
99704 Coleus saxicola Gürke;岩生鞘蕊花■☆
99705 Coleus scandens Gürke = Englerastrum scandens (Gürke) Alston ■☆
99706 Coleus schimperi Vatke = Plectranthus semayatensis Cufod. ■☆
99707 Coleus schlechteri (T. C. E. Fr.) Robyns et Lebrun = Plectranthus schlechteri (T. C. E. Fr.) Hutch. et Dandy ■☆
99708 Coleus schweinfurthii Briq. = Solenostemon latifolius (Hochst. ex Benth.) J. K. Morton ■☆
99709 Coleus schweinfurthii Vatke;施韦鞘蕊花■☆
99710 Coleus scoposus C. H. Wright;花茎鞘蕊花■☆
99711 Coleus scutellarioides (L.) Benth.;五彩苏(彩色紫苏,彩苏,彩叶草,彩叶洋紫苏,黄芩香茶菜,金耳环,锦紫苏,盆上金耳环,小彩叶紫苏,小鞘蕊花,小纹草,小洋紫苏,洋紫苏);Coleus, Common Coleus, Dwarf Coleus, Flame Nettle, Garish Coleus, Garish Flamenettle, Painted Nettle, Trailing Coleus ■
99712 Coleus scutellarioides (L.) Benth. var. crispipilus (Merr.) H. Keng;小五彩苏(假紫苏,金耳环,金钱炮,苛留香,兰屿小鞘蕊花,盆上金耳环,五色草,小彩色紫苏,小彩叶紫苏,小洋紫苏,洋紫苏);Small Skullcaplike Coleus ■
99713 Coleus scutellarioides (L.) Benth. var. crispipilus (Merr.) H. Keng = Coleus formosanus Hayata ■
99714 Coleus seretii De Wild. = Plectranthus seretii (De Wild.) Vollesen ■☆
99715 Coleus serrulatus Robyns = Plectranthus serrulatus (Robyns) Troupin et Ayob. ■☆
99716 Coleus shirensis Gürke = Solenostemon autrani (Briq.) J. K. Morton ■☆
99717 Coleus somalensis S. Moore = Plectranthus lanuginosus (Hochst. ex Benth.) Agnew ■☆
99718 Coleus speciosus Baker f. = Plectranthus barbatus Andréws ■☆

99719 Coleus spicatus Benth. = Plectranthus caninus Roth ■☆
99720 Coleus spicatus Benth. var. rondinella Spreng. = Plectranthus ornatus Codd ■☆
99721 Coleus splendidus A. Chev. = Solenostemon latifolius (Hochst. ex Benth.) J. K. Morton ■☆
99722 Coleus stachyoides (Oliv.) E. A. Bruce = Plectranthus stachyoides Oliv. ■☆
99723 Coleus subscandens Gürke;亚攀缘鞘蕊花■☆
99724 Coleus subulatus Robyns;钻形鞘蕊花■☆
99725 Coleus succulentus Pax = Plectranthus pseudomarrubioides R. H. Willemse ■☆
99726 Coleus tenuicaulis Hook. f. = Plectranthus tenuicaulis (Hook. f.) J. K. Morton ■☆
99727 Coleus tenuiflorus Vatke = Plectranthus tenuiflorus (Vatke) Agnew ■☆
99728 Coleus ternatus A. Chev.;三出鞘蕊花■☆
99729 Coleus tetensis Baker = Plectranthus tetensis (Baker) Agnew ■☆
99730 Coleus tetragonus (Gürke) Robyns et Lebrun = Plectranthus tetragonus Gürke ■☆
99731 Coleus thyrsiflorus Lebrun et L. Touss. = Solenostemon thyrsiflorus (Lebrun et L. Touss.) Troupin ■☆
99732 Coleus thyrsoideus Baker = Plectranthus thyrsoideus (Baker) B. Mathew ■☆
99733 Coleus toroensis S. Moore;托罗鞘蕊花■☆
99734 Coleus tricholobus Gürke;毛片鞘蕊花■☆
99735 Coleus trichophorus Briq.;毛梗鞘蕊花■☆
99736 Coleus tuberosus A. Rich. = Plectranthus edulis (Vatke) Agnew ■☆
99737 Coleus uliginosus T. C. E. Fr.;沼泽鞘蕊花■☆
99738 Coleus umbrosus Vatke = Pycnostachys umbrosa (Vatke) Perkins ■☆
99739 Coleus urticifolius (Hook. f.) Roberty = Pycnostachys meyeri Gürke ■☆
99740 Coleus variifolius De Wild.;异叶鞘蕊花■☆
99741 Coleus vestitus Baker = Plectranthus barbatus Andréws ■☆
99742 Coleus viridis Briq.;绿鞘蕊花■☆
99743 Coleus welwitschii Briq.;韦尔鞘蕊花■☆
99744 Coleus wittei Robyns;维特鞘蕊花■☆
99745 Coleus wugensis Gürke;武加鞘蕊花■☆
99746 Coleus wulfenioides Diels = Orthosiphon rubicundus (D. Don) Benth. ■
99747 Coleus wulfenioides Diels = Orthosiphon wulfenioides (Diels) Hand. -Mazz. ■
99748 Coleus xanthanthus C. Y. Wu et Y. C. Huang;黄鞘蕊花;Yellow Flamenettle,Yellowflower Coleus,Yellow-flowered Coleus ●
99749 Coleus zatarhendi (Forssk.) Benth. = Plectranthus aegyptiacus (Forssk.) C. Chr. ■●☆
99750 Colicodendron Mart. = Capparis L. ●
99751 Colignonia Endl. (1837);大苞茉莉属■☆
99752 Colignonia parviflora (Kunth) Choisy;大苞茉莉■☆
99753 Colima (Ravenna) Aarón Rodr. et Ortiz-Cat. (2003);科里马鸢尾属■☆
99754 Colima (Ravenna) Aarón Rodr. et Ortiz-Cat. = Nemastylis Nutt. ■☆
99755 Colinil Adans. = Cracca Benth. (保留属名)●☆
99756 Colinil Adans. = Tephrosia Pers. (保留属名)●■
99757 Coliquea Bibra = Chusquea Kunth ●☆
99758 Coliquea Steud. ex Bibra = Chusquea Kunth ●☆
99759 Colla Raf. = Calla L. ■
99760 Collabiopsis S. S. Ying = Collabium Blume ■
99761 Collabiopsis S. S. Ying(1977);假吻兰属■
99762 Collabiopsis assamica (Hook. f.) S. S. Ying = Collabium assamicum (Hook. f.) Seidenf. ■
99763 Collabiopsis chinensis (Rolfe) S. S. Ying;乌来假吻兰■
99764 Collabiopsis chinensis (Rolfe) S. S. Ying = Collabium chinense (Rolfe) Ts. Tang et F. T. Wang ■
99765 Collabiopsis delavayi (Gagnep.) S. S. Ying = Collabium delavayi (Gagnep.) Seidenf. ■
99766 Collabiopsis delavayi (Gagnep.) Seidenf. = Collabium formosanum Hayata ■
99767 Collabiopsis formosana (Hayata) S. S. Ying;台湾假吻兰■
99768 Collabiopsis formosana (Hayata) S. S. Ying = Collabium formosanum Hayata ■
99769 Collabiopsis uraiensis (Fukuy.) S. S. Ying = Collabium chinense (Rolfe) Ts. Tang et F. T. Wang ■
99770 Collabium Blume(1825);吻兰属(柯丽白兰属);Collabium ■
99771 Collabium assamicum (Hook. f.) Seidenf. = Chrysoglossum assamicum Hook. f. ■
99772 Collabium balansae (Gagnep.) Ts. Tang et F. T. Wang = Collabium chinense (Rolfe) Ts. Tang et F. T. Wang ■
99773 Collabium chinense (Rolfe) Ts. Tang et F. T. Wang;吻兰(柯丽白兰);China Collabium,Chinese Collabium ■
99774 Collabium delavayi (Gagnep.) Seidenf.;南方吻兰■
99775 Collabium formosanum Hayata;台湾吻兰(金唇兰,台湾柯丽白兰);Taiwan Collabium ■
99776 Collabium nebulosum Blume;宽叶吻兰;Broadleaf Collabium ■☆
99777 Collabium uraiense Fukuy. = Collabium chinense (Rolfe) Ts. Tang et F. T. Wang ■
99778 Colladea Pers. = Colladoa Cav. ■
99779 Colladea Pers. = Ischaemum L. ■
99780 Colladoa Cav. = Ischaemum L. ■
99781 Colladonia DC. = Heptaptera Margot et Reut. ■☆
99782 Colladonia DC. = Perlebia DC. ■☆
99783 Colladonia Spreng. = Palicourea Aubl. ●☆
99784 Collaea Bert. ex Colla = Ardisia Sw. (保留属名)●■
99785 Collaea DC. = Galactia P. Browne ■
99786 Collaea Endl. = Collea Lindl. (废弃属名)■
99787 Collaea Endl. = Pelexia Poit. ex Lindl. (保留属名)■
99788 Collaea Spreng. = Chrysanthellum Rich. ex Pers. ■☆
99789 Collandra Lem. = Columnea L. ●■☆
99790 Collandra Lem. = Dalbergaria Tussac ●☆
99791 Collania Herb. = Bomarea Mirb. ■☆
99792 Collania Herb. = Wichuraea M. Roem. ■☆
99793 Collania Schult. et Schult. f. = Urceolina Rchb. (保留属名)■☆
99794 Collania Schult. f. = Urceolina Rchb. (保留属名)■☆
99795 Collare-stuartense Senghas et Bockemühl = Odontoglossum Kunth ■
99796 Collea Lindl. (废弃属名) = Pelexia Poit. ex Lindl. (保留属名)■
99797 Collema Adans. ex DC. = Goodenia Sm. ●■☆
99798 Collema Anderson ex DC. = Goodenia Sm. ●■☆
99799 Collema Anderson ex R. Br. = Goodenia Sm. ●■☆
99800 Collenucia Chiov. = Jatropha L. (保留属名)●■
99801 Collenucia paradoxa Chiov. = Jatropha paradoxa (Chiov.) Chiov. ●☆
99802 Colleteria David W. Taylor = Psychotria L. (保留属名)●

99803 Colleteria David W. Taylor(2003);多米尼加茜属●☆
99804 Colletia Comm. ex Juss.(1789)(保留属名);筒萼木属(科力木属,克莱梯木属);Colletia ●☆
99805 Colletia Endl. = Coletia Vell. ■☆
99806 Colletia Endl. = Mayaca Aubl. ■☆
99807 Colletia Juss. = Colletia Comm. ex Juss.(保留属名)●☆
99808 Colletia Scop.(废弃属名)= Celtis L. ●
99809 Colletia Scop.(废弃属名)= Colletia Comm. ex Juss.(保留属名)●☆
99810 Colletia armata Miers = Colletia hystrix Clos ●☆
99811 Colletia cruciata Gillies et Hook. = Colletia paradoxa (Spreng.) Escal. ●☆
99812 Colletia hystrix Clos;智利筒萼木(豪猪科力木,克莱梯木);Anchor Plant ●☆
99813 Colletia hystrix Clos 'Rosea';粉花智利筒萼木●☆
99814 Colletia paradoxa (Spreng.) Escal.;筒萼木(科力木);Anchor Bush,Anchor Plant ●☆
99815 Colletia spinosa J. F. Gmel.;秘鲁筒萼木(美丽科力木);Cylinder Colletia ●☆
99816 Colletoecema E. Petit.(1963);基茜树属●☆
99817 Colletoecema dewevrei (De Wild.) E. M. Petit;基茜树●☆
99818 Colletogyne Buchet(1939);黏蕊南星属■☆
99819 Colletogyne perrieri Buchet;黏蕊南星■☆
99820 Collignonia Endl. = Colignonia Endl. ■☆
99821 Colliguaja Molina(1782);考利桂属●☆
99822 Colliguaja odorifera Molina;考利桂;Colliguaji Bark ●☆
99823 Collinaria Ehrh. = Koeleria Pers. ■
99824 Collinaria Ehrh. = Poa L. ■
99825 Collinia (Liebm.) Liebm. ex Oerst. = Chamaedorea Willd.(保留属名)●☆
99826 Collinia (Liebm.) Oerst. = Chamaedorea Willd.(保留属名)●☆
99827 Collinia (Mart.) Liebm. = Collinia (Mart.) Liebm. ex Oerst. ●☆
99828 Collinia (Mart.) Liebm. ex Oerst. = Chamaedorea Willd.(保留属名)●☆
99829 Collinia Liebm. = Collinia (Mart.) Liebm. ex Oerst. ●☆
99830 Collinia Raf. = Collinsia Nutt. ■☆
99831 Collinia elegans Liebm. = Chamaedorea elegans Mart. ●☆
99832 Colliniana Raf. = Collinsia Nutt. ■☆
99833 Collinsia Nutt.(1817);科林花属(锦龙花属,柯林草属);Collinsia ■☆
99834 Collinsia bicolor Benth. = Collinsia heterophylla Buist ex Graham ●☆
99835 Collinsia bicolor DC. = Collinsia heterophylla Buist ex Graham ●☆
99836 Collinsia grandiflora Lindl.;大花科林花(大锦龙花);Blue Lips,Blue-lips ■☆
99837 Collinsia grandiflora Lindl. var. carminea Hort.;红色大花科林花■☆
99838 Collinsia heterophylla Buist ex Graham;二色科林花;China Houses,Chinese Houses,Chinese-houses,Purple Chinese Houses ●☆
99839 Collinsia parviflora Lindl.;小花科林花;Maiden Blue-eyed Mary ●☆
99840 Collinsia verna Nutt.;东部科林花;Blue-eyed Mary,Eastern Blue-eyed-mary,Spring Blue-eyed-mary ●☆
99841 Collinsia violacea Nutt.;堇色科林花;Violet Collinsia ●☆
99842 Collinsiana Raf. = Collinsia Nutt. ■☆
99843 Collinsonia L.(1753);二蕊紫苏属;Horse Balm ■☆
99844 Collinsonia canadensis L.;二蕊紫苏;Canada Horse-balm,Citronella,Hardback,Heal-all,Horse Balm,Horsebalm,Horseweed,Knob-root,Knobweed,Northern Horse-balm,Ox Balm,Ox-balm,Richleaf,Richweed,Stone Root,Stoneroot,Stone-root ●☆
99845 Collinsonia tenella (Pursh) Piper;矮二蕊紫苏●☆
99846 Colliquaja Augier = Colliguaja Molina ●☆
99847 Collococcus P. Browne = Cordia L.(保留属名)●
99848 Colloea Endl. = Collea Lindl.(废弃属名)■
99849 Colloea Endl. = Pelexia Poit. ex Lindl.(保留属名)■
99850 Collomia Nutt.(1818);胶壁籽属(粘胶花属);Collomia ■☆
99851 Collomia Sieber ex Steud. = Collomia Nutt. ■☆
99852 Collomia Sieber ex Steud. = Felicia Cass.(保留属名)●■
99853 Collomia bicolor Brand;二色黏胶花(二色粘胶花)■☆
99854 Collomia biflora Brand;二花胶壁籽(二花粘胶花)■☆
99855 Collomia cavanillesii Hook. et Arn.;胶壁籽(粘胶花)■☆
99856 Collomia coccinea Lehm. = Collomia biflora Brand ■☆
99857 Collomia coccinea Lehm. ex Benth. = Collomia biflora Brand ■☆
99858 Collomia grandiflora Douglas ex Lindl.;大花胶壁籽(大花粘胶花);Bigflower Gilia ■☆
99859 Collomia linearis Nutt.;线叶胶壁籽(线叶黏胶花);Collomia,Narrow-leaved Mountain Trumpet ■☆
99860 Collomiastrum (Brand) S. L. Welsh = Gilia Ruiz et Pav. ■●☆
99861 Collophora Mart. = Couma Aubl. ●☆
99862 Collospermum Skottsb.(1934);胶籽花属■☆
99863 Collospermum hastatum (Colenso) Skottsb.;胶籽花■☆
99864 Collospermum microspermum (Colenso et pro parte) Skottsb.;小籽胶籽花■☆
99865 Collospermum montanum (Seem.) Skottsb.;山地胶籽花■☆
99866 Collotapalus P. Br. = Cecropia Loefl.(保留属名)●☆
99867 Collotapalus P. Br. = Coilotapalus P. Browne(废弃属名)●☆
99868 Collyris Vahl = Dischidia R. Br. ●■
99869 Collyris Wahl = Dischidia R. Br. ●■
99870 Collyris minor Vahl = Dischidia nummularia R. Br. ■
99871 Colmeiroa F. Muell. = Corokia A. Cunn. ●☆
99872 Colmeiroa F. Muell. = Paracorokia M. Kral ●☆
99873 Colmeiroa Reut. = Flueggea Willd. ●
99874 Colmeiroa Reut. = Securinega Comm. ex Juss.(保留属名)●☆
99875 Colobachne P. Beauv. = Alopecurus L. ■
99876 Colobachne gerardii (All.) Link = Alopecurus alpinus Vill. ■☆
99877 Colobandra Bartl. = Hemigenia R. Br. ●☆
99878 Colobanthera Humbert(1923);平托菊属■☆
99879 Colobanthera waterlotii Humbert;平托菊■☆
99880 Colobanthium (Rchb.) G. Taylor = Sphenopholis Scribn. ■☆
99881 Colobanthium Rchb. = Avellinia Parl. ■☆
99882 Colobanthus (Trin.) Spach = Sphenopholis Scribn. ■☆
99883 Colobanthus Bartl.(1830);密缀属■☆
99884 Colobanthus Trin. = Colobanthium Rchb. ■☆
99885 Colobanthus crassifolius Hook. f. = Colobanthus quitensis (Kunth) Bartl. ■☆
99886 Colobanthus quitensis (Kunth) Bartl.;密缀■☆
99887 Colobatus Walp. = Buchenroedera Eckl. et Zeyh. ■
99888 Colobatus Walp. = Colobotus E. Mey. ■
99889 Colobium Roth = Leontodon L.(保留属名)■☆
99890 Colobocarpos Esser et Welzen = Croton L. ●
99891 Colobogyne Gagnep. = Acmella Rich. ex Pers. ■
99892 Colobogynium Schott = Schismatoglottis Zoll. et Moritzi ■
99893 Colobopetalum Post et Kuntze = Kolobopetalum Engl. ■☆
99894 Colobotus E. Mey. = Buchenroedera Eckl. et Zeyh. ■

99895 Colobotus ochreatus E. Mey.;鞘托叶密缀■☆
99896 Colocasia Link(废弃属名)=Colocasia Schott(保留属名)■
99897 Colocasia Link(废弃属名)=Richardia L. ■
99898 Colocasia Link(废弃属名)=Zantedeschia Spreng.(保留属名)■
99899 Colocasia Schott(1832)(保留属名);芋属;Colocasia,Dashen,Elephant's Ear,Elephant's-ear,Taro ■
99900 Colocasia aegyptiaca Samp. =Colocasia esculenta (L.) Schott ■
99901 Colocasia aethiopica (L.) Link =Zantedeschia aethiopica (L.) Spreng. ■
99902 Colocasia affinis Schott;近缘芋■☆
99903 Colocasia affinis Schott var. jenningsii (Veitch) Engl.;紫叶芋(车轮芋);Black Caladium ■☆
99904 Colocasia antiquorum Schott =Colocasia esculenta (L.) Schott ■
99905 Colocasia antiquorum Schott et Endl. =Colocasia esculenta (L.) Schott ■
99906 Colocasia antiquorum Schott et Endl. var. esculenta Seem. =Colocasia esculenta (L.) Schott ■
99907 Colocasia antiquorum Schott f. purpurea Makino =Colocasia tonoimo Nakai ■
99908 Colocasia antiquorum Schott var. esculenta Engl. =Colocasia esculenta (L.) Schott ■
99909 Colocasia cucullata Schott =Alocasia cucullata (Lour.) Schott et Endl. ●■
99910 Colocasia esculenta (L.) Schott;芋(白芋,槟榔芋,独皮叶,蹲鸱,旱芋,红广菜,红花野芋,红芋,红芋荷,接骨草,莒,老芋,连禅芋,梠芋,毛芋,青皮叶,青芋,水芋,土芝,香梗芋艿,野山芋,野芋,野芋荷,野芋艿,野芋头,芋根,芋荷,芋茇,芋魁,芋苗,芋奶,芋艿,芋头,真芋,紫芋);Coco,Coco Yam,Cocoyam,Coco-yam,Dasheen,Dashen,Eddo,Eddoes,Elephant Ears,Elephant's Ear,Elephant's-ear,Green Taro,Imo,Kalo,Keladi,Old Coco Yam,Old Coco-yam,Scratch Coco,Talas,Tanya,Taro ■
99911 Colocasia esculenta (L.) Schott 'Fontanesii';范塔尼斯芋■☆
99912 Colocasia esculenta (L.) Schott 'Illustris';紫斑芋■☆
99913 Colocasia esculenta (L.) Schott 'Rosea';粉芋■☆
99914 Colocasia esculenta (L.) Schott var. antiquorum (Schott) C. E. Hubb. et Rehder;槟榔芋■
99915 Colocasia esculenta (L.) Schott var. antiquorum (Schott) C. E. Hubb. et Rehder =Colocasia antiquorum Schott ■
99916 Colocasia esculenta (L.) Schott var. antiquorum (Schott) C. E. Hubb. et Rehder =Colocasia esculenta (L.) Schott ■
99917 Colocasia fallax Schott;假芋(山芋,野芋头);Falsetaro ■
99918 Colocasia formosana Hayata;台芋(红芋,兰屿芋,山芋,台湾青芋);Konishi Taro,Taiwan Taro ■
99919 Colocasia gigantea (Blume) Hook. f.;大野芋(白芋,滴水芋,莲芋,露芋,山野芋,水芋,抬板蕉,抬板七,土芝,象耳芋,印度芋);Dashen,Giant Taro ■
99920 Colocasia gigantea Schott =Colocasia gigantea (Blume) Hook. f. ■
99921 Colocasia heterochroma H. Li et Z. X. Wei;异色芋;Differentcolor Taro ■
99922 Colocasia indica (Lour.) Hassk.;印度绿芋(印度芋);Green Taro ■☆
99923 Colocasia indica Engl. =Colocasia gigantea (Blume) Hook. f. ■
99924 Colocasia jenningsii Veitch =Colocasia affinis Schott var. jenningsii (Veitch) Engl. ■☆
99925 Colocasia kerrii Gagnep. =Colocasia fallax Schott ■
99926 Colocasia konishii Hayata;小西氏芋■
99927 Colocasia konishii Hayata =Colocasia formosana Hayata ■
99928 Colocasia kotoensis Hayata;红头芋(兰屿芋,小西氏芋);Kato Taro ■
99929 Colocasia kotoensis Hayata =Schismatoglottis kotoensis (Hayata) T. C. Huang,J. L. Hsiao et H. Y. Yeh ■
99930 Colocasia macrorrhiza Schott =Alocasia macrorrhiza Schott ■
99931 Colocasia mucronata Kunth =Alocasia macrorrhiza Schott ■
99932 Colocasia neogyineensis André =Schismatoglottis novo-guineensis (André) N. E. Br. ■
99933 Colocasia odora Brongn. =Alocasia macrorrhiza Schott ■
99934 Colocasia pumila Kunth =Gonatanthus pumilus (D. Don) Engl. et Krause ■
99935 Colocasia rugosa Kunth =Alocasia cucullata (Lour.) Schott et Endl. ●■
99936 Colocasia tonoimo Nakai;紫芋(东南菜,广菜,老虎广菜,野芋头,芋头花);Dashen,Purple Taro ■
99937 Colocasia vivipara Thwaites =Remusatia vivipara (Lodd.) Schott ■
99938 Colocasiaceae Vines =Araceae Juss.(保留科名)■●
99939 Colococca Raf. =Collococcus P. Browne ●
99940 Colococca Raf. =Cordia L.(保留属名)●
99941 Colocrater T. Durand et Jacks. =Calocrater K. Schum. ●☆
99942 Colocynthis Mill.(废弃属名)=Citrullus Schrad. ex Eckl. et Zeyh.(保留属名)■
99943 Colocynthis amarissima Schrad. =Citrullus lanatus (Thunb.) Matsum. et Nakai ■
99944 Colocynthis citrullus (L.) Kuntze =Citrullus lanatus (Thunb.) Matsum. et Nakai ■
99945 Colocynthis citrullus (L.) Kuntze var. fistulosus (Stocks) Chakr. =Praecitrullus fistulosus (Stocks) Pangalo ■☆
99946 Colocynthis naudinianus (Sond.) Kuntze =Acanthosicyos naudinianus (Sond.) C. Jeffrey ■☆
99947 Colocynthis vulgaris Schrad. =Citrullus colocynthis (L.) Schrad. ■☆
99948 Cologania Kunth =Amphicarpaea Elliott ex Nutt.(保留属名)■
99949 Cologania Kunth(1824);热美两型豆属■☆
99950 Cologania affinis M. Martens et Galeotti;近缘热美两型豆■☆
99951 Cologania capitata Rose;头状热美两型豆■☆
99952 Cologania cordata Fearing ex McVaugh;心形热美两型豆■☆
99953 Cologania erecta Rose;直立热美两型豆■☆
99954 Cologania grandiflora Rose;大花热美两型豆■☆
99955 Cologania parviflora V. M. Badillo;小花热美两型豆■☆
99956 Cologania pulchella Kunth;美丽热美两型豆■☆
99957 Cologyne Griff. =Coelogyne Lindl. ■
99958 Cololobus H. Rob.(1994);短瓣斑鸠菊属●☆
99959 Cololobus hatschbachii H. Rob.;短瓣斑鸠菊●☆
99960 Colomandra Neck. =Aiouea Aubl. ●☆
99961 Colombiana Ospina =Pleurothallis R. Br. ■☆
99962 Colombobalanus Nixon et Crepet =Trigonobalanus Forman ●
99963 Colombobalanus excelsa (Lozano,Hern. Cam. et Henao) Nixon et Crepet =Trigonobalanus excelsa Lozano,Hern. Cam. et Henao ●☆
99964 Colona Cav.(1798);一担柴属(泡火绳属);Colona ●
99965 Colona floribunda (Wall. ex Kurz) Craib;一担柴(大毛叶子红绳,大泡火绳,柯榔木,野火绳);Manyflower Colona,Multiflorous Colona ●
99966 Colona floribunda (Wall.) Craib =Colona floribunda (Wall. ex Kurz) Craib ●
99967 Colona sinica Hu =Colona thorelii (Gagnep.) Burret ●

99968 Colona thorelii (Gagnep.) Burret;狭叶一担柴(柯仑木);Narrowleaf Colona,Thorel Colona ●
99969 Colonna J. St. -Hil. = Colona Cav. ●
99970 Colonnea Endl. = Calonnea Buc' hoz ■
99971 Colonnea Endl. = Gaillardia Foug. ■
99972 Colophonia Comm. ex Kunth = Canarium L. ●
99973 Colophonia Post et Kuntze = Ipomoea L. (保留属名)●■
99974 Colophonia Post et Kuntze = Kolofonia Raf. ●■
99975 Colophospermum J. Kirk ex Benth. = Colophospermum J. Kirk ex J. Léonard ●☆
99976 Colophospermum J. Kirk ex J. Léonard(1949);可乐豆属●☆
99977 Colophospermum J. Léonard = Colophospermum J. Kirk ex Benth. ●☆
99978 Colophospermum Kirk ex J. Léonard = Colophospermum J. Kirk ex Benth. ●☆
99979 Colophospermum mopane (J. Kirk ex Benth.) J. Kirk ex J. Léonard = Colophospermum mopane (J. Kirk ex Benth.) J. Léonard ●☆
99980 Colophospermum mopane (J. Kirk ex Benth.) J. Léonard;可乐豆;Balsam Tree,Mopane ●☆
99981 Coloptera J. M. Coult. et Rose = Cymopterus Raf. ■☆
99982 Coloradoa Boissev. et C. Davidson = Sclerocactus Britton et Rose ●☆
99983 Coloradoa mesae-verdae Boissev. et C. Davidson = Sclerocactus mesae-verdae (Boissev. et C. Davidson) L. D. Benson ●☆
99984 Colostephanus Harv. = Cynanchum L. ●■
99985 Colpias E. Mey. ex Benth. (1836);鞘玄参属●☆
99986 Colpias mollis E. Mey. ex Benth.;鞘玄参●☆
99987 Colpodium Trin. (1820);鞘柄茅属(拟沿沟草属,小沿沟草属)■
99988 Colpodium altaicum Trin.;柔毛小沿沟草■
99989 Colpodium altaicum Trin. = Paracolpodium altaicum (Trin.) Tzvelev ■
99990 Colpodium araraticum (Lipsky) Woronow;亚拉腊鞘柄茅■☆
99991 Colpodium bulbosum Trin. = Catabrosella humilis (M. Bieb.) Tzvelev ■
99992 Colpodium chionogeiton (Pilg.) Tzvelev;雪白鞘柄茅■☆
99993 Colpodium chrysanthum Woronow;黄花鞘柄茅■☆
99994 Colpodium colchicum (Albov) Woronow;黑海鞘柄茅■☆
99995 Colpodium drakensbergense Hedberg et I. Hedberg;德拉肯斯鞘柄茅■☆
99996 Colpodium fibrosum Trautv.;纤维鞘柄茅■☆
99997 Colpodium filifolium Trin. = Puccinellia filifolia (Trin.) Tzvelev ■
99998 Colpodium hedbergii (Melderis) Tzvelev;赫德鞘柄茅■☆
99999 Colpodium himalaicum (Hook. f.) Bor = Catabrosella himalaica (Hook. f.) Tzvelev ■☆
100000 Colpodium himalaicum Bor = Paracolpodium altaicum (Trin.) Tzvelev subsp. leucolepis (Nevski) Tzvelev ■
100001 Colpodium humile (M. Bieb.) Griseb.;矮小沿沟草■
100002 Colpodium humile (M. Bieb.) Griseb. = Catabrosella humilis (M. Bieb.) Tzvelev ■
100003 Colpodium leucolepis (Nevski) Tzvelev = Paracolpodium altaicum (Trin.) Tzvelev subsp. leucolepis (Nevski) Tzvelev ■☆
100004 Colpodium leucolepis Nevski;高山小沿沟草■
100005 Colpodium leucolepis Nevski = Paracolpodium altaicum (Trin.) Tzvelev subsp. leucolepis (Nevski) Tzvelev ■
100006 Colpodium nutans (Stapf) Bor = Hyalopoa nutans (Stapf) Alexeev ■☆
100007 Colpodium oreades (Peter) E. B. Alexeev = Colpodium chionogeiton (Pilg.) Tzvelev ■☆
100008 Colpodium ornatum Nevski;装饰鞘柄茅■☆
100009 Colpodium parviflorum Boiss. et Buhse;小花鞘柄茅■☆
100010 Colpodium ponticum Medvyedev;蓬特鞘柄茅■☆
100011 Colpodium pusillum Nees = Pentaschistis pusilla (Nees) H. P. Linder ■☆
100012 Colpodium tibeticum Bor;藏小沿沟草■
100013 Colpodium tibeticum Bor = Paracolpodium tibeticum (Bor) E. B. Alexeev ■☆
100014 Colpodium vahlianum (Liebm.) Nevski;瓦氏小沿沟草■☆
100015 Colpodium variegatum Boiss. ex Griseb.;斑点拟沿沟草■☆
100016 Colpodium versicolor (Steven) Woronow;变色拟沿沟草■☆
100017 Colpodium villosum Bor = Colpodium leucolepis Nevski ■
100018 Colpodium villosum Bor = Paracolpodium altaicum (Trin.) Tzvelev subsp. leucolepis (Nevski) Tzvelev ■
100019 Colpodium wallichii (Stapf) Bor;沃利克小沿沟草(瓦氏小沿沟草)■
100020 Colpogyne B. L. Burtt(1971);沟果苣苔属■☆
100021 Colpogyne betsiliensis (Humbert) B. L. Burtt;沟果苣苔■☆
100022 Colpoon P. J. Bergius = Osyris L. ●
100023 Colpoon P. J. Bergius(1767);檀枣属●☆
100024 Colpoon compressum P. J. Bergius = Osyris compressa (P. J. Bergius) A. DC. ●☆
100025 Colpoon speciosum (A. W. Hill) Bean = Osyris speciosa (A. W. Hill) J. C. Manning et Goldblatt ●☆
100026 Colpophyllos Trew = Ellisia L. (保留属名)■☆
100027 Colpothrinax Griseb. et H. Wendl. (1883);瓶棕属(瓶榈属,鞘扇榈属,桶棕属)●☆
100028 Colpothrinax cookii Read;瓶棕●☆
100029 Colpothrinax wrightii Griseb. et H. Wendl. ex Siebert et Voss;古巴瓶棕;Pot-bellied Palm ●☆
100030 Colquhounia Wall. (1822);火把花属;Colquhounia,Torchflower ●
100031 Colquhounia coccinea Wall.;深红火把花;Coccineus Colquhounia,Darkred Torchflower ●
100032 Colquhounia coccinea Wall. var. mollis (Schltdl.) Prain;火把花(密蒙花,炮仗花,细羊巴巴花);Pubescent Coccineus Colquhounia,Pubescent Colquhounia ●
100033 Colquhounia coccinea Wall. var. vestita Prain = Colquhounia vestita Wall. ●
100034 Colquhounia compta W. W. Sm.;金江火把花(金江炮仗花);Jinjiang Colquhounia, Magnific Torchflower, Thick-stemmed Colquhounia ●
100035 Colquhounia compta W. W. Sm. var. mekongensis (W. W. Sm.) Kudo;沧江火把花;Mekong Jinjiang Colquhounia ●
100036 Colquhounia decora Diels = Colquhounia seguinii Vaniot ●
100037 Colquhounia elegans Wall.;秀丽火把花(秀丽炮仗花);Elegant Colquhounia,Elegant Torchflower ●
100038 Colquhounia elegans Wall. var. pauciflora Prain = Colquhounia seguinii Vaniot ●
100039 Colquhounia elegans Wall. var. tenuiflora (Hook. f.) Prain;细花秀丽火把花(碎密花,细花火把花,细棉花,杂密花);Thinflower Colquhounia ●
100040 Colquhounia elegans Wall. var. typica Prain = Colquhounia elegans Wall. ●

100041 Colquhounia fluminis H. Lév. = Colquhounia seguinii Vaniot ●

100042 Colquhounia mekongensis W. W. Sm. = Colquhounia compta W. W. Sm. var. mekongensis (W. W. Sm.) Kudo ●

100043 Colquhounia mollis Schltdl. = Colquhounia coccinea Wall. var. mollis (Schltdl.) Prain ●

100044 Colquhounia seguinii Vaniot;藤状火把花(过接桥,苦梅叶,藤火把花,藤状炮仗花,小红花);Seguin Colquhounia, Seguin Torchflower ●

100045 Colquhounia seguinii Vaniot var. pilosa Rehder;长毛藤状火把花(长毛火把花);Pilose Seguin Colquhounia ●

100046 Colquhounia tenuiflora Hook. f. = Colquhounia elegans Wall. var. tenuiflora (Hook. f.) Prain ●

100047 Colquhounia tomentosa Houllet = Colquhounia coccinea Wall. var. mollis (Schltdl.) Prain ●

100048 Colquhounia vestita Wall.;白毛火把花(白毛炮仗花);Clothed Colquhounia, Whitehair Colquhounia, Whitehair Torchflower ●

100049 Colquhounia vestita Wall. = Colquhounia coccinea Wall. var. mollis (Schltdl.) Prain ●

100050 Colquhounia vestita Wall. var. rugosa C. B. Clarke ex Prain = Colquhounia coccinea Wall. var. mollis (Schltdl.) Prain ●

100051 Colsmannia Lehm. = Onosma L. ■

100052 Colubrina Brongn. = Colubrina Rich. ex Brongn.(保留属名)●

100053 Colubrina Friche-Joset et Montandon = Colubrina Rich. ex Brongn.(保留属名)●

100054 Colubrina Friche-Joset et Montandon = Polygonum L.(保留属名)■●

100055 Colubrina Montandon = Bistorta (L.) Scop. ■☆

100056 Colubrina Montandon = Colubrina Rich. ex Brongn.(保留属名)●

100057 Colubrina Montandon = Polygonum L.(保留属名)■●

100058 Colubrina Rich. ex Brongn.(1826)(保留属名);蛇藤属(滨枣属);Colubrina, Snakevine ●

100059 Colubrina arborescens (Mill.) Sarg.;大海蛇藤;Snake Bark ●

100060 Colubrina articulata (Capuron) Figueiredo;关节蛇藤●☆

100061 Colubrina asiatica (L.) Brongn.;蛇藤(亚洲滨枣);Asian Colubrina, Asian Nakedwood, Asian Snakevine, Asiatic Colubrina, Lather Leaf, Latherleaf ●

100062 Colubrina asiatica (L.) Brongn. var. subpubescens (Pit.) M. C. Johnst. = Colubrina javanica Miq. ●

100063 Colubrina bartramia (L.) Merr.;马来亚蛇藤●☆

100064 Colubrina decipiens (Baill.) Capuron;马岛蛇藤●☆

100065 Colubrina elliptica (Sw.) Brizicky et Stern;卷叶蛇藤●☆

100066 Colubrina ferruginosa Brongn.;锈毛蛇藤(铁锈色蛇藤,锈色蛇藤);West Indian Greenheart ●☆

100067 Colubrina ferruginosa Brongn. = Colubrina arborescens (Mill.) Sarg. ●

100068 Colubrina glandulosa Perkins;腺蛇藤;Saguaragy Bark ●☆

100069 Colubrina humbertii (H. Perrier) Capuron;亨伯特蛇藤●☆

100070 Colubrina javanica Miq.;毛蛇藤;Pubescent Colubrina, Pubescent Snakevine ●

100071 Colubrina macrocarpa (Cav.) G. Don;大果蛇藤●☆

100072 Colubrina nicholsonii A. E. van Wyk et Schrire;尼克尔森蛇藤●☆

100073 Colubrina oppositifofia Brongn. ex H. Mann;对叶蛇藤●☆

100074 Colubrina pubescens Kurz = Colubrina javanica Miq. ●

100075 Colubrina pubescens Kurz var. subpubescens Pit. = Colubrina javanica Miq. ●

100076 Colubrina reclinata Brongn. = Colubrina elliptica (Sw.) Brizicky et Stern ●☆

100077 Colubrina rufa Reissek;浅红蛇藤●☆

100078 Colubrina texensis A. Gray;北美蛇藤(得克萨斯蛇藤,得克萨斯野咖啡,得州蛇藤);Texas Colubrina ●☆

100079 Columbaria J. Presl et C. Presl = Scabiosa L. ●■

100080 Columbea Salisb. = Araucaria Juss. ●

100081 Columbia Pers. = Colona Cav. ●

100082 Columbia floribunda Kurz = Colona floribunda (Wall.) Craib ●

100083 Columbia floribunda Wall. ex Kurz = Colona floribunda (Wall. ex Kurz) Craib ●

100084 Columbia thorelii Gagnep. = Colona thorelii (Gagnep.) Burret ●

100085 Columbiadoria G. L. Nesom(1991);溪黄花属■☆

100086 Columbiadoria hallii (A. Gray) G. L. Nesom;溪黄花;Columbia River goldenrod ■☆

100087 Columbra Comm. ex Endl. = Cocculus DC.(保留属名)●

100088 Columella Comm. ex DC. = Pavonia Cav.(保留属名)●■☆

100089 Columella Lour.(废弃属名) = Cayratia Juss.(保留属名)●

100090 Columella Lour.(废弃属名) = Columellia Ruiz et Pav.(保留属名)●☆

100091 Columella Vahl = Columellia Ruiz et Pav.(保留属名)●☆

100092 Columella Vahl = Pisonia L. ●

100093 Columella Vell. = Pisonia L. ●

100094 Columella ciliifera Merr. = Cayratia ciliifera (Merr.) Chun ●

100095 Columella corniculam (Benth.) Merr. = Cayratia corniculata (Benth.) Gagnep. ●

100096 Columella geniculata (Blume) Merr. = Cayratia geniculata (Blume) Gagnep. ●

100097 Columella japonica (Thunb.) Merr. = Cayratia japonica (Thunb.) Gagnep. ●

100098 Columella japonica Merr. = Cayratia japonica (Thunb.) Gagnep. ●

100099 Columella oligocarpa (H. Lév. et Vaniot) Rehder = Cayratia oligocarpa Gagnep. ●

100100 Columella oligocarpa (H. Lév. et Vaniot) Rehder = Cayratia oligocarpa (H. Lév. et Vaniot) Gagnep. ●

100101 Columella pedata Lour. = Cayratia pedata (Lam.) Juss. ex Gagnep. ●

100102 Columella tenuifolia (Wight et Arn.) Merr. = Cayratia japonica (Thunb.) Gagnep. ●

100103 Columella tenuifolia Merr. = Cissus tenuifolia (Wight et Arn.) F. Heyne ex Planch. ●☆

100104 Columellaceae Dulac = Euphorbiaceae Juss.(保留科名) + Buxaceae Dumort.(保留科名)●■

100105 Columellea Jacq. = Nestlera Spreng. ■☆

100106 Columellea biennis Jacq. = Nestlera biennis (Jacq.) Spreng. ■☆

100107 Columellia Ruiz et Pav.(1794)(保留属名);弯药树属●☆

100108 Columellia oblonga Ruiz et Pav.;弯药树●☆

100109 Columelliaceae D. Don(1828)(保留科名);弯药树科●☆

100110 Columnea L.(1753);金鱼苣苔属(金鱼花属);Columnea ●■☆

100111 Columnea arguta Morton;锐齿金鱼花●☆

100112 Columnea chinensis Osbeck = Limnophila chinensis (Osbeck) Merr. ■

100113 Columnea crassifolia Brongn.;厚叶金鱼花●☆

100114 Columnea glabra Oerst.;光金鱼藤;Glabrous Columnea ●☆

100115 Columnea gloriosa Sprague;大红金鱼藤(大红金鱼花,大红鲸鱼花,金鱼花);Columnea, Cost Rica Columnea, Goldfish Plant, Scarlet Columnea ■☆

100116 Columnea heterophylla Roxb. = Limnophila heterophylla (Roxb.) Benth. ■
100117 Columnea hirta Klotzsch et Hanst.;短毛金鱼花●☆
100118 Columnea linearis Oerst.;细叶金鱼藤●☆
100119 Columnea longifolia L. = Artanema longifolium (L.) Vatke ■☆
100120 Columnea magnifica Klotzsch et Hanst. ex Oerst.;壮大金鱼藤(鲸鱼花);Robust Columnea ●☆
100121 Columnea microphylla Klotzsch et Hanst. = Columnea microphylla Klotzsch et Hanst. ex Oerst. ■☆
100122 Columnea microphylla Klotzsch et Hanst. ex Oerst.;小叶金鱼花(纽扣叶鲸鱼花);Small-leaf Columnea ■☆
100123 Columnea microphylla Klotzsch et Hanst. ex Oerst. 'Variegata';彩纹小叶金鱼花(斑叶小叶金鱼花)■☆
100124 Columnea oerstediana Klotzsch ex Oerst.;欧氏金鱼藤;Oersted Columnea ●☆
100125 Columnea sanguinea Hanst.;血红金鱼藤●☆
100126 Columnea schiedeana Schltdl.;希氏金鱼藤;Schiede Columnea ●☆
100127 Columnea stenophylla Standl. = Columnea crassifolia Brongn. ●☆
100128 Columnea tulae Urb.;图拉金鱼藤●☆
100129 Columnea tulae Urb. var. flava Urb.;浅黄图拉金鱼藤●☆
100130 Columnea tulae Urb. var. rubra Urb.;红花图拉金鱼藤●☆
100131 Coluppa Adans. = Gomphrena L. ●■
100132 Coluria R. Br. (1823);无尾果属;Coluria, Clove-root Plant ■
100133 Coluria elegans Cardot = Coluria longifolia Maxim. ■
100134 Coluria elegans Cardot var. imbricata Cardot = Coluria longifolia Maxim. ■
100135 Coluria geoides Ledeb.;苦鲁■☆
100136 Coluria henryi Batalin;大头叶无尾果;Henry Taillessfruit ■
100137 Coluria henryi Batalin var. grandiflora Cardot = Coluria henryi Batalin ■
100138 Coluria henryi Batalin var. pluriflora Cardot = Coluria henryi Batalin ■
100139 Coluria longifolia Maxim.;无尾果(长叶无尾果);Longleaf Taillessfruit ■
100140 Coluria longifolia Maxim. = Acomastylis elata (Royle) F. Bolle var. leiocarpa (Evans) F. Bolle ■
100141 Coluria longifolia Maxim. f. uniflora T. C. Ku;单花无尾果;Oneflower Taillessfruit ■
100142 Coluria longifolia Maxim. f. uniflora T. C. Ku = Coluria longifolia Maxim. ■
100143 Coluria oligocarpa (J. Krause) F. Bolle;汶川无尾果;Fewfruit Taillessfruit ■
100144 Coluria omeiensis T. C. Ku;峨眉无尾果;Emei Taillessfruit ■
100145 Coluria omeiensis T. C. Ku var. nanzhengensis Te T. Yu et T. C. Ku;光柱无尾果;Nanzheng Taillessfruit ■
100146 Coluria potertilloides R. Br.;委陵无尾果■☆
100147 Coluria purdomii (N. E. Br.) Evans = Coluria longifolia Maxim. ■
100148 Colus Raeusch. = Coleus Lour. ●■
100149 Colutea L. (1753);膀胱豆属(气囊豆属,鱼鳔槐属);Bladder Senna, Bladdersenna, Bladder-senna, Soundbean ●
100150 Colutea abyssinica Kunth et Bouché;阿比西尼亚膀胱豆●☆
100151 Colutea abyssinica Kunth et Bouché var. gillettii Browicz;吉莱特膀胱豆●☆
100152 Colutea acutifolia Shap.;尖叶膀胱豆●☆
100153 Colutea arborescens L.;鱼鳔槐(灯笼槐);Bastard Senna, Bladder Senna, Bladdersenna, Bladder-senna, Common Bladder, Common Bladder Senna, Common Bladdersenna, Common Bladder-senna, Sene, Senna, Soundbean ●
100154 Colutea arborescens L. 'Bullata';皱叶鱼鳔槐●☆
100155 Colutea arborescens L. 'Variegata';斑叶鱼鳔槐●☆
100156 Colutea arborescens L. subsp. atlantica (Browicz) Ponert = Colutea atlantica Browicz ●☆
100157 Colutea arborescens L. var. affinis Pomel = Colutea atlantica Browicz ●☆
100158 Colutea arborescens L. var. atrocalyx Maire = Colutea atlantica Browicz ●☆
100159 Colutea arborescens L. var. brevidentata Murb. = Colutea atlantica Browicz ●☆
100160 Colutea arborescens L. var. bullata Rehder = Colutea arborescens L. 'Bullata' ●☆
100161 Colutea arborescens L. var. melanocalyx (Boiss. et Heldr.) Maire = Colutea atlantica Browicz ●☆
100162 Colutea arborescens L. var. nepalensis (Sims) Baker = Colutea nepalensis Sims ●
100163 Colutea arborescens L. var. parvifolia Faure et Maire = Colutea atlantica Browicz ●☆
100164 Colutea armata Hemsl. et Lace;具刺膀胱豆●☆
100165 Colutea armena Boiss. et Huet;亚美尼亚膀胱豆●☆
100166 Colutea atlantica Browicz;大西洋膀胱豆●☆
100167 Colutea atlantica Browicz var. longeracemosa (Sennen) Browicz = Colutea atlantica Browicz ●☆
100168 Colutea brevialata Lange;短翼膀胱豆●☆
100169 Colutea buhsei (Boiss.) Shap.;布氏膀胱豆●☆
100170 Colutea canescens Shap.;灰膀胱豆●☆
100171 Colutea cilicica Boiss. et Balansa;小亚细亚鱼鳔槐●☆
100172 Colutea cruenta Aiton = Colutea orientalis Lam. ●☆
100173 Colutea delavayi Franch.;膀胱豆(命子花,无路花);Bladderbean, Delavay Bladdersenna ●
100174 Colutea frutescens L. = Sutherlandia frutescens (L.) R. Br. ●☆
100175 Colutea gracilis Freyn et Sint. ex Freyn;纤细膀胱豆●☆
100176 Colutea herbacea L. = Lessertia herbacea (L.) Druce ■☆
100177 Colutea hybrida Shap.;杂种膀胱豆●☆
100178 Colutea istria Mill.;中东鱼鳔槐●☆
100179 Colutea jarmnolenkoi Shap.;雅氏膀胱豆●☆
100180 Colutea kopetdaghensis B. Fedtsch.;科佩特膀胱豆●☆
100181 Colutea linearis Thunb. = Lessertia linearis (Thunb.) DC. ■☆
100182 Colutea media Willd.;红花鱼鳔槐(红花豆鱼鳔槐,杂种鱼鳔槐,中型灯笼槐);Average Bladder-senna, Hybrid Soundbean, Orange Bladder-senna ●
100183 Colutea media Willd. 'Copper Beauty';铜色美人红花鱼鳔槐●☆
100184 Colutea mesantha Shap. ex Ali = Colutea paulsenii Freyn subsp. mesantha (Shap. ex Ali) Ali ●☆
100185 Colutea nepalensis Sims;尼泊尔鱼鳔槐;Nepal Bladdersenna, Soundbean ●
100186 Colutea obtusata Thunb. = Lessertia obtusata (Thunb.) DC. ■☆
100187 Colutea orientalis Lam.;东方鱼鳔槐(东方膀胱豆,土耳其气囊豆);Oriental Bladder Senna, Oriental Bladdersenna, Oriental Bladder-senna ●☆
100188 Colutea paulsenii Freyn;帕氏膀胱豆●☆
100189 Colutea paulsenii Freyn et Sint. = Colutea paulsenii Freyn ●☆
100190 Colutea paulsenii Freyn subsp. mesantha (Shap. ex Ali) Ali;间

花鱼鳔槐●☆
100191 Colutea paulsenii Freyn var. mesantha (Shap. ex Ali) Browicz = Colutea paulsenii Freyn subsp. mesantha (Shap. ex Ali) Ali ●☆
100192 Colutea perennans Jacq. = Lessertia perennans (Jacq.) DC. ■☆
100193 Colutea procumbens Mill. = Lessertia procumbens (Mill.) DC. ■☆
100194 Colutea prostrata Thunb. = Lessertia prostrata (Thunb.) DC. ●☆
100195 Colutea pubescens Thunb. = Lessertia pubescens (Thunb.) DC. ●☆
100196 Colutea vesicaria Thunb. = Lessertia vesicaria (Thunb.) DC. ■☆
100197 Coluteastrum Fabr. (废弃属名) = Lessertia DC. (保留属名) ●■☆
100198 Coluteastrum Heist. ex Fabr. = Lessertia DC. (保留属名) ●■☆
100199 Coluteastrum Möhring ex Kuntze = Lessertia DC. (保留属名) ●■☆
100200 Coluteastrum benguellense (Baker) Hiern = Lessertia benguellensis Baker f. ■☆
100201 Coluteocarpus Boiss. (1841); 鳔果芥属(鱼鳔果芥属) ■☆
100202 Coluteocarpus vesicaria (L.) Holmb.; 鳔果芥■☆
100203 Colutia Medik. = Sutherlandia R. Br. (保留属名) ●☆
100204 Colvillea Bojer = Colvillea Bojer ex Hook. ●☆
100205 Colvillea Bojer ex Hook. (1834); 异凤凰木属(垂花楹属) ●☆
100206 Colvillea racemosa Bojer ex Hook.; 异凤凰木●☆
100207 Colymbada Hill = Centaurea L. (保留属名) ●■
100208 Colymbada balansae (Boiss.) Holub = Centaurea balansae Boiss. et Reut. ■☆
100209 Colymbada clementei (Boiss.) Holub = Centaurea clementei Boiss. ■☆
100210 Colymbada eryngioides (Lam.) Holub = Centaurea eryngioides Lam. ■☆
100211 Colymbada omphalotricha (Batt.) Holub = Centaurea pubescens Willd. subsp. omphalotricha Batt. ■☆
100212 Colymbea Steud. = Araucaria Juss. ●
100213 Colymbea Steud. = Columbea Salisb. ●
100214 Colyris Endl. = Collyris Vahl ●■
100215 Colyris Endl. = Dischidia R. Br. ●■
100216 Colythrum Schott = Esenbeckia Kunth ●☆
100217 Comaceae Dulac = Tamaricaceae Link (保留科名) ●■
100218 Comacephalus Klotzsch = Eremia D. Don ●☆
100219 Comacephalus incurvus Klotzsch = Erica eriocephala Lam. ●☆
100220 Comachlinium Scheidw. et Planch. = Dyssodia Cav. ■☆
100221 Comaclinium Scheidw. et Planch. (1852); 山橙菊属■☆
100222 Comaclinium Scheidw. et Planch. = Dyssodia Cav. ■☆
100223 Comaclinium montanum (Benth.) Strother; 山橙菊■☆
100224 Comacum Adans. = Myristica Gronov. (保留属名) ●
100225 Comandra Nutt. (1818); 毛蕊木属(假柳穿鱼属); Commandra ●☆
100226 Comandra livida Richardson; 北方毛蕊木; Northern Commandra ●☆
100227 Comandra pallida A. DC.; 苍白毛蕊木; Bastard Toadflax ●☆
100228 Comandra richardsiana Fernald = Comandra umbellata (L.) Nutt. ●☆
100229 Comandra umbellata (L.) Nutt.; 伞毛蕊木; Bastard Toadflax, Bastard-toadflax, False Toadflax, Star Toadflax, Toadflax ●☆
100230 Comandraceae Nickrent et Der; 毛蕊木科●☆
100231 Comanthera L. B. Sm. = Syngonanthus Ruhland ■☆
100232 Comanthosphace S. Moore (1877); 绵穗苏属(天人草属); Comanthosphace ■
100233 Comanthosphace formosana Ohwi; 台湾绵穗苏(白木草, 台湾白笏草, 台湾白木草) ■
100234 Comanthosphace japonica (Miq.) S. Moore ex Hook. f. = Leucosceptrum japonicum (Miq.) Kitam. et Murata ■
100235 Comanthosphace japonica (Miq.) S. Moore f. barbinervis (Miq.) Koidz. = Leucosceptrum japonicum (Miq.) Kitam. et Murata f. barbinerve (Miq.) Kitam. et Murata ■☆
100236 Comanthosphace nanchuanensis C. Y. Wu et H. W. Li; 南川绵穗苏; Nanchuan Comanthosphace, S. Sichuan Comanthosphace ■
100237 Comanthosphace nepalensis Kitam. et Murata = Leucosceptrum canum Sm. ●
100238 Comanthosphace ningpoensis (Hemsl.) Hand. -Mazz.; 绵穗苏(半边苏, 火胡麻, 野苏, 野鱼香); Ningpo Comanthosphace ■
100239 Comanthosphace ningpoensis (Hemsl.) Hand. -Mazz. var. stellipiloides C. Y. Wu; 绒毛绵穗苏(石荠苎); Hair Ningpo Comanthosphace ■
100240 Comanthosphace stellipila (Miq.) S. Moore = Leucosceptrum stellipilum (Miq.) Kitam. et Murata ●
100241 Comanthosphace stellipila (Miq.) S. Moore var. japonica (Miq.) Matsum. et Koidz. = Comanthosphace japonica (Miq.) S. Moore ex Hook. f. ■
100242 Comanthosphace stellipila (Miq.) S. Moore var. tosaensis (Makino ex Koidz.) Makino = Leucosceptrum stellipilum (Miq.) Kitam. et Murata var. radicans (Honda) T. Yamaz. et Murata ●☆
100243 Comanthosphace sublanceolata (Miq.) S. Moore = Comanthosphace ningpoensis (Hemsl.) Hand. -Mazz. ■
100244 Comanthosphace tosaensis Makino ex Koidz. = Leucosceptrum stellipilum (Miq.) Kitam. et Murata var. radicans (Honda) T. Yamaz. et Murata ●☆
100245 Comarella Rydb. = Potentilla L. ■●
100246 Comarobatia Greene = Rubus L. ●■
100247 Comaropsis Rich. = Waldsteinia Willd. ■
100248 Comaropsis sibirica (Tratt.) Ser. = Waldsteinia ternata (Stephan) Fritsch ■
100249 Comarostaphylis Zucc. (1837); 夏冬青属●☆
100250 Comarostaphylis diversifolia (Parry) Greene; 夏冬青; Summer Holly ●☆
100251 Comarostaphylis diversifolia Greene = Comarostaphylis diversifolia (Parry) Greene ●☆
100252 Comarostaphylos Zucc. = Comarostaphylis Zucc. ●☆
100253 Comarouna Carrière = Dipteryx Schreb. (保留属名) ●☆
100254 Comarum L. (1753); 沼委陵菜属; Cinquefoil, Marsh Cinquefoil ●■
100255 Comarum L. = Pancovia Heist. ex Fabr. (废弃属名) ●☆
100256 Comarum L. = Potentilla L. ■●
100257 Comarum palustre L.; 沼委陵菜; Bog Strawberry, Bogberry, Cowberry, Marsh Cinquefoil, Marsh Five-finger, Northern Cinquefoil, Purple Cinquefoil ●■
100258 Comarum palustre L. var. villosum Pers. = Comarum palustre L. ●■
100259 Comarum salesovianum (Stephan) Asch. et Graebn.; 西北沼委陵菜(白花沼委陵菜, 红茎委陵菜); Shrubby Cinquefoil, Shrubby Marsh Cinquefoil ●■
100260 Comarum salesovii (Steph. ex Willd.) Bunge = Comarum salesovianum (Stephan) Asch. et Graebn. ●■
100261 Comarum salesovii Bunge = Comarum salesovianum (Stephan) Asch. et Graebn. ●■

100262 Comaspermum Pers. = Comesperma Labill. ●☆
100263 Comastoma (Wettst.) Toyok. (1961);喉毛花属(喉花草属,喉花属,喉毛草属);Comastoma, Throathair ■
100264 Comastoma (Wettst.) Toyok. = Gentiana L. ■
100265 Comastoma Toyok. = Comastoma (Wettst.) Toyok. ■
100266 Comastoma arrectum (Franch.) Holub = Comastoma pulmonarium (Turcz.) Toyok. ■
100267 Comastoma beesianum (W. W. Sm.) Holub = Comastoma traillianum (Forrest) Holub ■
100268 Comastoma cyananthiflorum (Franch. ex Hemsl.) Holub;蓝钟喉毛花;Bluebell Comastoma, Bluebell Throathair ■
100269 Comastoma cyananthiflorum (Franch. ex Hemsl.) Holub var. acutifolium Ma et H. W. Li ex T. N. Ho;尖叶蓝钟喉毛花(尖叶喉毛花);Sharpleaf Bluebell Comastoma, Sharpleaf Bluebell Throathair ■
100270 Comastoma disepalum H. W. Li ex T. N. Ho;二萼喉毛花(三萼喉毛花);Disepal Comastoma, Disepal Throathair ■
100271 Comastoma falcatum (Turcz. ex Kar. et Kir.) Toyok.;镰萼喉毛花(镰萼假龙胆,镰萼龙胆);Sicklecalyx Comastoma, Sicklecalyx Throathair ■
100272 Comastoma henryi (Hemsl.) Holub;鄂西喉毛花;W. Hubei Comastoma, W. Hubei Throathair ■
100273 Comastoma limprichtii (Grüning) Toyok. = Comastoma polycladum (Diels et Gilg) T. N. Ho ■
100274 Comastoma maclearenii (Harry Sm.) Holub = Comastoma cyananthiflorum (Franch. ex Hemsl.) Holub ■
100275 Comastoma muliense (C. Marquand) T. N. Ho;木里喉毛花;Muli Comastoma, Muli Throathair ■
100276 Comastoma pedunculatum (Royle ex D. Don) Holub;长梗喉毛花(长柄喉毛花);Longstalk Comastoma, Longstalk Throathair ■
100277 Comastoma polycladum (Diels et Gilg) T. N. Ho;皱边喉毛花(林氏龙胆,皱萼喉毛花);Wrinkleedge Comastoma, Wrinkleedge Throathair ■
100278 Comastoma pulmonarium (Turcz.) Toyok.;喉毛花(喉花草);Lungwort Comastoma, Lungwort Throathair ■
100279 Comastoma pulmonarium (Turcz.) Toyok. subsp. arrectum (Franch.) Harry Sm. = Comastoma pulmonarium (Turcz.) Toyok. ■
100280 Comastoma pulmonarium (Turcz.) Toyok. subsp. sectum (Satake) Toyok.;刚毛喉毛花■☆
100281 Comastoma pulmonarium (Turcz.) Toyok. subsp. sectum (Satake) Toyok. f. albiflorum Hid. Takah.;白花刚毛喉毛花■☆
100282 Comastoma stellariifolium (Franch. ex Hemsl.) Holub;纤枝喉毛花;Thinbranch Comastoma, Thinbranch Throathair ■
100283 Comastoma stellariifolium (Franch.) Holub = Comastoma stellariifolium (Franch. ex Hemsl.) Holub ■
100284 Comastoma tenellum (Rottb.) Toyok.;柔弱喉毛花(柔弱喉花草);Tender Comastoma, Tender Throathair ■
100285 Comastoma traillianum (Forrest) Holub;高杯喉毛花;Wineglass Comastoma, Wineglass Throathair ■
100286 Comatocroton H. Karst. = Croton L. ●
100287 Comatoglossum (Comatoglosum) H. Karst. et Triana = Talisia Aubl. ●☆
100288 Comatoglossum H. Karst. et Triana = Talisia Aubl. ●☆
100289 Combera Sandwith (1936);库默茄属■☆
100290 Combera minima (Reiche) Sandwith;库默茄■☆
100291 Combera minima Sandwith = Combera minima (Reiche) Sandwith ■☆
100292 Combesia A. Rich. = Crassula L. ●■☆
100293 Comborhiza Anderb. et K. Bremer (1991);粗根鼠麴木属●☆
100294 Comborhiza longipes (K. Bremer) Anderb. et K. Bremer;长梗粗根鼠麴木●☆
100295 Comborhiza virgata (N. E. Br.) Anderb. et K. Bremer;粗根鼠麴木●☆
100296 Combretaceae R. Br. (1810)(保留科名);使君子科;Combretum Family, Myrobalan Family ●
100297 Combretocarpus Hook. f. (1865);风车果属●☆
100298 Combretocarpus motleyi Hook. f.;风车果●☆
100299 Combretodendron A. Chev. = Combretodendron A. Chev. ex Exell ●☆
100300 Combretodendron A. Chev. = Petersianthus Merr. ●☆
100301 Combretodendron A. Chev. ex Exell = Petersianthus Merr. ●☆
100302 Combretodendron A. Chev. ex Exell (1909);风车玉蕊属●☆
100303 Combretodendron africanum (Welw. ex Benth. et Hook. f.) Exell = Petersianthus macrocarpus (P. Beauv.) Liben ●☆
100304 Combretodendron macrocarpum (P. Beauv.) Keay;大果风车玉蕊●☆
100305 Combretodendron macrocarpum (P. Beauv.) Keay = Petersianthus macrocarpus (P. Beauv.) Liben ●☆
100306 Combretodendron quadrialatum (Merr.) Merr.;四翅风车玉蕊●☆
100307 Combretodendron viridiflorum A. Chev. = Petersianthus macrocarpus (P. Beauv.) Liben ●☆
100308 Combretopsis K. Schum. = Lophopyxis Hook. f. ●☆
100309 Combretum Loefl. (1758)(保留属名);风车子属(风车藤属,藤诃子属);Combretum, Indian Gum, Redwithe, Windmill ●
100310 Combretum abbreviatum Engl. = Combretum paniculatum Vent. ●☆
100311 Combretum abercornense Exell = Combretum collinum Fresen. subsp. elgonense (Exell) Okafor ●☆
100312 Combretum aculeatum Vent.;皮刺风车子●☆
100313 Combretum acutifolium Exell;尖叶风车子●☆
100314 Combretum acutum M. A. Lawson;尖风车子●☆
100315 Combretum adenogonium Steud. ex A. Rich.;腺角风车子●☆
100316 Combretum affine De Wild. = Combretum holstii Engl. ●☆
100317 Combretum afzelii Engl. et Diels = Combretum conchipetalum Engl. et Diels ●☆
100318 Combretum albidiflorum Engl. et Diels = Combretum adenogonium Steud. ex A. Rich. ●☆
100319 Combretum albopunctatum Suess.;白斑风车子●☆
100320 Combretum album De Wild. = Combretum collinum Fresen. subsp. gazense (Swynn. et Baker f.) Okafor ●☆
100321 Combretum alfredii Hance;风车子(华风车子,清凉树,使君子藤,水番桃,四角风);Alfred Combretum, Alfred Windmill ●
100322 Combretum altum Guillaumin et Perr. ex DC.;小花风车子(小翅风车子);Little-flower Combretum ●☆
100323 Combretum altum Perr. = Combretum micranthum G. Don ●☆
100324 Combretum anacardifolium Engl. = Combretum psidioides Welw. subsp. dinteri (Schinz) Exell ●☆
100325 Combretum angolense M. A. Lawson;安哥拉风车子●☆
100326 Combretum angustifolium De Wild. = Combretum platypetalum Welw. ex M. A. Lawson subsp. oatesii (Rolfe) Exell ●☆
100327 Combretum angustipetalum Chiov.;狭瓣风车子●☆
100328 Combretum anisopterum Welw. ex M. A. Lawson = Pteleopsis anisoptera (Welw. ex M. A. Lawson) Engl. et Diels ●☆
100329 Combretum ankolense Bagsh. et Baker f. = Combretum molle R.

Br. ex G. Don ●☆
100330 Combretum antunesii Engl. et Diels = Combretum zeyheri Sond. ●☆
100331 Combretum aphanopetalum Engl. et Diels;隐瓣风车子●☆
100332 Combretum apiculatum Sond.;细尖风车子●☆
100333 Combretum apiculatum Sond. subsp. boreale Exell = Combretum apiculatum Sond. ●☆
100334 Combretum apiculatum Sond. subsp. leutweinii (Schinz) Exell;洛伊特风车子●☆
100335 Combretum arbuscula Engl. et Diels = Combretum molle R. Br. ex G. Don ●☆
100336 Combretum arenarium Portères = Combretum lineare Keay ●☆
100337 Combretum arengense Sim = Combretum molle R. Br. ex G. Don ●☆
100338 Combretum argyrochryseum Engl. et Gilg = Combretum platypetalum Welw. ex M. A. Lawson subsp. baumii (Engl. et Gilg) Exell ●☆
100339 Combretum argyrotrichum Welw. ex M. A. Lawson;银毛风车子●☆
100340 Combretum atelanthum Diels = Combretum molle R. Br. ex G. Don ●☆
100341 Combretum atropurpureum Engl. et Diels = Combretum mannii M. A. Lawson ex Engl. et Diels ●☆
100342 Combretum augustinum Diels = Combretum adenogonium Steud. ex A. Rich. ●☆
100343 Combretum aureonitens Engl. et Gilg;黄光风车子●☆
100344 Combretum auriculatum C. Y. Wu et T. Z. Hsu;耳叶风车子;Auriculate Combretum,Auriculate Windmill ●
100345 Combretum auriculatum C. Y. Wu et T. Z. Hsu = Combretum wallichii DC. ●
100346 Combretum auriculatum Engl. et Diels;耳形风车子●☆
100347 Combretum auriculatum Engl. et Diels var. longispicatum De Wild. = Combretum auriculatum Engl. et Diels ●☆
100348 Combretum bajonense Sim = Combretum collinum Fresen. subsp. gazense (Swynn. et Baker f.) Okafor ●☆
100349 Combretum basarense Engl. = Combretum adenogonium Steud. ex A. Rich. ●☆
100350 Combretum batesii Exell = Combretum fuscum Planch. ex Benth. ●☆
100351 Combretum baumii Engl. et Gilg = Combretum platypetalum Welw. ex M. A. Lawson subsp. baumii (Engl. et Gilg) Exell ●☆
100352 Combretum bequaertii De Wild. = Combretum lukafuense De Wild. ●☆
100353 Combretum binderianum Kotschy = Combretum collinum Fresen. subsp. binderanum (Kotschy) Okafor ●☆
100354 Combretum bipindense Engl. et Diels;比平迪风车子●☆
100355 Combretum borumense Engl. et Diels = Combretum hereroense Schinz ■☆
100356 Combretum bosoi De Wild. = Combretum longipilosum Engl. et Diels ●☆
100357 Combretum brachypetalum R. E. Fr. = Combretum psidioides Welw. ●☆
100358 Combretum bracteatum (M. A. Lawson) Engl. et Diels;具苞风车子●☆
100359 Combretum bracteosum (Hochst.) Brandis;疗呃果;Hiccup Nut ●☆
100360 Combretum bracteosum Brandis ex Engl. = Combretum bracteosum (Hochst.) Brandis ●☆
100361 Combretum bragae Engl. = Combretum zeyheri Sond. ●☆
100362 Combretum bricchettii Engl. et Diels = Combretum molle R. Br. ex G. Don ●☆
100363 Combretum brosigianum Engl. et Diels = Combretum collinum Fresen. subsp. suluense (Engl. et Diels) Okafor ●☆
100364 Combretum bruchhausenianum Engl. et Diels = Combretum hereroense Schinz var. parvifolium (Engl.) Wickens ●☆
100365 Combretum bruneelii De Wild. = Combretum paniculatum Vent. ●☆
100366 Combretum bucciniflorum Exell = Combretum bracteatum (M. A. Lawson) Engl. et Diels ●☆
100367 Combretum buchananii Engl. et Diels = Combretum apiculatum Sond. ●☆
100368 Combretum bulongense De Wild. = Combretum subglabratum De Wild. ●☆
100369 Combretum burttii Exell = Combretum collinum Fresen. subsp. taborense (Engl.) Okafor ●☆
100370 Combretum bussei Engl. et Diels = Combretum constrictum (Benth.) M. A. Lawson ●☆
100371 Combretum butayei De Wild. = Combretum celastroides Welw. ex M. A. Lawson subsp. laxiflorum (Welw. ex M. A. Lawson) Exell ●☆
100372 Combretum buvumense Baker f. = Combretum paniculatum Vent. ●☆
100373 Combretum cabrae De Wild. et T. Durand = Combretum cinereopetalum Engl. et Diels ●☆
100374 Combretum caffrum (Eckl. et Zeyh.) Kuntze;卡菲风车子●☆
100375 Combretum calobotrys Engl. et Diels = Combretum mannii M. A. Lawson ex Engl. et Diels ●☆
100376 Combretum calocarpum Gilg ex Dinter = Combretum zeyheri Sond. ●☆
100377 Combretum calvescens Exell = Combretum conchipetalum Engl. et Diels ●☆
100378 Combretum camporum Engl.;弯风车子●☆
100379 Combretum capitatum De Wild. et Exell;头状风车子●☆
100380 Combretum capituliflorum Fenzl ex Schweinf.;杯花风车子●☆
100381 Combretum carvalhoi Engl. = Combretum paniculatum Vent. ●☆
100382 Combretum cataractarum Diels = Combretum mossambicense (Klotzsch) Engl. ●☆
100383 Combretum caudatisepalum Exell et J. G. Garcia;尾萼风车子●☆
100384 Combretum celastroides Welw. ex M. A. Lawson;南蛇藤风车子●☆
100385 Combretum celastroides Welw. ex M. A. Lawson subsp. laxiflorum (Welw. ex M. A. Lawson) Exell;疏松南蛇藤风车子●☆
100386 Combretum celastroides Welw. ex M. A. Lawson subsp. orientale Exell;东方南蛇藤风车子●☆
100387 Combretum chevalieri Diels = Combretum adenogonium Steud. ex A. Rich. ●☆
100388 Combretum chinense Roxb. = Combretum griffithii Van Heurck et Müll. Arg. ●
100389 Combretum chionanthoides Engl. et Diels;雪花风车子●☆
100390 Combretum chlorocarpum Exell = Combretum engleri Schinz ●☆
100391 Combretum cinereopetalum Engl. et Diels;灰瓣风车子●☆
100392 Combretum cinnabarinum Engl. et Diels;朱红风车子●☆
100393 Combretum clarense Jongkind;克莱尔风车子●☆
100394 Combretum coccineum (Aubl.) Engl. et Diels;亮红风车子●☆
100395 Combretum coccineum (Sonn.) Lam. = Combretum coccineum

(Aubl.) Engl. et Diels ●☆

100396 Combretum cognatum Diels;共生风车子●☆

100397 Combretum collinum Fresen.;小丘风车子;Hill Combretum, Hill-growing Combretum, Hill-growing Windmill ●☆

100398 Combretum collinum Fresen. subsp. binderanum (Kotschy) Okafor;双皮小丘风车子●☆

100399 Combretum collinum Fresen. subsp. dumetorum (Exell) Okafor = Combretum dumetorum Exell ●☆

100400 Combretum collinum Fresen. subsp. elgonense (Exell) Okafor;埃尔贡风车子●☆

100401 Combretum collinum Fresen. subsp. gazense (Swynn. et Baker f.) Okafor;加兹风车子●☆

100402 Combretum collinum Fresen. subsp. geitonophyllum (Diels) Okafor;邻叶风车子●☆

100403 Combretum collinum Fresen. subsp. hypopilinum (Diels) Okafor;背毛风车子●☆

100404 Combretum collinum Fresen. subsp. ondongense (Engl. et Diels) Okafor;翁氏风车子●☆

100405 Combretum collinum Fresen. subsp. suluense (Engl. et Diels) Okafor;苏卢风车子●☆

100406 Combretum collinum Fresen. subsp. taborense (Engl.) Okafor;泰伯风车子●☆

100407 Combretum comosum G. Don;簇毛风车子●☆

100408 Combretum conchipetalum Engl. et Diels;贝壳风车子●☆

100409 Combretum confertum (Benth.) M. A. Lawson;密集风车子●☆

100410 Combretum confusum Merr. et Rolfe = Combretum sundaicum Miq. ●

100411 Combretum congolanum Liben;刚果风车子●☆

100412 Combretum constrictum (Benth.) M. A. Lawson;缢缩风车子●☆

100413 Combretum constrictum (Benth.) M. A. Lawson var. somalense Pamp. = Combretum constrictum (Benth.) M. A. Lawson ●☆

100414 Combretum constrictum (Benth.) M. A. Lawson var. tomentellum Engl. = Combretum constrictum (Benth.) M. A. Lawson ●☆

100415 Combretum contractum Engl. et Diels;紧缩风车子●☆

100416 Combretum copaliferum Chiov. = Combretum schumannii Engl. ●☆

100417 Combretum cordifolium Engl. ex De Wild. = Combretum comosum G. Don ●☆

100418 Combretum coriaceum Schinz = Combretum collinum Fresen. ●☆

100419 Combretum coriifolium Engl. et Diels = Combretum fuscum Planch. ex Benth. ●☆

100420 Combretum crotonoides Hutch. et Dalziel = Combretum collinum Fresen. ●☆

100421 Combretum cufodontii Chiov. = Combretum hereroense Schinz var. parvifolium (Engl.) Wickens ●☆

100422 Combretum cuspidatum Planch. ex Benth.;骤尖风车子●☆

100423 Combretum cyclocarpum Chiov. = Combretum collinum Fresen. subsp. binderanum (Kotschy) Okafor ●☆

100424 Combretum cyclophyllum Steud. = Combretum latifolium Blume ●

100425 Combretum dalzielii Hutch. = Combretum adenogonium Steud. ex A. Rich. ●☆

100426 Combretum dasystachyum Kurz = Combretum griffithii Van Heurck et Müll. Arg. ●

100427 Combretum decandrum Roxb. = Combretum roxburghii Spreng. ●

100428 Combretum dekindtianum Exell = Combretum molle R. Br. ex G. Don ●☆

100429 Combretum demeusei De Wild.;德氏风车子(德穆风车子)●☆

100430 Combretum denhardtiorum Engl. et Diels = Combretum aculeatum Vent. ●☆

100431 Combretum deserti Engl. = Combretum molle R. Br. ex G. Don ●☆

100432 Combretum detinens Dinter = Combretum mossambicense (Klotzsch) Engl. ●☆

100433 Combretum didymostachys Engl. et Diels = Combretum exalatum Engl. ●☆

100434 Combretum dilembense De Wild. = Combretum zeyheri Sond. ●☆

100435 Combretum dinteri Schinz = Combretum psidioides Welw. subsp. dinteri (Schinz) Exell ●☆

100436 Combretum dipterum Welw. = Pteleopsis diptera (Welw.) Engl. et Diels ●☆

100437 Combretum distillatorium Blanco = Combretum punctatum Blume subsp. squamosum (Roxb. ex G. Don) Exell ●

100438 Combretum dolichopetalum Engl. et Diels;长瓣风车子(金风藤)●☆

100439 Combretum dolichopodum Gilg ex Engl.;长足风车子●☆

100440 Combretum dumetorum Exell;灌丛风车子●☆

100441 Combretum duparquetiana Baill. = Combretum mannii M. A. Lawson ex Engl. et Diels ●☆

100442 Combretum edwardsii Exell;爱德华兹风车子●☆

100443 Combretum eilkerianum Schinz = Combretum hereroense Schinz ■☆

100444 Combretum eke Exell = Combretum longipilosum Engl. et Diels ●☆

100445 Combretum elaeagnoides Klotzsch;胡颓子风车子●☆

100446 Combretum elgonense Exell = Combretum collinum Fresen. subsp. elgonense (Exell) Okafor ●☆

100447 Combretum elliotii Engl. et Diels = Combretum nigricans Lepr. ex Guillaumin et Perr. var. elliotii (Engl. et Diels) Aubrév. ●☆

100448 Combretum ellipticum Sim = Combretum molle R. Br. ex G. Don ●☆

100449 Combretum elmeri Merr.;埃默风车子●☆

100450 Combretum engleri Schinz;恩格勒风车子●☆

100451 Combretum erlangerianum Engl. = Combretum hereroense Schinz var. parvifolium (Engl.) Wickens ●☆

100452 Combretum erosum Jongkind;啮蚀状风车子●☆

100453 Combretum erythrophloeum Gilg et Ledermann;红皮风车子●☆

100454 Combretum erythrophyllum (Burch.) Sond.;红叶风车子;River Bushwillow ●☆

100455 Combretum erythrophyllum (Burch.) Sond. var. obscurum Van Heurck et Müll. Arg. = Combretum erythrophyllum (Burch.) Sond. ●☆

100456 Combretum erythrophyllum Sond. = Combretum erythrophyllum (Burch.) Sond. ●☆

100457 Combretum exalatum Engl.;无翅风车子●☆

100458 Combretum excelsum Keay = Combretum pecoense Exell ●☆

100459 Combretum exellii Jongkind;埃克塞尔风车子●☆

100460 Combretum extensum Roxb. ex G. Don = Combretum latifolium Blume ●

100461 Combretum eylesii Exell = Combretum collinum Fresen. subsp. gazense (Swynn. et Baker f.) Okafor ●☆

100462 Combretum falcatum (Welw. ex Hiern) Jongkind;镰形风车子●☆

100463 Combretum ferrugineum A. Rich. = Combretum molle R. Br. ex G. Don ●☆

100464 Combretum fischeri Engl. = Combretum collinum Fresen. subsp.

suluense (Engl. et Diels) Okafor ●☆
100465 Combretum flammeum (Welw. ex M. A. Lawson) Hiern = Combretum racemosum P. Beauv. ●☆
100466 Combretum flaviflorum Exell = Combretum collinum Fresen. subsp. hypopilinum (Diels) Okafor ●☆
100467 Combretum formosum Griff. = Combretum latifolium Blume ●
100468 Combretum fragrans F. Hoffm. = Combretum adenogonium Steud. ex A. Rich. ●☆
100469 Combretum frommii Engl. = Combretum collinum Fresen. subsp. binderanum (Kotschy) Okafor ●☆
100470 Combretum frommii Gilg ex Engl.;弗罗姆风车子●☆
100471 Combretum fruticosum (Loefl.) Stuntz;灌木风车子●☆
100472 Combretum fulvum Keay;黄褐风车子●☆
100473 Combretum fuscum Planch. ex Benth.;棕色风车子●☆
100474 Combretum gabonense Exell;加蓬风车子●☆
100475 Combretum gallabatense Schweinf. = Combretum rochetianum A. Rich. ex A. Juss. ●☆
100476 Combretum galpinii Engl. et Diels = Combretum molle R. Br. ex G. Don ●☆
100477 Combretum gazense Swynn. et Baker f. = Combretum collinum Fresen. subsp. gazense (Swynn. et Baker f.) Okafor ●☆
100478 Combretum geitonophyllum Diels = Combretum collinum Fresen. subsp. geitonophyllum (Diels) Okafor ●☆
100479 Combretum gentilii De Wild. = Combretum bracteatum (M. A. Lawson) Engl. et Diels ●☆
100480 Combretum germainii Liben = Combretum fuscum Planch. ex Benth. ●☆
100481 Combretum ghasalense Engl. et Diels;加沙尔风车子;Ghasal Combretum,Ghasal Windmill ●☆
100482 Combretum ghasalense Engl. et Diels = Combretum adenogonium Steud. ex A. Rich. ●☆
100483 Combretum ghesquierei Liben;盖斯基埃风车子●☆
100484 Combretum gillettianum Liben;吉莱特风车子●☆
100485 Combretum giorgii De Wild. et Exell = Combretum padoides Engl. et Diels ●☆
100486 Combretum glandulosum F. Hoffm. = Combretum zeyheri Sond. ●☆
100487 Combretum glomeruliflorum Sond. = Combretum erythrophyllum (Burch.) Sond. ●☆
100488 Combretum glutinosum Perr. ex DC.;黏性风车子●☆
100489 Combretum gnidioides Engl. et Diels = Combretum platypetalum Welw. ex M. A. Lawson subsp. baumii (Engl. et Gilg) Exell ●☆
100490 Combretum goetzei Engl. et Diels;格兹风车子●☆
100491 Combretum goetzenianum Diels = Combretum collinum Fresen. subsp. taborense (Engl.) Okafor ●☆
100492 Combretum gondense F. Hoffm. = Combretum molle R. Br. ex G. Don ●☆
100493 Combretum goossensii De Wild. et Exell;古森斯风车子●☆
100494 Combretum gossweileri Exell;戈斯风车子●☆
100495 Combretum grandiflorum G. Don;大花藤诃子(大花风车藤);Large-flowered Combretum ●
100496 Combretum grandifolium F. Hoffm. = Combretum psidioides Welw. ●☆
100497 Combretum grandifolium F. Hoffm. var. eickii Engl. et Diels = Combretum psidioides Welw. ●☆
100498 Combretum grandifolium F. Hoffm. var. retusa? = Combretum psidioides Welw. ●☆
100499 Combretum greenwayi Exell = Combretum hereroense Schinz var. volkensii (Engl.) Wickens ●☆
100500 Combretum griffithii Van Heurck et Müll. Arg.;西南风车子;Griffith Combretum,Griffith Windmill ●
100501 Combretum griffithii Van Heurck et Müll. Arg. var. yunnanense (Exell) Turland et C. Chen;云南风车子;Yunnan Combretum, Yunnan Windmill ●
100502 Combretum griseiflorum S. Moore = Combretum collinum Fresen. ●☆
100503 Combretum grotei Exell = Combretum hereroense Schinz subsp. grotei (Exell) Wickens ●☆
100504 Combretum gueinzii Sond. = Combretum molle R. Br. ex G. Don ●☆
100505 Combretum gueinzii Sond. subsp. splendens (Engl.) Exell ex Brenan = Combretum molle R. Br. ex G. Don ●☆
100506 Combretum gueinzii Sond. var. holosericeum (Sond.) Exell ex Burtt Davy et Hoyle = Combretum molle R. Br. ex G. Don ●☆
100507 Combretum gueinzii Sond. var. holosericeum (Sond.) Exell ex Rendle = Combretum molle R. Br. ex G. Don ●☆
100508 Combretum harrisii Wickens;哈里斯风车子●☆
100509 Combretum haullevilleanum De Wild.;豪尔风车子●☆
100510 Combretum haynesianum Diels = Combretum adenogonium Steud. ex A. Rich. ●☆
100511 Combretum hensii Engl. et Diels;亨斯风车子●☆
100512 Combretum hensii Engl. et Diels var. pyriforme (De Wild.) Exell = Combretum hensii Engl. et Diels ●☆
100513 Combretum herbaceum G. Don = Combretum sericeum G. Don ●☆
100514 Combretum hereroense Schinz;草本风车子■☆
100515 Combretum hereroense Schinz subsp. grotei (Exell) Wickens;格罗特风车子●☆
100516 Combretum hereroense Schinz var. parvifolium (Engl.) Wickens;小叶草本风车子●☆
100517 Combretum hereroense Schinz var. villosissimum Engl. et Diels = Combretum hereroense Schinz ■☆
100518 Combretum hereroense Schinz var. volkensii (Engl.) Wickens;福尔风车子●☆
100519 Combretum hildebrandtii Engl. = Combretum illairii Engl. ●☆
100520 Combretum hispidum M. A. Lawson = Combretum comosum G. Don ●☆
100521 Combretum hobol Engl. et Diels = Combretum molle R. Br. ex G. Don ●☆
100522 Combretum hockii De Wild. = Combretum haullevilleanum De Wild. ●☆
100523 Combretum holosericeum Sond. = Combretum molle R. Br. ex G. Don ●☆
100524 Combretum holstii Engl.;霍尔风车子●☆
100525 Combretum holtzii Diels = Combretum molle R. Br. ex G. Don ●☆
100526 Combretum homalioides Hutch. et Dalziel;平滑风车子●☆
100527 Combretum homblei De Wild. = Combretum padoides Engl. et Diels ●☆
100528 Combretum horsfieldii Miq. = Combretum latifolium Blume ●
100529 Combretum houyanum Mildbr. = Combretum longispicatum (Engl.) Engl. et Diels ●☆
100530 Combretum hypopilinum Diels = Combretum collinum Fresen. subsp. hypopilinum (Diels) Okafor ●☆
100531 Combretum illairii Engl.;希尔德风车子●☆
100532 Combretum imberbe Wawra;无须风车子●☆

100533 Combretum imberbe Wawra var. dielsii Engl. = Combretum imberbe Wawra ●☆

100534 Combretum imberbe Wawra var. petersii (Klotzsch) Engl. et Diels = Combretum imberbe Wawra ●☆

100535 Combretum imberbe Wawra var. truncatum (M. A. Lawson) Burtt Davy = Combretum imberbe Wawra ●☆

100536 Combretum incertum Hand. -Mazz. = Combretum wallichii DC. ●

100537 Combretum indicum (L.) DeFilipps;印度风车子●☆

100538 Combretum indicum (L.) Jongkind = Combretum indicum (L.) DeFilipps ●☆

100539 Combretum indicum (L.) Jongkind = Quisqualis indica L. ●

100540 Combretum inflatum Jongkind;膨胀风车子●☆

100541 Combretum infundibuliforme Engl. = Combretum constrictum (Benth.) M. A. Lawson ●☆

100542 Combretum insculptum Engl. et Diels = Combretum molle R. Br. ex G. Don ●☆

100543 Combretum insigne Van Heurck et Müll. Arg. = Combretum pilosum Roxb. ●

100544 Combretum insulare Engl. et Diels = Combretum cuspidatum Planch. ex Benth. ●☆

100545 Combretum ischnothyrsum Engl. et Diels = Combretum mossambicense (Klotzsch) Engl. ●☆

100546 Combretum itsoghense Pellegr. = Combretum fuscum Planch. ex Benth. ●☆

100547 Combretum jacquini Griseb. f. ovatifolium (Pohl) Eichler = Chrysostachys ovatifolia Pohl ●☆

100548 Combretum kabadense Exell = Combretum collinum Fresen. subsp. elgonense (Exell) Okafor ●☆

100549 Combretum kachinense King et Prain = Sarcosperma kachinense (King et Prain) Exell ●

100550 Combretum kachinense King et Prain ex Prain = Sarcosperma kachinense (King et Prain) Exell ●

100551 Combretum kamatutu De Wild. = Combretum adenogonium Steud. ex A. Rich. ●☆

100552 Combretum karaguense Engl. et Diels = Combretum collinum Fresen. subsp. binderanum (Kotschy) Okafor ●☆

100553 Combretum kasaiense Liben;开赛风车子●☆

100554 Combretum katangense De Wild. = Combretum platypetalum Welw. ex M. A. Lawson ●☆

100555 Combretum kerengense Diels = Combretum collinum Fresen. subsp. suluense (Engl. et Diels) Okafor ●☆

100556 Combretum kerstingii Engl. et Diels = Combretum collinum Fresen. subsp. geitonophyllum (Diels) Okafor ●☆

100557 Combretum kilossanum Engl. et Diels = Combretum adenogonium Steud. ex A. Rich. ●☆

100558 Combretum kirkii M. A. Lawson;柯克风车子●☆

100559 Combretum klotzschii Welw. ex Lawson = Combretum mannii M. A. Lawson ex Engl. et Diels ●☆

100560 Combretum kraussii Hochst.;克劳斯风车子●☆

100561 Combretum kwangsiense H. L. Li = Combretum alfredii Hance ●

100562 Combretum kwebense N. E. Br. = Combretum apiculatum Sond. subsp. leutweinii (Schinz) Exell ●☆

100563 Combretum kwinkiti De Wild. = Combretum psidioides Welw. ●☆

100564 Combretum laboniense M. B. Moss = Combretum collinum Fresen. subsp. elgonense (Exell) Okafor ●☆

100565 Combretum laeteviride Engl. et Gilg = Combretum collinum Fresen. ●☆

100566 Combretum lamprocarpum Diels = Combretum collinum Fresen. subsp. geitonophyllum (Diels) Okafor ●☆

100567 Combretum landanaense De Wild. et Exell = Combretum holstii Engl. ●☆

100568 Combretum lasiocarpum Engl. et Diels;毛果风车子●☆

100569 Combretum lasiopetalum Engl. et Diels = Combretum pentagonum M. A. Lawson ●☆

100570 Combretum latialatum Engl. ex Engl. et Diels;宽翅风车子●☆

100571 Combretum latifolium Blume;阔叶风车子(大叶地耳);Broadleaf Combretum, Broadleaf Windmill, Broad-leaved Combretum ●

100572 Combretum laurentii De Wild. = Combretum platypterum (Welw.) Hutch. et Dalziel ●☆

100573 Combretum laurifolium Engl. = Combretum fuscum Planch. ex Benth. ●☆

100574 Combretum lawsonianum A. Chev. = Combretum dolichopetalum Engl. et Diels ●☆

100575 Combretum lawsonianum Engl. et Diels = Combretum platypterum (Welw.) Hutch. et Dalziel ●☆

100576 Combretum laxiflorum Welw. ex M. A. Lawson = Combretum celastroides Welw. ex M. A. Lawson subsp. laxiflorum (Welw. ex M. A. Lawson) Exell ●☆

100577 Combretum lebrunii Exell = Combretum capituliflorum Fenzl ex Schweinf. ●☆

100578 Combretum lecananthum Engl. et Diels = Combretum nigricans Lepr. ex Guillaumin et Perr. var. elliotii (Engl. et Diels) Aubrév. ●☆

100579 Combretum lecardii Engl. et Diels;莱卡德风车子●☆

100580 Combretum leiophyllum Diels = Combretum fuscum Planch. ex Benth. ●☆

100581 Combretum lemairei De Wild. et Exell = Combretum paniculatum Vent. ●☆

100582 Combretum leonense Engl. et Diels = Combretum platypetalum Welw. ex M. A. Lawson subsp. baumii (Engl. et Gilg) Exell ●☆

100583 Combretum lepidiflorum Exell = Combretum longipilosum Engl. et Diels ●☆

100584 Combretum lepidotum A. Rich. = Combretum molle R. Br. ex G. Don ●☆

100585 Combretum lepidotum C. Presl = Combretum punctatum Blume subsp. squamosum (Roxb. ex G. Don) Exell ●

100586 Combretum letestui Exell = Combretum homalioides Hutch. et Dalziel ●☆

100587 Combretum leucanthum Van Heurck et Müll. Arg. = Combretum latifolium Blume ●

100588 Combretum leuconeurum Gilg et Ledermann;白脉风车子●☆

100589 Combretum leuconili Schweinf. = Combretum aculeatum Vent. ●☆

100590 Combretum leutweinii Schinz = Combretum apiculatum Sond. subsp. leutweinii (Schinz) Exell ●☆

100591 Combretum ligustrifolium Engl. et Diels ex Baker f. = Combretum erythrophyllum (Burch.) Sond. ●☆

100592 Combretum lindense Exell et Mildbr.;林德风车子●☆

100593 Combretum lineare Keay;线形风车子●☆

100594 Combretum linyenense Hand. -Mazz. = Combretum wallichii DC. ●

100595 Combretum lisowskii Jongkind;利索风车子●☆

100596 Combretum littoreum (Engl.) Engl. et Diels = Combretum falcatum (Welw. ex Hiern) Jongkind ●☆

100597 Combretum lomuense Sim = Combretum microphyllum Klotzsch ●☆

100598 Combretum longipilosum Engl. et Diels;长毛风车子●☆

100599 Combretum longispicatum (Engl.) Engl. et Diels;长穗风车子●☆

100600 Combretum lopolense Engl. et Diels = Combretum zeyheri Sond. ●☆

100601 Combretum louisii Liben;路易斯风车子●☆

100602 Combretum lucidum E. Mey. ex Drège = Combretum kraussii Hochst. ●☆

100603 Combretum lukafuense De Wild.;卢卡夫风车子●☆

100604 Combretum luluense Exell = Combretum mannii M. A. Lawson ex Engl. et Diels ●☆

100605 Combretum luxenii Exell;卢森风车子●☆

100606 Combretum lydenburgianum Engl. et Diels = Combretum erythrophyllum (Burch.) Sond. ●☆

100607 Combretum maclaudii Aubrév.;马克洛风车子●☆

100608 Combretum macrocarpum P. Beauv. = Petersianthus macrocarpus (P. Beauv.) Liben ●☆

100609 Combretum macrophyllum Roxb. = Combretum latifolium Blume ●

100610 Combretum macrostigmateum Engl. et Diels = Combretum schumannii Engl. ●☆

100611 Combretum makindense Gilg ex Engl. = Combretum collinum Fresen. subsp. suluense (Engl. et Diels) Okafor ●☆

100612 Combretum mannii M. A. Lawson ex Engl. et Diels;曼氏风车子●☆

100613 Combretum marchettii Chiov. = Combretum collinum Fresen. subsp. binderanum (Kotschy) Okafor ●☆

100614 Combretum mardaf Chiov. = Combretum angustipetalum Chiov. ●☆

100615 Combretum marginatum Engl. et Diels;具边风车子●☆

100616 Combretum mayumbense Exell = Combretum cuspidatum Planch. ex Benth. ●☆

100617 Combretum mechowianum O. Hoffm. = Combretum collinum Fresen. subsp. gazense (Swynn. et Baker f.) Okafor ●☆

100618 Combretum mechowianum O. Hoffm. subsp. gazense (Swynn. et Baker f.) P. A. Duvign. = Combretum collinum Fresen. subsp. gazense (Swynn. et Baker f.) Okafor ●☆

100619 Combretum mechowianum O. Hoffm. subsp. taborense (Engl.) P. A. Duvign. = Combretum collinum Fresen. subsp. taborense (Engl.) Okafor ●☆

100620 Combretum megalocarpum Exell ex Brenan = Combretum zeyheri Sond. ●☆

100621 Combretum melchiorianum H. J. P. Winkl. = Combretum illairii Engl. ●☆

100622 Combretum menyhartii Engl. et Diels = Combretum kirkii M. A. Lawson ●☆

100623 Combretum meruense Engl. = Combretum illairii Engl. ●☆

100624 Combretum micranthum G. Don = Combretum altum Guillaumin et Perr. ex DC. ●☆

100625 Combretum microlepidotum Engl. = Combretum molle R. Br. ex G. Don ●☆

100626 Combretum micropetalum Llanos = Combretum latifolium Blume ●

100627 Combretum microphyllum Klotzsch;小叶风车子;Burning Bush, Flame Creeper ●☆

100628 Combretum migeodii Exell = Combretum mossambicense (Klotzsch) Engl. ●☆

100629 Combretum mildbraedii Hutch. et Dalziel = Combretum demeusei De Wild. ●☆

100630 Combretum millerianum Burtt Davy = Combretum collinum Fresen. ●☆

100631 Combretum minimipetalum Chiov. = Combretum molle R. Br. ex G. Don ●☆

100632 Combretum minutiflorum Exell = Combretum padoides Engl. et Diels ●☆

100633 Combretum mittuense Engl. et Diels = Combretum cinereopetalum Engl. et Diels ●☆

100634 Combretum moggii Exell;莫诺风车子●☆

100635 Combretum molle (Klotzsch) Engl. et Diels = Combretum pisoniiflorum (Klotzsch) Engl. ●☆

100636 Combretum molle R. Br. = Combretum molle R. Br. ex G. Don ●☆

100637 Combretum molle R. Br. ex G. Don;毛风车子;Hairy Combretum ●☆

100638 Combretum monticola Engl. et Gilg;山地风车子●☆

100639 Combretum mooreanum Exell;穆尔风车子●☆

100640 Combretum mortehanii De Wild. et Exell;莫特汉风车子●☆

100641 Combretum mossambicense (Klotzsch) Engl.;莫桑比克风车子●☆

100642 Combretum mucronatum Schumach. et Thonn.;短尖风车子●☆

100643 Combretum mucronatum Schumach. et Thonn. f. acutum (M. A. Lawson) Roberty = Combretum conchipetalum Engl. et Diels ●☆

100644 Combretum mucronatum Schumach. et Thonn. f. cuspidatum (Planch. ex Benth.) Roberty = Combretum cuspidatum Planch. ex Benth. ●☆

100645 Combretum multiflorum Pamp. = Combretum contractum Engl. et Diels ●☆

100646 Combretum multinervium Exell;多脉风车子●☆

100647 Combretum multispicatum Engl. et Diels = Combretum adenogonium Steud. ex A. Rich. ●☆

100648 Combretum mussaendiflorum Engl. et Diels = Combretum falcatum (Welw. ex Hiern) Jongkind ●☆

100649 Combretum muvarzense Exell = Combretum collinum Fresen. subsp. elgonense (Exell) Okafor ●☆

100650 Combretum mweroense Baker;姆韦鲁风车子●☆

100651 Combretum myrtifolium M. A. Lawson = Pteleopsis myrtifolia (M. A. Lawson) Engl. et Diels ●☆

100652 Combretum myrtillifolium Engl. = Combretum engleri Schinz ●☆

100653 Combretum nanum Buch. -Ham.;矮小风车子●☆

100654 Combretum ndjoleense Jongkind;恩乔莱风车子●☆

100655 Combretum nelsonii Dümmer;纳尔逊风车子●☆

100656 Combretum nelsonii Dümmer = Combretum kraussii Hochst. ●☆

100657 Combretum nervosum Engl. et Diels = Combretum platypterum (Welw.) Hutch. et Dalziel ●☆

100658 Combretum nigricans Lepr. ex Guillaumin et Perr.;变黑风车子●☆

100659 Combretum nigricans Lepr. ex Guillaumin et Perr. var. elliotii (Engl. et Diels) Aubrév.;埃利风车子●☆

100660 Combretum nitrophilum Gilg et Ledermann ex Mildbr.;喜碱风车子●☆

100661 Combretum nyikae Engl. = Combretum molle R. Br. ex G. Don ●☆

100662 Combretum nyikae Engl. var. boehmii? = Combretum molle R. Br. ex G. Don ●☆

100663 Combretum oatesii Rolfe = Combretum platypetalum Welw. ex M. A. Lawson subsp. oatesii (Rolfe) Exell ●☆

100664 Combretum obanense (Baker f.) Hutch. et Dalziel = Combretum marginatum Engl. et Diels ●☆

100665 Combretum oblongum F. Hoffm. = Combretum zeyheri Sond. ●☆

100666 Combretum obovatum F. Hoffm.;倒卵风车子●☆

100667 Combretum obtusatum Engl. et Diels = Combretum molle R. Br. ex G. Don ●☆

100668 Combretum odontopetalum Engl. et Diels = Combretum zeyheri Sond. ●☆

100669 Combretum olivaceum Engl.;橄榄绿风车子●☆

100670 Combretum oliverianum Engl. = Combretum collinum Fresen. subsp. suluense (Engl. et Diels) Okafor ●☆

100671 Combretum oliviforme A. C. Chao = Combretum sundaicum Miq. ●

100672 Combretum oliviforme A. C. Chao var. yaxianense Y. R. Ling = Combretum sundaicum Miq. ●

100673 Combretum oliviforme H. C. Chao = Combretum sundaicum Miq. ●

100674 Combretum oliviforme H. C. Chao var. yaxianense Y. R. Ling;崖县风车子;Oliveform Combretum, Yaxian Windmill ●

100675 Combretum omahekae Gilg et Dinter = Combretum psidioides Welw. ●☆

100676 Combretum ondongense Engl. et Diels = Combretum collinum Fresen. subsp. ondongense (Engl. et Diels) Okafor ●☆

100677 Combretum orophilum Liben = Combretum fuscum Planch. ex Benth. ●☆

100678 Combretum ovale G. Don = Combretum aculeatum Vent. ●☆

100679 Combretum ovalifolium Roxb.;卵叶藤诃子●

100680 Combretum oxystachyum Welw. ex M. A. Lawson;尖穗风车子●☆

100681 Combretum padoides Engl. et Diels;微花风车子●☆

100682 Combretum paniculatum Vent.;圆锥风车子●☆

100683 Combretum paniculatum Vent. subsp. microphyllum (Klotzsch) Wickens = Combretum microphyllum Klotzsch ●☆

100684 Combretum paniculatum Vent. var. vanderystii De Wild. = Combretum paniculatum Vent. ●☆

100685 Combretum paradoxum Welw. ex M. A. Lawson;奇异风车子●☆

100686 Combretum parviflorum Rchb. = Combretum micranthum G. Don ●☆

100687 Combretum parvifolium Dinter = Combretum engleri Schinz ●☆

100688 Combretum parvifolium Engl. = Combretum hereroense Schinz var. parvifolium (Engl.) Wickens ●☆

100689 Combretum parvulum Engl. et Diels;较小风车子●☆

100690 Combretum passargei Engl. et Diels = Combretum goetzei Engl. et Diels ●☆

100691 Combretum patelliforme Engl. et Diels = Combretum celastroides Welw. ex M. A. Lawson ●☆

100692 Combretum paucinervium Engl. et Diels;少脉风车子●☆

100693 Combretum paucinervium Engl. et Diels var. obanense Baker f. = Combretum marginatum Engl. et Diels ●☆

100694 Combretum pecoense Exell;佩科风车子●☆

100695 Combretum pellegrinianum Exell = Combretum falcatum (Welw. ex Hiern) Jongkind ●☆

100696 Combretum pellucidum Exell = Combretum mucronatum Schumach. et Thonn. ●☆

100697 Combretum pengheense De Wild. et Exell = Combretum marginatum Engl. et Diels ●☆

100698 Combretum pentagonum M. A. Lawson;五角风车子●☆

100699 Combretum petersii Klotzsch = Combretum imberbe Wawra ●☆

100700 Combretum petitianum A. Rich. = Combretum molle R. Br. ex G. Don ●☆

100701 Combretum petrophilum Retief;喜岩风车子●☆

100702 Combretum pilosum Roxb.;柔毛风车子(长毛风车子,风车子树,康柏树);Pilose Combretum, Pilose Windmill ●

100703 Combretum pincianum Hook. = Combretum paniculatum Vent. ●☆

100704 Combretum pisoniiflorum (Klotzsch) Engl.;腺果藤花风车子●☆

100705 Combretum platycarpum Engl. et Diels = Combretum zeyheri Sond. ●☆

100706 Combretum platypetalum Welw. ex M. A. Lawson;宽瓣风车子●☆

100707 Combretum platypetalum Welw. ex M. A. Lawson subsp. baumii (Engl. et Gilg) Exell;鲍姆风车子●☆

100708 Combretum platypetalum Welw. ex M. A. Lawson subsp. oatesii (Rolfe) Exell;奥茨风车子●☆

100709 Combretum platypetalum Welw. ex M. A. Lawson subsp. virgatum (Welw. ex M. A. Lawson) Exell;条纹风车子●☆

100710 Combretum platyphyllum Van Heurck et Müll. Arg. = Combretum latifolium Blume ●

100711 Combretum platypterum (Welw.) Hutch. et Dalziel = Cacoucia platyptera Hemsl. ●☆

100712 Combretum platypterum (Welw.) Hutch. et Dalziel = Combretum bracteatum (M. A. Lawson) Engl. et Diels ●☆

100713 Combretum poggei Engl. et Diels = Combretum hensii Engl. et Diels ●☆

100714 Combretum polyanthum Jongkind;多花风车子●☆

100715 Combretum polystictum Welw. ex Hiern = Combretum camporum Engl. ●☆

100716 Combretum polystictum Welw. ex Hiern var. undulatum Hiern = Combretum camporum Engl. ●☆

100717 Combretum populifolium Engl. et Diels = Combretum collinum Fresen. subsp. binderanum (Kotschy) Okafor ●☆

100718 Combretum porphyrobotrys Engl. et Diels = Combretum mannii M. A. Lawson ex Engl. et Diels ●☆

100719 Combretum porphyrolepis Engl. et Diels = Combretum hereroense Schinz ■☆

100720 Combretum praecox De Wild. = Combretum platypetalum Welw. ex M. A. Lawson subsp. baumii (Engl. et Gilg) Exell ●☆

100721 Combretum primigenum Marloth ex Engl. = Combretum imberbe Wawra ●☆

100722 Combretum psammophilum Engl. et Diels = Combretum collinum Fresen. subsp. taborense (Engl.) Okafor ●☆

100723 Combretum psidioides Welw.;短瓣风车子●☆

100724 Combretum psidioides Welw. subsp. dinteri (Schinz) Exell;丁特风车子●☆

100725 Combretum psidioides Welw. subsp. glabrum Exell;光滑短瓣风车子●☆

100726 Combretum psidioides Welw. subsp. psilophyllum Wickens;裸叶风车子●☆

100727 Combretum punctatum A. Rich. = Combretum molle R. Br. ex G. Don ●☆

100728 Combretum punctatum Blume;盾鳞风车子;Dotted Combretum, Peltatescaled Windmill, Punctate Combretum ●

100729 Combretum punctatum Blume subsp. squamosum (Roxb. ex G. Don) Exell;水密花(盾鳞风车子)●

100730 Combretum purpurascens Hand. -Mazz. = Combretum wallichii DC. ●

100731 Combretum purpureiflorum Engl.;紫花风车子●☆

100732 Combretum purpureum Vahl = Combretum coccineum (Sonn.) Lam. ●☆

100733 Combretum pynaertii De Wild. = Combretum holstii Engl. ●☆

100734 Combretum pyriforme De Wild. = Combretum hensii Engl. et Diels ●☆

100735 Combretum quadrangulare Kurz;四角风车子(四棱鸭跖草);Quadrangular Combretum,Quadrangular Windmill ●☆

100736 Combretum quangense Engl. et Diels = Combretum mossambicense (Klotzsch) Engl. ●☆

100737 Combretum quartinianum A. Rich. = Combretum molle R. Br. ex G. Don ●☆

100738 Combretum quirirense Engl. et Diels = Combretum psidioides Welw. subsp. dinteri (Schinz) Exell ●☆

100739 Combretum rabiense Jongkind;拉比风车子●☆

100740 Combretum racemosum P. Beauv.;总花风车子●☆

100741 Combretum racemosum P. Beauv. var. flammeum Welw. ex M. A. Lawson = Combretum racemosum P. Beauv. ●☆

100742 Combretum raimbaultii Heckel = Combretum micranthum G. Don ●☆

100743 Combretum ramosissimum Engl. et Diels = Combretum paniculatum Vent. ●☆

100744 Combretum rautanenii Engl. et Diels = Combretum hereroense Schinz ■☆

100745 Combretum reticulatum Fresen. = Combretum molle R. Br. ex G. Don ●☆

100746 Combretum rhodanthum Engl. et Diels = Combretum comosum G. Don ●☆

100747 Combretum rhodesicum Baker f. = Combretum hereroense Schinz ■☆

100748 Combretum rhombifolium Exell = Combretum fuscum Planch. ex Benth. ●☆

100749 Combretum richardianum Van Heurck et Müll. Arg. = Combretum molle R. Br. ex G. Don ●☆

100750 Combretum riggenbachianum Gilg et Ledermann;里根巴赫风车子●☆

100751 Combretum rigidifolium Welw. ex Hiern = Combretum mossambicense (Klotzsch) Engl. ●☆

100752 Combretum riparium Sond. = Combretum erythrophyllum (Burch.) Sond. ●☆

100753 Combretum ritschardii De Wild. et Exell = Combretum collinum Fresen. subsp. gazense (Swynn. et Baker f.) Okafor ●☆

100754 Combretum robustum Jongkind;粗壮风车子●☆

100755 Combretum robynsii Exell;罗宾斯风车子●☆

100756 Combretum rochetianum A. Rich. ex A. Juss.;罗歇风车子●☆

100757 Combretum rotundifolium Roxb. = Combretum latifolium Blume ●

100758 Combretum roxburghii Spreng.;十蕊风车子(十字风车子);Roxburgh Combretum,Roxburgh Windmill ●

100759 Combretum rubriflorum De Wild. = Combretum haullevilleanum De Wild. ●☆

100760 Combretum rueppellianum A. Rich. = Combretum molle R. Br. ex G. Don ●☆

100761 Combretum salicifolium E. Mey. ex Hook. = Combretum caffrum (Eckl. et Zeyh.) Kuntze ●☆

100762 Combretum sambesiacum Engl. et Diels = Combretum hereroense Schinz ■☆

100763 Combretum sankisiense De Wild. = Combretum zeyheri Sond. ●☆

100764 Combretum sapinii De Wild. = Combretum haullevilleanum De Wild. ●☆

100765 Combretum scandens Liben;攀缘风车子●☆

100766 Combretum schelei Engl. = Combretum molle R. Br. ex G. Don ●☆

100767 Combretum schimperianum A. Rich. = Combretum molle R. Br. ex G. Don ●☆

100768 Combretum schinzii Engl. ex Engl. et Diels = Combretum collinum Fresen. ●☆

100769 Combretum schliebenii Exell et Mildbr. = Combretum illairii Engl. ●☆

100770 Combretum schumannii Engl.;舒曼风车子(非洲风车子);Schuman Combretum,Schuman Windmill ●☆

100771 Combretum schweinfurthii Engl. et Diels;施韦风车子●☆

100772 Combretum sennii Chiov. = Combretum exalatum Engl. ●☆

100773 Combretum seretii De Wild. = Combretum paniculatum Vent. ●☆

100774 Combretum seretii De Wild. var. grandiflora = Combretum paniculatum Vent. ●☆

100775 Combretum sericeum (Walp.) Wall. ex C. B. Clarke = Getonia floribunda Roxb. ●

100776 Combretum sericeum G. Don;绢毛风车子●☆

100777 Combretum sericogyne Engl. et Diels = Combretum falcatum (Welw. ex Hiern) Jongkind ●☆

100778 Combretum sericogyne Engl. et Diels var. glabrescens De Wild. = Combretum falcatum (Welw. ex Hiern) Jongkind ●☆

100779 Combretum simulans Portères;相似风车子●☆

100780 Combretum singidense Exell = Combretum collinum Fresen. subsp. gazense (Swynn. et Baker f.) Okafor ●☆

100781 Combretum sinuatipetalum De Wild. = Combretum zeyheri Sond. ●☆

100782 Combretum smeathmannii G. Don = Combretum mucronatum Schumach. et Thonn. ●☆

100783 Combretum sokodense Engl. = Combretum molle R. Br. ex G. Don ●☆

100784 Combretum somalense Engl. et Diels = Combretum molle R. Br. ex G. Don ●☆

100785 Combretum sonderi Gerrard ex Harv. = Combretum erythrophyllum (Burch.) Sond. ●☆

100786 Combretum sordidum Exell;暗色风车子●☆

100787 Combretum spinosum G. Don = Combretum paniculatum Vent. ●☆

100788 Combretum splendens Engl. = Combretum molle R. Br. ex G. Don ●☆

100789 Combretum squamosum Roxb. ex G. Don = Combretum punctatum Blume subsp. squamosum (Roxb. ex G. Don) Exell ●

100790 Combretum squamosum Roxb. ex G. Don var. dissitum Craib = Combretum punctatum Blume subsp. squamosum (Roxb. ex G. Don) Exell ●

100791 Combretum squamosum Roxb. ex G. Don var. luzonicum C. Presl = Combretum punctatum Blume subsp. squamosum (Roxb. ex G. Don) Exell ●

100792 Combretum stefaninianum Pamp. = Combretum aculeatum Vent. ●☆

100793 Combretum stenanthoides Mildbr. = Combretum xanthothyrsum Engl. et Diels ●☆

100794 Combretum stenanthum Diels = Combretum chionanthoides Engl. et Diels ●☆

100795 Combretum stenophyllum R. E. Fr. = Combretum platypetalum Welw. ex M. A. Lawson subsp. oatesii (Rolfe) Exell ●☆

100796 Combretum stocksii Sprague;斯托克斯风车子●☆

100797 Combretum struempellianum Gilg et Ledermann ex Engl. = Combretum collinum Fresen. ●☆

100798 Combretum subglabratum De Wild.;近光风车子●☆
100799 Combretum subglomeruliflorum De Wild. = Combretum gossweileri Exell ●☆
100800 Combretum sublancifolium Chiov. = Combretum molle R. Br. ex G. Don ●☆
100801 Combretum subscabrum De Wild. = Combretum haullevilleanum De Wild. ●☆
100802 Combretum subvernicosum Engl. et Diels = Combretum adenogonium Steud. ex A. Rich. ●☆
100803 Combretum suluense Engl. et Diels = Combretum collinum Fresen. subsp. suluense (Engl. et Diels) Okafor ●☆
100804 Combretum sundaicum DC. = Combretum sundaicum Miq. ●
100805 Combretum sundaicum Miq.;榄形风车子(马来风车子);Oliveform Combretum, Oliveform Windmill, Olive-like Combretum ●
100806 Combretum taborense Engl. = Combretum collinum Fresen. subsp. taborense (Engl.) Okafor ●☆
100807 Combretum taitense Engl. et Diels = Combretum exalatum Engl. ●☆
100808 Combretum tanaense Clark;塔纳风车子●☆
100809 Combretum tavetense Diels = Combretum exalatum Engl. ●☆
100810 Combretum tenuifolium Exell = Combretum cuspidatum Planch. ex Benth. ●☆
100811 Combretum tenuipes Engl. et Diels = Combretum padoides Engl. et Diels ●☆
100812 Combretum tenuipetiolatum Wickens;细柄风车子●☆
100813 Combretum tenuispicatum Engl. = Combretum molle R. Br. ex G. Don ●☆
100814 Combretum ternifolium Engl. et Diels = Combretum adenogonium Steud. ex A. Rich. ●☆
100815 Combretum tessmannii Gilg ex Engl.;泰斯曼风车子●☆
100816 Combretum tetragonum M. A. Lawson = Combretum pisoniiflorum (Klotzsch) Engl. ●☆
100817 Combretum tetrandrum Exell = Meiostemon tetrandrus (Exell) Exell et Stace ●☆
100818 Combretum tetraphyllum Diels = Combretum adenogonium Steud. ex A. Rich. ●☆
100819 Combretum theuschii O. Hoffm. = Combretum zeyheri Sond. ●☆
100820 Combretum thonneri De Wild. = Combretum paniculatum Vent. ●☆
100821 Combretum thonneri De Wild. var. laurentii? = Combretum paniculatum Vent. ●☆
100822 Combretum tibatiense Gilg et Ledermann ex Mildbr.;蒂巴蒂风车子●☆
100823 Combretum tinctorium Welw. ex M. A. Lawson = Combretum zeyheri Sond. ●☆
100824 Combretum tisserantii Exell = Combretum schweinfurthii Engl. et Diels ●☆
100825 Combretum tomentosum G. Don;绒毛风车子●☆
100826 Combretum transvaalense Schinz = Combretum hereroense Schinz ■☆
100827 Combretum transvaalense Schinz var. villosissimum (Engl. et Diels) Burtt Davy = Combretum hereroense Schinz ■☆
100828 Combretum trichanthum Fresen. = Combretum molle R. Br. ex G. Don ●☆
100829 Combretum trichanthum Fresen. var. angustifolium Fiori = Combretum molle R. Br. ex G. Don ●☆
100830 Combretum trichanthum Fresen. var. petitianum (A. Rich.) Fiori = Combretum molle R. Br. ex G. Don ●☆
100831 Combretum trichopetalum Engl. = Combretum mossambicense (Klotzsch) Engl. ●☆
100832 Combretum trothae Engl. et Diels = Combretum celastroides Welw. ex M. A. Lawson subsp. orientale Exell ●☆
100833 Combretum truncatum Welw. ex M. A. Lawson = Combretum imberbe Wawra ●☆
100834 Combretum turbinatum F. Hoffm. = Combretum platypetalum Welw. ex M. A. Lawson subsp. oatesii (Rolfe) Exell ●☆
100835 Combretum ukambense Engl. = Combretum mossambicense (Klotzsch) Engl. ●☆
100836 Combretum ulugurense Engl. et Diels = Combretum molle R. Br. ex G. Don ●☆
100837 Combretum umbricola Engl.;荫地风车子●☆
100838 Combretum undulato-marginatum De Wild. et Exell = Combretum capituliflorum Fenzl ex Schweinf. ●☆
100839 Combretum undulatum Engl. et Diels = Combretum adenogonium Steud. ex A. Rich. ●☆
100840 Combretum unyorense Bagsh. et Baker f. = Combretum paniculatum Vent. ●☆
100841 Combretum usaramense Engl. = Combretum hereroense Schinz ■☆
100842 Combretum vanderystii De Wild. = Combretum psidioides Welw. subsp. psilophyllum Wickens ●☆
100843 Combretum velutinum (S. Moore) Engl. et Diels = Combretum mooreanum Exell ●☆
100844 Combretum velutinum DC. = Combretum molle R. Br. ex G. Don ●☆
100845 Combretum verticillatum Engl. et Diels = Combretum collinum Fresen. subsp. hypopilinum (Diels) Okafor ●☆
100846 Combretum virgatum Welw. ex M. A. Lawson = Combretum platypetalum Welw. ex M. A. Lawson subsp. virgatum (Welw. ex M. A. Lawson) Exell ●☆
100847 Combretum viscosum Exell;黏风车子●☆
100848 Combretum volkensii Engl. = Combretum hereroense Schinz var. volkensii (Engl.) Wickens ●☆
100849 Combretum wakefieldii Engl. = Combretum pentagonum M. A. Lawson ●☆
100850 Combretum wallichii DC.;石风车子(凌云风车子,牛板筋,瓦氏风车子,紫风车子);Stone Windmill, Wallich Combretum ●
100851 Combretum wallichii DC. var. griffithii (Van Heurck et Müll. Arg.) M. G. Gangop. et Chakrab. = Combretum griffithii Van Heurck et Müll. Arg. ●
100852 Combretum wallichii DC. var. pubinerve C. Y. Wu = Combretum wallichii DC. ●
100853 Combretum wallichii DC. var. pubinerve C. Y. Wu ex T. Z. Hsu;毛脉石风车子;Hairynerved Wallich Combretum ●
100854 Combretum wallichii DC. var. yunnanense (Exell) M. G. Gangop. et Chakrab. = Combretum griffithii Van Heurck et Müll. Arg. var. yunnanense (Exell) Turland et C. Chen ●
100855 Combretum wattii Exell;瓦特风车子●☆
100856 Combretum welwitschii Engl. et Diels = Combretum molle R. Br. ex G. Don ●☆
100857 Combretum wightianum Wall. ex Wight et Arn. = Combretum latifolium Blume ●
100858 Combretum wittei Exell = Combretum haullevilleanum De Wild. ●☆
100859 Combretum woodii Dümmer = Combretum kraussii Hochst. ●☆

100860 Combretum xanthothyrsum Engl. et Diels;黄序风车子●☆
100861 Combretum youngii Exell = Combretum pecoense Exell ●☆
100862 Combretum yuankiangense C. C. Huang et S. C. Huang ex T. Z. Hsu;元江风车子;Yuanjiang Windmill ●
100863 Combretum yuankiangense C. C. Huang et S. C. Huang ex T. Z. Hsu = Combretum griffithii Van Heurck et Müll. Arg. ●
100864 Combretum yunnanense Exell = Combretum griffithii Van Heurck et Müll. Arg. var. yunnanense (Exell) Turland et C. Chen ●
100865 Combretum zastrowii Dinter = Combretum platypetalum Welw. ex M. A. Lawson ●☆
100866 Combretum zechii Diels = Combretum adenogonium Steud. ex A. Rich. ●☆
100867 Combretum zeyheri Sond.;大果风车子(蔡赫风车子);Large-fruited Bushwillow ●☆
100868 Comera Furtado = Calamus L. ●
100869 Comesperma Labill. (1806);雅志属●☆
100870 Comesperma Labill. = Bredemeyera Willd. ●☆
100871 Comesperma ericinum DC.;绵毛雅志;Heath Milkwort ●☆
100872 Comesperma volubile Labill.;扭雅志;Love Creeper, Love-creeper ●☆
100873 Cometes L. (1767);彗星花属■☆
100874 Cometes abyssinica R. Br. ex Wall.;阿比西尼亚彗星花■☆
100875 Cometes alterniflora L. = Cometes surattensis L. ■☆
100876 Cometes surattensis L.;彗星花■☆
100877 Cometia Thouars ex Baill. = Drypetes Vahl + Thecacoris A. Juss. ●
100878 Cometia Thouars ex Baill. = Drypetes Vahl ●
100879 Cometia thouarsii Baill. = Drypetes thouarsii (Baill.) Leandri ●☆
100880 Comeurya Baill. = Dracontomelon Blume ●
100881 Cominia P. Browne = Allophylus L. ●
100882 Cominia Raf. = Rhus L. ●
100883 Cominsia Hemsl. (1891);长瓣竹芋属■☆
100884 Cominsia gigantea K. Schum.;大长瓣竹芋■☆
100885 Cominsia guppyi Hemsl.;长瓣竹芋■☆
100886 Cominsia minor Valeton;小长瓣竹芋■☆
100887 Cominsia rubra Valeton;红长瓣竹芋■☆
100888 Comiphyton Floret(1974);加蓬红树属●☆
100889 Comiphyton gabonense Floret;加蓬红树●☆
100890 Commarum Schrank = Comarum L. ●■
100891 Commarum Schrank = Potentilla L. ■●
100892 Commelina L. (1753);鸭跖草属;Day Flower, Dayflower, Dayflower, Spiderflower, Spider-wort, Widow's-tears ■
100893 Commelina Plumier ex L. = Commelina L. ■
100894 Commelina acutispatha De Wild.;尖苞鸭跖草■☆
100895 Commelina aequinoctiale P. Beauv. = Aneilema aequinoctiale (P. Beauv.) G. Don ■☆
100896 Commelina aethiopica C. B. Clarke;埃塞俄比亚鸭跖草■☆
100897 Commelina africana L.;非洲鸭跖草■☆
100898 Commelina africana L. f. glabrata Chiov. = Commelina arenicola Faden ■☆
100899 Commelina africana L. f. pilosa Chiov. = Commelina arenicola Faden ■☆
100900 Commelina africana L. var. barberae (C. B. Clarke) C. B. Clarke;巴尔鸭跖草■☆
100901 Commelina africana L. var. boehmiana (K. Schum.) Brenan = Commelina africana L. var. krebsiana (Kunth) C. B. Clarke ■☆
100902 Commelina africana L. var. brevipila Brenan = Commelina africana L. var. krebsiana (Kunth) C. B. Clarke ■☆
100903 Commelina africana L. var. circinnata Chiov. = Commelina arenicola Faden ■☆
100904 Commelina africana L. var. diffusa Brenan;松散非洲鸭跖草■☆
100905 Commelina africana L. var. glabriuscula (Norl.) Brenan = Commelina africana L. var. lancispatha C. B. Clarke ■☆
100906 Commelina africana L. var. krebsiana (Kunth) C. B. Clarke;克雷布斯非洲鸭跖草■☆
100907 Commelina africana L. var. lancispatha C. B. Clarke;细苞非洲鸭跖草■☆
100908 Commelina africana L. var. mannii (C. B. Clarke) Brenan;曼氏非洲鸭跖草■☆
100909 Commelina africana L. var. milleri Brenan = Commelina africana L. var. krebsiana (Kunth) C. B. Clarke ■☆
100910 Commelina africana L. var. polyclada Welw. ex C. B. Clarke = Commelina africana L. var. krebsiana (Kunth) C. B. Clarke ■☆
100911 Commelina africana L. var. villosior (C. B. Clarke) Brenan = Commelina africana L. var. krebsiana (Kunth) C. B. Clarke ■☆
100912 Commelina agraria Kunth = Commelina nudiflora L. ■
100913 Commelina albescens Hassk.;白鸭跖草■☆
100914 Commelina albescens Hassk. var. hirsutissima Chiov. = Commelina albescens Hassk. ■☆
100915 Commelina albescens Hassk. var. occidentalis C. B. Clarke = Commelina erecta L. subsp. livingstonii (C. B. Clarke) J. K. Morton ■☆
100916 Commelina albiflora Faden;白花鸭跖草■☆
100917 Commelina ambigua P. Beauv. = Palisota ambigua (P. Beauv.) C. B. Clarke ■☆
100918 Commelina amphibia A. Chev. = Commelina congesta C. B. Clarke ■☆
100919 Commelina amphibia A. Chev. var. hirsuta? = Commelina congesta C. B. Clarke ■☆
100920 Commelina amplexicaulis Hassk.;抱茎鸭跖草■☆
100921 Commelina angolensis C. B. Clarke = Commelina africana L. ■☆
100922 Commelina angustifolia Michx.;狭叶鸭跖草■☆
100923 Commelina angustifolia Michx. = Commelina erecta L. ■☆
100924 Commelina angustissima K. Schum.;狭鸭跖草■☆
100925 Commelina aquatica J. K. Morton = Commelina diffusa Burm. f. ■
100926 Commelina arenicola Faden;沙生鸭跖草■☆
100927 Commelina ascendens Morton;上升鸭跖草■☆
100928 Commelina aspera Benth.;粗糙鸭跖草■☆
100929 Commelina aurantiiflora Faden;橙花鸭跖草■☆
100930 Commelina auriculata Blume;耳苞鸭跖草(耳叶鸭跖草,蓬莱鸭跖草);Earbract Dayflower ■
100931 Commelina auriculata Blume f. albiflora Tawada;白花耳苞鸭跖草■☆
100932 Commelina axillaris L. = Amischophacelus axillaris (L.) R. S. Rao et Kammathy ■☆
100933 Commelina axillaris L. = Cyanotis axillaris (L.) Sweet ■
100934 Commelina baidoensis Chiov. = Commelina somalensis Chiov. ■☆
100935 Commelina bainesii C. B. Clarke = Commelina bracteosa Hassk. ■☆
100936 Commelina bakueana A. Chev. = Commelina africana L. ■☆
100937 Commelina barbata Lam. var. villosior C. B. Clarke = Commelina africana L. var. krebsiana (Kunth) C. B. Clarke ■☆
100938 Commelina beccariana Martelli = Commelina africana L. ■☆
100939 Commelina bella Oberm.;雅致鸭跖草■☆
100940 Commelina benghalensis L.;饭包草(大叶兰花,大叶兰花草,

淡竹叶,火柴头,火炭头,兰花菜,卵叶鸭跖草,马耳草,千日晒,圆叶鸭跖草,竹菜,竹叶菜,竹叶草,竹竹菜,竹仔菜,竹子菜,竹子草);Bengal Dayflower,Jio ■

100941 Commelina benghalensis L. subsp. hirsuta (C. B. Clarke) J. K. Morton = Commelina benghalensis L. var. hirsuta C. B. Clarke ■☆

100942 Commelina benghalensis L. var. hirsuta C. B. Clarke;粗毛饭包草■☆

100943 Commelina beniniensis P. Beauv. = Aneilema beniniense (P. Beauv.) Kunth ■☆

100944 Commelina bequaertii De Wild.;贝卡尔鸭跖草■☆

100945 Commelina bianoensis De Wild.;比亚诺鸭跖草■☆

100946 Commelina boehmiana K. Schum. = Commelina africana L. var. krebsiana (Kunth) C. B. Clarke ■☆

100947 Commelina boissieriana C. B. Clarke;布瓦西耶鸭跖草■☆

100948 Commelina bracteolata Lam. = Murdannia spirata (L.) Brückn. ■

100949 Commelina bracteosa Hassk.;多苞鸭跖草■☆

100950 Commelina buchananii C. B. Clarke;布坎南鸭跖草■☆

100951 Commelina cameroonensis J. K. Morton;喀麦隆鸭跖草■☆

100952 Commelina canescens Vahl = Commelina benghalensis L. ■

100953 Commelina capitata Benth.;头状鸭跖草■☆

100954 Commelina caroliniana Willd. ex Kunth;卡罗林鸭跖草;Dayflower ■☆

100955 Commelina carsonii C. B. Clarke;卡森鸭跖草■☆

100956 Commelina cavaleriei H. Lév. = Commelina benghalensis L. ■

100957 Commelina chantransia Roem. et Schult. = Floscopa africana (P. Beauv.) C. B. Clarke ■☆

100958 Commelina claessensii De Wild.;克莱森斯鸭跖草■☆

100959 Commelina clarkeana K. Schum.;克拉克鸭跖草■☆

100960 Commelina coelestis Willd.;墨西哥鸭跖草;Carolina Dayflower,Commelina,Dayllower,Mexican Dayflower,Mexican Dayflower ■☆

100961 Commelina communis L.;鸭跖草(鼻斫草,碧蝉花,碧蝉蛇,碧蟾蜍,碧竹草,碧竹子,翠娥眉,翠蝴蝶,淡竹叶,淡竹叶菜,地地藕,靛青花草,鹅儿菜,耳环草,哥哥啼草,挂兰青,桂竹草,鸡冠菜,鸡舌草,兰花草,兰紫草,蓝姑草,蓝蝴蝶,蓝花菜,蓝花姑娘,蓝花水竹草,蓝石竹,菱角伞,露草,帽子花,牛耳朵草,笪竹花,青耳环花,三筴子菜,三荚菜,三荚子菜,三角菜,水浮草,水竹草,水竹叶草,水竹子,倭青草,鸦雀草,鸭脚板草,鸭脚草,鸭脚青,鸭鹊草,鸭食草,鸭仔草,鸭子菜,野靛青,萤火虫草,竹根菜,竹管草,竹鸡草,竹鸡苋,竹夹菜,竹夹草,竹剪草,竹节菜,竹节草,竹壳菜,竹叶菜,竹叶草,竹叶活血丹,竹叶兰,竹叶青,竹叶青菜,竹叶水草);Asiatic Dayflower,Common Day Flower,Common Dayflower,Dayflower ■

100962 Commelina communis L. 'Aureo-striata';金线鸭跖草■

100963 Commelina communis L. f. alba Ti Chen = Commelina communis L. f. albiflora Makino ■

100964 Commelina communis L. f. albiflora Makino;普通白花鸭跖草;White Dayflower ■

100965 Commelina communis L. f. ciliata (Masam.) Murata = Commelina communis L. f. ciliata Pennell ■☆

100966 Commelina communis L. f. ciliata Pennell;缘毛鸭跖草■☆

100967 Commelina communis L. f. miranda Hiyama;奇异鸭跖草■☆

100968 Commelina communis L. var. angustifolia Nakai = Commelina communis L. var. ludens (Miq.) C. B. Clarke ■☆

100969 Commelina communis L. var. angustifolia Nakai = Commelina communis L. ■

100970 Commelina communis L. var. hortensis Makino;庭园鸭跖草■☆

100971 Commelina communis L. var. hortensis Makino f. candida Hiyama;白花庭园鸭跖草■☆

100972 Commelina communis L. var. ludens (Miq.) C. B. Clarke;细叶鸭跖草■☆

100973 Commelina communis L. var. ludens (Miq.) C. B. Clarke = Commelina communis L. ■

100974 Commelina condensata C. B. Clarke = Commelina congesta C. B. Clarke ■☆

100975 Commelina congesta C. B. Clarke;密集鸭跖草■☆

100976 Commelina conspicua Blume = Dictyospermum conspicuum (Blume) Hassk. ■

100977 Commelina corbisieri De Wild.;科比西尔鸭跖草■☆

100978 Commelina cordifolia A. Rich. = Commelina africana L. ■☆

100979 Commelina coreana H. Lév. et Vaniot = Commelina communis L. ■

100980 Commelina covaleriei H. Lév. = Commelina benghalensis L. ■

100981 Commelina crassicaulis C. B. Clarke;粗茎鸭跖草■☆

100982 Commelina crispa L.;皱叶鸭跖草■☆

100983 Commelina crispa Wooton = Commelina erecta L. ■☆

100984 Commelina cristata L. = Cyanotis cristata (L.) D. Don ex Sweet ■

100985 Commelina cufodontii Chiov.;卡佛鸭跖草■☆

100986 Commelina cuneata C. B. Clarke;楔形鸭跖草■☆

100987 Commelina debilis Ledeb. = Commelina communis L. ■

100988 Commelina demissa C. B. Clarke;下垂鸭跖草■☆

100989 Commelina dianthifolia Delile;石竹叶节节草;Birdbill Dayflower,Western Dayllower ■☆

100990 Commelina diffusa Burm. f.;节节草(白竹仔菜,白竹仔草,翠娥眉,翠蝴蝶,笪竹花,黄花草,倭青草,鸭跖草,竹菜,竹蒿草,竹节菜,竹节草,竹节花,竹筋草,竹叶花,竹仔菜);Dayflower,Diffuse Dayflower,Pond Grass ■

100991 Commelina diffusa Burm. f. = Commelina nudiflora L. ■

100992 Commelina diffusa Burm. f. subsp. aquatica (J. K. Morton) J. K. Morton = Commelina diffusa Burm. f. ■

100993 Commelina diffusa Burm. f. subsp. montana J. K. Morton;山地节节草■☆

100994 Commelina diffusa Burm. f. subsp. scandens (Welw. ex C. B. Clarke) Oberm.;攀缘鸭跖草■☆

100995 Commelina diffusa Burm. f. var. gigas (Small) Faden;大节节草(大鸭跖草)■☆

100996 Commelina dinteri Mildbr. = Commelina africana L. var. lancispatha C. B. Clarke ■☆

100997 Commelina dubia Jacq. = Heteranthera dubia (Jacq.) MacMill. ■☆

100998 Commelina echinosperma K. Schum.;刺子鸭跖草■☆

100999 Commelina echinulata J. -P. Lebrun et Taton;小刺鸭跖草■☆

101000 Commelina eckloniana Kunth;埃氏鸭跖草■☆

101001 Commelina edulis A. Rich. = Commelina africana L. ■☆

101002 Commelina edulis Stokes = Murdannia edulis (Stokes) Faden ■

101003 Commelina elata Vahl = Murdannia japonica (Thunb.) Faden ■

101004 Commelina elegans Kunth = Commelina erecta L. ■☆

101005 Commelina elegantula K. Schum.;雅丽鸭跖草■☆

101006 Commelina elgonensis Bullock;埃尔贡鸭跖草■☆

101007 Commelina elliotii C. B. Clarke et Rendle = Commelina africana L. var. lancispatha C. B. Clarke ■☆

101008 Commelina erecta L.;直立鸭跖草(雅致鸭跖草);Dayflower,Erect Dayflower,Narrow-leaved Dayflower,Whitemouth Dayflower,

White-mouth Dayflower ■☆
101009 Commelina erecta L. f. crispa (Wooton) Fernald = Commelina erecta L. ■☆
101010 Commelina erecta L. f. intercursa Fernald = Commelina erecta L. ■☆
101011 Commelina erecta L. subsp. livingstonii (C. B. Clarke) J. K. Morton;利文斯顿鸭跖草■☆
101012 Commelina erecta L. subsp. maritima (J. K. Morton) J. K. Morton;滨海鸭跖草■☆
101013 Commelina erecta L. var. angustifolia (Michx.) Fernald = Commelina erecta L. ■☆
101014 Commelina erecta L. var. deamiana Fernald = Commelina erecta L. ■☆
101015 Commelina erecta L. var. greenei Fassett = Commelina erecta L. var. deamiana Fernald ■☆
101016 Commelina filifolia K. Schum.;丝叶鸭跖草■☆
101017 Commelina firma Rendle;丝鸭跖草■☆
101018 Commelina flava Salisb. = Commelina africana L. ■☆
101019 Commelina fluviatilis Brenan;河岸鸭跖草■☆
101020 Commelina foliacea Chiov.;叶状鸭跖草■☆
101021 Commelina foliacea Chiov. subsp. amplexicaulis Faden;抱茎叶状鸭跖草■☆
101022 Commelina forskalaei Vahl var. ramulosa C. B. Clarke = Commelina ramulosa H. Perrier ■☆
101023 Commelina forskallii Hochst. ex C. B. Clarke;福斯科尔鸭跖草;Rat's Ear ●☆
101024 Commelina forskallii Hochst. ex C. B. Clarke var. hirsutula C. B. Clarke = Commelina forskallii Hochst. ex C. B. Clarke ●☆
101025 Commelina forskallii Hochst. ex C. B. Clarke var. major Chiov. = Commelina forskallii Hochst. ex C. B. Clarke ●☆
101026 Commelina frutescens Faden;灌木鸭跖草●☆
101027 Commelina gambiae C. B. Clarke;冈比亚鸭跖草;Gambian Dayflower ■☆
101028 Commelina gambiae C. B. Clarke = Commelina nigritana Benth. var. gambiae (C. B. Clarke) Brenan ■☆
101029 Commelina gerrardii C. B. Clarke = Commelina erecta L. ■☆
101030 Commelina gerrardii C. B. Clarke subsp. maritima J. K. Morton = Commelina erecta L. subsp. maritima (J. K. Morton) J. K. Morton ■☆
101031 Commelina gigas Small = Commelina diffusa Burm. f. var. gigas (Small) Faden ■☆
101032 Commelina giorgii De Wild.;乔治鸭跖草■☆
101033 Commelina gourmaca A. Chev. = Commelina nigritana Benth. ■☆
101034 Commelina gourmaensis A. Chev. = Commelina nigritana Benth. ■☆
101035 Commelina graminifolia Sessé et Moc.;禾叶鸭跖草■☆
101036 Commelina grandis Brenan;大鸭跖草■☆
101037 Commelina grossa C. B. Clarke;粗鸭跖草■☆
101038 Commelina guineensis Hua = Commelina erecta L. ■☆
101039 Commelina herbacea Roxb. = Murdannia japonica (Thunb.) Faden ■
101040 Commelina heudelotii C. B. Clarke = Commelina congesta C. B. Clarke ■☆
101041 Commelina hirsuta Hochst. = Cyanotis barbata D. Don ■
101042 Commelina hockii De Wild.;霍克鸭跖草■☆
101043 Commelina holubii C. B. Clarke;霍勒布鸭跖草■☆
101044 Commelina homblei De Wild.;洪布勒鸭跖草■☆
101045 Commelina hookeri Dietr. = Murdannia simplex (Vahl) Brenan ■
101046 Commelina huillensis C. B. Clarke;威拉鸭跖草■☆
101047 Commelina humblotii H. Perrier;洪布鸭跖草■☆
101048 Commelina imberbis Ehrenb. ex Hassk.;无须鸭跖草■☆
101049 Commelina imberbis Ehrenb. ex Hassk. var. loandensis C. B. Clarke = Commelina petersii Hassk. subsp. loandensis (C. B. Clarke) Faden ■☆
101050 Commelina involucrosa A. Rich.;内卷鸭跖草■☆
101051 Commelina irumuensis De Wild.;伊鲁姆鸭跖草■☆
101052 Commelina japonica Thunb. = Murdannia japonica (Thunb.) Faden ■
101053 Commelina kabarensis De Wild.;卡巴雷鸭跖草■☆
101054 Commelina kagerensis J. -P. Lebrun et Taton;卡盖拉鸭跖草■☆
101055 Commelina kapiriensis De Wild.;卡皮里鸭跖草■☆
101056 Commelina karooica C. B. Clarke = Commelina africana L. var. barberae (C. B. Clarke) C. B. Clarke ■☆
101057 Commelina karooica C. B. Clarke var. barberae? = Commelina africana L. var. barberae (C. B. Clarke) C. B. Clarke ■☆
101058 Commelina kilimandscharica K. Schum. = Commelina benghalensis L. var. hirsuta C. B. Clarke ■☆
101059 Commelina kirkii C. B. Clarke;柯克鸭跖草■☆
101060 Commelina kisantuensis De Wild.;基桑图鸭跖草■☆
101061 Commelina kotschyi Hassk.;科奇鸭跖草■☆
101062 Commelina kotschyi Hassk. = Commelina imberbis Ehrenb. ex Hassk. ■☆
101063 Commelina krebsiana (Kunth) C. B. Clarke var. glabriuscula Norl. = Commelina africana L. var. lancispatha C. B. Clarke ■☆
101064 Commelina krebsiana Kunth = Commelina africana L. var. krebsiana (Kunth) C. B. Clarke ■☆
101065 Commelina kurzii C. B. Clarke = Commelina undulata R. Br. ■
101066 Commelina lagosensis C. B. Clarke;拉各斯鸭跖草■☆
101067 Commelina lagosensis C. B. Clarke var. subglabra A. Chev. = Commelina lagosensis C. B. Clarke ■☆
101068 Commelina lateriticola A. Chev. = Commelina schweinfurthii C. B. Clarke ■☆
101069 Commelina latifolia C. B. Clarke = Commelina imberbis Ehrenb. ex Hassk. ■☆
101070 Commelina latifolia Hochst. ex A. Rich.;宽叶鸭跖草■☆
101071 Commelina lineolata Blume = Murdannia japonica (Thunb.) Faden ■
101072 Commelina livingstonii C. B. Clarke = Commelina erecta L. subsp. livingstonii (C. B. Clarke) J. K. Morton ■☆
101073 Commelina longicapsa C. B. Clarke;长果鸭跖草■☆
101074 Commelina longicaulis Jacq. = Commelina diffusa Burm. f. ■
101075 Commelina longifolia (Hook.) Spreng. = Murdannia simplex (Vahl) Brenan ■
101076 Commelina loureirii Kunth = Commelina diffusa Burm. f. ■
101077 Commelina ludens Miq. = Commelina communis L. ■
101078 Commelina lugardii Bullock;卢格德鸭跖草■☆
101079 Commelina lutea Moench = Commelina africana L. ■☆
101080 Commelina luteiflora De Wild.;黄花鸭跖草■☆
101081 Commelina lyallii (C. B. Clarke) Chiov. = Commelina africana L. var. mannii (C. B. Clarke) Brenan ■☆
101082 Commelina lyallii (C. B. Clarke) H. Perrier = Commelina africana L. var. mannii (C. B. Clarke) Brenan ■☆
101083 Commelina macrospatha Gilg et Lederm. ex Mildbr.;大苞鸭跖草■☆

101084 Commelina macrosperma Morton;大籽鸭跖草■☆

101085 Commelina maculata Edgew.;地地藕(小竹叶菜);Maculate Dayflower ■

101086 Commelina madagascarica C. B. Clarke;马岛鸭跖草■☆

101087 Commelina mannii C. B. Clarke = Commelina africana L. var. mannii (C. B. Clarke) Brenan ■☆

101088 Commelina mannii C. B. Clarke var. lyallii C. B. Clarke = Commelina africana L. var. mannii (C. B. Clarke) Brenan ■☆

101089 Commelina mascarenica C. B. Clarke;马斯克林鸭跖草■☆

101090 Commelina medica Lour. = Murdannia medica (Lour.) D. Y. Hong ■

101091 Commelina melanorrhiza Faden;黑根鸭跖草■☆

101092 Commelina membranacea Robyns;膜质鸭跖草■☆

101093 Commelina mensensis Schweinf.;芒斯鸭跖草■☆

101094 Commelina merkeri K. Schum.;默克鸭跖草■☆

101095 Commelina microspatha K. Schum.;小苞鸭跖草■☆

101096 Commelina modesta Oberm.;适度鸭跖草■☆

101097 Commelina montana K. Schum. ex Engl.;山地鸭跖草■☆

101098 Commelina montigena H. Perrier;山生鸭跖草■☆

101099 Commelina multicaulis Hochst. ex Clarke = Commelina albescens Hassk. ■☆

101100 Commelina nairobiensis Faden;内罗比鸭跖草■☆

101101 Commelina nana Roxb. = Murdannia spirata (L.) Brückn. ■

101102 Commelina neurophylla C. B. Clarke;脉叶鸭跖草■☆

101103 Commelina nigritana Benth.;尼格里塔鸭跖草;African Dayflower ■☆

101104 Commelina nigritana Benth. var. gambiae (C. B. Clarke) Brenan = Commelina gambiae C. B. Clarke ■☆

101105 Commelina nudiflora L. = Murdannia nudiflora (L.) Brenan ■

101106 Commelina nyasensis C. B. Clarke;尼亚斯鸭跖草■☆

101107 Commelina obliqua Buch. -Ham. ex D. Don = Commelina paludosa Blume ■

101108 Commelina obliqua Buch. -Ham. ex D. Don var. mathewii C. B. Clarke = Commelina undulata R. Br. ■

101109 Commelina obliqua Buch. -Ham. ex D. Don var. viscida C. B. Clarke = Commelina maculata Edgew. ■

101110 Commelina obscura K. Schum.;隐匿鸭跖草■☆

101111 Commelina pallida De Wild.;苍白鸭跖草■☆

101112 Commelina paludosa Blume;沼泽鸭跖草(大苞地地藕,大苞鸭跖草,大鸭跖草,大叶鸭跖草,大叶竹仔,大叶竹仔菜,大竹叶菜,凤眼灵芝,竹叶茶);Bigbract Dayflower ■

101113 Commelina paludosa Blume f. pedunculate Qaiser et Jafri;梗花大苞鸭跖草■☆

101114 Commelina paludosa Blume var. mathewii (C. B. Clarke) R. S. Rao et Kammathy = Commelina undulata R. Br. ■

101115 Commelina paludosa Blume var. viscida (C. B. Clarke) R. S. Rao et Kammathy = Commelina maculata Edgew. ■

101116 Commelina paniculata (Benth.) Roberty = Polyspatha paniculata Benth. ●☆

101117 Commelina paniculata Vahl = Aneilema forskalii Kunth ■☆

101118 Commelina petersii Hassk.;彼得斯鸭跖草■☆

101119 Commelina petersii Hassk. subsp. loandensis (C. B. Clarke) Faden;罗安达鸭跖草■☆

101120 Commelina phaeochaeta Chiov.;褐毛鸭跖草■☆

101121 Commelina pilosissima Hutch.;多毛鸭跖草■☆

101122 Commelina polhillii Faden et Alford;普尔鸭跖草■☆

101123 Commelina poligama Blanco;碧蝉花■☆

101124 Commelina praecox T. C. E. Fr.;早鸭跖草■☆

101125 Commelina pseudopurpurea Faden;假紫鸭跖草■☆

101126 Commelina pseudoscaposa De Wild.;假花茎鸭跖草■☆

101127 Commelina purpurea C. B. Clarke ex Rendle;紫鸭跖草■☆

101128 Commelina pycnospatha Brenan;密苞鸭跖草■☆

101129 Commelina pynaertii De Wild.;皮那鸭跖草■☆

101130 Commelina pyrrhoblepharis Hassk.;火红脉鸭跖草■☆

101131 Commelina pyrrhoblepharis Hassk. var. glabra Pic. Serm.;光滑鸭跖草■☆

101132 Commelina quarrei De Wild.;卡雷鸭跖草■☆

101133 Commelina radicans Spreng. = Murdannia nudiflora (L.) Brenan ■

101134 Commelina ramulosa H. Perrier;多枝鸭跖草■☆

101135 Commelina reptans Brenan;匍匐鸭跖草■☆

101136 Commelina reygaertii De Wild.;赖氏鸭跖草■☆

101137 Commelina rhodesica Norl.;罗得西亚鸭跖草■☆

101138 Commelina robynsi De Wild.;罗宾斯鸭跖草■☆

101139 Commelina rogersii Burtt Davy;罗杰斯鸭跖草■☆

101140 Commelina ruandensis De Wild.;卢旺达鸭跖草■☆

101141 Commelina rufociliata C. B. Clarke;红睫毛鸭跖草■☆

101142 Commelina rupicola Font Quer;岩生鸭跖草■☆

101143 Commelina sabatieri C. B. Clarke;萨巴蒂耶尔鸭跖草■☆

101144 Commelina salicifolia Roxb. = Commelina diffusa Burm. f. ■

101145 Commelina saxatilis H. Perrier;岩地鸭跖草■☆

101146 Commelina saxosa De Wild.;岩栖鸭跖草■☆

101147 Commelina scaberrima Blume = Rhopalephora scaberrima (Blume) Faden ■

101148 Commelina scandens Welw. ex C. B. Clarke = Commelina diffusa Burm. f. subsp. scandens (Welw. ex C. B. Clarke) Oberm. ■☆

101149 Commelina scapiflora Roxb. = Murdannia edulis (Stokes) Faden ■

101150 Commelina scaposa C. B. Clarke ex De Wild. et T. Durand;花茎鸭跖草■☆

101151 Commelina schimperiana Hochst. ex Clarke = Commelina albescens Hassk. ■☆

101152 Commelina schliebenii Mildbr.;施利本鸭跖草■☆

101153 Commelina schweinfurthii C. B. Clarke;施韦鸭跖草■☆

101154 Commelina secundiflora Blume = Pollia secundiflora (Blume) Bakh. f. ■

101155 Commelina simplex Vahl = Murdannia simplex (Vahl) Brenan ■

101156 Commelina sinica (Ker Gawl.) Roem. et Schult. = Murdannia simplex (Vahl) Brenan ■

101157 Commelina somalensis Chiov.;索马里鸭跖草■☆

101158 Commelina spectabilis C. B. Clarke;壮观鸭跖草■☆

101159 Commelina spectabilis C. B. Clarke var. ramosa;分枝壮观鸭跖草■☆

101160 Commelina sphaerosperma C. B. Clarke = Commelina erecta L. subsp. livingstonii (C. B. Clarke) J. K. Morton ■☆

101161 Commelina spirata L. = Murdannia spirata (L.) Brückn. ■

101162 Commelina stefaniniana Chiov.;斯特鸭跖草■☆

101163 Commelina stolzii Mildbr.;斯托尔兹鸭跖草■☆

101164 Commelina striata Edgew. = Commelina undulata R. Br. ■

101165 Commelina subalbescens Berhaut = Commelina erecta L. subsp. livingstonii (C. B. Clarke) J. K. Morton ■☆

101166 Commelina subcucullata C. B. Clarke;僧帽状鸭跖草■☆

101167 Commelina subscabrifolia De Wild.;亚糙叶鸭跖草■☆

101168 Commelina subulata Roth;钻形鸭跖草■☆

101169 Commelina suffruticosa Blume;大叶鸭跖草;Bigleaf Dayflower ■
101170 Commelina sulcata Benth. = Commelina erecta L. ■☆
101171 Commelina sylvatica De Wild.;森林鸭跖草■☆
101172 Commelina taiwaniana S. S. Ying = Commelina auriculata Blume ■
101173 Commelina thomasii Hutch. = Commelina acutispatha De Wild. ■☆
101174 Commelina trachysperma Chiov.;糙籽鸭跖草■☆
101175 Commelina triangulispatha Mildbr.;三棱苞鸭跖草■☆
101176 Commelina trilobosperma K. Schum.;三裂籽鸭跖草■☆
101177 Commelina tuberosa Forssk. = Aneilema forskalii Kunth ■☆
101178 Commelina tuberosa L.;块茎鸭跖草;Tuber Dayflower ■☆
101179 Commelina tuberosa L. = Murdannia edulis (Stokes) Faden ■
101180 Commelina tuberosa Lour. = Murdannia edulis (Stokes) Faden ■
101181 Commelina umbellata Thonn. = Commelina erecta L. ■☆
101182 Commelina umbellata Thonn. var. gambiae (C. B. Clarke) J. K. Morton = Commelina nigritana Benth. var. gambiae (C. B. Clarke) Brenan ■☆
101183 Commelina umbrosa Vahl = Aneilema umbrosum (Vahl) Kunth ■☆
101184 Commelina uncata C. B. Clarke;具钩鸭跖草■☆
101185 Commelina undulata R. Br.;波缘鸭跖草(波叶鸭跖菜,波叶鸭跖草);Wavemargin Dayflower ■
101186 Commelina undulata R. Br. = Commelina erecta L. ■☆
101187 Commelina vaginata L. = Murdannia vaginata (L.) Brückn. ■
101188 Commelina vanderystii De Wild.;范德鸭跖草■☆
101189 Commelina velutina Mildbr.;短绒毛鸭跖草■☆
101190 Commelina venusta C. B. Clarke = Commelina erecta L. ■☆
101191 Commelina vermoesenii De Wild.;韦尔蒙森鸭跖草■☆
101192 Commelina violacea C. B. Clarke = Commelina subulata Roth ■☆
101193 Commelina virginica L.;北美鸭跖草;Dayflower, Virginia Dayflower ■☆
101194 Commelina vogelii C. B. Clarke = Commelina erecta L. ■☆
101195 Commelina vogelii C. B. Clarke var. angustior? = Commelina erecta L. ■☆
101196 Commelina weimarckiana Norl. = Commelina eckloniana Kunth ■☆
101197 Commelina welwitschii C. B. Clarke = Commelina africana L. ■☆
101198 Commelina zambesica C. B. Clarke;赞比西鸭跖草■☆
101199 Commelina zenkeri C. B. Clarke;岑克尔鸭跖草■☆
101200 Commelinaceae Mirb. (1804)(保留科名);鸭跖草科;Dayflower Family, Spiderwort Family ■
101201 Commelinaceae R. Br. = Commelinaceae Mirb.(保留科名)●■
101202 Commelinantia Tharp = Tinantia Scheidw.(保留属名)■☆
101203 Commelinantia anomala (Torr.) Tharp = Tinantia anomala (Torr.) C. B. Clarke ■☆
101204 Commelinidium Stapf = Acroceras Stapf ■
101205 Commelinidium gabunense (Hack.) Stapf = Acroceras gabunense (Hack.) Clayton ■☆
101206 Commelinidium mayumbense (Franch.) Stapf = Acroceras gabunense (Hack.) Clayton ■☆
101207 Commelinidium nervosum Stapf = Acroceras gabunense (Hack.) Clayton ■☆
101208 Commelinopsis Pichon = Commelina L. ■
101209 Commelyna Endl. = Commelina L. ■
101210 Commelyna Hoffmanns. ex Endl. = Commelina L. ■
101211 Commercona Sonn. = Barringtonia J. R. Forst. et G. Forst.(保留属名)●
101212 Commerconia F. Muell. = Commersonia J. R. Forst. et G. Forst. ●
101213 Commerconia F. Muell. ex Tate = Commersonia J. R. Forst. et G. Forst. ●
101214 Commersia Thouars = Commersorchis Thouars ■
101215 Commersia Thouars = Dendrobium Sw.(保留属名)■
101216 Commersona Sonn. = Barringtonia J. R. Forst. et G. Forst.(保留属名)●
101217 Commersona Sonn. = Mitraria J. F. Gmel.(废弃属名)●☆
101218 Commersonia Comm. ex Juss. = Polycardia Juss. ●☆
101219 Commersonia J. R. Forst. et G. Forst. (1775);山麻树属;Commersonia ●
101220 Commersonia bartramia (L.) Merr.;山麻树(大叶麻木,红山麻,阔叶山麻树);Bartram Commersonia, Broadleaf Commersonia, Common Commersonia ●
101221 Commersonia echinata J. R. Forst. et G. Forst. = Commersonia bartramia (L.) Merr. ●
101222 Commersonia echinata J. R. Forst. et G. Forst. var. platyphylla (Andréws) Gagnep. = Commersonia bartramia (L.) Merr. ●
101223 Commersonia platyphylla Andréws = Commersonia bartramia (L.) Merr. ●
101224 Commersophylis Thouars = Bulbophyllum Thouars(保留属名)■
101225 Commersorchis Thouars = Dendrobium Sw.(保留属名)■
101226 Commia Ham. ex Meisn. = Aporusa Blume ●
101227 Commia Lour. = Excoecaria L. ●
101228 Commia cochinchinensis Lour. = Excoecaria agallocha L. ●
101229 Commianthus Benth. = Retiniphyllum Humb. et Bonpl. ●☆
101230 Commicarpus Standl. (1909);黏腺果属;Mucilagefruit ●
101231 Commicarpus Standl. = Boerhavia L. ■
101232 Commicarpus africanus (Lour.) Dandy = Boerhavia diffusa L. ■
101233 Commicarpus africanus (Lour.) Dandy var. sinuato-lobatus (Chiov.) Cufod. = Commicarpus sinuatus Meikle ●☆
101234 Commicarpus ambiguus Meikle;可疑黏腺果●☆
101235 Commicarpus boissieri (Heimerl) Cufod.;布瓦西耶黏腺果●☆
101236 Commicarpus chinensis (L.) Heimerl;中华黏腺果(华黄细心);China Mucilagefruit ●
101237 Commicarpus chinensis (L.) Heimerl = Commicarpus lantsangensis D. Q. Lu ●
101238 Commicarpus chinensis (L.) Heimerl subsp. natalensis Meikle;纳塔尔黏腺果●☆
101239 Commicarpus commersonii (Baill.) Cavaco = Commicarpus plumbagineus (Cav.) Standl. ●☆
101240 Commicarpus commersonii Cavaco = Commicarpus plumbagineus (Cav.) Standl. ●☆
101241 Commicarpus decipiens Meikle;迷惑黏腺果●☆
101242 Commicarpus ehrenbergii Täckh. et Boulos = Commicarpus sinuatus Meikle ●☆
101243 Commicarpus fallacissimus (Heimerl) Heimerl ex Oberm., Schweick. et I. Verd.;含糊黏腺果●☆
101244 Commicarpus fallacissimus (Heimerl) Pohnert = Commicarpus fallacissimus (Heimerl) Heimerl ex Oberm., Schweick. et I. Verd. ●☆
101245 Commicarpus fallacissimus (Heimerl) Pohnert f. pilosus Heimerl = Commicarpus pilosus (Heimerl) Meikle ●☆
101246 Commicarpus fruticosus Pohnert;灌丛黏腺果●☆
101247 Commicarpus grandiflorus (A. Rich.) Standl.;大花黏腺果●☆
101248 Commicarpus greenwayi Meikle;格林韦黏腺果●☆
101249 Commicarpus helenae (Roem. et Schult.) Meikle;海氏黏腺果(海伦娜黏腺果)●☆

101250 Commicarpus helenae (Roem. et Schult.) Meikle var. barbatus Meikle;髯毛黏腺果●☆
101251 Commicarpus helenae (Schult.) Meikle = Commicarpus helenae (Roem. et Schult.) Meikle ●☆
101252 Commicarpus hiranensis Thulin;希兰黏腺果●☆
101253 Commicarpus lantsangensis D. Q. Lu;澜沧黏腺果;Lancang Mucilagefruit ●
101254 Commicarpus mistus Thulin;丝线黏腺果●☆
101255 Commicarpus montanus Miré et H. Gillet et Quézel;山地黏腺果●☆
101256 Commicarpus montanus Miré et H. Gillet et Quézel var. obovatus Miré = Commicarpus montanus Miré et H. Gillet et Quézel ●☆
101257 Commicarpus parviflorus Thulin;小花黏腺果●☆
101258 Commicarpus pedunculosus (A. Rich.) Cufod.;梗花黏腺果●☆
101259 Commicarpus pentandrus (Burch.) Heimerl;五蕊黏腺果●☆
101260 Commicarpus pilosus (Heimerl) Meikle;疏毛黏腺果●☆
101261 Commicarpus plumbagineus (Cav.) Standl.;白花丹黏腺果●☆
101262 Commicarpus plumbagineus (Cav.) Standl. var. trichocarpus (Heimerl) Meikle;毛果黏腺果●☆
101263 Commicarpus ramosissimus Thulin;多枝黏腺果●☆
101264 Commicarpus raynalii J. -P. Lebrun et Meikle;雷纳尔黏腺果●☆
101265 Commicarpus reniformis (Chiov.) Cufod.;肾形黏腺果●☆
101266 Commicarpus scandens (L.) Standl.;攀缘黏腺果;Climbing wortclub ●☆
101267 Commicarpus sinuatus Meikle;深波黏腺果●☆
101268 Commicarpus somalensis (Chiov.) J. -P. Lebrun et Stork = Acleisanthes somalensis (Chiov.) R. A. Levin ■☆
101269 Commicarpus squarrosus (Heimerl) Standl.;粗鳞黏腺果●☆
101270 Commicarpus stellatus (Wight) Berhaut = Commicarpus helenae (Roem. et Schult.) Meikle ●☆
101271 Commicarpus stellatus sensu F. G. Bhopal et M. N. Choudhri = Commicarpus boissieri (Heimerl) Cufod. ●☆
101272 Commicarpus stenocarpus (Chiov.) Cufod.;窄果黏腺果●☆
101273 Commicarpus transvaalensis Gand. = Commicarpus pentandrus (Burch.) Heimerl ●☆
101274 Commicarpus verticillatus (Poir.) Standl. = Commicarpus plumbagineus (Cav.) Standl. ●☆
101275 Commicarpus verticillatus (Poir.) Standl. var. glandulosus (Franch.) Cufod. = Commicarpus plumbagineus (Cav.) Standl. ●☆
101276 Commicarpus verticillatus (Poir.) Standl. var. puberulus Hutch. et E. A. Bruce = Commicarpus mistus Thulin ●☆
101277 Commicarpus verticillatus sensu Heimerl = Commicarpus helenae (Roem. et Schult.) Meikle ●☆
101278 Commidendron Lem. = Commidendrum Burch. ex DC. ●☆
101279 Commidendrum Burch. ex DC. (1833);胶菀木属●☆
101280 Commidendrum DC. = Commidendrum Burch. ex DC. ●☆
101281 Commidendrum burchellii Hemsl.;布尔胶菀木●☆
101282 Commidendrum rotundifolium (Roxb.) DC.;圆叶胶菀木●☆
101283 Commidendrum spurium (G. Forst.) DC.;胶菀木●☆
101284 Commilobium Benth. = Pterodon Vogel ■☆
101285 Commiphora Jacq. (1797)(保留属名);没药属;Myrrh, Myrrh Tree, Myrrhtree ●
101286 Commiphora abyssinica (O. Berg) Engl. = Commiphora kua (R. Br. ex Royle) Vollesen ●☆
101287 Commiphora abyssinica Engl.;阿比西尼亚没药;Abyssinian Myrrh, Abyssinian Myrrh Tree ●☆
101288 Commiphora acuminata Mattick;渐尖没药●☆
101289 Commiphora acutifoliolata Mattick = Commiphora engleri Guillaumin ●☆
101290 Commiphora africana (A. Rich.) Engl.;非洲没药;African Myrrh, African Myrrh Tree ●☆
101291 Commiphora africana (A. Rich.) Engl. var. glaucidula (Engl.) J. B. Gillett;灰绿非洲没药●☆
101292 Commiphora africana (A. Rich.) Engl. var. oblongifoliolata (Engl.) J. B. Gillett;矩圆叶非洲没药●☆
101293 Commiphora africana (A. Rich.) Engl. var. ramosissima (Oliv.) Engl.;多枝非洲没药●☆
101294 Commiphora africana (A. Rich.) Engl. var. rubriflora (Engl.) Wild;红花非洲没药●☆
101295 Commiphora africana (A. Rich.) Engl. var. tubuk (Sprague) J. B. Gillett = Commiphora africana (A. Rich.) Engl. ●☆
101296 Commiphora africanum Endl. = Commiphora africana (A. Rich.) Engl. ●☆
101297 Commiphora agallocha Engl. = Commiphora madagascariensis Jacq. ●☆
101298 Commiphora agar Chiov. = Commiphora alata Chiov. ●☆
101299 Commiphora airica A. Chev. = Commiphora quadricincta Schweinf. ●☆
101300 Commiphora alata Chiov.;翅没药●☆
101301 Commiphora alaticaulis J. B. Gillett et Vollesen;翅茎没药●☆
101302 Commiphora albiflora Engl. = Commiphora gileadensis (L.) C. Chr. ●☆
101303 Commiphora allophylla Sprague = Commiphora kataf (Forssk.) Engl. ●☆
101304 Commiphora anacardiifolia Dinter et Engl.;腰果叶没药●☆
101305 Commiphora ancistrophora Chiov. = Commiphora gileadensis (L.) C. Chr. ●☆
101306 Commiphora anfractuosa Chiov. = Commiphora gileadensis (L.) C. Chr. ●☆
101307 Commiphora anglosomaliae Chiov. = Commiphora serrulata Engl. ●☆
101308 Commiphora angolensis Engl.;安哥拉没药●☆
101309 Commiphora angustefoliolata Mendes;窄小叶没药●☆
101310 Commiphora antunesii Engl.;安图内思没药●☆
101311 Commiphora ararobba Engl. = Commiphora kerstingii Engl. ●☆
101312 Commiphora arenaria Thulin;沙地没药●☆
101313 Commiphora arussensis Engl. = Commiphora schimperi (O. Berg) Engl. ●☆
101314 Commiphora assaortensis Chiov. = Commiphora kua (R. Br. ex Royle) Vollesen ●☆
101315 Commiphora atramentaria Chiov. = Commiphora kua (R. Br. ex Royle) Vollesen ●☆
101316 Commiphora baluensis Engl.;巴卢没药●☆
101317 Commiphora benadirensis Mattei = Commiphora africana (A. Rich.) Engl. ●☆
101318 Commiphora berberidifolia Engl. = Commiphora glandulosa Schinz ●☆
101319 Commiphora betschuanica Engl. = Commiphora schimperi (O. Berg) Engl. ●☆
101320 Commiphora boehmii Engl. = Commiphora mollis (Oliv.) Engl. ●☆
101321 Commiphora boiviniana Engl. = Commiphora edulis (Klotzsch) Engl. subsp. boiviniana (Engl.) J. B. Gillett ●☆
101322 Commiphora boiviniana Engl. var. crenata = Commiphora edulis

(Klotzsch) Engl. subsp. boiviniana (Engl.) J. B. Gillett ●☆
101323 Commiphora boranensis Vollesen;博兰没药●☆
101324 Commiphora brevicalyx H. Perrier;短萼没药●☆
101325 Commiphora bricchettii Chiov. = Lannea obovata (Hook. f. ex Oliv.) Engl. ●☆
101326 Commiphora bruceae Chiov. = Commiphora kua (R. Br. ex Royle) Vollesen ●☆
101327 Commiphora buraensis Engl. = Commiphora schimperi (O. Berg) Engl. ●☆
101328 Commiphora caerulea Burtt;天蓝没药●☆
101329 Commiphora calciicola Engl. = Commiphora africana (A. Rich.) Engl. ●☆
101330 Commiphora campestris Engl.;田野没药●☆
101331 Commiphora campestris Engl. subsp. glabrata (Engl.) J. B. Gillett;光滑田野没药●☆
101332 Commiphora campestris Engl. subsp. shinyangensis J. B. Gillett;希尼安加没药●☆
101333 Commiphora campestris Engl. subsp. wajirensis J. B. Gillett;瓦吉尔没药●☆
101334 Commiphora campestris Engl. var. heterophylla (Engl.) J. B. Gillett;互叶田野没药●☆
101335 Commiphora candidula Sprague = Commiphora kua (R. Br. ex Royle) Vollesen ●☆
101336 Commiphora capensis (Sond.) Engl.;好望角没药●☆
101337 Commiphora capuronii Bard. -Vauc.;凯普伦没药●☆
101338 Commiphora caryifolia Oliv. = Commiphora woodii Engl. ●☆
101339 Commiphora cassan Chiov. = Commiphora gileadensis (L.) C. Chr. ●☆
101340 Commiphora cerasiformis Chiov. = Commiphora sphaerocarpa Chiov. ●☆
101341 Commiphora chariensis A. Chev.;沙里没药●☆
101342 Commiphora chevalieri Engl.;舍瓦利耶没药●☆
101343 Commiphora chiovendana J. B. Gillett ex Thulin;基奥文达没药●☆
101344 Commiphora chlorocarpa Engl. = Commiphora edulis (Klotzsch) Engl. ●☆
101345 Commiphora ciliata Vollesen;缘毛没药●☆
101346 Commiphora cinerea Engl. = Commiphora mollis (Oliv.) Engl. ●☆
101347 Commiphora confusa Vollesen;混乱没药●☆
101348 Commiphora coriacea Engl. = Commiphora myrrha (T. Nees) Engl. ●
101349 Commiphora cornii Chiov. = Commiphora cyclophylla Chiov. ●☆
101350 Commiphora coronillifolia Chiov. = Commiphora gileadensis (L.) C. Chr. ●☆
101351 Commiphora corrugata J. B. Gillett et Vollesen;皱褶没药●☆
101352 Commiphora crassispina Sprague = Commiphora samharensis Schweinf. ●☆
101353 Commiphora crenatolobata Chiov. = Commiphora truncata Engl. ●☆
101354 Commiphora crenatoserrata Engl.;圆齿没药●☆
101355 Commiphora crenulata A. Terracc. ex Chiov. = Commiphora kua (R. Br. ex Royle) Vollesen ●☆
101356 Commiphora cuneaphylla Chiov. ex Guid.;楔叶没药●☆
101357 Commiphora cuspidata Chiov. = Commiphora myrrha (T. Nees) Engl. ●
101358 Commiphora cyclophylla Chiov.;圆叶没药●☆
101359 Commiphora dalzielii Hutch.;达尔齐尔没药●☆
101360 Commiphora dancaliensis Chiov. = Commiphora kua (R. Br. ex Royle) Vollesen ●☆
101361 Commiphora danduensis J. B. Gillett = Commiphora samharensis Schweinf. ●☆
101362 Commiphora dekindtiana Engl. = Commiphora mollis (Oliv.) Engl. ●☆
101363 Commiphora dinteri Engl.;丁特没药●☆
101364 Commiphora discolor Mendes;异色没药●☆
101365 Commiphora dulcis Engl. = Commiphora edulis (Klotzsch) Engl. ●☆
101366 Commiphora dulcis Engl. = Commiphora saxicola Engl. ●☆
101367 Commiphora edulis (Klotzsch) Engl.;甜没药●☆
101368 Commiphora edulis (Klotzsch) Engl. subsp. boiviniana (Engl.) J. B. Gillett;博伊文没药●☆
101369 Commiphora edulis (Klotzsch) Engl. subsp. holosericea (Engl.) J. B. Gillett;全毛甜没药●☆
101370 Commiphora ellenbeckii Engl. = Commiphora kua (R. Br. ex Royle) Vollesen ●☆
101371 Commiphora ellisiae Vollesen = Commiphora sphaerophylla Chiov. ●☆
101372 Commiphora eminii Engl.;埃明没药●☆
101373 Commiphora eminii Engl. subsp. trifoliolata (Engl.) J. B. Gillett;三小叶甜没药●☆
101374 Commiphora eminii Engl. subsp. zimmermannii (Engl.) J. B. Gillett;齐默尔曼没药●☆
101375 Commiphora engleri Guillaumin;恩格勒没药●☆
101376 Commiphora enneaphylla Chiov.;九叶没药●☆
101377 Commiphora erlangeriana Engl.;厄兰格没药●☆
101378 Commiphora erosa Vollesen;啮蚀状没药●☆
101379 Commiphora erythraea (Ehrenb.) Engl. = Commiphora kataf (Forssk.) Engl. ●☆
101380 Commiphora erythraea (Ehrenb.) Engl. var. glabrescens Engl. = Commiphora gorinii Chiov. ●☆
101381 Commiphora erythraea Engl.;红色没药●☆
101382 Commiphora fischeri Engl. = Commiphora mossambicensis (Oliv.) Engl. ●☆
101383 Commiphora flabellulifera Chiov. = Commiphora schimperi (O. Berg) Engl. ●☆
101384 Commiphora flaviflora Engl. = Commiphora kua (R. Br. ex Royle) Vollesen ●☆
101385 Commiphora foliacea Sprague;多叶没药●☆
101386 Commiphora foliolosa (Hiern) K. Schum. = Haplocoelum foliolosum (Hiern) Bullock ●☆
101387 Commiphora fragariifolia Mattick = Commiphora stolonifera Burtt ●☆
101388 Commiphora fulvotomentosa Engl.;褐绒毛没药●☆
101389 Commiphora gallaensis (Engl.) Engl. = Commiphora baluensis Engl. ●☆
101390 Commiphora giessii J. J. A. van der Walt;吉斯没药●☆
101391 Commiphora gileadensis (L.) C. Chr.;麦加没药(爱伦堡没药树);Balm of Gilead, Mecca Balsam, Mecca Myrrh ●☆
101392 Commiphora gileadensis (L.) C. Chr. var. pubescens (Stocks) J. B. Gillett;毛麦加没药●☆
101393 Commiphora gileadensis (L.) M. R. Almeida = Commiphora gileadensis (L.) C. Chr. ●☆
101394 Commiphora gillettii Chiov. = Commiphora gileadensis (L.) C.

Chr. ●☆

101395 Commiphora glabrata Engl. = Commiphora campestris Engl. subsp. glabrata (Engl.) J. B. Gillett ●☆

101396 Commiphora glandulosa Schinz；腺没药；Tall Common Corkwood, Tall Firethorn Corkwood ●☆

101397 Commiphora glaucescens Engl.；灰没药●☆

101398 Commiphora gorinii Chiov.；戈林没药●☆

101399 Commiphora gowlello Sprague = Commiphora kua (R. Br. ex Royle) Vollesen ●☆

101400 Commiphora gracilifrondosa Dinter ex J. J. A. van der Walt；多细叶没药●☆

101401 Commiphora gracilispina J. B. Gillett = Commiphora kua (R. Br. ex Royle) Vollesen ●☆

101402 Commiphora grandifolia Engl.；大花没药●☆

101403 Commiphora grosswelleri Engl. = Commiphora angolensis Engl. ●☆

101404 Commiphora guidottii Chiov.；索马里没药●☆

101405 Commiphora habessinica (O. Berg) Engl. = Commiphora kua (R. Br. ex Royle) Vollesen ●☆

101406 Commiphora habessinica (O. Berg) Engl. var. grossedentata Chiov. = Commiphora myrrha (T. Nees) Engl. ●

101407 Commiphora habessinica (O. Berg) Engl. var. simplicifolia Schweinf. = Commiphora kua (R. Br. ex Royle) Vollesen ●☆

101408 Commiphora harveyi (Engl.) Engl.；铜纸树；Bronze Paper Tree ●☆

101409 Commiphora harveyi Engl. = Commiphora harveyi (Engl.) Engl. ●☆

101410 Commiphora hereroensis Schinz = Commiphora glaucescens Engl. ●☆

101411 Commiphora heterophylla Engl. = Commiphora campestris Engl. var. heterophylla (Engl.) J. B. Gillett ●☆

101412 Commiphora hildebrandtii (Engl.) Engl.；希尔德没药●☆

101413 Commiphora hildebrandtii (Engl.) Engl. var. gallaensis Engl. = Commiphora baluensis Engl. ●☆

101414 Commiphora hirtella Chiov. = Commiphora sphaerocarpa Chiov. ●☆

101415 Commiphora hollisii Burtt Davy = Commiphora schimperi (O. Berg) Engl. ●☆

101416 Commiphora holosericea Engl. = Commiphora edulis (Klotzsch) Engl. subsp. holosericea (Engl.) J. B. Gillett ●☆

101417 Commiphora holstii Engl. = Combretum aculeatum Vent. ●☆

101418 Commiphora holtziana Engl.；霍尔茨没药●☆

101419 Commiphora holtziana Engl. = Commiphora kataf (Forssk.) Engl. ●☆

101420 Commiphora holtziana Engl. subsp. microphylla J. B. Gillett = Commiphora kataf (Forssk.) Engl. ●☆

101421 Commiphora hornbyi Burtt；赫恩比没药●☆

101422 Commiphora horrida Chiov.；多刺没药●☆

101423 Commiphora humbertii H. Perrier；亨伯特没药●☆

101424 Commiphora incisa Chiov. = Commiphora kua (R. Br. ex Royle) Vollesen ●☆

101425 Commiphora iringensis Engl. = Commiphora mollis (Oliv.) Engl. ●☆

101426 Commiphora kaokoensis Swanepoel；卡奥科没药●☆

101427 Commiphora karibensis Wild；卡里巴没药●☆

101428 Commiphora kataf (Forssk.) Engl.；卡塔夫没药(阿拉伯没药)；Opopanax ●☆

101429 Commiphora kataf (Forssk.) Engl. subsp. turkanaensis J. B. Gillett = Commiphora kataf (Forssk.) Engl. ●☆

101430 Commiphora kataf Engl. = Commiphora kataf (Forssk.) Engl. ●☆

101431 Commiphora kerstingii Engl.；克斯廷没药●☆

101432 Commiphora kilimandscharica Engl. = Lepidotrichilia volkensii (Gürke) Leroy ●☆

101433 Commiphora krausei Engl. = Commiphora mollis (Oliv.) Engl. ●☆

101434 Commiphora kua (R. Br. ex Royle) Vollesen；苦没药●☆

101435 Commiphora kua (R. Br. ex Royle) Vollesen var. gowlello (Sprague) J. B. Gillett = Commiphora kua (R. Br. ex Royle) Vollesen ●☆

101436 Commiphora kucharii Thulin；库哈尔没药●☆

101437 Commiphora kwebensis N. E. Br. = Commiphora angolensis Engl. ●☆

101438 Commiphora kyimbilensis Engl. = Commiphora eminii Engl. subsp. zimmermannii (Engl.) J. B. Gillett ●☆

101439 Commiphora lacerata Thulin；撕裂没药●☆

101440 Commiphora lamii H. Perrier；拉姆没药●☆

101441 Commiphora lasiodisca H. Perrier；毛盘没药●☆

101442 Commiphora laxiflora Engl. = Commiphora engleri Guillaumin ●☆

101443 Commiphora leandriana H. Perrier；利安没药●☆

101444 Commiphora ledermannii Engl. = Commiphora pedunculata (Kotschy et Peyr.) Engl. ●☆

101445 Commiphora lindensis Engl. = Commiphora kua (R. Br. ex Royle) Vollesen ●☆

101446 Commiphora loandensis Engl. = Commiphora africana (A. Rich.) Engl. ●☆

101447 Commiphora lobatospathulata J. B. Gillett ex Thulin；裂苞没药●☆

101448 Commiphora longebracteata Engl. = Commiphora angolensis Engl. ●☆

101449 Commiphora longipedicellata Vollesen = Commiphora paolii Chiov. ●☆

101450 Commiphora lugardae N. E. Br. = Commiphora glandulosa Schinz ●☆

101451 Commiphora lughensis Chiov. = Commiphora cyclophylla Chiov. ●☆

101452 Commiphora macrophylla J. B. Gillett；大叶没药●☆

101453 Commiphora madagascariensis Jacq.；马岛没药(印度百代留，印度没药)●☆

101454 Commiphora mahafaliensis Capuron；马哈法里没药●☆

101455 Commiphora marchandii Engl.；马尔尚没药●☆

101456 Commiphora marlothii Engl.；马洛斯没药●☆

101457 Commiphora merkeri Engl.；东非没药●☆

101458 Commiphora microcarpa Chiov. = Commiphora gileadensis (L.) C. Chr. ●☆

101459 Commiphora mildbraedii Engl.；米尔德没药●☆

101460 Commiphora missionis Chiov. = Commiphora eminii Engl. subsp. zimmermannii (Engl.) J. B. Gillett ●☆

101461 Commiphora mollis (Oliv.) Engl.；柔软没药●☆

101462 Commiphora mollissima Engl. = Commiphora pedunculata (Kotschy et Peyr.) Engl. ●☆

101463 Commiphora molmol (Engl.) Engl. = Commiphora myrrha (T. Nees) Engl. ●

101464 Commiphora mombassensis Engl.；蒙巴萨没药●☆

101465 Commiphora monoica Vollesen；同株没药●☆

101466 Commiphora montana Engl. = Commiphora mollis (Oliv.)

Engl. ●☆

101467 Commiphora morogorensis Engl. = Commiphora edulis (Klotzsch) Engl. ●☆

101468 Commiphora mossambicensis (Oliv.) Engl.;莫桑比克没药●☆

101469 Commiphora mossamedensis Mendes;莫萨梅迪没药●☆

101470 Commiphora mukul (Hook. ex Stocks) Engl. = Commiphora wightii (Arn.) Bhandari ●

101471 Commiphora mukul (Stocks) Engl. = Commiphora wightii (Arn.) Bhandari ●

101472 Commiphora multifoliolata J. B. Gillett ex Thulin;多小叶没药●☆

101473 Commiphora multijuga (Hiern) K. Schum.;多对没药●☆

101474 Commiphora myrrha (T. Nees) Engl.;没药(明没药,末药);Common Myrrh,Common Myrrh Tree,Myrrh,Myrrh Tree,Myrrhtree ●

101475 Commiphora myrrha (T. Nees) Engl. var. molmol Engl. = Commiphora myrrha (T. Nees) Engl. ●

101476 Commiphora myrrha Engl. = Commiphora myrrha (T. Nees) Engl. ●

101477 Commiphora namaensis Schinz;纳马没药●☆

101478 Commiphora ndemfi Engl. = Commiphora mollis (Oliv.) Engl. ●☆

101479 Commiphora neglecta I. Verd.;忽视没药●☆

101480 Commiphora neumannii Engl. = Commiphora schimperi (O. Berg) Engl. ●☆

101481 Commiphora nigrescens Engl. = Commiphora angolensis Engl. ●☆

101482 Commiphora nkolola Engl. = Commiphora africana (A. Rich.) Engl. ●☆

101483 Commiphora oblanceolata Schinz;倒披针没药●☆

101484 Commiphora oblongifolia J. B. Gillett;矩圆叶没药●☆

101485 Commiphora obovata Chiov.;倒卵没药●☆

101486 Commiphora ogadensis Chiov. = Commiphora hildebrandtii (Engl.) Engl. ●☆

101487 Commiphora oliveri Engl. = Commiphora angolensis Engl. ●☆

101488 Commiphora opobalsamum (L.) Engl.;阿拉伯没药;Mecca Balsam,Mecca Myrrh,Mecca Myrrh Tree ●☆

101489 Commiphora opobalsamum (L.) Engl. = Commiphora gileadensis (L.) C. Chr. ●☆

101490 Commiphora opobalsamum (L.) Engl. var. ehrenbergianum (O. Berg) Engl. = Commiphora gileadensis (L.) C. Chr. ●☆

101491 Commiphora opobalsamum (L.) Engl. var. induta (Hutch.) Sprague ex Bruce = Commiphora gileadensis (L.) C. Chr. ●☆

101492 Commiphora opobalsamum Engl. = Commiphora opobalsamum (L.) Engl. ●☆

101493 Commiphora orbicularis Engl.;圆形没药●☆

101494 Commiphora ovalifolia J. B. Gillett;卵叶没药●☆

101495 Commiphora palmatifoliolata Chiov. = Commiphora africana (A. Rich.) Engl. ●☆

101496 Commiphora paolii Chiov.;保尔没药●☆

101497 Commiphora parvifructa J. B. Gillett;小果没药●☆

101498 Commiphora pedunculata (Kotschy et Peyr.) Engl.;梗花没药●☆

101499 Commiphora pervilleana Engl.;佩尔没药●☆

101500 Commiphora pilosa (Engl.) Engl. = Commiphora africana (A. Rich.) Engl. ●☆

101501 Commiphora pilosa (Engl.) Engl. var. glaucidula Engl. = Commiphora africana (A. Rich.) Engl. var. glaucidula (Engl.) J. B. Gillett ●☆

101502 Commiphora pilosa (Engl.) Engl. var. meyeri-johannis Engl. = Commiphora africana (A. Rich.) Engl. ●☆

101503 Commiphora pilosa (Engl.) Engl. var. oblongifoliolata Engl. = Commiphora africana (A. Rich.) Engl. var. oblongifoliolata (Engl.) J. B. Gillett ●☆

101504 Commiphora pilosa (Engl.) Engl. var. venosa Mattick = Commiphora africana (A. Rich.) Engl. var. rubriflora (Engl.) Wild ●☆

101505 Commiphora pilosissima Engl. = Commiphora edulis (Klotzsch) Engl. subsp. holosericea (Engl.) J. B. Gillett ●☆

101506 Commiphora playfairii (Hook. f. ex Oliv.) Engl. var. benadirensis Chiov. = Commiphora myrrha (T. Nees) Engl. ●

101507 Commiphora porensis Engl. = Lannea schweinfurthii (Engl.) Engl. ●☆

101508 Commiphora pruinosa Engl. = Commiphora glaucescens Engl. ●☆

101509 Commiphora pseudopaolii J. B. Gillett = Commiphora kataf (Forssk.) Engl. ●☆

101510 Commiphora pterocarpa H. Perrier;翅果没药●☆

101511 Commiphora pubescens Engl.;印度柔毛没药●☆

101512 Commiphora puguensis Engl. = Commiphora eminii Engl. subsp. zimmermannii (Engl.) J. B. Gillett ●☆

101513 Commiphora pyracanthoides Engl.;火棘没药●☆

101514 Commiphora pyracanthoides Engl. subsp. glandulosa (Schinz) Wild = Commiphora glandulosa Schinz ●☆

101515 Commiphora quadricincta Schweinf.;四被没药●☆

101516 Commiphora quercifoliola J. B. Gillett ex Thulin;栎叶没药●☆

101517 Commiphora rangeana Engl. = Commiphora capensis (Sond.) Engl. ●☆

101518 Commiphora reghinii Chiov. = Euphorbia jatrophoides Pax ●☆

101519 Commiphora rehmannii Engl. = Commiphora angolensis Engl. ●☆

101520 Commiphora resiniflua Martelli = Commiphora schimperi (O. Berg) Engl. ●☆

101521 Commiphora retifolia Chiov. = Commiphora erlangeriana Engl. ●☆

101522 Commiphora riparia Engl. = Commiphora mildbraedii Engl. ●☆

101523 Commiphora rivae Engl. = Commiphora myrrha (T. Nees) Engl. ●

101524 Commiphora robecchii Engl. = Commiphora rostrata Engl. ●☆

101525 Commiphora rosifolia Engl. = Commiphora pedunculata (Kotschy et Peyr.) Engl. ●☆

101526 Commiphora rotundifolia Dinter et Engl. = Commiphora namaensis Schinz ●☆

101527 Commiphora roxburghii (Arn.) Engl. var. serratifolia Haines = Commiphora madagascariensis Jacq. ●☆

101528 Commiphora roxburghii (Stocks) Engl. = Commiphora wightii (Arn.) Bhandari ●

101529 Commiphora roxburghii Engl. = Commiphora wightii (Arn.) Bhandari ●

101530 Commiphora ruahensis Mattick = Commiphora glandulosa Schinz ●☆

101531 Commiphora rubriflora Engl. = Commiphora africana (A. Rich.) Engl. var. rubriflora (Engl.) Wild ●☆

101532 Commiphora rugosa Engl. = Commiphora africana (A. Rich.) Engl. ●☆

101533 Commiphora ruquietiana Dinter et Engl. = Commiphora capensis (Sond.) Engl. ●☆

101534 Commiphora ruspolii Chiov.;鲁斯波利没药●☆

101535 Commiphora salubris Engl. = Commiphora kua (R. Br. ex Royle) Vollesen ●☆

101536 Commiphora sambesiaca Engl. = Commiphora africana (A. Rich.) Engl. ●☆

101537 Commiphora samharensis Schweinf.;塞姆哈尔没药●☆

101538 Commiphora samharensis Schweinf. subsp. terebinthina (Vollesen) J. B. Gillett = Commiphora samharensis Schweinf. ●☆

101539 Commiphora savoiae Chiov. = Commiphora edulis (Klotzsch) Engl. subsp. boiviniana (Engl.) J. B. Gillett ●☆

101540 Commiphora saxicola Engl.;岩生没药●☆

101541 Commiphora scaberula Engl. = Commiphora edulis (Klotzsch) Engl. subsp. boiviniana (Engl.) J. B. Gillett ●☆

101542 Commiphora schefferi Engl. = Commiphora campestris Engl. ●☆

101543 Commiphora schimperi (O. Berg) Engl.;欣珀没药●☆

101544 Commiphora schlechteri Engl.;施莱没药●☆

101545 Commiphora seineri Engl. = Commiphora glandulosa Schinz ●☆

101546 Commiphora sennii Chiov.;森恩没药●☆

101547 Commiphora serrata Engl.;粗齿没药●☆

101548 Commiphora serrata Engl. var. multipinnata = Commiphora serrata Engl. ●☆

101549 Commiphora serrulata Engl.;细齿没药●☆

101550 Commiphora serrulata Engl. var. tenuipes;细梗细齿没药●☆

101551 Commiphora sessiliflora Vollesen = Commiphora guidottii Chiov. ●☆

101552 Commiphora simplicifolia H. Perrier;单叶没药●☆

101553 Commiphora sinuata H. Perrier;深波没药●☆

101554 Commiphora somalensis Engl. = Commiphora kataf (Forssk.) Engl. ●☆

101555 Commiphora spathulata Mattick;匙形没药●☆

101556 Commiphora sphaerocarpa Chiov.;球果没药●☆

101557 Commiphora sphaerophylla Chiov.;球叶没药●☆

101558 Commiphora spinulosa J. B. Gillett ex Thulin;小刺没药●☆

101559 Commiphora spondioides Engl. = Commiphora zanzibarica (Baill.) Engl. ●☆

101560 Commiphora staphyleifolia Chiov.;省沽油没药●☆

101561 Commiphora stellatopubescens J. B. Gillett ex Thulin;短柔毛没药●☆

101562 Commiphora stocksiana (Engl.) Engl.;柔毛没药●☆

101563 Commiphora stolonifera Burtt;匍匐没药●☆

101564 Commiphora stolzii Engl. = Commiphora mossambicensis (Oliv.) Engl. ●☆

101565 Commiphora stuhlmannii Engl. = Commiphora mollis (Oliv.) Engl. ●☆

101566 Commiphora subglauca Engl. = Sclerocarya birrea (A. Rich.) Hochst. subsp. caffra (Sond.) Kokwaro ●☆

101567 Commiphora subsessilifolia Engl. = Commiphora kua (R. Br. ex Royle) Vollesen ●☆

101568 Commiphora suckertiana Chiov. = Commiphora gileadensis (L.) C. Chr. ●☆

101569 Commiphora sulcata Chiov.;纵沟没药●☆

101570 Commiphora swynnertonii Burtt;斯温纳顿没药●☆

101571 Commiphora taborensis Engl. = Lannea humilis (Oliv.) Engl. ●☆

101572 Commiphora tenuipetiolata Engl.;细柄没药●☆

101573 Commiphora tenuis Vollesen = Commiphora gurreh Engl. ●☆

101574 Commiphora tephrodes Chiov. = Commiphora hildebrandtii (Engl.) Engl. ●☆

101575 Commiphora terebinthina Vollesen = Commiphora samharensis Schweinf. ●☆

101576 Commiphora thermitaria Lisowski, Malaisse et Symoens = Commiphora glandulosa Schinz ●☆

101577 Commiphora tomentosa Engl. = Lannea rivae (Chiov.) Sacleux ●☆

101578 Commiphora torrei Mendes = Commiphora fulvotomentosa Engl. ●☆

101579 Commiphora trollii Mattick = Commiphora edulis (Klotzsch) Engl. ●☆

101580 Commiphora trothae Engl. = Commiphora schimperi (O. Berg) Engl. ●☆

101581 Commiphora truncata Engl.;平截没药●☆

101582 Commiphora tubuk Sprague = Commiphora africana (A. Rich.) Engl. ●☆

101583 Commiphora ugogensis Engl.;热非没药●☆

101584 Commiphora ulugurensis Engl.;乌卢古尔没药●☆

101585 Commiphora unilobata J. B. Gillett et Vollesen;单浅裂没药●☆

101586 Commiphora velutina Chiov. = Commiphora gileadensis (L.) C. Chr. ●☆

101587 Commiphora viminea Burtt Davy;软枝没药●☆

101588 Commiphora virgata Engl.;条纹没药●☆

101589 Commiphora voense Engl. = Platycelyphium voense (Engl.) Wild ■☆

101590 Commiphora welwitschii Engl. = Commiphora mollis (Oliv.) Engl. ●☆

101591 Commiphora wightii (Arn.) Bhandari;印度没药(罗氏没药,摩库尔没药);Guggul,India Myrrhtree ●

101592 Commiphora wildii Merxm.;瓦尔德没药●☆

101593 Commiphora woodii Engl.;伍得没药●☆

101594 Commiphora zanzibarica (Baill.) Engl.;赞比亚没药●☆

101595 Commiphora zanzibarica (Baill.) Engl. var. elongata Engl. = Commiphora zanzibarica (Baill.) Engl. ●☆

101596 Commiphora zimmermannii Engl. = Commiphora eminii Engl. subsp. zimmermannii (Engl.) J. B. Gillett ●☆

101597 Commirhoea Miers = Chrysochlamys Poepp. ●☆

101598 Commitheca Bremek. (1940);胶囊茜属●☆

101599 Commitheca letestuana N. Hallé = Pauridiantha letestuana (N. Hallé) Ntore et Dessein ●☆

101600 Commitheca liebrechtsiana (De Wild. et T. Durand) Bremek. = Pauridiantha liebrechtsiana (De Wild. et T. Durand) Ntore et Dessein ●☆

101601 Comocarpa Rydb. = Potentilla L. ■●

101602 Comocladia P. Browne(1756);毛枝漆属●☆

101603 Comocladia dentata Jacq.;齿状毛枝漆●☆

101604 Comocladiaceae Martinov = Anacardiaceae R. Br. (保留科名)●

101605 Comolia DC. (1828);腺海棠属●☆

101606 Comolia sertularia Triana;腺海棠●☆

101607 Comoliopsis Wurdack(1984);类腺海棠属●☆

101608 Comoliopsis neblinae Wurdack;类腺海棠●☆

101609 Comomyrsine Hook. f. = Cybianthus Mart. (保留属名)●☆

101610 Comomyrsine Hook. f. = Weigeltia A. DC. ●☆

101611 Comoneura Pierre = Strombosia Blume ●☆

101612 Comoneura Pierre ex Engl. = Strombosia Blume ●☆

101613 Comopyena Kuntze = Pycnocoma Benth. ●☆

101614 Comopyrum (Jaub. et Spach) Á. Löve = Aegilops L. (保留属名)■

101615 Comopyrum Á. Löve = Aegilops L. (保留属名)■

101616 Comopyrum comosum (Sibth. et Sm.) Á. Löve = Aegilops comosa Sibth. et Sm. ■☆

101617 Comoranthus Knobl. (1934);毛花木犀属●☆
101618 Comoranthus madagascariensis H. Perrier;马岛毛花木犀●☆
101619 Comoranthus minor H. Perrier;小毛花木犀●☆
101620 Comoroa Oliv. = Teclea Delile(保留属名)●☆
101621 Comosperma Poir. = Comesperma Labill. ●☆
101622 Comospermum Rauschert(1982);毛籽吊兰属■☆
101623 Comospermum platypetalum (Masam.) Rauschert;宽瓣毛籽吊兰■☆
101624 Comospermum yedoense (Maxim. ex Franch. et Sav.) Rauschert;毛籽吊兰■☆
101625 Comostemum Nees = Androtrichum (Brongn.) Brongn. ■☆
101626 Comparettia Poepp. et Endl. (1836);考姆兰属■☆
101627 Comparettia coccinea Lindl.;朱红考姆兰■☆
101628 Comparettia falcata Poepp. et Endl.;粉红考姆兰■☆
101629 Comparettia macroplectron Rchb. f. et Triana;大距考姆兰■☆
101630 Comparettia rosea Lindl. = Comparettia falcata Poepp. et Endl. ■☆
101631 Comperia C. Koch = Orchis L. ■
101632 Comperia K. Koch = Orchis L. ■
101633 Comperia K. Koch(1849);康珀兰属■☆
101634 Comperia comperiana (Steven) Asch. et Graebn.;康珀兰;Komper's Orchid ■☆
101635 Comperia taurica K. Koch;达乌尔康珀兰☆
101636 Comphoropsis Moq. = Camphoropsis Moq. ex Pfeiff. ●■
101637 Comphoropsis Moq. = Nanophyton Less. ●■
101638 Comphrena Aubl. = Gomphrena L. ●■
101639 Complaya Strother = Sphagneticola O. Hoffm. ■☆
101640 Compositae Adans. = Asteraceae Bercht. et J. Presl(保留科名)●■
101641 Compositae Adans. = Compositae Giseke(保留科名)●■
101642 Compositae Giseke(1792)(保留科名);菊科;Aster Family, Composite Family, Composites, Daisy Family ●■
101643 Compositae Giseke(保留科名) = Asteraceae Bercht. et J. Presl(保留科名)●■
101644 Compsanthus Spreng. = Compsoa D. Don(废弃属名)■
101645 Compsanthus Spreng. = Tricyrtis Wall. (保留属名)■
101646 Compsanthus maculatus Spreng. = Tricyrtis maculata (D. Don) J. F. Macbr. ■
101647 Compsoa D. Don(废弃属名) = Tricyrtis Wall. (保留属名)■
101648 Compsoa maculata D. Don = Tricyrtis pilosa Wall. ■
101649 Compsoaceae Horan. = Colchicaceae DC. (保留科名)■
101650 Compsoneura (A. DC.) Warb. (1896);饰脉树属(聚脉树属)●☆
101651 Compsoneura Warb. = Compsoneura (A. DC.) Warb. ●☆
101652 Compsoneura costaricensis Warb.;饰脉树●☆
101653 Compsos maculata D. Don = Tricyrtis pilosa Wall. ■
101654 Comptonanthus B. Nord. = Ifloga Cass. ■☆
101655 Comptonanthus brachypterus (O. Hoffm. ex Zahlbr.) B. Nord. = Lasiopogon brachypterus O. Hoffm. ex Zahlbr. ■☆
101656 Comptonanthus molluginoides (DC.) B. Nord. = Ifloga molluginoides (DC.) Hilliard ■☆
101657 Comptonanthus subcarnosus B. Nord. = Lasiopogon debilis (Thunb.) Hilliard ■☆
101658 Comptonella Baker f. (1921);肖长苞杨梅属(肖香蕨木属)●☆
101659 Comptonella albiflora Baker f.;白花肖长苞杨梅●☆
101660 Comptonella microcarpa (Perkins) T. G. Hartley;小果肖长苞杨梅●☆
101661 Comptonella sessilifoliola (Guillaumin) T. G. Hartley;无柄肖长苞杨梅●☆
101662 Comptonia Banks ex Gaertn. = Myrica L. ●
101663 Comptonia L'Hér. = Comptonia L'Hér. ex Aiton ●☆
101664 Comptonia L'Hér. ex Aiton(1789);长苞杨梅属(香蕨木属);Sweet Fern, Sweet Fern Shrub ●☆
101665 Comptonia asplenifolia Banks = Comptonia peregrina (L.) J. M. Coult. ●☆
101666 Comptonia ceterach Mirb. = Comptonia peregrina (L.) J. M. Coult. ●☆
101667 Comptonia peregrina (L.) J. M. Coult.;长苞杨梅(香蕨木,洋香蕨木);Sweet Fern, Wavy Hair-grass ●☆
101668 Comptonia peregrina (L.) J. M. Coult. var. asplenifolia (L.) Fernald = Comptonia peregrina (L.) J. M. Coult. ●☆
101669 Comptonia peregrina (L.) J. M. Coult. var. tomentosa A. Chev. = Comptonia peregrina (L.) J. M. Coult. ●☆
101670 Comularia Pichon = Hunteria Roxb. ●
101671 Comularia camerunensis (K. Schum. ex Hallier f.) Pichon = Hunteria camerunensis K. Schum. ex Hallier f. ●☆
101672 Comus Salisb. = Leopoldia Parl. (保留属名)■☆
101673 Comus Salisb. = Muscari Mill. ■☆
101674 Conami Aubl. = Phyllanthus L. ●■
101675 Conamomum Ridl. = Amomum Roxb. (保留属名)■
101676 Conandrium (K. Schum.) Mez(1902);锥蕊紫金牛属●☆
101677 Conandrium Mez = Conandrium (K. Schum.) Mez ●☆
101678 Conandrium finisterrae Mez;锥蕊紫金牛●☆
101679 Conandrium polyanthum Mez;多花锥蕊紫金牛●☆
101680 Conandron Siebold et Zucc. (1843);苦苣苔属;Conandron ■
101681 Conandron ramondioides Siebold et Zucc.;苦苣苔(水鳖草,一张白,钻骨草);Common Conandron ■
101682 Conandron ramondioides Siebold et Zucc. f. leucanthus (Nakai) Okuyama;白花苦苣苔草■
101683 Conandron ramondioides Siebold et Zucc. f. pilosum (Makino) Ohwi = Conandron ramondioides Siebold et Zucc. var. pilosum Makino ■☆
101684 Conandron ramondioides Siebold et Zucc. var. pilosum Makino;毛脉苦苣苔草■☆
101685 Conandron ramondioides Siebold et Zucc. var. ryukyuense Masam. = Conandron ramondioides Siebold et Zucc. var. taiwanense Masam. ■
101686 Conandron ramondioides Siebold et Zucc. var. taiwanense Masam.;台湾苦苣苔草■
101687 Conandron ramondioides Siebold et Zucc. var. taiwanense Masam. = Conandron ramondioides Siebold et Zucc. ■
101688 Conanthera Ruiz et Pav. (1802);锥药花属■☆
101689 Conanthera albiflora Cham. et Schltdl.;白锥药花■☆
101690 Conanthera bifolia Ruiz et Pav.;双叶锥药花■☆
101691 Conanthera minima Grau;小锥药花■☆
101692 Conantheraceae Endl. = Tecophilaeaceae Leyb. (保留科名)■☆
101693 Conantheraceae Hook. f. = Tecophilaeaceae Leyb. (保留科名)■☆
101694 Conanthes Raf. = Pitcairnia L'Hér. (保留属名)■☆
101695 Conanthodium A. Gray = Helichrysum Mill. (保留属名)●■
101696 Conanthus S. Watson = Nama L. (保留属名)■
101697 Conceveiba Aubl. (1775);康斯大戟属●☆
101698 Conceveiba africana D. W. Thomas = Conceveiba macrostachys Breteler ●☆
101699 Conceveiba africana D. W. Thomas = Neoboutonia mannii

Benth. ●☆

101700 Conceveiba leptostachys Breteler;细穗康斯大戟●☆

101701 Conceveiba macrostachys Breteler;大穗康斯大戟●☆

101702 Conceveibastrum (Müll. Arg.) Pax et K. Hoffm. = Conceveiba Aubl. ●☆

101703 Conceveibastrum Pax et K. Hoffm. = Conceveiba Aubl. ●☆

101704 Conceveibum A. Rich. ex A. Juss. = Aparisthmium Endl. ●☆

101705 Conceveibum A. Rich. ex A. Juss. = Conceveiba Aubl. ●☆

101706 Conchidium Griff. (1851);蛤兰属■

101707 Conchidium Griff. = Eria Lindl. (保留属名)■

101708 Conchidium japonicum (Maxim.) S. C. Chen et J. J. Wood;高山蛤兰■

101709 Conchidium muscicola (Lindl.) Rauschert;网鞘蛤兰■

101710 Conchidium pusillum Griff.;蛤兰■

101711 Conchidium pusillum Griff. = Eria pusilla (Griff.) Lindl. ■

101712 Conchidium rhomboidale (Ts. Tang et F. T. Wang) S. C. Chen et J. J. Wood;菱唇蛤兰■

101713 Conchidium sinicum Lindl. = Conchidium pusillum Griff. ■

101714 Conchidium sinicum Lindl. = Eria sinica (Lindl.) Lindl. ■

101715 Conchium Sm. = Hakea Schrad. ●☆

101716 Conchium aciculare Vent. = Hakea sericea Schrad. et J. C. Wendl. ●☆

101717 Conchium drupaceum C. F. Gaertn. = Hakea drupacea (C. F. Gaertn.) Roem. et Schult. ●☆

101718 Conchocarpus Mikan = Angostura Roem. et Schult. ●☆

101719 Conchochilus Hassk. = Appendicula Blume ■

101720 Conchopetalum Radlk. (1887);壳瓣花属●☆

101721 Conchopetalum brachysepalum Capuron;壳瓣花●☆

101722 Conchopetalum madagascariense Radlk.;马岛壳瓣花●☆

101723 Conchophyllum Blume = Dischidia R. Br. ●■

101724 Conchophyllum Blume(1827);壳叶萝藦属☆

101725 Conchophyllum maximum Kraenzl. = Hoya maxima Teijsm. et Binn. ●☆

101726 Concilium Raf. = Lightfootia L'Hér. ■●

101727 Concocidium Romowicz et Szlach. = Oncidium Sw. (保留属名) ■☆

101728 Condaea Steud. = Condea Adans. (废弃属名)●■

101729 Condaea Steud. = Hyptis Jacq. (保留属名)●■

101730 Condalia Cav. (1799)(保留属名);康达木属;Condalia ●☆

101731 Condalia Ruiz et Pav. (废弃属名) = Coccocypselum P. Browne (保留属名)●☆

101732 Condalia Ruiz et Pav. (废弃属名) = Condalia Cav. (保留属名)●☆

101733 Condalia ericoides (A. Gray) M. C. Johnst.;野猪康达木;Javelina Bush ●☆

101734 Condalia globosa I. M. Johnst.;球状康达木;Bitter Condalia, Bitter Snakewood ●☆

101735 Condalia lycioides (A. Gray) Weberb.;枸杞状康达木;Southwestern Condalia ●☆

101736 Condalia mexicana Schltdl.;墨西哥康达木;Mexican Bloewood Condalia ●☆

101737 Condalia obovata Hook.;倒卵叶康达木;Bloewood Condalia ●☆

101738 Condalia obovata Hook. var. edwardsiana Cory;爱德华倒卵叶康达木;Edwards Condalia ●☆

101739 Condalia obtusifolia (Hook. ex Torr. et A. Gray) Weberb.;钝叶康达木;Lotebush Condalia ●☆

101740 Condalia paradoxa Spreng. = Colletia paradoxa (Spreng.) Escal. ●☆

101741 Condalia spathulata A. Gray;匙叶康达木;Knife-leaf Condalia ●☆

101742 Condalia velutina I. M. Johnst.;短毛康达木(绿康达木);Green Condalia ●☆

101743 Condalia viridis I. M. Johnst. var. reedii Cory;里德绿康达木;Reed's Green Condalia ●☆

101744 Condalia warnockii M. C. Johnst.;瓦氏康达木;Mexican Crucillo, Warnock's Javelina Bush ●☆

101745 Condaliopsis (Weberb.) Suess. = Condalia Cav. (保留属名)●☆

101746 Condaminea DC. (1830);安第斯茜属●☆

101747 Condaminea angustifolia Rusby;窄叶安第斯茜●☆

101748 Condaminea breviflora Standl.;短花安第斯茜●☆

101749 Condaminea corymbosa DC.;安第斯茜●☆

101750 Condea Adans. (废弃属名) = Hyptis Jacq. (保留属名)●■

101751 Condgiea Baill. ex Tiegh. = Klainedoxa Pierre ex Engl. ●☆

101752 Condylago Luer(1982);节瘤兰属■☆

101753 Condylago rodrigoi Luer.;节瘤兰■☆

101754 Condylicarpus Steud. = Tordylium L. ■☆

101755 Condylidium R. M. King et H. Rob. (1972);狭管尖泽兰属■●☆

101756 Condylidium iresinoides (Kunth) R. M. King et H. Rob.;狭管尖泽兰■☆

101757 Condylocarpon Desf. (1822);瘤果夹竹桃属●☆

101758 Condylocarpon guyanense Desf.;瘤果夹竹桃●☆

101759 Condylocarpon occidentale Markgr.;西方瘤果夹竹桃●☆

101760 Condylocarpon reticulatum Ducke;网状瘤果夹竹桃●☆

101761 Condylocarpus Hoffm. = Tordylium L. ■☆

101762 Condylocarpus K. Schum. = Condylocarpon Desf. ●☆

101763 Condylocarpus Salisb. ex Lamb. = Sequoia Endl. (保留属名)●

101764 Condylocarya Bess. ex Endl. = Rapistrum Crantz(保留属名)■☆

101765 Condylopodium R. M. King et H. Rob. (1972);微腺修泽兰属●☆

101766 Condylopodium cuatrecasasii R. M. King et H. Rob.;微腺修泽兰●☆

101767 Condylostylis Piper = Vigna Savi(保留属名)■

101768 Confluaceae Dulac = Globulariaceae DC. (保留科名)●■☆

101769 Conforata Caesalp. ex Fourr. = Achillea L. ■

101770 Conforata Fourr. = Achillea L. ■

101771 Congdonia Jeps. = Declieuxia Kunth ■☆

101772 Congdonia Jeps. = Sedum L. ●■

101773 Congdonia Müll. Arg. = Declieuxia Kunth ■☆

101774 Congea Roxb. (1820);绒苞藤属(五翅藤属);Congea ●

101775 Congea azurea Wall. = Congea tomentosa Roxb. ●

101776 Congea chinensis Moldenke;华绒苞藤;China Congea, Chinese Congea ●

101777 Congea tomentosa C. P'ei = Congea chinensis Moldenke ●

101778 Congea tomentosa Roxb.;绒苞藤;Lluvia De Orquideas, Tomentose Congea ●

101779 Congea tomentosa Roxb. var. oblongbifolia Schauer = Congea tomentosa Roxb. ●

101780 Congea vestita Griff.;康吉木;Covered Congea ●☆

101781 Conghas Wall. = Schleichera Willd. (保留属名)●☆

101782 Conghas Wall. ex Hiern = Schleichera Willd. (保留属名)●☆

101783 Congolanthus A. Raynal(1968);康吉龙胆属■☆

101784 Congolanthus longidens (N. E. Br.) A. Raynal;康吉龙胆■☆

101785 Coniandra Eckl. et Zeyh. = Kedrostis Medik. ■☆

101786 Coniandra Schrad. = Kedrostis Medik. ■☆

101787 Coniandra Schrad. ex Eckl. et Zeyh. = Kedrostis Medik. ■☆

101788 Coniandra africana (L.) Sond. = Kedrostis africana (L.)

Cogn. ■☆

101789 Coniandra digitata (Thunb.) Sond. = Kedrostis africana (L.) Cogn. ■☆

101790 Coniandra dissecta Schrad. = Kedrostis africana (L.) Cogn. ■☆

101791 Coniandra glauca Schrad. = Kedrostis africana (L.) Cogn. ■☆

101792 Coniandra grossulariifolia E. Mey. ex Arn. = Kedrostis africana (L.) Cogn. ■☆

101793 Coniandra molle (Kunze) Sond. = Kedrostis nana (Lam.) Cogn. ■☆

101794 Coniandra pinnatisecta Schrad. = Kedrostis africana (L.) Cogn. ■☆

101795 Coniandra punctulata Sond. = Kedrostis africana (L.) Cogn. ■☆

101796 Coniandra thunbergii Sond. = Kedrostis nana (Lam.) Cogn. ■☆

101797 Coniandra zeyheri Schrad. = Kedrostis nana (Lam.) Cogn. var. zeyheri (Schrad.) A. Meeuse ■☆

101798 Conicosia N. E. Br. (1925);锥果玉属■☆

101799 Conicosia affinis N. E. Br. = Conicosia elongata (Haw.) N. E. Br. ■☆

101800 Conicosia alborosea L. Bolus = Conicosia pugioniformis (L.) N. E. Br. subsp. alborosea (L. Bolus) Ihlenf. et Gerbaulet ■☆

101801 Conicosia australis L. Bolus = Conicosia pugioniformis (L.) N. E. Br. subsp. muiri (N. E. Br.) Ihlenf. et Gerbaulet ■☆

101802 Conicosia bijlii N. E. Br. = Conicosia pugioniformis (L.) N. E. Br. subsp. muiri (N. E. Br.) Ihlenf. et Gerbaulet ■☆

101803 Conicosia brevicaulis (Haw.) Schwantes = Conicosia pugioniformis (L.) N. E. Br. ■☆

101804 Conicosia capensis N. E. Br. = Conicosia pugioniformis (L.) N. E. Br. ■☆

101805 Conicosia communis N. E. Br. = Conicosia pugioniformis (L.) N. E. Br. ■☆

101806 Conicosia coruscans (Haw.) Schwantes = Conicosia pugioniformis (L.) N. E. Br. ■☆

101807 Conicosia elongata (Haw.) N. E. Br.;长锥果玉■☆

101808 Conicosia fusiformis (Haw.) N. E. Br. = Conicosia elongata (Haw.) N. E. Br. ■☆

101809 Conicosia muirii N. E. Br. = Conicosia pugioniformis (L.) N. E. Br. subsp. muiri (N. E. Br.) Ihlenf. et Gerbaulet ■☆

101810 Conicosia pugioniformis (L.) N. E. Br.;狭叶锥果玉;Narrow-leaved Iceplant ■☆

101811 Conicosia pugioniformis (L.) N. E. Br. subsp. alborosea (L. Bolus) Ihlenf. et Gerbaulet;白红狭叶锥果玉■☆

101812 Conicosia pugioniformis (L.) N. E. Br. subsp. muiri (N. E. Br.) Ihlenf. et Gerbaulet;缪里锥果玉■☆

101813 Conicosia pulliloba N. E. Br. = Conicosia pugioniformis (L.) N. E. Br. ■☆

101814 Conicosia robusta N. E. Br. = Conicosia elongata (Haw.) N. E. Br. ■☆

101815 Conilaria Raf. = Lecokia DC. ■☆

101816 Conimitella Rydb. = Heuchera L. ■☆

101817 Coniogeton Blume = Buchanania Spreng. ●

101818 Coniogeton arborescens Blume = Buchanania arborescens (Blume) Blume ●

101819 Conioneura Pierre ex Engl. = Strombosia Blume ●☆

101820 Conioselinum Fisch. ex Hoffm. (1814);山芎属(川芎属,滇芎属,弯柱芎属);Hemlock Parsley, Hemlockparsley ■

101821 Conioselinum Hoffm. = Conioselinum Fisch. ex Hoffm. ■

101822 Conioselinum boreale Schischk.;北方山芎■☆

101823 Conioselinum chinense (L.) Britton, Sterns et Poggenb.;山芎;Chinese Hemlock-parsley, Hemlockparsley, Hemlock-parsley ■

101824 Conioselinum czernaevia Fisch. et C. A. Mey. = Czernaevia laevigata Turcz. ■

101825 Conioselinum filicinum (H. Wolff) H. Hara;蕨山芎■☆

101826 Conioselinum filicinum (H. Wolff) H. Hara f. maritimum H. Hara = Conioselinum chinense (L.) Britton, Sterns et Poggenb. ■

101827 Conioselinum kamtschaticum Rupr.;勘察加山芎■☆

101828 Conioselinum kamtschaticum Rupr. = Conioselinum chinense (L.) Britton, Sterns et Poggenb. ■

101829 Conioselinum latifolium Rupr.;宽叶山芎■☆

101830 Conioselinum longifolium Turcz.;长叶山芎■☆

101831 Conioselinum morrisonense Hayata;台湾山芎(玉山弯柱芎,玉山芎);Taiwan Hemlockparsley ■

101832 Conioselinum nematophyllum Pimenov et Kljuykov = Ligusticum nematophyllum (Pimenov et Kljuykov) F. T. Fu et M. F. Watson ■

101833 Conioselinum nipponicum Hara = Conioselinum chinense (L.) Britton, Sterns et Poggenb. ■

101834 Conioselinum pinnatifolium (Korovin) Schischk.;羽叶山芎■☆

101835 Conioselinum pumilum Rose = Conioselinum chinense (L.) Britton, Sterns et Poggenb. ■

101836 Conioselinum tataricum Hoffm. = Conioselinum vaginatum (Spreng.) Thell. ■

101837 Conioselinum univittatum Turcz. ex Kar. et Kir.;单带山芎(川芎)■

101838 Conioselinum univittatum Turcz. ex Kar. et Kir. = Conioselinum vaginatum (Spreng.) Thell. ■

101839 Conioselinum vaginatum (Spreng.) Thell.;鞘山芎(滇前胡,山芎,新疆藁本);Sheath Hemlockparsley ■

101840 Conioselinum victoris Schischk.;维多利亚山芎■☆

101841 Coniothele DC. = Blennosperma Less. ■☆

101842 Coniphylis Thouars = Bulbophyllum Thouars(保留属名)■

101843 Conirostrum Dulac = Brassica L. ■●

101844 Conisa Desf. ex Steud. = Conyza Less.(保留属名)■

101845 Conium L. (1753);毒参属(毒胡萝蔔属);Conium, Hemlock, Poisonhemlock ■

101846 Conium africanum L. = Capnophyllum africanum (L.) W. D. J. Koch ■☆

101847 Conium dichotomum Desf. = Krubera peregrinum (L.) Hoffm. ■☆

101848 Conium fontanum Hilliard et B. L. Burtt;泉毒参■☆

101849 Conium fontanum Hilliard et B. L. Burtt var. alticola Hilliard et B. L. Burtt;高原毒参■☆

101850 Conium fontanum Hilliard et B. L. Burtt var. silvaticumHilliard et B. L. Burtt;森林毒参■☆

101851 Conium jacquinii (DC.) D. Dietr. = Dasispermum suffruticosum (P. J. Bergius) B. L. Burtt ■☆

101852 Conium maculatum L.;毒参(斑药芹,欧毒芹,芹叶钩吻);Bad Man's Oatmeal, Beaver Poison, Break-your-mother's-heart, Bunks, Caise, Cakeseed, Cakezie, California Fern, Cambuck, Carrot, Cartwheel, Cashes, Caxes, Dead Man's Flourish, Devil's Flower, Devil's Oats, Eldrot, Ever, Gixy Gix, Gypsy Curtains, Gypsy Flower, Gypsy's Curtains, Hare's Parsley, Hech-How, Hemlock, Herb Bennett, Hever, Honiton Lace, Humlick, Humlock, Humly, Kaka, Kakezie, Kecksy, Keeks, Keicer Keice, Keik-kecksy, Keish, Kelk, Kelk-kecksy, Kesh, Kewsies, Kex, Kexies, Kicksy, Kix, Kous, Koushe, Koushle, Kricksies, Lace-flower, Lady's Lace, Lady's Needlework,

Mother-die, Mushquash Root, Na-How, Nebraska Fern, Nosebleed, Poison Fool's Parsley, Poison Hemlock, Poison Parsley, Poisonhemlock, Poison-hemlock, Scabby Hands, Snakeweed, Spotted Cowbane, Spotted Hemlock, Spotted Parsley, St. Bennet's Herb, Wode Whistle, Wode-whistle ■

101853 Conium rugosum Thunb. = Anginon rugosum (Thunb.) Raf. ■☆

101854 Conium sphaerocarpum Hilliard et B. L. Burtt;球果毒参■☆

101855 Conium suffruticosum P. J. Bergius = Dasispermum suffruticosum (P. J. Bergius) B. L. Burtt ■☆

101856 Conium tenuifolium Vahl = Itasina filifolia (Thunb.) Raf. ■☆

101857 Coniza Neck. = Conyza Less. (保留属名) ■

101858 Connaraceae R. Br. (1818) (保留科名);牛栓藤科;Connarus Family ●

101859 Connaropsis Planch. ex Hook. f. = Sarcotheca Blume ●☆

101860 Connarus L. (1753);牛栓藤属;Connarus ●

101861 Connarus africanus G. Mey. = Connarus africanus Lam. ●☆

101862 Connarus africanus Lam.;非洲牛栓藤●☆

101863 Connarus congolanus G. Schellenb.;刚果牛栓藤●☆

101864 Connarus djalonensis A. Chev. = Connarus africanus Lam. ●☆

101865 Connarus duparquetianus Baill. = Jollydora duparquetiana (Baill.) Pierre ●☆

101866 Connarus englerianus Gilg = Connarus griffonianus Baill. ●☆

101867 Connarus fernandesianus Exell et Mendonça = Connarus griffonianus Baill. ●☆

101868 Connarus floribundus Schumach. et Thonn. = Connarus thonningii (DC.) G. Schellenb. ●☆

101869 Connarus florulentus Hiern = Connarus griffonianus Baill. ●☆

101870 Connarus gabonensis Lemmens;加蓬牛栓藤●☆

101871 Connarus griffonianus Baill.;格里牛栓藤●☆

101872 Connarus griffonianus Baill. var. subsericeus (G. Schellenb.) Troupin = Connarus griffonianus Baill. ●☆

101873 Connarus guianensis Lamb. ex DC.;圭亚那牛栓藤;Zebra Wood ●☆

101874 Connarus hainanensis Merr. = Connarus paniculatus Roxb. ●

101875 Connarus incurvatus G. Schellenb. = Connarus griffonianus Baill. ●☆

101876 Connarus lambertii (DC.) Sagot;兰伯牛栓藤●☆

101877 Connarus libericus Stapf = Rourea thomsonii (Baker) Jongkind ●☆

101878 Connarus longistipitatus Gilg;长柄牛栓藤●☆

101879 Connarus longistipulatus Gossw. et Mendonça = Connarus longistipitatus Gilg ●☆

101880 Connarus luluensis Gilg = Connarus griffonianus Baill. ●☆

101881 Connarus mannii Baker = Cnestis mannii (Baker) G. Schellenb. ●☆

101882 Connarus microphyllus Hook. et Arn. = Rourea microphylla (Hook. et Arn.) Planch. ●

101883 Connarus mildbraedii G. Schellenb. = Connarus longistipitatus Gilg ●☆

101884 Connarus monocarpus L.;单果牛栓藤●☆

101885 Connarus nemorosus Vahl = Connarus thonningii (DC.) G. Schellenb. ●☆

101886 Connarus nigrensis Gilg = Connarus africanus Lam. ●☆

101887 Connarus obovatus G. Schellenb. = Connarus griffonianus Baill. ●☆

101888 Connarus orientalis G. Schellenb. = Connarus griffonianus Baill. ●☆

101889 Connarus paniculatus Roxb.;牛栓藤;Paniculate Connarus ●

101890 Connarus pentagynus Lam. = Agelaea pentagyna (Lam.) Baill. ●☆

101891 Connarus planchonianus Schellenb.;普氏牛栓藤●☆

101892 Connarus pseudoracemosus Gilg = Cnestis mannii (Baker) G. Schellenb. ●☆

101893 Connarus puberulus G. Schellenb. = Connarus griffonianus Baill. ●☆

101894 Connarus pubescens Baker = Rourea thomsonii (Baker) Jongkind ●☆

101895 Connarus punctulatus Hiern = Agelaea pentagyna (Lam.) Baill. ●☆

101896 Connarus reynoldsii Stapf = Rourea solanderi Baker ●☆

101897 Connarus roxburghii Hook. et Arn. = Rourea minor (Gaertn.) Leenh. ●

101898 Connarus santaloides Vahl = Rourea minor (Gaertn.) Leenh. ●

101899 Connarus sapinii De Wild. = Connarus griffonianus Baill. ●☆

101900 Connarus sapinii G. Schellenb. = Connarus congolanus G. Schellenb. ●☆

101901 Connarus sericeus G. Schellenb. = Connarus griffonianus Baill. ●☆

101902 Connarus smeathmannii (DC.) Planch.;斯米牛栓藤●☆

101903 Connarus staudtii Gilg;施陶牛栓藤●☆

101904 Connarus stuhlmannianus Gilg = Connarus longistipitatus Gilg ●☆

101905 Connarus subsericeus G. Schellenb. = Connarus griffonianus Baill. ●☆

101906 Connarus thomsonii Baker = Rourea thomsonii (Baker) Jongkind ●☆

101907 Connarus thonningii (DC.) G. Schellenb.;通宁牛栓藤●☆

101908 Connarus tonkinensis Lecomte = Connarus paniculatus Roxb. ●

101909 Connarus triangularis G. Schellenb. = Connarus griffonianus Baill. ●☆

101910 Connarus venosus Smeathman = Connarus africanus Lam. ●☆

101911 Connarus villosiflorus Gilg = Connarus griffonianus Baill. ●☆

101912 Connarus vrydaghii Troupin = Connarus longistipitatus Gilg ●☆

101913 Connarus yunnanensis Schellenb.;云南牛栓藤;Yunnan Connarus ●

101914 Connellia N. E. Br. (1901);点头凤梨属■☆

101915 Connellia augustae N. E. Br.;窄点头凤梨■☆

101916 Conobaea Bert. ex Steud. = Muehlenbeckia Meisn. (保留属名) ●☆

101917 Conobea Aubl. (1775);双唇婆婆纳属■☆

101918 Conobea alata Graham;翅双唇婆婆纳■☆

101919 Conobea aquatica Aubl.;双唇婆婆纳■☆

101920 Conobea borealis Spreng.;北方双唇婆婆纳■☆

101921 Conobea indica Spreng.;印度双唇婆婆纳■☆

101922 Conobea multifida (Michx.) Benth. = Leucospora multifida (Michx.) Nutt. ■☆

101923 Conobea ovata Spreng. ex Schrank;卵形双唇婆婆纳■☆

101924 Conobea polystachya (Brandegee) Minod;多穗双唇婆婆纳■☆

101925 Conocalpis Bojer ex Decne. = Gymnema R. Br. ●

101926 Conocalyx Benoist(1967);锥萼爵床属☆

101927 Conocalyx laxus Benoist;锥萼爵床☆

101928 Conocarpus Adans. = Leucadendron R. Br. (保留属名) ●

101929 Conocarpus Adans. = Protea L. (废弃属名) ●☆

101930 Conocarpus L. (1753);锥果藤属(圆锥果属,锥果木属)●☆

101931 Conocarpus acuminatus Roxb. ex DC. = Anogeissus acuminata (Roxb. ex DC.) Guillaumin et al. ●◇

101932 Conocarpus acuminatus Roxb. ex DC. var. lanceolata Wall. ex C. B. Clarke =Anogeissus acuminata (Roxb. ex DC.) Guillaumin et al. ●◇

101933 Conocarpus erectus L.;直立锥果藤(直立圆锥果);Button Mangrove,Button Wood,Buttonwood ●☆

101934 Conocarpus lancifolius Engl. ex Engl. et Diels;剑叶锥果藤●☆

101935 Conocarpus latifolius DC. =Anogeissus latifolius (DC.) Bedd. ●☆

101936 Conocarpus leiocarpa DC. =Anogeissus leiocarpa (DC.) Guillaumin et Perr. ●☆

101937 Conocarpus parviflorus Hochst. =Anogeissus leiocarpa (DC.) Guillaumin et Perr. ●☆

101938 Conocarpus racemosus L. =Laguncularia racemosa (L.) C. F. Gaertn. ●☆

101939 Conocarpus schimperi Hochst. =Anogeissus leiocarpa (DC.) Guillaumin et Perr. ●☆

101940 Conocephalopsis Kuntze =Conocephalus Blume ●

101941 Conocephalus Blume =Poikilospermum Zipp. ex Miq. ●

101942 Conocephalus lanceolatus Trécul =Poikilospermum lanceolatum (Trécul) Merr. ●

101943 Conocephalus naucleiflorus Roxb. ex Lindl. =Poikilospermum naucleiflorum (Roxb. ex Lindl.) Chew ●

101944 Conocephalus niveus Wight =Debregeasia longifolia (Burm. f.) Wedd. ●

101945 Conocephalus sinensis C. H. Wright =Poikilospermum suaveolens (Blume) Merr. ●

101946 Conocephalus suaveolens Blume =Poikilospermum suaveolens (Blume) Merr. ●

101947 Conocliniopsis R. M. King et H. Rob. (1972);齿缘柄泽兰属●☆

101948 Conocliniopsis prasiifolia (DC.) R. M. King et H. Rob.;齿缘柄泽兰■☆

101949 Conoclinium DC. (1836);破坏草属■

101950 Conoclinium DC. =Eupatorium L. ■●

101951 Conoclinium betonicifolium (Mill.) R. M. King et H. Rob.;石蚕叶破坏草;Betony-leaf mistflower ■☆

101952 Conoclinium betonicifolium (Mill.) R. M. King et H. Rob. var. integrifolium (A. Gray) T. F. Patt.;全叶破坏草■☆

101953 Conoclinium betonicum DC. =Conoclinium betonicifolium (Mill.) R. M. King et H. Rob. ■☆

101954 Conoclinium betonicum DC. var. integrifolium A. Gray =Conoclinium betonicifolium (Mill.) R. M. King et H. Rob. var. integrifolium (A. Gray) T. F. Patt. ■☆

101955 Conoclinium coelestinum (L.) DC.;破坏草(大泽兰,黑头草,解放草,马鹿草,紫茎泽兰);Blue Boneset,Blue Mistflower,Eupatorium,Hardy Ageratum,Mist Flower,Mistflower,Mist-flower,Skyblue Bogorchid,Wild Ageratum ■

101956 Conoclinium coelestinum (L.) DC. =Eupatorium coelestinum L. ■

101957 Conoclinium dichotomum Chapm. =Conoclinium coelestinum (L.) DC. ■

101958 Conoclinium dissectum A. Gray;掌叶破坏草(格氏泽兰);Boothill Eupatorium,Palm-leaf Mistflower,Throughwort ■☆

101959 Conoclinium greggii (A. Gray) Small =Conoclinium dissectum A. Gray ■☆

101960 Conoclinium greggii (A. Gray) Small =Eupatorium greggii A. Gray ■☆

101961 Conoclinium integrifolium (A. Gray) Small =Conoclinium betonicifolium (Mill.) R. M. King et H. Rob. var. integrifolium (A. Gray) T. F. Patt. ■☆

101962 Conogyne (R. Br.) Spach =Grevillea R. Br. ex Knight(保留属名)●

101963 Conohoria Aubl. (废弃属名)=Rinorea Aubl. (保留属名)●

101964 Conomitra Fenzl =Glossonema Decne. ■☆

101965 Conomitra Fenzl(1839);锥帽萝藦属■☆

101966 Conomitra linearis Fenzl;锥帽萝藦■☆

101967 Conomorpha A. DC. =Cybianthus Mart. (保留属名)●☆

101968 Conophallus Schott =Amorphophallus Blume ex Decne. (保留属名)■●

101969 Conopharyngia G. Don =Pandaca Noronha ex Thouars ●

101970 Conopharyngia G. Don =Tabernaemontana L. ●

101971 Conopharyngia angolensis (Stapf) Stapf =Tabernaemontana pachysiphon Stapf ●☆

101972 Conopharyngia bequaertii De Wild. =Tabernaemontana stapfiana Britten ●☆

101973 Conopharyngia brachyantha (Stapf) Stapf =Tabernaemontana brachyantha Stapf ●☆

101974 Conopharyngia chippii Stapf =Tabernaemontana africana Hook. ●☆

101975 Conopharyngia coffeoides (Bojer ex A. DC.) Summerh. =Tabernaemontana coffeoides Bojer ex A. DC. ●☆

101976 Conopharyngia contorta (Stapf) Stapf =Tabernaemontana contorta Stapf ●☆

101977 Conopharyngia crassa (Benth.) Stapf =Tabernaemontana crassa Benth. ●☆

101978 Conopharyngia cumminsii Stapf =Tabernaemontana pachysiphon Stapf ●☆

101979 Conopharyngia durissima (Stapf) Stapf =Tabernaemontana crassa Benth. ●☆

101980 Conopharyngia elegans (Stapf) Stapf;雅致锥帽萝藦;Toad Tree ●☆

101981 Conopharyngia elegans (Stapf) Stapf =Tabernaemontana elegans Stapf ●☆

101982 Conopharyngia holstii (K. Schum.) Stapf =Tabernaemontana pachysiphon Stapf ●☆

101983 Conopharyngia humilis Chiov. =Ephippiocarpa humilis (Chiov.) Boiteau ●☆

101984 Conopharyngia johnstonii Stapf =Tabernaemontana stapfiana Britten ●☆

101985 Conopharyngia johnstonii Stapf var. grandiflora Markgr. =Tabernaemontana stapfiana Britten ●☆

101986 Conopharyngia jollyana Stapf =Tabernaemontana crassa Benth. ●☆

101987 Conopharyngia longiflora (Benth.) Stapf =Tabernaemontana africana Hook. ●☆

101988 Conopharyngia macrosiphon Schellenb. =Tabernaemontana crassa Benth. ●☆

101989 Conopharyngia pachysiphon (Stapf) Stapf =Tabernaemontana pachysiphon Stapf ●☆

101990 Conopharyngia penduliflora (K. Schum.) Stapf =Tabernaemontana penduliflora K. Schum. ●☆

101991 Conopharyngia retusa (Lam.) G. Don =Tabernaemontana retusa (Lam.) Palacky ●☆

101992 Conopharyngia rutshurensis De Wild. =Tabernaemontana ventricosa Hochst. ex A. DC. ●☆

101993 Conopharyngia smithii (Stapf) Stapf = Tabernaemontana crassa Benth. ●☆

101994 Conopharyngia smithii (Stapf) Stapf var. brevituba De Wild. = Tabernaemontana crassa Benth. ●☆

101995 Conopharyngia stapfiana (Britten) Stapf = Tabernaemontana stapfiana Britten ●☆

101996 Conopharyngia stenosiphon (Stapf) Stapf = Tabernaemontana stenosiphon Stapf ●☆

101997 Conopharyngia thonneri (Stapf) Stapf var. demeusei De Wild. = Tabernaemontana crassa Benth. ●☆

101998 Conopharyngia thonneri (Stapf) Stapf var. lescrauwaetii De Wild. = Tabernaemontana crassa Benth. ●☆

101999 Conopharyngia thonneri (T. Durand et De Wild. ex Stapf) Stapf = Tabernaemontana crassa Benth. ●☆

102000 Conopharyngia usambarensis (K. Schum. ex Engl.) Stapf = Tabernaemontana ventricosa Hochst. ex A. DC. ●☆

102001 Conopharyngia ventricosa (Hochst. ex A. DC.) Stapf = Tabernaemontana ventricosa Hochst. ex A. DC. ●☆

102002 Conopholis Wallr. (1825);锥鳞叶属;Cancer-root ■☆

102003 Conopholis americana (L.) Wallr.;美洲锥鳞叶;American Squawroot, Cancer Corn, Squaw Root, Squawroot, Squaw-root ■☆

102004 Conopholis ludoviciana (Nutt.) A. W. Wood = Orobanche ludoviciana Nutt. ■☆

102005 Conophora (DC.) Nieuwl. = Arnoglossum Raf. ■☆

102006 Conophora (DC.) Nieuwl. = Mesadenia Raf. ■☆

102007 Conophora Nieuwl. = Arnoglossum Raf. ■☆

102008 Conophora Nieuwl. = Mesadenia Raf. ■☆

102009 Conophora atriplicifolia (L.) Nieuwl. = Arnoglossum atriplicifolium (L.) H. Rob. ■☆

102010 Conophora diversifolia (Torr. et A. Gray) Nieuwl. = Arnoglossum diversifolium (Torr. et A. Gray) H. Rob. ■☆

102011 Conophora floridana (A. Gray) Nieuwl. = Arnoglossum floridanum (A. Gray) H. Rob. ■☆

102012 Conophora ovata (Walter) Nieuwl. = Arnoglossum ovatum (Walter) H. Rob. ■☆

102013 Conophora reniformis (Hook.) Nieuwl. = Arnoglossum reniforme (Hook.) H. Rob. ■☆

102014 Conophora similis (Small) Nieuwl. = Arnoglossum atriplicifolium (L.) H. Rob. ■☆

102015 Conophora tuberosa (Nutt.) Nieuwl. = Arnoglossum plantagineum Raf. ■☆

102016 Conophyllum Schwantes = Mitrophyllum Schwantes ●☆

102017 Conophyllum Schwantes(1928);玉条草属●☆

102018 Conophyllum angustifolium L. Bolus = Mitrophyllum clivorum (N. E. Br.) Schwantes ●☆

102019 Conophyllum articulatum L. Bolus = Mitrophyllum dissitum (N. E. Br.) Schwantes ●☆

102020 Conophyllum brevisepalum L. Bolus = Mitrophyllum grande N. E. Br. ●☆

102021 Conophyllum carterianum L. Bolus = Mitrophyllum grande N. E. Br. ●☆

102022 Conophyllum compactum L. Bolus = Mitrophyllum clivorum (N. E. Br.) Schwantes ●☆

102023 Conophyllum cuspidatum L. Bolus = Mitrophyllum grande N. E. Br. ●☆

102024 Conophyllum globosum L. Bolus = Mitrophyllum globosum (L. Bolus) Ihlenf. ●☆

102025 Conophyllum gracile Schwantes = Mitrophyllum abbreviatum L. Bolus ●☆

102026 Conophyllum grande (N. E. Br.) L. Bolus;始祖鸟●☆

102027 Conophyllum hallii L. Bolus = Mitrophyllum clivorum (N. E. Br.) Schwantes ●☆

102028 Conophyllum herrei L. Bolus = Mitrophyllum clivorum (N. E. Br.) Schwantes ●☆

102029 Conophyllum latibracteatum L. Bolus = Mitrophyllum grande N. E. Br. ●☆

102030 Conophyllum nanum L. Bolus = Diplosoma retroversum (Kensit) Schwantes ■☆

102031 Conophyllum obtusipetalum L. Bolus = Mitrophyllum dissitum (N. E. Br.) Schwantes ●☆

102032 Conophyllum ripense L. Bolus = Mitrophyllum clivorum (N. E. Br.) Schwantes ●☆

102033 Conophyllum tenuifolium (L.) Schwantes;玉条草●☆

102034 Conophyllum vanheerdei L. Bolus = Mitrophyllum clivorum (N. E. Br.) Schwantes ●☆

102035 Conophyta Schum. ex Hook. f. = Thonningia Vahl ■☆

102036 Conophyton Haw. = Conophytum N. E. Br. ■☆

102037 Conophytum N. E. Br. (1922);肉锥花属(厚锥花属);Pebble Plant ■☆

102038 Conophytum absimile L. Bolus f. majus? = Conophytum bilobum (Marloth) N. E. Br. ■☆

102039 Conophytum absimile L. Bolus f. umbrosum? = Conophytum bilobum (Marloth) N. E. Br. ■☆

102040 Conophytum absimile L. Bolus var. absimile? = Conophytum bilobum (Marloth) N. E. Br. ■☆

102041 Conophytum absimile L. Bolus var. major? = Conophytum bilobum (Marloth) N. E. Br. ■☆

102042 Conophytum acutum L. Bolus;尖肉锥花■☆

102043 Conophytum admiraalii L. Bolus = Conophytum jucundum (N. E. Br.) N. E. Br. ■☆

102044 Conophytum advenum N. E. Br. = Conophytum piluliforme (N. E. Br.) N. E. Br. ■☆

102045 Conophytum aequale L. Bolus = Conophytum bilobum (Marloth) N. E. Br. ■☆

102046 Conophytum aequatum L. Bolus = Conophytum pageae (N. E. Br.) N. E. Br. ■☆

102047 Conophytum aggregatum (Haw. ex N. E. Br.) N. E. Br. = Conophytum piluliforme (N. E. Br.) N. E. Br. ■☆

102048 Conophytum albertense (N. E. Br.) N. E. Br. = Conophytum truncatum (Thunb.) N. E. Br. ■☆

102049 Conophytum albescens N. E. Br. = Conophytum bilobum (Marloth) N. E. Br. ■☆

102050 Conophytum albifissum Tischer = Conophytum minimum (Haw.) N. E. Br. ■☆

102051 Conophytum albiflorum (Rawé) S. A. Hammer;立雏■☆

102052 Conophytum altile (N. E. Br.) N. E. Br. = Conophytum ficiforme (Haw.) N. E. Br. ■☆

102053 Conophytum altum L. Bolus = Conophytum bilobum (Marloth) N. E. Br. subsp. altum (L. Bolus) S. A. Hammer ■☆

102054 Conophytum altum L. Bolus var. plenum? = Conophytum bilobum (Marloth) N. E. Br. subsp. altum (L. Bolus) S. A. Hammer ■☆

102055 Conophytum ampliatum L. Bolus = Conophytum bilobum (Marloth) N. E. Br. ■☆

102056 Conophytum amplum L. Bolus = Conophytum bilobum (Marloth) N. E. Br. ■☆

102057 Conophytum andausanum N. E. Br. = Conophytum bilobum (Marloth) N. E. Br. ■☆

102058 Conophytum andausanum N. E. Br. var. immaculatum L. Bolus = Conophytum bilobum (Marloth) N. E. Br. ■☆

102059 Conophytum angelicae (Dinter et Schwantes) N. E. Br.;安杰利卡肉锥花■☆

102060 Conophytum angustum L. Bolus = Conophytum bilobum (Marloth) N. E. Br. ■☆

102061 Conophytum angustum N. E. Br. = Conophytum bilobum (Marloth) N. E. Br. ■☆

102062 Conophytum anjametae de Boer = Conophytum violaciflorum Schick et Tischer ■☆

102063 Conophytum anomalum L. Bolus = Conophytum bilobum (Marloth) N. E. Br. ■☆

102064 Conophytum apertum Tischer = Conophytum bilobum (Marloth) N. E. Br. subsp. altum (L. Bolus) S. A. Hammer ■☆

102065 Conophytum apiatum (N. E. Br.) N. E. Br. = Conophytum bilobum (Marloth) N. E. Br. ■☆

102066 Conophytum apiculatum N. E. Br. = Conophytum bilobum (Marloth) N. E. Br. ■☆

102067 Conophytum approximatum Lavis = Conophytum bilobum (Marloth) N. E. Br. ■☆

102068 Conophytum archeri Lavis;阿谢尔肉锥花■☆

102069 Conophytum archeri Lavis = Conophytum piluliforme (N. E. Br.) N. E. Br. ■☆

102070 Conophytum archeri Lavis var. stayneri L. Bolus = Conophytum truncatum (Thunb.) N. E. Br. subsp. viridicatum (N. E. Br.) S. A. Hammer ■☆

102071 Conophytum areolatum Littlew. = Conophytum pellucidum Schwantes ■☆

102072 Conophytum asperulum L. Bolus;粗糙肉锥花■☆

102073 Conophytum asperulum L. Bolus var. asperulum? = Conophytum bilobum (Marloth) N. E. Br. ■☆

102074 Conophytum asperulum L. Bolus var. brevistylum? = Conophytum bilobum (Marloth) N. E. Br. ■☆

102075 Conophytum assimile (N. E. Br.) N. E. Br. = Conophytum ficiforme (Haw.) N. E. Br. ■☆

102076 Conophytum astylum L. Bolus = Conophytum pellucidum Schwantes var. cupreatum (Tischer) S. A. Hammer ■☆

102077 Conophytum auctum N. E. Br. f. approximatum (Lavis) Rawé = Conophytum bilobum (Marloth) N. E. Br. ■☆

102078 Conophytum auctum N. E. Br. f. auctum? = Conophytum bilobum (Marloth) N. E. Br. ■☆

102079 Conophytum australe L. Bolus = Conophytum bilobum (Marloth) N. E. Br. ■☆

102080 Conophytum avenantii L. Bolus = Conophytum jucundum (N. E. Br.) N. E. Br. subsp. fragile (Tischer) S. A. Hammer ■☆

102081 Conophytum barbatum L. Bolus = Conophytum obscurum N. E. Br. subsp. barbatum (L. Bolus) S. A. Hammer ■☆

102082 Conophytum barkerae L. Bolus = Conophytum bilobum (Marloth) N. E. Br. ■☆

102083 Conophytum batesii N. E. Br. = Conophytum minimum (Haw.) N. E. Br. ■☆

102084 Conophytum beekenkampianum Tischer = Conophytum bilobum (Marloth) N. E. Br. ■☆

102085 Conophytum bicarinatum L. Bolus;双棱肉锥花■☆

102086 Conophytum bilobum (Marloth) N. E. Br.;少将肉锥花(少将)■☆

102087 Conophytum bilobum (Marloth) N. E. Br. subsp. altum (L. Bolus) S. A. Hammer;淡雪(淡春,高大少将肉锥花)■☆

102088 Conophytum bilobum (Marloth) N. E. Br. subsp. claviferens S. A. Hammer;棍棒肉锥花■☆

102089 Conophytum bilobum (Marloth) N. E. Br. subsp. gracilistylum (L. Bolus) S. A. Hammer;细柱肉锥花■☆

102090 Conophytum bilobum (Marloth) N. E. Br. var. elishae (N. E. Br.) S. A. Hammer;式典■☆

102091 Conophytum bilobum (Marloth) N. E. Br. var. linearilucidum (L. Bolus) S. A. Hammer;亮线少将肉锥花■☆

102092 Conophytum blandum L. Bolus;光滑肉锥花■☆

102093 Conophytum bolusiae Schwantes;博卢斯肉锥花■☆

102094 Conophytum boreale L. Bolus = Conophytum lithopsoides L. Bolus subsp. boreale (L. Bolus) S. A. Hammer ■☆

102095 Conophytum braunsii Tischer = Conophytum minutum (Haw.) N. E. Br. ■☆

102096 Conophytum breve N. E. Br.;短肉锥花■☆

102097 Conophytum breve N. E. Br. var. minor L. Bolus = Conophytum breve N. E. Br. ■☆

102098 Conophytum breve N. E. Br. var. minutiflorum (Schwantes) Rawé = Conophytum pageae (N. E. Br.) N. E. Br. ■☆

102099 Conophytum breve N. E. Br. var. swanepoelii Rawé = Conophytum pageae (N. E. Br.) N. E. Br. ■☆

102100 Conophytum breve N. E. Br. var. vanzylii (Lavis) Rawé = Conophytum calculus (A. Berger) N. E. Br. subsp. vanzylii (Lavis) S. A. Hammer ■☆

102101 Conophytum brevilineatum Tischer = Conophytum minimum (Haw.) N. E. Br. ■☆

102102 Conophytum brevipes L. Bolus = Conophytum wettsteinii (A. Berger) N. E. Br. ■☆

102103 Conophytum brevipetalum Lavis = Conophytum piluliforme (N. E. Br.) N. E. Br. ■☆

102104 Conophytum brevisectum L. Bolus = Conophytum bilobum (Marloth) N. E. Br. ■☆

102105 Conophytum brevitubum Lavis = Conophytum truncatum (Thunb.) N. E. Br. ■☆

102106 Conophytum brownii Tischer = Conophytum ectypum N. E. Br. subsp. brownii (Tischer) S. A. Hammer ■☆

102107 Conophytum brunneum S. A. Hammer;褐色肉锥花■☆

102108 Conophytum burgeri L. Bolus;伯格肉锥花■☆

102109 Conophytum buysianum A. R. Mitch. et S. A. Hammer = Conophytum reconditum A. R. Mitch. subsp. buysianum (A. R. Mitch. et S. A. Hammer) S. A. Hammer ■☆

102110 Conophytum calculus (A. Berger) N. E. Br.;石灰肉锥花■☆

102111 Conophytum calculus (A. Berger) N. E. Br. subsp. vanzylii (Lavis) S. A. Hammer;万齐肉锥花■☆

102112 Conophytum calculus (A. Berger) N. E. Br. var. komkansicum (L. Bolus) Rawé = Conophytum calculus (A. Berger) N. E. Br. ■☆

102113 Conophytum calculus (A. Berger) N. E. Br. var. protusum L. Bolus = Conophytum pageae (N. E. Br.) N. E. Br. ■☆

102114 Conophytum calitzdorpense L. Bolus = Conophytum truncatum (Thunb.) N. E. Br. ■☆

102115 Conophytum candelabriforme de Boer = Conophytum minimum (Haw.) N. E. Br. ■☆

102116 Conophytum candidum L. Bolus = Conophytum meyeri N. E. Br. ■☆

102117 Conophytum carolii Lavis;卡罗尔肉锥花■☆

102118 Conophytum catervum (N. E. Br.) N. E. Br. = Conophytum truncatum (Thunb.) N. E. Br. subsp. viridicatum (N. E. Br.) S. A. Hammer ■☆

102119 Conophytum cauliferum N. E. Br.;茎生肉锥花■☆

102120 Conophytum cauliferum N. E. Br. = Conophytum bilobum (Marloth) N. E. Br. ■☆

102121 Conophytum cauliferum N. E. Br. var. lekkersingense L. Bolus = Conophytum bilobum (Marloth) N. E. Br. ■☆

102122 Conophytum ceresianum L. Bol.;云映玉■☆

102123 Conophytum ceresianum L. Bolus = Conophytum obcordellum (Haw.) N. E. Br. var. ceresianum (L. Bolus) S. A. Hammer ■☆

102124 Conophytum ceresianum L. Bolus var. divergens (L. Bolus) Rawé = Conophytum obcordellum (Haw.) N. E. Br. var. ceresianum (L. Bolus) S. A. Hammer ■☆

102125 Conophytum chauviniae (Schwantes) S. A. Hammer;豪氏肉锥花■☆

102126 Conophytum chloratum Tischer = Conophytum ectypum N. E. Br. ■☆

102127 Conophytum christiansenianum L. Bolus = Conophytum bilobum (Marloth) N. E. Br. ■☆

102128 Conophytum cibdelum N. E. Br. = Conophytum truncatum (Thunb.) N. E. Br. ■☆

102129 Conophytum cinereum Lavis = Conophytum bilobum (Marloth) N. E. Br. ■☆

102130 Conophytum circumpunctatum Schick et Tischer = Conophytum wettsteinii (A. Berger) N. E. Br. ■☆

102131 Conophytum citrinum L. Bolus = Conophytum bilobum (Marloth) N. E. Br. ■☆

102132 Conophytum clarum N. E. Br. = Conophytum uviforme (Haw.) N. E. Br. ■☆

102133 Conophytum clavatum L. Bolus = Conophytum obscurum N. E. Br. ■☆

102134 Conophytum colorans Lavis = Conophytum uviforme (Haw.) N. E. Br. ■☆

102135 Conophytum complanatum L. Bolus = Conophytum truncatum (Thunb.) N. E. Br. subsp. viridicatum (N. E. Br.) S. A. Hammer ■☆

102136 Conophytum compressum N. E. Br.;世尊■☆

102137 Conophytum compressum N. E. Br. = Conophytum bilobum (Marloth) N. E. Br. ■☆

102138 Conophytum comptonii N. E. Br.;康普顿肉锥花■☆

102139 Conophytum concavum L. Bolus;凹肉锥花■☆

102140 Conophytum concinnum Schwantes = Conophytum flavum N. E. Br. ■☆

102141 Conophytum concordans G. D. Rowley;聚心形肉锥花■☆

102142 Conophytum conformale N. E. Br. = Conophytum pageae (N. E. Br.) N. E. Br. ■☆

102143 Conophytum conradii L. Bolus = Conophytum bilobum (Marloth) N. E. Br. ■☆

102144 Conophytum convexum L. Bolus = Conophytum bilobum (Marloth) N. E. Br. ■☆

102145 Conophytum corculum Schwantes = Conophytum meyeri N. E. Br. ■☆

102146 Conophytum cordatum Schick et Tischer;心形肉锥花■☆

102147 Conophytum cordatum Schick et Tischer = Conophytum bilobum (Marloth) N. E. Br. ■☆

102148 Conophytum cordatum Schick et Tischer var. macrostigma L. Bolus = Conophytum bilobum (Marloth) N. E. Br. ■☆

102149 Conophytum coriaceum L. Bolus = Conophytum bilobum (Marloth) N. E. Br. ■☆

102150 Conophytum corniferum Schick et Tischer;小迪■☆

102151 Conophytum corniferum Schick et Tischer = Conophytum bilobum (Marloth) N. E. Br. subsp. altum (L. Bolus) S. A. Hammer ■☆

102152 Conophytum crassum L. Bolus = Conophytum bilobum (Marloth) N. E. Br. ■☆

102153 Conophytum craterulum Tischer = Conophytum velutinum Schwantes ■☆

102154 Conophytum creperum N. E. Br. = Conophytum obcordellum (Haw.) N. E. Br. ■☆

102155 Conophytum cubicum Pavelka;管状肉锥花■☆

102156 Conophytum cuneatum Tischer = Conophytum halenbergense (Dinter et Schwantes) N. E. Br. ■☆

102157 Conophytum cupreatum Tischer = Conophytum pellucidum Schwantes var. cupreatum (Tischer) S. A. Hammer ■☆

102158 Conophytum cupreiflorum Tischer;铜花肉锥花■☆

102159 Conophytum curtum L. Bolus;碧天玉■☆

102160 Conophytum curtum L. Bolus = Conophytum bilobum (Marloth) N. E. Br. ■☆

102161 Conophytum cylindratum Schwantes = Conophytum roodiae N. E. Br. subsp. cylindratum (Schwantes) Smale ■☆

102162 Conophytum cylindratum Schwantes var. primosii (Lavis) Rawé = Conophytum roodiae N. E. Br. subsp. cylindratum (Schwantes) Smale ■☆

102163 Conophytum declinatum L. Bolus;七星座■☆

102164 Conophytum declinatum L. Bolus = Conophytum obcordellum (Haw.) N. E. Br. ■☆

102165 Conophytum decoratum N. E. Br. = Conophytum uviforme (Haw.) N. E. Br. subsp. decoratum (N. E. Br.) S. A. Hammer ■☆

102166 Conophytum dedicatum N. E. Br. = Conophytum minimum (Haw.) N. E. Br. ■☆

102167 Conophytum dennisii N. E. Br. = Conophytum bilobum (Marloth) N. E. Br. ■☆

102168 Conophytum depressum Lavis;凹陷肉锥花■☆

102169 Conophytum devium G. D. Rowley;荒野肉锥花■☆

102170 Conophytum difforme L. Bolus = Conophytum bilobum (Marloth) N. E. Br. ■☆

102171 Conophytum dilatatum Tischer = Conophytum bilobum (Marloth) N. E. Br. ■☆

102172 Conophytum discrepans G. D. Rowley f. discrepans = Conophytum maughanii N. E. Br. subsp. latum (Tischer) S. A. Hammer ■☆

102173 Conophytum discrepans G. D. Rowley f. rubrum (Tischer) G. D. Rowley = Conophytum maughanii N. E. Br. subsp. latum (Tischer) S. A. Hammer ■☆

102174 Conophytum dispar N. E. Br. = Conophytum truncatum (Thunb.) N. E. Br. subsp. viridicatum (N. E. Br.) S. A. Hammer ■☆

102175 Conophytum dissimile L. Bolus = Conophytum bilobum (Marloth) N. E. Br. ■☆

102176 Conophytum distans L. Bolus = Conophytum bilobum (Marloth) N. E. Br. ■☆

102177 Conophytum distinctum Tischer = Conophytum ectypum N. E.

Br. subsp. sulcatum (L. Bolus) S. A. Hammer ■☆
102178 Conophytum divaricatum N. E. Br. = Conophytum bilobum (Marloth) N. E. Br. ■☆
102179 Conophytum divergens L. Bolus = Conophytum obcordellum (Haw.) N. E. Br. var. ceresianum (L. Bolus) S. A. Hammer ■☆
102180 Conophytum diversum N. E. Br. = Conophytum bilobum (Marloth) N. E. Br. ■☆
102181 Conophytum dolomiticum Tischer = Conophytum bilobum (Marloth) N. E. Br. ■☆
102182 Conophytum doornense N. E. Br. = Conophytum breve N. E. Br. ■☆
102183 Conophytum durnale N. E. Br. = Conophytum bilobum (Marloth) N. E. Br. subsp. altum (L. Bolus) S. A. Hammer ■☆
102184 Conophytum ecarinatum L. Bolus;无棱肉锥花■☆
102185 Conophytum ecarinatum L. Bolus = Conophytum bilobum (Marloth) N. E. Br. ■☆
102186 Conophytum ecarinatum L. Bolus var. angustum? = Conophytum bilobum (Marloth) N. E. Br. ■☆
102187 Conophytum ecarinatum L. Bolus var. candidum (L. Bolus) Rawé = Conophytum meyeri N. E. Br. ■☆
102188 Conophytum ecarinatum L. Bolus var. mutabile? = Conophytum bilobum (Marloth) N. E. Br. ■☆
102189 Conophytum ectypum N. E. Br.;雕饰肉锥花■☆
102190 Conophytum ectypum N. E. Br. subsp. brownii (Tischer) S. A. Hammer;布朗肉锥花■☆
102191 Conophytum ectypum N. E. Br. subsp. cruciatum S. A. Hammer;十字肉锥花■☆
102192 Conophytum ectypum N. E. Br. subsp. sulcatum (L. Bolus) S. A. Hammer;纵沟肉锥花■☆
102193 Conophytum ectypum N. E. Br. var. brownii (Tischer) Tischer = Conophytum ectypum N. E. Br. subsp. brownii (Tischer) S. A. Hammer ■☆
102194 Conophytum ectypum N. E. Br. var. limbatum (N. E. Br.) Tischer = Conophytum ectypum N. E. Br. ■☆
102195 Conophytum ectypum N. E. Br. var. tischleri (Schwantes) Tischer = Conophytum ectypum N. E. Br. ■☆
102196 Conophytum edithiae N. E. Br. = Conophytum subfenestratum Schwantes ■☆
102197 Conophytum edwardii Schwantes = Conophytum piluliforme (N. E. Br.) N. E. Br. subsp. edwardii (Schwantes) S. A. Hammer ■☆
102198 Conophytum edwardsiae Lavis = Conophytum luckhoffii Lavis ■☆
102199 Conophytum edwardsiae Lavis var. albiflorum Rawé = Conophytum albiflorum (Rawé) S. A. Hammer ■☆
102200 Conophytum elegans N. E. Br. = Conophytum pellucidum Schwantes ■☆
102201 Conophytum elishae (N. E. Br.) N. E. Br. = Conophytum bilobum (Marloth) N. E. Br. var. elishae (N. E. Br.) S. A. Hammer ■☆
102202 Conophytum ellipticum Tischer = Conophytum flavum N. E. Br. subsp. novicium (N. E. Br.) S. A. Hammer ■☆
102203 Conophytum elongatum Schick et Tischer = Conophytum hians N. E. Br. ■☆
102204 Conophytum etaylorii Schwantes = Conophytum piluliforme (N. E. Br.) N. E. Br. ■☆
102205 Conophytum excisum L. Bolus = Conophytum bilobum (Marloth) N. E. Br. ■☆
102206 Conophytum exiguum N. E. Br. = Conophytum saxetanum (N. E. Br.) N. E. Br. ■☆
102207 Conophytum exsertum N. E. Br. = Conophytum bilobum (Marloth) N. E. Br. ■☆
102208 Conophytum extractum Tischer = Conophytum meyeri N. E. Br. ■☆
102209 Conophytum fenestratum N. E. Br.;肉锥花(秋想)■☆
102210 Conophytum fenestratum Schwantes = Conophytum pellucidum Schwantes ■☆
102211 Conophytum fenestriferum N. E. Br. = Conophytum ficiforme (Haw.) N. E. Br. ■☆
102212 Conophytum fibuliforme (Haw.) N. E. Br.;纽扣肉锥花■☆
102213 Conophytum ficiforme (Haw.) N. E. Br.;无花果肉锥花;Cone Plant,Pebble Plant ■☆
102214 Conophytum ficiforme (Haw.) N. E. Br. var. placitum (N. E. Br.) Rawé = Conophytum ficiforme (Haw.) N. E. Br. ■☆
102215 Conophytum flavum N. E. Br.;玉彦■☆
102216 Conophytum flavum N. E. Br. subsp. novicium (N. E. Br.) S. A. Hammer;鲜玉彦■☆
102217 Conophytum flavum N. E. Br. var. luteum (N. E. Br.) Boom = Conophytum flavum N. E. Br. ■☆
102218 Conophytum forresteri L. Bolus = Conophytum pageae (N. E. Br.) N. E. Br. ■☆
102219 Conophytum fossulatum Tischer = Conophytum ficiforme (Haw.) N. E. Br. ■☆
102220 Conophytum fragile Tischer = Conophytum jucundum (N. E. Br.) N. E. Br. subsp. fragile (Tischer) S. A. Hammer ■☆
102221 Conophytum framesii Lavis = Conophytum uviforme (Haw.) N. E. Br. ■☆
102222 Conophytum franciscii L. Bolus = Conophytum uviforme (Haw.) N. E. Br. ■☆
102223 Conophytum francoiseae (S. A. Hammer) S. A. Hammer;法兰西斯肉锥花■☆
102224 Conophytum fraternum (N. E. Br.) N. E. Br.;兄弟肉锥花■☆
102225 Conophytum fraternum (N. E. Br.) N. E. Br. var. leptanthum (L. Bolus) L. Bolus = Conophytum jucundum (N. E. Br.) N. E. Br. subsp. marlothii (N. E. Br.) S. A. Hammer ■☆
102226 Conophytum friedrichiae (Dinter) Schwantes;弗里德利希肉锥花■☆
102227 Conophytum frutescens Schwantes;寂光■☆
102228 Conophytum fulleri L. Bolus;富勒肉锥花■☆
102229 Conophytum furcatum N. E. Br. = Conophytum bilobum (Marloth) N. E. Br. ■☆
102230 Conophytum geminum N. E. Br. = Conophytum bilobum (Marloth) N. E. Br. ■☆
102231 Conophytum geometricum Lavis = Conophytum violaciflorum Schick et Tischer ■☆
102232 Conophytum germanum N. E. Br. = Conophytum obcordellum (Haw.) N. E. Br. ■☆
102233 Conophytum geyeri L. Bolus = Conophytum jucundum (N. E. Br.) N. E. Br. ■☆
102234 Conophytum giftbergense Tischer;翠黛■☆
102235 Conophytum giftbergense Tischer = Conophytum obcordellum (Haw.) N. E. Br. ■☆
102236 Conophytum glabrum Tischer = Conophytum minutum (Haw.) N. E. Br. ■☆
102237 Conophytum glaucum N. E. Br. = Conophytum bilobum (Marloth) N. E. Br. ■☆

102238 Conophytum globosum (N. E. Br.) N. E. Br.;球形肉锥花■☆
102239 Conophytum globosum (N. E. Br.) N. E. Br. var. vanbredae (L. Bolus) Rawé = Conophytum globosum (N. E. Br.) N. E. Br. ■☆
102240 Conophytum globuliforme Schick et Tischer = Conophytum meyeri N. E. Br. ■☆
102241 Conophytum gonapense L. Bolus = Conophytum bilobum (Marloth) N. E. Br. ■☆
102242 Conophytum gonapense L. Bolus var. numeesicum? = Conophytum bilobum (Marloth) N. E. Br. ■☆
102243 Conophytum gothicum Tischer = Conophytum hians N. E. Br. ■☆
102244 Conophytum gracile N. E. Br.;纤细肉锥花■☆
102245 Conophytum gracile N. E. Br. = Conophytum bilobum (Marloth) N. E. Br. subsp. altum (L. Bolus) S. A. Hammer ■☆
102246 Conophytum gracile N. E. Br. var. majusculum L. Bolus = Conophytum bilobum (Marloth) N. E. Br. subsp. altum (L. Bolus) S. A. Hammer ■☆
102247 Conophytum graciliramosum L. Bolus = Conophytum bilobum (Marloth) N. E. Br. subsp. altum (L. Bolus) S. A. Hammer ■☆
102248 Conophytum gracilistylum (L. Bolus) N. E. Br. = Conophytum bilobum (Marloth) N. E. Br. subsp. gracilistylum (L. Bolus) S. A. Hammer ■☆
102249 Conophytum graessneri Tischer;小公女■☆
102250 Conophytum graessneri Tischer = Conophytum saxetanum (N. E. Br.) N. E. Br. ■☆
102251 Conophytum grandiflorum L. Bolus = Conophytum bilobum (Marloth) N. E. Br. ■☆
102252 Conophytum gratum (N. E. Br.) N. E. Br.;愉悦肉锥花(可爱肉锥花)■☆
102253 Conophytum gratum (N. E. Br.) N. E. Br. = Conophytum jucundum (N. E. Br.) N. E. Br. ■☆
102254 Conophytum gratum (N. E. Br.) N. E. Br. subsp. marlothii (N. E. Br.) S. A. Hammer = Conophytum jucundum (N. E. Br.) N. E. Br. subsp. marlothii (N. E. Br.) S. A. Hammer ■☆
102255 Conophytum gratum N. E. Br.;雨月■☆
102256 Conophytum gregale N. E. Br. = Conophytum bilobum (Marloth) N. E. Br. ■☆
102257 Conophytum halenbergense (Dinter et Schwantes) N. E. Br.;哈伦肉锥花■☆
102258 Conophytum halenbergense N. E. Br.;中将姬■☆
102259 Conophytum hallii L. Bolus = Conophytum roodiae N. E. Br. ■☆
102260 Conophytum hansii N. E. Br. = Conophytum angelicae (Dinter et Schwantes) N. E. Br. ■☆
102261 Conophytum haramoepense (L. Bolus) G. D. Rowley = Conophytum lydiae (H. Jacobsen) G. D. Rowley ■☆
102262 Conophytum haramoepense L. Bolus = Conophytum marginatum Lavis subsp. haramoepense (L. Bolus) S. A. Hammer ■☆
102263 Conophytum helenae Rawé = Conophytum tantillum N. E. Br. subsp. heleniae (Rawé) S. A. Hammer ■☆
102264 Conophytum herreanthus S. A. Hammer;赫勒肉锥花■☆
102265 Conophytum herreanthus S. A. Hammer subsp. rex S. A. Hammer;国王肉锥花■☆
102266 Conophytum herrei Schwantes = Conophytum minusculum (N. E. Br.) N. E. Br. ■☆
102267 Conophytum hians N. E. Br.;开裂肉锥花■☆
102268 Conophytum hians N. E. Br. var. acuminatum L. Bolus = Conophytum hians N. E. Br. ■☆
102269 Conophytum hillii L. Bolus = Conophytum uviforme (Haw.) N. E. Br. ■☆
102270 Conophytum hirtum Schwantes;小笛■☆
102271 Conophytum hirtum Schwantes = Conophytum hians N. E. Br. ■☆
102272 Conophytum igniflorum?;圣园■☆
102273 Conophytum impressum Tischer = Conophytum obcordellum (Haw.) N. E. Br. ■☆
102274 Conophytum inclusum L. Bolus = Conophytum bilobum (Marloth) N. E. Br. ■☆
102275 Conophytum incurvum N. E. Br.;内折肉锥花■☆
102276 Conophytum incurvum N. E. Br. = Conophytum bilobum (Marloth) N. E. Br. ■☆
102277 Conophytum incurvum N. E. Br. var. leucanthum (Lavis) Tischer = Conophytum bilobum (Marloth) N. E. Br. ■☆
102278 Conophytum indefinitum L. Bolus = Conophytum bilobum (Marloth) N. E. Br. ■☆
102279 Conophytum indutum L. Bolus = Conophytum obscurum N. E. Br. ■☆
102280 Conophytum inornatum N. E. Br.;无饰肉锥花■☆
102281 Conophytum insigne L. Bolus = Conophytum bilobum (Marloth) N. E. Br. ■☆
102282 Conophytum intermedium L. Bolus = Conophytum loeschianum Tischer ■☆
102283 Conophytum intrepidum L. Bolus = Conophytum bolusiae Schwantes ■☆
102284 Conophytum jacobsenianum Tischer = Conophytum jucundum (N. E. Br.) N. E. Br. ■☆
102285 Conophytum johannis-winkleri (Dinter et Schwantes) N. E. Br. = Conophytum pageae (N. E. Br.) N. E. Br. ■☆
102286 Conophytum joubertii Lavis;朱伯特肉锥花■☆
102287 Conophytum jucundum (N. E. Br.) N. E. Br.;惬意肉锥花■☆
102288 Conophytum jucundum (N. E. Br.) N. E. Br. subsp. fragile (Tischer) S. A. Hammer;脆惬意肉锥花■☆
102289 Conophytum jucundum (N. E. Br.) N. E. Br. subsp. marlothii (N. E. Br.) S. A. Hammer;马洛斯肉锥花■☆
102290 Conophytum jucundum (N. E. Br.) N. E. Br. subsp. ruschii (Schwantes) S. A. Hammer;鲁施肉锥花■☆
102291 Conophytum julii Schwantes;明镜玉■☆
102292 Conophytum julii Schwantes ex Jacobsen = Conophytum uviforme (Haw.) N. E. Br. ■☆
102293 Conophytum khamiesbergense (L. Bolus) Schwantes;卡米肉锥花■☆
102294 Conophytum klaverense N. E. Br. = Conophytum obcordellum (Haw.) N. E. Br. ■☆
102295 Conophytum klinghardtense Rawé;克林肉锥花■☆
102296 Conophytum klinghardtense Rawé subsp. baradii (Rawé) S. A. Hammer;巴拉德肉锥花■☆
102297 Conophytum klipbokbergense L. Bolus = Conophytum bilobum (Marloth) N. E. Br. ■☆
102298 Conophytum komkansicum L. Bolus = Conophytum calculus (A. Berger) N. E. Br. ■☆
102299 Conophytum koupense Tischer = Conophytum truncatum (Thunb.) N. E. Br. subsp. viridicatum (N. E. Br.) S. A. Hammer ■☆
102300 Conophytum kubusbergense Tischer = Conophytum bilobum (Marloth) N. E. Br. ■☆
102301 Conophytum labiatum Tischer = Conophytum pageae (N. E. Br.) N. E. Br. ■☆
102302 Conophytum labyrintheum (N. E. Br.) N. E. Br. = Conophytum

minimum (Haw.) N. E. Br. ■☆
102303 Conophytum labyrintheum N. E. Br.;延历肉锥花■☆
102304 Conophytum lacteum L. Bolus = Conophytum bilobum (Marloth) N. E. Br. ■☆
102305 Conophytum laetum L. Bolus = Conophytum meyeri N. E. Br. ■☆
102306 Conophytum laetum L. Bolus var. extractum (Tischer) Rawé = Conophytum meyeri N. E. Br. ■☆
102307 Conophytum lambertense Schick et Tischer = Conophytum obcordellum (Haw.) N. E. Br. ■☆
102308 Conophytum lambertense Schick et Tischer var. conspicuum Rawé = Conophytum obcordellum (Haw.) N. E. Br. ■☆
102309 Conophytum lambertense Schick et Tischer var. rolfii (de Boer) Rawé = Conophytum obcordellum (Haw.) N. E. Br. subsp. rolfii (de Boer) S. A. Hammer ■☆
102310 Conophytum largum L. Bolus = Conophytum bilobum (Marloth) N. E. Br. ■☆
102311 Conophytum latum L. Bolus = Conophytum bilobum (Marloth) N. E. Br. ■☆
102312 Conophytum lavisianum L. Bolus = Conophytum bilobum (Marloth) N. E. Br. ■☆
102313 Conophytum lavranosii Rawé;拉夫拉诺斯肉锥花■☆
102314 Conophytum lavranosii Rawé var. cuneatum Rawé = Conophytum taylorianum (Dinter et Schwantes) N. E. Br. ■☆
102315 Conophytum lavranosii Rawé var. lavranosii Rawé = Conophytum taylorianum (Dinter et Schwantes) N. E. Br. ■☆
102316 Conophytum laxipetalum N. E. Br. = Conophytum bilobum (Marloth) N. E. Br. ■☆
102317 Conophytum leightoniae L. Bolus = Conophytum piluliforme (N. E. Br.) N. E. Br. ■☆
102318 Conophytum leipoldtii N. E. Br. = Conophytum minusculum (N. E. Br.) N. E. Br. subsp. leipoldtii (N. E. Br.) S. A. Hammer ■☆
102319 Conophytum lekkersingense L. Bolus = Conophytum bilobum (Marloth) N. E. Br. ■☆
102320 Conophytum leopardinum L. Bolus = Conophytum meyeri N. E. Br. ■☆
102321 Conophytum leptanthum L. Bolus = Conophytum jucundum (N. E. Br.) N. E. Br. subsp. marlothii (N. E. Br.) S. A. Hammer ■☆
102322 Conophytum leucanthum Lavis = Conophytum bilobum (Marloth) N. E. Br. ■☆
102323 Conophytum leucanthum Lavis var. multipetalum L. Bolus = Conophytum bilobum (Marloth) N. E. Br. ■☆
102324 Conophytum leviculum (N. E. Br.) N. E. Br.;椿姬■☆
102325 Conophytum leviculum (N. E. Br.) N. E. Br. = Conophytum minimum (Haw.) N. E. Br. ■☆
102326 Conophytum leviculum N. E. Br. = Conophytum leviculum (N. E. Br.) N. E. Br. ■☆
102327 Conophytum lilianum Littlew. = Conophytum pellucidum Schwantes var. lilianum (Littlew.) S. A. Hammer ■☆
102328 Conophytum limbatum N. E. Br. = Conophytum ectypum N. E. Br. ■☆
102329 Conophytum lindenianum Lavis et S. A. Hammer = Conophytum tantillum N. E. Br. subsp. lindenianum (Lavis et S. A. Hammer) S. A. Hammer ■☆
102330 Conophytum linearilucidum L. Bolus = Conophytum bilobum (Marloth) N. E. Br. var. linearilucidum (L. Bolus) S. A. Hammer ■☆
102331 Conophytum literatum N. E. Br. = Conophytum minimum (Haw.) N. E. Br. ■☆
102332 Conophytum lithopsoides L. Bolus;生石花肉锥花■☆
102333 Conophytum lithopsoides L. Bolus subsp. boreale (L. Bolus) S. A. Hammer;北方肉锥花■☆
102334 Conophytum litorale L. Bolus = Conophytum uviforme (Haw.) N. E. Br. ■☆
102335 Conophytum littlewoodii L. Bolus = Conophytum marginatum Lavis subsp. littlewoodii (L. Bolus) S. A. Hammer ■☆
102336 Conophytum loeschianum Tischer;勒施肉锥花■☆
102337 Conophytum longibracteatum L. Bolus;长苞肉锥花■☆
102338 Conophytum longifissum Tischer = Conophytum obcordellum (Haw.) N. E. Br. ■☆
102339 Conophytum longipetalum L. Bolus = Conophytum pageae (N. E. Br.) N. E. Br. ■☆
102340 Conophytum longistylum N. E. Br. = Conophytum jucundum (N. E. Br.) N. E. Br. ■☆
102341 Conophytum longitubum L. Bolus = Conophytum truncatum (Thunb.) N. E. Br. subsp. viridicatum (N. E. Br.) S. A. Hammer ■☆
102342 Conophytum longum N. E. Br.;长肉锥花■☆
102343 Conophytum lucipunctum N. E. Br. = Conophytum subfenestratum Schwantes ■☆
102344 Conophytum luckhoffii Lavis;吕克霍夫肉锥花■☆
102345 Conophytum luckhoffii Lavis var. angustipetalum L. Bolus = Conophytum luckhoffii Lavis ■☆
102346 Conophytum luisae Schwantes;珠贝玉■☆
102347 Conophytum luiseae Schwantes = Conophytum bilobum (Marloth) N. E. Br. subsp. altum (L. Bolus) S. A. Hammer ■☆
102348 Conophytum luiseae Schwantes var. papillatum L. Bolus = Conophytum bilobum (Marloth) N. E. Br. subsp. altum (L. Bolus) S. A. Hammer ■☆
102349 Conophytum luteolum L. Bolus;淡黄肉锥花■☆
102350 Conophytum luteolum L. Bolus = Conophytum flavum N. E. Br. subsp. novicium (N. E. Br.) S. A. Hammer ■☆
102351 Conophytum luteolum L. Bolus var. macrostigma? = Conophytum flavum N. E. Br. subsp. novicium (N. E. Br.) S. A. Hammer ■☆
102352 Conophytum luteopurpureum N. E. Br. = Conophytum flavum N. E. Br. ■☆
102353 Conophytum luteum N. E. Br. = Conophytum flavum N. E. Br. ■☆
102354 Conophytum lydiae (H. Jacobsen) G. D. Rowley;利迪亚肉锥花■☆
102355 Conophytum macrostigma (L. Bolus) Schwantes = Conophytum bilobum (Marloth) N. E. Br. ■☆
102356 Conophytum marginatum Lavis;具边肉锥花■☆
102357 Conophytum marginatum Lavis subsp. haramoepense (L. Bolus) S. A. Hammer;哈拉肉锥花■☆
102358 Conophytum marginatum Lavis subsp. littlewoodii (L. Bolus) S. A. Hammer;利特尔伍德肉锥花■☆
102359 Conophytum marginatum Lavis var. haramoepense (L. Bolus) Rawé = Conophytum marginatum Lavis subsp. haramoepense (L. Bolus) S. A. Hammer ■☆
102360 Conophytum marginatum Lavis var. littlewoodii (L. Bolus) Rawé = Conophytum marginatum Lavis subsp. littlewoodii (L. Bolus) S. A. Hammer ■☆
102361 Conophytum markoetterae Schwantes = Conophytum bilobum (Marloth) N. E. Br. ■☆
102362 Conophytum marlothii N. E. Br. = Conophytum jucundum (N. E. Br.) N. E. Br. subsp. marlothii (N. E. Br.) S. A. Hammer ■☆

102363 Conophytum maughanii N. E. Br. ;莫恩肉锥花■☆

102364 Conophytum maughanii N. E. Br. subsp. armeniacum S. A. Hammer;亚美尼亚肉锥花■☆

102365 Conophytum maughanii N. E. Br. subsp. latum (Tischer) S. A. Hammer;侧肉锥花■☆

102366 Conophytum maximum Tischer = Conophytum jucundum (N. E. Br.) N. E. Br. ■☆

102367 Conophytum meleagris L. Bolus = Conophytum uviforme (Haw.) N. E. Br. ■☆

102368 Conophytum membranaceum L. Bolus = Conophytum breve N. E. Br. ■☆

102369 Conophytum meridianum L. Bolus;南方肉锥花■☆

102370 Conophytum meridianum L. Bolus var. meridianum? = Conophytum pellucidum Schwantes var. cupreatum (Tischer) S. A. Hammer ■☆

102371 Conophytum meridianum L. Bolus var. pulverulentum? = Conophytum pellucidum Schwantes var. cupreatum (Tischer) S. A. Hammer ■☆

102372 Conophytum meyerae Schwantes = Conophytum bilobum (Marloth) N. E. Br. ■☆

102373 Conophytum meyerae Schwantes f. alatum Tischer = Conophytum bilobum (Marloth) N. E. Br. ■☆

102374 Conophytum meyerae Schwantes f. apiculatum (N. E. Br.) Tischer = Conophytum bilobum (Marloth) N. E. Br. ■☆

102375 Conophytum meyerae Schwantes f. asperulum (L. Bolus) H. Jacobsen = Conophytum bilobum (Marloth) N. E. Br. ■☆

102376 Conophytum meyerae Schwantes f. pole-evansii (N. E. Br.) Tischer = Conophytum bilobum (Marloth) N. E. Br. ■☆

102377 Conophytum meyeri N. E. Br. ;神铃■☆

102378 Conophytum meyeri N. E. Br. f. semilunulum (Tischer) Rawé = Conophytum meyeri N. E. Br. ■☆

102379 Conophytum meyeri N. E. Br. var. globuliforme (Schick et Tischer) Rawé = Conophytum meyeri N. E. Br. ■☆

102380 Conophytum meyeri N. E. Br. var. quinarium L. Bolus = Conophytum meyeri N. E. Br. ■☆

102381 Conophytum meyeri N. E. Br. var. ramosum (Lavis) Rawé = Conophytum meyeri N. E. Br. ■☆

102382 Conophytum microstoma L. Bolus = Conophytum meyeri N. E. Br. ■☆

102383 Conophytum middlemostii L. Bolus = Conophytum jucundum (N. E. Br.) N. E. Br. subsp. fragile (Tischer) S. A. Hammer ■☆

102384 Conophytum minimum (Haw.) N. E. Br. ;小肉锥花■☆

102385 Conophytum minusculum (N. E. Br.) N. E. Br. ;翠卵(纳言)■☆

102386 Conophytum minusculum (N. E. Br.) N. E. Br. f. reticulatum (L. Bolus) Raweex G. D. Rowley = Conophytum minusculum (N. E. Br.) N. E. Br. ■☆

102387 Conophytum minusculum (N. E. Br.) N. E. Br. f. roseum (G. D. Rowley) G. D. Rowley = Conophytum minusculum (N. E. Br.) N. E. Br. ■☆

102388 Conophytum minusculum (N. E. Br.) N. E. Br. subsp. leipoldtii (N. E. Br.) S. A. Hammer;莱波尔德肉锥花■☆

102389 Conophytum minusculum (N. E. Br.) N. E. Br. var. paucilineatum Rawé = Conophytum minusculum (N. E. Br.) N. E. Br. ■☆

102390 Conophytum minusculum (N. E. Br.) N. E. Br. var. roseum (G. D. Rowley) Tischer = Conophytum minusculum (N. E. Br.) N. E. Br. ■☆

102391 Conophytum minutiflorum (Schwantes) N. E. Br. = Conophytum pageae (N. E. Br.) N. E. Br. ■☆

102392 Conophytum minutum (Haw.) N. E. Br. ;微小肉锥花(清姬)■☆

102393 Conophytum minutum (Haw.) N. E. Br. f. sellatum (Tischer) Rawé = Conophytum minutum (Haw.) N. E. Br. ■☆

102394 Conophytum minutum (Haw.) N. E. Br. var. laxum Lavis = Conophytum minutum (Haw.) N. E. Br. ■☆

102395 Conophytum minutum (Haw.) N. E. Br. var. nudum (Tischer) Boom;裸碧玉(群碧玉)■☆

102396 Conophytum minutum (Haw.) N. E. Br. var. pearsonii (N. E. Br.) Boom;皮尔逊肉锥花■☆

102397 Conophytum minutum (Haw.) N. E. Br. var. sellatum (Tischer) Boom = Conophytum minutum (Haw.) N. E. Br. ■☆

102398 Conophytum minutum N. E. Br. ;群碧玉(碧玉)■☆

102399 Conophytum mirabile A. R. Mitch. et S. A. Hammer;奇异肉锥花■☆

102400 Conophytum misellum N. E. Br. = Conophytum saxetanum (N. E. Br.) N. E. Br. ■☆

102401 Conophytum miserum N. E. Br. = Conophytum hians N. E. Br. ■☆

102402 Conophytum modestum L. Bolus = Conophytum quaesitum (N. E. Br.) N. E. Br. ■☆

102403 Conophytum morganii Lavis = Conophytum truncatum (Thunb.) N. E. Br. ■☆

102404 Conophytum muiri N. E. Br. ;七小町■☆

102405 Conophytum muirii N. E. Br. = Conophytum truncatum (Thunb.) N. E. Br. subsp. viridicatum (N. E. Br.) S. A. Hammer ■☆

102406 Conophytum multicolor Tischer = Conophytum obcordellum (Haw.) N. E. Br. ■☆

102407 Conophytum multipunctatum Tischer = Conophytum truncatum (Thunb.) N. E. Br. ■☆

102408 Conophytum mundum N. E. Br. ;阿娇■☆

102409 Conophytum mundum N. E. Br. = Conophytum obcordellum (Haw.) N. E. Br. ■☆

102410 Conophytum muscosipapillatum Lavis;舞子■☆

102411 Conophytum namibense N. E. Br. = Conophytum saxetanum (N. E. Br.) N. E. Br. ■☆

102412 Conophytum namiesicum L. Bolus = Conophytum calculus (A. Berger) N. E. Br. subsp. vanzylii (Lavis) S. A. Hammer ■☆

102413 Conophytum nanum Tischer = Conophytum meyeri N. E. Br. ■☆

102414 Conophytum nelianum Schwantes = Conophytum bilobum (Marloth) N. E. Br. ■☆

102415 Conophytum nevillei (N. E. Br.) N. E. Br. = Conophytum obcordellum (Haw.) N. E. Br. ■☆

102416 Conophytum nevillei N. E. Br. ;内侍■☆

102417 Conophytum niveum L. Bolus = Conophytum meyeri N. E. Br. ■☆

102418 Conophytum noctiflorum (L. Bolus) G. D. Rowley = Conophytum maughanii N. E. Br. subsp. latum (Tischer) S. A. Hammer ■☆

102419 Conophytum noisabisense L. Bolus = Conophytum bilobum (Marloth) N. E. Br. ■☆

102420 Conophytum nordenstamii L. Bolus = Conophytum jucundum (N. E. Br.) N. E. Br. subsp. fragile (Tischer) S. A. Hammer ■☆

102421 Conophytum notabile N. E. Br. ;显突肉锥花■☆

102422 Conophytum notabile N. E. Br. = Conophytum frutescens Schwantes ■☆

102423 Conophytum notatum N. E. Br. ;初音■☆

102424 Conophytum notatum N. E. Br. = Conophytum minimum (Haw.) N. E. Br. ■☆

102425 Conophytum novellum N. E. Br. = Conophytum truncatum (Thunb.) N. E. Br. subsp. viridicatum (N. E. Br.) S. A. Hammer ■☆

102426 Conophytum novicium N. E. Br. = Conophytum flavum N. E. Br. subsp. novicium (N. E. Br.) S. A. Hammer ■☆

102427 Conophytum nudum Tischer;赤映玉■☆

102428 Conophytum nudum Tischer = Conophytum minutum (Haw.) N. E. Br. var. nudum (Tischer) Boom ■☆

102429 Conophytum nutaboiense Tischer = Conophytum bilobum (Marloth) N. E. Br. ■☆

102430 Conophytum obconellum (Haw.) Schwantes = Conophytum obcordellum (Haw.) N. E. Br. ■☆

102431 Conophytum obcordellum (Haw.) N. E. Br.;倒心形肉锥花■☆

102432 Conophytum obcordellum (Haw.) N. E. Br. f. declinatum (L. Bolus) Tischer = Conophytum obcordellum (Haw.) N. E. Br. ■☆

102433 Conophytum obcordellum (Haw.) N. E. Br. f. multicolor (Tischer) Tischer = Conophytum obcordellum (Haw.) N. E. Br. ■☆

102434 Conophytum obcordellum (Haw.) N. E. Br. f. mundum (N. E. Br.) Rawé = Conophytum obcordellum (Haw.) N. E. Br. ■☆

102435 Conophytum obcordellum (Haw.) N. E. Br. f. picturatum (N. E. Br.) Rawé = Conophytum obcordellum (Haw.) N. E. Br. ■☆

102436 Conophytum obcordellum (Haw.) N. E. Br. f. picturatum (N. E. Br.) Tischer = Conophytum obcordellum (Haw.) N. E. Br. ■☆

102437 Conophytum obcordellum (Haw.) N. E. Br. f. stayneri (L. Bolus) Rawé = Conophytum obcordellum (Haw.) N. E. Br. ■☆

102438 Conophytum obcordellum (Haw.) N. E. Br. f. ursprungianum (Tischer) Rawé = Conophytum obcordellum (Haw.) N. E. Br. ■☆

102439 Conophytum obcordellum (Haw.) N. E. Br. subsp. rolfii (de Boer) S. A. Hammer;罗尔夫肉锥花■☆

102440 Conophytum obcordellum (Haw.) N. E. Br. subsp. stenandrum (L. Bolus) S. A. Hammer;狭蕊肉锥花■☆

102441 Conophytum obcordellum (Haw.) N. E. Br. var. ceresianum (L. Bolus) S. A. Hammer;塞里斯肉锥花■☆

102442 Conophytum obcordellum (Haw.) N. E. Br. var. germanum (N. E. Br.) Rawé = Conophytum obcordellum (Haw.) N. E. Br. ■☆

102443 Conophytum obcordellum (Haw.) N. E. Br. var. parvipetalum (N. E. Br.) Tischer = Conophytum obcordellum (Haw.) N. E. Br. ■☆

102444 Conophytum obcordellum N. E. Br.;白眉玉(玉彦)■☆

102445 Conophytum obmetale (N. E. Br.) N. E. Br. = Conophytum minimum (Haw.) N. E. Br. ■☆

102446 Conophytum obovatum Lavis;倒卵肉锥花■☆

102447 Conophytum obovatum Lavis = Conophytum globosum (N. E. Br.) N. E. Br. ■☆

102448 Conophytum obovatum Lavis var. obtusum L. Bolus = Conophytum globosum (N. E. Br.) N. E. Br. ■☆

102449 Conophytum obscurum N. E. Br.;滴翠玉■☆

102450 Conophytum obscurum N. E. Br. subsp. barbatum (L. Bolus) S. A. Hammer;髯毛肉锥花■☆

102451 Conophytum obscurum N. E. Br. subsp. vitreopapillum (Rawé) S. A. Hammer;透明滴翠玉■☆

102452 Conophytum obscurum N. E. Br. var. alticola L. Bolus = Conophytum bolusiae Schwantes ■☆

102453 Conophytum obscurum N. E. Br. var. puberulum L. Bolus = Conophytum bolusiae Schwantes ■☆

102454 Conophytum obtusum N. E. Br.;钝肉锥花■☆

102455 Conophytum obtusum N. E. Br. = Conophytum bilobum (Marloth) N. E. Br. ■☆

102456 Conophytum obtusum N. E. Br. var. amplum (L. Bolus) Rawé = Conophytum bilobum (Marloth) N. E. Br. ■☆

102457 Conophytum occultum L. Bol.;王宫殿■☆

102458 Conophytum occultum L. Bolus = Conophytum uviforme (Haw.) N. E. Br. subsp. decoratum (N. E. Br.) S. A. Hammer ■☆

102459 Conophytum odoratum (N. E. Br.) N. E. Br. = Conophytum ficiforme (Haw.) N. E. Br. ■☆

102460 Conophytum odoratum N. E. Br.;青春玉■☆

102461 Conophytum orbicum N. E. Br. ex Tischer = Conophytum jucundum (N. E. Br.) N. E. Br. ■☆

102462 Conophytum orientale L. Bolus = Conophytum truncatum (Thunb.) N. E. Br. ■☆

102463 Conophytum oripictum N. E. Br. = Conophytum pageae (N. E. Br.) N. E. Br. ■☆

102464 Conophytum ornatum Lavis;上腊■☆

102465 Conophytum ornatum Lavis = Conophytum flavum N. E. Br. ■☆

102466 Conophytum ovatum L. Bolus = Conophytum bilobum (Marloth) N. E. Br. ■☆

102467 Conophytum ovigerum Schwantes;倾国■☆

102468 Conophytum ovigerum Schwantes = Conophytum meyeri N. E. Br. ■☆

102469 Conophytum pageae (N. E. Br.) N. E. Br.;微花肉锥花■☆

102470 Conophytum pageae (N. E. Br.) N. E. Br. var. albiflorum Rawé = Conophytum pageae (N. E. Br.) N. E. Br. ■☆

102471 Conophytum pageae (N. E. Br.) N. E. Br. var. pygmaeum (Schick et Tischer) Rawé = Conophytum breve N. E. Br. ■☆

102472 Conophytum pallidum (N. E. Br.) N. E. Br. = Conophytum ficiforme (Haw.) N. E. Br. ■☆

102473 Conophytum papillatum L. Bolus = Conophytum meyeri N. E. Br. ■☆

102474 Conophytum parcum N. E. Br. = Conophytum minimum (Haw.) N. E. Br. ■☆

102475 Conophytum pardicolor Tischer = Conophytum pellucidum Schwantes ■☆

102476 Conophytum pardivisum Tischer = Conophytum uviforme (Haw.) N. E. Br. ■☆

102477 Conophytum parviflorum N. E. Br. = Conophytum obcordellum (Haw.) N. E. Br. ■☆

102478 Conophytum parviflorum N. E. Br. var. impressum (Tischer) Tischer = Conophytum obcordellum (Haw.) N. E. Br. ■☆

102479 Conophytum parvimarinum L. Bolus = Conophytum hians N. E. Br. ■☆

102480 Conophytum parvipetalum (N. E. Br.) N. E. Br. = Conophytum obcordellum (Haw.) N. E. Br. ■☆

102481 Conophytum parvipunctum Tischer = Conophytum truncatum (Thunb.) N. E. Br. ■☆

102482 Conophytum parvulum L. Bolus = Conophytum bilobum (Marloth) N. E. Br. ■☆

102483 Conophytum paucipunctum Tischer = Conophytum breve N. E. Br. ■☆

102484 Conophytum pauperae L. Bolus = Conophytum pageae (N. E. Br.) N. E. Br. ■☆

102485 Conophytum pauxillum (N. E. Br.) N. E. Br. = Conophytum minimum (Haw.) N. E. Br. ■☆

102486 Conophytum pauxillum N. E. Br.;大纳言(细玉)■☆

102487 Conophytum pearsonii N. E. Br. ;银星肉锥花(凤雏玉)■☆

102488 Conophytum pearsonii N. E. Br. = Conophytum minutum (Haw.) N. E. Br. var. pearsonii (N. E. Br.) Boom ■☆

102489 Conophytum pearsonii N. E. Br. var. latisectum L. Bolus = Conophytum minutum (Haw.) N. E. Br. var. pearsonii (N. E. Br.) Boom ■☆

102490 Conophytum pearsonii N. E. Br. var. minor? = Conophytum minutum (Haw.) N. E. Br. var. pearsonii (N. E. Br.) Boom ■☆

102491 Conophytum peersii Lavis;水晶玉■☆

102492 Conophytum peersii Lavis = Conophytum truncatum (Thunb.) N. E. Br. ■☆

102493 Conophytum peersii Lavis var. multipunctatum (Tischer) Rawé = Conophytum truncatum (Thunb.) N. E. Br. ■☆

102494 Conophytum pellucidum Schwantes;透明肉锥花■☆

102495 Conophytum pellucidum Schwantes var. cupreatum (Tischer) S. A. Hammer;铜色肉锥花■☆

102496 Conophytum pellucidum Schwantes var. lilianum (Littlew.) S. A. Hammer;百合肉锥花■☆

102497 Conophytum pellucidum Schwantes var. neohallii S. A. Hammer;新霍尔肉锥花■☆

102498 Conophytum pellucidum Schwantes var. terrestre (Tischer) S. A. Hammer = Conophytum pellucidum Schwantes var. terricolor (Tischer) Littlew. ex S. A. Hammer ■☆

102499 Conophytum pellucidum Schwantes var. terricolor (Tischer) Littlew. ex S. A. Hammer;宝槌(土色肉锥花)■☆

102500 Conophytum percrassum Schick et Tischer = Conophytum flavum N. E. Br. ■☆

102501 Conophytum permaculatum Tischer = Conophytum truncatum (Thunb.) N. E. Br. ■☆

102502 Conophytum perpusillum (Haw.) N. E. Br. = Conophytum minimum (Haw.) N. E. Br. ■☆

102503 Conophytum perpusillum N. E. Br. ;晓山■☆

102504 Conophytum petraeum N. E. Br. = Conophytum minimum (Haw.) N. E. Br. ■☆

102505 Conophytum philippii L. Bolus = Conophytum bilobum (Marloth) N. E. Br. ■☆

102506 Conophytum phoeniceum S. A. Hammer;紫红肉锥花■☆

102507 Conophytum pictum (N. E. Br.) N. E. Br. = Conophytum minimum (Haw.) N. E. Br. ■☆

102508 Conophytum pictum N. E. Br. ;中纳言(青光玉)■☆

102509 Conophytum picturatum N. E. Br. = Conophytum obcordellum (Haw.) N. E. Br. ■☆

102510 Conophytum pilansii Lavis;翠光玉(不死鸟,静明玉)■☆

102511 Conophytum pillansii Lavis = Conophytum subfenestratum Schwantes ■☆

102512 Conophytum piluliforme (N. E. Br.) N. E. Br. ;都鸟■☆

102513 Conophytum piluliforme (N. E. Br.) N. E. Br. subsp. edwardii (Schwantes) S. A. Hammer;爱德华肉锥花■☆

102514 Conophytum piluliforme (N. E. Br.) N. E. Br. var. advenum (N. E. Br.) Rawé = Conophytum piluliforme (N. E. Br.) N. E. Br. ■☆

102515 Conophytum piluliforme (N. E. Br.) N. E. Br. var. brevipetalum (Lavis) Rawé = Conophytum piluliforme (N. E. Br.) N. E. Br. ■☆

102516 Conophytum piriforme L. Bolus = Conophytum bilobum (Marloth) N. E. Br. ■☆

102517 Conophytum pisinnum (N. E. Br.) N. E. Br. = Conophytum truncatum (Thunb.) N. E. Br. subsp. viridicatum (N. E. Br.) S. A. Hammer ■☆

102518 Conophytum pisinnum N. E. Br. ;云母绘■☆

102519 Conophytum placitum (N. E. Br.) N. E. Br. = Conophytum ficiforme (Haw.) N. E. Br. ■☆

102520 Conophytum placitum (N. E. Br.) N. E. Br. var. pubescens Littlew. = Conophytum ficiforme (Haw.) N. E. Br. ■☆

102521 Conophytum plenum N. E. Br. = Conophytum bilobum (Marloth) N. E. Br. ■☆

102522 Conophytum pluriforme L. Bolus = Conophytum bilobum (Marloth) N. E. Br. ■☆

102523 Conophytum poellnitzianum Schwantes = Conophytum pageae (N. E. Br.) N. E. Br. ■☆

102524 Conophytum pole-evansii N. E. Br. ;明珍■☆

102525 Conophytum pole-evansii N. E. Br. = Conophytum bilobum (Marloth) N. E. Br. ■☆

102526 Conophytum polulum N. E. Br. = Conophytum minimum (Haw.) N. E. Br. ■☆

102527 Conophytum polyandrum Lavis = Conophytum velutinum Schwantes subsp. polyandrum (Lavis) S. A. Hammer ■☆

102528 Conophytum praecinctum N. E. Br. = Conophytum minimum (Haw.) N. E. Br. ■☆

102529 Conophytum praecox N. E. Br. ;浜千鸟■☆

102530 Conophytum praecox N. E. Br. = Conophytum fraternum (N. E. Br.) N. E. Br. ■☆

102531 Conophytum praegratum Tischer = Conophytum jucundum (N. E. Br.) N. E. Br. ■☆

102532 Conophytum praeparvum N. E. Br. = Conophytum uviforme (Haw.) N. E. Br. ■☆

102533 Conophytum praeparvum N. E. Br. var. roseum Lavis = Conophytum uviforme (Haw.) N. E. Br. ■☆

102534 Conophytum primosii Lavis = Conophytum roodiae N. E. Br. subsp. cylindratum (Schwantes) Smale ■☆

102535 Conophytum productum L. Bolus = Conophytum pageae (N. E. Br.) N. E. Br. ■☆

102536 Conophytum prolongatum L. Bolus = Conophytum uviforme (Haw.) N. E. Br. ■☆

102537 Conophytum prospersum N. E. Br. = Conophytum flavum N. E. Br. ■☆

102538 Conophytum proximum L. Bolus = Conophytum bilobum (Marloth) N. E. Br. ■☆

102539 Conophytum puberulum Lavis = Conophytum meyeri N. E. Br. ■☆

102540 Conophytum pubescens (Tischer) G. D. Rowley;短柔毛肉锥花■☆

102541 Conophytum pubicalyx Lavis;毛萼肉锥花■☆

102542 Conophytum pulchellum Tischer;群萤■☆

102543 Conophytum pulchellum Tischer = Conophytum obscurum N. E. Br. ■☆

102544 Conophytum pumilum N. E. Br. = Conophytum breve N. E. Br. ■☆

102545 Conophytum purpusii (Schwantes) N. E. Br. = Conophytum truncatum (Thunb.) N. E. Br. ■☆

102546 Conophytum purpusii N. E. Br. ;若鲇玉■☆

102547 Conophytum pusillum (N. E. Br.) N. E. Br. = Conophytum minimum (Haw.) N. E. Br. ■☆

102548 Conophytum pusillum N. E. Br. ;璎珞■☆

102549 Conophytum pygmaeum Schick et Tischer = Conophytum breve N. E. Br. ■☆

102550 Conophytum quaesitum (N. E. Br.) N. E. Br. ;蝶羽玉■☆

102551 Conophytum quaesitum (N. E. Br.) N. E. Br. var. rostratum (Tischer) S. A. Hammer;喙蝶羽玉■☆
102552 Conophytum quartziticum Tischer = Conophytum quaesitum (N. E. Br.) N. E. Br. ■☆
102553 Conophytum radiatum Tischer = Conophytum minimum (Haw.) N. E. Br. ■☆
102554 Conophytum ramosum Lavis;白鸠(雏鸠)■☆
102555 Conophytum ramosum Lavis = Conophytum meyeri N. E. Br. ■☆
102556 Conophytum rarum N. E. Br. = Conophytum jucundum (N. E. Br.) N. E. Br. ■☆
102557 Conophytum rauhii Tischer = Conophytum uviforme (Haw.) N. E. Br. subsp. rauhii (Tischer) S. A. Hammer ■☆
102558 Conophytum rawei G. D. Rowley = Conophytum longum N. E. Br. ■☆
102559 Conophytum recisum N. E. Br. = Conophytum bilobum (Marloth) N. E. Br. ■☆
102560 Conophytum reconditum A. R. Mitch.;隐蔽肉锥花■☆
102561 Conophytum reconditum A. R. Mitch. subsp. buysianum (A. R. Mitch. et S. A. Hammer) S. A. Hammer;布伊隐蔽肉锥花■☆
102562 Conophytum renniei Lavis = Conophytum truncatum (Thunb.) N. E. Br. ■☆
102563 Conophytum renominatum G. D. Rowley = Conophytum friedrichiae (Dinter) Schwantes ■☆
102564 Conophytum reticulatum L. Bolus = Conophytum minusculum (N. E. Br.) N. E. Br. ■☆
102565 Conophytum reticulatum L. Bolus f. roseum G. D. Rowley = Conophytum minusculum (N. E. Br.) N. E. Br. ■☆
102566 Conophytum retusum N. E. Br. = Conophytum meyeri N. E. Br. ■☆
102567 Conophytum ricardianum Loesch et Tischer;理查德肉锥花■☆
102568 Conophytum ricardianum Loesch et Tischer subsp. rubriflorum Tischer;红花理查德肉锥花■☆
102569 Conophytum robustum Tischer = Conophytum jucundum (N. E. Br.) N. E. Br. ■☆
102570 Conophytum rolfii de Boer = Conophytum obcordellum (Haw.) N. E. Br. subsp. rolfii (de Boer) S. A. Hammer ■☆
102571 Conophytum roodiae N. E. Br.;鲁迪亚肉锥花■☆
102572 Conophytum roodiae N. E. Br. subsp. corrugatum Smale;皱褶肉锥花■☆
102573 Conophytum roodiae N. E. Br. subsp. cylindratum (Schwantes) Smale;柱形肉锥花■☆
102574 Conophytum roodiae N. E. Br. subsp. sanguineum (S. A. Hammer) Smale;血红鲁迪亚肉锥花■☆
102575 Conophytum rooipanense L. Bolus = Conophytum uviforme (Haw.) N. E. Br. ■☆
102576 Conophytum roseolineatum Tischer = Conophytum minimum (Haw.) N. E. Br. ■☆
102577 Conophytum rostratum Tischer = Conophytum quaesitum (N. E. Br.) N. E. Br. var. rostratum (Tischer) S. A. Hammer ■☆
102578 Conophytum rubricarinatum Tischer = Conophytum loeschianum Tischer ■☆
102579 Conophytum rubristylosum Tischer = Conophytum flavum N. E. Br. subsp. novicium (N. E. Br.) S. A. Hammer ■☆
102580 Conophytum rubrolineatum Rawé = Conophytum swanepoelianum Rawé subsp. rubrolineatum (Rawé) S. A. Hammer ■☆
102581 Conophytum rubroniveum L. Bolus = Conophytum roodiae N. E. Br. ■☆
102582 Conophytum rubrum L. Bolus = Conophytum piluliforme (N. E. Br.) N. E. Br. subsp. edwardii (Schwantes) S. A. Hammer ■☆
102583 Conophytum rufescens N. E. Br. = Conophytum maughanii N. E. Br. ■☆
102584 Conophytum rugosum S. A. Hammer;皱缩肉锥花■☆
102585 Conophytum rugosum S. A. Hammer subsp. sanguineum? = Conophytum roodiae N. E. Br. subsp. sanguineum (S. A. Hammer) Smale ■☆
102586 Conophytum ruschii Schwantes = Conophytum jucundum (N. E. Br.) N. E. Br. subsp. ruschii (Schwantes) S. A. Hammer ■☆
102587 Conophytum ruschii Schwantes var. obtusipetalum L. Bolus = Conophytum jucundum (N. E. Br.) N. E. Br. subsp. ruschii (Schwantes) S. A. Hammer ■☆
102588 Conophytum salmonicolor L. Bolus = Conophytum frutescens Schwantes ■☆
102589 Conophytum saxetanum (N. E. Br.) N. E. Br.;大纳马兰肉锥花■☆
102590 Conophytum saxetanum (N. E. Br.) N. E. Br. f. hallianum G. D. Rowley = Conophytum saxetanum (N. E. Br.) N. E. Br. ■☆
102591 Conophytum saxetanum (N. E. Br.) N. E. Br. var. loeschianum (Tischer) Rawé = Conophytum loeschianum Tischer ■☆
102592 Conophytum saxetanum (N. E. Br.) N. E. Br. var. misellum (N. E. Br.) Rawé = Conophytum saxetanum (N. E. Br.) N. E. Br. ■☆
102593 Conophytum schickianum Tischer = Conophytum pageae (N. E. Br.) N. E. Br. ■☆
102594 Conophytum schlechteri Schwantes;施莱肉锥花■☆
102595 Conophytum schwantesii G. D. Rowley = Conophytum friedrichiae (Dinter) Schwantes ■☆
102596 Conophytum scitulum (N. E. Br.) N. E. Br. = Conophytum minimum (Haw.) N. E. Br. ■☆
102597 Conophytum scitulum N. E. Br. = Conophytum minimum (Haw.) N. E. Br. ■☆
102598 Conophytum sellatum Tischer = Conophytum minutum (Haw.) N. E. Br. ■☆
102599 Conophytum semilunulum Tischer = Conophytum meyeri N. E. Br. ■☆
102600 Conophytum semivestitum L. Bolus;半被肉锥花■☆
102601 Conophytum senarium L. Bolus = Conophytum marginatum Lavis subsp. haramoepense (L. Bolus) S. A. Hammer ■☆
102602 Conophytum signatum (N. E. Br.) N. E. Br. = Conophytum minimum (Haw.) N. E. Br. ■☆
102603 Conophytum simile N. E. Br. = Conophytum bilobum (Marloth) N. E. Br. ■☆
102604 Conophytum simplum N. E. Br. = Conophytum bilobum (Marloth) N. E. Br. ■☆
102605 Conophytum singulare G. D. Rowley = Conophytum caroli Lavis ■☆
102606 Conophytum sitzlerianum Schwantes = Conophytum bilobum (Marloth) N. E. Br. ■☆
102607 Conophytum smithersii L. Bolus = Conophytum bilobum (Marloth) N. E. Br. ■☆
102608 Conophytum sororium N. E. Br. = Conophytum bilobum (Marloth) N. E. Br. ■☆
102609 Conophytum spathulatum (L. Bolus) G. D. Rowley = Conophytum lydiae (H. Jacobsen) G. D. Rowley ■☆
102610 Conophytum speciosum Tischer = Conophytum jucundum (N. E. Br.) N. E. Br. subsp. ruschii (Schwantes) S. A. Hammer ■☆

102611 Conophytum spectabile Lavis = Conophytum obcordellum (Haw.) N. E. Br. ■☆

102612 Conophytum spirale N. E. Br. = Conophytum truncatum (Thunb.) N. E. Br. ■☆

102613 Conophytum springbokense N. E. Br.;樱贝(宝贝草,春雨)■☆

102614 Conophytum springbokense N. E. Br. = Conophytum bilobum (Marloth) N. E. Br. ■☆

102615 Conophytum stegmannianum L. Bolus = Conophytum truncatum (Thunb.) N. E. Br. ■☆

102616 Conophytum stenandrum L. Bolus = Conophytum obcordellum (Haw.) N. E. Br. subsp. stenandrum (L. Bolus) S. A. Hammer ■☆

102617 Conophytum stephanii Schwantes;美冠肉锥花■☆

102618 Conophytum stephanii Schwantes subsp. abductum S. A. Hammer = Conophytum stephanii Schwantes ■☆

102619 Conophytum steytlervillense Tischer = Conophytum truncatum (Thunb.) N. E. Br. ■☆

102620 Conophytum stipitatum L. Bolus = Conophytum uviforme (Haw.) N. E. Br. ■☆

102621 Conophytum strictum L. Bolus;刚直肉锥花■☆

102622 Conophytum strictum L. Bolus = Conophytum bilobum (Marloth) N. E. Br. ■☆

102623 Conophytum strictum L. Bolus var. inaequale? = Conophytum bilobum (Marloth) N. E. Br. ■☆

102624 Conophytum stylosum (N. E. Br.) Tischer = Conophytum bilobum (Marloth) N. E. Br. ■☆

102625 Conophytum subacutum L. Bolus = Conophytum bilobum (Marloth) N. E. Br. ■☆

102626 Conophytum subconfusum Tischer = Conophytum piluliforme (N. E. Br.) N. E. Br. ■☆

102627 Conophytum subcylindricum L. Bolus = Conophytum bilobum (Marloth) N. E. Br. ■☆

102628 Conophytum subfenestratum Schwantes;翠星■☆

102629 Conophytum subglobosum Tischer = Conophytum truncatum (Thunb.) N. E. Br. ■☆

102630 Conophytum subincanum Tischer = Conophytum uviforme (Haw.) N. E. Br. subsp. subincanum (Tischer) S. A. Hammer ■☆

102631 Conophytum subrisum (N. E. Br.) N. E. Br. = Conophytum pageae (N. E. Br.) N. E. Br. ■☆

102632 Conophytum subtenue L. Bolus = Conophytum bilobum (Marloth) N. E. Br. ■☆

102633 Conophytum subterraneum Smale et Jacobs;地下肉锥花■☆

102634 Conophytum subtile N. E. Br. = Conophytum breve N. E. Br. ■☆

102635 Conophytum sulcatum L. Bolus = Conophytum ectypum N. E. Br. subsp. sulcatum (L. Bolus) S. A. Hammer ■☆

102636 Conophytum supremum L. Bolus = Conophytum bilobum (Marloth) N. E. Br. ■☆

102637 Conophytum swanepoelianum Rawé;斯旺肉锥花■☆

102638 Conophytum swanepoelianum Rawé subsp. proliferans S. A. Hammer;多育肉锥花■☆

102639 Conophytum swanepoelianum Rawé subsp. rubrolineatum (Rawé) S. A. Hammer;红线肉锥花■☆

102640 Conophytum tantillum N. E. Br. subsp. amicorum S. A. Hammer et Barnhill;可爱肉锥花■☆

102641 Conophytum tantillum N. E. Br. subsp. heleniae (Rawé) S. A. Hammer;海伦娜肉锥花■☆

102642 Conophytum tantillum N. E. Br. subsp. lindenianum (Lavis et S. A. Hammer) S. A. Hammer;林登肉锥花■☆

102643 Conophytum taylorianum (Dinter et Schwantes) N. E. Br.;泰勒肉锥花■☆

102644 Conophytum tectum N. E. Br. = Conophytum bilobum (Marloth) N. E. Br. ■☆

102645 Conophytum teguliflorum Tischer = Conophytum frutescens Schwantes ■☆

102646 Conophytum tenuisectum L. Bolus = Conophytum pageae (N. E. Br.) N. E. Br. ■☆

102647 Conophytum terrestre Tischer = Conophytum pellucidum Schwantes var. terrestre (Tischer) S. A. Hammer ■☆

102648 Conophytum terricolor Tischer = Conophytum pellucidum Schwantes var. terricolor (Tischer) Littlew. ex S. A. Hammer ■☆

102649 Conophytum tetracarpum Lavis = Conophytum flavum N. E. Br. ■☆

102650 Conophytum thudichumii L. Bolus = Conophytum pageae (N. E. Br.) N. E. Br. ■☆

102651 Conophytum tinctum Lavis = Conophytum flavum N. E. Br. ■☆

102652 Conophytum tischeri Schick;天使■☆

102653 Conophytum tischeri Schick = Conophytum velutinum Schwantes ■☆

102654 Conophytum tischleri Schwantes = Conophytum ectypum N. E. Br. ■☆

102655 Conophytum translucens N. E. Br. = Conophytum truncatum (Thunb.) N. E. Br. ■☆

102656 Conophytum triebneri Schwantes = Conophytum marginatum Lavis subsp. haramoepense (L. Bolus) S. A. Hammer ■☆

102657 Conophytum truncatellum (Haw.) N. E. Br.;小红翠玉■☆

102658 Conophytum truncatum (Thunb.) N. E. Br.;红翠玉■☆

102659 Conophytum truncatum (Thunb.) N. E. Br. f. parvipunctum (Tischer) Tischer = Conophytum truncatum (Thunb.) N. E. Br. ■☆

102660 Conophytum truncatum (Thunb.) N. E. Br. f. rennei (Lavis) Tischer = Conophytum truncatum (Thunb.) N. E. Br. ■☆

102661 Conophytum truncatum (Thunb.) N. E. Br. subsp. viridicatum (N. E. Br.) S. A. Hammer;绿翠玉■☆

102662 Conophytum truncatum (Thunb.) N. E. Br. var. brevitubum (Lavis) Tischer = Conophytum truncatum (Thunb.) N. E. Br. ■☆

102663 Conophytum truncatum (Thunb.) N. E. Br. var. parvipunctum (Tischer) Tischer = Conophytum truncatum (Thunb.) N. E. Br. ■☆

102664 Conophytum truncatum (Thunb.) N. E. Br. var. wiggettiae (N. E. Br.) Rawé;维格肉锥花■☆

102665 Conophytum tubatum Tischer = Conophytum minutum (Haw.) N. E. Br. ■☆

102666 Conophytum tumidum N. E. Br. = Conophytum bilobum (Marloth) N. E. Br. ■☆

102667 Conophytum tumidum N. E. Br. var. asperulum L. Bolus = Conophytum bilobum (Marloth) N. E. Br. ■☆

102668 Conophytum turrigerum (N. E. Br.) N. E. Br.;春侍玉■☆

102669 Conophytum turrigerum N. E. Br. = Conophytum turrigerum (N. E. Br.) N. E. Br. ■☆

102670 Conophytum udabibense Loesch et Tischer = Conophytum pageae (N. E. Br.) N. E. Br. ■☆

102671 Conophytum umdausense L. Bolus = Conophytum bilobum (Marloth) N. E. Br. ■☆

102672 Conophytum ursprungianum Tischer = Conophytum obcordellum (Haw.) N. E. Br. ■☆

102673 Conophytum ursprungianum Tischer var. stayneri L. Bolus = Conophytum obcordellum (Haw.) N. E. Br. ■☆

102674 Conophytum uvaeforme N. E. Br.;萤光玉■☆

102675 Conophytum uviforme (Haw.) N. E. Br.;葡萄肉锥花■☆

102676 Conophytum uviforme (Haw.) N. E. Br. f. framesii (Lavis) Tischer = Conophytum uviforme (Haw.) N. E. Br. ■☆

102677 Conophytum uviforme (Haw.) N. E. Br. f. meleagris (L. Bolus) Tischer = Conophytum uviforme (Haw.) N. E. Br. ■☆

102678 Conophytum uviforme (Haw.) N. E. Br. subsp. decoratum (N. E. Br.) S. A. Hammer;装饰肉锥花■☆

102679 Conophytum uviforme (Haw.) N. E. Br. subsp. rauhii (Tischer) S. A. Hammer;劳氏肉锥花■☆

102680 Conophytum uviforme (Haw.) N. E. Br. subsp. subincanum (Tischer) S. A. Hammer;灰毛肉锥花■☆

102681 Conophytum uviforme (Haw.) N. E. Br. var. clarum (N. E. Br.) Rawé = Conophytum uviforme (Haw.) N. E. Br. ■☆

102682 Conophytum uviforme (Haw.) N. E. Br. var. litorale (L. Bolus) Rawé = Conophytum uviforme (Haw.) N. E. Br. ■☆

102683 Conophytum uviforme (Haw.) N. E. Br. var. occultum (L. Bolus) Rawé = Conophytum uviforme (Haw.) N. E. Br. subsp. decoratum (N. E. Br.) S. A. Hammer ■☆

102684 Conophytum uviforme (Haw.) N. E. Br. var. subincanum (Tischer) Rawé = Conophytum uviforme (Haw.) N. E. Br. subsp. subincanum (Tischer) S. A. Hammer ■☆

102685 Conophytum vagum N. E. Br. = Conophytum minimum (Haw.) N. E. Br. ■☆

102686 Conophytum vanbredai L. Bolus = Conophytum globosum (N. E. Br.) N. E. Br. ■☆

102687 Conophytum vanheerdei Tischer;黑尔德肉锥花■☆

102688 Conophytum vanrhynsdorpense Schwantes = Conophytum uviforme (Haw.) N. E. Br. ■☆

102689 Conophytum vanzylii Lavis = Conophytum calculus (A. Berger) N. E. Br. subsp. vanzylii (Lavis) S. A. Hammer ■☆

102690 Conophytum variabile L. Bolus = Conophytum bilobum (Marloth) N. E. Br. ■☆

102691 Conophytum varians L. Bolus = Conophytum uviforme (Haw.) N. E. Br. subsp. decoratum (N. E. Br.) S. A. Hammer ■☆

102692 Conophytum velutinum Schwantes;雏鸠(雏鸟)■☆

102693 Conophytum velutinum Schwantes subsp. polyandrum (Lavis) S. A. Hammer;多花雏鸠■☆

102694 Conophytum velutinum Schwantes var. craterulum (Tischer) Rawé = Conophytum velutinum Schwantes ■☆

102695 Conophytum verrucosum (Lavis) G. D. Rowley;多疣肉锥花■☆

102696 Conophytum vescum N. E. Br. = Conophytum saxetanum (N. E. Br.) N. E. Br. ■☆

102697 Conophytum victoris Lavis = Conophytum pageae (N. E. Br.) N. E. Br. ■☆

102698 Conophytum villetii L. Bolus = Conophytum pageae (N. E. Br.) N. E. Br. ■☆

102699 Conophytum violaciflorum Schick et Tischer;明窗玉■☆

102700 Conophytum virens L. Bolus = Conophytum ectypum N. E. Br. ■☆

102701 Conophytum viride Tischer = Conophytum joubertii Lavis ■☆

102702 Conophytum viridicatum (N. E. Br.) N. E. Br.;乙彦■☆

102703 Conophytum viridicatum (N. E. Br.) N. E. Br. = Conophytum truncatum (Thunb.) N. E. Br. subsp. viridicatum (N. E. Br.) S. A. Hammer ■☆

102704 Conophytum viridicatum (N. E. Br.) N. E. Br. var. pisinnum (N. E. Br.) Rawé = Conophytum truncatum (Thunb.) N. E. Br. subsp. viridicatum (N. E. Br.) S. A. Hammer ■☆

102705 Conophytum viridicatum (N. E. Br.) N. E. Br. var. punctatum N. E. Br. = Conophytum truncatum (Thunb.) N. E. Br. subsp. viridicatum (N. E. Br.) S. A. Hammer ■☆

102706 Conophytum viridicatum N. E. Br. = Conophytum viridicatum (N. E. Br.) N. E. Br. ■☆

102707 Conophytum vitreopapillum Rawé = Conophytum obscurum N. E. Br. subsp. vitreopapillum (Rawé) S. A. Hammer ■☆

102708 Conophytum vlakmynense L. Bolus = Conophytum bilobum (Marloth) N. E. Br. ■☆

102709 Conophytum wagneriorum Schwantes = Conophytum truncatum (Thunb.) N. E. Br. ■☆

102710 Conophytum wettsteinii (A. Berger) N. E. Br.;小槌(黄花小槌)■☆

102711 Conophytum wettsteinii (A. Berger) N. E. Br. subsp. fragile (Tischer) S. A. Hammer = Conophytum jucundum (N. E. Br.) N. E. Br. subsp. fragile (Tischer) S. A. Hammer ■☆

102712 Conophytum wettsteinii (A. Berger) N. E. Br. subsp. francoiseae S. A. Hammer = Conophytum francoiseae (S. A. Hammer) S. A. Hammer ■☆

102713 Conophytum wettsteinii (A. Berger) N. E. Br. subsp. ruschii (Schwantes) S. A. Hammer = Conophytum jucundum (N. E. Br.) N. E. Br. subsp. ruschii (Schwantes) S. A. Hammer ■☆

102714 Conophytum wettsteinii (A. Berger) N. E. Br. var. oculatum L. Bolus = Conophytum jucundum (N. E. Br.) N. E. Br. subsp. ruschii (Schwantes) S. A. Hammer ■☆

102715 Conophytum wettsteinii (A. Berger) N. E. Br. var. speciosum (Tischer) Tischer = Conophytum jucundum (N. E. Br.) N. E. Br. subsp. ruschii (Schwantes) S. A. Hammer ■☆

102716 Conophytum wettsteinii N. E. Br. = Conophytum wettsteinii (A. Berger) N. E. Br. ■☆

102717 Conophytum wiesemannianum Schwantes = Conophytum fulleri L. Bolus ■☆

102718 Conophytum wiggettiae N. E. Br. = Conophytum truncatum (Thunb.) N. E. Br. var. wiggettiae (N. E. Br.) Rawé ■☆

102719 Conophytum wittebergense de Boer = Conophytum minimum (Haw.) N. E. Br. ■☆

102720 Conopodium W. D. J. Koch(1824)(保留属名);锥足芹属(锥足草属);Earth-nut,Hognut,Pignut ■☆

102721 Conopodium bourgaei Coss.;布尔热锥足芹■☆

102722 Conopodium bunioides (Boiss.) Calest.;布留芹状锥足芹■☆

102723 Conopodium bunioides (Boiss.) Calest. subsp. atlantis (Humbert et Maire) Molero;亚特兰大锥足芹■☆

102724 Conopodium bunioides (Boiss.) Calest. var. atlantis Humbert et Maire = Conopodium bunioides (Boiss.) Calest. subsp. atlantis (Humbert et Maire) Molero ■☆

102725 Conopodium cyminum Benth. et Hook. = Sphallerocarpus gracilis (Trevir.) Koso-Pol. ■

102726 Conopodium glaberrimum (Desf.) Engstrand;光滑锥足芹■☆

102727 Conopodium majus (Gouan) Loret;锥足芹;Arnut,Bald Man's Bread,Briza Jocks,Cain-and-abel,Catnut,Cuckoo Potato,Cuckoo Potatoes,Curlans,Curly Nuts,Devil's Bread,Devil's Oatmeal,Dothering Jockies,Earth Chestnut,Earth Nut,Earthnut,Earth-nut,Fairy Potato,Fairy Potatoes,Farenut,Fern Nut,Gernut,Gourlins,Gowlins,Ground Nut,Grove-nut,Grunnut,Hare Nut,Harenut,Hawk Nut,Hawk-nut,Heare-nut,Hog Nut,Hognut,Hornicks Hornecks,Jack-durnals,Jack-jennets,Jacky-jurnals,Jocky Jurnals,Jur-nut,Kellas,Kelly,Killimore,Knotty Meal,Lousy Arnot,Lousy Arnut,Lucy

Arnut, Lucyarnut, Meat Nut, Peg-nut, Pig Nut, Pignut, Scabby Hands, St. Anthony's Nut, Swinebread, Trembling Jockies, Truffle, Underground Nut, Varenut, Yarnut, Yernut Yennut, Yethnut, Yornut, Yowe Yornut, Yowe-yornut, Yowie Yorlin, Yowie-yorlin ■☆
102728 Conopodium majus (Gouan) Loret = Bunium majus Gouan ■☆
102729 Conopodium marianum Lange;玛利亚锥足芹■☆
102730 Conopodium setaceum (Schrenk) Korovin. = Scaligeria setacea (Schrenk) Korovin ■
102731 Conopodium smyrnioides (H. Wolff) M. Hiroe = Changium smyrnioides H. Wolff ■
102732 Conopsidium Wallr. = Platanthera Rich.(保留属名)■
102733 Conoria Juss. = Rinorea Aubl.(保留属名)●
102734 Conosapium Müll. Arg. = Sapium Jacq.(保留属名)●
102735 Conosilene (Rohrb.) Fourr. = Pleconax Raf. ■
102736 Conosilene (Rohrb.) Fourr. = Silene L.(保留属名)■
102737 Conosilene Fourr. = Pleconax Raf. ■
102738 Conosilene Fourr. = Silene L.(保留属名)■
102739 Conosiphon Poepp. = Sphinctanthus Benth. ●☆
102740 Conosiphon Poepp. et Endl. = Sphinctanthus Benth. ●☆
102741 Conospermum Sm. (1798);烟木属;Smoke Bush ●☆
102742 Conospermum burgessiorum L. A. S. Johnson et McGill.;布格烟木●☆
102743 Conospermum longifolium Sm.;长叶烟木;Smoke Bush ●☆
102744 Conospermum stoechadis Endl.;烟木;Smoke Bush ●☆
102745 Conospermum taxifolium C. F. Gaertn.;紫杉叶烟木;Smoke Bush ●☆
102746 Conospermum tenuifolium Sieber ex Roem. et Schult.;薄叶烟木●☆
102747 Conospermum teretifolium R. Br.;蜘蛛烟木;Spider Smokebush ●☆
102748 Conostalix (Kraenzl.) Brieger = Dendrobium Sw.(保留属名)■
102749 Conostalix (Kraenzl.) Brieger = Eria Lindl.(保留属名)■
102750 Conostalix (Schltr.) Brieger = Dendrobium Sw.(保留属名)■
102751 Conostegia D. Don(1823);锥被野牡丹属■☆
102752 Conostegia xalapensis (Bonpl.) D. Don;锥被野牡丹■☆
102753 Conostemum Kunth = Androtrichum (Brongn.) Brongn. ■☆
102754 Conostemum Kunth = Comostemum Nees ■☆
102755 Conostephiopsis Stschegl. = Conostephium Benth. ●☆
102756 Conostephium Benth. (1837);锥花石南属(梭花石南属)●☆
102757 Conostephium minus Lindl.;小锥花石南●☆
102758 Conostephium nitens B. D. Jacks.;光亮锥花石南●☆
102759 Conostephium roei Benth.;锥花石南●☆
102760 Conostomium (Stapf) Cufod. (1948);锥口茜属■☆
102761 Conostomium brevirostrum Bremek. = Pentanopsis fragrans Rendle ■☆
102762 Conostomium camptopodum Bremek. = Conostomium kenyense Bremek. ■☆
102763 Conostomium fasciculatum (Hiern) Cufod. = Conostomium longitubum (Beck) Cufod. ■☆
102764 Conostomium floribundum Agnew = Conostomium kenyense Bremek. ■☆
102765 Conostomium gazense Verdc.;加兹锥口茜■☆
102766 Conostomium hispidulum Bremek. = Conostomium longitubum (Beck) Cufod. ■☆
102767 Conostomium kenyense Bremek.;肯尼亚锥口茜■☆
102768 Conostomium kenyense Bremek. var. subglabrum? = Conostomium kenyense Bremek. ■☆
102769 Conostomium longitubum (Beck) Cufod.;长管锥口茜■☆
102770 Conostomium microcarpum Bremek. = Conostomium kenyense Bremek. ■☆
102771 Conostomium natalense (Hochst.) Bremek.;纳塔尔锥口茜■☆
102772 Conostomium natalense (Hochst.) Bremek. var. glabrum Bremek.;光滑锥口茜■☆
102773 Conostomium natalense (Hochst.) Bremek. var. hirsuta Baer = Conostomium natalense (Hochst.) Bremek. ■☆
102774 Conostomium natalense (Hochst.) Bremek. var. ovalifolium Bremek.;卵叶纳塔尔锥口茜■☆
102775 Conostomium natalense (Hochst.) Bremek. var. tomentellum Bremek.;绒毛锥口茜■☆
102776 Conostomium quadrangulare (Rendle) Cufod.;四棱锥口茜■☆
102777 Conostomium rhynchothecum (K. Schum.) Cufod. = Conostomium longitubum (Beck) Cufod. ■☆
102778 Conostomium rotatum (Baker) Cufod. = Conostomium longitubum (Beck) Cufod. ■☆
102779 Conostomium squarrosum Bremek. = Pentanopsis fragrans Rendle ■☆
102780 Conostomium zoutpansbergense (Bremek.) Bremek.;佐特锥口茜■☆
102781 Conostylidaceae (Pax) Takht. = Haemodoraceae R. Br.(保留科名)■☆
102782 Conostylidaceae Takht. (1987);锥柱草科(叉毛草科)■☆
102783 Conostylidaceae Takht. = Haemodoraceae R. Br.(保留科名)■☆
102784 Conostylis R. Br. (1810);锥柱草属(叉毛草属)■☆
102785 Conostylis albicans A. Cunn. ex Benth.;白锥柱草■☆
102786 Conostylis albicans Benth. = Conostylis albicans A. Cunn. ex Benth. ■☆
102787 Conostylis americana Pursh;美洲锥柱草■☆
102788 Conostylis angustifolia Hopper;窄叶锥柱草■☆
102789 Conostylis argentea (J. W. Green) Hopper;银锥柱草■☆
102790 Conostylis aurea Lindl.;黄锥柱草■☆
102791 Conostylis canescens (Lindl.) F. Muell.;灰锥柱草;Red Bogles, Red Bugles ■☆
102792 Conostylis canescens F. Muell. = Conostylis canescens (Lindl.) F. Muell. ■☆
102793 Conostylis micrantha Hopper;小花锥柱草■☆
102794 Conostylis robusta Diels ex Diels et E. Pritz.;粗壮锥柱草■☆
102795 Conostylus Pohl ex A. DC. = Conomorpha A. DC. ●☆
102796 Conothamnus Lindl. (1839);锥灌桃金娘属●☆
102797 Conothamnus aureus Domin;黄锥灌桃金娘●☆
102798 Conothamnus trinervis Lindl.;锥灌桃金娘●☆
102799 Conotrichia A. Rich. = Manettia Mutis ex L.(保留属名)●■☆
102800 Conradia Mart. = Gesneria L. ●☆
102801 Conradia Nutt. = Macranthera Nutt. ex Benth. ■☆
102802 Conradia Raf. = Tofieldia Huds. ■
102803 Conradina A. Gray(1870);假迷迭香属(康拉德草属)●☆
102804 Conradina canescens A. Gray;灰假迷迭香;False rosemary ●☆
102805 Conradina verticillata Jennison;轮生假迷迭香(康拉德草);Cumberland Rosemary ●☆
102806 Conringia Adans. = Conringia Heist. ex Fabr. ■
102807 Conringia Fabr. = Conringia Heist. ex Fabr. ■
102808 Conringia Heist. ex Fabr. (1759);线果芥属(肋果芥属,肋果荠属,四棱芥属);Conringia, Hare's-ear, Hare's-ear Mustard, Ribsilique ■

102809 Conringia austriaca (Jacq.) Sweet;南方线果芥■☆

102810 Conringia orientalis (L.) Andrz. = Conringia orientalis (L.) Dum. Cours. ■☆

102811 Conringia orientalis (L.) Dum. Cours.;东方线果芥;Hare-ear Mustard, Hare's Ear, Hare's Ear Cabbage, Hare's Ear Mustard, Hare's-ear Cabbage, Hare's-ear Mustard, Treacle Hare's-ear, Treacle Mustard ■☆

102812 Conringia perfoliata (C. A. Mey.) N. Busch;穿叶线果芥■☆

102813 Conringia perfoliata Link = Conringia orientalis (L.) Dum. Cours. ■☆

102814 Conringia persica Boiss.;波斯线果芥■☆

102815 Conringia planisiliqua Fisch. et C. A. Mey.;线果芥(四棱芥);Flatpodded Conringia, Ribsilique ■

102816 Conringia ramosa Boiss. = Conringia persica Boiss. ■☆

102817 Consana Adans. = Subularia L. ■☆

102818 Consolea Lem. (1862);康氏掌属■☆

102819 Consolea Lem. = Opuntia Mill. ●

102820 Consolea acaulis (Ekman et Werderm.) F. M. Knuth;无茎康氏掌■☆

102821 Consolea corallicola Small;佛罗里达康氏掌;Florida Semaphore Cactus, Semaphore Cactus ■☆

102822 Consolea leucacantha Lem.;白刺康氏掌■☆

102823 Consolea macracantha A. Berger;大花康氏掌■☆

102824 Consolea microcarpa E. F. Anderson;小果康氏掌■☆

102825 Consolea rubescens Lem.;红康氏掌;Road Kill Cactus ●☆

102826 Consolea spinosissima Lem.;大刺康氏掌;Florida Semaphore Cactus ●☆

102827 Consolida (DC.) Gray(1821);飞燕草属;Consolida, Larkspur ■

102828 Consolida (DC.) Opiz = Consolida (DC.) Gray ■

102829 Consolida (DC.) Opiz = Delphinium L. ■

102830 Consolida Gilib. = Symphytum L. ■

102831 Consolida Gray = Consolida (DC.) Gray ■

102832 Consolida Riv. ex Rupp. = Symphytum L. ■

102833 Consolida ajacis (L.) Schur;飞燕草(矮飞燕草,彩雀,翠雀,蝴蝶花,琉璃飞燕草,千鸟草,疏花翠雀花,硬飞燕草);Annual Delphinium, Blue Butterfly, Branching Larkspur, Dolphin-flower, Doubtful Knight's Spur, Doubtful Knight's-spur, Field Larkspur, Forking Larkspur, Granny's Bonnet, Granny's Bonnets, Granny's Nightcap, Hyacinth-flowered Larkspur, Jacob's Ladder, King's Consound, Knight's Spur, Lark's Claws, Lark's Heel, Lark's Toe, Larkspur, Noah's Ark, Oriental Knight's-spur, Rocket Consolida, Rocket Larkspur ■

102834 Consolida ambigua (L.) P. W. Ball et Heywood;迷惑飞燕草;Garden Larkspur, Rocket Larkspur ■☆

102835 Consolida flava (DC.) Schrödinger;黄飞燕草■☆

102836 Consolida hispanica (Costa) Greuter et Burdet = Consolida orientalis (J. Gay) Schrödinger ■☆

102837 Consolida mauritanica (Coss.) Munz;毛里塔尼亚飞燕草■☆

102838 Consolida orientalis (J. Gay) Schrödinger;东方翠雀花;Eastern Larkspur, Oriental Larkspur ■☆

102839 Consolida pubescens (DC.) Soó;毛飞燕草;Hairy Knight's-spur, Knight's-spur ■☆

102840 Consolida pubescens Soó = Consolida pubescens (DC.) Soó ■☆

102841 Consolida regalis Gray;高贵飞燕草;Field Larkspur, Forking Larkspur, Garden Rocket, Knight's-spur, Rocket Larkspur, Royal Knight's-spur ■☆

102842 Consolida rugulosa (Boiss.) Schrödinger;凸脉飞燕草(飞燕草,皱叶翠雀花);Rugulose Consolida, Rugulose Cenesolida ■

102843 Consolida tenuissima (Sibth. et Sm.) Soó;长飞燕草;Long Knight's-spur ■☆

102844 Consoligo (DC.) Opiz = Adonis L. (保留属名)■

102845 Constancea B. G. Baldwin(2000);绵菊木属●☆

102846 Constancea nevinii (A. Gray) B. G. Baldwin;绵菊木●☆

102847 Constantia Barb. Rodr. (1877);孔唐兰属■☆

102848 Constantia rupestris Barb. Rodr.;孔唐兰■☆

102849 Contarena Adans. = Corymbium L. ■●☆

102850 Contarenia Vand. = ? Alectra Thunb. ■

102851 Contortaceae Dulac = Convolvulaceae Juss. + Cuscutaecae Dumort. ●■

102852 Contortuplicata Medik. = Astragalus L. ●■

102853 Contrarenia J. St. -Hil. = ? Alectra Thunb. ■

102854 Contrarenia J. St. -Hil. = Contarenia Vand. ■

102855 Conuleum A. Rich. = Siparuna Aubl. ●☆

102856 Convallaria L. (1753);铃兰属(草玉铃属);ヨキミカゲソウ属;Lily of the Valley, Lily-of-the-valley, Valley Lily ■

102857 Convallaria biflora Walter = Polygonatum biflorum (Walter) Elliott ■☆

102858 Convallaria bifolia L. = Maianthemum bifolium (L.) F. W. Schmidt ■

102859 Convallaria cirrhifolia Wall. = Polygonatum cirrhifolium (Wall.) Royle ■

102860 Convallaria fruticosa L. = Cordyline fruticosa (L.) A. Chev. ●

102861 Convallaria japonica L. f. = Ophiopogon japonicus (L. f.) Ker Gawl. ■

102862 Convallaria japonica L. f. var. minor Thunb. = Ophiopogon japonicus (L. f.) Ker Gawl. ■

102863 Convallaria keiskei Miq.;伊藤氏铃兰(君影草)■☆

102864 Convallaria keiskei Miq. = Convallaria majalis L. ■

102865 Convallaria keiskei Miq. var. trifolla Y. C. Chu et al. = Convallaria majalis L. ■

102866 Convallaria latifolia Jacq. = Polygonatum latifolium (Jacq.) Desf. ■☆

102867 Convallaria majalis L.;铃兰(草寸香,草玉兰,草玉铃,君影草,铃铛花,芦藜草,芦藜花,鹿铃,鹿铃草,鹿铃花,糜子菜,扫帚糜子,香水花,小芦藜,小芦铃);Conval Lily, Convally, Dangle Bells, Dangling Bells, European Lily of the Valley, European Lily-of-the-valley, Fairy Bells, Glovewort, Great Park Lily, Great Solomon's Seal, Innocent, Jacob's Ladder, Jacob's Tears, Ladder To Heaven, Ladder-to-heaven, Lady's Tears, Lilly-convally, Lily of the Valley, Lily-confancy, Lily-constancy, Lily-of-the-valley, Linen Buttons, Linen-buttons, Liricon-fancy, Liricum-fancy, Liry-confancy, Little White Bells, Male Lily, May Lily, May-blossom, Mayflower Lily, May-flower Lily, May-lily, Mugget, Muguet, Valley Lily, Valleys, Virgin's Tears, White Bells, Wood Lily ■

102868 Convallaria majalis L. 'Flore Pleno';重瓣铃兰■☆

102869 Convallaria majalis L. 'Fortin's Giant';福庭大花铃兰■☆

102870 Convallaria majalis L. var. manshurica Kom. = Convallaria majalis L. ■

102871 Convallaria majalis L. var. montana (Raf.) H. E. Ahles;美洲铃兰;American Lily-of the-valley ■☆

102872 Convallaria montana Raf. = Convallaria majalis L. var. montana (Raf.) H. E. Ahles ■☆

102873 Convallaria multiflora Ball. = Polygonatum odoratum (Mill.) Druce ■

102874 Convallaria odorata Mill. = Polygonatum odoratum (Mill.) Druce ■

102875 Convallaria oppositifolia Wall. = Polygonatum oppositifolium (Wall.) Royle ■

102876 Convallaria polygonatum L. = Polygonatum odoratum (Mill.) Druce ■

102877 Convallaria pubescens Willd. = Polygonatum pubescens (Willd.) Pursh ■☆

102878 Convallaria racemosa Forssk. = Sansevieria forskaoliana (Schult. f.) Hepper et J. R. I. Wood ■☆

102879 Convallaria racemosa L. = Maianthemum racemosum (L.) Link ■☆

102880 Convallaria rosea Ledeb. = Polygonatum roseum (Ledeb.) Kunth ■

102881 Convallaria spicata Thunb. = Liriope spicata (Thunb.) Lour. ■

102882 Convallaria stellata L. = Maianthemum stellatum (L.) Link ■☆

102883 Convallaria transcaucasica Utkin ex Grossh.;外高加索铃兰■☆

102884 Convallaria tricolor L.;三色铃兰;Dwarf Morning-glory, Three-coloured Bindweed ■☆

102885 Convallaria trifolia L. = Maianthemum trifolium (L.) Slobada ■

102886 Convallaria umbellulata Michx. = Clintonia umbellulata (Michx.) Morong ■☆

102887 Convallaria verticillata L. = Polygonatum verticillatum (L.) All. ■

102888 Convallariaceae Horan. (1834);铃兰科;Morning-glory Family ■

102889 Convallariaceae Horan. = Liliaceae Juss. (保留科名) ■●

102890 Convallariaceae L. = Convallariaceae Horan. ■

102891 Convallariaceae L. = Ruscaceae M. Roem. (保留科名) ●

102892 Convolvulaceae Juss. (1789) (保留科名);旋花科;Bindweed Family, Glorybind Family, Morning Glory Family, Morningglory Family, Morning-glory Family ●■

102893 Convolvulaster Fabr. = Convolvulus L. ■●

102894 Convolvuloides Moench(废弃属名) = Ipomoea L. (保留属名) ●■

102895 Convolvuloides Moench(废弃属名) = Pharbitis Choisy(保留属名) ■

102896 Convolvuloides elongata Moench = Ipomoea sibirica (L.) Pers. ■

102897 Convolvuloides elongata Moench = Merremia sibirica (L.) Hallier f. ■

102898 Convolvuloides leucosperma Moench = Ipomoea purpurea (L.) Roth ■

102899 Convolvuloides palmata Moench = Ipomoea pes-tigridis L. ■

102900 Convolvuloides purpurea Moench = Ipomoea purpurea (L.) Roth ■

102901 Convolvuloides triloba Moench = Ipomoea hederacea Jacq. ■

102902 Convolvuloides triloba Moench = Ipomoea nil (L.) Roth ■

102903 Convolvulus L. (1753);旋花属(鼓子花属,三色牵牛属);Bindweed, Bind-weed, Convolvulus, Field Bindweed, Glorybind, Morning Glory ■●

102904 Convolvulus acetosellifolius Desr. = Merremia hederacea (Burm. f.) Hallier f. ■

102905 Convolvulus acetosifolius Turcz. = Calystegia hederacea Wall. ex Roxb. ■

102906 Convolvulus acetosifolius Vahl = Ipomoea imperati (Vahl) Griseb. ■

102907 Convolvulus acicularis Vatke = Convolvulus sericophyllus T. Anderson ■☆

102908 Convolvulus aculeatus L. = Ipomoea alba L. ■

102909 Convolvulus acuminatus Vahl = Ipomoea indica (Burm.) Merr. ■

102910 Convolvulus agrestis (Hochst. ex Schweinf.) Hallier f. = Convolvulus siculus L. subsp. agrestis (Hochst. ex Schweinf.) Verdc. ■☆

102911 Convolvulus albivenius Lindl. = Ipomoea albivenia (Lindl.) Sweet ■☆

102912 Convolvulus alceifolius Lam. = Convolvulus capensis Burm. f. ■☆

102913 Convolvulus alsinoides L. = Evolvulus alsinoides (L.) L. ●■

102914 Convolvulus althaeoides L.;地中海旋花(蜀葵叶旋花);Mallow Bindweed, Mallow-leaved Bindweed, Mediterranean Glorybind ■☆

102915 Convolvulus althaeoides L. subsp. elegantissimus (Mill.) Quézel et Santa = Convolvulus elegantissimus Mill. ■☆

102916 Convolvulus althaeoides L. subsp. tenuissimus (Sibth. et Sm.) Batt. = Convolvulus elegantissimus Mill. ■☆

102917 Convolvulus althaeoides L. var. albidiflorus Braun-Blanq. et Maire = Convolvulus althaeoides L. ■☆

102918 Convolvulus althaeoides L. var. angustisectus Pamp. = Convolvulus althaeoides L. ■☆

102919 Convolvulus althaeoides L. var. dissectus Faure et Maire = Convolvulus althaeoides L. ■☆

102920 Convolvulus althaeoides L. var. jolyi Sauvage et Vindt = Convolvulus althaeoides L. ■☆

102921 Convolvulus althaeoides L. var. pedatus Choisy = Convolvulus althaeoides L. ■☆

102922 Convolvulus althaeoides L. var. repandus Faure et Maire = Convolvulus althaeoides L. ■☆

102923 Convolvulus althaeoides L. var. scandens Maire et Sennen = Convolvulus althaeoides L. ■☆

102924 Convolvulus altheaoides Thunb. = Convolvulus thunbergii Roem. et Schult. ■☆

102925 Convolvulus ambigens House = Convolvulus arvensis L. ■

102926 Convolvulus ammannii Desr.;银灰旋花(阿氏旋花,彩木,小旋花);Ammann Glorybind, Sivery-grey Glorybind ■

102927 Convolvulus anceps L. = Convolvulus turpethum L. ■

102928 Convolvulus angolensis Baker = Convolvulus sagittatus Thunb. ■☆

102929 Convolvulus angularis Burm. f. = Merremia vitifolia (Burm. f.) Hallier f. ■

102930 Convolvulus arborescens Humb. et Bonpl. ex Willd. = Ipomoea arborescens (Humb. et Bonpl. ex Willd.) G. Don ●☆

102931 Convolvulus argillicola Pilg.;白土旋花■☆

102932 Convolvulus armatus Delile = Convolvulus hystrix Vahl ■☆

102933 Convolvulus arvensis L.;田旋花(白花藤,车子蔓,打碗花,扶田秧,扶秧苗,扶秧田,箭叶旋花,拉拉菀,面根藤,曲节藤,三齿草藤,田福花,小旋花,燕子草,野牵牛,野旋花,中国旋花);Banebind, Barbine, Barweed, Bearbind, Bearbine, Beddywind, Bedwind, Bedwine, Bellbind, Bellwind, Beswlne, Bethroot, Bettywind, Billy Clippe, Billy Clipper, Bind-corn, Bindweed, Bine Lily, Binf-lily, Bithwind, Bithwine, Bithywind, Bunwede, China Glorybind, Common Bindweed, Corn Bind, Corn Lily, Corn-bin, Cornbind, Corn-bind, Corn-bine, Creeping Jenny, Deer's-foot Bindweed, Devil's Entrails, Devil's Garter, Devil's Garters, Devil's Guts, Devil's Nightcap, Devil's Twine, Devil-weed, Dodder, Dralyer, Earwig, Europe Glorybind, European Glorybind, Fairy's Umbrella, Fairy's Umbrellas, Fairy's Winecup, Fairy's Winecups, Field Bindweed, Field Morning Glory, Granny's Nightcap, Ground Ivy, Ground Lily, Gypsy's Hat, Hedge

Bells, Hellweed, Jack-run-in-the-country, Kettle Smock, Kettle-smocks, Lady's Nightcap, Lady's Smock, Lady's Sunshade, Lady's Umbrella, Laplove, Lesser Bindweed, Lily, Lily Bind, Lily-bind, Morning Glory, Old Man's Nightcap, Parasol, Ragged Shirt, Reedbind, Robin Run-the-dyke, Robin-run-in-the-field, Ropewind, Sheepbine, Shirt, Shirts-and-shimmies, Small Bindweed, Sunshade, Tare, Thunder Flower, Wandering Willy, Waywind, Weedwind, Wheatbine, White Smock, Widdy-wine, Widwind, Willow-wind, Willywind, With-vine, Withwind, Withwine, Withybind, Withyvine, Withyweed, Withywind, Withywine, Withywing, Young Man's Death ■

102934 Convolvulus arvensis L. subsp. hortensis (Batt.) Maire = Convolvulus tricolor L. ■☆

102935 Convolvulus arvensis L. var. angustatus Ledeb. = Convolvulus arvensis L. ■

102936 Convolvulus arvensis L. var. aphacifolius Pomel = Convolvulus arvensis L. ■

102937 Convolvulus arvensis L. var. biflorus Pau = Convolvulus arvensis L. ■

102938 Convolvulus arvensis L. var. crassifolius Choisy = Convolvulus arvensis L. ■

102939 Convolvulus arvensis L. var. filicaulis Pomel = Convolvulus arvensis L. ■

102940 Convolvulus arvensis L. var. linearifolius Choisy = Convolvulus arvensis L. ■

102941 Convolvulus arvensis L. var. minutus Maire = Convolvulus arvensis L. ■

102942 Convolvulus arvensis L. var. obtusifolius Choisy = Convolvulus arvensis L. ■

102943 Convolvulus arvensis L. var. paui Maire = Convolvulus arvensis L. ■

102944 Convolvulus arvensis L. var. sagittatus Ledeb. = Convolvulus arvensis L. ■

102945 Convolvulus arvensis L. var. sagittifolius Turcz. = Convolvulus arvensis L. ■

102946 Convolvulus arvensis L. var. trigonophyllus Maire = Convolvulus arvensis L. ■

102947 Convolvulus asarifolius Desr. = Ipomoea asarifolia (Desr.) Roem. et Schult. ■☆

102948 Convolvulus asarifolius Salisb. = Calystegia soldanella (L.) R. Br. ■

102949 Convolvulus aschersonii Engl.;阿舍森旋花■☆

102950 Convolvulus askabadensis Bornm. et Sint. = Convolvulus pseudocantabrica Schrenk ■

102951 Convolvulus atropurpureus Wall. = Argyreia pierreana Bois ●

102952 Convolvulus auricomus (A. Rich.) Bhandari;金毛旋花■☆

102953 Convolvulus austro-aegyptiacus Abdallah et Sa'ad;南埃及旋花■☆

102954 Convolvulus batatas L. = Ipomoea batatas (L.) Lam. ■

102955 Convolvulus bicolor Vahl = Hewittia malabarica (L.) Suresh ■

102956 Convolvulus bidentatus Bernh. ex C. Krauss;双齿旋花■☆

102957 Convolvulus bifidus Vell. = Jacquemontia ovalifolia (Vahl) Hallier f. ■☆

102958 Convolvulus biflorus L. = Ipomoea biflora (L.) Pers. ■

102959 Convolvulus bilobatus Roxb. = Ipomoea pes-caprae (L.) R. Br. ■

102960 Convolvulus binectariferus Wall. ex Roxb. = Lepistemon binectarifer (Wall.) Kuntze ■

102961 Convolvulus bornmuelleri Hausskn. = Convolvulus deserti Hochst. et Steud. ex Baker et Rendle ■☆

102962 Convolvulus bracteatus Vahl = Hewittia malabarica (L.) Suresh ■

102963 Convolvulus brasiliensis L. = Ipomoea pes-caprae (L.) R. Br. subsp. brasiliensis (L.) Ooststr. ■

102964 Convolvulus brasiliensis L. = Ipomoea pes-caprae (L.) R. Br. ■

102965 Convolvulus brevipes Pomel = Convolvulus supinus Coss. et Kralik ■☆

102966 Convolvulus bryoneae-folius Sims;硬毛旋花■

102967 Convolvulus bullerianus Rendle = Convolvulus natalensis Bernh. ex Krauss ■☆

102968 Convolvulus bussei Pilg.;布瑟旋花●☆

102969 Convolvulus caespitosus Roxb. = Merremia hirta (L.) Merr. ■

102970 Convolvulus cairicus L. = Ipomoea cairica (L.) Sweet ■

102971 Convolvulus calvertii Boiss.;卡费氏旋花;Calvert Glorybind ■☆

102972 Convolvulus calycinus E. Mey. ex Drège = Convolvulus natalensis Bernh. ex Krauss ■☆

102973 Convolvulus calycinus Roxb. = Ipomoea biflora (L.) Pers. ■

102974 Convolvulus calycinus Roxb. = Ipomoea sinensis (Desr.) Choisy ■

102975 Convolvulus calystegioides Choisy = Calystegia hederacea Wall. ex Roxb. ■

102976 Convolvulus campanulatus Spreng. = Stictocardia tiliifolia (Desr.) Hallier f. ●■

102977 Convolvulus campanulatus Zopr. = Stictocardia tiliifolia (Desr.) Hallier f. ●■

102978 Convolvulus canariensis L.;加那利旋花■☆

102979 Convolvulus candicans Soland. ex Sims = Ipomoea batatas (L.) Lam. ■

102980 Convolvulus cantabrica L.;坎塔布连山旋花■☆

102981 Convolvulus cantabrica L. var. mazicum (Emb. et Maire) Font Quer = Convolvulus mazicum Emb. et Maire ■☆

102982 Convolvulus cantabricus L. subsp. atlantis Emb. = Convolvulus mazicum Emb. et Maire ■☆

102983 Convolvulus cantabricus L. subsp. mazicum (Emb. et Maire) Maire = Convolvulus mazicum Emb. et Maire ■☆

102984 Convolvulus capensis Burm. f.;好望角旋花■☆

102985 Convolvulus capensis Burm. f. var. bowieanus (Rendle) A. Meeuse = Convolvulus capensis Burm. f. ■☆

102986 Convolvulus capensis Burm. f. var. plicatus (Desr.) Baker = Convolvulus capensis Burm. f. ■☆

102987 Convolvulus capitatus Desr. = Jacquemontia tamnifolia (L.) Griseb. ■☆

102988 Convolvulus capitatus Vahl = Argyreia capitiformis (Poir.) Ooststr. ●

102989 Convolvulus capitiformis Poir. = Argyreia capitiformis (Poir.) Ooststr. ●

102990 Convolvulus capituliferus Franch. = Convolvulus rhyniospermus Hochst. ex Choisy ■☆

102991 Convolvulus capituliferus Franch. var. foliaceus Verdc. = Convolvulus rhyniospermus Hochst. ex Choisy ■☆

102992 Convolvulus capituliferus Franch. var. suberectus? = Convolvulus rhyniospermus Hochst. ex Choisy ■☆

102993 Convolvulus cephalanthus Wall. = Lepistemon binectarifer (Wall.) Kuntze ■

102994 Convolvulus charmelii Sennen et Mauricio = Convolvulus valentinus Cav. subsp. suffruticosus (Desf.) Maire ●☆

102995 Convolvulus chinensis Ker Gawl. = Convolvulus arvensis L. ■

102996 Convolvulus chryseides (Ker Gawl.) Spreng. = Merremia hederacea (Burm. f.) Hallier f. ■

102997 Convolvulus cirrhosus R. Br. = Convolvulus arvensis L. ■

102998 Convolvulus cneorus L.;银旋花(银毛旋花);Bush Morning Glory, Shrubby Bindweed, Silver Bush Morning Glory, Silver Glorybind, Silverbush, Silvery Bindweed ●☆

102999 Convolvulus cneorus L. subsp. latifolius (Rchb.) Sa'ad;宽叶银旋花●☆

103000 Convolvulus coccinea Salisb. = Quamoclit coccinea (L.) Moench ■

103001 Convolvulus colubrinus Blanco = Ipomoea turbinata Lag. ■

103002 Convolvulus congestus (R. Br.) Spreng. = Ipomoea indica (Burm.) Merr. ■

103003 Convolvulus congestus R. Br. = Convolvulus auricomus (A. Rich.) Bhandari ■☆

103004 Convolvulus copticus L. = Ipomoea coptica (L.) Roth ex Roem. et Schult. ■☆

103005 Convolvulus cordifolius Thunb. = Convolvulus farinosus L. ■☆

103006 Convolvulus corymbosus L. = Turbina corymbosa (L.) Raf. ■☆

103007 Convolvulus crispus Thunb. = Ipomoea crispa (Thunb.) Hallier f. ■☆

103008 Convolvulus cujanensis Bowdich = Merremia aegyptia (L.) Urb. ■☆

103009 Convolvulus cupanianus Tod. = Convolvulus tricolor L. subsp. cupanianus (Tod.) Cavara et Grande ■☆

103010 Convolvulus cupanianus Tod. var. guttatus Batt. et Maire = Convolvulus tricolor L. subsp. cupanianus (Tod.) Cavara et Grande ■☆

103011 Convolvulus cymosus Desr. = Merremia umbellata (L.) Hallier f. ■

103012 Convolvulus cymosus Roem. et Schult. = Bonamia thunbergiana (Roem. et Schult.) F. N. Williams ●☆

103013 Convolvulus dahuricus Herb.;达呼里旋花;Dahurian Glorybind ■☆

103014 Convolvulus dahuricus Herb. = Calystegia pellita (Ledeb.) G. Don ■

103015 Convolvulus dahuricus Herb. = Calystegia sepium (L.) R. Br. subsp. spectabilis Brummitt ■

103016 Convolvulus defloratus Choisy = Ipomoea polymorpha Roem. et Schult. ■

103017 Convolvulus dentatus Vahl = Merremia hederacea (Burm. f.) Hallier f. ■

103018 Convolvulus denticulatus Desr. = Ipomoea littoralis (L.) Blume ■

103019 Convolvulus deserti Hochst. et Steud. ex Baker et Rendle;荒漠旋花■☆

103020 Convolvulus dianthoides Kar. et Kir. = Convolvulus pseudocantabrica Schrenk ■

103021 Convolvulus dichrous Roem. et Schult. = Ipomoea dichroa Choisy ■☆

103022 Convolvulus dinteri Pilg. = Convolvulus ocellatus Hook. f. ■☆

103023 Convolvulus dissectus Jacq. = Merremia dissecta (Jacq.) Hallier f. ■

103024 Convolvulus distillatorius Blanco = Merremia similis Elmer ■

103025 Convolvulus divaricatus Regel et Schmalh.;叉开旋花■☆

103026 Convolvulus diversifolius Schumach. et Thonn. = Ipomoea marginata (Desr.) Verdc. ■

103027 Convolvulus dregeanus Choisy = Convolvulus liniformis Rendle ■☆

103028 Convolvulus dryadum Maire;林旋花■☆

103029 Convolvulus dryadum Maire var. tazzekkensis Sauvage et Vindt = Convolvulus dryadum Maire ■☆

103030 Convolvulus durandoi Pomel;杜朗多旋花■☆

103031 Convolvulus edulis Thunb. = Ipomoea batatas (L.) Lam. ■

103032 Convolvulus edulis Thunb. ex Murray = Ipomoea batatas (L.) Lam. ■

103033 Convolvulus elarishensis Boulos = Convolvulus lanatus Vahl ■☆

103034 Convolvulus elegantissimus Mill.;雅致旋花■☆

103035 Convolvulus erinaceus Ledeb.;具刺旋花■☆

103036 Convolvulus eriocarpus (R. Br.) Spreng. = Ipomoea eriocarpa R. Br. ■

103037 Convolvulus erubescens Sims;变红旋花;Pinkflower Bindweed ■☆

103038 Convolvulus erythrocarpus Wall. = Argyreia wallichii Choisy ●

103039 Convolvulus evolvuloides Desf. = Convolvulus humilis Jacq. ■☆

103040 Convolvulus falkia Jacq. = Convolvulus capensis Burm. f. ■☆

103041 Convolvulus farinosus L.;被粉旋花■☆

103042 Convolvulus fatmensis Kunze;法蒂玛旋花■☆

103043 Convolvulus faurotii Franch. = Convolvulus auricomus (A. Rich.) Bhandari ■☆

103044 Convolvulus filiformis Thunb. = Convolvulus capensis Burm. f. ■☆

103045 Convolvulus flavescens D. Dietr. = Lepistemon binectarifer (Wall.) Kuntze ■

103046 Convolvulus flavus Willd. = Merremia hederacea (Burm. f.) Hallier f. ■

103047 Convolvulus flexuosus Pomel = Convolvulus siculus L. ■☆

103048 Convolvulus floridus L. f.;繁花旋花●☆

103049 Convolvulus floridus L. f. var. angustifolius Pit. = Convolvulus floridus L. f. ●☆

103050 Convolvulus floridus L. f. var. densiflorus Christ = Convolvulus floridus L. f. ●☆

103051 Convolvulus floridus L. f. var. virgatus (Webb et Berthel.) Mend. -Heuer = Convolvulus floridus L. f. ●☆

103052 Convolvulus forskalei Delile = Convolvulus lanatus Vahl ■☆

103053 Convolvulus fruticosus Pall.;灌木旋花;Fruticose Glorybind ●

103054 Convolvulus fruticosus Pall. f. tianschanica Palib. = Convolvulus tragacanthoides Turcz. ■

103055 Convolvulus fruticulosus Desr.;亚灌木旋花●☆

103056 Convolvulus fruticulosus Desr. var. glabrior Sa'ad = Convolvulus fruticulosus Desr. ●☆

103057 Convolvulus fruticulosus Desr. var. glandulosus (Webb) Sa'ad = Convolvulus glandulosus (Webb) Hallier ■☆

103058 Convolvulus galpinii C. H. Wright;盖尔旋花■☆

103059 Convolvulus gemellus Burm. f. = Merremia gemella (Burm. f.) Hallier f. ■

103060 Convolvulus gemellus Burm. f. = Merremia hederacea (Burm. f.) Hallier f. ■

103061 Convolvulus gharbensis Batt. et Pit.;盖尔比旋花■☆

103062 Convolvulus gilbertii Sebsebe;吉尔伯特旋花■☆

103063 Convolvulus glandulosus (Webb) Hallier;具腺旋花■☆

103064 Convolvulus glomeratus Choisy;团集旋花■☆

103065 Convolvulus gortschakovii Schrenk;鹰爪柴(郭氏木旋花,铁猫刺,鹰爪);Gortschakov Glorybind ●

103066 Convolvulus gracilis (R. Br.) Spreng. = Ipomoea littoralis (L.) Blume ■

103067 Convolvulus grandiflorus Jacq. = Ipomoea violacea L. ■

103068 Convolvulus grandiflorus L. f. = Ipomoea violacea L. ■
103069 Convolvulus guineensis Schumach. = Jacquemontia tamnifolia (L.) Griseb. ■☆
103070 Convolvulus hadramauticus Baker = Merremia somalensis (Vatke) Hallier f. ■☆
103071 Convolvulus hallierianus Schulze-Menz = Convolvulus sagittatus Thunb. ■☆
103072 Convolvulus hamadae (Vved.) Petr.;哈马达旋花■☆
103073 Convolvulus hardwickii Spreng. = Ipomoea biflora (L.) Pers. ■
103074 Convolvulus hardwickii Spreng. = Ipomoea sinensis (Desr.) Choisy ■
103075 Convolvulus hastatus Desr. = Xenostegia tridentata (L.) D. F. Austin et Staples ■
103076 Convolvulus hastatus Thunb. = Convolvulus bidentatus Bernh. ex C. Krauss ■☆
103077 Convolvulus hederaceus L. = Ipomoea nil (L.) Roth ■
103078 Convolvulus heterotrichus Maire = Convolvulus prostratus Forssk. ■☆
103079 Convolvulus hildebrandtii Vatke;希尔德旋花■☆
103080 Convolvulus hirsutus Steven;毛旋花■☆
103081 Convolvulus hirtellus Hallier f. = Convolvulus sagittatus Thunb. ■☆
103082 Convolvulus hirtus L. = Merremia hirta (L.) Merr. ■
103083 Convolvulus hispidus Vahl = Ipomoea eriocarpa R. Br. ■
103084 Convolvulus holosericeus M. Bieb.;密毛旋花■☆
103085 Convolvulus huillensis (Baker) Rendle = Convolvulus sagittatus Thunb. ■☆
103086 Convolvulus humilis Jacq.;低矮旋花■☆
103087 Convolvulus hyoscyamoides Vatke = Astripomoea hyoscyamoides (Vatke) Verdc. ■☆
103088 Convolvulus hystrix Vahl;豪猪旋花■☆
103089 Convolvulus hystrix Vahl f. inermis Chiov. = Convolvulus hystrix Vahl ■☆
103090 Convolvulus ifniensis Caball. = Convolvulus trabutianus Schweinf. et Muschl. ■☆
103091 Convolvulus imperati Vahl = Ipomoea imperati (Vahl) Griseb. ■
103092 Convolvulus inconspicuus Hallier f. = Convolvulus capensis Burm. f. ■☆
103093 Convolvulus indicus Burm. = Ipomoea indica (Burm.) Merr. ■
103094 Convolvulus involucellatus Klotzsch = Convolvulus rhyniospermus Hochst. ex Choisy ■☆
103095 Convolvulus japonicus Choisy;日本旋花;California Glorybind, Japan Glorybind ■☆
103096 Convolvulus japonicus Thunb. = Calystegia hederacea Wall. ex Roxb. ■
103097 Convolvulus jefferyi Verdc.;杰弗里旋花■☆
103098 Convolvulus keniensis Standl. = Convolvulus kilimandschari Engl. ■☆
103099 Convolvulus kilimandschari Engl.;基利旋花■☆
103100 Convolvulus kilimandschari Engl. var. glabratus Hallier f. = Convolvulus kilimandschari Engl. ■☆
103101 Convolvulus korolkovii Regel et Schmalh.;科罗氏旋花■☆
103102 Convolvulus krauseanus Regelet Schmalh.;克氏旋花■☆
103103 Convolvulus lanatus Vahl;绵毛旋花■☆
103104 Convolvulus lanuginosus Desr.;多绵毛旋花■☆
103105 Convolvulus lanuginosus Desr. var. argenteus Choisy = Convolvulus lanuginosus Desr. ■☆
103106 Convolvulus lanuginosus Desr. var. villosus Boiss. = Convolvulus lanuginosus Desr. ■☆
103107 Convolvulus lapathifolius Spreng. = Merremia hederacea (Burm. f.) Hallier f. ■
103108 Convolvulus leucochnous Benoist = Convolvulus pitardii Batt. ■☆
103109 Convolvulus leucotrichus Pomel = Convolvulus supinus Coss. et Kralik ■☆
103110 Convolvulus lineatus L.;线叶旋花;Linearleaf Glorybind ■
103111 Convolvulus lineatus L. var. minutus Maire et Weiller = Convolvulus lineatus L. ■
103112 Convolvulus lineatus L. var. pentapetaloides Batt. = Convolvulus lineatus L. ■
103113 Convolvulus linifolius L. = Evolvulus alsinoides (L.) L. ●■
103114 Convolvulus liniformis Rendle = Convolvulus dregeanus Choisy ■☆
103115 Convolvulus littoralis L. = Ipomoea imperati (Vahl) Griseb. ■
103116 Convolvulus littoralis L. = Ipomoea littoralis (L.) Blume ■
103117 Convolvulus littoralis L. = Ipomoea stolonifera (Cirillo) J. F. Gmel. ■
103118 Convolvulus littoralis Vatke = Convolvulus rhyniospermus Hochst. ex Choisy ■☆
103119 Convolvulus loureiri G. Don = Calystegia hederacea Wall. ex Roxb. ■
103120 Convolvulus lycioides Boiss.;枸杞旋花■☆
103121 Convolvulus macrocarpus L. = Operculina macrocarpa (L.) Urb. ■☆
103122 Convolvulus maireanus Pamp.;迈雷旋花■☆
103123 Convolvulus major Gilib. = Ipomoea purpurea (L.) Roth ■
103124 Convolvulus malabaricus L. = Hewittia malabarica (L.) Suresh ■
103125 Convolvulus marginatus Desr. = Ipomoea marginata (Desr.) Verdc. ■
103126 Convolvulus maritimus Desr. = Ipomoea pes-caprae (L.) R. Br. ■
103127 Convolvulus marittimus Lam. = Calystegia soldanella (L.) R. Br. ■
103128 Convolvulus maroccanus Batt. = Convolvulus tricolor L. ■☆
103129 Convolvulus martiniciensis Jacq. = Aniseia martinicensis (Jacq.) Choisy ■☆
103130 Convolvulus massonii A. Dietr.;马森旋花■☆
103131 Convolvulus mauritanicus Boiss. = Convolvulus sabatius Viv. subsp. mauritanicus (Boiss.) Murb. ■☆
103132 Convolvulus mauritanicus Boiss. = Convolvulus sabatius Viv. ■☆
103133 Convolvulus maximus L. f. = Ipomoea marginata (Desr.) Verdc. ■
103134 Convolvulus mazicum Emb. et Maire;马兹旋花■☆
103135 Convolvulus mazicum Emb. et Maire var. atlantis (Emb.) Sauvage et Vindt = Convolvulus mazicum Emb. et Maire ■☆
103136 Convolvulus mechoacanus Vand.;米却肯旋花■☆
103137 Convolvulus medium L. = Merremia medium (L.) Hallier f. ■☆
103138 Convolvulus meonanthus Hoffmanns. et Link;细旋花■☆
103139 Convolvulus michelsonii Petr.;米氏旋花■☆
103140 Convolvulus microphyllus Sieber ex Spreng. = Convolvulus prostratus Forssk. ■☆
103141 Convolvulus microphyllus Sieber ex Spreng. var. heterotrichus (Maire) Maire = Convolvulus prostratus Forssk. ■☆
103142 Convolvulus microphyllus Sieber ex Spreng. var. orreanus (Murb.) Maire = Convolvulus prostratus Forssk. ■☆

103143 Convolvulus microphyllus Sieber var. longipes Maire = Convolvulus deserti Hochst. et Steud. ex Baker et Rendle ■☆

103144 Convolvulus minor Gilib. = Convolvulus tricolor L. ■☆

103145 Convolvulus mollis Burm. f. = Argyreia mollis (Burm. f.) Choisy ●

103146 Convolvulus mucronatus Engl. = Seddera suffruticosa (Schinz) Hallier f. ●☆

103147 Convolvulus multifidus Hallier f. = Convolvulus ocellatus Hook. f. ■☆

103148 Convolvulus multifidus Thunb. ;多裂旋花■☆

103149 Convolvulus muricatus L. = Ipomoea turbinata Lag. ■

103150 Convolvulus nashii House = Calystegia sepium (L.) R. Br. ■

103151 Convolvulus natalensis Bernh. ex Krauss;纳他尔旋花■☆

103152 Convolvulus natalensis Bernh. ex Krauss var. angustifolia C. H. Wright = Convolvulus natalensis Bernh. ex Krauss ■☆

103153 Convolvulus natalensis Bernh. ex Krauss var. integrifolia C. H. Wright = Convolvulus natalensis Bernh. ex Krauss ■☆

103154 Convolvulus natalensis Bernh. ex Krauss var. transvaalensis (Schltr.) A. Meeuse = Convolvulus natalensis Bernh. ex Krauss ■☆

103155 Convolvulus nervosus Burm. f. = Argyreia nervosa (Burm. f.) Bojer ●

103156 Convolvulus nil L. = Ipomoea nil (L.) Roth ■

103157 Convolvulus nolaniflorus Zipp. ex Span. = Ipomoea polymorpha Roem. et Schult. ■

103158 Convolvulus nummularius L. = Evolvulus nummularius (L.) L. ■

103159 Convolvulus obscurus L. = Ipomoea obscura (L.) Ker Gawl. ■

103160 Convolvulus obtectus Wall. = Argyreia mollis (Burm. f.) Choisy ●

103161 Convolvulus ocellatus Hook. f. ;单眼旋花■☆

103162 Convolvulus ocellatus Hook. f. var. ornatus (Engl.) A. Meeuse = Convolvulus ocellatus Hook. f. ■☆

103163 Convolvulus ochraceus Lindl. = Ipomoea ochracea (Lindl.) G. Don ■☆

103164 Convolvulus oenotherae Vatke = Ipomoea oenotherae (Vatke) Hallier f. ■☆

103165 Convolvulus oenotheroides L. f. = Ipomoea oenotheroides (L. f.) Raf. ex Hallier f. ■☆

103166 Convolvulus oleifolius Desr. ;木犀榄叶旋花■☆

103167 Convolvulus oleifolius Desr. var. angustifolius Bég. et Vacc. = Convolvulus oleifolius Desr. ■☆

103168 Convolvulus olgae Regel et Schmalh. ;奥氏旋花■☆

103169 Convolvulus ornatus Engl. = Convolvulus ocellatus Hook. f. ■☆

103170 Convolvulus ovalifolius Vahl = Jacquemontia ovalifolia (Vahl) Hallier f. ■☆

103171 Convolvulus panduratus L. ;琴叶旋花(提琴叶旋花)■☆

103172 Convolvulus paniculatus A. Rich. = Convolvulus farinosus L. ■☆

103173 Convolvulus paniculatus L. = Ipomoea mauritiana Jacq. ■

103174 Convolvulus parasiticus Kunth = Ipomoea parasitica (Kunth) G. Don ■☆

103175 Convolvulus parviflorus Spreng. = Convolvulus prostratus Forssk. ■☆

103176 Convolvulus parviflorus Vahl = Jacquemontia paniculata (Burm. f.) Hallier f. ■☆

103177 Convolvulus pellitus Ledeb. = Calystegia pellita (Ledeb.) G. Don ■

103178 Convolvulus pellitus Ledeb. f. anestius Fernald = Calystegia pubescens Lindl. ■

103179 Convolvulus peltatus L. = Merremia peltata (L.) Merr. ■☆

103180 Convolvulus penicillatus A. Rich. = Convolvulus farinosus L. ■☆

103181 Convolvulus pennatus Desr. = Ipomoea quamoclit L. ■

103182 Convolvulus pennatus Desr. = Quamoclit pennata (Desr.) Bojer ■

103183 Convolvulus pentanthus Jacq. = Jacquemontia pentantha (Jacq.) G. Don ■☆

103184 Convolvulus pentapetaloides L. ;五瓣旋花■☆

103185 Convolvulus pentaphyllus L. = Merremia aegyptia (L.) Urb. ■☆

103186 Convolvulus perfoliatus Schumach. et Thonn. = Ipomoea involucrata P. Beauv. ■☆

103187 Convolvulus perraudieri Coss. = Convolvulus fruticulosus Desr. ●☆

103188 Convolvulus persicus L. ;波斯旋花■☆

103189 Convolvulus pes-caprae L. = Ipomoea pes-caprae (L.) R. Br. ■

103190 Convolvulus pes-tigridis (L.) Spreng. = Ipomoea pes-tigridis L. ■

103191 Convolvulus phillipsiae Baker = Astripomoea malvacea (Klotzsch) A. Meeuse var. volkensii (Dammer) Verdc. ■☆

103192 Convolvulus phyllosepalus Hallier f. = Convolvulus sagittatus Thunb. ■☆

103193 Convolvulus pileatus (Roxb.) Spreng. = Ipomoea pileata Roxb. ■

103194 Convolvulus pilosellifolius Desr. ;毛叶旋花■☆

103195 Convolvulus pilosellifolius Desr. var. orreanus Murb. = Convolvulus pilosellifolius Desr. ■☆

103196 Convolvulus pilosus Roxb. = Ipomoea dichroa Choisy ■☆

103197 Convolvulus pitardii Batt. ;皮塔德旋花■☆

103198 Convolvulus pitardii Batt. var. leucochnous (Benoist) Maire = Convolvulus pitardii Batt. ■☆

103199 Convolvulus plantagineus Choisy = Ipomoea simplex Thunb. ■☆

103200 Convolvulus plebeius (R. Br.) Spreng. = Ipomoea biflora (L.) Pers. ■

103201 Convolvulus plicatus Desr. = Convolvulus capensis Burm. f. ■☆

103202 Convolvulus pluricaulis Choisy = Convolvulus prostratus Forssk. ■☆

103203 Convolvulus pluricaulis Choisy var. longipes Maire = Convolvulus prostratus Forssk. ■☆

103204 Convolvulus prostratus Forssk. ;平卧旋花■☆

103205 Convolvulus prostratus Forssk. var. deserti (Hochst. et Steud. ex Baker et Rendle) Parmar = Convolvulus deserti Hochst. et Steud. ex Baker et Rendle ■☆

103206 Convolvulus pseudocantabrica Schrenk; 直立旋花; Erect Glorybind ■

103207 Convolvulus pseudocantabrica Schrenk subsp. dianthoides Vved. = Convolvulus pseudocantabrica Schrenk ■

103208 Convolvulus pseudocantabricus Schrenk subsp. dianthoides (Kar. et Kir.) Vved. = Convolvulus pseudocantabrica Schrenk ■

103209 Convolvulus pseudoscammonia K. Koch;假矢叶旋花■☆

103210 Convolvulus pungens Kar. et Kir. = Convolvulus gortschakovii Schrenk ●

103211 Convolvulus purga Wender. ;泻薯(旋花)■☆

103212 Convolvulus purpureus L. = Ipomoea purpurea (L.) Roth ■

103213 Convolvulus pycnanthus Hochst. ex Choisy = Jacquemontia tamnifolia (L.) Griseb. ■☆

103214 Convolvulus quamoclit Spreng. = Quamoclit pennata (Desr.) Bojer ■

103215 Convolvulus quinatus (R. Br.) Spreng. = Merremia quinata (R. Br.) Ooststr. ■

103216 Convolvulus quinqueflorus Vahl = Convolvulus farinosus L. ■☆

103217 Convolvulus racemosus Roem. et Schult. = Ipomoea sumatrana

(Blume) Ooststr. ■
103218 Convolvulus radicans Thunb. = Ipomoea imperati (Vahl) Griseb. ■
103219 Convolvulus randii Rendle = Convolvulus ocellatus Hook. f. ■☆
103220 Convolvulus refractus Pomel = Convolvulus siculus L. subsp. elongatus Batt. ■☆
103221 Convolvulus reniformis (R. Br.) Poir. = Calystegia soldanella (L.) R. Br. ■
103222 Convolvulus reniformis Roxb. = Merremia emarginata (Burm. f.) Hallier f. ■
103223 Convolvulus repens L. = Calystegia sepium (L.) R. Br. ■
103224 Convolvulus repens Vahl = Ipomoea aquatica Forssk. ■
103225 Convolvulus retans L. = Merremia hirta (L.) Merr. ■
103226 Convolvulus rhynchophyllus Baker ex Hallier f. = Convolvulus sagittatus Thunb. ■☆
103227 Convolvulus rhyniospermus Hochst. ex Choisy;锉籽旋花■☆
103228 Convolvulus rnaritimus Desr. = Ipomoea pes-caprae (L.) R. Br. ■
103229 Convolvulus robertianus Spreng. = Ipomoea polymorpha Roem. et Schult. ■
103230 Convolvulus roxburghii Wall. = Argyreia roxburghii (Wall.) Arn. ex Choisy ●
103231 Convolvulus ruber Vahl = Ipomoea setifera Poir. ■☆
103232 Convolvulus ruspolii Dammer ex Hallier f.;鲁斯波利旋花■☆
103233 Convolvulus ruspolii Dammer ex Hallier f. var. pilosa Sebsebe;疏毛旋花■☆
103234 Convolvulus sabatius Viv.;北非旋花(摩洛哥旋花);Ground Morning Glory, Morocco Glorybind ■☆
103235 Convolvulus sabatius Viv. subsp. mauritanicus (Boiss.) Murb.;毛里塔尼亚旋花■☆
103236 Convolvulus sabatius Viv. var. atlanticus Ball = Convolvulus sabatius Viv. ■☆
103237 Convolvulus sagittatus Thunb.;箭叶旋花■☆
103238 Convolvulus sagittatus Thunb. var. abyssinicus (Hallier f.) Rendle = Convolvulus sagittatus Thunb. ■☆
103239 Convolvulus sagittatus Thunb. var. aschersonii (Engl.) Verdc. = Convolvulus aschersonii Engl. ■☆
103240 Convolvulus sagittatus Thunb. var. graminifolius (Hallier f.) Baker et C. H. Wright ex A. Meeuse = Convolvulus sagittatus Thunb. ■☆
103241 Convolvulus sagittatus Thunb. var. grandiflorus (Hallier f.) A. Meeuse = Convolvulus sagittatus Thunb. ■☆
103242 Convolvulus sagittatus Thunb. var. grandiflorus Hallier f. = Convolvulus sagittatus Thunb. ■☆
103243 Convolvulus sagittatus Thunb. var. hirtellus (Hallier f.) A. Meeuse = Convolvulus sagittatus Thunb. ■☆
103244 Convolvulus sagittatus Thunb. var. latifolius C. H. Wright = Convolvulus sagittatus Thunb. ■☆
103245 Convolvulus sagittatus Thunb. var. linearifolius (Hallier f.) Baker et C. H. Wright = Convolvulus sagittatus Thunb. ■☆
103246 Convolvulus sagittatus Thunb. var. linearifolius (Hallier f.) Baker et C. H. Wright ex A. Meeuse = Convolvulus sagittatus Thunb. ■☆
103247 Convolvulus sagittatus Thunb. var. namaquensis A. Meeuse = Convolvulus sagittatus Thunb. ■☆
103248 Convolvulus sagittatus Thunb. var. parviflorus Hallier f. = Convolvulus sagittatus Thunb. ■☆
103249 Convolvulus sagittatus Thunb. var. phyllosepalus (Hallier f.) A. Meeuse = Convolvulus sagittatus Thunb. ■☆
103250 Convolvulus sagittatus Thunb. var. phyllosepalus Hallier f. = Convolvulus sagittatus Thunb. ■☆
103251 Convolvulus sagittatus Thunb. var. subcordata (Hallier f.) Baker = Convolvulus sagittatus Thunb. ■☆
103252 Convolvulus sagittatus Thunb. var. ulosepalus (Hallier f.) Verdc. = Convolvulus sagittatus Thunb. ■☆
103253 Convolvulus sagittatus Thunb. var. villosus (Hallier f.) Rendle = Convolvulus sagittatus Thunb. ■☆
103254 Convolvulus sagittifolius (Fisch.) T. Liou et Y. Ling = Convolvulus arvensis L. ■
103255 Convolvulus scammonia L.;矢叶旋花;Levant Scammony, Scammony, Scammony Bindweed, Scammony Glorybind, Syrian Bindweed ■☆
103256 Convolvulus scammonia L. = Calystegia hederacea Wall. ex Roxb. ■
103257 Convolvulus scammonia Lour. = Calystegia hederacea Wall. ex Roxb. ■
103258 Convolvulus scandens J. König ex Milne = Hewittia malabarica (L.) Suresh ■
103259 Convolvulus scandens Milne = Hewittia malabarica (L.) Suresh ■
103260 Convolvulus schimperi Engl. = Convolvulus kilimandschari Engl. ■☆
103261 Convolvulus schweinfurthii Engl. = Convolvulus farinosus L. ■☆
103262 Convolvulus scoparius L. f.;帚枝旋花■☆
103263 Convolvulus scopulatus Thulin;岩栖旋花■☆
103264 Convolvulus semidigynus Roxb. = Bonamia semidigyna (Roxb.) Hallier f. ●☆
103265 Convolvulus senegambiae Spreng. = Bonamia thunbergiana (Roem. et Schult.) F. N. Williams ●☆
103266 Convolvulus sepium L. = Calystegia sepium (L.) R. Br. ■
103267 Convolvulus sepium L. f. coloratus Lange = Calystegia sepium (L.) R. Br. ■
103268 Convolvulus sepium L. f. malacophyllus Fernald = Calystegia sepium (L.) R. Br. ■
103269 Convolvulus sepium L. var. americanus Sims = Calystegia sepium (L.) R. Br. ■
103270 Convolvulus sepium L. var. communis R. M. Tryon = Calystegia sepium (L.) R. Br. ■
103271 Convolvulus sepium L. var. fraterniflorus Mack. et Bush = Calystegia silvatica (Kit.) Griseb. ■☆
103272 Convolvulus sepium L. var. pubescens (A. Gray) Fernald = Calystegia spithamaea (L.) Pursh ■☆
103273 Convolvulus sepium L. var. repens (L.) A. Gray = Calystegia sepium (L.) R. Br. ■
103274 Convolvulus ser Spreng. = Ipomoea biflora (L.) Pers. ■
103275 Convolvulus sericeus L. = Argyreia mollis (Burm. f.) Choisy ●
103276 Convolvulus sericocephalus Juz.;绢毛头旋花■☆
103277 Convolvulus sericophyllus T. Anderson;绢毛叶旋花■☆
103278 Convolvulus sessiliflorus (Roth) Spreng. = Ipomoea eriocarpa R. Br. ■
103279 Convolvulus setosus (Ker Gawl.) Spreng. = Ipomoea setosa Ker Gawl. ■
103280 Convolvulus sibiricus L. = Merremia sibirica (L.) Hallier f. ■
103281 Convolvulus siculus L.;西西里旋花■☆
103282 Convolvulus siculus L. subsp. agrestis (Hochst. ex Schweinf.)

Verdc. ;野生旋花■☆

103283 Convolvulus siculus L. subsp. elongatus Batt. ;伸长旋花■☆

103284 Convolvulus siculus L. subsp. pseudosiculus (Cav.) Fiori = Convolvulus siculus L. subsp. elongatus Batt. ■☆

103285 Convolvulus siculus L. var. flexuosus (Pomel) Batt. = Convolvulus siculus L. ■☆

103286 Convolvulus sidifolius Kunth = Turbina corymbosa (L.) Raf. ■☆

103287 Convolvulus sinensis Desr. = Ipomoea biflora (L.) Pers. ■

103288 Convolvulus sinensis L. = Ipomoea biflora (L.) Pers. ■

103289 Convolvulus sinuatus Petagna = Ipomoea imperati (Vahl) Griseb. ■

103290 Convolvulus soldanella L. = Calystegia soldanella (L.) R. Br. ex Roem. et Schult. ■

103291 Convolvulus soldanellus L. = Calystegia soldanella (L.) R. Br. ■

103292 Convolvulus somalensis Franch. = Convolvulus sericophyllus T. Anderson ■☆

103293 Convolvulus somalensis Vatke = Merremia somalensis (Vatke) Hallier f. ■☆

103294 Convolvulus speciosus L. f. = Argyreia nervosa (Burm. f.) Bojer ●

103295 Convolvulus sphaerocephalus Roxb. = Argyreia pierreana Bois ●

103296 Convolvulus sphaerophorus Baker = Convolvulus rhyniospermus Hochst. ex Choisy ■☆

103297 Convolvulus spicifolius Desr. = Convolvulus lineatus L. ■

103298 Convolvulus spinifer Popov = Convolvulus tragacanthoides Turcz. ■

103299 Convolvulus spinosus Bunge = Convolvulus tragacanthoides Turcz. ■

103300 Convolvulus spinosus Desr. = Convolvulus fruticosus Pall. ●

103301 Convolvulus spithamaeus L. = Calystegia spithamaea (L.) Pursh ■☆

103302 Convolvulus spithamaeus L. subsp. stans (Michx.) Wherry = Calystegia spithamaea (L.) Pursh ■☆

103303 Convolvulus spithamaeus L. var. pubescens (A. Gray) Fernald = Calystegia spithamaea (L.) Pursh ■☆

103304 Convolvulus spithamaeus L. var. stans (Michx.) Farw. = Calystegia spithamaea (L.) Pursh ■☆

103305 Convolvulus splendens Hornem. = Argyreia splendens (Roxb.) Sweet ●

103306 Convolvulus stachydifolius Choisy;穗叶旋花■☆

103307 Convolvulus stachydifolius Choisy var. villosus Hallier f. = Convolvulus stachydifolius Choisy ■☆

103308 Convolvulus stans Michx. = Calystegia spithamaea (L.) Pursh ■☆

103309 Convolvulus stenocladus Chiov. ;狭枝旋花■☆

103310 Convolvulus steppicola Hand. -Mazz. ;草坡旋花;Grasslaving Glorybind ■

103311 Convolvulus steudneri Engl. ;斯托德旋花■☆

103312 Convolvulus stipulatus Desr. = Ipomoea coptica (L.) Roth ex Roem. et Schult. ■☆

103313 Convolvulus stolonifer Cirillo = Ipomoea stolonifera (Cirillo) J. F. Gmel. ■

103314 Convolvulus stoloniferus Cirillo = Ipomoea imperati (Vahl) Griseb. ■

103315 Convolvulus strigosus Wall. = Argyreia capitiformis (Poir.) Ooststr. ●

103316 Convolvulus subauriculatus (Burch.) Linding. ;耳形旋花■☆

103317 Convolvulus subhirsutus Regel et Schmalh. ;假毛旋花■☆

103318 Convolvulus sublobatus L. f. = Hewittia malabarica (L.) Suresh ■

103319 Convolvulus subseticeus Schrenk;亚刚毛旋花■☆

103320 Convolvulus subspathulatus Vatke;亚匙形旋花■☆

103321 Convolvulus suffruticosus Desf. = Convolvulus valentinus Cav. subsp. suffruticosus (Desf.) Maire ●☆

103322 Convolvulus suffruticosus Desf. var. ovatus Andr. = Convolvulus valentinus Cav. subsp. suffruticosus (Desf.) Maire ●☆

103323 Convolvulus supinus Coss. et Kralik;仰卧形旋花■☆

103324 Convolvulus supinus Coss. et Kralik subsp. brevipes (Pomel) Quézel et Santa = Convolvulus supinus Coss. et Kralik ■☆

103325 Convolvulus supinus Coss. et Kralik var. atrichogynus Maire et Wilczek = Convolvulus supinus Coss. et Kralik ■☆

103326 Convolvulus supinus Coss. et Kralik var. leucotrichus (Pomel) Batt. = Convolvulus supinus Coss. et Kralik ■☆

103327 Convolvulus supinus Coss. et Kralik var. sulphurescens Maire et Wilczek = Convolvulus supinus Coss. et Kralik ■☆

103328 Convolvulus supinus Coss. et Kralik var. tripolitanus Borzí et Mattei = Convolvulus supinus Coss. et Kralik ■☆

103329 Convolvulus tauricus (Bornm.) Juz. ;克里木旋花■☆

103330 Convolvulus tenuissimus Sibth. et Sm. = Convolvulus elegantissimus Mill. ■☆

103331 Convolvulus thomsonii Baker = Convolvulus sagittatus Thunb. ■☆

103332 Convolvulus thonningii Schumach. et Thonn. = Ipomoea coptica (L.) Roth ex Roem. et Schult. ■☆

103333 Convolvulus thunbergianus Roem. et Schult. = Bonamia thunbergiana (Roem. et Schult.) F. N. Williams ●☆

103334 Convolvulus thunbergii Hallier f. = Convolvulus multifidus Thunb. ■☆

103335 Convolvulus thunbergii Roem. et Schult. ;通贝里旋花■☆

103336 Convolvulus tiliifolius Desr. = Stictocardia tiliifolia (Desr.) Hallier f. ●■

103337 Convolvulus trabutianus Schweinf. et Muschl. ;特拉布特旋花■☆

103338 Convolvulus tragacanthoides Turcz. ;刺旋花(木旋花);Spiny Glorybind ■

103339 Convolvulus transvaalensis Schltr. = Convolvulus natalensis Bernh. ex Krauss ■☆

103340 Convolvulus tricolor L. ;三色旋花(彩旋花,旋花);Convolvulus, Dwarf Glorybind, Dwarf Morning Glory, Dwarf Morning-glory, Dwart Morning Glory, Three-coloured Bindweed ■☆

103341 Convolvulus tricolor L. 'Blue Flash';蓝光三色旋花■☆

103342 Convolvulus tricolor L. 'Flying Saucers';飞碟三色旋花■☆

103343 Convolvulus tricolor L. subsp. cupanianus (Tod.) Cavara et Grande;库潘旋花■☆

103344 Convolvulus tricolor L. subsp. hortensis (Batt.) Maire = Convolvulus tricolor L. ■☆

103345 Convolvulus tricolor L. subsp. meonanthus (Hoffmanns. et Link) Maire = Convolvulus meonanthus Hoffmanns. et Link ■☆

103346 Convolvulus tricolor L. subsp. pentapetaloides (L.) O. Bolòs et Vigo = Convolvulus pentapetaloides L. ■☆

103347 Convolvulus tricolor L. var. guttatus Batt. et Maire = Convolvulus tricolor L. ■☆

103348 Convolvulus tricolor L. var. heterocalyx Maire = Convolvulus tricolor L. ■☆

103349 Convolvulus tricolor L. var. maroccanus (Batt.) Maire = Convolvulus tricolor L. ■☆

103350 Convolvulus tricolor L. var. pseudotricolor (Bertol.) Fiori = Convolvulus tricolor L. ■☆

103351 Convolvulus tricolor L. var. pseudotricolor (Viv.) Fiori = Convolvulus tricolor L. ■☆
103352 Convolvulus tricolor L. var. quadricolor Batt. et Maire = Convolvulus tricolor L. ■☆
103353 Convolvulus tridentatus L. = Merremia tridentata (L.) Hallier f. ■
103354 Convolvulus tridentatus L. = Xenostegia tridentata (L.) D. F. Austin et Staples ■
103355 Convolvulus trifidus Kunth = Ipomoea trifida (Kunth) G. Don ■☆
103356 Convolvulus trilobus (L.) Desr. = Ipomoea triloba L. ■
103357 Convolvulus trilobus Thunb. = Ipomoea ficifolia Lindl. ■☆
103358 Convolvulus trinervis Thunb. = Tripterospermum japonicum (Siebold et Zucc.) Maxim. ■
103359 Convolvulus tuba Schltdl. = Ipomoea macrantha Roem. et Schult. ■
103360 Convolvulus tuba Schltdl. = Ipomoea violacea L. ■
103361 Convolvulus tuberculatus Desr. = Ipomoea cairica (L.) Sweet ■
103362 Convolvulus turpethum L.;盒果藤(红薯藤,宽筋藤,软筋藤,松筋藤,印度牵牛,紫翅藤);Boxfruitvine, Foully Operculina, St. Thomas Lidpod ■
103363 Convolvulus turpethum L. = Operculina turpetha (L.) Silva Manso ■
103364 Convolvulus ulosepalus Hallier f. = Convolvulus sagittatus Thunb. ■☆
103365 Convolvulus umbellatus L. = Merremia umbellata (L.) Hallier f. ■
103366 Convolvulus undulatus Cav. = Convolvulus humilis Jacq. ■☆
103367 Convolvulus valentinus Cav.;强壮旋花■☆
103368 Convolvulus valentinus Cav. subsp. suffruticosus (Desf.) Maire;半灌木旋花●☆
103369 Convolvulus valentinus Cav. var. adpressipilis Maire et Wilczek = Convolvulus valentinus Cav. ■☆
103370 Convolvulus valentinus Cav. var. debilis Sennen et Mauricio = Convolvulus valentinus Cav. ■☆
103371 Convolvulus valentinus Cav. var. embergeri Sauvage et Vindt = Convolvulus valentinus Cav. ■☆
103372 Convolvulus valentinus Cav. var. melillensis Pau = Convolvulus valentinus Cav. ■☆
103373 Convolvulus valentinus Cav. var. oranensis Pomel = Convolvulus valentinus Cav. ■☆
103374 Convolvulus valentinus Cav. var. simulans Maire = Convolvulus valentinus Cav. ■☆
103375 Convolvulus valentinus Cav. var. sulfureus Batt. = Convolvulus valentinus Cav. ■☆
103376 Convolvulus valentinus Cav. var. transfretanus Pau et Font Quer = Convolvulus valentinus Cav. ■☆
103377 Convolvulus valentinus Cav. var. transiens Maire et Wilczek = Convolvulus valentinus Cav. ■☆
103378 Convolvulus venosus Desr. = Ipomoea venosa (Desr.) Roem. et Schult. ■☆
103379 Convolvulus ventricosus Silva Manso;汤氏盒果藤;St. Thomas Lidpod ■☆
103380 Convolvulus verdcourtianus Sebsebe;韦尔德旋花■☆
103381 Convolvulus verrucosus (Blume) D. Dietr. = Ipomoea marginata (Desr.) Verdc. ■
103382 Convolvulus vidalii Pau;韦达尔旋花■☆
103383 Convolvulus violaceus Vahl = Jacquemontia paniculata (Burm. f.) Hallier f. ■☆
103384 Convolvulus vitifolius Burm. f. = Merremia vitifolia (Burm. f.) Hallier f. ■
103385 Convolvulus vollesenii Sebsebe;福勒森旋花■☆
103386 Convolvulus volubilis Link;缠绕旋花■☆
103387 Convolvulus wallichianus Spreng.;沃利克旋花;Wallich's Bindweed ■☆
103388 Convolvulus wallichianus Spreng. = Calystegia hederacea Wall. ex Roxb. ■
103389 Convolvulus wightii Wall. = Ipomoea wightii (Wall.) Choisy ■☆
103390 Convolvulus zernyi Schulze-Menz = Convolvulus bussei Pilg. ●☆
103391 Convovlulus strigosus Wall. = Argyreia capitiformis (Poir.) Ooststr. ●
103392 Conysa Adans. = Conyza Less.(保留属名)■
103393 Conysa Burm. f. = Conyza Less.(保留属名)■
103394 Conysa anthelmintica L. = Vernonia anthelmintica (L.) Willd. ■
103395 Conysa attenuate Wall. = Vernonia attenuata (Wall.) DC. ■
103396 Conysa extensa Wall. = Vernonia extensa (Wall.) DC. ●
103397 Conysa patula Dryand. = Vernonia patula (Dryand.) Merr. ■
103398 Conysa saligna Wall. = Vernonia saligna (Wall.) DC. ■
103399 Conystylus Pritz. = Ascochilopsis Carr ■☆
103400 Conyza Hill = Conyza Less.(保留属名)■
103401 Conyza L.(废弃属名)= Conyza Less.(保留属名)■
103402 Conyza Less.(1832)(保留属名);白酒草属(假蓬属);Conyza, Fleabane, Horseweed ■
103403 Conyza absinthifolia DC. = Conyza stricta Willd. ex DC. var. pinnatifida (D. Don) Kitam. ■
103404 Conyza abyssinica Sch. Bip. ex A. Rich.;阿比西尼亚白酒草■☆
103405 Conyza adolfi-friderici (Muschl.) Wild = Conyza vernonioides (Sch. Bip. ex A. Rich.) Wild ■☆
103406 Conyza adolfi-friderici Muschl. = Conyza vernonioides (Sch. Bip. ex A. Rich.) Wild ■☆
103407 Conyza aegyptiaca (L.) Aiton;埃及白酒草(埃及假蓬);Egypt Conyza ■
103408 Conyza aegyptiaca (L.) Aiton var. lineariloba (DC.) O. Hoffm. = Conyza aegyptiaca (L.) Aiton ■
103409 Conyza ageratoides DC.;藿香蓟白酒草●☆
103410 Conyza alata (D. Don) Roxb. = Laggera alata (D. Don) Sch. Bip. ex Oliv. ■
103411 Conyza alata Roxb. = Laggera alata (D. Don) Sch. Bip. ex Oliv. ■
103412 Conyza albida Spreng. = Conyza sumatrensis (Retz.) E. Walker ■
103413 Conyza altaica DC. = Brachyactis ciliata Ledeb. ■
103414 Conyza ambigua DC. = Conyza bonariensis (L.) Cronquist ■
103415 Conyza andringitrana Humbert;安德林吉特拉山白酒草●☆
103416 Conyza androrangensis Humbert;安德鲁兰加白酒草●☆
103417 Conyza anthelmintica L. = Vernonia anthelmintica (L.) Willd. ■
103418 Conyza apiculata Hutch. et M. B. Moss = Conyza ruwenzoriensis (S. Moore) R. E. Fr. ■☆
103419 Conyza arabica Willd. = Blumea decurrens (Vahl) Merxm. ■☆
103420 Conyza argentea Wall. = Duhaldea chinensis DC. ●■
103421 Conyza argentea Wall. = Inula cappa (Buch.-Ham.) DC. ●■
103422 Conyza asteroides DC. = Conyza japonica (Thunb.) Less. ■
103423 Conyza attenuata DC.;渐狭白酒草■☆
103424 Conyza attenuata DC. var. hispidula;硬毛白酒草●☆
103425 Conyza attenuata Wall. = Vernonia attenuata (Wall.) DC. ■
103426 Conyza auriculifera R. E. Fr.;耳状白酒草●☆
103427 Conyza aurita L. f. = Pseudoconyza viscosa (Mill.) D'Arey ■☆

103428 Conyza axillaris Lam. = Blumea axillaris (Lam.) DC. ■
103429 Conyza baccharis Mill. = Pluchea baccharis (Mill.) Pruski ■☆
103430 Conyza bakeri Humbert;贝克白酒草■☆
103431 Conyza balsamifera L. = Blumea balsamifera (L.) DC. ■
103432 Conyza bampsiana (Lisowski) Lisowski;邦氏白酒草■☆
103433 Conyza bifoliatus Walter = Sericocarpus tortifolius (Michx.) Nees ■☆
103434 Conyza bilbaoana Remy = Conyza floribunda Kunth ■☆
103435 Conyza blanda Wall. = Vernonia blanda (Wall.) DC. ●■
103436 Conyza blinii H. Lév.;熊胆草(矮脚苦蒿,虎胆草,金蒿枝,金龙胆草,劲直假蓬,苦艾,苦草,苦丁,苦蒿,苦蒿尖,苦龙胆,苦龙胆草,刘寄奴,龙胆草,龙胆蒿,毛苦蒿,细苦蒿,油蒿,鱼胆草);Blin Conyza,Blin's Conyza ■
103437 Conyza bojeri DC. = Pluchea bojeri (DC.) Humbert ■☆
103438 Conyza bonariensis (L.) Cronquist;香丝草(火苗草,美洲假蓬,蓑衣草,消息草,小白菊,小加蓬,小山艾,野地黄菊,野塘蒿,野桐蒿);Argentine Fleabane,Asthmaweed,Bona Conyza,Buenos Aires Conyza ■
103439 Conyza bonariensis (L.) Cronquist var. leiotheca (S. F. Blake) Cuatrecasas = Conyza floribunda Kunth ■☆
103440 Conyza boranensis (S. Moore) Cufod.;博兰白酒草■☆
103441 Conyza bovei DC. = Blumea bovei (DC.) Vatke ■☆
103442 Conyza brevipetiolata (Muschl.) Lisowski = Conyza vernonioides (Sch. Bip. ex A. Rich.) Wild ■☆
103443 Conyza britannica (L.) Rupr. = Inula britannica L. ■
103444 Conyza cafra DC. = Blumea cafra (DC.) O. Hoffm. ■☆
103445 Conyza calycina Cav. = Phagnalon calycinum (Cav.) DC. ■☆
103446 Conyza canadensis (L.) Cronquist;小蓬草(臭艾,飞蓬,加拿大飞蓬,加拿大蓬,苦蒿,破布艾,祁州一枝蒿,蛇舌草,小白酒草,小飞蓬,小山艾,鱼胆草,鱼肥草,竹叶艾);Bitterweed,Blood Staunch,Butterweed,Canada Gentian,Canadian Fleabane,Canadian Horseweed,Cobbler's Pegs,Colt's Tail,Colt's Tails,Cow's Tail,Cow's Tails,Deadweed,Fireweed,Fleabane,Fleawort,Hog Weed,Hogweed,Horse Weed,Horseweed,Horseweed Fleabane,Horseweed Fleahane,Mare's Tail,Mare's Tails,Prideweed,Scabious ■
103447 Conyza canadensis (L.) Cronquist var. glabrata (A. Gray) Cronquist = Conyza canadensis (L.) Cronquist ■
103448 Conyza canadensis (L.) Cronquist var. gracilis K. M. Liu;细叶小蓬草■
103449 Conyza canadensis (L.) Cronquist var. pusilla (Nutt.) Cronquist;光茎飞蓬■
103450 Conyza canadensis (L.) Cronquist var. pusilla (Nutt.) Cronquist = Conyza canadensis (L.) Cronquist ■
103451 Conyza canadensis (L.) Cronquist var. pusilla (Nutt.) Cronquist = Erigeron pusillus Nutt. ■
103452 Conyza canariensis Willd. = Allagopappus canariensis (Willd.) Greuter ■☆
103453 Conyza candolleana Boiss. = Pluchea wallichiana DC. ■☆
103454 Conyza canescens L. f. = Vernonia capensis (Houtt.) Druce ■☆
103455 Conyza cappa Buch. -Ham. ex D. Don = Duhaldea chinensis DC. ●■
103456 Conyza carolinensis J. Jacq. = Pluchea carolinensis (J. Jacq.) G. Don ●
103457 Conyza chilensis Spreng.;智利白酒草■☆
103458 Conyza chinensis L. = Vernonia cinerea (L.) Less. ■
103459 Conyza chinensis Lour. = Blumea lanceolaria (Roxb.) Druce ■
103460 Conyza chrysocoma (DC.) Vatke = Conyza stricta Willd. ex DC. ■
103461 Conyza chrysocomoides Desf. = Nolletia chrysocomoides (Desf.) Cass. ■●☆
103462 Conyza cinerea L. = Cyanthillium cinereum (L.) H. Rob. ■
103463 Conyza cinerea L. = Vernonia cinerea (L.) Less. ■
103464 Conyza clarenceana (Hook. f.) Oliv. et Hiern;克拉伦斯白酒草■☆
103465 Conyza conspicua Wall. = Aster albescens (DC.) Wall. ex Hand. -Mazz. ●
103466 Conyza coulteri A. Gray = Laennecia coulteri (A. Gray) G. L. Nesom ■☆
103467 Conyza crispata Vahl = Blumea crispata (Vahl) Merxm. ■☆
103468 Conyza crispata Vahl = Laggera pterodonta (DC.) Benth. ■
103469 Conyza dasycoma Miq. var. pinnatifida Miq. = Blumea densiflora DC. ■
103470 Conyza decurrens L. = Neojeffreya decurrens (L.) Cabrera ■☆
103471 Conyza dentata Blanco = Blumea lacera (Burm. f.) DC. ■
103472 Conyza dentata Blanco = Duhaldea chinensis DC. ●■
103473 Conyza dentata Blanco = Inula cappa (Buch. -Ham.) DC. ●■
103474 Conyza dunniana H. Lév. = Conyza blinii H. Lév. ■
103475 Conyza echioides A. Rich. = Conyza aegyptiaca (L.) Aiton ■
103476 Conyza ellisii Baker = Conyza neocandolleana Humbert ●☆
103477 Conyza eriophora Wall. = Duhaldea chinensis DC. ●■
103478 Conyza eriophora Wall. = Inula cappa (Buch. -Ham.) DC. ●■
103479 Conyza eriophylla (A. Gray) Cronquist = Laennecia eriophylla (A. Gray) G. L. Nesom ■☆
103480 Conyza eupatorioides Wall. = Inula eupatorioides DC. ●
103481 Conyza extensa Wall. = Vernonia extensa (Wall.) DC. ●
103482 Conyza feae (Bég.) Wild;费厄白酒草■☆
103483 Conyza filaginoides (DC.) Hieron.;似絮菊假蓬■☆
103484 Conyza filaginoides (DC.) Hieron. = Laennecia filaginoides DC. ■☆
103485 Conyza fismlosa Roxb. = Blumea fistulosa (Roxb.) Kurz ■
103486 Conyza flabellata Mesfin;扇状白酒草■☆
103487 Conyza flexilis DC. = Psiadia lucida (Cass.) Drake ●☆
103488 Conyza floribunda Kunth = Conyza albida Spreng. ■
103489 Conyza floribunda Kunth = Conyza sumatrensis (Retz.) E. Walker ■
103490 Conyza fruticulosa O. Hoffm.;灌木状白酒草■☆
103491 Conyza gallianii Chiov.;加利安白酒草■☆
103492 Conyza gigantea O. Hoffm.;巨大白酒草■☆
103493 Conyza glabrescens Pax;渐光白酒草■☆
103494 Conyza gouanii (L.) Willd.;古安白酒草■☆
103495 Conyza heudelotii Oliv. et Hiern = Microglossa pyrifolia (Lam.) Kuntze ●
103496 Conyza hieracifolia Spreng. = Blumea hieraciifolia (D. Don) DC. ■
103497 Conyza hirtella DC.;多毛白酒草●☆
103498 Conyza hochstetteri Sch. Bip. ex A. Rich.;霍赫白酒草■☆
103499 Conyza hochstetteri Sch. Bip. ex A. Rich. var. agrestis Vatke = Conyza gouanii (L.) Willd. ■☆
103500 Conyza hochstetteri Sch. Bip. ex A. Rich. var. glabra A. Rich. = Conyza gouanii (L.) Willd. ■☆
103501 Conyza hochstetteri Sch. Bip. ex A. Rich. var. silvestris Vatke = Conyza tigrensis Oliv. et Hiern ■☆
103502 Conyza hochstteteri Sch. Bip. ex A. Rich. var. montana Vatke = Conyza variegata Sch. Bip. ex A. Rich. ■☆

103503 Conyza hypoleuca A. Rich.;里白白酒草■☆
103504 Conyza iliensis Trautv. = Psychrogeton nigromontanus (Boiss. et Buhse) Grierson ■
103505 Conyza incana (Vahl) Willd.;灰毛白酒草■☆
103506 Conyza incisa Aiton = Conyza ulmifolia (Burm. f.) Kuntze ■☆
103507 Conyza ivifolia (L.) Less. = Conyza scabrida DC. ■☆
103508 Conyza ivifolia Burm. f.;奥尔巴尼白酒草;Albany Gall-sick Bush,Oven Bush ■☆
103509 Conyza japonica (Thunb.) Less.;白酒草(白桦白酒草,白酒棵,白酒香,喉痛草,假蓬,酒香草,酒药草,日本假蓬,山地菊,银钮子,鱼腥草);Japan Conyza,Japanese Conyza ■
103510 Conyza kotschyi (Sch. Bip. ex Schweinf. et Asch.) Sch. Bip. ex Schweinf. et Asch. = Pentanema indicum (L.) Y. Ling ■
103511 Conyza lacera Burm. f. = Blumea lacera (Burm. f.) DC. ■
103512 Conyza laciniata Roxb. = Blumea laciniata (Roxb.) DC. ■
103513 Conyza lanceolaria Roxb. = Blumea lanceolaria (Roxb.) Druce ■
103514 Conyza lanuginosa Wall. = Duhaldea chinensis DC. ●■
103515 Conyza lanuginosa Wall. = Inula cappa (Buch.-Ham.) DC. ●■
103516 Conyza leucantha (D. Don) Ludlow et Raven;黏毛白酒草(白花白酒草,假蓬,粘毛假蓬);Whiteflower Conyza ■
103517 Conyza leucodasys Miq. = Conyza bonariensis (L.) Cronquist ■
103518 Conyza leucophylla Sch. Bip. ex A. Rich. = Conyza incana (Vahl) Willd. ■☆
103519 Conyza limosa O. Hoffm.;湿地白酒草■☆
103520 Conyza lineariloba (O. Hoffm.) DC. = Conyza aegyptiaca (L.) Aiton ■
103521 Conyza lineariloba DC. = Conyza aegyptiaca (L.) Aiton ■
103522 Conyza linifolia (Willd.) Täckh. = Conyza bonariensis (L.) Cronquist ■
103523 Conyza linifolia L.;亚麻叶白酒草■☆
103524 Conyza linifolia L. = Sericocarpus linifolius Britton, Sterns et Poggenb. ■☆
103525 Conyza longipedunculata Klatt;长梗白酒草●☆
103526 Conyza lyrata Kunth = Pseudoconyza viscosa (Mill.) D'Arey ■☆
103527 Conyza macrorrhiza Sch. Bip. ex A. Rich. = Conyza stricta Willd. ex DC. ■
103528 Conyza madagascariensis Lam. = Psiadia altissima (DC.) Drake ●☆
103529 Conyza mairei H. Lév. = Conyza stricta Willd. ex DC. var. pinnatifida (D. Don) Kitam. ■
103530 Conyza mandrarensis Humbert;曼德拉白酒草●☆
103531 Conyza megensis F. G. Davies;梅加白酒草●☆
103532 Conyza messeri Pic. Serm.;梅瑟白酒草●☆
103533 Conyza mildbraedii (Muschl.) Robyns = Conyza limosa O. Hoffm. ■☆
103534 Conyza miniata Klatt = Psiadia leucophylla (Baker) Humbert ●☆
103535 Conyza modatensis Sch. Bip. = Pluchea dioscoridis (L.) DC. ●☆
103536 Conyza mollis H. Lév. = Anaphalis bulleyana (Jeffrey) C. C. Chang ■
103537 Conyza montevidensis Spreng. = Baccharis pingraea DC. ●☆
103538 Conyza montigena S. Moore;山生白酒草■☆
103539 Conyza muliensis Y. L. Chen;木里白酒草;Muli Conyza ■
103540 Conyza multicaulis DC. = Conyza japonica (Thunb.) Less. ■
103541 Conyza nana Sch. Bip. ex Oliv. et Hiern;矮白酒草■☆
103542 Conyza natalensis Sch. Bip. = Blumea cafra (DC.) O. Hoffm. ■☆
103543 Conyza naudinii Bonnet = Conyza bonariensis (L.) Cronquist ■
103544 Conyza neglecta R. E. Fr.;忽视白酒草■☆
103545 Conyza neocandolleana Humbert;新康氏白酒草●☆
103546 Conyza neriifolia (L.) L'Hér. ex Steud. = Brachylaena neriifolia (L.) R. Br. ●☆
103547 Conyza newii Oliv. et Hiern;纽白酒草■☆
103548 Conyza obscura DC.;隐匿白酒草■☆
103549 Conyza odontophylla Boiss. = Pluchea arguta Boiss. ■☆
103550 Conyza odontoptera Webb = Blumea crispata (Vahl) Merxm. ■☆
103551 Conyza odorata L. = Pluchea odorata (L.) Cass. ●■☆
103552 Conyza pallidiflora R. E. Fr.;苍白花白酒草■☆
103553 Conyza pannosa Webb;毡状白酒草■☆
103554 Conyza parva Cronquist;小白酒草■☆
103555 Conyza parva Cronquist = Conyza canadensis (L.) Cronquist ■
103556 Conyza parvicapitulata Lisowski = Conyza gouanii (L.) Willd. ■☆
103557 Conyza patula Dryand. = Vernonia patula (Dryand.) Merr. ■
103558 Conyza pauciflora Willd. = Vernonia galamensis (Cass.) Less. ■☆
103559 Conyza pectinata Sch. Bip. ex Oliv. et Hiern;篦状白酒草■☆
103560 Conyza pedunculata (Oliv.) Wild = Conyza boranensis (S. Moore) Cufod. ■☆
103561 Conyza perennis Hand.-Mazz.;宿根白酒草;Perennial Conyza ■
103562 Conyza perrieri Humbert;佩里耶白酒草■☆
103563 Conyza persicariifolia (Benth.) Oliv. et Hiern = Conyza attenuata DC. ■☆
103564 Conyza persicifolia (Benth.) Oliv. et Hiern = Conyza attenuata DC. ■☆
103565 Conyza pinifolia Lam. = Vernonia capensis (Houtt.) Druce ■☆
103566 Conyza pinnata (L. f.) Kuntze;羽状白酒草■☆
103567 Conyza pinnatifida (Thunb.) Less.;羽裂白酒草■☆
103568 Conyza pinnatifida Buch.-Ham. ex Roxb. = Conyza stricta Willd. ex DC. var. pinnatifida (D. Don) Kitam. ■
103569 Conyza pinnatifida Dunn = Conyza blinii H. Lév. ■
103570 Conyza pinnatifida Franch. = Conyza blinii H. Lév. ■
103571 Conyza pinnatilobata DC. = Conyza pinnata (L. f.) Kuntze ■☆
103572 Conyza podocephala DC.;梗头白酒草■☆
103573 Conyza pubescens DC. = Brachyactis pubescens (DC.) Aitch. et C. B. Clarke ■
103574 Conyza pulsatilloides O. Hoffm.;白头翁白酒草■☆
103575 Conyza pusilla Houtt.;微小白酒草■☆
103576 Conyza pycnostachya Michx. = Pterocaulon pycnostachyum (Michx.) Elliott ■☆
103577 Conyza pyrifolia Lam. = Microglossa pyrifolia (Lam.) Kuntze ●
103578 Conyza pyrrhopappa Sch. Bip. ex A. Rich.;安哥拉白酒草■☆
103579 Conyza pyrrhopappa Sch. Bip. ex A. Rich. subsp. oblongifolia (O. Hoffm.) Wild = Conyza pyrrhopappa Sch. Bip. ex A. Rich. ■☆
103580 Conyza ramosissima Cronquist;铺散白酒草;Dwarf Fleabane, Spreading Fleabane ■☆
103581 Conyza redolens Willd. = Pterocaulon redolens (G. Forst.) Fern.-Vill. ■
103582 Conyza repanda Roxb. = Blumea repanda (Roxb.) Hand.-Mazz. ■
103583 Conyza rhizocephala Rupr. = Inula rhizocephala Schrenk ■
103584 Conyza riparia Blume = Blumea riparia (Blume) DC. ■
103585 Conyza roylei DC. = Brachyactis roylei (DC.) Wendelbo ■
103586 Conyza rubricaulis R. E. Fr. = Conyza clarenceana (Hook. f.) Oliv. et Hiern ■☆
103587 Conyza rupestris L. = Phagnalon rupestre (L.) DC. ■☆

103588 Conyza ruwenzoriensis (S. Moore) R. E. Fr. = Carduus ruwenzoriensis S. Moore ■☆
103589 Conyza sagittalis Lam. = Pluchea sagittalis (Lam.) Cabrera ■
103590 Conyza salicina Rupr. = Inula salicina L. ■
103591 Conyza saligna Wall. = Vernonia saligna (Wall.) DC. ■
103592 Conyza salsoloides Turcz. = Inula salsoloides (Turcz.) Ostenf. ●■
103593 Conyza sarmentosa Humbert;蔓茎白酒草■☆
103594 Conyza saxatilis L. = Phagnalon saxatile (L.) Cass. ■☆
103595 Conyza scabrida DC.;微糙白酒草■☆
103596 Conyza schiedeana (Less.) Cronquist = Laennecia schiedeana (Less.) G. L. Nesom ■☆
103597 Conyza schimperi Sch. Bip. ex A. Rich.;欣珀白酒草■☆
103598 Conyza schimperi Sch. Bip. ex A. Rich. subsp. longipapposa R. E. Fr. = Conyza schimperi Sch. Bip. ex A. Rich. ■☆
103599 Conyza senegalensis Willd. = Pseudoconyza viscosa (Mill.) D'Arcy ■☆
103600 Conyza sennii Chiov.;森恩白酒草■☆
103601 Conyza serratifolia Baker = Conyza attenuata DC. ■☆
103602 Conyza setschwanica Hand. -Mazz. = Blumea aromatica DC. ■
103603 Conyza sordida L. = Phagnalon sordidum (L.) Rchb. ■☆
103604 Conyza spartioides O. Hoffm. = Nidorella spartioides (O. Hoffm.) Cronquist ■☆
103605 Conyza spinosa Sch. Bip. ex Oliv. et Hiern;具刺白酒草■☆
103606 Conyza squamata Spreng. = Aster squamatus (Spreng.) Hieron. ■☆
103607 Conyza squamata Spreng. = Symphyotrichum subulatum (Michx.) G. L. Nesom var. squamatum (Spreng.) S. D. Sundb. ■☆
103608 Conyza steudelii Sch. Bip. ex A. Rich.;斯托白酒草■☆
103609 Conyza stricta Wall. = Conyza stricta Willd. ex DC. ■
103610 Conyza stricta Wall. ex DC. = Conyza japonica (Thunb.) Less. ■
103611 Conyza stricta Willd. = Conyza stricta Willd. ex DC. ■
103612 Conyza stricta Willd. ex DC.;劲直白酒草(劲直假蓬);Strict Conyza ■
103613 Conyza stricta Willd. ex DC. var. pinnatifida (D. Don) Kitam.;羽裂劲直白酒草(羽裂白酒草);Pinnatifid Conyza ■
103614 Conyza stricta Willd. ex DC. var. pinnatifida (D. Don) Kitam. = Conyza stricta Willd. ex DC. ■
103615 Conyza stricta Willd. var. pinnatifida (D. Don) Kitam. = Conyza stricta Willd. ex DC. ■
103616 Conyza subscaposa O. Hoffm.;亚花茎白酒草■☆
103617 Conyza sumatrensis (Retz.) E. Walker;苏门答腊白酒草(苏门白酒草,野茼蒿,野筒蒿,竹叶艾);Asthmaweed, Guernsey Fleabane, Sumatra Conyza, Sumatra Fleabane ■
103618 Conyza sumatrensis (Retz.) E. Walker = Conyza albida Spreng. ■
103619 Conyza syringaeifolia Meyen et Walp. = Microglossa pyrifolia (Lam.) Kuntze ●
103620 Conyza theodori R. E. Fr. = Conyza clarenceana (Hook. f.) Oliv. et Hiern ■☆
103621 Conyza tigrensis Oliv. et Hiern;蒂格雷白酒草■☆
103622 Conyza tigrensis Oliv. et Hiern var. erythrolepis Chiov. = Conyza tigrensis Oliv. et Hiern ■☆
103623 Conyza tigrensis Oliv. et Hiern var. pratensis (Vatke) Engl. = Conyza tigrensis Oliv. et Hiern ■☆
103624 Conyza tigrensis Oliv. et Hiern var. sylvestris (Vatke) Engl. = Conyza tigrensis Oliv. et Hiern ■☆
103625 Conyza tomentosa Burm. f. = Capelio tomentosa (Burm. f.) B. Nord. ■☆
103626 Conyza transvaalensis Bremek. = Conyza aegyptiaca (L.) Aiton ■
103627 Conyza triloba Decne. = Conyza stricta Willd. ex DC. ■
103628 Conyza ulmifolia (Burm. f.) Kuntze;榆叶白酒草■☆
103629 Conyza umbrosa Ker. et Kir. = Brachyactis roylei (DC.) Wendelbo ■
103630 Conyza urticifolia (Baker) Humbert;荨麻叶白酒草■☆
103631 Conyza varia (Webb) Wild;变异白酒草■☆
103632 Conyza variegata Sch. Bip. ex A. Rich.;斑叶白酒草■☆
103633 Conyza variegata Sch. Bip. ex A. Rich. var. pseudohochstetteri Chiov.;假霍赫白酒草■☆
103634 Conyza vatkeana Oliv. et Hiern = Conyza stricta Willd. ex DC. ■
103635 Conyza velutina H. Lév. = Blumea lacera (Burm. f.) DC. ■
103636 Conyza velutina H. Lév. et Vaniot = Blumea lacera (Burm. f.) DC. ■
103637 Conyza vernonicifolia Wall. ex DC. = Conyza japonica (Thunb.) Less. ■
103638 Conyza vernonioides (Sch. Bip. ex A. Rich.) Wild;斑鸠菊白酒草■☆
103639 Conyza vernonioides (Sch. Bip. ex A. Rich.) Wild subsp. arborea (R. E. Fr.) Lisowski = Conyza vernonioides (Sch. Bip. ex A. Rich.) Wild ■☆
103640 Conyza vernonioides (Sch. Bip. ex A. Rich.) Wild subsp. inuloides (O. Hoffm.) Wild = Conyza vernonioides (Sch. Bip. ex A. Rich.) Wild ■☆
103641 Conyza veronicifolia Wall. ex DC. = Conyza japonica (Thunb.) Less. ■
103642 Conyza viguieri Humbert;维基耶白酒草●☆
103643 Conyza virgata DC. = Conyza neocandolleana Humbert ●☆
103644 Conyza virgata DC. var. ellisii (Baker) Humbert = Conyza neocandolleana Humbert ●☆
103645 Conyza viscidula Wall. = Conyza leucantha (D. Don) Ludlow et Raven ■
103646 Conyza viscosa Mill. = Pseudoconyza viscosa (Mill.) D'Arey ■☆
103647 Conyza volkameriifolia Wall. = Vernonia volkameriifolia (Wall.) DC. ●
103648 Conyza volkensii O. Hoffm. = Conyza steudelii Sch. Bip. ex A. Rich. ■☆
103649 Conyza welwitschii (S. Moore) Wild;韦尔白酒草●☆
103650 Conyza wittei Robyns = Conyza vernonioides (Sch. Bip. ex A. Rich.) Wild ■☆
103651 Conyzanthus Tamamsch. (1959);科尼花属☆
103652 Conyzanthus Tamamsch. = Aster L. ●■
103653 Conyzanthus graminifolius (Spreng.) Tamamsch.;禾叶科尼花☆
103654 Conyzanthus squamatus (Spreng.) Tamamsch.;科尼花☆
103655 Conyzella Fabr. = Conyza Less. (保留属名)■
103656 Conyzella Rupr. = Erigeron L. ■●
103657 Conyzoides DC. = Carpesium L. ■
103658 Conyzoides Fabr. = Erigeron L. ■●
103659 Conyzoides Tourn. ex DC. = Carpesium L. ■
103660 Conzattia Rose(1909);黄花苏木属●☆
103661 Conzattia arborea Rose;黄花苏木●☆
103662 Conzattia multiflora Standl.;多黄花苏木●☆
103663 Coockia Batsch = Clausena Burm. f. ●
103664 Coockia Batsch = Cookia Sonn. ●
103665 Cookia J. F. Gmel. = Pimelea Banks ex Gaertn. (保留属名)●☆

103666 Cookia Sonn. = Clausena Burm. f. ●

103667 Cookia anisumolens Blanco = Clausena anisumolens (Blanco) Merr. ●

103668 Cookia wampi Blanco = Clausena lansium (Lour.) Skeels ●

103669 Cooktownia D. L. Jones(1997);澳昆兰属■☆

103670 Coombea P. Royen = Medicosma Hook. f. ●☆

103671 Cooperia Herb. (1836);夜星花属(雨百合属);Rain Lily ■☆

103672 Cooperia Herb. = Zephyranthes Herb. (保留属名)■

103673 Cooperia chlorosolen Herb. = Zephyranthes chlorosolen (Herb.) D. Dietr. ■☆

103674 Cooperia drummondii Herb.;夜星花;Evening Star, Rain Lily ■☆

103675 Cooperia drummondii Herb. = Zephyranthes chlorosolen (Herb.) D. Dietr. ■☆

103676 Cooperia jonesii Cory = Zephyranthes jonesii (Cory) Traub ■☆

103677 Cooperia kansensis W. Stevens = Zephyranthes chlorosolen (Herb.) D. Dietr. ■☆

103678 Cooperia pedunculata Herb. = Zephyranthes drummondii D. Don ■☆

103679 Cooperia smallii Alexander = Zephyranthes smallii (Alexander) Traub ■☆

103680 Cooperia traubii W. Hayw. = Zephyranthes traubii (W. Hayw.) Moldenke ■☆

103681 Coopernookia Carolin(1968);库珀草海桐属●☆

103682 Coopernookia barbata (R. Br.) Carolin;库珀草海桐●☆

103683 Copaiba Adans. = Copaifera L. (保留属名)●☆

103684 Copaiba Mill(废弃属名) = Copaifera L. (保留属名)●☆

103685 Copaiba arnoldiana De Wild. et T. Durand = Guibourtia arnoldiana (De Wild. et T. Durand) J. Léonard ●☆

103686 Copaiba coleosperma (Benth.) Kuntze = Guibourtia coleosperma (Benth.) J. Léonard ●☆

103687 Copaiba conjugata (Bolle) Kuntze = Guibourtia conjugata (Bolle) J. Léonard ●☆

103688 Copaiba langsdorfii (Desf.) Kuntze = Copaifera langsdorffii Desf. ●☆

103689 Copaiba mopane (J. Kirk ex Benth.) Kuntze = Colophospermum mopane (J. Kirk ex Benth.) J. Kirk ex J. Léonard ●☆

103690 Copaica T. Durand et Jacks. = Copaifera L. (保留属名)●☆

103691 Copaifera L. (1762)(保留属名);古巴香脂树属(柯比胶树属,香脂树属,香脂苏木属);Balsam-tree, Copaifera, Copal Tree, Copal-tree, Ironwood ●☆

103692 Copaifera baumiana Harms;鲍姆古巴香脂树●☆

103693 Copaifera carrissoana Exell = Guibourtia carrissoana (Exell) J. Léonard ●☆

103694 Copaifera coleosperma Benth. = Guibourtia coleosperma (Benth.) J. Léonard ●☆

103695 Copaifera conjugata (Bolle) Milne-Redh. = Guibourtia conjugata (Bolle) J. Léonard ●☆

103696 Copaifera copallifera (Benn.) Milne-Redh.;西非昂香脂树(塞拉里昂香脂树);Copaifera, Gum Copal, Red Gum, Sierra Leone Copal Tree, Sierra Leone Copal-tree, Sierra Leone Gum Copal, Yellow Gum ●☆

103697 Copaifera copallifera (Benn.) Milne-Redh. = Guibourtia copallifera Benn. ●☆

103698 Copaifera demeusei Harms;刚果香脂苏木●☆

103699 Copaifera demeusei Harms = Guibourtia demeusei (Harms) J. Léonard ●☆

103700 Copaifera dinklagei Harms = Guibourtia dinklagei (Harms) J. Léonard ●☆

103701 Copaifera duckei Dwyer;达凯香脂苏木●☆

103702 Copaifera eguminosae?;荚果香脂苏木●☆

103703 Copaifera ehie A. Chev. = Guibourtia ehie (A. Chev.) J. Léonard ●☆

103704 Copaifera gorskia Schinz = Guibourtia conjugata (Bolle) J. Léonard ●☆

103705 Copaifera gorskiana Benth. = Guibourtia conjugata (Bolle) J. Léonard ●☆

103706 Copaifera gossweileri Exell = Guibourtia carrissoana (Exell) J. Léonard var. gossweileri? ●☆

103707 Copaifera guibourtiana Benth. = Guibourtia copallifera Benn. ●☆

103708 Copaifera guyanensis Desf.;圭亚那香脂苏木●☆

103709 Copaifera langsdorffii Desf.;朗氏香脂苏木●☆

103710 Copaifera laurentii De Wild. = Guibourtia demeusei (Harms) J. Léonard ●☆

103711 Copaifera letestui (Pellegr.) Pellegr.;加蓬香脂苏木●☆

103712 Copaifera letestui (Pellegr.) Pellegr. = Sindoropsis letestui (Pellegr.) J. Léonard ●☆

103713 Copaifera mannii Baill. = Prioria mannii (Baill.) Breteler ●☆

103714 Copaifera martii Hayne;马氏香脂苏木●☆

103715 Copaifera mildbraedii Harms;米尔香脂苏木●☆

103716 Copaifera mopane J. Kirk ex Benth.;罗得西亚香脂树;Balsam Tree, Mopane, Rhodesian Ironwood, Turpentine Tree ●☆

103717 Copaifera mopane J. Kirk ex Benth. = Hardwickia mopane (J. Kirk ex Benth.) Breteler ●☆

103718 Copaifera multijuga Hayne;多对香脂苏木;Copaiba Balsam, Copaiba Oil ●☆

103719 Copaifera officinalis (Jacq.) L.;药用古巴香脂树(柯伯胶树,药用香脂苏木);Copaiba, Copaiba Balsam, Copaiba Copal Tree, Copaiba Copal-tree ●☆

103720 Copaifera officinalis L. = Copaifera officinalis (Jacq.) L. ●☆

103721 Copaifera religiosa J. Léonard;神圣香脂苏木●☆

103722 Copaifera reticulata Ducke;网脉香脂苏木●☆

103723 Copaifera salikounda Heckel;西非香脂苏木;Bubinga ●☆

103724 Copaifera tessmannii Harms = Guibourtia tessmannii (Harms) J. Léonard ●☆

103725 Copaifera vuilletiana A. Chev. = Guibourtia copallifera Benn. ●☆

103726 Copaifera vuilletii A. Chev. = Guibourtia copallifera Benn. ●☆

103727 Copaiva Jacq. (废弃属名) = Copaifera L. (保留属名)●☆

103728 Copaiva coleosperma (Benth.) Britton = Guibourtia coleosperma (Benth.) J. Léonard ●☆

103729 Copedesma Gleason = Miconia Ruiz et Pav. (保留属名)●☆

103730 Copernicia Mart. = Copernicia Mart. ex Endl. ●☆

103731 Copernicia Mart. ex Endl. (1837);哥白尼棕属(巴西蜡棕属,杯形花属,粗柄扇椰子属,科布榈属,蜡榈属,蜡棕属);Caranda Palm, Copernicia, Copernicus Palm, Wax Palm ●☆

103732 Copernicia alba Morong;白巴西蜡棕(白腊棕,腊棕);White Copernicia ●☆

103733 Copernicia baileyana León;壮蜡棕;Yarey, Yarey Hembra, Yarreyon ●☆

103734 Copernicia cerifera (Arruda) Mart. = Copernicia prunifera (Mill.) H. E. Moore ●☆

103735 Copernicia hospita Mart.;奇异哥白尼棕(奇异蜡棕)●☆

103736 Copernicia macroglossa H. Wendl.;大舌蜡棕(巴西蜡棕,蜡棕);Carnaba ●☆

103737 Copernicia prunifera (Mill.) H. E. Moore;裙蜡棕(巴西蜡

棕);Brazilian Wax Palm, Carnauba Palm, Carnauba Wax, Carnauba Wax Palm, Cuban Petticoat Palm, Jata De Guanbacoa ●☆
103738 Copernicia torreana Leon;托里氏蜡棕●☆
103739 Copiapoa Britton et Rose(1922);龙爪球属(龙爪玉属,南美仙人球属);Copiapoa ●
103740 Copiapoa bridgesii (Pfeiff.) Backeb.;舞龙球(舞龙丸);Copiapoa De Bridges ●☆
103741 Copiapoa calderana F. Ritter;帝龙冠●☆
103742 Copiapoa carrizalensis F. Ritter;黑闪玉●☆
103743 Copiapoa chaniaralensis F. Ritter;加奈留玉●☆
103744 Copiapoa cinerascens Britton et Rose;龙牙玉●☆
103745 Copiapoa cinerea (Phil.) Britton et Rose;黑王球(黑天球,黑王丸,黑王玉);Blackspines Copiapoa, Copiapoa De Philippi, Grey Copiapoa ●☆
103746 Copiapoa cinerea (Phil.) Britton et Rose subsp. columna-alba (F. Ritter) D. R. Hunt;孤龙玉(孤龙丸)●☆
103747 Copiapoa cinerea Britton et Rose var. albispina F. Ritter;白刺龙爪球(白刺黑王丸,白刺龙爪玉)●☆
103748 Copiapoa cinerea Britton et Rose var. dealbata (F. Ritter) Backeb. = Copiapoa dealbata F. Ritter ●☆
103749 Copiapoa coquimbana (Karw. ex Rümpler) Britton et Rose;龙爪球(龙爪玉);Coquimbano, Dragonclaw Copiapoa ●☆
103750 Copiapoa coquimbana (Karw. ex Rümpler) Britton et Rose var. wagenknechtii F. Ritter;和严玉●☆
103751 Copiapoa cuprea F. Ritter;君光球(君光丸)●☆
103752 Copiapoa dealbata F. Ritter;黑士冠●☆
103753 Copiapoa dura F. Ritter;铜罗球(铜罗丸)●☆
103754 Copiapoa echinoides Britton et Rose;龙魔玉●☆
103755 Copiapoa ferox Lembcke et Backeb.;猛虎玉●☆
103756 Copiapoa gigantea Backeb.;雷血球(雷血丸)●☆
103757 Copiapoa grandiflora F. Ritter;秋霜玉●☆
103758 Copiapoa haseltoniana Backeb.;豹髯玉●☆
103759 Copiapoa humilis (Phil.) Hutchison;公子球(公子丸);Humildito ●☆
103760 Copiapoa krainziana F. Ritter ex Backeb.;稀翁玉●☆
103761 Copiapoa lembckei Backeb.;冥王球(冥王丸)●☆
103762 Copiapoa longistaminea F. Ritter;鬼神龙●☆
103763 Copiapoa malletiana (Lem.) Backeb.;豪枪球(豪枪丸);Copiapoa De Carrizal ●☆
103764 Copiapoa marginata (Salm-Dyck) Britton et Rose;龙鳞球(龙鳞玉,龙爪仙人掌)●☆
103765 Copiapoa megarhiza Britton et Rose;虎髯玉●☆
103766 Copiapoa montana F. Ritter ex Doweld;松风玉;Bajotierra ●☆
103767 Copiapoa pepiniana (Lem. ex Salm-Dyck) Backeb.;海龙玉●☆
103768 Copiapoa pseudocoquimbana F. Ritter;假龙爪球●☆
103769 Copiapoa rubriflora F. Ritter;赤鬼玉●☆
103770 Copiapoa streptocaulon (Hook.) F. Ritter;黑云城●☆
103771 Copiapoa taltalensis (Werderm.) Looser;黑王殿;Quisco Del Desierto ●☆
103772 Copiapoa tenuissima F. Ritter;鱼鳞玉●☆
103773 Copioglossa Miers = Ruellia L. ■●
103774 Copisma E. Mey. = Rhynchosia Lour.(保留属名)●■
103775 Copisma falcatum E. Mey. = Rhynchosia minima (L.) DC. var. falcata (E. Mey.) Verdc. ■☆
103776 Copisma nitidum E. Mey. = Rhynchosia nitida (E. Mey.) Steud. ■☆
103777 Copisma paniculata E. Mey. = Rhynchosia totta (Thunb.) DC. ■☆
103778 Copisma pilosum E. Mey. = Rhynchosia totta (Thunb.) DC. ■☆
103779 Copisma rotundifolium E. Mey. = Rhynchosia rotundifolia (E. Mey.) Steud. ■☆
103780 Coppenaia Dumort. = Oncidium Sw.(保留属名)■☆
103781 Coppoleria Todaro = Vicia L. ■
103782 Coprosma J. R. Forst. et G. Forst. (1775);异味树属(臭味木属,染料木属,污生境属);Coprosma ●☆
103783 Coprosma acerosa A. Cunn.;针叶异味树(针叶污生境);Sand Coprosma ●☆
103784 Coprosma baueri Endl. = Coprosma repens Hook. f. ●☆
103785 Coprosma brunnea Cockayne;亚高山异味树(亚高山污生境)●☆
103786 Coprosma foetidissima A. Cunn.;烈异味树;Stinkwood ●☆
103787 Coprosma hirtella Labill.;糙叶异味树(糙叶污生境)●☆
103788 Coprosma kawakamii Hayata = Lonicera kawakamii (Hayata) Maxim. ●
103789 Coprosma kirkii Cheeseman;凯克异味树(凯克污生境,狭叶染料木)●☆
103790 Coprosma kirkii Cheeseman 'Variegata';银边异味树(斑叶凯克污生境,银边狭叶染料木)●☆
103791 Coprosma lucida G. Forst.;亮叶异味树(亮叶污生境);Karamu ●☆
103792 Coprosma macrocarpa Cheeseman;大果异味树(大果污生境)●☆
103793 Coprosma marginata (Salm-Dyck) Britton et Rose;龙爪仙人掌●
103794 Coprosma petriei Cheeseman;帕氏异味树(帕氏污生境)●☆
103795 Coprosma prisca W. R. B. Oliv.;古老异味树(古老污生境)●☆
103796 Coprosma propinqua A. Cunn.;角异味树(角污生境)●☆
103797 Coprosma quadrifida (Labill.) B. L. Rob.;刺异味树(刺污生境)●☆
103798 Coprosma repens Hook. f.;匍匐异味树(宽叶染料木,匍匐污生境);Creeping Mirrorplant, Looking-glass Plant, Mirror Bush, Mirror Plant, Taupata, Tree Bedstraw ●☆
103799 Coprosma repens Hook. f. 'Marble Queen';白斑皇后匍匐异味树(白斑皇后匍匐污生境)●☆
103800 Coprosma repens Hook. f. 'Painter Palette';调色板匍匐异味树(调色板匍匐污生境)●☆
103801 Coprosma repens Hook. f. 'Picturata';美景匍匐异味树(斑叶宽叶染料木,美景匍匐污生境)●☆
103802 Coprosma repens Hook. f. 'Variegata';斑叶匍匐异味树(斑叶匍匐污生境)●☆
103803 Coprosma repens Hook. f. 'Yvonne';伊冯匍匐异味树(伊冯匍匐污生境)●☆
103804 Coprosma rhamnoides A. Cunn.;鼠李异味树(鼠李污生境)●☆
103805 Coprosma rigida Cheeseman;硬叶异味树(硬叶污生境)●☆
103806 Coprosma robusta M. Raoul;健壮异味树(健壮污生境);Karamu ●☆
103807 Coprosma rugosa Cheeseman;矮生异味树(矮生污生境)●☆
103808 Coprosma virescens Petrie;小叶异味树(小叶污生境)●☆
103809 Coprosmanthus (Torr.) Kunth = Smilax L. ●
103810 Coprosmanthus Kunth = Smilax L. ●
103811 Coprosmanthus ecirrhatus (S. Watson) Chapm. = Smilax ecirrhata S. Watson ●☆
103812 Coprosmanthus herbaceus (L.) Kunth = Smilax herbacea L. ●☆
103813 Coprosmanthus herbaceus (L.) Kunth var. ecirratus Engelm. ex Kunth = Smilax ecirrhata S. Watson ●☆
103814 Coprosmanthus lasioneura (Hook.) Kunth = Smilax lasioneura

Hook. ●☆

103815 Coprosmanthus lasioneuron (Hook.) Kunth = Smilax lasioneura Hook. ●☆

103816 Coprosmanthus peduncularis (Muhl. ex Willd.) Kunth = Smilax herbacea L. ●☆

103817 Coprosmanthus tamnifolius (Michx.) Kunth = Smilax pseudochina Lour. ●☆

103818 Coptaceae Á. Löve et D. Löve = Ranunculaceae Juss. (保留科名)●■

103819 Coptidaceae Á. Löve et D. Löve = Ranunculaceae Juss. (保留科名)●■

103820 Coptidium (Prantl) Á. Löve et D. Löve ex Tzvelev = Ranunculus L. ■

103821 Coptidium (Prantl) Beurl. ex Rydb. = Ranunculus L. ■

103822 Coptidium (Prantl) Tzvelev = Ranunculus L. ■

103823 Coptidium Nyman = Ranunculus L. ■

103824 Coptis Salisb. (1807);黄连属;Cankerberry, Coptide, Goldthread ■

103825 Coptis anemonifolia Siebold et Zucc.;菊叶黄连■☆

103826 Coptis brachypetala Siebold et Zucc.;芹叶黄连■☆

103827 Coptis brachypetala Siebold et Zucc. var. pygmaca Miq.;细叶黄连■☆

103828 Coptis chinensis Franch.;黄连(川连,鸡爪黄连,鸡爪连,王连,味连,仙姑草,支连);China Goldthread, Chinese Goldthread ■

103829 Coptis chinensis Franch. var. angustiloba W. Y. Kong;狭裂黄连■

103830 Coptis chinensis Franch. var. brevisepala W. T. Wang;短萼黄连(土黄连);Shortsepal Goldthread ■

103831 Coptis chinensis Franch. var. omeiensis F. H. Chen = Coptis omeiensis (F. H. Chen) C. Y. Cheng ■

103832 Coptis deltoidea C. Y. Cheng et P. K. Hsiao;三角叶黄连(峨眉黄连,峨眉家连,雅连);Deltaleaf Goldthread, Deltoid Goldthread ■

103833 Coptis groenlandica (Oeder) Fernald = Coptis trifolia (L.) Salisb. ■☆

103834 Coptis gulinensis;古蔺黄连■☆

103835 Coptis japonica Makino;日本黄连;Japan Goldthread ■☆

103836 Coptis japonica Makino var. disecta Nakai;深裂黄连■☆

103837 Coptis japonica Makino var. major Satake;大黄连■☆

103838 Coptis morii Hayata = Coptis quinquefolia Miq. ■

103839 Coptis occidentalis (Nutt.) Torr. et A. Gray;西方黄连■☆

103840 Coptis omeiensis (F. H. Chen) C. Y. Cheng;峨眉黄连(峨眉野连,凤尾连,岩黄连,野黄连);Emei Goldthread, Omei Goldthread ■

103841 Coptis omeiensis (F. H. Chen) C. Y. Cheng var. stolonifera S. L. Zhang;草黄连■

103842 Coptis ospriocarpa Brühl = Souliea vaginata (Maxim.) Franch. ■

103843 Coptis quinquefolia Miq.;五叶黄连(台湾黄连);Fiveleaf Goldthread ■

103844 Coptis quinquefolia Miq. f. ramosa Makino = Coptis quinquefolia Miq. ■

103845 Coptis quinquesecta W. T. Wang;五裂黄连(台湾黄连,五加叶黄连,五叶黄连);Fivelobe Goldthread, Fivesplit Goldthread ■

103846 Coptis teeta Wall.;云南黄连(滴胆芝,黄连,鸡脚黄连,水莲,王连,云连,支连);Yunnan Goldthread ■

103847 Coptis teeta Wall. var. chinensis (Franch.) Finet et Gagnep. = Coptis chinensis Franch. ■

103848 Coptis teeta Wall. var. chinensis (Franch.) Finet et Gagnep. subvar. rhizomatosa H. Lév. = Coptis chinensis Franch. ■

103849 Coptis teetoides C. Y. Cheng = Coptis teeta Wall. ■

103850 Coptis trifolia (L.) Salisb.;三叶黄连;Canker Root, Canker-root, Golden Thread, Goldenroot, Goldthread, Mouthroot, Savoyana, Three-leaf Goldthread, Three-leaved Gold-thread, Vegetable Gold, Yellow Snakeroot ■☆

103851 Coptis trifolia (L.) Salisb. subsp. groenlandica (Oeder) Hultén = Coptis trifolia (L.) Salisb. ■☆

103852 Coptis trifolia (L.) Salisb. var. groenlandica (Oeder) Fassett = Coptis trifolia (L.) Salisb. ■☆

103853 Coptis trifoliolata (Makino) Makino;三小叶黄连■☆

103854 Coptocheile Hoffmanns. = ? Gesneria L. ●☆

103855 Coptocheile Hoffmanns. ex L. = ? Gesneria L. ●☆

103856 Coptophyllum Gardner(废弃属名) = Coptophyllum Korth. (保留属名)■☆

103857 Coptophyllum Korth. (1851)(保留属名);裂叶茜属■☆

103858 Coptophyllum bracteatum Korth.;裂叶茜■☆

103859 Coptophyllum pilosum Miq.;毛裂叶茜■☆

103860 Coptophyllum reptans (Backer ex Bremek.) Bakh. f.;匍匐裂叶茜■☆

103861 Coptosapelta Korth. (1851)(保留属名);流苏子属;Coptosapelta ●

103862 Coptosapelta diffusa (Champ. ex Benth.) Steenis;流苏子(臭沙藤,凉藤,棉花藤,棉坡藤,棉丝藤,棉藤,棉絮藤,牛老药,牛老药藤,瓢箪藤,千叶藤,伤药藤,上树逼,乌龙藤,小青藤,苎丝藤);Ditfuse Coptosapelta, Thysanospermum ●

103863 Coptospelta K. Schum. = Coptosapelta Korth. (保留属名)●

103864 Coptospelta T. Durand et Jacks. = Coptosapelta Korth. (保留属名)●

103865 Coptosperma Hook. f. (1873);裂籽茜属●☆

103866 Coptosperma Hook. f. = Tarenna Gaertn. ●

103867 Coptosperma graveolens (S. Moore) Degreef;臭裂籽茜●☆

103868 Coptosperma graveolens (S. Moore) Degreef subsp. arabicum (Cufod.) Degreef;阿拉伯臭裂籽茜●☆

103869 Coptosperma graveolens (S. Moore) Degreef var. impolitum (Bridson) Degreef;暗色臭裂籽茜●☆

103870 Coptosperma kibuwae (Bridson) Degreef;基布瓦裂籽茜●☆

103871 Coptosperma littorale (Hiern) Degreef;滨海裂籽茜●☆

103872 Coptosperma madagascariensis J. G. Garcia = Coptosperma nigrescens Hook. f. ●☆

103873 Coptosperma neurophyllum (S. Moore) Degreef;脉叶裂籽茜●☆

103874 Coptosperma nigrescens Hook. f.;裂籽茜●☆

103875 Coptosperma peteri (Bridson) Degreef;彼得裂籽茜●☆

103876 Coptosperma rhodesiacum (Bremek.) Degreef;罗得西亚裂籽茜●☆

103877 Coptosperma rhodesiacum (Bremek.) Degreef = Tarenna zimbabwensis Bridson ●☆

103878 Coptosperma somaliense Degreef;索马里裂籽茜●☆

103879 Coptosperma supra-axillare (Hemsl.) Degreef;腋生裂籽茜●☆

103880 Coptosperma wajirense (Bridson) Degreef;瓦吉尔裂籽茜●☆

103881 Coptosperma zygoon (Bridson) Degreef;烈味裂籽茜●☆

103882 Coptosperma zygoon (Bridson) Degreef = Tarenna zygoon Bridson ●☆

103883 Coquebertia Brongn. = Zollernia Wied-Neuw. et Nees ●☆

103884 Coralliokyphos H. Fleischm. et Rech. = Moerenhoutia Blume ■☆

103885 Coralliorrhiza Asch. = Corallorhiza Gagnebin(保留属名)■

103886 Corallobotrys Hook. f. = Agapetes D. Don ex G. Don ●

103887 Corallocarpus Welw. ex Benth. et Hook. f. (1867);珊瑚果属■☆

103888 Corallocarpus Welw. ex Hook. f. = Corallocarpus Welw. ex

Benth. et Hook. f. ■☆

103889 Corallocarpus bainesii (Hook. f.) A. Meeuse;贝恩斯珊瑚果■☆

103890 Corallocarpus bequaertii De Wild. = Corallocarpus epigaeus (Rottler) C. B. Clarke ■☆

103891 Corallocarpus boehmii (Cogn.) C. Jeffrey;贝姆珊瑚果■☆

103892 Corallocarpus brevipedunculatus Gilg = Corallocarpus schimperi (Naudin) Hook. f. ■☆

103893 Corallocarpus bussei Gilg = Corallocarpus bainesii (Hook. f.) A. Meeuse ●☆

103894 Corallocarpus congolensis Cogn. = Corallocarpus welwitschii (Naudin) Hook. f. ex Welw. ■☆

103895 Corallocarpus conocarpus (Dalzell et Gibson) Hook. f. ex Clarke;束果珊瑚果■☆

103896 Corallocarpus corallinus (Naudin) Cogn. = Corallocarpus epigaeus (Rottler) C. B. Clarke ■☆

103897 Corallocarpus courbonii (Naudin) Cogn. = Corallocarpus schimperi (Naudin) Hook. f. ■☆

103898 Corallocarpus dinteri Cogn. = Corallocarpus bainesii (Hook. f.) A. Meeuse ●☆

103899 Corallocarpus dissectus Cogn.;深裂珊瑚果■☆

103900 Corallocarpus ehrenbergii (Schweinf.) Hook. f. = Corallocarpus schimperi (Naudin) Hook. f. ■☆

103901 Corallocarpus elegans Gilg = Corallocarpus schimperi (Naudin) Hook. f. ■☆

103902 Corallocarpus ellipticus Chiov.;椭圆珊瑚果■☆

103903 Corallocarpus epigaeus (Rottler) C. B. Clarke = Corallocarpus epigaeus (Rottler) Hook. f. ex C. B. Clarke ■☆

103904 Corallocarpus epigaeus (Rottler) Hook. f. ex C. B. Clarke;地生珊瑚果■☆

103905 Corallocarpus epigaeus Benth. et Hook. f. = Corallocarpus epigaeus (Rottler) Hook. f. ex C. B. Clarke ■☆

103906 Corallocarpus erostris (Schweinf.) Hook. f. = Corallocarpus schimperi (Naudin) Hook. f. ■☆

103907 Corallocarpus fenzlii Hook. f. = Corallocarpus epigaeus (Rottler) C. B. Clarke ■☆

103908 Corallocarpus gijef (J. F. Gmel.) Hook. f. = Kedrostis gijef (J. F. Gmel.) C. Jeffrey ■☆

103909 Corallocarpus gilgianus Cogn. = Corallocarpus welwitschii (Naudin) Hook. f. ex Welw. ■☆

103910 Corallocarpus glaucicaulis Dinter et Gilg ex Dinter = Corallocarpus welwitschii (Naudin) Hook. f. ex Welw. ■☆

103911 Corallocarpus glomeruliflorus (Deflers) Cogn.;团花珊瑚果■☆

103912 Corallocarpus gracilipes (Naudin) Cogn. = Corallocarpus epigaeus (Rottler) Hook. f. ex C. B. Clarke ■☆

103913 Corallocarpus grevei (Rabenant.) Rabenant.;格雷弗珊瑚果■☆

103914 Corallocarpus harmsii A. Zimm. = Kedrostis gijef (J. F. Gmel.) C. Jeffrey ■☆

103915 Corallocarpus hildebrandtii Gilg = Corallocarpus epigaeus (Rottler) C. B. Clarke ■☆

103916 Corallocarpus leiocarpus Gilg = Diplocyclos decipiens (Hook. f.) C. Jeffrey ■☆

103917 Corallocarpus longiracemosus Gilg = Corallocarpus schimperi (Naudin) Hook. f. ■☆

103918 Corallocarpus palmatus Cogn. = Corallocarpus epigaeus (Rottler) Hook. f. ex C. B. Clarke ■☆

103919 Corallocarpus pedunculosus (Naudin) Cogn. = Corallocarpus schimperi (Naudin) Hook. f. ■☆

103920 Corallocarpus perrieri (Rabenant.) Rabenant.;佩里耶珊瑚果■☆

103921 Corallocarpus poissonii Cogn. = Corallocarpus bainesii (Hook. f.) A. Meeuse ●☆

103922 Corallocarpus pseudogijef Gilg = Kedrostis pseudogijef (Gilg) C. Jeffrey ■☆

103923 Corallocarpus scaber Dinter et Gilg ex Dinter = Corallocarpus welwitschii (Naudin) Hook. f. ex Welw. ■☆

103924 Corallocarpus schimperi (Naudin) Hook. f.;欣珀珊瑚果■☆

103925 Corallocarpus schinzii Cogn.;欣兹珊瑚果■☆

103926 Corallocarpus sphaerocarpus Cogn. = Corallocarpus bainesii (Hook. f.) A. Meeuse ●☆

103927 Corallocarpus subhastatus Cogn. = Corallocarpus schinzii Cogn. ■☆

103928 Corallocarpus tavetensis Gilg = Corallocarpus epigaeus (Rottler) C. B. Clarke ■☆

103929 Corallocarpus tenuissimus Buscal. et Muschl.;极细珊瑚果■☆

103930 Corallocarpus triangularis Cogn.;三角珊瑚果■☆

103931 Corallocarpus velutinus (Dalzell et Gibson) Hook. f. ex Clarke = Corallocarpus schimperi (Naudin) Hook. f. ■☆

103932 Corallocarpus welwitschii (Naudin) Hook. f. ex Welw.;韦尔珊瑚果■☆

103933 Corallocarpus wildii C. Jeffrey;维尔德珊瑚果■☆

103934 Corallodendron Kuntze = Erythrina L. ●■

103935 Corallodendron Mill. = Erythrina L. ●■

103936 Corallodendron Tourn. ex Rupp. = Erythrina L. ●■

103937 Corallodiscus Batalin (1892);珊瑚苣苔属(珊瑚盘属);Corallodiscus ■

103938 Corallodiscus bullatus (Craib) B. L. Burtt;泡状珊瑚苣苔草;Bullate Corallodiscus ■

103939 Corallodiscus bullatus (Craib) B. L. Burtt = Corallodiscus lanuginosus (Wall. ex A. DC.) B. L. Burtt ■

103940 Corallodiscus conchifolius Batalin;小石花;Stone Flower ■

103941 Corallodiscus cordatulus (Craib) B. L. Burtt;珊瑚苣苔(滴滴花,翻魂草,瓜米还阳,虎耳还魂草,还魂草,还阳草,九倒生,牛耳草,铜钱还阳,岩白菜);Common Corallodiscus ■

103942 Corallodiscus cordatulus (Craib) B. L. Burtt = Corallodiscus lanuginosus (Wall. ex R. Br.) B. L. Burtt ■

103943 Corallodiscus cotinifolius W. T. Wang;瑶山苣苔草;Yaoshan Corallodiscus ■

103944 Corallodiscus flabellatus (Craib) B. L. Burtt;石花(扁叶珊瑚盘,生扯拢,石胆草,石荷叶,石蝴蝶,石莲花,石指甲,岩指甲,镇心草);Fan-shaped Corallodiscus,Stone Corallodiscus ■

103945 Corallodiscus flabellatus (Craib) B. L. Burtt = Corallodiscus lanuginosus (Wall. ex A. DC.) B. L. Burtt ■

103946 Corallodiscus flabellatus (Craib) B. L. Burtt var. leiocalyx W. T. Wang;光萼石花;Smoothcalyx Fan-shaped Corallodiscus ■

103947 Corallodiscus flabellatus (Craib) B. L. Burtt var. leiocalyx W. T. Wang = Corallodiscus lanuginosus (Wall. ex A. DC.) B. L. Burtt ■

103948 Corallodiscus flabellatus (Craib) B. L. Burtt var. luteus (Craib) K. Y. Pan;黄花石花;Yellow Corallodiscus ■

103949 Corallodiscus flabellatus (Craib) B. L. Burtt var. luteus (Craib) K. Y. Pan = Corallodiscus lanuginosus (Wall. ex A. DC.) B. L. Burtt ■

103950 Corallodiscus flabellatus (Craib) B. L. Burtt var. puberulus K. Y. Pan;锈毛石花;Rusthair Corallodiscus ■

103951 Corallodiscus flabellatus (Craib) B. L. Burtt var. puberulus K.

Y. Pan = Corallodiscus lanuginosus (Wall. ex A. DC.) B. L. Burtt ■

103952 Corallodiscus flabellatus (Craib) B. L. Burtt var. sericeus (Craib) K. Y. Pan;绢毛石花(绢毛石胆草,石莲花,石芮);Silky Corallodiscus ■

103953 Corallodiscus flabellatus (Craib) B. L. Burtt var. sericeus (Craib) K. Y. Pan = Corallodiscus lanuginosus (Wall. ex A. DC.) B. L. Burtt ■

103954 Corallodiscus forrestii (Anthony) B. L. Burtt = Corallodiscus conchifolius Batalin ■

103955 Corallodiscus grandis (Craib) B. L. Burtt;大叶珊瑚苣苔草;Bigleaf Corallodiscus ■☆

103956 Corallodiscus grandis (Craib) B. L. Burtt = Corallodiscus kingianus (Craib) B. L. Burtt ■

103957 Corallodiscus kingianus (Craib) B. L. Burtt;卷丝苣苔(卷丝苦苣苔,卷丝珊瑚苣苔);King Corallodiscus ■

103958 Corallodiscus labordei (Craib) B. L. Burtt = Corallodiscus lanuginosus (Wall. ex R. Br.) B. L. Burtt ■

103959 Corallodiscus lanuginosus (Wall. ex A. DC.) B. L. Burtt = Corallodiscus lanuginosus (Wall. ex R. Br.) B. L. Burtt ■

103960 Corallodiscus lanuginosus (Wall. ex R. Br.) B. L. Burtt;西藏珊瑚苣苔草;Xizang Corallodiscus ■

103961 Corallodiscus lineatus (Craib) B. L. Burtt = Corallodiscus lanuginosus (Wall. ex R. Br.) B. L. Burtt ■

103962 Corallodiscus luteus (Craib) B. L. Burtt = Corallodiscus lanuginosus (Wall. ex R. Br.) B. L. Burtt ■

103963 Corallodiscus mengtzeanus (Craib) B. L. Burtt = Corallodiscus lanuginosus (Wall. ex R. Br.) B. L. Burtt ■

103964 Corallodiscus patens (Craib) B. L. Burtt;多花珊瑚苣苔草;Flowery Corallodiscus ■

103965 Corallodiscus patens (Craib) B. L. Burtt = Corallodiscus lanuginosus (Wall. ex R. Br.) B. L. Burtt ■

103966 Corallodiscus plicatus (Franch.) B. L. Burtt;长柄珊瑚苣苔草;Longstipal Corallodiscus ■

103967 Corallodiscus plicatus (Franch.) B. L. Burtt = Corallodiscus lanuginosus (Wall. ex R. Br.) B. L. Burtt ■

103968 Corallodiscus plicatus (Franch.) B. L. Burtt var. lineatus (Craib) K. Y. Pan;短柄珊瑚苣苔■

103969 Corallodiscus plicatus (Franch.) B. L. Burtt var. lineatus (Craib) K. Y. Pan = Corallodiscus lanuginosus (Wall. ex R. Br.) B. L. Burtt ■

103970 Corallodiscus sericeus (Craib) B. L. Burtt = Corallodiscus flabellatus (Craib) B. L. Burtt var. sericeus (Craib) K. Y. Pan ■

103971 Corallodiscus sericeus (Craib) B. L. Burtt = Corallodiscus lanuginosus (Wall. ex R. Br.) B. L. Burtt ■

103972 Corallodiscus taliensis (Craib) B. L. Burtt;大理珊瑚苣苔草;Dali Corallodiscus ■

103973 Corallodiscus taliensis (Craib) B. L. Burtt = Corallodiscus lanuginosus (Wall. ex R. Br.) B. L. Burtt ■

103974 Corallophyllum Kumh = Lennoa Lex. ■☆

103975 Corallorhiza Châtel. = Corallorhiza Gagnebin(保留属名)■

103976 Corallorhiza Gagnebin(1755)(保留属名)('Corallorrhiza');珊瑚兰属;Chichen's Toes, Corallorhiza, Coralroot, Coral-root, Coralroot Orchid ■

103977 Corallorhiza Nutt. = Corallorhiza Gagnebin(保留属名)■

103978 Corallorhiza arizonica S. Watson = Hexalectris spicata (Walter) Barnhart var. arizonica (S. Watson) Catling et V. S. Engel ■☆

103979 Corallorhiza bentleyi Freudenst.;本氏珊瑚兰;Bentley's Coral-root ■☆

103980 Corallorhiza bigelovii S. Watson = Corallorhiza striata Lindl. var. vreelandii (Rydb.) L. O. Williams ■☆

103981 Corallorhiza corallorhiza (L.) H. Karst. = Corallorhiza trifida Chatel. ■

103982 Corallorhiza grab-hamii Cockerell = Corallorhiza maculata (Raf.) Raf. var. occidentalis (Lindl.) Ames ■☆

103983 Corallorhiza grandiflora A. Rich. et Galeotti = Hexalectris grandiflora (A. Rich. et Galeotti) L. O. Williams ■☆

103984 Corallorhiza hortensis Suksd. = Corallorhiza wisteriana Conrad ■☆

103985 Corallorhiza indica Lindl. = Oreorchis indica (Lindl.) Hook. f. ■

103986 Corallorhiza innata R. Br. = Corallorhiza trifida Chatel. ■

103987 Corallorhiza innata R. Br. var. virescens Farr = Corallorhiza trifida Chatel. ■

103988 Corallorhiza leimbachiana Suksd. = Corallorhiza maculata (Raf.) Raf. var. occidentalis (Lindl.) Ames ■☆

103989 Corallorhiza macraei A. Gray = Corallorhiza striata Lindl. ■☆

103990 Corallorhiza maculata (Raf.) Raf.;斑唇珊瑚兰;Large Coralroot, Spotted Coralroot, Spotted coral-root, Spottedlip Corallorhiza, Summer Coralroot ■☆

103991 Corallorhiza maculata (Raf.) Raf. f. flavida (M. Peck) Farw. = Corallorhiza maculata (Raf.) Raf. var. occidentalis (Lindl.) Ames ■☆

103992 Corallorhiza maculata (Raf.) Raf. f. intermedia (Farw.) Farw. = Corallorhiza maculata (Raf.) Raf. var. occidentalis (Lindl.) Ames ■☆

103993 Corallorhiza maculata (Raf.) Raf. f. punicea (Bartlett) Weath. et J. Adams = Corallorhiza maculata (Raf.) Raf. var. occidentalis (Lindl.) Ames ■☆

103994 Corallorhiza maculata (Raf.) Raf. subsp. mertensiana (Bong.) Calder et R. L. Taylor = Corallorhiza mertensiana Bong. ■☆

103995 Corallorhiza maculata (Raf.) Raf. subsp. occidentalis (Lindl.) Cockerell = Corallorhiza maculata (Raf.) Raf. var. occidentalis (Lindl.) Ames ■☆

103996 Corallorhiza maculata (Raf.) Raf. var. flavida (M. Peck) Cockerell = Corallorhiza maculata (Raf.) Raf. var. occidentalis (Lindl.) Ames ■☆

103997 Corallorhiza maculata (Raf.) Raf. var. fusca Bartlett = Corallorhiza maculata (Raf.) Raf. var. occidentalis (Lindl.) Ames ■☆

103998 Corallorhiza maculata (Raf.) Raf. var. immaculata M. Peck = Corallorhiza maculata (Raf.) Raf. var. occidentalis (Lindl.) Ames ■☆

103999 Corallorhiza maculata (Raf.) Raf. var. intermedia Farw. = Corallorhiza maculata (Raf.) Raf. var. occidentalis (Lindl.) Ames ■☆

104000 Corallorhiza maculata (Raf.) Raf. var. occidentalis (Lindl.) Ames;西方斑唇珊瑚兰;Spotted Coralroot, Summer Coralroot ■☆

104001 Corallorhiza maculata (Raf.) Raf. var. punicea Bartlett = Corallorhiza maculata (Raf.) Raf. var. occidentalis (Lindl.) Ames ■☆

104002 Corallorhiza mertensiana Bong.;梅尔滕珊瑚兰;Merten's Coral-root, Western Coral-root ■☆

104003 Corallorhiza micrantha Chapm. = Corallorhiza odontorhiza (Willd.) Poir. ■☆

104004 Corallorhiza multiflora Nutt. = Corallorhiza maculata (Raf.) Raf. ■☆

104005 Corallorhiza multiflora Nutt. var. flavida Peck = Corallorhiza maculata (Raf.) Raf. var. occidentalis (Lindl.) Ames ■☆

104006 Corallorhiza multiflora Nutt. var. occidentalis Lindl. = Corallorhiza maculata (Raf.) Raf. var. occidentalis (Lindl.) Ames ■☆

104007 Corallorhiza multiflora Nutt. var. sulphurea Suksd. = Corallorhiza maculata (Raf.) Raf. ■☆

104008 Corallorhiza ochroleuca Rydb. = Corallorhiza striata Lindl. var. vreelandii (Rydb.) L. O. Williams ■☆

104009 Corallorhiza odontorhiza (Willd.) Poir.;齿根珊瑚兰(小花珊瑚兰);Autumn Coralroot, Autumn Coral-root, Coral, Coral Root, Crawley, Dragon Claw, Fall Coralroot, Fall Coral-root, Late Coral Root, Late Coralroot, Small-flowered Coralroot, Small-flowered Coral-root ■☆

104010 Corallorhiza odontorhiza (Willd.) Poir. var. pringlei (Greenm.) Freudenst.;普氏齿根珊瑚兰■☆

104011 Corallorhiza patens Lindl. = Oreorchis patens (Lindl.) Lindl. ■

104012 Corallorhiza pringlei Greenm. = Corallorhiza odontorhiza (Willd.) Poir. var. pringlei (Greenm.) Freudenst. ■☆

104013 Corallorhiza pringlei Greenm. = Corallorhiza odontorhiza (Willd.) Poir. ■☆

104014 Corallorhiza purpurea L. O. Williams = Corallorhiza mertensiana Bong. ■☆

104015 Corallorhiza spicata (Walter) Tidestr. = Hexalectris spicata (Walter) Barnhart ■☆

104016 Corallorhiza striata Lindl.;条纹珊瑚兰;Hooded Coralroot, Striped Coralroot, Striped Coral-root ■☆

104017 Corallorhiza striata Lindl. var. flavida Todsen et T. A. Todsen = Corallorhiza striata Lindl. var. vreelandii (Rydb.) L. O. Williams ■☆

104018 Corallorhiza striata Lindl. var. vreelandii (Rydb.) L. O. Williams;弗里兰珊瑚兰■☆

104019 Corallorhiza trifida Chatel.;珊瑚兰;Chichen's Toes, Coralroot Orchid, Coral-root Orchid, Early Coralroot, Early Coral-root, Hooded Coralroot, Northern Coralroot, Northern Coral-root, Striped Coralroot, Trifid Corallorhiza, Yellow Coralroot ■

104020 Corallorhiza vancouveriana Finet = Corallorhiza maculata (Raf.) Raf. ■☆

104021 Corallorhiza verna Nutt. = Corallorhiza trifida Chatel. ■

104022 Corallorhiza vreelandii Rydb. = Corallorhiza striata Lindl. var. vreelandii (Rydb.) L. O. Williams ■☆

104023 Corallorhiza wisteriana Conrad;维氏珊瑚兰;Coral Root, Wister's coral-root, Wister's Coralroot ■☆

104024 Corallorhiza wyomingensis Hellm. et K. Hellm. = Corallorhiza trifida Chatel. ■

104025 Corallorrhiza Gagnebin = Corallorhiza Gagnebin(保留属名)■

104026 Corallorrhiza Rupp. ex Gagnebin. = Corallorhiza Gagnebin(保留属名)■

104027 Corallorrhiza trifida Chatel. = Corallorhiza trifida Chatel. ■

104028 Corallospartium J. B. Armstr. (1881);珊瑚雀枝属●☆

104029 Corallospartium crassicaule (Hook. f.) J. B. Armstr.;珊瑚雀枝;Coral Broom ●☆

104030 Coralluma Schrank ex Haw. = Caralluma R. Br. ■

104031 Coralorhiza Raf. = Corallorhiza Gagnebin(保留属名)■

104032 Corbassona Aubrév. = Niemeyera F. Muell. (保留属名)●☆

104033 Corbichonia Scop. (1777);多瓣粟草属■●☆

104034 Corbichonia decumbens (Forssk.) Exell;多瓣粟草■☆

104035 Corbichonia rubriviolacea (Friedrich) C. Jeffrey;红多瓣粟草■☆

104036 Corbularia Salisb. = Narcissus L. ■

104037 Corbularia monophylla Durieu = Narcissus cantabricus DC. subsp. monophyllus (Durieu) A. Fern. ■☆

104038 Corbularia obesa Salisb. = Narcissus obesus Salisb. ■☆

104039 Corchoropsis Siebold et Zucc. (1843);田麻属;Corchoropsis ■●

104040 Corchoropsis crenata Siebold et Zucc.;圆齿田麻(田麻)■☆

104041 Corchoropsis crenata Siebold et Zucc. = Corchoropsis tomentosa (Thunb.) Makino ■

104042 Corchoropsis crenata Siebold et Zucc. var. hupehensis Pamp.;光果圆齿田麻(光果田麻)■

104043 Corchoropsis intermedia Nakai = Corchoropsis tomentosa (Thunb.) Makino ■

104044 Corchoropsis psilocarpa Harms et Loes. = Corchoropsis crenata Siebold et Zucc. var. hupehensis Pamp. ■

104045 Corchoropsis psilocarpa Harms et Loes. ex Gilg et Loes. = Corchoropsis tomentosa (Thunb.) Makino var. psilocarpa (Harms et Loes.) C. Y. Wu et Y. Tang ■

104046 Corchoropsis tomentosa (Thunb.) Makino;田麻(地构叶,毛果田麻,野花生);Tomentose Corchoropsis ■

104047 Corchoropsis tomentosa (Thunb.) Makino f. glabrescens (Nakai) H. Hara;光田麻■☆

104048 Corchoropsis tomentosa (Thunb.) Makino f. glabrescens (Nakai) H. Hara = Corchoropsis tomentosa (Thunb.) Makino ■

104049 Corchoropsis tomentosa (Thunb.) Makino var. glabrescens Nakai = Corchoropsis tomentosa (Thunb.) Makino ■

104050 Corchoropsis tomentosa (Thunb.) Makino var. micropetala Y. T. Chang = Corchoropsis tomentosa (Thunb.) Makino ■

104051 Corchoropsis tomentosa (Thunb.) Makino var. psilocarpa (Harms et Loes.) C. Y. Wu et Y. Tang;光果田麻;Glabrousfruit Corchoropsis ■

104052 Corchoropsis tomentosa (Thunb.) Makino var. tomentosicarpa P. L. Chiu et G. R. Zhong = Corchoropsis tomentosa (Thunb.) Makino ■

104053 Corchoropsis tomentosa Makino var. micropetala Y. T. Zhang = Corchoropsis crenata Siebold et Zucc. ■☆

104054 Corchoropsis tomentosa Makino var. psilocarpa (Harms et Loes.) C. Y. Wu et Y. Tang = Corchoropsis crenata Siebold et Zucc. var. hupehensis Pamp. ■

104055 Corchoropsis tomentosa Makino var. tomentosicarpa P. L. Chiu et G. R. Zhong = Corchoropsis crenata Siebold et Zucc. ■☆

104056 Corchorus L. (1753);黄麻属;Gunny, Jute ■●

104057 Corchorus acutangulus Lam.;假黄麻(假麻区,麻瓯,甜麻,野黄麻,针筒草);Acuteangular Jute ●■

104058 Corchorus acutangulus Lam. = Corchorus aestuans L. ■

104059 Corchorus aestuans L.;甜麻(假黄麻,假麻区,绳黄麻,水丁香,藤肥皂,铁茵陈,野黄麻,针筒草,针筒麻);Round-podded Jute, Sweet Jute ■

104060 Corchorus aestuans L. var. brevicaulis (Hosok.) Tang S. Liu et H. S. Lo;短茎甜麻(短茎绳黄麻);Shortstem Jute ■

104061 Corchorus africanus Bari;非洲黄麻■☆

104062 Corchorus angolensis Exell et Mendonça;安哥拉黄麻■☆

104063 Corchorus antichorus Raeusch. = Corchorus depressus (L.) Stocks ■☆

104064 Corchorus argillicola Moeaha et P. Winter;白土黄麻■☆

104065 Corchorus asplenifolius Burch.;铁线蕨叶黄麻■☆

104066 Corchorus asplenifolius E. Mey. ex Harv. et Sond. = Corchorus trilocularis L. ■

104067 Corchorus brachycarpus Guillaumin et Perr. = Corchorus fascicularis Lam. ■☆
104068 Corchorus brevicaulis Hosok. = Corchorus aestuans L. var. brevicaulis (Hosok.) Tang S. Liu et H. S. Lo ■
104069 Corchorus brevicornutus Vollesen;短角黄麻■☆
104070 Corchorus bricchettii Weim. = Corchorus cinerascens Deflers ■☆
104071 Corchorus burmanii DC. = Corchorus tridens L. ■☆
104072 Corchorus capsularis L.;黄麻(大麻,苦麻,绿麻,络麻,麻皮,牛泥茨,牛泥刺,三珠草,天紫苏,印度麻,圆果黄麻);Indian Grass,Jute,Roundpod Jute,Round-podded Jute ■
104073 Corchorus catharticus Blanco = Corchorus olitorius L. ■
104074 Corchorus cavaleriei H. Lév. = Helicteres glabriuscula Wall. ex Mast. ●
104075 Corchorus cinerascens Deflers;变灰黄麻■☆
104076 Corchorus confusus Wild;混乱黄麻■☆
104077 Corchorus cordifolius Salisb. = Corchorus capsularis L. ■
104078 Corchorus decemangularis Roxb. = Corchorus olitorius L. ■
104079 Corchorus depressus (L.) Stocks;平展黄麻■☆
104080 Corchorus discolor N. E. Br. = Corchorus junodii (Schinz) N. E. Br. ■☆
104081 Corchorus echinatus Hochst. ex Garcke = Corchorus pseudocapsularis Schweinf. ■☆
104082 Corchorus erinaceus Weim. = Corchorus cinerascens Deflers ■☆
104083 Corchorus fascicularis Lam.;簇生黄麻■☆
104084 Corchorus fuscus Roxb. = Corchorus aestuans L. ■
104085 Corchorus gillettii Bari;吉莱特黄麻■☆
104086 Corchorus gracilis R. Br. = Corchorus trilocularis L. ■
104087 Corchorus hirsutus L. var. stenophyllus K. Schum. = Corchorus cinerascens Deflers ■☆
104088 Corchorus hirtus L.;毛黄麻■☆
104089 Corchorus hirtus L. var. orinocensis (Kunth) Schum. = Corchorus hirtus L. ■☆
104090 Corchorus hochstetteri Milne-Redh. = Corchorus pseudocapsularis Schweinf. ■☆
104091 Corchorus japonicus Thunb. = Kerria japonica (L.) DC. ●
104092 Corchorus junodii (Schinz) N. E. Br.;朱诺德黄麻■☆
104093 Corchorus kirkii N. E. Br.;柯克黄麻■☆
104094 Corchorus longipedunculatus Mast.;长花梗黄麻■☆
104095 Corchorus malchairii De Wild. = Corchorus olitorius L. var. malchairii (De Wild.) R. Wilczek ■☆
104096 Corchorus marua Buch.-Ham. = Corchorus capsularis L. ■
104097 Corchorus merxmuelleri Wild;梅尔黄麻■☆
104098 Corchorus microphyllus Fresen. = Corchorus depressus (L.) Stocks ■☆
104099 Corchorus mucilagineus Gibbs = Corchorus asplenifolius Burch. ■☆
104100 Corchorus muricatus Hochst. ex A. Rich. = Corchorus schimperi Cufod. ■☆
104101 Corchorus muricatus Schumach. et Thonn.;粗糙黄麻■☆
104102 Corchorus oenotheroides H. Lév. = Indigofera squalida Prain ■
104103 Corchorus olitorius L.;长蒴黄麻(长果黄麻,斗鹿,山麻,台湾黄麻);Banji Pat,Bush Okra,Jews Mallow,Jew's Mallow,Jews-mallow,Jute,Long-fruited Jute,Nalta Jute,Potherb Jute,Tossa Jute ■
104104 Corchorus olitorius L. var. incisifolius Asch. et Schweinf.;锐裂长蒴黄麻■☆
104105 Corchorus olitorius L. var. malchairii (De Wild.) R. Wilczek;马尔黄麻■☆
104106 Corchorus oppositiflorus Hassk. = Corchorus aestuans L. ■
104107 Corchorus orinocensis Kunth = Corchorus hirtus L. ■☆
104108 Corchorus parvifolius Sebsebe;小叶黄麻■☆
104109 Corchorus pinnatipartitus Wild;羽裂黄麻■☆
104110 Corchorus polygonatum H. Lév. = Tricyrtis pilosa Wall. ■
104111 Corchorus pongolensis Burtt Davy et Greenway = Corchorus kirkii N. E. Br. ■☆
104112 Corchorus prostratus Royle = Corchorus depressus (L.) Stocks ■☆
104113 Corchorus psammophilus Codd;喜沙黄麻■☆
104114 Corchorus pseudocapsularis Schweinf.;假裂果黄麻■☆
104115 Corchorus pseudoolitorius A. K. Islam et Zaid;假长蒴黄麻■☆
104116 Corchorus pyriformis?;梨形黄麻■☆
104117 Corchorus quadrangularis G. Don;四棱黄麻■☆
104118 Corchorus quadrangularis J. A. Schmidt = Corchorus trilocularis L. ■
104119 Corchorus quinquelocularis Moench. = Corchorus olitorius L. ■
104120 Corchorus quinquenervis Hochst. ex A. Rich. = Corchorus urticifolius Wight et Arn. ■☆
104121 Corchorus saxatilis Wild;岩生黄麻■☆
104122 Corchorus scandens Thunb. = Rhodotypos scandens (Thunb.) Makino ●
104123 Corchorus schimperi Cufod.;欣珀黄麻■☆
104124 Corchorus senegalensis Juss. ex Steud. = Corchorus tridens L. ■☆
104125 Corchorus serrata Thunb. = Zelkova serrata (Thunb.) Makino ●
104126 Corchorus serrifolius Burch. = Corchorus asplenifolius Burch. ■☆
104127 Corchorus serrifolius DC. = Corchorus trilocularis L. ■
104128 Corchorus siliquosus L.;长荚果黄麻■☆
104129 Corchorus somalicus Gand. = Corchorus trilocularis L. ■
104130 Corchorus stenophyllus (K. Schum.) Weim. = Corchorus cinerascens Deflers ■☆
104131 Corchorus sulcatus I. Verd.;纵沟黄麻■☆
104132 Corchorus tomentosus Thunb. = Corchoropsis tomentosa (Thunb.) Makino ■
104133 Corchorus tridens L.;三齿黄麻;Tridens Corchorus ■☆
104134 Corchorus triflorus Bojer = Corchorus trilocularis L. ■
104135 Corchorus trilocularis L.;三室黄麻■
104136 Corchorus urticifolius Wight et Arn.;荨麻叶黄麻■☆
104137 Corchorus velutinus Wild;短绒毛黄麻■☆
104138 Corculum Stuntz = Antigonon Endl. ●■
104139 Corda St.-Lag. = Cordia L.(保留属名)●
104140 Cordaea Spreng. = Cyamopsis DC. ■
104141 Cordanthera L. O. Williams.(1941);心药兰属■☆
104142 Cordanthera andina L. O. Williams;心药兰■☆
104143 Cordeauxia Hemsl.(1907);野合豆属■☆
104144 Cordeauxia edulis Hemsl.;野合豆;Jeheb Nut,Ye'eb Nut,Yeheh Nut ■☆
104145 Cordemoya Baill.(1861);科尔大戟属●☆
104146 Cordemoya acuminata Baill.;科尔大戟●☆
104147 Cordia L.(1753)(保留属名);破布木属(破布子属);Cordia ●
104148 Cordia abyssinica R. Br.;苏丹破布木(埃塞俄比亚破布木,埃塞破布木);E. Africa Cordia,Mukumari,Sudan Teak ●☆
104149 Cordia abyssinica R. Br. = Cordia africana Lam. ●☆
104150 Cordia africana Lam.;非洲破布木;Mukumari,Muringa ●☆
104151 Cordia alba (Jacq.) Roem. et Schult.;白破布木;White Cordia ●☆
104152 Cordia alba Roem. et Schult. = Cordia alba (Jacq.) Roem. et

Schult. ●☆

104153 Cordia alliodora (Ruiz et Pav.) Oken;南美破布木(蒜味破布木,蒜叶破布木);Capa Prieto,Cyp,Cypre,Ecuador Laurel,Laurel Negro,Onion Cordia,Salmwood ●☆

104154 Cordia aspera G. Forst. subsp. kanehirai (Hayata) H. Y. Liu = Cordia kanehirai Hayata ●

104155 Cordia aurantiaca Baker;奥兰特破布木●☆

104156 Cordia bakeri Britten = Cordia monoica Roxb. ●☆

104157 Cordia batesii Wernham = Cordia aurantiaca Baker ●☆

104158 Cordia bequaertii De Wild. = Cordia guineensis Thonn. ●☆

104159 Cordia boissieri A. DC.;得州破布木(布西破布木,得克萨斯破布木);Anacahuita,Texas Olive ●☆

104160 Cordia caffra Sond.;南非破布木●☆

104161 Cordia candidissima A. Chev. = Cordia platythyrsa Baker ●☆

104162 Cordia chisimajensis Chiov. = Cordia crenata Delile subsp. meridionalis Warfa ●☆

104163 Cordia chrysocarpa Baker = Cordia millenii Baker ●☆

104164 Cordia cochinchinensis Gagnep.;越南破布木;Cochinchina Cordia,Cochin-China Cordia,Vietnam Cordia ●

104165 Cordia crenata Delile;圆齿破布木●☆

104166 Cordia crenata Delile subsp. meridionalis Warfa;南方破布木●☆

104167 Cordia crenata Delile subsp. shinyangensis Verdc.;希尼安加破布木●☆

104168 Cordia cumingiana Vidal;吕宋破布木●

104169 Cordia cumingiana Vidal = Cordia aspera G. Forst. subsp. kanehirai (Hayata) H. Y. Liu ●

104170 Cordia cumingiana Vidal = Cordia kanehirai Hayata ●

104171 Cordia curassavica (Jacq.) Roem. et Schult.;加勒比破布木●☆

104172 Cordia cylindrostachya Roem. et Schult.;桂花破布木●☆

104173 Cordia dentata Poir.;尖齿破布木●☆

104174 Cordia dewevrei De Wild. et T. Durand;德韦破布木●☆

104175 Cordia dichotoma G. Forst.;破布木(白茶,风筝子,狗屎木,破布子,青桐翠木,青桐木);Bird-lime Tree,Dichotomous Cordia,Fragrant Manjack,Sebastan Plum Cordia ●

104176 Cordia dioica A. DC. = Cordia monoica Roxb. ●☆

104177 Cordia dodecandra DC.;十二雄蕊破布木;Zirieote ●☆

104178 Cordia dodecandra Sessé et Moc. = Cordia dodecandra DC. ●☆

104179 Cordia dusenii Gürke = Cordia aurantiaca Baker ●☆

104180 Cordia ellenbeckii Gürke;埃伦破布木●☆

104181 Cordia faulknerae Verdc.;福克纳破布木●☆

104182 Cordia fissistyla Vollesen;半裂柱破布木●☆

104183 Cordia forcans Johnst.;二叉破布木;Fork Cordia,Furcate Cordia ●

104184 Cordia gerascanthus Jacq. = Cordia gerascanthus L. ●☆

104185 Cordia gerascanthus L.;拉美破布木(委内瑞拉破布木);Eucador Laurel,Prince Wood,Princewood,Salmwood,Spanish Elm ●☆

104186 Cordia gharaf (Forssk.) Ehrenb. ex Asch. = Cordia sinensis Lam. ●

104187 Cordia gilletii De Wild.;吉勒特破布木●☆

104188 Cordia glabrata A. DC.;脱毛破布木●☆

104189 Cordia globosa Andrieux ex DC.;球形破布木●☆

104190 Cordia goeldiana Huber;亚马逊破布木(龙凤檀,南美胡桃);Frei Jo,Freijo,South American Walnut ●☆

104191 Cordia goetzei Gürke;格兹破布木●☆

104192 Cordia goossensii De Wild. = Cordia aurantiaca Baker ●☆

104193 Cordia grandicalyx Oberm.;大萼破布木●☆

104194 Cordia guineensis Thonn.;几内亚破布木●☆

104195 Cordia guineensis Thonn. subsp. mutica Verdc.;无尖破布木●☆

104196 Cordia heudelotii Baker = Cordia senegalensis Juss. ●☆

104197 Cordia holstii Gürke = Cordia africana Lam. ●☆

104198 Cordia incana Royle = Cordia vestita (DC.) Hook. f. et Thomson ●☆

104199 Cordia indica Lam. = Cordia dichotoma G. Forst. ●

104200 Cordia irvingii Baker = Cordia millenii Baker ●☆

104201 Cordia johnsonii Baker = Cordia guineensis Thonn. ●☆

104202 Cordia kabarensis De Wild. = Cordia monoica Roxb. ●☆

104203 Cordia kanehirai Hayata;台湾破布木(金平氏破布木,金平氏破布子,吕宋破布木,破布子,邱氏破布木);Cuming Cordia,Taiwan Cordia ●

104204 Cordia kanehirai Hayata = Cordia aspera G. Forst. subsp. kanehirai (Hayata) H. Y. Liu ●

104205 Cordia kirkii Baker = Cordia pilosissima Baker ●☆

104206 Cordia liebrechtsiana De Wild. et T. Durand = Cordia millenii Baker ●☆

104207 Cordia longipes Baker = Cordia millenii Baker ●☆

104208 Cordia longipetiolata Warfa;长柄破布木●☆

104209 Cordia lowriana Brandis = Cordia crenata Delile ●☆

104210 Cordia lowryana J. S. Mill.;劳里破布木●☆

104211 Cordia lutea Lam.;黄破布木●☆

104212 Cordia macleodii (Griff.) Hook. f. et Thomson;马氏破布木●☆

104213 Cordia macleodii Hook. f. et Thomson = Cordia macleodii (Griff.) Hook. f. et Thomson ●☆

104214 Cordia mairei Humbert;迈雷破布木●☆

104215 Cordia mannii C. H. Wright = Cordia senegalensis Juss. ●☆

104216 Cordia millenii Baker;米氏破布木●☆

104217 Cordia monoica Roxb. = Cordia ovalis R. Br. ex A. DC. ●☆

104218 Cordia myxa L.;毛叶破布木(胶果木,破布乌,破布子,破故子,破果子,破子,亚述);Assyrian Plum,Hairleaf Cordia,Hair-leaved Cordia,Safistan,Sebestan Plum,Selu,Sudan Teak ●☆

104219 Cordia nevillii Alston = Cordia quercifolia Klotzsch ●☆

104220 Cordia obliqua Willd.;斜叶破布木(倾斜破布木);Clammy Cherry,Oblique Leaf Cordia ●☆

104221 Cordia oblongifolia Hochst. ex DC. = Cordia sinensis Lam. ●

104222 Cordia obovata Baker = Cordia monoica Roxb. ●☆

104223 Cordia obtusa Balf. f.;钝破布木●☆

104224 Cordia odorata Gürke = Cordia platythyrsa Baker ●☆

104225 Cordia ovalis R. Br. ex A. DC.;卵叶破布木;Sandpaper Tree ●☆

104226 Cordia ovalis R. Br. ex A. DC. = Cordia monoica Roxb. ●☆

104227 Cordia parviflora Ortega;小花破布木;Littleleaf Cordia ●☆

104228 Cordia parvifolia A. DC.;小叶破布木;Little Leaf Cordia ●☆

104229 Cordia peteri Verdc.;彼得破布木●☆

104230 Cordia pilosissima Baker;多毛破布木●☆

104231 Cordia platythyrsa Baker;聚伞破布木●☆

104232 Cordia populifolia Baker = Cordia platythyrsa Baker ●☆

104233 Cordia quarensis Gürke;夸伦破布木●☆

104234 Cordia quarensis Gürke = Cordia monoica Roxb. ●☆

104235 Cordia quercifolia Klotzsch;栎叶破布木●☆

104236 Cordia ravae Chiov. = Cordia goetzei Gürke ●☆

104237 Cordia reticulata Roth = Cordia sinensis Lam. ●

104238 Cordia retusa Vahl = Carmona retusa (Vahl) Masam. ●

104239 Cordia rogersii Hutch.;罗杰斯破布木●☆

104240 Cordia rothii Roem. et Schult. = Cordia gharaf (Forssk.) Ehrenb. ex Asch. ●

104241 Cordia rothii Roem. et Schult. = Cordia sinensis Lam. ●

104242 Cordia rubra Hiern = Cordia monoica Roxb. ●☆
104243 Cordia salicifolia Cham.;柳叶破布木●☆
104244 Cordia schatziana J. S. Mill.;沙茨破布木●☆
104245 Cordia sebestena L.;黏液破布木(塞贝破布木);Geiger Tree, Largeleaf Geigertree, Scarlet Cordia, Scarlet Cordia-tree ●☆
104246 Cordia senegalensis Juss.;塞内加尔破布木●☆
104247 Cordia sinensis Lam.;中华破布木;Chinese Cordia ●
104248 Cordia somaliensis Baker;索马里破布木●☆
104249 Cordia spinescens L.;刺状破布木●☆
104250 Cordia stuhlmannii Gürke;斯图尔曼破布木●☆
104251 Cordia subcordata Lam.;橙花破布木(心叶破布木);Indopacific Strand, Kou, Mareer, Marer, Orangeflower Cordia, Subcordate Cordia ●
104252 Cordia subopposita DC. = Cordia sinensis Lam. ●
104253 Cordia suckertii Chiov.;祖克特破布木●☆
104254 Cordia suckertii Chiov. var. exasperata Verdc. = Cordia suckertii Chiov. ●☆
104255 Cordia thyrsiflora Siebold et Zucc. = Ehretia acuminata R. Br. var. obovata (Lindl.) I. M. Johnst. ●
104256 Cordia thyrsiflora Siebold et Zucc. = Ehretia acuminata R. Br. ●
104257 Cordia thyrsiflora Siebold et Zucc. = Ehretia thyrsiflora (Siebold et Zucc.) Nakai ●
104258 Cordia tisserantii Aubrév. = Cordia uncinulata De Wild. ●☆
104259 Cordia torrei E. S. Martins;托雷破布木●☆
104260 Cordia trichocladophylla Verdc.;毛枝破布木●☆
104261 Cordia trichotoma Vell. ex Steud.;三出破布木;Lauro Cordia, Lauro Pardo ●
104262 Cordia ubanghensis A. Chev. = Cordia africana Lam. ●☆
104263 Cordia ugandensis S. Moore = Cordia millenii Baker ●☆
104264 Cordia uncinulata De Wild.;钩破布木●☆
104265 Cordia unyorensis Stapf = Cordia millenii Baker ●☆
104266 Cordia venosa Hemsl. = Clerodendrum cyrtophyllum Turcz. ●
104267 Cordia verbenacea DC.;马鞭菊破布木●☆
104268 Cordia vestita (DC.) Hook. f. et Thomson;包被破布木●☆
104269 Cordia vignei Hutch. et Dalziel;维涅破布木●☆
104270 Cordia warneckei Baker et C. H. Wright = Cordia guineensis Thonn. ●☆
104271 Cordia zedambae Martelli = Cordia crenata Delile ●☆
104272 Cordiaceae R. Br. ex Dumort. (1829)(保留科名);破布木科(破布树科)●■
104273 Cordiaceae R. Br. ex Dumort.(保留科名) = Boraginaceae Juss.(保留科名)■●
104274 Cordiaceae R. Br. ex Dumort.(保留科名) = Ehretiaceae Mart.(保留科名)●
104275 Cordiada Vell. = Cordia L.(保留属名)●
104276 Cordiera A. Rich. = Alibertia A. Rich. ex DC. ●☆
104277 Cordiera A. Rich. ex DC. = Alibertia A. Rich. ex DC. ●☆
104278 Cordiglottis J. J. Sm. (1922);心舌兰属■☆
104279 Cordiglottis filiformis (Hook. f.) Garay;线形心舌兰■☆
104280 Cordiglottis major (Carr) Garay;大心舌兰■☆
104281 Cordiglottis multicolor (Ridl.) Garay;多色心舌兰■☆
104282 Cordiglottis westenenki J. J. Sm.;心舌兰■☆
104283 Cordiopsis Desv. = Cordia L.(保留属名)●
104284 Cordiopsis Desv. ex Ham. = Cordia L.(保留属名)●
104285 Cordisepalum Verdc. (1971);心萼旋花属●☆
104286 Cordisepalum thorelii (Gagnep.) Verdc.;心萼旋花●☆
104287 Cordobia Nied. (1912);克尔金虎尾属●☆
104288 Cordobia argentea Nied.;克尔金虎尾●☆
104289 Cordula Raf.(废弃属名) = Paphiopedilum Pfitzer(保留属名)■
104290 Cordula appletoniana (Gower) Rolfe = Paphiopedilum appletonianum (Gower) Rolfe ■
104291 Cordula bellatula (Rchb. f.) Rolfe = Paphiopedilum bellatulum (Rchb. f.) Stein ■
104292 Cordula boxallii (Rchb. f.) Rolfe = Paphiopedilum villosum Lindl. var. boxallii (Rchb. f.) Pfitzer ■
104293 Cordula charlesworthii (Rolfe) Rolfe = Paphiopedilum charlesworthii (Rolfe) Pfitzer ■
104294 Cordula concolor (Lindl. ex Bateman) Rolfe = Paphiopedilum concolor (Bateman) Pfitzer ■
104295 Cordula concolor (Lindl.) Rolfe = Paphiopedilum concolor (Bateman) Pfitzer ■
104296 Cordula esquirolei (Schltr.) Hu = Paphiopedilum hirsutissimum (Lindl. ex Hook.) Stein ■
104297 Cordula gratrixiana (Rolfe) Rolfe = Paphiopedilum gratrixianum Rolfe ■
104298 Cordula hirsutissima (Lindl. ex Hook.) Rolfe = Paphiopedilum hirsutissimum (Lindl. ex Hook.) Stein ■
104299 Cordula insignis (Lindl.) Raf. = Paphiopedilum insigne (Lindl.) Pfitzer ■
104300 Cordula insignis (Wall. ex Lindl.) Rafinesque = Paphiopedilum insigne (Lindl.) Pfitzer ■
104301 Cordula parishii (Rchb. f.) Rolfe = Paphiopedilum parishii (Rchb. f.) Pfitzer ■
104302 Cordula purpurata (Lindl.) Rolfe = Paphiopedilum purpuratum (Lindl.) Stein ■
104303 Cordula spiceriana (Rchb. f.) Rolfe = Paphiopedilum spicerianum (Rchb. f.) Pfitzer ■
104304 Cordula venustum (Wall. ex Sims) Rolfe = Paphiopedilum venustum (Wall. ex Sims) Pfitzer ■
104305 Cordula villosum (Lindl.) Rolfe = Paphiopedilum villosum (Lindl.) Stein ■
104306 Cordyla Blume = Nervilia Comm. ex Gaudich.(保留属名)■
104307 Cordyla Blume = Roptrostemon Blume ■
104308 Cordyla Lour. (1790);棒状苏木属●☆
104309 Cordyla Lour. = Dupuya J. H. Kirkbr. ●☆
104310 Cordyla Post et Kuntze = Cordula Raf. ■
104311 Cordyla Post et Kuntze = Paphiopedilum Pfitzer(保留属名)■
104312 Cordyla africana Lour.;非洲棒状苏木●☆
104313 Cordyla densiflora Milne-Redh.;密花棒状苏木●☆
104314 Cordyla discolor Blume = Nervilia plicata (Andréws) Schltr. ■
104315 Cordyla haraka Capuron = Dupuya haraka (Capuron) J. H. Kirkbr. ●☆
104316 Cordyla madagascariensis R. Vig.;马岛棒状苏木●☆
104317 Cordyla madagascariensis R. Vig. = Dupuya madagascariensis (R. Vig.) J. H. Kirkbr. ●☆
104318 Cordyla pinnata (A. Rich.) Milne-Redh. = Cordyla pinnata (Lepr. ex A. Rich.) Milne-Redh. ●☆
104319 Cordyla pinnata (Lepr. ex A. Rich.) Milne-Redh.;棒状苏木(羽叶棒状苏木);Bush Mango ●☆
104320 Cordyla richardii Milne-Redh.;理查德棒状苏木●☆
104321 Cordyla richardii Planch. = Cordyla richardii Milne-Redh. ●☆
104322 Cordyla somalensis J. B. Gillett;索马里棒状苏木●☆
104323 Cordyla somalensis J. B. Gillett subsp. littoralis;滨海棒状苏木●☆
104324 Cordylanthus Blume = Homalium Jacq. ●

104325 Cordylanthus Nutt. ex Benth. (1846)(保留属名);棒花列当属(棒花参属)■☆

104326 Cordylanthus filifolius Nutt. ex Benth.;棒花列当(棒花参)■☆

104327 Cordylanthus maritimus;海滨棒花列当(海滨棒花参);Sattmarsh Club-flower ■☆

104328 Cordylanthus wrightii A. Gray;赖特棒花列当■☆

104329 Cordylestylis Falc. = Goodyera R. Br. ■

104330 Cordylia Pers. = Cordyla Lour. ●☆

104331 Cordyline Adans.(废弃属名) = Cordyline Comm. ex R. Br.(保留属名)●

104332 Cordyline Adans.(废弃属名) = Sansevieria Thunb.(保留属名)■

104333 Cordyline Comm. ex Juss. = Cordyline Comm. ex R. Br.(保留属名)●

104334 Cordyline Comm. ex R. Br. (1810)(保留属名);朱蕉属;Cabbage Palm, Cabbage Tree, Club Palm, Cordyline, Dracaena, Dracaena Palm, Ti Plant ●

104335 Cordyline Fabr. = Dracaena Vand. ex L. ●■

104336 Cordyline Royen ex Adans. = Sansevieria Thunb.(保留属名)■

104337 Cordyline 'Glanca';耐荫朱蕉●☆

104338 Cordyline australis (G. Forst.) Endl. = Cordyline australis (G. Forst.) Hook. f. ●

104339 Cordyline australis (G. Forst.) Hook. f.;大朱蕉(剑叶朱蕉,巨朱蕉,南朱蕉,香朱蕉,香棕榈兰,新西兰朱蕉);Cabbage Palm, Cabbage Tree, Cabbage-tree, Club Palm, Cordyline, Dracaena-palm, Giant Dracaena, Green Dracaena, New Zealand Cabbage Palm, New Zealand Cabbage Tree, New Zealand Cabbage-tree, New Zealand Dracaena, Palm Lily, Ti Kouka, Tikouka ●

104340 Cordyline australis (G. Forst.) Hook. f. 'Arbertii';阿尔伯迪新西兰朱蕉●☆

104341 Cordyline australis (G. Forst.) Hook. f. 'Atropurpurea';深紫叶新西兰朱蕉●☆

104342 Cordyline australis (G. Forst.) Hook. f. 'Purpurea';紫叶新西兰朱蕉●☆

104343 Cordyline australis (G. Forst.) Hook. f. 'Veitchii';维茨新西兰朱蕉●☆

104344 Cordyline banksii Hook. f.;班克斯朱蕉●☆

104345 Cordyline baueri Hook. f.;鲍氏朱蕉;Bauer's dracaena ●☆

104346 Cordyline betschleriana Goepp. = Dracaena concinna Kunth ●☆

104347 Cordyline beuckelaerii K. Koch = Cordyline banksii Hook. f. ●☆

104348 Cordyline cannifolia R. Br.;美人蕉叶朱蕉●☆

104349 Cordyline congesta Endl. = Cordyline stricta Endl. ●

104350 Cordyline fruticosa (L.) A. Chev.;朱蕉(红铁树,红叶铁树,假槟榔树,铁连草,铁树,牙竹麻,朱竹);Cabbage Tree, Fruticose Cordyline, Fruticose Coriaria, Fruticose Dracaena, Goodluck Plant, Hawaiian Ti Plant, Tanket, Ti, Ti Plant, Ti Tree ●

104351 Cordyline fruticosa (L.) A. Chev. 'Baptisii';巴氏朱蕉●

104352 Cordyline fruticosa (L.) A. Chev. 'Imperialis';帝王朱蕉●

104353 Cordyline fruticosa (L.) A. Chev. 'Tricolor';七彩朱蕉●☆

104354 Cordyline fruticosa (L.) A. Chev. = Cordyline terminalis (L.) Kunth ●

104355 Cordyline fruticosa (L.) Göpp. = Cordyline fruticosa (L.) A. Chev. ●

104356 Cordyline guineensis (L.) Britton ex Small = Sansevieria hyacinthoides (L.) Druce ■☆

104357 Cordyline haageana K. Koch;哈吉朱蕉■☆

104358 Cordyline indivisa (G. Forst.) Endl. = Cordyline indivisa (G. Forst.) Kunth ●☆

104359 Cordyline indivisa (G. Forst.) Kunth;高山朱蕉(蓝朱蕉,裂叶龙血树,新西兰铁树);Blue Dracaena, Broad-leaved Cabbage-tree Mountain Cabbage Tree, Cabbage Palm, Cabbage Tree, Dracena, Fountain Dracaena, Palm Lily, Tie Palm, Toi ●☆

104360 Cordyline indivisa Kunth = Cordyline indivisa (G. Forst.) Kunth ●☆

104361 Cordyline maculata Planch. = Dracaena elliptica Thunb. var. maculata Roxb. ●☆

104362 Cordyline petiolaris (Domin) Pedley;阔叶朱蕉;Btoad-leaf Palm Lily ●☆

104363 Cordyline rubra Hügel = Cordyline rubra Hügel ex Kunth ●☆

104364 Cordyline rubra Hügel ex Kunth;红铁树●☆

104365 Cordyline rumphii Hook. = Dracaena aletriformis (Haw.) Bos ●☆

104366 Cordyline stricta Endl.;剑叶朱蕉(澳大利亚朱蕉,澳洲朱蕉,红剑叶朱蕉,剑叶铁树,剑叶万年青,细叶朱蕉,小叶铁树,小朱蕉);Australia Dracaena, Narrow-leaved Palm Lily, Slender Palm Lily ●

104367 Cordyline terminalis (L.) Kunth;顶生朱蕉(红竹,红竹叶,千年木,铁树,五彩铁树,朱蕉,朱竹);Boundary Mark, Common Dracaena, Good-luck Plant, Red Dracaena, Terminal Dracaena, Ti Plant ●

104368 Cordyline terminalis (L.) Kunth 'Aichiaka';亮叶朱蕉●

104369 Cordyline terminalis (L.) Kunth 'Alba Rosea';乳红朱蕉●

104370 Cordyline terminalis (L.) Kunth 'Amacilis';锦朱蕉(彩叶朱蕉)●

104371 Cordyline terminalis (L.) Kunth 'Baptistii';斜纹朱蕉●

104372 Cordyline terminalis (L.) Kunth 'Bella';细叶朱蕉●☆

104373 Cordyline terminalis (L.) Kunth 'Bicolor';二色朱蕉●

104374 Cordyline terminalis (L.) Kunth 'Crystal';翡翠朱蕉●☆

104375 Cordyline terminalis (L.) Kunth 'Fire fovntain';火焰朱蕉●☆

104376 Cordyline terminalis (L.) Kunth 'Goshikiba';五彩朱蕉(锦叶朱蕉)●

104377 Cordyline terminalis (L.) Kunth 'Hakuba';白马朱蕉●☆

104378 Cordyline terminalis (L.) Kunth 'Metallica';金泽朱蕉●

104379 Cordyline terminalis (L.) Kunth 'Minima Tricolor';三色姬朱蕉●

104380 Cordyline terminalis (L.) Kunth 'Minimus';姬朱蕉●

104381 Cordyline terminalis (L.) Kunth 'Neger';浓紫朱蕉●☆

104382 Cordyline terminalis (L.) Kunth 'Pulchella';艳美朱蕉●

104383 Cordyline terminalis (L.) Kunth 'Red Edge';红边朱蕉●

104384 Cordyline terminalis (L.) Kunth 'Rosea';粉叶朱蕉●

104385 Cordyline terminalis (L.) Kunth 'Rubra';红叶朱蕉●

104386 Cordyline terminalis (L.) Kunth 'Rubrostriata';红条朱蕉●

104387 Cordyline terminalis (L.) Kunth 'Ti';绿叶朱蕉●☆

104388 Cordyline terminalis (L.) Kunth 'Youmeninshiki';锦翠朱蕉●☆

104389 Cordyline terminalis (L.) Kunth = Cordyline fruticosa (L.) A. Chev. ●

104390 Cordyline terminalis (L.) Kunth var. ferra (L.) Baker = Cordyline fruticosa (L.) A. Chev. ●

104391 Cordyline terminalis (L.) Kunth var. ferrea (L.) Baker = Cordyline fruticosa (L.) A. Chev. ●

104392 Cordyline terminalis Kunth = Cordyline terminalis (L.) Kunth ●

104393 Cordyloblaste Hensch. ex Moritzi = Symplocos Jacq. ●

104394 Cordyloblaste Moritzi = Symplocos Jacq. ●

104395 Cordyloblaste confusa (Brand) Hatus. = Symplocos sonoharae Koidz. ●

104396 Cordylocarpus Desf. (1798);北非棒果芥属■☆

104397 Cordylocarpus muricatus Desf.;北非棒果芥■☆
104398 Cordylocarpus muricatus Desf. var. leiocarpus Faure et Maire = Cordylocarpus muricatus Desf. ■☆
104399 Cordylocarpus muricatus Desf. var. trichocarpus Faure et Maire = Cordylocarpus muricatus Desf. ■☆
104400 Cordylocarpus pumilus Sennen et Mauricio = Cordylocarpus muricatus Desf. ■☆
104401 Cordylocarya Besser ex DC. = Rapistrum Crantz(保留属名)■☆
104402 Cordylogne Lindl. = Cordylogyne E. Mey. ■☆
104403 Cordylogyne E. Mey. (1838);棒蕊萝藦属■☆
104404 Cordylogyne argillicola Dinter = Periglossum angustifolium Decne. ■☆
104405 Cordylogyne globosa E. Mey.;棒蕊萝藦■☆
104406 Cordylogyne kassnerianum (Schltr.) Eyles = Periglossum mackenii Harv. ■☆
104407 Cordylogyne mossambicense (Schltr.) Eyles = Periglossum mackenii Harv. ■☆
104408 Cordylophorum (Nutt. ex Torr. et A. Gray) Rydb. = Epilobium L. ■
104409 Cordylophorum Rydb. = Epilobium L. ■
104410 Cordylostylis Post et Kuntze = Cordylestylis Falc. ■
104411 Cordylostylis Post et Kuntze = Goodyera R. Br. ■
104412 Coreanomecon Nakai = Chelidonium L. ■
104413 Coreanomecon Nakai = Hylomecon Maxim. ■
104414 Coreanomecon Nakai(1935);荷青花白屈菜属■☆
104415 Coreanomecon hylomeconoides Nakai;荷青花白屈菜■☆
104416 Coreanomecon hylomeconoides Nakai = Chelidonium hylomeconoides (Nakai) Ohwi ■☆
104417 Coredia Hook. f. = Cordia L. (保留属名)●
104418 Corellia A. M. Powell = Perityle Benth. ●■☆
104419 Corema Bercht. et J. Presl = Sarothamnus Wimm. (保留属名)●☆
104420 Corema D. Don(1826);岩帚兰属(丛枝木属,帚高兰属)●☆
104421 Corema album (L.) D. Don;白岩帚兰●☆
104422 Corema conradii (Torr.) Loudon;岩帚兰●☆
104423 Coreocarpus Benth. (1844);虫籽菊属(虫子菊属)■☆
104424 Coreocarpus arizonicus (A. Gray) S. F. Blake;虫籽菊■☆
104425 Coreopis Gunn = Coreopsis L. ●■
104426 Coreopsidaceae Link = Asteraceae Bercht. et J. Presl(保留科名)●■
104427 Coreopsidaceae Link = Compositae Giseke(保留科名)●■
104428 Coreopsidaceae Link;金鸡菊科●■
104429 Coreopsis L. (1753);金鸡菊属;Coreopsis, Sea Dahlia, Tickseed, Tickweed ●■
104430 Coreopsis abyssinica Sch. Bip. ex Walp. = Bidens camporum (Hutch.) Mesfin ■☆
104431 Coreopsis abyssinica Sch. Bip. ex Walp. f. latisecta Vatke = Bidens rueppellii (Sch. Bip. ex Walp.) Sherff ■☆
104432 Coreopsis abyssinica Sch. Bip. ex Walp. var. bipinnato-partita Chiov. = Bidens setigera (Sch. Bip. ex Vatke) Sherff subsp. bipinnato-partita (Chiov.) Mesfin ■☆
104433 Coreopsis abyssinica Sch. Bip. ex Walp. var. glabrior Oliv. et Hiern = Bidens camporum (Hutch.) Mesfin ■☆
104434 Coreopsis alternifolia L. = Verbesina alternifolia (L.) Britton ex Kearney ■☆
104435 Coreopsis ambacensis Hiern = Bidens steppia (Steetz) Sherff. ■☆
104436 Coreopsis arenicola S. Moore = Bidens arenicola (S. Moore) T. G. J. Rayner ■☆
104437 Coreopsis aristosa Michx. = Bidens aristosa (Michx.) Britton ■☆
104438 Coreopsis aristosa Michx. var. mutica A. Gray = Bidens aristosa (Michx.) Britton ■☆
104439 Coreopsis artemisiifolia Jacq. = Cosmos sulphureus Cav. ■
104440 Coreopsis asperata Hutch. et Dalziel = Bidens asperata (Hutch. et Dalziel) Sherff ■☆
104441 Coreopsis aspilioides Baker = Aspilia mossambicensis (Oliv.) Wild ■☆
104442 Coreopsis atkinsoniana Douglas ex Lindl. = Coreopsis tinctoria Nutt. ■
104443 Coreopsis aurea Aiton = Bidens aurea (Aiton) Sherff ■☆
104444 Coreopsis auriculata L.;耳裂金鸡菊;Coreopsis, Eared Coreopsis, Mouse Ear Coreopsis ■☆
104445 Coreopsis auriculata L. 'Superba';华丽耳裂金鸡菊■☆
104446 Coreopsis badia Sherff = Bidens barteri (Oliv. et Hiern) T. G. J. Rayner ■☆
104447 Coreopsis barteri Oliv. et Hiern = Bidens barteri (Oliv. et Hiern) T. G. J. Rayner ■☆
104448 Coreopsis basalis (A. Dietr.) S. F. Blake;金鸡菊(基生叶金鸡菊,金鸡兰);Golden Wave, Goldenwave Coreopsis ■
104449 Coreopsis basalis (A. Dietr.) S. F. Blake var. wrightii (A. Gray) S. F. Blake = Coreopsis basalis (A. Dietr.) S. F. Blake ■
104450 Coreopsis baumii O. Hoffm. = Bidens baumii (O. Hoffm.) Sherff ■☆
104451 Coreopsis beguinotii Chiov. = Bidens pachyloma (Oliv. et Hiern) Cufod. ■☆
104452 Coreopsis bella Hutch. = Bidens rueppellii (Sch. Bip. ex Walp.) Sherff ■☆
104453 Coreopsis bigelovii (A. Gray) Voss;毕氏金鸡菊■☆
104454 Coreopsis bipinnatus Cav. = Cosmos bipinnatus Cav. ■
104455 Coreopsis biternata Lour. = Bidens biternata (Lour.) Merr. et Sherff ex Sherff ■
104456 Coreopsis borianiana Sch. Bip. ex Schweinf. et Asch. = Bidens borianiana (Sch. Bip. ex Schweinf. et Asch.) Cufod. ■☆
104457 Coreopsis borianiana Sch. Bip. ex Schweinf. et Asch. var. cannabina? = Bidens borianiana (Sch. Bip. ex Schweinf. et Asch.) Cufod. ■☆
104458 Coreopsis borianiana Sch. Bip. ex Schweinf. et Asch. var. multiplex Sherff = Bidens borianiana (Sch. Bip. ex Schweinf. et Asch.) Cufod. ■☆
104459 Coreopsis buchingeri Sch. Bip. ex Schweinf. et Asch. = Bidens ternata (Chiov.) Sherff var. vatkei (Sherff) Mesfin ■☆
104460 Coreopsis buchneri Klatt = Bidens buchneri (Klatt) Sherff ■☆
104461 Coreopsis californica (Nutt.) H. Sharsm.;加州金鸡菊■☆
104462 Coreopsis californica (Nutt.) H. Sharsm. subsp. newberryi (A. Gray) E. Murray = Coreopsis californica (Nutt.) H. Sharsm. ■☆
104463 Coreopsis californica (Nutt.) H. Sharsm. var. newberryi (A. Gray) E. B. Sm. = Coreopsis californica (Nutt.) H. Sharsm. ■☆
104464 Coreopsis calliopsidea (DC.) A. Gray;莫哈维金鸡菊;Mohave Coreopsis ■☆
104465 Coreopsis callosa Sch. Bip. = Bidens pachyloma (Oliv. et Hiern) Cufod. ■☆
104466 Coreopsis camporum Hutch. = Bidens camporum (Hutch.) Mesfin ■☆
104467 Coreopsis cardaminifolia (DC.) Torr. et A. Gray = Coreopsis tinctoria Nutt. ■
104468 Coreopsis cardaminifolia Torr. et A. Gray;心叶金鸡菊;

Heartleaf Coreopsis ■☆

104469 Coreopsis cardaminifolia Torr. et A. Gray = Coreopsis tinctoria Nutt. ■

104470 Coreopsis chevalieri Hoffm. et Muschl. = Bidens borianiana (Sch. Bip. ex Schweinf. et Asch.) Cufod. ■☆

104471 Coreopsis chippii M. B. Moss = Bidens chippii (M. B. Moss) Mesfin ■☆

104472 Coreopsis chrysantha Vatke = Bidens ternata (Chiov.) Sherff var. vatkei (Sherff) Mesfin ■☆

104473 Coreopsis chrysantha Vatke var. simplicifolia? = Bidens ternata (Chiov.) Sherff ■☆

104474 Coreopsis chrysopterocarpa Chiov. = Bidens borianiana (Sch. Bip. ex Schweinf. et Asch.) Cufod. ■☆

104475 Coreopsis coriacea O. Hoffm. = Bidens buchneri (Klatt) Sherff ■☆

104476 Coreopsis cosmophylla Sherff = Bidens ochracea (O. Hoffm.) Sherff ■☆

104477 Coreopsis crassifolia Aiton = Coreopsis lanceolata L. ■

104478 Coreopsis crataegifolia O. Hoffm. = Bidens kilimandscharica (O. Hoffm.) Sherff ■☆

104479 Coreopsis discoidea Torr. et A. Gray = Bidens discoidea (Torr. et A. Gray) Britton ■☆

104480 Coreopsis douglasii (DC.) H. M. Hall;道格拉斯金鸡菊;Douglas Coreopsis ■☆

104481 Coreopsis douglasii Hall. = Coreopsis douglasii (DC.) H. M. Hall ■☆

104482 Coreopsis drummondii (G. Don) Torr. et A. Gray;金鸡兰;Golden Wave ■

104483 Coreopsis drummondii (G. Don) Torr. et A. Gray = Coreopsis basalis (A. Dietr.) S. F. Blake ■

104484 Coreopsis elgonensis Sherff = Bidens elgonensis (Sherff) Agnew ■☆

104485 Coreopsis ellenbeckii O. Hoffm. = Bidens macroptera (Sch. Bip. ex Chiov.) Mesfin ■☆

104486 Coreopsis elliotii S. Moore = Bidens elliotii (S. Moore) Sherff ■☆

104487 Coreopsis exaristata O. Hoffm. = Bidens schimperi Sch. Bip. ex Walp. ■☆

104488 Coreopsis exaristata O. Hoffm. var. gracilior = Bidens taylori (S. Moore) Sherff ■☆

104489 Coreopsis exilis Sherff = Bidens oblonga (Sherff) Wild ■☆

104490 Coreopsis falcata F. E. Boynton = Coreopsis gladiata Walter ■☆

104491 Coreopsis feruloides Sherff = Bidens rueppellii (Sch. Bip. ex Walp.) Sherff ■☆

104492 Coreopsis filifolia Hook. = Thelesperma filifolium (Hook.) A. Gray ■☆

104493 Coreopsis fischeri O. Hoffm. = Bidens fischeri (O. Hoffm.) Sherff ■☆

104494 Coreopsis floridana E. B. Sm. = Coreopsis gladiata Walter ■☆

104495 Coreopsis formosa Bonato = Cosmos bipinnatus Cav. ■

104496 Coreopsis frondosa O. Hoffm. = Bidens magnifolia Sherff ■☆

104497 Coreopsis galericulata Sherff = Guizotia scabra (Vis.) Chiov. ■☆

104498 Coreopsis gigantea (Kellogg) H. M. Hall;高大金鸡菊■☆

104499 Coreopsis giorgii Sherff = Bidens grantii (Oliv.) Sherff ■☆

104500 Coreopsis gladiata Walter;腺点金鸡菊■☆

104501 Coreopsis gladiata Walter var. linifolia (Nutt.) Cronquist = Coreopsis gladiata Walter ■☆

104502 Coreopsis glaucescens Oliv. et Hiern = Bidens rueppellii (Sch. Bip. ex Walp.) Sherff ■☆

104503 Coreopsis goffardii Sherff = Bidens oblonga (Sherff) Wild ■☆

104504 Coreopsis grandiflora Hogg 'Baby Sun';初阳金鸡菊■

104505 Coreopsis grandiflora Hogg 'Bandengold';冰金大花金鸡菊■

104506 Coreopsis grandiflora Hogg 'Double New Gold';重瓣黄金鸡菊■

104507 Coreopsis grandiflora Hogg 'Sunburst';旭日金鸡菊■

104508 Coreopsis grandiflora Hogg = Coreopsis grandiflora Hogg ex Sweet ●■

104509 Coreopsis grandiflora Hogg ex Sweet;大花金鸡菊(大花波斯菊);Bigflower Coreopsis, Big-flower Tickseed, Coreopsis, Large Headed Tickseed, Large-flowered Tickseed, Tickseed, Tickweed ●■

104510 Coreopsis grandiflora Hogg ex Sweet var. harveyana (A. Gray) Sherff;异鳞金鸡菊;Big-flower Tickseed, Large-flowered Tickseed ■☆

104511 Coreopsis grandiflora Hogg var. harveyana (A. Gray) Sherff = Coreopsis grandiflora Hogg ●■

104512 Coreopsis grandiflora Hogg var. longipes (Hook.) Torr. et A. Gray = Coreopsis grandiflora Hogg ●■

104513 Coreopsis grandiflora Hogg var. saxicola (Alexander) E. B. Sm. = Coreopsis grandiflora Hogg ●■

104514 Coreopsis grantii Oliv. = Bidens grantii (Oliv.) Sherff ■☆

104515 Coreopsis guineensis Oliv. et Hiern = Bidens borianiana (Sch. Bip. ex Schweinf. et Asch.) Cufod. ■☆

104516 Coreopsis hamiltonii (Elmer) H. Sharsm.;哈米尔顿金鸡菊■☆

104517 Coreopsis helianthoides Beadle = Coreopsis gladiata Walter ■☆

104518 Coreopsis heterocarpa Chiov. = Bidens prestinaria (Sch. Bip. ex Walp.) Cufod. ■☆

104519 Coreopsis heterogyna Fernald = Coreopsis lanceolata L. ■

104520 Coreopsis heterolepis Sherff = Coreopsis grandiflora Hogg ex Sweet var. harveyana (A. Gray) Sherff ■☆

104521 Coreopsis holstii O. Hoffm. = Bidens holstii (O. Hoffm.) Sherff ■☆

104522 Coreopsis injucunda Sherff = Bidens steppia (Steetz) Sherff. ■☆

104523 Coreopsis insecta S. Moore = Bidens kirkii (Oliv. et Hiern) Sherff ■☆

104524 Coreopsis involucrata Nutt. = Bidens polylepis S. F. Blake ■☆

104525 Coreopsis involucrata Sch. Bip. ex Walp. = Bidens pachyloma (Oliv. et Hiern) Cufod. ■☆

104526 Coreopsis isokoensis Sherff = Bidens oligoflora (Klatt) Wild ■☆

104527 Coreopsis jacksonii S. Moore = Guizotia jacksonii (S. Moore) J. Baagoe ■☆

104528 Coreopsis jacksonii S. Moore var. arthrochaeta Sherff = Guizotia jacksonii (S. Moore) J. Baagoe ■☆

104529 Coreopsis kilimandscharica O. Hoffm. = Bidens kilimandscharica (O. Hoffm.) Sherff ■☆

104530 Coreopsis kirkii Oliv. et Hiern = Bidens kirkii (Oliv. et Hiern) Sherff ■☆

104531 Coreopsis lanceolata L.;剑叶金鸡菊(除虫菊,大金鸡菊,大蛇目菊,狭叶金鸡菊,线叶金鸡菊);Coreopsis, Lanceleaf Coreopsis, Lance-leaf Tickseed, Lance-leaved Tickseed, Lance-leaved-tickseed, Long-stalk Tickseed, Sand Coreopsis, Sand Tickseed Lance Coreopsis, Tickseed Coreopsis ■

104532 Coreopsis lanceolata L. var. villosa Michx.;柔毛剑叶金鸡菊;Lance-leaf Tickseed, Long-stalk Tickseed, Sand Coreopsis, Sand Tickseed ■☆

104533 Coreopsis lanceolata L. var. villosa Michx. = Coreopsis lanceolata L. ■

104534 Coreopsis leptoglossa Sherff = Bidens kilimandscharica (O.

Hoffm.) Sherff ■☆
104535 Coreopsis leucantha L. = Bidens pilosa L. ■
104536 Coreopsis linearifolia Oliv. et Hiern = Bidens ugandensis (S. Moore) Sherff ■☆
104537 Coreopsis linifolia Nutt. = Coreopsis gladiata Walter ■☆
104538 Coreopsis longifolia Small = Coreopsis gladiata Walter ■☆
104539 Coreopsis lupulina O. Hoffm. = Bidens pinnatipartita (O. Hoffm.) Wild ■☆
104540 Coreopsis macrantha Sch. Bip. = Bidens macroptera (Sch. Bip. ex Chiov.) Mesfin ■☆
104541 Coreopsis macroptera Sch. Bip. ex Chiov. = Bidens macroptera (Sch. Bip. ex Chiov.) Mesfin ■☆
104542 Coreopsis major Walter;大叶金鸡菊;Bigleaf Tickseed, Greater Coreopsis, Greater Tickseed, Largeleaf Tickseed, Wood Tickseed ■
104543 Coreopsis major Walter var. rigida (Nutt.) F. E. Boynton = Coreopsis major Walter ■
104544 Coreopsis major Walter var. stellata (Nutt.) B. L. Rob. = Coreopsis major Walter ■
104545 Coreopsis maritima (Nutt.) Hook. f.;滨海金鸡菊;Sea Coreopsis, Sea Dahlia, Winter Marguerite ●■☆
104546 Coreopsis mattfeldii Sherff = Bidens oligoflora (Klatt) Wild ■☆
104547 Coreopsis microglossa Sherff = Bidens negriana (Sherff) Cufod. ■☆
104548 Coreopsis mildbraedii Muschl. = Bidens elliotii (S. Moore) Sherff ■☆
104549 Coreopsis mitis Michx. = Bidens mitis (Michx.) Sherff ■☆
104550 Coreopsis monticola (Hook. f.) Oliv. = Bidens mannii T. G. J. Rayner ■☆
104551 Coreopsis monticola (Hook. f.) Oliv. var. pilosa Hutch. et Dalziel = Bidens mannii T. G. J. Rayner ■☆
104552 Coreopsis morotonensis Sherff = Bidens elgonensis (Sherff) Agnew ■☆
104553 Coreopsis multiflora Sherff = Bidens steppia (Steetz) Sherff. ■☆
104554 Coreopsis negriana Sherff = Bidens negriana (Sherff) Cufod. ■☆
104555 Coreopsis neumannii Sherff = Bidens ternata (Chiov.) Sherff ■☆
104556 Coreopsis oblonga Sherff = Bidens oblonga (Sherff) Wild ■☆
104557 Coreopsis occidentalis (Hutch. et Dalziel) C. D. Adams = Bidens occidentalis (Hutch. et Dalziel) Mesfin ■☆
104558 Coreopsis ochracea O. Hoffm. = Bidens ochracea (O. Hoffm.) Sherff ■☆
104559 Coreopsis ochracea O. Hoffm. var. lugardii Sherff = Bidens ugandensis (S. Moore) Sherff ■☆
104560 Coreopsis ochraceoides Sherff = Bidens ochracea (O. Hoffm.) Sherff ■☆
104561 Coreopsis odora Sherff = Bidens odora (Sherff) T. G. J. Rayner ■☆
104562 Coreopsis oligantha Klatt = Bidens oligoflora (Klatt) Wild ■☆
104563 Coreopsis oligoflora Klatt = Bidens oligoflora (Klatt) Wild ■☆
104564 Coreopsis pachyloma Oliv. et Hiern = Bidens pachyloma (Oliv. et Hiern) Cufod. ■☆
104565 Coreopsis pachyloma Oliv. et Hiern var. inanis Sherff = Bidens pachyloma (Oliv. et Hiern) Cufod. ■☆
104566 Coreopsis palmata Nutt.;掌叶金鸡菊;Finger Coreopsis, Finger Tickseed, Prairie Coreopsis, Prairie Tickseed, Stiff Tickseed ■☆
104567 Coreopsis parviflora Jacq. = Cosmos parviflorus (Jacq.) Pers. ■☆
104568 Coreopsis pinnatipartita O. Hoffm. = Bidens pinnatipartita (O. Hoffm.) Wild ■☆
104569 Coreopsis prestinaria Sch. Bip. ex Walp. = Bidens prestinaria (Sch. Bip. ex Walp.) Cufod. ■☆
104570 Coreopsis prestinariaeformis (Vatke) Cufod. var. incisa Sherff = Bidens prestinaria (Sch. Bip. ex Walp.) Cufod. ■☆
104571 Coreopsis prestinariaeformis Vatke = Bidens prestinaria (Sch. Bip. ex Walp.) Cufod. ■☆
104572 Coreopsis pubescens Elliott;毛金鸡菊;Star Tickseed ■☆
104573 Coreopsis pubescens Elliott var. debilis (Sherff) E. B. Sm. = Coreopsis pubescens Elliott ■☆
104574 Coreopsis pubescens Elliott var. robusta A. Gray ex Eames = Coreopsis pubescens Elliott ■☆
104575 Coreopsis pulchella O. Hoffm. = Bidens microphylla Sherff ■☆
104576 Coreopsis quarrei Sherff = Bidens grantii (Oliv.) Sherff ■☆
104577 Coreopsis rosea Nutt.;粉金鸡菊;Pink Tickseed, Rose Coreopsis ■☆
104578 Coreopsis rueppellii Sch. Bip. ex Walp. = Bidens rueppellii (Sch. Bip. ex Walp.) Sherff ■☆
104579 Coreopsis rueppellii Sch. Bip. f. angustisecta Chiov. = Bidens rueppellii (Sch. Bip. ex Walp.) Sherff ■☆
104580 Coreopsis rueppellii Sch. Bip. var. incisior Sherff = Bidens rueppellii (Sch. Bip. ex Walp.) Sherff ■☆
104581 Coreopsis rueppellii Sch. Bip. var. simplicifolia (Vatke) Chiov. = Bidens ternata (Chiov.) Sherff ■☆
104582 Coreopsis ruwenzoriensis S. Moore = Bidens buchneri (Klatt) Sherff ■☆
104583 Coreopsis saxicola Alexander = Coreopsis grandiflora Hogg ●■
104584 Coreopsis scabrifolia Sherff = Bidens baumii (O. Hoffm.) Sherff ■☆
104585 Coreopsis schimperi O. Hoffm. = Bidens carinata Cufod. ex Mesfin ■☆
104586 Coreopsis schlechteri (Sherff) Burtt Davy = Bidens kirkii (Oliv. et Hiern) Sherff ■☆
104587 Coreopsis seretii De Wild. = Bidens buchneri (Klatt) Sherff ■☆
104588 Coreopsis setigera Sch. Bip. ex Walp. = Bidens setigera (Sch. Bip. ex Walp.) Sherff ■☆
104589 Coreopsis simplicifolia (Vatke) Engl. = Bidens ternata (Chiov.) Sherff ■☆
104590 Coreopsis speciosa Hiern = Bidens buchneri (Klatt) Sherff ■☆
104591 Coreopsis stenophylla F. E. Boynton = Coreopsis tinctoria Nutt. ■
104592 Coreopsis steppia Steetz = Bidens steppia (Steetz) Sherff. ■☆
104593 Coreopsis stillmanii (A. Gray) S. F. Blake;斯提波斯菊;Stillman's Daisy ■☆
104594 Coreopsis stuhlmannii O. Hoffm. = Bidens buchneri (Klatt) Sherff ■☆
104595 Coreopsis tannensis G. Forst. ex Spreng. = Glossocardia bidens (Retz.) Veldkamp ■
104596 Coreopsis taylori S. Moore = Bidens taylori (S. Moore) Sherff ■☆
104597 Coreopsis ternata Chiov. = Bidens ternata (Chiov.) Sherff ■☆
104598 Coreopsis tinctoria Nutt.;两色金鸡菊(波斯菊,金钱聚菊,金钱梅,痢疾草,蛇目菊,小波斯菊,园庭金鸡菊,紫心梅);Annual Coreopsis, Calliopsis, Coreopsis, Garden Coreopsis, Garden Tickseed, Golden Coreopsis, Golden Tickseed, Plains Coreopsis, Plains Tickseed, Tickseed ■
104599 Coreopsis tinctoria Nutt. 'Golden Crown';金冠两色金鸡菊■
104600 Coreopsis tinctoria Nutt. f. atropurpurea (Hook.) Fernald = Coreopsis tinctoria Nutt. ■
104601 Coreopsis tinctoria Nutt. var. atkinsoniana (Douglas ex Lindl.)

H. M. Parker ex E. B. Sm. = Coreopsis tinctoria Nutt. ■

104602 Coreopsis tinctoria Nutt. var. atropurpurea Hook.;暗紫两色金鸡菊■☆

104603 Coreopsis tinctoria Nutt. var. imminuta Sherff = Coreopsis tinctoria Nutt. ■

104604 Coreopsis tinctoria Nutt. var. purpurea Hook.;紫花两色金鸡菊■☆

104605 Coreopsis tinctoria Nutt. var. similis (F. E. Boynton) H. M. Parker ex E. B. Sm. = Coreopsis tinctoria Nutt. ■

104606 Coreopsis togoensis Sherff = Bidens borianiana (Sch. Bip. ex Schweinf. et Asch.) Cufod. ■☆

104607 Coreopsis trichosperma Michx. = Bidens coronata (L.) Britton ■☆

104608 Coreopsis trichosperma Michx. = Bidens trichosperma (Michx.) Britton ■☆

104609 Coreopsis trichosperma Michx. var. tenuiloba A. Gray = Bidens coronata (L.) Britton ■☆

104610 Coreopsis tripartita M. B. Moss = Bidens chippii (M. B. Moss) Mesfin ■☆

104611 Coreopsis tripteris L.;三叶金鸡菊;Tall Coreopsis, Tall Tickseed, Threeleaf Coreopsis ■

104612 Coreopsis tripteris L. var. deamii Standl. = Coreopsis tripteris L. ■

104613 Coreopsis tripteris L. var. smithii Sherff = Coreopsis tripteris L. ■

104614 Coreopsis ugandensis S. Moore = Bidens ugandensis (S. Moore) Sherff ■☆

104615 Coreopsis ulugurica Gilli;乌卢古尔金鸡菊■☆

104616 Coreopsis verticillata L.;轮叶金鸡菊;Coreopsis, Fern-leaf Coreopsis, Threadleaf Coreopsis, Verticillate Coreopsis, Whorled Coreopsis ■

104617 Coreopsis vulgaris Sherff = Bidens steppia (Steetz) Sherff. ■☆

104618 Coreopsis whytei S. Moore = Bidens pinnatipartita (O. Hoffm.) Wild ■☆

104619 Coreopsoides Moench = Coreopsis L. ●■

104620 Coreosma Spach = Ribes L. ●

104621 Coreosma americana (Mill.) Nieuwl. = Ribes americanum Mill. ●

104622 Coreosma americana Nieuwl. = Ribes americanum Mill. ●

104623 Coreosma florida (L'Hér.) Spach = Ribes americanum Mill. ●

104624 Coreosma forida Spach = Ribes americanum Mill. ●

104625 Coreosma longifolia Lunell = Ribes odoratum H. L. Wendl. ●

104626 Coreosma odorata (H. L. Wendl.) Nieuwl. = Ribes odoratum H. L. Wendl. ●

104627 Coreosma odorata Nieuwl. = Ribes odoratum H. L. Wendl. ●

104628 Coreosma tristis (Pall.) Lunell = Ribes triste Pall. ●

104629 Coreosma tristis Lunell = Ribes triste Pall. ●

104630 Coresantha Alef. = Iris L. ■

104631 Coresanthe Baker = Coresantha Alef ■

104632 Coreta P. Browne = Corchorus L. ■●

104633 Corethamnium R. M. King et H. Rob. (1978);展瓣亮泽兰属●☆

104634 Corethamnium chocoensis R. M. King et H. Rob.;展瓣亮泽兰■☆

104635 Corethrodendron Fisch. et Basiner = Hedysarum L. (保留属名)●■

104636 Corethrodendron Fisch. et Basiner (1845);扫帚木属(山竹子属)●

104637 Corethrodendron fruticosum (Pall.) B. H. Choi et H. Ohashi;灌木扫帚木(山竹子)●

104638 Corethrodendron fruticosum (Pall.) B. H. Choi et H. Ohashi var. mongolicum (Turcz.) Turcz. ex Kitag.;蒙古山竹子●

104639 Corethrodendron krassnovii (B. Fedtsch.) B. H. Choi et H. Ohashi;帕米尔山竹子●

104640 Corethrodendron lignosum (Trautv.) L. R. Xu et B. H. Choi;木山竹子●

104641 Corethrodendron lignosum (Trautv.) L. R. Xu et B. H. Choi var. laeve (Maxim.) L. R. Xu et B. H. Choi;塔落山竹子●

104642 Corethrodendron multijugum (Maxim.) B. H. Choi et H. Ohashi;红花山竹子●

104643 Corethrodendron scoparium (Fisch. et C. A. Mey.) Fisch. et Basiner;细枝山竹子●

104644 Corethrodendron scoparium (Fisch. et C. A. Mey.) Fisch. et Basiner = Hedysarum scoparium Fisch. et C. A. Mey. ●

104645 Corethrodendron scoparium Fisch. et Basiner = Hedysarum scoparium Fisch. et C. A. Mey. ●

104646 Corethrogyne DC. (1836);沙紫菀属;Sandaster ●■☆

104647 Corethrogyne DC. = Lessingia Cham. ■☆

104648 Corethrogyne californica DC. = Corethrogyne filaginifolia (Hook. et Arn.) Nutt. ●■☆

104649 Corethrogyne californica DC. var. lyonii S. F. Blake = Corethrogyne filaginifolia (Hook. et Arn.) Nutt. ●■☆

104650 Corethrogyne californica DC. var. obovata (Benth.) Kuntze = Corethrogyne filaginifolia (Hook. et Arn.) Nutt. ●■☆

104651 Corethrogyne detonsa Greene = Hazardia detonsa (Greene) Greene ●☆

104652 Corethrogyne filaginifolia (Hook. et Arn.) Nutt.;沙紫菀;California Aster, Common Sand Aster ●■☆

104653 Corethrogyne filaginifolia (Hook. et Arn.) Nutt. var. bernardina (Abrams) H. M. Hall = Corethrogyne filaginifolia (Hook. et Arn.) Nutt. ●■☆

104654 Corethrogyne filaginifolia (Hook. et Arn.) Nutt. var. brevicula (Greene) Canby = Corethrogyne filaginifolia (Hook. et Arn.) Nutt. ●■☆

104655 Corethrogyne filaginifolia (Hook. et Arn.) Nutt. var. californica (DC.) Saroyan = Corethrogyne filaginifolia (Hook. et Arn.) Nutt. ●■☆

104656 Corethrogyne filaginifolia (Hook. et Arn.) Nutt. var. glomerata H. M. Hall = Corethrogyne filaginifolia (Hook. et Arn.) Nutt. ●■☆

104657 Corethrogyne filaginifolia (Hook. et Arn.) Nutt. var. hamiltonensis D. D. Keck = Corethrogyne filaginifolia (Hook. et Arn.) Nutt. ●■☆

104658 Corethrogyne filaginifolia (Hook. et Arn.) Nutt. var. incana (Lindl.) Canby = Corethrogyne filaginifolia (Hook. et Arn.) Nutt. ●■☆

104659 Corethrogyne filaginifolia (Hook. et Arn.) Nutt. var. latifolia H. M. Hall = Corethrogyne filaginifolia (Hook. et Arn.) Nutt. ●■☆

104660 Corethrogyne filaginifolia (Hook. et Arn.) Nutt. var. linifolia H. M. Hall = Corethrogyne filaginifolia (Hook. et Arn.) Nutt. ●■☆

104661 Corethrogyne filaginifolia (Hook. et Arn.) Nutt. var. peirsonii Canby = Corethrogyne filaginifolia (Hook. et Arn.) Nutt. ●■☆

104662 Corethrogyne filaginifolia (Hook. et Arn.) Nutt. var. pinetorum I. M. Johnst. = Corethrogyne filaginifolia (Hook. et Arn.) Nutt. ●■☆

104663 Corethrogyne filaginifolia (Hook. et Arn.) Nutt. var. rigida A. Gray = Corethrogyne filaginifolia (Hook. et Arn.) Nutt. ●■☆

104664 Corethrogyne filaginifolia (Hook. et Arn.) Nutt. var. robusta Greene = Corethrogyne filaginifolia (Hook. et Arn.) Nutt. ●■☆

104665 Corethrogyne filaginifolia (Hook. et Arn.) Nutt. var. sessilis (Greene) Canby = Corethrogyne filaginifolia (Hook. et Arn.) Nutt. ●■☆

104666 Corethrogyne filaginifolia (Hook. et Arn.) Nutt. var. virgata (Benth.) A. Gray = Corethrogyne filaginifolia (Hook. et Arn.) Nutt. ●■☆
104667 Corethrogyne filaginifolia (Hook. et Arn.) Nutt. var. viscidula (Greene) D. D. Keck = Corethrogyne filaginifolia (Hook. et Arn.) Nutt. ●■☆
104668 Corethrogyne incana (Lindl.) Nutt. = Corethrogyne filaginifolia (Hook. et Arn.) Nutt. ●■☆
104669 Corethrogyne leucophylla (Lindl.) Jeps. = Corethrogyne filaginifolia (Hook. et Arn.) Nutt. ●■☆
104670 Corethrogyne linifolia (H. M. Hall) Ferris = Corethrogyne filaginifolia (Hook. et Arn.) Nutt. ●■☆
104671 Corethrogyne obovata Benth. = Corethrogyne filaginifolia (Hook. et Arn.) Nutt. ●■☆
104672 Corethrogyne sessilis Greene = Corethrogyne filaginifolia (Hook. et Arn.) Nutt. ●■☆
104673 Corethrostyles Benth. et Hook. f. = Corethrostylis Endl. ●☆
104674 Corethrostyles Endl. = Corethrostylis Endl. ●☆
104675 Corethrostylis Endl. = Lasiopetalum Sm. ●☆
104676 Corethrum Vahl = Botelua Lag. ■
104677 Corethrum Vahl = Bouteloua Lag.(保留属名)■
104678 Corethrum Vahl(1810);扫帚禾属■☆
104679 Corethrum bromoides Vahl;扫帚禾■☆
104680 Coriaceae J. Agardh = Primulaceae Batsch ex Borkh.(保留科名)●■
104681 Coriandraceae Burnett = Apiaceae Lindl.(保留科名)●■
104682 Coriandraceae Burnett = Umbelliferae Juss.(保留科名)■●
104683 Coriandraceae Burnett;芫荽科■☆
104684 Coriandropsis H. Wolff = Coriandrum L. ■
104685 Coriandropsis H. Wolff(1921);拟芫荽属■☆
104686 Coriandropsis syriaca H. Wolff;拟芫荽■☆
104687 Coriandrum L. (1753);芫荽属;Coriander ■
104688 Coriandrum cicuta Crantz = Conium maculatum L. ■
104689 Coriandrum maculatum (L.) Roth = Conium maculatum L. ■
104690 Coriandrum sativum L.;芫荽(白卯段花,胡菜,胡荽,葫荽,满天星,松须菜,荽,葰,莞荽,莞蓑,香菜,香茜,香荽,延荽,芫菜,芫茜,莚葛草,莚荽菜,园荽,蒝荽);Cellender, Chinese Parsley, Cilantro, Col, Colander, Coliander, Coriander, Dhani ■
104691 Coriaria L. (1753);马桑属;Coriaria ●
104692 Coriaria Niss. ex L. = Coriaria L. ●
104693 Coriaria arborea Linds.;树马桑(乔木马桑);Tree Tutu ●☆
104694 Coriaria intermedia Matsum.;台湾马桑(马桑);Chinese Coriaria, Taiwan Coriaria ●
104695 Coriaria intermedia Matsum. = Coriaria japonica A. Gray ●☆
104696 Coriaria japonica A. Gray;日本马桑(毒空木,日本毒空木);Japan Coriaria, Japanese Coriaria ●☆
104697 Coriaria japonica A. Gray subsp. intermedia (Matsum.) S. F. Huang et T. C. Huang = Coriaria intermedia Matsum. ●
104698 Coriaria japonica A. Gray subsp. intermedia (Matsum.) S. F. Huang et T. C. Huang = Coriaria japonica A. Gray ●☆
104699 Coriaria kweichowensis Hu = Coriaria nepalensis Wall. ●
104700 Coriaria myrtifolia L.;欧马桑(番樱桃叶马桑,香叶木马桑);French Sumac, Mediterranean Coriaria, Myrtle Coriaria, Myrtle Eoriaria, Myrtleleaf Coriaria, Redoul ●☆
104701 Coriaria nepalensis Wall.;马桑(蛤蟆树,黑果果,黑虎大王,黑龙须,红马桑,红娘子,蓝蛇风,马鞍子,马桑紫,闹鱼儿,尼泊尔马桑,千年红,上天梯,水马桑,乌龙须,野马桑,鱼尾草,紫桑,醉鱼草,醉鱼儿);Chinese Coriaria, Mussoorie-berry, Nepal Coriaria, Tanner's Tree ●
104702 Coriaria ruscifolia L.;新西兰马桑●☆
104703 Coriaria sarmentosa G. Forst.;蔓茎马桑●☆
104704 Coriaria sinica Maxim. = Coriaria nepalensis Wall. ●
104705 Coriaria summicola Hayata = Coriaria intermedia Matsum. ●
104706 Coriaria terminalis Hemsl.;顶序马桑(草马桑);Sikkim Coriaria ●■
104707 Coriaria terminalis Hemsl. var. xanthocarpa Rehder et E. H. Wilson;黄果顶序马桑(黄果草马桑);Yellow-fruited Sikkim Coriaria ●■
104708 Coriaria terminalis Hemsl. var. xanthocarpa Rehder et E. H. Wilson = Coriaria terminalis Hemsl. ●■
104709 Coriaria thymifolia Humb. et Bonpl. ex Willd.;百里香叶马桑●☆
104710 Coriariaceae DC. (1824)(保留科名);马桑科;Coriaria Family ●
104711 Coriariaceae DC.(保留科名) = Primulaceae Batsch ex Borkh.(保留科名)●■
104712 Coridaceae J. Agardh.;麝香报春科(麝香草科)●☆
104713 Coridaceae J. Agardh. = Myrsinaceae R. Br.(保留科名)●
104714 Coridaceae J. Agardh. = Primulaceae Batsch ex Borkh.(保留科名)●■
104715 Coridochloa Nees = Alloteropsis J. Presl ex C. Presl ■
104716 Coridochloa Nees ex Graham = Alloteropsis J. Presl ex C. Presl ■
104717 Coridochloa cimicina (L.) Nees = Alloteropsis cimicina (L.) Stapf ■
104718 Coridothymus Rchb. f. = Thymbra L. ●☆
104719 Coridothymus capitatus (L.) Rchb. f. = Thymbra capitata (L.) Cav. ■☆
104720 Coriflora W. A. Weber = Clematis L. ●■
104721 Coriflora hirsutissima (Pursh) W. A. Weber = Clematis hirsutissima Pursh ■☆
104722 Coriflora viorna (L.) W. A. Weber = Clematis viorna L. ■☆
104723 Corilus Nocca = Corylus L. ●
104724 Corindum Adans. = Paullinia L. ●☆
104725 Corindum Mill. = Cardiospermum L. ■
104726 Corindum Tourn. ex Medik. = Cardiospermum L. ■
104727 Coringia J. Presl et C. Presl = Conringia Heist. ex Fabr. ■
104728 Corinocarpus Poir. = Corynocarpus J. R. Forst. et G. Forst. ●☆
104729 Coriocarpus Pax et K. Hoffm. = Coreocarpus Benth. ■☆
104730 Corion Hoffmanns. et Link = Bifora Hoffm.(保留属名)■☆
104731 Corion Mitch. = Spergularia (Pers.) J. Presl et C. Presl(保留属名)■
104732 Coriophyllus Rydb. = Cymopterus Raf. ■☆
104733 Coriospermum Post et Kuntze = Corispermum L. ■
104734 Coris L. (1753);麝香报春属(考丽草属)●☆
104735 Coris Tourn. ex L. = Coris L. ●☆
104736 Coris monspeliensis L.;麝香报春■☆
104737 Coris monspeliensis L. subsp. maroccana (Murb.) Greuter et Burdet;摩洛哥麝香草■☆
104738 Coris monspeliensis L. subsp. syrtica (Murb.) Masclans;瑟尔特麝香报春■☆
104739 Coris monspeliensis L. var. longinqua Airy Shaw;长麝香报春■☆
104740 Coris monspeliensis L. var. longispina Murb. = Coris monspeliensis L. ■☆
104741 Coris monspeliensis L. var. maroccana Murb. = Coris monspeliensis L. subsp. maroccana (Murb.) Greuter et Burdet ■☆
104742 Coris monspeliensis L. var. syrtica Murb. = Coris monspeliensis

L. subsp. syrtica (Murb.) Masclans ■☆
104743 Corisanthera C. B. Clarke = Corysanthera Wall. ex Benth. ●
104744 Corisanthera C. B. Clarke = Rhynchotechum Blume ●
104745 Corisanthes Steud. = Criosanthes Raf. ■
104746 Corispermaceae Link = Amaranthaceae Juss.(保留科名)●■
104747 Corispermaceae Link = Chenopodiaceae Vent.(保留科名)●■
104748 Corispermum (B. Juss.) ex L. = Corispermum L. ■
104749 Corispermum B. Juss. ex L. = Corispermum L. ■
104750 Corispermum L. (1753);虫实属;Bugseed,Tickseed,Tick-seed ■
104751 Corispermum algidum Iljin;喜冰虫实■☆
104752 Corispermum altaicum Iljin;阿尔泰虫实■☆
104753 Corispermum americanum (Nutt.) Nutt.;美洲虫实;American Bugseed,Bugseed ■☆
104754 Corispermum aralo-caspicum Iljin;里海虫实■☆
104755 Corispermum candelabrum Iljin;烛台虫实(乌丹虫实);Candelabra Tickseed,Candlestick Tickseed ■
104756 Corispermum canescens Kit. ex Schult.;灰白虫实■☆
104757 Corispermum canesces Kit.;灰虫实■☆
104758 Corispermum chinganicum Iljin;兴安虫实;Chingan Tickseed,Xing'an Tickseed ■
104759 Corispermum chinganicum Iljin var. microcarpum Iljin;小果兴安虫实■
104760 Corispermum chinganicum Iljin var. microcarpum Iljin = Corispermum chinganicum Iljin ■
104761 Corispermum chinganicum Iljin var. stellipile C. P. Tsien et C. G. Ma;毛果兴安虫实(毛果虫实);Hairfruit Xing'an Tickseed,Hairyfruit Chingan Tickseed ■
104762 Corispermum confertum Bunge;密穗虫实;Confert Tickseed ■
104763 Corispermum crassifolium Turcz.;厚叶虫实■☆
104764 Corispermum declinatum Stephan ex Iljin var. tylocarpum (Hance) C. P. Tsien et C. G. Ma = Corispermum tylocarpum Hance ■
104765 Corispermum declinatum Stephan ex Steven;绳虫实;Declinate Tickseed,Rope Tickseed ■
104766 Corispermum declinatum Stephan ex Steven var. tylocarpum (Hance) C. P. Tsien et C. G. Ma = Corispermum tylocarpum Hance ■
104767 Corispermum dilutum (Kitag.) C. P. Tsien et C. G. Ma;辽西虫实;Liaoxi Tickseed,Pale Tickseed ■
104768 Corispermum dilutum (Kitag.) C. P. Tsien et C. G. Ma var. hebecarpum C. P. Tsien et C. G. Ma;毛果辽西虫实;Hairfruit Liaoxi Tickseed,Hairyfruit Pale Tickseed ■
104769 Corispermum dilutum (Kitag.) C. P. Tsien et C. G. Ma var. hebecarpum C. P. Tsien et C. G. Ma = Corispermum dilutum (Kitag.) C. P. Tsien et C. G. Ma ■
104770 Corispermum dutreuilii Iljin;粗喙虫实;Thickbill Tickseed,Thickrostrum Tickseed ■
104771 Corispermum elongatum Bunge ex Maxim.;长穗虫实;Elongate Tickseed,Longspike Tickseed ■
104772 Corispermum elongatum Bunge ex Maxim. var. latifolium Bunge;宽叶长穗虫实■
104773 Corispermum elongatum Bunge ex Maxim. var. stellatopilosum Wang-wei et P. Y. Fu;星毛虫实(毛果长穗虫实);Starhair Tickseed ■
104774 Corispermum emarginatum Rydb. = Corispermum villosum Rydb. ■☆
104775 Corispermum erosum Iljin;啮蚀状虫实■☆
104776 Corispermum falcatum Iljin;镰叶虫实;Falcate Tickseed,Sickleleaf Tickseed ■
104777 Corispermum filifolium C. A. Mey.;线叶虫实;Linearleaf Tickseed ■☆
104778 Corispermum flrxuosum Wang-Wei et P. Y. Fu;屈枝虫实;Bentbranch Tickseed ■
104779 Corispermum flrxuosum Wang-Wei et P. Y. Fu var. leiocarpum Wang-Wei et P. Y. Fu;光果屈枝虫实;Smootjfruit Bentbranch Tickseed ■
104780 Corispermum gelidum Iljin;寒地虫实■☆
104781 Corispermum gmelinii Bunge = Corispermum declinatum Stephan ex Steven var. tylocarpum (Hance) C. P. Tsien et C. G. Ma ■
104782 Corispermum gmelinii Bunge = Corispermum tylocarpum Hance ■
104783 Corispermum heptapotamicum Iljin;中亚虫实;Centrol Asia Tickseed ■
104784 Corispermum hilariae Iljin;伊拉虫实■☆
104785 Corispermum huanghoense C. P. Tsien et C. G. Ma;黄河虫实(施氏虫实);Huangho Tickseed,Yellow River Tickseed ■
104786 Corispermum hyssopifolium L.;神香草叶虫实(虫实,斐梭浦叶虫实);Hyssop-leaved Bugseed,Hyssop-leaved Tickseed ■☆
104787 Corispermum hyssopifolium L. sensu Maihle ? et Blackw. = Corispermum villosum Rydb. ■☆
104788 Corispermum hyssopifolium L. var. americanum Nutt. = Corispermum americanum (Nutt.) Nutt. ■☆
104789 Corispermum hyssopifolium L. var. emarginatum (Rydb.) B. Boivin = Corispermum villosum Rydb. ■☆
104790 Corispermum hyssopifolium L. var. leptopterum Asch. = Corispermum pallasii Steven ■☆
104791 Corispermum imbricatum A. Nelson = Corispermum americanum (Nutt.) Nutt. ■☆
104792 Corispermum komarovli Iljin;科马罗夫虫实■☆
104793 Corispermum korovinii Iijin;科罗温虫实■☆
104794 Corispermum krylovii Iljin;克雷罗夫虫实■☆
104795 Corispermum ladakhianum Grey-Wilson et Wadhwa = Corispermum tibeticum Iljin ■
104796 Corispermum laxiflorum Schrenk;疏花虫实■☆
104797 Corispermum lehmannianum Bunge;倒披针叶虫实(早熟虫实);Oblanceolata Tickseed ■
104798 Corispermum lepidocarpum Grubov;鳞果虫实;Scalefruit Tickseed,Scalyfruit Tickseed ■
104799 Corispermum lepidocarpum Grubov var. kokonoricum R. F. Huang;青海虫实;Qinghai Tickseed ■
104800 Corispermum leptopterum (Asch.) Iljin = Corispermum pallasii Steven ■☆
104801 Corispermum lhasaense C. P. Tsien et C. G. Ma;拉萨虫实;Lasa Tickseed ■
104802 Corispermum macrocarpum Bunge = Corispermum macrocarpum Bunge ex Maxim. ■
104803 Corispermum macrocarpum Bunge ex Maxim.;大果虫实;Bigfruit Tickseed ■
104804 Corispermum macrocarpum Bunge ex Maxim. var. microstachyum P. Y. Fu et W. Wang = Corispermum macrocarpum Bunge ex Maxim. ■
104805 Corispermum macrocarpum Bunge ex Maxim. var. rubrum P. Y. Fu et W. Wang = Corispermum macrocarpum Bunge ex Maxim. ■
104806 Corispermum macrocarpum Bunge f. elongatum (Fuh et Wang-Wei) Kitag. = Corispermum macrocarpum Bunge var. elongatum P. Y. Fu et Wang-Wei ■
104807 Corispermum macrocarpum Bunge var. elongatum P. Y. Fu et Wang-Wei = Corispermum macrocarpum Bunge var. rubrum P. Y. Fu

et Wang-Wei ■
104808 Corispermum macrocarpum Bunge var. elongatum Wang-wei et Fuh = Corispermum candelabrum Iljin ■
104809 Corispermum macrocarpum Bunge var. microstachyum P. Y. Fu et Wang-Wei;小穗虫实■
104810 Corispermum macrocarpum Bunge var. microstachyum P. Y. Fu et Wang-Wei = Corispermum macrocarpum Bunge var. rubrum P. Y. Fu et Wang-Wei ■
104811 Corispermum macrocarpum Bunge var. rubrum P. Y. Fu et Wang-Wei;毛大果虫实(红虫实,毛果虫实);Hairfruit Bigfruit Tickseed,Hairy Bigfruit Tickseed ■
104812 Corispermum marginale Rydb. = Corispermum americanum (Nutt.) Nutt. ■☆
104813 Corispermum marschallii Steven;马氏虫实■☆
104814 Corispermum mongolicum Iljin;蒙古虫实;Mongol Tickseed, Mongolian Tickseed ■
104815 Corispermum navicula Mosyakin;舟状虫实;Boat-shaped Bugseed ■☆
104816 Corispermum nitidum Kit. ex Schult.;光虫实;Shiny Bugseed, Slender Bugseed ■☆
104817 Corispermum nitidum Kit. ex Schult. = Corispermum americanum (Nutt.) Nutt. ■☆
104818 Corispermum ochotense Ignatov;阿拉斯加虫实;Alaskan Bugseed,Okhotian Bugseed ■☆
104819 Corispermum orientale Lam.;东方虫实;Oriental Tickseed ■
104820 Corispermum orientale Lam. Maihle and Blackw. = Corispermum pallasii Steven ■☆
104821 Corispermum orientale Lam. var. emarginatum (Rydb.) J. F. Macbr. = Corispermum villosum Rydb. ■☆
104822 Corispermum orientale Lam. var. emarginatum (Rydb.) J. F. Macbr. = Corispermum pallasii Steven ■☆
104823 Corispermum pacificum Mosyakin;太平洋虫实;Pacific Bugseed ■☆
104824 Corispermum pallasii Steven;帕拉斯虫实;Bugseed, Pallas Bugseeed,Siberian Bugseed ■☆
104825 Corispermum pallidum Mosyakin;苍白虫实;Pale Bugseed ■☆
104826 Corispermum pamiricum Iljin;帕米尔虫实;Pamir Tickseed ■
104827 Corispermum pamiricum Iljin var. pilocarpum C. P. Tsien et C. G. Ma;毛果帕米尔虫实;Hairfruit Pamir Tickseed,Hairyfruit Pamir Tickseed ■
104828 Corispermum papillosum (Kuntze) Iljin;乳突虫实■☆
104829 Corispermum patelliforme Iljin;碟果虫实;Dishfruit Tickseed, Patelliform Tickseed ■
104830 Corispermum patelliforme Iljin var. pelviforme H. C. Fu et Z. Y. Chu;盆果虫实;Pelviform Tickseed ■
104831 Corispermum platypterum Kitag.;宽翅虫实(鳞虫实);Broadwing Tickseed ■
104832 Corispermum praecox C. P. Tsien et C. G. Ma;早熟虫实;Early-ripe Tickseed,Henan Tickseed ■
104833 Corispermum pseudofalcatum C. P. Tsien et C. G. Ma;假镰叶虫实;False Sickleleaf Tickseed,Falsefalcate Tickseed ■
104834 Corispermum puberulum Iljin;软毛虫实;Puberulous Tickseed ■
104835 Corispermum puberulum Iljin = Corispermum candelabrum Iljin ■
104836 Corispermum puberulum Iljin var. ellipsocarpum C. P. Tsien et C. G. Ma;光果软毛虫实;Smoothfruit Tickseed ■
104837 Corispermum puberulum Iljin var. ellipsocarpum C. P. Tsien et C. G. Ma = Corispermum puberulum Iljin ■
104838 Corispermum pungens Vahl = Agriophyllum squarrosum (L.) Moq. ■
104839 Corispermum redowskii Fisch.;列多夫斯基虫实■☆
104840 Corispermum retortum Wang-Wei et P. Y. Fu;扭果虫实;Twisted Tickseed,Twistedfruit Tickseed ■
104841 Corispermum rostratum A. I. Baranov et Skvortsov;喙虫实;Beak Tickseed ■
104842 Corispermum rostratum A. I. Baranov et Skvortsov ex W. Wang = Corispermum tylocarpum Hance ■
104843 Corispermum rostratum A. I. Baranov et Skvortsov ex W. Wang = Corispermum declinatum Stephan ex Steven var. tylocarpum (Hance) C. P. Tsien et C. G. Ma ■
104844 Corispermum sibiricum Iljin;西伯利亚虫实;Siberia Tickseed ■
104845 Corispermum sibiricum Iljin subsp. baicalense Iljin = Corispermum pallasii Steven ■☆
104846 Corispermum simplicissimum Lunell = Corispermum americanum (Nutt.) Nutt. ■☆
104847 Corispermum squarrosum L.;糙虫实■☆
104848 Corispermum squarrosum L. = Agriophyllum squarrosum (L.) Moq. ■
104849 Corispermum stauntonii Moq.;华虫实(施氏虫实);China Tickseed,Staunton Tickseed ■
104850 Corispermum stenolepis Kitag.;细苞虫实;Narrowbract Tickseed,Stenobracteoid Tickseed ■
104851 Corispermum stenolepis Kitag. f. psilocarpum (Kitag.) Kitag. = Corispermum stenolepis Kitag. var. psilocarpum Kitag. ■
104852 Corispermum stenolepis Kitag. var. psilocarpum Kitag.;光果细苞虫实;Nakedfruit Tickseed,Smoothfruit Stenobracteoid Tickseed ■
104853 Corispermum stenolepis Kitag. var. psilocarpum Kitag. = Corispermum stenolepis Kitag. ■
104854 Corispermum thelelegium Kitag. = Corispermum candelabrum Iljin ■
104855 Corispermum thelelegium Kitag. var. dilutum Kitag. = Corispermum dilutum (Kitag.) C. P. Tsien et C. G. Ma ■
104856 Corispermum tibeticum Iljin;藏虫实;Tibet Tickseed, Xizang Tickseed ■
104857 Corispermum tibeticum Iljin var. pilocarpum R. F. Huang;毛果西藏虫实;Hairfruit Xizang Tickseed ■
104858 Corispermum tylocarpum Hance;毛果绳虫实(喙虫实,瘤果虫实);Hairyfruit Tickseed ■
104859 Corispermum tylocarpum Hance = Corispermum declinatum Stephan ex Steven var. tylocarpum (Hance) C. P. Tsien et C. G. Ma ■
104860 Corispermum ulopterum Fenzl;卷翅虫实■☆
104861 Corispermum villosum Rydb.;毛虫实;Common Bugseed,Hairy Bugseed ■☆
104862 Corispermum welshii Mosyakin;威尔士虫实;Welsh's Bugseed ■☆
104863 Coristospermum Bertol.(1838);臭虫草属■☆
104864 Coristospermum Bertol. = Ligusticum L. ■
104865 Corium Post et Kuntze = Bifora Hoffm.(保留属名)■☆
104866 Corium Post et Kuntze = Corion Hoffmanns. et Link ■☆
104867 Corium Post et Kuntze = Corion Mitch. ■
104868 Corium Post et Kuntze = Spergularia (Pers.) J. Presl et C. Presl (保留属名)■
104869 Corizospermum Zipp. ex Blume = Casearia Jacq. ●
104870 Cormigonus Raf. = Bikkia Reinw.(保留属名)●☆
104871 Cormonema Reissek = Colubrina Rich. ex Brongn.(保留属名)●
104872 Cormonema Reissek ex Endl. = Colubrina Rich. ex Brongn.(保

留属名)●
104873 Cormus Spach = Sorbus L. ●
104874 Cormus yunnanensis (Franch.) Koidz. = Malus yunnanensis (Franch.) C. K. Schneid. ●
104875 Cormylus Raf. = Hedyotis L. (保留属名)●■
104876 Corna Noronha = Avicennia L. ●
104877 Cornacchinia Endl. = Baeolepis Decne. ex Moq. ■☆
104878 Cornacchinia Savi = Clerodendrum L. ●■
104879 Cornacchinia fragiformis Savi = Clerodendrum acerbianum (Vis.) Benth. ●☆
104880 Cornaceae Bercht. et J. Presl(1825)(保留科名);山茱萸科(四照花科);Dogwood Family ●
104881 Cornaceae Dumort. = Cornaceae Bercht. et J. Presl(保留科名)●■
104882 Cornachina Endl. = Baeolepis Decne. ex Moq. ■☆
104883 Cornachina Endl. = Cornacchinia Endl. ■☆
104884 Cornelia Ard. = Ammannia L. ■
104885 Cornelia Rydb. = Chamaepericlymenum Asch. et Graebn. ■
104886 Cornella canadensis (L.) Rydb. = Chamaepericlymenum canadense (L.) Asch. et Graebn. ■
104887 Cornella canadensis (L.) Rydb. = Cornus canadensis L. ■
104888 Cornera Furtado = Calamus L. ●
104889 Cornera Furtado(1955);科纳棕属(可耐拉棕属);Cornera ●☆
104890 Cornera conirostris (Becc.) Furtado;锥喙科纳棕(圆锥喙可耐拉棕);Conical-beaked Cornera ●☆
104891 Cornera lobbiana (Becc.) Furtado;洛卜科纳棕(洛卜可耐拉棕);Lobb Cornera ●☆
104892 Cornera pycnocarpa Furtado;密果科纳棕(密果可耐拉棕);Dense-fruited Cornera ●☆
104893 Corneria A. V. Bobrov et Melikyan = Calamus L. ●
104894 Corneria A. V. Bobrov et Melikyan = Cornera Furtado ●☆
104895 Cornicina Boiss. = Anthyllis L. ■☆
104896 Cornicina hamosa (Desf.) Boiss. = Hymenocarpos hamosus (Desf.) Vis. ■☆
104897 Cornicina loeflingii Boiss. = Hymenocarpos cornicinus (L.) Vis. ■☆
104898 Cornicina lotoides (L.) Boiss. = Hymenocarpos lotoides (L.) Vis. ■☆
104899 Cornidia Ruiz et Pav. = Hydrangea L. ●
104900 Corniola Adans. = Genista L. ●
104901 Corniveum Nienwi. = Dicentra Bernh. (保留属名)■
104902 Cornthamnus (Koch) C. Presl = Genista L. ●
104903 Cornucopiae L. (1753);角刀草属■☆
104904 Cornucopiae cucullatum L.;角刀草■☆
104905 Cornucopiae hyemalis Walter = Agrostis hyemalis (Walter) Britton, Sterns et Poggenb. ■☆
104906 Cornucopiae perennans Walter = Agrostis perennans (Walter) Tuck. ■☆
104907 Cornuella Pierre = Chrysophyllum L. ●
104908 Cornukaempferia Mood et K. Larsen(1997);考尔姜属■☆
104909 Cornulaca Delile(1813);单刺蓬属(单刺花属,单刺属);Cornulaca ●■
104910 Cornulaca alaschanica C. P. Tsien et G. L. Chu;阿拉善单刺蓬(阿拉善单刺,阿拉善单刺花);Alashan Cornulaca ●◇
104911 Cornulaca aucheri Moq.;奥切尔单刺蓬●☆
104912 Cornulaca ehrenbergii Asch.;爱伦堡单刺蓬●☆
104913 Cornulaca korshinskyi Litv.;考尔单刺蓬●☆
104914 Cornulaca monacantha Delile;单刺蓬●☆
104915 Cornulaca monacantha Delile var. diacantha Maire = Cornulaca monacantha Delile ●☆
104916 Cornulus Fabr. = Swida Opiz ●
104917 Cornus L. (1753);山茱萸属(梾木属);Bunchberry, Cornel, Dogwood ●
104918 Cornus Salisb. = Muscari Mill. ■☆
104919 Cornus Spach = Sorbus L. ●
104920 Cornus alba L.;红瑞木(红瑞山茱萸,凉子木);Ivy Halo Dogwood, Red-barked Dogwood, Siberian Dogwood, Slavin's Dogwood, Tartar Dogwood, Tatarian Dogwood, White Dogwood ●
104921 Cornus alba L. 'Argenteo-Marginata';白边红瑞木;Tatarian Dogwood, Variegated Tartarian Dogwood ●
104922 Cornus alba L. 'Atrosanguinea' = Cornus alba L. 'Sibirica' ●
104923 Cornus alba L. 'Aurea';黄叶红瑞木;Tatarian Dogwood ●
104924 Cornus alba L. 'Elegantissima';雅致红瑞木●☆
104925 Cornus alba L. 'Gouchartii';高查尔迪红瑞木(花叶红瑞木)●
104926 Cornus alba L. 'Ivory Halo';银环红瑞木●
104927 Cornus alba L. 'Kesselringii';凯瑟琳红瑞木(紫枝红瑞木)●
104928 Cornus alba L. 'Sibirica Red Gnome';西伯利亚矮红瑞木;Siberian Dogwood ●☆
104929 Cornus alba L. 'Sibirica Variegata';西伯利亚斑叶红瑞木●
104930 Cornus alba L. 'Sibirica';西伯利亚红瑞木;Westonbirt Dogwood ●
104931 Cornus alba L. 'Spaethii';黄阔边红瑞木(史佩斯红瑞木)●
104932 Cornus alba L. 'Touch of Elegance';斑叶鞑靼红瑞木;Variegated Tatarian Dogwood ●☆
104933 Cornus alba L. = Cornus stolonifera Michx. ●☆
104934 Cornus alba L. = Swida alba (L.) Opiz ●
104935 Cornus alba L. = Swida macrophylla (Wall.) Soják ●
104936 Cornus alba L. subsp. stolonifera (Michx.) Wangerin = Cornus stolonifera Michx. ●☆
104937 Cornus alba L. var. argenteo-marginata Rehder = Cornus alba L. 'Argenteo-Marginata' ●
104938 Cornus alba L. var. baileyi (J. M. Coult. et W. H. Evans) B. Boivin = Cornus stolonifera Michx. ●☆
104939 Cornus alba L. var. californica (C. A. Mey.) B. Boivin = Cornus stolonifera Michx. ●☆
104940 Cornus alba L. var. interior (Rydb.) B. Boivin = Cornus stolonifera Michx. ●☆
104941 Cornus alba L. var. sibirica Loudon = Cornus alba L. 'Sibirica' ●
104942 Cornus alba L. var. spaethii Wittm. = Cornus alba L. 'Spaethii' ●
104943 Cornus alpina W. P. Fang et W. K. Hu = Cornus macrophylla Wall. ●
104944 Cornus alpina W. P. Fang et W. K. Hu = Swida alpina (W. P. Fang et W. K. Hu) W. P. Fang et W. K. Hu ●
104945 Cornus alsophila W. W. Sm. = Cornus hemsleyi C. K. Schneid. et Wangerin ●
104946 Cornus alsophila W. W. Sm. = Swida alsophila (W. W. Sm.) Holub ●
104947 Cornus alternifolia L. f.;互叶梾木(北美灯台树,拉美灯台树);Alternateleaf Dogwood, Alternate-leaved Dogwood, Blue Dogwood, Green Osier, Pagoda Dogwood, Pigeon Berry ●☆
104948 Cornus alternifolia L. f. 'Argentea';银斑互叶梾木(银白斑纹互叶梾木);Variegated Pagoda Tree ●☆
104949 Cornus amomum Mill.;熊果梾木(北美山茱萸花);Blueberry, Female Dogwood, Kinnikinnik, Knob-styled Dogwood, Pale Dogwood,

Red American Osier, Red Dogwood, Red Willow, Rose Willow, Silky Cornel, Silky Dogwood, Silky-dogwood, Swamp Dogwood ●☆

104950 Cornus amomum Mill. subsp. obliqua (Raf.) J. S. Wilson = Cornus amomum Mill. var. schuetzeana (C. A. Mey.) Rickett ●☆

104951 Cornus amomum Mill. var. schuetzeana (C. A. Mey.) Rickett; 蓝熊果梾木; Blue-fruited Dogwood, Silky Dogwood ●☆

104952 Cornus angustata (Chun) T. R. Dudley = Cornus elliptica (Pojark.) Q. Y. Xiang et Boufford ●

104953 Cornus angustata (Chun) T. R. Dudley subsp. angustata (Chun) Q. Y. Xiang = Cornus elliptica (Pojark.) Q. Y. Xiang et Boufford ●

104954 Cornus angustata (Chun) T. R. Dudley var. angustata (Chun) W. P. Fang = Cornus elliptica (Pojark.) Q. Y. Xiang et Boufford ●

104955 Cornus angustata (Chun) T. R. Dudley var. hypoleuca H. Lév. = Cornus elliptica (Pojark.) Q. Y. Xiang et Boufford ●

104956 Cornus angustata (Chun) T. R. Dudley var. mollis Rehder = Cornus elliptica (Pojark.) Q. Y. Xiang et Boufford ●

104957 Cornus aspera Wangerin = Cornus bretschneideri L. Henry ●

104958 Cornus aspera Wangerin = Swida bretschneideri (J. Henry) Soják ●

104959 Cornus asperifolia Michx.; 糙叶梾木; Rough Dogwood, Roughleaf Dogwood ●☆

104960 Cornus australis C. A. Mey.; 南方梾木●☆

104961 Cornus austrosinensis W. P. Fang et W. K. Hu; 华南梾木; S. China Dogwood, South China Dogwood ●

104962 Cornus austrosinensis W. P. Fang et W. K. Hu = Swida austrosinensis (W. P. Fang et W. K. Hu) W. P. Fang et W. K. Hu ●

104963 Cornus baileyi Coult. et Evans; 贝利红瑞木(贝蕾红瑞木); Bailey Redtwig Dogwood, Bailey Red-twig Dogwood ●☆

104964 Cornus baileyi J. M. Coult. et W. H. Evans = Cornus stolonifera Michx. ●☆

104965 Cornus boliviana J. F. Macbr. = Cornus peruviana J. F. Macbr. ●☆

104966 Cornus brachypoda C. A. Mey.; 短梗山茱萸●☆

104967 Cornus brachypoda C. A. Mey. = Cornus macrophylla Wall. ●

104968 Cornus brachypoda C. A. Mey. = Swida macrophylla (Wall.) Soják ●

104969 Cornus brachypoda Miq. = Cornus controversa Hemsl. ex Prain ●

104970 Cornus brachypoda Miq. = Swida controversa (Hemsl.) Soják ●

104971 Cornus bretschneideri L. Henry = Swida bretschneideri (L. Henry) Soják ●

104972 Cornus bretschneideri L. Henry var. crispa W. P. Fang et W. K. Hu; 卷毛沙梾; Crisped Bretschneider Dogwood ●

104973 Cornus bretschneideri L. Henry var. crispa W. P. Fang et W. K. Hu = Swida bretschneideri (L. Henry) Soják var. crispa (W. P. Fang et W. K. Hu) W. P. Fang et W. K. Hu ●

104974 Cornus bretschneideri L. Henry var. gracilis Wangerin; 细梗沙梾; Slenderstalk Dogwood ●

104975 Cornus bretschneideri L. Henry var. gracilis Wangerin = Cornus bretschneideri L. Henry ●

104976 Cornus bretschneideri L. Henry var. gracilis Wangerin = Swida bretschneideri (J. Henry) Soják var. gracilis (Wangerin) W. K. Hu ●

104977 Cornus california C. A. Mey.; 加州梾木; Creek Dogwood ●☆

104978 Cornus californica C. A. Mey. = Cornus stolonifera Michx. ●☆

104979 Cornus canadensis L. = Chamaepericlymenum canadense (L.) Asch. et Graebn. ■

104980 Cornus canadensis L. f. dutillyi Lepage = Chamaepericlymenum canadense (L.) Asch. et Graebn. ■

104981 Cornus canadensis L. f. elongata M. Peck = Chamaepericlymenum canadense (L.) Asch. et Graebn. ■

104982 Cornus canadensis L. var. dutillyi (Lepage) B. Boivin = Chamaepericlymenum canadense (L.) Asch. et Graebn. ■

104983 Cornus canadensis L. var. intermedia Farr = Chamaepericlymenum canadense (L.) Asch. et Graebn. ■

104984 Cornus candidissima Marshall = Cornus racemosa Lam. ●☆

104985 Cornus capitata Wall. = Dendrobenthamia capitata (Wall.) Hutch. ●

104986 Cornus capitata Wall. ex Roxb. = Benthamia fragifera Lindl. ●

104987 Cornus capitata Wall. ex Roxb. = Dendrobenthamia capitata (Wall. ex Roxb.) Hutch. ●

104988 Cornus capitata Wall. subsp. angustata (Chun) Q. Y. Xiang = Cornus elliptica (Pojark.) Q. Y. Xiang et Boufford ●

104989 Cornus capitata Wall. subsp. brevipedunculata (W. P. Fang et Y. T. Hsieh) Q. Y. Xiang = Cornus capitata Wall. ●

104990 Cornus capitata Wall. subsp. emeiensis (W. P. Fang et Y. T. Hsieh) Q. Y. Xiang = Cornus capitata Wall. ●

104991 Cornus capitata Wall. var. angustata (Chun) W. P. Fang = Cornus elliptica (Pojark.) Q. Y. Xiang et Boufford ●

104992 Cornus capitata Wall. var. hypoleuca H. Lév. = Cornus elliptica (Pojark.) Q. Y. Xiang et Boufford ●

104993 Cornus capitata Wall. var. molis Rehder = Dendrobenthamia angustata (Chun) W. P. Fang var. molis (Rehder) W. P. Fang ●

104994 Cornus chinensis Wangerin; 川鄂山茱萸; China Dogwood, China Macrocarpium, Chinese Cornel, Chinese Dogwood, Chinese Macrocarpium ●

104995 Cornus chinensis Wangerin f. jinyangense (W. K. Hu) W. K. Hu = Cornus chinensis Wangerin ●

104996 Cornus chinensis Wangerin f. longipedunculata (W. P. Fang et W. K. Hu) W. P. Fang et W. K. Hu = Cornus chinensis Wangerin ●

104997 Cornus chinensis Wangerin f. microcarpa (W. K. Hu) W. K. Hu = Cornus chinensis Wangerin ●

104998 Cornus circinata L'Her. = Cornus rugosa Lam. ●☆

104999 Cornus conerlea Lam. = Cornus amomum Mill. ●☆

105000 Cornus controversa Hemsl. = Cornus controversa Hemsl. ex Prain ●

105001 Cornus controversa Hemsl. ex C. K. Schneid. var. alpina Wangerin = Swida controversa (Hemsl. ex Prain) Soják var. alpina (Wangerin) H. Hara ex Noshiro ●☆

105002 Cornus controversa Hemsl. ex Prain 'Pagoda'; 塔灯台树●

105003 Cornus controversa Hemsl. ex Prain 'Variegata'; 银边灯台树(斑叶灯台树); Variegata Giant Dogwood, Wedding-cake-tree ●

105004 Cornus controversa Hemsl. ex Prain = Bothrocaryum controversum (Hemsl. ex Prain) Pojark. ●

105005 Cornus controversa Hemsl. ex Prain var. angustifolia Wangerin = Bothrocaryum controversum (Hemsl. ex Prain) Pojark. ●

105006 Cornus controversa Hemsl. ex Prain var. angustifolia Wangerin = Cornus controversa Hemsl. ex Prain ●

105007 Cornus controversa Hemsl. ex Prain var. angustifolia Wangerin = Cornus controversa Hemsl. ●

105008 Cornus controversa Hemsl. ex Prain var. shikokumontana Hiyama = Swida controversa (Hemsl. ex Prain) Soják var. shikokumontana (Hiyama) H. Hara ex Noshiro ●☆

105009 Cornus controversa Hemsl. ex Prain var. variegata Rehder = Cornus controversa Hemsl. ex Prain 'Variegata' ●

105010 Cornus coreana Wangerin = Swida coreana (Wangerin) Soják ●

105011 Cornus corynostylis Koehne = Cornus macrophylla Wall. ●

105012 Cornus corynostylis Koehne = Swida macrophylla (Wall.) Soják ●
105013 Cornus crispula Hance = Cornus macrophylla Wall. ●
105014 Cornus crispula Hance = Swida macrophylla (Wall.) Soják ●
105015 Cornus daijinensis W. P. Fang et W. K. Hu = Cornus schindleri Wangerin ●
105016 Cornus daijinensis W. P. Fang et W. K. Hu = Swida daijinensis (W. P. Fang et W. K. Hu) W. P. Fang et W. K. Hu ●◇
105017 Cornus drummondii C. A. Mey.;糙叶山茱萸;Northern Roughleaf Dogwood,Roughleaf Dogwood,Rough-leaved Dogwood ●☆
105018 Cornus elliptica (Pojark.) Q. Y. Xiang et Boufford;尖叶四照花●
105019 Cornus excelsa Kunth;高梾木●☆
105020 Cornus ferruginea Y. C. Wu = Cornus hongkongensis Hemsl. subsp. ferruginea (Y. C. Wu) Q. Y. Xiang ●
105021 Cornus ferruginea Y. C. Wu = Cornus hongkongensis Hemsl. ●
105022 Cornus ferruginea Y. C. Wu = Dendrobenthamia ferruginea (Y. C. Wu) W. P. Fang ●
105023 Cornus florida L.;佛罗里达四照花(北美山茱萸,大花四照花,多花梾木,佛罗里达黄杨,佛罗里达梾木,佛州四照花,花梾木,箭木,美山茱萸,群花梾木,山茱萸,伪黄杨木,御膳橘);American Dogwood,Boxwood,Dog Tree,Eastern Flowering Dogwood,Flowering Dogwood, Great-flowered Cornel, Green Osier, New England Boxwood, Red Berry, Redberry, Virginian Boxwood, Virginian Dogwood ●☆
105024 Cornus florida L.'Apple Blossom';苹果花佛州四照花●☆
105025 Cornus florida L.'Cherokee Chief';深玫瑰多花梾木;Deep Rosy Flowering Dogwood ●☆
105026 Cornus florida L.'Cherokee Princess';白花多花梾木;White Flowering Dogwood ●☆
105027 Cornus florida L.'Cloud Nine';美丽多花梾木;Beautiful Flowering Dogwood ●☆
105028 Cornus florida L.'Pendula';垂枝多花梾木;Weeping Flowering Dogwood ●☆
105029 Cornus florida L.'Rainbow';鲜黄叶多花梾木;Yellowleaf Flowering Dogwood ●☆
105030 Cornus florida L.'Rubra';玫瑰梾木;Rose Flowering Dogwood ●☆
105031 Cornus florida L.'Spring Song';春之歌佛州四照花●☆
105032 Cornus florida L.'Welchii';彩叶多花梾木(韦尔奇佛州四照花);Tricolor Flowering Dogwood ●☆
105033 Cornus florida L.'White Cloud';白云佛州四照花●☆
105034 Cornus florida L. = Benthamidia florida (L.) Spach ●☆
105035 Cornus florida L. var. pendula Dippel = Cornus florida L. 'Pendula' ●☆
105036 Cornus florida L. var. pluribracteata Rehder;多苞佛罗里达四照花●☆
105037 Cornus florida L. var. rubra West. = Cornus florida L. 'Rubra' ●☆
105038 Cornus florida L. var. xanthocarpa Rehder;黄果佛罗里达四照花●☆
105039 Cornus foemina Mill.;沼泽梾木;Gray Dogwood,Stiff Dogwood,Swamp Dogwood ●☆
105040 Cornus foemina Mill. = Cornus racemosa Lam. ●☆
105041 Cornus foemina Mill. subsp. racemosa (Lam.) J. S. Wilson = Cornus racemosa Lam. ●☆
105042 Cornus fordii Hemsl. = Cornus wilsoniana Wangerin ●
105043 Cornus fordii Hemsl. = Swida wilsoniana (Wangerin) Soják ●
105044 Cornus fulvescens W. P. Fang et W. K. Hu = Cornus schindleri Wangerin ●
105045 Cornus fulvescens W. P. Fang et W. K. Hu = Swida fulvescens (W. P. Fang et W. K. Hu) W. P. Fang et W. K. Hu ●
105046 Cornus gharaf Forssk. = Cordia gharaf (Forssk.) Ehrenb. ex Asch. ●
105047 Cornus gigantea (Hand. -Mazz.) Tardieu = Cornus hongkongensis Hemsl. subsp. gigantea (Hand. -Mazz.) Q. Y. Xiang ●
105048 Cornus hemsleyi C. K. Schneid. et Wangerin = Swida hemsleyi (C. K. Schneid. et Wangerin) Soják ●
105049 Cornus hemsleyi C. K. Schneid. et Wangerin var. gracilipes W. P. Fang et W. K. Hu = Swida hemsleyi (C. K. Schneid. et Wangerin) Soják var. gracilipes (W. P. Fang et W. K. Hu) W. P. Fang et W. K. Hu ●
105050 Cornus hemsleyi C. K. Schneid. et Wangerin var. gracilipes W. P. Fang et W. K. Hu = Cornus hemsleyi C. K. Schneid. et Wangerin ●
105051 Cornus hemsleyi C. K. Schneid. et Wangerin var. longistyla W. P. Fang et W. K. Hu = Swida hemsleyi (C. K. Schneid. et Wangerin) Soják var. longistyla (W. P. Fang et W. K. Hu) W. P. Fang et W. K. Hu ●
105052 Cornus hemsleyi C. K. Schneid. et Wangerin var. longistyla W. P. Fang et W. K. Hu = Cornus hemsleyi C. K. Schneid. et Wangerin ●
105053 Cornus henryi Hemsl. = Swida walteri (Wangerin) Soják ●
105054 Cornus henryi Hemsl. ex Wangerin = Cornus walteri Wangerin ●
105055 Cornus hessei Koehne;东北亚山茱萸●☆
105056 Cornus hongkongensis Hemsl.;香港四照花(山荔枝);Hongkong Dendrobenthamia, Hongkong Dogwood, Hongkong Four-involucre ●
105057 Cornus hongkongensis Hemsl. = Benthamidia japonica (Siebold et Zucc.) H. Hara ●
105058 Cornus hongkongensis Hemsl. = Dendrobenthamia hongkongensis (Hemsl.) Hutch. ●
105059 Cornus hongkongensis Hemsl. subsp. elegans (W. P. Fang et Y. T. Hsieh) Q. Y. Xiang;秀丽四照花;Beautiful Dogwood, Elegant Dendrobenthamia, Elegant Four-involucre ●
105060 Cornus hongkongensis Hemsl. subsp. ferruginea (Y. C. Wu) Q. Y. Xiang;褐毛四照花;Brownhair Four-involucre, Rusty-coloured Dendrobenthamia, Rusty-hairs Dogwood ●
105061 Cornus hongkongensis Hemsl. subsp. gigantea (Hand. -Mazz.) Q. Y. Xiang;大型香港四照花●
105062 Cornus hongkongensis Hemsl. subsp. melanotricha (Pojark.) Q. Y. Xiang;黑毛四照花●
105063 Cornus hongkongensis Hemsl. subsp. tonkinensis (W. P. Fang) Q. Y. Xiang;东京四照花(西南四照花);Tonkin Dendrobenthamia, Tonkin Dogwood, Tonkin Four-involucre ●
105064 Cornus hongkongensis Hemsl. var. gigantea Hand. -Mazz. = Cornus hongkongensis Hemsl. subsp. gigantea (Hand. -Mazz.) Q. Y. Xiang ●
105065 Cornus hongkongensis Hemsl. var. gigantea Hand. -Mazz. = Dendrobenthamia gigantea (Hand. -Mazz.) W. P. Fang ●
105066 Cornus hongkongensis Hemsl. var. jinyunensis (W. P. Fang et W. K. Hu) Q. Y. Xiang = Cornus hongkongensis Hemsl. subsp. melanotricha (Pojark.) Q. Y. Xiang ●
105067 Cornus hybrida Hort.;杂种梾木;Stellar Hybrids Dogwood ●☆
105068 Cornus iberica Woronow;伊比利亚肖梾木●☆
105069 Cornus ignorata K. Koch = Bothrocaryum controversum (Hemsl. ex Prain) Pojark. ●

105070 Cornus ignorata Shiras. = Cornus brachypoda C. A. Mey. ●
105071 Cornus instolonea A. Nelson = Cornus stolonifera Michx. ●☆
105072 Cornus interior (Rydb.) N. Petersen = Cornus stolonifera Michx. ●☆
105073 Cornus japonica DC. = Benthamidia japonica (Siebold et Zucc.) H. Hara ●
105074 Cornus japonica DC. = Dendrobenthamia japonica (DC.) W. P. Fang ●
105075 Cornus japonica DC. = Viburnum japonicum (Thunb.) Spreng. ●
105076 Cornus koehneana Wangerin = Swida koehneana (Wangerin) Soják ●
105077 Cornus kousa Buerger = Dendrobenthamia japonica (DC.) W. P. Fang var. chineneis (Osborn) W. P. Fang ●
105078 Cornus kousa Buerger ex Hance = Benthamidia japonica (Siebold et Zucc.) H. Hara ●
105079 Cornus kousa Buerger ex Hance subsp. chinensis (Osborn) Q. Y. Xiang = Benthamidia japonica (Siebold et Zucc.) H. Hara var. chinensis (Osborn) H. Hara ●
105080 Cornus kousa Buerger ex Hance subsp. chinensis (Osborn) Q. Y. Xiang = Dendrobenthamia japonica (DC.) W. P. Fang var. chineneis (Osborn) W. P. Fang ●
105081 Cornus kousa Buerger ex Hance var. chinensis Osborn = Benthamidia japonica (Siebold et Zucc.) H. Hara var. chinensis (Osborn) H. Hara ●
105082 Cornus kousa Buerger ex Hance var. chinensis Osborn = Dendrobenthamia japonica (DC.) W. P. Fang var. chineneis (Osborn) W. P. Fang ●
105083 Cornus kousa Buerger ex Miq. = Dendrobenthamia japonica (DC.) W. P. Fang ●
105084 Cornus kousa Buerger ex Miq. = Dendrobenthamia japonica (DC.) W. P. Fang var. chineneis (Osborn) W. P. Fang ●
105085 Cornus kousa Hance ex Diels = Dendrobenthamia japonica (DC.) W. P. Fang var. chinensis (Osborn) W. P. Fang ●
105086 Cornus kousa Hance ex Diels subsp. chinensis (Osborn) Q. Y. Xiang = Dendrobenthamia japonica (DC.) W. P. Fang var. chineneis (Osborn) W. P. Fang ●
105087 Cornus kousa Hance ex Diels var. angustata Chun = Cornus elliptica (Pojark.) Q. Y. Xiang et Boufford ●
105088 Cornus kousa Hance ex Diels var. angustata Chun = Dendrobenthamia angustata (Chun) W. P. Fang ●
105089 Cornus kousa Hance ex Diels var. chinensis Osborn = Cornus kousa Buerger ex Hance subsp. chinensis (Osborn) Q. Y. Xiang ●
105090 Cornus kousa Hance ex Diels var. chinensis Osborn = Dendrobenthamia japonica (DC.) W. P. Fang var. chineneis (Osborn) W. P. Fang ●
105091 Cornus kousa Hance ex Diels var. leucotricha (W. P. Fang et Y. T. Hsieh) Q. Y. Xiang = Cornus kousa Buerger ex Hance subsp. chinensis (Osborn) Q. Y. Xiang ●
105092 Cornus kousa Hance ex Diels var. yaeyamensis Hatus. = Benthamidia japonica (Siebold et Zucc.) H. Hara ●
105093 Cornus kousa Rehder = Dendrobenthamia multinervosa (Pojark.) W. P. Fang ●
105094 Cornus kweichowensis H. L. Li = Cornus wilsoniana Wangerin ●
105095 Cornus liangkwangensis W. P. Fang et W. K. Hu;两广梾木;Liangguang Dogwood ●
105096 Cornus lixianensis W. P. Fang et W. K. Hu = Cornus schindleri Wangerin ●
105097 Cornus lixianensis W. P. Fang et W. K. Hu = Swida schindleri (Wangerin) Soják var. lixianensis (W. P. Fang et W. K. Hu) W. P. Fang et W. K. Hu ●
105098 Cornus longipedunculata W. P. Fang et W. K. Hu = Cornus macrophylla Wall. ●
105099 Cornus longipedunculata W. P. Fang et W. K. Wu = Swida macrophylla (Wall.) Soják ●
105100 Cornus longipetiolata Hayata = Bothrocaryum controversum (Hemsl. ex Prain) Pojark. ●
105101 Cornus longipetiolata Hayata = Cornus macrophylla Wall. ●
105102 Cornus longipetiolata Hayata = Swida macrophylla (Wall.) Soják ●
105103 Cornus lowriana Brandis = Cordia dichotoma G. Forst. ●
105104 Cornus macrophylla Wall. = Cornus brachypoda C. A. Mey. ●
105105 Cornus macrophylla Wall. = Swida controversa (Hemsl. ex Prain) Soják ●
105106 Cornus macrophylla Wall. = Swida macrophylla (Wall.) Soják ●
105107 Cornus macrophylla Wall. ex Roxb.;大叶山茱萸●☆
105108 Cornus macrophylla Wall. var. stracheyi C. B. Clarke;密毛梾木●
105109 Cornus malifolia W. P. Fang et W. K. Hu = Cornus schindleri Wangerin ●
105110 Cornus malifolia W. P. Fang et W. K. Hu = Swida poliophylla (C. K. Schneid. et Wangerin) Soják var. malifolia (W. P. Fang et W. K. Wu) W. P. Fang et W. K. Wu ●
105111 Cornus mas L.;欧亚山茱萸(地中海梾木,欧洲山茱萸,欧茱萸);Cornel, Cornel Cherry, Cornelia Tree, Cornelian Cherry, Cornelian Cherry Dogwood, Cornelian Dogwood, Cornelian-cherry Dogwood, Long Cherry, Male Cornel, Male Dogwood, Sorbet, Tame Cornel ●☆
105112 Cornus mas L. 'Aurea';金叶欧洲山茱萸;Yellow-leaved Cornelissen Dogw ●☆
105113 Cornus mas L. 'Aureoelegantissima';金边欧洲山茱萸(黄雅致欧茱萸)●☆
105114 Cornus mas L. 'Elegantissima' = Cornus mas L. 'Aureoelegantissima' ●☆
105115 Cornus mas L. 'Golden Glory';金色荣耀欧洲山茱萸●☆
105116 Cornus mas L. 'Macrocarpa';大果欧洲山茱萸●☆
105117 Cornus mas L. 'Pyramidalis';塔形欧洲山茱萸●☆
105118 Cornus mas L. 'Variegata';白边欧洲山茱萸(银边欧茱萸)●☆
105119 Cornus mascula Hort. = Cornus mas L. ●☆
105120 Cornus mascula L.;雄性梾木●☆
105121 Cornus melanosticta C. Y. Wu;黑线梾木;Blackline Dogwood ●
105122 Cornus mombeigii Hemsl. = Swida monbeigii (Hemsl.) Soják ●
105123 Cornus mombeigii Hemsl. subsp. crassa W. P. Fang et W. K. Hu = Swida monbeigii (Hemsl.) Soják var. crassa (W. P. Fang et W. K. Hu) W. P. Fang et W. K. Hu ●
105124 Cornus monbeigii Hemsl. = Cornus schindleri Wangerin ●
105125 Cornus monbeigii Hemsl. subsp. crassa W. P. Fang et W. K. Hu = Cornus schindleri Wangerin ●
105126 Cornus monbeigii Hemsl. subsp. popolufolia W. P. Fang et W. K. Hu = Swida monbeigii (Hemsl.) Soják var. popolufolia (W. P. Fang et W. K. Hu) W. P. Fang et W. K. Hu ●
105127 Cornus monbeigii Hemsl. subsp. populifolia W. P. Fang et W. K. Hu = Cornus schindleri Wangerin ●
105128 Cornus monbeigii Hemsl. subsp. populifolia W. P. Fang et W. K. Hu = Swida monbeigii (Hemsl.) Soják var. popolufolia (W. P. Fang et W. K. Hu) W. P. Fang et W. K. Hu ●

105129 Cornus multinervosa (Pojark.) Q. Y. Xiang = Dendrobenthamia multinervosa (Pojark.) W. P. Fang ●

105130 Cornus nuttallii Audubon;太平洋楝木(美国西部四照花,太平洋四照花);Canadian Dogwood, Flowering Dogwood, Mountain Dogwood, Nuttall's Dogwood, Pacific Dogwood, Pacific Flowering Dogwood, Western Flowering Dogwood ●☆

105131 Cornus nuttallii Audubon 'Colrigo Giant';直立太平洋楝木;Columbia R. Giant Dogwood, Erect Pacific Dogwood ●☆

105132 Cornus nuttallii Audubon 'Goldspot';黄斑太平洋楝木(金斑太平洋楝木);Goldspot Pacific Dogwood ●☆

105133 Cornus obliqua Raf.;斜叶楝木;Pale Dogwood, Silky Dogwood ●☆

105134 Cornus obliqua Raf. = Cornus amomum Mill. var. schuetzeana (C. A. Mey.) Rickett ●☆

105135 Cornus obliqua Raf. = Cornus amomum Mill. ●☆

105136 Cornus obliqua Willd. = Cordia myxa L. ●☆

105137 Cornus obliqua Willd. var. tomentosa (Wall.) Kazmi = Cordia myxa L. ●☆

105138 Cornus oblonga Wall. = Swida oblonga (Wall.) Soják ●

105139 Cornus oblonga Wall. ex Roxb.;矩圆楝木●☆

105140 Cornus oblonga Wall. f. pilosula H. L. Li = Swida oblonga (Wall.) Soják var. griffithii (C. B. Clarke) W. K. Hu ●

105141 Cornus oblonga Wall. var. glabrescena W. P. Fang et W. K. Hu = Swida oblonga (Wall.) Soják var. glabrescena (W. P. Fang et W. K. Hu) W. P. Fang et W. K. Hu ●

105142 Cornus oblonga Wall. var. griffithii C. B. Clarke = Swida oblonga (Wall.) Soják var. griffithii (C. B. Clarke) W. K. Hu ●

105143 Cornus obovata Thunb. = Bothrocaryum controversum (Hemsl. ex Prain) Pojark. ●

105144 Cornus obovata Thunb. = Cornus controversa Hemsl. ex Prain ●

105145 Cornus obovata Thunb. = Swida controversa (Hemsl. ex Prain) Soják ●

105146 Cornus officinalis Siebold et Zucc.;山茱萸(红枣皮,鸡足,寇思,魃实,肉枣,山萸,山萸肉,石枣,实枣儿,实枣儿树,鼠矢,蜀酸枣,蜀枣,药枣,野春桂,萸肉,枣皮,茱萸肉);Common Macrocarpium, Dogwood, Japanese Cornal, Japanese Cornal Dogwood, Japanese Cornelian Cherry, Japanese Cornelian-cherry, Japanese Dogwood, Medical Dogwood, Medicinal Cornel ●

105147 Cornus officinalis Siebold et Zucc. var. koreana Kitam.;朝鲜山茱萸●☆

105148 Cornus oligophlebia Merr. = Swida oligophlebia (Merr.) W. K. Hu ●

105149 Cornus paniculata Buch.-Ham. ex D. Don = Cornus oblonga Wall. ex Roxb. ●☆

105150 Cornus paniculata Buch.-Ham. ex D. Don = Swida oblonga (Wall.) Soják ●

105151 Cornus paniculata L'Her. = Cornus racemosa Lam. ●☆

105152 Cornus papillosa W. P. Fang et W. K. Hu = Swida papillosa (W. P. Fang et W. K. Hu) W. P. Fang et W. K. Hu ●

105153 Cornus parviflora S. S. Chien = Swida parviflora (S. S. Chien) Holub ●

105154 Cornus paucinervis Hance = Cornus quinquenervis Franch. ●

105155 Cornus paucinervis Hance = Swida paucinervis (Hance) Soják ●

105156 Cornus peruviana J. F. Macbr.;秘鲁山茱萸●☆

105157 Cornus poliophylla C. K. Schneid. et Wangerin = Cornus schindleri Wangerin subsp. poliophylla (C. K. Schneid. et Wangerin) Q. Y. Xiang ●

105158 Cornus poliophylla C. K. Schneid. et Wangerin = Swida poliophylla (C. K. Schneid. et Wangerin) Soják ●

105159 Cornus poliophylla C. K. Schneid. et Wangerin var. microphylla L. C. Wang et X. G. Sun = Cornus schindleri Wangerin subsp. poliophylla (C. K. Schneid. et Wangerin) Q. Y. Xiang ●

105160 Cornus poliophylla C. K. Schneid. et Wangerin var. microphylla L. G. Wang et X. G. Sun;小叶黑椋子●

105161 Cornus poliophylla C. K. Schneid. et Wangerin var. praelonga W. P. Fang et W. K. Hu = Cornus schindleri Wangerin ●

105162 Cornus poliophylla C. K. Schneid. et Wangerin var. prelonga W. P. Fang et W. K. Hu = Swida poliophylla (C. K. Schneid. et Wangerin) Soják var. praelonga (W. P. Fang et W. K. Hu) W. P. Fang et W. K. Hu ●

105163 Cornus polyantha W. P. Fang et W. K. Hu = Cornus hemsleyi C. K. Schneid. et Wangerin ●

105164 Cornus polyantha W. P. Fang et W. K. Hu = Swida polyantha (W. P. Fang et W. K. Hu) W. P. Fang et W. K. Hu ●

105165 Cornus priceae Small = Cornus drummondii C. A. Mey. ●☆

105166 Cornus pumila Koehne;矮生楝木;Dwarf Redtip Dogwood, Dwarf Red-tipped Dogwood ●☆

105167 Cornus purpusii Koehne = Cornus amomum Mill. var. schuetzeana (C. A. Mey.) Rickett ●☆

105168 Cornus quinquenervis Franch. = Swida paucinervis (Hance) Soják ●

105169 Cornus racemosa Lam.;圆锥楝木;Gray Dogwood, Northern Swamp Dogwood, Panicled Dogwood ●☆

105170 Cornus racemosa Lam. = Cornus foemina Mill. ●☆

105171 Cornus rugosa Lam.;圆叶楝木;Roundleaf Dogwood, Round-leafed Dogwood ●☆

105172 Cornus sanguinea L. 'Widwinter Fire';隆冬之火欧洲红瑞木●☆

105173 Cornus sanguinea L. 'Winter Beauty';冬美人欧洲红瑞木●☆

105174 Cornus sanguinea L. = Bothrocaryum controversum (Hemsl. ex Prain) Pojark. ●

105175 Cornus sanguinea L. = Swida controversa (Hemsl. ex Prain) Soják ●

105176 Cornus sanguinea L. = Swida macrophylla (Wall.) Soják ●

105177 Cornus sanguinea L. = Swida sanguinea (L.) Opiz ●

105178 Cornus sanguinea L. var. australis C. A. Mey. = Cornus australis C. A. Mey. ●☆

105179 Cornus sanguinea Thunb. = Cornus controversa Hemsl. ex Prain ●

105180 Cornus scabrida Franch. = Cornus schindleri Wangerin ●

105181 Cornus scabrida Franch. = Swida scabrida (Franch.) Holub ●

105182 Cornus schindleri Wangerin = Swida schindleri (Wangerin) Soják ●

105183 Cornus schindleri Wangerin subsp. poliophylla (C. K. Schneid. et Wangerin) Q. Y. Xiang = Swida poliophylla (C. K. Schneid. et Wangerin) Soják ●

105184 Cornus sericea L. 'Flaviramea' = Cornus stolonifera Michx. 'Flaviramea' ●☆

105185 Cornus sericea L. = Cornus amomum Mill. ●☆

105186 Cornus sericea L. = Cornus stolonifera Michx. ●☆

105187 Cornus sericea L. subsp. stolonifera (Michx.) Fosberg = Cornus stolonifera Michx. ●☆

105188 Cornus sericea L. var. interior (Rydb.) H. St. John = Cornus stolonifera Michx. ●☆

105189 Cornus sericea L. var. schuetzeana C. A. Mey. = Cornus amomum Mill. var. schuetzeana (C. A. Mey.) Rickett ●☆

105190 Cornus sessilis Torr. ex Durand;无柄梾木●☆
105191 Cornus sibirica Lodd. = Cornus tatarica Mill. ●☆
105192 Cornus sinensis Pojark. = Dendrobenthamia angustata (Chun) W. P. Fang ●
105193 Cornus stolonifera Michx.;偃伏梾木;American Dogwood, Kinnikinnik, Red Brush, Red Osier Dogwood, Red Twig Dogwood, Red Willow, Red-osier Dogwood, Red-stem Dogwood, Redtwig Dogwood, Sericeous Dogwood, Squawbusb, Squaw-bush, Yellowtwig Dogwood ●☆
105194 Cornus stolonifera Michx. 'Flaviramea';黄枝偃伏梾木;Yellow Twig Dogwood ●☆
105195 Cornus stolonifera Michx. 'Isanti';伊善提偃伏梾木●☆
105196 Cornus stolonifera Michx. 'Sunshine';阳光偃伏梾木;Gold-leaved Red Osier Dogwood ●☆
105197 Cornus stolonifera Michx. var. baileyi (J. M. Coult. et W. H. Evans) Drescher = Cornus stolonifera Michx. ●☆
105198 Cornus stolonifera Michx. var. glaviamea Rehder;金枝偃伏梾木;Goldentwig Dogwood ●☆
105199 Cornus stolonifera Michx. var. interior (Rydb.) H. St. John = Cornus stolonifera Michx. ●☆
105200 Cornus stracheyi (C. B. Clarke) Hemsl. = Cornus macrophylla Wall. var. stracheyi C. B. Clarke ●
105201 Cornus stricta Lam.;刚直梾木;Swamp Dogwood ●☆
105202 Cornus subumbellata Komatsu;库页梾木●☆
105203 Cornus suecica A. Gray = Chamaepericlymenum canadense (L.) Asch. et Graebn. ■
105204 Cornus suecica L.;瑞典梾木(瑞典草茱萸);Dwarf Cornel, Dwarf Honeysuckle, Honeysuckle, Lapland Cornel, Northern Dwarf Cornel, Plant of Gluttony ●☆
105205 Cornus taiwanensis Kaneh. = Cornus macrophylla Wall. ●
105206 Cornus taiwanensis Kaneh. = Swida macrophylla (Wall.) Soják ●
105207 Cornus tatarica Mill.;鞑靼梾木;Siberian Dogwood, Tatar Dogwood ●☆
105208 Cornus thelicanis Lebas = Swida macrophylla (Wall.) Soják ●
105209 Cornus tonkinensis (W. P. Fang) Tardieu = Cornus hongkongensis Hemsl. subsp. tonkinensis (W. P. Fang) Q. Y. Xiang ●
105210 Cornus tonkinensis (W. P. Fang) Tardieu = Dendrobenthamia tonkinensis W. P. Fang ●
105211 Cornus ulotricha C. K. Schneid. et Wangerin = Swida ulotricha (C. K. Schneid. et Wangerin) Soják ●
105212 Cornus ulotricha C. K. Schneid. et Wangerin var. leptophylla W. K. Hu ex P. C. Li = Cornus ulotricha C. K. Schneid. et Wangerin ●
105213 Cornus unalaschkensis Ledeb. = Chamaepericlymenum canadense (L.) Asch. et Graebn. ■
105214 Cornus volkensii Harms = Afrocrania volkensii (Harms) Hutch. ●☆
105215 Cornus walteri Wangerin = Swida walteri (Wangerin) Soják ●
105216 Cornus walteri Wangerin var. conferrtiflora W. P. Fang et W. K. Hu = Swida walteri (Wangerin) Soják ●
105217 Cornus walteri Wangerin var. confertiflora W. P. Fang et W. K. Hu = Cornus walteri Wangerin ●
105218 Cornus walteri Wangerin var. insignis W. P. Fang et W. K. Hu = Cornus walteri Wangerin ●
105219 Cornus walteri Wangerin var. insignis W. P. Fang et W. K. Hu = Swida walteri (Wangerin) Soják ●
105220 Cornus wilsoniana Wangerin = Swida wilsoniana (Wangerin) Soják ●
105221 Cornus xanthotricha W. P. Fang et W. K. Hu = Cornus schindleri Wangerin ●
105222 Cornus xanthotricha W. P. Fang et W. K. Hu = Swida monbeigii (Hemsl.) Soják var. xanthotricha (W. P. Fang et W. K. Hu) W. P. Fang et W. K. Hu ●
105223 Cornus yaeyamensis Hatus. = Benthamidia japonica (Siebold et Zucc.) H. Hara ●
105224 Cornus yunnanensis H. L. Li = Cornus walteri Wangerin ●
105225 Cornus yunnanensis H. L. Li = Swida walteri (Wangerin) Soják ●
105226 Cornus yunnanensis Koidz. = Malus yunnanensis (Franch.) C. K. Schneid. ●
105227 Cornuta L. = Cornus L. ●
105228 Cornutia Burm. f. = Premna L. (保留属名) ●■
105229 Cornutia L. (1753);墨蓝花属(科努草属);Cornutia ■☆
105230 Cornutia corymbosa Burm. f. = Premna corymbosa (Burm. f.) Rottl. et Willd. ●
105231 Cornutia corymbosa Burm. f. = Premna serratifolia L. ●
105232 Cornutia grandiflora Steud.;大花墨蓝花(大花科努草);Mexican-blue ●☆
105233 Cornutia grandifolia (Schltdl. et Cham.) Schauer;大叶墨蓝花(大叶科努草)●☆
105234 Cornutia pyramidata L.;三棱墨蓝花(三棱科努草)●☆
105235 Cornutia quinata Lour. = Vitex quinata (Lour.) F. N. Williams ●
105236 Corocephalus D. Dietr. = Conocephalus Blume ●
105237 Corocephalus D. Dietr. = Poikilospermum Zipp. ex Miq. ●
105238 Corokia A. Cunn. (1839);宿萼果属(假醉鱼草属,克劳凯奥属,克罗开木属,秋叶果属);Whakataka ●☆
105239 Corokia buddleioides A. Cunn.;宿萼果(假醉鱼草,克劳凯奥);Corokio ●☆
105240 Corokia cotoneaster Raoul;栒子状宿萼果(细枝克劳凯奥,栒子状假醉鱼草);Corokia, Wire Netting Bush, Wire-netting Bush ●☆
105241 Corokia macrocarpa Kirk;大果宿萼果(大果克劳凯奥);Whakataka ●☆
105242 Corokia virgata Turrill;帚状宿萼果(橙果假醉鱼草,帚状克劳凯奥)●☆
105243 Corokia virgata Turrill 'Bronze King';青铜色之王帚状宿萼果(青铜色之王帚状克劳凯奥)●☆
105244 Corokia virgata Turrill 'Cheesemanii';小叶帚状宿萼果(小叶帚状克劳凯奥)●☆
105245 Corokia virgata Turrill 'Frosted Chocolate';巧克力帚状宿萼果(巧克力帚状克劳凯奥)●☆
105246 Corokia virgata Turrill 'Red Wonder';红色奇迹帚状宿萼果(红色奇迹帚状克劳凯奥)●☆
105247 Corokia virgata Turrill 'Yellow Wonder';黄色奇迹帚状宿萼果(黄色奇迹帚状克劳凯奥)●☆
105248 Corokia whiteana L. S. Sm.;白花宿萼果(白花克劳凯奥)●☆
105249 Corokiaceae Kapil ex Takht.;宿萼果科●☆
105250 Corokiaceae Kapil ex Takht. = Argophyllaceae Takht. ●☆
105251 Corokiaceae Kapil ex Takht. = Corsiaceae Becc. (保留科名) ■
105252 Corollonema Schltr. (1914);冠丝萝藦属☆
105253 Corollonema boliviense Schltr.;冠丝萝藦☆
105254 Corona Fisch. ex Graham = Fritillaria L. ■
105255 Corona leucantha Fisch. ex Graham = Fritillaria leucantha Fisch ex Schult. f. ■☆
105256 Coronanthera Vieill. ex C. B. Clarke (1883);冠药苣苔属●☆
105257 Coronanthera grandis G. W. Gillett;大冠药苣苔●☆
105258 Coronanthera pulchra C. B. Clarke;美丽冠药苣苔●☆

105259 Coronaria Guett. = Lychnis L.(废弃属名)■
105260 Coronaria Guett. = Silene L.(保留属名)■
105261 Coronaria coriacea (Moench) Schischk. = Lychnis coronaria (L.) Desr. ■
105262 Coronaria coriacea (Moench) Schischk. = Silene coronaria (L.) Clairv. ■
105263 Coronaria coriacea (Moench) Schischk. et Gorschk. = Lychnis coronaria (L.) Desr. ■
105264 Coronaria flos-cuculi (L.) A. Br. = Lychnis flos-cuculi L. ■☆
105265 Coronaria flos-cuculi (L.) A. Br. = Silene flos-cuculi (L.) Greuter et Burdet ■☆
105266 Corone Hoffmanns. ex Steud. = Silene L.(保留属名)■
105267 Coronidium Paul G. Wilson(2008);尖鳞菊属☆
105268 Coronilla Ehrh. = Coronilla L.(保留属名)●■
105269 Coronilla L.(1753)(保留属名);小冠花属;Coronilla,Crown Vetch,Crownvetch,Scorpion's Senna,Scorpionsenna,Scorpion-vetch ●■
105270 Coronilla aculeata Willd. = Sesbania bispinosa (Jacq.) W. Wight ■
105271 Coronilla arenivaga Pau = Coronilla repanda (Poir.) Guss. ●☆
105272 Coronilla argentea Burm. f. = Lessertia argentea Harv. ■☆
105273 Coronilla atlantica (Boiss. et Reut.) Boiss. = Securigera atlantica Boiss. et Reut. ■☆
105274 Coronilla balansae Boiss.;巴拉小冠花●☆
105275 Coronilla buxifolia Hance;黄杨叶小冠花(厦门小冠花,小冠花);Boxleaf Coronilla,Box-leaved Coronilla ●
105276 Coronilla coronata L.;小冠花●☆
105277 Coronilla cretica L.;克里特小冠花●☆
105278 Coronilla dura (Cav.) Boiss. = Coronilla repanda (Poir.) Guss. subsp. dura (Cav.) Cout. ●☆
105279 Coronilla emeroides Boiss. et Spruner;假蝎子旃那●☆
105280 Coronilla emeroides Boiss. et Spruner = Coronilla emerus L. ●
105281 Coronilla emerus L.;蝎子旃那;Scorpion Senna,Scorpionsenna Coronilla,Scorpion-senna Coronilla ●
105282 Coronilla emerus L. = Hippocrepis emerus (L.) Lassen ●
105283 Coronilla emerus L. subsp. emeroides (Boiss. et Spruner) Holmboe = Coronilla emeroides Boiss. et Spruner ●☆
105284 Coronilla glauca L.;蓝绿小冠花(粉绿小冠花,灰小冠花)●☆
105285 Coronilla glauca L. var. pentaphylloides Rouy = Coronilla valentina L. subsp. glauca (L.) Batt. ●☆
105286 Coronilla hirsuta DC. = Lessertia argentea Harv. ■☆
105287 Coronilla hyrcana Prilipko;西加小冠花●☆
105288 Coronilla juncea L.;灯心草小冠花●☆
105289 Coronilla juncea L. subsp. pomelii Batt.;波梅尔小冠花●☆
105290 Coronilla juncea L. subsp. ramosissima Ball = Coronilla ramosissima (Ball) Ball ●☆
105291 Coronilla juncea L. var. pomelii (Batt.) Batt. = Coronilla juncea L. subsp. pomelii Batt. ●☆
105292 Coronilla latifolia Jav.;大叶小冠花●☆
105293 Coronilla minima L.;侏儒小冠花●☆
105294 Coronilla minima L. subsp. clusii (Dufour) Murb. = Coronilla minima L. subsp. lotoides (Koch) Nyman ●☆
105295 Coronilla minima L. subsp. lotoides (Koch) Nyman;君迁子小冠花●☆
105296 Coronilla minima L. var. clusii (Dufour) Batt. = Coronilla minima L. ●☆
105297 Coronilla minima L. var. fruticans Burnat = Coronilla minima L. ●☆
105298 Coronilla minima L. var. mairei Hrabetova = Coronilla minima L. ●☆
105299 Coronilla montana Scop.;山地小冠花;Mountain Coronilla ●☆
105300 Coronilla orientalis Mill.;东方小冠花●☆
105301 Coronilla parviflora Willd.;小花小冠花●☆
105302 Coronilla pentaphylla Desf. = Coronilla valentina L. subsp. pentaphylla (Desf.) Batt ●☆
105303 Coronilla pubescens Schumach. et Thonn.;短柔毛小冠花●☆
105304 Coronilla pulchra Ball = Coronilla viminalis Salisb. ●☆
105305 Coronilla ramosissima (Ball) Ball;多枝小冠花●☆
105306 Coronilla ramosissima (Ball) Ball var. ifniensis (Caball.) Font Quer = Coronilla ramosissima (Ball) Ball ●☆
105307 Coronilla repanda (Poir.) Guss.;浅波状小冠花●☆
105308 Coronilla repanda (Poir.) Guss. subsp. dura (Cav.) Cout.;硬浅波状小冠花●☆
105309 Coronilla repanda (Poir.) Guss. var. arenivaga (Pau) Maire = Coronilla repanda (Poir.) Guss. ●☆
105310 Coronilla scorpioides (L.) W. D. J. Koch;蝎尾小冠花;Annual Scorpion-vetch,Yellow Crownvetch ●☆
105311 Coronilla scorpioides St.-Lag. = Coronilla scorpioides (L.) W. D. J. Koch ●☆
105312 Coronilla securidaca L. = Securigera securidaca (L.) Degen. et Dorf. ■☆
105313 Coronilla sericea Willd. = Sesbania sericea (Willd.) Link ■☆
105314 Coronilla sesban Willd. = Sesbania sesban (L.) Merr. ●
105315 Coronilla somalensis Thulin = Securigera somalensis (Thulin) Lassen ●☆
105316 Coronilla thymifolia Burm. f.;百里香叶小冠花●☆
105317 Coronilla valentina L.;巴伦西亚小冠花;Bastard Senna,Mediterranean Crownvetch,Shrubby Scorpion-vetch ●☆
105318 Coronilla valentina L. subsp. glauca (L.) Batt.;灰绿小冠花●☆
105319 Coronilla valentina L. subsp. pentaphylla (Desf.) Batt.;五叶灰绿小冠花●☆
105320 Coronilla valentina L. subsp. speciosa (Hrabetova) Greuter et Burdet;美丽灰绿小冠花●☆
105321 Coronilla valentina L. var. pentaphylloides (Rouy) Asch. et Graebn. = Coronilla valentina L. ●☆
105322 Coronilla valentina L. var. polyarthra Maire = Coronilla valentina L. ●☆
105323 Coronilla valentina L. var. zaianica Emb. et Maire = Coronilla valentina L. ●☆
105324 Coronilla varia L.;绣球小冠花(变色小冠花,变异小冠花,多变小冠花,小冠花);Axseed,Axwort,Changeable Scorpionsenna,Crown Vetch,Crownvetch,Crown-vetch,Crown-vetch Coronilla,Devil's Shoestrings,Purple Crown Vetch,Purple Crown-vetch,Russian Clover,Scorpion Senna ●
105325 Coronilla varia L. = Securigera varia (L.) Lassen ●
105326 Coronilla viminalis Salisb.;柳条小冠花●☆
105327 Coronillaceae Martinov = Fabaceae Lindl.(保留科名)●■
105328 Coronillaceae Martinov = Leguminosae Juss.(保留科名)●■
105329 Coronocarpus Schumach.(1827);冠果菊属■☆
105330 Coronocarpus Schumach. et Thonn. = Aspilia Thouars ■☆
105331 Coronocarpus Schumach. et Thonn. = Coronocarpus Schumach. ■☆
105332 Coronocarpus gayanus Benth. var. peduncularis ? = Aspilia helianthoides (Schumach. et Thonn.) Oliv. et Hiern subsp. ciliata (Schumach.) C. D. Adams ■☆

105333 Coronocarpus helianthoides Schumach. et Thonn. = Aspilia helianthoides (Schumach. et Thonn.) Oliv. et Hiern ■☆

105334 Coronocarpus kotschyi (Sch. Bip. ex Hochst.) Benth. = Aspilia kotschyi (Sch. Bip.) Oliv. ■☆

105335 Coronopus Mill.(废弃属名) = Coronopus Zinn(保留属名)■

105336 Coronopus Mill.(废弃属名) = Plantago L. ■●

105337 Coronopus Rchb. = Plantago L. ■●

105338 Coronopus Zinn(1757)(保留属名);臭荠属(滨芥属,肾果荠属,硬果芥属,硬果荠属);Swine Cress,Swine-cress,Swine's-cress,Wart Cress,Wartcress ■

105339 Coronopus Zinn(保留属名) = Lepidium L. ■

105340 Coronopus didymus (L.) Sm.;臭荠(臭滨芥,肾果荠);Lesser Swinecress,Lesser Swine-cress,Lesser Swine's Cress,Slender Wart Cress,Slender Wart-cress,Swine Cress,Swine Wartcress,Swine's Cress,Swine's-cress,Wartcress,Wart-cress ■

105341 Coronopus didymus (L.) Sm. = Lepidium didymum L. ■

105342 Coronopus englerianus Muschl. = Coronopus integrifolius (DC.) Spreng. ■

105343 Coronopus integrifolius (DC.) Prantl = Coronopus integrifolius (DC.) Spreng. ■

105344 Coronopus integrifolius (DC.) Spreng. = Lepidium englerianum (Muschl.) Al-Shehbaz ■

105345 Coronopus lepidioides (Coss. et Durieu) Kuntze = Lepidium lepidioides (Coss. et Durieu) Al-Shehbaz ■☆

105346 Coronopus lepidioides (Coss. et Durieu) Kuntze var. garamas Maire = Lepidium lepidioides (Coss. et Durieu) Al-Shehbaz ■☆

105347 Coronopus linoides (DC.) Spreng. = Coronopus integrifolius (DC.) Spreng. ■

105348 Coronopus niloticus (Delile) Spreng.;尼罗河臭荠■☆

105349 Coronopus niloticus (Delile) Spreng. subsp. lepidioides (Coss. et Durieu) Quézel = Lepidium lepidioides (Coss. et Durieu) Al-Shehbaz ■☆

105350 Coronopus procumbens Gilib.;匍匐臭荠;Creeping Wart Cress,Creeping Wartcress ■☆

105351 Coronopus procumbens Gilib. = Lepidium squamatum Forssk. ■☆

105352 Coronopus squamatus (Forssk.) Asch.;鳞臭荠;Greater Swinecress,Hartshorn,Herb Eve,Herb Ive,Herb Ivy,Hog-grass,Lesser Swine's Cress,Ribwort,Sow Grass,Star of the Earth,Swine Cress,Swinecress,Swine's Cress,Swine's-cress,Wart Cress,Wartcress,Wart-cress ■☆

105353 Coronopus squamatus (Forssk.) Asch. subsp. conradi Muschl. = Lepidium squamatum Forssk. ■☆

105354 Coronopus squamatus (Forssk.) Asch. subsp. verrucarius (Muschl.) Maire = Lepidium squamatum Forssk. ■☆

105355 Coronopus verrucarius Muschl. et Thell. = Lepidium squamatum Forssk. ■☆

105356 Coronopus violaceus (Munby) Kuntze = Lepidium violaceum (Munby) Al-Shehbaz ■☆

105357 Coronopus violaceus (Munby) Kuntze var. condensatus Maire = Lepidium violaceum (Munby) Al-Shehbaz ■☆

105358 Coronopus wrightii H. Hara = Coronopus integrifolius (DC.) Spreng. ■

105359 Coronopus wrightii H. Hara = Lepidium englerianum (Muschl.) Al-Shehbaz ■

105360 Coropsis Adans. = Coreopsis L. ●■

105361 Corothamnus (W. D. J. Koch) C. Presl = Corothamnus C. Presl ●☆

105362 Corothamnus C. Presl = Cytisus Desf.(保留属名)●

105363 Corothamnus C. Presl = Genista L. ●

105364 Corothamnus C. Presl(1845);乌灌豆属●☆

105365 Corothamnus diffusus C. Presl;乌灌豆●☆

105366 Coroya Pierre = Dalbergia L. f.(保留属名)●

105367 Coroya Pierre(1899);肖黄檀属●☆

105368 Coroya dialoides Pierre;肖黄檀●☆

105369 Corozo Jacq. ex Giseke = Elaeis Jacq. ●

105370 Corpodetes Rchb. = Stenomesson Herb. ■☆

105371 Corpuscularia Schwantes(1926);白绒玉属●☆

105372 Corpuscularia angustifolia (L. Bolus) H. E. K. Hartmann;窄叶白绒玉●☆

105373 Corpuscularia angustipetala (Lavis) H. E. K. Hartmann;窄瓣白绒玉●☆

105374 Corpuscularia britteniae (L. Bolus) H. E. K. Hartmann;布里滕白绒玉●☆

105375 Corpuscularia buchubergense Dinter et Schwantes ex Range = Antimima buchubergensis (Dinter) H. E. K. Hartmann ■☆

105376 Corpuscularia cymbiformis (Haw.) Schwantes;船状白绒玉●☆

105377 Corpuscularia dolomitica (Dinter) Schwantes = Antimima dolomitica (Dinter) H. E. K. Hartmann ■☆

105378 Corpuscularia gracilis H. E. K. Hartmann;纤细白绒玉●☆

105379 Corpuscularia lehmannii (Eckl. et Zeyh.) Schwantes;莱曼白绒玉●☆

105380 Corpuscularia molle (Aiton) Schwantes = Malephora mollis (Aiton) N. E. Br. ■☆

105381 Corpuscularia perdiantha Tischer = Nelia schlechteri Schwantes ■☆

105382 Corpuscularia quartzitica (Dinter) Schwantes = Antimima quarzitica (Dinter) H. E. K. Hartmann ■☆

105383 Corpuscularia taylori (N. E. Br.) Schwantes;泰勒白绒玉●☆

105384 Corpuscularia thunbergii (Haw.) Schwantes = Malephora thunbergii (Haw.) Schwantes ■☆

105385 Corpusoularia Schwantes = Delosperma N. E. Br. ●☆

105386 Corraea Sm. = Correa Andréws(保留属名)●☆

105387 Correa Andréws(1798)(保留属名);考来木属(澳吊钟属);Correa,Australian Fuchsia ●☆

105388 Correa Becerra = Dialium L. ●☆

105389 Correa aemula (Lindl.) F. Muell.;毛考来木;Hairy Correa ●☆

105390 Correa alba Andréws;白花考来木●☆

105391 Correa backhousiana Hook.;密枝考来木(紫花澳吊钟)●☆

105392 Correa baeuerlenii F. Muell.;厨师帽考来木;Chef's Cap Correa ●☆

105393 Correa lawrenciana Hook.;树考来木;Mountain Correa,Tree Correa ●☆

105394 Correa pulchella Lindl.;美丽考来木(光叶澳吊钟);Australian Fuchsia ●☆

105395 Correa reflexa (Labill.) Vent.;多变考来木(丽花澳吊钟);Native Fuchsia ●☆

105396 Correa schlechtendalii Behr;灰叶考来木●☆

105397 Correaceae J. Agardh = Rutaceae Juss.(保留科名)●■

105398 Correaea T. Post et Kuntze = Ouratea Aubl.(保留属名)●

105399 Correas Hoffmanns. = Correa Andréws(保留属名)●☆

105400 Correia Vand.(废弃属名) = Correa Andréws(保留属名)●☆

105401 Correia Vand.(废弃属名) = Ouratea Aubl.(保留属名)●

105402 Correia Vell.(废弃属名) = Correa Andréws(保留属名)●☆

105403 Correia Vell.(废弃属名) = Ouratea Aubl.(保留属名)●

105404 Correllia A. M. Powell(1973);科雷尔菊属☆
105405 Correllia montana A. M. Powell;科雷尔菊☆
105406 Correlliana D'Arcy = Cybianthus Mart.(保留属名)●☆
105407 Correorchis Szlach.(2008);智利柱穗兰属■☆
105408 Correorchis Szlach. = Chloraea Lindl. ■☆
105409 Corrigiola Kuntze = Illecebrum L. ■☆
105410 Corrigiola L.(1753);互叶指甲草属;Corrigiole,Knotgrass,Strapwort ■☆
105411 Corrigiola albella Forssk. = Paronychia arabica (L.) DC. ■☆
105412 Corrigiola barotsensis Wild = Corrigiola paniculata Peter ■☆
105413 Corrigiola capensis Willd.;好望角互叶指甲草■☆
105414 Corrigiola capensis Willd. subsp. africana (Turrill) Chaudhri;非洲互叶指甲草■☆
105415 Corrigiola drymarioides Baker f.;荷莲豆状互叶指甲草■☆
105416 Corrigiola litoralis L.;互叶指甲草;Knot-grass,Strapwort ■☆
105417 Corrigiola litoralis L. f. typica Graebn. = Corrigiola litoralis L. ■☆
105418 Corrigiola litoralis L. subsp. africana Turrill = Corrigiola capensis Willd. subsp. africana (Turrill) Chaudhri ■☆
105419 Corrigiola litoralis L. subsp. foliosa (Pérez Lara) Devesa = Corrigiola litoralis L. subsp. perez-larae Chaudhri et al. ■☆
105420 Corrigiola litoralis L. subsp. perez-larae Chaudhri et al.;佩雷互叶指甲草■☆
105421 Corrigiola litoralis L. subsp. telephiifolia (Pourr.) Briq. = Corrigiola telephiifolia Pourr. ■☆
105422 Corrigiola litoralis L. var. perennans Chaudhri;多年指甲草■☆
105423 Corrigiola litoralis L. var. purpurascens Giraudias = Corrigiola litoralis L. ■☆
105424 Corrigiola paniculata Peter;圆锥互叶指甲草■☆
105425 Corrigiola psammatrophoides Baker = Corrigiola litoralis L. ■☆
105426 Corrigiola repens Forssk. = Polycarpaea repens (Forssk.) Asch. et Schweinf. ■☆
105427 Corrigiola russelliana A. Chev. = Corrigiola litoralis L. ■☆
105428 Corrigiola telephiifolia Pourr.;疣叶指甲草■☆
105429 Corrigiola telephiifolia Pourr. subsp. paronychioides Emb. = Corrigiola telephiifolia Pourr. ■☆
105430 Corrigiola telephiifolia Pourr. var. paronychioides (Emb.) Maire = Corrigiola telephiifolia Pourr. ■☆
105431 Corrigiolaceae (Dumort.) Dumort. = Caryophyllaceae Juss.(保留科名)■●
105432 Corrigiolaceae Dumort. = Caryophyllaceae Juss.(保留科名)■●
105433 Corroea Paxton = Correa Andréws(保留属名)●☆
105434 Corryocactus Britton et Rose(1920);恐龙角属●☆
105435 Corryocactus aureus (F. A. C. Weber) Hutchison;金黄恐龙角●☆
105436 Corryocactus brachypetalus Britton et Rose;恐龙阁●☆
105437 Corryocactus brevistylus Britton et Rose;新绿阁;Guacalla ●☆
105438 Corryocactus melanotrichus (K. Schum.) Britton et Rose;百万灯●☆
105439 Corryocactus pulquinensis Cardenas;破天荒●☆
105440 Corryocactus tarijensis Cárdenas;观音柱●☆
105441 Corryocactus tenuiculus (Backeb.) Hutchison;细恐龙角●☆
105442 Corryocereus Frič et Kreuz. = Corryocactus Britton et Rose ●☆
105443 Corsia Becc.(1878);腐蛛草属■☆
105444 Corsia acuminata L. O. Williams;渐尖腐蛛草■☆
105445 Corsia cordata Schltr.;心形腐蛛草■☆
105446 Corsia purpurata L. O. Williams;紫腐蛛草■☆
105447 Corsiaceae Becc.(1878)(保留科名);腐蛛草科(白玉簪科,美丽腐草科,美丽腐生草科)■
105448 Corsiopsis D. X. Zhang,R. M. K. Saunders et C. M. Hu(1999);类腐蛛草属(白玉簪属);Corsiopsis ■★
105449 Corsiopsis chinensis D. X. Zhang,R. M. K. Saunders et C. M. Hu;类腐蛛草(白玉簪);Corsiopsis ■
105450 Cortaderia Stapf(1897)(保留属名);蒲苇属(银芦属);Pampas Grass ■
105451 Cortaderia argentea Stapf = Cortaderia selloana (Schult. et Schult. f.) Asch. et Graebn. ■
105452 Cortaderia dioeca (Speg.) Speg. = Cortaderia selloana (Schult. et Schult. f.) Asch. et Graebn. ■
105453 Cortaderia jubata (Lemoine ex Carrière) Stapf;紫蒲苇;Purple Pampas Grass ■☆
105454 Cortaderia jubata (Lemoine) Stapf = Cortaderia jubata (Lemoine ex Carrière) Stapf ■☆
105455 Cortaderia quila Stapf;奎拉蒲苇■☆
105456 Cortaderia selloana (Schult. et Schult. f.) Asch. et Graebn.;蒲苇(白金芦,白银芦,德国蒲苇,乌拉圭蒲苇,银芦);Pampas Grass,Pampas-grass,Sellon Pampas Grass,Uruguayan Pampas Grass,White Pampas Grass ■
105457 Cortaderia selloana (Schult. et Schult. f.) Asch. et Graebn. = Gynerium argenteum Nees ■☆
105458 Cortaderia selloana Asch. et Graebn. = Cortaderia selloana (Schult. et Schult. f.) Asch. et Graebn. ■
105459 Cortaderia sellosa (Schult. et Schult. f.) Asch. et Graebn. 'Aureolineata';金边蒲苇(金边银芦)■☆
105460 Cortaderia sellosa (Schult. et Schult. f.) Asch. et Graebn. 'Silver Comet';白边蒲苇(白边银芦)■☆
105461 Cortaderia sellosa (Schult. et Schult. f.) Asch. et Graebn. 'Sunningdale Silver';乳白穗蒲苇(乳白穗银芦)■☆
105462 Cortaderia sellosa (Schult. et Schult. f.) Asch. et Graebn. Gold Band = Cortaderia sellosa (Schult. et Schult. f.) Asch. et Graebn. 'Aureolineata' ■☆
105463 Cortesia Cav.(1798);澳大利亚紫草属;Australian Fuchsia ■☆
105464 Cortesia cuneifolia Cav.;澳大利亚紫草■☆
105465 Corthumia Rchb. = Pelargonium L'Hér. ex Aiton ●■
105466 Corthusa Rchb. = Cortusa L. ■
105467 Cortia DC.(1830);喜峰芹属(郑栓果芹属);Lovebeecelery ■
105468 Cortia candollii (DC.) Leute = Selinum candollei DC. ■
105469 Cortia depressa (D. Don) C. Norman;喜峰芹;Lovebeecelery ■
105470 Cortia elata Edgew. = Ligusticum elatum (Edgew.) C. B. Clarke ■
105471 Cortia hookeri C. B. Clarke = Cortiella hookeri (C. B. Clarke) C. Norman ■
105472 Cortia lindleyi DC. = Cortia depressa (D. Don) C. Norman ■
105473 Cortia nepalensis C. Norman = Cortia depressa (D. Don) C. Norman ■
105474 Cortia oreomyrrhiformis Farille et S. B. Malla = Cortia depressa (D. Don) C. Norman ■
105475 Cortia papyraceum (C. B. Clarke) Leute = Selinum papyraceum C. B. Clarke ■☆
105476 Cortia schmidii Nasir;巴基斯坦喜峰芹■☆
105477 Cortia striata (DC.) Leute = Ligusticum striatum Wall. ex DC. ■
105478 Cortia vaginata Edgew. = Selinum vaginatum (Edgew.) C. B. Clarke ■☆
105479 Cortia wallichiana (DC.) Leute = Selinum candollei DC. ■
105480 Cortia wallichiana (DC.) Leute = Selinum wallichianum (DC.) Raizada et H. O. Saxena ■

105481 Cortiella C. Norman(1937);栓果芹属;Cortiella ■★

105482 Cortiella caespitosa R. H. Shan et M. L. Sheh;宽叶栓果芹;Broadleaf Cortiella ■

105483 Cortiella cauwetmarciana Farille et S. B. Malla = Cortiella hookeri (C. B. Clarke) C. Norman ■

105484 Cortiella cortioides (C. Norman) M. F. Watson;锡金栓果芹■

105485 Cortiella glacialis Bonner = Cortiella hookeri (C. B. Clarke) C. Norman ■

105486 Cortiella hedinii (Diels) C. Norman = Pleurospermum hedinii Diels ■

105487 Cortiella hookeri (C. B. Clarke) C. Norman;栓果芹;Hooker Cortiella ■

105488 Cortusa L. (1753);假报春属(假报春花属);Bear's-ear Sanicle, Cortusa ■

105489 Cortusa altaica Losinsk.;阿尔泰假报春;Altai Cortusa ■

105490 Cortusa amurensis Fed.;阿穆尔假报春■☆

105491 Cortusa brotheri Pax ex Lipsky;布拉泽假报春■☆

105492 Cortusa gmelinii Gaertn. = Androsace gmelinii (Gaertn.) Roem. et Schult. ■

105493 Cortusa gmelinii L. = Androsace gmelinii (Gaertn.) Roem. et Schult. ■

105494 Cortusa himalaica Losinsk. = Cortusa brotheri Pax ex Lipsky ■☆

105495 Cortusa matthioli L. f. brotheri (Pax ex Lipsky) R. Knuth = Cortusa brotheri Pax ex Lipsky ■☆

105496 Cortusa matthiolii L.;假报春(马氏假报春);Matthiol Cortusa, Mountain Sanicle ■

105497 Cortusa matthiolii L. f. pekingensis V. A. Rich. = Cortusa matthiolii L. subsp. pekingensis (V. A. Rich.) Kitag. ■

105498 Cortusa matthiolii L. subsp. pekinensis (V. A. Rich.) Kitag. var. jozana (Miyabe et Tatew.) H. Hara ex Ohwi = Cortusa matthiolii L. subsp. pekinensis (V. A. Rich.) Kitag. var. sachalinensis (Losinsk.) T. Yamaz. ■☆

105499 Cortusa matthiolii L. subsp. pekinensis (V. A. Rich.) Kitag. var. sachalinensis (Losinsk.) T. Yamaz.;库页假报春■☆

105500 Cortusa matthiolii L. subsp. pekinensis (V. A. Rich.) Kitag. var. yezoensis Miyabe et Tatew. = Cortusa matthiolii L. subsp. pekinensis (V. A. Rich.) Kitag. var. sachalinensis (Losinsk.) T. Yamaz. ■☆

105501 Cortusa matthiolii L. subsp. pekingensis (V. A. Rich.) Kitag.;河北假报春(北京假报春,假报春,京报春);Beijing Cortusa, Hebei Cortusa, Peking Cortusa ■

105502 Cortusa matthiolii L. var. yezoensis (Miyabe et Tatew.) H. Hara = Cortusa matthiolii L. subsp. pekinensis (V. A. Rich.) Kitag. var. sachalinensis (Losinsk.) T. Yamaz. ■☆

105503 Cortusa pekinensis (A. Rich.) Losinsk. = Cortusa matthiolii L. subsp. pekingensis (V. A. Rich.) Kitag. ■

105504 Cortusa sachalinensis Losinsk. = Cortusa matthiolii L. subsp. pekinensis (V. A. Rich.) Kitag. var. sachalinensis (Losinsk.) T. Yamaz. ■☆

105505 Cortusa sibirica Andrz. ex Besser;西伯利亚假报春■☆

105506 Cortusa stenocalyx Maxim.;狭萼假报春■☆

105507 Cortusa turkestanica Losinsk.;土耳其斯坦假报春■☆

105508 Cortusina Eckl. et Zeyh. = Pelargonium L'Hér. ex Aiton ●■

105509 Corunastylis Fitzg. = Anticheirostylis Fitzg. ■☆

105510 Corunastylis Fitzg. = Genoplesium R. Br. ■☆

105511 Corunostylis T. Post et Kuntze = Corunastylis Fitzg. ■☆

105512 Corvina B. D. Jacks. = Muricaria Desv. ■☆

105513 Corvina Steud. = Muricaria Desv. ■☆

105514 Corvinia Stadtm. ex Willem. = Litchi Sonn. ●

105515 Corvisartia Mérat = Inula L. ●■

105516 Corvisartia helenium Mérat = Inula helenium L. ■

105517 Corya Raf. = Carya Nutt. (保留属名)●

105518 Coryanthes Hook. (1831);吊桶兰属(科瑞安兰属,盔兰属,帽花兰属,头盔兰属);Bucket Orchid, Helmet Flower ■☆

105519 Coryanthes Lam. = Coryanthes Hook. ■☆

105520 Coryanthes Schltr. = Coryanthes Hook. ■☆

105521 Coryanthes macrantha (Hook.) Hook.;大花吊桶兰;Large-flower Bucket Orchid ■☆

105522 Coryanthes maculata Hook.;吊桶兰;Common Bucket Orchid ■☆

105523 Coryanthes speciosa Hook.;美丽吊桶兰;Beautiful Coryanthes ■☆

105524 Corybas Salisb. (1807);铠兰属(盔兰属);Corybas, Helmet Orchid ■

105525 Corybas fanjingshanensis Y. X. Xiong;梵净山铠兰■

105526 Corybas himalaicus (King et Pantl.) Schltr.;杉林溪铠兰■

105527 Corybas macranthus (Hook. f.) Rchb. f.;大花铠兰■☆

105528 Corybas pruinosums (A. Cunn.) Rchb. f.;澳洲铠兰;Australian Helmet Orchid ■☆

105529 Corybas purpureus J. Joseph et Yogan. = Corybas himalaicus (King et Pantl.) Schltr. ■

105530 Corybas shanlinshiensis W. M. Lin, T. C. Hsu et T. P. Lin = Corybas himalaicus (King et Pantl.) Schltr. ■

105531 Corybas sinii Ts. Tang et F. T. Wang;铠兰(辛氏铠兰,辛氏盔兰);Common Corybas ■

105532 Corybas taiwanensis T. P. Lin et S. Y. Leu;台湾铠兰(红盔兰);Taiwan Corybas, Taiwan Helmet Orchid ■

105533 Corybas taliensis Ts. Tang et F. T. Wang;大理铠兰;Dali Corybas, Dali Helmet Orchid, Tali Helmet Orchid ■

105534 Corycarpus Spreng. = Diarrhena P. Beauv. (保留属名)■

105535 Corycarpus Zea ex Spreng. = Diarrhena P. Beauv. (保留属名)■

105536 Corycium Sw. (1800);蜜兰属■☆

105537 Corycium alticola Parkman et Schelpe;高原蜜兰■☆

105538 Corycium bicolorum (Thunb.) Sw.;二色蜜兰■☆

105539 Corycium bifidum Sond.;二裂蜜兰■☆

105540 Corycium carnosum (Lindl.) Rolfe;肉质蜜兰■☆

105541 Corycium crispum (Thunb.) Sw.;皱波蜜兰■☆

105542 Corycium deflexum (Bolus) Rolfe;外折蜜兰■☆

105543 Corycium dracomontanum Parkman et Schelpe;德拉科蜜兰■☆

105544 Corycium excisum Lindl.;缺刻蜜兰■☆

105545 Corycium flanaganii (Bolus) Kurzweil et H. P. Linder;弗拉纳根蜜兰■☆

105546 Corycium ligulatum Rchb. f. = Corycium bifidum Sond. ■☆

105547 Corycium magnum (Rchb. f.) Rolfe = Pterygodium magnum Rchb. f. ■☆

105548 Corycium microglossum Lindl.;小舌蜜兰■☆

105549 Corycium nigrescens Sond.;黑蜜兰■☆

105550 Corycium orobanchoides (L. f.) Sw.;列当蜜兰■☆

105551 Corycium rubiginosum (Sond. ex Bolus) Rolfe = Evotella rubiginosa (Sond. ex Bolus) Kurzweil et H. P. Linder ■☆

105552 Corycium tricuspidatum Bolus;三尖蜜兰■☆

105553 Corycium venosum (Lindl.) Rolfe = Ceratandra venosa (Lindl.) Schltr. ■☆

105554 Corycium vestitum Sw. = Corycium orobanchoides (L. f.) Sw. ■☆

105555 Coryda B. D. Jacks. = Cordyla Lour. ●☆

105556 Corydalaceae Vest = Fumariaceae Marquis(保留科名)■☆
105557 Corydalis DC. (1805)(保留属名);紫堇属(延胡索属);Corydalis, Rock Harlequin, Yanhusuo ■
105558 Corydalis Medik.(废弃属名)= Corydalis DC.(保留属名)■
105559 Corydalis Medik.(废弃属名)= Cysticapnos Mill.(废弃属名)■
105560 Corydalis Vent. = Capnoides Mill.(废弃属名)■
105561 Corydalis Vent. = Corydalis DC.(保留属名)■
105562 Corydalis acropteryx Fedde;顶冠黄堇(假顶冠黄堇,松潘黄堇);Roofwing Corydalis ■
105563 Corydalis acuminata Franch.;川东紫堇(地丁,苦地丁);Acuminate Corydalis ■
105564 Corydalis acuminata Franch. subsp. hupehensis C. Y. Wu;湖北紫堇;Hubei Corydalis ■
105565 Corydalis acuminata Franch. subsp. hupehensis C. Y. Wu = Corydalis acuminata Franch. ■
105566 Corydalis adiantifolia Hook. f. et Thomson;铁线蕨叶黄堇;Adiantumleaf Corydalis ■
105567 Corydalis adoxifolia C. Y. Wu;东义紫堇;Muskrootleaf Corydalis ■
105568 Corydalis adrienii Prain;美丽紫堇(美紫堇);Pretty Corydalis ■
105569 Corydalis adrienii Prain = Corydalis melanochlora Maxim. ■
105570 Corydalis adrienii Prain var. forrestii Fedde = Corydalis adrienii Prain ■
105571 Corydalis adrienii Prain var. forrestii Fedde = Corydalis melanochlora Maxim. ■
105572 Corydalis adunca Maxim.;灰绿黄堇(早生紫堇,黄草花,入夏蒿);Greygreen Corydalis, Greyishgreen Corydalis ■
105573 Corydalis adunca Maxim. subsp. microsperma Lidén et Z. Y. Su;滇西灰绿黄堇;Smallseed Greygreen Corydalis ■
105574 Corydalis adunca Maxim. subsp. microsperma Lidén et Z. Y. Su = Corydalis adunca Maxim. ■
105575 Corydalis adunca Maxim. subsp. scaphopetala (Fedde) C. Y. Wu et Z. Y. Su;帚枝灰绿黄堇■
105576 Corydalis adunca Maxim. subsp. scaphopetala (Fedde) C. Y. Wu et Z. Y. Su = Corydalis adunca Maxim. ■
105577 Corydalis adunca Maxim. var. humilis Maxim. = Corydalis adunca Maxim. ■
105578 Corydalis aeaeae X. F. Gao;艳巫岛紫堇■
105579 Corydalis aeditua Lidén et Z. Y. Su;湿崖紫堇■
105580 Corydalis aegopodioides H. Lév. et Vaniot = Corydalis temulifolia Franch. subsp. aegopodioides (H. Lév. et Vaniot) C. Y. Wu ■
105581 Corydalis aitchisonii Popov;埃奇紫堇■☆
105582 Corydalis alaschanica (Maxim.) Peshkova;贺兰山延胡索(贺兰山稀花紫堇);Helanshan Corydalis ■
105583 Corydalis albicaulis Franch. = Corydalis adunca Maxim. ■
105584 Corydalis albicaulis Franch. var. latiloba Franch. = Corydalis latiloba (Franch.) Hand. -Mazz. ■
105585 Corydalis alburyi Ludlow = Corydalis latiflora Hook. f. et Thomson subsp. gerdae (Fedde) Lidén ex C. Y. Wu, H. Chuang et Z. Y. Su ■
105586 Corydalis alburyi Ludlow et Stearn = Corydalis latiflora Hook. f. et Thomson ■
105587 Corydalis alexeenkoana N. Busch;阿来紫堇■☆
105588 Corydalis alpestris C. A. Mey.;高山延胡索(少花延胡索)■
105589 Corydalis alpestris C. A. Mey. var. bayeriana (Rupr.) Popov = Corydalis tianzhuensis M. S. Yan et Ching J. Wang ■
105590 Corydalis alpigena C. Y. Wu et H. Chuang = Corydalis trachycarpa Maxim. var. leucostachya (C. Y. Wu et H. Chuang) C. Y. Wu ■
105591 Corydalis altaica (Ledeb.) Besser = Bromus japonicus Thunb. ■
105592 Corydalis altaica (Ledeb.) Besser = Corydalis pauciflora (Stephan ex Willd.) Pers. ■
105593 Corydalis amabilis Migo = Corydalis decumbens (Thunb.) Pers. ■
105594 Corydalis ambigua Cham. et Schltdl.;迷延胡索(滴金卵,东北延胡索,虾夷延胡索,玄胡,玄胡索,延胡索,元胡);Amur Corydalis, Northeastern Corydalis ■
105595 Corydalis ambigua Cham. et Schltdl. = Corydalis yanhusuo W. T. Wang ex Z. Y. Su et C. Y. Wu ■
105596 Corydalis ambigua Cham. et Schltdl. f. albiflora Tatew.;白花迷延胡索■☆
105597 Corydalis ambigua Cham. et Schltdl. f. dentata Y. H. Chou = Corydalis fumariifolia Maxim. ■
105598 Corydalis ambigua Cham. et Schltdl. f. fumariifolia (Maxim.) Kitag. = Corydalis fumariifolia Maxim. ■
105599 Corydalis ambigua Cham. et Schltdl. f. lineariloba Maxim.;线裂迷延胡索(线裂东北延胡索);Linearilobe Corydalis ■
105600 Corydalis ambigua Cham. et Schltdl. f. multifida Y. H. Chou = Corydalis fumariifolia Maxim. ■
105601 Corydalis ambigua Cham. et Schltdl. f. pectinata Kom.;栉裂迷延胡索(栉齿东北延胡索,栉裂东北延胡索);Pectinate Corydalis ■
105602 Corydalis ambigua Cham. et Schltdl. f. rotundiloba Maxim.;圆裂迷延胡索(圆裂东北延胡索);Rotundlobe Corydalis ■
105603 Corydalis ambigua Cham. et Schltdl. f. rotundiloba Maxim. = Corydalis fumariifolia Maxim. ■
105604 Corydalis ambigua Cham. et Schltdl. var. amurensis Maxim. = Corydalis repens Mandl et Muehld. ■
105605 Corydalis ambigua Cham. et Schltdl. var. amurensis Maxim. = Corydalis caudata (Lam.) Pers. ■
105606 Corydalis ambigua Cham. et Schltdl. var. amurensis Maxim. = Corydalis fumariifolia Maxim. ■
105607 Corydalis ambigua Cham. et Schltdl. var. angustifolia Yatabe;狭叶迷延胡索■☆
105608 Corydalis ambigua Cham. et Schltdl. var. glabra Takeda;光秃迷延胡索(光秃东北延胡索);Glabrous Northeastern Corydalis ■
105609 Corydalis ambigua Cham. et Schltdl. var. lineariloba Maxim. = Corydalis ambigua Cham. et Schltdl. f. lineariloba Maxim. ■
105610 Corydalis ampelos Lidén et Z. Y. Su;攀缘黄堇■
105611 Corydalis amphipogon Lidén;文县紫堇■
105612 Corydalis amplisepala Z. Y. Su et Lidén;圆萼紫堇■
105613 Corydalis anaginova Lidén et Z. Y. Su;藏中黄堇;Central Xizang Corydalis ■
105614 Corydalis ananke Lidén;齿瓣紫堇■
105615 Corydalis anethifolia C. Y. Wu et Z. Y. Su;莳萝叶紫堇(窄叶紫堇);Dillleaf Corydalis ■
105616 Corydalis angusta Z. Y. Su et Lidén;细距紫堇■
105617 Corydalis angustiflora C. Y. Wu;狭花紫堇;Narrowflower Corydalis ■
105618 Corydalis angustifolia (M. Bieb.) DC.;狭叶紫堇■☆
105619 Corydalis anthocrene Lidén et J. Van de Veire;泉涌花紫堇■
105620 Corydalis anthriscifolia Franch.;峨参叶紫堇;Cherviileaf Corydalis ■
105621 Corydalis apletonii Hutch.;念果紫堇■
105622 Corydalis appendiculata Hand. -Mazz.;小距紫堇(小草乌,雪

山一枝蒿);Apependiculate Corydalis,Littlespur Corydalis ■
105623 Corydalis aquilegioides Z. Y. Su;假耧斗菜紫堇(假耧斗菜);False Columbine Corydalis ■
105624 Corydalis arctica Popov;北极紫堇■☆
105625 Corydalis arctica Popov =Corydalis pauciflora (Stephan) Pers. ■
105626 Corydalis aspleniifolia Lidén et Z. Y. Su =Corydalis ternatifolia C. Y. Wu,Z. Y. Su et Lidén ■
105627 Corydalis asterostigma H. Lév.;贵州紫堇;Guizhou Corydalis ■
105628 Corydalis asterostigma H. Lév. =Corydalis duclouxii H. Lév. et Vaniot ■
105629 Corydalis astragalina Hook. f. et Thomson =Corydalis stricta Steph. ex DC. ■
105630 Corydalis atuntsuensis W. W. Sm.;阿墩紫堇(刺毛黄堇,粗毛黄堇);Adun Corydalis ■
105631 Corydalis aurea Willd.;北美延胡索;Golden Corydalis,Scrambled Eggs ■☆
105632 Corydalis aurea Willd. subsp. occidentalis (Engelm. ex A. Gray) G. B. Ownbey =Corydalis curvisiliqua (A. Gray) A. Gray ■☆
105633 Corydalis aurea Willd. var. australis Chapm. =Corydalis micrantha (Engelm. ex A. Gray) A. Gray subsp. australis (Chapm.) G. B. Ownbey ■☆
105634 Corydalis aurea Willd. var. curvisiliqua A. Gray =Corydalis curvisiliqua (A. Gray) A. Gray ■☆
105635 Corydalis aurea Willd. var. micrantha Engelm. ex A. Gray =Corydalis micrantha (Engelm. ex A. Gray) A. Gray ■☆
105636 Corydalis aurea Willd. var. speciosa (Maxim.) Regel =Corydalis speciosa Maxim. ■
105637 Corydalis auricilla Lidén et Z. Y. Su;高黎贡山黄堇■
105638 Corydalis auriculata Lidén et Z. Y. Su;耳柄紫堇;Earstalk Corydalis ■
105639 Corydalis balansae Prain;北越紫堇(臭草,黄连,鸡屎草,台湾黄堇);Acorn Corydalis,Taiwan Corydalis ■
105640 Corydalis balfouriana Diels;直梗紫堇(苍山紫堇,天葵叶紫堇);Balfour Corydalis ■
105641 Corydalis balfouriana Diels =Corydalis oxypetala Franch. subsp. balfouriana (Diels) Lidén ■
105642 Corydalis balfouriana Diels =Corydalis pseudoadoxa C. Y. Wu et H. Chuang ■
105643 Corydalis balfouriana Diels var. pseudoadoxa C. Y. Wu et H. Chuang =Corydalis pseudoadoxa C. Y. Wu et H. Chuang ■
105644 Corydalis balfouriana Diels var. pseudoboxa C. Y. Wu et H. Chuang;藏天葵叶紫堇■
105645 Corydalis balsamiflora Prain =Corydalis flexuosa Franch. subsp. balsamiflora (Prain) C. Y. Wu ■
105646 Corydalis barbisepala Hand. -Mazz. et Fedde;髯萼紫堇(髯萼黄堇);Beardsepal Corydalis ■
105647 Corydalis benecincta W. W. Sm.;囊距紫堇(美国紫堇,紫堇);Bagspur Corydalis,Succatespur Corydalis ■
105648 Corydalis benecincta W. W. Sm. subsp. trilobipetala (Hand. -Mazz.) Lidén =Corydalis trilobipetala Hand. -Mazz. ■
105649 Corydalis bibracteolata Z. Y. Su;梗苞黄堇;Stipebract Corydalis ■
105650 Corydalis bijiangensis C. Y. Wu et H. Chuang;碧江黄堇;Bijiang Corydalis ■
105651 Corydalis bimaculata C. Y. Wu et T. Y. Shu;双斑黄堇;Bimaculate Corydalis,Twinspot Corydalis ■
105652 Corydalis binderae Fedde =Corydalis melanochlora Maxim. ■
105653 Corydalis binderae Fedde =Corydalis pseudohamata Fedde ■
105654 Corydalis binderae Fedde subsp. pseudohamata (Fedde) Z. Y. Su =Corydalis hamata Franch. ■
105655 Corydalis binderae Fedde subsp. pseudohamata (Fedde) Z. Y. Su =Corydalis pseudohamata Fedde ■
105656 Corydalis bokuensis L. H. Zhou =Corydalis tianzhuensis M. S. Yan et Ching J. Wang ■
105657 Corydalis borii C. E. C. Fisch.;那加黄堇;Bor Corydalis ■
105658 Corydalis boweri Hemsl.;金球黄堇(东丝儿,东丝勒,黄花紫堇,金球紫堇,圆锥黄堇);Goldball Corydalis,Golden Sphaerical Corydalis ■
105659 Corydalis boweri Hemsl. =Corydalis mucronifera Maxim. ■
105660 Corydalis bowes-lyonii D. G. Long =Corydalis crispa Prain ■
105661 Corydalis brachyceras Lidén et J. Van de Veire;江达紫堇■
105662 Corydalis brachystyla H. Koidz. =Corydalis orthopoda Hayata ■
105663 Corydalis bracteata (Stephan ex Willd.) Pers.;苞叶延胡索(对叶延胡索,栉苞延胡索);Bracteate Corydalis ■
105664 Corydalis brandegei S. Watson =Corydalis caseana A. Gray subsp. brandegei (S. Watson) G. B. Ownbey ■☆
105665 Corydalis brevipedunculata (Z. Y. Su) Z. Y. Su et Lidén;短轴黄堇(短轴臭黄堇);Shortpeduncle Corydalis ■
105666 Corydalis brevirostrata C. Y. Wu et Z. Y. Su;蔓生黄堇(短喙黄堇);Shotbeak Corydalis,Vine Corydalis ■
105667 Corydalis brevirostrata C. Y. Wu et Z. Y. Su subsp. tibetica (Maxim.) Lidén;西藏蔓生黄堇■
105668 Corydalis brunneovaginata Fedde;褐鞘紫堇;Brownsheath Corydalis ■
105669 Corydalis bucharica Popov;布哈尔紫堇■☆
105670 Corydalis bulbifera C. Y. Wu;鳞叶紫堇;Bulbiferous Corydalis,Scaleleaf Corydalis ■
105671 Corydalis bulbilligera C. Y. Wu;巫溪紫堇;Wuxi Corydalis ■
105672 Corydalis bulbosa (L.) DC. =Corydalis solida (L.) Clairv. ■
105673 Corydalis bulbosa (L.) DC. =Corydalis yanhusuo W. T. Wang ex Z. Y. Su et C. Y. Wu ■
105674 Corydalis bulbosa (L.) DC. f. ternata Nakai =Corydalis ternata (Nakai) Nakai ■
105675 Corydalis bulbosa (L.) DC. var. remota (Fisch. ex Maxim.) Nakai =Corydalis turtschaninovii Besser ■
105676 Corydalis bulbosa (L.) DC. var. remota (Maxim.) Nakai =Corydalis turtschaninovii Besser ■
105677 Corydalis bulbosa (L.) DC. var. remota (Maxim.) Nakai f. ternata Nakai =Corydalis ternata (Nakai) Nakai ■
105678 Corydalis bulleyana Diels;齿冠金钩如意草(齿冠紫堇,滇西紫堇);Bulley Corydalis,Bulley Dali Corydalis,Bulley Tali Corydalis ■
105679 Corydalis bulleyana Diels subsp. muliensis Lidén et Z. Y. Su;木里齿冠紫堇(木里滇西紫堇);Muli Corydalis ■
105680 Corydalis bungeana Turcz.;地丁草(布氏紫堇,地丁,地丁紫堇,苦地丁,苦丁,彭氏紫堇,小鸡菜,紫花地丁,紫堇,紫堇地丁);Bunge Corydalis ■
105681 Corydalis bungeana Turcz. var. odontopetala Hemsl. =Corydalis bungeana Turcz. ■
105682 Corydalis buschii Nakai;东紫堇;Busch Corydalis ■
105683 Corydalis cabulica Gilli =Corydalis ledebouriana Kar. et Kir. ■
105684 Corydalis cachemiriana Royle =Corydalis pachycentra Franch. ■
105685 Corydalis caespitosa C. Y. Wu;丛生黄堇;Clustered Corydalis ■
105686 Corydalis caespitosa C. Y. Wu =Corydalis feddeana H. Lév. et Fedde ■
105687 Corydalis calcicola W. W. Sm.;灰岩紫堇(丽江马尾黄连);

Calcicolous Corydalis, Limestone Corydalis ■
105688 Corydalis calcicola W. W. Sm. var. szechuanica Fedde;四川灰岩紫堇;Sichuan Corydalis, Szechuan Corydalis ■
105689 Corydalis calcicola W. W. Sm. var. szechuanica Fedde = Corydalis trachycarpa Maxim. ■
105690 Corydalis calycosa H. Chuang;显萼紫堇■
105691 Corydalis calycosa H. Chuang = Corydalis flexuosa Franch. subsp. gemmipara (H. Chuang) C. Y. Wu ■
105692 Corydalis campestris (Britton) J. Buchholz et E. J. Palmer = Corydalis micrantha (Engelm. ex A. Gray) A. Gray subsp. australis (Chapm.) G. B. Ownbey ■☆
105693 Corydalis campulicarpa Hayata;台东紫堇(弯果黄堇);Taidong Corydalis ■
105694 Corydalis campulicarpa Hayata = Corydalis ophiocarpa Hook. f. et Thomson ■
105695 Corydalis canadensis Goldie = Dicentra canadensis (Goldie) Walp. ■☆
105696 Corydalis capillaris (Makino) Takeda = Corydalis lineariloba Siebold et Zucc. var. capillaris (Makino) Ohwi ■☆
105697 Corydalis capillipes Franch. = Corydalis orthoceras Siebold et Zucc. ■☆
105698 Corydalis capitata X. F. Gao;头花紫堇■
105699 Corydalis capnoides (L.) Pers.;方茎黄堇(山黄堇,山紫堇,真堇);True Corydalis ■
105700 Corydalis capnoides (L.) Pers. var. tibetica Maxim. = Corydalis brevirostrata C. Y. Wu et Z. Y. Su ■
105701 Corydalis caput-medusae Z. Y. Su et Lidén;泸定紫堇■
105702 Corydalis carinata Lidén et Z. Y. Su;龙骨籽紫堇■
105703 Corydalis caseana A. Gray;凯斯紫堇;Case's Fireweed, Fitweed ■☆
105704 Corydalis caseana A. Gray subsp. brachycarpa (Rydb.) G. B. Ownbey;短果凯斯紫堇■☆
105705 Corydalis caseana A. Gray subsp. brandegei (S. Watson) G. B. Ownbey;布朗紫堇■☆
105706 Corydalis caseana A. Gray subsp. cusickii (S. Watson) G. B. Ownbey;库西克紫堇■☆
105707 Corydalis caseana A. Gray subsp. hastata (Rydb.) G. B. Ownbey;戟形凯斯紫堇■☆
105708 Corydalis caseana A. Gray var. cusickii (S. Watson) C. L. Hitchc. = Corydalis caseana A. Gray subsp. cusickii (S. Watson) G. B. Ownbey ■☆
105709 Corydalis caseana A. Gray var. hastata (Rydb.) C. L. Hitchc. = Corydalis caseana A. Gray subsp. hastata (Rydb.) G. B. Ownbey ■☆
105710 Corydalis cashmeriana Royle;克什米尔紫堇;Kashmir Corydalis ■
105711 Corydalis cashmeriana Royle subsp. brevicornu (Prain) D. G. Long = Corydalis jigmei C. E. C. Fisch. et Kaul ■
105712 Corydalis cashmeriana Royle subsp. longicalcarata (D. G. Long) Lidén;少花克什米尔紫堇■
105713 Corydalis cashmeriana Royle var. brevicornu Prain = Corydalis ecristata (Prain) D. G. Long ■
105714 Corydalis cashmeriana Royle var. brevicornu Prain = Corydalis pseudoadoxa C. Y. Wu et H. Chuang ■
105715 Corydalis cashmeriana Royle var. ecristata Prain = Corydalis ecristata (Prain) D. G. Long ■
105716 Corydalis casimiriana Duthie et Prain ex Prain;铺散黄堇(大花长梗黄堇); Bigflower Longstalk Corydalis, Diffuse Corydalis, Spreading Corydalis ■
105717 Corydalis casimiriana Duthie et Prain ex Prain var. meeboldii Fedde = Corydalis cornuta Royle ■
105718 Corydalis cataractarum Lidén;飞流紫堇■
105719 Corydalis caucasica DC.;高加索紫堇;Caucasia Corydalis ■☆
105720 Corydalis caudata (Lam.) Pers.;小药八旦子(北京元胡,匍匐延胡索,全叶土延胡,全叶延胡索,苏延胡,土元胡,元胡);Thinanther Corydalis ■
105721 Corydalis cava (L.) Schweigg. et Körte;凹陷紫堇(大块茎延胡索,瓶紫堇);Bulbous Corydalis, Hollow-root, Hollowroot Birthwort ■☆
105722 Corydalis cava Schweigg. = Corydalis cava (L.) Schweigg. et Körte ■☆
105723 Corydalis cavaleriei H. Lév. = Corydalis balansae Prain ■
105724 Corydalis cavei D. G. Long = Corydalis longipes DC. var. pubescens (C. Y. Wu et H. Chuang) C. Y. Wu ■
105725 Corydalis chamdoensis C. Y. Wu et H. Chuang;昌都紫堇;Changdu Corydalis ■
105726 Corydalis chanbaishanensis M. L. Zhang et Y. W. Wang;长白山黄堇;Changbaishan Corydalis ■
105727 Corydalis chanetii H. Lév. = Corydalis wilfordii Regel ■
105728 Corydalis chanetii H. Lév. et Fedde = Corydalis wilfordii Regel ■
105729 Corydalis changuensis D. G. Long;显囊黄堇■
105730 Corydalis cheilanthifolia Hemsl.;地柏枝(地白子,地黄连,雀雀菜,石菜子,碎米蕨叶黄堇);China Corydalis, Chinese Corydalis, Fern-leaved Corydalis ■
105731 Corydalis cheilosticta Z. Y. Su et Lidén;斑花紫堇■
105732 Corydalis cheilosticta Z. Y. Su et Lidén subsp. borealis Lidén et Z. Y. Su;北邻斑花紫堇■
105733 Corydalis cheirifolia Franch.;掌叶紫堇;Palmateleaf Corydalis, Palmleaf Corydalis ■
105734 Corydalis chelidoniifolia H. Lév. = Corydalis sheareri S. Moore ■
105735 Corydalis chelidonium Fedde;白屈菜紫堇;Celandine Corydalis ■
105736 Corydalis chelidonium Fedde = Corydalis stenantha Franch. ■
105737 Corydalis chinensis Franch. = Corydalis edulis Maxim. ■
105738 Corydalis chingii Fedde;甘肃紫堇;Gansu Corydalis, Kansu Corydalis ■
105739 Corydalis chingii Fedde var. shansiensis W. T. Wang ex C. Y. Wu et Z. Y. Su;大花甘肃紫堇;Bigflower Gansu Corydalis ■
105740 Corydalis chingii Fedde var. shansiensis W. T. Wang ex C. Y. Wu et Z. Y. Su = Corydalis chingii Fedde ■
105741 Corydalis chionophila Czerniak.;喜雪紫堇■☆
105742 Corydalis chosenensis Ohwi = Corydalis buschii Nakai ■
105743 Corydalis chrysosphaera C. Marquand et Airy Shaw = Corydalis boweri Hemsl. ■
105744 Corydalis clarkei Prain;卡那克黄堇(卡拉黄堇);C. B. Clarke Corydalis ■☆
105745 Corydalis claviculata DC.;小棒紫堇;Climbing Corydalis, Climbing Fumitory, Hen's Foot ■☆
105746 Corydalis claviculata DC. = Ceratocapnos claviculata (L.) Lidén ■☆
105747 Corydalis clematis H. Lév. = Corydalis davidii Franch. ■
105748 Corydalis clematis H. Lév. = Corydalis longicornu Franch. ■
105749 Corydalis clematis H. Lév. et Fedde;开阳黄堇;Clematis Corydalis ■
105750 Corydalis concinna C. Y. Wu et H. Chuang;优雅黄堇(大花黄堇);Elegant Corydalis ■
105751 Corydalis concinna C. Y. Wu et H. Chuang = Corydalis

pseudocristata Fedde ■
105752 Corydalis conorhiza Ledeh.;束根紫堇■☆
105753 Corydalis conspersa Maxim.;斑花黄堇(广布紫堇,密花黄堇,密花紫堇);Denseflower Corydalis,Spoptflower Corydalis ■
105754 Corydalis cornuta Royle;角状黄堇;Horn Corydalis ■
105755 Corydalis cornuta Royle var. meeboldii Fedde = Corydalis cornuta Royle ■
105756 Corydalis cornutior (C. Marquand et Airy Shaw) C. Y. Wu et T. Y. Su;长柄黄堇;Longstipe Corydalis ■
105757 Corydalis cornutior (C. Marquand et Airy Shaw) C. Y. Wu et Z. Y. Su = Corydalis dubia Prain ■
105758 Corydalis corymbosa C. Y. Wu et T. Y. Shu;伞花黄堇;Corymb Corydalis,Corymbose Corydalis ■
105759 Corydalis crassicalcarata C. Y. Wu et H. Chuang = Corydalis eugeniae Fedde ■
105760 Corydalis crassifolia Royle;厚叶紫堇■☆
105761 Corydalis crassirhizomata (C. Y. Wu) C. Y. Wu;粗颈紫堇;Thickroot Corydalis ■
105762 Corydalis crassirhizomata (C. Y. Wu) C. Y. Wu = Corydalis weigoldii Fedde ■
105763 Corydalis crispa Prain;皱波黄堇(齿冠皱波黄堇,多毛皱波黄堇,光棱皱波黄堇,无棱皱波黄堇,小锥花黄堇);Crisped Corydalis, False Thyres Corydalis, Manyhair Crisped Corydalis, Smoothangular Crisped Corydalis, Walton Crisped Corydalis, Wrinklewave Corydalis ■
105764 Corydalis crispa Prain subsp. laeviangula (C. Y. Wu et H. Chuang) Lidén et Z. Y. Su;光棱皱波黄堇■
105765 Corydalis crispa Prain var. laeviangula C. Y. Wu et H. Chuang = Corydalis crispa Prain subsp. laeviangula (C. Y. Wu et H. Chuang) Lidén et Z. Y. Su ■
105766 Corydalis crispa Prain var. laeviangula C. Y. Wu et H. Chuang = Corydalis crispa Prain ■
105767 Corydalis crispa Prain var. setulosa C. Y. Wu et H. Chuang = Corydalis crispa Prain ■
105768 Corydalis crispa Prain var. waltoni Fedde = Corydalis crispa Prain ■
105769 Corydalis crista-galli Maxim.;鸡冠黄堇;Comb Corydalis ■
105770 Corydalis cristata Maxim.;具冠黄堇(秦岭紫堇);Crested Corydalis ■
105771 Corydalis cristata Maxim. = Corydalis trisecta Franch. ■
105772 Corydalis cristata Maxim. var. pseudoflaccida Fedde = Corydalis dajingensis C. Y. Wu et Z. Y. Su ■
105773 Corydalis cristata Maxim. var. ramosa C. Y. Wu et H. Chuang;眉县具冠黄堇;Meixian Crested Corydalis ■
105774 Corydalis cristata Maxim. var. ramosa C. Y. Wu et H. Chuang = Corydalis trisecta Franch. ■
105775 Corydalis crystallina Engelm. ex A. Gray;北美紫堇;Mealy Corydalis ■☆
105776 Corydalis curvicalcarata Miyabe et Kudo;弯距紫堇■☆
105777 Corydalis curvicalcarata Miyabe et Kudo = Corydalis gigantea Trautv. et C. A. Mey. ■
105778 Corydalis curviflora Maxim. ex Hemsl.;曲花紫堇(螺样花,洛阳花,弯花紫堇);Curvedflower Corydalis,Curveflower Corydalis ■
105779 Corydalis curviflora Maxim. ex Hemsl. subsp. altecristata (C. Y. Wu et H. Chuang) C. Y. Wu. = Corydalis cytisiflora (Fedde) Lidén ex C. Y. Wu,H. Chuang et Z. Y. Su subsp. altecristata (C. Y. Wu et H. Chuang) Lidén ■
105780 Corydalis curviflora Maxim. ex Hemsl. subsp. alteocristata (C. Y. Wu et H. Chuang) C. Y. Wu;高冠曲花紫堇■
105781 Corydalis curviflora Maxim. ex Hemsl. subsp. minuticristata (Fedde) C. Y. Wu;直距曲花紫堇■
105782 Corydalis curviflora Maxim. ex Hemsl. subsp. minuticristata (Fedde) C. Y. Wu = Corydalis cytisiflora (Fedde) Lidén ex C. Y. Wu,H. Chuang et Z. Y. Su subsp. minuticristata (Fedde) Lidén ■
105783 Corydalis curviflora Maxim. ex Hemsl. subsp. pseudosmith (Fedde) C. Y. Wu;流苏曲花紫堇■
105784 Corydalis curviflora Maxim. ex Hemsl. subsp. pseudosmithii (Fedde) C. Y. Wu = Corydalis cytisiflora (Fedde) Lidén ex C. Y. Wu,H. Chuang et Z. Y. Su subsp. pseudosmithii (Fedde) Lidén ■
105785 Corydalis curviflora Maxim. ex Hemsl. subsp. rosthornii (Fedde) C. Y. Wu;具爪曲花紫堇(具爪弯花紫堇,弯花紫堇);Rosthorn Curvedflower Corydalis ■
105786 Corydalis curviflora Maxim. ex Hemsl. var. alteocristata C. Y. Wu et H. Chuang = Corydalis curviflora Maxim. ex Hemsl. subsp. alteocristata (C. Y. Wu et H. Chuang) C. Y. Wu ■
105787 Corydalis curviflora Maxim. ex Hemsl. var. giraldii Fedde = Corydalis shensiana Lidén ■
105788 Corydalis curviflora Maxim. ex Hemsl. var. minuticristata Fedde = Corydalis curviflora Maxim. ex Hemsl. subsp. minuticristata (Fedde) C. Y. Wu ■
105789 Corydalis curviflora Maxim. ex Hemsl. var. pseudosmithii Fedde = Corydalis cytisiflora (Fedde) Lidén ex C. Y. Wu,H. Chuang et Z. Y. Su subsp. pseudosmithii (Fedde) Lidén ■
105790 Corydalis curviflora Maxim. ex Hemsl. var. rosthornii Fedde = Corydalis curviflora Maxim. ex Hemsl. subsp. rosthornii (Fedde) C. Y. Wu ■
105791 Corydalis curviflora Maxim. ex Hemsl. var. smithii Fedde = Corydalis cytisiflora (Fedde) Lidén ex C. Y. Wu,H. Chuang et Z. Y. Su ■
105792 Corydalis curviflora Maxim. ex Hemsl. var. trifida W. T. Wang;裂苞曲花紫堇■
105793 Corydalis curviflora Maxim. ex Hemsl. var. trifida W. T. Wang ex C. Y. Wu et H. Chuang = Corydalis curviflora Maxim. ex Hemsl. ■
105794 Corydalis curviflora Maxim. var. giraldii Fedde = Corydalis shensiana Lidén ■
105795 Corydalis curvisiliqua (A. Gray) A. Gray;弯果紫堇;Corydalis ■☆
105796 Corydalis curvisiliqua (A. Gray) A. Gray subsp. grandibracteata (Fedde) G. B. Ownbey;大苞弯果紫堇■☆
105797 Corydalis curvisiliqua (A. Gray) A. Gray var. grandibracteata Fedde = Corydalis curvisiliqua (A. Gray) A. Gray subsp. grandibracteata (Fedde) G. B. Ownbey ■☆
105798 Corydalis cytisiflora (Fedde) Lidén ex C. Y. Wu,H. Chuang et Z. Y. Su;金雀花黄堇(金雀花紫堇);Broomflower Corydalis ■
105799 Corydalis cytisiflora (Fedde) Lidén ex C. Y. Wu,H. Chuang et Z. Y. Su subsp. altecristata (C. Y. Wu et H. Chuang) Lidén;高冠金雀花紫堇■
105800 Corydalis cytisiflora (Fedde) Lidén ex C. Y. Wu,H. Chuang et Z. Y. Su subsp. minuticristata (Fedde) Lidén;直距金雀花黄堇■
105801 Corydalis cytisiflora (Fedde) Lidén ex C. Y. Wu,H. Chuang et Z. Y. Su subsp. pseudosmithii (Fedde) Lidén;流苏金雀花紫堇■
105802 Corydalis dajingensis C. Y. Wu et Z. Y. Su;大金紫堇;Dajin Corydalis ■
105803 Corydalis darwasica Regel ex Prain;达尔瓦斯紫堇■☆

105804 Corydalis dasyptera Maxim.;迭裂黄堇(黄连,鸡爪黄连);Repeatedsplit Corydalis,Thickwing Corydalis ■
105805 Corydalis dasyptera Maxim. var. tenuiflora C. Y. Wu et T. Y. Shu;狭花迭裂黄堇(隆思,木冬欧蒿,细花迭裂黄堇)■
105806 Corydalis daucifolia H. Lév. = Corydalis cheilanthifolia Hemsl. ■
105807 Corydalis daucifolia H. Lév. et Vaniot = Corydalis cheilanthifolia Hemsl. ■
105808 Corydalis davidii Franch.;南黄堇(百脉根,断肠草,何及南博,黄断肠草,老龙草,南黄紫堇,牛角花,山香,水黄连,土黄芩,小牛角草,啄木冠草);David Corydalis ■
105809 Corydalis debilis Edgew. = Corydalis cornuta Royle ■
105810 Corydalis decumbens (Thunb.) Pers.;夏天无(洞里仙,伏地延胡索,伏茎紫堇,伏生紫堇,落水珠,无柄紫堇,夏天棕,野延胡,一粒金丹);Decumbent Corydalis,Nothing in Summer ■
105811 Corydalis decumbens (Thunb.) Pers. = Corydalis buschii Nakai ■
105812 Corydalis decumbens (Thunb.) Pers. f. albescens (Takeda) Ohwi;白花夏天无■☆
105813 Corydalis deflexi-calcarata C. Y. Wu;拟昌都紫堇;False Changdu Corydalis ■
105814 Corydalis deflexi-calcarata C. Y. Wu = Corydalis trachycarpa Maxim. ■
105815 Corydalis degensis C. Y. Wu et H. Chuang;德格紫堇;Deg Corydalis ■
105816 Corydalis delavayi Franch.;苍山黄堇(苍山紫堇,丽江黄堇,丽江紫堇,马尾黄连);Delavay Corydalis ■
105817 Corydalis delavayi Franch. var. euryphylla Fedde;宽叶苍山黄堇;Broadleaf Corydalis ■
105818 Corydalis delavayi Franch. var. euryphylla Fedde = Corydalis mayae Hand.-Mazz. ■
105819 Corydalis delavayi Franch. var. stenophylla Fedde;狭叶苍山黄堇;Narrowleaf Corydalis ■
105820 Corydalis delavayi Franch. var. stenophylla Fedde = Corydalis mayae Hand.-Mazz. ■
105821 Corydalis delavayi Franch. var. stenophylla Fedde = Corydalis mayae Hand.-Mazz. var. stenophylla (Fedde) C. Y. Wu ■
105822 Corydalis delicatula D. G. Long;娇嫩黄堇;Delicate Corydalis ■
105823 Corydalis delphinioides Fedde;飞燕黄堇(翠雀状紫堇,飞燕紫堇,假飞燕草);Larkspur Corydalis,Larkspurlike Corydalis ■
105824 Corydalis densispica C. Y. Wu;密穗黄堇;Densespike Corydalis,Spikate Corydalis ■
105825 Corydalis denticulatobracteata Fedde = Corydalis hookeri Prain ■
105826 Corydalis diffusa Lidén;展枝黄堇■
105827 Corydalis dingdonis Airy Shaw;多雄黄堇■
105828 Corydalis diphylla Wall.;尼泊尔延胡索(双叶紫堇)■☆
105829 Corydalis dolichocentra Z. Y. Su et Lidén;雅曲距紫堇■
105830 Corydalis dongchuanensis Z. Y. Su et Lidén;东川紫堇■
105831 Corydalis dorjii D. G. Long;不丹紫堇;Dorj Corydalis ■
105832 Corydalis drakeana Prain;短爪黄堇(悬果黄堇);Drake Corydalis,Shortspur Corydalis ■
105833 Corydalis drakeana Prain var. tibetica C. Y. Wu et H. Chuang;西藏短爪黄堇;Tibet Shortspur Corydalis,Xizang Drake Corydalis ■
105834 Corydalis drakeana Prain var. tibetica C. Y. Wu et H. Chuang = Corydalis pseudodrakeana Lidén ■
105835 Corydalis dubia Prain;稀花黄堇(稀花紫堇,稀毛黄堇);Laxflower Corydalis,Laxity-hair Corydalis ■
105836 Corydalis duclouxii H. Lév. et Vaniot;师宗紫堇(地锦苗,断肠草,金钩如意草,如意草,水黄连,水金钩如意,无冠金钩如意草,五味草,紫堇);Cristateless Dali Corydalis,Cristateless Tali Corydalis,Ducloux Corydalis ■
105837 Corydalis dulongjiangensis H. Chuang;独龙江紫堇;Dulongjiang Corydalis ■
105838 Corydalis ecalcarata (Z. Y. Su) Y. H. Zhang;无距小花黄堇;Spurless Corydalis ■
105839 Corydalis ecalcarata (Z. Y. Su) Y. H. Zhang = Corydalis balansae Prain ■
105840 Corydalis eccremocarpa Franch.;悬果紫堇■☆
105841 Corydalis eccremocarpa W. W. Sm. = Corydalis drakeana Prain ■
105842 Corydalis echinocarpa Franch. = Corydalis sheareri S. Moore ■
105843 Corydalis ecristata (Prain) D. G. Long;无冠紫堇(雀子都,无冠克什米尔紫堇);Crestless Corydalis,Cristateless Kashmir Corydalis ■
105844 Corydalis ecristata (Prain) D. G. Long subsp. longicarcarata (D. G. Long) C. Y. Wu;长距无冠紫堇;Longspur Crestless Corydalis ■
105845 Corydalis ecristata (Prain) D. G. Long var. longicarcarata D. G. Long = Corydalis ecristata (Prain) D. G. Long subsp. longicarcarata (D. G. Long) C. Y. Wu ■
105846 Corydalis edulis Maxim.;紫堇(赤芹,楚葵,断肠草,麦黄草,闷头花,起贫草,蜀堇,水卜菜,水葡菜,苔菜,小柄紫堇,蝎子花,野花生,紫芹);Common Corydalis,Eatable Corydalis,Microstalk Corydalis ■
105847 Corydalis edulis Maxim. var. cicutariifolia Fedde = Corydalis racemosa (Thunb.) Pers. ■
105848 Corydalis eduloides Fedde = Corydalis decumbens (Thunb.) Pers. ■
105849 Corydalis eduloides Fedde var. haimensis Fedde = Corydalis decumbens (Thunb.) Pers. ■
105850 Corydalis elata Bureau et Franch.;高茎紫堇;Tall Corydalis ■
105851 Corydalis elata Bureau et Franch. subsp. ecristata C. Y. Wu = Corydalis harrysmithii Lidén et Z. Y. Su ■
105852 Corydalis elata Bureau et Franch. subsp. ecristata Lidén;无冠高茎紫堇;Crestless Tall Corydalis ■
105853 Corydalis elegans Wall. = Corydalis clarkei Prain ■☆
105854 Corydalis elegans Wall. ex Hook. et Thomson;幽雅黄堇■
105855 Corydalis ellipticarpa C. Y. Wu et Z. Y. Su;椭果黄堇;Ellipticfruit Corydalis ■
105856 Corydalis ellipticarpa C. Y. Wu et Z. Y. Su var. taipaica C. Y. Wu;陕西椭果黄堇;Shaanxi Ellipticfruit Corydalis ■
105857 Corydalis ellipticarpa C. Y. Wu et Z. Y. Su var. taipaica C. Y. Wu = Corydalis jingyuanensis C. Y. Wu et H. Chuang ■
105858 Corydalis emanuelii C. A. Mey.;埃玛紫堇■☆
105859 Corydalis enantiophylla Lidén;对叶紫堇■
105860 Corydalis erdelii Zucc.;埃尔紫堇■☆
105861 Corydalis erythrocarpa H. Lév. = Dactylicapnos torulosa (Hook. f. et Thomson) Hutch. ■
105862 Corydalis erythrocarpa H. Lév. = Dicentra torulosa Hook. f. et Thomson ■
105863 Corydalis esquirolii H. Lév. et Fedde;籽纹紫堇(埃氏紫堇,高山羊不吃,山香);Esquirol Corydalis ■
105864 Corydalis eugeniae Fedde;粗距紫堇(粗毛黄堇,康定紫堇);Thickspur Corydalis,Toughhair Corydalis ■
105865 Corydalis eugeniae Fedde subsp. fissibracteata (Fedde) Lidén ex C. Y. Wu,H. Chuang et Z. Y. Su = Corydalis eugeniae Fedde ■
105866 Corydalis eugeniae Fedde var. fissibracteata (Fedde) Lidén;裂

苞粗距紫堇(裂苞条裂紫堇)■

105867 Corydalis eugeniae Fedde var. fissibracteata Fedde = Corydalis eugeniae Fedde var. fissibracteata (Fedde) Lidén ■

105868 Corydalis fangshanensis W. T. Wang;房山紫堇(石黄连,土黄连);Fangshan Corydalis ■

105869 Corydalis fargesii Franch.;北岭黄堇(倒卵果紫堇,南黄紫堇);Farges Corydalis ■

105870 Corydalis feddeana H. Lév. et Fedde;大海黄堇(断肠草);Fedde Corydalis ■

105871 Corydalis feddei H. Lév. ex Fedde = Corydalis feddeana H. Lév. et Fedde ■

105872 Corydalis fedtschenkoana Regel;天山囊果紫堇(囊果紫堇);Fedtschenko Corydalis ■

105873 Corydalis filisecta C. Y. Wu;丝叶紫堇;Filiform-leaf Corydalis, Silkleaf Corydalis ■

105874 Corydalis filistipes Nakai;丝梗紫堇■☆

105875 Corydalis fimbripetala Ludlow et Stearn;流苏瓣缘黄堇(流苏黄堇);Fimbriate-petal Corydalis, Tasselpeta Corydalis ■

105876 Corydalis fimbripetala Ludlow et Stearn = Corydalis inopinata Prain ex Fedde ■

105877 Corydalis flabellata Edgew.;扇叶黄堇;Fanleaf Corydalis ■

105878 Corydalis flaccida Hook. f. et Thomson;裂冠紫堇(裂冠黄堇,柔弱紫堇);Flaccid Corydalis, Frail Corydalis ■

105879 Corydalis flavifibrillosa C. Y. Wu = Corydalis flexuosa Franch. subsp. pseudoheterocentra (Fedde) Lidén ■

105880 Corydalis flavula (Raf.) DC. = Fumaria flavula Raf. ■☆

105881 Corydalis flexuosa Franch.;穆坪紫堇;Flaxuose Corydalis, Muping Corydalis ■

105882 Corydalis flexuosa Franch. f. bulbillifera C. Y. Wu;珠芽穆坪紫堇;Bulbil Flaxuose Corydalis ■

105883 Corydalis flexuosa Franch. subsp. balsamiflora (Prain) C. Y. Wu;香花紫堇;Fragrant Muping Corydalis ■

105884 Corydalis flexuosa Franch. subsp. gemmipara (H. Chuang) C. Y. Wu;显芽紫堇■

105885 Corydalis flexuosa Franch. subsp. gemmipara (H. Chuang) C. Y. Wu = Corydalis calycosa H. Chuang ■

105886 Corydalis flexuosa Franch. subsp. kuanhsienensis C. Y. Wu;灌县紫堇;Guanxian Muping Corydalis ■

105887 Corydalis flexuosa Franch. subsp. kuanhsienensis C. Y. Wu = Corydalis calycosa H. Chuang ■

105888 Corydalis flexuosa Franch. subsp. microflora (C. Y. Wu et H. Chuang) C. Y. Wu;小花穆坪紫堇;Smallflower Muping Corydalis ■

105889 Corydalis flexuosa Franch. subsp. microflora (C. Y. Wu et H. Chuang) C. Y. Wu = Corydalis microflora (C. Y. Wu et H. Chuang) Z. Y. Su et Lidén ■

105890 Corydalis flexuosa Franch. subsp. mucronipetala (C. Y. Wu et H. Chuang) C. Y. Wu;尖突穆坪紫堇;Mucronipetale Muping Corydalis ■

105891 Corydalis flexuosa Franch. subsp. mucronipetala (C. Y. Wu et H. Chuang) C. Y. Wu = Corydalis mucronipetala (C. Y. Wu et H. Chuang) Lidén et Z. Y. Su ■

105892 Corydalis flexuosa Franch. subsp. omeiana (C. Y. Wu et H. Chuang) C. Y. Wu;金顶紫堇(淡蓝断肠草,断肠草,蓝芹续草,紫断肠草);Emei Muping Corydalis ■

105893 Corydalis flexuosa Franch. subsp. omeiana (C. Y. Wu et H. Chuang) C. Y. Wu = Corydalis omeiana (C. Y. Wu et H. Chuang) Z. Y. Su et Lidén ■

105894 Corydalis flexuosa Franch. subsp. pinnatibracteata (C. Y. Wu et H. Chuang) C. Y. Wu;羽苞穆坪紫堇;Pinnatebract Muping Corydalis ■

105895 Corydalis flexuosa Franch. subsp. pinnatibracteata (C. Y. Wu et H. Chuang) C. Y. Wu = Corydalis calycosa H. Chuang ■

105896 Corydalis flexuosa Franch. subsp. pseudoheterocentra (Fedde) Lidén ex C. Y. Wu;黄根紫堇(黄花草);Yellowroot Muping Corydalis ■

105897 Corydalis flexuosa Franch. subsp. pseudoheterocentra (Fedde) Lidén = Corydalis flexuosa Franch. subsp. pseudoheterocentra (Fedde) Lidén ex C. Y. Wu ■

105898 Corydalis flexuosa Franch. var. microflora C. Y. Wu et H. Chuang = Corydalis microflora (C. Y. Wu et H. Chuang) Z. Y. Su et Lidén ■

105899 Corydalis flexuosa Franch. var. microflora C. Y. Wu et H. Chuang = Corydalis flexuosa Franch. subsp. microflora (C. Y. Wu et H. Chuang) C. Y. Wu ■

105900 Corydalis flexuosa Franch. var. mucronipetala C. Y. Wu et H. Chuang = Corydalis mucronipetala (C. Y. Wu et H. Chuang) Lidén et Z. Y. Su ■

105901 Corydalis flexuosa Franch. var. mucronipetala C. Y. Wu et H. Chuang = Corydalis flexuosa Franch. subsp. mucronipetala (C. Y. Wu et H. Chuang) C. Y. Wu ■

105902 Corydalis flexuosa Franch. var. omeiana C. Y. Wu et H. Chuang = Corydalis omeiana (C. Y. Wu et H. Chuang) Z. Y. Su et Lidén ■

105903 Corydalis flexuosa Franch. var. omeiana C. Y. Wu et H. Chuang = Corydalis flexuosa Franch. subsp. omeiana (C. Y. Wu et H. Chuang) C. Y. Wu ■

105904 Corydalis flexuosa Franch. var. pinnatibracteata C. Y. Wu et H. Chuang = Corydalis flexuosa Franch. subsp. pinnatibracteata (C. Y. Wu et H. Chuang) C. Y. Wu ■

105905 Corydalis flexuosa Franch. var. pinnatibracteata C. Y. Wu et H. Chuang = Corydalis calycosa H. Chuang ■

105906 Corydalis fluminicola W. W. Sm. = Corydalis hamata Franch. ■

105907 Corydalis foetida C. Y. Wu et Z. Y. Su;臭黄堇(断肠草);Stink Corydalis ■

105908 Corydalis foetida C. Y. Wu et Z. Y. Su var. brevipedunculata Z. Y. Su;短轴臭黄堇;Short-stalked Stink Corydalis ■

105909 Corydalis foetida C. Y. Wu et Z. Y. Su var. brevipedunculata Z. Y. Su = Corydalis brevipedunculata (Z. Y. Su) Z. Y. Su et Lidén ■

105910 Corydalis foliaceobracteata C. Y. Wu et Z. Y. Su;叶苞紫堇;Leafybract Corydalis ■

105911 Corydalis formosana Hayata = Corydalis heterocarpa Siebold et Zucc. var. koidzumiana (Ohwi) Ohwi ■

105912 Corydalis formosana Hayata var. microphylla Sasaki = Corydalis pallida (Thunb.) Pers. ■

105913 Corydalis formosana Hayata var. microphylla Sasaki = Corydalis wilfordii Regel ■

105914 Corydalis franchetiana Prain;春丕黄堇;Franchet Corydalis ■

105915 Corydalis fukuharae Lidén;福原紫堇■☆

105916 Corydalis fumaria H. Lév. et Vaniot = Corydalis racemosa (Thunb.) Pers. ■

105917 Corydalis fumariifolia Maxim.;堇叶延胡索;Fumariaeleaf Corydalis, Fumitoryleaf Corydalis ■

105918 Corydalis fumariifolia Maxim. subsp. azurea Lidén et Zetterl.;阿摺紫堇■☆

105919 Corydalis fumariifolia Maxim. var. incisa Popov;栉苞堇叶延胡

索;Incised Fumitoryleaf Corydalis ■
105920 Corydalis gamosepala Maxim.;北京延胡索(山延胡索);Beijing Corydalis ■
105921 Corydalis gaoxinfeniae Lidén;柄苞黄堇■
105922 Corydalis gebleri Ledeb. = Corydalis capnoides (L.) Pers. ■
105923 Corydalis gemmipara H. Chuang = Corydalis calycosa H. Chuang ■
105924 Corydalis gemmipara H. Chuang = Corydalis flexuosa Franch. subsp. gemmipara (H. Chuang) C. Y. Wu ■
105925 Corydalis geocarpa Harry Sm. ex Lidén;弯柄紫堇;Bowstalk Corydalis ■
105926 Corydalis geocarpa Harry Sm. ex Lidén = Corydalis dajingensis C. Y. Wu et Z. Y. Su ■
105927 Corydalis gerdae Fedde = Corydalis latiflora Hook. f. et Thomson subsp. gerdae (Fedde) Lidén ex C. Y. Wu, H. Chuang et Z. Y. Su ■
105928 Corydalis gerdae Fedde = Corydalis latiflora Hook. f. et Thomson ■
105929 Corydalis gigantea Trautv. et C. A. Mey.;巨紫堇;Giant Corydalis ■
105930 Corydalis gigantea Trautv. et C. A. Mey. = Corydalis macrantha (Regel) Popov ■
105931 Corydalis gigantea Trautv. et C. A. Mey. var. amurensis Regel = Corydalis gigantea Trautv. et C. A. Mey. ■
105932 Corydalis gigantea Trautv. et C. A. Mey. var. genuina Regel = Corydalis gigantea Trautv. et C. A. Mey. ■
105933 Corydalis gigantea Trautv. et C. A. Mey. var. macrantha Regel = Corydalis gigantea Trautv. et C. A. Mey. ■
105934 Corydalis giraldii Fedde;小花宽瓣黄堇;Girald Corydalis ■
105935 Corydalis glareosa Sommier et H. Lév. = Corydalis alpestris C. A. Mey. ■
105936 Corydalis glaucescens Regel;新疆元胡(粉绿延胡索,灰叶延胡索,灰叶元胡,元胡);Glaucescent Corydalis, Greyleaf Corydalis ■
105937 Corydalis glaucissima Lidén et Z. Y. Su;苍白紫堇■
105938 Corydalis glycyphyllos Fedde;甘草叶紫堇(甜叶紫堇);Liquoriceleaf Corydalis, Sweetleaf Corydalis ■
105939 Corydalis gortschakovii Schrenk;新疆黄堇(高山黄堇,高山紫堇,戈氏紫堇);Gortschakov Corydalis ■
105940 Corydalis gortschakovii Schrenk = Corydalis moorcroftiana Wall. ■
105941 Corydalis gortschakovii Schrenk = Corydalis thyrsiflora Prain ■☆
105942 Corydalis gortschakovii Schrenk subsp. onobrychis (Fedde) Wendelbo = Corydalis onobrychis Fedde ■
105943 Corydalis govaniana Wall.;库莽黄堇(高文紫堇);Govan Corydalis ■
105944 Corydalis gracilipes S. Moore = Corydalis decumbens (Thunb.) Pers. ■
105945 Corydalis gracilis Franch. = Corydalis gracillima C. Y. Wu ex Govaerts ■
105946 Corydalis gracillima C. Y. Wu = Corydalis casimiriana Duthie et Prain ex Prain ■
105947 Corydalis gracillima C. Y. Wu ex Govaerts;纤细黄堇(纤细紫堇,小黄断肠草);Thinnest Corydalis ■
105948 Corydalis gracillima C. Y. Wu ex Govaerts var. microcarcarata H. Chuang;小距纤细黄堇;Smallspur Thinnest Corydalis ■
105949 Corydalis graminea Prain = Corydalis polygalina Hook. f. et Thomson ■
105950 Corydalis grandiflora C. Y. Wu et Z. Y. Su;丹巴黄堇(川西黄堇);Danba Corydalis ■
105951 Corydalis grubovii Michajlova = Corydalis stricta Steph. ex DC. ■
105952 Corydalis gymnopoda Z. Y. Su et Lidén;寡叶裸茎紫堇■
105953 Corydalis gyrophylla Lidén;裸茎延胡索;Nakestem Corydalis ■
105954 Corydalis halei (Small) Fernald et B. G. Schub. = Corydalis micrantha (Engelm. ex A. Gray) A. Gray ■☆
105955 Corydalis halei (Small) Fernald et Schub. = Corydalis micrantha (Engelm. ex A. Gray) A. Gray subsp. australis (Chapm.) G. B. Ownbey ■☆
105956 Corydalis halleri Willd. = Corydalis solida (L.) Clairv. ■☆
105957 Corydalis hamata Franch.;钩距黄堇(分枝钩距黄堇,钩状黄堇,钩状紫堇,溪畔紫堇);Branch Hookspur Corydalis, Hookspur Corydalis, Streamside Corydalis ■
105958 Corydalis hamata Franch. = Corydalis conspersa Maxim. ■
105959 Corydalis hamata Franch. = Corydalis glycyphyllos Fedde ■
105960 Corydalis hamata Franch. = Corydalis melanochlora Maxim. ■
105961 Corydalis hamata Franch. var. ramosa Z. Y. Su = Corydalis hamata Franch. ■
105962 Corydalis handel-mazzettii Fedde = Corydalis racemosa (Thunb.) Pers. ■
105963 Corydalis hannae J. Buchholz = Corydalis trachycarpa Maxim. ■
105964 Corydalis hannae Kanitz = Corydalis calcicola W. W. Sm. ■
105965 Corydalis harrysmithii Lidén et Z. Y. Su;康定紫堇■
105966 Corydalis hebephylla C. Y. Wu et Z. Y. Su;毛被黄堇;Obtuseleaf Corydalis ■
105967 Corydalis hebephylla C. Y. Wu et Z. Y. Su var. glabrescens C. Y. Wu et Z. Y. Su;假毛被黄堇(钝叶微毛紫堇)■
105968 Corydalis hebephylla C. Y. Wu et Z. Y. Su var. glabrescens C. Y. Wu et Z. Y. Su = Corydalis sigmantha Z. Y. Su et C. Y. Wu ■
105969 Corydalis hemidicentra Hand.-Mazz.;半荷包紫堇(三叶紫堇);Halfpouch Corydalis, Three-leaf Corydalis ■
105970 Corydalis hemsleyana Franch. ex Prain;巴东紫堇(异齿紫堇);Badong Corydalis, Hemsley Corydalis ■
105971 Corydalis hendersonii Hemsl.;尼泊尔黄堇(矮紫堇,亨氏黄堇,尼泊尔紫堇);Henderson Corydalis, Nepal Corydalis ■
105972 Corydalis hendersonii Hemsl. var. alto-cristata C. Y. Wu et Z. Y. Su;高冠尼泊尔黄堇;Highcristate Henderson Corydalis ■
105973 Corydalis hepaticifolia C. Y. Wu et T. Y. Shu;假獐耳紫堇;Hepaticaleaf Corydalis, Mossleaf Corydalis ■
105974 Corydalis heracleifolia C. Y. Wu et Z. Y. Su;独活叶紫堇;Cowparsnipleaf Corydalis ■
105975 Corydalis heterocarpa (Durieu) Ball = Ceratocapnos heterocarpa Durieu ■☆
105976 Corydalis heterocarpa Siebold et Zucc.;异果黄堇(宽果黄堇,阔果紫堇,阔叶紫堇);Differfruit Corydalis ■
105977 Corydalis heterocarpa Siebold et Zucc. var. brachystyla (Koidz.) Ohwi;短穗异果黄堇■☆
105978 Corydalis heterocarpa Siebold et Zucc. var. japonica (Franch. et Sav.) Ohwi;日本异果黄堇■
105979 Corydalis heterocarpa Siebold et Zucc. var. japonica Ohwi = Corydalis heterocarpa Siebold et Zucc. ■
105980 Corydalis heterocarpa Siebold et Zucc. var. koidzumiana (Ohwi) Ohwi = Corydalis orthopoda Hayata ■
105981 Corydalis heterocarpa Siebold et Zucc. var. simadae Ohwi = Corydalis heterocarpa Siebold et Zucc. var. japonica (Franch. et Sav.) Ohwi ■
105982 Corydalis heterocentra Diels;异心紫堇;Differcentre Corydalis, Heterocentre Corydalis ■
105983 Corydalis heterocentra Diels = Corydalis flexuosa Franch. subsp. pseudoheterocentra (Fedde) Lidén ■

105984 Corydalis heterocentra Diels = Corydalis petrophila Franch. ■

105985 Corydalis heterodonta H. Lév. et Fedde;异齿紫堇;Differtooth Corydalis ■

105986 Corydalis heterothylax C. Y. Wu ex Z. Y. Su et Lidén;异距紫堇■

105987 Corydalis holopetala Diels;同瓣黄堇;Copetal Corydalis, Equalpetal Corydalis ■

105988 Corydalis hondoensis Ohwi = Corydalis pallida (Thunb.) Pers. var. tenuis Yatabe ■

105989 Corydalis hondoensis Ohwi = Corydalis speciosa Maxim. ■

105990 Corydalis hongbashanensis Lidén et Y. W. Wang;洪坝山紫堇■

105991 Corydalis hookeri Prain;拟锥花黄堇(齿苞黄堇);Hooker Corydalis,Tooth-bract Corydalis ■

105992 Corydalis hsiaowutaishanensis T. P. Wang;五台山延胡索;Wutaishan Corydalis ■

105993 Corydalis humicola Hand. -Mazz.;湿生紫堇;Wetfoot Corydalis ■

105994 Corydalis humilis O. U. Oh et Y. S. Kim;矮生延胡索;Dwarf Corydalis ■

105995 Corydalis humosa Migo;土元胡;Humus Corydalis ■

105996 Corydalis humosa Migo = Corydalis caudata (Lam.) Pers. ■

105997 Corydalis hupehensis C. Y. Wu ex Z. Zheng = Corydalis acuminata Franch. subsp. hupehensis C. Y. Wu ■

105998 Corydalis imbricata Z. Y. Su et Lidén;银瑞;Imbricate Corydalis ■

105999 Corydalis impatiens (Pall.) Fisch.;赛北紫堇(断肠草);Impatient Corydalis ■

106000 Corydalis impatiens (Pall.) Fisch. = Corydalis pseudoimpatiens Fedde ■

106001 Corydalis impatiens (Pall.) Fisch. var. maxima Michajlova = Corydalis pseudoimpatiens Fedde ■

106002 Corydalis impatiens (Pall.) Fisch. var. minima Michajlova = Corydalis impatiens (Pall.) Fisch. ■

106003 Corydalis impatiens (Pallas) Fisch. var. maxima Michajlova = Corydalis pseudoimpatiens Fedde ■

106004 Corydalis incisa (Thunb.) Pers.;刻叶紫堇(地锦苗,断肠草,裂苞紫堇,烫伤草,天奎草,羊不吃,羊不吃草,野黄连,紫花鱼灯草,紫堇);Gapleaf Corydalis,Incised Corydalis ■

106005 Corydalis incisa (Thunb.) Pers. f. bicolor Hayashi;二色刻叶紫堇■☆

106006 Corydalis incisa (Thunb.) Pers. f. candida Hiyama;白色刻叶紫堇■☆

106007 Corydalis incisa (Thunb.) Pers. f. liuchiuensis Nakai = Corydalis incisa (Thunb.) Pers. ■

106008 Corydalis incisa (Thunb.) Pers. f. pallescens Makino = Corydalis incisa (Thunb.) Pers. ■

106009 Corydalis incisa (Thunb.) Pers. var. alba S. Y. Wang = Corydalis incisa (Thunb.) Pers. ■

106010 Corydalis incisa (Thunb.) Pers. var. koreana Fedde = Corydalis incisa (Thunb.) Pers. ■

106011 Corydalis incisa (Thunb.) Pers. var. pseudomakinoana Fedde = Corydalis incisa (Thunb.) Pers. ■

106012 Corydalis incisa (Thunb.) Pers. var. tschekiangensis Fedde = Corydalis incisa (Thunb.) Pers. ■

106013 Corydalis inconspicua Bunge ex Ledeb.;小株紫堇(二色堇);Indistinct Corydalis ■

106014 Corydalis inopinata Prain ex Fedde;卡惹拉黄堇;Sudden Corydalis,Unexpected Corydalis ■

106015 Corydalis inopinata Prain ex Fedde var. glabra C. Y. Wu et Z. Y. Su;无毛卡惹拉黄堇;Glabrous Sudden Corydalis, Glabrous Unexpected Corydalis ■

106016 Corydalis inopinata Prain ex Fedde var. glabra C. Y. Wu et Z. Y. Su = Corydalis inopinata Prain ex Fedde ■

106017 Corydalis intermedia (L.) Merat;中型紫堇■☆

106018 Corydalis iochanensis H. Lév.;药山紫堇;Yaoshan Corydalis ■

106019 Corydalis ischnosiphon Lidén et Z. Y. Su;瘦距紫堇■

106020 Corydalis ivaschkeviczii Aparina = Corydalis watanabei Kitag. ■

106021 Corydalis japonica Makino;日本紫堇■☆

106022 Corydalis japonica Makino = Corydalis ophiocarpa Hook. f. et Thomson ■

106023 Corydalis japonica Siebold ex Miq. = Corydalis incisa (Thunb.) Pers ■

106024 Corydalis jigmei C. E. C. Fisch. et Kaul;藏南紫堇;S. Xizang Corydalis,South Tibet Corydalis ■

106025 Corydalis jingyuanensis C. Y. Wu et H. Chuang;泾源紫堇;Jingyuan Corydalis ■

106026 Corydalis jiulongensis Z. Y. Su et Lidén;九龙黄堇■

106027 Corydalis juncea Wall.;裸茎黄堇;Juncus Corydalis,Nakedstem Corydalis ■

106028 Corydalis kailiensis Z. Y. Su;凯里紫堇;Kaili Corydalis ■

106029 Corydalis kansuana Fedde = Corydalis chingii Fedde ■

106030 Corydalis kareliniana Pritz. = Corydalis inconspicua Bunge ex Ledeb. ■

106031 Corydalis kareliniana Pritz. ex Walp. = Corydalis inconspicua Bunge ex Ledeb. ■

106032 Corydalis kaschgarica Rupr.;喀什黄堇(喀什紫堇);Kashi Corydalis ■

106033 Corydalis kelungensis Hayata = Corydalis decumbens (Thunb.) Pers. ■

106034 Corydalis kiautschouensis Poelln.;胶州延胡索(老鼠屎,山东延胡索,山东紫堇);Jiaozhou Corydalis ■

106035 Corydalis kingdonis Airy Shaw;墨脱黄堇(多雄黄堇);Kingdon Corydalis ■

106036 Corydalis kingii Prain;帕里紫堇(帕里黄堇);King Corydalis ■

106037 Corydalis kingii Prain var. megalantha C. Y. Wu et Z. Y. Su;大花帕里紫堇;Bigflower King Corydalis ■

106038 Corydalis kingii Prain var. megalantha C. Y. Wu et Z. Y. Su = Corydalis lasiocarpa Lidén et Z. Y. Su ■

106039 Corydalis kingii Prain var. minuticalcata C. Y. Wu;小距帕里紫堇;Smallspur Corydalis ■

106040 Corydalis kiukiangensis C. Y. Wu,Z. Y. Su et Lidén;俅江紫堇;Qiujiang Corydalis ■

106041 Corydalis koidzumiana Ohwi = Corydalis formosana Hayata ■

106042 Corydalis koidzumiana Ohwi = Corydalis orthopoda Hayata ■

106043 Corydalis kokiana Hand. -Mazz.;狭距紫堇;Kok Corydalis, Narrowspur Corydalis ■

106044 Corydalis kokiana Hand. -Mazz. var. micrantha C. Y. Wu et H. Chuang = Corydalis minutiflora C. Y. Wu ■

106045 Corydalis kokiana Hand. -Mazz. var. robusta C. Y. Wu = Corydalis nigro-apiculata C. Y. Wu ■

106046 Corydalis kokiana Hand. -Mazz. var. robusta C. Y. Wu et H. Chuang = Corydalis kokiana Hand. -Mazz. ■

106047 Corydalis kolpakovskiana Regel = Corydalis glaucescens Regel ■

106048 Corydalis kolpakovskiana Regel var. hennigii Fedde = Corydalis glaucescens Regel ■

106049 Corydalis krasnovii Michajlova;南疆黄堇;S. Xinjiang Corydalis ■

106050 Corydalis kuruchuensis Lidén;库如措紫堇■

106051 Corydalis kushiroensis Fukuhara;钏路紫堇■☆

106052 Corydalis laelia Prain;高冠黄堇(翅冠黄堇);Longcomb Corydalis,Winged Corolla Corydalis ■

106053 Corydalis laelia Prain subsp. bhutanica D. G. Long = Corydalis laelia Prain ■

106054 Corydalis lagochila Lidén et Z. Y. Su;兔唇紫堇■

106055 Corydalis lasiocarpa Lidén et Z. Y. Su;毛果紫堇;Cottonyfruit Corydalis ■

106056 Corydalis lathyrophylla C. Y. Wu;长冠紫堇;Vetchingleaf Corydalis ■

106057 Corydalis lathyrophylla C. Y. Wu subsp. dawuensis Lidén;道孚长冠紫堇■

106058 Corydalis latiflora Hook. f. et Thomson;宽花紫堇;Broadflower Corydalis ■

106059 Corydalis latiflora Hook. f. et Thomson subsp. gerdae (Fedde) Lidén ex C. Y. Wu, H. Chuang et Z. Y. Su;西藏宽花紫堇(不丹紫堇);Bhutan Corydalis, Xizang Broadflower Corydalis ■

106060 Corydalis latiflora Hook. f. et Thomson subsp. gerdae (Fedde) Lidén ex C. Y. Wu, H. Chuang et Z. Y. Su = Corydalis latiflora Hook. f. et Thomson ■

106061 Corydalis latiloba (Franch.) Hand. -Mazz.;宽裂黄堇(岩黄连,岩连);Broadlobe Corydalis ■

106062 Corydalis latiloba (Franch.) Hand. -Mazz. subsp. wumungensis C. Y. Wu et Z. Y. Su;乌蒙黄堇(豆瓣鹿含);Wumeng Broadlobe ■

106063 Corydalis latiloba (Franch.) Hand. -Mazz. var. tibetica Z. Y. Su et Lidén;西藏宽裂黄堇;Xizang Broadlobe Corydalis ■

106064 Corydalis laucheana Fedde;松潘黄堇(曲瓣紫堇,紫苞黄堇);Songpan Corydalis ■

106065 Corydalis laxiflora Lidén;疏花黄堇■

106066 Corydalis ledebouriana Kar. et Kir.;薯根延胡索(对叶延胡索,对叶元胡,对叶紫堇,元胡);Yamroot Corydalis ■

106067 Corydalis leptocarpa Hook. f. et Thomson;细果紫堇(泰国紫堇);Smallfruit Corydalis ■

106068 Corydalis leucanthema C. Y. Wu;粉叶紫堇(白断肠草,假苁蓉);Whiteflower Corydalis ■

106069 Corydalis leucostachya C. Y. Wu et H. Chuang = Corydalis trachycarpa Maxim. var. leucostachya (C. Y. Wu et H. Chuang) C. Y. Wu ■

106070 Corydalis lhasaensis C. Y. Wu et Z. Y. Su;拉萨黄堇(洛隆紫堇,无冠细叶黄堇);Lasa Corydalis ■

106071 Corydalis lhorongensis C. Y. Wu et H. Chuang;洛隆紫堇;Lhorong Corydalis, Luolong Corydalis ■

106072 Corydalis liana Lidén et Z. Y. Su;绕曲黄堇■

106073 Corydalis lichuanensis Z. Zhang = Corydalis hemsleyana Franch. ex Prain ■

106074 Corydalis lidenii Z. Y. Su;积鳞紫堇■

106075 Corydalis linarioides Maxim. var. fissibracteata Fedde = Corydalis eugeniae Fedde ■

106076 Corydalis lineariloba Siebold et Zucc.;线裂紫堇■☆

106077 Corydalis lineariloba Siebold et Zucc. f. pectinata (Kom.) Kitag. = Corydalis fumariifolia Maxim. ■

106078 Corydalis lineariloba Siebold et Zucc. var. capillaris (Makino) Ohwi;发状线裂紫堇■☆

106079 Corydalis lineariloba Siebold et Zucc. var. fumariifolia (Maxim.) Kitag. = Corydalis fumariifolia Maxim. ■

106080 Corydalis lineariloba Siebold et Zucc. var. micrantha Ohwi = Corydalis repens Mandl et Muehld. ■

106081 Corydalis lineariloba Siebold et Zucc. var. papillata (Ohwi) Ohwi = Corydalis repens Mandl et Muehld. ■

106082 Corydalis lineariloba Siebold et Zucc. var. papilligera (Ohwi) Ohwi ex Akiyama = Corydalis papilligera Ohwi ■☆

106083 Corydalis linearioides Maxim.;条裂黄堇(条裂紫堇,铜棒锤,铜锤紫堇);Linearsegmented Corydalis, Toadflaxlike Corydalis ■

106084 Corydalis linearioides Maxim. var. fissibracteata Fexide = Corydalis eugeniae Fedde var. fissibracteata (Fedde) Lidén ■

106085 Corydalis linearis C. Y. Wu;线叶黄堇;Linearleaf Corydalis ■

106086 Corydalis linjiangensis Z. Y. Su ex Lidén;临江延胡索;Linjiang Corydalis ■

106087 Corydalis linstowiana Fedde;变根紫堇(断肠草,康定紫堇,水黄连);Linstow Corydalis ■

106088 Corydalis livida Maxim.;红花紫堇;Redflower Corydalis ■

106089 Corydalis livida Maxim. var. denticulato-cristata Z. Y. Su;齿冠红花紫堇;Toothedcrown Redflower Corydalis ■

106090 Corydalis lofouensis H. Lév. = Corydalis balansae Prain ■

106091 Corydalis longibracteata Ludlow et Stearn;长苞紫堇;Longbract Corydalis ■

106092 Corydalis longicalcarata H. Chuang et Z. Y. Su;长距紫堇(断肠草,高山羊不吃);Longspur Corydalis ■

106093 Corydalis longicalcarata H. Chuang et Z. Y. Su var. multipinnata Z. Y. Su;多裂长距紫堇■

106094 Corydalis longicalcarata H. Chuang et Z. Y. Su var. non-saccata Z. Y. Su;无囊长距紫堇■

106095 Corydalis longicornu Franch. = Corydalis davidii Franch. ■

106096 Corydalis longiflora (Willd.) Pers. = Corydalis schanginii (Pall.) B. Fedtsch. ■

106097 Corydalis longiflora (Willd.) Pers. var. caudata (Lam.) DC. = Corydalis caudata (Lam.) Pers. ■

106098 Corydalis longiflora Pers. var. caudata (Lam.) DC. = Corydalis caudata (Lam.) Pers. ■

106099 Corydalis longipes DC.;长梗黄堇;Longpedicel Corydalis, Longstalk Corydalis ■

106100 Corydalis longipes DC. = Corydalis pseudolongipes Lidén ■

106101 Corydalis longipes DC. var. burkillii Fedde = Corydalis pseudolongipes Lidén ■

106102 Corydalis longipes DC. var. chumbica Prain ex W. W. Sm. = Corydalis cornuta Royle ■

106103 Corydalis longipes DC. var. megalantha H. Chuang = Corydalis casimiriana Duthie et Prain ex Prain ■

106104 Corydalis longipes DC. var. megalantha H. Chuang = Corydalis rubrisepala Lidén subsp. zhuangiana Lidén ■

106105 Corydalis longipes DC. var. phallutiana Fedde = Corydalis pseudolongipes Lidén ■

106106 Corydalis longipes DC. var. pubescens (C. Y. Wu et H. Chuang) C. Y. Wu;毛长梗黄堇(毛黄堇,聂拉木黄堇);Cave Corydalis, Hairy Corydalis, Pubescent Corydalis, Pubescent Longstalk Corydalis ■

106107 Corydalis longipes DC. var. smithii Fedde = Corydalis pseudolongipes Lidén ■

106108 Corydalis longistyla Z. Y. Su et Lidén;长柱黄堇■

106109 Corydalis longkiensis C. Y. Wu, Lidén et Z. Y. Su;龙溪紫堇;Longxi Corydalis ■

106110 Corydalis lophophora Lidén et Z. Y. Su;齿冠紫堇■

106111 Corydalis lopinensis Franch.;罗平山黄堇;Luoping Corydalis ■

106112 Corydalis lowndesii Lidén;齿瓣黄堇■

106113 Corydalis lowndesii Lidén = Corydalis polygalina Hook. f. et

Thomson var. micrantha C. Y. Wu ■
106114 Corydalis ludlowii Stearn;单叶紫堇;Ludlow Corydalis,Single-leaf Corydalis ■
106115 Corydalis lupinoides C. Marquand et Airy Shaw;米林紫堇;Lupinlike Corydalis,Milin Corydalis ■
106116 Corydalis luquanensis H. Chuang;禄劝黄堇;Luquan Corydalis ■
106117 Corydalis lutea (L.) DC.;黄色紫堇(黄堇,欧黄堇,深色黄堇);Fingers-and-thumbs,Fumitory,Haliwort,Lady's Pincushion,Mother-of-thousands,Pincushion,Rock Fumewort,Yellow Birthwort,Yellow Corydalis,Yellow Fumrrory ■☆
106118 Corydalis lutea (L.) DC. = Pseudofumaria lutea Medik. ■☆
106119 Corydalis lutescens C. Y. Wu ex H. Chuang;岩石紫堇■
106120 Corydalis maackii Rupr. ex Trautv. = Corydalis speciosa Maxim. ■
106121 Corydalis macrantha (Regel) Popov;大花紫堇(大花巨紫堇);Largeflower Corydalis ■
106122 Corydalis macrantha (Regel) Popov = Corydalis gigantea Trautv. et C. A. Mey. ■
106123 Corydalis macrocalyx Litv.;大萼紫堇■☆
106124 Corydalis macrocentra Regel;大刺紫堇■☆
106125 Corydalis madida Lidén et Z. Y. Su;喜湿紫堇■
106126 Corydalis mairei H. Lév.;会泽紫堇(滇东紫堇);East Yunnan Corydalis,Huize Corydalis ■
106127 Corydalis mairei H. Lév. var. megalantha C. Y. Wu;大花会泽紫堇;Bigflower Huize Corydalis ■
106128 Corydalis mairei H. Lév. var. megalantha C. Y. Wu = Corydalis pseudomairei C. Y. Wu ex Z. Y. Su et Lidén ■
106129 Corydalis makinoana Matsum. = Corydalis ophiocarpa Hook. f. et Thomson ■
106130 Corydalis marschalliana Pers.;马查紫堇■☆
106131 Corydalis martinii H. Lév. et Vaniot = Corydalis temulifolia Franch. subsp. aegopodioides (H. Lév. et Vaniot) C. Y. Wu ■
106132 Corydalis maximowicziana Nakai = Corydalis speciosa Maxim. ■
106133 Corydalis mayae Hand.-Mazz.;马牙黄堇;Maya Corydalis ■
106134 Corydalis mayae Hand.-Mazz. var. stenophylla (Fedde) C. Y. Wu;狭叶马牙黄堇;Narrowleaf Maya Corydalis ■
106135 Corydalis mayae Hand.-Mazz. var. stenophylla (Fedde) C. Y. Wu = Corydalis mayae Hand.-Mazz. ■
106136 Corydalis mediterranea Z. Y. Su et Lidén;中国紫堇■
106137 Corydalis megalantha C. Y. Wu = Corydalis concinna C. Y. Wu et H. Chuang ■
106138 Corydalis megalantha C. Y. Wu = Corydalis pseudocristata Fedde ■
106139 Corydalis megalantha C. Y. Wu var. laevis C. Y. Wu et H. Chuang = Corydalis pseudocristata Fedde ■
106140 Corydalis megalosperma Z. Y. Su;少子黄堇;Largeseed Corydalis ■
106141 Corydalis meifolia Wall.;细叶黄堇(寸冬欧蒿,细叶紫堇,小叶紫堇);Narrowleaf Corydalis,Spigneleaf Corydalis ■
106142 Corydalis meifolia Wall. = Corydalis pulchella Franch. ■☆
106143 Corydalis meifolia Wall. var. cornutior C. Marquand et Airy Shaw = Corydalis dubia Prain ■
106144 Corydalis meifolia Wall. var. ecristata C. Y. Wu et Z. Y. Su = Corydalis lhasaensis C. Y. Wu et Z. Y. Su ■
106145 Corydalis meifolia Wall. var. sikkimensis Prain = Corydalis stracheyi Duthie ex Prain ■
106146 Corydalis melanochlora Maxim.;暗绿紫堇;Darkgreen Corydalis ■
106147 Corydalis melanochlora Maxim. var. pallescens Maxim. = Corydalis scaberula Maxim. ■
106148 Corydalis mianningensis C. Y. Wu = Corydalis schweriniana Fedde ■
106149 Corydalis micrantha (Engelm. ex A. Gray) A. Gray;细小花紫堇;Slender Corydalis,Slender Fumewort,Small-flowered Corydalis ■☆
106150 Corydalis micrantha (Engelm. ex A. Gray) A. Gray subsp. australis (Chapm.) G. B. Ownbey;南方细小花紫堇■☆
106151 Corydalis micrantha (Engelm. ex A. Gray) A. Gray subsp. texensis G. B. Ownbey;得州细小花紫堇■☆
106152 Corydalis micrantha (Engelm. ex A. Gray) A. Gray var. australis (Chapm.) Shinners = Corydalis micrantha (Engelm. ex A. Gray) A. Gray subsp. australis (Chapm.) G. B. Ownbey ■☆
106153 Corydalis micrantha (Engelm. ex A. Gray) A. Gray var. texensis (G. B. Ownbey) Shinners = Corydalis micrantha (Engelm. ex A. Gray) A. Gray subsp. texensis G. B. Ownbey ■☆
106154 Corydalis microflora (C. Y. Wu et H. Chuang) Z. Y. Su et Lidén;叶状苞紫堇■
106155 Corydalis micropoda Franch. = Corydalis edulis Maxim. ■
106156 Corydalis microsperma Lidén;小籽紫堇■
106157 Corydalis mienningensis C. Y. Wu = Corydalis schweriniana Fedde ■
106158 Corydalis mildbraedii Fedde = Corydalis cornuta Royle ■
106159 Corydalis minutiflora C. Y. Wu;小花紫堇(小花狭距紫堇);Miniflower Corydalis ■
106160 Corydalis mira (Batalin) C. Y. Wu et H. Chuang;疆堇;Surprise Corydalis ■
106161 Corydalis mira Batalin = Corydalis mira (Batalin) C. Y. Wu et H. Chuang ■
106162 Corydalis mitae Kitag. = Corydalis latiflora Hook. f. et Thomson subsp. gerdae (Fedde) Lidén ex C. Y. Wu,H. Chuang et Z. Y. Su ■
106163 Corydalis mitae Kitam. = Corydalis latiflora Hook. f. et Thomson ■
106164 Corydalis montana Engelm. = Corydalis curvisiliqua (A. Gray) A. Gray ■☆
106165 Corydalis moorcroftiana Wall.;革吉黄堇(藏西黄堇);Geji Corydalis,West Tibet Corydalis ■
106166 Corydalis moorcroftiana Wall. = Corydalis clarkei Prain ■☆
106167 Corydalis moorcroftiana Wall. = Corydalis gortschakovii Schrenk ■
106168 Corydalis moupinensis Franch.;尿罐草(宝兴黄堇,断肠草);Niaoguan Corydalis ■
106169 Corydalis mucronata Franch.;突尖紫堇;Mucronate Corydalis ■
106170 Corydalis mucronifera Maxim.;尖突黄堇(扁柄黄堇,东丝勒,冬司,短尖黄堇,黄花紫堇);Flatstiped Corydalis,Shortine Corydalis ■
106171 Corydalis mucronipetala (C. Y. Wu et H. Chuang) Lidén et Z. Y. Su;天全紫堇■
106172 Corydalis muliensis C. Y. Wu et Z. Y. Su;木里黄堇;Muli Corydalis ■
106173 Corydalis multiflora Michajlova = Corydalis gigantea Trautv. et C. A. Mey. ■
106174 Corydalis multisecta C. Y. Wu et H. Chuang;多裂紫堇;Manysplit Corydalis ■
106175 Corydalis multisecta C. Y. Wu et H. Chuang = Corydalis oxypetala Franch. ■
106176 Corydalis myriophylla Lidén;富叶紫堇■
106177 Corydalis nakaii Ishid. = Corydalis ternata (Nakai) Nakai ■
106178 Corydalis nakaii Ishidoya = Corydalis ternata (Nakai) Nakai ■
106179 Corydalis nana Royle;矬紫堇■

106180 Corydalis nana Royle var. jacquemontii Fedde = Corydalis stracheyi Duthie ex Prain ■
106181 Corydalis nanwutaishanensis Z. Y. Su et Lidén;南五台山紫堇■
106182 Corydalis napuligera C. Y. Wu;细花黄堇;Thinflower Corydalis ■
106183 Corydalis napuligera C. Y. Wu = Corydalis lupinoides C. Marquand et Airy Shaw ■
106184 Corydalis nematopoda Lidén et Z. Y. Su;线基紫堇■
106185 Corydalis nemoralis C. Y. Wu et H. Chuang;林生紫堇;Forest Corydalis,Jungle Corydalis ■
106186 Corydalis nepalensis Kitam.;矮紫堇■☆
106187 Corydalis nepalensis Kitam. = Corydalis hendersonii Hemsl. ■
106188 Corydalis nevskii Popov;奈氏紫堇■☆
106189 Corydalis nigro-apiculata C. Y. Wu;黑顶黄堇;Blacktine Corydalis,Blacktop Corydalis ■
106190 Corydalis nigro-apiculata C. Y. Wu var. erosipetala C. Y. Wu et H. Chuang;心瓣黑顶黄堇;Gnawedpetal Corydalis,Heartpetal Corydalis ■
106191 Corydalis nigroapiculata var. erosipetala C. Y. Wu et H. Chuang = Corydalis nigro-apiculata C. Y. Wu ■
106192 Corydalis nivalis (L.) Pers. = Corydalis alpestris C. A. Mey. ■
106193 Corydalis nobilis (L.) Pers.;阿山黄堇■
106194 Corydalis nobilis (L.) Pers. = Corydalis solida (L.) Clairv. ■
106195 Corydalis nubicola Z. Y. Su et Lidén;凌云紫堇■
106196 Corydalis nudicaulis Regel;裸茎紫堇■☆
106197 Corydalis ochotensis Turcz.;黄紫堇(黄龙脱壳,气草,疏花黄堇);Ochotsk Corydalis ■
106198 Corydalis ochotensis Turcz. f. raddeana (Regel) Nakai = Corydalis raddeana Regel ■
106199 Corydalis ochotensis Turcz. var. pedunculata Nakai = Corydalis raddeana Regel ■
106200 Corydalis ochotensis Turcz. var. raddeana (Regel) Nakai = Corydalis raddeana Regel ■
106201 Corydalis ochroleuca Koch;乳黄堇;Pale Corydalis ■☆
106202 Corydalis ochroleuca Koch = Pseudofumaria alba (Mill.) Lidén ■☆
106203 Corydalis octocornuta C. Y. Wu = Corydalis trachycarpa Maxim. var. octocornuta (C. Y. Wu) C. Y. Wu ■
106204 Corydalis odontostigma Fedde = Corydalis adunca Maxim. ■
106205 Corydalis oldhamii Koidz. = Corydalis heterocarpa Siebold et Zucc. ■
106206 Corydalis oligantha Ludlow et Stearn;少花紫堇;Fewflower Corydalis ■
106207 Corydalis oligosperma C. Y. Wu et T. Y. Shu;稀子黄堇(棉子黄堇);Fewseed Corydalis,Oligospermous Corydalis ■
106208 Corydalis omeiana (C. Y. Wu et H. Chuang) Z. Y. Su et Lidén = Corydalis flexuosa Franch. subsp. omeiana (C. Y. Wu et H. Chuang) C. Y. Wu ■
106209 Corydalis omphalocarpa Hayata = Corydalis balansae Prain ■
106210 Corydalis onobrychis Fedde;假驴豆■
106211 Corydalis onobrychis Fedde = Corydalis gortschakovii Schrenk ■
106212 Corydalis onobrychoides Fedde = Corydalis gortschakovii Schrenk ■
106213 Corydalis onobrychoides Fedde = Corydalis moorcroftiana Wall. ■
106214 Corydalis ophiocarpa Hook. f. et Thomson;蛇果黄堇(断肠草,扭果黄堇,日本紫堇,弯果黄堇,小前胡);Snakefruit Corydalis ■
106215 Corydalis oreocoma Lidén et Z. Y. Su;线足紫堇■
106216 Corydalis orthocarpa Hayata = Corydalis formosana Hayata ■
106217 Corydalis orthoceras Siebold et Zucc.;直角紫堇■☆
106218 Corydalis orthopoda Hayata;密花黄堇(小泉紫堇);Flowery Corydalis ■
106219 Corydalis osmastonii Fedde = Corydalis mira (Batalin) C. Y. Wu et H. Chuang ■
106220 Corydalis oxalidifolia Ludlow et Stearn;假酢浆草■
106221 Corydalis oxypetala Franch.;尖瓣紫堇;Tinepetal Corydalis ■
106222 Corydalis oxypetala Franch. subsp. balfouriana (Diels) Lidén;小花尖瓣紫堇■
106223 Corydalis pachycentra Franch.;浪穹紫堇;Thickspur Corydalis ■
106224 Corydalis pachypoda (Franch.) Hand.-Mazz.;粗梗黄堇(粗梗紫堇,马尾连,土黄连);Thickpedicelled Corydalis,Thickstalk Corydalis ■
106225 Corydalis paczoskii N. Busch;帕氏紫堇■☆
106226 Corydalis paeoniifolia (Steph.) Pers.;芍药叶紫堇■☆
106227 Corydalis pallida (Thunb.) Pers.;黄堇(黄花地丁,黄紫堇,鸡粪草,鸡爪莲,菊花黄连,千人耳子,山黄堇,深山黄堇,水黄连,土黄连,细深山黄堇,岩黄连,野芹菜,珠果黄堇);Pall Corydalis,Yellowflower Corydalis ■
106228 Corydalis pallida (Thunb.) Pers. var. chanetii (H. Lév.) Govaerts;河北黄堇■
106229 Corydalis pallida (Thunb.) Pers. var. chanetii (H. Lév.) Govaerts = Corydalis speciosa Maxim. ■
106230 Corydalis pallida (Thunb.) Pers. var. chanetii (H. Lév.) Govaerts = Corydalis wilfordii Regel ■
106231 Corydalis pallida (Thunb.) Pers. var. microphylla (Sasaki) Ohwi = Corydalis wilfordii Regel ■
106232 Corydalis pallida (Thunb.) Pers. var. platycarpa Maxim. ex Palib. = Corydalis heterocarpa Siebold et Zucc. ■
106233 Corydalis pallida (Thunb.) Pers. var. ramosissima Kom. = Corydalis pallida (Thunb.) Pers. ■
106234 Corydalis pallida (Thunb.) Pers. var. ramosissima Kom. = Corydalis speciosa Maxim. ■
106235 Corydalis pallida (Thunb.) Pers. var. sparsimamma (Ohwi) Ohwi;凹子黄堇■
106236 Corydalis pallida (Thunb.) Pers. var. sparsimamma (Ohwi) Ohwi = Corydalis wilfordii Regel ■
106237 Corydalis pallida (Thunb.) Pers. var. speciosa (Maxim.) Kom. = Corydalis speciosa Maxim. ■
106238 Corydalis pallida (Thunb.) Pers. var. tenuis Yatabe;细深山黄堇■
106239 Corydalis pallida (Thunb.) Pers. var. tenuis Yatabe = Corydalis pallida (Thunb.) Pers. ■
106240 Corydalis pallida (Thunb.) Pers. var. zhejiangensis Y. H. Zhang;浙江黄堇;Zhejiang Corydalis ■
106241 Corydalis pallida (Thunb.) Pers. var. zhejiangensis Y. H. Zhang = Corydalis wilfordii Regel ■
106242 Corydalis pallidiflora (Rupr.) N. Busch;苍白花紫堇■☆
106243 Corydalis panda Lidén et Y. W. Wang;熊猫紫堇(熊猫之友)■
106244 Corydalis paniculata C. Y. Wu et H. Chuang;散穗黄堇;Paniculate Corydalis ■
106245 Corydalis paniculata C. Y. Wu et H. Chuang = Corydalis hookeri Prain ■
106246 Corydalis paniculigera Regel et Schmalh. ex Regel;帕米尔黄堇;Pamir Corydalis ■
106247 Corydalis papilligera Ohwi;乳突紫堇■☆
106248 Corydalis papillipes C. Y. Wu;聂拉木黄堇;Nielamu Corydalis,

Papillastalk Corydalis ■
106249 Corydalis papillipes C. Y. Wu = Corydalis cavei D. G. Long ■
106250 Corydalis papillipes C. Y. Wu = Corydalis crispa Prain ■
106251 Corydalis papillosa Z. Y. Su et Lidén;冕宁紫堇■
106252 Corydalis parviflora Z. Y. Su et Lidén; 贵州黄堇; Guizhou Corydalis ■
106253 Corydalis pauciflora (Stephan ex Willd.) Pers. var. alaschanica Maxim. = Corydalis alaschanica (Maxim.) Peshkova ■
106254 Corydalis pauciflora (Stephan ex Willd.) Pers. var. albiflora A. E. Porsild = Corydalis pauciflora (Stephan) Pers. ■
106255 Corydalis pauciflora (Stephan) Pers.;少花延胡索(少花紫堇);Fewflower Corydalis ■
106256 Corydalis pauciflora (Stephan) Pers. = Bromus japonicus Thunb. ■
106257 Corydalis pauciflora (Stephan) Pers. = Corydalis alpestris C. A. Mey. ■
106258 Corydalis pauciflora (Stephan) Pers. = Corydalis diphylla Wall. ■☆
106259 Corydalis pauciflora (Stephan) Pers. var. alaschanica Maxim. = Corydalis alaschanica (Maxim.) Peschkova ■
106260 Corydalis pauciflora (Stephan) Pers. var. alpestris ? = Corydalis alpestris C. A. Mey. ■
106261 Corydalis pauciflora (Stephan) Pers. var. aquilegiifolia DC. = Corydalis pauciflora (Stephan) Pers. ■
106262 Corydalis pauciflora (Stephan) Pers. var. holanschanica Fedde = Corydalis alaschanica (Maxim.) Peschkova ■
106263 Corydalis pauciflora (Stephan) Pers. var. latiloba Maxim. = Corydalis tangutica Peschkova subsp. bullata (Lidén) Z. Y. Su ■
106264 Corydalis pauciflora (Stephan) Pers. var. latiloba Maxim. = Corydalis tangutica Peschkova ■
106265 Corydalis pauciflora (Stephan) Pers. var. latiloba Maxim. = Corydalis alpestris C. A. Mey. ■
106266 Corydalis pauciflora (Stephan) Pers. var. nivalis ? = Corydalis alpestris C. A. Mey. ■
106267 Corydalis peltata Lidén et Z. Y. Su; 盾萼紫堇; Peltatesepal Corydalis ■
106268 Corydalis persica Cham. et Schltdl.;波斯紫堇■
106269 Corydalis petrodoxa Lidén et Z. Y. Su;喜石黄堇■
106270 Corydalis petrophila Franch.;岩生紫堇(椭果紫堇);Rocky Corydalis, Saxicolous Corydalis ■
106271 Corydalis petrophila Franch. = Corydalis smithiana Fedde ■
106272 Corydalis physocarpa Cambess. = Corydalis crassifolia Royle ■☆
106273 Corydalis pingwuensis C. Y. Wu;平武紫堇(断肠草,飞燕草,蓝花紫堇);Pingwu Corydalis ■
106274 Corydalis pinnata Lidén et Z. Y. Su; 羽叶紫堇; Pinnateleaf Corydalis ■
106275 Corydalis pinnatibracteata Y. W. Wang;羽苞黄堇■
106276 Corydalis platycarpa (Maxim.) Makino = Corydalis heterocarpa Siebold et Zucc. ■
106277 Corydalis platycarpa Makino = Corydalis heterocarpa Siebold et Zucc. ■
106278 Corydalis polygalina Hook. f. et Thomson;远志黄堇(远志紫堇);Milkwort Corydalis ■
106279 Corydalis polygalina Hook. f. et Thomson var. micrantha C. Y. Wu;小花远志黄堇; Littleflower Milkwort Corydalis, Smallflower Milkwort Corydalis ■
106280 Corydalis polyphylla Hand.-Mazz.;多叶紫堇;Leafy Corydalis, Polyphyllous Corydalis ■
106281 Corydalis popovii Nevski ex Popov;白距紫堇■☆
106282 Corydalis porphyrantha C. Y. Wu;紫花紫堇;Purple Corydalis ■
106283 Corydalis potaninii Maxim.;半裸茎黄堇;Potanin Corydalis ■
106284 Corydalis praecipitorum C. Y. Wu, Z. Y. Su et Lidén;峭壁紫堇; Precipice Corydalis ■
106285 Corydalis prainiana Kanodia et S. K. Mukerjee = Corydalis casimiriana Duthie et Prain ex Prain ■
106286 Corydalis prattii Franch.;草甸黄堇;Pratt Corydalis ■
106287 Corydalis procera Lidén et Z. Y. Su;白花紫堇■
106288 Corydalis pseudacropteryx Fedde = Corydalis acropteryx Fedde ■
106289 Corydalis pseudasterostigma Fedde = Corydalis duclouxii H. Lév. et Vaniot ■
106290 Corydalis pseudasterostigma Fedde = Corydalis taliensis Franch. ■
106291 Corydalis pseudoadoxa C. Y. Wu et H. Chuang;波密紫堇(藏天葵叶紫堇,天葵叶紫堇);Bomi Corydalis, False Muskroot Corydalis ■
106292 Corydalis pseudoadunca Popov;假钩状紫堇■☆
106293 Corydalis pseudo-alpestris Popov; 假高山延胡索; False Alp Corydalis ■
106294 Corydalis pseudo-asterostigma Fedde;假星头紫堇;Falseasterostigma Corydalis ■☆
106295 Corydalis pseudobalfouriana Lidén et Z. Y. Su; 弯梗紫堇; Curvestalk Corydalis ■
106296 Corydalis pseudobarbisepala Fedde;假髯萼紫堇(假髯萼黄堇);False Beardsepal Corydalis ■
106297 Corydalis pseudoclematis Fedde = Corydalis davidii Franch. ■
106298 Corydalis pseudocristata Fedde; 美花黄堇; False Crested Corydalis ■
106299 Corydalis pseudocrithmifolia Jafri = Corydalis tibetica Hook. f. et Thomson ■
106300 Corydalis pseudodensispica Z. Y. Su et Lidén;假密穗黄堇■
106301 Corydalis pseudodrakeana Lidén; 甲格黄堇; False Drake Corydalis ■
106302 Corydalis pseudofargesii H. Chuang;假北岭黄堇;False Farges Corydalis ■
106303 Corydalis pseudofilisecta Lidén et Z. Y. Su;假丝叶紫堇;False Silkleaf Corydalis ■
106304 Corydalis pseudofluminicola Fedde;假溪畔紫堇(假多叶黄堇,拟溪边黄堇)■
106305 Corydalis pseudofluminicola Fedde = Corydalis corymbosa C. Y. Wu et T. Y. Shu ■
106306 Corydalis pseudohamata Fedde;川北钩距黄堇;False Hookspur Corydalis ■
106307 Corydalis pseudohamata Fedde = Corydalis hamata Franch. ■
106308 Corydalis pseudoheterocentra Fedde = Corydalis flexuosa Franch. subsp. pseudoheterocentra (Fedde) Lidén ex C. Y. Wu ■
106309 Corydalis pseudoheterocentra Fedde = Corydalis flexuosa Franch. subsp. pseudoheterocentra (Fedde) Lidén ■
106310 Corydalis pseudoimpatiens Fedde;假塞北紫堇;False Impatiens Corydalis ■
106311 Corydalis pseudoincisa C. Y. Wu, Z. Y. Su et Lidén;假刻叶紫堇;False Gapleaf Corydalis ■
106312 Corydalis pseudojuncea Ludlow et Stearn;拟裸茎黄堇;False Juncus Corydalis, False Nakedstem Corydalis ■
106313 Corydalis pseudolluminicola Fedde;假多叶黄堇;False Leafy Corydalis ■
106314 Corydalis pseudolongipes Lidén; 短腺黄堇; False Longstalk

Corydalis ■

106315 Corydalis pseudomairei C. Y. Wu ex Z. Y. Su et Lidén;会泽大花紫堇(大花会泽紫堇)■

106316 Corydalis pseudomicrophylla Z. Y. Su;假小叶黄堇;Falselittleleaf Corydalis ■

106317 Corydalis pseudomucronata C. Y. Wu;长突尖紫堇■

106318 Corydalis pseudomucronata C. Y. Wu Z. Y. Su et Lidén var. cristata C. Y. Wu = Corydalis amplisepala Z. Y. Su et Lidén ■

106319 Corydalis pseudorupestris Lidén et Z. Y. Su;短葶黄堇(岩黄连);Shortscape Corydalis ■

106320 Corydalis pseudoscaberula Lidén et Z. Y. Su = Corydalis scaberula Maxim. ■

106321 Corydalis pseudoschlechteriana Fedde = Corydalis atuntsuensis W. W. Sm. ■

106322 Corydalis pseudoschlechteriana Fedde = Corydalis eugeniae Fedde ■

106323 Corydalis pseudosibirica Lidén et Z. Y. Su;假北紫堇■

106324 Corydalis pseudothyrsiflora C. Y. Wu et T. Y. Shu = Corydalis crispa Prain ■

106325 Corydalis pseudotomentella Fedde = Corydalis balansae Prain ■

106326 Corydalis pseudotongolensis Lidén;假全冠黄堇;False Tongol Corydalis ■

106327 Corydalis pseudoweigoldii Z. Y. Su;假川西紫堇;Weigold Corydalis ■

106328 Corydalis psudomucronata C. Y. Wu,Z. Y. Su et Lidén;长尖突紫堇(断肠草,牛尿草,野指甲花)■

106329 Corydalis psudomucronata C. Y. Wu, Z. Y. Su et Lidén var. cristata C. Y. Wu;圆萼长尖突紫堇(圆萼紫堇,老老嫩)■

106330 Corydalis pterophora Ohwi = Corydalis speciosa Maxim. ■

106331 Corydalis pterygopetala Hand. -Mazz. ;翅瓣黄堇(翅瓣紫堇,断肠草);Winged Petal Corydalis, Wingpetal Corydalis ■

106332 Corydalis pterygopetala Hand. -Mazz. var. divaricata Z. Y. Su et Lidén;展枝翅瓣黄堇■

106333 Corydalis pterygopetala Hand. -Mazz. var. ecristata H. Chuang;无冠翅瓣黄堇;Crownless Wingpetal Corydalis ■

106334 Corydalis pterygopetala Hand. -Mazz. var. megalantha (Diels) Lidén et Z. Y. Su;大花翅瓣黄堇■

106335 Corydalis pterygopetala Hand. -Mazz. var. parviflora Lidén;小花翅瓣黄堇■

106336 Corydalis pubescens C. Y. Wu et H. Chuang = Corydalis longipes DC. var. pubescens (C. Y. Wu et H. Chuang) C. Y. Wu ■

106337 Corydalis pubicaulis C. Y. Wu et H. Chuang;毛茎紫堇;Hairstem Corydalis ■

106338 Corydalis pulchella Franch. = Corydalis adrienii Prain ■

106339 Corydalis pulchella Franch. = Corydalis melanochlora Maxim. ■

106340 Corydalis punicea C. Y. Wu ex Govaerts;玫瑰红堇(红花紫堇);Pinkflower Corydalis, Redflower Corydalis ■

106341 Corydalis punicea C. Y. Wu ex Govaerts = Corydalis livida Maxim. ■

106342 Corydalis purpureocalcarata C. Y. Wu et Z. Y. Su;紫距黄堇;Purplespur Corydalis ■

106343 Corydalis purpureocalcarata C. Y. Wu et Z. Y. Su = Corydalis stracheyi Duthie ex Prain ■

106344 Corydalis pycnopus Lidén;巨萼紫堇■

106345 Corydalis pygmaea C. Y. Wu et Z. Y. Su;矮黄堇;Dwarf Corydalis, Short Corydalis ■

106346 Corydalis qinghaiensis Z. Y. Su et Lidén;青海黄堇;Qinghai Corydalis ■

106347 Corydalis quadriflora Hand. -Mazz. = Corydalis trifoliolata Franch. ■

106348 Corydalis quantmeyeriana Fedde;掌苞紫堇;Palmbract Corydalis ■

106349 Corydalis quinquefoliolata Ludlow et Stearn;朗县黄堇;Fivefoliolate Corydalis, Langxian Corydalis ■

106350 Corydalis racemosa (Thunb.) Pers. ;小花黄堇(白刺梨果,白断肠草,断肠草,粪桶草,黄荷包牡丹,黄花地锦苗,黄花鱼灯草,黄堇,石莲,水黄连,虾子草,小花紫堇,烟紫堇,野水芹,鱼子草);Fumitory Corydalis, Raceme Corydalis, Racemose Corydalis ■

106351 Corydalis racemosa (Thunb.) Pers. = Corydalis bungeana Turcz. ■

106352 Corydalis racemosa (Thunb.) Pers. var. ecalcarata Z. Y. Su = Corydalis balansae Prain ■

106353 Corydalis racemosa (Thunb.) Pers. var. ecalcarata Z. Y. Su = Corydalis ecalcarata (Z. Y. Su) Y. H. Zhang ■

106354 Corydalis racemosa Bunge = Corydalis bungeana Turcz. ■

106355 Corydalis raddeana Regel;小黄紫堇(黄花地丁,蔓黄堇);Radde Corydalis, Small Ochotsk Corydalis ■

106356 Corydalis radicans Hand. -Mazz. ;裂瓣紫堇;Radicat Corydalis, Rooting Corydalis ■

106357 Corydalis ramosa Hook. f. et Thomson = Corydalis stracheyi Duthie ex Prain ■

106358 Corydalis redowskii Fedde;列氏紫堇■

106359 Corydalis regia Z. Y. Su et Lidén;高雅紫堇■

106360 Corydalis remota Fisch. ex Maxim. = Corydalis turtschaninovii Besser ■

106361 Corydalis remota Fisch. ex Maxim. f. haitaoensis (Y. H. Chou et C. Q. Xu) C. Y. Wu et Z. Y. Su = Corydalis gamosepala Maxim. ■

106362 Corydalis remota Fisch. ex Maxim. f. heteroclita (K. T. Fu) C. Y. Wu et Z. Y. Su = Corydalis gamosepala Maxim. ■

106363 Corydalis remota Fisch. ex Maxim. f. lineariloba (Maxim.) C. Y. Wu et Z. Y. Su = Corydalis remota Fisch. ex Maxim. var. lineariloba Maxim. ■

106364 Corydalis remota Fisch. ex Maxim. f. nonapiculata (Ohwi) C. Y. Wu et Z. Y. Su = Corydalis gamosepala Maxim. ■

106365 Corydalis remota Fisch. ex Maxim. f. papillosa (Kitag.) C. Y. Wu et Z. Y. Su = Corydalis turtschaninovii Besser subsp. vernyi (Franch. et Sav.) Lidén ■

106366 Corydalis remota Fisch. ex Maxim. f. punctata Skvortsov;斑叶延胡索;Punctate Corydalis ■☆

106367 Corydalis remota Fisch. ex Maxim. var. fumariifolia (Maxim.) Kom. = Corydalis fumariifolia Maxim. ■

106368 Corydalis remota Fisch. ex Maxim. var. heteroclita K. T. Fu;山延胡索(异叶齿瓣延胡索);Differentleaf Corydalis ■

106369 Corydalis remota Fisch. ex Maxim. var. heteroclita K. T. Fu = Corydalis gamosepala Maxim. ■

106370 Corydalis remota Fisch. ex Maxim. var. lineariloba Maxim. ;线齿瓣延胡索(狭裂延胡索,线叶齿瓣延胡索,元胡);Linearlobe Corydalis ■

106371 Corydalis remota Fisch. ex Maxim. var. lineariloba Maxim. = Corydalis turtschaninovii Besser ■

106372 Corydalis remota Fisch. ex Maxim. var. papillosa (Kitag.) Baranov et Skvortsov;瘤叶延胡索;Papillalaef Corydalis ■

106373 Corydalis remota Fisch. ex Maxim. var. papillosa (Kitag.) Baranov et Skvortzov = Corydalis turtschaninovii Besser subsp. vernyi

(Franch. et Sav.) Lidén ■

106374 Corydalis remota Fisch. ex Maxim. var. papillosa (Kitag.) Baranov et Skvortsov = Corydalis turtschaninovii Besser ■

106375 Corydalis remota Fisch. ex Maxim. var. pectinata Kom. = Corydalis fumariifolia Maxim. ■

106376 Corydalis remota Fisch. ex Maxim. var. punctata Skvortzov = Corydalis turtschaninovii Besser ■

106377 Corydalis remota Fisch. ex Maxim. var. rotundiloba Maxim. = Corydalis turtschaninovii Besser ■

106378 Corydalis remota Fisch. ex Maxim. var. ternata (Nakai) Makino = Corydalis ternata (Nakai) Nakai ■

106379 Corydalis remota Fisch. ex Maxim. var. ternata Makino = Corydalis ternata (Nakai) Nakai ■

106380 Corydalis repens Mandl et Muehld.;全叶延胡索(苏延胡,土玄胡,土延胡,土元胡,玄胡);Creeping Corydalis, Entireleaf Corydalis ■

106381 Corydalis repens Mandl et Muehld. var. humosoides Y. H. Zhang = Corydalis caudata (Lam.) Pers. ■

106382 Corydalis repens Mandl et Muehld. var. jiangsuensis Y. H. Zhang = Corydalis caudata (Lam.) Pers. ■

106383 Corydalis repens Mandl et Muehld. var. manshurica Skvortzov = Corydalis repens Mandl et Muehld. ■

106384 Corydalis repens Mandl et Muehld. var. pubescens Skvortzov = Corydalis repens Mandl et Muehld. ■

106385 Corydalis repens Mandl et Muehld. var. watanabei (Kitag.) Y. H. Chou;角瓣延胡索■

106386 Corydalis repens Mandl et Muehld. var. watanabei (Kitag.) Y. H. Chou = Corydalis watanabei Kitag. ■

106387 Corydalis retingensis Ludlow;囊果紫堇;Corydalis, Sacfruit Corydalis ■

106388 Corydalis rheinbabeniana Fedde;扇苞黄堇;Fanbract Corydalis, Fanshaped-bract Corydalis ■

106389 Corydalis rheinbabeniana Fedde var. leioneura H. Chuang;无毛扇苞黄堇;Smooth Fanbract Corydalis ■

106390 Corydalis rockiana C. Y. Wu, Z. Y. Su et Lidén = Corydalis nigro-apiculata C. Y. Wu ■

106391 Corydalis rockii Fedde = Corydalis petrophila Franch. ■

106392 Corydalis rorida H. Chuang;露点紫堇;Dew Corydalis ■

106393 Corydalis rosea Maxim. = Corydalis livida Maxim. ■

106394 Corydalis rosea Maxim. = Corydalis punicea C. Y. Wu ex Govaerts ■

106395 Corydalis rosea Steud.;粉红紫堇■☆

106396 Corydalis roseotincta C. Y. Wu et H. Chuang;拟鳞叶紫堇;False Scaleleaf Corydalis ■

106397 Corydalis roseotincta C. Y. Wu et H. Chuang = Corydalis melanochlora Maxim. ■

106398 Corydalis rostellata Lidén;具喙黄堇■

106399 Corydalis rotundiloba (Maxim.) C. Y. Wu et Z. Y. Su;圆齿瓣延胡索■

106400 Corydalis rubrisepala Lidén subsp. zhuangiana Lidén;西藏红萼黄堇■

106401 Corydalis rupifraga C. Y. Wu et Z. Y. Su;石隙紫堇;Rockgap Corydalis ■

106402 Corydalis saccata Z. Y. Su et Lidén;囊瓣延胡索;Sacpetal Corydalis ■

106403 Corydalis saltatoria W. W. Sm.;中缅黄堇■

106404 Corydalis sarcolepis Lidén et Z. Y. Su;肉鳞紫堇■

106405 Corydalis saxicola Bunting;石生黄堇(白蓬紫堇,黄连,鸡爪连,菊花黄连,石生黄连,土黄连,岩胡,岩黄堇,岩黄连,岩连);Lithological Corydalis, Rockliving Corydalis ■

106406 Corydalis saxicola Bunting var. pasuiflora C. Y. Wu et T. Y. Shu;岩黄堇(岩川芎)■

106407 Corydalis scaberula Maxim.;粗糙黄堇(粗糙紫堇,粗毛黄堇);Rugged Corydalis, Scabrate Corydalis ■

106408 Corydalis scaberula Maxim. var. glabra Z. C. Zuo et L. H. Zhou;无毛粗糙黄堇■

106409 Corydalis scaberula Maxim. var. purpurescens C. Y. Wu;紫花粗糙黄堇;Purpleflower Rugged Corydalis, Purpleflower Scabrate Corydalis ■

106410 Corydalis scaberula Maxim. var. ramifera C. Y. Wu et H. Chuang;分枝粗糙黄堇;Branchy Rugged Corydalis ■

106411 Corydalis scaberula Maxim. var. ramifera C. Y. Wu et H. Chuang = Corydalis scaberula Maxim. ■

106412 Corydalis scaphopetala Fedde = Corydalis adunca Maxim. subsp. scaphopetala (Fedde) C. Y. Wu et Z. Y. Su ■

106413 Corydalis scaphopetala Fedde = Corydalis adunca Maxim. ■

106414 Corydalis schanginii (Pall.) B. Fedtsch.;长距元胡(长花延胡索,新疆元胡,元胡);Longspur Corydalis, Schangin Corydalis ■

106415 Corydalis schelesnowiana Regel et Schmalh.;塞氏紫堇■☆

106416 Corydalis schistostigma X. F. Gao;裂柱紫堇■

106417 Corydalis schlagintweitii Fedde = Corydalis stricta Stephan ex DC. ■

106418 Corydalis schlechteriana Fedde = Corydalis linearioides Maxim. ■

106419 Corydalis schochii Fedde = Corydalis duclouxii H. Lév. et Vaniot ■

106420 Corydalis schochii Fedde = Corydalis taliensis Franch. ■

106421 Corydalis schusteriana Fedde;甘洛紫堇;Schuster Corydalis ■

106422 Corydalis schusteriana Fedde var. crassirhizomata C. Y. Wu;粗茎紫堇■

106423 Corydalis schusteriana Fedde var. crassirhizomata C. Y. Wu = Corydalis weigoldii Fedde ■

106424 Corydalis schusteriana Fedde var. crassirhizomata C. Y. Wu = Corydalis crassirhizomata (C. Y. Wu) C. Y. Wu ■

106425 Corydalis schweriniana Fedde;巧家紫堇(冕宁紫堇);Sehwerin Corydalis ■

106426 Corydalis scouleri Hook.;斯库勒紫堇■☆

106427 Corydalis semenovii Regel et Herder;中亚紫堇(天山黄堇,天山紫堇);Semenov Corydalis ■

106428 Corydalis semiaquilegiifolia C. Y. Wu et H. Chuang;天葵叶紫堇■

106429 Corydalis semiaquilegiifolia C. Y. Wu et H. Chuang = Corydalis pseudoadoxa C. Y. Wu et H. Chuang ■

106430 Corydalis sempervirens (L.) Pers.;常绿紫堇;Harlequin-flower, Pale Corydalis, Pink and Yellow Corydalis, Pink Corydalis, Rock Harlequin ■☆

106431 Corydalis sewerzovii Regel;大苞延胡索(谢氏紫堇);Bigbract Corydalis, Sewerzov Corydalis ■

106432 Corydalis sheareri S. Moore;地锦苗(大流尿草,大羊不吃草,断肠草,飞菜,高山羊不吃,荷包牡丹,红花鸡距草,护心胆,尖距紫堇,苦心胆,鹿耳草,牛奶七,牛屎草,芹菜,三月烂,山芹菜,蛇含七,铁板道人);Sharpspur Corydalis, Shearer Corydalis ■

106433 Corydalis sheareri S. Moore var. bulbillifera Hand. -Mazz.;珠芽地锦苗(一串金丹,珠芽紫堇);Bulbilferous Corydalis ■

106434 Corydalis sheareri S. Moore var. changyangensis Fedde =

Corydalis sheareri S. Moore ■

106435 Corydalis shennongensis H. Chuang; 鄂西黄堇; Shenlong Corydalis, Shennong Corydalis ■

106436 Corydalis shensiana Lidén;陕西紫堇(长距曲花紫堇,秦岭弯花紫堇);Shaanxi Corydalis ■

106437 Corydalis sherriffii Ludlow;巴嘎紫堇;Sherriff Corydalis ■

106438 Corydalis shimienensis C. Y. Wu et Z. Y. Su;石棉紫堇(倒地掬,断肠草);Shimian Corydalis ■

106439 Corydalis siamensis Craib;泰国紫堇;Siam Corydalis ■☆

106440 Corydalis siamensis Craib = Corydalis leptocarpa Hook. f. et Thomson ■

106441 Corydalis sibirica (L. f.) Pers.;北紫堇■

106442 Corydalis sibirica (L. f.) Pers. = Corydalis solida (L.) Clairv. ■

106443 Corydalis sibirica (L. f.) Pers. subsp. elata Lidén = Corydalis pseudosibirica Lidén et Z. Y. Su ■

106444 Corydalis sibirica (L. f.) Pers. subsp. impatiens (Pall.) A. Gubanov = Corydalis impatiens (Pall.) Fisch. ■

106445 Corydalis sibirica (L. f.) Pers. var. impatiense (Pall.) Regel = Corydalis impatiens (Pall.) Fisch. ■

106446 Corydalis sibirica Pers. = Corydalis longipes DC. var. pubescens (C. Y. Wu et H. Chuang) C. Y. Wu ■

106447 Corydalis sibirica Pers. = Corydalis longipes DC. ■

106448 Corydalis sibirica Pers. var. impatiens (Pall.) Regel = Corydalis impatiens (Pall.) Fisch. ■

106449 Corydalis sigmantha Z. Y. Su et C. Y. Wu;甘南紫堇;Gannan Corydalis ■

106450 Corydalis sigmoides C. Y. Wu et H. Chuang;宝兴黄堇■

106451 Corydalis sigmoides C. Y. Wu et H. Chuang = Corydalis linearis C. Y. Wu ■

106452 Corydalis smithiana Fedde;箐边紫堇;Smith Corydalis ■

106453 Corydalis solida (L.) Clairv.;多花延胡索(阿尔泰黄堇,阿山黄堇,阿山紫堇,北紫堇,哈勒氏紫堇,山延胡索,西伯利亚紫堇);Bird-in-a-bush, Bird-on-a-thorn, Bulbous Fumitory, Kalmuk Corydalis, Noble Corydalis, North Corydalis, Purple Corydalls, Siberian Corydalis, Spring Fumewort ■

106454 Corydalis solida (L.) Clairv. 'G. P. Baker' = Corydalis solida (L.) Clairv. 'George Baker' ■☆

106455 Corydalis solida (L.) Clairv. 'George Baker';乔治·贝克多花延胡索■☆

106456 Corydalis solida (L.) Clairv. subsp. densiflora (C. Presl) Hayek;密花紫堇■☆

106457 Corydalis solida (L.) Clairv. subsp. remota (Fisch. ex Maxim.) Korsh. = Corydalis turtschaninovii Besser ■

106458 Corydalis solida (L.) Clairv. subsp. remota (Maxim.) Korsh. = Corydalis turtschaninovii Besser ■

106459 Corydalis solida (L.) Clairv. var. bracteosa Batt. et Trab. = Corydalis solida (L.) Clairv. subsp. densiflora (C. Presl) Hayek ■☆

106460 Corydalis sophronitis Z. Y. Su et Lidén;石渠黄堇■

106461 Corydalis souliei Franch. = Corydalis calcicola W. W. Sm. ■

106462 Corydalis souliei Franch. = Corydalis trachycarpa Maxim. ■

106463 Corydalis sparsimamma Ohwi = Corydalis pallida (Thunb.) Pers. var. sparsimamma (Ohwi) Ohwi ■

106464 Corydalis sparsimamma Ohwi = Corydalis wilfordii Regel ■

106465 Corydalis spathulata Prain ex Craib;匙苞黄堇(匙苞紫堇,东丝儿);Spathulate Corydalis, Spoonbract Corydalis ■

106466 Corydalis speciosa Maxim.;珠果黄堇(河北黄堇,胡黄堇,狭裂珠果黄堇);Beautiful Corydalis ■

106467 Corydalis speciosa Maxim. var. pterophora (Ohwi) Ohwi = Corydalis speciosa Maxim. ■

106468 Corydalis speciosa Maxim. var. ramosissima (Kom.) Kitag. = Corydalis speciosa Maxim. ■

106469 Corydalis spinulosa H. Chuang;刺毛黄堇;Spine Corydalis ■

106470 Corydalis spinulosa H. Chuang = Corydalis atuntsuensis W. W. Sm. ■

106471 Corydalis squamigera Z. Y. Su;具鳞黄堇;Squamate Corydalis ■

106472 Corydalis squamigera Z. Y. Su = Corydalis ellipticarpa C. Y. Wu et Z. Y. Su ■

106473 Corydalis stenantha Franch.;洱源紫堇(白屈莱状紫堇,地锦苗,金钩如意草,水金钩如意,五味草,紫堇);Eryuan Corydalis ■

106474 Corydalis stenantha Franch. = Corydalis duclouxii H. Lév. et Vaniot ■

106475 Corydalis stewartii Fedde = Corydalis cornuta Royle ■

106476 Corydalis stolonifera Lidén;匍匐茎紫堇■

106477 Corydalis stracheyi Duthie ex Prain;折曲黄堇;Flex Corydalis, Strachey Corydalis ■

106478 Corydalis stracheyi Duthie ex Prain var. ecristata Prain;无冠折曲黄堇;Crownless Flex Corydalis ■

106479 Corydalis stracheyioides Fedde; 变冠黄堇; Like Strachey Corydalis ■☆

106480 Corydalis stracheyioides Fedde = Corydalis crispa Prain ■

106481 Corydalis straminea Maxim. = Corydalis stramineoides C. Y. Wu et Z. Y. Su ■

106482 Corydalis straminea Maxim. ex Hemsl.;草黄堇(草黄花紫堇);Herb Corydalis, Strawcoloured Corydalis ■

106483 Corydalis straminea Maxim. ex Hemsl. var. megacalyx Z. Y. Su;大萼草黄堇■

106484 Corydalis straminea Maxim. var. megacalyx Z. Y. Su = Corydalis straminea Maxim. ex Hemsl. var. megacalyx Z. Y. Su ■

106485 Corydalis stramineoides C. Y. Wu et Z. Y. Su;索县黄堇;Suoxian Corydalis ■

106486 Corydalis streptocarpa Maxim. = Corydalis ophiocarpa Hook. f. et Thomson ■

106487 Corydalis striatocarpa H. Chuang; 纹果紫堇; Striatefruit Corydalis ■

106488 Corydalis stricta Steph. ex DC.;直茎黄堇(劲直黄堇,玉门透骨草,直立黄堇,直立紫堇);Erect Corydalis, Strict Corydalis ■

106489 Corydalis stricta Stephan ex DC. subsp. holosepala Michajlova = Corydalis stricta Steph. ex DC. ■

106490 Corydalis stricta Stephan ex DC. subsp. spathosepala Michajlova = Corydalis stricta Steph. ex DC. ■

106491 Corydalis stricta Stephan ex DC. var. potaninii Fedde = Corydalis stricta Steph. ex DC. ■

106492 Corydalis suaveolens Hance = Corydalis sheareri S. Moore ■

106493 Corydalis susannae Lidén;幽溪紫堇■

106494 Corydalis suzhiyunii Lidén;茎节生根紫堇■

106495 Corydalis taipaishanica H. Chuang; 太白紫堇; Taibaishan Corydalis ■

106496 Corydalis taipaishanica H. Chuang = Corydalis jingyuanensis C. Y. Wu et H. Chuang ■

106497 Corydalis taitoensis Hayata = Corydalis balansae Prain ■

106498 Corydalis taiwanensis Ohwi = Corydalis pallida (Thunb.) Pers. ■

106499 Corydalis taiwanensis Ohwi = Corydalis wilfordii Regel ■

106500 Corydalis taliensis Franch.;金钩如意草(大理紫堇,大理紫堇草,地锦苗,断肠草,金钩黄堇,金钩如玉草,苦地丁,如意草,水

黄连,水金钩如意,水晶金钩如意草,五味草);Dali Corydalis,Tali Corydalis ■

106501 Corydalis taliensis Franch. var. bulleyana (Diels) C. Y. Wu et H. Chuang = Corydalis bulleyana Diels ■

106502 Corydalis taliensis Franch. var. ecristata Hand. -Mazz. = Corydalis duclouxii H. Lév. et Vaniot ■

106503 Corydalis taliensis Franch. var. ecristata Hand. -Mazz. = Corydalis stenantha Franch. ■

106504 Corydalis taliensis Franch. var. patentillifolia C. Y. Wu et H. Chuang;禄春金钩如意草(绿春金钩如意草)■

106505 Corydalis taliensis Franch. var. siamensis (Craib) H. Chuang = Corydalis leptocarpa Hook. f. et Thomson ■

106506 Corydalis tangutica Peschkova;唐古特延胡索;Tangut Corydalis ■

106507 Corydalis tangutica Peschkova subsp. bullata (Lidén) Z. Y. Su;长轴唐古特延胡索;Bullate Tangut Corydalis ■

106508 Corydalis tashiroi Makino;台湾黄堇;Taiwan Corydalis ■

106509 Corydalis tashiroi Makino = Corydalis balansae Prain ■

106510 Corydalis temolana C. Y. Wu et H. Chuang;黄绿紫堇;Yellow-green Corydalis ■

106511 Corydalis temulifolia Franch.;大叶紫堇(城口紫堇,断肠草,冷草,闷头花,山臭草);Bigleaf Corydalis,Largeleaf Corydalis ■

106512 Corydalis temulifolia Franch. subsp. aegopodioides (H. Lév. et Vaniot) C. Y. Wu;鸡血七(断肠草,人血七)■

106513 Corydalis tenantha Franch. = Corydalis duclouxii H. Lév. et Vaniot ■

106514 Corydalis tenella Kar. et Kir. = Corydalis inconspicua Bunge ex Ledeb. ■

106515 Corydalis tenerrima C. Y. Wu, Z. Y. Su et Lidén;柔弱黄堇;Weak Corydalis ■

106516 Corydalis tenuicalcarata C. Y. Wu et T. Y. Shu;细距黄堇;Finespur Corydalis ■

106517 Corydalis tenuicalcarata C. Y. Wu et T. Y. Shu = Corydalis chamdoensis C. Y. Wu et H. Chuang ■

106518 Corydalis tenuipes Lidén et Z. Y. Su;细柄黄堇■

106519 Corydalis ternata (Nakai) Nakai;三裂延胡索(三出延胡索,中井黄堇);Nakai Corydalis ■

106520 Corydalis ternata (Nakai) Nakai f. yanhusuo (Y. H. Chou et C. C. Hsu) Y. C. Zhu = Corydalis yanhusuo W. T. Wang ex Z. Y. Su et C. Y. Wu ■

106521 Corydalis ternatifolia C. Y. Wu, Z. Y. Su et Lidén;神农架紫堇;Ternateleaf Corydalis ■

106522 Corydalis thalictrifolia Franch. = Corydalis saxicola Bunting ■

106523 Corydalis thalictrifolia Jameson ex Regel = Corydalis cornuta Royle ■

106524 Corydalis thyrsiflora Prain;锥花黄堇;Thyrse Corydalis ■☆

106525 Corydalis thyrsiflora Prain = Corydalis hookeri Prain ■

106526 Corydalis thyrsiflora Prain var. minor C. Y. Wu;小锥花黄堇;Small Thyrse Corydalis ■

106527 Corydalis tianshanica Lidén;天山黄堇■

106528 Corydalis tianzhuensis M. S. Yan et C. J. Wang subsp. bullata Lidén = Corydalis tangutica Peschkova subsp. bullata (Lidén) Z. Y. Su ■

106529 Corydalis tianzhuensis M. S. Yan et Ching J. Wang;天祝黄堇;Tianzhu Corydalis ■

106530 Corydalis tianzhuensis M. S. Yan et Ching J. Wang subsp. bullata Lidén = Corydalis tangutica Peschkova subsp. bullata (Lidén) Z. Y. Su ■

106531 Corydalis tibetica Hook. f. et Thomson;西藏黄堇;Xizang Corydalis ■

106532 Corydalis tibetica Hook. f. et Thomson = Corydalis pachypoda (Franch.) Hand. -Mazz. ■

106533 Corydalis tibetica Hook. f. et Thomson = Corydalis tibeto-alpina C. Y. Wu et Z. Y. Su ■

106534 Corydalis tibetica Hook. f. et Thomson var. pachypoda Franch. = Corydalis pachypoda (Franch.) Hand. -Mazz. ■

106535 Corydalis tibeto-alpina C. Y. Wu et Z. Y. Su;西藏高山紫堇(西藏对叶黄堇,西藏高山黄堇,西藏紫堇);Tibet Corydalis, Xizang Alp Corydalis ■

106536 Corydalis tibeto-oppositifolia C. Y. Wu et T. Y. Shu;西藏对叶黄堇;Xizang Oppositeleaf Corydalis ■

106537 Corydalis tomentella Franch.;毛黄堇(干岩矸,千岩堇,绒毛黄堇,三尖刀,土黄芩,岩黄连);Hair Corydalis ■

106538 Corydalis tomentosa N. E. Br. = Corydalis tomentella Franch. ■

106539 Corydalis tongolensis Franch.;全冠黄堇(东谷黄堇);Tongo Corydalis, Tongol Corydalis ■

106540 Corydalis tongolensis Franch. = Corydalis pseudotongolensis Lidén ■

106541 Corydalis trachycarpa Maxim.;糙果紫堇;Coarsefruit Corydalis, Roughfruit Corydalis ■

106542 Corydalis trachycarpa Maxim. = Corydalis calcicola W. W. Sm. ■

106543 Corydalis trachycarpa Maxim. var. leucostachya (C. Y. Wu et H. Chuang) C. Y. Wu;白穗紫堇(高山紫堇);Alp Corydalis, Whitestachys Corydalis ■

106544 Corydalis trachycarpa Maxim. var. nana C. Y. Wu et H. Chuang;小糙果紫堇■

106545 Corydalis trachycarpa Maxim. var. nana C. Y. Wu et H. Chuang = Corydalis trachycarpa Maxim. ■

106546 Corydalis trachycarpa Maxim. var. octocornuta (C. Y. Wu) C. Y. Wu;淡花紫堇(淡花黄堇,淡黄花黄堇);Eightcornut Corydalis, Octocrown Corydalis ■

106547 Corydalis trifoliolata Franch.;三裂紫堇(三裂黄堇);Trifoliolate Corydalis, Trileaf Corydalis ■

106548 Corydalis trigibbosa H. Chuang;三囊紫堇■

106549 Corydalis trigibbosa H. Chuang = Corydalis petrophila Franch. ■

106550 Corydalis trilobipetala Hand. -Mazz.;三裂瓣紫堇(裂瓣紫堇);Trilobatepetal Corydalis ■

106551 Corydalis trisecta Franch.;秦岭紫堇;Trisect Corydalis ■

106552 Corydalis triternata Franch. = Corydalis triternatifolia C. Y. Wu ■

106553 Corydalis triternatifolia C. Y. Wu;重三出黄堇;Triternateleaf Corydalis ■

106554 Corydalis tsangensis Lidén et Z. Y. Su;藏紫堇;Zang Corydalis ■

106555 Corydalis tsariensis Ludlow = Corydalis dubia Prain ■

106556 Corydalis tsariensis Ludlow et Stearn = Corydalis dubia Prain ■

106557 Corydalis tsayulensis C. Y. Wu et H. Chuang;察隅紫堇;Chayu Corydalis ■

106558 Corydalis tuberipisiformis Z. Y. Su;豌豆根紫堇;Pearoot Corydalis ■

106559 Corydalis tuberipisiformis Z. Y. Su = Corydalis dajingensis C. Y. Wu et Z. Y. Su ■

106560 Corydalis tuberosa DC.;块茎紫堇;Holewort, Hollow-root, Hollow-wort, Tuberose Corydalis ■☆

106561 Corydalis tuberosa L. = Corydalis cava (L.) Schweigg. et Körte ■☆

106562 Corydalis turczaninovii Besser = Corydalis turtschaninovii Besser ■

106563 Corydalis turczaninovii Besser subsp. vernyi (Franch. et Sav.) Lidén = Corydalis turtschaninovii Besser subsp. vernyi (Franch. et Sav.) Lidén ■

106564 Corydalis turtschaninovii Besser;齿瓣延胡索(东北延胡索,东北元胡,兰花菜,兰雀花,蓝花菜,蓝雀花,山延胡索,土元胡,狭裂延胡索,线齿瓣延胡索,线叶齿瓣延胡索,延胡索);Toothedpetal Corydalis,Turtschaninov Corydalis ■

106565 Corydalis turtschaninovii Besser f. fumariifolia (Maxim.) Y. H. Chou = Corydalis fumariifolia Maxim. ■

106566 Corydalis turtschaninovii Besser f. fumariifolia Y. C. Zhu = Corydalis fumariifolia Maxim. ■

106567 Corydalis turtschaninovii Besser f. haitaoensis Y. H. Chou et Ch. Q. Xu = Corydalis gamosepala Maxim. ■

106568 Corydalis turtschaninovii Besser f. lineariloba Kitag.;线裂齿瓣延胡索■

106569 Corydalis turtschaninovii Besser f. multisecta P. Y. Fu;多裂齿瓣延胡索;Multifid Turtschaninov Corydalis ■

106570 Corydalis turtschaninovii Besser f. yanhusuo Y. H. Chou et C. C. Hsu = Corydalis yanhusuo W. T. Wang ex Z. Y. Su et C. Y. Wu ■

106571 Corydalis turtschaninovii Besser subsp. vernyi (Franch. et Sav.) Lidén;少花齿瓣延胡索■

106572 Corydalis turtschaninovii Besser var. non-apiculata Ohwi = Corydalis gamosepala Maxim. ■

106573 Corydalis turtschaninovii Besser var. papillata (Ohwi) Ohwi = Corydalis repens Mandl et Muehld. ■

106574 Corydalis turtschaninovii Besser var. papillata Ohwi = Corydalis repens Mandl et Muehld. ■

106575 Corydalis turtschaninovii Besser var. papillosa Kitag. = Corydalis turtschaninovii Besser subsp. vernyi (Franch. et Sav.) Lidén ■

106576 Corydalis turtschaninovii Besser var. papillosa Kitag. = Corydalis turtschaninovii Besser ■

106577 Corydalis turtschaninovii Besser var. ternata (Nakai) Ohwi = Corydalis ternata (Nakai) Nakai ■

106578 Corydalis turtschaninovii Besser var. ternata Ohwi = Corydalis turtschaninovii Besser subsp. vernyi (Franch. et Sav.) Lidén ■

106579 Corydalis uranoscopa Lidén;立花黄堇■

106580 Corydalis urbaniana Fedde;紫苞黄堇;Violetbract Corydalis ■

106581 Corydalis urbaniana Fedde = Corydalis laucheana Fedde ■

106582 Corydalis urosepala Fedde;尾萼紫堇■☆

106583 Corydalis ussuriensis Aparina;吉林延胡索■

106584 Corydalis uvaria Lidén et Z. Y. Su;圆根紫堇;Grape Corydalis ■

106585 Corydalis vaginans Royle;具鞘紫堇■☆

106586 Corydalis varicolor C. Y. Wu;变色紫堇■

106587 Corydalis varicolor C. Y. Wu = Corydalis nigro-apiculata C. Y. Wu ■

106588 Corydalis vermicularis Lidén et Z. Y. Su = Corydalis brevirostrata C. Y. Wu et Z. Y. Su ■

106589 Corydalis verna Z. Y. Su et Lidén;春花紫堇■

106590 Corydalis vernyi Franch. et Sav. = Corydalis turtschaninovii Besser subsp. vernyi (Franch. et Sav.) Lidén ■

106591 Corydalis virginea Lidén et Z. Y. Su;腋含珠紫堇■

106592 Corydalis vivipara Fedde;胎生紫堇;Viviparous Corydalis,Vivipary Corydalis ■

106593 Corydalis wandoensis Y. Lee = Corydalis turtschaninovii Besser subsp. vernyi (Franch. et Sav.) Lidén ■

106594 Corydalis wardii C. Marquart et Airy Shaw = Corydalis dingdonis Airy Shaw ■

106595 Corydalis wardii W. W. Sm = Corydalis calcicola W. W. Sm. ■

106596 Corydalis wardii W. W. Sm.;滇西紫堇;Ward Corydalis ■

106597 Corydalis washingtoniana Fedde = Corydalis aurea Willd. ■☆

106598 Corydalis watanabei Kitag. = Corydalis repens Mandl et Muehld. var. watanabei (Kitag.) Y. H. Chou ■

106599 Corydalis weigoldii Fedde;川西紫堇;W. Sichuan Corydalis ■

106600 Corydalis weisiensis H. Chuang;维西黄堇;Weixi Corydalis ■

106601 Corydalis weisiensis H. Chuang = Corydalis lopinensis Franch. ■

106602 Corydalis wilfordii Regel;阜平黄堇;Fuping Corydalis ■

106603 Corydalis wilfordii Regel var. japonica Franch. et Sav. = Corydalis heterocarpa Siebold et Zucc. ■

106604 Corydalis wilsonii N. E. Br.;川鄂黄堇(川鄂紫堇,金花草,土黄连,岩黄连);E. H. Wilson Corydalis ■

106605 Corydalis wilsonii N. E. Br. = Corydalis latiloba (Franch.) Hand. -Mazz. ■

106606 Corydalis wumengensis C. Y. Wu;岩连(岩黄连,岩莲);Wumeng Corydalis ■

106607 Corydalis wuzhengyiana Z. Y. Su et Lidén;齿苞黄堇;Zhengyi Corydalis ■

106608 Corydalis yanhusuo W. T. Wang = Corydalis yanhusuo W. T. Wang ex Z. Y. Su et C. Y. Wu ■

106609 Corydalis yanhusuo W. T. Wang ex Z. Y. Su et C. Y. Wu;延胡索(玄胡索,延胡,元胡,元胡索);Yahusuo ■

106610 Corydalis yaoi Lidén et Z. Y. Su;覆鳞紫堇■

106611 Corydalis yargongensis C. Y. Wu;雅江紫堇;Yajiang Corydalis ■

106612 Corydalis yui Lidén;瘤籽黄堇;Yu Corydalis ■

106613 Corydalis yunnanensis Franch.;滇黄堇(黄水金钩如意);Yunnan Corydalis ■

106614 Corydalis yunnanensis Franch. var. megalantha Diels;大花滇黄堇;Bigflower Yunnan Corydalis,Largeflower Yunnan Corydalis ■

106615 Corydalis yunnanensis Franch. var. megalantha Diels = Corydalis pterygopetala Hand. -Mazz. var. megalantha (Diels) Lidén et Z. Y. Su ■

106616 Corydalis zadoiensis L. H. Zhou;杂多紫堇(扎多紫堇);Zaduo Corydalis ■

106617 Corydalis zambuii C. E. C. Fisch. et Kaul = Corydalis conspersa Maxim. ■

106618 Corydalis zeaensis Michajlova = Corydalis gigantea Trautv. et C. A. Mey. ■

106619 Corydalis zeaensis Michajlova = Corydalis macrantha (Regel) Popov ■

106620 Corydalis zhongdianensis C. Y. Wu et Lidén;中甸黄堇;Zhongdian Corydalis ■

106621 Corydallis Asch. = Corydalis DC. (保留属名)■

106622 Corydandra Rchb. = Galeandra Lindl. et Bauer ■☆

106623 Corylaceae Mirb. (1815)(保留科名);榛科(榛木科);Filbert Family,Hazal Family,Hazelnut Family ●

106624 Corylaceae Mirb. (保留科名) = Betulaceae Gray(保留科名)●

106625 Corylopasania (Hickel et A. Camus) Nakai = Lithocarpus Blume ●

106626 Corylopasania (Hickel et A. Camus) Nakai = Pasania (Miq.) Oerst. ●

106627 Corylopasania tubulosa (Hickel et A. Camus) Nakai = Lithocarpus tubulosus (Hickel et A. Camus) A. Camus ●

106628 Corylopsis Siebold et Zucc. (1836);蜡瓣花属(瑞木属);Corylopsis,Waxpetal,Winter Hazel,Winterhazel,Winter-hazel ●

106629 Corylopsis alnifolia (H. Lév.) C. K. Schneid.;桤叶蜡瓣花;Alderleaf Waxpetal,Alderleaf Winterhazel,Alder-leaved Winterhazel ●

106630 Corylopsis brevistyla Hung T. Chang;短柱蜡瓣花;Shortstyle Waxpetal,Shortstyle Winterhazel,Short-styled Winterhazel ●

106631 Corylopsis calcicola C. Y. Wu;灰岩蜡瓣花;Limestone Waxpetal ●

106632 Corylopsis cavaleriei H. Lév. = Corylopsis multiflora Hance ●

106633 Corylopsis cordata Merr. = Corylopsis multiflora Hance var. cordata (Merr.) Hung T. Chang ●

106634 Corylopsis cordata Merr. ex H. L. Li = Corylopsis multiflora Hance var. cordata (Merr.) Hung T. Chang ●

106635 Corylopsis cordata Merr. ex H. L. Li = Corylopsis multiflora Hance ●

106636 Corylopsis coreana Uyeki ex Hatus.;朝鲜蜡瓣花●☆

106637 Corylopsis glabrescens Franch. et Sav.;沧江蜡瓣花(香蜡瓣花);Fragrant Winterhazel,Fragrant Witer Hazel,Fragrant Witerhazel ●☆

106638 Corylopsis glabrescens Franch. et Sav. var. gotoana (Makino) T. Yamanaka = Corylopsis gotoana Makino ●☆

106639 Corylopsis glabrescens Franch. et Sav. var. gotoana (Makino) T. Yamanaka f. pubescens (Nakai) T. Yamanaka = Corylopsis gotoana Makino var. pubescens (Nakai) T. Yamaz. ●☆

106640 Corylopsis glandulifera Hemsl.;腺果蜡瓣花(腺蜡瓣花);Glandbearing Cryptocarya, Glandbearing Waxpetal, Glandular Winterhazel,Glandulifering Winterhazel ●

106641 Corylopsis glandulifera Hemsl. var. hypoglauca (W. C. Cheng) Hung T. Chang;灰白蜡瓣花;Glaucousback Waxpetal,Glaucousback Winterhazel ●

106642 Corylopsis glandulifera Hemsl. var. hypoglauca (W. C. Cheng) Hung T. Chang = Corylopsis glandulifera Hemsl. ●

106643 Corylopsis glaucescens Hand. -Mazz.;怒江蜡瓣花(沧江蜡瓣花,香蜡瓣花);Fragrant Waxpetal,Fragrant Winterhazel,Glaucescent Winterhazel,Nujiang Winterhazel ●

106644 Corylopsis gotoana Makino;高山蜡瓣花●☆

106645 Corylopsis gotoana Makino f. pubescens (Nakai) H. Ohba = Corylopsis gotoana Makino var. pubescens (Nakai) T. Yamaz. ●☆

106646 Corylopsis gotoana Makino var. coreana (Uyeki) T. Yamaz. = Corylopsis coreana Uyeki ex Hatus. ●☆

106647 Corylopsis gotoana Makino var. pubescens (Nakai) T. Yamaz.;毛高山蜡瓣花●☆

106648 Corylopsis griffithii Hemsl.;紫果蜡瓣花;Griffith Waxpetal ●

106649 Corylopsis henryi Hemsl.;鄂西蜡瓣花(小扇木);Henry Waxpetal,Henry Winterhazel ●

106650 Corylopsis himalayana Griff.;西域蜡瓣花(喜马拉雅蜡瓣花);Himalayan Waxpetal ●

106651 Corylopsis hypoglauca W. C. Cheng = Corylopsis glandulifera Hemsl. ●

106652 Corylopsis hypoglauca W. C. Cheng = Corylopsis glandulifera Hemsl. var. hypoglauca (W. C. Cheng) Hung T. Chang ●

106653 Corylopsis hypoglauca W. C. Cheng var. glaucescens W. C. Cheng = Corylopsis glandulifera Hemsl. ●

106654 Corylopsis macrostachya Pamp. = Sinowilsonia henryi Hemsl. ●

106655 Corylopsis matsudae Kaneh. et Sasaki = Corylopsis pauciflora Siebold et Zucc. ●

106656 Corylopsis matsudai Kaneh. et Sasaki;台湾蜡瓣花(小叶瑞木);Matsuda Winterhazel,Taiwan Winterhazel ●

106657 Corylopsis matsudai Kaneh. et Sasaki = Corylopsis pauciflora Siebold et Zucc. ●

106658 Corylopsis microcarpa Hung T. Chang;小果蜡瓣花;Littlefruit Winterhazel, Smallfruit Waxpetal, Smallfruit Winterhazel, Smallfruited Winterhazel ●

106659 Corylopsis multiflora Hance;瑞木(大果蜡瓣花,多花蜡瓣花,峨眉蜡瓣花,假榛,朴扇木);Manyflower Waxpetal, Manyflower Winterhazel,Multiflorous Winterhazel ●

106660 Corylopsis multiflora Hance var. cordata (Merr. ex H. L. Li) Hung T. Chang;心叶瑞木;Cordateleaf Waxpetal,Cordateleaf Winterhazel ●

106661 Corylopsis multiflora Hance var. cordata (Merr. ex H. L. Li) Hung T. Chang = Corylopsis multiflora Hance ●

106662 Corylopsis multiflora Hance var. cordata (Merr.) Hung T. Chang = Corylopsis multiflora Hance var. cordata (Merr. ex H. L. Li) Hung T. Chang ●

106663 Corylopsis multiflora Hance var. nivea Hung T. Chang;白背瑞木(白背蜡瓣花);Snow-white Winterhazel ●

106664 Corylopsis multiflora Hance var. parvifolia Hung T. Chang;小叶瑞木(小叶大果蜡瓣花);Small-leaf Winterhazel ●

106665 Corylopsis multiflora Hance var. parvifolia Hung T. Chang = Corylopsis multiflora Hance ●

106666 Corylopsis multiflora sensu Hung T. Chang = Corylopsis multiflora Hance var. cordata (Merr.) Hung T. Chang ●

106667 Corylopsis obovata Hung T. Chang;贵州蜡瓣花(黔蜡瓣花);Obovate Waxpetal,Obovate Winterhazel ●

106668 Corylopsis omeiensis Yen C. Yang;峨眉蜡瓣花;Emei Mountain Winterhazel,Emei Waxpetal,Emei Winterhazel ●

106669 Corylopsis pauciflora Siebold et Zucc.;少花蜡瓣花(寡叶蜡瓣花,少花瑞木,疏花蜡瓣花,疏花瑞木,小叶瑞木);Buttercup Waxpetal,Buttercup Winter Hazel,Buttercup Winterhazel,Buttercup Winter-hazel,Winter Hazel ●

106670 Corylopsis platypetala Rehder et E. H. Wilson;阔瓣蜡瓣花(阔蜡瓣花);Broad Waxpetal, Broadpetal Winterhazel, Broad-petaled Winterhazel ●

106671 Corylopsis platypetala Rehder et E. H. Wilson var. levis Rehder et E. H. Wilson;川西阔瓣蜡瓣花(宽瓣蜡瓣花,秃蜡瓣花);West Sichuan Broadpetal Winterhazel ●

106672 Corylopsis platypetala Rehder et E. H. Wilson var. levis Rehder et E. H. Wilson = Corylopsis platypetala Rehder et E. H. Wilson ●

106673 Corylopsis polyneura H. L. Li = Corylopsis glaucescens Hand. -Mazz. ●

106674 Corylopsis rotundifolia Hung T. Chang;圆叶蜡瓣花;Roundleaf Waxpetal,Roundleaf Winterhazel,Round-leaved Winterhazel ●

106675 Corylopsis sinensis Hemsl.;蜡瓣花(华蜡瓣花,连合子,连核梅,支那水木,中华蜡瓣花);China Waxpetal, Chinese Winter Hazel, Chinese Winterhazel, Chinese Winter-hazel, Chinese Witch Hazel ●

106676 Corylopsis sinensis Hemsl. 'Spring Purple';春紫蜡瓣花●

106677 Corylopsis sinensis Hemsl. f. veitchiana (Bean) B. D. Morley et J. M. Chao = Corylopsis veitchiana Bean ●

106678 Corylopsis sinensis Hemsl. var. calvescens Rehder et E. H. Wilson;秃蜡瓣花(变秃蜡瓣花);Glabrescent Waxpetal, Glabrescent Winterhazel ●

106679 Corylopsis sinensis Hemsl. var. glandulifera (Hemsl.) Rehder et E. H. Wilson = Corylopsis glandulifera Hemsl. ●

106680 Corylopsis sinensis Hemsl. var. parvifolia Hung T. Chang;小叶蜡瓣花(小蜡瓣花);Small-leaf Winterhazel ●

106681 Corylopsis spicata Hemsl. = Corylopsis sinensis Hemsl. ●

106682 Corylopsis spicata Siebold et Zucc.;穗状蜡瓣花(长穗蜡瓣花,土佐水木);Cowslip Bush,Spike Winter Hazel,Spike Winter-hazel ●☆

106683 Corylopsis stelligera Guillaumin;星毛蜡瓣花;Stellatehair Waxpetal,Stellatehair Winterhazel,Stellate-haired Winterhazel ●

106684 Corylopsis stenopetala Hayata;台湾瑞木;Taiwan Winterhazel ●

106685 Corylopsis stenopetala Hayata = Corylopsis multiflora Hance ●

106686 Corylopsis trabeculosa Hu et W. C. Cheng;俅江蜡瓣花;Qiujiang Waxpetal,Qiujiang Winterhazel,Trabeculate Winterhazel ●

106687 Corylopsis veitchiana Bean;红药蜡瓣花;Veitch Waxpetal,Veitch Winterhazel,Veitch's Winter Hazel ●

106688 Corylopsis velutina Hand. -Mazz.;绒毛蜡瓣花;Velutinous Waxpetal,Velvety Winterhazel ●

106689 Corylopsis willmottiae Rehder et E. H. Wilson;四川蜡瓣花;Willmott Waxpetal,Willmott Winter Hazel,Willmott Winterhazel,Willmott Winter-hazel ●

106690 Corylopsis willmottiae Rehder et E. H. Wilson = Corylopsis sinensis Hemsl. ●

106691 Corylopsis willmottiae Rehder et E. H. Wilson var. chekiangensis Hung T. Cheng = Corylopsis glandulifera Hemsl. ●

106692 Corylopsis willmottiae Rehder et E. H. Wilson var. chekiangensis W. C. Cheng = Corylopsis glandulifera Hemsl. ●

106693 Corylopsis wilsonii Hemsl. = Corylopsis multiflora Hance ●

106694 Corylopsis yui Hu et W. C. Cheng;长穗蜡瓣花(疏花蜡瓣花);Yu Waxpetal,Yu Winterhazel ●

106695 Corylopsis yunnanensis Diels;云南蜡瓣花(滇蜡瓣花);Yunnan Waxpetal,Yunnan Winterhazel ●

106696 Corylus L.(1753);榛属(榛木属);Cobnut,Cobnuts,Filbert,Filberts,Hazel,Hazelnut,Hazelnuts ●

106697 Corylus americana Marshall = Corylus americana Walter ●☆

106698 Corylus americana Walter;美洲榛;American Filbert,American Hazel,American Hazelnut,American Wild Hazel,Hazelnut ●☆

106699 Corylus americana Walter f. missouriensis (A. DC.) Fernald = Corylus americana Walter ●☆

106700 Corylus americana Walter var. altior Fernald = Corylus americana Walter ●☆

106701 Corylus americana Walter var. indehiscens E. J. Palmer et Steyerm. = Corylus americana Walter ●☆

106702 Corylus americana Walter var. missourensis A. DC. = Corylus americana Walter ●☆

106703 Corylus avellana L.;欧洲榛(褐色榛,欧榛,欧洲榛子,西洋榛子,洋榛子);Aglet,Aglet-tree,Baa Lambs Tails,Baa-lambs,Baa-lambs' Tails,Baccy Lambs,Barcelona Nut,Beard-tree,Brown Shillers,Brown-shillers,Carskin,Cat O' nine Tails,Cat's Tails,Cat's Tall,Cats-and-kittens,Cob Nut,Cobbedy-cut,Cobbly-cut,Cobnut,Cobnut Hazel,Common Filbert,Common Hazel,Contorted Filbert,Corkscrew Hazel,Cowll,Crack Nut,Crack-nut,European Filbert,European Hazel,European Hazelnut,Filbeard,Filberd,Filbert,Filbert Nut-tree,Filbord,Foxtail,Fussy Cats,Halenut,Hales,Halse,Harry Lauder's walking stick,Hasill Tree,Haskett,Haul,Hazel,Hazel Catkins,Hazelnut,Hazzle,Hedge Nut,Hessel,Hezzle,Hole,Kitten's Tails,Kittens' tails,Lambkins,Lamb's Tails,Leemers,Nut Bush,Nut Hall,Nut Halse,Nut Palm,Nut Rags,Nut Stowel,Nut Stowell,Nut Tree,Nuttall,Palm,Pussy Cat's Tail,Pussy Cat's Tails,Pussy-cats,Rags,Sheep's Tails,Stock Nut,Victor Nut,Victor-nut,Witch Halse,Wood-nut ●

106704 Corylus avellana L. 'Aurea';金叶欧洲榛●☆

106705 Corylus avellana L. 'Contorta';扭枝欧洲榛(扭枝欧榛);Corkscrew Hazel,Twisted Hazel ●☆

106706 Corylus avellana L. 'Pendula';垂枝欧洲榛●☆

106707 Corylus avellana L. var. atropurpurea Kirchin;紫色欧洲榛●☆

106708 Corylus avellana L. var. aurea Kirchin = Corylus avellana L. 'Aurea' ●☆

106709 Corylus avellana L. var. davurica Ledeb. = Corylus heterophylla Fisch. ex Trautv. ●

106710 Corylus avellana L. var. fusco-rubra Dippel;赤褐欧洲榛●☆

106711 Corylus avellana L. var. laciniata Kirchin;条裂欧洲榛●☆

106712 Corylus avellana L. var. pendula Goeshke ? = Corylus avellana L. 'Pendula' ●☆

106713 Corylus avellana L. var. pontica ? = Corylus pontica K. Koch ●☆

106714 Corylus brevituba Kom.;短管榛●☆

106715 Corylus californica (A. DC.) Rose = Corylus cornuta Marshall subsp. californica (A. DC.) E. Murray ●☆

106716 Corylus chinensis Franch.;华榛(鸡栗子,山白果,榛树);China Filbert,Chinese Filbert,Chinese Hazel,Chinese Hazelnut ●◇

106717 Corylus chinensis Franch. var. brevilimba Hu ex T. Hong et J. W. Li;钟苞榛;Shortbract Hazelnut ●☆

106718 Corylus chinensis Franch. var. macrocarpa Hu = Corylus chinensis Franch. ●◇

106719 Corylus colchica Albov;黑海榛●☆

106720 Corylus colurna L.;土耳其榛;Beaked Hazelnut,Turkish Filbert,Turkish Hazel ●☆

106721 Corylus colurna L. var. chinensis (Franch.) Burkill = Corylus chinensis Franch. ●◇

106722 Corylus colurna L. var. chinensis Burkill = Corylus chinensis Franch. ●◇

106723 Corylus cornuta Du Roi ex Steud. = Corylus cornuta Marshall ●☆

106724 Corylus cornuta Marshall;欧洲尖果榛(高加索榛,角状榛,土耳其榛);Beaked Hazel,Beaked Hazelnut,Caucasian Filbert,Constantinople Hazel,Tree Hazelnut,Turkish Filbert,Turkish Hazel,Turkish Hazelnut ●☆

106725 Corylus cornuta Marshall subsp. californica (A. DC.) E. Murray;加州榛;California Filbert,California Hazel,California Hazelnut ●☆

106726 Corylus cornuta Marshall var. californica (A. DC.) Sharp = Corylus cornuta Marshall subsp. californica (A. DC.) E. Murray ●☆

106727 Corylus cornuta Marshall var. glandulosa B. Boivin = Corylus cornuta Marshall subsp. californica (A. DC.) E. Murray ●☆

106728 Corylus cornuta Marshall var. megaphylla Vict. et J. Rousseau = Corylus cornuta Marshall ●☆

106729 Corylus davidiana (Decne.) Baill. = Ostryopsis davidiana Decne. ●

106730 Corylus davidiana Baill. = Ostryopsis davidiana Decne. ●

106731 Corylus fargesii C. K. Schneid.;披针叶榛(绒苞榛);Farges Filbert,Farges Hazel ●

106732 Corylus fargesii C. K. Schneid. var. latifolia T. Hong et J. W. Li;宽叶绒苞榛(宽叶榛,山白果);Broadleaf Filbert ●

106733 Corylus ferox Wall.;刺榛(滇刺榛,山板栗);Himalayas Filbert,Tibetan Hazel ●

106734 Corylus ferox Wall. var. thibetica (Batalin) Franch.;腺毛刺榛(藏刺榛,藏榛,猴板栗树,西藏榛树);Tibet Filbert,Tibetan Filbert ●

106735 Corylus formosana Hayata;台湾榛;Taiwan Filbert ●

106736 Corylus heterophylla Fisch. ex Trautv.;榛(榧子,和尚头,毛榛,平榛,日本榛,山白果,山反栗,榛柴棵子,榛栗,榛树,榛子);Chinese Hazel,Siberia Filbert,Siberian Filbert,Siberian Hazel,Siberian Hazelnut,Thunberg Filbert ●

106737 Corylus heterophylla Fisch. ex Trautv. f. brevituba (Kom.) Kitag.;短苞毛榛●
106738 Corylus heterophylla Fisch. ex Trautv. var. cristagalli Burkill = Corylus heterophylla Fisch. ex Trautv. var. sutchuenensis Franch. ●
106739 Corylus heterophylla Fisch. ex Trautv. var. japonica Koidz. = Corylus heterophylla Fisch. ex Trautv. var. thunbergii Blume ●
106740 Corylus heterophylla Fisch. ex Trautv. var. sutchuenensis Franch.;川榛;Sichuan Filbert,Szechwan Filbert ●
106741 Corylus heterophylla Fisch. ex Trautv. var. thunbergii Blume;日本榛●
106742 Corylus heterophylla Fisch. ex Trautv. var. thunbergii Blume = Corylus heterophylla Fisch. ex Trautv. ●
106743 Corylus heterophylla Fisch. ex Trautv. var. yezoensis Koidz.;北海道榛●☆
106744 Corylus heterophylla Fisch. ex Trautv. var. yunnanensis Franch. = Corylus yunnanensis (Franch.) A. Camus ●
106745 Corylus heterophylla Fisch. var. yunnanensis Franch. = Corylus yunnanensis (Franch.) A. Camus ●
106746 Corylus jacquemontii Decne. = Corylus colurna L. ●☆
106747 Corylus jasquemontii Decne. = Corylus wangii Hu ●
106748 Corylus kweichouensis Hu;贵州榛;Guizhou Hazelnut ●
106749 Corylus kweichouensis Hu = Corylus heterophylla Fisch. ex Trautv. var. sutchuenensis Franch. ●
106750 Corylus kweichouensis Hu var. brevipes W. J. Liang;短柄贵州榛(短柄川榛);Shortstalk Guizhou Hazelnut ●
106751 Corylus mandshurica Maxim.;毛榛(胡榛子,火榛子,角榛,毛榛子,小榛树);Manchurian Hazel,Manshurian Filbert ●
106752 Corylus mandshurica Maxim. = Corylus sieboldiana Blume var. mandshurica (Maxim.) C. K. Schneid. ●
106753 Corylus mandshurica Maxim. ex Rupr. = Corylus mandshurica Maxim. ●
106754 Corylus mandshurica Maxim. f. glandulosa S. L. Tung;腺毛毛榛;Glandulose Filbert ●
106755 Corylus mandshurica Maxim. var. fargesii (Franch.) Burkill = Corylus fargesii C. K. Schneid. ●
106756 Corylus mandshurica Maxim. var. fargesii Burkill = Corylus fargesii C. K. Schneid. ●
106757 Corylus maxima Mill.;大榛(马氏榛,南欧榛树);Filbert,Giant Filbert,Kentish Cob,Lambert Nut,Lambert's Filbert ●☆
106758 Corylus maxima Mill. 'Atropurpurea' = Corylus maxima Mill. 'Purpurea' ●☆
106759 Corylus maxima Mill. 'Contorta';扭曲榛;Corkscrew Hazel ●☆
106760 Corylus maxima Mill. 'Purpurea';紫叶大榛(紫红色大榛);Purple Nut,Purple-leaf Filbert ●☆
106761 Corylus maxima Mill. var. purpurea Rehder = Corylus maxima Mill. 'Purpurea' ●☆
106762 Corylus papyracea Hickel = Corylus chinensis Franch. ●◇
106763 Corylus pontica Dochmahl = Corylus avellana L. ●
106764 Corylus pontica K. Koch = Corylus colurna L. ●☆
106765 Corylus rostrata Aiton = Corylus colurna L. ●☆
106766 Corylus rostrata Aiton = Corylus cornuta Marshall ●☆
106767 Corylus rostrata Aiton var. californica A. DC. = Corylus cornuta Marshall subsp. californica (A. DC.) E. Murray ●☆
106768 Corylus rostrata Aiton var. fargesii Franch. = Corylus fargesii C. K. Schneid. ●
106769 Corylus rostrata Aiton var. mandshurica (Maxim.) Regel = Corylus mandshurica Maxim. ●
106770 Corylus rostrata Aiton var. tracyi Jeps. = Corylus cornuta Marshall subsp. californica (A. DC.) E. Murray ●☆
106771 Corylus rostrata Dippel;喙榛;Beaked Hazel ●☆
106772 Corylus rostrata Dippel var. californica A. DC.;加州喙榛;Californian Hazel ●☆
106773 Corylus rostrata Dippel var. fargesii Franch. = Corylus fargesii C. K. Schneid. ●
106774 Corylus rostrata Dippel var. mandshurica (Maxim.) Regel = Corylus mandshurica Maxim. ●
106775 Corylus rostrata Dippel var. sieboldiana Maxim. = Corylus sieboldiana Blume ●
106776 Corylus rostrata Marshall var. tracyi Jeps. = Corylus cornuta Marshall subsp. californica (A. DC.) E. Murray ●☆
106777 Corylus sieboldiana Blume;西氏榛(日本榛,席氏榛);Japanese Filbert,Japanese Hazel,Japanese Hazelnut,Siebold Filbert ●
106778 Corylus sieboldiana Blume f. mitis (Maxim.) Sugim.;乳白西氏榛●☆
106779 Corylus sieboldiana Blume var. brevirostris C. K. Schneid.;短角西氏榛(短角毛榛)●☆
106780 Corylus sieboldiana Blume var. mandshurica (Maxim.) C. K. Schneid. = Corylus mandshurica Maxim. ●
106781 Corylus sieboldiana Blume var. mandshurica (Maxim.) C. K. Schneid. f. longissimorostris M. Kikuchi;长喙西氏榛;Manchu Filbert ●☆
106782 Corylus sieboldiana Blume var. mitis (Maxim.) Nakai = Corylus sieboldiana Blume f. mitis (Maxim.) Sugim. ●☆
106783 Corylus thibetica Batalin = Corylus ferox Wall. var. thibetica (Batalin) Franch. ●
106784 Corylus wangii Hu;维西榛(卵叶尖齿刺榛);Jacquemont Hazelnut,Wang Filbert ●
106785 Corylus wulingensis Q. X. Liu et C. M. Zhang;武陵榛;Wuling Hazelnut ●
106786 Corylus yunnanensis (Franch.) A. Camus;滇榛(猴核桃);Yunnan Filbert ●
106787 Corymbia K. D. Hill et L. A. S. Johnson = Eucalyptus L'Hér. ●
106788 Corymbia K. D. Hill et L. A. S. Johnson(1995);伞房花桉属(科林比亚属,柠檬桉属);Blood Wood,Ghost Gum,Gum,Eucalypt ●
106789 Corymbia aparrerinja K. D. Hill et L. A. S. Johnson;中澳伞房花桉(中澳科林比亚);Central Australian Ghost Gum ●☆
106790 Corymbia bleeseri (Blakely) K. D. Hill et L. A. S. Johnson;光皮红木桉;Smooth-stemmed Bloodwood ●
106791 Corymbia calophylla (Lindl.) K. D. Hill et L. A. S. Johnson;美叶伞房花桉●☆
106792 Corymbia calophylla (Lindl.) K. D. Hill et L. A. S. Johnson = Eucalyptus calophylla Lindl. ●
106793 Corymbia citriodora (Hook.) K. D. Hill et L. A. S. Johnson;柠檬桉(白树,留香久,柠檬科林比亚,柠檬伞房花桉,柠檬香桉树,香桉,油桉树);Lemon Eucalyptus,Lemon Scented Gum,Lemon-scented Gum,Lemon-scented Spotted Gum ●
106794 Corymbia citriodora (Hook.) K. D. Hill et L. A. S. Johnson = Eucalyptus citriodora Hook. ●
106795 Corymbia confertiflora (Kippist ex F. Muell.) K. D. Hill et L. A. S. Johnson;密花桉●☆
106796 Corymbia dichromophloia (F. Muell.) K. D. Hill et L. A. S. Johnson;二色伞房花桉(二色桉)●☆
106797 Corymbia dichromophloia (F. Muell.) K. D. Hill et L. A. S. Johnson = Eucalyptus dichromophloia F. Muell. ●☆

106798 Corymbia eximia (Schauer) K. D. Hill et L. A. S. Johnson;曲叶伞房花桉(曲叶科林比亚,特桉);Yellow Bloodwood ●

106799 Corymbia eximia (Schauer) K. D. Hill et L. A. S. Johnson = Eucalyptus eximia Schauer ●

106800 Corymbia ficifolia (F. Muell.) K. D. Hill et L. A. S. Johnson;红花伞房花桉(红花桉,红花科林比亚,美丽桉,美丽花桉);Crimson-flower Eucalyptus, Flowering Gum, Red Flowering Gum, Red Gum, Redflower Gum, Red-flowered Wood, Red-flowering Gum, Scarlet Eucalyptus, Scarlet Flowering Gum, Scarlet Gum, Scarlet-flowered Gum, Scarlet-flowering Gum ●☆

106801 Corymbia ficifolia (F. Muell.) K. D. Hill et L. A. S. Johnson = Eucalyptus ficifolia F. Muell. ●☆

106802 Corymbia grandifolia (R. Br. ex Benth.) K. D. Hill et L. A. S. Johnson;大叶伞房花桉●☆

106803 Corymbia grandifolia (R. Br. ex Benth.) K. D. Hill et L. A. S. Johnson = Eucalyptus grandifolia R. Br. ex Benth. ●☆

106804 Corymbia gummifera (Gaertn. ex Gaertn.) K. D. Hill et L. A. S. Johnson = Eucalyptus gummifera (Gaertn.) Hochr. ●

106805 Corymbia gummifera (Gaertn.) K. D. Hill et L. A. S. Johnson;伞房花桉(红血桉,科林比亚,树胶桉);Bloodwood, Cedar Gum, Cider Gum, Red Bloodwood, Sandhill Blackbuut, Tasmanian Cider Tree, Victoria Bloodwood ●☆

106806 Corymbia gummifera (Gaertn.) K. D. Hill et L. A. S. Johnson = Eucalyptus gummifera (Gaertn.) Hochr. ●

106807 Corymbia intermedia (R. T. Baker) K. D. Hill et L. A. S. Johnson;桃红木桉;Pink Bloodwood ●

106808 Corymbia intermedia (R. T. Baker) K. D. Hill et L. A. S. Johnson = Eucalyptus intermedia R. T. Baker ●

106809 Corymbia maculata (Hook.) K. D. Hill et L. A. S. Johnson;斑皮伞房花桉(斑桉,斑皮桉,斑皮科林比亚,花皮桉);Spotted Gum ●☆

106810 Corymbia maculata (Hook.) K. D. Hill et L. A. S. Johnson = Eucalyptus maculata Hook. ●

106811 Corymbia nesophila (Blakely) K. D. Hill et L. A. S. Johnson;海岛桉;Island Bloodwood ●

106812 Corymbia nesophila (Blakely) K. D. Hill et L. A. S. Johnson = Eucalyptus nesophila Blakeley ●

106813 Corymbia pachycarpa K. D. Hill et L. A. S. Johnson;毛果伞房花桉●☆

106814 Corymbia papuana (F. Muell.) K. D. Hill et L. A. S. Johnson;鬼伞房花桉(巴布亚桉,鬼桉,鬼胶树);Ghost Gum, Ghost Gum-bark White ●☆

106815 Corymbia papuana (F. Muell.) K. D. Hill et L. A. S. Johnson = Eucalyptus papuana F. Muell. ●☆

106816 Corymbia peltata (Benth.) K. D. Hill et L. A. S. Johnson;盾叶桉●☆

106817 Corymbia peltata (Benth.) K. D. Hill et L. A. S. Johnson = Eucalyptus peltata Benth. ●☆

106818 Corymbia polycarpa (F. Muell.) K. D. Hill et L. A. S. Johnson;多果伞房花桉●☆

106819 Corymbia polycarpa (F. Muell.) K. D. Hill et L. A. S. Johnson = Eucalyptus polycarpa F. Muell. ●☆

106820 Corymbia ptychocarpa (F. Muell.) K. D. Hill et L. A. S. Johnson;皱果伞房花桉(皱果桉,皱果科林比亚,皱皮桉);Red Bloodwood, Swamp Bloodwood ●☆

106821 Corymbia ptychocarpa (F. Muell.) K. D. Hill et L. A. S. Johnson = Eucalyptus ptychocarpa F. Muell. ●☆

106822 Corymbia tessellaris (F. Muell.) K. D. Hill et L. A. S. Johnson;白皮伞房花桉(白皮科林比亚,方格皮桉,方块皮桉);Garbeen, Moreton Bay Ash ●☆

106823 Corymbia tessellaris (F. Muell.) K. D. Hill et L. A. S. Johnson = Eucalyptus tessellaris F. Muell. ●

106824 Corymbia torelliana (F. Muell.) K. D. Hill et L. A. S. Johnson;托里伞房花桉(速生科林比亚,托里桉);Cadaga, Cadagi ●☆

106825 Corymbia torelliana (F. Muell.) K. D. Hill et L. A. S. Johnson = Eucalyptus torelliana F. Muell. ●

106826 Corymbiferae Juss. = Asteraceae Bercht. et J. Presl(保留科名)●■

106827 Corymbiferae Juss. = Compositae Giseke(保留科名)●■

106828 Corymbis Lindl. = Corymborkis Thouars ■

106829 Corymbis Rchb. f. = Corymborkis Thouars ■

106830 Corymbis Thouars = Corymborkis Thouars ■

106831 Corymbis bolusiana Schltr. = Corymborkis corymbis Thouars ■☆

106832 Corymbis corymbosa Ridl. = Corymborkis corymbis Thouars ■☆

106833 Corymbis disticha Lindl. = Corymborkis corymbis Thouars ■☆

106834 Corymbis leptantha Kraenzl. ex Engl. = Corymborkis corymbis Thouars ■☆

106835 Corymbis polystachya (Sw.) Benth. ex Fawc. = Tropidia polystachya (Sw.) Ames ■☆

106836 Corymbis thouarsii Rchb. f. = Corymborkis corymbis Thouars ■☆

106837 Corymbis veratrifolia (Reinw.) Rchb. f. = Corymborkis veratrifolia (Reinw.) Blume ■

106838 Corymbis welwitschii Rchb. f. = Corymborkis corymbis Thouars ■☆

106839 Corymbium L. (1753);绣球菊属■●☆

106840 Corymbium africanum L.;非洲绣球菊■☆

106841 Corymbium africanum L. var. fourcadei (Hutch.) Weitz;富尔卡德绣球菊■☆

106842 Corymbium africanum L. var. gramineum (Burm. f.) Weitz;禾叶非洲绣球菊■☆

106843 Corymbium africanum L. var. scabridum (P. J. Bergius) Weitz;微糙绣球菊■☆

106844 Corymbium congestum E. Mey. ex DC.;密集绣球菊■☆

106845 Corymbium cymosum E. Mey. ex DC.;聚伞绣球菊■☆

106846 Corymbium elsiae Weitz;埃尔斯绣球菊■☆

106847 Corymbium enerve Markötter;无脉绣球菊■☆

106848 Corymbium filiforme L. f. = Corymbium africanum L. var. gramineum (Burm. f.) Weitz ■☆

106849 Corymbium fourcadei Hutch. = Corymbium africanum L. var. fourcadei (Hutch.) Weitz ■☆

106850 Corymbium glabrum L.;光绣球菊■☆

106851 Corymbium glabrum L. var. rogersii (Markötter) Weitz;罗杰斯绣球菊■☆

106852 Corymbium gramineum Burm. f. = Corymbium africanum L. var. gramineum (Burm. f.) Weitz ■☆

106853 Corymbium harveyanum Markötter = Corymbium glabrum L. ■☆

106854 Corymbium hirsutum Eckl. ex DC. = Corymbium villosum L. f. ■☆

106855 Corymbium hirtum Thunb. = Corymbium villosum L. f. ■☆

106856 Corymbium latifolium Harv. = Corymbium glabrum L. ■☆

106857 Corymbium laxum Compton;疏松绣球菊■☆

106858 Corymbium laxum Compton subsp. bolusii Weitz;博卢斯绣球菊■☆

106859 Corymbium luteum E. Mey. ex DC. = Corymbium africanum L. var. gramineum (Burm. f.) Weitz ■☆

106860 Corymbium nervosum Thunb. = Corymbium glabrum L. ■☆

106861 Corymbium rogersii Markötter = Corymbium glabrum L. var. rogersii (Markötter) Weitz ■☆
106862 Corymbium salteri Markötter = Corymbium glabrum L. ■☆
106863 Corymbium scabridum P. J. Bergius = Corymbium africanum L. var. scabridum (P. J. Bergius) Weitz ■☆
106864 Corymbium scabrum L.;粗糙绣球菊■☆
106865 Corymbium scabrum L. var. filiforme (L. f.) Thunb. = Corymbium africanum L. var. gramineum (Burm. f.) Weitz ■☆
106866 Corymbium scabrum L. var. scabrum ? = Corymbium africanum L. var. scabridum (P. J. Bergius) Weitz ■☆
106867 Corymbium theileri Markötter;泰勒绣球菊■☆
106868 Corymbium villosum L. f.;长柔毛绣球菊■☆
106869 Corymborchis Thouars = Corymborkis Thouars ■
106870 Corymborchis polystachya (Sw.) Kuntze = Tropidia polystachya (Sw.) Ames ■☆
106871 Corymborkis Thouars (1809);管花兰属;Corymborkis, Tubeorchis ■
106872 Corymborkis assamica Blume = Corymborkis veratrifolia (Reinw.) Blume ■
106873 Corymborkis bolusiana Schltr. = Corymborkis corymbis Thouars ■☆
106874 Corymborkis corymbis Thouars;马岛管花兰■☆
106875 Corymborkis corymbosa (Ridl.) Kuntze = Corymborkis corymbis Thouars ■☆
106876 Corymborkis flava (Sw.) Kuntze;淡黄管花兰;Yellowish Corymborkis ■☆
106877 Corymborkis forcipigera (Rchb. f.) L. O. Williams;钳叉状管花兰;Forkedlike Corymborkis ■☆
106878 Corymborkis minima P. J. Cribb;极小管花兰■☆
106879 Corymborkis sakisimensis Fukuy. = Corymborkis veratrifolia (Reinw.) Blume ■
106880 Corymborkis subdensa (Schltr.) Masam.;亚密管花兰■☆
106881 Corymborkis thouarsii Blume = Corymborkis corymbis Thouars ■☆
106882 Corymborkis veratri Blume;管花兰;Common Corymborkis ■☆
106883 Corymborkis veratrifolia (Reinw.) Blume;藜芦叶管花兰(管花兰);Common Corymborkis, Common Tubeorchis ■
106884 Corymborkis welwitschii (Rchb. f.) Kuntze = Corymborkis corymbis Thouars ■☆
106885 Corymbostachys Lindau = Anisostachya Nees ●■
106886 Corymbostachys Lindau = Justicia L. ●■
106887 Corymbula Raf. = Polygala L. ●■
106888 Corynabutilon (K. Schum.) Kearney = Abutilon Mill. ●■
106889 Corynabutilon (K. Schum.) Kearney(1949);棒苘麻属●☆
106890 Corynabutilon bicolor (Phil. ex K. Schum.) Kearney;二色棒苘麻●☆
106891 Corynabutilon viride (Phil.) A. E. Martic.;绿花棒苘麻●☆
106892 Corynabutilon vitifolium (Cav.) Kearney;葡萄叶棒苘麻●☆
106893 Corynaea Hook. f. (1856);安第斯菰属■☆
106894 Corynaea crassa Hook. f.;安第斯菰■☆
106895 Corynandra Schrad. = Cleome L. ●■
106896 Corynanthe Welw. (1869);宾树属(棒花属,柯楠属)●☆
106897 Corynanthe bequaertii De Wild. = Pausinystalia macroceras (K. Schum.) Pierre ex Beille ●☆
106898 Corynanthe brachythyrsus K. Schum. = Pausinystalia brachythyrsum (K. Schum.) W. Brandt ●☆
106899 Corynanthe dolichocarpa W. Brandt = Pausinystalia talbotii Wernham ●☆
106900 Corynanthe gabonensis A. Chev. = Pausinystalia macroceras (K. Schum.) Pierre ex Beille ●☆
106901 Corynanthe ituriense De Wild. = Pausinystalia lane-poolei (Hutch.) Hutch. ex Lane-Poole subsp. ituriense (De Wild.) Stoffelen et Robbr. ●☆
106902 Corynanthe johimbe K. Schum.;育亨宾树●☆
106903 Corynanthe johimbe K. Schum. = Pausinystalia johimbe (K. Schum.) Pierre ex Beille ●☆
106904 Corynanthe lane-poolei Hutch. = Pausinystalia lane-poolei (Hutch.) Hutch. ex Lane-Poole ●☆
106905 Corynanthe macroceras K. Schum. = Pausinystalia macroceras (K. Schum.) Pierre ex Beille ●☆
106906 Corynanthe mayumbensis (R. D. Good) Raym. -Hamet ex N. Hallé;大角宾树●☆
106907 Corynanthe moebiusii W. Brandt = Pausinystalia talbotii Wernham ●☆
106908 Corynanthe pachyceras K. Schum.;粗角宾树●☆
106909 Corynanthe paniculata Welw.;圆锥宾树●☆
106910 Corynanthe tenuis W. Brandt = Pausinystalia talbotii Wernham ●☆
106911 Corynanthelium Kunze = Mikania Willd. (保留属名)■
106912 Corynanthera J. W. Green(1979);棒药桃金娘属●☆
106913 Corynanthera flava J. W. Green;棒药桃金娘●☆
106914 Corynanthes Schltdl. = Coryanthes Hook. ■☆
106915 Corynelia Rchb. = Corynella DC. ■☆
106916 Corynella DC. (1825);小棒豆属■☆
106917 Corynella gracilis Griseb.;纤细小棒豆■☆
106918 Corynella immarginata C. Wright = Sauvallella immarginata (C. Wright) Rydb. ■☆
106919 Corynella paucifolia DC.;寡叶小棒豆■☆
106920 Corynella polyantha DC.;多花小棒豆■☆
106921 Corynelobos R. Roem. = Brassica L. ■●
106922 Corynelobos R. Roem. ex Willk. = Brassica L. ■●
106923 Corynemyrtus (Kiaersk.) Mattos = Myrtus L. ●
106924 Corynephorus P. Beauv. (1812)(保留属名);棒芒草属;Club Awn Grass, Grey Hair-grass, Club Awn-grass, Clubawngrass ■☆
106925 Corynephorus articulatus (Desf.) P. Beauv.;关节棒芒草■☆
106926 Corynephorus articulatus (Desf.) P. Beauv. subsp. fasciculatus (Boiss. et Reut.) Husn. = Corynephorus divaricatus (Pourr.) Breistr. ■☆
106927 Corynephorus articulatus (Desf.) P. Beauv. subsp. macrantherus (Boiss. et Reut.) Maire;大药关节棒芒草■☆
106928 Corynephorus articulatus (Desf.) P. Beauv. subsp. oranensis (Murb.) Maire et Weiller = Corynephorus oranensis Murb. ■☆
106929 Corynephorus articulatus (Desf.) P. Beauv. var. gracilis (Guss.) Coss. et Durieu = Corynephorus divaricatus (Pourr.) Breistr. ■☆
106930 Corynephorus articulatus (Desf.) P. Beauv. var. intermedius Maire = Corynephorus articulatus (Desf.) P. Beauv. ■☆
106931 Corynephorus articulatus P. Beauv. = Corynephorus articulatus (Desf.) P. Beauv. ■☆
106932 Corynephorus canescens (L.) P. Beauv.;灰白棒芒草;Gray Clubawn Grass, Gray Hair Grass, Grey Hair-grass, Hoary Clubawngrass ■☆
106933 Corynephorus canescens P. Beauv. = Corynephorus canescens (L.) P. Beauv. ■☆
106934 Corynephorus divaricatus (Pourr.) Breistr.;叉开棒芒草■☆
106935 Corynephorus divaricatus (Pourr.) Breistr. subsp. macrantherus

(Boiss. et Reut.) Paunero = Corynephorus articulatus (Desf.) P. Beauv. subsp. macrantherus (Boiss. et Reut.) Maire ■☆
106936 Corynephorus gracilis Rouy;纤细棒芒草■☆
106937 Corynephorus macrantherus Boiss. et Reut. = Corynephorus articulatus (Desf.) P. Beauv. subsp. macrantherus (Boiss. et Reut.) Maire ■☆
106938 Corynephorus oranensis Murb.;奥兰棒芒草■☆
106939 Corynephyllum Rose = Sedum L. ●■
106940 Corynephyllum Rose(1905);棒叶景天属■☆
106941 Corynephyllum viride Rose;棒叶景天■☆
106942 Corynitis Spreng. = Corynella DC. ■☆
106943 Corynocarpaceae Engl.(1897)(保留科名);棒果木科(棒果科,毛利果科)●☆
106944 Corynocarpus J. R. Forst. et G. Forst.(1775);棒果木属(棒果属,卡拉卡属,毛利果属);Karaka Nut ●☆
106945 Corynocarpus laevigata J. R. Forst. et G. Forst.;棒果木(毛利果,平滑卡拉卡);Karaka,Karaka Nut,New Zealand Laurel ●☆
106946 Corynocarpus rupestris Guymer;岩生卡拉卡●☆
106947 Corynolobus Post et Kuntze = Brassica L. ■●
106948 Corynolobus Post et Kuntze = Corynelobos R. Roem. ●■
106949 Corynophallus Schott = Amorphophallus Blume ex Decne.(保留属名)■●
106950 Corynophallus afzelii Schott = Amorphophallus aphyllus (Hook.) Hutch. ■☆
106951 Corynophallus angolensis (Welw. ex Schott) Kuntze = Amorphophallus angolensis (Welw. ex Schott) N. E. Br. ■☆
106952 Corynophallus gratus (Schott) Kuntze = Amorphophallus abyssinicus (A. Rich.) N. E. Br. ■☆
106953 Corynophallus maximus (Engl.) Kuntze = Amorphophallus maximus (Engl.) N. E. Br. ■☆
106954 Corynophorus Kunth = Corynephorus P. Beauv.(保留属名)■☆
106955 Corynopuntia F. M. Knuth = Grusonia F. Rchb. ex K. Schum. ■☆
106956 Corynopuntia F. M. Knuth = Opuntia Mill. ●
106957 Corynopuntia F. M. Knuth(1936);棍棒仙人掌属■☆
106958 Corynopuntia clavata (Engelm.) F. M. Knuth;豆麒麟■☆
106959 Corynopuntia clavata (Engelm.) F. M. Knuth = Grusonia clavata (Engelm.) H. Rob. ■☆
106960 Corynopuntia grahamii (Engelm.) F. M. Knuth = Grusonia grahamii (Engelm.) H. Rob. ■☆
106961 Corynopuntia invicta (Brandegee) F. M. Knuth;武者团扇■☆
106962 Corynopuntia parishii (Orcutt) F. M. Knuth = Grusonia parishii (Orcutt) Pinkava ■☆
106963 Corynopuntia pulchella (Engelm.) F. M. Knuth = Grusonia pulchella (Engelm.) H. Rob. ■☆
106964 Corynopuntia schottii (Engelm.) F. M. Knuth = Grusonia schottii (Engelm.) H. Rob. ■☆
106965 Corynopuntia stanlyi (Engelm. ex B. D. Jacks.) F. M. Knuth = Grusonia emoryi (Engelm.) Pinkava ■☆
106966 Corynopuntia stanlyi (Engelm. ex B. D. Jacks.) F. M. Knuth var. kunzei (Rose) Backeb. = Grusonia kunzei (Rose) Pinkava ■☆
106967 Corynopuntia stanlyi (Engelm. ex B. D. Jacks.) F. M. Knuth var. parishii (Orcutt) Backeb. = Grusonia parishii (Orcutt) Pinkava ■☆
106968 Corynopuntia stanlyi (Engelm. ex B. D. Jacks.) F. M. Knuth var. wrightiana (E. M. Baxter) Backeb. = Grusonia kunzei (Rose) Pinkava ■☆
106969 Corynopuntia stanlyi Engelm. ex B. D. Jacks. var. parishii (Orcutt) L. D. Benson = Grusonia parishii (Orcutt) Pinkava ■☆
106970 Corynopuntia vilis (Rose) F. M. Knuth;棍棒仙人掌(姬武者)■☆
106971 Corynosicyos F. Muell. = Cucumeropsis Naudin ■☆
106972 Corynostigma C. Presl = Ludwigia L. ●■
106973 Corynostylis Mart.(1824);盘柱堇属■☆
106974 Corynostylis hybanthus Mart. et Zucc.;盘柱堇■☆
106975 Corynostylus T. Post et Kuntze = Anticheirostylis Fitzg. ■☆
106976 Corynostylus T. Post et Kuntze = Corunastylis Fitzg. ■☆
106977 Corynostylus T. Post et Kuntze = Genoplesium R. Br. ■☆
106978 Corynotheca F. Muell. = Corynotheca F. Muell. ex Benth. ■☆
106979 Corynotheca F. Muell. ex Benth.(1878);棒室吊兰属■☆
106980 Corynotheca acanthoclada (F. Muell.) Benth.;刺枝棒室吊兰■☆
106981 Corynotheca lateriflora (R. Br.) Benth.;棒室吊兰■☆
106982 Corynotheca micrantha Druce;小花棒室吊兰■☆
106983 Corynula Hook. f. = Leptostigma Arn. ■☆
106984 Corypha L.(1753);贝叶棕属(贝叶属,大叶棕属,顶榈属,金丝榈属,金丝葵属,团扇葵属,行李榈属,行李叶椰子属);Corypha,Cowryleafpalm,Talipot Palm ●
106985 Corypha australis R. Br. = Livistona australis (R. Br.) Mart. ●☆
106986 Corypha elata Roxb.;高贝叶棕(高行李叶椰子,马来扇叶椰子);Buri Palm,Gebang,Gebang Palm ●☆
106987 Corypha elata Roxb. = Corypha utan Lam. ●☆
106988 Corypha minor Jacq. = Sabal minor (Jacq.) Pers. ●
106989 Corypha obliqua W. Bartram = Serenoa repens (Bartram) Small ●☆
106990 Corypha palma W. Bartram = Sabal palmetto (Walter) Lodd. ex Roem. et Schult. f. ●
106991 Corypha palmetto Walter = Sabal palmetto (Walter) Lodd. ex Roem. et Schult. f. ●
106992 Corypha pilearia Lour. = Licuala spinosa Thunb. ●
106993 Corypha pumila Walter = Sabal minor (Jacq.) Pers. ●
106994 Corypha repens W. Bartram = Serenoa repens (Bartram) Small ●☆
106995 Corypha rotundifolia Lam. = Livistona rotundifolia (Lam.) Mart. ●
106996 Corypha saribus Lour. = Livistona saribus (Lour.) Merr. ex A. Chev. ●◇
106997 Corypha thebaica L. = Hyphaene thebaica (L.) Mart. ●☆
106998 Corypha umbraculifera Jacq. = Sabal palmetto (Walter) Lodd. ex Roem. et Schult. f. ●
106999 Corypha umbraculifera L.;贝叶棕(贝多罗,伞形行李叶椰子,团扇葵,锡兰行李叶椰子,行李叶椰子);Corypha,Cowryleafpalm,Fan Palm,Talipot Palm ●
107000 Corypha utan Lam.;金丝榈(金丝葵);Gebang,Gebang Palm ●☆
107001 Coryphaceae Schultz Sch. = Arecaceae Bercht. et J. Presl(保留科名)●
107002 Coryphaceae Schultz Sch. = Palmae Juss.(保留科名)●
107003 Coryphadenia Morley = Votomita Aubl. ●☆
107004 Coryphantha (Engelm.) Lem.(1868)(保留属名);菠萝球属(顶花球属);Coryphantha,Pincushion Cactus,Red Cactus ●■
107005 Coryphantha Lem. = Coryphantha (Engelm.) Lem.(保留属名)●■
107006 Coryphantha albicolumnaria (Hester) Dale = Coryphantha sneedii (Britton et Rose) A. Berger ■☆
107007 Coryphantha alversonii (J. M. Coult.) Orcutt;安氏菠萝球;Cushion Foxtail Cactus ■☆

107008 Coryphantha andreae (J. A. Purpus et Boed.) A. Berger;巨象球;Giant Coryphantha ■☆
107009 Coryphantha bumamma (Ehrenb.) Britton et Rose;天司球■☆
107010 Coryphantha calipensis Bravo;壮农■☆
107011 Coryphantha calochlora Boed. = Coryphantha nickelsiae (K. Brandegee) Britton et Rose ■☆
107012 Coryphantha chaffeyi (Britton et Rose) Fosberg;查菲菠萝球■☆
107013 Coryphantha chlorantha (Engelm.) Britton et Rose;绿花菠萝球■☆
107014 Coryphantha clavata (Schum.) Backeb.;针刺仙人球■☆
107015 Coryphantha compacta Britton et Rose;千头仙人球■
107016 Coryphantha cornifera (DC.) Lem.;狮子奋迅(顶花仙人球,玉狮子);Rhinoceros Cactus ■
107017 Coryphantha cornifera (DC.) Lem. var. echinus (Engelm.) L. D. Benson = Coryphantha echinus (Engelm.) Britton et Rose ■☆
107018 Coryphantha dasyacantha (Engelm.) Orcutt;毛花菠萝球;Big Bend Cactus ■☆
107019 Coryphantha dasyacantha (Engelm.) Orcutt var. varicolor (Tiegel) L. D. Benson = Coryphantha tuberculosa (Engelm.) A. Berger ■☆
107020 Coryphantha duncanii (Hester) L. D. Benson;邓肯菠萝球;Duncan's Pincushion Cactus ■☆
107021 Coryphantha echinoidea Britton et Rose;针鼠丸(赤目狮子)■☆
107022 Coryphantha echinus (Engelm.) Britton et Rose;海胆菠萝球;Prickly Beehive Cactus, Rhinocerous Cactus, Sea-urchin Cactus ■☆
107023 Coryphantha elephantidens (Lem.) Lem.;象牙球(象牙丸,象牙仙人球);Elephants Tooth, Elephanttooth Coryphantha ■
107024 Coryphantha emskoetteriana (Quehl) A. Berger = Coryphantha robertii A. Berger ■☆
107025 Coryphantha engelmannii Cory = Coryphantha robustispina (Schott ex Engelm.) Britton et Rose ■☆
107026 Coryphantha erecta (Lem.) Lem.;直立仙人球(杨贵妃)■
107027 Coryphantha glanduligera (Otto et A. Dietr.) Lem.;具腺菠萝球■☆
107028 Coryphantha hesteri Y. Wright;赫氏菠萝球;Hester's Foxtail Cactus, Hester's Pincushion Cactus ■☆
107029 Coryphantha laui Bremer = Coryphantha nickelsiae (K. Brandegee) Britton et Rose ■☆
107030 Coryphantha longicornis Boed.;长角菠萝球■☆
107031 Coryphantha macromeris (Engelm.) Lem.;乳突菠萝球;Long Mamma, Nipple Beehive Cactus ■☆
107032 Coryphantha macromeris (Engelm.) Lem. subsp. runyonii (Britton et Rose) N. P. Taylor = Coryphantha macromeris (Engelm.) Lem. ■☆
107033 Coryphantha macromeris (Engelm.) Lem. var. runyonii (Britton et Rose) L. D. Benson = Coryphantha macromeris (Engelm.) Lem. ■☆
107034 Coryphantha maiz-tablasensis Fritz Schwarz;黑象球■☆
107035 Coryphantha minima Baird;袖珍仙人球;Nellie's Pincushion Cactus ■☆
107036 Coryphantha missouriensis (Sweet) Britton et Rose;密苏里菠萝球;Missouri Foxtail Cactus ■☆
107037 Coryphantha missouriensis (Sweet) Britton et Rose var. caespitosa (Engelm.) L. D. Benson = Coryphantha missouriensis (Sweet) Britton et Rose ■☆
107038 Coryphantha missouriensis (Sweet) Britton et Rose var. marstonii (Clover) L. D. Benson = Coryphantha vivipara (Nutt.) Britton et Rose ■☆
107039 Coryphantha missouriensis (Sweet) Britton et Rose var. robustior (Engelm.) L. D. Benson = Coryphantha missouriensis (Sweet) Britton et Rose ■☆
107040 Coryphantha muehlbaueriana Boed. = Coryphantha robertii A. Berger ■☆
107041 Coryphantha muehlenpfordtii (Poselg.) Britton et Rose = Coryphantha robustispina (Schott ex Engelm.) Britton et Rose ■☆
107042 Coryphantha neglecta Bremer = Coryphantha nickelsiae (K. Brandegee) Britton et Rose ■☆
107043 Coryphantha nellieae Croizat = Coryphantha minima Baird ■☆
107044 Coryphantha neoscheeri Backeb. = Coryphantha robustispina (Schott ex Engelm.) Britton et Rose ■☆
107045 Coryphantha nickelsiae (K. Brandegee) Britton et Rose;尼克菠萝球;Nickel's Coryphantha, Nickels' Pincushion Cactus ■☆
107046 Coryphantha orcuttii (Boed.) Zimmerman = Coryphantha sneedii (Britton et Rose) A. Berger ■☆
107047 Coryphantha organensis Zimmerman = Coryphantha sneedii (Britton et Rose) A. Berger ■☆
107048 Coryphantha ottonis Lem.;印度菠萝球;Indian Head ■☆
107049 Coryphantha pallida Britton et Rose;金环蚀■
107050 Coryphantha palmeri Britton et Rose;帕氏菠萝球■☆
107051 Coryphantha pectinata (Engelm.) Britton et Rose = Coryphantha echinus (Engelm.) Britton et Rose ■☆
107052 Coryphantha pirtlei Werderm. = Coryphantha macromeris (Engelm.) Lem. ■☆
107053 Coryphantha poselgeriana (Dietr.) Britton et Rose;大祥冠(宝珠)■
107054 Coryphantha pycnacantha (Mart.) Lem.;菠萝球;Densespine Coryphantha, Dense-spines Coryphantha ■
107055 Coryphantha radians (DC.) Britton et Rose;玉狮子(顶花仙人球)■
107056 Coryphantha radians (DC.) Britton et Rose = Coryphantha cornifera Lem. ■
107057 Coryphantha radiosa (Engelm.) Rydb.;大疣仙人球■
107058 Coryphantha ramillosa Cutak;刺仙人球;Big Bend Cory Cactus ■☆
107059 Coryphantha recurvata (Engelm.) Britton et Rose;丽阳球(丽阳丸);Golden Chested Beehive Cactus, Santa Cruz Beehive, Santa Cruz Beehive Cactus ■
107060 Coryphantha reduncuspina Boed.;钩刺菠萝球■☆
107061 Coryphantha robbinsorum (W. H. Earle) Zimmerman;罗宾菠萝球;Cochise Foxtail cactus, Cochise Pincushion Cactus ●☆
107062 Coryphantha robertii A. Berger;罗氏菠萝球■☆
107063 Coryphantha robustispina (Schott ex Engelm.) Britton et Rose;粗刺菠萝球;Pineapple Cactus ■☆
107064 Coryphantha runyonii Britton et Rose = Coryphantha macromeris (Engelm.) Lem. ■☆
107065 Coryphantha scheeri Lem. var. robustispina (Schott ex Engelm.) L. D. Benson = Coryphantha robustispina (Schott ex Engelm.) Britton et Rose ■☆
107066 Coryphantha scheeri Lem. var. valida (Schott ex Engelm.) L. D. Benson = Coryphantha robustispina (Schott ex Engelm.) Britton et Rose ■☆
107067 Coryphantha sneedii (Britton et Rose) A. Berger;地毯菠萝球;Carpet Foxtail cactus ■☆
107068 Coryphantha sneedii (Britton et Rose) A. Berger var. leei (Rose

et Boed.) L. D. Benson = Coryphantha sneedii (Britton et Rose) A. Berger ■☆

107069 Coryphantha strobiliformis (Scheer) Moran var. orcuttii (Boed.) L. D. Benson = Coryphantha sneedii (Britton et Rose) A. Berger ■☆

107070 Coryphantha sulcata (Engelm.) Britton et Rose;具槽菠萝球;Finger Cactus,Nipple Cactus,Pineapple Cactus ■☆

107071 Coryphantha sulcata (Engelm.) Britton et Rose var. nickelsiae (K. Brandegee) L. D. Benson = Coryphantha nickelsiae (K. Brandegee) Britton et Rose ■☆

107072 Coryphantha tuberculosa (Engelm.) A. Berger;白柱菠萝球;Cob cactus,White-column Foxtail Cactus ■☆

107073 Coryphantha varicolor Tiegel = Coryphantha tuberculosa (Engelm.) A. Berger ■☆

107074 Coryphantha vivipara (Engelm.) Britton et Rose var. buoflama P. C. Fisch. = Coryphantha chlorantha (Engelm.) Britton et Rose ■☆

107075 Coryphantha vivipara (Engelm.) Britton et Rose var. deserti (Engelm.) T. Marshall = Coryphantha chlorantha (Engelm.) Britton et Rose ■☆

107076 Coryphantha vivipara (Nutt.) Britton et Rose;横网(北极丸);Beehive Cactus,Cushion Cactus,Pincushion Cactus ■☆

107077 Coryphantha vivipara (Nutt.) Britton et Rose var. alversonii (J. M. Coult.) L. D. Benson = Coryphantha alversonii (J. M. Coult.) Orcutt ■☆

107078 Coryphantha vivipara (Nutt.) Britton et Rose var. arizonica (Engelm.) T. Marshall = Coryphantha vivipara (Nutt.) Britton et Rose ■☆

107079 Coryphantha vivipara (Nutt.) Britton et Rose var. bisbeeana (Orcutt) L. D. Benson = Coryphantha vivipara (Nutt.) Britton et Rose ■☆

107080 Coryphantha vivipara (Nutt.) Britton et Rose var. kaibabensis P. C. Fisch. = Coryphantha vivipara (Nutt.) Britton et Rose ■☆

107081 Coryphantha vivipara (Nutt.) Britton et Rose var. neomexicana (Engelm.) Backeb. = Coryphantha vivipara (Nutt.) Britton et Rose ■☆

107082 Coryphantha vivipara (Nutt.) Britton et Rose var. radiosa (Engelm.) Backeb. = Coryphantha vivipara (Nutt.) Britton et Rose ■☆

107083 Coryphantha vivipara (Nutt.) Britton et Rose var. rosea (Clokey) L. D. Benson = Coryphantha vivipara (Nutt.) Britton et Rose ■☆

107084 Coryphantha werdermannii Boed.;精美球■☆

107085 Coryphomia Rojas = Copernicia Mart. ex Endl. ●☆

107086 Coryphomia Rojas(1918);阿根廷蜡棕属●☆

107087 Coryphomia tectorum Rojas;阿根廷蜡棕●☆

107088 Coryphothamnus Steyerm. (1965);头灌茜属●☆

107089 Coryphothamnus auyantepuiensis (Steyerm.) Steyerm.;头灌茜●☆

107090 Corysadenia Griff. = Illigera Blume ●■

107091 Corysanthera Decne. ex Regel = Heppiella Regel ■☆

107092 Corysanthera Endl. = Rhynchotechum Blume ●

107093 Corysanthera Regel = Corysanthera Wall. ex Benth. ●

107094 Corysanthera Wall. = Corysanthera Wall. ex Benth. ●

107095 Corysanthera Wall. = Rhynchotechum Blume ●

107096 Corysanthera Wall. ex Benth. = Rhynchotechum Blume ●

107097 Corysanthera Wall. ex Endl. = Rhynchotechum Blume ●

107098 Corysanthera elliptica Wall. ex D. Dietr. = Rhynchotechum ellipticum (Wall. ex D. Dietr.) A. DC. ●

107099 Corysanthes R. Br. = Corybas Salisb. ■

107100 Corysanthes himalaica King et Pantl. = Corybas himalaicus (King et Pantl.) Schltr. ■

107101 Corythacanthus Nees = Clistax Mart. ☆

107102 Corythanthes Lem. = Coryanthes Hook. ■☆

107103 Corythea S. Watson = Acalypha L. ●■

107104 Corytholobium Benth. = Securidaca L. (保留属名)●

107105 Corytholoblum Mart. ex Benth. = Securidaca L. (保留属名)●

107106 Corytholoma (Benth.) Decne. = Rechsteineria Regel(保留属名)■☆

107107 Corytholoma (Benth.) Decne. = Sinningia Nees ●■☆

107108 Corytholoma Decne. = Rechsteineria Regel(保留属名)■☆

107109 Corythophora R. Knuth(1939);头梗玉蕊属●☆

107110 Corythophora alta R. Knuth;头梗玉蕊●☆

107111 Corytoplectus Oerst. (1858);奥氏苣苔属■☆

107112 Corytoplectus Oerst. = Alloplectus Mart. (保留属名)●■☆

107113 Corytoplectus capitatus Oerst.;奥氏苣苔■☆

107114 Coryzadenia Griff. = Illigera Blume ●■

107115 Cosaria J. F. Gmel. = Dorstenia L. ■●☆

107116 Cosaria J. F. Gmel. = Kosaria Forssk. ■●☆

107117 Cosbaea Lem. = Schisandra Michx. (保留属名)●

107118 Cosbaea coccinea Lem. = Kadsura coccinea (Lem.) A. C. Sm. ●

107119 Coscinium Colebr. (1821);箣藤属(南洋药藤属)●☆

107120 Coscinium blumeanum Miers;布氏箣藤(布卢姆箣藤);False Calumba Root ●☆

107121 Coscinium colaniae Gagnep. = Pericampylus glaucus (Lam.) Merr. ●

107122 Coscinium fenestratum Colebr.;膜孔箣藤●☆

107123 Coscinium usitatum Pierre;普通箣藤●☆

107124 Cosmanthus Nolte ex A. DC. = Phacelia Juss. ■☆

107125 Cosmarium Dulac = Adonis L. (保留属名)■

107126 Cosmea Willd. = Cosmos Cav. ■

107127 Cosmea bipinnatus Hort. = Cosmos bipinnatus Cav. ■

107128 Cosmelia R. Br. (1810);笔管石南属●☆

107129 Cosmelia rubra R. Br.;笔管石南●☆

107130 Cosmia Dombey ex Juss. = Calandrinia Kunth(保留属名)■☆

107131 Cosmianthemum Bremek. (1960);秋英爵床属●■

107132 Cosmianthemum guangxiense H. S. Lo et D. Fang;广西秋英爵床■

107133 Cosmianthemum knoxifolium (C. B. Clarke) B. Hansen;节叶秋英爵床●

107134 Cosmianthemum longiflorum D. Fang et H. S. Lo;长花秋英爵床■

107135 Cosmianthemum viriduliflorum (C. Y. Wu et H. S. Lo) H. S. Lo;海南秋英爵床(琼紫叶)■

107136 Cosmibuena Ruiz et Pav. (1802)(保留属名);长管栀子属●☆

107137 Cosmibuena acuminata Ruiz et Pav.;渐尖长管栀子●☆

107138 Cosmibuena arborea Standl.;树状长管栀子●☆

107139 Cosmibuena grandiflora Rusby;大花长管栀子●☆

107140 Cosmibuena latifolia Klotzsch ex Walp.;宽叶长管栀子●☆

107141 Cosmibuena macrocarpa Klotzsch ex Walp.;大果长管栀子●☆

107142 Cosmibuena obtusifolia Ruiz et Pav.;钝叶长管栀子●☆

107143 Cosmibuena triflora Klotzsch;三花长管栀子●☆

107144 Cosmidium Nutt. = Thelesperma Less. ■●☆

107145 Cosmidium burridgeanum Regel = Thelesperma burridgeanum (Regel) S. F. Blake ■☆

107146 Cosmidium simplicifolium A. Gray = Thelesperma simplicifolium

(A. Gray) A. Gray ■☆
107147 Cosmiusa Alef. = Parochetus Buch. -Ham. ex D. Don ■
107148 Cosmiza Raf. (废弃属名) = Polypompholyx Lehm. (保留属名) ■
107149 Cosmiza Raf. (废弃属名) = Utricularia L. ■
107150 Cosmocalyx Standl. (1930); 齐萼茜属●☆
107151 Cosmocalyx spectabilis Standl.; 齐萼茜●☆
107152 Cosmoneuron Pierre = Octoknema Pierre ●☆
107153 Cosmoneuron Pierre = Strombosia Blume ●☆
107154 Cosmoneuron klaineanum Pierre = Octoknema klaineana Pierre ●☆
107155 Cosmoneuron klaineanum Pierre = Strombosia grandifolia Hook. f. ●☆
107156 Cosmophyllum C. Koch = Podachaenium Benth. ex Oerst. ●☆
107157 Cosmophyllum K. Koch = Podachaenium Benth. ex Oerst. ●☆
107158 Cosmos Cav. (1791); 秋英属(波斯菊属,大波斯菊属,秋樱属); Cosmos, Mexican Aster ■
107159 Cosmos atrosanguineus (Hook.) Voss = Cosmos atrosanguineus (Ortega) Voss ■☆
107160 Cosmos atrosanguineus (Ortega) Voss; 血紫秋英(紫红秋英); Chocolate Cosmos, Chocolate Plant ■☆
107161 Cosmos bipinnatus Cav.; 秋英(八瓣梅,波斯菊,大波斯菊,非洲菊,红菊,水茼蒿); Common Cosmos, Common Garden Cosmos, Cosmos, Garden Cosmos, Mexican Aster, Purple Aster, Purple Mexican Aster, Tall Cosmos ■
107162 Cosmos bipinnatus Cav. 'Candy Stripe'; 彩纹秋英■☆
107163 Cosmos bipinnatus Cav. 'Sea Shells'; 海贝秋英■☆
107164 Cosmos bipinnatus Cav. var. albiflorus Hort.; 白花秋英■☆
107165 Cosmos bipinnatus Cav. var. grandiflorus Hort.; 大花秋英■☆
107166 Cosmos bipinnatus Cav. var. purpureus Hort.; 紫花秋英■☆
107167 Cosmos caudatus Kunth; 尾状秋英■☆
107168 Cosmos diversifolius Otto; 异叶秋英(小波斯菊,异叶波斯菊); Black Cosmos ■☆
107169 Cosmos formosa Bonato = Cosmos bipinnatus Cav. ■
107170 Cosmos parviflorus (Jacq.) Pers.; 小花秋英; Cosmos ■☆
107171 Cosmos peucedanifolius Wedd.; 前胡叶秋英■☆
107172 Cosmos sulphureus Cav.; 黄秋英(黄波斯菊,硫华菊,硫黄菊); Cosmos, Klondike Cosmos, Orange Cosmos, Sulphur Cosmos, Yellow Cosmos ■
107173 Cosmos tinctoris Nutt. = Coreopsis tinctoria Nutt. ■
107174 Cosmostigma Wight (1834); 荟蔓藤属(阔柱藤属); Cosmostigma ●
107175 Cosmostigma hainanense Tsiang; 荟蔓藤; Hainan Cosmostigma ●
107176 Cossignea Willd. = Cossinia Comm. ex Lam. ●☆
107177 Cossignia Comm. ex Lam. = Cossinia Comm. ex Lam. ●☆
107178 Cossignya Baker = Cossinia Comm. ex Lam. ●☆
107179 Cossinia Comm. ex Lam. (1786); 澳木患属●☆
107180 Cossinia australiana S. T. Reynolds; 澳木患●☆
107181 Cossonia Durieu = Raffenaldia Godr. ■☆
107182 Cossonia africana Durieu = Raffenaldia primuloides Godr. ■☆
107183 Cossonia africana Durieu var. lutea Maire = Raffenaldia primuloides Godr. ■☆
107184 Cossonia africana Durieu var. violacea Maire = Raffenaldia primuloides Godr. ■☆
107185 Cossonia intermedia Coss. = Raffenaldia primuloides Godr. ■☆
107186 Cossonia platycarpa Coss. = Raffenaldia platycarpa (Coss.) Stapf ■☆
107187 Costa Vell. = Galipea Aubl. ●☆
107188 Costaceae (Meisn.) Nakai = Zingiberaceae Martinov(保留科名) ■
107189 Costaceae (Meisn.) Nakai(1941); 闭鞘姜科■
107190 Costaceae Nakai = Costaceae (Meisn.) Nakai ■
107191 Costaceae Nakai = Zingiberaceae Martinov(保留科名) ■
107192 Costaea A. Rich. = Purdiaea Planch. ●☆
107193 Costaea Post et Kuntze = Agropyron Gaertn. ■
107194 Costaea Post et Kuntze = Costa Vell. ●☆
107195 Costaea Post et Kuntze = Costia Willk. ■
107196 Costaea Post et Kuntze = Galipea Aubl. ●☆
107197 Costaea Post et Kuntze = Iris L. ■
107198 Costantina Bullock = Lygisma Hook. f. ■
107199 Costantina inflexa (Costantin) Bullock = Lygisma inflexum (Costantin) Kerr ■
107200 Costarica L. D. Gomez = Sicyos L. ■
107201 Costaricaea Schltr. = Hexisea Lindl. (废弃属名) ■☆
107202 Costaricaea Schltr. = Scaphyglottis Poepp. et Endl. (保留属名) ■☆
107203 Costea A. Rich. = Costaea A. Rich. ●☆
107204 Costea A. Rich. = Purdiaea Planch. ●☆
107205 Costera J. J. Sm. (1910); 腺叶莓属●☆
107206 Costera borneensis J. J. Sm.; 腺叶莓●☆
107207 Costera lucida (Merr.) Airy Shaw et J. J. Sm.; 光亮腺叶莓●☆
107208 Costera macrantha Argent; 大花腺叶莓●☆
107209 Costera ovalifolia J. J. Sm.; 卵叶腺叶莓●☆
107210 Costia Willk. (1860) = Iris L. ■
107211 Costia Willk. (1958) = Agropyron Gaertn. ■
107212 Costia orientalis (L.) Willk. = Eremopyrum orientale (L.) Jaub. et Spach ■
107213 Costularia C. B. Clarke = Costularia C. B. Clarke ex Dyer ■☆
107214 Costularia C. B. Clarke ex Dyer(1898); 细脉莎草属■☆
107215 Costularia baronii C. B. Clarke; 细脉莎草■☆
107216 Costularia brevicaulis C. B. Clarke; 短茎细脉莎草■☆
107217 Costularia brevifolia Cherm.; 短叶细脉莎草■☆
107218 Costularia breviseta Raynal; 短刚毛细脉莎草■☆
107219 Costularia leucocarpa H. Pfeiff.; 白果细脉莎草■☆
107220 Costularia microcarpa (Cherm.) Kük.; 小果细脉莎草■☆
107221 Costus L. (1753); 闭鞘姜属(巴西掖姜花属,广商陆属,鞘姜属); Spiral Flag, Spiral Ginger, Spiralflag ■
107222 Costus adolfi-friderici Loes.; 弗里德里西闭鞘姜■☆
107223 Costus afer Ker Gawl.; 非洲闭鞘姜■☆
107224 Costus albus A. Chev. ex Koechlin = Costus dubius (Afzel.) K. Schum. ■☆
107225 Costus anomocalyx K. Schum. = Costus afer Ker Gawl. ■☆
107226 Costus arabicus Vell.; 阿拉伯闭鞘姜■☆
107227 Costus barbatus Suess.; 毛闭鞘姜; Spiral Ginger ■☆
107228 Costus bicolor J. Braun et K. Schum.; 二色闭鞘姜■☆
107229 Costus bingervillensis A. Chev. = Costus afer Ker Gawl. ■☆
107230 Costus chinensis T. L. Wu et Senjen = Costus lacerus Gagnep. ■
107231 Costus deistelii K. Schum.; 戴氏闭鞘姜■☆
107232 Costus dewevrei De Wild. et T. Durand; 德韦闭鞘姜■☆
107233 Costus dinklagei K. Schum.; 丁克闭鞘姜■☆
107234 Costus discolor Roscoe; 异色闭鞘姜■☆
107235 Costus dubius (Afzel.) K. Schum.; 不定闭鞘姜■☆
107236 Costus dubius K. Schum = Costus dubius (Afzel.) K. Schum. ■☆
107237 Costus edulis De Wild. et T. Durand = Costus dubius (Afzel.) K. Schum. ■☆

107238 Costus englerianus K. Schum. = Paracostus englerianus (K. Schum.) Engl. ■☆

107239 Costus fimbriatus Pellegr. = Costus ligularis Baker ■☆

107240 Costus foliaceus Lock et A. D. Poulsen;多叶闭鞘姜■☆

107241 Costus formosanus Nakai = Costus speciosus (König) Sm. ■

107242 Costus gabonensis Koechlin;加蓬闭鞘姜■☆

107243 Costus giganteus Welw. ex Ridl.;巨大闭鞘姜■☆

107244 Costus igneus N. E. Br.;巴西闭鞘姜(巴西福神草)■☆

107245 Costus insularis A. Chev. = Costus afer Ker Gawl. ■☆

107246 Costus lacerus Gagnep.;莴笋花;China Spiralflag ■

107247 Costus lateriflorus Baker;侧花闭鞘姜■☆

107248 Costus ledermannii Loes.;莱德闭鞘姜■☆

107249 Costus letestui Pellegr.;莱泰斯图闭鞘姜■☆

107250 Costus ligularis Baker;舌状闭鞘姜■☆

107251 Costus littoralis K. Schum.;滨海闭鞘姜■☆

107252 Costus macranthus K. Schum.;大花闭鞘姜■☆

107253 Costus maculatus Roscoe;斑点闭鞘姜■☆

107254 Costus malortieanus H. Wendl.;绒叶闭鞘姜;Spiral Ginger ■☆

107255 Costus megalobracta K. Schum.;大苞闭鞘姜■☆

107256 Costus nemotrichus K. Schum.;丝毛闭鞘姜■☆

107257 Costus nepalensis Roscoe = Costus speciosus (König) Sm. ■

107258 Costus ngouniensis Pellegr. = Costus ligularis Baker ■☆

107259 Costus nudicaulis Baker;裸茎闭鞘姜■☆

107260 Costus oblitterans K. Schum. = Costus afer Ker Gawl. ■☆

107261 Costus oblongus S. Q. Tong;长圆闭鞘姜;Oblong Spiralflag ■

107262 Costus phaeotrichus Loes.;褐毛闭鞘姜■☆

107263 Costus pictus D. Don ex Lindl.;花叶闭鞘姜;Spiral Ginger ■☆

107264 Costus pleiostachyum K. Schum. = Zingiber pleiostachyum K. Schum. ■

107265 Costus productus Gleason ex Maas;伸展闭鞘姜;Spiral Ginger ■☆

107266 Costus pterometra K. Schum. = Costus afer Ker Gawl. ■☆

107267 Costus pulcherrimus A. Chev. = Costus deistelii K. Schum. ■☆

107268 Costus sarmentosus Bojer;蔓茎闭鞘姜■☆

107269 Costus scaber Ruiz et Pav.;洋闭鞘姜■☆

107270 Costus schlechteri H. Winkl.;施莱闭鞘姜■☆

107271 Costus speciosus (König) Sm.;闭鞘姜(白石笋,福神草,广东商陆,广商陆,姜商陆,绢毛鸢尾,老妈妈拐棍,山冬笋,水蕉花,水莲花,象甘蔗,樟柳头);Canereed Spiral Flag, Canereed Spiralflag, Malay Ginger ■

107272 Costus speciosus (König) Sm. var. angustifolius ? = Costus speciosus (König) Sm. ■

107273 Costus speciosus (König) Sm. var. formosanus (Nakai) S. S. Ying = Costus speciosus (König) Sm. ■

107274 Costus speciosus (König) Sm. var. hirsutus Blume = Costus speciosus (König) Sm. ■

107275 Costus speciosus (König) Sm. var. leocalyx (K. Schum) Nakai = Costus speciosus (König) Sm. ■

107276 Costus speciosus (König) Sm. var. nepalensis ? = Costus speciosus (König) Sm. ■

107277 Costus spectabilis (Fenzl) K. Schum.;壮观闭鞘姜■☆

107278 Costus spiralis Roscoe;螺旋闭鞘姜■☆

107279 Costus subbiflorus K. Schum.;亚双花闭鞘姜■☆

107280 Costus talbotii Ridl.;塔尔博特闭鞘姜■☆

107281 Costus tonkinensis Gagnep.;光叶闭鞘姜;Tonkin Spiralflag ■

107282 Costus trachyphyllus K. Schum. = Costus dubius (Afzel.) K. Schum. ■☆

107283 Costus ubangiensis Gagnep.;乌班吉闭鞘姜■☆

107284 Costus ulugurensis K. Schum. = Costus sarmentosus Bojer ■☆

107285 Costus villosissimus Jacq.;柔毛闭鞘姜■☆

107286 Costus violaceus Koechlin;堇色闭鞘姜■☆

107287 Costus viridis S. Q. Tong;绿苞闭鞘姜;Greenbract Spiralflag ■

107288 Costus zechii K. Schum. = Costus dubius (Afzel.) K. Schum. ■☆

107289 Costus zerumbert Pers. = Alpinia zerumbet (Pers.) B. L. Burtt et R. M. Sm. ■

107290 Cota J. Gay = Cota J. Gay ex Guss. ■☆

107291 Cota J. Gay ex Guss. (1845);全黄菊属■☆

107292 Cota J. Gay ex Guss. = Anthemis L. ■

107293 Cota tinctoria (L.) J. Gay = Anthemis tinctoria L. ■

107294 Cota tinctoria (L.) J. Gay = Cota tinctoria (L.) J. Gay ex Guss. ■☆

107295 Cota tinctoria (L.) J. Gay ex Guss.;全黄菊;Golden Marguerite, Yellow Chamomile ■☆

107296 Cotema Britton et P. Wilson = Spirotecoma (Baill.) Dalla Torre et Harms ●☆

107297 Cotinus Mill. (1754);黄栌属;Smoke Bush, Smoke Tree, Smoketree, Smoke-tree, Smoke-wood ●

107298 Cotinus americana Nutt. = Cotinus obovata Raf. ●

107299 Cotinus cinerea F. A. Barkley = Cotinus coggygria Scop. var. cinerea Engl. ●

107300 Cotinus coggygria Scop.;黄栌(黄道栌,栌木,月亮柴);Aaron's-beard, Common Smoke Tree, Common Smoketree, Common Smoke-tree, Eurasian Smoke Bush, European Smoketree, Fustic, Green Smokebush, Hungarian, Hungarian Fustic, Indian Sumac, Indian Sumach, Purple Fringe, Scotino, Smoke Bush, Smoke Plant, Smoke Tree, Smokebush, Smoketree, Smoke-tree, Venetian Sumach, Venus' Sumach, White Olive, Wig Tree, Wigtree, Wig-tree, Young Fustic, Zante Fustic, Zante Wood ●

107301 Cotinus coggygria Scop. 'Atropurpureus' = Cotinus coggygria Scop. var. purpureus Rehder ●

107302 Cotinus coggygria Scop. 'Blazeaway';连发黄栌●☆

107303 Cotinus coggygria Scop. 'Flame';火焰黄栌;Common Smoke Bush, Common Smoke Tree ●☆

107304 Cotinus coggygria Scop. 'Nordine Red';紫红黄栌(紫叶黄栌);Purple-leaved Smokebush ●☆

107305 Cotinus coggygria Scop. 'Notcutt's Variety';紫晕黄栌●☆

107306 Cotinus coggygria Scop. 'Royal Purple';蓝紫黄栌(品紫黄栌);Purple Smoke Bush ●☆

107307 Cotinus coggygria Scop. 'Velvet Cloak';丝绒披风黄栌;Select Purple Smoketree ●☆

107308 Cotinus coggygria Scop. var. chengkouensis Y. T. Wu;城口黄栌;Chengkou Smoketree ●

107309 Cotinus coggygria Scop. var. cinerea Engl.;红叶黄栌(红叶,黄道栌,黄栌,灰毛黄栌);Ash-colored Smoke Tree, Ash-colored Smoketree, Red Leaf ●

107310 Cotinus coggygria Scop. var. glaucophylla C. Y. Wu;粉背黄栌;Glaucousback Smoketree, Glaucousleaf Smoketree ●

107311 Cotinus coggygria Scop. var. pendulus Dippel;垂枝黄栌;Pendulous Smoketree ●

107312 Cotinus coggygria Scop. var. pubescens Engl.;毛黄栌(栌木,毛叶黄栌,柔毛黄栌,岩棕树);Pubescent Smoketree ●

107313 Cotinus coggygria Scop. var. purpureus Rehder;紫叶黄栌;Burning Bush, Purple Smoke Bush, Purple Smoketree ●

107314 Cotinus nana W. W. Sm.;矮黄栌;Dwarf Smoketree, Dwarf Smoke-tree ●

107315 Cotinus obovata Raf.;美洲黄栌(北美黄栌,美国黄栌,美果黄栌);America Smoketree,American fustic Chittam,American Smoke Tree,American Smoketree,Chittam,Chittam Wood,Chittamwood,Yellowwood ●

107316 Cotinus szechuanensis Pénzes;四川黄栌;Sichuan Smoketree,Sichuan Smoke-tree ●

107317 Cotonea Raf. = Cotoneaster Medik. ●

107318 Cotoneaster J. B. Ehrh. = Cotoneaster Medik. ●

107319 Cotoneaster Medik.(1789);栒子属(铺地蜈蚣属);Cotoneaster,Quince-leaved Medlar,Rose-box ●

107320 Cotoneaster Rupp. = Cotoneaster Medik. ●

107321 Cotoneaster 'Hessei';黑塞栒子;Cotoneaster,Hessei ●☆

107322 Cotoneaster acuminatus Lindl.;尖叶栒子(灰栒子);Sharpleaf Cotoneaster,Sharp-leaf Cotoneaster ●

107323 Cotoneaster acuminatus Lindl. var. lucidus (Schltdl.) L. T. Lu = Cotoneaster lucidus Schltdl. ●

107324 Cotoneaster acuminatus Lindl. var. prostratus Hook. ex Decne. = Cotoneaster horizontalis Decne. ●

107325 Cotoneaster acutifolius Turcz.;灰栒子;Beijing Cotoneaster,Cotoneaster,Peking Cotoneaster ●

107326 Cotoneaster acutifolius Turcz. = Cotoneaster villosulus (Rehder et E. H. Wilson) Flinck et Hylmö ●

107327 Cotoneaster acutifolius Turcz. f. glabriusculus Hurus. = Cotoneaster hurusawaianus G. Klotz ●

107328 Cotoneaster acutifolius Turcz. f. glabriusculus Hurus. = Cotoneaster acutifolius Turcz. ●

107329 Cotoneaster acutifolius Turcz. var. ambiguus (Rehder et E. H. Wilson) Hurus. = Cotoneaster ambiguus Rehder et E. H. Wilson ●

107330 Cotoneaster acutifolius Turcz. var. glabricalyx Hurus.;光萼灰栒子●

107331 Cotoneaster acutifolius Turcz. var. laetevirens Rehder et E. H. Wilson = Cotoneaster acutifolius Turcz. ●

107332 Cotoneaster acutifolius Turcz. var. lucidus (Schltdl.) L. T. Lu;甘南灰栒子●

107333 Cotoneaster acutifolius Turcz. var. pekiennsis Koehne = Cotoneaster acutifolius Turcz. ●

107334 Cotoneaster acutifolius Turcz. var. villosulus Rehder et E. H. Wilson;密毛灰栒子●

107335 Cotoneaster acutifolius Turcz. var. villosulus Rehder et E. H. Wilson = Cotoneaster villosulus (Rehder et E. H. Wilson) Flinck et Hylmö ●

107336 Cotoneaster adpressus Boiss.;匍匐栒子(伏栒子,匍匐灰栒子);Creeping Cotoneaster ●

107337 Cotoneaster affinis Lindl.;藏边栒子;Brownberry Cotoneaster,Brown-berry Cotoneaster,Purple-berry Cotoneaster ●

107338 Cotoneaster allochroa Pojark.;异花栒子●

107339 Cotoneaster ambiguus Rehder et E. H. Wilson;四川栒子(川康栒子);Doubtful Cotoneaster ●

107340 Cotoneaster amoenus E. H. Wilson;美丽栒子(可爱栒子,马蝗果);Beautiful Cotoneaster ●

107341 Cotoneaster amoenus E. H. Wilson = Cotoneaster franchetii Boiss. ●

107342 Cotoneaster angustifolius Franch. = Pyracantha angustifolia (Franch.) C. K. Schneid. ●

107343 Cotoneaster angustus (Te T. Yu) G. Klotz = Cotoneaster salicifolius Franch. var. angustus Te T. Yu ●

107344 Cotoneaster antoninae A. V. Vassil.;安氏栒子●☆

107345 Cotoneaster apiculatus Rehder et E. H. Wilson;细尖栒子(尖叶栒);Cranberry Cotoneaster,Cran-berry Cotoneaster ●

107346 Cotoneaster applanatus Duthie ex Veitch = Cotoneaster dielsianus Pritz. ex Diels ●

107347 Cotoneaster arbusculus G. Klotz = Cotoneaster glaucophyllus Franch. var. meiophyllus W. W. Sm. ●

107348 Cotoneaster argenteus G. Klotz;凸尖栒子(银毛栒子)●

107349 Cotoneaster argenteus G. Klotz = Cotoneaster buxifolius Wall. ex Lindl. ●

107350 Cotoneaster astrophorus J. Fryer et E. C. Nelson = Cotoneaster poluninii G. Klotz ●

107351 Cotoneaster atlanticus G. Klotz = Cotoneaster granatensis Boiss. ●☆

107352 Cotoneaster atropurpureus Flinck et B. Hylmö;暗紫栒子;Purple-flowered Cotoneaster ●☆

107353 Cotoneaster bacillaris Lindl. var. affinis (Lindl.) Hook. f. = Cotoneaster affinis Lindl. ●

107354 Cotoneaster bacillaris Wall. ex Lindl.;喜马拉雅栒子;Open-fruited Cotoneaster ●☆

107355 Cotoneaster bacillaris Wall. ex Lindl. var. affinis Hook. f. = Cotoneaster affinis Lindl. ●

107356 Cotoneaster bakeri G. Klotz;毛萼栒子(贝克栒子,柔毛栒子);Baker Cotoneaster ●

107357 Cotoneaster bakeri G. Klotz = Cotoneaster acuminatus Lindl. ●

107358 Cotoneaster blinii H. Lév. = Photinia blinii (H. Lév.) Rehder ●

107359 Cotoneaster bodinieri H. Lév. = Docynia delavayi (Franch.) C. K. Schneid. ●

107360 Cotoneaster borealichinensis (Hurus.) Hurus. = Cotoneaster submultiflorus Popov ●

107361 Cotoneaster brevirameus Rehder et E. H. Wilson = Cotoneaster buxifolius Wall. ex Lindl. ●

107362 Cotoneaster bullatus Bois;泡叶栒子;Hollyberry Cotoneaster,Holly-berry Cotoneaster ●☆

107363 Cotoneaster bullatus Bois 'Firebird';火鸟泡叶栒子;Hollyberry Cotoneaster ●

107364 Cotoneaster bullatus Bois f. floribundus (Stapf) Rehder;多花泡叶栒子;Manyflower Hollyberry Cotoneaster ●

107365 Cotoneaster bullatus Bois f. floribundus (Stapf) Rehder et E. H. Wilson = Cotoneaster bullatus Bois f. floribundus (Stapf) Rehder ●

107366 Cotoneaster bullatus Bois var. macrophyllus Rehder et E. H. Wilson;大叶泡叶栒子(大泡叶栒子);Bullate Cotoneaster,Largeleaf Hollyberry Cotoneaster ●

107367 Cotoneaster buxifolius Bois f. vellaeus Franch. = Cotoneaster buxifolius Wall. ex Lindl. ●

107368 Cotoneaster buxifolius Lindl. f. cochleatus Franch. = Cotoneaster microphyllus Wall. ex Lindl. var. cochleatus (Franch.) Rehder et E. H. Wilson ●

107369 Cotoneaster buxifolius Lindl. f. melanotrichus Franch. = Cotoneaster microphyllus Wall. ex Lindl. ●

107370 Cotoneaster buxifolius Wall. ex Lindl.;黄杨叶栒子(车轮棠);Boxleaved Cotoneaster,Box-leaved Cotoneaster ●

107371 Cotoneaster buxifolius Wall. ex Lindl. f. cochleatus Franch. = Cotoneaster microphyllus Wall. ex Lindl. var. cochleatus (Franch.) Rehder et E. H. Wilson ●

107372 Cotoneaster buxifolius Wall. ex Lindl. f. cochleatus Franch. = Cotoneaster microphyllus Wall. ex Lindl. ●

107373 Cotoneaster buxifolius Wall. ex Lindl. f. vellaeus Franch. =

Cotoneaster buxifolius Wall. ex Lindl. var. vellaeus (Franch.) G. Klotz ●

107374 Cotoneaster buxifolius Wall. ex Lindl. f. vellaeus Franch. = Cotoneaster poluninii G. Klotz ●

107375 Cotoneaster buxifolius Wall. ex Lindl. var. cochleatus (Franch.) Rehder et E. H. Wilson = Cotoneaster microphyllus Wall. ex Lindl. ●

107376 Cotoneaster buxifolius Wall. ex Lindl. var. cochleatus (Franch.) Rehder et E. H. Wilson = Cotoneaster microphyllus Wall. ex Lindl. var. cochleatus (Franch.) Rehder et E. H. Wilson ●

107377 Cotoneaster buxifolius Wall. ex Lindl. var. marginatus Loudon;多花黄杨叶栒子;Manyflower Boxleaved Cotoneaster ●

107378 Cotoneaster buxifolius Wall. ex Lindl. var. melanotichus (Franch.) Hand. -Mazz. = Cotoneaster microphyllus Wall. ex Lindl. ●

107379 Cotoneaster buxifolius Wall. ex Lindl. var. rockii (G. Klotz) L. T. Lu et A. R. Brach;西南黄杨叶栒子●

107380 Cotoneaster buxifolius Wall. ex Lindl. var. vellaeus (Franch.) G. Klotz;小叶黄杨叶栒子(小黄杨叶栒子);Little Boxleaved Cotoneaster ●

107381 Cotoneaster buxifolius Wall. ex Lindl. var. vellaeus (Franch.) G. Klotz = Cotoneaster buxifolius Wall. ex Lindl. ●

107382 Cotoneaster buxifolius Wall. ex Lindl. var. vellaeus (Franch.) G. Klotz = Cotoneaster poluninii G. Klotz ●

107383 Cotoneaster calocarpus (Rehder et E. H. Wilson) Flinck et Hylmö = Cotoneaster multiflorus Bunge var. calocarpus Rehder et E. H. Wilson ●

107384 Cotoneaster camilli-schneideri Pojark.;肉苞栒子(卡氏栒子)●

107385 Cotoneaster cashmiriensis Klotz;克什米尔栒子;Kashmir Cotoneaster ●

107386 Cotoneaster cavei G. Klotz = Cotoneaster nitidus Jacq. var. parvifolius (Te T. Yu) Te T. Yu ●

107387 Cotoneaster chengkangensis Te T. Yu;镇康栒子;Zhenkang Cotoneaster ●

107388 Cotoneaster cochleatus (Franch.) G. Klotz = Cotoneaster cashmiriensis Klotz ●

107389 Cotoneaster cochleatus (Franch.) G. Klotz = Cotoneaster microphyllus Wall. ex Lindl. ●

107390 Cotoneaster cochleatus (Franch.) G. Klotz = Cotoneaster microphyllus Wall. ex Lindl. var. cochleatus (Franch.) Rehder et E. H. Wilson ●

107391 Cotoneaster cochleatus (Franch.) G. Klotz f. melanotrichus (Franch.) G. Klotz = Cotoneaster microphyllus Wall. ex Lindl. ●

107392 Cotoneaster congestus Baker;比利牛斯栒子;Congested Cotoneaster ●☆

107393 Cotoneaster congestus Baker = Cotoneaster microphyllus Wall. ex Lindl. var. glacialis Hook. f. ●

107394 Cotoneaster congestus Baker = Cotoneaster microphyllus Wall. ex Lindl. ●

107395 Cotoneaster conspicuus (Messel) Messel = Cotoneaster microphyllus Wall. ex Lindl. var. conspicuus Messel ●

107396 Cotoneaster conspicuus (Messel) Messel et C. Marquand ex G. Klotz;猩红果栒子(大果栒子);Tibetan Cotoneaster ●

107397 Cotoneaster conspicuus C. Marquand = Cotoneaster conspicuus (Messel) Messel et C. Marquand ex G. Klotz ●

107398 Cotoneaster conspicuus C. Marquand ex G. Klotz = Cotoneaster conspicuus (Messel) Messel et C. Marquand ex G. Klotz ●

107399 Cotoneaster conspicuus C. Marquand var. decorus Russell = Cotoneaster conspicuus (Messel) Messel et C. Marquand ex G. Klotz ●

107400 Cotoneaster conspicuus C. Marquand var. nanus G. Klotz = Cotoneaster conspicuus (Messel) Messel et C. Marquand ex G. Klotz ●

107401 Cotoneaster conspicuus Comber = Cotoneaster conspicuus (Messel) Messel et C. Marquand ex G. Klotz ●

107402 Cotoneaster conspicuus Comber ex Marquand = Cotoneaster conspicuus (Messel) Messel et C. Marquand ex G. Klotz ●

107403 Cotoneaster conspicuus Comber ex Marquand var. decorus Russell = Cotoneaster conspicuus (Messel) Messel et C. Marquand ex G. Klotz ●

107404 Cotoneaster conspicuus Comber ex Marquand var. nanus G. Klotz = Cotoneaster conspicuus (Messel) Messel et C. Marquand ex G. Klotz ●

107405 Cotoneaster conspicuus Messel = Cotoneaster microphyllus Wall. ex Lindl. var. conspicuus Messel ●

107406 Cotoneaster conspicuus Messel var. decorus Russell = Cotoneaster microphyllus Wall. ex Lindl. var. conspicuus Messel ●

107407 Cotoneaster cooperi C. Marquand = Cotoneaster bacillaris Wall. ex Lindl. ●☆

107408 Cotoneaster cordifolius G. Klotz = Cotoneaster nitidus Jacq. var. parvifolius (Te T. Yu) Te T. Yu ●

107409 Cotoneaster coreanus H. Lév. = Symplocos coreana (H. Lév.) Ohwi ●☆

107410 Cotoneaster coreanus H. Lév. = Symplocos paniculata (Thunb.) Miq. ●

107411 Cotoneaster coriaceus Franch.;厚叶栒子(野苦梨);Coriaceous Cotoneaster ●

107412 Cotoneaster cornifolius Flinck et Hylmö = Cotoneaster obscurus Rehder et E. H. Wilson var. cornifolius Rehder et E. H. Wilson ●

107413 Cotoneaster crenulatus (D. Don) K. Koch = Pyracantha crenulata (D. Don) M. Roem. ●

107414 Cotoneaster crenulatus K. Koch = Pyracantha crenulata (D. Don) M. Roem. ●

107415 Cotoneaster cuspidatus C. Marquand ex J. Fryer et B. Hylmö;红叶栒子●☆

107416 Cotoneaster dammeri C. K. Schneid.;矮生栒子(矮栒子);Bearberry Cotoneaster, Bear-berry Cotoneaster ●

107417 Cotoneaster dammeri C. K. Schneid. 'Coral Beauty';红珊瑚矮栒子;Coral Beauty Cotoneaster ●☆

107418 Cotoneaster dammeri C. K. Schneid. 'Eichholz';小叶矮栒子●☆

107419 Cotoneaster dammeri C. K. Schneid. 'Major';大叶矮栒子●☆

107420 Cotoneaster dammeri C. K. Schneid. var. radicans (Dammer) C. K. Schneid.;长柄矮生栒子;Longstalk Bearberry Cotoneaster ●

107421 Cotoneaster davidianus Hort. = Cotoneaster horizontalis Decne. ●

107422 Cotoneaster delavayanus G. Klotz;滇西北栒子(滇西栒子,铺茎栒子);Delavay Cotoneaster ●

107423 Cotoneaster dielsianus Pritz. = Cotoneaster dielsianus Pritz. ex Diels ●

107424 Cotoneaster dielsianus Pritz. ex Diels;木帚栒子(狄氏栒子,地柲椤树,茅铁香,木帚子,石板柴);Diels Cotoneaster, Diels' Cotoneaster, Woodenbroom Cotoneaster ●

107425 Cotoneaster dielsianus Pritz. var. elegans Rehder et E. H. Wilson;小叶木帚栒子(小叶木帚子);Littleleaf Diels Cotoneaster ●

107426 Cotoneaster difficilis G. Klotz = Cotoneaster gracilis Rehder et E. H. Wilson var. difficilis (G. Klotz) L. T. Lu ●

107427 Cotoneaster difficilis G. Klotz = Cotoneaster gracilis Rehder et E. H. Wilson ●

107428 Cotoneaster dissimilis G. Klotz;陕西栒子(脱毛栒子)●

107429 Cotoneaster distichus Lange = Cotoneaster nitidus Jacq. ●

107430 Cotoneaster distichus Lange var. duthieanus C. K. Schneid. = Cotoneaster duthieanus (C. K. Schneid.) G. Klotz ●

107431 Cotoneaster distichus Lange var. duthieanus C. K. Schneid. = Cotoneaster nitidus Jacq. var. duthieanus (C. K. Schneid.) Te T. Yu ●

107432 Cotoneaster distichus Lange var. parvifolius Te T. Yu = Cotoneaster nitidus Jacq. var. parvifolius (Te T. Yu) Te T. Yu ●

107433 Cotoneaster distichus Lange var. perpusillus (C. K. Schneid.) C. K. Schneid. = Cotoneaster horizontalis Decne. var. perpusillus C. K. Schneid. ●

107434 Cotoneaster distichus Lange var. verruculosus (Diels) Te T. Yu = Cotoneaster verruculosus Diels ●

107435 Cotoneaster divaricatus Rehder et E. H. Wilson;散生栒子(散枝栒子,张枝栒子);Spreading Cotoneaster ●

107436 Cotoneaster dokeriensis G. Klotz;毛瓣栒子●

107437 Cotoneaster duthieanus (C. K. Schneid.) G. Klotz = Cotoneaster nitidus Jacq. var. duthieanus (C. K. Schneid.) Te T. Yu ●

107438 Cotoneaster elatus G. Klotz = Cotoneaster microphyllus Wall. ex Lindl. ●

107439 Cotoneaster elegans Flinck et Hylmö = Cotoneaster dielsianus Pritz. var. elegans Rehder et E. H. Wilson ●

107440 Cotoneaster ellipticus Loudon;椭圆栒子;Lindley's Cotoneaster ●☆

107441 Cotoneaster esquirolii H. Lév. = Photinia esquirolii (H. Lév.) Rehder ●

107442 Cotoneaster fangianus E. S. Yu;方氏栒子(恩施栒子);Fang's Cotoneaster ●

107443 Cotoneaster floccosus (Rehder et E. H. Wilson) Flinck et Hylmö = Cotoneaster salicifolius Franch. ●

107444 Cotoneaster fontanesii Spach = Cotoneaster granatensis Boiss. ●☆

107445 Cotoneaster fontanesii Spach var. soongoricus Regel = Cotoneaster songoricus (Regel et Herder) Popov ●

107446 Cotoneaster formosanus Hayata = Pyracantha koidzumii (Hayata) Rehder ●

107447 Cotoneaster forrestii G. Klotz = Cotoneaster nitidus Jacq. var. duthieanus (C. K. Schneid.) Te T. Yu ●

107448 Cotoneaster foveolatus Rehder et E. H. Wilson;麻核栒子(网脉灰栒子);Glossy Cotoneaster ●

107449 Cotoneaster foveolatus Rehder et E. H. Wilson = Cotoneaster moupinensis Franch. ●

107450 Cotoneaster franchetii Bois var. cinerascens Rehder = Cotoneaster franchetii Boiss. ●

107451 Cotoneaster franchetii Boiss.;西南栒子(佛氏栒子,弗朗奇迪,马蝗果,美丽栒子);Franchet Cotoneaster, Franchet's Cotoneaster, Orange Cotoneaster ●

107452 Cotoneaster frigidus Lindl. = Cotoneaster frigidus Wall. ex Lindl. ●

107453 Cotoneaster frigidus Lindl. var. affinis (Lindl.) Wenz. = Cotoneaster affinis Lindl. ●

107454 Cotoneaster frigidus Wall. ex Lindl.;耐寒栒子;Himalayan Cotoneaster, Himalayan Tree Cotoneaster, Tree Cotoneaster ●

107455 Cotoneaster frigidus Wall. ex Lindl. 'Fructu Luteo';黄果耐寒栒子●☆

107456 Cotoneaster frigidus Wall. ex Lindl. 'Notcutt's Variety';大叶耐寒栒子●☆

107457 Cotoneaster frigidus Wall. ex Lindl. var. affinis Wenz. = Cotoneaster affinis Lindl. ●

107458 Cotoneaster fulvidus (W. W. Sm.) G. Klotz = Cotoneaster hebephyllus Diels var. fulvidus W. W. Sm. ●

107459 Cotoneaster giraldii Flinck et Hylmö ex G. Klotz = Cotoneaster hebephyllus Diels ●

107460 Cotoneaster glabratus Rehder et E. H. Wilson;光叶栒子;Smooth Cotoneaster ●

107461 Cotoneaster glacialis (Hook. f. ex Wenz.) Panigrahi et Arv. Kumar = Cotoneaster microphyllus Wall. ex Lindl. var. glacialis Hook. f. ●

107462 Cotoneaster glaucophyllus Franch.;粉叶栒子(野山楂);Brightbead Cotoneaster, Bright-bead Cotoneaster, Glaucosleaf Cotoneaster, Godalming Cotoneaster ●

107463 Cotoneaster glaucophyllus Franch. f. serotinus (Hutch.) Stapf;多花粉叶栒子;Manyflower Glaucosleaf Cotoneaster ●

107464 Cotoneaster glaucophyllus Franch. var. meiophyllus W. W. Sm.;小叶粉叶栒子;Small-leaf Glaucosleaf Cotoneaster ●

107465 Cotoneaster glaucophyllus Franch. var. serotinus (Hutch.) L. T. Lu et A. R. Brach = Cotoneaster glaucophyllus Franch. f. serotinus (Hutch.) Stapf ●

107466 Cotoneaster glaucophyllus Franch. var. vestius W. W. Sm.;毛萼粉叶栒子;Hairycalyx Glaucosleaf Cotoneaster ●

107467 Cotoneaster glomerulatus W. W. Sm.;球花栒子;Ballflower Cotoneaster, Globular-flowered Cotoneaster, Glomerule Cotoneaster ●

107468 Cotoneaster gracilis Rehder et E. H. Wilson;细弱栒子(细弱灰栒子,细枝栒子); Slendertwig Cotoneaster, Slender-twigged Cotoneaster ●

107469 Cotoneaster gracilis Rehder et E. H. Wilson var. difficilis (G. Klotz) L. T. Lu;小叶细弱栒子●

107470 Cotoneaster granatensis Boiss.;大西洋栒子●☆

107471 Cotoneaster handel-mazzettii G. Klotz;西昌栒子●

107472 Cotoneaster harrovianus E. H. Wilson;蒙自栒子(华西栒子,爬山虎,铺地蜈蚣,野苦梨);Mengzi Cotoneaster ●

107473 Cotoneaster harrysmithii Flinck et Hylmö;丹巴栒子;Harrysmith Cotoneaster ●

107474 Cotoneaster hebephyllus Diels;钝叶栒子(云南栒子);Cherryred Cotoneaster, Cherry-red Cotoneaster ●

107475 Cotoneaster hebephyllus Diels var. fulvidus W. W. Sm.;黄毛钝叶栒子;Yellower Cherryred Cotoneaster ●

107476 Cotoneaster hebephyllus Diels var. incanus W. W. Sm.;灰毛钝叶栒子;Hoary Cherryred Cotoneaster ●

107477 Cotoneaster hebephyllus Diels var. majusculus W. W. Sm.;大果钝叶栒子;Bigfruit Cherryred Cotoneaster ●

107478 Cotoneaster hebephyllus Diels var. monopyrenus W. W. Sm. = Cotoneaster hebephyllus Diels ●

107479 Cotoneaster henryanus (C. K. Schneid.) Rehder et E. H. Wilson = Cotoneaster salicifolius Franch. ●

107480 Cotoneaster henryanus (C. K. Schneid.) Rehder et E. H. Wilson = Cotoneaster salicifolius Franch. var. henryanus (C. K. Schneid.) Te T. Yu ●

107481 Cotoneaster himalayensis Zabel = Cotoneaster frigidus Wall. ex Lindl. ●

107482 Cotoneaster hissaricus Pojark. = Cotoneaster ignotus G. Klotz ●☆

107483 Cotoneaster hjelmqvistii Flinck et B. Hylmö = Cotoneaster horizontalis Decne. 'Robustus' ●☆

107484 Cotoneaster hodjingensis G. Klotz;鹤庆栒子;Heqing Cotoneaster ●

107485 Cotoneaster hodjingensis G. Klotz = Cotoneaster buxifolius Wall.

ex Lindl. ●

107486 Cotoneaster horizontalis Decne.;平枝栒子(矮红子,白马骨,被告惹,高山带子,平枝灰栒子,平枝铺地蜈蚣,铺地蜈蚣,铺地栒子,山头姑娘,水莲沙,栒刺木,栒刺子,岩楞子);Fishbone Cotoneaster,Herringbone Cotoneaster,Rock Cotoneaster,Rockspray Cotoneaster,Wall Cotoneaster,Wall Spray,Wall-spray ●

107487 Cotoneaster horizontalis Decne.'Ascendens';幸运星平枝栒子 ●☆

107488 Cotoneaster horizontalis Decne.'Robustus';粗壮平枝栒子;Hjelrnquvist's Cotoneaster ●☆

107489 Cotoneaster horizontalis Decne. var. adpressus (Bois) C. K. Schneid. = Cotoneaster adpressus Boiss. ●

107490 Cotoneaster horizontalis Decne. var. perpusillus C. K. Schneid.;小叶平枝栒子(矮红子,地红子,小叶平枝灰栒子,小叶栒刺木);Compact Rockspray,Littleleaf Rock Cotoneaster ●

107491 Cotoneaster humifusus Duthie ex Veitch = Cotoneaster dammeri C. K. Schneid. ●

107492 Cotoneaster hupehensis Rehder et E. H. Wilson = Cotoneaster silvestrii Pamp. ●

107493 Cotoneaster hurusawaianus G. Klotz;梨果栒子(河北栒子)●

107494 Cotoneaster hurusawaianus G. Klotz = Cotoneaster acutifolius Turcz. ●

107495 Cotoneaster hylmoei Flinck et J. Fryer = Cotoneaster salicifolius Franch. var. rugosus (Pritz.) Rehder et E. H. Wilson ●

107496 Cotoneaster ignotus G. Klotz;毛枝栒子;Black-grape Cotoneaster,Round-leaved Cotoneaster ●☆

107497 Cotoneaster improvisus G. Klotz = Cotoneaster chengkangensis Te T. Yu ●

107498 Cotoneaster incanus (W. W. Sm.) G. Klotz = Cotoneaster hebephyllus Diels var. incanus W. W. Sm. ●

107499 Cotoneaster insculptus Diels;怒江栒子(陷脉栒子);Engraved Cotoncaster ●

107500 Cotoneaster insculptus Diels = Cotoneaster franchetii Boiss. ●

107501 Cotoneaster insignius Pojark. = Cotoneaster ellipticus Loudon ●☆

107502 Cotoneaster insolitus G. Klotz;黑山门栒子●

107503 Cotoneaster insolitus G. Klotz = Cotoneaster buxifolius Wall. ex Lindl. var. rockii (G. Klotz) L. T. Lu et A. R. Brach ●

107504 Cotoneaster integerrimus Medik.;全缘栒子(欧洲灰栒子,全缘栒子木,全缘叶栒子);Common Cotoneaster,Entire-leaved Cotoneaster,European Cotoneaster ●

107505 Cotoneaster integerrimus Medik. var. frnigro Medik. = Cotoneaster melanocarpus Fisch. ex Loudon ●

107506 Cotoneaster integerrimus Medik. var. uniflorus (Bunge) C. K. Schneid. = Cotoneaster uniflorus Bunge ●

107507 Cotoneaster integrifolius (Roxb.) G. Klotz = Cotoneaster microphyllus Wall. ex Lindl. ●

107508 Cotoneaster integrifolius (Roxb.) G. Klotz = Cotoneaster microphyllus Wall. ex Lindl. var. thymifolius (Baker) Koehne ●

107509 Cotoneaster kangtinensis G. Klotz = Cotoneaster silvestrii Pamp. ●

107510 Cotoneaster kansuensis E. H. Wilson;甘肃栒子●

107511 Cotoneaster kaschkarowii Pojark.;巴塘栒子;Batang Cotoneaster ●

107512 Cotoneaster kinishii Hayata et Hylmö;纸叶栒子●

107513 Cotoneaster koidzumii Hayata = Pyracantha koidzumii (Hayata) Rehder ●

107514 Cotoneaster kongboensis G. Klotz;康布栒子●

107515 Cotoneaster kongboensis G. Klotz = Cotoneaster acuminatus Lindl. ●

107516 Cotoneaster konishii Hayata;小西氏铁杪椤(高山铁树,马太鞍栒子,台湾铺地蜈蚣)●

107517 Cotoneaster konishii Hayata = Cotoneaster acutifolius Turcz. ●

107518 Cotoneaster kudoi Masam. = Cotoneaster konishii Hayata ●

107519 Cotoneaster kweitschoviensis G. Klotz = Cotoneaster dammeri C. K. Schneid. ●

107520 Cotoneaster lacteus W. W. Sm.;乳白花栒子(刚毛栒子,乳白栒子,团花栒子);Late Cotoneaster,Lecteous Cotoneaster,Milkflower Cotoneaster,Parney Cotoneaster ●

107521 Cotoneaster laetevirens (Rehder et E. H. Wilson) G. Klotz = Cotoneaster microphyllus Wall. ex Lindl. ●

107522 Cotoneaster laetevirens (Rehder et E. H. Wilson) G. Klotz = Cotoneaster acutifolius Turcz. ●

107523 Cotoneaster lanatus Jacq. = Cotoneaster buxifolius Wall. ex Lindl. ●

107524 Cotoneaster lanatus Otto = Cotoneaster buxifolius Wall. ex Lindl. var. marginatus Loudon ●

107525 Cotoneaster langei G. Klotz;中甸栒子;Chungtien Cotoneaster,Zhongdian Cotoneaster ●

107526 Cotoneaster lidjiangensis G. Klotz;丽江栒子;Lijiang Cotoneaster ●

107527 Cotoneaster lidjiangensis G. Klotz = Cotoneaster buxifolius Wall. ex Lindl. ●

107528 Cotoneaster lindleyi Steud. = Cotoneaster ellipticus Loudon ●☆

107529 Cotoneaster linearifolius (G. Klotz) G. Klotz = Cotoneaster microphyllus Wall. ex Lindl. var. thymifolius (Baker) Koehne ●

107530 Cotoneaster linearifolius (G. Klotz) G. Klotz = Cotoneaster microphyllus Wall. ex Lindl. ●

107531 Cotoneaster lucidus Schltdl.;光亮叶栒子(贝加尔栒子);Hodge Cotoneaster,Shiny Cotoneaster ●

107532 Cotoneaster lucidus Schltdl. = Cotoneaster acutifolius Turcz. var. lucidus (Schltdl.) L. T. Lu ●

107533 Cotoneaster ludlowii G. Klotz;鲁氏栒子(单核栒子,西藏栒子);Ludlow Cotoneaster ●

107534 Cotoneaster magnificus J. Fryer et B. Hylmö = Cotoneaster multiflorus Bunge ●

107535 Cotoneaster mairei H. Lév. = Cotoneaster franchetii Boiss. ●

107536 Cotoneaster mairei H. Lév. var. albiflorus H. Lév. = Cotoneaster franchetii Boiss. ●

107537 Cotoneaster majuscul us (W. W. Sm.) G. Klotz = Cotoneaster hebephyllus Diels var. majusculus W. W. Sm. ●

107538 Cotoneaster marginatus Lindl. ex Loudon = Cotoneaster buxifolius Wall. ex Lindl. var. marginatus Loudon ●

107539 Cotoneaster marginatus Lindl. ex Schltdl.;边境栒子(喜马拉雅栒子)●

107540 Cotoneaster marginatus Lindl. ex Schltdl. = Cotoneaster buxifolius Wall. ex Lindl. ●

107541 Cotoneaster megalocarpus Popov;大果栒子;Big-leaved Cotoneaster ●

107542 Cotoneaster meiophyllus (W. W. Sm.) G. Klotz = Cotoneaster glaucophyllus Franch. var. meiophyllus W. W. Sm. ●

107543 Cotoneaster melanocarpus Fisch. ex Loudon;黑果栒子(黑果灰栒子,黑果栒子木);Black Cotoneaster,Black-berried Eotoneaster,Blackfruited Cotoneaster,Black-fruited Cotoneaster ●

107544 Cotoneaster melanocarpus Fisch. ex Loudon var. typicus Schneid. = Cotoneaster melanocarpus Fisch. ex Loudon ●

107545 Cotoneaster melanocarpus Loudon = Cotoneaster melanocarpus Fisch. ex Loudon ●

107546 Cotoneaster melanotrichus (Franch.) G. Klotz = Cotoneaster microphyllus Wall. ex Lindl. ●

107547 Cotoneaster microcarpus (Rehder et E. H. Wilson) Flinck et Hylmö = Cotoneaster songoricus (Regel et Herder) Popov var. microcarpus (Rehder et E. H. Wilson) G. Klotz ●

107548 Cotoneaster microphyllus Wall. = Cotoneaster microphyllus Wall. ex Lindl. ●

107549 Cotoneaster microphyllus Wall. ex Lindl.;小叶栒子(大泡叶栒子,刀口药,地锅巴,地锅粑,钝叶栒子,黑牛筋,耐冬果,铺地蜈蚣,铁杪椤树,狭叶栒子,小黑牛筋);Ampfield Cotoncaster, Littleleaf Cotoneaster, Rockspray, Rockspray Cotoneaster, Rock-spray Cotoneaster, Rose Box, Small-leaved Cotoneaster, Thyme-leaved Cotoneaster, Wallspray ●

107550 Cotoneaster microphyllus Wall. ex Lindl. f. linearifolius G. Klotz = Cotoneaster microphyllus Wall. ex Lindl. var. thymifolius (Baker) Koehne ●

107551 Cotoneaster microphyllus Wall. ex Lindl. f. linearifolius G. Klotz = Cotoneaster microphyllus Wall. ex Lindl. ●

107552 Cotoneaster microphyllus Wall. ex Lindl. f. melanoticha (Franch.) Hand.-Mazz.;黑毛小叶栒子(耐冬果)●

107553 Cotoneaster microphyllus Wall. ex Lindl. f. melanoticha (Franch.) Hand.-Mazz. = Cotoneaster microphyllus Wall. ex Lindl. ●

107554 Cotoneaster microphyllus Wall. ex Lindl. var. buxifolas Dippel = Cotoneaster buxifolius Wall. ex Lindl. ●

107555 Cotoneaster microphyllus Wall. ex Lindl. var. buxifolia Dippel f. lanatus Dippel = Cotoneaster buxifolius Wall. ex Lindl. var. marginatus Loudon ●

107556 Cotoneaster microphyllus Wall. ex Lindl. var. cochleat us (Franch.) Rehder et E. H. Wilson = Cotoneaster microphyllus Wall. ex Lindl. ●

107557 Cotoneaster microphyllus Wall. ex Lindl. var. cochleatus (Franch.) Rehder et E. H. Wilson;白毛小叶栒子;White-hair Cotoneaster ●

107558 Cotoneaster microphyllus Wall. ex Lindl. var. conspicuus Messel;大果小叶栒子(白毛小叶栒子,大果栒子,螺卷栒子,美丽栒子);Bigfruit Cotoneaster, Bigfruit Rockspray Cotoneaster, Thyme-leaved Cotoneaster ●

107559 Cotoneaster microphyllus Wall. ex Lindl. var. conspicuus Messel = Cotoneaster conspicuus (Messel) Messel et C. Marquand ex G. Klotz ●

107560 Cotoneaster microphyllus Wall. ex Lindl. var. glacialis Hook. f.;无毛小叶栒子;Hairless Rockspray Cotoneaster ●

107561 Cotoneaster microphyllus Wall. ex Lindl. var. glacialis Hook. f. = Cotoneaster microphyllus Wall. ex Lindl. ●

107562 Cotoneaster microphyllus Wall. ex Lindl. var. melanotrichus (Franch.) Rehder et E. H. Wilson = Cotoneaster microphyllus Wall. ex Lindl. ●

107563 Cotoneaster microphyllus Wall. ex Lindl. var. nivalis G. Klotz = Cotoneaster microphyllus Wall. ex Lindl. var. glacialis Hook. f. ●

107564 Cotoneaster microphyllus Wall. ex Lindl. var. nivalis G. Klotz = Cotoneaster microphyllus Wall. ex Lindl. ●

107565 Cotoneaster microphyllus Wall. ex Lindl. var. rotundifolius (Wall. ex Lindl.) Wenz. = Cotoneaster rotundifolius Wall. ex Lindl. ●

107566 Cotoneaster microphyllus Wall. ex Lindl. var. rotundifolius Wenz. = Cotoneaster rotundifolius Wall. ex Lindl. ●

107567 Cotoneaster microphyllus Wall. ex Lindl. var. thymifolius (Baker) Koehne;细小叶栒子(细叶小叶栒子)●

107568 Cotoneaster microphyllus Wall. ex Lindl. var. thymifolius (Baker) Koehne = Cotoneaster microphyllus Wall. ex Lindl. ●

107569 Cotoneaster microphyllus Wall. ex Lindl. var. uva-ursi Lindl. = Cotoneaster rotundifolius Wall. ex Lindl. ●

107570 Cotoneaster microphyllus Wall. ex Lindl. var. vellaeus (Franch.) Rehder et E. H. Wilson = Cotoneaster poluninii G. Klotz ●

107571 Cotoneaster microphyllus Wall. ex Lindl. var. vellaeus (Franch.) Rehder et E. H. Wilson = Cotoneaster buxifolius Wall. ex Lindl. ●

107572 Cotoneaster microphyllus Wall. ex Lindl. var. vellaeus Rehder et E. H. Wilson = Cotoneaster buxifolius Wall. ex Lindl. f. vellaeus Franch. ●

107573 Cotoneaster mongolicus Pojark.;蒙古栒子;Mongol Cotoneaster, Mongolian Cotoneaster ●

107574 Cotoneaster monopyrenus (W. W. Sm.) Flinck et Hylmö = Cotoneaster hebephyllus Diels ●

107575 Cotoneaster morrisonensis Hayata;台湾栒子(玉山铺地蜈蚣);Morrison Cotoneaster, Taiwan Cotoneaster ●

107576 Cotoneaster moupinensis Franch.;穆坪栒子(宝兴栒子);Moupin Cotoneaster, Muping Cotoneaster ●

107577 Cotoneaster moupinensis Franch. = Cotoneaster bullatus Bois ●☆

107578 Cotoneaster moupinensis Franch. f. floribundus Stapf = Cotoneaster bullatus Bois f. floribundus (Stapf) Rehder ●

107579 Cotoneaster mucronatus Franch.;短尖头栒子;Mucronate Cotoneaster ●

107580 Cotoneaster mucronatus Franch. = Cotoneaster acuminatus Lindl. ●

107581 Cotoneaster muliensis G. Klotz;木里栒子;Muli Cotoneaster ●

107582 Cotoneaster muliensis G. Klotz = Cotoneaster sherriffii G. Klotz ●

107583 Cotoneaster multiflorus Bunge;水栒子(多花灰栒子,多花栒子,灰栒子,香李,栒子木);Flowery Cotoneaster, Manyflower Cotoneaster, Manyflowered Cotoneaster, Many-flowered Cotoneaster, Multiflorous Cotoneaster, Water Cotoneaster ●

107584 Cotoneaster multiflorus Bunge var. atropurpureus Te T. Yu;紫果水栒子(紫果栒子);Purplefruit Manyflower Cotoneaster ●

107585 Cotoneaster multiflorus Bunge var. calocarpus Rehder et E. H. Wilson;大果水栒子(大实水栒子);Bigfruit Manyflower Cotoneaster ●

107586 Cotoneaster multiflorus Bunge var. typicus Hurus. = Cotoneaster multiflorus Bunge ●

107587 Cotoneaster nanshan ?;早生矮栒子;Dwarf Cotoneaster ●☆

107588 Cotoneaster nanus (G. Klotz) G. Klotz = Cotoneaster conspicuus (Messel) Messel et C. Marquand ex G. Klotz ●

107589 Cotoneaster nepalensis André = Cotoneaster acuminatus Lindl. ●

107590 Cotoneaster niger (Wahlb.) Fr. = Cotoneaster melanocarpus Fisch. ex Loudon ●

107591 Cotoneaster niger (Wahlb.) Fr. var. acutifolius Wenz. = Cotoneaster acutifolius Turcz. ●

107592 Cotoneaster niger Fr. = Cotoneaster melanocarpus Fisch. ex Loudon ●

107593 Cotoneaster nitens Rehder et E. H. Wilson;光泽栒子(亮叶栒子);Few-flowered Cotoneaster, Pinkblush Cotoneaster, Shining Cotoneaster ●

107594 Cotoneaster nitidifolius C. Marquand;亮叶栒子;Shiningleaf Cotoneaster, Shining-leaved Cotoneaster ●

107595 Cotoneaster nitidus Jacq.;两列栒子(两列枝栒子);Distichous Cotoneaster ●

107596 Cotoneaster nitidus Jacq. subsp. cavei (G. Klotz) H. Ohashi = Cotoneaster nitidus Jacq. var. parvifolius (Te T. Yu) Te T. Yu ●
107597 Cotoneaster nitidus Jacq. subsp. cavei (Klotz) Ohashi = Cotoneaster nitidus Jacq. var. parvifolius (Te T. Yu) Te T. Yu ●
107598 Cotoneaster nitidus Jacq. subsp. taylorii (Te T. Yu) H. Ohashi = Cotoneaster taylorii Te T. Yu ●
107599 Cotoneaster nitidus Jacq. var. duthieanus (C. K. Schneid.) Te T. Yu;大叶两列栒子;Largeleaf Distichous Cotoneaster ●
107600 Cotoneaster nitidus Jacq. var. parvifolius (Te T. Yu) Te T. Yu;小叶两列栒子;Littleleaf Distichous Cotoneaster ●
107601 Cotoneaster notabilis G. Klotz;密柔毛栒子(显著栒子)●
107602 Cotoneaster notabilis G. Klotz = Cotoneaster rubens W. W. Sm. ●
107603 Cotoneaster nummularius Fisch. et C. A. Mey.;圆板栒子●☆
107604 Cotoneaster nummularius Fisch. et C. A. Mey. var. ovalifolius Boiss. = Cotoneaster songoricus (Regel et Herder) Popov ●
107605 Cotoneaster nummularius Fisch. et C. A. Mey. var. racemiflorus (Desf.) Wenz. = Cotoneaster granatensis Boiss. ●☆
107606 Cotoneaster nummularius Fisch. et C. A. Mey. var. soongoricus Regel et Herder = Cotoneaster songoricus (Regel et Herder) Popov ●
107607 Cotoneaster obscurus Rehder et E. H. Wilson;暗红栒子(暗红果栒子);Bloodberry Cotoneaster, Blood-berry Cotoneaster, Dartford Cotuneaster ●
107608 Cotoneaster obscurus Rehder et E. H. Wilson var. cornifolius Rehder et E. H. Wilson;大叶暗红栒子;Bigleaf Bloodberry Cotoneaster ●
107609 Cotoneaster obscurus Rehder et E. H. Wilson var. cornifolius Rehder et E. H. Wilson = Cotoneaster foveolatus Rehder et E. H. Wilson ●
107610 Cotoneaster obtusus Wall. = Cotoneaster bacillaris Wall. ex Lindl. ●☆
107611 Cotoneaster oliganthus Pojark.;少花栒子;Fewflower Cotoneaster, Few-flowered Cotoneaster ●
107612 Cotoneaster oligocarpus C. K. Schneid.;少果栒子;Fewfruit Cotoneaster ●
107613 Cotoneaster orientalis A. Kern. = Cotoneaster melanocarpus Fisch. ex Loudon ●
107614 Cotoneaster ottoschwarzii G. Klotz = Cotoneaster acutifolius Turcz. ●
107615 Cotoneaster pannosus Franch.;毡毛栒子;Silverleaf Cotoneaster, Silverleaf Cotoneaster, Silver-leaved Cotoneaster ●
107616 Cotoneaster pannosus Franch. var. robustior W. W. Sm.;大叶毡毛栒子;Large Silverleaf Cotoneaster ●
107617 Cotoneaster peduncularis Boiss. = Cotoneaster melanocarpus Fisch. ex Loudon ●
107618 Cotoneaster pekinensis (Koehne) Zabel = Cotoneaster acutifolius Turcz. ●
107619 Cotoneaster pekinensis Zabel;北京栒子●
107620 Cotoneaster permutatus G. Klotz = Cotoneaster conspicuus (Messel) Messel et C. Marquand ex G. Klotz ●
107621 Cotoneaster perpusillus (C. K. Schneid.) Flinck et Hylmö = Cotoneaster horizontalis Decne. var. perpusillus C. K. Schneid. ●
107622 Cotoneaster pleuriflorus G. Klotz;侧花栒子(微凸栒子);Pleuriflower Cotoneaster ●
107623 Cotoneaster pluriflorus G. Klotz = Cotoneaster conspicuus (Messel) Messel et C. Marquand ex G. Klotz ●
107624 Cotoneaster pojarkovae Zakirov;波氏栒子●☆
107625 Cotoneaster poluninii G. Klotz;绒毛细叶栒子;Polunin's Cotoneaster ●
107626 Cotoneaster potaninii Pojark. = Cotoneaster songoricus (Regel et Herder) Popov var. microcarpus (Rehder et E. H. Wilson) G. Klotz ●
107627 Cotoneaster prostratus Baker = Cotoneaster rotundifolius Wall. ex Lindl. ●
107628 Cotoneaster prostratus Baker var. lanatus (Dippel) Rehder = Cotoneaster buxifolius Wall. ex Lindl. var. marginatus Loudon ●
107629 Cotoneaster przewalskii Pojark. = Cotoneaster multiflorus Bunge var. calocarpus Rehder et E. H. Wilson ●
107630 Cotoneaster pseudoambiguus J. Fryer et B. Hylmö = Cotoneaster ambiguus Rehder et E. H. Wilson ●
107631 Cotoneaster pseudomultiflorus Popov;假多花栒子●☆
107632 Cotoneaster pyracanthus (L.) Spach = Pyracantha coccinea M. Roem. ●☆
107633 Cotoneaster racemiflorus (Desf.) Bosse var. desfontainii Regel = Cotoneaster granatensis Boiss. ●☆
107634 Cotoneaster racemiflorus (Desf.) Bosse var. nummularius (Fisch. et C. A. Mey.) Regel = Cotoneaster granatensis Boiss. ●☆
107635 Cotoneaster racemiflorus (Desf.) Bosse var. tomentellus Maire = Cotoneaster granatensis Boiss. ●☆
107636 Cotoneaster racemiflorus (Desf.) C. Koch = Cotoneaster racemiflorus (Desf.) K. Koch ●☆
107637 Cotoneaster racemiflorus (Desf.) K. Koch;总花栒子(圆锥栒子);Redbead Cotoneaster ●☆
107638 Cotoneaster racemiflorus (Desf.) K. Koch var. microcarpus Rehder et E. H. Wilson = Cotoneaster songoricus (Regel et Herder) Popov var. microcarpus (Rehder et E. H. Wilson) G. Klotz ●
107639 Cotoneaster racemiflorus (Desf.) K. Koch var. ovalifolius (Boiss.) Hurus. = Cotoneaster songoricus (Regel et Herder) Popov ●
107640 Cotoneaster racemiflorus (Desf.) K. Koch var. songoricus Regel et Herder = Cotoneaster songoricus (Regel et Herder) Popov ●
107641 Cotoneaster racemiflorus (Desf.) K. Koch var. soongoricus (Regel et Herder) Schneid. = Cotoneaster songoricus (Regel et Herder) Popov ●
107642 Cotoneaster racemiflorus (Desf.) K. Koch var. soongoricus (Regel et Herder) C. K. Schneid. = Cotoneaster songoricus (Regel et Herder) Popov ●
107643 Cotoneaster racemiflorus (Desf.) K. Koch var. veitchii Rehder et E. H. Wilson = Cotoneaster silvestrii Pamp. ●
107644 Cotoneaster radicans (Schneid.) Klotz = Cotoneaster dammeri C. K. Schneid. var. radicans (Dammer) C. K. Schneid. ●
107645 Cotoneaster radicans Dammer = Cotoneaster dammeri C. K. Schneid. var. radicans (Dammer) C. K. Schneid. ●
107646 Cotoneaster reflexus Carrière = Cotoneaster multiflorus Bunge ●
107647 Cotoneaster rehderi Pojark. = Cotoneaster bullatus Bois var. macrophyllus Rehder et E. H. Wilson ●
107648 Cotoneaster reticulatus Rehder et E. H. Wilson;网脉栒子;Jetbead Cotoneaster, Netted Cotoneaster, Netveined Cotoneaster ●
107649 Cotoneaster rhytidophyllus Rehder et E. H. Wilson;麻叶栒子;Orangebead Cotoneaster, Orange-beaded Cotoneaster ●
107650 Cotoneaster roborowskii Pojark.;罗氏栒子(梨果栒子)●
107651 Cotoneaster rockii G. Klotz;川藏栒子●
107652 Cotoneaster rockii G. Klotz = Cotoneaster buxifolius Wall. ex Lindl. ●
107653 Cotoneaster rockii G. Klotz = Cotoneaster buxifolius Wall. ex Lindl. var. rockii (G. Klotz) L. T. Lu et A. R. Brach ●
107654 Cotoneaster rokujodaisanensis Hayata;乐山铺地蜈蚣●

107655 Cotoneaster rokujodaisanensis Hayata = Cotoneaster morrisonensis Hayata ●

107656 Cotoneaster rosea Edgew.;粉红花栒子●☆

107657 Cotoneaster rotundifolius Lindl. = Cotoneaster nitidus Jacq. ●

107658 Cotoneaster rotundifolius Lindl. var. lanatus (Dippel) C. K. Schneid. = Cotoneaster buxifolius Wall. ex Lindl. var. marginatus Loudon ●

107659 Cotoneaster rotundifolius Wall. ex Lindl.;圆叶栒子;Redbox Cotoneaster,Red-box Cotoneaster,Roundleaf Cotoneaster ●

107660 Cotoneaster rotundifolius Wall. ex Lindl. var. lanatus (Dippel) C. K. Schneid. = Cotoneaster buxifolius Wall. ex Lindl. var. marginatus Loudon ●

107661 Cotoneaster rubens W. W. Sm.;红花栒子;Redflower Cotoneaster,Red-flowered Cotoneaster ●

107662 Cotoneaster rubens W. W. Sm. var. miniatus Te T. Yu;小叶红花栒子;Littleleaf Redflower Cotoneaster ●

107663 Cotoneaster rubens W. W. Sm. var. miniatus Te T. Yu = Cotoneaster buxifolius Wall. ex Lindl. ●

107664 Cotoneaster rugosus E. Pritz. = Cotoneaster salicifolius Franch. var. rugosus (Pritz.) Rehder et E. H. Wilson ●

107665 Cotoneaster rugosus E. Pritz. var. henryanus C. K. Schneid. = Cotoneaster salicifolius Franch. var. henryanus (C. K. Schneid.) Te T. Yu ●

107666 Cotoneaster rugosus W. W. Sm. var. typicus Schneid. = Cotoneaster salicifolius Franch. var. rugosus (Pritz.) Rehder et E. H. Wilson ●

107667 Cotoneaster rupestris Charles = Cotoneaster nitidus Jacq. ●

107668 Cotoneaster salicifolius Franch.;柳叶栒子(把把柴,翻白柴,木帚子,山米麻);Willowleaf Cotoneaster,Willow-leaf Cotoneaster,Willow-leaved Cotoneaster ●

107669 Cotoneaster salicifolius Franch. 'Autumn Fire' = Cotoneaster salicifolius Franch. 'Herbstfeuer' ●☆

107670 Cotoneaster salicifolius Franch. 'Herbstfeuer';球水柳叶栒子●☆

107671 Cotoneaster salicifolius Franch. 'Repens';匍匐柳叶栒子;Willowleaf Cotoneaster ●☆

107672 Cotoneaster salicifolius Franch. var. angustus Te T. Yu;窄柳叶栒子(狭叶栒子,窄叶柳叶栒子);Narrow Willowleaf Cotoneaster ●

107673 Cotoneaster salicifolius Franch. var. floccosus Rehder et E. H. Wilson = Cotoneaster salicifolius Franch. ●

107674 Cotoneaster salicifolius Franch. var. henryanus (C. K. Schneid.) Te T. Yu;大柳叶栒子(大叶柳叶栒子,大叶栒子);Henry Willowleaf Cotoneaster ●

107675 Cotoneaster salicifolius Franch. var. rugosus (Pritz.) Rehder et E. H. Wilson;皱柳叶栒子(小叶山米麻,皱叶柳叶栒子,皱叶栒子);Wrinkly Willowleaf Cotoneaster ●

107676 Cotoneaster sanguineus Te T. Yu;血色栒子;Sanguine Cotoneaster ●

107677 Cotoneaster saxatilis Pojark.;岩生栒子●☆

107678 Cotoneaster schantungensis G. Klotz;山东栒子●

107679 Cotoneaster schlechtendalii G. Klotz;灰绿栒子(钝尖栒子)●

107680 Cotoneaster schlechtendalii G. Klotz = Cotoneaster microphyllus Wall. ex Lindl. ●

107681 Cotoneaster schlechtendalii G. Klotz = Cotoneaster sherriffii G. Klotz ●

107682 Cotoneaster serotinus Hutch.;灰毛栒子(多花粉叶栒子)●☆

107683 Cotoneaster serotinus Hutch. = Cotoneaster glaucophyllus Franch. f. serotinus (Hutch.) Stapf ●

107684 Cotoneaster sherriffii G. Klotz;康巴栒子;Kangba Cotoneaster,Sherriff Cotoneaster ●

107685 Cotoneaster sikangensis Flinck et Hylmö;西康栒子;Xikang Cotoneaster ●

107686 Cotoneaster silvestrii Pamp.;华中栒子(鄂栒子,湖北栒子);Central China Cotoneaster,Forest Cotoneaster,Silvestri Cotoneaster ●

107687 Cotoneaster simonsii Baker;西蒙氏栒子(西蒙斯栒子);Himalayan Cotoneaster,Khasia Berry,Simons Cotoneaster,Simons' Cotoneaster ●☆

107688 Cotoneaster songoricus (Regel et Herder) Popov;准噶尔栒子(准噶尔总花栒子);Dzungar Cotoneaster,Dzungaria Cotoneaster,Songar Cotoneaster,Sungari Redbead Cotoneaster ●

107689 Cotoneaster songoricus (Regel et Herder) Popov var. microcarpus (Rehder et E. H. Wilson) G. Klotz;小果准噶尔栒子;Small-fruit Songar Cotoneaster ●

107690 Cotoneaster splendens Flinck;康定栒子(亮叶栒子);Showy Cotoneaster ●

107691 Cotoneaster staintonii G. Klotz;斯氏栒子(尼泊尔栒子,尼东栒子);Stainton's Cotoneaster ●

107692 Cotoneaster sternianus (Turrill) Boom;斯腾栒子;Stern's Cotoneaster ●☆

107693 Cotoneaster strigosus G. Klotz = Cotoneaster chengkangensis Te T. Yu ●

107694 Cotoneaster suavis Pojark.;甜栒子●

107695 Cotoneaster suavis Pojark. = Cotoneaster songoricus (Regel et Herder) Popov ●

107696 Cotoneaster subadpressus Te T. Yu;高山栒子;Alpine Cotoneaster ●

107697 Cotoneaster submultiflorus Popov;毛叶水栒子(毛叶栒子);Hairyleaf Cotoneaster,Hairy-leaved Cotoneaster ●

107698 Cotoneaster symonsii Loudon ex Koehne = Cotoneaster horizontalis Decne. ●

107699 Cotoneaster taitoensis Hayata = Pyracantha koidzumii (Hayata) Rehder ●

107700 Cotoneaster taoensis G. Klotz = Cotoneaster adpressus Boiss. ●

107701 Cotoneaster tauricus Pojark.;克里木栒子;Klimu Cotoneaster ●☆

107702 Cotoneaster taylorii Te T. Yu;藏南栒子;S. Xizang Cotoneaster,Taylor Cotoneaster ●

107703 Cotoneaster tenuipes Rehder et E. H. Wilson;细枝栒子(细梗栒子);Slender Cotoneaster,Thinbrabch Cotoneaster ●

107704 Cotoneaster thymifolius Baker = Cotoneaster microphyllus Wall. ex Lindl. ●

107705 Cotoneaster thymifolius Baker = Cotoneaster microphyllus Wall. ex Lindl. var. thymifolius (Baker) Koehne ●

107706 Cotoneaster thymifolius Baker var. cochleatus (Franch.) Franch. = Cotoneaster microphyllus Wall. ex Lindl. var. cochleatus (Franch.) Rehder et E. H. Wilson ●

107707 Cotoneaster tibeticus G. Klotz;西藏栒子(拉萨栒子);Xizang Cotoneaster ●

107708 Cotoneaster tibeticus G. Klotz = Cotoneaster songoricus (Regel et Herder) Popov ●

107709 Cotoneaster tomentellus Pojark. = Cotoneaster songoricus (Regel et Herder) Popov ●

107710 Cotoneaster tomentosus Lindl.;毛栒子;Brickberry Cotoneaster ●☆

107711 Cotoneaster transens G. Klotz = Cotoneaster glaucophyllus Franch. ●

107712 Cotoneaster tumeticus Pojark.;土默特栒子●

107713 Cotoneaster tumeticus Pojark. = Cotoneaster mongolicus Pojark. ●

107714 Cotoneaster turbinetus Craib;陀螺果栒子(陀螺栒子);Brightberry Cotoneaster,Bright-berry Cotoneaster ●

107715 Cotoneaster uniflorus Bunge;单花栒子;Singleberry Cotoneaster,Single-berry Cotoneaster,Singleflower Cotoneaster ●

107716 Cotoneaster veitchii (Rehder et E. H. Wilson) G. Klotz = Cotoneaster silvestrii Pamp. ●

107717 Cotoneaster vernae C. K. Schneid.;雪山栒子●

107718 Cotoneaster verruculosus Diels;疣枝栒子;Scarletbead Cotoneaster,Scarlet-beaded Cotoneaster ●

107719 Cotoneaster vestitus (W. W. Sm.) Flinck et Hylmö = Cotoneaster glaucophyllus Franch. var. vestius W. W. Sm. ●

107720 Cotoneaster villosulus (Rehder et E. H. Wilson) Flinck et Hylmö;毛灰栒子(河北栒子,密毛灰栒子,细柔毛栒子);Hairy Beijing Cotoneaster,Hairy Peking Cotoneaster,Lleyn Cotoneaster ●

107721 Cotoneaster villosulus (Rehder et E. H. Wilson) Flinck et Hylmö = Cotoneaster acutifolius Turcz. var. villosulus Rehder et E. H. Wilson ●

107722 Cotoneaster vilmorinianus G. Klotz;宿萼栒子(维西栒子)●

107723 Cotoneaster vulgaris Lindl. = Cotoneaster integerrimus Medik. ●

107724 Cotoneaster vulgaris Lindl. var. melanocarpus (Lodd.) Ledeb. = Cotoneaster melanocarpus Fisch. ex Loudon ●

107725 Cotoneaster vulgaris Lindl. var. uniflorus (Bunge) Regel = Cotoneaster uniflorus Bunge ●

107726 Cotoneaster wardii W. W. Sm.;白毛栒子(瓦德栒子);Ward Cotoneaster,Whitehair Cotoneaster ●

107727 Cotoneaster watereri Exell;沃特尔栒子;Waterer's Cotoneaster ●☆

107728 Cotoneaster wilsonii Nakai;威氏栒子●☆

107729 Cotoneaster zabelii C. K. Schneid.;西北栒子(担棍子,土兰条,杂氏栒子,札氏栒子);Cherry-berried Cotoneaster,Cherryberry Cotoneaster,Cherry-berry Cotoneaster,Cherry-red Cotoneaster,NW. China Cotoneaster ●

107730 Cotoneaster zabelii C. K. Schneid. var. miniatus Rehder et E. H. Wilson = Cotoneaster zabelii C. K. Schneid. ●

107731 Cotoneaster zayulensis G. Klotz;札尤栒子(札尤路栒子);Zhayou Cotoneaster ●

107732 Cotoneaster zayulensis G. Klotz = Cotoneaster songoricus (Regel et Herder) Popov ●

107733 Cotopaxia Mathias et Constance(1952);哥伦比亚草属☆

107734 Cotopaxia asplundii Mathias et Constance;哥伦比亚草☆

107735 Cottaea Endl. = Cottea Kunth ■

107736 Cottea Kunth(1829);寇蒂禾属■

107737 Cottea pappophoroides Kunth;寇蒂禾■☆

107738 Cottendorfia Schult. et Schult. f. (1830);卡田凤梨属(卡田道夫属)■☆

107739 Cottendorfia Schult. f. = Cottendorfia Schult. et Schult. f. ■☆

107740 Cottendorfia neogranatensis Baker;卡田凤梨■☆

107741 Cottetia Gand. = Rosa L. ●

107742 Cottonia Wight(1851);科顿兰属■☆

107743 Cottonia championii Lindl. = Diploprora championii (Lindl.) Hook. f. ■

107744 Cottonia championii Lindl. ex Benth. = Diploprora championii (Lindl. ex Benth.) Hook. f. ■

107745 Cottonia macroatachya Wight;科顿兰■☆

107746 Cottsia Dubard et Dop = Janusia A. Juss. ex Endl. ●☆

107747 Cotula L. (1753);山芫荽属(铜扣菊属,芫荽属);Brass Buttons,Brassbuttons,Buttonweed ■

107748 Cotula abyssinica Sch. Bip. ex A. Rich.;阿比西尼亚山芫荽■☆

107749 Cotula abyssinica Sch. Bip. ex A. Rich. var. nana Sch. Bip. = Cotula abyssinica Sch. Bip. ex A. Rich. ■☆

107750 Cotula abyssinica Sch. Bip. ex A. Rich. var. sessilis Hedberg;无梗阿比西尼亚山芫荽■☆

107751 Cotula andreae (E. Phillips) Bremer et Humphries;安氏山芫荽■☆

107752 Cotula anthemoides L.;芫荽菊(山芫荽);Canaomile Brassbuttons ■

107753 Cotula anthemoides L. = Grangea maderaspatana (L.) Poir. ■

107754 Cotula aurea L.;黄山芫荽■☆

107755 Cotula australis (Sieber ex Spreng.) Hook. f.;澳洲山芫荽;Annual Buttonweed,Australian Waterbuttons ■☆

107756 Cotula australis (Spreng.) Hook. f. = Cotula australis (Sieber ex Spreng.) Hook. f. ■☆

107757 Cotula barbata DC.;髯毛山芫荽;Pincushion Plant ■☆

107758 Cotula bicolor Roth = Dichrocephala auriculata (Thunb.) Druce ■

107759 Cotula bicolor Roth = Dichrocephala integrifolia (L.) Kuntze ■

107760 Cotula bipinnata Thunb.;双羽山芫荽■☆

107761 Cotula bracteolata E. Mey. ex DC.;小苞片山芫荽■☆

107762 Cotula burchellii DC.;伯切尔山芫荽■☆

107763 Cotula chinensis Kitam. = Cotula hemisphaerica Wall. ■

107764 Cotula chrysanthemifolia Blume = Dichrocephala chrysanthemifolia (Blume) DC. ■

107765 Cotula cinerea Delile;灰色山芫荽■☆

107766 Cotula coronopifolia L.;芥叶山芫荽(铜扣菊);Brass Buttons,Brass-buttons,Buttonweed,Common Brassbuttons,Cotoneaster,Duck's Eyes,Water Buttons,Waterbuttons ■

107767 Cotula cryptocephala Sch. Bip. ex A. Rich.;隐头山芫荽■☆

107768 Cotula dichrocephala Sch. Bip. ex A. Rich. = Cotula anthemoides L. ■

107769 Cotula dielsii Muschl.;迪尔斯山芫荽■☆

107770 Cotula dioica Hook. f.;无毛山芫荽;Hairless Leptinella ■☆

107771 Cotula duckittiae (L. Bolus) K. Bremer et Humphries;达克山芫荽■☆

107772 Cotula eckloniana (DC.) Levyns;埃氏山芫荽■☆

107773 Cotula filifolia Thunb.;线叶山芫荽■☆

107774 Cotula globifera Thunb. = Oncosiphon piluliferum (L. f.) Källersjö ■☆

107775 Cotula grandis L. = Plagius grandis (L.) Alavi et Heywood ■☆

107776 Cotula hemisphaerica (Roxb.) Wall. ex Benth. et Hook. f.;山芫荽;Hemisphaerical Brassbuttons ■

107777 Cotula hemisphaerica Wall. = Cotula hemisphaerica (Roxb.) Wall. ex Benth. et Hook. f. ■

107778 Cotula heterocarpa DC.;异果山芫荽■☆

107779 Cotula hispida (DC.) Harv.;硬毛山芫荽■☆

107780 Cotula latifolia Pers. = Dichrocephala auriculata (Thunb.) Druce ■

107781 Cotula laxa DC.;疏松山芫荽■☆

107782 Cotula lineariloba (DC.) Hilliard;线裂山芫荽■☆

107783 Cotula loganii Hutch.;洛根山芫荽■☆

107784 Cotula macroglossa Bolus ex Schltr.;大舌山芫荽■☆

107785 Cotula maderaspatana (L.) Willd. = Grangea maderaspatana (L.) Poir. ■

107786 Cotula mariae K. Bremer et Humphries;玛利亚山芫荽■☆

107787 Cotula melaleuca Bolus;黑白山芫荽■☆
107788 Cotula membranifolia Hilliard;膜叶山芫荽■☆
107789 Cotula mexicana (DC.) Cabrera;墨西哥山芫荽;Mexican Brassbuttons ■☆
107790 Cotula microcephala DC. = Cotula anthemoides L. ■
107791 Cotula microglossa (DC.) O. Hoffm. et Kuntze ex Kuntze;小舌山芫荽■☆
107792 Cotula minima (L.) Willd. = Centipeda minima (L.) A. Braun et Asch. ■
107793 Cotula minima (L.) Willd. = Centipeda orbicularis Lour. ■
107794 Cotula montana Compton;山地山芫荽■☆
107795 Cotula multifida DC. = Cotula villosa DC. ■☆
107796 Cotula myriophylloides Harv.;多叶山芫荽■☆
107797 Cotula nigellifolia (DC.) K. Bremer et Humphries;黑种草叶山芫荽■☆
107798 Cotula nigellifolia (DC.) K. Bremer et Humphries var. tenuior (DC.) Herman;瘦山芫荽■☆
107799 Cotula nudicaulis Thunb.;裸茎山芫荽■☆
107800 Cotula orbicularis Lour. = Centipeda minima (L.) A. Braun et Asch. ■
107801 Cotula orbicularis Lour. = Centipeda orbicularis Lour. ■
107802 Cotula oxyodonta DC. = Cotula bipinnata Thunb. ■☆
107803 Cotula paludosa Hilliard;沼泽山芫荽■☆
107804 Cotula paradoxa Schinz;奇异山芫荽■☆
107805 Cotula pedicellata Compton;梗花山芫荽■☆
107806 Cotula pedunculata (Schltr.) E. Phillips;序花山芫荽■☆
107807 Cotula pilulifera L. f. = Oncosiphon piluliferum (L. f.) Källersjö ■☆
107808 Cotula pterocarpa DC.;翅果山芫荽■☆
107809 Cotula pubescens Desf. = Aaronsohnia pubescens (Desf.) K. Bremer et Humphries ■☆
107810 Cotula purpurea ?;紫山芫荽;Cotula ■☆
107811 Cotula pusilla Thunb.;微小山芫荽■☆
107812 Cotula quinquefida Thunb. = Pentzia quinquefida (Thunb.) Less. ■☆
107813 Cotula quinqueloba L. f. = Lidbeckia quinqueloba (L. f.) Cass. ●☆
107814 Cotula radicalis (Killick et C. Claassen) Hilliard et B. L. Burtt;辐射山芫荽■☆
107815 Cotula sericea L. f.;绢毛山芫荽■☆
107816 Cotula sericea Thunb. = Cotula lineariloba (DC.) Hilliard ■☆
107817 Cotula socialis Hilliard;群生山芫荽■☆
107818 Cotula sonchifolia M. Bieb. = Dichrocephala integrifolia (L. f.) Kuntze ■
107819 Cotula sororia DC.;堆积山芫荽■☆
107820 Cotula sphaeranthus Link. = Grangea maderaspatana (L.) Poir. ■
107821 Cotula squalida Hook. f.;新西兰山芫荽;Leptinella, New Zealand Brass Buttons ■☆
107822 Cotula tanacetifolia L. = Oncosiphon schlechteri (Bolus ex Schltr.) Källersjö ■☆
107823 Cotula tanacetifolia L. = Oncosiphon suffruticosum (L.) Källersjö ■☆
107824 Cotula tenella E. Mey. ex DC.;柔软山芫荽■☆
107825 Cotula thunbergii Harv.;通贝里山芫荽■☆
107826 Cotula turbinata L.;陀螺山芫荽■☆
107827 Cotula umbellata L. f. = Schistostephium umbellatum (L. f.) Bremer et Humphries ●☆
107828 Cotula villosa DC. = Cotula australis (Sieber ex Spreng.) Hook. f. ■☆
107829 Cotula vulgaris Levyns;普通山芫荽■☆
107830 Cotula zeyheri Fenzl;泽赫山芫荽■☆
107831 Cotulina Pomel = Cenocline K. Koch ■
107832 Cotulina Pomel = Cotula L. ■
107833 Cotulina Pomel = Matricaria L. ■
107834 Cotulina aurea (L.) Pomel = Cotula aurea L. ■☆
107835 Cotylanthera Blume(1826);杯药草属(杯蕊草属,杯蕊属);Cotylanthera ■
107836 Cotylanthera paucisquama C. B. Clarke;杯药草;Cotylanthera, Fewscale Cotylanthera ■
107837 Cotylanthera tenuis Blume;细杯药草;Thin Brassbuttons ■☆
107838 Cotylanthera yunnanensis W. W. Sm. = Cotylanthera paucisquama C. B. Clarke ■
107839 Cotylaria Raf. = Cotyledon L. ●■☆
107840 Cotyledon L. (1753);长筒莲属(圣塔属,银波锦属);Cotyledon ●■☆
107841 Cotyledon Tourn. ex L. = Cotyledon L. ●■☆
107842 Cotyledon acaulon Walther ex Steud. = Cotyledon papillaris L. f. ●☆
107843 Cotyledon adscendens R. A. Dyer;上举长筒莲●☆
107844 Cotyledon affinis (Schrenk) Maxim. = Pseudosedum affine (Schrenk) A. Berger ●■
107845 Cotyledon alstonii Schönland et Baker f. = Adromischus alstonii (Schönland et Baker f.) C. A. Sm. ■☆
107846 Cotyledon alternans Haw. = Adromischus maculatus (Salm-Dyck) Lem. ■☆
107847 Cotyledon arborescens Mill. = Crassula arborescens (Mill.) Willd. ●■☆
107848 Cotyledon attenuata H. Lindb. = Pistorinia attenuata (H. Lindb.) Greuter ●☆
107849 Cotyledon attenuata H. Lindb. subsp. mairei H. Lindb. = Pistorinia attenuata (H. Lindb.) Greuter subsp. mairei Greuter ●☆
107850 Cotyledon attenuata H. Lindb. var. maculata Maire = Pistorinia attenuata (H. Lindb.) Greuter ●☆
107851 Cotyledon attenuata H. Lindb. var. purpurea Maire = Pistorinia attenuata (H. Lindb.) Greuter subsp. mairei (H. Lindb.) Greuter ●☆
107852 Cotyledon ausana Dinter;白蝶长筒莲■☆
107853 Cotyledon ausana Dinter = Cotyledon orbiculata L. ●☆
107854 Cotyledon barbeyi Schweinf. ex Baker;长柔毛长筒莲●☆
107855 Cotyledon barbeyi Schweinf. ex Penz. = Cotyledon barbeyi Schweinf. ex Baker ●☆
107856 Cotyledon beckeri Schönland et Baker f. = Cotyledon velutina Hook. f. ●☆
107857 Cotyledon bolusii Schönland = Adromischus caryophyllaceus (Burm. f.) Lem. ■●☆
107858 Cotyledon bolusii Schönland var. karroensis ? = Adromischus triflorus (L. f.) A. Berger ■☆
107859 Cotyledon brachyantha (Coss.) Maire = Pistorinia brachyantha Coss. ●☆
107860 Cotyledon brachyantha (Coss.) Maire var. aurea Maire = Pistorinia brachyantha Coss. ●☆
107861 Cotyledon brachyantha (Coss.) Maire var. ochroleuca Maire = Pistorinia brachyantha Coss. ●☆
107862 Cotyledon brachyantha (Coss.) Maire var. purpurea Maire = Pistorinia brachyantha Coss. ●☆

107863 Cotyledon brachyantha (Coss.) Maire var. versicolor Maire = Pistorinia brachyantha Coss. ●☆
107864 Cotyledon breviflora (Boiss.) Maire;短花长筒莲●☆
107865 Cotyledon breviflora (Boiss.) Maire = Pistorinia breviflora Boiss. ●☆
107866 Cotyledon breviflora (Boiss.) Maire subsp. intermedia (Boiss. et Reut.) Maire = Pistorinia breviflora Boiss. subsp. intermedia (Boiss. et Reut.) Greuter et Burdet ●☆
107867 Cotyledon breviflora (Boiss.) Maire subsp. salzmannii ? = Pistorinia breviflora Boiss. ●☆
107868 Cotyledon breviflora (Boiss.) Maire var. flava Maire = Pistorinia breviflora Boiss. ●☆
107869 Cotyledon breviflora (Boiss.) Maire var. flaviflora Batt. = Pistorinia breviflora Boiss. ●☆
107870 Cotyledon breviflora (Boiss.) Maire var. rhodantha Maire = Pistorinia breviflora Boiss. ●☆
107871 Cotyledon breviflora (Boiss.) Maire var. rubella Batt. = Pistorinia breviflora Boiss. ●☆
107872 Cotyledon breviflora (Boiss.) Maire var. subbrachyantha Maire = Pistorinia breviflora Boiss. ●☆
107873 Cotyledon breviflora (Boiss.) Maire var. variegata Gatt. et Maire = Pistorinia breviflora Boiss. ●☆
107874 Cotyledon breviflora (Boiss.) Maire var. xanthantha Maire = Pistorinia breviflora Boiss. ●☆
107875 Cotyledon buchholziana Schuldt et Stephens = Tylecodon buchholzianus (Schuldt et Stephens) Toelken ●☆
107876 Cotyledon cacalioides L. f. = Tylecodon cacalioides (L. f.) Toelken ●☆
107877 Cotyledon campanulata Marloth;风铃草状长筒莲●☆
107878 Cotyledon canaliculata Haw. = Cotyledon orbiculata L. var. oblonga (Haw.) DC. ●☆
107879 Cotyledon canalifolia Haw. = Cotyledon orbiculata L. var. oblonga (Haw.) DC. ●☆
107880 Cotyledon caryophyllacea Burm. f. = Adromischus caryophyllaceus (Burm. f.) Lem. ■●☆
107881 Cotyledon chloroleuca Dinter ex Friedrich = Tylecodon racemosus (Harv.) Toelken ●☆
107882 Cotyledon clavifolia Haw. = Adromischus cristatus (Haw.) Lem. var. clavifolius (Haw.) Toelken ■☆
107883 Cotyledon cooperi Baker = Adromischus cooperi (Baker) A. Berger ■
107884 Cotyledon cooperi Baker var. immaculata Schönland et Baker f. = Adromischus cooperi (Baker) A. Berger ■
107885 Cotyledon coruscans Haw. = Cotyledon orbiculata L. var. oblonga (Haw.) DC. ●☆
107886 Cotyledon cossoniana Ball = Pistorinia brachyantha Coss. ●☆
107887 Cotyledon crassifolia Haw. = Cotyledon orbiculata L. var. oblonga (Haw.) DC. ●☆
107888 Cotyledon crassifolia Salisb. = Adromischus hemisphaericus (L.) Lem. ■☆
107889 Cotyledon cristata Haw. = Adromischus cristatus (Haw.) Lem. ■☆
107890 Cotyledon cuneata Thunb.;楔形长筒莲●☆
107891 Cotyledon cuneiformis Haw. = Cotyledon orbiculata L. var. oblonga (Haw.) DC. ●☆
107892 Cotyledon curviflora Sims = Tylecodon grandiflorus (Burm. f.) Toelken ●☆
107893 Cotyledon deasii Schönland = Cotyledon cuneata Thunb. ●☆
107894 Cotyledon decussata Sims;对生长筒莲(宝塔草)●☆
107895 Cotyledon decussata Sims = Cotyledon orbiculata L. ●☆
107896 Cotyledon decussata Sims var. dielsii Schltr. ex Poelln. = Cotyledon orbiculata L. ●☆
107897 Cotyledon decussata Sims var. flavida (Fourc.) Poelln. = Cotyledon orbiculata L. var. oblonga (Haw.) DC. ●☆
107898 Cotyledon decussata Sims var. hinrichseniana H. Jacobsen = Cotyledon orbiculata L. ●☆
107899 Cotyledon decussata Sims var. rubra Poelln. = Cotyledon orbiculata L. var. oblonga (Haw.) DC. ●☆
107900 Cotyledon dichotoma Haw. = Tylecodon reticulatus (L. f.) Toelken ●☆
107901 Cotyledon dinteri Baker f. = Tylecodon wallichii (Harv.) Toelken subsp. ecklonianus (Harv.) Toelken ●☆
107902 Cotyledon eckloniana Harv. = Tylecodon wallichii (Harv.) Toelken subsp. ecklonianus (Harv.) Toelken ●☆
107903 Cotyledon elata Haw. = Cotyledon orbiculata L. ●☆
107904 Cotyledon engleri A. Berger et Dinter = Cotyledon orbiculata L. ●☆
107905 Cotyledon erubescens (Maxim.) Franch. et Sav. = Orostachys spinosa (L.) Sweet ■
107906 Cotyledon fascicularis Aiton = Tylecodon paniculatus (L. f.) Toelken ●☆
107907 Cotyledon filicaulis Eckl. et Zeyh. = Adromischus filicaulis (Eckl. et Zeyh.) C. A. Sm. ■☆
107908 Cotyledon fimbriata Turcz. = Orostachys fimbriata (Turcz.) A. Berger ■
107909 Cotyledon fimbriata Turcz. var. ramosissima (Maxim.) Maxim. = Orostachys fimbriata (Turcz.) A. Berger ■
107910 Cotyledon flanaganii Schönland et Baker f. = Cotyledon orbiculata L. var. flanaganii (Schönland et Baker f.) Toelken ●☆
107911 Cotyledon flanaganii Schönland et Baker f. var. karroensis ? = Cotyledon orbiculata L. ●☆
107912 Cotyledon flavida Fourc. = Cotyledon orbiculata L. var. oblonga (Haw.) DC. ●☆
107913 Cotyledon fragilis R. A. Dyer = Tylecodon fragilis (R. A. Dyer) Toelken ●☆
107914 Cotyledon fusiformis Rolfe = Adromischus filicaulis (Eckl. et Zeyh.) C. A. Sm. ■☆
107915 Cotyledon gaditana (Boiss.) Pau = Umbilicus gaditanus Boiss. ■☆
107916 Cotyledon gaditana (Boiss.) Pau subsp. fontqueri Maire et Sennen = Umbilicus gaditanus Boiss. ■☆
107917 Cotyledon galpinii Schönland et Baker f. = Cotyledon orbiculata L. var. oblonga (Haw.) DC. ●☆
107918 Cotyledon glutinosa Schönland = Cotyledon papillaris L. f. ●☆
107919 Cotyledon gracilis Haw.;纤细长筒莲●☆
107920 Cotyledon gracilis Haw. = Cotyledon papillaris L. f. ●☆
107921 Cotyledon grandiflora Burm. f. = Tylecodon grandiflorus (Burm. f.) Toelken ●☆
107922 Cotyledon hallii Toelken = Tylecodon hallii (Toelken) Toelken ■☆
107923 Cotyledon hemispherica L. = Adromischus hemisphaericus (L.) Lem. ■☆
107924 Cotyledon herrei W. F. Barker = Adromischus marianiae (Marloth) A. Berger var. immaculatus Uitewaal ■☆

107925 Cotyledon heterophylla Schönland = Cotyledon tomentosa Harv. subsp. ladismithiensis (Poelln.) Toelken ●☆

107926 Cotyledon hirtifolia W. F. Barker = Tylecodon hirtifolius (W. F. Barker) Toelken ●☆

107927 Cotyledon hispanica L. subsp. cossoniana Ball = Pistorinia brachyantha Coss. ●☆

107928 Cotyledon hispanica L. var. flaviflora Maire = Pistorinia attenuata (H. Lindb.) Greuter ●☆

107929 Cotyledon hispanica L. var. maculata Maire = Pistorinia attenuata (H. Lindb.) Greuter ●☆

107930 Cotyledon hispanica L. var. purpurea Maire = Pistorinia attenuata (H. Lindb.) Greuter subsp. mairei (H. Lindb.) Greuter ●☆

107931 Cotyledon hispanica L. var. salzmannii (Boiss.) Ball = Pistorinia breviflora Boiss. ●☆

107932 Cotyledon horizontalis Guss. = Umbilicus horizontalis (Guss.) DC. ■☆

107933 Cotyledon horizontalis Guss. var. micranthus Pamp. = Umbilicus horizontalis (Guss.) DC. ■☆

107934 Cotyledon humilis Marloth = Adromischus humilis (Marloth) Poelln. ■☆

107935 Cotyledon insignis N. E. Br. = Kalanchoe elizae A. Berger ■☆

107936 Cotyledon integra Medik. = Kalanchoe integra (Medik.) Kuntze ■

107937 Cotyledon intermedia (Boiss.) Bornm. = Umbilicus intermedius Boiss. ■☆

107938 Cotyledon jacobseniana Poelln. = Cotyledon papillaris L. f. ●☆

107939 Cotyledon jasminiflora Salm-Dyck = Adromischus caryophyllaceus (Burm. f.) Lem. ■☆

107940 Cotyledon laciniata L. = Kalanchoe ceratophylla Haw. ■

107941 Cotyledon laciniata L. = Kalanchoe laciniata (L.) DC. ■

107942 Cotyledon ladismithensis Poelln. = Cotyledon tomentosa Harv. subsp. ladismithiensis (Poelln.) Toelken ●☆

107943 Cotyledon ladysmithiensis Poelln.;莱城圣塔●☆

107944 Cotyledon lanceolata Forssk. = Kalanchoe lanceolata (Forssk.) Pers. ■☆

107945 Cotyledon leucantha Ledeb. = Orostachys thyrsiflora Fisch. ■

107946 Cotyledon leucophylla C. A. Sm. = Cotyledon orbiculata L. var. oblonga (Haw.) DC. ●☆

107947 Cotyledon leucothrix (C. A. Sm.) Fourc. = Tylecodon leucothrix (C. A. Sm.) Toelken ●☆

107948 Cotyledon lievenii Ledeb. = Pseudosedum lievenii (Ledeb.) A. Berger ●■

107949 Cotyledon luteosquamata Poelln.;黄鳞长筒莲●☆

107950 Cotyledon luteosquamata Poelln. = Tylecodon pearsonii (Schönland) Toelken ●☆

107951 Cotyledon macrantha A. Berger var. virescens (Schönland et Baker f.) Poelln. = Cotyledon orbiculata L. var. oblonga (Haw.) DC. ●☆

107952 Cotyledon macrantha L.;大花长筒莲;Bigflower Cotyledon, Cotyledon ●☆

107953 Cotyledon maculata Salm-Dyck = Adromischus maculatus (Salm-Dyck) Lem. ■☆

107954 Cotyledon malacophylla Pall. = Orostachys malacophylla (Pall.) Fisch. ■

107955 Cotyledon mammillaris L. f. = Adromischus mammillaris (L. f.) Lem. ■

107956 Cotyledon marianiae Marloth = Adromischus marianiae (Marloth) A. Berger ■☆

107957 Cotyledon marlothii Schönland = Adromischus filicaulis (Eckl. et Zeyh.) C. A. Sm. subsp. marlothii (Schönland) Toelken ■☆

107958 Cotyledon meyeri Harv. = Cotyledon papillaris L. f. ●☆

107959 Cotyledon minuta Kom. = Orostachys minuta (Kom.) A. Berger ■

107960 Cotyledon mollis Dinter = Tylecodon paniculatus (L. f.) Toelken ●☆

107961 Cotyledon mollis Schönland = Cotyledon velutina Hook. f. ●☆

107962 Cotyledon mucizonia Ortega = Sedum mucizonia (Ortega) Raym. -Hamet ■☆

107963 Cotyledon mucizonia Ortega subsp. abylaea Font Quer et Maire = Sedum mucizonia (Ortega) Raym. -Hamet subsp. abylaeum (Font Quer et Maire) Spring. ■☆

107964 Cotyledon mucizonia Ortega var. glabra Braun-Blanq. et Maire = Sedum mucizonia (Ortega) Raym. -Hamet ■☆

107965 Cotyledon mucizonia Ortega var. hispida (Lam.) Pérez Lara = Sedum mucizonia (Ortega) Raym. -Hamet ■☆

107966 Cotyledon mucizonia Ortega var. parviflora Pau = Sedum mucizonia (Ortega) Raym. -Hamet ■☆

107967 Cotyledon mucronata Lam. = Cotyledon orbiculata L. ●☆

107968 Cotyledon muirii Schönland = Cotyledon papillaris L. f. ●☆

107969 Cotyledon nana Marloth = Adromischus humilis (Marloth) Poelln. ■☆

107970 Cotyledon nana N. E. Br. = Adromischus nanus (N. E. Br.) Poelln. ■☆

107971 Cotyledon nussbaumeriana Poelln. = Adromischus cristatus (Haw.) Lem. var. clavifolius (Haw.) Toelken ■☆

107972 Cotyledon obermeyeriana Poelln. = Cotyledon orbiculata L. var. oblonga (Haw.) DC. ●☆

107973 Cotyledon oblonga Haw. = Cotyledon orbiculata L. var. oblonga (Haw.) DC. ●☆

107974 Cotyledon occultans Toelken = Tylecodon occultans (Toelken) Toelken ●☆

107975 Cotyledon oppositifolia Ledeb. ex Nordm.;对叶长筒莲;Lamb's Tails ●☆

107976 Cotyledon orbiculata L.;圆叶长筒莲(圣塔);Cotyledon, Pig's Ear, Roudleaf Cotyledon, Round-leaved Navelwort, Round-leaved Navel-wort ●☆

107977 Cotyledon orbiculata L. var. ausana (Dinter) H. Jacobsen = Cotyledon orbiculata L. ●☆

107978 Cotyledon orbiculata L. var. dactylopsis Toelken;指状圆叶长筒莲●☆

107979 Cotyledon orbiculata L. var. dinteri H. Jacobsen = Cotyledon orbiculata L. ●☆

107980 Cotyledon orbiculata L. var. elata (Haw.) DC. = Cotyledon orbiculata L. ●☆

107981 Cotyledon orbiculata L. var. engleri (Dinter et A. Berger) Dinter = Cotyledon orbiculata L. ●☆

107982 Cotyledon orbiculata L. var. flanaganii (Schönland et Baker f.) Toelken;弗拉纳根长筒莲●☆

107983 Cotyledon orbiculata L. var. higginsiae H. Jacobsen = Cotyledon orbiculata L. ●☆

107984 Cotyledon orbiculata L. var. hinrichseniana H. Jacobsen = Cotyledon orbiculata L. ●☆

107985 Cotyledon orbiculata L. var. oblonga (Haw.) DC.;矩圆叶长筒莲;Pig's Ear ●☆

107986 Cotyledon orbiculata L. var. obovata DC. = Cotyledon orbiculata L. ●☆

107987 Cotyledon orbiculata L. var. oophylla Dinter = Cotyledon orbiculata L. ●☆

107988 Cotyledon orbiculata L. var. ramosa (Haw.) DC. = Cotyledon orbiculata L. ●☆

107989 Cotyledon orbiculata L. var. rotundifolia DC. = Cotyledon orbiculata L. ●☆

107990 Cotyledon orbiculata L. var. spuria (L.) Toelken;可疑长筒莲●☆

107991 Cotyledon orbiculata L. var. viridis Dinter ex Range = Cotyledon orbiculata L. ●☆

107992 Cotyledon oreades (Decne.) C. B. Clarke = Sedum oreades (Decne.) Raym. -Hamet ■

107993 Cotyledon ovata Haw. = Cotyledon orbiculata L. ●☆

107994 Cotyledon ovata Mill. = Crassula ovata (Mill.) Druce ●☆

107995 Cotyledon paniculata L. f. = Tylecodon paniculatus (L. f.) Toelken ●☆

107996 Cotyledon paniculata O. Fedtsch. et B. Fedtsch. = Tylecodon paniculatus (L. f.) Toelken ●☆

107997 Cotyledon papillaris L. f.;乳突长筒莲●☆

107998 Cotyledon papillaris L. f. var. glutinosa (Schönland) Poelln. = Cotyledon papillaris L. f. ●☆

107999 Cotyledon papillaris L. f. var. robusta Schönland et Baker f. = Cotyledon papillaris L. f. ●☆

108000 Cotyledon papillaris L. f. var. subundulata Poelln. = Cotyledon papillaris L. f. ●☆

108001 Cotyledon papillaris L. f. var. tricuspidata (Haw.) DC. = Cotyledon orbiculata L. ●☆

108002 Cotyledon parvula Burch. = Tylecodon reticulatus (L. f.) Toelken ●☆

108003 Cotyledon pearsonii Schönland = Tylecodon hallii (Toelken) Toelken ■☆

108004 Cotyledon pendens Van Jaarsv.;彭达长筒莲●☆

108005 Cotyledon phillipsiae Marloth = Adromischus phillipsiae (Marloth) Poelln. ■☆

108006 Cotyledon pillansii Schönland = Cotyledon cuneata Thunb. ●☆

108007 Cotyledon pinnata Lam. = Bryophyllum pinnatum (Lam.) Oken ■

108008 Cotyledon pinnata Lam. = Kalanchoe pinnata (Lam.) Pers. ■

108009 Cotyledon praealta (Brot.) Mariz = Umbilicus heylandianus Webb et Berthel. ●☆

108010 Cotyledon procurva N. E. Br. = Adromischus triflorus (L. f.) A. Berger ■☆

108011 Cotyledon pseudogracilis Poelln. = Cotyledon papillaris L. f. ●☆

108012 Cotyledon purpurea Thunb. = Cotyledon orbiculata L. var. spuria (L.) Toelken ●☆

108013 Cotyledon pygmaea W. F. Barker = Tylecodon pygmaeus (W. F. Barker) Toelken ●☆

108014 Cotyledon pygmaeus W. F. Barker var. tenuis Toelken = Tylecodon tenuis (Toelken) Bruyns ●☆

108015 Cotyledon racemosa E. Mey.;总状长筒莲●☆

108016 Cotyledon racemosa E. Mey. ex Harv. = Tylecodon hallii (Toelken) Toelken ■☆

108017 Cotyledon racemosa Harv. = Tylecodon racemosus (Harv.) Toelken ●☆

108018 Cotyledon ramosa Haw. = Cotyledon orbiculata L. ●☆

108019 Cotyledon ramosissima Salm-Dyck ex Haw. = Cotyledon woodii Schönland et Baker f. ●☆

108020 Cotyledon reticulata L. f. = Tylecodon reticulatus (L. f.) Toelken ●☆

108021 Cotyledon reticulata Thunb. = Tylecodon reticulatus (L. f.) Toelken ●☆

108022 Cotyledon rhombifolia Haw. var. spathulata N. E. Br. ex Marloth = Adromischus trigynus (Burch.) Poelln. ■☆

108023 Cotyledon rotundifolia Haw. = Adromischus hemisphaericus (L.) Lem. ■☆

108024 Cotyledon rubrovenosa Dinter = Tylecodon rubrovenosus (Dinter) Toelken ●☆

108025 Cotyledon rudatisii Poelln. = Cotyledon orbiculata L. var. oblonga (Haw.) DC. ●☆

108026 Cotyledon salmiana Poelln. var. woodii (Schönland et Baker f.) Poelln. = Cotyledon woodii Schönland et Baker f. ●☆

108027 Cotyledon salzmannii (Willk.) Font Quer = Pistorinia attenuata (H. Lindb.) Greuter ●☆

108028 Cotyledon schonlandii E. Phillips = Adromischus cristatus (Haw.) Lem. var. schonlandii (E. Phillips) Toelken ■☆

108029 Cotyledon simensis Britten = Rosularia semiensis (J. Gay ex A. Rich.) H. Ohba ■☆

108030 Cotyledon similis Toelken = Tylecodon similis (Toelken) Toelken ●☆

108031 Cotyledon simulans Schönland ex Poelln. = Cotyledon orbiculata L. var. oblonga (Haw.) DC. ●☆

108032 Cotyledon simulans Schönland ex Poelln. var. spathulata ? = Cotyledon orbiculata L. var. oblonga (Haw.) DC. ●☆

108033 Cotyledon singularis R. A. Dyer = Tylecodon singularis (R. A. Dyer) Toelken ●☆

108034 Cotyledon sinus-alexandri Poelln. = Tylecodon schaeferianus (Dinter) Toelken ●☆

108035 Cotyledon spathulata (DC.) Poir. = Kalanchoe integra (Medik.) Kuntze ■

108036 Cotyledon spathulata (DC.) Poir. = Sedum oreades (Decne.) Raym. -Hamet ■

108037 Cotyledon spinosa L. = Orostachys spinosa (L.) Sweet ■

108038 Cotyledon spuria L. = Cotyledon orbiculata L. var. spuria (L.) Toelken ●☆

108039 Cotyledon striata Hutchison = Tylecodon striatus (Hutchison) Toelken ●☆

108040 Cotyledon sturmiana Poelln. = Cotyledon barbeyi Schweinf. ex Baker ●☆

108041 Cotyledon sulphurea Toelken = Tylecodon sulphureus (Toelken) Toelken ●☆

108042 Cotyledon swartbergensis Poelln. = Tylecodon leucothrix (C. A. Sm.) Toelken ●☆

108043 Cotyledon tardiflora Bonpl. = Tylecodon paniculatus (L. f.) Toelken ●☆

108044 Cotyledon teretifolia Lam. = Cotyledon papillaris L. f. ●☆

108045 Cotyledon teretifolia Thunb.;筒叶长筒莲(佐保姬)●☆

108046 Cotyledon teretifolia Thunb. = Cotyledon campanulata Marloth ●☆

108047 Cotyledon thyrsiflora (Fisch.) Maxim. = Orostachys thyrsiflora Fisch. ■

108048 Cotyledon tomentosa Harv.;毛长筒莲;Bear's Paw ●☆

108049 Cotyledon tomentosa Harv. subsp. ladismithiensis (Poelln.) Toelken;莱迪史密斯长筒莲●☆

108050 Cotyledon transvaalensis Guillaumet = Cotyledon barbeyi Schweinf. ex Baker ●☆

108051 Cotyledon tricuspidata Haw. = Cotyledon orbiculata L. ●☆

108052 Cotyledon triflora L. f. = Adromischus triflorus (L. f.) A.

Berger ■☆

108053 Cotyledon trigyna Burch. = Adromischus trigynus (Burch.) Poelln. ■☆

108054 Cotyledon tuberculosa Lam. = Tylecodon grandiflorus (Burm. f.) Toelken ●☆

108055 Cotyledon turkestanica (Regel et Winkl.) O. Fedtsch. et B. Fedtsch. = Rosularia turkestanica (Regel et Winkl.) A. Berger ■

108056 Cotyledon umbilicus-veneris L. = Umbilicus rupestris (Salisb.) Dandy ■☆

108057 Cotyledon umbilicus-veneris L. subsp. erecta (Desf.) Batt. = Umbilicus gaditanus Boiss. ■☆

108058 Cotyledon umbilicus-veneris L. subsp. horizontalis (Guss.) Batt. = Umbilicus horizontalis (Guss.) DC. ■☆

108059 Cotyledon umbilicus-veneris L. subsp. patens (Pomel) Batt. = Umbilicus patens Pomel ■☆

108060 Cotyledon umbilicus-veneris L. subsp. pendulina (DC.) Batt. = Umbilicus rupestris (Salisb.) Dandy ■☆

108061 Cotyledon umbilicus-veneris L. var. amphitropa Batt. = Umbilicus horizontalis (Guss.) DC. ■☆

108062 Cotyledon umbilicus-veneris L. var. deflexa (Pomel) Batt. = Umbilicus rupestris (Salisb.) Dandy ■☆

108063 Cotyledon umbilicus-veneris L. var. fontqueri (Maire et Sennen) Maire = Atriplex prostrata DC. ●

108064 Cotyledon umbilicus-veneris L. var. gaditana (Boiss.) Batt. = Atriplex prostrata DC. ●

108065 Cotyledon umbilicus-veneris L. var. gigantea (Batt.) Maire = Atriplex prostrata DC. ●

108066 Cotyledon umbilicus-veneris L. var. horizontalis (Guss.) Lowe = Umbilicus horizontalis (Guss.) DC. ■☆

108067 Cotyledon umbilicus-veneris L. var. intermedia (Boiss.) Maire = Umbilicus rupestris (Salisb.) Dandy ■☆

108068 Cotyledon umbilicus-veneris L. var. micrantha (Pamp.) Maire = Umbilicus rupestris (Salisb.) Dandy ■☆

108069 Cotyledon umbilicus-veneris L. var. patula (Pomel) Maire = Umbilicus rupestris (Salisb.) Dandy ■☆

108070 Cotyledon umbilicus-veneris L. var. pomelii Maire = Atriplex prostrata DC. ●

108071 Cotyledon umbilicus-veneris L. var. purpurea Maire = Atriplex prostrata DC. ●

108072 Cotyledon umbilicus-veneris L. var. suberecta Maire = Umbilicus horizontalis (Guss.) DC. ■☆

108073 Cotyledon umbilicus-veneris L. var. subhorizontalis Maire et Weiller = Umbilicus rupestris (Salisb.) Dandy ■☆

108074 Cotyledon umbilicus-veneris L. var. tuberosa L. = Umbilicus rupestris (Salisb.) Dandy ■☆

108075 Cotyledon undulata Haw.;银波锦(边草,脐状瓦松,银冠);Silver Crown,Silver-crown,Silver-ruffles ●☆

108076 Cotyledon undulata Haw. = Cotyledon orbiculata L. var. oblonga (Haw.) DC. ●☆

108077 Cotyledon undulata Haw. var. mucronata (Lam.) Poelln. = Cotyledon orbiculata L. ●☆

108078 Cotyledon ungulata Lam. = Cotyledon orbiculata L. ●☆

108079 Cotyledon velutina Hook. f.;短绒毛长筒莲●☆

108080 Cotyledon velutina Hook. f. var. beckeri (Schönland et Baker f.) Schönland = Cotyledon velutina Hook. f. ●☆

108081 Cotyledon ventricosa Burm. f. = Tylecodon ventricosus (Burm. f.) Toelken ●☆

108082 Cotyledon ventricosa Burm. f. var. alpina Harv. = Tylecodon ventricosus (Burm. f.) Toelken ●☆

108083 Cotyledon verea Jacq. = Kalanchoe crenata (Andréws) Haw. ■☆

108084 Cotyledon virescens Schönland et Baker f. = Cotyledon orbiculata L. var. oblonga (Haw.) DC. ●☆

108085 Cotyledon viridiflora Toelken = Tylecodon viridiflorus (Toelken) Toelken ●☆

108086 Cotyledon viridis Haw. = Cotyledon orbiculata L. var. oblonga (Haw.) DC. ●☆

108087 Cotyledon wallichii Harv. = Tylecodon wallichii (Harv.) Toelken ●☆

108088 Cotyledon weikensii Schönland;威氏长筒莲●☆

108089 Cotyledon whiteae Schönland et Baker f. = Cotyledon orbiculata L. var. oblonga (Haw.) DC. ●☆

108090 Cotyledon wickensii Schönland = Cotyledon barbeyi Schweinf. ex Baker ●☆

108091 Cotyledon wickensii Schönland var. glandulosa Poelln. = Cotyledon barbeyi Schweinf. ex Baker ●☆

108092 Cotyledon woodii Schönland et Baker f.;伍得长筒莲●☆

108093 Cotyledon zeyheri Harv. = Adromischus cristatus (Haw.) Lem. var. zeyheri (Harv.) Toelken ■☆

108094 Cotyledon zuluensis Schönland ex Poelln. = Cotyledon orbiculata L. var. oblonga (Haw.) DC. ●☆

108095 Cotyledonaceae Martinov = Crassulaceae J. St. -Hil.(保留科名)●■

108096 Cotylelobiopsis Heim = Copaifera L.(保留属名)●☆

108097 Cotylelobium Pierre(1890);杯裂香属●☆

108098 Cotylelobium lewisianum (Trimen ex Hook. f.) P. S. Ashton;刘易斯杯裂香木●☆

108099 Cotylelobium melanoxylon Pierre;黑木杯裂香●☆

108100 Cotylephora Meisn. = Neesia Blume(保留属名)●☆

108101 Cotylina Post et Kuntze = Cotula L. ■

108102 Cotylina Post et Kuntze = Cotulina Pomel ■

108103 Cotyliphyllum Link = Cotyledon L. ●■☆

108104 Cotyliscus Desv. = Coronopus Zinn(保留属名)■

108105 Cotyliscus Desv. = Lepidium L. ■

108106 Cotylodiscus Radlk. = Plagioscyphus Radlk. ●☆

108107 Cotylodiscus stelechanthus Radlk. = Plagioscyphus stelechanthus (Radlk.) Capuron ●☆

108108 Cotylolabium Garay = Stenorrhynchos Rich. ex Spreng. ■☆

108109 Cotylolobiopsis Post et Kuntze = Copaifera L.(保留属名)●☆

108110 Cotylolobiopsis Post et Kuntze = Cotylelobiopsis Heim ●☆

108111 Cotylolobium Post et Kuntze = Cotylelobium Pierre ●☆

108112 Cotylonia C. Norman = Dickinsia Franch. ■★

108113 Cotylonia bracteata C. Norman. = Dickinsia hydrocotyloides Franch. ■

108114 Cotylonychia Stapf = Pentadiplandra Baill. ●☆

108115 Cotylonychia Stapf(1908);杯距梧桐属●☆

108116 Cotylonychia chevalieri Stapf;杯距梧桐●☆

108117 Cotylophyllum Post et Kuntze = Cotyliphyllum Link ●■☆

108118 Cotylophyllum Post et Kuntze = Umbilicus DC. ■☆

108119 Cotyloplecta Alef. = Hibiscus L.(保留属名)●■

108120 Coublandia Aubl.(废弃属名) = Lonchocarpus Kunth(保留属名)●■☆

108121 Coublandia Aubl.(废弃属名) = Muellera L. f.(保留属名)●■☆

108122 Coudenbergia Marchal = Pentapanax Seem. ●

108123 Couepia Aubl.(1775);库佩果属●☆

108124 Couepia edulis (Prance) Prance;库佩果;Castanha De Curia ●☆
108125 Coula Baill. (1862);柯拉铁青树属●☆
108126 Coula cabrae De Wild. et T. Durand = Coula edulis Baill. ●☆
108127 Coula edulis Baill.;柯拉铁青树;African Walnut,Gaboon Nut ●☆
108128 Coula utilis S. Moore = Coula edulis Baill. ●☆
108129 Coulaceae Tiegh. = Erythropalaceae Planch. ex Miq. (保留科名)●
108130 Coulaceae Tiegh. = Olacaceae R. Br. (保留科名)●
108131 Coulaceae Tiegh. ex Bullock = Erythropalaceae Planch. ex Miq. (保留科名)●
108132 Coulaceae Tiegh. ex Bullock = Olacaceae R. Br. (保留科名)●
108133 Coulejia Dennst. = Antidesma L. ●
108134 Coulterella Tiegh. = Pterocephalus Vaill. ex Adans. ●■
108135 Coulterella Vasey et Rose ex O. Hoffm. = Coulterella Vasey et Rose ■☆
108136 Coulterella Vasey et Rose(1890);盘头菊属■☆
108137 Coulterella capitata Vasey et Rose;盘头菊■☆
108138 Coulteria Kunth = Caesalpinia L. ●
108139 Coulteria africana Guillaumin et Perr. = Prosopis africana (Guillaumin et Perr.) Taub. ●☆
108140 Coulteria tinctoria Kunth = Caesalpinia spinosa (Molina) Kuntze ●☆
108141 Coulterina Kuntze = Physaria (Nutt. ex Torr. et A. Gray) A. Gray ■
108142 Coulterophytum B. L. Rob. (1892);考特草属☆
108143 Coulterophytum brevipes J. M. Coult. et Rose;短梗考特草☆
108144 Coulterophytum laxum B. L. Rob.;松散考特草☆
108145 Coulterophytum macrophyllum J. M. Coult. et Rose;大叶考特草☆
108146 Coulterophytum pubescens J. M. Coult. et Rose;毛考特草☆
108147 Couma Aubl. (1775);牛奶木属;Couma Rubber ●☆
108148 Couma guatemalensis Standl.;危地马拉牛奶木●☆
108149 Couma guianensis Aubl.;圭亚那牛奶木●☆
108150 Couma macrocarpa Barb. Rodr.;大果牛奶木;Sorva ●☆
108151 Couma pentaphylla Huber;五叶牛奶木●☆
108152 Couma rigida Müll. Arg.;坚硬牛奶木●☆
108153 Couma utilis Müll. Arg.;良木牛奶木;Cow Tree,Milk Tree ●☆
108154 Coumarouna Aubl. (废弃属名) = Dipteryx Schreb. (保留属名)●☆
108155 Coumarouna odorata Aubl. = Dipteryx odorata Willd. ●☆
108156 Coupla G. Don = Goupia Aubl. ●☆
108157 Coupoui Aubl. = Duroia L. f. (保留属名)●☆
108158 Coupuia Raf. = Duroia L. f. (保留属名)●☆
108159 Coupuya Raf. = Duroia L. f. (保留属名)●☆
108160 Couralia Splitg. = Potamoxylon Raf. ●☆
108161 Couralia Splitg. = Tabebuia Gomes ex DC. ●☆
108162 Courantia Lem. = Echeveria DC. ●■☆
108163 Couratari Aubl. (1775);纤皮玉蕊属●☆
108164 Couratari asterotricha Prance;星毛纤皮玉蕊●☆
108165 Couratari coriacea Mart. ex O. Berg;革质纤皮玉蕊●☆
108166 Couratari fagifolia (Miq.) Eyma;水青冈叶纤皮玉蕊●☆
108167 Couratari gloriosa Sandwith;格勒纤皮玉蕊●☆
108168 Couratari guianensis Aubl.;圭亚那纤皮玉蕊●☆
108169 Couratari multiflora Eyma;多花纤皮玉蕊●☆
108170 Couratari oblongifolia Ducke et R. Knuth;长椭圆叶纤皮玉蕊●☆
108171 Couratari stellata A. C. Sm.;星芒纤皮玉蕊●☆
108172 Courbari Adans. = Courbaril Mill. ●
108173 Courbari Adans. = Hymenaea L. ●
108174 Courbaril Mill. = Hymenaea L. ●
108175 Courbaril Plum. ex Endl. = Courbaril Mill. ●
108176 Courbonia Brongn. = Maerua Forssk. ●☆
108177 Courbonia brevipilosa Gilg = Maerua decumbens (Brongn.) DeWolf ●☆
108178 Courbonia bussei Gilg et Gilg-Ben. = Maerua edulis (Gilg et Gilg-Ben.) DeWolf ●☆
108179 Courbonia calothamna Gilg et Gilg-Ben. = Maerua edulis (Gilg et Gilg-Ben.) DeWolf ●☆
108180 Courbonia camporum Gilg et Gilg-Ben. = Maerua edulis (Gilg et Gilg-Ben.) DeWolf ●☆
108181 Courbonia decumbens Brongn. = Maerua decumbens (Brongn.) DeWolf ●☆
108182 Courbonia edulis Gilg et Gilg-Ben. = Maerua edulis (Gilg et Gilg-Ben.) DeWolf ●☆
108183 Courbonia glauca (Klotzsch) Gilg et Gilg-Ben. = Maerua edulis (Gilg et Gilg-Ben.) DeWolf ●☆
108184 Courbonia nummularifolia Mattei = Maerua subcordata (Gilg) DeWolf ●☆
108185 Courbonia prunicarpa Gilg et Gilg-Ben. = Maerua edulis (Gilg et Gilg-Ben.) DeWolf ●☆
108186 Courbonia pseudopetalosa Gilg et Gilg-Ben. = Maerua pseudopetalosa (Gilg et Gilg-Ben.) DeWolf ●☆
108187 Courbonia subcordata Gilg = Maerua subcordata (Gilg) DeWolf ●☆
108188 Courbonia tubulosa Gilg et Gilg-Ben. = Maerua subcordata (Gilg) DeWolf ●☆
108189 Courbonia virgata Brongn. = Maerua pseudopetalosa (Gilg et Gilg-Ben.) DeWolf ●☆
108190 Courimari Aubl. = Sloanea L. ●
108191 Couringia Adans. = Conringia Heist. ex Fabr. ■
108192 Courondi Adans. (废弃属名) = Salacia L. (保留属名)●
108193 Couroupita Aubl. (1775);炮弹树属(炮弹果属);Cannonball Tree,Cannon-ball Tree ●☆
108194 Couroupita amazonica R. Knuth;亚马孙炮弹果●☆
108195 Couroupita guianensis Aubl.;炮弹树(安贵玉蕊,圭亚那炮弹果,贵安玉蕊,炮弹果);Cannonball Tree,Cannon-ball Tree ●☆
108196 Couroupita peruviana O. Berg;秘鲁炮弹果●☆
108197 Courrantia Sch. Bip. = Matricaria L. ■
108198 Coursetia DC. (1825);婴帽豆属●☆
108199 Coursetia glandulifera (Benth.) J. F. Macbr.;婴帽豆;Coursetia ●☆
108200 Coursiana Homolle = Schismatoclada Baker ■☆
108201 Coursiana Homolle(1942);库斯茜属■☆
108202 Coursiana homolleana Cavaco;库斯茜■☆
108203 Courtenia R. Br. = Cola Schott et Endl. (保留属名)●☆
108204 Courtenia triloba R. Br. = Cola triloba (R. Br.) K. Schum. ●☆
108205 Courtoisia Nees = Courtoisina Soják ■
108206 Courtoisia Rchb. = Collomia Nutt. ■☆
108207 Courtoisia Rchb. = Phlox L. ■
108208 Courtoisia assimilis (Steud.) C. B. Clarke = Courtoisina assimilis (Steud.) Maquet ■☆
108209 Courtoisia assimilis (Steud.) Maquet = Courtoisina assimilis (Steud.) Maquet ■☆
108210 Courtoisia cyperoides (Roxb.) Nees = Courtoisina cyperoides (Roxb.) Soják ■
108211 Courtoisia cyperoides (Roxb.) Nees var. africana C. B. Clarke

= Courtoisina cyperoides (Roxb.) Soják ■
108212 Courtoisia cyperoides Nees = Courtoisina cyperoides (Roxb.) Soják ■
108213 Courtoisina Soják(1980);翅鳞莎属;Courtoisia,Courtoisina ■
108214 Courtoisina assimilis (Steud.) Maquet;阿西翅鳞莎■☆
108215 Courtoisina cyperoides (A. Dietr.) Soják;翅鳞莎;Flatsedge Courtoisia,Flatsedge-like Courtoisia ■
108216 Courtoisina cyperoides (Roxb.) Soják = Courtoisina cyperoides (A. Dietr.) Soják ■
108217 Cousinia Cass. (1827);刺头菊属(假蓟属,栲新菊属);Cousinia,Vinegentian ●■
108218 Cousinia abbreviata Tscherneva;缩短刺头菊■☆
108219 Cousinia abolinii Kult. ex Tscherneva;阿伯刺头菊■☆
108220 Cousinia adenophora Juz.;腺梗刺头菊■☆
108221 Cousinia affinis Schrenk;近缘刺头菊;Related Cousinia,Related Vinegentian ■
108222 Cousinia agelocephala Tscherneva;头刺头菊■☆
108223 Cousinia alaica Juz. ex Tscherneva;阿赖刺头菊■☆
108224 Cousinia alata Schrenk;翼茎刺头菊;Wingstem Cousinia,Wingstem Vinegentian ■
108225 Cousinia albertii Regel et Schmalh.;阿尔伯特刺头菊■☆
108226 Cousinia albiflora (Bornm. et Sint.) Bornm.;白花刺头菊■☆
108227 Cousinia alpestris Bornm. ex Juz.;高山刺头菊■☆
108228 Cousinia alpina Bunge;山地刺头菊■☆
108229 Cousinia ambigens Juz.;含糊刺头菊■☆
108230 Cousinia amoena C. Winkl.;秀丽刺头菊■☆
108231 Cousinia androssovii Juz.;安德罗索夫刺头菊■☆
108232 Cousinia angusticeps Juz.;窄头刺头菊■☆
108233 Cousinia anomala Franch.;异常刺头菊■☆
108234 Cousinia antonowii C. Winkl.;安托刺头菊■☆
108235 Cousinia apiculata Tscherneva;细尖刺头菊■☆
108236 Cousinia arachnoidea Fisch. et C. A. Mey.;蛛毛刺头菊■☆
108237 Cousinia arctioides Schrenk;拟北极刺头菊■☆
108238 Cousinia armena Takht.;亚美尼亚刺头菊■☆
108239 Cousinia astracanica (Spreng.) Tamamsch.;伏尔加刺头菊;Volga Cousinia,Volga Vinegentian ■☆
108240 Cousinia aurea C. Winkl.;黄刺头菊■☆
108241 Cousinia badghysi Kult.;巴德刺头菊■☆
108242 Cousinia batalinii C. Winkl.;巴塔林刺头菊■☆
108243 Cousinia bobrovii Juz.;鲍勃刺头菊■☆
108244 Cousinia bonvalotii Franck;鲍恩刺头菊■☆
108245 Cousinia botschantzevii Juz. ex Tscherneva;包兹刺头菊■☆
108246 Cousinia brachyptera DC.;短翅刺头菊■☆
108247 Cousinia bungeana Regel et Schmalh.;布氏刺头菊■☆
108248 Cousinia buphtalmoides Regel;牛眼菊状刺头菊■☆
108249 Cousinia caespitosa C. Winkl.;丛生刺头菊;Caespitose Cousinia,Caespitose Vinegentian ■
108250 Cousinia calva Juz.;光秃刺头菊■☆
108251 Cousinia campyloraphis Tscherneva;弯刺头菊■☆
108252 Cousinia candicans Juz.;纯白刺头菊■☆
108253 Cousinia centauroides Fisch. et C. A. Mey.;矢车菊状刺头菊■☆
108254 Cousinia chaetocephala Kult.;毛头刺头菊■☆
108255 Cousinia chlorantha Kult.;绿花刺头菊■☆
108256 Cousinia chlorocephala C. A. Mey.;绿头刺头菊■☆
108257 Cousinia chrysantha Kult.;金花刺头菊■☆
108258 Cousinia coerulea Kuic ex Tscherneva;青蓝刺头菊■☆
108259 Cousinia congesta Bunge;密集刺头菊■☆
108260 Cousinia coronata Franck;冠状刺头菊■☆
108261 Cousinia corymbosa C. Winkl.;伞序刺头菊■☆
108262 Cousinia cynaroides (M, Bieb.) C. A. Mey.;菜蓟刺头菊■☆
108263 Cousinia darwasica Winkl.;达尔瓦斯刺头菊■☆
108264 Cousinia decurrentifolia Juz. ex Tschemcva;下延叶刺头菊■☆
108265 Cousinia dichotoma Bunge;二歧刺头菊■☆
108266 Cousinia dichromata Kult.;二色刺头菊■☆
108267 Cousinia dimoana Kult.;迪茂刺头菊■☆
108268 Cousinia dissecta Kar. et Kir.;深裂刺头菊;Dissect Cousinia,Dissect Vinegentian ■
108269 Cousinia dissecta Kar. et Kir. = Cousinia alata Schrenk ■
108270 Cousinia dissectifolia Kult.;深裂叶刺头菊■☆
108271 Cousinia divaricata C. Winkl.;叉开刺头菊■☆
108272 Cousinia dolichoclada Juz.;长枝刺头菊■☆
108273 Cousinia dolicholepis Schrenk;长鳞刺头菊■☆
108274 Cousinia dolichophylla Kult.;长叶刺头菊■☆
108275 Cousinia dubia Popov.;可疑刺头菊■☆
108276 Cousinia egregia Juz.;优秀刺头菊■☆
108277 Cousinia erectispina Tscherneva;中亚直刺刺头菊■☆
108278 Cousinia eriophora Regel et Schmalh. = Schmalhausenia nidulans (Regel) Petr. ■
108279 Cousinia eriotricha Juz.;红毛刺头菊■☆
108280 Cousinia erivanensis Bornm.;埃里温刺头菊■☆
108281 Cousinia eryngioides Boiss.;刺芹刺头菊■☆
108282 Cousinia eugenii Kult.;欧根刺头菊■☆
108283 Cousinia falconeri Hook. f.;穗花刺头菊;Spikeflower Cousinia,Spikeflower Vinegentian ■
108284 Cousinia fallax C. Winkl.;迷惑刺头菊■☆
108285 Cousinia fascicularis Juz.;簇生刺头菊■☆
108286 Cousinia fedtschenkoana Bornm.;范氏刺头菊■☆
108287 Cousinia ferganensis Bornm.;费尔干刺头菊■☆
108288 Cousinia ferruginea Kult.;锈色刺头菊■☆
108289 Cousinia fetissowii C. Winkl.;费季索娃刺头菊■☆
108290 Cousinia franchetii C. Winkl.;弗朗刺头菊■☆
108291 Cousinia glabriseta Kult.;无毛刺头菊■☆
108292 Cousinia glandulosa Kult.;多腺刺头菊■☆
108293 Cousinia glaphyrocephala Juz. ex Tscherneva;空头刺头菊■☆
108294 Cousinia glochidiata Kult.;钩毛刺头菊■☆
108295 Cousinia gnezdilloi Tscherneva;格涅刺头菊■☆
108296 Cousinia gontscharowii Juz.;高恩恰洛夫刺头菊■☆
108297 Cousinia grandifolia Kult.;大叶刺头菊■☆
108298 Cousinia grigoriewii Juz.;格里刺头菊■☆
108299 Cousinia grisea Kult.;灰刺头菊■☆
108300 Cousinia hamadae Juz.;哈马达刺头菊■☆
108301 Cousinia hastifolia C. Winkl.;戟叶刺头菊■☆
108302 Cousinia hilariae Kult.;希拉里刺头菊■☆
108303 Cousinia hohenackeri Fisch. et C. A. Mey.;豪氏刺头菊■☆
108304 Cousinia horridula Juz.;小刺刺头菊■☆
108305 Cousinia hypopolia Bornm. et Sint.;背毛刺头菊■☆
108306 Cousinia hystrix C. A. Mey.;豪猪刺头菊■☆
108307 Cousinia iljinii Takht.;伊尔金刺头菊■☆
108308 Cousinia integrifolia Franch.;全叶刺头菊■☆
108309 Cousinia karatavica Regel et Schmalh.;卡拉塔夫刺头菊■☆
108310 Cousinia knotringiae Bornm.;克诺氏刺头菊■☆
108311 Cousinia kokanica Regel et Schmalh.;浩罕刺头菊■☆
108312 Cousinia komarowii (Kuntze) C. Winkl.;科马罗夫刺头菊■☆
108313 Cousinia korolkovii Regel et Schmalh.;科罗尔科夫刺头菊■☆

108314 Cousinia korshinskyi C. Winkl. ;考尔刺头菊■☆
108315 Cousinia krauseana Regel et Schmalh. ;克劳斯刺头菊■☆
108316 Cousinia kuekenthalii Bornm. ;屈肯刺头菊■☆
108317 Cousinia laetevirens C. Winkl. ;鲜绿刺头菊■☆
108318 Cousinia lanata C. Winkl. ;绵毛刺头菊■☆
108319 Cousinia laniceps Juz. ;绵毛梗刺头菊■☆
108320 Cousinia lappacea Schrenk;拟钩毛刺头菊■☆
108321 Cousinia lasiophylla C. Shih;丝毛刺头菊;Flossleaf Cousinia, Flossleaf Vinegentian ■
108322 Cousinia lasiosiphon Juz. ;毛管刺头菊■☆
108323 Cousinia leiocephala (Regel) Juz. ;光苞刺头菊; Flathead Cousinia, Flathead Vinegentian ■
108324 Cousinia leptacantha (Bornm.) Juz. ;细刺刺头菊■☆
108325 Cousinia leptocampyla Bornm. ;细弯刺头菊■☆
108326 Cousinia leptocephala Fisch. et C. A. Mey. ;细头刺头菊■☆
108327 Cousinia leptoclada Kult. ;细枝刺头菊■☆
108328 Cousinia leptocladoides Tscherneva;拟细刺刺头菊■☆
108329 Cousinia leucantha Bornm. et Sint. ;白刺刺头菊■☆
108330 Cousinia linczewskii Juz. ;林契刺头菊■☆
108331 Cousinia litvinovii Kult. ex Juz. ;里特刺头菊■☆
108332 Cousinia lomakinii C. Winkl. ;洛马金刺头菊■☆
108333 Cousinia lyrata Bunge;大头羽裂刺头菊■☆
108334 Cousinia macrocephala C. A. Mey. ;大头刺头菊■☆
108335 Cousinia macroptera C. A. Mey. ;大翅刺头菊■☆
108336 Cousinia magnifica Juz. ;华丽刺头菊■☆
108337 Cousinia maracandica Juz. ;马拉坎达刺头菊■☆
108338 Cousinia margaritae Kult. ;马尔刺头菊■☆
108339 Cousinia medians Juz. ;中间刺头菊■☆
108340 Cousinia microcarpa Boiss. ;小果刺头菊■☆
108341 Cousinia microcephala C. A. Mey. ;小头刺头菊■☆
108342 Cousinia minkwitziae Bornm. ;明克刺头菊■☆
108343 Cousinia minuta Boiss. ;微小刺头菊■☆
108344 Cousinia mogoltavica Tschernevaet Wed. ;莫戈尔塔夫刺头菊■☆
108345 Cousinia mollis Schrenk;柔软刺头菊■☆
108346 Cousinia mucida Kult. ;黏刺头菊■☆
108347 Cousinia multiloba DC. ;多裂刺头菊■☆
108348 Cousinia neglecta Juz. ;忽视刺头菊■☆
108349 Cousinia newesskiana C. Winkl. ;纽氏刺头菊■☆
108350 Cousinia ninae Juz. ;尼娜刺头菊■☆
108351 Cousinia olgae Regel et Schmalh. ;奥氏刺头菊■☆
108352 Cousinia omphalodes Tscherneva;脐状刺头菊■☆
108353 Cousinia onopordinides Ledeb. ;大翅蓟刺头菊■☆
108354 Cousinia oopoda Juz. ;卵足刺头菊■☆
108355 Cousinia oreodoxa Bornm. et Sint. ;山景刺头菊■☆
108356 Cousinia oreoxerophila Kult. ;旱生刺头菊■☆
108357 Cousinia orientalis (Adams) K. Koch;东方刺头菊■☆
108358 Cousinia orthacantha Tscherneva;直刺刺头菊■☆
108359 Cousinia ortholepis Juz. ex Tscherneva;直鳞刺头菊■☆
108360 Cousinia ovczinnikovii Tscherneva;奥夫钦尼科夫刺头菊■☆
108361 Cousinia oxiana Tscherneva;阿穆达尔刺头菊■☆
108362 Cousinia oxytoma Rech. f. ;尖刺头菊■☆
108363 Cousinia pannosa C. Winkl. ;毛毡刺头菊■☆
108364 Cousinia pannosiformis Tscherneva;毡状刺头菊■☆
108365 Cousinia pauciramosa Kult. ;少枝刺头菊■☆
108366 Cousinia peduncularis Juz. ex Tscherneva;梗花刺头菊■☆
108367 Cousinia pentacantha Regel et Schmalh. ;五花刺头菊■☆
108368 Cousinia pentacanthoides Juz. ex Tscherneva;拟五花刺头菊■☆
108369 Cousinia platylepis Schrenk ex Fisch. , C. A. Mey. et Avé-Lall. ; 宽苞刺头菊;Broadbract Cousinia, Broadbract Vinegentian ■
108370 Cousinia platystegia Tscherneva;平盖刺头菊■☆
108371 Cousinia podophylla Tscherneva;足叶刺头菊■☆
108372 Cousinia polycephala Rupr. ;多头刺头菊(多花刺头菊); Manyhead Cousinia, Manyhead Vinegentian ■
108373 Cousinia praestans Tscherneva et Wed. ;优越刺头菊■☆
108374 Cousinia princeps Franch. ;帝王刺头菊■☆
108375 Cousinia proxima Juz. ;鹿刺头菊■☆
108376 Cousinia psammophila Kult. ;喜沙刺头菊■☆
108377 Cousinia pseudoaffinis Kult. ;假近缘刺头菊■☆
108378 Cousinia pseudolanata Popov. ex Tscherneva;假绵毛刺头菊■☆
108379 Cousinia pseudomollis C. Winkl. ;略柔软刺头菊■☆
108380 Cousinia pterolepida Kult. ;翅鳞刺头菊■☆
108381 Cousinia pulchella Bunge;亚美刺头菊■☆
108382 Cousinia pulchra C. Winkl. ;美丽刺头菊■☆
108383 Cousinia pungens Juz. ;多刺刺头菊■☆
108384 Cousinia purpurea C. A. Mey. ;紫刺头菊■☆
108385 Cousinia pusilla C. Winkl. ;弱小刺头菊■☆
108386 Cousinia pygmaea C. Winkl. ;小刺头菊■☆
108387 Cousinia raddeana C. Winkl. ;拉德刺头菊■☆
108388 Cousinia radians Bunge;辐射刺头菊■☆
108389 Cousinia ramulosa Rech. f. ;多枝刺头菊■☆
108390 Cousinia rava C. Winkl. ;拉瓦刺头菊■☆
108391 Cousinia refracta Juz. ;反折刺头菊■☆
108392 Cousinia resinosa Juz. ;树脂刺头菊■☆
108393 Cousinia rhodantha Kult. ;粉红花刺头菊■☆
108394 Cousinia rigida Kult. ;硬刺头菊■☆
108395 Cousinia rosea Kult. ;粉红刺头菊■☆
108396 Cousinia rotundifolia C. Winkl. ;圆叶刺头菊■☆
108397 Cousinia rubiginosa Kult. ;锈红刺头菊■☆
108398 Cousinia scabrida Juz. ;微糙刺头菊■☆
108399 Cousinia schischkinii Juz. ;希施刺头菊■☆
108400 Cousinia sclerolepis C. Shih;硬苞刺头菊;Hardbract Cousinia, Hardbract Vinegentian ■
108401 Cousinia sclerophylla Juz. ;硬叶刺头菊■☆
108402 Cousinia semidecurrens C. Winkl. ;半下延刺头菊■☆
108403 Cousinia semilacera Juz. ;半撕裂刺头菊■☆
108404 Cousinia sewertzowii Regel var. leiocephala Regel = Cousinia leiocephala (Regel) Juz. ■
108405 Cousinia sororia Juz. ;堆头刺头菊■☆
108406 Cousinia speciosa C. Winkl. ;艳丽刺头菊■☆
108407 Cousinia spiridonovii Juz. ;斯皮里刺头菊■☆
108408 Cousinia splendida C. Winkl. ;闪光刺头菊■☆
108409 Cousinia sporadocephala Juz. ;散头刺头菊■☆
108410 Cousinia stahliana Bornm. et Gauba;斯达刺头菊■☆
108411 Cousinia stellaris Bornm. ;星状刺头菊■☆
108412 Cousinia stenophylla Kult. ;狭叶刺头菊■☆
108413 Cousinia stephanophora C. Winkl. ;美冠刺头菊■☆
108414 Cousinia stricta Tscherneva;刚直刺头菊■☆
108415 Cousinia strobilocephala Tscherneva et Vved. ;球头刺头菊■☆
108416 Cousinia subappendiculata Kult. ;小附属物刺头菊■☆
108417 Cousinia subcandicans Tscherneva;亚白刺头菊■☆
108418 Cousinia submutica Franch. ;无尖刺头菊■☆
108419 Cousinia sylvicola Bunge;西尔维亚刺头菊■☆
108420 Cousinia syrdariensis Kult. ;锡尔达里亚刺头菊■☆
108421 Cousinia talassica (Kult.) Juz. ;塔拉斯刺头菊■☆

108422 Cousinia tamarae Juz.;塔马拉刺头菊■☆
108423 Cousinia tenella Fisch. et C. A. Mey.;细弱刺头菊■
108424 Cousinia tenuisecta Juz.;细齿刺头菊■☆
108425 Cousinia thomsonii C. B. Clarke;毛苞刺头菊(绵刺头菊);Thomson Cousinia,Thomson Vinegentian ■
108426 Cousinia tianschanica Kult.;天山刺头菊;Tianshan Cousinia ■☆
108427 Cousinia tomentella C. Winkl.;绒毛刺头菊■☆
108428 Cousinia trachyphylla Juz.;糙叶刺头菊■☆
108429 Cousinia transiliensis Juz.;外伊犁刺头菊■☆
108430 Cousinia transoxana Tscherneva;外阿穆达尔刺头菊■☆
108431 Cousinia trautvetteri Regel = Alfredia nivea Kar. et Kir. ■
108432 Cousinia triceps Kult.;三头刺头菊■☆
108433 Cousinia trichopora Kult.;毛梗刺头菊■☆
108434 Cousinia triflora Schrenk;三花刺头菊■☆
108435 Cousinia turcomanica C. Winkl.;土库曼刺头菊■☆
108436 Cousinia turkestanica (Regel) Juz.;土耳其斯坦刺头菊■☆
108437 Cousinia ulotoma Bornm.;卷片刺头菊■☆
108438 Cousinia umblicata Juz.;脐刺头菊■☆
108439 Cousinia umbrosa Bunge;耐荫刺头菊■☆
108440 Cousinia vavilovii Kult.;瓦韦罗夫刺头菊■☆
108441 Cousinia verticillaris Bunge;轮生刺头菊■☆
108442 Cousinia vicaria Kult.;替代刺头菊■☆
108443 Cousinia vvedenskyl Tscherneva;韦氏刺头菊■☆
108444 Cousinia waldheimiana Bornm.;瓦尔刺头菊■☆
108445 Cousinia wolgensis C. A. Mey. ex DC. = Cousinia astracanica (Spreng.) Tamamsch. ■☆
108446 Cousinia wolgensis C. A. Mey. ex DC. var. affinis Regel = Cousinia affinis Schrenk ■
108447 Cousinia wolgensis C. A. Mey. var. affinis Regel = Cousinia affinis Schrenk ■
108448 Cousinia xanthina Bornm.;黄色刺头菊■☆
108449 Cousinia xanthiocephala Tscherneva;黄头刺头菊■☆
108450 Cousiniopsis Nevski(1937);蓝刺菊属■☆
108451 Cousiniopsis atractylodes (C. Winkl.) Nevski;蓝刺菊■☆
108452 Coussapoa Aubl. (1775);糙麻树属●☆
108453 Coussapoa asperifolia Trécul;糙麻树●☆
108454 Coussarea Aubl. (1775);热美茜属●☆
108455 Coussarea acuminata (Ruiz et Pav.) Zappi;渐尖热美茜●☆
108456 Coussarea albescens Müll. Arg.;渐白热美茜●☆
108457 Coussarea americana (L.) M. Gómez;美洲热美茜●☆
108458 Coussarea brevicaulis K. Krause;短茎热美茜●☆
108459 Coussarea chiriquiensis (Dwyer) C. M. Taylor = Psychotria nebulosa K. Krause ●☆
108460 Coussarea grandis Müll. Arg.;大热美茜●☆
108461 Coussarea nebulosa Dwyer = Psychotria nebulosa K. Krause ●☆
108462 Coutaportla Urb. (1923);美茜树属●☆
108463 Coutaportla Urb. = Portlandia P. Browne ●☆
108464 Coutaportla ghiesbreghtiana (Baill.) Urb.;美茜树●☆
108465 Coutarea Aubl. (1775);南美茜属(库塔茜属)●■☆
108466 Coutarea hexandra (Jacq.) Schum.;南美茜●☆
108467 Coutarea latifolia Moc. et Sessé ex DC.;宽花库塔茜(墨西哥南美茜)■☆
108468 Coutarea pterosperma Standl.;翅籽南美茜(翅籽库塔茜)■☆
108469 Coutareaceae Martinov = Rubiaceae Juss.(保留科名)●■
108470 Couthovia A. Gray = Neuburgia Blume ●☆
108471 Coutinia Vell.(废弃属名) = Aspidosperma Mart. et Zucc.(保留属名)●☆
108472 Coutiria Willis = Coutinia Vell.(废弃属名)●☆
108473 Coutoubaea Ham. = Coutoubea Aubl. ■☆
108474 Coutoubea Aubl. (1775);库塔龙胆属(库塔草属)■☆
108475 Coutoubea spicata Aubl.;穗花库塔龙胆■☆
108476 Coutoubeaceae Martinov = Gentianaceae Juss.(保留科名)●■
108477 Coutubea Steud. = Coutoubea Aubl. ■☆
108478 Covalia Rchb. = Covolia Neck. ex Raf. ●■
108479 Covalia Rchb. = Spermacoce L. ●■
108480 Covelia Endl. = Covalia Rchb. ●■
108481 Covellia Gasp. = Ficus L. ●
108482 Covellia cunia (Buch.-Ham. ex Roxb.) Miq. = Ficus semicostata (Buch.-Ham. ex Sm.) F. M. Bailey ●
108483 Covellia cunia (Buch.-Ham.) Miq. = Ficus semicordata Buch.-Ham. ex Sm. ●
108484 Covellia cyrtophylla Wall. ex Miq. = Ficus cyrtophylla (Wall. ex Miq.) Miq. ●
108485 Covellia glomerata (Roxb.) Miq. = Ficus racemosa L. ●
108486 Covellia hispida (L. f.) Miq. = Ficus hispida L. f. ●
108487 Covellia prostrata Miq. = Ficus prostrata Wall. ex Miq. ●
108488 Covellia prostrata Wall. ex Miq. = Ficus prostrata Wall. ex Miq. ●
108489 Covilhamia Korth. = Stixis Lour. ●
108490 Covillea Vail = Larrea Cav.(保留属名)●☆
108491 Covola Medik. = Salvia L. ●■
108492 Covolia Neck. ex Raf. = Spermacoce L. ●■
108493 Covolvulus tridentata L. = Xenostegia tridentata (L.) D. F. Austin et Staples ■
108494 Cowania D. Don = Purshia DC. ex Poir. ●☆
108495 Cowania D. Don ex Okamoto et Taylor = Cowania D. Don ●☆
108496 Cowania D. Don(1824);考恩蔷薇属●☆
108497 Cowania mexicana D. Don;墨西哥考恩蔷薇;Cliffrose,Ouininc Bush,Stansbury Cliffrose ●☆
108498 Cowellocassia Britton = Cassia L.(保留属名)●■
108499 Cowellocassia Britton = Crudia Schreb.(保留属名)●☆
108500 Cowiea Wernham = Hypobathrum Blume ●☆
108501 Cowiea Wernlmm = Petunga DC. ●☆
108502 Coxella Cheeseman et Hemsl. = Aciphylla J. R. Forst. et G. Forst. ■☆
108503 Coxia Endl. = Lysimachia L. ●■
108504 Crabbea Harv. (1838)(保留属名);克拉布爵床属■☆
108505 Crabbea acaulis N. E. Br.;无茎克拉布爵床■☆
108506 Crabbea albolutea Thulin;白黄克拉布爵床■☆
108507 Crabbea angustifolia Nees;窄叶克拉布爵床■☆
108508 Crabbea cirsioides (Nees) Nees = Crabbea hirsuta Harv. ■☆
108509 Crabbea galpinii C. B. Clarke;盖尔克拉布爵床■☆
108510 Crabbea hirsuta Harv.;粗毛克拉布爵床■☆
108511 Crabbea hirsuta Harv. var. somalensis Lindau = Crabbea hirsuta Harv. ■☆
108512 Crabbea kaessneri S. Moore;卡斯纳克拉布爵床■
108513 Crabbea longipes Mildbr.;长梗克拉布爵床■☆
108514 Crabbea migiurtina (Chiov.) Thulin;米朱蒂克拉布爵床■☆
108515 Crabbea nana Nees;矮小克拉布爵床■☆
108516 Crabbea ovalifolia Ficalho et Hiern = Crabbea nana Nees ■☆
108517 Crabbea pedunculata N. E. Br. = Crabbea nana Nees ■☆
108518 Crabbea pinnatifida Thulin;羽裂克拉布爵床■☆
108519 Crabbea reticulata C. B. Clarke = Crabbea velutina S. Moore ■☆
108520 Crabbea robusta N. E. Br. = Crabbea hirsuta Harv. ■☆
108521 Crabbea thymifolia (Chiov.) Thulin;百里香叶克拉布爵床■☆

108522 Crabbea undulatifolia Engl. = Crabbea angustifolia Nees ■☆
108523 Crabbea velutina S. Moore;短绒毛克拉布爵床■☆
108524 Crabowskia G. Don = Grabowskia Schltdl. ●☆
108525 Cracca Benth. (1853)(保留属名);大巢菜属●☆
108526 Cracca Benth. (保留属名) = Coursetia DC. ●☆
108527 Cracca Benth. ex Oerst. = Cracca Benth. (保留属名)●☆
108528 Cracca Hill = Vicia L. ■
108529 Cracca L. (废弃属名) = Cracca Benth. (保留属名)●☆
108530 Cracca L. (废弃属名) = Tephrosia Pers. (保留属名)●■
108531 Cracca acaciifolia (Baker) Kuntze = Tephrosia acaciifolia Welw. ex Baker ■☆
108532 Cracca bracteolata (Guillaumin et Perr.) Kuntze var. microfoliata Pires de Lima = Tephrosia reptans Baker var. microfoliata (Pires de Lima) Brummitt ■☆
108533 Cracca capensis (Pers.) Kuntze = Tephrosia capensis (Jacq.) Pers. ■☆
108534 Cracca glandulifera Benth.;大巢菜●☆
108535 Cracca latidens Small = Tephrosia virginiana (L.) Pers. ■☆
108536 Cracca lupinifolia (DC.) Kuntze = Tephrosia lupinifolia DC. ●☆
108537 Cracca mohrii Rydb. = Tephrosia virginiana (L.) Pers. ■☆
108538 Cracca purpurea L. = Tephrosia purpurea (L.) Pers. ●■
108539 Cracca semiglabra (Sond.) Kuntze = Tephrosia semiglabra Sond. ●☆
108540 Cracca villosa L. = Tephrosia villosa (L.) Pers. ■
108541 Cracca virginiana L. = Tephrosia virginiana (L.) Pers. ■☆
108542 Craccina Steven = Astragalus L. ●■
108543 Cracosna Gagnep. (1929);老挝龙胆属■☆
108544 Cracosna carinata (Dop) Thiv;龙骨老挝龙胆■☆
108545 Cracosna gracilis (Dop) Thiv;纤细老挝龙胆■☆
108546 Cracosna xyridiformis Gagnep.;老挝龙胆■☆
108547 Craepalia Schrank = Lolium L. ■
108548 Craepaloprumnon H. Karst. = Xylosma G. Forst. (保留属名)●
108549 Crafordia Raf. = Tephrosia Pers. (保留属名)●■
108550 Craibella R. M. K. Saunders, Y. C. F. Su et Chalermglin(2004);泰国番荔枝属●☆
108551 Craibia Dunn = Craibia Harms et Dunn ●☆
108552 Craibia Harms et Dunn(1911);克来豆属●☆
108553 Craibia affinis (De Wild.) De Wild.;近缘克来豆●☆
108554 Craibia atlantica Dunn;大西洋克来豆●☆
108555 Craibia bequaertii De Wild. = Millettia psilopetala Harms ●☆
108556 Craibia brevicaudata (Vatke) Dunn;短尾克来豆●☆
108557 Craibia brevicaudata (Vatke) Dunn subsp. baptistarum (Büttner) J. B. Gillett;染色克来豆●☆
108558 Craibia brevicaudata (Vatke) Dunn subsp. burttii (Baker f.) J. B. Gillett;伯特克来豆●☆
108559 Craibia brevicaudata (Vatke) Dunn subsp. schliebenii (Harms) J. B. Gillett;施利本克来豆●☆
108560 Craibia brownii Dunn;布朗克来豆●☆
108561 Craibia burttii Baker f. = Craibia brevicaudata (Vatke) Dunn subsp. burttii (Baker f.) J. B. Gillett ●☆
108562 Craibia elliotii Dunn = Craibia brownii Dunn ●☆
108563 Craibia filipes Dunn var. macrantha Pellegr. = Craibia macrantha (Pellegr.) J. B. Gillett ●☆
108564 Craibia gazensis (Baker f.) Baker f. = Craibia brevicaudata (Vatke) Dunn subsp. baptistarum (Büttner) J. B. Gillett ●☆
108565 Craibia grandiflora (Micheli) Baker f.;大花克来豆●☆
108566 Craibia laurentii (De Wild.) De Wild.;洛朗克来豆●☆
108567 Craibia lujae De Wild.;卢亚克来豆●☆
108568 Craibia macrantha (Pellegr.) J. B. Gillett;非洲大花克来豆●☆
108569 Craibia mildbreadii Harms = Craibia grandiflora (Micheli) Baker f. ●☆
108570 Craibia schliebenii Harms = Craibia brevicaudata (Vatke) Dunn subsp. schliebenii (Harms) J. B. Gillett ●☆
108571 Craibia simplex Dunn;简单克来豆●☆
108572 Craibia utilis M. B. Moss = Craibia laurentii (De Wild.) De Wild. ●☆
108573 Craibia zimmermannii (Harms) Dunn;齐默尔曼克来豆●☆
108574 Craibiodendron W. W. Sm. (1911);金叶子属(假木荷属,克檑树属,泡花树属);Craibiodendron, Goldleaf ●
108575 Craibiodendron calyculata (L.) Moench = Cassandra calyculata (L.) Moench ●
108576 Craibiodendron forrestii W. W. Sm.;怒江泡花树;Forrest Goldleaf, Nujiang Craibiodendron ●
108577 Craibiodendron forrestii W. W. Sm. = Quercus rehderiana Hand. -Mazz. ●
108578 Craibiodendron henryi W. W. Sm.;柳叶金叶子(毒药树,柳叶假木荷,柳叶泡花树);Henry Craibiodendron, Henry Goldleaf ●
108579 Craibiodendron kwangtungense S. Y. Hu = Craibiodendron scleranthum (Dop) Judd var. kwangtungense (S. Y. Hu) Judd ●
108580 Craibiodendron kwangtungense S. Y. Hu var. frutescens S. Y. Hu = Craibiodendron scleranthum (Dop) Judd var. kwangtungense (S. Y. Hu) Judd ●
108581 Craibiodendron mannii W. W. Sm. = Craibiodendron henryi W. W. Sm. ●
108582 Craibiodendron scleranthum (Dop) Judd;硬花金叶子;Hardflower Craibiodendron, Hardflower Goldleaf, Hard-flowered Craibiodendron ●
108583 Craibiodendron scleranthum (Dop) Judd var. kwangtungense (S. Y. Hu) Judd;广东金叶子(独牛角,广东假吊钟,广东假木荷,广东克檑木,广东泡花树,红皮紫陵,碎骨红);Guangdong Craibiodendron, Guangdong Goldleaf, Kwangtung Craibiodendron ●
108584 Craibiodendron shanicum W. W. Sm. = Craibiodendron stellatum (Pierre) W. W. Sm. ●
108585 Craibiodendron stellatum (Pierre) W. W. Sm.;金叶子(粗糠树,狗脚草,厚皮金叶子,火炭木,假木荷,老火树,美娥,泡花树,三百棒,小栗叶,星芒克檑木);Stellate Craibiodendron, Stellate Goldleaf ●
108586 Craibiodendron stellatum (Pierre) W. W. Sm. = Craibiodendron scleranthum (Dop) Judd var. kwangtungense (S. Y. Hu) Judd ●
108587 Craibiodendron yunnanense W. W. Sm.;云南金叶子(补骨灵,滇假木荷,毒羊叶,疯姑娘,狗脚草,果母,假吊钟,金叶子,紧羊叶,劳伤叶,麻虱子,马虱子,马虱子草,马虱子树,美娥,闹羊花,泡花树,细叶子,云南假木荷,云南克雷木,云南克檑木,云南泡花树);Yunnan Craibiodendron, Yunnan Goldleaf ●
108588 Craigia W. W. Sm. et Evans(1921);滇桐属;Craigia ●★
108589 Craigia kwangsiensis J. R. Xue;桂滇桐;Guangxi Craigia ●◇
108590 Craigia yunnanensis W. W. Sm. et Evans;滇桐;Yunnan Craigia ●◇
108591 Crambe L. (1753);两节荠属(甘比菜属,海边芥蓝属,海甘蓝属);Colewort, Crambe, Kale, Sea-kale ■
108592 Crambe abyssinica Hochst. ex R. E. Fr.;阿比西尼亚心叶海甘蓝■☆
108593 Crambe abyssinica Hochst. ex R. E. Fr. = Crambe hispanica L. ■☆
108594 Crambe abyssinica Hochst. ex R. E. Fr. var. meyeri O. E. Schulz

= Crambe hispanica L. ■☆
108595 Crambe arborea H. Christ;北方两节荠■☆
108596 Crambe arborea H. Christ var. indivisa Svent. = Crambe arborea H. Christ ■☆
108597 Crambe armena N. Busch;亚美尼亚荠■☆
108598 Crambe aspera M. Bieb.;糙海甘蓝■☆
108599 Crambe cordifolia Steven;心叶两节荠(心叶海甘蓝);Greater Sea-kale ■☆
108600 Crambe cordifolia Steven subsp. kotschyana (Boiss.) Jafri;科奇两节荠■☆
108601 Crambe cordifolia Steven subsp. kotschyana (Boiss.) Jafri = Crambe kotschyana Boiss. ■
108602 Crambe cordifolia Steven var. kotschyana (Boiss.) O. E. Schulz = Crambe kotschyana Boiss. ■
108603 Crambe cordifolia Steven var. kotschyana (Boiss.) O. E. Schulz = Crambe cordifolia Steven subsp. kotschyana (Boiss.) Jafri ■☆
108604 Crambe edentula Fisch. et C. A. Mey. ex Kar.;无齿两节荠■☆
108605 Crambe filiformis Jacq.;丝状两节荠■☆
108606 Crambe fruticosa L. f.;灌丛两节荠■☆
108607 Crambe gibberosa Rupr.;浅囊两节荠■☆
108608 Crambe gigantea (Ceballos et Ortuno) Bramwell;巨大两节荠■☆
108609 Crambe glabrata DC. = Crambe hispanica L. ■☆
108610 Crambe glauca ?;灰蓝两节荠;Billy Buttons ■☆
108611 Crambe gomeraea H. Christ;戈梅拉两节荠■☆
108612 Crambe gordjaginii Sprygin et Popov;高尔海甘蓝■☆
108613 Crambe grandiflora DC.;大花海甘蓝;Bigflower Colewort ■☆
108614 Crambe hispanica Ball = Crambe filiformis Jacq. ■☆
108615 Crambe hispanica L.;西班牙两节荠■☆
108616 Crambe hispanica L. subsp. glabrata (DC.) Cout.;光滑西班牙两节荠■☆
108617 Crambe hispanica L. var. glabrata (DC.) Coss. = Crambe hispanica L. ■☆
108618 Crambe juncea M. Bieb.;灯心草两节荠■☆
108619 Crambe kilimandscharica O. E. Schulz;基利两节荠■☆
108620 Crambe koktebelica (Junge) N. Busch;科克两节荠■☆
108621 Crambe kotschyana Boiss.;两节荠;Kotschy Colewort ■
108622 Crambe kotschyana Boiss. = Crambe cordifolia Steven var. kotschyana (Boiss.) O. E. Schulz ■
108623 Crambe kralikii Coss.;克拉利克两节荠■☆
108624 Crambe kralikii Coss. var. garamas Maire = Crambe kralikii Coss. ■☆
108625 Crambe laevigata H. Christ;光滑两节荠■☆
108626 Crambe litwinowii Grossh.;利特氏海甘蓝(李文氏海甘蓝);Litwinov Colewort ■☆
108627 Crambe maritima L.;海甘蓝(浜菜,滨菜);Cole, Colewort, Common Colewort, Sea Cabbage, Sea Colewort, Sea Kale, Sea Keele, Sea-kale, Strand Cabbage ■☆
108628 Crambe orientalis L.;东方海甘蓝■☆
108629 Crambe pinnatifida R. Br.;羽裂海甘蓝;Pinnatifid Colewort ■☆
108630 Crambe pontica Steven ex Rupr.;黑海海甘蓝■☆
108631 Crambe pritzelii Bolle;普里特两节荠■☆
108632 Crambe reniformis Desf. = Crambe filiformis Jacq. ■☆
108633 Crambe scaberrima Bramwell;粗糙两节荠■☆
108634 Crambe schugnana Korsh.;舒格南两节荠■☆
108635 Crambe scoparia Svent.;帚状两节荠■☆
108636 Crambe sinuato-dentata F. Petri;深波齿两节荠■☆
108637 Crambe steveniana Rupr.;司梯氏海甘蓝;Steven Colewort ■☆
108638 Crambe strigosa L'Hér.;糙伏毛两节荠■☆
108639 Crambe strigosa L'Hér. var. gigantea Ceballos et Ortuno = Crambe gigantea (Ceballos et Ortuno) Bramwell ■☆
108640 Crambe sventenii Bramwell et Sunding;斯文顿两节荠■☆
108641 Crambe tataria Sebeok;鞑靼海甘蓝;Tatar Colewort, Tatarian Bread ■☆
108642 Crambe teretifolia Batt. et Trab. = Crambella teretifolia (Batt. et Trab.) Maire ■☆
108643 Crambe tetuanensis Pit. = Crambe hispanica L. subsp. glabrata (DC.) Cout. ■☆
108644 Crambella Maire(1924);小两节荠属(柱叶荠属)■☆
108645 Crambella teretifolia (Batt. et Trab.) Maire;小两节荠■☆
108646 Crambella teretifolia (Batt.) Maire = Crambella teretifolia (Batt. et Trab.) Maire ■☆
108647 Crameria Murr. = Krameria L. ex Loefl. ●■☆
108648 Crangonorchis D. L. Jones et M. A. Clem. (2002);澳洲虾兰属■☆
108649 Crangonorchis D. L. Jones et M. A. Clem. = Pterostylis R. Br. (保留属名)■☆
108650 Cranichis Sw. (1788);宝石兰属;Helmet Orchid, Jewel Orchid ■☆
108651 Cranichis gracilis L. O. Williams;小宝石兰■☆
108652 Cranichis luteola Sw. = Polystachya concreta (Jacq.) Garay et H. R. Sweet ■
108653 Cranichis micrantha Spreng. = Prescottia oligantha (Sw.) Lindl. ■☆
108654 Cranichis muscosa Sw.;宝石兰;Jewel Orchid ■☆
108655 Cranichis nudifolia (Lour.) Pers. = Galeola nudifolia Lour. ■
108656 Cranichis oligantha Sw. = Prescottia oligantha (Sw.) Lindl. ■☆
108657 Craniolaria L. (1753);长管角胡麻属■☆
108658 Craniolaria annua L.;一年长管角胡麻■☆
108659 Craniolaria fragrans Schenk;香长管角胡麻■☆
108660 Craniolaria fruticosa L.;灌木状长管角胡麻■☆
108661 Craniospermum Lehm. (1818);颅果草属;Craniospermum, Headfruit ■
108662 Craniospermum canescens DC.;灰颅果草■☆
108663 Craniospermum echioides (Schrenk) Bunge = Craniospermum mongolicum I. M. Johnst. ■
108664 Craniospermum hirsutum DC.;毛颅果草■☆
108665 Craniospermum mongolicum I. M. Johnst.;颅果草;Echiumlike Craniospermum, Headfruit ■
108666 Craniospermum subfloccosum Krylov;卷毛颅果草■
108667 Craniotome Rchb. (1825);簇序草属(颅萼草属);Craniotome ■
108668 Craniotome furcata (Link) Kuntze;簇序草;Furcate Craniotome ■
108669 Craniotome versicolor Rchb.;变色簇序草;Versicolor Craniotome ■
108670 Craniotome versicolor Rchb. = Craniotome furcata (Link) Kuntze ■
108671 Cranocarpus Benth. (1859);巴西盔豆属■☆
108672 Cranocarpus gracilis Afr. Fern. et P. Bezerra;细巴西盔豆■☆
108673 Cranocarpus martii Benth.;巴西盔豆■☆
108674 Crantzia DC. = Cranzia Schreb. ●
108675 Crantzia DC. = Toddalia Juss. (保留属名)●
108676 Crantzia Lag. ex DC. = Conringia Adans. + Moricandia DC. ■☆
108677 Crantzia Nutt. = Crantziola F. Muell. ■☆
108678 Crantzia Nutt. = Lilaeopsis Greene ■☆
108679 Crantzia Scop. (废弃属名) = Alloplectus Mart. (保留属名)●■☆

108680 Crantzia Sw. = Buxus L. ●
108681 Crantzia Sw. = Tricera Schreb. ●
108682 Crantzia Vell. = Centratherum Cass. ■☆
108683 Crantzia asiatica Kuntze = Toddalia asiatica (L.) Lam. ●
108684 Crantziola F. Muell. = Lilaeopsis Greene ■☆
108685 Crantziola F. Muell. ex Koso-Pol. = Lilaeopsis Greene ■☆
108686 Crantziola Koso-Pol. = Hydrocotyle L. ■
108687 Cranzia J. F. Gmel. = Buxus L. ●
108688 Cranzia J. F. Gmel. = Crantzia Sw. ●
108689 Cranzia Schreb. = Toddalia Juss. (保留属名)●
108690 Cranzia angolensis Hiern = Vepris hiernii Gereau ●☆
108691 Crasanloma D. Dietr. = Geissoloma Lindl. ex Kunth ●☆
108692 Craspedia G. Forst. (1786);金杖球属(金绣球属)■☆
108693 Craspedia globosa (Benth.) Benth.;槌金杖球;Drumsticks ■☆
108694 Craspedia incana Cockayne et Allan;白毛金杖球■☆
108695 Craspedia uniflora G. Forst.;单花金杖球;Billy Buttons ■☆
108696 Craspedolepis Steud. = Restio Rottb. (保留属名)■☆
108697 Craspedolepis verreauxii Steud. = Restio filiformis Poir. ■☆
108698 Craspedolobium Harms (1921);巴豆藤属;Craspedolobium, Crotonvine ●★
108699 Craspedolobium schochii Harms = Craspedolobium unijugum (Gagnep.) Z. Wei et Pedley ●
108700 Craspedolobium unijugum (Gagnep.) Z. Wei et Pedley;巴豆藤(黑藤,禄劝鸡血藤,三叶藤,铁藤,铁血藤,血藤);Schoch Craspedolobium, Schoch Crotonvine ●
108701 Craspedorhachis Benth. (1882);流苏舌草属■☆
108702 Craspedorhachis africana Benth.;非洲流苏舌草■☆
108703 Craspedorhachis digitata Kupicha et Cope;指裂流苏舌草■☆
108704 Craspedorhachis menyharthii Hack. = Leptochloa uniflora Hochst. ex A. Rich. ■☆
108705 Craspedorhachis rhodesiaca Rendle var. gracilior C. E. Hubb. = Craspedorhachis rhodesiana Rendle ■☆
108706 Craspedorhachis rhodesiana Rendle;罗得西亚流苏舌草■☆
108707 Craspedorhachis sarmentosa (Hack.) Pilg. = Willkommia sarmentosa Hack. ■☆
108708 Craspedorhachis uniflora (Hochst. ex A. Rich.) Chippind. = Leptochloa uniflora Hochst. ex A. Rich. ■☆
108709 Craspedospermum Airy Shaw = Craspidospermum Bojer ex DC. ●☆
108710 Craspedospermum Bojer ex DC. = Craspidospermum Bojer ex DC. ●☆
108711 Craspedostoma Domke = Gnidia L. ●☆
108712 Craspedostoma Domke(1934);缘口香属●☆
108713 Craspedostoma linoides (Wikstr.) Domke;缘口香●☆
108714 Craspedostoma pubescens (P. J. Bergius) Domke;毛缘口香●☆
108715 Craspedum Lour. = Elaeocarpus L. ●
108716 Craspidospermum Bojer ex DC. (1844);轮生夹竹桃属●☆
108717 Craspidospermum verticillatum Bojer ex A. DC.;轮生夹竹桃●☆
108718 Craspidospermum verticillatum Bojer ex A. DC. var. petiolare A. DC. = Craspidospermum verticillatum Bojer ex A. DC. ●☆
108719 Craspidospermum verticillatum Bojer ex A. DC. var. sessile Markgr. = Craspidospermum verticillatum Bojer ex A. DC. ●☆
108720 Crassangis Thouars = Angraecum Bory ■
108721 Crassina Scepin(废弃属名) = Zinnia L. (保留属名)●■
108722 Crassipes Swallen = Sclerochloa P. Beauv. ■
108723 Crassocephalum Moench(1794)(废弃属名);野茼蒿属(木耳菜属,昭和草属);Velvet Plant ■
108724 Crassocephalum Moench(废弃属名) = Gynura Cass. (保留属名)■●
108725 Crassocephalum afromontanum R. E. Fr. = Crassocephalum montuosum (S. Moore) Milne-Redh. ■☆
108726 Crassocephalum amplexicaulis (Oliv. et Hiern) S. Moore = Gynura amplexicaulis Oliv. et Hiern ■☆
108727 Crassocephalum auriformis S. Moore = Gynura scandens O. Hoffm. ■☆
108728 Crassocephalum baoulense (Hutch. et Dalziel) Milne-Redh.;巴乌莱野茼蒿■☆
108729 Crassocephalum behmianum (Muschl.) S. Moore = Crassocephalum ducis-aprutii (Chiov.) S. Moore ■☆
108730 Crassocephalum bojeri (DC.) Robyns = Solanecio angulatus (Vahl) C. Jeffrey ■☆
108731 Crassocephalum bumbense S. Moore = Crassocephalum montuosum (S. Moore) Milne-Redh. ■☆
108732 Crassocephalum butagensis (Muschl.) S. Moore = Crassocephalum montuosum (S. Moore) Milne-Redh. ■☆
108733 Crassocephalum cernuum (L. f.) Moench = Crassocephalum rubens (Juss. ex Jacq.) S. Moore ■
108734 Crassocephalum cernuum Moench = Crassocephalum rubens (Juss. ex Jacq.) S. Moore ■
108735 Crassocephalum coeruleum (O. Hoffm.) R. E. Fr.;天蓝野茼蒿■☆
108736 Crassocephalum crepidioides (Benth.) S. Moore;野茼蒿(安南草,冬风菜,飞机菜,革命菜,假茼蒿,满天飞,山茼蒿,昭和草);Hawksbeard Velvetplant, Redflower Ragleaf, Thickhead ■
108737 Crassocephalum diversifolium Hiern = Crassocephalum crepidioides (Benth.) S. Moore ■
108738 Crassocephalum ducis-aprutii (Chiov.) S. Moore;阿普鲁特野茼蒿■☆
108739 Crassocephalum effusum (Mattf.) C. Jeffrey;开展野茼蒿■☆
108740 Crassocephalum flavum Decne. = Senecio flavus (Decne.) Sch. Bip. ■☆
108741 Crassocephalum goetzenii (O. Hoffm.) S. Moore;格兹野茼蒿■☆
108742 Crassocephalum gossweileri S. Moore = Emilia gossweileri (S. Moore) C. Jeffrey ■☆
108743 Crassocephalum gracile (Hook. f.) Milne-Redh. ex Guinea;纤细野茼蒿■☆
108744 Crassocephalum guineense C. D. Adams;几内亚野茼蒿■☆
108745 Crassocephalum heteromorphum (Hutch. et B. L. Burtt) C. Jeffrey = Crassocephalum radiatum S. Moore ■☆
108746 Crassocephalum kassneri S. Moore = Crassocephalum kassnerianum (Muschl.) Lisowski ■☆
108747 Crassocephalum kassnerianum (Muschl.) Lisowski;卡斯纳野茼蒿■☆
108748 Crassocephalum libericum S. Moore;离生野茼蒿■☆
108749 Crassocephalum longirameum S. Moore = Emilia longiramea (S. Moore) C. Jeffrey ■☆
108750 Crassocephalum luteum (Humb.) Humb. = Crassocephalum montuosum (S. Moore) Milne-Redh. ■☆
108751 Crassocephalum macropappum (Sch. Bip. ex A. Rich.) S. Moore;大毛野茼蒿■☆
108752 Crassocephalum mannii (Hook. f.) Milne-Redh. = Solanecio mannii (Hook. f.) C. Jeffrey ■☆
108753 Crassocephalum montuosum (S. Moore) Milne-Redh.;山区野茼蒿■☆

108754 Crassocephalum multicorymbosum (Klatt) S. Moore = Solanecio mannii (Hook. f.) C. Jeffrey ■☆
108755 Crassocephalum notonioides S. Moore = Kleinia abyssinica (A. Rich.) A. Berger var. hildebrandtii (Vatke) C. Jeffrey ■☆
108756 Crassocephalum paludum C. Jeffrey;沼泽野茼蒿■☆
108757 Crassocephalum picridifolium (DC.) S. Moore;直梗野茼蒿■☆
108758 Crassocephalum radiatum S. Moore;辐射野茼蒿■☆
108759 Crassocephalum rubens (Juss. ex Jacq.) S. Moore;昭和草(红三七)■
108760 Crassocephalum rubens (Juss. ex Jacq.) S. Moore var. sarcobasis (DC.) C. Jeffrey et Beentje;肉基昭和草■
108761 Crassocephalum ruwenzoriensis S. Moore = Gynura scandens O. Hoffm. ■☆
108762 Crassocephalum sarcobasis (DC.) S. Moore = Crassocephalum rubens (Juss. ex Jacq.) S. Moore var. sarcobasis (DC.) C. Jeffrey et Beentje ■
108763 Crassocephalum scandens (O. Hoffm.) Hiern = Gynura scandens O. Hoffm. ■☆
108764 Crassocephalum sonchifolium (L.) Less. = Emilia sonchifolia (L.) DC. ex Wight ■
108765 Crassocephalum splendens C. Jeffrey;光亮野茼蒿■☆
108766 Crassocephalum subscandens (Hochst. ex A. Rich.) S. Moore = Solanecio angulatus (Vahl) C. Jeffrey ■☆
108767 Crassocephalum togoense C. D. Adams;多哥野茼蒿■☆
108768 Crassocephalum torreanum Lisowski;托尔野茼蒿■☆
108769 Crassocephalum uvens (Hiern) S. Moore;葡萄野茼蒿■☆
108770 Crassocephalum vitellinum (Benth.) S. Moore;蛋黄色野茼蒿■☆
108771 Crassopetalum Northrop = Crossopetalum P. Browne ●☆
108772 Crassopetalum Northrop = Myginda Jacq. ●☆
108773 Crassouvia Comm. ex DC. = Bryophyllum Salisb. ■
108774 Crassouvia Comm. ex DC. = Crassuvia Comm. ex Lam. ●■
108775 Crassula L. (1753);青锁龙属(厚叶属);Crassula, Letter Flower, Pygmyweed ●■☆
108776 Crassula abyssinica A. Rich. = Crassula alba Forssk. ■☆
108777 Crassula abyssinica A. Rich. var. vaginata (Eckl. et Zeyh.) Engl. = Crassula vaginata Eckl. et Zeyh. ■☆
108778 Crassula acinaciformis Schinz;长刀形青锁龙●☆
108779 Crassula acuminata E. Mey. ex Drège = Crassula flanaganii Schönland et Baker f. ●☆
108780 Crassula acutifolia Lam. = Crassula tetragona L. subsp. acutifolia (Lam.) Toelken ●☆
108781 Crassula acutifolia Lam. var. densifolia (Harv.) Schönland = Crassula tetragona L. ●☆
108782 Crassula acutifolia Lam. var. harveyi Schönland = Crassula tetragona L. subsp. lignescens Toelken ●☆
108783 Crassula acutifolia Lam. var. radicans Harv. = Crassula tetragona L. subsp. acutifolia (Lam.) Toelken ●☆
108784 Crassula acutifolia Lam. var. typica Schönland = Crassula tetragona L. subsp. acutifolia (Lam.) Toelken ●☆
108785 Crassula adscendens Thunb.;上举青锁龙●☆
108786 Crassula aitonii Britten et Baker f. = Crassula cordata Thunb. ■☆
108787 Crassula alata (Viv.) A. Berger = Tillaea alata Viv. ■
108788 Crassula alata (Viv.) A. Berger subsp. pharnaceoides (Fisch. et C. A. Mey.) Wickens et Bywater;线叶粟草青锁龙■☆
108789 Crassula alata (Viv.) A. Berger subsp. pharnaceoides (Fisch. et C. A. Mey.) Wickens et Bywater = Tillaea alata Viv. ■
108790 Crassula alata (Viv.) A. Berger var. trichopoda (Fenzl) Post = Crassula alata (Viv.) A. Berger ■
108791 Crassula alba Forssk.;白青锁龙■☆
108792 Crassula alba Forssk. var. pallida Toelken;苍白青锁龙■☆
108793 Crassula alba Forssk. var. parvisepala (Schönland) Toelken;小瓣青锁龙■☆
108794 Crassula albanensis Schönland = Crassula capitella Thunb. ■☆
108795 Crassula albertiniae Schönland = Crassula capensis (L.) Baill. var. albertiniae (Schönland) Toelken ■☆
108796 Crassula albicaulis Harv. = Crassula expansa Dryand. ■☆
108797 Crassula albiflora Sims = Crassula dejecta Jacq. ■☆
108798 Crassula albiflora Sims var. minor Schönland = Crassula dejecta Jacq. ■☆
108799 Crassula alcicornis Schönland;尖角青锁龙■☆
108800 Crassula aliciae Raym. -Hamet = Kungia aliciae (Raym. -Hamet) K. T. Fu ■
108801 Crassula aliciae Raym. -Hamet = Orostachys aliciae (Raym. -Hamet) H. Ohba ■
108802 Crassula aloides N. E. Br. = Crassula acinaciformis Schinz ●☆
108803 Crassula alooides Dryand. = Crassula hemisphaerica Thunb. ■☆
108804 Crassula alpestris Thunb. subsp. massonii (Britten et Baker f.) Toelken = Crassula alpestris Thunb. ■☆
108805 Crassula alpina (Eckl. et Zeyh.) Walp. = Crassula umbellata Thunb. ■☆
108806 Crassula alsinoides (Hook. f.) Engl. = Crassula pellucida L. subsp. alsinoides (Hook. f.) Toelken ■☆
108807 Crassula alstonii Marloth;阿尔斯顿青锁龙■☆
108808 Crassula alternifolia L.;异叶青锁龙■☆
108809 Crassula alticola R. Fern.;高原青锁龙■☆
108810 Crassula ammophila Toelken;喜沙青锁龙■☆
108811 Crassula anguina Harv. = Crassula muscosa L. ●☆
108812 Crassula anomala Schönland et Baker f. = Crassula atropurpurea (Haw.) D. Dietr. var. anomala (Schönland et Baker f.) Toelken ●☆
108813 Crassula anthurus E. Mey. ex Drège = Crassula perforata Thunb. ●☆
108814 Crassula aphylla Schönland et Baker f.;无叶青锁龙■☆
108815 Crassula aquatica (L.) Schönland = Tillaea aquatica L. ■
108816 Crassula arborea L. = Portulacaria afra Jacq. ●☆
108817 Crassula arborea Medik. = Crassula arborescens (Mill.) Willd. ●■☆
108818 Crassula arborescens (Mill.) Willd.;亚高青锁龙(花月,玉树);Beestebul, Jade Plant, Jade Tree, Silver Dollar Plant, Silver Jade Plant ●■☆
108819 Crassula arborescens (Mill.) Willd. subsp. undulatifolia Toelken;波叶亚高青锁龙●☆
108820 Crassula archeri Compton = Crassula pyramidalis Thunb. ●☆
108821 Crassula arenicola Toelken = Crassula cymosa P. J. Bergius ■☆
108822 Crassula argentea L. f. = Crassula ovata Druce ●☆
108823 Crassula argentea Thunb. = Crassula ovata (Mill.) Druce ●☆
108824 Crassula argyrophylla Diels ex Schönland et Baker f. = Crassula swaziensis Schönland ■☆
108825 Crassula argyrophylla Diels ex Schönland et Baker f. var. ramosa Schönland = Crassula swaziensis Schönland ■☆
108826 Crassula argyrophylla Diels ex Schönland et Baker f. var. swaziensis (Schönland) Schönland = Crassula swaziensis Schönland ■☆
108827 Crassula aristata Schönland = Crassula bergioides Harv. ■☆
108828 Crassula arta Schönland;龙宫城■☆

108829 Crassula arta Schönland = Crassula deceptor Schönland et Baker f. ●☆
108830 Crassula articulata Zuccagni = Crassula ovata (Mill.) Druce ●☆
108831 Crassula atropurpurea (Haw.) D. Dietr.;深紫青锁龙●☆
108832 Crassula atropurpurea (Haw.) D. Dietr. = Crassula obliqua Haw. ●☆
108833 Crassula atropurpurea (Haw.) D. Dietr. var. anomala (Schönland et Baker f.) Toelken;异常青锁龙●☆
108834 Crassula atropurpurea (Haw.) D. Dietr. var. cultriformis (Friedrich) Toelken;刀形深紫青锁龙■☆
108835 Crassula atropurpurea (Haw.) D. Dietr. var. muirii (Schönland) R. Fern.;缪里青锁龙■☆
108836 Crassula atropurpurea (Haw.) D. Dietr. var. rubella (Compton) Toelken = Crassula atropurpurea (Haw.) D. Dietr. var. muirii (Schönland) R. Fern. ■☆
108837 Crassula atropurpurea (Haw.) D. Dietr. var. watermeyeri (Compton) Toelken;沃特迈耶青锁龙■☆
108838 Crassula atrosanguinea Beauverd = Crassula alba Forssk. var. parvisepala (Schönland) Toelken ■☆
108839 Crassula aurosensis Dinter = Crassula exilis Harv. subsp. sedifolia (N. E. Br.) Toelken ●☆
108840 Crassula ausensis Hutchison;奥斯青锁龙■☆
108841 Crassula ausensis Hutchison subsp. giessii (Friedrich) Toelken;吉斯青锁龙■☆
108842 Crassula avasimontana Dinter = Crassula capitella Thunb. subsp. nodulosa (Schönland) Toelken ■☆
108843 Crassula bakeri Schönland = Crassula grisea Schönland ■☆
108844 Crassula barbata Thunb.;髯毛青锁龙(月光)■☆
108845 Crassula barklyana Schönland = Crassula setulosa Harv. var. rubra (N. E. Br.) G. D. Rowley ■☆
108846 Crassula barklyi N. E. Br.;圆柱青锁龙(玉椿);Rattlesnake Tail,Wurmplakkie ■☆
108847 Crassula bartlettii Schönland = Crassula capensis (L.) Baill. var. albertiniae (Schönland) Toelken ■☆
108848 Crassula basutica Schönland = Crassula dependens Bolus ●☆
108849 Crassula bergioides Harv.;田繁缕青锁龙■☆
108850 Crassula bibracteata Haw. = Crassula tetragona L. subsp. acutifolia (Lam.) Toelken ●☆
108851 Crassula biconvexa (Eckl. et Zeyh.) Harv. = Crassula pubescens Thunb. ●☆
108852 Crassula biconvexa Haw. = Crassula fascicularis Lam. ■☆
108853 Crassula biplanata Haw.;扁青锁龙●☆
108854 Crassula bloubergensis R. A. Dyer = Crassula setulosa Harv. ■☆
108855 Crassula bolusii Hook. = Crassula exilis Harv. subsp. cooperi (Regel) Toelken ●☆
108856 Crassula brachypetala Drège ex Harv. = Crassula pellucida L. subsp. brachypetala (Drège ex Harv.) Toelken ●☆
108857 Crassula brachyphylla Adamson = Crassula decumbens Thunb. var. brachyphylla (Adamson) Toelken ●☆
108858 Crassula brachystachya Toelken;短穗青锁龙●☆
108859 Crassula brevifolia (Eckl. et Zeyh.) Schönland = Crassula decumbens Thunb. var. brachyphylla (Adamson) Toelken ●☆
108860 Crassula brevifolia Harv.;短叶青锁龙●☆
108861 Crassula brevifolia Harv. subsp. psammophila Toelken;喜沙短叶青锁龙●☆
108862 Crassula brevistyla Baker f. = Crassula capitella Thunb. subsp. meyeri (Harv.) Toelken ■☆
108863 Crassula browniana Burtt Davy = Crassula expansa Dryand. subsp. filicaulis (Haw.) Toelken ■☆
108864 Crassula bullulata Haw. = Crassula flava L. ●☆
108865 Crassula burmanniana (Eckl. et Zeyh.) D. Dietr. = Crassula flava L. ●☆
108866 Crassula caerulata J. F. Gmel. = Crassula crenulata Thunb. ■☆
108867 Crassula caffra L.;开菲尔青锁龙●☆
108868 Crassula campestris (Eckl. et Zeyh.) Endl. ex Walp.;田野青锁龙●☆
108869 Crassula campestris (Eckl. et Zeyh.) Endl. ex Walp. f. compacta Schönland = Crassula campestris (Eckl. et Zeyh.) Endl. ex Walp. ●☆
108870 Crassula campestris (Eckl. et Zeyh.) Endl. ex Walp. f. laxa Schönland = Crassula campestris (Eckl. et Zeyh.) Endl. ex Walp. ●☆
108871 Crassula campestris (Eckl. et Zeyh.) Endl. ex Walp. subsp. pharnaceoides (Fisch. et C. A. Mey.) Toelken = Crassula alata (Viv.) A. Berger subsp. pharnaceoides (Fisch. et C. A. Mey.) Wickens et Bywater ■☆
108872 Crassula campestris (Eckl. et Zeyh.) Endl. ex Walp. subsp. rhodesica (Merxm.) R. Fern. = Crassula rhodesica (Merxm.) Wickens et M. Bywater ■☆
108873 Crassula canescens (Haw.) Schult. = Crassula nudicaulis L. ■☆
108874 Crassula canescens (Haw.) Schult. var. angustifolia (Eckl. et Zeyh.) Harv. = Crassula nudicaulis L. ■☆
108875 Crassula capensis (L.) Baill.;好望角青锁龙●■☆
108876 Crassula capensis (L.) Baill. var. albertiniae (Schönland) Toelken;艾伯蒂尼亚青锁龙■☆
108877 Crassula capensis (L.) Baill. var. promontorii (Schönland et Baker f.) Toelken;普罗青锁龙■☆
108878 Crassula capillacea E. Mey. ex Drège = Crassula filiformis (Eckl. et Zeyh.) D. Dietr. ■☆
108879 Crassula capitata Lam. = Crassula subulata L. ■☆
108880 Crassula capitata Lodd. = Crassula fascicularis Lam. ■☆
108881 Crassula capitella Thunb.;小头青锁龙■☆
108882 Crassula capitella Thunb. subsp. enantiophylla (Baker f.) Toelken = Crassula capitella Thunb. subsp. nodulosa (Schönland) Toelken ■☆
108883 Crassula capitella Thunb. subsp. meyeri (Harv.) Toelken;迈尔青锁龙■☆
108884 Crassula capitella Thunb. subsp. nodulosa (Schönland) Toelken;长萼青锁龙■☆
108885 Crassula capitella Thunb. subsp. thyrsiflora (Thunb.) Toelken;锥花青锁龙;Aanteel-poprosie ■☆
108886 Crassula capitella Thunb. subsp. thyrsiflora (Thunb.) Toelken = Crassula thyrsiflora Thunb. ■☆
108887 Crassula centauroides L. = Crassula strigosa L. ■☆
108888 Crassula centauroides L. var. marginalis (Dryand.) Harv. = Crassula pellucida L. subsp. marginalis (Dryand.) Toelken ■☆
108889 Crassula cephalophora Thunb. = Crassula nudicaulis L. ■☆
108890 Crassula cephalophora Thunb. var. basutica Schönland = Crassula nudicaulis L. ■☆
108891 Crassula cephalophora Thunb. var. dubia (Schönland) Schönland = Crassula cotyledonis Thunb. ■☆
108892 Crassula cephalophora Thunb. var. tayloriae (Schönland) Schönland = Crassula cotyledonis Thunb. ■☆
108893 Crassula cephalophora Thunb. var. thunbergii Schönland = Crassula nudicaulis L. ■☆

108894 Crassula chloraeflora (Haw.) D. Dietr. = Crassula dichotoma L. ●☆

108895 Crassula ciliata L.;睫毛青锁龙■☆

108896 Crassula ciliata L. var. acutifolia E. Mey. ex Drège = Crassula vaginata Eckl. et Zeyh. ■☆

108897 Crassula cinerea Friedrich;灰色青锁龙■☆

108898 Crassula clavata N. E. Br.;棍棒青锁龙■☆

108899 Crassula clavifolia Harv. = Crassula atropurpurea (Haw.) D. Dietr. ●☆

108900 Crassula clavifolia Harv. var. muirii Schönland = Crassula atropurpurea (Haw.) D. Dietr. var. muirii (Schönland) R. Fern. ■☆

108901 Crassula coccinea L.;杜神刀●☆

108902 Crassula coerulescens Schönland = Crassula nemorosa (Eckl. et Zeyh.) Endl. ex Walp. ■☆

108903 Crassula cogmansensis (Kuntze) K. Schum. = Crassula subaphylla (Eckl. et Zeyh.) Harv. ■☆

108904 Crassula columella Marloth et Schönland;小柱青锁龙;Silinderplakkie ■☆

108905 Crassula columnaris Thunb.;柱状青锁龙(丽人);Koesnaatjie ■☆

108906 Crassula columnaris Thunb. subsp. prolifera Friedrich;多育青锁龙■☆

108907 Crassula columnaris Thunb. var. elongata E. Mey. ex Drège = Crassula columnaris Thunb. subsp. prolifera Friedrich ■☆

108908 Crassula commutata Friedrich = Crassula rupestris Thunb. subsp. commutata (Friedrich) Toelken ●☆

108909 Crassula compacta Schönland;紧密青锁龙●☆

108910 Crassula compacta Schönland var. elatior Baker f. = Crassula compacta Schönland ●☆

108911 Crassula comptonii Hutchison et Pillans = Crassula namaquensis Schönland et Baker f. subsp. comptonii (Hutch. et Pillans) Toelken ■☆

108912 Crassula confusa Schönland et Baker f. = Crassula nemorosa (Eckl. et Zeyh.) Endl. ex Walp. ■☆

108913 Crassula congesta N. E. Br.;粗三角青锁龙●☆

108914 Crassula congesta N. E. Br. subsp. laticephala (Schönland) Toelken;宽头粗三角青锁龙●☆

108915 Crassula conjuncta N. E. Br. = Crassula perforata Thunb. ●☆

108916 Crassula connivens Schönland = Crassula tetragona L. subsp. connivens (Schönland) Toelken ●☆

108917 Crassula conspicua Haw. = Crassula tomentosa Thunb. ■☆

108918 Crassula cooperi Regel;库珀青锁龙(乙姬)●☆

108919 Crassula cooperi Regel = Crassula exilis Harv. subsp. cooperi (Regel) Toelken ●☆

108920 Crassula cooperi Regel var. subnodulosa R. Fern.;亚节库珀青锁龙●☆

108921 Crassula corallina Thunb.;白妙■☆

108922 Crassula corallina Thunb. subsp. macrorrhiza Toelken;大根白妙■☆

108923 Crassula cordata Thunb.;心形青锁龙■☆

108924 Crassula cordifolia Baker;心叶青锁龙■☆

108925 Crassula cornuta Schönland et Baker f.;梦殿■☆

108926 Crassula cornuta Schönland et Baker f. = Crassula deceptor Schönland et Baker f. ●☆

108927 Crassula corpuscularioptsis Boom = Crassula elegans Schönland et Baker f. ■☆

108928 Crassula corymbulosa Link et Otto = Crassula capitella Thunb. subsp. thyrsiflora (Thunb.) Toelken ■☆

108929 Crassula corymbulosa Link et Otto var. cordata Schönland = Crassula capitella Thunb. subsp. thyrsiflora (Thunb.) Toelken ■☆

108930 Crassula corymbulosa Link et Otto var. lanceolata Schönland = Crassula capitella Thunb. subsp. thyrsiflora (Thunb.) Toelken ■☆

108931 Crassula corymbulosa Link et Otto var. major Schönland = Crassula capitella Thunb. subsp. thyrsiflora (Thunb.) Toelken ■☆

108932 Crassula corymbulosa Link et Otto var. typica Schönland = Crassula capitella Thunb. subsp. thyrsiflora (Thunb.) Toelken ■☆

108933 Crassula cotyledon Jacq. = Crassula arborescens (Mill.) Willd. ●■☆

108934 Crassula cotyledonifolia Salisb. = Crassula arborescens (Mill.) Willd. ●■☆

108935 Crassula cotyledonis Thunb.;瓦松青锁龙■☆

108936 Crassula crassiflora (Kuntze) K. Schum. = Crassula vaginata Eckl. et Zeyh. ■☆

108937 Crassula cremnophila Van Jaarsv. et A. E. van Wyk;悬崖青锁龙■☆

108938 Crassula crenatifolia Baker f. = Crassula umbraticola N. E. Br. ■☆

108939 Crassula crenulata Thunb.;细圆齿青锁龙■☆

108940 Crassula cultrata L.;刀形青锁龙;Plakkiebos ■☆

108941 Crassula cultrata L. var. typica Schönland = Crassula cultrata L. ■☆

108942 Crassula cultriformis Friedrich = Crassula atropurpurea (Haw.) D. Dietr. var. cultriformis (Friedrich) Toelken ■☆

108943 Crassula cultriformis Friedrich var. robusta ? = Crassula atropurpurea (Haw.) D. Dietr. var. cultriformis (Friedrich) Toelken ■☆

108944 Crassula curta N. E. Br. = Crassula setulosa Harv. var. rubra (N. E. Br.) G. D. Rowley ■☆

108945 Crassula curta N. E. Br. var. rubra ? = Crassula setulosa Harv. var. rubra (N. E. Br.) G. D. Rowley ■☆

108946 Crassula cyclophylla Schönland et Baker f. = Crassula spathulata Thunb. ●☆

108947 Crassula cylindrica Schönland = Crassula pyramidalis Thunb. ●☆

108948 Crassula cymbiformis Toelken;船状青锁龙■☆

108949 Crassula cymosa P. J. Bergius;聚伞青锁龙■☆

108950 Crassula dasyphylla Harv. = Crassula corallina Thunb. ■☆

108951 Crassula debilis Thunb. = Crassula thunbergiana Schult. ■☆

108952 Crassula deceptor Schönland et Baker f.;雅儿姿●☆

108953 Crassula deceptrix Schönland = Crassula deceptor Schönland et Baker f. ●☆

108954 Crassula decidua Schönland;脱落青锁龙;Norsveld Plakkie ●☆

108955 Crassula decipiens N. E. Br. = Crassula tecta Thunb. ■☆

108956 Crassula decumbens (Willd.) Harv. = Crassula thunbergiana Schult. ■☆

108957 Crassula decumbens Thunb.;大花青锁龙;Scilly Pygmyweed ●☆

108958 Crassula decumbens Thunb. var. brachyphylla (Adamson) Toelken;短叶大花青锁龙●☆

108959 Crassula dejecta Jacq.;白花青锁龙■☆

108960 Crassula deltoidea Thunb.;三角青锁龙;Gruisplakkie, Silver Beads ■☆

108961 Crassula deminuta Diels = Crassula setulosa Harv. var. deminuta (Diels) Toelken ■☆

108962 Crassula densa N. E. Br. = Crassula elegans Schönland et Baker f. ■☆

108963 Crassula densifolia Harv. = Crassula tetragona L. ●☆

108964 Crassula dentata Thunb.;尖齿青锁龙■☆

108965 Crassula dentata Thunb. var. minor Harv. = Crassula dentata Thunb. ■☆
108966 Crassula dependens Bolus;悬垂青锁龙●☆
108967 Crassula depressa (Eckl. et Zeyh.) Toelken;凹陷青锁龙●☆
108968 Crassula dewinteri Friedrich = Crassula pubescens Thunb. ●☆
108969 Crassula diabolica N. E. Br. = Crassula pellucida L. subsp. brachypetala (Drège ex Harv.) Toelken ●☆
108970 Crassula diaphana Drège ex Harv. = Crassula strigosa L. ■☆
108971 Crassula dichotoma L.;二歧青锁龙●☆
108972 Crassula dielsii Schönland = Crassula dentata Thunb. ■☆
108973 Crassula diffusa Dryand.;松散青锁龙●☆
108974 Crassula dinteri Schönland = Crassula elegans Schönland et Baker f. ■☆
108975 Crassula divaricata Eckl. et Zeyh. = Crassula muricata Thunb. ■☆
108976 Crassula dodii Schönland et Baker f.;多德青锁龙●☆
108977 Crassula drakensbergensis Schönland = Crassula vaginata Eckl. et Zeyh. ■☆
108978 Crassula dregeana Harv. = Crassula obovata Haw. var. dregeana (Harv.) Toelken ■☆
108979 Crassula dregei (Harv.) Schönland = Crassula pellucida L. subsp. brachypetala (Drège ex Harv.) Toelken ●☆
108980 Crassula dubia Schönland = Crassula cotyledonis Thunb. ■☆
108981 Crassula ecklonii D. Dietr. = Crassula depressa (Eckl. et Zeyh.) Toelken ●☆
108982 Crassula eendornensis Dinter = Crassula tomentosa Thunb. var. glabrifolia (Harv.) G. D. Rowley ■☆
108983 Crassula elata N. E. Br. = Crassula capitella Thunb. subsp. nodulosa (Schönland) Toelken ■☆
108984 Crassula elatinoides (Eckl. et Zeyh.) Friedrich;高青锁龙●☆
108985 Crassula elegans Schönland et Baker f.;光绿青锁龙;Elegant Crassula ■☆
108986 Crassula elegans Schönland et Baker f. subsp. namibensis (Friedrich) Toelken;纳米比亚青锁龙●☆
108987 Crassula ellenbeckiana Schönland = Crassula alba Forssk. ■☆
108988 Crassula elongata Schönland = Crassula pellucida L. subsp. brachypetala (Drège ex Harv.) Toelken ●☆
108989 Crassula elsieae Toelken;埃尔西青锁龙■☆
108990 Crassula enantiophylla Baker f. = Crassula capitella Thunb. subsp. nodulosa (Schönland) Toelken ■☆
108991 Crassula engleri Schönland = Crassula montana Thunb. ●☆
108992 Crassula ericoides Haw.;石南状青锁龙●☆
108993 Crassula ericoides Haw. subsp. tortuosa Toelken;扭曲青锁龙●☆
108994 Crassula ernestii Schönland et Baker f. = Crassula lanuginosa Harv. var. pachystemon (Schönland et Baker f.) Toelken ●☆
108995 Crassula erosula N. E. Br. = Crassula subacaulis Schönland et Baker f. subsp. erosula (N. E. Br.) Toelken ■☆
108996 Crassula erubescens Bullock = Crassula granvikii Mildbr. ■☆
108997 Crassula exilis Harv.;瘦小青锁龙●☆
108998 Crassula exilis Harv. subsp. cooperi (Regel) Toelken = Crassula cooperi Regel ●☆
108999 Crassula exilis Harv. subsp. picturata (Boom) G. D. Rowley;色彩青锁龙●☆
109000 Crassula exilis Harv. subsp. sedifolia (N. E. Br.) Toelken;景天叶青锁龙●☆
109001 Crassula expansa Dryand. = Crassula parviflora E. Mey. ex Drège ■☆
109002 Crassula expansa Dryand. subsp. filicaulis (Haw.) Toelken;线茎青锁龙■☆
109003 Crassula expansa Dryand. subsp. fragilis (Baker) Toelken;脆青锁龙■☆
109004 Crassula expansa Dryand. subsp. pyrifolia (Compton) Toelken;梨叶青锁龙●☆
109005 Crassula expansa Dryand. var. longifolia R. Fern. = Crassula expansa Dryand. ■☆
109006 Crassula falcata H. Wendl. = Crassula perfoliata L. var. minor (Haw.) G. D. Rowley ●☆
109007 Crassula falcata J. C. Wendl. = Crassula perfoliata L. var. minor (Haw.) G. D. Rowley ●☆
109008 Crassula fallax Friedrich;迷惑青锁龙■☆
109009 Crassula falx Linding. = Crassula perfoliata L. var. minor (Haw.) G. D. Rowley ●☆
109010 Crassula fascicularis Lam.;簇生青锁龙■☆
109011 Crassula fastigiata Schönland = Crassula subulata L. var. fastigiata (Schönland) Toelken ■☆
109012 Crassula fergusoniae Schönland = Crassula pubescens Thunb. ●☆
109013 Crassula fergusoniae Schönland f. major ? = Crassula pubescens Thunb. ●☆
109014 Crassula filamentosa Schönland = Crassula lanceolata (Eckl. et Zeyh.) Endl. ex Walp. ■☆
109015 Crassula filicaulis Haw. = Crassula expansa Dryand. subsp. filicaulis (Haw.) Toelken ■☆
109016 Crassula filicaulis Haw. = Crassula expansa Dryand. ■☆
109017 Crassula filiformis (Eckl. et Zeyh.) D. Dietr.;线形青锁龙■☆
109018 Crassula flabellifolia Harv. = Crassula umbella Jacq. ●☆
109019 Crassula flanaganii Schönland et Baker f.;弗拉纳根青锁龙●☆
109020 Crassula flava L.;黄青锁龙●☆
109021 Crassula flavovirens Pillans = Crassula brevifolia Harv. ●☆
109022 Crassula foveata Van Jaarsv.;浅凹青锁龙●☆
109023 Crassula fragilis Baker = Crassula expansa Dryand. subsp. filicaulis (Haw.) Toelken ■☆
109024 Crassula fragilis Baker var. suborbicularis R. Fern. = Crassula expansa Dryand. subsp. filicaulis (Haw.) Toelken ■☆
109025 Crassula fragilis Schönland = Crassula pubescens Thunb. ●☆
109026 Crassula fragillima Dinter = Crassula brevifolia Harv. ●☆
109027 Crassula fruticulosa L.;灌木状青锁龙●☆
109028 Crassula furcata (Eckl. et Zeyh.) Endl. ex Walp. = Crassula ericoides Haw. ●☆
109029 Crassula fusca Herre;棕色青锁龙●☆
109030 Crassula galunkensis Engl. = Crassula volkensii Engl. ■☆
109031 Crassula garibina Marloth et Schönland subsp. glabra Toelken;光滑青锁龙●☆
109032 Crassula gentianoides Lam. = Crassula dichotoma L. ●☆
109033 Crassula giessii Friedrich = Crassula ausensis Hutchison subsp. giessii (Friedrich) Toelken ■☆
109034 Crassula gifbergensis Friedrich = Crassula atropurpurea (Haw.) D. Dietr. var. watermeyeri (Compton) Toelken ■☆
109035 Crassula gillii Schönland = Crassula montana Thunb. subsp. quadrangularis (Schönland) Toelken ■☆
109036 Crassula glabra Haw. = Crassula glomerata P. J. Bergius ●☆
109037 Crassula glabrifolia Harv. = Crassula tomentosa Thunb. var. glabrifolia (Harv.) G. D. Rowley ■☆
109038 Crassula glauca Schönland = Crassula cordata Thunb. ■☆
109039 Crassula globifera (Sims.) Spreng. = Crassula capensis (L.) Baill. ●■☆

109040 Crassula globosa N. E. Br. = Crassula elegans Schönland et Baker f. ■☆

109041 Crassula globularioides Britten;球花木青锁龙●☆

109042 Crassula globularioides Britten f. longiciliata R. Fern.;长缘毛球花木青锁龙●☆

109043 Crassula globularioides Britten f. pilosa R. Fern.;疏毛青锁龙●☆

109044 Crassula globularioides Britten subsp. argyrophylla (Schönland et Baker f.) Toelken = Crassula swaziensis Schönland ■☆

109045 Crassula glomerata P. J. Bergius;团集青锁龙●☆

109046 Crassula glomerata P. J. Bergius var. patens Eckl. et Zeyh. = Crassula glomerata P. J. Bergius ●☆

109047 Crassula grammanthoides (Schönland) Toelken;凸纹青锁龙●☆

109048 Crassula granvikii Mildbr.;格兰维克青锁龙■☆

109049 Crassula griquaensis Schönland = Crassula dependens Bolus ●☆

109050 Crassula grisea Schönland;灰青锁龙■☆

109051 Crassula guchabensis Merxm. = Crassula capitella Thunb. subsp. nodulosa (Schönland) Toelken ■☆

109052 Crassula guilelmi-trollii Stopp = Crassula hirsuta Schönland et Baker f. ■☆

109053 Crassula hallii Adcock = Crassula sericea Schönland ■☆

109054 Crassula harveyi Britten et Baker f. = Crassula dependens Bolus ●☆

109055 Crassula harveyi Britten et Baker f. var. dependens (Bolus) Schönland = Crassula dependens Bolus ●☆

109056 Crassula harveyi Britten et Baker f. var. intermedia Schönland = Crassula dependens Bolus ●☆

109057 Crassula harveyi Britten et Baker f. var. typica Schönland = Crassula dependens Bolus ●☆

109058 Crassula hedbergii Wickens et M. Bywater;赫德青锁龙■☆

109059 Crassula helmsii (Kirk) Cockayne;澳洲青锁龙;Australian Swamp Stonecrop,New Zealand Pygmyweed ●☆

109060 Crassula hemisphaerica Thunb.;半球青锁龙■☆

109061 Crassula hemisphaerica Thunb. var. foliosa Schönland = Crassula hemisphaerica Thunb. ■☆

109062 Crassula hemisphaerica Thunb. var. recurva Schönland = Crassula capitella Thunb. ■☆

109063 Crassula hemisphaerica Thunb. var. typica Schönland = Crassula hemisphaerica Thunb. ■☆

109064 Crassula herrei Friedrich = Crassula nudicaulis L. var. herrei (Friedrich) Toelken ■☆

109065 Crassula heterotricha Schinz = Crassula perfoliata L. var. heterotricha (Schinz) Toelken ■☆

109066 Crassula hirsuta Schönland et Baker f.;多毛青锁龙■☆

109067 Crassula hirta Thunb. = Crassula nudicaulis L. ■☆

109068 Crassula hirta Thunb. var. dyeri Schönland = Crassula nudicaulis L. ■☆

109069 Crassula hirtipes Harv.;毛梗青锁龙■☆

109070 Crassula hispida (Haw.) D. Dietr. = Crassula mesembryanthoides (Haw.) D. Dietr. subsp. hispida (Haw.) Toelken ■☆

109071 Crassula hofmeyeriana Dinter = Crassula ausensis Hutchison ■☆

109072 Crassula holstii Pax ex Engl.;霍尔青锁龙■☆

109073 Crassula hottentotta Marloth et Schönland = Crassula sericea Schönland var. hottentotta (Marloth et Schönland) Toelken ■☆

109074 Crassula humbertii Desc.;亨伯特青锁龙■☆

109075 Crassula humilis N. E. Br. = Crassula elegans Schönland et Baker f. ■☆

109076 Crassula hystrix Schönland = Crassula hirtipes Harv. ■☆

109077 Crassula ihlenfeldtii Friedrich = Crassula grisea Schönland ■☆

109078 Crassula imbricata Burm. f. = Crassula muscosa L. ●☆

109079 Crassula impressa (Haw.) D. Dietr. = Crassula capitella Thunb. ■☆

109080 Crassula inaequalis Schönland = Crassula acinaciformis Schinz ●☆

109081 Crassula inamoena N. E. Br. = Crassula subacaulis Schönland et Baker f. subsp. erosula (N. E. Br.) Toelken ■☆

109082 Crassula inandensis Schönland et Baker f.;伊南德青锁龙■☆

109083 Crassula inanis Thunb.;空青锁龙■☆

109084 Crassula incana (Eckl. et Zeyh.) Harv. = Crassula subaphylla (Eckl. et Zeyh.) Harv. ■☆

109085 Crassula inchangensis Engl. = Crassula obovata Haw. ■☆

109086 Crassula indica Decne. = Sinocrassula indica (Decne.) A. Berger ■

109087 Crassula intermedia Schönland;间型青锁龙■☆

109088 Crassula interrupta Drège ex Harv. = Crassula tomentosa Thunb. var. glabrifolia (Harv.) G. D. Rowley ■☆

109089 Crassula interrupta Drège ex Harv. var. glabrifolia (Harv.) Schönland = Crassula tomentosa Thunb. var. glabrifolia (Harv.) G. D. Rowley ■☆

109090 Crassula involucrata Schönland = Crassula pellucida L. subsp. brachypetala (Drège ex Harv.) Toelken ●☆

109091 Crassula jacobseniana Poelln. = Crassula ericoides Haw. ●☆

109092 Crassula jasminiana Haw. ex Sims = Crassula obtusa Haw. ■☆

109093 Crassula johannis-winkleri Linding. = Crassula perfoliata L. var. coccinea (Sweet) G. D. Rowley ●☆

109094 Crassula karasana Friedrich = Crassula ausensis Hutchison ■☆

109095 Crassula klinghardtensis Schönland = Crassula sericea Schönland ■☆

109096 Crassula kuhnii Schönland = Crassula biplanata Haw. ●☆

109097 Crassula lactea Sol.;乳白青锁龙(洛东);Krysna Crassula, Taylor's Parches ●☆

109098 Crassula lambertiana Schönland et Baker f. = Crassula oblanceolata Schönland et Baker f. ■☆

109099 Crassula lanceolata (Eckl. et Zeyh.) Endl. ex Walp.;剑形青锁龙●☆

109100 Crassula lanceolata (Eckl. et Zeyh.) Endl. ex Walp. = Crassula schimperi Fisch. et C. A. Mey. f. filamentosa (Schönland) R. Fern. ■☆

109101 Crassula lanceolata (Eckl. et Zeyh.) Endl. ex Walp. subsp. denticulata (Brenan) Toelken;细齿剑形青锁龙■☆

109102 Crassula lanceolata (Eckl. et Zeyh.) Endl. ex Walp. subsp. transvaalensis (Kuntze) Toelken;德兰士瓦青锁龙■☆

109103 Crassula langebergensis Schönland = Crassula decumbens Thunb. ●☆

109104 Crassula lanuginosa Harv.;多绵毛青锁龙●☆

109105 Crassula lanuginosa Harv. var. pachystemon (Schönland et Baker f.) Toelken;粗冠青锁龙●☆

109106 Crassula lasiantha Drège ex Harv.;毛花青锁龙■☆

109107 Crassula latibracteata Toelken;宽苞青锁龙■☆

109108 Crassula laticephala Schönland = Crassula congesta N. E. Br. subsp. laticephala (Schönland) Toelken ●☆

109109 Crassula latispathulata Schönland et Baker f. = Crassula spathulata Thunb. ●☆

109110 Crassula laxa Schönland = Crassula dependens Bolus ●☆

109111 Crassula leachii R. Fern.;利奇青锁龙■☆

109112 Crassula leipoldtii Schönland et Baker f. = Crassula decumbens Thunb. ●☆

109113 Crassula lettyae E. Phillips = Crassula barbata Thunb. ■☆

109114 Crassula leucantha Schönland et Baker f. = Crassula multiflora Schönland et Baker f. subsp. leucantha (Schönland et Baker f.) Toelken ■☆

109115 Crassula levynsiae Adamson = Crassula natans Thunb. ■☆

109116 Crassula liebuschiana Engl. = Crassula globularioides Britten ●☆

109117 Crassula lignosa Burtt Davy = Crassula sarcocaulis Eckl. et Zeyh. ●☆

109118 Crassula limosa Schönland = Crassula papillosa Schönland et Baker f. ■☆

109119 Crassula lineolata Dryand. = Crassula pellucida L. subsp. marginalis (Dryand.) Toelken ■☆

109120 Crassula lingua D. Dietr. = Crassula nudicaulis L. ■☆

109121 Crassula linguifolia Haw. = Crassula tomentosa Thunb. ■☆

109122 Crassula liquiritiodora Dinter = Crassula elegans Schönland et Baker f. ■☆

109123 Crassula littlewoodii Friedrich = Crassula ausensis Hutchison ■☆

109124 Crassula littoralis (Eckl. et Zeyh.) Endl. et Walp. = Crassula muscosa L. ●☆

109125 Crassula loganiana Compton = Crassula subaphylla (Eckl. et Zeyh.) Harv. ■☆

109126 Crassula longistyla Schönland = Crassula obovata Haw. var. dregeana (Harv.) Toelken ■☆

109127 Crassula loriformis Schönland et Baker f. = Crassula umbella Jacq. ●☆

109128 Crassula lucens Gram = Crassula ovata (Mill.) Druce ●☆

109129 Crassula lucida Lam. = Crassula spathulata Thunb. ●☆

109130 Crassula luederitzii Schönland;吕德里茨青锁龙■☆

109131 Crassula lutea (Schönland) Friedrich = Crassula namaquensis Schönland et Baker f. subsp. lutea (Schönland) Toelken ■☆

109132 Crassula lycopodioides Lam. = Crassula muscosa L. ●☆

109133 Crassula lycopodioides Lam. var. obtusifolia Harv. = Crassula muscosa L. var. obtusifolia (Harv.) G. D. Rowley ●☆

109134 Crassula lycopodioides Lam. var. pseudolycopodioides (Dinter) Walther ex Jacobsen = Crassula muscosa L. ●☆

109135 Crassula macowaniana Schönland et Baker f.;麦克欧文青锁龙■☆

109136 Crassula macowaniana Schönland et Baker f. var. crassifolia Schönland = Crassula macowaniana Schönland et Baker f. ■☆

109137 Crassula macowanii Scott-Elliot = Crassula tetragona L. subsp. acutifolia (Lam.) Toelken ●☆

109138 Crassula macrantha (Hook. f.) Diels et E. Pritz. = Crassula decumbens Thunb. ●☆

109139 Crassula marchandii Friedrich;马尔尚青锁龙■☆

109140 Crassula margaritifera (Eckl. et Zeyh.) Harv. = Crassula mollis Thunb. ■☆

109141 Crassula marginalis Dryand. = Crassula pellucida L. subsp. marginalis (Dryand.) Toelken ■☆

109142 Crassula marginata Thunb. = Crassula pellucida L. ■☆

109143 Crassula mariae Raym.-Hamet = Crassula capitella Thunb. subsp. nodulosa (Schönland) Toelken ■☆

109144 Crassula maritima Schönland = Crassula expansa Dryand. subsp. filicaulis (Haw.) Toelken ■☆

109145 Crassula marlothii Schönland = Crassula dentata Thunb. ■☆

109146 Crassula marnierana Huber et Jacobsen = Crassula rupestris Thunb. subsp. marnierana (H. E. Huber et Jacobsen) Toelken ●☆

109147 Crassula massonii Britten et Baker f. = Crassula alpestris Thunb. subsp. massonii (Britten et Baker f.) Toelken ■☆

109148 Crassula massonioides Diels = Crassula compacta Schönland ●☆

109149 Crassula media (Haw.) D. Dietr. = Crassula fascicularis Lam. ■☆

109150 Crassula merxmuelleri Friedrich = Crassula sericea Schönland var. hottentotta (Marloth et Schönland) Toelken ■☆

109151 Crassula mesembrianthemoides Dinter et Berger;银箭●☆

109152 Crassula mesembrianthemopsis Dinter = Crassula mesembrianthemoides Dinter et Berger ●☆

109153 Crassula mesembrianthoides Dinter et A. Berger = Crassula elegans Schönland et Baker f. subsp. namibensis (Friedrich) Toelken ●☆

109154 Crassula mesembrianthoides Schönland et Baker f. = Crassula elegans Schönland et Baker f. ■☆

109155 Crassula mesembryanthoides (Haw.) D. Dietr. = Crassula mesembrianthemoides Dinter et Berger ●☆

109156 Crassula mesembryanthoides (Haw.) D. Dietr. subsp. hispida (Haw.) Toelken;硬毛青锁龙■☆

109157 Crassula meyeri Harv. = Crassula capitella Thunb. subsp. meyeri (Harv.) Toelken ■☆

109158 Crassula micans Vahl ex Baill.;弱光泽青锁龙■☆

109159 Crassula micrantha Schönland = Crassula atropurpurea (Haw.) D. Dietr. ●☆

109160 Crassula milfordiae Byles = Crassula setulosa Harv. var. rubra (N. E. Br.) G. D. Rowley ■☆

109161 Crassula milleriana Burtt Davy = Crassula alba Forssk. ■☆

109162 Crassula minima Thunb. = Crassula dentata Thunb. ■☆

109163 Crassula minuta Toelken;微小青锁龙■☆

109164 Crassula minutiflora Schönland et Baker f. = Crassula thunbergiana Schult. subsp. minutiflora (Schönland et Baker f.) Toelken ■☆

109165 Crassula mitrata Friedrich = Crassula columnaris Thunb. ■☆

109166 Crassula mollis Thunb.;柔软青锁龙●☆

109167 Crassula mongolica Franch. = Tillaea mongolica (Franch.) S. H. Fu ■

109168 Crassula montana Thunb.;高山青锁龙●☆

109169 Crassula montana Thunb. subsp. quadrangularis (Schönland) Toelken;四棱青锁龙■☆

109170 Crassula monticola N. E. Br. = Crassula rupestris Thunb. ●☆

109171 Crassula montis-draconis Dinter = Crassula brevifolia Harv. ●☆

109172 Crassula mossii Schönland = Crassula compacta Schönland ●☆

109173 Crassula mucronata Keissl. = Crassula southii Schönland ■☆

109174 Crassula multicava Lem.;鸣户(多孔神刀,猪耳朵);Cape Province Pygmyweed ●☆

109175 Crassula multicava Lem. subsp. floribunda Friedrich ex Toelken;繁花鸣户●☆

109176 Crassula multiceps Harv.;多头青锁龙■☆

109177 Crassula multiflora Schönland et Baker f.;多花青锁龙■☆

109178 Crassula multiflora Schönland et Baker f. subsp. leucantha (Schönland et Baker f.) Toelken;白色多花青锁龙■☆

109179 Crassula muricata Thunb.;粗糙青锁龙■☆

109180 Crassula muscosa L.;青锁龙;Lizard's Tail, Rattail Crassula, Watch Chain, Watch Chain Crassula ●☆

109181 Crassula muscosa L. var. obtusifolia (Harv.) G. D. Rowley;钝叶青锁龙●☆

109182 Crassula muscosa L. var. parvula (Eckl. et Zeyh.) Toelken;较小青锁龙●☆

109183 Crassula muscosa L. var. polpodacea (Eckl. et Zeyh.) G. D.

Rowley;粟米草青锁龙●☆

109184 Crassula muscosa L. var. rigida Toelken = Crassula muscosa L. var. obtusifolia (Harv.) G. D. Rowley ●☆

109185 Crassula muscosa L. var. sinuata Toelken = Crassula muscosa L. var. polpodacea (Eckl. et Zeyh.) G. D. Rowley ●☆

109186 Crassula namaquensis Schönland et Baker f.;纳马夸青锁龙■☆

109187 Crassula namaquensis Schönland et Baker f. subsp. comptonii (Hutch. et Pillans) Toelken;康氏纳马夸青锁龙■☆

109188 Crassula namaquensis Schönland et Baker f. subsp. lutea (Schönland) Toelken;黄纳马夸青锁龙■☆

109189 Crassula namaquensis Schönland et Baker f. var. brevifolia Schönland = Crassula namaquensis Schönland et Baker f. ■☆

109190 Crassula namaquensis Schönland et Baker f. var. lutea Schönland = Crassula namaquensis Schönland et Baker f. subsp. lutea (Schönland) Toelken ■☆

109191 Crassula namibensis Friedrich = Crassula elegans Schönland et Baker f. subsp. namibensis (Friedrich) Toelken ●☆

109192 Crassula nana Schönland et Baker f. = Crassula umbellata Thunb. ■☆

109193 Crassula natalensis Schönland;纳塔尔青锁龙■☆

109194 Crassula natalensis Schönland var. mosii ? = Crassula capitella Thunb. subsp. meyeri (Harv.) Toelken ■☆

109195 Crassula natans Thunb.;浮水青锁龙■☆

109196 Crassula natans Thunb. f. amphibia (Harv.) Schönland = Crassula natans Thunb. ■☆

109197 Crassula natans Thunb. f. filiformis (Eckl. et Zeyh.) Schönland = Crassula natans Thunb. var. minus (Eckl. et Zeyh.) G. D. Rowley ■☆

109198 Crassula natans Thunb. f. fluitans (Eckl. et Zeyh.) Schönland = Crassula natans Thunb. ■☆

109199 Crassula natans Thunb. f. obovata (Eckl. et Zeyh.) Schönland = Crassula natans Thunb. ■☆

109200 Crassula natans Thunb. f. parvifolia Schönland = Crassula natans Thunb. ■☆

109201 Crassula natans Thunb. subsp. filiformis (Eckl. et Zeyh.) Friedrich = Crassula natans Thunb. var. minus (Eckl. et Zeyh.) G. D. Rowley ■☆

109202 Crassula natans Thunb. var. filiformis (Eckl. et Zeyh.) Toelken = Crassula natans Thunb. var. minus (Eckl. et Zeyh.) G. D. Rowley ■☆

109203 Crassula natans Thunb. var. minus (Eckl. et Zeyh.) G. D. Rowley;小浮水青锁龙■☆

109204 Crassula nealeana Higgins = Crassula perforata Thunb. ●☆

109205 Crassula neglecta Schult. = Crassula cordata Thunb. ■☆

109206 Crassula nemorosa (Eckl. et Zeyh.) Endl. ex Walp.;森林青锁龙●☆

109207 Crassula nitida Schönland = Crassula ovata (Mill.) Druce ●☆

109208 Crassula nivalis (Eckl. et Zeyh.) Endl. et Walp. = Crassula nemorosa (Eckl. et Zeyh.) Endl. ex Walp. ■☆

109209 Crassula nodulosa Schönland = Crassula capitella Thunb. subsp. nodulosa (Schönland) Toelken ■☆

109210 Crassula nodulosa Schönland f. rhodesica R. Fern. = Crassula capitella Thunb. subsp. nodulosa (Schönland) Toelken ■☆

109211 Crassula nodulosa Schönland var. longisepala R. Fern. = Crassula capitella Thunb. subsp. nodulosa (Schönland) Toelken ■☆

109212 Crassula nuda Compton = Crassula capitella Thunb. subsp. thyrsiflora (Thunb.) Toelken ■☆

109213 Crassula nudicaulis L.;裸茎青锁龙●☆

109214 Crassula nudicaulis L. var. glabra Schönland = Crassula nudicaulis L. ■☆

109215 Crassula nudicaulis L. var. herrei (Friedrich) Toelken;赫勒青锁龙■☆

109216 Crassula nudicaulis L. var. platyphylla (Harv.) Toelken;宽叶裸茎青锁龙●☆

109217 Crassula nummularifolia Baker = Crassula pellucida L. subsp. alsinoides (Hook. f.) Toelken ■☆

109218 Crassula nyikensis Baker f. = Crassula globularioides Britten ●☆

109219 Crassula oblanceolata Schönland et Baker f.;倒披针形青锁龙■☆

109220 Crassula obliqua Haw. = Crassula atropurpurea (Haw.) D. Dietr. ●☆

109221 Crassula obliqua Sol.;燕子掌●☆

109222 Crassula obliqua Sol. = Crassula ovata (Mill.) Druce ●☆

109223 Crassula obovata Haw.;倒卵青锁龙■☆

109224 Crassula obovata Haw. var. dregeana (Harv.) Toelken;德雷青锁龙■☆

109225 Crassula obtusa Haw.;钝青锁龙■☆

109226 Crassula obvallata L. = Crassula nudicaulis L. ■☆

109227 Crassula odoratissima Andréws = Crassula fascicularis Lam. ■☆

109228 Crassula orbicularis L.;圆盘青锁龙■☆

109229 Crassula ovata (Mill.) Druce;玉树(翡翠木,银白青锁龙);Crassula, Friendship Tree, Jade Plant, Jade Tree, Japanese Rubber Plant, Money Plant, Money Tree ●☆

109230 Crassula ovata Druce = Crassula ovata (Mill.) Druce ●☆

109231 Crassula pachyphylla Schönland = Crassula congesta N. E. Br. ●☆

109232 Crassula pachystemon Schönland et Baker f. = Crassula lanuginosa Harv. var. pachystemon (Schönland et Baker f.) Toelken ●☆

109233 Crassula pageae Toelken;纸青锁龙■☆

109234 Crassula pallens Schönland et Baker f.;变苍白青锁龙■☆

109235 Crassula pallida Baker = Crassula perfoliata L. ●☆

109236 Crassula paniculata (Haw.) D. Dietr.;圆锥青锁龙;Butter Tree ●☆

109237 Crassula paniculata (Haw.) D. Dietr. = Crassula capitella Thunb. ■☆

109238 Crassula papillosa Schönland et Baker f.;乳头青锁龙■☆

109239 Crassula parviflora E. Mey. ex Drège = Crassula expansa Dryand. ■☆

109240 Crassula parvipetala Schönland = Crassula tenuipedicellata Schönland et Baker f. ■☆

109241 Crassula parvisepala Schönland = Crassula sarcocaulis Eckl. et Zeyh. ●☆

109242 Crassula parvula (Eckl. et Zeyh.) Endl. ex Walp. = Crassula muscosa L. var. parvula (Eckl. et Zeyh.) Toelken ●☆

109243 Crassula patens (Eckl. et Zeyh.) Endl. et Walp. = Crassula dentata Thunb. ■☆

109244 Crassula patersoniae Schönland = Crassula perforata Thunb. ●☆

109245 Crassula pearsonii Schönland = Crassula brevifolia Harv. ●☆

109246 Crassula pectinata Conrath = Crassula capitella Thunb. subsp. nodulosa (Schönland) Toelken ■☆

109247 Crassula peculiaris (Toelken) Toelken et Wickens;特殊青锁龙■☆

109248 Crassula peglerae Schönland = Crassula obovata Haw. ■☆

109249 Crassula pellucida L.;透明青锁龙■☆

109250 Crassula pellucida L. subsp. alsinoides (Hook. f.) Toelken;繁

缕青锁龙■☆
109251 Crassula pellucida L. subsp. brachypetala (Drège ex Harv.) Toelken;短瓣青锁龙●☆
109252 Crassula pellucida L. subsp. marginalis (Dryand.) Toelken;边生青锁龙■☆
109253 Crassula pellucida L. subsp. spongiosa Toelken;海绵青锁龙■☆
109254 Crassula peltata ?;盾状青锁龙;Pagoda Plant ■☆
109255 Crassula pentandra (Royle ex Edgew.) Schönland;五蕊青锁龙■☆
109256 Crassula pentandra (Royle ex Edgew.) Schönland = Tillaea schimperi (C. A. Mey.) M. G. Gilbert, H. Ohba et K. T. Fu ■
109257 Crassula pentandra (Royle ex Edgew.) Schönland var. denticulata Brenan = Crassula lanceolata (Eckl. et Zeyh.) Endl. ex Walp. subsp. denticulata (Brenan) Toelken ■☆
109258 Crassula perfilata Scop. = Crassula perforata Thunb. ●☆
109259 Crassula perfoliata L. var. albiflora Harv. = Crassula perfoliata L. ●☆
109260 Crassula perfoliata L. var. coccinea (Sweet) G. D. Rowley;红花串钱景天●☆
109261 Crassula perfoliata L. var. falcata (J. C. Wendl.) Toelken = Crassula perfoliata L. var. minor (Haw.) G. D. Rowley ●☆
109262 Crassula perfoliata L. var. heterotricha (Schinz) Toelken;异毛串钱景天■☆
109263 Crassula perfoliata L. var. miniata Toelken = Crassula perfoliata L. var. coccinea (Sweet) G. D. Rowley ●☆
109264 Crassula perfoliata L. var. minor (Haw.) G. D. Rowley;神刀(宝刀,神刀草);Aeroplane Propeller, Propeller Plant, Sickle Plant ●☆
109265 Crassula perforata L. f. = Crassula perforata Thunb. ●☆
109266 Crassula perforata Thunb.;串钱景天(星乙女);Button on a String, Sosatieplakkie, String of Buttons ●☆
109267 Crassula perfossa Lam. = Crassula perforata Thunb. ●☆
109268 Crassula petraea Schönland = Crassula exilis Harv. ●☆
109269 Crassula petrogeton Endl. ex Walp. = Crassula dentata Thunb. ■☆
109270 Crassula pharnaceoides Fisch. et C. A. Mey. = Crassula alata (Viv.) A. Berger subsp. pharnaceoides (Fisch. et C. A. Mey.) Wickens et Bywater ■☆
109271 Crassula pharnaceoides Fisch. et C. A. Mey. = Tillaea alata Viv. ■
109272 Crassula pharnaceoides Fisch. et C. A. Mey. subsp. rhodesica Merxm. = Crassula rhodesica (Merxm.) Wickens et M. Bywater ■☆
109273 Crassula picturata Boom = Crassula exilis Harv. subsp. picturata (Boom) G. D. Rowley ●☆
109274 Crassula pinnata L. f. = Bryophyllum pinnatum (L. f.) Oken ■
109275 Crassula planifolia Schönland;平花青锁龙■☆
109276 Crassula platyphylla Harv. = Crassula nudicaulis L. var. platyphylla (Harv.) Toelken ●☆
109277 Crassula portulacea Lam. = Crassula ovata (Mill.) Druce ●☆
109278 Crassula profusa Hook. f. = Crassula pellucida L. subsp. marginalis (Dryand.) Toelken ■☆
109279 Crassula promontorii Schönland et Baker f. = Crassula capensis (L.) Baill. var. promontorii (Schönland et Baker f.) Toelken ■☆
109280 Crassula propinqua (Eckl. et Zeyh.) Endl. ex Walp. = Crassula muscosa L. var. obtusifolia (Harv.) G. D. Rowley ●☆
109281 Crassula prostrata E. Mey. ex Drège = Crassula pellucida L. subsp. brachypetala (Drège ex Harv.) Toelken ●☆
109282 Crassula prostratum Thunb. = Crassula expansa Dryand. ■☆
109283 Crassula pruinosa L.;粉景天青锁龙;Skurwemannetjie ●☆
109284 Crassula pseudohemisphaerica Friedrich;假半球青锁龙■☆
109285 Crassula pseudolycopodioides Dinter et Schinz = Crassula muscosa L. ●☆
109286 Crassula pubescens (Eckl. et Zeyh.) Walp. = Crassula strigosa L. ■☆
109287 Crassula pubescens Thunb.;短柔毛青锁龙;Jersey Pigmyweed, Red Carpet ●☆
109288 Crassula pubescens Thunb. subsp. radicans (Haw.) Toelken = Crassula pubescens Thunb. ●☆
109289 Crassula pulchella Dryand.;美丽青锁龙■☆
109290 Crassula punctata L.;斑点青锁龙■☆
109291 Crassula punctulata Schönland et Baker f. = Crassula biplanata Haw. ●☆
109292 Crassula pustulata Toelken;泡青锁龙■☆
109293 Crassula pyramidalis Thunb.;绿塔青锁龙(绿塔)●☆
109294 Crassula pyramidalis Thunb. var. ramosa Schönland = Crassula pyramidalis Thunb. ●☆
109295 Crassula pyrifolia Compton = Crassula expansa Dryand. subsp. pyrifolia (Compton) Toelken ●☆
109296 Crassula quadrangula (Eckl. et Zeyh.) Endl. et Walp. = Crassula pyramidalis Thunb. ●☆
109297 Crassula quadrangularis Schönland = Crassula montana Thunb. subsp. quadrangularis (Schönland) Toelken ■☆
109298 Crassula quadrifida Baker f. = Crassula multicava Lem. ●☆
109299 Crassula radicans (Haw.) D. Dietr. = Crassula pubescens Thunb. subsp. radicans (Haw.) Toelken ●☆
109300 Crassula radicans (Haw.) D. Dietr. var. fastigiata Schönland = Crassula pubescens Thunb. ●☆
109301 Crassula radicans (Haw.) D. Dietr. var. phillipsii Schönland = Crassula pubescens Thunb. ●☆
109302 Crassula radicans (Haw.) D. Dietr. var. typica Schönland = Crassula pubescens Thunb. subsp. radicans (Haw.) Toelken ●☆
109303 Crassula ramosa Thunb. = Crassula subulata L. ■☆
109304 Crassula ramuliflora Link et Otto = Crassula obovata Haw. ■☆
109305 Crassula ramuliflora Link et Otto var. bolusii Schönland = Crassula obovata Haw. ■☆
109306 Crassula ramuliflora Link et Otto var. flanaganii Schönland = Crassula obovata Haw. ■☆
109307 Crassula ramuliflora Link et Otto var. rattrayi Schönland = Crassula obovata Haw. ■☆
109308 Crassula ramuliflora Link et Otto var. simii Schönland = Crassula obovata Haw. ■☆
109309 Crassula ramuliflora Link et Otto var. stachyera (Eckl. et Zeyh.) Schönland = Crassula obovata Haw. ■☆
109310 Crassula ramuliflora Link et Otto var. transvaalensis Schönland = Crassula setulosa Harv. ■☆
109311 Crassula ramuliflora Link et Otto var. typica Schönland = Crassula obovata Haw. ■☆
109312 Crassula rauhii Friedrich = Crassula subacaulis Schönland et Baker f. subsp. erosula (N. E. Br.) Toelken ■☆
109313 Crassula recurva N. E. Br. = Crassula alba Forssk. ■☆
109314 Crassula recurva N. E. Br. = Crassula helmsii (Kirk) Cockayne ●☆
109315 Crassula rehmannii Baker f. = Crassula cotyledonis Thunb. ■☆
109316 Crassula remota Schönland = Crassula subaphylla (Eckl. et Zeyh.) Harv. ■☆
109317 Crassula retroflexa Thunb. = Crassula dichotoma L. ●☆
109318 Crassula retrorsa Hutch. = Crassula vaginata Eckl. et Zeyh. ■☆

109319 Crassula reversisetosa Bitter = Crassula obovata Haw. ■☆
109320 Crassula revolvens Haw.;外卷青锁龙■☆
109321 Crassula rhodesica (Merxm.) Wickens et M. Bywater;罗得西亚青锁龙■☆
109322 Crassula rhodogyna Friedrich = Crassula capitella Thunb. subsp. thyrsiflora (Thunb.) Toelken ■☆
109323 Crassula rhomboidea N. E. Br. = Crassula deltoidea Thunb. ■☆
109324 Crassula rivularis (Peter) Hutch. et E. A. Bruce = Crassula granvikii Mildbr. ■☆
109325 Crassula robusta Toelken = Crassula tetragona L. subsp. robusta (Toelken) Toelken ●☆
109326 Crassula rogersii Schönland;罗杰斯青锁龙■☆
109327 Crassula roggeveldii Schönland;罗格青锁龙■☆
109328 Crassula rosularis Haw. = Crassula orbicularis L. ■☆
109329 Crassula rotundifolia Haw. = Kalanchoe rotundifolia (Haw.) Haw. ■☆
109330 Crassula rubella Compton = Crassula atropurpurea (Haw.) D. Dietr. var. muirii (Schönland) R. Fern. ■☆
109331 Crassula rubescens Schönland et Baker = Crassula natalensis Schönland ■☆
109332 Crassula rubescens Schönland et Baker f. var. intermedia Schönland = Crassula natalensis Schönland ■☆
109333 Crassula rubescens Schönland et Baker f. var. laxa Schönland = Crassula natalensis Schönland ■☆
109334 Crassula rubicunda Drège ex Harv. = Crassula alba Forssk. ■☆
109335 Crassula rubicunda Drège ex Harv. var. flexuosa Schönland = Crassula alba Forssk. ■☆
109336 Crassula rubicunda Drège ex Harv. var. hispida Schönland = Crassula alba Forssk. ■☆
109337 Crassula rubicunda Drège ex Harv. var. lydenburgensis Schönland = Crassula alba Forssk. var. parvisepala (Schönland) Toelken ■☆
109338 Crassula rubicunda Drège ex Harv. var. milleriana (Burtt Davy) Schönland = Crassula alba Forssk. ■☆
109339 Crassula rubicunda Drège ex Harv. var. parvisepala Schönland = Crassula alba Forssk. var. parvisepala (Schönland) Toelken ■☆
109340 Crassula rubicunda Drège ex Harv. var. rubicunda R. A. Dyer = Crassula alba Forssk. ■☆
109341 Crassula rubicunda Drège ex Harv. var. similis (Baker f.) Schönland = Crassula alba Forssk. var. parvisepala (Schönland) Toelken ■☆
109342 Crassula rubicunda Drège ex Harv. var. subglabra Schönland = Crassula alba Forssk. ■☆
109343 Crassula rubicunda Drège ex Harv. var. typica Schönland = Crassula alba Forssk. ■☆
109344 Crassula rubricaulis Eckl. et Zeyh.;红茎青锁龙■☆
109345 Crassula rubricaulis Eckl. et Zeyh. var. muirii Schönland = Crassula rubricaulis Eckl. et Zeyh. ■☆
109346 Crassula rudis Schönland et Baker f. = Crassula tetragona L. subsp. rudis (Schönland et Baker f.) Toelken ●☆
109347 Crassula rudolfii Schönland et Baker f.;鲁道夫青锁龙■☆
109348 Crassula rufo-punctata Schönland = Crassula capitella Thunb. ■☆
109349 Crassula rupestris Thunb.;岩地青锁龙;Inrygertjie,Rosary Plant,Sosaties ●☆
109350 Crassula rupestris Thunb. subsp. commutata (Friedrich) Toelken;变异岩地青锁龙●☆
109351 Crassula rupestris Thunb. subsp. marnierana (H. E. Huber et Jacobsen) Toelken;马尔岩地青锁龙●☆
109352 Crassula rustii Schönland = Crassula subulata L. ■☆
109353 Crassula sarcocaulis Eckl. et Zeyh.;肉茎神刀(肉果景天);Bonsai Crassula ●☆
109354 Crassula sarcocaulis Eckl. et Zeyh. subsp. rupicola Toelken;岩生肉茎神刀■☆
109355 Crassula sarcocaulis Eckl. et Zeyh. var. milanjiana R. Fern. = Crassula sarcocaulis Eckl. et Zeyh. subsp. rupicola Toelken ●☆
109356 Crassula sarcocaulis Eckl. et Zeyh. var. scaberula Harv. = Crassula sarcocaulis Eckl. et Zeyh. ●☆
109357 Crassula sarcolipes Harv. = Crassula strigosa L. ■☆
109358 Crassula sarmentosa Harv.;蔓茎青锁龙■☆
109359 Crassula sarmentosa Harv. var. integrifolia Toelken;全缘叶蔓茎青锁龙■☆
109360 Crassula saxifraga Harv.;岩栖青锁龙■☆
109361 Crassula scabra L.;糙青锁龙■☆
109362 Crassula scabrella Haw. = Crassula pruinosa L. ●☆
109363 Crassula scalaris Schönland et Baker f. = Crassula tomentosa Thunb. var. glabrifolia (Harv.) G. D. Rowley ■☆
109364 Crassula scheppigiana Diels = Crassula setulosa Harv. ■☆
109365 Crassula schimperi C. A. Mey. = Tillaea schimperi (C. A. Mey.) M. G. Gilbert,H. Ohba et K. T. Fu ■
109366 Crassula schimperi Fisch. et C. A. Mey.;欣珀青锁龙■☆
109367 Crassula schimperi Fisch. et C. A. Mey. f. filamentosa (Schönland) R. Fern. = Crassula lanceolata (Eckl. et Zeyh.) Endl. ex Walp. ■☆
109368 Crassula schimperi Fisch. et C. A. Mey. f. transvaalensis (Kuntze) R. Fern. = Crassula lanceolata (Eckl. et Zeyh.) Endl. ex Walp. subsp. transvaalensis (Kuntze) Toelken ■☆
109369 Crassula schimperi Fisch. et C. A. Mey. subsp. transvaalensis (Kuntze) R. Fern. = Crassula lanceolata (Eckl. et Zeyh.) Endl. ex Walp. subsp. transvaalensis (Kuntze) Toelken ■☆
109370 Crassula schimperi Fisch. et C. A. Mey. var. denticulata (Brenan) R. Fern. = Crassula lanceolata (Eckl. et Zeyh.) Endl. ex Walp. subsp. denticulata (Brenan) Toelken ■☆
109371 Crassula schimperi Fisch. et C. A. Mey. var. illecebroides (Welw. ex Hiern) G. D. Rowley = Crassula lanceolata (Eckl. et Zeyh.) Endl. ex Walp. ■☆
109372 Crassula schimperi Fisch. et C. A. Mey. var. lanceolata (Eckl. et Zeyh.) Toelken = Crassula lanceolata (Eckl. et Zeyh.) Endl. ex Walp. subsp. denticulata (Brenan) Toelken ■☆
109373 Crassula schimperi Fisch. et C. A. Mey. var. transvaalensis (Kuntze) R. Fern. = Crassula lanceolata (Eckl. et Zeyh.) Endl. ex Walp. subsp. transvaalensis (Kuntze) Toelken ■☆
109374 Crassula schmidtii Regel;施密特青锁龙(筑羽根)●☆
109375 Crassula schoenlandii H. Jacobsen = Crassula elegans Schönland et Baker f. ■☆
109376 Crassula schweinfurthii De Wild. = Crassula vaginata Eckl. et Zeyh. ■☆
109377 Crassula scleranthoides Burm. f. = Crassula glomerata P. J. Bergius ●☆
109378 Crassula scutellaria Burm. f. = Polyscias scutellaria (Burm. f.) Fosberg ●
109379 Crassula sebaeoides (Eckl. et Zeyh.) Toelken;小黄管青锁龙■☆
109380 Crassula sediflora (Eckl. et Zeyh.) Endl. et Walp. = Crassula tenuifolia Schönland ■☆
109381 Crassula sediflora (Eckl. et Zeyh.) Endl. et Walp. var.

laxifoliosa Schönland = Crassula obovata Haw. ■☆
109382 Crassula sedifolia N. E. Br. = Crassula exilis Harv. subsp. sedifolia (N. E. Br.) Toelken ●☆
109383 Crassula sedoides Mill. = Crassula orbicularis L. ■☆
109384 Crassula selago Dinter = Crassula lanceolata (Eckl. et Zeyh.) Endl. ex Walp. subsp. transvaalensis (Kuntze) Toelken ■☆
109385 Crassula semiorbicularis Eckl. et Zeyh. = Crassula columnaris Thunb. subsp. prolifera Friedrich ■☆
109386 Crassula septas Thunb. = Crassula capensis (L.) Baill. ●■☆
109387 Crassula septas Thunb. var. leipoldtii Schönland = Crassula capensis (L.) Baill. ●■☆
109388 Crassula sericea Schönland var. hottentotta (Marloth et Schönland) Toelken;豪顿青锁龙■☆
109389 Crassula sericea Schönland var. velutina (Friedrich) Toelken;短绒毛青锁龙■☆
109390 Crassula serpentaria Schönland;蛇药青锁龙■☆
109391 Crassula sessilifolia Baker f. = Crassula natalensis Schönland ■☆
109392 Crassula setigera (Eckl. et Zeyh.) Schönland = Crassula tomentosa Thunb. ■☆
109393 Crassula setulosa Harv.;细刚毛青锁龙■☆
109394 Crassula setulosa Harv. f. latipetala R. Fern. = Crassula setulosa Harv. ■☆
109395 Crassula setulosa Harv. var. basutica Schönland = Crassula setulosa Harv. ■☆
109396 Crassula setulosa Harv. var. curta (N. E. Br.) Schönland = Crassula setulosa Harv. var. rubra (N. E. Br.) G. D. Rowley ■☆
109397 Crassula setulosa Harv. var. deminuta (Diels) Toelken;缩小青锁龙■☆
109398 Crassula setulosa Harv. var. lanceolata Schönland = Crassula setulosa Harv. ■☆
109399 Crassula setulosa Harv. var. longiciliata Toelken;长缘毛细刚毛青锁龙■☆
109400 Crassula setulosa Harv. var. ovata Schönland = Crassula setulosa Harv. ■☆
109401 Crassula setulosa Harv. var. ramosa Schönland = Crassula setulosa Harv. ■☆
109402 Crassula setulosa Harv. var. robusta Schönland = Crassula setulosa Harv. ■☆
109403 Crassula setulosa Harv. var. rubra (N. E. Br.) G. D. Rowley;红细刚毛青锁龙■☆
109404 Crassula sieberiana Druce;西伯尔青锁龙;Siberian Pygmyweed ■☆
109405 Crassula similis Baker f. = Crassula alba Forssk. var. parvisepala (Schönland) Toelken ■☆
109406 Crassula simulans Schönland;相似青锁龙■☆
109407 Crassula sladenii Schönland;斯莱登青锁龙●☆
109408 Crassula smutsii Schönland = Crassula atropurpurea (Haw.) D. Dietr. ●☆
109409 Crassula socialis Schönland;莲座青锁龙●☆
109410 Crassula southii Schönland;索斯青锁龙■☆
109411 Crassula southii Schönland subsp. sphaerocephala Toelken;球头青锁龙■☆
109412 Crassula spathulata Thunb.;匙形青锁龙●☆
109413 Crassula spectabilis Schönland = Crassula vaginata Eckl. et Zeyh. ■☆
109414 Crassula sphaeritis Harv. = Crassula subulata L. ■☆
109415 Crassula spicata Thunb. = Crassula capitella Thunb. ■☆
109416 Crassula squamulosa Schltdl. = Crassula pruinosa L. ●☆
109417 Crassula stachyera Eckl. et Zeyh. = Crassula obovata Haw. ■☆
109418 Crassula stachyera Eckl. et Zeyh. var. pulchella Harv. = Crassula setulosa Harv. ■☆
109419 Crassula stachyera Eckl. et Zeyh. var. rotundifolia Harv. = Crassula obovata Haw. ■☆
109420 Crassula stewartiae Burtt Davy = Crassula alba Forssk. ■☆
109421 Crassula streyi Toelken;施特赖青锁龙;Stonecrop ■☆
109422 Crassula strigosa L.;糙伏毛青锁龙■☆
109423 Crassula suavis Friedrich = Crassula sericea Schönland ■☆
109424 Crassula subacaulis Schönland et Baker f. subsp. erosula (N. E. Br.) Toelken;啮蚀状近无叶青锁龙■☆
109425 Crassula subaphylla (Eckl. et Zeyh.) Harv.;近无叶青锁龙■☆
109426 Crassula subaphylla (Eckl. et Zeyh.) Harv. var. puberula ? = Crassula subaphylla (Eckl. et Zeyh.) Harv. ■☆
109427 Crassula subaphylla (Eckl. et Zeyh.) Harv. var. virgata (Harv.) Toelken;条纹近无叶青锁龙■☆
109428 Crassula subaphylla (Eckl. et Zeyh.) Harv. var. virgata (Harv.) Toelken = Crassula virgata Harv. ■☆
109429 Crassula subbifaria Schönland = Crassula capitella Thunb. ■☆
109430 Crassula subincana (Haw.) D. Dietr. = Crassula mollis Thunb. ■☆
109431 Crassula subsessilis W. F. Barker = Crassula tetragona L. subsp. connivens (Schönland) Toelken ●☆
109432 Crassula subulata L.;钻头青锁龙■☆
109433 Crassula subulata L. var. fastigiata (Schönland) Toelken;帚状钻头青锁龙■☆
109434 Crassula subulata L. var. hispida Toelken;硬毛钻头青锁龙■☆
109435 Crassula sulcata (Haw.) D. Dietr. = Crassula nudicaulis L. ■☆
109436 Crassula sulphurea Kunze = Crassula vaginata Eckl. et Zeyh. ■☆
109437 Crassula susannae Rauh et Friedrich;苏珊娜青锁龙■☆
109438 Crassula swaziensis Schönland;斯威士青锁龙■☆
109439 Crassula swaziensis Schönland f. argyrophylla (Diels ex Schönland et Baker f.) R. Fern. = Crassula swaziensis Schönland ■☆
109440 Crassula swaziensis Schönland f. brevipilosa R. Fern. = Crassula swaziensis Schönland ■☆
109441 Crassula swaziensis Schönland subsp. brachycarpa R. Fern. = Crassula swaziensis Schönland ■☆
109442 Crassula swaziensis Schönland var. gurnensis R. Fern. = Crassula swaziensis Schönland ■☆
109443 Crassula sylvatica Licht. ex Schult. = Crassula strigosa L. ■☆
109444 Crassula tabularis Dinter;扁平青锁龙■☆
109445 Crassula tayloriae Schönland = Crassula cotyledonis Thunb. ■☆
109446 Crassula tecta Thunb.;小野衣■☆
109447 Crassula tenuicaulis Schönland;细茎青锁龙■☆
109448 Crassula tenuifolia Schönland = Crassula sediflora (Eckl. et Zeyh.) Endl. et Walp. ■☆
109449 Crassula tenuipedicellata Schönland et Baker f.;细花梗青锁龙■☆
109450 Crassula tenuis Wolley-Dod = Crassula umbellata Thunb. ■☆
109451 Crassula teres Marloth = Crassula barklyi N. E. Br. ■☆
109452 Crassula tetragona L.; 四角青锁龙(桃源境); Crassula, Miniature Pine Tree ●☆
109453 Crassula tetragona L. subsp. acutifolia (Lam.) Toelken;尖叶四角青锁龙●☆
109454 Crassula tetragona L. subsp. connivens (Schönland) Toelken;靠合四角青锁龙●☆

109455 Crassula tetragona L. subsp. lignescens Toelken;木质四角青锁龙●☆

109456 Crassula tetragona L. subsp. robusta (Toelken) Toelken;粗壮四角青锁龙●☆

109457 Crassula tetragona L. subsp. rudis (Schönland et Baker f.) Toelken;粗糙四角青锁龙●☆

109458 Crassula thorncroftii Burtt Davy = Crassula expansa Dryand. subsp. fragilis (Baker) Toelken ■☆

109459 Crassula thunbergiana Schult.;桑伯格青锁龙■☆

109460 Crassula thunbergiana Schult. subsp. minutiflora (Schönland et Baker f.) Toelken;微花桑伯格青锁龙■☆

109461 Crassula thyrsiflora Thunb. = Crassula capitella Thunb. subsp. thyrsiflora (Thunb.) Toelken ■☆

109462 Crassula tibestica Miré et Quézel = Crassula schimperi Fisch. et C. A. Mey. ■☆

109463 Crassula tillaea Lest. -Garl.;丛生东爪草;Moss Pygmyweed, Mossy Stonecrop ■☆

109464 Crassula tomentosa Thunb.;绒毛东爪草■☆

109465 Crassula tomentosa Thunb. var. glabrifolia (Harv.) G. D. Rowley;光叶绒毛东爪草■☆

109466 Crassula tomentosa Thunb. var. interrupta (Drège ex Harv.) Toelken = Crassula tomentosa Thunb. var. glabrifolia (Harv.) G. D. Rowley ■☆

109467 Crassula tomentosa Thunb. var. setigera (Eckl. et Zeyh.) Schönland = Crassula tomentosa Thunb. ■☆

109468 Crassula torquata Baker f. = Crassula cultrata L. ■☆

109469 Crassula trachysantha (Eckl. et Zeyh.) Harv. = Crassula mesembryanthoides (Haw.) D. Dietr. ●☆

109470 Crassula transvaalensis (Kuntze) K. Schum. = Crassula lanceolata (Eckl. et Zeyh.) Endl. ex Walp. subsp. transvaalensis (Kuntze) Toelken ■☆

109471 Crassula triebneri Schönland ex H. Jacobsen = Crassula capitella Thunb. subsp. thyrsiflora (Thunb.) Toelken ■☆

109472 Crassula turrita Thunb. = Crassula capitella Thunb. subsp. thyrsiflora (Thunb.) Toelken ■☆

109473 Crassula turrita Thunb. var. latifolia Harv. = Crassula capitella Thunb. ■☆

109474 Crassula turrita Thunb. var. rosea Haw. = Crassula capitella Thunb. subsp. thyrsiflora (Thunb.) Toelken ■☆

109475 Crassula tysonii Schönland = Crassula pellucida L. subsp. brachypetala (Drège ex Harv.) Toelken ●☆

109476 Crassula umbella Jacq.;伞花青锁龙●☆

109477 Crassula umbellata Thunb.;小伞青锁龙■☆

109478 Crassula umbellata Thunb. var. nana (Schönland et Baker f.) Schönland = Crassula umbellata Thunb. ■☆

109479 Crassula umbraticola N. E. Br.;荫蔽青锁龙■☆

109480 Crassula undata Haw. = Crassula dejecta Jacq. ■☆

109481 Crassula undulata Haw. = Crassula dejecta Jacq. ■☆

109482 Crassula uniflora Schönland = Crassula expansa Dryand. subsp. filicaulis (Haw.) Toelken ■☆

109483 Crassula vaginata Eckl. et Zeyh.;具鞘青锁龙■☆

109484 Crassula vaginata Eckl. et Zeyh. subsp. minuta Toelken;小具鞘青锁龙■☆

109485 Crassula vaginata Eckl. et Zeyh. var. hispida Keissl. = Crassula vaginata Eckl. et Zeyh. ■☆

109486 Crassula vaginata Eckl. et Zeyh. var. laxa Keissl. = Crassula vaginata Eckl. et Zeyh. ■☆

109487 Crassula vaginata Eckl. et Zeyh. var. parviflora Keissl. = Crassula vaginata Eckl. et Zeyh. ■☆

109488 Crassula vaillantii (Willd.) Roth;迭罗汉●☆

109489 Crassula velutina Friedrich = Crassula sericea Schönland var. velutina (Friedrich) Toelken ■☆

109490 Crassula versicolor Burch. ex Ker Gawl. = Crassula coccinea L. ●☆

109491 Crassula vestita Thunb.;包被青锁龙■☆

109492 Crassula virgata Harv. = Crassula subaphylla (Eckl. et Zeyh.) Harv. var. virgata (Harv.) Toelken ■☆

109493 Crassula volkensii Engl.;福尔青锁龙■☆

109494 Crassula watermeyeri Compton = Crassula atropurpurea (Haw.) D. Dietr. var. watermeyeri (Compton) Toelken ■☆

109495 Crassula whiteheadii Harv.;怀特黑德青锁龙■☆

109496 Crassula whyteana Schönland = Crassula globularioides Britten ●☆

109497 Crassula wilmsii Diels = Crassula alba Forssk. var. parvisepala (Schönland) Toelken ■☆

109498 Crassula woodii Schönland = Crassula expansa Dryand. subsp. fragilis (Baker) Toelken ■☆

109499 Crassula wrightiana Bullock = Crassula granvikii Mildbr. ■☆

109500 Crassula yunnanensis Franch. = Sinocrassula yunnanensis (Franch.) A. Berger ■

109501 Crassula zeyheriana Schönland = Crassula thunbergiana Schult. ■☆

109502 Crassula zimmermannii Engl. = Crassula expansa Dryand. subsp. fragilis (Baker) Toelken ■☆

109503 Crassula zombensis Baker f.;宗巴青锁龙■☆

109504 Crassulaceae DC. = Crassulaceae J. St. -Hil. (保留科名)●■

109505 Crassulaceae J. St. -Hil. (1805)(保留科名);景天科;Crassula Family, Orpine Family, Stonecrop Family ●■

109506 Crassularia Hochst. ex Schweinf. = Crassula L. ●■☆

109507 Crassuvia Comm. ex Lam. = Kalanchoe Adans. ●■

109508 Crataegosorbus Makino = Sorbus L. ●

109509 Crataegus L. (1753);山楂属;Hawthorn, May Flower, May Thorn, Ornamental Thorn, Thorn, Thornapple ●

109510 Crataegus Tourn. ex L. = Crataegus L. ●

109511 Crataegus aboriginum Sarg. = Crataegus chrysocarpa Ashe var. aboriginum (Sarg.) Kruschke ●☆

109512 Crataegus acanthacolonensis Laughlin = Crataegus calpodendron (Ehrh.) Medik. ●☆

109513 Crataegus acutifolia Sarg.;尖叶山楂;Hawthorn ●☆

109514 Crataegus acutifolia Sarg. = Crataegus crus-galli L. ●

109515 Crataegus acutifolia Sarg. var. insignis (Sarg.) Palmer = Crataegus crus-galli L. ●

109516 Crataegus acutiserrata Kruschke;尖齿山楂●☆

109517 Crataegus albicans Ashe = Crataegus mollis (Torr. et A. Gray) Scheele ●☆

109518 Crataegus algens Beadle = Crataegus crus-galli L. ●

109519 Crataegus alnifolia Siebold et Zucc. = Aria alnifolia (Siebold et Zucc.) Decne. ●

109520 Crataegus alnifolia Siebold et Zucc. = Sorbus alnifolia (Siebold et Zucc.) K. Koch ●

109521 Crataegus alnorum Sarg. = Crataegus schuettei Ashe ●☆

109522 Crataegus altaica (Loudon) Lange;阿尔泰山楂;Altai Hawthorn, Altai Mountain Hawthorn ●

109523 Crataegus altaica (Loudon) Lange var. villosa (Rupr.) Lange = Crataegus maximowiczii C. K. Schneid. ●

109524 Crataegus ambigua Becker;可疑山楂●☆
109525 Crataegus ambrosia Sarg. = Crataegus succulenta Schrad. ex Link ●
109526 Crataegus anomala Sarg.;畸山楂;Anomalous Hawthorn ●☆
109527 Crataegus apiifolia Medik.;欧芹叶山楂;Parsley Hawthorn, Parsley Hawtitorn, Parsley-leafed Thorn ●☆
109528 Crataegus apiomorpha Sarg.;纵纹山楂;Fort Sheridan Hawthorn ●☆
109529 Crataegus apiomorpha Sarg. f. paucispina (Sarg.) Kruschke = Crataegus apiomorpha Sarg. ●☆
109530 Crataegus apiomorpha Sarg. var. cyanophylla (Sarg.) Kruschke = Crataegus apiomorpha Sarg. ●☆
109531 Crataegus aquilonaris Sarg.;北部山楂;Northern Hawthorn ●☆
109532 Crataegus ardula Sarg. = Crataegus succulenta Schrad. ex Link ●
109533 Crataegus argyi H. Lév. et Vaniot = Crataegus cuneata Siebold et Zucc. ●
109534 Crataegus arkansana Sarg. = Crataegus mollis (Torr. et A. Gray) Scheele ●☆
109535 Crataegus armena Pojark.;杏黄山楂●☆
109536 Crataegus arnoldiana Sarg.;阿诺德山楂;Arnold Hawthorn ●
109537 Crataegus arnoldiana Sarg. = Crataegus submollis Sarg. ●☆
109538 Crataegus aronia Decne.;扶移山楂●☆
109539 Crataegus aspera Sarg. = Crataegus pruinosa (H. L. Wendl.) K. Koch ●☆
109540 Crataegus asperata Sarg. = Crataegus lucorum Sarg. ●☆
109541 Crataegus atrorubens Ashe;黑红山楂;Hawthorn ●☆
109542 Crataegus atrosanguinea Pojark.;暗紫山楂●☆
109543 Crataegus aulica Sarg. = Crataegus pedicellata Sarg. ●☆
109544 Crataegus aurantia Pojark.;橘红山楂;Orangered Hawthorn, Orange-red Hawthorn ●
109545 Crataegus azarolus L.;南欧山楂;Azarole, Azarole Hawthorn, Mediterranean Medlar ●☆
109546 Crataegus azarolus L. var. aronia ? = Crataegus azarolus L. ●☆
109547 Crataegus balkwillii Sarg. = Crataegus scabrida Sarg. ●☆
109548 Crataegus baroussana Eggl.;墨西哥山楂;Tejocote ●☆
109549 Crataegus barrettiana Sarg. = Crataegus crus-galli L. ●
109550 Crataegus basilica Beadle = Crataegus schuettei Ashe ●☆
109551 Crataegus beata Sarg.;邓氏山楂;Dunbar's Hawthorn ●☆
109552 Crataegus beckeriana Pojark.;白氏山楂(白克氏山楂);Becker Hawthorn ●☆
109553 Crataegus beipiaogensis S. L. Tung et X. J. Tian;北票山楂;Beipiao Hawthorn ●
109554 Crataegus beipiaogensis S. L. Tung et X. J. Tian = Crataegus maximowiczii C. K. Schneid. ●
109555 Crataegus berberifolia Torr. et Gray;小檗叶山楂;Barberry Hawthorn, Barberryleaf Hawthorn, Bigtree Hawthorn ●☆
109556 Crataegus bibas Lour. = Eriobotrya japonica (Thunb. ex Murray) Lindl. ●
109557 Crataegus bicknellii Eggl. = Crataegus florifera Sarg. ●☆
109558 Crataegus bodinieri H. Lév. = Crataegus scabrifolia (Franch.) Rehder ●
109559 Crataegus brachyacantha Sarg. et Engelm.;蓝果山楂;Blue Hawthorn, Blueberry, Hawthorn ●☆
109560 Crataegus brachyphylla Sarg. = Crataegus mollis (Torr. et A. Gray) Scheele ●☆
109561 Crataegus brainderi Sarg.;布雷纳德氏山楂;Brainerd Hawthorn ●☆
109562 Crataegus brainerdii Sarg. var. asperifolia (Sarg.) Eggl. = Crataegus scabrida Sarg. ●☆
109563 Crataegus brainerdii Sarg. var. cyclophylla (Sarg.) E. J. Palmer = Crataegus scabrida Sarg. ●☆
109564 Crataegus brainerdii Sarg. var. egglestonii (Sarg.) B. L. Rob. = Crataegus scabrida Sarg. ●☆
109565 Crataegus brainerdii Sarg. var. scabrida (Sarg.) Eggl. = Crataegus scabrida Sarg. ●☆
109566 Crataegus brevispina Kuntze = Crataegus monogyna Jacq. ●☆
109567 Crataegus brockwayae Sarg. = Crataegus douglasii Lindl. ●
109568 Crataegus brunetiana Sarg. = Crataegus chrysocarpa Ashe ●
109569 Crataegus brunetiana Sarg. var. fernaldii (Sarg.) E. J. Palmer = Crataegus chrysocarpa Ashe ●
109570 Crataegus bushii Sarg. = Crataegus crus-galli L. ●
109571 Crataegus caesariata Sarg. = Crataegus lumaria Ashe ●☆
109572 Crataegus caliciglabrata Schuette = Crataegus chrysocarpa Ashe var. phoenicea E. J. Palmer ●☆
109573 Crataegus calpodendron (Ehrh.) Medik.;梨山楂;Pear Hawthorn, Sugar Hawthorn, Urn-tree Hawthorn ●☆
109574 Crataegus calpodendron (Ehrh.) Medik. = Crataegus macracantha G. Lodd. ex Loudon ●☆
109575 Crataegus calpodendron (Ehrh.) Medik. var. gigantea Kruschke = Crataegus calpodendron (Ehrh.) Medik. ●☆
109576 Crataegus calpodendron (Ehrh.) Medik. var. globosa (Sarg.) E. J. Palmer = Crataegus calpodendron (Ehrh.) Medik. ●☆
109577 Crataegus calpodendron (Ehrh.) Medik. var. hispida (Sarg.) E. J. Palmer = Crataegus calpodendron (Ehrh.) Medik. ●☆
109578 Crataegus calpodendron (Ehrh.) Medik. var. hispidula (Sarg.) E. J. Palmer = Crataegus calpodendron (Ehrh.) Medik. ●☆
109579 Crataegus calpodendron (Ehrh.) Medik. var. microcarpa (Chapm.) E. J. Palmer = Crataegus calpodendron (Ehrh.) Medik. ●☆
109580 Crataegus calpodendron (Ehrh.) Medik. var. mollicula (Sarg.) E. J. Palmer = Crataegus calpodendron (Ehrh.) Medik. ●☆
109581 Crataegus calpodendron (Ehrh.) Medik. var. obesa (Ashe) E. J. Palmer = Crataegus calpodendron (Ehrh.) Medik. ●☆
109582 Crataegus canadensis Sarg.;加拿大山楂;Canada Hawthorn ●☆
109583 Crataegus canbyi Sarg. = Crataegus crus-galli L. ●
109584 Crataegus carrierei Vauvel;卡里山楂(卡里埃氏山楂);Carriere Hawthorn ●☆
109585 Crataegus caucasica K. Koch;高加索山楂●☆
109586 Crataegus cavaleriei H. Lév. = Photinia komarovii (H. Lév. et Vaniot) L. T. Lu et C. L. Li ●
109587 Crataegus cavaleriei H. Lév. et Vaniot = Malus sieboldii (Regel) Rehder ●
109588 Crataegus cavaleriei H. Lév. et Vaniot = Malus toringo (Siebold) Siebold ex de Vriese ●
109589 Crataegus celsa Sarg. = Crataegus florifera Sarg. ●☆
109590 Crataegus chadsfordiana Sarg. = Crataegus macrosperma Ashe ●☆
109591 Crataegus champlainensis Sarg.;香普兰山楂;Lake Champlain Hawthorn ●☆
109592 Crataegus chantcha H. Lév. = Crataegus cuneata Siebold et Zucc. ●
109593 Crataegus chapmanii (Beadle) Ashe = Crataegus calpodendron (Ehrh.) Medik. ●☆
109594 Crataegus cherokeensis Sarg. = Crataegus crus-galli L. ●
109595 Crataegus chippewaensis Sarg. = Crataegus fulleriana Sarg. ●☆

109596 Crataegus chitaensis Sarg. = Crataegus dahurica Koehne ex C. K. Schneid. ●

109597 Crataegus chlorosarca Maxim.;绿肉山楂(黑果山楂,库页山楂,绿果山楂);Blackfruit Hawthorn, Black-fruited Hawthorn, Greenfleshy Hawthorn ●

109598 Crataegus chrysocarpa Ashe;黄果山楂(刺山楂);Fireberry Hawthorn, Golden-fruit Hawthorn, Roundleaf Hawthorn, Round-leaved Hawthorn ●

109599 Crataegus chrysocarpa Ashe var. aboriginum (Sarg.) Kruschke;土著黄果山楂;Fireberry Hawthorn ●☆

109600 Crataegus chrysocarpa Ashe var. bicknellii (Eggl.) E. J. Palmer = Crataegus succulenta Schrad. ex Link ●

109601 Crataegus chrysocarpa Ashe var. caesariata (Sarg.) E. J. Palmer = Crataegus lumaria Ashe ●☆

109602 Crataegus chrysocarpa Ashe var. longiacuminata Kruschke = Crataegus chrysocarpa Ashe ●

109603 Crataegus chrysocarpa Ashe var. phoenicea E. J. Palmer;紫红山楂;Fireberry Hawthorn ●☆

109604 Crataegus chrysocarpa Ashe var. phoenicea E. J. Palmer = Crataegus rotundifolia Borkh. ●☆

109605 Crataegus chrysocarpa Ashe var. rotundifolia (Moench) Sarg. = Crataegus chrysocarpa Ashe ●

109606 Crataegus chungtienensis W. W. Sm.;中甸山楂(野山楂);Zhongdian Hawthorn ●

109607 Crataegus cibaria Beadle = Crataegus mollis (Torr. et A. Gray) Scheele ●☆

109608 Crataegus coccinata Sarg. = Crataegus chrysocarpa Ashe var. aboriginum (Sarg.) Kruschke ●☆

109609 Crataegus coccinea L.;美洲山楂(梨果山楂,裂果山楂);Biltmore Hawthorn, Copenhagen Hawthorn, Pear-fruited Gockspur-thorn, Scarlet Haw, Scarlet Hawthorn, Thicket Hawthorn ●☆

109610 Crataegus coccinea L. = Crataegus pedicellata Sarg. ●☆

109611 Crataegus coccinea L. var. mollis Torr. et A. Gray = Crataegus mollis (Torr. et A. Gray) Scheele ●☆

109612 Crataegus coccinioides Ashe;堪萨斯山楂;Kansas Hawthorn, Late-flowered Cockspur-thorn, Mississippi Hawberry ●☆

109613 Crataegus cocksii Sarg. = Crataegus crus-galli L. ●

109614 Crataegus collicola Ashe = Crataegus disperma Ashe ●☆

109615 Crataegus collina Chapm. = Crataegus punctata Jacq. ●

109616 Crataegus collina Chapm. var. collicola (Ashe) E. J. Palmer = Crataegus disperma Ashe ●☆

109617 Crataegus collina Chapm. var. collicola (Ashe) E. J. Palmer = Crataegus collina Chapm. ●

109618 Crataegus collina Chapm. var. secta (Sarg.) E. J. Palmer = Crataegus collina Chapm. ●

109619 Crataegus collina Chapm. var. sordida (Sarg.) E. J. Palmer = Crataegus collina Chapm. ●

109620 Crataegus collina Chapm. var. succincta (Sarg.) E. J. Palmer = Crataegus collina Chapm. ●

109621 Crataegus columbiana Howell; 哥伦比亚山楂; Columbia Hawthorn ●☆

109622 Crataegus columbiana J. T. Howell var. chrysocarpa (Ashe) Dorn = Crataegus chrysocarpa Ashe ●

109623 Crataegus columbiana J. T. Howell var. occidentalis (Britton) Dorn = Crataegus macracantha G. Lodd. ex Loudon var. occidentalis (Britton) Eggl. ●☆

109624 Crataegus confragosa Sarg. = Crataegus fulleriana Sarg. ●☆

109625 Crataegus congesta Sarg. = Crataegus pruinosa (H. L. Wendl.) K. Koch ●☆

109626 Crataegus cordata Elliott = Crataegus phaenopyrum Borkh. ●☆

109627 Crataegus coreana H. Lév. = Crataegus pinnatifida Bunge var. psilosa C. K. Schneid. ●

109628 Crataegus corusca Sarg.;亮枝山楂;Shining-branch Hawthorn ●☆

109629 Crataegus corusca Sarg. var. gigantea Kruschke = Crataegus corusca Sarg. ●☆

109630 Crataegus corusca Sarg. var. hillii (Sarg.) Kruschke = Crataegus corusca Sarg. ●☆

109631 Crataegus crassifolia Sarg. = Crataegus dodgei Ashe ●☆

109632 Crataegus crawfordiana Sarg. = Crataegus pruinosa (H. L. Wendl.) K. Koch ●☆

109633 Crataegus crenulata (D. Don) Roxb. = Pyracantha crenulata (D. Don) M. Roem. ●

109634 Crataegus crenulata Roxb. = Pyracantha crenulata (D. Don) M. Roem. ●

109635 Crataegus crudelis Sarg.;劣味山楂;Cruel Hawthorn ●☆

109636 Crataegus crus-galli L.;鸡脚山楂(鸡距山楂,重齿叶山楂);Cockspur Hawthorn, Cockspur Thorn, Cockspur-thorn, Hog-apple, Newcastle-thorn, Thornless Cockspur Hawthorn ●

109637 Crataegus crus-galli L. 'Inermis';无刺鸡脚山楂●☆

109638 Crataegus crus-galli L. = Crataegus hannibalensis E. J. Palmer ●☆

109639 Crataegus crus-galli L. var. barrettiana (Sarg.) E. J. Palmer = Crataegus crus-galli L. ●

109640 Crataegus crus-galli L. var. bellica (Sarg.) E. J. Palmer = Crataegus crus-galli L. ●

109641 Crataegus crus-galli L. var. capillata Sarg. = Crataegus crus-galli L. ●

109642 Crataegus crus-galli L. var. crus-galli f. truncata (Sarg.) E. J. Palmer = Crataegus crus-galli L. ●

109643 Crataegus crus-galli L. var. exigua (Sarg.) Eggl. = Crataegus crus-galli L. ●

109644 Crataegus crus-galli L. var. leptophylla (Sarg.) E. J. Palmer = Crataegus crus-galli L. ●

109645 Crataegus crus-galli L. var. macra (Beadle) E. J. Palmer = Crataegus crus-galli L. ●

109646 Crataegus crus-galli L. var. oblongata Sarg. = Crataegus crus-galli L. ●

109647 Crataegus crus-galli L. var. pachyphylla (Sarg.) E. J. Palmer = Crataegus crus-galli L. ●

109648 Crataegus crus-galli L. var. pyracanthifolia Aiton = Crataegus crus-galli L. ●

109649 Crataegus crus-galli L. var. salicifolia (Medik.) Wood;柳叶鸡脚山楂●☆

109650 Crataegus cuneata Siebold et Zucc.;野山楂(鼻涕团,大红子,浮萍果,红果子,猴楂,候抓子,栌,毛枣子,牧虎梨,南山楂,山梨,山里果树,山里红,山栌子,山楂,小叶山楂,药山栌);Nippon Hawthorn ●

109651 Crataegus cuneata Siebold et Zucc. f. lutea Matsum. ex Koidz.;黄果野山楂;Yellow-fruited Nippon Hawthorn ●☆

109652 Crataegus cuneata Siebold et Zucc. f. pleniflora S. X. Qian;重瓣野山楂;Doubleflower Nippon Hawthorn ●

109653 Crataegus cuneata Siebold et Zucc. f. pleniflora S. X. Qian = Crataegus cuneata Siebold et Zucc. ●

109654 Crataegus cuneata Siebold et Zucc. f. tangchungchangii (F. P. Metcalf) Y. T. Chang = Crataegus cuneata Siebold et Zucc. var.

tangchungchangii (F. P. Metcalf) T. C. Ku et Spongberg ●

109655 Crataegus cuneata Siebold et Zucc. f. xanthocarpa Nakai = Crataegus cuneata Siebold et Zucc. f. lutea Matsum. ex Koidz. ●☆

109656 Crataegus cuneata Siebold et Zucc. var. shangnanensis L. Mao et T. C. Cui;匍匐野山楂;Dwarf Nippon Hawthorn,Shannan Hawthorn ●

109657 Crataegus cuneata Siebold et Zucc. var. shangnanensis L. Mao et T. C. Cui = Crataegus cuneata Siebold et Zucc. ●

109658 Crataegus cuneata Siebold et Zucc. var. tangchungchangii (F. P. Metcalf) T. C. Ku et Spongberg;小叶野山楂●

109659 Crataegus cuneiformis (Marshall) Eggl. = Crataegus disperma Ashe ●☆

109660 Crataegus curvisepala Lindm.;东欧山楂●☆

109661 Crataegus cuspidata Spach = Sorbus cuspidata (Spach) Hedl. ●

109662 Crataegus dahurica Koehne ex C. K. Schneid.;光叶山楂;Dahur Hawthorn,Dahuria Hawthorn,Dahurian Hawthorn ●

109663 Crataegus dahurica Koehne ex C. K. Schneid. var. laevicalyx (J. X. Huang,L. Y. Sun et T. J. Feng) T. C. Ku et Spongberg;光萼山楂;Laevicalyx Hawthorn,Smooth-calyx Hawthorn ●

109664 Crataegus danielsii E. J. Palmer = Crataegus crus-galli L. ●

109665 Crataegus delosii Sarg. = Crataegus dodgei Ashe ●☆

109666 Crataegus deltoides Ashe = Crataegus pruinosa (H. L. Wendl.) K. Koch ●☆

109667 Crataegus denaria Beadle = Crataegus crus-galli L. ●

109668 Crataegus desueta Sarg.;纽约山楂;New York Hawthorn ●☆

109669 Crataegus desueta Sarg. var. wausaukiensis Kruschke = Crataegus desueta Sarg. ●☆

109670 Crataegus dilatata Sarg.;阔叶山楂;Apple-leaf Hawthorn, Broadleaf Hawthorn ●☆

109671 Crataegus dipyrena Pojark.;二籽山楂●☆

109672 Crataegus disjuncta Sarg. = Crataegus dissona Sarg. ●☆

109673 Crataegus disjuncta Sarg. = Crataegus pruinosa (H. L. Wendl.) K. Koch ●☆

109674 Crataegus disperma Ashe;撒布山楂;Hawthorn, Spreading Hawthorn ●☆

109675 Crataegus disperma Ashe var. peoriensis (Sarg.) Kruschke = Crataegus disperma Ashe ●☆

109676 Crataegus dissona Sarg.;西北部山楂;Northern Hawthorn ●☆

109677 Crataegus dissona Sarg. var. bellula (Sarg.) Kruschke = Crataegus dissona Sarg. ●☆

109678 Crataegus distincta Kruschke;显著山楂;Distinct Hawthorn ●☆

109679 Crataegus divida Sarg. = Crataegus macracantha G. Lodd. ex Loudon var. occidentalis (Britton) Eggl. ●☆

109680 Crataegus dodgei Ashe;道奇山楂;Dodge's Hawthorn ●☆

109681 Crataegus dodgei Ashe var. flavida (Sarg.) P. G. Sm. et J. B. Phipps = Crataegus dodgei Ashe ●☆

109682 Crataegus dodgei Ashe var. lumaria (Ashe) Sarg. = Crataegus lumaria Ashe ●☆

109683 Crataegus dodgei Ashe var. rotundata (Sarg.) Kruschke = Crataegus dodgei Ashe ●☆

109684 Crataegus douglasii Lindl.;道格拉斯山楂;Black Hawberry, Black Hawthorn,Douglas Hawthorn,River Hawthorn ●

109685 Crataegus dsungarica Zabel;北亚山楂(天山山楂,准噶尔山楂)●☆

109686 Crataegus dunbarii Sarg. = Crataegus scabrida Sarg. ●☆

109687 Crataegus durobrivensis Sarg.;锐齿山楂●☆

109688 Crataegus ellwangeriana Sarg.;猩红梨果山楂;Ellwanger Hawthorn,Scarlet Hawthorn ●☆

109689 Crataegus ellwangeriana Sarg. = Crataegus pedicellata Sarg. var. ellwangeriana (Sarg.) Eggl. ●☆

109690 Crataegus ellwangeriana Sarg. var. sinistra (Beadle) E. J. Palmer = Crataegus ellwangeriana Sarg. ●☆

109691 Crataegus engelmannii Sarg.;恩格尔曼山楂;Barberry-leaved Hawthorn ●☆

109692 Crataegus engelmannii Sarg. f. nuda E. J. Palmer = Crataegus crus-galli L. ●

109693 Crataegus eriantha Pojark.;毛花山楂●☆

109694 Crataegus eriocarpa Pomel = Crataegus laciniata Ucria ●☆

109695 Crataegus evansiana Sarg. = Crataegus margaretta Ashe ●☆

109696 Crataegus faxoni Sarg. = Crataegus chrysocarpa Ashe ●

109697 Crataegus faxoni Sarg. var. durifructa Kruschke = Crataegus chrysocarpa Ashe ●

109698 Crataegus faxoni Sarg. var. praecoqua (Sarg.) Kruschke = Crataegus chrysocarpa Ashe ●

109699 Crataegus faxoni Sarg. var. praetermissa (Sarg.) E. J. Palmer = Crataegus chrysocarpa Ashe ●

109700 Crataegus fecunda Sarg.;多育山楂;Hawthorn ●☆

109701 Crataegus fecunda Sarg. = Crataegus crus-galli L. ●

109702 Crataegus ferrissii Ashe = Crataegus schuettei Ashe ●☆

109703 Crataegus ferta Sarg. = Crataegus succulenta Schrad. ex Link ●

109704 Crataegus filipes Ashe = Crataegus irrasa Sarg. ●☆

109705 Crataegus fischeri C. K. Schneid. = Crataegus songorica K. Koch ●

109706 Crataegus flabellata (Bosc ex Spach) K. Koch var. grayana (Eggl.) E. J. Palmer = Crataegus flabellata (Bosc ex Spach) Rydb. ●☆

109707 Crataegus flabellata (Bosc ex Spach) Rydb.;扇叶山楂;Fanleaf Hawthorn,Fan-leaf Hawthorn,Fan-shaped Hawthorn ●☆

109708 Crataegus flabellata sensu Gleason et Cronquist = Crataegus macrosperma Ashe ●☆

109709 Crataegus flava Aiton;黄山楂(黄果山楂);Summer Hawthorn, Yellow Haw,Yellow Hawthorn,Yellow-fruited Hawthorn ●☆

109710 Crataegus flavida Sarg. = Crataegus dodgei Ashe ●☆

109711 Crataegus florifera Sarg.;花山楂●☆

109712 Crataegus florifera Sarg. var. celsa (Sarg.) Kruschke = Crataegus florifera Sarg. ●☆

109713 Crataegus florifera Sarg. var. mortonis (Laughlin) Kruschke = Crataegus succulenta Schrad. ex Link ●

109714 Crataegus florifera Sarg. var. shirleyensis (Sarg.) Kruschke = Crataegus florifera Sarg. ●☆

109715 Crataegus florifera Sarg. var. virilis (Sarg.) Kruschke = Crataegus succulenta Schrad. ex Link ●

109716 Crataegus fluviatilis Sarg.;河旁山楂●☆

109717 Crataegus fluviatilis Sarg. = Crataegus roanensis Ashe var. fluviatilis (Sarg.) Kruschke ●☆

109718 Crataegus fontanesiana (Spach) Steud. = Crataegus calpodendron (Ehrh.) Medik. ●☆

109719 Crataegus formosa Sarg. = Crataegus pruinosa (H. L. Wendl.) K. Koch ●☆

109720 Crataegus franklinensis Sarg. = Crataegus pruinosa (H. L. Wendl.) K. Koch ●☆

109721 Crataegus fretalis Sarg. = Crataegus macrosperma Ashe ●☆

109722 Crataegus fulleriana (Sarg.) Kruschke = Crataegus fulleriana Sarg. ●☆

109723 Crataegus fulleriana Sarg.;富勒山楂;Fuller's Hawthorn ●☆

109724 Crataegus fulleriana Sarg. var. chippewaensis (Sarg.) Kruschke

= Crataegus fulleriana Sarg. ●☆
109725 Crataegus fulleriana Sarg. var. gigantea Kruschke = Crataegus fulleriana Sarg. ●☆
109726 Crataegus fulleriana Sarg. var. magniflora (Sarg.) E. J. Palmer = Crataegus fulleriana Sarg. ●☆
109727 Crataegus fulleriana Sarg. var. miranda (Sarg.) Kruschke = Crataegus fulleriana Sarg. ●☆
109728 Crataegus fulleriana Sarg. var. miranda (Sarg.) Kruschke f. magniflora (Sarg.) Kruschke = Crataegus fulleriana Sarg. ●☆
109729 Crataegus gattingeri Ashe;加氏山楂;Hawthorn ●☆
109730 Crataegus gattingeri Ashe = Crataegus pruinosa (H. L. Wendl.) K. Koch ●☆
109731 Crataegus gattingeri Ashe var. rigida E. J. Palmer = Crataegus pruinosa (H. L. Wendl.) K. Koch ●☆
109732 Crataegus gaudens Sarg. = Crataegus pruinosa (H. L. Wendl.) K. Koch ●☆
109733 Crataegus gemmosa Sarg. = Crataegus succulenta Schrad. ex Link ●
109734 Crataegus georgiana Sarg. = Crataegus pruinosa (H. L. Wendl.) K. Koch ●☆
109735 Crataegus glabra Thunb. = Photinia glabra (Thunb.) Maxim. ●
109736 Crataegus glauca Wall. ex G. Don = Stranvaesia nussia (Buch.-Ham. ex D. Don) Decne. ●
109737 Crataegus globosa Sarg. = Crataegus calpodendron (Ehrh.) Medik. ●☆
109738 Crataegus granatensis Boiss.;格拉山楂●☆
109739 Crataegus gravida Beadle = Crataegus mollis (Torr. et A. Gray) Scheele ●☆
109740 Crataegus gravis Ashe = Crataegus prona Sarg. ●☆
109741 Crataegus grayana Eggl. = Crataegus flabellata (Bosc ex Spach) Rydb. ●☆
109742 Crataegus greggiana Eggl.;格雷格山楂;Gregg Hawthorn ●☆
109743 Crataegus grignoniensis Mouill.;丰花山楂;Hawthorn ●☆
109744 Crataegus habereri Sarg. = Crataegus pedicellata Sarg. ●☆
109745 Crataegus hadleyana Sarg. = Crataegus scabrida Sarg. ●☆
109746 Crataegus hannibalensis E. J. Palmer;哈尼巴尔山楂;Hawthorn ●☆
109747 Crataegus hannibalensis E. J. Palmer = Crataegus crus-galli L. ●
109748 Crataegus harbisonii Beadle;哈比森山楂;Harbison Hawthorn ●☆
109749 Crataegus harveyana Sarg.;哈维山楂;Hawthorn ●☆
109750 Crataegus heldreichii Boiss.;黑氏山楂;Heldreich Hawthorn ●☆
109751 Crataegus henryi Dunn = Crataegus scabrifolia (Franch.) Rehder ●
109752 Crataegus heterophylla Stev.; 互叶山楂; Various-leaved Hawthorn ●☆
109753 Crataegus hillii Sarg. = Crataegus corusca Sarg. ●☆
109754 Crataegus holmesiana Ashe;霍尔曼山楂;Holmes' Hawthorn, Red Haw ●☆
109755 Crataegus holmesiana Ashe var. amicta (Ashe) E. J. Palmer = Crataegus holmesiana Ashe ●☆
109756 Crataegus holmesiana Ashe var. chippewaensis (Sarg.) E. J. Palmer = Crataegus fulleriana Sarg. ●☆
109757 Crataegus holmesiana Ashe var. magniflora (Sarg.) E. J. Palmer = Crataegus fulleriana Sarg. ●☆
109758 Crataegus holmesiana Ashe var. villipes Ashe = Crataegus holmesiana Ashe ●☆
109759 Crataegus horridula Sarg. = Crataegus pruinosa (H. L. Wendl.) K. Koch ●☆
109760 Crataegus hupehensis Sarg.;湖北山楂(大山枣,猴楂子,山楂,酸枣);Hubei Hawthorn, Hupeh Hawthorn ●
109761 Crataegus hupehensis Sarg. var. flavida S. Y. Wang;黄果湖北山楂;Yellow-fruit Hubei Hawthorn ●
109762 Crataegus hupehensis Sarg. var. flavida S. Y. Wang = Crataegus hupehensis Sarg. ●
109763 Crataegus illecebrosa Sarg. = Crataegus fulleriana Sarg. ●☆
109764 Crataegus illuminata Sarg. = Crataegus chrysocarpa Ashe ●
109765 Crataegus improvisa Sarg. = Crataegus scabrida Sarg. ●☆
109766 Crataegus incerta Sarg. = Crataegus florifera Sarg. ●☆
109767 Crataegus incisa Sarg. = Crataegus dissona Sarg. ●☆
109768 Crataegus indica L. = Rhaphiolepis indica (L.) Lindl. ex Ker ●
109769 Crataegus induta Sarg. = Crataegus mollis (Torr. et A. Gray) Scheele ●☆
109770 Crataegus insolens Sarg. = Crataegus lucorum Sarg. ●☆
109771 Crataegus integrifolia Roxb. = Cotoneaster microphyllus Wall. ex Lindl. var. thymifolius (Baker) Koehne ●
109772 Crataegus integrifolia Roxb. = Cotoneaster microphyllus Wall. ex Lindl. ●
109773 Crataegus intricata Lange = Crataegus coccinea L. ●☆
109774 Crataegus invisa Sarg. = Crataegus mollis (Torr. et A. Gray) Scheele ●☆
109775 Crataegus irrasa Sarg.;布兰山楂;Blanchard's Hawthorn ●☆
109776 Crataegus irrasa Sarg. var. blanchardii (Sarg.) Eggl. = Crataegus irrasa Sarg. ●☆
109777 Crataegus iterata Sarg. = Crataegus scabrida Sarg. ●☆
109778 Crataegus jackii Sarg. = Crataegus chrysocarpa Ashe var. aboriginum (Sarg.) Kruschke ●☆
109779 Crataegus jesupii Sarg.;杰瑟普山楂;Jesup's Hawthorn ●☆
109780 Crataegus jozana C. K. Schneid.;虾夷山楂●☆
109781 Crataegus kansuensis E. H. Wilson;甘肃山楂(面旦子,山楂,野山楂);Chinese Hawthorn, Gansu Hawthorn, Kansu Hawthorn ●
109782 Crataegus kelloggii Sarg.;凯洛格山楂;Hawthorn ●☆
109783 Crataegus komarovii Sarg. = Malus komarovii (Sarg.) Rehder ●◇
109784 Crataegus kulingensis Sarg. = Crataegus cuneata Siebold et Zucc. ●
109785 Crataegus kyrtostyla Fingerh. ex Schltdl.;弯柱山楂●☆
109786 Crataegus lacera Sarg. = Crataegus mollis (Torr. et A. Gray) Scheele ●☆
109787 Crataegus laciniata Borkh. = Crataegus oxyacantha L. ●☆
109788 Crataegus laciniata Steven; 东方山楂 (深裂叶山楂); Hawthorn, Oriental Hawthorn, Oriental Thorn, Silver Hawthorn ●☆
109789 Crataegus laevicalyx J. X. Huang, L. Y. Sun et T. J. Feng = Crataegus dahurica Koehne ex C. K. Schneid. var. laevicalyx (J. X. Huang, L. Y. Sun et T. J. Feng) T. C. Ku et Spongberg ●
109790 Crataegus laevigata (Poir.) DC.;平滑山楂(刺山楂,钝裂叶山楂,多刺山楂,光滑山楂,尖刺山楂,锐刺山楂,西洋山楂,英国山楂); English Hawthorn, European Hawthorn, Hawthorn, May Bush, May-bush, Midland Hawthorn, Quick, Quick Set Thorn, Smooth Hawthorn, White Thorn, Whitethorn, Woodland Hawthorn ●☆
109791 Crataegus laevigata DC. 'Gireeoudii';吉瑞奥迪平滑山楂●☆
109792 Crataegus laevigata DC. 'Paul's Scarlet';猩红平滑山楂(宝罗红钝裂叶山楂);Paul's Scarlet Hawthorn ●☆
109793 Crataegus laevigata DC. 'Plena';白花重瓣平滑山楂●☆
109794 Crataegus laevigata DC. 'Punicea';石榴红钝裂叶山楂●☆
109795 Crataegus laevigata DC. 'Rosea Flore Pleno';粉红重瓣平滑山

楂●☆

109796 Crataegus lambertiana K. Koch;兰伯特山楂;Lambert Hawthorn ●☆

109797 Crataegus lanuginosa Sarg.;绵毛山楂;Woolly Hawthorn ●☆

109798 Crataegus laurentiana Sarg. var. brunetiana (Sarg.) Kruschke = Crataegus chrysocarpa Ashe ●

109799 Crataegus laurentiana Sarg. var. dissimilifolia Kruschke = Crataegus chrysocarpa Ashe ●

109800 Crataegus lavallei Sarg.; 红蕊山楂 (拉氏山楂); Lavelle Hawthorn ●☆

109801 Crataegus lavallei Sarg. 'Carrièrei';卡里埃拉氏山楂●☆

109802 Crataegus laxiflora Sarg. = Crataegus succulenta Schrad. ex Link ●

109803 Crataegus lecta Sarg. = Crataegus pruinosa (H. L. Wendl.) K. Koch var. virella (Ashe) Kruschke ●☆

109804 Crataegus leiophylla Sarg. = Crataegus pruinosa (H. L. Wendl.) K. Koch var. leiophylla (Sarg.) J. B. Phipps ●☆

109805 Crataegus lenta Ashe = Crataegus holmesiana Ashe ●☆

109806 Crataegus letchworthiana Sarg. = Crataegus pedicellata Sarg. ●☆

109807 Crataegus limaria Sarg. = Crataegus mollis (Torr. et A. Gray) Scheele ●☆

109808 Crataegus limatula Sarg. = Crataegus nitidula Sarg. ●☆

109809 Crataegus lucorum Sarg.;小树林山楂;Grove Hawthorn ●☆

109810 Crataegus lumaria Ashe;圆叶山楂;Round-leaved Hawthorn ●☆

109811 Crataegus lyi H. Lév. = Symplocos paniculata (Thunb.) Miq. ●

109812 Crataegus mackenziei Sarg. = Crataegus pruinosa (H. L. Wendl.) K. Koch var. virella (Ashe) Kruschke ●☆

109813 Crataegus mackenziei Sarg. var. aspera (Sarg.) E. J. Palmer = Crataegus pruinosa (H. L. Wendl.) K. Koch ●☆

109814 Crataegus mackenziei Sarg. var. bracteata (Sarg.) E. J. Palmer = Crataegus pruinosa (H. L. Wendl.) K. Koch ●☆

109815 Crataegus mackenzii Sarg. = Crataegus pruinosa (H. L. Wendl.) K. Koch ●☆

109816 Crataegus mackenzii Sarg. var. aspera (Sarg.) E. J. Palmer = Crataegus pruinosa (H. L. Wendl.) K. Koch ●☆

109817 Crataegus mackenzii Sarg. var. bracteata (Sarg.) E. J. Palmer = Crataegus pruinosa (H. L. Wendl.) K. Koch ●☆

109818 Crataegus macracantha G. Lodd. ex Loudon; 大花山楂; Hawthorn, Long-spiked Thorn, Scarlet Haw ●☆

109819 Crataegus macracantha G. Lodd. ex Loudon var. colorado (Ashe) Kruschke = Crataegus succulenta Schrad. ex Link ●

109820 Crataegus macracantha G. Lodd. ex Loudon var. divida (Sarg.) Kruschke = Crataegus macracantha G. Lodd. ex Loudon var. occidentalis (Britton) Eggl. ●☆

109821 Crataegus macracantha G. Lodd. ex Loudon var. occidentalis (Britton) Eggl.;西部大花山楂;Hawthorn ●☆

109822 Crataegus macracantha G. Lodd. ex Loudon var. pertomentosa (Ashe) Kruschke;毛大花山楂;Hawthorn ●☆

109823 Crataegus macrosperma Ashe;大籽山楂;Big-fruit Hawthorn, Large-seeded Hawthorn ●☆

109824 Crataegus macrosperma Ashe var. acutiloba (Sarg.) Eggl.;锐裂叶大籽山楂●☆

109825 Crataegus macrosperma Ashe var. acutiloba (Sarg.) Eggl. = Crataegus macrosperma Ashe ●☆

109826 Crataegus macrosperma Ashe var. eganii (Ashe) Kruschke = Crataegus macrosperma Ashe ●☆

109827 Crataegus macrosperma Ashe var. matura (Sarg.) Eggl. = Crataegus macrosperma Ashe ●☆

109828 Crataegus macrosperma Ashe var. pastora (Sarg.) Eggl. = Crataegus macrosperma Ashe ●☆

109829 Crataegus macrosperma Ashe var. pentandra (Sarg.) Eggl. = Crataegus macrosperma Ashe ●☆

109830 Crataegus macrosperma Ashe var. roanensis (Ashe) E. J. Palmer = Crataegus macrosperma Ashe ●☆

109831 Crataegus mansfieldensis Sarg. = Crataegus irrasa Sarg. ●☆

109832 Crataegus margaretta Ashe; 玛格蕾塔山楂; Hawthorn, Margarett's Hawthorn ●☆

109833 Crataegus margaretta Ashe var. angustifolia E. J. Palmer = Crataegus margaretta Ashe ●☆

109834 Crataegus margaretta Ashe var. brownii (Britton) Sarg. = Crataegus margaretta Ashe ●☆

109835 Crataegus margaretta Ashe var. meiophylla (Sarg.) E. J. Palmer = Crataegus margaretta Ashe ●☆

109836 Crataegus marshallii Eggl.;欧芹山楂;Parsley Hawthorn ●☆

109837 Crataegus matura Sarg.;成熟山楂;Mature Hawthorn ●☆

109838 Crataegus maura L. f. = Crataegus monogyna Jacq. ●☆

109839 Crataegus maximowiczii C. K. Schneid.; 毛山楂; Maximowicz Hawthorn ●

109840 Crataegus maximowiczii C. K. Schneid. var. ninganensis S. Q. Nie et B. J. Jen = Crataegus maximowiczii C. K. Schneid. ●

109841 Crataegus melanocarpa M. Bieb. = Crataegus pentagyna Waldst. et Kit. ●☆

109842 Crataegus mercerensis Sarg. = Crataegus chrysocarpa Ashe ●

109843 Crataegus merita Sarg. = Crataegus apiomorpha Sarg. ●☆

109844 Crataegus mexicana D. Don = Crataegus pubescens Steud. ●☆

109845 Crataegus meyeri Pojark.;密毛山楂●☆

109846 Crataegus microphylla K. Koch;小叶山楂●☆

109847 Crataegus minutiflora Sarg. = Crataegus dodgei Ashe ●☆

109848 Crataegus miranda Sarg. = Crataegus fulleriana Sarg. ●☆

109849 Crataegus mohrii Beadle = Crataegus crus-galli L. ●

109850 Crataegus mollis (Torr. et A. Gray) Scheele;柔毛山楂(皱叶山楂);American Hawthorn, Arnold Hawthorn, Down Hawthorn, Downy Hawthorn, Red Hawthorn, Summer Haw, Turkey Apple, Woolly Thorn ●☆

109851 Crataegus mollis (Torr. et A. Gray) Scheele f. dumetosa (Sarg.) E. J. Palmer = Crataegus mollis (Torr. et A. Gray) Scheele ●☆

109852 Crataegus mollis (Torr. et A. Gray) Scheele var. dumetosa (Sarg.) Kruschke = Crataegus mollis (Torr. et A. Gray) Scheele ●☆

109853 Crataegus mollis (Torr. et A. Gray) Scheele var. gigantea Kruschke = Crataegus mollis (Torr. et A. Gray) Scheele ●☆

109854 Crataegus mollis (Torr. et A. Gray) Scheele var. sera (Sarg.) Eggl. = Crataegus mollis (Torr. et A. Gray) Scheele ●☆

109855 Crataegus mollis (Torr. et A. Gray) Scheele var. sera (Sarg.) Eggl. f. mecocantha Kruschke = Crataegus mollis (Torr. et A. Gray) Scheele ●☆

109856 Crataegus mollis Scheele = Crataegus mollis (Torr. et A. Gray) Scheele ●☆

109857 Crataegus monogyna Jacq.;单柱山楂(单花柱山楂,单子山楂,金钩如意草,普通山楂,英国山楂);Abbespine, Albaspyne, Albespeine, Aubespyne, Awglen, Azzy-tree, Beead-and-cider, Bird Eagle, Bird Eagles, Bird Pears, Bird's Cherries, Bird's Cherry, Bird's Eegle, Bird's Egg, Bird's Eggle, Bird's Eggs, Bird's Meat, Blossom, Bread-and-cheese Tree, Bread-and-cider, Bull Haw, Bull Haws, Bulls, Butter Haw, Butter-and-bread, Cat Haw, Cat Haws, Chaws,

Cheese-and-bread, Chucky-cheese, Chueky Cheese, Common Hawthorn, Cuckoo's Beads, Cuckoo's Bread-and-cheese Tree, Eggers, Eggle, Eggs-eggs, Eglon Eglet, English Hawthorn, Fattahs, Frith, Gazel, Gazle, God's Meat, Gypsy Nut, Gypsy Nuts, Hab-nabs, Hag, Hag Bush, Hag Haw, Hag Tree, Hagag, Hagberry, Haggas, Haggil, Hag-haw, Hagthorn, Hague, Hahs-bush, Haig, Haigh, Hales, Half-and-half, Halve, Harbs, Harsy, Harves, Harvies, Haves, Haw, Haw Bush, Haw Gaw, Haw Tree, Hawberry, Hawthorn, Haythorn, Hazel, Hazle, Hedge Thorn, Hedgespecks, Heethen-berry, Heg-peg, Heg-peg Bush, Hep-thorn, Herbs, Hip Haw, Hipperty Haw, Hog Haw, Hogail, Hogarve, Hogasses, Hogazel, Hog-berry, Hoggan, Hoggin, Hog-gosse, Hog-haghes, Hog-haw, Hog-hazel, Holy Innocents, Howes, Johnny Mac Gorey, Lady's Meat, Mahaw, May, May Bread-and-cheese Bush, May Bread-and-cheese Tree, May Thorn, May Tree, Mayberry, May-bush, May-flower, May-fruit, Maythorn, May-thorn, May-tree, Moon Flower, Mother-will-die, Oneseed Hawthorn, One-seeded Hawthorn, Orglon, Peggall-bush, Peggle, Peggy-ailes, Peggyiles, Pig Haw, Pig's Ailes, Pigales, Pig-ales, Pigall, Pigaul, Pig-berry, Pighau, Pig's Hales, Pig's Haws, Pig's Heels, Pig's Hells, Pig's Isles, Pig's Pear, Pig's Pears, Pigshell, Pigsy Pear, Pigsy-pears, Pill's Heel, Pixy Pear, Pixy-pear, Quer Citron Oak, Quick, Quick Wood, Quicken, Quickset, Quickthorn, Quickthorn Quickset, Sates, Scrag, Scrog, Scrog-bush, Sgeach, Shiggy, Single-seed Hawthorn, Skayug, Skeeog, Skeg, Snowflake, Thornberries, Whicks, White Thorn, Whitethorn, Wibrow-wobrow, Wlbrow ●☆

109858 Crataegus monogyna Jacq. 'Biflora';二花英国山楂(双花山楂);Glastonbury Thorn ●☆

109859 Crataegus monogyna Jacq. 'Praecox' = Crataegus monogyna Jacq. 'Biflora' ●☆

109860 Crataegus monogyna Jacq. 'Stricta';柱状英国山楂●☆

109861 Crataegus monogyna Jacq. subsp. brevispina (Kunze) Franco = Crataegus monogyna Jacq. ●☆

109862 Crataegus monogyna Jacq. var. lasiocarpa (Lange) K. I. Chr. = Crataegus monogyna Jacq. ●☆

109863 Crataegus monogyna Jacq. var. praecox?; 早花单柱山楂; Glastonbury Thorn, Greens, Holy Thorn ●☆

109864 Crataegus multiflora Ashe;多花山楂;Inkberry Hawthorn ●☆

109865 Crataegus neofluvialis Ashe = Crataegus succulenta Schrad. ex Link ●

109866 Crataegus nevadensis K. I. Chr. = Cotoneaster granatensis Boiss. ●☆

109867 Crataegus nigra Pall. ex Steud. = Crataegus oxyacantha L. ●☆

109868 Crataegus nigra Waldst. et Kit.; 匈牙利山楂(黑山楂); European Black Hawthorn, Hungarian Thorn ●☆

109869 Crataegus ninganensis S. Q. Nie et B. J. Jen;宁安山楂;Ning'an Hawthorn ●

109870 Crataegus nitida (Engelm.) Sarg.; 光亮山楂; Hawthorn, Ontario Hawthorn ●☆

109871 Crataegus nitidula Sarg.;安大略山楂;Ontario Hawthorn ●☆

109872 Crataegus nitidula Sarg. var. limatula (Sarg.) Kruschke = Crataegus nitidula Sarg. ●☆

109873 Crataegus nitidula Sarg. var. macrocarpa Kruschke = Crataegus nitidula Sarg. ●☆

109874 Crataegus nitidula Sarg. var. recedens (Sarg.) Kruschke = Crataegus nitidula Sarg. ●☆

109875 Crataegus noelensis Sarg. = Crataegus mollis (Torr. et A. Gray) Scheele ●☆

109876 Crataegus nuda Sarg.;裸山楂;Hawthorn ●☆

109877 Crataegus oakesiana Eggl. = Crataegus irrasa Sarg. ●☆

109878 Crataegus occidentalis Britton = Crataegus macracantha G. Lodd. ex Loudon var. occidentalis (Britton) Eggl. ●☆

109879 Crataegus opaca Hook. et Arn. ex Hook.;暗山楂;Mayhaw ●☆

109880 Crataegus operta Ashe = Crataegus crus-galli L. ●

109881 Crataegus oresbia W. W. Sm.; 滇西山楂; W. Yunnan Hawthorn, West Yunnan Hawthorn ●

109882 Crataegus orientalis K. I. Chr. = Crataegus laciniata Ucria ●☆

109883 Crataegus orientalis K. I. Chr. subsp. presliana K. I. Chr. = Crataegus laciniata Ucria ●☆

109884 Crataegus orientalis Pall. = Crataegus laciniata Steven ●☆

109885 Crataegus oxyacantha L. 'Paulii';泡利刺山楂●☆

109886 Crataegus oxyacantha L. = Crataegus laevigata (Poir.) DC. ●☆

109887 Crataegus oxyacantha L. subsp. maura (L. f.) Maire = Crataegus monogyna Jacq. ●☆

109888 Crataegus oxyacantha L. subsp. monogyna (Jacq.) Rouy et Camus = Crataegus monogyna Jacq. ●☆

109889 Crataegus oxyacantha L. subsp. oxyacanthoides (Thuill.) Maire = Crataegus laevigata (Poir.) DC. ●☆

109890 Crataegus oxyacantha L. var. brevispina (Kuntze) Dippel = Crataegus laevigata (Poir.) DC. ●☆

109891 Crataegus oxyacantha L. var. ciliata Maire = Crataegus monogyna Jacq. ●☆

109892 Crataegus oxyacantha L. var. coriacea Maire = Crataegus laevigata (Poir.) DC. ●☆

109893 Crataegus oxyacantha L. var. fallax Maire = Crataegus monogyna Jacq. ●☆

109894 Crataegus oxyacantha L. var. heterophylla (Flüggé) Maire = Crataegus laevigata (Poir.) DC. ●☆

109895 Crataegus oxyacantha L. var. hirsuta Boiss. = Crataegus monogyna Jacq. ●☆

109896 Crataegus oxyacantha L. var. maura (L. f.) Batt. = Crataegus laevigata (Poir.) DC. ●☆

109897 Crataegus oxyacantha L. var. miniata Maire = Crataegus laevigata (Poir.) DC. ●☆

109898 Crataegus oxyacantha L. var. monogyna (Jacq.) Batt. = Crataegus monogyna Jacq. ●☆

109899 Crataegus oxyacantha L. var. paulii (Rehder) Rehder = Crataegus monogyna Jacq. ●☆

109900 Crataegus oxyacantha L. var. pinnatifida Regel = Crataegus pinnatifida Bunge ●

109901 Crataegus oxyacantha L. var. saccardyana Maire = Crataegus laevigata (Poir.) DC. ●☆

109902 Crataegus oxyacantha L. var. stenoloba Maire = Crataegus monogyna Jacq. ●☆

109903 Crataegus oxyacantha L. var. supravillosa Maire = Crataegus granatensis Boiss. ●☆

109904 Crataegus oxyacantha L. var. triloba (Poir.) Batt. = Crataegus azarolus L. ●☆

109905 Crataegus oxyacantha Scop. = Crataegus monogyna Jacq. ●☆

109906 Crataegus oxyacanthoides Thuill. = Crataegus laevigata (Poir.) DC. ●☆

109907 Crataegus pallasii Griseb.;帕拉斯山楂●☆

109908 Crataegus palliata Sarg. = Crataegus crus-galli L. ●

109909 Crataegus palmeri Sarg.;帕默山楂;Hawthorn ●☆

109910 Crataegus palmeri Sarg. = Crataegus crus-galli L. ●

109911 Crataegus parea Ashe = Crataegus flabellata (Bosc ex Spach) Rydb. ●☆
109912 Crataegus pausiaca Ashe = Crataegus disperma Ashe ●☆
109913 Crataegus pedicellata Sarg.;梗山楂;Scarlet Hawthorn ●☆
109914 Crataegus pedicellata Sarg. = Crataegus coccinea L. ●☆
109915 Crataegus pedicellata Sarg. var. albicans (Ashe) E. J. Palmer = Crataegus mollis (Torr. et A. Gray) Scheele ●☆
109916 Crataegus pedicellata Sarg. var. assurgens (Sarg.) E. J. Palmer = Crataegus pedicellata Sarg. ●☆
109917 Crataegus pedicellata Sarg. var. caesa (Ashe) Kruschke = Crataegus pedicellata Sarg. ●☆
109918 Crataegus pedicellata Sarg. var. ellwangeriana (Sarg.) Eggl.;埃尔梗山楂;Scarlet Hawthorn ●☆
109919 Crataegus pedicellata Sarg. var. ellwangeriana (Sarg.) Eggl. = Crataegus ellwangeriana Sarg. ●☆
109920 Crataegus pedicellata Sarg. var. ellwangeriana (Sarg.) Eggl. f. assurgens (Sarg.) Kruschke = Crataegus pedicellata Sarg. var. ellwangeriana (Sarg.) Eggl. ●☆
109921 Crataegus pedicellata Sarg. var. robesoniana (Sarg.) E. J. Palmer = Crataegus pedicellata Sarg. var. ellwangeriana (Sarg.) Eggl. ●☆
109922 Crataegus pedicellata Sarg. var. sertata (Sarg.) Kruschke = Crataegus pedicellata Sarg. ●☆
109923 Crataegus pennsylvanica Ashe; 宾州山楂; Pennsylvania Hawthorn ●☆
109924 Crataegus pentagyna Waldst. et Kit.;五蕊山楂(五子山楂,紫果山楂);Chinese Hawthorn ●☆
109925 Crataegus peoriensis Sarg. = Crataegus disperma Ashe ●☆
109926 Crataegus perampla Sarg. = Crataegus pruinosa (H. L. Wendl.) K. Koch ●☆
109927 Crataegus permixta E. J. Palmer = Crataegus crus-galli L. ●
109928 Crataegus persimilis Sarg. 'Prunifolia';桃叶山楂(樱叶山楂);Broad-leaved Cockspur-thorn ●☆
109929 Crataegus persimilis Sarg. = Crataegus nuda Sarg. ●☆
109930 Crataegus pertomentosa Ashe = Crataegus macracantha G. Lodd. ex Loudon var. pertomentosa (Ashe) Kruschke ●☆
109931 Crataegus phaenopyrum Borkh.;华盛顿山楂(心叶山楂);Washington Hawthorn, Washington Thorn, Washington's Hawthorn, Washington's-thorn Washington-thorn ●☆
109932 Crataegus pinguis Sarg. = Crataegus scabrida Sarg. ●☆
109933 Crataegus pinnatifida Bunge;山楂(赤瓜木,猴栌,猴楂,猴樝,檕梅,裂叶山楂,山梨红,山里红,山楂扣,棠棣子);China Hawthorn, Chinese Hawthorn ●
109934 Crataegus pinnatifida Bunge var. korolkowi Y. Yabe = Crataegus pinnatifida Bunge var. major N. E. Br. ●
109935 Crataegus pinnatifida Bunge var. korolkowii (Asch. et Graebn.) Y. Yabe = Crataegus pinnatifida Bunge var. major N. E. Br. ●
109936 Crataegus pinnatifida Bunge var. major N. E. Br.;山里红(大果山楂,黑山楂,红果,红果子,猴栌,猴楂,猴栌,檕梅,辽山楂,毛枣子,茅栌,牧虎梨,南山楂,机,机子,山查,山果子,山梨,山里果子,山里红果,山栌,山楂,柿栌子,鼠查,酸查,酸里红,酸梅子,酸枣,酸楂,唐棣子,棠棣,棠梂,棠梂子,小叶山楂,羊梂,羊梂子,药山楂,野山楂,映山红果);Chinese Hawthorn, Large Chinese Hawthorn, Major Chinese Hawthorn, Red Hawthorn ●
109937 Crataegus pinnatifida Bunge var. psilosa C. K. Schneid.;无毛山楂(长毛山楂,柔毛山楂);Hairless Chinese Hawthorn ●
109938 Crataegus pinnatifida Bunge var. songarica Dippel = Crataegus pinnatifida Bunge ●
109939 Crataegus pinnatifida Bunge var. typica Schneid. = Crataegus pinnatifida Bunge ●
109940 Crataegus pisifera Sarg. = Crataegus succulenta Schrad. ex Link ●
109941 Crataegus placens Sarg. = Crataegus mollis (Torr. et A. Gray) Scheele ●☆
109942 Crataegus platycarpa Sarg.;宽果山楂;Hawthorn ●☆
109943 Crataegus platycarpa Sarg. = Crataegus pruinosa (H. L. Wendl.) K. Koch ●☆
109944 Crataegus pontica K. Koch = Crataegus azarolus L. ●☆
109945 Crataegus pringlei Sarg.;普氏山楂;Pringle's Hawthorn ●☆
109946 Crataegus pringlei Sarg. var. exclusa (Sarg.) Eggl. = Crataegus pringlei Sarg. ●☆
109947 Crataegus pringlei Sarg. var. lobulata (Sarg.) Eggl. = Crataegus pringlei Sarg. ●☆
109948 Crataegus prona Sarg.;伊州山楂;Illinois Hawthorn ●☆
109949 Crataegus pruinosa (H. L. Wendl.) K. Koch;霜果山楂(棘刺山楂);Frosted Hawthorn, Waxy-fruit Thorn ●☆
109950 Crataegus pruinosa (H. L. Wendl.) K. Koch var. brachypoda (Sarg.) E. J. Palmer = Crataegus dissona Sarg. ●☆
109951 Crataegus pruinosa (H. L. Wendl.) K. Koch var. congesta (Sarg.) J. B. Phipps = Crataegus pruinosa (H. L. Wendl.) K. Koch ●☆
109952 Crataegus pruinosa (H. L. Wendl.) K. Koch var. delawarensis (Sarg.) E. J. Palmer = Crataegus dissona Sarg. ●☆
109953 Crataegus pruinosa (H. L. Wendl.) K. Koch var. dissona (Sarg.) Eggl. = Crataegus dissona Sarg. ●☆
109954 Crataegus pruinosa (H. L. Wendl.) K. Koch var. grandiflora Kruschke f. mecocantha Kruschke = Crataegus pruinosa (H. L. Wendl.) K. Koch ●☆
109955 Crataegus pruinosa (H. L. Wendl.) K. Koch var. grandiflora Kruschke = Crataegus pruinosa (H. L. Wendl.) K. Koch ●☆
109956 Crataegus pruinosa (H. L. Wendl.) K. Koch var. latisepala (Sarg.) Eggl. = Crataegus pruinosa (H. L. Wendl.) K. Koch ●☆
109957 Crataegus pruinosa (H. L. Wendl.) K. Koch var. leiophylla (Sarg.) J. B. Phipps; 光叶霜果山楂(粉被山楂); Frosted Hawthorn ●☆
109958 Crataegus pruinosa (H. L. Wendl.) K. Koch var. pachypoda (Sarg.) E. J. Palmer = Crataegus pruinosa (H. L. Wendl.) K. Koch ●☆
109959 Crataegus pruinosa (H. L. Wendl.) K. Koch var. rugosa (Ashe) Kruschke = Crataegus pruinosa (H. L. Wendl.) K. Koch var. leiophylla (Sarg.) J. B. Phipps ●☆
109960 Crataegus pruinosa (H. L. Wendl.) K. Koch var. virella (Ashe) Kruschke;浅绿霜果山楂;Frosted Hawthorn ●☆
109961 Crataegus prunifolia (Poir.) Pers. = Crataegus persimilis Sarg. 'Prunifolia' ●☆
109962 Crataegus prunifolia Bosc = Crataegus crus-galli L. ●
109963 Crataegus przewalskii Pojark.;西湾山楂●
109964 Crataegus pseudoambigua Pojark.;疑似山楂●☆
109965 Crataegus pseudoheterophylla Pojark.;紫蕊山楂●☆
109966 Crataegus pseudomelanocarpa Popov;拟黑果山楂●☆
109967 Crataegus pubescens Steud.; 短毛山楂 (短毛山里红); Mexican Hawthorn, Stone Plum ●☆
109968 Crataegus pubescens Steud. stipulata Stapf; 托叶短毛山楂; Manzanilla ●☆
109969 Crataegus pulcherrima Ashe;美丽山楂;Beautiful Hawthorn ●☆

109970 Crataegus punctata Jacq.; 斑点山楂(斑山楂); Dotted Hawthorn, Dotted Thorn, Large-fruit Thorn, Punctuate Hawthorn, Whitehaw ●
109971 Crataegus punctata Jacq. 'Aurea';黄果斑点山楂●☆
109972 Crataegus punctata Jacq. 'Ohio Pioneer';俄亥俄斑点山楂●☆
109973 Crataegus punctata Jacq. = Crataegus collina Chapm. ●
109974 Crataegus punctata Jacq. f. aurea (Aiton) Rehder = Crataegus punctata Jacq. ●
109975 Crataegus punctata Jacq. f. canescens (Britton) Kruschke = Crataegus punctata Jacq. ●
109976 Crataegus punctata Jacq. f. intermedia Kruschke = Crataegus punctata Jacq. ●
109977 Crataegus punctata Jacq. var. aurea Aiton = Crataegus punctata Jacq. ●
109978 Crataegus punctata Jacq. var. canescens Britton = Crataegus punctata Jacq. ●
109979 Crataegus punctata Jacq. var. microphylla Sarg. = Crataegus punctata Jacq. ●
109980 Crataegus punctata Jacq. var. pausiaca (Ashe) E. J. Palmer = Crataegus disperma Ashe ●☆
109981 Crataegus purpurea Bosc var. altaica Loudon = Crataegus altaica (Loudon) Lange ●
109982 Crataegus purpurea Bosc. ex DC. = Crataegus dahurica Koehne ex C. K. Schneid. ●
109983 Crataegus putnamiana Sarg. = Crataegus pedicellata Sarg. ●☆
109984 Crataegus pyracantha (L.) Medik. = Pyracantha atalantioides (Hance) Stapf ●
109985 Crataegus pyracantha Hemsl. = Pyracantha atalantioides (Hance) Stapf ●
109986 Crataegus pyracantha Hemsl. var. crenulata (D. Don) Loudon = Pyracantha crenulata (D. Don) M. Roem. ●
109987 Crataegus pyracantha Hemsl. var. crenulata Loudon = Pyracantha crenulata (D. Don) M. Roem. ●
109988 Crataegus pyracanthoides (Aiton) Beadle = Crataegus crus-galli L. ●
109989 Crataegus randiana Sarg. = Crataegus macrosperma Ashe ●☆
109990 Crataegus recedens Sarg. = Crataegus nitidula Sarg. ●☆
109991 Crataegus regalis Beadle = Crataegus crus-galli L. ●
109992 Crataegus regalis Beadle var. paradoxa (Sarg.) E. J. Palmer = Crataegus crus-galli L. ●
109993 Crataegus relicta Sarg. = Crataegus dissona Sarg. ●☆
109994 Crataegus remotilobata Raikova ex Popov; 裂叶山楂; Remotelobated Hawthorn, Remote-lobated Hawthorn, Remote-lobed Hawthorn ●
109995 Crataegus reverchonii Sarg.;勒韦雄山楂(簕氏山楂);Hawthorn, Reverchon Hawthorn ●☆
109996 Crataegus rivularis Nutt. ex Torr. et Gray;河岸山楂;Hawberry, River Hawthorn ●☆
109997 Crataegus rivularis W. H. Brewer et S. Watson = Crataegus douglasii Lindl. ●
109998 Crataegus roanensis Ashe = Crataegus macrosperma Ashe ●☆
109999 Crataegus roanensis Ashe var. fluviatilis (Sarg.) Kruschke = Crataegus fluviatilis Sarg. ●☆
110000 Crataegus roanensis Ashe var. heidelbergensis (Sarg.) Kruschke = Crataegus macrosperma Ashe ●☆
110001 Crataegus robesoniana Sarg.;罗伯逊山楂;Robeson Hawthorn ●☆
110002 Crataegus rotundata Sarg. = Crataegus dodgei Ashe ●☆
110003 Crataegus rotundifolia Borkh.;西方圆叶山楂(凤凰山楂); Eastern Fire-berry Hawthorn, Fireberry Hawthorn ●☆
110004 Crataegus rotundifolia Moench = Crataegus chrysocarpa Ashe ●
110005 Crataegus rubra Lour. = Rhaphiolepis indica (L.) Lindl. ex Ker ●
110006 Crataegus rufula Sarg.;五月山楂;May Hawthorn ●☆
110007 Crataegus rugosa Ashe = Crataegus pruinosa (H. L. Wendl.) K. Koch var. leiophylla (Sarg.) J. B. Phipps ●☆
110008 Crataegus rugosa Ashe = Crataegus pruinosa (H. L. Wendl.) K. Koch ●☆
110009 Crataegus rupicola Sarg.;岩生山楂;Hawthorn ●☆
110010 Crataegus rutila Sarg. = Crataegus succulenta Schrad. ex Link ●
110011 Crataegus sabineana Ashe = Crataegus crus-galli L. ●
110012 Crataegus saeva Sarg. = Crataegus succulenta Schrad. ex Link ●
110013 Crataegus saligna Greene;柳叶山楂;Willow Hawthorn ●☆
110014 Crataegus sanguinea Pall.;辽宁山楂(白槎子,白海棠,红果山楂,辽东山楂,面果果,山楂);Red Hawthorn, Redhaw Hawthorn, Red-haw Hawthorn ●
110015 Crataegus sanguinea Pall. = Crataegus kansuensis E. H. Wilson ●
110016 Crataegus sanguinea Pall. var. glabra Maxim. = Crataegus dahurica Koehne ex C. K. Schneid. ●
110017 Crataegus sanguinea Pall. var. incisa Regel = Crataegus altaica (Loudon) Lange ●
110018 Crataegus sanguinea Pall. var. inermis Kar. et Kir. = Crataegus altaica (Loudon) Lange ●
110019 Crataegus sanguinea Pall. var. villosa Rupr. = Crataegus maximowiczii C. K. Schneid. ●
110020 Crataegus sargentii Beadle;萨金特山楂●☆
110021 Crataegus scabrida Sarg.;粗糙山楂;Niagara Hawthorn, Rough Hawthorn ●☆
110022 Crataegus scabrida Sarg. var. asperifolia (Sarg.) Kruschke = Crataegus scabrida Sarg. ●☆
110023 Crataegus scabrida Sarg. var. balkwillii (Sarg.) J. B. Phipps = Crataegus scabrida Sarg. ●☆
110024 Crataegus scabrida Sarg. var. cyclophylla (Sarg.) Kruschke = Crataegus scabrida Sarg. ●☆
110025 Crataegus scabrida Sarg. var. dunbarii (Sarg.) Kruschke = Crataegus scabrida Sarg. ●☆
110026 Crataegus scabrida Sarg. var. egglestoni (Sarg.) Kruschke = Crataegus scabrida Sarg. ●☆
110027 Crataegus scabrida Sarg. var. hadleyana (Sarg.) J. B. Phipps = Crataegus scabrida Sarg. ●☆
110028 Crataegus scabrida Sarg. var. honesta (Sarg.) Kruschke = Crataegus scabrida Sarg. ●☆
110029 Crataegus scabrida Sarg. var. improvisa (Sarg.) J. B. Phipps = Crataegus scabrida Sarg. ●☆
110030 Crataegus scabrifolia (Franch.) Rehder;云南山楂(大果山楂,山林果,山楂,酸冷果);Yunnan Hawthorn ●
110031 Crataegus schizophylla Eggl. = Crataegus crus-galli L. ●
110032 Crataegus schraderiana Ledeb.; 灰毛山楂(施氏山楂); Schrader Hawthorn ●☆
110033 Crataegus schuettei Ashe;许氏山楂;Schuette's Hawthorn ●☆
110034 Crataegus schuettei Ashe var. basilica (Beadle) J. B. Phipps = Crataegus schuettei Ashe ●☆
110035 Crataegus schuettei Ashe var. cuneata Kruschke = Crataegus schuettei Ashe ●☆
110036 Crataegus schuettei Ashe var. ferrissii (Ashe) Kruschke = Crataegus schuettei Ashe ●☆

110037 Crataegus schuettei Ashe var. gigantea Kruschke = Crataegus schuettei Ashe ●☆
110038 Crataegus serratifolia Desf. = Photinia serratifolia (Desf.) Kalkman ●
110039 Crataegus shandongensis F. Z. Li et W. D. Peng;山东山楂;Shandong Hawthorn ●
110040 Crataegus shenxiensis Pojark.;陕西山楂;Shaanxi Hawthorn, Shensi Hawthorn ●
110041 Crataegus shinnersii Kruschke = Crataegus crus-galli L. ●
110042 Crataegus shirleyensis Sarg. = Crataegus florifera Sarg. ●☆
110043 Crataegus sicca Sarg. = Crataegus margaretta Ashe ●☆
110044 Crataegus sicca Sarg. var. glabrifolia (Sarg.) E. J. Palmer = Crataegus margaretta Ashe ●☆
110045 Crataegus signata Beadle = Crataegus crus-galli L. ●
110046 Crataegus silvestris Sarg. = Crataegus florifera Sarg. ●☆
110047 Crataegus smithii (DC.) Chalon;史密斯山楂;Hybrid Thorn, Red Mexican Thorn ●☆
110048 Crataegus songorica K. Koch;准噶尔山楂;Dzungar Hawthorn, Songor Hawthorn, Xinjiang Hawthorn ●
110049 Crataegus spathulata H. Lév. = Crataegus cuneata Siebold et Zucc. ●
110050 Crataegus spathulata Michx.;糖山楂;Sugar Hawthorn, Littlehip Hawthorn ●☆
110051 Crataegus spathulata Pursh = Crataegus flava Aiton ●☆
110052 Crataegus sphaenophylla Pojark.;楔叶山楂;Wedge-leaved Hawthorn ●☆
110053 Crataegus spicata Lam. = Amelanchier spicata (Lam.) K. Koch ●
110054 Crataegus spthulata Michx.;匙形山楂;Littlehip Hawthorn, Pasture Hawthorn, Small-fruit Hawthorn ●☆
110055 Crataegus stephanostyla H. Lév. et Vaniot = Crataegus cuneata Siebold et Zucc. ●
110056 Crataegus stevenii Pojark.;司梯氏山楂;Steven Hawthorn ●☆
110057 Crataegus stipulacea Lodd. = Crataegus pubescens Steud. ●☆
110058 Crataegus stipulosa (Kunth) Steud.;托叶山楂●☆
110059 Crataegus stipulosa Steud. = Crataegus stipulosa (Kunth) Steud. ●☆
110060 Crataegus submollis Sarg.;魁北克山楂(羽裂叶山楂);Hairy Cockspur-thorn, Quebec Hawthorn, Red Haw, Velvety Hawthorn ●☆
110061 Crataegus subpilosa Sarg. = Crataegus crus-galli L. ●
110062 Crataegus subrotundifolia Sarg. = Crataegus chrysocarpa Ashe var. aboriginum (Sarg.) Kruschke ●☆
110063 Crataegus succulenta Link = Crataegus succulenta Schrad. ex Link ●
110064 Crataegus succulenta Link var. pertomentosa (Ashe) E. J. Palmer = Crataegus macracantha G. Lodd. ex Loudon ●☆
110065 Crataegus succulenta Schrad. ex Link;多浆山楂(肉质山楂);Caughnawaga Thorn, Fleshy Hawthorn, Long-spine Hawthorn, Red Haw, Round-fruited Cockspur-thorn, Succulent Hawthorn ●
110066 Crataegus succulenta Schrad. ex Link var. gemmosa (Sarg.) Kruschke = Crataegus succulenta Schrad. ex Link ●
110067 Crataegus succulenta Schrad. ex Link var. laxiflora (Sarg.) Kruschke = Crataegus succulenta Schrad. ex Link ●
110068 Crataegus succulenta Schrad. ex Link var. macracantha (Lodd.) Eggl. = Crataegus macracantha G. Lodd. ex Loudon ●☆
110069 Crataegus succulenta Schrad. ex Link var. michiganensis (Ashe) E. J. Palmer = Crataegus succulenta Schrad. ex Link ●
110070 Crataegus succulenta Schrad. ex Link var. neofluvialis (Ashe) E. J. Palmer = Crataegus succulenta Schrad. ex Link ●
110071 Crataegus succulenta Schrad. ex Link var. occidentalis Britton = Crataegus macracantha G. Lodd. ex Loudon var. occidentalis (Britton) Eggl. ●☆
110072 Crataegus succulenta Schrad. ex Link var. pertomentosa (Ashe) E. J. Palmer = Crataegus macracantha G. Lodd. ex Loudon var. pertomentosa (Ashe) Kruschke ●☆
110073 Crataegus succulenta Schrad. ex Link var. pisifera (Sarg.) Kruschke = Crataegus succulenta Schrad. ex Link ●
110074 Crataegus succulenta Schrad. ex Link var. rutila (Sarg.) Kruschke = Crataegus succulenta Schrad. ex Link ●
110075 Crataegus sylvestris Sarg. = Crataegus florifera Sarg. ●☆
110076 Crataegus szovitsii Pojark.;绍氏山楂●☆
110077 Crataegus tanacetifolia Pers.;艾菊叶山楂;Syrian Hawberry, Tansy-leafed Hawthorn, Tansy-leafed Thorn ●☆
110078 Crataegus tangchungchangii F. P. Metcalf = Crataegus cuneata Siebold et Zucc. var. tangchungchangii (F. P. Metcalf) T. C. Ku et Spongberg ●
110079 Crataegus tangchungchangii Metcalfe;福建山楂●
110080 Crataegus tantula Sarg. = Crataegus crus-galli L. ●
110081 Crataegus taquetii H. Lév. = Malus sieboldii (Regel) Rehder ●
110082 Crataegus taquetii H. Lév. = Malus toringo (Siebold) Siebold ex de Vriese ●
110083 Crataegus tatnalliana Sarg. = Crataegus pennsylvanica Ashe ●☆
110084 Crataegus taurica Pojark.;克里木山楂●
110085 Crataegus tenuifolia Britton. = Malus komarovii (Sarg.) Rehder ●◇
110086 Crataegus texana Buckley;得克萨斯山楂;Hawthorn, Texas Hawthorn ●☆
110087 Crataegus tianschanica Pojark.;天山山楂●☆
110088 Crataegus tomentosa L.;梨形山楂;Pear Thorn ●☆
110089 Crataegus tortilis Ashe = Crataegus schuettei Ashe ●☆
110090 Crataegus tracyi Ashe;山地山楂;Mountain Hawthorne ●☆
110091 Crataegus transcaspica Pojark.;里海山楂●☆
110092 Crataegus triflora Chapm.;三花山楂;Threeflower Hawthorn ●☆
110093 Crataegus triloba Poir. = Crataegus azarolus L. ●☆
110094 Crataegus triumphalis Sarg. = Crataegus crus-galli L. ●
110095 Crataegus turcomanica Pojark.;土库曼山楂●☆
110096 Crataegus turkestanica Pojark.;土耳其斯坦山楂;Turkestan Hawthorn ●☆
110097 Crataegus ucrainica Pojark.;乌克兰山楂;Ukraine Hawthorn ●☆
110098 Crataegus uniflora Muenchh.;单花山楂;Dwarf Hawthorn, Oneflower Hawthorn, One-flower Hawthorn ●☆
110099 Crataegus uniqua Sarg. = Crataegus crus-galli L. ●
110100 Crataegus vallicola Sarg. = Crataegus crus-galli L. ●
110101 Crataegus venulosa Sarg. = Crataegus succulenta Schrad. ex Link ●
110102 Crataegus verruculosa Sarg.;小疣山楂●☆
110103 Crataegus vicinalis Beadle = Crataegus pruinosa (H. L. Wendl.) K. Koch ●☆
110104 Crataegus villipes (Ashe) Ashe = Crataegus holmesiana Ashe ●☆
110105 Crataegus villosa Thunb. = Photinia villosa (Thunb.) DC. ●
110106 Crataegus virella Ashe = Crataegus pruinosa (H. L. Wendl.) K. Koch var. virella (Ashe) Kruschke ●☆
110107 Crataegus viridis L.;绿山楂;Green Hawthorn, Southern Hawthorn ●☆
110108 Crataegus virilis Sarg. = Crataegus succulenta Schrad. ex Link ●
110109 Crataegus vittata Ashe = Crataegus apiomorpha Sarg. ●☆

110110 Crataegus volgensis Pojark.;伏尔加山楂;Volga Hawthorn ●☆
110111 Crataegus wattiana Hemsl. et Lace;瓦特山楂(山东山楂);Watt Hawthorn ●
110112 Crataegus wattiana Hemsl. et Lace = Crataegus kansuensis E. H. Wilson ●
110113 Crataegus wattiana Hemsl. et Lace var. incisa (Regel) C. K. Schneid. = Crataegus altaica (Loudon) Lange ●
110114 Crataegus wilsonii Sarg.;华东山楂(华中山楂);Central China Hawthorn, E. H. Wilson Hawthorn, Wilson Hawthorn ●
110115 Crataegus wisconsinensis Kruschke = Crataegus florifera Sarg. ●☆
110116 Crataegus zangezura Pojark.;赞格祖尔山楂●☆
110117 Crataeva L. = Crateva L. ●
110118 Crataeva adansonii DC. = Crateva trifoliata (Roxb.) B. S. Sun ●
110119 Crataeva adansonii DC. subsp. odora (Ham.) Jacobs = Crateva trifoliata (Roxb.) B. S. Sun ●
110120 Crataeva adansonii DC. subsp. trifoliata (Roxb.) Jacobs = Crateva trifoliata (Roxb.) B. S. Sun ●
110121 Crataeva marnelos L. = Aegle marmelos (L.) Corrêa ex Roxb. ●
110122 Cratalaria duboisii H. Lév. = Polygala arillata Buch.-Ham. ex D. Don ●
110123 Crateola Raf. = ? Hemigraphis Nees ■
110124 Crateranthus Baker f. (1913);杯花玉蕊属●☆
110125 Crateranthus congolensis Lecomte;刚果杯花玉蕊●☆
110126 Crateranthus letestuii Lecomte;莱泰斯图杯花玉蕊●☆
110127 Crateranthus talbotii Baker f.;塔尔博特杯花玉蕊●☆
110128 Crateria Pers. = Casearia Jacq. ●
110129 Crateria Pers. = Chaetocrater Ruiz et Pav. ●
110130 Craterianthus Valeton ex K. Heyne = Pellacalyx Korth. ●
110131 Cratericarpium Spach = Boisduvalia Spach ■☆
110132 Cratericarpium Spach = Oenothera L. ●■
110133 Crateriphytum Scheff. ex Koord. = Neuburgia Blume ●☆
110134 Craterispermum Benth. (1849);杯籽茜属(杯籽属)●☆
110135 Craterispermum angustifolium De Wild. et T. Durand = Craterispermum schweinfurthii Hiern ●☆
110136 Craterispermum aristatum Wernham;具芒杯籽茜●☆
110137 Craterispermum caudatum Hutch.;尾状杯籽茜●☆
110138 Craterispermum cerinanthum Hiern;蜡黄花杯籽茜●☆
110139 Craterispermum congolanum De Wild. et T. Durand = Craterispermum schweinfurthii Hiern ●☆
110140 Craterispermum dewevrei De Wild. et T. Durand;德韦杯籽茜●☆
110141 Craterispermum goossensii De Wild. = Craterispermum schweinfurthii Hiern ●☆
110142 Craterispermum grumileoides K. Schum.;类九节杯籽茜●☆
110143 Craterispermum inquisitorium Wernham var. longepedunculatum Good;长花梗杯籽茜●☆
110144 Craterispermum laurinum (Poir.) Benth.;劳津杯籽茜●☆
110145 Craterispermum laurinum Benth. = Craterispermum laurinum (Poir.) Benth. ●☆
110146 Craterispermum ledermannii K. Krause;莱德杯籽茜●☆
110147 Craterispermum longipedunculatum Verdc.;长梗杯籽茜●☆
110148 Craterispermum montanum Hiern;山地杯籽茜●☆
110149 Craterispermum reticulatum De Wild. = Craterispermum schweinfurthii Hiern ●☆
110150 Craterispermum schweinfurthii Hiern;狭叶杯籽●☆
110151 Crateritecoma Lindl. = Craterotecoma Mart. ex DC. ●☆
110152 Crateritecoma Lindl. = Lundia DC. (保留属名)●☆
110153 Craterocapsa Hilliard et B. L. Burtt(1973);杯囊桔梗属☆
110154 Craterocapsa congesta Hilliard et B. L. Burtt;密集杯囊桔梗☆
110155 Craterocapsa insizwae (Zahlbr.) Hilliard et B. L. Burtt = Craterocapsa montana (A. DC.) Hilliard et B. L. Burtt ■☆
110156 Craterocapsa montana (A. DC.) Hilliard et B. L. Burtt;山地杯囊桔梗■☆
110157 Craterocoma Mart. ex DC. = Lundia DC. (保留属名)●☆
110158 Craterogyne Lanj. = Dorstenia L. ■●☆
110159 Craterogyne africana (Baill.) Lanj. = Dorstenia africana (Baill.) C. C. Berg ●☆
110160 Craterogyne dorstenioides (Engl.) Lanj. = Dorstenia dorstenioides (Engl.) Hijman et C. C. Berg ■☆
110161 Craterogyne kameruniana (Engl.) Lanj. = Dorstenia kameruniana Engl. ■☆
110162 Craterogyne oligogyna (Pellegr.) Lanj. = Dorstenia oligogyna (Pellegr.) Berg ●☆
110163 Craterosiphon Engl. et Gilg(1894);宽管瑞香属●☆
110164 Craterosiphon beniense Domke;贝尼宽管瑞香●☆
110165 Craterosiphon devredii A. Robyns;德夫雷宽管瑞香●☆
110166 Craterosiphon louisii R. Wilczek ex A. Robyns;路易斯宽管瑞香●☆
110167 Craterosiphon micranthum A. Robyns;小花宽管瑞香●☆
110168 Craterosiphon montanum Domke;山地宽管瑞香●☆
110169 Craterosiphon pseudoscandens Domke;假攀缘宽管瑞香●☆
110170 Craterosiphon quarrei Staner;卡雷宽管瑞香●☆
110171 Craterosiphon scandens Engl. et Gilg;攀缘宽管瑞香●☆
110172 Craterosiphon schmitzii A. Robyns;施密茨宽管瑞香●☆
110173 Craterostemma K. Schum. = Brachystelma R. Br. (保留属名)■
110174 Craterostemma schinzii K. Schum. = Brachystelma schinzii (K. Schum.) N. E. Br. ■☆
110175 Craterostigma Hochst. (1841);杯柱玄参属■☆
110176 Craterostigma alatum Hepper;具翅杯柱玄参■☆
110177 Craterostigma auriculifolium Benth. et Hook. f. = Craterostigma pumilum Hochst. ■☆
110178 Craterostigma boranense Chiov.;博兰杯柱玄参■☆
110179 Craterostigma capitatum Hepper;头状杯柱玄参■☆
110180 Craterostigma chironioides S. Moore = Crepidorhopalon chironioides (S. Moore) Eb. Fisch. ■☆
110181 Craterostigma crassifolium Engl. = Lindernia crassifolia (Engl.) Eb. Fisch. ■☆
110182 Craterostigma goetzei Engl. = Crepidorhopalon goetzei (Engl.) Eb. Fisch. ■☆
110183 Craterostigma gracile Pilg. = Crepidorhopalon gracilis (Pilg.) Eb. Fisch. ■☆
110184 Craterostigma guineense Hepper = Crepidorhopalon gracilis (Pilg.) Eb. Fisch. ■☆
110185 Craterostigma hirsutum S. Moore;粗毛杯柱玄参■☆
110186 Craterostigma lanceolatum (Engl.) Skan;披针形杯柱玄参■☆
110187 Craterostigma latibracteatum Skan = Crepidorhopalon latibracteatus (Skan) Eb. Fisch. ■☆
110188 Craterostigma lindernioides E. A. Bruce = Lindernia oliverana Dandy ■☆
110189 Craterostigma linearifolia Engl. = Lindernia linearifolia (Engl.) Eb. Fisch. ■☆
110190 Craterostigma longicarpum Hepper;长果杯柱玄参■☆
110191 Craterostigma monroi S. Moore = Lindernia monroi (S. Moore) Eb. Fisch. ■☆
110192 Craterostigma nanum (Benth.) Engl. = Craterostigma plantagi-

neum Hochst. ■☆

110193 Craterostigma nanum (Benth.) Engl. var. elatior Oliv. = Craterostigma wilmsii Engl. ■☆

110194 Craterostigma nanum (Benth.) Engl. var. lanceolatum Engl. = Craterostigma lanceolatum (Engl.) Skan ■☆

110195 Craterostigma ndassekerense Engl. = Craterostigma hirsutum S. Moore ■☆

110196 Craterostigma plantagineum Hochst.;车前杯柱玄参;Blue Gem ■☆

110197 Craterostigma pumilum Hochst.;矮小杯柱玄参■☆

110198 Craterostigma purpureum Lebrun et L. Touss.;紫杯柱玄参■☆

110199 Craterostigma schweinfurthii (Oliv.) Engl. = Crepidorhopalon schweinfurthii (Oliv.) Eb. Fisch. ■☆

110200 Craterostigma smithii S. Moore;史密斯杯柱玄参■☆

110201 Craterostigma welwitschii Engl. = Crepidorhopalon welwitschii (Engl.) Eb. Fisch. ■☆

110202 Craterostigma wilmsii Engl.;维尔姆斯杯柱玄参■☆

110203 Craterotecoma Mart. ex DC. = Lundia DC.(保留属名)●☆

110204 Craterotecoma Mart. ex Meisn. = Lundia DC.(保留属名)●☆

110205 Crateva L.(1753);鱼木属;Crateva, Fishwood, Garlic Pear, Medlar-Hawthorn ●

110206 Crateva adansonii DC.;阿当松鱼木●☆

110207 Crateva adansonii DC. subsp. formosensis Jacobs = Crateva formosensis (Jacobs) B. S. Sun ●

110208 Crateva adansonii DC. subsp. odora (Ham.) Jacobs;芳香鱼木●☆

110209 Crateva adansonii DC. subsp. trifoliata (Roxb.) Jacobs = Crateva trifoliata (Roxb.) B. S. Sun ●

110210 Crateva avicularis Burch. = Maerua cafra (DC.) Pax ●☆

110211 Crateva cafra Burch. = Maerua cafra (DC.) Pax ●☆

110212 Crateva capparoides Andréws = Ritchiea capparoides (Andréws) Britten ●☆

110213 Crateva erythrocarpa Gagnep. = Crateva trifoliata (Roxb.) B. S. Sun ●

110214 Crateva excelsa Bojer;高大鱼木●☆

110215 Crateva falcata (Lour.) DC.;广东鱼木(鸡爪菜,镰叶鱼木);Falcate Crateva ●

110216 Crateva formosensis (Jacobs) B. S. Sun;台湾鱼木(鱼木);Spider Tree, Taiwan Adanson Caper, Taiwan Crateva, Taiwan Fishwood ●

110217 Crateva fragrans Sims = Ritchiea capparoides (Andréws) Britten ●☆

110218 Crateva greveana Baill.;格雷鱼木●☆

110219 Crateva humblotii (Baill.) Hadj-Moust.;洪布鱼木●☆

110220 Crateva lophosperma Kurz = Crateva magna (Lour.) DC. ●

110221 Crateva lophosperma Kurz = Crateva nurvala Buch.-Ham. ●

110222 Crateva magna (Lour.) DC.;沙梨木(巨大鱼木)●

110223 Crateva membranifolia Miq. = Crateva religiosa G. Forst. ●

110224 Crateva nurvala Buch.-Ham.;刺籽鱼木(刺子鱼木,沙梨木);Sandpear Fishwood, Tubercledseed Crateva, Tuberculate-seeded Crateva ●

110225 Crateva nurvala Buch.-Ham. = Crateva magna (Lour.) DC. ●

110226 Crateva obovata Vahl;倒卵鱼木●☆

110227 Crateva religiosa G. Forst.;鱼木(鹅脚木,牛角歪,千斤藤,千金藤,三脚鳖);Barna Tree, Japanese Stephania, Spider Tree ●

110228 Crateva religiosa G. Forst. = Crateva formosensis (Jacobs) B. S. Sun ●

110229 Crateva religiosa G. Forst. = Crateva unilocularis Buch.-Ham. ●

110230 Crateva roxburghii R. Br.;罗氏鱼木●☆

110231 Crateva roxburghii R. Br. var. erythrocarpa (Gagnep.) Gagnep. = Crateva trifoliata (Roxb.) B. S. Sun ●

110232 Crateva simplicifolia J. S. Mill.;单叶鱼木●☆

110233 Crateva trifoliata (Roxb.) B. S. Sun;钝叶鱼木(赤果鱼木);Bluntleaf Fishwood, Three-leaves Taiwan Crateva, Trifoliate Crateva ●

110234 Crateva trifoliata (Roxb.) B. S. Sun var. macrosperma S. M. Hwang;大籽鱼木(四方灯盏);Big-seed Crateva ●

110235 Crateva unilocularia Buch.-Ham.;树头菜(鹅脚木,虎王,鸡爪菜,苦洞树,龙头花,四方灯盏,鱼木);Barna, Crateva, Single Fishwood, Unilocular Crateva ●

110236 Cratoehwilia Neck. = Clutia L. ■☆

110237 Cratoxylon Blume = Cratoxylum Blume ●

110238 Cratoxylum Blume(1823);黄牛木属(九芎木属,山竹子属);Cratoxylum, Oxwood ●

110239 Cratoxylum acuminatum Merr.;尖叶黄牛木●☆

110240 Cratoxylum arborescens (Vahl) Blume;树状黄牛木;Geroggong ●☆

110241 Cratoxylum arborescens Blume = Cratoxylum blancoi Blume ●☆

110242 Cratoxylum arborescens Blume var. borneense A. C. Church et P. F. Stevens;博尔纳黄牛木●☆

110243 Cratoxylum arboreum Elmer;菲律宾黄牛木●☆

110244 Cratoxylum biflorum (Lam.) Turcz. = Cratoxylum cochinchinense (Lour.) Blume ●

110245 Cratoxylum biflorum Turcz. = Cratoxylum cochinchinense (Lour.) Blume ●

110246 Cratoxylum blancoi Blume;乔状黄牛木(树黄牛木,树状黄牛木);Genonggang ●☆

110247 Cratoxylum carneum Kurz. = Cratoxylum cochinchinense (Lour.) Blume ●

110248 Cratoxylum chinense Merr. = Cratoxylum cochinchinense (Lour.) Blume ●

110249 Cratoxylum chinense Merr. = Hypericum chinense L. ●

110250 Cratoxylum clandestinum Blume;隐匿黄牛木●☆

110251 Cratoxylum cochinchinense (Lour.) Blume;黄牛木(穿破石,狗牙茶,狗牙木,狗芽木,海牙茶,海芽茶,何线藤,黄尝,黄金桂,黄牛茶,黄丝鸡兰,黄芽木,节节花,九芽木,满天红,梅低优,雀笼木,山狗芽,水杧果,土苏木,畏芝,越南黄牛木,鹧鸪木);Cochin-China Cratoxylum, Common Oxwood, Shortstyle Cratoxylum, Tree Avens, Yoke-wood Tree ●

110252 Cratoxylum cochinchinense Blume = Cratoxylum formosum (Jack) Benth. et Hook. f. ex Dyer ●

110253 Cratoxylum cuneatum Miq.;楔形黄牛木●☆

110254 Cratoxylum dasyphyllum Hand.-Mazz. = Cratoxylum formosum (Jack) Benth. et Hook. f. ex Dyer subsp. pruniflorum (Kurz) Gogelein ●

110255 Cratoxylum floribundum F. Vill.;繁花黄牛木●☆

110256 Cratoxylum formosum (Jack) Benth. et Hook. f. ex Dyer;越南黄牛木(多花黄牛木,九芎木,毛叶黄牛木);Pretty Cratoxylum, Thickleaf Cratoxylum, Viet Nam Cratoxylum, Vietnam Cratoxylum, Yoke-wood Tree ●

110257 Cratoxylum formosum (Jack) Benth. et Hook. f. ex Dyer = Elodea formosa Jack ●

110258 Cratoxylum formosum (Jack) Benth. et Hook. f. ex Dyer subsp. pruniflorum (Kurz) Gogelein;李花黄牛木(红芽木,红眼树,黄浆果,黄浆树,苦沉茶,苦丁茶,毛叶黄牛木,牛丁角,酸浆树,土茶,樱桃叶黄牛木,樱叶苦丁茶);Bitter Tea Tree, Kuding Cratoxylum, Kudingcha, Plumleaf Cratoxylum, Plumleaf Oxwood ●

110259 Cratoxylum glaucum Korth.;灰绿黄牛木●☆

110260 Cratoxylum harmandii Pierre;柬埔寨黄牛木●☆

110261 Cratoxylum hornschuchii Blume;爪哇黄牛木●☆

110262 Cratoxylum hypericinum Merr. = Hornschuchia hypericina Blume ●☆

110263 Cratoxylum hypoleuca Elmer;里白黄牛木●☆

110264 Cratoxylum lanceolatum Miq. = Cratoxylum cochinchinense (Lour.) Blume ●

110265 Cratoxylum ligustrinum (Spach) Blume = Cratoxylum cochinchinense (Lour.) Blume ●

110266 Cratoxylum ligustrinum Blume = Cratoxylum cochinchinense (Lour.) Blume ●

110267 Cratoxylum maingayi Dyer;马来黄牛木●☆

110268 Cratoxylum microphyllum Miq.;小叶黄牛木●☆

110269 Cratoxylum myrtifolium Blume;香桃木叶黄牛木●☆

110270 Cratoxylum neriifolium Kurz;缅甸黄牛木●☆

110271 Cratoxylum parvifolium Merr.;非洲小叶黄牛木●☆

110272 Cratoxylum pentadelphum Turcz.;五冠黄牛木●☆

110273 Cratoxylum petiolatum Blume = Cratoxylum cochinchinense (Lour.) Blume ●

110274 Cratoxylum polyanthum Korth. = Cratoxylum cochinchinense (Lour.) Blume ●

110275 Cratoxylum polyanthum Korth. var. ligustrinum Dyer = Cratoxylum cochinchinense (Lour.) Blume ●

110276 Cratoxylum polystachyum Turcz.;多穗黄牛木●☆

110277 Cratoxylum procerum Diels;高大黄牛木●☆

110278 Cratoxylum pruniflorum (Kurz) Kurz. = Cratoxylum formosum (Jack) Benth. et Hook. f. ex Dyer subsp. pruniflorum (Kurz) Gogelein ●

110279 Cratoxylum pruniflorum Kurz = Cratoxylum formosum (Jack) Benth. et Hook. f. ex Dyer subsp. pruniflorum (Kurz) Gogelein ●

110280 Cratoxylum punctulatum Elmer ex Merr. = Cratoxylum blancoi Blume ●☆

110281 Cratoxylum racemosum Blume;总花黄牛木●☆

110282 Cratoxylum subglaucum Merr.;粉绿黄牛木●☆

110283 Cratoxylum sumatranum (Jack) Blume;苏门答腊黄牛木;Sumatra Cratoxylum,Sumatra Oxwood ●☆

110284 Cratoxylum thorelii Pierre ex Gagnep.;托雷尔黄牛木●☆

110285 Cratoxylun pruniflorum Kurz = Cratoxylum formosum (Jack) Benth. et Hook. f. ex Dyer subsp. pruniflorum (Kurz) Gogelein ●

110286 Cratylia Mart. ex Benth. (1837);克拉豆属●☆

110287 Cratylia floribunda Benth.;繁花克拉豆●☆

110288 Cratylia nuda Tul.;裸克拉豆●☆

110289 Cratylia pauciflora Harms;少花克拉豆●☆

110290 Cratystylis S. Moore(1905);束柱菊属■☆

110291 Cratystylis conocephala (F. Muell.) S. Moore;束柱菊■☆

110292 Crawfurdia Wall. (1826);蔓龙胆属;Crawfurdia,Vinegentian ■

110293 Crawfurdia Wall. = Gentiana L. ■

110294 Crawfurdia Wall. = Tripterospermum Blume ■

110295 Crawfurdia angustata C. B. Clarke;大花蔓龙胆;Bigflower Vinegentian ■

110296 Crawfurdia blumei G. Don;布氏蔓龙胆■☆

110297 Crawfurdia bulleyana Forrest = Crawfurdia campanulacea Wall. et Griff. ex C. B. Clarke ■

110298 Crawfurdia campanulacea Wall. et Griff. ex C. B. Clarke;云南蔓龙胆(双蝴蝶);Yunnan Crawfurdia,Yunnan Vinegentian ■

110299 Crawfurdia chinensis Migo = Tripterospermum chinense (Migo) Harry Sm. ■

110300 Crawfurdia coerulea Hand.-Mazz. = Tripterospermum coeruleum (Hand.-Mazz. ex Harry Sm.) Harry Sm. ■

110301 Crawfurdia cordata (Marquart) Hand.-Mazz. = Tripterospermum cordatum (C. Marquand) Harry Sm. ■

110302 Crawfurdia cordifolia Gogelein = Tripterospermum cordifolium (Gogelein) Satake ■

110303 Crawfurdia crawfurdioides (C. Marquand) Harry Sm.;裂萼蔓龙胆;Splitcalyx Vinegentian ■

110304 Crawfurdia crawfurdioides (C. Marquand) Harry Sm. var. iochroa (C. Marquand) C. J. Wu;根茎蔓龙胆■

110305 Crawfurdia crawfurdioides (C. Marquand) Harry Sm. var. macrophylla Marquart = Crawfurdia crawfurdioides (C. Marquand) Harry Sm. ■

110306 Crawfurdia delavayi Franch.;披针叶蔓龙胆;Delavay Vinegentian ■

110307 Crawfurdia dimidiata (C. Marquand) Harry Sm.;半侧蔓龙胆;Oneside Vinegentian ■

110308 Crawfurdia fasciculata Wall. = Tripterospermum filicaule (Hemsl.) Harry Sm. ■

110309 Crawfurdia forrestii Harry Sm. = Crawfurdia crawfurdioides (C. Marquand) Harry Sm. ■

110310 Crawfurdia gracilipes Harry Sm.;细柄蔓龙胆;Smallstipe Vinegentian ■

110311 Crawfurdia iochroa (C. Marquand) Hand.-Mazz. = Crawfurdia crawfurdioides (C. Marquand) Harry Sm. var. iochroa (C. Marquand) C. J. Wu ■

110312 Crawfurdia japonica Siebold et Zucc. = Tripterospermum japonicum (Siebold et Zucc.) Maxim. ■

110313 Crawfurdia japonica Siebold et Zucc. var. luteoviridis (C. B. Clarke) C. B. Clarke = Tripterospermum volubile (D. Don) H. Hara ■

110314 Crawfurdia japonica Siebold et Zucc. var. taiwanense Masam. = Tripterospermum taiwanense (Masam.) Satake ■

110315 Crawfurdia japonica Siebold et Zucc. var. tenuis ? = Tripterospermum japonicum (Siebold et Zucc.) Maxim. ■

110316 Crawfurdia lanceolata Hayata = Tripterospermum lanceolatum (Hayata) Hara et Satake ■

110317 Crawfurdia lobatilimba W. L. Cheng;裂膜蔓龙胆■

110318 Crawfurdia luteoviridis C. B. Clarke = Tripterospermum pallidum Harry Sm. ■

110319 Crawfurdia luteoviridis C. B. Clarke = Tripterospermum volubile (D. Don) H. Hara ■

110320 Crawfurdia luzonensis Vidal = Tripterospermum luzonense (Vidal) Eggl. ■

110321 Crawfurdia maculaticaulis C. Y. Wu ex C. J. Wu;斑茎蔓龙胆;Blotchstem Vinegentian ■

110322 Crawfurdia nienkui (C. Marquand) Chun = Tripterospermum nienkui (C. Marquand) C. J. Wu ■

110323 Crawfurdia nyingchiensis K. Yao et W. L. Cheng;林芝蔓龙胆;Linzhi Vinegentian ■

110324 Crawfurdia parvifolia (Hayata) Hayata = Tripterospermum microphyllum Harry Sm. ■

110325 Crawfurdia pricei (C. Marquand) Harry Sm.;福建蔓龙胆(蝴蝶草);Fujian Vinegentian ■

110326 Crawfurdia pterygocalyx Hemsl. = Pterygocalyx volubilis Maxim. ■

110327 Crawfurdia puberula C. B. Clarke;毛叶蔓龙胆;Hairleaf Vinegentian ■

110328 Crawfurdia semialata (C. Marquand) Harry Sm.;直立蔓龙胆;Erect Vinegentian ■

110329 Crawfurdia sessiliflora (C. Marquand) Harry Sm.;无柄蔓龙胆;Sessile Vinegentian ■

110330 Crawfurdia speciosa Wall.;穗序蔓龙胆;Spike Vinegentian ■

110331 Crawfurdia speciosa Wall. = Crawfurdia sessiliflora (C. Marquand) Harry Sm. ■

110332 Crawfurdia taiwanense Masam. = Tripterospermum taiwanense (Masam.) Satake ■

110333 Crawfurdia thibetica Franch.;四川蔓龙胆;Xizang Vinegentian ■

110334 Crawfurdia trailliana Forrest = Crawfurdia angustata C. B. Clarke ■

110335 Crawfurdia trinervis Makino = Tripterospermum japonicum (Siebold et Zucc.) Maxim. ■

110336 Crawfurdia tsangshaoensis C. J. Wu;苍山蔓龙胆;Cangshan Vinegentian ■

110337 Crawfurdia volubilis (Maxim.) Makino = Pterygocalyx volubilis Maxim. ■

110338 Crawfurdia wardii C. Marquand = Crawfurdia speciosa Wall. ■

110339 Creaghia Scort. = Mussaendopsis Baill. ●■☆

110340 Creaghiella Stapf = Anerincleistus Korth. ●☆

110341 Creatantha Standl. = Isertia Schreb. ●☆

110342 Cremanium D. Don = Miconia Ruiz et Pav.(保留属名)●☆

110343 Cremanthodium Benth.(1873);垂头菊属;Cremanthodium, Nutantdaisy ■

110344 Cremanthodium acernuum R. D. Good = Cremanthodium smithianum (Hand.-Mazz.) Hand.-Mazz. ■

110345 Cremanthodium angustifolium W. W. Sm.;狭叶垂头菊(垂头菊);Narrowleaf Cremanthodium, Narrowleaf Nutantdaisy ■

110346 Cremanthodium angustifolium W. W. Sm. var. roseum Hand.-Mazz.;红花狭叶垂头菊■

110347 Cremanthodium arnicoides (DC. ex Royle) R. D. Good;宽舌垂头菊(革叶垂头菊);Broatongue Nutantdaisy, Leatheryleaf Cremanthodium ■

110348 Cremanthodium atrocapitatum R. D. Good;黑垂头菊■

110349 Cremanthodium atroviolaceum (Franch.) R. D. Good = Ligularia atroviolacea (Franch.) Hand.-Mazz. ■

110350 Cremanthodium bhutanicum Ludlow;不丹垂头菊;Bhutan Cremanthodium, Bhutan Nutantdaisy ■

110351 Cremanthodium botrycephalum S. W. Liu;总状垂头菊;Raceme Cremanthodium, Raceme Nutantdaisy ■

110352 Cremanthodium brachychaetum C. C. Chang;短缨垂头菊(短冠垂头菊);Shorttassell Cremanthodium, Shorttassell Nutantdaisy ■

110353 Cremanthodium brunneopilosum S. W. Liu;褐毛垂头菊(点头菊);Brownhair Cremanthodium, Brownhair Nutantdaisy ■

110354 Cremanthodium bulbilliferum W. W. Sm.;珠芽垂头菊;Bulbil Cremanthodium, Bulbil Nutantdaisy ■

110355 Cremanthodium bupleurifolium W. W. Sm.;柴胡叶垂头菊;Thorowaxleaf Cremanthodium, Thorowaxleaf Nutantdaisy ■

110356 Cremanthodium calcieola W. W. Sm.;长鞘垂头菊;Longsheath Cremanthodium ■

110357 Cremanthodium calotum Diels = Doronicum thibetanum Cavill. ■

110358 Cremanthodium campanulatum (Franch.) Diels;钟花垂头菊;Campanulate Cremanthodium, Campanulate Nutantdaisy ■

110359 Cremanthodium campanulatum (Franch.) Diels ex H. Lév. var. pinnatiseetum Ludlow = Cremanthodium pinnatiseetum (Ludlow) Y. L. Chen et S. W. Liu ■

110360 Cremanthodium campanulatum (Franch.) Diels var. brachytrichum Y. Ling et S. W. Liu;短毛钟花垂头菊;Shorthair Calcicole Nutantdaisy, Shorthair Campanulate Cremanthodium ■

110361 Cremanthodium campanulatum (Franch.) Diels var. flavidum S. W. Liu et T. N. Ho;黄苞钟花垂头菊(黄苞垂头菊);Yellow Campanulate Cremanthodium ■

110362 Cremanthodium campanulatum (Franch.) Diels var. pinnatisectum Ludlow = Cremanthodium pinnatisectum (Ludlow) Y. L. Chen et S. W. Liu ■

110363 Cremanthodium chungtienense Y. Ling et S. W. Liu;中甸垂头菊;Zhongdian Cremanthodium, Zhongdian Nutantdaisy ■

110364 Cremanthodium citriflorum R. D. Good;柠檬色垂头菊;Orangeflower Cremanthodium, Orangeflower Nutantdaisy ■

110365 Cremanthodium comptum W. W. Sm. = Cremanthodium humile Maxim. ■

110366 Cremanthodium conaense S. W. Liu;错那垂头菊;Cuona Cremanthodium ■

110367 Cremanthodium cordatum S. W. Liu;心叶垂头菊■

110368 Cremanthodium coriaceum S. W. Liu;革叶垂头菊;Cremanthodium, Leatheryleaf Nutantdaisy ■

110369 Cremanthodium cremanthodioides (Hand.-Mazz.) R. D. Good = Ligularia cremanthodioides Hand.-Mazz. ■

110370 Cremanthodium cucullatum Y. Ling et S. W. Liu;兜鞘垂头菊;Poketlike Cremanthodium, Poketlike Nutantdaisy ■

110371 Cremanthodium cuculliferum W. W. Sm. = Cremanthodium discoideum Maxim. ■

110372 Cremanthodium cyclaminanthum Hand.-Mazz.;仙客来垂头菊;Cyclamenlike Cremanthodium, Cyclamenlike Nutantdaisy ■

110373 Cremanthodium cymosum Hand.-Mazz. = Ligularia cymosa (Hand.-Mazz.) S. W. Liu ■

110374 Cremanthodium daochengense Y. Ling et S. W. Liu;稻城垂头菊;Daocheng Cremanthodium, Daocheng Nutantdaisy ■

110375 Cremanthodium deasyi Hemsl. = Cremanthodium nanum (Decne.) W. W. Sm. ■

110376 Cremanthodium decaisnei C. B. Clarke;喜马拉雅垂头菊(须弥垂头菊);Himalayan Cremanthodium, Himalayas Nutantdaisy ■

110377 Cremanthodium decaisnei C. B. Clarke = Cremanthodium smithianum (Hand.-Mazz.) Hand.-Mazz. ■

110378 Cremanthodium decaisnei C. B. Clarke f. clarkei R. D. Good = Cremanthodium decaisnei C. B. Clarke ■

110379 Cremanthodium decaisnei C. B. Clarke f. sinense R. D. Good = Cremanthodium decaisnei C. B. Clarke ■

110380 Cremanthodium delavayi (Franch.) Diels ex H. Lév.;大理垂头菊;Delavay Cremanthodium, Delavay Nutantdaisy ■

110381 Cremanthodium dicoideum Maxim. subsp. ramosum Y. Ling = Cremanthodium ellisii (Hook. f.) Kitam. var. ramosum (Y. Ling) Y. Ling et S. W. Liu ■

110382 Cremanthodium discoideum Maxim.;盘花垂头菊;Discoid Cremanthodium, Discoid Nutantdaisy ■

110383 Cremanthodium dissectum Grierson;细裂垂头菊;Dissected Cremanthodium, Dissected Nutantdaisy ■

110384 Cremanthodium ellisii (Hook. f.) Kitam.;车前叶垂头菊(点头菊,块根垂头菊);Planain Cremanthodium, Planain Nutantdaisy, Plantainshaped Cremanthodium ■

110385 Cremanthodium ellisii (Hook. f.) Kitam. var. ramosum (Y. Ling) Y. Ling et S. W. Liu;祁连垂头菊;Qilianshan Cremanthodium, Qilianshan Nutantdaisy ■

110386 Cremanthodium ellisii (Hook. f.) Kitam. var. roseum (Hand.-

Mazz.) S. W. Liu;红舌垂头菊;Rose Cremanthodium ■

110387 Cremanthodium ellisii (Hook. f.) S. W. Liu = Cremanthodium petiolatum S. W. Liu ■

110388 Cremanthodium farreri W. W. Sm.;红花垂头菊;Redflower Cremanthodium,Redflower Nutantdaisy ■

110389 Cremanthodium fletcheri (Hemsl.) Hemsl. = Cremanthodium ellisii (Hook. f.) Kitam. ■

110390 Cremanthodium forrestii Jeffrey;矢叶垂头菊;Forrest Cremanthodium,Forrest Nutantdaisy ■

110391 Cremanthodium glandulipilosum Y. L. Chen ex S. W. Liu;腺毛垂头菊;Glandhair Cremanthodium,Glandhair Nutantdaisy ■

110392 Cremanthodium glaucum Hand.-Mazz.;灰绿垂头菊;Greygreen Cremanthodium,Greygreen Nutantdaisy ■

110393 Cremanthodium goringense (Hemsl.) Hemsl. = Cremanthodium ellisii (Hook. f.) Kitam. ■

110394 Cremanthodium gracillimum W. W. Sm. = Cremanthodium rhodocephalum Diels ■

110395 Cremanthodium gypsophilum R. D. Good = Cremanthodium principis (Franch.) R. D. Good ■

110396 Cremanthodium hederifolium (Dunn) C. C. Chang = Sinosenecio hederifolius (Dümmer) B. Nord. ■

110397 Cremanthodium helianthus (Franch.) W. W. Sm.;向日垂头菊;Sunflower Cremanthodium,Sunflower Nutantdaisy ■

110398 Cremanthodium heterocephalum Y. L. Chen;异首垂头菊(异头垂头菊);Heterohead Cremanthodium,Heterohead Nutantdaisy ■

110399 Cremanthodium hirtiflorum S. W. Liu;毛花垂头菊;Hairflower Cremanthodium,Hairflower Nutantdaisy ■

110400 Cremanthodium hirtiflorum S. W. Liu = Cremanthodium spathulifolium S. W. Liu ■

110401 Cremanthodium hookeri C. B. Clarke = Ligularia hookeri (C. B. Clarke) Hand.-Mazz. ■

110402 Cremanthodium hookeri C. B. Clarke var. polycephalum R. D. Good = Ligularia hookeri (C. B. Clarke) Hand.-Mazz. ■

110403 Cremanthodium humile Maxim.;矮垂头菊(小垂头菊);Low Cremanthodium,Low Nutantdaisy ■

110404 Cremanthodium laciniatum Y. Ling et Y. L. Chen ex S. W. Liu;条裂垂头菊;Laciniate Cremanthodium,Laciniate Nutantdaisy ■

110405 Cremanthodium larium Hand.-Mazz = Cremanthodium campanulatum (Franch.) Diels ■

110406 Cremanthodium limprichtii Diels ex Limpr. = Cremanthodium potaninii C. Winkl. ■

110407 Cremanthodium lineare Maxim.;条叶垂头菊(线叶垂头菊);Linearleaf Cremanthodium,Linearleaf Nutantdaisy ■

110408 Cremanthodium lineare Maxim. var. eligulatum Y. Ling et S. W. Liu;无舌条叶垂头菊(舌线叶垂头菊);Tongueless Cremanthodium,Tongueless Nutantdaisy ■

110409 Cremanthodium lineare Maxim. var. roseum Hand.-Mazz.;红花条叶垂头菊;Rose Linearleaf Cremanthodium, Rose Linearleaf Nutantdaisy ■

110410 Cremanthodium lingulatum S. W. Liu;舌叶垂头菊;Tongueleaf Cremanthodium,Tongueleaf Nutantdaisy ■

110411 Cremanthodium lobatum Grierson = Cremanthodium forrestii Jeffrey ■

110412 Cremanthodium microcephalum Hand.-Mazz. = Ligularia microcephala (Hand.-Mazz.) Hand.-Mazz. ■

110413 Cremanthodium microglossum S. W. Liu;小舌垂头菊■

110414 Cremanthodium microphyllum S. W. Liu;小叶垂头菊;Smallleaf Cremanthodium,Smallleaf Nutantdaisy ■

110415 Cremanthodium nakaoi Kitam. = Cremanthodium oblongatum C. B. Clarke ■

110416 Cremanthodium nanum (Decne.) W. W. Sm.;小垂头菊;Small Cremanthodium,Small Nutantdaisy ■

110417 Cremanthodium nepalense Kitam.; 尼泊尔垂头菊; Nepal Cremanthodium,Nepal Nutantdaisy ■

110418 Cremanthodium nervosum S. W. Liu;显脉垂头菊;Nervose Cremanthodium,Nervose Nutantdaisy ■

110419 Cremanthodium nobile (Franch.) Diels ex H. Lév.;壮观垂头菊(北欧垂头菊);Noble Cremanthodium,Noble Nutantdaisy ■

110420 Cremanthodium oblongatum C. B. Clarke; 矩叶垂头菊; Oblongateleaf Cremanthodium,Oblongateleaf Nutantdaisy ■

110421 Cremanthodium oblongatum C. B. Clarke var. villosior C. B. Clarke = Cremanthodium ellisii (Hook. f.) Kitam. ■

110422 Cremanthodium obovatum Y. Ling et S. W. Liu;硕首垂头菊;Obovate Cremanthodium,Obovate Nutantdaisy ■

110423 Cremanthodium palmatum Benth.; 掌叶垂头菊; Palmleaf Cremanthodium,Palmleaf Nutantdaisy ■

110424 Cremanthodium palmatum Benth. subsp. benthami R. D. Good = Cremanthodium palmatum Benth. ■

110425 Cremanthodium palmatum Benth. subsp. rhodocephalum (Diels) R. D. Good = Cremanthodium rhodocephalum Diels ■

110426 Cremanthodium palmatum Benth. var. rhodocephalum (Diels) R. D. Good = Cremanthodium rhodocephalum Diels ■

110427 Cremanthodium petiolatum S. W. Liu;长柄垂头菊;Longstiped Cremanthodium,Longstiped Nutantdaisy ■

110428 Cremanthodium phoenicochaetum (Franch.) R. D. Good = Ligularia phoenicochaeta (Franch.) S. W. Liu ■

110429 Cremanthodium phyllodineum S. W. Liu;叶状柄垂头菊;Leaflikestiped Cremanthodium,Leaflikestiped Nutantdaisy ■

110430 Cremanthodium pilosum S. W. Liu;黄毛垂头菊■

110431 Cremanthodium pinnatifidum Benth.;羽裂垂头菊;Pinnatifid Cremanthodium,Pinnatifid Nutantdaisy ■

110432 Cremanthodium pinnatisectum (Ludlow) Y. L. Chen et S. W. Liu;裂叶垂头菊;Pinnatisect Cremanthodium,Pinnatisect Nutantdaisy ■

110433 Cremanthodium plantagineum Maxim. = Cremanthodium ellisii (Hook. f.) Kitam. ■

110434 Cremanthodium plantagineum Maxim. f. albidum Good = Cremanthodium ellisii (Hook. f.) Kitam. ■

110435 Cremanthodium plantagineum Maxim. f. goringense (Hemsl.) R. D. Good = Cremanthodium ellisii (Hook. f.) Kitam. ■

110436 Cremanthodium plantagineum Maxim. f. roseum Hand.-Mazz. = Cremanthodium ellisii (Hook. f.) Kitam. var. roseum (Hand.-Mazz.) S. W. Liu ■

110437 Cremanthodium plantagineum Maxim. var. maximowiczii (Franch.) Aswal = Cremanthodium ellisii (Hook. f.) Kitam. ■

110438 Cremanthodium plantagineum Maxim. var. ramosum (Y. Ling) Y. Ling et S. W. Liu = Cremanthodium ellisii (Hook. f.) Kitam. var. ramosum (Y. Ling) Y. Ling et S. W. Liu ■

110439 Cremanthodium plantaginifolium (Franch.) R. D. Good = Ligularia virgaurea (Maxim.) Mattf. ex Rehder et Kobuski ■

110440 Cremanthodium plantaginifolium (Franch.) R. D. Good subsp. franchettii f. winkleri Good = Ligularia liatroides (C. Winkl.) Hand.-Mazz. ■

110441 Cremanthodium plantaginifolium (Franch.) R. D. Good subsp. oligocephalum R. D. Good = Ligularia virgaurea (Maxim.) Mattf. ex

Rehder et Kobuski var. oligocephala (R. D. Good) S. W. Liu ■

110442 Cremanthodium pleurocaule (Franch.) R. D. Good = Ligularia pleurocaulis (Franch.) Hand. -Mazz. ■

110443 Cremanthodium potaninii C. Winkl.; 戟叶垂头菊; Potanin Nutantdaisy, Potanin's Cremanthodium ■

110444 Cremanthodium prattii (Hemsl.) R. D. Good; 长舌垂头菊; Longtongue Cremanthodium, Longtongue Nutantdaisy ■

110445 Cremanthodium principis (Franch.) R. D. Good; 方叶垂头菊; Squwreleaf Cremanthodium, Squwreleaf Nutantdaisy ■

110446 Cremanthodium pseudo-oblongatum R. D. Good; 无毛垂头菊; Hairless Cremanthodium, Hairless Nutantdaisy ■

110447 Cremanthodium pteridophyllum Y. L. Chen = Ligularia paradoxa Hand. -Mazz. ■

110448 Cremanthodium puberulum S. W. Liu; 毛叶垂头菊; Hairleaf Cremanthodium, Hairleaf Nutantdaisy ■

110449 Cremanthodium pulchrum R. D. Good; 美丽垂头菊; Beautiful Cremanthodium, Beautiful Nutantdaisy ■

110450 Cremanthodium purpureifolium Kitam.; 紫叶垂头菊; Purpleleaf Cremanthodium, Purpleleaf Nutantdaisy ■

110451 Cremanthodium reniforme (DC.) Benth.; 肾叶垂头菊(垂头菊); Reniform Cremanthodium, Reniform Nutantdaisy ■

110452 Cremanthodium retusum (DC.) R. D. Good = Ligularia retusa DC. ■

110453 Cremanthodium retusum (Wall. ex Hook. f.) Good = Ligularia retusa DC. ■

110454 Cremanthodium rhodocephalum Diels; 长柱垂头菊(红头垂头菊); Redhead Cremanthodium, Redhead Nutantdaisy ■

110455 Cremanthodium rumicifolium (Drumm.) R. D. Good = Ligularia rumicifolia (Drumm.) S. W. Liu ■

110456 Cremanthodium sagittifolium Y. Ling et Y. L. Chen ex S. W. Liu; 箭叶垂头菊; Arrowleaf Cremanthodium, Arrowleaf Nutantdaisy ■

110457 Cremanthodium sherriffii H. R. Fletcher = Cremanthodium rhodoceghalum Diels ■

110458 Cremanthodium sino-oblongatum Good; 铲叶垂头菊; Spadeleaf Cremanthodium, Spadeleaf Nutantdaisy ■

110459 Cremanthodium smithianum (Hand. -Mazz.) Hand. -Mazz.; 紫茎垂头菊; Purplestem Cremanthodium, Purplestem Nutantdaisy ■

110460 Cremanthodium spathulifolium S. W. Liu; 匙叶垂头菊; Spoonleaf Cremanthodium, Spoonleaf Nutantdaisy ■

110461 Cremanthodium stenactinium Diels ex H. Limpr.; 膜苞垂头菊(丛叶垂头菊,阔叶垂头菊,千穷娃); Membranebract Cremanthodium, Membranebract Nutantdaisy ■

110462 Cremanthodium stenactinium Diels ex H. Limpr. var. evillosum Hand. -Mazz. = Cremanthodium stenactinium Diels ex H. Limpr. ■

110463 Cremanthodium stenoglossum Y. Ling et S. W. Liu; 狭舌垂头菊(线舌垂头菊); Narrowtongue Cremanthodium, Narrowtongue Nutantdaisy ■

110464 Cremanthodium suave W. W. Sm.; 木里垂头菊; Muli Cremanthodium, Muli Nutantdaisy ■

110465 Cremanthodium thomsonii C. B. Clarke; 叉舌垂头菊; Forktogue Cremanthodium, Forktogue Nutantdaisy ■

110466 Cremanthodium trilobum S. W. Liu; 裂舌垂头菊; Threelobe Cremanthodium, Threelobe Nutantdaisy ■

110467 Cremanthodium variifolium Good; 变叶垂头菊; Variantleaf Cremanthodium, Variantleaf Nutantdaisy ■

110468 Cremanthodium virgaurea (Maxim.) Hand. -Mazz. = Ligularia virgaurea (Maxim.) Mattf. ex Rehder et Kobuski ■

110469 Cremanthodium wardii W. W. Sm. = Cremanthodium campanulatum (Franch.) Diels ■

110470 Cremanthodium yadongense S. W. Liu; 亚东垂头菊; Yadong Cremanthodium, Yadong Nutantdaisy ■

110471 Cremaspora Benth. (1849); 悬籽茜属●☆

110472 Cremaspora africana Benth. = Cremaspora triflora (Thonn.) K. Schum. ●☆

110473 Cremaspora bocandeana Webb = Cremaspora triflora (Thonn.) K. Schum. ●☆

110474 Cremaspora congesta Baill. = Polysphaeria multiflora Hiern ■☆

110475 Cremaspora glabra Wernham = Argocoffeopsis pulchella (K. Schum.) Robbr. ●☆

110476 Cremaspora heterophylla Didr. = Cremaspora triflora (Thonn.) K. Schum. ●☆

110477 Cremaspora heterophylla K. Schum. = Cremaspora triflora (Thonn.) K. Schum. ●☆

110478 Cremaspora microcarpa (DC.) Baill. = Cremaspora triflora (Thonn.) K. Schum. ●☆

110479 Cremaspora thomsonii Hiern; 托马森悬籽茜●☆

110480 Cremaspora triflora (Thonn.) K. Schum.; 三花悬籽茜●☆

110481 Cremaspora wernhamiana Hutch. et Dalziel = Tricalysia wernhamiana (Hutch. et Dalziel) Keay ●☆

110482 Cremastogyne (H. Winkl.) Czerep. = Alnus Mill. ●

110483 Cremastogyne (H. Winkl.) De Moor (1955); 悬蕊桤属●

110484 Cremastogyne ferdinandi-coburgii (C. K. Schneid.) Czerep.; 悬蕊桤●☆

110485 Cremastogyne lanata (Duthie ex Bean) Czerep.; 绵毛悬蕊桤●☆

110486 Cremastogyne longipes Czerep.; 长梗悬蕊桤●☆

110487 Cremastopus Paul G. Wilson (1962); 悬足葫芦属■☆

110488 Cremastopus minimus (S. Watson) Paul G. Wilson; 小悬足葫芦■☆

110489 Cremastopus rostratus Paul G. Wilson; 悬足葫芦■☆

110490 Cremastosciadium Rech. f. = Eriocycla Lindl. ■

110491 Cremastosperma R. E. Fr. (1930); 焰子木属●☆

110492 Cremastosperma brevipes (DC.) R. E. Fr.; 短梗焰子木●☆

110493 Cremastosperma gracilipes R. E. Fr.; 细梗焰子木●☆

110494 Cremastosperma longicuspe R. E. Fr.; 焰子木●☆

110495 Cremastosperma monospermum (Rusby) R. E. Fr.; 单籽焰子木●☆

110496 Cremastosperma stenophyllum Pirie; 窄叶焰子木●☆

110497 Cremastostemon Jacq. (1809); 悬蕊千屈菜属●☆

110498 Cremastostemon Jacq. = Olinia Thunb. ●☆

110499 Cremastostemon capensis Jacq. = Olinia capensis (Jacq.) Klotzsch ●☆

110500 Cremastra Lindl. (1833); 杜鹃兰属(马鞭兰属,毛慈姑属); Cremastra, Cuckoo-orchis ■

110501 Cremastra aphylla Yukawa; 无叶杜鹃兰■☆

110502 Cremastra appendiculata (D. Don) Makino; 杜鹃兰(白地栗,白毛姑,朝天一柱香,处姑,大白芨,鬼灯檠,金灯,金灯花,鹿蹄草,马鞭兰,马笃七,毛慈姑,毛姑,茅慈姑,泥宾子,泥滨子,泥冰子,人头七,人头芪,三道箍,三道圈,三七笋,僧帽杜鹃兰,山茨菇,山茨菰,山慈姑,算盘七,朱姑); Appendiculate Cremastra, Common Cuckoo-orchis ■

110503 Cremastra appendiculata (D. Don) Makino var. miyabei T. Inoue = Cremastra aphylla Yukawa ■☆

110504 Cremastra appendiculata (D. Don) Makino var. triloba (Hayata) S. S. Ying = Cremastra appendiculata (D. Don) Makino ■

110505 Cremastra appendiculata (D. Don) Makino var. variabilis (Blume) I. D. Lund f. viridiflora (Honda) Honda;绿花多变杜鹃兰■☆

110506 Cremastra appendiculata (D. Don) Makino var. variabilis (Blume) I. D. Lund;翅柱杜鹃兰(多变杜鹃兰)■

110507 Cremastra appendiculata (D. Don) Makino var. variabilis (Blume) I. D. Lund = Cremastra appendiculata (D. Don) Makino ■

110508 Cremastra appendiculata (D. Don) Makino var. viridiflora (Honda) Aver. = Cremastra appendiculata (D. Don) Makino var. variabilis (Blume) I. D. Lund ■

110509 Cremastra bifolia C. L. Tso = Cremastra appendiculata (D. Don) Makino ■

110510 Cremastra guizhouensis Q. H. Chen et S. C. Chen;贵州杜鹃兰;Guizhou Cremastra ■

110511 Cremastra lanceolata (Kraenzl.) Schltr. = Cremastra appendiculata (D. Don) Makino ■

110512 Cremastra lanceolata (Kraenzl.) Schltr. = Cremastra appendiculata (D. Don) Makino var. variabilis (Blume) I. D. Lund ■

110513 Cremastra mitrata A. Gray = Cremastra appendiculata (D. Don) Makino var. variabilis (Blume) I. D. Lund ■

110514 Cremastra mitrata A. Gray = Cremastra appendiculata (D. Don) Makino ■

110515 Cremastra triloba Hayata = Cremastra appendiculata (D. Don) Makino ■

110516 Cremastra unguiculata (Finet) Finet;斑叶杜鹃兰;Spotleaf Cremastra,Spotleaf Cuckoo-orchis ■

110517 Cremastra variabilis (Blume) Nakai = Cremastra appendiculata (D. Don) Makino var. variabilis (Blume) I. D. Lund ■

110518 Cremastra variabilis (Blume) Nakai = Cremastra appendiculata (D. Don) Makino ■

110519 Cremastra variabilis (Blume) Nakai var. miyabei T. Inoue = Cremastra aphylla Yukawa ■☆

110520 Cremastra variabilis (Blume) Nakai var. viridiflora Honda = Cremastra appendiculata (D. Don) Makino var. variabilis (Blume) I. D. Lund ■

110521 Cremastra wallichiana Lindl. = Cremastra appendiculata (D. Don) Makino ■

110522 Cremastus Miers = Arrabidaea DC. ●☆

110523 Crematomia Miers = Bourreria P. Browne(保留属名)●☆

110524 Crematomia Miers = Morelosia Lex. ●☆

110525 Cremersia Feuillet et L. E. Skog(2003);克里苣苔属■☆

110526 Cremersia platula Feuillet et L. E. Skog;克里苣苔■☆

110527 Cremnobates Ridl. = Schizomeria D. Don ●☆

110528 Cremnophila Rose = Sedum L. ●■

110529 Cremnophila Rose(1905);悬崖景天属■☆

110530 Cremnophila nutans Rose;悬崖景天■☆

110531 Cremnophyton Brullo et Pavone(1987);岩石藜属●☆

110532 Cremnophyton lanfrancoi Brullo et Pavone;岩石藜●☆

110533 Cremnothamnus Puttock(1854);黄花鼠麹木属●☆

110534 Cremnothamnus thomsonii (F. Muell.) Puttock;黄花鼠麹木●☆

110535 Cremobotrys Beer = Billbergia Thunb. ■

110536 Cremobotrys Beer = Eucallias Raf. ■

110537 Cremocarpon Baill. = Cremocarpon Boiv. ex Baill. ●☆

110538 Cremocarpon Boiv. ex Baill. (1879);悬果茜属●☆

110539 Cremocarpon boivinianum Baill.;悬果茜●☆

110540 Cremocarpus K. Schum. = Cremocarpon Boiv. ex Baill. ●☆

110541 Cremocephallum Miq. = Cremocephalum Cass. ■●

110542 Cremocephalum Cass. = Crassocephalum Moench(废弃属名)■

110543 Cremocephalum Cass. = Gynura Cass.(保留属名)■●

110544 Cremocephalum cernuum Cass. = Crassocephalum rubens (Juss. ex Jacq.) S. Moore ■

110545 Cremochilus Turcz. = Siphocampylus Pohl ■●☆

110546 Cremolobus DC. (1821);双钱荠属■☆

110547 Cremolobus bolivianus Britton.;玻利维亚双钱荠■☆

110548 Cremolobus chilensis DC.;智利双钱荠■☆

110549 Cremolobus parviflorus Wedd.;小花双钱荠■☆

110550 Cremolobus pinnatifidus Hook.;羽裂双钱荠■☆

110551 Cremolobus pubescens Hook.;毛双钱荠■☆

110552 Cremolobus stenophyllus Muschl.;窄叶双钱荠■☆

110553 Cremophyllum Scheidw. = Dalechampia L. ●

110554 Cremopyrum Schur = Agropyron Gaertn. ■

110555 Cremopyrum Schur = Eremopyrum (Ledeb.) Jaub. et Spach ■

110556 Cremosperma Benth. (1846);悬子苣苔属■☆

110557 Cremosperma album C. V. Morton;白悬子苣苔■☆

110558 Cremosperma auriculatum C. V. Morton;小耳悬子苣苔■☆

110559 Cremosperma filicifolium L. P. Kvist et L. E. Skog;线叶悬子苣苔■☆

110560 Cremosperma maculatum L. E. Skog;斑点悬子苣苔■☆

110561 Cremosperma monticola C. V. Morton;山生悬子苣苔■☆

110562 Cremosperma parviflorum C. V. Morton;小花悬子苣苔■☆

110563 Cremosperma rotundatum C. V. Morton;圆叶悬子苣苔■☆

110564 Cremosperma serratum C. V. Morton;齿叶悬子苣苔■☆

110565 Cremosperma sylvaticum C. V. Morton;林地悬子苣苔■☆

110566 Cremospermopsis L. E. Skog et L. P. Kvist(2002);拟悬子苣苔属■●☆

110567 Cremospermopsis cestroides (Fritsch) L. E. Skog et L. P. Kvist;拟悬子苣苔■●☆

110568 Cremospermopsis parviflora L. E. Skog et L. P. Kvist;小花拟悬子苣苔■●☆

110569 Cremospora Post et Kuntze = Cremaspora Benth. ●☆

110570 Cremostachys Tul. = Galearia Zoll. et Moritzi(保留属名)●☆

110571 Crena Scop. = Crenea Aubl. ●☆

110572 Crenaea Schreb. = Crenea Aubl. ●☆

110573 Crenamon Raf. = ? Leontodon L.(保留属名) + Picris L. ■

110574 Crenamum Adans. = Crepis L. + Picris L. ■

110575 Crenamum Adans. = Helminthotheca Zinn ■☆

110576 Crenea Aubl. (1775);巴西千屈菜属●☆

110577 Crenea maritima Aubl.;巴西千屈菜●☆

110578 Crenias A. Spreng. (1827);巴西川苔草属■☆

110579 Crenias A. Spreng. = Mniopsis Mart. ■☆

110580 Crenias scopulorum A. Spreng.;巴西川苔草■☆

110581 Crenidium Haegi(1981);澳大利亚茄属☆

110582 Crenidium spinescens Haegi;澳大利亚茄☆

110583 Crenosciadium Boiss. et Heldr. (1849);纹伞芹属■☆

110584 Crenosciadium Boiss. et Heldr. = Opopanax W. D. J. Koch ■☆

110585 Crenosciadium siifolium Boiss. et Heldr.;纹伞芹■☆

110586 Crenularia Boiss. = Aethionema R. Br. ■☆

110587 Crenulluma Plowes = Caralluma R. Br. ■

110588 Creochiton Blume(1831);肉被野牡丹属●☆

110589 Creochiton bibracteata Blume;肉被野牡丹●☆

110590 Creochiton diptera Elmer;双翅肉被野牡丹●☆

110591 Creochiton monticola (Ridl.) Veldkamp;山地肉被野牡丹●☆

110592 Creochiton rosea Merr.;粉红肉被野牡丹●☆

110593 Creocome Kunae = Oreocome Edgew. ■☆

110594 Creocome Kunae = Selinum L.(保留属名)■
110595 Creodus Lour. = Chloranthus Sw. ■●
110596 Creolobus Lilja = Mentzelia L. ●■☆
110597 Crepalia Steud. = Craepalia Schrank ■
110598 Crepalia Steud. = Lolium L. ■
110599 Crepidaria Haw. = Pedilanthus Neck. ex Poit.(保留属名)●
110600 Crepidaria Haw. = Tithymalus Mill.(废弃属名)●
110601 Crepidiastrum Nakai = Ixeris (Cass.) Cass. ■
110602 Crepidiastrum Nakai(1920);假还阳参属(假还羊参属,假黄鹌菜属);Crepidiastrum ●■
110603 Crepidiastrum × surugense (Hisauti) Yonek.;骏河湾假还阳参■☆
110604 Crepidiastrum ameristophyllum (Nakai) Nakai;小笠原假还阳参■☆
110605 Crepidiastrum chelidoniifolium (Makino) J. H. Pak et Kawano;白屈菜叶假还阳参■☆
110606 Crepidiastrum denticulatum (Houtt.) J. H. Pak et Kawano;细齿假还阳参■☆
110607 Crepidiastrum denticulatum (Houtt.) J. H. Pak et Kawano f. pallescens (Momiy. et Tuyama) Yonek.;灰白屈菜叶假还阳参■☆
110608 Crepidiastrum denticulatum (Houtt.) J. H. Pak et Kawano f. pinnatipartitum (Makino) Sennikov;羽裂白屈菜叶假还阳参(羽裂黄瓜菜)■
110609 Crepidiastrum denticulatum (Houtt.) J. H. Pak et Kawano f. pinnatipartitum (Makino) Sennikov = Paraixeris pinnatipartita (Makino) Tzvelev ■
110610 Crepidiastrum grandicollum (Koidz.) Nakai;粗颈假还阳参■☆
110611 Crepidiastrum keiskeanum (Maxim.) Nakai;伊藤假还阳参■☆
110612 Crepidiastrum keiskeanum (Maxim.) Nakai f. pinnatilobum Hisauti;羽裂伊藤假还阳参■☆
110613 Crepidiastrum koshunense (Hayata) Nakai = Crepidiastrum lanceolatum (Houtt.) Nakai ■
110614 Crepidiastrum koshunense (Hayata) Nakai var. taiwananum (Nakai) Gogelein = Crepidiastrum taiwanianum Nakai ●■
110615 Crepidiastrum lanceolatum (Houtt.) Nakai;假还阳参(花莲假黄鹌菜,细叶假黄鹌菜);Lanceolate Cuckoo-orchis ■
110616 Crepidiastrum lanceolatum (Houtt.) Nakai f. batakanensis (Kitam.) Kitam.;花莲假黄鹌菜(半羽假还阳参)■
110617 Crepidiastrum lanceolatum (Houtt.) Nakai f. batakensis (Kitam.) Kitam. = Crepidiastrum lanceolatum (Houtt.) Nakai ■
110618 Crepidiastrum lanceolatum (Houtt.) Nakai f. pinnatilobum (Maxim.) Nakai;羽裂假还阳参■
110619 Crepidiastrum lanceolatum (Houtt.) Nakai f. pinnatilobum (Maxim.) Nakai = Crepidiastrum lanceolatum (Houtt.) Nakai ■
110620 Crepidiastrum lanceolatum (Houtt.) Nakai var. batakanense (Kitam.) Nemoto = Crepidiastrum lanceolatum (Houtt.) Nakai ■
110621 Crepidiastrum lanceolatum (Houtt.) Nakai var. daitoense (Tawada) Hatus.;大东假还阳参■☆
110622 Crepidiastrum linguifolium (A. Gray) Nakai;舌叶假还阳参■☆
110623 Crepidiastrum platyphyllum (Franch. et Sav.) Kitam.;宽叶假还阳参■☆
110624 Crepidiastrum quercus (H. Lév. et Vaniot) Nakai = Crepidiastrum lanceolatum (Houtt.) Nakai ■
110625 Crepidiastrum quercus (H. Lév. et Vaniot) Nakai = Crepidiastrum lanceolatum (Houtt.) Nakai f. pinnatilobum (Maxim.) Nakai ■
110626 Crepidiastrum sonchifolium (Bunge) J. H. Pak et Kawano;苣叶假还阳参■☆
110627 Crepidiastrum taiwanianum Nakai;台湾假还阳参(假还阳参,台湾假黄鹌菜);Taiwan Crepidiastrum ●■
110628 Crepidiastrum yoshinoi (Makino) J. H. Pak et Kawano;吉野氏假还阳参■☆
110629 Crepidifolium Sennikov(2007);还阳参叶菊属■☆
110630 Crepidispermum Fr. = Hieracium L. ■
110631 Crepidium Blume = Malaxis Sol. ex Sw. ■
110632 Crepidium Blume(1825);沼兰属■
110633 Crepidium Tausch = Crepis L. ■
110634 Crepidium Tausch = Endoptera DC. ■
110635 Crepidium acuminatum (D. Don) Szlach.;浅裂沼兰;Lobed Addermonth Orchid,Lobed Bogorchis ■
110636 Crepidium bahanense (Hand.-Mazz.) S. C. Chen et J. J. Wood;云南沼兰;Yunnan Addermonth Orchid,Yunnan Bogorchis ■
110637 Crepidium bancanoides (Ames) Szlach.;兰屿沼兰(裂唇软叶兰);Lanyu Addermonth Orchid,Lanyu Bogorchis ■
110638 Crepidium biauritum (Lindl.) Szlach.;二耳沼兰;Twoear Addermonth Orchid,Twoear Bogorchis ■
110639 Crepidium bilobum (Lindl.) Szlach. = Crepidium acuminatum (D. Don) Szlach. ■
110640 Crepidium calophyllum (Rchb. f.) Szlach.;美叶沼兰;Fairleaf Addermonth Orchid,Spiffyleaf Bogorchis ■
110641 Crepidium concavum (Seidenf.) Szlach.;凹唇沼兰;Concavelip Addermonth Orchid,Concavelip Bogorchis ■
110642 Crepidium finetii (Gagnepa.) S. C. Chen et J. J. Wood;二脊沼兰;Finet Addermonth Orchid,Finet Bogorchis ■
110643 Crepidium glaucum Nutt. = Crepis runcinata (E. James) Torr. et A. Gray subsp. glauca (Nutt.) Babc. et Stebbins ■☆
110644 Crepidium hainanense (Ts. Tang et F. T. Wang) S. C. Chen et J. J. Wood;海南沼兰;Hainan Addermonth Orchid,Hainan Bogorchis ■
110645 Crepidium insulare (Ts. Tang et F. T. Wang) S. C. Chen et J. J. Wood;琼岛沼兰;Island Addermonth Orchid,Island Bogorchis ■
110646 Crepidium khasianum (Hook. f.) Szlach.;细茎沼兰;Khans Addermonth Orchid,Khans Bogorchis ■
110647 Crepidium mackinnonii (Duthie) Szlach.;铺叶沼兰;Mackinnon Addermonth Orchid,Mackinnon Bogorchis ■
110648 Crepidium matsudae (Yamam.) Szlach.;鞍唇沼兰(凹唇软叶兰);Matsuda Addermonth Orchid,Matsuda Bogorchis ■
110649 Crepidium ophrydis (J. König) M. A. Clem. et D. L. Jones = Dienia ophrydis (J. König) Ormerod et Seidenf. ■
110650 Crepidium orbiculare (W. W. Sm. et Jeffrey) Seidenf.;齿唇沼兰;Orbicular Addermonth Orchid,Orbicular Bogorchis ■
110651 Crepidium ovalisepalum (J. J. Sm.) Szlach.;卵萼沼兰;Oosepal Addermonth Orchid,Oosepal Bogorchis ■
110652 Crepidium purpureum (Lindl.) Szlach.;深裂沼兰(红花沼兰,紫花软叶兰);Purplered Addermonth Orchid, Purplered Bogorchis ■
110653 Crepidium ramosii (Ames) Szlach.;心唇沼兰(圆唇软叶兰,圆唇小柱兰);Heartlip Addermonth Orchid,Heartlip Bogorchis ■
110654 Crepidium sichuanicum (Ts. Tang et F. T. Wang) S. C. Chen et J. J. Wood;四川沼兰;Sichuan Orchid ■
110655 Crepidocarpus Klotzsch ex Bocck. = Scirpus L.(保留属名)■
110656 Crepidopsis Arv.-Touv. = Hieracium L. ■
110657 Crepidopteris Benth. = Crepidotropis Walp. ■☆
110658 Crepidopteris Benth. = Dioclea Kunth ■☆
110659 Crepidorhopalon Eb. Fisch.(1989);肖蝴蝶草属■☆

110660 Crepidorhopalon Eb. Fisch. = Torenia L. ■
110661 Crepidorhopalon affinis (De Wild.) Eb. Fisch. et Govaerts;近缘肖蝴蝶草■☆
110662 Crepidorhopalon alatocalycinus Eb. Fisch.;翅萼肖蝴蝶草■☆
110663 Crepidorhopalon bifolius (Skan) Eb. Fisch.;双叶肖蝴蝶草■☆
110664 Crepidorhopalon bifolius (V. Naray.) Eb. Fisch. = Crepidorhopalon bifolius (Skan) Eb. Fisch. ■☆
110665 Crepidorhopalon chironioides (S. Moore) Eb. Fisch.;圣诞果肖蝴蝶草■☆
110666 Crepidorhopalon damblonii (P. A. Duvign.) Eb. Fisch. = Crepidorhopalon tenuis (S. Moore) Eb. Fisch. ■☆
110667 Crepidorhopalon debilis (Skan) Eb. Fisch.;弱小肖蝴蝶草■☆
110668 Crepidorhopalon debilis (V. Naray.) Eb. Fisch. = Crepidorhopalon debilis (Skan) Eb. Fisch. ■☆
110669 Crepidorhopalon goetzei (Engl.) Eb. Fisch.;格兹肖蝴蝶草■☆
110670 Crepidorhopalon gracilis (Pilg.) Eb. Fisch.;纤细肖蝴蝶草■☆
110671 Crepidorhopalon gracilis (Pilg.) Eb. Fisch. = Torenia ledermannii Hepper ■☆
110672 Crepidorhopalon hartlii Eb. Fisch.;哈特尔肖蝴蝶草■☆
110673 Crepidorhopalon hepperi Eb. Fisch.;赫佩肖蝴蝶草■☆
110674 Crepidorhopalon insularis (Skan) Eb. Fisch. = Crepidorhopalon rupestris (Engl.) Eb. Fisch. ■☆
110675 Crepidorhopalon involucratus (Philcox) Eb. Fisch.;总苞肖蝴蝶草■☆
110676 Crepidorhopalon latibracteatus (Skan) Eb. Fisch.;宽苞肖蝴蝶草■☆
110677 Crepidorhopalon latibracteatus (Skan) Eb. Fisch. subsp. parviflorus (Philcox) Eb. Fisch. = Crepidorhopalon parviflorus (Philcox) Eb. Fisch. ■☆
110678 Crepidorhopalon latibracteatus (V. Naray.) Eb. Fisch. = Crepidorhopalon latibracteatus (Skan) Eb. Fisch. ■☆
110679 Crepidorhopalon laxiflorus Eb. Fisch.;疏花肖蝴蝶草■☆
110680 Crepidorhopalon malaissei Eb. Fisch.;马莱泽肖蝴蝶草■☆
110681 Crepidorhopalon membranocalycinus Eb. Fisch.;膜萼肖蝴蝶草■☆
110682 Crepidorhopalon parviflorus (Philcox) Eb. Fisch.;小花肖蝴蝶草■☆
110683 Crepidorhopalon perennis (P. A. Duvign.) Eb. Fisch.;多年生肖蝴蝶草■☆
110684 Crepidorhopalon robynsii Eb. Fisch.;罗宾斯肖蝴蝶草■☆
110685 Crepidorhopalon rupestris (Engl.) Eb. Fisch.;岩生肖蝴蝶草■☆
110686 Crepidorhopalon rupestris (Engl.) Eb. Fisch. = Lindernia insularis Skan ■☆
110687 Crepidorhopalon schweinfurthii (Oliv.) Eb. Fisch.;施韦肖蝴蝶草■☆
110688 Crepidorhopalon spicatus (Engl.) Eb. Fisch.;长穗肖蝴蝶草■☆
110689 Crepidorhopalon symoensii Eb. Fisch.;西莫肖蝴蝶草■☆
110690 Crepidorhopalon tanzanicus Eb. Fisch.;坦桑尼亚肖蝴蝶草■☆
110691 Crepidorhopalon tenuifolius (Philcox) Eb. Fisch.;细叶肖蝴蝶草■☆
110692 Crepidorhopalon tenuis (S. Moore) Eb. Fisch.;细肖蝴蝶草■☆
110693 Crepidorhopalon uvens (Hiern) Eb. Fisch.;尤瓦肖蝴蝶草■☆
110694 Crepidorhopalon welwitschii (Engl.) Eb. Fisch.;韦尔肖蝴蝶草■☆
110695 Crepidorhopalon whytei (Skan) Eb. Fisch.;怀特肖蝴蝶草■☆
110696 Crepidorhopalon whytei (V. Naray.) Eb. Fisch. = Crepidorhopalon whytei (Skan) Eb. Fisch. ■☆
110697 Crepidospermum Benth. et Hook. f. = Crepidospermum Hook. f. ●☆
110698 Crepidospermum Benth. et Hook. f. = Hieracium L. ■
110699 Crepidospermum Hook. f. (1862);鞋籽橄榄属●☆
110700 Crepidospermum cuneifolium (Cuatrec.) Daly;楔叶鞋籽橄榄●☆
110701 Crepidospermum guyanense Marchand ex Triana et Planch.;圭亚那鞋籽橄榄●☆
110702 Crepidospermum sprucei Hook. f.;鞋籽橄榄●☆
110703 Crepidotropis Walp. = Dioclea Kunth ■☆
110704 Crepinella (Marchal) ex Oliver = Schefflera J. R. Forst. et G. Forst. (保留属名)●
110705 Crepinella Marchal = Schefflera J. R. Forst. et G. Forst. (保留属名)●
110706 Crepinia Gand. = Rosa L. ●
110707 Crepinia Rchb. = Crepis L. ■
110708 Crepinia Rchb. = Pterotheca Cass. ■
110709 Crepinodendron Pierre = Micropholis (Griseb.) Pierre ●☆
110710 Crepis L. (1753);还阳参属(还羊参属,驴打滚草属);Dandelion, Hawk's Beard, Hawksbeard, Hawks-beard, Hawk's-beard ■
110711 Crepis L. = Youngia Cass. ■
110712 Crepis abietina Boiss. et Balansa ex Boiss. = Prenanthes abietina (Boiss. et Balansa) Kirp. ■☆
110713 Crepis abyssinica Sch. Bip. = Crepis rueppellii Sch. Bip. ■☆
110714 Crepis acaulis (Roxb.) Hook. f. = Launaea acaulis (Roxb.) Babc. ex Kerr ■
110715 Crepis achyrophoroides Vatke;拟猫儿菊还阳参■☆
110716 Crepis aculeata (DC.) Boiss.;皮刺还阳参■☆
110717 Crepis acuminata Nutt.;锐尖还阳参;Longleaf Hawksbeard, Sharp Hawksbeard, Tapertip Hawksbeard ■☆
110718 Crepis acuminata Nutt. subsp. pluriflora Babc. et Stebbins = Crepis acuminata Nutt. ■☆
110719 Crepis acuminata Nutt. var. intermedia (A. Gray) Jeps. = Crepis intermedia A. Gray ■☆
110720 Crepis acuminata Nutt. var. pleurocarpa (A. Gray) Jeps. = Crepis pleurocarpa A. Gray ■☆
110721 Crepis aegyptiaca (Schweinf.) Täckh. et Boulos = Heteroderis pusilla (Boiss.) Boiss. ■☆
110722 Crepis alaica Krasch. et Popov;红齿还阳参■
110723 Crepis albida Vill.;白还阳参■☆
110724 Crepis albida Vill. var. minor Willk. = Crepis albida Vill. ■☆
110725 Crepis albiflora Babc.;白花还阳参■☆
110726 Crepis alikeri Tamamsch.;阿氏还阳参■☆
110727 Crepis alpina L.;高山还阳参■☆
110728 Crepis altissima Balb. = Tolpis virgata (Desf.) Bertol. ■☆
110729 Crepis ambacensis Hiern = Lactuca ambacensis (Hiern) C. Jeffrey ■☆
110730 Crepis ambigua A. Gray = Hieracium fendleri Sch. Bip. ■☆
110731 Crepis ambigua Balb. = Tolpis virgata (Desf.) Bertol. ■☆
110732 Crepis amplexifolia (Godr.) Willk.;褶叶还阳参■☆
110733 Crepis andersonii A. Gray = Crepis runcinata (E. James) Torr. et A. Gray subsp. andersonii (A. Gray) Babc. et Stebbins ■☆
110734 Crepis angustata Rydb. = Crepis acuminata Nutt. ■☆
110735 Crepis arenaria (Pomel) Pomel;沙地还阳参■☆
110736 Crepis arenaria (Pomel) Pomel subsp. suberostris (Batt.) Greuter;木栓质还阳参■☆
110737 Crepis aspera L.;粗糙还阳参■☆
110738 Crepis astrachanica Steven ex De Moor;阿斯特拉罕还阳参;

Astrachan Hawksbeard ■☆
110739 Crepis atribarba A. Heller;暗毛还阳参;Dark Hawksbeard, Slender Hawksbeard ■☆
110740 Crepis atripappa Babc. = Youngia stebbinsiana S. Y. Hu ■
110741 Crepis aurea (L.) Carrière;金黄还阳参(黄花还阳参,金黄色还阳参);Golden Hawksbeard, Yellow Crepis ■☆
110742 Crepis aurea (L.) Carrière var. crocea Froelich ex DC. = Crepis crocea (Lam.) Babc. ■
110743 Crepis baicalensis Ledeb. = Youngia tenuicaulis (Babc. et Stebbins) De Moor ■
110744 Crepis baicalensis Ledeb. = Youngia tenuifolia (Willd.) Babc. et Stebbins ■
110745 Crepis bakeri Greene;贝克还阳参;Baker's Hawksbeard ■☆
110746 Crepis bakeri Greene subsp. cusickii (Eastw.) Babc. et Stebbins;库氏还阳参;Cusick's Hawksbeard ■☆
110747 Crepis bakeri Greene subsp. idahoensis Babc. et Stebbins;爱达荷还阳参;Idaho Hawksbeard ■☆
110748 Crepis balliana Babc.;博利还阳参■☆
110749 Crepis barbata L. = Tolpis barbata (L.) Gaertn. ■☆
110750 Crepis barberi Greenm. = Crepis runcinata (E. James) Torr. et A. Gray subsp. barberi (Greenm.) Babc. et Stebbins ■☆
110751 Crepis bhotaniea Hutch. = Dubyaea bhotanica (Hutch.) C. Shih ■
110752 Crepis biennis L.;二年生还阳参(粗糙还阳参);Biennial Hawksbeard, Rough Hawksbeard, Rough Hawk's-beard ■☆
110753 Crepis bifurcata (Babc. et Stebbins) Hand. -Mazz. = Youngia bifurcata Babc. et Stebbins ■
110754 Crepis blinii H. Lév. = Youngia blinii (H. Lév.) Lauener ■
110755 Crepis blinii H. Lév. = Youngia fusca (Babc.) Babc. et Stebbins ■
110756 Crepis bodinieri H. Lév.;果山还阳参(波氏还羊参,波氏还阳参,黑果还阳参,还阳参);Bodinier Hawksbeard ■
110757 Crepis boekiana Diels = Youngia heterophylla (Hemsl.) Babc. et Stebbins ■
110758 Crepis bonii Gagnep. = Ixeris polycephala Cass. ■
110759 Crepis bruceae Babc. = Crepis newii Oliv. et Hiern ■☆
110760 Crepis bulbosa (L.) Tausch = Sonchus bulbosus (L.) N. Kilian et Greuter ■☆
110761 Crepis bumbensis Hiern subsp. itakensis (Babc.) Babc. = Crepis newii Oliv. et Hiern ■☆
110762 Crepis bungei Ledeb. ex DC.;邦奇还阳参■☆
110763 Crepis burejensis Eastw.;布列亚山还阳参■☆
110764 Crepis bursifolia L.;意大利还阳参;Italian Hawksbeard ■☆
110765 Crepis cameroonica Babc. ex Hutch. et Dalziel = Crepis newii Oliv. et Hiern subsp. oliveriana (Kuntze) C. Jeffrey et Beentje ■☆
110766 Crepis canariensis (Sch. Bip.) Babc.;加那利还阳参■☆
110767 Crepis capillaris (L.) Wallr.;纤细还阳参(发还阳参,强壮还阳参);Hawk's-beard, Slender Hawksbeard, Smooth Hawksbeard, Smooth Hawk's-beard ■☆
110768 Crepis carbonaria Sch. Bip.;卡勃还阳参(卡勃还羊参)■☆
110769 Crepis caucasica C. A. Mey.;高加索还阳参■☆
110770 Crepis caucasigena De Moor;高加索产还阳参■☆
110771 Crepis chanetii H. Lév. = Tephroseris subdentata (Bunge) Holub ■
110772 Crepis charbonnelii H. Lév. = Mulgedium tataricum (L.) DC. ■
110773 Crepis chloroclada Collett et Hemsl.;竹叶还阳参(小竹叶防风)■☆
110774 Crepis chloroclada Collett et Hemsl. = Crepis lignea (Vaniot) Babc. ■
110775 Crepis chrysantha (Ledeb.) Turcz.;黄花还阳参(黄还阳参,金黄还阳参);Goldenyellow Hawksbeard ■
110776 Crepis cichorioides Hiern = Lactuca cichorioides (Hiern) C. Jeffrey ■☆
110777 Crepis ciliata K. Koch;缘毛还阳参■☆
110778 Crepis cineripappa Babc. = Youngia cineripappa (Babc.) Babc. et Stebbins ■
110779 Crepis claryi Batt.;克莱里还阳参■☆
110780 Crepis clausonis (Pomel) Batt.;克劳森还阳参■☆
110781 Crepis congoensis Babc. = Crepis hypochaeridea (DC.) Thell. ■☆
110782 Crepis cooperi A. Gray = Crepis capillaris (L.) Wallr. ■☆
110783 Crepis corniculata Regel et Schmalh. ex Regel;圆锥还阳参■☆
110784 Crepis coronopifolia Desf. = Tolpis coronopifolia (Desf.) Biv. ■☆
110785 Crepis crocea (Lam.) Babc.;北方还阳参(还羊参,还阳参,黄花还羊参,黄花还阳参,驴打滚儿,驴打滚儿草,屠还阳参,小苦荬);Common Hawksbeard ■
110786 Crepis cusickii Eastw. = Crepis bakeri Greene subsp. cusickii (Eastw.) Babc. et Stebbins ■☆
110787 Crepis czuensis Serg.;丘还阳参■☆
110788 Crepis darvazica Krasch.;中亚还阳参(达地还阳参,新疆还阳参)■
110789 Crepis depressa Hook. f. et Thomson = Soroseris glomerata (Decne.) Stebbins ■
110790 Crepis depressa Hook. f. et Thomson = Youngia depressa (Hook. f. et Thomson) Babc. et Stebbins ■
110791 Crepis disciformis Mattf. = Syncalathium disciforme (Mattf.) Y. Ling ■
110792 Crepis divaricata (Lowe) F. W. Schultz;叉开还阳参■☆
110793 Crepis djimilensis K. Koch;德吉米尔还阳参■☆
110794 Crepis dubyaea (C. B. Clarke) Marquart et Shaw = Dubyaea bhotanica (Hutch.) C. Shih ■
110795 Crepis eharbonnelii H. Lév. = Mulgedium tataricum (L.) DC. ■
110796 Crepis elegans Hook.;雅致还阳参;Elegant Hawksbeard ■☆
110797 Crepis ellenbeckii R. E. Fr. = Crepis carbonaria Sch. Bip. ■☆
110798 Crepis elongata Babc.;藏滇还阳参(长茎还羊参,长茎还阳参,独花蒲公英,还阳参,天竺参,铁刷把,万丈深,西藏还羊参,西藏还阳参,有根无叶,竹叶青)■
110799 Crepis ephemera Hiern = Tolpis capensis (L.) Sch. Bip. ■☆
110800 Crepis ephemeroides S. Moore = Tolpis capensis (L.) Sch. Bip. ■☆
110801 Crepis eritreensis Babc. = Crepis foetida L. ■☆
110802 Crepis exilis Osterh. = Crepis atribarba A. Heller ■☆
110803 Crepis exilis Osterh. subsp. originalis Babc. et Stebbins = Crepis atribarba A. Heller ■☆
110804 Crepis faureliana Maire;福雷尔还阳参■☆
110805 Crepis filiformis Viv. = Crepis senecioides Delile subsp. filiformis (Viv.) Alavi ■☆
110806 Crepis flexuosa (Ledeb.) C. B. Clarke;弯茎还阳参;Flexuouse Hawksbeard ■
110807 Crepis flexuosa (Ledeb.) C. B. Clarke var. tenuifolia C. H. An;细叶还阳参■
110808 Crepis foetida L.;臭味还阳参;Roadside Hawksbeard, Stink Hawksbeard, Stinking Hawksbeard, Stinking Hawk's-beard ■☆
110809 Crepis foliosa Babc.;多叶还阳参■☆

110810 Crepis formosana Hayata = Youngia japonica (L.) DC. subsp. formosana (Hayata) Kitam. ■
110811 Crepis formosana Hayata = Youngia japonica (L.) DC. ■
110812 Crepis forskalii Babc. = Crepis rueppellii Sch. Bip. ■☆
110813 Crepis friesii Babc.;弗里斯还阳参■☆
110814 Crepis fusca Babc. = Youngia fusca (Babc.) Babc. et Stebbins ■
110815 Crepis geisseana Phil. = Malacothrix clevelandii A. Gray ■☆
110816 Crepis gillii S. Moore = Soroseris gillii (S. Moore) Stebbins ■
110817 Crepis gillii S. Moore var. bellidifolia Hand. -Mazz. = Soroseris glomerata (Decne.) Stebbins ■
110818 Crepis gillii S. Moore var. erysimoides Hand. -Mazz. = Soroseris erysimoides (Hand. -Mazz.) C. Shih ■
110819 Crepis gillii S. Moore var. hirsuta J. Anthony = Soroseris hirsuta (J. Anthony) C. Shih ■
110820 Crepis glabra Boiss.;光还阳参■☆
110821 Crepis glandulosissima R. E. Fr. = Crepis carbonaria Sch. Bip. ■☆
110822 Crepis glareosa Piper = Crepis modocensis Greene subsp. glareosa (Piper) Babc. et Stebbins ■☆
110823 Crepis glomerata (Decne.) Benth. et Hook. f. = Soroseris glomerata (Decne.) Stebbins ■
110824 Crepis glomerata (Decne.) Benth. et Hook. f. var. porphyrea C. Marquand et Shaw = Syncalathium porphyreum (C. Marquand et Airy Shaw) Y. Ling ■
110825 Crepis glomerata (Decne.) C. B. Clarke = Youngia depressa (Hook. f. et Thomson) Babc. et Stebbins ■
110826 Crepis glomerata Decne. var. porphyrea C. Marquand et Airy Shaw = Syncalathium porphyreum (C. Marquand et Airy Shaw) Y. Ling ■
110827 Crepis gmelinii (L.) Tausch;格氏还阳参■☆
110828 Crepis gmelinii (L.) Tausch var. grandiflora Tausch. = Crepis crocea (Lam.) Babc. ■
110829 Crepis gmelinii Schult. var. grandifolia Tausch = Crepis crocea (Lam.) Babc. ■
110830 Crepis gmelinii Schult. var. grandifolia Tausch = Crepis pallasii (Pall.) Turcz. ■
110831 Crepis gossweileri S. Moore;戈斯还阳参■☆
110832 Crepis gracilipes Hook. f. = Youngia gracilipes (Hook. f.) Babc. et Stebbins ■
110833 Crepis gracilis Hook. f. et Hook. f. ex C. B. Clarke = Youngia stebbinsiana S. Y. Hu ■
110834 Crepis graminifolia Ledeb. = Ixeridium graminifolium (Ledeb.) Tzvelev ■
110835 Crepis gymnopus Koidz.;裸足还阳参■☆
110836 Crepis henryi Diels = Youngia henryi (Diels) Babc. et Stebbins ■
110837 Crepis heterophylla Hemsl. = Youngia heterophylla (Hemsl.) Babc. et Stebbins ■
110838 Crepis hieracium H. Lév. = Faberia sinensis Hemsl. ■
110839 Crepis hirsuta Pomel = Crepis vesicaria L. subsp. taraxacifolia (Thuill.) Schinz et Keller ■☆
110840 Crepis hispanica Pau = Crepis pulchra L. ■☆
110841 Crepis hokkaidoensis Babc.;北海道还阳参■☆
110842 Crepis hookeriana Ball;胡克还阳参■☆
110843 Crepis hookeriana Ball var. aspera Emb. = Crepis hookeriana Ball ■☆
110844 Crepis hookeriana Ball var. balliana Emb. et Maire = Crepis hookeriana Ball ■☆
110845 Crepis hookeriana C. B. Clarke = Soroseris hookeriana (C. B. Clarke) Stebbins ■
110846 Crepis hookeriana Oliv. et Hiern = Crepis newii Oliv. et Hiern subsp. oliveriana (Kuntze) C. Jeffrey et Beentje ■☆
110847 Crepis humilis Fisch. ex Herder = Crepis nana Richardson ■
110848 Crepis hypochaeridea (DC.) Thell.;猫儿菊还阳参■☆
110849 Crepis hypochaeridea (DC.) Thell. subsp. brevicaulis Babc. = Crepis hypochaeridea (DC.) Thell. ■☆
110850 Crepis hypochaeridea (DC.) Thell. subsp. rhodesica Babc. = Crepis hypochaeridea (DC.) Thell. ■☆
110851 Crepis hypochaeridea (DC.) Thell. var. genuina Thell. = Crepis hypochaeridea (DC.) Thell. ■☆
110852 Crepis hypochaeridea (DC.) Thell. var. junodiana Thell. = Crepis hypochaeridea (DC.) Thell. ■☆
110853 Crepis hypochaeridea (DC.) Thell. var. woodii Thell. = Crepis hypochaeridea (DC.) Thell. ■☆
110854 Crepis incana Ledeb.;粉花还阳参;Northern Hawk's-beard, Pink Dandelion ■☆
110855 Crepis incarnata Tausch;肉色还阳参■☆
110856 Crepis integra (Thunb.) Miq. = Crepidiastrum lanceolatum (Houtt.) Nakai ■
110857 Crepis integra Miq. = Crepidiastrum platyphyllum (Franch. et Sav.) Kitam. ■☆
110858 Crepis integra Miq. var. pinnatiloba Maxim. = Crepidiastrum lanceolatum (Houtt.) Nakai f. pinnatilobum (Maxim.) Nakai ■
110859 Crepis integra Miq. var. pinnatiloba Maxim. = Crepidiastrum lanceolatum (Houtt.) Nakai ■
110860 Crepis integrifolia C. Shih;全叶还阳参;Entireleaf Hawksbeard ■
110861 Crepis intermedia A. Gray;灰岩还阳参;Limestone Hawksbeard, Small-flower Hawksbeard ■☆
110862 Crepis intermedia A. Gray var. pleurocarpa (A. Gray) A. Gray = Crepis pleurocarpa A. Gray ■☆
110863 Crepis iringensis Babc. = Crepis newii Oliv. et Hiern subsp. oliveriana (Kuntze) C. Jeffrey et Beentje ■☆
110864 Crepis itakensis Babc. = Crepis newii Oliv. et Hiern ■☆
110865 Crepis japonica (L.) Benth. = Youngia japonica (L.) DC. ■
110866 Crepis japonica (L.) Benth. f. foliosa Matsuda = Youngia rosthornii (Diels) Babc. et Stebbins ■
110867 Crepis japonica (L.) Benth. subsp. genuina (Hochr.) Hochr. = Youngia japonica (L.) DC. ■
110868 Crepis japonica (L.) Benth. subsp. longiflora (Babc. et Stebbins) Hand. -Mazz. = Youngia longiflora (Babc. et Stebbins) C. Shih ■
110869 Crepis japonica (L.) Benth. var. elstonii Hochr. = Youngia pseudosenecio (Vaniot) C. Shih ■
110870 Crepis japonica (L.) Benth. var. genuina Hochr. = Youngia japonica (L.) DC. ■
110871 Crepis karakuschensis De Moor;卡拉库还阳参■☆
110872 Crepis karelinii Popov et Schischk. ex De Moor;乌恰还阳参(广布还阳参);Wuqia Hawksbeard ■
110873 Crepis keniensis (R. E. Fr.) Babc. = Crepis newii Oliv. et Hiern subsp. oliveriana (Kuntze) C. Jeffrey et Beentje ■☆
110874 Crepis khorassanica Boiss.;浩拉山还阳参■☆
110875 Crepis kilimandscharica O. Hoffm. = Crepis newii Oliv. et Hiern subsp. oliveriana (Kuntze) C. Jeffrey et Beentje ■☆
110876 Crepis kilimandscharica O. Hoffm. var. keniensis R. E. Fr. = Crepis newii Oliv. et Hiern subsp. oliveriana (Kuntze) C. Jeffrey et Beentje ■☆

110877 Crepis kilimandscharica O. Hoffm. var. meruensis R. E. Fr. = Crepis newii Oliv. et Hiern subsp. oliveriana (Kuntze) C. Jeffrey et Beentje ■☆

110878 Crepis koshunensis Hayata = Crepidiastrum lanceolatum (Houtt.) Nakai ■

110879 Crepis kotschyana Boiss.;科奇还阳参■☆

110880 Crepis lactea Lipsch.;红花还阳参(紫花还阳参);Redflower Hawksbeard ■

110881 Crepis laevigata (Blume) Sch. Bip. ex Zoll. = Ixeridium laevigatum (Blume) C. Shih ■

110882 Crepis lanceolata Sch. Bip. = Crepidiastrum lanceolatum (Houtt.) Nakai ■

110883 Crepis lanceolata Sch. Bip. var. pinnatiloba (Maxim.) Makino = Crepidiastrum lanceolatum (Houtt.) Nakai f. pinnatilobum (Maxim.) Nakai ■

110884 Crepis lanceolata Sch. Bip. var. pinnatiloba (Maxim.) Makino = Crepidiastrum lanceolatum (Houtt.) Nakai ■

110885 Crepis libyca (Pamp.) Shab.;利比亚还阳参■☆

110886 Crepis lignea (Vaniot) Babc.;绿茎还阳参(马尾参,奶浆参,刷把细辛,铁扫把,土麻黄,万丈深,细草,细叶万丈深,竹叶青);Woody Hawksbeard ■

110887 Crepis litardierei Emb.;利塔还阳参■☆

110888 Crepis longipes Hemsl. = Youngia longipes (Hemsl.) Babc. et Stebbins ■

110889 Crepis lyrata (L.) Froel.;琴叶还阳参■

110890 Crepis lyrata (Poir.) Benth. ex C. B. Clarke = Youngia japonica (L.) DC. ■

110891 Crepis macrophylla Desf. = Crepis vesicaria L. ■

110892 Crepis mairei H. Lév. = Youngia mairei (H. Lév.) Babc. et Stebbins ■

110893 Crepis marschallii F. Schultz;马氏还阳参■☆

110894 Crepis melanthera C. H. An;黑药还阳参■

110895 Crepis meruensis (R. E. Fr.) Babc. = Crepis newii Oliv. et Hiern subsp. oliveriana (Kuntze) C. Jeffrey et Beentje ■☆

110896 Crepis micrantha De Moor; 小花还阳参; Smallflower Hawksbeard ■☆

110897 Crepis mildbraedii Babc. = Crepis newii Oliv. et Hiern ■☆

110898 Crepis minuta Kitam. = Crepis lactea Lipsch. ■

110899 Crepis miyabei Tatew. et Kitam. = Crepis hokkaidoensis Babc. ■☆

110900 Crepis modocensis Greene;莫道克还阳参;Modoc Hawksbeard, Siskiyou Hawksbeard ■☆

110901 Crepis modocensis Greene subsp. glareosa (Piper) Babc. et Stebbins;石砾还阳参■☆

110902 Crepis modocensis Greene subsp. rostrata (Coville) Babc. et Stebbins;喙还阳参■☆

110903 Crepis modocensis Greene subsp. subacaulis (Kellogg) Babc. et Stebbins;近无茎还阳参■☆

110904 Crepis mollis Asch.;软毛还阳参(柔色还阳参);Northern Hawksbeard, Soft Hawksbeard, Softcolor Hawksbeard ■☆

110905 Crepis monticola Coville;山生还阳参;Mountain Hawksbeard ■☆

110906 Crepis multicaulis Ledeb.;多茎还阳参;Manystem Hawksbeard ■

110907 Crepis multicaulis Ledeb. subsp. congesta (Regel et Herder) Babc. = Crepis multicaulis Ledeb. ■

110908 Crepis multicaulis Ledeb. subsp. congesta (Regel) Babc. = Crepis multicaulis Ledeb. ■

110909 Crepis multicaulis Ledeb. subsp. genuina (Regel) Babc. = Crepis multicaulis Ledeb. ■

110910 Crepis multicaulis Ledeb. subsp. subintegrifolia Tolm. et Rchb. = Crepis multicaulis Ledeb. ■

110911 Crepis multicaulis Ledeb. var. congesta Regel = Crepis multicaulis Ledeb. ■

110912 Crepis multicaulis Ledeb. var. congesta Regel et Herder = Crepis multicaulis Ledeb. ■

110913 Crepis multicaulis Ledeb. var. genuina Regel = Crepis multicaulis Ledeb. ■

110914 Crepis multicaulis Ledeb. var. laxa Regel = Crepis multicaulis Ledeb. ■

110915 Crepis multicaulis Ledeb. var. laxa Regel et Herder = Crepis multicaulis Ledeb. ■

110916 Crepis nana Richardson;矮小还阳参(矮还阳参,小还阳参);Dwarf Alpine Hawksbeard, Dwarf Hawksbeard, Tiny Hawksbeard ■

110917 Crepis nana Richardson subsp. ramosa Babc. = Crepis nana Richardson ■

110918 Crepis nana Richardson subsp. typica Babc. = Crepis nana Richardson ■

110919 Crepis nana Richardson var. lyratifolia (Turcz.) Hultén = Crepis nana Richardson ■

110920 Crepis napifera (Franch.) Babc.;芜菁还阳参(抽葶还阳参,大一支箭,丽江一支箭,肉根还阳参,万丈深,芜菁还羊参,一支箭);Turnip-shaped Hawksbeard ■

110921 Crepis newii Oliv. et Hiern;纽还阳参■☆

110922 Crepis newii Oliv. et Hiern subsp. bumbensis (Hiern) Babc. = Crepis hypochaeridea (DC.) Thell. ■☆

110923 Crepis newii Oliv. et Hiern subsp. greenwayi Babc. = Crepis newii Oliv. et Hiern ■☆

110924 Crepis newii Oliv. et Hiern subsp. itakensis (Babc.) Babc. = Crepis newii Oliv. et Hiern ■☆

110925 Crepis newii Oliv. et Hiern subsp. kundensis (Babc.) Babc. = Crepis hypochaeridea (DC.) Thell. ■☆

110926 Crepis newii Oliv. et Hiern subsp. mbuluensis Babc. = Crepis newii Oliv. et Hiern ■☆

110927 Crepis newii Oliv. et Hiern subsp. nyasensis Babc. = Crepis newii Oliv. et Hiern ■☆

110928 Crepis newii Oliv. et Hiern subsp. oliveriana (Kuntze) C. Jeffrey et Beentje;奥里弗还阳参■☆

110929 Crepis nicaeensis Balb. ex Pers.; 法国还阳参; French Hawksbeard, Turkish Hawksbeard ■☆

110930 Crepis nigrescens Puttocke;黑色还阳参■☆

110931 Crepis nigricans Viv.;浅黑还阳参■☆

110932 Crepis noronhaea Jenkins;诺罗尼亚还阳参■☆

110933 Crepis nudiflora Viv. = Crepis senecioides Delile ■☆

110934 Crepis occidentalis Nutt.; 西方还阳参; Gray Hawksbeard, Largeflower Hawksbeard, Western Hawksbeard ■☆

110935 Crepis occidentalis Nutt. subsp. costata (A. Gray) Babc. et Stebbins;中脉还阳参■☆

110936 Crepis occidentalis Nutt. subsp. pumila (Rydb.) Babc. et Stebbins;小还阳参■☆

110937 Crepis occidentalis Nutt. var. costata A. Gray = Crepis occidentalis Nutt. subsp. costata (A. Gray) Babc. et Stebbins ■☆

110938 Crepis occidentalis Nutt. var. crinita A. Gray = Crepis monticola Coville ■☆

110939 Crepis occidentalis Nutt. var. gracilis D. C. Eaton = Crepis atribarba A. Heller ■☆

110940 Crepis occidentalis Nutt. var. subacaulis Kellogg = Crepis

modocensis Greene subsp. subacaulis (Kellogg) Babc. et Stebbins ■☆
110941 Crepis oliveriana (Kuntze) C. Jeffrey = Crepis newii Oliv. et Hiern subsp. oliveriana (Kuntze) C. Jeffrey et Beentje ■☆
110942 Crepis oreades Schrenk;山地还阳参(中山还阳参)■
110943 Crepis paleacea Diels = Youngia paleacea (Diels) Babc. et Stebbins ■
110944 Crepis pallasii (Pall.) Turcz. = Crepis crocea (Lam.) Babc. ■
110945 Crepis pallasii Turcz. = Crepis crocea (Lam.) Babc. ■
110946 Crepis paludosa Moench;沼生还阳参(沼泽还阳参);Marsh Hawksbeard,Marsh Hawk's-beard ■☆
110947 Crepis pannonica (Jacq.) K. Koch;潘城还阳参;Pasture Hawksbeard ■☆
110948 Crepis pannonica K. Koch = Crepis pannonica (Jacq.) K. Koch ■☆
110949 Crepis parva (Babc. et Stebbins) Hand. -Mazz. = Youngia parva Babc. et Stebbins ■
110950 Crepis parviflora Desf. = Crepis micrantha De Moor ■☆
110951 Crepis patula Poir.;张口还阳参■☆
110952 Crepis phoenix Dunn;万丈深(岔子菜,还阳参,马尾参,奶浆参,奶浆柴胡,瘦地草,细防风,小黏连,竹叶参,竹叶青,竹叶万丈深)■
110953 Crepis pleurocarpa A. Gray;脉果还阳参;Naked Hawksbeard ■☆
110954 Crepis polytricha (Ledeb.) Turcz.;毛还阳参■
110955 Crepis pontica C. A. Mey.;蓬特还阳参■☆
110956 Crepis praemorsa Tausch;残还阳参;Leafless Hawk's-beard ■☆
110957 Crepis pratensis C. Shih;草甸还阳参;Grassland Hawksbeard ■
110958 Crepis pratti Babc. = Youngia prattii (Babc.) Babc. et Stebbins ■
110959 Crepis prenanthoides Hemsl. = Paraprenanthes prenanthoides (Hemsl.) C. Shih ■
110960 Crepis primulifolia Hook. f. ex Benth. et Hook. f. = Youngia cineripappa (Babc.) Babc. et Stebbins ■
110961 Crepis pseudonaniformis C. Shih;长苞还阳参;Longbract Hawksbeard ■
110962 Crepis pseudovirens H. Lév. = Ixeridium gramineum (Fisch.) Tzvelev ■
110963 Crepis pulcherrima Fisch. ex Link = Youngia tenuicaulis (Babc. et Stebbins) De Moor ■
110964 Crepis pulcherrima Fisch. ex Link = Youngia tenuifolia (Willd.) Babc. et Stebbins ■
110965 Crepis pulchra L.;美还阳参;Beautiful Hawksbeard,Hawks Beard,Smallflower Hawksbeard ■☆
110966 Crepis pulchra L. = Phaecasium lampsanoides Cass. ■☆
110967 Crepis pulchra L. subsp. africana Babc.;非洲美还阳参■☆
110968 Crepis pulchra L. var. valentina Willk. = Crepis pulchra L. ■☆
110969 Crepis pumila Rydb. = Crepis occidentalis Nutt. subsp. pumila (Rydb.) Babc. et Stebbins ■☆
110970 Crepis pusilla (Sommier) Merxm.;微小还阳参■☆
110971 Crepis racemifera Hook. f. = Youngia racemifera (Hook. f.) Babc. et Stebbins ■
110972 Crepis radicata Forssk. = Picris asplenioides L. ■☆
110973 Crepis radicata Forssk. var. kralikii (Pomel) Murb. = Picris asplenioides L. ■☆
110974 Crepis radicata Forssk. var. nudiflora (Viv.) Pamp. = Crepis senecioides Delile subsp. nudiflora (Viv.) Alavi ■☆
110975 Crepis ramosissima d'Urv.;多分枝还阳参■☆
110976 Crepis rapunculoides Dunn = Youngia racemifera (Hook. f.) Babc. et Stebbins ■
110977 Crepis rhagadioloides L. = Picris rhagadioloides (L.) Desf. ■☆
110978 Crepis rifana Maire et Sennen;里夫还阳参■☆
110979 Crepis rigescens Diels;还阳参(岔子菜,川滇还羊参,滇川还阳参,独花蒲公英,马尾参,奶浆柴胡,瘦地草,天竹参,细防风,小黏连,竹叶青);Rigescent Hawksbeard ■
110980 Crepis rigescens Diels subsp. lignescens Babc. = Crepis rigescens Diels ■
110981 Crepis rigescens Diels subsp. typica Babc. = Crepis rigescens Diels ■
110982 Crepis rosthornii Diels = Youngia rosthornii (Diels) Babc. et Stebbins ■
110983 Crepis rostrata Coville = Crepis modocensis Greene subsp. rostrata (Coville) Babc. et Stebbins ■☆
110984 Crepis rosularis Diels = Soroseris glomerata (Decne.) Stebbins ■
110985 Crepis rubra L.;倒披针叶还阳参(红花还阳参,红还阳参);Pink Hawk's-beard,Red Hawksbeard ■☆
110986 Crepis rueppellii Sch. Bip.;吕埃还阳参■☆
110987 Crepis rueppellii Sch. Bip. subsp. ugandensis (Babc.) Babc. = Crepis rueppellii Sch. Bip. ■☆
110988 Crepis rueppellii Sch. Bip. var. centrali-africana R. E. Fr. = Crepis friesii Babc. ■☆
110989 Crepis rueppellii Sch. Bip. var. somalensis R. E. Fr. = Crepis rueppellii Sch. Bip. ■☆
110990 Crepis runcinata (E. James) Torr. et A. Gray;倒齿还阳参;Dandelion Hawksbeard, Fiddleleaf Hawksbeard, Naked-stem Hawksbeard,Scapose Hawksbeard ■☆
110991 Crepis runcinata (E. James) Torr. et A. Gray subsp. andersonii (A. Gray) Babc. et Stebbins;安氏还阳参;Anderson's Hawksbeard ■☆
110992 Crepis runcinata (E. James) Torr. et A. Gray subsp. barberi (Greenm.) Babc. et Stebbins;巴伯还阳参;Barber's Hawksbeard ■☆
110993 Crepis runcinata (E. James) Torr. et A. Gray subsp. glauca (Nutt.) Babc. et Stebbins;光滑还阳参;Smooth Hawksbeard ■☆
110994 Crepis runcinata (E. James) Torr. et A. Gray subsp. hallii Babc. et Stebbins; 豪尔还阳参; Hall's Hawksbeard, Meadow Hawksbeard ■☆
110995 Crepis runcinata (E. James) Torr. et A. Gray subsp. hispidulosa (Howell ex Rydb.) Babc. et Stebbins;毛倒齿还阳参■☆
110996 Crepis runcinata (E. James) Torr. et A. Gray var. andersonii (A. Gray) Cronquist = Crepis runcinata (E. James) Torr. et A. Gray subsp. andersonii (A. Gray) Babc. et Stebbins ■☆
110997 Crepis runcinata (E. James) Torr. et A. Gray var. glauca (Nutt.) B. Boivin = Crepis runcinata (E. James) Torr. et A. Gray subsp. glauca (Nutt.) Babc. et Stebbins ■☆
110998 Crepis runcinata (E. James) Torr. et A. Gray var. hispidulosa Howell ex Rydb. = Crepis runcinata (E. James) Torr. et A. Gray subsp. hispidulosa (Howell ex Rydb.) Babc. et Stebbins ■☆
110999 Crepis ruprechtii Boiss. = Crepis sibirica L. ■
111000 Crepis sahendi Boiss. et Buhse;萨亨迪还阳参■☆
111001 Crepis salzmannii Babc.;萨尔还阳参■☆
111002 Crepis sancta (L.) Babc. = Crepis sancta (L.) Bornm. ■☆
111003 Crepis sancta (L.) Babc. subsp. bifida ? = Crepis sancta (L.) Bornm. ■☆
111004 Crepis sancta (L.) Bornm.;神圣还阳参■☆
111005 Crepis sancta (L.) K. Maly = Crepis sancta (L.) Bornm. ■☆
111006 Crepis scaposa C. C. Chang = Youngia szechuanica (Soderb.) S. Y. Hu ■
111007 Crepis scaposa R. E. Fr. = Crepis carbonaria Sch. Bip. ■☆

111008　Crepis schimperi (Sch. Bip. ex A. Richardson) Schweinf. = Crepis foetida L. ■☆

111009　Crepis schultzii (Hochst. ex A. Richardson) Vatke;舒尔茨还阳参■☆

111010　Crepis scopulorum Coville = Crepis modocensis Greene ■☆

111011　Crepis senecioides Delile;千里光还阳参■☆

111012　Crepis senecioides Delile subsp. filiformis (Viv.) Alavi;丝形还阳参■☆

111013　Crepis senecioides Delile subsp. nudiflora (Viv.) Alavi;裸花还阳参■☆

111014　Crepis seselifolia Rydb. = Crepis acuminata Nutt. ■☆

111015　Crepis setigera Scott = Youngia blinii (H. Lév.) Lauener ■

111016　Crepis setigera Scott ex W. W. Sm. = Youngia blinii (H. Lév.) Lauener ■

111017　Crepis setosa Hallier f.;刺毛还阳参;Bristly Hawksbeard, Bristly Hawk's-beard, Hawk's Beard, Setose Hawksbeard ■☆

111018　Crepis shawanensis C. Shih;沙湾还阳参;Shawan Hawksbeard ■

111019　Crepis sibirica L.;西伯利亚还阳参;Siberia Hawksbeard ■

111020　Crepis simulatrix Babc. = Youngia simulatrix (Babc.) Babc. et Stebbins ■

111021　Crepis sinuata Lam. = Picris sinuata (Lam.) Lack ■☆

111022　Crepis smithiana Hand. -Mazz. = Youngia simulatrix (Babc.) Babc. et Stebbins ■

111023　Crepis sorocephala Hemsl. = Soroseris glomerata (Decne.) Stebbins ■

111024　Crepis spathulata Guss.;匙形还阳参■☆

111025　Crepis stenoma Turcz. = Youngia stenoma (Turcz.) Ledeb. ■

111026　Crepis stenoma Turcz. ex DC. = Youngia stenoma (Turcz.) Ledeb. ■

111027　Crepis stolonifera H. Lév. = Paraixeris humifusa (Dunn) C. Shih ■

111028　Crepis suberostris Batt. = Crepis arenaria (Pomel) Pomel subsp. suberostris (Batt.) Greuter ■☆

111029　Crepis suberostris Batt. subsp. arenaria (Pomel) Babc. = Crepis arenaria (Pomel) Pomel ■☆

111030　Crepis subscaposa Collett et Hemsl.;抽茎还阳参■

111031　Crepis suffruticosa Babc. = Crepis newii Oliv. et Hiern subsp. oliveriana (Kuntze) C. Jeffrey et Beentje ■☆

111032　Crepis swynnertonii S. Moore = Crepis newii Oliv. et Hiern ■☆

111033　Crepis szechuanica Soderb. = Youngia szechuanica (Soderb.) S. Y. Hu ■

111034　Crepis taquetii (H. Lév. et Vaniot) H. Lév. = Youngia japonica (L.) DC. ■

111035　Crepis taraxacifolia Thuill. = Crepis vesicaria L. subsp. taraxacifolia (Thuill.) Schinz et Keller ■☆

111036　Crepis taraxacifolia Thuill. subsp. myriocephala (Coss. et Durieu) H. Lindb. = Crepis vesicaria L. subsp. myriocephala (Coss. et Durieu) Babc. ■☆

111037　Crepis taraxacifolia Thuill. subsp. stellata Ball = Crepis vesicaria L. subsp. stellata (Ball) Babc. ■☆

111038　Crepis taraxacifolia Thuill. subsp. tingitana (Salzm.) Batt. = Crepis salzmannii Babc. ■☆

111039　Crepis taraxacifolia Thuill. subsp. vesicaria (L.) Batt. = Crepis vesicaria L. ■

111040　Crepis taraxacifolia Thuill. var. hiemalis DC. = Crepis vesicaria L. subsp. taraxacifolia (Thuill.) Schinz et Keller ■☆

111041　Crepis taraxacifolia Thuill. var. hirsuta (Pomel) Batt. = Crepis vesicaria L. subsp. taraxacifolia (Thuill.) Schinz et Keller ■☆

111042　Crepis taraxacifolia Thuill. var. libyca Pamp. = Crepis libyca (Pamp.) Shab. ■☆

111043　Crepis taraxacifolia Thuill. var. myriocephala Coss. et Durieu = Crepis vesicaria L. subsp. myriocephala (Coss. et Durieu) Babc. ■☆

111044　Crepis taraxacifolia Thuill. var. numidica (Pomel) Batt. = Crepis vesicaria L. subsp. taraxacifolia (Thuill.) Schinz et Keller ■☆

111045　Crepis taraxacifolia Thuill. var. spathulata (Guss.) Pamp. = Crepis spathulata Guss. ■☆

111046　Crepis taraxacifolia Thuill. var. stellata (Ball) Ball = Crepis vesicaria L. subsp. stellata (Ball) Babc. ■☆

111047　Crepis taraxacifolia Thuill. var. sulphurea Maire et Wilczek = Crepis vesicaria L. subsp. stellata (Ball) Babc. ■☆

111048　Crepis taraxacifolia Thuill. var. vesicaria (L.) Barratte = Crepis vesicaria L. ■

111049　Crepis taraxacoides Desf. = Crepis vesicaria L. subsp. taraxacifolia (Thuill.) Schinz et Keller ■☆

111050　Crepis tectorum L.;窄叶还阳参(屋根草,屋生还阳参);Hawk's Beard, Hawk's-beard, Narrowleaf Hawksbeard, Narrow-leaved Hawk's-beard ■

111051　Crepis tectorum L. var. gracilis Wallr.;细屋根草■

111052　Crepis tectorum L. var. melancephala Ledeb.;羽叶屋根草■

111053　Crepis tenerrima (Sch. Bip.) R. E. Fr.;极细还阳参■☆

111054　Crepis tenuifolia Willd. = Youngia tenuicaulis (Babc. et Stebbins) De Moor ■

111055　Crepis tenuifolia Willd. = Youngia tenuifolia (Willd.) Babc. et Stebbins ■

111056　Crepis tenuifolia Willd. subsp. tenuicaulis (Babc. et Stebbins) Hand. -Mazz. = Youngia tenuicaulis (Babc. et Stebbins) De Moor ■

111057　Crepis tenuifolia Willd. subsp. tenuicaulis (Babc. et Stebbins) Hand. -Mazz. = Youngia tenuicaulis (Babc. et Stebbins) Czerep. ■

111058　Crepis thomsonii Babc. = Crepis foetida L. ■☆

111059　Crepis tianshanica C. Shih;天山还阳参;Tianshan Hawksbeard ■

111060　Crepis tibetica Babc. = Crepis elongata Babc. ■

111061　Crepis tingitana Ball;丹吉尔还阳参■☆

111062　Crepis tingitana Ball var. ramosissima Maire = Crepis tingitana Ball ■☆

111063　Crepis tsarongensis (W. W. Sm.) Anthony var. chimiliensis (W. W. Sm.) Anthony = Dubyaea tsarongensis (W. W. Sm.) Stebbins ■

111064　Crepis tsarongensis (W. W. Sm.) J. Anthony = Dubyaea tsarongensis (W. W. Sm.) Stebbins ■

111065　Crepis tsarongensis (W. W. Sm.) J. Anthony var. chimiliensis (W. W. Sm.) J. Anthony = Dubyaea tsarongensis (W. W. Sm.) Stebbins ■

111066　Crepis turcomanica Krasch.;土库曼还阳参■☆

111067　Crepis turczaninowii C. A. Mey. = Crepis crocea (Lam.) Babc. ■

111068　Crepis turczaninowii C. A. Mey. ex Turcz. = Crepis crocea (Lam.) Babc. ■

111069　Crepis ugandensis Babc. = Crepis rueppellii Sch. Bip. ■☆

111070　Crepis umbrella Franch. = Stebbinsia umbrella (Franch.) Lipsch. ■

111071　Crepis urundica Babc.;乌隆迪还阳参■☆

111072　Crepis vaniotii H. Lév. = Ixeridium gramineum (Fisch.) Tzvelev ■

111073　Crepis vesicaria L.;膀胱还阳参;Beaked Hawksbeard, Beaked Hawk's-beard, Bladder Hawksbeard, Weedy Hawksbeard ■

111074 Crepis vesicaria L. subsp. andryaloides (Lowe) Babc.;毛托菊还阳参■☆
111075 Crepis vesicaria L. subsp. haenseleri (Boiss.) P. D. Sell = Crepis vesicaria L. subsp. taraxacifolia (Thuill.) Schinz et Keller ■☆
111076 Crepis vesicaria L. subsp. myriocephala (Coss. et Durieu) Babc.;多头还阳参■☆
111077 Crepis vesicaria L. subsp. stellata (Ball) Babc.;星状还阳参■☆
111078 Crepis vesicaria L. subsp. taraxacifolia (Thuill.) Schinz et Keller;北非还阳参■☆
111079 Crepis vesicaria L. var. intybacea (Brot.) Fiori = Crepis vesicaria L. subsp. taraxacifolia (Thuill.) Schinz et Keller ■☆
111080 Crepis vesicaria L. var. laciniata (Willk.) Emb. et Maire = Crepis vesicaria L. subsp. taraxacifolia (Thuill.) Schinz et Keller ■☆
111081 Crepis vesicaria L. var. libyca (Pamp.) Maire et Weiller = Crepis libyca (Pamp.) Shab. ■☆
111082 Crepis vesicaria L. var. longiseta Maire = Crepis vesicaria L. ■
111083 Crepis vesicaria L. var. myriocephala (Coss. et Durieu) Maire = Crepis vesicaria L. subsp. myriocephala (Coss. et Durieu) Babc. ■☆
111084 Crepis vesicaria L. var. pectinata (Willk.) Maire = Crepis vesicaria L. ■
111085 Crepis vesicaria L. var. ramosissima Maire = Crepis vesicaria L. ■
111086 Crepis vesicaria L. var. recognita (Hallier f.) Maire = Crepis vesicaria L. ■
111087 Crepis vesicaria L. var. tangerina (Pau) Maire = Crepis vesicaria L. subsp. stellata (Ball) Babc. ■☆
111088 Crepis vesicaria L. var. taraxacifolia (Thuill.) B. Boivin;具喙膀胱还阳参;Beaked Hawksbeard ■☆
111089 Crepis virens L. = Crepis capillaris (L.) Wallr. ■☆
111090 Crepis virgata Desf. = Tolpis virgata (Desf.) Bertol. ■☆
111091 Crepis willdenowii De Moor;魏氏还阳参■☆
111092 Crepis willemetioides Boiss.;鳞果苣还阳参■☆
111093 Crepis wilsonii Babc. = Youngia wilsonii (Babc.) Babc. et Stebbins ■
111094 Crepis xylorrhiza Sch. Bip.;木根还阳参■☆
111095 Crepis yunnanensis Babc. = Youngia paleacea (Diels) Babc. et Stebbins ■
111096 Crepis zacintha (L.) Babc.;斑纹还阳参;Striped Hawksbeard ■☆
111097 Crepula Hill = Cirsium Mill. ■
111098 Crepula Noronha = Phrynium Willd. (保留属名) ■
111099 Creranobates Ridl. = Schizomeria D. Don ●☆
111100 Crescentia L. (1753);葫芦树属(炮弹果属,蒲瓜树属);Calabash Tree, Calabashtree, Calabash-tree ●
111101 Crescentia alata Kunth;十字架树(叉叶木,叉叶树,具翅炮弹果,三叉木);Croos Tree, Morrito, Trifoliate Calabash-tree ●
111102 Crescentia amazonica Ducke;安巴卡葫芦树●☆
111103 Crescentia cucurbitana L.;葫芦炮弹果●☆
111104 Crescentia cujete L.;炮弹果(叉叶树,红椤,葫芦树,炮弹树,瓢瓜木,蒲瓜树);Calabash, Calabash Tree, Calabashtree, Calabash-tree, Common Calabash Tree, Tree Calabash ●
111105 Crescentia omieta L.;葫芦树●☆
111106 Crescentia pinnata Jacq. = Kigelia africana (Lam.) Benth. ●☆
111107 Crescentia trifolia Blanco = Crescentia alata Kunth ●
111108 Crescentiaceae Dumort.;葫芦树科(炮弹果科)●
111109 Crescentiaceae Dumort. = Bignoniaceae Juss. (保留科名) ●■
111110 Creslobus Lilja = Mentzelia L. ●■☆
111111 Cressa L. (1753);克里特旋花属■☆
111112 Cressa arabica Forssk. = Seddera arabica (Forssk.) Choisy ●☆
111113 Cressa cretica L.;克里特旋花■☆
111114 Cressa cretica L. var. salina J. A. Schmidt = Cressa cretica L. ■☆
111115 Cressaceae Raf. = Convolvulaceae Juss. (保留科名) ●■
111116 Cressaria Raf. = Cressa L. ■☆
111117 Creusa P. V. Heath = Crassula L. ●■☆
111118 Cribbia Senghas (1985);克里布兰属■☆
111119 Cribbia brachyceras (Summerh.) Senghas;短角克里布兰■☆
111120 Cribbia confusa P. J. Cribb;混乱克里布兰■☆
111121 Cribbia pendula la Croix et P. J. Cribb;下垂克里布兰■☆
111122 Cribbia thomensis la Croix et P. J. Cribb;托芒斯克里布兰■☆
111123 Criciuma Soderstr. et Londoño (1987);环草属■☆
111124 Criciuma asymmetrica Soderstr. et Londono;环草■☆
111125 Crimaea Vassilcz. = Medicago L. (保留属名) ●■
111126 Crinaceae Vest = Amaryllidaceae J. St. -Hil. (保留科名) ●■
111127 Crinaceae Vest = Gramineae Juss. (保留科名) ■●
111128 Crinaceae Vest = Poaceae Barnhart (保留科名) ■●
111129 Crinipes Hochst. (1855);毛发草属■☆
111130 Crinipes abyssinicus (Hochst. ex A. Rich.) Hochst.;阿比西尼亚毛发草■☆
111131 Crinipes gynoglossa Gooss. = Styppeiochloa gynoglossa (Gooss.) De Winter ■☆
111132 Crinipes longifolius C. E. Hubb.;长叶毛发草■☆
111133 Crinipes longipes (Stapf et C. E. Hubb.) C. E. Hubb. = Nematopoa longipes (Stapf et C. E. Hubb.) C. E. Hubb. ■☆
111134 Crinissa Rchb. = Pyrrhopappus DC. (保留属名) ■☆
111135 Crinita Houtt. = Pavetta L. ●
111136 Crinita Moench = Crinitaria Cass. ■●☆
111137 Crinita Moench = Linosyris Cass. ■
111138 Crinita capensis Houtt. = Pavetta capensis (Houtt.) Bremek. ●☆
111139 Crinitaria Cass. (1825);毛麻菀属■●☆
111140 Crinitaria Cass. = Aster L. ●■
111141 Crinitaria Cass. = Linosyris Cass. ■
111142 Crinitaria biflora (L.) Cass. = Galatella biflora (L.) Nees ■
111143 Crinitaria viscidiflora Hook. = Chrysothamnus viscidiflorus (Hook.) Nutt. ●☆
111144 Crinitaris tatarica (Less.) Novopokr. = Linosyris tatarica (Less.) C. A. Mey. ■
111145 Crinodendron Molina (1782);智利灯笼树属(百合木属);Lantern Tree ●☆
111146 Crinodendron hookerianum Gay;智利灯笼树(红百合木);Chile Lantern Tree, Lantern Tree, Lantern-tree, Vinegar Weed ●☆
111147 Crinodendron patagua Molina;毛智利灯笼树●☆
111148 Crinodendrum Juss. = Crinodendron Molina ●☆
111149 Crinonia Banks ex Tul. = Hedycarya J. R. Forst. et G. Forst. ●☆
111150 Crinonia Blume = Pholidota Lindl. ex Hook. ■
111151 Crinopsis Herb. = Crinum L. ■
111152 Crinum L. (1753);文珠兰属(文殊兰属);Cape Lily, Crinum, Crinum Lily, Crinum-Lilies, Spider Lily, String-lily, Swamp-lily, Veld Lily ■
111153 Crinum abyssinicum Hochst = Crinum abyssinicum Hochst. ex A. Rich. ■☆
111154 Crinum abyssinicum Hochst. ex A. Rich.;阿比尼西亚文珠兰■☆
111155 Crinum acaule Baker;无茎文珠兰■☆
111156 Crinum africanum L. = Agapanthus africanus (L.) Hoffmanns. ■☆
111157 Crinum amabile Donn;美丽文殊兰(红花文殊兰,美丽文珠

兰,优美文殊兰);Sumatra Crinum ■☆

111158 Crinum americanum L.;美洲文殊兰(北美文殊兰,美国文殊兰);American Crinum,Florida Crinum,Florida Swamp-lily,Southern Swamp Crinum,Southern Swamp Lily,Swamp Crinum ■☆

111159 Crinum americanum L. var. traubii (Moldenke) L. S. Hannibal;特劳布文殊兰■☆

111160 Crinum angolense (Baker) Benth. ex Baker = Ammocharis angolensis (Baker) Milne-Redh. et Schweick. ■☆

111161 Crinum angustifolium Herb. ex Steud. = Crinum asisticum L. ■

111162 Crinum angustifolium Houtt. = Polianthes tuberosa L. ■

111163 Crinum angustifolium L. f. = Cyrtanthus angustifolius (L. f.) W. T. Aiton ■☆

111164 Crinum angustifolium R. Br.;狭叶文殊兰■☆

111165 Crinum angustifolium Tate = Crinum flaccidum Herb. ■☆

111166 Crinum angustum Roxb.;紫文殊兰■☆

111167 Crinum asiaticum L.;亚洲文殊兰(扁担叶,东亚文殊兰,海带七,海蕉,罗裙带,牛黄散,千层喜,秦琼剑,裙带草,十八学士,水蕉,水笑草,万年青,文兰树,文殊兰,文珠兰,腰带七,引水蕉,玉带风,郁蕉,郁金叶);Asia Crinum,Grand Crinum,Poison Bulb,Poisonbulb,Tree Crinum ■

111168 Crinum asiaticum L. var. anomalum Herb.;奇形东亚文殊兰(畸形文殊兰);Anomalous Crinum ■☆

111169 Crinum asiaticum L. var. declinatum Herb.;下弯东亚文殊兰(垂文殊兰,文珠兰);Declined Crinum ■☆

111170 Crinum asiaticum L. var. japonicum Baker;日本文殊兰(文珠兰);Japan Crinum,Japanese Crinum ■☆

111171 Crinum asiaticum L. var. sinicum (Roxb. ex Herb.) Baker;文殊兰(白花石蒜,扁担叶,海带七,海蕉,罗裙带,牛黄伞,牛黄散,千层喜,秦琼剑,裙带草,十八学士,水蕉,水笑草,万年青,万寿兰,文兰树,文珠兰,腰带七,引水蕉,玉带风,郁蕉,郁金叶,中国文殊兰,朱兰);China Crinum,China Grand Crinum,Chinese Crinum,Grand Crinum,St. Johns Lily ■

111172 Crinum asiaticum L. var. sinicum (Roxb. ex Herb.) Baker = Crinum asiaticum L. ■

111173 Crinum bambusetum Nordal et Sebsebe;邦布塞特文殊兰■☆

111174 Crinum baumii Harms;鲍姆文殊兰■☆

111175 Crinum baumii Harms = Ammocharis baumii (Harms) Milne-Redh. et Schweick. ■☆

111176 Crinum bequaertii De Wild. = Crinum jagus (J. Thomps.) Dandy ■☆

111177 Crinum biflorum Baker;双花文殊兰■☆

111178 Crinum boehmii Baker = Crinum ornatum (L. f. ex Aiton) Bury ■☆

111179 Crinum broussonetii (A. DC.) Herb.;布鲁文殊兰■☆

111180 Crinum bulbispermum (Burm. f.) Milne-Redh. et Schweick.;鳞茎文殊兰(长叶文殊兰,鳞子文殊兰);Hardy Crinum,Orange River Lily ■☆

111181 Crinum bulbispermum (Burm. f.) Milne-Redh. et Schweick. = Crinum powellii Baker ■☆

111182 Crinum buphanoides Welw. ex Baker;石蒜文殊兰■☆

111183 Crinum campanulatum Herb.;风铃草状文殊兰■☆

111184 Crinum capense Herb. = Crinum longifolium (L.) Thunb. ■☆

111185 Crinum congolense De Wild. = Crinum jagus (J. Thomps.) Dandy ■☆

111186 Crinum corradii Chiov. = Crinum macowanii Baker ■☆

111187 Crinum crassicaule Baker;粗茎文殊兰■☆

111188 Crinum crispum E. Phillips = Crinum lugardiae N. E. Br. ■☆

111189 Crinum curvifolium Baker = Ammocharis angolensis (Baker) Milne-Redh. et Schweick. ■☆

111190 Crinum definum Ker Gawl.;内生文殊兰■☆

111191 Crinum delagoense I. Verd. = Crinum stuhlmannii Baker ■☆

111192 Crinum distichum Herb.;二列文殊兰■☆

111193 Crinum esquirolii H. Lév. = Crinum latifolium L. ■

111194 Crinum falcatum Jacq. = Ammocharis longifolia (L.) M. Roem. ■☆

111195 Crinum flaccidum Herb.;柔弱文殊兰;Darling Lily ■☆

111196 Crinum forbesii (Lindl.) Schult. et Schult. f. = Crinum paludosum I. Verd. ■☆

111197 Crinum giesii Lehmiller;吉斯文殊兰■☆

111198 Crinum giganteum Andréws;大花文殊兰■☆

111199 Crinum giganteum Andréws = Crinum jagus (J. Thomps.) Dandy ■☆

111200 Crinum gigas Nakai;大文殊兰■☆

111201 Crinum glaucum A. Chev.;灰绿文殊兰■☆

111202 Crinum gouwsii Traub = Crinum macowanii Baker ■☆

111203 Crinum graminicola I. Verd.;草莺文殊兰■☆

111204 Crinum harmsii Baker;哈姆斯文殊兰■☆

111205 Crinum herbertianum Schult. f. = Crinum americanum L. ■☆

111206 Crinum herbertianum Wall. = Crinum zeylanicum (L.) L. ■☆

111207 Crinum heterostylum Bullock = Ammocharis angolensis (Baker) Milne-Redh. et Schweick. ■☆

111208 Crinum humile A. Chev. = Crinum nubicum Hannibal ■☆

111209 Crinum imbricatum Baker = Crinum moorei Hook. f. ■☆

111210 Crinum jagus (J. Thomps.) Dandy;沼地文殊兰;Spider Lily,Swamp Lily ■☆

111211 Crinum johnstonii Baker = Crinum macowanii Baker ■☆

111212 Crinum kirkii Baker;柯克文殊兰;Kirk Crinum,Pyjama Lily ■☆

111213 Crinum kirkii Baker var. reductum ? = Crinum kirkii Baker ■☆

111214 Crinum latifolium L.;西南文殊兰(西南文珠兰);Broadleaf Crinum,Large-leaved Crinum ■

111215 Crinum latifolium L. var. zeylanicum (L.) Hook. f. = Crinum zeylanicum (L.) L. ■☆

111216 Crinum laurentii T. Durand et De Wild. = Crinum jagus (J. Thomps.) Dandy ■☆

111217 Crinum longifolium (L.) Thunb.;长叶文殊兰;Hardy Crinum ■☆

111218 Crinum longifolium (L.) Thunb. = Ammocharis longifolia (L.) M. Roem. ■☆

111219 Crinum longifolium Roxb. = Crinum bulbispermum (Burm. f.) Milne-Redh. et Schweick. ■☆

111220 Crinum longitubum Pax;长管文殊兰■☆

111221 Crinum loureiri M. Roem.;劳瑞氏文殊兰;Loureir Crinum ■

111222 Crinum lugardiae N. E. Br.;多叶文殊兰■☆

111223 Crinum macowanii Baker;波叶文殊兰■☆

111224 Crinum macowanii Baker subsp. confusum I. Verd. = Crinum macowanii Baker ■☆

111225 Crinum makoyanum Hort. = Crinum moorei Hook. f. ■☆

111226 Crinum maritimum Siebold ex Nakai = Crinum asiaticum L. var. japonicum Baker ■☆

111227 Crinum massaianum (L. Linden et Rodigas) N. E. Br. = Crinum kirkii Baker ■☆

111228 Crinum menyharthii Baker;迈尼哈尔特文殊兰■☆

111229 Crinum minimum Milne-Redh.;微小文殊兰■☆

111230 Crinum moorei Hook. f.;宽叶文殊兰(长颈文殊兰,紫花文殊

兰);Longneck Crinum ■☆
111231 Crinum natans Baker;浮水文殊兰■☆
111232 Crinum nerinoides Baker = Ammocharis nerinoides (Baker) Lehmiller ■☆
111233 Crinum nubicum Hannibal;云雾文殊兰■☆
111234 Crinum obliquum L. f. = Cyrtanthus obliquus (L. f.) W. T. Aiton ■☆
111235 Crinum occiduale R. A. Dyer = Crinum lugardiae N. E. Br. ■☆
111236 Crinum octobris Nakai et Tuyama;十月文殊兰■☆
111237 Crinum ornatum (L. f. ex Aiton) Bury var. letifolium Herb. = Crinum latifolium L. ■
111238 Crinum paludosum I. Verd.;沼泽文殊兰■☆
111239 Crinum papillosum Nordal;乳头文殊兰■☆
111240 Crinum parvibulbosum Dinter ex Overkott;小鳞茎文殊兰■☆
111241 Crinum parvum Baker;小文殊兰■☆
111242 Crinum pauciflorum Baker;少花文殊兰■☆
111243 Crinum pedicellatum Pax = Crinum macowanii Baker ■☆
111244 Crinum pedunculatum R. Br.;花柄文殊兰;Swamp Lily ■☆
111245 Crinum piliferum Nordal;纤毛文殊兰■☆
111246 Crinum podophyllum Baker = Crinum jagus (J. Thomps.) Dandy ■☆
111247 Crinum poggei Pax;波格文殊兰■☆
111248 Crinum politifolium R. Wahlstr.;亮叶文殊兰■☆
111249 Crinum polyphyllum Baker = Crinum lugardiae N. E. Br. ■☆
111250 Crinum powellii Baker;鲍威尔文殊兰(鲍氏文殊兰);Cape Coast Lily, Crinum Lily, Powell Crinum, Swamp Lily ■☆
111251 Crinum purpurascens Herb.;淡紫色文殊兰■☆
111252 Crinum purpurascens Herb. var. angustilobium De Wild. = Crinum purpurascens Herb. ■☆
111253 Crinum rattrayii Hort. = Crinum glaucum A. Chev. ■☆
111254 Crinum rautanenianum Schinz;劳塔宁文殊兰■☆
111255 Crinum riparium Herb. = Crinum longifolium (L.) Thunb. ■☆
111256 Crinum samueli Worsley;萨姆埃尔文殊兰■☆
111257 Crinum sanderianum Baker;桑德文殊兰■☆
111258 Crinum sanderianum Baker = Crinum broussonetii (A. DC.) Herb. ■☆
111259 Crinum schimperi Vatke ex K. Schum. = Crinum abyssinicum Hochst. ex A. Rich. ■☆
111260 Crinum schmidtii Regel = Crinum moorei Hook. f. ■☆
111261 Crinum sinicum Roxb. ex Herb. = Crinum asiaticum L. var. sinicum (Roxb. ex Herb.) Baker ■
111262 Crinum sinicum Roxb. ex Herb. = Crinum asiaticum L. ■
111263 Crinum somalense Chiov. = Crinum stuhlmannii Baker ■☆
111264 Crinum stenophyllum Baker;窄叶文殊兰■☆
111265 Crinum strictum Herb. = Crinum americanum L. ■☆
111266 Crinum strictum Herb. var. traubii Moldenke = Crinum americanum L. var. traubii (Moldenke) L. S. Hannibal ■☆
111267 Crinum stuhlmannii Baker;斯图尔曼文殊兰■☆
111268 Crinum subcernuum Baker;俯垂文殊兰■☆
111269 Crinum superbum Roxb. = Crinum amabile Donn ■☆
111270 Crinum tanganyikense Baker = Crinum ornatum (L. f. ex Aiton) Bury ■☆
111271 Crinum tenellum L. f. = Strumaria tenella (L. f.) Snijman ■☆
111272 Crinum texanum L. S. Hannibal = Crinum americanum L. ■☆
111273 Crinum uniflorum F. Muell.;单花文殊兰■☆
111274 Crinum variabile (Jacq.) Herb.;易变文殊兰■☆
111275 Crinum vassei Bois;瓦塞文殊兰■☆
111276 Crinum verdoorniae Lehmiller;韦尔文殊兰■☆
111277 Crinum wallichianum M. Roem. = Crinum zeylanicum (L.) L. ■☆
111278 Crinum walteri Overkott = Crinum minimum Milne-Redh. ■☆
111279 Crinum yuccaeflorum Salisb. = Crinum broussonetii (A. DC.) Herb. ■☆
111280 Crinum zeylanicum (L.) L.;锡兰文殊兰;Ceylon Swamplily ■☆
111281 Crinum zeylanicum L. = Crinum zeylanicum (L.) L. ■☆
111282 Crioceras Pierre(1897);羊角夹竹桃属●☆
111283 Crioceras dipladeniiflorus (Stapf) K. Schum.;羊角夹竹桃●☆
111284 Crioceras longiflorus Pierre = Crioceras dipladeniiflorus (Stapf) K. Schum. ●☆
111285 Criogenes Salisb. = Cypripedium L. ■
111286 Criosanthes Raf. = Cypripedium L. ■
111287 Criosanthes arietina (R. Br.) House = Cypripedium arietinum R. Br. ■☆
111288 Criosophila Post et Kuntze = Cryosophila Blume ●☆
111289 Criptangis Thouars = Angraecum Bory ■
111290 Criptina Raf. = Crypta Nutt. ■
111291 Criptina Raf. = Elatine L. ■
111292 Criptophylis Thouars = Bulbophyllum Thouars(保留属名)■
111293 Criscia Katinas(1994);橙花钝柱菊属●☆
111294 Criscia stricta (Spreng.) Katinas;橙花钝柱菊■☆
111295 Crispaceae Dulac = Balsaminaceae A. Rich.(保留科名)■
111296 Crispiloba Steenis(1984);二籽假海桐属●☆
111297 Crispiloba disperma (S. Moore) Steenis;二籽假海桐●☆
111298 Cristaria Cav. (1799)(保留属名);冠毛锦葵属■●☆
111299 Cristaria Sonn.(废弃属名) = Combretum Loefl.(保留属名)●
111300 Cristaria Sonn.(废弃属名) = Cristaria Cav.(保留属名)■●☆
111301 Cristaria australis Phil.;澳洲冠毛锦葵■☆
111302 Cristaria coccinea Sonn. = Combretum coccineum (Sonn.) Lam. ●☆
111303 Cristaria elegans Gay;雅致冠毛锦葵■☆
111304 Cristaria glabra Phil.;无毛冠毛锦葵■☆
111305 Cristaria glandulosa Phil.;多腺冠毛锦葵■☆
111306 Cristaria glaucophylla Cav.;灰叶冠毛锦葵■☆
111307 Cristaria grandidentata Phil. ex Baker f.;大齿冠毛锦葵■☆
111308 Cristaria grandiflora Phil.;大花冠毛锦葵■☆
111309 Cristaria heterophylla (Cav.) Hook. et Arn.;异叶冠毛锦葵■☆
111310 Cristaria microptera Phil. ex Baker f.;小翅冠毛锦葵■☆
111311 Cristaria multifida Cav.;多裂冠毛锦葵■☆
111312 Cristaria multiflora Gay;多花冠毛锦葵■☆
111313 Cristaria pilosa Phil.;绒毛冠毛锦葵■☆
111314 Cristaria trifida Phil.;三裂冠毛锦葵■☆
111315 Cristatella Nutt. (1834);冠毛山柑属■☆
111316 Cristatella erosa Nutt.;冠毛山柑■☆
111317 Cristatella jamesii Torr. et A. Gray = Polanisia jamesii (Torr. et A. Gray) H. H. Iltis ■☆
111318 Cristella Raf. = Cristatella Nutt. ■☆
111319 Cristesion Raf. = Hordeum L. ■
111320 Cristesion jubatum (L.) Nevski = Hordeum jubatum L. ■
111321 Cristonia J. H. Ross = Bossiaea Vent. ●☆
111322 Cristonia J. H. Ross(2001);可利豆属●☆
111323 Critamus Besser = Falcaria Fabr.(保留属名)■
111324 Critamus Hoffm. = Apium L. ■
111325 Critesia Raf. = Salsola L. ●■
111326 Critesion Raf. (1819);芒麦草属■☆

111327 Critesion Raf. = Hordeum L. ■
111328 Critesion bogdanii (Wilensky) Á. Löve = Hordeum bogdanii Wilensky ■
111329 Critesion brevisubulatum (Trin.) Á. Löve = Hordeum brevisubulatum (Trin.) Link ■
111330 Critesion brevisubulatum (Trin.) Link subsp. nevskianum (Bowden) Á. Löve = Hordeum brevisubulatum (Trin.) Link var. nevskianum (Bowden) Tzvelev ■
111331 Critesion brevisubulatum (Trin.) Link subsp. turkestanicum Á. Löve = Hordeum brevisubulatum (Trin.) Link subsp. turkestanicum Tzvelev ■
111332 Critesion bulbosum (L.) Á. Löve = Hordeum bulbosum L. ■
111333 Critesion californicum (Covas et Stebbins) Á. Löve subsp. sibiricum Á. Löve = Hordeum roshevitzii Bowden ■
111334 Critesion glaucum (Steud.) Á. Löve = Hordeum glaucum Steud. ■☆
111335 Critesion hystrix (Roth) Á. Löve = Hordeum geniculatum All. ■☆
111336 Critesion jubatum (L.) Nevski = Hordeum jubatum L. ■
111337 Critesion marinum (Huds.) Á. Löve = Hordeum marinum Huds. ■☆
111338 Critesion murinum (L.) Á. Löve = Hordeum murinum L. ■☆
111339 Critesion nevskianum (Bowden) Tzvelev = Hordeum brevisubulatum (Trin.) Link var. nevskianum (Bowden) Tzvelev ■
111340 Critesion roshevitzii (Bowden) Tzvelev = Hordeum roshevitzii Bowden ■
111341 Critesion turkestanicum Tzvelev = Hordeum brevisubulatum (Trin.) Link subsp. turkestanicum Tzvelev ■
111342 Critesium Endl. = Critesion Raf. ■☆
111343 Crithmum L. (1753);海茴香属;Rock Samphire,Samphire ■☆
111344 Crithmum canariense Cav. = Crithmum maritimum L. ■☆
111345 Crithmum latifolium L. f. = Astydamia canariensis DC. ■☆
111346 Crithmum latifolium L. f. = Astydamia latifolia (L. f.) Kuntze ■☆
111347 Crithmum maritimum L.;海茴香;Camphire,Creevereegh,Crestmarine,Passper,Peter's Cress,Pierce-stone,Rock Samphire,Rock Sampier,Samfer,Samphire,Sampkins,Sea Fennel,Sea Samphire,Sea-fennel,Semper,Shamsher ■☆
111348 Crithmum mediterraneum M. Bieb. = Cenolophium denudatum (Hornem.) Tutin ■
111349 Critho E. Mey. = Hordeum L. ■
111350 Crithodium Link = Triticum L. ■
111351 Crithodium monococcum (L.) Á. Löve = Triticum monococcum L. ■
111352 Crithopsis Jaub. et Spach(1851);类大麦属■☆
111353 Crithopsis delileana (Schult.) Roshev.;类大麦■☆
111354 Crithopyrum Hort. Prag. ex Steud. = Agropyron Gaertn. ■
111355 Crithopyrum Steud. = Elymus L. ■
111356 Critonia Cass. = Vernonia Schreb. (保留属名) ●■
111357 Critonia Gaertn. = Kalmia L. ●
111358 Critonia P. Browne(1756);亮泽兰属●☆
111359 Critonia chrysocephala (Klatt) R. M. King et H. Rob.;金头亮泽兰●☆
111360 Critonia elliptica Raf. ex DC.;椭圆亮泽兰●☆
111361 Critonia eriocarpa (B. L. Rob. et Greenm.) R. M. King et H. Rob.;毛果亮泽兰●☆
111362 Critonia heteroneura Ernst;异脉亮泽兰●☆
111363 Critonia laurifolia (B. L. Rob.) R. M. King et H. Rob.;桂叶亮泽兰●☆
111364 Critonia macropoda DC.;大足亮泽兰●☆
111365 Critonia megaphylla (Baker) R. M. King et H. Rob.;大叶亮泽兰●☆
111366 Critonia morifolia (Mill.) R. M. King et H. Rob.;桑叶亮泽兰●☆
111367 Critonia tenuifolia (Kunth) V. M. Badillo;细叶亮泽兰●☆
111368 Critoniadelphus R. M. King et H. Rob. (1971);腺果亮泽兰属●☆
111369 Critoniadelphus microdon (B. L. Rob.) R. M. King et H. Rob.;小齿腺果亮泽兰●☆
111370 Critoniadelphus nubigenus (Benth.) R. M. King et H. Rob.;腺果亮泽兰●☆
111371 Critoniella R. M. King et H. Rob. (1975);柔柱亮泽兰属■●☆
111372 Critoniella acuminata (Kunth) R. M. King et H. Rob.;渐尖柔柱亮泽兰●☆
111373 Critoniella lebrijensis (B. L. Rob.) R. M. King et H. Rob.;柔柱亮泽兰●☆
111374 Critoniella tenuifolia (Kunth) R. M. King et H. Rob.;细叶柔柱亮泽兰●☆
111375 Critoniopsis Sch. Bip. (1863);腺瓣落苞菊属●■☆
111376 Critoniopsis Sch. Bip. = Vernonia Schreb. (保留属名) ●■
111377 Critoniopsis angusta (Gleason) H. Rob.;窄腺瓣落苞菊●☆
111378 Critoniopsis boliviana (Britton) H. Rob.;玻利维亚腺瓣落苞菊●☆
111379 Critoniopsis floribunda (Kunth) H. Rob.;多花腺瓣落苞菊●☆
111380 Critoniopsis foliosa (Benth.) H. Rob.;多叶腺瓣落苞菊●☆
111381 Critoniopsis leiocarpa (DC.) H. Rob.;光果腺瓣落苞菊●☆
111382 Critoniopsis oolepis (S. F. Blake) H. Rob.;卵鳞腺瓣落苞菊●☆
111383 Critoniopsis pallida (Cuatrec.) H. Rob.;苍白腺瓣落苞菊●☆
111384 Critoniopsis salicifolia (DC.) H. Rob.;柳叶腺瓣落苞菊●☆
111385 Critoniopsis suaveolens (Kunth) H. Rob.;香腺瓣落苞菊●☆
111386 Critoniopsis tomentosa (La Llave et Mexia) H. Rob.;毛腺瓣落苞菊●☆
111387 Critoniopsis uniflora (Sch. Bip.) H. Rob.;单花腺瓣落苞菊●☆
111388 Crlnissa Rchb. = Pyrrhopappus DC. (保留属名) ■☆
111389 Crlnita Houtr. = Pavetta L. ●
111390 Crnciata Gilib. = Gentiana L. ■
111391 Croaspila Raf. = Chaerophyllum L. ■
111392 Croatiella E. G. Gonç. = Asterostigma Fisch. et C. A. Mey. ■☆
111393 Crobylanthe Bremek. (1940);辫花茜属●☆
111394 Crobylanthe pellacalyx (Ridl.) Bremek.;辫花茜●☆
111395 Crocaceae Vest = Iridaceae Juss. (保留科名) ■●
111396 Crocanthemum Spach = Halimium (Dunal) Spach ●☆
111397 Crocanthemum Spach(1836);拟番红花属;Frostwort ●☆
111398 Crocanthemum bicknellii (Fernald) Barnhart = Helianthemum bicknellii Fernald ●☆
111399 Crocanthemum canadense (L.) Britton = Helianthemum canadense (L.) Michx. ●☆
111400 Crocanthemum carolinianum Spach;拟番红花●☆
111401 Crocanthemum majus sensu Britton = Helianthemum bicknellii Fernald ●☆
111402 Crocanthus Klotzsch ex Klatt = Crocosmia Planch. ■
111403 Crocanthus L. Bolus = Malephora N. E. Br. ■☆
111404 Crocanthus croceus (Jacq.) L. Bolus = Malephora crocea (Jacq.) Schwantes ■☆
111405 Crocanthus luteolus (Haw.) L. Bolus = Malephora luteola (Haw.) Schwantes ■☆

111406 Crocanthus mossambicensis Klotzsch ex Klatt = Crocosmia aurea (Pappe ex Hook.) Planch. ■

111407 Crocanthus purpureo-croceus (Haw.) L. Bolus = Malephora purpureo-crocea (Haw.) Schwantes ■☆

111408 Crocanthus thunbergii (Haw.) L. Bolus = Malephora thunbergii (Haw.) Schwantes ■☆

111409 Crocaria Noronha = Microcos Burm. ex L. ●

111410 Crocidium Hook. (1834);腋绒菊属■☆

111411 Crocidium multicaule Hook.;腋绒菊■☆

111412 Crocion Nieuwl. = Viola L. ■●

111413 Crocion Nieuwl. et Kaczm. = Viola L. ■●

111414 Crociris Schur = Crocus L. ■

111415 Crociseris (Rchb.) Fourr. = Senecio L. ■●

111416 Crociseris Fourr. = Senecio L. ■●

111417 Crockeria Greene ex A. Gray = Lasthenia Cass. ■☆

111418 Crockeria chrysantha Greene ex A. Gray = Lasthenia chrysantha (Greene ex A. Gray) Greene ■☆

111419 Crococylum Steud. = Crocoxylon Eckl. et Zeyh. ●☆

111420 Crocodeilanthe Rchb. f. = Pleurothallis R. Br. ■☆

111421 Crocodeilanthe Rchb. f. et Warsz. = Plazia Ruiz et Pav. ●☆

111422 Crocodilina Bubani = Atractylis L. ■☆

111423 Crocodilium Hill = Centaurea L. (保留属名)●■

111424 Crocodilodes Adans. (废弃属名) = Berkheya Ehrh. (保留属名)●■☆

111425 Crocodilodes andongensis Hiern = Berkheya angolensis O. Hoffm. ■☆

111426 Crocodilodes angolensis (O. Hoffm.) Hiern = Berkheya angolensis O. Hoffm. ■☆

111427 Crocodilodes annectens (Harv.) Kuntze = Berkheya annectens Harv. ■☆

111428 Crocodilodes antunesii (O. Hoffm.) Hiern = Hirpicium antunesii (O. Hoffm.) Rössler ■☆

111429 Crocodilodes bipinnatifidum (Harv.) Kuntze = Berkheya bipinnatifida (Harv.) Rössler ■☆

111430 Crocodilodes carlinopsis (Welw. ex O. Hoffm.) Hiern = Berkheya carlinopsis Welw. ex O. Hoffm. ■☆

111431 Crocodilodes coriaceum (Harv.) Kuntze = Berkheya coriacea Harv. ■☆

111432 Crocodilodes dregei (Harv.) Kuntze = Berkheya dregei Harv. ■☆

111433 Crocodilodes eryngiifolium (Less.) Kuntze = Heterorhachis aculeata (Burm. f.) Rössler ●☆

111434 Crocodilodes gorterioides (Oliv. et Hiern) Kuntze = Hirpicium gazanioides (Harv.) Rössler ■☆

111435 Crocodilodes gracilis (O. Hoffm.) Hiern = Hirpicium gracile (O. Hoffm.) Rössler ■☆

111436 Crocodilodes harveyanum Kuntze = Berkheya seminivea Harv. et Sond. ■☆

111437 Crocodilodes palmatum (Thunb.) Kuntze = Heterorhachis aculeata (Burm. f.) Rössler ●☆

111438 Crocodilodes pinnatum (Thunb.) Kuntze = Heterorhachis aculeata (Burm. f.) Rössler ●☆

111439 Crocodilodes radula (Harv.) Kuntze = Berkheya radula (Harv.) De Wild. ■☆

111440 Crocodilodes seminiveum (DC.) Kuntze = Berkheya bipinnatifida (Harv.) Rössler ■☆

111441 Crocodilodes setiferum (DC.) Kuntze = Berkheya setifera DC. ■☆

111442 Crocodilodes speciosum (DC.) Kuntze = Berkheya speciosa (DC.) O. Hoffm. ■☆

111443 Crocodilodes spekeanum (Oliv.) Kuntze = Berkheya spekeana Oliv. ■☆

111444 Crocodilodes subulatum (Harv.) Kuntze = Berkheya subulata Harv. ■☆

111445 Crocodilodes umbellatum (DC.) Kuntze = Berkheya umbellata DC. ■☆

111446 Crocodilodes welwitschii (O. Hoffm.) Hiern = Berkheya welwitschii O. Hoffm. ■☆

111447 Crocodilodes zeyheri (Oliv. et Hiern) Kuntze = Berkheya zeyheri Oliv. et Hiern ■☆

111448 Crocodiloides B. D. Jacks. = Crocodilodes Adans. (废弃属名)●■☆

111449 Crocodylium Hill = Centaurea L. (保留属名)●■

111450 Crocodylium Hill = Crocodylium Vaill. ■☆

111451 Crocodylium Vaill. = Centaurea L. (保留属名)●■

111452 Crocodylium Vaill. = Crocodilium Hill ●■

111453 Crocopsis Pax = Stenomesson Herb. ■☆

111454 Crocosma Klatt = Crocosmia Planch. ■

111455 Crocosmia Planch. (1851-1852);雄黄兰属(臭藏红花属,观音兰属,香鸢尾属);Coppertip, Crocosmia, Falling Star, Montbretia, Pleated Leaves ■

111456 Crocosmia ambongensis (H. Perrier) Goldblatt et J. C. Manning;马岛雄黄兰■☆

111457 Crocosmia aurea (Hook.) Planch. = Crocosmia aurea (Pappe ex Hook.) Planch. ■

111458 Crocosmia aurea (Pappe ex Hook.) Planch.;金黄雄黄兰(桧叶水仙,金黄臭藏红花,金黄火星花);Coppertip, Falling Star, Falling Stars, Golden Coppertip ■

111459 Crocosmia aurea (Pappe ex Hook.) Planch. subsp. pauciflora (Milne-Redh.) Goldblatt;少花金黄雄黄兰■☆

111460 Crocosmia aurea (Pappe ex Hook.) Planch. var. maculata Baker;斑点金黄雄黄兰■☆

111461 Crocosmia cinnabarina (Pax) de Vos = Crocosmia aurea (Pappe ex Hook.) Planch. subsp. pauciflora (Milne-Redh.) Goldblatt ■☆

111462 Crocosmia crocosmiiflora (Burb. et Dean) N. E. Br. = Crocosmia crocosmiiflora (Nicholson) N. E. Br. ■

111463 Crocosmia crocosmiiflora (Lemoine ex Anonym.) N. E. Br. = Crocosmia crocosmiiflora (Nicholson) N. E. Br. ■

111464 Crocosmia crocosmiiflora (Nicholson) N. E. Br.;雄黄兰(标竿花,倒挂金钩,观音兰,黄大蒜,火星花,火焰兰,姬桧扇水仙,扭子药,鸢尾兰);Common Crocosmia, Crocosmia, Hybrid Montebretia, Montbretia ■

111465 Crocosmia fucata (Herb.) M. P. de Vos;着色雄黄兰■☆

111466 Crocosmia hybrida Hort.;杂种雄黄兰;Coppertip, Falling Stars ■☆

111467 Crocosmia maculata (Baker) N. E. Br. = Crocosmia aurea (Pappe ex Hook.) Planch. var. maculata Baker ■☆

111468 Crocosmia masonorum (L. Bolus) N. E. Br.;马氏雄黄兰;Crocosmia, Giant Montbretia, Golden Swan ■☆

111469 Crocosmia mathewsiana (L. Bolus) Goldblatt;马修斯雄黄兰■☆

111470 Crocosmia paniculata (Klatt) Goldblatt;锥序雄黄兰;Aunt-eliza ■☆

111471 Crocosmia pauciflora Milne-Redh. = Crocosmia aurea (Pappe ex Hook.) Planch. subsp. pauciflora (Milne-Redh.) Goldblatt ■☆

111472 Crocosmia pearsei Oberm. ;皮尔斯雄黄兰■☆

111473 Crocosmia pottsii (Baker) N. E. Br. = Crocosmia pottsii (Macnab ex Baker) N. E. Br. ■

111474 Crocosmia pottsii (Macnab ex Baker) N. E. Br. ;射干鸢尾(火焰兰,帕氏火星花,普氏鸢尾);Potts Flower,Pott's Montbretia ■

111475 Crocoxylon Eckl. et Zeyh. (1835);番红花卫矛属●☆

111476 Crocoxylon Eckl. et Zeyh. = Elaeodendron J. Jacq. ●☆

111477 Crocoxylon croceum (Thunb.) N. Robson = Elaeodendron croceum (Thunb.) DC. ●☆

111478 Crocoxylon excelsum Eckl. et Zeyh. = Elaeodendron zeyheri Spreng. ex Turcz. ●☆

111479 Crocoxylon transvaalense (Burtt Davy) N. Robson = Elaeodendron transvaalense (Burtt Davy) R. H. Archer ●☆

111480 Crocus L. (1753);番红花属(藏红花属);Crocus,Saffron ■

111481 Crocus adamii J. Gay;阿达姆番红花■☆

111482 Crocus aerius Herb. ;悬垂番红花■☆

111483 Crocus alatavicus Regel et Semen. ;白番红花(阿拉套番红花);Alata Crocus,Alata Mountain Crocus ■

111484 Crocus ancyrensis Maw;安塞里番红花;Golden Bunch ■☆

111485 Crocus angustifolius Weston;狭叶番红花■☆

111486 Crocus angustifolius Weston = Crocus susianus Ker Gawl. ■☆

111487 Crocus artvinensis (I. Phil.) Grossh. ;阿尔特温番红花■☆

111488 Crocus asturicus Herb. ;西班牙番红花■☆

111489 Crocus atlanticus Pomel = Crocus nevadensis Amo et Campo ■☆

111490 Crocus aureus Sibth. et Sm. = Crocus flavus Weston ■☆

111491 Crocus aureus Sibth. et Sm. = Crocus moesiacus Ker Gawl. ■☆

111492 Crocus autranii Albov;奥特番红花■☆

111493 Crocus balansae J. Gay = Crocus olivieri J. Gay subsp. balansae (J. Gay ex Maw) B. Mathew ■☆

111494 Crocus balansae J. Gay ex Maw = Crocus olivieri J. Gay subsp. balansae (J. Gay ex Maw) B. Mathew ■☆

111495 Crocus banaticus Heuff. ;匈牙利番红花■☆

111496 Crocus banaticus J. Gay = Crocus byzantinus Ker Gawl. ■☆

111497 Crocus baylopiorum B. Mathew;蓝色番红花■☆

111498 Crocus biflorus Mill. ; 双花番红花; Scotch Crocus, Silvery Crocus ■☆

111499 Crocus biflorus Mill. subsp. alexandri (Nicic ex Velen.) B. Mathew;亚历山大双花番红花■☆

111500 Crocus biflorus Mill. subsp. pulchricolor (Herb.) B. Mathew;艳色双花番红花■☆

111501 Crocus biflorus Mill. var. argenteus Sabine;银叶双花番红花■☆

111502 Crocus biflorus Mill. var. parkinsonii Sabine; 白金番红花; White Parkinson Crocus ■☆

111503 Crocus biflorus Mill. var. pusillus (Ten.) Baker = Crocus biflorus Mill. ■☆

111504 Crocus biflorus Mill. var. weldenii Baker;韦尔登番红花;Scotch Crocus,Slaty-purple Crocus ■☆

111505 Crocus boryi J. Gay;波里番红花■☆

111506 Crocus boulosii Greuter;布洛番红花■☆

111507 Crocus byzantinus Ker Gawl. ;比赞番红花■☆

111508 Crocus cancellatus Herb. ;格纹番红花■☆

111509 Crocus candidus Boiss. = Crocus fleischeri J. Gay ■☆

111510 Crocus candidus Clarke;纯白番红花■☆

111511 Crocus capensis Burm. f. = Romulea rosea (L.) Eckl. ■☆

111512 Crocus cartwrightianus Herb. ;卡氏番红花■☆

111513 Crocus cashmerianus Royle = Crocus sativus L. ■

111514 Crocus caspicus Fisch. et C. A. Mey. ;里海番红花■☆

111515 Crocus chrysanthus (Herb.) Herb. ;金黄番红花(菊黄);Golden Crocus,Golden-winter-crocus Crocus ■☆

111516 Crocus chrysanthus Herb. = Crocus chrysanthus (Herb.) Herb. ■☆

111517 Crocus chrysanthus Herb. subsp. multifolius Papan. et Zacharof;多叶番红花■☆

111518 Crocus chrysanthus Herb. var. blue-bird Hort. ;蓝鸟番红花;Blue-bird-crocus ■☆

111519 Crocus chrysanthus Herb. var. blupeter Hort. ;蓝彼得番红花;Blue Peter Crocus ■☆

111520 Crocus chrysanthus Herb. var. bowles Hort. ; 金盏番红花;Golden-bowl-crocus Crocus ■☆

111521 Crocus chrysanthus Herb. var. gibsy-girl Hort. ;吉普赛番红花;Gipsygirl Crocus ■☆

111522 Crocus chrysanthus Herb. var. ladykiller Hort. ;褐瓣番红花;Ladykiller Crocus ■☆

111523 Crocus clusii J. Gay = Crocus serotinus Salisb. subsp. clusii (J. Gay) Matthews ■☆

111524 Crocus clusii J. Gay var. mauritii Maire et Sennen = Crocus serotinus Salisb. subsp. clusii (J. Gay) Matthews ■☆

111525 Crocus clusii J. Gay var. xanthostylus Maire = Crocus serotinus Salisb. subsp. clusii (J. Gay) Matthews ■☆

111526 Crocus corsicus Vanucchi = Crocus corsicus Vanucchi ex Maw ■☆

111527 Crocus corsicus Vanucchi ex Maw;科尔西卡番红花■☆

111528 Crocus cvijici Kosanin;克氏番红花■☆

111529 Crocus dalmaticus Vis. ;达尔马特番红花■☆

111530 Crocus dianthus K. Koch = Crocus cancellatus Herb. ■☆

111531 Crocus etruscus Maw;冬番红花;Winter Crocus ■☆

111532 Crocus etruscus Maw 'Zwanenburg';茨仁堡冬番红花■☆

111533 Crocus fimbriatus Lapeyr. = Crocus nudiflorus Sm. ■☆

111534 Crocus flavus Haw. = Crocus flavus Weston ■☆

111535 Crocus flavus Weston; 鲜黄番红花; Golden Crocus, Yellow Crocus ■☆

111536 Crocus fleischeri J. Gay;弗氏番红花■☆

111537 Crocus gargaricus Herb. ;土耳其番红花■☆

111538 Crocus goulimyi Turrill;吉利米番红花■☆

111539 Crocus hadriaticus Herb. ;地中海番红花■☆

111540 Crocus heuffelianus Herb. ; 亥氏番红花; Heuffel Saffron, Heuffel's Saffron ■☆

111541 Crocus heuffelianus Herb. = Crocus banaticus Heuff. ■☆

111542 Crocus imperati Ten. ;艾佩雷特番红花(帝王番红花);Early Crocus,Imperate Crocus ■☆

111543 Crocus imperati Ten. 'De Jager';猎手帝王番红花■☆

111544 Crocus iridiflorus Heuff. ex Rchb. = Crocus banaticus J. Gay ■☆

111545 Crocus italicus Gaudin = Crocus biflorus Mill. ■☆

111546 Crocus karsianus Fomin;卡尔斯番红花■☆

111547 Crocus kirkii Maw = Crocus candidus Clarke ■☆

111548 Crocus korolkowii Maw et Regel;科洛番红花;Celandine Crocus ■☆

111549 Crocus kotschyanus K. Koch;科奇番红花;Kotchy's Crocus, Kotschy Crocus ■☆

111550 Crocus kotschyanus K. Koch = Crocus zonatus Gay ■☆

111551 Crocus laevigatus Bory et Chaub. ;平滑番红花■☆

111552 Crocus longiflorus Raf. ;长花番红花■☆

111553 Crocus maesiacus Ker Gawl. ;番黄花■☆

111554 Crocus malyi Vis. ;玛利番红花■☆

111555 Crocus medius Balb. ;中间番红花(裂柱番红花)■☆

111556 Crocus michelsonii B. Fedtsch. ;米氏番红花■☆
111557 Crocus minimus Ten. ;小番红花;Tiny Crocus ■☆
111558 Crocus moesiacus Ker Gawl. ;金番红花(番黄花,金黄番红花);Dutch Yellow Crocus,Goldenyellow Crocus,Yellow Crocus ■☆
111559 Crocus multifidus Ramond = Crocus nudiflorus Sm. ■☆
111560 Crocus multifidus Rchb. = Crocus speciosus M. Bieb. ■☆
111561 Crocus neapolitanus Ten. = Crocus imperati Ten. ■☆
111562 Crocus nevadensis Amo et Campo;内华达番黄花■☆
111563 Crocus nevadensis Amo et Campo var. atlanticus (Pomel) Pau et Font Quer = Crocus nevadensis Amo et Campo ■☆
111564 Crocus nivalis Bory et Chaub. = Crocus sieberi J. Gay ■☆
111565 Crocus niveus Bowles;洁净番红花■☆
111566 Crocus nudiflorus Sm. ;光花番红花(长管番红花);Autumn Crocus,Naked Boys,Naked Ladies,Saffron ■☆
111567 Crocus ochroleucus Boiss. et Gaillon;浅黄番红花■☆
111568 Crocus odorus Biv. = Crocus longiflorus Raf. ■☆
111569 Crocus officinalis Martyn = Crocus sativus L. ■
111570 Crocus olivieri J. Gay;奥氏番红花■☆
111571 Crocus olivieri J. Gay subsp. balansae (J. Gay ex Maw) B. Mathew;西亚番红花■☆
111572 Crocus orphanidis Hook. f. = Crocus tournefortii J. Gay ■☆
111573 Crocus pallasii Goldb. ;帕拉氏番红花;Pallas Saffron ■☆
111574 Crocus penicillatus Steud. = Crocus fleischeri J. Gay ■☆
111575 Crocus penicillatus Steud. ex Baker = Crocus fleischeri J. Gay ■☆
111576 Crocus pestalozzae Boiss. ;佩氏番红花■☆
111577 Crocus pulchellus Herb. ;艳丽番红花;Hairy Crocus ■☆
111578 Crocus pulchellus Herb. = Crocus speciosus M. Bieb. ■☆
111579 Crocus purpureus ? = Crocus vernus Wulfen ■☆
111580 Crocus pusillus Ten. = Crocus biflorus Mill. ■☆
111581 Crocus pyrenaeus Herb. = Crocus nudiflorus Sm. ■☆
111582 Crocus reticulatus Steven ex Adam;网脉番红花;Reticulate Saffron ■☆
111583 Crocus revolutus Haw. = Crocus susianus Ker Gawl. ■☆
111584 Crocus roopiae Woronow;鲁普番红花■☆
111585 Crocus salzmannii J. Gay = Crocus serotinus Salisb. subsp. salzmannii (J. Gay) Matthews ■☆
111586 Crocus salzmannii J. Gay = Crocus serotinus Salisb. ■☆
111587 Crocus salzmannii J. Gay var. coloratus Maire = Crocus serotinus Salisb. subsp. salzmannii (J. Gay) Matthews ■☆
111588 Crocus salzmannii J. Gay var. pallidus Maire = Crocus serotinus Salisb. subsp. salzmannii (J. Gay) Matthews ■☆
111589 Crocus sativus L. ;番红花(泊夫蓝,藏红花,蕃红花,红花,撒馥兰,西藏红花,西红花,咱法兰,咱夫兰);Crocus,Safforne,Saffron,Saffron Crocus ■
111590 Crocus scharojanii Rupr. ;沙罗番红花■☆
111591 Crocus schimperi J. Gay = Crocus cancellatus Herb. ■☆
111592 Crocus schimperi J. Gay ex Baker = Crocus cancellatus Herb. ■☆
111593 Crocus serotinus Bert. ? = Crocus longiflorus Raf. ■☆
111594 Crocus serotinus Salisb. ;迟番红花;Salzmann Crocus ■☆
111595 Crocus serotinus Salisb. subsp. clusii (J. Gay) Matthews;克鲁斯番红花(克路西番红花);Cluse Crocus ■☆
111596 Crocus serotinus Salisb. subsp. salzmannii (J. Gay) Matthews;索尔曼番红花;Salzmann Crocus ■☆
111597 Crocus sieberi J. Gay;西伯番红花;Sieber Crocus,Sieber's Crocus ■☆
111598 Crocus sieberi J. Gay 'Bowles White';鲍威尔斯白西伯番红花■☆
111599 Crocus sieberi J. Gay 'Hubert Edelsten';休伯特宝石西伯番红花■☆
111600 Crocus sieberi J. Gay subsp. atticus (Boiss. et Orph.) B. Mathew;雅致西伯番红花■☆
111601 Crocus smyrnensis Poech = Crocus fleischeri J. Gay ■☆
111602 Crocus speciosus Griseb. = Crocus pulchellus Herb. ■☆
111603 Crocus speciosus M. Bieb. ;美丽番红花;Bieberstein's Crocus,Pretty Crocus ■☆
111604 Crocus speciosus M. Bieb. 'Conqueror';征服者美丽番红花■☆
111605 Crocus speciosus M. Bieb. 'Oxinian';牛津人美丽番红花■☆
111606 Crocus speciosus M. Bieb. var. aitchisonii Hort. ;大花美丽番红花;Largeflower Pretty Crocus ■☆
111607 Crocus speciosus Wilson = Crocus nudiflorus Sm. ■☆
111608 Crocus spruneri Boiss. = Crocus cancellatus Herb. ■☆
111609 Crocus spruneri Boiss. et Heldr. = Crocus cancellatus Herb. ■☆
111610 Crocus stellaris Haw. ;星形番红花■☆
111611 Crocus sublimis Herb. = Crocus sieberi J. Gay ■☆
111612 Crocus susianus Ker Gawl. ;苏萨番红花(高加索番红花,金线番红花,狭叶番红花);Cloth of Gold,Cloth of Gold Crocus,Cloth-of-gold-crocus ■☆
111613 Crocus susianus Ker Gawl. = Crocus angustifolius Weston ■☆
111614 Crocus suworowianus K. Koch;苏氏番红花■☆
111615 Crocus tauricus Steven ex Nyman;克里木番红花■☆
111616 Crocus tingitanus Herb. = Crocus salzmannii J. Gay ■☆
111617 Crocus tomasinianus Herb. ;托马西尼番红花;Early Crocus,Tomasinini Crocus,Woodland Crocus ■☆
111618 Crocus tomasinianus Herb. 'Ruby Giant';红宝石托马西尼番红花■☆
111619 Crocus tomasinianus Herb. 'Whitewell Purple';怀城紫托马西尼番红花■☆
111620 Crocus tomasinianus Herb. var. ruby-giant Hort. ;红宝石番红花;Ruby Giangt Crocus ■☆
111621 Crocus tournefortii J. Gay;图氏番红花(希腊番红花)■☆
111622 Crocus triflorus Burm. f. = Romulea triflora (Burm. f.) N. E. Br. ■☆
111623 Crocus vallicola Herb. ;谷番红花■☆
111624 Crocus variegatus Hooper et Hornem. ;花斑番红花■☆
111625 Crocus vernus (L.) Hill;春番红花(春郁金香,春郁香,番紫花,荷兰番红花);Common Crocus,Dutch Crocus,Spring Crocus,Spring Saffron ■☆
111626 Crocus vernus Wulfen 'Jeanned Arc';弧光春番红花■☆
111627 Crocus vernus Wulfen 'Pickwick';匹克威克春番红花■☆
111628 Crocus vernus Wulfen 'Prinses Juliana';朱丽公主春番红花■☆
111629 Crocus vernus Wulfen 'Purpureus Grandiflorus';大紫花春番红花■☆
111630 Crocus vernus Wulfen 'Queen of the Blues';布鲁斯皇后春番红花■☆
111631 Crocus vernus Wulfen 'Remembrance';回想春番红花■☆
111632 Crocus vernus Wulfen 'Vabguard';先锋春番红花■☆
111633 Crocus vernus Wulfen = Crocus vernus (L.) Hill ■☆
111634 Crocus versicolor Barcelo;变色番红花;Versicolorous Crocus ■☆
111635 Crocus zonatus Gay;虎斑番红花■☆
111636 Crocus zonatus Gay = Crocus kotschyanus K. Koch ■☆
111637 Crocyllis E. Mey. ex Hook. f. (1873);南非茜属●☆
111638 Crocyllis anthospermoides E. Mey. ex K. Schum. ;南非茜●☆
111639 Crocyllis intricatissima Dinter = Gaillonia crocyllis (Sond.) Thulin ■☆

111640 Crodisperrna Poit. ex Cass. = Wulffia Neck. ex Cass. ■☆
111641 Croftia King et Prain = Pommereschea Wittm. ■
111642 Croftia Small = Carlowrightia A. Gray(保留属名)☆
111643 Croftia spectabilis King et Prain = Pommereschea spectabilis (King et Prain) K. Schum. ■
111644 Croixia Pierre = Palaquium Blanco ●
111645 Croizatia Steyerm. (1952);克罗大戟属☆
111646 Croizatia neotropica Steyerm.;克罗大戟☆
111647 Crolocos Raf. = Salvia L. ●■
111648 Crolocos Raf. = Stiefia Medik. ●■
111649 Cromapanax Grierson. (1991);克罗参属☆
111650 Cromapanax lobatus Grierson;克罗参☆
111651 Cromidon Compton(1931);苞萼玄参属■☆
111652 Cromidon confusum Hilliard;混乱苞萼玄参■☆
111653 Cromidon corrigioloides (Rolfe) Compton;互叶指甲草苞萼玄参■☆
111654 Cromidon decumbens (Thunb.) Hilliard;外倾苞萼玄参■☆
111655 Cromidon dregei Hilliard;德雷苞萼玄参■☆
111656 Cromidon gracile Hilliard;纤细苞萼玄参■☆
111657 Cromidon hamulosum (E. Mey.) Hilliard;具钩苞萼玄参■☆
111658 Cromidon microechinos Hilliard;小刺苞萼玄参■☆
111659 Cromidon minutum (Rolfe) Hilliard;小苞萼玄参■☆
111660 Cromidon plantaginis (L. f.) Hilliard;车前苞萼玄参■☆
111661 Cromidon pusillum (Rössler) Hilliard;微小苞萼玄参■☆
111662 Cromidon varicalyx Hilliard;杂萼苞萼玄参■☆
111663 Croninia J. M. Powell(1857);沙鞭石南属●☆
111664 Croninia kingiana (F. Muell.) J. M. Powell;沙鞭石南●☆
111665 Cronquistia R. M. King = Carphochaete A. Gray ■☆
111666 Cronquistia R. M. King(1968);长芒菊属■☆
111667 Cronquistia pringlei (S. Watson) R. M. King;长芒菊■☆
111668 Cronquistianthus R. M. King et H. Rob. (1972);圆苞亮泽兰属●☆
111669 Cronquistianthus leucophyllus (Kunth) R. M. King et H. Rob.;白叶圆苞亮泽兰●☆
111670 Cronquistianthus niveus (Kunth) R. M. King et H. Rob.;圆苞亮泽兰●☆
111671 Cronquistianthus rugosus (Kunth) R. M. King et H. Rob.;褶皱圆苞亮泽兰●☆
111672 Cronyxium Raf. = Lloydia Salisb. ex Rchb. (保留属名)■
111673 Crookea Small = Hypericum L. ■●
111674 Croomia Torr. (1840);黄精叶钩吻属(金刚大属);Croomia ■
111675 Croomia Torr. ex Torr. et A. Gray = Croomia Torr. ■
111676 Croomia heterosepala (Baker) Okuyama;异萼黄精叶钩吻■☆
111677 Croomia japonica Miq.;黄精叶钩吻(金刚大);Japan Croomia, Japanese Croomia ■
111678 Croomia pauciflora (Nutt.) Torr.;少花黄精叶钩吻■☆
111679 Croomiaceae Nakai = Stemonaceae Caruel(保留科名)■
111680 Croomiaceae Nakai;黄精叶钩吻科(金刚大科)■
111681 Croptilon Raf. (1837);划雏菊属;Scratchdaisy ■☆
111682 Croptilon Raf. = Haplopappus Cass. (保留属名)■●☆
111683 Croptilon divaricatum (Nutt.) Raf.;纤细划雏菊;Slender Scratchdaisy ■☆
111684 Croptilon divaricatum (Nutt.) Raf. var. graniticum (E. B. Sm.) Shinners = Croptilon hookerianum (Torr. et A. Gray) House var. graniticum (E. B. Sm.) E. B. Sm. ■☆
111685 Croptilon divaricatum (Nutt.) Raf. var. hirtellum (Shinners) Shinners = Croptilon rigidifolium (E. B. Sm.) E. B. Sm. ■☆
111686 Croptilon divaricatum (Nutt.) Raf. var. hookerianum (Torr. et A. Gray) Shinners = Croptilon hookerianum (Torr. et A. Gray) House ■☆
111687 Croptilon hookerianum (Torr. et A. Gray) House;胡克划雏菊(虎克划雏菊);Hooker's Scratchdaisy ■☆
111688 Croptilon hookerianum (Torr. et A. Gray) House var. graniticum (E. B. Sm.) E. B. Sm.;花岗岩划雏菊■☆
111689 Croptilon hookerianum (Torr. et A. Gray) House var. validum (Rydb.) E. B. Sm.;强壮划雏菊■☆
111690 Croptilon rigidifolium (E. B. Sm.) E. B. Sm.;硬叶划雏菊;Stiff-leaf Scratchdaisy ■☆
111691 Crosapila Raf. = Chaerophyllum L. ■
111692 Crosperma Raf. = Amianthium A. Gray(保留属名)■☆
111693 Crossandra Salisb. (1805);十字爵床属(半边黄属,鸟尾花属);Crossandra ●
111694 Crossandra acutiloba Vollesen;尖裂十字爵床■☆
111695 Crossandra afromontana Mildbr. = Stenandrium afromontanum (Mildbr.) Vollesen ■☆
111696 Crossandra albolineata Benoist;白线十字爵床■☆
111697 Crossandra angolensis S. Moore;安哥拉十字爵床■☆
111698 Crossandra arenicola Vollesen;沙生十字爵床■☆
111699 Crossandra armandii Benoist;阿尔芒十字爵床■☆
111700 Crossandra axillaris Nees = Crossandra infundibuliformis (L.) Nees ●
111701 Crossandra baccarinii Fiori;巴卡林十字爵床■☆
111702 Crossandra benoistii Vollesen;本诺十字爵床■☆
111703 Crossandra brachstachys (Franch.) Lindau = Crossandra infundibuliformis (L.) Nees subsp. brachystachys (Franch.) Napper ●☆
111704 Crossandra buntingii S. Moore = Stenandrium buntingii (S. Moore) Vollesen ■☆
111705 Crossandra cephalostachya Mildbr. = Crossandra tridentata Lindau ■☆
111706 Crossandra cinnabarina Vollesen;朱红十字爵床●☆
111707 Crossandra citrina Benoist var. pilosa Benoist = Crossandra pilosa (Benoist) Vollesen ●☆
111708 Crossandra citrina Benoist var. subintegra Benoist = Crossandra strobilifera (Lam.) Benoist ●☆
111709 Crossandra cloiselii S. Moore;克卢塞尔十字爵床■☆
111710 Crossandra cloiselii S. Moore var. brevis Benoist = Crossandra strobilifera (Lam.) Benoist ●☆
111711 Crossandra crocea S. Moore = Crossandra infundibuliformis (L.) Nees subsp. crocea (S. Moore) Napper ●☆
111712 Crossandra elatior S. Moore = Stenandrium guineense (Nees) Vollesen ■☆
111713 Crossandra flava Hook.;黄十字爵床;Yellow Crossandra, Yellow-flowered Crossandra ●☆
111714 Crossandra flavicaulis Vollesen;黄茎十字爵床●☆
111715 Crossandra friesiorum Mildbr.;弗里斯十字爵床●☆
111716 Crossandra fruticulosa Lindau;灌木状十字爵床●☆
111717 Crossandra gabonica Benoist = Stenandrium gabonicum (Benoist) Vollesen ■☆
111718 Crossandra gossweileri S. Moore = Stenandrium gabonicum (Benoist) Vollesen ■☆
111719 Crossandra grandidieri (Baill.) Benoist;格朗十字爵床●☆
111720 Crossandra greenstockii S. Moore;格林斯十字爵床●☆
111721 Crossandra guineensis Nees = Stenandrium guineense (Nees)

Vollesen ■☆
111722 Crossandra horrida Vollesen;多刺十字爵床●☆
111723 Crossandra humbertii Benoist;亨伯特十字爵床●☆
111724 Crossandra infundibuliformis (L.) Nees;十字爵床(半边黄,橙色单药花,鸟尾花);Crossandra,Fire Cracker Flower,Firecracker Flower,Funnel-shaped Crossandra ●
111725 Crossandra infundibuliformis (L.) Nees subsp. boranensis Vollesen;博兰十字爵床●☆
111726 Crossandra infundibuliformis (L.) Nees subsp. brachystachys (Franch.) Napper;短穗十字爵床●☆
111727 Crossandra infundibuliformis (L.) Nees subsp. crocea (S. Moore) Napper;镉黄十字爵床●☆
111728 Crossandra infundibuliformis (L.) Nees subsp. eglandulosa Vollesen;无腺十字爵床●☆
111729 Crossandra infundibuliformis (L.) Nees var. brachystachys Franch. = Crossandra infundibuliformis (L.) Nees subsp. brachystachys (Franch.) Napper ●☆
111730 Crossandra infundibuliformis Nees = Crossandra infundibuliformis (L.) Nees ●
111731 Crossandra isaloensis Vollesen;伊萨卢十字爵床●☆
111732 Crossandra jashii Lindau = Crossandra puberula Klotzsch ●☆
111733 Crossandra leucodonta Vollesen;白齿十字爵床■☆
111734 Crossandra longehirsuta Vollesen;长毛十字爵床■☆
111735 Crossandra longipes S. Moore;长梗十字爵床●☆
111736 Crossandra longispica Benoist;长穗十字爵床●☆
111737 Crossandra madagascariensis T. Anderson = Crossandra stenandrium (Nees) Lindau ●☆
111738 Crossandra massaica Mildbr.;马萨十字爵床●☆
111739 Crossandra mucronata Lindau;钝尖十字爵床■☆
111740 Crossandra multidentata Vollesen;多齿十字爵床●☆
111741 Crossandra nilotica Oliv.;尼罗河十字爵床(黄鸟尾花,尼罗鸟尾花);Nile Crossandra ●☆
111742 Crossandra nilotica Oliv. subsp. massaica (Mildbr.) Napper = Crossandra massaica Mildbr. ●☆
111743 Crossandra nilotica Oliv. var. acuminata C. B. Clarke = Crossandra infundibuliformis (L.) Nees subsp. brachystachys (Franch.) Napper ●☆
111744 Crossandra nilotica Oliv. var. acuminata Lindau = Crossandra nilotica Oliv. ●☆
111745 Crossandra nobilis Benoist;名贵十字爵床●☆
111746 Crossandra obanensis Heine;奥班十字爵床●☆
111747 Crossandra parvifolia Lindau = Crossandra spinosa Beck ■☆
111748 Crossandra pilosa (Benoist) Vollesen;疏毛十字爵床●☆
111749 Crossandra pinguior S. Moore;肥厚十字爵床●☆
111750 Crossandra poissonii Benoist;普瓦松十字爵床●☆
111751 Crossandra praecox Vollesen;早十字爵床■☆
111752 Crossandra puberula Klotzsch;微毛十字爵床●☆
111753 Crossandra puberula Klotzsch var. smithii C. B. Clarke = Crossandra massaica Mildbr. ●☆
111754 Crossandra pubescens Klotzsch = Crossandra puberula Klotzsch ●☆
111755 Crossandra pungens Lindau;辛辣十字爵床;Pungent Crossandra ●☆
111756 Crossandra pyrophila Vollesen;喜炎十字爵床■☆
111757 Crossandra quadridentata Benoist;四齿十字爵床■☆
111758 Crossandra raripila Benoist;稀毛十字爵床●☆
111759 Crossandra rhynchocarpa (Klotzsch) Cufod. = Crossandra nilotica Oliv. ●☆
111760 Crossandra rupestris Benoist;岩生十字爵床●☆
111761 Crossandra smithii S. Moore = Crossandra nilotica Oliv. ●☆
111762 Crossandra spinescens Dunkley;小刺十字爵床■☆
111763 Crossandra spinosa Beck;大刺十字爵床■☆
111764 Crossandra stenandrium (Nees) Lindau;狭蕊十字爵床■☆
111765 Crossandra stenostachya (Lindau) C. B. Clarke;细穗十字爵床●☆
111766 Crossandra stenostachya (Lindau) C. B. Clarke var. somalensis Fiori = Crossandra stenostachya (Lindau) C. B. Clarke ●☆
111767 Crossandra strobilifera (Lam.) Benoist;球果十字爵床■☆
111768 Crossandra strobilifera (Lam.) Benoist var. brevis (Benoist) Benoist = Crossandra strobilifera (Lam.) Benoist ●☆
111769 Crossandra subacaulis C. B. Clarke;近无茎十字爵床■☆
111770 Crossandra sulphurea Lindau;硫黄十字爵床■☆
111771 Crossandra talbotii S. Moore = Stenandrium talbotii (S. Moore) Vollesen ■☆
111772 Crossandra thomensis Milne-Redh. = Stenandrium thomense (Milne-Redh.) Vollesen ■☆
111773 Crossandra tridentata Lindau;三齿十字爵床■☆
111774 Crossandra undulifolia Salisb.;台湾十字爵床(鸟尾花草);Taiwan Crossandra ●
111775 Crossandra undulifolia Salisb. = Crossandra infundibuliformis (L.) Nees ●
111776 Crossandra usambarensis Mildbr. = Stenandrium warneckei (S. Moore) Vollesen ■☆
111777 Crossandra vestita Benoist;包被十字爵床●☆
111778 Crossandra warneckei S. Moore = Stenandrium warneckei (S. Moore) Vollesen ■☆
111779 Crossandra zuluensis W. T. Vos et T. J. Edwards;祖卢十字爵床■☆
111780 Crossandrella C. B. Clarke(1906);小十字爵床属■☆
111781 Crossandrella adamii Heine;阿达姆小十字爵床■☆
111782 Crossandrella dusenii (Lindau) S. Moore;小十字爵床■☆
111783 Crossandrella laxispicata C. B. Clarke = Crossandrella dusenii (Lindau) S. Moore ■☆
111784 Crossangis Schltr. = Diaphananthe Schltr. ■☆
111785 Crosslandia W. Fitzg. (1918);克罗莎草属■☆
111786 Crosslandia setifolia W. Fitzg.;克罗莎草■☆
111787 Crossocephalum Britten(1901);须头草属■☆
111788 Crossocephalum crepidioides (Benth.) S. Moore;须头草■☆
111789 Crossocoma Hook. = Crossosoma Nutt. ●☆
111790 Crossoglossa Dressler et Dodson = Microstylis (Nutt.) Eaton (保留属名)■☆
111791 Crossolepis Benth. = Angianthus J. C. Wendl. (保留属名)■●☆
111792 Crossolepis Benth. = Gnephosis Cass. ■☆
111793 Crossolepis Less. = Gnephosis Cass. ■☆
111794 Crossolepsis N. T. Burb. = Crossolepis Less. ■☆
111795 Crossoliparis Marg. = Liparis Rich. (保留属名)■
111796 Crossonephelis Baill. = Glenniea Hook. f. ●☆
111797 Crossonephelis adamii Fouilloy = Glenniea adamii (Fouilloy) Leenh. ●☆
111798 Crossonephelis africanus (Radlk.) Leenh. = Glenniea africana (Radlk.) Leenh. ●☆
111799 Crossonephelis oblongus Capuron ex Fouilloy = Glenniea africana (Radlk.) Leenh. ●☆
111800 Crossonephelis unijugatus (Pellegr.) Leenh. = Glenniea

unijugata (Pellegr.) Leenh. ●☆
111801 Crossopetalon Adans. = Crossopetalum P. Browne ●☆
111802 Crossopetalum P. Browne(1756);缨瓣属●☆
111803 Crossopetalum Roth = Gentiana L. ■
111804 Crossopetalum Roth = Gentianopsis Ma ■
111805 Crossopetalum gaumeri (Loes.) Lundell;高梅缨瓣●☆
111806 Crossophora Link = Chrozophora A. Juss.(保留属名)●
111807 Crossophrys Klotzsch = Clethra Gronov. ex L. ●
111808 Crossopteryx Fenzl(1839);缨翼茜属■☆
111809 Crossopteryx africana Baill. = Crossopteryx febrifuga (Afzel. ex G. Don) Benth. ■☆
111810 Crossopteryx africana Baill. = Crossopteryx kotschyana Fenzl ■☆
111811 Crossopteryx africana K. Schum. = Crossopteryx kotschyana Fenzl ■☆
111812 Crossopteryx febrifuga (Afzel. ex G. Don) Benth.;解热缨翼茜●☆
111813 Crossopteryx febrifuga Benth. = Crossopteryx febrifuga (Afzel. ex G. Don) Benth. ■☆
111814 Crossopteryx kotschyana Fenzl;科奇缨翼茜■☆
111815 Crossopteryx kotschyana Fenzl = Crossopteryx febrifuga (Afzel. ex G. Don) Benth. ■☆
111816 Crossosoma Nutt.(1848);流苏亮籽属(穗子属)●☆
111817 Crossosoma bigelovii S. Watson;毕氏流苏亮籽;Ragged Rock Flower ●☆
111818 Crossosoma californica Nutt.;流苏亮籽●☆
111819 Crossosomataceae Engl.(1897)(保留科名);流苏亮籽科(燧体木科)●☆
111820 Crossosperma T. G. Hartley(1997);茎花芸香属●☆
111821 Crossostemma Planch. ex Benth.(1849);十字西番莲属●☆
111822 Crossostemma laurifolium Planch. ex Benth.;十字西番莲●☆
111823 Crossostephium Less.(1831);芙蓉菊属(蕲艾属,千年艾属,玉芙蓉属);Crossostephium,Lotusdaisy ●
111824 Crossostephium artemisioides Less. = Crossostephium chinense (L.) Makino ●
111825 Crossostephium californicum (Less.) Rydb. = Artemisia californica Less. ●☆
111826 Crossostephium chinense (L.) Makino;芙蓉菊(白艾,白香菊,芙蓉,老人花,蕲艾,千年艾,香菊,玉芙蓉);China Lotusdaisy,Chinese Crossostephium,Moxa ●
111827 Crossostephium insulare Rydb. = Artemisia nesiotica P. H. Raven ■
111828 Crossostigma Spach = Epilobium L. ■
111829 Crossostigma Spach(1835);缨柱柳叶菜属■☆
111830 Crossostigma lindleyi Spach;缨柱柳叶菜■☆
111831 Crossostoma Spach = Scaevola L.(保留属名)●■
111832 Crossostyles Benth. et Hook. f. = Crossostylis J. R. Forst. et G. Forst. ●☆
111833 Crossostylis J. R. Forst. et G. Forst.(1775);缨柱红树属●☆
111834 Crossostylis biflora J. R. Forst. et G. Forst.;双花缨柱红树●☆
111835 Crossostylis grandiflora Brongn. et Gris;大花缨柱红树●☆
111836 Crossostylis multiflora Brongn. et Gris ex Pancher et Sebert;多花缨柱红树●☆
111837 Crossothamnus R. M. King et H. Rob.(1972);腺果修泽兰属(缨灌菊属)●☆
111838 Crossothamnus weberbaueri (Hieron.) R. M. King et H. Rob.;腺果修泽兰●☆
111839 Crossotoma (G. Don) Spach = Trichoneura Andersson ■☆
111840 Crossotropis Stapf = Trichoneura Andersson ■☆
111841 Crossotropis arenaria (Hochst. et Steud. ex Steud.) Rendle = Trichoneura mollis (Kunth) Ekman ■☆
111842 Crossotropis eleusinoides Rendle = Trichoneura eleusinoides (Rendle) Ekman ■☆
111843 Crossotropis grandiglumis (Nees) Rendle var. minor Rendle = Trichoneura grandiglumis (Nees) Ekman var. minor (Rendle) Chippind. ■☆
111844 Crossyne Salisb.(1866);红斑石蒜属■☆
111845 Crossyne Salisb. = Boophone Herb. ■☆
111846 Crossyne Salisb. = Buphane Herb. ■☆
111847 Crossyne ciliaris Salisb.;缘毛红斑石蒜■☆
111848 Crossyne flava (W. F. Barker ex Snijman) D. Müll. -Doblies et U. Müll. -Doblies;黄花红斑石蒜(黄斑石蒜)■☆
111849 Crossyne guttata (L.) D. Müll. -Doblies et U. Müll. -Doblies;红斑石蒜■☆
111850 Crotalaria Dill. ex L. = Crotalaria L. ●■
111851 Crotalaria L.(1753);猪屎豆属(野百合属);Crotalaria,Rattlebox,Rattle-box,Rattlepod ●■
111852 Crotalaria abbreviata Baker f.;缩短猪屎豆●☆
111853 Crotalaria abscondita Welw. ex Baker;隐匿猪屎豆●☆
111854 Crotalaria abyssinica D. Dietr. = Crotalaria impressa Nees ex Walp. ■☆
111855 Crotalaria acervata Baker f. = Crotalaria subcapitata De Wild. ■☆
111856 Crotalaria acicularis Buch. -Ham. ex Benth.;针状猪屎豆(圆叶野百合,针叶状铃豆,针状野百合,针状叶猪屎豆);Needlelike Rattlebox,Needleshaped Rattlebox ■
111857 Crotalaria aculeata De Wild.;皮刺猪屎豆●☆
111858 Crotalaria aculeata De Wild. subsp. claessensii (De Wild.) Polhill;克莱森斯猪屎豆■☆
111859 Crotalaria aculeata De Wild. var. claessensii (De Wild.) R. Wilczek = Crotalaria aculeata De Wild. subsp. claessensii (De Wild.) Polhill ■☆
111860 Crotalaria acuminata DC. = Bolusia acuminata (DC.) Polhill ■☆
111861 Crotalaria acuminata G. Don = Crotalaria verrucosa L. ■
111862 Crotalaria adamii R. Wilczek;亚当斯猪屎豆■☆
111863 Crotalaria adamsonii Baker f.;亚当森猪屎豆■☆
111864 Crotalaria adenocarpoides Taub.;腺果猪屎豆■☆
111865 Crotalaria aegyptiaca Benth.;埃及猪屎豆■☆
111866 Crotalaria africana Buscal. et Muschl. = Crotalaria oocarpa Baker subsp. microcarpa Milne-Redh. ■☆
111867 Crotalaria afrocentralis Polhill;中非猪屎豆■☆
111868 Crotalaria agatiflora Schweinf.;金花猪屎豆(田菁猪屎豆);Bird Flower,Canary Bird Bush,Canary-bird Bush,Lion's Claw,Lion's Claws ●☆
111869 Crotalaria agatiflora Schweinf. subsp. engleri (Harms ex Baker) Polhill;恩格勒猪屎豆■☆
111870 Crotalaria agatiflora Schweinf. subsp. erlangeri Baker f.;厄兰格猪屎豆■☆
111871 Crotalaria agatiflora Schweinf. subsp. imperialis (Taub.) Polhill;壮丽猪屎豆■☆
111872 Crotalaria agatiflora Schweinf. subsp. vaginifera Polhill;具鞘猪屎豆■☆
111873 Crotalaria akoensis Hayata = Crotalaria chinensis L. ■
111874 Crotalaria alata Buch. -Ham. ex D. Don;翅托叶猪屎豆(翅托叶野百合,翼柄野百合);Wingstipule Rattlebox ■
111875 Crotalaria alata Buch. -Ham. ex Roxb. = Crotalaria alata Buch. -

Ham. ex D. Don ■

111876 Crotalaria alata H. Lév. = Crotalaria alata Buch. -Ham. ex D. Don ■

111877 Crotalaria albertiana Baker f. = Crotalaria brevidens Benth. var. intermedia (Kotschy) Polhill ■☆

111878 Crotalaria albicaulis Franch. ;白茎响铃豆■☆

111879 Crotalaria albida K. Heyne ex Roth;响铃豆(摆子药,狗响铃,猴丝草,黄花地丁,马口铃,响铃草,小响铃,野豌豆);Diabolo Rattlebox,Whitish Rattlebox ■

111880 Crotalaria albida K. Heyne ex Roth var. gengmaensis (Z. Wei et C. Y. Yang) C. Chen et J. Q. Li;耿马猪屎豆;Gengma Rattlebox ■

111881 Crotalaria alexandri Baker f. ;亚历山大猪屎豆■☆

111882 Crotalaria alleniii I. Verd. = Crotalaria orientalis Burtt Davy ex I. Verd. subsp. allenii (I. Verd.) Polhill et A. Schreib. ■☆

111883 Crotalaria allophylla Thulin;无叶猪屎豆●☆

111884 Crotalaria alticola Polhill;高原猪屎豆●☆

111885 Crotalaria amadiensis De Wild. = Crotalaria glauca Willd. ■☆

111886 Crotalaria amoena Welw. ex Baker;秀丽猪屎豆●☆

111887 Crotalaria amplexicaulis L. = Rafnia amplexicaulis (L.) Thunb. ■☆

111888 Crotalaria anagyroides Kunth = Crotalaria micans Link ■

111889 Crotalaria andringitrensis R. Vig. ;安德林吉特拉山猪屎豆●☆

111890 Crotalaria androyensis R. Vig. ;安德罗猪屎豆●☆

111891 Crotalaria angulicaulis Harms;棱茎猪屎豆●☆

111892 Crotalaria angulosa Lam. = Crotalaria verrucosa L. ■

111893 Crotalaria angustissima E. Mey. = Crotalaria excisa (Thunb.) Baker f. subsp. namaquensis Polhill ●☆

111894 Crotalaria anisophylla (Hiern) Welw. ex Baker f. ;异叶猪屎豆●☆

111895 Crotalaria ankaizinensis R. Vig. ;安凯济纳猪屎豆●☆

111896 Crotalaria ankaranensis M. Pelt. = Crotalaria bernieri Baill. ■☆

111897 Crotalaria ankaratrana R. Vig. ;安卡拉特拉猪屎豆●☆

111898 Crotalaria anningensis X. Y. Zhu et Y. F. Du;安宁猪屎豆■

111899 Crotalaria annua Milne-Redh. ;一年猪屎豆●☆

111900 Crotalaria anomala R. Vig. ;异常猪屎豆●☆

111901 Crotalaria anthylloides D. Don = Crotalaria calycina Schrank ■

111902 Crotalaria anthyllopsis Welw. ex Baker;绒毛花猪屎豆●☆

111903 Crotalaria antunesii Baker f. ;安图内思猪屎豆●☆

111904 Crotalaria apiculata Schinz = Crotalaria platysepala Harv. ■☆

111905 Crotalaria arcuata Polhill;拱猪屎豆●☆

111906 Crotalaria arenaria Benth. ;沙地猪屎豆●☆

111907 Crotalaria argentea Jacq. = Argyrolobium argenteum (Jacq.) Eckl. et Zeyh. ●☆

111908 Crotalaria argenteotomentosa R. Wilczek;白绒毛猪屎豆●☆

111909 Crotalaria argenteotomentosa R. Wilczek subsp. dolosa Polhill;假白绒毛猪屎豆●☆

111910 Crotalaria argyraea Welw. ex Baker;银色猪屎豆●☆

111911 Crotalaria argyrolobioides Baker;银豆猪屎豆●☆

111912 Crotalaria arushae Milne-Redh. ex Polhill;阿鲁沙猪屎豆●☆

111913 Crotalaria arvensis Klotzsch = Crotalaria podocarpa DC. ■☆

111914 Crotalaria aspalathoides Lam. ;芳香木猪屎豆●☆

111915 Crotalaria assamica Benth. ;大猪屎豆(大金不换,大猪屎青,马铃根,山豆根,十字珍珠草,通心草,通心蓉,凸尖野百合,凸类野百合,野靛叶,自消容,自消融);Assam Crotalaria,Assam Rattlebox,Indian Rattlebox ■

111916 Crotalaria assurgens Polhill;上升猪屎豆●☆

111917 Crotalaria athroophylla I. Verd. = Crotalaria pallidicaulis Harms ■☆

111918 Crotalaria atrorubens Hochst. ex Benth. ;暗红猪屎豆●☆

111919 Crotalaria aurantiaca Baker = Crotalaria laburnifolia L. subsp. petiolaris (Franch.) Polhill ■☆

111920 Crotalaria aurea Dinter ex Baker f. ;黄猪屎豆●☆

111921 Crotalaria australis (Baker f.) Baker f. ex I. Verd. = Crotalaria laburnifolia L. subsp. australis (Baker f.) Polhill ■☆

111922 Crotalaria axillaris Aiton;腋生猪屎豆●☆

111923 Crotalaria axilliflora Baker f. ;腋花猪屎豆●☆

111924 Crotalaria axillifloroides Baker f. ex R. Wilczek;拟腋花猪屎豆●☆

111925 Crotalaria axillifloroides Baker f. ex R. Wilczek var. gracilis R. Wilczek = Crotalaria subtilis Polhill ●☆

111926 Crotalaria azaisii Sacleux = Crotalaria quartiniana A. Rich. ■☆

111927 Crotalaria azurea Eckl. et Zeyh. = Lotononis azurea (Eckl. et Zeyh.) Benth. ■☆

111928 Crotalaria bagamoyoensis Baker f. = Crotalaria laburnoides Klotzsch ■☆

111929 Crotalaria bakeriana Rossberg;贝克猪屎豆☆

111930 Crotalaria ballyi Polhill;博利猪屎豆●☆

111931 Crotalaria bamendae Hepper;巴门达猪屎豆●☆

111932 Crotalaria barkae Schweinf. ;巴克猪屎豆●☆

111933 Crotalaria barkae Schweinf. subsp. cordisepala Polhill;心萼猪屎豆●☆

111934 Crotalaria barkae Schweinf. subsp. teitensis (Sacleux) Polhill;泰塔猪屎豆●☆

111935 Crotalaria barkae Schweinf. subsp. zimmermannii (Baker f.) Polhill;齐默尔曼猪屎豆●☆

111936 Crotalaria barnabassii Dinter ex Baker f. var. cunenensis Torre = Crotalaria ulbrichiana Harms ●☆

111937 Crotalaria baumii Harms;鲍姆猪屎豆☆

111938 Crotalaria becquetii R. Wilczek;贝凯猪屎豆●☆

111939 Crotalaria becquetii R. Wilczek subsp. turgida Polhill;膨胀猪屎豆☆

111940 Crotalaria belckii Schinz = Crotalaria podocarpa DC. ■☆

111941 Crotalaria benadirensis Chiov. ;贝纳迪尔猪屎豆●☆

111942 Crotalaria benghalensis Lam. = Crotalaria juncea L. ■

111943 Crotalaria benguellensis Baker f. ;本格拉猪屎豆■☆

111944 Crotalaria benguellensis Baker f. var. bailundensis Torre = Crotalaria benguellensis Baker f. ■☆

111945 Crotalaria beniensis De Wild. = Crotalaria recta Steud. ex A. Rich. ■☆

111946 Crotalaria bequaertii Baker f. ;贝卡尔猪屎豆■☆

111947 Crotalaria bequaertii Baker f. var. pubescens R. Wilczek = Crotalaria bequaertii Baker f. ■☆

111948 Crotalaria bernieri Baill. ;伯尼尔猪屎豆■☆

111949 Crotalaria berteriana DC. ;伯氏猪屎豆;Berteron's Rattlebox☆

111950 Crotalaria bialata Roxb. = Crotalaria alata Buch. -Ham. ex D. Don ■

111951 Crotalaria bialata Schrenk;翼茎野百合;Wingstem Rattlebox ■

111952 Crotalaria bianoensis P. A. Duvign. ex Timp. = Crotalaria pseudodiloloensis R. Wilczek ■☆

111953 Crotalaria bicolor I. M. Johnst. = Crotalaria quangensis Taub. ■●☆

111954 Crotalaria bidiei Gamble = Crotalaria alata Buch. -Ham. ex D. Don ■

111955 Crotalaria biflora L. ;双花猪屎豆;Twoflower Rattlebox☆

111956 Crotalaria blanda Polhill;光滑猪屎豆●☆

111957 Crotalaria bodinieri H. Lév. = Crotalaria ferruginea Graham ex

Benth. ■
111958 Crotalaria boehmii Taub.;贝姆猪屎豆■☆
111959 Crotalaria bondii Baker f. ex Torre;邦德猪屎豆●☆
111960 Crotalaria bongensis Baker f.;邦戈猪屎豆●☆
111961 Crotalaria bongensis Baker f. var. camerunensis ? = Crotalaria bongensis Baker f. ●☆
111962 Crotalaria bongensis Baker f. var. shirensis ? = Crotalaria shirensis (Baker f.) Milne-Redh. ■☆
111963 Crotalaria boranica Harms ex Baker f.;博兰猪屎豆■☆
111964 Crotalaria boranica Harms ex Baker f. subsp. trichocarpa Polhill;毛果博兰猪屎豆■☆
111965 Crotalaria bosseri M. Pelt.;博瑟猪屎豆●☆
111966 Crotalaria boutiqueana R. Wilczek;布蒂克猪屎豆●☆
111967 Crotalaria brachycarpa (Benth.) Burtt Davy ex I. Verd.;短果猪屎豆■☆
111968 Crotalaria bracteata Roxb. ex DC.;苞叶猪屎豆(大苞叶猪屎豆,毛果猪屎豆);Bracteate Rattlebox,Hairfruit Rattlebox ■●
111969 Crotalaria brevicornuta Polhill;短角猪屎豆☆
111970 Crotalaria brevidens Benth.;埃塞俄比亚猪屎豆;Ethiopian Rattlebox ■☆
111971 Crotalaria brevidens Benth. var. dorumaensis (R. Wilczek) Polhill;扎伊尔猪屎豆■☆
111972 Crotalaria brevidens Benth. var. intermedia (Kotschy) Polhill;全叶埃塞俄比亚猪屎豆;Ethiopian Rattlebox ■☆
111973 Crotalaria brevidens Benth. var. parviflora (Baker f.) Polhill;小花埃塞俄比亚猪屎豆■☆
111974 Crotalaria breviflora Champ. = Crotalaria sessiliflora L. ■
111975 Crotalaria breviflora DC.;短花猪屎豆;Shortflower Rattlebox ■☆
111976 Crotalaria brevipes Champ. = Crotalaria sessiliflora L. ■
111977 Crotalaria brevipes Champ. ex Benth. = Crotalaria sessiliflora L. ■
111978 Crotalaria breyeri N. E. Br. = Crotalaria monteiroi Taub. ex Baker f. var. galpinii Burtt Davy ex I. Verd. ■☆
111979 Crotalaria buchananii Baker f. = Crotalaria alexandri Baker f. ■☆
111980 Crotalaria burkeana Benth.;伯克猪屎豆■☆
111981 Crotalaria burkeana Benth. var. sparsipila Harv. = Crotalaria burkeana Benth. ■☆
111982 Crotalaria burmanni DC. = Crotalaria assamica Benth. ■
111983 Crotalaria burttii Baker f.;伯特猪屎豆■☆
111984 Crotalaria callensii R. Wilczek;卡伦斯猪屎豆●☆
111985 Crotalaria calliantha Polhill;美花猪屎豆●☆
111986 Crotalaria calva R. Vig.;光秃猪屎豆●☆
111987 Crotalaria calycina Schrank;长萼猪屎豆(长萼野百合,猪铃草);Longcalyx Rattlebox,Longsepaled Rattlebox ■
111988 Crotalaria camerounensis Polhill = Crotalaria vagans Polhill ■☆
111989 Crotalaria campestris Polhill;田野猪屎豆☆
111990 Crotalaria camptosepala Thulin;萼猪屎豆☆
111991 Crotalaria cannabina Schweinf. ex Baker f. = Crotalaria ochroleuca G. Don ■
111992 Crotalaria capensis Jacq.;好望角猪屎豆;Cape Laburnum ●☆
111993 Crotalaria capensis Jacq. var. obscura E. Mey. = Crotalaria capensis Jacq. ●☆
111994 Crotalaria capillipes Polhill;发梗猪屎豆☆
111995 Crotalaria capitata Baker = Crotalaria mairei H. Lév. ■
111996 Crotalaria capitata Benth. ex Baker = Crotalaria mairei H. Lév. ■
111997 Crotalaria capitata Lam. = Liparia umbellifera Thunb. ■☆
111998 Crotalaria capituliformis R. Wilczek = Crotalaria quangensis Taub. var. capituliformis (R. Wilczek) Polhill ■☆
111999 Crotalaria capuronii M. Pelt.;凯普伦猪屎豆■☆
112000 Crotalaria carinata Steud. ex A. Rich. = Crotalaria senegalensis (Pers.) Bacle ex DC. ■☆
112001 Crotalaria carrissoana Torre;卡里索猪屎豆●☆
112002 Crotalaria carsonii Baker f.;卡森猪屎豆■☆
112003 Crotalaria carsonioides R. Wilczek;拟卡森猪屎豆■☆
112004 Crotalaria cataractarum Baker f. = Crotalaria flavicarinata Baker f. ■☆
112005 Crotalaria catatii Drake = Argyrolobium catatii (Drake) M. Peltier ●☆
112006 Crotalaria caudata Welw. ex Baker;尾状猪屎豆■☆
112007 Crotalaria ceciliae I. Verd. = Crotalaria capensis Jacq. ●☆
112008 Crotalaria cephalotes Steud. ex A. Rich.;大头猪屎豆■☆
112009 Crotalaria cephalotes Steud. ex A. Rich. var. moeroensis Baker f. = Crotalaria cephalotes Steud. ex A. Rich. ■☆
112010 Crotalaria cernua Schinz = Crotalaria sphaerocarpa Perr. ex DC. ■☆
112011 Crotalaria cernua Schinz f. latifolia ? = Crotalaria sphaerocarpa Perr. ex DC. ■☆
112012 Crotalaria chiayiana Y. C. Liu et F. Y. Lu;红花假地蓝■
112013 Crotalaria chinensis L.;中国猪屎豆(华野百合,华猪屎豆,台湾野百合);China Rattlebox,Chinese Rattlebox ■
112014 Crotalaria chirindae Baker f.;奇林达猪屎豆■☆
112015 Crotalaria chondrocarpa Polhill;骨果猪屎豆●☆
112016 Crotalaria chrysochlora Baker f. ex Harms;金芽猪屎豆■☆
112017 Crotalaria chrysotricha Polhill;金毛猪屎豆■☆
112018 Crotalaria cinerea Burtt Davy ex I. Verd. = Crotalaria globifera E. Mey. ■☆
112019 Crotalaria cistoides Welw. ex Baker;岩蔷薇猪屎豆●☆
112020 Crotalaria cistoides Welw. ex Baker subsp. orientalis Polhill;东方岩蔷薇猪屎豆■☆
112021 Crotalaria citriocolorata Baker f. = Crotalaria ukambensis Vatke ■☆
112022 Crotalaria claessensii De Wild. = Crotalaria aculeata De Wild. subsp. claessensii (De Wild.) Polhill ■☆
112023 Crotalaria cleomifolia Welw. ex Baker;白花菜猪屎豆●☆
112024 Crotalaria cleomifolia Welw. ex Baker var. kassneri Baker f. = Crotalaria cleomifolia Welw. ex Baker ●☆
112025 Crotalaria cleomifolia Welw. ex Baker var. seretii De Wild. = Crotalaria cleomifolia Welw. ex Baker ●☆
112026 Crotalaria cleomoides Klotzsch = Crotalaria trichotoma Bojer ■
112027 Crotalaria cobalticola P. A. Duvign. et Plancke;科博尔特猪屎豆■☆
112028 Crotalaria collina Polhill;山丘猪屎豆■☆
112029 Crotalaria colorata Schinz;着色猪屎豆■☆
112030 Crotalaria colorata Schinz subsp. erecta (Schinz) Polhill;直立着色猪屎豆■☆
112031 Crotalaria coluteoides Lam. = Hypocalyptus coluteoides (Lam.) R. Dahlgren ●☆
112032 Crotalaria comosa Baker;簇毛猪屎豆■☆
112033 Crotalaria concinna Polhill;整洁猪屎豆■☆
112034 Crotalaria confertiflora Polhill;密花猪屎豆■☆
112035 Crotalaria confusa Hepper;混乱猪屎豆■☆
112036 Crotalaria congesta Polhill;团集猪屎豆■☆
112037 Crotalaria congoensis Baker f. = Crotalaria florida Welw. ex Baker var. congolensis (Baker f.) R. Wilczek ■☆
112038 Crotalaria cordata Welw. ex Baker;心形猪屎豆■☆

112039 Crotalaria cordifolia L. = Hypocalyptus sophoroides (P. J. Bergius) Baill. ●☆
112040 Crotalaria cornetii Taub. et Dewèvre;科尔内猪屎豆■☆
112041 Crotalaria corymbosa Torre;伞序猪屎豆■☆
112042 Crotalaria coursii M. Pelt.;库尔斯猪屎豆●☆
112043 Crotalaria crebra Polhill;密集猪屎豆●☆
112044 Crotalaria crepitans Hutch. = Crotalaria recta Steud. ex A. Rich. ■☆
112045 Crotalaria crispata F. Muell. ex Benth.;皱波猪屎豆■☆
112046 Crotalaria cuneiformis Lam. = Rafnia capensis (L.) Schinz subsp. ovata (P. J. Bergius) G. J. Campb. et B. -E. van Wyk ●☆
112047 Crotalaria cunninghamii R. Br.;澳洲猪屎豆(坎宁猪屎豆); Greenbird Flower ●☆
112048 Crotalaria cupricola Leteint.;喜铜猪屎豆●☆
112049 Crotalaria cuspidata Taub.;骤尖猪屎豆■☆
112050 Crotalaria cyanea Baker;蓝色猪屎豆■☆
112051 Crotalaria cyanoxantha R. Vig.;蓝黄猪屎豆■☆
112052 Crotalaria cylindrica A. Rich.;柱形猪屎豆■☆
112053 Crotalaria cylindrica A. Rich. subsp. afrorientalis Polhill;东非柱形猪屎豆■☆
112054 Crotalaria cylindrocarpa DC.;柱果猪屎豆■☆
112055 Crotalaria cylindroclados Baker f. et Martin = Crotalaria glauca Willd. ■☆
112056 Crotalaria cylindrostachys Welw. ex Baker;柱序猪屎豆■☆
112057 Crotalaria cytisoides Hils. et Bojer ex Benth. = Crotalaria tanety Du Puy, Labat et H. E. Ireland ■☆
112058 Crotalaria cytisoides Roxb. ex DC.;金雀猪屎豆(黄雀儿)■
112059 Crotalaria cytisoides Roxb. ex DC. = Crotalaria psoralioides D. Don ■
112060 Crotalaria cytisoides Roxb. ex DC. = Priotropis cytisoides (Roxb. ex DC.) Wight et Arn. ■
112061 Crotalaria damarensis Engl.;达马尔猪屎豆■☆
112062 Crotalaria damarensis Engl. var. maraisiana Torre = Crotalaria podocarpa DC. ■☆
112063 Crotalaria dasyclada Polhill;毛枝猪屎豆■☆
112064 Crotalaria debilis Polhill;弱小猪屎豆■☆
112065 Crotalaria decaryana R. Vig.;德卡里猪屎豆●☆
112066 Crotalaria decasperma Nair = Crotalaria mysorensis Roth ■
112067 Crotalaria decora Polhill;装饰猪屎豆■☆
112068 Crotalaria deflersii Schweinf.;德弗莱尔猪屎豆●☆
112069 Crotalaria deightonii Hepper;戴顿猪屎豆●☆
112070 Crotalaria densicephala Welw. ex Baker;密头猪屎豆■☆
112071 Crotalaria densiflora De Wild. = Crotalaria cuspidata Taub. ■☆
112072 Crotalaria depressa Polhill;凹陷猪屎豆■☆
112073 Crotalaria descampsii Micheli;德康猪屎豆●☆
112074 Crotalaria deserticola Taub. ex Baker f.;荒漠猪屎豆■☆
112075 Crotalaria deserticola Taub. ex Baker f. var. robusta Polhill;粗壮荒漠猪屎豆■☆
112076 Crotalaria dewildemaniana R. Wilczek;德怀尔德曼猪屎豆■☆
112077 Crotalaria dewildemaniana R. Wilczek subsp. oxyrhyncha Polhill;尖喙荒漠猪屎豆■☆
112078 Crotalaria diffusa E. Mey. = Crotalaria excisa (Thunb.) Baker f. ■☆
112079 Crotalaria dilatata Polhill;膨大猪屎豆■☆
112080 Crotalaria diloloensis Baker f. = Crotalaria graminicola Taub. ex Baker f. ■☆
112081 Crotalaria diloloensis Baker f. var. prostrata R. Wilczek = Crotalaria graminicola Taub. ex Baker f. ■☆
112082 Crotalaria diminuta Polhill;缩小猪屎豆■☆
112083 Crotalaria dinteri Schinz;丁特猪屎豆■☆
112084 Crotalaria diosmifolia Benth.;逸香木叶●☆
112085 Crotalaria distans Benth.;远离猪屎豆■☆
112086 Crotalaria distans Benth. subsp. macrotropis (Baker f.) Polhill;大棱远离猪屎豆■☆
112087 Crotalaria distans Benth. subsp. mediocris Polhill;中位猪屎豆■☆
112088 Crotalaria divaricato-ramosa De Wild. = Crotalaria cephalotes Steud. ex A. Rich. ■☆
112089 Crotalaria dolichantha Polhill;长花猪屎豆■☆
112090 Crotalaria dolichonyx Baker f. et Martin;长刺猪屎豆■☆
112091 Crotalaria doniana Baker;唐猪屎豆■☆
112092 Crotalaria drummondii Milne-Redh. = Crotalaria scassellatii Chiov. ■☆
112093 Crotalaria dubia De Wild. = Crotalaria prolongata Baker ■☆
112094 Crotalaria dubia Graham ex Benth.;卵苞猪屎豆; Doubt Rattlebox, Ovate-bract Rattlebox ■
112095 Crotalaria duboisii R. Wilczek;杜氏猪屎豆■☆
112096 Crotalaria duboisii R. Wilczek subsp. mutica Polhill;无尖猪屎豆■☆
112097 Crotalaria dumosa Franch.;灌丛猪屎豆●☆
112098 Crotalaria dura J. M. Wood et M. S. Evans;硬猪屎豆■☆
112099 Crotalaria dura J. M. Wood et M. S. Evans subsp. mozambica Polhill;莫桑比克猪屎豆■☆
112100 Crotalaria durandiana R. Wilczek;杜朗猪屎豆■☆
112101 Crotalaria duvigneaudii Timp.;迪维尼奥猪屎豆■☆
112102 Crotalaria ebenoides (Guillaumin et Perr.) Walp.;黑檀猪屎豆●☆
112103 Crotalaria ecklonis Harv. = Lotononis stenophylla (Eckl. et Zeyh.) B. -E. van Wyk ■☆
112104 Crotalaria effusa E. Mey.;开展猪屎豆■☆
112105 Crotalaria egregia Polhill;优秀猪屎豆■☆
112106 Crotalaria elata Welw. ex Baker = Crotalaria lachnophora A. Rich. ■☆
112107 Crotalaria elisabethae Baker f.;艾利萨猪屎豆■☆
112108 Crotalaria elliptica Roxb.;双子野百合■
112109 Crotalaria elliptica Roxb. = Crotalaria uncinella Lam. subsp. elliptica (Roxb.) Polhill ■
112110 Crotalaria elliptica Roxb. = Crotalaria uncinella Lam. ■
112111 Crotalaria elongata Thunb.;伸长猪屎豆■☆
112112 Crotalaria emarginata Bojer ex Benth.;微缺猪屎豆■☆
112113 Crotalaria emarginella Vatke;无边猪屎豆■☆
112114 Crotalaria emirnensis Benth.;埃米猪屎豆●☆
112115 Crotalaria endlichii Harms = Crotalaria oocarpa Baker ■☆
112116 Crotalaria engleri Harms ex Baker = Crotalaria agatiflora Schweinf. subsp. engleri (Harms ex Baker) Polhill ■☆
112117 Crotalaria ephemera Polhill;短命猪屎豆■☆
112118 Crotalaria eremicola Baker f.;沙生猪屎豆●☆
112119 Crotalaria eremicola Baker f. subsp. parviflora Polhill;小花沙生猪屎豆■☆
112120 Crotalaria ericoides Torre;石南状猪屎豆■☆
112121 Crotalaria erisemoides Ficalho et Hiern = Lotononis calycina (E. Mey.) Benth. ■☆
112122 Crotalaria erlangeri (Baker f.) Harms ex Hutch. et E. A. Bruce = Crotalaria agatiflora Schweinf. subsp. erlangeri Baker f. ■☆
112123 Crotalaria ervoides Welw. ex Baker = Bolusia ervoides (Welw.

ex Baker) Torre ■☆

112124 Crotalaria erythrophleba Welw. ex Baker;红脉猪屎豆■☆

112125 Crotalaria esquirolii H. Lév. = Crotalaria tetragona Roxb. ex Andréws ■

112126 Crotalaria eurycalyx Polhill;良萼猪屎豆■☆

112127 Crotalaria exaltata Polhill;极高猪屎豆■☆

112128 Crotalaria excisa (Thunb.) Baker f.;缺刻猪屎豆■☆

112129 Crotalaria excisa (Thunb.) Baker f. subsp. namaquensis Polhill;纳马夸猪屎豆●☆

112130 Crotalaria exelliana R. Wilczek;埃克塞尔猪屎豆●☆

112131 Crotalaria exilipes Polhill;弱梗猪屎豆■☆

112132 Crotalaria exilis Polhill;瘦小猪屎豆■☆

112133 Crotalaria eximia Polhill;优异猪屎豆■☆

112134 Crotalaria falcata Vahl ex DC. = Crotalaria pallida Aiton var. obovata (G. Don) Polhill ■☆

112135 Crotalaria fallax Chiov.;迷惑猪屎豆■☆

112136 Crotalaria fastigiata E. Mey. = Lotononis fastigiata (E. Mey.) B. -E. van Wyk ●☆

112137 Crotalaria fenarolii Torre;费纳罗利猪屎豆●☆

112138 Crotalaria ferruginea Graham ex Benth.;假地蓝(大响铃豆,地响铃,狗响铃,荷承草,荷猪草,黄花野百合,假地豆,假地兰,假地蓝猪屎豆,假花生,铃铃草,马铃草,马响铃,马小莲,肾气草,响亮草,响铃草,响铃子,小狗响铃,野花生,野毛豆,野豌豆,猪屎豆);Rustcolor Rattlebox,Rustcolored Rattlebox ■

112139 Crotalaria ferruginea Graham ex Benth. var. chiayiana (Y. C. Liu et F. Y. Lu) S. S. Ying = Crotalaria chiayiana Y. C. Liu et F. Y. Lu ■

112140 Crotalaria ferruginea Graham ex Benth. var. pilosissima Benth. ex Baker = Crotalaria ferruginea Graham ex Benth. ■

112141 Crotalaria fertilis Delile = Crotalaria pallida Aiton ●■

112142 Crotalaria fiherenensis R. Vig.;马岛猪屎豆●☆

112143 Crotalaria filicaulis Welw. ex Baker;线茎猪屎豆■☆

112144 Crotalaria filicaulis Welw. ex Baker var. grandiflora Polhill;大花线茎猪屎豆■☆

112145 Crotalaria filicauloides R. Wilczek = Crotalaria filicaulis Welw. ex Baker var. grandiflora Polhill ■☆

112146 Crotalaria filifolia De Wild. = Crotalaria durandiana R. Wilczek ■☆

112147 Crotalaria flavicarinata Baker f.;黄棱猪屎豆■☆

112148 Crotalaria fleckii Schinz = Crotalaria damarensis Engl. ■☆

112149 Crotalaria flexuosa Baker = Crotalaria podocarpa DC. ■☆

112150 Crotalaria floribunda Lodd. = Wiborgia obcordata (P. J. Bergius) Thunb. ■☆

112151 Crotalaria florida Welw. ex Baker;繁花猪屎豆■☆

112152 Crotalaria florida Welw. ex Baker var. congolensis (Baker f.) R. Wilczek;刚果猪屎豆■☆

112153 Crotalaria florida Welw. ex Baker var. monosperma (De Wild.) R. Wilczek;单籽猪屎豆■☆

112154 Crotalaria forbesii Baker = Crotalaria virgulata Klotzsch subsp. forbesii (Baker) Polhill ■☆

112155 Crotalaria forbesii Baker var. vanmelii R. Wilczek = Crotalaria virgulata Klotzsch subsp. pauciflora (Baker) Polhill ●☆

112156 Crotalaria formosana Matsum. ex T. Ito et Matsum. = Crotalaria albida K. Heyne ex Roth ■

112157 Crotalaria formosana T. Ito = Crotalaria albida K. Heyne ex Roth ■

112158 Crotalaria friesii I. Verd.;弗里斯猪屎豆■☆

112159 Crotalaria fruticosa Mill. = Crotalaria sagittalis L. ■☆

112160 Crotalaria fulgida Baker = Crotalaria podocarpa DC. ■☆

112161 Crotalaria fulva Roxb.;暗黄猪屎豆■☆

112162 Crotalaria fulvella Merxm. = Crotalaria abbreviata Baker f. ●☆

112163 Crotalaria furfuracea Boiss. = Crotalaria persica (Burm. f.) Merr. ■☆

112164 Crotalaria fwamboensis Baker f. = Crotalaria subcapitata De Wild. var. fwamboensis (Baker f.) Polhill ■☆

112165 Crotalaria gambica Taub. = Crotalaria perrottetii DC. ●☆

112166 Crotalaria gazensis Baker f.;加兹猪屎豆■☆

112167 Crotalaria gazensis Baker f. subsp. herbacea Polhill;草本猪屎豆■☆

112168 Crotalaria gengmaensis Z. Wei et C. Y. Yang = Crotalaria albida K. Heyne ex Roth var. gengmaensis (Z. Wei et C. Y. Yang) C. Chen et J. Q. Li ■

112169 Crotalaria genistoides Lam. = Liparia genistoides (Lam.) A. L. Schutte ■☆

112170 Crotalaria genistoides Willd. = Crotalaria willdenowiana DC. ■☆

112171 Crotalaria germainii R. Wilczek;杰曼猪屎豆■☆

112172 Crotalaria gillettii Polhill;吉莱特猪屎豆■☆

112173 Crotalaria glabripedicellata R. Wilczek;光梗猪屎豆■☆

112174 Crotalaria glauca Willd. = Crotalaria cylindroclados Baker f. et Martin ■☆

112175 Crotalaria glauca Willd. var. anisophylla Hiern = Crotalaria anisophylla (Hiern) Welw. ex Baker f. ●☆

112176 Crotalaria glauca Willd. var. beniensis De Wild. = Crotalaria glauca Willd. ■☆

112177 Crotalaria glauca Willd. var. elliotii Baker f. = Crotalaria glauca Willd. ■☆

112178 Crotalaria glauca Willd. var. mildbraedii Baker f. = Crotalaria glauca Willd. ■☆

112179 Crotalaria glauca Willd. var. welwitschii Baker f. = Crotalaria glauca Willd. ■☆

112180 Crotalaria glaucifolia Baker;灰绿叶猪屎豆■☆

112181 Crotalaria glaucoides Baker f. = Crotalaria paludosa A. Chev. ■☆

112182 Crotalaria globifera E. Mey.;球形猪屎豆■☆

112183 Crotalaria globifera E. Mey. var. brachycarpa Benth. = Crotalaria brachycarpa (Benth.) Burtt Davy ex I. Verd. ■☆

112184 Crotalaria globifera E. Mey. var. congolensis Baker f. = Crotalaria florida Welw. ex Baker var. congolensis (Baker f.) R. Wilczek ■☆

112185 Crotalaria globifera E. Mey. var. glabra Harv. = Crotalaria globifera E. Mey. ■☆

112186 Crotalaria globifera E. Mey. var. pubescens ? = Crotalaria globifera E. Mey. ■☆

112187 Crotalaria gnidioides R. Wilczek;格尼瑞香猪屎豆●☆

112188 Crotalaria goetzei Harms;三叶猪屎豆(三叶菽麻)●☆

112189 Crotalaria goodiiformis Vatke;托马森猪屎豆●☆

112190 Crotalaria goreensis Guillaumin et Perr.;戈雷猪屎豆●☆

112191 Crotalaria gracilis (Guillaumin et Perr.) Walp. = Crotalaria perrottetii DC. ●☆

112192 Crotalaria gracillima Klotzsch = Crotalaria hyssopifolia Klotzsch ●☆

112193 Crotalaria grahamiana Wight et Arn.;格雷厄姆猪屎豆●☆

112194 Crotalaria graminicola Taub. ex Baker f.;草莺猪屎豆■☆

112195 Crotalaria grandibracteata Taub.;大苞猪屎豆■☆

112196 Crotalaria grandiflora Zoll. = Crotalaria tetragona Roxb. ex Andréws ■

112197 Crotalaria grandistipulata Harms;大托叶猪屎豆■☆

112198 Crotalaria grantiana Harv. = Crotalaria virgulata Klotzsch subsp. grantiana (Harv.) Polhill ●☆
112199 Crotalaria grantii Baker = Crotalaria polysperma Kotschy ■☆
112200 Crotalaria grata Polhill;可爱猪屎豆■☆
112201 Crotalaria greenwayi Baker f.;格林韦猪屎豆■☆
112202 Crotalaria grevei Drake;格雷弗猪屎豆●☆
112203 Crotalaria griquensis L. Bolus;格里夸猪屎豆●☆
112204 Crotalaria griseofusca Baker f.;棕色猪屎豆■☆
112205 Crotalaria gweloensis (Baker f.) Milne-Redh. = Crotalaria variegata Welw. ex Baker ■☆
112206 Crotalaria gymnocalyx Baker = Crotalaria natalitia Meisn. ■☆
112207 Crotalaria hainaensis C. C. Huang;海南猪屎豆(海南野百合);Hainan Rattlebox ■
112208 Crotalaria harmsiana Taub. = Crotalaria caudata Welw. ex Baker ■☆
112209 Crotalaria haumaniana R. Wilczek;豪曼猪屎豆■☆
112210 Crotalaria helenae Buscal. et Muschl. = Crotalaria laburnifolia L. ■☆
112211 Crotalaria hemsleyi Milne-Redh.;昂斯莱猪屎豆●☆
112212 Crotalaria heqingensis C. Y. Yang = Crotalaria yunnanensis Franch. var. heqingensis (C. Y. Yang) C. Chen et J. Q. Li ■
112213 Crotalaria heterotricha Polhill;异毛猪屎豆■☆
112214 Crotalaria hildebrandtii Baill. = Crotalaria xanthoclada Bojer ex Benth. ■☆
112215 Crotalaria hildebrandtii Vatke = Crotalaria axillaris Aiton ●☆
112216 Crotalaria hirsutissima Schinz = Crotalaria podocarpa DC. ■☆
112217 Crotalaria hirta Willd.;多毛猪屎豆■☆
112218 Crotalaria hislopii Corbishley = Crotalaria anisophylla (Hiern) Welw. ex Baker f. ●☆
112219 Crotalaria hispida Schinz = Crotalaria steudneri Schweinf. ■☆
112220 Crotalaria hoffmannii R. Wilczek;豪夫曼猪屎豆■☆
112221 Crotalaria hoffmannii R. Wilczek var. glabra Polhill;光豪夫曼猪屎豆■☆
112222 Crotalaria holoptera Welw. ex Baker;全翅猪屎豆■☆
112223 Crotalaria homblei De Wild. = Crotalaria lachnophora A. Rich. ■☆
112224 Crotalaria horrida Polhill;多刺猪屎豆■☆
112225 Crotalaria huillensis Taub.;威拉猪屎豆■☆
112226 Crotalaria huillensis Taub. subsp. zambesiaca Polhill;赞比西猪屎豆■☆
112227 Crotalaria huillensis Taub. var. cacondensis Baker f. ex Torre;卡孔达猪屎豆■☆
112228 Crotalaria humbertiana M. Pelt.;亨氏猪屎豆●☆
112229 Crotalaria humbertii R. Vig.;亨伯特猪屎豆●☆
112230 Crotalaria humifusa Graham ex Benth.;匍地猪屎豆■
112231 Crotalaria humilis Eckl. et Zeyh.;低矮猪屎豆■☆
112232 Crotalaria hypargyrea Chiov.;下银猪屎豆■☆
112233 Crotalaria hyssopifolia Klotzsch;神香草叶猪屎豆●☆
112234 Crotalaria ibityensis R. Vig. et Humbert;伊比提猪屎豆●☆
112235 Crotalaria imbricata L. = Amphithalea imbricata (L.) Druce ■☆
112236 Crotalaria imperialis Taub. = Crotalaria agatiflora Schweinf. subsp. imperialis (Taub.) Polhill ■☆
112237 Crotalaria impressa Nees ex Walp.;凹猪屎豆■☆
112238 Crotalaria incana L.;圆叶猪屎豆(光叶猪屎豆,恒春野百合);Hoary Rattlebox,Roundleaf Rattlebox,Shackshack Crataloria ■
112239 Crotalaria incana L. f. glabrescens R. Wilczek = Crotalaria incana L. ■
112240 Crotalaria incana L. f. lanata R. Wilczek = Crotalaria incana L. subsp. purpurascens (Lam.) Milne-Redh. ■☆
112241 Crotalaria incana L. subsp. purpurascens (Lam.) Milne-Redh.;紫圆叶猪屎豆■☆
112242 Crotalaria incana L. subsp. purpurascens (Lam.) Milne-Redh. = Crotalaria incana L. ■
112243 Crotalaria incrassifolia Polhill;厚叶猪屎豆■☆
112244 Crotalaria inflexa Polhill;内折猪屎豆■☆
112245 Crotalaria inhabilis I. Verd. = Crotalaria monteiroi Taub. ex Baker f. var. galpinii Burtt Davy ex I. Verd. ■☆
112246 Crotalaria inopinata (Harms) Polhill;意外猪屎豆■☆
112247 Crotalaria insignis Polhill;显著猪屎豆■☆
112248 Crotalaria intermedia Kotschy;间型猪屎豆;Slender-leaf Cretalaria ■☆
112249 Crotalaria intermedia Kotschy = Crotalaria brevidens Benth. var. intermedia (Kotschy) Polhill ■☆
112250 Crotalaria intermedia Kotschy = Crotalaria ochroleuca G. Don ■
112251 Crotalaria intermedia Kotschy var. abyssinica Taub. ex Engl. = Crotalaria brevidens Benth. ■☆
112252 Crotalaria intermedia Kotschy var. dorumaensis R. Wilczek = Crotalaria brevidens Benth. var. dorumaensis (R. Wilczek) Polhill ■☆
112253 Crotalaria intermedia Kotschy var. parviflora Baker f. = Crotalaria brevidens Benth. var. parviflora (Baker f.) Polhill ■☆
112254 Crotalaria intonsa Polhill;须毛猪屎豆■☆
112255 Crotalaria intricata Thulin;缠绕猪屎豆■☆
112256 Crotalaria involutifolia Polhill;内卷叶猪屎豆■☆
112257 Crotalaria inyangensis Polhill;伊尼扬加猪屎豆■☆
112258 Crotalaria ionoptera Polhill;堇翅猪屎豆■☆
112259 Crotalaria iringana Harms;伊林加猪屎豆■☆
112260 Crotalaria isaloensis R. Vig.;伊萨卢猪屎豆●☆
112261 Crotalaria jacksonii Baker f.;杰克逊猪屎豆■☆
112262 Crotalaria jamesii Oliv. = Crotalaria dumosa Franch. ●☆
112263 Crotalaria jianfengensis C. Y. Yang;尖峰猪屎豆;Jianfeng Rattlebox ■
112264 Crotalaria jinpingensis C. Y. Yang = Crotalaria prostrata Rottler ex Willd. ■
112265 Crotalaria jinpingensis C. Y. Yang = Crotalaria prostrata Rottler ex Willd. var. jinpingensis (C. Y. Yang) C. Y. Yang ■
112266 Crotalaria johannis Torre;约翰猪屎豆■☆
112267 Crotalaria johnstonii Baker;约翰斯顿猪屎豆■☆
112268 Crotalaria jubae Polhill;朱巴猪屎豆●☆
112269 Crotalaria juncea L.;菽麻(大狗响铃,大响铃,赫麻,太阳麻,印度麻,自消容);Bombay Hemp,East Indies Hemp,False Hemp,Indian Hemp,Madras Hemp,Rush-like Rattlebox,San Hemp,Sann Hemp,Sun Hemp,Sun-hemp,Sunn Crotalaria,Sunn Hemp,Sunn Rattlebox ■
112270 Crotalaria junodiana Baker f. = Crotalaria laburnoides Klotzsch ■☆
112271 Crotalaria kambolensis Baker f.;坎布尔猪屎豆■☆
112272 Crotalaria kandoensis Baker f.;坎多猪屎豆■☆
112273 Crotalaria kapiriensis De Wild.;卡皮里猪屎豆■☆
112274 Crotalaria karagwensis Taub.;赞比亚猪屎豆■☆
112275 Crotalaria karongensis Baker = Crotalaria senegalensis (Pers.) Bacle ex DC. ■☆
112276 Crotalaria kasaiensis R. Wilczek = Crotalaria sapinii De Wild. subsp. kasaiensis (R. Wilczek) Polhill ■☆
112277 Crotalaria kasikiensis Baker f. = Crotalaria argyrolobioides Baker ●☆
112278 Crotalaria kassneri Baker f.;卡斯纳猪屎豆■☆

112279 Crotalaria katangensis Dewèvre;加丹加猪屎豆■☆
112280 Crotalaria katongaensis R. Wilczek = Crotalaria sparsifolia Baker ■☆
112281 Crotalaria kawakamii Hayata = Crotalaria chinensis L. ■
112282 Crotalaria kelaensis Baker f.;凯拉猪屎豆●☆
112283 Crotalaria keniensis Baker f.;肯尼亚猪屎豆■☆
112284 Crotalaria kibaraensis R. Wilczek;基巴拉猪屎豆●☆
112285 Crotalaria kigesiensis Baker f. = Crotalaria deserticola Taub. ex Baker f. ■☆
112286 Crotalaria kilimandscharica Taub. = Crotalaria natalitia Meisn. ■☆
112287 Crotalaria kipiriensis R. Wilczek = Crotalaria florida Welw. ex Baker var. monosperma (De Wild.) R. Wilczek ■☆
112288 Crotalaria kirkii Baker;柯克猪屎豆■☆
112289 Crotalaria kundelunguensis Baker f.;昆德龙古猪屎豆●☆
112290 Crotalaria kurzii Baker ex Kurz;薄叶猪屎豆■●
112291 Crotalaria kwengeensis R. Wilczek;昆盖猪屎豆●☆
112292 Crotalaria kwengeensis R. Wilczek var. parviflora Polhill;小花昆盖猪屎豆■☆
112293 Crotalaria kyimbilae Harms = Crotalaria nyikensis Baker ■☆
112294 Crotalaria laburnifolia L.;金莲花叶猪屎豆■☆
112295 Crotalaria laburnifolia L. subsp. australis (Baker f.) Polhill;南方金莲花叶猪屎豆■☆
112296 Crotalaria laburnifolia L. subsp. petiolaris (Franch.) Polhill;柄叶金莲花叶猪屎豆■☆
112297 Crotalaria laburnifolia L. subsp. tenuicarpa Polhill;细果金莲花叶猪屎豆■☆
112298 Crotalaria laburnoides Klotzsch;金莲花状猪屎豆■☆
112299 Crotalaria laburnoides Klotzsch var. nudicarpa Polhill;裸果金莲花叶猪屎豆■☆
112300 Crotalaria lachnocarpa Hochst. ex Baker = Crotalaria lachnophora A. Rich. ■☆
112301 Crotalaria lachnocarpoides Engl.;拟毛果猪屎豆■☆
112302 Crotalaria lachnoclada Harms = Crotalaria densicephala Welw. ex Baker ■☆
112303 Crotalaria lachnophora A. Rich.;毛梗猪屎豆■☆
112304 Crotalaria lachnosema Stapf;绵毛猪屎豆■☆
112305 Crotalaria laevigata Lam.;平滑猪屎豆●☆
112306 Crotalaria lanata Thunb. = Xiphotheca fruticosa (L.) A. L. Schutte et B. -E. van Wyk ■☆
112307 Crotalaria lanceolata E. Mey.;长果猪屎豆(长叶猪屎豆,披针叶猪屎豆);Lance-leaf Crotalaria, Lanceleaf Rattlebox, Lanceolate Rattlebox, Longfruit Rattlebox ■
112308 Crotalaria lanceolata E. Mey. subsp. contigua Polhill;邻近猪屎豆■☆
112309 Crotalaria lanceolata E. Mey. subsp. exigua Polhill;弱小长果猪屎豆■☆
112310 Crotalaria lanceolata E. Mey. var. malangensis Baker f. = Crotalaria subcapitata De Wild. ■☆
112311 Crotalaria lancifoliolata Torre;披针形猪屎豆■☆
112312 Crotalaria lasiocarpa Polhill;毛果猪屎豆■☆
112313 Crotalaria lathouwersii Baker f. = Crotalaria dolichonyx Baker f. et Martin ■☆
112314 Crotalaria lathyroides Guillaumin et Perr.;山黧豆状猪屎豆■☆
112315 Crotalaria latifoliolata (De Wild.) R. Wilczek = Crotalaria laburnoides Klotzsch ■☆
112316 Crotalaria lawalreeana R. Wilczek;拉瓦尔猪屎豆●☆
112317 Crotalaria laxa Franch. = Crotalaria emarginella Vatke ■☆
112318 Crotalaria laxiflora Baker;疏花猪屎豆■☆
112319 Crotalaria laxiflora Baker var. acuta Polhill;尖疏花猪屎豆■☆
112320 Crotalaria leandriana M. Pelt.;利安猪屎豆●☆
112321 Crotalaria lebeckioides Bond;南非针叶豆猪屎豆●☆
112322 Crotalaria lebrunii Baker f.;勒布伦猪屎豆■☆
112323 Crotalaria ledermannii Baker f.;莱德猪屎豆■☆
112324 Crotalaria lenticula E. Mey. = Lotononis lenticula (E. Mey.) Benth. ■☆
112325 Crotalaria leonardiana Timp.;莱奥猪屎豆■☆
112326 Crotalaria lepidissima Baker f.;多鳞猪屎豆■☆
112327 Crotalaria leprieurii Guillaumin et Perr.;莱普里厄猪屎豆■☆
112328 Crotalaria leptocarpa Baker f.;细果猪屎豆■☆
112329 Crotalaria leptocarpa Baker f. subsp. aberrans Polhill;异常细果猪屎豆■☆
112330 Crotalaria leptocarpa Baker f. subsp. contracta Polhill;紧缩细果猪屎豆■☆
112331 Crotalaria leptopoda Harms = Crotalaria virgulata Klotzsch subsp. pauciflora (Baker) Polhill ●☆
112332 Crotalaria leschenaultii DC. = Crotalaria spectabilis Roth ■
112333 Crotalaria leucoclada Baker;白枝猪屎豆■☆
112334 Crotalaria leucotricha Baker = Crotalaria johnstonii Baker ■☆
112335 Crotalaria limosa Polhill;湿地猪屎豆■☆
112336 Crotalaria lindneri Schinz = Crotalaria damarensis Engl. ■☆
112337 Crotalaria linearifolia De Wild. = Crotalaria leprieurii Guillaumin et Perr. ■☆
112338 Crotalaria linearifoliolata Chiov.;线托叶猪屎豆■☆
112339 Crotalaria lineata Thunb. = Eriosema squarrosum (Thunb.) Walp. ■☆
112340 Crotalaria linifolia L. f.;线叶猪屎豆(假花生,密叶猪屎豆,条叶猪屎豆,线叶野百合,响铃草,小苦参,狭叶猪屎豆);Linearleaf Rattlebox ■
112341 Crotalaria linifolia L. f. var. pygmea Gogelein = Crotalaria linifolia L. f. ■
112342 Crotalaria linifolia L. f. var. stenophylla (Vogel) C. Y. Yang;窄叶猪屎豆(条叶猪屎豆,狭线叶猪屎豆,狭叶猪屎豆);Narrowleaf Rattlebox ■
112343 Crotalaria linifolia L. f. var. stenophylla C. Y. Yang = Crotalaria linifolia L. f. var. stenophylla (Vogel) C. Y. Yang ■
112344 Crotalaria lisowskii Polhill;利索猪屎豆■☆
112345 Crotalaria loandae Baker f.;罗安达猪屎豆■☆
112346 Crotalaria loandae Baker f. var. annua Torre = Crotalaria loandae Baker f. ■☆
112347 Crotalaria lonchophylla Hand. -Mazz. = Crotalaria ferruginea Graham ex Benth. ■
112348 Crotalaria longiclavata Polhill;长棒猪屎豆■☆
112349 Crotalaria longidens Burtt Davy ex I. Verd.;长齿猪屎豆■☆
112350 Crotalaria longifoliolata De Wild. = Crotalaria glaucifolia Baker ■☆
112351 Crotalaria longipedunculata De Wild. ex R. Wilczek = Crotalaria subcapitata De Wild. ■☆
112352 Crotalaria longipedunculata De Wild. ex R. Wilczek f. glabra R. Wilczek = Crotalaria subcapitata De Wild. ■☆
112353 Crotalaria longirostrata Hook. et Arn.;长喙猪屎豆;Longbeak Rattlebox ■☆
112354 Crotalaria longistyla Baker f. = Crotalaria virgulata Klotzsch subsp. longistyla (Baker f.) Polhill ●☆
112355 Crotalaria longithyrsa Baker f.;长序猪屎豆■☆
112356 Crotalaria longithyrsa Baker f. var. latifolia R. Wilczek;宽叶猪

屎豆■☆
112357 Crotalaria lotoides Benth.;君迁子猪屎豆■☆
112358 Crotalaria lotononis Welw. ex Baker = Crotalaria glaucifolia Baker ■☆
112359 Crotalaria lugardiorum Bullock = Crotalaria karagwensis Taub. ■☆
112360 Crotalaria lukafuensis De Wild.;卢卡夫猪屎豆■☆
112361 Crotalaria lukuluensis Baker f. = Crotalaria subcapitata De Wild. ■☆
112362 Crotalaria lukwangulensis Harms;卢夸古尔猪屎豆●☆
112363 Crotalaria lunaris L. = Argyrolobium lunare (L.) Druce ●☆
112364 Crotalaria lunata Bedd. ex Polhill;新月猪屎豆■☆
112365 Crotalaria lundensis Torre;隆德猪屎豆■☆
112366 Crotalaria luniemuensis Baker f. ex R. Wilczek = Crotalaria subcapitata De Wild. ■☆
112367 Crotalaria luteo-rubella Baker = Crotalaria pervillei Baill. ●☆
112368 Crotalaria luxenii Baker f.;卢森猪屎豆■☆
112369 Crotalaria luxurians Benth. = Crotalaria medicaginea Lam. var. luxurians (Benth.) Baker ■●
112370 Crotalaria macrantha Polhill;大花猪屎豆■☆
112371 Crotalaria macrocalyx Benth.;大萼猪屎豆■☆
112372 Crotalaria macrocarpa E. Mey.;大果猪屎豆■☆
112373 Crotalaria macrophylla Weinm. = Crotalaria spectabilis Roth ■
112374 Crotalaria macrophylla Willd. = Flemingia congesta Roxb. ex W. T. Aiton ●
112375 Crotalaria macrophylla Willd. = Flemingia macrophylla (Willd.) Kuntze ex Merr. ●
112376 Crotalaria macropoda Baker = Crotalaria laevigata Lam. ●☆
112377 Crotalaria macrostachya Sond. = Crotalaria globifera E. Mey. ■☆
112378 Crotalaria macrostyla D. Don = Campylotropis macrostyla (D. Don) Schindl. ●☆
112379 Crotalaria macrotropis Baker f. = Crotalaria distans Benth. subsp. macrotropis (Baker f.) Polhill ■☆
112380 Crotalaria madecassa R. Vig. = Crotalaria uncinella Lam. ■
112381 Crotalaria madurensis Wight;马都拉猪屎豆;Madura Rattlebox ■☆
112382 Crotalaria mahafalensis R. Vig.;马哈法尔猪屎豆●☆
112383 Crotalaria mairei H. Lév.;头花猪屎豆(大丁草,大丁香,地草果,鸡儿头,蓝花水豌豆);Capitate Rattlebox,Head Rattlebox ■
112384 Crotalaria mairei H. Lév. var. pubescens C. Chen et J. Q. Li;短毛头花猪屎豆(短头花猪屎豆);Pubescent Head Rattlebox ■
112385 Crotalaria malaissei Polhill;马莱泽猪屎豆●☆
112386 Crotalaria malangensis Baker f. = Crotalaria quangensis Taub. var. malangensis (Baker f.) Polhill ●☆
112387 Crotalaria malindiensis Polhill;马林迪猪屎豆●☆
112388 Crotalaria mandrarensis R. Vig.;曼德拉猪屎豆●☆
112389 Crotalaria manongarivensis R. Vig.;马农加猪屎豆●☆
112390 Crotalaria marginata N. E. Br. = Rhynchotropis marginata (N. E. Br.) J. B. Gillett ■☆
112391 Crotalaria maritima Chapm.;海边猪屎豆■☆
112392 Crotalaria marlothii Engl. = Crotalaria damarensis Engl. ■☆
112393 Crotalaria massaiensis Taub.;微花猪屎豆■☆
112394 Crotalaria maxillaris Klotzsch = Crotalaria laburnoides Klotzsch ■☆
112395 Crotalaria maxillaris Klotzsch var. latifoliolata De Wild. = Crotalaria laburnoides Klotzsch ■☆
112396 Crotalaria medicaginea Lam.;假苜蓿;Alfafa-like Rattlebox,Sham Medic ■●
112397 Crotalaria medicaginea Lam. var. luxurians (Benth.) Baker;大叶假苜蓿;Largeleaf Alfafa-like Rattlebox ■●
112398 Crotalaria megapteryx Baker f. et Martin = Crotalaria chrysochlora Baker f. ex Harms ■☆
112399 Crotalaria melanocalyx Polhill;黑萼猪屎豆■☆
112400 Crotalaria mendesii Torre;门代斯猪屎豆●☆
112401 Crotalaria mendoncae Torre;门东萨猪屎豆●☆
112402 Crotalaria mesopontica Taub. f. glabrescens R. Wilczek = Crotalaria mesopontica Taub. subsp. glabrescens (R. Wilczek) Milne-Redh. ■☆
112403 Crotalaria mesoponticoides R. Wilczek = Crotalaria subcapitata De Wild. ■☆
112404 Crotalaria meyeriana Steud.;迈尔猪屎豆■☆
112405 Crotalaria micans Link;三尖叶猪屎豆(黄野百合,黄猪屎豆,美洲野百合,闪光猪屎豆);Anagyris-like Rattle-box,Bright Rattlebox,Caracas Rattlebox,Three-sharp-leaves Rattlebox ■
112406 Crotalaria micheliana R. Wilczek;米歇尔猪屎豆■☆
112407 Crotalaria microcarpa Hochst. ex Benth.;小果猪屎豆■☆
112408 Crotalaria microcarpa Hochst. ex Benth. var. dawei Baker f. = Crotalaria microcarpa Hochst. ex Benth. ■☆
112409 Crotalaria microcarpa Hochst. ex Benth. var. sudanica Baker f. = Crotalaria microcarpa Hochst. ex Benth. ■☆
112410 Crotalaria microcereus Timp. = Crotalaria duboisii R. Wilczek ■☆
112411 Crotalaria microphylla Vahl;小叶猪屎豆■☆
112412 Crotalaria microthamnus Robyns ex R. Wilczek;矮小猪屎豆■☆
112413 Crotalaria mildbraedii Baker f.;米尔德猪屎豆■☆
112414 Crotalaria milneana R. Wilczek;米尔恩猪屎豆■☆
112415 Crotalaria minima Baker f. = Crotalaria spinosa Hochst. ex Benth. ■☆
112416 Crotalaria minor C. H. Wright = Lotus mlanjeanus J. B. Gillett ■☆
112417 Crotalaria minutiflora Baker f. = Crotalaria massaiensis Taub. ■☆
112418 Crotalaria minutissima Baker f.;极小猪屎豆■☆
112419 Crotalaria miranda Milne-Redh.;奇异猪屎豆■☆
112420 Crotalaria mitwabaensis Timp. = Crotalaria lawalreeana R. Wilczek ●☆
112421 Crotalaria mocubensis Polhill;莫库巴猪屎豆●☆
112422 Crotalaria modesta Polhill;适度猪屎豆■☆
112423 Crotalaria mokoroensis R. Wilczek = Crotalaria axilliflora Baker f. ●☆
112424 Crotalaria mollii Polhill;柔软猪屎豆■☆
112425 Crotalaria mollis E. Mey. = Crotalaria meyeriana Steud. ■☆
112426 Crotalaria mollis E. Mey. var. erecta Schinz = Crotalaria colorata Schinz subsp. erecta (Schinz) Polhill ■☆
112427 Crotalaria mollis Weinm. = Crotalaria verrucosa L. ■
112428 Crotalaria mongaensis Baker f. = Crotalaria cuspidata Taub. ■☆
112429 Crotalaria monophylla Germish.;单叶猪屎豆■☆
112430 Crotalaria monosperma De Wild. = Crotalaria florida Welw. ex Baker var. monosperma (De Wild.) R. Wilczek ■☆
112431 Crotalaria montana A. Rich. = Crotalaria incana L. subsp. purpurascens (Lam.) Milne-Redh. ■☆
112432 Crotalaria montana B. Heyne ex Roth;山地猪屎豆☆
112433 Crotalaria montana B. Heyne ex Roth var. angustifolia (Gagnep.) Niyomdham;狭叶山地猪屎豆☆
112434 Crotalaria montana Roxb. = Crotalaria albida K. Heyne ex Roth ■
112435 Crotalaria monteiroi Taub. ex Baker f.;蒙泰鲁猪屎豆●☆
112436 Crotalaria monteiroi Taub. ex Baker f. var. galpinii Burtt Davy ex I. Verd.;盖尔猪屎豆■☆

112437 Crotalaria mortonii Hepper;莫顿猪屎豆■☆
112438 Crotalaria morumbensis Baker f.;莫卢猪屎豆●☆
112439 Crotalaria mossambicensis Klotzsch = Crotalaria lanceolata E. Mey. ■
112440 Crotalaria mucronata Desv. = Crotalaria pallida Aiton var. obovata (G. Don) Polhill ■☆
112441 Crotalaria mucronata Desv. = Crotalaria pallida Aiton ●■
112442 Crotalaria mudugensis Thulin;穆杜格猪屎豆●☆
112443 Crotalaria multicolor Merxm. = Crotalaria pallidicaulis Harms ■☆
112444 Crotalaria mundyi Baker f. = Crotalaria distans Benth. subsp. macrotropis (Baker f.) Polhill ■☆
112445 Crotalaria mutabilis Schinz = Crotalaria podocarpa DC. ■☆
112446 Crotalaria mysorensis Roth;褐毛猪屎豆;Brownhair Rattlebox ■
112447 Crotalaria nana Burm. f.;小猪屎豆(小野百合);Small Rattlebox ■
112448 Crotalaria nana Burm. f. var. patula Baker = Crotalaria nana Burm. f. var. patula Graham ex Baker ■
112449 Crotalaria nana Burm. f. var. patula Graham ex Baker;座地猪屎豆;Dwarf Rattlebox,Spreading Rattlebox ■
112450 Crotalaria natalensis Baker f.;纳塔尔猪屎豆■☆
112451 Crotalaria natalitia Meisn.;纳塔利特猪屎豆■☆
112452 Crotalaria natalitia Meisn. var. procumbens Baker f. = Crotalaria rhodesiae Baker f. ■☆
112453 Crotalaria natalitia Meisn. var. pseudorhodesiae Merxm. = Crotalaria rhodesiae Baker f. ■☆
112454 Crotalaria natalitia Meisn. var. sengensis Baker f. = Crotalaria prittwitzii Baker f. ■☆
112455 Crotalaria neglecta Wight et Arn. = Crotalaria medicaginea Lam. ■
112456 Crotalaria nematophylla Baker f.;蠕虫叶猪屎豆■☆
112457 Crotalaria newtoniana Torre;纽敦猪屎豆■☆
112458 Crotalaria nicholsonii Baker f. = Crotalaria subcapitata De Wild. ■☆
112459 Crotalaria nigrescens Chiov. = Crotalaria cylindrica A. Rich. ■☆
112460 Crotalaria nigricans Baker;变黑猪屎豆■☆
112461 Crotalaria nigricans Baker var. erecta Milne-Redh.;直立变黑猪屎豆■☆
112462 Crotalaria nogalensis Chiov. = Crotalaria emarginella Vatke ■☆
112463 Crotalaria noldeae Rossberg = Crotalaria adamsonii Baker f. ■☆
112464 Crotalaria nubica Benth. = Crotalaria sphaerocarpa Perr. ex DC. ■☆
112465 Crotalaria nuda Polhill;裸猪屎豆■☆
112466 Crotalaria nudiflora Polhill;裸花猪屎豆■☆
112467 Crotalaria nutans Welw. ex Baker = Crotalaria sphaerocarpa Perr. ex DC. ■☆
112468 Crotalaria nyikensis Baker;尼卡猪屎豆■☆
112469 Crotalaria obcordata P. J. Bergius = Wiborgia obcordata (P. J. Bergius) Thunb. ■☆
112470 Crotalaria obovata G. Don = Crotalaria pallida Aiton var. obovata (G. Don) Polhill ■☆
112471 Crotalaria obovata G. Don = Crotalaria pallida Aiton ●■
112472 Crotalaria obscura DC.;模糊猪屎豆■☆
112473 Crotalaria occidentalis Hepper;西方猪屎豆■☆
112474 Crotalaria occulta Graham ex Benth.;紫花猪屎豆;Purple Rattlebox,Purpleflower Rattlebox ■
112475 Crotalaria ochroleuca G. Don;狭叶猪屎豆;Narrowleaf Rattlebox,Slender Leaf Rattlebox,Yellowish-white Rattlebox ■
112476 Crotalaria oligosperma Polhill;寡籽猪屎豆■☆
112477 Crotalaria oligostachya Baker;寡穗猪屎豆■☆
112478 Crotalaria oocarpa Baker;卵果猪屎豆■☆
112479 Crotalaria oocarpa Baker subsp. microcarpa Milne-Redh.;小卵果猪屎豆■☆
112480 Crotalaria oosterboschiana Timp. = Crotalaria oxyphylloides R. Wilczek ■☆
112481 Crotalaria oreadum Baker f. = Crotalaria subcapitata De Wild. subsp. oreadum (Baker f.) Polhill ■☆
112482 Crotalaria orientalis Burtt Davy ex I. Verd.;东方猪屎豆■☆
112483 Crotalaria orientalis Burtt Davy ex I. Verd. subsp. allenii (I. Verd.) Polhill et A. Schreib.;阿伦猪屎豆■☆
112484 Crotalaria orthoclada Baker = Crotalaria xanthoclada Bojer ex Benth. ■☆
112485 Crotalaria orthoclada Welw. ex Baker;直枝猪屎豆■☆
112486 Crotalaria ovalis Pursh;椭圆叶猪屎豆■☆
112487 Crotalaria ovata Polhill;卵叶猪屎豆■☆
112488 Crotalaria oxyphylla Harms;尖叶猪屎豆■☆
112489 Crotalaria oxyphylloides R. Wilczek;拟尖叶猪屎豆■☆
112490 Crotalaria oxyptera E. Mey. = Lotononis oxyptera (E. Mey.) Benth. ●☆
112491 Crotalaria pallida Aiton;猪屎豆(白猪屎豆,大马铃,大眼蓝,黄野百合,水蓼竹,土沙苑子,野花生,野黄豆,猪屎青);Pallid Rattlebox,Pallid Rattle-box,Smooth Rattlebox,Striped Crotalaria ●■
112492 Crotalaria pallida Aiton var. obovata (G. Don) Polhill;倒卵叶猪屎豆;Smooth Rattlebox ■☆
112493 Crotalaria pallida Aiton var. obovata (G. Don) Polhill = Crotalaria pallida Aiton ●■
112494 Crotalaria pallidicaulis Harms;苍白茎猪屎豆■☆
112495 Crotalaria paludosa A. Chev. = Crotalaria glaucoides Baker f. ■☆
112496 Crotalaria paniculata Willd.;圆锥猪屎豆;Paniculate Rattlebox ■☆
112497 Crotalaria paolii Cufod. = Crotalaria massaiensis Taub. ■☆
112498 Crotalaria paraspartea Polhill;假鹰爪豆■☆
112499 Crotalaria parsonsii Baker f. = Crotalaria leprieurii Guillaumin et Perr. ■☆
112500 Crotalaria parvifolia Thunb. = Amphithalea parvifolia (Thunb.) A. L. Schutte ●☆
112501 Crotalaria parvula Beck = Crotalaria dumosa Franch. ●☆
112502 Crotalaria passargei Taub. = Crotalaria confusa Hepper ■☆
112503 Crotalaria passerinoides Taub.;雀猪屎豆■☆
112504 Crotalaria patula Polhill;张开猪屎豆■☆
112505 Crotalaria pauciflora Baker = Crotalaria virgulata Klotzsch subsp. pauciflora (Baker) Polhill ●☆
112506 Crotalaria paulitschkei Baker f. = Crotalaria dumosa Franch. ●☆
112507 Crotalaria pearsonii Baker f.;皮尔逊猪屎豆■☆
112508 Crotalaria pechueliana Schinz = Crotalaria argyraea Welw. ex Baker ●☆
112509 Crotalaria peguana Benth. ex Baker;庇古猪屎豆(薄叶猪屎豆);Thinleaf Rattlebox ■
112510 Crotalaria peguana Benth. ex Baker = Crotalaria kurzii Baker ex Kurz ■●
112511 Crotalaria peguana Benth. ex Baker var. qiubeiensis (C. Y. Yang) C. Chen et J. Q. Li = Crotalaria kurzii Baker ex Kurz ■●
112512 Crotalaria peltieri Polhill;盾状猪屎豆●☆
112513 Crotalaria pentaphylla Baker f.;五叶猪屎豆■☆
112514 Crotalaria pentheri Gand. = Crotalaria globifera E. Mey. ■☆
112515 Crotalaria peregrina Polhill;外来猪屎豆■☆

112516 Crotalaria perplexa E. Mey. = Lotononis perplexa (E. Mey.) Eckl. et Zeyh. ■☆
112517 Crotalaria perrieri R. Vig.;佩里耶猪屎豆■☆
112518 Crotalaria perrottetii DC.;佩罗猪屎豆●☆
112519 Crotalaria persica (Burm. f.) Merr.;波斯猪屎豆●☆
112520 Crotalaria pervillei Baill.;佩尔猪屎豆●☆
112521 Crotalaria peschiana P. A. Duvign. et Timp.;佩施猪屎豆●☆
112522 Crotalaria petiolaris Franch. = Crotalaria laburnifolia L. subsp. petiolaris (Franch.) Polhill ■☆
112523 Crotalaria petiolaris Franch. var. australis Baker f. = Crotalaria laburnifolia L. subsp. australis (Baker f.) Polhill ■☆
112524 Crotalaria petiolata Vogel ex Walp.;柄叶猪屎豆■☆
112525 Crotalaria petitiana (A. Rich.) Walp.;佩蒂蒂猪屎豆■☆
112526 Crotalaria phillipsiae Baker;菲利猪屎豆■☆
112527 Crotalaria phylicoides Wild;菲利木猪屎豆●☆
112528 Crotalaria phylloloba Harms;浅裂叶猪屎豆■☆
112529 Crotalaria phyllostachys Baker;叶穗猪屎豆■☆
112530 Crotalaria piedboeufii R. Wilczek = Crotalaria kambolensis Baker f. ■☆
112531 Crotalaria pilifera Klotzsch = Crotalaria podocarpa DC. ■☆
112532 Crotalaria pilosa Thunb. = Crotalaria obscura DC. ■☆
112533 Crotalaria pilosa Thunb. var. collina Eckl. et Zeyh. = Crotalaria obscura DC. ■☆
112534 Crotalaria pilosa Thunb. var. riparia Eckl. et Zeyh. = Crotalaria obscura DC. ■☆
112535 Crotalaria pilosiflora Baker;毛花猪屎豆■☆
112536 Crotalaria pilosissima Miq. = Crotalaria ferruginea Graham ex Benth. ■
112537 Crotalaria platycalyx Steud. ex Baker = Crotalaria quartiniana A. Rich. ■☆
112538 Crotalaria platysepala Harv.;宽萼猪屎豆■☆
112539 Crotalaria pleiophylla Polhill;多叶猪屎豆■☆
112540 Crotalaria podocarpa DC.;柄果猪屎豆■☆
112541 Crotalaria podocarpa DC. subsp. flexuosa (Baker) Baker f. = Crotalaria podocarpa DC. ■☆
112542 Crotalaria poecilantha Polhill;杂色猪屎豆●☆
112543 Crotalaria poissonii R. Vig.;普瓦松猪屎豆●☆
112544 Crotalaria polhillii Thulin;普尔猪屎豆■☆
112545 Crotalaria poliochlora Harms;灰绿猪屎豆■☆
112546 Crotalaria polyantha Taub.;多花猪屎豆■☆
112547 Crotalaria polycarpa Benth. = Crotalaria sphaerocarpa Perr. ex DC. subsp. polycarpa (Benth.) Hepper ■☆
112548 Crotalaria polychotoma Taub. = Crotalaria microcarpa Hochst. ex Benth. ■☆
112549 Crotalaria polychroma Polhill;多色猪屎豆■☆
112550 Crotalaria polyclados Welw. ex Baker = Crotalaria cyanea Baker ■☆
112551 Crotalaria polygaloides Welw. ex Baker;远志猪屎豆●☆
112552 Crotalaria polygaloides Welw. ex Baker subsp. orientalis Polhill;东方远志猪屎豆●☆
112553 Crotalaria polysperma Kotschy;多籽猪屎豆■☆
112554 Crotalaria polysperma Kotschy subsp. stewartii (Baker) Baker f. = Crotalaria polysperma Kotschy ■☆
112555 Crotalaria polytricha Polhill;密毛猪屎豆■☆
112556 Crotalaria praecox Milne-Redh. = Crotalaria graminicola Taub. ex Baker f. ■☆
112557 Crotalaria preladoi Baker f.;普拉多猪屎豆●☆
112558 Crotalaria prittwitzii Baker f.;普里特猪屎豆■☆
112559 Crotalaria procumbens C. Sm.;平铺猪屎豆■☆
112560 Crotalaria prolifera E. Mey. = Lotononis prolifera (E. Mey.) B.-E. van Wyk ■☆
112561 Crotalaria prolongata Baker;延长猪屎豆■☆
112562 Crotalaria prostrata Rottler ex Willd.;俯伏猪屎豆;Protracte Rattlebox ■
112563 Crotalaria prostrata Rottler ex Willd. var. jinpingensis (C. Y. Yang) C. Y. Yang;金平猪屎豆;Jinping Rattlebox ■
112564 Crotalaria prostrata Rottler ex Willd. var. jinpingensis (C. Y. Yang) C. Y. Yang = Crotalaria prostrata Rottler ex Willd. ■
112565 Crotalaria prostrata Roxb. = Crotalaria prostrata Rottler ex Willd. ■
112566 Crotalaria prostrata Roxb. ex D. Don = Crotalaria prostrata Rottler ex Willd. ■
112567 Crotalaria protensa Welw. ex Baker;伸展猪屎豆■☆
112568 Crotalaria psammophila Harms;喜沙猪屎豆■☆
112569 Crotalaria pseudo-alexandri R. Wilczek;假亚历山大猪屎豆■☆
112570 Crotalaria pseudodescampsii Baker f. = Crotalaria lukafuensis De Wild. ■☆
112571 Crotalaria pseudo-eriosema Vatke = Crotalaria vasculosa Wall. ex Benth. ■☆
112572 Crotalaria pseudoflorida R. Wilczek = Crotalaria kambolensis Baker f. ■☆
112573 Crotalaria pseudonatalitia R. Wilczek = Crotalaria prittwitzii Baker f. ■☆
112574 Crotalaria pseudopodocarpa R. E. Fr. = Crotalaria cylindrocarpa DC. ■☆
112575 Crotalaria pseudoseretii R. Wilczek;假赛雷猪屎豆■☆
112576 Crotalaria pseudospartium Baker f.;假绳索猪屎豆■☆
112577 Crotalaria psoralioides D. Don = Prioptropis cytisoides (Roxb. ex DC.) Wight et Arn. ■
112578 Crotalaria psoraloides Lam. = Eriosema psoraloides (Lam.) G. Don ■☆
112579 Crotalaria pterocalyx Harms;翅萼猪屎豆■☆
112580 Crotalaria pteropoda Balf. f.;翅足猪屎豆■☆
112581 Crotalaria pterospartioides Torre;索翅猪屎豆■☆
112582 Crotalaria pudica Polhill;羞涩猪屎豆■☆
112583 Crotalaria pulchella Andréws = Lebeckia cytisoides Thunb. ■☆
112584 Crotalaria purpurascens Lam. = Crotalaria incana L. subsp. purpurascens (Lam.) Milne-Redh. ■☆
112585 Crotalaria purpurascens Lam. = Crotalaria incana L. ■
112586 Crotalaria purpurea Vent. = Hypocalyptus coluteoides (Lam.) R. Dahlgren ●☆
112587 Crotalaria purpureolineata Baker f. = Crotalaria brevidens Benth. var. intermedia (Kotschy) Polhill ■☆
112588 Crotalaria pycnostachya Benth.;密穗猪屎豆■☆
112589 Crotalaria pycnostachya Benth. subsp. donaldsonii (Baker f.) Polhill;唐纳森猪屎豆■☆
112590 Crotalaria pycnostachya Benth. var. donaldsonii Baker f. = Crotalaria pycnostachya Benth. subsp. donaldsonii (Baker f.) Polhill ■☆
112591 Crotalaria pygmaea Polhill;微小猪屎豆■☆
112592 Crotalaria qiubeiensis C. Y. Yang;邱北猪屎豆;Qiubei Rattlebox ■
112593 Crotalaria qiubeiensis C. Y. Yang = Crotalaria kurzii Baker ex Kurz ■●

112594 Crotalaria quangensis Taub.;热非猪屎豆■●☆

112595 Crotalaria quangensis Taub. var. capituliformis (R. Wilczek) Polhill;小头猪屎豆■☆

112596 Crotalaria quangensis Taub. var. malangensis (Baker f.) Polhill;马兰加猪屎豆●☆

112597 Crotalaria quarrei Baker f.;卡雷猪屎豆●☆

112598 Crotalaria quarrei Baker f. var. longipes Polhill;长梗猪屎豆■☆

112599 Crotalaria quartiniana A. Rich.;夸尔廷猪屎豆■☆

112600 Crotalaria quinata E. Mey. = Lotononis acutiflora Benth. ■☆

112601 Crotalaria quinquefolia L.;热带五叶猪屎豆■☆

112602 Crotalaria raffillii Milne-Redh. = Crotalaria rosenii (Pax) Milne-Redh. ex Polhill ●☆

112603 Crotalaria ramosissima Baker = Crotalaria huillensis Taub. ■☆

112604 Crotalaria randii Baker f. = Crotalaria alexandri Baker f. ■☆

112605 Crotalaria rangei Harms;兰格猪屎豆■☆

112606 Crotalaria rathjensiana O. Schwartz = Crotalaria emarginella Vatke ■☆

112607 Crotalaria reclinata Polhill;拱垂猪屎豆■☆

112608 Crotalaria recta Steud. ex A. Rich.;直立猪屎豆■☆

112609 Crotalaria recta Steud. ex A. Rich. subsp. simplex ? = Crotalaria recta Steud. ex A. Rich. ■☆

112610 Crotalaria recta Steud. ex A. Rich. var. katangensis Robyns ex R. Wilczek = Crotalaria recta Steud. ex A. Rich. ■☆

112611 Crotalaria reflexa Thunb. = Xiphotheca reflexa (Thunb.) A. L. Schutte et B. -E. van Wyk ■☆

112612 Crotalaria renieriana R. Wilczek;雷尼尔猪屎豆■☆

112613 Crotalaria reniformis Lam. = Rafnia amplexicaulis (L.) Thunb. ■☆

112614 Crotalaria reptans Taub.;匍匐猪屎豆■☆

112615 Crotalaria retusa L.;吊裙草(凹叶野百合,凹猪屎豆);Hangskirt Rattlebox, Yellow-flowering Rattlebox ■

112616 Crotalaria rhizoclada Polhill;根枝猪屎豆■☆

112617 Crotalaria rhodesiae Baker f.;罗得西亚猪屎豆■☆

112618 Crotalaria rhynchocarpa Polhill;喙果猪屎豆■☆

112619 Crotalaria rigidula Baker f. = Crotalaria monteiroi Taub. ex Baker f. var. galpinii Burtt Davy ex I. Verd. ■☆

112620 Crotalaria ringoetii Baker f.;林戈猪屎豆■☆

112621 Crotalaria riparia Polhill;河岸猪屎豆■☆

112622 Crotalaria robinsoniana Torre = Crotalaria germainii R. Wilczek ■☆

112623 Crotalaria robynsii R. Wilczek = Crotalaria kambolensis Baker f. ■☆

112624 Crotalaria rogersii Baker f.;罗杰斯猪屎豆■☆

112625 Crotalaria rogersii Baker f. f. kilwaensis R. Wilczek = Crotalaria rogersii Baker f. ■☆

112626 Crotalaria rosenii (Pax) Milne-Redh. ex Polhill;罗森猪屎豆●☆

112627 Crotalaria rotundicarinata Baker f. = Crotalaria goetzei Harms ●☆

112628 Crotalaria roxburghiana DC. = Crotalaria calycina Schrank ■

112629 Crotalaria rubiginosa Willd.;锈红猪屎豆■☆

112630 Crotalaria rufescens Franch. = Crotalaria ferruginea Graham ex Benth. ■

112631 Crotalaria rufocarpa Gilli = Crotalaria vasculosa Wall. ex Benth. ■☆

112632 Crotalaria rufocaulis Gilli;浅红茎猪屎豆■☆

112633 Crotalaria rupicola Baker f.;岩生猪屎豆■☆

112634 Crotalaria ruspoliana Chiov.;鲁斯波利猪屎豆■☆

112635 Crotalaria sacculata Chiov.;小囊猪屎豆■☆

112636 Crotalaria sagittalis L.;翼茎猪屎豆;Arrowhead Rattle-box, Rattlebox, Weedy Rattle-box ■☆

112637 Crotalaria sagittalis L. var. blumeriana Senn = Crotalaria sagittalis L. ■☆

112638 Crotalaria sagittalis L. var. fruticosa (Mill.) Fawc. et Rendle = Crotalaria sagittalis L. ■☆

112639 Crotalaria sagittalis L. var. oblonga Michx. = Crotalaria sagittalis L. ■☆

112640 Crotalaria saharae Coss.;左原猪屎豆■☆

112641 Crotalaria saltiana Andr.;非洲猪屎豆;African Rattlebox ☆

112642 Crotalaria saltiana Prain ex King = Crotalaria pallida Aiton ●■

112643 Crotalaria sapinii De Wild.;萨潘猪屎豆■☆

112644 Crotalaria sapinii De Wild. subsp. kasaiensis (R. Wilczek) Polhill;开赛猪屎豆■☆

112645 Crotalaria scassellatii Chiov.;斯卡猪屎豆■☆

112646 Crotalaria schimperi A. Rich. = Crotalaria incana L. ■

112647 Crotalaria schinzii Baker f.;欣兹猪屎豆■☆

112648 Crotalaria schlechteri Baker f.;施莱猪屎豆■☆

112649 Crotalaria schliebenii Polhill;施利本猪屎豆■☆

112650 Crotalaria schmitzii R. Wilczek;施密茨猪屎豆■☆

112651 Crotalaria schultzei Harms = Crotalaria colorata Schinz subsp. erecta (Schinz) Polhill ■☆

112652 Crotalaria schweinfurthii Deflers = Lotus garcinii DC. ■☆

112653 Crotalaria semperflorens Vent.;速生猪屎豆●☆

112654 Crotalaria senegalensis (Pers.) Bacle ex DC.;塞内加尔猪屎豆■☆

112655 Crotalaria senegalensis (Pers.) Bacle ex DC. var. carinata (Steud. ex A. Rich.) Baker f. = Crotalaria senegalensis (Pers.) Bacle ex DC. ■☆

112656 Crotalaria sengae Baker f. ex R. Wilczek = Crotalaria axilliflora Baker f. ●☆

112657 Crotalaria sengensis Baker f.;森加猪屎豆●☆

112658 Crotalaria sennii Chiov. = Crotalaria boranica Harms ex Baker f. subsp. trichocarpa Polhill ■☆

112659 Crotalaria seretii De Wild. = Crotalaria cuspidata Taub. ■☆

112660 Crotalaria sericea Burm. f. = Crotalaria assamica Benth. ■

112661 Crotalaria sericea Retz. = Crotalaria spectabilis Roth ■

112662 Crotalaria sericea Willd. = Crotalaria juncea L. ■

112663 Crotalaria sericifolia Harms;绢毛叶猪屎豆■☆

112664 Crotalaria sericifolia Harms var. gweleonsis Baker f. = Crotalaria variegata Welw. ex Baker ■☆

112665 Crotalaria serpens E. Mey. = Lotononis hirsuta (Thunb.) D. Dietr. ■☆

112666 Crotalaria serpentinicola Leteint. et Polhill;蛇纹岩猪屎豆■☆

112667 Crotalaria sessiliflora L.;无梗猪屎豆(倒挂山芝麻,佛指花,佛指甲,狗铃草,兰花野百合,兰花猪屎豆,蓝花野百合,蓝花猪屎豆,狸豆,农吉利,山油麻,鼠蛋草,鼠蛋叶,细叶芝麻铃,响铃草,羊屎蛋,野百合,野花生,野芝麻,芝麻响铃铃,猪铃草,紫花野百合);Pediselless Rattlebox, Sessileflower Rattlebox, Wild Lily ■

112668 Crotalaria sessilis De Wild.;无柄猪屎豆■☆

112669 Crotalaria shirensis (Baker f.) Milne-Redh.;希尔猪屎豆■☆

112670 Crotalaria sidamaensis Chiov. = Crotalaria ruspoliana Chiov. ■☆

112671 Crotalaria similis Hemsl.;屏东猪屎豆(鹅銮鼻野百合);Pingdong Rattlebox, Similar Rattlebox ■

112672 Crotalaria simplex A. Rich. = Crotalaria recta Steud. ex A. Rich. ■☆

112673 Crotalaria simulans Milne-Redh.;相似猪屎豆■☆

112674 Crotalaria sinensis J. F. Gmel. = Crotalaria chinensis L. ■

112675 Crotalaria singulifloroides R. Wilczek;单花猪屎豆■☆

112676 Crotalaria somalensis Chiov.;索马里猪屎豆■☆

112677 Crotalaria sparsiflora E. Mey. = Lotononis sparsiflora (E. Mey.) B. -E. van Wyk ■☆

112678 Crotalaria sparsifolia Baker;稀叶猪屎豆■☆

112679 Crotalaria spartea Baker;鹰爪猪屎豆■☆

112680 Crotalaria spartioides DC.;拟鹰爪猪屎豆■☆

112681 Crotalaria spathulato-foliolata Torre;匙叶猪屎豆■☆

112682 Crotalaria spectabilis Roth;柔毛猪屎豆(大托叶猪屎豆,美丽猪屎豆,丝毛野百合,紫花野百合);Beautiful Rattlebox, Largestipule Rattlebox, Showy Crotalaria, Showy Rattlebox ■

112683 Crotalaria sphaerocarpa Perr. ex DC.;非洲球果猪屎豆■☆

112684 Crotalaria sphaerocarpa Perr. ex DC. subsp. polycarpa (Benth.) Hepper;多球果猪屎豆■☆

112685 Crotalaria sphaerocarpa Perr. ex DC. var. angustifolia Hochst. ex Kuntze = Crotalaria sphaerocarpa Perr. ex DC. ■☆

112686 Crotalaria sphaerocarpa Perr. ex DC. var. grandiflora Schweinf. ex Baker f. = Crotalaria sphaerocarpa Perr. ex DC. ■☆

112687 Crotalaria sphaerocarpa Perr. ex DC. var. lanceolata Schinz = Crotalaria sphaerocarpa Perr. ex DC. ■☆

112688 Crotalaria sphaerocarpa Perr. ex DC. var. polycarpa (Benth.) Kuntze = Crotalaria sphaerocarpa Perr. ex DC. subsp. polycarpa (Benth.) Hepper ■☆

112689 Crotalaria spinosa Hochst. ex Benth.;具刺猪屎豆■☆

112690 Crotalaria spinosa Hochst. ex Benth. subsp. aculeata (De Wild.) Baker f. = Crotalaria aculeata De Wild. ●☆

112691 Crotalaria spinosa Hochst. ex Benth. var. macrocarpa Baker f. = Crotalaria kapiriensis De Wild. ■☆

112692 Crotalaria spinosa Hochst. ex Benth. var. schlechteri Baker f. = Crotalaria eremicola Baker f. ●☆

112693 Crotalaria splendens Vogel = Crotalaria uncinella Lam. subsp. elliptica (Roxb.) Polhill ■

112694 Crotalaria splendens Vogel = Crotalaria uncinella Lam. ■

112695 Crotalaria squarrosa Schinz = Crotalaria sphaerocarpa Perr. ex DC. ■☆

112696 Crotalaria staneriana Baker f.;斯塔内猪屎豆●☆

112697 Crotalaria stenocladon Baker f. = Crotalaria graminicola Taub. ex Baker f. ■☆

112698 Crotalaria stenophylla Bojer = Crotalaria diosmifolia Benth. ●☆

112699 Crotalaria stenophylla Eckl. et Zeyh. = Lotononis stenophylla (Eckl. et Zeyh.) B. -E. van Wyk ■☆

112700 Crotalaria stenophylla Vogel = Crotalaria linifolia L. f. var. stenophylla (Vogel) C. Y. Yang ■

112701 Crotalaria stenopoda Baker f.;窄足猪屎豆■☆

112702 Crotalaria stenoptera Welw. ex Baker;窄翅猪屎豆■☆

112703 Crotalaria stenoptera Welw. ex Baker var. latifolia Baker f. = Crotalaria stenoptera Welw. ex Baker ■☆

112704 Crotalaria stenorhampha Harms;狭猪屎豆■☆

112705 Crotalaria stenothyrsa Taub.;窄序猪屎豆■☆

112706 Crotalaria steudneri Schweinf.;斯托德猪屎豆■☆

112707 Crotalaria stewartii Baker = Crotalaria polysperma Kotschy ■☆

112708 Crotalaria stipulacea Roxb. = Crotalaria mysorensis Roth ■

112709 Crotalaria stipularia Desv.;托叶猪屎豆■☆

112710 Crotalaria stolzii (Baker f.) Milne-Redh.;斯托尔兹猪屎豆■☆

112711 Crotalaria streptorrhyncha Milne-Redh.;扭喙猪屎豆■☆

112712 Crotalaria striata DC. = Crotalaria pallida Aiton ●■

112713 Crotalaria stuhlmannii Taub.;斯图尔曼猪屎豆■☆

112714 Crotalaria stuhlmannii Taub. var. acuticarinata Polhill;尖棱斯图尔曼猪屎豆■☆

112715 Crotalaria stuhlmannii Taub. var. crassicarpa Polhill;粗果斯图尔曼猪屎豆■☆

112716 Crotalaria subcaespitosa Polhill;丛生猪屎豆■☆

112717 Crotalaria subcalvata Polhill;近光猪屎豆■☆

112718 Crotalaria subcapitata De Wild.;亚头状猪屎豆■☆

112719 Crotalaria subcapitata De Wild. = Crotalaria lanceolata E. Mey. var. malangensis Baker f. ■☆

112720 Crotalaria subcapitata De Wild. subsp. oreadum (Baker f.) Polhill;山地亚头状猪屎豆■☆

112721 Crotalaria subcapitata De Wild. var. fwamboensis (Baker f.) Polhill;富瓦姆波猪屎豆■☆

112722 Crotalaria subdisperma Baker f. = Crotalaria hyssopifolia Klotzsch ●☆

112723 Crotalaria subsessilis Harms;近无柄猪屎豆■☆

112724 Crotalaria subspicata Polhill;穗状猪屎豆■☆

112725 Crotalaria subtilis Polhill;纤细猪屎豆●☆

112726 Crotalaria subumbellata Torre = Crotalaria elisabethae Baker f. ■☆

112727 Crotalaria sylvicola Baker f.;西尔维亚猪屎豆■☆

112728 Crotalaria symoensiana Timp. = Crotalaria annua Milne-Redh. ●☆

112729 Crotalaria szemaoensis Gagnep.;思茅猪屎豆(小扁豆);Simao Rattlebox, Simao Rattle-box ■

112730 Crotalaria szemaoensis Gagnep. = Crotalaria psoralioides D. Don ■

112731 Crotalaria szemaoensis Gagnep. = Priotropis cytisoides (Roxb. ex DC.) Wight et Arn. ■

112732 Crotalaria tabularis Baker f.;扁平猪屎豆■☆

112733 Crotalaria tanety Du Puy, Labat et H. E. Ireland = Crotalaria cytisoides Hils. et Bojer ex Benth. ■☆

112734 Crotalaria tanganyikensis Baker f. = Crotalaria steudneri Schweinf. ■☆

112735 Crotalaria teitensis Sacleux = Crotalaria barkae Schweinf. subsp. teitensis (Sacleux) Polhill ●☆

112736 Crotalaria teixeirae Torre;特谢拉猪屎豆●☆

112737 Crotalaria tenuiflora Steud.;瘦叶猪屎豆■☆

112738 Crotalaria tenuifolia Roxb. ex DC. = Crotalaria juncea L. ■

112739 Crotalaria tenuifolia Roxb. ex Hornem. = Crotalaria juncea L. ■

112740 Crotalaria tenuipedicellata Baker f.;细花梗猪屎豆■☆

112741 Crotalaria tenuirama Welw. ex Baker;细枝猪屎豆■☆

112742 Crotalaria tenuirostrata Polhill;细喙猪屎豆■☆

112743 Crotalaria tenuis Baker = Crotalaria xanthoclada Bojer ex Benth. ■☆

112744 Crotalaria teretifolia Milne-Redh.;圆柱叶猪屎豆■☆

112745 Crotalaria tetragona Roxb. ex Andréws;四棱猪屎豆(化金丹);Fourangular Rattlebox, Fourridgy Rattlebox ■

112746 Crotalaria tetraptera Torre;四翅猪屎豆■☆

112747 Crotalaria thaumasiophylla Harms = Crotalaria variegata Welw. ex Baker ■☆

112748 Crotalaria thomasii Harms;托马斯猪屎豆■☆

112749 Crotalaria thomensis Baker f. = Crotalaria trichotoma Bojer ■

112750 Crotalaria thomsonii Oliv. = Crotalaria goodiiformis Vatke ●☆

112751 Crotalaria tiantaiensis Yan C. Jiang et al.;天台猪屎豆■

112752 Crotalaria tigrensis Baker = Crotalaria cylindrica A. Rich. ■☆

112753 Crotalaria tomentosa Thunb.;绒毛猪屎豆■☆

112754 Crotalaria torrei Polhill;托雷猪屎豆■☆

112755 Crotalaria trachycarpa Taub. ex Baker f. = Crotalaria barkae

Schweinf. ●☆
112756 Crotalaria trichopoda E. Mey. = Lotononis glabra (Thunb.) D. Dietr. ■☆
112757 Crotalaria trichotoma Bojer;光萼猪屎豆(光萼野百合,苦罗豆,南美猪屎豆,桑岛猪屎豆,西印度猪屎豆);Glabroussepal Rattlebox, Velvet Rattlebox, West Indian Rattlebox, Zanzibar Rattlebox ■
112758 Crotalaria triflora L. = Rafnia triflora (L.) Thunb. ■☆
112759 Crotalaria trifoliastrum Willd. = Crotalaria medicaginea Lam. ■
112760 Crotalaria trifoliastrum Willd. = Crotalaria pallida Aiton var. obovata (G. Don) Polhill ■☆
112761 Crotalaria trifoliolata Baker f.;三小叶猪屎豆■☆
112762 Crotalaria trinervia Polhill;三脉猪屎豆■☆
112763 Crotalaria triquatra Dalzell;砂地野百合■
112764 Crotalaria triquetra Dalzell var. garambiensis Y. C. Liu et F. Y. Lu = Crotalaria triquetra Dalzell ■
112765 Crotalaria tristis Polhill;暗淡猪屎豆■☆
112766 Crotalaria truncata E. Mey. = Lotononis umbellata (L.) Benth. ●☆
112767 Crotalaria truncata Schinz = Crotalaria sphaerocarpa Perr. ex DC. ■☆
112768 Crotalaria tuberosa Buch.-Ham. ex D. Don = Eriosema himalaicum Ohashi ■
112769 Crotalaria ugandensis Baker f. = Crotalaria vasculosa Wall. ex Benth. ■☆
112770 Crotalaria uguenensis Taub.;乌古猪屎豆●☆
112771 Crotalaria ukambensis Vatke;乌卡猪屎豆■☆
112772 Crotalaria ukingensis Harms;尤金猪屎豆●☆
112773 Crotalaria uliginosa C. C. Huang;湿生猪屎豆;Humid Rattlebox, Swamp Rattlebox ■
112774 Crotalaria umbellifera R. E. Fr.;伞花猪屎豆■☆
112775 Crotalaria uncinata Welw. ex Baker;具钩猪屎豆■☆
112776 Crotalaria uncinella Lam.;球果猪屎豆(钩状猪屎豆);Ballfruit Rattlebox, Hooked Rattlebox ■
112777 Crotalaria uncinella Lam. subsp. elliptica (Roxb.) Polhill;椭圆球果猪屎豆■
112778 Crotalaria uncinella Lam. var. senegalensis Pers. = Crotalaria senegalensis (Pers.) Bacle ex DC. ■☆
112779 Crotalaria unicaulis Bullock;单茎猪屎豆■☆
112780 Crotalaria uniflora Baker = Crotalaria occidentalis Hepper ■☆
112781 Crotalaria usaramoensis Baker f. = Crotalaria trichotoma Bojer ■
112782 Crotalaria vagans Polhill;漫游猪屎豆■☆
112783 Crotalaria valida Baker;刚直猪屎豆■☆
112784 Crotalaria vallicola Baker f.;河谷猪屎豆●☆
112785 Crotalaria vandenbrandii R. Wilczek;范登布兰德猪屎豆■☆
112786 Crotalaria vanderystii R. Wilczek;范德猪屎豆■☆
112787 Crotalaria vanmeelii R. Wilczek;范米尔猪屎豆●☆
112788 Crotalaria variegata Welw. ex Baker;变色猪屎豆■☆
112789 Crotalaria vasculosa Wall. ex Benth.;瓶状猪屎豆■☆
112790 Crotalaria vatkeana Engl.;瓦特凯猪屎豆●☆
112791 Crotalaria verdcourtii Polhill;韦尔德猪屎豆■☆
112792 Crotalaria verrucosa L.;多疣猪屎豆(大叶野百合,多疣野百合);Verrucose Rattlebox, Warty Rattlebox ■
112793 Crotalaria versicolor Baker = Crotalaria polychroma Polhill ■☆
112794 Crotalaria versicolor E. Mey. = Lotononis decumbens (Thunb.) B.-E. van Wyk ■●☆
112795 Crotalaria vexillata E. Mey. = Lotononis prostrata (L.) Benth. ■☆
112796 Crotalaria vialis Milne-Redh.;路边猪屎豆■☆
112797 Crotalaria villosa Thunb.;长柔毛猪屎豆■☆
112798 Crotalaria virgata Thunb. = Rafnia angulata Thunb. subsp. thunbergii (Harv.) G. J. Campb. et B.-E. van Wyk ●☆
112799 Crotalaria virgulata Klotzsch;灌木猪屎豆;Thicket Rattlebox ●☆
112800 Crotalaria virgulata Klotzsch subsp. forbesii (Baker) Polhill;福布斯猪屎豆■☆
112801 Crotalaria virgulata Klotzsch subsp. grantiana (Harv.) Polhill;格氏灌木猪屎豆;Grant's Rattlebox ●☆
112802 Crotalaria virgulata Klotzsch subsp. longistyla (Baker f.) Polhill;长柱灌木猪屎豆●☆
112803 Crotalaria virgulata Klotzsch subsp. pauciflora (Baker) Polhill;少花灌木猪屎豆●☆
112804 Crotalaria virgulta Spreng.;条纹猪屎豆●☆
112805 Crotalaria virgultalis Burch. ex DC.;小条纹猪屎豆●☆
112806 Crotalaria vogelii Benth. = Crotalaria leprieurii Guillaumin et Perr. ■☆
112807 Crotalaria volubilis Thunb.;缠结猪屎豆■☆
112808 Crotalaria warkeri Arn.;沃氏猪屎豆;Warker Rattlebox ■☆
112809 Crotalaria welwitschii Baker;韦氏猪屎豆■☆
112810 Crotalaria wilczekiana Timp.;维尔切克猪屎豆●☆
112811 Crotalaria wildemanii Baker f. et Martin = Crotalaria mildbraedii Baker f. ■☆
112812 Crotalaria willdenowiana DC.;威尔猪屎豆■☆
112813 Crotalaria winkleri Baker f. = Crotalaria polysperma Kotschy ■☆
112814 Crotalaria wissmannii O. Schwartz = Crotalaria aegyptiaca Benth. ■☆
112815 Crotalaria wittei Baker f. = Crotalaria chrysochlora Baker f. ex Harms ■☆
112816 Crotalaria xanthoclada Bojer ex Benth.;黄枝猪屎豆■☆
112817 Crotalaria xanthoclada Bojer ex Benth. var. stolzii Baker f. = Crotalaria stolzii (Baker f.) Milne-Redh. ■☆
112818 Crotalaria xassenguensis Torre = Crotalaria quangensis Taub. ■●☆
112819 Crotalaria yaihsienensis T. C. Chen;崖州猪屎豆(崖州野百合);Yaihsien Rattlebox, Yaxian Rattlebox, Yazhou Rattlebox ■
112820 Crotalaria youngii Baker f.;扬氏猪屎豆■☆
112821 Crotalaria yuanjiangensis C. Y. Yang;元江猪屎豆;Yuanjiang Rattlebox ■●
112822 Crotalaria yuanjiangensis C. Y. Yang = Crotalaria medicaginea Lam. ■
112823 Crotalaria yunnanensis Franch.;云南猪屎豆;Yunnan Rattlebox ■
112824 Crotalaria yunnanensis Franch. var. heqingensis (C. Y. Yang) C. Chen et J. Q. Li;鹤庆猪屎豆;Heqing Rattlebox ■
112825 Crotalaria zanzibarica Benth. = Crotalaria trichotoma Bojer ■
112826 Crotalaria zimmermannii Baker f. = Crotalaria barkae Schweinf. subsp. zimmermannii (Baker f.) Polhill ●☆
112827 Crotalopsis Michx. ex DC. = Baptisia Vent. ■☆
112828 Crotolaria Neck. = Crotalaria L. ●■
112829 Croton L. (1753);巴豆属;Croton, Fever Bark ●
112830 Croton aceroides Radcl.-Sm.;槭巴豆●☆
112831 Croton acuminatum Thunb. = Mallotus japonicus (L. f.) Müll. Arg. ●
112832 Croton alabamense E. A. Sm. ex Chapm.;阿州巴豆;Alabama Croton ●☆
112833 Croton albicans (Blume) Rchb. f. et Zoll. = Sumbaviopsis albicans (Blume) J. J. Sm. ●
112834 Croton alienum Pax;外来巴豆●☆

112835 Croton amabile Müll. Arg. = Croton gratissimum Burch. ●☆
112836 Croton angolense Müll. Arg. ;安哥拉巴豆●☆
112837 Croton antunesii Pax = Croton gratissimum Burch. ●☆
112838 Croton argenteum L. ;银巴豆;Silver July Croton ●☆
112839 Croton aromaticum L. ;香巴豆●☆
112840 Croton asperifolium Pax = Croton sylvaticum Hochst. ex Krauss ●☆
112841 Croton aubrevillei J. Léonard;奥布巴豆●☆
112842 Croton balsamiferum Müll. Arg. ; 海滨巴豆; Seaside Sage, Yellow Balsam ●☆
112843 Croton barotsense Gibbs = Croton leuconeurus Pax ●☆
112844 Croton billbergianum Müll. Arg. ;比尔巴豆●☆
112845 Croton bonplandianum Baill. ;邦氏巴豆;Bonpland's Croton ●☆
112846 Croton brieyi De Wild. ;布里巴豆●☆
112847 Croton bukobensis Pax = Croton sylvaticum Hochst. ex Krauss ●☆
112848 Croton butaguensis De Wild. = Croton macrostachyus Hochst. ex Delile ●☆
112849 Croton cajucarum Benth. ;卡朱巴豆●☆
112850 Croton californicum Müll. Arg. ;加州巴豆●☆
112851 Croton capense L. f. = Jatropha capensis (L. f.) Sond. ●☆
112852 Croton capitatum Michx. ; 毛巴豆; Hogwort, Hog-wort, Woolly Croton ●☆
112853 Croton cascarilloides Raeusch. ;银叶巴豆(里白巴豆,邱氏巴豆,柿糊,叶下白);Cuming Croton,Silverleaf Gentian ●
112854 Croton cascarilloides Raeusch. f. pilosum Y. T. Zhang;毛银叶巴豆;Pilose Silverleaf Gentian ●
112855 Croton caudatiformis Hand. -Mazz. = Croton euryphyllum W. W. Sm. ●
112856 Croton caudatum Geiseler;尾叶巴豆(卵叶巴豆,毛尾叶巴豆,尾状巴豆);Ovateleaf Gentian Croton,Tailed Croton ●
112857 Croton caudatum Geiseler var. tomentosum Hook. f. ;毛尾叶巴豆(毛叶巴豆);Tomentose Ovateleaf Croton ●
112858 Croton cavaleriei Gagnep. = Croton euryphyllum W. W. Sm. ●
112859 Croton chevalieri Beille = Crotonogyne chevalieri (Beille) Keay ●☆
112860 Croton chinense Benth. = Croton crassifolium Geiseler ●
112861 Croton chinense Geiseler = Mallotus paxii Pamp. ●
112862 Croton chunianum Croizat;光果巴豆;Chun Croton ●
112863 Croton ciliatoglanduliferum Ortega;毛腺巴豆●☆
112864 Croton cliffordii Hutch. et E. A. Bruce = Croton somalense Vatke ex Pax ●☆
112865 Croton collenettei Hutch. et Dalziel = Croton dispar N. E. Br. ●☆
112866 Croton columnaris Airy Shaw;柱状巴豆●☆
112867 Croton confertus Baker;密集巴豆●☆
112868 Croton congense De Wild. ;康格巴豆●☆
112869 Croton congestum Lour. = Xylosma congesta (Lour.) Merr. ●
112870 Croton courtetii Beille = Astraea lobata (L.) Klotzsch ●☆
112871 Croton crassifolium Geiseler;鸡骨香(矮脚猪,驳骨消,地灵香,滚地龙,过山香,黄牛香,鸡脚香,木沉香,山豆根,透地龙,土沉香);Cockbone's Aroma,Thickleaf Croton,Thick-leaved Croton ●
112872 Croton crispatum Thulin;皱波巴豆●☆
112873 Croton cumingii Müll. Arg. = Croton cascarilloides Raeusch. ●
112874 Croton damayeshu Y. T. Chang; 大麻叶巴豆; Damayeshu, Hempleaf Croton,Hemp-leaved Croton ●
112875 Croton decorsei Beille = Astraea lobata (L.) Klotzsch ●☆
112876 Croton dictyophlebodes Radcl. -Sm. ;网脉巴豆●☆
112877 Croton dinklagei Pax et K. Hoffm. = Croton nigritanum Scott-Elliot ●☆
112878 Croton dioicum Cav. ;异株巴豆●☆
112879 Croton dispar N. E. Br. ;异型巴豆●☆
112880 Croton draco Schltdl. ;龙血巴豆●
112881 Croton draconoides Müll. Arg. ;拟龙血巴豆●☆
112882 Croton duclouxii Gagnep. = Croton yunnanense W. W. Sm. ●
112883 Croton dybowskii Hutch. ;迪布巴豆●☆
112884 Croton eleuteria Bennet; 苦香树; Bastard Jesuit's Bark, Cascarilla,Seaside Balsam,Sweet-wood ●☆
112885 Croton elliotianum Engl. ex Pax = Croton megalocarpum Hutch. ●☆
112886 Croton elliottii Chapm. ;埃氏巴豆(埃利巴豆)●☆
112887 Croton elskensi De Wild. = Croton sylvaticum Hochst. ex Krauss ●☆
112888 Croton eluteria (L.) Sw. ;西印度巴豆;Cascarilla Bark ●☆
112889 Croton eluteria (L.) Wright = Croton eluteria (L.) Sw. ●☆
112890 Croton erythrochilum Müll. Arg. ;红唇巴豆●☆
112891 Croton euryphyllum W. W. Sm. ;石山巴豆(宽叶巴豆);Broad-leaved Croton,Rockhill Croton ●
112892 Croton ferrugineum Kunth;锈色巴豆●☆
112893 Croton fragile Kunth;脆巴豆●☆
112894 Croton fruticulosum Müll. Arg. ; 灌木巴豆; Bush Croton, Encinilla,Hierba Loca ●☆
112895 Croton glandulosum L. ; 腺点巴豆; Sand Croton, Tooth-leaved Croton,Tropic Croton,Vente Conmigo ●☆
112896 Croton glandulosum L. var. septentrionalis Müll. Arg. ;隔腺点巴豆; Sand Croton, Tooth-leaved Croton, Tropical Croton, Vente Conmigo ●☆
112897 Croton gossweileri Hutch. ;戈斯巴豆●☆
112898 Croton gossypiifolium Vahl;流肢巴豆●
112899 Croton gratissimum Burch. ;可爱巴豆●☆
112900 Croton guatemalense Lotsy;危地马拉巴豆●☆
112901 Croton gubouga S. Moore;南非巴豆●☆
112902 Croton gubouga S. Moore = Croton megalobotrys Müll. Arg. ●☆
112903 Croton guerzesiense Beille ex A. Chev. = Croton macrostachyus Hochst. ex Delile ●☆
112904 Croton hainanense Merr. et F. P. Metcalf = Croton laui Merr. et F. P. Metcalf ●◇
112905 Croton hancei Benth. ; 香港巴豆; Hance Croton, Hongkong Croton ●
112906 Croton hastatum L. = Tragia plukenetii Radcl. -Sm. ●☆
112907 Croton haumanianus J. Léonard;豪曼巴豆●☆
112908 Croton hibiscifolium Kunth;木槿叶巴豆●
112909 Croton hirtus L'Hér. ;硬毛巴豆●
112910 Croton hookeri Croizat;羊痂树;Hooker Croton ●
112911 Croton howii Merr. et Chun ex Y. T. Chang; 宽昭巴豆; How Croton,Kuanzhao Croton ●
112912 Croton humile L. ;矮巴豆●☆
112913 Croton inhambanense Radcl. -Sm. ;伊尼扬巴内巴豆●☆
112914 Croton integrifolium Pax;全叶巴豆●☆
112915 Croton jansii J. Léonard;简斯巴豆●☆
112916 Croton japonicum Thunb. = Mallotus japonicus (L. f.) Müll. Arg. ●
112917 Croton jatrophoides Pax;麻疯树巴豆●☆
112918 Croton joufra Roxb. ;长果巴豆;Long-fruited Croton ●
112919 Croton kamerunicum Pax et K. Hoffm. = Croton leuconeurus Pax ●☆

112920 Croton kongense Gagnep.;越南巴豆(假弹草,孔巴豆,银叶巴豆);Kong Croton, Tonkin Croton, Viatnam, Viet Nam Croton, Vietnam Croton ●

112921 Croton kroneanum Miq. = Croton crassifolium Geiseler ●

112922 Croton kwangsiense Croizat = Croton lachnocarpum Benth. var. kwangsiense (Croizat) H. S. Kiu ●

112923 Croton kwangsiense Croizat = Croton lachnocarpum Benth. ●

112924 Croton kwebensis N. E. Br. = Croton menyharthii Pax ●☆

112925 Croton laccifer Blanco = Macaranga tanarius (L.) Müll. Arg. ●

112926 Croton laccifer L.;虫胶巴豆●☆

112927 Croton lachnocarpum Benth.;毛果巴豆(狗屎藤,山辣蓼,山辣子,桃叶双眼龙,细叶双眼龙,下山虎,小叶双眼龙,巡山虎);Hainryfruit Croton, Hainry-fruited Croton ●

112928 Croton lachnocarpum Benth. var. kwangsiense (Croizat) H. S. Kiu;黄毛果巴豆(野巴豆)●

112929 Croton laciniatistylum J. Léonard;撕裂巴豆●☆

112930 Croton laechleri F. Muell.;秘鲁巴豆(莱克巴豆)●☆

112931 Croton laevigatum Vahl;光叶巴豆(抱龙,圆叶巴豆);Glabrous Croton, Nitidleaf Croton, Smooth-leaved Croton ●

112932 Croton lanuginosum K. Schum. ex Schweinf. = Chrozophora plicata (Vahl) A. Juss. ex Spreng. ●☆

112933 Croton laui Merr. et F. P. Metcalf;海南巴豆;Hainan Croton ●◇

112934 Croton lehmbachii Hutch. = Croton longiracemosum Hutch. ●☆

112935 Croton leonense Hutch.;莱昂巴豆●☆

112936 Croton leptostachyum Kunth;细穗花巴豆●☆

112937 Croton leuconeurus Pax;白脉巴豆●☆

112938 Croton leuconeurus Pax subsp. mossambicense Radcl.-Sm.;莫桑比克巴豆●☆

112939 Croton levatii Guillaumin;列氏巴豆●☆

112940 Croton limitincolum Croizat; 疏齿巴豆; Poortooth Croton, Remote-toothed Croton, Soóse-toothed Croton ●

112941 Croton lindheimerianus E. Scheele; 林氏巴豆; Lindheimer's Croton ●☆

112942 Croton lobatus Forssk. = Jatropha glauca Vahl ●☆

112943 Croton lobatus L. = Astraea lobata (L.) Klotzsch ●☆

112944 Croton longipedicellatum J. Léonard;长梗巴豆●☆

112945 Croton longipedicellatum J. Léonard subsp. austrotanzanicum Radcl.-Sm.;南坦桑尼亚巴豆●☆

112946 Croton longipedicellatum J. Léonard var. glabrescens Radcl.-Sm.;渐光巴豆●☆

112947 Croton longiracemosum Hutch.;长序巴豆●☆

112948 Croton loukandense Pellegr.;路坎德巴豆●☆

112949 Croton loukandense Pellegr. = Croton longiracemosum Hutch. ●☆

112950 Croton lucidum L.;光泽巴豆●☆

112951 Croton lundellii Standl.;伦德巴豆●☆

112952 Croton macrostachyus Hochst. ex A. Rich.;大穗巴豆(长穗巴豆);Baccanisa, Longispiked Croton, Long-spiked Croton ●

112953 Croton malambo H. Karst.;马拉巴豆;Malambo ●☆

112954 Croton mangelong Y. T. Chang;曼哥龙巴豆;Mangelong Croton ●

112955 Croton mayumbense J. Léonard;马永巴巴豆●☆

112956 Croton mearnsi De Wild. = Croton leuconeurus Pax ●☆

112957 Croton megalobotrys Müll. Arg.;热非大穗巴豆;Musine ●☆

112958 Croton megalocarpoides Friis et M. G. Gilbert;拟大果巴豆●☆

112959 Croton megalocarpum Hutch.;大果巴豆;Musine ●☆

112960 Croton membranaceus Müll. Arg.;膜质巴豆●☆

112961 Croton menyharthii Pax;迈尼哈尔特巴豆●☆

112962 Croton merrillianum Croizat;厚叶巴豆;Merrill's Croton ●

112963 Croton michauxii G. L. Webster;米氏巴豆;Rushfoil ●☆

112964 Croton microbotryus Pax = Croton gratissimum Burch. ●☆

112965 Croton monanthogynum Michx.;单籽巴豆(单花巴豆);One-seed Croton, Prairie Tea, Prairie-tea ●☆

112966 Croton montanum Willd. = Mallotus philippensis (Lam.) Müll. Arg. ●

112967 Croton mooriae Greenway ex Burtt Davy et Hoyle = Croton penduliflorum Hutch. ●☆

112968 Croton multiglandulosum Reinw. ex Blume = Melanolepis multiglandulosa (Reinw. ex Blume) Rchb. f. et Zoll. ●

112969 Croton nepetifolium Baill.;荆芥叶巴豆●☆

112970 Croton nigritanum Scott-Elliot;尼格里塔巴豆●☆

112971 Croton niloticus Müll. Arg. = Neoboutonia melleri (Müll. Arg.) Prain ●☆

112972 Croton niveus Jacq.;雪白巴豆;Copalehi Bark ●☆

112973 Croton nudifolium Baker et Hutch. = Croton nigritanum Scott-Elliot ●☆

112974 Croton obliquifolium Vis. = Chrozophora plicata (Vahl) A. Juss. ex Spreng. ●☆

112975 Croton obliquum Vahl = Chrozophora obliqua (Vahl) A. Juss. ex Spreng. ●☆

112976 Croton oblongifolium Delile = Chrozophora oblongifolia (Delile) A. Juss. ex Spreng. ●☆

112977 Croton oblongifolium Roxb. = Croton laevigatum Vahl ●

112978 Croton oligandrus Pierre ex Hutch.;寡蕊巴豆●☆

112979 Croton olivaceum Y. T. Chang et P. T. Li;榄绿巴豆;Olivaceous Croton, Olive Croton ●

112980 Croton oppositifolium Geiseler = Mallotus oppositifolius (Geiseler) Müll. Arg. ●☆

112981 Croton oxypetalum Müll. Arg. = Croton sylvaticum Hochst. ex Krauss ●☆

112982 Croton palanostigma Klotzsch;枝梗巴豆●☆

112983 Croton panamense Klotzsch;巴拿马巴豆●☆

112984 Croton paniculatum Lam. = Mallotus paniculatus (Lam.) Müll. Arg. ●

112985 Croton penduliflorum Hutch.;垂花巴豆●☆

112986 Croton peraeruginosum Croizat;铜绿巴豆●☆

112987 Croton perrottetianum Baill. = Astraea lobata (L.) Klotzsch ●☆

112988 Croton philippense Lam. = Mallotus philippensis (Lam.) Müll. Arg. ●

112989 Croton pictus Lodd. = Codiaeum variegatum (L.) Blume ●

112990 Croton pierrei Gagnep. = Croton cascarilloides Raeusch. ●

112991 Croton plicatum Vahl = Chrozophora plicata (Vahl) A. Juss. ex Spreng. ●☆

112992 Croton poggei Pax;波格巴豆●☆

112993 Croton polystachyum Hook. et Arn. = Croton cascarilloides Raeusch. ●

112994 Croton polytrichum Pax;密毛巴豆;Manyhairs Croton ●☆

112995 Croton polytrichum Pax subsp. brachystachys Radcl.-Sm.;短穗多毛巴豆●☆

112996 Croton pseudoniloticum De Wild.;假尼罗河巴豆●☆

112997 Croton pseudopulchellum Pax;美丽巴豆●☆

112998 Croton punctatum Lour. = Croton cascarilloides Raeusch. ●

112999 Croton purpurascens Y. T. Chang;淡紫毛巴豆(淡紫巴豆);Lilac Croton, Purplehair Croton, Purplish Croton ●

113000 Croton pynaertii De Wild.;皮那巴豆●☆

113001 Croton pyrifolium Müll. Arg.;梨叶巴豆●☆

113002 Croton reflexifolium Kunth;反曲叶巴豆●☆

113003 Croton repandus Willd. = Mallotus repandus (Willd.) Müll. Arg. ●

113004 Croton repens Schltdl.;匍匐巴豆●☆

113005 Croton rivularis Müll. Arg.;溪边巴豆●☆

113006 Croton roxburghii N. P. Balakr. = Croton laevigatum Vahl ●

113007 Croton rubinoense Aubrév. = Croton penduliflorum Hutch. ●☆

113008 Croton ruizianum Müll. Arg.;鲁伊斯巴豆●☆

113009 Croton sacaquinba Croizat;萨卡巴豆●☆

113010 Croton salutaris Casar.;健身巴豆●☆

113011 Croton sanguifluum Kunth = Croton gossypiifolium Vahl ●

113012 Croton scheffleri Pax;谢夫勒巴豆●☆

113013 Croton schiedeanum Schltdl.;希特巴豆●☆

113014 Croton schimperianum Hochst. ex;欣珀巴豆●☆

113015 Croton scouleri Hook. f.;加拉帕戈斯巴豆●☆

113016 Croton sebiferum L. = Sapium sebiferum (L.) Roxb. ●

113017 Croton sebiferum L. = Triadica sebifera (L.) Small ●

113018 Croton seineri Pax = Croton leuconeurus Pax ●☆

113019 Croton senegalense Lam. = Chrozophora senegalensis (Lam.) A. Juss. ex Spreng. ●☆

113020 Croton serratum (Turcz.) Hochst. ex Baill. = Caperonia serrata (Turcz.) C. Presl ■☆

113021 Croton setigerum Hook.;刚毛巴豆;Dove-weed ●☆

113022 Croton setigerum Hook. = Eremocarpus setigerus (Hook.) Benth. ■☆

113023 Croton siraki Siebold et Zucc. = Mallotus japonicus (L. f.) Müll. Arg. ●

113024 Croton siraki Siebold et Zucc. = Sapium japonicum (Siebold et Zucc.) Pax et K. Hoffm. ●

113025 Croton somalense Vatke ex Pax;索马里巴豆●☆

113026 Croton sordidum Benth.;暗巴豆●☆

113027 Croton sparsiflorum Morong;散花巴豆●☆

113028 Croton sphaerocarpum Kuntze = Croton sylvaticum Hochst. ex Krauss ●☆

113029 Croton spinosus Forssk. = Jatropha spinosa Vahl ●☆

113030 Croton steenkampianum Gerstner;斯滕巴豆●☆

113031 Croton stelluliferum Hutch.;星状巴豆●☆

113032 Croton stuhlmannii Pax = Croton sylvaticum Hochst. ex Krauss ●☆

113033 Croton sublyratum Kurz;泰国巴豆(近琴巴豆)●☆

113034 Croton swynnertonii S. Moore = Tannodia swynnertonii (S. Moore) Prain ■☆

113035 Croton sylvaticum Hochst. ex Krauss;林地巴豆●☆

113036 Croton tchibangense Pellegr.;奇班加巴豆●☆

113037 Croton texense (Klotzsch) Müll. Arg.;得州巴豆;Dove-weed, Hogwort, Skunk Weed, Skunk-weed, Texas Croton ●☆

113038 Croton texense Klotzsch = Croton texense (Klotzsch) Müll. Arg. ●☆

113039 Croton tiglium L.;巴豆(八百力,巴仁,巴菽,巴霜刚子,大叶双眼龙,挡蛇剑,毒鱼子,独行千里,刚子,红子仁,江子,老阳子,銮豆,猛子树,双眼龙,双眼虾);Burging Croton, Croton, Croton-oil Plant, Physic-nut, Purging Croton ●

113040 Croton tiglium L. var. xiaopadou Y. T. Chang et S. Z. Huang;小巴豆;Little Croton ●

113041 Croton tiliifolium Müll. Arg. var. aromaticum Lam. = Mallotus tiliifolius (Blume) Müll. Arg. ●

113042 Croton tinctorium L.;染色巴豆;Turnsole ●☆

113043 Croton tomentosum Lour. = Croton crassifolium Geiseler ●

113044 Croton tonkinense Gagnep. = Croton kongense Gagnep. ●

113045 Croton torreyanum Müll. Arg.;榧巴豆●☆

113046 Croton trinitatis Millsp.;路旁巴豆;Roadside Croton ●☆

113047 Croton trinotatum Millsp.;三点巴豆●☆

113048 Croton tuberculatum Bunge = Speranskia tuberculata (Bunge) Baill. ■

113049 Croton urens L. = Tragia plukenetii Radcl. -Sm. ●☆

113050 Croton urticifolium Y. T. Chang et Q. H. Chen;荨麻叶巴豆;Nettle-leaf Croton, Nettle-leaved Croton ●

113051 Croton urticifolium Y. T. Chang et Q. H. Chen var. dui Y. T. Chang;孟连巴豆;Menglian Nettle-leaf Croton ●

113052 Croton urucurana Baill.;乌鲁巴豆●☆

113053 Croton variegatum L. = Codiaeum variegatum (L.) Blume ●

113054 Croton verdickii De Wild. = Croton sylvaticum Hochst. ex Krauss ●☆

113055 Croton villosus Forssk. = Jatropha pelargoniifolia Courbon ●☆

113056 Croton wellensii De Wild.;韦伦斯巴豆●☆

113057 Croton welwitschianum Müll. Arg. = Croton gratissimum Burch. ●☆

113058 Croton willdenowii G. L. Webster;威尔巴豆;Rushfoil ●☆

113059 Croton xalapense Kunth;沙拉巴豆●☆

113060 Croton yanhuii Y. T. Chang;延辉巴豆;Yanhui Croton ●

113061 Croton yunnanense W. W. Sm.;云南巴豆(滇巴豆);Yunnan Croton ●

113062 Croton yunnanense W. W. Sm. var. megadentum W. T. Wang;大齿云南巴豆(大齿滇巴豆);Bigtooth Yunnan Croton ●

113063 Croton zambesicus Müll. Arg. = Croton gratissimum Burch. ●☆

113064 Croton zehntneri Pax et K. Hoffm.;曾内巴豆●☆

113065 Crotonaceae J. Agardh = Euphorbiaceae Juss.(保留科名)●■

113066 Crotonaceae J. Agardh;巴豆科●

113067 Crotonanthus Klotzsch ex Schltdl. = Croton L. ●

113068 Crotonogyne Müll. Arg. (1864);虫蕊大戟属●☆

113069 Crotonogyne angustifolia Pax;窄叶虫蕊大戟●☆

113070 Crotonogyne caterviflora N. E. Br.;簇花巴豆●☆

113071 Crotonogyne chevalieri (Beille) Keay;舍瓦利耶虫蕊大戟●☆

113072 Crotonogyne gabunensis Pax;加蓬大戟●☆

113073 Crotonogyne giorgii De Wild.;乔治虫蕊大戟●☆

113074 Crotonogyne ikelembensis Prain = Crotonogyne poggei Pax ●☆

113075 Crotonogyne impedita Prain;累赘巴豆●☆

113076 Crotonogyne lasiocarpa Prain;毛果虫蕊大戟●☆

113077 Crotonogyne laurentii De Wild. = Crotonogyne poggei Pax ●☆

113078 Crotonogyne ledermanniana (Pax et K. Hoffm.) Pax et K. Hoffm.;莱德虫蕊大戟●☆

113079 Crotonogyne manniana Müll. Arg.;曼氏虫蕊大戟●☆

113080 Crotonogyne manniana Müll. Arg. subsp. congolensis J. Léonard;刚果虫蕊大戟●☆

113081 Crotonogyne parvifolia Prain;小叶虫蕊大戟●☆

113082 Crotonogyne poggei Pax;波格虫蕊大戟●☆

113083 Crotonogyne preussii Pax;普罗伊斯虫蕊大戟●☆

113084 Crotonogyne sapinii De Wild. = Crotonogyne poggei Pax ●☆

113085 Crotonogyne strigosa Prain;糙伏毛虫蕊大戟●☆

113086 Crotonogyne thonneri De Wild. = Crotonogyne poggei Pax ●☆

113087 Crotonogyne zenkeri Pax;岑克尔虫蕊大戟●☆

113088 Crotonogynopsis Pax(1899);乌桑巴拉大戟属☆

113089 Crotonogynopsis akeassii J. Léonard;阿克斯乌桑巴拉大戟☆

113090 Crotonogynopsis usambarica Pax;乌桑巴拉大戟☆

113091 Crotonopsis Michx. (1803);拟巴豆属●☆

113092 Crotonopsis Michx. = Croton L. ●
113093 Crotonopsis linearis Michx.;拟巴豆●☆
113094 Crotonopsis linearis Michx. = Croton michauxii G. L. Webster ●☆
113095 Crototerum Desv. ex Baill. = Adriana Gaudich. ●☆
113096 Crototerum Desv. ex Baill. = Trachycaryon Klotzsch ●☆
113097 Crotularia Medik. = Crotalaria L. ●■
113098 Crotularius Medik. = Crotalaria L. ●■
113099 Croum Gled. = Ervum L. ■
113100 Croum Pfeiff. = Ervum L. ■
113101 Croum Pfeiff. = Lens Mill. + Vicia L. ■
113102 Crowea Sm. (1798);异蜡花木属●☆
113103 Crowea exalata F. Muell.;无翅异蜡花木●☆
113104 Crowea saligna Andréws;曲叶异蜡花木●☆
113105 Crozophora A. Juss. = Chrozophora A. Juss. (保留属名)●
113106 Crozophyla Raf. = Codiaeum A. Juss. (保留属名)●
113107 Crrptothladia (Bunge) M. J. Cannon = Morina L. ■
113108 Cruciaceae Dulac = Brassicaceae Burnett(保留科名)■●
113109 Cruciaceae Dulac = Cruciferae Juss. (保留科名)■●
113110 Crucianella L. (1753);十字叶属(长柱花属);Crosswort ●■☆
113111 Crucianella aegyptiaca L.;埃及十字叶■☆
113112 Crucianella angustifolia L.;窄十字叶;Narrowleaf Crucianella ■☆
113113 Crucianella angustifolia L. var. chabertii (Gand.) Nyman = Crucianella angustifolia L. ■☆
113114 Crucianella bou-arfae Andr. = Crucianella hirta Pomel ■☆
113115 Crucianella chabertii Gand. = Crucianella angustifolia L. ■☆
113116 Crucianella ciliata Lam.;睫毛十字叶■☆
113117 Crucianella herbacea Forssk. = Crucianella aegyptiaca L. ■☆
113118 Crucianella hirta Pomel;硬毛十字叶■☆
113119 Crucianella latifolia L.;宽叶十字叶■☆
113120 Crucianella maritima L.;滨海十字叶■☆
113121 Crucianella mucronata Roth = Crucianella angustifolia L. ■☆
113122 Crucianella patula L.;张开十字叶■☆
113123 Crucianella pentandra Dufour ex Roem. et Schult. = Crucianella patula L. ■☆
113124 Crucianella rupestris Guss. = Crucianella maritima L. ■☆
113125 Crucianella spicata Lam. = Crucianella angustifolia L. ■☆
113126 Crucianella squarrosa Sennen et Mauricio = Crucianella patula L. ■☆
113127 Crucianella stylosa Trin. = Phuopsis stylosa (Trin.) Hook. f. ■
113128 Cruciata Gilib. = Gentiana L. ■
113129 Cruciata Gilib. = Tretorhiza Adans. ■
113130 Cruciata Mill. (1754);十字茜属;Crosswort ■☆
113131 Cruciata Mill. = Galium L. ■●
113132 Cruciata Tourn. ex Adans. = Galium L. ■●
113133 Cruciata articulata (L.) Ehrend.;关节十字茜■☆
113134 Cruciata glabra (L.) Ehrend.;光滑十字茜■☆
113135 Cruciata laevipes Opiz;赫尔松十字茜;Cherson Bedstraw, Cross Wort, Crosswort, Mugwort ■☆
113136 Cruciata laevipes Opiz = Galium chersonense Roem. et Schult. ■☆
113137 Cruciata pedemontana (Bellardi) Ehrend.;佩德十字茜;Piedmont Bedstraw ■☆
113138 Crucicaryum O. Brand(1929);十字果紫草属☆
113139 Crucicaryum papuanum O. Brand;十字果紫草☆
113140 Cruciella Leschen. ex DC. = Xanthosia Rudge ■☆
113141 Crucifera brassica Krause = Brassica oleracea L. ■
113142 Cruciferae Adans = Brassicaceae Burnett(保留科名)■●
113143 Cruciferae Adans = Cruciferae Juss. (保留科名)■●
113144 Cruciferae Juss. (1789)(保留科名);十字花科;Cabbage Family, Mustard Family ■●
113145 Cruciferae Juss. (保留科名) = Brassicaceae Burnett(保留科名)■●
113146 Crucihimalaya Al-Shehbaz, O' Kane et R. A. Price(1999);须弥芥属■
113147 Crucihimalaya axillaris (Hook. f. et Thomson) Al-Shehbaz, O' Kane et R. A. Price;腋花须弥芥■
113148 Crucihimalaya himalaica (Edgew.) Al-Shehbaz, O' Kane et R. A. Price;须弥芥(喜马拉雅鼠耳芥);Himalayan Mouseear Cress, Himalayas Mouseear Cress ■
113149 Crucihimalaya kneuckeri (Bornm.) Al-Shehbaz, O' Kane et R. A. Price;科氏须弥芥■☆
113150 Crucihimalaya lasiocarpa (Hook. f. et Thomson) Al-Shehbaz, O' Kane et R. A. Price;毛果须弥芥(粗根鼠耳芥);Roughroot Mouseear Cress ■
113151 Crucihimalaya mollissima (C. A. Mey.) Al-Shehbaz, O' Kane et R. A. Price;柔毛须弥芥(柔毛鼠耳芥);Softhair Mouseear Cress ■
113152 Crucihimalaya mongolica (Botsch.) Al-Shehbaz, O' Kane et R. A. Price;蒙古须弥芥■☆
113153 Crucihimalaya ovczinnikovii (Botsch.) Al-Shehbaz, O' Kane et R. A. Price;中亚须弥芥■☆
113154 Crucihimalaya stricta (Cambess.) Al-Shehbaz, O' Kane et R. A. Price;直须弥芥■
113155 Crucihimalaya wallichii (Hook. f. et Thomson) Al-Shehbaz, O' Kane et R. A. Price;卵叶须弥芥■
113156 Crucita L. = Cruzeta Loefl. ●■
113157 Crucita L. = Iresine P. Browne(保留属名)●■
113158 Cruciundula Raf. = Thlaspi L. ■
113159 Cruckshanksia Hook. (废弃属名) = Balbisia Cav. (保留属名)●☆
113160 Cruckshanksia Hook. (废弃属名) = Cruckshanksia Hook. et Arn. (保留属名)●☆
113161 Cruckshanksia Hook. et Arn. (1833)(保留属名);克鲁茜属●☆
113162 Cruckshanksia Miers(1826) = Solenomelus Miers ■☆
113163 Cruckshanksia capitata Phil.;头状克鲁茜☆
113164 Cruckshanksia chrysantha Phil.;金花克鲁茜☆
113165 Cruckshanksia densifolia Phil.;密叶克鲁茜☆
113166 Cruckshanksia hymenodon Hook. et Arn.;克鲁茜☆
113167 Cruckshanksia macrantha Phil.;大花克鲁茜☆
113168 Cruckshanksia montana Clos;山地克鲁茜☆
113169 Cruckshanksia paradoxa Phil.;奇异克鲁茜☆
113170 Cruckshanksia verticillata Phil.;轮生克鲁茜☆
113171 Cruddasia Prain = Ophrestia H. M. L. Forbes ●■
113172 Crudea K. Schum. = Crudia Schreb. (保留属名)●☆
113173 Crudia Schreb. (1789)(保留属名);库地苏木属●☆
113174 Crudia curtisii Prain ex King;库地苏木●☆
113175 Crudia gabonensis Pierre ex Harms;加蓬库地苏木●☆
113176 Crudia gossweileri Baker f.;戈斯库地苏木●☆
113177 Crudia harmsiana De Wild.;哈姆斯库地苏木●☆
113178 Crudia harmsiana De Wild. var. velutina J. Léonard;短绒毛库地苏木●☆
113179 Crudia klainei Pierre ex De Wild.;克莱恩库地苏木●☆
113180 Crudia laurentii De Wild.;洛朗库地苏木●☆
113181 Crudia ledermannii Harms;莱德库地苏木●☆
113182 Crudia michelsonii J. Léonard;米歇尔松库地苏木●☆
113183 Crudia monophylla Harms = Haplormosia monophylla (Harms)

Harms ●☆
113184 Crudia senegalensis Planch. ex Benth.;塞内加尔库地苏木●☆
113185 Crudia zenkeri Harms ex De Wild. = Crudia gabonensis Pierre ex Harms ●☆
113186 Crudya Batsch = Crudia Schreb.(保留属名)●☆
113187 Cruicita L. = Iresine P. Browne(保留属名)●■
113188 Cruikshanksia Benth. et Hook. f. = Balbisia Cav.(保留属名)●☆
113189 Cruikshanksia Benth. et Hook. f. = Cruckshanksia Hook. et Arn.(保留属名)●☆
113190 Cruikshanksia Rchb. = Solenomelus Miers ■☆
113191 Crula Nieuwl. = Acer L. ●
113192 Crula grisea (Franch.) Nieuwl. = Acer griseum (Franch.) Pax ●
113193 Crula henryi (Pax) Nieuwl. = Acer henryi Pax ●
113194 Crula mandshurica (Maxim.) Nieuwl. = Acer mandshuricum Maxim. ●
113195 Crula nikoense (Maxim.) Nieuwl. = Acer nikoense Maxim. ●
113196 Crula sutchuenensis (Franch.) Nieuwl. = Acer sutchuenense Franch. ●
113197 Crula triflora (Kom.) Nieuwl. = Acer triflorum Kom. ●
113198 Crumenaria Mart. (1826);袋鼠李属●☆
113199 Crumenaria choretroides Mart. ex Reissek;袋鼠李●☆
113200 Cruminium Desv. = Centrosema (DC.) Benth.(保留属名)●■☆
113201 Crunocallis Rydb. (1906);球茎水繁缕属■☆
113202 Crunocallis Rydb. = Montia L. ■☆
113203 Crunocallis chamissoi (Ledeb. ex Spreng.) Rydb. = Montia chamissoi (Ledeb. ex Spreng.) Greene ■☆
113204 Crupina (Pers.) Cass. = Crupina (Pers.) DC. ■
113205 Crupina (Pers.) DC. (1810);半毛菊属(谷粒菊属);Crupina ■
113206 Crupina Cass. = Crupina (Pers.) DC. ■
113207 Crupina DC. = Crupina (Pers.) DC. ■
113208 Crupina crupinastrum (Moris) Vis.;地中海半毛菊■☆
113209 Crupina crupinastrum Vis. = Crupina crupinastrum (Moris) Vis. ■☆
113210 Crupina intermedia (Mutel) Walp.;间型半毛菊■☆
113211 Crupina vulgaris Cass. = Crupina vulgaris Pers. ex Cass. ■
113212 Crupina vulgaris Cass. subsp. crupinastrum (Moris) Batt. = Crupina crupinastrum (Moris) Vis. ■☆
113213 Crupina vulgaris Pers. = Crupina vulgaris Pers. ex Cass. ■
113214 Crupina vulgaris Pers. ex Cass.;半毛菊(欧谷粒菊);Bearded Creeper, Common Crupina, Crupina ■
113215 Crupinastrum Schur = Serratula L. ■
113216 Crusea A. Rich. = Chione DC. ■☆
113217 Crusea Cham. et Schltdl. (1830);克吕兹茜属■☆
113218 Crusea Cham. ex DC. = Mitracarpus Zucc. ex Schult. et Schult. f. ■
113219 Crusea brachyphylla Cham. et Schltdl.;克吕兹茜■☆
113220 Crusea glaucescens E. Mey. ex Harv. = Pentanisia prunelloides (Klotzsch ex Eckl. et Zeyh.) Walp. ■☆
113221 Crusea lanceolata E. Mey. ex Hochst. = Pentanisia prunelloides (Klotzsch ex Eckl. et Zeyh.) Walp. ■☆
113222 Crusea variabilis E. Mey. ex Harv. = Pentanisia prunelloides (Klotzsch ex Eckl. et Zeyh.) Walp. ■☆
113223 Cruzea A. Rich. = Chione DC. ■☆
113224 Cruzea A. Rich. = Crusea A. Rich. ■☆
113225 Cruzeta Loefl. = Iresine P. Browne(保留属名)●■
113226 Cruzia Phil. = Scutellaria L. ●■
113227 Cruzita L. = Cruzeta Loefl. ●■
113228 Cruzita L. = Iresine P. Browne(保留属名)●■
113229 Cryanthemum Kamelin = Tanacetum L. ■●
113230 Crybe Lindl. (1836);南美白芨属;Crybe ■☆
113231 Crybe rosea Lindl.;南美白芨;Crybe ■☆
113232 Cryophytum N. E. Br. = Gasoul Adans. ■
113233 Cryophytum N. E. Br. = Mesembryanthemum L.(保留属名)■●
113234 Cryophytum acuminatum L. Bolus = Mesembryanthemum guerichianum Pax ■☆
113235 Cryophytum alatum L. Bolus = Mesembryanthemum guerichianum Pax ■☆
113236 Cryophytum alkalifugum Dinter ex Range = Mesembryanthemum barklyi N. E. Br. ■☆
113237 Cryophytum aureum L. Bolus = Mesembryanthemum excavatum L. Bolus ■☆
113238 Cryophytum barklyi (N. E. Br.) N. E. Br. ex L. Bolus = Mesembryanthemum barklyi N. E. Br. ■☆
113239 Cryophytum burchellii N. E. Br. = Mesembryanthemum aitonis Jacq. ■☆
113240 Cryophytum calycinum L. Bolus = Mesembryanthemum guerichianum Pax ■☆
113241 Cryophytum carinatum L. Bolus = Mesembryanthemum guerichianum Pax ■☆
113242 Cryophytum clandestinum (Haw.) L. Bolus = Mesembryanthemum aitonis Jacq. ■☆
113243 Cryophytum clavatum L. Bolus = Mesembryanthemum eurystigmatum Gerbault ■☆
113244 Cryophytum cleistum L. Bolus = Mesembryanthemum nodiflorum L. ■☆
113245 Cryophytum conjectum N. E. Br. = Mesembryanthemum aitonis Jacq. ■☆
113246 Cryophytum crassifolium L. Bolus = Mesembryanthemum guerichianum Pax ■☆
113247 Cryophytum crassipes L. Bolus = Mesembryanthemum fastigiatum Thunb. ■☆
113248 Cryophytum crystallinum (L.) N. E. Br. = Mesembryanthemum crystallinum L. ■
113249 Cryophytum dejagerae L. Bolus = Mesembryanthemum stenandrum (L. Bolus) L. Bolus ■☆
113250 Cryophytum excavatum (L. Bolus) L. Bolus = Mesembryanthemum excavatum L. Bolus ■☆
113251 Cryophytum fenchelii (Schinz) N. E. Br. = Mesembryanthemum guerichianum Pax ■☆
113252 Cryophytum framesii L. Bolus = Mesembryanthemum guerichianum Pax ■☆
113253 Cryophytum framesii L. Bolus var. laxum ? = Mesembryanthemum guerichianum Pax ■☆
113254 Cryophytum galpinii L. Bolus = Mesembryanthemum stenandrum (L. Bolus) L. Bolus ■☆
113255 Cryophytum gibbosum N. E. Br. = Mesembryanthemum nodiflorum L. ■☆
113256 Cryophytum glaucum Dinter = Mesembryanthemum barklyi N. E. Br. ■☆
113257 Cryophytum grandiflorum Dinter et Schwantes ex Range = Mesembryanthemum guerichianum Pax ■☆
113258 Cryophytum grandifolium (Schinz) Dinter et Schwantes = Mesembryanthemum guerichianum Pax ■☆
113259 Cryophytum guerichianum (Pax) Schwantes = Mesembryanthemum

gerichianum Pax ■☆

113260 Cryophytum intermedium L. Bolus = Mesembryanthemum longistylum DC. ■☆

113261 Cryophytum karrooicum L. Bolus = Mesembryanthemum guerichianum Pax ■☆

113262 Cryophytum latisepalum L. Bolus = Mesembryanthemum guerichianum Pax ■☆

113263 Cryophytum liebendalense (L. Bolus) J. W. Ingram = Mesembryanthemum stenandrum (L. Bolus) L. Bolus ■☆

113264 Cryophytum lineare L. Bolus = Mesembryanthemum longistylum DC. ■☆

113265 Cryophytum longipapillatum L. Bolus = Mesembryanthemum stenandrum (L. Bolus) L. Bolus ■☆

113266 Cryophytum maxwellii L. Bolus = Mesembryanthemum guerichianum Pax ■☆

113267 Cryophytum nanum N. E. Br. = Mesembryanthemum crystallinum L. ■

113268 Cryophytum neglectum N. E. Br. = Mesembryanthemum longistylum DC. ■☆

113269 Cryophytum neilsoniae L. Bolus = Mesembryanthemum guerichianum Pax ■☆

113270 Cryophytum nodiflorum (L.) L. Bolus = Mesembryanthemum nodiflorum L. ■☆

113271 Cryophytum parvum L. Bolus = Mesembryanthemum crystallinum L. ■

113272 Cryophytum paulum N. E. Br. = Mesembryanthemum aitonis Jacq. ■☆

113273 Cryophytum pentagonum L. Bolus = Mesembryanthemum guerichianum Pax ■☆

113274 Cryophytum planum L. Bolus = Mesembryanthemum aitonis Jacq. ■☆

113275 Cryophytum rogersii L. Bolus = Mesembryanthemum nodiflorum L. ■☆

113276 Cryophytum roseum L. Bolus = Mesembryanthemum guerichianum Pax ■☆

113277 Cryophytum sessiliflorum (Aiton) N. E. Br. = Cleretum papulosum (L. f.) L. Bolus ■☆

113278 Cryophytum sessiliflorum L. Bolus = Mesembryanthemum guerichianum Pax ■☆

113279 Cryophytum sessiliforum (Aiton) N. E. Br. var. luteum (Haw.) H. Jacobsen = Cleretum papulosum (L. f.) L. Bolus ■☆

113280 Cryophytum setosum L. Bolus = Mesembryanthemum guerichianum Pax ■☆

113281 Cryophytum squamulosum Dinter = Mesembryanthemum barklyi N. E. Br. ■☆

113282 Cryophytum squamulosum L. Bolus = Mesembryanthemum guerichianum Pax ■☆

113283 Cryophytum stenandrum L. Bolus = Mesembryanthemum stenandrum (L. Bolus) L. Bolus ■☆

113284 Cryophytum suaveolens (L. Bolus) J. W. Ingram = Phyllobolus lignescens (L. Bolus) Gerbaulet ●☆

113285 Cryophytum subulatum L. Bolus = Mesembryanthemum guerichianum Pax ■☆

113286 Cryophytum suffruticosum L. Bolus = Phyllobolus suffruticosus (L. Bolus) Gerbaulet ●☆

113287 Cryophytum truncatum L. Bolus = Mesembryanthemum guerichianum Pax ■☆

113288 Cryophytum velutinum L. Bolus = Mesembryanthemum barklyi N. E. Br. ■☆

113289 Cryophytum wilmaniae L. Bolus = Mesembryanthemum stenandrum (L. Bolus) L. Bolus ■☆

113290 Cryosophila Blume(1838);根刺棕属(叉刺棕属,刺根榈属,根刺椰子属,克利索榈属);Root Spine Palm,Rootspine Palm ●☆

113291 Cryosophila albida Bartlett;白根刺棕●☆

113292 Cryosophila argentea Bartlett;中美根刺棕;Escoba Palm ●☆

113293 Cryosophila warscewiczii (H. Wendl.) Bartlett;根刺棕(华西根刺棕,华西威根刺棕);Root Spine Palm ●☆

113294 Cryphaea Buch. -Ham. = Chloranthus Sw. ■●

113295 Cryphaea Buch. -Ham. = Peperidia Rchb. ■●

113296 Cryphaea erecta Buch. -Ham. = Chloranthus erectus (Buch. -Ham.) Verdc. ■

113297 Cryphia R. Br. = Prostanthera Labill. ●☆

113298 Cryphiacanthus Nees = Ruellia L. ■●

113299 Cryphiantha Eckl. et Zeyh. = Amphithalea Eckl. et Zeyh. ■☆

113300 Cryphiantha imbricata Eckl. et Zeyh. = Amphithalea micrantha (E. Mey.) Walp. ■☆

113301 Cryphiospermum P. Beauv. = Enydra Lour. ■

113302 Cryphiospermum abyssinicum (Sch. Bip.) Sch. Bip. = Micractis bojeri DC. ■☆

113303 Crypsinna E. Fourn. = Muhlenbergia Schreb. ■

113304 Crypsinna E. Fourn. ex Benth. = Muhlenbergia Schreb. ■

113305 Crypsis Aiton(1789)(保留属名);隐花草属(扎股草属);Crypsis,Prickle Grass,Pricklegrass,Prickle-grass ■

113306 Crypsis aculeata (L.) Aiton;隐花草(扎股草,扎屁股草);Common Crypsis,Common Pricklegrass,Prickle-grass ■

113307 Crypsis alopecuroides (Piller et Mitterp.) Schrad.;看麦娘隐花草■☆

113308 Crypsis alopecuroides (Piller et Mitterp.) Schrad. subsp. brachystachys (C. Presl) Trab. = Crypsis alopecuroides (Piller et Mitterp.) Schrad. ■☆

113309 Crypsis alopecuroides (Piller et Mitterp.) Schrad. var. nigricans (Guss.) Coss. et Durieu = Crypsis alopecuroides (Piller et Mitterp.) Schrad. ■☆

113310 Crypsis alopecuroides Guss. ex Schult.;狐尾隐花草;Foxtail Pricklegrass ■☆

113311 Crypsis arenaria Desf. = Phleum arenarium L. ■☆

113312 Crypsis borszczowii Regel;鲍尔隐花草■☆

113313 Crypsis dura Boiss. = Urochondra setulosa (Trin.) C. E. Hubb. ■☆

113314 Crypsis nigricans Guss. = Crypsis alopecuroides (Piller et Mitterp.) Schrad. ■☆

113315 Crypsis niliaca Fig. et De Not. = Crypsis schoenoides (L.) Lam. ■

113316 Crypsis schoenoides (L.) Lam.;蔺状隐花草;Bogrushlike Crypsis,Bogrushlike Pricklegrass,False-timothy,Pickle Grass,Swamp Pickle Grass,Swamp Pricklegrass ■

113317 Crypsis setulosa (Trin.) Mez = Urochondra setulosa (Trin.) C. E. Hubb. ■☆

113318 Crypsis turkestanica Eig;突厥隐花草■

113319 Crypsis vaginiflora (Forssk.) Opiz;鞘隐花草;Modest Prickle Grass ■☆

113320 Crypsocalyx Endl. = Chrysocalyx Guill. et Perr. ■☆

113321 Crypsocalyx Endl. = Crotalaria L. ●■

113322 Crypsophila Benth. et Hook. f. = Cryosophila Blume ●☆

113323 Crypta Nutt. = Elatine L. ■
113324 Cryptaceae Raf. = Elatinaceae Dumort.(保留科名)■
113325 Cryptadenia Meisn.(1841);隐腺瑞香属●☆
113326 Cryptadenia Meisn. = Lachnaea L. ●☆
113327 Cryptadenia breviflora Meisn. = Lachnaea grandiflora (L. f.) Baill. ●☆
113328 Cryptadenia filicaulis Meisn. = Lachnaea filicaulis (Meisn.) Beyers ●☆
113329 Cryptadenia grandiflora (L. f.) Meisn.;大花隐腺瑞香●☆
113330 Cryptadenia grandiflora (L. f.) Meisn. = Lachnaea grandiflora (L. f.) Baill. ●☆
113331 Cryptadenia grandiflora (L. f.) Meisn. var. latifolia Meisn. = Lachnaea grandiflora (L. f.) Baill. ●☆
113332 Cryptadenia laxa C. H. Wright = Lachnaea laxa (C. H. Wright) Beyers ●☆
113333 Cryptadenia uniflora (L.) Meisn. = Lachnaea uniflora (L.) Crantz ●☆
113334 Cryptadia Lindl. ex Endl. = Gymnarrhena Desf. ■☆
113335 Cryptandra Sm.(1798);缩苞木属;Cryptandra ●☆
113336 Cryptandra albicans F. Muell.;灰白缩苞木●☆
113337 Cryptandra alpina Hook. f.;高山缩苞木;Alpine Cryptandra ●☆
113338 Cryptandra amara Sm.;苦味缩苞木;Bitter Cryptandra ●☆
113339 Cryptandra ericoides Sm.;尖苞缩苞木;Acuminate-bract Cryptandra ●☆
113340 Cryptandra floribunda Steud.;多花缩苞木●☆
113341 Cryptandra propinqua A. Cunn. ex Fenzl;线叶缩苞木;Linearleaf Cryptandra ●☆
113342 Cryptandra spinescens DC.;阔苞缩苞木;Broad-bract Cryptandra ●☆
113343 Cryptandraceae Barldey = Rhamnaceae Juss.(保留科名)●
113344 Cryptangium Schrad. ex Nees = Lagenocarpus Nees ■☆
113345 Cryptanopsis Ule = Orthophytum Beer ■☆
113346 Cryptantha G. Don = Cryptantha Lehm. ex G. Don ■☆
113347 Cryptantha Lehm. = Cryptantha Lehm. ex G. Don ■☆
113348 Cryptantha Lehm. ex Fisch. et C. A. Mey. = Cryptantha Lehm. ex G. Don ■☆
113349 Cryptantha Lehm. ex Fisch. et C. A. Mey. = Eritrichium Schrad. ex Gaudin ■
113350 Cryptantha Lehm. ex G. Don(1837);秘花草属;Cryptantha ■☆
113351 Cryptantha affinis Greene;近缘秘花草■☆
113352 Cryptantha alpicola Cronquist;高山秘花草■☆
113353 Cryptantha angustifolia Greene;窄叶秘花草■☆
113354 Cryptantha axillaris Reiche;腋花秘花草■☆
113355 Cryptantha breviflora Payson;短花秘花草■☆
113356 Cryptantha cinerea (Greene) Cronquist;灰秘花草(詹氏秘花草);James' Cryptantha ■☆
113357 Cryptantha hispida (Phil.) Reiche;硬毛秘花草■☆
113358 Cryptanthe Benth. et Hook. f. = Cryptantha Lehm. ex G. Don ■☆
113359 Cryptanthela Gagnep. = Argyreia Lour. ●
113360 Cryptanthemis Rupp = Rhizanthella R. S. Rogers ■☆
113361 Cryptanthopsis Ule = Orthophytum Beer ■☆
113362 Cryptanthus Nutt. ex Moq. = Aphanisma Nutt. ■☆
113363 Cryptanthus Osbeck(废弃属名) = Cryptanthus Otto et A. Dietr.(保留属名)■☆
113364 Cryptanthus Otto et A. Dietr.(1757)(保留属名);姬凤梨属(锦纹凤梨属,迷你凤梨属,无柄凤梨属,小凤梨属,小型凤梨属,隐花凤梨属,隐花小凤兰属,隐花属);Cryptanthus, Earth Star, Earth Stars, Starfish ■☆
113365 Cryptanthus acaulis (Lindl.) Beer;姬凤梨(无茎隐花凤梨,小菠萝);Green Earth Star, Star Bromelia, Starfish plant, Stenless Cryptanthus ■☆
113366 Cryptanthus acaulis (Lindl.) Beer 'Ruber';红叶姬凤梨(红叶小菠萝)■☆
113367 Cryptanthus acaulis (Lindl.) Beer var. argenteus Beer;银叶姬凤梨■☆
113368 Cryptanthus acaulis Beer = Cryptanthus acaulis (Lindl.) Beer ■☆
113369 Cryptanthus acaulis Beer var. ruber Beer = Cryptanthus acaulis Beer 'Ruber' ■☆
113370 Cryptanthus bahianus L. B. Sm.;巴氏姬凤梨■☆
113371 Cryptanthus bahianus L. B. Sm. 'Blonde';铜色姬凤梨■☆
113372 Cryptanthus beuckeri E. Morren;小花姬凤梨■☆
113373 Cryptanthus bivittafus Regel 'Tricolor';红边凤梨■☆
113374 Cryptanthus bivittatus (Hook.) Regel;双条带姬凤梨(绒叶姬凤梨,绒叶小凤梨,纵纹姬凤梨);Earth Star, Twovittae Cryptanthus ■☆
113375 Cryptanthus bivittatus Regel 'Lueddemannii';绒叶姬凤梨■☆
113376 Cryptanthus bivittatus Regel 'Minor';玫红姬凤梨■☆
113377 Cryptanthus bivittatus Regel 'Pink Starlight';粉绒姬凤梨(粉绒小凤梨)■☆
113378 Cryptanthus bivittatus Regel 'Ruby';暗红小凤梨■☆
113379 Cryptanthus bivittatus Regel = Cryptanthus bivittatus (Hook.) Regel ■☆
113380 Cryptanthus bromelioides Otto et A. Dietr.;大姬凤梨(长叶小菠萝,长叶小凤梨,隐花凤梨);Rainbow Star ■☆
113381 Cryptanthus bromelioides Otto et A. Dietr. 'Tricolor';三色隐花凤梨(长叶小菠萝,三色小凤梨)■☆
113382 Cryptanthus bromelioides Otto et A. Dietr. var. tricolor M. B. Foster = Cryptanthus bromelioides Otto et A. Dietr. 'Tricolor' ■☆
113383 Cryptanthus chinensis Osbeck = Clerodendrum chinense (Osbeck) Mabb. ●
113384 Cryptanthus diversifolius Beer;异叶小菠萝■☆
113385 Cryptanthus fosterianus L. B. Sm.;福斯特姬凤梨■☆
113386 Cryptanthus maritimus L. B. Sm.;海滨姬凤梨■☆
113387 Cryptanthus zonatus (Vis.) Beer;环带姬凤梨(虎纹小菠萝,环条带姬凤梨);Zonate Cryptanthus ■☆
113388 Cryptanthus zonatus (Vis.) Regel = Cryptanthus zonatus (Vis.) Beer ■☆
113389 Cryptanthus zonatus Beer 'Zebrinus';银纹环带姬凤梨■☆
113390 Cryptanthus zonatus Beer var. fuscus Mez;褐色环带姬凤梨■☆
113391 Cryptaria Raf. = Crypta Nutt. ■
113392 Cryptaria Raf. = Elatine L. ■
113393 Cryptarrhena R. Br.(1816);藏蕊兰属■☆
113394 Cryptarrhena lunata R. Br.;藏蕊兰■☆
113395 Cryptella Raf. = Crypta Nutt. ■
113396 Cryptella Raf. = Elatine L. ■
113397 Crypteronia Blume(1827);隐翼木属(隐翼属);Crypteronia ●
113398 Crypteronia cumingii Endl.;库氏隐翼木(库氏隐翼)●☆
113399 Crypteronia glabra (Wall.) Blume = Crypteronia paniculata Blume ●◇
113400 Crypteronia griffithii C. B. Clarke;格氏隐翼木●☆
113401 Crypteronia paniculata Blume;隐翼木(隐翼);Paniculate Crypteronia ●◇
113402 Crypteroniaceae A. DC.(1868)(保留科名);隐翼科;

Crypteronia Family ●
113403 Crypterpis Thouars = Goodyera R. Br. ■
113404 Crypterpis Thouars = Platylepis A. Rich.(保留属名)■☆
113405 Cryptina Raf. = Crypta Nutt. ■
113406 Cryptina Raf. = Elatine L. ■
113407 Cryptobasis Nevski = Iris L. ■
113408 Cryptocalyx Benth. = Lippia L. ●■☆
113409 Cryptocapnos Rech. f. (1968);垫状烟堇属■☆
113410 Cryptocapnos chasmophyticus Rech. f.;垫状烟堇■☆
113411 Cryptocaria Raf. = Cryptocarya R. Br.(保留属名)●
113412 Cryptocarpa Steud. = Acicarpha Juss. ■☆
113413 Cryptocarpa Steud. = Cryptocarpha Cass. ■☆
113414 Cryptocarpa Tayl. ex Tul. = Tristicha Thouars ■☆
113415 Cryptocarpha Cass. = Acicarpha Juss. ■☆
113416 Cryptocarpum (Dunal) Wijk et al. = Solanum L. ●■
113417 Cryptocarpus Kunth(1817);微花茉莉属●☆
113418 Cryptocarpus Post et Kuntze = Cryptocarpa Tayl. ex Tul. ●☆
113419 Cryptocarpus Post et Kuntze = Tristicha Thouars ■☆
113420 Cryptocarpus capitatus S. Watson = Pisonia capitata (S. Watson) Standl. ●☆
113421 Cryptocarpus pyriformis Kunth;微花茉莉■☆
113422 Cryptocarya R. Br. (1810)(保留属名);厚壳桂属(芳香厚壳桂属,佳叶樟属,拉文萨拉属);Cryptocarya,Thickshellcassia ●
113423 Cryptocarya acuminata Sim = Cryptocarya woodii Engl. ●☆
113424 Cryptocarya acutifolia H. W. Li;尖叶厚壳桂;Acuteleaf Cryptocarya,Acuteleaf Thickshellcassia,Acute-leaved Cryptocarya ●◇
113425 Cryptocarya agathophylla van der Werff = Agathophyllum aromaticum Willd. ●☆
113426 Cryptocarya amygdalina Nees;杏仁厚壳桂;Amygdaline Cryptocarya,Amygdaline Thickshellcassia ●
113427 Cryptocarya andersonii King ex Hook. f. = Alseodaphne andersonii (King ex Hook. f.) Kosterm. ●
113428 Cryptocarya angustifolia E. Mey. ex Meisn.;窄叶厚壳桂●☆
113429 Cryptocarya aromatica (Becc.) Kosterm.;芳香厚壳桂;Massoy Bark ●☆
113430 Cryptocarya australis (Hook.) Benth.;灰厚壳桂●☆
113431 Cryptocarya austrokweichouensis X. H. Song;黔南厚壳桂;S. Guizhou Cryptocarya, S. Guizhou Thickshellcassia, South Guizhou Cryptocarya ●
113432 Cryptocarya bourdilloni Gamble;布隆迪厚壳桂●☆
113433 Cryptocarya brachythyrsa H. W. Li;短序厚壳桂;Shortthyrse Cryptocarya,Shortthyrse Thickshellcassia,Short-thyrsed Cryptocarya ●
113434 Cryptocarya brassii C. K. Allen;巴西厚壳桂●☆
113435 Cryptocarya calcicola H. W. Li;岩生厚壳桂;Calicicolous Cryptocarya,Calicicolous Thickshellcassia ●
113436 Cryptocarya chinensis (Hance) Hemsl.;厚壳桂(华厚壳桂,山饼斗,铜锣桂,香果,香花桂,硬壳槁);Chinese Cryptocarya, Chinese Thickshellcassia ●
113437 Cryptocarya chingii W. C. Cheng;硬壳桂(芳果槁,流鼻槁,流涎槁,平阳厚壳桂,仁昌桂,仁昌厚壳桂,仁昌硬壳桂,硬壳槁,硬壳果);Ching Cryptocarya,Ching Thickshellcassia ●
113438 Cryptocarya concinna Hance;黄果厚壳桂(长果厚壳桂,大香叶树,海南厚壳桂,黄果,黄果桂,生虫树,石楠,土楠,香港厚壳桂);Konishi Cryptocarya, Yellowfruit Cryptocarya, Yellowfruit Thickshellcassia,Yellow-fruited Cryptocarya ●
113439 Cryptocarya cordata Allen;心形厚壳桂●☆
113440 Cryptocarya corrugata C. T. White et W. D. Francis;高大厚壳桂;Corduroy,Oak Walnut ●☆
113441 Cryptocarya densiflora Blume;丛花厚壳桂(白面槁,丛花桂,大果铜锣桂,密花厚壳桂,平滑厚壳桂,硬壳槁);Denseflower Cryptocarya, Denseflower Thickshellcassia, Densi-flowered Cryptocarya ●
113442 Cryptocarya depauperata H. W. Li;贫花厚壳桂;Depauperate Cryptocarya,Fewflower Cryptocarya,Poorflower Thickshellcassia ●
113443 Cryptocarya dolichocarpa Kosterm. = Aspidostemon dolichocarpum (Kosterm.) Rohwer ●☆
113444 Cryptocarya elliotii Kosterm. = Aspidostemon parvifolium (Scott-Elliot) van der Werff ●☆
113445 Cryptocarya elliptifolia Merr.;菲律宾厚壳桂(大果厚壳桂,菲岛厚壳桂)●
113446 Cryptocarya glabella Domin;光滑厚壳桂;Poison Walnut ●☆
113447 Cryptocarya glaucescens R. Br.;灰绿厚壳桂;Brown Beech, Jackwood ●☆
113448 Cryptocarya glaucosepala Scott-Elliot = Beilschmiedia madagascariensis (Baill.) Kosterm. ●☆
113449 Cryptocarya grandifolia Stapf ex Engl. = Beilschmiedia grandifolia (Stapf) Robyns et R. Wilczek ●☆
113450 Cryptocarya guianensis Meisn.;圭亚那厚壳桂●☆
113451 Cryptocarya hainanensis Merr.;海南厚壳桂;Hainan Cryptocarya,Hainan Thickshellcassia ●
113452 Cryptocarya hornet Gillespie;玻利尼西亚厚壳桂●☆
113453 Cryptocarya howii C. K. Allen = Cryptocarya metcalfiana Allen ●
113454 Cryptocarya humbertiana Kosterm. = Aspidostemon humbertianum (Kosterm.) Rohwer ●☆
113455 Cryptocarya impressiaervia H. W. Li.;钝叶厚壳桂(大叶乌面槁,钝叶桂,那果);Obtuseleaf Cryptocarya, Obtuseleaf Thickshellcassia,Obtuse-leaved Cryptocarya ●
113456 Cryptocarya konishii Hayata = Cryptocarya concinna Hance ●
113457 Cryptocarya kwangtungensis Hung T. Chang;广东厚壳桂;Guangdong Cryptocarya, Guangdong Thickshellcassia, Kwangtung Thickshellcassia ●
113458 Cryptocarya laevigata Blume;亮叶厚壳桂;Glossy Laurel ●☆
113459 Cryptocarya laevigata Elmer = Cryptocarya densiflora Blume ●
113460 Cryptocarya latifolia Sond.;阔叶厚壳桂;Ntonga Nut ●☆
113461 Cryptocarya laui Merr. et F. P. Metcalf = Cryptocarya chingii W. C. Cheng ●
113462 Cryptocarya leiana C. K. Allen;鸡卵槁(黎厚壳桂);Egg Thickshellcassia,Lei Cryptocarya ●
113463 Cryptocarya lenticellata Lecomte = Cryptocarya concinna Hance ●
113464 Cryptocarya liebertiana Engl.;利伯特厚壳桂●☆
113465 Cryptocarya louvelii Danguy;卢氏厚壳桂●☆
113466 Cryptocarya lyoniifolia S. K. Lee et F. N. Wei;南烛厚壳桂;Pittea Leaf Cryptocarya ●☆
113467 Cryptocarya maclurei Merr.;白背厚壳桂(白叶厚壳桂);Maclure Cryptocarya,Maclure Thickshellcassia ●
113468 Cryptocarya maculata H. W. Li;斑果厚壳桂;Maculate-fruited Cryptocarya,Spoted Fruit Cryptocarya,Spottedfruit Thickshellcassia ●
113469 Cryptocarya merrilliana C. K. Allen = Cryptocarya chingii W. C. Cheng ●
113470 Cryptocarya metcalfiana Allen;长序厚壳桂(麦桂,麦氏厚壳桂,小果厚壳桂);F. P. Metcalf Cryptocarya, F. P. Metcalf Thickshellcassia,Long Inflorescence Cryptocarya,Metcalf Cryptocarya ●
113471 Cryptocarya microcarpa F. N. Wei;小果厚壳桂;Smallfruit Thickshellcassia ●

113472 Cryptocarya microcarpa F. N. Wei = Cryptocarya concinna Hance ●
113473 Cryptocarya moschata Nees et Mart.;麝香厚壳桂(巴西厚壳桂);Brazil Nutmeg,Brazilian Nutmeg ●☆
113474 Cryptocarya murrayi F. Muell.;大叶厚壳桂●☆
113475 Cryptocarya myrtifolia Stapf;香桃木叶厚壳桂●☆
113476 Cryptocarya obovata R. Br.;臭厚壳桂●☆
113477 Cryptocarya obtusifolia Merr. = Cryptocarya impressiaervia H. W. Li. ●
113478 Cryptocarya parvifolia (Scott-Elliot) Kosterm. = Aspidostemon parvifolium (Scott-Elliot) van der Werff ●☆
113479 Cryptocarya percoriacea Kosterm. = Aspidostemon percoriaceum (Kosterm.) Rohwer ●☆
113480 Cryptocarya perrieri Danguy;白厚壳桂●☆
113481 Cryptocarya perrieri Danguy = Aspidostemon perrieri (Danguy) Rohwer ●☆
113482 Cryptocarya pleurosperma C. T. White et W. D. Francis;侧厚壳桂●☆
113483 Cryptocarya reticulata Yen C. Yang = Cryptocarya yaanica N. Chao ●
113484 Cryptocarya rolletii H. Wang et H. Zhu = Cryptocarya hainanensis Merr. ●
113485 Cryptocarya rubra (Molina) Skeels;红厚壳桂●☆
113486 Cryptocarya siebertiana Burtt Davy = Cryptocarya liebertiana Engl. ●☆
113487 Cryptocarya sutherlandii Stapf = Cryptocarya wyliei Stapf ●☆
113488 Cryptocarya transvaalensis Burtt Davy;德兰士瓦厚壳桂●☆
113489 Cryptocarya trianthera Kosterm. = Aspidostemon trianthera (Kosterm.) Rohwer ●☆
113490 Cryptocarya tsangii Nakai;红柄厚壳桂(怀德厚壳桂);Redstalk Cryptocarya,Redstalk Thickshellcassia,Tsang Cryptocarya ●
113491 Cryptocarya vacciniifolia Stapf = Cryptocarya myrtifolia Stapf ●☆
113492 Cryptocarya wightiana Thwaites;斯里兰卡厚壳桂●☆
113493 Cryptocarya woodii Engl.;伍得厚壳桂●☆
113494 Cryptocarya wyliei Stapf;怀利厚壳桂●☆
113495 Cryptocarya yaanica N. Chao = Cryptocarya yaanica N. Chao ex H. W. Li et al. ●
113496 Cryptocarya yaanica N. Chao ex H. W. Li et al.;雅安厚壳桂;Ya'an Cryptocarya,Ya'an Thickshellcassia ●
113497 Cryptocarya yunnanensis H. W. Li;云南厚壳桂;Yunnan Cryptocarya,Yunnan Thickshellcassia ●
113498 Cryptocarynaceae J. Agardh = Araceae Juss.(保留科名)■●
113499 Cryptocentrum Benth. (1880);隐距兰属■☆
113500 Cryptocentrum flavum Schltr.;黄隐距兰■☆
113501 Cryptocentrum gracilipes Schltr.;细梗隐距兰■☆
113502 Cryptocentrum jamesonii Benth.;隐距兰■☆
113503 Cryptocentrum latifolium Schltr.;宽叶隐距兰■☆
113504 Cryptocentrum minus Schltr.;小隐距兰■☆
113505 Cryptoceras Schott et Kotschy = Corydalis DC.(保留属名)■
113506 Cryptocereus Alexander = Selenicereus (A. Berger) Britton et Rose ●
113507 Cryptocereus Alexander(1950);隐柱昙花属(隐柱天轮柱属)■☆
113508 Cryptocereus anthonyanus Alexander;隐柱昙花(安氏隐柱)■☆
113509 Cryptochaete Ralmondi ex Herrera = Laccopetalum Ulbr. ■☆
113510 Cryptochilos Spreng. = Cryptochilus Wall. ■
113511 Cryptochilus Wall. (1824);宿苞兰属;Cryptochilus ■
113512 Cryptochilus farreri Schltr. = Cryptochilus luteus Lindl. ■
113513 Cryptochilus luteus Lindl.;宿苞兰;Common Cryptochilus ■
113514 Cryptochilus roseus (Lindl.) S. C. Chen et J. J. Wood;玫瑰宿苞兰(玫瑰毛兰);Rose Eria,Rose Hairorchis ■
113515 Cryptochilus sanguineus Wall.;红花宿苞兰(血红宿苞兰);Bloodred Cryptochilus,Red Cryptochilus ■
113516 Cryptochloa Swallen(1942);隐藏禾属■☆
113517 Cryptochloa concinna (Hook. f.) Swallen;隐藏禾■☆
113518 Cryptochloa granulifera Swallen;腺隐藏禾■☆
113519 Cryptochloa strictiflora (E. Fourn.) Swallen;直花隐藏禾■☆
113520 Cryptochloris Benth. = Tetrapogon Desf. ■☆
113521 Cryptocodon Fed. (1957);隐钟草属■☆
113522 Cryptocodon Fed. = Asyneuma Griseb. et Schenk ■
113523 Cryptocodon monocephalus (Trautv.) Fed.;隐钟草■☆
113524 Cryptocorynaceae J. G. Agardh = Araceae Juss.(保留科名)■●
113525 Cryptocoryne Fisch. = Cryptocoryne Fisch. ex Wydler ●■
113526 Cryptocoryne Fisch. et C. A. Mey. = Cryptocoryne Fisch. ex Wydler ●■
113527 Cryptocoryne Fisch. ex Wydler(1830);隐棒花属;Cryptocoryne ●■
113528 Cryptocoryne beckettii Thwaites ex Trimen;贝氏隐棒花;Beckett Cryptocoryne,Beckett's Water Trumpet ■☆
113529 Cryptocoryne ciliata (Roxb.) Fisch. ex Schott = Cryptocoryne ciliata Blume ■☆
113530 Cryptocoryne ciliata Blume;纤毛隐棒花(睫苞隐棒花,三角椒草);Ciliate Cryptocoryne ■☆
113531 Cryptocoryne cordata Griff.;心叶隐棒花;Cordate-leaf Cryptocoryne ■☆
113532 Cryptocoryne crispatula Engl. = Cryptocoryne retrospiralis (Roxb.) Fisch. ex Wydler ■
113533 Cryptocoryne drymorrhisa Zipp. ex Schott = Cryptocoryne ciliata Blume ■☆
113534 Cryptocoryne elata Griff. = Cryptocoryne ciliata Blume ■☆
113535 Cryptocoryne grandis Ridl.;大叶隐棒花;Large-leaf Cryptocoryne ■☆
113536 Cryptocoryne grandis Ridl. = Cryptocoryne cordata Griff. ■☆
113537 Cryptocoryne griffithii Schottky;格里菲斯隐棒花■☆
113538 Cryptocoryne johorensis Engl.;马来隐棒花■☆
113539 Cryptocoryne kwangsiensis H. Li;广西隐棒花;Guangxi Cryptocoryne ■
113540 Cryptocoryne retrospiralis (Roxb.) Fisch. ex Wydler;旋苞隐棒花;Retrospiral Cryptocoryne ■
113541 Cryptocoryne sinensis Merr.;隐棒花(沙滩草,岩榄);China Cryptocoryne,Chinese Cryptocoryne ■
113542 Cryptocoryne spiralis (Retz.) Wydler;螺旋隐棒花(花椒草);Indian Ipecacuanha ■☆
113543 Cryptocoryne spiralis Fisch. ex Wydler = Cryptocoryne spiralis (Retz.) Wydler ■☆
113544 Cryptocoryne walkeri Schott;斯里兰卡隐棒花■☆
113545 Cryptocoryne willissii Engl.;威氏隐棒花;Wendt's Water Trumpet ■☆
113546 Cryptocoryne yunnanensis H. Li;八仙过海;Yunnan Cryptocoryne ■
113547 Cryptodia Sch. Bip. = Cryptadia Lindl. ex Endl. ■☆
113548 Cryptodia Sch. Bip. = Gymnarrhena Desf. ■☆
113549 Cryptodiscus Schrenk = Neocryptodiscus Hedge et Lamond ■☆

113550 Cryptodiscus Schrenk ex Fisch. et C. A. Mey. = Prangos Lindl. ■☆
113551 Cryptodiscus Schrenk(1841);隐盘芹属;Cryptodiscus ■
113552 Cryptodiscus ammophilus Bunge;喜沙隐盘芹■☆
113553 Cryptodiscus arenarius Schischk.;沙地隐盘芹■☆
113554 Cryptodiscus cachroides Schrenk;绵果隐盘芹(隐盘芹);Common Cryptodiscus ■
113555 Cryptodiscus didymus (Regel) Korovin;双生隐盘芹;Pairing Cryptodiscus,Twin Cryptodiscus ■
113556 Cryptodiscus didymus (Regel) Korovin = Prangos didyma (Regel) Pimenov et V. N. Tikhom. ■
113557 Cryptoglochin Heuff. = Carex L. ■
113558 Cryptoglottis Blume = Podochilus Blume ■
113559 Cryptogyne Cass.(废弃属名) = Cryptogyne Hook. f.(保留属名)●☆
113560 Cryptogyne Cass.(废弃属名) = Eriocephalus L. ●☆
113561 Cryptogyne Hook. f.(1876)(保留属名);隐蕊榄属●☆
113562 Cryptogyne Hook. f.(保留属名) = Sideroxylon L. ●☆
113563 Cryptogyne gerrardiana Hook. f. = Sideroxylon gerrardianum (Hook. f.) Aubrév. ●☆
113564 Cryptolappa (A. Jussieu) Kuntze = Camarea A. St. -Hll. ●☆
113565 Cryptolappa Kuntze = Camarea A. St. -Hll. ●☆
113566 Cryptolepis R. Br.(1810);白叶藤属(半架牛属,隐鳞藤属);Cryptolepis ●
113567 Cryptolepis R. Br. = Pentopetia Decne. ■☆
113568 Cryptolepis acutifolia (Sond.) N. E. Br. = Cryptolepis oblongifolia (Meisn.) Schltr. ●☆
113569 Cryptolepis albicans Jum. et H. Perrier = Pentopetia albicans (Jum. et H. Perrier) Klack. ■☆
113570 Cryptolepis angolensis Welw. ex Hiern = Cryptolepis oblongifolia (Meisn.) Schltr. ●☆
113571 Cryptolepis apiculata K. Schum.;细尖白叶藤●☆
113572 Cryptolepis arenicola Dinter = Cryptolepis oblongifolia (Meisn.) Schltr. ●☆
113573 Cryptolepis barteri K. Schum. = Cryptolepis sanguinolenta (Lindl.) Schltr. ●☆
113574 Cryptolepis baumii N. E. Br. = Cryptolepis oblongifolia (Meisn.) Schltr. ●☆
113575 Cryptolepis baumii Schltr. = Ectadiopsis producta (N. E. Br.) Bullock ●☆
113576 Cryptolepis brazzaei Baill. = Cryptolepis oblongifolia (Meisn.) Schltr. ●☆
113577 Cryptolepis buchananii Roem. et Schult.;古钩藤(白都宗,白浆藤,白马连鞍,白叶藤,半架牛,布坎南白叶藤,大暗消,大叶白藤,大叶白叶藤,断肠草,个卜汁,扣过怀,老鸦咀,奶浆藤,牛挂脖子藤,牛角藤,牛奶藤,羊嘛,羊排果);Ancient Hookime,Buchanan Cryptolepis ●
113578 Cryptolepis buxifolia Chiov. = Cryptolepis oblongifolia (Meisn.) Schltr. ●☆
113579 Cryptolepis capensis Schltr.;好望角白叶藤●☆
113580 Cryptolepis cryptolepidioides (Schltr.) Bullock;隐鳞白叶藤●☆
113581 Cryptolepis debeerstii De Wild. = Cryptolepis oblongifolia (Meisn.) Schltr. ●☆
113582 Cryptolepis decidua (Planch. ex Benth.) N. E. Br.;脱落白叶藤●☆
113583 Cryptolepis delagoensis Schltr.;迪拉果白叶藤●☆
113584 Cryptolepis eburnea (Pichon) Venter;象牙白白叶藤●☆
113585 Cryptolepis edithae (Hance) Benth. et Hook. f. ex Maxim. = Cryptolepis sinensis (Lour.) Merr. ●
113586 Cryptolepis edithae Benth. et Hook. f. ex Maxim. = Cryptolepis sinensis (Lour.) Merr. ●
113587 Cryptolepis elegans Wall. = Cryptolepis sinensis (Lour.) Merr. ●
113588 Cryptolepis elegans Wall. ex G. Don = Cryptolepis sinensis (Lour.) Merr. ●
113589 Cryptolepis elliotii Schltr. = Cryptolepis oblongifolia (Meisn.) Schltr. ●☆
113590 Cryptolepis gillettii Hutch. et E. A. Bruce;吉莱特白叶藤●☆
113591 Cryptolepis gossweileri S. Moore;戈斯白叶藤●☆
113592 Cryptolepis grevei Baill. = Pentopetia grevei (Baill.) Venter ■☆
113593 Cryptolepis hensii N. E. Br. = Cryptolepis oblongifolia (Meisn.) Schltr. ●☆
113594 Cryptolepis hypoglauca K. Schum.;粉绿背白叶藤●☆
113595 Cryptolepis laurentii De Wild.;洛朗白叶藤●☆
113596 Cryptolepis laxa Baill.;疏松白叶藤●☆
113597 Cryptolepis linearis N. E. Br. = Cryptolepis oblongifolia (Meisn.) Schltr. ●☆
113598 Cryptolepis macrophylla (Radcl. -Sm.) Venter;大叶白叶藤●☆
113599 Cryptolepis microphylla Baill.;小叶白叶藤●☆
113600 Cryptolepis migiurtina Chiov.;米朱蒂白叶藤●☆
113601 Cryptolepis monteiroae Oliv. = Stomatostemma monteiroae (Oliv.) N. E. Br. ■☆
113602 Cryptolepis myrtifolia (Baill.) Schltr. = Cryptolepis oblongifolia (Meisn.) Schltr. ●☆
113603 Cryptolepis nigritana (Benth.) N. E. Br. = Cryptolepis oblongifolia (Meisn.) Schltr. ●☆
113604 Cryptolepis oblongifolia (Meisn.) Schltr.;矩圆白叶藤●☆
113605 Cryptolepis oblongifolia (Meisn.) Schltr. = Cryptolepis sanguinolenta (Lindl.) Schltr. ●☆
113606 Cryptolepis obtusa N. E. Br.;钝白叶藤●☆
113607 Cryptolepis orbicularis Chiov.;圆白叶藤●☆
113608 Cryptolepis producta N. E. Br. = Ectadiopsis producta (N. E. Br.) Bullock ●☆
113609 Cryptolepis ruspolii Chiov.;鲁斯波利白叶藤●☆
113610 Cryptolepis sanguinolenta (Lindl.) Schltr.;血红白叶藤●☆
113611 Cryptolepis scandens (K. Schum.) Schltr. = Cryptolepis oblongifolia (Meisn.) Schltr. ●☆
113612 Cryptolepis sinensis (Lour.) Merr.;白叶藤(飞扬藤,红丝线,红藤仔,扛棺回,抗棺回,篱尾蛇,淋汁藤,鸟仔藤,牛蹄藤,七娘藤,铁边,脱皮藤,蜈蚣草,隐鳞藤);China Cryptolepis,Chinese Cryptolepis ●
113613 Cryptolepis sinensis (Lour.) Merr. subsp. africana Bullock;非洲白叶藤●☆
113614 Cryptolepis sizenandii Rolfe = Cryptolepis oblongifolia (Meisn.) Schltr. ●☆
113615 Cryptolepis socotrana (Balf. f.) Venter = Cochlanthus socotranus Balf. f. ●☆
113616 Cryptolepis somaliensis Venter et Thulin;索马里白叶藤●☆
113617 Cryptolepis stefaninii Chiov.;斯特白叶藤●☆
113618 Cryptolepis suffruticosa (K. Schum.) N. E. Br. = Cryptolepis oblongifolia (Meisn.) Schltr. ●☆
113619 Cryptolepis transvaalensis Schltr. = Cryptolepis cryptolepidioides (Schltr.) Bullock ●☆
113620 Cryptolepis triangularis N. E. Br.;三棱白叶藤●☆
113621 Cryptolepis triangularis N. E. Br. = Cryptolepis sanguinolenta (Lindl.) Schltr. ●☆

113622 Cryptolepis welwitschii (Baill.) Schltr. = Cryptolepis oblongifolia (Meisn.) Schltr. ●☆

113623 Cryptolepis welwitschii (Baill.) Schltr. var. luteola Hiern = Cryptolepis oblongifolia (Meisn.) Schltr. ●☆

113624 Cryptolluma Plowes = Boucerosia Wight et Arn. ■☆

113625 Cryptolluma Plowes(1995);食萝藦属■☆

113626 Cryptolluma edulis (Edgew.) Plowes = Caralluma edulis (Edgew.) Benth. ■☆

113627 Cryptolobus Endl. = Cryptolepis R. Br. ●

113628 Cryptolobus Meisn. ex Steud. = Cryptolepis R. Br. ●

113629 Cryptolobus Spreng. = Amphicarpaea Elliott + Vigna Savi(保留属名)■

113630 Cryptolobus Spreng. = Voandzeia Thouars(废弃属名)■

113631 Cryptoloma Hanst. = Isoloma Decne. ●■☆

113632 Cryptoloma Hanst. = Kohleria Regel ●■☆

113633 Cryptomeria D. Don(1838);柳杉属;Chinese Cedar,Cryptomeria, Japan Cedar,Japanese Cedar,Japanese Red-cedar ●

113634 Cryptomeria araucarioides Henkel et Hochst. = Cryptomeria japonica (Thunb. ex L. f.) D. Don 'Compactoglobosa' ●

113635 Cryptomeria elegans Jacob-Makoy = Cryptomeria japonica (Thunb. ex L. f.) D. Don 'Elegans' ●

113636 Cryptomeria fortunei Hooibr. ex Otto et A. Dietr. = Cryptomeria japonica (Thunb. ex L. f.) D. Don var. sinensis Siebold et Zucc. ●

113637 Cryptomeria fortunei Hooibr. ex Otto et A. Dietr. f. kawaii (Hayata) W. C. Cheng et H. P. Tsui = Cryptomeria japonica (Thunb. ex L. f.) D. Don var. sinensis Miq. ●

113638 Cryptomeria japonica (L. f.) D. Don 'Albospica' = Cryptomeria japonica (Thunb. ex L. f.) D. Don 'Albospica' ●☆

113639 Cryptomeria japonica (L. f.) D. Don 'Araucarioides' = Cryptomeria japonica (Thunb. ex L. f.) D. Don 'Araucarioides' ●

113640 Cryptomeria japonica (L. f.) D. Don 'Bandai' = Cryptomeria japonica (Thunb. ex L. f.) D. Don 'Bandai' ●☆

113641 Cryptomeria japonica (L. f.) D. Don 'Bandai-sugi' = Cryptomeria japonica (Thunb. ex L. f.) D. Don 'Bandai-sugi' ●☆

113642 Cryptomeria japonica (L. f.) D. Don 'Compactoglobosa' = Cryptomeria japonica (Thunb. ex L. f.) D. Don 'Compactoglobosa' ●

113643 Cryptomeria japonica (L. f.) D. Don 'Compressa' = Cryptomeria japonica (Thunb. ex L. f.) D. Don 'Compressa' ●☆

113644 Cryptomeria japonica (L. f.) D. Don 'Cristata' = Cryptomeria japonica (Thunb. ex L. f.) D. Don 'Cristata' ●☆

113645 Cryptomeria japonica (L. f.) D. Don 'Dacrydioides' = Cryptomeria japonica (Thunb. ex L. f.) D. Don 'Dacrydioides' ●

113646 Cryptomeria japonica (L. f.) D. Don 'Elegans Compacta' = Cryptomeria japonica (Thunb. ex L. f.) D. Don 'Elegans Compacta' ●☆

113647 Cryptomeria japonica (L. f.) D. Don 'Elegans Nana' = Cryptomeria japonica (Thunb. ex L. f.) D. Don 'Elegans Nana' ●☆

113648 Cryptomeria japonica (L. f.) D. Don 'Elegans' = Cryptomeria japonica (Thunb. ex L. f.) D. Don 'Elegans' ●

113649 Cryptomeria japonica (L. f.) D. Don 'Globosa Nana' = Cryptomeria japonica (Thunb. ex L. f.) D. Don 'Globosa Nana' ●☆

113650 Cryptomeria japonica (L. f.) D. Don 'Lobbii Nana' = Cryptomeria japonica (Thunb. ex L. f.) D. Don 'Lobbii Nana' ●

113651 Cryptomeria japonica (L. f.) D. Don 'Lycopodioides' = Cryptomeria japonica (Thunb. ex L. f.) D. Don 'Lycopodioides' ●☆

113652 Cryptomeria japonica (L. f.) D. Don 'Nana' = Cryptomeria japonica (Thunb. ex L. f.) D. Don 'Nana' ●☆

113653 Cryptomeria japonica (L. f.) D. Don 'Pygmaea' = Cryptomeria japonica (Thunb. ex L. f.) D. Don 'Pygmaea' ●

113654 Cryptomeria japonica (L. f.) D. Don 'Sekkansugi' = Cryptomeria japonica (Thunb. ex L. f.) D. Don 'Sekkansugi' ●☆

113655 Cryptomeria japonica (L. f.) D. Don 'Spiralis' = Cryptomeria japonica (Thunb. ex L. f.) D. Don 'Spiralis' ●

113656 Cryptomeria japonica (L. f.) D. Don 'Tansu' = Cryptomeria japonica (Thunb. ex L. f.) D. Don 'Tansu' ●☆

113657 Cryptomeria japonica (L. f.) D. Don 'Unciata' = Cryptomeria japonica (Thunb. ex L. f.) D. Don 'Unciata' ●☆

113658 Cryptomeria japonica (L. f.) D. Don 'Vilmoriniana' = Cryptomeria japonica (Thunb. ex L. f.) D. Don 'Vilmoriniana' ●

113659 Cryptomeria japonica (L. f.) D. Don 'Yoshino' = Cryptomeria japonica (Thunb. ex L. f.) D. Don 'Yoshino' ●☆

113660 Cryptomeria japonica (L. f.) D. Don 'Yuantouliusha' = Cryptomeria japonica (Thunb. ex L. f.) D. Don 'Yuantouliusha' ●

113661 Cryptomeria japonica (L. f.) D. Don = Cryptomeria japonica (Thunb. ex L. f.) D. Don ●

113662 Cryptomeria japonica (L. f.) D. Don f. albo-variegata Dallim. et Jacks. = Cryptomeria japonica (Thunb. ex L. f.) D. Don f. albo-variegata Dallim. et Jacks. ●

113663 Cryptomeria japonica (L. f.) D. Don f. araucarioides (Henkel et Hochst.) Beissn. = Cryptomeria japonica (Thunb. ex L. f.) D. Don 'Compactoglobosa' ●

113664 Cryptomeria japonica (L. f.) D. Don f. clathrata Sugim. = Cryptomeria japonica (Thunb. ex L. f.) D. Don f. clathrata Sugim. ●☆

113665 Cryptomeria japonica (L. f.) D. Don f. compacta (Beissn.) Beissn. = Cryptomeria japonica (Thunb. ex L. f.) D. Don f. compacta (Beissn.) Beissn. ●

113666 Cryptomeria japonica (L. f.) D. Don f. compactoglobosa F. H. Chen = Cryptomeria japonica (Thunb. ex L. f.) D. Don 'Compactoglobosa' ●

113667 Cryptomeria japonica (L. f.) D. Don f. dacrydioides (Carrière) Rehder = Cryptomeria japonica (Thunb. ex L. f.) D. Don 'Dacrydioides' ●

113668 Cryptomeria japonica (L. f.) D. Don f. elegans (Jacob-Makoy) Beissn. = Cryptomeria japonica (Thunb. ex L. f.) D. Don 'Elegans' ●

113669 Cryptomeria japonica (L. f.) D. Don f. fasciata Dallim. et Jacks. = Cryptomeria japonica (Thunb. ex L. f.) D. Don f. fasciata Dallim. et Jacks. ●

113670 Cryptomeria japonica (L. f.) D. Don f. husari-sugi Dallim. et Jacks. = Cryptomeria japonica (Thunb. ex L. f.) D. Don f. husari-sugi Dallim. et Jacks. ●

113671 Cryptomeria japonica (L. f.) D. Don f. kusari-sugi Dallim. et Jacks. = Cryptomeria japonica (Thunb. ex L. f.) D. Don f. kusari-sugi Dallim. et Jacks. ●

113672 Cryptomeria japonica (L. f.) D. Don f. lobbii (Carrière) Beissn. = Cryptomeria japonica (Thunb. ex L. f.) D. Don f. lobbii (Carrière) Beissn. ●

113673 Cryptomeria japonica (L. f.) D. Don f. nana (Carrière) Beissn. = Cryptomeria japonica (Thunb. ex L. f.) D. Don f. nana (Carrière) Beissn. ●

113674 Cryptomeria japonica (L. f.) D. Don f. pungens (Carrière) Beissn. = Cryptomeria japonica (Thunb. ex L. f.) D. Don f. pungens (Carrière) Beissn. ●

113675 Cryptomeria japonica (L. f.) D. Don f. radicans (Nakai) Sugim. et Muroi = Cryptomeria japonica (Thunb. ex L. f.) D. Don var. radicans Nakai ●☆

113676 Cryptomeria japonica (L. f.) D. Don f. spiralis (Siebold) Rehder = Cryptomeria japonica (Thunb. ex L. f.) D. Don 'Spiralis' ●

113677 Cryptomeria japonica (L. f.) D. Don f. variegata Dallim. et Jacks. = Cryptomeria japonica (Thunb. ex L. f.) D. Don f. variegata Dallim. et Jacks. ●

113678 Cryptomeria japonica (L. f.) D. Don var. araucarioides Henkel et Hochst. = Cryptomeria japonica (Thunb. ex L. f.) D. Don 'Compactoglobosa' ●

113679 Cryptomeria japonica (L. f.) D. Don var. araucarioides Siebold = Cryptomeria japonica (Thunb. ex L. f.) D. Don 'Compactoglobosa' ●

113680 Cryptomeria japonica (L. f.) D. Don var. dacrydioides Carrière = Cryptomeria japonica (Thunb. ex L. f.) D. Don 'Dacrydioides' ●

113681 Cryptomeria japonica (L. f.) D. Don var. elegans Mast. = Cryptomeria japonica (Thunb. ex L. f.) D. Don 'Elegans' ●

113682 Cryptomeria japonica (L. f.) D. Don var. fortunei Henry = Cryptomeria japonica (Thunb. ex L. f.) D. Don var. sinensis Miq. ●

113683 Cryptomeria japonica (L. f.) D. Don var. radicans Nakai = Cryptomeria japonica (Thunb. ex L. f.) D. Don var. radicans Nakai ●☆

113684 Cryptomeria japonica (L. f.) D. Don var. vilmoriniana Hornibr. = Cryptomeria japonica (Thunb. ex L. f.) D. Don 'Vilmoriniana' ●

113685 Cryptomeria japonica (Thunb. ex L. f.) D. Don;日本柳杉(孔雀松,柳杉,日本杉,山杉,杉,倭木);Common Cryptomeria, Cryptomeria, Japan Cedar, Japan Cryptomeria, Japanese Cedar, Japanese Cryptomeria, Japanese Red Cedar, Japanese Red-cedar, Peacock Pine, Sugi ●

113686 Cryptomeria japonica (Thunb. ex L. f.) D. Don 'Albospica';白穗柳杉●☆

113687 Cryptomeria japonica (Thunb. ex L. f.) D. Don 'Araucarioides';短叶柳杉(短叶孔雀杉,绫杉,猿猴杉);Shortleaf Cryptomeria ●

113688 Cryptomeria japonica (Thunb. ex L. f.) D. Don 'Bandai';班代日本柳杉●☆

113689 Cryptomeria japonica (Thunb. ex L. f.) D. Don 'Bandai-sugi';丛生日本柳杉●☆

113690 Cryptomeria japonica (Thunb. ex L. f.) D. Don 'Compactoglobosa';圆球日本柳杉(圆球柳杉,猿臂柳杉,猿猴杉);Compact Cryptomeria ●

113691 Cryptomeria japonica (Thunb. ex L. f.) D. Don 'Compressa';紫叶矮日本柳杉●☆

113692 Cryptomeria japonica (Thunb. ex L. f.) D. Don 'Cristata';鸡冠叶柳杉(鸡冠柳杉,鸡冠日本柳杉);Cockscomb Japanese Cedar, Cristate Cryptomeria ●☆

113693 Cryptomeria japonica (Thunb. ex L. f.) D. Don 'Dacrydioides';鳞叶日本柳杉(老人杉,鳞叶柳杉,软枝柳杉,仙人杉);Scaleleaf Cryptomeria ●

113694 Cryptomeria japonica (Thunb. ex L. f.) D. Don 'Elegans Compacta';密枝日本柳杉●☆

113695 Cryptomeria japonica (Thunb. ex L. f.) D. Don 'Elegans Nana';矮扁叶柳杉;Elegans Cryptomeria ●☆

113696 Cryptomeria japonica (Thunb. ex L. f.) D. Don 'Elegans';扁叶柳杉(扁叶日本柳杉,密枝柳杉,唐杉,优美日本柳杉);Plume Cedar, Plume Cryptomeria ●

113697 Cryptomeria japonica (Thunb. ex L. f.) D. Don 'Globosa Nana';矮球柳杉;Dwarf Cryptomeria ●☆

113698 Cryptomeria japonica (Thunb. ex L. f.) D. Don 'Lobbii Nana';洛布矮柳杉;Lobb Dwarf Cryptomeria ●

113699 Cryptomeria japonica (Thunb. ex L. f.) D. Don 'Lycopodioides';锚杉●☆

113700 Cryptomeria japonica (Thunb. ex L. f.) D. Don 'Nana';矮生日本柳杉(万代杉)●☆

113701 Cryptomeria japonica (Thunb. ex L. f.) D. Don 'Pygmaea';矮柳杉(塔状日本柳杉);Dwarf Cryptomeria ●

113702 Cryptomeria japonica (Thunb. ex L. f.) D. Don 'Sekkansugi';黄叶日本柳杉●☆

113703 Cryptomeria japonica (Thunb. ex L. f.) D. Don 'Spiralis';旋叶日本柳杉(龙卷杉,锁杉,旋叶柳杉);Grannies Ringlets ●

113704 Cryptomeria japonica (Thunb. ex L. f.) D. Don 'Tansu';坦苏日本柳杉●☆

113705 Cryptomeria japonica (Thunb. ex L. f.) D. Don 'Uncinata';钩杉●☆

113706 Cryptomeria japonica (Thunb. ex L. f.) D. Don 'Vilmoriniana';千头柳杉(矮球日本柳杉,亮叶日本柳杉);Dwarf Japanese Cedar, Vilmorin Cryptomeria ●

113707 Cryptomeria japonica (Thunb. ex L. f.) D. Don 'Yoshino';和纸日本柳杉●☆

113708 Cryptomeria japonica (Thunb. ex L. f.) D. Don 'Yuantouliusha';圆头柳杉;Roundhead Cryptomeria ●

113709 Cryptomeria japonica (Thunb. ex L. f.) D. Don f. albo-variegata Dallim. et Jacks.;花叶柳杉●

113710 Cryptomeria japonica (Thunb. ex L. f.) D. Don f. caespitosa Sugim.;狮子杉(万吉杉)●

113711 Cryptomeria japonica (Thunb. ex L. f.) D. Don f. clathrata Sugim.;格子日本柳杉●☆

113712 Cryptomeria japonica (Thunb. ex L. f.) D. Don f. compacta (Beissn.) Beissn.;密锥柳杉●

113713 Cryptomeria japonica (Thunb. ex L. f.) D. Don f. fasciata Dallim. et Jacks.;丛枝柳杉●

113714 Cryptomeria japonica (Thunb. ex L. f.) D. Don f. husari-sugi Dallim. et Jacks.;曲枝柳杉●

113715 Cryptomeria japonica (Thunb. ex L. f.) D. Don f. kusari-sugi Dallim. et Jacks.;侏儒柳杉●

113716 Cryptomeria japonica (Thunb. ex L. f.) D. Don f. lobbii (Carrière) Beissn.;洛布柳杉●

113717 Cryptomeria japonica (Thunb. ex L. f.) D. Don f. nana (Carrière) Beissn.;伏地柳杉(万代杉)●

113718 Cryptomeria japonica (Thunb. ex L. f.) D. Don f. pungens (Carrière) Beissn.;锐叶柳杉●

113719 Cryptomeria japonica (Thunb. ex L. f.) D. Don f. variegata Dallim. et Jacks.;黄叶柳杉●

113720 Cryptomeria japonica (Thunb. ex L. f.) D. Don var. radicans Nakai;芦生杉●☆

113721 Cryptomeria japonica (Thunb. ex L. f.) D. Don var. sinensis Miq.;柳杉(宝树,长叶孔雀松,长叶柳杉,华杉,孔雀杉,沙罗树,湿沙天树,小果柳杉,云南柳杉,中国柳杉);China Cedar, China Cryptomeria, Chinese Cedar, Chinese Cryptomeria, Peacock Fir, Yunnan Cryptomeria ●

113722 Cryptomeria japonica D. Don = Cryptomeria japonica (Thunb. ex L. f.) D. Don ●

113723 Cryptomeria kawaii Hayata = Cryptomeria japonica (Thunb. ex L. f.) D. Don var. sinensis Miq. ●

113724 Cryptomeria mairei (H. Lév.) Nakai = Cryptomeria japonica (Thunb. ex L. f.) D. Don var. sinensis Miq. ●

113725 Cryptomeriaceae Gorozh. = Cryptomeriaceae Hayata ●

113726 Cryptomeriaceae Gorozh. = Cupressaceae Gray(保留科名)●
113727 Cryptomeriaceae Gorozh. = Taxodiaceae Saporta(保留科名)●
113728 Cryptomeriaceae Hayata = Cupressaceae Gray(保留科名)●
113729 Cryptomeriaceae Hayata = Taxodiaceae Saporta(保留科名)●
113730 Cryptomeriaceae Hayata;柳杉科●
113731 Cryptonema Turcz. = Burmannia L. ■
113732 Cryptonema Turcz. = Nephrocoelium Turcz. ■
113733 Cryptopera oliveriana Rchb. f. = Eulophia oliveriana (Rchb. f.) Bolus ■☆
113734 Cryptopetalon Cass. = Pectis L. ■☆
113735 Cryptopetalum Hook. et Arn. = Lepuropetalon Elliott ■☆
113736 Cryptophaeseolus Kuntze = Canavalia Adans. (保留属名)●■
113737 Cryptophila W. Wolf = Monotropsis Schwein. ●☆
113738 Cryptophoranthus Barb. Rodr. (1881);窗兰属(萼包兰属);Window Bearing Orchid, Window Orchid ■☆
113739 Cryptophoranthus Barb. Rodr. = Pleurothallis R. Br. ■☆
113740 Cryptophoranthus atropurpureus (Lindl.) Rolfe;暗紫红窗兰;Darkpurpleflower Window Orchid ■☆
113741 Cryptophoranthus dayanus Rolfe;戴氏窗兰■☆
113742 Cryptophoranthus lepidotus L. O. Williams;鳞窗兰;Common Window Orchid ■☆
113743 Cryptophragmia Benth. et Hook. f. = Cryptophragmium Nees ■☆
113744 Cryptophragmium Nees = Gymnostachyum Nees ■
113745 Cryptophragmium Nees(1832);小苞爵床属■☆
113746 Cryptophragmium affine Kuntze;近缘小苞爵床■☆
113747 Cryptophragmium axillare Nees;叶花小苞爵床■☆
113748 Cryptophragmium canescens Nees;银灰小苞爵床■☆
113749 Cryptophragmium latifolium Dalzell;宽叶小苞爵床■☆
113750 Cryptophragmium paniculatum Kuntze;圆锥小苞爵床■☆
113751 Cryptophragmium polyanthum Kuntze;多花小苞爵床■☆
113752 Cryptophragmium sanguinolentum Nees = Gymnostachyum sanguinolentum (Vahl) T. Anderson ■
113753 Cryptophragmium tomentosum Kuntze;毛小苞爵床■☆
113754 Cryptophragmium tonkinense Benoist;东京小苞爵床■☆
113755 Cryptophragmum sanguinolentum (Vahl) Nees = Gymnostachyum sanguinolentum (Vahl) T. Anderson ■
113756 Cryptophysa Standl. et J. F. Macbr. = Conostegia D. Don ■☆
113757 Cryptopleura Nutt. = Troximon Gaertn. ■☆
113758 Cryptopleura californica Nutt. = Agoseris heterophylla (Nutt.) Greene var. cryptopleura Greene ■☆
113759 Cryptopodium Schrad. ex Nees = Scleria P. J. Bergius ■
113760 Cryptopus Lindl. (1824);隐足兰属■☆
113761 Cryptopus brachiatus H. Perrier;短隐足兰■☆
113762 Cryptopus dissectus (Bosser) Bosser;深裂隐足兰■☆
113763 Cryptopus elatus (Thouars) Lindl.;隐足兰■☆
113764 Cryptopus elatus (Thouars) Lindl. subsp. dissectus Bosser = Cryptopus brachiatus H. Perrier ■☆
113765 Cryptopus paniculatus H. Perrier;圆锥隐足兰■☆
113766 Cryptopylos Garay(1972);隐口兰属■☆
113767 Cryptopylos clausus (J. J. Sm.) Garay;隐口兰■☆
113768 Cryptopyrum Heynh. = Elymus L. ■
113769 Cryptopyrum Heynh. = Triticum L. ■
113770 Cryptorhiza Urb. = Pimenta Lindl. ●☆
113771 Cryptorrhynchus Nevski = Astragalus L. ●■
113772 Cryptorrhynchus Nevski(1937);隐喙豆属●☆
113773 Cryptorrhynchus aemulans Nevski;隐喙豆●☆
113774 Cryptosaccus Rchb. f. = Leochilus Knowles et Westc. ■☆
113775 Cryptosanus Scheidw. = Cryptosaccus Rchb. f. ■☆
113776 Cryptosanus Scheidw. = Leochilus Knowles et Westc. ■☆
113777 Cryptosema Meisn. = Jansonia Kippist ■☆
113778 Cryptosepalum Benth. (1865);隐萼豆属(垂籽树属)●☆
113779 Cryptosepalum ambamense Letouzey;安巴姆隐萼豆●☆
113780 Cryptosepalum bequaertii De Wild. = Cryptosepalum katangense (De Wild.) J. Léonard ●☆
113781 Cryptosepalum bifolium De Wild. = Cryptosepalum maraviense Oliv. ●☆
113782 Cryptosepalum boehmii Harms = Cryptosepalum maraviense Oliv. ●☆
113783 Cryptosepalum busseanum Harms = Cryptosepalum maraviense Oliv. ●☆
113784 Cryptosepalum congolanum (De Wild.) J. Léonard;刚果隐萼豆●☆
113785 Cryptosepalum crassiusculum P. A. Duvign. = Cryptosepalum maraviense Oliv. ●☆
113786 Cryptosepalum curtisiorum I. M. Johnst. = Cryptosepalum maraviense Oliv. ●☆
113787 Cryptosepalum dasycladum Harms = Cryptosepalum maraviense Oliv. ●☆
113788 Cryptosepalum debeerstii De Wild. = Cryptosepalum maraviense Oliv. ●☆
113789 Cryptosepalum delevoyi De Wild. = Cryptosepalum maraviense Oliv. ●☆
113790 Cryptosepalum diphyllum P. A. Duvign.;二叶隐萼豆●☆
113791 Cryptosepalum elegans Letouzey = Cryptosepalum ambamense Letouzey ●☆
113792 Cryptosepalum elegans P. A. Duvign. = Cryptosepalum maraviense Oliv. ●☆
113793 Cryptosepalum exfoliatum De Wild. subsp. puberulum P. A. Duvign. et Brenan;微毛隐萼豆●☆
113794 Cryptosepalum exfoliatum De Wild. subsp. suffruticans (P. A. Duvign.) P. A. Duvign. et Brenan;亚灌木隐萼豆●☆
113795 Cryptosepalum exfoliatum De Wild. var. fruticosum (Hutch.) P. A. Duvign. et Brenan;灌木状隐萼豆■☆
113796 Cryptosepalum exfoliatum De Wild. var. fruticosum (Hutch.) P. A. Duvign. et Brenan = Cryptosepalum fruticosum Hutch. ■☆
113797 Cryptosepalum exfoliatum De Wild. var. pubescens P. A. Duvign. et Brenan;短柔毛隐萼豆●☆
113798 Cryptosepalum fruticosum Hutch. = Cryptosepalum exfoliatum De Wild. var. fruticosum (Hutch.) P. A. Duvign. et Brenan ■☆
113799 Cryptosepalum hockii De Wild. = Cryptosepalum maraviense Oliv. ●☆
113800 Cryptosepalum katangense (De Wild.) J. Léonard;加丹加隐萼豆●☆
113801 Cryptosepalum maraviense Oliv.;美丽隐萼豆●☆
113802 Cryptosepalum mimosoides Welw. ex Oliv.;含羞草隐萼豆●☆
113803 Cryptosepalum minutifolium (A. Chev.) Hutch. et Dalziel;微花隐萼豆●☆
113804 Cryptosepalum pellegrinianum (J. Léonard) J. Léonard;佩尔格兰隐萼豆●☆
113805 Cryptosepalum pulchellum Harms = Cryptosepalum maraviense Oliv. ●☆
113806 Cryptosepalum robynsii De Wild. et Staner = Cryptosepalum maraviense Oliv. ●☆
113807 Cryptosepalum staudtii Harms;斯氏隐萼豆●☆

113808 Cryptosepalum subelegans P. A. Duvign. = Cryptosepalum maraviense Oliv. ●☆
113809 Cryptosepalum tetraphyllum (Hook. f.) Benth.;四叶隐萼豆●☆
113810 Cryptosepalum verdickii De Wild. = Cryptosepalum maraviense Oliv. ●☆
113811 Cryptospermum Steud. = Cryptotaenia DC.(保留属名)■
113812 Cryptospermum Steud. = Cyrtospermum Raf. ex DC. ■
113813 Cryptospermum Young = Opercularia Gaertn. ■☆
113814 Cryptospermum Young ex Pers. = Opercularia Gaertn. ■☆
113815 Cryptospora Kar. et Kir. (1842);隐籽芥属(隐子芥属);Cryptospora ■
113816 Cryptospora dentata Freyn et Sint. = Torularia dentata (Freyn et Sint.) Kitam. ■☆
113817 Cryptospora falcata Kar. et Kir.;隐籽芥(隐子芥);Falcate Cryptospora ■
113818 Cryptospora omissa Botsch. = Cryptospora falcata Kar. et Kir. ■
113819 Cryptostachys Steud. = Sporobolus R. Br. ■
113820 Cryptostegia R. Br. (1820);桉叶藤属(隐冠藤属);Cryptostegia, India Rubber Vine, Madagascar Rubber, Rubber Vine ●
113821 Cryptostegia glaberrima Hochr. = Cryptostegia madagascariensis Bojer ex Decne. ●☆
113822 Cryptostegia grandiflora R. Br. = Cryptostegia grandiflora R. Br. ex Lindl. ●
113823 Cryptostegia grandiflora R. Br. ex Lindl.;桉叶藤(大花桉叶藤,大花胶藤);India Rubber Vine, Largeflower Cryptostegia, Palay Rubber, Palay Rubbervine, Pink Allemande, Purple Allamanda, Rubber Vine ●
113824 Cryptostegia madagascariensis Bojer = Cryptostegia madagascariensis Bojer ex Decne. ●☆
113825 Cryptostegia madagascariensis Bojer ex Decne.;马岛桉叶藤(马达加斯加胶藤,马达加斯加橡胶树);Madagascar Cryptostegia, Madagascar Rubbervine, Rubber Vine ●☆
113826 Cryptostegia madagascariensis Bojer ex Decne. var. glaberrima (Hochr.) Marohasy et P. I. Forst. = Cryptostegia madagascariensis Bojer ex Decne. ●☆
113827 Cryptostegia madagascariensis Bojer ex Decne. var. septentrionalis Marohasy et P. I. Forst. = Cryptostegia madagascariensis Bojer ex Decne. ●☆
113828 Cryptostegiaceae Hayata = Apocynaceae Juss.(保留科名)●■
113829 Cryptostemma R. Br. ex W. T. Aiton = Arctotheca J. C. Wendl. ■☆
113830 Cryptostemma calendula (L.) Druce = Arctotheca calendula (L.) Levyns ■☆
113831 Cryptostemma calendulaceum (L.) R. Br. = Arctotheca calendula (L.) Levyns ■☆
113832 Cryptostemma forbesianum (DC.) Harv. = Arctotheca forbesiana (DC.) K. Lewin ■☆
113833 Cryptostemma niveum (L. f.) Nicolson = Arctotheca populifolia (P. J. Bergius) Norl. ■☆
113834 Cryptostemon F. Muell. et Miq. (1856);隐蕊桃金娘属●☆
113835 Cryptostemon F. Muell. ex Miq. = Darwinia Rudge ●☆
113836 Cryptostemon ericaeus F. Muell. et Miq.;隐蕊桃金娘●☆
113837 Cryptostemon fascicularis F. Muell. ex Miq.;簇生隐蕊桃金娘●☆
113838 Cryptostephane Sch. Bip. = Dicoma Cass. ●☆
113839 Cryptostephanus Welw. ex Baker(1878);隐冠石蒜属■☆
113840 Cryptostephanus densiflorus Welw. ex Baker;密花隐冠石蒜■☆
113841 Cryptostephanus haemanthoides Pax;血红隐冠石蒜■☆
113842 Cryptostephanus herrei F. M. Leight. = Cyrtanthus herrei (F. M. Leight.) R. A. Dyer ■☆
113843 Cryptostephanus merenskyanus Dinter et G. M. Schulze = Cyrtanthus herrei (F. M. Leight.) R. A. Dyer ■☆
113844 Cryptostoma D. Dietr. = Cryptostomum Schreb. ●☆
113845 Cryptostomum Schreb. = Moutabea Aubl. ●☆
113846 Cryptostylis R. Br. (1810);隐柱兰属;Cryptostylis ■
113847 Cryptostylis alismatifolia F. Muell. = Cryptostylis arachnites (Blume) Blume ■
113848 Cryptostylis arachnites (Blume) Blume;隐柱兰(红唇隐柱兰,满绿隐柱兰,美唇隐柱兰,蜘蛛样隐柱兰);Cryptostylis, Spiderlike Cryptostylis ■
113849 Cryptostylis arachnites (Blume) Blume var. taiwaniana (Masam.) S. S. Ying = Cryptostylis taiwaniana Masam. ■
113850 Cryptostylis arachnites (Blume) Hassk. = Cryptostylis arachnites (Blume) Blume ■
113851 Cryptostylis arachnites (Blume) Hassk. var. philippinensis (Schltr.) S. S. Ying = Cryptostylis taiwaniana Masam. ■
113852 Cryptostylis arachnites (Blume) Hassk. var. taiwaniana (Masam.) S. S. Ying = Cryptostylis taiwaniana Masam. ■
113853 Cryptostylis erecta R. Br.;直立隐柱兰;Erect Cryptostylis ■☆
113854 Cryptostylis erythroglossa Hayata = Cryptostylis arachnites (Blume) Blume ■
113855 Cryptostylis fulva Schltr. = Cryptostylis arachnites (Blume) Blume ■
113856 Cryptostylis fulva Schltr. var. subregularis Schltr. = Cryptostylis arachnites (Blume) Blume ■
113857 Cryptostylis leptochila F. Muell. ex Benth.;薄唇隐柱兰;Thinlip Cryptostylis ■☆
113858 Cryptostylis papuana Schltr. = Cryptostylis arachnites (Blume) Blume ■
113859 Cryptostylis philippinensis Schltr. = Cryptostylis arachnites (Blume) Hassk. ■
113860 Cryptostylis philippinensis Schltr. = Cryptostylis taiwaniana Masam. ■
113861 Cryptostylis stenochila Schltr. = Cryptostylis arachnites (Blume) Blume ■
113862 Cryptostylis subulata (Labill.) Rchb. f.;钻形隐柱兰;Awl-shaped Cryptostylis, Large Tongue Orchid ■☆
113863 Cryptostylis taiwaniana Masam.;台湾隐柱兰(蓬莱隐柱兰);Taiwan Cryptostylis ■
113864 Cryptostylis taiwaniana Masam. = Cryptostylis arachnites (Blume) Hassk. var. taiwaniana (Masam.) S. S. Ying ■
113865 Cryptostylis vitiensis Schltr. = Cryptostylis arachnites (Blume) Blume ■
113866 Cryptostylis walkerae (Wight) Blume = Cryptostylis arachnites (Blume) Blume ■
113867 Cryptostylis zeylanica (Lindl.) Blume = Cryptostylis arachnites (Blume) Blume ■
113868 Cryptotaenia DC. (1829)(保留属名);鸭儿芹属■
113869 Cryptotaenia africana (Hook. f.) Drude;非洲鸭儿芹■☆
113870 Cryptotaenia calycina C. C. Towns.;萼状鸭儿芹■☆
113871 Cryptotaenia calycina C. C. Towns. var. dissecta ?;深裂萼状鸭儿芹■☆
113872 Cryptotaenia canadensis (L.) DC.;加拿大鸭儿芹(北美鸭儿芹,鸭儿芹);Canada Ducklingcelery, Canadian Honewort, Honewort, Japan Parsley, Mitsuba, Mitzuba, White Chervil, Wild Chervil ■

113873 Cryptotaenia canadensis DC. = Cryptotaenia canadensis (L.) DC. ■
113874 Cryptotaenia canadensis DC. = Cryptotaenia japonica Hassk. ■
113875 Cryptotaenia canadensis DC. subsp. japonica (Hassk.) Hand. -Mazz. = Cryptotaenia japonica Hassk. ■
113876 Cryptotaenia canadensis DC. var. japonica (Hassk.) Makino f. dissecta (Y. Yabe) Makino = Cryptotaenia japonica Hassk. ■
113877 Cryptotaenia canadensis DC. var. japonica f. dissecta (Y. Yabe) Makino = Cryptotaenia japonica Hassk. f. dissecta (Y. Yabe) Hara ■
113878 Cryptotaenia elegans Bolle;雅致鸭儿芹■☆
113879 Cryptotaenia japonica Hassk.;鸭儿芹(大鸭脚板,当归,鹅脚板,赴鱼,红鸭脚板,起莫,三石,三叶,三叶芹,水白芷,水芹菜,鸭脚板,鸭脚板草,鸭脚板芹,牙痛草,野芹菜,野蜀葵);Japan Ducklingcelery,Mitsu-ba ■
113880 Cryptotaenia japonica Hassk. f. atropurpurea (Makino) Ohwi = Cryptotaenia japonica Hassk. f. atropurpurea (Makino) Ohwi ex H. Hara ■
113881 Cryptotaenia japonica Hassk. f. atropurpurea (Makino) Ohwi ex H. Hara;紫鸭儿芹■
113882 Cryptotaenia japonica Hassk. f. atropurpurea (Makino) Ohwi ex H. Hara = Cryptotaenia japonica Hassk. ■
113883 Cryptotaenia japonica Hassk. f. dissecta (Y. Yabe) H. Hara;深裂鸭儿芹■
113884 Cryptotaenia japonica Hassk. f. dissecta (Y. Yabe) H. Hara = Cryptotaenia japonica Hassk. ■
113885 Cryptotaenia japonica Hassk. f. pinnatisecta S. L. Liou;羽裂鸭儿芹;Pinnatifid Ducklingcelery ■
113886 Cryptotaenia japonica Hassk. f. warabiana (Makino) H. Hara = Cryptotaenia japonica Hassk. ■
113887 Cryptotaenia japonica Hassk. var. atropurpurea Makino = Cryptotaenia japonica Hassk. ■
113888 Cryptotaenia japonica Hassk. var. dissecta Y. Yabe = Cryptotaenia japonica Hassk. f. dissecta (Y. Yabe) Hara ■
113889 Cryptotaenia japonica Hassk. var. dissecta Y. Yabe = Cryptotaenia japonica Hassk. ■
113890 Cryptotaenia japonica Hassk. var. warabiana Makino = Cryptotaenia japonica Hassk. ■
113891 Cryptotaenia polygama C. C. Towns.;多籽鸭儿芹■☆
113892 Cryptotaeniopsis Dunn = Pternopetalum Franch. ■
113893 Cryptotaeniopsis Dunn(1902);拟鸭儿芹属■☆
113894 Cryptotaeniopsis affinis H. Wolff = Pternopetalum delicatulum (H. Wolff) Hand. -Mazz. ■
113895 Cryptotaeniopsis botrychioides Dunn = Pternopetalum botrychioides (Dunn) Hand. -Mazz. ■
113896 Cryptotaeniopsis cardiocarpa (Franch.) Dunn = Pternopetalum cardiocarpum (Franch.) Hand. -Mazz. ■
113897 Cryptotaeniopsis cuneifolia H. Wolff = Pternopetalum molle (Franch.) Hand. -Mazz. ■
113898 Cryptotaeniopsis davidii (Franch.) H. Wolff = Pternopetalum davidii Franch. ■
113899 Cryptotaeniopsis decipiens C. Norman = Pternopetalum kiangsiense (H. Wolff) Hand. -Mazz. ■
113900 Cryptotaeniopsis decipiens C. Norman = Pternopetalum trichomanifolium (Franch.) Hand. -Mazz. ■
113901 Cryptotaeniopsis delavayi (Franch.) Dunn = Pternopetalum delavayi (Franch.) Hand. -Mazz. ■
113902 Cryptotaeniopsis delavayi Dunn;拟鸭儿芹■☆
113903 Cryptotaeniopsis delavayi Dunn = Pternopetalum delavayi (Franch.) Hand. -Mazz. ■
113904 Cryptotaeniopsis delicatula (H. Wolff) H. Wolff = Pternopetalum delicatulum (H. Wolff) Hand. -Mazz. ■
113905 Cryptotaeniopsis filicina (Franch.) H. Boissieu = Pternopetalum filicinum (Franch.) Hand. -Mazz. ■
113906 Cryptotaeniopsis gracillima H. Wolff = Chamarea gracillima (H. Wolff) B. L. Burtt ■☆
113907 Cryptotaeniopsis gracillima H. Wolff = Pternopetalum gracillimum (H. Wolff) Hand. -Mazz. ■
113908 Cryptotaeniopsis kiangsiensis H. Wolff = Pternopetalum trichomanifolium (Franch.) Hand. -Mazz. ■
113909 Cryptotaeniopsis leptophylla Dunn = Pternopetalum leptophyllum (Dunn) Hand. -Mazz. ■
113910 Cryptotaeniopsis mollis (Franch.) Dunn = Pternopetalum molle (Franch.) Hand. -Mazz. ■
113911 Cryptotaeniopsis nudicaulis H. Boissieu = Pternopetalum nudicaule (H. Boissieu) Hand. -Mazz. ■
113912 Cryptotaeniopsis rosthornii (Diels) H. Wolff = Pternopetalum rosthornii (Diels) Hand. -Mazz. ■
113913 Cryptotaeniopsis sinensis (Franch.) H. Wolff = Pternopetalum sinense (Franch.) Hand. -Mazz. ■
113914 Cryptotaeniopsis tanakae (Franch. et Sav.) H. Boissieu = Pternopetalum tanakae (Franch. et Sav.) Hand. -Mazz. ■
113915 Cryptotaeniopsis trichomanifolia (Franch.) H. Boissieu = Pternopetalum trichomanifolium (Franch.) Hand. -Mazz. ■
113916 Cryptotaeniopsis trichomanifolia (Franch.) H. Wolff = Pternopetalum trichomanifolium (Franch.) Hand. -Mazz. ■
113917 Cryptotaeniopsis viridis C. Norman = Pternopetalum leptophyllum (Dunn) Hand. -Mazz. ■
113918 Cryptotaeniopsis vulgaris Dunn = Pternopetalum vulgare (Dunn) Hand. -Mazz. ■
113919 Cryptotaeniopsis wolffiana Fedde ex H. Wolff = Pternopetalum wolffianum (Fedde) Hand. -Mazz. ■
113920 Cryptotenia Raf. = Cryptotaenia DC.(保留属名)■
113921 Cryptotheca Blume = Ammannia L. ■
113922 Cryptothladia (Bunge) M. J. Cannon = Morina L. ■
113923 Cryptothladia chinensis M. J. Cannon = Morina chinensis (Batalin ex Diels) P. Y. Pai ■
113924 Cryptothladia chlorantha (Diels) M. J. Cannon = Morina chlorantha Diels ■
113925 Cryptothladia kokonorica (K. S. Hao) M. J. Cannon = Morina kokonorica K. S. Hao ■
113926 Cryptothladia ludlowii M. J. Cannon = Morina ludlowii (M. J. Cannon) D. Y. Hong et F. Barrie ■
113927 Cryptothladia polyphylla (Wall. ex DC.) M. J. Cannon = Morina polyphylla Wall. ex DC. ■
113928 Cryptotonia Tausch = Cryptotaenia DC.(保留属名)■
113929 Crypturus Link = Lolium L. ■
113930 Crypturus Trin. = Lolium L. ■
113931 Crypwcarya lenticellata Lecomte = Cryptocarya concinna Hance ●
113932 Crysophila Benth. et Hook. f. = Cryosophila Blume ●☆
113933 Crystallopollen Steetz = Vernonia Schreb.(保留属名)●■
113934 Crystallopollen angustifolium Steetz = Vernonia rhodanthoidea Muschl. ■☆
113935 Crystallopollen angustifolium Steetz f. vulgaris ? = Vernonia rhodanthoidea Muschl. ■☆

113936 Crystallopollen angustifolium Steetz var. chlorolepis ? = Vernonia steetziana Oliv. et Hiern ●☆
113937 Crystallopollen latifolium Steetz = Vernonia petersii Oliv. et Hiern ex Oliv. ●☆
113938 Csapodya Borhidi = Deppea Cham. et Schltdl. ●☆
113939 Cszernaevia Endl. = Angelica L. ■
113940 Cszernaevia Endl. = Archangelica Hoffm. ■
113941 Cszernaevia Endl. = Czernaevia Turcz. ■
113942 Ctenadena Prokh. = Euphorbia L. ●■
113943 Ctenanthe Bichl. = Myrosma L. f. ■☆
113944 Ctenanthe Eichler (1884);栉花芋属(栉花小芭蕉属);Etenanthe ■☆
113945 Ctenanthe lubbersiana Eichler ex Petersen;栉花芋■☆
113946 Ctenanthe lubbersiana Eichler ex Petersen 'Happy Dream';蔓栉花芋(蔓斑竹芋)■☆
113947 Ctenanthe oppenheimiana (E. Morren) K. Schum.;奥氏栉花芋■☆
113948 Ctenanthe oppenheimiana (E. Morren) K. Schum. 'Tricolor';三色奥氏栉花芋(锦竹芋)■☆
113949 Ctenanthe oppenheimiana K. Schum. = Ctenanthe oppenheimiana (E. Morren) K. Schum. ■☆
113950 Ctenanthe setosa Eichler;毛柄栉花芋(毛柄银羽竹芋)■☆
113951 Ctenanthe setosa Eichler 'Greystar';银叶栉花芋(银叶竹芋)■☆
113952 Ctenardisia Ducke(1930);栉花紫金牛属●☆
113953 Ctenardisia speciosa Ducke;栉花紫金牛●☆
113954 Ctenium Panz. (1813)(保留属名);栉茅属■☆
113955 Ctenium aromaticum (Walter) Wood;牙栉茅;Toothache Grass ■☆
113956 Ctenium camposum A. Chev. = Ctenium newtonii Hack. ■☆
113957 Ctenium canescens Benth.;灰白栉茅■☆
113958 Ctenium concinnum Nees;整洁栉茅■☆
113959 Ctenium concinnum Nees var. minus Pilg. = Ctenium somalense (Chiov.) Chiov. ■☆
113960 Ctenium elegans Kunth;雅致栉茅■☆
113961 Ctenium ledermannii Pilg.;莱德栉茅■☆
113962 Ctenium longiglume Kupicha ex Longhi-Wagner et Cope;长颖栉茅■☆
113963 Ctenium minus (Pilg.) Clayton = Ctenium somalense (Chiov.) Chiov. ■☆
113964 Ctenium newtonii Hack.;纽敦栉茅■☆
113965 Ctenium newtonii Hack. var. annuum J. -P. Lebrun;一年纽敦栉茅■☆
113966 Ctenium nubicum De Not.;云雾栉茅■☆
113967 Ctenium nubicum De Not. var. somalense Chiov. = Ctenium somalense (Chiov.) Chiov. ■☆
113968 Ctenium rupestre J. A. Schmidt = Enteropogon rupestris (J. A. Schmidt) A. Chev. ■☆
113969 Ctenium schweinfurthii Pilg. = Ctenium newtonii Hack. ■☆
113970 Ctenium serpentinum Steud. = Ctenium elegans Kunth ■☆
113971 Ctenium somalense (Chiov.) Chiov.;索马里栉茅■☆
113972 Ctenium villosum Berhaut;长柔毛栉茅■☆
113973 Ctenocladium Airy Shaw = Dorstenia L. ■●☆
113974 Ctenocladus Engl. = Ctenocladium Airy Shaw ■●☆
113975 Ctenocladus mildbraedii Engl. = Dorstenia psilurus Welw. ■☆
113976 Ctenodaucus Pomel = Daucus L. ■
113977 Ctenodaucus virgatus (Poir.) Pomel = Daucus virgatus (Poir.) Maire ■☆
113978 Ctenodon Baill. = Aeschynomene L. ●■
113979 Ctenolepis Hook. f. (1867);梳鳞葫芦属■☆
113980 Ctenolepis cerasiformis (Stocks) Hook. f.;梳鳞葫芦;Ctenolepis ■☆
113981 Ctenolophon Oliv. (1873);垂籽树属(垂子树属)●☆
113982 Ctenolophon englerianus Mildbr.;垂籽树●☆
113983 Ctenolophon grandifolius Oliv.;大叶垂籽树●☆
113984 Ctenolophon parvifolius Oliv.;小叶垂籽树●☆
113985 Ctenolophonaceae Exell et Mendonça = Linaceae DC. ex Perleb (保留科名)■●
113986 Ctenolophonaceae Exell et Mendonça(1951);垂籽树科●☆
113987 Ctenomeria Harv. (1842);篦大戟属●☆
113988 Ctenomeria capensis (Thunb.) Harv. ex Sond.;篦大戟●☆
113989 Ctenomeria cordata Harv. = Ctenomeria capensis (Thunb.) Harv. ex Sond. ●☆
113990 Ctenomeria schlechteri (Pax) Prain = Ctenomeria capensis (Thunb.) Harv. ex Sond. ●☆
113991 Ctenopaepale Bremek. = Strobilanthes Blume ●■
113992 Ctenophrynium K. Schum. = Saranthe (Regel et Körn.) Eichler ■☆
113993 Ctenophyllum Rydb. = Astragalus L. ●■
113994 Ctenopsis De Not. (1848);篦茅属■☆
113995 Ctenopsis De Not. = Vulpia C. C. Gmel. ■
113996 Ctenopsis Naudin = Blastania Kotschy et Peyr. ■☆
113997 Ctenopsis Naudin = Ctenolepis Hook. f. ■☆
113998 Ctenopsis cynosuroides (Desf.) Paunero ex Romero García;篦茅■☆
113999 Ctenopsis cynosuroides (Desf.) Paunero ex Romero García = Festuca cynosuroides Desf. ■☆
114000 Ctenopsis pectinella (Delile) De Not.;小篦茅■☆
114001 Ctenopsis pectinella (Delile) De Not. var. delileana Maire et Weiller = Ctenopsis pectinella (Delile) De Not. ■☆
114002 Ctenopsis pectinella (Delile) De Not. var. kralikiana Hack. = Ctenopsis pectinella (Delile) De Not. ■☆
114003 Ctenopsis pectinella (Delile) De Not. var. pubescens Pamp. = Ctenopsis pectinella (Delile) De Not. ■☆
114004 Ctenorchis K. Schum. = Angraecum Bory ■
114005 Ctenorchis pectinata (Thouars) K. Schum. = Angraecum pectinatum Thouars ■☆
114006 Ctenosachna Post et Kuntze = Ktenosachne Steud. ■☆
114007 Ctenosachna Post et Kuntze = Prionanthium Desv. ■☆
114008 Ctenosachna Post et Kuntze = Rostraria Trin. ■☆
114009 Ctenosperma F. Muell. ex Pfeiff. = Brachycome Cass. ●■☆
114010 Ctenosperma Hook. f. = Cotula L. ■
114011 Ctenospermum Lehm. ex T. Post et Kuntze = Ktenospermum Lehm. ●☆
114012 Ctenospermum Lehm. ex T. Post et Kuntze = Pectocarya DC. ex Meisn. ●☆
114013 Ctenospermum T. Post et Kuntze = Ktenospermum Lehm. ●☆
114014 Ctenospermum T. Post et Kuntze = Pectocarya DC. ex Meisn. ●☆
114015 Ctinogyne K. Schum. = Marantochloa Brongn. ex Gris ■☆
114016 Cuapidaria (DC.) Besser = Acachmena H. P. Fuchs(废弃属名)●■
114017 Cuapidaria (DC.) Besser = Erysimum L. ■●
114018 Cuatrecasanthus H. Rob. (1989);单花落苞菊属●■☆
114019 Cuatrecasanthus H. Rob. = Vernonia Schreb. (保留属名)●■

114020 Cuatrecasanthus flexipappus (Gleason) H. Rob.;单花落苞菊■☆
114021 Cuatrecasasia Standl. = Cuatrecasasiodendron Standl. et Steyerm. ●☆
114022 Cuatrecasasiella H. Rob. (1985);对叶紫绒草属■☆
114023 Cuatrecasasiella argentina (Cabrera) H. Rob.;对叶紫绒草■☆
114024 Cuatrecasasiodendron Standl. et Steyerm. (1964);夸特木属●☆
114025 Cuatrecasasiodendron Steyerm. = Cuatrecasasiodendron Standl. et Steyerm. ●☆
114026 Cuatrecasasiodendron colombianum Standl. et Steyerm.;夸特木●☆
114027 Cuatrecasea Dugand = Iriartella H. Wendl. ●☆
114028 Cuatresia Hunz. (1977);酸浆茄属●☆
114029 Cuatresia cuspidata (Dunal) Hunz.;酸浆茄●☆
114030 Cuba Scop. = Tachigalia Aubl. ●☆
114031 Cubacroton Alain(1961);古巴巴豆属●☆
114032 Cubacroton maestrensis Alain;古巴巴豆●☆
114033 Cubaea Schreb. = Tachigalia Aubl. ●☆
114034 Cubanola Aiello(1979);古巴茜属●☆
114035 Cubanola daphnoides (Graham) Aiello;古巴茜●☆
114036 Cubanola domingensis (Britton) Aiello;多明古巴茜●☆
114037 Cubanthus (Boiss.) Millsp. (1913);古巴花属■☆
114038 Cubanthus Millsp. = Cubanthus (Boiss.) Millsp. ■☆
114039 Cubanthus brittoni Millsp.;古巴花■☆
114040 Cubanthus linearifolius Millsp.;线叶古巴花;Cussonia ■☆
114041 Cubeba Raf. = Litsea Lam. (保留属名)●
114042 Cubeba Raf. = Piper L. ●■
114043 Cubeba clusii Miq. = Piper guineense Schumach. et Thonn. ●☆
114044 Cubelium Raf. = Hybanthus Jacq. (保留属名)●■
114045 Cubelium Raf. ex Britton et A. Br. = Hybanthus Jacq. (保留属名)●■
114046 Cubelium concolor (T. F. Forst.) Raf. = Hybanthus concolor (T. F. Forst.) Spreng. ●☆
114047 Cubilia Blume(1849);南洋丹属●☆
114048 Cubilia cubili (Blanco) Adelb.;南洋丹;Kubili Nut ●☆
114049 Cubincola Urb. = Cneorum L. ●☆
114050 Cubitanthus Barringer(1984);肘花苣苔属■☆
114051 Cubitanthus alatus (Cham. et Schltdl.) Barringer;具翅肘花苣苔■☆
114052 Cubospermum Lour. = Ludwigia L. ●■
114053 Cucholzia philoxeroides Mart. = Alternanthera philoxeroides (Mart.) Griseb. ■
114054 Cuchumatanea Seid. et Beaman(1966);危地马拉菊属■☆
114055 Cuchumatanea steyermarkii Seid. et Beaman;危地马拉菊■☆
114056 Cucifera Delile = Hyphaene Gaertn. ●☆
114057 Cucubalus L. (1753);狗筋蔓属;Berry Catchftly, Bladder Campion, Bladdercampion, Cucubalus ■
114058 Cucubalus L. = Silene L. (保留属名)■
114059 Cucubalus acaulis L. = Silene acaulis L. ■☆
114060 Cucubalus baccifer L.;狗筋蔓(白牛膝,长深根,称筋散,抽筋草,大被单草,大鹅肠菜,大种鹅肠菜,大种鹅儿肠,高果果鸟,狗夺子,九股牛,九股牛膝,舒筋草,水筋骨,太极草,小被单草,小九古牛,小九股牛);Berry Catchfly, Berry-bearing Campion, Berry-bearing Chickweed, Bladdercampion ■
114061 Cucubalus baccifer L. = Silene baccifera (L.) Roth ■
114062 Cucubalus baccifer L. var. angustifolius L. H. Zhou;窄叶狗筋蔓;Narrowleaf Bladdercampion ■
114063 Cucubalus baccifer L. var. angustifolius L. H. Zhou = Silene baccifera (L.) Roth ■
114064 Cucubalus baccifer L. var. cavaleriei H. Lév. = Cucubalus baccifer L. ■
114065 Cucubalus baccifer L. var. cavaleriei H. Lév. = Silene baccifera (L.) Roth ■
114066 Cucubalus baccifer L. var. japonicus Miq.;日本狗筋蔓(大鹅肠菜,大鸡肠草,鹅儿肠,和筋草);Japan Bladdercampion ■
114067 Cucubalus baccifer L. var. japonicus Miq. = Cucubalus baccifer L. ■
114068 Cucubalus baccifer L. var. japonicus Miq. = Silene baccifera (L.) Roth var. japonica (Miq.) H. Ohashi et H. Nakai ■
114069 Cucubalus baccifer L. var. japonicus Miq. = Silene baccifera (L.) Roth ■
114070 Cucubalus behen L. = Silene vulgaris (Moench) Garcke ■
114071 Cucubalus fruticulosus Pall. = Silene altaica Pers. ■
114072 Cucubalus japonicus (Miq.) Vorosch. = Silene baccifera (L.) Roth var. japonica (Miq.) H. Ohashi et H. Nakai ■
114073 Cucubalus latifolius Mill. = Silene vulgaris (Moench) Garcke ■
114074 Cucubalus niveus Nutt. = Silene nivea (Nutt.) Muhl. ex Otth ■☆
114075 Cucubalus otites L. = Silene otites (L.) Wibel ■
114076 Cucubalus polypetalus Walter = Silene polypetala (Walter) Fernald et B. G. Schub. ■☆
114077 Cucubalus stellatus L. = Silene stellata (L.) W. T. Aiton ■☆
114078 Cucubalus venosus Gilib. = Silene vulgaris (Moench) Garcke ■
114079 Cucubalus wolgensis Willd. = Silene wolgensis (Willd.) Besser ex Spreng. ■
114080 Cucularia Raf. = Dicentra Bernh. (保留属名)■
114081 Cuculina Raf. = Catasetum Rich. ex Kunth ■☆
114082 Cucullangis Thouars = Angraecum Bory ■
114083 Cucullaria Endl. = Cucularia Raf. ■
114084 Cucullaria Endl. = Dicentra Bernh. (保留属名)■
114085 Cucullaria Fabr. = Lychnis L. (废弃属名)■
114086 Cucullaria Kramer ex Kuntze = Callipeltis Steven ■☆
114087 Cucullaria Kramer ex Schreb. = Vochysia Aubl. (保留属名)●☆
114088 Cucullaria Kuntze = Callipeltis Steven ■☆
114089 Cucullifera Nees = Cannomois P. Beauv. ex Desv. ■☆
114090 Cucullifera dura Nees = Chondropetalum nudum Rottb. ■☆
114091 Cuculligera Mast. = Cannomois P. Beauv. ex Desv. ■☆
114092 Cucumella Chiov. (1929);小香瓜属■☆
114093 Cucumella bryoniifolia (Merxm.) C. Jeffrey = Cucumis bryoniifolia (Merxm.) Ghebret. et Thulin ■☆
114094 Cucumella cinerea (Cogn.) C. Jeffrey = Cucumis cinerea (Cogn.) Ghebret. et Thulin ■☆
114095 Cucumella engleri (Gilg) C. Jeffrey = Cucumis engleri (Gilg) Ghebret. et Thulin ■☆
114096 Cucumella jeffreyana J. H. Kirkbr. = Cucumis kirkbridei Ghebret. et Thulin ■☆
114097 Cucumella kelleri (Cogn.) C. Jeffrey = Cucumis kelleri (Cogn.) Ghebret. et Thulin ■☆
114098 Cucumella reticulata R. Fern. et A. Fern. = Cucumis reticulatus (A. Fern. et R. Fern.) Ghebret. et Thulin ■☆
114099 Cucumella robecchii Chiov.;小香瓜■☆
114100 Cucumella robecchii Chiov. = Cucumis kelleri (Cogn.) Ghebret. et Thulin ■☆
114101 Cucumeria Luer = Pleurothallis R. Br. ■☆
114102 Cucumeroides Gaertn. = Trichosanthes L. ■●
114103 Cucumeropsis Naudin(1866);热非葫芦属■☆

114104 Cucumeropsis edulis (Hook. f.) Cogn. = Cucumeropsis mannii Naudin ■☆
114105 Cucumeropsis edulis (Hook. f.) Cogn. = Momordica charantia L. ■
114106 Cucumeropsis mannii Naudin;热非葫芦;Egusi ■☆
114107 Cucumeropsis mannii Naudin = Momordica charantia L. ■
114108 Cucumis L. (1753);黄瓜属(甜瓜属,香瓜属);Cucumber, Cucumis, Melon, Muskmelon ■
114109 Cucumis abyssinicus A. Rich. = Cucumis ficifolia A. Rich. ■☆
114110 Cucumis acidus Jacq. = Cucumis melo L. subsp. agrestis (Naudin) Pangalo ■
114111 Cucumis acidus Jacq. = Cucumis melo L. var. agrestis Naudin ■
114112 Cucumis acidus Jacq. = Cucumis melo L. ■
114113 Cucumis aculeata Cogn.;皮刺瓜■☆
114114 Cucumis acutangula L. = Luffa acutangula (L.) Roxb. ■
114115 Cucumis acutangula Wall. = Luffa cylindrica (L.) M. Roem. ■☆
114116 Cucumis africanus L. f.;非洲黄瓜■☆
114117 Cucumis africanus L. f. = Momordica charantia L. ■
114118 Cucumis africanus L. f. var. echinatus Herm. = Cucumis africanus L. f. ■☆
114119 Cucumis africanus L. f. var. zeyheri (Sond.) Burtt Davy = Cucumis zeyheri Sond. ■☆
114120 Cucumis angolensis Hook. f. ex Cogn.;安哥拉香瓜■☆
114121 Cucumis angolensis Hook. f. ex Cogn. = Cucumis sagittata Peyr. ■☆
114122 Cucumis anguria L.;牙买加香瓜(西印度香瓜);Anguria, Bur Cucumber, Bur Gherkin, Gereken, Gherkin, Jerusalem Cucumber, West Indian Gherkin, Wild Cucumber ■☆
114123 Cucumis anguria L. var. longaculeatus J. H. Kirkbr.;纳米比亚香瓜;West Indian Gherkin ■☆
114124 Cucumis anguria L. var. longipes (Hook. f.) A. Meeuse = Cucumis anguria L. ■☆
114125 Cucumis arenaria Schrad. = Cucumis africanus L. f. ■☆
114126 Cucumis argyi H. Lév. = Momordica charantia L. ■
114127 Cucumis asper Cogn.;粗糙黄瓜■☆
114128 Cucumis bisexualis A. M. Lu et G. C. Wang ex A. M. Lu et Z. Y. Zhang;小马泡(马包,小马包);Bisexual Cucumis, Bisexual Muskmelon ■
114129 Cucumis bisexualis A. M. Lu et G. C. Wang ex A. M. Lu et Zhi Y. Zhang = Cucumis melo L. ■
114130 Cucumis bryoniifolia (Merxm.) Ghebret. et Thulin;泻根黄瓜■☆
114131 Cucumis bryoniifolia (Merxm.) H. Schaef. = Cucumis bryoniifolia (Merxm.) Ghebret. et Thulin ■☆
114132 Cucumis callosus (Rottler) Cogn. et Harms = Cucumis melo L. subsp. agrestis (Naudin) Pangalo ■
114133 Cucumis callosus Cogn. et Harms = Cucumis melo L. var. agrestis Naudin ■
114134 Cucumis carolinus J. H. Kirkbr.;卡罗来纳黄瓜■☆
114135 Cucumis cecili N. E. Br. = Cucumis oreosyce H. Schaef. ■☆
114136 Cucumis chate Hasselq.;毛叶香瓜;Hairy Cucumber, Round-leaved Egyptian Melon ■
114137 Cucumis chate Hasselq. = Cucumis melo L. ■
114138 Cucumis chrysocomus Schumach. var. echinophorus (Naudin) Hiern = Cucumis prophetarum L. subsp. dissectus (Naudin) C. Jeffrey ■☆
114139 Cucumis cinereus (Cogn.) Ghebret. et Thulin;灰瓜■☆
114140 Cucumis citrullus (L.) Ser. = Citrullus lanatus (Thunb.) Matsum. et Nakai ■
114141 Cucumis clavipetiolatus (J. H. Kirkbr.) Ghebret. et Thulin;棒梗瓜■☆
114142 Cucumis cognatus Fenzl ex Cogn. = Cucumis melo L. ■
114143 Cucumis cogniauxianus Dinter ex Cogn. et Harms = Cucumis sagittata Peyr. ■☆
114144 Cucumis conomon Thunb. = Cucumis melo L. var. conomon (Thunb.) Makino ■
114145 Cucumis courtoisii H. Lév. = Thladiantha nudiflora Hemsl. ex Forbes et Hemsl. ■
114146 Cucumis dinteri Cogn. = Cucumis sagittata Peyr. ■☆
114147 Cucumis dipsaceus Ehrenb. ex Spach.;刺猬香瓜;Hedgehog Gourd ■☆
114148 Cucumis dipsaceus Spach. = Cucumis dipsaceus Ehrenb. ex Spach. ■☆
114149 Cucumis dissectifolius Naudin = Cucumis myriocarpus Naudin ■☆
114150 Cucumis dudaim L.;橘味香瓜;Dudaim Melon, Lemon-scented Cucumber, Queen Anne's Pocket Melon ■☆
114151 Cucumis dudaim L. = Cucumis melo L. ■
114152 Cucumis engleri (Gilg) Ghebret. et Thulin;恩格勒瓜■☆
114153 Cucumis esculentus Salisb. = Cucumis sativus L. ■
114154 Cucumis ficifolius A. Rich.;榕叶香瓜■☆
114155 Cucumis figarei Delile ex Naudin = Cucumis ficifolia A. Rich. ■☆
114156 Cucumis figarei Delile ex Naudin var. cyrtopodus Naudin = Cucumis ficifolia A. Rich. ■☆
114157 Cucumis figarei Delile ex Naudin var. dissectus Naudin = Cucumis prophetarum L. subsp. dissectus (Naudin) C. Jeffrey ■☆
114158 Cucumis figarei Delile ex Naudin var. echinophorus Naudin = Cucumis prophetarum L. subsp. dissectus (Naudin) C. Jeffrey ■☆
114159 Cucumis figarei Delile ex Naudin var. ficifolius Naudin = Cucumis pustulata Naudin ex Hook. f. ■☆
114160 Cucumis figarei Delile ex Naudin var. microphyllus Naudin = Cucumis ficifolia A. Rich. ■☆
114161 Cucumis flexuosa L. = Cucumis melo L. var. flexuosa (L.) Naudin ■☆
114162 Cucumis globosa C. Jeffrey;球形香瓜■☆
114163 Cucumis gossweileri Norman = Cucumis hirsuta Sond. ■☆
114164 Cucumis halabarda Chiov. = Cucumis prophetarum L. subsp. dissectus (Naudin) C. Jeffrey ■☆
114165 Cucumis hardwickii Royle = Cucumis sativus L. var. hardwickii (Royle) Gabaev ■
114166 Cucumis hastata Thulin;戟形香瓜■☆
114167 Cucumis heptadactylus Naudin;七指香瓜■☆
114168 Cucumis hirsuta Sond.;粗毛香瓜■☆
114169 Cucumis homblei De Wild. ex Cogn. = Cucumis hirsuta Sond. ■☆
114170 Cucumis hookeri Naudin = Cucumis africanus L. f. ■☆
114171 Cucumis hystrix Chakr.;野黄瓜(老鼠瓜,鸟苦瓜,酸黄瓜);Porcupine Cucumis ■
114172 Cucumis insignis C. Jeffrey;显著瓜■☆
114173 Cucumis integrifolia Roxb. = Gymnopetalum integrifolium (Roxb.) Kurz ■
114174 Cucumis jeffreyanus Thulin;杰弗里黄瓜■☆
114175 Cucumis kalahariensis A. Meeuse;卡拉哈利黄瓜■☆
114176 Cucumis kelleri (Cogn.) Ghebret. et Thulin;开乐瓜■☆
114177 Cucumis kelleri (Cogn.) H. Schaef. = Cucumis kelleri (Cogn.) Ghebret. et Thulin ■☆
114178 Cucumis kirkbridei Ghebret. et Thulin;克尔瓜■☆

114179 Cucumis laciniosus Eckl. ex Schrad. = Citrullus lanatus (Thunb.) Matsum. et Nakai ■

114180 Cucumis laevigatus Chiov. = Cucumis melo L. ■

114181 Cucumis leptodermis Schweick. = Cucumis myriocarpus Naudin subsp. leptodermis (Schweick.) C. Jeffrey et P. Halliday ■☆

114182 Cucumis lineatus Bosc = Luffa cylindrica (L.) M. Roem. ■☆

114183 Cucumis longipes Hook. f. = Cucumis anguria L. ■☆

114184 Cucumis lyratus Zimm. = Cucumis prophetarum L. subsp. dissectus (Naudin) C. Jeffrey ■☆

114185 Cucumis maderaspatanus L. = Mukia maderaspatana (L.) M. Roem. ■

114186 Cucumis mairei H. Lév. = Lagenaria siceraria (Molina) Standl. ■

114187 Cucumis megacarpus G. Don = Luffa cylindrica (L.) M. Roem. ■☆

114188 Cucumis melo Blanco = Cucumis trigonus Roxb. ■☆

114189 Cucumis melo L.;甜瓜(白兰瓜,穿肠瓜,粪甜瓜,甘瓜,果瓜,哈密瓜,华莱士瓜,黄瓜仔,梨瓜,蜜糖埕,熟瓜,香瓜);Cantaloupe,Melon,Milion,Musk Melon,Musk Million,Muskmelon,Netted Melon,Quash,Squash,Sweet Melon ■

114190 Cucumis melo L. = Cucumis sativus L. ■

114191 Cucumis melo L. f. variegatus Makino;岛瓜■☆

114192 Cucumis melo L. f. viridis Makino;腌瓜■☆

114193 Cucumis melo L. subsp. agrestis (Naudin) Greb. = Cucumis melo L. subsp. agrestis (Naudin) Pangalo ■

114194 Cucumis melo L. subsp. agrestis (Naudin) Pangalo;马泡瓜(甜瓜菜瓜);Field Muskmelon ■

114195 Cucumis melo L. var. aestivalis Fil.;夏甜瓜■

114196 Cucumis melo L. var. agrestis Naudin = Cucumis melo L. subsp. agrestis (Naudin) Pangalo ■

114197 Cucumis melo L. var. albida Makino;白兰瓜■

114198 Cucumis melo L. var. autumnalis (Thunb.) Makino;秋甜瓜■

114199 Cucumis melo L. var. cantalupensis Naudin;罗马甜瓜;Cantaloup Melon,Cantaloupe,Rock Melon ■☆

114200 Cucumis melo L. var. chito Naudin;杧果瓜;Garden Lemon,Mango Melon,Vegetable Orange ■☆

114201 Cucumis melo L. var. conomon (Thunb.) Makino;菜瓜(白瓜,梢瓜,稍瓜,生瓜,腌瓜,羊角瓜,越瓜);Common Muskmelon,Oriental Pickling Cucumis,Oriental Pickling Melon,Queen Anne's Pocket Melon ■

114202 Cucumis melo L. var. conomon (Thunb.) Makino = Cucumis melo L. var. utilissimus (Roxb.) Duthie et Fuller 'Albus' ■

114203 Cucumis melo L. var. conomon (Thunb.) Makino = Cucumis melo L. var. utilissimus (Roxb.) Duthie et Fuller ■

114204 Cucumis melo L. var. conomon (Thunb.) Makino f. albus Makino = Cucumis melo L. var. conomon (Thunb.) Makino ■

114205 Cucumis melo L. var. conomon Makino = Cucumis melo L. var. conomon (Thunb.) Makino ■

114206 Cucumis melo L. var. cultus Kurz. = Cucumis melo L. ■

114207 Cucumis melo L. var. dudaim Naudin = Cucumis dudaim L. ■☆

114208 Cucumis melo L. var. flava Makino;黄皮小瓜■☆

114209 Cucumis melo L. var. flexuosa (L.) Naudin;羊角瓜;Serpent Melon,Snake Cucumber,Snake Melon ■☆

114210 Cucumis melo L. var. flexuosus Naudin = Cucumis melo L. var. flexuosa (L.) Naudin ■☆

114211 Cucumis melo L. var. hibermus Fil.;冬甜瓜■

114212 Cucumis melo L. var. inodorus Naudin;西班牙瓜;Neapolitan Melon,Spanish Melon,Winter Melon ■☆

114213 Cucumis melo L. var. makuwa Makino;马氏甜瓜(香瓜)■☆

114214 Cucumis melo L. var. reticulatus Naudin;网香瓜;Netted Melon,Nutmeg Melon,Rock Melon ■☆

114215 Cucumis melo L. var. scandens Naudin;哈密瓜;Netted Melon ■☆

114216 Cucumis melo L. var. utilissimus (Roxb.) Duthie et Fuller 'Albus' = Cucumis melo L. var. conomon (Thunb.) Makino ■

114217 Cucumis melo L. var. utilissimus (Roxb.) Duthie et Fuller = Cucumis melo L. var. conomon (Thunb.) Makino ■

114218 Cucumis membranifolius Hook. f. = Cucumis oreosyce H. Schaef. ■☆

114219 Cucumis merxmuelleri Suess. = Cucumis myriocarpus Naudin ■☆

114220 Cucumis messorius (C. Jeffrey) Ghebret. et Thulin;密苏里瓜■☆

114221 Cucumis metuliferus E. Mey. ex Naudin;非洲角瓜;African Horned Cucumber, Horned Cucumber, Horned Melon, Kiwano, Prickly Cucumber ■☆

114222 Cucumis metuliferus Naudin = Cucumis metuliferus E. Mey. ex Naudin ■☆

114223 Cucumis muricatus Willd. = Cucumis sativus L. ■

114224 Cucumis muriculatus Chakr. = Cucumis hystrix Chakr. ■

114225 Cucumis myriocarpus Naudin;南非甜瓜(蜜果甜瓜);Gooseberry Gourd ■☆

114226 Cucumis myriocarpus Naudin subsp. leptodermis (Schweick.) C. Jeffrey et P. Halliday;薄皮南非甜瓜■☆

114227 Cucumis naudinianus Sond. = Acanthosicyos naudinianus (Sond.) C. Jeffrey ■☆

114228 Cucumis nigristriatus Zimm. = Cucumis prophetarum L. subsp. dissectus (Naudin) C. Jeffrey ■☆

114229 Cucumis oreosyce H. Schaef.;非洲山瓜■☆

114230 Cucumis parvifolius Cogn. = Oreosyce africana Hook. f. ■☆

114231 Cucumis prophetarum L.;球瓜(阿比西尼亚柔刺瓜);Globe Cucumber ■☆

114232 Cucumis prophetarum L. subsp. dissectus (Naudin) C. Jeffrey;深裂球瓜■☆

114233 Cucumis prophetarum L. subsp. zeyheri (Sond.) C. Jeffrey = Cucumis zeyheri Sond. ■☆

114234 Cucumis pubituberculatus Thulin;多疣瓜■☆

114235 Cucumis pustulatus Hook. f. = Cucumis pustulata Naudin ex Hook. f. ■☆

114236 Cucumis pustulatus Hook. f. var. echinophorus A. Terracc. = Cucumis prophetarum L. ■☆

114237 Cucumis pustulatus Naudin ex Hook. f.;泡状甜瓜■☆

114238 Cucumis quintanilhae R. Fern. et A. Fern.;金塔利尼亚瓜■☆

114239 Cucumis reticulatus (A. Fern. et R. Fern.) Ghebret. et Thulin;网瓜■☆

114240 Cucumis reticulatus (R. Fern. et A. Fern.) H. Schaef. = Cucumis reticulata (A. Fern. et R. Fern.) Ghebret. et Thulin ■☆

114241 Cucumis rigidus E. Mey. ex Sond.;硬瓜■☆

114242 Cucumis rostratus J. H. Kirkbr.;喙瓜■☆

114243 Cucumis sacleuxii Pailleux et Bois;萨克勒瓜■☆

114244 Cucumis sagittatus Peyr.;安哥拉瓜■☆

114245 Cucumis sativus L.;黄瓜(白瓜,刺瓜,蒴瓜,吊瓜,胡瓜,勤瓜,青瓜,甜瓜,王瓜);Bottle, Clinger, Conger, Congo, Conkers, Cowcumber, Cow-cummer, Cucumber, Cucummer, Garden Cucumber, Gherkin, Pumpion ■

114246 Cucumis sativus L. var. anglicus Bailey;英国黄瓜(无刺黄瓜);English Forcing Cucumis ■☆

114247 Cucumis sativus L. var. falcatus Gabaev;镰形黄瓜■☆

114248 Cucumis sativus L. var. hardwickii (Royle) Alef. = Cucumis sativus L. var. hardwickii (Royle) Gabaev ■

114249 Cucumis sativus L. var. hardwickii (Royle) Gabaev;西南野黄瓜;Hardwick Cucumis ■

114250 Cucumis sativus L. var. sikkimensis Hook.;印度黄瓜;Sikkim Cucumber ■☆

114251 Cucumis sativus L. var. sqummosus Gabaev;鳞皮黄瓜■☆

114252 Cucumis sativus L. var. tuberculatus Gabaev;花皮黄瓜■☆

114253 Cucumis sativus L. var. usambarensis A. Zimm. = Cucumis sacleuxii Pailleux et Bois ■☆

114254 Cucumis seretii De Wild. = Cucumis hirsuta Sond. ■☆

114255 Cucumis seretioides Suess. = Cucumis hirsuta Sond. ■☆

114256 Cucumis sonderi Cogn. = Cucumis hirsuta Sond. ■☆

114257 Cucumis subsericeus Hook. f. = Cucumis oreosyce H. Schaef. ■☆

114258 Cucumis subsericeus Hook. f. = Oreosyce africana Hook. f. ■☆

114259 Cucumis thulinianus J. H. Kirkbr.;图林瓜■☆

114260 Cucumis trigonus Benth. = Cucumis callosus Cogn. et Harms ■

114261 Cucumis trigonus Roxb.;三角香瓜(三角形瓜,野苦瓜);Trigona Cucumber ■☆

114262 Cucumis trigonus Roxb. = Cucumis melo L. var. agrestis Naudin ■

114263 Cucumis turbinatus Roxb. = Cucumis callosus Cogn. et Harms ■

114264 Cucumis umbrosus A. Meeuse et Strey = Cucumis bryoniifolia (Merxm.) Ghebret. et Thulin ■☆

114265 Cucumis utilissimus Roxb. = Cucumis melo L. var. utilissimus (Roxb.) Duthie et Fuller ■

114266 Cucumis utilissimus Roxb. = Cucumis melo L. ■

114267 Cucumis vulgaris E. H. L. Krause = Citrullus lanatus (Thunb.) Matsum. et Nakai ■

114268 Cucumis welwitschii Cogn. = Cucumis hirsuta Sond. ■☆

114269 Cucumis wildemanianus Cogn. = Cucumis hirsuta Sond. ■☆

114270 Cucumis zeyheri Sond.;蔡赫瓜■☆

114271 Cucurbita L. (1753);南瓜属;Gourd, Marrow, Pumpkin, Squash, Winter Squash ■

114272 Cucurbita acutangula (L.) Blume = Luffa acutangula (L.) Roxb. ■

114273 Cucurbita argyresperma K. Koch;银籽南瓜;Cushaw, Pumpkin, Silverseed Gourd ■☆

114274 Cucurbita caffra Eckl. et Zeyh. = Citrullus lanatus (Thunb.) Matsum. et Nakai ■

114275 Cucurbita citrullus L. = Citrullus lanatus (Thunb.) Matsum. et Nakai ■

114276 Cucurbita digitata A. Gray;指叶南瓜;Finger-leaved Gourd ■☆

114277 Cucurbita ficifolia Bouché;马拉巴尔瓜;Bush Pumpkin, Cidra, Figleaf Gourd, Fig-leaf Gourd, Malabar Gourd, Malabar Squash ■☆

114278 Cucurbita foetidissima Kunth;水牛瓜(臭瓜);American Pumpkin, Buffalo Gourd, Chilicote, Foetid Wild Gourd, Missouri Gourd, Wild Pumpkin ■☆

114279 Cucurbita hispida Thunb. = Benincasa hispida (Thunb. ex A. Murray) Cogn. ■

114280 Cucurbita hispida Thunb. = Benincasa hispida (Thunb.) Cogn. ■

114281 Cucurbita hispida Thunb. ex A. Murray = Benincasa hispida (Thunb. ex A. Murray) Cogn. ■

114282 Cucurbita lagenaria L. = Lagenaria siceraria (Molina) Standl. ■

114283 Cucurbita leucantha Duchesne = Lagenaria siceraria (Molina) Standl. ■

114284 Cucurbita leucantha Duchesne ex Lam. = Lagenaria siceraria (Molina) Standl. ■

114285 Cucurbita leucantha Lam. = Lagenaria siceraria (Molina) Standl. ■

114286 Cucurbita maxima Duche. = Cucurbita maxima Duchesne ex Lam. ■

114287 Cucurbita maxima Duchesne = Cucurbita maxima Duchesne ex Lam. ■

114288 Cucurbita maxima Duchesne ex Lam.;笋瓜(北瓜,大瓜,番瓜,番南瓜,蕃南瓜,饭瓜,搅丝瓜,金瓜,栗南瓜,南瓜,桃南瓜,西印度南瓜,荀瓜,印度南瓜,玉瓜);Autumn and Winter Squash, Autumn Squash, Elephant Pumpkin, Giant Pumpkin, Great Squash, Hubbard Squash, Melon Pumpkin, Million, Pumpkin, Spanish Gourd, Turban Squash, Winter Squash, Yellow Field Pumpkin ■

114289 Cucurbita melanosperma A. Braun = Cucurbita ficifolia Bouché ■☆

114290 Cucurbita melopepo Lour. = Cucurbita moschata Duchesne ■

114291 Cucurbita mixta Pangalo;混杂瓜■☆

114292 Cucurbita moschata (Duchesne ex Lam.) Duchesne ex Poir.;南瓜(北瓜,冬瓜,番瓜,番瓜藤,番南瓜,番蒲,翻瓜,蕃瓜,饭瓜,伏瓜,红南瓜,金冬瓜,金瓜,癞瓜,蓝藤,老缅瓜,麦瓜,蛮南瓜,美国金瓜,美国南瓜,盘肠草,倭瓜,窝瓜,香蒲,中国南瓜);Barbary Squash, Canada Crook-neck Squash, China Squash, Crook-neck Squash, Cushaw, Cushaw Squash, Musky Gourd, Pumpkin, Pumpkin Winter Squash, Seminole Pumpkin, Squash, Winter Crookneck Squash, Winter Squash ■

114293 Cucurbita moschata (Duchesne ex Lam.) Duchesne ex Poir. var. lufiiformis H. Hara;丝瓜状南瓜■☆

114294 Cucurbita moschata (Duchesne ex Lam.) Duchesne ex Poir. var. meloniiformis (Carrière) Makino;毛壳南瓜(金瓜,南瓜,倭瓜)■☆

114295 Cucurbita moschata (Duchesne ex Lam.) Duchesne ex Poir. var. toonas Makino;光壳南瓜(番金瓜,番南瓜,牛腿南瓜)■☆

114296 Cucurbita moschata (Duchesne) Poir. = Cucurbita moschata (Duchesne ex Lam.) Duchesne ex Poir. ■

114297 Cucurbita moschata Duchesne = Cucurbita moschata (Duchesne ex Lam.) Duchesne ex Poir. ■

114298 Cucurbita moschata Duchesne ex Lam. = Cucurbita moschata (Duchesne ex Lam.) Duchesne ex Poir. ■

114299 Cucurbita palmata S. Watson;掌状南瓜;Coyote Melon ■☆

114300 Cucurbita pepo L.;西葫芦(吊瓜,鼎足瓜,冬瓜,番瓜,红南瓜,皎瓜,搅瓜,金瓜,看瓜,美洲南瓜,南瓜,气豆,水葫芦,笋瓜,桃南瓜,倭瓜);Cocozelle, Courgette, Field Pumpkin, Marrow, Pear Gourd, Pepita, Pomion, Pompion, Pumpkin, Scalloped Summer Squash, Squash, Summer and Autumn Pumpkins, Summer Squash, Vegetable Marrow, Vegetable Spaghetti, Yellow-flowered Gourd ■

114301 Cucurbita pepo L. 'Melopepo';扁西葫芦;Bush Pumpkin, Custard Squash, Pattypan Squash, Patty-pan Squash, Scallop Gourd ■☆

114302 Cucurbita pepo L. var. akoda Makino;桃南瓜(吊瓜,鼎足瓜,红南瓜,金瓜,看瓜)■

114303 Cucurbita pepo L. var. condensa ?;密集西葫芦;Summer Squash ■☆

114304 Cucurbita pepo L. var. giraumontia ?;日劳南瓜;Marrow Squash, Marrow-type Pumpkin, Vegetable Marrow ■

114305 Cucurbita pepo L. var. kintoga Makino;金冬瓜(北瓜,金瓜)■

114306 Cucurbita pepo L. var. maxima (Duchesne) Delile = Cucurbita maxima Duchesne ■

114307 Cucurbita pepo L. var. melopepo Duchesne ex Lam. = Cucurbita pepo L. ' Melopepo' ■☆

114308 Cucurbita pepo L. var. moschata Duchesne ex Lam. = Cucurbita moschata (Duchesne ex Lam.) Duchesne ex Poir. ■

114309 Cucurbita pepo L. var. moschata Duchesne ex Lam. = Cucurbita moschata Duchesne ■

114310 Cucurbita pepo L. var. ovifera (L.) Alef.;观赏瓜(观赏南瓜,金瓜,卵北瓜,珠瓜);Egg Squash, Marrow, Ornamental Pepo, Yellow-flowered Gourds ■☆

114311 Cucurbita pepo L. var. pafisson ?;飞碟瓜■☆

114312 Cucurbita siceraria Molina = Lagenaria siceraria (Molina) Standl. ■

114313 Cucurbitaceae Juss. (1789)(保留科名);葫芦科(瓜科,南瓜科);Gourd Family, White Bryony Family ■●

114314 Cucurbitella Walp. (1846);小南瓜属■☆

114315 Cucurbitella asperata (Gillies ex Hook. et Arn.) Walp.;小南瓜■☆

114316 Cucurbitula (M. Roem.) Kuntze = Blastania Kotschy et Peyr. ■☆

114317 Cucurbitula (M. Roem.) Post et Kuntze = Zehneria Endl. ■

114318 Cudicia Buch. -Ham. ex G. Don = Pottsia Hook. et Arn. + Parsonsia R. Br. ●

114319 Cudrania Trécul(1847)(保留属名);柘树属(葨芝属,柘属);Cudrania ●

114320 Cudrania Trécul(保留属名) = Maclura Nutt. (保留属名)●

114321 Cudrania amboinensis (Blume) Miq. = Maclura amboinensis Blume ●

114322 Cudrania bodinieri H. Lév. = Capparis cantoniensis Lour. ●

114323 Cudrania cochinchinensis (Lour.) Corner = Cratoxylum cochinchinense (Lour.) Blume ●

114324 Cudrania cochinchinensis (Lour.) Kudo et Masam. = Maclura cochinchinensis (Lour.) Corner ●

114325 Cudrania cochinchinensis (Lour.) Kudo et Masam. var. gerontogea (Siebold et Zucc.) Kudo et Masam.;香港柘(凹头葨芝,刺格仔,大丁黄);Emarginate-leaf Silkwormthorn, Hongkong Cudrania ●

114326 Cudrania cochinchinensis (Lour.) Kudo et Masam. var. gerontogea (Siebold et Zucc.) Kudo et Masam. = Maclura cochinchinensis (Lour.) Corner var. gerontogea (Siebold et Zucc.) H. Ohashi ●

114327 Cudrania cochinchinensis (Lour.) Kudo et Masam. var. gerontogea (Siebold et Zucc.) Kudo et Masam. = Maclura cochinchinensis (Lour.) Corner ●

114328 Cudrania cochinchinensis (Lour.) Kudo et Masam. var. gerontogea Kudo et Masam. = Cudrania cochinchinensis (Lour.) Kudo et Masam. ●

114329 Cudrania fruticosa (Roxb.) Wight ex Kurz = Maclura fruticosa (Roxb.) Corner ●

114330 Cudrania grandifolia Merr. = Cudrania amboinensis (Blume) Miq. ●

114331 Cudrania grandifolia Merr. = Maclura amboinensis Blume ●

114332 Cudrania integra F. T. Wang et Ts. Tang;全缘柘;Entire Cudrania ●

114333 Cudrania integra F. T. Wang et Ts. Tang = Cudrania cochinchinensis (Lour.) Kudo et Masam. ●

114334 Cudrania integra F. T. Wang et Ts. Tang = Maclura cochinchinensis (Lour.) Corner ●

114335 Cudrania javanensis Trécul = Cudrania cochinchinensis (Lour.) Kudo et Masam. ●

114336 Cudrania javanensis Trécul = Maclura cochinchinensis (Lour.) Corner ●

114337 Cudrania jingdongensis S. S. Chang = Cudrania amboinensis (Blume) Miq. ●

114338 Cudrania jingdongensis S. S. Chang = Maclura amboinensis Blume ●

114339 Cudrania jinghongensis S. S. Chang = Cudrania pubescens Trécul ●

114340 Cudrania jinghongensis S. S. Chang = Maclura pubescens (Trécul) Z. K. Zhou et M. G. Gilbert ●

114341 Cudrania obovata Trécul = Maclura cochinchinensis (Lour.) Corner ●

114342 Cudrania pubescens Miq. = Cudrania pubescens Trécul ●

114343 Cudrania pubescens Trécul = Maclura pubescens (Trécul) Z. K. Zhou et M. G. Gilbert ●

114344 Cudrania rectispina Hance = Maclura cochinchinensis (Lour.) Corner ●

114345 Cudrania tricuspidata (Carrière) Bureau = Maclura tricuspidata Carrière ●

114346 Cudrania tricuspidata (Carrière) Bureau ex Lavallée = Maclura tricuspidata Carrière ●

114347 Cudrania tricuspidata Bureau = Maclura tricuspidata Carrière ●

114348 Cudrania triloba Hance = Maclura tricuspidata Carrière ●

114349 Cudranus Kuntze = Cudrania Trécul(保留属名)●

114350 Cudranus Miq. = Cudrania Trécul(保留属名)●

114351 Cudranus Rumph. ex Miq. = Cudrania Trécul(保留属名)●

114352 Cudranus pubescens Miq. = Cudrania pubescens Trécul ●

114353 Cudranus triloba Hance = Cudrania tricuspidata (Carrière) Bureau ex Lavallée ●

114354 Cuellara Pers. = Cuellaria Ruiz et Pav. ●

114355 Cuellaria Ruiz et Pav. = Gilibertia J. F. Gmel. ●

114356 Cuenotia Rizzini(1956);巴东北爵床属☆

114357 Cuenotia speciosa Rizzini;巴东北爵床☆

114358 Cuepia J. F. Gmel. = Couepia Aubl. ●☆

114359 Cuervea Triana ex Miers(1872);膜杯卫矛属●☆

114360 Cuervea isangiensis (De Wild.) N. Hallé;膜杯卫矛●☆

114361 Cuervea macrophylla (Vahl) R. Wilczek ex N. Hallé;大叶膜杯卫矛●☆

114362 Cufodontia Woodson = Aspidosperma Mart. et Zucc. (保留属名)●☆

114363 Cuiavus Trew = Psidium L. ●

114364 Cuiete Adans. = Crescentia L. ●

114365 Cuiete Mill. = Crescentia L. ●

114366 Cuitlanzina Lindl. = Cuitlauzina La Llave et Lex. ■

114367 Cuitlanzina Roeper = Cuitlauzina La Llave et Lex. ■

114368 Cuitlanzina Roeper = Odontoglossum Kunth ■

114369 Cuitlauzina La Llave et Lex. = Odontoglossum Kunth ■

114370 Cuitlauzinia Rchb. = Cuitlauzina La Llave et Lex. ■

114371 Cujunia Alef. = Vicia L. ■

114372 Culcasia P. Beauv. (1803)(保留属名);库卡芋属■☆

114373 Culcasia afzelii Schott = Cercestis afzelii Schott ■☆

114374 Culcasia angolensis Welw. ex Schott;库卡芋■☆

114375 Culcasia annetii Ntépé-Nyamè;非洲库卡芋■☆

114376 Culcasia barombensis N. E. Br. = Culcasia angolensis Welw. ex Schott ■☆

114377 Culcasia bequaertii De Wild. = Culcasia sapinii De Wild. ■☆

114378 Culcasia caudata Engl.;尾状库卡芋■☆

114379 Culcasia dinklagei Engl.;丁克库卡芋■☆
114380 Culcasia engleriana A. Chev. = Culcasia striolata Engl. ■☆
114381 Culcasia falcifolia Engl.;镰叶库卡芋■☆
114382 Culcasia glandulosa Hepper;具腺库卡芋■☆
114383 Culcasia gracilis N. E. Br. = Culcasia scandens P. Beauv. ■☆
114384 Culcasia insulana N. E. Br.;海岛库卡芋■☆
114385 Culcasia kasaiensis De Wild. = Culcasia tenuifolia Engl. ■☆
114386 Culcasia lanceolata Engl.;披针形库卡芋■☆
114387 Culcasia lancifolia N. E. Br. = Culcasia scandens P. Beauv. ■☆
114388 Culcasia liberica N. E. Br.;离生库卡芋■☆
114389 Culcasia longevaginata Engl. = Culcasia striolata Engl. ■☆
114390 Culcasia loukandensis Pellegr.;路坎德库卡芋■☆
114391 Culcasia mannii (Hook. f.) Engl.;曼氏库卡芋■☆
114392 Culcasia obliquifolia Engl.;斜叶库卡芋■☆
114393 Culcasia orientalis Mayo;东方库卡芋■☆
114394 Culcasia panduriformis Engl. et K. Krause;琴形库卡芋■☆
114395 Culcasia parviflora N. E. Br.;小花库卡芋■☆
114396 Culcasia piperoides A. Chev. = Culcasia parviflora N. E. Br. ■☆
114397 Culcasia pynaertii De Wild. = Culcasia dinklagei Engl. ■☆
114398 Culcasia reticulata Voss = Culcasia mannii (Hook. f.) Engl. ■☆
114399 Culcasia rotundifolia Bogner;圆叶库卡芋■☆
114400 Culcasia sanagensis Ntepe-Nyame;萨纳加库卡芋■☆
114401 Culcasia sapinii De Wild. = Culcasia seretii De Wild. ■☆
114402 Culcasia saxatilis A. Chev.;岩生库卡芋■☆
114403 Culcasia scandens P. Beauv.;攀缘库卡芋■☆
114404 Culcasia scandens P. Beauv. = Culcasia gracilis N. E. Br. ■☆
114405 Culcasia scandens P. Beauv. = Culcasia lancifolia N. E. Br. ■☆
114406 Culcasia scandens P. Beauv. f. ovatifolia Engl. = Culcasia falcifolia Engl. ■☆
114407 Culcasia seretii De Wild.;赛雷库卡芋■☆
114408 Culcasia striolata Engl. = Culcasia longevaginata Engl. ■☆
114409 Culcasia tenuifolia Engl.;细叶库卡芋■☆
114410 Culcasia tubulifera Engl. = Culcasia insulana N. E. Br. ■☆
114411 Culcitium Bonpl. (1808);垂绒菊属■☆
114412 Culcitium Bonpl. = Senecio L. ■●
114413 Culcitium Humb. et Bonpl. = Culcitium Bonpl. ■☆
114414 Culcitium canescens Humb. et Bonpl.;灰色垂绒菊■☆
114415 Culcitium discolor Herrera;异色垂绒菊■☆
114416 Culcitium ferrugineum Klatt;锈色垂绒菊■☆
114417 Culcitium longifolium Turcz.;长叶垂绒菊■☆
114418 Culcitium nitidum Spreng.;光亮垂绒菊■☆
114419 Culcitum N. T. Burb. = Culcitium Bonpl. ■☆
114420 Culhamia Forssk. = Sterculia L. ●
114421 Culhamia simplex (L.) Nakai = Firmiana simplex (L.) W. Wight ●
114422 Cullay Molina ex Steud. = Quillaja Molina ●☆
114423 Cullen Medik. (1787);热带补骨脂属●■
114424 Cullen Medik. = Psoralea L. ●■
114425 Cullen americanum (L.) Rydb.;美洲热带补骨脂;American Scurfpea ●☆
114426 Cullen biflorum (Harv.) C. H. Stirt.;双花热带补骨脂■☆
114427 Cullen corylifolium (L.) Medik.;补骨脂(补骨鸱,川故子,和兰苋,黑故纸,黑故子,胡故子,胡韭子,怀故子,吉故子,马骝姜,婆固脂,破故纸,破故子,榛叶热带补骨脂);Bawchan Seed, Malaysian Scurfpea, Malaytea Scurfpea, Scurf Pea, Scurfpea ■
114428 Cullen holubii (Burtt Davy) C. H. Stirt.;霍勒布热带补骨脂■☆
114429 Cullen obtusifolium (DC.) C. H. Stirt. = Cullen tomentosum (Thunb.) J. W. Grimes ■☆
114430 Cullen plicatum (Delile) C. H. Stirt.;折扇热带补骨脂■☆
114431 Cullen tomentosum (Thunb.) J. W. Grimes;绒毛热带补骨脂■☆
114432 Cullenia Wight = Durio Adans. ●
114433 Cullenia Wight(1851);卡伦木棉属●☆
114434 Cullenia excelsa Wight;卡伦木棉●☆
114435 Cullmannia Distefano = Peniocereus (A. Berger) Britton et Rose ●
114436 Cullmannia Distefano = Wilcoxia Britton et Rose ■☆
114437 Cullomia Juss. = Collomia Nutt. ■☆
114438 Cullumia R. Br. (1813);帚叶联苞菊属■☆
114439 Cullumia aculeata (Houtt.) Rössler;皮刺帚叶联苞菊■☆
114440 Cullumia aculeata (Houtt.) Rössler var. sublanata (DC.) Rössler;亚绵毛帚叶联苞菊■☆
114441 Cullumia adnata DC. = Cullumia setosa (L.) R. Br. var. adnata (DC.) Harv. ■☆
114442 Cullumia bisulca (Thunb.) Less.;二沟帚叶联苞菊■☆
114443 Cullumia carlinoides DC.;刺苞菊状帚叶联苞菊■☆
114444 Cullumia ciliaris (L.) R. Br.;缘毛帚叶联苞菊■☆
114445 Cullumia ciliaris (L.) R. Br. subsp. angustifolia (Hutch.) Rössler;窄叶缘毛帚叶联苞菊■☆
114446 Cullumia ciliaris (L.) R. Br. var. angustifolia Hutch. = Cullumia ciliaris (L.) R. Br. subsp. angustifolia (Hutch.) Rössler ■☆
114447 Cullumia cirsioides DC.;蓟帚叶联苞菊■☆
114448 Cullumia decurrens Less.;下延帚叶联苞菊■☆
114449 Cullumia floccosa E. Mey. ex DC.;丛卷毛帚叶联苞菊■☆
114450 Cullumia hispida (L. f.) Less. = Cullumia aculeata (Houtt.) Rössler ■☆
114451 Cullumia massoni S. Moore = Cullumia pectinata (Thunb.) Less. ■☆
114452 Cullumia micracantha DC.;小刺联苞菊■☆
114453 Cullumia obovata E. Mey. ex DC. = Cullumia micracantha DC. ■☆
114454 Cullumia patula (Thunb.) Less.;张开联苞菊■☆
114455 Cullumia patula (Thunb.) Less. subsp. uncinata Rössler;具钩帚叶联苞菊■☆
114456 Cullumia pectinata (Thunb.) Less.;篦状帚叶联苞菊■☆
114457 Cullumia rigida DC.;坚挺帚叶联苞菊■☆
114458 Cullumia setosa (L.) R. Br.;刚毛帚叶联苞菊■☆
114459 Cullumia setosa (L.) R. Br. var. adnata (DC.) Harv.;贴生帚叶联苞菊■☆
114460 Cullumia setosa (L.) R. Br. var. microcephala Rössler;小头帚叶联苞菊■☆
114461 Cullumia setosa Sieber ex DC. = Polyarrhena reflexa (L.) Cass ●☆
114462 Cullumia squarrosa (L.) R. Br.;粗鳞帚叶联苞菊■☆
114463 Cullumia stricta Compton = Cullumia sulcata (Thunb.) Less. ■☆
114464 Cullumia sublanata DC. = Cullumia aculeata (Houtt.) Rössler var. sublanata (DC.) Rössler ■☆
114465 Cullumia sulcata (Thunb.) Less.;纵沟帚叶联苞菊■☆
114466 Cullumia sulcata (Thunb.) Less. var. intercedens Rössler;中间帚叶联苞菊■☆
114467 Cullumiopsis Drake = Dicoma Cass. ●☆
114468 Cullumiopsis Drake = Macledium Cass. ●☆
114469 Cullumiopsis grandidieri Drake = Macledium grandidieri (Drake) S. Ortiz ●☆
114470 Cultridendris Thouars = Dendrobium Sw. (保留属名)■
114471 Cultridendris Thouars = Polystachya Hook. (保留属名)■

114472 Cuma Post et Kuntze = Couma Aubl. ●☆

114473 Cumarinia (Knuth) Buxb. (1951);薰大将属■☆

114474 Cumarinia (Knuth) Buxb. = Coryphantha (Engelm.) Lem. (保留属名)●■

114475 Cumarinia Buxb. = Coryphantha (Engelm.) Lem. (保留属名)●■

114476 Cumarinia Buxb. = Cumarinia (Knuth) Buxb. ■☆

114477 Cumarinia odorata (Boed.) Buxb.;薰大将■☆

114478 Cumarouma Steud. = Cumaruna J. F. Gmel. ●☆

114479 Cumaruma Steud. = Cumaruna J. F. Gmel. ●☆

114480 Cumaruna J. F. Gmel. = Dipteryx Schreb. (保留属名)●☆

114481 Cumaruna Kuntze = Coumarouna Aubl. (废弃属名)●☆

114482 Cumaruna Kuntze = Dipteryx Schreb. (保留属名)●☆

114483 Cumbalu Adans. = Catalpa Scop. ●

114484 Cumbata Raf. = Rubus L. ●■

114485 Cumbea Wight et Arn. = Cumbia Buch. -Ham. ●☆

114486 Cumbia Buch. -Ham. = Careya Roxb. (保留属名)●☆

114487 Cumbula Steud. = Cumbulu Adans. ●

114488 Cumbulu Adans. = Gmelina L. ●

114489 Cumetea Raf. = Myrcia DC. ex Guill. ●☆

114490 Cumingia Kunth = Conanthera Ruiz et Pav. ■☆

114491 Cumingia Kunth = Cummingia D. Don(废弃属名)■☆

114492 Cumingia Vidal(1885)(保留属名);卡明木棉属●☆

114493 Cumingia Vidal(保留属名) = Camptostemon Mast. ●☆

114494 Cumingia philippinensis S. Vidal;卡明木棉●☆

114495 Cuminia B. D. Jacks. = Cuminum L. ■

114496 Cuminia Colla(1835);马岛塔花属●☆

114497 Cuminia fernandezia Colla;马岛塔花●☆

114498 Cuminoides Fabr. = Lagoecia L. ■☆

114499 Cuminoides Moench = Lagoecia L. ■☆

114500 Cuminoides Tourn. = Lagoecia L. ■☆

114501 Cuminoides Tourn. ex Moench = Lagoecia L. ■☆

114502 Cuminum L. (1753);孜然芹属(小茴香属);Cumin, Cuminum, Cummin ■

114503 Cuminum cyminum L.;孜然芹(枯茗,香旱芹,孜然);Cumin, Cymous Cuminum ■

114504 Cuminum maroccanum P. H. Davis et Hedge = Ammodaucus leucotrichus Coss. et Durieu ■☆

114505 Cuminum nigrum L.;黑孜然芹■☆

114506 Cuminum setifolium (Boiss.) Koso-Pol.;中亚孜然芹■☆

114507 Cuminum sinense (Dunn) M. Hiroe = Carlesia sinensis Dunn ■

114508 Cummin Hill = Cuminum L. ■

114509 Cummingia D. Don(废弃属名) = Conanthera Ruiz et Pav. ■☆

114510 Cummingia D. Don(废弃属名) = Cumingia Vidal(保留属名)●☆

114511 Cumminsia King ex Prain = Meconopsis R. Vig. ■

114512 Cumulopuntia F. Ritter = Opuntia Mill. ●

114513 Cuncea Buch. -Ham. ex D. Don = Knoxia L. ■

114514 Cunibalu B. D. Jacks. = Cumbulu Adans. ●

114515 Cunibalu B. D. Jacks. = Gmelina L. ●

114516 Cuniculotinus Urbatsch, R. P. Roberts et Neubig(2005);兔黄花属(禾状兔黄花属);Rock Goldenrod ●☆

114517 Cuniculotinus gramineus (H. M. Hall) Urbatsch, R. P. Roberts et Neubig;兔黄花(禾状兔黄花);Charleston Rabbitbrush, Panamint Rock Goldenrod ●☆

114518 Cunigunda Bubani = Eupatorium L. ■●

114519 Cunila D. Royen ex L. = Cunila L. (保留属名)●☆

114520 Cunila L. (1759)(保留属名);岩薄荷属;Stone Mint, Stone-mint ●☆

114521 Cunila L. ex Mill. (废弃属名) = Cunila L. (保留属名)●☆

114522 Cunila Mill. = Cunila L. (保留属名)●☆

114523 Cunila Mill. = Sideritis L. ■●

114524 Cunila lythrifolia Benth.;千屈菜叶岩薄荷■☆

114525 Cunila mariana L. = Cunila origanoides (L.) Britton ■☆

114526 Cunila nepalensis D. Don = Mosla dianthera (Buch. -Ham. ex Roxb.) Maxim. ■

114527 Cunila origanoides (L.) Britton;牛至岩薄荷;American Dittany, Dittany ■☆

114528 Cunila pulegioides L. = Hedeoma pulegioides (L.) Pers. ■☆

114529 Cunila spicata Benth.;穗花岩薄荷■☆

114530 Cunina Clos = Nertera Banks ex Gaertn. (保留属名)■

114531 Cunina Gay = Nertera Banks ex Gaertn. (保留属名)■

114532 Cunninghamia R. Br. (1873)(保留属名);杉木属;China Fir, Chinafir, China-fir, Chinese Fir ●★

114533 Cunninghamia R. Br. ex Rich. = Cunninghamia R. Br. (保留属名)●★

114534 Cunninghamia R. Br. ex Rich. et A. Rich. = Cunninghamia R. Br. (保留属名)●★

114535 Cunninghamia Schreb. (废弃属名) = Malanea Aubl. ●☆

114536 Cunninghamia chinensis de Vos = Cunninghamia lanceolata (Lamb.) Hook. ●

114537 Cunninghamia kawakamii Hayata = Cunninghamia konishii Hayata ●

114538 Cunninghamia konishii Hayata = Cunninghamia lanceolata (Lamb.) Hook. var. konishii (Hayata) Fujita ●

114539 Cunninghamia lanceolata (Lamb.) Hook.;杉木(椵,椵木,刺杉,福州杉,广东杉,广叶杉,黄杉木,琉球杉,木头树,泡杉,千把刀,檠木,沙木,沙树,杉,杉树,黏木,天蜈蚣,西杉木,香木,香杉,真杉木,正木,正杉);China Fir, China-fir, Chinese Cedar, Chinese Cunninghamia, Chinese Fir, Common China Fir ●

114540 Cunninghamia lanceolata (Lamb.) Hook. 'Glauca';灰叶杉木(白皇油杉,白叶杉,深绿叶杉木,油杉);Bluecolored Chinese Fir, Glaucous China-fir ●

114541 Cunninghamia lanceolata (Lamb.) Hook. 'Mollifolia';软叶杉木(柔叶杉木);Flaccidleaf Chinese Fir ●

114542 Cunninghamia lanceolata (Lamb.) Hook. f. glauca (Dallim. et Jacks.) S. Y. Hu = Cunninghamia lanceolata (Lamb.) Hook. 'Glauca' ●

114543 Cunninghamia lanceolata (Lamb.) Hook. var. corticosa Z. Y. Que et J. X. Li = Cunninghamia lanceolata (Lamb.) Hook. ●

114544 Cunninghamia lanceolata (Lamb.) Hook. var. glauca (Dallim. et Jacks.) Dallim. et Jacks. = Cunninghamia lanceolata (Lamb.) Hook. 'Glauca' ●

114545 Cunninghamia lanceolata (Lamb.) Hook. var. glauca (Dallim. et Jacks.) Dallim. et Jacks. = Cunninghamia lanceolata (Lamb.) Hook. ●

114546 Cunninghamia lanceolata (Lamb.) Hook. var. glauca Sasaki;白叶杉(白皇油杉)●

114547 Cunninghamia lanceolata (Lamb.) Hook. var. kawakamii (Hayata) Fujita = Cunninghamia konishii Hayata ●

114548 Cunninghamia lanceolata (Lamb.) Hook. var. konishii (Hayata) Fujita;台湾杉木(福州杉,广东杉,峦大杉,香杉);China-fir, Konishi China Fir, Konishi China-fir, Konishi-fir, Luanta-fir ●

114549 Cunninghamia lanceolata (Lamb.) Hook. var. konishii

(Hayata) Hayata = Cunninghamia lanceolata (Lamb.) Hook. var. konishii (Hayata) Fujita ●
114550 Cunninghamia lanceolata (Lamb.) Hook. var. zhaotongensis Dallim. et D. Yan;昭通杉木;Zhaotong Chinese Fir ●
114551 Cunninghamia sinensis R. Br. ex Rich. = Cunninghamia lanceolata (Lamb.) Hook. ●
114552 Cunninghamia sinensis R. Br. ex Rich. var. glauca Dallim. et Jacks. = Cunninghamia lanceolata (Lamb.) Hook. 'Glauca' ●
114553 Cunninghamia sinensis R. Br. ex Rich. var. glauca Dallim. et Jacks. = Cunninghamia lanceolata (Lamb.) Hook. ●
114554 Cunninghamia sinensis R. Br. ex Rich. var. probfera Lemée et H. Lév. = Cunninghamia lanceolata (Lamb.) Hook. ●
114555 Cunninghamia unicanaliculata D. Y. Wang et H. L. Liu;德昌杉木;Unicanaliculate China Fir ●
114556 Cunninghamia unicanaliculata D. Y. Wang et H. L. Liu = Cunninghamia lanceolata (Lamb.) Hook. ●
114557 Cunninghamia unicanaliculata D. Y. Wang et H. L. Liu var. pyramidalis D. Y. Wang et H. L. Liu;米德杉木;Mide China fir ●
114558 Cunninghamia unicanaliculata D. Y. Wang et H. L. Liu var. pyramidalis D. Y. Wang et H. L. Liu = Cunninghamia lanceolata (Lamb.) Hook. ●
114559 Cunninghamiaceae Hayata = Cupressaceae Gray(保留科名)●
114560 Cunninghamiaceae Siebold et Zucc. = Cupressaceae Gray(保留科名)●
114561 Cunninghamiaceae Siebold et Zucc. = Taxodiaceae Saporta(保留科名)●
114562 Cunninghamiaceae Zucc. = Taxodiaceae Saporta(保留科名)●
114563 Cunonia L. (1759)(保留属名);火把树属(匙木属,库诺尼属)●☆
114564 Cunonia Mill. (废弃属名) = Cunonia L. (保留属名)●☆
114565 Cunonia Mill. (废弃属名) = Gladiolus L. ■
114566 Cunonia capensis L.;好望角火把树(匙木,好望角库诺尼);African Red Alder, Butterknife Bush, Spoon Bush ●☆
114567 Cunoniaceae R. Br. (1814)(保留科名);火把树科(常绿棱枝树科,角瓣木科,库诺尼科,南蔷薇科,轻木科)●☆
114568 Cunto Adans. = Acronychia J. R. Forst. et G. Forst. (保留属名)●
114569 Cunuria Baill. = Micrandra Benth. (保留属名)●☆
114570 Cupadessa Hassk. = Cipadessa Blume ●
114571 Cupameni Adans. = Acalypha L. ●■
114572 Cupamenis Raf. = Cupameni Adans. ●■
114573 Cupania (Plum.) ex L. = Cupania L. ●☆
114574 Cupania L. (1753);库潘树属;Cupania, Guara, Lobinlly Tree ●☆
114575 Cupania Plum. ex L. = Cupania L. ●☆
114576 Cupania chapelieriana Cambess. = Tina chapelieriana (Cambess.) Kalkman ●☆
114577 Cupania dissitiflora Baker = Tinopsis dissitiflora (Baker) Capuron ●☆
114578 Cupania ferruginea Baker = Laccodiscus ferrugineus (Baker) Radlk. ●☆
114579 Cupania isomera Baker = Neotina isoneura (Radlk.) Capuron ●☆
114580 Cupania pentapetala (Roxb.) Wight et Arn. = Mischocarpus pentapetalus (Roxb.) Radlk. ●
114581 Cupania sapida (König) Oken;库潘树(阿开木,美味阿开木,西非荔枝果);Ackee, Akee, Akee Apple ●☆
114582 Cupania sapida (König) Oken = Blighia sapida K. König ●☆
114583 Cupania thouarsiana Cambess. = Tina thouarsiana (Cambess.) Capuron ●☆
114584 Cupaniopsis Radlk. (1879);拟库潘树属(库帕尼奥属,拟火把树属)●☆
114585 Cupaniopsis anacardioides (A. Rich.) Radlk.;美丽拟库潘树(美丽库帕尼奥);Carrot Wood, Carrotwood, Tuckeroo ●☆
114586 Cupaniopsis angustifolia Radlk.;窄叶拟库潘树●☆
114587 Cupaniopsis leptobotrys Radlk.;细穗拟库潘树●☆
114588 Cupaniopsis macrocarpa Radlk.;大果拟库潘树●☆
114589 Cupaniopsis multidens Radlk.;多齿拟库潘树●☆
114590 Cupaniopsis punctulata Radlk.;斑点拟库潘树●☆
114591 Cupaniopsis reticulata Merr. et L. M. Perry;网状拟库潘树●☆
114592 Cupaniopsis sylvatica Guillaumin;林地拟库潘树●☆
114593 Cuparilla Raf. = Acacia Mill. (保留属名)●■
114594 Cuphaea Moench = Cuphea Adans. ex P. Browne ●■
114595 Cuphea Adans. ex P. Browne(1756);萼距花属(花柳属,克非亚属,雪茄花属);Cuphea ●■
114596 Cuphea P. Browne = Cuphea Adans. ex P. Browne ●■
114597 Cuphea aequipetala Cav.;同瓣萼距花■☆
114598 Cuphea balsamona Cham. et Schltdl.;香膏萼距花;Balsam Cuphea ●
114599 Cuphea carthagenensis (Jacq.) J. F. Macbr.;克非亚草;Colombian Waxweed ■
114600 Cuphea cyanea DC. = Cuphea cyanea Moc. et Sessé ■☆
114601 Cuphea cyanea Moc. et Sessé;堇色萼距花(雪茄花);Violet Cuphea ■☆
114602 Cuphea eminens Planch. et Linden = Cuphea micropetala Kunth ●
114603 Cuphea glutinosa Cham. et Schltdl.;黏萼距花;Sticky Waxweed ■☆
114604 Cuphea hookeriana Walp.;胡克萼距花(萼距花);Bat Face, Bat-faced Cuphea, Heather, Hooker Cuphea, Red Cuphea, St. Peter Plant ●
114605 Cuphea hyssopifolia Kunth;狭叶萼距花(海索草叶萼距花,神香草叶萼距花,细叶萼距花,细叶雪茄花);Barbados Heather, False Heather, Mexican Heather ●☆
114606 Cuphea ignea A. DC.;萼距花(火红萼距花,雪茄花,烟花,黏毛萼距花);Brodspur Cuphea, Cigar Flower, Cigar Plant, Cigarette Plant, Cigar-flower, Cuphea, Firecracker Plant, Fire-cracker Plant, Mexican Cigar Plant ●
114607 Cuphea ignea A. DC. = Cuphea platycentra Lem. ●
114608 Cuphea jorullensis Lindl. = Cuphea micropetala Kunth ●
114609 Cuphea lanceolata Aiton;披针叶萼距花;Lanceoleaf Cuphea ●
114610 Cuphea llavea Lindl. = Cuphea hookeriana Walp. ●
114611 Cuphea micropetala Kunth;小瓣萼距花(微瓣雪茄花);Big Cigar, Tinypetal Cuphea ●
114612 Cuphea mimuloides Cham. et Schltdl.;沟酸浆萼距花●☆
114613 Cuphea miniata Brongn.;朱红萼距花;Cinnabar Cuphea ●☆
114614 Cuphea petiolata (L.) Koehne;黏毛萼距花;Blue Waxweed, Clammy Cuphea, Petioled Cuphea ●
114615 Cuphea petiolata (L.) Koehne = Cuphea viscosissima Jacq. ●
114616 Cuphea pinetorum Benth.;松林萼距花●☆
114617 Cuphea platycentra Lem. = Cuphea ignea A. DC. ●
114618 Cuphea procumbens Cav.;平卧萼距花(匍匐萼距花,偃伏雪茄花);Creeping Waxweed, Fire Cracker, Procumbent Cuphea ●
114619 Cuphea procumbens Ortega = Cuphea procumbens Cav. ●
114620 Cuphea purpurea Lem.;紫萼距花;Firefly ■☆
114621 Cuphea viscosa Rose;黏瓣萼距花■☆
114622 Cuphea viscosissima Jacq. = Cuphea petiolata (L.) Koehne ●

114623 Cupheanthus Seem. (1865);弯花桃金娘属●☆
114624 Cupheanthus microphyllus Guillaumin;小叶弯花桃金娘●☆
114625 Cupheanthus neocaledonicus Seem.;弯花桃金娘●☆
114626 Cuphocarpus Decne. et Planch. (1854);弯果五加属●☆
114627 Cuphocarpus Decne. et Planch. = Polyscias J. R. Forst. et G. Forst. ●
114628 Cuphocarpus aculeatus Decne. et Planch.;弯果五加●☆
114629 Cuphocarpus briquetianus Bernardi = Polyscias briquetiana (Bernardi) Lowry et G. Plunkett ●☆
114630 Cuphocarpus humbertianus Bernardi = Polyscias humbertiana (Bernardi) Lowry et G. Plunkett ●☆
114631 Cuphocarpus leandrianus Bernardi = Polyscias leandriana (Bernardi) Lowry et G. Plunkett ●☆
114632 Cuphoea Brongn. ex Neumann = Cuphea Adans. ex P. Browne ●■
114633 Cuphonotus O. E. Schulz(1933);驼缘荠属■☆
114634 Cupi Adans. = Chomelia L. (废弃属名)●☆
114635 Cupi Adans. = Rondeletia L. ●
114636 Cupi Adans. = Tarenna Gaertn. ●
114637 Cupia (Schult.) DC. = Anomanthodia Hook. f. ●☆
114638 Cupia DC. = Cupi Adans. ●
114639 Cupia mollissima Hook. et Arn. = Tarenna mollissima (Hook. et Arn.) Rob. ●
114640 Cupidone Lem. = Catananche L. ■☆
114641 Cupidonia Bubani = Catananche L. ■☆
114642 Cupirana Miers = Coupoui Aubl. ●☆
114643 Cupirana Miers = Duroia L. f. (保留属名)●☆
114644 Cuprespinnata J. Nelson = Taxodium Rich. ●
114645 Cupressaceae Gray(1822)(保留科名);柏科;Cypress Family, Juniper Family, Redwood Family ●
114646 Cupressaceae Neger = Cupressaceae Gray(保留科名)●
114647 Cupressaceae Rich. ex Bartl. = Cupressaceae Gray(保留科名)●
114648 Cupresstellata J. Nelson = Fitzmya Hook. f. ex Lindl. ●☆
114649 Cupressus L. (1753);柏属(柏木属);Cypress, True Cypress ●
114650 Cupressus abramsiana C. B. Wolf;艾布拉姆柏;Abrams Cypress ●☆
114651 Cupressus abramsiana C. B. Wolf = Cupressus goveniana Gordon ex Lindl. ●
114652 Cupressus arizonica Greene;绿干柏(美洲柏木,亚利桑那柏木,亚利桑那丝杉);Arizona Cypress, Arizona Smooth Cypress, Cedro Blanco, Cuyamaca Cypress, Rough-bark, Rough-barked Arizona Cypress ●
114653 Cupressus arizonica Greene 'Blue Ice';蓝叶绿干柏●☆
114654 Cupressus arizonica Greene subsp. stephensonii (C. B. Wolf) A. E. Murray;光皮绿干柏●☆
114655 Cupressus arizonica Greene var. bonita Lemmon = Cupressus arizonica Greene ●
114656 Cupressus arizonica Greene var. glabra (Sudw.) Little = Cupressus arizonica Greene ●
114657 Cupressus arizonica Greene var. glabra (Sudw.) Little = Cupressus glabra Sudw. ●☆
114658 Cupressus arizonica Greene var. nevadensis (Abrams) Little = Cupressus arizonica Greene ●
114659 Cupressus arizonica Greene var. stephensonii (C. B. Wolf) Little = Cupressus arizonica Greene ●
114660 Cupressus atlantica Gaussen;大西洋柏木●☆
114661 Cupressus austrotibetica Silba = Cupressus duclouxiana Hickel ●
114662 Cupressus bakeri Jeps.;巴克柏木;Baker Cypress, Modoc Cypress, Sikiyou Cypress ●☆
114663 Cupressus bakeri Jeps. subsp. matthewsii C. B. Wolf = Cupressus bakeri Jeps. ●☆
114664 Cupressus benthamii Endl.;本瑟姆柏木(本瑟柏木,鞭人柏木);Bentham Portuguese Cypress ●
114665 Cupressus benthamii Endl. var. arizonica (Greene) Mast. = Cupressus arizonica Greene ●
114666 Cupressus californica Carrière = Cupressus glabra Sudw. ●☆
114667 Cupressus californica Carrière = Cupressus goveniana Gordon ex Lindl. ●
114668 Cupressus cashmeriana Royle ex Carrière;克什米尔柏木;Bhutan Cypress, Kashmir Cypress ●☆
114669 Cupressus chengiana S. Y. Hu;岷江柏木;Cheng Cypress ●◇
114670 Cupressus chengiana S. Y. Hu var. jiangeensis (N. Chao) C. T. Kuan = Cupressus jiangeensis N. Chao ●
114671 Cupressus chengiana S. Y. Hu var. kansouensis Silba = Cupressus chengiana S. Y. Hu ●◇
114672 Cupressus chengiana S. Y. Hu var. wenchuanhsiensis Silba = Cupressus chengiana S. Y. Hu ●◇
114673 Cupressus columnaris G. Forst. = Araucaria columnaris (G. Forst.) Hook. ●☆
114674 Cupressus corneyana Hort. = Cupressus lusitanica Mill. ●
114675 Cupressus disticha L. = Taxodium distichum (L.) Rich. ●
114676 Cupressus disticha L. var. imbricaria Nutt. = Taxodium distichum (L.) Rich. var. imbricatum (Nutt.) Croom ●
114677 Cupressus disticha L. var. nutans Aiton = Taxodium ascendens Brongn. 'Nutans' ●
114678 Cupressus disticha L. var. nutans Aiton = Taxodium distichum (L.) Rich. var. imbricatum (Nutt.) Croom ●
114679 Cupressus dupreziana Camus;阿尔及利亚柏木;Algeria Cypress, Duprez Cypress ●☆
114680 Cupressus duclouxiana Hickel;干香柏(扁柏,冲天柏,滇柏,干柏杉,云南柏);Bhutan Cypress, Chinese Cypress, Ducloux Cypress, Yunnan Cypress ●
114681 Cupressus duclouxiana Hickel = Cupressus chengiana S. Y. Hu ●◇
114682 Cupressus duclouxiana Hickel = Cupressus torulosa D. Don ●
114683 Cupressus dupreziana A. Camus = Tassilicyparis dupreziana (A. Camus) A. V. Bobrov et Melikyan ●☆
114684 Cupressus expansa Targ. Tozz. = Cupressus sempervirens L. var. horizontalis (Mill.) Gordon ●☆
114685 Cupressus expansa Targ. Tozz. ex Steud. = Cupressus sempervirens L. var. horizontalis (Mill.) Gordon ●☆
114686 Cupressus fallax Franco = Cupressus chengiana S. Y. Hu ●◇
114687 Cupressus fomosensis (Matsum.) Henry = Chamaecyparis fomosensis Matsum. ●◇
114688 Cupressus forbesii Jeps. = Cupressus guadalupensis S. Watson var. forbesii (Jeps.) Little ●☆
114689 Cupressus formosensis (Matsum.) A. Henry = Chamaecyparis formosensis Matsum. ●◇
114690 Cupressus funebris Endl.;柏木(白木树,柏,柏木树,柏青树,柏树,柏香树,扁柏,垂柏,垂丝柏,垂枝柏,吊柏,黄柏,密密柏,密密松,扫帚柏,宋柏,香柏,香柏树,香扁柏,璎珞柏);China Mountain Cypress, Chinese Funeral-cypress, Chinese Weeping Cypress, Chinese-weeping Cypress, Funeral Cypress, Funereal Cypress, Mourning Cypress, Weeping Cypress, White Fir ●
114691 Cupressus gigantea W. C. Cheng et L. K. Fu;巨柏(雅鲁藏布江

柏木);Big Cypress ●◇

114692 Cupressus glabra Sudw.;光皮柏木(光皮亚利桑那柏木);Arizona Cypress, Smooth Arizona Cypress, Smooth Cypress, Smooth-bark Arizona Cypress ●☆

114693 Cupressus glabra Sudw. 'Hodginsii';霍金斯柏木;Hodgin's Arizona Cypress ●☆

114694 Cupressus glabra Sudw. = Cupressus arizonica Greene ●

114695 Cupressus glauca Lam.;印度柏木(平滑柏木);Blue Gowen Cypress ●☆

114696 Cupressus glauca Lam. = Cupressus lusitanica Mill. ●

114697 Cupressus goveniana Gordon = Cupressus goveniana Gordon ex Lindl. ●

114698 Cupressus goveniana Gordon ex Lindl.;加利福尼亚柏木(戈温柏木);Califolian Cypress, Dwarf Cypress, Goven Cypress, Gowen Cypress, Mendocino Cypress, Santa Cruz Cypress ●

114699 Cupressus goveniana Gordon ex Lindl. 'Glauca';灰蓝叶加利福尼亚柏木;Blue Gowen Cypress ●☆

114700 Cupressus goveniana Gordon ex Lindl. var. abramsiana (C. B. Wolf) Little = Cupressus goveniana Gordon ex Lindl. ●

114701 Cupressus goveniana Gordon ex Lindl. var. pigmaea Lemmon = Cupressus goveniana Gordon ex Lindl. ●

114702 Cupressus guadalupensis S. Watson;瓜达罗浦柏木;Gaudalup Cypress, Guadalupe Cypress, Tecate Cypress ●☆

114703 Cupressus guadalupensis S. Watson var. forbesii (Jeps.) Little;福布斯柏木;Tecate Cypress ●☆

114704 Cupressus guadalupensis S. Watson var. forbesii (Jeps.) Little = Cupressus forbesii Jeps. ●☆

114705 Cupressus hodginsii Dunn = Fokienia hodginsii (Dunn) A. Henry et H. H. Thomas ●◇

114706 Cupressus horizontalis Mill. = Cupressus sempervirens L. var. horizontalis (Mill.) Gordon ●☆

114707 Cupressus horizontalis Mill. = Cupressus sempervirens L. ●

114708 Cupressus japonica Thunb. ex L. f. = Cryptomeria japonica (Thunb. ex L. f.) D. Don ●

114709 Cupressus jiangeensis N. Chao;剑阁柏木;Jiange Cypress ●

114710 Cupressus knightiana Perry ex Gordon = Cupressus benthamii Endl. ●

114711 Cupressus lawsoniana A. Murray = Chamaecyparis lawsoniana (A. Murray) Parl. ●

114712 Cupressus lawsoniana A. Murray bis = Chamaecyparis lawsoniana (A. Murray) Parl. ●

114713 Cupressus lindleyi Klotzsch ex Endl. subsp. benthamii (Endl.) Silba = Cupressus benthamii Endl. ●

114714 Cupressus lusitanica Mill.;墨西哥柏木(白粉柏,墨西哥柏,葡萄牙柏木,速生柏);Cedar of Goa, Cedar-of-goa, Lusitanian Cypress, Mexican Cypress, Mexico Cypress, Portuguese Cypress ●

114715 Cupressus lusitanica Mill. subsp. arizonica (Greene) Maire = Cupressus arizonica Greene ●

114716 Cupressus lusitanica Mill. subsp. benthamii (Endl.) Franco = Cupressus benthamii Endl. ●

114717 Cupressus lusitanica Mill. subsp. mexicana (Koch) Maire = Cupressus lusitanica Mill. ●

114718 Cupressus lusitanica Mill. var. benthamii (Endl.) Carrière = Cupressus benthamii Endl. ●

114719 Cupressus lusitanica Mill. var. benthamii Carrière = Cupressus benthamii Endl. ●

114720 Cupressus lusitanica Mill. var. knightiana Rehder;奈特柏木;Knight Portuguese Cypress ●☆

114721 Cupressus macnabiana A. Murray;加州柏木;MacNab Cypress, MacNab's Cypress ●☆

114722 Cupressus macnabiana A. Murray bis var. bakeri (Jeps.) Jeps. = Cupressus bakeri Jeps. ●☆

114723 Cupressus macrocarpa Hartw.;大果柏木(大果柏);Bigfruit Cypress, Monterey Cedar, Monterey Cypress, Monterrey Cypress ●☆

114724 Cupressus macrocarpa Hartw. 'Aurea Saligna';金叶大果柏木●☆

114725 Cupressus macrocarpa Hartw. 'Brunniana';直枝大果柏木●☆

114726 Cupressus macrocarpa Hartw. 'Coneybearii Aurea' = Cupressus macrocarpa Hartw. 'Aurea Saligna' ●☆

114727 Cupressus macrocarpa Hartw. 'Donard Gold';唐纳德金大果柏木●☆

114728 Cupressus macrocarpa Hartw. 'Greenstead Magnificent';矮生蓝大果柏木●☆

114729 Cupressus macrocarpa Hartw. 'Horizontalis';平冠大果柏木●☆

114730 Cupressus macrocarpa Hartw. var. lutea Dicksons;黄叶大果柏木●☆

114731 Cupressus mairei H. Lév. = Cryptomeria japonica (Thunb. ex L. f.) D. Don var. sinensis Miq. ●

114732 Cupressus mas Garsault = Cupressus sempervirens L. var. horizontalis (Mill.) Gordon ●☆

114733 Cupressus nevadensis Abrams = Cupressus arizonica Greene ●

114734 Cupressus nootkatensis D. Don = Chamaecyparis nootkatensis (D. Don) Spach ●

114735 Cupressus obtusa (Siebold et Zucc.) F. Muell. = Chamaecyparis obtusa (Siebold et Zucc.) Siebold et Zucc. ex Endl. ●

114736 Cupressus obtusa (Siebold et Zucc.) Mast. = Chamaecyparis obtusa (Siebold et Zucc.) Siebold et Zucc. ex Endl. ●

114737 Cupressus obtusa (Siebold et Zucc.) Mast. f. formosana (Hayata) Clinton-Baker = Chamaecyparis obtusa (Siebold et Zucc.) Siebold et Zucc. ex Endl. var. formosana (Hayata) Rehder ●

114738 Cupressus obtusa (Siebold et Zucc.) Mast. f. formosana (Hayata) Clinton-Baker = Chamaecyparis obtusa (Siebold et Zucc.) Siebold et Zucc. ex Endl. ●

114739 Cupressus obtusa (Siebold et Zucc.) Mast. var. formosana (Hayata) Dallim. et Jacks. = Chamaecyparis obtusa (Siebold et Zucc.) Siebold et Zucc. ex Endl. var. formosana (Hayata) Rehder ●

114740 Cupressus obtusa (Siebold et Zucc.) Mast. var. formosana (Hayata) Dallim. et A. B. Jacks. = Chamaecyparis obtusa (Siebold et Zucc.) Siebold et Zucc. ex Endl. ●

114741 Cupressus pendula Abel = Cupressus funebris Endl. ●

114742 Cupressus pigmaea (Lemmon) Sarg. = Cupressus goveniana Gordon ex Lindl. ●

114743 Cupressus pisifera (Siebold et Zucc.) Koch = Chamaecyparis pisifera (Siebold et Zucc.) Siebold et Zucc. ex Endl. ●

114744 Cupressus pisifera (Siebold et Zucc.) Koch f. squarrosa (Zucc.) Mast. = Chamaecyparis pisifera (Siebold et Zucc.) Siebold et Zucc. ex Endl. 'Squarrosa' ●

114745 Cupressus pisifera (Siebold et Zucc.) Koch var. filifera (Veitch) H. Lév. = Chamaecyparis pisifera (Siebold et Zucc.) Siebold et Zucc. ex Endl. 'Filifera' ●

114746 Cupressus pisifera (Siebold et Zucc.) Koch var. plumosa (Carrière) Geerinck = Chamaecyparis pisifera (Siebold et Zucc.) Siebold et Zucc. ex Endl. 'Plumosa' ●

114747 Cupressus pisifera (Siebold et Zucc.) Koch var. plumosa H. Lév. = Chamaecyparis pisifera (Siebold et Zucc.) Siebold et Zucc.

ex Endl. 'Plumosa' ●

114748 Cupressus pisifera (Siebold et Zucc.) Koch var. squarrosa (Zucc.) Keng = Chamaecyparis pisifera (Siebold et Zucc.) Siebold et Zucc. ex Endl. 'Squarrosa' ●

114749 Cupressus pygmaea (Lemmon) Sarg.; 矮柏木; Mendocino Cypress ●☆

114750 Cupressus pyramidalis Targ. Tozz. = Cupressus sempervirens L. var. stricta Aiton ●☆

114751 Cupressus sargentii Jeps.; 萨金特柏木; Sargent Cypress ●☆

114752 Cupressus sargentii Jeps. var. duttonii Jeps. = Cupressus sargentii Jeps. ●☆

114753 Cupressus sempervirens L.; 地中海柏木(柏,常绿柏,鳞皮柏,西洋柏木,洋柏,意大利柏); Common Cypress, Funeral Cypress, Gopherwood, Italian Cypress, Italy Cypress, Mediterranean Cypress, Roman Cypress ●

114754 Cupressus sempervirens L. 'Karoonda'; 卡荣德地中海柏木●☆

114755 Cupressus sempervirens L. 'Stricta'; 窄冠地中海柏木●☆

114756 Cupressus sempervirens L. 'Swane's Golden'; 天鹅金地中海柏木; Swane's Golden Pencil-pine ●☆

114757 Cupressus sempervirens L. = Cupressus duclouxiana Hickel ●

114758 Cupressus sempervirens L. var. atlantica (Gaussen) Silba = Cupressus atlantica Gaussen ●☆

114759 Cupressus sempervirens L. var. femina ? = Cupressus sempervirens L. var. stricta Aiton ●☆

114760 Cupressus sempervirens L. var. horizontalis (Mill.) Gordon; 匍匐地中海柏木; Horizontal Italian Cypress, Spreading Roman Cypress ●☆

114761 Cupressus sempervirens L. var. horizontalis (Mill.) Gordon = Cupressus sempervirens L. ●

114762 Cupressus sempervirens L. var. horizontalis Gordon = Cupressus sempervirens L. var. horizontalis (Mill.) Gordon ●☆

114763 Cupressus sempervirens L. var. indica Parl.; 亮绿地中海柏木(亮叶地中海柏木); Bright-green Italian Cypress ●☆

114764 Cupressus sempervirens L. var. numidica Trab. = Cupressus sempervirens L. ●

114765 Cupressus sempervirens L. var. pyramidalis (Targ. Tozz.) Nyman; 塔形地中海柏木; Pyramidal Italian Cypress ●☆

114766 Cupressus sempervirens L. var. pyramidalis (Targ. Tozz.) Nyman = Cupressus sempervirens L. var. stricta Aiton ●☆

114767 Cupressus sempervirens L. var. pyramidalis (Targ. Tozz.) Nyman = Cupressus sempervirens L. ●

114768 Cupressus sempervirens L. var. stricta Aiton; 柱形地中海柏木; Columar Italian Cypress, Upright Roman Cypress ●☆

114769 Cupressus sempervirens L. var. stricta Aiton = Cupressus sempervirens L. ●

114770 Cupressus stephensonii C. B. Wolf = Cupressus arizonica Greene ●

114771 Cupressus thyoides L. = Chamaecyparis thyoides (L.) Britton, Sterns et Poggenb. ●

114772 Cupressus tongmaiensis Silba = Cupressus torulosa D. Don ●

114773 Cupressus tongmaiensis Silba var. ludlowii Silba = Cupressus torulosa D. Don ●

114774 Cupressus tonkinensis Silba = Cupressus torulosa D. Don ●

114775 Cupressus torulosa D. Don; 西藏柏木(干柏杉,西藏柏,喜马拉雅柏,喜马拉雅柏木); Bhutan Cypress, Himalaya Cypress, Himalayan Cypress, Himalayas Cypress, Yunan Cypress ●

114776 Cupressus torulosa D. Don 'Cashmeriana' = Cupressus cashmeriana Royle ex Carrière ●☆

114777 Cupressus torulosa D. Don = Cupressus chengiana S. Y. Hu ●◇

114778 Cuprestellata Carrière = Cupresstellata J. Nelson ●☆

114779 Cupuia Raf. = Coupoui Aubl. ●☆

114780 Cupuia Raf. = Duroia L. f. (保留属名) ●☆

114781 Cupulaceae Dulac = Cupuliferae A. Rich. ●

114782 Cupulanthus Hutch. (1964); 盆花豆属■☆

114783 Cupulanthus bracteolosus (F. Muell.) Hutch.; 盆花豆■☆

114784 Cupularia Godr. et Gren. = Dittrichia Greuter ■☆

114785 Cupularia Godr. et Gren. = Inula L. ●■

114786 Cupularia Godr. et Gren. ex Godr. = Dittrichia Greuter ■☆

114787 Cupularia Godr. et Gren. ex Godr. = Inula L. ●■

114788 Cupularia viscosa Gren. et Godr. = Dittrichia viscosa (L.) Greuter ■

114789 Cupuliferae A. Rich. = Betulaceae Gray (保留科名) + Corylaceae Mirb. (保留科名) + Fagaceae Dumort. (保留科名) ●

114790 Cupuliferae A. Rich. = Betulaceae Gray (保留科名) + Fagaceae Dumort. (保留科名) ●

114791 Cupulissa Raf. (废弃属名) = Anemopaegma Mart. ex Meisn. (保留属名) ●☆

114792 Cupuya Raf. = Coupoui Aubl. ●☆

114793 Cupuya Raf. = Duroia L. f. (保留属名) ●☆

114794 Curanga Juss. (1807); 苦味草属■☆

114795 Curanga Juss. = Picria Lour. ■

114796 Curanga amara Juss.; 苦味草■☆

114797 Curanga amara Juss. = Picria felterrae Lour. ■

114798 Curanga bivalvis Druce = Besleria bivalvis L. f. ■☆

114799 Curanga felterrae (Lour.) Merr. = Picria felterrae Lour. ■

114800 Curania Roem. et Schult. = Curanga Juss. ■☆

114801 Curare Kunth ex Humb. = Strychnos L. ●

114802 Curarea Barneby et Krukoff (1971); 箭毒藤属●☆

114803 Curarea candicans (Rich.) Barneby et Krukoff; 白箭毒藤■☆

114804 Curarea tecunarum Barneby et Krukoff; 巴西箭毒藤■☆

114805 Curarea toxicofera (Wedd.) Barneby et Krukoff; 箭毒藤■☆

114806 Curatari J. F. Gmel. = Couratari Aubl. ●☆

114807 Curataria Spreng. = Couratari Aubl. ●☆

114808 Curatella Loefl. (1758); 拭戈木属(库拉五桠果木属); Curatella ●☆

114809 Curatella americana L.; 拭戈木(库拉五桠果木); American Curatella ●☆

114810 Curbaril Post et Kuntze = Courbaril Mill. ●

114811 Curbaril Post et Kuntze = Hymenaea L. ●

114812 Curcas Adans. = Jatropha L. (保留属名) ●■

114813 Curcubitella Walp. = Cucurbitella Walp. ■☆

114814 Curculigo Gaertn. (1788); 仙茅属; Curculigo ■

114815 Curculigo baguirmiensis A. Chev. = Hypoxis angustifolia Lam. ■☆

114816 Curculigo borneensis Merr.; 婆罗洲仙茅■☆

114817 Curculigo brevifolia Dryand. = Curculigo orchioides Gaertn. ■

114818 Curculigo breviscapa S. C. Chen; 短葶仙茅(大莎草); Shortscap Curculigo ■

114819 Curculigo capitulata (Lour.) Kuntze; 大叶仙茅(船仔草,船子草,大白芨,大地棕,大仙茅,花叶,假槟榔树,山棕,松兰,头花仙茅,土七厘丹,岩棕,野棕,竹灵芝,棕参); Largeleaf Curculigo, Palm-grass, Weevil Plant ■

114820 Curculigo capitulata (Lour.) Kuntze = Molineria capitulata (Lour.) Herb. ■

114821 Curculigo crassifolia (Baker) Hook. f.; 绒叶仙茅; Thickleaf Curculigo ■

114822 Curculigo djalonensis A. Chev. = Hypoxis angustifolia Lam. ■☆

114823 Curculigo ensifolia R. Br. = Curculigo orchioides Gaertn. ■

114824 Curculigo fuziwarae Gogelein = Curculigo capitulata (Lour.) Kuntze ■

114825 Curculigo gallabatensis Schweinf. ex Baker = Curculigo pilosa (Schumach. et Thonn.) Engl. ■☆

114826 Curculigo gallabatensis Schweinf. var. major Baker = Curculigo pilosa (Schumach. et Thonn.) Engl. subsp. major (Baker) Wiland ■☆

114827 Curculigo glabrescens (Ridl.) Merr.;光叶仙茅(无毛仙茅);Glabrous Curculigo ■

114828 Curculigo gracilis (Wall. ex Kurz) Hook. f.;疏花仙茅(小棕包);Laxflower Curculigo ■

114829 Curculigo graminifolia Nimmo = Hypoxis aurea Lour. ■

114830 Curculigo latifolia Dryand. = Molineria latifolia (Dryand. ex W. T. Aiton) Herb. ex Kurz ■☆

114831 Curculigo latifolia Dryand. ex W. T. Aiton = Molineria latifolia (Dryand. ex W. T. Aiton) Herb. ex Kurz ■☆

114832 Curculigo latifolia Dryand. ex W. T. Aiton var. glabrescens Ridl. = Curculigo glabrescens (Ridl.) Merr. ■

114833 Curculigo latifolia Dryand. var. glabrescens Ridl. = Curculigo glabrescens (Ridl.) Merr. ■

114834 Curculigo malabarica Wight = Curculigo orchioides Gaertn. ■

114835 Curculigo minor Guinea = Curculigo pilosa (Schumach. et Thonn.) Engl. subsp. minor (Guinea) Wiland ■☆

114836 Curculigo multiflora Zimudzi = Curculigo pilosa (Schumach. et Thonn.) Engl. subsp. minor (Guinea) Wiland ■☆

114837 Curculigo namaquensis Baker = Empodium namaquensis (Baker) M. F. Thomps. ■☆

114838 Curculigo orchioides Gaertn.;仙茅(白仙茅,地棕,冬虫草,独角丝茅,独角仙茅,独脚黄茅,独脚丝茅,独脚仙茅,独毛,独茅,独足绿茅,番龙草,风苔草,海南参,黄茅参,假虫草,尖刀草,冷饭草,茅爪子,盘棕,蟠龙草,平肝薯,婆罗门参,千年棕,山党参,山兰花,山棕,天棕,土白芍,仙毛,仙茅参,小地棕,小棕包,小棕苞,芽爪子,鹧鸪苁);Common Curculigo,Curculigo ■

114839 Curculigo orchioides Gaertn. var. minor Benth. = Curculigo orchioides Gaertn. ■

114840 Curculigo pilosa (Schumach. et Thonn.) Engl.;疏毛仙茅■☆

114841 Curculigo pilosa (Schumach. et Thonn.) Engl. subsp. major (Baker) Wiland;大疏毛仙茅■☆

114842 Curculigo pilosa (Schumach. et Thonn.) Engl. subsp. minor (Guinea) Wiland;小疏毛仙茅■☆

114843 Curculigo plicata (Thunb.) Dryand. = Empodium plicatum (Thunb.) Garside ■☆

114844 Curculigo plicata (Thunb.) Dryand. var. barberae Baker = Empodium flexile (Nel) M. F. Thomps. ex Snijman ■☆

114845 Curculigo recurvata Dryand. = Curculigo capitulata (Lour.) Kuntze ■

114846 Curculigo recurvata Dryand. = Molineria capitulata (Lour.) Herb. ■

114847 Curculigo recurvata W. T. Aiton = Curculigo capitulata (Lour.) Kuntze ■

114848 Curculigo senporeiensis Gogelein = Curculigo glabrescens (Ridl.) Merr. ■

114849 Curculigo sinensis S. C. Chen;中华仙茅;China Curculigo, Chinese Curculigo ■

114850 Curculigo strobiliformis D. Fang et D. H. Qin = Curculigo capitulata (Lour.) Kuntze ■

114851 Curculigo veratrifolia (Willd.) Baker = Empodium veratrifolium (Willd.) M. F. Thomps. ■☆

114852 Curcuma L. (1753)(保留属名);姜黄属(郁金属);Curcuma, Hidden Lily, Hidden-lily, Turmeric ■

114853 Curcuma aeruginosa Roxb. = Curcuma zedoaria (Christm.) Roscoe ●■

114854 Curcuma albicoma S. Q. Tong = Curcuma sichuanensis X. X. Chen ■

114855 Curcuma alismatifolia Gagnep.;泽泻叶姜黄;Pineapple Ginger ■☆

114856 Curcuma amada Roxb.;杧果姜;Mango Ginger ■☆

114857 Curcuma amarissima Roscoe;极苦姜黄(味极苦姜黄)■

114858 Curcuma angustifolia Roxb.;狭叶姜黄;Bombay Arrowroot, East Indian Arrowroot, East Indies Arrowroot, Indian Arrowroot, Tikor ■

114859 Curcuma aromatica Salisb.;郁金(姜黄,毛郁金,温郁金);Aromatic Turmeric ■

114860 Curcuma aromatica Salisb. 'Wenyujin' = Curcuma wenyujin Y. H. Chen et C. Ling ■

114861 Curcuma australis ?;南方姜黄■☆

114862 Curcuma caesia Roxb. = Curcuma phaeocaulis Valeton ■

114863 Curcuma chuanyujin C. K. Hsieh et H. Zhang = Curcuma kwangsiensis S. G. Lee et C. F. Liang ■

114864 Curcuma domestica Valeton = Curcuma longa L. ■

114865 Curcuma elata Roxb.;高姜黄(大莪术)■☆

114866 Curcuma exigua N. Liu;细莪术■

114867 Curcuma flaviflora S. Q. Tong;黄花姜黄;Yellowflower Turmeric ■

114868 Curcuma kwangsiensis S. G. Lee et C. F. Liang;广西莪术(白丝郁金,莪术,桂莪术,黄郁,马蒁,毛莪术,玉金);Guangxi Turmeric, Kwangsi Turmeric ■

114869 Curcuma kwangsiensis S. G. Lee et C. F. Liang var. affinis Y. H. Chen;紫脉莪术■☆

114870 Curcuma kwangsiensis S. G. Lee et C. F. Liang var. puberula Y. H. Chen;软毛莪术(毛莪术)■☆

114871 Curcuma longa L.;姜黄(白丝郁金,宝鼎香,黄姜,黄丝郁金,黄郁,马蒁,毛姜黄,乙金,玉金,郁金);Common Turmeric, Curry, Cypress of Babylon, Cypress of India, Cypress of Malabar, Indian Saffron, Long-rooted Turmeric, Turmeric, Turn-Merick ■

114872 Curcuma pallida Lour. = Curcuma phaeocaulis Valeton ■

114873 Curcuma pallida Lour. = Curcuma zedoaria (Christm.) Roscoe ●■

114874 Curcuma petiolata Roxb.;皇后姜黄(马来亚姜黄);Queen Lily ■☆

114875 Curcuma phaeocaulis Valeton;莪术(白丝郁金,臭屎姜,淡蓝姜黄,莪蒁,风姜,广茂,广术,黑褐姜黄,黑姜,黑心姜,黄郁,兰姜,蓝姜,蓝心姜,绿姜,马蒁,缅郁金,蓬莪,蓬莪茂,蓬莪术,蓬莪蒁,蓬术,蓬蒁,羌七,山姜黄,蒁药,文术,乌姜,玉金);Black Zedoary ■

114876 Curcuma roscoeana Wall.;马来姜黄■☆

114877 Curcuma rotunda L. = Boesenbergia rotunda (L.) Mansf. ■

114878 Curcuma sichuanensis X. X. Chen;川郁金(白顶姜黄,白丝郁金,川莪术,土文术);Sichuan Queen Lily ■

114879 Curcuma viridifolia Roxb.;绿叶姜黄(二黄)■

114880 Curcuma wenyujin Y. H. Chen et C. Ling;温郁金(白丝郁金,莪术,黄郁,姜黄,马蒁,温莪术,温州蓬莪茂,玉金)■

114881 Curcuma xanthorrhiza Roxb.;黄红姜黄(黄根姜黄,印尼莪术);Yellow-red Turmeric ■

114882 Curcuma yunnanensis N. Liu et S. J. Chen;顶花莪术;Yunnan Turmeric ■

114883 Curcuma zanthorrhiza Roxb.;印尼莪术(印度莪术)■
114884 Curcuma zedoaria (Christm.) Roscoe;黄莪术(臭尿姜,莪术,莪莛,姜黄,蓬莪茂,蓬莪术,山姜黄);Banquet Herb, Round Zedoary, Zedary, Zedoary, Zeodary, Zodoary, Zodoary Turmeric ●■
114885 Curcuma zedoaria (Christm.) Roscoe = Curcuma phaeocaulis Valeton ■
114886 Curcumaceae Dumort. = Zingiberaceae Martinov(保留科名)■
114887 Curcumorpha A. S. Rao et D. M. Verma = Boesenbergia Kuntze ■
114888 Curcumorpha longiflora (Wall.) A. S. Rao et D. M. Verma = Boesenbergia longiflora (Wall.) Kuntze ■
114889 Curima O. F. Cook = Aiphanes Willd. ●☆
114890 Curimari Post et Kuntze = Courimari Aubl. ●
114891 Curimari Post et Kuntze = Sloanea L. ●
114892 Curinila Raf. = Leptadenia R. Br. ●☆
114893 Curinila Roem. et Schult. = Leptadenia R. Br. ●☆
114894 Curinila Schult. = Leptadenia R. Br. ●☆
114895 Curio P. V. Heath = Senecio L. ■●
114896 Curitiba Salywon et Landrum = Eugenia L. ●
114897 Curmeria Linden et André = Homalomena Schott ■
114898 Curnilla Raf. = Curinila Roem. et Schult. ●☆
114899 Curnilla Raf. = Leptadenia R. Br. ●☆
114900 Curnmingia D. Don = Conanthera Ruiz et Pav. ■☆
114901 Curondia Raf. = Salacia L. (保留属名)●
114902 Curraniodendron Merr. = Quintinia A. DC. ●☆
114903 Curroria Planch. ex Benth. (1849);库萝藦属●☆
114904 Curroria decidua Planch. ex Benth.;脱落库萝藦●☆
114905 Curroria decidua Planch. ex Benth. subsp. gillettii (Hutch. et E. A. Bruce) Bullock = Cryptolepis gillettii Hutch. et E. A. Bruce ●☆
114906 Curroria macrophylla Radcl. -Sm. = Cryptolepis macrophylla (Radcl. -Sm.) Venter ●☆
114907 Curroria migiurtina (Chiov.) Bullock = Cryptolepis migiurtina Chiov. ●☆
114908 Curroria volubilis (Schltr.) Bullock = Buckollia volubilis (Schltr.) Venter et R. L. Verh. ●☆
114909 Cursonia Nutt. = Onoseris Willd. ●■☆
114910 Curtia Cham. et Schltdl. (1826);库尔特龙胆属■☆
114911 Curtia gentianoides Cham. et Schltdl.;库尔特龙胆■☆
114912 Curtia obtusifolia Knobl.;钝叶库尔特龙胆■☆
114913 Curtia tenuifolia Knobl.;细叶库尔特龙胆■☆
114914 Curtia verticillaris Knobl.;轮生库尔特龙胆■☆
114915 Curtisia Aiton(1789)(保留属名);南非茱萸属(短山茱萸属,山茱萸树属);Assegai Tree ●☆
114916 Curtisia Schreb. (废弃属名) = Curtisia Aiton(保留属名)●☆
114917 Curtisia Schreb. (废弃属名) = Zanthoxylum L. ●
114918 Curtisia dentata (Burm. f.) C. A. Sm.;南非茱萸(短山茱萸,山茱萸树);Assagai Wood, Assegai Tree ●☆
114919 Curtisia faginea Aiton = Curtisia dentata (Burm. f.) C. A. Sm. ●☆
114920 Curtisiaceae (Harms) Takht. = Cornaceae Bercht. et J. Presl (保留科名)●■
114921 Curtisiaceae Takht. (1987);南非茱萸科(菲茱萸科,柯茱萸科,山茱萸树科)●☆
114922 Curtisiaceae Takht. = Cornaceae Bercht. et J. Presl(保留科名) ●■
114923 Curtisina Ridl. = Dacryodes Vahl ●☆
114924 Curtogyne Haw. = Crassula L. ●■☆
114925 Curtogyne albiflora (Sims) Eckl. et Zeyh. = Crassula dejecta Jacq. ■☆
114926 Curtogyne burmanniana Eckl. et Zeyh. = Crassula flava L. ●☆
114927 Curtogyne dejecta (Jacq.) DC. = Crassula dejecta Jacq. ■☆
114928 Curtogyne flava (L.) Eckl. et Zeyh. = Crassula flava L. ●☆
114929 Curtogyne undata (Haw.) Haw. = Crassula dejecta Jacq. ■☆
114930 Curtogyne undosa Haw. = Crassula dejecta Jacq. ■☆
114931 Curtogyne undulata (Haw.) Haw. = Crassula dejecta Jacq. ■☆
114932 Curtolsia Endl. = Collomia Nutt. ■☆
114933 Curtolsia Endl. = Courtoisia Rchb. ■☆
114934 Curtonus N. E. Br. = Crocosmia Planch. ■
114935 Curtonus paniculatus (Klatt) N. E. Br. = Crocosmia paniculata (Klatt) Goldblatt ■☆
114936 Curtopogon P. Beauv. = Aristida L. ■
114937 Curupira G. A. Black(1948);巴西铁青树属●☆
114938 Curupira tefeensis G. A. Black;巴西铁青树●☆
114939 Curupita Post et Kuntze = Couroupita Aubl. ●☆
114940 Cururu Mill. = Paullinia L. ●☆
114941 Curvangis Thouars = Angraecum Bory ■
114942 Curvangis Thouars = Jumellea Schltr. ■☆
114943 Curvophylis Thouars = Bulbophyllum Thouars(保留属名)■
114944 Cuscatlania Standl. (1923);纤苞茉莉属■☆
114945 Cuscatlania vulcanicola Standl.;纤苞茉莉■☆
114946 Cuscuaria Schott = Scindapsus Schott ■
114947 Cuscuta L. (1753);菟丝子属;Devil's Bit Guts, Devil's Guts, Dodder, Scald ■
114948 Cuscuta abyssinica A. Rich.;阿比西尼亚菟丝子■☆
114949 Cuscuta abyssinica A. Rich. var. ghindensis Yunck. = Cuscuta somaliensis Yunck. ■☆
114950 Cuscuta acuminata Pomel = Cuscuta approximata Bab. ■
114951 Cuscuta aegyptiaca Trab.;埃及菟丝子■☆
114952 Cuscuta africana Willd.;非洲菟丝子■☆
114953 Cuscuta alba J. Presl et C. Presl;白菟丝子■☆
114954 Cuscuta alpestris Fourc. = Cuscuta africana Willd. ■☆
114955 Cuscuta americana L.;美洲菟丝子;Love-vine ■☆
114956 Cuscuta americana Thunb. = Cuscuta africana Willd. ■☆
114957 Cuscuta anguina Edgew. = Cuscuta reflexa Roxb. var. anguina (Edgew.) C. B. Clarke ■
114958 Cuscuta angulata Engelm.;棱角菟丝子■☆
114959 Cuscuta appendiculata Engelm.;附属物菟丝子■☆
114960 Cuscuta appendiculata Engelm. var. macroflora Yunck. = Cuscuta appendiculata Engelm. ■☆
114961 Cuscuta approximata Bab.;杯花菟丝子(杯状菟丝子);Alfalfa Dodder, Cupulate Dodder ■
114962 Cuscuta approximata Bab. = Cuscuta europaea L. ■
114963 Cuscuta approximata Bab. var. urceolata (Kuntze) Yunck. = Cuscuta approximata Bab. ■
114964 Cuscuta arabica Fresen. = Cuscuta pedicellata Ledeb. ■☆
114965 Cuscuta araratica Butkov;亚拉腊菟丝子■☆
114966 Cuscuta arvensis Beyr. ex Engelm. = Cuscuta campestris Yunck. ■
114967 Cuscuta arvensis Beyr. ex Engelm. var. calycina Engl. = Cuscuta campestris Yunck. ■
114968 Cuscuta arvensis Beyr. ex Hook. = Cuscuta pentagona Engelm. ■☆
114969 Cuscuta astyla Engelm. = Cuscuta japonica Choisy ■
114970 Cuscuta astyla Engelm. = Cuscuta monogyna Vahl ■
114971 Cuscuta atlantica Trab. = Cuscuta approximata Bab. ■
114972 Cuscuta australis R. Br.;南方菟丝子(大豆菟丝子,飞扬藤,金线藤,南菟丝子,女萝,欧洲菟丝子,菟丝子);European

Dodder, Southern Dodder ■

114973 Cuscuta babylonica Aucher ex Choisy;巴比伦菟丝子■☆

114974 Cuscuta balansae Boiss. et Reut. = Cuscuta planiflora Ten. ■☆

114975 Cuscuta balansae Boiss. et Reut. var. mossamedensis (Hiern) Yunck. = Cuscuta planiflora Ten. ■☆

114976 Cuscuta bifurcata Yunck.;双叉菟丝子■☆

114977 Cuscuta blepharolepis Welw. ex Hiern;毛鳞菟丝子■☆

114978 Cuscuta boldinghii Urb.;鲍氏菟丝子;Boldingh's Dodder ■☆

114979 Cuscuta brevistyla A. Braun = Cuscuta planiflora Ten. ■☆

114980 Cuscuta brevistyla A. Braun ex A. Rich.;短柱菟丝子■☆

114981 Cuscuta brevistyla A. Rich. = Cuscuta brevistyla A. Braun ex A. Rich. ■☆

114982 Cuscuta bucharica Palib.;布哈尔菟丝子■☆

114983 Cuscuta californica Choisy;加州菟丝子;California Dodder ■☆

114984 Cuscuta callinema Butkov;美茎菟丝子■☆

114985 Cuscuta callosa Pomel = Cuscuta microcephala Pomel ■☆

114986 Cuscuta campestris Yunck.;原野菟丝子(平原菟丝子,田间菟丝子);Field Dodder, Yellow Dodder ■

114987 Cuscuta campestris Yunck. = Cuscuta pentagona Engelm. ■☆

114988 Cuscuta capensis Choisy = Cuscuta africana Willd. ■☆

114989 Cuscuta capitata Roxb.;头状菟丝子;Capitate Dodder ■☆

114990 Cuscuta cassytoides Engelm. = Cuscuta cassytoides Nees ex Engelm. ■☆

114991 Cuscuta cassytoides Nees = Cuscuta cassytoides Nees ex Engelm. ■☆

114992 Cuscuta cassytoides Nees ex Engelm.;无根藤菟丝子;African Dodder ■☆

114993 Cuscuta cephalanthi Engelm.;风箱树菟丝子;Buttonbush Dodder ■☆

114994 Cuscuta chinensis Lam.;菟丝子(禅真,缠豆藤,缠龙子,缠丝蔓,赤网,豆寄生,豆马黄,豆须子,复实,狐丝,黄腊须,黄乱丝,黄萝子,黄丝,黄丝草,黄丝藤,黄藤子,黄弯子,黄湾子,黄网子,火焰草,鸡血藤,金黄丝子,金丝草,金丝藤,雷真子,龙须子,萝丝子,麻棱丝,蒙,莫娘藤,女萝,盘死豆,山麻子,丝子,蕬,唐,吐丝子,吐血丝,吐血丝子,兔儿须,兔丘,兔丝子,兔蕬,菟累,菟芦,菟缕,菟丝,王女,无根草,无根藤,无娘藤,无娘藤米米,无叶藤,鸮萝,野狐浆草,野狐丝,中国菟丝子);China Dodder, Chinese Dodder ■

114995 Cuscuta colorans Maxim. = Cuscuta japonica Choisy ■

114996 Cuscuta compacta C. Juss.;紧凑菟丝子;Compact Dodder, Dodder, Love Vine, Love-vine ■☆

114997 Cuscuta convallariiflora Pavlov;环花菟丝子■☆

114998 Cuscuta cordofana (Engelm.) Yunck. = Cuscuta australis R. Br. ■

114999 Cuscuta coryli Engelm.;榛菟丝子;Hazel Dodder ■☆

115000 Cuscuta cucullata Yunck. = Cuscuta gerrardii Baker ■☆

115001 Cuscuta cupulata Engelm. = Cuscuta approximata Bab. ■

115002 Cuscuta curta (Engelm.) Rydb. = Cuscuta megalocarpa Rydb. ■☆

115003 Cuscuta cuspidata Engelm.;尖尾菟丝子;Cusp Dodder ■☆

115004 Cuscuta cuspidata Pomel = Cuscuta scabrella Trab. ■☆

115005 Cuscuta denticulata Mart. ex Engelm.;细齿菟丝子;Toothed Dodder ■☆

115006 Cuscuta engelmanni Korsh.;长柱菟丝子■☆

115007 Cuscuta epilinum Weihe;亚麻菟丝子(寄亚麻无根草);Flax Dodder, Wild Flax ■

115008 Cuscuta epithymum (L.) L.;百里香菟丝子(附生菟丝子);Adder's Cotton, Alfalfa, Beggarweed, Claver Devil, Clover Dodder, Common Dodder, Devil's Guts, Devil's Net, Devil's Thread, Devil's Threads, Devil's-guts, Dodder, Dodder-of-thyme, Dother, Epiphany, Fairy Hair, Flax Dodder, Hairweed, Hairy Bind, Hale-weed, Hellbind, Hellweed, Lady's Lace, Lady's Laces, Lesser Dodder, Love Vine, Maiden's Hair, Mulberry, Podder, Pother, Red Tangle, Scald, Scaldweed, Strangle Tare, Strangleweed, Wicked Tree ■☆

115009 Cuscuta epithymum (L.) L. subsp. jahandiezii Trab. = Cuscuta obtusata (Engelm.) Trab. ■☆

115010 Cuscuta epithymum (L.) L. subsp. macrostemon Trab.;大冠菟丝子■☆

115011 Cuscuta epithymum (L.) L. subsp. obtusata (Engelm.) Trab. = Cuscuta obtusata (Engelm.) Trab. ■☆

115012 Cuscuta epithymum (L.) L. subsp. planiflora (Ten.) Rouy = Cuscuta planiflora Ten. ■☆

115013 Cuscuta epithymum (L.) L. subsp. trifolii (Bab.) P. Fourn. = Cuscuta epithymum (L.) L. ■☆

115014 Cuscuta epithymum (L.) L. var. alba (J. Presl et C. Presl) Trab. = Cuscuta epithymum (L.) L. ■☆

115015 Cuscuta epithymum (L.) L. var. biloba Trab. = Cuscuta epithymum (L.) L. ■☆

115016 Cuscuta epithymum (L.) L. var. brevistyla (A. Braun) Jahand. et Maire = Cuscuta brevistyla A. Rich. ■☆

115017 Cuscuta epithymum (L.) L. var. bullata Batt. = Cuscuta epithymum (L.) L. ■☆

115018 Cuscuta epithymum (L.) L. var. calycina (Webb) Trab. = Cuscuta epithymum (L.) L. ■☆

115019 Cuscuta epithymum (L.) L. var. decipiens Maire = Cuscuta epithymum (L.) L. ■☆

115020 Cuscuta epithymum (L.) L. var. godronii (Des Moul.) Rouy = Cuscuta planiflora Ten. ■☆

115021 Cuscuta epithymum (L.) L. var. macrostemon (Trab.) Maire = Cuscuta planiflora Ten. ■☆

115022 Cuscuta epithymum (L.) L. var. microcephala (Pomel) Trab. = Cuscuta microcephala Pomel ■☆

115023 Cuscuta epithymum (L.) L. var. rubella Engelm. = Cuscuta epithymum (L.) L. ■☆

115024 Cuscuta epithymum (L.) L. var. subobtusata Maire et Sauvage = Cuscuta epithymum (L.) L. ■☆

115025 Cuscuta epithymum (L.) L. var. subulata (Tineo) Trab. = Cuscuta epithymum (L.) L. ■☆

115026 Cuscuta epithymum (L.) L. var. tenorei Engelm. = Cuscuta planiflora Ten. ■☆

115027 Cuscuta epithymum (L.) L. var. trifolii Bab. = Cuscuta epithymum (L.) L. ■☆

115028 Cuscuta epithymum (L.) L. var. withaniae Maire = Cuscuta epithymum (L.) L. ■☆

115029 Cuscuta epithymum (L.) Murray = Cuscuta epithymum (L.) L. ■☆

115030 Cuscuta epithymum Murray = Cuscuta epithymum (L.) L. ■☆

115031 Cuscuta europaea L.;欧洲菟丝子(大菟丝子,金灯藤,苜蓿菟丝子,欧菟丝子);Beggarweed, Big Dodder, Devil's Guts, Europe Dodder, Greater Dodder, Hailweed, Hairweed, Hale-weed, Podder, Pother, Strangle Tare ■

115032 Cuscuta europaea L. var. indica Engelm. = Cuscuta europaea L. ■

115033 Cuscuta ferganensis Butkov;费尔干菟丝子■☆

115034 Cuscuta flava Siev. ex Ledeb. = Cuscuta lupuliformis Krock. ■

115035 Cuscuta formosana Hayata = Cuscuta japonica Choisy var. formosana (Hayata) Yunck. ■

115036 Cuscuta gerrardii Baker;杰勒德菟丝子■☆

115037 Cuscuta gigantea Griff.;巨大菟丝子(高大菟丝子)■☆

115038 Cuscuta glomerata Choisy;球花菟丝子(美洲菟丝子);Rope Dodder ■☆

115039 Cuscuta grandiflora Kunth = Cuscuta reflexa Roxb. ■

115040 Cuscuta gronovii Willd. ex Roem. et Schult.;普通菟丝子;Common Dodder, Gronovius Dodder, Love Dodder, Love Vine, Love-vine, Scald-weed, Swamp Dodder ■☆

115041 Cuscuta gronovii Willd. ex Roem. et Schult. var. curta Engelm. = Cuscuta megalocarpa Rydb. ■☆

115042 Cuscuta gronovii Willd. ex Roem. et Schult. var. latiflora Engelm. = Cuscuta gronovii Willd. ex Roem. et Schult. ■☆

115043 Cuscuta gronovii Willd. ex Roem. et Schult. var. saururi (Engelm.) MacMill. = Cuscuta gronovii Willd. ex Roem. et Schult. ■☆

115044 Cuscuta hyalina Roth;透明菟丝子■☆

115045 Cuscuta hyalina Roth var. nubica Yunck. = Cuscuta hyalina Roth ■☆

115046 Cuscuta hygrophilae H. Pearson = Cuscuta australis R. Br. ■

115047 Cuscuta indecora Choisy;大苜蓿菟丝子;Large Alfalfa Dodder, Love Vine, Love-vine ■☆

115048 Cuscuta indecora Choisy var. neuropetala (Engelm.) Hitchc. = Cuscuta indecora Choisy ■☆

115049 Cuscuta indica (Engelm.) Petr. ex Butkov;印度菟丝子;Indian Dodder ■☆

115050 Cuscuta japonica Choisy;金灯藤(大粒菟丝子,大菟丝子,飞来花,飞来藤,红无根藤,红雾水藤,黄丝藤,金灯笼,金镫藤,金丝草,金丝藤,毛芽藤,日本菟丝子,山老虎,天蓬草,兔蕬,菟丝子,无根草,无根藤,无量藤,无娘藤,无头藤,雾水藤);Japan Dodder, Japanese Dodder ■

115051 Cuscuta japonica Choisy f. viridicaulis (Honda) Sugim.;绿茎金灯藤■☆

115052 Cuscuta japonica Choisy var. fissistyla Engelm.;川西金灯藤;W. Sichuan Dodder ■

115053 Cuscuta japonica Choisy var. formosana (Hayata) Yunck.;台湾菟丝子;Taiwan Dodder ■

115054 Cuscuta japonica Choisy var. paniculata Engelm. = Cuscuta japonica Choisy ■

115055 Cuscuta japonica Choisy var. thyrsoidea Engelm. = Cuscuta japonica Choisy ■

115056 Cuscuta karatavica Pavlov;卡拉塔夫菟丝子■☆

115057 Cuscuta kawakamii Hayata = Cuscuta australis R. Br. ■

115058 Cuscuta keetii Schltr. = Cuscuta africana Willd. ■☆

115059 Cuscuta kilimanjari Oliv.;基利曼菟丝子■☆

115060 Cuscuta kilimanjari Oliv. var. major Verdc.;大基利曼菟丝子■☆

115061 Cuscuta kotschyana Boiss.;考奇菟丝子■☆

115062 Cuscuta kotschyana Boiss. = Cuscuta europaea L. ■

115063 Cuscuta lehmanniana Bunge;莱曼菟丝子■☆

115064 Cuscuta letourneuxii Trab. = Cuscuta planiflora Ten. ■☆

115065 Cuscuta lophosepala Butkov;冠萼菟丝子■☆

115066 Cuscuta lupuliformis Krock.;啤酒花菟丝子;Hop-shaped Dodder ■

115067 Cuscuta macrantha G. Don = Cuscuta reflexa Roxb. ■

115068 Cuscuta macrolepis R. C. Fang et S. H. Huang;大鳞菟丝子;Bigdcale Dodder ■

115069 Cuscuta madagascarensis Yunck. = Cuscuta planiflora Ten. var. madagascarensis (Yunck.) Verdc. ■☆

115070 Cuscuta major Gilib.;大菟丝子;Teddy-bear Plant ■☆

115071 Cuscuta major J. Bauhin = Cuscuta europaea L. ■

115072 Cuscuta maritima Makino;海滨菟丝子■☆

115073 Cuscuta maritima Makino = Cuscuta chinensis Lam. ■

115074 Cuscuta maroccana Trab.;摩洛哥菟丝子■☆

115075 Cuscuta mearnsii Yunck. = Cuscuta planiflora Ten. var. madagascarensis (Yunck.) Verdc. ■☆

115076 Cuscuta medicaginis C. H. Wright = Cuscuta suaveolens Ser. ■☆

115077 Cuscuta megalocarpa Rydb.;大果菟丝子;Big-fruit Dodder ■☆

115078 Cuscuta microcephala Pomel;小头菟丝子■☆

115079 Cuscuta millettii Hook. et Arn. = Cuscuta australis R. Br. ■

115080 Cuscuta monogyna Vahl;单柱菟丝子(榆树菟丝子);Unistyle Dodder ■

115081 Cuscuta monogyna Vahl = Cuscuta lupuliformis Krock. ■

115082 Cuscuta monogyna Vahl var. blancheana (Desm.) Maire = Cuscuta monogyna Vahl ■

115083 Cuscuta monogyna Vahl var. tenuicaulis Maire et Sennen = Cuscuta monogyna Vahl ■

115084 Cuscuta monogyna Vahl var. vahlianaMaire = Cuscuta monogyna Vahl ■

115085 Cuscuta natalensis Baker;纳塔尔菟丝子■☆

115086 Cuscuta ndorensis Schweinf. = Cuscuta kilimanjari Oliv. ■☆

115087 Cuscuta nitida Choisy;光亮菟丝子■☆

115088 Cuscuta nivea M. A. Garcia = Cuscuta scabrella Trab. ■☆

115089 Cuscuta notochlaena A. Chev. = Cuscuta planiflora Ten. ■☆

115090 Cuscuta obtusata (Engelm.) Trab.;钝菟丝子■☆

115091 Cuscuta obtusiflora Kunth var. australis (R. Br.) Engelm. = Cuscuta australis R. Br. ■

115092 Cuscuta obtusiflora Kunth var. australis Engelm. = Cuscuta australis R. Br. ■

115093 Cuscuta obtusiflora Kunth var. cordofana Engelm. = Cuscuta australis R. Br. ■

115094 Cuscuta ormosana Hayata = Cuscuta japonica Choisy var. formosana (Hayata) Yunck. ■

115095 Cuscuta palaestina Boiss.;燥地菟丝子■☆

115096 Cuscuta pamirica Butkov;帕米尔菟丝子■☆

115097 Cuscuta paradoxa Raf.;奇异菟丝子;Love-vine ■☆

115098 Cuscuta pedicellata Ledeb.;具梗菟丝子■☆

115099 Cuscuta pentagona Engelm.;五角菟丝子;Field Dodder, Five-angled Dodder, Knotweed Dodder, Pentagon Dodder, Smartweed Dodder ■☆

115100 Cuscuta pentagona Engelm. var. calycina Engelm. = Cuscuta campestris Yunck. ■

115101 Cuscuta pentagona Engelm. var. calycina Engelm. = Cuscuta pentagona Engelm. ■☆

115102 Cuscuta pentagona Engelm. var. calycina Engelm. = Cuscuta polygonorum Engelm. ■☆

115103 Cuscuta pentagona Engelm. var. subulata Yunck. = Cuscuta campestris Yunck. ■

115104 Cuscuta planiflora Ten.;由花菟丝子■☆

115105 Cuscuta planiflora Ten. subsp. approximata (Bab.) H. Lindb. = Cuscuta approximata Bab. ■

115106 Cuscuta planiflora Ten. subsp. macrostemon Trab. = Cuscuta planiflora Ten. ■☆

115107 Cuscuta planiflora Ten. var. abyssinica (A. Rich.) Verdc. = Cuscuta abyssinica A. Rich. ■☆

115108 Cuscuta planiflora Ten. var. algeriana Yunck. = Cuscuta planiflora Ten. ■☆
115109 Cuscuta planiflora Ten. var. approximata (Bab.) Engelm. = Cuscuta approximata Bab. ■
115110 Cuscuta planiflora Ten. var. callosa (Pomel) Batt. = Cuscuta planiflora Ten. ■☆
115111 Cuscuta planiflora Ten. var. deserti Trab. = Cuscuta planiflora Ten. ■☆
115112 Cuscuta planiflora Ten. var. holstii Baker et Rendle;霍尔菟丝子■☆
115113 Cuscuta planiflora Ten. var. madagascarensis (Yunck.) Verdc.;马岛尔菟丝子■☆
115114 Cuscuta planiflora Ten. var. mossamedensis Hiern = Cuscuta planiflora Ten. ■☆
115115 Cuscuta planiflora Ten. var. sicula (Tineo) Trab. = Cuscuta planiflora Ten. ■☆
115116 Cuscuta polygonorum Engelm.;荨麻菟丝子;Knotweed Dodder, Smartweed Dodder ■☆
115117 Cuscuta pretoriana Yunck. = Cuscuta planiflora Ten. ■☆
115118 Cuscuta pulchella Engelm.;美丽菟丝子■☆
115119 Cuscuta pulchella Engelm. var. afghana ? = Cuscuta pulchella Engelm. ■☆
115120 Cuscuta pulchella Engelm. var. altaica ? = Cuscuta capitata Roxb. ■☆
115121 Cuscuta pulverulenta Sennen et Mauricio = Cuscuta scabrella Trab. ■☆
115122 Cuscuta racemosa Mart. = Cuscuta suaveolens Ser. ■☆
115123 Cuscuta reflexa Roxb.;大花菟丝子(红无娘藤,黄藤草,金丝藤,卷瓣菟丝子,蛇系腰,无根花,无娘藤,云南菟丝子,展瓣菟丝子);Giant Dodder, Reflexed Dodder, Yunnan Dodder ■
115124 Cuscuta reflexa Roxb. var. anguina (Edgew.) C. B. Clarke;短柱头菟丝子;Shortstigma Dodder ■
115125 Cuscuta reflexa Roxb. var. anguina (Edgew.) C. B. Clarke = Cuscuta reflexa Roxb. var. brachystigma Engelm. ■
115126 Cuscuta reflexa Roxb. var. anguina (Edgew.) Yunck. = Cuscuta reflexa Roxb. var. anguina (Edgew.) C. B. Clarke ■
115127 Cuscuta reflexa Roxb. var. brachystigma Engelm. = Cuscuta reflexa Roxb. var. anguina (Edgew.) C. B. Clarke ■
115128 Cuscuta reflexa Roxb. var. densiflora Benth. = Cuscuta japonica Choisy ■
115129 Cuscuta reflexa Roxb. var. grandiflora Engelm. = Cuscuta reflexa Roxb. ■
115130 Cuscuta rhodesiana Yunck. = Cuscuta planiflora Ten. var. approximata (Bab.) Engelm. ■
115131 Cuscuta rostrata Shuttlew. ex Engelm. et A. Gray;具喙菟丝子;Beaked Dodder ■☆
115132 Cuscuta sandwichiana Choisy;桑威奇菟丝子■☆
115133 Cuscuta santapaui Banerji et Sitesh Das = Cuscuta reflexa Roxb. var. brachystigma Engelm. ■
115134 Cuscuta scabrella Trab.;略粗糙菟丝子■☆
115135 Cuscuta scandens Brot.;攀缘菟丝子■☆
115136 Cuscuta schlechteri Yunck.;施莱菟丝子■☆
115137 Cuscuta somaliensis Yunck.;索马里菟丝子■☆
115138 Cuscuta stapfiana Palib.;斯氏菟丝子■☆
115139 Cuscuta stenantha Trab. = Cuscuta microcephala Pomel ■☆
115140 Cuscuta stenocalycina Palib.;狭萼菟丝子■☆
115141 Cuscuta suaveolens Ser.;流苏菟丝子;Chile Dodder, Fringed Dodder ■☆
115142 Cuscuta tasmanica Engelm.;木麻黄菟丝子;Casuarina Cuscuta ■☆
115143 Cuscuta tianschanica Palib.;天山菟丝子;Tianshan Cuscuta ■
115144 Cuscuta tianschanica Palib. = Cuscuta monogyna Vahl ■
115145 Cuscuta timorensis Engelm. = Cuscuta cassytoides Engelm. ■☆
115146 Cuscuta trifolii Bab.;三叶菟丝子;Clover Dodder ■☆
115147 Cuscuta trifolii Bab. = Cuscuta epithymum Murray ■☆
115148 Cuscuta triumvirati Lange subsp. jahandiezii (Maire) Dobignard;贾汉菟丝子■☆
115149 Cuscuta triumvirati Lange var. jahandiezii Maire = Cuscuta triumvirati Lange subsp. jahandiezii (Maire) Dobignard ■☆
115150 Cuscuta upcraftii Pearson = Cuscuta japonica Choisy var. fissistyla Engelm. ■
115151 Cuscuta urceolata Kuntze = Cuscuta approximata Bab. ■
115152 Cuscuta verrucosa Engelm. = Cuscuta reflexa Roxb. ■
115153 Cuscuta violacea Rajput et Syeda;堇色菟丝子■☆
115154 Cuscutaceae Bercht. et J. Presl = Convolvulaceae Juss.(保留科名)●■
115155 Cuscutaceae Dumort.(1829)(保留科名);菟丝子科;Cuscuta Family, Dodder Family ■
115156 Cuscutaceae Dumort.(保留科名) = Convolvulaceae Juss.(保留科名)●■
115157 Cuscutina Pfeiff. = Cuscuta L. ■
115158 Cusickia M. E. Jones = Lomatium Raf. ■☆
115159 Cusickiella Rollins(1988);库西葶苈属■☆
115160 Cusickiella douglasii (A. Gray) Rollins;道格拉斯库西葶苈■☆
115161 Cusickiella quadricostata (Rollins) Rollins;库西葶苈■☆
115162 Cuspa Humb. = Rinorea Aubl.(保留属名)●
115163 Cusparia D. Dietr. = Bauhinia L. ●
115164 Cusparia D. Dietr. = Casparia Kunth ●
115165 Cusparia Humb. = Angostura Roem. et Schult. ●☆
115166 Cusparia Humb. ex DC. = Angostura Roem. et Schult. ●☆
115167 Cusparia Humb. ex R. Br.(1807);库柏属(西花椒属)●☆
115168 Cusparia Humb. ex R. Br. = Angostura Roem. et Schult. ●☆
115169 Cusparia trifoliata Engl.;三叶库柏(安古斯塔树)●☆
115170 Cuspariaceae J. Agardh = Rutaceae Juss.(保留科名)●■
115171 Cuspariaceae Tratt. = Rutaceae Juss.(保留科名)●■
115172 Cuspidaria (DC.) Besser(废弃属名) = Acachmena H. P. Fuchs(废弃属名)●■
115173 Cuspidaria (DC.) Besser(废弃属名) = Cuspidaria DC.(保留属名)●☆
115174 Cuspidaria (DC.) Besser(废弃属名) = Erysimum L. ■●
115175 Cuspidaria DC.(1838)(保留属名);尖紫葳属●☆
115176 Cuspidaria Link = Erysimum L. ■●
115177 Cuspidaria angustidens DC.;尖紫葳●☆
115178 Cuspidaria cordata A. Mattos;心形尖紫葳●☆
115179 Cuspidia Gaertn.(1791);尖头联苞菊属(杯头联苞菊属)■☆
115180 Cuspidia cernua (L. f.) B. L. Burtt;杯头联苞菊■☆
115181 Cuspidia cernua (L. f.) B. L. Burtt subsp. annua (Less.) Rössler;一年生杯头联苞菊■☆
115182 Cuspidocarpus Sperm. = Micromeria Benth.(保留属名)■●
115183 Cussambium Buch.-Ham. = Schleichera Willd.(保留属名)●☆
115184 Cussambium Lam.(废弃属名) = Schleichera Willd.(保留属名)●☆
115185 Cussapoa Post et Kuntze = Coussapoa Aubl. ●☆
115186 Cussarea Post et Kuntze = Coussarea Aubl. ●☆

115187 Cussetia M. Kato = Terniola Tul. ■
115188 Cussetia M. Kato(2006);印度支那川苔草属■☆
115189 Cusso Bruce = Brayera Kunth ■●☆
115190 Cussonia Thunb. (1780);甘蓝树属(黑五加属,库松木属);Cabbage Tree,Umbrella Tree ●☆
115191 Cussonia Thunb. = Schefflera J. R. Forst. et G. Forst. (保留属名)●
115192 Cussonia angolensis (Seem.) Hiern;安哥拉甘蓝树●☆
115193 Cussonia arborea Hochst. ex A. Rich.;北方甘蓝树●☆
115194 Cussonia arenicola Strey;沙生甘蓝树●☆
115195 Cussonia bancoensis Aubrév. et Pellegr.;邦克甘蓝树●☆
115196 Cussonia barteri Seem. = Cussonia arborea Hochst. ex A. Rich. ●☆
115197 Cussonia bequaerti De Wild. = Cussonia holstii Harms ex Engl. ●☆
115198 Cussonia bojeri Seem. = Schefflera bojeri (Seem.) R. Vig. ●☆
115199 Cussonia brieyi De Wild.;布里甘蓝树●☆
115200 Cussonia buchananii Harms = Schefflera umbellifera (Sond.) Baill. ●☆
115201 Cussonia calophylla Miq. = Cussonia spicata Thunb. ●☆
115202 Cussonia capuroniana Bernardi = Schefflera capuroniana (Bernardi) Bernardi ●☆
115203 Cussonia capuroniana Bernardi var. bracteolata Bernardi = Schefflera bracteolifera Frodin ●☆
115204 Cussonia corbisieri De Wild.;科比西尔甘蓝树●☆
115205 Cussonia delevoyi De Wild. = Cussonia arborea Hochst. ex A. Rich. ●☆
115206 Cussonia djalonensis A. Chev. = Cussonia arborea Hochst. ex A. Rich. ●☆
115207 Cussonia fosbergiana Bernardi = Schefflera fosbergiana (Bernardi) Bernardi ●☆
115208 Cussonia fraxinifolia Baker = Polyscias fraxinifolia Harms ●☆
115209 Cussonia gamtoosensis Strey;加姆图斯甘蓝树●☆
115210 Cussonia gerrardii Seem. = Seemannaralia gerrardii (Seem.) Harms ●☆
115211 Cussonia hamata Harms = Cussonia arborea Hochst. ex A. Rich. ●☆
115212 Cussonia herteri ?;赫氏甘蓝树(赫脱库松木)●☆
115213 Cussonia holstii Harms ex Engl.;小穗甘蓝树●☆
115214 Cussonia holstii Harms ex Engl. var. tomentosa Tennant = Cussonia holstii Harms ex Engl. ●☆
115215 Cussonia homblei De Wild. = Cussonia arborea Hochst. ex A. Rich. ●☆
115216 Cussonia jatrophoides Hutch. et E. A. Bruce;麻疯甘蓝树●☆
115217 Cussonia kirkii Seem.;甘蓝树;Cabbage Tree,Umbrella Tree ●☆
115218 Cussonia kirkii Seem. = Cussonia arborea Hochst. ex A. Rich. ●☆
115219 Cussonia kraussii Hochst. = Cussonia spicata Thunb. ●☆
115220 Cussonia laciniata Harms = Cussonia arborea Hochst. ex A. Rich. ●☆
115221 Cussonia lanceolata Harms;剑叶甘蓝树●☆
115222 Cussonia longipedicellata Lecomte = Schefflera longipedicellata (Lecomte) Bernardi ●☆
115223 Cussonia longissima Hutch. et Dalziel = Cussonia arborea Hochst. ex A. Rich. ●☆
115224 Cussonia lukwangulensis Tennant = Schefflera lukwangulensis (Tennant) Bernardi ●☆
115225 Cussonia microstachys Harms = Cussonia holstii Harms ex Engl. ●☆
115226 Cussonia monophylla Baker = Schefflera monophylla (Baker) Bernardi ●☆
115227 Cussonia myriantha Baker = Schefflera myriantha (Baker) Drake ●☆
115228 Cussonia natalensis Sond.;纳塔尔甘蓝树●☆
115229 Cussonia nicholsonii Strey;尼克尔森甘蓝树●☆
115230 Cussonia nigerica Hutch. = Cussonia arborea Hochst. ex A. Rich. ●☆
115231 Cussonia paniculata E. Mey. = Cussonia paniculata Eckl. et Zeyh. ●☆
115232 Cussonia paniculata Eckl. et Zeyh.;高原甘蓝树;Highveld Cabbage Tree,Little Cabbage Tree ●☆
115233 Cussonia paniculata Eckl. et Zeyh. subsp. sinuata (Reyneke et Kok) De Winter;深波甘蓝树●☆
115234 Cussonia paniculata Eckl. et Zeyh. var. sinuata Reyneke et Kok = Cussonia paniculata Eckl. et Zeyh. subsp. sinuata (Reyneke et Kok) De Winter ●☆
115235 Cussonia quarrei De Wild. = Cussonia corbisieri De Wild. ●☆
115236 Cussonia racemosa Baker = Schefflera macerosa Bernardi ●☆
115237 Cussonia sessilis Lebrun;无柄甘蓝树●☆
115238 Cussonia sphaerocephala Strey;球头甘蓝树●☆
115239 Cussonia spicata Thunb.;穗花甘蓝树(粗根树,甘蓝树,穗花黑五加);Common Cabbage Tree ●☆
115240 Cussonia thyrsiflora Thunb.;聚伞甘蓝树●☆
115241 Cussonia tisserantii Aubrév. et Pellegr. = Cussonia arborea Hochst. ex A. Rich. ●☆
115242 Cussonia transvaalensis Reyneke;德兰士瓦甘蓝树●☆
115243 Cussonia triptera Colla = Cussonia spicata Thunb. ●☆
115244 Cussonia umbellifera Sond. = Schefflera umbellifera (Sond.) Baill. ●☆
115245 Cussonia umbellifera Sond. var. buchananii (Harms) Tennant = Schefflera umbellifera (Sond.) Baill. ●☆
115246 Cussonia zimmermannii Harms;齐默尔曼甘蓝树●☆
115247 Cussonia zuluensis Strey;祖卢甘蓝树●☆
115248 Cussutha Benth. et Hook. f. = Cassutha Des Moul. ■
115249 Cussutha Benth. et Hook. f. = Cuscuta L. ■
115250 Custenia Steud. = Cussutha Benth. et Hook. f. ■
115251 Custinia Neck. = Salacia L. (保留属名)●
115252 Cutandia Wilk. (1860);海滨草属;Cutandia ■☆
115253 Cutandia dichotoma (Forssk.) Trab.;二歧海滨草■☆
115254 Cutandia divaricata (Desf.) Benth.;叉开海滨草■☆
115255 Cutandia divaricata (Desf.) Benth. var. laxiflora Cavara et Trotter = Cutandia divaricata (Desf.) Benth. ■☆
115256 Cutandia incrassata (Lam.) Benth. = Vulpiella tenuis (Tineo) Kerguélen ■☆
115257 Cutandia maritima (L.) Barbey = Cutandia maritima (L.) Benth. ex Barley ■☆
115258 Cutandia maritima (L.) Benth. ex Barbey;海滨草;Sea-coast Cutandia ■☆
115259 Cutandia memphitica (Spreng.) K. Richt.;孟斐斯海滨草;Memphisgrass ■☆
115260 Cutandia memphitica (Spreng.) K. Richt. var. dichotoma (Forssk.) Batt. et Trab. = Cutandia dichotoma (Forssk.) Trab. ■☆
115261 Cutandia memphitica K. Richt. = Cutandia memphitica (Spreng.) K. Richt. ■☆
115262 Cutarea J. St. -Hil. = Coutarea Aubl. ●■☆
115263 Cutaria Brign. = Coutarea Aubl. ●■☆

115264 Cutaria Brign. = Cutarea J. St. -Hil. ●■☆
115265 Cuthbertia Small = Callisia Loefl. ■☆
115266 Cuthbertia Small = Phyodina Raf. ■☆
115267 Cuthbertia graminea Small = Callisia graminea (Small) G. C. Tucker ■☆
115268 Cuthbertia ornata Small = Callisia ornata (Small) G. C. Tucker ■☆
115269 Cuthbertia rosea (Vent.) Small = Callisia rosea (Vent.) D. R. Hunt ■☆
115270 Cutlera Raf. = Gentiana L. ■
115271 Cutsis Burns-Bal., E. W. Greenw. et Gonzales = Dichromanthus Garay ■☆
115272 Cuttera Raf. = Cutlera Raf. ■
115273 Cuttsia F. Muell. (1865);卡茨鼠刺属●☆
115274 Cuttsia viburnea F. Muell.;卡茨鼠刺●☆
115275 Cutubaea Post et Kuntze = Cutubea J. St. -Hil. ■☆
115276 Cutubea J. St. -Hil. = Coutoubea Aubl. ■☆
115277 Cuveraca Jones = Cedrela P. Browne ●
115278 Cuviera DC. (1807)(保留属名);居维叶茜草属;Cuviera ■☆
115279 Cuviera Koeler(废弃属名) = Cuviera DC. (保留属名)■☆
115280 Cuviera Koeler(废弃属名) = Hordelymus (Jess.) Jess. ex Harz ■☆
115281 Cuviera acutiflora DC.;尖居维叶茜草■☆
115282 Cuviera africana Spreng. = Cuviera acutiflora DC. ■☆
115283 Cuviera angolensis Welw. ex K. Schum. = Cuviera longiflora Hiern ■☆
115284 Cuviera australis K. Schum. = Lagynias lasiantha (Sond.) Bullock ■☆
115285 Cuviera bolo Aubrév. et Pellegr. = Robynsia glabrata Hutch. ■☆
115286 Cuviera calycosa Wernham;多萼居维叶茜草■☆
115287 Cuviera cienkowskii (Schweinf.) Roberty = Fadogia cienkowskii Schweinf. ●☆
115288 Cuviera djalonensis A. Chev. = Cuviera macroura K. Schum. ■☆
115289 Cuviera latior Wernham;宽居维叶茜草■☆
115290 Cuviera latior Wernham var. hispidula N. Hallé;细毛居维叶茜草■☆
115291 Cuviera ledermannii K. Krause;莱德居维叶茜草■☆
115292 Cuviera leniochlamys K. Schum.;被居维叶茜草■☆
115293 Cuviera letestui Pellegr.;莱泰斯图居维叶茜草■☆
115294 Cuviera longiflora Hiern;长居维叶茜草■☆
115295 Cuviera macroura K. Schum.;大尾居维叶茜草■☆
115296 Cuviera migeodii Verdc.;米容德居维叶茜草■☆
115297 Cuviera minor (Wernham) Verdc. = Globulostylis minor Wernham ■☆
115298 Cuviera minor C. H. Wright = Cuviera nigrescens (Scott-Elliot ex Oliv.) Wernham ■☆
115299 Cuviera nigrescens (Scott-Elliot ex Oliv.) Wernham;变黑居维叶茜草■☆
115300 Cuviera physinodes K. Schum.;肿节居维叶茜草■☆
115301 Cuviera pierrei N. Hallé;皮埃尔居维叶茜草■☆
115302 Cuviera plagiophylla K. Schum. = Cuviera subuliflora Benth. ■☆
115303 Cuviera schliebenii Verdc.;施利本居维叶茜草■☆
115304 Cuviera semsei Verdc.;塞姆居维叶茜草■☆
115305 Cuviera subuliflora Benth.;钻叶居维叶茜草■☆
115306 Cuviera talbotii (Wernham) Verdc. = Globulostylis talbotii Wernham ■☆
115307 Cuviera tomentosa Verdc.;毛居维叶茜草■☆
115308 Cuviera trichostephana K. Schum. = Cuviera nigrescens (Scott-Elliot ex Oliv.) Wernham ■☆
115309 Cuviera truncata Hutch. et Dalziel;截居维叶茜草■☆
115310 Cuviera uncinula N. Hallé;钩居维叶茜草■☆
115311 Cuviera wernhamii Cheek = Globulostylis minor Wernham ■☆
115312 Cwangayana Rauschert = Acanthophora Merr. ●■
115313 Cwangayana Rauschert = Aralia L. ●■
115314 Cwangayana Rauschert = Neoacanthophora Bennet ●■
115315 Cyamopsis DC. (1826);瓜儿豆属(瓜胶豆属);Guar, Cyamopsis ■
115316 Cyamopsis dentata (N. E. Br.) Torre;瓜儿豆■☆
115317 Cyamopsis psoraloides (Lam.) DC. = Cyamopsis tetragonoloba (L.) Taub. ■
115318 Cyamopsis psoraloides DC. = Cyamopsis tetragonoloba (L.) Taub. ■
115319 Cyamopsis senegalensis Guillaumin et Perr.;塞内加尔瓜儿豆■☆
115320 Cyamopsis senegalensis Guillaumin et Perr. var. stenophylla Bonnet = Cyamopsis senegalensis Guillaumin et Perr. ■☆
115321 Cyamopsis serrata Schinz;齿叶瓜儿豆■☆
115322 Cyamopsis stenophylla (Bonnet) Chev. = Cyamopsis senegalensis Guillaumin et Perr. ■☆
115323 Cyamopsis tetragonoloba (L.) Taub.;四棱瓜儿豆(瓜儿豆);Cluster Bean Guvar, Guar, Gwar ■
115324 Cyamus Sm. = Nelumbo Adans. ■
115325 Cyanaeorchis Barb. Rodr. (1877);巴西青兰属■☆
115326 Cyanaeorchis arundinae Barb. Rodr.;巴西青兰■☆
115327 Cyanaeorchis minor Schltr.;小巴西青兰■☆
115328 Cyanandrium Stapf(1895);蓝蕊野牡丹属☆
115329 Cyanandrium glabrum M. P. Nayar;无毛蓝蕊野牡丹☆
115330 Cyanandrium guttatum Stapf;蓝蕊野牡丹☆
115331 Cyananthaceae J. Agardh = Campanulaceae Juss. (保留科名)■●
115332 Cyananthus Griff. = Stauranthera Benth. ■
115333 Cyananthus Miers = Burmannia L. ■
115334 Cyananthus Raf. (废弃属名) = Centaurea L. (保留属名)●■
115335 Cyananthus Raf. (废弃属名) = Cyananthus Miers ■
115336 Cyananthus Raf. (废弃属名) = Cyananthus Wall. ex Benth. (保留属名)■
115337 Cyananthus Raf. (废弃属名) = Cyanus P. Mill. (废弃属名)●■
115338 Cyananthus Wall. ex Benth. (1836)(保留属名);蓝钟花属;Bluebell, Bluebellflower, Cyananthus, Trailing Bell-flower ■
115339 Cyananthus albiflorus D. F. Chamb. = Cyananthus flavus C. Marquand subsp. montanus (C. Y. Wu) D. Y. Hong et L. M. Ma ■
115340 Cyananthus argenteus C. Marquand;总花蓝钟花(补草根,马鬃参,小白锦,银叶蓝钟花);Argent Bluebellflower, Raceme Bluebell ■
115341 Cyananthus argenteus C. Marquand = Cyananthus longiflorus Franch. ■
115342 Cyananthus barbatus Franch. = Cyananthus delavayi Franch. ■
115343 Cyananthus chungdianensis C. Y. Wu;中甸蓝钟花;Chungdian Bluebellflower, Zhongdian Bluebell, Zhongdian Bluebellflower ■
115344 Cyananthus chungdianensis C. Y. Wu = Cyananthus formosus Diels ■
115345 Cyananthus cordifolius Duthie;心叶蓝钟花;Cordateleaf Bluebellflower, Heartleaf Bluebell ■
115346 Cyananthus cronquistii Shrestha = Cyananthus hookeri C. B. Clarke ■
115347 Cyananthus delavayi Franch.;细叶蓝钟花(小菱叶蓝钟花);Delavay Bluebell, Delavay Bluebellflower ■

115348 Cyananthus dolichosceles C. Marquand; 川西蓝钟花; W. Sichuan Bluebell, W. Sichuan Bluebellflower, West Szechuan Bluebellflower ■

115349 Cyananthus fasciculatus C. Marquand; 束花蓝钟花; Fasciculate Bluebell, Fasciculate Bluebellflower ■

115350 Cyananthus flavus C. Marquand; 黄钟花(丽江黄钟花); Yellow Bluebell, Yellow Bluebellflower ■

115351 Cyananthus flavus C. Marquand subsp. montanus (C. Y. Wu) D. Y. Hong et L. M. Ma; 白钟花(山地蓝钟花); Montanous Bluebell, Montanous Bluebellflower ■

115352 Cyananthus flavus C. Marquand var. glaber C. Y. Wu; 光叶黄钟花; Glabrous Yellow Bluebell, Glabrousleaf Bluebellflower ■

115353 Cyananthus formosus Diels; 美丽蓝钟花; Beautiful Bluebell, Beautiful Bluebellflower ■

115354 Cyananthus forrestii Diels = Cyananthus inflatus Hook. f. et Thomson ■

115355 Cyananthus hookeri C. B. Clarke; 蓝钟花; Hooker Bluebell, Hooker Bluebellflower ■

115356 Cyananthus hookeri C. B. Clarke var. densus C. B. Clarke; 密枝蓝钟花 ■

115357 Cyananthus hookeri C. B. Clarke var. densus C. Marquand = Cyananthus hookeri C. B. Clarke ■

115358 Cyananthus hookeri C. B. Clarke var. grandiflorus C. Marquand; 长柔毛蓝钟花 ■

115359 Cyananthus hookeri C. B. Clarke var. grandiflorus C. Marquand = Cyananthus hookeri C. B. Clarke ■

115360 Cyananthus hookeri C. B. Clarke var. hispidus Franch. = Cyananthus hookeri C. B. Clarke ■

115361 Cyananthus hookeri C. B. Clarke var. levicalyx Y. S. Lian; 光萼蓝钟花; Smoothcalyx Bluebell, Smoothcalyx Bluebellflower ■

115362 Cyananthus hookeri C. B. Clarke var. levicalyx Y. S. Lian = Cyananthus hookeri C. B. Clarke ■

115363 Cyananthus hookeri C. B. Clarke var. levicaulis Franch.; 光茎蓝钟花; Glabrousstem Bluebell, Glabrousstem Bluebellflower ■

115364 Cyananthus hookeri C. B. Clarke var. levicaulis Franch. = Cyananthus hookeri C. B. Clarke ■

115365 Cyananthus incanus Hook. f. et Thomson; 灰毛蓝钟花(矮小蓝钟花,草补药,蔓茎蓝钟花,小白棉); Decumbent Bluebellflower, Greyhair Bluebell, Small Bluebellflower ■

115366 Cyananthus incanus Hook. f. et Thomson subsp. orientalis K. Shrestha = Cyananthus incanus Hook. f. et Thomson ■

115367 Cyananthus incanus Hook. f. et Thomson subsp. petiolatus (Franch.) D. Y. Hong et L. M. Ma; 毛叶蓝钟花(黄白花蓝钟花); Hairyleaf Bluebell, Hairyleaf Bluebellflower, Hoary Bluebellflower ■

115368 Cyananthus incanus Hook. f. et Thomson var. decumbens Y. S. Lian = Cyananthus incanus Hook. f. et Thomson ■

115369 Cyananthus incanus Hook. f. et Thomson var. leiocalyx Franch. = Cyananthus macrocalyx Franch. ■

115370 Cyananthus incanus Hook. f. et Thomson var. leiocalyx Franch. = Cyananthus leiocalyx (Franch.) Cowan ■

115371 Cyananthus incanus Hook. f. et Thomson var. leiocalyx Franch. = Cyananthus incanus Hook. f. et Thomson ■

115372 Cyananthus incanus Hook. f. et Thomson var. parvus C. Marquand = Cyananthus incanus Hook. f. et Thomson ■

115373 Cyananthus incanus Hook. f. et Thomson var. rufus Franch.; 粗茎蓝钟花 ■

115374 Cyananthus incanus Hook. f. et Thomson var. sylvestris C. Marquand; 长柄蓝钟花 ■

115375 Cyananthus incanus Hook. f. et Thomson var. trichocalyx Franch. = Cyananthus incanus Hook. f. et Thomson ■

115376 Cyananthus inflatus Hook. f. et Thomson; 胀萼蓝钟花; Inflatecalyx Bluebell, Inflated Bluebellflower ■

115377 Cyananthus inflatus Hook. f. et Thomson var. tenuis Franch. = Cyananthus inflatus Hook. f. et Thomson ■

115378 Cyananthus insignis Grahame = Cyananthus lobatus Wall. ex Benth. ■

115379 Cyananthus leiocalyx (Franch.) Cowan; 光蓝钟花 ■

115380 Cyananthus leiocalyx (Franch.) Cowan = Cyananthus macrocalyx Franch. ■

115381 Cyananthus leiocalyx (Franch.) Cowan subsp. lucidus K. Shrestha = Cyananthus macrocalyx Franch. ■

115382 Cyananthus lichiangensis W. W. Sm.; 丽江蓝钟花; Lijiang Bluebell, Lijiang Bluebellflower, Likiang Bluebellflower ■

115383 Cyananthus lobatus Wall. ex Benth.; 裂叶蓝钟花(浅裂蓝钟花); Lobedleaf Bluebell, Lobedleaf Bluebellflower ■

115384 Cyananthus lobatus Wall. ex Benth. var. farreri C. Marquand = Cyananthus lobatus Wall. ex Benth. ■

115385 Cyananthus longiflorus Franch.; 长花蓝钟花; Longflower Bluebell, Longflower Bluebellflower ■

115386 Cyananthus macrocalyx Franch.; 大萼蓝钟花(光萼蓝钟花,脉萼蓝钟花); Bigcalyx Bluebell, Bigcalyx Bluebellflower, Nervedcalyx Bluebellflower, Pilose Bluebellflower, Smoothcalyx Bluebellflower, Yellowpurple Bluebellflower ■

115387 Cyananthus macrocalyx Franch. subsp. spathulifolius (Nannf.) K. Shrestha; 匙叶蓝钟花 ■

115388 Cyananthus macrocalyx Franch. var. flavopurpureus C. Marquand; 黄紫花蓝钟花 ■

115389 Cyananthus macrocalyx Franch. var. flavopurpureus C. Marquand = Cyananthus macrocalyx Franch. ■

115390 Cyananthus macrocalyx Franch. var. pilosus C. Marquand; 毛蓝钟花 ■

115391 Cyananthus macrocalyx Franch. var. pilosus C. Marquand = Cyananthus macrocalyx Franch. ■

115392 Cyananthus mairei (H. Lév.) Cowan = Cyananthus flavus C. Marquand subsp. montanus (C. Y. Wu) D. Y. Hong et L. M. Ma ■

115393 Cyananthus mairei (H. Lév.) Cowan = Cyananthus montanus C. Y. Wu ■

115394 Cyananthus mairei H. Lév. = Codonopsis bulleyana Forrest ex Diels ■

115395 Cyananthus microphyllus Edgew.; 小叶蓝钟花; Smallleaf Bluebell, Smallleaf Bluebellflower ■

115396 Cyananthus microrhombeus C. Y. Wu; 小菱叶蓝钟花; Littlerhomboidleaf Bluebellflower ■

115397 Cyananthus microrhombeus C. Y. Wu = Cyananthus delavayi Franch. ■

115398 Cyananthus microrhombeus C. Y. Wu var. delavayi Franch.; 光萼小菱叶蓝钟花; Glabrouscalyx Littlerhomboidleaf Bluebellflower ■

115399 Cyananthus microrhombeus C. Y. Wu var. leiocalyx C. Y. Wu = Cyananthus delavayi Franch. ■

115400 Cyananthus montanus C. Y. Wu = Cyananthus flavus C. Marquand subsp. montanus (C. Y. Wu) D. Y. Hong et L. M. Ma ■

115401 Cyananthus neglectus C. Marquand = Cyananthus incanus Hook. f. et Thomson subsp. petiolatus (Franch.) D. Y. Hong et L. M. Ma ■

115402 Cyananthus neglectus C. Marquand = Cyananthus petiolatus Franch. ■
115403 Cyananthus nepalensis Kitam. = Cyananthus microphyllus Edgew. ■
115404 Cyananthus nepalensis Kitam. = Cyananthus sherriffii Cowan ■
115405 Cyananthus neurocalyx C. Y. Wu;脉萼蓝钟花■
115406 Cyananthus neurocalyx C. Y. Wu = Cyananthus macrocalyx Franch. ■
115407 Cyananthus obtusilobus C. Marquand = Cyananthus longiflorus Franch. ■
115408 Cyananthus pedunculatus C. B. Clarke; 有梗蓝钟花; Pedunculate Bluebell, Pedunculate Bluebellflower ■
115409 Cyananthus petiolatus Franch. = Cyananthus incanus Hook. f. et Thomson subsp. petiolatus (Franch.) D. Y. Hong et L. M. Ma ■
115410 Cyananthus petiolatus Franch. var. pilifolius (C. Y. Wu) Y. S. Lian = Cyananthus incanus Hook. f. et Thomson subsp. petiolatus (Franch.) D. Y. Hong et L. M. Ma ■
115411 Cyananthus pilifolius C. Y. Wu = Cyananthus incanus Hook. f. et Thomson subsp. petiolatus (Franch.) D. Y. Hong et L. M. Ma ■
115412 Cyananthus pilifolius C. Y. Wu = Cyananthus petiolatus Franch. var. pilifolius (C. Y. Wu) Y. S. Lian ■
115413 Cyananthus pilifolius C. Y. Wu var. minor C. Y. Wu = Cyananthus incanus Hook. f. et Thomson subsp. petiolatus (Franch.) D. Y. Hong et L. M. Ma ■
115414 Cyananthus pilifolius C. Y. Wu var. minor C. Y. Wu = Cyananthus petiolatus Franch. var. pilifolius (C. Y. Wu) Y. S. Lian ■
115415 Cyananthus pilifolius C. Y. Wu var. pallidocoeruleus C. Y. Wu = Cyananthus petiolatus Franch. ■
115416 Cyananthus pilifolius C. Y. Wu var. pallidocoeruleus C. Y. Wu = Cyananthus incanus Hook. f. et Thomson subsp. petiolatus (Franch.) D. Y. Hong et L. M. Ma ■
115417 Cyananthus pilosus (C. Marquand) K. K. Shrestha = Cyananthus macrocalyx Franch. ■
115418 Cyananthus pseudoinflatus P. C. Tsoong; 短毛蓝钟花; Bluebellflower, Shorthaired Bluebell ■
115419 Cyananthus pseudoinflatus P. C. Tsoong = Cyananthus inflatus Hook. f. et Thomson ■
115420 Cyananthus sericeus Y. S. Lian;绢毛蓝钟花;Sericeus Bluebell, Sericeus Bluebellflower ■
115421 Cyananthus sherriffii Cowan; 杂毛蓝钟花; Sherriff Bluebell, Sherriff Bluebellflower ■
115422 Cyananthus spathulifolius Nannf. = Cyananthus macrocalyx Franch. subsp. spathulifolius (Nannf.) K. Shrestha ■
115423 Cyananthus umbrosus Griff. = Stauranthera umbrosa (Griff.) C. B. Clarke ■
115424 Cyananthus wardii C. Marquand;棕毛蓝钟花■
115425 Cyananthus wardii C. Marquand = Cyananthus macrocalyx Franch. var. pilosus C. Marquand ■
115426 Cyanastraceae Engl. (1900)(保留科名);蓝星科■☆
115427 Cyanastraceae Engl. (保留科名) = Tecophilaeaceae Leyb. (保留科名)■☆
115428 Cyanastrum Cass. = Volutarella Cass. ■☆
115429 Cyanastrum Oliv. (1891);蓝星属■☆
115430 Cyanastrum bussei Engl. = Kabuyea hostifolia (Engl.) Brummitt ■☆
115431 Cyanastrum cordifolium Oliv.;心叶蓝星■☆
115432 Cyanastrum goetzeanum Engl.;蓝星■☆
115433 Cyanastrum hockii De Wild. = Cyanastrum johnstonii Baker ■☆
115434 Cyanastrum hostifolium Engl. = Kabuyea hostifolia (Engl.) Brummitt ■☆
115435 Cyanastrum johnstonii Baker;约翰斯顿蓝星■☆
115436 Cyanastrum johnstonii Baker var. cuneifolium S. Carter = Cyanastrum johnstonii Baker ■☆
115437 Cyanastrum verdickii De Wild. = Cyanastrum johnstonii Baker ■☆
115438 Cyanea Gaudich. (1829);蓝桔梗属●☆
115439 Cyanea acuminata Hillebr.;渐尖蓝桔梗●☆
115440 Cyanea angustifolia Hillebr.;窄叶蓝桔梗●☆
115441 Cyanea arborea Hillebr.;乔木蓝桔梗●☆
115442 Cyanea aspera A. Gray;粗糙蓝桔梗●☆
115443 Cyanea bicolor H. St. John;二色蓝桔梗●☆
115444 Cyanea elliptica (Rock) Lammers;椭圆蓝桔梗●☆
115445 Cyanea glabra (E. Wimm.) H. St. John;光蓝桔梗●☆
115446 Cyanea linearifolia Rock;线叶蓝桔梗●☆
115447 Cyanea longiflora (Wawra) Lammers, Givnish et Sytsma;长花蓝桔梗●☆
115448 Cyanea mannii Hillebr.;曼氏蓝桔梗●☆
115449 Cyanea megacarpa (Rock) Rock;大果蓝桔梗●☆
115450 Cyanea multispicata H. Lév.;多穗蓝桔梗●☆
115451 Cyanea obtusa Hillebr.;钝蓝桔梗●☆
115452 Cyanea parvifolia (C. N. Forbes) Lammers, Givnish et Sytsma;小叶蓝桔梗●☆
115453 Cyanea pinnatifida (Cham.) E. Wimm.;羽裂蓝桔梗●☆
115454 Cyanea platyphylla Hillebr.;宽叶蓝桔梗●☆
115455 Cyanea pulchra Rock;美丽蓝桔梗●☆
115456 Cyanea salicina H. Lév.;柳叶蓝桔梗●☆
115457 Cyanea sessilifolia (O. Deg.) Lammers;无柄蓝桔梗●☆
115458 Cyanea sylvestris A. Heller;林地蓝桔梗●☆
115459 Cyanella L. (1754);蓝蒂可花属■☆
115460 Cyanella Royen ex L. = Cyanella L. ■☆
115461 Cyanella alba L. f.;白蓝蒂可花■☆
115462 Cyanella alba L. f. subsp. flavescens J. C. Manning;黄白蓝蒂可花■☆
115463 Cyanella aquatica Oberm. ex G. A. M. Scott;水蓝蒂可花■☆
115464 Cyanella capensis L. = Cyanella hyacinthoides L. ■☆
115465 Cyanella cygnea G. Scott;普通蓝蒂可花■☆
115466 Cyanella hyacinthoides L.;好望角蓝蒂可花■☆
115467 Cyanella krauseana Dinter et G. M. Schulze = Cyanella ramosissima (Engl. et K. Krause) Engl. et K. Krause ■☆
115468 Cyanella lineata Burch. = Cyanella lutea L. f. ■☆
115469 Cyanella lutea L. f.;黄蓝蒂可花■☆
115470 Cyanella lutea L. f. var. angustifolia Schinz = Cyanella lutea L. f. ■☆
115471 Cyanella lutea L. f. var. rosea Baker = Cyanella lutea L. f. ■☆
115472 Cyanella orchidiformis Jacq.;兰状蓝蒂可花■☆
115473 Cyanella pentheri Zahlbr. = Cyanella hyacinthoides L. ■☆
115474 Cyanella racemosa Schinz = Cyanella lutea L. f. ■☆
115475 Cyanella ramosissima (Engl. et K. Krause) Engl. et K. Krause;多枝蓝蒂可花■☆
115476 Cyanellaceae Salisb. = Tecophilaeaceae Leyb. (保留科名)■☆
115477 Cyanitis Reinw. = Dichroa Lour. ●
115478 Cyanitis chinensis Hook. f. et Thomson = Dichroa febrifuga Lour. ●
115479 Cyanitis sylvatica Reinw. = Dichroa febrifuga Lour. ●
115480 Cyanitis versicolor Hook. f. et Thomson = Dichroa febrifuga

Lour. ●

115481 Cyanixia Goldblatt et J. C. Manning = Babiana Ker Gawl. ex Sims(保留属名)■☆

115482 Cyanobotrys Zucc. = Muellera L. f.(保留属名)●■☆

115483 Cyanocarpus F. M. Bailey = Helicia Lour. ●

115484 Cyanococcus (A. Gray) Rydb. = Vaccinium L. ●

115485 Cyanococcus canadensis (Kalm ex A. Rich.) Rydb. = Vaccinium myrtilloides Michx. ●☆

115486 Cyanococcus corymbosus (L.) Rydb. = Vaccinium corymbosum L. ●☆

115487 Cyanococcus cuthbertii Small = Vaccinium corymbosum L. ●☆

115488 Cyanocoecus Rydb. = Vaccinium L. ●

115489 Cyanodaphne Blume = Dehaasia Blume ●

115490 Cyanoneuron C. Tange(1998);蓝脉茜属●☆

115491 Cyanophyllum Naudin = Miconia Ruiz et Pav.(保留属名)●☆

115492 Cyanopis Blume = Cyanthillium Blume ■

115493 Cyanopis Blume = Vernonia Schreb.(保留属名)●■

115494 Cyanopis Steud. = Cyanopsis Cass. ■☆

115495 Cyanopis Steud. = Volutarella Cass. ■☆

115496 Cyanopngon Welw. ex C. B. Clarke = Cyanotis D. Don(保留属名)■

115497 Cyanopsis Cass. = Volutaria Cass. ■☆

115498 Cyanopsis Endl. = Cyanopis Blume ●■

115499 Cyanopsis Endl. = Vernonia Schreb.(保留属名)●■

115500 Cyanopsis muricata (L.) Dostál = Volutaria muricata (L.) Maire ■☆

115501 Cyanorchis Thouars = Cyanorkis Thouars ■

115502 Cyanorchis Thouars = Phaius Lour. ■

115503 Cyanorchis Thouars ex Steud. = Phaius Lour. ■

115504 Cyanorkis Thouars = Cyanorchis Thouars ■

115505 Cyanoseris (Koch) Schur = Lactuca L. ■

115506 Cyanoseris Schur = Lactuca L. ■

115507 Cyanospermum Wight et Arn. = Rhynchosia Lour.(保留属名)●■

115508 Cyanostegia Turcz.(1849);蓝被草属●☆

115509 Cyanostegia angustifolia Turcz.;窄叶蓝被草●☆

115510 Cyanostegia cyanocalyx (F. Muell.) C. A. Gardner;蓝萼蓝被草●☆

115511 Cyanostegia intermedia Turcz.;间型蓝被草●☆

115512 Cyanostegia lanceolata Turcz.;披针叶蓝被草●☆

115513 Cyanostegia microphylla S. Moore;小叶蓝被草●☆

115514 Cyanostremma Benth. ex Hook. et Arn. = Calopogonium Desv. ●

115515 Cyanothamnus Lindl. = Boronia Sm. ●☆

115516 Cyanothyrsus Harms = Daniellia Benn. ●☆

115517 Cyanothyrsus klainei Pierre = Daniellia klainei (Pierre) De Wild. ●☆

115518 Cyanothyrsus mortehanii De Wild. = Daniellia mortehanii De Wild. ●☆

115519 Cyanothyrsus oblongus (Oliv.) Harms = Daniellia oblonga Oliv. ●☆

115520 Cyanothyrsus soyauxii Harms = Daniellia soyauxii (Harms) Rolfe ●☆

115521 Cyanotis D. Don(1825)(保留属名);蓝耳草属(露水草属,鸭舌疝属,银毛冠属);Blueeargrass,Cyanotis ■

115522 Cyanotis Miers = Burmannia L. ■

115523 Cyanotis abyssinica A. Rich. = Cyanotis barbata D. Don ■

115524 Cyanotis akeassii Brenan;阿克斯耳草■☆

115525 Cyanotis angusta C. B. Clarke;狭蓝耳草■☆

115526 Cyanotis arachnoidea C. B. Clarke;蛛丝毛蓝耳草(换肺散,鸡出头草,鸡冠参,蓝耳草,露水草,鸭脚菜,鸭舌疝,珍珠露水草,蛛毛蓝耳草,紫背鹿衔草);Spiderweb Blueeargrass,Spiderweb Cyanotis ■

115527 Cyanotis arachnoidea C. B. Clarke var. pilosa Brenan;疏毛蓝耳草■☆

115528 Cyanotis axillaris (L.) D. Don ex Sweet;鞘花蓝耳草(鞘苞花);Common Amischophacelus ■

115529 Cyanotis axillaris (L.) Sweet = Amischophacelus axillaris (L.) R. S. Rao et Kammathy ■☆

115530 Cyanotis axillaris (L.) Sweet = Cyanotis axillaris (L.) D. Don ex Sweet ■

115531 Cyanotis barbata D. Don = Cyanotis vaga (Lour.) Roem. et Schult. ■

115532 Cyanotis bodinieri H. Lév. et Vaniot;大蓝耳草(老来红)■☆

115533 Cyanotis bodinieri H. Lév. et Vaniot = Cyanotis arachnoidea C. B. Clarke ■

115534 Cyanotis bulbifera Hutch. = Cyanotis angusta C. B. Clarke ■☆

115535 Cyanotis bulbosa H. Lév. = Cyanotis vaga (Lour.) Roem. et Schult. ■

115536 Cyanotis caespitosa Kotschy et Peyr.;丛生蓝耳草■☆

115537 Cyanotis capitata (Blume) C. B. Clarke = Belosynapsis ciliata (Blume) R. S. Rao ■

115538 Cyanotis cavaleriei H. Lév. et Vaniot = Cyanotis cristata (L.) D. Don ■

115539 Cyanotis ciliata (Blume) Bakh. f. = Belosynapsis ciliata (Blume) R. S. Rao ■

115540 Cyanotis cristata (L.) D. Don = Cyanotis cristata (L.) D. Don ex Sweet ■

115541 Cyanotis cristata (L.) D. Don ex Sweet;四孔草(蛇通管,竹夹草);Cristate Blueeargrass,Cristate Cyanotis ■

115542 Cyanotis cristata (L.) Schult. f. = Cyanotis cristata (L.) D. Don ex Sweet ■

115543 Cyanotis cupricola P. A. Duvign.;喜铜蓝耳草■☆

115544 Cyanotis deightonii C. B. Clarke = Cyanotis longifolia Benth. ■☆

115545 Cyanotis deightonii Hutch. = Cyanotis longifolia Benth. ■☆

115546 Cyanotis dybowskii Hua;迪布蓝耳草■☆

115547 Cyanotis fasciculata Roem. et Schult. = Cyanotis vaga (Lour.) Roem. et Schult. ■

115548 Cyanotis flexuosa C. B. Clarke;之字蓝耳草■☆

115549 Cyanotis geniculata C. B. Clarke = Cyanotis loureiriana (Roem. et Schult.) Merr. ■

115550 Cyanotis grandidieri H. Perrier;格朗蓝耳草■☆

115551 Cyanotis hepperi Brenan;赫佩蓝耳草■☆

115552 Cyanotis hirsuta Fisch. et E. Mey. = Cyanotis barbata D. Don ■

115553 Cyanotis homblei De Wild.;洪布勒蓝耳草■☆

115554 Cyanotis kawakamii Hayata = Belosynapsis ciliata (Blume) R. S. Rao ■

115555 Cyanotis kawakamii Hayata = Belosynapsis kawakamii (Hayata) C. I. Peng et Y. J. Chen ■

115556 Cyanotis kewensis C. B. Clarke;红背草(邱园蓝耳草,邱园蓝花草);Teddy-bear Vine ■☆

115557 Cyanotis labordei H. Lév. et Vaniot = Cyanotis arachnoidea C. B. Clarke ■

115558 Cyanotis lanata Benth.;绵毛蓝耳草■☆

115559 Cyanotis lanata Benth. var. gracilis Schnell = Cyanotis longifolia

Benth. var. gracilis (Schnell) Schnell ■☆

115560 Cyanotis lanata Benth. var. rubescens (A. Chev.) Schnell = Cyanotis lanata Benth. ■☆

115561 Cyanotis lapidosa E. Phillips;石砾蓝耳草■☆

115562 Cyanotis longifolia Benth.;长叶蓝花草■☆

115563 Cyanotis longifolia Benth. subsp. deightonii (Hutch.) J. K. Morton = Cyanotis longifolia Benth. ■☆

115564 Cyanotis longifolia Benth. var. albolanescens Schnell = Cyanotis longifolia Benth. ■☆

115565 Cyanotis longifolia Benth. var. deightonii (Hutch.) Schnell = Cyanotis longifolia Benth. ■☆

115566 Cyanotis longifolia Benth. var. fonensis Schnell = Cyanotis longifolia Benth. var. rupicola (Schnell) Schnell ■☆

115567 Cyanotis longifolia Benth. var. gracilis (Schnell) Schnell;纤细长叶蓝花草■☆

115568 Cyanotis longifolia Benth. var. maliensis Schnell = Cyanotis longifolia Benth. var. rupicola (Schnell) Schnell ■☆

115569 Cyanotis longifolia Benth. var. pseudorupicola Schnell = Cyanotis longifolia Benth. var. rupicola (Schnell) Schnell ■☆

115570 Cyanotis longifolia Benth. var. rupicola (Schnell) Schnell;岩地蓝花草■☆

115571 Cyanotis loureiriana (Roem. et Schult.) Merr.;沙地蓝耳草;Desert Blueeargrass,Desert Cyanotis ■

115572 Cyanotis mannii C. B. Clarke = Cyanotis barbata D. Don ■

115573 Cyanotis minima De Wild.;极小蓝耳草■☆

115574 Cyanotis nobilis Hassk. = Cyanotis vaga (Lour.) Roem. et Schult. ■

115575 Cyanotis nodiflora (Lam.) Kunth = Cyanotis speciosa (L. f.) Hassk. ■☆

115576 Cyanotis pachyrrhiza Oberm.;粗根蓝耳草■☆

115577 Cyanotis paludosa Brenan;沼泽蓝耳草■☆

115578 Cyanotis parasitica Hochst. ex Hassk. = Cyanotis barbata D. Don ■

115579 Cyanotis pauciflora A. Rich. = Cyanotis barbata D. Don ■

115580 Cyanotis pilosa Roem. et Schult. = Cyanotis arachnoidea C. B. Clarke ■

115581 Cyanotis polyrrhiza Hochst. ex Hassk.;多根蓝耳草■☆

115582 Cyanotis racemosa C. B. Clarke = Cyanotis cristata (L.) D. Don ■

115583 Cyanotis robusta Oberm.;粗壮蓝耳草■☆

115584 Cyanotis rubescens A. Chev. = Cyanotis lanata Benth. ■☆

115585 Cyanotis rupicola Schnell = Cyanotis longifolia Benth. var. rupicola (Schnell) Schnell ■☆

115586 Cyanotis scaberula Hutch.;粗糙蓝耳草■☆

115587 Cyanotis somaliensis C. B. Clarke;毛蓝耳草(长毛鸭跖草,毛蓝花草,索马里蓝耳草,银毛冠);Furry Kittens,Pussy Ears ■

115588 Cyanotis speciosa (L. f.) Hassk.;美丽蓝耳草■☆

115589 Cyanotis vaga (Lour.) Roem. et Schult.;蓝耳草(勾蚕贝,勾蛋贝,鸡冠参,假苍贝母,苦籽,露水草,如意草,土贝母,鸭舌疝);Common Blueeargrass,Common Cyanotis ■

115590 Cyanotris Raf.(废弃属名)= Camassia Lindl.(保留属名)■☆

115591 Cyanotris Raf.(废弃属名)= Zigadenus Michx. ■

115592 Cyanotris scilloides Raf. = Camassia scilloides (Raf.) Cory ■☆

115593 Cyanthillium Blume = Vernonia Schreb.(保留属名)●■

115594 Cyanthillium Blume(1826);夜香牛属■

115595 Cyanthillium cinereum (L.) H. Rob.;夜香牛(大号一枝香,返魂香,拐棍参,还魂香,寄色草,假咸虾,假咸虾花,染色草,伤寒草,缩盖斑鸠菊,消山虎,星拭草,夜牵牛,夜香草,一枝香,枝香草);Ashycoloured Ironweed,Little Ironweed,Spice Ox in Night ■

115596 Cyanthillium cinereum (L.) H. Rob. = Vernonia cinerea (L.) Less. ■

115597 Cyanthillium cordifolium (Benth. ex Oliv.) H. Rob. = Gutenbergia cordifolia Benth. ex Oliv. ■☆

115598 Cyanthillium polytrichotoma (Wech.) H. Rob. = Gutenbergia polytrichotoma Wech. ■☆

115599 Cyanthillium stelluliferum (Benth.) H. Rob. = Vernonia stellulifera (Benth.) C. Jeffrey ■☆

115600 Cyanus Juss. = Centaurea L.(保留属名)●■

115601 Cyanus P. Mill.(废弃属名)= Centaurea L.(保留属名)●■

115602 Cyanus segetum Hill. = Centaurea cyanus L. ■

115603 Cyathanthera Pohl = Miconia Ruiz et Pav.(保留属名)●☆

115604 Cyathanthera Puttock = Miconia Ruiz et Pav.(保留属名)●☆

115605 Cyathanthus Engl. = Scyphosyce Baill. ●☆

115606 Cyathella Decne. = Cynanchum L. ●■

115607 Cyathella bojeriana Decne. = Cynanchum bojerianum (Decne.) Choux ●☆

115608 Cyathella callialata (Buch. -Ham. ex Wight) C. Y. Wu et D. Z. Li = Cynanchum callialatum Buch. -Ham. ex Wight ■

115609 Cyathella cathayensis (Tsiang et H. D. Zhang) C. Y. Wu et D. Z. Li = Cynanchum acutum L. subsp. sibiricum (Willd.) Rech. f. ■

115610 Cyathella corymbosa (Wight) C. Y. Wu et D. Z. Li = Cynanchum corymbosum Wight ■

115611 Cyathella formosana (Maxim.) C. Y. Wu et D. Z. Li = Cynanchum formosanum (Maxim.) Hemsl. ex Forbes et Hemsl. ●■

115612 Cyathella formosana Maxim. var. ovalifolia (Tsiang et P. T. Li) C. Y. Wu et D. Z. Li = Cynanchum formosanum (Maxim.) Hemsl. ex Forbes et Hemsl. ●■

115613 Cyathella insulana (Hance) Hemsl. var. lineare Tsiang et H. D. Zhang = Cynanchum insulanum (Hance) Hemsl. var. lineare (Tsiang et H. D. Zhang) Tsiang et H. D. Zhang ■

115614 Cyathella insulana (Hance) Tsiang et H. D. Zhang = Cynanchum insulanum (Hance) Hemsl. ■

115615 Cyathella insulana (Hance) Tsiang et H. D. Zhang var. linearis Tsiang ex H. D. Zhang = Cynanchum insulanum (Hance) Hemsl. var. lineare (Tsiang et H. D. Zhang) Tsiang et H. D. Zhang ■

115616 Cyathella insulana Tsiang et H. D. Zhang = Cynanchum insulanum (Hance) Hemsl. ■

115617 Cyathella kwangsiensis (Tsiang et H. D. Zhang) C. Y. Wu et D. Z. Li = Cynanchum kwangsiense Tsiang et H. D. Zhang ■

115618 Cyathella mucronata Decne. = Cynanchum obovatum (Decne.) Choux ●☆

115619 Cyathella otophylla (C. K. Schneid.) C. Y. Wu et D. Z. Li = Cynanchum otophyllum C. K. Schneid. ■

115620 Cyathella purpurea (Pall.) C. Y. Wu et D. Z. Li = Cynanchum purpureum (Pall.) K. Schum. ■

115621 Cyathella repanda Decne. = Cynanchum repandum (Decne.) K. Schum. ●☆

115622 Cyathella wallichii (Wight) C. Y. Wu et D. Z. Li = Cynanchum wallichii Wight ■

115623 Cyathidium Lindl. = Saussurea DC.(保留属名)●■

115624 Cyathidium Lindl. ex Royle = Saussurea DC.(保留属名)●■

115625 Cyathidium taraxacifolium Lindl. ex Royle = Saussurea taraxacifolia Wall. ex DC. ■

115626 Cyathiscus Tiegh. = Baratranthus (Korth.) Miq. ●☆

115627 Cyathobasis Aellen(1949);鞘叶藜属●☆

115628 Cyathobasis fruticulosa (Bunge) Aellen;鞘叶藜■☆

115629 Cyathocalyx Champ. ex Hook. f. et Thomson(1855);杯萼木属(杯萼树属,杯萼藤属)●
115630 Cyathocalyx bancana Boerl.;邦卡杯萼木●☆
115631 Cyathocephalum Nakai = Ligularia Cass.(保留属名)■
115632 Cyathocephalum angustum Nakai = Ligularia angusta (Nakai) Kitam. ■☆
115633 Cyathocephalum schmidtii (Maxim.) Nakai = Ligularia schmidtii (Maxim.) Makino ■
115634 Cyathochaeta Nees(1846);刚毛莎属■☆
115635 Cyathochaeta australis C. A. Gardner;南方刚毛莎■☆
115636 Cyathochaeta diandra (R. Br.) Nees;双蕊刚毛莎■☆
115637 Cyathochaeta diandra Nees = Cyathochaeta diandra (R. Br.) Nees ■☆
115638 Cyathochaete Benth. = Cyathochaeta Nees ■☆
115639 Cyathocline Cass.(1829);杯菊属;Cupdaisy,Cyathocline ■
115640 Cyathocline lyrata Cass. = Cyathocline purpurea (Buch. -Ham. ex D. Don) Kuntze ■
115641 Cyathocline purpurea (Buch. -Ham. ex D. Don) Kuntze;杯菊(红蒿枝,小艾,小红蒿);Purple Cupdaisy,Purple Cyathocline ■
115642 Cyathocnemis Klotzsch = Begonia L. ●■
115643 Cyathocoma Nees = Tetraria P. Beauv. ■☆
115644 Cyathocoma Nees(1834);杯毛莎草属■☆
115645 Cyathocoma bachmannii (Kük.) C. Archer;巴克曼杯毛莎草■☆
115646 Cyathocoma ecklonii Nees;埃氏杯毛莎草■☆
115647 Cyathocoma hexandra (Nees) Browning;六蕊杯毛莎草■☆
115648 Cyathocoma nigrovaginata Nees = Tetraria nigrovaginata (Nees) C. B. Clarke ■☆
115649 Cyathodes Labill.(1805);核果尖苞木属(杯果尖苞木属,杜松石南属)●☆
115650 Cyathodes Labill. = Styphelia (Sol. ex G. Forst.) Sm. ●☆
115651 Cyathodes colensoi Hook. f.;核果尖苞木(杜松石南)●☆
115652 Cyathodiscus Hochst. = Peddiea Harv. ex Hook. ●☆
115653 Cyathoglottis Poepp. et Endl. = Sobralia Ruiz et Pav. ■☆
115654 Cyathogyne Müll. Arg.(1864);肖囊大戟属●☆
115655 Cyathogyne Müll. Arg. = Thecacoris A. Juss. ●☆
115656 Cyathogyne bussei Pax = Thecacoris spathulifolia (Pax) Léandri ●☆
115657 Cyathogyne dewevrei Pax = Cyathogyne viridis Müll. Arg. subsp. dewevrei (Pax) J. Léonard ●☆
115658 Cyathogyne grandifolia Pax et K. Hoffm. = Thecacoris grandifolia (Pax et K. Hoffm.) Govaerts ●☆
115659 Cyathogyne preussii Pax = Cyathogyne viridis Müll. Arg. ●☆
115660 Cyathogyne spathulifolia Pax = Thecacoris spathulifolia (Pax) Léandri ●☆
115661 Cyathogyne usambarensis (Verdc.) J. Léonard;乌桑巴拉肖囊大戟●☆
115662 Cyathogyne viridis Müll. Arg.;绿肖囊大戟●☆
115663 Cyathogyne viridis Müll. Arg. subsp. dewevrei (Pax) J. Léonard;德韦肖囊大戟●☆
115664 Cyathogyne viridis Müll. Arg. subsp. glabra J. Léonard;光滑绿肖囊大戟●☆
115665 Cyathogyne viridis Müll. Arg. var. preussii (Pax) Pax = Cyathogyne viridis Müll. Arg. ●☆
115666 Cyathogyne viridis Müll. Arg. var. subintegra Pax et K. Hoffm. = Cyathogyne viridis Müll. Arg. ●☆
115667 Cyathomiscus Turcz. = Marianthus Hügel ex Endl. ●☆
115668 Cyathomone S. F. Blake(1923);杯冠菊属●☆
115669 Cyathomone sodiroi (Hieron.) S. F. Blake;杯冠菊■☆
115670 Cyathopappus F. Muell. = Gnephosis Cass. ■☆
115671 Cyathopappus Sch. Bip. = Elytropappus Cass. ●☆
115672 Cyathopappus metalasioides Sch. Bip. = Elytropappus hispidus (L. f.) Druce ●☆
115673 Cyathophora Raf. = Euphorbia L. ●■
115674 Cyathophylla Bocquet et Strid(1986);杯叶花属(杯叶石竹属)■☆
115675 Cyathophylla chlorifolia (Poir.) Bocquet et Strid;杯叶花■☆
115676 Cyathopsis Brongn. et Gris(1864);新喀岛尖苞木属(拟杜松石南属)●☆
115677 Cyathopsis floribunda Brongn. et Gris;新喀岛尖苞木●☆
115678 Cyathopus Stapf(1895);锡金杯禾属(杯禾属)■
115679 Cyathopus sikkimensis Stapf;锡金杯禾■
115680 Cyathorhachis Nees ex Steud. = Polytoca R. Br. ■
115681 Cyathorhachis Steud. = Polytoca R. Br. ■
115682 Cyathorhaehls Nees ex Steud. = Polytoca R. Br. ■
115683 Cyathoselinum Benth.(1867);杯蛇床属■☆
115684 Cyathoselinum tomentosum (Vis.) B. D. Jacks.;杯蛇床■☆
115685 Cyathospermum Wall. ex D. Don = Gardneria Wall. ex Roxb. ●
115686 Cyathostegia (Benth.) Schery(1950);杯豆属■☆
115687 Cyathostegia weberbaueri (Harms) Schery;杯豆■☆
115688 Cyathostelma E. Fourn.(1885);叉萝藦属■☆
115689 Cyathostelma furcatum E. Fourn.;叉萝藦■☆
115690 Cyathostemma Griff.(1854);杯冠木属;Cyathostemma ●
115691 Cyathostemma vietnamense Ban = Cyathostemma yunnanense Hu ●◇
115692 Cyathostemma yunnanense Hu;杯冠木(云南杯冠木);Yunnan Cyathostemma ●◇
115693 Cyathostemon Turcz.(1852);杯蕊桃金娘属●☆
115694 Cyathostemon Turcz. = Baeckea L. ●
115695 Cyathostemon tenuifolius Turcz.;杯蕊桃金娘●☆
115696 Cyathostyles Schott ex Meisn. = Cyphomandra Mart. ex Sendtn. ●■
115697 Cyathula Blume(1826)(保留属名);杯苋属(川牛膝属);Cyathula ■
115698 Cyathula Lour.(废弃属名)= Achyranthes L.(保留属名)■
115699 Cyathula Lour.(废弃属名)= Cyathula Blume(保留属名)■
115700 Cyathula achyranthoides (Kunth) Moq.;牛膝杯苋■☆
115701 Cyathula albida Lopr. = Cyathula cylindrica Moq. ■☆
115702 Cyathula angustifolia Moq. = Kyphocarpa angustifolia (Moq.) Lopr. ■☆
115703 Cyathula braunii Gilg ex Schinz;布劳恩杯苋■☆
115704 Cyathula capitata (Wall.) Moq.;头花杯苋(白牛膝,川牛膝,麻牛膝,头花蒽草,头序杯苋,头状杯苋);Capitate Cyathula ■
115705 Cyathula capitata (Wall.) Moq. = Cyathula officinalis K. C. Kuan ■
115706 Cyathula cordifolia Chiov. = Cyathula polycephala Baker ■☆
115707 Cyathula coriacea Schinz;革质杯苋■☆
115708 Cyathula crispa Schinz = Cyathula lanceolata Schinz ■☆
115709 Cyathula cylindrica Moq.;柱杯苋■☆
115710 Cyathula cylindrica Moq. var. abbreviata Suess.;缩短柱杯苋■☆
115711 Cyathula cylindrica Moq. var. mannii (Baker) Suess. = Cyathula cylindrica Moq. ■☆
115712 Cyathula cylindrica Moq. var. orbicularis Suess. = Cyathula cylindrica Moq. ■☆
115713 Cyathula deserti (N. E. Br.) Suess. = Cyathula lanceolata

Schinz ■☆
115714 Cyathula distorta (Hiern) C. B. Clarke = Cyathula cylindrica Moq. ■☆
115715 Cyathula echinulata Hauman = Cyathula polycephala Baker ■☆
115716 Cyathula erinacea Schinz;刺杯苋■☆
115717 Cyathula geminata (Thonn.) Moq. = Cyathula achyranthoides (Kunth) Moq. ■☆
115718 Cyathula globulifera Moq. = Cyathula uncinulata (Schrad.) Schinz ■☆
115719 Cyathula gregorii (S. Moore) Schinz;格雷戈尔杯苋■☆
115720 Cyathula hereroensis Schinz = Cyathula lanceolata Schinz ■☆
115721 Cyathula kilimandscharica Suess. et Beyerle = Cyathula orthacantha (Hochst. ex Asch.) Schinz ■☆
115722 Cyathula lanceloata Schinz var. merkeri (Gilg) Schinz = Cyathula lanceolata Schinz ■☆
115723 Cyathula lanceolata Schinz;剑叶杯苋■☆
115724 Cyathula lanceolata Schinz var. scabrida ? = Cyathula lanceolata Schinz ■☆
115725 Cyathula lindaviana Lopr. = Sericocomopsis hildebrandtii Schinz ■☆
115726 Cyathula mannii Baker = Cyathula cylindrica Moq. ■☆
115727 Cyathula merkeri Gilg = Cyathula lanceolata Schinz ■☆
115728 Cyathula merkeri Gilg var. strigosa Suess. = Cyathula lanceolata Schinz ■☆
115729 Cyathula mollis C. C. Towns.;柔软杯苋■☆
115730 Cyathula natalensis Sond.;纳塔尔杯苋■☆
115731 Cyathula officinalis K. C. Kuan;川牛膝(白牛膝,川膝,大牛膝,拐牛膝,龙牛膝,毛牛膝,米牛膝,肉牛膝,天全牛膝,甜川牛膝,甜牛膝);Medicinal Cyathula ■
115732 Cyathula orthacantha (Hochst. ex Asch.) Schinz;直刺杯苋■☆
115733 Cyathula orthacanthoides Suess. = Cyathula orthacantha (Hochst. ex Asch.) Schinz ■☆
115734 Cyathula paniculata Hauman = Cyathula coriacea Schinz ■☆
115735 Cyathula pedicellata C. B. Clarke = Cyathula prostrata (L.) Blume var. pedicellata (C. B. Clarke) Cavaco ■☆
115736 Cyathula pobeguinii Jacq. -Fél.;波别杯苋■☆
115737 Cyathula polycephala Baker;多头杯苋■☆
115738 Cyathula prostrata (L.) Blume;杯苋(拔子弹草,假川牛膝,镜面草,牛奶藤,蛇草,蛇见怕,蛇惊慌,细样倒扣草,细叶蛇总管,小马鞭草);Pastureweed,Prostrate Cyathula ■
115739 Cyathula prostrata (L.) Blume f. pedicellata (C. B. Clarke) Hauman = Cyathula prostrata (L.) Blume var. pedicellata (C. B. Clarke) Cavaco ■☆
115740 Cyathula prostrata (L.) Blume var. grandiflora Suess. = Pupalia micrantha Hauman ■☆
115741 Cyathula prostrata (L.) Blume var. pedicellata (C. B. Clarke) Cavaco;梗花杯苋■☆
115742 Cyathula schimperiana Moq. = Cyathula cylindrica Moq. ■☆
115743 Cyathula schimperiana Moq. subvar. subfusca Suess. = Cyathula cylindrica Moq. ■☆
115744 Cyathula schimperiana Moq. var. tomentosa Suess. = Cyathula polycephala Baker ■☆
115745 Cyathula semirosulata Masam.;半莲座杯苋;Semirosulate Cyathula ■☆
115746 Cyathula spathulata Schinz = Cyathula natalensis Sond. ■☆
115747 Cyathula spathulifolia Lopr. = Cyathula natalensis Sond. ■☆
115748 Cyathula strigosa Suess. = Cyathula lanceolata Schinz ■☆
115749 Cyathula tomentosa (Roth) Moq.;绒毛杯苋(藏牛膝,川牛膝,毛杯苋);Tomentose Cyathula ■
115750 Cyathula tomentosa (Roth) Moq. = Cyathula officinalis K. C. Kuan ■
115751 Cyathula uncinulata (Schrad.) Schinz;钩杯苋■☆
115752 Cyathula uncinulata (Schrad.) Schinz var. pleiocephala Suess. = Cyathula uncinulata (Schrad.) Schinz ■☆
115753 Cyatochaete Kük. = Cyathochaeta Nees ■☆
115754 Cybanthus Post et Kuntze = Cybianthus Mart. (保留属名)●☆
115755 Cybbanthera Buch. -Ham. ex D. Don = Limnophila R. Br. (保留属名)■
115756 Cybbanthera connata Buch. -Ham. ex D. Don = Limnophila connata (Buch. -Ham. ex D. Don) Hand. -Mazz. ■
115757 Cybebus Garay(1978);哥伦比亚兰属■☆
115758 Cybebus grandis Garay;哥伦比亚兰■☆
115759 Cybela Falc ex Lindl. = Herminium L. ■
115760 Cybele Falc. = Herminium L. ■
115761 Cybele Salisb. = Stenocarpus R. Br. (保留属名)●☆
115762 Cybele Salisb. ex Knight(废弃属名) = Stenocarpus R. Br. (保留属名)●☆
115763 Cybelion Spreng. = Ionopsis Kunth ■☆
115764 Cybianthopsis (Mez) Lundell = Cybianthus Mart. (保留属名)●☆
115765 Cybianthus Mart. (1831)(保留属名);立方花属●☆
115766 Cybianthus angustifolius A. DC.;窄叶立方花●☆
115767 Cybianthus foliosus Rusby;多叶立方花●☆
115768 Cybianthus fuscus Mart.;褐立方花●☆
115769 Cybianthus longifolius Miq.;长叶立方花●☆
115770 Cybianthus microbotrys A. DC.;小穗立方花●☆
115771 Cybianthus montanus (Lundell) G. Agostini;山地立方花●☆
115772 Cybianthus multiflorus (A. C. Sm.) G. Agostini;多花立方花●☆
115773 Cybianthus multipunctatus A. DC.;多斑立方花●☆
115774 Cybianthus nitidus Miq.;光亮立方花●☆
115775 Cybianthus parviflorus C. Muell.;小花立方花●☆
115776 Cybianthus philippinensis Hook. f. ex S. Vidal;菲律宾立方花●☆
115777 Cybianthus sylvaticus (Gleason) G. Agostini;林地立方花●☆
115778 Cybianthus verticillatus (Vell.) G. Agostini;轮生立方花●☆
115779 Cybianthus viridiflorus A. C. Sm.;绿花立方花●☆
115780 Cybiostigma Turcz. = Ayenia L. ●☆
115781 Cybistax Mart. = Cybistax Mart. ex Meisn. ●☆
115782 Cybistax Mart. ex Meisn. (1840);艳阳花属●☆
115783 Cybistax antisyphilitica (Mart.) Mart. ex DC.;艳阳花●☆
115784 Cybistax donnell-smithii (Rose) Seibert;金艳阳花;Gold Tree, Prima Vera, Primavera, Sunshine Tree ●☆
115785 Cybistax donnell-smithii (Rose) Siebert = Tabebuia donnell-smithii Rose ●☆
115786 Cybistax longifolia (L.) Milne-Redh. et Schweick.;长叶艳阳花●☆
115787 Cybistetes Milne-Redh. et Schweick. (1939);非洲大球石蒜属■☆
115788 Cybistetes herrei (F. M. Leight.) D. Müll. -Doblies et U. Müll. -Doblies = Ammocharis longifolia (L.) M. Roem. ■☆
115789 Cybistetes longifolia (L.) Milne-Redh. et Schweick.;非洲大球石蒜■☆
115790 Cybistetes longifolia (L.) Milne-Redh. et Schweick. = Ammocharis longifolia (L.) M. Roem. ■☆
115791 Cybostigma Post et Kuntze = Ayenia L. ●☆

115792 Cybostigma Post et Kuntze = Cybiostigma Turcz. ●☆

115793 Cycadaceae Pers. (1807)(保留科名);苏铁科;Cycad Family, Cycas Family ●

115794 Cycas L. (1753);苏铁属;Bread Palm, Conehead, Cycad, Cycas, Funeral Palm, Sago Cycas, Sago Palm ●

115795 Cycas acuminatissima Hung T. Chang et al. = Cycas segmentifida D. Yue Wang et C. Y. Deng ●

115796 Cycas armata Miq. = Cycas rumphii Miq. ●◇

115797 Cycas armstrongii Miq.;阿姆斯特朗苏铁●☆

115798 Cycas armstrongii Miq. = Cycas circinalis L. ●◇

115799 Cycas arnhemica K. D. Hill;阿纳姆苏铁●☆

115800 Cycas baguanheensis L. K. Fu et S. Z. Cheng;把关河苏铁;Baguanhe Cycas, Bakuanho Cycas ●

115801 Cycas baguanheensis L. K. Fu et S. Z. Cheng = Cycas panzhihuaensis L. Zhou et S. Y. Yang ●◇

115802 Cycas balansae Warb.;宽叶苏铁●

115803 Cycas basaltica C. A. Gardner;玄武岩苏铁●☆

115804 Cycas bougainvilleana K. D. Hill;布干维尔苏铁●☆

115805 Cycas brevipinnata Hung T. Chang et al. = Cycas miquelii Warb. ●

115806 Cycas cairnsiana F. Muell.;凯恩苏铁●☆

115807 Cycas calcicola Maconochie;银叶苏铁●☆

115808 Cycas canalis K. D. Hill;龙骨叶苏铁●☆

115809 Cycas celebica Miq. = Cycas circinalis L. ●◇

115810 Cycas changjiangensis N. Liu;葫芦苏铁(佛肚苏铁);Changjiang Cycas ●

115811 Cycas chevalieri Leand. = Cycas balansae Warb. ●

115812 Cycas circinalis L.;拳叶苏铁(凤尾蕉,卷圈苏铁,铁树果,西米苏铁,旋叶苏铁,掌叶苏铁);Crozier Cycas, Eastindian Cycad, False Sago, Fern Cycas, Fern Palm, Queen Sago, Sago Cycad, Sago Palm ●◇

115813 Cycas circinalis L. sensu Roxb. = Cycas rumphii Miq. ●◇

115814 Cycas circinalis L. subsp. thouarsii (Gaudich.) Engl. = Cycas thouarsii Gaudich. ●

115815 Cycas circinalis L. subsp. vera var. pectinata (Griff.) Schuster = Cycas pectinata Buch. -Ham. ●◇

115816 Cycas circinalis L. var. pectinata (Griff.) Schuster = Cycas pectinata Buch. -Ham. ●◇

115817 Cycas comorensis Bruant = Cycas thouarsii Gaudich. ●

115818 Cycas conferta Chirgwin;密叶苏铁●☆

115819 Cycas debaoensis Y. C. Zhong et C. J. Chen;德保苏铁;Debao Cycas ●

115820 Cycas diannanensis Z. T. Guan et G. D. Tao;滇南苏铁●

115821 Cycas diannanensis Z. T. Guan et G. D. Tao = Cycas taiwaniana Carruth. ●◇

115822 Cycas dilatata Griff. = Cycas pectinata Buch. -Ham. ●◇

115823 Cycas elongata (Leandri) D. Yue Wang;越南篦齿苏铁●☆

115824 Cycas fairylakea D. Yue Wang;仙湖苏铁(广东苏铁)●

115825 Cycas fairylakea D. Yue Wang = Cycas taiwaniana Carruth. ●◇

115826 Cycas ferruginea F. N. Wei;绣毛苏铁●

115827 Cycas formosana Gogelein = Cycas taiwaniana Carruth. ●◇

115828 Cycas furfuracea W. Fitzg.;软鳞苏铁●☆

115829 Cycas glauca Miq. = Cycas rumphii Miq. ●◇

115830 Cycas guizhouensis K. M. Lan et R. F. Zou;贵州苏铁;Guizhou Cycas ●

115831 Cycas guizhouensis K. M. Lan et R. F. Zou = Cycas szechuanensis C. Y. Cheng, W. C. Cheng et L. K. Fu ●◇

115832 Cycas guizhouensis K. M. Lan et R. F. Zou f. tenuifolia C. Y. Deng et D. Yue Wang ex Z. T. Guan, L. Zhou et K. S. Hsu = Cycas guizhouensis K. M. Lan et R. F. Zou ●

115833 Cycas hainanensis C. J. Chen ex C. Y. Cheng, W. C. Cheng et L. K. Fu;海南苏铁(刺柄苏铁);Hainan Cycas ●◇

115834 Cycas hamelini J. Schust. = Cycas circinalis L. ●◇

115835 Cycas hongheensis S. Y. Yang et S. L. Yang ex D. Yue Wang;灰干苏铁(红河苏铁)●

115836 Cycas immersa Craib = Cycas siamensis Miq. ●◇

115837 Cycas inermis Lour. = Cycas revoluta Thunb. ●◇

115838 Cycas inermis Oudem.;无刺苏铁●☆

115839 Cycas inermis Oudem. = Cycas revoluta Thunb. ●◇

115840 Cycas intermedia B. S. Williams = Cycas siamensis Miq. ●◇

115841 Cycas javana (Miq.) de Laub.;爪哇苏铁●☆

115842 Cycas jenkinsiana Griff. = Cycas pectinata Buch. -Ham. ●◇

115843 Cycas lane-poolei C. A. Gardner;兰普乐苏铁●☆

115844 Cycas lingshuiensis G. A. Fu;念珠苏铁;Lingshui Cycas ●

115845 Cycas longiconifera Hung T. Chang et al. = Cycas segmentifida D. Yue Wang et C. Y. Deng ●

115846 Cycas longipetiolula D. Yue Wang;长柄叉叶苏铁●

115847 Cycas longipetiolula D. Yue Wang = Cycas multipinnata C. J. Chen et S. Y. Yang ●

115848 Cycas longisporophylla F. N. Wei = Cycas miquelii Warb. ●

115849 Cycas longlinensis Hung T. Chang et Y. C. Zhong = Cycas segmentifida D. Yue Wang et C. Y. Deng ●

115850 Cycas macrocarpa Griff. = Cycas rumphii Miq. ●◇

115851 Cycas madagascariensis Miq. = Cycas thouarsii Gaudich. ●

115852 Cycas media R. Br.;大子苏铁(澳洲苏铁,臭苏铁,大籽苏铁,智利苏铁);Bigseed Cycas, Nut Palm, Zamia Palm ●☆

115853 Cycas megacarpa K. D. Hill;大果苏铁●

115854 Cycas micholitzii Dyer;叉叶苏铁(龙口苏铁);Forkleaf Cycas, Micholitz Cycas ●◇

115855 Cycas micholitzii Dyer f. distichus Z. T. Guan = Cycas micholitzii Dyer ●◇

115856 Cycas micholitzii Dyer f. stonensis (S. L. Yang) Z. T. Guan = Cycas micholitzii Dyer ●◇

115857 Cycas micholitzii Dyer var. simplicipinna Smitinand = Cycas balansae Warb. ●

115858 Cycas micholitzii Dyer var. stonensis S. L. Yang = Cycas micholitzii Dyer ●◇

115859 Cycas miquelii Warb.;石山苏铁●

115860 Cycas multifida Hung T. Chang et Y. C. Zhong = Cycas segmentifida D. Yue Wang et C. Y. Deng ●

115861 Cycas multifrondis D. Yue Wang;多羽叉叶苏铁●

115862 Cycas multifrondis D. Yue Wang = Cycas micholitzii Dyer ●◇

115863 Cycas multiovula D. Yue Wang;多胚苏铁●

115864 Cycas multiovula D. Yue Wang = Cycas szechuanensis C. Y. Cheng, W. C. Cheng et L. K. Fu ●◇

115865 Cycas multipinnata C. J. Chen et S. Y. Yang;多歧苏铁(独把铁,独脚铁);Many-pinnate Cycas, Multipinnate Cycas ●

115866 Cycas neocaledonica Linden = Cycas circinalis L. ●◇

115867 Cycas ophiolitica K. D. Hill;蛇纹岩苏铁●

115868 Cycas orientis K. D. Hill;东部苏铁●☆

115869 Cycas palmatifida Hung T. Chang et al. = Cycas balansae Warb. ●

115870 Cycas panzhihuaensis L. Zhou et S. Y. Yang;攀枝花苏铁;Panzhihua Cycas, Tukou Cycas ●◇

115871 Cycas parvula S. L. Yang ex D. Yue Wang;元江苏铁●

115872 Cycas parvula S. L. Yang ex D. Yue Wang = Cycas balansae Warb. ●

115873 Cycas pectinata Blume = Cycas rumphii Miq. ●◇

115874 Cycas pectinata Buch. -Ham.；篦齿苏铁(篦叶苏铁，凤尾蕉)；Nepal Cycas ●◇

115875 Cycas pectinata Buch. -Ham. f. hongheensis (S. Y. Yang et S. L. Yang ex D. Yue Wang) Z. T. Guan = Cycas hongheensis S. Y. Yang et S. L. Yang ex D. Yue Wang ●

115876 Cycas pectinata Griff. = Cycas pectinata Buch. -Ham. ●◇

115877 Cycas pectinata Griff. = Cycas taiwaniana Carruth. ●◇

115878 Cycas pectinata Griff. subsp. manhaoensis C. Chen et P. Yun = Cycas taiwaniana Carruth. ●◇

115879 Cycas platyphylla K. D. Hill；阔叶苏铁●☆

115880 Cycas pluma Hort. = Cycas circinalis L. ●◇

115881 Cycas recurvata Blume = Cycas rumphii Miq. ●◇

115882 Cycas recurvata Blume ex Schuster = Cycas rumphii Miq. ●◇

115883 Cycas revoluta Blume = Cycas rumphii Miq. ●◇

115884 Cycas revoluta Thunb.；苏铁(避火蕉，避火树，番蕉，凤皇蛋，凤尾，凤尾蕉，凤尾松，凤尾棕，凤竹，金边凤尾，苏铁树，梭罗，铁甲松，铁蕉，铁树)；Cycas，Japanese Sago Cycad，Japanese Sago Plum，Sago，Sago Cycad，Sago Cycas，Sago Palm，Sago Plum，Sago-plum ●◇

115885 Cycas revoluta Thunb. var. inermis Miq. = Cycas revoluta Thunb. ●◇

115886 Cycas revoluta Thunb. var. taiwaniana (Carruth.) J. Schust. = Cycas taiwaniana Carruth. ●◇

115887 Cycas rumphii A. DC. = Cycas circinalis L. ●◇

115888 Cycas rumphii Miq.；华南苏铁(刺叶苏铁，刺针苏铁，华东苏铁，龙尾苏铁，印度苏铁)；Indian Cycas，Rumph Cycas，S. China Cycas ●◇

115889 Cycas rumphii Miq. = Cycas siamensis Miq. ●◇

115890 Cycas rumphii Miq. var. bifida Dyer = Cycas micholitzii Dyer ●◇

115891 Cycas rumphii Roxb. = Cycas circinalis L. ●◇

115892 Cycas segmentifida D. Yue Wang et C. Y. Deng；叉苞苏铁●

115893 Cycas septemsperma Hung T. Chang et al. = Cycas miquelii Warb. ●

115894 Cycas sexseminifera F. N. Wei = Cycas miquelii Warb. ●

115895 Cycas shiwandashanica Hung T. Chang et Y. C. Zhong；十万大山苏铁●

115896 Cycas shiwandashanica Hung T. Chang et Y. C. Zhong = Cycas balansae Warb. ●

115897 Cycas siamensis Miq.；云南苏铁(凤凰蛋，凤尾蕉，凤尾松，孔雀抱蛋，台湾凤尾蕉，泰国苏铁，铁树，暹罗苏铁，象尾菜)；Siam Cycas，Siamese Cycas ●◇

115898 Cycas siamensis Miq. subsp. balansae (Warb.) J. Schust. = Cycas balansae Warb. ●

115899 Cycas silvestris K. D. Hill；密林苏铁●☆

115900 Cycas simplicipinna (Smitinand) K. D. Hill；单羽苏铁●

115901 Cycas simplicipinna (Smitinand) K. D. Hill = Cycas balansae Warb. ●

115902 Cycas speciosa D. Don = Cycas rumphii Miq. ●◇

115903 Cycas sphaerica Roxb. = Cycas circinalis L. ●◇

115904 Cycas spiniformis J. Y. Liang = Cycas miquelii Warb. ●

115905 Cycas squamosa Lodd. = Cycas circinalis L. ●◇

115906 Cycas squarrosa Steud. = Cycas circinalis L. ●◇

115907 Cycas sundaica Miq. = Cycas rumphii Miq. ●◇

115908 Cycas sundaica Miq. ex Schuster = Cycas rumphii Miq. ●◇

115909 Cycas szechuanensis C. Y. Cheng，W. C. Cheng et L. K. Fu；四川苏铁；Sichuan Cycas，Szechuan Cycas ●◇

115910 Cycas taitungensis C. F. Shen，K. D. Hill，C. H. Tsou et C. J. Chen；台东苏铁；Taidong Cycas ●

115911 Cycas taiwaniana Carrière = Cycas taiwaniana Carruth. ●◇

115912 Cycas taiwaniana Carruth.；台湾苏铁(海铁鸡，铁海鸥，肖楠木屋)；Taiwan Cycas ●◇

115913 Cycas tanqingii D. Yue Wang；潭清苏铁●

115914 Cycas tanqingii D. Yue Wang = Cycas balansae Warb. ●

115915 Cycas thouarsii Gaudich.；光果苏铁(光果凤尾蕉，托氏苏铁)；Cycad，Thouars Cycas ●

115916 Cycas thouarsii R. Br. = Cycas circinalis L. ●◇

115917 Cycas timorensis Miq. = Cycas rumphii Miq. ●◇

115918 Cycas wadei Merr.；韦德苏铁●☆

115919 Cycas wallichii Miq. = Cycas pectinata Buch. -Ham. ●◇

115920 Cycas wallichii Miq. = Cycas rumphii Miq. ●◇

115921 Cycas wendlandii Sander. = Cycas circinalis L. ●◇

115922 Cycas xilignensis Hung T. Chang et Y. C. Zhong = Cycas segmentifida D. Yue Wang et C. Y. Deng ●

115923 Cycca Batsch = Cicca L. ●

115924 Cyclacanthus S. Moore(1921)；环刺爵床属☆

115925 Cyclacanthus coccineus S. Moore；环刺爵床☆

115926 Cyclachaena Fresen. = Euphrosyne DC. ■☆

115927 Cyclachaena ambrosiifolia (A. Gray) Rydb. = Hedosyne ambrosiifolia (A. Gray) Strother ■☆

115928 Cyclachaena xanthifolia (Nutt.) Fresen. = Iva xanthifolia Nutt. ●☆

115929 Cycladenia Benth. (1849)；环腺夹竹桃属●☆

115930 Cycladenia humilis Benth.；环腺夹竹桃●☆

115931 Cyclamen L. (1753)；仙客来属(萝卜海棠属)；Alpine Violet，Cyclamen，Persian Violet，Sowbread ■

115932 Cyclamen abchasicum (Medw.) Kolak. ex Pobed.；阿扎仙客来■☆

115933 Cyclamen adzharicum Pobed. = Cyclamen abchasicum(Medw.) Kolak. ex Pobed. ■☆

115934 Cyclamen africanum Boiss. et Reut.；非洲仙客来■☆

115935 Cyclamen africanum Boiss. et Reut. subsp. saldense (Pomel) Batt. = Cyclamen africanum Boiss. et Reut. ■☆

115936 Cyclamen alpinum Sprenger = Cyclamen trochopteranthum O. Schwarz ■☆

115937 Cyclamen atkinsii Moore；阿氏仙客来■☆

115938 Cyclamen caucasicum Willd. ex Boiss. = Cyclamen coum Rchb. var. caucasicum (K. Koch) Meikle ■☆

115939 Cyclamen cilicicum Boiss. et Heldr.；西里西亚仙客来(土耳其仙客来)■☆

115940 Cyclamen circassicum Pobed.；切尔卡西亚仙客来■☆

115941 Cyclamen clusii Lindl. = Cyclamen europaeum L. ■☆

115942 Cyclamen coum Mill.；春仙客来(科姆仙客来，小花仙客来，早花仙客来)；Eastern Cyclamen，Spring Cyclamen，Spring-cyclamen ■☆

115943 Cyclamen coum Mill. f. albissimum R. H. Bailey；白花早花仙客来■☆

115944 Cyclamen coum Mill. var. caucasicum (K. Koch) Meikle；高加索早花仙客来■☆

115945 Cyclamen coum Rchb. var. caucasicum (K. Koch) Meikle = Cyclamen coum Mill. var. caucasicum (K. Koch) Meikle ■☆

115946 Cyclamen creticum Hildebrandt；白香仙客来■☆

115947 Cyclamen cyprium Unger et Kotschy;塞浦路斯仙客来■☆
115948 Cyclamen elegans Boiss. et Buhse;雅致仙客来■☆
115949 Cyclamen europaeum L.;欧洲仙客来(烈香仙客来,仙客来);Bleeding Nun, Cyclamen, Earth Apple, European Cyclamen, Rape Violet, Slite, Snitchback, Sowbread ■☆
115950 Cyclamen europaeum L. = Cyclamen purpurascens Mill. ■☆
115951 Cyclamen fatrense Halda et Soják = Cyclamen europaeum L. ■☆
115952 Cyclamen graecum Link;细网纹仙客来■☆
115953 Cyclamen hederifolium Aiton;角叶仙客来(地中海仙客来,那不仙客来);Cyclamen, Ivyleaved Cyclamen, Neapolitan Cyclamen, Shooting Star, Sow Bread, Sowbread ■☆
115954 Cyclamen hederifolium Aiton f. albiflorum (Jord.) Grey-Wilson;白花角叶仙客来■☆
115955 Cyclamen hederifolium Sibth. et Sm. = Cyclamen persicum Mill. ■
115956 Cyclamen ibericum Steven ex Boiss. = Cyclamen vernum Sw. ■☆
115957 Cyclamen indicum L. = Cyclamen persicum Mill. ■
115958 Cyclamen latifolium Sibth. et Sm.;广叶仙客来■☆
115959 Cyclamen latifolium Sibth. et Sm. = Cyclamen persicum Mill. ■
115960 Cyclamen libanoticum Hildebrandt;黎巴嫩仙客来■☆
115961 Cyclamen mirabile Hildebrandt;齿瓣仙客来(奇异仙客来)■☆
115962 Cyclamen neapolitanum Ten. = Cyclamen hederifolium Aiton ■☆
115963 Cyclamen orbiculatum Mill. = Cyclamen coum Mill. ■☆
115964 Cyclamen parviflorum Pobed.;小花仙客来■☆
115965 Cyclamen persicum Mill.;仙客来(萝卜海棠,兔耳花,兔子花,一品冠,印度仙客来);Common Cyclamen, Cyclamen, Florist's Cyclamen, Ivyleaf Cyclamen, Ivy-leaved Cyclamen, Persian Cyclamen, Persian Violet, Sowbread ■
115966 Cyclamen persicum Mill. 'Esmeralda';埃斯米拉达仙客来■☆
115967 Cyclamen persicum Mill. 'Pearl Ware';珍珠波仙客来■☆
115968 Cyclamen persicum Mill. 'Renown';名望仙客来■☆
115969 Cyclamen persicum Mill. 'Seentsation';芳香仙客来■☆
115970 Cyclamen ponticum (Albov) Pobed.;蓬特仙客来■☆
115971 Cyclamen pseudibericum Hildebrandt;白口仙客来■☆
115972 Cyclamen punicum Pomel = Cyclamen persicum Mill. ■
115973 Cyclamen purpurascens Mill. = Cyclamen europaeum L. ■☆
115974 Cyclamen repandum Sm.;波叶仙客来(波缘仙客来);Spring Cyclamen ■☆
115975 Cyclamen repandum Sm. var. baborense (Debussche) Quézel = Cyclamen repandum Sm. ■☆
115976 Cyclamen rohlfsianum Asch.;天竺葵叶仙客来■☆
115977 Cyclamen romanum Griseb. = Cyclamen repandum Sm. ■☆
115978 Cyclamen saldense Pomel = Cyclamen africanum Boiss. et Reut. ■☆
115979 Cyclamen somalense Thulin et Warfa;索马里仙客来■☆
115980 Cyclamen trochopteranthum O. Schwarz;扭花仙客来■☆
115981 Cyclamen vernum Sw.;高加索仙客来;Caucasia Cyclamen ■☆
115982 Cyclaminos Heldr. = Cyclamen L. ■
115983 Cyclaminum Bubani = Cyclamen L. ■
115984 Cyclaminus Asch. = Cyclamen L. ■
115985 Cyclaminus Haller = Cyclamen L. ■
115986 Cyclandra Lauterb. = Ternstroemia Mutis ex L. f. (保留属名)●
115987 Cyclandrophora Hassk. = Atuna Raf. ●☆
115988 Cyclanthaceae Dumort. = Cyclanthaceae Poit. ex A. Rich. (保留科名)●■
115989 Cyclanthaceae Poit. ex A. Rich. (1824)(保留科名);巴拿马草科(环花科);Cyclanthus Family ■●
115990 Cyclanthera Schrad. (1831);小雀瓜属(辣子瓜属);Cyclanthera ■
115991 Cyclanthera brachystachya (Ser.) Cogn.;短穗小雀瓜(辣子瓜)■☆
115992 Cyclanthera digitata Arn. = Cyclanthera pedata (L.) Schrad. ■
115993 Cyclanthera elastica Hort. = Cyclanthera explodens Naudin ■☆
115994 Cyclanthera explodens Naudin = Cyclanthera brachystachya (Ser.) Cogn. ■☆
115995 Cyclanthera pedata (L.) Schrad.;小雀瓜(辣子瓜);Pedate Cyclanthera ■
115996 Cyclantheraceae Lilja = Cucurbitaceae Juss. (保留科名)●■
115997 Cyclantheropsis Harms(1896);拟小雀瓜属●■☆
115998 Cyclantheropsis madagascariensis Rabenant.;马岛拟小雀瓜●☆
115999 Cyclantheropsis occidentalis Gilg et Mildbr.;西部拟小雀瓜■☆
116000 Cyclantheropsis parviflora (Cogn.) Harms;拟小雀瓜■☆
116001 Cyclanthus Poit. (1822);环花草属■☆
116002 Cyclanthus Poit. ex A. Rich. = Cyclanthus Poit. ■☆
116003 Cyclanthus Poit. ex Spreng. = Cyclanthus Poit. ■☆
116004 Cyclanthus bipartitus Poit.;环花草■☆
116005 Cyclas Schreb. = Apalatoa Aubl. (废弃属名)●☆
116006 Cyclas Schreb. = Crudia Schreb. (保留属名)●☆
116007 Cyclea Arn. = Cyclea Arn. ex Wight ●■
116008 Cyclea Arn. ex Wight(1840);轮环藤属(银不换属);Cyclea, Ringvine ●■
116009 Cyclea atjehensis Forman;亚齐轮环藤■☆
116010 Cyclea barbata Miers;毛叶轮环藤(大叶金锁匙,金线风,九条牛,毛参箕藤,散血丹,银不换,银锁匙,猪肠换);Barbate Cyclea, Barbate Ringvine ■
116011 Cyclea burmanii Miers;布满轮环藤■
116012 Cyclea ciliata Craib = Cyclea barbata Miers ■
116013 Cyclea debiliflora Miers;纤花轮环藤;Fineflower Ringvine ■
116014 Cyclea deltoidea Miers;三角轮环藤■
116015 Cyclea deltoidea Miers = Cyclea hypoglauca (Schauer) Diels ●
116016 Cyclea densiflora (Gogelein) Y. C. Tang et H. C. Lo = Cyclea gracillima Diels ●
116017 Cyclea gracillima Diels;纤细轮环藤(密花轮环藤,土防己);Fine Ringvine, Slender Cyclea ●
116018 Cyclea hainanensis Merr. = Cyclea polypetala Dunn ●■
116019 Cyclea hypoglauca (Schauer) Diels;粉叶轮环藤(百解藤,铲鸡藤,穿山龙,二十四风藤,粉背轮环藤,蛤仔藤,黑皮蛇,金钱风,金钥匙,凉粉藤,青藤子,山豆根,山苦参,乌皮龙,银锁匙,有毛粪箕笃);Glaucousleaf Cyclea, Glaucous-leaved Cyclea, Paleback Ringvine ●
116020 Cyclea insularis (Makino) Hatus.;海岛轮环藤(海岛锡生藤,兰屿土防己,土防己);Insular Cyclea, Island Ringvine ●■
116021 Cyclea insularis (Makino) Hatus. var. guangxiensis H. S. Lo;黔桂轮环藤;Guangxi Island Ringvine ■
116022 Cyclea longgangensis J. Y. Luo;弄岗轮环藤;Nonggang Cyclea, Nonggang Ringvine ●■
116023 Cyclea madagascariensis Baker;马达加斯加轮环藤■☆
116024 Cyclea meeboldii Diels;云南轮环藤;Yunnan Cyclea, Yunnan Ringvine ●■
116025 Cyclea migoana Gogelein;福建轮环藤;Fujian Ringvine ■
116026 Cyclea migoana Gogelein = Cyclea hypoglauca (Schauer) Diels ●
116027 Cyclea migoana Yamam. = Cyclea hypoglauca (Schauer) Diels ●
116028 Cyclea ochiaiana (Gogelein) S. F. Huang et T. C. Huang = Cyclea ochiaiana (Yamam.) S. F. Huang et T. C. Huang ●
116029 Cyclea ochiaiana (Yamam.) S. F. Huang et T. C. Huang;台湾

轮环藤(台湾土防己,台湾锡生藤);Taiwan Cissampelos,Taiwan Cyclea ●

116030 Cyclea peltata Hook. f. et Thomson;盾叶轮环藤■☆

116031 Cyclea polypetala Dunn;多瓣轮环藤(百解藤,海南轮环藤,棵叶黑,离瓣轮环藤,铁藤,须龙藤,银不换);Iron Ringvine,Iron Vine,Manypetal Cyclea,Polypetal Cyclea ●■

116032 Cyclea racemosa Oliv.;轮环藤(滚天龙,淮通,金蚂蝗,良藤,青藤,青藤细辛,山豆根,铁石鞭,小青藤香);Racemose Cyclea,Racemose Ringvine ●■

116033 Cyclea racemosa Oliv. var. emeiensis H. C. Lo et S. Y. Zhao;峨眉轮环藤;Emei Cyclea ●■

116034 Cyclea racemosa Oliv. var. emeiensis H. C. Lo et S. Y. Zhao = Cyclea racemosa Oliv. ●■

116035 Cyclea racemosa Oliv. var. longgangensis J. Y. Luo = Cyclea longgangensis J. Y. Luo ●■

116036 Cyclea sutchuenensis Gagnep.;四川轮环藤(光叶金钥匙,金钱风,良藤);Sichuan Cyclea,Sichuan Ringvine,Szechuan Cyclea,Szechwan Cyclea ●■

116037 Cyclea suwhuenensis Gagnep. var. sessilis Y. C. Wu = Cyclea sutchuenensis Gagnep. ●■

116038 Cyclea tonkinensis Gagnep.;南轮环藤(槟榔花,小花轮环藤,越南轮环藤);Tonkin Cyclea,Tonkin Ringvine ■

116039 Cyclea wallichii Diels = Cyclea barbata Miers ■

116040 Cyclea wattii Diels;西南轮环藤;Southwest Cyclea,Watt Cyclea,Watt Ringvine ●

116041 Cyclium Steud. = Cycnium E. Mey. ex Benth. ■●☆

116042 Cyclobalanopsis (Endl.) Oerst. = Cyclobalanopsis Oerst.(保留属名)●

116043 Cyclobalanopsis (Endl.) Oerst. = Quercus L. ●

116044 Cyclobalanopsis Oerst.(1867)(保留属名);青冈属(椆属,青冈栎属,青刚栎属);Cyclobalanopsis,Oak,Qinggang ●

116045 Cyclobalanopsis Oerst.(保留属名)= Quercus L. ●

116046 Cyclobalanopsis acuta (Thunb. ex A. Murray) Oerst. = Quercus acuta Thunb. ex A. Murray ●☆

116047 Cyclobalanopsis acuta (Thunb. ex A. Murray) Oerst. f. lanceolata (Hatus.) Honda = Quercus acuta Thunb. f. lanceolata Hatus. ●☆

116048 Cyclobalanopsis acuta (Thunb. ex A. Murray) Oerst. var. paucidentata (Franch. ex Nakai) J. C. Liao = Cyclobalanopsis sessilifolia (Blume) Schottky ●

116049 Cyclobalanopsis acuta (Thunb.) Oerst. var. paucidentata (Franch. ex Nakai) J. C. Liao = Cyclobalanopsis sessilifolia (Blume) Schottky ●

116050 Cyclobalanopsis albicaulis (Chun et W. C. Ko) Y. C. Hsu et H. Wei Jen;白枝青冈(白枝椆);Whitebranch Oak,White-branched Cyclobalanopsis,Whitetwig Qinggang ●◇

116051 Cyclobalanopsis angustinii (Skan) Schottky = Cyclobalanopsis angustinii (Skan) Schottky ●

116052 Cyclobalanopsis angustinii (V. Naray.) Schottky = Cyclobalanopsis augustinii (Skan) Schottky ●

116053 Cyclobalanopsis angustissima (Makino) Kudo et Masam. ex Kudo = Cyclobalanopsis salicina (Blume) Oerst. ●

116054 Cyclobalanopsis annulata (Sm.) Oerst.;轮环青冈(环青冈);Annular Cyclobalanopsis,Annular Oak,Annular Qinggang ●

116055 Cyclobalanopsis argyrotricha (A. Camus) Chun et Y. T. Chang ex Y. C. Hsu et H. Wei Jen;贵州青冈;Argyrotrichia Oak,Argyrotrichia Qinggang,Guizhou Oak,Guizhou Qinggang,Silver-haired Cyclobalanopsis ●◇

116056 Cyclobalanopsis augustinii (Skan) Schottky;窄叶青冈(扫把椆);Narrowleaf Oak,Narrowleaf Qinggang,Narrow-leaved Cyclobalanopsis ●

116057 Cyclobalanopsis augustinii (Skan) Schottky var. nigrinux (Hu) M. Deng et Z. K. Zhou;黑果窄叶青冈●

116058 Cyclobalanopsis augustinii Schottky = Cyclobalanopsis augustinii (Skan) Schottky ●

116059 Cyclobalanopsis austrocochinchinensis (Hickel et A. Camus) Hjelmq.;越南青冈;Cochin-China Cyclobalanopsis,Indochinese Oak,Vietnam Qinggang ●

116060 Cyclobalanopsis austro-glauca Y. T. Chang ex Y. C. Hsu et H. Wei Jen;滇南青冈;South Yunnan Cyclobalanopsis,South Yunnan Oak,South Yunnan Qinggang ●

116061 Cyclobalanopsis austro-yunnanensis Hu = Cyclobalanopsis fleuryi (Hickel et A. Camus) Chun ●

116062 Cyclobalanopsis bambusifolia (Hance) Chun ex Y. C. Hsu et H. Wei Jen = Cyclobalanopsis neglecta Schottky ●

116063 Cyclobalanopsis bapouensis H. Li et Y. C. Hsu;巴坡青冈;Bapo Cyclobalanopsis,Bapo Oak ●

116064 Cyclobalanopsis bella (Chun et Tsiang) Chun ex Y. C. Hsu et H. Wei Jen;槟榔青冈(槟榔椆);Beautiful Cyclobalanopsis,Beautiful Oak,Beautiful Qinggang ●

116065 Cyclobalanopsis blakei (R. H. Shan) Schottky;栎子青冈(栎子椆);Blake Cyclobalanopsis,Blake Oak,Blake Qinggang ●

116066 Cyclobalanopsis breviradiata W. C. Cheng;短星毛青冈;Short-radiate Cyclobalanopsis,Short-starhair Oak,Short-starhair Qinggang ●

116067 Cyclobalanopsis breviradiata W. C. Cheng = Cyclobalanopsis oxyodon (Miq.) Oerst. ●

116068 Cyclobalanopsis breviradiata W. C. Cheng ex Y. C. Hsu et H. Wei Jen = Cyclobalanopsis oxyodon (Miq.) Oerst. ●

116069 Cyclobalanopsis camusae (Trel. ex Hickel et A. Camus) Y. C. Hsu et H. Wei Jen;法斗青冈(法斗椆);Camus Cyclobalanopsis,Fadou Oak,Fadou Qinggang ●

116070 Cyclobalanopsis championii (Benth.) Oerst. = Cyclobalanopsis championii (Benth.) Oerst. ex Schottky ●

116071 Cyclobalanopsis championii (Benth.) Oerst. ex Schottky;岭南青冈(岭南椆,岭南青刚栎);Champion Cyclobalanopsis,Champion Oak,Champion Qinggang,Hongkong Oak ●

116072 Cyclobalanopsis chapensis (Hickel et A. Camus) Y. C. Hsu et H. Wei Jen;扁果青冈;Flatfruit Oak,Flatfruit Qinggang,Flat-fruited Cyclobalanopsis,Flat-fruited Oak ●

116073 Cyclobalanopsis chevalieri (Hickel et A. Camus) Y. C. Hsu et H. Wei Jen;黑果青冈;Chevalier Oak,Blackfruit Qinggang,Chevalier Cyclobalanopsis ●

116074 Cyclobalanopsis chingsiensis (Y. T. Chang) Y. T. Chang;靖西青冈(靖西椆,靖西栎);Jingxi Cyclobalanopsis,Jingxi Oak,Jingxi Qinggang ●

116075 Cyclobalanopsis chrysocalyx (Hickel et A. Camus) Hjelmq.;毛斗青冈;Hairycapsule Oak,Hairy-cupuled Cyclobalanopsis,Hairycupuled Qinggang ●

116076 Cyclobalanopsis chungii (F. P. Metcalf) Y. C. Hsu et H. Wei Jen = Cyclobalanopsis chungii (F. P. Metcalf) Y. C. Hsu et H. Wei Jen ex Q. F. Zheng ●

116077 Cyclobalanopsis chungii (F. P. Metcalf) Y. C. Hsu et H. Wei Jen ex Q. F. Zheng;福建青冈;Chung Cyclobalanopsis,Chung Oak,Chung Qinggang ●

116078 Cyclobalanopsis damingshanensis S. K. Lee;大明山青冈;

Damingshan Cyclobalanopsis, Damingshan Oak, Damingshan Qinggang ●

116079 Cyclobalanopsis delavayi (Franch.) Schottky;黄毛青冈(滇黄栎,黄背叶青冈,黄椆,黄栎,黄栗树,黄青冈,黄桐);Delavay Cyclobalanopsis,Delavay Oak,Yellowhair Qinggang ●

116080 Cyclobalanopsis delicatula (Chun et Tsiang) Y. C. Hsu et H. Wei Jen;上思青冈;Cheeful Qinggang, Delicate Cyclobalanopsis, Delight Oak,Shangsi Oak ●

116081 Cyclobalanopsis dinghuensis (C. C. Huang) Y. C. Hsu et H. Wei Jen;鼎湖青冈(鼎湖椆);Dinghu Oak, Dinghu Qinggang, Dinghushan Cyclobalanopsis ●

116082 Cyclobalanopsis disciformis (Chun et Tsiang) Y. C. Hsu et H. Wei Jen;碟斗青冈(碟叶青冈);Dishcupula Oak, Dishcupula Qinggang,Dish-cupule Cyclobalanopsis,Dish-cupule Oak ●

116083 Cyclobalanopsis dongfangensis (C. C. Huang, F. W. Xing et Z. X. Li) Y. T. Chang;东方青冈;Oriental Oak,Oriental Qinggang ●

116084 Cyclobalanopsis dulongensis H. Li et Y. C. Hsu;独龙青冈;Dulong Cyclobalanopsis,Dulong Oak ●

116085 Cyclobalanopsis dulongensis H. Li et Y. C. Hsu = Cyclobalanopsis gambleana (A. Camus) Y. C. Hsu et H. Wei Jen ●

116086 Cyclobalanopsis edithae (Skan) Schottky;华南青冈(华南椆);S. China Qinggang, South China Cyclobalanopsis, South China Oak ●

116087 Cyclobalanopsis edithae (V. Naray.) Schottky = Cyclobalanopsis edithae (Skan) Schottky ●

116088 Cyclobalanopsis elevaticostata Q. F. Zheng;突脉青冈;Covexvein Oak, Elevated-midrib Cyclobalanopsis, Projectingvein Qinggang ●

116089 Cyclobalanopsis faadoouensis Hu = Cyclobalanopsis camusae (Trel. ex Hickel et A. Camus) Y. C. Hsu et H. Wei Jen ●

116090 Cyclobalanopsis fengii Hu et W. C. Cheng;冯氏槠;Feng's Cyclobalanopsis,Feng's Oak,Feng's Qinggang ●

116091 Cyclobalanopsis fengii Hu et W. C. Cheng = Cyclobalanopsis lamellosa (Sm.) Oerst. ●

116092 Cyclobalanopsis fleuryi (Hickel et A. Camus) Chun;饭甑青冈(饭甑椆);Fleury Cyclobalanopsis, Fleury Oak, Ricesteamer Qinggang ●

116093 Cyclobalanopsis fuhsingensis (Y. T. Chang) Y. T. Chang = Cyclobalanopsis xanthotricha (A. Camus) Y. C. Hsu et H. Wei Jen ●

116094 Cyclobalanopsis fuhsingensis (Y. T. Chang) Y. T. Chang ex Y. C. Hsu et H. Wei Jen = Cyclobalanopsis xanthotricha (A. Camus) Y. C. Hsu et H. Wei Jen ●

116095 Cyclobalanopsis fuliginosa (Chun et W. C. Ko) Y. Y. Luo et R. J. Wang;污托栎;Fuliginos Oak ●

116096 Cyclobalanopsis fulviseriaca Y. C. Hsu et D. M. Wang = Cyclobalanopsis lungmaiensis Hu ●

116097 Cyclobalanopsis gambleana (A. Camus) Y. C. Hsu et H. Wei Jen;毛蔓青冈;Gamble Cyclobalanopsis, Gamble Oak, Gamble Qinggang ●

116098 Cyclobalanopsis gilva (Blume) Oerst.;赤皮青冈(赤柯,赤皮,赤皮椆,橹樫,石槠);Red Bark Oak, Red-bark Cyclobalanopsis, Redbark Oak,Redbark Qinggang ●

116099 Cyclobalanopsis gilva (Blume) Oerst. = Quercus gilva Blume ●

116100 Cyclobalanopsis glauca (Thunb. ex A. Murray) Oerst.;青冈(白校欑,椆,椆树,大叶青冈,谷园青刚栎,九层,九桧,青冈栎,青栲,实椆,铁椆,铁栎,铁青冈,小栗子树,校欑,崖青冈树,槠);Grey-blue Cyclobalanopsis, Japanese Blue Oak, Qinggang, Ring Cupped Oak,Ring-cupped Oak ●

116101 Cyclobalanopsis glauca (Thunb. ex A. Murray) Oerst. = Quercus glauca Thunb. ●

116102 Cyclobalanopsis glauca (Thunb. ex A. Murray) Oerst. f. gracilis (Rehder et E. H. Wilson) Y. T. Chang = Cyclobalanopsis gracilis (Rehder et E. H. Wilson) W. C. Cheng et T. Hong ●

116103 Cyclobalanopsis glauca (Thunb. ex A. Murray) Oerst. var. gracilis (Rehder et E. H. Wilson) Y. T. Chang = Cyclobalanopsis gracilis (Rehder et E. H. Wilson) W. C. Cheng et T. Hong ●

116104 Cyclobalanopsis glauca (Thunb. ex A. Murray) Oerst. var. kuyuensis (J. C. Liao) J. C. Liao;谷园青冈栎(谷园青刚栎,台湾铁椆);Guyuan Cyclobalanopsis ●

116105 Cyclobalanopsis glauca (Thunb. ex A. Murray) Oerst. var. kuyuensis (J. C. Liao) J. C. Liao = Cyclobalanopsis glauca (Thunb. ex A. Murray) Oerst. ●

116106 Cyclobalanopsis glauca (Thunb. ex A. Murray) Oerst. var. stricta Makino;纹叶青冈栎●☆

116107 Cyclobalanopsis glauca (Thunb.) Oerst. = Cyclobalanopsis glauca (Thunb. ex A. Murray) Oerst. ●

116108 Cyclobalanopsis glauca (Thunb.) Oerst. var. gracilis (Rehder et E. H. Wilson) Y. T. Chang = Cyclobalanopsis gracilis (Rehder et E. H. Wilson) W. C. Cheng et T. Hong ●

116109 Cyclobalanopsis glaucoides Schottky;滇青冈(滇椆,滇青栎);Schottky Cyclobalanopsis, Schottky Oak, Yunnan Oak, Yunnan Qinggang ●

116110 Cyclobalanopsis globosa W. F. Lin et T. Liu;圆果青冈栎(圆果椆,圆果青刚栎);Globose Glans Cyclobalanopsis, Globose Glans Oak ●

116111 Cyclobalanopsis globosa W. F. Lin et T. Liu f. chiapautaiensis (J. C. Liao) J. C. Liao;佳保台圆果青冈栎(佳保台圆果青刚栎);Jiabaotai Cyclobalanopsis ●

116112 Cyclobalanopsis gracilis (Rehder et E. H. Wilson) W. C. Cheng et T. Hong;细叶青冈(小叶青冈栎);Slender Oak, Slender Qinggang, Slender-leaved Cyclobalanopsis, Thinleaf Oak, Thinleaf Qinggang ●

116113 Cyclobalanopsis helferiana (A. DC.) Oerst.;毛枝青冈;Hairybranch Oak, Hairybranch Qinggang, Hairy-branched Cyclobalanopsis,Hairy-branched Oak ●

116114 Cyclobalanopsis hui (Chun) Chun ex Y. C. Hsu et H. Wei Jen;雷公青冈(雷公椆);Hu's Cyclobalanopsis, Hu's Oak, Thundergot Qinggang ●

116115 Cyclobalanopsis hunanensis (Hand. -Mazz.) W. C. Cheng et T. Hong;湖南青冈;Hunan Cyclobalanopsis, Hunan Oak, Hunan Qinggang ●

116116 Cyclobalanopsis hunanensis (Hand. -Mazz.) W. C. Cheng et T. Hong = Cyclobalanopsis gilva (Blume) Oerst. ●

116117 Cyclobalanopsis hypargyrea (Seemen) Y. C. Hsu et H. Wei Jen = Cyclobalanopsis multinervis W. C. Cheng et T. Hong ●

116118 Cyclobalanopsis hypophaea (Hayata) Kudo;绒毛青冈(灰背栎,灰背石栎,灰绒椆,绒毛栎);Darkgray-leaved Cyclobalanopsis, Downy Qinggang,Sea-buckthorn Leaf Tanoak ●

116119 Cyclobalanopsis jenseniana (Hand. -Mazz.) W. C. Cheng et T. Hong = Cyclobalanopsis jenseniana (Hand. -Mazz.) W. C. Cheng et T. Hong ex Q. F. Zheng ●

116120 Cyclobalanopsis jenseniana (Hand. -Mazz.) W. C. Cheng et T. Hong ex Q. F. Zheng;大叶青冈(大叶椆);Bigleaf Qinggang, Largeleaf Oak,Large-leaved Cyclobalanopsis,Large-leaved Oak ●

116121 Cyclobalanopsis jinpinensis Y. C. Hsu et H. Wei Jen;金平青冈;Jinping Cyclobalanopsis,Jinping Oak,Jinping Qinggang ●

116122 Cyclobalanopsis kerrii (Craib) Hu;毛叶青冈(理博树,平脉椆);Hairleaf Cyclobalanopsis,Hairleaf Qinggang,Kerr Oak ●

116123 Cyclobalanopsis kiukiangensis Y. T. Chang ex Y. C. Hsu et H. Wei Jen;俅江青冈;Qiujiang Cyclobalanopsis, Qiujiang Oak, Qiujiang Qinggang ●

116124 Cyclobalanopsis kontumensis (A. Camus) Y. C. Hsu et H. Wei Jen = Cyclobalanopsis saravanensis (A. Camus) Hjelmq. ●

116125 Cyclobalanopsis kontumensis (A. Camus) Y. C. Hsu et H. Wei Jen ex Y. T. Chang = Cyclobalanopsis saravanensis (A. Camus) Hjelmq. ●

116126 Cyclobalanopsis kouangsiensis (A. Camus) Y. C. Hsu et H. Wei Jen;广西青冈;Guangxi Cyclobalanopsis, Guangxi Oak, Guangxi Qinggang,Kwangsi Cyclobalanopsis,Kwangsi Oak ●◇

116127 Cyclobalanopsis koumeii Hu = Cyclobalanopsis chapensis (Hickel et A. Camus) Y. C. Hsu et H. Wei Jen ●

116128 Cyclobalanopsis lamelloides (C. C. Huang) Y. T. Chang;拟薄片青冈;Laminate-like Cyclobalanopsis ●

116129 Cyclobalanopsis lamelloides (C. C. Huang) Y. T. Chang = Cyclobalanopsis lamellosa (Sm.) Oerst. ●

116130 Cyclobalanopsis lamellosa (Sm.) Oerst.;薄片青冈(薄片椆,大铁椆树,茅青冈树);Laminate Cyclobalanopsis, Laminate Oak, Slice Qinggang ●

116131 Cyclobalanopsis lineata (Blume) Oerst. var. lobbii (Hook. f. et Thomson ex Wenz.) Schottky = Cyclobalanopsis lobbii (Hook. f. et Thomson ex Wenz.) Y. C. Hsu et H. Wei Jen ●

116132 Cyclobalanopsis litoralis Chun et P. C. Tam = Cyclobalanopsis litoralis Chun et P. C. Tam ex Y. C. Hsu et H. Wei Jen ●

116133 Cyclobalanopsis litoralis Chun et P. C. Tam ex Y. C. Hsu et H. Wei Jen;尖峰青冈(海南青冈);Jianfengling Cyclobalanopsis, Littoral Oak,Plage Qinggang ●

116134 Cyclobalanopsis litseoides (Dunn) Y. C. Hsu et H. Wei Jen;木姜叶青冈;Litsea-leaved Cyclobalanopsis, Litsea-leaved Oak, Litseleaf Oak,Litseleaf Qinggang ●

116135 Cyclobalanopsis lobbii (Etting) Y. C. Hsu et H. Wei Jen = Cyclobalanopsis lobbii (Hook. f. et Thomson ex Wenz.) Y. C. Hsu et H. Wei Jen ●

116136 Cyclobalanopsis lobbii (Hook. f. et Thomson ex Wenz.) Y. C. Hsu et H. Wei Jen;滇西青冈;Lobb Oak, W. Yunnan Cyclobalanopsis,W. Yunnan Qinggang ●

116137 Cyclobalanopsis longifolia Y. C. Hsu et Q. Z. Dong;长叶青冈;Longleaf Oak,Longleaf Qinggang,Long-leaved Cyclobalanopsis ●

116138 Cyclobalanopsis longifolia Y. C. Hsu et Q. Z. Dong = Cyclobalanopsis lungmaiensis Hu ●

116139 Cyclobalanopsis longinux (Hayata) Schottky;长果青冈(椆子,锥果椆,锥果栎);Long Glans Oak, Longfruit Oak, Longfruit Qinggang,Long-fruited Cyclobalanopsis,Long-nut Oak ●

116140 Cyclobalanopsis longinux (Hayata) Schottky var. kanehirai (Nakai) J. C. Liao = Cyclobalanopsis longinux (Hayata) Schottky var. kuoi J. C. Liao ●

116141 Cyclobalanopsis longinux (Hayata) Schottky var. kuoi J. C. Liao;无粉锥果栎(郭氏锥果栎,无粉锥果椆);Kuo's Longfruit Oak,Kuo's Longfruit Qinggang ●

116142 Cyclobalanopsis longinux (Hayata) Schottky var. kuoi J. C. Liao = Cyclobalanopsis longinux (Hayata) Schottky ●

116143 Cyclobalanopsis longinux (Hayata) Schottky var. lativiolaciifolia J. C. Liao;紫背锥果栎●

116144 Cyclobalanopsis longinux (Hayata) Schottky var. lativiolaciifolia J. C. Liao = Cyclobalanopsis longinux (Hayata) Schottky ●

116145 Cyclobalanopsis longinux (Hayata) Schottky var. pseudomyrsinifolia (Hayata) J. C. Liao = Cyclobalanopsis longinux (Hayata) Schottky ●

116146 Cyclobalanopsis lungmaiensis Hu;龙迈青冈;Longmai Cyclobalanopsis,Longmai Oak,Longmai Qinggang ●

116147 Cyclobalanopsis meihuashanensis Q. F. Zheng;梅花山青冈;Meihuashan Cyclobalanopsis,Meihuashan Oak,Meihuashan Qinggang ●

116148 Cyclobalanopsis meihuashanensis Q. F. Zheng = Cyclobalanopsis obovatifolia (C. C. Huang) Y. C. Hsu et H. Wei Jen ●

116149 Cyclobalanopsis miyagii (Koidz.) Kudo et Masam. = Quercus miyagii Koidz. ●☆

116150 Cyclobalanopsis morii (Hayata) Schottky;台湾青冈(赤椆,赤柯,森氏栎,校櫟);Mori Cyclobalanopsis, Mori Oak, Red Oak, Taiwan Qinggang ●

116151 Cyclobalanopsis motuoensis (C. C. Huang) Y. C. Hsu et H. Wei Jen;墨脱青冈;Motuo Cyclobalanopsis,Motuo Oak,Motuo Qinggang ●

116152 Cyclobalanopsis multinervis W. C. Cheng et T. Hong;多脉青冈(多脉青冈栎,粉背青冈);Manyveins Oak, Multiveined Cyclobalanopsis,Veiny Qinggang ●

116153 Cyclobalanopsis myrsinifolia (Blume) Oerst.;小叶青冈(黑椆,黑栎,面槠,青椆,青栲,铁槠);Bamboo-leaved Oak,Japanese Oak, Littleleaf Qinggang, Myrsina-leaved Cyclobalanopsis, Myrsineleaf Oak,Myrsine-leaf Oak ●

116154 Cyclobalanopsis myrsinifolia (Blume) Oerst. = Quercus myrsinifolia Blume ●

116155 Cyclobalanopsis nanchuanica (C. C. Huang) Y. T. Chang = Cyclobalanopsis gambleana (A. Camus) Y. C. Hsu et H. Wei Jen ●

116156 Cyclobalanopsis neglecta Schottky;竹叶青冈(红椆,铁槠,竹叶,竹叶青冈栎);Bamboo Leaf Oak,Bambooleaf Oak,Bambooleaf Qinggang,Bamboo-leaved Cyclobalanopsis,Bamboo-leaved Oak ●

116157 Cyclobalanopsis nengpulaensis H. Li et Y. C. Hsu;能铺拉青冈;Nengpula Qinggang ●

116158 Cyclobalanopsis nengpulaensis H. Li et Y. C. Hsu = Cyclobalanopsis fleuryi (Hickel et A. Camus) Chun ●

116159 Cyclobalanopsis nigrinervis Hu = Cyclobalanopsis chevalieri (Hickel et A. Camus) Y. C. Hsu et H. Wei Jen ●

116160 Cyclobalanopsis nigrinervis Hu = Cyclobalanopsis lamellosa (Sm.) Oerst. ●

116161 Cyclobalanopsis nigrinux Hu = Cyclobalanopsis chevalieri (Hickel et A. Camus) Y. C. Hsu et H. Wei Jen ●

116162 Cyclobalanopsis ningangensis W. C. Cheng et Y. C. Hsu;宁冈青冈;Ninggang Cyclobalanopsis,Ninggang Oak,Ninggang Qinggang ●

116163 Cyclobalanopsis nubium (Hand. -Mazz.) Chun ex Q. F. Zheng = Cyclobalanopsis sessilifolia (Blume) Schottky ●

116164 Cyclobalanopsis obovatifolia (C. C. Huang) Y. C. Hsu et H. Wei Jen;倒卵叶青冈;Obvateleaf Oak, Obvateleaf Qinggang, Obvate-leaved Cyclobalanopsis,Obvate-leaved Oak ●

116165 Cyclobalanopsis oxyodon (Miq.) Oerst.;曼青冈(曼椆);Sharpdent Oak,Sharpdent Qinggang,Sharp-dented Cyclobalanopsis, Sharp-dented Oak ●

116166 Cyclobalanopsis oxyodon (Miq.) Oerst. var. tomentosa Hu = Cyclobalanopsis gambleana (A. Camus) Y. C. Hsu et H. Wei Jen ●

116167 Cyclobalanopsis pachyloma (Seemen) Schottky;毛果青冈(赤椆,红校櫟,金斗椆,卷斗栎);Hairhull Qinggang,Revolute Cupule

Oak, Thik-leaf Oak, Thik-leaved Cyclobalanopsis ●

116168 Cyclobalanopsis pachyloma (Seemen) Schottky var. mubianensis (Y. C. Hsu et H. Wei Jen) C. C. Huang = Cyclobalanopsis pachyloma (Seemen) Schottky ●

116169 Cyclobalanopsis pachyloma (Seemen) Schottky var. mubianensis Y. C. Hsu et H. Wei Jen;睦边青冈;Mubian Cyclobalanopsis, Mubian Oak, Mubian Qinggang ●

116170 Cyclobalanopsis pachyloma (Seemen) Schottky var. mubianensis Y. C. Hsu et H. Wei Jen = Cyclobalanopsis pachyloma (Seemen) Schottky ●

116171 Cyclobalanopsis pachyloma (Seemen) Schottky var. tomentosicupula (Hayata) J. C. Liao = Cyclobalanopsis pachyloma (Seemen) Schottky ●

116172 Cyclobalanopsis patelliformis (Chun) Y. C. Hsu et H. Wei Jen;托盘青冈(托盘椆);Dish Oak, Dish-shaped Oak, Patelliform Cyclobalanopsis, Salver Qinggang ●

116173 Cyclobalanopsis paucidentata (Franch. ex Nakai) Kudo et Masam. = Cyclobalanopsis sessilifolia (Blume) Schottky ●

116174 Cyclobalanopsis paucidentata (Franch.) Kudo et Masam.;毽子栎;Shuttle-cock Oak ●

116175 Cyclobalanopsis paucidentata (Franch.) Kudo et Masam. = Cyclobalanopsis sessilifolia (Blume) Schottky ●

116176 Cyclobalanopsis pentacycla (Y. T. Chang) Y. T. Chang ex Y. C. Hsu et H. Wei Jen;五环青冈(五环椆);Fivering Qinggang, Pentacyclic Cyclobalanopsis, Pentacyclic Oak ●

116177 Cyclobalanopsis phanera (Chun) Y. C. Hsu et H. Wei Jen;亮叶青冈;Brightleaf Qinggang, Cyclobalanopsis, Obvious Cyclobalanopsis, Shinyleaf Oak, Shiny-leaved Oak ●

116178 Cyclobalanopsis pinbianensis Y. C. Hsu et H. Wei Jen;屏边青冈;Pingbian Cyclobalanopsis, Pingbian Oak, Pingbian Qinggang ●

116179 Cyclobalanopsis pinbianensis Y. C. Hsu et H. Wei Jen = Cyclobalanopsis jenseniana (Hand. -Mazz.) W. C. Cheng et T. Hong ex Q. F. Zheng ●

116180 Cyclobalanopsis poilanei (Hickel et A. Camus) Hjelmq.;黄背青冈;Poilane Cyclobalanopsis, Poilane Oak, Yellowback Qinggang ●

116181 Cyclobalanopsis pseudoglauca Y. K. Li et X. M. Wang;长叶粉背青冈;False Blue Japanese Oak, False Grey-blue Cyclobalanopsis, Longleaf Whiteback Qinggang ●

116182 Cyclobalanopsis pseudoglauca Y. K. Li et X. M. Wang = Cyclobalanopsis gracilis (Rehder et E. H. Wilson) W. C. Cheng et T. Hong ●

116183 Cyclobalanopsis pypargyrea (Seem.) Y. C. Hsu et H. Wei Jen = Cyclobalanopsis multinervis W. C. Cheng et T. Hong ●

116184 Cyclobalanopsis repandifolia (J. C. Liao) J. C. Liao;波叶栎(波叶椆)●

116185 Cyclobalanopsis repandifolia (J. C. Liao) J. C. Liao = Cyclobalanopsis glauca (Thunb. ex A. Murray) Oerst. ●

116186 Cyclobalanopsis rex (Hemsl.) Schottky;大果青冈(大果椆);Bigfruit Qinggang, Big-fruited Cyclobalanopsis, Big-fruited Oak, Largefruit Oak ●◇

116187 Cyclobalanopsis salicina (Blume) Oerst.;柳叶青冈(白背栎,柳栎,柳叶槠,竹野青冈栎);Willow Oak, Willowleaf Cyclobalanopsis, Willowleaf Qinggang, Willow-like Oak ●

116188 Cyclobalanopsis salicina (Blume) Oerst. = Quercus salicina Blume ●

116189 Cyclobalanopsis saravanensis (A. Camus) Hjelmq.;薄叶青冈;Kontum Cyclobalanopsis, Thinleaf Oak, Thinleaf Qinggang ●

116190 Cyclobalanopsis semiserrata (Roxb.) Oerst.;无齿青冈●

116191 Cyclobalanopsis semiserratoides Y. C. Hsu et H. Wei Jen;拟半齿青冈(无齿青冈);Semitooth Cyclobalanopsis, Semitooth Oak, Toothless Qinggang ●

116192 Cyclobalanopsis semiserratoides Y. C. Hsu et H. Wei Jen = Cyclobalanopsis semiserrata (Roxb.) Oerst. ●

116193 Cyclobalanopsis sessilifolia (Blume) Schottky;云山青冈(毽子椆,毽子栎,云山椆);Sessile Qinggang, Sessile-leaf Oak, Yunshan Cyclobalanopsis, Yunshan Oak ●

116194 Cyclobalanopsis sessilifolia (Blume) Schottky = Quercus sessilifolia Blume ●

116195 Cyclobalanopsis shennongii (C. C. Huang et S. H. Fu) Y. C. Hsu et H. Wei Jen;神农青冈(神农栎);Shennong Cyclobalanopsis, Shennong Oak, Shennong Qinggang ●

116196 Cyclobalanopsis shennongii (C. C. Huang et S. H. Fu) Y. C. Hsu et H. Wei Jen = Cyclobalanopsis gracilis (Rehder et E. H. Wilson) W. C. Cheng et T. Hong ●

116197 Cyclobalanopsis shiangpyngensis Hu = Cyclobalanopsis chapensis (Hickel et A. Camus) Y. C. Hsu et H. Wei Jen ●

116198 Cyclobalanopsis sichourensis Hu;西畴青冈;Xichou Cyclobalanopsis, Xichou Oak, Xichou Qinggang ●◇

116199 Cyclobalanopsis stenophylla (Blume) Schottky var. stenophylloides (Hayata) J. C. Liao = Cyclobalanopsis stenophylloides (Hayata) Kudo et Masam. ex Kudo ●

116200 Cyclobalanopsis stenophylloides (Hayata) Kudo et Masam. ex Kudo;台湾窄叶青冈(狭叶椆,狭叶高山栎,狭叶栎);Arishan Oak, Narrow-leaved Cyclobalanopsis, Narrow-leaved Oak, Taiwan Narrowleaf Qinggang, Taiwan Oak ●

116201 Cyclobalanopsis stewardiana (A. Camus) Y. C. Hsu et H. Wei Jen;褐叶青冈;Brownleaf Qinggang, Steward Cyclobalanopsis, Steward Oak ●

116202 Cyclobalanopsis stewardiana (A. Camus) Y. C. Hsu et H. Wei Jen var. longicaudata Y. C. Hsu et al.;长尾青冈;Longtail Qinggang, Longtail Steward Cyclobalanopsis, Longtail Steward Oak ●

116203 Cyclobalanopsis stewardiana (A. Camus) Y. C. Hsu et H. Wei Jen var. longicaudata Y. C. Hsu et al. = Cyclobalanopsis stewardiana (A. Camus) Y. C. Hsu et H. Wei Jen ●

116204 Cyclobalanopsis subhinoidea (Chun et W. C. Ko) Y. C. Hsu et H. Wei Jen ex Y. T. Chang;平脉青冈(海南青冈,鹿茸椆,鹿茸青冈);Antlerpilose Qinggang, Hainan Oak, Pilose-antlered Cyclobalanopsis ●

116205 Cyclobalanopsis takaoyamensis (Makino) Kudo et Masam. = Quercus × takaoyamensis Makino ●☆

116206 Cyclobalanopsis tenuicupula Y. C. Hsu et H. Wei Jen;薄斗青冈;Thincupula Oak, Thincupula Qinggang, Thin-cupule Cyclobalanopsis ●

116207 Cyclobalanopsis ternaticupula (Hayata) Kudo = Lithocarpus hancei (Benth.) Rehder ●

116208 Cyclobalanopsis ternaticupula (Hayata) Kudo f. arisanensis (Hayata) Kudo = Lithocarpus hancei (Benth.) Rehder ●

116209 Cyclobalanopsis ternaticupula (Hayata) Kudo f. arisanensis (Hayata) Kaneh. = Lithocarpus hancei (Benth.) Rehder ●

116210 Cyclobalanopsis thorelii (Hickel et A. Camus) Hu;厚缘青冈;Thickdge Qinggang, Thorel Cyclobalanopsis, Thorel Oak ●

116211 Cyclobalanopsis tiaoloshanica (Chun et W. C. Ko) Y. C. Hsu et H. Wei Jen;吊罗山青冈;Diaoluoshan Cyclobalanopsis, Diaoluoshan Oak, Diaoluoshan Qinggang ●

116212 Cyclobalanopsis tomentosinervis Y. C. Hsu et H. Wei Jen;毛脉青冈;Hairyvein Oak,Hairyvein Qinggang,Hairy-veined Cyclobalanopsis ●

116213 Cyclobalanopsis xanthotricha (A. Camus) Y. C. Hsu et H. Wei Jen;思茅青冈;Simao Cyclobalanopsis,Simao Oak,Simao Qinggang ●

116214 Cyclobalanopsis xiangxiensis C. J. Qi et Q. Z. Lin;湘西青冈;Xiangxi Cyclobalanopsis,Xiangxi Qinggang ●

116215 Cyclobalanopsis xiangxiensis C. J. Qi et Q. Z. Lin = Cyclobalanopsis kiukiangensis Y. T. Chang ex Y. C. Hsu et H. Wei Jen ●

116216 Cyclobalanopsis xizangensis Y. C. Hsu et H. Wei Jen;西藏青冈;Xizang Cyclobalanopsis,Xizang Oak,Xizang Qinggang ●

116217 Cyclobalanopsis xizangensis Y. C. Hsu et H. Wei Jen = Cyclobalanopsis kiukiangensis Y. T. Chang ex Y. C. Hsu et H. Wei Jen ●

116218 Cyclobalanopsis yingjiangensis Y. C. Hsu et Q. Z. Dong;盈江青冈;Yingjiang Cyclobalanopsis,Yingjiang Oak,Yingjiang Qinggang ●

116219 Cyclobalanopsis yonganensis (L. Lin et C. C. Huang) Y. C. Hsu et H. Wei Jen;永安青冈;Yong'an Cyclobalanopsis,Yong'an Oak,Yong'an Qinggang ●

116220 Cyclobalanus (Endl.) Oerst. (1867);红肉杜属●☆

116221 Cyclobalanus (Endl.) Oerst. = Lithocarpus Blume ●

116222 Cyclobalanus Oerst. = Cyclobalanus (Endl.) Oerst. ●☆

116223 Cyclobalanus hancei (Benth.) Oerst. = Lithocarpus hancei (Benth.) Rehder ●

116224 Cyclobalanus ternaticupula (Hayata) Nakai = Lithocarpus hancei (Benth.) Rehder ●

116225 Cyclobothra D. Don = Calochortus Pursh ■☆

116226 Cyclobothra D. Don ex Sweet = Calochortus Pursh ■☆

116227 Cyclobothra alba Benth. = Calochortus albus (Benth.) Douglas ex Benth. ■☆

116228 Cyclobothra coerulea Kellogg = Calochortus coeruleus (Kellogg) S. Watson ■☆

116229 Cyclobothra monophylla Lindl. = Calochortus monophyllus (Lindl.) Lem. ■☆

116230 Cyclobothra pulchella Benth. = Calochortus pulchellus (Benth.) A. W. Wood ■☆

116231 Cyclocampe Benth. et Hook. f. = Lophoschoenus Stapf ■☆

116232 Cyclocampe Steud. = Schoenus L. ■

116233 Cyclocarpa Afzel. ex Baker = Cyclocarpa Afzel. ex Urb. ■☆

116234 Cyclocarpa Afzel. ex Urb. (1884);球豆属■☆

116235 Cyclocarpa Miq. = Cyclocampe Steud. ■

116236 Cyclocarpa Miq. = Schoenus L. ■

116237 Cyclocarpa stellaris Afzel. ex Baker;球豆■☆

116238 Cyclocarpus Jungh. = Evodia J. R. Forst. et G. Forst. ●

116239 Cyclocarya Iljinsk. (1953);青钱柳属;Cyclocarya ●★

116240 Cyclocarya paliurus (Batalin) Iljinsk.;青钱柳(大叶水化香,麻柳,青钱李,山沟树,山化树,山麻柳,甜茶树,摇钱树,一串钱);Cyclocarya,Diskfruit Wingnut,Round Wingfruit Cyclocarya,Roundwingfruit Cyclocarya,Round-wing-fruited Cyclocarya ●

116241 Cyclocarya paliurus (Batalin) Iljinsk. var. micropaliurus (P. C. Tsoong) P. S. Hsu et al. = Cyclocarya paliurus (Batalin) Iljinsk. ●

116242 Cyclocheilaceae Marais = Orobanchaceae Vent. (保留科名) ●■

116243 Cyclocheilaceae Marais(1981);盘果木科(圆唇花科)●☆

116244 Cyclocheilon Oliv. (1895);盘果木属(圆唇花属)●☆

116245 Cyclocheilon eriantherum (Vatke) Engl. = Asepalum eriantherum (Vatke) Marais ●☆

116246 Cyclocheilon eriantherum (Vatke) Engl. var. decurrens Chiov. = Asepalum eriantherum (Vatke) Marais ●☆

116247 Cyclocheilon kelleri Engl.;凯乐盘果木●☆

116248 Cyclocheilon minutibracteolatum Engl. = Asepalum eriantherum (Vatke) Marais ●☆

116249 Cyclocheilon physocalyx Chiov.;囊果盘果木●☆

116250 Cyclocheilon somalense Oliv. var. kelleri (Engl.) Stapf = Cyclocheilon kelleri Engl. ●☆

116251 Cyclocheilon somaliense Oliv.;盘果木●☆

116252 Cyclochilus Post et Kuntze = Cyclocheilon Oliv. ●☆

116253 Cyclocodon Griff. (1858);土党参属(轮钟草属,轮钟花属)■

116254 Cyclocodon Griff. = Campanumoea Blume ■

116255 Cyclocodon Griff. = Codonopsis Wall. ex Roxb. ■

116256 Cyclocodon adnatus Griff. = Campanumoea lancifolia (Roxb.) Merr. ■

116257 Cyclocodon celebicus (Blume) D. Y. Hong;小叶轮钟草;Smallleaf Campanumoea,Smallleaf Leopard ■

116258 Cyclocodon lancifolius (Roxb.) Kurz;台湾土党参(轮钟花)■

116259 Cyclocodon lancifolius (Roxb.) Kurz = Campanumoea lancifolia (Roxb.) Merr. ■

116260 Cyclocodon lancifolius (Roxb.) Moeliono = Cyclocodon lancifolius (Roxb.) Kurz ■

116261 Cyclocodon parviflorus (Wall. ex A. DC.) Hook. f. et Thomson = Campanumoea parviflora (Wall. ex A. DC.) Benth. et Hook. f. ■

116262 Cyclocodon parviflorus (Wall.) Hook. f. et Thomson = Campanumoea parviflora (Wall. ex A. DC.) Benth. et Hook. f. ■

116263 Cyclocodon truncatus (Wall. ex A. DC.) Hook. f. et Thomson = Cyclocodon lancifolius (Roxb.) Kurz ■

116264 Cyclocodon truncatus (Wall.) Hook. f. et Thomson = Campanumoea lancifolia (Roxb.) Merr. ■

116265 Cyclocotyla Stapf(1908);环杯夹竹桃属●☆

116266 Cyclocotyla congolensis Stapf;刚果环杯夹竹桃●☆

116267 Cyclocotyla oligosperma Wernham = Cyclocotyla congolensis Stapf ●☆

116268 Cyclodiscus K. Schum. = Cylicodiscus Harms ●☆

116269 Cyclodiscus Klotzsch = Apama Lam. ●

116270 Cyclodiscus Klotzsch = Munnickia Blume ex Rchb. ●

116271 Cyclodiscus Klotzsch = Thottea Rottb. ●

116272 Cyclodon Small = Vincetoxicum Wolf ●■

116273 Cyclogyne Benth. = Swainsona Salisb. ●■☆

116274 Cyclogyne Benth. ex Lindl. = Swainsona Salisb. ●■☆

116275 Cyclolepis Gillies ex D. Don(1832);脱叶菊属●☆

116276 Cyclolepis Moq. = Cycloloma Moq. ■☆

116277 Cyclolepis Moq. = Petermarmia Rchb. ■☆

116278 Cyclolepis genistoides Gillies ex D. Don;脱叶菊●☆

116279 Cyclolepis platyphylla (Michx.) Moq. = Cycloloma atriplicifolium (Spreng.) J. M. Coult. ■☆

116280 Cyclolepsis Endl. = Cyclolepis Gillies ex D. Don ●☆

116281 Cyclolobium Benth. (1837);环裂豆属●☆

116282 Cyclolobium clausseni Benth.;环裂豆●☆

116283 Cyclolobium vecchii A. Samp. ex Hoehne;韦氏环裂豆●☆

116284 Cycloloma Moq. (1840);环翅藜属(环翅萼藜属);Winged Pigweed ■☆

116285 Cycloloma atriplicifolium (Spreng.) J. M. Coult. = Cycloloma atriplicifoliura (Roth) J. M. Coult. ■☆

116286 Cycloloma atriplicifoliura (Roth) J. M. Coult.;环翅藜;Tumble Ringwing,Winged Pigweed ■☆

116287 Cycloloma platyphyllum (Michx.) Moq. = Cycloloma atriplicifolium (Spreng.) J. M. Coult. ■☆
116288 Cyclomorium Walp. = Desmodium Desv. (保留属名)●■
116289 Cyclonema Hochst. = Clerodendrum L. ●■
116290 Cyclonema Hochst. = Rotheca Raf. ●☆
116291 Cyclonema discolor Klotzsch = Rotheca myricoides (Hochst.) Steane et Mabb. var. discolor (Klotzsch) Verdc. ●☆
116292 Cyclonema hirsutum Hochst. = Rotheca hirsuta (Hochst.) R. Fern. ●☆
116293 Cyclonema mucronatum Klotzsch = Karomia tettensis (Klotzsch) R. Fern. ●☆
116294 Cyclonema myricoides (Hochst.) Hochst. = Rotheca myricoides (Hochst.) Steane et Mabb. ●☆
116295 Cyclonema serratum Hochst.;具齿环翅藜●☆
116296 Cyclonema spinescens Klotzsch = Karomia tettensis (Klotzsch) R. Fern. ●☆
116297 Cyclonema spinescens Oliv. = Clerodendrum uncinatum Schinz ●☆
116298 Cyclonema sylvaticum Hochst. = Rotheca myricoides (Hochst.) Steane et Mabb. ●☆
116299 Cyclonema tettensis Klotzsch = Karomia tettensis (Klotzsch) R. Fern. ●☆
116300 Cyclonema triphyllum Harv. = Rotheca hirsuta (Hochst.) R. Fern. ●☆
116301 Cyclopappus Cass. ex Sch. Bip. = Asteraceae Bercht. et J. Presl (保留科名)●■
116302 Cyclopappus Cass. ex Sch. Bip. = Compositae Giseke(保留科名)●■
116303 Cyclophyllum Hook. f. (1873);圆叶茜属●☆
116304 Cyclophyllum fragrans (Schltr. et K. Krause) Mouly;香圆叶茜●☆
116305 Cyclophyllum longiflorum (Valeton) A. P. Davis et Ruhsam;长花圆叶茜●☆
116306 Cyclophyllum sessilifolium (A. Gray) A. C. Sm. et S. P. Darwin;无梗圆叶茜●☆
116307 Cyclopia Vent. (1808);南非蜜茶属●☆
116308 Cyclopia alopecuroides A. L. Schutte;看麦娘南非蜜茶●☆
116309 Cyclopia alpina A. L. Schutte;高山南非蜜茶●☆
116310 Cyclopia ashtonii Hofmeyr et E. Phillips = Cyclopia bowieana Harv. ●☆
116311 Cyclopia aurea Fourc. = Cyclopia intermedia E. Mey. ●☆
116312 Cyclopia aurescens Kies var. glauca ? = Cyclopia buxifolia (Burm. f.) Kies ●☆
116313 Cyclopia bolusii Hofmeyr et E. Phillips;博卢斯南非蜜茶●☆
116314 Cyclopia bowieana Harv.;博韦南非蜜茶●☆
116315 Cyclopia brachypoda Benth. = Cyclopia sessiliflora Eckl. et Zeyh. ●☆
116316 Cyclopia brachypoda Benth. var. intermedia (E. Mey.) Hofmeyr et E. Phillips = Cyclopia intermedia E. Mey. ●☆
116317 Cyclopia burtonii Hofmeyr et E. Phillips;伯顿南非蜜茶●☆
116318 Cyclopia buxifolia (Burm. f.) Kies;黄杨叶南非蜜茶●☆
116319 Cyclopia capensis T. M. Salter = Cyclopia galioides (P. J. Bergius) DC. ●☆
116320 Cyclopia cordifolia Benth. = Cyclopia latifolia DC. ●☆
116321 Cyclopia dregeana Kies = Cyclopia buxifolia (Burm. f.) Kies ●☆
116322 Cyclopia falcata (Harv.) Kies;镰形南非蜜茶●☆
116323 Cyclopia falcata (Harv.) Kies var. ovata Kies = Cyclopia buxifolia (Burm. f.) Kies ●☆
116324 Cyclopia filiformis Kies;线形南非蜜茶●☆
116325 Cyclopia galioides (P. J. Bergius) DC.;好望角南非蜜茶●☆
116326 Cyclopia genistoides (L.) R. Br.;金雀南非蜜茶●☆
116327 Cyclopia genistoides (L.) R. Br. var. heterophylla Harv. = Cyclopia genistoides (L.) R. Br. ●☆
116328 Cyclopia genistoides (L.) R. Br. var. ovalifolia Kies = Cyclopia alpina A. L. Schutte ●☆
116329 Cyclopia genistoides (L.) R. Br. var. teretifolia (Eckl. et Zeyh.) Kies = Cyclopia genistoides (L.) R. Br. ●☆
116330 Cyclopia glabra (Hofmeyr et E. Phillips) A. L. Schutte;光滑南非蜜茶●☆
116331 Cyclopia grandiflora A. DC. = Cyclopia subternata Vogel ●☆
116332 Cyclopia heterophylla Eckl. et Zeyh. = Cyclopia genistoides (L.) R. Br. ●☆
116333 Cyclopia intermedia E. Mey.;间型南非蜜茶●☆
116334 Cyclopia laricina E. Mey. = Cyclopia maculata (Andréws) Kies ●☆
116335 Cyclopia latifolia DC.;宽叶南非蜜茶●☆
116336 Cyclopia laxiflora Benth.;疏花南非蜜茶●☆
116337 Cyclopia longifolia Vogel;长叶南非蜜茶●☆
116338 Cyclopia maculata (Andréws) Kies;斑点南非蜜茶●☆
116339 Cyclopia meyeriana Walp.;迈尔南非蜜茶●☆
116340 Cyclopia montana Hofmeyr et E. Phillips;山地南非蜜茶●☆
116341 Cyclopia montana Hofmeyr et E. Phillips var. glabra ? = Cyclopia glabra (Hofmeyr et E. Phillips) A. L. Schutte ●☆
116342 Cyclopia plicata Kies;折叠南非蜜茶●☆
116343 Cyclopia pubescens Eckl. et Zeyh.;短柔毛南非蜜茶●☆
116344 Cyclopia sessiliflora E. Mey. = Cyclopia meyeriana Walp. ●☆
116345 Cyclopia sessiliflora Eckl. et Zeyh.;无花梗南非蜜茶●☆
116346 Cyclopia squamosa A. L. Schutte;多鳞南非蜜茶●☆
116347 Cyclopia subternata Vogel;近三出南非蜜茶●☆
116348 Cyclopia subternata Vogel var. laxiflora (Benth.) Kies = Cyclopia laxiflora Benth. ●☆
116349 Cyclopia tenuifolia Lehm. = Cyclopia maculata (Andréws) Kies ●☆
116350 Cyclopia teretifolia Eckl. et Zeyh. = Cyclopia genistoides (L.) R. Br. ●☆
116351 Cyclopia vogelii Harv. var. brachypoda (Benth.) Harv. = Cyclopia sessiliflora Eckl. et Zeyh. ●☆
116352 Cyclopia vogelii Harv. var. falcata Harv. = Cyclopia falcata (Harv.) Kies ●☆
116353 Cyclopia vogelii Harv. var. intermedia (E. Mey.) Harv. = Cyclopia intermedia E. Mey. ●☆
116354 Cyclopia vogelii Harv. var. laxiflora (Benth.) Harv. = Cyclopia laxiflora Benth. ●☆
116355 Cyclopia vogelii Harv. var. subternata (Vogel) Harv. = Cyclopia subternata Vogel ●☆
116356 Cyclopis Guill. = Cyclolepis Gillies ex D. Don ●☆
116357 Cyclopogon C. Presl(1827);萼基毛兰属(环毛兰属)■☆
116358 Cyclopogon americana (C. Schweinf. et Garay) Burns = Manniella americana C. Schweinf. et Garay ■☆
116359 Cyclopogon cranichoides (Griseb.) Schltr.;萼基毛兰■☆
116360 Cyclopogon elatus (Sw.) Schltr.;大萼基毛兰■☆
116361 Cycloptera (R. Br.) Spach = Grevillea R. Br. ex Knight(保留属名)●
116362 Cycloptera Endl. = Cyclopogon C. Presl ■☆

116363 Cyclopterygium Hochst. = Schouwia DC. (保留属名) ■☆
116364 Cycloptychis E. Mey. = Cycloptychis E. Mey. ex Sond. ■☆
116365 Cycloptychis E. Mey. ex Arn. = Cycloptychis E. Mey. ex Sond. ■☆
116366 Cycloptychis E. Mey. ex Sond. (1841);南非褶芥属■☆
116367 Cycloptychis polygaloides Sond. = Heliophila nubigena Schltr. ■☆
116368 Cycloptychis virgata (Thunb.) E. Mey. ex Sond. = Heliophila maraisiana Al-Shehbaz et Mummenhoff ■☆
116369 Cycloptychis virgata E. Mey.;南非褶芥■☆
116370 Cyclorhiza M. L. Sheh et R. H. Shan (1980);环根芹属; Cyclorhiza ■★
116371 Cyclorhiza major (M. L. Sheh et R. H. Shan) M. L. Shen = Cyclorhiza peucedanifolia (Franch.) Constance ■
116372 Cyclorhiza peucedanifolia (Franch.) Constance;南竹叶环根芹;Bigger Cyclorhiza ■
116373 Cyclorhiza waltonii (H. Wolff) M. L. Sheh et R. H. Shan;环根芹;Walton Cyclorhiza ■
116374 Cyclorhiza waltonii (H. Wolff) M. L. Sheh et R. H. Shan var. major M. L. Sheh et R. H. Shan = Cyclorhiza major (M. L. Sheh et R. H. Shan) M. L. Sheh ■
116375 Cyclorhiza waltonii (H. Wolff) M. L. Sheh et R. H. Shan var. major M. L. Sheh et R. H. Shan = Cyclorhiza peucedanifolia (Franch.) Constance ■
116376 Cyclosanthes Poepp. = Cyclanthus Poit. ex A. Rich. ■☆
116377 Cyclosia Klotzsch = Mormodes Lindl. ■☆
116378 Cyclospathe O. F. Cook = Pseudophoenix H. Wendl. ex Sarg. (废弃属名) ●☆
116379 Cyclospathe O. F. Cook = Sargentia S. Watson(保留属名) ●☆
116380 Cyclospermum Caruel = Ciclospermum Lag. ■
116381 Cyclospermum Lag. (1821) ('Ciclospermum') (保留属名);细叶旱芹属(圆果旱芹属) ■
116382 Cyclospermum leptophyllum (Pers.) Sprague = Apium leptophyllum (Pers.) F. Muell. ex Benth. ■
116383 Cyclospermum leptophyllum (Pers.) Sprague = Cyclospermum leptophyllum (Pers.) Sprague ex Britton et P. Wilson ■
116384 Cyclospermum leptophyllum (Pers.) Sprague ex Britton et P. Wilson;细叶旱芹(薄叶芹菜,纤叶芹,圆果旱芹);Marsh Parsley, Thinleaf Celery ■
116385 Cyclospermum leptophyllum (Pers.) Sprague ex Britton et P. Wilson = Apium leptophyllum (Pers.) F. Muell. ex Benth. ■
116386 Cyclostachya Reeder et C. Reeder(1963);匍匐圆穗草属■☆
116387 Cyclostachya stolonifera (Scribn.) Reeder et C. Reeder;匍匐圆穗草■☆
116388 Cyclostegia Benth. = Elsholtzia Willd. ●■
116389 Cyclostegia strobilifera Benth. = Elsholtzia strobilifera Benth. ■
116390 Cyclostemon Blume = Drypetes Vahl ●
116391 Cyclostemon afzelii Pax = Drypetes afzelii (Pax) Hutch. ●☆
116392 Cyclostemon aquifolium Scott-Elliot = Drypetes madagascariensis (Lam.) Humbert et Leandri ●☆
116393 Cyclostemon argutus Müll. Arg. = Drypetes arguta (Müll. Arg.) Hutch. ●☆
116394 Cyclostemon bipindensis Pax = Drypetes bipindensis (Pax) Hutch. ●☆
116395 Cyclostemon cumingii Baill. = Drypetes cumingii (Baill.) Pax et K. Hoffm. ●
116396 Cyclostemon cuspidatum Blume = Aphananthe cuspidata (Blume) Planch. ●
116397 Cyclostemon dinklagei Pax = Drypetes dinklagei (Pax) Hutch. ●☆
116398 Cyclostemon floribundus Müll. Arg. = Drypetes floribunda (Müll. Arg.) Hutch. ●☆
116399 Cyclostemon gabonensis Pierre ex Hutch. = Drypetes gabonensis (Pierre ex Hutch.) Hutch. ●☆
116400 Cyclostemon gilgianus Pax = Drypetes gilgiana (Pax) Pax et K. Hoffm. ●☆
116401 Cyclostemon glaber Pax = Drypetes glabra (Pax) Hutch. ●☆
116402 Cyclostemon glomeratus Müll. Arg. = Drypetes fernandopoana Brenan ●☆
116403 Cyclostemon griffithii Hook. f. = Drypetes indica (Müll. Arg.) Pax et K. Hoffm. ●◇
116404 Cyclostemon henriquesii Pax = Drypetes henriquesii (Pax) Hutch. ●☆
116405 Cyclostemon hieranensis Hayata = Drypetes indica (Müll. Arg.) Pax et K. Hoffm. ●◇
116406 Cyclostemon indicus Müll. Arg. = Drypetes indica (Müll. Arg.) Pax et K. Hoffm. ●◇
116407 Cyclostemon iwahigensis Elmer = Drypetes littoralis (C. B. Rob.) Merr. ●
116408 Cyclostemon karapinensis Hayata = Drypetes indica (Müll. Arg.) Pax et K. Hoffm. ●◇
116409 Cyclostemon klaineanum Pierre = Drypetes klaineana (Pierre) Breteler ●☆
116410 Cyclostemon laciniatus Pax = Drypetes laciniata (Pax) Hutch. ●☆
116411 Cyclostemon lancifolius Hook. f. = Drypetes indica (Müll. Arg.) Pax et K. Hoffm. ●◇
116412 Cyclostemon leonensis Pax = Drypetes inaequalis Hutch. ●☆
116413 Cyclostemon littoralis C. B. Rob. = Drypetes littoralis (C. B. Rob.) Merr. ●
116414 Cyclostemon magnistipulus Pax = Drypetes magnistipula (Pax) Hutch. ●☆
116415 Cyclostemon major Pax = Drypetes natalensis (Harv.) Hutch. ●☆
116416 Cyclostemon mildbraedii Pax = Drypetes mildbraedii (Pax) Hutch. ●☆
116417 Cyclostemon mindorensis Merr. = Drypetes littoralis (C. B. Rob.) Merr. ●
116418 Cyclostemon mottikoro Léandri = Drypetes aylmeri Hutch. et Dalziel ●☆
116419 Cyclostemon natalensis Harv. = Drypetes natalensis (Harv.) Hutch. ●☆
116420 Cyclostemon occidentalis Müll. Arg. = Drypetes occidentalis (Müll. Arg.) Hutch. ●☆
116421 Cyclostemon parvifolius Müll. Arg. = Drypetes parvifolia (Müll. Arg.) Pax et K. Hoffm. ●☆
116422 Cyclostemon preussii Pax = Drypetes preussii (Pax) Hutch. ●☆
116423 Cyclostemon principum Müll. Arg. = Drypetes principum (Müll. Arg.) Hutch. ●☆
116424 Cyclostemon staudtii Pax = Drypetes staudtii (Pax) Hutch. ●☆
116425 Cyclostemon stipularis Müll. Arg. = Drypetes stipularis (Müll. Arg.) Hutch. ●☆
116426 Cyclostemon tessmannianus Pax = Drypetes tessmanniana (Pax) Pax et K. Hoffm. ●☆
116427 Cyclostemon ugandensis Rendle = Drypetes ugandensis (Rendle) Hutch. ●☆
116428 Cyclostemon usambaricus Pax = Drypetes usambarica (Pax) Hutch. ●☆

116429 Cyclostemon yamadae Kaneh. et Sasaki = Drypetes littoralis (C. B. Rob.) Merr. ●
116430 Cyclostigma Hochst. ex Endl. = Voacanga Thouars ●
116431 Cyclostigma Klotzsch = Croton L. ●
116432 Cyclostigma Phil. = Leptoglossis Benth. ■☆
116433 Cyclostigma natalense (Hochst.) Hochst. = Voacanga thouarsii Roem. et Schult. ●☆
116434 Cyclotaxis Boiss. = Scandix L. ■
116435 Cycloteria Stapf = Coelorachis Brongn. ■
116436 Cycloteria Stapf = Coelorhachis Brongn. + Rhytachne Desv. ex Ham. ■
116437 Cyclotheca Moq. = Gyrostemon Desf. ●☆
116438 Cyclotrichium (Boiss.) Manden. et Scheng. (1953);环毛草属●☆
116439 Cyclotrichium Manden. et Scheng. = Cyclotrichium (Boiss.) Manden. et Scheng. ●☆
116440 Cyclotrichium floridum (Boiss.) Manden. et Scheng.;佛罗里达环毛草●☆
116441 Cyclotrichium glabrescens (Boiss. et Kotsch ex Rech. f.) Leblebici;渐光环毛草●☆
116442 Cyclotrichium leucotrichum (Stapf) Leblebici;白毛环毛草●☆
116443 Cyclotrichium longiflorum Leblebici;长花环毛草●☆
116444 Cyclotrichium niveum (Boiss.) Manden. et Scheng.;雪白环毛草●☆
116445 Cycnia Griff. = ? Parinari Aubl. ●☆
116446 Cycnia Lindl. = Prinsepia Royle ●
116447 Cycniopsis Engl. (1905);拟鹅参属■☆
116448 Cycniopsis humifusa (Forssk.) Engl.;拟鹅参■☆
116449 Cycniopsis humifusa (Forssk.) Engl. f. hispida Fiori = Cycniopsis humifusa (Forssk.) Engl. ■☆
116450 Cycniopsis humifusa (Forssk.) Engl. var. parviflora Pax = Cycniopsis humilis (Hochst. ex Benth.) Backlund, Asfaw Hunde et E. Langström ■☆
116451 Cycniopsis humilis (Hochst. ex Benth.) Backlund, Asfaw Hunde et E. Langström;低矮拟鹅参■☆
116452 Cycniopsis minima Engl. = Cycniopsis humifusa (Forssk.) Engl. ■☆
116453 Cycniopsis obtusifolia Skan = Cycniopsis humifusa (Forssk.) Engl. ■☆
116454 Cycnium E. Mey. = Cycnium E. Mey. ex Benth. ■●☆
116455 Cycnium E. Mey. ex Benth. (1836);鹅参属■●☆
116456 Cycnium adoense Benth. et Hook. f. = Cycnium adonense E. Mey. ex Benth. ■☆
116457 Cycnium adonense E. Mey. ex Benth.;阿多鹅参■☆
116458 Cycnium adonense E. Mey. ex Benth. subsp. camporum (Engl.) O. J. Hansen;弯阿多鹅参■☆
116459 Cycnium adonense E. Mey. ex Benth. var. adscendens Oliv. = Cycnium adonense E. Mey. ex Benth. ■☆
116460 Cycnium amaniense Engl.;阿马尼鹅参■☆
116461 Cycnium angolense (Engl.) O. J. Hansen;安哥拉鹅参■☆
116462 Cycnium aquaticum Engl. = Cycnium tubulosum (L. f.) Engl. ■☆
116463 Cycnium bequaertii De Wild. = Cycnium herzfeldianum (Vatke) Engl. ■☆
116464 Cycnium brachycalyx Schweinf.;短萼鹅参■☆
116465 Cycnium breviflorum Ghaz.;短花鹅参■☆
116466 Cycnium brevifolium De Wild. = Cycnium tubulosum (L. f.) Engl. subsp. montanum (N. E. Br.) O. J. Hansen ■☆
116467 Cycnium bricchetii Engl. = Cycnium volkensii Engl. ■☆
116468 Cycnium cameronianum (Oliv.) Engl.;卡梅伦鹅参■☆
116469 Cycnium camporum Engl. = Cycnium adonense E. Mey. ex Benth. subsp. camporum (Engl.) O. J. Hansen ■☆
116470 Cycnium carvalhoi Engl. = Jamesbrittenia carvalhoi (Engl.) Hilliard ■☆
116471 Cycnium chevalieri Diels;舍瓦利耶鹅参■☆
116472 Cycnium claessensii De Wild. = Cycnium tubulosum (L. f.) Engl. subsp. montanum (N. E. Br.) O. J. Hansen ■☆
116473 Cycnium dewevrei De Wild. et T. Durand = Cycnium adonense E. Mey. ex Benth. subsp. camporum (Engl.) O. J. Hansen ■☆
116474 Cycnium elskensii De Wild. = Cycnium tubulosum (L. f.) Engl. subsp. montanum (N. E. Br.) O. J. Hansen ■☆
116475 Cycnium erectum Rendle;直立鹅参■☆
116476 Cycnium filicalyx (E. A. Bruce) O. J. Hansen;丝萼鹅参■☆
116477 Cycnium fruticans Engl. = Cycnium erectum Rendle ■☆
116478 Cycnium gallaense Engl. = Cycnium volkensii Engl. ■☆
116479 Cycnium hamatum Engl. et Gilg = Cycnium tubulosum (L. f.) Engl. ■☆
116480 Cycnium herzfeldianum (Vatke) Engl.;赫茨菲尔德鹅参■☆
116481 Cycnium herzfeldianum (Vatke) Engl. f. holstii Engl. ex Engl. = Cycnium herzfeldianum (Vatke) Engl. ■☆
116482 Cycnium herzfeldianum (Vatke) Engl. var. subauriculata ? = Cycnium herzfeldianum (Vatke) Engl. ■☆
116483 Cycnium heuglinii (Hochst. ex Schweinf.) Engl. = Cycnium tubulosum (L. f.) Engl. ■☆
116484 Cycnium humifusum (Forssk.) Engl. = Cycniopsis humifusa (Forssk.) Engl. ■☆
116485 Cycnium huttoniae Hiern = Cycnium racemosum Benth. ■☆
116486 Cycnium jamesii (Skan) O. J. Hansen;詹姆斯鹅参■☆
116487 Cycnium jamesii (V. Naray.) O. J. Hansen = Cycnium jamesii (Skan) O. J. Hansen ■☆
116488 Cycnium kraussianum Benth. = Cycnium racemosum Benth. ■☆
116489 Cycnium longiflorum Eckl. et Zeyh. = Cycnium adonense E. Mey. ex Benth. ■☆
116490 Cycnium meyeri-johannis (Engl.) Engl. = Cycnium recurvum (Oliv.) Engl. ■☆
116491 Cycnium meyeri-johannis Engl. = Cycnium herzfeldianum (Vatke) Engl. ■☆
116492 Cycnium paucidentatum (Engl.) Engl. = Cycnium tubulosum (L. f.) Engl. subsp. montanum (N. E. Br.) O. J. Hansen ■☆
116493 Cycnium pentheri Gand. = Cycnium adonense E. Mey. ex Benth. ■☆
116494 Cycnium petunioides Hutch. = Cycnium adonense E. Mey. ex Benth. subsp. camporum (Engl.) O. J. Hansen ■☆
116495 Cycnium pratense Engl. = Striga forbesii Benth. ■☆
116496 Cycnium questieauxianum De Wild. = Cycnium tubulosum (L. f.) Engl. ■☆
116497 Cycnium racemosum Benth.;总花鹅参■☆
116498 Cycnium rectum Gand. = Cycnium adonense E. Mey. ex Benth. ■☆
116499 Cycnium recurvum (Oliv.) Engl.;反折鹅参■☆
116500 Cycnium rubrifolium Engl. = Cycnium tubulosum (L. f.) Engl. subsp. montanum (N. E. Br.) O. J. Hansen ■☆
116501 Cycnium sandersonii Harv. = Cycnium racemosum Benth. ■☆
116502 Cycnium serratum (Klotzsch) Engl. = Cycnium tubulosum (L. f.) Engl. ■☆

116503 Cycnium serratum (Klotzsch) Engl. f. paucidentatum Engl. = Cycnium tubulosum (L. f.) Engl. subsp. montanum (N. E. Br.) O. J. Hansen ■☆
116504 Cycnium strictum Engl. = Striga latericea Vatke ■☆
116505 Cycnium suffruticosum Engl. = Cycnium veronicifolium (Vatke) Engl. subsp. suffruticosum (Engl.) O. J. Hansen ●☆
116506 Cycnium tenuisectum (Standl.) O. J. Hansen = Cycnium recurvum (Oliv.) Engl. ■☆
116507 Cycnium tomentosum Engl. = Cycnium erectum Rendle ■☆
116508 Cycnium tubatum Benth. = Harveya speciosa Bernh. ■☆
116509 Cycnium tubulosum (L. f.) Engl.;管状鹅参■☆
116510 Cycnium tubulosum (L. f.) Engl. = Cycnium aquaticum Engl. ■☆
116511 Cycnium tubulosum (L. f.) Engl. subsp. montanum (N. E. Br.) O. J. Hansen;山地管状鹅参■☆
116512 Cycnium verdickii De Wild. = Cycnium adonense E. Mey. ex Benth. ■☆
116513 Cycnium veronicifolium (Vatke) Engl.;婆婆纳叶鹅参■☆
116514 Cycnium veronicifolium (Vatke) Engl. subsp. suffruticosum (Engl.) O. J. Hansen;亚灌木鹅参●☆
116515 Cycnium volkensii Engl.;沃尔鹅参■☆
116516 Cycnoches Lindl. (1832);天鹅兰属(肉唇兰属);Swan Neck, Swan Neck Orchid, Swan Orchid, Swan-orchid, Swan-plant ■☆
116517 Cycnoches chlorochilon Klotzsch;绿舌天鹅兰(绿舌肉唇兰)■☆
116518 Cycnoches egertoniana Bateman;艾氏天鹅兰(艾氏肉唇兰);Egerton Swan Orchid ■☆
116519 Cycnoches egertoniana Bateman var. aurea (Lindl.) P. H. Allen;黄艾氏天鹅兰(黄艾氏肉唇兰);Golden Egerton Swan Orchid ■☆
116520 Cycnoches egertoniana Bateman var. dianae (Rchb. f.) P. H. Allen;丹氏肉唇兰(丹氏天鹅兰);Dian Egerton Swan Orchid ■☆
116521 Cycnoches maculata Lindl.;斑唇天鹅兰(斑唇肉唇兰);Spotted Swan Orchid ■☆
116522 Cycnoches pentadactylon Lindl.;五指天鹅兰(五指肉唇兰);Fivefinger Swan Orchid ■☆
116523 Cycnogeton Endl. = Triglochin L. ■
116524 Cycnopodium Naudin = Graffenrieda DC. ●☆
116525 Cycnoseris Endl. = Hypochaeris L. ■
116526 Cycoctonum Post et Kuntze = Cynoctonum J. F. Gmel. ■
116527 Cycoctonum Post et Kuntze = Mitreola L. ■
116528 Cydenis Sallab. = Narcissus L. ■
116529 Cydista Miers(1863);优紫葳属●☆
116530 Cydista aequinoctialis Miers;优紫葳;Garlic Vine ●☆
116531 Cydista diversifolia Miers;异叶优紫葳●☆
116532 Cydonia Mill. (1754);榅桲属;Common Quince, Quince ●
116533 Cydonia Tourn. ex Mill. = Cydonia Mill. ●
116534 Cydonia cathayensis Hemsl. = Chaenomeles cathayensis (Hemsl.) C. K. Schneid. ●
116535 Cydonia delavayi (Franch.) Cardot = Docynia delavayi (Franch.) C. K. Schneid. ●
116536 Cydonia delavayi Cardot = Docynia delavayi (Franch.) C. K. Schneid. ●
116537 Cydonia indica (Wall.) Spach = Docynia indica (Wall.) Decne. ●
116538 Cydonia indica Spach = Docynia indica (Wall.) Decne. ●
116539 Cydonia japonica (Thunb.) Pers. = Chaenomeles japonica (Thunb.) Lindl. ex Spach ●
116540 Cydonia japonica (Thunb.) Pers. var. cathayensis (Hemsl.) Cardot = Chaenomeles cathayensis (Hemsl.) C. K. Schneid. ●
116541 Cydonia japonica (Thunb.) Pers. var. lagenaria (Loisel.) Makino = Chaenomeles speciosa (Sweet) Nakai ●
116542 Cydonia japonica Pers. = Chaenomeles japonica (Thunb.) Lindl. ex Spach ●
116543 Cydonia lagenaria Loisel. = Chaenomeles speciosa (Sweet) Nakai ●
116544 Cydonia maliformis Mill. = Cydonia oblonga Mill. ●
116545 Cydonia maulei T. Moore = Cydonia japonica (Thunb.) Pers. ●
116546 Cydonia oblonga Mill.;榅桲(榠栌,木梨,木李蛮栌,土木瓜);Common Quince, Coynes, Maule's Quince, Portuguese Quince, Quince, Quitte ●
116547 Cydonia oblonga Mill. var. maliformis C. K. Schneid.;苹果形榅桲;Apple Quince ●
116548 Cydonia oblonga Mill. var. pyriformis Rehder;梨形榅桲●
116549 Cydonia sinensis Thouin = Chaenomeles sinensis (Thouin) Koehne ●
116550 Cydonia sinensis Thouin = Pseudocydonia sinensis C. K. Schneid. ●
116551 Cydonia speciosa Sweet = Chaenomeles speciosa (Sweet) Nakai ●
116552 Cydonia vulgaris Pers. = Cydonia oblonga Mill. ●
116553 Cydoniaceae Schnizl. = Rosaceae Juss. (保留科名)●■
116554 Cydostigma Klotzsch ex Seem. = Croton L. ●
116555 Cylastis Raf. = Rubus L. ●■
116556 Cylastis pubescens (Raf.) W. A. Weber = Rubus pubescens Raf. ●☆
116557 Cylastis saxatilis (L.) Á. Löve = Rubus saxatilis L. ●■
116558 Cylbanida Noronha ex Tul. = Pittosporum Banks ex Gaertn. (保留属名)●
116559 Cylicadenia Lem. = Odontadenia Benth. ●☆
116560 Cylichnanthus Dulac = Dianthus L. ■
116561 Cylichnium Dulac = Gaudinia P. Beauv. ■☆
116562 Cylichnium Mizush. = Gaudinia P. Beauv. ■☆
116563 Cylicodaphne Nees = Litsea Lam. (保留属名)●
116564 Cylicodaphne akoensis (Hayata) Nakai = Litsea akoensis Hayata ●
116565 Cylicodaphne garciae (Vidal) Nakai = Litsea garciae Vidal ●
116566 Cylicodaphne hayatae (Kaneh.) Nakai = Litsea hayatae Kaneh. ●
116567 Cylicodiscus Harms(1897);圆盘豆属(轮盘豆属)●☆
116568 Cylicodiscus battiscombei Baker f. = Newtonia paucijuga (Harms) Brenan ●☆
116569 Cylicodiscus gabunensis (Taub.) Harms;加蓬圆盘豆木;Africa Greenheart, Denya, Okan ●☆
116570 Cylicodiscus gabunensis Harms = Cylicodiscus gabunensis (Taub.) Harms ●☆
116571 Cylicodiscus paucijugus (Harms) Verdc. = Newtonia paucijuga (Harms) Brenan ●☆
116572 Cylicomorpha Urb. (1901);叉刺番瓜树属(非洲番瓜树属)●☆
116573 Cylicomorpha parviflora Urb.;叉刺番瓜树(非洲番瓜树)●☆
116574 Cylicomorpha solmsii (Urb.) Urb.;索尔叉刺番瓜树(索尔非洲番瓜树)●☆
116575 Cylindrachne Rchb. = Cylindrocline Cass. ●☆
116576 Cylindria Lour. = Chionanthus L. ●
116577 Cylindrilluma Plowes = Caralluma R. Br. ■
116578 Cylindrilluma Plowes(1886);沟梗水牛角属■☆
116579 Cylindrocarpa Regel(1877);柱果桔梗属■☆

116580 Cylindrocarpa sewerzowi Regel;柱果桔梗■☆
116581 Cylindrochilus Thwaites = Thrixspermum Lour. ■
116582 Cylindrocline Cass. (1817);绵背菊属●☆
116583 Cylindrocline commersonii Cass.;绵背菊●☆
116584 Cylindrokelupha Hutch. = Cylindrokelupha Kosterm. ●
116585 Cylindrokelupha Kosterm. (1954);棋子豆属(柱可卢法属);Chessbean,Cylindrokelupha ●
116586 Cylindrokelupha Kosterm. = Archidendron F. Muell. ●
116587 Cylindrokelupha alternifoliolata T. L. Wu;长叶棋子豆;Alternateleaf Chessbean, Longleaf Cylindrokelupha, Long-leaved Cylindrokelupha ●
116588 Cylindrokelupha alternifoliolata T. L. Wu = Archidendron alternifoliolatum (T. L. Wu) I. C. Nielsen ●
116589 Cylindrokelupha balansae (Oliv.) Kosterm.;锈毛棋子豆(马蛋果,马粪树);Balansa Cylindrokelupha,Rusthair Chessbean ●
116590 Cylindrokelupha balansae (Oliv.) Kosterm. = Archidendron balansae (Oliv.) I. C. Nielsen ●
116591 Cylindrokelupha chevalieri Kosterm.;坛腺棋子豆;Chevalier's Cylindrokelupha,Chevalier Chessbean ●
116592 Cylindrokelupha chevalieri Kosterm. = Archidendron chevalieri (Kosterm.) I. C. Nielsen ●
116593 Cylindrokelupha dalatensis (Kosterm.) T. L. Wu;显脉棋子豆;Dalat Cylindrokelupha, Distinctvein Cylindrokelupha, Nervose Chessbean ●
116594 Cylindrokelupha eberhardtii (I. C. Nielsen) T. L. Wu = Archidendron eberhardtii I. C. Nielsen ●
116595 Cylindrokelupha eberhardtii (Nielsen) T. L. Wu;大棋子豆;Eberhardt Cylindrokelupha,Large Chessbean ●
116596 Cylindrokelupha glabrifolia T. L. Wu;光叶棋子豆;Glabrous-leaved Cylindrokelupha, Smoothleaf Chessbean, Smoothleaf Cylindrokelupha ●
116597 Cylindrokelupha glabrifolia T. L. Wu = Archidendron alternifoliolatum (T. L. Wu) I. C. Nielsen ●
116598 Cylindrokelupha glabrifolia T. L. Wu = Cylindrokelupha alternifoliolata T. L. Wu ●
116599 Cylindrokelupha kerrii (Gagnep.) T. L. Wu;碟腺棋子豆;Dishgland Chessbean,Kerr Cylindrokelupha ●
116600 Cylindrokelupha kerrii (Gagnep.) T. L. Wu = Archidendron kerrii (Gagnep.) I. C. Nielsen ●
116601 Cylindrokelupha laoticum (Gagnep.) C. Chen et H. Sun;老挝棋子豆;Laos Cylindrokelupha ●
116602 Cylindrokelupha laoticum (Gagnep.) C. Chen et H. Sun = Archidendron laoticum (Gagnep.) I. C. Nielsen ●
116603 Cylindrokelupha macrophylla T. L. Wu;大叶棋子豆;Bigleaf Cylindrokelupha ●
116604 Cylindrokelupha macrophylla T. L. Wu = Archidendron eberhardtii I. C. Nielsen ●
116605 Cylindrokelupha macrophylla T. L. Wu = Cylindrokelupha eberhardtii (Nielsen) T. L. Wu ●
116606 Cylindrokelupha robinsonii (Gagnep.) Kosterm.;棋子豆(广西棋子豆);Robinson Chessbean,Robinson Cylindrokelupha ●
116607 Cylindrokelupha robinsonii (Gagnep.) Kosterm. = Archidendron robinsonii (Gagnep.) I. C. Nielsen ●
116608 Cylindrokelupha tonkinensis (I. C. Nielsen) T. L. Wu;绢毛棋子豆;Silkyhair Chessbean,Tonkin Cylindrokelupha ●
116609 Cylindrokelupha tonkinensis (I. C. Nielsen) T. L. Wu = Archidendron tonkinense I. C. Nielsen ●
116610 Cylindrokelupha turgida (Merr.) T. L. Wu;大叶合欢(鼎湖合欢,两广合欢,胀荚合欢);Inflated Cylindrokelupha, Swollen Cylindrokelupha,Turgid Chessbean ●
116611 Cylindrokelupha turgida (Merr.) T. L. Wu = Archidendron turgidum (Merr.) I. C. Nielsen ●
116612 Cylindrokelupha xichouensis C. Chen et H. Sun = Archidendron xichouensis (C. Chen et H. Sun) T. L. Wu ●
116613 Cylindrokelupha yunnanensis (Kosterm.) T. L. Wu;云南棋子豆;Yunnan Chessbean,Yunnan Cylindrokelupha ●
116614 Cylindrokelupha yunnanensis (Kosterm.) T. L. Wu = Archidendron kerrii (Gagnep.) I. C. Nielsen ●
116615 Cylindrolepis Boeck. = Mariscus Gaertn. ■
116616 Cylindrolobus (Blume) Brieger = Eria Lindl. (保留属名)■
116617 Cylindrolobus Blume = Eria Lindl. (保留属名)■
116618 Cylindrolobus Blume(1828);柱兰属■
116619 Cylindrolobus bambusifolius (Lindl.) Brieger = Callostylis bambusifolia (Lindl.) S. C. Chen et J. J. Wood ■
116620 Cylindrolobus cristatus (Rolfe) S. C. Chen et J. J. Wood;鸡冠柱兰■
116621 Cylindrolobus marginatus (Rolfe) S. C. Chen et J. J. Wood;柱兰(棒茎);Marginate Eria,Marginate Hairorchis ■
116622 Cylindrolobus tenuicaulis (S. C. Chen et Z. H. Tsi) S. C. Chen et J. J. Wood;细茎柱兰(细茎毛兰);Thinstem Eria,Thinstem Hairorchis ■
116623 Cylindrophyllum Schwantes(1927);筒叶玉属●☆
116624 Cylindrophyllum comptonii L. Bolus;康普顿筒叶玉●☆
116625 Cylindrophyllum dichroum (Rolfe) Schwantes = Ruschia dichroa (Rolfe) L. Bolus ●☆
116626 Cylindrophyllum dyeri L. Bolus = Cylindrophyllum calamiforme (L.) Schwantes ●☆
116627 Cylindrophyllum hallii L. Bolus;霍尔筒叶玉●☆
116628 Cylindrophyllum obsubulatum (Haw.) Schwantes;倒钻形筒叶玉●☆
116629 Cylindrophyllum tugwelliae L. Bolus;特格筒叶玉●☆
116630 Cylindropsis Pierre(1898);柱状夹竹桃属●☆
116631 Cylindropsis parvifolia Pierre;柱状夹竹桃●☆
116632 Cylindropsis talbotii Wernham = Cylindropsis parvifolia Pierre ●☆
116633 Cylindropsis togolana Hallier f. = Landolphia togolana (Hallier f.) Pichon ●☆
116634 Cylindropsis watsoniana (Roxb.) Hallier f. = Landolphia watsoniana Roxb. ●☆
116635 Cylindropuntia (Engelm.) F. M. Knuth = Opuntia Mill. ●
116636 Cylindropuntia (Engelm.) F. M. Knuth(1930);圆筒仙人掌属;Cholla ■☆
116637 Cylindropuntia (Engelm.) Frič et Schelle ex Kreuz. = Opuntia Mill. ●
116638 Cylindropuntia abyssi (Hester) Backeb.;桃花圆筒仙人掌;Peach Springs Cholla ■☆
116639 Cylindropuntia acanthocarpa (Engelm. et J. M. Bigelow) F. M. Knuth;尖果圆筒仙人掌;Buckhorn Cholla ■☆
116640 Cylindropuntia acanthocarpa (Engelm. et J. M. Bigelow) F. M. Knuth var. coloradensis (L. D. Benson) Pinkava;鹿角尖果圆筒仙人掌;Buckhorn Cholla ■☆
116641 Cylindropuntia acanthocarpa (Engelm. et J. M. Bigelow) F. M. Knuth var. major (Engelm.) Pinkava;大尖果圆筒仙人掌;Major Cholla ■☆
116642 Cylindropuntia acanthocarpa (Engelm. et J. M. Bigelow) F. M.

Knuth var. thornberi (Thornber et Bonker) Backeb.;陶尔圆筒仙人掌;Thornber Cholla ■☆

116643 Cylindropuntia acanthocarpa (Engelm. et J. M. Bigelow) F. M. Knuth var. major (Engelm. et J. M. Bigelow) L. D. Benson = Cylindropuntia acanthocarpa (Engelm. et J. M. Bigelow) F. M. Knuth var. major (Engelm.) Pinkava ■☆

116644 Cylindropuntia acanthocarpa (Engelm. et J. M. Bigelow) F. M. Knuth var. ramosa (Peebles) Backeb. = Cylindropuntia acanthocarpa (Engelm. et J. M. Bigelow) F. M. Knuth var. major (Engelm.) Pinkava ■☆

116645 Cylindropuntia acanthocarpa (Engelm. et J. M. Bigelow) F. M. Knuth var. ramosa Peebles = Cylindropuntia acanthocarpa (Engelm. et J. M. Bigelow) F. M. Knuth var. major (Engelm.) Pinkava ■☆

116646 Cylindropuntia acanthocarpa (Engelm. et J. M. Bigelow) F. M. Knuth = Opuntia acanthocarpa Engelm. et J. M. Bigelow ■☆

116647 Cylindropuntia arbuscula (Engelm.) F. M. Knuth;圆筒仙人掌;Pencil Cholla ■☆

116648 Cylindropuntia bigelovii (Engelm.) F. M. Knuth;松岚;Teddy Bear Cholla,Teddy-bear Cholla ■☆

116649 Cylindropuntia caerulescens (Griffiths) F. M. Knuth;蓝圆筒仙人掌■☆

116650 Cylindropuntia californica (Torr. et A. Gray) F. M. Knuth;加州圆筒仙人掌;California Cholla,Snake Cholla ■☆

116651 Cylindropuntia californica (Torr. et A. Gray) F. M. Knuth var. parkeri (J. M. Coult.) Pinkava;帕克圆筒仙人掌■☆

116652 Cylindropuntia cholla (F. A. C. Weber) F. M. Knuth;瘤珊瑚;Chain-link Cholla ■☆

116653 Cylindropuntia davisii (Engelm. et J. M. Bigelow) F. M. Knuth;戴维斯圆筒仙人掌;Davis Cholla ■☆

116654 Cylindropuntia echinocarpa (Engelm. et J. M. Bigelow) F. M. Knuth;刺果圆筒仙人掌;Golden Cholla,Silver Cholla ■☆

116655 Cylindropuntia echinocarpa (Engelm. et J. M. Bigelow) F. M. Knuth var. robustior J. M. Coult. = Cylindropuntia acanthocarpa (Engelm. et J. M. Bigelow) F. M. Knuth var. major (Engelm.) Pinkava ■☆

116656 Cylindropuntia fosbergii (C. B. Wolf) Rebman,M. A. Baker et Pinkava;马森圆筒仙人掌;Mason Valley Cholla ■☆

116657 Cylindropuntia fulgida (Engelm.) F. M. Knuth;鳞团扇;Boxing Glove Cholla,Chain-fruit Cholla,Smooth Chain Fruit Cholla ■☆

116658 Cylindropuntia fulgida (Engelm.) F. M. Knuth var. mamillata (Schott ex Engelm.) Backeb.;乳突圆筒仙人掌■☆

116659 Cylindropuntia ganderi (C. B. Wolf) Rebman et Pinkava;甘德圆筒仙人掌;Gander Cholla,Gander's Cholla ■☆

116660 Cylindropuntia hualpaensis Backeb. = Cylindropuntia whipplei (Engelm. et J. M. Bigelow) F. M. Knuth ■☆

116661 Cylindropuntia imbricata (Haw.) F. M. Knuth;鬼子角;Tree Cholla ■☆

116662 Cylindropuntia imbricata (Haw.) F. M. Knuth var. argentea (M. S. Anthony) Backeb.;银白鬼子角■☆

116663 Cylindropuntia kelvinensis (V. E. Grant et K. A. Grant) P. V. Heath;克文圆筒仙人掌■☆

116664 Cylindropuntia kleiniae (DC.) F. M. Knuth;克雷恩圆筒仙人掌;Klein Cholla ■☆

116665 Cylindropuntia leptocaulis (DC.) F. M. Knuth;细茎圆筒仙人掌;Christmas Cactus, Desert Christmas Cactus, Desert Christmas Cholla,Pencil Cholla ■☆

116666 Cylindropuntia leptocaulis (DC.) F. M. Knuth var. brevispina (Engelm.) F. M. Knuth = Cylindropuntia leptocaulis (DC.) F. M. Knuth ■☆

116667 Cylindropuntia molesta (Brandegee) F. M. Knuth;吴竹■☆

116668 Cylindropuntia multigeniculata (Clokey) Backeb.;多曲圆筒仙人掌;Blue Diamond Cholla ■☆

116669 Cylindropuntia munzii (C. B. Wolf) Backeb.;蒙茨圆筒仙人掌;Munz Cholla ■☆

116670 Cylindropuntia prolifera (Engelm.) F. M. Knuth;沿海圆筒仙人掌;Coastal Cholla,Jumping Cholla ■☆

116671 Cylindropuntia ramosissima (Engelm.) F. M. Knuth;钻石圆筒仙人掌;Diamond Cholla,Pencil Cholla ■☆

116672 Cylindropuntia spinosior (Engelm.) F. M. Knuth;细长圆筒仙人掌;Cane Cholla,Walkingstick Cactus ■☆

116673 Cylindropuntia tetracantha (Toumey) F. M. Knuth;四刺圆筒仙人掌■☆

116674 Cylindropuntia tunicata (Lehm.) F. M. Knuth;鞘圆筒仙人掌;Sheathed Cholla ■☆

116675 Cylindropuntia versicolor (Engelm. ex J. M. Coult.) F. M. Knuth;珊瑚圆筒仙人掌;Staghorn Cholla ■☆

116676 Cylindropuntia whipplei (Engelm. et J. M. Bigelow) F. M. Knuth;惠普尔圆筒仙人掌;Whipple Cholla ■☆

116677 Cylindropuntia whipplei (Engelm. et J. M. Bigelow) F. M. Knuth var. enodis (Peebles) Backeb. = Cylindropuntia whipplei (Engelm. et J. M. Bigelow) F. M. Knuth ■☆

116678 Cylindropuntia whipplei (Engelm. et J. M. Bigelow) F. M. Knuth var. enodis Peebles = Cylindropuntia whipplei (Engelm. et J. M. Bigelow) F. M. Knuth ■☆

116679 Cylindropuntia wigginsii (L. D. Benson) H. Rob. = Cylindropuntia echinocarpa (Engelm. et J. M. Bigelow) F. M. Knuth ■☆

116680 Cylindropuntia wolfii (L. D. Benson) M. A. Baker;沃尔夫圆筒仙人掌;Wolf Cholla ■☆

116681 Cylindropus Nees = Scleria P. J. Bergius ■

116682 Cylindropyrum (Jaub. et Spach) Á. Löve = Aegilops L. (保留属名)■

116683 Cylindrorebutia Frič et Kreuz. = Rebutia K. Schum. ●

116684 Cylindrosolen Kuntze = Cylindrosolenium Lindau ■☆

116685 Cylindrosolenium Lindau(1897);筒爵床属■☆

116686 Cylindrosolenium sprucei Lindau;筒爵床■☆

116687 Cylindrosorus Benth. = Angianthus J. C. Wendl. (保留属名)■●☆

116688 Cylindrosperma Ducke = Microplumeria Baill. ●☆

116689 Cyliodaphne garciae (Vidal) Nakai = Litsea garciae Vidal ●

116690 Cylipogon Raf. = Dalea L. (保留属名)●■☆

116691 Cylista Aiton(废弃属名) = Paracalyx Ali ■☆

116692 Cylista Aiton(废弃属名) = Rhynchosia Lour. (保留属名)●■

116693 Cylista albiflora Sims = Rhynchosia hirta (Andréws) Meikle et Verdc. ■☆

116694 Cylista argentea Eckl. et Zeyh. = Rhynchosia leucoscias Benth. ex Harv. ■☆

116695 Cylista microphylla Chiov. = Paracalyx microphyllus (Chiov.) Ali ■☆

116696 Cylista nogalensis Chiov. = Paracalyx nogalensis (Chiov.) Ali ■☆

116697 Cylista preussii Harms = Rhynchosia preussii (Harms) Taub. ex Harms ■☆

116698 Cylista pycnostachya DC. = Rhynchosia pycnostachya (DC.) Meikle ■☆

116699 Cylista scariosa Roxb. = Paracalyx scariosus (Roxb.) Ali ■☆

116700 Cylista somalorum Vierh. = Paracalyx somalorum (Vierh.) Ali ■☆
116701 Cylista villosa Aiton = Rhynchosia hirta (Andréws) Meikle et Verdc. ■☆
116702 Cylixylon Llanos = Gymnanthera R. Br. ●
116703 Cylizoma Neck. = Derris Lour.(保留属名)●
116704 Cyllenium Schott = Biarum Schott(保留属名)■☆
116705 Cylopogon Post et Kuntze = Cylipogon Raf. ●■☆
116706 Cylopogon Post et Kuntze = Dalea L.(保留属名)●■☆
116707 Cymapleura Post et Kuntze = Kymapleura Nutt. ■☆
116708 Cymapleura Post et Kuntze = Troximon Gaertn. ■☆
116709 Cymaria Benth.(1830);歧伞花属(伞荆芥属);Cymaria ●
116710 Cymaria acuminata Decne.;长柄歧伞花;Acuminate Cymaria ●
116711 Cymaria dichotoma Benth.;歧伞花;Dichotomous Cymaria ●
116712 Cymaria elongata Benth.;长歧伞花;Elongated Cymaria ●
116713 Cymation Spreng. = Lichtensteinia Willd.(废弃属名)■☆
116714 Cymation Spreng. = Ornithoglossum Salisb. ■☆
116715 Cymation undulatum (Willd.) Spreng. = Ornithoglossum undulatum Sweet ■☆
116716 Cymatocarpus O. E. Schulz(1924);歧果芥属■☆
116717 Cymatocarpus grossheimii N. Busch;格罗歧果芥■☆
116718 Cymatocarpus heterophyllus (Popov) N. Busch;互叶歧果芥■☆
116719 Cymatocarpus pilosissimus (Trautv.) O. E. Schulz;多毛歧果芥■☆
116720 Cymatochloa Schltdl. = Paspalum L. ■
116721 Cymatoptera Turcz. = Menonvillea R. Br. ex DC. ■●☆
116722 Cymba Dulac = Tofieldia Huds. ■
116723 Cymba Noronha = Agalmyla Blume ●☆
116724 Cymbachne Retz. = Rottboellia L. f.(保留属名)■
116725 Cymbachne amplectens (Nees) Roberty = Diheteropogon amplectens (Nees) Clayton ■☆
116726 Cymbachne amplectens (Nees) Roberty subvar. heteropogonoides Roberty = Diheteropogon amplectens (Nees) Clayton ■☆
116727 Cymbachne fastigiata (Sw.) Roberty = Andropogon fastigiatus Sw. ■☆
116728 Cymbachne filifolia (Nees) Roberty = Diheteropogon filifolius (Nees) Clayton ■☆
116729 Cymbachne guineensis (Schumach.) Roberty = Andropogon gayanus Kunth ■☆
116730 Cymbachne textilis (Welw. ex Rendle) Roberty = Andropogon textilis Welw. ex Rendle ■☆
116731 Cymbaecarpa Cav. = Coreopsis L. ●■
116732 Cymbalaria Hill(1756);假金鱼草属(蔓柳穿鱼属,铙钹花属,铙钹藤属);Basket Ivy, Ivy-leaved Toadflax, Kenilworth Ivy ■☆
116733 Cymbalaria Medik. = Linaria Mill. ■
116734 Cymbalaria aequitriloba (Viv.) A. Chev.;杂色假金鱼草;Variegated Creeping Charlie ■☆
116735 Cymbalaria hepaticifolia Wettst.;科西嘉假金鱼;Corsican Toadflax ■☆
116736 Cymbalaria muralis P. Gaertn., B. Mey. et Scherb.;假金鱼草(蔓柳穿鱼,梅花草,铙绂花);Aaron's Beard, Aaron's-beard, Bastard Navelwort, Bunny Rabbits, Bunny Rabbit's Mouth, Church Bells, Climbing Sailor, Coliseum-ivy, Colosseum Ivy, Creeping Jenny, Creeping Sailor, Fleas-and-lice, Hen-and-chickens, Hundreds-and-thousands, Ivy Leaved Toadflax, Ivy-leaved Antirrhinum, Ivy-leaved Snapdragon, Ivy-leaved Toadflax, Ivywort, Kenilworth Ivy, Kenilworth-Ivy, Lavender Snips, Monkey Jaw, Monkey Jaws, Monkey Mouth, Mother of Thousands, Mother-of-millions, Mother-of-thousands, Nanny Goat's Mouth, Oxford Ivy, Oxford Weed, Pedlar's Basket, Pennywort, Pickpocket, Rabbit Flower, Rabbit's Mouth, Rambling Sailor, Roving Jenny, Snapdragon, Thousand-flower, Thread-of-life, Underground Ivy, Wall Toadflax, Wandering Jack, Wandering Jew, Wandering Sailor, Wan-dering Sailor ■☆
116737 Cymbalaria muralis P. Gaertn., B. Mey. et Scherb. = Antirrhinum cymbalaria L. ■☆
116738 Cymbalaria pallida Wettst.;意大利假金鱼草;Italian Toadflax ■☆
116739 Cymbalaria spuria (L.) P. Gaertn., B. Mey. et Schreb. = Kickxia spuria (L.) Dumort. ■☆
116740 Cymbalariella Nappi = Saxifraga L. ■
116741 Cymbalina Raf. = Cymbalaria Hill ■☆
116742 Cymbanthaceae Salisb. = Melanthiaceae Batsch ex Borkh.(保留科名)■
116743 Cymbanthelia Andersson = Cymbopogon Spreng. ■
116744 Cymbanthes Salisb. = Androcymbium Willd. ■☆
116745 Cymbaria L.(1753);芯芭属(大黄花属);Cymbaria ■
116746 Cymbaria borysthenica Pall.;第聂伯芯芭■☆
116747 Cymbaria dahurica L.;达乌里芯芭(白蒿茶,大黄花,芯芭,芯玛芭,兴安芯芭);Dahur Cymbaria, Dahuria Cymbaria ■
116748 Cymbaria dahurica L. var. aspera Franch.;粗糙达乌里芯芭■
116749 Cymbaria linearifolia K. S. Hao = Cymbaria mongolica Maxim. ■
116750 Cymbaria mongolica Maxim.;蒙古芯芭(光药大黄花);Mongol Cymbaria, Mongolian Cymbaria ■
116751 Cymbia (Torr. et A. Gray) Standl. = Krigia Schreb.(保留属名)■☆
116752 Cymbia Standl. = Krigia Schreb.(保留属名)■☆
116753 Cymbia occidentalis (Nutt.) Standl. = Krigia occidentalis Nutt. ■☆
116754 Cymbicarpos Steven = Astragalus L. ●■
116755 Cymbidiella Rolfe(1918);马岛兰属(小建兰属);Cymbidiella ■☆
116756 Cymbidiella falcigera (Rchb. f.) Garay;簇生马岛兰■☆
116757 Cymbidiella flabellata (Thouars) Rolfe;马岛兰;Common Cymbidiella ■☆
116758 Cymbidiella flabellata Rolfe = Cymbidiella flabellata (Thouars) Rolfe ■☆
116759 Cymbidiella humblotii (Rolfe) Rolfe;洪氏马岛兰;Humblot Cymbidiella ■☆
116760 Cymbidiella humblotii (Rolfe) Rolfe = Cymbidiella falcigera (Rchb. f.) Garay ■☆
116761 Cymbidiella humblotii Rolfe = Cymbidiella humblotii (Rolfe) Rolfe ■☆
116762 Cymbidiella pardalina (Rchb. f.) Garay;豹斑马岛兰■☆
116763 Cymbidiella perrieri Schltr.;佩里耶马岛兰;Perrier Cymbidiella ■☆
116764 Cymbidiella perrieri Schltr. = Cymbidiella flabellata (Thouars) Rolfe ■☆
116765 Cymbidiella rhodochila (Rolfe) Rolfe = Cymbidiella pardalina (Rchb. f.) Garay ■☆
116766 Cymbidiella rhodochila Rolfe;红唇马岛兰;Redlip Cymbidiella ■☆
116767 Cymbidiopsis H. J. Chowdhery = Cymbidium Sw. ■
116768 Cymbidiopsis H. J. Chowdhery(2009);南亚兰属■☆
116769 Cymbidium Sw.(1799);兰属(蕙兰属,圃兰属);Cymbidium,

Cymbidium Orchid, Orchis ■
116770 Cymbidium × iansonii Rolfe = Cymbidium lowianum (Rchb. f.) Rchb. f. var. iansonii (Rolfe) P. J. Cribb et Du Puy ■
116771 Cymbidium aberrans (Finet) Schltr. = Cymbidium macrorhizum Lindl. f. aberrans (Schltr.) Hid. Takah. et Ohba ■☆
116772 Cymbidium aculeatum (L. f.) Sw. = Eulophia aculeata (L. f.) Spreng. ■☆
116773 Cymbidium acutum Ridl.; 寒凤兰■☆
116774 Cymbidium acutum Ridl. var. vernale Makino; 春凤兰■☆
116775 Cymbidium adenoglossum Lindl. = Eulophia adenoglossa (Lindl.) Rchb. f. ■☆
116776 Cymbidium aestivum Z. J. Liu et S. C. Chen; 夏凤兰■
116777 Cymbidium albojucundissimum Hayata; 白花报岁兰■
116778 Cymbidium albojucundissimum Hayata = Cymbidium sinense (Jacks. ex Andréws) Willd. ■
116779 Cymbidium albomarginatum Makino; 白边兰(古今轮)■☆
116780 Cymbidium alborubens Makino = Cymbidium dayanum Rchb. f. ■
116781 Cymbidium aloifolium (L.) Sw.; 纹瓣兰(大剑兰, 剑兰, 芦荟叶兰, 硬叶吊兰); Aloeleaf Cymbidium, Imitating Cymbidium, Linepetal Orchis ■☆
116782 Cymbidium angolense Rchb. f. = Eulophia angolensis (Rchb. f.) Summerh. ■☆
116783 Cymbidium aphyllum Ames et Schltr. = Cymbidium macrorrhizum Lindl. ■
116784 Cymbidium appendiculatum D. Don = Cremastra appendiculata (D. Don) Makino ■
116785 Cymbidium arrogans Hayata = Cymbidium ensifolium (L.) Sw. ■
116786 Cymbidium aspidistrifolium Fukuy. = Cymbidium lancifolium Hook. var. aspidistrifolium (Fukuy.) S. S. Ying ■
116787 Cymbidium aspidistrifolium Fukuy. = Cymbidium lancifolium Hook. ■
116788 Cymbidium atropurpureum (Lindl.) Rolfe; 椰香兰■
116789 Cymbidium atropurpureum (Lindl.) Rolfe = Cymbidium aloifolium (L.) Sw. ■☆
116790 Cymbidium atropurpureum (Lindl.) Rolfe var. olivaceum J. J. Sm. = Cymbidium atropurpureum (Lindl.) Rolfe ■
116791 Cymbidium babae (Kudo ex Masam.) Masam. = Cymbidium cochleare Lindl. ■
116792 Cymbidium bambusifolium Fowlie = Cymbidium lancifolium Hook. var. aspidistrifolium (Fukuy.) S. S. Ying ■
116793 Cymbidium bambusifolium Fowlie = Cymbidium lancifolium Hook. ■
116794 Cymbidium bambusifolium Fowlie, Mark et C. N. Ho = Cymbidium lancifolium Hook. var. aspidistrifolium (Fukuy.) S. S. Ying ■
116795 Cymbidium baoshanense F. Y. Liu et H. Perner; 保山兰■
116796 Cymbidium bicolor Lindl.; 南亚硬叶兰; Bicolor Cymbidium ■
116797 Cymbidium bicolor Lindl. subsp. obtusum Du Puy et P. J. Cribb; 硬叶兰(硬叶吊兰); Hardleaf Bicolor Orchis, Hardleaf Cymbidium ■
116798 Cymbidium bicolor Lindl. subsp. obtusum Du Puy et P. J. Cribb = Cymbidium mannii Rchb. f. ■
116799 Cymbidium boreale Sw. = Calypso bulbosa (L.) Oakes ■
116800 Cymbidium buchananii Rchb. f. = Eulophia foliosa (Lindl.) Bolus ■☆
116801 Cymbidium calcaratum Schltr. = Oeceoclades calcarata (Schltr.) Garay et P. Taylor ■☆
116802 Cymbidium canaliculatum R. Br.; 纵沟兰(澳洲建兰)■☆
116803 Cymbidium carnosum Griff. = Cymbidium cyperifolium Wall. ex Lindl. ■
116804 Cymbidium cerinum Schltr. = Cymbidium faberi Rolfe ■
116805 Cymbidium changningense Z. J. Liu et S. C. Chen; 昌宁兰■
116806 Cymbidium chawalongense C. L. Long, H. Li et Z. L. Dao = Cymbidium floribundum Lindl. ■
116807 Cymbidium chinense Heynh. = Cymbidium sinense (Jacks. ex Andréws) Willd. ■
116808 Cymbidium chuan-lan C. Chow = Cymbidium goeringii (Rchb. f.) Rchb. f. ■
116809 Cymbidium cochleare Lindl.; 垂花兰(莎草兰, 香莎草兰); Nutateflower Cymbidium, Nutateflower Orchis ■
116810 Cymbidium concinnum Z. J. Liu et S. C. Chen; 丽花兰■
116811 Cymbidium crassifolium Wall. = Cymbidium aloifolium (L.) Sw. ■☆
116812 Cymbidium crinum Schltr. = Cymbidium faberi Rolfe ■
116813 Cymbidium cyperifolium Wall. ex Lindl.; 莎叶兰(套叶兰); Flatsedgeleaf Cymbidium, Sedgeleaf Orchis ■
116814 Cymbidium cyperifolium Wall. ex Lindl. var. szechuanicum (Y. S. Wu et S. C. Chen) S. C. Chen et Z. J. Liu; 送春; Sichuan Cymbidium, Sichuan Orchis, Szechuan Cymbidium ■
116815 Cymbidium dayanum Rchb. f.; 冬凤兰(东凤兰, 凤兰, 立奇兰); Pendulous Cymbidium, Pendulous Orchis ■
116816 Cymbidium dayanum Rchb. f. subsp. leachianum (Rchb. f.) S. S. Ying = Cymbidium dayanum Rchb. f. ■
116817 Cymbidium dayanum Rchb. f. var. albiflorum S. S. Ying = Cymbidium dayanum Rchb. f. ■
116818 Cymbidium dayanum Rchb. f. var. austrojaponicum Tuyama = Cymbidium dayanum Rchb. f. ■
116819 Cymbidium dayanum Rchb. f. var. leachianum (Rchb. f.) S. S. Ying = Cymbidium dayanum Rchb. f. ■
116820 Cymbidium defoliatum Y. S. Wu et S. C. Chen; 落叶兰; Defoliate Cymbidium, Defoliate Orchis ■
116821 Cymbidium dependens Lodd. = Cirrhaea dependens (Lodd.) Rchb. f. ■☆
116822 Cymbidium devonianum Paxton; 福兰(德氏凤兰); Rigid Cymbidium ■
116823 Cymbidium eburneum Lindl.; 独占春(象牙白花兰); Ivorywhite Cymbidium, Monopolize Orchis ■
116824 Cymbidium eburneum Lindl. var. austrojaponicum (Tuyama) Hiroe = Cymbidium dayanum Rchb. f. ■
116825 Cymbidium eburneum Lindl. var. longzhouense Z. J. Liu et S. C. Chen; 龙州兰■
116826 Cymbidium eburneum Lindl. var. parishi (Rchb. f.) Hook. f.; 大独占春(大雪兰); Parish Cymbidium ■
116827 Cymbidium elegans Lindl.; 莎草兰(长叶兰); Elegant Orchis, Longleaf Cymbidium ■
116828 Cymbidium elegans Lindl. var. lushuiense (Z. J. Liu, S. C. Chen et X. C. Shi) Z. J. Liu et S. C. Chen; 泸水兰■
116829 Cymbidium ensifolium (L.) Sw.; 建兰(八月兰, 报春兰, 草兰, 春兰, 大菁兰, 大青兰, 官兰, 官兰花, 红丝毛草, 建兰花, 焦叶兰, 骏河兰, 兰草, 兰花, 兰蕙花, 青兰, 秋兰, 山兰花, 烧刃兰, 烧叶兰, 四季兰, 土续断, 夏兰, 燕草, 幽兰, 玉叶兰, 芝兰); Sword Orchis, Swordleaf Cymbidium ■
116830 Cymbidium ensifolium (L.) Sw. f. arcuatum T. K. Yen = Cymbidium ensifolium (L.) Sw. ■
116831 Cymbidium ensifolium (L.) Sw. f. falcatum T. K. Yen =

Cymbidium ensifolium (L.) Sw. ■
116832 Cymbidium ensifolium (L.) Sw. f. flaccidior Makino;漳兰(雌兰)■
116833 Cymbidium ensifolium (L.) Sw. subsp. haematodes (Lindl.) Du Puy et P. J. Cribb = Cymbidium haematodes Lindl. ■
116834 Cymbidium ensifolium (L.) Sw. var. haematodes (Lindl.) Trimen = Cymbidium haematodes Lindl. ■
116835 Cymbidium ensifolium (L.) Sw. var. misericors (Hayata) T. P. Lin = Cymbidium ensifolium (L.) Sw. ■
116836 Cymbidium ensifolium (L.) Sw. var. misericors (Hayata) Tang S. Liu et H. J. Su;素心兰(焦尾兰)■
116837 Cymbidium ensifolium (L.) Sw. var. munronianum King et Pantl. = Cymbidium sinense (Jacks. ex Andréws) Willd. ■
116838 Cymbidium ensifolium (L.) Sw. var. rubrigemmum (Hayata) Tang S. Liu et H. J. Su;四季兰■
116839 Cymbidium ensifolium (L.) Sw. var. rubrigemmum (Hayata) Tang S. Liu et H. J. Su = Cymbidium ensifolium (L.) Sw. ■
116840 Cymbidium ensifolium (L.) Sw. var. striatum Lindl. = Cymbidium ensifolium (L.) Sw. ■
116841 Cymbidium ensifolium (L.) Sw. var. susin T. K. Yen = Cymbidium ensifolium (L.) Sw. ■
116842 Cymbidium ensifolium (L.) Sw. var. susin T. K. Yen f. arcuatum T. K. Yen = Cymbidium ensifolium (L.) Sw. ■
116843 Cymbidium ensifolium (L.) Sw. var. suxin T. K. Yen;素心建兰;Susin Cymbidium ■
116844 Cymbidium ensifolium (L.) Sw. var. xiphiifolium (Lindl.) S. S. Ying = Cymbidium ensifolium (L.) Sw. ■
116845 Cymbidium ensifolium (L.) Sw. var. yakibaran (Makino) Y. S. Wu et S. C. Chen;焦叶兰;Yakibaran Cymbidium ■
116846 Cymbidium ensifolium (L.) Sw. var. yakibaran (Makino) Y. S. Wu et S. C. Chen = Cymbidium ensifolium (L.) Sw. ■
116847 Cymbidium erythraeum Lindl.;长叶兰;Longleaf Cymbidium, Longleaf Orchis ■
116848 Cymbidium erythraeum Lindl. var. flavum (Z. J. Liu et J. Yong Zhang) Z. J. Liu, S. C. Chen et P. J. Cribb;黄花长叶兰■
116849 Cymbidium erythrostylum Rolfe;红柱兰■☆
116850 Cymbidium evrardii Guillaumin = Coelogyne assamica Linden et Rchb. f. ■
116851 Cymbidium faberi Rolfe;蕙兰(长叶兰,化气兰,九华兰,九节兰,九子兰,土百部,线兰,一茎九花,一茎九华);Faber Cymbidium, Faber Orchis ■
116852 Cymbidium faberi Rolfe f. viridiflorum S. S. Ying = Cymbidium faberi Rolfe ■
116853 Cymbidium faberi Rolfe var. omeiense (Y. S. Wu et S. C. Chen) Y. S. Wu et S. C. Chen;峨眉蕙兰(峨眉春蕙);Emei Cymbidium, Emei Orchis, Omei Cymbidium ■
116854 Cymbidium faberi Rolfe var. omeiense (Y. S. Wu et S. C. Chen) Y. S. Wu et S. C. Chen = Cymbidium omeiense Y. S. Wu et S. C. Chen ■
116855 Cymbidium faberi Rolfe var. szechuanicum (Y. S. Wu et S. C. Chen) Y. S. Wu et S. C. Chen = Cymbidium cyperifolium Wall. ex Lindl. var. szechuanicum (Y. S. Wu et S. C. Chen) S. C. Chen et Z. J. Liu ■
116856 Cymbidium falcatum (Thunb.) Sw. = Cephalanthera falcata (Thunb. ex A. Murray) Blume ■
116857 Cymbidium finlaysonianum Lindl.;垂花剑兰■☆
116858 Cymbidium finlaysonianum Wall. ex Lindl. var. atropurpureum (Lindl.) Veitch = Cymbidium atropurpureum (Lindl.) Rolfe ■
116859 Cymbidium flabellatum Spreng. = Cymbidiella flabellata (Thouars) Rolfe ■☆
116860 Cymbidium flaccidum Schltr. = Cymbidium bicolor Lindl. subsp. obtusum Du Puy et P. J. Cribb ■
116861 Cymbidium flaccidum Schltr. = Cymbidium mannii Rchb. f. ■
116862 Cymbidium flavum Z. J. Liu et J. Yong Zhang = Cymbidium erythraeum Lindl. var. flavum (Z. J. Liu et J. Yong Zhang) Z. J. Liu, S. C. Chen et P. J. Cribb ■
116863 Cymbidium floribundum Lindl.;多花兰(长寿兰,红兰,蕙兰,金棱边,金棱边兰,金龙边,九头兰,六月兰,蜜蜂兰,牛角七,山慈姑,石羊果,台兰,夏兰,小蜜蜂兰);Dwarf Cymbidium, Flowery Orchis, Manyflower Cymbidium ■
116864 Cymbidium floribundum Lindl. var. pumilum (Rolfe) Y. S. Wu et S. C. Chen = Cymbidium floribundum Lindl. ■
116865 Cymbidium formosanum Hayata = Cymbidium goeringii (Rchb. f.) Rchb. f. ■
116866 Cymbidium formosanum Hayata f. albiflorum S. S. Ying = Cymbidium goeringii (Rchb. f.) Rchb. f. ■
116867 Cymbidium formosanum Hayata var. gracillimum (Fukuy.) Tang S. Liu et H. J. Su = Cymbidium serratum Schltr. ■
116868 Cymbidium formosanum Hayata var. gracillimum (Fukuy.) Tang S. Liu et H. J. Su = Cymbidium goeringii (Rchb. f.) Rchb. f. var. serratum (Schltr.) Y. S. Wu et S. C. Chen ■
116869 Cymbidium formosanum Lindl. f. albiflorum S. S. Ying = Cymbidium goeringii (Rchb. f.) Rchb. f. ■
116870 Cymbidium forrestii Rolfe = Cymbidium goeringii (Rchb. f.) Rchb. f. ■
116871 Cymbidium fragrans Salisb. = Cymbidium sinense (Jacks. ex Andréws) Willd. ■
116872 Cymbidium fukienense T. K. Yen = Cymbidium faberi Rolfe ■
116873 Cymbidium gaoligongense Z. J. Liu et J. Yong Zhang;金蝉兰;Gaoligong Cymbidium ■
116874 Cymbidium giganteum Lindl. = Cymbidium iridioides D. Don ■
116875 Cymbidium giganteum Wall. = Cymbidium iridioides D. Don ■
116876 Cymbidium giganteum Wall. ex Lindl. = Cymbidium iridioides D. Don ■
116877 Cymbidium giganteum Wall. ex Lindl. var. hookerianum (Rchb. f.) Bois = Cymbidium hookerianum Rchb. f. ■
116878 Cymbidium giganteum Wall. ex Lindl. var. lowianum Rchb. f. = Cymbidium lowianum (Rchb. f.) Rchb. f. ■
116879 Cymbidium giganteum Wall. ex Lindl. var. wilsonii Rolfe ex Cook = Cymbidium wilsonii (Rolfe ex Cook) Rolfe ■
116880 Cymbidium goeringii (Rchb. f.) Rchb. f.;春兰(草兰,草素,吊花兰,朵朵香,兰花,日本春兰,山花,山兰,双飞燕,丝兰,台湾春兰,台湾兰);Goering Cymbidium, Spring Cymbidium, Spring Orchis ■
116881 Cymbidium goeringii (Rchb. f.) Rchb. f. f. albiflorum (S. S. Ying) S. S. Ying = Cymbidium goeringii (Rchb. f.) Rchb. f. ■
116882 Cymbidium goeringii (Rchb. f.) Rchb. f. var. angustatum F. Maek. = Cymbidium goeringii (Rchb. f.) Rchb. f. var. serratum (Schltr.) Y. S. Wu et S. C. Chen ■
116883 Cymbidium goeringii (Rchb. f.) Rchb. f. var. angustatum F. Maek. = Cymbidium goeringii (Rchb. f.) Rchb. f. var. gracillimum (Fukuy.) Tang S. Liu et H. J. Su ■
116884 Cymbidium goeringii (Rchb. f.) Rchb. f. var. formosanum (Hayata) S. S. Ying f. albiflorum S. S. Ying = Cymbidium goeringii

(Rchb. f.) Rchb. f. ■

116885 Cymbidium goeringii (Rchb. f.) Rchb. f. var. formosanum (Hayata) S. S. Ying = Cymbidium goeringii (Rchb. f.) Rchb. f. ■

116886 Cymbidium goeringii (Rchb. f.) Rchb. f. var. gracillimum (Fukuy.) Govaerts = Cymbidium serratum Schltr. ■

116887 Cymbidium goeringii (Rchb. f.) Rchb. f. var. gracillimum (Fukuy.) Tang S. Liu et H. J. Su;线叶春兰(细叶春兰);Linearleaf Cymbidium ■

116888 Cymbidium goeringii (Rchb. f.) Rchb. f. var. longibracteatum (Y. S. Wu et S. C. Chen) Y. S. Wu et S. C. Chen = Cymbidium tortisepalum Fukuy. var. longibracteatum (Y. S. Wu et S. C. Chen) S. C. Chen et Z. J. Liu ■

116889 Cymbidium goeringii (Rchb. f.) Rchb. f. var. papyfiflorum Y. S. Wu = Cymbidium goeringii (Rchb. f.) Rchb. f. ■

116890 Cymbidium goeringii (Rchb. f.) Rchb. f. var. serratum (Schltr.) Y. S. Wu et S. C. Chen = Cymbidium goeringii (Rchb. f.) Rchb. f. var. gracillimum (Fukuy.) Tang S. Liu et H. J. Su ■

116891 Cymbidium goeringii (Rchb. f.) Rchb. f. var. serratum (Schltr.) Y. S. Wu et S. C. Chen = Cymbidium serratum Schltr. ■

116892 Cymbidium goeringii (Rchb. f.) Rchb. f. var. tortisepalum (Fukuy.) Y. S. Wu et S. C. Chen f. albiflorum (S. S. Ying) S. S. Ying = Cymbidium goeringii (Rchb. f.) Rchb. f. var. tortisepalum (Fukuy.) Y. S. Wu et S. C. Chen ■

116893 Cymbidium goeringii (Rchb. f.) Rchb. f. var. tortisepalum (Fukuy.) Y. S. Wu et S. C. Chen = Cymbidium tortisepalum Fukuy. ■

116894 Cymbidium goeringii (Rchb. f.) Rchb. f. var. tortisepalum f. albiflorum S. S. Ying = Cymbidium goeringii (Rchb. f.) Rchb. f. var. tortisepalum (Fukuy.) Y. S. Wu et S. C. Chen ■

116895 Cymbidium goeringii (Rchb. f.) Rchb. f. var. tortisepalum f. albiflorum (S. S. Ying) S. S. Ying = Cymbidium goeringii (Rchb. f.) Rchb. f. var. tortisepalum (Fukuy.) Y. S. Wu et S. C. Chen ■

116896 Cymbidium gracillimum Fukuy. = Cymbidium goeringii (Rchb. f.) Rchb. f. var. serratum (Schltr.) Y. S. Wu et S. C. Chen ■

116897 Cymbidium gracillimum Fukuy. = Cymbidium serratum Schltr. ■

116898 Cymbidium grandiflorum Griff. = Cymbidium hookerianum Rchb. f. ■

116899 Cymbidium grandiflorum Griff. var. kalawense Colyear = Cymbidium lowianum (Rchb. f.) Rchb. f. var. iansonii (Rolfe) P. J. Cribb et Du Puy ■

116900 Cymbidium gyokuchin Makino;鱼魫兰(鱼魫,玉魫兰)■

116901 Cymbidium gyokuchin Makino = Cymbidium ensifolium (L.) Sw. ■

116902 Cymbidium gyokuchin Makino var. arrogans (Hayata) S. S. Ying = Cymbidium ensifolium (L.) Sw. ■

116903 Cymbidium gyokuchin Makino var. soshin Makino = Cymbidium ensifolium (L.) Sw. ■

116904 Cymbidium haematodes Lindl.;秋墨兰■

116905 Cymbidium hookerianum Rchb. f.;虎头兰(蝉兰,青蝉);Bigflower Cymbidium,Tiger Orchis ■

116906 Cymbidium hookerianum Rchb. f. var. lowianum (Rchb. f.) Y. S. Wu et S. C. Chen = Cymbidium lowianum (Rchb. f.) Rchb. f. ■

116907 Cymbidium hoosai Makino = Cymbidium sinense (Jacks. ex Andréws) Willd. ■

116908 Cymbidium humblotii Rolfe = Cymbidiella falcigera (Rchb. f.) Garay ■☆

116909 Cymbidium hyacinthinum Sm. = Bletilla striata (Thunb. ex A. Murray) Rchb. f. ■

116910 Cymbidium hybridum Hort.;东亚兰(绿花大花惠兰,绿花红唇,绿珍珠)■

116911 Cymbidium hyemale Muhl. ex Willd. = Aplectrum hyemale (Muhl. ex Willd.) Torr. ■☆

116912 Cymbidium illiberale Hayata = Cymbidium floribundum Lindl. ■

116913 Cymbidium imbricatum (Hook.) Roxb. = Pholidota imbricata Hook. ■

116914 Cymbidium imbricatum Roxb. = Pholidota imbricata Hook. ■

116915 Cymbidium induratifolium Z. J. Liu et J. N. Zhang;硬叶夏兰;Cymbidium,Hardleaf Orchis ■

116916 Cymbidium insigne Rolfe;美花兰(安南建兰,独占春);Remarkable Cymbidium,Spiffyflower Orchis ■

116917 Cymbidium iridifolium Roxb. = Oberonia iridifolia Roxb. ex Lindl. ■

116918 Cymbidium iridifolium Roxb. = Oberonia mucronata (D. Don) Ormerod et Seidenf. ■

116919 Cymbidium iridioides D. Don;黄蝉兰;Iris-like Cymbidium,Iris-like Orchis ■

116920 Cymbidium ixioides D. Don = Spathoglottis ixioides (D. Don) Lindl. ■

116921 Cymbidium javanicum Blume;无齿兔耳兰;Java Cymbidium ■☆

116922 Cymbidium javanicum Blume = Cymbidium lancifolium Hook. ■

116923 Cymbidium javanicum Blume var. aspidistrifolium (Fukuy.) F. Maek. = Cymbidium lancifolium Hook. ■

116924 Cymbidium javanicum Blume var. aspidistrifolium (Fukuy.) F. Maek. = Cymbidium lancifolium Hook. var. aspidistrifolium (Fukuy.) S. S. Ying ■

116925 Cymbidium javanicum Blume var. aspidistrifolium (Fukuy.) F. Maekawa = Cymbidium lancifolium Hook. ■

116926 Cymbidium kanran Makino;寒兰;Cold Orchis,Kanran,Smoothlip Cymbidium,Winter Cymbidium ■

116927 Cymbidium kanran Makino f. purpureoviridescens Makino;更纱寒兰■☆

116928 Cymbidium kanran Makino f. purpurescens Makino;紫花瓯兰(紫寒兰);Purple-flowered Cold Orchis ■

116929 Cymbidium kanran Makino f. rubescens Makino;红寒兰■☆

116930 Cymbidium kanran Makino f. viridescens Makino;绿花瓯兰(青寒兰);Green-flowered Cold Orchis ■

116931 Cymbidium kanran Makino var. aestivale Y. S. Wu = Cymbidium kanran Makino ■

116932 Cymbidium kanran Makino var. babae (Kudo ex Masam.) S. S. Ying = Cymbidium cochleare Lindl. ■

116933 Cymbidium kanran Makino var. latifolium Makino;大叶寒兰■☆

116934 Cymbidium kanran Makino var. misericors (Hayata) S. S. Ying = Cymbidium ensifolium (L.) Sw. ■

116935 Cymbidium kanran Makino var. purpureohiemale (Hayata) S. S. Ying = Cymbidium kanran Makino ■

116936 Cymbidium koran Makino;小兰■☆

116937 Cymbidium koran Makino = Cymbidium ensifolium (L.) Sw. ■

116938 Cymbidium lancifolium Hook.;竹柏兰(地青梅,搜山虎,兔耳兰,无齿兔耳兰,续筋草);Java Cymbidium,Lanceolate Cymbidium,Rabbitear Orchis ■

116939 Cymbidium lancifolium Hook. f. aspidistrifolium (Fukuy.) T. P. Lin = Cymbidium lancifolium Hook. ■

116940 Cymbidium lancifolium Hook. f. aspidistrifolium (Fukuy.) T. P. Lin = Cymbidium lancifolium Hook. var. aspidistrifolium

(Fukuy.) S. S. Ying ■
116941 Cymbidium lancifolium Hook. var. aspidistrifolium (Fukuy.) S. S. Ying;绿花竹柏兰;Green-flowered Rabbitear Orchis ■
116942 Cymbidium lancifolium Hook. var. aspidistrifolium (Fukuy.) S. S. Ying = Cymbidium lancifolium Hook. ■
116943 Cymbidium lancifolium Hook. var. papuanum (Schltr.) S. S. Ying = Cymbidium lancifolium Hook. ■
116944 Cymbidium lancifolium Hook. var. syunitianum (Fukuy.) S. S. Ying;大竹柏兰■
116945 Cymbidium lancifolium Hook. var. syunitianum (Fukuy.) S. S. Ying = Cymbidium lancifolium Hook. ■
116946 Cymbidium leachianum Rchb. f.;立奇兰■
116947 Cymbidium leachianum Rchb. f. = Cymbidium dayanum Rchb. f. ■
116948 Cymbidium lianpan Ts. Tang et F. T. Wang ex Y. S. Wu = Cymbidium tortisepalum Fukuy. ■
116949 Cymbidium linearisepalum Gogelein = Cymbidium kanran Makino ■
116950 Cymbidium linearisepalum Gogelein f. atropurpureum Gogelein = Cymbidium kanran Makino ■
116951 Cymbidium linearisepalum Gogelein f. atropurpureum Yamam. = Cymbidium kanran Makino ■
116952 Cymbidium linearisepalum Gogelein f. atrovirens Gogelein = Cymbidium kanran Makino ■
116953 Cymbidium linearisepalum Gogelein f. atrovirens Yamam. = Cymbidium kanran Makino ■
116954 Cymbidium linearisepalum Gogelein var. atropurpureum (Gogelein) Masam. = Cymbidium kanran Makino ■
116955 Cymbidium linearisepalum Gogelein var. atrovirens (Yamam.) Masam. = Cymbidium kanran Makino ■
116956 Cymbidium linearisepalum Yamam. = Cymbidium kanran Makino ■
116957 Cymbidium loise-chauvierii Hort. = Cymbidiella pardalina (Rchb. f.) Garay ■☆
116958 Cymbidium longibracteatum Y. S. Wu et S. C. Chen = Cymbidium goeringii (Rchb. f.) Rchb. f. var. longibracteatum (Y. S. Wu et S. C. Chen) Y. S. Wu et S. C. Chen ■
116959 Cymbidium longibracteatum Y. S. Wu et S. C. Chen = Cymbidium tortisepalum Fukuy. var. longibracteatum (Y. S. Wu et S. C. Chen) S. C. Chen et Z. J. Liu ■
116960 Cymbidium longibracteatum Y. S. Wu et S. C. Chen var. flaccidifolium Y. S. Wu = Cymbidium tortisepalum Fukuy. var. longibracteatum (Y. S. Wu et S. C. Chen) S. C. Chen et Z. J. Liu ■
116961 Cymbidium longibracteatum Y. S. Wu et S. C. Chen var. rubisepalum Y. S. Wu = Cymbidium tortisepalum Fukuy. var. longibracteatum (Y. S. Wu et S. C. Chen) S. C. Chen et Z. J. Liu ■
116962 Cymbidium longibracteatum Y. S. Wu et S. C. Chen var. tonghaiense Y. S. Wu = Cymbidium tortisepalum Fukuy. var. longibracteatum (Y. S. Wu et S. C. Chen) S. C. Chen et Z. J. Liu ■
116963 Cymbidium longibracteatum Y. S. Wu et S. C. Chen var. tortisepalum (Fukuy.) Y. S. Wu = Cymbidium tortisepalum Fukuy. ■
116964 Cymbidium longifolium D. Don = Cymbidium elegans Lindl. ■
116965 Cymbidium longifolium D. Don = Cymbidium erythraeum Lindl. ■
116966 Cymbidium longipes Z. J. Liu et J. N. Zhang;长柄兰;Longstalk Cymbidium, Longstalk Orchis ■
116967 Cymbidium lowianum (Rchb. f.) Rchb. f.;碧玉兰(村葱慈姑);Jade Cymbidium, Jade Orchis ■
116968 Cymbidium lowianum (Rchb. f.) Rchb. f. var. changningense X. M. Xu = Cymbidium changningense Z. J. Liu et S. C. Chen ■
116969 Cymbidium lowianum (Rchb. f.) Rchb. f. var. iansonii (Rolfe) P. J. Cribb et Du Puy;浅斑碧玉兰■
116970 Cymbidium lowianum (Rchb. f.) Rchb. f. var. kalawense (Colyear) Govaerts = Cymbidium lowianum (Rchb. f.) Rchb. f. var. iansonii (Rolfe) P. J. Cribb et Du Puy ■
116971 Cymbidium lushuiense Z. J. Liu, S. C. Chen et X. C. Shi = Cymbidium elegans Lindl. var. lushuiense (Z. J. Liu, S. C. Chen et X. C. Shi) Z. J. Liu et S. C. Chen ■
116972 Cymbidium maclehoseae S. Y. Hu = Cymbidium lancifolium Hook. ■
116973 Cymbidium macrorhizon Lindl. = Cymbidium macrorrhizum Lindl. ■
116974 Cymbidium macrorhizum Lindl. f. aberrans (Schltr.) Hid. Takah. et Ohba = Cymbidium macrorrhizum Lindl. f. aberrans (Schltr.) Hid. Takah. et Ohba ■☆
116975 Cymbidium macrorrhizum Lindl.;大根兰;Largeroot Orchis, Saprophylic Cymbidium ■
116976 Cymbidium macrorrhizum Lindl. f. aberrans (Schltr.) Hid. Takah. et Ohba;白大根兰■☆
116977 Cymbidium madidum Lindl.;潮湿兰■☆
116978 Cymbidium maguanense F. Y. Liu;马关兰(象牙白);Maguan Cymbidium, Maguan Orchis ■
116979 Cymbidium maguanense F. Y. Liu = Cymbidium mastersii Griff. ex Lindl. ■
116980 Cymbidium mandaianum Gower = Cymbidium lowianum (Rchb. f.) Rchb. f. var. iansonii (Rolfe) P. J. Cribb et Du Puy ■
116981 Cymbidium mannii Rchb. f. = Cymbidium aloifolium (L.) Sw. ■☆
116982 Cymbidium mannii Rchb. f. = Cymbidium bicolor Lindl. subsp. obtusum Du Puy et P. J. Cribb ■
116983 Cymbidium mastersii Griff. = Cymbidium mastersii Griff. ex Lindl. ■
116984 Cymbidium mastersii Griff. ex Lindl.;大雪兰;Heavysnow Cymbidium, Heavysnow Orchis ■
116985 Cymbidium micans Schauer = Cymbidium ensifolium (L.) Sw. ■
116986 Cymbidium micranthum Z. J. Liu et S. C. Chen;细花兰■
116987 Cymbidium micromeson Lindl. = Cymbidium mastersii Griff. ex Lindl. ■
116988 Cymbidium misericors Hayata = Cymbidium ensifolium (L.) Sw. ■
116989 Cymbidium misericors Hayata var. oreophyllum (Hayata) Hayata = Cymbidium kanran Makino ■
116990 Cymbidium moschatum (Buch. -Ham.) Willd. = Dendrobium moschatum (Buch. -Ham.) Sw. ■
116991 Cymbidium multiradicatum Z. J. Liu et S. C. Chen;多根兰■
116992 Cymbidium nagifolium Masam. = Cymbidium lancifolium Hook. ■
116993 Cymbidium nanulum Y. S. Wu et S. C. Chen;珍珠矮兰;Dwarf-pear Cymbidium, Dwarf-pear Orchis ■
116994 Cymbidium nigrovenium Z. J. Liu et J. N. Zhang;黑脉寒兰;Blackvein Cymbidium, Blackvein Orchis ■
116995 Cymbidium nigrovenium Z. J. Liu et J. N. Zhang = Cymbidium kanran Makino ■
116996 Cymbidium nipponicum (Franch. et Sav.) Makino = Cymbidium macrorrhizum Lindl. ■
116997 Cymbidium nipponicum (Franch. et Sav.) Makino f. sagamiense

(Nakai) Sugim. = Cymbidium macrorrhizum Lindl. ■

116998 Cymbidium nitidum Roxb. = Coelogyne punctulata Lindl. ■

116999 Cymbidium nitidum Wall. ex D. Don = Coelogyne nitida (Wall. ex D. Don) Lindl. ■

117000 Cymbidium niveo-marginatum Makino;玉花兰■

117001 Cymbidium odontorhizon Willd. = Corallorhiza odontorhiza (Willd.) Poir. ■☆

117002 Cymbidium oiwakense Hayata = Cymbidium faberi Rolfe ■

117003 Cymbidium omeiense Y. S. Wu et S. C. Chen = Cymbidium faberi Rolfe var. omeiense (Y. S. Wu et S. C. Chen) Y. S. Wu et S. C. Chen ■

117004 Cymbidium oreophyllum Hayata = Cymbidium kanran Makino ■

117005 Cymbidium papuanum Schltr. = Cymbidium lancifolium Hook. ■

117006 Cymbidium paucifolium Z. J. Liu et S. C. Chen;少叶硬叶兰■

117007 Cymbidium pedicellatum (L. f.) Sw. = Eulophia aculeata (L. f.) Spreng. ■☆

117008 Cymbidium pendulum (Roxb.) Sw.;硬叶吊兰(大剑兰,大凉药,大甩头,倒吊兰,吊兰,吊兰子,虎头兰,剑兰,石吊兰,树茭瓜,卧吊兰,硬叶兰);Rigidleaf Cymbidium ■

117009 Cymbidium pendulum (Roxb.) Sw. = Cymbidium aloifolium (L.) Sw. ■☆

117010 Cymbidium pendulum (Roxb.) Sw. = Cymbidium bicolor Lindl. subsp. obtusum Du Puy et P. J. Cribb ■

117011 Cymbidium pendulum (Roxb.) Sw. var. atropurpureum Lindl. = Cymbidium atropurpureum (Lindl.) Rolfe ■

117012 Cymbidium pendulum Sw. = Cymbidium pendulum (Roxb.) Sw. ■

117013 Cymbidium pendulum Sw. var. purpureum W. Watson = Cymbidium atropurpureum (Lindl.) Rolfe ■

117014 Cymbidium pictum R. Br. = Geodorum densiflorum (Lam.) Schltr. ■

117015 Cymbidium plicatum Harv. ex Lindl. = Eulophia aculeata (L. f.) Spreng. ■☆

117016 Cymbidium poilanei Gagnep. = Cymbidium dayanum Rchb. f. ■

117017 Cymbidium pomporenium Z. J. Liu et J. N. Zhang;显脉四季兰;Distinctvein Cymbidium ■

117018 Cymbidium praecox (Sm.) Lindl. = Pleione praecox (Sm.) D. Don ■

117019 Cymbidium prompovenium Z. J. Liu et J. N. Zhang = Cymbidium ensifolium (L.) Sw. ■

117020 Cymbidium pseudovirens Schltr. = Cymbidium goeringii (Rchb. f.) Rchb. f. ■

117021 Cymbidium pulchellum (Salisb.) Sw. = Calopogon tuberosus (L.) Britton, Sterns et Poggenb. ■☆

117022 Cymbidium pumilum Rolfe;金棱边兰(长寿兰,金龙边)■

117023 Cymbidium pumilum Rolfe = Cymbidium floribundum Lindl. ■

117024 Cymbidium purpureohiemale Hayata = Cymbidium kanran Makino ■

117025 Cymbidium qiubeiense K. M. Feng et H. Li;邱北冬蕙兰(邱北蕙兰);Qiubei Cymbidium, Qiubei Orchis ■

117026 Cymbidium quinquelobum Z. J. Liu et S. C. Chen = Cymbidium wenshanense Y. S. Wu et F. Y. Liu var. quinquelobum (Z. J. Liu et S. C. Chen) Z. J. Liu, S. C. Chen et P. J. Cribb ■

117027 Cymbidium recurvatum Z. J. Liu, S. C. Chen et P. J. Cribb;长茎兔耳兰■

117028 Cymbidium rhizomatosum Z. J. Liu et S. C. Chen;二叶兰■

117029 Cymbidium rhodochilum Rolfe = Cymbidiella pardalina (Rchb. f.) Garay ■☆

117030 Cymbidium rhodochilum Rolfe = Cymbidiella rhodochila Rolfe ■☆

117031 Cymbidium rigidum Z. J. Liu et S. C. Chen = Cymbidium devonianum Paxton ■

117032 Cymbidium rubrigemmum Hayata = Cymbidium ensifolium (L.) Sw. ■

117033 Cymbidium scabroserrulatum Makino = Cymbidium faberi Rolfe ■

117034 Cymbidium schroederi Rolfe;薛氏兰■

117035 Cymbidium serratum Schltr.;豆瓣兰(细叶春兰)■

117036 Cymbidium serratum Schltr. = Cymbidium goeringii (Rchb. f.) Rchb. f. var. serratum (Schltr.) Y. S. Wu et S. C. Chen ■

117037 Cymbidium siamense Rolfe ex Downie = Cymbidium haematodes Lindl. ■

117038 Cymbidium sichuanicum Z. J. Liu et S. C. Chen;川西兰■

117039 Cymbidium simlans Rolfe = Cymbidium aloifolium (L.) Sw. ■☆

117040 Cymbidium simonsianum King et Pantl. = Cymbidium dayanum Rchb. f. ■

117041 Cymbidium simulans Rolfe = Cymbidium aloifolium (L.) Sw. ■☆

117042 Cymbidium sinense (Jacks. ex Andréws) Willd.;墨兰(白花报岁兰,报岁兰,春兰,丰岁兰);China Orchis, Chinese Cymbidium, Dark-purple Cymbidium ■

117043 Cymbidium sinense (Jacks. ex Andréws) Willd. f. albojucundissimum (Hayata) Fukuy. = Cymbidium sinense (Jacks. ex Andréws) Willd. ■

117044 Cymbidium sinense (Jacks. ex Andréws) Willd. f. aureomarginatum T. K. Yen = Cymbidium sinense (Jacks. ex Andréws) Willd. ■

117045 Cymbidium sinense (Jacks. ex Andréws) Willd. f. hakuran Makino;大明兰■☆

117046 Cymbidium sinense (Jacks. ex Andréws) Willd. f. margicoloratum (Hayata) Fukuy. = Cymbidium sinense (Jacks. ex Andréws) Willd. ■

117047 Cymbidium sinense (Jacks. ex Andréws) Willd. f. pallidiflorum S. S. Ying = Cymbidium sinense (Jacks. ex Andréws) Willd. ■

117048 Cymbidium sinense (Jacks. ex Andréws) Willd. f. taiwanianum (S. S. Ying) S. S. Ying = Cymbidium sinense (Jacks. ex Andréws) Willd. ■

117049 Cymbidium sinense (Jacks. ex Andréws) Willd. f. taiwanianum S. S. Ying = Cymbidium sinense (Jacks. ex Andréws) Willd. ■

117050 Cymbidium sinense (Jacks. ex Andréws) Willd. f. viridiflorum T. K. Yen = Cymbidium sinense (Jacks. ex Andréws) Willd. ■

117051 Cymbidium sinense (Jacks. ex Andréws) Willd. var. albojucundissimum (Hayata) Masam.;白墨兰(白花丰岁兰);Whiteflower Chinese Cymbidium ■

117052 Cymbidium sinense (Jacks. ex Andréws) Willd. var. albojucundissimum (Hayata) Masam. = Cymbidium sinense (Jacks. ex Andréws) Willd. ■

117053 Cymbidium sinense (Jacks. ex Andréws) Willd. var. album T. K. Yen f. viridiflorum T. K. Yen = Cymbidium sinense (Jacks. ex Andréws) Willd. ■

117054 Cymbidium sinense (Jacks. ex Andréws) Willd. var. album T. K. Yen = Cymbidium sinense (Jacks. ex Andréws) Willd. ■

117055 Cymbidium sinense (Jacks. ex Andréws) Willd. var. autumnale Y. S. Wu = Cymbidium sinense (Jacks. ex Andréws) Willd. ■

117056 Cymbidium sinense (Jacks. ex Andréws) Willd. var. bellum T. K. Yen = Cymbidium sinense (Jacks. ex Andréws) Willd. ■

117057 Cymbidium sinense (Jacks. ex Andréws) Willd. var. haematodes

(Lindl.) Z. J. Liu et S. C. Chen = Cymbidium haematodes Lindl. ■

117058 Cymbidium sinense (Jacks. ex Andréws) Willd. var. margicoloratum Hayata;彩边墨兰;Margincolored Cymbidium ■

117059 Cymbidium sinense (Jacks. ex Andréws) Willd. var. margicoloratum Hayata = Cymbidium sinense (Jacks. ex Andréws) Willd. ■

117060 Cymbidium sinense (Jacks. ex Andréws) Willd. var. pallidiflorum S. S. Ying = Cymbidium sinense (Jacks. ex Andréws) Willd. ■

117061 Cymbidium sinense (Jacks. ex Andréws) Willd. var. taiwanianum S. S. Ying = Cymbidium sinense (Jacks. ex Andréws) Willd. ■

117062 Cymbidium sinokanran T. K. Yen = Cymbidium kanran Makino ■

117063 Cymbidium sinokanran T. K. Yen var. atropurpureum T. K. Yen = Cymbidium kanran Makino ■

117064 Cymbidium speciosissimum D. Don = Coelogyne cristata Lindl. ■

117065 Cymbidium striatum (Thunb.) Swartz = Bletilla striata (Thunb. ex A. Murray) Rchb. f. ■

117066 Cymbidium strictum D. Don = Coelogyne stricta (D. Don) Schltr. ■

117067 Cymbidium suavissimum Sander ex C. H. Curtis;果香兰;Fragrant Cymbidium, Fragrant Orchis, Sweet Cymbidium ■

117068 Cymbidium sumbidium Rolfe ex Downie = Cymbidium dayanum Rchb. f. ■

117069 Cymbidium sundaicum Schltr. = Cymbidium haematodes Lindl. ■

117070 Cymbidium sundaicum Schltr. var. estriatum Schltr. = Cymbidium haematodes Lindl. ■

117071 Cymbidium sutepense Rolfe ex Downie = Cymbidium dayanum Rchb. f. ■

117072 Cymbidium syringodorum Griff. = Cymbidium eburneum Lindl. ■

117073 Cymbidium syunitianum Fukuy. = Cymbidium lancifolium Hook. var. syunitianum (Fukuy.) S. S. Ying ■

117074 Cymbidium syunitianum Fukuy. = Cymbidium lancifolium Hook. ■

117075 Cymbidium szechuanensis S. Y. Hu = Cymbidium macrorrhizum Lindl. ■

117076 Cymbidium szechuanicum Y. S. Wu et S. C. Chen = Cymbidium cyperifolium Wall. ex Lindl. var. szechuanicum (Y. S. Wu et S. C. Chen) S. C. Chen et Z. J. Liu ■

117077 Cymbidium szechuanicum Y. S. Wu et S. C. Chen = Cymbidium faberi Rolfe var. szechuanicum (Y. S. Wu et S. C. Chen) Y. S. Wu et S. C. Chen ■

117078 Cymbidium tabulare (L. f.) Sw. = Eulophia tabularis (L. f.) Bolus ■☆

117079 Cymbidium tentyozanense Masam. = Cymbidium goeringii (Rchb. f.) Rchb. f. ■

117080 Cymbidium teretipetiolatum Z. J. Liu et S. C. Chen;奇瓣红春素■

117081 Cymbidium tigrinum E. C. Parish ex Hook.;斑舌兰;Spottongue Cymbidium, Spottongue Orchis ■

117082 Cymbidium tortisepalum Fukuy.;菅草兰(莲瓣兰,苗栗素心兰);Twistsepal Cymbidium ■

117083 Cymbidium tortisepalum Fukuy. = Cymbidium goeringii (Rchb. f.) Rchb. f. var. tortisepalum (Fukuy.) Y. S. Wu et S. C. Chen ■

117084 Cymbidium tortisepalum Fukuy. f. albiflorum S. S. Ying = Cymbidium goeringii (Rchb. f.) Rchb. f. var. tortisepalum (Fukuy.) Y. S. Wu et S. C. Chen ■

117085 Cymbidium tortisepalum Fukuy. f. albiflorum S. S. Ying = Cymbidium tortisepalum Fukuy. ■

117086 Cymbidium tortisepalum Fukuy. var. longibracteatum (Y. S. Wu et S. C. Chen) S. C. Chen et Z. J. Liu;春剑;Longbract Spring Orchis, Long-bracted Cymbidium ■

117087 Cymbidium tortisepalum Fukuy. var. viridiflorum S. S. Ying = Cymbidium tortisepalum Fukuy. ■

117088 Cymbidium tortisepalum Fukuy. var. viridiflorum S. S. Ying = Cymbidium faberi Rolfe ■

117089 Cymbidium tosyaense Masam. = Cymbidium kanran Makino ■

117090 Cymbidium tracyanum L. Castle;西藏虎头兰(虎头兰);Tibet Cymbidium, Tracy Orchis, Xizang Cymbidium ■

117091 Cymbidium tsukengensis C. Chow = Cymbidium tortisepalum Fukuy. ■

117092 Cymbidium umbellatum Spreng. = Bulbophyllum longiflorum Thouars ■☆

117093 Cymbidium undulatum (Sw.) Sw. = Trichocentrum undulatum (Sw.) Ackerman et M. W. Chase ■☆

117094 Cymbidium uniflorum T. K. Yen = Cymbidium goeringii (Rchb. f.) Rchb. f. ■

117095 Cymbidium ustulatum Bolus = Acrolophia ustulata (Bolus) Schltr. et Bolus ■☆

117096 Cymbidium verecundum (Salisb.) Sw. = Bletia purpurea (Lam.) DC. ■☆

117097 Cymbidium virens Rchb. f. = Cymbidium goeringii (Rchb. f.) Rchb. f. ■

117098 Cymbidium virescens Lindl. = Cymbidium goeringii (Rchb. f.) Rchb. f. ■

117099 Cymbidium viridiflorum Griff. = Cymbidium cyperifolium Wall. ex Lindl. ■

117100 Cymbidium wenshanense Y. S. Wu et F. Y. Liu;文山红柱兰(文山虎头兰);Wenshan Cymbidium, Wenshan Orchis ■

117101 Cymbidium wenshanense Y. S. Wu et F. Y. Liu var. quinquelobum (Z. J. Liu et S. C. Chen) Z. J. Liu, S. C. Chen et P. J. Cribb;五裂红柱兰■

117102 Cymbidium wilsonii (Rolfe ex Cook) Rolfe;滇南虎头兰(短叶虎头兰,兰草,兰花,细叶土兰);E. H. Wilson Cymbidium, E. H. Wilson Orchis ■

117103 Cymbidium xiphiifolium Lindl. = Cymbidium ensifolium (L.) Sw. ■

117104 Cymbidium yakibaran Makino;蕉叶兰■

117105 Cymbidium yakibaran Makino = Cymbidium ensifolium (L.) Sw. ■

117106 Cymbidium yongfuense Z. J. Liu et J. N. Zhang;线叶建兰;Yongfu Cymbidium, Yongfu Orchis ■

117107 Cymbidium yongfuense Z. J. Liu et J. N. Zhang = Cymbidium ensifolium (L.) Sw. ■

117108 Cymbidium yunnanense Schltr. = Cymbidium goeringii (Rchb. f.) Rchb. f. ■

117109 Cymbiglossum Halb. = Lemboglossum Halb. ■

117110 Cymbispatha Pichon = Tradescantia L. ■

117111 Cymbocarpa Miers(1840);舟果水玉簪属■☆

117112 Cymbocarpa refracta Miers;舟果水玉簪;Cat's Milk ■☆

117113 Cymbocarpum DC. = Cymbocarpum DC. ex C. A. Mey. ■☆

117114 Cymbocarpum DC. ex C. A. Mey. (1831);舟果芹属■☆

117115 Cymbocarpum anethoides DC.;舟果芹■☆

117116 Cymbochasma (Endl.) Klokov et Zoz = Cymbaria L. ■

117117 Cymbochasma (Endl.) Klokov et Zoz(1839);舟口玄参属■☆

117118 Cymbochasma Endl. = Cymbaria L. ■
117119 Cymbochasma borystheaica (Pall.) Klokov et Zoz;舟口玄参☆
117120 Cymboglossum (J. J. Sm.) Brieger = Eria Lindl.(保留属名)■
117121 Cymbolaena Smoljan. (1955);长柱紫绒草属■☆
117122 Cymbolaena griffithii (A. Gray) Wagenitz;长柱紫绒草■☆
117123 Cymbolaena longifolia (Boiss. et Reut.) Smoljan. = Cymbolaena griffithii (A. Gray) Wagenitz ■☆
117124 Cymbonotus Cass. (1825);澳洲熊耳菊属■☆
117125 Cymbonotus preissianus Steetz;澳洲熊耳菊■☆
117126 Cymbopappus B. Nord. (1976);舟冠菊属●☆
117127 Cymbopappus adenosolen (Harv.) B. Nord.;腺管舟冠菊■☆
117128 Cymbopappus hilliardiae B. Nord.;希利亚德舟冠菊■☆
117129 Cymbopappus lasiopodus (Hutch.) B. Nord. = Cymbopappus piliferus (Thell.) B. Nord. ■☆
117130 Cymbopappus piliferus (Thell.) B. Nord.;纤毛舟冠菊■☆
117131 Cymbopetalum Benth. (1860);舟瓣花属●☆
117132 Cymbopetalum odoratissimum Barb. Rodr.;巴西舟瓣花●☆
117133 Cymbopetalum penduliflorum (Dunal) Baill.;垂花舟瓣花●☆
117134 Cymbopetalum penduliflorum Baill. = Cymbopetalum penduliflorum (Dunal) Baill. ●☆
117135 Cymbophyllum F. Muell. = Veronica L. ■
117136 Cymbopogon Spreng. (1815);香茅属(橘草属);Cirtonella, Lemon Grass, Lemongrass, Oil Grass ■
117137 Cymbopogon acutispathaceus De Wild. = Hyparrhenia variabilis Stapf ■☆
117138 Cymbopogon afronardus Stapf = Cymbopogon nardus (L.) Rendle ■
117139 Cymbopogon ambiguus A. Camus;含糊香茅■☆
117140 Cymbopogon andongensis Rendle = Hyparrhenia andongensis (Rendle) Stapf ■☆
117141 Cymbopogon annamensis (A. Camus) A. Camus;圆基香茅■
117142 Cymbopogon arriani (Edgew.) Aitch. = Cymbopogon jwarancusus (Jones) Schult. subsp. olivieri (Boiss.) Soenarko ■
117143 Cymbopogon arriani (Edgew.) Aitch. = Cymbopogon olivieri (Boiss.) Bor ■
117144 Cymbopogon arundinaceus (Roxb.) Schult. = Themeda arundinacea (Roxb.) Ridl. ■
117145 Cymbopogon auritus B. S. Sun;长耳香茅■
117146 Cymbopogon bassacensis A. Camus = Cymbopogon annamensis (A. Camus) A. Camus ■
117147 Cymbopogon bequaertii De Wild. = Hyparrhenia gossweileri Stapf ■☆
117148 Cymbopogon bombycinus (R. Br.) Domin;丝质香茅■☆
117149 Cymbopogon bracteata (Humb. et Bonpl. ex Willd.) Hitchc. = Hyparrhenia bracteata (Humb. et Bonpl. ex Willd.) Stapf ■
117150 Cymbopogon caesius (Nees ex Hook. et Arn.) Stapf;青香茅(橘香草,枯香草,香花草);Green Lemongrass, Grey Lemongrass ■
117151 Cymbopogon caesius (Nees ex Hook. et Arn.) Stapf subsp. giganteus (Chiov.) Sales;大青香茅■☆
117152 Cymbopogon chrysargyreus Stapf = Hyparrhenia nyassae (Rendle) Stapf ■☆
117153 Cymbopogon citratus (DC.) Stapf;香茅(大风茅,风茅,风茅草,姜巴草,姜巴茅,姜芭果,姜芭茅,姜草,茅草茶,茅香草,柠檬草,柠檬茅,柠檬香茅,牛腿芒,香巴茅,香芭茅,香麻,香茅草);Lemon Grass, Lemongrass, Lemon-grass, Malabar Oil, Serai ■
117154 Cymbopogon claessensii Robyns = Cymbopogon nardus (L.) Rendle ■
117155 Cymbopogon commutatus (Steud.) Stapf;变异香茅■☆
117156 Cymbopogon confertiflorus (Steud.) Stapf var. traninhensis A. Camus = Cymbopogon traninhensis (A. Camus) Soenarko ■
117157 Cymbopogon cyanescens Stapf = Hyparrhenia cyanescens (Stapf) Stapf ■☆
117158 Cymbopogon densiflorus (Steud.) Stapf;密花香茅;Denseflower Lemongrass ■☆
117159 Cymbopogon densiflorus Stapf = Cymbopogon densiflorus (Steud.) Stapf ■☆
117160 Cymbopogon dieterlenii Stapf ex E. Phillips;迪氏香茅■☆
117161 Cymbopogon diplandrus (Hack.) De Wild. = Hyparrhenia diplandra (Hack.) Stapf ■
117162 Cymbopogon distans (Nees ex Steud.) W. Watson;芸香草(臭草,韭叶芸香草,茅草筋骨,山茅草,射香草,麝香草,石灰草,细叶茅草,香茅草,香茅筋骨草,小香茅草,野芸香草,诸葛草);Remote Lemongrass, Rue Lemongrass ■
117163 Cymbopogon divaricatus Stapf = Cymbopogon commutatus (Steud.) Stapf ■☆
117164 Cymbopogon eberhardtii A. Camus = Hyparrhenia diplandra (Hack.) Stapf ■
117165 Cymbopogon eugenolatus L. Liou;香酚草;Henol Lemongrass ■
117166 Cymbopogon excavatus (Hochst.) Stapf ex Burtt Davy = Cymbopogon caesius (Nees ex Hook. et Arn.) Stapf ■
117167 Cymbopogon fibrosus B. S. Sun;纤鞘香茅■
117168 Cymbopogon filipendulus (Hochst.) Rendle = Hyparrhenia filipendula (Hochst.) Stapf ■
117169 Cymbopogon filipendulus (Hochst.) Rendle var. thwaitesii (Hochst.) Hand. -Mazz.;吊丝云香草■
117170 Cymbopogon flexuasus (Steud.) W. Watson = Cymbopogon flexuosus (Nees ex Steud.) W. Watson ■
117171 Cymbopogon flexuosus (Nees ex Steud.) W. Watson;曲序香茅(东印度柠檬草,蜿蜒香茅);Flexspike Lemongrass, Malabar Oil ■
117172 Cymbopogon flexuosus (Nees ex Steud.) W. Watson var. microstachys (Hook. f.) Bor = Cymbopogon microstachys (Hook. f.) Soenarko ■
117173 Cymbopogon flexuosus (Nees ex Steud.) W. Watson var. microstachys (Hook. f.) Bor. = Cymbopogon microstachys (Hook. f.) Soenarko ■
117174 Cymbopogon foliosus (Humb. et Bonpl. ex Willd.) Roem. et Schult. = Hyparrhenia bracteata (Humb. et Bonpl. ex Willd.) Stapf ■
117175 Cymbopogon gazensis Rendle = Hyparrhenia gazensis (Rendle) Stapf ■☆
117176 Cymbopogon gidarba (Buch. -Ham. ex Steud.) A. Camus var. burmanicus Bor;缅甸浅囊香茅■
117177 Cymbopogon giganteus Chiov. = Cymbopogon caesius (Nees ex Hook. et Arn.) Stapf subsp. giganteus (Chiov.) Sales ■☆
117178 Cymbopogon giganteus Chiov. var. densiflorus (Steud.) Chiov. = Cymbopogon densiflorus (Steud.) Stapf ■☆
117179 Cymbopogon giganteus Chiov. var. inermis Clayton = Cymbopogon caesius (Nees ex Hook. et Arn.) Stapf subsp. giganteus (Chiov.) Sales ■☆
117180 Cymbopogon giganteus Chiov. var. madagascariensis A. Camus = Cymbopogon caesius (Nees ex Hook. et Arn.) Stapf ■
117181 Cymbopogon goeringii (Steud.) A. Camus;橘草(桔草,五香草,香茅,香茅草,野香茅);Goering Lemongrass ■
117182 Cymbopogon goeringii (Steud.) A. Camus = Cymbopogon tortilis (J. Presl) Hitchc. var. goeringii (Steud.) Hand. -Mazz. ■

117183 Cymbopogon goeringii (Steud.) A. Camus var. hongkongensis Soenarko = Cymbopogon goeringii (Steud.) A. Camus ■

117184 Cymbopogon hamatulus (Hook. et Arn.) A. Camus = Cymbopogon tortilis (J. Presl) A. Camus ■

117185 Cymbopogon hamatulus (Nees ex Hook. et Arn.) A. Camus;扭鞘香茅(臭草,韭叶芸香草,括花草,扭曲香茅,生姜草,细香草,野香茅,芸香草);Tweaksheath Lemongrass,Twisted Lemongrass ■

117186 Cymbopogon hookeri (Munro ex Hack.) Stapf ex Bor = Andropogon munroi C. B. Clarke ■

117187 Cymbopogon iwarancusus Schult. = Andropogon iwarancusa Roxb. ■☆

117188 Cymbopogon jinshaensis R. Zhang et C. H. Li;金沙香茅;Jinsha Lemongrass ■

117189 Cymbopogon jinshaensis R. Zhang et C. H. Li = Cymbopogon tortilis (J. Presl) A. Camus ■

117190 Cymbopogon jwarancusus (Jones) Schult.;辣薄荷草(喜马拉雅香茅,伊瓦须芒草);Hotmint Lemongrass,Iwarancusa Grass ■

117191 Cymbopogon jwarancusus (Jones) Schult. = Andropogon jwarancusa Roxb. ■☆

117192 Cymbopogon jwarancusus (Jones) Schult. subsp. olivieri (Boiss.) Soenarko;西亚香茅;W. Asia Lemongrass ■

117193 Cymbopogon jwarancusys (Jones) Schult. subsp. olivieri (Boiss.) Soenarko = Cymbopogon olivieri (Boiss.) Bor ■

117194 Cymbopogon kapandensis De Wild. = Hyparrhenia diplandra (Hack.) Stapf ■

117195 Cymbopogon khasianus (Munro ex Hack.) Stapf ex Bor;卡西香茅(卡西山香茅);Khas Lemongrass ■

117196 Cymbopogon khasianus (Munro ex Hack.) Stapf ex Bor var. nagensis Bor = Cymbopogon traninhensis (A. Camus) Soenarko ■

117197 Cymbopogon ladakhensis B. K. Gupta = Cymbopogon jwarancusus (Jones) Schult. subsp. olivieri (Boiss.) Soenarko ■

117198 Cymbopogon ladakhensis B. K. Gupta = Cymbopogon olivieri (Boiss.) Bor ■

117199 Cymbopogon lanceifolius L. Liou;披针叶香茅;Lanceleaf Lemongrass ■

117200 Cymbopogon lepidus (Nees) Chiov. = Hyparrhenia cymbaria (L.) Stapf ■☆

117201 Cymbopogon liangshanensis L. Liou;凉山香茅;Liangshan Lemongrass ■

117202 Cymbopogon marginatus (Steud.) Stapf ex Burtt Davy;具边香茅■☆

117203 Cymbopogon martinianus Schult. = Cymbopogon martinii (Roxb.) W. Watson ■

117204 Cymbopogon martinii (Roxb.) W. Watson;鲁沙香茅(红秆草,马丁香茅,香茅,鱼沙香茅);Geranium Oil,Ginger Grass,Martin Lemongrass,Palina-rosa,Palma Rosa,Palmerosa Oil Grass,Rosha,Rusha Grass,Sweet Calamus ■

117205 Cymbopogon martinii (Roxb.) W. Watson 'Motia';玫瑰草■☆

117206 Cymbopogon martinii (Roxb.) W. Watson 'Sofia';索菲亚香茅■☆

117207 Cymbopogon martinii (Roxb.) W. Watson var. anna-mensis A. Camus = Cymbopogon annamensis (A. Camus) A. Camus ■

117208 Cymbopogon martinii (Roxb.) W. Watson var. motia ? = Cymbopogon martinii (Roxb.) W. Watson 'Motia' ■☆

117209 Cymbopogon martinii (Roxb.) W. Watson var. sofia B. K. Gupta = Cymbopogon martinii (Roxb.) W. Watson 'Sofia' ■☆

117210 Cymbopogon martinii (Roxb.) W. Watson var. sofia B. K. Gupta = Cymbopogon martinii (Roxb.) W. Watson ■

117211 Cymbopogon melanocarpus (Elliott) Spreng. = Heteropogon melanocarpus (Elliott) Benth. ■

117212 Cymbopogon melanocarpus Spreng. = Heteropogon melanocarpus (Elliott) Benth. ■

117213 Cymbopogon micratherus Pilg. = Hyparrhenia dregeana (Nees) Stapf ex Stent ■☆

117214 Cymbopogon microstachys (Hook. f.) Soenarko;细穗香茅;Smallspike Lemongrass ■

117215 Cymbopogon minor B. S. Sun et R. Zhang ex S. M. Phillips et H. Peng;细小香茅■

117216 Cymbopogon modicus De Wild. = Hyparrhenia hirta (L.) Stapf ■☆

117217 Cymbopogon motia B. K. Gupta = Cymbopogon martinii (Roxb.) W. Watson ■

117218 Cymbopogon nardus (L.) Rendle;亚香茅(枫茅,金橘草,精香茅,香茅,香水茅);Citronella Grass,Citronella Oil,Citronella-grass,Mana Grass,Nard Grass,Nard-grass,Nardus Lemongrass,Second Lemongrass ■

117219 Cymbopogon nardus (L.) Rendle subsp. hamatulus (Hook. et Arn.) Rendle = Cymbopogon tortilis (J. Presl) A. Camus ■

117220 Cymbopogon nardus (L.) Rendle var. confertiflorus (Steud.) Bor = Cymbopogon nardus (L.) Rendle ■

117221 Cymbopogon nardus (L.) Rendle var. goeringii (Steud.) Rendle = Cymbopogon goeringii (Steud.) A. Camus ■

117222 Cymbopogon nardus (L.) Rendle var. stracheyi Hook. f. = Cymbopogon pospischilii (K. Schum.) C. E. Hubb. ■

117223 Cymbopogon nardus (L.) Rendle var. tortilis (J. Presl) Merr. ex Griff. et al. = Cymbopogon hamatulus (Nees ex Hook. et Arn.) A. Camus ■

117224 Cymbopogon nervosus B. S. Sun;多脉香茅■

117225 Cymbopogon obtectus S. T. Blake;被覆香茅;Silky-heads ■☆

117226 Cymbopogon olivieri (Boiss.) Bor = Cymbopogon jwarancusus (Jones) Schult. subsp. olivieri (Boiss.) Soenarko ■

117227 Cymbopogon pachnodes (Trin.) W. Watson = Cymbopogon martinii (Roxb.) W. Watson ■

117228 Cymbopogon parkeri Stapf = Cymbopogon commutatus (Steud.) Stapf ■☆

117229 Cymbopogon patens ?;铺展香茅■☆

117230 Cymbopogon pendulus (Nees ex Steud.) W. Watson;垂序香茅■

117231 Cymbopogon phoenix Rendle = Hyparrhenia diplandra (Hack.) Stapf ■

117232 Cymbopogon pilosovaginatus De Wild. = Hyparrhenia bracteata (Humb. et Bonpl. ex Willd.) Stapf ■

117233 Cymbopogon plurinodis (Stapf) Stapf ex Burtt Davy = Cymbopogon pospischilii (K. Schum.) C. E. Hubb. ■

117234 Cymbopogon pospischilii (K. Schum.) C. E. Hubb.;喜马拉雅香茅;Himalayas Lemongrass ■

117235 Cymbopogon princeps Stapf = Hyparrhenia dybowskii (Franch.) Roberty ■☆

117236 Cymbopogon procerus (R. Br.) Domin;高大香茅■☆

117237 Cymbopogon prolixus (Stapf) E. Phillips;伸展香茅■☆

117238 Cymbopogon proximus (A. Rich.) Stapf var. sennarensis (Hochst.) Drar = Cymbopogon schoenanthus (L.) Spreng. subsp. proximus (Hochst. ex A. Rich.) Maire et Weiller ■☆

117239 Cymbopogon proximus (Hochst. ex A. Rich.) Chiov.;近轴香茅■☆

117240 Cymbopogon refractus (R. Br.) A. Camus;反折香茅;Barbwire Grass ■☆

117241 Cymbopogon refractus A. Camus = Cymbopogon refractus (R. Br.) A. Camus ■☆

117242 Cymbopogon rufus (Nees) Rendle var. major Rendle = Hyparrhenia rufa (Nees) Stapf ■☆

117243 Cymbopogon scabrimarginatus De Wild. = Hyparrhenia collina (Pilg.) Stapf ■☆

117244 Cymbopogon schimperi (Hochst. ex A. Rich.) Rendle = Hyparrhenia schimperi (Hochst. ex A. Rich.) Andersson ex Stapf ■☆

117245 Cymbopogon schoenanthus (L.) Spreng.;蔺花香茅(茅香,签草香茅);Canal Grass,Schoenus-flower Lemongrass ■

117246 Cymbopogon schoenanthus (L.) Spreng. = Cymbopogon commutatus (Steud.) Stapf ■☆

117247 Cymbopogon schoenanthus (L.) Spreng. = Cymbopogon jwarancusus (Jones) Schult. subsp. olivieri (Boiss.) Soenarko ■

117248 Cymbopogon schoenanthus (L.) Spreng. = Cymbopogon olivieri (Boiss.) Bor ■

117249 Cymbopogon schoenanthus (L.) Spreng. subsp. laniger (Hook.) Maire et Weiller = Cymbopogon schoenanthus (L.) Spreng. ■

117250 Cymbopogon schoenanthus (L.) Spreng. subsp. proximus (Hochst. ex A. Rich.) Maire et Weiller;近基香茅■☆

117251 Cymbopogon schoenanthus (L.) Spreng. var. stypticus (Welw.) Rendle = Cymbopogon stypticus (Welw.) Fritsch ■☆

117252 Cymbopogon sennaarensis Chiov.;信浓香茅■☆

117253 Cymbopogon stolzii Pilg. = Hyparrhenia pilgerana C. E. Hubb. ■☆

117254 Cymbopogon stracheyi (Hook. f.) Raizada et Jain = Cymbopogon pospischilii (K. Schum.) C. E. Hubb. ■

117255 Cymbopogon stypticus (Welw.) Fritsch;收敛香茅■☆

117256 Cymbopogon subcordatifolius De Wild. = Diheteropogon amplectens (Nees) Clayton ■☆

117257 Cymbopogon tenuis Gilli = Hyparrhenia mobukensis (Chiov.) Chiov. ■☆

117258 Cymbopogon tibeticus Bor;藏香茅;Xizang Lemongrass ■

117259 Cymbopogon tibeticus Bor = Andropogon munroi C. B. Clarke ■

117260 Cymbopogon tortilis (J. Presl) A. Camus = Cymbopogon hamatulus (Nees ex Hook. et Arn.) A. Camus ■

117261 Cymbopogon tortilis (J. Presl) A. Camus subsp. goeringii (Steud.) Koyama = Cymbopogon goeringii (Steud.) A. Camus ■

117262 Cymbopogon tortilis (J. Presl) A. Camus var. goeringii (Steud.) Hand. -Mazz. = Cymbopogon goeringii (Steud.) A. Camus ■

117263 Cymbopogon tortilis (J. Presl) Hitchc. subsp. goeringii (Steud.) T. Koyama = Cymbopogon tortilis (J. Presl) Hitchc. var. goeringii (Steud.) Hand. -Mazz. ■

117264 Cymbopogon tortilis (J. Presl) Hitchc. var. goeringii (Steud.) Hand. -Mazz. = Cymbopogon goeringii (Steud.) A. Camus ■

117265 Cymbopogon traninhensis (A. Camus) Soenarko;横香茅■

117266 Cymbopogon travancorensis Bor = Cymbopogon flexuosus (Nees ex Steud.) W. Watson ■

117267 Cymbopogon tungmaiensis L. Liou;通麦香茅;Tongmai Lemongrass ■

117268 Cymbopogon umbrosus (Hochst.) Pilg. = Hyparrhenia umbrosa (Hochst.) Andersson ex Clayton ■☆

117269 Cymbopogon validus (Stapf) Stapf ex Burtt Davy = Cymbopogon nardus (L.) Rendle ■

117270 Cymbopogon validus (Stapf) Stapf ex Burtt Davy var. lysocladus Stapf = Cymbopogon nardus (L.) Rendle ■

117271 Cymbopogon vanderystii De Wild. = Hyparrhenia nyassae (Rendle) Stapf ■☆

117272 Cymbopogon welwitschii Rendle = Hyparrhenia welwitschii (Rendle) Stapf ■☆

117273 Cymbopogon winterianus Jowitt;枫茅(文氏香茅,爪哇香茅);Maple Lemongrass ■

117274 Cymbopogon xichangensis R. Zhang et B. S. Sun;西昌香茅;Xichang Lemongrass ■

117275 Cymbosema Benth. (1840);淡红豆属■☆

117276 Cymbosema roseum Benth.;淡红豆■☆

117277 Cymbosepalum Baker = Haematoxylum L. ●

117278 Cymbosepalum Baker(1895);舟萼豆属●☆

117279 Cymbosepalum baronii Baker;舟萼豆●☆

117280 Cymboseris Boiss. = Phaecasium Cass. ■

117281 Cymbosetaria Schweick. = Setaria P. Beauv. (保留属名) ■

117282 Cymbosetaria sagittifolia (A. Rich.) Schweick. = Setaria sagittifolia (A. Rich.) Walp. ■☆

117283 Cymbostemon Spach = Illicium L. ●

117284 Cymburus Raf. = Elytraria Michx. (保留属名) ●☆

117285 Cymburus Salisb. = Stachytarpheta Vahl(保留属名) ■●

117286 Cymburus urticaetolius Salisb. = Stachytarpheta urticifolia (Salisb.) Sims ●☆

117287 Cymelonema C. Presl = Urophyllum Jack ex Wall. ●

117288 Cymicifuga Rchb. = Cimicifuga L. ●■

117289 Cyminon St. -Lag. = Cuminum L. ■

117290 Cyminosma Gaertn. = Acronychia J. R. Forst. et G. Forst. (保留属名) ●

117291 Cyminosma pedunculata DC. = Acronychia pedunculata (L.) Miq. ●

117292 Cyminum Boiss. = Microsciadium Boiss. ■☆

117293 Cyminum Hill = Cuminum L. ■

117294 Cyminum Post et Kuntze = Cuminoides Moench ■☆

117295 Cyminum Post et Kuntze = Lagoecia L. ■☆

117296 Cymodocea K. D. König(1805)(保留属名);丝粉藻属(海神草属);Manateagrass ■

117297 Cymodocea acaulis Peter = Cymodocea serrulata (R. Br.) Asch. et Magnus ■☆

117298 Cymodocea aequorea J. König = Cymodocea nodosa (Ucria) Asch. ■☆

117299 Cymodocea asiatica Makino = Cymodocea serrulata (R. Br.) Asch. et Magnus ■☆

117300 Cymodocea ciliata (Forssk.) Ehrenb. ex Asch. = Thalassodendron ciliatum (Forssk.) Hartog ■☆

117301 Cymodocea filiformis (Kurtz) Correll = Syringodium filiforme Kurtz ■☆

117302 Cymodocea isoetifolia Asch. = Syringodium isoetifolium (Asch.) Dandy ■

117303 Cymodocea major (Cavolini) Grande = Cymodocea nodosa (Ucria) Asch. ■☆

117304 Cymodocea manatorum Asch. = Syringodium filiforme Kurtz ■☆

117305 Cymodocea nodosa (Ucria) Asch.;多节丝粉藻■☆

117306 Cymodocea rotundata Asch. et Schweinf. = Cymodocea rotundata Ehrenb. et Hemprich ex Asch. et Schweinf. ■

117307 Cymodocea rotundata Ehrenb. et Hemprich ex Asch. = Cymodocea rotundata Ehrenb. et Hemprich ex Asch. et Schweinf. ■

117308 Cymodocea rotundata Ehrenb. et Hemprich ex Asch. et

Schweinf.;丝粉藻;Manateagrass,Round Manateagrass ■
117309 Cymodocea serrulata (R. Br.) Asch. et Magnus;琉球丝粉藻■☆
117310 Cymodoceaceae N. Taylor = Cymodoceaceae Vines(保留科名)■
117311 Cymodoceaceae Vines(1895)(保留科名);丝粉藻科(海参草科,绿粉藻科);Manatee-grass Family ■
117312 Cymonamia (Roberty) Roberty = Bonamia Thouars(保留属名)●☆
117313 Cymonamia Roberty = Bonamia Thouars(保留属名)●☆
117314 Cymonetra Roberty = Gilletiodendron Vermoesen ●☆
117315 Cymonetra glandulosa (Portères) Roberty = Gilletiodendron glandulosum (Portères) J. Léonard ●☆
117316 Cymophora B. L. Rob. (1907);银光菊属;Cymophora ■☆
117317 Cymophora B. L. Rob. = Tridax L. ■●
117318 Cymophora accedens (S. F. Blake) B. L. Turner et A. M. Powell;银光菊;Cymophora ■☆
117319 Cymophora pringlei B. L. Rob.;普雷银光菊■☆
117320 Cymophyllus Mack. (1913);波叶莎草属■☆
117321 Cymophyllus Mack. ex Britton = Cymophyllus Mack. ■☆
117322 Cymophyllus Mack. ex Britton et A. Br. = Cymophyllus Mack. ■☆
117323 Cymophyllus fraseri (Andrews) Mack. = Cymophyllus fraserianus (Ker Gawl.) Kartesz et Gandhi ■☆
117324 Cymophyllus fraserianus (Ker Gawl.) Kartesz et Gandhi;波叶莎草;Fraser's Sedge ■☆
117325 Cymopterus Raf. (1819);聚散翼属■☆
117326 Cymopterus littoralis J. G. Cooper et A. Gray = Glehnia littoralis F. Schmidt ex Miq. ■
117327 Cymopterus longipes S. Watson;长柄聚散翼(长柄聚伞翼)■☆
117328 Cymopterus watsonii (J. M. Coult. et Rose) M. E. Jones;聚散翼(聚伞翼)■☆
117329 Cymothoe Airy Shaw = Costera J. J. Sm. ●☆
117330 Cyna Lour. = Styrax L. ●
117331 Cynamonum Deniker = Cinnamomum Schaeff. (保留属名)●
117332 Cynanchaceae G. Mey. = Apocynaceae Juss. (保留科名)●■
117333 Cynanchica Fourr. = Asperula L. (保留属名)■
117334 Cynanchum L. (1753);鹅绒藤属(白前属,牛皮消属);Mosquitotrap,Swallowwort,Swallow Wort,Swallow-wort ●■
117335 Cynanchum absconditum Liede;隐匿鹅绒藤●☆
117336 Cynanchum abyssinicum Decne.;阿比西尼亚鹅绒藤●☆
117337 Cynanchum abyssinicum Decne. var. tomentosum Oliv. = Cynanchum abyssinicum Decne. ●☆
117338 Cynanchum aculeatum (Desc.) Liede et Meve;皮刺鹅绒藤●☆
117339 Cynanchum acuminatifolium Hemsl.;潮风草(白薇,尖叶白前,小葛瓢);Acuminate Mosquitotrap,Acuminate Swallowwort ■
117340 Cynanchum acuminatifolium Hemsl. = Cynanchum ascyrifolium (Franch. et Sav.) Matsum. ■
117341 Cynanchum acuminatifolium Hemsl. = Vincetoxicum acuminatum Decne. ■
117342 Cynanchum acuminatum (Decne.) Matsum. = Cynanchum acuminatifolium Hemsl. ■
117343 Cynanchum acuminatum Royle ex Wight var. amamianum (Hatus.) T. Yamaz. = Cynanchum boudieri H. Lév. et Vaniot ■
117344 Cynanchum acuminatum Thunb. = Pentatropis capensis (L. f.) Bullock ■☆
117345 Cynanchum acutum L.;尖牛皮消(尖白前);Stranglewort ■☆
117346 Cynanchum acutum L. subsp. sibiricum (Willd.) Rech. f.;西伯利亚牛皮消(戟叶鹅绒藤,戟叶牛皮消,沙牛皮消);Siberia Mosquitotrap,Siberia Swallowwort ■
117347 Cynanchum acutum L. var. fissum (Pomel) Batt. = Cynanchum acutum L. ■☆
117348 Cynanchum acutum L. var. longifolium (Mart.) Ledeb. = Cynanchum acutum L. subsp. sibiricum (Willd.) Rech. f. ■
117349 Cynanchum adalinae (K. Schum.) K. Schum.;热非杯冠藤●☆
117350 Cynanchum adalinae (K. Schum.) K. Schum. subsp. mannii (Scott-Elliot) Bullock;曼氏杯冠藤■☆
117351 Cynanchum adriaticum R. Beck;德里白前;Mosquitotrap,Swallowwort ■☆
117352 Cynanchum aequilongum Choux = Cynanchum implicatum (Jum. et H. Perrier) Jum. et H. Perrier ■☆
117353 Cynanchum affine Hemsl. = Cynanchum mooreanum Hemsl. ■
117354 Cynanchum africanum (L.) Hoffmanns.;非洲杯冠藤●☆
117355 Cynanchum africanum R. Br. var. crassifolium N. E. Br. = Cynanchum africanum (L.) Hoffmanns. ●☆
117356 Cynanchum alatum Buch.-Ham. ex Wight et Arn.;翅果杯冠藤;Wing Mosquitotrap,Wingedfruit Swallowwort ■
117357 Cynanchum albowianum Kusn.;阿氏杯冠藤■☆
117358 Cynanchum altiscandens K. Schum.;攀缘鹅绒藤■☆
117359 Cynanchum ambositrense Choux = Cynanchum mahafalense Jum. et H. Perrier ■☆
117360 Cynanchum ambovombense (Liede) Liede et Meve;安布文贝鹅绒藤■☆
117361 Cynanchum ampanihense Jum. et H. Perrier;马达加斯加鹅绒藤■☆
117362 Cynanchum amphibolum Schneid. = Cynanchum boudieri H. Lév. et Vaniot ■
117363 Cynanchum ampibolum C. K. Schneid. = Cynanchum boudieri H. Lév. et Vaniot ■
117364 Cynanchum amplexicaule (Siebold et Zucc.) Hemsl.;合掌消(抱茎白前,合掌草,神仙对坐草,土胆草,野豆蕉,野荚豆,硬皮草);Amplexicaul Mosquitotrap,Amplexicaul Swallowwort ■
117365 Cynanchum amplexicaule (Siebold et Zucc.) Hemsl. = Vincetoxicum amplexicaule Siebold et Zucc. ■
117366 Cynanchum amplexicaule (Siebold et Zucc.) Hemsl. f. castaneum (Makino) Ohwi = Vincetoxicum amplexicaule Siebold et Zucc. f. castaneum (Makino) Kitag. ■
117367 Cynanchum amplexicaule (Siebold et Zucc.) Hemsl. f. castaneum (Makino) Ohwi = Cynanchum amplexicaule (Siebold et Zucc.) Hemsl. ■
117368 Cynanchum amplexicaule (Siebold et Zucc.) Hemsl. var. castaneum Makino;紫花合掌消(合掌草,合掌消,甜胆草,土胆草,硬皮草);Purpleflower Amplexicaul Swallowwort,Purpleflower Mosquitotrap ■
117369 Cynanchum amplexicaule (Siebold et Zucc.) Hemsl. var. castaneum Makino = Vincetoxicum amplexicaule Siebold et Zucc. f. castaneum (Makino) Kitag. ■
117370 Cynanchum amplexicaule (Siebold et Zucc.) Hemsl. var. castaneum Makino = Cynanchum amplexicaule (Siebold et Zucc.) Hemsl. ■
117371 Cynanchum analamazaotrense Choux;阿纳拉马鹅绒藤●☆
117372 Cynanchum andringitrense Choux;安德林吉特拉山鹅绒藤●☆
117373 Cynanchum anthonyanum Hand.-Mazz.;小叶鹅绒藤(滇白前);Smallleaf Mosquitotrap,Smallleaf Swallowwort ■
117374 Cynanchum anthopotamicum Hand.-Mazz. = Tylophora anthopotamica (Hand.-Mazz.) Tsiang et H. T. Zhang ●

117375 Cynanchum antsiranense (Meve et Liede) Liede et Meve;安齐朗鹅绒藤●☆

117376 Cynanchum aphyllum (Thunb.) Schltr. = Sarcostemma viminale (L.) R. Br. ■

117377 Cynanchum appendiculatopsis Liede;拟附属物鹅绒藤●☆

117378 Cynanchum appendiculatum Choux;附属物鹅绒藤●☆

117379 Cynanchum arboreum Forssk. = Leptadenia arborea (Forssk.) Schweinf. ●☆

117380 Cynanchum arenarium Jum. et H. Perrier;沙地鹅绒藤●☆

117381 Cynanchum arnottianum Wight = Vincetoxicum arnottianum (Wight) Wight ■☆

117382 Cynanchum ascyrifolium (Franch. et Sav.) Matsum. var. calcareum (H. Ohashi) T. Yamaz. = Vincetoxicum calcareum (H. Ohashi) Akasawa ■☆

117383 Cynanchum ascyrifolium (Franch. ex Sav.) Matsum. = Cynanchum acuminatifolium Hemsl. ■

117384 Cynanchum ascyrifolium (Franch. ex Sav.) Matsum. = Vincetoxicum acuminatum Decne. ■

117385 Cynanchum atratum Bunge;白薇(白马薇,白马尾,白幕,白前,白尾,百荡草,变色白薇,春草,骨美,九牛力,苦胆草,拉瓜瓢,老瓜瓢,老君须,老水牛瓢,龙胆白薇,马尾白薇,芒草,牛角胆草,牛皮消,荞麦细辛,三百根,山白薇,山烟根子,实白薇,薇草,羊奶子,硬白薇,知微老,直立白薇);Blackend Mosquitotrap, Blackend Swallowwort ■

117386 Cynanchum atratum Bunge = Vincetoxicum atratum (Bunge) C. Morren et Decne. ■

117387 Cynanchum atratum Bunge f. multinerve (Franch. et Sav.) T. Yamaz.;多脉白薇■☆

117388 Cynanchum atratum Bunge f. viridescens H. Hara;绿白薇■☆

117389 Cynanchum auriculatum Hemsl. = Cynanchum auriculatum Royle ex Wight ●■

117390 Cynanchum auriculatum Royle ex Wight;牛皮消(白何首乌,白首乌,耳叶牛皮消,飞来鹤,隔山锹,隔山撬,隔山消,何首乌,剪蛇珠,老牛瓢,奶浆藤,牛皮冻,瓢瓢藤,七股莲,山步虎,土白蔹,万世竹,羊角藤,野番薯,野红苕);Auriculate Mosquitotrap, Auriculate Swallowwort ●■

117391 Cynanchum auriculatum Royle ex Wight var. amaninatum (Hatus.) T. Yamaz. = Cynanchum boudieri H. Lév. et Vaniot ■

117392 Cynanchum auriculatum Royle ex Wight var. taiwanianum (T. Yamaz.) F. Y. Lu et C. H. Ou = Cynanchum boudieri H. Lév. et Vaniot ■

117393 Cynanchum austrokiusianum Koidz. = Vincetoxicum austrokiusianum (Koidz.) Kitag. ■☆

117394 Cynanchum balense Liede = Pentarrhinum balense (Liede) Liede ■☆

117395 Cynanchum balfourianum (Schltr.) Tsiang et H. D. Zhang;椭圆叶白前;Balfour Mosquitotrap, Oblongleaf Swallowwort ■

117396 Cynanchum balfourianum (Schltr.) Tsiang et H. D. Zhang = Cynanchum forrestii Schltr. ■

117397 Cynanchum baronii Choux;巴隆白前■☆

117398 Cynanchum batangense P. T. Li;巴塘白前;Batang Mosquitotrap, Swallowwort ■

117399 Cynanchum bicampanulatum M. G. Gilbert et P. T. Li;钟冠白前;Bellcrown Mosquitotrap, Swallowwort ■

117400 Cynanchum biondioides W. T. Wang ex Tsiang et P. T. Li;秦岭藤白前;Biondia Mosquitotrap, Chinling Swallowwort, Qinling Swallowwort ■

117401 Cynanchum bisinuatum Jum. et H. Perrier;双深波杯冠藤●☆

117402 Cynanchum blyttioides Liede;布吕特萝藦杯冠藤●☆

117403 Cynanchum bodinieri Schltr. ex H. Lév. = Cynanchum officinale (Hemsl.) Tsiang et H. D. Zhang ●■

117404 Cynanchum boerhavifolium Hook. et Arn.;智利鹅绒藤●☆

117405 Cynanchum boissieri Kusn.;布瓦西耶杯冠藤■

117406 Cynanchum bojerianum (Decne.) Choux;博耶尔杯冠藤●☆

117407 Cynanchum bosseri Liede;博瑟杯冠藤■☆

117408 Cynanchum boudieri H. Lév. et Vaniot;折冠牛皮消(薄叶牛皮消);Boudier Swallowwort ■

117409 Cynanchum boudieri H. Lév. et Vaniot subsp. caudatum (Miq.) P. T. Li, M. G. Gilbert et W. D. Stevens = Cynanchum caudatum (Miq.) Maxim. ■☆

117410 Cynanchum boveanum Decne. = Glossonema boveanum (Decne.) Decne. ■☆

117411 Cynanchum brevicoronatum M. G. Gilbert et P. T. Li;短冠豹药藤;Shortcrown Mosquitotrap, Swallowwort ■

117412 Cynanchum bungei Decne.;白首乌(柏氏白前,大根牛皮消,地葫芦,和尚乌,戟叶牛皮消,山东何首乌,山葫芦,泰山白首乌,泰山何首乌,野山药);Bunge Mosquitotrap, Bunge Swallowwort ●■

117413 Cynanchum calcareum H. Ohashi = Vincetoxicum calcareum (H. Ohashi) Akasawa ■☆

117414 Cynanchum callialatum Buch. -Ham. ex Wight;美翼杯冠藤(萝藦藤);Beatifulwing Mosquitotrap, Beatifulwing Swallowwort ■

117415 Cynanchum canescens (Willd.) K. Schum.;灰绿白前;Mosquitotrap, Swallowwort ■

117416 Cynanchum canescens (Willd.) K. Schum. = Vincetoxicum canescens (Willd.) Decne. ■

117417 Cynanchum capense L. f. = Pentatropis capensis (L. f.) Bullock ■☆

117418 Cynanchum capense R. Br. = Cynanchum obtusifolium L. f. ●☆

117419 Cynanchum capense Thunb. = Cynanchum ellipticum (Harv.) R. A. Dyer ●☆

117420 Cynanchum cathayense Tsiang et H. D. Zhang;羊角子草;Hastate Mosquitotrap, Hastate Swallowwort ■

117421 Cynanchum cathayense Tsiang et H. D. Zhang = Cynanchum acutum L. subsp. sibiricum (Willd.) Rech. f. ■

117422 Cynanchum caudatum (Miq.) Maxim. = Cynanchum maximowiczii Pobed. ■☆

117423 Cynanchum caudatum (Miq.) Maxim. var. tanzawamontanum Kigawa;丹泽山鹅绒藤●☆

117424 Cynanchum chekiangense M. Cheng ex Tsiang et P. T. Li;蔓剪草(蔓白薇,四叶对剪草,浙江白前);Chekiang Swallowwort, Zhejiang Mosquitotrap, Zhejiang Swallowwort ■

117425 Cynanchum chinense R. Br.;鹅绒藤(牛皮消,羊奶角角,祖子花);China Mosquitotrap, Chinese Swallow Wort, Chinese Swallowwort ■

117426 Cynanchum chirindense S. Moore;奇林达鹅绒藤●☆

117427 Cynanchum clavidens N. E. Br. = Cynanchum hastifolium K. Schum. subsp. clavidens (N. E. Br.) Liede ●☆

117428 Cynanchum clavidens N. E. Br. subsp. hastifolium (N. E. Br.) Liede = Cynanchum hastifolium K. Schum. ●☆

117429 Cynanchum compactum Choux;紧密鹅绒藤■☆

117430 Cynanchum complexum N. E. Br. = Cynanchum mossambicense K. Schum. ■☆

117431 Cynanchum congolense De Wild. = Cynanchum adalinae (K. Schum.) K. Schum. ●☆

117432 Cynanchum cordifolium (E. Mey.) D. Dietr. = Schizoglossum cordifolium E. Mey. ■☆
117433 Cynanchum corymbosum Wight;藤刺瓜(刺瓜,乳蚕,小刺瓜,野苦瓜,叶苦瓜);Corymbose Mosquitotrap,Corymbose Swallowwort ■
117434 Cynanchum crassiantherae Liede;粗药鹅绒藤●☆
117435 Cynanchum crassifolium Hatus. = Cynanchum formosanum (Maxim.) Hemsl. ex Forbes et Hemsl. ●■
117436 Cynanchum crassifolium R. Br. = Cynanchum africanum (L.) Hoffmanns. ●☆
117437 Cynanchum crassipedicellatum Meve et Liede;粗梗鹅绒藤■☆
117438 Cynanchum cucullatum N. E. Br.;僧帽状鹅绒藤■☆
117439 Cynanchum danguyanum Choux;当吉鹅绒藤●☆
117440 Cynanchum decaisnianum Desc.;德凯纳鹅绒藤●☆
117441 Cynanchum decaisnianum Desc. var. longicoronae Liede;长冠德凯纳鹅绒藤●☆
117442 Cynanchum decaryi Choux;德卡里鹅绒藤●☆
117443 Cynanchum decipiens C. K. Schneid.;豹药藤(川白前,西川白前,西川鹅绒藤);Deceiving Mosquitotrap,Deceiving Swallowwort ●■
117444 Cynanchum decorsei (Costantin et Gallaud) Liede et Meve;德科斯鹅绒藤■☆
117445 Cynanchum defoliascens K. Schum. = Blyttia fruticulosa (Decne.) D. V. Field ■☆
117446 Cynanchum dehoideum Hance = Cynanchum chinense R. Br. ■
117447 Cynanchum dehoideum Hook. f. = Cynanchum otophyllum C. K. Schneid. ■
117448 Cynanchum dentatum K. Schum. = Pentarrhinum somaliense (N. E. Br.) Liede ●☆
117449 Cynanchum descoingsii Rauh;德斯鹅绒藤■☆
117450 Cynanchum dinklagei Schltr. ex Mildbr. = Cynanchum adalinae (K. Schum.) K. Schum. subsp. mannii (Scott-Elliot) Bullock ●☆
117451 Cynanchum doianum Koidz. = Vincetoxicum doianum (Koidz.) Kitag. ■☆
117452 Cynanchum dregeanum Decne. = Cynanchum obtusifolium L. f. ●☆
117453 Cynanchum dubium Kitag. = Cynanchum paniculatum (Bunge) Kitag. ex H. Hara ■
117454 Cynanchum duclouxii M. G. Gilbert et P. T. Li;小花杯冠藤;Littleflower Mosquitotrap,Smallflower Swallowwort ■
117455 Cynanchum edule Jum. et H. Perrier = Cynanchum gerrardii (Harv.) Liede ●☆
117456 Cynanchum elachistemmoides (Liede et Meve) Liede et Meve;微冠鹅绒藤■☆
117457 Cynanchum ellipticum (Harv.) R. A. Dyer;椭圆鹅绒藤●☆
117458 Cynanchum erythranthum Jum. et H. Perrier;红花杯冠藤●☆
117459 Cynanchum eurychitoides (K. Schum.) K. Schum.;拟宽鹅绒藤●☆
117460 Cynanchum eurychiton (Decne.) K. Schum.;宽鹅绒藤●☆
117461 Cynanchum excelsum Desf. = Cynanchum acutum L. ■☆
117462 Cynanchum extensum Jacq. = Pergularia daemia (Forssk.) Chiov. ■☆
117463 Cynanchum falcatum Hutch. et E. A. Bruce;镰状杯冠藤●☆
117464 Cynanchum filliforme L. f. = Schizoglossum filiforme (L. f.) Druce ■☆
117465 Cynanchum fimbricoronum P. T. Li;流苏杯冠藤■☆
117466 Cynanchum fissum Pomel = Cynanchum acutum L. ■☆
117467 Cynanchum floribundum R. Br.;繁花杯冠藤(多花牛皮消)■☆
117468 Cynanchum fordii Hemsl.;山白前;Ford Mosquitotrap,Ford Swallowwort ■
117469 Cynanchum formosanum (Maxim.) Hemsl. ex Forbes et Hemsl.;台湾杯冠藤(台湾白薇,台湾牛皮消);Formosan Swallowwort,Taiwan Mosquitotrap,Taiwan Swallowwort ●■
117470 Cynanchum formosanum (Maxim.) Hemsl. ex Forbes et Hemsl. var. ovalifolium Tsiang et P. T. Li;卵叶杯冠藤;Ovateleaf Mosquitotrap,Ovateleaf Swallowwort ●■
117471 Cynanchum formosanum (Maxim.) Hemsl. ex Forbes et Hemsl. var. ovalifolium Tsiang et P. T. Li = Cynanchum formosanum (Maxim.) Hemsl. ex Forbes et Hemsl. ●■
117472 Cynanchum formosanum Maxim. = Cynanchum formosanum (Maxim.) Hemsl. ex Forbes et Hemsl. ●■
117473 Cynanchum forrestii Schltr.;大理白前(白龙须,白薇,狗毒,群虎草,蛇辣子,搜山虎,小白薇);Forrest Mosquitotrap,Forrest Swallowwort ■
117474 Cynanchum forrestii Schltr. var. balfourianum Schltr. = Cynanchum balfourianum (Schltr.) Tsiang et H. D. Zhang ■
117475 Cynanchum forrestii Schltr. var. balfourianum Schltr. = Cynanchum forrestii Schltr. ■
117476 Cynanchum forrestii Schltr. var. stenolobum Tsiang et H. D. Zhang;石棉白前;Mosquitotrap,Swallowwort ■
117477 Cynanchum forrestii Schltr. var. stenolobum Tsiang et H. D. Zhang = Cynanchum forrestii Schltr. ■
117478 Cynanchum franchetii Nakai = Vincetoxicum sublanceolatum (Miq.) Maxim. var. macranthum Maxim. ■☆
117479 Cynanchum fraternum N. E. Br. = Cynanchum heteromorphum Vatke ●☆
117480 Cynanchum fruticulosum Decne. = Blyttia fruticulosa (Decne.) D. V. Field ■☆
117481 Cynanchum gerrardii (Harv.) Liede;杰勒德鹅绒藤●☆
117482 Cynanchum giraldii Schltr.;峨眉牛皮消(峨眉白前);Girald Mosquitotrap,Girald Swallowwort ●■
117483 Cynanchum glabrum Nakai = Vincetoxicum glabrum (Nakai) Kitag. ■☆
117484 Cynanchum glabrum Nakai f. viridescens Murata = Vincetoxicum glabrum (Nakai) Kitag. f. viridescens (Murata) Sugim. ■☆
117485 Cynanchum glabrum Nakai var. rotundifolium Honda = Vincetoxicum glabrum (Nakai) Kitag. var. rotundifolium (Honda) Sugim. ■☆
117486 Cynanchum glaucescens (Decne.) Hand. -Mazz.;白前(打狗耳,鹅白前,鹅管白前,咳药,沙消,石蓝,水竹消,溪瓢羹,狭叶牛皮消,消结草,芫花叶白前,羊奶子,竹叶白前);Glaucescent Mosquitotrap,Glaucescent Swallowwort ●■
117487 Cynanchum glaucum Wall. = Vincetoxicum canescens (Willd.) Decne. ■
117488 Cynanchum glaucum Wall. ex Wight;蓝绿白前;Mosquitotrap,Swallowwort ■
117489 Cynanchum glaucum Wall. ex Wight = Cynanchum canescens (Willd.) K. Schum. ■
117490 Cynanchum gonoloboides Schltr. = Pentarrhinum gonoloboides (Schltr.) Liede ●☆
117491 Cynanchum gossweileri S. Moore = Schizostephanus gossweileri (S. Moore) Liede ■☆
117492 Cynanchum gracilipes Tsiang et H. D. Zhang;细梗白前;Slenderstalk Swallowwort,Thinstalk Mosquitotrap ■
117493 Cynanchum gracilipes Tsiang et H. D. Zhang = Cynanchum

taihangense Tsiang et H. D. Zhang ■

117494 Cynanchum gracillimum Wall. ex Wight = Adelostemma gracillimum (Wall. ex Wight) Hook. f. ■

117495 Cynanchum graminiforme Liede;禾状鹅绒藤■☆

117496 Cynanchum grandidieri Liede et Meve;格兰鹅绒藤●☆

117497 Cynanchum grandifolium Hemsl.;大叶鹅绒藤■☆

117498 Cynanchum grandifolium Hemsl. = Vincetoxicum macrophyllum Siebold et Zucc. ■☆

117499 Cynanchum grandifolium Hemsl. var. nikoense (Maxim.) Ohwi;日光鹅绒藤(日光大叶白前)■☆

117500 Cynanchum grandifolium Hemsl. var. nikoense (Maxim.) Ohwi = Vincetoxicum macrophyllum Siebold et Zucc. var. nikoense Maxim. ■☆

117501 Cynanchum hamsimai P. T. Li = Cynanchum formosanum (Maxim.) Hemsl. ex Forbes et Hemsl. ●■

117502 Cynanchum hancockianum (Maxim.) Iljinski = Cynanchum mongolicum (Maxim.) Hemsl. ■

117503 Cynanchum hancockianum (Maxim.) Iljinski f. angustifolium K. T. Chow;狭叶华北白前(刺瓜)■

117504 Cynanchum hardyi Liede et Meve;哈迪鹅绒藤■☆

117505 Cynanchum hastatum Lam. = Cynanchum bungei Decne. ●■

117506 Cynanchum hastatum Pers. = Leptadenia hastata (Pers.) Decne. ●☆

117507 Cynanchum hastifolium K. Schum.;戟叶鹅绒藤●☆

117508 Cynanchum hastifolium K. Schum. subsp. clavidens (N. E. Br.) Liede;棒齿鹅绒藤●☆

117509 Cynanchum hastifolium N. E. Br. = Cynanchum hastifolium K. Schum. ●☆

117510 Cynanchum hatusimai P. T. Li = Cynanchum formosanum (Maxim.) Hemsl. ex Forbes et Hemsl. ●■

117511 Cynanchum helicoideum Choux = Cynanchum madagascariense K. Schum. ■☆

117512 Cynanchum henryi Warb. ex Schltr. et Diels = Biondia henryi (Warb. ex Schltr. et Diels) Tsiang et P. T. Li ●

117513 Cynanchum heteromorphum Vatke;异形鹅绒藤●☆

117514 Cynanchum heterophyllum Delile = Leptadenia arborea (Forssk.) Schweinf. ●☆

117515 Cynanchum heydei Hook. f.;西藏鹅绒藤;Xizang Mosquitotrap,Xizang Swallowwort ■

117516 Cynanchum holstii (K. Schum.) K. Schum. = Cynanchum abyssinicum Decne. ●☆

117517 Cynanchum humbert-capuronii Liede et Meve;亨凯鹅绒藤■☆

117518 Cynanchum humbertii Choux = Cynanchum ampanihense Jum. et H. Perrier ■☆

117519 Cynanchum hydrophilum Tsiang et H. D. Zhang;水白前;Water Mosquitotrap,Waterloving Swallowwort ■

117520 Cynanchum ikema Ohwi = Cynanchum caudatum (Miq.) Maxim. ■☆

117521 Cynanchum implicatum (Jum. et H. Perrier) Jum. et H. Perrier;纠缠杯冠藤■☆

117522 Cynanchum inamoenum (Maxim.) Loes. = Vincetoxicum inamoenum Maxim. ■

117523 Cynanchum inamoenum (Maxim.) Loes. ex Gilg et Loes.;竹灵消(白龙须,白薇,川白薇,大羊角瓢,恶牛皮消,九连台,老君须,牛角风,瓢儿瓜,婆婆衣,婆婆针线包,绒针,犀角细辛,雪里蟠桃,正骨草,直立白前);Unpleasant Mosquitotrap,Unpleasant Swallowwort ■

117524 Cynanchum inodorum Lour. = Gymnema inodorum (Lour.) Decne. ●

117525 Cynanchum insigne (N. E. Br.) Liede et Meve;显著杯冠藤■☆

117526 Cynanchum insulanum (Hance) Hemsl.;海南杯冠藤;Hainan Mosquitotrap,Hainan Swallowwort ■

117527 Cynanchum insulanum (Hance) Hemsl. var. lineare (Tsiang et H. D. Zhang) Tsiang et H. D. Zhang;线叶杯冠藤;Linearleaf Mosquitotrap,Linearleaf Swallowwort ■

117528 Cynanchum intermedium N. E. Br. = Cynanchum africanum (L.) Hoffmanns. ●☆

117529 Cynanchum itremense Liede;伊特雷穆杯冠藤■☆

117530 Cynanchum jacquemontianum Decne.;雅克蒙杯冠藤■☆

117531 Cynanchum japonicum C. Morren et Decne. = Vincetoxicum japonicum C. Morren et Decne. ■☆

117532 Cynanchum japonicum C. Morren et Decne. var. albiflorum (Franch. et Sav.) H. Hara;白花日本白前■☆

117533 Cynanchum japonicum C. Morren et Decne. var. albiflorum (Franch. et Sav.) H. Hara f. puncticulatum (Koidz.) Ohwi = Vincetoxicum japonicum C. Morren et Decne. f. puncticlatum (Koidz.) Kitag. ■☆

117534 Cynanchum japonicum C. Morren et Decne. var. albiflorum (Franch. et Sav.) H. Hara = Vincetoxicum japonicum C. Morren et Decne. var. albiflorum (Franch. et Sav.) Kitag. ■☆

117535 Cynanchum japonicum C. Morren et Decne. var. puncticulatum (Koidz.) H. Hara;斑点日本白前■☆

117536 Cynanchum japonicum C. Morren et Decne. var. puncticulatum (Koidz.) H. Hara = Vincetoxicum japonicum C. Morren et Decne. f. puncticlatum (Koidz.) Kitag. ■☆

117537 Cynanchum japonicum E. Morren et Decne.;日本白前(白薇);Japan Mosquitotrap ■☆

117538 Cynanchum japonicum E. Morren et Decne. var. purpurascens Maxim.;变紫日本白前(紫色白前)■

117539 Cynanchum juliani-marnieri Desc.;朱利安鹅绒藤■☆

117540 Cynanchum jumellei Choux;朱迈尔鹅绒藤■☆

117541 Cynanchum junciforme (Decne.) Liede;灯芯草鹅绒藤■☆

117542 Cynanchum kaschgaricum Y. X. Liou;阿克苏牛皮消;Akesu Swallowwort ■

117543 Cynanchum katoi Ohwi f. albescens (H. Hara) Ohwi = Vincetoxicum katoi (Ohwi) Kitag. f. albescens (H. Hara) Kitag. ■☆

117544 Cynanchum katoi Ohwi var. albescens H. Hara = Vincetoxicum katoi (Ohwi) Kitag. f. albescens (H. Hara) Kitag. ■☆

117545 Cynanchum kindonwardii Tsiang;宁蒗杯冠藤;Ninglang Mosquitotrap,Ninglang Swallowwort ■

117546 Cynanchum kintungense Tsiang;景东杯冠藤;Jingdong Mosquitotrap,Jingdong Swallowwort ■

117547 Cynanchum kiusianum Nakai = Vincetoxicum macrophyllum Siebold et Zucc. ■☆

117548 Cynanchum komarovii Iljinski;老瓜头(黑老鸦脖子,黑牛心朴子,黑心朴子,芦蕊草,牛心朴子);Komalov Mosquitotrap,Komalov Swallowwort ■

117549 Cynanchum komarovii Iljinski = Cynanchum mongolicum (Maxim.) Hemsl. ■

117550 Cynanchum krameri (Franch. et Sav.) Matsum. = Vincetoxicum krameri Franch. et Sav. ■☆

117551 Cynanchum kwangsiense Tsiang et H. D. Zhang;广西杯冠藤;Guangxi Mosquitotrap,Guangxi Swallowwort,Kwangsi Swallowwort ■

117552 Cynanchum laeve (Michx.) Pers.;平滑杯冠藤;Angle-pod,

Blue Vine, Climbing Milkweed, Sand Vine ■☆

117553 Cynanchum lanceolatum Poir. = Leptadenia hastata (Pers.) Decne. ■☆

117554 Cynanchum lancifolium Schumach. = Leptadenia hastata (Pers.) Decne. ■☆

117555 Cynanchum lanhsuense T. Yamaz.;兰屿牛皮消(兰屿白薇);Lanyu Swallowwort ■

117556 Cynanchum lateriflorum (Hemsl.) Kitag. = Cynanchum mongolicum (Maxim.) Hemsl. ■

117557 Cynanchum laxum Bartl.;松散杯冠藤●☆

117558 Cynanchum lecomtei Choux;勒孔特杯冠藤●☆

117559 Cynanchum ledermannii Schltr.;莱德杯冠藤●☆

117560 Cynanchum lenewtonii Liede;莱牛顿杯冠藤■☆

117561 Cynanchum leucanthum (K. Schum.) K. Schum.;白花杯冠藤■☆

117562 Cynanchum leveilleanum Schltr. ex H. Lév. = Cynanchum verticillatum Hemsl. ●■

117563 Cynanchum lightii Dunn = Cynanchum glaucescens (Decne.) Hand. -Mazz. ●■

117564 Cynanchum likiangense W. T. Wang ex Tsiang et P. T. Li;丽江牛皮消■

117565 Cynanchum likiangense W. T. Wang ex Tsiang et P. T. Li = Cynanchum lysimachioides Tsiang et P. T. Li ■

117566 Cynanchum limprichtii Schltr.;康定白前;Kangding Mosquitotrap, Limprichet Swallowwort ■

117567 Cynanchum limprichtii Schltr. = Cynanchum forrestii Schltr. ■

117568 Cynanchum lineare N. E. Br.;线形杯冠藤■☆

117569 Cynanchum linearifolium Hemsl. = Cynanchum stauntonii (Decne.) Schltr. ex H. Lév. ●■

117570 Cynanchum linearisepalum P. T. Li;线萼白前;Linearsepal Mosquitotrap, Linearsepal Swallowwort ■

117571 Cynanchum liukiuense Warb.;琉球牛皮消■☆

117572 Cynanchum longifolium Mart. = Cynanchum acutum L. subsp. sibiricum (Willd.) Rech. f. ■

117573 Cynanchum longipedunculatum M. G. Gilbert et P. T. Li;短柱豹药藤(长柱豹药藤);Longstyle Mosquitotrap, Longstyle Swallowwort ■

117574 Cynanchum longipes N. E. Br.;长梗杯冠藤■☆

117575 Cynanchum louiseae Kartesz et Gandhi;路易斯白前;Black Swallowwort, Black Swallow-wort, Louise's Swallow-wort ■☆

117576 Cynanchum lysimachioides Tsiang et P. T. Li;白牛皮消(丽江牛皮消);Lijiang Mosquitotrap, Lijiang Swallowwort, Likiang Swallowwort, Loosestrifeleaf Mosquitotrap, Loosestrifeleaf Swallowwort ■

117577 Cynanchum macinense A. Chev. = Cynanchum hastifolium K. Schum. ●☆

117578 Cynanchum macranthum (Maxim.) Nakai = Vincetoxicum sublanceolatum (Miq.) Maxim. var. macranthum Maxim. ■☆

117579 Cynanchum macranthum (Maxim.) Nakai var. dickinsii (Franch. et Sav.) Ohwi = Vincetoxicum sublanceolatum (Miq.) Maxim. var. auriculatum Franch. et Sav. ■☆

117580 Cynanchum macranthum Jum. et H. Perrier;大花白前■☆

117581 Cynanchum macrolobum Jum. et H. Perrier;大裂白前■☆

117582 Cynanchum macrophyllum Thunb. = Cynanchum grandifolium Hemsl. ■☆

117583 Cynanchum madagascariense K. Schum.;马岛白前■☆

117584 Cynanchum madecassum Desc. = Cynanchum arenarium Jum. et H. Perrier ●☆

117585 Cynanchum magnificum Nakai = Vincetoxicum magnificum (Nakai) Kitag. ■☆

117586 Cynanchum mahafalense Jum. et H. Perrier;马哈法尔鹅绒藤■☆

117587 Cynanchum mairei Schltr. ex H. Lév. = Cynanchum alatum Buch. -Ham. ex Wight et Arn. ■

117588 Cynanchum mandshuricum (Hance) Hemsl. = Cynanchum versicolor Bunge ●■

117589 Cynanchum mandshuricum Hemsl. = Cynanchum versicolor Bunge ●■

117590 Cynanchum mannii (Scott-Elliot) N. E. Br. = Cynanchum adalinae (K. Schum.) K. Schum. subsp. mannii (Scott-Elliot) Bullock ●☆

117591 Cynanchum mariense (Meve et Liede) Liede et Meve;马里安鹅绒藤■☆

117592 Cynanchum masoalense Choux;马苏阿拉鹅绒藤■☆

117593 Cynanchum matsumurae T. Yamaz. = Tylophora matsumurae (T. Yamaz.) T. Yamash. et Tateishi ■☆

117594 Cynanchum mauritianum Lam. = Camptocarpus mauritianus (Lam.) Decne. ●☆

117595 Cynanchum maximoviczii Pobed. = Cynanchum caudatum (Miq.) Maxim. ■☆

117596 Cynanchum maximowiczianum (Warb.) Nakai = Cynanchum sublanceolatum (Miq.) Matsum. ■

117597 Cynanchum maximowiczii Pobed.;生马牛皮消(白兔藿,牛皮消,尾状牛皮消);Mosquitotrap, Swallowwort, Swallow-wort ■☆

117598 Cynanchum megalanthum M. G. Gilbert et P. T. Li;大花刺瓜;Bigflower Swallowwort, Mosquitotrap ■

117599 Cynanchum membranaceum (Liede et Meve) Liede et Meve;膜质鹅绒藤●☆

117600 Cynanchum mensense Schweinf. ex K. Schum. = Cynanchum altiscandens K. Schum. ■☆

117601 Cynanchum messeri (Buchenau) Jum. et H. Perrier;梅瑟鹅绒藤●☆

117602 Cynanchum meyeri (Decne.) Schltr.;迈尔鹅绒藤●☆

117603 Cynanchum microstegium K. Schum. = Blyttia fruticulosa (Decne.) D. V. Field ■☆

117604 Cynanchum minutiflorum K. Schum. = Cynanchum schistoglossum Schltr. ■☆

117605 Cynanchum molle (E. Mey.) D. Dietr. = Anisotoma cordifolia Fenzl ■☆

117606 Cynanchum mongolicum (Maxim.) Hemsl.;华北白前(侧花徐长卿,对叶草,阔叶徐长卿,老瓜头,牛心朴,牛心朴子,牛心秧,瓢菜);Hancock Mosquitotrap, Hancock Swallowwort, Mongolian Mosquitotrap, Mongolian Swallowwort ■

117607 Cynanchum mongolicum (Maxim.) Hemsl. = Cynanchum hancockianum (Maxim.) Iljinski ■

117608 Cynanchum mongolicum (Maxim.) K. Schum. = Cynanchum mongolicum (Maxim.) Hemsl. ■

117609 Cynanchum mongolicum (Maxim.) Kom. = Cynanchum mongolicum (Maxim.) Hemsl. ■

117610 Cynanchum mooreanum Hemsl.;毛白前(白地牛,白毛藤,老君须,龙胆白前);Moore Mosquitotrap, Moore Swallowwort ■

117611 Cynanchum moramangense Choux;莫拉芒鹅绒藤■☆

117612 Cynanchum moratii Liede;莫拉特鹅绒藤■☆

117613 Cynanchum mossambicense K. Schum.;莫桑比克白前■☆

117614 Cynanchum muliense Tsiang;木里白前;Muli Mosquitotrap, Muli Swallowwort ■

117615 Cynanchum muliense Tsiang = Cynanchum forrestii Schltr. ■

117616 Cynanchum multinerve (Franch. et Sav. r) Matsum. = Cynanchum atratum Bunge ■

117617 Cynanchum multinerve (Franch. et Sav.) Matsum. = Cynanchum atratum Bunge f. multinerve (Franch. et Sav.) T. Yamaz. ■☆

117618 Cynanchum multinerve (Franch. et Sav.) Matsum. var. kiyohikoanum (Honda) Ohwi = Vincetoxicum multinerve Franch. et Sav. var. kiyohikoanum (Honda) Kitag. ■☆

117619 Cynanchum napiferum Choux;芜菁鹅绒藤■☆

117620 Cynanchum natalitium Schltr.;纳塔利特鹅绒藤■☆

117621 Cynanchum nematostemma Liede;虫冠鹅绒藤■☆

117622 Cynanchum nigrum (L.) Pers. = Cynanchum louiseae Kartesz et Gandhi ■☆

117623 Cynanchum nikoense (Maxim.) Makino = Cynanchum grandifolium Hemsl. ■☆

117624 Cynanchum nikoense (Maxim.) Makino = Vincetoxicum macrophyllum Siebold et Zucc. var. nikoense Maxim. ■☆

117625 Cynanchum nipponicum Matsum. = Vincetoxicum nipponicum (Matsum.) Kitag. ■☆

117626 Cynanchum nipponicum Matsum. f. abukumense (Koidz.) H. Hara = Vincetoxicum glabrum (Nakai) Kitag. ■☆

117627 Cynanchum nipponicum Matsum. f. abukumense (Koidz.) H. Hara = Vincetoxicum nipponicum (Matsum.) Kitag. f. abukumense (Koidz.) Kitag. ■☆

117628 Cynanchum nipponicum Matsum. f. rotundifolium (Honda) T. Yamaz. = Vincetoxicum glabrum (Nakai) Kitag. var. rotundifolium (Honda) Sugim. ■☆

117629 Cynanchum nipponicum Matsum. var. glabrum (Nakai) H. Hara f. viridescens (Murata) Murata = Vincetoxicum glabrum (Nakai) Kitag. f. viridescens (Murata) Sugim. ■☆

117630 Cynanchum nipponicum Matsum. var. rotundifolium (Honda) Murata = Vincetoxicum glabrum (Nakai) Kitag. var. rotundifolium (Honda) Sugim. ■☆

117631 Cynanchum nodosum (Jum. et H. Perrier) Desc. = Cynanchum arenarium Jum. et H. Perrier ●☆

117632 Cynanchum obovatum (Decne.) Choux;倒卵叶鹅绒藤●☆

117633 Cynanchum obscurum K. Schum. = Cynanchum polyanthum K. Schum. ■☆

117634 Cynanchum obtusifolium L. f.;钝叶鹅绒藤●☆

117635 Cynanchum obtusifolium L. f. var. pilosum Schltr. = Cynanchum obtusifolium L. f. ●☆

117636 Cynanchum odoratissimum Lour. = Telosma cordata (Burm. f.) Merr. ●

117637 Cynanchum officinale (Hemsl.) Tsiang et H. D. Zhang;朱砂藤(白薮,赤芍,湖北白前,桔梗,青阳参,藤白芍,托腰散,野红薯藤,朱砂莲);Medicinal Swallowwort, Sinnabar Mosquitotrap ●■

117638 Cynanchum omissum Bullock = Fockea angustifolia K. Schum. ●☆

117639 Cynanchum orangeanum (Schltr.) N. E. Br.;奥兰鹅绒藤■☆

117640 Cynanchum otophyllum C. K. Schneid.;青羊参(白岑,白薮,白芪,白芍,白石参,白首乌,白药,地藕,毒狗药,对节参,鹅绒藤,奶参,奶浆草,奶浆藤,闹狗药,牛尾参,千年生,青阳参,青洋参,小白薮,小绿牛角藤);Auricledleaf Mosquitotrap, Auricledleaf Swallowwort ■

117641 Cynanchum pachycladon Choux;粗枝鹅绒藤●☆

117642 Cynanchum paniculatum (Bunge) Kitag. = Cynanchum paniculatum (Bunge) Kitag. ex H. Hara ■

117643 Cynanchum paniculatum (Bunge) Kitag. ex H. Hara;徐长卿(白细辛,别仙踪,察察竹,刁竹,钓鱼竿,对节莲,对叶莲,对月草,对月莲,谷茬细辛,鬼督邮,黑薇,尖刀儿苗,九头狮,九头狮子草,老君须,痢止草,寥刁竹,寮刁竹,了刁竹,料刁竹,料吊,铃柴胡,柳叶细辛,柳枝癀,千云竹,三百根,山刁竹,上天梯,蛇草,蛇利草,蛇山草,生竹,石下长卿,天竹,铜锣草,土细辛,蜈蚣草,溪柳,线香草,香摇边,逍遥竹,小对叶草,牙蛀消,摇边竹,摇竹消,遥竹逍,瑶山竹,药王,一枝箭,一枝香,英雄草,獐耳草,中心草,竹叶细辛);Paniculate Mosquitotrap, Paniculate Swallowwort, Xu Changqing ■

117644 Cynanchum paniculatum (Bunge) Kitag. ex H. Hara = Vincetoxicum pycnostelma Kitag. ■

117645 Cynanchum paniculatum (Bunge) Kitag. ex H. Hara var. latifolium (Makino) H. Hara;宽叶徐长卿■☆

117646 Cynanchum paniculatum (Bunge) Kitag. ex H. Hara var. latifolium (Makino) H. Hara = Vincetoxicum pycnostelma Kitag. f. latifolium (Makino) Kitag. ■☆

117647 Cynanchum papillatum Choux;乳突鹅绒藤■☆

117648 Cynanchum papillosum Weim. = Cynanchum chirindense S. Moore ●☆

117649 Cynanchum pendulum Poir.;下垂鹅绒藤●☆

117650 Cynanchum perrieri Choux;佩里耶杯冠藤■☆

117651 Cynanchum petrense Hemsl. et Lace;岩生鹅绒藤●☆

117652 Cynanchum phillipsonianum Liede et Meve;菲利鹅绒藤■☆

117653 Cynanchum pilosum R. Br. = Cynanchum africanum (L.) Hoffmanns. ●☆

117654 Cynanchum pingshanicum M. G. Gilbert et P. T. Li;平山白前;Pingshan Mosquitotrap, Pingshan Swallowwort ■

117655 Cynanchum pleianthum K. Schum. = Cynanchum heteromorphum Vatke ●☆

117656 Cynanchum polyanthum K. Schum.;多花杯冠藤■☆

117657 Cynanchum praecox Schltr. ex S. Moore;早花杯冠藤■☆

117658 Cynanchum pubescens Bunge = Cynanchum chinense R. Br. ■

117659 Cynanchum purpurascens (C. Morren et Decne.) Matsum. = Vincetoxicum × purpurascens C. Morren et Decne. ■☆

117660 Cynanchum purpureum (Pall.) K. Schum.;紫花杯冠藤(紫花白前,紫花牛皮消);Purple Mosquitotrap, Purpleflower Swallowwort ■

117661 Cynanchum pycnoneuroides Choux;密脉杯冠藤■☆

117662 Cynanchum pygmaeum Schltr. = Cynanchum praecox Schltr. ex S. Moore ■☆

117663 Cynanchum radiatum Jum. et H. Perrier;辐射鹅绒藤●☆

117664 Cynanchum rauhianum Desc.;劳氏鹅绒藤■☆

117665 Cynanchum repandum (Decne.) K. Schum.;浅波状杯冠藤●☆

117666 Cynanchum riparium Tsiang et H. D. Zhang;荷花柳;Lotus Mosquitotrap, Riverbank Swallowwort ■

117667 Cynanchum rockii M. G. Gilbert et P. T. Li;高冠白前;Rock Mosquitotrap ■

117668 Cynanchum roseum Chiov. = Tylophora heterophylla A. Rich. ●☆

117669 Cynanchum roseum R. Br. = Cynanchum purpureum (Pall.) K. Schum. ■

117670 Cynanchum rossicum (Kleopow) Borhidi;欧洲白前;European Swallow-wort ■☆

117671 Cynanchum rossii Rauh;罗斯鹅绒藤■☆

117672 Cynanchum rotundifolium Thunb. ex Decne. = Cynanchum africanum (L.) Hoffmanns. ●☆

117673 Cynanchum rungweense Bullock;伦圭鹅绒藤■☆

117674 Cynanchum rusillonii Hochr. = Cynanchum junciforme (Decne.) Liede ■☆

117675 Cynanchum saccatum W. T. Wang ex Tsiang et P. T. Li;西藏牛皮消;Cystoid Mosquitotrap,Tibet Swallowwort ■

117676 Cynanchum saccatum W. T. Wang ex Tsiang et P. T. Li = Cynanchum auriculatum Royle ex Wight ●■

117677 Cynanchum sarcostemmatoides K. Schum. = Cynanchum gerrardii (Harv.) Liede ●☆

117678 Cynanchum schistoglossum Schltr.;微花白前■☆

117679 Cynanchum schmalhausenii Kusn.;施马白前■☆

117680 Cynanchum sessiliflorum (Decne.) Liede;无花梗白前■☆

117681 Cynanchum sibiricum (L.) R. Br. = Cynanchum thesioides (Freyn) K. Schum. ■

117682 Cynanchum sibiricum (L.) R. Br. var. australe (Maxim.) Maxim. ex Kom. = Cynanchum thesioides (Freyn) K. Schum. ■

117683 Cynanchum sibiricum (L.) R. Br. var. gracilentum Nakai et Kitag. = Cynanchum thesioides (Freyn) K. Schum. ■

117684 Cynanchum sibiricum (L.) R. Br. var. latifolium Kitag. = Cynanchum thesioides (Freyn) K. Schum. ■

117685 Cynanchum sibiricum R. Br. = Cynanchum thesioides (Freyn) K. Schum. ■

117686 Cynanchum sibiricum R. Br. var. austale Maxim. ex Kom. = Cynanchum thesioides (Freyn) K. Schum. var. australe (Maxim.) Tsiang et P. T. Li ■

117687 Cynanchum sibiricum R. Br. var. gracilentum Nakai et Kitag. = Cynanchum thesioides (Freyn) K. Schum. ■

117688 Cynanchum sibiricum R. Br. var. latifolium Katag. = Cynanchum thesioides (Freyn) K. Schum. var. australe (Maxim.) Tsiang et P. T. Li ■

117689 Cynanchum sibiricum Willd. = Cynanchum acutum L. subsp. sibiricum (Willd.) Rech. f. ■

117690 Cynanchum sinoracemosum M. G. Gilbert et P. T. Li;尖叶杯冠藤;Mosquitotrap,Sharpleaf Swallowwort,Swallowwort ■

117691 Cynanchum somaliense (N. E. Br.) N. E. Br. = Pentarrhinum somaliense (N. E. Br.) Liede ●☆

117692 Cynanchum stauntonii (Decne.) Schltr. ex H. Lév.;柳叶白前(白前,草白前,打狗耳,鹅白前,鹅管白前,江杨柳,酒叶草,咳药,石蓝,水豆粘,水杨柳,斯氏牛皮消,西河柳,狭叶牛皮消,竹叶白前);Willowleaf Mosquitotrap,Willowleaf Swallowwort ●■

117693 Cynanchum stauntonii Hand. -Mazz. = Cynanchum stauntonii (Decne.) Schltr. ex H. Lév. ●■

117694 Cynanchum stenophyllum Hemsl.;狭叶白前;Narrowleaf Mosquitotrap,Narrowleaf Swallowwort ■

117695 Cynanchum steppicola Hand. -Mazz.;卵叶白前;Ovateleaf Mosquitotrap,Ovateleaf Swallowwort ■

117696 Cynanchum steppicola Hand. -Mazz. = Cynanchum forrestii Schltr. ■

117697 Cynanchum subcoriaceum Schltr. = Cynanchum repandum (Decne.) K. Schum. ●☆

117698 Cynanchum sublanceolatum (Miq.) Matsum.;镇江白前;Lanceolate Swallowwort,Zhenjiang Mosquitotrap ■

117699 Cynanchum sublanceolatum (Miq.) Matsum. = Vincetoxicum sublanceolatum (Miq.) Maxim. ■

117700 Cynanchum sublanceolatum (Miq.) Matsum. f. albiflorum (Franch. et Sav.) T. Yamaz.;日本娃儿藤●☆

117701 Cynanchum sublanceolatum (Miq.) Matsum. f. albiflorum (Franch. et Sav.) T. Yamaz. = Vincetoxicum sublanceolatum (Miq.) Maxim. var. albiflorum (Franch. et Sav.) Kitag. ●☆

117702 Cynanchum sublanceolatum (Miq.) Matsum. var. albiflorum (Franch. et Sav.) H. Hara = Vincetoxicum sublanceolatum (Miq.) Maxim. var. albiflorum (Franch. et Sav.) Kitag. ●☆

117703 Cynanchum sublanceolatum (Miq.) Matsum. var. auriculatum (Franch. et Sav.) Matsum.;耳状日本娃儿藤●☆

117704 Cynanchum sublanceolatum (Miq.) Matsum. var. auriculatum (Franch. et Sav.) Matsum. = Vincetoxicum sublanceolatum (Miq.) Maxim. var. auriculatum Franch. et Sav. ■☆

117705 Cynanchum sublanceolatum (Miq.) Matsum. var. kinokuniense T. Yamaz.;木国鹅绒藤●☆

117706 Cynanchum sublanceolatum (Miq.) Matsum. var. macranthum (Maxim.) Matsum.;大花日本娃儿藤■☆

117707 Cynanchum sublanceolatum (Miq.) Matsum. var. macranthum (Maxim.) Matsum. f. yesoense (Nakai) H. Hara = Vincetoxicum sublanceolatum (Miq.) Maxim. var. macranthum Maxim. f. yesoense (Nakai) Kitag. ■☆

117708 Cynanchum sublanceolatum (Miq.) Matsum. var. macranthum (Maxim.) Matsum. f. auriculatum (Franch. et Sav.) T. Yamaz. = Vincetoxicum sublanceolatum (Miq.) Maxim. var. auriculatum Franch. et Sav. ■☆

117709 Cynanchum sublanceolatum (Miq.) Matsum. var. macranthum (Maxim.) Matsum. = Vincetoxicum sublanceolatum (Miq.) Maxim. var. macranthum Maxim. ■☆

117710 Cynanchum sublanceolatum (Miq.) Matsum. var. obtusulum (Franch. et Sav.;钝叶镇江白前■

117711 Cynanchum sublanceolatum (Miq.) Matsum. var. obtusulum (Franch. et Sav.) Matsum. = Vincetoxicum sublanceolatum (Miq.) Maxim. ■

117712 Cynanchum subtilis Liede;纤细鹅绒藤■☆

117713 Cynanchum szechuanense Tsiang et H. D. Zhang;四川鹅绒藤(白花四川鹅绒藤);Sichuan Mosquitotrap,Sichuan Swallowwort,Szechuan Swallowwort ■

117714 Cynanchum szechuanense Tsiang et H. D. Zhang var. albescens Tsiang et H. D. Zhang = Cynanchum szechuanense Tsiang et H. D. Zhang ■

117715 Cynanchum taihangense Tsiang et H. D. Zhang;太行白前;Taihang Mosquitotrap,Taihang Swallowwort ■

117716 Cynanchum tailandense R. T. Li;泰国白前;Tailand Swallowwort ■☆

117717 Cynanchum taiwanianum T. Yamaz.;薄叶牛皮消(薄叶白薇);Taqiwan Swallowwort ■

117718 Cynanchum taiwanianum T. Yamaz. = Cynanchum auriculatum Royle ex Wight var. amaninatum (Hatus.) T. Yamaz. ■

117719 Cynanchum taiwanianum T. Yamaz. = Cynanchum boudieri H. Lév. et Vaniot ■

117720 Cynanchum tetrapterum (Turcz.) R. A. Dyer = Sarcostemma viminale (L.) R. Br. ■

117721 Cynanchum thesioides (Freyn) K. Schum.;地梢瓜(地瓜瓢,地梢花,浮瓢棵,老瓜瓢,女青,沙奶草,沙奶奶,西伯利亚白前,细叶白前,小丝瓜,羊不奶棵,羊奶草);Bastardtoadflax-like Swallowwort,Thesionlike Mosquitotrap ●■

117722 Cynanchum thesioides (Freyn) K. Schum. var. australe (Maxim.) Tsiang et P. T. Li;雀瓢(地瓜瓜,马奶草,省瓜);Southern Mosquitotrap,Southern Swallowwort ■

117723 Cynanchum thesioides (Freyn) K. Schum. var. australe (Maxim.) Tsiang et P. T. Li = Cynanchum thesioides (Freyn) K. Schum. ■

117724 Cynanchum toliari Liede et Meve;托里鹅绒藤●☆

117725 Cynanchum trifurcatum Schltr. = Pentarrhinum somaliense (N. E. Br.) Liede ●☆
117726 Cynanchum tsaratananense Choux;察拉塔纳纳鹅绒藤●☆
117727 Cynanchum tsiangii P. T. Li;贵州白前;Guizhou Mosquitotrap, Guizhou Swallowwort ■
117728 Cynanchum tsiangii P. T. Li = Tylophora tsiangii (P. T. Li) M. G. Gilbert, W. D. Stevens et P. T. Li ●
117729 Cynanchum tylophoroideum Schltr. ex H. Lév. = Cynanchum mooreanum Hemsl. ■
117730 Cynanchum utriculosum Costantin;越南白前;Vietnam Mosquitotrap, Vietnam Swallowwort ■
117731 Cynanchum validum N. E. Br.;刚直鹅绒藤●☆
117732 Cynanchum vernyi (Franch. et Sav.) Matsum. = Vincetoxicum vernyi Franch. et Sav. ■☆
117733 Cynanchum verrucosum (Desc.) Liede et Meve;多疣鹅绒藤■☆
117734 Cynanchum versicolor Bunge;变色白前(白花牛皮消,白龙须,白马尾,白微,白尾,半蔓白薇,春草,骨美,龙胆白薇,蔓生白前,蔓生白薇,芒草,薇草);Vorsicolorous Mosquitotrap, Vorsicolorous Swallowwort ●■
117735 Cynanchum verticillatum Hemsl.;轮叶白前(细蓼仔,细叶蓼);Verticillate Mosquitotrap, Verticillate Swallowwort ●■
117736 Cynanchum verticillatum Hemsl. var. arenicola Tsiang et H. D. Zhang ex Tsiang et P. T. Li = Cynanchum verticillatum Hemsl. ●■
117737 Cynanchum verticillatum Hemsl. var. arenicola Tsiang et H. D. Zhang = Cynanchum verticillatum Hemsl. ●■
117738 Cynanchum verticillatum Hemsl. var. arenicola Tsiang et H. T. Zhang ex Tsiang et P. T. Li;富宁白前;Funing Mosquitotrap, Funing Swallowwort ■
117739 Cynanchum villosum T. Yamaz. = Tylophora matsumurae (T. Yamaz.) T. Yamash. et Tateishi ■☆
117740 Cynanchum viminale (L.) L. = Sarcostemma viminale (L.) R. Br. ■
117741 Cynanchum viminale (L.) L. subsp. crassicaule Liede et Meve;粗茎鹅绒藤■☆
117742 Cynanchum viminale (L.) L. subsp. stippitaceum (Forssk.) Meve et Liede = Sarcostemma viminale (L.) R. Br. subsp. stipitaceum (Forssk.) Meve et Liede ■☆
117743 Cynanchum vincetoxicum (L.) Pers.;催吐白前(催吐白薇,药用白前);Emetic Mosquitotrap, Swallow-wort, Vincetoxicum, White Swallowwort, White Swallow-wort ■
117744 Cynanchum virens (E. Mey.) D. Dietr.;绿鹅绒藤■☆
117745 Cynanchum volubile (Maxim.) Hemsl.;蔓白前;Swallowwort, Twine Mosquitotrap ■
117746 Cynanchum volubile sensu Courtois = Cynanchum sublanceolatum (Miq.) Matsum. ■
117747 Cynanchum wallichii Wight;昆明杯冠藤(断节参,对节参,金线壶芦,昆明白前,青洋参,团花奶浆根);Kunming Mosquitotrap, Wallich Swallowwort ■
117748 Cynanchum wangii P. T. Li et W. Kittr.;启无白前;Wang Mosquitotrap, Wang Swallowwort ■
117749 Cynanchum welwitschii Schltr. et Rendle = Cynanchum polyanthum K. Schum. ■☆
117750 Cynanchum wilfordii (Maxim.) Hemsl. = Cynanchum wilfordii (Maxim.) Hook. f. ■
117751 Cynanchum wilfordii (Maxim.) Hook. f.;隔山消(白奶奶,白首乌,豆角蛤蝌,隔山牛皮消,隔山撬,过山飘,戟叶牛皮消,无梁藤);Wilfoed Swallowwort, Wilford Mosquitotrap ■
117752 Cynanchum wilfordii (Maxim.) Hook. f. var. amamianum Hatus. = Cynanchum boudieri H. Lév. et Vaniot ■
117753 Cynanchum yamanakae Ohwi et H. Ohashi = Vincetoxicum yamanakae (Ohwi et H. Ohashi) H. Ohashi ■☆
117754 Cynanchum yonakuniense Hatus. = Vincetoxicum yonakuniense (Hatus.) T. Yamash. et Tateishi ■☆
117755 Cynanchum yunnanense Anthony = Cynanchum anthonyanum Hand. -Mazz. ■
117756 Cynanchum yunnanense H. Lév. = Paederia yunnanensis (H. Lév.) Rehder ●
117757 Cynanchum zeyheri Schltr.;泽赫鹅绒藤●☆
117758 Cynapium Bubani = Aethusa L. ■☆
117759 Cynapium Nutt. (1840);犬足芹属■☆
117760 Cynapium Nutt. = Ligusticum L. ■
117761 Cynapium Nutt. ex Torr. et A. Gray = Cynapium Nutt. ■☆
117762 Cynapium Nutt. ex Torr. et A. Gray = Ligusticum L. ■
117763 Cynapium Rupr. = Aethusa L. ■☆
117764 Cynapium apiifolium Nutt. ex Torr. et A. Gray;犬足芹■☆
117765 Cynara L. (1753);菜蓟属;Cardoon ■
117766 Cynara Vaill. ex L. = Cynara L. ■
117767 Cynara acaulis L. = Dittrichia viscosa (L.) Greuter ■
117768 Cynara baetica (Spreng.) Pau;伯蒂卡菜蓟■☆
117769 Cynara baetica (Spreng.) Pau subsp. maroccana Wiklund;摩洛哥菜蓟■☆
117770 Cynara cardunculus L.;刺苞菜蓟(刺菜蓟,大叶菜蓟,西班牙菜蓟);Artichoke, Artichoke Thistle, Cardoon ■
117771 Cynara cardunculus L. subsp. flavescens Wiklund;浅黄菜蓟■☆
117772 Cynara cardunculus L. var. altilis DC. = Cynara cardunculus L. ■
117773 Cynara cardunculus L. var. elata Cavara = Cynara cardunculus L. ■
117774 Cynara cardunculus L. var. ferocissima Lowe = Cynara cardunculus L. ■
117775 Cynara cardunculus L. var. inermis DC. = Cynara cardunculus L. ■
117776 Cynara cardunculus L. var. scolymus (L.) Fiori = Cynara scolymus L. ■
117777 Cynara cardunculus L. var. sylvestris (Lam.) Fiori = Cynara cardunculus L. ■
117778 Cynara cornigera Lindl.;角状菜蓟■☆
117779 Cynara cyrenaica Maire et Weiller;昔兰尼菜蓟■☆
117780 Cynara glomerata Thunb. = Platycarpha glomerata (Thunb.) Less. ■☆
117781 Cynara horrida Aiton = Cynara cardunculus L. ■
117782 Cynara humilis L.;低矮菜蓟■☆
117783 Cynara humilis L. var. leucantha (Coss.) Cout. = Cynara humilis L. ■☆
117784 Cynara humilis L. var. reflexa Batt. = Cynara humilis L. ■☆
117785 Cynara humilis L. var. walliana Maire = Cynara humilis L. ■☆
117786 Cynara hystrix Ball = Cynara baetica (Spreng.) Pau ■☆
117787 Cynara scolymus L.;菜蓟(朝蓟,朝鲜蓟,洋蓟);Artichoke, Cardoon, Chards, Common Artichoke, French Artichoke, Globe Artichoke, Globe Cardoon, Green Artichoke, Heartychoke ■
117788 Cynara scolymus L. = Cynara cardunculus L. ■
117789 Cynara sibthorpiana Boiss. et Heldr. = Cynara cornigera Lindl. ■☆
117790 Cynara sibthorpiana Boiss. et Heldr. var. elata Bég. et Vacc. = Cynara cornigera Lindl. ■☆
117791 Cynara sibthorpiana Boiss. et Heldr. var. mauginiana Pamp. =

Cynara cornigera Lindl. ■☆

117792 Cynara tournefortii Boiss. et Reut. = Arcyna tournefortii (Boiss. et Reut.) Wiklund ■☆

117793 Cynaraceae Burnett = Asteraceae Bercht. et J. Presl(保留科名)●■

117794 Cynaraceae Burnett = Compositae Giseke(保留科名)●■

117795 Cynaraceae Juss. = Asteraceae Bercht. et J. Presl(保留科名)●■

117796 Cynaraceae Juss. = Compositae Giseke(保留科名)●■

117797 Cynaraceae Lindl. = Asteraceae Bercht. et J. Presl(保留科名)●■

117798 Cynaraceae Spenn. = Asteraceae Bercht. et J. Presl(保留科名)●■

117799 Cynaraceae Spenn. = Compositae Giseke(保留科名)●■

117800 Cynaroides (Boiss. ex Walp.) Dostál = Centaurea L. (保留属名)●■

117801 Cynaropsis Kuntze = Cynara L. ■

117802 Cynaropsis Kuntze = Silybum Vaill. (保留属名)■

117803 Cynarospermum Vollesen = Blepharis Juss. ●■

117804 Cynarospermum Vollesen(1999);印度百簕花属■☆

117805 Cyne Danser(1929);犬寄生属●☆

117806 Cyne banahaensis Danser;犬寄生●☆

117807 Cynocardamum Webb et Berthel. = Lepidium L. ■

117808 Cynocrambaceae Meisn. = Theligonaceae Dumort. (保留科名)■

117809 Cynocrambaceae Nees = Rubiaceae Juss. (保留科名)●■

117810 Cynocrambaceae Nees = Theligonaceae Dumort. (保留科名)■

117811 Cynocrambe Gagnep. = Theligonum L. ■

117812 Cynocrambe Hill = Mercurialis L. ■

117813 Cynocrambe formosana Ohwi = Theligonum formosanum (Ohwi) Ohwi et Tang S. Liu ■

117814 Cynocrambe japonica (Okubo et Makino) Makino = Theligonum japonicum Okubo et Makino ■

117815 Cynocrambe macrantha Poulsen = Theligonum macranthum Franch. ■

117816 Cynoctonum E. Mey. = Cynanchum L. ●■

117817 Cynoctonum J. F. Gmel. = Mitreola L. ■

117818 Cynoctonum acuminatum Benth. = Cynanchum adalinae (K. Schum.) K. Schum. subsp. mannii (Scott-Elliot) Bullock ●☆

117819 Cynoctonum alatum Decne. = Cynanchum alatum Buch. -Ham. ex Wight et Arn. ■

117820 Cynoctonum bojerianum Decne. = Cynanchum bojerianum (Decne.) Choux ●☆

117821 Cynoctonum brownii Meisn. = Cynanchum obtusifolium L. f. ●☆

117822 Cynoctonum callialata Decne. = Cynanchum callialatum Buch. -Ham. ex Wight ■

117823 Cynoctonum callialatum (Buch. -Ham. ex Wight) Decne. = Cynanchum callialatum Buch. -Ham. ex Wight ■

117824 Cynoctonum capense (L. f.) E. Mey. = Pentatropis capensis (L. f.) Bullock ■☆

117825 Cynoctonum capense (R. Br.) E. Mey. = Cynanchum obtusifolium L. f. ●☆

117826 Cynoctonum corymbosum (Wight) Decne. = Cynanchum corymbosum Wight ■

117827 Cynoctonum corymbosum Decne. = Cynanchum corymbosum Wight ■

117828 Cynoctonum crassifolium E. Mey. = Cynanchum africanum (L.) Hoffmanns. ●☆

117829 Cynoctonum formosanum Maxim. = Cynanchum formosanum (Maxim.) Hemsl. ex Forbes et Hemsl. ●■

117830 Cynoctonum insulanum Hance = Cynanchum insulanum (Hance) Hemsl. ■

117831 Cynoctonum meyeri Decne. = Cynanchum meyeri (Decne.) Schltr. ●☆

117832 Cynoctonum mitreola (L.) Britton = Mitreola petiolata (J. F. Gmel.) Torr. et A. Gray ■

117833 Cynoctonum molle E. Mey. = Anisotoma cordifolia Fenzl ■☆

117834 Cynoctonum obovatum Decne. = Cynanchum obovatum (Decne.) Choux ●☆

117835 Cynoctonum oldenlandioides (Wall. ex DC.) B. L. Rob. = Mitreola petiolata (J. F. Gmel.) Torr. et A. Gray ■

117836 Cynoctonum paniculatum (Wall. ex G. Don) B. L. Rob. = Mitreola petiolata (J. F. Gmel.) Torr. et A. Gray ■

117837 Cynoctonum pedicellatum (Benth.) B. L. Rob. = Mitreola pedicellata Benth. ■

117838 Cynoctonum pedicellatum (Benth.) B. L. Rob. = Mitreola petiolata (J. F. Gmel.) Torr. et A. Gray ■

117839 Cynoctonum petiolatum J. F. Gmel. = Mitreola petiolata (J. F. Gmel.) Torr. et A. Gray ■

117840 Cynoctonum pilosum E. Mey. = Cynanchum africanum (L.) Hoffmanns. ●☆

117841 Cynoctonum purpureum Pobed. = Cynanchum purpureum (Pall.) K. Schum. ■

117842 Cynoctonum repandum (Decne.) Decne. = Cynanchum repandum (Decne.) K. Schum. ●☆

117843 Cynoctonum roseum (R. Br.) Decne. = Cynanchum purpureum (Pall.) K. Schum. ■

117844 Cynoctonum roseum Decne. = Cynanchum purpureum (Pall.) K. Schum. ■

117845 Cynoctonum succulentum R. W. Long = Mitreola petiolata (J. F. Gmel.) Torr. et A. Gray ■

117846 Cynoctonum virens E. Mey. = Cynanchum virens (E. Mey.) D. Dietr. ■☆

117847 Cynoctonum wallichii (Wight) Decne. = Cynanchum wallichii Wight ■

117848 Cynoctonum wallichii Decne. = Cynanchum wallichii Wight ■

117849 Cynoctonum wilfordii Maxim. = Cynanchum wilfordii (Maxim.) Hook. f. ■

117850 Cynodendron Baehni = Chrysophyllum L. ●

117851 Cynodon Pers. = Cynodon Rich. (保留属名)■

117852 Cynodon Rich. (1805)(保留属名);狗牙根属(绊根草属);Bermuda-grass, Dogstooth Grass, Dogtoothgrass, Star Grass ■

117853 Cynodon aethiopicus Clayton et Harlan;埃塞俄比亚狗牙根;Ethiopian Dogstooth Grass ■☆

117854 Cynodon affinis Caro et E. A. Sánchez = Cynodon dactylon (L.) Pers. var. affinis (Caro et E. A. Sánchez) Romero Zarco ■

117855 Cynodon arcuatus J. Presl = Cynodon radiatus Roth ex Roem. et Schult. ■

117856 Cynodon arcuatus J. Presl et C. Presl = Cynodon radiatus Roth ex Roem. et Schult. ■

117857 Cynodon ciliaris (L.) Raspail = Eragrostis ciliaris (L.) R. Br. ■

117858 Cynodon curtipendulus (Michx.) Raspail = Bouteloua curtipendula (Michx.) Torr. ■

117859 Cynodon dactylon (L.) Pers.;狗牙根(巴根草,绊根草,草皮子,动地虎,狗牙草,鸡肠草,马鞭子草,马根子草,马挽手,牛马根,爬根草,铺地草,堑头草,铁线草,铜丝金,蟋蟀草,咸沙草,行仪芝);Bahama Grass, Bermuda Grass, Bermudagrass, Bermuda-grass, Couch, Couch Grass, Dhob, Dhub, Dogtoothgrass, Doob,

Kweek, Kweek Grass, Star Grass ■

117860 Cynodon dactylon (L.) Pers. subsp. nipponicus (Ohwi) T. Koyama = Cynodon dactylon (L.) Pers. var. nipponicus Ohwi ■☆

117861 Cynodon dactylon (L.) Pers. var. affinis (Caro et E. A. Sánchez) Romero Zarco = Cynodon dactylon (L.) Pers. ■

117862 Cynodon dactylon (L.) Pers. var. biflorus Merino; 双花狗牙根; Biflower Dogtoothgrass, Twoflower Dogtoothgrass ■

117863 Cynodon dactylon (L.) Pers. var. glabratus (Steud.) Chiov. = Cynodon dactylon (L.) Pers. ■

117864 Cynodon dactylon (L.) Pers. var. hirsutissimus (Litard. et Maire) Maire = Cynodon dactylon (L.) Pers. ■

117865 Cynodon dactylon (L.) Pers. var. intermedius (Rang. et Tadulingham) C. E. C. Fischer = Cynodon radiatus Roth ex Roem. et Schult. ■

117866 Cynodon dactylon (L.) Pers. var. intermedius (Rang. et Tadulingham) C. E. C. Fischer = Cynodon arcuatus J. Presl et C. Presl ■

117867 Cynodon dactylon (L.) Pers. var. nipponicus Ohwi; 日本狗牙根■☆

117868 Cynodon decipiens Caro et E. A. Sánchez; 迷惑狗牙根■☆

117869 Cynodon fransvaalensis Burtt Davy; 乌干达狗牙根; Masindi Grass, Uganda Grass ■☆

117870 Cynodon glabratus Steud. = Cynodon dactylon (L.) Pers. ■

117871 Cynodon hirsutus Stent; 粗毛狗牙根■☆

117872 Cynodon intermedius Rang. et Tadulingham = Cynodon arcuatus J. Presl ■

117873 Cynodon intermedius Rang. et Tadulingham = Cynodon radiatus Roth ex Roem. et Schult. ■

117874 Cynodon magennisii Hurcombe; 马氏狗牙根; Magennis' Dogstooth Grass ■☆

117875 Cynodon nlemfuensis Vanderyst; 非洲狗牙根; African Bermudagrass ■☆

117876 Cynodon nlemfuensis Vanderyst var. robustus Clayton et Harlan; 粗壮非洲狗牙根■☆

117877 Cynodon occidentalis Willd. ex Steud. = Cynodon dactylon (L.) Pers. ■

117878 Cynodon plectostachyus (K. Schum.) Pilg.; 织穗狗牙根; Stargrass ■☆

117879 Cynodon plectostachyus (K. Schum.) Pilg. var. ruspolianus (Chiov.) Chiov. = Cynodon plectostachyus (K. Schum.) Pilg. ■☆

117880 Cynodon polevansii Stent; 埃文斯狗牙根■☆

117881 Cynodon radiatus Roth ex Roem. et Schult.; 弯穗狗牙根(恒春狗牙根,宽叶绊根草); Bentspike Dogtoothgrass ■

117882 Cynodon ruspolianus Chiov. = Cynodon plectostachyus (K. Schum.) Pilg. ■☆

117883 Cynodon tener J. Presl = Eustachys tener (J. Presl) A. Camus ■

117884 Cynodon ternatus A. Rich. = Digitaria ternata (A. Rich.) Stapf ■☆

117885 Cynodon ternatus Hochst. ex A. Rich. = Digitaria ternata (Hochst. ex A. Rich.) Stapf ex Dyer ■

117886 Cynodon transvaalensis Burtt Davy; 德兰士瓦狗牙根; African Dogstooth Grass, Masindi, Uganda Grass ■☆

117887 Cynodontaceae Link = Gramineae Juss. (保留科名) ■●

117888 Cynodontaceae Link = Poaceae Barnhart (保留科名) ■●

117889 Cynogeton Kunth = Cycnogeton Endl. ■

117890 Cynogeton Kunth = Triglochin L. ■

117891 Cynoglossaceae Döll = Boraginaceae Juss. (保留科名) ■●

117892 Cynoglossopsis Brand (1931); 拟琉璃草属■☆

117893 Cynoglossopsis latifolia (Hochst. ex A. Rich.) Brand; 拟琉璃草■☆

117894 Cynoglossopsis somaliensis Riedl; 索马里拟琉璃草■☆

117895 Cynoglossospermum Kuntze (1891) = Echinospermum Sw. ■

117896 Cynoglossospermum Kuntze (1898) = Eritrichium Schrad. ex Gaudin ■

117897 Cynoglossospermum Siegesb. = Echinospermum Sw. ■

117898 Cynoglossospermum Siegesb. ex Kuntze = Echinospermum Sw. ■

117899 Cynoglossospermum deflexum Kuntze = Eritrichium deflexum (Wahlenb.) Y. S. Lian et J. Q. Wang ■

117900 Cynoglossum L. (1753); 琉璃草属(倒提壶属,狗舌草属); Chinese Forget-me-not, Hound's Tongue, Houndstongue, Hound's-tongue ■

117901 Cynoglossum abyssinicum Hochst. ex Schweinf. = Cynoglossum coeruleum A. DC. ■☆

117902 Cynoglossum aequinoctiale T. C. E. Fr.; 昼夜琉璃草■☆

117903 Cynoglossum afrocaeruleum (R. R. Mill) Riedl = Cynoglossum coeruleum A. DC. ■☆

117904 Cynoglossum alpestre Ohwi; 高山倒提壶■

117905 Cynoglossum alpinum (Brand) B. L. Burtt; 高山琉璃草■☆

117906 Cynoglossum alpinum Riedl = Cynoglossum alpinum (Brand) B. L. Burtt ■☆

117907 Cynoglossum alticola Hilliard et B. L. Burtt; 高原琉璃草■☆

117908 Cynoglossum amabile Stapf et J. R. Drumm.; 倒提壶(大肥根,附地菜,狗尿蓝布裙,狗尿蓝花,狗舌草,狗舌花,狗屎花,狗屎蓝花,狗屎萝卜,鸡爪参,接骨草,兰花参,拦路虎,蓝布裙,蓝狗屎花,蓝花参,蓝花叶,莲子叶,六肥根,六月肥,龙须草,绿花心,绿花叶,牛舌头草,牛舌头花,七星剑,七星箭,贴骨散,铁骨散,一把抓,中国勿忘草); China Houndstongue, Chinese Forgetmenot, Chinese Forget-me-not, Chinese Hound's-tongue ■

117909 Cynoglossum amabile Stapf et J. R. Drumm. 'Firmament'; 天蓝倒提壶■

117910 Cynoglossum amabile Stapf et J. R. Drumm. f. leucanthum X. D. Dong; 白花倒提壶; Whiteflower China Houndstongue ■

117911 Cynoglossum amabile Stapf et J. R. Drumm. f. ruberum X. D. Dong; 红花倒提壶; Redflower China Houndstongue ■

117912 Cynoglossum amabile Stapf et J. R. Drumm. var. pauciglochidiatum Y. L. Liu; 滇西琉璃草(滇西倒提壶); W. Yunnan China Houndstongue ■

117913 Cynoglossum amplifolium A. DC.; 大叶琉璃草■☆

117914 Cynoglossum amplifolium A. DC. f. macrocarpum Brand = Cynoglossum amplifolium A. DC. var. subalpinum (T. C. E. Fr.) Verdc. ■☆

117915 Cynoglossum amplifolium A. DC. var. subalpinum (T. C. E. Fr.) Verdc.; 亚高山琉璃草■☆

117916 Cynoglossum arundanum Coss. = Cynoglossum cheirifolium L. subsp. heterocarpum (Kunze) Maire ■☆

117917 Cynoglossum asperrimum Nakai = Paracynoglossum asperrimum (Nakai) Popov ■☆

117918 Cynoglossum asperrimum Nakai var. tosaense (Nakai) H. Hara = Cynoglossum asperrimum Nakai ■☆

117919 Cynoglossum asperrimum Nakai var. yesoense Nakai = Cynoglossum asperrimum Nakai ■☆

117920 Cynoglossum atlanticum Murb. = Cynoglossum creticum Mill. ■☆

117921 Cynoglossum austroafricanum Hilliard et B. L. Burtt; 南非琉璃草■☆

117922 Cynoglossum austroafricanum Weim. = Cynoglossum austroafricanum Hilliard et B. L. Burtt ■☆

117923 Cynoglossum bequaertii De Wild. = Cynoglossum amplifolium A. DC. var. subalpinum (T. C. E. Fr.) Verdc. ■☆

117924 Cynoglossum bequaertii De Wild. = Cynoglossum amplifolium A. DC. ■☆

117925 Cynoglossum birkinshawii J. S. Mill.;伯金肖琉璃草■☆

117926 Cynoglossum boreale Fernald;北方倒提壶;Northern Houndstongue,Northern Wild Comfrey,Wild Comfrey ■☆

117927 Cynoglossum canescens Willd. = Cynoglossum lanceolatum Forssk. ■

117928 Cynoglossum cavaleriei H. Lév. = Antiotrema dunnianum (Diels) Hand. -Mazz. ■

117929 Cynoglossum cernuum Baker;俯垂琉璃草■☆

117930 Cynoglossum cheirifolium L.;掌叶琉璃草■☆

117931 Cynoglossum cheirifolium L. subsp. heterocarpum (Kunze) Maire;异果掌叶琉璃草■☆

117932 Cynoglossum cheirifolium L. var. antiatlanticum Molero et J. M. Monts. = Cynoglossum cheirifolium L. ■☆

117933 Cynoglossum cheirifolium L. var. arundanum (Coss.) Maire = Cynoglossum cheirifolium L. subsp. heterocarpum (Kunze) Maire ■☆

117934 Cynoglossum cheirifolium L. var. controversum (Sennen) Maire = Cynoglossum cheirifolium L. ■☆

117935 Cynoglossum cheirifolium L. var. gomaricum Font Quer = Cynoglossum cheirifolium L. ■☆

117936 Cynoglossum cheirifolium L. var. heterocarpum Kunze = Cynoglossum cheirifolium L. subsp. heterocarpum (Kunze) Maire ■☆

117937 Cynoglossum cheirifolium L. var. lasianthum Murb. = Cynoglossum cheirifolium L. ■☆

117938 Cynoglossum cheirifolium L. var. tubuliflorum Maire = Cynoglossum cheirifolium L. ■☆

117939 Cynoglossum cheranganienese Verdc.;切兰加尼琉璃草■☆

117940 Cynoglossum clandestinum Desf.;隐匿琉璃草■☆

117941 Cynoglossum coeruleum A. DC.;天蓝琉璃草■☆

117942 Cynoglossum coeruleum A. DC. subsp. kenyense Verdc.;肯尼亚琉璃草■☆

117943 Cynoglossum coeruleum A. DC. subsp. latifolium Verdc.;宽叶天蓝琉璃草■☆

117944 Cynoglossum coeruleum A. DC. var. johnstonii (Baker) Baker et C. H. Wright;约翰斯顿琉璃草■☆

117945 Cynoglossum coeruleum A. DC. var. mannii (Baker et C. H. Wright) Verdc.;曼氏琉璃草■☆

117946 Cynoglossum coeruleum A. DC. var. winkleri Brand = Cynoglossum coeruleum A. DC. subsp. kenyense Verdc. ■☆

117947 Cynoglossum controversum Sennen = Cynoglossum cheirifolium L. ■☆

117948 Cynoglossum creticum Mill.;杂斑倒提壶(克里特琉璃草)■☆

117949 Cynoglossum creticum Mill. var. atlanticum Murb. = Cynoglossum creticum Mill. ■☆

117950 Cynoglossum creticum Mill. var. doumerguei Sennen et Mauricio = Cynoglossum creticum Mill. ■☆

117951 Cynoglossum creticum Mill. var. pictum (Aiton) Maire = Cynoglossum creticum Mill. ■☆

117952 Cynoglossum densefoliatum Chiov.;密叶倒提壶■☆

117953 Cynoglossum denticulatum A. DC. = Cynoglossum glochidiatum (Wall. ex Benth.) Kazmi ■

117954 Cynoglossum denticulatum A. DC. = Cynoglossum wallichii G. Don ■

117955 Cynoglossum diffusum Roxb. = Bothriospermum tenellum (Hornem.) Fisch. et C. A. Mey. ■

117956 Cynoglossum divaricatum Steph. ex Lehm.;大果琉璃草(大赖鸡毛子,大赖毛七,大赖毛子,大粘柒子,倒提壶,琉璃草,粘染子,展枝倒提壶);Divaricate Houndstongue ■

117957 Cynoglossum dunnianum Diels = Antiotrema dunnianum (Diels) Hand. -Mazz. ■

117958 Cynoglossum edgeworthii A. DC. = Cynoglossum glochidiatum (Wall. ex Benth.) Kazmi ■

117959 Cynoglossum edgeworthii A. DC. = Cynoglossum wallichii G. Don ■

117960 Cynoglossum enerve Turcz. = Cynoglossum hispidum Thunb. ■☆

117961 Cynoglossum formosanum Nakai;台湾倒提壶(台湾琉璃草,台湾小花琉璃草)■

117962 Cynoglossum formosanum Nakai = Cynoglossum lanceolatum Forssk. var. formosanum (Nakai) H. Hara ■

117963 Cynoglossum formosanum Nakai = Cynoglossum lanceolatum Forssk. ■

117964 Cynoglossum formosanum Nakai f. albiflorum Masam.;白花台湾倒提壶■☆

117965 Cynoglossum furcatum Wall.;琉璃草(大琉璃草,狗屎花,火草,捆仙绳,拦路虎,绿花莱,母猪油子,牛舌头草,青菜参,生扯拢,贴骨草,贴骨散,铁板道,铁道板,铁箍散,锡兰倒提壶,小生地,粘娘娘,猪尾巴);Ceylon Hound's Tongue,Ceylon Houndstongue ■

117966 Cynoglossum furcatum Wall. var. villosulum (Nakai) Riedl;短毛琉璃草■

117967 Cynoglossum gansuense Y. L. Liu;甘青琉璃草(甘肃琉璃草);Gansu Houndstongue,Kansu Houndstongue ■

117968 Cynoglossum germanicum Jacq.;德国琉璃草(德国倒提壶);German Houndstongue, Green Hound's Tongue, Green Hound's-tongue,Green-leaved Hound's-tongue ■☆

117969 Cynoglossum glochidianum Wall. ex Benth. = Cynoglossum glochidiatum (Wall. ex Benth.) Kazmi ■

117970 Cynoglossum glochidiatum (Wall. ex Benth.) Kazmi;倒钩琉璃草;Glochidiate Houndstongue,Prickly Hound's Tongue ■

117971 Cynoglossum grande Douglas ex Lehm.;大琉璃草;Hound's-tongue ■☆

117972 Cynoglossum gymnandrum (Coss.) Greuter et Burdet;裸蕊琉璃草■☆

117973 Cynoglossum hedbergiorum Riedl;赫德琉璃草■☆

117974 Cynoglossum hirsutum Jacq. = Cynoglossum lanceolatum Forssk. ■

117975 Cynoglossum hirsutum Thunb. = Cynoglossum lanceolatum Forssk. ■

117976 Cynoglossum hispidum Thunb.;硬毛琉璃草■☆

117977 Cynoglossum hochstetteri Vatke ex Engl. = Cynoglossopsis latifolia (Hochst. ex A. Rich.) Brand ■☆

117978 Cynoglossum holosericeum Steven;全毛琉璃草■☆

117979 Cynoglossum intermedium Fresen.;间型琉璃草■☆

117980 Cynoglossum intermedium Fresen. = Paracaryum intermedium (Fresen.) Lipsky ■☆

117981 Cynoglossum inyangense E. S. Martins;伊尼扬加琉璃草■☆

117982 Cynoglossum japonicum Thunb. ex A. Murray = Omphalodes japonica (Thunb. ex A. Murray) Maxim. ■☆

117983 Cynoglossum johnstonii Baker = Cynoglossum coeruleum A. DC. var. johnstonii (Baker) Baker et C. H. Wright ■☆

117984 Cynoglossum karamojense Verdc.;卡拉莫贾琉璃草■☆

117985 Cynoglossum laevigatum L. f. = Rindera tetraspis Pall. ■

117986 Cynoglossum lanceolatum Forssk.;小花琉璃草(半边龙,大号疟草,鹤虱,母一条根,披针叶琉璃草,破布草,破布粘,台湾倒提壶,小花倒提壶,牙痈草,一条龙);Smallflower Houndstongue ■

117987 Cynoglossum lanceolatum Forssk. subsp. geometricum (Baker et C. H. Wright) Brand = Cynoglossum coeruleum A. DC. var. mannii (Baker et C. H. Wright) Verdc. ■☆

117988 Cynoglossum lanceolatum Forssk. var. formosanum (Nakai) H. Hara = Cynoglossum formosanum Nakai ■

117989 Cynoglossum lancifolium Hook. f. = Cynoglossum amplifolium A. DC. var. subalpinum (T. C. E. Fr.) Verdc. ■☆

117990 Cynoglossum laxum G. Don = Hackelia uncinata (Royle ex Benth.) C. E. C. Fisch. ■

117991 Cynoglossum linifolium (L.) Moench = Omphalodes linifolia (L.) Moench ■☆

117992 Cynoglossum linifolium L. = Omphalodes linifolia (L.) Moench ■☆

117993 Cynoglossum longepetiolatum De Wild. = Cynoglossum amplifolium A. DC. ■☆

117994 Cynoglossum lowryanum J. S. Mill.;劳里琉璃草■☆

117995 Cynoglossum macrocalycinum Riedl;大萼琉璃草■

117996 Cynoglossum macrophyllum Royle ex Benth. = Hackelia macrophylla (Brand) I. M. Johnst. ■☆

117997 Cynoglossum mannii Baker et C. H. Wright = Cynoglossum coeruleum A. DC. var. mannii (Baker et C. H. Wright) Verdc. ■☆

117998 Cynoglossum mathezii Greuter et Burdet;马泰琉璃草■☆

117999 Cynoglossum micranthum Desf. = Cynoglossum lanceolatum Forssk. ■

118000 Cynoglossum micranthum Hook. f. et Thomson ex Clarke = Cynoglossum wallichii G. Don ■

118001 Cynoglossum micranthum Poir. = Cynoglossum lanceolatum Forssk. ■

118002 Cynoglossum microglochin Benth.;小毛琉璃草;Smallbristle Hound's Tongue ■☆

118003 Cynoglossum monophlebium Baker;单脉琉璃草■☆

118004 Cynoglossum montanum L.;山地琉璃草■☆

118005 Cynoglossum montanum L. var. alpinum Brand = Cynoglossum alpinum (Brand) B. L. Burtt ■☆

118006 Cynoglossum nervosum Benth. ex Hook. f.;喜马拉雅琉璃草;Himalayan Hound's Tongue, Himalayan Hound's-tongue ■☆

118007 Cynoglossum obtusicalyx Retief et A. E. van Wyk;钝萼琉璃草■☆

118008 Cynoglossum officinale L.;红花琉璃草(新疆倒提壶,药用倒提壶,药用琉璃草);Common Hound's Tongue, Common Hound's-tongue, Dog's Tongue, Gypsy Flower, Gypsyflower, Gypsy-flower, Hound's Mie, Hound's Mile, Hound's Piss, Hound's Tongue, Hound's-tongue, Little Burdock, Navelwort, Rats-and-mice, Redflower Houndstongue, Rose Noble, Scald-head, Sticky-buds ■

118009 Cynoglossum officinale L. f. bicolor (Willd.) Lehm. = Cynoglossum officinale L. ■

118010 Cynoglossum pictum Aiton;着色琉璃草■☆

118011 Cynoglossum pictum Aiton = Cynoglossum creticum Mill. ■☆

118012 Cynoglossum pitardianum Greuter et Burdet;皮塔德琉璃草■☆

118013 Cynoglossum prostratum D. Don = Bothriospermum tenellum (Hornem.) Fisch. et C. A. Mey. ■

118014 Cynoglossum racemosum Roxb. = Cynoglossum lanceolatum Forssk. ■

118015 Cynoglossum roylei Wall. = Hackelia uncinata (Royle ex Benth.) C. E. C. Fisch. ■

118016 Cynoglossum roylei Wall. ex G. Don = Hackelia uncinata (Royle ex Benth.) C. E. C. Fisch. ■

118017 Cynoglossum schlagintweitii (Brand) Kazmi;西藏琉璃草;Xizang Houndstongue ■

118018 Cynoglossum seravschanicum (B. Fedtsch.) Popov;塞拉夫琉璃草■☆

118019 Cynoglossum stylosa Kar. et Kir. = Lindelofia stylosa (Kar. et Kir.) Brand ■

118020 Cynoglossum stylosum Kar. et Kir. = Lindelofia stylosa (Kar. et Kir.) Brand ■

118021 Cynoglossum subalpinum T. C. E. Fr. = Cynoglossum amplifolium A. DC. var. subalpinum (T. C. E. Fr.) Verdc. ■☆

118022 Cynoglossum tianschanicum Popov;天山琉璃草;Tianshan Houndstongue ■☆

118023 Cynoglossum triste Diels;心叶琉璃草;Heartleaf Houndstongue ■

118024 Cynoglossum tsaratananense J. S. Mill.;察拉塔纳纳琉璃草■☆

118025 Cynoglossum tubiflorum (Murb.) Greuter et Burdet;管花琉璃草■☆

118026 Cynoglossum ukaguruense Verdc.;乌卡古鲁琉璃草■☆

118027 Cynoglossum uncinatum Benth. = Hackelia uncinata (Royle ex Benth.) C. E. C. Fisch. ■

118028 Cynoglossum uncinatum Royle ex Benth. = Hackelia uncinata (Royle ex Benth.) C. E. C. Fisch. ■

118029 Cynoglossum uncinatum Royle ex Benth. var. laxiforum Royle ex Benth. = Hackelia macrophylla (Brand) I. M. Johnst. ■☆

118030 Cynoglossum villosulum Nakai = Cynoglossum zeylanicum (Vahl ex Hornem.) Thunb. ex Lehm. ■☆

118031 Cynoglossum virginianum L.;北方琉璃草;Giant Forget-me-not, Northern Wild Comfrey, Virginia Hound's-tongue, Wild Comfrey ■☆

118032 Cynoglossum virginianum L. var. boreale (Fernald) Cooperr. = Cynoglossum boreale Fernald ■☆

118033 Cynoglossum viridiflorum Pall. ex Lehm.;绿花琉璃草(绿花倒提壶);Greenflower Houndstongue ■

118034 Cynoglossum wallichii G. Don;西南琉璃草;Wallich Houndstongue ■

118035 Cynoglossum wallichii G. Don var. glochidiatum (Wall. ex Benth.) Kazmi = Cynoglossum glochidiatum (Wall. ex Benth.) Kazmi ■

118036 Cynoglossum watieri (Batt. et Maire) Braun-Blanq. et Maire;瓦捷琉璃草■☆

118037 Cynoglossum wildii E. S. Martins;维尔德琉璃草■☆

118038 Cynoglossum zeylanicum (Vahl ex Hornem.) Thunb. ex Lehm.;锡兰琉璃草(琉璃草)■☆

118039 Cynoglossum zeylanicum (Vahl ex Hornem.) Thunb. ex Lehm. = Cynoglossum furcatum Wall. ■

118040 Cynoglossum zeylanicum (Vahl ex Hornem.) Thunb. ex Lehm. f. albiflorum H. Hara;白花锡兰琉璃草■☆

118041 Cynoglossum zeylanicum (Vahl ex Hornem.) Thunb. ex Lehm. var. villosulum (Nakai) Ohwi = Cynoglossum furcatum Wall. var. villosulum (Nakai) Riedl ■

118042 Cynoglossum zeylanicum (Vahl ex Hornem.) Thunb. ex Lehm. var. villosulum (Nakai) Ohwi = Cynoglossum zeylanicum (Vahl ex Hornem.) Thunb. ex Lehm. ■☆

118043 Cynoglottis (Gusul.) Vural et Kit Tan(1983);欧洲狗舌草属■☆

118044 Cynoglottis barrelieri (All.) Vural et Kit Tan = Anchusa barrelieri Vitman ■☆

118045 Cynomarathrum Nutt. = Lomatium Raf. ■☆
118046 Cynomarathrum Nutt. ex J. M. Coult. et Rose = Lomatium Raf. ■☆
118047 Cynometra L. (1753);茎花豆属(喃果苏木属,喃喃果属)●☆
118048 Cynometra abrahamii Du Puy et R. Rabev.;亚伯拉罕茎花豆●☆
118049 Cynometra afzelii Oliv. = Hymenostegia afzelii (Oliv.) Harms ■☆
118050 Cynometra alexandri C. H. Wright;乌干达茎花豆(乌干达喃果苏木);Ironwood,Muhimbi ●☆
118051 Cynometra ananta Hutch. et Dalziel;假凤梨喃果苏木●☆
118052 Cynometra ankaranensis Dupuy et R. Rabev.;安卡兰茎花豆●☆
118053 Cynometra aubrevillei Pellegr. = Hymenostegia neoaubrevillei J. Léonard ■☆
118054 Cynometra aurita R. Vig.;耳茎花豆●☆
118055 Cynometra bequaertii De Wild. = Normandiodendron bequaertii (De Wild.) J. Léonard ■☆
118056 Cynometra bipetala Pellegr. = Hymenostegia pellegrinii (A. Chev.) J. Léonard ■☆
118057 Cynometra brachyrachis Harms;短轴茎花豆(短轴喃果苏木)●☆
118058 Cynometra brachyura Harms = Hymenostegia brachyura (Harms) J. Léonard ■☆
118059 Cynometra capuronii Du Puy et R. Rabev.;凯普伦喃果苏木●☆
118060 Cynometra cauliflora L.;茎花豆;Lamuta,Nam-Nam ●☆
118061 Cynometra cauliflora Wall. = Cynometra cauliflora L. ●☆
118062 Cynometra citrina (Taub.) Harms = Zenkerella citrina Taub. ■☆
118063 Cynometra claessensii De Wild. = Normandiodendron bequaertii (De Wild.) J. Léonard ■☆
118064 Cynometra cloiselii Drake = Cynometra commersoniana Baill. ●☆
118065 Cynometra commersoniana Baill.;科梅逊茎花豆(科梅逊喃果苏木)●☆
118066 Cynometra congensis De Wild.;刚果茎花豆(刚果喃果苏木)●☆
118067 Cynometra dauphinensis Dupuy et R. Rabev.;多芬茎花豆(多芬喃果苏木)●☆
118068 Cynometra djumaensis De Wild. = Aphanocalyx djumaensis (De Wild.) J. Léonard ■☆
118069 Cynometra egregia Bullock = Zenkerella egregia J. Léonard ●☆
118070 Cynometra engleri Harms;恩格勒茎花豆(恩格勒喃果苏木)●☆
118071 Cynometra escherichii Harms = Gilletiodendron escherichii (Harms) J. Léonard ●☆
118072 Cynometra felicis (A. Chev.) Pellegr. = Hymenostegia felicis (A. Chev.) J. Léonard ■☆
118073 Cynometra filifera Harms;丝茎花豆(丝喃果苏木)●☆
118074 Cynometra gilletii De Wild. = Cynometra sessiliflora Harms ●☆
118075 Cynometra gillmanii J. Léonard;吉尔曼茎花豆(吉尔曼喃果苏木)●☆
118076 Cynometra glabra De Wild. = Lebruniodendron leptanthum (Harms) J. Léonard ●☆
118077 Cynometra glabra R. Vig. = Cynometra commersoniana Baill. ●☆
118078 Cynometra glandulosa Portères = Gilletiodendron glandulosum (Portères) J. Léonard ●☆
118079 Cynometra greenwayi Brenan;格林韦茎花豆(格林韦喃果苏木)●☆
118080 Cynometra grotei Harms = Zenkerella capparidacea (Taub.) J. Léonard subsp. grotei (Harms) Temu ■☆
118081 Cynometra hedinii A. Chev. = Aphanocalyx hedinii (A. Chev.) Wieringa ■☆
118082 Cynometra kisantuensis Vermoesen ex De Wild. = Gilletiodendron kisantuense (Vermoesen ex De Wild.) J. Léonard ●☆
118083 Cynometra koko De Wild. = Lebruniodendron leptanthum (Harms) J. Léonard ●☆
118084 Cynometra laurentii De Wild. = Cynometra sessiliflora Harms var. laurentii (De Wild.) J. -P. Lebrun ●☆
118085 Cynometra laxiflora Benth. = Hymenostegia laxiflora (Benth.) Harms ■☆
118086 Cynometra leonensis Hutch. et Dalziel;莱昂茎花豆(莱昂喃果苏木)●☆
118087 Cynometra leptantha Harms = Lebruniodendron leptanthum (Harms) J. Léonard ●☆
118088 Cynometra letestui (Pellegr.) J. Léonard;莱泰斯图茎花豆(喃果苏木)●☆
118089 Cynometra longepedicellata Harms;长花梗茎花豆(喃果苏木)●☆
118090 Cynometra longituba Harms = Plagiosiphon longitubus (Harms) J. Léonard ■☆
118091 Cynometra lujae De Wild.;卢杰茎花豆(卢杰喃喃果)●☆
118092 Cynometra lyallii Baker;莱尔茎花豆(莱尔喃果苏木)●☆
118093 Cynometra madagascariensis Baill.;马岛茎花豆(喃果苏木)●☆
118094 Cynometra malaccensis Meeuwen;马六喃果苏木茎花豆(喃果苏木)●☆
118095 Cynometra mannii Oliv.;曼氏茎花豆(喃果苏木)●☆
118096 Cynometra megalophylla Harms;大叶茎花豆(喃果苏木)●☆
118097 Cynometra michelsonii J. Léonard;米歇尔松茎花豆(喃果苏木)●☆
118098 Cynometra mildbraedii Harms = Gilletiodendron mildbraedii (Harms) Vermoesen ●☆
118099 Cynometra multijuga Harms = Plagiosiphon multijugus (Harms) J. Léonard ■☆
118100 Cynometra nyangensis Pellegr.;尼扬加茎花豆(喃果苏木)●☆
118101 Cynometra oddonii De Wild.;奥顿茎花豆(喃果苏木)●☆
118102 Cynometra pachycarpa A. Chev. = Brachystegia mildbraedii Harms ●☆
118103 Cynometra palustris J. Léonard;沼泽茎花豆(喃果苏木)●☆
118104 Cynometra pedicellata De Wild.;梗花茎花豆(喃果苏木)●☆
118105 Cynometra pervilleana Baill.;佩尔茎花豆(喃果苏木)●☆
118106 Cynometra pierreana Harms = Gilletiodendron pierreanum (Harms) J. Léonard ●☆
118107 Cynometra pinnata Lour. = Ormosia pinnata (Lour.) Merr. ●
118108 Cynometra purpureo-caerulea Baker f. = Normandiodendron bequaertii (De Wild.) J. Léonard ■☆
118109 Cynometra rubriflora De Wild. = Normandiodendron bequaertii (De Wild.) J. Léonard ■☆
118110 Cynometra sakalava Du Puy et R. Rabev.;萨卡拉瓦茎花豆●☆
118111 Cynometra sanagaensis Aubrév.;萨纳加茎花豆●☆
118112 Cynometra sankuruensis Vermoesen = Cynometra alexandri C. H. Wright ●☆
118113 Cynometra schlechteri Harms;施莱茎花豆(施莱喃果苏木)●☆
118114 Cynometra sessiliflora Harms;无花梗茎花豆(无花梗喃果苏木)●☆
118115 Cynometra sessiliflora Harms var. laurentii (De Wild.) J. -P. Lebrun;洛朗茎花豆●☆
118116 Cynometra tetraphylla Hook. f. = Cryptosepalum tetraphyllum (Hook. f.) Benth. ●☆
118117 Cynometra ulugurensis Harms;乌卢古尔茎花豆(乌卢古尔喃果苏木)●☆
118118 Cynometra vogelii Hook. f.;沃格尔茎花豆(沃格尔喃果苏木)●☆

118119 Cynometra webberi Baker f. ;肯尼亚茎花豆(肯尼亚喃果苏木)●☆
118120 Cynomora R. Hedw. = Cynometra L. ●☆
118121 Cynomorbium Opiz = Hericinia Fourr. ■
118122 Cynomorbium Opiz = Pfundia Opiz ex Nevski ■
118123 Cynomoriaceae (Agardh) Lindl. = Cynomoriaceae Endl. ex Lindl.(保留科名)■
118124 Cynomoriaceae Endl. ex Lindl.(1833)(保留科名);锁阳科;Cynomorium Family ■
118125 Cynomoriaceae Lindl. = Balanophoraceae Rich.(保留科名)●■
118126 Cynomoriaceae Lindl. = Cynomoriaceae Endl. ex Lindl.(保留科名)■
118127 Cynomorium L.(1753);锁阳属;Cynomorium ■
118128 Cynomorium coccineum Boiss. = Cynomorium songaricum Rupr. ■
118129 Cynomorium coccineum L. ;欧锁阳■☆
118130 Cynomorium coccineum L. = Cynomorium songaricum Rupr. ■
118131 Cynomorium coccineum L. subsp. songaricum (Rupr.) J. Léonard = Cynomorium songaricum Rupr. ■
118132 Cynomorium purpureum Rupr. = Cynomorium coccineum L. ■☆
118133 Cynomorium songaricum Rupr. ;锁阳(不老药,地毛球,绯红锁阳,黄骨狼,琐阳,锁严子,锁燕,铁棒锤,锈铁棒,锈铁锤,雪央,羊锁不拉,朱红锁阳);Dzungar Cynomorium,Scarlet Cynomorium,Songaria Cynomorium ■
118134 Cynomyrtus Scriv. = Rhodomyrtus (DC.) Rchb. ●
118135 Cynopaema Lunell = Apocynum L. ●■
118136 Cynophalla J. Presl = Capparis L. ●
118137 Cynopoa Ehrh. = Agropyron Gaertn. ■
118138 Cynopoa Ehrh. = Elymus L. ■
118139 Cynopsole Endl. = Balanophora J. R. Forst. et G. Forst. ■
118140 Cynopsole elongata (Blume) Endl. ex Jacks. = Balanophora elongata Blume ■
118141 Cynorchis Thouars = Cynorkis Thouars ■☆
118142 Cynorhiza Eckl. et Zeyh.(1837);狗根草属■☆
118143 Cynorhiza Eckl. et Zeyh. = Peucedanum L. ■
118144 Cynorhiza alta Eckl. et Zeyh. ;翘狗根草■☆
118145 Cynorhiza montana Eckl. et Zeyh. ;山地狗根草■☆
118146 Cynorhiza sulcata Eckl. et Zeyh. ;狗根草■☆
118147 Cynorkis Thouars(1809);西澳兰属■☆
118148 Cynorkis alborubra Schltr. ;白红西澳兰■☆
118149 Cynorkis ampullifera H. Perrier;瓶形西澳兰■☆
118150 Cynorkis anacamptoides Kraenzl. ;倒距西澳兰■☆
118151 Cynorkis anacamptoides Kraenzl. var. ecalcarata P. J. Cribb;无距西澳兰■☆
118152 Cynorkis andohahelensis H. Perrier;安杜哈赫尔西澳兰■☆
118153 Cynorkis andringitrana Schltr. ;安德林吉特拉山西澳兰■☆
118154 Cynorkis angustipetala Ridl. ;狭瓣西澳兰■☆
118155 Cynorkis anisoloba Summerh. ;不等裂西澳兰■☆
118156 Cynorkis aphylla Schltr. ;无叶西澳兰■☆
118157 Cynorkis barlaea Schltr. = Cynorkis anacamptoides Kraenzl. ■☆
118158 Cynorkis baronii Rolfe;巴龙西澳兰■☆
118159 Cynorkis bathiei Schltr. ;巴西西澳兰■☆
118160 Cynorkis betsileensis Kraenzl. ;贝齐尔西澳兰■☆
118161 Cynorkis bimaculata (Ridl.) H. Perrier;双斑西澳兰■☆
118162 Cynorkis boinana Schltr. ;博伊纳西澳兰■☆
118163 Cynorkis bosseriana Szlach. = Cynorkis buchananii Rolfe ■☆
118164 Cynorkis brachyceras Schltr. ;马岛短角西澳兰■☆
118165 Cynorkis brachystachya Bosser;短穗西澳兰■☆
118166 Cynorkis brauniana Kraenzl. ;布劳恩西澳兰■☆
118167 Cynorkis braunii Kraenzl. = Cynorkis buchwaldiana Kraenzl. subsp. braunii (Kraenzl.) Summerh. ■☆
118168 Cynorkis brevicalcar P. J. Cribb;短距西澳兰■☆
118169 Cynorkis brevicornu Ridl. ;短角西澳兰■☆
118170 Cynorkis buchananii Rolfe;布坎南西澳兰■☆
118171 Cynorkis buchwaldiana Kraenzl. ;布赫西澳兰■☆
118172 Cynorkis buchwaldiana Kraenzl. subsp. braunii (Kraenzl.) Summerh. ;布朗西澳兰■☆
118173 Cynorkis buchwaldiana Kraenzl. subsp. nyassana (Schltr.) Summerh. = Cynorkis buchananii Rolfe ■☆
118174 Cynorkis cardiophylla Schltr. ;心叶西澳兰■☆
118175 Cynorkis catatii Bosser;卡他西澳兰■☆
118176 Cynorkis clarae Geerinck;克拉拉西澳兰■☆
118177 Cynorkis coccinelloides Schltr. ;绯红西澳兰■☆
118178 Cynorkis compacta Rchb. f. ;紧密西澳兰■☆
118179 Cynorkis confusa H. Perrier;混乱西澳兰■☆
118180 Cynorkis cuneilabia Schltr. ;楔形西澳兰■☆
118181 Cynorkis debilis (Hook. f.) Summerh. ;弱小西澳兰■☆
118182 Cynorkis decolorata Schltr. = Cynorkis fastigiata Thouars ■☆
118183 Cynorkis elata Rolfe;高西澳兰■☆
118184 Cynorkis elegans Rchb. f. ;雅致西澳兰■☆
118185 Cynorkis exilis Schltr. = Cynorkis schlechterii H. Perrier ■☆
118186 Cynorkis fallax Schltr. = Cynorkis flexuosa Lindl. ■☆
118187 Cynorkis fastigiata Thouars;帚状西澳兰■☆
118188 Cynorkis fastigiata Thouars var. decolorata (Schltr.) H. Perrier = Cynorkis fastigiata Thouars ■☆
118189 Cynorkis fastigiata Thouars var. diplorhyncha (Schltr.) H. Perrier = Cynorkis fastigiata Thouars ■☆
118190 Cynorkis fastigiata Thouars var. hygrophila (Schltr.) H. Perrier = Cynorkis fastigiata Thouars ■☆
118191 Cynorkis fastigiata Thouars var. laggiarae (Schltr.) H. Perrier = Cynorkis fastigiata Thouars ■☆
118192 Cynorkis filiformis (Kraenzl.) H. Perrier = Cynorkis papillosa (Ridl.) Summerh. ■☆
118193 Cynorkis flabellifera H. Perrier;扇状西澳兰■☆
118194 Cynorkis flexuosa Lindl. ;曲折西澳兰■☆
118195 Cynorkis formosa Bosser;美丽西澳兰■☆
118196 Cynorkis gabonensis Summerh. ;加蓬西澳兰■☆
118197 Cynorkis galeata Rchb. f. ;盔形西澳兰■☆
118198 Cynorkis gibbosa Ridl. ;浅囊西澳兰■☆
118199 Cynorkis gigas Schltr. ;巨大西澳兰■☆
118200 Cynorkis glandulosa L. Bolus = Cynorkis kassneriana Kraenzl. ■☆
118201 Cynorkis glandulosa Ridl. ;具腺西澳兰■☆
118202 Cynorkis globifera H. Perrier;小球西澳兰■☆
118203 Cynorkis globosa Schltr. ;球形西澳兰■☆
118204 Cynorkis graminea (Thouars) Schltr. ;禾叶西澳兰■☆
118205 Cynorkis gymnadenoides Schltr. = Cynorkis anacamptoides Kraenzl. ■☆
118206 Cynorkis gymnochiloides (Schltr.) H. Perrier;裸唇兰西澳兰■☆
118207 Cynorkis hanningtonii Rolfe = Cynorkis debilis (Hook. f.) Summerh. ■☆
118208 Cynorkis henrici Schltr. ;昂里克西澳兰■☆
118209 Cynorkis heterochroma Schltr. = Cynorkis ridleyi T. Durand et Schinz ■☆
118210 Cynorkis hispidula Ridl. ;细毛西澳兰■☆
118211 Cynorkis hologlossa Schltr. ;全舌西澳兰■☆

118212 Cynorkis humbertii Bosser;亨伯特西澳兰■☆

118213 Cynorkis hygrophila Schltr. = Cynorkis fastigiata Thouars ■☆

118214 Cynorkis inversa Schltr. = Cynorkis bathiei Schltr. ■☆

118215 Cynorkis johnsonii Rolfe = Cynorkis debilis (Hook. f.) Summerh. ■☆

118216 Cynorkis jumelleana Schltr.;朱迈尔西澳兰■☆

118217 Cynorkis kassneriana Kraenzl.;卡斯纳西澳兰■☆

118218 Cynorkis kassneriana Kraenzl. subsp. tenuior Summerh.;细卡斯纳西澳兰■☆

118219 Cynorkis kirkii Rolfe;柯克西澳兰■☆

118220 Cynorkis laeta Schltr.;愉悦西澳兰■☆

118221 Cynorkis laggiarae Schltr. = Cynorkis fastigiata Thouars ■☆

118222 Cynorkis lancilabia Schltr.;披针形西澳兰■☆

118223 Cynorkis latipetala H. Perrier;阔瓣西澳兰■☆

118224 Cynorkis lilacina Ridl.;紫丁香西澳兰■☆

118225 Cynorkis lisowskii Szlach. = Cynorkis debilis (Hook. f.) Summerh. ■☆

118226 Cynorkis longifolia (Lindl.) Schltr. = Cynorkis graminea (Thouars) Schltr. ■☆

118227 Cynorkis macloughlinii L. Bolus = Stenoglottis woodii Schltr. ■☆

118228 Cynorkis marojejyensis Bosser;马罗西澳兰■☆

118229 Cynorkis melinantha Schltr.;蜜花西澳兰■☆

118230 Cynorkis minuticalcar Toill. -Gen. et Bosser;微西澳兰■☆

118231 Cynorkis monadenia H. Perrier;单腺西澳兰■☆

118232 Cynorkis morlandii Rolfe = Cynorkis kirkii Rolfe ■☆

118233 Cynorkis muscicola Bosser;苔地西澳兰■☆

118234 Cynorkis nigrescens Schltr. = Cynorkis baronii Rolfe ■☆

118235 Cynorkis nutans H. Perrier;俯垂西澳兰■☆

118236 Cynorkis nyassana Schltr. = Cynorkis buchananii Rolfe ■☆

118237 Cynorkis obcordata (Willemet) Schltr. = Cynorkis fastigiata Thouars ■☆

118238 Cynorkis oblonga Schltr. = Cynorkis kirkii Rolfe ■☆

118239 Cynorkis occidentalis (Lindl.) T. Durand et Schinz = Habenaria occidentalis (Lindl.) Summerh. ■☆

118240 Cynorkis ochroglossa Schltr.;美舌西澳兰■☆

118241 Cynorkis ochyrae Szlach. et Olszewski;奥吉拉西澳兰■☆

118242 Cynorkis orchioides Schltr.;兰状西澳兰■☆

118243 Cynorkis papilio Bosser;蝶形西澳兰■☆

118244 Cynorkis papillosa (Ridl.) Summerh.;乳头西澳兰■☆

118245 Cynorkis parva Summerh. = Habenaria parva (Summerh.) Summerh. ■☆

118246 Cynorkis pauciflora Rolfe = Cynorkis baronii Rolfe ■☆

118247 Cynorkis perrieri Schltr.;佩里耶西澳兰■☆

118248 Cynorkis petiolata H. Perrier;柄叶西澳兰■☆

118249 Cynorkis peyrotii Bosser;佩罗西澳兰■☆

118250 Cynorkis platyclinoides Kraenzl. = Cynorkis anacamptoides Kraenzl. ■☆

118251 Cynorkis pleistadenia (Rchb. f.) Schltr.;多腺西澳兰■☆

118252 Cynorkis praecox Schltr. = Cynorkis purpurascens Thouars ■☆

118253 Cynorkis pseudorolfei H. Perrier;假罗尔夫西澳兰■☆

118254 Cynorkis purpurascens Thouars;浅紫西澳兰■☆

118255 Cynorkis purpurea (Thouars) Kraenzl.;紫西澳兰■☆

118256 Cynorkis quinqueloba H. Perrier;五浅裂西澳兰■☆

118257 Cynorkis quinquepartita H. Perrier;五深裂西澳兰■☆

118258 Cynorkis ridleyi T. Durand et Schinz;里德利西澳兰■☆

118259 Cynorkis rolfei Hochr.;罗尔夫西澳兰■☆

118260 Cynorkis rosellata (Thouars) Bosser;玫瑰西澳兰■☆

118261 Cynorkis rungweensis Schltr.;伦圭西澳兰■☆

118262 Cynorkis rupicola Schltr. = Cynorkis kassneriana Kraenzl. ■☆

118263 Cynorkis sacculata Schltr.;小囊西澳兰■☆

118264 Cynorkis sagittata H. Perrier;箭头西澳兰■☆

118265 Cynorkis sambiranoensis Schltr.;桑比拉诺西澳兰■☆

118266 Cynorkis saxicola Schltr.;岩生西澳兰■☆

118267 Cynorkis schlechterii H. Perrier;施莱西澳兰■☆

118268 Cynorkis similis Schltr. = Cynorkis graminea (Thouars) Schltr. ■☆

118269 Cynorkis sororia Schltr.;堆积西澳兰■☆

118270 Cynorkis souegesii Bosser et Veyret;苏埃热西澳兰■☆

118271 Cynorkis spatulata H. Perrier;匙形西澳兰■☆

118272 Cynorkis stenoglossa Kraenzl.;窄舌西澳兰■☆

118273 Cynorkis stolonifera (Schltr.) Schltr.;匍匐西澳兰■☆

118274 Cynorkis subtilis Bosser;纤细西澳兰■☆

118275 Cynorkis summerhayesiana Geerinck;萨默海斯西澳兰■☆

118276 Cynorkis sylvatica Bosser;森林西澳兰■☆

118277 Cynorkis symoensii Geerinck et Tournay;西莫西澳兰■☆

118278 Cynorkis tenella Ridl.;柔弱西澳兰■☆

118279 Cynorkis tenerrima (Ridl.) Kraenzl.;极细西澳兰■☆

118280 Cynorkis tenuicalcar Schltr.;细距西澳兰■☆

118281 Cynorkis tristis Bosser;暗淡西澳兰■☆

118282 Cynorkis tryphioides Schltr.;全毛兰西澳兰■☆

118283 Cynorkis uncata (Rolfe) Kraenzl.;具钩西澳兰■☆

118284 Cynorkis uncinata H. Perrier;钩西澳兰■☆

118285 Cynorkis uniflora Lindl.;单花西澳兰■☆

118286 Cynorkis usambarae Rolfe;乌桑巴拉西澳兰■☆

118287 Cynorkis verrucosa Bosser;多疣西澳兰■☆

118288 Cynorkis villosa Rolfe ex Hook. f.;长柔毛西澳兰■☆

118289 Cynorkis violacea Schltr.;堇色西澳兰■☆

118290 Cynorkis volkensii Kraenzl. = Cynorkis pleistadenia (Rchb. f.) Schltr. ■☆

118291 Cynorkis zaratananae Schltr.;萨拉坦西澳兰■☆

118292 Cynorrhiza Eckl. et Zeyh. = Peucedanum L. ■

118293 Cynorrhynchium Mitch. = Mimulus L. ●■

118294 Cynosbata (DC.) Rchb. = Pelargonium L'Hér. ex Aiton ●■

118295 Cynosbata Rchb. = Pelargonium L'Hér. ex Aiton ●■

118296 Cynosciadium DC. (1829);犬伞芹属■☆

118297 Cynosciadium digitatum DC.;犬伞芹■☆

118298 Cynosciadium pinnatum DC. = Limnosciadium pinnatum (DC.) Mathias et Constance ■☆

118299 Cynosorchis Thouars = Cynorkis Thouars ■☆

118300 Cynosorchis calanthoides Kraenzl. = Cynorkis purpurascens Thouars ■☆

118301 Cynosorchis chinensis Rolfe = Amitostigma gracile (Blume) Schltr. ■

118302 Cynosorchis gracilis (Blume) Kraenzl. = Amitostigma gracile (Blume) Schltr. ■

118303 Cynosorchis grandiflora Ridl. = Cynorkis uniflora Lindl. ■☆

118304 Cynosorchis micrantha Schltr. = Cynorkis anacamptoides Kraenzl. ■☆

118305 Cynosuraceae Link = Gramineae Juss.(保留科名)■●

118306 Cynosuraceae Link = Poaceae Barnhart(保留科名)■●

118307 Cynosurus L. (1753);洋狗尾草属;Dog's-tail, Dogstail Grass, Dog's-tail Grass, Dog's-tail-grass, Dogtailgrass, Silky Bent Grass ■

118308 Cynosurus aegyptius L. = Dactyloctenium aegyptium (L.) Willd. ■

118309 Cynosurus aurasiacus Murb. = Cynosurus elegans Desf. ■☆
118310 Cynosurus aureus L. = Lamarckia aurea (L.) Moench ■☆
118311 Cynosurus balansae Coss. et Durieu;巴兰萨洋狗尾草■☆
118312 Cynosurus callitrichus Barbey = Cynosurus coloratus Lehm. ex Nees ■☆
118313 Cynosurus coloratus Lehm. ex Nees;澳非洋狗尾草■☆
118314 Cynosurus coracanus L. = Eleusine coracana (L.) Gaertn. ■
118315 Cynosurus cristatus L.;洋狗尾草;Crested Dog's-tail, Crested Dogstail Grass, Crested Dog's-tail Grass, Crested Dogtail, Crested Dogtailgrass, Dog's-tail, Dogstail Grass, Wimble-straw ■
118316 Cynosurus cristatus L. subsp. polybracteatus (Poir.) Trab. = Cynosurus polybracteatus Poir. ■☆
118317 Cynosurus cristatus L. var. polybracteatus (Poir.) Coss. et Durieu = Cynosurus polybracteatus Poir. ■☆
118318 Cynosurus durus L. = Sclerochloa dura (L.) P. Beauv. ■
118319 Cynosurus echinatus L.;刺洋狗尾草;Bristly Dogstail Grass, Dogstail Grass, Hedgehog Dog's-tail, Hedgehog Dogtailgrass, Rough Dogstail, Rough Dog's-tail ■☆
118320 Cynosurus echinatus L. subsp. hystrix (Pomel) Trab. = Cynosurus echinatus L. ■☆
118321 Cynosurus echinatus L. var. hystrix (Pomel) Maire = Cynosurus echinatus L. ■☆
118322 Cynosurus echinatus L. var. tangerinus Pau = Cynosurus echinatus L. ■☆
118323 Cynosurus effusus Link;开展洋狗尾草■☆
118324 Cynosurus effusus Link var. gracilis (Moris) Kerguélen = Cynosurus effusus Link ■☆
118325 Cynosurus elegans Desf.;美丽洋狗尾草;Lovely Dogtailgrass ■☆
118326 Cynosurus elegans Desf. subsp. aurasiacus (Murb.) Maire = Cynosurus elegans Desf. ■☆
118327 Cynosurus elegans Desf. subsp. obliquatus (Link) Trab. = Cynosurus effusus Link ■☆
118328 Cynosurus elegans Desf. var. obliquatus (Link) Trab. = Cynosurus effusus Link ■☆
118329 Cynosurus falcatus (L. f.) Thunb. = Harpochloa falx (L. f.) Kuntze ■☆
118330 Cynosurus floccifolius Forssk. = Eleusine floccifolia (Forssk.) Spreng. ■☆
118331 Cynosurus hystrix Pomel = Cynosurus echinatus L. ■☆
118332 Cynosurus indicus L. = Eleusine indica (L.) Gaertn. ■
118333 Cynosurus junceus Murb.;灯心草状狗尾草■☆
118334 Cynosurus lagopoides Burm. f. = Coelachyrum lagopoides (Burm. f.) Senaratna ■☆
118335 Cynosurus lima L. = Wangenheimia lima (L.) Trin. ■☆
118336 Cynosurus monostachyos Vahl = Enteropogon monostachyus (Vahl) K. Schum ■☆
118337 Cynosurus obligatus Link = Cynosurus effusus Link ■☆
118338 Cynosurus paspaloides Vahl = Eustachys paspaloides (Vahl) Lanza et Mattei ■☆
118339 Cynosurus peltieri Maire;盾状洋狗尾草■☆
118340 Cynosurus phleoides Desf. = Rostraria phleoides (Desf.) Holub ■☆
118341 Cynosurus polybracteatus Poir.;多苞洋狗尾草■☆
118342 Cynosurus retroflexus Vahl = Dinebra retroflexa (Forssk. ex Vahl) Panz. ■
118343 Cynosurus retroflexus Vahl = Dinebra retroflexa (Vahl) Panz. ■☆
118344 Cynosurus tenerrimus Hornem. = Leptochloa chinensis (L.) Nees ■
118345 Cynosurus tenerrimus Hornem. = Leptochloa panicea (Retz.) Ohwi ■
118346 Cynosurus tristachyus Lam. = Eleusine tristachya (Lam.) Lam ■☆
118347 Cynosurus uniolae L. f. = Tribolium uniolae (L. f.) Renvoize ■☆
118348 Cynosurus virgata L. = Leptochloa virgata (L.) P. Beauv. ■☆
118349 Cynotis Hoffmanns. = Arctotheca J. C. Wendl. ■☆
118350 Cynotis Hoffmanns. = Cryptostemma R. Br. ex W. T. Aiton ■☆
118351 Cynotis Hoffmanns. = Odontoptera Cass. ●■☆
118352 Cynotoxicum Vell. (1829);毒犬藤属●☆
118353 Cynoxylon (Raf.) Small = Benthamidia Spach ●☆
118354 Cynoxylon (Raf.) Small = Cornus L. ●
118355 Cynoxylon Raf. = Cornus L. ●
118356 Cynoxylon capitata (Wall.) Nakai = Cornus capitata Wall. ex Roxb. ●
118357 Cynoxylon capitata (Wall.) Nakai = Dendrobenthamia capitata (Wall.) Hutch. ●
118358 Cynoxylon capitatum (Wall.) Nakai = Cornus capitata Wall. ●
118359 Cynoxylon ellipticum Pojark. = Cornus elliptica (Pojark.) Q. Y. Xiang et Boufford ●
118360 Cynoxylon ellipticum Pojark. = Dendrobenthamia angustata (Chun) W. P. Fang ●
118361 Cynoxylon ferrugineum (Y. C. Wu) Hara = Dendrobenthamia ferruginea (Y. C. Wu) W. P. Fang ●
118362 Cynoxylon ferrugineum (Y. C. Wu) Pojark. = Cornus hongkongensis Hemsl. ●
118363 Cynoxylon ferrugineum (Y. C. Wu) Pojark. = Cornus hongkongensis Hemsl. subsp. ferruginea (Y. C. Wu) Q. Y. Xiang ●
118364 Cynoxylon floridum (Dippel) Moldenke = Benthamidia florida (L.) Spach ●☆
118365 Cynoxylon floridum (Dippel) Moldenke = Cornus florida L. ●☆
118366 Cynoxylon glabriusculum Pojark. = Cornus capitata Wall. ●
118367 Cynoxylon hongkongense (Hemsl.) Nakai = Cornus hongkongensis Hemsl. ●
118368 Cynoxylon hongkongense (Hemsl.) Nakai = Dendrobenthamia hongkongensis (Hemsl.) Hutch. ●
118369 Cynoxylon japonica (Siebold et Zucc.) Nakai = Dendrobenthamia japonica (DC.) W. P. Fang ●
118370 Cynoxylon japonica Siebold et Zucc. = Benthamidia japonica (Siebold et Zucc.) H. Hara ●
118371 Cynoxylon kousa Nakai = Dendrobenthamia japonica (DC.) W. P. Fang ●
118372 Cynoxylon melanotrichum Pojark. = Cornus hongkongensis Hemsl. subsp. melanotricha (Pojark.) Q. Y. Xiang ●
118373 Cynoxylon melanotrichum Pojark. = Dendrobenthamia melanotricha (Pojark.) W. P. Fang ●
118374 Cynoxylon multinervosa Pojark. = Dendrobenthamia multinervosa (Pojark.) W. P. Fang ●
118375 Cynoxylon multinervosum Pojark. = Cornus multinervosa (Pojark.) Q. Y. Xiang ●
118376 Cynoxylon pseudokousa Pojark. = Cornus kousa Buerger ex Hance subsp. chinensis (Osborn) Q. Y. Xiang ●
118377 Cynoxylon pseudokousa Pojark. = Dendrobenthamia japonica (DC.) W. P. Fang var. chineneis (Osborn) W. P. Fang ●
118378 Cynoxylon sinense Nakai = Cornus kousa Buerger ex Hance subsp. chinensis (Osborn) Q. Y. Xiang ●

118379 Cynoxylon sinense Nakai = Dendrobenthamia japonica (DC.) W. P. Fang var. chineneis (Osborn) W. P. Fang ●
118380 Cynoxylon sinensis Pojark. = Dendrobenthamia angustata (Chun) W. P. Fang ●
118381 Cynoxylon yunnanense Pojark. = Cornus capitata Wall. ●
118382 Cynoxylon yunnanense Pojark. = Dendrobenthamia capitata (Wall.) Hutch. ●
118383 Cynoxylum Pluk. = Nyssa L. ●
118384 Cynoxylum florida Raf. = Cornus florida L. ●☆
118385 Cynthia D. Don = Krigia Schreb. (保留属名) ■☆
118386 Cynthia dandelion (L.) DC. = Krigia dandelion (L.) Nutt. ■☆
118387 Cynthia montana (Michx.) Standl. = Krigia montana (Michx.) Nutt. ■☆
118388 Cynthia virginica (L.) D. Don = Krigia biflora (Walter) S. F. Blake ■☆
118389 Cynthia virginica (L.) D. Don ex DC. = Krigia biflora (Walter) S. F. Blake ■☆
118390 Cynthia viridis Standl. = Krigia biflora (Walter) S. F. Blake ■☆
118391 Cynura d'Orb. = Gynura Cass. (保留属名) ■●
118392 Cyparissia Hoffmanns. = Callitris Vent. ●
118393 Cypella Herb. (1826); 杯鸢花属; Cypella ■☆
118394 Cypella Klatt = Marica Ker Gawl. ■☆
118395 Cypella drummondii Graham = Alophia drummondii (Graham) R. C. Foster ■☆
118396 Cypella drummondii Graham = Herbertia lahue (Molina) Goldblatt ■☆
118397 Cypella elegans Speg.; 杯鸢花 ■☆
118398 Cypella herbertii Hook.; 赫氏杯鸢花 ■☆
118399 Cypella plumbea Lindl.; 铅色杯鸢花 ■☆
118400 Cypellium Desv. = Styrax L. ●
118401 Cypeola maritima L. = Lobularia maritima (L.) Desv. ■
118402 Cypeola minor L. = Alyssum minus (L.) Rothm. ■
118403 Cyperaceae Juss. (1789) (保留科名); 莎草科; Sedge Family ■
118404 Cyperella Kramer = Luzula DC. (保留属名) ■
118405 Cyperella MacMill. = Luzula DC. (保留属名) ■
118406 Cyperochloa Lazarides et L. Watson (1987); 苔草禾属 ■☆
118407 Cyperochloa hirsuta Lazarides et L. Watson; 苔草禾 ■☆
118408 Cyperoides Ség. = Carex L. ■
118409 Cyperorchis Blume = Cymbidium Sw. ■
118410 Cyperorchis babae Kudo ex Masam. = Cymbidium cochleare Lindl. ■
118411 Cyperorchis cochlearis (Lindl.) Benth. = Cymbidium cochleare Lindl. ■
118412 Cyperorchis eburnea (Lindl.) Schltr. = Cymbidium eburneum Lindl. ■
118413 Cyperorchis elegans (Lindl.) Blume = Cymbidium elegans Lindl. ■
118414 Cyperorchis gigantea (Wall. ex Lindl.) Schltr. = Cymbidium iridioides D. Don ■
118415 Cyperorchis gigantea Schltr. = Cymbidium iridioides D. Don ■
118416 Cyperorchis grandiflora (Griff.) Schltr. = Cymbidium hookerianum Rchb. f. ■
118417 Cyperorchis insignis (Rolfe) Schltr. = Cymbidium insigne Rolfe ■
118418 Cyperorchis longifolia (D. Don) Schltr. = Cymbidium erythraeum Lindl. ■
118419 Cyperorchis lowiana (Rchb. f.) Schltr. = Cymbidium lowianum (Rchb. f.) Rchb. f. ■
118420 Cyperorchis mastersii (Griff. ex Lindl.) Benth. = Cymbidium mastersii Griff. ■
118421 Cyperorchis mastersii (Griff.) Benth. = Cymbidium mastersii Griff. ex Lindl. ■
118422 Cyperorchis schroederi (Rolfe) Schltr. = Cymbidium schroederi Rolfe ■
118423 Cyperorchis tigrina (E. C. Parish ex Hook.) Schltr. = Cymbidium tigrinum E. C. Parish ex Hook. ■
118424 Cyperorchis tracyana (L. Castle) Schltr. = Cymbidium tracyanum L. Castle ■
118425 Cyperorchis wallichii Blume = Cymbidium cyperifolium Wall. ex Lindl. ■
118426 Cyperorchis wilsonii (Rolfe ex Cook) Schltr. = Cymbidium wilsonii (Rolfe ex Cook) Rolfe ■
118427 Cyperus (Griseb.) C. B. Clarke = Cyperus L. ■
118428 Cyperus L. (1753); 莎草属; Cypress Grass, Cypressgrass, Cypress-grass, Flat Sedge, Flatsedge, Flat-sedge, Galingale, Sedge, Umbrella Sedge, Umbrella-sedge ■
118429 Cyperus × amuricocompressus T. Koyama = Cyperus amuricus Maxim. ■
118430 Cyperus × amuricoides T. Koyama; 拟阿穆尔莎草 ■☆
118431 Cyperus × babae T. Koyama; 马场莎草 ■☆
118432 Cyperus × borealikiiensis T. Koyama; 北纪伊莎草 ■☆
118433 Cyperus × condensatus T. Koyama; 密集莎草 ■☆
118434 Cyperus × kadzusensis T. Koyama; 上总莎草 ■☆
118435 Cyperus × mesochorus Geise; 杂种草原莎草; Midland Sand Sedge ■☆
118436 Cyperus × mihasii T. Koyama; 三阶莎草 ■☆
118437 Cyperus × ogawae T. Koyama; 小川莎草 ■☆
118438 Cyperus × yamamotoi T. Koyama; 山本莎草 ■☆
118439 Cyperus acicularis (Nees) Steud. = Cyperus odoratus L. ■
118440 Cyperus acicularis (Nees) Steud. = Torulinium odoratum (L.) S. S. Hooper ■
118441 Cyperus actinostachys Ridl. = Cyperus tenax Boeck. ■☆
118442 Cyperus acuminatus Roxb. ex C. B. Clarke; 线叶莎草; Tapeleaf Galingale, Umbrella Sedge ■
118443 Cyperus acuminatus Torr. et Hook. = Cyperus acuminatus Roxb. ex C. B. Clarke ■
118444 Cyperus acuminatus Torr. et Hook. var. cyrtolepis (Torr. et Hook.) Kük. = Cyperus acuminatus Roxb. ex C. B. Clarke ■
118445 Cyperus acuticarinatus Kük. = Pycreus acuticarinatus (Kük.) Cherm. ■☆
118446 Cyperus adansonii C. B. Clarke = Cyperus jeminicus Rottb. ■☆
118447 Cyperus adenophorus Schrad. et Nees; 巴西莎草; Brazillian Galingale ■☆
118448 Cyperus adoensis Hochst. ex A. Rich. = Cyperus rigidifolius Steud. ■☆
118449 Cyperus aegyptiacus Gloxin = Cyperus capitatus Vand. ■☆
118450 Cyperus aequalis Vahl = Cyperus prolifer Lam. ■☆
118451 Cyperus aequalis Vahl var. subaequalis (Baker) Cherm. = Cyperus subaequalis Baker ■☆
118452 Cyperus afroalpinus Lye; 非洲亚高山莎草 ■☆
118453 Cyperus afrodunensis Lye; 非洲砂丘莎草 ■☆
118454 Cyperus afroechinatus Lye = Kyllinga echinata S. S. Hooper ■☆
118455 Cyperus afromontanus Lye; 非洲山地莎草 ■☆
118456 Cyperus afropumilus (Lye) ye = Kyllinga afropumila Lye ■☆
118457 Cyperus afzelii Boeck. = Pycreus capillifolius (A. Rich.) C. B.

Clarke ■☆

118458 Cyperus aggregatus (Willd.) Endl.;集生莎草■☆

118459 Cyperus alatus (Nees) F. Muell. = Kyllinga alata Nees ■☆

118460 Cyperus alba-purpureus (Lye) Lye = Kyllinga alba-purpurea Lye ■☆

118461 Cyperus albiceps Ridl. = Kyllinga albiceps (Ridl.) Rendle ■☆

118462 Cyperus albiflorus Cherm. = Cyperus phaeolepis Cherm. ■☆

118463 Cyperus albogracilis (Lye) Lye = Kyllinga albogracilis Lye ■☆

118464 Cyperus albomarginatus (Mart. et Schrad. ex Nees) Steud. = Cyperus flavicomus Michx. ■☆

118465 Cyperus albomarginatus (Nees) Steud. var. sabulosus (Mart. et Schrad. ex Nees) Kük. = Cyperus flavicomus Michx. ■☆

118466 Cyperus albomarginatus (Nees) Steud. var. tenuis (Boeck.) Kük. = Pycreus macrostachyos (Lam.) J. Raynal var. tenuis (Boeck.) Wickens ■☆

118467 Cyperus albopilosus (C. B. Clarke) Kük. = Mariscus albopilosus C. B. Clarke ■☆

118468 Cyperus albopurpureus Cherm.;浅紫莎草■☆

118469 Cyperus albo-sanguineus Kük. = Mariscus alba-sanguineus (Kük.) Napper ■☆

118470 Cyperus albostriatus Schrad.;白脉莎草■☆

118471 Cyperus albostriatus Schrad. 'Variegatus';白纹白脉莎草■☆

118472 Cyperus algeriensis Väre et Kukkonen = Cyperus conglomeratus Rottb. ■☆

118473 Cyperus alopecuroides Rottb.;狐尾莎草;Foxtail Flatsedge ■☆

118474 Cyperus alpestris K. Schum. = Mariscus alpestris (K. Schum.) C. B. Clarke ■☆

118475 Cyperus alterniflorus Schwein. = Cyperus schweinitzii Torr. ■☆

118476 Cyperus alternifolius L.;伞莎草(风车草,异叶莎草);Sedge, Umbrella Flat Sedge, Umbrella Flatsedge, Umbrella Flat-sedge, Umbrella Galingale, Umbrella Palm, Umbrella Plant, Umbrella Sedge, Umbrella-grass, Umbrella-plant ■

118477 Cyperus alternifolius L. subsp. flabelliformis (Rottb.) Kük.;鞭伞莎草(风车草,旱伞草,九龙吐珠,伞莎草);Fanshaped Umbrellasedge, Umbrella Flat Sedge, Umbrella-plant, Windmill Cypressgrass ■

118478 Cyperus alternifolius L. subsp. flabelliformis (Rottb.) Kük. 'Variegatus';斑叶风车草(风车草)■

118479 Cyperus alternifolius L. subsp. flabelliformis (Rottb.) Kük. = Cyperus involucratus Rottb. ■

118480 Cyperus alternifolius L. var. flabelliformis (Rottb.) T. Koyama = Cyperus alternifolius L. subsp. flabelliformis (Rottb.) Kük. ■

118481 Cyperus alternifolius L. var. gracilis Hort.;细弱莎草;Slender Umbrella Galingale ■

118482 Cyperus alternifolius L. var. macrostachys Robyns et Tournay = Cyperus involucratus Rottb. ■

118483 Cyperus alternifolius L. var. obtusangulus (Boeck.) T. Koyama = Cyperus alternifolius L. subsp. flabelliformis (Rottb.) Kük. ■

118484 Cyperus amabilis Vahl;秀丽莎草■☆

118485 Cyperus amabilis Vahl var. capitatus Chiov.;头状秀丽莎草■☆

118486 Cyperus amabilis Vahl var. macrostachyus (Boeck.) Kük. = Cyperus amabilis Vahl ■☆

118487 Cyperus amabilis Vahl var. subacaulis Kük. = Cyperus amabilis Vahl ■☆

118488 Cyperus amauropus Steud. = Mariscus amauropus (Steud.) Cufod. ■☆

118489 Cyperus ambongensis Boeck. = Cyperus obtusiflorus Vahl ■☆

118490 Cyperus amplissimus Steud. = Cyperus prolixus Kunth ■☆

118491 Cyperus amuricus Maxim.;阿穆尔莎草(灯台草,黑龙江莎草,三棱草,三楞草);Amur Cypressgrass, Amur Galingale, Asian Flatsedge ■

118492 Cyperus amuricus Maxim. var. japonicus Kük. = Cyperus microiria Steud. ■

118493 Cyperus amuricus Maxim. var. japonicus Miq.;日本莎草■☆

118494 Cyperus amuricus Maxim. var. teztori (Miq.) Kük. = Cyperus microiria Steud. ■

118495 Cyperus anceps Liebm. = Cyperus thyrsiflorus Jungh. ■☆

118496 Cyperus andongensis Ridl. = Cyperus tenax Boeck. ■☆

118497 Cyperus angolensis Boeck.;安哥拉莎草■☆

118498 Cyperus angulatus Nees = Pycreus unioloides (R. Br.) Urb. ■

118499 Cyperus angustifolius Nees = Cyperus cuspidatus Kunth ■

118500 Cyperus angustifolius Schumach. et Thonn.;窄叶莎草■☆

118501 Cyperus ankaizinensis Cherm.;安凯济纳莎草■☆

118502 Cyperus ankaratrensis Cherm.;安卡拉特拉莎草■☆

118503 Cyperus apricus Ridl.;向阳莎草■☆

118504 Cyperus arenarius Retz.;沙地莎草■☆

118505 Cyperus arenicola Steud. = Cyperus reflexus Vahl ■☆

118506 Cyperus argenteus Ridl. = Cyperus niveus Retz. var. leucocephalus (Kunth) Fosberg ■☆

118507 Cyperus argyraeus Steud. = Pycreus macrostachyos (Lam.) J. Raynal ■☆

118508 Cyperus aristatus Rottb. = Cyperus squarrosus L. ■☆

118509 Cyperus aristatus Rottb. = Mariscus aristatus (Rottb.) Ts. Tang et F. T. Wang ■

118510 Cyperus aristatus Rottb. subsp. hamulosus (M. Bieb.) Asch. et Graebn. = Mariscus hamulosus (M. Bieb.) S. S. Hooper ■☆

118511 Cyperus aristatus Rottb. var. floribundus E. G. Camus = Mariscus aristatus (Rottb.) Ts. Tang et F. T. Wang ■

118512 Cyperus aristatus Rottb. var. hamulosus (M. Bieb.) Boeck. = Mariscus hamulosus (M. Bieb.) S. S. Hooper ■☆

118513 Cyperus aristatus Rottb. var. inflexus (Muhl.) Kük. = Mariscus aristatus (Rottb.) Ts. Tang et F. T. Wang ■

118514 Cyperus aristatus Rottb. var. pitardii (Trab. ex Pit.) Maire = Mariscus hamulosus (M. Bieb.) S. S. Hooper ■☆

118515 Cyperus aristatus Rottb. var. runyonii O'Neill = Mariscus aristatus (Rottb.) Ts. Tang et F. T. Wang ■

118516 Cyperus aristatus Rottb. var. semiglobosus Kük. = Mariscus squarrosus (L.) C. B. Clarke ■☆

118517 Cyperus aromaticus (Ridl.) Mattf. et Kük. = Kyllinga erecta Schumach. var. polyphylla (Willd. ex Kunth) S. S. Hooper ■☆

118518 Cyperus aromaticus (Ridl.) Mattf. et Kük. var. elatior (Kunth) Kük. = Kyllinga elatior Kunth ■☆

118519 Cyperus aromaticus (Ridl.) Mattf. et Kük. var. elatus (Steud.) Kük. = Kyllinga melanosperma Nees var. elata (Steud.) J.-P. Lebrun et Stork ■☆

118520 Cyperus articulatus L.;节莎草;Adrue, Jointed Flatsedge ■☆

118521 Cyperus articulatus L. var. multiflorus Kük. = Cyperus articulatus L. ■☆

118522 Cyperus articulatus L. var. nodosus (Willd.) Kük. = Cyperus articulatus L. ■☆

118523 Cyperus ascolepidioides (Cherm.) Kük. = Kyllinga alba Nees subsp. ascolepidioides (Cherm.) Lye ■☆

118524 Cyperus asper (Liebm.) O'Neill = Cyperus mutisii (Kunth) Andersson ■☆

118525 Cyperus asperrimus Liebm. = Cyperus manimae Kunth var. asperrimus (Liebm.) Kük. ■☆

118526 Cyperus asperrimus Liebm. var. multiflorus Liebm. = Cyperus manimae Kunth var. asperrimus (Liebm.) Kük. ■☆

118527 Cyperus assimilis Steud. = Courtoisina assimilis (Steud.) Maquet ■☆

118528 Cyperus assimilis Steud. var. depressa ? = Courtoisina assimilis (Steud.) Maquet ■☆

118529 Cyperus atribulbus Kük. = Pycreus atribulbus (Kük.) Napper ■☆

118530 Cyperus atronervatus Boeck. = Pycreus atronervatus (Boeck.) C. B. Clarke ■☆

118531 Cyperus atronitens Hochst. = Pycreus niger (Ruiz et Pav.) Cufod. subsp. elegantulus (Steud.) Lye ■☆

118532 Cyperus atrorubidus (Nelmes) Raymond = Pycreus atrorubidus Nelmes ■☆

118533 Cyperus atrosanguineus Hochst. ex Steud. = Cyperus rigidifolius Steud. ■☆

118534 Cyperus atroviridis C. B. Clarke;暗绿莎草■☆

118535 Cyperus aucheri Jaub. et Spach;奥切尔莎草■☆

118536 Cyperus aurantiacus Kunth = Cyperus amabilis Vahl ■☆

118537 Cyperus aureoalatus Lye;黄翅莎草■☆

118538 Cyperus aureo-brunneus C. B. Clarke = Cyperus denudatus L. f. var. aureo-brunneus (C. B. Clarke) Kük. ■☆

118539 Cyperus aureo-stramineus Mattf. et Kük. = Kyllinga chrysantha K. Schum. ■☆

118540 Cyperus aureovillosus (Lye) Lye = Kyllinga aureovillosa Lye ■☆

118541 Cyperus aureus Kunth = Cyperus amabilis Vahl ■☆

118542 Cyperus aureus Kunth var. aurantiacus (Kunth) Boeck. = Cyperus amabilis Vahl ■☆

118543 Cyperus aureus Kunth var. macrostachyus Boeck. = Cyperus amabilis Vahl ■☆

118544 Cyperus aureus Kunth var. oligostachyus (Kunth) Boeck. = Cyperus amabilis Vahl ■☆

118545 Cyperus aureus Ten. = Cyperus esculentus L. ■☆

118546 Cyperus auricomus Sieber ex Spreng. = Cyperus digitatus Roxb. var. auricomus (Sieber ex Spreng.) Kük. ■☆

118547 Cyperus autumnalis Vahl = Cyperus haspan L. ■

118548 Cyperus babakan Steud.;刺鳞莎草■

118549 Cyperus babakensis Steud. ex Miq. = Cyperus babakan Steud. ■

118550 Cyperus badius Desf. = Cyperus longus L. subsp. badius (Desf.) Asch. et Graebn. ■☆

118551 Cyperus balbisii Kunth = Cyperus sphacelatus Rottb. ■☆

118552 Cyperus baldwinii Torr. = Cyperus croceus Vahl ■☆

118553 Cyperus bancanus Miq. = Mariscus trialatus (Boeck.) Ts. Tang et F. T. Wang ■

118554 Cyperus baoulensis Kük.;巴乌莱莎草■☆

118555 Cyperus baronii C. B. Clarke;巴龙莎草■☆

118556 Cyperus baronii C. B. Clarke var. densus C. B. Clarke ex Cherm. = Cyperus baronii C. B. Clarke ■☆

118557 Cyperus baronii C. B. Clarke var. mannii (C. B. Clarke) Kük. = Cyperus mannii C. B. Clarke ■☆

118558 Cyperus bellus Kunth;雅致莎草■☆

118559 Cyperus benadirensis Chiov. = Cyperus poecilus C. B. Clarke ■☆

118560 Cyperus bequaertii (Cherm.) Robyns et Tournay = Mariscus ferrugineoviridis (C. B. Clarke) Cherm. ■☆

118561 Cyperus betafensis Cherm.;贝塔夫莎草■☆

118562 Cyperus bidentatus Vahl = Mariscus ligularis (L.) Urb. ■☆

118563 Cyperus bifax C. B. Clarke;二裂莎草■☆

118564 Cyperus bifolius Lye;二叶莎草■☆

118565 Cyperus bipartitus Torr.;纤细莎草;Nut Grass, Shining Flat Sedge, Slender Flat Sedge ■☆

118566 Cyperus blandus Kunth = Cyperus marginatus Thunb. ■☆

118567 Cyperus blysmoides C. B. Clarke = Cyperus bulbosus Vahl var. spicatus Boeck. ■☆

118568 Cyperus blysmoides Hochst. = Cyperus bulbosus Vahl var. spicatus Boeck. ■☆

118569 Cyperus boehmii Boeck. = Cyperus tenax Boeck. var. pseudocastaneus (Kük.) Kük. ■☆

118570 Cyperus boreobellus Lye;北美莎草■☆

118571 Cyperus brevibracteatus (Domin) Domin;短苞叶莎草■☆

118572 Cyperus brevifolioides Delahoussaye et Thieret = Kyllinga gracillima Miq. ■☆

118573 Cyperus brevifolius (Rottb.) Endl. ex Hassk. = Kyllinga brevifolia Rottb. ■

118574 Cyperus brevifolius (Rottb.) Endl. ex Hassk. var. stellulatus J. V. Suringar = Kyllinga brevifolia Rottb. var. stellulata (J. V. Suringar) Ts. Tang et F. T. Wang ■

118575 Cyperus brevifolius (Rottb.) Hassk. = Kyllinga brevifolia Rottb. ■

118576 Cyperus brevifolius (Rottb.) Hassk. f. pumilus Suringar = Kyllinga brevifolia Rottb. var. pumila (J. V. Suringar) Ts. Tang et F. T. Wang ■

118577 Cyperus brevifolius (Rottb.) Hassk. subsp. intricatus (Cherm.) Lye = Kyllinga brevifolia Rottb. subsp. intricata (Cherm.) J.-P. Lebrun et Stork ■☆

118578 Cyperus brevifolius (Rottb.) Hassk. subsp. leiolepis (Franch. et Sav.) T. Koyama = Cyperus brevifolius (Rottb.) Hassk. var. leiolepis (Franch. et Sav.) T. Koyama ■

118579 Cyperus brevifolius (Rottb.) Hassk. subsp. luridus (Kük.) Lye = Kyllinga brevifolia Rottb. subsp. lurida (Kük.) Lye ■☆

118580 Cyperus brevifolius (Rottb.) Hassk. Vahl = Mariscus javanicus (Houtt.) Merr. et F. P. Metcalf ■

118581 Cyperus brevifolius (Rottb.) Hassk. var. leiolepis (Franch. et Sav.) T. Koyama f. macrolepis H. Hara ex T. Koyama = Kyllinga brevifolia Rottb. subsp. leiolepis (Franch. et Sav.) T. Koyama f. macrolepis T. Koyama ■☆

118582 Cyperus brevifolius (Rottb.) Hassk. var. leiolepis (Franch. et Savigny) H. Hara = Kyllinga gracillima Miq. ■☆

118583 Cyperus brevifolius (Rottb.) Hassk. var. stellulatus Suringar = Kyllinga brevifolia Rottb. var. stellulata (J. V. Suringar) Ts. Tang et F. T. Wang ■

118584 Cyperus brevifolius Hassk. var. gracillimus (Miq.) Kük. = Kyllinga brevifolia Rottb. var. leiolepis (Franch. et Sav.) H. Hara ■

118585 Cyperus brevis Boeck.;短莎草■☆

118586 Cyperus brizaeus J. Presl et C. Presl = Cyperus polystachyus (Rottb.) P. Beauv. ■

118587 Cyperus brizaeus Vahl = Cyperus planifolius Rich. ■☆

118588 Cyperus brunneofibrosus Lye;褐纤维莎草■☆

118589 Cyperus brunneo-vaginatus Boeck. = Cyperus marginatus Thunb. ■☆

118590 Cyperus brunneus Sw. = Cyperus planifolius Rich. ■☆

118591 Cyperus buchananii Boeck. = Cyperus esculentus L. ■☆

118592 Cyperus buchholzii Boeck. = Cyperus laxus Lam. subsp. buchholzii (Boeck.) Lye ■☆

118593 Cyperus buckleyi Britton ex J. M. Coult. = Cyperus spectabilis Spreng. ■☆
118594 Cyperus buettneri Boeck.;比特纳莎草■☆
118595 Cyperus bulbiferus A. Dietr. = Cyperus bulbosus Vahl ■☆
118596 Cyperus bulbipes Mattf. et Kük. = Kyllinga crassipes Boeck. ■☆
118597 Cyperus bulbocaulis (Hochst. ex A. Rich.) Boeck. = Mariscus bulbocaulis Hochst. ex A. Rich. ■☆
118598 Cyperus bulbocaulis (Hochst. ex A. Rich.) Boeck. var. atrosanguineus (Hochst. ex A. Rich.) Kük. = Mariscus bulbocaulis Hochst. ex A. Rich. var. atrosanguineus (Hochst. ex A. Rich.) C. B. Clarke ■☆
118599 Cyperus bulbosus Vahl;鳞茎莎草■☆
118600 Cyperus bulbosus Vahl subsp. giolii (Chiov.) Kük. = Cyperus giolii Chiov. ■☆
118601 Cyperus bulbosus Vahl var. flavus Chiov.;黄莎草■☆
118602 Cyperus bulbosus Vahl var. grandibulbosus (C. B. Clarke) Chiov. = Cyperus grandibulbosus C. B. Clarke ■☆
118603 Cyperus bulbosus Vahl var. longebracteatus A. Terracc.;长苞黄莎草■☆
118604 Cyperus bulbosus Vahl var. melanolepis Kük.;黑鳞黄莎草■☆
118605 Cyperus bulbosus Vahl var. spicatus Boeck.;穗状黄莎草■☆
118606 Cyperus bushii Britton = Cyperus lupulinus (Spreng.) Marcks ■☆
118607 Cyperus californicus S. Watson = Cyperus odoratus L. ■
118608 Cyperus californicus S. Watson = Torulinium odoratum (L.) S. S. Hooper ■
118609 Cyperus cancellatus Ridl. = Cyperus haspan L. ■
118610 Cyperus capensis (Steud.) Endl.;好望角莎草■☆
118611 Cyperus capillifolius A. Rich. = Pycreus capillifolius (A. Rich.) C. B. Clarke ■☆
118612 Cyperus capitatus Vand.;头状莎草■☆
118613 Cyperus caricinus D. Don = Carex filicina Nees ex Wight ■
118614 Cyperus cartilagineus (K. Schum.) Mattf. et Kük. var. angustatus Peter et Kük. = Kyllinga chrysantha K. Schum. var. comosipes (Mattf. et Kük.) J. -P. Lebrun et Stork ■☆
118615 Cyperus castaneobellus Lye;栗莎草■☆
118616 Cyperus cataractarum (C. B. Clarke) Kük. = Pycreus cataractarum C. B. Clarke ■☆
118617 Cyperus cayennensis (Lam.) Britton = Cyperus aggregatus (Willd.) Endl. ■☆
118618 Cyperus cephalanthus Torr. et Hook.;头花莎草■☆
118619 Cyperus chaetophyllus (Chiov.) Kük. = Mariscus chaetophyllus Chiov ■☆
118620 Cyperus chamaecephalus Cherm.;矮头莎草■☆
118621 Cyperus chermezonianus Robyns et Tournay = Mariscus foliosus C. B. Clarke ■☆
118622 Cyperus chevalieri Kük. = Mariscus stolonifer C. B. Clarke ■☆
118623 Cyperus chlorostachys Boeck. = Pycreus macrostachyos (Lam.) J. Raynal ■☆
118624 Cyperus chlorotropis (Steud.) Mattf. et Kük. = Kyllinga chlorotropis Steud. ■☆
118625 Cyperus chordorrhizus Chiov.;索根莎草■☆
118626 Cyperus chrysanthus Boeck. = Pycreus chrysanthus (Boeck.) C. B. Clarke ■☆
118627 Cyperus chrysocephalus (K. Schum.) Kük. = Mariscus chrysocephalus K. Schum. ■☆
118628 Cyperus ciliato-pilosus Mattf. et Kük. = Kyllinga platyphylla K. Schum. ex C. B. Clarke ■☆
118629 Cyperus clandestinus Steud. = Ficinia clandestina (Steud.) Boeck. ■☆
118630 Cyperus clavinux C. B. Clarke;棒莎草■☆
118631 Cyperus cognatus Kunth = Cyperus rupestris Kunth ■☆
118632 Cyperus coloratus Vahl = Mariscus dubius (Rottb.) Kük. ex C. E. C. Fisch. subsp. coloratus (Vahl) Lye ■☆
118633 Cyperus coloratus Vahl var. longinux Kük. = Mariscus alba-sanguineus (Kük.) Napper ■☆
118634 Cyperus commixtus Kük.;混合莎草■☆
118635 Cyperus comosipes Mattf. et Kük. = Kyllinga chrysantha K. Schum. var. comosipes (Mattf. et Kük.) J. -P. Lebrun et Stork ■☆
118636 Cyperus compactus Lam. = Cyperus niveus Retz. var. leucocephalus (Kunth) Fosberg ■☆
118637 Cyperus compactus Lam. = Cyperus obtusiflorus Vahl ■☆
118638 Cyperus compactus Lam. var. macrostachys Graebn. = Cyperus obtusiflorus Vahl var. macrostachys (Graebn.) Robyns et Tournay ■☆
118639 Cyperus compactus Lam. var. tenerior C. B. Clarke = Cyperus niveus Retz. var. tisserantii (Cherm.) Lye ■☆
118640 Cyperus compactus Retz. = Mariscus compactus (Retz.) Druce ■
118641 Cyperus compactus Retz. var. macrostachys (Boeck.) Kük. = Mariscus compactus (Retz.) Druce var. macrostachys (Boeck.) F. C. How ex Ts. Tang et F. T. Wang ■
118642 Cyperus compressus L.;扁穗莎草(木虱草,沙田草,天打锤);Coco Grass,Compress Cypressgrass,Compress Galingale,Compressed Flat-sedge ■
118643 Cyperus compressus L. var. pectiniformis (Roem. et Schult.) C. B. Clarke = Cyperus compressus L. ■
118644 Cyperus compressus L. var. triqueter Boeck. = Cyperus mutisii (Kunth) Andersson ■☆
118645 Cyperus congensis C. B. Clarke;康格莎草■☆
118646 Cyperus congestus Vahl;团集莎草;Clustered Flatsedge ■☆
118647 Cyperus congestus Vahl var. brevis (Boeck.) Kük. = Cyperus brevis Boeck. ■☆
118648 Cyperus congestus Vahl var. glanduliferus (C. B. Clarke) Kük. = Cyperus congestus Vahl ■☆
118649 Cyperus congestus Vahl var. grandiceps Kük. = Cyperus congestus Vahl ■☆
118650 Cyperus congestus Vahl var. parishii (Britton ex Parish) Kük. = Cyperus parishii Britton ex Parish ■☆
118651 Cyperus congestus Vahl var. pseudonatalensis Kük. = Cyperus congestus Vahl ■☆
118652 Cyperus conglomeratus Rottb.;聚集莎草■☆
118653 Cyperus conglomeratus Rottb. f. strictus Kük. = Cyperus algeriensis Väre et Kukkonen ■☆
118654 Cyperus conglomeratus Rottb. var. arenarius (Retz.) Coss. = Cyperus arenarius Retz. ■☆
118655 Cyperus conglomeratus Rottb. var. aucheri (Jaub. et Spach) C. B. Clarke = Cyperus aucheri Jaub. et Spach ■☆
118656 Cyperus conglomeratus Rottb. var. effusus (Rottb.) Coss. et Durieu;开展莎草■☆
118657 Cyperus conglomeratus Rottb. var. ensifolius (Nees) Kük.;密叶聚集莎草■☆
118658 Cyperus conglomeratus Rottb. var. minor Boeck.;较小聚集莎草■☆
118659 Cyperus conglomeratus Rottb. var. multiculmis (Boeck.) Kük. = Cyperus jeminicus Rottb. ■☆

118660 Cyperus controversus (Steud.) Mattf. et Kük. var. subexalatus (C. B. Clarke) Kük. = Kyllinga welwitschii Ridl. ■☆

118661 Cyperus cooperi (C. B. Clarke) K. Schum. = Cyperus congestus Vahl ■☆

118662 Cyperus corymbosus Rottb.;伞序莎草■☆

118663 Cyperus corymbosus Rottb. var. subnodosus (Nees et Meyen) Kük. = Cyperus articulatus L. ■☆

118664 Cyperus costatus Mattf. et Kük. = Kyllinga nervosa Steud. ■☆

118665 Cyperus crassipes Vahl;粗梗莎草■☆

118666 Cyperus crassivaginatus Lye;厚鞘莎草■☆

118667 Cyperus cremeomariscus Lye;悬垂莎草■☆

118668 Cyperus cristatus (Kunth) Mattf. et Kük.;冠状莎草■☆

118669 Cyperus cristatus (Kunth) Mattf. et Kük. = Kyllinga alba Nees ■☆

118670 Cyperus cristatus (Kunth) Mattf. et Kük. subsp. ascolepidioides (Cherm.) Kük. = Kyllinga alba Nees subsp. ascolepidioides (Cherm.) Lye ■☆

118671 Cyperus cristatus (Kunth) Mattf. et Kük. var. exaltatus Merxm. = Kyllinga alba Nees ■☆

118672 Cyperus cristatus (Kunth) Mattf. et Kük. var. nigritanus (C. B. Clarke) Kük. = Kyllinga alba Nees subsp. nigritana (C. B. Clarke) J. -P. Lebrun et Stork ■☆

118673 Cyperus croceus Vahl;鲍尔温莎草;Baldwin's Sedge ■☆

118674 Cyperus cruentus Retz. = Pycreus sanguinolentus (Vahl) Nees ■

118675 Cyperus crustaceus Raymond = Pycreus acuticarinatus (Kük.) Cherm. ■☆

118676 Cyperus cuanzensis Ridl. = Pycreus smithianus (Ridl.) C. B. Clarke ■☆

118677 Cyperus cupreus J. Presl et C. Presl = Cyperus erythrorhizos Muhl. ■☆

118678 Cyperus cuspidatus Baker = Cyperus betafensis Cherm. ■☆

118679 Cyperus cuspidatus Kunth;长尖莎草(长叶莎草,碎米香附);Cuspidate Cypressgrass,Cuspidate Galingale ■

118680 Cyperus cuspidatus Kunth f. angustifolius (Nees) Kük. = Cyperus cuspidatus Kunth ■

118681 Cyperus cyclostachyus Griseb. = Cyperus croceus Vahl ■☆

118682 Cyperus cylindricus (Elliott) Britton = Cyperus retrorsus Chapm. ■☆

118683 Cyperus cylindricus Chapm. = Cyperus ovatus Baldwin ■☆

118684 Cyperus cyperinus (Retz.) Suringar = Mariscus cyperinus (Retz.) Vahl ■

118685 Cyperus cyperinus (Retz.) Suringar var. bengalensis (C. B. Clarke) Kük. = Mariscus cyperinus Vahl var. bengalensis C. B. Clarke ■

118686 Cyperus cyperinus (Vahl) Suringar = Mariscus cyperinus Vahl ■

118687 Cyperus cyperinus Ohwi = Mariscus umbellatus Vahl var. microstachys (Kük.) E. G. Camus ■

118688 Cyperus cyperoides (L.) Kuntze;普通莎草■☆

118689 Cyperus cyperoides (L.) Kuntze = Mariscus sumatrensis (Retz.) J. Raynal ■

118690 Cyperus cyperoides (L.) Kuntze = Mariscus umbellatus Vahl ■

118691 Cyperus cyperoides (L.) Kuntze f. longipedunculatus Kük. = Mariscus steudelianus (Boeck.) Cufod. ■☆

118692 Cyperus cyperoides (L.) Kuntze subsp. cyperinus (Retz.) Kük. = Mariscus cyperinus (Retz.) Vahl ■

118693 Cyperus cyperoides (L.) Kuntze subsp. cyperinus Kük. = Mariscus cyperinus Vahl ■

118694 Cyperus cyperoides (L.) Kuntze subsp. flavus Lye;黄普通莎草■☆

118695 Cyperus cyperoides (L.) Kuntze subsp. macrocarpus (Kunth) Lye = Mariscus macrocarpus Kunth ■☆

118696 Cyperus cyperoides (L.) Kuntze subsp. pseudoflavus (Kük.) Lye;拟黄普通莎草■☆

118697 Cyperus cyperoides (L.) Kuntze var. aureus Peter ex Kük. = Mariscus sumatrensis (Retz.) J. Raynal ■

118698 Cyperus cyperoides (L.) Kuntze var. microstachys Kük. = Mariscus sumatrensis (Retz.) J. Raynal var. microstachys (Kük.) L. K. Dai ■

118699 Cyperus cyperoides (L.) Kuntze var. microstachys Kük. = Mariscus umbellatus Vahl var. microstachys (Kük.) E. G. Camus ■

118700 Cyperus cyperoides (L.) Kuntze var. nossibeensis (Steud.) Kük. = Mariscus sumatrensis (Retz.) J. Raynal ■

118701 Cyperus cyperoides (L.) Kuntze var. polyphyllus (Steud.) Kük. = Mariscus sumatrensis (Retz.) J. Raynal ■

118702 Cyperus cyperoides (L.) Kuntze var. subcompositus (C. B. Clarke) Kük. = Mariscus sumatrensis (Retz.) J. Raynal var. subcompositus (C. B. Clarke) Karthik. ■

118703 Cyperus cyplindrostachys Boeck. = Mariscus umbellatus Vahl ■

118704 Cyperus cyrtolepis Torr. et Hook. = Cyperus acuminatus Roxb. ex C. B. Clarke ■

118705 Cyperus cyrtolepis Torr. et Hook. var. denticarinatus (Britton) Britton = Cyperus acuminatus Roxb. ex C. B. Clarke ■

118706 Cyperus dactyliformis Boeck. = Mariscus gueinzii C. B. Clarke ■☆

118707 Cyperus debilissimus Baker;瘦小莎草■☆

118708 Cyperus debilissimus Baker var. calochrous (Cherm.) Cherm. = Cyperus debilissimus Baker ■☆

118709 Cyperus debilissimus Baker var. triqueter Cherm. = Cyperus debilissimus Baker ■☆

118710 Cyperus deciduus Boeck.;脱落莎草■☆

118711 Cyperus decurvatus (C. B. Clarke) C. Archer et Goetgh. = Cyperus indecorus Kunth var. decurvatus (C. B. Clarke) Kük. ■☆

118712 Cyperus deeringianus Britton et Small = Cyperus ovatus Baldwin ■☆

118713 Cyperus delavayi (C. B. Clarke) Kük. = Pycreus delavayi C. B. Clarke ■

118714 Cyperus delavayi Kük. = Pycreus delavayi C. B. Clarke ■

118715 Cyperus dense-spicatus Hayata = Cyperus imbricatus Retz. var. dense-spicatus (Hayata) Ohwi ■

118716 Cyperus dense-spicatus Hayata = Cyperus imbricatus Retz. var. elongatus (Boeck.) T. Koyama ■

118717 Cyperus densibulbosus Lye;密鳞茎莎草■☆

118718 Cyperus densicaespitosus Mattf. et Kük. = Kyllinga pumila Michx. ■☆

118719 Cyperus densicaespitosus Mattf. et Kük. var. major (Nees) Kük. = Kyllinga robusta Boeck. ■☆

118720 Cyperus densicaespitosus Mattf. et Kük. var. rigidulus (Steud.) Kük. = Kyllinga pumila Michx. ■☆

118721 Cyperus densispicatus Cherm. = Cyperus grandis C. B. Clarke ■☆

118722 Cyperus densus Link = Pycreus lanceolatus (Poir.) C. B. Clarke ■☆

118723 Cyperus dentatus Torr.;齿莎草;Toothed Cyperus ■☆

118724 Cyperus dentoniae G. C. Tucker;登顿莎草■☆

118725 Cyperus denudatus L. f.;裸露莎草■☆

118726 Cyperus denudatus L. f. f. subdelicatulus Kük. = Cyperus denudatus L. f. ■☆

118727 Cyperus denudatus L. f. var. aureo-brunneus (C. B. Clarke) Kük.;黄褐莎草■☆

118728 Cyperus denudatus L. f. var. delicatulus C. B. Clarke = Cyperus denudatus L. f. ■☆

118729 Cyperus denudatus L. f. var. lucenti-nigricans (K. Schum.) Kük.;变黑裸露莎草■☆

118730 Cyperus denudatus L. f. var. sphaerospermus (Schrad.) Kük. = Cyperus sphaerospermus Schrad. ■☆

118731 Cyperus depauperatus Vahl = Eleocharis retroflexa (Poir.) Urb. ■☆

118732 Cyperus deremensis K. Schum. = Cyperus renschii Boeck. ■☆

118733 Cyperus dewildeorum (J. Raynal) Lye = Pycreus dewildeorum J. Raynal ■☆

118734 Cyperus diandrus Torr.;双雄莎草;Low Cyperus, Two-stamened Cyperus, Umbrella Flat Sedge, Umbrella Sedge ■☆

118735 Cyperus diandrus Torr. var. castaneus S. Watson = Cyperus niger Ruiz et Pav. ■☆

118736 Cyperus diaphanus Schrad. ex Roem. et Schult. = Cyperus diaphanus Schrad. ex Schult. ■

118737 Cyperus diaphanus Schrad. ex Schult.;东北扁莎■

118738 Cyperus diaphanus Schrad. ex Schult. = Pycreus diaphanus (Schrad. ex Schult.) S. S. Hooper et T. Koyama ■

118739 Cyperus diaphanus Schrad. ex Schult. var. setiformis (Korsh.) Kern = Cyperus diaphanus Schrad. ex Roem. et Schult. ■

118740 Cyperus dichromenaeformis Kunth var. major Boeck. = Cyperus mapanioides C. B. Clarke ■☆

118741 Cyperus dichromus C. B. Clarke;二色莎草■☆

118742 Cyperus dichroostachyus Hochst. ex A. Rich.;二色穗莎草■☆

118743 Cyperus dichrostachyus Kük. = Cyperus duclouxii E. G. Camus ■

118744 Cyperus difformis L.;异花莎草(碱草,密穗莎草,球穗莎草,三方草,水蜈蚣,王母钗,咸草,香附,异型莎草);Difformed Galingale, Dirty Dora, Globose Head Hat-sedge, Variable Flatsedge ■

118745 Cyperus difformis L. var. humilis Debeaux;矮异型莎草;Dwarf Difformed Galingale ■

118746 Cyperus difformis L. var. subdecompositus Kük. = Cyperus difformis L. ■

118747 Cyperus diffusus L. f. turgidulus Suringar = Mariscus trialatus (Boeck.) Ts. Tang et F. T. Wang ■

118748 Cyperus diffusus L. subsp. bancanus (Miq.) Kük. = Mariscus trialatus (Boeck.) Ts. Tang et F. T. Wang ■

118749 Cyperus diffusus Vahl;多脉莎草;Diffuse Cypressgrass, Diffuse Galingale, Dwarf Umbrella Grass ■

118750 Cyperus diffusus Vahl = Cyperus laxus Lam. ■☆

118751 Cyperus diffusus Vahl subsp. bancanus (Miq.) Kük. = Mariscus trialatus (Boeck.) Ts. Tang et F. T. Wang ■

118752 Cyperus diffusus Vahl subsp. buchholzii (Boeck.) Kük. = Cyperus laxus Lam. subsp. buchholzii (Boeck.) Lye ■☆

118753 Cyperus diffusus Vahl subsp. sylvestris (Ridl.) Kük. = Cyperus laxus Lam. subsp. sylvestris (Ridl.) Lye ■☆

118754 Cyperus diffusus Vahl var. latifolius L. K. Dai;宽叶多脉莎草;Broadleaf Diffuse Cypressgrass ■

118755 Cyperus diffusus Vahl var. multispicatus S. M. Hwang;多穗莎草;Mantspike Diffuse Cypressgrass ■

118756 Cyperus digitatus Roxb.;长小穗莎草(恒春莎草);Longspikelet Cypressgrass, Longspikelet Galingale ■

118757 Cyperus digitatus Roxb. var. auricomus (Sieber ex Spreng.) Kük.;金毛莎草■☆

118758 Cyperus digitatus Roxb. var. laxiflorus L. K. Dai;少花穗莎草;Few-flower Spike Galingale, Fewspike Cypressgrass ■

118759 Cyperus digitatus Roxb. var. pingbienensis L. K. Dai;屏边莎草;Pingbian Cypressgrass, Pingbian Galingale, Pingpien Galingale ■

118760 Cyperus dilatatus Schumach. et Thonn.;指裂莎草■☆

118761 Cyperus dilutus Vahl = Mariscus compactus (Retz.) Druce ■

118762 Cyperus dilutus Vahl var. macrostachys Boeck. = Mariscus compactus (Retz.) Druce var. macrostachys (Boeck.) F. C. How ex Ts. Tang et F. T. Wang ■

118763 Cyperus dipsaceus Liebm.;赖氏莎草;Teasel-flatsedge ■☆

118764 Cyperus dissitiflorus Nees ex Torr. = Cyperus thyrsiflorus Jungh. ■☆

118765 Cyperus distans L. f.;疏穗莎草(落滥草);Laxspiculate Cypressgrass, Laxspiculate Galingale ■

118766 Cyperus distans L. f. = Mariscus rubrotinctus Cherm. ■☆

118767 Cyperus distans L. f. subsp. longibracteatus (Cherm.) Lye = Mariscus longibracteatus Cherm. ■☆

118768 Cyperus distans L. f. var. kilimandscharica K. Schum. = Cyperus atroviridis C. B. Clarke ■☆

118769 Cyperus distans L. f. var. mucronatus Berhaut = Cyperus congensis C. B. Clarke ■☆

118770 Cyperus distans L. f. var. niger C. B. Clarke = Mariscus rubrotinctus Cherm. ■☆

118771 Cyperus distans L. f. var. rubrotinctus (Cherm.) Lye = Mariscus rubrotinctus Cherm. ■☆

118772 Cyperus distinctus Steud.;离生莎草■☆

118773 Cyperus diurensis Boeck. = Mariscus diurensis (Boeck.) C. B. Clarke ■☆

118774 Cyperus dives Delile;富有莎草■☆

118775 Cyperus drummondii Torr. et Hook.;德拉蒙德莎草■☆

118776 Cyperus dubius Rottb. = Mariscus dubius (Rottb.) Kük. ■

118777 Cyperus dubius Rottb. subsp. coloratus (Vahl) Lye = Mariscus dubius (Rottb.) Kük. ex C. E. C. Fisch. subsp. coloratus (Vahl) Lye ■☆

118778 Cyperus dubius Rottb. var. caespitosus Boeck. = Mariscus circumclusus C. B. Clarke ■☆

118779 Cyperus dubius Rottb. var. coloratus (Vahl) Kük. = Mariscus dubius (Rottb.) Kük. ex C. E. C. Fisch. subsp. coloratus (Vahl) Lye ■☆

118780 Cyperus ducis Buscal. et Muschl. = Cyperus kirkii C. B. Clarke ■☆

118781 Cyperus duclouxii E. G. Camus;云南莎草;Ducloux Cypressgrass, Ducloux Galingale ■

118782 Cyperus durus Kunth;硬莎草■☆

118783 Cyperus dwarkensis Sahni et H. B. Naithani = Pycreus dwarkensis (Sahni et H. B. Naithani) S. S. Hooper ■☆

118784 Cyperus echinatus (L.) A. W. Wood;刺莎草;Globe Flat Sedge, Hedgehog Club Rush, Teasel sedge ■☆

118785 Cyperus echinatus (L.) A. W. Wood var. multiflora Chapm. = Cyperus croceus Vahl ■☆

118786 Cyperus echinolepis T. Koyama = Lipocarpha albiceps Ridl. ■☆

118787 Cyperus edulis Dinter = Cyperus usitatus Burch. ■☆

118788 Cyperus effusus Rottb. = Cyperus conglomeratus Rottb. var. effusus (Rottb.) Coss. et Durieu ■☆

118789 Cyperus eggersii Boeck. = Cyperus odoratus L. ■

118790 Cyperus eggersii Boeck. = Torulinium odoratum (L.) S. S.

Hooper ■

118791 Cyperus ehrenbergianus Boeck. = Cyperus mutisii (Kunth) Andersson ■☆

118792 Cyperus elatior Boeck.;较高莎草■☆

118793 Cyperus elatus L.;高黄翅莎草■

118794 Cyperus elatus Rottb. = Cyperus distans L. f. ■

118795 Cyperus elegans L.;雅丽莎草■☆

118796 Cyperus elegantulus Steud. = Pycreus niger (Ruiz et Pav.) Cufod. subsp. elegantulus (Steud.) Lye ■☆

118797 Cyperus elephantinus (C. B. Clarke) Kük.;象莎草■☆

118798 Cyperus eleusinoides Kunth;穇穗莎草(三角草,土三棱);Yardgrass Galingale, Yardgrass-like Cypressgrass ■

118799 Cyperus eleusinoides Kunth var. dinklageanus Kük. = Cyperus congensis C. B. Clarke ■☆

118800 Cyperus eleusinoides Kunth var. subprolixus Kük. = Cyperus nutans Vahl subsp. subprolixus (Kük.) T. Koyama ■

118801 Cyperus elongatus Steud. = Pycreus intactus (Vahl) J. Raynal ■☆

118802 Cyperus endlichii Kük.;恩德莎草■☆

118803 Cyperus engelmannii Steud.;恩格莎草■☆

118804 Cyperus engelmannii Steud. = Cyperus odoratus L. ■

118805 Cyperus engelmannii Steud. = Torulinium odoratum (L.) S. S. Hooper ■

118806 Cyperus eragrostis Hell. var. flaccidus Boeck. = Pycreus sanguinolentus (Vahl) Nees f. flaccidus (Boeck.) Cufod. ■☆

118807 Cyperus eragrostis Lam. = Pycreus sanguinolentus (Vahl) Nees ex C. B. Clarke ■

118808 Cyperus eragrostis Vahl = Pycreus sanguinolentus (Vahl) Nees ex C. B. Clarke ■

118809 Cyperus eragrostis Vahl f. melanocephalus Suringar = Pycreus sanguinolentus (Vahl) Nees ex C. B. Clarke f. melanocephalus (Miq.) L. K. Dai ■

118810 Cyperus eragrostis Vahl f. rubromarginatus Kük. = Pycreus sanguinolentus (Vahl) Nees ex C. B. Clarke f. rubromarginatus (Schrenk) L. K. Dai ■

118811 Cyperus eragrostis Vahl var. humilis Miq. = Pycreus sanguinolentus (Vahl) Nees ex C. B. Clarke f. humilis (Miq.) L. K. Dai ■

118812 Cyperus erectus (Schumach.) Mattf. et Kük. = Kyllinga erecta Schumach. ■☆

118813 Cyperus erectus (Schumach.) Mattf. et Kük. subsp. albescens (Lye) Lye = Kyllinga erecta Schumach. subsp. albescens Lye ■☆

118814 Cyperus erectus (Schumach.) Mattf. et Kük. var. aurata (Nees) Kük. = Kyllinga erecta Schumach. ■☆

118815 Cyperus erectus (Schumach.) Mattf. et Kük. var. intricatus (Cherm.) Kük. = Kyllinga brevifolia Rottb. subsp. intricata (Cherm.) J. -P. Lebrun et Stork ■☆

118816 Cyperus erectus (Schumach.) Mattf. et Kük. var. luridus (Kük.) Kük. = Kyllinga brevifolia Rottb. subsp. lurida (Kük.) Lye ■☆

118817 Cyperus erectus (Schumach.) Mattf. et Kük. var. pleiocarpus Kük. = Kyllinga erecta Schumach. ■☆

118818 Cyperus erinaceus (Ridl.) Kük. = Sphaerocyperus erinaceus (Ridl.) Lye ■☆

118819 Cyperus erythraeus Schrad. = Pycreus sanguinolentus (Vahl) Nees ■

118820 Cyperus erythrorhizos Muhl.;红根莎草;Nutgrass, Redroot Flatsedge, Red-rooted Flatsedge, Red-rooted Sedge ■☆

118821 Cyperus erythrorhizos Muhl. var. cupreus (J. Presl et C. Presl) Kük. = Cyperus erythrorhizos Muhl. ■☆

118822 Cyperus esculentus L.;铁荸荠(地栗,黄莎草,油莎草,油莎豆);Chufa, Chufa Flat Sedge, Chufa Nut, Earth Almond, Edible Cyperus, Field Nut Sedge, Flat Sedge, Rush Nut, Tiger Nut, Tigernut, Yellow Nut Grass, Yellow Nut Sedge, Yellow Nutsedge, Zulu Nut ■☆

118823 Cyperus esculentus L. = Cyperus aureus Ten. ■☆

118824 Cyperus esculentus L. f. angustispicatus (Britton) Fernald = Cyperus esculentus L. var. leptostachyus Boeck. ■☆

118825 Cyperus esculentus L. f. angustispicatus (Britton) Fernald = Cyperus esculentus L. ■☆

118826 Cyperus esculentus L. var. angustispicatus Britton = Cyperus esculentus L. var. leptostachyus Boeck. ■☆

118827 Cyperus esculentus L. var. cyclolepis Kük. = Cyperus esculentus L. ■☆

118828 Cyperus esculentus L. var. heermannii (Buckley) Britton;黑尔曼铁荸荠■☆

118829 Cyperus esculentus L. var. leptostachyus Boeck.;细穗铁荸荠;Field Nut Sedge, Flat Sedge, Yellow Nut Sedge ■☆

118830 Cyperus esculentus L. var. leptostachyus Boeck. = Cyperus esculentus L. ■☆

118831 Cyperus esculentus L. var. lutesceus (Torr. et Hook.) Kük. = Cyperus esculentus L. var. macrostachyus Boeck. ■☆

118832 Cyperus esculentus L. var. macrostachyus Boeck.;大穗铁荸荠■☆

118833 Cyperus esculentus L. var. macrostachyus Boeck. = Cyperus esculentus L. ■☆

118834 Cyperus esculentus L. var. nervoso-striatus (Turrill) Kük. = Cyperus esculentus L. ■☆

118835 Cyperus esculentus L. var. phymatodes (Muhl.) Kük. = Cyperus esculentus L. var. leptostachyus Boeck. ■☆

118836 Cyperus esculentus L. var. rufus Chiov.;浅红莎草■☆

118837 Cyperus esculentus L. var. sativus Boeck.;油莎豆(油莎草,栽培油莎草);Chufa, Earth Almond, Rush Nul, Tiger Nut, Yellow Nutsedge, Zulu Nul ■

118838 Cyperus esculentus L. var. sativus Boeck. = Cyperus esculentus L. ■☆

118839 Cyperus esphacelatus Kük.;无毒莎草■☆

118840 Cyperus eurystachys Ridl. = Mariscus eurystachys (Ridl.) C. B. Clarke ■☆

118841 Cyperus exaltatus Retz.;高秆莎草(无翅莎草);Tallculm Cypressgrass, Tallculm Galingale ■

118842 Cyperus exaltatus Retz. subsp. iwasakii (Makino) T. Koyama = Cyperus exaltatus Retz. var. iwasakii (Makino) T. Koyama ■☆

118843 Cyperus exaltatus Retz. var. dives (Delile) C. B. Clarke = Cyperus dives Delile ■☆

118844 Cyperus exaltatus Retz. var. hainanensis L. K. Dai;海南高秆莎草;Hainan Tallculm Cypressgrass, Hainan Tallculm Galingale ■

118845 Cyperus exaltatus Retz. var. iwasakii (Makino) T. Koyama;岩崎高秆莎草■☆

118846 Cyperus exaltatus Retz. var. megalanthus Kük.;长穗高秆莎草;Longspike Tallculm Cypressgrass, Longspike Tallculm Galingale ■

118847 Cyperus exaltatus Retz. var. tenuispicatus L. K. Dai;广东高秆莎草;Guangdong Tallculm Cypressgrass, Guangdong Tallculm Galingale ■

118848 Cyperus eximius (C. B. Clarke) Mattf. et Kük. = Kyllinga eximia C. B. Clarke ■☆

118849 Cyperus eximius (C. B. Clarke) Mattf. et Kük. var. kelleri (C.

B. Clarke) Kük. = Kyllinga eximia C. B. Clarke var. kelleri ? ■☆
118850 Cyperus extremiorientalis Ohwi = Cyperus pygmaeus Rottb. ■
118851 Cyperus falcatus Boeck.;镰状莎草■☆
118852 Cyperus fascicularis Poir. = Pycreus polystachyos (Rottb.) P. Beauv. ■
118853 Cyperus fastigiatus Rottb.;帚状莎草■☆
118854 Cyperus fendlerianus Boeck. var. debilis (Britton) Kük. = Cyperus sphaerolepis Boeck. ■☆
118855 Cyperus fendlerianus Boeck. var. leucolepis (Boeck.) Kük. = Cyperus sphaerolepis Boeck. ■☆
118856 Cyperus fenzelianus K. Schum. = Cyperus rotundus L. var. tuberosus (Rottb.) Kük. ■☆
118857 Cyperus fenzelianus Steud. = Cyperus longus L. var. pallidus Boeck. ■☆
118858 Cyperus fenzelianus Steud. var. badiiformis Chiov. = Cyperus longus L. f. badiiformis (Chiov.) Kük. ■☆
118859 Cyperus ferax L. = Torulinium ferax (Rich.) Urb. ■
118860 Cyperus ferax Rich. = Cyperus odoratus L. ■
118861 Cyperus ferax Rich. = Torulinium ferax (Rich.) Ham. ■
118862 Cyperus ferax Rich. = Torulinium odoratum (L.) S. S. Hooper ■
118863 Cyperus ferax subsp. engelmannii (Steud.) Kük. = Cyperus odoratus L. ■
118864 Cyperus ferax subsp. speciosus (Vahl) Kük. = Cyperus odoratus L. ■
118865 Cyperus ferrugineoviridis (C. B. Clarke) Kük. = Mariscus ferrugineoviridis (C. B. Clarke) Cherm. ■☆
118866 Cyperus ferrugineoviridis (C. B. Clarke) Kük. var. distantiformis Kük. = Mariscus rubrotinctus Cherm. ■☆
118867 Cyperus ferruginescens Boeck.;锈色莎草■☆
118868 Cyperus ferruginescens Boeck. = Cyperus odoratus L. ■
118869 Cyperus ferruginescens Boeck. = Torulinium odoratum (L.) S. S. Hooper ■
118870 Cyperus ferrugineus Poir. = Pycreus intactus (Vahl) J. Raynal ■☆
118871 Cyperus ferrugineus Poir. var. baroni (C. B. Clarke) Kük. = Pycreus intactus (Vahl) J. Raynal ■☆
118872 Cyperus fertilis Boeck.;多产莎草■☆
118873 Cyperus fibrillosus Kük. = Pycreus fibrillosus (Kük.) Cherm. ■☆
118874 Cyperus fibrillosus Kük. var. scaettae (Cherm.) Kük. = Pycreus fibrillosus (Kük.) Cherm. ■☆
118875 Cyperus filicinus Vahl;冬青叶莎草■☆
118876 Cyperus filiculmis Vahl = Cyperus lupulinus (Spreng.) Marcks ■☆
118877 Cyperus filiculmis Vahl var. grayi (Torr.) Boeck. = Cyperus grayi Torr. ■☆
118878 Cyperus filiculmis Vahl var. macilentus Fernald = Cyperus lupulinus (Spreng.) Marcks subsp. macilentus (Fernald) Marcks ■☆
118879 Cyperus filiculmis Vahl var. macilentus Fernald = Cyperus lupulinus (Spreng.) Marcks ■☆
118880 Cyperus filiculmis Vahl var. oblitus Fernald et Griscom = Cyperus grayi Torr. ■☆
118881 Cyperus filiformis Sw.;线形莎草■☆
118882 Cyperus fimbristyloides T. Koyama = Alinula paradoxa (Cherm.) Goetgh. et Vorster ■☆
118883 Cyperus firmipes (C. B. Clarke) Kük. = Mariscus firmipes C. B. Clarke ■☆
118884 Cyperus fischerianus G. W. Schimp. ex A. Rich.;菲氏莎草■☆
118885 Cyperus fischerianus G. W. Schimp. ex A. Rich. var. ugandensis Lye;乌干达菲氏莎草■☆
118886 Cyperus fissus Steud.;半裂莎草■☆
118887 Cyperus flabelliformis Rottb. = Cyperus alternifolius L. subsp. flabelliformis (Rottb.) Kük. ■
118888 Cyperus flabelliformis Rottb. = Cyperus involucratus Rottb. ■
118889 Cyperus flaccidus R. Br.;柔软莎草■☆
118890 Cyperus flavescens L.;淡黄莎草;Umbrella Sedge ■☆
118891 Cyperus flavescens L. = Pycreus flavescens (L.) P. Beauv. ex Rchb. ■☆
118892 Cyperus flavescens L. f. rubromarginatus Schrenk = Pycreus sanguinolentus (Vahl) Nees ex C. B. Clarke f. rubromarginatus (Schrenk) L. K. Dai ■
118893 Cyperus flavescens L. subsp. microglumis (Lye) Lye = Pycreus flavescens (L.) P. Beauv. ex Rchb. subsp. microglumis Lye ■☆
118894 Cyperus flavescens L. subsp. tanaensis (Kük.) Lye = Pycreus flavescens (L.) P. Beauv. ex Rchb. subsp. tanaensis (Kük.) Lye ■☆
118895 Cyperus flavescens L. var. castaneus (Lye) Lye = Pycreus flavescens (L.) P. Beauv. ex Rchb. ■☆
118896 Cyperus flavescens L. var. poiformis (Pursh) Fernald = Cyperus flavescens L. ■☆
118897 Cyperus flavicomus Michx.;黄冠莎草;Nut Sedge ■☆
118898 Cyperus flavidus (Retz.) T. Koyama;浅黄莎草■☆
118899 Cyperus flavidus Retz. = Cyperus tenuispica Steud. ■
118900 Cyperus flavidus Retz. = Pycreus flavidus (Retz.) T. Koyama ■
118901 Cyperus flavissimus Schrad. = Cyperus niveus Retz. var. flavissimus (Schrad.) Lye ■☆
118902 Cyperus flavoculmis Lye;粗秆莎草■☆
118903 Cyperus flavomariscus Griseb. var. peduncularis Britton = Cyperus dentoniae G. C. Tucker ■☆
118904 Cyperus flavus (Britton) Kük. var. peduncularis (Britton) Kük. = Cyperus dentoniae G. C. Tucker ■☆
118905 Cyperus flavus (Vahl) Nees = Cyperus aggregatus (Willd.) Endl. ■☆
118906 Cyperus flavus (Vahl) Nees var. aggregatus (Willd.) Kük. = Cyperus aggregatus (Willd.) Endl. ■☆
118907 Cyperus flavus (Vahl) Nees var. laevis Kük. = Cyperus aggregatus (Willd.) Endl. ■☆
118908 Cyperus flavus Ridl. = Mariscus macrocarpus Kunth ■☆
118909 Cyperus floribundus (Kük.) R. Carter et S. D. Jones;单花莎草■☆
118910 Cyperus floridanus Britton ex Small = Cyperus filiformis Sw. ■☆
118911 Cyperus fluminalis Ridl. = Pycreus fluminalis (Ridl.) Troupin ■☆
118912 Cyperus fluviaticus (Torr.) A. Gray var. yagara (Ohwi) T. Koyama. = Bolboschoenus yagara (Ohwi) Y. C. Yang et M. Zhan ■
118913 Cyperus foliaceus C. B. Clarke;密叶莎草■☆
118914 Cyperus foliosus K. Schum. = Mariscus foliosus C. B. Clarke ■☆
118915 Cyperus foliosus Willd. ex Kunth = Pycreus intactus (Vahl) J. Raynal ■☆
118916 Cyperus fortunei Steud. = Cyperus malaccensis Lam. ■
118917 Cyperus fraternus Kunth = Cyperus reflexus Vahl ■☆
118918 Cyperus fresenii Steud. = Cyperus dichroostachyus Hochst. ex A. Rich. ■☆
118919 Cyperus friesii Kük. = Mariscus amauropus (Steud.) Cufod. var. friesii (Kük.) Cufod. ■☆
118920 Cyperus fucatus Boeck. = Mariscus albescens Gaudich. ■
118921 Cyperus fugax Liebm.;早萎莎草■☆

118922 Cyperus fulgens C. B. Clarke;光亮莎草■☆
118923 Cyperus fulgens C. B. Clarke var. contractus Kük.;紧缩莎草■☆
118924 Cyperus fulvescens Liebm. = Cyperus esculentus L. var. leptostachyus Boeck. ■☆
118925 Cyperus fulvus Ridl. = Pycreus nitidus (Lam.) J. Raynal ■☆
118926 Cyperus fuscescens Link;浅棕色莎草■☆
118927 Cyperus fuscovaginatus Kük.;污鞘莎草■☆
118928 Cyperus fuscus L.;褐穗莎草(密穗莎草,三棱草);Black Cyperus, Brown Flatsedge, Brown Galingale, Fuscous Cypressgrass, Fuscous Galingale ■
118929 Cyperus fuscus L. f. pallescens Husn.;绿白莎草;Pallid-green Galingale ■
118930 Cyperus fuscus L. f. pallescens Husn. = Cyperus fuscus L. ■
118931 Cyperus fuscus L. f. virescens (Hoffm.) Vahl;北褐穗莎草莎草;North Galingale ■
118932 Cyperus fuscus L. var. virescens (Hoffm.) Coss. et Durieu = Cyperus fuscus L. f. virescens (Hoffm.) Vahl ■
118933 Cyperus fuscus L. var. virescens (Hoffm.) Coss. et Durieu = Cyperus fuscus L. ■
118934 Cyperus geminiflora (Steud.) Wickens;对花莎草■☆
118935 Cyperus giganteus Vahl;墨西哥莎草;Mexican Papyrus ■☆
118936 Cyperus gigantobulbes Lye;巨莎草■☆
118937 Cyperus ginge Welw. = Cyperus involucratus Rottb. ■
118938 Cyperus giolii Chiov.;焦尔莎草■☆
118939 Cyperus giolii Chiov. var. latifolius ?;宽叶焦尔莎草■☆
118940 Cyperus giolii Chiov. var. nogalensis (Chiov.) Kük.;诺加尔莎草■☆
118941 Cyperus glaber L.;光莎草■☆
118942 Cyperus glaucophyllus Boeck.;灰绿莎草■☆
118943 Cyperus glaucoviridis Boeck. = Mariscus ligularis (L.) Urb. ■☆
118944 Cyperus globifer (C. B. Clarke) Lye = Mariscus globifer C. B. Clarke ■☆
118945 Cyperus globosus All. = Pycreus flavidus (Retz.) T. Koyama ■
118946 Cyperus globosus All. = Pycreus globosus (All.) Rchb. ■
118947 Cyperus globosus All. f. fuscoater (Meinsh.) Kük. = Cyperus flavidus Retz. ■
118948 Cyperus globosus All. f. minimus Kük. = Pycreus flavidus (Retz.) T. Koyama var. minimus (Kük.) L. K. Dai ■
118949 Cyperus globosus All. f. minimus Kük. = Pycreus globosus (All.) Rchb. var. minimvs (Kük.) Ts. Tang et F. T. Wang ■
118950 Cyperus globosus All. var. nilagiricus (Hochst. ex Steud.) C. B. Clarke = Pycreus flavidus (Retz.) T. Koyama var. nilagiricus (Hochst. ex Steud.) C. Y. Wu ex Karthik. ■
118951 Cyperus globosus All. var. nilagiricus (Hochst. ex Steud.) C. B. Clarke = Cyperus flavidus (Retz.) T. Koyama ■☆
118952 Cyperus globosus All. var. nilagiricus (Hochst. ex Steud.) C. B. Clarke = Pycreus globosus (All.) Rchb. var. nilagiricus (Hochst.) C. B. Clarke ■
118953 Cyperus globosus All. var. nilagiricus (Hochst.) C. B. Clarke = Pycreus globosus (All.) Rchb. var. nilagiricus (Hochst.) C. B. Clarke ■
118954 Cyperus globosus All. var. strictus (Roxb.) C. B. Clarke = Pycreus globosus (All.) Rchb. var. strictus (Roxb.) C. B. Clarke ■
118955 Cyperus globulosus Aubl. var. pseudofiliculmis Kük. = Cyperus croceus Vahl ■☆
118956 Cyperus globulosus Aubl. var. robustus (Boeck.) Shinners = Cyperus croceus Vahl ■☆
118957 Cyperus glomeratus L.;头状穗莎草(聚穗莎草,球穗莎草,球形莎草,三轮草,水莎草,头穗莎草,团花莎草,喂香壶,状元花);Glomerate Cypressgrass, Glomerate Galingale ■
118958 Cyperus gossweileri Kük. = Pycreus pubescens Turrill ■☆
118959 Cyperus graciliculmis Lye;细秆莎草■☆
118960 Cyperus gracilinux C. B. Clarke = Cyperus dilatatus Schumach. et Thonn. ■☆
118961 Cyperus gracilinux C. B. Clarke var. platyphylla ? = Cyperus dilatatus Schumach. et Thonn. ■☆
118962 Cyperus gracilis R. Br.;细小莎草;Slimjim Flatsedge ■☆
118963 Cyperus gracillimus (Chiov.) Kük. = Pycreus gracillimus Chiov. ■☆
118964 Cyperus gracillimus Miq. = Kyllinga brevifolia Rottb. var. leiolepis (Franch. et Sav.) H. Hara ■
118965 Cyperus grandibulbosus C. B. Clarke;大球莎草■☆
118966 Cyperus grandibulbosus C. B. Clarke var. amplus Kük.;膨大莎草■☆
118967 Cyperus grandis C. B. Clarke;大莎草■☆
118968 Cyperus grantii C. B. Clarke = Cyperus tenax Boeck. ■☆
118969 Cyperus gratus C. B. Clarke = Cyperus cuspidatus Kunth ■
118970 Cyperus grayi Torr.;格雷莎草;Gray's Sedge ■☆
118971 Cyperus grayoides Mohlenbr.;假格雷莎草;Mohlenbrock's Sedge ■☆
118972 Cyperus groteanus Kük. = Mariscus taylori C. B. Clarke var. groteanus (Kük.) Napper ■☆
118973 Cyperus guineensis Nelmes = Cyperus tenuiculmis Boeck. var. guineensis (Nelmes) S. S. Hooper ■☆
118974 Cyperus gypsophilus Lye;喜钙莎草■☆
118975 Cyperus haematocephalus C. B. Clarke;血头莎草■☆
118976 Cyperus haenkei J. Presl et C. Presl = Cyperus odoratus L. ■
118977 Cyperus haenkei J. Presl et C. Presl = Torulinium odoratum (L.) S. S. Hooper ■
118978 Cyperus hakonensis Franch. et Sav. = Cyperus flaccidus R. Br. ■☆
118979 Cyperus halei Torr. ex Britton = Cyperus erythrorhizos Muhl. ■☆
118980 Cyperus hamulosus M. Bieb.;钩莎草■☆
118981 Cyperus hansenii Britton = Cyperus strigosus L. ■☆
118982 Cyperus haspan L.;畦畔莎草(鸡屎青,三棱草);Asidefield Cypressgrass, Asidefield Galingale, Paddyfield Flat-sedge, Sharp Edge Sedge ■
118983 Cyperus haspan L. subsp. juncoides (Lam.) Kük. = Cyperus haspan L. ■
118984 Cyperus haspan L. var. americanus Boeck. = Cyperus haspan L. ■
118985 Cyperus haspan L. var. flaccidissimus Kük.;软秆莎草■
118986 Cyperus haspan L. var. flaccidissimus Kük. = Cyperus haspan L. ■
118987 Cyperus haspan L. var. microhaspan Makino;小畦畔莎草■☆
118988 Cyperus haspan L. var. sphaerospermoides (Cherm.) Cherm. = Cyperus denudatus L. f. ■☆
118989 Cyperus haspan L. var. transiens (Cherm.) Cherm. = Cyperus denudatus L. f. ■☆
118990 Cyperus haspan L. var. tuberiferus T. Koyama;瘤畦畔莎草■☆
118991 Cyperus haspanoides C. B. Clarke = Cyperus denudatus L. f. ■☆
118992 Cyperus heermannii Buckley = Cyperus esculentus L. var. heermannii (Buckley) Britton ■☆
118993 Cyperus hemisphaericus Boeck. = Mariscus hemisphaericus (Boeck.) C. B. Clarke ■☆
118994 Cyperus hensii T. Durand et Schinz;亨斯莎草■☆

118995 Cyperus heudelotii C. B. Clarke = Cyperus maculatus Boeck. ■☆

118996 Cyperus hildebrandtii Boeck. = Mariscus hemisphaericus (Boeck.) C. B. Clarke ■☆

118997 Cyperus hildebrandtii K. Schum. = Pycreus hildebrandtii (K. Schum.) C. B. Clarke ■☆

118998 Cyperus hirsutus P. J. Bergius = Fuirena hirsuta (P. J. Bergius) P. L. Forbes ■☆

118999 Cyperus hirtellus (Chiov.) Kük. = Mariscus hirtellus Chiov. ■☆

119000 Cyperus hirtus Thunb. = Fimbristylis squarrosa (Poir.) Vahl ■

119001 Cyperus hochstetteri Krauss var. tenuis Boeck. = Pycreus macrostachyos (Lam.) J. Raynal var. tenuis (Boeck.) Wickens ■☆

119002 Cyperus holostigma C. B. Clarke ex Schweinf. = Cyperus schinzii Boeck. ■☆

119003 Cyperus holstii Kük.;霍尔莎草■☆

119004 Cyperus houghtonii Torr.;霍顿莎草;Houghton's Cyperus, Houghton's Flat Sedge, Houghton's Flatsedge ■☆

119005 Cyperus huarmensis (Kunth) M. C. Johnst. = Cyperus odoratus L. ■

119006 Cyperus huarmensis (Kunth) M. C. Johnst. = Torulinium odoratum (L.) S. S. Hooper ■

119007 Cyperus huillensis Ridl. = Cyperus sphaerospermus Schrad. ■☆

119008 Cyperus huillensis Ridl. var. aphyllus ? = Cyperus sphaerospermus Schrad. ■☆

119009 Cyperus humboldtianus Schult. = Pycreus humboldtianus (Schult.) Cufod. ■☆

119010 Cyperus humboldtianus Schult. = Pycreus lanceolatus (Poir.) C. B. Clarke ■☆

119011 Cyperus hyalinus Vahl = Queenslandiella hyalina (Vahl) F. Ballard ■☆

119012 Cyperus hylaeus Ridl. = Cyperus renschii Boeck. ■☆

119013 Cyperus hylophilus Cherm.;喜盐莎草■☆

119014 Cyperus hypopitys G. C. Tucker;松林莎草;Pinewoods Sedge ■☆

119015 Cyperus hystricinus Fernald;豪猪莎草;Bristly Flatsedge, Nut Grass ■☆

119016 Cyperus imbricatus Retz.;迭穗莎草(叠穗莎草,覆瓦状莎草);Imbricate Cypressgrass, Imbricate Galingale ■

119017 Cyperus imbricatus Retz. var. capitatus (Boeck.) Kük. = Cyperus imbricatus Retz. ■

119018 Cyperus imbricatus Retz. var. dense-spicatus (Hayata) Ohwi;密迭穗莎草(大密穗莎草);Densespike Imbricate Galingale ■

119019 Cyperus imbricatus Retz. var. dense-spicatus (Hayata) Ohwi = Cyperus imbricatus Retz. ■

119020 Cyperus imbricatus Retz. var. elongatus (Boeck.) L. K. Dai;大密穗莎草(士林莎草);Elongate Imbricate Galingale ■

119021 Cyperus imbricatus Retz. var. elongatus (Boeck.) T. Koyama = Cyperus imbricatus Retz. var. elongatus (Boeck.) L. K. Dai ■

119022 Cyperus imbricatus Retz. var. multiflorus Kük. = Cyperus imbricatus Retz. ■

119023 Cyperus imbricatus Retz. var. multiflorus Kük. = Cyperus imbricatus Retz. var. dense-spicatus (Hayata) Ohwi ■

119024 Cyperus immensus C. B. Clarke = Cyperus dives Delile ■☆

119025 Cyperus immensus C. B. Clarke var. petherickii (C. B. Clarke) Kük. = Cyperus dives Delile ■☆

119026 Cyperus impubes Steud. = Mariscus richardii Steud. ■☆

119027 Cyperus inauratus (Nees) Mattf. et Kük. = Kyllinga inaurata Nees ex Boeck. var. laevicarinatus Kük. ■☆

119028 Cyperus inconspicuus Liebm. = Cyperus fugax Liebm. ■☆

119029 Cyperus indecorus Kunth;装饰莎草■☆

119030 Cyperus indecorus Kunth var. decurvatus (C. B. Clarke) Kük.;下延莎草■☆

119031 Cyperus indecorus Kunth var. inflatus (C. B. Clarke) Kük.;膨胀莎草■☆

119032 Cyperus indecorus Kunth var. namaquensis Kük.;纳马夸莎草■☆

119033 Cyperus inflexus Muhl. = Mariscus aristatus (Rottb.) Ts. Tang et F. T. Wang ■

119034 Cyperus ingratus Kunth = Cyperus sphaerospermus Schrad. ■☆

119035 Cyperus intactus Vahl = Pycreus intactus (Vahl) J. Raynal ■☆

119036 Cyperus intermedius Steud. = Pycreus intermedius (Steud.) C. B. Clarke ■☆

119037 Cyperus intermedius Steud. var. tenuis Boeck. = Pycreus intermedius (Steud.) C. B. Clarke f. tenuis (Boeck.) Cufod. ■☆

119038 Cyperus inundatus Roxb. = Juncellus serotinus (Rottb.) C. B. Clarke var. inundatus (Roxb.) L. K. Dai ■

119039 Cyperus involucratus Poir. = Cyperus alternifolius L. ■

119040 Cyperus involucratus Rottb. = Cyperus alternifolius L. subsp. flabelliformis (Rottb.) Kük. ■

119041 Cyperus iria L.;碎米莎草(见骨草,荆三棱,米莎草,三方草,三棱草,三楞草,三轮草,莎草,四方草,野席草);Crushed-rice Flat-sedge, Rice Cypressgrass, Rice Galingale, Ricefield Flatsedge, Umbrella Sedge ■

119042 Cyperus iria L. var. microiria (Steud.) Franch. et Sav. = Cyperus microiria Steud. ■

119043 Cyperus iria L. var. parviflorus (Nees) Miq. = Cyperus microiria Steud. ■

119044 Cyperus iria L. var. rectangularis Kük. = Cyperus microiria Steud. ■

119045 Cyperus isocladus Kunth = Cyperus prolifer Lam. ■☆

119046 Cyperus iwasakii Makino = Cyperus exaltatus Retz. var. iwasakii (Makino) T. Koyama ■☆

119047 Cyperus javanicus Houtt. = Mariscus javanicus (Houtt.) Merr. et F. P. Metcalf ■

119048 Cyperus jeminicus Rottb.;杰敏莎草■☆

119049 Cyperus jeminicus Rottb. var. spicatus (Boeck.) Chiov. = Cyperus bulbosus Vahl var. spicatus Boeck. ■☆

119050 Cyperus juncoides Lam. = Cyperus haspan L. ■

119051 Cyperus kabarensis Cherm.;卡巴雷莎草■☆

119052 Cyperus kaessneri C. B. Clarke;卡斯纳莎草■☆

119053 Cyperus kalli (Forssk.) Murb. = Cyperus capitatus Vand. ■☆

119054 Cyperus karisimbiensis (Cherm.) Kük. = Mariscus karisimbiensis Cherm. ■☆

119055 Cyperus kasamensis Podlech;卡萨马莎草■☆

119056 Cyperus keniensis Kük.;肯尼亚莎草■☆

119057 Cyperus kernianus Ohwi et T. Koyama = Cyperus sesquiflorus (Torr.) Mattf. et Kük. var. cylindricus (Nees) Kük. ■

119058 Cyperus kernianus Ohwi et T. Koyama = Kyllinga odorata Vahl ■☆

119059 Cyperus kerstenii Boeck. = Mariscus kerstenii (Boeck.) C. B. Clarke ■☆

119060 Cyperus kilimandscharicus Kük.;基利莎草■☆

119061 Cyperus kirkii C. B. Clarke;柯克莎草■☆

119062 Cyperus kivuensis Cherm.;基伍莎草■☆

119063 Cyperus kleinianus Hochst. ex Steud. = Courtoisina cyperoides (Roxb.) Soják ■

119064 Cyperus kottensis Cherm. = Cyperus tonkinensis C. B. Clarke var. baikiei (C. B. Clarke) S. S. Hooper ■☆

119065 Cyperus kwaleensis Lye;夸莱莎草■☆
119066 Cyperus kyllingaeformis Lye;水蜈蚣莎草■☆
119067 Cyperus kyllingia Endl. = Kyllinga monocephala Rottb. ■
119068 Cyperus kyllingia Endl. = Kyllinga nemoralis (J. R. Forst. et G. Forst.) Dandy ex Hutch. et Dalziel ■
119069 Cyperus kyllingioides Vahl = Mariscus dubius (Rottb.) Kük. ex C. E. C. Fisch. ■
119070 Cyperus labiatus Peter = Cyperus latifolius Poir. ■☆
119071 Cyperus lacteus Steud. = Cyperus obtusiflorus Vahl ■☆
119072 Cyperus laetus J. Presl et C. Presl var. cephalanthus (Torr. et Hook.) Kük. = Cyperus cephalanthus Torr. et Hook. ■☆
119073 Cyperus laevigatus L.;平滑莎草;Makalao ■☆
119074 Cyperus laevigatus L. subsp. albidus (Vahl) Maire et Weiller;白色平滑莎草■☆
119075 Cyperus laevigatus L. var. albidus Vahl = Cyperus laevigatus L. subsp. albidus (Vahl) Maire et Weiller ■☆
119076 Cyperus laevigatus L. var. subaphyllus (Boeck.) Kük. = Cyperus laevigatus L. ■☆
119077 Cyperus lanceolatus Poir. = Pycreus lanceolatus (Poir.) C. B. Clarke ■☆
119078 Cyperus lanceolatus Poir. var. compositus J. Presl et C. Presl = Pycreus lanceolatus (Poir.) C. B. Clarke ■☆
119079 Cyperus lanceus Thunb. = Pycreus nitidus (Lam.) J. Raynal ■☆
119080 Cyperus lanceus Thunb. var. angustifolius Ridl. = Pycreus macranthus (Boeck.) C. B. Clarke var. angustifolius (Ridl.) C. B. Clarke ex Rendle ■☆
119081 Cyperus lanceus Thunb. var. macrostachya Kunth = Pycreus macranthus (Boeck.) C. B. Clarke ■☆
119082 Cyperus lanceus Thunb. var. melanopus Boeck. = Pycreus nitidus (Lam.) J. Raynal ■☆
119083 Cyperus lapidicola Kük. = Cyperus semitrifidus Schrad. ■☆
119084 Cyperus lateriflorus Torr. = Cyperus difformis L. ■
119085 Cyperus lateriticus J. Raynal;砖红莎草■☆
119086 Cyperus latespicatus Boeck. = Pycreus diaphanus (Schrad. ex Roem. et Schult.) S. S. Hooper et T. Koyama ex S. S. Hooper ■
119087 Cyperus latespicatus Boeck. = Pycreus diaphanus (Schrad. ex Schult.) S. S. Hooper et T. Koyama ■
119088 Cyperus latespicatus Boeck. = Pycreus latespicatus (Boeck.) C. B. Clarke ■
119089 Cyperus latespicatus Boeck. var. gracilescens Kük. = Pycreus pseudolatespicatus L. K. Dai ■
119090 Cyperus latifolius Poir.;宽叶莎草■☆
119091 Cyperus latifolius Poir. var. angustifolius Hochst. = Cyperus latifolius Poir. ■☆
119092 Cyperus latifolius Poir. var. austro-africanus Kük. = Cyperus latifolius Poir. ■☆
119093 Cyperus latispicatus Boeck. var. setiformis (Korsh.) T. Koyama = Cyperus diaphanus Schrad. ex Roem. et Schult. ■
119094 Cyperus latispicatus Boeck. var. setiformis (Korsh.) T. Koyama = Pycreus diaphanus (Schrad. ex Roem. et Schult.) S. S. Hooper et T. Koyama ex S. S. Hooper ■
119095 Cyperus laxespicatus Kük.;稀穗莎草■☆
119096 Cyperus laxus Lam.;疏松莎草■☆
119097 Cyperus laxus Lam. subsp. buchholzii (Boeck.) Lye;布赫疏松莎草■☆
119098 Cyperus laxus Lam. subsp. sylvestris (Ridl.) Lye;林地疏松莎草■☆
119099 Cyperus ledermannii (Kük.) Hooper = Cyperus niveus Retz. var. ledermannii (Kük.) Lye ■☆
119100 Cyperus lentiginosus Millsp. et Chase;雀斑莎草■☆
119101 Cyperus leptocladus Kunth;细枝莎草■☆
119102 Cyperus leptolepis Peter ex Kük.;细鳞莎草■☆
119103 Cyperus leptophyllus Hochst. ex Boeck. var. friesii (Kük.) Kük. = Mariscus amauropus (Steud.) Cufod. var. friesii (Kük.) Cufod. ■☆
119104 Cyperus leptorhachis Mattf. et Kük. = Kyllinga debilis C. B. Clarke ■☆
119105 Cyperus leucocephalus Nees = Cyperus pulchellus R. Br. ■☆
119106 Cyperus leucolepis Boeck. = Cyperus sphaerolepis Boeck. ■☆
119107 Cyperus leucolepis J. Carey ex C. B. Clarke = Cyperus pumilus L. ■
119108 Cyperus leucoloma Nees = Isolepis leucoloma (Nees) C. Archer ■☆
119109 Cyperus ligularis L. = Mariscus ligularis (L.) Urb. ■☆
119110 Cyperus limosus Maxim. = Juncellus limosus (Maxim.) C. B. Clarke ■
119111 Cyperus linearispiculatus L. K. Dai;线状穗莎草;Linearspike Cypressgrass,Linearspike Galingale ■
119112 Cyperus lipocarpha T. Koyama = Lipocarpha chinensis (Osbeck) J. Kern. ■
119113 Cyperus litoreus (C. B. Clarke) Britton = Cyperus ovatus Baldwin ■☆
119114 Cyperus locuples C. B. Clarke = Cyperus sphacelatus Rottb. ■☆
119115 Cyperus longibracteatus (Cherm.) Kük. = Mariscus longibracteatus Cherm. ■☆
119116 Cyperus longibracteatus (Cherm.) Kük. var. niger (C. B. Clarke) Lye = Mariscus rubrotinctus Cherm. ■☆
119117 Cyperus longibracteatus (Cherm.) Kük. var. rubrotinctus ? = Mariscus rubrotinctus Cherm. ■☆
119118 Cyperus longifolius Poir.;长叶莎草■☆
119119 Cyperus longifolius Poir. var. condensatus Cherm. = Cyperus longifolius Poir. ■☆
119120 Cyperus longi-involucratus Lye;长总苞莎草■☆
119121 Cyperus longispicatus Norton = Torulinium odoratum (L.) S. S. Hooper ■
119122 Cyperus longispiculata Muasya et D. A. Simpson;西方长尖莎草■☆
119123 Cyperus longistolon Peter et Kük. = Pycreus longistolon (Peter et Kük.) Napper ■☆
119124 Cyperus longistolon Peter et Kük. subsp. atrofuscus (Lye) Lye = Pycreus longistolon (Peter et Kük.) Napper subsp. atrofuscus Lye ■☆
119125 Cyperus longus L.;长莎草;Cypress-root,Galangale,Galingale,Galingale Flat Sedge,Long Cyperus,Sweet Cypress,Sweet Galingale ■☆
119126 Cyperus longus L. f. badiiformis (Chiov.) Kük.;栗形莎草■☆
119127 Cyperus longus L. subsp. badius (Desf.) Asch. et Graebn.;栗色莎草■☆
119128 Cyperus longus L. subsp. badius (Desf.) Asch. et Graebn. = Cyperus badius Desf. ■☆
119129 Cyperus longus L. var. adoensis Boeck. = Cyperus rigidifolius Steud. ■☆
119130 Cyperus longus L. var. badius (Desf.) J. Gay = Cyperus longus L. subsp. badius (Desf.) Asch. et Graebn. ■☆
119131 Cyperus longus L. var. gracillimus (Chiov.) Cufod.;细长莎草■☆

119132 Cyperus longus L. var. pallescens (Desf.) Coss. et Durieu = Cyperus longus L. ■☆

119133 Cyperus longus L. var. pallidus Boeck.;苍白长莎草■☆

119134 Cyperus longus L. var. tenuiflorus (Rottb.) Boeck.;细花长莎草■☆

119135 Cyperus longus L. var. tenuiflorus (Rottb.) Kük. = Cyperus longus L. var. tenuiflorus (Rottb.) Boeck. ■☆

119136 Cyperus louisianensis Thieret = Cyperus sanguinolentus Vahl ■

119137 Cyperus lucenti-nigricans K. Schum. = Cyperus denudatus L. f. var. lucenti-nigricans (K. Schum.) Kük. ■☆

119138 Cyperus lucidulus C. B. Clarke = Cyperus zollingeri Steud. ■

119139 Cyperus lupulinus (Spreng.) Marcks;草原莎草(纤弱莎草);Great Plains Flat Sedge, Nut Sedge, Sand Cyperus, Sand Sedge, Slender Gyperus ■☆

119140 Cyperus lupulinus (Spreng.) Marcks subsp. macilentus (Fernald) Marcks;细草原莎草;Great Plains Flat Sedge, Sand Cyperus, Slender Sand Sedge ■☆

119141 Cyperus luteo-stramineus Mattf. et Kük. = Kyllinga exigua Boeck. ■☆

119142 Cyperus luteus Boeck. = Mariscus foliosus C. B. Clarke ■☆

119143 Cyperus luzulae (L.) Retz.;地杨梅莎草■☆

119144 Cyperus luzuliformis Boeck. = Mariscus luzuliformis (Boeck.) C. B. Clarke ■☆

119145 Cyperus macer (Kunth) K. Schum. = Cyperus cyperoides (L.) Kuntze subsp. pseudoflavus (Kük.) Lye ■☆

119146 Cyperus macilentus (Fernald) E. P. Bicknell = Cyperus lupulinus (Spreng.) Marcks subsp. macilentus (Fernald) Marcks ■☆

119147 Cyperus macranthus Boeck. = Pycreus macranthus (Boeck.) C. B. Clarke ■☆

119148 Cyperus macranthus Boeck. f. acuticarinatus (Kük.) Kük. = Pycreus acuticarinatus (Kük.) Cherm. ■☆

119149 Cyperus macrocarpus (Kunth) Boeck.;大果莎草■☆

119150 Cyperus macrocephalus Liebm. = Cyperus odoratus L. ■

119151 Cyperus macrocephalus Liebm. = Torulinium odoratum (L.) S. S. Hooper ■

119152 Cyperus macropus Boeck. = Mariscus circumclusus C. B. Clarke ■☆

119153 Cyperus macrorrhizus Nees;大根莎草■☆

119154 Cyperus macrostachyos Lam. = Pycreus macrostachyos (Lam.) J. Raynal ■☆

119155 Cyperus macrostachyos Lam. subsp. tremulus (Poir.) Lye = Pycreus macrostachyos (Lam.) J. Raynal ■☆

119156 Cyperus maculatus Boeck.;斑点莎草■☆

119157 Cyperus maculatus Boeck. var. contractus Cherm. = Cyperus maculatus Boeck. ■☆

119158 Cyperus madagascariensis (Willd.) Roem. et Schult.;马岛莎草■☆

119159 Cyperus maderaspatanus Willd. = Mariscus maderaspatanus (Willd.) Napper ■☆

119160 Cyperus malaccensis Lam.;茳芏碱草(碱草,茳芏,卤草);Malacca Cypressgrass, Malacca Galingale ■

119161 Cyperus malaccensis Lam. subsp. brevifolius (Boeck.) T. Koyama = Cyperus malaccensis Lam. subsp. monophyllus (Vahl) T. Koyama ■

119162 Cyperus malaccensis Lam. subsp. monophyllus (Vahl) T. Koyama;短叶茳芏(单叶碱草,茳芏,席草,咸水草);Shortleaf Galingale, Shortleaf Malaca Cypressgrass ■

119163 Cyperus malaccensis Lam. var. brevifolius Boeck. = Cyperus malaccensis Lam. subsp. monophyllus (Vahl) T. Koyama ■

119164 Cyperus manimae Kunth;马尼玛莎草■☆

119165 Cyperus manimae Kunth var. asperrimus (Liebm.) Kük.;粗糙莎草■☆

119166 Cyperus mannii C. B. Clarke;曼氏莎草■☆

119167 Cyperus mapanioides C. B. Clarke;擂鼓艻莎草■☆

119168 Cyperus mapanioides C. B. Clarke var. major (Boeck.) Kük. = Cyperus mapanioides C. B. Clarke ■☆

119169 Cyperus maranguensis K. Schum.;马兰古莎草■☆

119170 Cyperus maranguensis K. Schum. var. ferrugineoviridis C. B. Clarke = Mariscus ferrugineoviridis (C. B. Clarke) Cherm. ■☆

119171 Cyperus margaritaceus Vahl;珍珠莎草■☆

119172 Cyperus margaritaceus Vahl var. pseudoniveus (Boeck.) C. B. Clarke = Cyperus margaritaceus Vahl ■☆

119173 Cyperus margaritaceus Vahl var. tisserantii (Cherm.) Kük. = Cyperus niveus Retz. var. tisserantii (Cherm.) Lye ■☆

119174 Cyperus marginatus Thunb.;具边莎草■☆

119175 Cyperus marginellus Nees = Cyperus pilosus Vahl ■

119176 Cyperus maritimus Poir. = Cyperus angolensis Boeck. ■☆

119177 Cyperus marlothii Boeck.;马洛斯莎草■☆

119178 Cyperus martindalei Britton = Cyperus filiculmis Vahl ■☆

119179 Cyperus mediorubescens Hayata = Cyperus imbricatus Retz. ■

119180 Cyperus meeboldii Kük. var. gigas Berhaut = Cyperus clavinux C. B. Clarke ■☆

119181 Cyperus melanacme (Nelmes) Raymond = Pycreus melanacme Nelmes ■☆

119182 Cyperus melanocephalus Miq. = Pycreus sanguinolentus (Vahl) Nees ex C. B. Clarke f. melanocephalus (Miq.) L. K. Dai ■

119183 Cyperus melanosperma (Nees) Suringar = Kyllinga melanosperma Nees ■

119184 Cyperus melanospermus (Nees) Suringar = Kyllinga melanosperma Nees ■

119185 Cyperus melanospermus (Nees) Suringar subsp. elatus (Steud.) Lye = Kyllinga melanosperma Nees var. elata (Steud.) J. -P. Lebrun et Stork ■☆

119186 Cyperus melanospermus (Nees) Suringar var. hexalatus (Lye) Lye = Kyllinga melanosperma Nees var. hexalata Lye ■☆

119187 Cyperus melanostachyus Kunth = Cyperus niger Ruiz et Pav. ■☆

119188 Cyperus merkeri C. B. Clarke = Cyperus rotundus L. subsp. merkeri (C. B. Clarke) Kük. ■☆

119189 Cyperus merxmuelleri (Podlech) Lye = Kyllinga merxmuelleri Podlech ■☆

119190 Cyperus metzii (Hochst. ex Steud.) Mattf. et Kük. = Kyllinga squamulata Thonn. ex Vahl ■

119191 Cyperus metzii (Hochst.) Mattf. et Kük. = Kyllinga squamulata Thonn. ex Vahl ■

119192 Cyperus meyerianus Kunth;迈尔莎草;Meyen's Flatsedge ■☆

119193 Cyperus micans Kunth = Pycreus intactus (Vahl) J. Raynal ■☆

119194 Cyperus michauxianus Schult. = Cyperus odoratus L. ■

119195 Cyperus michauxianus Schult. = Torulinium odoratum (L.) S. S. Hooper ■

119196 Cyperus michelianus (L.) Link;旋鳞莎草(附心草,护儿草,护心草);Michel Cypressgrass, Michel Galingale ■

119197 Cyperus michelianus (L.) Link subsp. pacificus (Ohwi) T. Koyama = Cyperus pacificus (Ohwi) Ohwi ■☆

119198 Cyperus michelianus (L.) Link subsp. pygmaeus (Rottb.)

Asch. et Graebn. = Cyperus pygmaeus Rottb. ■

119199 Cyperus michelianus (L.) Link subsp. pygmaeus (Rottb.) Asch. et Graebn. var. nipponicus (Franch. et Sav.) Kük. = Cyperus nipponicus Franch. et Sav. ■

119200 Cyperus michelianus (L.) Link var. pacificus Ohwi = Cyperus pacificus (Ohwi) Ohwi ■☆

119201 Cyperus microaureus Lye = Alinula peteri (Kük.) Goetgh. et Vorster ■☆

119202 Cyperus microbolbos C. B. Clarke = Mariscus microbolbos (C. B. Clarke) Vorster ■☆

119203 Cyperus microbracteatus (Lye) Lye = Kyllinga microbracteata Lye ■☆

119204 Cyperus microbulbosus (Lye) Lye = Kyllinga microbulbosa Lye ■☆

119205 Cyperus microcarpus Boeck. = Cyperus haspan L. ■

119206 Cyperus microcristatus Lye;小冠莎草■☆

119207 Cyperus microglumis Simpson;小颖莎草■☆

119208 Cyperus microiria Steud.;具芒碎米莎草(黄颖莎草);Awned Cypressgrass,Awned Galingale ■

119209 Cyperus microlepis Boeck. = Cyperus submicrolepis Kük. ■☆

119210 Cyperus microstachyus Vahl = Cyperus amabilis Vahl ■☆

119211 Cyperus microstylis (C. B. Clarke) Mattf. et Kük. = Kyllinga microstyla C. B. Clarke ■☆

119212 Cyperus microumbellatus Lye;小伞莎草■☆

119213 Cyperus minimus K. Schum. = Pycreus minimus (K. Schum.) C. B. Clarke ■☆

119214 Cyperus minutulus K. Schum. = Pycreus hildebrandtii (K. Schum.) C. B. Clarke ■☆

119215 Cyperus minutus (C. B. Clarke) Kük.;微小莎草■☆

119216 Cyperus minutus Roth = Isolepis marginata (Thunb.) A. Dietr. ■☆

119217 Cyperus molliglumis Cherm.;毛颖莎草■☆

119218 Cyperus mollipes (C. B. Clarke) Kük. var. bulbocaulis (Boeck.) Kük. = Mariscus amomodorus (K. Schum.) Cufod. var. bulbocaulis (Boeck.) Cufod. ■☆

119219 Cyperus mollipes (C. B. Clarke) Kük. var. paolii (Chiov.) Kük. = Mariscus amomodorus (K. Schum.) Cufod. var. paolii (Chiov.) Cufod. ■☆

119220 Cyperus monocephalus (Rottb.) F. Muell.;单头莎草;Sharp Edge Sedge ■☆

119221 Cyperus monophyllus Vahl = Cyperus malaccensis Lam. subsp. monophyllus (Vahl) T. Koyama ■

119222 Cyperus monostachyos L. = Abildgaardia ovata (Burm. f.) Král ■

119223 Cyperus monostachyus (L.) Vahl = Fimbristylis monostachys (L.) Hassk. ■

119224 Cyperus monostachyus L. = Fimbristylis ovata (Burm. f.) J. Kern ■

119225 Cyperus monostigma C. B. Clarke = Cyperus clavinux C. B. Clarke ■☆

119226 Cyperus monroviensis Boeck. = Cyperus tenax Boeck. ■☆

119227 Cyperus morandinii Pic. Serm.;莫兰莎草■☆

119228 Cyperus mortonii (S. S. Hooper) Lye = Pycreus mortonii S. S. Hooper ■☆

119229 Cyperus mossii Turrill;莫西莎草■☆

119230 Cyperus mucronatus (L.) Mabille = Cyperus capitatus Vand. ■☆

119231 Cyperus mudugensis D. A. Simpson;穆杜格莎草■☆

119232 Cyperus muelleri Boeck.;米勒莎草■☆

119233 Cyperus multiglumis Turrill = Cyperus semitrifidus Schrad. ■☆

119234 Cyperus mundtii (Nees) Kunth = Pycreus mundtii Nees ■☆

119235 Cyperus muricatus Kük. = Pycreus muricatus (Kük.) Napper ■☆

119236 Cyperus mutisii (Kunth) Andersson;穆蒂斯莎草■☆

119237 Cyperus mutisii (Kunth) Andersson var. asper (Liebm.) Kük. = Cyperus mutisii (Kunth) Andersson ■☆

119238 Cyperus mutisii (Kunth) Andersson var. contractus Kük. = Cyperus mutisii (Kunth) Andersson ■☆

119239 Cyperus mutisii (Kunth) Andersson var. semitribrachiatus (Boeck.) Kük. = Cyperus mutisii (Kunth) Andersson ■☆

119240 Cyperus mwinilungensis Podlech;穆维尼莎草■☆

119241 Cyperus mwinilungensis Podlech var. maior ?;大穆维尼莎草■☆

119242 Cyperus myrmecias Ridl. = Mariscus myrmecias (Ridl.) C. B. Clarke ■☆

119243 Cyperus nanellus Ts. Tang et F. T. Wang;汾河莎草;Fenhe Cypressgrass,Fenhe Galingale ■

119244 Cyperus natalensis Hochst. ex Krauss;纳塔尔莎草■☆

119245 Cyperus natalensis Hochst. ex Krauss var. longibracteatus C. B. Clarke = Cyperus natalensis Hochst. ex Krauss ■☆

119246 Cyperus nemoralis Cherm.;森林莎草■☆

119247 Cyperus neotropicalis Alain = Lipocarpha maculata (Michx.) Torr. ■☆

119248 Cyperus nervoso-striatus Turrill = Cyperus esculentus L. ■☆

119249 Cyperus neurotropis Steud. = Pycreus sanguinolentus (Vahl) Nees f. neurotropis (Steud.) Cufod. ■☆

119250 Cyperus niger Ruiz et Pav.;黑色莎草■☆

119251 Cyperus niger Ruiz et Pav. var. castaneus (S. Watson) Kük. = Cyperus niger Ruiz et Pav. ■☆

119252 Cyperus niger Ruiz et Pav. var. rivularis (Kunth) V. E. Grant = Cyperus bipartitus Torr. ■☆

119253 Cyperus nigricans Steud. = Pycreus nigricans (Steud.) C. B. Clarke ■☆

119254 Cyperus nigricans Steud. var. firmior Kük. = Pycreus nigricans (Steud.) C. B. Clarke var. firmior (Kük.) Cherm. ■☆

119255 Cyperus nigripes (C. B. Clarke) Kük. = Kyllinga nigripes C. B. Clarke ■☆

119256 Cyperus nigrofuscus L. K. Dai;黑穗莎草;Blackspike Cypressgrass,Nigrescentspike Galingale ■

119257 Cyperus nigrofuscus L. K. Dai f. pallescens Huanet;线白穗莎草■

119258 Cyperus nigrofuscus L. K. Dai f. virescens (Hoffm.) Vahl;北莎草■

119259 Cyperus niigatensis Ohwi;新泻莎草■☆

119260 Cyperus nilagiricus Hochst. ex Steud. = Pycreus flavidus (Retz.) T. Koyama var. nilagiricus (Hochst. ex Steud.) C. Y. Wu ex Karthik. ■

119261 Cyperus nilagiricus Hochst. ex Steud. = Pycreus globosus (All.) Rchb. var. nilagiricus (Hochst.) C. B. Clarke ■

119262 Cyperus nipponicus Franch. et Sav.;白鳞莎草;Whitescale Cypressgrass,Whitescale Galingale ■

119263 Cyperus nipponicus Franch. et Sav. subsp. niigatensis (Ohwi) T. Koyama = Cyperus niigatensis Ohwi ■☆

119264 Cyperus nipponicus Franch. et Sav. subsp. spiralis (Ohwi) T. Koyama = Cyperus nipponicus Franch. et Sav. var. spiralis Ohwi ■☆

119265 Cyperus nipponicus Franch. et Sav. var. spiralis Ohwi;螺旋白鳞莎草■☆

119266 Cyperus nitens Retz. = Pycreus pumilus (L.) Domin ■

119267 Cyperus nitens Vahl = Pycreus pumilus (L.) Domin ■

119268 Cyperus nitidus Lam. = Pycreus nitidus (Lam.) J. Raynal ■☆
119269 Cyperus niveoides C. B. Clarke;雪莎草■☆
119270 Cyperus niveus Retz.;南莎草;Snowwhite Cypressgrass, Snowwhite Galingale ■
119271 Cyperus niveus Retz. var. flavissimus (Schrad.) Lye;黄南莎草■☆
119272 Cyperus niveus Retz. var. ledermannii (Kük.) Lye;莱德南莎草■☆
119273 Cyperus niveus Retz. var. leucocephalus (Kunth) Fosberg;白头南莎草■☆
119274 Cyperus niveus Retz. var. polyphyllus Boeck.;多叶南莎草■☆
119275 Cyperus niveus Retz. var. tisserantii (Cherm.) Lye;蒂斯朗特南莎草■☆
119276 Cyperus nodosus Willd. = Cyperus articulatus L. ■☆
119277 Cyperus nodosus Willd. var. subnodosus (Nees et Meyen) Boeck. = Cyperus articulatus L. ■☆
119278 Cyperus nubianus Gand. = Cyperus proteinolepis Steud. ■☆
119279 Cyperus nubicus C. B. Clarke = Cyperus rotundus L. ■
119280 Cyperus nudicaulis Poir. = Anosporum pectinatus (Vahl) Lye ■☆
119281 Cyperus nudiculmis Sieber ex C. B. Clarke = Cyperus schimperianus Steud. ■☆
119282 Cyperus nutans Vahl;垂穗莎草(十瘳楼,十廖楼);Nodding Galingale ■
119283 Cyperus nutans Vahl subsp. eleusinoides (Kunth) Koyama = Cyperus eleusinoides Kunth ■
119284 Cyperus nutans Vahl subsp. subprolixus (Kük.) T. Koyama;点头莎草■
119285 Cyperus nutans Vahl var. eleusinoides (Kunth) Haines = Cyperus eleusinoides Kunth ■
119286 Cyperus nutans Vahl var. subprolixus (Kük.) T. Koyama = Cyperus nutans Vahl subsp. subprolixus (Kük.) T. Koyama ■
119287 Cyperus nyassensis (Podlech) Lye = Mariscus nyasensis Podlech ■☆
119288 Cyperus obbiadensis Chiov.;奥比亚德莎草■☆
119289 Cyperus obliquus Nees = Cyperus pilosus Vahl var. obliquus (Nees) C. B. Clarke ■
119290 Cyperus oblongo-incrassatus Kük. = Mariscus taylori C. B. Clarke ■☆
119291 Cyperus oblongus (C. B. Clarke) Kük.;矩圆莎草■☆
119292 Cyperus oblongus (C. B. Clarke) Kük. = Kyllinga nervosa Steud. subsp. oblonga (C. B. Clarke) J.-P. Lebrun et Stork ■☆
119293 Cyperus oblongus (C. B. Clarke) Kük. subsp. flavus (C. B. Clarke) Lye = Kyllinga nervosa Steud. var. flava (C. B. Clarke) Lye ■☆
119294 Cyperus oblongus (C. B. Clarke) Kük. subsp. nervosus (Steud.) Lye = Kyllinga nervosa Steud. ■☆
119295 Cyperus oblongus (C. B. Clarke) Kük. var. ruwenzoriensis ? = Kyllinga nervosa Steud. var. ruwenzoriensis (C. B. Clarke) Lye ■☆
119296 Cyperus obsoletenervosus Peter et Kük. = Mariscus obsoletenervosus (Peter et Kük.) Greenway ■☆
119297 Cyperus obtusatus (J. Presl) Mattf. et Kük. var. africanus Kük. = Kyllinga erecta Schumach. var. africana (Kük.) S. S. Hooper ■☆
119298 Cyperus obtusiflorus Vahl;钝花莎草■☆
119299 Cyperus obtusiflorus Vahl = Cyperus niveus Retz. var. leucocephalus (Kunth) Fosberg ■☆
119300 Cyperus obtusiflorus Vahl var. flavissimus (Schrad.) Boeck.;黄钝花莎草■☆
119301 Cyperus obtusiflorus Vahl var. ledermannii Kük. = Cyperus niveus Retz. var. ledermannii (Kük.) Lye ■☆
119302 Cyperus obtusiflorus Vahl var. macrostachys (Graebn.) Robyns et Tournay;大穗莎草■☆
119303 Cyperus obtusiflorus Vahl var. sphaerocephalus (Vahl) Kük. = Cyperus obtusiflorus Vahl var. flavissimus (Schrad.) Boeck. ■☆
119304 Cyperus occidentalis Torr. = Cyperus erythrorhizos Muhl. ■☆
119305 Cyperus ochreatus Boeck. = Cyperus mutisii (Kunth) Andersson ■☆
119306 Cyperus ochrocarpus K. Schum. = Cyperus renschii Boeck. ■☆
119307 Cyperus ochrocephalus (Boeck.) C. B. Clarke = Cyperus rhynchosporoides Kük. ■☆
119308 Cyperus ochroleucus Boeck.;白绿莎草■☆
119309 Cyperus odorata Hell. subsp. cylindrica (Nees) T. Koyama = Kyllinga odorata Vahl ■☆
119310 Cyperus odorata Hell. var. cylindrica (Nees) Merr. = Kyllinga odorata Vahl ■☆
119311 Cyperus odoratus L. = Torulinium odoratum (L.) S. S. Hooper ■
119312 Cyperus odoratus L. var. acicularis (Schrad. ex Nees) O'Neill = Torulinium odoratum (L.) S. S. Hooper ■
119313 Cyperus odoratus L. var. engelmannii (Steud.) R. Carter, S. D. Jones et Wipff = Torulinium odoratum (L.) S. S. Hooper ■
119314 Cyperus odoratus L. var. engelmannii (Steud.) R. Carter, S. D. Jones et Wipff = Cyperus odoratus L. ■
119315 Cyperus odoratus L. var. squarrosus (Britton) Gilly = Torulinium odoratum (L.) S. S. Hooper ■
119316 Cyperus ohwii Kük.;大井氏莎草■☆
119317 Cyperus oligostachyus Kunth = Cyperus amabilis Vahl ■☆
119318 Cyperus onivensis Cherm. = Cyperus nemoralis Cherm. ■☆
119319 Cyperus orthostachyus Franch. et Sav.;毛笠莎草(牛草,三轮草,水蜈蚣);Upright-spiked Cypressgrass, Upright-spiked Galingale ■
119320 Cyperus orthostachyus Franch. et Sav. var. longibracteatus L. K. Dai;长苞三轮草;Longbract Upright-spiked Cypressgrass, Longbract Upright-spiked Galingale ■
119321 Cyperus orthostachyus Franch. et Sav. var. robustus Hara;粗壮毛笠莎草■
119322 Cyperus ossicaulis Lye;茎莎草■☆
119323 Cyperus ovatus Baldwin;卵莎草■☆
119324 Cyperus overlaetii (Cherm. ex S. S. Hooper et J. Raynal) Lye = Pycreus overlaetii Cherm. ex S. S. Hooper et J. Raynal ■☆
119325 Cyperus ovularis (Michx.) Torr. = Cyperus echinatus (L.) A. W. Wood ■☆
119326 Cyperus ovularis (Michx.) Torr. var. sphaericus Boeck. = Cyperus echinatus (L.) A. W. Wood ■☆
119327 Cyperus ovularis Boeck. = Cyperus cyperoides (L.) Kuntze subsp. flavus Lye ■☆
119328 Cyperus ovularis Boeck. var. americanus Boeck. = Cyperus echinatus (L.) A. W. Wood ■☆
119329 Cyperus ovularis Boeck. var. sphaericus Boeck. = Cyperus echinatus (L.) A. W. Wood ■☆
119330 Cyperus ovularis Boeck. var. wolfii (A. W. Wood) Kük. = Cyperus echinatus (L.) A. W. Wood ■☆
119331 Cyperus owahuensis Nees = Mariscus javanicus (Houtt.) Merr. et F. P. Metcalf ■
119332 Cyperus owanii Boeck.;欧文莎草;Owan's Flatsedge ■☆
119333 Cyperus oxycarioides Britton = Cyperus odoratus L. ■
119334 Cyperus oxycarioides Britton = Torulinium odoratum (L.) S. S. Hooper ■

119335 Cyperus pachyrhizus Nees;粗根莎草■☆
119336 Cyperus pachystylus (Kük.) Kük. = Kyllinga pachystyla Kük. ■☆
119337 Cyperus pacificus (Ohwi) Ohwi;太平洋莎草■☆
119338 Cyperus pagotii (J. Raynal) Lye = Pycreus pagotii J. Raynal ■☆
119339 Cyperus pallens (Liebm.) Standl. et Steyerm. = Cyperus thyrsiflorus Jungh. ■☆
119340 Cyperus pallescens Desf. = Cyperus longus L. ■☆
119341 Cyperus pallidicolor (Kük.) G. C. Tucker;苍白莎草■☆
119342 Cyperus pangorei Rottb.;红翅莎草;Redwing Cypressgrass, Redwing Galingale ■
119343 Cyperus pannonicus Jacq. = Juncellus pannonicus (Jacq.) C. B. Clarke ■
119344 Cyperus paolii Chiov. = Mariscus amomodorus (K. Schum.) Cufod. var. paolii (Chiov.) Cufod. ■☆
119345 Cyperus pappii Gand.;保普莎草■☆
119346 Cyperus papyroides Poir. = Cyperus prolifer Lam. ■☆
119347 Cyperus papyrus L.;纸莎草(纸蔺);Egyptian Paper Plant, Egyptian Paper Reed, Egyptian Paper Rush, Egyptian Paper-reed, Egyptian Papyrus, Egyptian Reed, Giant Papyrus, Paper Reed, Paper Sedge, Papyrus, Papyrus Plant ■
119348 Cyperus papyrus L. 'Nana';矮生纸莎草■☆
119349 Cyperus papyrus L. subsp. antiquorum (Willd.) Chiov. = Cyperus papyrus L. ■
119350 Cyperus papyrus L. subsp. nyassicus Chiov. = Cyperus papyrus L. ■
119351 Cyperus papyrus L. subsp. zairensis (Chiov.) Kük.;扎伊尔莎草■☆
119352 Cyperus parishii Britton ex Parish;帕尔什莎草■☆
119353 Cyperus parryi Britton ex C. B. Clarke = Cyperus spectabilis Spreng. ■☆
119354 Cyperus parvinux C. B. Clarke = Cyperus rupestris Kunth var. parvinux (C. B. Clarke) Kük. ■☆
119355 Cyperus patens Vahl = Pycreus pumilus (L.) Domin ■
119356 Cyperus pauper A. Rich. = Pycreus pauper (A. Rich.) C. B. Clarke ■☆
119357 Cyperus pectinatus Roxb. = Cyperus compressus L. ■
119358 Cyperus pectinatus Vahl = Anosporum pectinatus (Vahl) Lye ■☆
119359 Cyperus pectiniformis Roem. et Schult. = Cyperus compressus L. ■
119360 Cyperus pedunculatus (R. Br.) J. Kern = Remirea maritima Aubl. ■
119361 Cyperus pendulus Cherm.;下垂莎草■☆
119362 Cyperus pennatus Lam. = Mariscus javanicus (Houtt.) Merr. et F. P. Metcalf ■
119363 Cyperus permacer C. B. Clarke;瘦弱莎草■☆
119364 Cyperus permutatus Boeck. = Pycreus permutatus (Boeck.) Napper ■☆
119365 Cyperus persquarrosus T. Koyama = Lipocarpha nana (A. Rich.) Cherm. ■☆
119366 Cyperus peruvianus (Lam.) F. N. Williams = Kyllinga peruviana Lam. ■☆
119367 Cyperus peteri Kük. ex Peter = Kyllinga peteri (Kük. ex Peter) Lye ■☆
119368 Cyperus petersianus Boeck. = Cyperus involucratus Rottb. ■
119369 Cyperus petherickii C. B. Clarke = Cyperus dives Delile ■☆
119370 Cyperus phaeolepis Cherm.;马达加斯加莎草;Madagascar Flatsedge ■☆
119371 Cyperus phaeolepis Cherm. var. albiflorus (Cherm.) Cherm. = Cyperus phaeolepis Cherm. ■☆
119372 Cyperus phaeorrhizus K. Schum. = Cyperus denudatus L. f. ■☆
119373 Cyperus phaeorrhizus K. Schum. var. pallidus Gilli = Cyperus denudatus L. f. ■☆
119374 Cyperus phaeorrhizus K. Schum. var. princeae (C. B. Clarke) Kük. = Cyperus denudatus L. f. ■☆
119375 Cyperus phillipsiae (C. B. Clarke) Kük. = Mariscus phillipsiae C. B. Clarke ■☆
119376 Cyperus phymatodes Muhl. = Cyperus esculentus L. var. leptostachyus Boeck. ■☆
119377 Cyperus pilosulus K. Schum. = Mariscus pilosulus (K. Schum.) C. B. Clarke ■☆
119378 Cyperus pilosus Vahl;毛轴莎草(三角草,三棱草);Fuzzy Flatsedge, Pilose Cypressgrass, Pilose Galingale ■
119379 Cyperus pilosus Vahl var. oblicuus (Nees) C. B. Clarke;白花毛轴莎草;Oblique Pilose Cypressgrass, Oblique Pilose Galingale ■
119380 Cyperus pilosus Vahl var. oblicuus (Nees) C. B. Clarke = Cyperus pilosus Vahl ■
119381 Cyperus pilosus Vahl var. pauciflorus L. K. Dai;少花毛轴莎草;Fewflower Pilose Galingale ■
119382 Cyperus pilosus Vahl var. purpurascens L. K. Dai;紫穗毛轴莎草;Purple-spike Galingale ■
119383 Cyperus pinguis (C. B. Clarke) Mattf. et Kük. = Kyllinga elatior Kunth ■☆
119384 Cyperus planifolius Rich.;平叶莎草■☆
119385 Cyperus plankii Britton = Cyperus croceus Vahl ■☆
119386 Cyperus plantaginifolius Cherm.;车前叶莎草■☆
119387 Cyperus plantaginifolius Cherm. var. minor Cherm. = Cyperus plantaginifolius Cherm. ■☆
119388 Cyperus platycaulis Baker;扁茎莎草■☆
119389 Cyperus platycaulis Baker var. platycaulis = Cyperus denudatus L. f. var. lucenti-nigricans (K. Schum.) Kük. ■☆
119390 Cyperus platystylis R. Br.;宽柱莎草■
119391 Cyperus plukenetii Fernald;普拉莎草;Plukenet's Sedge, Plukenet's Umbrella Sedge ■☆
119392 Cyperus pluricephalus Lye;多头莎草■☆
119393 Cyperus plurinervosus Bodard;密脉莎草■☆
119394 Cyperus podocarpus Boeck.;足果莎草■☆
119395 Cyperus poecilus C. B. Clarke;杂色莎草■☆
119396 Cyperus poecilus C. B. Clarke var. evolutus Kük. = Cyperus poecilus C. B. Clarke ■☆
119397 Cyperus poiformis Pursh = Cyperus flavescens L. ■☆
119398 Cyperus pollardii Britton = Cyperus ovatus Baldwin ■☆
119399 Cyperus polyphyllus Vahl;多叶莎草■
119400 Cyperus polyphyllus Vahl = Cyperus bulbosus Vahl ■☆
119401 Cyperus polystachyos Rottb. = Pycreus polystachyos (Rottb.) P. Beauv. ■
119402 Cyperus polystachyos Rottb. subsp. laxiflorus (Benth.) Lye = Pycreus polystachyos (Rottb.) P. Beauv. var. laxiflorus (Benth.) C. B. Clarke ■☆
119403 Cyperus polystachyos Rottb. var. baroni C. B. Clarke = Pycreus intactus (Vahl) J. Raynal ■☆
119404 Cyperus polystachyos Rottb. var. brevispiculatus K. C. How = Pycreus polystachyos (Rottb.) P. Beauv. var. brevispiculatus (F. C. How) L. K. Dai ■

119405 Cyperus polystachyos Rottb. var. ferrugineus (Poir.) Boeck. = Pycreus intactus (Vahl) J. Raynal ■☆

119406 Cyperus polystachyos Rottb. var. laxiflorus Benth. = Pycreus polystachyos (Rottb.) P. Beauv. var. laxiflorus (Benth.) C. B. Clarke ■☆

119407 Cyperus polystachyos Rottb. var. leptostachyus Boeck. = Cyperus polystachyus (Rottb.) P. Beauv. ■

119408 Cyperus polystachyos Rottb. var. macrostachyos Boeck. = Cyperus filicinus Vahl ■☆

119409 Cyperus polystachyos Rottb. var. micans (Kunth) C. B. Clarke = Pycreus intactus (Vahl) J. Raynal ■☆

119410 Cyperus polystachyus (Rottb.) P. Beauv.;多枝扁莎草(多枝扁莎,三方草,水草);Branchy Flatsedge, Branchy Pycreus, Manyspike Galingale, Many-spiked Umbrella Sedge, Numerous-branched Flat-sedge ■

119411 Cyperus polystachyus Rottb. = Pycreus polystachyus (Rottb.) P. Beauv. ■

119412 Cyperus prasinus Kunth = Cyperus leptocladus Kunth ■☆

119413 Cyperus pratensis Boeck. = Mariscus pratensis (Boeck.) Cufod. ■☆

119414 Cyperus pratensis Boeck. var. laxa C. B. Clarke = Mariscus pratensis (Boeck.) Cufod. ■☆

119415 Cyperus pratensis Boeck. var. radiatus C. B. Clarke = Mariscus pratensis (Boeck.) Cufod. ■☆

119416 Cyperus prieurianus (Steud.) Koyama = Lipocarpha prieuriana Steud. ■☆

119417 Cyperus princeae C. B. Clarke = Cyperus denudatus L. f. ■☆

119418 Cyperus pringlei Britton = Cyperus tetragonus Elliott ■☆

119419 Cyperus procerus Rottb.;拟毛轴莎草■

119420 Cyperus procerus Rottb. var. stenanthus Kük.;细花拟毛轴莎草■

119421 Cyperus procerus Rottb. var. stenanthus Kük. = Cyperus procerus Rottb. ■

119422 Cyperus prolifer Lam.;多育莎草(微小莎草);Dwarf Papyrus, Miniature Flatsedge ■☆

119423 Cyperus prolixus Kunth;莎草蚊子;Mosquito Flatsedge ■☆

119424 Cyperus proteinolepis Steud.;良鳞枝莎草■☆

119425 Cyperus pseudocallistus Kük. = Mariscus vestitus (Hochst. ex Krauss) C. B. Clarke ■☆

119426 Cyperus pseudodiaphanus (S. S. Hooper) Lye = Pycreus pseudodiaphanus S. S. Hooper ■☆

119427 Cyperus pseudohildebrandtii Kük. = Pycreus hildebrandtii (K. Schum.) C. B. Clarke ■☆

119428 Cyperus pseudokyllingioides Kük. = Courtoisina cyperoides (A. Dietr.) Soják ■

119429 Cyperus pseudokyllingioides Kük. = Courtoisina cyperoides (Roxb.) Soják ■

119430 Cyperus pseudokyllingioides Kük. var. africanus Kük. = Courtoisina cyperoides (Roxb.) Soják ■

119431 Cyperus pseudokyllingoides Kük. = Courtoisia cyperoides Nees ■

119432 Cyperus pseudoleptocladus Kük.;假细枝莎草■☆

119433 Cyperus pseudomarginatus Dinter;假边莎草■☆

119434 Cyperus pseudoniveus Boeck. = Cyperus margaritaceus Vahl ■☆

119435 Cyperus pseudosomaliensis Kük. ex Kük. = Mariscus somaliensis C. B. Clarke ■☆

119436 Cyperus pseudosphacelatus Chiov. = Cyperus esphacelatus Kük. ■☆

119437 Cyperus pseudostrigosus Steud. = Torulinium odoratum (L.) S. S. Hooper ■

119438 Cyperus pseudothyrsiflorus (Kük.) R. Carter et S. D. Jones;假伞莎草■☆

119439 Cyperus pseudovegetus Steud. var. arenicola (Steud.) Kük. = Cyperus reflexus Vahl ■☆

119440 Cyperus pseudovestitus (C. B. Clarke) Kük.;假覆被伞莎草■☆

119441 Cyperus pseudovestitus (C. B. Clarke) Kük. var. perrieri (Cherm.) Kük.;佩里耶莎草■☆

119442 Cyperus psilostachys (C. B. Clarke) Kük. = Mariscus psilostachys C. B. Clarke ■☆

119443 Cyperus pubens Kük. = Mariscus pubens (Kük.) Podlech ■☆

119444 Cyperus pubescens Steud. = Bulbostylis puberula (Poir.) C. B. Clarke ■

119445 Cyperus pulchellus R. Br.;非洲小莎草■☆

119446 Cyperus pulcher Thunb.;美丽莎草■☆

119447 Cyperus pulcherrimus Willd. ex Kunth;极美莎草■☆

119448 Cyperus pumilus L. = Pycreus pumilus (L.) Domin ■

119449 Cyperus pumilus L. var. membranaceus ? = Pycreus pumilus (L.) Domin ■

119450 Cyperus pumilus L. var. muticus (Boeck.) C. B. Clarke = Pycreus pumilus (L.) Domin ■

119451 Cyperus pumilus L. var. patens (Vahl) Kük. = Pycreus pumilus (L.) Domin ■

119452 Cyperus pungens Boeck. var. multiculmis ? = Cyperus jeminicus Rottb. ■☆

119453 Cyperus pungens Boeck. var. tenuis ? = Cyperus macrorrhizus Nees ■☆

119454 Cyperus purpurascens Vahl = Cyperus planifolius Rich. ■☆

119455 Cyperus purpureoglandulosus Mattf. et Kük. = Kyllinga sphaerocephala Boeck. ■☆

119456 Cyperus purpureoviridis Lye;紫绿莎草■☆

119457 Cyperus purpureus Boeck. = Cyperus schinzii Boeck. ■☆

119458 Cyperus pustulatus Vahl;泡状莎草■☆

119459 Cyperus pustulatus Vahl var. debilis Kük.;弱小泡状莎草■☆

119460 Cyperus pustulatus Vahl var. djalonensis (A. Chev.) Kük. = Cyperus pustulatus Vahl ■☆

119461 Cyperus pygmaeus Rottb.;矮莎草;Low Cypressgrass, Low Galingale ■

119462 Cyperus pygmaeus Rottb. = Cyperus michelianus (L.) Link subsp. pygmaeus (Rottb.) Asch. et Graebn. ■

119463 Cyperus quadriflorus Boeck. = Mariscus richardii Steud. ■☆

119464 Cyperus racemosus Poir. = Cyperus alternifolius L. ■

119465 Cyperus radians Nees et Meyen = Mariscus radians (Nees et Meyen) Ts. Tang et F. T. Wang ■

119466 Cyperus radians Nees et Meyen ex Kunth = Mariscus radians (Nees et Meyen) Ts. Tang et F. T. Wang ■

119467 Cyperus radians Nees et Meyen ex Nees = Mariscus radians (Nees et Meyen) Ts. Tang et F. T. Wang ■

119468 Cyperus radiatus Vahl = Cyperus imbricatus Retz. ■

119469 Cyperus radiatus Vahl var. capitatus Boeck. = Cyperus imbricatus Retz. ■

119470 Cyperus radiatus Vahl var. elongatus Boeck. = Cyperus imbricatus Retz. var. elongatus (Boeck.) L. K. Dai ■

119471 Cyperus rarissimus Steud. = Bulbostylis rarissima (Steud.) C. B. Clarke ■☆

119472 Cyperus recurvispicatus Lye;曲穗莎草■☆

119473 Cyperus recurvus Vahl = Cyperus cuspidatus Kunth ■

119474 Cyperus reflexus Vahl = Cyperus refractus Engelm. ex Boeck. ■☆

119475 Cyperus reflexus Vahl var. fraternus (Kunth) Kuntze = Cyperus reflexus Vahl ■☆

119476 Cyperus refractus Engelm. ex Boeck.;反折莎草;Teasel Nut Sedge ■☆

119477 Cyperus regiomontanus Britton var. pallens (Liebm.) Kük. = Cyperus thyrsiflorus Jungh. ■☆

119478 Cyperus rehmannianus (C. B. Clarke) Boeck. ex Kuntze = Pycreus flavescens (L.) P. Beauv. ex Rchb. ■☆

119479 Cyperus rehmii Merxm.;雷姆莎草■☆

119480 Cyperus remotispicatus S. S. Hooper;寡穗莎草■☆

119481 Cyperus remotus (C. B. Clarke) Kük. = Mariscus remotus C. B. Clarke ■☆

119482 Cyperus renschii Boeck.;伦施莎草■☆

119483 Cyperus renschii Boeck. var. scabrida Lye;微糙莎草■☆

119484 Cyperus repens Elliott = Cyperus esculentus L. var. leptostachyus Boeck. ■☆

119485 Cyperus retroflexus Buckley;反曲莎草■☆

119486 Cyperus retrofractus (L.) Torr. var. hystricinus (Fernald) Kük. = Cyperus hystricinus Fernald ■☆

119487 Cyperus retrorsus Chapm.;倒向莎草■☆

119488 Cyperus retrorsus Chapm. var. curtisii (C. B. Clarke) Kük. = Cyperus ovatus Baldwin ■☆

119489 Cyperus retrorsus Chapm. var. deeringianus (Britton et Small) Fernald et Griscom = Cyperus ovatus Baldwin ■☆

119490 Cyperus retrorsus Chapm. var. robustus Kük. = Cyperus croceus Vahl ■☆

119491 Cyperus retzii Nees = Cyperus rotundus L. ■

119492 Cyperus rhynchosporoides Kük.;刺子莞莎草■☆

119493 Cyperus richardii Steud. = Kyllinga bulbosa P. Beauv. ■☆

119494 Cyperus richardii Steud. var. angustior (C. B. Clarke) Kük. = Kyllinga bulbosa P. Beauv. ■☆

119495 Cyperus ridleyi Mattf. et Kük. = Kyllinga pauciflora Ridl. ■☆

119496 Cyperus rigidifolius Steud.;挺叶莎草■☆

119497 Cyperus rivularis Kunth;溪畔莎草;River Cyperus ■☆

119498 Cyperus rivularis Kunth = Cyperus bipartitus Torr. ■☆

119499 Cyperus rivularis Kunth f. elutus (C. B. Clarke) Kük. = Cyperus bipartitus Torr. ■☆

119500 Cyperus robinsonii Podlech;鲁滨逊莎草■☆

119501 Cyperus robustus Kunth = Cyperus drummondii Torr. et Hook. ■☆

119502 Cyperus rothianus Roem. et Schult. = Isolepis marginata (Thunb.) A. Dietr. ■☆

119503 Cyperus rotundus L.;香附子(草头香,出莎,地糕草,地沟草,地贯草,地久姜,地韭姜,地藾根,地毛,吊马鬃,隔夜抽,姑娘草,藒,藒侯,侯莎,回头青,荆三棱,苦羌头,辣姜草,雷公草,雷公头,米珠子,目萃哆,雀头香,三棱草,莎草,莎随,山莎,鼠莎,水莎,缩缩草,土香草,香附,香头草,小三棱,野韭菜,野韭姜,夜夜青,猪荸荠,猪通草茹,猪鬃草);Coco Grass, Coco-grass, Nut Grass, Nutgrass, Nut-grass, Nutgrass Cypressgrass, Nut-grass Flat Sedge, Nutgrass Flatsedge, Nut-grass Flat-sedge, Nutgrass Galingale, Purple Nutgrass, Purple Nut-sedge ■

119504 Cyperus rotundus L. f. contractus Kük. = Cyperus rotundus L. ■

119505 Cyperus rotundus L. subsp. merkeri (C. B. Clarke) Kük.;迈尔香附子■☆

119506 Cyperus rotundus L. subsp. retzii (Nees) Kük. = Cyperus rotundus L. ■

119507 Cyperus rotundus L. subsp. tuberosus (Rottb.) Kük. = Cyperus tuberosus Rottb. ■

119508 Cyperus rotundus L. var. alpinus Chiov. = Cyperus rotundus L. ■

119509 Cyperus rotundus L. var. comosus (Sibth. et Sm.) Batt. et Trab. = Cyperus rotundus L. ■

119510 Cyperus rotundus L. var. fenzelianus (Steud.) Habashy = Cyperus rotundus L. ■

119511 Cyperus rotundus L. var. nubicus (C. B. Clarke) Kük. = Cyperus rotundus L. ■

119512 Cyperus rotundus L. var. platystachys C. B. Clarke = Cyperus rotundus L. var. tuberosus (Rottb.) Kük. ■☆

119513 Cyperus rotundus L. var. quimoyensis L. K. Dai;金门莎草;Jinmen Galingale ■

119514 Cyperus rotundus L. var. quimoyensis L. K. Dai = Cyperus rotundus L. ■

119515 Cyperus rotundus L. var. taylorii (C. B. Clarke) Kük.;泰勒莎草■☆

119516 Cyperus rotundus L. var. tetrastachyos (Desf.) Trab. = Cyperus rotundus L. ■

119517 Cyperus rotundus L. var. tuberosus (Rottb.) Kük.;块状香附子■☆

119518 Cyperus rotundus L. var. yoshinagae (Ohwi) Ohwi;吉永香附子■☆

119519 Cyperus rubicundus Vahl;稍红莎草■☆

119520 Cyperus rubicundus Vahl var. longimucronatus (Kük.) Cufod. = Cyperus rubicundus Vahl ■☆

119521 Cyperus rufescens Torr. et Hook. var. denticarinatus Britton = Cyperus acuminatus Roxb. ex C. B. Clarke ■

119522 Cyperus ruficomus Buckley = Cyperus esculentus L. var. macrostachyus Boeck. ■☆

119523 Cyperus rufostriatus C. B. Clarke ex Cherm.;红纹莎草■☆

119524 Cyperus rupestris Kunth;岩生莎草■☆

119525 Cyperus rupestris Kunth var. parvinux (C. B. Clarke) Kük.;小果岩生莎草■☆

119526 Cyperus rusbyi Britton = Cyperus sphaerolepis Boeck. ■☆

119527 Cyperus sabulicola Ridl. = Cyperus tenax Boeck. ■☆

119528 Cyperus sabulosus (Mart. et Schrad. ex Nees) Steud. = Cyperus flavicomus Michx. ■☆

119529 Cyperus sanguinolentus Vahl = Pycreus sanguinolentus (Vahl) Nees ex C. B. Clarke ■

119530 Cyperus sanguinolentus Vahl f. flaccidus Boeck. = Pycreus sanguinolentus (Vahl) Nees f. flaccidus (Boeck.) Cufod. ■☆

119531 Cyperus sanguinolentus Vahl f. humilis (Miq.) Kük. = Pycreus sanguinolentus (Vahl) Nees ex C. B. Clarke f. humilis (Miq.) L. K. Dai ■

119532 Cyperus sanguinolentus Vahl f. melanocephalus (Miq.) Kük. = Pycreus sanguinolentus (Vahl) Nees ex C. B. Clarke f. melanocephalus (Miq.) L. K. Dai ■

119533 Cyperus sanguinolentus Vahl f. neurotropis (Steud.) Kük. = Pycreus sanguinolentus (Vahl) Nees f. neurotropis (Steud.) Cufod. ■☆

119534 Cyperus sanguinolentus Vahl f. rubromarginatus (Schrenk.) Kük. = Pycreus sanguinolentus (Vahl) Nees ex C. B. Clarke f. rubromarginatus (Schrenk) L. K. Dai ■

119535 Cyperus sanguinolentus Vahl f. spectabilis (Makino) Ohwi;雅观莎草■☆

119536 Cyperus sanguinolentus Vahl subsp. nairobiensis (Lye) Lye = Pycreus sanguinolentus (Vahl) Nees subsp. nairobiensis Lye ex C.

B. Clarke ■☆

119537 Cyperus sanguinolentus Vahl var. nipponicus Ohwi = Cyperus sanguinolentus Vahl ■

119538 Cyperus sanguinolentus Vahl var. spectabilis Makino = Cyperus sanguinolentus Vahl f. spectabilis (Makino) Ohwi ■☆

119539 Cyperus sanguinolentus Vahl var. uniceps C. B. Clarke = Pycreus mundtii Nees ■☆

119540 Cyperus scaberrimus Nees = Cyperus spectabilis Spreng. ■☆

119541 Cyperus scabricaulis Lye;糙茎莎草■☆

119542 Cyperus scariosus R. Br.;干膜莎草■☆

119543 Cyperus schimperianus Steud.;欣珀莎草■☆

119544 Cyperus schinzii Boeck.;欣兹莎草■☆

119545 Cyperus schlechteri C. B. Clarke;施莱莎草■☆

119546 Cyperus schoenoides Griseb. = Cyperus capitatus Vand. ■☆

119547 Cyperus schweinfurthianus Boeck. = Cyperus tenuiculmis Boeck. var. schweinfurthianus (Boeck.) S. S. Hooper ■☆

119548 Cyperus schweinfurthii (Chiov.) Kük. = Mariscus schweinfurthii Chiov. ■☆

119549 Cyperus schweinitzii Torr.;史威尼兹莎草;Great Plains Sand Sedge, Nut Sedge, Schweinitz's Flatsedge, Schweinitz's Cyperus, Schweinitz's Flat Sedge ■☆

119550 Cyperus schweinitzii Torr. var. debilis Britton = Cyperus sphaerolepis Boeck. ■☆

119551 Cyperus scirpoides Vahl = Cyperus crassipes Vahl ■☆

119552 Cyperus scleropodus Chiov. = Mariscus scleropodus (Chiov.) Cufod. ■☆

119553 Cyperus semitribrachiatus Boeck. = Cyperus mutisii (Kunth) Andersson ■☆

119554 Cyperus semitrifidus Schrad.;半三裂莎草■☆

119555 Cyperus semitrifidus Schrad. var. apricus (Ridl.) Kük. = Cyperus apricus Ridl. ■☆

119556 Cyperus semitrifidus Schrad. var. multiglumis (Turrill) Kük. = Cyperus semitrifidus Schrad. ■☆

119557 Cyperus senegalensis (C. B. Clarke) Mattf. et Kük. = Kyllinga erecta Schumach. var. polyphylla (Willd. ex Kunth) S. S. Hooper ■☆

119558 Cyperus serotinus Rottb. = Juncellus serotinus (Rottb.) C. B. Clarke ■

119559 Cyperus serotinus Rottb. f. depauperatus Kük. = Juncellus serotinus (Rottb.) C. B. Clarke f. depauperatus (Kük.) L. K. Dai ■

119560 Cyperus serotinus Rottb. var. inundatus (Roxb.) Kük. = Juncellus serotinus (Rottb.) C. B. Clarke var. inundatus (Roxb.) L. K. Dai ■

119561 Cyperus serpens Cherm.;蛇形莎草■☆

119562 Cyperus sesquiflorus (Torr.) Mattf. et Kük. = Kyllinga odorata Vahl ■☆

119563 Cyperus sesquiflorus (Torr.) Mattf. et Kük. f. spinulosus Kük. = Kyllinga odorata Vahl ■☆

119564 Cyperus sesquiflorus (Torr.) Mattf. et Kük. subsp. appendiculatus (K. Schum.) Lye = Kyllinga appendiculata K. Schum. ■☆

119565 Cyperus sesquiflorus (Torr.) Mattf. et Kük. subsp. cylindrica (Nees) T. Koyama = Kyllinga odorata Vahl ■☆

119566 Cyperus sesquiflorus (Torr.) Mattf. et Kük. subsp. cylindricus (Nees) T. Koyama = Kyllinga cylindrica Nees ■

119567 Cyperus sesquiflorus (Torr.) Mattf. et Kük. subsp. cylindricus (Nees) T. Koyama = Cyperus sesquiflorus (Torr.) Mattf. et Kük. var. cylindricus (Nees) Kük. ■

119568 Cyperus sesquiflorus (Torr.) Mattf. et Kük. var. cylindricus (Nees) Kük. = Kyllinga cylindrica Nees ■

119569 Cyperus sesquiflorus (Torr.) Mattf. et Kük. var. cylindricus (Nees) Kük. = Kyllinga odorata Vahl ■☆

119570 Cyperus sesquiflorus (Torr.) Mattf. et Kük. var. major (C. B. Clarke) Kük. = Kyllinga appendiculata K. Schum. ■☆

119571 Cyperus sesquiflorus (Torr.) Mattf. et Kük. var. subtriceps (Nees) T. Koyama = Kyllinga odorata Vahl ■☆

119572 Cyperus sesquiflorus (Torr.) Mattf. et Kük. var. subtriceps (Nees) T. Koyama = Cyperus sesquiflorus (Torr.) Mattf. et Kük. var. cylindricus (Nees) Kük. ■

119573 Cyperus setaceus Retz. = Heleocharis chaetaria Roem. et Schult. ■

119574 Cyperus setiformis Korsh. = Cyperus diaphanus Schrad. ex Roem. et Schult. ■

119575 Cyperus setiformis Korsh. = Pycreus setiformis (Korsh.) Nakai ■

119576 Cyperus setigerus Torr. et Hook.;刚毛莎草;Bristled Umbrella Sedge ■☆

119577 Cyperus sexangularis Nees;六棱莎草■☆

119578 Cyperus shandongense F. Z. Li;山东白鳞莎草;Shandong Galingale ■

119579 Cyperus shimadae Ohwi = Cyperus radians Nees et Meyen ex Kunth ■

119580 Cyperus sieberianus (Nees ex Steud.) K. Schum. var. polyphylla (Steud.) K. Schum. = Mariscus steudelianus (Boeck.) Cufod. ■☆

119581 Cyperus simaoensis Y. Y. Qian;思茅莎草■

119582 Cyperus sinensis Debeaux = Mariscus radians (Nees et Meyen) Ts. Tang et F. T. Wang ■

119583 Cyperus sinensis Debeaux var. floribundus E. G. Camus = Mariscus radians (Nees et Meyen) Ts. Tang et F. T. Wang var. floribundus (E. G. Camus) S. M. Huang ■

119584 Cyperus smithianus Ridl. = Pycreus smithianus (Ridl.) C. B. Clarke ■☆

119585 Cyperus smithii McLean = Cyperus leptocladus Kunth ■☆

119586 Cyperus socialis C. B. Clarke = Mariscus socialis (C. B. Clarke) S. S. Hooper ■☆

119587 Cyperus solidus Kunth;污浊莎草■☆

119588 Cyperus solidus Kunth var. elatior ? = Cyperus solidus Kunth ■☆

119589 Cyperus somalicus Gand. = Cyperus niveus Retz. var. leucocephalus (Kunth) Fosberg ■☆

119590 Cyperus somalidunensis Lye;索马里砂丘莎草■☆

119591 Cyperus somaliensis C. B. Clarke;索马里莎草■☆

119592 Cyperus sonderi J. A. Schmidt = Pycreus polystachyos (Rottb.) P. Beauv. ■

119593 Cyperus soongoricus Kar. et Kir.;准噶尔莎草■

119594 Cyperus sorostachys Boeck. = Cyperus pulchellus R. Br. ■☆

119595 Cyperus soyauxii Boeck. = Mariscus soyauxii (Boeck.) C. B. Clarke ■☆

119596 Cyperus soyauxii Boeck. subsp. pallescens Lye;变苍白莎草■☆

119597 Cyperus spathaceus L. = Dulichium arundinaceum (L.) Britton ■☆

119598 Cyperus speciosus Vahl = Cyperus odoratus L. ■

119599 Cyperus speciosus Vahl = Torulinium odoratum (L.) S. S. Hooper ■

119600 Cyperus speciosus Vahl var. ferruginescens (Boeck.) Britton = Torulinium odoratum (L.) S. S. Hooper ■

119601 Cyperus speciosus Vahl var. squarrosus Britton = Torulinium odoratum (L.) S. S. Hooper ■

119602 Cyperus spectabilis Spreng. ;美花莎草;Showy Sedge ■☆
119603 Cyperus spectabilis Spreng. var. coarctatus Boeck. = Cyperus manimae Kunth ■☆
119604 Cyperus spectabilis Spreng. var. filiformis Boeck. = Cyperus spectabilis Spreng. ■☆
119605 Cyperus spectabilis Spreng. var. parryi (C. B. Clarke) Kük. = Cyperus spectabilis Spreng. ■☆
119606 Cyperus spectabilis Spreng. var. scaberrimus (Nees) Boeck. = Cyperus spectabilis Spreng. ■☆
119607 Cyperus sphacelatus Rottb. ;毒莎草■☆
119608 Cyperus sphaeranthelus Chiov. ;球花莎草■☆
119609 Cyperus sphaerocephalus Vahl = Cyperus obtusiflorus Vahl var. flavissimus (Schrad.) Boeck. ■☆
119610 Cyperus sphaerocephalus Vahl var. leucocephalus Kunth = Cyperus niveus Retz. var. leucocephalus (Kunth) Fosberg ■☆
119611 Cyperus sphaerolepis Boeck. ;球鳞莎草;Rusby's Sedge ■☆
119612 Cyperus sphaerospermoides Cherm. = Cyperus denudatus L. f. ■☆
119613 Cyperus sphaerospermus Schrad. ;球籽莎草■☆
119614 Cyperus squarrosus L. ;叉开莎草■☆
119615 Cyperus squarrosus L. var. parvus Britton = Cyperus odoratus L. ■
119616 Cyperus squarrosus L. var. parvus Britton = Torulinium odoratum (L.) S. S. Hooper ■
119617 Cyperus stenolepis Torr. = Cyperus strigosus L. ■☆
119618 Cyperus steudelianus Boeck. = Mariscus steudelianus (Boeck.) Cufod. ■☆
119619 Cyperus stoloniferus Retz. ; 粗 根 茎 莎 草; Stolon-bearing Cypressgrass, Stolon-bearing Galingale ■
119620 Cyperus stramineo-ferrugineus Kük. = Mariscus stramineo-ferrugineus (Kük.) Napper ■☆
119621 Cyperus striatulus Vahl = Cyperus obtusiflorus Vahl ■☆
119622 Cyperus strictus Roxb. = Pycreus flavidus (Retz.) T. Koyama var. strictus (Roxb.) C. Y. Wu ex Karthik. ■
119623 Cyperus strictus Roxb. = Pycreus globosus (All.) Rchb. var. strictus (Roxb.) C. B. Clarke ■
119624 Cyperus strigosus L. ; 糙 伏 毛 莎 草; False Nut Sedge, False Nutgrass, False Nutsedge, Straw-colored Cyperus, Umbrella Sedge ■☆
119625 Cyperus strigosus L. f. robustior Kunth = Cyperus strigosus L. ■☆
119626 Cyperus strigosus L. var. capitatus Boeck. = Cyperus strigosus L. ■☆
119627 Cyperus strigosus L. var. gracilis Britton = Cyperus lentiginosus Millsp. et Chase ■☆
119628 Cyperus strigosus L. var. hansenii (Britton) Kük. = Cyperus strigosus L. ■☆
119629 Cyperus strigosus L. var. multiflorus Geise = Cyperus strigosus L. ■☆
119630 Cyperus strigosus L. var. robustior (Kunth) Britton = Cyperus strigosus L. ■☆
119631 Cyperus strigosus L. var. robustior Britton = Cyperus strigosus L. ■☆
119632 Cyperus strigosus L. var. stenolepis (Torr.) Kük. = Cyperus strigosus L. ■☆
119633 Cyperus stuhlmannii C. B. Clarke ex K. Schum. = Cyperus usitatus Burch. var. stuhlmannii (C. B. Clarke ex K. Schum.) Lye ■☆
119634 Cyperus subaequalis Baker;近对称莎草■☆
119635 Cyperus subambiguus Kük. var. pallidicolor Kük. = Cyperus pallidicolor (Kük.) G. C. Tucker ■☆
119636 Cyperus subaphyllus Boeck. = Cyperus laevigatus L. ■☆
119637 Cyperus subintermedius Kük. = Pycreus intermedius (Steud.) C. B. Clarke ■☆
119638 Cyperus subintermedius Kük. f. tenuis (Boeck.) Kük. = Pycreus intermedius (Steud.) C. B. Clarke f. tenuis (Boeck.) Cufod. ■☆
119639 Cyperus sublimis (C. B. Clarke) Dandy = Mariscus sublimis C. B. Clarke ■☆
119640 Cyperus sublimis (C. B. Clarke) Dandy var. subglobosus (Kük.) Robyns et Tournay = Mariscus sublimis C. B. Clarke ■☆
119641 Cyperus submacropus Kük. ;亚大足莎草■☆
119642 Cyperus submacropus Kük. var. albescens Chiov. = Mariscus macropus C. B. Clarke var. albescens (Chiov.) Cufod. ■☆
119643 Cyperus submaculatus T. Koyama = Lipocarpha chinensis (Osbeck) J. Kern. ■
119644 Cyperus submicrolepis Kük. ;亚小鳞莎草■☆
119645 Cyperus subnodosus Nees et Meyen = Cyperus articulatus L. ■☆
119646 Cyperus subparadoxus Kük. = Alinula paradoxa (Cherm.) Goetgh. et Vorster ■☆
119647 Cyperus subtenax Kük. ;黏莎草■☆
119648 Cyperus subtilis (Kük.) Väre et Kukkonen;纤弱莎草■☆
119649 Cyperus subtrigonus (C. B. Clarke) Kük. = Pycreus subtrigonus C. B. Clarke ■☆
119650 Cyperus subulatus S. V. Meyen var. confertus Benth. = Cyperus bulbosus Vahl var. spicatus Boeck. ■☆
119651 Cyperus subumbellatus Kük. = Cyperus cyperoides (L.) Kuntze subsp. flavus Lye ■☆
119652 Cyperus subumbellatus Kük. = Mariscus sublimis C. B. Clarke ■☆
119653 Cyperus subxerophilus Kük. = Cyperus xerophilus Cherm. ■☆
119654 Cyperus sulcinux C. B. Clarke = Pycreus sulcinux (C. B. Clarke) C. B. Clarke ■
119655 Cyperus sylvestris Ridl. = Cyperus laxus Lam. subsp. sylvestris (Ridl.) Lye ■☆
119656 Cyperus sylvicola Ridl. = Cyperus renschii Boeck. ■☆
119657 Cyperus szechuanensis T. Koyama; 四 川 莎 草; Sichuan Cypressgrass, Sichuan Galingale ■
119658 Cyperus tabularis Schrad. ;扁平莎草■☆
119659 Cyperus tanaensis Kük. = Pycreus flavescens (L.) P. Beauv. ex Rchb. subsp. tanaensis (Kük.) Lye ■☆
119660 Cyperus tanganyicanus (Kük.) Lye;坦噶尼喀莎草■☆
119661 Cyperus tanzaniae (Lye) Lye = Kyllinga tanzaniae Lye ■☆
119662 Cyperus taylorii C. B. Clarke = Cyperus rotundus L. var. taylorii (C. B. Clarke) Kük. ■☆
119663 Cyperus tegetiformis C. B. Clarke = Cyperus malaccensis Lam. subsp. brevifolius (Boeck.) T. Koyama ■
119664 Cyperus tenax Boeck. ;强黏莎草■☆
119665 Cyperus tenax Boeck. var. monroviensis (Boeck.) Kük. = Cyperus tenax Boeck. ■☆
119666 Cyperus tenax Boeck. var. pseudocastaneus (Kük.) Kük. ;假栗色莎草■☆
119667 Cyperus tenellus J. Presl et C. Presl = Cyperus fugax Liebm. ■☆
119668 Cyperus tenellus L. f. ;柔弱莎草■☆
119669 Cyperus tenellus L. f. var. gracilis Nees = Cyperus tenellus L. f. ■☆
119670 Cyperus teneriffae Poir. = Cyperus rubicundus Vahl ■☆
119671 Cyperus teneriffae Poir. var. longimucronatus Kük. = Cyperus rubicundus Vahl ■☆
119672 Cyperus teneristolon Mattf. et Kük. = Kyllinga pulchella Kunth ■☆

119673 Cyperus teneristolon Mattf. et Kük. var. robustior (Kük.) Kük. = Kyllinga pulchella Kunth var. robustior (Kük.) Podlech ■☆
119674 Cyperus tenuiculmis Boeck.;细秆穗莎草(棱穗莎草,四棱莎草,四棱穗莎草);Tenuousculm Galingale ■
119675 Cyperus tenuiculmis Boeck. var. guineensis (Nelmes) S. S. Hooper;几内亚莎草■☆
119676 Cyperus tenuiculmis Boeck. var. longiramulosus (Kük.) Meneses;长枝细秆穗莎草■☆
119677 Cyperus tenuiculmis Boeck. var. schweinfurthianus (Boeck.) S. S. Hooper;施韦莎草■☆
119678 Cyperus tenuiflorus Rottb. = Cyperus longus L. var. tenuiflorus (Rottb.) Boeck. ■☆
119679 Cyperus tenuiflorus Rottb. var. gracillimus Chiov. = Cyperus longus L. var. gracillimus (Chiov.) Cufod. ■☆
119680 Cyperus tenuifolius (Steud.) Dandy = Kyllinga pumila Michx. ■☆
119681 Cyperus tenuifolius (Steud.) Dandy = Kyllinga tenuifolia Steud. ■☆
119682 Cyperus tenuifolius L. K. Dai;疏鳞莎草;Tenuous Leaf Galingale,Thinleaf Cypressgrass ■
119683 Cyperus tenuis Sw. = Mariscus flabelliformis Kunth ■☆
119684 Cyperus tenuis Sw. var. aximensis (C. B. Clarke) Kük. = Mariscus flabelliformis Kunth var. aximensis (C. B. Clarke) S. S. Hooper ■☆
119685 Cyperus tenuis Sw. var. lentiginosus (Millsp. et Chase) Kük. = Cyperus lentiginosus Millsp. et Chase ■☆
119686 Cyperus tenuis Sw. var. luridus (C. B. Clarke) Kük. = Mariscus luridus C. B. Clarke ■☆
119687 Cyperus tenuispica Steud.;窄穗莎草(窄翅莎草);Narrowspiculate Galingale,Narrowspike Cypressgrass ■
119688 Cyperus teretifolius A. Rich. = Cyperus laevigatus L. ■☆
119689 Cyperus terminalis Steud. = Pycreus pumilus (L.) Domin ■
119690 Cyperus tetragonus Elliott;四角莎草■☆
119691 Cyperus tetragonus Elliott var. pringlei (Britton) Kük. = Cyperus tetragonus Elliott ■☆
119692 Cyperus tetrastachyos Desf. = Cyperus rotundus L. ■
119693 Cyperus textilis Thunb.;编织莎草■☆
119694 Cyperus thorncroftii McClean = Cyperus kirkii C. B. Clarke ■☆
119695 Cyperus thunbergii Vahl;通贝里莎草■☆
119696 Cyperus thyrsiflorus Jungh.;聚伞莎草■☆
119697 Cyperus tisserantii Cherm. = Cyperus niveus Retz. var. tisserantii (Cherm.) Lye ■☆
119698 Cyperus toisensis Kuntze = Cyperus solidus Kunth ■☆
119699 Cyperus tomatophyllus Engl.;毛叶莎草■☆
119700 Cyperus tonkinensis C. B. Clarke;东京莎草■☆
119701 Cyperus tonkinensis C. B. Clarke var. baikiei (C. B. Clarke) S. S. Hooper;拜克东京莎草■☆
119702 Cyperus trachynotus Torr. = Cyperus elegans L. ■☆
119703 Cyperus transitorius Kük. = Kyllinga pulchella Kunth ■☆
119704 Cyperus tremulus Poir. = Pycreus macrostachyos (Lam.) J. Raynal ■☆
119705 Cyperus trialatus (Boeck.) Kern = Mariscus trialatus (Boeck.) Ts. Tang et F. T. Wang ■
119706 Cyperus tribrachiatus (Liebm.) Kük. = Cyperus thyrsiflorus Jungh. ■☆
119707 Cyperus triceps (Rottb.) Engl. = Kyllinga triceps Rottb. ■
119708 Cyperus triceps Endl. = Kyllinga tenuifolia Steud. ■☆
119709 Cyperus triceps Endl. var. ciliata (Boeck.) Kük. = Kyllinga welwitschii Ridl. ■☆
119710 Cyperus triceps Endl. var. obtusiflorus (Boeck.) Kük. = Kyllinga odorata Vahl ■☆
119711 Cyperus triceps Nees = Cyperus manimae Kunth var. asperrimus (Liebm.) Kük. ■☆
119712 Cyperus triflorus L. = Abildgaardia triflora (L.) Abeyw. ■☆
119713 Cyperus trinervis R. Br.;三脉莎草;Australian Flatsedge ■☆
119714 Cyperus triqueter Boeck. = Cyperus amabilis Vahl ■☆
119715 Cyperus tristis Kunth;暗淡莎草■☆
119716 Cyperus truncatus Turcz. = Cyperus orthostachyus Franch. et Sav. ■
119717 Cyperus tuberosus Pursh = Cyperus esculentus L. var. leptostachyus Boeck. ■☆
119718 Cyperus tuberosus Rottb.;假香附子(粗根茎莎草);Tuberous Cypressgrass,Tuberous Galingale ■
119719 Cyperus tuberosus Rottb. = Cyperus rotundus L. var. tuberosus (Rottb.) Kük. ■☆
119720 Cyperus turgidulus C. B. Clarke = Mariscus trialatus (Boeck.) Ts. Tang et F. T. Wang ■
119721 Cyperus turrillii Kük.;图尔香附子■☆
119722 Cyperus ugogensis Peter et Kük. = Kyllinga ugogensis (Peter et Kük.) Lye ■☆
119723 Cyperus umbellatus (Rottb.) Benth. f. cyperinus (Retz.) C. B. Clarke = Mariscus cyperinus (Retz.) Vahl ■
119724 Cyperus umbellatus Benth. = Mariscus cyperinus Vahl ■
119725 Cyperus umbellatus Benth. f. cyperinus C. B. Clarke = Mariscus cyperinus Vahl ■
119726 Cyperus umbellatus Vahl = Mariscus umbellatus Vahl ■
119727 Cyperus umbellatus Vahl var. cylindrostachys C. B. Clarke = Mariscus umbellatus Vahl var. evolutior (C. B. Clarke) Ts. Tang et F. T. Wang ■
119728 Cyperus umbilensis Boeck. = Mariscus umbilensis (Boeck.) C. B. Clarke ■☆
119729 Cyperus uncinatus C. B. Clarke = Cyperus cuspidatus Kunth ■
119730 Cyperus uncinatus C. B. Clarke var. gratus (C. B. Clarke) Kük. = Cyperus cuspidatus Kunth ■
119731 Cyperus uncinatus Poir. = Cyperus cuspidatus Kunth ■
119732 Cyperus undulatus Kük.;波状莎草■☆
119733 Cyperus uniflorus Torr. et Hook. = Cyperus floribundus (Kük.) R. Carter et S. D. Jones ■☆
119734 Cyperus uniflorus Torr. et Hook. var. floribundus Kük. = Cyperus floribundus (Kük.) R. Carter et S. D. Jones ■☆
119735 Cyperus uniflorus Torr. et Hook. var. floribundus Kük. = Cyperus retroflexus Buckley ■☆
119736 Cyperus uniflorus Torr. et Hook. var. pseudothyrsiflorus Kük. = Cyperus pseudothyrsiflorus (Kük.) R. Carter et S. D. Jones ■☆
119737 Cyperus uniflorus Torr. et Hook. var. retroflexus (Buckley) Kük. = Cyperus retroflexus Buckley ■☆
119738 Cyperus unioloides R. Br. = Pycreus unioloides (R. Br.) Urb. ■
119739 Cyperus unistamen T. Koyama = Lipocarpha nana (A. Rich.) Cherm. ■☆
119740 Cyperus usitatus Burch.;习见莎草■☆
119741 Cyperus usitatus Burch. subsp. palmatus Lye;掌裂莎草■☆
119742 Cyperus usitatus Burch. var. stuhlmannii (C. B. Clarke ex K. Schum.) Lye;斯图尔曼莎草■☆
119743 Cyperus vaginatissimus K. Schum. = Mariscus kerstenii

(Boeck.) C. B. Clarke ■☆

119744 Cyperus variegatus Boeck. var. atrosanguineus (Hochst. ex A. Rich.) Boeck. = Mariscus bulbocaulis Hochst. ex A. Rich. var. atrosanguineus (Hochst. ex A. Rich.) C. B. Clarke ■☆

119745 Cyperus vegetus Salzm. ex Steud. = Pycreus sanguinolentus (Vahl) Nees ex C. B. Clarke ■

119746 Cyperus verrucinux C. B. Clarke = Cyperus denudatus L. f. ■☆

119747 Cyperus vestitus Hochst. ex Krauss = Mariscus vestitus (Hochst. ex Krauss) C. B. Clarke ■☆

119748 Cyperus vestitus Hochst. ex Krauss var. pseudocallistus (Kük.) Kük. = Mariscus vestitus (Hochst. ex Krauss) C. B. Clarke ■☆

119749 Cyperus vexillatus Peter ex Kük.;旗瓣莎草■☆

119750 Cyperus virens Michx.;绿花莎草;Green Flat Sedge ■☆

119751 Cyperus virens Michx. subsp. drummondii (Torr. et Hook.) T. Koyama = Cyperus drummondii Torr. et Hook. ■☆

119752 Cyperus virens Michx. var. brittonii C. B. Clarke = Cyperus distinctus Steud. ■☆

119753 Cyperus virens Michx. var. drummondii (Torr. et Hook.) Kük. = Cyperus drummondii Torr. et Hook. ■☆

119754 Cyperus virens Michx. var. robustus (Kunth) Kük. = Cyperus drummondii Torr. et Hook. ■☆

119755 Cyperus waillyi (Cherm.) Lye = Pycreus waillyi Cherm. ■☆

119756 Cyperus washingtonensis Gand. = Cyperus erythrorhizos Muhl. ■☆

119757 Cyperus webbianus Steud. = Cyperus sexangularis Nees ■☆

119758 Cyperus welwitschii (Ridl.) Lye = Kyllinga welwitschii Ridl. ■☆

119759 Cyperus wightii Hance = Cyperus zollingeri Steud. ■

119760 Cyperus wilmsii Gand.;维尔姆斯莎草■☆

119761 Cyperus winkleri Britton et Small = Cyperus ovatus Baldwin ■☆

119762 Cyperus wittei Cherm.;维特莎草■☆

119763 Cyperus wolfii A. W. Wood = Cyperus echinatus (L.) A. W. Wood ■☆

119764 Cyperus wrightii Britton = Cyperus dipsaceus Liebm. ■☆

119765 Cyperus xerophilus Cherm.;旱生莎草■☆

119766 Cyperus zairensis Chiov. = Cyperus papyrus L. subsp. zairensis (Chiov.) Kük. ■☆

119767 Cyperus zambesiensis C. B. Clarke;赞比西莎草■☆

119768 Cyperus zanzibarensis C. B. Clarke = Cyperus pulchellus R. Br. ■☆

119769 Cyperus zollingeri Steud.;四棱穗莎草;Zollinger Cypressgrass, Zollinger Galingale ■

119770 Cyperus zollingeri Steud. var. longiramulosus Kük. = Cyperus tenuiculmis Boeck. var. longiramulosus (Kük.) Meneses ■☆

119771 Cyperus zollingeri Steud. var. permacer (C. B. Clarke) Kük. = Cyperus permacer C. B. Clarke ■☆

119772 Cyperus zollingeri Steud. var. robusta K. Schum. = Cyperus holstii Kük. ■☆

119773 Cyperus zollingeri Steud. var. schweinfurthianus (Boeck.) Kük. = Cyperus tenuiculmis Boeck. var. schweinfurthianus (Boeck.) S. S. Hooper ■☆

119774 Cyperus zollingeriana (Boeck.) T. Koyama = Lipocarpha microcephala (R. Br.) Kunth ■

119775 Cyperus zollingerioides C. B. Clarke;拟佐林格莎草■☆

119776 Cyperus zonatissimus (Cherm.) Kük. = Pycreus zonatus Cherm. ■☆

119777 Cyperus zonatus Kük.;带莎草■☆

119778 Cyphacanthus Leonard(1953);弯刺爵床属■☆

119779 Cyphacanthus atopus Léonard;弯刺爵床■☆

119780 Cyphadenia Post et Kuntze = Chrysactinia A. Gray ●☆

119781 Cyphadenia Post et Kuntze = Kyphadenia Sch. Bip. ex O. Hoffm. ●☆

119782 Cyphaea Lem. = Cuphea Adans. ex P. Browne ●■

119783 Cyphanthe Raf. = Orobanche L. ■

119784 Cyphanthera Miers = Anthocercis Labill. ●■☆

119785 Cyphanthera Miers(1853);驼药茄属■☆

119786 Cyphanthera albicans Miers;驼药茄■☆

119787 Cyphanthera microphylla Miers;小叶驼药茄■☆

119788 Cyphanthera tomentosa (Benth.) Miers;毛驼药茄■☆

119789 Cyphea Post et Kuntze = Cuphea Adans. ex P. Browne ●■

119790 Cypheanthus Post et Kuntze = Cupheanthus Seem. ●☆

119791 Cyphella Post et Kuntze = Cypella Herb. ■☆

119792 Cyphia P. J. Bergius(1767);驼曲草属(腔柱草属)■☆

119793 Cyphia alba N. E. Br.;白驼曲草(白腔柱草)■☆

119794 Cyphia alba N. E. Br. f. purpurea E. Wimm. = Cyphia alba N. E. Br. ■☆

119795 Cyphia angustifolia C. Presl ex Eckl. et Zeyh.;窄叶驼曲草■☆

119796 Cyphia antunesii Engl. = Cyphia lasiandra Diels ■☆

119797 Cyphia aspergilloides E. Wimm.;帚状驼曲草■☆

119798 Cyphia aspergilloides E. Wimm. var. brevipes;短梗帚状驼曲草■☆

119799 Cyphia assimilis Sond.;相似驼曲草■☆

119800 Cyphia assimilis Sond. var. latifolia E. Phillips = Cyphia phillipsii E. Wimm. ■☆

119801 Cyphia basiloba E. Wimm.;基裂驼曲草■☆

119802 Cyphia bechuanensis Bremek. et Oberm. = Cyphia stenopetala Diels ■☆

119803 Cyphia bolusii E. Phillips;博卢斯驼曲草■☆

119804 Cyphia brachyandra Thulin;短蕊驼曲草■☆

119805 Cyphia brevifolia Thulin;短叶驼曲草■☆

119806 Cyphia brummittii Thulin;布鲁米特驼曲草■☆

119807 Cyphia bulbosa (L.) P. J. Bergius;腔柱草■☆

119808 Cyphia bulbosa (L.) P. J. Bergius var. acocksii E. Wimm.;阿氏驼曲草■☆

119809 Cyphia bulbosa (L.) P. J. Bergius var. orientalis E. Phillips = Cyphia linarioides C. Presl ■☆

119810 Cyphia cacondensis Good = Cyphia lasiandra Diels ■☆

119811 Cyphia campestris C. Presl;田野驼曲草■☆

119812 Cyphia campestris C. Presl var. nudiuscula E. Wimm.;稍裸驼曲草■☆

119813 Cyphia comptonii Bond;康普顿驼曲草■☆

119814 Cyphia crenata (Thunb.) C. Presl;圆齿驼曲草■☆

119815 Cyphia crenata (Thunb.) C. Presl var. angustifolia E. Wimm.;窄叶圆齿驼曲草■☆

119816 Cyphia decora Thulin;装饰驼曲草■☆

119817 Cyphia deltoidea E. Wimm.;三角驼曲草■☆

119818 Cyphia dentariifolia C. Presl;齿叶驼曲草■☆

119819 Cyphia dentata E. Wimm.;具齿驼曲草■☆

119820 Cyphia digitata (Thunb.) Willd.;指裂驼曲草■☆

119821 Cyphia digitata (Thunb.) Willd. subsp. gracilis E. Wimm.;纤细驼曲草■☆

119822 Cyphia digitata (Thunb.) Willd. var. tomentosa (C. Presl) Sond. = Cyphia digitata (Thunb.) Willd. ■☆

119823 Cyphia digitata (Thunb.) Willd. var. trimera E. Wimm. = Cyphia digitata (Thunb.) Willd. ■☆

119824 Cyphia eckloniana C. Presl;埃氏驼曲草■☆

119825 Cyphia elata Harv.;高驼曲草■☆
119826 Cyphia elata Harv. f. decurrens E. Wimm. = Cyphia elata Harv. var. gerrardii (Harv.) E. Wimm. ■☆
119827 Cyphia elata Harv. f. depauperata E. Wimm. = Cyphia revoluta E. Wimm. ■☆
119828 Cyphia elata Harv. f. flanagana E. Wimm. = Cyphia revoluta E. Wimm. ■☆
119829 Cyphia elata Harv. f. latifolia E. Wimm. = Cyphia elata Harv. var. globularis E. Wimm. ■☆
119830 Cyphia elata Harv. f. thodeana E. Wimm. = Cyphia elata Harv. var. gerrardii (Harv.) E. Wimm. ■☆
119831 Cyphia elata Harv. f. truncata E. Wimm. = Cyphia elata Harv. var. glabra ? ■☆
119832 Cyphia elata Harv. var. gerrardii (Harv.) E. Wimm.;杰勒德驼曲草■☆
119833 Cyphia elata Harv. var. glabra ?;光滑高驼曲草■☆
119834 Cyphia elata Harv. var. globularis E. Wimm.;小球驼曲草■☆
119835 Cyphia elata Harv. var. oblongifolia (Sond. et Harv.) E. Phillips = Cyphia oblongifolia Sond. et Harv. ■☆
119836 Cyphia erecta De Wild.;直立驼曲草■☆
119837 Cyphia erecta De Wild. f. minor ? = Cyphia erecta De Wild. ■☆
119838 Cyphia erecta De Wild. var. ufipana E. Wimm. = Cyphia erecta De Wild. ■☆
119839 Cyphia erecta De Wild. var. witteana E. Wimm.;维特驼曲草■☆
119840 Cyphia exelliana E. Wimm. = Cyphia lasiandra Diels ■☆
119841 Cyphia floribunda E. Wimm. = Cyphia lasiandra Diels ■☆
119842 Cyphia galpinii E. Wimm.;盖尔驼曲草■☆
119843 Cyphia gamopetala P. A. Duvign. et Denaeyer;瓣驼曲草■☆
119844 Cyphia georgica E. Wimm.;乔治驼曲草■☆
119845 Cyphia gerrardii Harv. = Cyphia elata Harv. var. gerrardii (Harv.) E. Wimm. ■☆
119846 Cyphia glabra E. Wimm.;光滑驼曲草■☆
119847 Cyphia glandulifera Hochst. ex A. Rich.;腺体驼曲草■☆
119848 Cyphia glandulifera Hochst. ex A. Rich. f. obovatifolia E. Wimm. = Cyphia glandulifera Hochst. ex A. Rich. ■☆
119849 Cyphia glandulifera Hochst. ex A. Rich. f. subcuspidata E. Wimm. = Cyphia glandulifera Hochst. ex A. Rich. ■☆
119850 Cyphia heterophylla C. Presl;互叶腔柱草■☆
119851 Cyphia incisa (Thunb.) Willd.;锐裂驼曲草■☆
119852 Cyphia incisa (Thunb.) Willd. var. bracteata E. Phillips;具苞腔柱草■☆
119853 Cyphia incisa (Thunb.) Willd. var. lyrata E. Wimm.;大头羽裂驼曲草■☆
119854 Cyphia incisa (Thunb.) Willd. var. sinuata E. Wimm.;深波驼曲草■☆
119855 Cyphia lasiandra Diels;毛蕊腔柱草■☆
119856 Cyphia latipetala C. Presl;阔瓣驼曲草■☆
119857 Cyphia linarioides C. Presl;柳穿鱼驼曲草■☆
119858 Cyphia longiflora Schltr.;长花腔柱草■☆
119859 Cyphia longifolia N. E. Br.;长叶腔柱草■☆
119860 Cyphia longifolia N. E. Br. var. baurii E. Phillips = Cyphia longifolia N. E. Br. ■☆
119861 Cyphia longilobata E. Phillips;长裂腔柱草■☆
119862 Cyphia longipedicellata E. Wimm.;长花梗驼曲草■☆
119863 Cyphia longipetala C. Presl;长瓣腔柱草■☆
119864 Cyphia maculosa E. Phillips;斑点腔柱草■☆
119865 Cyphia mazoensis S. Moore;马索腔柱草■☆
119866 Cyphia mazoensis S. Moore f. angustior E. Wimm. = Cyphia mazoensis S. Moore ■☆
119867 Cyphia mazoensis S. Moore var. stellaris E. Wimm. = Cyphia mazoensis S. Moore ■☆
119868 Cyphia natalensis E. Phillips;纳塔尔驼曲草■☆
119869 Cyphia nyasica Baker = Cyphia lasiandra Diels ■☆
119870 Cyphia nyikensis Thulin;尼卡驼曲草■☆
119871 Cyphia oblongifolia Sond. et Harv.;矩圆叶驼曲草■☆
119872 Cyphia oligotricha Schltr.;寡毛驼曲草■☆
119873 Cyphia pectinata E. Wimm.;篦状驼曲草■☆
119874 Cyphia persicifolia E. Mey.;桃叶驼曲草■☆
119875 Cyphia peteriana E. Wimm. = Cyphia lasiandra Diels ■☆
119876 Cyphia phillipsii E. Wimm.;菲利驼曲草■☆
119877 Cyphia phyteuma (L.) Willd.;牧根草驼曲草■☆
119878 Cyphia phyteuma (L.) Willd. var. ciliata E. Wimm.;缘毛驼曲草■☆
119879 Cyphia phyteuma (L.) Willd. var. grandidentata E. Wimm.;大齿驼曲草■☆
119880 Cyphia psilostemon E. Wimm.;裸冠驼曲草■☆
119881 Cyphia ramosa E. Wimm.;分枝驼曲草■☆
119882 Cyphia ranunculifolia E. Wimm.;毛茛叶驼曲草■☆
119883 Cyphia reducta E. Wimm.;退缩驼曲草■☆
119884 Cyphia regularis E. Wimm. = Cyphia erecta De Wild. ■☆
119885 Cyphia revoluta E. Wimm.;外卷驼曲草■☆
119886 Cyphia rhodesiaca E. Wimm. = Cyphia erecta De Wild. ■☆
119887 Cyphia richardsiae E. Wimm.;理查兹驼曲草■☆
119888 Cyphia rivularis E. Wimm. = Cyphia mazoensis S. Moore ■☆
119889 Cyphia rogersii S. Moore;罗杰斯驼曲草■☆
119890 Cyphia rogersii S. Moore subsp. winteri E. Wimm.;温特驼曲草■☆
119891 Cyphia rupestris E. Wimm.;岩生驼曲草■☆
119892 Cyphia salicifolia C. Presl = Cyphia sylvatica Eckl. var. salicifolia (C. Presl) E. Wimm. ■☆
119893 Cyphia salteri E. Wimm.;索尔特驼曲草■☆
119894 Cyphia scandens De Wild. = Cyphia erecta De Wild. ■☆
119895 Cyphia schlechteri E. Phillips;施莱驼曲草■☆
119896 Cyphia smithiae L. Bolus = Cyphia crenata (Thunb.) C. Presl ■☆
119897 Cyphia stenodonta E. Wimm.;窄齿驼曲草■☆
119898 Cyphia stenopetala Diels;窄瓣驼曲草■☆
119899 Cyphia stenopetala Diels var. johannesburgensis E. Wimm. = Cyphia stenopetala Diels ■☆
119900 Cyphia stenophylla E. Wimm.;狭叶驼曲草■☆
119901 Cyphia stephensii E. Wimm. = Cyphia bulbosa (L.) P. J. Bergius var. acocksii E. Wimm. ■☆
119902 Cyphia subscandens E. Wimm. = Cyphia mazoensis S. Moore ■☆
119903 Cyphia subtubulata E. Wimm.;稍弯驼曲草■☆
119904 Cyphia sylvatica Eckl.;林地驼曲草■☆
119905 Cyphia sylvatica Eckl. var. graminea E. Wimm.;禾状林地驼曲草■☆
119906 Cyphia sylvatica Eckl. var. salicifolia (C. Presl) E. Wimm.;柳叶林地驼曲草■☆
119907 Cyphia tenera Diels;极细驼曲草■☆
119908 Cyphia tomentosa C. Presl = Cyphia digitata (Thunb.) Willd. ■☆
119909 Cyphia tortilis N. E. Br.;螺旋状驼曲草■☆
119910 Cyphia transvaalensis E. Phillips;德兰士瓦驼曲草■☆
119911 Cyphia tricuspis E. Wimm.;三尖驼曲草■☆
119912 Cyphia triphylla E. Phillips;三叶驼曲草■☆

119913 Cyphia tysonii E. Phillips;泰森驼曲草■☆
119914 Cyphia undulata Eckl.;波状驼曲草■☆
119915 Cyphia undulata Eckl. var. linarioides (C. Presl) E. Wimm. = Cyphia linarioides C. Presl ■☆
119916 Cyphia volubilis (Burm. f.) Willd.;缠绕驼曲草■☆
119917 Cyphia volubilis (Burm. f.) Willd. var. banksiana E. Wimm.;班克斯驼曲草■☆
119918 Cyphia volubilis (Burm. f.) Willd. var. intermedia E. Wimm. = Cyphia longipetala C. Presl ■☆
119919 Cyphia volubilis (Burm. f.) Willd. var. latipetala (C. Presl) E. Wimm. = Cyphia latipetala C. Presl ■☆
119920 Cyphia volubilis (Burm. f.) Willd. var. longipes E. Wimm. = Cyphia angustiloba C. Presl ex Eckl. et Zeyh. ■☆
119921 Cyphia wilmsiana Diels = Cyphia elata Harv. var. gerrardii (Harv.) E. Wimm. ■☆
119922 Cyphia zernyana E. Wimm. = Cyphia lasiandra Diels ■☆
119923 Cyphia zeyheriana C. Presl;泽耶尔驼曲草■☆
119924 Cyphia zeyheriana C. Presl var. eckloniana (C. Presl) Sond. = Cyphia eckloniana C. Presl ■☆
119925 Cyphiaceae A. DC. (1839);驼曲草科(腔柱草科)■☆
119926 Cyphiaceae A. DC. = Campanulaceae Juss.(保留科名)■●
119927 Cyphiaceae A. DC. = Cyphocarpaceae Miers ■☆
119928 Cyphiella (Presl) Spach = Cyphia P. J. Bergius ■☆
119929 Cyphiella (Presl) Spach = Cyphopsis Kuntze ■☆
119930 Cyphisia Rizzini = Justicia L. ●■
119931 Cyphium J. F. Gmel. = Cyphia P. J. Bergius ■☆
119932 Cyphocalyx C. Presl = Aspalathus L. ●☆
119933 Cyphocalyx C. Presl(1846);弯萼豆属●☆
119934 Cyphocalyx Gagnep. = Trungboa Rauschert ●☆
119935 Cyphocalyx aridus C. Presl = Aspalathus arida E. Mey. ●☆
119936 Cyphocalyx major C. Presl;弯萼豆●☆
119937 Cyphocardamum Hedge(1968);阿富汗白花芥属■☆
119938 Cyphocardamum aretioides Hedge;阿富汗白花芥■☆
119939 Cyphocarpa (Fenzl) Lopr. = Kyphocarpa (Fenzl ex Endl.) Lopr. ■☆
119940 Cyphocarpa Lopr. (1899);弯果草属■☆
119941 Cyphocarpa gregorii S. Moore = Cyathula gregorii (S. Moore) Schinz ■☆
119942 Cyphocarpa hildebrandtii (Schinz) C. B. Clarke = Sericocomopsis hildebrandtii Schinz ■☆
119943 Cyphocarpa kilimandscharica Suess. et Beyerle = Cyathula orthacantha (Hochst. ex Asch.) Schinz ■☆
119944 Cyphocarpa kuhlweiniana Peter = Cyathula lanceolata Schinz ■☆
119945 Cyphocarpa kuhlweiniana Peter var. melanacantha ? = Cyathula lanceolata Schinz ■☆
119946 Cyphocarpa orthacantha (Hochst. ex Asch.) C. B. Clarke = Cyathula orthacantha (Hochst. ex Asch.) Schinz ■☆
119947 Cyphocarpa orthacanthoides Suess. = Cyathula orthacantha (Hochst. ex Asch.) Schinz ■☆
119948 Cyphocarpa pallida (S. Moore) C. B. Clarke = Sericocomopsis pallida (S. Moore) Schinz ■☆
119949 Cyphocarpa quadrangula (Engl.) C. B. Clarke = Nelsia quadrangula (Engl.) Schinz ■☆
119950 Cyphocarpa trichinioides (Fenzl) Lopr.;弯果草■☆
119951 Cyphocarpa welwitschii (Baker) C. B. Clarke = Nelsia quadrangula (Engl.) Schinz ■☆
119952 Cyphocarpaceae Miers = Campanulaceae Juss.(保留科名)■●
119953 Cyphocarpaceae Miers = Cypripediaceae Lindl. ■
119954 Cyphocarpaceae Miers(1848);弯果草科■☆
119955 Cyphocarpaceae Reveal et Hoogland = Campanulaceae Juss.(保留科名)■●
119956 Cyphocarpus Miers(1848);弯果桔梗属(弯果草属)■☆
119957 Cyphocarpus Post et Kuntze = Cuphocarpus Decne. et Planch. ●☆
119958 Cyphocarpus Post et Kuntze = Polyscias J. R. Forst. et G. Forst. ●
119959 Cyphocarpus rigescens Miers;弯果桔梗■☆
119960 Cyphochilus Schltr. = Appendicula Blume ■
119961 Cyphochlaena Hack. (1901);驼蜀黍属■☆
119962 Cyphochlaena madagascariensis Hack.;驼蜀黍■☆
119963 Cyphokentia Brongn. (1873);赛佛棕属(粉蕊椰属,粉雄椰属,瓶棕属,弯堪蒂榈属);Cyphokentia ●☆
119964 Cyphokentia balansae Brongn.;赛佛棕●☆
119965 Cyphokentia macrostachya Brongn.;大穗赛佛棕●☆
119966 Cypholepis Chiov. = Coelachyrum Hochst. et Nees ■☆
119967 Cypholepis yemenica (Schweinf.) Chiov. = Coelachyrum yemenicum (Schweinf.) S. M. Phillips ■☆
119968 Cypholophus Wedd. (1854);疣冠麻属(瘤冠麻属,隆冠麻属);Cypholophus,Tumorcomb ●
119969 Cypholophus moluccanus (Blume) Miq.;疣冠麻(瘤冠麻,隆冠麻);Molucca Cypholophus,Tumorcomb ●
119970 Cypholoron Dodson et Dressler(1972);驼兰属■☆
119971 Cypholoron Dressler et Dodson = Cypholoron Dodson et Dressler ■☆
119972 Cypholoron frigidum Dodson et Dressler;驼兰■☆
119973 Cyphomandra Mart. ex Sendtn. (1845);树番茄属;Tamarillo, Tree Tomato,Treetomato,Tree-tomato ●■
119974 Cyphomandra betacea (Cav.) Sendtn.;树番茄(缅茄,木立蕃茄,树西红柿);Beet-like Mandrake, Tamarilla, Tamarillo, Tree Tomato,Treetomato,Tree-tomato,Vegetable Mercury ●
119975 Cyphomandra betacea Cav. = Cyphomandra betacea (Cav.) Sendtn. ●
119976 Cyphomandra betacea Sendtn. = Cyphomandra betacea (Cav.) Sendtn. ●
119977 Cyphomandra crassicaulis ? = Cyphomandra betacea (Cav.) Sendtn. ●
119978 Cyphomandra crassifolia (Ortega) Kuntze = Cyphomandra betacea (Cav.) Sendtn. ●
119979 Cyphomandra procera Wawra = Cyphomandra betacea (Cav.) Sendtn. ●
119980 Cyphomattia Boiss. = Rindera Pall. ■
119981 Cyphomeris Standl. (1911);瘤果茉莉属(歪果茉莉属)■☆
119982 Cyphomeris Standl. = Boerhavia L. ■
119983 Cyphomeris crassifolia (Standl.) Standl.;厚叶瘤果茉莉■☆
119984 Cyphomeris gypsophiloides (M. Martens et Galeotti) Standl.;石头花瘤果茉莉■☆
119985 Cyphonanthus Zuloaga et Morrone = Panicum L. ■
119986 Cyphonanthus Zuloaga et Morrone(2007);弯花黍属■☆
119987 Cyphonema Herb. = Cyrtanthus Aiton(保留属名)■☆
119988 Cyphophoenix H. Wendl. ex Benth. et Hook. f. (1883);膨颈椰属(大洋洲刺葵属,加罗林椰树,弯棕属,瘤茎椰子属,弯曲扇榈属)●☆
119989 Cyphophoenix H. Wendl. ex Hook. f. = Cyphophoenix H. Wendl. ex Benth. et Hook. f. ●☆
119990 Cyphophoenix carolinensis Kaneh. et Hatus.;卡罗来纳膨颈椰●☆
119991 Cyphophoenix elegans Benth. et Hook. f.;优雅膨颈椰●☆

119992 Cyphophoenix fulcita Benth. et Hook. f.;膨颈椰●☆
119993 Cyphopsis Kuntze = Cyphia P. J. Bergius ■☆
119994 Cyphorima Raf. = Lithospermum L. ■
119995 Cyphosperma H. Wendl. = Cyphosperma H. Wendl. ex Benth. et Hook. f. ●☆
119996 Cyphosperma H. Wendl. ex Benth. et Hook. f.(1883);肿瘤椰属(啮籽棕属)●☆
119997 Cyphosperma H. Wendl. ex Hook. f. = Cyphosperma H. Wendl. ex Benth. et Hook. f. ●☆
119998 Cyphosperma balansae H. Wendl. ex Benth. et Hook. f.;肿瘤椰●☆
119999 Cyphostemma (Planch.) Alston = Cissus L. ●
120000 Cyphostemma (Planch.) Alston(1931);树葡萄属(葡萄瓮属)●■☆
120001 Cyphostemma Alston = Cyphostemma (Planch.) Alston ●■☆
120002 Cyphostemma abercornense Wild et R. B. Drumm.;阿伯康膨颈椰●☆
120003 Cyphostemma adamii Desc.;亚当斯树葡萄●☆
120004 Cyphostemma adenanthum (Fresen.) Desc.;腺花树葡萄●☆
120005 Cyphostemma adenocarpum (Gilg et M. Brandt) Desc.;腺果树葡萄●☆
120006 Cyphostemma adenocaule (Steud. ex A. Rich.) Desc. ex Wild et R. B. Drumm. subsp. pulverulentum Verdc.;细粉腺茎树葡萄●☆
120007 Cyphostemma adenocaule (Steud. ex A. Rich.) Desc. ex Wild et R. B. Drumm.;腺茎树葡萄●☆
120008 Cyphostemma adenocaule (Steud. ex A. Rich.) Desc. ex Wild et R. B. Drumm. var. eglandulosum (Dewit) Desc. = Cyphostemma adenocaule (Steud. ex A. Rich.) Desc. ex Wild et R. B. Drumm. ●☆
120009 Cyphostemma adenocaule (Steud. ex A. Rich.) Desc. ex Wild et R. B. Drumm. var. pubescens (Dewit) Desc. = Cyphostemma adenocaule (Steud. ex A. Rich.) Desc. ex Wild et R. B. Drumm. ●☆
120010 Cyphostemma adenocaulis (Steud. ex A. Rich.) Desc. ex Wild et R. B. Drumm. var. chiovendae Lanza = Cyphostemma adenocaule (Steud. ex A. Rich.) Desc. ex Wild et R. B. Drumm. ●☆
120011 Cyphostemma adenocephalum (Gilg et M. Brandt) Desc. = Cyphostemma dembianense (Chiov.) Vollesen ●☆
120012 Cyphostemma adenopodum (Sprague) Desc.;腺梗树葡萄●☆
120013 Cyphostemma agnus-castus (Planch.) Desc. = Cyphostemma bororense (Klotzsch) Desc. ex Wild et R. B. Drumm. ●☆
120014 Cyphostemma allophylloides (Gilg et M. Brandt) Desc.;异叶花葱树葡萄●☆
120015 Cyphostemma alnifolium (Schweinf. ex Planch.) Desc.;桤叶树葡萄●☆
120016 Cyphostemma amplexicaule Desc.;抱茎树葡萄●☆
120017 Cyphostemma amplexum (Baker) Desc. ex Wild et R. B. Drumm. = Cyphostemma bororense (Klotzsch) Desc. ex Wild et R. B. Drumm. ●☆
120018 Cyphostemma andongensis (Welw. ex Baker) Desc. = Cyphostemma chloroleucum (Welw. ex Baker) Desc. ex Wild et R. B. Drumm. ●☆
120019 Cyphostemma bainesii (Hook. f.) Desc.;白氏葡萄瓮;Gouty Vine ●☆
120020 Cyphostemma bambuseti (Gilg et M. Brandt) Desc. ex Wild et R. B. Drumm.;邦布塞特树葡萄●☆
120021 Cyphostemma bambuseti (Gilg et M. Brandt) Desc. ex Wild et R. B. Drumm. var. glandulosissima (Dewit) Desc.;具腺邦布塞特树葡萄●☆
120022 Cyphostemma barbosae Wild et R. B. Drumm.;巴尔博萨树葡萄●☆
120023 Cyphostemma biternata (Chiov.) Desc. = Cyphostemma ternato-multifidum (Chiov.) Desc. ●☆
120024 Cyphostemma bororense (Klotzsch) Desc. ex Wild et R. B. Drumm.;博罗雷树葡萄●☆
120025 Cyphostemma braunii (Gilg et M. Brandt) Desc.;布劳恩树葡萄●☆
120026 Cyphostemma brieyi (De Wild.) Compère;布里树葡萄●☆
120027 Cyphostemma buchananii (Planch.) Desc. ex Wild et R. B. Drumm.;布坎南树葡萄●☆
120028 Cyphostemma bullatum (Gilg et M. Brandt) Desc.;泡状树葡萄●☆
120029 Cyphostemma burgeri Vollesen;伯格树葡萄●☆
120030 Cyphostemma centrali-africanum (Gilg et R. E. Fr.) Desc. = Cyphostemma mildbraedii (Gilg et M. Brandt) Desc. ex Wild et R. B. Drumm. ●☆
120031 Cyphostemma chevalieri (Gilg et M. Brandt) Desc. = Cyphostemma flavicans (Baker) Desc. ●☆
120032 Cyphostemma chloroleucum (Welw. ex Baker) Desc. ex Wild et R. B. Drumm.;白树葡萄●☆
120033 Cyphostemma chrysadenium (Gilg) Desc.;金腺树葡萄●☆
120034 Cyphostemma cirrhosum (Thunb.) Desc. ex Wild et R. B. Drumm.;卷须树葡萄●☆
120035 Cyphostemma cirrhosum (Thunb.) Desc. ex Wild et R. B. Drumm. subsp. rhodesicum Wild et R. B. Drumm.;罗得西亚卷须树葡萄●☆
120036 Cyphostemma cirrhosum (Thunb.) Desc. ex Wild et R. B. Drumm. subsp. transvaalense (Szyszyl.) Wild et R. B. Drumm.;德兰士瓦树葡萄●☆
120037 Cyphostemma congestum (Baker) Desc. ex Wild et R. B. Drumm.;密集树葡萄●☆
120038 Cyphostemma congoense Desc.;刚果树葡萄●☆
120039 Cyphostemma cornigera Desc.;角状树葡萄●☆
120040 Cyphostemma crameranum (Schinz) Desc. = Cyphostemma currorii (Hook. f.) Desc. ●☆
120041 Cyphostemma crassiusculum (Baker) Desc.;厚叶树葡萄●☆
120042 Cyphostemma crinitum (Planch.) Desc.;长软毛树葡萄●☆
120043 Cyphostemma crithmifolium (Chiov.) Desc.;海茴香叶树葡萄●☆
120044 Cyphostemma crotalarioides (Planch.) Desc. ex Wild et R. B. Drumm.;猪屎豆树葡萄●☆
120045 Cyphostemma cryptoglandulosum Verdc.;隐腺树葡萄●☆
120046 Cyphostemma cuneatum (Gilg et M. Brandt) Desc.;楔形树葡萄●☆
120047 Cyphostemma currorii (Hook. f.) Desc.;库洛里树葡萄●☆
120048 Cyphostemma currorii (Hook. f.) Desc. = Cyphostemma crameranum (Schinz) Desc. ●☆
120049 Cyphostemma currorii (Hook. f.) Desc. = Cyphostemma macropus (Welw.) Desc. ●☆
120050 Cyphostemma curvipodum (Baker) Desc.;弯梗树葡萄●☆
120051 Cyphostemma cymosum (Schumach. et Thonn.) Desc.;聚伞树葡萄●☆
120052 Cyphostemma cymosum (Schumach. et Thonn.) Desc. subsp. orientale Verdc.;东方聚伞树葡萄●☆
120053 Cyphostemma cyphopetalum (Fresen.) Desc. ex Wild et R. B. Drumm.;曲瓣树葡萄●☆

120054 Cyphostemma cyphopetalum (Fresen.) Desc. ex Wild et R. B. Drumm. var. nodiglandulosum (T. C. E. Fr.) Verdc.;瘤腺曲瓣树葡萄●☆

120055 Cyphostemma dasycarpum Verdc.;毛果树葡萄●☆

120056 Cyphostemma dasypleurum (C. A. Sm.) J. J. M. van der Merwe;毛脉树葡萄●☆

120057 Cyphostemma decurrens (Gilg et M. Brandt) Desc. = Cyphostemma chloroleucum (Welw. ex Baker) Desc. ex Wild et R. B. Drumm. ●☆

120058 Cyphostemma delphinensis Desc.;德尔芬树葡萄●☆

120059 Cyphostemma dembianense (Chiov.) Vollesen;登比树葡萄●☆

120060 Cyphostemma descoingsii Lavie;德斯树葡萄●☆

120061 Cyphostemma digitatum (Forssk.) Desc.;指裂树葡萄●☆

120062 Cyphostemma duparquetii (Planch.) Desc.;迪帕树葡萄●☆

120063 Cyphostemma dysocarpum (Gilg et M. Brandt) Desc. subsp. glandulosissimum Verdc.;多腺树葡萄●☆

120064 Cyphostemma echinocarpum Desc.;刺果树葡萄●☆

120065 Cyphostemma egregium (Gilg) Desc. = Cyphostemma cyphopetalum (Fresen.) Desc. ex Wild et R. B. Drumm. ●☆

120066 Cyphostemma elephantopus Desc.;象足树葡萄●☆

120067 Cyphostemma eminii (Gilg) Desc.;埃明树葡萄●☆

120068 Cyphostemma engleri (Gilg) Desc.;恩格勒树葡萄●☆

120069 Cyphostemma erythraeae (Gilg et M. Brandt) Desc. = Cyphostemma molle (Steud. ex Baker) Desc. ●☆

120070 Cyphostemma erythrocephalum (Gilg et M. Brandt) Desc.;红头树葡萄●☆

120071 Cyphostemma flavicans (Baker) Desc.;浅黄树葡萄●☆

120072 Cyphostemma flaviflorum (Sprague) Desc.;黄花树葡萄●☆

120073 Cyphostemma fugosioides (Gilg) Desc. ex Wild et R. B. Drumm.;棉树葡萄●☆

120074 Cyphostemma fugosioides (Gilg) Desc. ex Wild et R. B. Drumm. = Cyphostemma gracillimoides (Dewit) Desc. ●☆

120075 Cyphostemma gallaense (Gilg et M. Brandt) Desc. = Cyphostemma cyphopetalum (Fresen.) Desc. ex Wild et R. B. Drumm. ●☆

120076 Cyphostemma gigantophyllum (Gilg et M. Brandt) Desc. ex Wild et R. B. Drumm.;巨叶树葡萄●☆

120077 Cyphostemma gilletii (De Wild. et T. Durand) Desc.;吉勒特树葡萄●☆

120078 Cyphostemma glandulosissimum (Gilg et M. Brandt) Desc. ex Wild et R. B. Drumm.;密腺树葡萄●☆

120079 Cyphostemma glanduloso-pilosa Desc.;腺毛树葡萄●☆

120080 Cyphostemma gracillimoides (Dewit) Desc. = Cyphostemma fugosioides (Gilg) Desc. ex Wild et R. B. Drumm. ●☆

120081 Cyphostemma gracillimum (Werderm.) Desc.;细长树葡萄●☆

120082 Cyphostemma grahamii Verdc.;格雷厄姆树葡萄●☆

120083 Cyphostemma grandistipulatum (Gilg et M. Brandt) Desc.;大托叶树葡萄●☆

120084 Cyphostemma graniticum (Wild et R. B. Drumm.) Wild et R. B. Drumm.;花岗岩树葡萄●☆

120085 Cyphostemma greenwayi Verdc.;格林韦树葡萄●☆

120086 Cyphostemma greveana Desc.;格雷弗树葡萄●☆

120087 Cyphostemma griseo-rubrum (Gilg et M. Brandt) Desc.;灰红树葡萄●☆

120088 Cyphostemma hardyi Retief;哈迪树葡萄●☆

120089 Cyphostemma haumanii (Dewit) Desc.;豪曼树葡萄●☆

120090 Cyphostemma hereroense (Schinz) Desc. ex Wild et R. B. Drumm.;赫雷罗树葡萄●☆

120091 Cyphostemma hermannioides Wild et R. B. Drumm. = Cyphostemma wittei (Staner) Wild et R. B. Drumm. ●☆

120092 Cyphostemma heterotrichum (Gilg et R. E. Fr.) Desc. ex Wild et R. B. Drumm.;异毛树葡萄●☆

120093 Cyphostemma hildebrandtii (Gilg) Desc. ex Wild et R. B. Drumm.;希尔德树葡萄●☆

120094 Cyphostemma hispidiflorum (C. A. Sm.) J. J. M. van der Merwe;毛花树葡萄●☆

120095 Cyphostemma homblei (De Wild.) Desc.;洪布勒树葡萄●☆

120096 Cyphostemma huillense (Exell et Mendonça) Desc.;威拉树葡萄●☆

120097 Cyphostemma humile (N. E. Br.) Desc. ex Wild et R. B. Drumm.;矮小树葡萄●☆

120098 Cyphostemma humile (N. E. Br.) Desc. ex Wild et R. B. Drumm. subsp. dolichopus (C. A. Sm.) Wild et R. B. Drumm.;长矮小树葡萄●☆

120099 Cyphostemma hypargyrea (Gilg) Desc. = Cyphostemma chloroleucum (Welw. ex Baker) Desc. ex Wild et R. B. Drumm. ●☆

120100 Cyphostemma hypoleucum (Harv.) Desc. ex Wild et R. B. Drumm.;白背树葡萄●☆

120101 Cyphostemma jaegeri (Gilg et M. Brandt) Desc. ex Wild et R. B. Drumm. = Cyphostemma kilimandscharicum (Gilg) Desc. ex Wild et R. B. Drumm. var. jaegeri (Gilg et M. Brandt) Verdc. ●☆

120102 Cyphostemma jatrophoides (Baker) Desc. = Cyphostemma junceum (Webb) Wild et R. B. Drumm. subsp. jatrophoides (Baker) Verdc. ●☆

120103 Cyphostemma jiguu Verdc. = Cyphostemma pachypus Verdc. ●☆

120104 Cyphostemma johannis (Exell et Mendonça) Desc.;约翰树葡萄●☆

120105 Cyphostemma junceum (Webb) Wild et R. B. Drumm.;灯心草树葡萄●☆

120106 Cyphostemma junceum (Webb) Wild et R. B. Drumm. subsp. jatrophoides (Baker) Verdc.;麻疯树葡萄●☆

120107 Cyphostemma junceum (Webb) Wild et R. B. Drumm. var. bambariense Desc.;班巴里树葡萄●☆

120108 Cyphostemma juttae (Dinter et Gilg ex Gilg et M. Brandt) Desc.;树葡萄(葡萄瓮,青紫葛,肉瓶树);Bastard Cobas, Baster Kobas, Basterkobas, Fleshy-trunk Treebine, Jutt Cissus, Tree Grape ●☆

120109 Cyphostemma juttae (Dinter et Gilg ex Gilg et M. Brandt) Desc. = Cissus juttae Dinter et Gilg ex Gilg et M. Brandt ●☆

120110 Cyphostemma juttae (Dinter et Gilg) Desc. = Cyphostemma juttae (Dinter et Gilg ex Gilg et M. Brandt) Desc. ●☆

120111 Cyphostemma kaessneri (Gilg et M. Brandt) Wild et R. B. Drumm. = Cyphostemma cyphopetalum (Fresen.) Desc. ex Wild et R. B. Drumm. ●☆

120112 Cyphostemma kaniamae (Dewit) Desc.;卡尼亚马树葡萄●☆

120113 Cyphostemma kaniamae (Dewit) Desc. subsp. tanzaniae Verdc.;坦桑尼亚树葡萄●☆

120114 Cyphostemma kapiriense (Dewit) Desc.;卡皮里树葡萄●☆

120115 Cyphostemma keilii (Gilg et M. Brandt) Desc.;凯尔树葡萄●☆

120116 Cyphostemma keniense (T. C. E. Fr.) Desc. = Cyphostemma knittelii (Gilg) Desc. ●☆

120117 Cyphostemma kilimandscharicum (Gilg) Desc. ex Wild et R. B. Drumm.;基利树葡萄●☆

120118 Cyphostemma kilimandscharicum (Gilg) Desc. ex Wild et R. B. Drumm. var. jaegeri (Gilg et M. Brandt) Verdc.;耶格树葡萄●☆

120119 Cyphostemma kirkianum (Planch.) Desc. ex Wild et R. B. Drumm.;柯克树葡萄●☆

120120 Cyphostemma kirkianum (Planch.) Desc. ex Wild et R. B. Drumm. subsp. trifoliolatum Verdc.;三叶柯克树葡萄●☆

120121 Cyphostemma knittelii (Gilg) Desc.;克尼特尔树葡萄●☆

120122 Cyphostemma kundelunguense Malaisse;昆德龙古树葡萄●☆

120123 Cyphostemma lageniflorum (Gilg et M. Brandt) Desc.;烧瓶树葡萄●☆

120124 Cyphostemma lanigerum (Harv.) Desc. ex Wild et R. B. Drumm.;绵毛树葡萄●☆

120125 Cyphostemma laza Desc.;拉扎树葡萄●☆

120126 Cyphostemma leandrii Desc.;利安树葡萄●☆

120127 Cyphostemma ledermannii (Gilg et M. Brandt) Desc.;莱德树葡萄●☆

120128 Cyphostemma letouzeyanum Desc.;勒图树葡萄●☆

120129 Cyphostemma leucorufescens Desc.;浅红树葡萄●☆

120130 Cyphostemma leucotrichum (Gilg et M. Brandt) Desc.;白毛树葡萄●☆

120131 Cyphostemma libenii (Dewit) Desc.;利本树葡萄●☆

120132 Cyphostemma luteum (Exell et Mendonça) Desc.;黄树葡萄●☆

120133 Cyphostemma lynesii (Dewit) Desc. ex Wild et R. B. Drumm.;莱恩斯树葡萄●☆

120134 Cyphostemma macrocarpa Desc.;大果树葡萄●☆

120135 Cyphostemma macropus (Welw.) Desc. = Cyphostemma currorii (Hook. f.) Desc. ●☆

120136 Cyphostemma macrothyrsum (Gilg) Desc. = Cyphostemma cyphopetalum (Fresen.) Desc. ex Wild et R. B. Drumm. ●☆

120137 Cyphostemma manikense (De Wild.) Desc. ex Wild et R. B. Drumm.;马尼科树葡萄●☆

120138 Cyphostemma mannii (Baker) Desc.;曼氏树葡萄●☆

120139 Cyphostemma maranguense (Gilg) Desc.;马兰古树葡萄●☆

120140 Cyphostemma marunguense (Dewit) Desc.;马龙古树葡萄●☆

120141 Cyphostemma masukuense (Baker) Desc. ex Wild et R. B. Drumm. subsp. ferrugineo-velutinum Verdc.;锈绒毛树葡萄●☆

120142 Cyphostemma masukuense (Baker) Desc. ex Wild et R. B. Drumm. subsp. nguruense Verdc.;恩古鲁树葡萄●☆

120143 Cyphostemma masukuense (Baker) Desc. ex Wild et R. B. Drumm. subsp. ulugurense Verdc.;乌卢古尔树葡萄●☆

120144 Cyphostemma meyeri-johannis (Gilg et M. Brandt) Verdc.;迈尔约翰树葡萄●☆

120145 Cyphostemma michelii (Dewit) Desc.;米歇尔树葡萄●☆

120146 Cyphostemma micradenium (Gilg et M. Brandt) Desc.;小腺树葡萄●☆

120147 Cyphostemma microdiptera (Baker) Desc.;二小翅树葡萄●☆

120148 Cyphostemma migiurtinorum (Chiov.) Desc.;米朱蒂树葡萄●☆

120149 Cyphostemma mildbraedii (Gilg et M. Brandt) Desc. ex Wild et R. B. Drumm.;米尔德树葡萄●☆

120150 Cyphostemma milleri Wild et R. B. Drumm.;米勒树葡萄●☆

120151 Cyphostemma molle (Steud. ex Baker) Desc.;柔软树葡萄●☆

120152 Cyphostemma montanum Wild et R. B. Drumm.;山地树葡萄●☆

120153 Cyphostemma muhuluense (Mildbr.) Desc.;穆胡卢树葡萄●☆

120154 Cyphostemma nanellum (Gilg et R. E. Fr.) Desc. ex Wild et R. B. Drumm.;极矮树葡萄●☆

120155 Cyphostemma natalitium (Szyszyl.) J. J. M. van der Merwe;蝙蝠树葡萄●☆

120156 Cyphostemma nieriense (T. C. E. Fr.) Desc. = Cyphostemma cyphopetalum (Fresen.) Desc. ex Wild et R. B. Drumm. ●☆

120157 Cyphostemma nigroglandulosum (Gilg et M. Brandt) Desc.;黑腺树葡萄●☆

120158 Cyphostemma niveum (Hochst. ex Schweinf.) Desc.;雪白树葡萄●☆

120159 Cyphostemma nodiglandulosum (T. C. E. Fr.) Desc. = Cyphostemma cyphopetalum (Fresen.) Desc. ex Wild et R. B. Drumm. var. nodiglandulosum (T. C. E. Fr.) Verdc. ●☆

120160 Cyphostemma obovato-oblongum (De Wild.) Desc. ex Wild et R. B. Drumm.;倒卵-矩圆树葡萄●☆

120161 Cyphostemma odontadenium (Gilg) Desc.;齿腺树葡萄●☆

120162 Cyphostemma oleraceum (Bolus) J. J. M. van der Merwe;蔬菜树葡萄●☆

120163 Cyphostemma omburense (Gilg et M. Brandt) Desc.;翁布尔树葡萄●☆

120164 Cyphostemma ornatum (A. Chev. ex Hutch. et Dalziel) Desc.;装饰树葡萄●☆

120165 Cyphostemma orondo (Gilg et M. Brandt) Desc. = Cyphostemma serpens (A. Rich.) Desc. ●☆

120166 Cyphostemma ouakense Desc. var. glandulosum ?;具腺树葡萄●☆

120167 Cyphostemma overlaetii (Dewit) Desc.;奥弗莱特树葡萄●☆

120168 Cyphostemma oxyphyllum (A. Rich.) Vollesen;尖叶树葡萄●☆

120169 Cyphostemma pachyanthum (Gilg et M. Brandt) Desc.;粗花树葡萄●☆

120170 Cyphostemma pachypus Desc.;粗足树葡萄●☆

120171 Cyphostemma pachypus Verdc. = Cyphostemma pachypus Desc. ●☆

120172 Cyphostemma pachyrrhachis (Gilg et M. Brandt) Desc. = Cyphostemma cyphopetalum (Fresen.) Desc. ex Wild et R. B. Drumm. ●☆

120173 Cyphostemma pannosum Vollesen;毡状树葡萄●☆

120174 Cyphostemma passargei (Gilg et M. Brandt) Desc.;帕萨树葡萄●☆

120175 Cyphostemma paucidentatum (Klotzsch) Desc. ex Wild et R. B. Drumm.;少齿树葡萄●☆

120176 Cyphostemma paucidentatum (Klotzsch) Desc. ex Wild et R. B. Drumm. subsp. zanzibaricum Verdc.;桑给巴尔树葡萄●☆

120177 Cyphostemma penduloides (Dewit) Desc. = Cyphostemma hildebrandtii (Gilg) Desc. ex Wild et R. B. Drumm. ●☆

120178 Cyphostemma pendulum (Welw. ex Baker) Desc.;下垂树葡萄●☆

120179 Cyphostemma perforatum (Louis ex Dewit) Desc.;穿孔树葡萄●☆

120180 Cyphostemma phyllomicron (Chiov.) Desc.;小树葡萄●☆

120181 Cyphostemma phyllomicron (Chiov.) Desc. f. eglandulosum ?;无腺树葡萄●☆

120182 Cyphostemma pobeguinianum Desc.;波别树葡萄●☆

120183 Cyphostemma pobeguinii A. Chev. = Cyphostemma junceum (Webb) Wild et R. B. Drumm. subsp. jatrophoides (Baker) Verdc. ●☆

120184 Cyphostemma psammophilum (Gilg et M. Brandt) Desc. = Cyphostemma setosum (Roxb.) Alston ●☆

120185 Cyphostemma pseudoburgeri Verdc.;假伯格树葡萄●☆

120186 Cyphostemma pseudomanikensis (Dewit) Desc. = Cyphostemma manikense (De Wild.) Desc. ex Wild et R. B. Drumm. ●☆

120187 Cyphostemma pseudoniveum (Gilg et M. Brandt) Desc. = Cyphostemma cyphopetalum (Fresen.) Desc. ex Wild et R. B. Drumm. ●☆

120188 Cyphostemma pseudorhodesiae (Dewit) Desc.;假罗得西亚树葡萄●☆

120189 Cyphostemma pseudoupembaense (Dewit) Desc.;假乌彭贝树葡萄●☆

120190 Cyphostemma puberulum (C. A. Sm.) Wild et R. B. Drumm.;微毛树葡萄●☆

120191 Cyphostemma pubescens (Dewit) Desc. var. eglanduloso-pubescens = Cyphostemma adenocaule (Steud. ex A. Rich.) Desc. ex Wild et R. B. Drumm. ●☆

120192 Cyphostemma pumila Desc.;低矮树葡萄●☆

120193 Cyphostemma quinatum (Dryand.) Desc. ex Wild et R. B. Drumm.;五出树葡萄●☆

120194 Cyphostemma reedii (Dewit) Desc. = Cyphostemma heterotrichum (Gilg et R. E. Fr.) Desc. ex Wild et R. B. Drumm. ●☆

120195 Cyphostemma rhodesiae (Gilg et M. Brandt) Desc. ex Wild et R. B. Drumm.;罗得西亚树葡萄●☆

120196 Cyphostemma richardsiae Wild et R. B. Drumm.;理查兹树葡萄●☆

120197 Cyphostemma ringoetii (De Wild.) Desc. = Cyphostemma obovato-oblongum (De Wild.) Desc. ex Wild et R. B. Drumm. ●☆

120198 Cyphostemma rivae (Gilg) Desc.;沟树葡萄●☆

120199 Cyphostemma robsonii Wild et R. B. Drumm.;罗伯逊树葡萄●☆

120200 Cyphostemma robynsii (Dewit) Desc.;罗宾斯树葡萄●☆

120201 Cyphostemma roseiglandulosa Desc.;粉红腺树葡萄●☆

120202 Cyphostemma rotundistipulatum Wild et R. B. Drumm.;圆托叶树葡萄●☆

120203 Cyphostemma rowlandii (Gilg et M. Brandt) Desc. = Cyphostemma adenopodum (Sprague) Desc. ●☆

120204 Cyphostemma ruacanense (Exell et Mendonça) Desc.;鲁阿卡纳树葡萄●☆

120205 Cyphostemma rubroglandulosum Retief et A. E. van Wyk;红腺树葡萄●☆

120206 Cyphostemma rubromarginatum (Gilg et M. Brandt) Desc.;红边树葡萄●☆

120207 Cyphostemma rubrosetosum (Gilg et M. Brandt) Desc.;红毛树葡萄●☆

120208 Cyphostemma rupicola (Gilg et M. Brandt) Desc.;岩地树葡萄●☆

120209 Cyphostemma sandersonii (Harv.) Desc.;桑德森树葡萄●☆

120210 Cyphostemma sarcospathulum (Chiov.) Desc.;肉苞树葡萄●☆

120211 Cyphostemma saxicola (Gilg et R. E. Fr.) Desc. ex Wild et R. B. Drumm.;岩生树葡萄●☆

120212 Cyphostemma scarlatinum (Gilg et M. Brandt) Desc.;斯卡拉特树葡萄●☆

120213 Cyphostemma schimperi (Hochst. ex Planch.) Desc. = Cyphostemma oxyphyllum (A. Rich.) Vollesen ●☆

120214 Cyphostemma schlechteri (Gilg et M. Brandt) Desc. ex Wild et R. B. Drumm.;施莱树葡萄●☆

120215 Cyphostemma schliebenii (Mildbr.) Desc.;施利本树葡萄●☆

120216 Cyphostemma schweinfurthii (Planch.) Desc. = Cyphostemma serpens (A. Rich.) Desc. ●☆

120217 Cyphostemma segmentatum (C. A. Sm.) J. J. M. van der Merwe;短裂树葡萄●☆

120218 Cyphostemma seitzianum (Gilg et M. Brandt) Desc. = Cyphostemma bainesii (Hook. f.) Desc. ●☆

120219 Cyphostemma serpens (A. Rich.) Desc.;蛇形树葡萄●☆

120220 Cyphostemma sesquipedale (Gilg) Desc. = Cyphostemma serpens (A. Rich.) Desc. ●☆

120221 Cyphostemma sessilifolium (Dewit) Desc.;无柄叶树葡萄●☆

120222 Cyphostemma setosum (Roxb.) Alston;刚毛树葡萄●☆

120223 Cyphostemma shinyangense Verdc.;希尼安加树葡萄●☆

120224 Cyphostemma simplicifolium A. Bjornstad;单叶树葡萄●☆

120225 Cyphostemma simulans (C. A. Sm.) Wild et R. B. Drumm.;相似树葡萄●☆

120226 Cyphostemma spinosopilosum (Gilg et M. Brandt) Desc.;刺毛树葡萄●☆

120227 Cyphostemma stefaninianum (Chiov.) Desc.;斯特树葡萄●☆

120228 Cyphostemma stenolobum (Welw. ex Baker) Desc. ex Wild et R. B. Drumm.;窄裂片树葡萄●☆

120229 Cyphostemma stenopodum (Gilg) Desc.;细梗树葡萄●☆

120230 Cyphostemma stipulaceum (Baker) Desc.;托叶树葡萄●☆

120231 Cyphostemma strigosum (Dewit) Desc.;糙伏毛树葡萄●☆

120232 Cyphostemma subciliatum (Baker) Desc. ex Wild et R. B. Drumm.;缘毛树葡萄●☆

120233 Cyphostemma subglaucescens (Planch.) Desc. = Cyphostemma subciliatum (Baker) Desc. ex Wild et R. B. Drumm. ●☆

120234 Cyphostemma sulcatum (C. A. Sm.) J. J. M. van der Merwe;纵沟树葡萄●☆

120235 Cyphostemma taborense Verdc.;泰伯树葡萄●☆

120236 Cyphostemma tenuipes (Gilg et R. E. Fr.) Desc. = Cyphostemma cyphopetalum (Fresen.) Desc. ex Wild et R. B. Drumm. ●☆

120237 Cyphostemma tenuissimum (Gilg et R. E. Fr.) Desc. ex Wild et R. B. Drumm.;极细树葡萄●☆

120238 Cyphostemma termetophilum (De Wild.) Desc. = Cyphostemma mildbraedii (Gilg et M. Brandt) Desc. ex Wild et R. B. Drumm. ●☆

120239 Cyphostemma ternato-multifidum (Chiov.) Desc.;多裂树葡萄●☆

120240 Cyphostemma ternatum (Forssk.) Desc.;三出树葡萄●☆

120241 Cyphostemma thomasii (Gilg et M. Brandt) Desc.;托马斯树葡萄●☆

120242 Cyphostemma tisserantii Desc.;蒂斯朗特树葡萄●☆

120243 Cyphostemma triumfettioides (Gilg et M. Brandt) Desc. = Cyphostemma rhodesiae (Gilg et M. Brandt) Desc. ex Wild et R. B. Drumm. ●☆

120244 Cyphostemma tsaratananensis Desc.;察拉塔纳纳树葡萄●☆

120245 Cyphostemma ukerewense (Gilg) Desc.;乌凯雷韦树葡萄●☆

120246 Cyphostemma ukerewense (Gilg) Desc. var. gabonicum Desc.;加蓬树葡萄●☆

120247 Cyphostemma unguiformifolium (C. A. Sm.) Desc. = Cyphostemma schlechteri (Gilg et M. Brandt) Desc. ex Wild et R. B. Drumm. ●☆

120248 Cyphostemma upembaensis (Dewit) Desc. = Cyphostemma zombense (Baker) Desc. ex Wild et R. B. Drumm. ●☆

120249 Cyphostemma urophyllum (Gilg et M. Brandt) Desc.;尾叶树葡萄●☆

120250 Cyphostemma uter (Exell et Mendonça) Desc.;囊树葡萄●☆

120251 Cyphostemma vandenbergheae Malaisse et Matamba;范登树葡萄●☆

120252 Cyphostemma vandenbrandeanum (Dewit) Desc. ex Wild et R. B. Drumm.;范登布兰德树葡萄●☆

120253 Cyphostemma vanderbenii (Dewit) Desc.;范德树葡萄●☆

120254 Cyphostemma vanmeelii (Lawalrée) Wild et R. B. Drumm.;范米尔树葡萄●☆

120255 Cyphostemma variifolia (Baker) Desc. = Cyphostemma crotalarioides (Planch.) Desc. ex Wild et R. B. Drumm. ●☆
120256 Cyphostemma villosicaule Verdc.;长毛茎树葡萄●☆
120257 Cyphostemma villosiglandulosum (Werderm.) Desc.;腺树葡萄●☆
120258 Cyphostemma violaceoglandulosum (Gilg) Desc.;堇腺树葡萄●☆
120259 Cyphostemma viscosum (Gilg et R. E. Fr.) Desc. ex Wild et R. B. Drumm.;黏树葡萄●☆
120260 Cyphostemma vogelii (Hook. f.) Desc.;沃格尔树葡萄●☆
120261 Cyphostemma vollesenii Verdc.;福勒森树葡萄●☆
120262 Cyphostemma waterlotii (A. Chev.) Desc.;瓦泰洛树葡萄●☆
120263 Cyphostemma wilmsii (Gilg et M. Brandt) Desc.;维尔姆斯树葡萄●☆
120264 Cyphostemma wittei (Staner) Wild et R. B. Drumm.;沃特树葡萄●☆
120265 Cyphostemma woodii (Gilg et M. Brandt) Desc.;伍得树葡萄●☆
120266 Cyphostemma zechianum (Gilg et M. Brandt) Desc.;策希树葡萄●☆
120267 Cyphostemma zimmermannii Verdc.;齐默尔曼树葡萄●☆
120268 Cyphostemma zombense (Baker) Desc. ex Wild et R. B. Drumm.;宗巴树葡萄●☆
120269 Cyphostigma Benth. (1882);驼柱姜属■☆
120270 Cyphostigma pulchellum Benth.;驼柱姜■☆
120271 Cyphostyla Gleason(1929);驼柱野牡丹属☆
120272 Cyphostyla hirsuta Gleason = Allomaieta hirsuta (Gleason) Lozano☆
120273 Cyphostyla strigosa Gleason;驼柱野牡丹☆
120274 Cyphostyla villosa Gleason;毛驼柱野牡丹☆
120275 Cyphotheca Diels(1932);药囊花属(瘤药花属,弯棕属);Cyphotheca ●★
120276 Cyphotheca hispida S. Y. Hu = Phyllagathis fengii C. Hansen ●
120277 Cyphotheca hispida S. Y. Hu = Phyllagathis hispida (S. Y. Hu) C. Y. Wu ●
120278 Cyphotheca montana Diels;药囊花(弯棕);Montane Cyphotheca ●◇
120279 Cyprianthe Spach = Ranunculus L. ■
120280 Cypringlea M. T. Strong = Scirpus L. (保留属名)■
120281 Cypringlea M. T. Strong(2003);墨西哥藨草属■☆
120282 Cyprinia Browicz(1966);鲤鱼萝藦属●☆
120283 Cyprinia gracilis (Boiss.) Browicz;鲤鱼萝藦●☆
120284 Cypripediaceae Lindl.;杓兰科■
120285 Cypripediaceae Lindl. = Orchidaceae Juss. (保留科名)■
120286 Cypripedilon St.-Lag. = Cypripedium L. ■
120287 Cypripedium L. (1753);杓兰属(喜普鞋兰属,仙履兰属);Cypripedium, Lady Slipper, Lady's Slipper, Lady's Slipper Orchid, Ladyslipper, Lady's-slipper, Lady's-slipper Orchid, Moccasin Flower, Moccasin-flower, Slipper Orchid ■
120288 Cypripedium × favillianum J. T. Curtis;法氏杓兰;Lady's-slipper ■☆
120289 Cypripedium × ventricosum Sw.;东北杓兰;NE. China Lady's Slipper, NE. China Ladyslipper ■
120290 Cypripedium acaule Aiton;粉红杓兰;Mocassin Flower, Moccasin-flower, Nerve-root, Pink Lady's Slipper, Pink Lady's-slipper, Pink Mocassin, Pink Moccasin Flower, Pink Moccasin-flower, Small Pink Lady's-slipper, Stemless lady's-slipper, Stemless Lady's Slipper, Valerian ■☆
120291 Cypripedium acaule Aiton f. albiflorum E. L. Rand et Redfield = Cypripedium acaule Aiton ■☆
120292 Cypripedium amesianum Schltr. = Cypripedium yunnanense Franch. ■
120293 Cypripedium andrewsii A. M. Fuller;安氏杓兰;Andrews' Lady's-slipper ■☆
120294 Cypripedium andrewsii A. M. Fuller var. favillianum (J. T. Curtis) B. Boivin = Cypripedium favillianum J. T. Curtis ■☆
120295 Cypripedium andrewsii A. M. Fuller var. landonii (Garay) B. Boivin = Cypripedium andrewsii A. M. Fuller ■☆
120296 Cypripedium appletonianum Gower = Paphiopedilum appletonianum (Gower) Rolfe ■
120297 Cypripedium argus Rchb. f. = Paphiopedilum argus (Rchb. f) Stein ■☆
120298 Cypripedium arietinum Franch. = Cypripedium plectrochilum Franch. ■
120299 Cypripedium arietinum R. Br.;羊角杓兰;Arietinous Lady's Slipper, Ram's Head Lady's Slipper, Ram's-head, Ram's-head Lady's Slipper, Ram's-head Lady's-slipper ■☆
120300 Cypripedium arietinum R. Br. = Cypripedium plectrochilum Franch. ■
120301 Cypripedium barbatum Lindl. = Paphiopedilum barbatum (Lindl.) Pfitzer ■☆
120302 Cypripedium bardolphianum W. W. Sm. et Farrer;无苞杓兰;Bractless Lady's Slipper, Bractless Ladyslipper ■
120303 Cypripedium bardolphianum W. W. Sm. et Farrer var. zhongdianense S. C. Chen;中甸杓兰;Zhongdian Lady's Slipper, Zhongdian Ladyslipper ■
120304 Cypripedium bardolphianum W. W. Sm. et Farrer var. zhongdianense S. C. Chen = Cypripedium forrestii P. J. Cribb ■
120305 Cypripedium bellatulum Rchb. f. = Paphiopedilum bellatulum (Rchb. f.) Stein ■
120306 Cypripedium bouffordianum Yong H. Zhang et H. Sun = Cypripedium guttatum Sw. ■
120307 Cypripedium boxallii Rchb. f. = Paphiopedilum villosum Lindl. var. boxallii (Rchb. f.) Pfitzer ■
120308 Cypripedium bulbosum L. = Calypso bulbosa (L.) Oakes ■
120309 Cypripedium bullenianum Rchb. f. var. appletonianum (Gower) Rolfe = Paphiopedilum appletonianum (Gower) Rolfe ■
120310 Cypripedium calceolus L.;杓兰(斑花兰,黄囊杓兰,履状杓兰,履状囊兰,鞋杓兰);Boots-and-shoes, Calcalary, Calscalary, European Lady's Slipper, European Ladyslipper, European Lady-slipper, European Lady's-slipper, Fingers-and-thumbs, Golden Slipper, Lady's Boots, Lady's Slipper Orchid, Large Yellow Lady's Slipper, Small Yellow Lady's Slipper, Yellow Lady Slipper, Yellow Lady's Slipper, Yellow Lady's Slipper Orchid, Yellow Lady's-slipper ■
120311 Cypripedium calceolus L. subsp. parviflorum (Salisb.) Hultén = Cypripedium calceolus L. var. parviflorum (Salisb.) Fernald ■☆
120312 Cypripedium calceolus L. var. parviflorum (Salisb.) Fernald;北美小花杓兰;Small Yellow Lady's Slipper ■☆
120313 Cypripedium calceolus L. var. planipetalum (Fernald) Vict. et J. Rousseau = Cypripedium parviflorum Salisb. var. pubescens (Willd.) O. W. Knight ■
120314 Cypripedium calceolus L. var. pubescens (Willd.) Correll;柔毛杓兰(短毛杓兰);Pubescent Lady's Slipper ■☆
120315 Cypripedium calceolus L. var. pubescens (Willd.) Correll = Cypripedium parviflorum Salisb. var. pubescens (Willd.) O. W. Knight ■

120316 Cypripedium calcicola Schltr.；褐花杓兰；Brownflower Lady's Slipper，Brownflower Ladyslipper ■

120317 Cypripedium californicum A. Gray；加州杓兰；California Lady's-slipper ■☆

120318 Cypripedium callosum Rchb. f. = Paphiopedilum callosum (Rchb. f.) Pfitzer ■☆

120319 Cypripedium calsicola Schltr. = Cypripedium smithii Schltr. ■

120320 Cypripedium candidum Muhl. ex Willd.；白杓兰；Small White Lady's Slipper，Small White Lady's-slipper，White Lady's-slipper ■☆

120321 Cypripedium cardiophyllum Franch. et Sav. = Cypripedium debile Rchb. f. ■

120322 Cypripedium cathayenum S. S. Chien = Cypripedium japonicum Thunb. ■

120323 Cypripedium chamberlainianum O' Brien = Paphiopedilum chamberlainianum (O' Brien) Pfitzer ■☆

120324 Cypripedium charlesworthii Rolfe = Paphiopedilum charlesworthii (Rolfe) Pfitzer ■

120325 Cypripedium cheniae Torelli = Cypripedium farreri W. W. Sm. ■

120326 Cypripedium chinense Franch. = Cypripedium henryi Rolfe ■

120327 Cypripedium ciliolare Rchb. f. = Paphiopedilum ciliolare (Rchb. f.) Stein ■☆

120328 Cypripedium compactum Schltr. = Cypripedium tibeticum King ex Hemsl. ■

120329 Cypripedium concolor Bateman = Paphiopedilum concolor (Bateman) Pfitzer ■

120330 Cypripedium concolor Lindl. = Paphiopedilum concolor (Bateman) Pfitzer ■

120331 Cypripedium concolor Rchb. f. = Paphiopedilum curtisii (Rchb. f.) Pfitzer ■☆

120332 Cypripedium cordigerum D. Don；白唇杓兰；Whitelip Lady's Slipper，Whitelip Ladyslipper ■

120333 Cypripedium corrugatum Franch. = Cypripedium tibeticum King ex Hemsl. ■

120334 Cypripedium corrugatum Franch. var. obesum Franch. = Cypripedium tibeticum King ex Hemsl. ■

120335 Cypripedium curtisii Rchb. f. = Paphiopedilum curtisii (Rchb. f.) Pfitzer ■☆

120336 Cypripedium daliense S. C. Chen et J. L. Wu；大理杓兰；Dali Slipper ■

120337 Cypripedium daliense S. C. Chen et J. L. Wu = Cypripedium lichiangense S. C. Chen et P. J. Cribb ■

120338 Cypripedium daliense S. C. Chen et J. L. Wu = Cypripedium margaritaceum Franch. ■

120339 Cypripedium daweishanense (S. C. Chen et Z. J. Liu) S. C. Chen et Z. J. Liu；大围山杓兰；Daweishan Lady's Slipper ■

120340 Cypripedium dayanum Rchb. f. = Paphiopedilum dayanum (Rchb. f.) Pfitzer ■☆

120341 Cypripedium debile Rchb. f.；对叶杓兰(二叶兰，小喜普鞋兰)；Oppositeleaf Lady's Slipper，Oppositeleaf Ladyslipper ■

120342 Cypripedium delenatii (Guillaumin) C. H. Curtis = Paphiopedilum delenatii Guillaumin ■

120343 Cypripedium ebracteatum Rolfe = Cypripedium fargesii Franch. ■

120344 Cypripedium elegans Rchb. f.；雅致杓兰；Elegant Lady's Slipper，Elegant Ladyslipper ■

120345 Cypripedium fairieanum Lindl. = Paphiopedilum fairieanum (Lindl.) Stein. ■☆

120346 Cypripedium fargesii Franch.；毛瓣杓兰(飞蛾七，花叶两块瓦，马蹄兰)；Hairpetal Lady's Slipper，Hairpetal Ladyslipper ■

120347 Cypripedium farreri W. W. Sm.；华西杓兰；Ferrer Lady's Slipper，Ferrer Ladyslipper ■

120348 Cypripedium fasciculatum Kellogg ex S. Watson；北美杓兰；Brownie Lady's-slipper，Clustered Lady's-slipper ■☆

120349 Cypripedium fasciolatum Franch.；大叶杓兰(灯笼花，鸡嗉子，蜈蚣七)；Bigleaf Ladyslipper，Brownie Lady's Slipper ■

120350 Cypripedium favillianum J. T. Curtis；法维尔杓兰；Lady's-slipper ■☆

120351 Cypripedium flavescens DC. = Cypripedium parviflorum Salisb. var. pubescens (Willd.) O. W. Knight ■

120352 Cypripedium flavum P. F. Hunt et Summerh.；黄花杓兰(淡黄杓兰)；Yellow Ladyslipper，Yellowflower Lady's Slipper，Yellowish Lady's Slipper ■

120353 Cypripedium formosanum Hayata；台湾杓兰(台湾袋唇兰，台湾喜普鞋兰，一点红)；Taiwan Lady's Slipper，Taiwan Ladyslipper ■

120354 Cypripedium forrestii P. J. Cribb；玉龙杓兰；Forrest Lady's Slipper，Forrest Ladyslipper ■

120355 Cypripedium franchetii E. H. Wilson；毛杓兰(兰竹参，龙舌箭，牌骨七，牌楼七)；Franchet Lady's Slipper，Hair Ladyslipper ■

120356 Cypripedium gratrixianum Mast. = Paphiopedilum gratrixianum Rolfe ■

120357 Cypripedium guttatum Sw.；紫点杓兰(斑杓兰，斑花杓兰，小口袋花，小米口袋花，紫斑杓兰)；Purplespot Lady's Slipper，Purplespot Ladyslipper，Spotted Lady's-slipper ■

120358 Cypripedium guttatum Sw. subsp. yatabeanum (Makino) Hultén = Cypripedium yatabeanum Makino ■

120359 Cypripedium guttatum Sw. var. segawae (Masam.) S. S. Ying = Cypripedium segawai Masam. ■

120360 Cypripedium guttatum Sw. var. wardii (Rolfe) P. Taylor = Cypripedium wardii Rolfe ■

120361 Cypripedium guttatum Sw. var. yatabeanum (Makino) Pfitzer = Cypripedium yatabeanum Makino ■

120362 Cypripedium haynaldianum Rchb. f. = Paphiopedilum haynaldianum (Rchb. f.) Pfitzer ■☆

120363 Cypripedium henryi Rolfe；绿花杓兰(灯盏七，金龙七，九根毛，龙蛇箭，野竹兰)；Green Ladyslipper，Henry Lady's Slipper ■

120364 Cypripedium himalaicum Rolfe ex Hemsl.；高山杓兰(狭萼杓兰，腰子花)；Himalayas Ladyslipper，Himalayan Lady's Slipper ■

120365 Cypripedium hirsutissimum Lindl. ex Hook. = Paphiopedilum hirsutissimum (Lindl. ex Hook.) Stein ■

120366 Cypripedium hirsutum Mill. = Cypripedium reginae L. ■

120367 Cypripedium hookerae Rchb. f. = Paphiopedilum hookerae (Rchb. f.) Pfitzer ■☆

120368 Cypripedium humile Salisb. = Cypripedium acaule Aiton ■☆

120369 Cypripedium insigne Lindl. = Paphiopedilum insigne (Lindl.) Pfitzer ■

120370 Cypripedium insigne Wall. ex Lindl. = Paphiopedilum insigne (Lindl.) Pfitzer ■

120371 Cypripedium japonicum Thunb.；扇脉杓兰(大对月草，飞蛾七，菊花双叶草，老虎七，连子七，日本喜普鞋兰，扇叶杓兰，扇叶还阳，扇子还阳，扇子七，双扇兰，熊谷草，一把伞，阴阳扇)；Japan Ladyslipper，Japanese Lady's Slipper ■

120372 Cypripedium japonicum Thunb. f. urasawae T. Koyama；浦泽杓兰■☆

120373 Cypripedium japonicum Thunb. var. formosanum (Hayata) S. S. Ying = Cypripedium formosanum Hayata ■

120374 Cypripedium japonicum Thunb. var. glabrum M. Suzuki;光扇脉杓兰■☆

120375 Cypripedium kentuckiense C. F. Reed;象牙莎草;Ivory Lady's-slipper,Purloined Slipper ■☆

120376 Cypripedium knightiae A. Nelson = Cypripedium fasciculatum Kellogg ex S. Watson ■☆

120377 Cypripedium laevigatum Benth. = Paphiopedilum philippinense (Rchb. f.) Pfitzer ■☆

120378 Cypripedium langrhoa Gattef. = Cypripedium fasciolatum Franch. ■

120379 Cypripedium langrhoa Gattef. ex Const. = Cypripedium fasciolatum Franch. ■

120380 Cypripedium lanuginosum Schltr. = Cypripedium tibeticum King ex Hemsl. ■

120381 Cypripedium lawrenceanum Rchb. f. = Paphiopedilum lawrenceanum (Rchb. f.) Pfitzer ■☆

120382 Cypripedium lentiginosum P. J. Cribb et S. C. Chen;长瓣杓兰■

120383 Cypripedium lichiangense S. C. Chen et P. J. Cribb;丽江杓兰;Lijiang Lady's Slipper,Lijiang Ladyslipper ■

120384 Cypripedium lichiangense S. C. Chen et P. J. Cribb subsp. lentiginosum (P. J. Cribb et S. C. Chen) Eccarius = Cypripedium lentiginosum P. J. Cribb et S. C. Chen ■

120385 Cypripedium lichiangense S. C. Chen et P. J. Cribb var. daweishanense S. C. Chen et Z. J. Liu = Cypripedium daweishanense (S. C. Chen et Z. J. Liu) S. C. Chen et Z. J. Liu ■

120386 Cypripedium lowii Lindl. = Paphiopedilum lowii (Lindl.) Pfitzer ■☆

120387 Cypripedium ludlowii P. J. Cribb;波密杓兰;Ludlow Lady's Slipper,Ludlow Ladyslipper ■

120388 Cypripedium luteum Franch. = Cypripedium flavum P. F. Hunt et Summerh. ■

120389 Cypripedium luteum Raf. = Cypripedium parviflorum Salisb. ■☆

120390 Cypripedium macranthos Sw. f. albiflorum (Makino) Ohwi = Cypripedium macranthum Sw. ■

120391 Cypripedium macranthos Sw. var. albiflorum Makino = Cypripedium macranthum Sw. ■

120392 Cypripedium macranthos Sw. var. himalaicum (Rolfe) Kraenzl. = Cypripedium himalaicum Rolfe ex Hemsl. ■

120393 Cypripedium macranthos Sw. var. taiwanianum F. Maek. = Cypripedium macranthum Sw. ■

120394 Cypripedium macranthos Sw. var. tibeticum (King ex Rolfe) Kraenzl. = Cypripedium tibeticum King ex Hemsl. ■

120395 Cypripedium macranthos Sw. var. villosum Hand. -Mazz. = Cypripedium franchetii E. H. Wilson ■

120396 Cypripedium macranthum Sw.;大花杓兰(杓兰,大花囊兰,大口袋花,敦盛草,黑驴蛋,牌楼七,奇莱喜普鞋兰,蜈蚣七);Bigflower Lady's Slipper,Bigflower Ladyslipper ■

120397 Cypripedium macranthum Sw. f. albiflorum (Makino) Ohwi;白大花杓兰■☆

120398 Cypripedium macranthum Sw. f. albiflorum (Makino) Ohwi = Cypripedium macranthum Sw. ■

120399 Cypripedium macranthum Sw. f. albiflorum Makino = Cypripedium macranthum Sw. ■

120400 Cypripedium macranthum Sw. var. flavum Mandl.;黄大花杓兰■☆

120401 Cypripedium macranthum Sw. var. hotei-atsumorianum Sadovsky = Cypripedium macranthum Sw. ■

120402 Cypripedium macranthum Sw. var. speciosum (Rolfe) Koidz.;雅大花杓兰■☆

120403 Cypripedium macranthum Sw. var. speciosum (Rolfe) Koidz. = Cypripedium macranthum Sw. ■

120404 Cypripedium macranthum Sw. var. speciosum (Rolfe) Koidz. f. albiflorum (Makino) Ohwi = Cypripedium macranthum Sw. f. albiflorum (Makino) Ohwi ■☆

120405 Cypripedium macranthum Sw. var. taiwanianum F. Maek. = Cypripedium macranthum Sw. ■

120406 Cypripedium macranthum Sw. var. tibeticum (King ex Rolfe) Kraenzl. = Cypripedium tibeticum King ex Hemsl. ■

120407 Cypripedium macranthum Sw. var. ventricosum Rchb. f.;长侧瓣大花杓兰■

120408 Cypripedium macranthum Sw. var. villosum Hand. -Mazz. = Cypripedium franchetii E. H. Wilson ■

120409 Cypripedium malopoense S. C. Chen et Z. J. Liu;麻栗坡杓兰;Malipo Lady's Slipper ■

120410 Cypripedium maranthum Sw. = Cypripedium tibeticum King ex Hemsl. ■

120411 Cypripedium margaritaceum Franch.;斑叶杓兰(花叶两块瓦,兰花双叶草,扇子还阳);Variegate Lady's Slipper,Variegate Ladyslipper ■

120412 Cypripedium margaritaceum Franch. var. fargesii (Franch.) Pfitzer = Cypripedium fargesii Franch. ■

120413 Cypripedium mastersianum Rchb. f. = Paphiopedilum mastersianum (Rchb. f.) Pfitzer ■

120414 Cypripedium micranthum Franch.;小花杓兰;Littleflower Lady's Slipper,Smallflower Ladyslipper ■

120415 Cypripedium montanum Douglas ex Lindl.;山地杓兰;Montane Lady's Slipper ■☆

120416 Cypripedium niveum Rchb. f. = Paphiopedilum niveum (Rchb. f.) Pfitzer ■☆

120417 Cypripedium nutans Schltr. = Cypripedium bardolphianum W. W. Sm. et Farrer ■

120418 Cypripedium orientale Spreng. = Cypripedium guttatum Sw. ■

120419 Cypripedium palangshanense Ts. Tang et F. T. Wang;巴郎山杓兰;Balangshan Lady's Slipper,Balangshan Ladyslipper ■

120420 Cypripedium parishii Rchb. f. = Paphiopedilum parishii (Rchb. f.) Pfitzer ■

120421 Cypripedium parviflorum Salisb.;黄色杓兰;American Valerian,Downy Lady's Slipper,Noah's Ark,Small Yellow Lady's-slipper,Small Yellow Lady's Slipper,Yellow Lady's Slipper,Yellow Lady's-slipper ■☆

120422 Cypripedium parviflorum Salisb. = Cypripedium calceolus L. ■

120423 Cypripedium parviflorum Salisb. var. makasin (Farw.) Sheviak;小黄色杓兰;Small Yellow Lady's-slipper ■☆

120424 Cypripedium parviflorum Salisb. var. planipetalum Fernald = Cypripedium parviflorum Salisb. var. pubescens (Willd.) O. W. Knight ■

120425 Cypripedium parviflorum Salisb. var. pubescens (Willd.) O. W. Knight = Cypripedium pubescens Willd. ■

120426 Cypripedium passerinum Richardson;雀蛋杓兰;Franklin's Lady's-slipper,Sparrow's-egg Lady's-slipper ■☆

120427 Cypripedium passerinum Richardson var. minganense Vict. = Cypripedium passerinum Richardson ■☆

120428 Cypripedium philippinense Rchb. f. = Paphiopedilum philippinense (Rchb. f.) Pfitzer ■☆

120429 Cypripedium planipetalum (Fernald) F. J. A. Morris et E. A. Eames = Cypripedium parviflorum Salisb. var. pubescens (Willd.) O. W. Knight ■

120430 Cypripedium plectrochilum Franch.;离萼杓兰;Dialysepalous Lady's Slipper, Discretesepal Ladyslipper ■

120431 Cypripedium prupramm Lindl. = Paphiopedilum purpuratum (Lindl.) Stein ■

120432 Cypripedium pubescens Willd.;软毛杓兰(北美毛杓兰,毛黄色杓兰);American Valerian, Downy Lady's Slipper, Large Yellow Lady's-slipper, Larger Yellow Lady's Slipper, Massl, Noah's Ark ■

120433 Cypripedium pubescens Willd. = Cypripedium calceolus L. ■

120434 Cypripedium pubescens Willd. = Cypripedium parviflorum Salisb. var. pubescens (Willd.) O. W. Knight ■

120435 Cypripedium pubescens Willd. var. makasin Farw. = Cypripedium parviflorum Salisb. var. makasin (Farw.) Sheviak ■☆

120436 Cypripedium pulchrum Ames et Schltr. = Cypripedium franchetii E. H. Wilson ■

120437 Cypripedium purpuratum Lindl. = Paphiopedilum purpuratum (Lindl.) Stein ■

120438 Cypripedium reginae L.;女王杓兰(华美杓兰,美丽杓兰);Queen Lady's Slipper, Queen Lady's-slipper, Royal Lady's-slipper, Showy Lady's-slipper, Showy Lady's Slipper, Showy Lady's Slipper Orchid, Whip-poor-Will's Shoe, Whip-poor-Will's Shoes ■

120439 Cypripedium reginae L. var. segawai (Masam.) S. S. Ying = Cypripedium segawai Masam. ■

120440 Cypripedium reginae Walter = Cypripedium flavum P. F. Hunt et Summerh. ■

120441 Cypripedium reginae Walter f. albolabium Fernald et B. G. Schub. = Cypripedium reginae L. ■

120442 Cypripedium reginae Walter var. segawae (Masam.) S. S. Ying = Cypripedium segawai Masam. ■

120443 Cypripedium ripedium var. segawai (Masam.) S. S. Ying = Cypripedium segawai Masam. ■

120444 Cypripedium rothschildianum Rchb. f. = Paphiopedilum rothschildianum (Rchb. f.) Pfitzer ■☆

120445 Cypripedium rubronerve Cavestro = Cypripedium franchetii E. H. Wilson ■

120446 Cypripedium segawai Masam.;宝岛杓兰(宝岛喜普鞋兰);Segawa Lady's Slipper, Segawa Ladyslipper ■

120447 Cypripedium shanxiense S. C. Chen;山西杓兰;Shanxi Lady's Slipper, Shanxi Ladyslipper ■

120448 Cypripedium sichuanense Perner;四川杓兰■

120449 Cypripedium sinicum Hance ex Rchb. f. = Paphiopedilum purpuratum (Lindl.) Stein ■

120450 Cypripedium smithii Schltr. = Cypripedium calcicola Schltr. ■

120451 Cypripedium speciosum Rolfe = Cypripedium macranthum Sw. var. speciosum (Rolfe) Koidz. ■☆

120452 Cypripedium speciosum Rolfe = Cypripedium macranthum Sw. ■

120453 Cypripedium spectabile Salisb. = Cypripedium reginae L. ■

120454 Cypripedium spectabile Sw. = Cypripedium reginae L. ■

120455 Cypripedium spicerianum Rchb. f. = Paphiopedilum spicerianum (Rchb. f.) Pfitzer ■

120456 Cypripedium stonei Hook. f. = Paphiopedilum stonei (Hook. f.) Pfitzer ■☆

120457 Cypripedium subtropicum S. C. Chen et K. Y. Lang;暖地杓兰;Subtropical Lady's Slipper, Subtropical Ladyslipper ■

120458 Cypripedium taibaiense G. H. Zhu et S. C. Chen;太白杓兰■

120459 Cypripedium taiwanianum Masam. = Cypripedium macranthum Sw. ■

120460 Cypripedium thunbergii Blume = Cypripedium macranthum Sw. ■

120461 Cypripedium tibeticum King ex Hemsl.;西藏杓兰(敦盛草,儿兰,绉杓兰);Xizang Lady's Slipper, Xizang Ladyslipper ■

120462 Cypripedium tonsum Rchb. f. = Paphiopedilum tonsum (Rchb. f.) Pfitzer ■☆

120463 Cypripedium veganum Cockerell, P. Barker et M. Barker = Cypripedium parviflorum Salisb. var. pubescens (Willd.) O. W. Knight ■

120464 Cypripedium ventricosum Sw.;膨杓兰■☆

120465 Cypripedium venustum Sims = Paphiopedilum venustum (Wall. ex Sims) Pfitzer ■

120466 Cypripedium venustum Wall. ex Sims = Paphiopedilum venustum (Wall. ex Sims) Pfitzer ■

120467 Cypripedium villosum Lindl. = Paphiopedilum villosum (Lindl.) Stein ■

120468 Cypripedium villosum Lindl. var. boxallii (Rchb. f.) Veitch = Paphiopedilum villosum Lindl. var. boxallii (Rchb. f.) Pfitzer ■

120469 Cypripedium wardii (Summerh.) C. Curtis = Paphiopedilum wardii Summerh. ■

120470 Cypripedium wardii Rolfe;宽口杓兰;Ward Lady's Slipper, Ward Ladyslipper ■

120471 Cypripedium wilsonii Rolfe = Cypripedium fasciolatum Franch. ■

120472 Cypripedium wumengense S. C. Chen;乌蒙杓兰;Wumeng Lady's Slipper, Wumeng Ladyslipper ■

120473 Cypripedium yatabeanum Makino;黄铃杓兰■

120474 Cypripedium yunnanense Franch.;云南杓兰;Yunnan Lady's Slipper, Yunnan Ladyslipper ■

120475 Cypripedium zhongdianensis Z. D. Fang = Cypripedium farreri W. W. Sm. ■

120476 Cypselea Turpin(1806);蜂箱草属■☆

120477 Cypselea humifuso Turpin;蜂箱草■☆

120478 Cypselocarpus F. Muell. (1873);蜂箱果属■☆

120479 Cypselocarpus haloragoides (Benth.) F. Muell.;蜂箱果●☆

120480 Cypselodontia DC. = Dicoma Cass. ●☆

120481 Cypselodontia eckloniana DC. = Dicoma picta (Thunb.) Druce ●☆

120482 Cypsophila P. Gaertn., B. Mey. et Scherb. = Gypsophila L. ■●

120483 Cyrbasium Endl. = Cristatella Nutt. ■☆

120484 Cyrenea F. Allam. (1770);昔兰尼禾属■☆

120485 Cyrilla Garden ex L. (1767);翅萼树属(翅萼木属,西里拉属);Leatherwood ●☆

120486 Cyrilla L. = Cyrilla Garden ex L. ●☆

120487 Cyrilla L'Hér. = Achimenes Pers. (保留属名)■☆

120488 Cyrilla racemiflora L.;翅萼树(翅萼木,西里拉,泽木);Leatherwood, Leatherwood Titi, Palo Colorado, Swamp Cyrilla, Titi, White Titi ●☆

120489 Cyrillaceae Endl. = Cyrillaceae Lindl. (保留科名)●☆

120490 Cyrillaceae Lindl. (1846)(保留科名);翅萼树科(翅萼木科,西里拉科)●☆

120491 Cyrillopsis Kuhlm. (1925);拟翅萼树属●☆

120492 Cyrillopsis micrantha (Steyerm.) P. E. Berry et N. Ramírez;小花拟翅萼树●☆

120493 Cyrillopsis paraensis Kuhlm.;拟翅萼树●☆

120494 Cyrilwhitea Ising = Bassia All. ■●

120495 Cyrilwhitea Ising = Sclerolaena R. Br. ●☆

120496 Cyrta Lour. = Styrax L. ●
120497 Cyrta agrestis Lour. = Styrax agrestis (Lour.) G. Don ●
120498 Cyrta japonica Miers = Styrax japonicus Siebold et Zucc. ●
120499 Cyrta suberifolia Miers = Styrax suberifolius Hook. et Arn. ●
120500 Cyrtacanthus Mart. ex Nees = Ruellia L. ■●
120501 Cyrtandra J. R. Forst. et G. Forst. (1775);浆果苣苔属(曲蕊花属,弯果苣苔属,弯蕊苣苔属,伪苦苣苔属);Cyrtandra, Leatherwood ●■
120502 Cyrtandra cumingii C. B. Clarke var. yaeyamae (Ohwi) Hatus. = Cyrtandra yaeyamae Ohwi ●☆
120503 Cyrtandra fortunei C. B. Clarke;曲蕊花;Fortune Cyrtandra ●
120504 Cyrtandra kotoensis Hosok. = Cyrtandra umbellifera Merr. ●
120505 Cyrtandra oblongifolia Benth. et Hook. f.;长圆叶浆果苣苔(长圆叶曲蕊花)●☆
120506 Cyrtandra pendula Blume;垂花浆果苣苔(垂花弯蕊苣苔);Nodding Cyrtandra ●☆
120507 Cyrtandra pritchardii Seem.;波瑞氏浆果苣苔(波瑞氏弯蕊苣苔);Pritchard Cyrtandra ●☆
120508 Cyrtandra umbellifera Merr.;浆果苣苔(雄胞囊草);Cyrtandra, Umbrella-flowered Cyrtandra ●
120509 Cyrtandra yaeyamae Ohwi;八重山浆果苣苔●☆
120510 Cyrtandraceae Jack = Gesneriaceae Rich. et Juss. (保留科名)■●
120511 Cyrtandroidea F. Br. = Cyrtandra J. R. Forst. et G. Forst. ●■
120512 Cyrtandromoea Zoll. (1855);囊萼花属■
120513 Cyrtandromoea grandiflora C. B. Clarke;大花囊萼花■
120514 Cyrtandromoea pterocaulis D. D. Tao et al.;翅茎囊萼花(云南囊萼花)■
120515 Cyrtandropsis C. B. Clarke ex DC. = Cyrtandra J. R. Forst. et G. Forst. ●■
120516 Cyrtandropsis C. B. Clarke ex DC. = Tetraphyllum C. B. Clarke ■☆
120517 Cyrtandropsis Lauterb. = Cyrtandra J. R. Forst. et G. Forst. ●■
120518 Cyrtanthaceae Salisb. = Amaryllidaceae J. St. -Hil. (保留科名)●■
120519 Cyrtanthaceae Salisb. = Gramineae Juss. (保留科名)■●
120520 Cyrtanthaceae Salisb. = Poaceae Barnhart(保留科名)■●
120521 Cyrtanthe F. M. Bailey ex T. Durand et B. D. Jacks. = Richea R. Br. (保留属名)●☆
120522 Cyrtanthe T. Durand et B. D. Jacks. = Richea R. Br. (保留属名)●☆
120523 Cyrtanthemum Oerst. = Besleria L. ●■☆
120524 Cyrtanthera Nees = Justicia L. ●■
120525 Cyrtanthera Nees(1847);珊瑚花属;Coralflower, Cyrtanthera ●■
120526 Cyrtanthera carnea (Lindl.) Bremek.;珊瑚花;Fleshy Coralflower, Fleshy Cyrtanthera ●■
120527 Cyrtantherella Oerst. (1854);小珊瑚花属■☆
120528 Cyrtantherella Oerst. = Jacobinia Nees ex Moric. (保留属名)●■☆
120529 Cyrtantherella Oerst. = Justicia L. ●■
120530 Cyrtantherella macrantha (Benth.) Oerst.;小珊瑚花■
120531 Cyrtantherella macrantha Oerst. = Cyrtantherella macrantha (Benth.) Oerst. ■
120532 Cyrtanthus Aiton(1789)(保留属名);曲花属(曲管花属);Cyrtanthus, Fire Lily, Ifafa Lily ■☆
120533 Cyrtanthus Aiton(保留属名) = Vallota Salisb. ex Herb. (保留属名)■☆
120534 Cyrtanthus Schreb. (废弃属名) = Cyrtanthus Aiton(保留属名)■☆
120535 Cyrtanthus Schreb. (废弃属名) = Posoqueria Aubl. ●☆
120536 Cyrtanthus × brownii Traub.;垂筒花■☆
120537 Cyrtanthus angustifolius (L. f.) W. T. Aiton;狭叶曲花■☆
120538 Cyrtanthus angustifolius Aiton = Cyrtanthus angustifolius (L. f.) W. T. Aiton ■☆
120539 Cyrtanthus attenuatus R. A. Dyer;渐狭曲花■☆
120540 Cyrtanthus balenii E. Phillips = Cyrtanthus galpinii Baker ■☆
120541 Cyrtanthus bicolor R. A. Dyer;二色曲花■☆
120542 Cyrtanthus brachyscyphus Baker;小花曲花(小曲管花);Dobo Lily, Littleflower Cyrtanthus ■☆
120543 Cyrtanthus brachysiphon Hilliard et B. L. Burtt;短管曲花■☆
120544 Cyrtanthus breviflorus Harv.;短花曲花(黄曲管花)■☆
120545 Cyrtanthus carneus Lindl.;肉色曲花■☆
120546 Cyrtanthus clavatus (L'Hér.) R. A. Dyer;棒状曲花;George Lily ■☆
120547 Cyrtanthus collinus Ker Gawl.;山丘曲花■☆
120548 Cyrtanthus contractus N. E. Br.;紧缩曲花■☆
120549 Cyrtanthus debilis Snijman;弱小曲花■☆
120550 Cyrtanthus elatus (Jacq.) Traub;漏斗曲管花;Scarborough Lily ■☆
120551 Cyrtanthus elatus (Jacq.) Traub = Vallota speciosa (L. f.) T. Durand et Schinz ■☆
120552 Cyrtanthus elliotii Baker = Cyrtanthus huttonii Baker ■☆
120553 Cyrtanthus erubescens Killick;变红曲花■☆
120554 Cyrtanthus falcatus R. A. Dyer;镰状曲花■☆
120555 Cyrtanthus fergusoniae L. Bolus;费格森曲花■☆
120556 Cyrtanthus flanaganii Baker;弗拉纳根曲花■☆
120557 Cyrtanthus flavus P. E. Barnes;黄曲花■☆
120558 Cyrtanthus galpinii Baker;盖尔曲花■☆
120559 Cyrtanthus guthrieae L. Bolus;格斯里曲花■☆
120560 Cyrtanthus herrei (F. M. Leight.) R. A. Dyer;赫勒曲花■☆
120561 Cyrtanthus huttonii Baker;赫顿曲花■☆
120562 Cyrtanthus inaequalis O'Brien;不等曲花■☆
120563 Cyrtanthus junodii P. Beauv.;朱诺德曲花■☆
120564 Cyrtanthus labiatus R. A. Dyer;唇状曲花■☆
120565 Cyrtanthus leptosiphon Snijman;细管曲花■☆
120566 Cyrtanthus leucanthus Schltr.;白花曲花■☆
120567 Cyrtanthus lutescens Herb. = Cyrtanthus ochroleucus (Herb.) Burch. ex Steud. ■☆
120568 Cyrtanthus lutescens Herb. var. cooperi Baker = Cyrtanthus mackenii Hook. f. subsp. cooperi (Baker) Snijman ■☆
120569 Cyrtanthus luteus Baker = Cyrtanthus breviflorus Harv. ■☆
120570 Cyrtanthus mackenii Hook. f.;马氏曲花(白垂筒花,白曲管花);White Ifafa Lily ■☆
120571 Cyrtanthus mackenii Hook. f. subsp. cooperi (Baker) Snijman;库珀曲花■☆
120572 Cyrtanthus mackenii Hook. f. var. cooperi (Baker) R. A. Dyer = Cyrtanthus mackenii Hook. f. subsp. cooperi (Baker) Snijman ■☆
120573 Cyrtanthus macowanii Baker;麦克欧文曲花■☆
120574 Cyrtanthus montanus R. A. Dyer;山地曲花■☆
120575 Cyrtanthus nutans R. A. Dyer;俯垂曲花■☆
120576 Cyrtanthus obliquus (L. f.) W. T. Aiton;斜叶曲花(扭叶曲管花)■☆
120577 Cyrtanthus obliquus Aiton = Cyrtanthus obliquus (L. f.) W. T. Aiton ■☆
120578 Cyrtanthus ochroleucus (Herb.) Burch. = Cyrtanthus ochroleu-

cus (Herb.) Burch. ex Steud. ■☆
120579 Cyrtanthus ochroleucus (Herb.) Burch. ex Steud.;黄垂筒花(黄曲花)■☆
120580 Cyrtanthus odorus Ker Gawl.;芳香曲花■☆
120581 Cyrtanthus pallidus Sims = Cyrtanthus ventricosus Willd. ■☆
120582 Cyrtanthus parviflorus Baker = Cyrtanthus brachyscyphus Baker ■☆
120583 Cyrtanthus pumilio Roem. et Schult. = Cyrtanthus clavatus (L'Hér.) R. A. Dyer ■☆
120584 Cyrtanthus purpureus (Aiton) Traub = Cyrtanthus elatus (Jacq.) Traub ■☆
120585 Cyrtanthus purpureus (Aiton) Traub = Vallota speciosa T. Durand et Schinz ■☆
120586 Cyrtanthus purpureus Herb. = Cyrtanthus elatus (Jacq.) Traub ■☆
120587 Cyrtanthus rectiflorus Baker = Cyrtanthus brachyscyphus Baker ■☆
120588 Cyrtanthus rhododactylus Stapf;粉红曲花■☆
120589 Cyrtanthus rotundilobus N. E. Br.;圆裂曲花■☆
120590 Cyrtanthus salmonoides P. R. O. Bally et S. Carter = Cyrtanthus sanguineus (Lindl.) Walp. subsp. salmonoides (P. R. O. Bally et S. Carter) Nordal ■☆
120591 Cyrtanthus sanguineus (Lindl.) Walp.;血红曲花■☆
120592 Cyrtanthus sanguineus (Lindl.) Walp. subsp. ballyi Nordal;博利曲花■☆
120593 Cyrtanthus sanguineus (Lindl.) Walp. subsp. minor Nordal;小血红曲花■☆
120594 Cyrtanthus sanguineus (Lindl.) Walp. subsp. salmonoides (P. R. O. Bally et S. Carter) Nordal;鲑色曲花■☆
120595 Cyrtanthus sanguineus (Lindl.) Walp. subsp. wakefieldii (Sealy) Nordal;韦克菲尔德曲花■☆
120596 Cyrtanthus sanguineus (Lindl.) Walp. var. wakefieldii Sealy = Cyrtanthus sanguineus (Lindl.) Walp. subsp. wakefieldii (Sealy) Nordal ■☆
120597 Cyrtanthus sanguineus Hook. = Cyrtanthus sanguineus (Lindl.) Walp. ■☆
120598 Cyrtanthus smithiae Watt ex Harv.;史密斯曲花■☆
120599 Cyrtanthus speciosus (L. f.) Traub = Cyrtanthus elatus (Jacq.) Traub ■☆
120600 Cyrtanthus spiralis Burch. ex Ker Gawl.;螺旋曲花■☆
120601 Cyrtanthus stayneri L. Bolus = Cyrtanthus suaveolens Schönland ■☆
120602 Cyrtanthus stenanthus Baker;狭花曲花■☆
120603 Cyrtanthus stenanthus Baker var. major R. A. Dyer;大狭花曲花■☆
120604 Cyrtanthus striatus Herb.;条纹曲花■☆
120605 Cyrtanthus suaveolens Schönland;香曲花■☆
120606 Cyrtanthus thorncroftii C. H. Wright;托恩曲花■☆
120607 Cyrtanthus tuckii Baker var. transvaalensis I. Verd.;德兰士瓦曲花■☆
120608 Cyrtanthus tuckii Baker var. viridilobus I. Verd.;浅裂曲花■☆
120609 Cyrtanthus uniflorus Ker Gawl. = Cyrtanthus clavatus (L'Hér.) R. A. Dyer ■☆
120610 Cyrtanthus ventricosus Willd.;偏肿曲花■☆
120611 Cyrtanthus welwitschii Hiern ex Baker;韦尔曲花■☆
120612 Cyrtidiorchis Rauschert(1982);弓兰属■☆
120613 Cyrtidiorchis alata (Lindl.) Rauschert;翅弓兰■☆
120614 Cyrtidiorchis triptera (Schltr.) Rauschert;三翅弓兰■☆
120615 Cyrtidium Schltr. = Cyrtidiorchis Rauschert ■☆
120616 Cyrtocarpa Kunth(1824);弓果漆属●☆
120617 Cyrtocarpa proceara Kunth;弓果漆(极高弓果漆);Chupadilla ●☆
120618 Cyrtoceras Benn. = Hoya R. Br. ●
120619 Cyrtochiloides N. H. Williams et M. W. Chase = Oncidium Sw. (保留属名)■☆
120620 Cyrtochilos Spreng. = Cyrtochilum Kunth ■☆
120621 Cyrtochilum Kunth = Oncidium Sw. (保留属名)■☆
120622 Cyrtochloa S. Dransf. (1851);弓竹属●☆
120623 Cyrtocladon Griff. = Homalomena Schott ■
120624 Cyrtococcum Stapf(1920);弓果黍属;Cyrtococcum ■
120625 Cyrtococcum accrescens (Trin.) Stapf = Cyrtococcum patens (L.) A. Camus var. latifolium (Honda) Ohwi ■
120626 Cyrtococcum bosseri A. Camus;博瑟弓果黍■☆
120627 Cyrtococcum chaetophoron (Roem. et Schult.) Dandy;刚毛弓果黍■☆
120628 Cyrtococcum multinode (Lam.) Clayton;多节弓果黍■☆
120629 Cyrtococcum muricatum (Retz.) Bor = Cyrtococcum patens (L.) A. Camus var. schmidtii (Hack.) A. Camus ■
120630 Cyrtococcum oxyphyllum (Hochst. ex Steud.) Stapf;尖叶弓果黍;Sharpleaf Cyrtococcum ■
120631 Cyrtococcum patens (L.) A. Camus;弓果黍;Spreading Cyrtococcum ■
120632 Cyrtococcum patens (L.) A. Camus var. latifolium (Honda) Ohwi;散穗弓果黍;Loose Spreading Cyrtococcum ■
120633 Cyrtococcum patens (L.) A. Camus var. latifolium (Honda) Ohwi = Cyrtococcum accrescens (Trin.) Stapf ■
120634 Cyrtococcum patens (L.) A. Camus var. schmidtii (Hack.) A. Camus;瘤穗弓果黍;Schmidt Spreading Cyrtococcum ■
120635 Cyrtococcum patens (L.) A. Camus var. schmidtii (Hack.) A. Camus = Cyrtococcum patens (L.) A. Camus ■
120636 Cyrtococcum patens (L.) A. Camus var. warburgii (Mez) Reeder = Cyrtococcum patens (L.) A. Camus ■
120637 Cyrtococcum pilipes (Nees et Arn. ex Büse) A. Camus = Cyrtococcum oxyphyllum (Hochst. ex Steud.) Stapf ■
120638 Cyrtococcum pilipes (Nees et Arn.) A. Camus;小叶弓果黍;Smallleaf Cyrtococcum ■
120639 Cyrtococcum pilipes (Nees et Arn.) A. Camus = Cyrtococcum oxyphyllum (Hochst. ex Steud.) Stapf ■
120640 Cyrtococcum radicans (Retz.) Stapf = Cyrtococcum patens (L.) A. Camus ■
120641 Cyrtococcum schmidtii (Hack.) Henrard = Cyrtococcum patens (L.) A. Camus var. schmidtii (Hack.) A. Camus ■
120642 Cyrtococcum setigerum (P. Beauv.) Stapf;非洲弓果黍;African Cyrtococcum ■☆
120643 Cyrtococcum trigonum (Retz.) A. Camus;三角弓果黍■☆
120644 Cyrtococcum warburgii (Mez) Stapf = Cyrtococcum patens (L.) A. Camus ■
120645 Cyrtococcus Willis = Cyrtococcum Stapf ■
120646 Cyrtocymura H. Rob. (1987);曲序斑鸠菊属■☆
120647 Cyrtocymura H. Rob. = Vernonia Schreb. (保留属名)●■
120648 Cyrtocymura scorpioides (Lamarck) H. Rob.;曲序斑鸠菊■☆
120649 Cyrtodeira Hanst. = Episcia Mart. ■☆
120650 Cyrtoglottis Schltr. = Podochilus Blume ■
120651 Cyrtoglottls Schltr. = Mormolyca Fenzl ■☆

120652 Cyrtogonone Prain(1911);弓大戟属●☆
120653 Cyrtogonone argentea (Pax) Prain;弓大戟●☆
120654 Cyrtogyma Post et Kuntze = Cyrtogyne Rchb. ●■☆
120655 Cyrtogyne Rchb. = Crassula L. ●■☆
120656 Cyrtogyne Rchb. = Curtogyne Haw. ●■☆
120657 Cyrtolepis Less. (1831);弓鳞菊属■☆
120658 Cyrtolepis Less. = Anacyclus L. ■☆
120659 Cyrtolepis alexandrina DC. = Anacyclus monanthos (L.) Thell. ■☆
120660 Cyrtolepis monantha Less. = Anacyclus alexandrinus Willd. ■☆
120661 Cyrtolobum R. Br. = Crotalaria L. ●■
120662 Cyrtonema Eckl. et Zeyh. = Kedrostis Medik. ■☆
120663 Cyrtonema Schrad. = Kedrostis Medik. ■☆
120664 Cyrtonema Schrad. ex Eckl. et Zeyh. = Kedrostis Medik. ■☆
120665 Cyrtonema convolvulaceum Fenzl = Kedrostis foetidissima (Jacq.) Cogn. ■☆
120666 Cyrtonema digitata (Thunb.) Schrad. = Kedrostis africana (L.) Cogn. ■☆
120667 Cyrtonema divergens A. Rich. = Kedrostis foetidissima (Jacq.) Cogn. ■☆
120668 Cyrtonema elegans Fenzl ex Cogn. = Kedrostis gijef (J. F. Gmel.) C. Jeffrey ■☆
120669 Cyrtonema foetens Hochst. ex Hook. f. = Kedrostis foetidissima (Jacq.) Cogn. ■☆
120670 Cyrtonema hirtella Hochst. = Kedrostis hirtella (Naudin) Cogn. ■☆
120671 Cyrtonema latiloba Schrad. = Kedrostis nana (Lam.) Cogn. ■☆
120672 Cyrtonema molle Kuntze = Kedrostis nana (Lam.) Cogn. ■☆
120673 Cyrtonema sphenoloba Schrad. = Kedrostis nana (Lam.) Cogn. var. zeyheri (Schrad.) A. Meeuse ■☆
120674 Cyrtonema triloba Schrad. = Kedrostis nana (Lam.) Cogn. ■☆
120675 Cyrtopera Lindl. = Eulophia R. Br. (保留属名)■
120676 Cyrtopera bicarinata Lindl. = Eulophia bicallosa (D. Don) P. F. Hunt et Summerh. ■
120677 Cyrtopera bituberculata Rolfe = Eulophia plantaginea (Thouars) Rolfe ex Hochr. ■☆
120678 Cyrtopera candida Lindl. = Eulophia bicallosa (D. Don) P. F. Hunt et Summerh. ■
120679 Cyrtopera cullenii Wight = Eulophia flava (Lindl.) Hook. f. ■
120680 Cyrtopera flava Lindl. = Eulophia flava (Lindl.) Hook. f. ■
120681 Cyrtopera flavopurpurea Rchb. f. = Eulophia flavopurpurea (Rchb. f.) Rolfe ■☆
120682 Cyrtopera flexuosa Rolfe = Polystachya dendrobiiflora Rchb. f. ■☆
120683 Cyrtopera foliosa Lindl. = Eulophia foliosa (Lindl.) Bolus ■☆
120684 Cyrtopera formosana Rolfe = Eulophia zollingeri (Rchb. f.) J. J. Sm. ■
120685 Cyrtopera holstiana Kraenzl. = Eulophia odontoglossa Rchb. f. ■☆
120686 Cyrtopera longifolia (Kunth) Rchb. f. = Eulophia alta (L.) Fawc. et Rendle ■☆
120687 Cyrtopera nuda (Lindl.) Rchb. f. = Eulophia spectabilis (Dennst.) Suresh ■
120688 Cyrtopera oliveriana Rchb. f. = Eulophia parviflora (Lindl.) A. V. Hall ■☆
120689 Cyrtopera papillosa Rolfe = Eulophia odontoglossa Rchb. f. ■☆
120690 Cyrtopera papuana Kraenzl. = Eulophia bicallosa (D. Don) P. F. Hunt et Summerh. ■
120691 Cyrtopera papuana Ridl. = Eulophia zollingeri (Rchb. f.) J. J. Sm. ■
120692 Cyrtopera pedicellata (L. f.) Lindl. = Eulophia aculeata (L. f.) Spreng. ■☆
120693 Cyrtopera plantaginea (Thouars) Lindl. = Eulophia plantaginea (Thouars) Rolfe ex Hochr. ■☆
120694 Cyrtopera rufa Thwaites = Eulophia zollingeri (Rchb. f.) J. J. Sm. ■
120695 Cyrtopera sanguinea Lindl. = Eulophia zollingeri (Rchb. f.) J. J. Sm. ■
120696 Cyrtopera shupangae Rchb. f. = Eulophia odontoglossa Rchb. f. ■☆
120697 Cyrtopera stolziana Kraenzl. = Eulophia angolensis (Rchb. f.) Summerh. ■☆
120698 Cyrtopera walleri Rchb. f. = Eulophia walleri (Rchb. f.) Kraenzl. ■☆
120699 Cyrtopera woodfordii (Sims) Lindl. = Eulophia alta (L.) Fawc. et Rendle ■☆
120700 Cyrtopera zollingeri Rchb. f. = Eulophia zollingeri (Rchb. f.) J. J. Sm. ■
120701 Cyrtophyllum Reinw. = Fagraea Thunb. ●
120702 Cyrtophyllum Reinw. ex Blume = Fagraea Thunb. ●
120703 Cyrtophyllum Reinw. ex Blume(1827);曲叶马钱属●☆
120704 Cyrtophyllum peregrinum Reinw. ex Blume;曲叶马钱●☆
120705 Cyrtopodium R. Br. (1813);曲足兰属;Bee-swarm Orchid ■☆
120706 Cyrtopodium andersonii (Lamb. ex Andr.) R. Br.;安氏曲足兰;Anderson Cyrtopodium ■☆
120707 Cyrtopodium bicolor Ridl. = Eulophia spectabilis (Dennst.) Suresh ■
120708 Cyrtopodium ecristatum Fernald = Pteroglossaspis ecristata (Fernald) Rolfe ■☆
120709 Cyrtopodium flavum (Lindl.) Benth. = Eulophia flava (Lindl.) Hook. f. ■
120710 Cyrtopodium paludicola Koehne;沼泽曲足兰;Swampy Cyrtopodium ■☆
120711 Cyrtopodium paranaense Schltr. = Cyrtopodium polyphyllum (Vell.) Pabst ex F. Barrios ■☆
120712 Cyrtopodium plantagineum (Thouars) Benth. = Eulophia plantaginea (Thouars) Rolfe ex Hochr. ■☆
120713 Cyrtopodium polyphyllum (Vell.) L. C. Menezes = Cyrtopodium polyphyllum (Vell.) Pabst ex F. Barrios ■☆
120714 Cyrtopodium polyphyllum (Vell.) Pabst ex F. Barrios;多叶曲足兰;Terrestrial Cowhorn Orchid ■☆
120715 Cyrtopodium punctatum (L.) Lindl.;斑点曲足兰;Bee-swarm Orchid, Cigar Orchid, Cowhorn Orchid, Punctate Cyrtopodium ■☆
120716 Cyrtopodium rufum (Thwaites) Trimen = Eulophia zollingeri (Rchb. f.) J. J. Sm. ■
120717 Cyrtopodium sanguineum (Lindl.) N. E. Br. = Eulophia zollingeri (Rchb. f.) J. J. Sm. ■
120718 Cyrtopodium strictum Griseb. = Pteroglossaspis ecristata (Fernald) Rolfe ■☆
120719 Cyrtopodium virescens Rchb. f. et Warm.;绿花曲足兰;Green-flower Cyrtopodium ■☆
120720 Cyrtopodium woodfordii Sims = Eulophia alta (L.) Fawc. et Rendle ■☆
120721 Cyrtopogon Spreng. = Aristida L. ■
120722 Cyrtopogon Spreng. = Curtopogon P. Beauv. ■
120723 Cyrtorchis Schltr. (1914);弯萼兰属■☆

120724 Cyrtorchis aberrans Mansf.;异常弯萼兰■☆
120725 Cyrtorchis acuminata (Rolfe) Schltr. = Cyrtorchis arcuata (Lindl.) Schltr. subsp. variabilis Summerh. ■☆
120726 Cyrtorchis arcuata (Lindl.) Schltr.;弯萼兰■☆
120727 Cyrtorchis arcuata (Lindl.) Schltr. subsp. leonensis Summerh.;莱昂弯萼兰■☆
120728 Cyrtorchis arcuata (Lindl.) Schltr. subsp. variabilis Summerh.;易变弯萼兰■☆
120729 Cyrtorchis arcuata (Lindl.) Schltr. subsp. whytei (Rolfe) Summerh.;怀特弯萼兰■☆
120730 Cyrtorchis arcuata (Lindl.) Schltr. var. variabilis (Summerh.) Geerinck = Cyrtorchis arcuata (Lindl.) Schltr. subsp. variabilis Summerh. ■☆
120731 Cyrtorchis aschersonii (Kraenzl.) Schltr.;阿舍森弯萼兰■☆
120732 Cyrtorchis bracteata Schltr. = Cyrtorchis arcuata (Lindl.) Schltr. ■☆
120733 Cyrtorchis brownii (Rolfe) Schltr. = Homocolleticon brownii (Rolfe) Szlach. et Olszewski ■☆
120734 Cyrtorchis crassifolia Schltr. = Homocolleticon crassifolia (Schltr.) Szlach. et Olszewski ■☆
120735 Cyrtorchis cufodontii Chiov. = Rangaeris amaniensis (Kraenzl.) Summerh. ■☆
120736 Cyrtorchis erythraeae (Rolfe) Schltr.;浅红弯萼兰■☆
120737 Cyrtorchis glaucifolia Summerh.;灰绿弯萼兰■☆
120738 Cyrtorchis hamata (Rolfe) Schltr.;顶钩弯萼兰■☆
120739 Cyrtorchis helicocalcar Bellone;距弯萼兰■☆
120740 Cyrtorchis henriquesiana (Ridl.) Rchb. f. = Homocolleticon henriquesiana (Ridl.) Szlach. et Olszewski ■☆
120741 Cyrtorchis injoloensis Schltr. = Homocolleticon injoloensis (De Wild.) Szlach. et Olszewski ■☆
120742 Cyrtorchis letouzeyi Szlach. et Olszewski;勒图弯萼兰■☆
120743 Cyrtorchis monteiroae (Rchb. f.) Schltr. = Homocolleticon monteriroae (Rchb. f.) Szlach. et Olszewski ■☆
120744 Cyrtorchis neglecta Summerh.;忽视弯萼兰■☆
120745 Cyrtorchis praetermissa Summerh. = Homocolleticon praetermissa (Summerh.) Szlach. et Olszewski ■☆
120746 Cyrtorchis praetermissa Summerh. subsp. zuluensis (E. R. Harrison) H. P. Linder = Homocolleticon praetermissa (Summerh.) Szlach. et Olszewski var. zuluensis (E. R. Harrison) Szlach. et Olszewski ■☆
120747 Cyrtorchis praetermissa Summerh. var. zuluensis E. R. Harrison = Cyrtorchis praetermissa Summerh. subsp. zuluensis (E. R. Harrison) H. P. Linder ■☆
120748 Cyrtorchis ringens (Rchb. f.) Summerh. = Homocolleticon ringens (Rchb. f.) Szlach. et Olszewski ■☆
120749 Cyrtorchis sedenii (Rchb. f.) Schltr. = Cyrtorchis arcuata (Lindl.) Schltr. subsp. variabilis Summerh. ■☆
120750 Cyrtorchis seretii (De Wild.) Schltr.;赛雷弯萼兰■☆
120751 Cyrtorchis submontana Stévart;亚山地弯萼兰■☆
120752 Cyrtorhyncha Nutt. = Cyrtorrhyncha Nutt. ex Torr. et A. Gray ■☆
120753 Cyrtorhyncha Nutt. = Ranunculus L. ■
120754 Cyrtorrhyncha Nutt. ex Torr. et A. Gray = Ranunculus L. ■
120755 Cyrtorrhyncha Nutt. ex Torr. et A. Gray(1838);曲喙毛茛属■☆
120756 Cyrtorrhyncha ranunculina Nutt.;曲喙毛茛■☆
120757 Cyrtosia Blume = Galeola Lour. ■
120758 Cyrtosia Blume(1825);肉果兰属;Cyrtosia ■
120759 Cyrtosia Lindl. = Eulophia R. Br.(保留属名)■
120760 Cyrtosia altissima Blume = Erythrorchis altissima (Blume) Blume ■
120761 Cyrtosia javanica Blume;肉果兰(爪哇山珊瑚);Java Cyrtosia ■
120762 Cyrtosia lindleyana Hook. f. et Thomson = Galeola lindleyana (Hook. f. et Thomson) Rchb. f. ■
120763 Cyrtosia nana (Rolfe ex Downie) Garay;矮小肉果兰;Dwarf Cyrtosia ■
120764 Cyrtosia septentrionalis (Rchb. f.) Garay;血红肉果兰(红果山珊瑚,山珊瑚,山珊瑚兰);Bloodred Cyrtosia ■
120765 Cyrtosiphonia Miq. = Rauvolfia L. ●
120766 Cyrtospadix C. Koch = Caladium Vent. ■
120767 Cyrtospadix K. Koch = Caladium Vent. ■
120768 Cyrtosperma Griff.(1851);曲籽芋属;Cyrtosperma ■
120769 Cyrtosperma afzelii (Schott) Engl. = Lasimorpha senegalensis Schott ■☆
120770 Cyrtosperma chamissonis Merr. = Cyrtosperma merkusii (Hassk.) Schott ■☆
120771 Cyrtosperma edule Schott = Cyrtosperma merkusii (Hassk.) Schott ■☆
120772 Cyrtosperma johstonii N. E. Br.;约氏曲籽芋■☆
120773 Cyrtosperma lasioides Griff.;曲籽芋;Common Cyrtosperma, Lasia-like Cyrtosperma ■
120774 Cyrtosperma merkusii (Hassk.) Schott;可食曲籽芋;Babai, Swamp Taro ■☆
120775 Cyrtosperma senegalense (Schott) Engl. = Lasimorpha senegalensis Schott ■☆
120776 Cyrtospermum Benth. = Campnosperma Thwaites(保留属名)●☆
120777 Cyrtospermum Raf. = Cryptotaenia DC.(保留属名)■
120778 Cyrtospermum Raf. ex DC. = Cryptotaenia DC.(保留属名)■
120779 Cyrtostachys Blume(1838);封蜡棕属(红柄椰属,红椰属,红椰子属,曲穗属,猩红椰属,猩红椰子属,猩猩椰子属);Sealing-wax Palm ●☆
120780 Cyrtostachys lakka Becc.;猩猩椰子(姬猩猩椰子);Rajah Sealing-wax Palm, Sealing-wax Palm ●☆
120781 Cyrtostachys lakka Becc. = Cyrtostachys renda Blume ●☆
120782 Cyrtostachys renda Blume;封蜡棕(赤轴椰子,猩猩椰子);Sumatra Sealing-wax Palm, Sealing-wax Palm, Sealing Wax Palm ●☆
120783 Cyrtostemma (Mert. et Koch) Spach = Scabiosa L. ●■
120784 Cyrtostemma Kunze = Clerodendrum L. ●■
120785 Cyrtostemma Spach = Scabiosa L. ●■
120786 Cyrtostylis R. Br.(1810);弯柱兰属■☆
120787 Cyrtostylis R. Br. = Acianthus R. Br. ■☆
120788 Cyrtostylis reniformis R. Br.;弯柱兰■☆
120789 Cyrtotropis Wall. = Apios Fabr.(保留属名)●
120790 Cyrtoxiphus Harms = Cylicodiscus Harms ●☆
120791 Cyrtoxiphus staudtii Harms = Cylicodiscus gabunensis Harms ●☆
120792 Cyssopetalum Turcz. = Oenanthe L. ■
120793 Cystacanthus T. Anderson = Phlogacanthus Nees ●■
120794 Cystacanthus T. Anderson(1867);鳔冠花属(鳔冠草属);Cystacanthus ●■
120795 Cystacanthus abbreviatus Craib = Phlogacanthus abbreviatus (Craib) Benoist ●
120796 Cystacanthus affinis W. W. Sm.;丽江鳔冠花(近缘鳔冠花);Affined Cystacanthus, Similar Cystacanthus ●
120797 Cystacanthus paniculatus T. Anderson;鳔冠花(鳔刺草);Paniculate Cystacanthus ●
120798 Cystacanthus yangtsekiangensis (H. Lév.) Rehder;金沙鳔冠

花(金江鳔冠花);Yangzijiang Cystacanthus ●
120799 Cystacanthus yunnanensis W. W. Sm.;滇鳔冠花(鳔冠花,蓝花楔);Yunnan Cystacanthus ●
120800 Cystanche Ledeb. = Cistanche Hoffmanns. et Link ■
120801 Cystanthe R. Br.(废弃属名)= Richea R. Br.(保留属名)●☆
120802 Cystibax Heynh. = Cybistax Mart. ex Meisn. ●☆
120803 Cysticapnos Mill.(废弃属名)= Corydalis DC.(保留属名)■
120804 Cysticorydalis Fedde = Corydalis DC.(保留属名)■
120805 Cysticorydalis Fedde ex Ikonn.(1936);鳔紫堇属■☆
120806 Cysticorydalis Fedde ex Ikonn. = Corydalis DC.(保留属名)■
120807 Cysticorydalis crassifolia (Royle) Fedde = Corydalis crassifolia Royle ■☆
120808 Cysticorydalis fedtschenkoana (Regel) Fedde = Corydalis fedtschenkoana Regel ■
120809 Cystidianthus Hassk.(1844);囊花萝藦属●☆
120810 Cystidianthus Hassk. = Physostelma Wight ●☆
120811 Cystidianthus campanulatus Hassk.;囊花萝藦●☆
120812 Cystidianthus laurifolius Blume;桂叶囊花萝藦●☆
120813 Cystidospermum Prokh.(1933);囊子大戟属●■☆
120814 Cystidospermum Prokh. = Euphorbia L. ●■
120815 Cystidospermum postii (Boiss.) Prokh.;囊子大戟■☆
120816 Cystistemon Post et Kuntze = Cystostemon Balf. f. ■☆
120817 Cystium (Steven) Steven = Astragalus L. ●■
120818 Cystium Steven = Astragalus L. ●■
120819 Cystocapnus Post et Kuntze = Corydalis DC.(保留属名)■
120820 Cystocarpum Benth. et Hook. = Vesicaria Tourn. ex Adans. ■☆
120821 Cystocarpus Lam. ex Post et Kuntze = Cystocarpum Benth. et Hook. ■☆
120822 Cystocarpus Lam. ex Post et Kuntze = Vesicaria Tourn. ex Adans. ■☆
120823 Cystochilum Barb. Rodr. = Cranichis Sw. ■☆
120824 Cystogyne Gasp. = Ficus L. ●
120825 Cystopora Lunell = Astragalus L. ●■
120826 Cystopora Lunell = Phaca L. ●■
120827 Cystopus Blume = Odontochilus Blume ■
120828 Cystopus Blume = Pristiglottis Cretz. et J. J. Sm. ■
120829 Cystopus brevistylis (Hook. f.) Kuntze = Odontochilus brevistylis Hook. f. ■
120830 Cystopus clarkei (Hook. f.) Kuntze = Odontochilus clarkei Hook. f. ■
120831 Cystopus crispus (Lindl.) Kuntze = Odontochilus crispus (Lindl.) Hook. f. ■
120832 Cystopus elwesii (C. B. Clarke ex Hook. f.) Kuntze = Odontochilus elwesii C. B. Clarke ex Hook. f. ■
120833 Cystopus elwesii (Clarke ex Hook. f.) Kuntze = Anoectochilus elwesii (C. B. Clarke ex Hook. f.) King et Pantl. ■
120834 Cystopus flavus (Benth. et Hook. f.) Kuntze = Odontochilus lanceolatus (Lindl.) Blume ■
120835 Cystopus humilis Fukuy. = Kuhlhasseltia yakushimensis (Yamam.) Ormerod ■
120836 Cystopus humilum Fukuy. = Vexillabium yakushimense (Yamam.) F. Maek. ■
120837 Cystopus lanceolatus (Lindl.) Kuntze = Odontochilus lanceolatus (Lindl.) Blume ■
120838 Cystopus pumilus (Hook. f.) Kuntze = Myrmechis pumila (Hook. f.) Ts. Tang et F. T. Wang ■
120839 Cystorchis Blume(1858);膀胱兰属■☆
120840 Cystorchis aphylla Ridl.;膀胱兰■☆
120841 Cystorchis fusca (Lindl.) Benth. et Hook. f. = Goodyera fusca (Lindl.) Hook. f. ■
120842 Cystorchis nebularum Hance = Goodyera foliosa (Lindl.) Benth. ex C. B. Clarke ■
120843 Cystostemma E. Fourn.(1885);囊冠萝藦属●■☆
120844 Cystostemma E. Fourn. = Oxypetalum R. Br.(保留属名)●■☆
120845 Cystostemma glandulosum Silveira;囊冠萝藦●■☆
120846 Cystostemon Balf. f.(1883);囊蕊紫草属■☆
120847 Cystostemon barbatus (Vaupel) A. G. Mill. et Riedl;髯毛囊蕊紫草■☆
120848 Cystostemon heliocharis (S. Moore) A. G. Mill. et Riedl;喜阳囊蕊紫草■☆
120849 Cystostemon hispidissimus (S. Moore) A. G. Mill. et Riedl;多硬毛囊蕊紫草■☆
120850 Cystostemon hispidissimus (S. Moore) A. G. Mill. et Riedl subsp. zambiensis A. G. Mill. et Riedl;赞比亚多毛囊蕊紫草■☆
120851 Cystostemon hispidus (Baker et C. H. Wright) A. G. Mill. et Riedl;硬毛囊蕊紫草■☆
120852 Cystostemon intricatus A. G. Mill. et Riedl;缠结囊蕊紫草■☆
120853 Cystostemon linearifolius E. S. Martins;线叶囊蕊紫草■☆
120854 Cystostemon macranthera (Gürke) A. G. Mill. et Riedl;大药囊蕊紫草■☆
120855 Cystostemon mechowii (Vaupel) A. G. Mill. et Riedl;梅休囊蕊紫草■☆
120856 Cystostemon medusa (Baker) A. G. Mill. et Riedl;梅杜萨囊蕊紫草■☆
120857 Cystostemon mwinilungensis E. S. Martins;穆维尼囊蕊紫草■☆
120858 Cystostemon somaliensis A. G. Mill. et Riedl;索马里囊蕊紫草■☆
120859 Cystostemon virescens A. G. Mill. et Riedl;绿囊蕊紫草■☆
120860 Cytharexylum Jacq. = Citharexylum L. ●☆
120861 Cytheraea (DC.) Wight et Arn. = Spondias L. ●
120862 Cytheraea Wight et Arn. = Spondias L. ●
120863 Cytherea Salisb. = Calypso Salisb.(保留属名)■
120864 Cytherea bulbosa (L.) House = Calypso bulbosa (L.) Oakes var. americana (R. Br.) Luer ■☆
120865 Cytherea bulbosa (L.) House = Calypso bulbosa (L.) Oakes ■
120866 Cytherea speciosa (Schltr.) Makino = Calypso bulbosa (L.) Oakes var. speciosa (Schltr.) Makino ■
120867 Cytheris Lindl. = Calanthe R. Br.(保留属名)■
120868 Cytheris Lindl. = Nephelaphyllum Blume ■
120869 Cythisus Schrank = Cytisus Desf.(保留属名)●
120870 Cyticus Link = Cythisus Schrank ●
120871 Cytinaceae (Brongn.) A. Rich. = Rafflesiaceae Dumort.(保留科名)■
120872 Cytinaceae A. Rich.(1824);簇花科(簇花草科,大花草科)■☆
120873 Cytinaceae A. Rich. = Rafflesiaceae Dumort.(保留科名)■
120874 Cytinaceae Brongn. = Rafflesiaceae Dumort.(保留科名)■
120875 Cytinus L.(1764)(保留属名);簇花属(簇花草属,大花草属);Cytinus ■☆
120876 Cytinus baroni Baker f.;巴龙簇花■☆
120877 Cytinus capensis Marloth;好望角簇花■☆
120878 Cytinus dioicus Juss. = Cytinus sanguineus (Thunb.) Fourc. ■☆
120879 Cytinus glandulosus Jum.;具腺簇花■☆
120880 Cytinus hypocistis (L.) L.;簇花■☆
120881 Cytinus hypocistis (L.) L. subsp. canariensis (Webb et Berthel.) G. Kunkel;加那利簇花■☆

120882 Cytinus hypocistis (L.) L. subsp. clusii Nyman;红簇花■☆

120883 Cytinus hypocistis (L.) L. subsp. kermesinus (Guss.) Arcang. = Cytinus hypocistis (L.) L. ■☆

120884 Cytinus hypocistis (L.) L. subsp. lutescens (Batt.) Maire;黄簇花■☆

120885 Cytinus hypocistis (L.) L. subsp. macranthus Wettst.;大花簇花■☆

120886 Cytinus hypocistis (L.) L. subsp. ochraceus (Guss.) Wettst. = Cytinus hypocistis (L.) L. ■☆

120887 Cytinus hypocistis (L.) L. subsp. orientalis Wettst.;东方簇花■☆

120888 Cytinus hypocistis (L.) L. var. gussonei Maire = Cytinus hypocistis (L.) L. ■☆

120889 Cytinus hypocistis (L.) L. var. striatus Maire = Cytinus hypocistis (L.) L. ■☆

120890 Cytinus ruber (Fourr.) Kom. = Cytinus hypocistis (L.) L. subsp. clusii Nyman ■☆

120891 Cytinus sanguineus (Thunb.) Fourc.;血红簇花■☆

120892 Cytinus visseri Burgoyne;维瑟簇花■☆

120893 Cytisanthus O. Lang = Genista L. ●

120894 Cytisanthus O. Lang(1843);欧金雀属●☆

120895 Cytisanthus aetnensis Biv. = Genista aetnensis (Biv.) DC. ●☆

120896 Cytiso-Genista Duhamel = Sarothamnus Wimm.(保留属名)●☆

120897 Cytisogenista Duhamel(废弃属名) = Sarothamnus Wimm.(保留属名)●☆

120898 Cytisophyllum O. Lang(1843);金雀儿叶属■☆

120899 Cytisophyllum sessilifolium (L.) O. Lang;金雀儿叶(无柄金雀花)●☆

120900 Cytisophyllum sessilifolium (L.) O. Lang = Cytisus sessilifolius L. ●☆

120901 Cytisopsis Jaub. et Spach(1844);西亚绒毛花属■☆

120902 Cytisopsis ahmedii (Batt. et Pit.) Lassen;非洲绒毛花■☆

120903 Cytisus Desf. (1798)(保留属名);金雀花属(金雀儿属);Broom ●

120904 Cytisus L.(废弃属名) = Cytisus Desf.(保留属名)●

120905 Cytisus aethiopicus L. = Melolobium aethiopicum (L.) Druce ■☆

120906 Cytisus albidus DC. = Chamaecytisus mollis (Cav.) Greuter et Burdet ●☆

120907 Cytisus albidus DC. var. ifnianus (Font Quer) Emb. et Maire = Chamaecytisus mollis (Cav.) Greuter et Burdet ●☆

120908 Cytisus albidus DC. var. tridentatus (Maire et Weiller) Maire = Chamaecytisus mollis (Cav.) Greuter et Burdet ●☆

120909 Cytisus albus Jacq.;白花金雀花(白花假金雀儿,白金雀儿,白雀花,西班牙金雀儿,西班牙金雀花);Pale Broom, Portuguese Broom, White Broom, White Portugal Broom, White Spanish Broom ●☆

120910 Cytisus albus Lam. = Cytisus multiflorus (L'Hér. ex Aiton) Sweet ●☆

120911 Cytisus alpinus Lam. = Laburnum alpinum J. Presl ●☆

120912 Cytisus arabicus Decne. = Argyrolobium arabicum (Decne.) Jaub. et Spach ●☆

120913 Cytisus arboreus (Desf.) DC.;乔木状金雀花●☆

120914 Cytisus arboreus (Desf.) DC. subsp. baeticus (Webb) Maire;伯蒂卡金雀花●☆

120915 Cytisus arboreus (Desf.) DC. subsp. macranthus (Ball) Maire = Cytisus arboreus (Desf.) DC. ●☆

120916 Cytisus arboreus (Desf.) DC. subsp. transiens (Maire) Maire;中间金雀花●☆

120917 Cytisus arboreus (Desf.) DC. var. africanus (Pau et Font Quer) Maire = Cytisus arboreus (Desf.) DC. ●☆

120918 Cytisus arboreus (Desf.) DC. var. ballianus (Briq.) Maire = Cytisus arboreus (Desf.) DC. ●☆

120919 Cytisus arboreus (Desf.) DC. var. extravagans Font Quer = Cytisus arboreus (Desf.) DC. ●☆

120920 Cytisus arboreus (Desf.) DC. var. haplotrichus Maire = Cytisus arboreus (Desf.) DC. ●☆

120921 Cytisus arboreus (Desf.) DC. var. intermedius (Emb. et Maire) Maire = Cytisus arboreus (Desf.) DC. ●☆

120922 Cytisus arboreus (Desf.) DC. var. leiocladus Maire = Cytisus arboreus (Desf.) DC. ●☆

120923 Cytisus arboreus (Desf.) DC. var. macranthus (Ball) Jahand. et Maire = Cytisus arboreus (Desf.) DC. ●☆

120924 Cytisus arboreus (Desf.) DC. var. tetuanensis Pau = Cytisus arboreus (Desf.) DC. ●☆

120925 Cytisus arboreus (Desf.) DC. var. transiens Maire = Cytisus arboreus (Desf.) DC. subsp. transiens (Maire) Maire ●☆

120926 Cytisus ardoinii E. Fourn.;法国金雀花(法国金雀儿)●☆

120927 Cytisus argenteus L. = Argyrolobium zanonii (Turra) P. W. Ball ●☆

120928 Cytisus austriacus L.;奥地利金雀花;Austrian Broom ●☆

120929 Cytisus baeticus (Webb) Steud. = Cytisus arboreus (Desf.) DC. subsp. baeticus (Webb) Maire ●☆

120930 Cytisus baeticus (Webb) Steud. var. africanus Pau et Font Quer = Cytisus arboreus (Desf.) DC. subsp. baeticus (Webb) Maire ●☆

120931 Cytisus baeticus (Webb) Steud. var. extravagans Font Quer = Cytisus arboreus (Desf.) DC. subsp. baeticus (Webb) Maire ●☆

120932 Cytisus baeticus (Webb) Steud. var. macranthus Ball = Cytisus arboreus (Desf.) DC. subsp. baeticus (Webb) Maire ●☆

120933 Cytisus balansae (Boiss.) Ball;巴兰萨金雀花●☆

120934 Cytisus balansae (Boiss.) Ball subsp. nevadensis Cantó et Rivas Mart. = Cytisus balansae (Boiss.) Ball var. galianoi (Talavera et Gibbs) G. López ●☆

120935 Cytisus balansae (Boiss.) Ball var. atlanticus Ball = Cytisus balansae (Boiss.) Ball ●☆

120936 Cytisus balansae (Boiss.) Ball var. galianoi (Talavera et Gibbs) G. López = Cytisus balansae (Boiss.) Ball ●☆

120937 Cytisus barbarus (Jahand. et Maire) Maire = Cytisus grandiflorus (Brot.) DC. ●☆

120938 Cytisus barbarus (Jahand. et Maire) Maire var. haplophyllus Maire et Sennen = Cytisus grandiflorus (Brot.) DC. ●☆

120939 Cytisus battandieri Maire;总序金雀花;Moroccan Broom, Pineapple Broom ●☆

120940 Cytisus battandieri Maire = Argyrocytisus battandieri (Maire) Raynaud ●☆

120941 Cytisus beanii Nicholson;美丽金雀花(长雀花,美丽金雀儿)●☆

120942 Cytisus blockii Krecz.;布劳氏金雀花;Block Broom ●☆

120943 Cytisus borysthenicus Grunner;第聂伯金雀花●☆

120944 Cytisus cajan L. = Cajanus cajan (L.) Millsp. ●

120945 Cytisus canariensis (L.) Kuntze = Chamaecytisus supinus (L.) Link ●☆

120946 Cytisus capense P. J. Bergius = Rafnia capensis (L.) Schinz ■☆

120947 Cytisus cincinnatus Ball = Adenocarpus cincinnatus (Ball) Maire ●☆

120948 Cytisus decumbens Spach;匍匐金雀花(匍匐金雀儿)●☆

120949 Cytisus flaccidus Royle = Argyrolobium flaccidum (Royle)

Jaub. et Spach ●☆

120950 Cytisus fontanesii Spach;丰塔纳金雀花●☆

120951 Cytisus fontanesii Spach subsp. incanus (Maire) H. Tahiri = Cytisus fontanesii Spach ●☆

120952 Cytisus fontanesii Spach subsp. plumosus (Boiss.) Nyman;羽状金雀花●☆

120953 Cytisus fontanesii Spach subsp. plumosus (Boiss.) Nyman = Cytisus fontanesii Spach ●☆

120954 Cytisus fontanesii Spach var. incanus Maire = Cytisus fontanesii Spach ●☆

120955 Cytisus fontanesii Spach var. platycarpus Doum. = Cytisus fontanesii Spach ●☆

120956 Cytisus fontanesii Spach var. plumosus (Boiss.) Willk. = Cytisus fontanesii Spach ●☆

120957 Cytisus glomeratus Bojer = Eriosema parviflorum E. Mey. ●☆

120958 Cytisus grandiflorus (Brot.) DC.;大花金雀儿;Woolly-podded Broom ●☆

120959 Cytisus grandiflorus (Brot.) DC. subsp. barbarus (Jahand. et Maire) Maire = Cytisus grandiflorus (Brot.) DC. ●☆

120960 Cytisus grandiflorus (Brot.) DC. subsp. haplophyllus (Maire et Sennen) Maire = Cytisus grandiflorus (Brot.) DC. ●☆

120961 Cytisus grandiflorus DC. = Cytisus grandiflorus (Brot.) DC. ●☆

120962 Cytisus hirsutissimus K. Koch;粗毛金雀花●☆

120963 Cytisus hirsutus L.;毛金雀花(毛金雀儿)●☆

120964 Cytisus hispidus Willd. = Ormocarpum sennoides (Willd.) DC. subsp. hispidum (Willd.) Brenan et J. Léonard ●☆

120965 Cytisus ifnianus Font Quer = Chamaecytisus mollis (Cav.) Greuter et Burdet ●☆

120966 Cytisus kewensis Bean;邱园白雀花(邱园金雀儿)●☆

120967 Cytisus laburnum L. = Laburnum anagyroides Medik. ●

120968 Cytisus leucanthus Waldst. et Kit. = Cytisus albus Jacq. ●☆

120969 Cytisus leucanthus Waldst. et Kit. = Cytisus multiflorus (L'Hér. ex Aiton) Sweet ●☆

120970 Cytisus lindemannii Krecz.;李曼氏金雀花;Lindemann Broom ●☆

120971 Cytisus linifolius (L.) Lam. = Teline linifolia (L.) Webb et Berthel. ●☆

120972 Cytisus litwinowii Krecz.;利特氏金雀花(李文氏金雀花);Litwinow Broom ●☆

120973 Cytisus mannii Hook. f. = Adenocarpus mannii (Hook. f.) Hook. f. ●☆

120974 Cytisus maurus Humbert et Maire;晚熟金雀花●☆

120975 Cytisus megalanthus (Pau et Font Quer) Font Quer = Cytisus striatus (Hill) Rothm. ●☆

120976 Cytisus mollis (Cav.) Pau = Chamaecytisus mollis (Cav.) Greuter et Burdet ●☆

120977 Cytisus monspessulanus L. = Genista monspessulana (L.) L. A. S. Johnson ●☆

120978 Cytisus monspessulanus L. = Teline monspessulana (L.) K. Koch ●☆

120979 Cytisus multiflorus (L'Hér. ex Aiton) Sweet;多花金雀儿(西班牙金雀花);Portuguese Broom, White Broom, White Spanish Broom ●☆

120980 Cytisus multiflorus (L'Hér. ex Aiton) Sweet = Cytisus multiflorus (L'Hér. ex Aiton) Sweet ●☆

120981 Cytisus multiflorus (L'Hér.) Sweet = Cytisus multiflorus (L'Hér. ex Aiton) Sweet ●☆

120982 Cytisus multiflorus Sweet = Cytisus multiflorus (L'Hér. ex Aiton) Sweet ●☆

120983 Cytisus nigricans L.;欧洲金雀花(变黑金雀儿);Black Broom, Blacken Cytisus, Spike Broom ●

120984 Cytisus osmarensis (Coss.) Ball;奥马金雀花(奥马金雀儿)●☆

120985 Cytisus osmarensis (Coss.) Ball = Teline osmarensis (Coss.) Gibbs et Dingwall ●☆

120986 Cytisus paszoskii Krecz.;帕曹氏金雀花●☆

120987 Cytisus persicus Burm. f. = Crotalaria persica (Burm. f.) Merr. ■☆

120988 Cytisus pinnatus L. = Pongamia pinnata (L.) Pierre ex Merr. ●

120989 Cytisus podolicus Blocki;波多尔金雀花●☆

120990 Cytisus polytrichus M. Bieb.;多毛金雀花●☆

120991 Cytisus praecox Wheeler;早开金雀花;Warminster Broom ●☆

120992 Cytisus procumbens Bojer ex Baker = Eriosema procumbens Benth. ex Baker ●☆

120993 Cytisus psoraloides L. = Indigofera psoraloides (L.) L. ■☆

120994 Cytisus purgans (L.) Boiss. = Genista scorpius (L.) DC. ●☆

120995 Cytisus purgans Spach;比利牛斯金雀花;Pyrenean Broom ●☆

120996 Cytisus purpureus Scop. = Chamaecytisus purpureus Link ●☆

120997 Cytisus ratisbonensis Schaeff.;德国金雀花;German Broom ●☆

120998 Cytisus rochelii Wierzb. ex Griseb. et Schenk;罗氏金雀花●☆

120999 Cytisus roseus Cambess. = Argyrolobium roseum (Cambess.) Jaub. et Spach ●☆

121000 Cytisus ruthenicus Fisch.;俄罗斯金雀花;Russia Broom ●☆

121001 Cytisus scoparius (L.) Link;金雀花(金雀儿);Andreas Broom, Banadle, Banathal, Bannal, Bannell, Basam, Basom, Bassam Broom, Beesom, Besom, Bream, Breem, Broom, Brush, Cat's Pea, Cat's Peas, Common Broom, European Broom, Genet, Golden Rod, Green Basom, Green Besom, Green Broom, Greenwood, Irish Broom, Lady's Slipper, Little Fair One, Scobe, Scotch Broom, Tree Trefoil, Woodwax ●

121002 Cytisus scoparius (L.) Link = Sarothamnus scoparius (L.) Wimm. ex K. Koch ●

121003 Cytisus scoparius (L.) Link subsp. maurus (Humbert et Maire) Talavera;海滨金雀儿●☆

121004 Cytisus scoparius (L.) Link subsp. maurus (Humbert et Maire) Talavera = Cytisus maurus Humbert et Maire ●☆

121005 Cytisus scoparius (L.) Link var. andreanus Dippel;安氏金雀花;Andrea's Broom, Scotch Broom ●☆

121006 Cytisus scoparius (L.) Link var. sulphureus Dippel;硫色金雀花;Moonlight Broom ●☆

121007 Cytisus scrobiszewskii Pacz.;司氏金雀花(司寇拜氏金雀花)●☆

121008 Cytisus segonnei Maire = Teline segonnei (Maire) Raynaud ●☆

121009 Cytisus sericeus Willd. = Mundulea sericea (Willd.) A. Chev. ●☆

121010 Cytisus sessiliflorus L. = Cytisophyllum sessilifolium (L.) O. Lang ●☆

121011 Cytisus sessilifolius L.;无梗叶金雀花●☆

121012 Cytisus skrobiszewskii Pacz.;斯克罗金雀花●☆

121013 Cytisus spachianus (Webb) Kuntze;香雀花;Easter Broom, Sweet Broom ●☆

121014 Cytisus spachianus C. Sm. = Cytisus spachianus (Webb) Kuntze ●☆

121015 Cytisus striatus (Hill) Rothm.;线纹金雀花;Hairy-fruited Broom, Striated Broom ●☆

121016 Cytisus supinus L. = Chamaecytisus supinus (L.) Link ●☆

121017 Cytisus supranubius Kuntze;耐寒金雀花(耐寒金雀儿);

Tenerife Broom ●☆
121018 Cytisus tomentosus Andréws = Argyrolobium tomentosum (Andréws) Druce ●☆
121019 Cytisus transiens (Maire) Talavera = Cytisus arboreus (Desf.) DC. subsp. transiens (Maire) Maire ●☆
121020 Cytisus tridentatus (L.) Vuk. = Pterospartum tridentatum (L.) Willk. ●☆
121021 Cytisus tridentatus (L.) Vuk. var. gomarica Emb. et Maire = Pterospartum tridentatum (L.) Willk. ●☆
121022 Cytisus tridentatus (L.) Vuk. var. lasianthus (Willk.) Briq. = Pterospartum tridentatum (L.) Willk. subsp. lasianthum (Spach) Talavera et Gibbs ●☆
121023 Cytisus triflorus L'Hér. = Cytisus villosus Pourr. ●☆
121024 Cytisus triflorus L'Hér. var. bidentatus Chabert = Cytisus villosus Pourr. ●☆
121025 Cytisus valdesii Talavera et Gibbs = Cytisus balansae (Boiss.) Ball ●☆
121026 Cytisus villosus J. Presl et C. Presl;密毛金雀花;Hairy Broom, Hairybroom ●☆
121027 Cytisus villosus Pourr. = Cytisus villosus J. Presl et C. Presl ●☆
121028 Cytisus virgatus Link;条纹金雀花●☆
121029 Cytisus volubilis Blanco = Cajanus crassus (Prain ex King) Maesen ●
121030 Cytisus wulffii Krecz.;武夫氏金雀花;Wulff Broom ●☆
121031 Cytisus zingeri (Nenukow ex Litv.) Krecz.;蔡格氏金雀花; Zinger Broom ●☆
121032 Cytogonidium B. G. Briggs et L. A. S. Johnson(1998);隐蕊帚灯草属■☆
121033 Cytogonidium leptocarpoides (Benth.) B. G. Briggs et L. A. S. Johnson;隐蕊帚灯草■☆
121034 Cytolanthera yunnanensis W. W. Sm. = Cotylanthera paucisquama C. B. Clarke ■
121035 Cyttaranthus J. Léonard(1955);隔花大戟属☆
121036 Cyttaranthus congolensis J. Léonard;隔花大戟☆
121037 Cyttarium Peterm. = Antennaria Gaertn. + Hehchrysum Mill. + Gnaphalium L. ■
121038 Czackia Andrz. = Paradisea Mazzuc.(保留属名)■☆
121039 Czeikia Ikonn. (2004);阿富汗石头花属■☆
121040 Czeikia Ikonn. = Gypsophila L. ■●
121041 Czekelia Schur = Muscari Mill. ■☆
121042 Czernaevia Turcz. = Angelica L. ■
121043 Czernaevia Turcz. = Czernaevia Turcz. ex Ledeb. ■
121044 Czernaevia Turcz. ex Ledeb. (1844);柳叶芹属;Willowcelery ■
121045 Czernaevia laevigata Turcz.;柳叶芹(叉子芹,鸡爪芹,小叶独活);Willowcelery ■
121046 Czernaevia laevigata Turcz. f. latipinna Y. C. Chu;宽叶柳叶芹; Broadleaf Willowcelery ■
121047 Czernaevia laevigata Turcz. f. latipinna Y. C. Chu = Czernaevia laevigata Turcz. ■
121048 Czernaevia laevigata Turcz. var. exatocarpa Y. C. Chu;无翼柳叶芹(无翅柳叶芹);Wingless Willowcelery ■
121049 Czerniaevia Ledeh. = Czernaevia Turcz. ex Ledeb. ■
121050 Czerniaevia Ledeh. = Deschampsia P. Beauv. ■
121051 Czerniaevia Turcz. ex Ledeb. = Deschampsia P. Beauv. ■
121052 Czerniajevia Turcz. = Czernaevia Turcz. ■
121053 Czerntajewia Post et Kuntze = Czerniaevia Ledeh. ■
121054 Czerntajewia Post et Kuntze = Deschampsia P. Beauv. ■
121055 Czernya C. Presl = Phragmites Adans. ■
121056 D'Ayena Monier ex Mill. = Ayenia L. ●☆
121057 Dabanus Kuntze = Pometia J. R. Forst. et G. Forst. ●
121058 Dabeocia C. Koch = Daboecia D. Don(保留属名)●☆
121059 Dabeocia K. Koch = Daboecia D. Don(保留属名)●☆
121060 Daboecia D. Don (1834)(保留属名);大宝石南属;St. Dabeoc's Heath ●☆
121061 Daboecia × scotica D. C. McClint.;苏格兰大宝石南●☆
121062 Daboecia × scotica D. C. McClint. 'Jack Drake';杰克·德雷克苏格兰大宝石南●☆
121063 Daboecia × scotica D. C. McClint. 'Silverwells';银泉苏格兰大宝石南●☆
121064 Daboecia × scotica D. C. McClint. 'William Buhanan';威廉·布坎南苏格兰大宝石南●☆
121065 Daboecia azorica Tutin et E. F. Warb.;红花大宝石南●☆
121066 Daboecia cantabrica (Huds.) K. Koch;大宝石南;Connemara Heath, Irish Heath, Irish Heather, Irish Worts, Saint Dabeoc's Heath, St. Dabeoc's Heath ●☆
121067 Daboecia cantabrica (Huds.) K. Koch 'Bicolor';双色大宝石南●☆
121068 Daboecia cantabrica (Huds.) K. Koch 'Praegerae';极品大宝石南●☆
121069 Daboecia cantabrica (Huds.) K. Koch 'Snowdrift';雪堆大宝石南●☆
121070 Daboecia polifolia D. Don = Daboecia cantabrica (Huds.) K. Koch ●☆
121071 Dachel Adans. = Phoenix L. ●
121072 Dachel Adans. = Phoenix L. + Elate L. ●
121073 Dacryanthus (Endl.) Spach = Dracophyllum Labill. ●☆
121074 Dacrycarpaceae A. V. Bobrov et Melikyan = Podocarpaceae Endl.(保留科名)●
121075 Dacrycarpaceae Melikyan et A. V. Bobrov = Podocarpaceae Endl.(保留科名)●
121076 Dacrycarpus (Endl.) de Laub. (1969);鸡毛松属;Kahikatea ●
121077 Dacrycarpus de Laub. = Dacrycarpus (Endl.) de Laub. ●
121078 Dacrycarpus dacrydioides (A. Rich.) de Laub. = Podocarpus dacrydioides Rich. ●☆
121079 Dacrycarpus imbricatus (Blume) de Laub.;鸡毛松(假柏木,岭南罗汉松,茂松,异叶罗汉松,爪哇罗汉松,爪哇松);Imbricate Podocarpus, Imbricate Yaccatree, Java Podocarpus, Javan Podocarpus ●◇
121080 Dacrycarpus imbricatus (Blume) de Laub. var. patulus de Laub.;阔鸡毛松(鸡毛松)●
121081 Dacrycarpus kawaii (Hayata) A. V. Bobrov et Melikyan = Dacrycarpus imbricatus (Blume) de Laub. var. patulus de Laub. ●
121082 Dacrydiaceae A. V. Bobrov et Melikyan;陆均松科●
121083 Dacrydium Lamb. = Dacrydium Sol. ex J. Forst. ●
121084 Dacrydium Sol. ex J. Forst. (1786);陆均松属(泪杉属); Dacrydium, Huon Pine, Huonpine, Rium ●
121085 Dacrydium Sol. ex Lamb. = Dacrydium Sol. ex J. Forst. ●
121086 Dacrydium balansae Brongn. et Griseb.;新喀里多尼亚陆均松;Balansa Dacrydium ●☆
121087 Dacrydium beccarii Parl.;加里曼丹陆均松;Beccar Dacrydium ●☆
121088 Dacrydium bidwillii Hook. f. ex Kirk;高山陆均松(沼生哈罗果松);Dog Pine, Mountain Pine, New Zealand Mountain Pine, Tarwood ●☆

121089 Dacrydium bidwillii Hook. f. ex Kirk = Halocarpus bidwillii (Hook. f. ex Kirk) Quinn ●☆
121090 Dacrydium biforme Pilg.;二型陆均松(二型泪柏,二型叶哈罗果松);Manoao,Pink Pine,Yellow Pine ●☆
121091 Dacrydium biforme Pilg. = Halocarpus biformis (Hook.) Quinn ●☆
121092 Dacrydium colensoi Hook. = Manoao colensoi (Hook.) Molloy ●☆
121093 Dacrydium comosum Corner;马来西亚陆均松;Malaysia Dacrydium,Malaysian Dacrydium ●☆
121094 Dacrydium cupressinum Sol. ex J. Forst.;柏木陆均松(类柏陆均松,新西兰陆均松);New Zealand Red Pine,New Zealand Rium,Red Pine,Rimu ●☆
121095 Dacrydium elatum (Roxb.) Wall. ex Hook.;巨陆均松(大陆均松);Sempilor,Tall Dacrydium ●
121096 Dacrydium elatum (Roxb.) Wall. ex Hook. = Dacrydium pierrei Hickel ●
121097 Dacrydium elatum Wall. = Dacrydium elatum (Roxb.) Wall. ex Hook. ●
121098 Dacrydium falciforme (Parl.) Pilg.;镰叶陆均松;Sickleleaf Dacrydium ●☆
121099 Dacrydium fonkii Benth.;智利陆均松;Chilean Rium Dacrydium ●☆
121100 Dacrydium franklinii Hook. f.;塔斯马尼亚陆均松;Huan Pine,Huon Dacrydium,Huon Pine,Tasmanian Teak ●☆
121101 Dacrydium intermedium Kirk;黄银陆均松;Yellow Silver Pine ●☆
121102 Dacrydium kirkii F. Muell. ex Parl.;柯克陆均松;Kirk Pine ●☆
121103 Dacrydium laxifolium Hook. f. = Lepidothamnus laxifolius (Hook. f.) Quinn ●☆
121104 Dacrydium lycopodioides Brongn. et Griseb.;石松叶陆均松;Lycopodium-like Dacrydium ●☆
121105 Dacrydium pectinatum Hickel;陆均松(红松,黄叶松,黄液松,金边柚,泪柏,泪杉,山松,卧子柏,卧子松);Pierre Dacrydium ●◇
121106 Dacrydium pierrei Hickel = Dacrydium elatum (Roxb.) Wall. ex Hook. ●
121107 Dacrydium xanthaudrum Pilg.;新几内亚陆均松;New Guinea Dacrydium ●☆
121108 Dacryodes Vahl(1810);蜡烛木属(蜡烛橄榄树属)●☆
121109 Dacryodes afzelii (Engl.) H. J. Lam;阿芙泽尔蜡烛木●☆
121110 Dacryodes bampsiana Pierlot;邦氏蜡烛木●☆
121111 Dacryodes buettneri (Engl.) H. J. Lam;中非蜡烛木●☆
121112 Dacryodes camerunensis Onana;喀麦隆蜡烛木●☆
121113 Dacryodes dahomensis (Engl.) H. J. Lam;达荷姆蜡烛木●☆
121114 Dacryodes edulis (G. Don) H. J. Lam;食果蜡烛木(可食蜡烛树,梨果橄榄);Bush Butter-tree ●☆
121115 Dacryodes edulis (G. Don) H. J. Lam var. parvicarpa Okafor;小果食果蜡烛木●☆
121116 Dacryodes excelsa Vahl;大蜡烛木;West India Elemi ●☆
121117 Dacryodes fraxinifolia (Engl.) H. J. Lam;白蜡叶蜡烛木●☆
121118 Dacryodes fusca (Engl.) H. J. Lam;棕色蜡烛木●☆
121119 Dacryodes heterotricha (Pellegr.) H. J. Lam;异毛蜡烛木●☆
121120 Dacryodes igaganga Aubrév. et Pellegr.;蜡烛木;Igaganga ●☆
121121 Dacryodes klaineana (Pierre) H. J. Lam;阿德蜡烛木●☆
121122 Dacryodes ledermannii (Engl.) H. J. Lam;莱德蜡烛木●☆
121123 Dacryodes leonardiana Pierlot;莱奥蜡烛木●☆
121124 Dacryodes letestui (Pellegr.) H. J. Lam;莫温多蜡烛木●☆
121125 Dacryodes macrophylla (Oliv.) H. J. Lam;大叶蜡烛木●☆
121126 Dacryodes normandii Aubrév. et Pellegr.;诺氏蜡烛木●☆
121127 Dacryodes osika (Guillaumin) H. J. Lam = Dacryodes yangambiensis Lam. ex Troupin ●☆
121128 Dacryodes pubescens (Vermoesen) H. J. Lam;柔毛蜡烛木●☆
121129 Dacryodes tessmannii (Engl.) H. J. Lam;泰斯曼蜡烛木●☆
121130 Dacryodes viridiflora (Engl.) H. J. Lam;绿花蜡烛木●☆
121131 Dacryodes yangambiensis Lam. ex Troupin = Dacryodes osika (Guillaumin) H. J. Lam ●☆
121132 Dacryodes zenkeri (Engl.) H. J. Lam = Dacryodes klaineana (Pierre) H. J. Lam ●☆
121133 Dacryotrichia Wild(1973);毛基黄属■☆
121134 Dacryotrichia robinsonii Wild;毛基黄■☆
121135 Dactilis Neck. = Dactylis L. ■
121136 Dactilon Vill.(废弃属名)= Cynodon Rich.(保留属名)■
121137 Dactimala Raf. = Chrysophyllum L. ●
121138 Dactiphyllon Raf. = Lupinaster Fabr. ■☆
121139 Dactiphyllum Raf. = Trifolium L. ■
121140 Dactychlaena Post et Kuntze = Dactylaena Schrad. ex Schnit. f. ■☆
121141 Dactyladenia Welw.(1859);指腺金壳果属●☆
121142 Dactyladenia Welw. = Acioa Aubl. ●☆
121143 Dactyladenia barteri (Hook. f. ex Oliv.) Prance et F. White;巴特指腺金壳果●☆
121144 Dactyladenia bellayana (Baill.) Prance et F. White;贝莱指腺金壳果●☆
121145 Dactyladenia buchneri (Engl.) Prance et Sothers;布赫纳指腺金壳果●☆
121146 Dactyladenia campestris (Engl.) Prance et F. White;田野指腺金壳果●☆
121147 Dactyladenia chevalieri (De Wild.) Prance et F. White;舍瓦利耶指腺金壳果●☆
121148 Dactyladenia cinerea (Engl. ex De Wild.) Prance et F. White;灰色指腺金壳果●☆
121149 Dactyladenia dewevrei (De Wild. et T. Durand) Prance et F. White;得威指腺金壳果●☆
121150 Dactyladenia dichotoma (De Wild.) Prance et F. White;二歧指腺金壳果●☆
121151 Dactyladenia dinklagei (Engl.) Prance et F. White;丁克指腺金壳果●☆
121152 Dactyladenia eketensis (De Wild.) Prance et F. White;埃凯特金壳果●☆
121153 Dactyladenia floretii Breteler;弗洛雷指腺金壳果●☆
121154 Dactyladenia floribunda Welw.;繁花指腺金壳果●☆
121155 Dactyladenia gilletii (De Wild.) Prance et F. White;吉勒特指腺金壳果●☆
121156 Dactyladenia hirsuta (A. Chev. ex De Wild.) Prance et F. White;粗毛指腺金壳果●☆
121157 Dactyladenia johnstonei (Hoyle) Prance et F. White;约翰斯顿指腺金壳果●☆
121158 Dactyladenia jongkindii Breteler;容金德指腺金壳果●☆
121159 Dactyladenia laevis (Pierre ex De Wild.) Prance et F. White;平滑指腺金壳果●☆
121160 Dactyladenia lehmbachii (Engl.) Prance et F. White;莱姆指腺金壳果●☆
121161 Dactyladenia letestui (Letouzey) Prance et F. White;莱泰斯图指腺金壳果●☆
121162 Dactyladenia librevillensis (Letouzey) Prance et F. White;利伯

维尔指腺金壳果●☆
121163 Dactyladenia lujae (De Wild.) Prance et F. White = Dactyladenia buchneri (Engl.) Prance et Sothers ●☆
121164 Dactyladenia mannii (Oliv.) Prance et F. White;曼氏指腺金壳果●☆
121165 Dactyladenia ndjoleensis Breteler;恩乔莱金壳果●☆
121166 Dactyladenia pallescens (Baill.) Prance et F. White;苍白指腺金壳果●☆
121167 Dactyladenia pierrei (De Wild.) Prance et F. White;皮埃尔指腺金壳果●☆
121168 Dactyladenia sapinii (De Wild.) Prance et F. White;萨潘指腺金壳果●☆
121169 Dactyladenia scabrifolia (Hua) Prance et F. White;糙叶指腺金壳果●☆
121170 Dactyladenia smeathmannii (Baill.) Prance et F. White;小叶指腺金壳果●☆
121171 Dactyladenia staudtii (Engl.) Prance et F. White;施陶指腺金壳果●☆
121172 Dactyladenia whytei (Stapf) Prance et F. White;怀特指腺金壳果●☆
121173 Dactylaea (Franch.) Farille = Sinocarum H. Wolff ex R. H. Shan et F. T. Pu ■★
121174 Dactylaea Fedde ex H. Wolff(1930);裂瓣芹属■
121175 Dactylaea H. Wolff = Sinocarum H. Wolff ex R. H. Shan et F. T. Pu ■★
121176 Dactylaea schizopetala (Franch.) Farille = Sinocarum schizopetalum (Franch.) H. Wolff ex R. H. Shan et F. T. Pu ■
121177 Dactylaea wolffiana Fedde ex H. Wolff = Sinocarum schizopetalum (Franch.) H. Wolff ex R. H. Shan et F. T. Pu var. bijiangense (S. L. Liou) X. T. Liu ■
121178 Dactylaea wolffiana Fedde ex H. Wolff = Sinocarum wolffianum (Fedde ex H. Wolff) R. H. Shan et F. T. Pu ■
121179 Dactylaena Schrad. ex Schult. f. (1829);指被山柑属■☆
121180 Dactylaena micrantha Schrad. ex Schult. f.;小花指被山柑■☆
121181 Dactylaena microphylla Eichler;小叶指被山柑■☆
121182 Dactylaena pauciflora Griseb.;少花指被山柑■☆
121183 Dactylanthaceae Takht. (1987);指花菰科(手指花科)■☆
121184 Dactylanthaceae Takht. = Balanophoraceae Rich. (保留科名)●■
121185 Dactylanthera Welw. (1859);指药藤黄属●☆
121186 Dactylanthera Welw. = Symphonia L. f. ●☆
121187 Dactylanthes Haw. = Euphorbia L. ●■
121188 Dactylanthes globosa Haw. = Euphorbia globosa (Haw.) Sims ■
121189 Dactylanthes hamata Haw. = Euphorbia hamata (Haw.) Sweet ●☆
121190 Dactylanthocactus Y. Ito = Notocactus (K. Schum.) A. Berger et Backeb. ■
121191 Dactylanthocactus Y. Ito = Parodia Speg. (保留属名)●
121192 Dactylanthus Hook. f. (1859);指花菰属(手指花属);Wood Rose ■☆
121193 Dactylanthus taylorii Hook. f.;指花菰;Wood Rose, Wooden Rose ■☆
121194 Dactylepia Raf. = Cuscuta L. ■
121195 Dactylethria Ehrh. = Digitalis L. ■
121196 Dactyliandra (Hook. f.) Hook. f. (1871);指蕊瓜属■☆
121197 Dactyliandra Hook. f. = Dactyliandra (Hook. f.) Hook. f. ■☆
121198 Dactyliandra luederitziana (Cogn.) Cogn. = Dactyliandra welwitschii Hook. f. ■☆
121199 Dactyliandra nigrescens C. Jeffrey = Dactyliandra stefaninii (Chiov.) C. Jeffrey ■☆
121200 Dactyliandra stefaninii (Chiov.) C. Jeffrey;斯氏指蕊瓜■☆
121201 Dactyliandra welwitschii Hook. f.;指蕊瓜■☆
121202 Dactylicapnos Wall. (废弃属名) = Dicentra Bernh. (保留属名)■
121203 Dactylicapnos burmanica (K. R. Stern) Lidén = Dicentra burmanica K. R. Stern ■
121204 Dactylicapnos gaoligongshanensis Lidén;滇西紫金龙(Dactylicapnos 是废弃属名,故此名称必须重组)■
121205 Dactylicapnos grandifoliolata Merr. = Dicentra grandifoliolata (Merr.) K. R. Stern ■
121206 Dactylicapnos leiosperma Lidén;平滑籽紫金龙(Dactylicapnos 是废弃属名,故此名称必须重组)■
121207 Dactylicapnos lichiangensis (Fedde) Hand. -Mazz. = Dicentra lichiangensis Fedde ■
121208 Dactylicapnos macrocapnos (Prain) Hutch. = Dicentra macrocapnos Prain ■
121209 Dactylicapnos macrocapnos Hutch. = Dicentra macrantha Oliv. ■☆
121210 Dactylicapnos multiflora Hu = Dicentra scandens (D. Don) Walp. ■
121211 Dactylicapnos roylei (Hook. f. et Thomson) Hutch. = Dicentra roylei Hook. f. et Thomson ■
121212 Dactylicapnos scandens (D. Don) Hutch. = Diclytra scandens D. Don ■☆
121213 Dactylicapnos schneideri (Fedde) Lidén = Dicentra schneideri Fedde ■
121214 Dactylicapnos thalictrifolia Wall. = Dicentra scandens (D. Don) Walp. ■
121215 Dactylicapnos thalictrifolia Wall. = Dicentra thalictrifolia (Wall.) Hook. f. et Thomson ■
121216 Dactylicapnos torulosa (Hook. f. et Thomson) Hutch. = Dicentra torulosa Hook. f. et Thomson ■
121217 Dactylicapnos wolfdietheri Fedde = Dactylicapnos torulosa (Hook. f. et Thomson) Hutch. ■
121218 Dactylicapnos wolfdietheri Fedde = Dicentra torulosa Hook. f. et Thomson ■
121219 Dactyliocapnos Spreng. = Dicentra Bernh. (保留属名)■
121220 Dactyliophora Tiegh. (1894);指梗寄生属●☆
121221 Dactyliophora basiflora Danser;基花指梗寄生●☆
121222 Dactyliophora verticillata Tiegh.;指梗寄生●☆
121223 Dactyliota (Blume) Blume = Hypenanthe (Blume) Blume ●
121224 Dactyliota (Blume) Blume = Medinilla Gaudich. ex DC. ●
121225 Dactyliota Blume = Hypenanthe (Blume) Blume ●
121226 Dactylis L. (1753);鸭茅属(鸡脚茅属);Cocksfoot, Cocksfoot-grass, Cock's-grass, Dactylis, Duckgrass, Orchard Grass, Orchardgrass, Orchard-grass ■
121227 Dactylis adscendens Schrad. ex Schult. = Tribolium ciliare (Stapf) Renvoize ■☆
121228 Dactylis altaica Besser = Dactylis glomerata L. ■
121229 Dactylis atlantica Sennen;大西洋鸭茅■
121230 Dactylis brevifolia (J. Konig ex Willd.) Nees ex Steud. = Aeluropus lagopoides (L.) Trin. ex Thwaites ■☆
121231 Dactylis brevifolia J. Konig ex Willd. = Aeluropus lagopoides (L.) Trin. ex Thwaites ■☆
121232 Dactylis cristata (L.) M. Bieb. = Koeleria cristata (L.) Pers. ■
121233 Dactylis glomerata L.;鸭茅(果园草,鸡脚草);Cocksfoot,

Cock's-foot, Cocksfoot-grass, Common Orchardgrass, Duckgrass, Orchard Grass, Orchardgrass, Orchard-grass ■

121234 Dactylis glomerata L. 'Variegata';银纹鸭茅■☆

121235 Dactylis glomerata L. subsp. altaica (Besser) Domin = Dactylis glomerata L. ■

121236 Dactylis glomerata L. subsp. aschersoniana (Graebn.) Thell. = Dactylis polygama Horv. ■☆

121237 Dactylis glomerata L. subsp. castellata Borrill et P. F. Parker;卡地鸭茅■☆

121238 Dactylis glomerata L. subsp. himalayensis Domin;喜马拉雅鸭茅■

121239 Dactylis glomerata L. subsp. himalayensis Domin = Dactylis glomerata L. ■

121240 Dactylis glomerata L. subsp. hispanica (Roth) Nyman;西班牙鸭茅■☆

121241 Dactylis glomerata L. subsp. mairei Stebbins et Zohary;迈雷鸭茅■☆

121242 Dactylis glomerata L. subsp. sinensis A. Camus = Dactylis glomerata L. ■

121243 Dactylis glomerata L. var. altaica (Besser) Keng = Dactylis glomerata L. ■

121244 Dactylis glomerata L. var. australis Willk. et Lange = Dactylis glomerata L. subsp. hispanica (Roth) Nyman ■☆

121245 Dactylis glomerata L. var. ciliata Peterm. = Dactylis glomerata L. ■

121246 Dactylis glomerata L. var. detonsa Fr. = Dactylis glomerata L. ■

121247 Dactylis glomerata L. var. hispanica (Roth) K. Koch = Dactylis glomerata L. subsp. hispanica (Roth) Nyman ■☆

121248 Dactylis glomerata L. var. maroccana Pau et Font Quer = Dactylis glomerata L. ■

121249 Dactylis glomerata L. var. melillensis Sennen = Dactylis glomerata L. ■

121250 Dactylis glomerata L. var. sibthorpii (Hack.) Boiss. = Dactylis glomerata L. ■

121251 Dactylis glomerata L. var. spicata Pamp. = Dactylis glomerata L. ■

121252 Dactylis glomerata L. var. spiciformis Hochr. = Dactylis glomerata L. ■

121253 Dactylis glomerata L. var. tetuanensis Maire = Dactylis glomerata L. ■

121254 Dactylis hirta Schrad. = Tribolium hispidum (Thunb.) Desv. ■☆

121255 Dactylis hispanica Roth = Dactylis glomerata L. subsp. hispanica (Roth) Nyman ■☆

121256 Dactylis hispida Thunb. = Tribolium hispidum (Thunb.) Desv. ■☆

121257 Dactylis lagopoides L. = Aeluropus lagopoides (L.) Trin. ex Thwaites ■☆

121258 Dactylis littoralis (Gouan) Willd. = Aeluropus littoralis (Gouan) Parl. ■☆

121259 Dactylis longifolia Schrad. = Tribolium hispidum (Thunb.) Desv. ■☆

121260 Dactylis maritima Curtis = Spartina maritima (Curtis) Fernald ■☆

121261 Dactylis paspaloides Willd. = Dinebra retroflexa (Forssk. ex Vahl) Panz. ■

121262 Dactylis paspaloides Willd. = Dinebra retroflexa (Vahl) Panz. ■

121263 Dactylis paucinervis Nees = Odyssea paucinervis (Nees) Stapf ■☆

121264 Dactylis polygama Horv.;杂性鸭茅;Ascherson's Orchardgrass, Polygamous Orchardgrass, Slender Cock's-foot ■☆

121265 Dactylis pungens Schreb. = Ammochloa pungens (Schreb.) Boiss. ■☆

121266 Dactylis repens Desf. = Aeluropus lagopoides (L.) Trin. ex Thwaites ■☆

121267 Dactylis smithii Link;史密斯鸭茅;Smith Orchardgrass ■☆

121268 Dactylis smithii Link subsp. marina (Borrill) P. F. Parker;海生斯密氏鸭茅■☆

121269 Dactylis spicata Willd. = Elytrophorus spicatus (Willd.) A. Camus ■

121270 Dactylis stricta Aiton = Spartina maritima (Curtis) Fernald ■☆

121271 Dactylis villosa Hook. f. = Aeluropus lagopoides (L.) Trin. ex Thwaites ■☆

121272 Dactylis villosa Thunb.;长柔毛鸭茅■☆

121273 Dactylis woronowii Ovcz.;沃氏鸭茅;Woronow Orchardgrass ■☆

121274 Dactyliscapnos B. D. Jacks. = Dicentra Bernh.(保留属名)■

121275 Dactylium Griff. = Erythropalum Blume ●

121276 Dactylium vagum Griff. = Erythropalum scandens Blume ●

121277 Dactylocardamum Al-Shehbaz(1989);秘鲁碎米荠属■☆

121278 Dactylocardamum imbricatifolium Al-Shehbaz;秘鲁碎米荠■☆

121279 Dactylocladus Oliv.(1895);钟康木属●☆

121280 Dactylocladus stenostachya Oliv.;钟康木;Jongkong Merubong Medang ●☆

121281 Dactyloctenium Willd.(1809);龙爪茅属;Button-grass, Craw Foot, Crowfootgrass ■

121282 Dactyloctenium aegypticum Willd. = Dactyloctenium aegyptium (L.) Willd. ■

121283 Dactyloctenium aegyptium (L.) P. Beauv. = Dactyloctenium aegyptium (L.) Willd. ■

121284 Dactyloctenium aegyptium (L.) Richt. = Dactyloctenium aegyptium (L.) P. Beauv. ■

121285 Dactyloctenium aegyptium (L.) Willd.;龙爪茅(鸭掌草);Crowfoot Grass, Durban Crowfootgrass, Egyptian Grass, Sudan Crowfoot Grass ■

121286 Dactyloctenium aegyptium (L.) Willd. var. aristatum (Link) A. Chev. = Dactyloctenium aristatum Link ■☆

121287 Dactyloctenium aristatum Link;具芒龙爪茅■☆

121288 Dactyloctenium australe Steud.;德班龙爪茅;Durban Grass ■☆

121289 Dactyloctenium bogdanii S. M. Phillips = Dactyloctenium geminatum Hack. ■☆

121290 Dactyloctenium capitatum A. Camus;头状龙爪茅■☆

121291 Dactyloctenium ciliare Chiov. = Dactyloctenium aegyptium (L.) Willd. ■

121292 Dactyloctenium falcatum (L. f.) Willd. = Harpochloa falx (L. f.) Kuntze ■☆

121293 Dactyloctenium figarei De Not. = Dactyloctenium aegyptium (L.) Willd. ■

121294 Dactyloctenium geminatum Hack.;双龙爪茅■☆

121295 Dactyloctenium giganteum B. S. Fisher et Schweick.;巨大龙爪茅■☆

121296 Dactyloctenium glabrum P. Beauv.;光龙爪茅■☆

121297 Dactyloctenium glaucophyllum Courbon = Dactyloctenium scindicum Boiss. ■☆

121298 Dactyloctenium glaucophyllum Courbon var. elongation Courbon = Dactyloctenium scindicum Boiss. ■☆

121299 Dactyloctenium glaucophyllum Courbon var. robustior Courbon = Dactyloctenium scindicum Boiss. ■☆

121300 Dactyloctenium glaucophyllum Courbon var. villosum Mattei = Dactyloctenium aristatum Link ■☆
121301 Dactyloctenium meridionale Ham. = Dactyloctenium aegyptium (L.) Willd. ■
121302 Dactyloctenium mpuetensis De Wild. = Dactyloctenium aegyptium (L.) Willd. ■
121303 Dactyloctenium mucronatum (Michx.) Willd. = Dactyloctenium aegyptium (L.) Willd. ■
121304 Dactyloctenium mucronatum Willd. = Dactyloctenium aegyptium (L.) P. Beauv. ■
121305 Dactyloctenium pilosum Stapf;疏毛龙爪茅■☆
121306 Dactyloctenium radulans P. Beauv.;刮刀龙爪茅;Buttongrass, Button-grass ■☆
121307 Dactyloctenium robecchii (Chiov.) Chiov.;罗贝克龙爪茅■☆
121308 Dactyloctenium scindicum Boiss.;灰绿龙爪茅■☆
121309 Dactyloctenium semipunctatum Courbon = Dactyloctenium aristatum Link ■☆
121310 Dactylodes Kuntze = Tripsacum L. ■
121311 Dactylodes Zanoni-Monti = Tripsacum L. ■
121312 Dactylodes Zanoni-Monti ex Kuntze = Tripsacum L. ■
121313 Dactylogramma Link = Muehlenbergia Schreb. ■
121314 Dactyloides Nieuwl. = Muscaria Haw. ■
121315 Dactyloides Nieuwl. = Saxifraga L. ■
121316 Dactylon Roem. et Schult. = Cynodon Rich.(保留属名)■
121317 Dactylon Roem. et Schult. = Dactilon Vill.(废弃属名)■
121318 Dactylon officinale Vill. = Cynodon dactylon (L.) Pers. ■
121319 Dactylopetalum Benth.(1859);指瓣树属●☆
121320 Dactylopetalum Benth. = Cassipourea Aubl. ●☆
121321 Dactylopetalum dinklagei Engl. = Cassipourea dinklagei (Engl.) Alston ●☆
121322 Dactylopetalum kamerunense Engl. = Cassipourea kamerunensis (Engl.) Alston ●☆
121323 Dactylopetalum mannii Hook. f. ex Oliv. = Cassipourea gummiflua Tul. var. mannii (Hook. f. ex Oliv.) J. Lewis ●☆
121324 Dactylopetalum parvifolium Scott-Elliot = Cassipourea afzelii (Oliv.) Alston ●☆
121325 Dactylopetalum sericeum Engl. = Cassipourea sericea (Engl.) Alston ●☆
121326 Dactylopetalum ugandense Stapf = Cassipourea gummiflua Tul. var. ugandensis (Stapf) J. Lewis ●☆
121327 Dactylopetalum verticillatum (N. E. Br.) Schinz = Cassipourea gummiflua Tul. var. verticillata (N. E. Br.) J. Lewis ●☆
121328 Dactylophora T. Durand et Jacks. = Dactyliophora Tiegh. ●☆
121329 Dactylophyllum (Benth.) Spach = Gilia Ruiz et Pav. ■●☆
121330 Dactylophyllum Spach = Gilia Ruiz et Pav. ■●☆
121331 Dactylophyllum Spach = Linanthus Benth. ■☆
121332 Dactylophyllum Spenn. = Potentilla L. ■●
121333 Dactylopsis N. E. Br.(1925);手指玉属■☆
121334 Dactylopsis digitata (Aiton) Gerbaulet subsp. littlewoodii (L. Bolus) Klak;里特手指玉■☆
121335 Dactylopsis digitata (Aiton) N. E. Br.;手指玉■☆
121336 Dactylopsis littlewoodii L. Bolus = Dactylopsis digitata (Aiton) Gerbaulet subsp. littlewoodii (L. Bolus) Klak ■☆
121337 Dactylorchis (Klinge) Verm.(1947);指兰属■☆
121338 Dactylorchis (Klinge) Verm. = Dactylorhiza Neck. ex Nevski (保留属名)■
121339 Dactylorchis (Klinge) Verm. = Orchis L. ■
121340 Dactylorchis aristata (Fisch. ex Lindl.) Verm. ex F. Maek. = Dactylorhiza aristata (Fisch. ex Lindl.) Soó ■
121341 Dactylorchis aristata (Fisch. ex Lindl.) Verm. f. punctata (Tatew.) F. Maek. = Dactylorhiza aristata (Fisch. ex Lindl.) Soó f. punctata (Tatew.) F. Maek. ex Toyok. ■☆
121342 Dactylorchis cruenta (O. F. Muell.) Verm. = Orchis cruenta O. F. Muell. ■
121343 Dactylorchis fuchsii (Druce) Verm. = Dactylorhiza fuchsii (Druce) Soó ■
121344 Dactylorchis fuchsii (Druce) Verm. = Orchis fuchsii Druce ■
121345 Dactylorchis incarnata (L.) Verm. = Orchis incarnata L. ■☆
121346 Dactylorchis latifolia (L.) Rathm. = Orchis latifolia L. ■
121347 Dactylorchis maculata (L.) Soó subsp. fuchsii (Druce) Hyl. = Dactylorhiza fuchsii (Druce) Soó ■
121348 Dactylorchis maculata (L.) Verm. = Orchis mascula L. ■☆
121349 Dactylorchis majalis (Rchb.) Verm. = Orchis majalis Rchb. ■☆
121350 Dactylorchis praetermissa (Druce) Verm. = Orchis praetermissa Druce ■☆
121351 Dactylorchis purpurella (T. Stephenson et T. A. Stephenson) Verm. = Orchis purpurella T. Stephenson et T. A. Stephenson ■☆
121352 Dactylorchis salina (Turcz. ex Lindl.) Verm. = Orchis latifolia L. ■
121353 Dactylorchis sambucina (L.) Verm. = Orchis sambucina L. ■☆
121354 Dactylorchis traunsteineri (Saut. ex Rchb.) Verm. = Orchis traunsteineri Saut. ex Rchb. ■☆
121355 Dactylorchis umbrosa (Kar. et Kir.) Wendelbo = Dactylorhiza umbrosa (Kar. et Kir.) Nevski ■
121356 Dactylorhiza (Neck. ex Nevski) Nevski = Orchis L. ■
121357 Dactylorhiza Neck. = Orchis L. ■
121358 Dactylorhiza Neck. ex Nevski(1937)(保留属名);掌根兰属(根爪兰属,肿根属);Marsh Orchid, Orchis, Salab-Misri, Salep ■
121359 Dactylorhiza aristata (Fisch. ex Lindl.) Soó;芒尖掌根兰(光掌根兰);Fischer's Orchid ■
121360 Dactylorhiza aristata (Fisch. ex Lindl.) Soó f. alba P. M. Br. = Dactylorhiza aristata (Fisch. ex Lindl.) Soó ■
121361 Dactylorhiza aristata (Fisch. ex Lindl.) Soó f. albiflora (Koidz.) F. Maek. ex Toyok.;白花光掌根兰■☆
121362 Dactylorhiza aristata (Fisch. ex Lindl.) Soó f. perbracteata (Lepage) Catling = Dactylorhiza aristata (Fisch. ex Lindl.) Soó ■
121363 Dactylorhiza aristata (Fisch. ex Lindl.) Soó f. punctata (Tatew.) F. Maek. ex Toyok.;斑光掌根兰■☆
121364 Dactylorhiza aristata (Fisch. ex Lindl.) Soó f. rosea P. M. Br. = Dactylorhiza aristata (Fisch. ex Lindl.) Soó ■
121365 Dactylorhiza aristata (Fisch. ex Lindl.) Soó var. kodiakensis Luer et G. M. Luer = Dactylorhiza aristata (Fisch. ex Lindl.) Soó ■
121366 Dactylorhiza atlantica Kreutz et Vlaciha = Dactylorhiza durandii (Boiss. et Reut.) M. Lainz ■☆
121367 Dactylorhiza battandieri Raynaud;巴坦掌根兰■☆
121368 Dactylorhiza chuhensis Renz et Taubenheim = Dactylorhiza umbrosa (Kar. et Kir.) Nevski ■
121369 Dactylorhiza cilicica (Klinge) P. F. Hunt et Summerh.;西里西亚掌根兰;Anatolian Marsh Orchid ■☆
121370 Dactylorhiza cordigera (Fr.) Soó;心掌根兰;Heart-shaped Orchid ■☆
121371 Dactylorhiza cruenta (O. F. Müll.) Soó = Dactylorhiza incarnata (L.) Soó subsp. cruenta (O. F. Müll.) P. D. Sell ■
121372 Dactylorhiza durandii (Boiss. et Reut.) M. Lainz;北非掌根兰■☆

121373 Dactylorhiza elata (Poir.) Soó;掌根兰;Robust Marsh Orchid ■☆
121374 Dactylorhiza elata (Poir.) Soó = Orchis elata Poir. ■☆
121375 Dactylorhiza elata (Poir.) Soó subsp. durandii (Boiss. et Reut.) Soó = Dactylorhiza durandii (Boiss. et Reut.) M. Lainz ■☆
121376 Dactylorhiza elata (Poir.) Soó subsp. mauritanica B. Baumann et H. Baumann = Dactylorhiza munbyana (Boiss. et Reut.) Aver. ■☆
121377 Dactylorhiza elata (Poir.) Soó var. durandii (Boiss. et Reut.) Landwehr = Dactylorhiza durandii (Boiss. et Reut.) M. Lainz ■☆
121378 Dactylorhiza elata (Poir.) Soó var. elongata (Maire) Raynaud = Orchis elata Poir. ■☆
121379 Dactylorhiza elata (Poir.) Soó var. munbyana (Boiss. et Reut.) Soó = Dactylorhiza munbyana (Boiss. et Reut.) Aver. ■☆
121380 Dactylorhiza foliosa (Lowe) Soó = Dactylorhiza foliosa (Sol. ex Lowe) Soó ■☆
121381 Dactylorhiza foliosa (Sol. ex Lowe) Soó;马德兰掌根兰;Madeiran Orchid ■☆
121382 Dactylorhiza fuchsii (Druce) Soó;紫斑掌裂兰(林地指兰,紫斑红门兰,紫斑叶红门兰);Common Spotted Orchid,Heath Orchid,Purplespot Orchis,Woodland Spotted Orchid ■
121383 Dactylorhiza fuchsii (Druce) Verm. = Orchis fuchsii Druce ■
121384 Dactylorhiza hatageterea (D. Don) Soó = Orchis latifolia L. ■
121385 Dactylorhiza hatagirea (D. Don) Soó;掌裂兰■
121386 Dactylorhiza hatagirea (D. Don) Soó = Orchis latifolia L. ■
121387 Dactylorhiza incarnata (L.) Soó = Orchis incarnata L. ■☆
121388 Dactylorhiza incarnata (L.) Soó subsp. africana (Klinge) H. Sund. = Dactylorhiza munbyana (Boiss. et Reut.) Aver. ■☆
121389 Dactylorhiza incarnata (L.) Soó subsp. cruenta (O. F. Müll.) P. D. Sell;紫点掌裂兰(紫点红门兰);Connaught Marsh Orchid,Purplespot Orchis ■
121390 Dactylorhiza incarnata (L.) Soó subsp. turkestanica (Klinge) H. Sund. = Dactylorhiza umbrosa (Kar. et Kir.) Nevski ■
121391 Dactylorhiza insularis (Sommier) Landwehr;海岛掌根兰■☆
121392 Dactylorhiza knorringiana (Kraenzl.) Ikonn. = Dactylorhiza umbrosa (Kar. et Kir.) Nevski ■
121393 Dactylorhiza kotschyi (Rchb. f.) P. F. Hunt et Summerh. = Dactylorhiza umbrosa (Kar. et Kir.) Nevski ■
121394 Dactylorhiza lapponica (Laest. ex Hartm.) Soó;拉普兰掌根兰;Lapland Marsh Orchid ■☆
121395 Dactylorhiza latifolia (L.) Rothm. = Orchis latifolia L. ■
121396 Dactylorhiza latifolia (L.) Soó = Orchis latifolia L. ■
121397 Dactylorhiza longebracteata (F. W. Schmidt) Holub = Dactylorhiza fuchsii (Druce) Verm. ■
121398 Dactylorhiza longebracteata (F. W. Schmidt) Holub = Orchis fuchsii Druce ■
121399 Dactylorhiza maculata (L.) Soó = Orchis mascula L. ■☆
121400 Dactylorhiza maculata (L.) Soó subsp. battandieri (Raynaud) Baumann et Künkele = Dactylorhiza battandieri Raynaud ■☆
121401 Dactylorhiza majalis (Rchb.) P. F. Hunt et Summerh. = Orchis majalis Rchb. ■☆
121402 Dactylorhiza majalis (Rchb.) P. F. Hunt et Summerh. var. ebudensis Wief. = Orchis traunsteineri Saut. ex Rchb. ■☆
121403 Dactylorhiza majalis (Rchb.) P. F. Hunt et Summerh. var. praetermissa (Druce) R. M. Bateman et Denholm = Orchis praetermissa Druce ■☆
121404 Dactylorhiza majalis (Rchb.) P. F. Hunt et Summerh. var. praetermissa (Druce) R. M. Bateman et Denholm = Orchis purpurella T. Stephenson et T. A. Stephenson ■☆
121405 Dactylorhiza majalis (Rchb.) P. F. Hunt et Summerh. var. traunsteinerioides (Pugsley) R. M. Bateman et Denholm = Orchis traunsteineri Saut. ex Rchb. ■☆
121406 Dactylorhiza markusii (Tineo) H. Baumann et Künkele = Dactylorhiza insularis (Sommier) Landwehr ■☆
121407 Dactylorhiza maurusia (Emb. et Maire) Raynaud;莫尔掌根兰■☆
121408 Dactylorhiza merovensis (Grossh.) Aver. = Dactylorhiza umbrosa (Kar. et Kir.) Nevski ■
121409 Dactylorhiza munbyana (Boiss. et Reut.) Aver.;芒比掌根兰■☆
121410 Dactylorhiza persica (Schltr.) Soó = Dactylorhiza umbrosa (Kar. et Kir.) Nevski ■
121411 Dactylorhiza praetermissa (Druce) Soó = Orchis praetermissa Druce ■☆
121412 Dactylorhiza pulchella (Druce) Aver. = Orchis purpurella T. Stephenson et T. A. Stephenson ■☆
121413 Dactylorhiza renzii Aver. = Dactylorhiza umbrosa (Kar. et Kir.) Nevski ■
121414 Dactylorhiza romana (Sebast.) Soó;罗马掌根兰■☆
121415 Dactylorhiza romana (Sebast.) Soó subsp. markusii (Tineo) Holub = Dactylorhiza insularis (Sommier) Landwehr ■☆
121416 Dactylorhiza romana (Sebast.) Soó subsp. siciliensis (Klinge) Soó = Dactylorhiza markusii (Tineo) H. Baumann et Künkele ■☆
121417 Dactylorhiza sambucina (L.) Soó = Orchis sambucina L. ■☆
121418 Dactylorhiza sanasunitensis (H. Fleischm.) Soó = Dactylorhiza umbrosa (Kar. et Kir.) Nevski ■
121419 Dactylorhiza sauna (Turcz. ex Lindl.) Soó = Orchis latifolia L. ■
121420 Dactylorhiza traunsteineri (Saut. ex Rchb.) Soó = Orchis traunsteineri Saut. ex Rchb. ■☆
121421 Dactylorhiza traunsteinerioides (Pugsley) R. M. Bateman et Denholm = Orchis traunsteineri Saut. ex Rchb. ■☆
121422 Dactylorhiza umbrosa (Kar. et Kir.) Nevski;耐荫掌根兰(阴生红门兰,阴生掌裂兰);Shady Orchis ■
121423 Dactylorhiza umbrosa (Kar. et Kir.) Nevski var. chuhensis (Renz et Taubenheim) Kreutz = Dactylorhiza umbrosa (Kar. et Kir.) Nevski ■
121424 Dactylorhiza umbrosa (Kar. et Kir.) Nevski var. knorringiana (Kraenzl.) Soó = Dactylorhiza umbrosa (Kar. et Kir.) Nevski ■
121425 Dactylorhiza umbrosa (Kar. et Kir.) Nevski var. longibracteata Renz = Dactylorhiza umbrosa (Kar. et Kir.) Nevski ■
121426 Dactylorhiza umbrosa (Kar. et Kir.) Nevski var. ochroleuca (Bornm.) Renz = Dactylorhiza umbrosa (Kar. et Kir.) Nevski ■
121427 Dactylorhiza vestita (Lag. et Rodr.) Aver.;包被掌根兰■☆
121428 Dactylorhiza viridis (L.) R. M. Bateman, Pridgeon et M. W. Chase;凹舌掌裂兰■
121429 Dactylorhiza viridis (L.) R. M. Bateman, Pridgeon et M. W. Chase = Coeloglossum viride (L.) Hartm. ■
121430 Dactylorhynchus Schltr. (1913);指喙兰属■☆
121431 Dactylorhynchus Schltr. = Bulbophyllum Thouars(保留属名)■
121432 Dactylorhynchus flavescens Schltr.;指喙兰■☆
121433 Dactylostalix Rchb. f. (1878);指脊兰属■☆
121434 Dactylostalix maculosa Miyabe et Kudo = Dactylostalix ringens Rchb. f. ■☆
121435 Dactylostalix ringens Rchb. f.;指脊兰■☆
121436 Dactylostalix ringens Rchb. f. f. punctata Miyabe et Tatew.;斑点指脊兰■☆
121437 Dactylostegium Nees = Dicliptera Juss. (保留属名)■
121438 Dactylostelma Schltr. (1895);指冠萝藦属☆

121439 Dactylostemon Klotzsch = Actinostemon Mart. ex Klotzsch ■☆
121440 Dactylostemon Klotzsch(1841);指蕊大戟属■☆
121441 Dactylostemon angustifolius Müll. Arg.;窄叶指蕊大戟■☆
121442 Dactylostemon australis Müll. Arg.;澳洲指蕊大戟■☆
121443 Dactylostemon brasiliensis Müll. Arg.;巴西指蕊大戟■☆
121444 Dactylostemon communis Müll. Arg.;普通指蕊大戟■☆
121445 Dactylostemon glabrescens Klotzsch;渐光指蕊大戟■☆
121446 Dactylostemon grandifolius Klotzsch;大叶指蕊大戟■☆
121447 Dactylostemon guianensis Klotzsch;圭亚那指蕊大戟■☆
121448 Dactylostemon lasiocarpus Klotzsch;光果指蕊大戟■☆
121449 Dactylostemon verticillatus Klotzsch;轮生指蕊大戟■☆
121450 Dactylostigma D. F. Austin = Hildebrandtia Vatke ex A. Braun. ●☆
121451 Dactylostigma D. F. Austin(1973);指柱旋花属■☆
121452 Dactylostigma linearifolia D. F. Austin = Hildebrandtia austinii Staples ●☆
121453 Dactylostyles Scheidw. = Zygostates Lindl. ■☆
121454 Dactylostylis Scheidw. = Zygostates Lindl. ■☆
121455 Dactylus Asch. = Cynodon Rich.(保留属名)■
121456 Dactylus Burm. f. = Microstegium Nees ■
121457 Dactylus Forssk. = Diospyros L. ●
121458 Dactymala Post et Kuntze = Chrysophyllum L. ●
121459 Dactymala Post et Kuntze = Dactimala Raf. ●
121460 Dactyphyllum Endl. = Dactiphyllum Raf. ■
121461 Dactyphyllum Endl. = Trifolium L. ■
121462 Dadia Vell.(1829);达蒂菊属■☆
121463 Dadjoua Parsa(1960);伊朗石竹属■☆
121464 Dadjoua pteranthoidea Parsa;伊朗石竹■☆
121465 Daedalacanthus T. Anderson = Eranthemum L. ●■
121466 Daedalacanthus nervosus (Vahl) T. Anderson = Eranthemum austrosinense H. S. Lo ■
121467 Daedalacanthus nervosus (Vahl) T. Anderson = Eranthemum pulchellum Andréws ●
121468 Daedalacanthus splendens T. Anderson = Eranthemum splendens (T. Anderson) Siebold et Voss ■
121469 Daemia Poir. = Doemia R. Br. ■☆
121470 Daemia Poir. = Pergularia L. ■☆
121471 Daemia R. Br = Doemia R. Br. ■☆
121472 Daemia aethiopica Decne. = Pergularia daemia (Forssk.) Chiov. ■☆
121473 Daemia barbata Klotzsch = Pergularia daemia (Forssk.) Chiov. subsp. barbata (Klotzsch) Goyder ■☆
121474 Daemia cordata (Forssk.) R. Br. = Pergularia tomentosa L. ●☆
121475 Daemia cordata (Forssk.) R. Br. var. schmidtiana (Pomel) Batt. = Pergularia tomentosa L. ●☆
121476 Daemia cordata R. Br. = Pergularia tomentosa L. ●☆
121477 Daemia cordifolia K. Schum. var. leiocarpa ? = Pergularia daemia (Forssk.) Chiov. subsp. garipensis (E. Mey.) Goyder ■☆
121478 Daemia extensa (Jacq.) Aiton f. = Pergularia daemia (Forssk.) Chiov. ■☆
121479 Daemia extensa (Jacq.) R. Br. = Pergularia daemia (Forssk.) Chiov. ■☆
121480 Daemia garipensis E. Mey. = Pergularia daemia (Forssk.) Chiov. subsp. garipensis (E. Mey.) Goyder ■☆
121481 Daemia glabra (Forssk.) Schult. = Pergularia glabra (Forssk.) Chiov. ■☆
121482 Daemia scandens G. Don = Pergularia daemia (Forssk.) Chiov. ■☆
121483 Daemia schmittiana Pomel = Pergularia tomentosa L. ●☆
121484 Daemia tomentosa (L.) Pomel = Pergularia tomentosa L. ●☆
121485 Daemonorops Blume ex Schult. f. = Daemonorops Blume ●
121486 Daemonorops Blume(1830);黄藤属(白藤属,红藤属,龙黄藤属,麒麟竭属,提摩藤属,小藤属);Devil Rattan, Devilrattan, Rattan, Yellowvine ●
121487 Daemonorops angustifolia Mart.;狭叶黄藤;Narrowleaf Devil Rattan, Slender Rotang Palm ●☆
121488 Daemonorops angustispatha Furtado;狭苞黄藤;Narrowspathe Devil Rattan, Rattan Palm ●☆
121489 Daemonorops brachystachys Furtado;短穗黄藤;Shortspik Devil Rattan ●☆
121490 Daemonorops calicarpa (Griff.) Mart.;秀果黄藤;Showyfruit Devil Rattan ●☆
121491 Daemonorops carcharodon Ridl. = Daemonorops angustifolia Mart. ●☆
121492 Daemonorops didymophylla Becc.;双叶黄藤(龙血黄藤);Didymophylla Devil Rattan ●☆
121493 Daemonorops draco (Willd.) Blume;龙血藤(海蜡,龙黄藤,木血竭,骐骥竭,麒麟竭,麒麟血,藤血竭,血竭);Sumatra-dragonsblood ●☆
121494 Daemonorops draconcella Becc.;含脂黄藤●☆
121495 Daemonorops fissa Blume;尖裂黄藤●☆
121496 Daemonorops geniculata (Griff.) Mart.;膝曲状黄藤;Kneed Devil Rattan ●☆
121497 Daemonorops grandis (Griff.) Mart.;大黄藤;Giant Devil Rattan, Giant Rotang Palm ●☆
121498 Daemonorops grandis Kurz = Daemonorops kurziana Becc. ●☆
121499 Daemonorops grandis Mart. = Daemonorops grandis (Griff.) Mart. ●☆
121500 Daemonorops hystrix Mart.;豪猪刺黄藤;Hystrix Devil Rattan ●☆
121501 Daemonorops imbellis Becc.;软弱黄藤;Feeble Devil Rattan ●☆
121502 Daemonorops intermedia Mart.;中间黄藤;Intermediate Devil Rattan ●☆
121503 Daemonorops javanica Furtado;爪哇黄藤●☆
121504 Daemonorops jenkensiana Mart.;长咀黄藤(长咀红藤)●☆
121505 Daemonorops kiahii Furtado;贾赫黄藤;Kiah Devil Rattan ●☆
121506 Daemonorops kunstleri Becc.;孔氏黄藤;Kunstler Devil Rattan ●☆
121507 Daemonorops kurziana Becc.;库氏黄藤;Dragon's Blood Palm ●☆
121508 Daemonorops laciniata Furtado;条裂黄藤;Laciniate Devil Rattan ●☆
121509 Daemonorops lasiospatha Furtado;毛苞黄藤;Hairyspathe Devil Rattan ●☆
121510 Daemonorops leptopa Mart.;细黄藤;Thin Devil Rattan ●☆
121511 Daemonorops lewisiana Mart.;路易斯黄藤;Lewis Devil Rattan ●☆
121512 Daemonorops longipes Mart.;长柄黄藤;Longstalk Devil Rattan ●☆
121513 Daemonorops macrophylla Becc.;大叶黄藤;Bigleaf Devil Rattan ●☆
121514 Daemonorops margaritae (Hance) Becc.;黄藤(白藤,赤藤,红藤,甲种黄藤,省藤,溪藤,正藤);Davil Devilrattan, Devil Rattan, Devilrattan, Margaret Rotang Palm, Margaric Devilrattan, Rattan, Yellow Rattan Palm, Yellowvine ●
121515 Daemonorops margaritae Becc. = Calamus orientalis C. E. Chang ●

121516 Daemonorops melanochaetes Blume;黑刚毛黄藤;Blacksetose Devil Rattan ●☆
121517 Daemonorops micracantha Becc.;小刺黄藤(小黄藤)●☆
121518 Daemonorops monticola Mart.;山生黄藤;Mountainliving Devil Rattan ●☆
121519 Daemonorops nurii Furtado;努尔黄藤;Nur Devil Rattan ●☆
121520 Daemonorops oligophylla Becc.;少叶黄藤;Fewleaf Devil Rattan ●☆
121521 Daemonorops periacantha Miq.;周刺黄藤;Suronded Devil Rattan ●☆
121522 Daemonorops propinqua Becc.;亲近黄藤;Affined Devil Rattan ●☆
121523 Daemonorops pseudosepala Becc.;假萼黄藤;False-sepal Devil Rattan ●☆
121524 Daemonorops sabut Becc.;萨布特黄藤;Sabut Devil Rattan ●☆
121525 Daemonorops scortechinii Becc.;斯考氏黄藤;Scortechin Devil Rattan ●☆
121526 Daemonorops sepala Becc.;萼片黄藤;Sepal Devil Rattan ●☆
121527 Daemonorops stipitata Furtado;有柄黄藤;Stalked Devil Rattan ●☆
121528 Daemonorops tabacina Becc.;烟草色黄藤;Tabaccooloured Devil Rattan ●☆
121529 Daemonorops verticillaris Mart.;轮生黄藤;Verticillate Devil Rattan ●☆
121530 Daenikera Hurl. et Stauffer(1957);达尼木属●☆
121531 Daenikera corallina Hurl. et Stauffer;达尼木●☆
121532 Daenikeranthus Baum.-Bod. = Dracophyllum Labill. ●☆
121533 Dahlbergia Raf. = Columnea L. ●■☆
121534 Dahlbergia Raf. = Dalbergaria Tussac ●☆
121535 Dahlgrenia Steyerm. = Dictyocaryum H. Wendl. ●☆
121536 Dahlgrenodendron J. J. M. van der Merwe et A. E. van Wyk = Cryptocarya R. Br.(保留属名)●
121537 Dahlgrenodendron J. J. M. van der Merwe et A. E. van Wyk (1988);纳塔尔樟属●☆
121538 Dahlgrenodendron natalense (J. H. Ross) J. J. M. van der Merwe et A. E. van Wyk;纳塔尔樟●☆
121539 Dahlia Cav.(1791);大丽花属;Dahlia,Pompon ■●
121540 Dahlia Thunb. = Trichocladus Pers. ●☆
121541 Dahlia × hortensis Guillaumin;大丽菊;Hybrids Dahlia ■☆
121542 Dahlia arborea Regel;高大丽花■☆
121543 Dahlia bidentifolia Salisb. = Dahlia coccinea Cav. ■
121544 Dahlia cervantesii (Sweet) Lag. ex DC. = Dahlia coccinea Cav. ■
121545 Dahlia cervantesii Lag. = Dahlia coccinea Cav. ■
121546 Dahlia cervantesii Lag. ex DC. = Dahlia coccinea Cav. ■
121547 Dahlia coccinea Cav.;圆叶大丽菊■
121548 Dahlia coccinea Cav. = Dahlia pinnata Cav. ■
121549 Dahlia crinita Thunb. = Trichocladus crinitus (Thunb.) Pers. ●☆
121550 Dahlia crocata Lag. = Dahlia pinnata Cav. ■
121551 Dahlia crocea Poir. = Dahlia coccinea Cav. ■
121552 Dahlia erecta Lag. = Dahlia pinnata Cav. ■
121553 Dahlia excelsa Benth.;平夹大丽花;Flat Tree Dahlia,Tree Dahlia ■☆
121554 Dahlia faurezii Hort.;法氏大丽花■☆
121555 Dahlia glabrata Lindl. = Dahlia merckii Lehm. ■☆
121556 Dahlia imperialis Roezl;垂头大丽花;Bell Tree,Bell Tree Dahlia,Candelabra Dahlia ■☆
121557 Dahlia jaurezii Hort.;华氏大丽花;Cactus Dahlia ■☆
121558 Dahlia maxonii Saff.;马氏大丽花■☆
121559 Dahlia merckii Lehm.;光滑大丽花;Bedding Dahlia ■☆
121560 Dahlia nana Andr. = Dahlia pinnata Cav. ■
121561 Dahlia pinnata Cav.;大丽花(大理花,大理菊,多变大丽花,红大丽花,金黄花,萝卜花,苕菊,天竺牡丹,西番莲,洋芍药,圆叶大丽花);Aztec Dahlia,Cocoxochitl,Common Dahlia,Dahlia,Dahlia Fair,Fire Dahlia,Garden Dahlia,Old Garden Dahlia,Pinnate Dahlia,Red Dahlia ■
121562 Dahlia pinnata Cav. = Dahlia × hortensis Guillaumin ■☆
121563 Dahlia purpurea Poir. = Dahlia pinnata Cav. ■
121564 Dahlia rosea Cav. = Dahlia pinnata Cav. ■
121565 Dahlia superflua Aiton = Dahlia pinnata Cav. ■
121566 Dahlia variabilis (Willd.) Desv. = Dahlia pinnata Cav. ■
121567 Dahlia variabilis Desv. = Dahlia pinnata Cav. ■
121568 Dahliaphyllum Constance et Breedlove(1994);大丽花叶属■☆
121569 Dahlstedtia Malme(1905);达氏豆属■☆
121570 Dahuronia Scop. = Licania Aubl. ●☆
121571 Dahuronia Scop. = Moquilea Aubl. ●☆
121572 Daiotyla Dressler = Chondrorhyncha Lindl. ■☆
121573 Dais L. = Dais Royen ex L. ●☆
121574 Dais Royen ex L.(1762);夏香属(篝火花属)●☆
121575 Dais cotinifolia L.;夏香;Pompon Bush,Pompon Tree ●☆
121576 Dais glaucescens Decne. ex C. A. Mey.;灰夏香●☆
121577 Dais madagascariensis Lam. = Gnidia daphnifolia L. f. ●☆
121578 Dais pubescens Lam. = Gnidia daphnifolia L. f. ●☆
121579 Daiswa Raf.(1838);蒴果重楼属■☆
121580 Daiswa Raf. = Paris L. ■
121581 Daiswa birmanica Takht. = Paris polyphylla Sm. var. yunnanensis (Franch.) Hand.-Mazz. ■
121582 Daiswa bockiana (Diels) Takht. = Paris polyphylla Sm. var. stenophylla Franch. ■
121583 Daiswa chinensis (Franch.) Takht. = Paris polyphylla Sm. var. chinensis (Franch.) H. Hara ■
121584 Daiswa chinensis (Franch.) Takht. subsp. brachysepala (Pamp.) Takht. = Paris polyphylla Sm. var. chinensis (Franch.) H. Hara ■
121585 Daiswa cronquistii Takht. = Paris cronquistii (Takht.) H. Li et Noltie ■
121586 Daiswa delavayi (Franch.) Takht. = Paris delavayi Franch. ■
121587 Daiswa dunniana (H. Lév.) Takht. = Paris dunniana H. Lév. ■
121588 Daiswa fargesii (Franch.) Takht. = Paris fargesii Franch. ■
121589 Daiswa fargesii (Franch.) Takht. var. brevipetalata T. C. Huang et K. C. Yang = Paris fargesii Franch. ■
121590 Daiswa fargesii (Franck) Takht. var. brevipetalata T. C. Huang et K. C. Yang = Paris fargesii Franch. ■
121591 Daiswa forrestii Takht. = Paris forrestii (Takht.) H. Li ■
121592 Daiswa hainanensis (Merr.) Takht. = Paris dunniana H. Lév. ■
121593 Daiswa hainanensis (Merr.) Takht. subsp. viemamensis Takht. = Paris vietnamensis (Takht.) H. Li ■
121594 Daiswa lancifolia (Hayata) Takht. = Paris polyphylla Sm. var. stenophylla Franch. ■
121595 Daiswa polyphylla (Sm.) Raf. = Paris polyphylla Sm. ■
121596 Daiswa pubescens (Hand.-Mazz.) Takht. = Paris mairei H. Lév. ■
121597 Daiswa thibetica (Franch.) Takht. = Paris thibetica Franch. ■
121598 Daiswa violacea (H. Lév.) Takht. = Paris mairei H. Lév. ■
121599 Daiswa yunnanensis (Franch.) Takht. = Paris polyphylla Sm.

var. yunnanensis (Franch.) Hand. -Mazz. ■
121600 Daknopholis Clayton(1967);咬鳞草属■☆
121601 Daknopholis boivinii (A. Camus) Clayton;咬鳞草■☆
121602 Dalanum Dostál = Galeopsis L. ■
121603 Dalbergaria Tussac = Alloplectus Mart.(保留属名)●■☆
121604 Dalbergaria Tussac = Columnea L. ●■☆
121605 Dalbergaria Tussac(1808-1813);达尔苣苔属●☆
121606 Dalbergaria phoenicea Tussac;达尔苣苔●☆
121607 Dalbergia L. f. (1782)(保留属名);黄檀属(檀属);Brazil Rosewood,Cocobolo,Nicaragua Wood,Palisander,Rosewood ●
121608 Dalbergia abrahamii Bosser et R. Rabev.;亚伯拉罕黄檀●☆
121609 Dalbergia acutifoliolata Mendonça et E. C. Sousa;尖托叶黄檀●☆
121610 Dalbergia adamii Berhaut;亚当黄檀●☆
121611 Dalbergia afzeliana G. Don;阿芙泽尔黄檀●☆
121612 Dalbergia afzeliana G. Don var. parvifolia Cronquist;小叶阿芙泽尔黄檀●☆
121613 Dalbergia afzelii Baker = Dalbergia afzeliana G. Don ●☆
121614 Dalbergia albiflora A. Chev. ex Hutch. et Dalziel;白花黄檀●☆
121615 Dalbergia albiflora A. Chev. ex Hutch. et Dalziel subsp. echinocarpa Stepper;刺果白花黄檀●☆
121616 Dalbergia altissima Baker f.;高大黄檀●☆
121617 Dalbergia ambongoensis Baill. = Dalbergia greveana Baill. ●☆
121618 Dalbergia andapensis Bosser et R. Rabev.;安达帕黄檀●☆
121619 Dalbergia arbutifolia Baker;浆果鹃叶黄檀●☆
121620 Dalbergia arbutifolia Baker subsp. aberrans Polhill;异常黄檀●☆
121621 Dalbergia arisanensis Hayata;阿里山黄檀;Alishan Rosewood ●
121622 Dalbergia armata E. Mey.;具刺黄檀●☆
121623 Dalbergia assamica Benth.;西南黄檀(思茅黄檀,秧青,紫花黄檀);Assam Rosewood,Purple-flower Rosewood,Twiner ●
121624 Dalbergia assamica Benth. = Dalbergia sericea G. Don ●
121625 Dalbergia aurea Bosser et R. Rabev.;金黄檀●☆
121626 Dalbergia bakeri Welw. ex Baker;贝克黄檀●☆
121627 Dalbergia bakeri Welw. ex Baker = Dalbergia foenumgraecum De Wild. ●☆
121628 Dalbergia bakeri Welw. ex Baker var. acutifoliolata P. Sousa;尖小叶贝克黄檀●☆
121629 Dalbergia balansae Prain;南岭黄檀(茶丫藤,黄类树,南岭檀,绳树,水相思,秧青);Balansa Rosewood,South China Rosewood ●
121630 Dalbergia barclayi Telfair ex Hook. = Mundulea barclayi (Telfair ex Hook.) R. Vig. ex Du Puy et Labat ●☆
121631 Dalbergia bariensis Pierre ex Prain;巴里黄檀;Asia Rosewood,Asiatic Rosewood ●☆
121632 Dalbergia baronii Baker;巴龙黄檀;Baron Rosewood ●☆
121633 Dalbergia bathiei R. Vig.;巴谢黄檀●☆
121634 Dalbergia benthamii Prain;两粤黄檀(蕉藤麻,两广黄檀,两粤檀,藤春,藤黄檀);Bentham Rosewood ●
121635 Dalbergia bequaertii De Wild. = Dalbergia nitidula Baker ●☆
121636 Dalbergia boehmii Taub.;贝姆黄檀●☆
121637 Dalbergia boehmii Taub. subsp. stuhlmannii (Taub.) Polhill;斯图尔曼黄檀●☆
121638 Dalbergia boinensis Jum. = Dalbergia trichocarpa Baker ●☆
121639 Dalbergia boivinii Baill. = Dalbergia hildebrandtii Vatke ●☆
121640 Dalbergia bojeri Drake;博耶尔黄檀●☆
121641 Dalbergia brachystachya Bosser et R. Rabev.;短穗黄檀●☆
121642 Dalbergia bracteolata Baker;小苞片黄檀●☆
121643 Dalbergia brevicaudata Vatke = Craibia brevicaudata (Vatke) Dunn ●☆
121644 Dalbergia brownei (Jacq.) Urb.;布朗黄檀●☆
121645 Dalbergia burmanica Prain;缅甸黄檀(倒钩刺);Burma Blackwood,Burma Rosewood ●
121646 Dalbergia campenonii Drake;康珀农黄檀●☆
121647 Dalbergia candenatensis (Dennst.) Prain;弯枝黄檀(扭黄檀);Candenat Rosewood,Curveshoot Rosewood,Tortuous Branch Rosewood ●
121648 Dalbergia capuronii Bosser et R. Rabev.;凯普伦黄檀●☆
121649 Dalbergia cavaleriei H. Lév. = Dalbergia stenophylla Prain ●
121650 Dalbergia cearensis Ducke;西阿拉黄檀(赛阿拉州黄檀,赛州黄檀);King Wood,Kingwood,Tulip Wood,Tulipwood ●☆
121651 Dalbergia chapelieri Baill.;沙普黄檀●☆
121652 Dalbergia chlorocarpa R. Vig.;绿果黄檀●☆
121653 Dalbergia cochinchinensis Pierre ex Laness.;交趾黄檀(印支黄檀);Phayung,Rose Black Wood,Siam Rosewood,Trac ●☆
121654 Dalbergia collettii Prain = Dalbergia yunnanensis Franch. var. collettii (Prain) Thoth. ●
121655 Dalbergia commiphoroides Baker f.;没药黄檀●☆
121656 Dalbergia commiphoroides Baker f. var. micrantha Chiov. = Dalbergia commiphoroides Baker f. ●☆
121657 Dalbergia congensis Baker f.;刚果黄檀●☆
121658 Dalbergia crispa Hepper;皱波黄檀●☆
121659 Dalbergia cultrata Graham = Dalbergia cultrata Graham ex Ralph ●
121660 Dalbergia cultrata Graham ex Benth. = Dalbergia cultrata Graham ex Ralph ●
121661 Dalbergia cultrata Graham ex Ralph;刀状黑黄檀(黑黄檀,小刀形黄檀)●
121662 Dalbergia dalzielii Baker f. ex Hutch. et Dalziel;达尔齐尔黄檀●☆
121663 Dalbergia decipularis Rizzini et A. Mattos;巴西黄檀;Brazil Tulipwood,Brazilian Tulipwood,Sebastiao-de-arruda,Tulipwood ●☆
121664 Dalbergia dekindtiana Harms = Dalbergia nitidula Baker ●☆
121665 Dalbergia delavayi Franch. = Cladrastis delavayi (Franch.) Prain ●
121666 Dalbergia delphinensis Bosser et R. Rabev.;德尔芬黄檀●☆
121667 Dalbergia densicoma Baill. = Dalbergia pervillei Vatke ●☆
121668 Dalbergia dinklagei Harms = Dalbergia oblongifolia G. Don ●☆
121669 Dalbergia discolor Blume ex Miq. = Dalbergia rimosa Roxb. ●
121670 Dalbergia dyeriana Prain ex Harms;大金刚藤黄檀(大金刚藤,土降香);Dyer Rosewood ●
121671 Dalbergia ealaensis De Wild.;埃阿拉黄檀●☆
121672 Dalbergia ecastophylla (L.) Taub.;伊卡托叶黄檀●☆
121673 Dalbergia elata Harms = Dalbergia boehmii Taub. ●☆
121674 Dalbergia emirnensis Benth.;埃米黄檀●☆
121675 Dalbergia eremicola Polhill;沙生黄檀●☆
121676 Dalbergia erubescens Bosser et R. Rabev.;变红黄檀●☆
121677 Dalbergia eurybothrya Drake = Dalbergia greveana Baill. ●☆
121678 Dalbergia ferruginea Roxb.;锈色黄檀(铁锈色黄檀);Rusty Rosewood ●☆
121679 Dalbergia fischeri Taub.;菲舍尔黄檀●☆
121680 Dalbergia floribunda Craib;繁花黄檀●☆
121681 Dalbergia florifera De Wild.;灰绿黄檀●☆
121682 Dalbergia foenumgraeca De Wild. = Dalbergia bakeri Welw. ex Baker ●☆
121683 Dalbergia frutescens (Vell.) Britton;灌木黄檀(粉木,黄檀木,玫瑰黑黄檀,绒毛黄檀,紫薇檀);Brazilian Tulipwood,

Jacaranda Rosa, Pau Rosa, Pinkwood ●☆

121684 Dalbergia fusca Pierre;黑黄檀(版纳黄檀);Black Rosewood ●

121685 Dalbergia fusca Pierre = Dalbergia cultrata Graham ex Benth. ●

121686 Dalbergia fusca Pierre var. enneandra S. Q. Zou et J. H. Liu = Dalbergia fusca Pierre ●

121687 Dalbergia fusca Pierre var. enneandra S. Q. Zou et J. H. Liu = Dalbergia cultrata Graham ex Benth. ●

121688 Dalbergia gautieri Bosser et R. Rabev.;戈捷黄檀●☆

121689 Dalbergia gentilii De Wild.;让蒂黄檀●☆

121690 Dalbergia gilbertii Cronquist;吉尔伯特黄檀●☆

121691 Dalbergia glaberrima Bosser et R. Rabev.;无毛滑黄檀●☆

121692 Dalbergia glandulosa Dunkley = Dalbergia martinii F. White ●☆

121693 Dalbergia glauca Kurz = Dalbergia obtusifolia (Baker) Prain ●

121694 Dalbergia glaucescens De Wild. = Dalbergia florifera De Wild. ●☆

121695 Dalbergia glaucocarpa Bosser et R. Rabev.;灰果黄檀●☆

121696 Dalbergia gloveri Q. Luke;格洛韦尔黄檀●☆

121697 Dalbergia gossweileri Baker f.;戈斯黄檀●☆

121698 Dalbergia granadillo Pittier;中美洲黄檀;Granadillo ●☆

121699 Dalbergia grandibracteata De Wild.;大苞黄檀●☆

121700 Dalbergia grandidieri Baill. = Dalbergia bracteolata Baker ●☆

121701 Dalbergia greveana Baill.;马达加斯加黄檀;Huanghuali ●☆

121702 Dalbergia hainanensis Merr. et Chun;海南黄檀(海南檀,花梨公,花梨木,牛筋树);Hainan Rosewood, Huanghuali ●

121703 Dalbergia hancei Benth.;藤黄檀(白鸡刺藤,梣果藤,大香藤,丁香柴,丁香藤,红香藤,黄龙脱衣,鸡踼香,橿树,降香,屈叶藤,藤檀,藤香,痛必灵,香藤刺,油香藤);Hance Rosewood, Scandent Rosewood, Vine Rosewood ●

121704 Dalbergia harmsiana De Wild. = Dalbergia boehmii Taub. ●☆

121705 Dalbergia henryana Prain;云南黄檀(亨利黄檀,蒙自黄檀);Henry Rosewood ●

121706 Dalbergia hepperi Jongkind;赫佩黄檀●☆

121707 Dalbergia heudelotii Stapf;厄德黄檀●☆

121708 Dalbergia hildebrandtii Vatke;希尔德黄檀●☆

121709 Dalbergia hircina Wall. = Dalbergia sericea G. Don ●

121710 Dalbergia hircine Buch. -Ham. ex Benth. = Dalbergia sericea G. Don ●

121711 Dalbergia hirticalyx Bosser et R. Rabev.;毛萼黄檀●☆

121712 Dalbergia horrida (Dennst.) Mabb.;喀拉拉黄檀●☆

121713 Dalbergia humbertii R. Vig.;亨伯特黄檀●☆

121714 Dalbergia hupeana Hance;黄檀(白檀,白檀树,不知春,扭杆树,水檀,檀,檀木,檀树,望水檀,硬檀树);Hubei Rosewood, Hupeh Rosewood ●

121715 Dalbergia hupeana Hance var. bauhiniifolia Pamp. = Dalbergia hupeana Hance ●

121716 Dalbergia ikopensis Jum. = Dalbergia greveana Baill. ●☆

121717 Dalbergia isaloensis R. Vig. = Dalbergia greveana Baill. ●☆

121718 Dalbergia jezoensis Maxim.;北海道黄檀;Yezo Rosewood ●☆

121719 Dalbergia jingxiensis S. Y. Liu;靖西黄檀●

121720 Dalbergia kamtschatica Maxim.;勘察加黄檀;Kamtschatka Rosewood ●☆

121721 Dalbergia kingiana Prain;滇南黄檀(金氏黄檀);King Rosewood, S. Yunnan Rosewood ●

121722 Dalbergia kisantuensis De Wild. et T. Durand;基桑图黄檀●☆

121723 Dalbergia lactea Vatke;乳白黄檀●☆

121724 Dalbergia lagosana Harms = Dalbergia rufa G. Don ●☆

121725 Dalbergia lanceolaria L. f.;披针黄檀(窄叶黄檀)●☆

121726 Dalbergia lanceolaria L. f. = Dalbergia assamica Benth. ●

121727 Dalbergia lanceolaria L. f. = Dalbergia balansae Prain ●

121728 Dalbergia lanceolaria L. f. var. assamica (Benth.) Thoth. = Dalbergia assamica Benth. ●

121729 Dalbergia lastoursvillensis Pellegr. = Dalbergia ngounyensis Pellegr. ●☆

121730 Dalbergia latifolia Roxb.;阔叶黄檀(东印度玫瑰木,广叶黄檀,马拉巴尔木,孟买黑木,孟买玫瑰木,印度黄檀,印度玫瑰木);Black Rosewood, Black Wood, Blackwood, Bombay Black Wood, Bombay Blackwood, Bombay Rosewood, East India Rosewood, East Indian Rosewood, India Rosewood, Indian Black Wood, Indian Rosewood, Malabar Black Wood, Malabar Blackwood, Malabar Rosewood ●☆

121731 Dalbergia laurentii De Wild. = Aganope lucida (Welw. ex Baker) Polhill ■☆

121732 Dalbergia laxiflora Micheli;疏花黄檀●☆

121733 Dalbergia lemurica Bosser et R. Rabev.;莱穆拉黄檀●☆

121734 Dalbergia liberiae Harms et Dinkl. ex Mildbr.;利比里亚黄檀●☆

121735 Dalbergia librevillensis Pellegr.;利伯维尔黄檀●☆

121736 Dalbergia louisii Cronquist;路易斯黄檀●☆

121737 Dalbergia louveli R. Vig.;卢氏黑黄檀●☆

121738 Dalbergia luluensis Harms = Dalbergia nitidula Baker ●☆

121739 Dalbergia macrocarpa Burtt Davy = Dalbergia afzeliana G. Don ●☆

121740 Dalbergia macrosperma Welw. ex Baker;大籽黄檀●☆

121741 Dalbergia madagascariensis Vatke;马岛黄檀●☆

121742 Dalbergia malangensis E. C. Sousa;马兰加黄檀●☆

121743 Dalbergia manongarivensis Bosser et R. Rabev.;卢赫梅罗黄檀●☆

121744 Dalbergia marginata Roxb. = Derris marginata (Roxb.) Benth. ●

121745 Dalbergia maritima R. Vig.;滨海黄檀●☆

121746 Dalbergia martinii F. White;马丁黄檀●☆

121747 Dalbergia masoalensis Bosser et R. Rabev.;马苏阿拉黄檀●☆

121748 Dalbergia mayumbensis Baker f.;马永巴黄檀●☆

121749 Dalbergia medicinalis De Wild. = Dalbergia nitidula Baker ●☆

121750 Dalbergia melanoxylon Guillaumin et Perr.;非洲黄檀(东非黑黄檀,非洲黑檀,刚果木,莫桑比克黑檀,塞内加尔黑檀,乌木黄檀);Africa Black Wood, Africa Rosewood, African Blackwood, African Ebony, African Rosewood, Congowood, Indian Blackwood, Indian Rosewood, Mozambique Ebony, Poyi, Rosewood, Senegal Ebony, Sudan Ebony ●

121751 Dalbergia mengsuoensis Y. Y. Qian;勐梭黄檀●

121752 Dalbergia micheliana De Wild. = Dalbergia bakeri Welw. ex Baker ●☆

121753 Dalbergia microcarpa R. Vig. = Dalbergia peltieri Bosser et R. Rabev. ●☆

121754 Dalbergia microcarpa Taub. ex Baker f. = Dalbergia microphylla Chiov. ●☆

121755 Dalbergia microphylla Chiov.;小叶黄檀●☆

121756 Dalbergia millettii Benth.;香港黄檀(港粤黄檀,孟葛藤,油香藤);Hongkong Rosewood, S. Guangdong Rosewood ●

121757 Dalbergia millettii Benth. = Dalbergia mimosoides Franch. ●

121758 Dalbergia millettii Benth. = Dalbergia stenophylla Prain ●

121759 Dalbergia millettii Benth. var. mimosoides (Franch.) Thoth. = Dalbergia mimosoides Franch. ●

121760 Dalbergia mimosoides Franch.;含羞草叶黄檀(鸡勾札,麦刺藤,象鼻藤,小黄檀);Mimoselike Rosewood, Mimose-like Rosewood, Trunk Rosewood ●

121761 Dalbergia mollis Bosser et R. Rabev.;柔软黄檀●☆

121762 Dalbergia monosperma Dalzell = Dalbergia candenatensis (Dennst.) Prain ●

121763 Dalbergia monticola Bosser et R. Rabev.;山生黄檀●☆

121764 Dalbergia mossambicensis Harms = Dalbergia nitidula Baker ●☆

121765 Dalbergia multijuga E. Mey.;多对黄檀●☆

121766 Dalbergia myriabotrys Baker = Dalbergia greveana Baill. ●☆

121767 Dalbergia nelsii Schinz = Philenoptera nelsii (Schinz) Schrire ●☆

121768 Dalbergia neoperrieri Bosser et R. Rabev.;新佩里耶黄檀●☆

121769 Dalbergia ngounyensis Pellegr.;加纳黄檀●☆

121770 Dalbergia nigra (Vell.) Allemão ex Benth.;巴西黑黄檀(钢琴木,黑黄檀,黑檀,紫薇);Bahia Rosewood,Bahia Rose-wood,Black Rosewood,Brazilian Rosewood,Jacaranda,Palisander,Pianowood,Rio Rosewood,Rosewood ●☆

121771 Dalbergia nigra (Vell.) Benth. = Dalbergia nigra (Vell.) Allemão ex Benth. ●☆

121772 Dalbergia nigrescens Kurz;黑色黄檀;Nigrescent Rosewood ●☆

121773 Dalbergia nitidula Baker;莫桑比克黄檀●☆

121774 Dalbergia noldeae Harms;诺尔德黄檀●☆

121775 Dalbergia normandii Bosser et R. Rabev.;诺曼德黄檀●☆

121776 Dalbergia oblongifolia G. Don;矩圆叶黄檀●☆

121777 Dalbergia obovata E. Mey.;倒卵黄檀●☆

121778 Dalbergia obtusa Lecomte = Dalbergia pervillei Vatke ●☆

121779 Dalbergia obtusifolia (Baker) Prain;钝叶黄檀(牛筋木,牛肋巴,铁刀木,紫梗树);Obtuse Leaf Rosewood,Obtuseleaf Rosewood,Obtuse-leaved Rosewood ●

121780 Dalbergia obtusifolia Prain = Dalbergia obtusifolia (Baker) Prain ●

121781 Dalbergia occulta Bosser et R. Rabev.;隐蔽黄檀●☆

121782 Dalbergia odorifera T. C. Chen;降香檀(番降,花梨,花梨母,花梨木,黄檀,鸡骨,降香,降香黄檀,降真,降真香,紫降香,紫藤香);Fragrant Rosewood,Scented Rosewood ●◇

121783 Dalbergia oligophylla Baker ex Hutch. et Dalziel;寡叶黄檀●☆

121784 Dalbergia oliveri Gamble ex Prain;缅甸红黄檀(奥氏黄檀);Burma Tulipwood ●☆

121785 Dalbergia orientalis Bosser et R. Rabev.;东方黄檀●☆

121786 Dalbergia ovata Graham ex Benth. var. obtusifolia Baker = Dalbergia obtusifolia (Baker) Prain ●

121787 Dalbergia pachycarpa (De Wild. et T. Durand) Ulbr. ex De Wild.;粗果黄檀●☆

121788 Dalbergia parviflora Roxb.;小花黄檀;Smallflower Rosewood ●

121789 Dalbergia peishaensis Chun et T. C. Chen;白沙黄檀(白沙檀);Baisha Rosewood ●☆

121790 Dalbergia peltieri Bosser et R. Rabev.;盾状黄檀●☆

121791 Dalbergia perrieri Jum. = Dalbergia greveana Baill. ●☆

121792 Dalbergia pervillei Vatke;佩尔黄檀●☆

121793 Dalbergia pinnata (Lour.) Prain;斜叶黄檀(罗望叶黄檀,罗望子叶黄檀,斜叶檀,羽叶檀);Oblique-leaflet Rosewood,Pinnate Rosewood,Slantingleaf Rosewood ●

121794 Dalbergia pluriflora Baker f.;多花黄檀●☆

121795 Dalbergia polyadelpha Prain;多体蕊黄檀(老秧草,绿叶玉蒿,云南黄檀);Polyadelphous Rosewood ●

121796 Dalbergia preussii Harms = Dalbergia saxatilis Hook. f. var. preussii (Harms) Cronquist ●☆

121797 Dalbergia pseudobaronii R. Vig.;假巴龙黄檀●☆

121798 Dalbergia pseudoviguieri Bosser et R. Rabev.;假维基耶黄檀●☆

121799 Dalbergia pterocarpiflora Baker = Dalbergia chapelieri Baill. ●☆

121800 Dalbergia pubescens Hook. f. = Dalbergia rufa G. Don ●☆

121801 Dalbergia purpurascens Baill.;紫黄檀●☆

121802 Dalbergia retusa Baill. = Dalbergia pervillei Vatke ●☆

121803 Dalbergia retusa Hemsl.;毒黄檀(黄檀,尼加拉瓜红木,微凹黄檀);Cocobolo,Granadillo,Nicaragua Rosewood,Palisandro ●

121804 Dalbergia richardii Baill. = Dalbergia bracteolata Baker ●☆

121805 Dalbergia rimosa Roxb.;多裂黄檀;Cleft Rosewood,Many-cleft Rosewood,Multifid Rosewood,Rimose Rosewood ●

121806 Dalbergia riparia (Mart.) Benth.;河岸黄檀●☆

121807 Dalbergia robusta Roxb. = Derris robusta (Roxb. ex DC.) Benth. ●

121808 Dalbergia robusta Roxb. ex DC. = Derris robusta (Roxb. ex DC.) Benth. ●

121809 Dalbergia rotundifolia Sond. = Pterocarpus rotundifolius (Sond.) Druce ●☆

121810 Dalbergia rubiginosa Roxb.;蔓黄檀(褐赤色藤黄檀,藤黄檀);Indian Rosewood ●

121811 Dalbergia rufa G. Don;浅红黄檀●☆

121812 Dalbergia rufotomentosa De Wild. = Dalbergia grandibracteata De Wild. ●☆

121813 Dalbergia rugosa Hepper;皱褶黄檀●☆

121814 Dalbergia ruwenzoriensis De Wild. = Dalbergia lactea Vatke ●☆

121815 Dalbergia sacerdotum Prain;上海黄檀;Shanghai Rosewood ●

121816 Dalbergia sapinii De Wild. = Dalbergia heudelotii Stapf ●☆

121817 Dalbergia saxatilis Hook. f.;岩生黄檀●☆

121818 Dalbergia saxatilis Hook. f. var. preussii (Harms) Cronquist;普罗伊斯黄檀●☆

121819 Dalbergia scandens Roxb. = Derris scandens (Roxb.) Benth. ●

121820 Dalbergia sciadendron Chiov. = Dalbergia lactea Vatke ●☆

121821 Dalbergia sericea Bojer = Mundulea sericea (Willd.) A. Chev. ●☆

121822 Dalbergia sericea G. Don;毛叶黄檀(绢毛黄檀);Hairleaf Rosewood,Sericeous-leaf Rosewood,Silky Rosewood ●

121823 Dalbergia sessiliflora Harms = Dalbergia obovata E. Mey. ●☆

121824 Dalbergia setifera Hutch. et Dalziel;刚毛黄檀●☆

121825 Dalbergia sissoo Roxb. = Dalbergia sissoo Roxb. ex DC. ●

121826 Dalbergia sissoo Roxb. ex DC.;印度黄檀(茶檀,印度檀);Indian Rosewood, Indiana Dalbergia, Sheesham, Shisham, Shisham Wood,Sisso,Sisso Rosewood,Sissoo,Sisu ●

121827 Dalbergia spruceana Benth.;亚马孙黄檀●☆

121828 Dalbergia stenocarpa Kurz var. typica R. Vig. = Dalbergia mollis Bosser et R. Rabev. ●☆

121829 Dalbergia stenophylla Prain;狭叶黄檀(贵州黄檀,黔黄檀,油香藤);Guizhou Rosewood,Narrow-leaflet Rosewood,Narrow-leaved Rosewood,Stenophyllous Rosewood ●

121830 Dalbergia stevensonii Standl.;史氏黄檀(伯利兹黄檀);Honduras Rosewood,Stevenson Rosewood ●☆

121831 Dalbergia stipulacea Roxb.;托叶黄檀;Stipulate Rosewood,Stipule Rosewood ●

121832 Dalbergia stocksii Benth. = Dalbergia melanoxylon Guillaumin et Perr. ●

121833 Dalbergia striata Bojer = Mundulea sericea (Willd.) A. Chev. subsp. madagascariensis Du Puy et Labat ●☆

121834 Dalbergia stuhlmannii Taub. = Dalbergia boehmii Taub. subsp. stuhlmannii (Taub.) Polhill ●☆

121835 Dalbergia suaresensis Baill.;苏亚雷斯黄檀●☆

121836 Dalbergia swynnertonii Baker f. = Dalbergia nitidula Baker ●☆

121837 Dalbergia szemaoensis Prain;思茅黄檀(酸香,秧青,紫梗树);

Simao Rosewood ●
121838 Dalbergia szemaoensis Prain = Dalbergia assamica Benth. ●
121839 Dalbergia tamarindifolia Roxb. = Dalbergia pinnata (Lour.) Prain ●
121840 Dalbergia teixeirae E. C. Sousa;特谢拉黄檀●☆
121841 Dalbergia tonkinensis Prain;越南黄檀(中越黄檀);Tonkin Rosewood,Vietnam Rosewood ●
121842 Dalbergia toroensis Baker f. = Dalbergia lactea Vatke ●☆
121843 Dalbergia torta Graham = Dalbergia candenatensis (Dennst.) Prain ●
121844 Dalbergia trichocarpa Baker;毛果黄檀●☆
121845 Dalbergia tricolor Drake;三色黄檀;Threecolor Rosewood ●☆
121846 Dalbergia tsaratananensis Bosser et R. Rabev.;察拉塔纳纳黄檀●☆
121847 Dalbergia tsoi Merr. et Chun;红果黄檀(红果檀,左氏黄檀);Redfruit Rosewood,Tso Rosewood ●
121848 Dalbergia tucurensis Donn. Sm.;危地马拉黄檀●☆
121849 Dalbergia uarandensis (Chiov.) Thulin;索马里黄檀●☆
121850 Dalbergia ugandensis Baker f. = Dalbergia lactea Vatke ●☆
121851 Dalbergia urschii Bosser et R. Rabev.;乌尔施黄檀●☆
121852 Dalbergia vacciniifolia Vatke;越橘黄檀●☆
121853 Dalbergia viguieri Bosser et R. Rabev.;维基耶黄檀●☆
121854 Dalbergia villosa Benth.;长柔毛黄檀●☆
121855 Dalbergia violacea (Vogel) Malme;堇紫色黄檀(紫堇黄檀,紫色黄檀);Violet Rosewood ●☆
121856 Dalbergia volubilis Roxb.;缠绕黄檀;Twining Rosewood ●☆
121857 Dalbergia xerophila Bosser et R. Rabev.;旱生黄檀●☆
121858 Dalbergia ximengensis Y. Y. Qian;西盟黄檀●
121859 Dalbergia yunnanensis Franch.;滇黔黄檀(虹香藤,秧青);Yunnan Rosewood ●
121860 Dalbergia yunnanensis Franch. var. collettii (Prain) Thoth.;高原黄檀(郭来得黄檀,昆明黄檀,酸香树,象鼻藤);Collett Rosewood,Highland Rosewood ●
121861 Dalbergia yunnanensis Franch. var. collettii (Prain) Thoth. = Dalbergia collettii Prain ●
121862 Dalbergiaceae Mart. = Fabaceae Lindl.(保留科名)●■
121863 Dalbergiaceae Mart. = Leguminosae Juss.(保留科名)●■
121864 Dalbergiella Baker f.(1928);小黄檀属●☆
121865 Dalbergiella gossweileri Baker f.;戈斯小黄檀●☆
121866 Dalbergiella nyassae Baker f.;尼亚萨小黄檀●☆
121867 Dalbergiella welwitschii (Baker) Baker f.;小黄檀●☆
121868 Dalea Cramer = Petalostemon Michx.(保留属名)■☆
121869 Dalea Gaertn. = Microdon Choisy ●☆
121870 Dalea Juss. = Dalea L.(保留属名)●■☆
121871 Dalea L.(1758)(保留属名);戴尔豆属(针叶豆属);Dalea ●■☆
121872 Dalea L. ex Juss. = Dalea L.(保留属名)●■☆
121873 Dalea Mill.(废弃属名) = Browallia L. ■☆
121874 Dalea Mill.(废弃属名) = Dalea L.(保留属名)●■☆
121875 Dalea P. Browne = Critonia P. Browne ●☆
121876 Dalea P. Browne = Eupatorium L. ■●
121877 Dalea albiflora A. Gray;白花戴尔豆;White Dalea ■☆
121878 Dalea alopecuroides Willd. = Dalea leporina (Aiton) Bullock ■☆
121879 Dalea argyraea A. Gray;银色戴尔豆(银二色戴尔豆);Silver Dalea ■☆
121880 Dalea bicolor Humb. et Bonpl.;二色戴尔豆;Baja Dalea,Blue Dalea,Silver Dalea ■☆
121881 Dalea bicolor Humb. et Bonpl. var. argyraea (A. Gray) Barneby = Dalea argyraea A. Gray ■☆
121882 Dalea candida Michx. ex Willd.;细戴尔豆(白瓣蕊豆);Slender White Prairie-clover,White Prairie Clover,White Prairie-clover ■☆
121883 Dalea candida Michx. ex Willd. var. oligophylla (Torr.) Shinners;寡叶戴尔豆;Slender White Prairie-clover,White Prairie-clover ■☆
121884 Dalea capitata S. Watson;橘色戴尔豆;Lemon Dalea ■☆
121885 Dalea carthagenensis (Jacq.) J. F. Macbr. var. barbata (Oerst.) Barneby;髯毛戴尔豆●☆
121886 Dalea enneandra Nutt.;九囊戴尔豆;Nine-anther Prairie-clover,Nine-anthered Prairie Clover,Sail-pod Dalea ■☆
121887 Dalea formosa Torr.;羽状戴尔豆(美丽戴尔豆);Feather Dalea,Feather Peabush,Feather Plume ■☆
121888 Dalea frutescens A. Gray;灌木戴尔豆;Black Dalea,Sierra Negra ●☆
121889 Dalea gattingeri (A. Heller) Barneby;嘎氏戴尔豆;Gattinger's Prairie Clover ■☆
121890 Dalea greggii A. Gray;格雷格戴尔豆;Spreading Dalea,Trailing Indigo Bush ■☆
121891 Dalea jamesii Torr. et A. Gray;詹氏戴尔豆;James' Prairie Clover☆
121892 Dalea lanata Spreng.;绵毛戴尔豆;Wooly Dalea☆
121893 Dalea leporina (Aiton) Bullock;狐尾戴尔豆;Foxtail Dalea,Hare's Foot Dalea ■☆
121894 Dalea lutea Willd.;黄戴尔豆;Yellow Dalea ■☆
121895 Dalea multiflora (Nutt.) Shinners;多花戴尔豆;Round-headed Prairie Clover ■☆
121896 Dalea mutabilis Willd.;变色戴尔豆■☆
121897 Dalea mutisii Kunth;穆氏戴尔豆■☆
121898 Dalea neomexicana (A. Gray) Cory;新墨西哥戴尔豆;Downy Prairie Dalea☆
121899 Dalea nigra M. Martens et Galeotti;黑戴尔豆☆
121900 Dalea oligophylla (Torr.) Shinners = Dalea candida Michx. ex Willd. var. oligophylla (Torr.) Shinners ■☆
121901 Dalea pulchella Moric.;美丽戴尔豆;Indigo Bush,Santa Catalina Prairie Clover☆
121902 Dalea purpurea Vent.;紫戴尔豆(紫瓣蕊豆);Purple Prairie Clover,Purple Prairie-clover,Violet Prairie-clover ■☆
121903 Dalea purpurea Vent. f. pubescens (A. Gray) Fassett = Dalea purpurea Vent. ■☆
121904 Dalea spinosa A. Gray;多刺戴尔豆;Smoke-tree ●☆
121905 Dalea tomentosa Willd.;柔毛戴尔豆;Wooly Dalea☆
121906 Dalea versicolor Zucc.;多色戴尔豆;Indigo Bush☆
121907 Dalea versicolor Zucc. var. sessilis (A. Gray) Barneby;无柄多色戴尔豆;Wislizenus Dalea☆
121908 Dalea villosa (Nutt.) Spreng.;长毛戴尔豆;Downy Prairie-clover,Silky Prairie-clover ■☆
121909 Dalea wrightii A. Gray;赖氏戴尔豆;Wright's Prairie Clover☆
121910 Daleaceae Bercht. et J. Presl = Fabaceae Lindl.(保留科名)●■
121911 Daleaceae Bercht. et J. Presl = Leguminosae Juss.(保留科名)●■
121912 Dalechampia L.(1753);黄蓉花属(化妆木属);Dalechampia ●
121913 Dalechampia Plum. ex L. = Dalechampia L. ●
121914 Dalechampia bidentata Blume;二齿黄蓉花;Twotooth Dalechampia ●

121915 Dalechampia bidentata Blume var. yunnanensis Pax et K. Hoffm.;黄蓉花;Yunnan Dalechampia ●

121916 Dalechampia bidentata Blume var. yunnanensis Pax et K. Hoffm. = Dalechampia bidentata Blume ●

121917 Dalechampia capensis A. Spreng.;好望角黄蓉花●☆

121918 Dalechampia capensis Sond. = Dalechampia scandens L. var. natalensis (Müll. Arg.) Pax et K. Hoffm. ●☆

121919 Dalechampia chevalieri Beille;舍瓦利耶黄蓉花●☆

121920 Dalechampia cordofana Hochst. ex Webb = Dalechampia scandens L. var. cordofana (Hochst. ex Webb) Müll. Arg. ●☆

121921 Dalechampia dioscoreifolia Poepp.;薯蓣叶黄蓉花;Costa Rican Butterfly Vine ●☆

121922 Dalechampia falcata Gagnep.;镰形黄蓉花●☆

121923 Dalechampia galpinii Pax;盖尔黄蓉花●☆

121924 Dalechampia hildebrandtii Pax = Dalechampia scandens L. var. hildebrandtii (Pax) Pax et K. Hoffm. ●☆

121925 Dalechampia ipomoeifolia Benth.;番薯叶黄蓉花●☆

121926 Dalechampia katangensis J. Léonard;加丹加黄蓉花●☆

121927 Dalechampia kirkii Prain = Dalechampia capensis A. Spreng. ●☆

121928 Dalechampia madagascariensis Pax et K. Hoffm.;马岛黄蓉花●☆

121929 Dalechampia magnoliifolia Müll. Arg.;木兰叶黄蓉花●☆

121930 Dalechampia natalensis Müll. Arg. = Dalechampia scandens L. var. natalensis (Müll. Arg.) Pax et K. Hoffm. ●☆

121931 Dalechampia parvifolia Lam. = Dalechampia scandens L. var. cordofana (Hochst. ex Webb) Müll. Arg. ●☆

121932 Dalechampia pavoniifolia (Chiov.) M. G. Gilbert;孔雀叶黄蓉花●☆

121933 Dalechampia roezliana Müll. Arg.;罗氏黄蓉花(凤蝶柏,化妆木)●☆

121934 Dalechampia scandens L.;藤状黄蓉花●☆

121935 Dalechampia scandens L. var. cordofana (Hochst. ex Webb) Müll. Arg.;塞内加尔黄蓉花●☆

121936 Dalechampia scandens L. var. hildebrandtii (Pax) Pax et K. Hoffm.;希尔德黄蓉花●☆

121937 Dalechampia scandens L. var. natalensis (Müll. Arg.) Pax et K. Hoffm.;纳塔尔黄蓉花●☆

121938 Dalechampia scandens L. var. parvifolia (Lam.) Müll. Arg. = Dalechampia scandens L. var. cordofana (Hochst. ex Webb) Müll. Arg. ●☆

121939 Dalechampia senegalensis A. Juss. ex Webb = Dalechampia scandens L. var. cordofana (Hochst. ex Webb) Müll. Arg. ●☆

121940 Dalechampia trifoliata Peter ex Verdc. et Greenway;三小叶黄蓉花●☆

121941 Dalechampia trifoliata Peter ex Verdc. et Greenway var. trifida Radcl. -Sm.;三裂黄蓉花●☆

121942 Dalechampia volubilis E. Mey. ex Prain = Dalechampia scandens L. var. natalensis (Müll. Arg.) Pax et K. Hoffm. ●☆

121943 Dalechampsia Post et Kuntze = Dalechampia L. ●

121944 Dalembertia Baill. (1858);达来大戟属☆

121945 Dalembertia hahniana Baill.;墨西哥达来大戟☆

121946 Dalembertia populifolia Baill.;达来大戟☆

121947 Dalembertia triangularis Müll. Arg.;三角达来大戟☆

121948 Dalenia Korth. (1844);达伦野牡丹属●☆

121949 Dalenia pulchra Korth.;达伦野牡丹●☆

121950 Dalhousiea Graham = Dalhousiea Wall. ex Benth. ■☆

121951 Dalhousiea Graham ex Benth. = Dalhousiea Wall. ex Benth. ■☆

121952 Dalhousiea Wall. ex Benth. (1838);光明豆属■☆

121953 Dalhousiea africana S. Moore;光明豆■☆

121954 Dalia Endl. = Dulia Adans. ●

121955 Dalia Endl. = Ledum L. ●

121956 Dalia St. -Lag. = Dalea L. (保留属名)●■☆

121957 Dalibarda L. = Rubus L. ●■

121958 Dalibarda calycina (Wall. ex D. Don) Ser. = Rubus calycinus Wall. ex D. Don ●■

121959 Dalibarda calycinus Ser. = Rubus calycinus Wall. ex D. Don ●■

121960 Dalibarda fragarioides Michx. = Waldsteinia fragarioides (Michx.) Tratt. ■☆

121961 Dalibarda repens L. = Rubus repens (L.) Kuntze ●☆

121962 Dalibarda ternata Steph. = Waldsteinia ternata (Stephan) Fritsch ■

121963 Dallachya F. Muell. (1875);肖猫乳属●☆

121964 Dallachya F. Muell. = Rhamnella Miq. ●

121965 Dallachya vitiensis (Benth.) F. Muell.;肖猫乳鼠李●☆

121966 Dallachya vitiensis F. Muell. = Dallachya vitiensis (Benth.) F. Muell. ●☆

121967 Dallwatsonia B. K. Simon(1992);昆士兰水禾属■☆

121968 Dalmatocytisus Trinajstić = Argyrolobium Eckl. et Zeyh. (保留属名)●☆

121969 Dalmatocytisus Trinajstić = Chamaecytisus Vis. ●☆

121970 Dalrympelea Roxb. (1820);戴尔省沽油属●☆

121971 Dalrympelea Roxb. = Turpinia Vent. (保留属名)●

121972 Dalrympelea pomifera Roxb.;戴尔省沽油●☆

121973 Dalrympelea pomifera Roxb. = Turpinia pomifera (Roxb.) DC. ●

121974 Dalrympelea pomifera Roxb. = Turpinia ternata Nakai ●

121975 Dalucum Adans. = Melica L. ■

121976 Dalzellia Hassk. = Belosynapsis Hassk. ■

121977 Dalzellia Hassk. = Cyanotis D. Don(保留属名)■

121978 Dalzellia Wight(1852);川藻属;Terniopsis ■

121979 Dalzellia sessilis (H. C. Chao) C. Cusset et G. Cusset;川藻(石蔓);Sessile Terniopsis ■

121980 Dalzielia Turrill(1916);达尔萝藦属■☆

121981 Dalzielia lanceolata Turrill;披针形达尔萝藦■☆

121982 Dalzielia oblanceolata Turrill;倒披针形达尔萝藦■☆

121983 Damapana Adans. (废弃属名) = Smithia Aiton(保留属名)●■

121984 Damapana aeschynomenoides (Welw. ex Baker) Kuntze = Kotschya aeschynomenoides (Welw. ex Baker) Dewit et P. A. Duvign. ●☆

121985 Damapana capitulifera (Welw. ex Baker) Kuntze = Kotschya capitulifera (Welw. ex Baker) Dewit et P. A. Duvign. ■☆

121986 Damapana strigosa (Benth.) Kuntze = Kotschya strigosa (Benth.) Dewit et P. A. Duvign. ■☆

121987 Damapana strobilantha (Welw. ex Baker) Kuntze = Kotschya strobilantha (Welw. ex Baker) Dewit et P. A. Duvign. ■☆

121988 Damasoniaceae Nakai = Alismataceae Vent. (保留科名)■

121989 Damasoniaceae Nakai;星果泽泻科■

121990 Damasonium Adans. = Alisma L. ■

121991 Damasonium Adans. = Limnocharis Bonpl. ■

121992 Damasonium Adans. = Ottelia Pers. ■

121993 Damasonium Mill. (1754);星果泽泻属(星果泻属);Damasonium, Starfruit, Star-Fruit, Thrumwort ■

121994 Damasonium Schreb. = Ottelia Pers. ■

121995 Damasonium alisma Mill.;星果泽泻;Starfruit, Star-fruit, Thrumwort ■

121996 Damasonium alisma Mill. subsp. polyspermum (Coss.) Maire;

多籽星果泽泻■☆

121997 Damasonium alisma Mill. subsp. stellatum (Thuill.) Maire = Damasonium alisma Mill. ■

121998 Damasonium alismoides (L.) R. Br. = Ottelia alismoides (L.) Pers. ■

121999 Damasonium bourgaei Coss. = Damasonium alisma Mill. subsp. bourgaei (Coss.) Maire ■☆

122000 Damasonium californicum Torr.;加州星果泽泻■☆

122001 Damasonium constrictum Juz.;缢缩星果泽泻■☆

122002 Damasonium indicum Willd. = Ottelia alismoides (L.) Pers. ■

122003 Damasonium minimum Lange = Damasonium polyspermum Coss. var. minimum (Lange) Coss. ■☆

122004 Damasonium polyspermum Coss. = Damasonium alisma Mill. subsp. polyspermum (Coss.) Maire ■☆

122005 Damasonium polyspermum Coss. var. medians Maire et Weiller = Damasonium alisma Mill. subsp. polyspermum (Coss.) Maire ■☆

122006 Damasonium polyspermum Coss. var. minimum (Lange) Coss. = Damasonium alisma Mill. subsp. polyspermum (Coss.) Maire ■☆

122007 Damasonium stellatum Rich. = Damasonium alisma Mill. ■

122008 Damasonium stellatum Thuill. = Damasonium alisma Mill. ■

122009 Damasonium ulvifolia Planch. = Ottelia ulvifolia (Planch.) Walp. ■☆

122010 Damatras Rchb. = Damatris Cass. ■☆

122011 Damatrias Rchb. = Damatris Cass. ■☆

122012 Damatris Cass. = Haplocarpha Less. ■☆

122013 Damburneya Raf. = Ocotea Aubl. ●☆

122014 Dameria Endl. = Dauceria Dennst. ●■

122015 Dameria Endl. = Embelia Burm. f.(保留属名)●■

122016 Damironia Cass. = Helipterum DC. ex Lindl. ■☆

122017 Dammara Gaertn. = Protium Burm. f.(保留属名)●

122018 Dammara Lam. = Agathis Salisb.(保留属名)●

122019 Dammara Link = Agathis Salisb.(保留属名)●

122020 Dammara alba Lam. = Agathis dammara (Lamb.) Rich. et A. Rich. ●

122021 Dammara alba Rumphius ex Blume. = Agathis dammara (Lamb.) Rich. et A. Rich. ●

122022 Dammara motleyi Parl. = Nageia motleyi (Parl.) de Laub. ●☆

122023 Dammaraceae Link = Arancariaceae Strasb. + Taxodiaceae Saporta(保留科名)●

122024 Dammaropsis Warb. = Ficus L. ●

122025 Dammera K. Schum. et Lanterb. = Licuala Thunb. ●

122026 Dammera Lauterb. et K. Schum. = Licuala Thunb. ●

122027 Damnacanthus C. F. Gaertn. (1805);虎刺属(伏牛花属);Damnacanthus, Tigerthorn ●

122028 Damnacanthus angustifolius Hayata;台湾虎刺(无刺伏牛花,细叶虎刺,窄叶岛虎刺);Narrow-leaved Damnacanthus, Narrow-leaved Tetraplasia, Taiwan Damnacanthus, Taiwan Tigerthorn ●

122029 Damnacanthus angustifolius Hayata = Damnacanthus giganteus (Makino) Nakai ●

122030 Damnacanthus angustifolius Hayata var. stenophyllus (Koidz.) Masam. = Damnacanthus angustifolius Hayata ●

122031 Damnacanthus biflorus (Rehder) Masam.;双花虎刺●☆

122032 Damnacanthus esquirolii H. Lév. = Carissa spinarum L. ●

122033 Damnacanthus formosanus (Nakai) Koidz. = Damnacanthus indicus (L.) C. F. Gaertn. ●

122034 Damnacanthus giganteus (Makino) Nakai;短刺虎刺(半球莲,长叶数珠根,长叶数珠树,黄鸡脚,黄鸡郎,黄鸡胖,鸡筋参,咳七风,树连藕,树莲藕,岩石羊);Shortspine Damnacanthus, Shortspine Tigerthorn, Short-spined Damnacanthus ●

122035 Damnacanthus giganteus (Makino) Nakai = Damnacanthus macrophyllus Siebold ex Miq. f. giganteus (Makino) T. Yamaz. ●

122036 Damnacanthus guangxiensis Y. Z. Ruan;广西虎刺;Guangxi Damnacanthus, Guangxi Tigerthorn ●

122037 Damnacanthus hainanensis (H. C. Lo) H. C. Lo ex Y. Z. Ruan;海南虎刺;Hainan Damnacanthus, Hainan Tigerthorn ●

122038 Damnacanthus henryi (H. Lév.) H. S. Lo;云桂虎刺(云贵虎刺);Henry Damnacanthus, Henry Tigerthorn ●

122039 Damnacanthus henryi (H. Lév.) H. S. Lo subsp. hainanensis H. C. Lo = Damnacanthus hainanensis (H. C. Lo) H. C. Lo ex Y. Z. Ruan ●

122040 Damnacanthus indicus (L.) C. F. Gaertn.;虎刺(白风珠,朝天刺,刺虎,大形虎刺,大叶虎刺,鹅嘴花,伏牛花,高骨老虎刺,隔虎刺花,红老鼠刺,红远志,虎刿,虎牙刺,虎叶,黄鸡兰,黄脚鸡,脚不踏,老虎刺,老鼠刺,老鼠枪,两面针,猫儿嚓,猫儿刺,鸟不踏,胖儿草,千金刺,千口针,雀不踏,蛇不过,寿庭木,寿星草,顺茶风,天针,铜针木,土鸡爪黄连,细花针,细木,小黄连,小叶刺风,绣花针,印度虎刺,针上叶);India Tigerthorn, Indian Damnacanthus ●

122041 Damnacanthus indicus (L.) C. F. Gaertn. f. microphyllus Makino;小叶虎刺;Smallleaf Tigerthorn ●☆

122042 Damnacanthus indicus (L.) C. F. Gaertn. f. minutispinus (Koidz.) Sugim. = Damnacanthus minutispinus Koidz. ●☆

122043 Damnacanthus indicus (L.) C. F. Gaertn. subsp. major (Siebold et Zucc.) T. Yamaz. = Damnacanthus major Siebold et Zucc. ●

122044 Damnacanthus indicus (L.) C. F. Gaertn. subsp. major (Siebold et Zucc.) T. Yamaz. = Damnacanthus indicus (L.) C. F. Gaertn. var. major (Siebold et Zucc.) Makino ●

122045 Damnacanthus indicus (L.) C. F. Gaertn. var. formosanus Nakai = Damnacanthus indicus (L.) C. F. Gaertn. ●

122046 Damnacanthus indicus (L.) C. F. Gaertn. var. giganteus Makino;巨虎刺●☆

122047 Damnacanthus indicus (L.) C. F. Gaertn. var. giganteus Makino = Damnacanthus giganteus (Makino) Nakai ●

122048 Damnacanthus indicus (L.) C. F. Gaertn. var. intermedius Matsum.;全叶虎刺(全叶印度虎刺)●☆

122049 Damnacanthus indicus (L.) C. F. Gaertn. var. lancifolius Makino;剑叶虎刺(剑叶印度虎刺,披针叶虎刺)●☆

122050 Damnacanthus indicus (L.) C. F. Gaertn. var. lancifolius Makino f. oblongus (Koidz.) Sugim. = Damnacanthus lancifolius (Makino) Koidz. var. oblongus Koidz. ●☆

122051 Damnacanthus indicus (L.) C. F. Gaertn. var. major (Siebold et Zucc.) Makino = Damnacanthus major Siebold et Zucc. ●

122052 Damnacanthus indicus (L.) C. F. Gaertn. var. major f. macrophylls (Siebold ex Miq.) Makino = Damnacanthus macrophyllus Siebold ex Miq. ●

122053 Damnacanthus indicus (L.) C. F. Gaertn. var. microphyllus (Makino) Makino ex Nakai = Damnacanthus indicus (L.) C. F. Gaertn. f. microphyllus Makino ●☆

122054 Damnacanthus indicus (L.) C. F. Gaertn. var. oblongus Koidz.;矩圆叶虎刺●☆

122055 Damnacanthus indicus (L.) C. F. Gaertn. var. okinawensis Hatus. = Damnacanthus okinawensis Hatus. ●☆

122056 Damnacanthus indicus (L.) C. F. Gaertn. var. ovatus Koidz.;

卵叶虎刺;Ovateleaf India Tigerthorn ●☆

122057 Damnacanthus indicus (L.) C. F. Gaertn. var. ovatus Koidz. = Damnacanthus indicus (L.) C. F. Gaertn. ●

122058 Damnacanthus indicus (L.) C. F. Gaertn. var. parvispinus Koidz. = Damnacanthus minutispinus Koidz. ●☆

122059 Damnacanthus indicus C. F. Gaertn. = Damnacanthus indicus (L.) C. F. Gaertn. ●

122060 Damnacanthus labordei (H. Lév.) H. S. Lo;柳叶虎刺(柳叶三角瓣花);Willowleaf Damnacanthus, Willowleaf Prismatomeris, Willowleaf Tigerthorn, Willow-leaved Damnacanthus, Willow-leaved Prismatomeris ●

122061 Damnacanthus lancifolius (Makino) Koidz. = Damnacanthus indicus (L.) C. F. Gaertn. var. lancifolius Makino ●☆

122062 Damnacanthus lancifolius (Makino) Koidz. var. oblongus Koidz. = Damnacanthus indicus (L.) C. F. Gaertn. var. oblongus Koidz. ●☆

122063 Damnacanthus lutchuensis (Koidz.) Hatus.;琉球虎刺●☆

122064 Damnacanthus macrophyllus Siebold ex Miq.;浙皖虎刺(大叶虎刺,浙江虎刺);Bigleaf Damnacanthus, Bigleaf Tigerthorn, Zhejiang Damnacanthus, Zhejiang Tigerthorn ●

122065 Damnacanthus macrophyllus Siebold ex Miq. f. giganteus (Makino) T. Yamaz. = Damnacanthus giganteus (Makino) Nakai ●

122066 Damnacanthus macrophyllus Siebold ex Miq. var. giganteus (Makino) Koidz. = Damnacanthus giganteus (Makino) Nakai ●

122067 Damnacanthus major Siebold et Zucc.;大卵叶虎刺(串珠虎刺,大形虎刺,大叶虎刺)●

122068 Damnacanthus major Siebold et Zucc. = Damnacanthus indicus (L.) C. F. Gaertn. var. major (Siebold et Zucc.) Makino ●

122069 Damnacanthus major Siebold et Zucc. var. lancifolius (Makino) Ohwi = Damnacanthus indicus (L.) C. F. Gaertn. var. lancifolius Makino ●☆

122070 Damnacanthus major Siebold et Zucc. var. macrophyllus (Siebold ex Miq.) Maxim. = Damnacanthus macrophyllus Siebold ex Miq. ●

122071 Damnacanthus major Siebold et Zucc. var. parvifolius Koidz. = Damnacanthus indicus (L.) C. F. Gaertn. var. intermedius Matsum. ●☆

122072 Damnacanthus major Siebold et Zucc. var. parvispinus (Koidz.) Koidz. = Damnacanthus minutispinus Koidz. ●☆

122073 Damnacanthus major Siebold et Zucc. var. parvispinus Koidz. = Damnacanthus macrophyllus Siebold ex Miq. ●

122074 Damnacanthus major Siebold et Zucc. var. submitis Maxim. et Regel = Damnacanthus macrophyllus Siebold ex Miq. ●

122075 Damnacanthus minutispinis Koidz. = Damnacanthus macrophyllus Siebold ex Miq. ●

122076 Damnacanthus moniliformis Koidz. = Damnacanthus macrophyllus Siebold ex Miq. ●

122077 Damnacanthus officinarum C. C. Huang;四川虎刺(古巴戟,土巴戟);Medicinal Damnacanthus, Medicinal Tigerthorn ●

122078 Damnacanthus okinawensis Hatus.;冲绳虎刺●☆

122079 Damnacanthus shanii K. Yao et M. B. Deng = Damnacanthus macrophyllus Siebold ex Miq. ●

122080 Damnacanthus stenophyllus Masam. = Damnacanthus angustifolius Hayata ●

122081 Damnacanthus subspinosus Hand. -Mazz. = Damnacanthus giganteus (Makino) Nakai ●

122082 Damnacanthus subspinosus Hand. -Mazz. var. salicifolius M. B. Deng et K. Yao = Damnacanthus macrophyllus Siebold ex Miq. ●

122083 Damnacanthus tashiroi Hayata = Damnacanthus biflorus (Rehder) Masam. ●☆

122084 Damnacanthus tsaii Hu;西南虎刺;H. T. Tsai Damnacanthus, H. T. Tsai Tigerthorn, Tsai Damnacanthus ●

122085 Damnamenia Given = Celmisia Cass. (保留属名)■☆

122086 Damnamenia Given(1973);漆光菊属■☆

122087 Damnamenia vernicosa (Hook. f.) Given;漆光菊■☆

122088 Damnxanthodium Strother(1987);金黄菊属■☆

122089 Damnxanthodium calvum (Greenm.) Strother;金黄菊■☆

122090 Dampiera R. Br. (1810);耳冠草海桐属●■☆

122091 Dampiera altissima F. Muell. ex Benth.;高耳冠草海桐■☆

122092 Dampiera bicolor de Vriese;二色耳冠草海桐■☆

122093 Dampiera canescens de Vriese;银灰耳冠草海桐■☆

122094 Dampiera cauloptera DC.;翅茎耳冠草海桐■☆

122095 Dampiera cinerea Ewart et O. B. Davies;灰耳冠草海桐■☆

122096 Dampiera cuneata R. Br.;楔形耳冠草海桐■☆

122097 Dampiera eriantha K. Krause;毛花耳冠草海桐■☆

122098 Dampiera eriocephala de Vriese;毛头耳冠草海桐■☆

122099 Dampiera fasciculata R. Br.;簇生耳冠草海桐■☆

122100 Dampiera ferruginea R. Br.;锈色耳冠草海桐■☆

122101 Dampiera fusca Rajput et Carolin;褐耳冠草海桐■☆

122102 Dampiera glabrescens Benth.;渐光耳冠草海桐■☆

122103 Dampiera humilis E. Pritz.;矮耳冠草海桐■☆

122104 Dampiera lanceolata A. Cunn. ex DC.;披针叶耳冠草海桐■☆

122105 Dampiera leptoclada Benth.;细枝耳冠草海桐■☆

122106 Dampiera linearis R. Br.;线形耳冠草海桐■☆

122107 Dampiera oligophylla Benth.;寡叶耳冠草海桐■☆

122108 Dampiera parvifolia R. Br.;小叶耳冠草海桐■☆

122109 Dampiera purpurea R. Br.;紫耳冠草海桐■☆

122110 Dampiera rotundifolia R. Br.;圆叶耳冠草海桐■☆

122111 Dampiera stenophylla K. Krause;窄叶耳冠草海桐■☆

122112 Dampiera tenuicaulis E. Pritz.;细茎耳冠草海桐■☆

122113 Dampiera tomentosa K. Krause;毛耳冠草海桐■☆

122114 Dampiera undulata R. Br.;波缘耳冠草海桐■☆

122115 Damrongia Kerr ex Craib = Chirita Buch. -Ham. ex D. Don ●■

122116 Damrongia kerrii (Craib) Pellegr. = Petrocosmea kerii Craib ■

122117 Danaa All. = Physospermum Cusson ex Juss. ■☆

122118 Danaa Colla = Senecio L. ■●

122119 Danaa gigantea Pau = Petroselinum crispum (Mill.) Nyman ex A. W. Hill ■

122120 Danaa verticillata (Waldst. et Kit.) Janch. = Physospermum verticillatum (Waldst. et Kit.) Vis. ■☆

122121 Danae Medik. (1787);大王桂属(亚历山大月桂属);Alexandrian Laurel ●☆

122122 Danae gayae Webb et Berthel. = Semele gayae (Webb) Svent. et G. Kunkel ■☆

122123 Danae laurus Medik. = Danae racemosa Moench ●☆

122124 Danae racemosa (L.) Moench;大王桂(亚历山大月桂,总状花假叶树);Alexandria Laurel, Alexandrian Laurel, Poet's Laurel ●☆

122125 Danae racemosa Moench = Danae racemosa (L.) Moench ●☆

122126 Danaidia Link = Danae Medik. ●☆

122127 Danais Comm. ex Vent. (1799);达奈茜属●☆

122128 Danais clematidea Drake = Danais volubilis Baker ●☆

122129 Danais fragrans (Comm. ex Lam.) Pers.;脆达奈茜●☆

122130 Danais latisepala Homolle = Danais volubilis Baker ●☆

122131 Danais longiflora Homolle = Danais volubilis Baker ●☆
122132 Danais volubilis Baker;达奈茜●☆
122133 Danatophorus Blume = Harpullia Roxb. ●
122134 Danbya Salisb. = Bomarea Mirb. ■☆
122135 Dancera Raf. = Clidemia D. Don ●☆
122136 Dandya H. E. Moore(1953);丹迪百合属■☆
122137 Dandya purpusii (Brandegee) H. E. Moore;丹迪百合■☆
122138 Dangervilla Ven. = Angostura Roem. et Schult. ●☆
122139 Danguya Benoist(1930);丹古爵床属☆
122140 Danguya pulchella Benoist;丹古爵床☆
122141 Danguyodrypetes Léandri(1939);丹古木属●☆
122142 Danguyodrypetes manongarivensis Léandri;丹古木●☆
122143 Danhatchia Garay et Christenson = Yoania Maxim. ■
122144 Danielia (DC.) Lem. = Crassula L. ●■☆
122145 Danielia Lem. = Crassula L. ●■☆
122146 Danielia Lem. = Rochea DC.(保留属名)●■☆
122147 Danielia Mello ex B. Verl. = Mansoa DC. ●☆
122148 Daniella Benn. = Daniellia Benn. ●☆
122149 Daniella Mello = Mansoa DC. ●☆
122150 Daniella Willis = Daniellia Benn. ●☆
122151 Daniellia Benn.(1854);丹尼尔苏木属(丹尼苏木属,西非苏木属);Bumbo ●☆
122152 Daniellia caillei A. Chev. = Daniellia thurifera Benn. ●☆
122153 Daniellia caudata Craib = Daniellia ogea (Harms) Rolfe ●☆
122154 Daniellia fosteri Craib = Daniellia ogea (Harms) Rolfe ●☆
122155 Daniellia klainei (Pierre) De Wild.;克氏丹尼尔苏木●☆
122156 Daniellia klainei Pierre ex De Wild. = Daniellia klainei (Pierre) De Wild. ●☆
122157 Daniellia laevis ?;平滑丹尼尔苏木●☆
122158 Daniellia mortehanii De Wild.;莫特汉丹尼尔苏木●☆
122159 Daniellia oblonga Oliv.;矩圆丹尼尔苏木●☆
122160 Daniellia ogea (Harms) Rolfe;干地丹尼尔苏木;Faro,Ogea ●☆
122161 Daniellia ogea Rolfe = Daniellia ogea (Harms) Rolfe ●☆
122162 Daniellia oliveri (Rolfe) Hutch. et Dalziel;奥氏丹尼尔苏木(奥氏西非苏木)●☆
122163 Daniellia oliveri Hutch. et Dalziel = Daniellia oliveri (Rolfe) Hutch. et Dalziel ●☆
122164 Daniellia punchii Craib = Daniellia ogea (Harms) Rolfe ●☆
122165 Daniellia pynaertii De Wild.;皮那丹尼尔苏木●☆
122166 Daniellia similis Craib = Daniellia ogea (Harms) Rolfe ●☆
122167 Daniellia soyauxii (Harms) Rolfe;索氏丹尼尔苏木●☆
122168 Daniellia soyauxii (Harms) Rolfe var. pilosa J. Léonard;疏毛丹尼尔苏木●☆
122169 Daniellia thurifera Benn.;乳香丹尼尔苏木(乳香西非苏木)●☆
122170 Daniellia thurifera Benn. var. chevalieri J. Léonard = Daniellia ogea (Harms) Rolfe ●☆
122171 Dankia Gagnep.(1939);丹基茶属●☆
122172 Dankia langbianensis Gagnep.;丹基茶●☆
122173 Dansera Steenis = Dialium L. ●☆
122174 Danserella Balle = Oncocalyx Tiegh. ●☆
122175 Dansiea Byrnes(1981);昆士兰使君子属●☆
122176 Dansiea elliptica Byrnes;昆士兰使君子●☆
122177 Dansiea grandiflora Pedley;大花昆士兰使君子●☆
122178 Danthia Steud. = Dantia Boehm. ●■
122179 Danthia Steud. = Ludwigia L. ●■
122180 Danthonia DC.(1805)(保留属名);扁芒草属(邓氏草属);Danthonia,Heath-grass,Sieglingia,Wallaby Grass,Wild Oatgrass ■
122181 Danthonia Lam. et DC. = Danthonia DC.(保留属名)■
122182 Danthonia abyssinica Hochst. ex A. Rich. = Crinipes abyssinicus (Hochst. ex A. Rich.) Hochst. ■☆
122183 Danthonia airoides Nees = Pentaschistis airoides (Nees) Stapf ■☆
122184 Danthonia andongensis Rendle = Alloeochaete andongensis (Rendle) C. E. Hubb. ■☆
122185 Danthonia angulata Nees = Pentaschistis tortuosa (Trin.) Stapf ■☆
122186 Danthonia angustifolia Nees = Pentaschistis pallida (Thunb.) H. P. Linder ■☆
122187 Danthonia angustifolia Nees var. micrathera ? = Pentaschistis glandulosa (Schrad.) H. P. Linder ■☆
122188 Danthonia anomala Steud. = Bromus pectinatus Thunb. ■
122189 Danthonia anthoxanthiformis Hochst. = Pentaschistis pictigluma (Steud.) Pilg. ■☆
122190 Danthonia arundinacea (P. J. Bergius) Schweick. = Merxmuellera arundinacea (P. J. Bergius) Conert ■☆
122191 Danthonia aureocephala J. G. Anderson = Merxmuellera aureocephala (J. G. Anderson) Conert ■☆
122192 Danthonia barbata Nees = Pentaschistis barbata (Nees) H. P. Linder ■☆
122193 Danthonia brachyacme Pilg. = Merxmuellera dura (Stapf) Conert ■☆
122194 Danthonia brachyphylla Stapf = Pseudopentameris brachyphylla (Stapf) Conert ■☆
122195 Danthonia buekeana Nees = Pentaschistis pallida (Thunb.) H. P. Linder ■☆
122196 Danthonia cachemyriana Hook. f. = Danthonia schneideri Pilg. ■
122197 Danthonia cachemyriana Jaub. et Spach;喀什米尔扁芒草■
122198 Danthonia cachemyriana Jaub. et Spach var. minor Hook. f. = Danthonia schneideri Pilg. ■
122199 Danthonia cachemyriana Jaub. et Spach var. minor Hook. f. = Danthonia cumminsii Hook. f. ■
122200 Danthonia calycina Roem. et Schult.;萼状扁芒草■☆
122201 Danthonia candida Hochst. ex Steud. = Rytidosperma subulata (A. Rich.) Cope ■☆
122202 Danthonia cincta Nees = Merxmuellera cincta (Nees) Conert ■☆
122203 Danthonia circinnata Steud. = Pentaschistis setifolia (Thunb.) McClean ■☆
122204 Danthonia cirrhata Hack.;卷须扁芒草■☆
122205 Danthonia cirrhulosa Nees = Pentaschistis cirrhulosa (Nees) H. P. Linder ■☆
122206 Danthonia collinita Nees = Pentaschistis triseta (Thunb.) Stapf ■☆
122207 Danthonia compressa Austin;普通扁芒草;Flattend Oat Grass, Flattened Wild-oat-grass ■☆
122208 Danthonia cumminsii Hook. f. = Danthonia schneideri Pilg. ■
122209 Danthonia davyi C. E. Hubb. = Rytidosperma davyi (C. E. Hubb.) Cope ■☆
122210 Danthonia decumbens (L.) DC.;斜生扁芒草;Common Heathgrass,Heath Grass,Heath-grass ■☆
122211 Danthonia decumbens (L.) DC. subsp. mauritanica (Maire) Dobignard;毛里塔尼亚扁芒草■☆
122212 Danthonia densifolia Nees = Pentaschistis densifolia (Nees) Stapf ■☆
122213 Danthonia denudata Nees = Pentaschistis curvifolia (Schrad.) Stapf ■☆

122214 Danthonia depressa Hochst. = Pentaschistis pictigluma (Steud.) Pilg. ■☆

122215 Danthonia disticha Nees = Rytidosperma disticha (Nees) Cope ■☆

122216 Danthonia distichophylla Lehm. = Pentameris distichophylla (Lehm.) Nees ■☆

122217 Danthonia drakensbergensis Schweick. = Merxmuellera drakensbergensis (Schweick.) Conert ■☆

122218 Danthonia dregeana (Nees) Steud. = Chaetobromus involucratus (Schrad.) Nees subsp. dregeanus (Nees) Verboom ■☆

122219 Danthonia dura Stapf = Merxmuellera dura (Stapf) Conert ■☆

122220 Danthonia elegans Nees = Pentaschistis elegans (Nees) Stapf ■☆

122221 Danthonia elephantina Nees = Merxmuellera arundinacea (P. J. Bergius) Conert ■☆

122222 Danthonia elongata Hochst. ex A. Rich. = Helictotrichon elongatum (Hochst. ex A. Rich.) C. E. Hubb. ■☆

122223 Danthonia eriostoma Nees = Pentaschistis eriostoma (Nees) Stapf ■☆

122224 Danthonia exilis Hook. f. = Danthonia cachemyriana Jaub. et Spach ■

122225 Danthonia fascicularis (Nees) Steud. = Merxmuellera stricta (Schrad.) Conert ■☆

122226 Danthonia filiformis Nees = Pentaschistis pallida (Thunb.) H. P. Linder ■☆

122227 Danthonia forskahlei (Vahl) Trin.;佛氏扁芒草■☆

122228 Danthonia forskalii (Vahl) R. Br. = Centropodia forskalii (Vahl) Cope ■☆

122229 Danthonia forskalii (Vahl) Trin. = Asthenatherum forskalii (Vahl) Nevski ■☆

122230 Danthonia forskalii (Vahl) Trin. = Centropodia forskalii (Vahl) Cope ■☆

122231 Danthonia forskalii (Vahl) Trin. = Danthonia forskahlii (Vahl) Trin. ■☆

122232 Danthonia fragilis Guinet et Sauvage = Centropodia fragilis (Guinet et Sauvage) Cope ■☆

122233 Danthonia glandulosa Schrad. = Pentaschistis glandulosa (Schrad.) H. P. Linder ■☆

122234 Danthonia glauca Nees = Centropodia glauca (Nees) Cope ■☆

122235 Danthonia glauca Nees var. lasiophylla Pilg. = Centropodia glauca (Nees) Cope ■☆

122236 Danthonia grandiflora Hochst. ex A. Rich. = Rytidosperma grandiflorum (Hochst. ex A. Rich.) S. M. Phillips ■☆

122237 Danthonia heptamera Nees = Pentaschistis heptamera (Nees) Stapf ■☆

122238 Danthonia heteropla Steud. = Pentaschistis patula (Nees) Stapf ■☆

122239 Danthonia hirsuta Nees = Pentaschistis aurea (Steud.) McClean subsp. pilosogluma (McClean) H. P. Linder ■☆

122240 Danthonia inermis Stapf = Schismus inermis (Stapf) C. E. Hubb. ■☆

122241 Danthonia intercepta (Nees) Steud. = Chaetobromus involucratus (Schrad.) Nees subsp. dregeanus (Nees) Verboom ■☆

122242 Danthonia intermedia Vasey;间型扁芒草■☆

122243 Danthonia involucrata (Schrad.) Schrad. = Chaetobromus involucratus (Schrad.) Nees ■☆

122244 Danthonia involuta Steud. = Pentaschistis pallescens (Schrad.) Stapf ■☆

122245 Danthonia jacquemontii Bor = Danthonia schneideri Pilg. ■

122246 Danthonia jacquemontii Bor var. minor ? = Danthonia cumminsii Hook. f. ■

122247 Danthonia juncea Trin. = Pseudopentameris macrantha (Schrad.) Conert ■☆

122248 Danthonia kuhlii Steud.;库尔扁芒草■☆

122249 Danthonia lanata (Schrad.) Schrad. = Merxmuellera rufa (Nees) Conert ■☆

122250 Danthonia lanata (Schrad.) Schrad. var. major Nees = Merxmuellera rufa (Nees) Conert ■☆

122251 Danthonia leptophylla A. Rich. = Tripogon leptophyllus (A. Rich.) Cufod. ■☆

122252 Danthonia lima Nees = Pentaschistis lima (Nees) Stapf ■☆

122253 Danthonia longearistata (A. Rich.) Engl. = Streblochaete longiarista (A. Rich.) Pilg. ■☆

122254 Danthonia longiglumis Nees = Pentameris longiglumis (Nees) Stapf ■☆

122255 Danthonia lupulina (Thunb.) P. Beauv. ex Roem. et Schult. = Merxmuellera lupulina (Thunb.) Conert ■☆

122256 Danthonia macowanii Stapf = Merxmuellera macowanii (Stapf) Conert ■☆

122257 Danthonia macrantha Schrad. = Pseudopentameris macrantha (Schrad.) Conert ■☆

122258 Danthonia macrocephala Stapf = Merxmuellera rufa (Nees) Conert ■☆

122259 Danthonia micrantha Trin. = Pentaschistis pallida (Thunb.) H. P. Linder ■☆

122260 Danthonia mossamedensis Rendle = Centropodia mossamedensis (Rendle) Cope ■☆

122261 Danthonia mutica Nees = Pentaschistis setifolia (Thunb.) McClean ■☆

122262 Danthonia nana Engl. = Pentaschistis pictigluma (Steud.) Pilg. ■☆

122263 Danthonia neuroelytrum Steud. = Arundinella setosa Trin. ■

122264 Danthonia nutans Nees = Pentaschistis tortuosa (Trin.) Stapf ■☆

122265 Danthonia obtusifolia Hochst. = Pseudopentameris obtusifolia (Hochst.) N. P. Barker ■☆

122266 Danthonia pallescens Schrad. = Pentaschistis pallescens (Schrad.) Stapf ■☆

122267 Danthonia papillosa Schrad. = Pentaschistis pallida (Thunb.) H. P. Linder ■☆

122268 Danthonia papposa Nees = Merxmuellera papposa (Nees) Conert ■☆

122269 Danthonia patula Nees = Pentaschistis patula (Nees) Stapf ■☆

122270 Danthonia patula Nees var. maior ? = Pentaschistis patula (Nees) Stapf ■☆

122271 Danthonia penicillata (Labill.) P. Beauv.;圆锥扁芒草■☆

122272 Danthonia penicillata P. Beauv. = Danthonia penicillata (Labill.) P. Beauv. ■☆

122273 Danthonia porosa Nees = Pentaschistis setifolia (Thunb.) McClean ■☆

122274 Danthonia propinqua Nees = Pentaschistis patula (Nees) Stapf ■☆

122275 Danthonia pumila Nees = Dregeochloa pumila (Nees) Conert ■☆

122276 Danthonia purpurea (L. f.) P. Beauv. = Karroochloa purpurea (L. f.) Conert et Türpe ■☆

122277 Danthonia purpurea P. Beauv. = Karroochloa purpurea (L. f.)

Conert et Türpe ■☆

122278 Danthonia racemosa R. Br. = Rytidosperma racemosum (R. Br.) Connor et Edgar ■☆

122279 Danthonia rangei Pilg. = Merxmuellera rangei (Pilg.) Conert ■☆

122280 Danthonia rigida Steud. = Pseudopentameris obtusifolia (Hochst.) N. P. Barker ■☆

122281 Danthonia rufa Nees = Merxmuellera rufa (Nees) Conert ■☆

122282 Danthonia rupestris Nees = Pentaschistis rupestris (Nees) Stapf ■☆

122283 Danthonia scabra Nees = Pentaschistis papillosa (Steud.) H. P. Linder ■☆

122284 Danthonia scabra Nees var. hirsuta ? = Pentaschistis rupestris (Nees) Stapf ■☆

122285 Danthonia schismoides Stapf ex Conert = Karroochloa schismoides (Stapf ex Conert) Conert et Türpe ■☆

122286 Danthonia schneideri Pilg.; 扁芒草(邓氏草); Schneider Oatgrass ■

122287 Danthonia schneideri Pilg. = Danthonia cumminsii Hook. f. ■

122288 Danthonia schneideri Pilg. var. minor (Hook. f.) Karthikeyan = Danthonia cumminsii Hook. f. ■

122289 Danthonia segetalis Hochst. = Pentaschistis trisetoides (Hochst. ex Steud.) Pilg. ■☆

122290 Danthonia setosa Nees = Karroochloa purpurea (L. f.) Conert et Türpe ■☆

122291 Danthonia speciosa Lehm. ex Nees = Pentameris macrocalycina (Steud.) Schweick. ■☆

122292 Danthonia spicata (L.) P. Beauv. ex Roem. et Schult.; 穗状扁芒草; Common Wild-oat-grass, Curly Oat Grass, Poverty Danthonia, Poverty Grass, Poverty Oat Grass ■☆

122293 Danthonia spicata (L.) P. Beauv. ex Roem. et Schult. var. longipila Scribn. et Merr. = Danthonia spicata (L.) P. Beauv. ex Roem. et Schult. ■☆

122294 Danthonia spicata (L.) P. Beauv. ex Roem. et Schult. var. pinetorum Piper = Danthonia spicata (L.) P. Beauv. ex Roem. et Schult. ■☆

122295 Danthonia stereophylla J. G. Anderson = Merxmuellera stereophylla (J. G. Anderson) Conert ■☆

122296 Danthonia streblochaeta Steud. = Streblochaete longiarista (A. Rich.) Pilg. ■☆

122297 Danthonia stricta Schrad. = Merxmuellera stricta (Schrad.) Conert ■☆

122298 Danthonia subulata A. Rich. = Rytidosperma subulata (A. Rich.) Cope ■☆

122299 Danthonia suffrutescens Stapf = Centropodia glauca (Nees) Cope ■☆

122300 Danthonia tenella Nees = Karroochloa tenella (Nees) Conert et Türpe ■☆

122301 Danthonia tenuiglumis Steud. = Crinipes abyssinicus (Hochst. ex A. Rich.) Hochst. ■☆

122302 Danthonia thermalis Scribn. = Danthonia spicata (L.) P. Beauv. ex Roem. et Schult. ■☆

122303 Danthonia thouarsii Nees = Pentameris thuarii P. Beauv. ■☆

122304 Danthonia thuarii (P. Beauv.) Desv. = Pentameris thuarii P. Beauv. ■☆

122305 Danthonia thunbergii Kunth = Pentaschistis triseta (Thunb.) Stapf ■☆

122306 Danthonia tortuosa Trin. = Pentaschistis tortuosa (Trin.) Stapf ■☆

122307 Danthonia trichopteryx Steud. = Trichopteryx dregeana Nees ■☆

122308 Danthonia trichotoma Nees = Pentaschistis aristidoides (Thunb.) Stapf ■☆

122309 Danthonia trisetoides Hochst. ex A. Rich. = Pentaschistis trisetoides (Hochst. ex Steud.) Pilg. ■☆

122310 Danthonia uberior Hochst. = Pentaschistis pictigluma (Steud.) Pilg. ■☆

122311 Danthonia viscidula Nees = Pentaschistis viscidula (Nees) Stapf ■☆

122312 Danthonia webbiana Steud. = Pentaschistis airoides (Nees) Stapf ■☆

122313 Danthonia zeyheriana Steud.; 泽赫扁芒草■☆

122314 Danthoniastrum (Holub) Holub = Metcaltia Conert ■☆

122315 Danthonidium C. E. Hubb. (1937); 印度扁芒草属■☆

122316 Danthonidium gammiei (Bhide) C. E. Hubb.; 印度扁芒草■☆

122317 Danthoniopsis Stapf(1916); 拟扁芒草属■☆

122318 Danthoniopsis acutigluma Chippind.; 尖颖拟扁芒草■☆

122319 Danthoniopsis anomala (C. E. Hubb. et Schweick.) Clayton = Danthoniopsis ramosa (Stapf) Clayton ■☆

122320 Danthoniopsis barbata (Nees) C. E. Hubb.; 髯毛拟扁芒草■☆

122321 Danthoniopsis chevalieri A. Camus et C. E. Hubb.; 舍瓦利耶拟扁芒草■☆

122322 Danthoniopsis chimanimaniensis (J. B. Phipps) Clayton; 奇马尼马尼拟扁芒草■☆

122323 Danthoniopsis dinteri (Pilg.) C. E. Hubb.; 丁特拟扁芒草■☆

122324 Danthoniopsis gossweileri Stapf = Danthoniopsis viridis (Rendle) C. E. Hubb. ■☆

122325 Danthoniopsis intermedia C. E. Hubb. = Danthoniopsis viridis (Rendle) C. E. Hubb. ■☆

122326 Danthoniopsis lignosa C. E. Hubb.; 木质拟扁芒草■☆

122327 Danthoniopsis minor Stapf et C. E. Hubb. = Danthoniopsis viridis (Rendle) C. E. Hubb. ■☆

122328 Danthoniopsis multinodis (C. E. Hubb.) Jacq. -Fél. = Loudetiopsis tristachyoides (Trin.) Conert ■☆

122329 Danthoniopsis occidentalis Jacq. -Fél. = Loudetiopsis occidentalis (Jacq. -Fél.) Clayton ■☆

122330 Danthoniopsis parva (J. B. Phipps) Clayton; 较小拟扁芒草■☆

122331 Danthoniopsis petiolata (J. B. Phipps) Clayton; 柄叶拟扁芒草■☆

122332 Danthoniopsis pobeguinii Jacq. -Fél. = Loudetiopsis pobeguinii (Jacq. -Fél.) Clayton ■☆

122333 Danthoniopsis pruinosa C. E. Hubb.; 白粉拟扁芒草■☆

122334 Danthoniopsis pruinosa C. E. Hubb. var. gracilis = Danthoniopsis pruinosa C. E. Hubb. ■☆

122335 Danthoniopsis purpurea (C. E. Hubb.) Jacq. -Fél. = Loudetiopsis tristachyoides (Trin.) Conert ■☆

122336 Danthoniopsis ramosa (Stapf) Clayton; 分枝拟扁芒草■☆

122337 Danthoniopsis scopulorum (J. B. Phipps) J. B. Phipps; 岩栖拟扁芒草■☆

122338 Danthoniopsis simulans (C. E. Hubb.) Clayton; 相似拟扁芒草■☆

122339 Danthoniopsis stocksii (Boiss.) C. E. Hubb.; 斯托克斯扁芒草■☆

122340 Danthoniopsis tristachyoides (Trin.) Jacq. -Fél. = Loudetiopsis tristachyoides (Trin.) Conert ■☆

122341 Danthoniopsis tuberculata (Stapf) Jacq. -Fél. = Loudetiopsis tristachyoides (Trin.) Conert ■☆

122342 Danthoniopsis viridis (Rendle) C. E. Hubb.;绿花拟扁芒草■☆
122343 Danthoniopsis westii J. B. Phipps = Danthoniopsis viridis (Rendle) C. E. Hubb. ■☆
122344 Danthorhiza Ten. = Helictotrichon Besser ex Schult. et Schult. f. ■
122345 Dantia Boehm. = Isnardia L. ●■
122346 Dantia Boehm. = Ludwigia L. ●■
122347 Dantia Lippi ex Choisy = Boerhavia L. ■
122348 Danubiunculus Sailer = Limosella L. ■
122349 Danzleria Bert. ex DC. = Diospyros L. ●
122350 Dapania Korth. (1854);五星藤属●☆
122351 Dapania macrophylla Knuth;大叶五星藤●☆
122352 Dapania pentandra Capuron;五蕊五星藤●☆
122353 Dapania racemosa Korth.;五星藤●☆
122354 Dapedostachys Börner = Carex L. ■
122355 Daphmanthus F. K. Ward = ? Daphne L. ●
122356 Daphnaceae Vent. = Thymelaea Mill.(保留属名)●■
122357 Daphnandra Benth. (1870);桂雄属(桂雄香属,花桂属,瑞香楠属)●☆
122358 Daphnandra micrantha Benth.;小花桂雄(小花瑞香楠)●☆
122359 Daphne L. (1753);瑞香属(芫花属);Bay, Daphne, Garland Flower, Mezercon, Spage Laurel ●
122360 Daphne acuminata Stocks = Daphne mucronata Royle ●☆
122361 Daphne acutiloba Rehder;尖瓣瑞香(尖裂瑞香);Sharplobed Daphne, Sharp-lobed Daphne ●
122362 Daphne alba (Hand. -Mazz.) Halda = Wikstroemia trichotoma (Thunb.) Makino ●
122363 Daphne albowiana Woronow ex Pobed.;阿氏瑞香●☆
122364 Daphne alpina L.;高山瑞香●☆
122365 Daphne altaica Pall.;阿尔泰瑞香(尖瓣瑞香,蒙古瑞香);Altai Daphne ●
122366 Daphne altaica Pall. subsp. fasciculiflora (T. Z. Hsu) Halda = Daphne altaica Pall. ●
122367 Daphne altaica Pall. var. longilobata Lecomte;狭瓣瑞香(长瓣瑞香);Longlobed Daphne ●
122368 Daphne altaica Pall. var. longilobata Lecomte = Daphne longilobata (Lecomte) Turrill ●
122369 Daphne alternifolia (Batalin) Halda = Wikstroemia alternifolia Batalin ●
122370 Daphne alternifolia (Batalin) Halda var. multiflora (Lecomte) Halda = Wikstroemia alternifolia Batalin ●
122371 Daphne ambigua Matsuda = Daphne gemmata E. Pritz. ex Diels ●
122372 Daphne angustifolia K. Koch = Daphne mucronata Royle ●☆
122373 Daphne angustiloba Rehder = Daphne tenuiflora Bureau et Franch. ●
122374 Daphne anhuiensis (D. C. Zhang et X. P. Zhang) Halda = Wikstroemia anhuiensis D. C. Zhang et X. P. Zhang ●
122375 Daphne aquilaria Blanco = Wikstroemia indica (L.) C. A. Mey. ●
122376 Daphne arbuscula Celak.;匈牙利瑞香●☆
122377 Daphne arisanensis Hayata;台湾瑞香(阿里山瑞香,白瑞香);Alishan Daphne, Taiwan Daphne ●
122378 Daphne aurantiaca Diels;橙花瑞香(橙黄瑞香,黄花瑞香,万年青,云南瑞香);Golden-flowered Daphne, Orange Daphne ●
122379 Daphne aurantiaca Diels var. calcicola (W. W. Sm.) Halda = Daphne aurantiaca Diels ●
122380 Daphne axillaris (Merr. et Chun) Chun et C. F. Wei;腋花瑞香(海南瑞香,腋生瑞香);Axillary Daphne ●
122381 Daphne axillaris (Merr. et Chun) Chun et C. F. Wei subsp. rhynchocarpa (C. Y. Chang) Halda = Daphne rhynchocarpa C. Yu Chang ●
122382 Daphne axilliflora (Keissl.) Pobed.;高加索腋花瑞香●☆
122383 Daphne baimashanensis (S. C. Huang) Halda = Wikstroemia baimashanensis S. C. Huang ●
122384 Daphne baimashanensis (S. C. Huang) Halda subsp. lungtzeensis (S. C. Huang) Halda = Wikstroemia lungtzeensis S. C. Huang ●
122385 Daphne balansae (Drake) Halda = Rhamnoneuron balansae (Drake) Gilg ●
122386 Daphne bholua Buch. -Ham. ex D. Don;藏东瑞香(亮叶瑞香,毛花瑞香);Hairflower Daphne, Lucid-leaved Daphne, Nepal Paper ●
122387 Daphne bholua Buch. -Ham. ex D. Don subsp. emeiensis (C. Y. Chang) Halda = Daphne emeiensis C. Yu Chang ●
122388 Daphne bholua Buch. -Ham. ex D. Don var. glacialis (W. W. Sm. et Cave) B. L. Burtt;落叶瑞香●
122389 Daphne blagayana Freyer;巴尔干瑞香●☆
122390 Daphne bodinieri H. Lév. = Daphne tangutica Maxim. ●
122391 Daphne brevipaniculata (Rehder) Halda = Wikstroemia micrantha Hemsl. ●
122392 Daphne brevituba H. F. Zhou ex C. Yu Chang;短管瑞香;Short-tube Daphne ●
122393 Daphne burkwoodii Turrill;伯氏瑞香;Burkwood Daphne ●☆
122394 Daphne burkwoodii Turrill 'Carole Mackie';卡罗尔伯氏瑞香;Carole Mackie Daphne ●☆
122395 Daphne burkwoodii Turrill 'Somerset';筋斗伯氏瑞香●☆
122396 Daphne calcicola W. W. Sm. = Daphne aurantiaca Diels ●
122397 Daphne canescens Wall. = Wikstroemia canescens (Wall.) Meisn. ●
122398 Daphne canescens Wikstr. subsp. capitatoracemosa (S. C. Huang) Halda = Wikstroemia capitatoracemosa S. C. Huang ●
122399 Daphne cannabina Lour. = Wikstroemia indica (L.) C. A. Mey. ●
122400 Daphne cannabina Lour. var. bholua (Buch. -Ham. ex D. Don) Keissl. = Daphne bholua Buch. -Ham. ex D. Don ●
122401 Daphne cannabina Lour. var. glacialis W. W. Sm. et Cave = Daphne bholua Buch. -Ham. ex D. Don var. glacialis (W. W. Sm. et Cave) B. L. Burtt ●
122402 Daphne cannabina Lour. var. kiusiana ? = Daphne kiusiana Miq. ●
122403 Daphne cannabina Wall. = Daphne papyracea Wall. ex Steud. ●
122404 Daphne cannabina Wall. var. bholua (Buch. -Ham. ex D. Don) Keissler = Daphne bholua Buch. -Ham. ex D. Don ●
122405 Daphne cannabina Wall. var. glacialis W. W. Sm. et Cave = Daphne bholua Buch. -Ham. ex D. Don var. glacialis (W. W. Sm. et Cave) B. L. Burtt ●
122406 Daphne capitata (Rehder) Halda = Wikstroemia capitata Rehder ●
122407 Daphne caucasica Pall.;高加索瑞香;Caucasian Daphne ●☆
122408 Daphne cavaleriei H. Lév. = Daphne papyracea Wall. ex Steud. ●
122409 Daphne chamaedaphne (Bunge) Halda = Wikstroemia chamaedaphne Meisn ●
122410 Daphne championii Benth.;长柱瑞香(白地菊,蒲银仔,小叶瑞香,野黄皮,一叶一枝花);Champion Daphne ●
122411 Daphne chinensis Spreng. = Daphne odora Thunb. ●
122412 Daphne chingshuishaniana S. S. Ying;清水山瑞香(高山瑞香)●

122413 Daphne chui (Merr.) Halda = Wikstroemia chuii Merr. ●
122414 Daphne circassica Woronow ex Pobed.;切尔卡西亚瑞香●☆
122415 Daphne clivicola Hand.-Mazz. = Daphne rosmarinifolia Rehder ●
122416 Daphne cneorum L.;松瑞香(南欧瑞香,欧洲瑞香);Garland Daphne,Garland Flower,Garland Flower Daphne,Mountain Widow-wail,Rock Daphne,Rose Daphne,Spurge Flax ●☆
122417 Daphne cneorum L. 'Eximia';卓越欧洲瑞香●☆
122418 Daphne collina Sm.;地中海瑞香●☆
122419 Daphne composita (L. f.) Gilg = Eriosolena composita (L. f.) Tiegh. ●
122420 Daphne delavayi (Lecomte) Halda = Wikstroemia delavayi Lecomte ●
122421 Daphne depauperata H. F. Zhou ex C. Yu Chang;少花瑞香;Fewflower Daphne ●
122422 Daphne dolichantha (Diels) Halda = Wikstroemia dolichantha Diels ●
122423 Daphne dolichantha (Diels) Halda var. effusa (Rehder) Halda = Wikstroemia dolichantha Diels ●
122424 Daphne emeiensis C. Yu Chang;峨眉瑞香;Emei Daphne ●
122425 Daphne erosiloba C. Yu Chang;啮蚀瓣瑞香;Premorse Daphne ●
122426 Daphne esquirolii H. Lév.;穗花瑞香(白脉瑞香,滇东瑞香);Esquirol Daphne ●
122427 Daphne esquirolii H. Lév. subsp. pedunculata (H. F. Zhou ex C. Y. Chang) Halda = Daphne penduculata H. F. Zhou ex C. Yu Chang ●
122428 Daphne fargesii (Domke) Halda = Wikstroemia fargesii (Lecomte) Domke ●
122429 Daphne fasciculiflora T. Z. Hsu;簇花瑞香;Fascicleflower Daphne ●
122430 Daphne fasciculiflora T. Z. Hsu = Daphne altaica Pall. ●
122431 Daphne feddei H. Lév.;滇瑞香(矮陀陀,川滇瑞香,短瓣瑞香,费氏瑞香,构皮,构皮岩托,构皮岩陀,桂花矮陀陀,桂花岩托,桂花岩陀,黄皮杜仲,黄山皮桃,黄山皮条,鸡蛋花,金腰带,梦花皮,山皮条,铜牛皮,万年青,万年青矮陀陀,小瑞香,小鼠皮,岩陀,野瑞香,月月绿);Shortpetal Daphne,Short-petaled Daphne ●
122432 Daphne feddei H. Lév. var. taliensis H. F. Zhou ex C. Yu Chang;大理瑞香;Dali Daphne ●
122433 Daphne flaviflora H. Winkl. = Daphne penicillata Rehder ●
122434 Daphne fortunei Benth. = Daphne championii Benth. ●
122435 Daphne fortunei Lindl. = Daphne genkwa Siebold et Zucc. ●
122436 Daphne gardneri Wall. = Edgeworthia gardneri (Wall.) Meisn. ●
122437 Daphne gemmata E. Pritz. ex Diels;川西瑞香(川西荛花);Budded Daphne,Gemmate Daphne,W. Sichuan Stringbush ●
122438 Daphne genkwa Siebold et Zucc.;芫花(败花,赤芫,大救驾,大米花,地棉花,丁香花,毒鱼,杜杭,杜芫,儿草,浮胀草,黄大戟,黄阳花,金腰带,九龙花,癞头花,老鼠花,闷头花,棉花条,南芫花,闹鱼花,泥秋树,泡米花,去水,石棉皮,蜀桑,铁牛皮,头疼花,头痛花,头痛皮,莞花,小叶金腰带,芫,芫花条,药鱼草,野丁香花,银腰带,鱼毒,杬,紫芫花);Lilac Daphne ●
122439 Daphne genkwa Siebold et Zucc. f. taitoensis Hamaya;台东瑞香●☆
122440 Daphne genkwa Siebold et Zucc. f. taitoensis Hamaya = Daphne genkwa Siebold et Zucc. ●
122441 Daphne genkwa Siebold et Zucc. subsp. jinzhaiensis (D. C. Zhang et J. Z. Shao) Halda = Daphne jinzhaiensis D. C. Zheng et J. Z. Shao ●
122442 Daphne genkwa Siebold et Zucc. subsp. leishanensis (H. F. Zhou ex C. Yu Chang) Halda = Daphne leishanensis H. F. Zhou ex C. Yu Chang ●◇
122443 Daphne genkwa Siebold et Zucc. var. fortunei (Lindl.) Franch. = Daphne genkwa Siebold et Zucc. ●
122444 Daphne giraldii Nitsche;黄瑞香(大救驾,黄狗皮,纪氏瑞香,金腰带,祖师麻,祖司麻);Girald Daphne ●
122445 Daphne glabra (W. C. Cheng) Halda = Wikstroemia glabra W. C. Cheng ●
122446 Daphne glabra (W. C. Cheng) Halda f. purpurea (W. C. Cheng) Halda = Wikstroemia glabra W. C. Cheng ●
122447 Daphne glomerata Lam.;团集瑞香●☆
122448 Daphne gnidium L.;白花欧瑞香(少叶瑞香);Garou Bush,Mediterranean Mezereon,Spurge Flax ●☆
122449 Daphne gnidium L. subsp. mauritanica (Nieto Fel.) Halda;毛里塔尼亚瑞香●☆
122450 Daphne gnidium L. var. lanata Faure et Maire = Daphne gnidium L. subsp. mauritanica (Nieto Fel.) Halda ●☆
122451 Daphne gnidium L. var. sericea Faure et Maire = Daphne gnidium L. subsp. mauritanica (Nieto Fel.) Halda ●☆
122452 Daphne gracilis E. Pritz.;小娃娃皮;Slender Daphne ●
122453 Daphne grayana Halda = Wikstroemia retusa A. Gray ●
122454 Daphne grueningana H. Winkl.;倒卵叶瑞香(天目瑞香);Obovateleaf Daphne ●◇
122455 Daphne hainanensis (Merr.) Halda = Wikstroemia hainanensis Merr. ●
122456 Daphne haoi (Domke) Halda = Wikstroemia haoii Domke ●
122457 Daphne hemsleyi Halda = Wikstroemia angustifolia Hemsl. ●
122458 Daphne holoserica (Diels) Hamaya;丝毛瑞香(绢毛瑞香);Woolly-silke Daphne,Woolly-silky Daphne ●
122459 Daphne holoserica (Diels) Hamaya var. thibetensis (Lecomte) Hamaya = Pentathymelaea thibetensis Lecomte ●
122460 Daphne holoserica (Diels) Hamaya var. wangeana Hamaya;少绢毛瑞香(少丝毛瑞香);Wang Daphne ●
122461 Daphne holosericea (Diels) Hamaya var. wangiana Hamaya = Daphne wangiana (Hamaya) Halda ●
122462 Daphne houtteana Lindl. et Paxton;紫叶瑞香●☆
122463 Daphne huidongensis (C. Y. Chang) Halda = Wikstroemia huidongensis C. Yu Chang ●
122464 Daphne hybrida Lindl. = Daphne odora Thunb. ●
122465 Daphne indica L. = Wikstroemia indica (L.) C. A. Mey. ●
122466 Daphne involucrata Wall. = Eriosolena composita (L. f.) Tiegh. ●
122467 Daphne japonica Siebold et Zucc. = Daphne pseudomezereum A. Gray ●
122468 Daphne japonica Thunb. = Daphne odora Thunb. ●
122469 Daphne japonica Thunb. = Daphne pseudomezereum A. Gray ●
122470 Daphne jasminea Sibth. et Sm.;小叶瑞香(希腊瑞香)●☆
122471 Daphne jezoensis Maxim.;北海道瑞香;Yezo Laurel ●☆
122472 Daphne jinyunensis C. Yu Chang;缙云瑞香;Jinyun Daphne ●
122473 Daphne jinyunensis C. Yu Chang var. pilostila C. Yu Chang;毛柱瑞香(金腰带,朦花);Pilostyle Laurel ●
122474 Daphne jinzhaiensis D. C. Zheng et J. Z. Shao;金寨瑞香;Jinzhai Daphne ●
122475 Daphne julia Koso-pol.;七月瑞香(朱丽瑞香)●☆
122476 Daphne kabylica Chabert = Daphne laureola L. ●☆
122477 Daphne kamtschalica Maxim.;勘察加瑞香;Kamtschatka Daphne ●☆

122478 Daphne kamtschatica Maxim. subsp. jezoensis (Maxim.) Vorosch. = Daphne jezoensis Maxim. ●☆

122479 Daphne kamtschatica Maxim. var. jezoensis (Maxim.) Ohwi = Daphne jezoensis Maxim. ●☆

122480 Daphne kamtschatica Maxim. var. rebunensis ? = Daphne jezoensis Maxim. ●☆

122481 Daphne karsana Nakai;长白瑞香●

122482 Daphne kiusiana Miq.;九州瑞香●

122483 Daphne kiusiana Miq. f. fasciata (T. Ito) H. Hara = Daphne kiusiana Miq. ●

122484 Daphne kiusiana Miq. var. atracaulis (Rehder) F. Maek.;紫茎瑞香(白花瑞香,大黄构,大金腰带,夺香花,金腰带,毛瑞香,蒙花,瑞香,野萝花,野梦花,贼腰带,紫枝瑞香);Black-caudexed Daphne,Taiwan Whiter Daphne ●

122485 Daphne kiusiana Miq. var. odora (Thunb.) Makino = Daphne odora Thunb. ●

122486 Daphne koreana Nakai;朝鲜瑞香(长白瑞香,辣根草);Korean Daphne ●

122487 Daphne kulingensis (Domke) Halda = Wikstroemia pilosa W. C. Cheng ●

122488 Daphne laciniata Lecomte;翼柄瑞香;Wingstipe Laurel ●

122489 Daphne laciniata Lecomte var. duclouxii Lecomte = Daphne tangutica Maxim. ●

122490 Daphne lamatsoensis (Hamaya) Halda = Wikstroemia lamatsoensis Hamaya ●

122491 Daphne lanceolata (Merr.) Halda = Wikstroemia lanceolata Merr. ●

122492 Daphne laureola L.;桂叶芫花(绿花欧瑞香,袂泽瑞香,欧瑞香,欧亚瑞香,紫欧瑞香);Copse Laurel,Fox Poison,Githcorn,Laurel,Laurel-wood,Laureole,Lauriel,Laury,Lawrell,Lorel,Loury,Lowry,Spurge Laurel,Spurge Laurel Daphne,Spurge Olive,Spurgelaurel,Spurge-laurel,Spurgeolive,Sturdy Lowries,Wood Laurel ●☆

122493 Daphne laureola L. subsp. latifolia (Coss.) Rivas Mart. = Daphne laureola L. ●☆

122494 Daphne laureola L. var. hosmariensis Ball = Daphne laureola L. ●☆

122495 Daphne laureola L. var. latifolia Coss. = Daphne laureola L. ●☆

122496 Daphne leishanensis H. F. Zhou ex C. Yu Chang;雷山瑞香;Leishan Daphne ●◇

122497 Daphne leptophylla (W. W. Sm.) Halda = Wikstroemia leptophylla W. W. Sm. ●

122498 Daphne leptophylla (W. W. Sm.) Halda var. atroviolacea (Hand.-Mazz.) Halda = Wikstroemia leptophylla W. W. Sm. var. atroviclacea Hand.-Mazz. ●

122499 Daphne leuconeura Rehder = Daphne esquirolii H. Lév. ●

122500 Daphne leuconeura Rehder var. mairei (Lecomte) Rehder et H. Lév. = Daphne esquirolii H. Lév. ●

122501 Daphne leuconeura Rehder var. mairei H. Lév. et Rehder = Daphne esquirolii H. Lév. ●

122502 Daphne liangii (Merr. et Chun) Halda = Wikstroemia liangii Merr. et Chun ●

122503 Daphne lichiangensis (W. W. Sm.) Halda = Wikstroemia lichiangensis W. W. Sm. ●

122504 Daphne ligustrina (Rehder) Halda = Wikstroemia ligustrina Rehder ●

122505 Daphne limprichtii H. Winkl.;铁牛皮;Limpricht Daphne ●

122506 Daphne linoides (Hemsl.) Halda = Wikstroemia linoides Hemsl. ex Forbes et Hemsl. ●

122507 Daphne longilobata (Lecomte) Turrill;长瓣瑞香(山地瑞香);Longpetal Daphne,Long-petaled Daphne ●

122508 Daphne longilobata (Lecomte) Turrill subsp. purpuras cens (S. C. Huang) Halda = Daphne purpurascens S. C. Huang ●

122509 Daphne longituba C. Yu Chang;长管瑞香;Longtube Daphne ●

122510 Daphne macrantha Ludlow;大花瑞香;Bigflower Daphne ●

122511 Daphne mairei H. Lév. = Daphne papyracea Wall. ex Steud. ●

122512 Daphne martinii H. Lév. = Daphne feddei H. Lév. ●

122513 Daphne mauritanica Nieto Fel. = Daphne gnidium L. subsp. mauritanica (Nieto Fel.) Halda ●☆

122514 Daphne mazelii Carrière = Daphne odora Thunb. ●

122515 Daphne mezereum L.;欧亚芫花(二月瑞香,密枝瑞香,欧瑞香,欧亚瑞香,紫欧瑞香);Daffany,Daffodil,Daphne,Daphne Cneorum,Dutch Mezerion,Dwarf Laurel,February Daphne,German Spurge Olive,Lady Laurel,Mazalium,Mazeerie,Mazell,Mezereon,Mezereon Spurge,Mezereon Tree,Mezereum,Mezereon Spurge,Mountain Pepper,Mysterious Plant,Paradise Plant,Red-berry Laurel,Spurge Flax,Spurge Laurel,Spurge Oil,Spurge Olive,Spurgeolive,Widow Wail,Wild Pepper ●

122516 Daphne micrantha (Hemsl.) Halda = Wikstroemia micrantha Hemsl. ●

122517 Daphne micrantha (Hemsl.) Halda subsp. longipaniculata (S. C. Huang) Halda = Wikstroemia longipaniculata S. C. Huang ●

122518 Daphne micrantha (Hemsl.) Halda subsp. paniculata (H. L. Li) Halda = Wikstroemia micrantha Hemsl. ●

122519 Daphne miyabeana Makino;宫部瑞香;Miyabe Daphne ●☆

122520 Daphne modesta Rehder;瘦叶瑞香(瘦叶荛花);Moderate Daphne,Narrow-leaf Daphne,Weakleaf Stringbush ●

122521 Daphne monnula (Hance) Halda = Wikstroemia monnula Hance ●

122522 Daphne monnula (Hance) Halda var. vaccinium (H. Lév.) Halda = Wikstroemia vaccinium (H. Lév.) Rehder ●

122523 Daphne monnula (Hance) Halda var. xiuningensis (D. C. Zhang et J. Z. Shao) Halda = Wikstroemia monnula Hance var. xiuningensis D. C. Zhang et J. Z. Shao ●

122524 Daphne mononectaria (Hayata) Halda = Wikstroemia mononectraria Hayata ●

122525 Daphne morrisonensis C. E. Chang;玉山瑞香;Yushan Daphne ●

122526 Daphne mucronata Royle;狭叶瑞香(渐尖瑞香);Acuminate Daphne ●☆

122527 Daphne myrtilloides Nitsche;乌饭瑞香●

122528 Daphne nana Tagawa;小芫花●

122529 Daphne neapolitana Lodd.;密叶瑞香●☆

122530 Daphne nutans (Champ. ex Benth.) Halda = Wikstroemia nutans Champ. ex Benth. ●

122531 Daphne nutans (Champ. ex Benth.) Halda var. brevior (Hand.-Mazz.) Halda = Wikstroemia nutans Champ. ex Benth. var. brevior Hand.-Mazz. ●

122532 Daphne odora Thunb.;瑞香(白瑞香,斗雪开,对雪丹,对雪开,夺皮香,夺香花,风流树,红总管,佳客,金腰带,锦被推,露甲,暖骨风,蓬莱花,蓬莱紫,千里香,瑞兰,山梦花,山棉皮,麝囊,沈丁花,沈丁香,世英,睡香,铁牛皮,雪地开花,雪冬花,雪冻花,雪花,野梦花,紫丁香,紫风流);Frangrant Daphne,Sweet Daphne,Variegated Winter Daphne,Whiter Daphne,Winter Daphne ●

122533 Daphne odora Thunb. 'Aureomarginata' = Daphne odora

Thunb. ' Variegata' ●
122534 Daphne odora Thunb. ' Marginata';金边瑞香;Variegata Sweet Daphne ●
122535 Daphne odora Thunb. ' Rubra';红花瑞香●
122536 Daphne odora Thunb. ' Variegata';花叶瑞香(金边瑞香)●
122537 Daphne odora Thunb. = Daphne kiusiana Miq. var. atracaulis (Rehder) F. Maek. ●
122538 Daphne odora Thunb. f. alba (Hemsl.) H. Hara;白花瑞香●☆
122539 Daphne odora Thunb. f. alba (Hemsl.) H. Hara = Daphne odora Thunb. ●
122540 Daphne odora Thunb. f. leucantha (Makino) Yong J. Li;朝鲜白花瑞香●☆
122541 Daphne odora Thunb. f. marginata Makino = Daphne odora Thunb. ●
122542 Daphne odora Thunb. f. marginata Makino = Daphne odora Thunb. ' Marginata' ●
122543 Daphne odora Thunb. f. rosacea (Makino) H. Hara;粉花瑞香(红花瑞香)●☆
122544 Daphne odora Thunb. f. rosacea (Makino) H. Hara = Daphne odora Thunb. ●
122545 Daphne odora Thunb. var. alba Hemsl. = Daphne odora Thunb. ●
122546 Daphne odora Thunb. var. atrocaulis Rehder = Daphne kiusiana Miq. var. atracaulis (Rehder) F. Maek. ●
122547 Daphne odora Thunb. var. crassiuscula Rehder;紫枝瑞香;Darkpurple Whiter Daphne ●
122548 Daphne odora Thunb. var. fasciata ? = Daphne kiusiana Miq. ●
122549 Daphne odora Thunb. var. kiusiana Keissl.;棕枝瑞香(铜牛皮)●
122550 Daphne odora Thunb. var. kiusiana Keissl. = Daphne kiusiana Miq. ●
122551 Daphne odora Thunb. var. leucantha Makino = Daphne odora Thunb. ●
122552 Daphne odora Thunb. var. leucantha Makino = Daphne odora Thunb. f. leucantha (Makino) Yong J. Li ●☆
122553 Daphne odora Thunb. var. marginata Miq. = Daphne odora Thunb. ●
122554 Daphne odora Thunb. var. marginata Thunb. = Daphne odora Thunb. f. marginata Makino ●
122555 Daphne odora Thunb. var. mazelii (Carrière) Hemsl. = Daphne odora Thunb. ●
122556 Daphne odora Thunb. var. rosacea Makino = Daphne odora Thunb. ●
122557 Daphne odora Thunb. var. taiwaniana Masam. = Daphne kiusiana Miq. var. atracaulis (Rehder) F. Maek. ●
122558 Daphne odora Thunb. var. variegata Bean = Daphne odora Thunb. ●
122559 Daphne odorata Lam. = Daphne cneorum L. ●☆
122560 Daphne oleoides Schreb.;多枝瑞香(齐墩果瑞香)●☆
122561 Daphne oleoides Schreb. var. atlantica Maire = Daphne oleoides Schreb. ●☆
122562 Daphne pachyphylla D. Fang;厚叶瑞香;Thickleaf Daphne ●
122563 Daphne pachyrachis (S. L. Tsai) Halda = Wikstroemia pachyrachis S. L. Tsai ●
122564 Daphne pampaninii (Rehder) Halda = Wikstroemia pampaninii Rehder ●
122565 Daphne pampaninii (Rehder) Halda subsp. cochlearifolia (S. C. Huang) Halda = Wikstroemia cochlearifolia S. C. Huang ●
122566 Daphne pampaninii (Rehder) Halda subsp. subcyclolepidota (L. P. Liu et Y. S. Lian) Halda = Wikstroemia subcyclolepidota L. P. Liu et Y. S. Lian ●
122567 Daphne papyracea Wall. ex Steud.;白瑞香(白芍,软皮树,小构皮,雪花构,纸用瑞香);Papery Daphne,White Daphne ●
122568 Daphne papyracea Wall. ex Steud. f. grandiflora Meisn. ex Diels = Daphne papyracea Wall. ex Steud. var. grandiflora (Meisn. ex Diels) C. Yu Chang ●
122569 Daphne papyracea Wall. ex Steud. subsp. jinyunensis (C. Yu Chang) Halda = Daphne jinyunensis C. Yu Chang ●
122570 Daphne papyracea Wall. ex Steud. subsp. yunnanensis (H. F. Zhou ex C. Yu Chang) H. F. Zhou ex Halda = Daphne yunnanensis H. F. Zhou ex C. Yu Chang ●
122571 Daphne papyracea Wall. ex Steud. var. crassiuscula Rehder;山辣子皮(麻树皮,毛花雪花构,小狗皮);Hairy-flower Papery Daphne ●
122572 Daphne papyracea Wall. ex Steud. var. duclouxii Lecomte;短柄白瑞香(云南雪花构);Shortstalk Daphne,Yunnan Papery Daphne ●
122573 Daphne papyracea Wall. ex Steud. var. grandiflora (Meisn. ex Diels) C. Yu Chang;大花白瑞香;Bigflower Papery Daphne ●
122574 Daphne papyracea Wall. ex Steud. var. longituba (C. Yu Chang) Halda = Daphne longituba C. Yu Chang ●
122575 Daphne papyracea Wall. ex Steud. var. ptilostyla (C. Yu Chang) Halda = Daphne jinyunensis C. Yu Chang var. pilostila C. Yu Chang ●
122576 Daphne papyracea Wall. ex Steud. var. xichouensis (H. F. Zhou ex C. Yu Chang) Halda = Daphne xichouensis H. F. Zhou ex C. Yu Chang ●
122577 Daphne papyrifera Buch. -Ham. ex D. Don = Daphne papyracea Wall. ex Steud. ●
122578 Daphne parviflora Halda = Wikstroemia sinoparviflora Yin Z. Wang et M. G. Gilbert ●
122579 Daphne paxiana (H. Winkl.) Halda = Wikstroemia paxiana H. Winkl. ●
122580 Daphne penduculata H. F. Zhou ex C. Yu Chang;长梗瑞香;Penduculate Daphne ●
122581 Daphne pendula Sm. = Eriosolena composita (L. f.) Tiegh. ●
122582 Daphne penicillata Rehder;岷江瑞香;Penicillate Daphne ●
122583 Daphne petraea Leyb.;意大利瑞香●☆
122584 Daphne petraea Leyb. ' Grandiflora';大花意大利瑞香●☆
122585 Daphne pontica L.;黑海瑞香(健壮瑞香)●☆
122586 Daphne pseudomezereum A. Gray;东北瑞香(假欧瑞香,拟欧瑞香)●
122587 Daphne pseudomezereum A. Gray f. atropurpurea Hiyama;深紫东北瑞香●☆
122588 Daphne pseudomezereum A. Gray subsp. jezoensis (Maxim.) Hamaya = Daphne jezoensis Maxim. ●☆
122589 Daphne pseudomezereum A. Gray var. koreana (Nakai) Hamaya = Daphne koreana Nakai ●
122590 Daphne purpurascens S. C. Huang;紫花瑞香;Purpleflower Daphne ●
122591 Daphne reginaldi-farreri Halda = Wikstroemia reginaldi-farreri (Halda) Yin Z. Wang et M. G. Gilbert ●
122592 Daphne rehderi Halda = Wikstroemia gracilis Hemsl. ●
122593 Daphne rehderi Halda subsp. techinensis (S. C. Huang) Halda = Wikstroemia techinensis S. C. Huang ●
122594 Daphne retusa Hemsl.;凹叶瑞香;Retuseleaf Daphne,Retuse-leaved Daphne ●

122595 Daphne rhynchocarpa C. Yu Chang;喙果瑞香;Rostratefruit Daphne ●
122596 Daphne rosmarinifolia Rehder;华瑞香;Rosmaryleaf Daphne, Rosmary-leaved Daphne ●
122597 Daphne rotundifolia L. f. = Wikstroemia indica (L.) C. A. Mey. ●
122598 Daphne salicina H. Lév. = Wikstroemia salicina H. Lév. et Vaniot ex Rehder ●
122599 Daphne scytophylla (Diels) Halda = Wikstroemia scytophylla Diels ●
122600 Daphne scytophylla (Diels) Halda subsp. mekongensis (W. W. Sm.) Halda = Wikstroemia delavayi Lecomte ●
122601 Daphne sericea Vahl;毛瑞香●☆
122602 Daphne shillong Banerji = Daphne sureil W. W. Sm. et Cave ●
122603 Daphne sinensis Lam. = Daphne odora Thunb. ●
122604 Daphne sophia Kolenicz.;索非亚瑞香;Sophia Daphne ●☆
122605 Daphne speciosissima Carrière = Daphne odora Thunb. ●
122606 Daphne squarrosa L. = Gnidia squarrosa (L.) Druce ●☆
122607 Daphne stenantha (Hemsl.) Halda = Wikstroemia monnula Hance ●
122608 Daphne stenophylla (E. Pritz. ex Diels) Halda = Wikstroemia stenophylla E. Pritz. ex Diels ●
122609 Daphne stenophylla (E. Pritz. ex Diels) Halda var. ziyangensis (C. Y. Yu) Halda = Wikstroemia stenophylla E. Pritz. ex Diels ●
122610 Daphne sureil W. W. Sm. et Cave;头序瑞香●
122611 Daphne szetschuanica H. Winkl. = Daphne tangutica Maxim. ●
122612 Daphne taiwanensis (C. E. Chang) Halda = Wikstroemia taiwanensis S. C. Chang ●
122613 Daphne taiwaniana (Masam.) Masam. = Daphne kiusiana Miq. var. atracaulis (Rehder) F. Maek. ●
122614 Daphne tangutica Maxim.;唐古特瑞香(甘肃瑞香,陕甘瑞香,陕西瑞香,四川瑞香);Tangut Daphne ●
122615 Daphne tangutica Maxim. var. retusa (Hemsl.) H. F. Chow = Daphne tangutica Maxim. ●
122616 Daphne tangutica Maxim. var. wilsonii (Rehder) H. F. Zhou ex C. Yu Chang;野梦花●
122617 Daphne tangutica Maxim. var. wilsonii (Rehder) H. F. Zhou ex C. Yu Chang = Daphne wilsonii Rehder ●
122618 Daphne taylorii Halda;西藏瑞香●
122619 Daphne tenuiflora Bureau et Franch.;细花瑞香(窄瓣瑞香);Thin-flower Daphne, Thin-flowered Daphne ●
122620 Daphne tenuiflora Bureau et Franch. var. legendrei (Lecomte) Hamaya;毛细花瑞香●
122621 Daphne tinifolia Sw.;截叶瑞香;Bonnace Tree, Burn-nose Tree ●☆
122622 Daphne transcaucasica Pobed.;外高加索瑞香●☆
122623 Daphne trichotoma (Thunb.) Halda = Wikstroemia trichotoma (Thunb.) Makino ●
122624 Daphne triflora Lour. = Daphne odora Thunb. ●
122625 Daphne tripartita H. F. Zhou ex C. Yu Chang;九龙瑞香;Jiulong Daphne ●
122626 Daphne vaccinium (H. Lév.) Halda = Wikstroemia vaccinium (H. Lév.) Rehder ●
122627 Daphne vaillantii Danguy = Daphne tangutica Maxim. ●
122628 Daphne vavaleriei H. Lév. = Daphne papyracea Wall. ex Steud. ●
122629 Daphne viridiflora Wall. = Wikstroemia indica (L.) C. A. Mey. ●
122630 Daphne wallichii Meisn. = Eriosolena composita (L. f.) Tiegh. ●
122631 Daphne wangiana (Hamaya) Halda;少丝瑞香●
122632 Daphne wilsonii Rehder;野萝花;E. H. Wilson Daphne ●
122633 Daphne wilsonii Rehder = Daphne tangutica Maxim. var. wilsonii (Rehder) H. F. Zhou ex C. Yu Chang ●
122634 Daphne xichouensis H. F. Zhou ex C. Yu Chang;西畴瑞香;Xichou Daphne ●
122635 Daphne yunnanensis H. F. Zhou ex C. Yu Chang;云南瑞香;Yunnan Daphne ●
122636 Daphne zhouana Halda = Wikstroemia linearifolia H. F. Zhou ex C. Yu Chang ●
122637 Daphnephyllum Hassk. = Daphniphyllum Blume ●
122638 Daphnicon Pohl = Tontelea Miers(保留属名)●☆
122639 Daphnidium Nees = Lindera Thunb.(保留属名)●
122640 Daphnidium bifarium Nees = Lindera nacusua (D. Don) Merr. ●
122641 Daphnidium caudatum Nees = Iteadaphne candata (Nees) H. W. Li ●
122642 Daphnidium caudatum Nees = Lindera caudata (Nees) Hook. f. ●
122643 Daphnidium cubeba Nees = Litsea cubeba (Lour.) Pers. ●
122644 Daphnidium elongatum Nees = Litsea elongata (Wall. ex Nees) Benth. et Hook. f. ●
122645 Daphnidium lancifolium Siebold et Zucc. = Litsea coreana H. Lév. ●
122646 Daphnidium pulcherrimum (Wall.) Nees = Lindera pulcherrima (Wall.) Benth. ex Hook. f. ●
122647 Daphnidium strychnifolium Siebold et Zucc. = Lindera aggregata (Sims) Kosterm. ●
122648 Daphnidium strychnifolium Siebold et Zucc. var. hemsleyanum (Diels) Nakai = Lindera pulcherrima (Wall.) Benth. ex Hook. f. var. hemsleiana (Diels) H. P. Tsui ●
122649 Daphnidium strynifolium Siebold et Zucc. = Lindera aggregata (Sims) Kosterm. ●
122650 Daphnidostaphylis Klotzsch = Arctostaphylos Adans.(保留属名)●☆
122651 Daphniluma Baill. = Pouteria Aubl. ●
122652 Daphnimorpha Nakai = Wikstroemia Endl.(保留属名)●
122653 Daphnimorpha Nakai(1937);肖瑞香属●☆
122654 Daphnimorpha capitellata (H. Hara) Nakai;小头肖瑞香●☆
122655 Daphnimorpha kudoi (Makino) Nakai;肖瑞香●☆
122656 Daphniphyllaceae Müll. Arg.(1869)(保留科名);虎皮楠科(交让木科);Daphniphyllum Family ●
122657 Daphniphyllopsis Kurz = Nyssa L. ●
122658 Daphniphyllopsis Kurz(1876);拟虎皮楠属●☆
122659 Daphniphyllopsis capitata Kurz;拟虎皮楠●☆
122660 Daphniphyllopsis capitata Kurz = Nyssa javanica (Blume) Wangerin ●
122661 Daphniphyllum Blume(1827);虎皮楠属(交让木属);Daphniphyllum, Tigernanmu ●
122662 Daphniphyllum africanum Müll. Arg. = Plagiostyles africana (Müll. Arg.) Prain ●☆
122663 Daphniphyllum amamiense Hurus. = Daphniphyllum teijsmannii Zoll. ex Kurz ●☆
122664 Daphniphyllum angustifolium Hutch.;狭叶虎皮楠;Narrowleaf Daphniphyllum, Narrowleaf Tigernanmu, Narrow-leaved Daphniphyllum ●
122665 Daphniphyllum atrobadium Croizat et F. P. Metcalf = Daphniphyllum paxianum K. Rosenthal ●

122666 Daphniphyllum beddomei Craib = Daphniphyllum paxianum K. Rosenthal ●

122667 Daphniphyllum bengalense Rosenthal = Daphniphyllum himalayense (Benth.) Müll. Arg. ●

122668 Daphniphyllum buergeri Müll. Arg. = Daphniphyllum teijsmannii Zoll. ex Kurz ●☆

122669 Daphniphyllum calycinum Benth.;牛耳枫(老虎耳,南岭虎皮楠,牛耳铃,土鸭胆子,羊屎子,猪肚果,猪颔木,猪仔木);Calyx-shaped Daphniphyllum,Calyx-shaped Tigernanmu ●

122670 Daphniphyllum candelabrum Croizat et F. P. Metcalf = Daphniphyllum calycinum Benth. ●

122671 Daphniphyllum candelabrum Croizat et F. P. Metcalf = Daphniphyllum paxianum K. Rosenthal ●

122672 Daphniphyllum cavaleriei H. Lév. = Nyssa sinensis Oliv. ●

122673 Daphniphyllum chartaceum K. Rosenthal;纸叶虎皮楠;Paper-leaf Daphniphyllum,Paper-leaf Tigernanmu,Papery Daphniphyllum ●

122674 Daphniphyllum chartaceum K. Rosenthal = Daphniphyllum himalayense (Benth.) Müll. Arg. ●

122675 Daphniphyllum crispifolium H. Keng = Daphniphyllum oldhamii (Hemsl.) K. Rosenthal ●

122676 Daphniphyllum divaricatum (C. C. Huang) J. X. Wang;叉柱虎皮楠;Divaricate Daphniphyllum,Divaricate Tigernanmu ●

122677 Daphniphyllum formosanum H. Keng = Daphniphyllum oblongum S. S. Chien subsp. salicifolium (S. S. Chien) J. X. Wang ●

122678 Daphniphyllum formosanum K. Rosenthal. = Daphniphyllum oldhamii (Hemsl.) K. Rosenthal ●

122679 Daphniphyllum glaucescens Blume;粉绿虎皮楠(虎皮楠,青黄刚树);Common Daphniphyllum,Common Tigernanmu ●

122680 Daphniphyllum glaucescens Blume = Daphniphyllum macropodum Miq. ●

122681 Daphniphyllum glaucescens Blume = Daphniphyllum oldhamii (Hemsl.) K. Rosenthal ●

122682 Daphniphyllum glaucescens Blume subsp. luzonense (Elmer) C. C. Huang;兰屿虎皮楠(菲律宾虎皮楠)■

122683 Daphniphyllum glaucescens Blume subsp. oldhamii (Hemsl.) C. C. Huang;奥氏虎皮楠(耿氏虎皮楠)■

122684 Daphniphyllum glaucescens Blume subsp. oldhamii (Hemsl.) T. C. Huang = Daphniphyllum oldhamii (Hemsl.) K. Rosenthal ●

122685 Daphniphyllum glaucescens Blume subsp. paxianum (K. Rosenthal) C. C. Huang = Daphniphyllum paxianum K. Rosenthal ●

122686 Daphniphyllum glaucescens Blume subsp. teijsmannii (Zoll. ex Kurz) T. C. Huang = Daphniphyllum teijsmannii Zoll. ex Kurz ●☆

122687 Daphniphyllum glaucescens Blume subsp. teijsmannii (Zoll. ex Kurz) T. C. Huang var. amamiense (Hurus.) T. C. Huang = Daphniphyllum teijsmannii Zoll. ex Kurz ●☆

122688 Daphniphyllum glaucescens Blume subsp. teijsmannii (Zoll. ex Kurz) T. C. Huang var. buergeri (Müll. Arg.) T. C. Huang = Daphniphyllum teijsmannii Zoll. ex Kurz ●☆

122689 Daphniphyllum glaucescens Blume subsp. teijsmannii (Zoll. ex Kurz) T. C. Huang var. iriomotense (Hurus.) T. C. Huang = Daphniphyllum teijsmannii Zoll. ex Kurz ●☆

122690 Daphniphyllum glaucescens Blume var. oldhamii Hemsl. = Daphniphyllum glaucescens Blume subsp. oldhamii (Hemsl.) C. C. Huang ■

122691 Daphniphyllum himalayense (Benth.) Müll. Arg.;喜马拉雅虎皮楠(西藏虎皮楠);Himalayan Daphniphyllum,Xizang Tigernanmu ●

122692 Daphniphyllum himalayense (Benth.) Müll. Arg. subsp. macropodum (Miq.) T. C. Huang = Daphniphyllum macropodum Miq. ●

122693 Daphniphyllum himalayense (Benth.) Müll. Arg. var. triangulatum C. C. Huang;角药虎皮楠;Triangulate Daphniphyllum,Triangulate Tigernanmu ●

122694 Daphniphyllum humile Maxim. = Daphniphyllum humile Maxim. ex Franch. et Sav. ●☆

122695 Daphniphyllum humile Maxim. ex Franch. et Sav.;矮交让木●☆

122696 Daphniphyllum kengii Hurus. = Daphniphyllum oblongum S. S. Chien subsp. salicifolium (S. S. Chien) J. X. Wang ●

122697 Daphniphyllum kengii Hurus. = Daphniphyllum oldhamii (Hemsl.) K. Rosenthal ●

122698 Daphniphyllum longeracemosum K. Rosenthal;长序虎皮楠(江西虎皮楠);Longiracemed Daphniphyllum,Longraceme Daphniphyllum,Longraceme Tigernanmu ●

122699 Daphniphyllum longistylum S. S. Chien = Daphniphyllum oldhamii (Hemsl.) K. Rosenthal var. longistylum (S. S. Chien) J. X. Wang ●

122700 Daphniphyllum macropodum Miq.;交让木(薄叶虎皮楠,薄叶交让木,长柄交让木,大柄虎皮楠,豆腐头,枸色子,枸血子,虎皮楠,画眉珠,山黄树,水红朴,五爪龙,猪仔木);Macropodous Daphniphyllum,Macropodous Tigernanmu,Sloumi ●

122701 Daphniphyllum macropodum Miq. f. variegatum (Bean) Rehder;斑点大柄虎皮楠●☆

122702 Daphniphyllum macropodum Miq. f. viridipes (Nakai) Ohwi;绿柄交让木●☆

122703 Daphniphyllum macropodum Miq. subsp. humile (Maxim. ex Franch. et Sav.) Hurus. = Daphniphyllum humile Maxim. ex Franch. et Sav. ●☆

122704 Daphniphyllum macropodum Miq. var. humile (Maxim. ex Franch. et Sav.) K. Rosenthal = Daphniphyllum macropodum Miq. subsp. humile (Maxim. ex Franch. et Sav.) Hurus. ●☆

122705 Daphniphyllum majus Müll. Arg.;大叶虎皮楠;Bigleaf Daphniphyllum,Big-leaved Daphniphyllum,Tigernanmu ●

122706 Daphniphyllum marchandii (H. Lév.) Croizat et F. P. Metcalf;两广虎皮楠;Marchand Daphniphyllum ●

122707 Daphniphyllum marchandii (H. Lév.) Croizat et F. P. Metcalf = Daphniphyllum macropodum Miq. ●

122708 Daphniphyllum membranaceum Hayata;台湾虎皮楠(膜叶虎皮楠);Taiwan Daphniphyllum ●

122709 Daphniphyllum membranaceum Hayata = Daphniphyllum macropodum Miq. ●

122710 Daphniphyllum oblongum S. S. Chien;长圆叶虎皮楠(矩叶虎皮楠,南宁虎皮楠,四川虎皮楠);Oblong Daphniphyllum,Oblong-leaf Daphniphyllum,Oblong-leaf Tigernanmu ●

122711 Daphniphyllum oblongum S. S. Chien = Daphniphyllum oldhamii (Hemsl.) K. Rosenthal ●

122712 Daphniphyllum oblongum S. S. Chien subsp. salicifolium (S. S. Chien) J. X. Wang;小叶虎皮楠(耿氏虎皮楠,兰屿虎皮楠,柳叶虎皮楠);Willow-leaf Daphniphyllum,Willow-leaf Tigernanmu ●

122713 Daphniphyllum oldhamii (Hemsl.) K. Rosenthal;虎皮楠(奥氏虎皮楠,俄氏虎皮楠,南方虎皮楠,南宁虎皮楠,四川虎皮楠,五蕊虎皮楠);Five Stamens Daphniphyllum,Oldham Daphniphyllum,Oldham Tigernanmu ●

122714 Daphniphyllum oldhamii (Hemsl.) K. Rosenthal subsp. salicifolium (S. S. Chien) J. X. Wang = Daphniphyllum oblongum S. S. Chien subsp. salicifolium (S. S. Chien) J. X. Wang ●

122715 Daphniphyllum oldhamii (Hemsl.) K. Rosenthal var. longistylum (S. S. Chien) J. X. Wang;长柱虎皮楠(广西虎皮楠);Longistyled Daphniphyllum, Longstyle Daphniphyllum, Long-style Oldham Daphniphyllum, Long-style Tigernanmu ●

122716 Daphniphyllum oldhamii (Hemsl.) K. Rosenthal var. oblongo-lanceolatum J. X. Wang;披针叶虎皮楠;Oblanceleaf Daphniphyllum, Oblanceleaf Tigernanmu ●

122717 Daphniphyllum paxianum K. Rosenthal;显脉虎皮楠(峨眉虎皮楠,海南虎皮楠,脉叶虎皮楠,土杧果,中叶羊屎);Hainan Daphniphyllum, Pax Daphniphyllum, Pax Tigernanmu ●

122718 Daphniphyllum pentandrum Hayata = Daphniphyllum oldhamii (Hemsl.) K. Rosenthal ●

122719 Daphniphyllum pentandrum Hayata var. okinawaense (Hurus.) Hurus. = Daphniphyllum teijsmannii Zoll. ex Kurz ●☆

122720 Daphniphyllum pentandrum Hayata var. oldhamii (Hemsl.) Hurus. = Daphniphyllum oldhamii (Hemsl.) K. Rosenthal ●

122721 Daphniphyllum pentandrum Hayata var. reticulatum (H. Keng) Hurus.;网脉虎皮楠;Reticulate Nerve Daphniphyllum ●

122722 Daphniphyllum reticulatum H. Keng = Daphniphyllum pentandrum Hayata var. reticulatum (H. Keng) Hurus. ●

122723 Daphniphyllum roxburghii Baill. = Daphniphyllum oldhamii (Hemsl.) K. Rosenthal ●

122724 Daphniphyllum roxburghii Baill. = Daphniphyllum teijsmannii Zoll. ex Kurz ●☆

122725 Daphniphyllum salicifolium S. S. Chien = Daphniphyllum oldhamii (Hemsl.) K. Rosenthal ●

122726 Daphniphyllum subverticillatum Merr.;假轮叶虎皮楠;False-verticillate Daphniphyllum, Subverticillate Daphniphyllum, Subverticillate Tigernanmu ●

122727 Daphniphyllum teijsmannii Zoll. ex Kurz;约翰逊虎皮楠●☆

122728 Daphniphyllum teijsmannii Zoll. ex Kurz var. amamiense (Hurus.) Hurus. = Daphniphyllum teijsmannii Zoll. ex Kurz ●☆

122729 Daphniphyllum teijsmannii Zoll. ex Kurz var. hisautii Hurus.;古沢虎皮楠●☆

122730 Daphniphyllum yunnanense C. C. Huang ex J. X. Wang;云南虎皮楠(大叶虎皮楠);Yunnan Daphniphyllum, Yunnan Tigernanmu ●

122731 Daphnitis Spreng. = Botryceras Willd. ●☆

122732 Daphnitis Spreng. = Laurophyllus Thunb. ●☆

122733 Daphnobryon Meisn. = Drapetes Banks ex Lam. ●☆

122734 Daphnophyllopsis Post et Kuntze = Daphniphyllopsis Kurz ●☆

122735 Daphnophyllopsis Post et Kuntze = Nyssa L. ●

122736 Daphnophyllum Post et Kuntze = Daphniphyllum Blume ●

122737 Daphnopsis Mart. (1824);拟瑞香属●☆

122738 Daphnopsis americana (Mill.) J. R. Johnst.;美洲拟瑞香●☆

122739 Daphnopsis microphylla (Kunth) Gilg;大叶拟瑞香●☆

122740 Daphonanthe Schrad. ex Nees = Calyptrocarya Nees ■☆

122741 Dapsilanthus B. G. Briggs et L. A. S. Johnson(1998);薄果草属■

122742 Dapsilanthus disjunctus (Mast.) B. G. Briggs et L. A. S. Johnson;薄果草■

122743 Darbya A. Gray = Buckleya Torr.(保留属名)●

122744 Darcya B. L. Turner et C. C. Cowan(1993);达西婆婆纳属■☆

122745 Darcya Hunz. = Darcyanthus Hunz. ■

122746 Darcya Hunz. = Physalis L. ■

122747 Darcya reliquiarum (D'Arcy) B. L. Turner et C. Cowan;达西婆婆纳■☆

122748 Darcyanthus Hunz. = Physalis L. ■

122749 Dardanis Raf. = Ferula L. ■

122750 Dardanis Raf. = Peucedanum L. ■

122751 Dargeria Decne. = Leptorhabdos Schrenk ■

122752 Dargeria Decne. ex Jacq. = Leptorhabdos Schrenk ■

122753 Darion Raf. = Kundmannia Scop. ■☆

122754 Darlingia F. Muell. (1866);达林木属●☆

122755 Darlingia ferruginea J. F. Bailey;锈色达林木●☆

122756 Darlingia spectatissima F. Muell.;达林木●☆

122757 Darlingtonia DC.(废弃属名) = Darlingtonia Torr.(保留属名) ■☆

122758 Darlingtonia Torr. (1851)(保留属名);眼镜蛇草属(加州瓶子草属);California Pitcher, California Pitcher Plant, California Pitcherplant, Pitcher Plants ■☆

122759 Darlingtonia californica Torr.;眼镜蛇草;California Pitcher Plant, Cobra Lily, Cobra Plant ■☆

122760 Darllngtonia Torr. = Styrax L. ●

122761 Darluca Raf. (1820) = Bouvardia Salisb. ●■☆

122762 Darluca Raf. (1838) = ? Evolvulus L. ●■

122763 Darmera Voss = Peltiphyllum Engl. ■☆

122764 Darmera Voss(1899);雨伞草属;Indian-rhubarb ■☆

122765 Darmera peltata (Benth.) Post et Kuntze = Darmera peltata (Torr. ex Benth.) Voss ■☆

122766 Darmera peltata (Torr. ex Benth.) Voss;雨伞草;Indian Rhubarb, Indian-rhubarb, Umbrella Plant ■☆

122767 Darniella Maire et Weiller = Salsola L. ●■

122768 Darniella cyrenaica Maire et Weiller = Salsola cyrenaica (Maire et Weiller) Brullo ■☆

122769 Darniella longifolia (Forssk.) Brullo = Salsola longifolia Forssk. ■☆

122770 Darniella schweinfurthii (Solms) Brullo = Salsola schweinfurthii Solms ■☆

122771 Darniella tunetana (Brullo) Brullo = Salsola tunetana Brullo ■☆

122772 Darniella zygophylla (Batt.) Trab. = Salsola zygophylla Batt. ■☆

122773 Dartus Lour. = Maesa Forssk. ●

122774 Dartus perlarius Lour. = Maesa perlaria (Lour.) Merr. ●

122775 Darwinia Dennst. = Darwinia Rudge ●☆

122776 Darwinia Dennst. = Litsea Lam.(保留属名)●

122777 Darwinia Raf. = Monoplectra Raf. ●■

122778 Darwinia Raf. = Sesbania Scop.(保留属名)●■

122779 Darwinia Rudge(1816);达尔文木属(达尔文属);Umbrella Plant ●☆

122780 Darwinia citriodora (Endl.) Benth.;柠檬达尔文木(橙香达氏木);Lemon-scented Darwinia ●☆

122781 Darwinia fascicularis Rudge;簇生达尔文木●☆

122782 Darwinia hookeriana Benth.;胡克达尔文木●☆

122783 Darwinia macrostegia (Turcz.) Benth.;长圆叶达尔文木;Mondurup Bell ●☆

122784 Darwinia oxylepis (Turcz.) N. G. Marchant et Keighery;垂花达尔文木●☆

122785 Darwinia taxifolia A. Cunn.;曲叶达尔文木●☆

122786 Darwinia uncinata (Schauer) F. Muell.;钩形达尔文木(西澳蜡花,玉梅);Geraldton Wax, Geraldton Wax Flower, Geraldton Waxflower ●☆

122787 Darwinia uncinata (Schauer) F. Muell. = Chamelaucium uncinatum Schauer ●☆

122788 Darwiniana Lindl. = Darwinia Dennst. ●

122789 Darwiniana Lindl. = Litsea Lam.(保留属名)●

122790 Darwiniella Braas et Lückel = Darwiniera Braas et Lückel ■☆

122791 Darwiniella Braas et Lückel = Stellilabium Schltr. ■☆
122792 Darwiniella Braas et Lückel = Trichoceros Kunth ■☆
122793 Darwiniella Speg. (1888);小达尔文兰属■☆
122794 Darwiniella antarctica Speg.;南极小达尔文兰■☆
122795 Darwiniella bergoldii (Garay et Dunst.) Braas et Lückel;小达尔文兰■☆
122796 Darwiniera Braas et Lückel = Stellilabium Schltr. ■☆
122797 Darwiniera Braas et Lückel(1982);达尔文兰属■☆
122798 Darwiniera bergoldii (Garay et Dunst.) Braas et Lückel;达尔文兰■☆
122799 Darwiniothamnus Harling(1962);达尔文菊属●☆
122800 Darwiniothamnus alternifolius Lawesson et Adsersen;达尔文菊●☆
122801 Darwiniothamnus lancifolius (Hook. f.) Harling;披针叶达尔文菊●☆
122802 Darwiniothamnus tenuifolius (Hook. f.) Harling;细叶达尔文菊●☆
122803 Darwynia Rchb. = Darwinia Rudge ●☆
122804 Dasanthera Raf. = Gerardia Benth. ■☆
122805 Dasanthera Raf. = Penstemon Schmidel ●■
122806 Dasianthera C. Presl = Scolopia Schreb. (保留属名)●
122807 Dasicephala Raf. = Cordia L. (保留属名)●
122808 Dasillipe Dubard = Madhuca Buch. -Ham. ex J. F. Gmel. ●
122809 Dasillipe pasquieri Dubard = Madhuca pasquieri (Dubard) H. J. Lam. ●◇
122810 Dasiogyne Raf. = Prosopis L. ●
122811 Dasiola Raf. = Festuca L. ■
122812 Dasiola Raf. = Vulpia C. C. Gmel. ■
122813 Dasiorima Raf. = Solidago L. ■
122814 Dasiphora Raf. = Potentilla L. ■●
122815 Dasiphora davurica (Nestl.) Kom. = Potentilla davurica Nestl. ●
122816 Dasiphora davurica (Nestl.) Kom. = Potentilla glabra Lodd. ●
122817 Dasiphora dryanthoides Juz. = Pentaphylloides dryanthoides (Juz.) Soják ●
122818 Dasiphora floribunda (Pursh) Raf. = Potentilla floribunda Pursh ●☆
122819 Dasiphora fruticosa (L.) Rydb. = Potentilla fruticosa L. ●
122820 Dasiphora fruticosa (L.) Rydb. var. veitchii (E. H. Wilson) Nakai = Potentilla glabra Lodd. var. veitchii (E. H. Wilson) Hand. -Mazz. ●
122821 Dasiphora mandshurica (Maxim.) Juz. = Potentilla fruticosa L. var. leucantha Makino ●
122822 Dasiphora mandshurica (Maxim.) Juz. = Potentilla glabra Lodd. var. mandshurica (Maxim.) Hand. -Mazz. ●
122823 Dasiphora parvifolia (Fisch. ex Lehm.) Juz. = Potentilla parvifolia Fisch. ●
122824 Dasiphora phyllocalyx Juz. = Pentaphylloides phyllocalyx (Juz.) Soják ●
122825 Dasiphora riparia Raf. = Potentilla fruticosa L. ●
122826 Dasiphora tenuifolia (Willd. ex Schltdl.) Y. C. Zhu = Potentilla tenuifolia Willd. ex Schltdl. ●☆
122827 Dasispermum Neck. ex Raf. (1840);毛籽芹属■☆
122828 Dasispermum Raf. = Dasispermum Neck. ex Raf. ■☆
122829 Dasispermum suffruticosum (P. J. Bergius) B. L. Burtt;毛籽芹■☆
122830 Dasistema Raf. = Dasistoma Raf. ■●☆
122831 Dasistemon Raf. = Aureolaria Raf. ■☆
122832 Dasistepha Raf. = Gentiana L. ■
122833 Dasistoma Raf. (1819);毛口列当属(毛口玄参属)■●☆
122834 Dasistoma aureum Raf.;毛口列当■●☆
122835 Dasistoma laevigatum (Raf.) Britton = Agalinis laevigata (Raf.) S. F. Blake ■☆
122836 Dasistoma macrophylla (Nutt.) Raf. = Afzelia macrophylla (Nutt.) Kuntze ●☆
122837 Dasistoma pedicularia (L.) Benth. = Aureolaria pedicularia (L.) Raf. ■☆
122838 Dasoclema J. Sinclair(1955);狭瓣玉盘属●☆
122839 Dasoclema marginalis (J. Scheff.) J. Sinclair;狭瓣玉盘●☆
122840 Dasouratea Tiegh. = Ouratea Aubl. (保留属名)●
122841 Dassovia Neck. = Asclepias L. ■
122842 Dastylepis Raf. = Cuscuta L. ■
122843 Dasurus Salisb. = Chamaelirium Willd. ■☆
122844 Dasurus Salisb. = Ophiostachys Delile ■☆
122845 Dasus Lour. (废弃属名) = Lasianthus Jack(保留属名)●
122846 Dasus verticillatus Lour. = Lasianthus verticillatus (Lour.) Merr. ●
122847 Dasyandantha H. Rob. (1993);毛瓣落苞菊属●☆
122848 Dasyandantha cuatrecasasiana (Aristeg.) H. Rob.;毛瓣落苞菊■☆
122849 Dasyanthera Rchb. = Dasianthera C. Presl ●
122850 Dasyanthera Rchb. = Scolopia Schreb. (保留属名)●
122851 Dasyanthes D. Don = Erica L. ●☆
122852 Dasyanthina H. Rob. (1993);毛瓣斑鸠菊属■☆
122853 Dasyanthina palustris (Gardner) H. Rob.;毛瓣斑鸠菊■☆
122854 Dasyanthina serrata (Less.) H. Rob.;齿叶毛瓣斑鸠菊■☆
122855 Dasyanthus Bubani = Gnaphalium L. ■
122856 Dasyanthus Bubani(1899);毛花鼠麴草属■☆
122857 Dasyanthus conglobatus Bubani;毛花鼠麴草(毛花菊)■☆
122858 Dasyaulus Thwaites = Madhuca Buch. -Ham. ex J. F. Gmel. ●
122859 Dasyaulus Thwaites = Payena A. DC. ●☆
122860 Dasycalyx F. Muell. (1859);毛萼苋属■☆
122861 Dasycalyx F. Muell. = Trichinium R. Br. ■●☆
122862 Dasycalyx zeyheri (Moq.) F. Muell.;毛萼苋■☆
122863 Dasycarpus Oerst. = Sloanea L. ●
122864 Dasycarya Liebm. = Cyrtocarpa Kunth ●☆
122865 Dasycephala (DC.) Benth. et Hook. f. = Diodia L. ■
122866 Dasycephala (DC.) Hook. f. = Diodia L. ■
122867 Dasycephala Benth. et Hook. f. = Diodia L. ■
122868 Dasycephala Borkh. ex Pfeiff. = Dasystephana Adans. ■
122869 Dasycephala Borkh. ex Pfeiff. = Gentiana L. ■
122870 Dasychloa Post et Kuntze = Danthonia DC. (保留属名)■
122871 Dasychloa Post et Kuntze = Dasyochloa Willd. ex Steud. ■
122872 Dasychloa Post et Kuntze = Sieglingia Bernh. (废弃属名)■
122873 Dasycoleum Turcz. = Chisocheton Blume ●
122874 Dasycondylus R. M. King et H. Rob. (1972);基节柄泽兰属●☆
122875 Dasycondylus debeauxii (B. L. Rob.) R. M. King et H. Rob.;基节柄泽兰●☆
122876 Dasydesmus Craib = Oreocharis Benth. (保留属名)■
122877 Dasydesmus Craib(1919);毛药苣苔属■
122878 Dasydesmus bodinieri (H. Lév.) Craib = Oreocharis bodinieri H. Lév. ■
122879 Dasydesmus bodinieri Craib = Oreocharis bodinieri H. Lév. ■
122880 Dasyglossum Königer et Schildh. = Odontoglossum Kunth ■
122881 Dasygyna Post et Kuntze = Dasiogyne Raf. ●
122882 Dasygyna Post et Kuntze = Prosopis L. ●

122883 Dasylepis Oliv.(1865);毛鳞大风子属●☆
122884 Dasylepis assinensis A. Chev. ex Hutch. et Dalziel = Dasylepis blackii (Oliv.) Chipp ●☆
122885 Dasylepis blackii (Oliv.) Chipp;布拉克毛鳞大风子●☆
122886 Dasylepis brevipedicellata Chipp = Dasylepis racemosa Oliv. ●☆
122887 Dasylepis burtt-davyi Edlin = Rawsonia burtt-davyi (Edlin) F. White ●☆
122888 Dasylepis eggelingii J. B. Gillett;埃格林大风子●☆
122889 Dasylepis integra Warb.;细叶毛鳞大风子●☆
122890 Dasylepis jansii Bamps = Dasylepis blackii (Oliv.) Chipp ●☆
122891 Dasylepis lasiocarpa Gilg = Dasylepis seretii De Wild. ●☆
122892 Dasylepis lebrunii Evrard = Dasylepis eggelingii J. B. Gillett ●☆
122893 Dasylepis leonensis (Oliv.) Warb. = Scottellia leonensis Oliv. ●☆
122894 Dasylepis leptophylla Gilg = Dasylepis integra Warb. ●☆
122895 Dasylepis racemosa Oliv.;总状毛鳞大风子●☆
122896 Dasylepis seretii De Wild.;赛雷毛鳞大风子●☆
122897 Dasylepis thomasii Obama et Breteler;托马斯毛鳞大风子●☆
122898 Dasylirion Zucc.(1838);毛百合属(锯齿龙属,有毛百合属);Bear Grass,Bear-grass,Hare's-foot Fern,Sotol,Tufted Lily ●☆
122899 Dasylirion acrotrichum Zucc.;尖毛百合;Brosh-tipped Sotol, Green Desert Spoon ●☆
122900 Dasylirion bigelovii Torr. = Nolina bigelovii S. Watson ■☆
122901 Dasylirion erumpens Torr. = Nolina erumpens (Torr.) S. Watson ■☆
122902 Dasylirion glaucophyllum Hook.;青绿毛百合;Blueleaf Sotol ●☆
122903 Dasylirion glaucum Carrière = Dasylirion glaucophyllum Hook. ●☆
122904 Dasylirion heteracanthum I. M. Johnst. = Dasylirion leiophyllum Engelm. ex Trel. ●☆
122905 Dasylirion leiophyllum Engelm. ex Trel.;平滑毛百合;Desert Candle,Smooth-leaved Sotol ●☆
122906 Dasylirion lindheimerianum Scheele = Nolina lindheimeriana (Scheele) S. Watson ■☆
122907 Dasylirion longifolium Zucc. = Nolina longifolia Hemsl. ●☆
122908 Dasylirion longissimum Lem.;墨西哥长毛百合;Mexican Grass Plant,Mexican Grass Tree,Toothless Sotol ●☆
122909 Dasylirion quadrangulatum S. Watson;墨西哥毛百合;Mexican Grass Tree☆
122910 Dasylirion texanum Scheele;得州毛百合(得州锯齿龙);Bear-grass,Green Sotol,Texas Sotol ■☆
122911 Dasylirion wheeleri S. Watson = Dasylirion wheeleri S. Watson ex Rothr. ●☆
122912 Dasylirion wheeleri S. Watson ex Rothr.;旱生毛百合;Desert Spoon,Sotol,Wheeler Sotol ●☆
122913 Dasyloma DC. = Oenanthe L. ■
122914 Dasyloma DC. = Seseli L. ■
122915 Dasyloma benghalense (Roxb.) DC. = Oenanthe benghalensis Benth. et Hook. f. ■
122916 Dasyloma glaucum DC. = Oenanthe benghalensis Benth. et Hook. f. ■
122917 Dasyloma japonicum Miq. = Oenanthe javanica (Blume) DC. ■
122918 Dasyloma javanicum (Blume) Miq. = Oenanthe javanica (Blume) DC. ■
122919 Dasyloma laciniatum Miq. = Oenanthe javanica (Blume) DC. ■
122920 Dasyloma latifolium Lindl. = Oenanthe javanica (Blume) DC. subsp. stolonifera (Roxb.) Murata ■
122921 Dasyloma subbipinnatum Miq. = Oenanthe javanica (Blume) DC. ■
122922 Dasymalla Endl. = Pityrodia R. Br. ●☆
122923 Dasymaschalon (Hook. f. et Thomson) Dalla Torre et Harms = Deppea Cham. et Schltdl. ●☆
122924 Dasymaschalon (Hook. f. et Thomson) Dalla Torre et Harms (1901);皂帽花属;Blackhatflower,Dasymaschalon ●
122925 Dasymaschalon Dalla Torre et Harms = Dasymaschalon (Hook. f. et Thomson) Dalla Torre et Harms ●
122926 Dasymaschalon glaucum Merr. et Chun = Dasymaschalon rostratum Merr. et Chun ●
122927 Dasymaschalon glaucum Merr. et Chun = Desmos rostratus (Merr. et Chun) P. T. Li ●
122928 Dasymaschalon macrocalyx Finet et Gagnep. = Desmos macrocalyx (Finet et Gagnep.) P. T. Li ●
122929 Dasymaschalon robinsonii Ast = Desmos robinsonii (Ast) P. T. Li ●
122930 Dasymaschalon rostratum Merr. et Chun = Desmos rostratus (Merr. et Chun) P. T. Li ●
122931 Dasymaschalon rostratum Merr. et Chun var. glaucum (Merr. et Chun) Ban = Dasymaschalon rostratum Merr. et Chun ●
122932 Dasymaschalon sootepense Craib = Desmos sootepensis (Craib) J. F. Maxwell ●
122933 Dasymaschalon trichophorum Merr.;皂帽花(毛皂帽花,乌木兰);Hanry Blackhatflower, Hanry Dasymaschalon, Lowflower Dasymaschalon ●
122934 Dasymaschalon yunnanense (Hu) Ban = Desmos yunnanense (Hu) P. T. Li ●
122935 Dasynema Schott = Sloanea L. ●
122936 Dasynotus I. M. Johnst.(1948);毛背紫草属■☆
122937 Dasynotus daubenmirei I. M. Johnst.;毛背紫草■☆
122938 Dasyochloa Rydb. = Erioneuron Nash ■☆
122939 Dasyochloa Willd. ex Rydb. = Erioneuron Nash ■☆
122940 Dasyochloa Willd. ex Steud. = Danthonia DC.(保留属名)■
122941 Dasyochloa Willd. ex Steud. = Sieglingia Bernh.(废弃属名)■
122942 Dasyochloa pulchella Willd. = Triodia pulchella Kunth ■☆
122943 Dasypetalum Pierre ex A. Chev. = Scottellia Oliv. ●☆
122944 Dasypetalum klaineanum Pierre ex A. Chev. = Scottellia klaineana Pierre ●☆
122945 Dasyphonion Raf. = Aristolochia L. ■●
122946 Dasyphora Post et Kuntze = Dasiphora Raf. ■●
122947 Dasyphora Post et Kuntze = Potentilla L. ■●
122948 Dasyphyllum Kunth(1818);毛叶刺菊木属●☆
122949 Dasyphyllum argenteum Kunth;毛叶刺菊木●☆
122950 Dasypoa Pilg. = Poa L. ■
122951 Dasypogon R. Br.(1810);毛瓣花属■☆
122952 Dasypogon bromeliifolius R. Br.;毛瓣花■☆
122953 Dasypogonaceae Dumort.(1892);毛瓣花科(多须草科)■☆
122954 Dasypogonaceae Dumort. = Xanthorrhoeaceae Dumort.(保留科名)●■☆
122955 Dasypogonia Rchb. = Dasypogon R. Br. ■☆
122956 Dasypyrum (Coss. et Durieu) Maire = Dasypyrum (Coss. et Durieu) T. Durand ■☆
122957 Dasypyrum (Coss. et Durieu) P. Candargy = Haynaldia Schur ■☆
122958 Dasypyrum (Coss. et Durieu) T. Durand(1888);簇毛麦属■☆
122959 Dasypyrum breviaristatum (H. Lindb.) Fred.;短芒簇毛麦■☆
122960 Dasypyrum hordeaceum (Coss. et Durieu) P. Candargy = Dasypyrum breviaristatum (H. Lindb.) Fred. ■☆
122961 Dasypyrum hordeaceum (Coss. et Durieu) P. Candargy var.

breviaristatum (H. Lindb.) Maire = Dasypyrum breviaristatum (H. Lindb.) Fred. ■☆
122962 Dasypyrum hordeaceum (Coss. et Durieu) P. Candargy var. velutinum (Maire) Maire = Dasypyrum breviaristatum (H. Lindb.) Fred. ■☆
122963 Dasypyrum villosum (L.) P. Candargy;簇毛麦;Mosquitograss ■☆
122964 Dasyranthus Raf. ex Steud. = Gnaphalium L. ■
122965 Dasys Lem. = Dasus Lour.(废弃属名)●
122966 Dasys Lem. = Lasianthus Jack(保留属名)●
122967 Dasyslachys Baker = Chlorophytum Ker Gawl. ■
122968 Dasyslachys Oersted = Chamaedorea Willd.(保留属名)●☆
122969 Dasyspermum Neck. = Conium L. + Tordylium L. + Ammi L. + Scandix L. ■
122970 Dasysphaera Volkens ex Gilg(1897);毛头苋属■☆
122971 Dasysphaera alternifolia Chiov.;互叶毛头苋■☆
122972 Dasysphaera breviflora C. C. Towns. = Dasysphaera hyposericea (Chiov.) C. C. Towns. ■☆
122973 Dasysphaera hyposericea (Chiov.) C. C. Towns.;短叶毛头苋■☆
122974 Dasysphaera prostrata (Volkens ex Gilg) Cavaco = Volkensinia prostrata (Volkens ex Gilg) Schinz ■☆
122975 Dasysphaera robecchii Lopr.;罗贝克毛头苋■☆
122976 Dasysphaera tomentosa Volkens ex Lopr.;毛头苋■☆
122977 Dasystachys Baker = Chlorophytum Ker Gawl. ■
122978 Dasystachys Oerst.(1859);毛穗棕属●☆
122979 Dasystachys Oerst. = Chamaedorea Willd.(保留属名)●☆
122980 Dasystachys africana (Baker) T. Durand et H. Durand = Chlorophytum africanum (Baker) Engl. ■☆
122981 Dasystachys atacorensis A. Chev. = Chlorophytum aureum Engl. ●☆
122982 Dasystachys aurea (Engl.) Baker = Chlorophytum aureum Engl. ●☆
122983 Dasystachys bequaertii De Wild. = Chlorophytum colubrinum (Welw. ex Baker) Engl. ■☆
122984 Dasystachys campanulata Baker = Chlorophytum colubrinum (Welw. ex Baker) Engl. ■☆
122985 Dasystachys colubrina Welw. ex Baker = Chlorophytum colubrinum (Welw. ex Baker) Engl. ■☆
122986 Dasystachys crassifolia Baker = Chlorophytum colubrinum (Welw. ex Baker) Engl. ■☆
122987 Dasystachys debilis Baker = Chlorophytum silvaticum Dammer ■☆
122988 Dasystachys deckeriana (Klotzsch) Oerst.;毛穗棕●☆
122989 Dasystachys decorata Baker = Chlorophytum colubrinum (Welw. ex Baker) Engl. ■☆
122990 Dasystachys drimiopsis (Baker) Baker ex Benth. et Hook. f. = Chlorophytum longifolium Schweinf. ex Baker ■☆
122991 Dasystachys falcata Baker = Chlorophytum longifolium Schweinf. ex Baker ■☆
122992 Dasystachys gracilis Baker = Chlorophytum silvaticum Dammer ■☆
122993 Dasystachys graminea A. Chev. = Chlorophytum affine Baker ■☆
122994 Dasystachys grantii Benth. = Chlorophytum africanum (Baker) Engl. ■☆
122995 Dasystachys grantii Benth. var. engleri Baker = Chlorophytum colubrinum (Welw. ex Baker) Engl. ■☆
122996 Dasystachys hockii De Wild. = Chlorophytum colubrinum (Welw. ex Baker) Engl. ■☆
122997 Dasystachys leptoneura C. H. Wright = Chlorophytum leptoneurum (C. H. Wright) Poelln. ■☆
122998 Dasystachys macinensis A. Chev. = Chlorophytum senegalense (Baker) Hepper ●☆
122999 Dasystachys marginata (Rendle) Baker = Chlorophytum africanum (Baker) Engl. ■☆
123000 Dasystachys melanocarpa Chiov. = Chlorophytum macrophyllum (A. Rich.) Asch. ■☆
123001 Dasystachys papillosa (Rendle) Baker = Chlorophytum longifolium Schweinf. ex Baker ■☆
123002 Dasystachys pleiostachya Baker = Chlorophytum colubrinum (Welw. ex Baker) Engl. ■☆
123003 Dasystachys polyphylla Baker = Chlorophytum suffruticosum Baker ●☆
123004 Dasystachys pulchella P. A. Duvign. et Dewit = Chlorophytum linearifolium Marais et Reilly ●☆
123005 Dasystachys senegalensis Baker = Chlorophytum senegalense (Baker) Hepper ●☆
123006 Dasystachys sombae A. Chev. = Chlorophytum aureum Engl. ●☆
123007 Dasystachys stenophylla R. E. Fr. = Chlorophytum colubrinum (Welw. ex Baker) Engl. ■☆
123008 Dasystachys verdickii De Wild. = Chlorophytum colubrinum (Welw. ex Baker) Engl. ■☆
123009 Dasystemon DC.(1828);毛蕊景天属■☆
123010 Dasystemon DC. = Crassula L. ●■☆
123011 Dasystemon calycina DC. = Crassula glomerata P. J. Bergius ●☆
123012 Dasystepha Post et Kuntze = Dasistepha Raf. ■
123013 Dasystepha Post et Kuntze = Gentiana L. ■
123014 Dasystephana Adans. = Gentiana L. ■
123015 Dasystephana andrewsii (Griseb.) Small = Gentiana andrewsii Griseb. ■☆
123016 Dasystephana grayi (Kusn.) Britton = Gentiana rubricaulis Schwein. ■☆
123017 Dasystephana linearis (Froel.) Britton = Gentiana linearis Froel. ■☆
123018 Dasystephana linearis Britton = Gentiana linearis Froel. ■☆
123019 Dasystephana sikokiana (Maxim.) Soják = Gentiana sikokiana Maxim. ■☆
123020 Dasystoma Raf. = Agalinis Raf.(保留属名)■☆
123021 Dasystoma Raf. = Dasistoma Raf. ■●☆
123022 Dasystoma Raf. ex Endl. = Aureolaria Raf. ■☆
123023 Dasytropis Urb.(1924);毛肋爵床属☆
123024 Dasytropis fragilis Urb.;毛肋爵床☆
123025 Dasyurus Post et Kuntze = Chamaelirium Willd. ■☆
123026 Dasyurus Post et Kuntze = Dasurus Salisb. ■☆
123027 Datisca L.(1753);疣柱花属(达麻属,达提斯加属,四数木属,野麻属);Datisca ●■☆
123028 Datisca cannabina L.;疣柱花(达蒂斯卡麻,野麻);Bastard Hemp,Datisca ●☆
123029 Datisca glomerata Baill.;加州疣柱花(加州麻,四数木);Durango Root ●☆
123030 Datisca hirta L. = Rhus hirta (L.) Sudw. ●☆
123031 Datisca nepalensis D. Don = Datisca cannabina L. ●☆
123032 Datiscaceae Bercht. et J. Presl = Datiscaceae Dumort.(保留科名)■●
123033 Datiscaceae Dumort.(1829)(保留科名);疣柱花科(达麻科,短序花科,四数木科,四数木科,野麻科);Datisca Family ■●

123034　Datiscaceae Lindl. = Datiscaceae Dumort.（保留科名）■●

123035　Datiscaceae R. Br. ex Lindl. = Datiscaceae Dumort.（保留科名）■●

123036　Datura L.（1753）；曼陀罗属；Angel's Trumpets, Datura, Jimsonweed, Jimson-weed, Thorn Apple, Thorn-apple ●■

123037　Datura alba Nees = Datura metel L. ●■

123038　Datura alba Rumph. ex Nees = Datura fastuosa L. ●■

123039　Datura alba Rumph. ex Ness = Datura metel L. ●■

123040　Datura arborea L. = Brugmansia arborea（L.）Steud. ●

123041　Datura arborea Ruiz et Pav. = Brugmansia candida Pers. ●☆

123042　Datura bertolonii Parl. ex Guss. = Datura stramonium L. var. inermis（Jacq.）Lundstr. ■

123043　Datura candida（Pers.）Saff. = Brugmansia candida Pers. ●☆

123044　Datura ceratocaulis Ortega；角茎曼陀罗■☆

123045　Datura cornigera Hook.；角曼陀罗■☆

123046　Datura cornigera Hook. = Brugmansia arborea（L.）Steud. ●

123047　Datura cornigera Hook. = Datura arborea L. ●

123048　Datura cornucopaea W. W. Sm. = Datura metel L. ●■

123049　Datura discolor Bernh.；紫斑曼陀罗；Purple-stained Toloache, Small Datura☆

123050　Datura fastuosa L. = Datura metel L. ●■

123051　Datura fastuosa L. var. alba（Nees）C. B. Clarke = Datura fastuosa L. ●■

123052　Datura fastuosa L. var. alba（Nees）C. B. Clarke = Datura metel L. ●■

123053　Datura fastuosa L. var. alba C. B. Clarke = Datura metel L. ●■

123054　Datura ferox L.；粗刺曼陀罗（多刺曼陀罗）；Angel's Trumpets ■

123055　Datura gardneri Hook. = Brugmansia suaveolens（Humb. et Bonpl. ex Willd.）Bercht. et C. Presl ●

123056　Datura hummatu Bernh. = Datura metel L. ●■

123057　Datura inermis Jacq. = Datura stramonium L. var. inermis（Jacq.）Lundstr. ■

123058　Datura inermis Juss. ex Jacq. = Datura stramonium L. ●■

123059　Datura inermis Juss. ex Jacq. = Datura stramonium L. var. inermis（Juss. ex Jacq.）Schinz et Thell. ■

123060　Datura innoxia Mill. = Datura inoxia Mill. ■

123061　Datura inoxia Mill.；毛曼陀罗（串筋花，风茄花，毛叶曼陀罗，洋金花）；Angel's Trumpet, Angel's-trumpet, Downy Thorn Apple, Hairleaf Datura, Hairy Datura, Hairy Jimsonweed, Pricklyburr ■

123062　Datura laevis L. f. = Datura stramonium L. ●■

123063　Datura laevis L. f. = Datura stramonium L. var. inermis（Jacq.）Lundstr. ■

123064　Datura lanosa Barclay ex Bye；绵毛曼陀罗■☆

123065　Datura metel L.；洋金花（白花曼陀罗，白曼陀罗，打破碗花，大喇叭花，大麻子花，大闹杨花，风麻花，风茄儿，风茄花，枫茄花，枫茄子，伏茄子，高大曼陀罗，狗核桃，关东大麻子花，广东闹羊花，胡茄，胡茄花，虎茄花，假荔枝，金盘托荔枝，金茄子，酒醉花，喇叭花，芳仙桃，六轴子，马兰花，曼多萝，曼陀罗，曼陀罗花，美丽曼陀罗，南洋金花，闹羊花，山大麻子，山茄花，山茄子，天茄子，万桃花，押不芦，羊惊花，洋大麻，洋大麻子花，洋喇叭花，重瓣曼陀罗，紫花曼陀罗，醉葡萄，醉仙桃）；Angel's Trumpet, Black Datura, Cornucopia Floripondia, Datura, Devil's Trumpet, Downy Thorn-apple, Hairy Apple, Hairy Thorn-apple, Hindu Datura, Horn of Plenty, Horn-of-plenty, Metel, Proud Datura, Thorn Apple ●■

123066　Datura metel L.'Floreplena'；重瓣洋金花●

123067　Datura metel L. = Datura fastuosa L. ●■

123068　Datura metel L. f. alba Chou = Datura metel L. ●■

123069　Datura metel L. sensu Fassett = Datura inoxia Mill. ■

123070　Datura meteloides DC. = Datura meteloides DC. ex Dunal ■☆

123071　Datura meteloides DC. ex Dunal；香曼陀罗（类米替曼陀罗）；Desert Trumpet Flower, Downy Apple, Downy Thorn-apple, Indian Apple, Ooze Apple, Sacred Datura, Thorn Apple, Toloache, Western Jimson ■☆

123072　Datura meteloides DC. ex Dunal = Datura inoxia Mill. ■

123073　Datura meteloides Dunal = Datura inoxia Mill. ■

123074　Datura nilhummatu Dunal = Datura metel L. ●■

123075　Datura quercifolia Kunth；栎叶曼陀罗■☆

123076　Datura sanguinea Ruiz et Pav.；红花曼陀罗；Angel's Tears, Angel's Trumpet ■☆

123077　Datura stramonium L.；曼陀罗（大喇叭花，番曼陀罗，枫茄花，佛茄儿，狗核桃，耗子阎王，美洲曼陀罗，闹羊花，欧曼陀罗，天麻花，万桃花，无刺曼陀罗，洋金花，野麻子，紫花曼陀罗，醉心花）；Angel's Trumpet, Apple of Peru, Common Thorn Apple, Datura, Deutery Dewtry, Devil's Apple, Devil's Herb, Devil's Trumpet, Dotroa, Fireweed, Goat Apple, Green Apple, Green Thorn-apple, Jamestown Lily, Jamestown Weed, Jimpson Weed, Jimson Weed, Jimsonweed, Jimson-weed, Jimson-weed Datura, Jimsy-weed, Mad Apple, Mad-apple, Magic Herb, Nightshade, Purple Apple, Purple Jimsonweed, Purple Stinkweed, Purple Thorn-apple, Purple-flowered Datura, Sopor, Stinkroot, Stinkweed, Stinkwort, Stramonium, Stramony, Thorn Apple, Thorn-apple, White Stinkweed ●■

123078　Datura stramonium L. f. tatura（L.）Danert = Datura tatula L. ●■

123079　Datura stramonium L. var. chalybea K. Koch = Datura stramonium L. ●■

123080　Datura stramonium L. var. inermis（Jacq.）Lundstr. = Datura stramonium L. var. inermis（Juss. ex Jacq.）Schinz et Thell. ■

123081　Datura stramonium L. var. inermis（Juss. ex Jacq.）Schinz et Thell.；无刺曼陀罗（洋金花）■

123082　Datura stramonium L. var. inermis（Juss. ex Jacq.）Schinz et Thell. = Datura stramonium L. ●■

123083　Datura stramonium L. var. tatula（L.）Torr.；番曼陀罗●■

123084　Datura stramonium L. var. tatula（L.）Torr. = Datura metel L. ●■

123085　Datura stramonium L. var. tatula（L.）Torr. = Datura stramonium L. ●■

123086　Datura stramonium L. var. tatula（Willd.）Clarke = Datura stramonium L. ●■

123087　Datura suaveolens Humb. et Bonpl. ex Willd. = Brugmansia suaveolens（Humb. et Bonpl. ex Willd.）Bercht. et C. Presl ●

123088　Datura tatula L. = Datura metel L. ●■

123089　Datura tatula L. = Datura stramonium L. ●■

123090　Datura vulgare Moench = Datura stramonium L. ●■

123091　Datura wallichii Dunal = Datura stramonium L. ●■

123092　Datura wrightii Regel；赖氏曼陀罗；Sacred Datura, South-westeru Thorn-apple, Wright's Jimson Weed ■☆

123093　Daturaceae Bercht. et J. Presl = Solanaceae Juss.（保留科名）●■

123094　Daturaceae Raf.；曼陀罗科●■

123095　Daturaceae Raf. = Solanaceae Juss.（保留科名）●■

123096　Daturicarpa Stapf = Tabernanthe Baill. ●☆

123097　Daturicarpa elliptica Stapf = Tabernanthe elliptica（Stapf）Leeuwenb. ●☆

123098　Daturicarpa firmula Stapf = Tabernanthe elliptica（Stapf）Leeuwenb. ●☆

123099　Daturicarpa lanceolata Stapf = Tabernanthe elliptica（Stapf）Leeuwenb. ●☆

123100 Daturicasoa Stapf = Tabernanthe Baill. ●☆

123101 Daubeninniopsis Rydb. = Sesbania Scop.（保留属名）●■

123102 Daubentonia DC. = Sesbania Scop.（保留属名）●■

123103 Daubentonia punicea DC. = Sesbania punicea Benth. ●☆

123104 Daubentoniopsis Rydb. = Sesbania Scop.（保留属名）●■

123105 Daubenya Lindl.（1835）；合花风信子属（多布尼草属）■☆

123106 Daubenya alba A. M. Van der Merwe；白合花风信子（白多布尼草）■☆

123107 Daubenya angustifolia（L. f.）A. M. Van der Merwe et J. C. Manning = Massonia echinata L. f. ■☆

123108 Daubenya aurea Lindl.；黄合花风信子（黄多布尼草）■☆

123109 Daubenya aurea Lindl. var. coccinea（Harv. ex Baker）Marloth = Daubenya aurea Lindl. ■☆

123110 Daubenya capensis（Schltr.）A. M. Van der Merwe et J. C. Manning；好望角合花风信子■☆

123111 Daubenya coccinea Harv. ex Baker = Daubenya aurea Lindl. ■☆

123112 Daubenya comata（Burch. ex Baker）J. C. Manning et A. M. Van der Merwe；束毛合花风信子■☆

123113 Daubenya fulva Lindl. = Daubenya aurea Lindl. ■☆

123114 Daubenya marginata（Willd. ex Kunth）J. C. Manning et A. M. Van der Merwe；具边合花风信子■☆

123115 Daubenya namaquensis（Schltr.）J. C. Manning et Goldblatt；纳马夸合花风信子■☆

123116 Daubenya stylosa（W. F. Barker）A. M. Van der Merwe et J. C. Manning = Amphisiphon stylosa W. F. Barker ■☆

123117 Daubenya zeyheri（Kunth）J. C. Manning et A. M. Van der Merwe；泽赫合花风信子■☆

123118 Daucaceae Dostal = Apiaceae Lindl.（保留科名）●■

123119 Daucaceae Dostal = Umbelliferae Juss.（保留科名）■●

123120 Daucaceae Martinov = Apiaceae Lindl.（保留科名）●■

123121 Daucaceae Martinov = Umbelliferae Juss.（保留科名）■●

123122 Daucalis Pomel = Caucalis L. ■☆

123123 Daucalis leptophylla（L.）Pomel = Torilis leptophylla（L.）Rchb. f. ■☆

123124 Dauceria Dennst. = Embelia Burm. f.（保留属名）●■

123125 Daucophyllum Rydb. = Daucophyllum（T. Nuttall ex J. Torrey et A. Gray）Rydb. ■☆

123126 Daucophyllum Rydb. = Musineon Raf. ex DC. ■☆

123127 Daucophyllum（T. Nuttall ex J. Torrey et A. Gray）Rydb.（1913）；萝卜叶属■☆

123128 Daucophyllum lineare Rydb.；萝卜叶■☆

123129 Daucophyllum tenuifolium Rydb.；细叶萝卜叶■☆

123130 Daucosma Engelm. et A. Gray = Discopleura DC. ■☆

123131 Daucosma Engelm. et A. Gray = Ptilimnium Raf. ■☆

123132 Daucosma Engelm. et A. Gray ex A. Gray = Daucosma Engelm. et A. Gray ■☆

123133 Daucosma Engelm. et A. Gray（1850）；萝卜芹属■☆

123134 Daucosma laciniata Engelm. et A. Gray；萝卜芹■☆

123135 Daucus L.（1753）；胡萝卜属；Corrot ■

123136 Daucus abyssinicus Hochst. ex A. Rich. = Daucus hochstetteri A. Br. ex Engl. ■☆

123137 Daucus alatus Poir. = Daucus carota L. subsp. commutatus（Paol.）Thell. ■☆

123138 Daucus aureus Desf.；黄胡萝卜■☆

123139 Daucus aureus Desf. var. tuberculatus Chabert = Daucus aureus Desf. ■☆

123140 Daucus capillifolius Gilli；发叶胡萝卜■☆

123141 Daucus carota L.；野胡萝卜（赤珊瑚，番萝卜，鹤虱，鹤虱草，鹤虱风，红菜头，红萝卜，胡萝卜，金笋，人参，山萝卜，野萝卜）；Bee's Nest，Bird's Nest，Bird's-nest，Carett，Caxes，Corrot，Corts，Crow's Nest，Curran-petris，Dauke，Devil's Plague，Fiddle，Hemp，Kager，Kaiyer，Keggers，Kex，Lady's Parasol，Mirrot，More，Pig's Parsley，Queen Anne's Lace，Queen Anne's-lace，Rantipole，Sea Carrot，Wild Carrot ■

123142 Daucus carota L. f. epurpuratus Farw. = Daucus carota L. ■

123143 Daucus carota L. glochidiatus ? = Daucus glochidiatus（Labill.）Fisch.，C. A. Mey. et Avé-Lall. ■☆

123144 Daucus carota L. subsp. azorica Franco = Daucus carota L. subsp. maximus（Desf.）Ball. ■☆

123145 Daucus carota L. subsp. commutatus（Paol.）Thell.；变异胡萝卜■☆

123146 Daucus carota L. subsp. commutatus Thell. = Daucus carota L. subsp. commutatus（Paol.）Thell. ■☆

123147 Daucus carota L. subsp. dentatus（Bertol.）Fiori = Daucus carota L. subsp. commutatus（Paol.）Thell. ■☆

123148 Daucus carota L. subsp. fontanesii Thell.；丰塔纳胡萝卜■☆

123149 Daucus carota L. subsp. gummifer（Syme）Hook. f.；产胶野胡萝卜；Sea Carrot ■☆

123150 Daucus carota L. subsp. hispanicus（Gouan）Thell.；西班牙胡萝卜■☆

123151 Daucus carota L. subsp. hispidus（Arcang.）Heywood = Daucus carota L. subsp. fontanesii Thell. ■☆

123152 Daucus carota L. subsp. maritimus（Lam.）Batt.；沿海野胡萝卜■☆

123153 Daucus carota L. subsp. mauritanicus（L.）Quézel et Santa = Daucus carota L. subsp. maximus（Desf.）Ball. ■☆

123154 Daucus carota L. subsp. maximus（Desf.）Ball.；大野胡萝卜■☆

123155 Daucus carota L. subsp. parviflorus（Desf.）Thell. = Daucus carota L. subsp. maritimus（Lam.）Batt. ■☆

123156 Daucus carota L. subsp. petroselinifolius Alleiz. = Daucus carota L. subsp. maritimus（Lam.）Batt. ■☆

123157 Daucus carota L. subsp. sativus（Hoffm.）Arcang. = Daucus carota L. var. sativa Hoffm. ■

123158 Daucus carota L. subsp. siculus（Tineo）Maire = Daucus carota L. ■

123159 Daucus carota L. var. brachycentrus Maire = Daucus carota L. ■

123160 Daucus carota L. var. excelsus Maire = Daucus carota L. ■

123161 Daucus carota L. var. herculeus（Pau）Maire = Daucus carota L. ■

123162 Daucus carota L. var. hipponensis Maire = Daucus carota L. ■

123163 Daucus carota L. var. maritimus（Lam.）Steud. = Daucus carota L. subsp. maritimus（Lam.）Batt. ■☆

123164 Daucus carota L. var. sativa Hoffm.；胡萝卜（丁香萝卜，鹤虱，红菜头，红芦菔，红萝卜，胡芦菔，胡萝菔，黄萝卜，金笋，野胡萝卜）；Corrot，Garden Carrot ■

123165 Daucus carota L. var. tenuiflorus Alleiz. = Daucus carota L. ■

123166 Daucus coptica（L.）Pers. = Trachyspermum ammi（L.）Sprague ■

123167 Daucus crinitus Desf.；长软毛胡萝卜■☆

123168 Daucus crinitus Desf. var. comosus Murb. = Daucus crinitus Desf. ■☆

123169 Daucus fontanesii Thell. = Daucus carota L. ■

123170 Daucus fontanesii Thell. var. melillensis Sennen = Daucus carota L. ■

123171 Daucus gingidium L. = Daucus carota L. ■
123172 Daucus gingidium L. = Daucus carota L. subsp. hispanicus (Gouan) Thell. ■☆
123173 Daucus gingidium L. subsp. fontanesii (Thell.) Onno = Daucus carota L. subsp. fontanesii Thell. ■☆
123174 Daucus gingidium L. subsp. gummifer (Lam.) Onno = Daucus carota L. subsp. hispanicus (Gouan) Thell. ■☆
123175 Daucus gingidium L. subsp. mauritanicus (L.) Onno = Daucus carota L. ■
123176 Daucus gingidium L. subsp. polygamus (Gouan) Onno = Daucus carota L. ■
123177 Daucus gingidium L. var. africanus Pau et Font Quer = Daucus carota L. ■
123178 Daucus gingidium L. var. hispidus (Desf.) Ball = Daucus carota L. subsp. fontanesii Thell. ■☆
123179 Daucus glaberrimus Desf. = Daucus carota L. subsp. maritimus (Lam.) Batt. ■☆
123180 Daucus glochidiatus (Labill.) Fisch., C. A. Mey. et Avé-Lall.; 钩毛胡萝卜; Australian Carrot ■☆
123181 Daucus gracilis Steinh.; 纤细胡萝卜■☆
123182 Daucus grandiflorus Desf. = Daucus muricatus (L.) L. ■☆
123183 Daucus gummifer Lam. = Daucus carota L. subsp. hispanicus (Gouan) Thell. ■☆
123184 Daucus guttatus Sibth. et Sm.; 斑点胡萝卜■☆
123185 Daucus herculeus Pau = Daucus carota L. subsp. maximus (Desf.) Ball. ■☆
123186 Daucus hispidissimus Sennen = Daucus muricatus (L.) L. ■☆
123187 Daucus hispidus Desf. = Daucus carota L. subsp. hispanicus (Gouan) Thell. ■☆
123188 Daucus hochstetteri A. Br. ex Engl.; 霍赫胡萝卜■☆
123189 Daucus jolensis Pomel = Daucus carota L. subsp. hispanicus (Gouan) Thell. ■☆
123190 Daucus laserpitioides DC. = Daucus virgatus (Poir.) Maire ■☆
123191 Daucus laserpitioides DC. var. apterus Batt. = Daucus virgatus (Poir.) Maire ■☆
123192 Daucus laserpitioides DC. var. stenopterus Batt. = Daucus virgatus (Poir.) Maire ■☆
123193 Daucus litoralis Sibth. et Sm.; 滨海胡萝卜■☆
123194 Daucus litoralis Sibth. et Sm. var. forsskalii Boiss. = Daucus litoralis Sibth. et Sm. ■☆
123195 Daucus mauritanicus L. = Daucus carota L. subsp. maximus (Desf.) Ball. ■☆
123196 Daucus mauritii Sennen = Daucus muricatus (L.) L. ■☆
123197 Daucus maximus Desf. = Daucus carota L. subsp. maximus (Desf.) Ball. ■☆
123198 Daucus micranthus Pomel = Daucus carota L. subsp. maritimus (Lam.) Batt. ■☆
123199 Daucus minusculus Pau = Pseudorlaya minuscula (Pau) M. Lainz ■☆
123200 Daucus muricatus (L.) L.; 短尖胡萝卜■☆
123201 Daucus muricatus (L.) L. var. erectus (Batt.) Maire = Daucus muricatus (L.) L. ■☆
123202 Daucus muricatus (L.) L. var. hispidissimus (Sennen) Maire = Daucus muricatus (L.) L. ■☆
123203 Daucus muricatus (L.) L. var. mauritii (Sennen) Maire = Daucus muricatus (L.) L. ■☆
123204 Daucus paralias Pomel = Daucus carota L. subsp. maritimus (Lam.) Batt. ■☆
123205 Daucus parviflorus Desf. = Daucus carota L. subsp. maritimus (Lam.) Batt. ■☆
123206 Daucus parviflorus Desf. subsp. breviumbellatus (Barratte) Le Houér. = Daucus carota L. subsp. maritimus (Lam.) Batt. ■☆
123207 Daucus parviflorus Desf. subsp. micranthus (Pomel) Le Houér. = Daucus carota L. subsp. maritimus (Lam.) Batt. ■☆
123208 Daucus parviflorus Desf. var. breviumbellatus Barratte = Daucus carota L. subsp. maritimus (Lam.) Batt. ■☆
123209 Daucus parviflorus Desf. var. glabra Le Houér. = Daucus carota L. subsp. maritimus (Lam.) Batt. ■☆
123210 Daucus parviflorus Desf. var. pubescens Le Houér. = Daucus carota L. subsp. maritimus (Lam.) Batt. ■☆
123211 Daucus pumilus (Gouan) Hoffmanns. et Link; 微小胡萝卜■☆
123212 Daucus pumilus (Gouan) Hoffmanns. et Link subsp. maritimus (Desf.) Maire = Daucus pumilus (Gouan) Hoffmanns. et Link ■☆
123213 Daucus pumilus (Gouan) Hoffmanns. et Link subsp. microcarpus (Loret et Barrandon) Maire = Daucus pumilus (Gouan) Hoffmanns. et Link ■☆
123214 Daucus pumilus (Gouan) Hoffmanns. et Link subsp. minusculus (Pau) Maire = Pseudorlaya minuscula (Pau) M. Lainz ■☆
123215 Daucus pusillus Michx.; 小胡萝卜; Beggar's Buttons, Beggar's Lice, Hemp, Queen Anne's Lace, Rattlesnake-weed, Small Wild Carrot, Wild Carrot ■☆
123216 Daucus reboudii Batt.; 雷博胡萝卜■☆
123217 Daucus sahariensis Murb.; 萨哈里胡萝卜■☆
123218 Daucus sahariensis Murb. var. elongatus Hochr. = Daucus sahariensis Murb. ■☆
123219 Daucus sativa (Hoffm.) Roehl = Daucus carota L. var. sativa Hoffm. ■
123220 Daucus serotinus Pomel = Daucus carota L. ■
123221 Daucus setifolius Desf.; 毛叶胡萝卜■☆
123222 Daucus setifolius Desf. subsp. tangerinus Sauvage; 丹吉尔胡萝卜■☆
123223 Daucus syrticus Murb.; 瑟尔特胡萝卜■☆
123224 Daucus tenuisectus Batt.; 细裂胡萝卜■☆
123225 Daucus virgatus (Poir.) Maire; 条纹胡萝卜■☆
123226 Daucus visnaga L. = Ammi visnaga (L.) Lam. ■
123227 Daumalia Airy Shaw = Urospermum Scop. ■☆
123228 Daumalia Arènes = Urospermum Scop. ■☆
123229 Daun-contu Adans(废弃属名) = Paederia L. (保留属名)●■
123230 Dauphinea Hedge(1983); 多芬草属●☆
123231 Dauphinea brevilabra Hedge; 多芬草●☆
123232 Dauresia B. Nord. et Pelser(2005); 白皮菊属●☆
123233 Dauresia alliariifolia (O. Hoffm.) B. Nord. et Pelser; 白皮菊■☆
123234 Dauthonia Link = Danthonia DC. (保留属名)■
123235 Dauventonia Rchb. = Daubentonia DC. ●■
123236 Davaea Gand. = Campanula L. ■●
123237 Davaella Gand. = Chondrilla L. ■
123238 Daveana T. Durand et Jacks. = Daveaua Willk. ex Mariz ■☆
123239 Daveaua Willk. = Daveaua Willk. ex Mariz ■☆
123240 Daveaua Willk. ex Mariz(1891)('Daveana'); 齿翅菊属■☆
123241 Daveaua anthemoides Mariz; 齿翅菊■☆
123242 Davejonesia M. A. Clem. = Dendrobium Sw. (保留属名)■
123243 Davejonesia M. A. Clem. = Dockrillia Brieger ■
123244 Davidia Baill. (1871); 珙桐属(珙垌属); Dove Tree, Dovetree, Dove-tree, Ghost-tree, Pocket-handkerchief Tree ●★

123245 Davidia involucrata Baill.;珙桐(空桐,山白果,水冬瓜,水梨子,土白果,中国鸽子树);China Dove Tree,Chinese Dove-tree,Dove Tree,Dovetree,Dove-tree,Ghost Tree,Ghost-tree,Handkerchief Tree,Pocket Handker Chief Tree,Pocket-handkerchief Tree ●◇

123246 Davidia involucrata Baill. subsp. vilmoriniana (Dode) Holub = Davidia involucrata Baill. var. vilmoriniana (Dode) Wangerin ●◇

123247 Davidia involucrata Baill. var. vilmoriniana (Dode) Wangerin;光叶珙桐;Virmorin Dovetree ●◇

123248 Davidia laeta Dode = Davidia involucrata Baill. var. vilmoriniana (Dode) Wangerin ●◇

123249 Davidia tibetana David = Davidia involucrata Baill. ●◇

123250 Davidia vilmoriniana Dode = Davidia involucrata Baill. var. vilmoriniana (Dode) Wangerin ●◇

123251 Davidiaceae (Harmus) H. L. Li = Cornaceae Bercht. et J. Presl (保留科名)●■

123252 Davidiaceae H. L. Li = Cornaceae Bercht. et J. Presl(保留科名)●■

123253 Davidiaceae Takht.;珙桐科;Davidia Family ●

123254 Davidiaceae Takht. = Cornaceae Bercht. et J. Presl(保留科名)●■

123255 Davidiaceae Takht. = Davidsoniaceae Bange ●☆

123256 Davidiaceae Takht. = Nyssaceae Juss. ex Dumort.(保留科名)●

123257 Davidsea Soderstr. et R. P. Ellis = Schizostachyum Nees ●

123258 Davidsea Soderstr. et R. P. Ellis(1988);戴维兹竹属(大维兹竹属)●☆

123259 Davidsea attenuata (Thwaites) Soderstr. et R. P. Ellis;戴维兹竹●☆

123260 Davidsonia F. Muell. (1867);澳楸属(大维逊李属)●☆

123261 Davidsonia pruriens F. Muell.;澳楸(大维逊李);Davidson's Plum ●☆

123262 Davidsoniaceae Bange = Cunoniaceae R. Br. (保留科名)●☆

123263 Davidsoniaceae Bange = Decaisneaceae Loconte ●

123264 Davidsoniaceae Bange(1952);澳楸科(大维逊李科)●☆

123265 Daviesia Poir. = Borya Labill. ■☆

123266 Daviesia Sm. (1798);澳苦豆属(苦豆属);Bacon And Eggs,Bitter Pea ●☆

123267 Daviesia arborea Scort.;乔木澳苦豆;Queenwood ●☆

123268 Daviesia latifolia R. Br.;药苦豆(广叶达比西亚);Hop Bitter Pea ●☆

123269 Daviesia mimosoides R. Br.;狭叶苦豆;Blunt-leaf Bitter Pea,Narrow-leaf Bitter Pea ●☆

123270 Daviesia squarrosa Sm.;细梗苦豆●☆

123271 Daviesia ulicifolia Andréws;刺苦豆;Gorse Bitter Pea ●☆

123272 Davilla Vand. (1788);达维木属●☆

123273 Davilla Vell. ex Vand. = Davilla Vand. ●☆

123274 Davilla angustifolia A. St. -Hil.;狭叶达维木●☆

123275 Davilla densiflora Triana et Planch.;密花达维木●☆

123276 Davilla grandifolia Moric. ex Eichler;大叶达维木●☆

123277 Davilla kunthii A. St. -Hil.;孔氏达维木●☆

123278 Davilla latifolia Casar.;阔叶达维木●☆

123279 Davilla lechleri Rusby = Tetracera parviflora (Rusby) Sleumer ●☆

123280 Davilla lucida C. Presl = Davilla kunthii A. St. -Hil. ●☆

123281 Davilla macrocarpa Eichler;大果达维木●☆

123282 Davilla microcalyx Herzog;小萼达维木●☆

123283 Davilla ovata C. Presl = Davilla kunthii A. St. -Hil. ●☆

123284 Davilla parviflora Rusby;小花达维木●☆

123285 Davilla rugosa Poir.;多皱达维木●☆

123286 Davilla sagraeana A. Rich.;古巴达维木●☆

123287 Davilla strigosa Kubitzki;巴西达维木●☆

123288 Davilla suaveolens Glaz.;芳香达维木●☆

123289 Davilla surinamensis Miq. = Davilla rugosa Poir. ●☆

123290 Davilla villosa Eichler;毛达维木●☆

123291 Davya DC. = Adelobotrys DC. ●☆

123292 Davya DC. = Meriania Sw. (保留属名)●☆

123293 Davya Moc. et Sessé ex DC. = Saurauia Willd. (保留属名)●

123294 Davyella Hack. = Neostapfia Burtt Davy ■☆

123295 Dawea Sprague ex Dawe = Warburgia Engl. (保留属名)●☆

123296 Dayaoshania W. T. Wang(1983);瑶山苣苔属;Dayaoshania ■★

123297 Dayaoshania cotinifolia W. T. Wang;瑶山苣苔;Dayaoshania ■

123298 Daydonia Britten = Anneslea Wall. (保留属名)●

123299 Dayena Adans. = Byttneria Loefl. (保留属名)●

123300 Dayena Adans. = Chaetaea Jacq. ●

123301 Dayenia Michx. ex Jaub. et Spach = Biebersteinia Stephan ex Fisch. ■

123302 Dayenia Mill. = Ayenia L. ●☆

123303 Dayia J. M. Porter = Gilia Ruiz et Pav. ■●☆

123304 Dazus Juss. = Dasus Lour. (废弃属名)●

123305 Dazus Juss. = Lasianthus Jack(保留属名)●

123306 Deamia Britton et Rose = Selenicereus (A. Berger) Britton et Rose ●

123307 Deamia Britton et Rose(1920);三棱尺属(迪姆属,龟甲仙人掌属)■☆

123308 Deamia testudo Britton et Rose;三棱尺(龟甲)■☆

123309 Deanea J. M. Coult. et Rose = Rhodosciadium S. Watson ■☆

123310 Deastella Loudon = Mimetes Salisb. ●☆

123311 Debeauxia Gand. = Cephalotos Adans. (废弃属名)●

123312 Debeauxia Gand. = Thymus L. ●

123313 Debesia Kuntze = Acrospira Welw. ex Baker ■☆

123314 Debesia Kuntze = Anthericum L. ■☆

123315 Debraea Roem. et Schult. = Erisma Rudge ●☆

123316 Debregeasia Gaudich. (1844);水麻属;Debregeasia,Waternettle ●

123317 Debregeasia atrata Gagnep. = Archiboehmeria atrata (Gagnep.) C. J. Chen ●◇

123318 Debregeasia bicolor (Roxb.) Wedd. = Debregeasia saeneb (Forssk.) Hepper et J. R. I. Wood ●

123319 Debregeasia bicolor (Roxb.) Wedd. = Debregeasia salicifolia (D. Don) Rendle ●

123320 Debregeasia ceylanica Hook. f. = Debregeasia wallichiana (Wedd.) Wedd. ●

123321 Debregeasia dichotoma (Blume) Wedd. = Debregeasia longifolia (Burm. f.) Wedd. ●

123322 Debregeasia edulis (Siebold et Zucc.) Wedd.;水麻(大水麻,冬里麻,红烟,尖麻,柳莓,水冬瓜,水麻柳,水麻桑,水麻叶,水麻子,水马桑,水苏麻);Edible Debregeasia,Yanagi ●

123323 Debregeasia elliptica C. J. Chen;椭圆叶水麻;Ellipticleaf Debregeasia,Ellipticleaf Waternettle ●

123324 Debregeasia hypoleuca (Steud.) Wedd. = Debregeasia saeneb (Forssk.) Hepper et J. R. I. Wood ●

123325 Debregeasia hypoleuca Wedd. = Debregeasia salicifolia (D. Don) Rendle ●

123326 Debregeasia japonica (Miq.) Koidz. = Debregeasia edulis (Siebold et Zucc.) Wedd. ●

123327 Debregeasia leucophylla Wedd. = Debregeasia wallichiana (Wedd.) Wedd. ●

123328 Debregeasia libera S. S. Chien et C. J. Chen;卵叶水麻;Ovate-leaved Debregeasia ●

123329 Debregeasia libera S. S. Chien et C. J. Chen = Debregeasia longifolia (Burm. f.) Wedd. ●

123330 Debregeasia longifolia (Burm. f.) Wedd.;长叶水麻(麻叶树,水珠麻);Longleaf Debregeasia, Longleaf Waternettle, Long-leaved Debregeasia ●

123331 Debregeasia obovata C. H. Wright = Oreocnide obovata (C. H. Wright) Merr. ●

123332 Debregeasia orientalis C. J. Chen;东方水麻(赤麻,冬里麻,柳眉莓,柳梅,沙连泡,水冬瓜,水麻,水麻柳,水麻桑,水麻叶,水马麻);Orintal Debregeasia, Waternettle ●

123333 Debregeasia saeneb (Forssk.) Hepper et J. R. I. Wood;柳叶水麻;Willowleaf Debregeasia, Willowleaf Waternettle ●

123334 Debregeasia salicifolia (D. Don) Rendle = Debregeasia saeneb (Forssk.) Hepper et J. R. I. Wood ●

123335 Debregeasia salicifolia Rendle = Debregeasia longifolia (Burm. f.) Wedd. ●

123336 Debregeasia spiculifera Merr. = Debregeasia squamata King ex Hook. ●

123337 Debregeasia squamata King ex Hook.;鳞片水麻(大血吉,卵叶水麻,山草麻,山野麻,山苎麻,野苎麻);Scale Debregeasia, Scale Waternettle ●

123338 Debregeasia velutina Gaudich. = Debregeasia longifolia (Burm. f.) Wedd. ●

123339 Debregeasia wallichiana (Wedd.) Wedd.;长序水麻;Wallich Debregeasia, Wallich Waternettle ●

123340 Decabelone Decne. = Tavaresia Welw. ex N. E. Br. ●■☆

123341 Decabelone barklyi Dyer = Tavaresia barklyi (Dyer) N. E. Br. ■☆

123342 Decabelone elegans Decne. = Tavaresia angolensis Welw. ■☆

123343 Decabelone grandiflora K. Schum. = Tavaresia barklyi (Dyer) N. E. Br. ■☆

123344 Decaceras Harv. = Anisotoma Fenzl ■☆

123345 Decaceras huttonii Harv. = Brachystelma huttonii (Harv.) N. E. Br. ■☆

123346 Decachaena (Hook.) Lindl. = Gaylussacia Kunth(保留属名) ●☆

123347 Decachaena (Torr. et A. Gray) Lindl. = Gaylussacia Kunth(保留属名)●☆

123348 Decachaena Torr. et A. Gray ex A. Gray = Gaylussacia Kunth (保留属名)●☆

123349 Decachaena (W. J. Hooker) Torrey et A. Gray ex Lindl. = Gaylussacia Kunth(保留属名)●☆

123350 Decachaena baccata (Wangenh.) Small = Gaylussacia baccata (Wangenh.) K. Koch ●☆

123351 Decachaeta DC. (1836);落冠毛泽兰属●☆

123352 Decachaeta Gardner = Ageratum L. ■●

123353 Decachaeta haenkeana DC.;落冠毛泽兰●☆

123354 Decachaeta longifolia Gardner;长叶落冠毛泽兰●☆

123355 Decachaeta ovatifolia (DC.) R. M. King et H. Rob.;卵叶落冠毛泽兰●☆

123356 Decadenium Raf. = Adenaria Kunth ■☆

123357 Decadia Lour. = Symplocos Jacq. ●

123358 Decadon G. Don = Decodon J. F. Gmel. ●☆

123359 Decadon G. Don = Nesaea Comm. ex Kunth(保留属名)■●☆

123360 Decadonia Raf. = Adenaria Kunth ■☆

123361 Decadonia Raf. = Decadenium Raf. ■☆

123362 Decadontia Griff. = Sphenodesme Jack ●

123363 Decagonocarpus Engl. (1874);十角芸香属●☆

123364 Decagonocarpus oppositifolius Engl.;十角芸香●☆

123365 Decaisnea Brongn. (废弃属名) = Decaisnea Hook f. et Thomson(保留属名)●

123366 Decaisnea Brongn. (废弃属名) = Prescottia Lindl. ■☆

123367 Decaisnea Hook. f. et Thomson(1855)(保留属名);猫儿子属(矮杞树属,猫儿屎属,猫耳屎属);Blue Bean Shrub, Decaisnea ●

123368 Decaisnea Lindl. = Tropidia Lindl. ■

123369 Decaisnea angulosa (Lindl.) Wall. = Tropidia angulosa (Lindl.) Blume ■

123370 Decaisnea fargesii Franch. = Decaisnea insignis (Griff.) Hook. f. et Thomson ●

123371 Decaisnea insignis (Griff.) Hook. f. et Thomson;猫儿子(矮杞树,齿果,鬼指梅,鸡肠子,老鸟吃,猫儿屎,猫耳屎,猫屎包,猫屎瓜,羊角立,羊角子,粘连子);Blue Bean Shrub, Farges Decaisnea ●

123372 Decaisneaceae Loconte = Lardizabalaceae R. Br. (保留科名)●

123373 Decaisneaceae Loconte;猫儿子科●

123374 Decaisnella Kuntze = Aquilaria Lam. (保留属名)●

123375 Decaisnella Kuntze = Gyrinops Gaertn. ●☆

123376 Decaisnella Kuntze = Gyrinopsis Decne. ●☆

123377 Decaisnina Tiegh. (1895);德卡寄生属●☆

123378 Decaisnina alata Tiegh.;翅德卡寄生●☆

123379 Decaisnina amplexans Tiegh.;德卡寄生●☆

123380 Decaisnina angustata (Barlow) Barlow;窄德卡寄生●☆

123381 Decaisnina parvifolia (Dans.) Barlow;小叶德卡寄生●☆

123382 Decaisnina triflora (Span.) Tiegh.;三花德卡寄生●☆

123383 Decaisnina triflora Tiegh. = Decaisnina triflora (Span.) Tiegh. ●☆

123384 Decaisnina viridis (Merr.) Barlow;绿花德卡寄生●☆

123385 Decalepidanthus Riedl(1963);十鳞草属●☆

123386 Decalepidanthus sericophyllus Riedl;十鳞草●☆

123387 Decalepis Boeck. = Boeckeleria T. Durand ■☆

123388 Decalepis Boeck. = Tetraria P. Beauv. ■☆

123389 Decalepis Wight et Arn. (1834);十鳞萝藦属☆

123390 Decalepis dregeana Boeck. = Tetraria nigrovaginata (Nees) C. B. Clarke☆

123391 Decalepis hamiltonii Wight et Arn.;十鳞萝藦☆

123392 Decaloba M. Roem. = Passiflora L. ●■

123393 Decaloba Raf. = Ipomoea L. (保留属名)●■

123394 Decalobanthus Ooststr. (1936);十裂花属■☆

123395 Decalobanthus sumatranus Ooststr.;十裂花■☆

123396 Decaloca F. Muell.;新几内亚尖苞木属●☆

123397 Decalophium Turcz. = Chamelaucium Desf. ●☆

123398 Decameria Welw. = Gardenia Ellis(保留属名)●

123399 Decamerium Nutt. = Gaylussacia Kunth(保留属名)●☆

123400 Decandolia Bastard = Agrostis L. (保留属名)■

123401 Decanema Decne. (1838);十蕊萝藦属■☆

123402 Decanema Decne. = Cynanchum L. ●■

123403 Decanema bojeriana Decne. = Cynanchum bojerianum (Decne.) Choux ●☆

123404 Decanema bojerianum Decne.;十蕊萝藦■☆

123405 Decanema grandiflorum Jum. et H. Perrier;大花十蕊萝藦■☆

123406 Decanema grandiflorum Jum. et H. Perrier = Cynanchum grandidieri Liede et Meve ●☆

123407 Decanemopsis Costantin et Gallaud = Sarcostemma R. Br. ■
123408 Decaneuron divergens DC. = Vernonia divergens (DC.) Edgew. ■
123409 Decaneuropsis H. Rob. et Skvarla = Vernonia Schreb. (保留属名)●■
123410 Decaneurum DC. = Centratherum Cass. ■☆
123411 Decaneurum DC. = Gymnanthemum Cass. ●■☆
123412 Decaneurum Sch. Bip. = Leucanthemella Tzvelev ■
123413 Decaneurum grande DC. = Vernonia colorata (Willd.) Drake subsp. grandis (DC.) C. Jeffrey ■☆
123414 Decapenta Raf. (1834) = Diodia L. ■
123415 Decapenta Raf. (1838) = Litsea Lam. (保留属名)●
123416 Decaphalangium Melch. = Clusia L. ●☆
123417 Decaprisma Raf. = Campanula L. ■●
123418 Decaptera Turcz. (1846);十翼芥属■☆
123419 Decaptera Turcz. = Menonvillea R. Br. ex DC. ■●☆
123420 Decaptera trifida Turcz.;十翼芥■☆
123421 Decaraphe Miq. = Melastoma L. ●■
123422 Decaraphe Miq. = Miconia Ruiz et Pav. (保留属名)●☆
123423 Decarinium Raf. = Croton L. ●
123424 Decarneria Welw. = Gardenia Ellis(保留属名)●
123425 Decarya Choux(1929);之形木属(德卡利木属)●☆
123426 Decarya madagascariensis Choux;之形木(德卡利木)●☆
123427 Decaryanthus Bonati(1927);德参属☆
123428 Decarydendron Danguy(1928);德卡瑞甜桂属●☆
123429 Decarydendron helenae Danguy;德卡瑞甜桂●☆
123430 Decarydendron lamii Cavaco;拉姆德卡瑞甜桂●☆
123431 Decarydendron perrieri Cavaco;佩里耶德卡瑞甜桂●☆
123432 Decarydendron ranomafanensis Lorence et Razafim.;拉努马法纳德卡瑞甜桂●☆
123433 Decaryella A. Camus(1931);小德卡草属■☆
123434 Decaryella madagascariensis A. Camus;小德卡草■☆
123435 Decaryia Choux = Decarya Choux ●☆
123436 Decaryia madagascariensis Choux = Decarya madagascariensis Choux ●☆
123437 Decaryochloa A. Camus(1946);德卡竹属●☆
123438 Decaryochloa diadelpha A. Camus;德卡竹●☆
123439 Decaschistia Wight et Arn. (1834);十裂葵属;Decaschistia ●■
123440 Decaschistia mouretii Gagnep.;中越十裂葵●
123441 Decaschistia mouretii Masam. var. nervifolia (Masam.) H. S. Kiu = Decaschistia nervifolia Masam. ●
123442 Decaschistia nervifolia Masam.;十裂葵●
123443 Decaspermum J. R. Forst. et G. Forst. (1775);子楝树属(十子木属,十子属);Decaspermum ●
123444 Decaspermum albociliatum Merr. et L. M. Perry;白毛子楝树;Whitehair Decaspermum,White-haired Decaspermum ●
123445 Decaspermum austro-hainanicum Hung T. Chang et R. H. Miao;琼南子楝树;Qiongnan Decaspermum, S. Hainan Decaspermum, South Hainan Decaspermum ●
123446 Decaspermum cambodianum Gagnep. = Decaspermum montanum Ridl. ●
123447 Decaspermum esquirolii (H. Lév.) Hung T. Chang et R. H. Miao = Decaspermum gracilentum (Hance) Merr. et L. M. Perry ●
123448 Decaspermum fruticosum J. R. Forst. et G. Forst.;灌木五瓣子楝树(米碎木,五瓣子楝树);Shrubby Decaspermum,Tailor Tree ●
123449 Decaspermum fruticosum J. R. Forst. et G. Forst. var. khasianum (Duthie) C. Y. Wu;狭叶米碎木●☆
123450 Decaspermum glabrum Hung T. Chang et R. H. Miao;秃子楝树;Bladseed Decaspermum, Glabrous Decaspermum, Smooth Decaspermum ●
123451 Decaspermum gracilentum (Hance) Merr. et L. M. Perry;子楝树(华夏子楝树,加入相,加入舅,米碎木,桑枝米碎木,山稔,山稔木,山桃稔,十子木,炸香棵);Chinese Decaspermum, Decaspermum,Esquirol Decaspermum,Slender Decaspermum ●
123452 Decaspermum hainanense (Merr.) Merr.;海南子楝树●
123453 Decaspermum hainanense (Merr.) Merr. = Pyrenocarpa hainanensis (Merr.) Hung T. Chang et R. H. Miao ●◇
123454 Decaspermum montanum Ridl.;柬埔寨子楝树(米花木,米碎叶,山大尼,山模,山齐,乌垒);Cambodia Decaspermum ●
123455 Decaspermum paniculatum (Lindl.) Kurz = Decaspermum parviflorum (Lam.) A. J. Scott ●
123456 Decaspermum paniculatum Kurz;圆锥子楝树(加入舅);Paniculate Decaspermum ●
123457 Decaspermum parviflorum (Lam.) A. J. Scott;五瓣子楝树●
123458 Decaspermum sericeum Hance;丝状子楝树●
123459 Decaspermum teretis Craven;圆枝子楝树●
123460 Decaspora R. Br. (1810);十子澳石南属●☆
123461 Decaspora R. Br. = Trochocarpa R. Br. ●☆
123462 Decaspora disticha R. Br.;十子澳石南●☆
123463 Decastelma Schltr. (1899);十冠萝藦属●☆
123464 Decastelma broadwayi Schltr.;十冠萝藦●☆
123465 Decastemon Klotzsch = Cleome L. ●■
123466 Decastemon Klotzsch(1861);十蕊白花菜属■☆
123467 Decastemon hirtus Klotzsch;毛十蕊白花菜■☆
123468 Decastemon hirtus Klotzsch = Cleome hirta (Klotzsch) Oliv. ■☆
123469 Decastemon zanzibaricus;十蕊白花菜■☆
123470 Decastemon zanzibaricus Klotzsch = Cleome strigosa (Bojer) Oliv. ■☆
123471 Decastia Raf. = Nigella L. ■
123472 Decastrophia Griff. = Erythropalum Blume ●
123473 Decastylocarpus Humb. (1923);十肋瘦片菊属■☆
123474 Decastylocarpus perrieri Humbert;十肋瘦片菊■☆
123475 Decateles Raf. = Bumelia Sw. (保留属名)●☆
123476 Decateles Raf. = Sideroxylon L. ●☆
123477 Decatoca F. Muell. (1889);十肋石南属●☆
123478 Decatoca spencerii F. Muell.;十肋石南●☆
123479 Decatropis Hook. f. (1862);十肋芸香属●☆
123480 Decatropis bicolor (Zucc.) Radlk.;二色十肋芸香●☆
123481 Decatropis bicolor Radlk. = Decatropis bicolor (Zucc.) Radlk. ●☆
123482 Decatropis coulteri Hook. f.;十肋芸香●☆
123483 Decavenia (Nakai) Koidz. = Pterostyrax Siebold et Zucc. ●
123484 Decavenia Koidz. = Pterostyrax Siebold et Zucc. ●
123485 Decazesia F. Muell. (1879);膜苞鼠麹草属■☆
123486 Decazesia hecatocephala F. Muell.;膜苞鼠麹草■☆
123487 Decazyx Pittier et S. F. Blake(1922);墨西哥芸香属●☆
123488 Decazyx esparzae F. Chiang;北美墨西哥芸香●☆
123489 Decazyx macrophyllus Pittier et S. F. Blake;墨西哥芸香●☆
123490 Deccania Tirveng. (1983);戴康茜属●☆
123491 Deccania pubescens (Roth) Tirveng.;戴康茜●☆
123492 Decemium Raf. = Hydrophyllum L. ■☆
123493 Decemium appendiculatum (Michx.) Small = Hydrophyllum appendiculatum Michx. ■☆
123494 Deceptor Seidenf. = Saccolabium Blume(保留属名)■

123495 Dechampsia Kunth = Deschampsia P. Beauv. ■
123496 Deckenia H. Wendl. = Deckenia H. Wendl. ex Seem. ●☆
123497 Deckenia H. Wendl. ex Seem. (1870);华丽桐属(华丽刺藤属,拟刺椰子属,塞舌尔王椰属);Deckenia Palm ●☆
123498 Deckenia nobilis H. Wendl.;华丽桐●☆
123499 Deckera Sch. Bip. = Picris L. ■
123500 Deckera aculeata Sch. Bip. = Helminthotheca aculeata (Vahl) Lack ■☆
123501 Deckera callosa Pomel = Helminthotheca aculeata (Vahl) Lack ■☆
123502 Deckera comosa (Boiss.) Batt. = Helminthotheca comosa (Boiss.) Lack ■☆
123503 Deckera glomerata Pomel = Helminthotheca glomerata (Pomel) Greuter ■☆
123504 Deckera montana Pomel = Helminthotheca aculeata (Vahl) Lack ■☆
123505 Deckera racemosa Pomel = Helminthotheca comosa (Boiss.) Lack ■☆
123506 Deckera racemosa Pomel var. rubiginosa (Pomel) Batt. = Helminthotheca comosa (Boiss.) Lack ■☆
123507 Deckera rubiginosa Pomel = Helminthotheca comosa (Boiss.) Lack ■☆
123508 Deckeria H. karst. = Iriartea Ruiz et Pav. ●☆
123509 Deckora Sch. Bip. = Picris L. ■
123510 Declieuxia Kunth(1819);戴克茜属■☆
123511 Declieuxia alba Zucc.;白戴克茜■☆
123512 Declieuxia brasiliensis Müll. Arg.;巴西戴克茜■☆
123513 Declieuxia coerulea Gardner;蓝戴克茜■☆
123514 Declieuxia foliosa DC.;多叶戴克茜■☆
123515 Declieuxia lancifolia J. H. Kirkbr.;披针叶戴克茜■☆
123516 Declieuxia latifolia Hochst.;宽叶戴克茜■☆
123517 Declieuxia latifolia Hochst. = Pentanisia prunelloides (Klotzsch ex Eckl. et Zeyh.) Walp. subsp. latifolia (Hochst.) Verdc. ■☆
123518 Declieuxia leiophylla Müll. Arg.;光叶戴克茜■☆
123519 Declieuxia mexicana DC.;墨西哥戴克茜■☆
123520 Declieuxia minutifolia Standl.;小叶戴克茜■☆
123521 Declieuxia mollis Zucc.;柔软戴克茜■☆
123522 Declieuxia prunelloides Klotzsch ex Eckl. et Zeyh. = Pentanisia prunelloides (Klotzsch ex Eckl. et Zeyh.) Walp. ■☆
123523 Declieuxia tenuiflora (Willd. ex Roem. et Schult.) Steyerm. et J. H. Kirkbr.;细花戴克茜■☆
123524 Declieuxia verticillata Müll. Arg.;轮生戴克茜■☆
123525 Decodon J. F. Gmel. (1791);敌克冬属;Decodon ●☆
123526 Decodon verticillatus (L.) Elliott;敌克冬;Swamp Loosestrife, Water Willow, Water-willow, Whorled Loosestrife ■☆
123527 Decodon verticillatus (L.) Elliott var. laevigatus Torr. et A. Gray;轮叶敌克冬;Swamp Loosestrife, Water-willow, Whorled Loosestrife ■☆
123528 Decodontia Haw. = Huernia R. Br. ■☆
123529 Decorima Raf. = Karwinskia Zucc. ●☆
123530 Decorsea Basse(1951);华美豆属●■☆
123531 Decorsea R. Vig. ex R. Vig. = Decorsea Basse ●■☆
123532 Decorsea dinteri (Harms) Verdc.;丁特华美豆■☆
123533 Decorsea galpinii (Burtt Davy) Verdc.;盖尔华美豆■☆
123534 Decorsea grandidieri (Baill.) R. Vig. ex M. Pelt.;格朗华美豆■☆
123535 Decorsea livida R. Vig.;铅色华美豆■☆
123536 Decorsea livida R. Vig. var. meridionalis R. Vig. = Decorsea meridionalis (R. Vig.) Du Puy et Labat ■☆
123537 Decorsea meridionalis (R. Vig.) Du Puy et Labat;南方华美豆■☆
123538 Decorsea schlechteri (Harms) Verdc.;施莱华美豆■☆
123539 Decorsella A. Chev. (1917);早裂堇属■☆
123540 Decorsella paradoxa A. Chev.;早裂堇■☆
123541 Decostea Ruiz et Pav. = Griselinia J. R. Forst. et G. Forst. ●☆
123542 Dectis Raf. = Commidendron Lem. ●☆
123543 Decumaria L. (1763);赤壁木属(赤壁草属,赤壁藤属,罩壁木属);Decumaria, Wood Vamp ●
123544 Decumaria barbata L.;毛赤壁木;Climbing Hydrangea ●
123545 Decumaria sinensis Oliv.;赤壁木(赤壁草,赤壁藤,钝叶冠盖藤,十出花,罩壁木);China Decumaria, Chinese Decumaria, Obtuse-leaved Pileostegia ●
123546 Decusocarpus de Laub. = Nageia Gaertn. (废弃属名)●
123547 Decussocarpus de Laub. = Nageia Gaertn. (废弃属名)●
123548 Decussocarpus de Laub. = Retrophyllum C. N. Page ●☆
123549 Decussocarpus fleuryi (Hickel) de Laub. = Nageia fleuryi (Hickel) de Laub. ●◇
123550 Decussocarpus maximus de Laub. = Nageia maxima (de Laub.) de Laub. ●☆
123551 Decussocarpus nagi (Thunb.) de Laub. = Nageia nagi (Thunb.) Kuntze ●
123552 Decussocarpus nagi (Thunb.) de Laub. var. formosensis (Dunn) Silba = Nageia nagi (Thunb.) Kuntze ●
123553 Decussocarpus wallichianus (C. Presl) de. Laub. = Nageia wallichiana (C. Presl) Kuntze ●
123554 Dedea Baill. = Quintinia A. DC. ●☆
123555 Dedeckera Reveal et J. T. Howell(1976);木黄蓼属(德克尔蓼属);Eureka, July Gold ●☆
123556 Dedeckera eurekensis Reveal et J. T. Howell;木黄蓼(德克尔蓼)●☆
123557 Deeringia Kuntze = Sison L. ■☆
123558 Deeringia R. Br. (1810);浆果苋属(地灵苋属,地苓苋属,纽藤属);Cladostachys ●■
123559 Deeringia amaranthoides (Lam.) Merr.;浆果苋(菜藤,地灵苋,地苓苋,纽藤,野苋);Frutescent Cladostachys, Amaranth Cladostachys ●■
123560 Deeringia amaranthoides (Lam.) Merr. = Cladostachys frutescens D. Don ●
123561 Deeringia baccata (Retz.) Moq. = Deeringia amaranthoides (Lam.) Merr. ●■
123562 Deeringia celosioides R. Br. = Cladostachys frutescens D. Don ●
123563 Deeringia celosioides R. Br. = Deeringia amaranthoides (Lam.) Merr. ●■
123564 Deeringia densiflora Cavaco;密花浆果苋●☆
123565 Deeringia humbertiana Cavaco;亨伯特浆果苋●☆
123566 Deeringia indica Zoll. ex Moq. var. pubescens Schinz = Deeringia polysperma (Roxb.) Moq. ■
123567 Deeringia madagascariensis Cavaco;马岛浆果苋●■
123568 Deeringia perrieriana Cavaco;佩里耶浆果苋●☆
123569 Deeringia polysperma (Roxb.) Moq.;白浆果苋(多籽地灵苋,多子浆果苋);Manyseed Cladostachys ■
123570 Deeringia polysperma (Roxb.) Moq. = Cladostachys polysperma (Roxb.) K. C. Kuan ■
123571 Deeringia polysperma (Roxb.) Moq. var. pubescens (Schinz)

Merr. = Deeringia polysperma (Roxb.) Moq. ■
123572 Deeringia spicata (Thouars) Schinz = Celosia spicata (Thouars) Spreng. ■☆
123573 Deeringiaceae J. Agardh = Amaranthaceae Juss.(保留科名)●■
123574 Deeringothamnus Small(1924);假泡泡果属●☆
123575 Deeringothamnus pulchellus Small;美丽假泡泡果;Beautiful Pawpaw ●☆
123576 Deeringothamnus rugelii (B. L. Rob.) Small;假泡泡果;Rugel's Pawpaw ●☆
123577 Deerlngia Kuntze = Cryptotaenia DC.(保留属名)■
123578 Deerlngia Kuntze = Deringa Adans.(废弃属名)■
123579 Defforgia Lam. = Forgesia Comm. ex Juss. ●☆
123580 Deflersia Gand. = Verbascum L. ■●
123581 Deflersia Schweinf. ex Penz.(1893);肖红果大戟属●☆
123582 Deflersia Schweinf. ex Penz. = Erythrococca Benth. ●☆
123583 Deflersia erythrococca Schweinf.;肖红果大戟●☆
123584 Degeneria I. W. Bailey et A. C. Sm.(1942);单心木兰属●☆
123585 Degeneria vitiensis I. W. Bailey et A. C. Sm.;单心木兰●☆
123586 Degeneriaceae I. W. Bailey = Degeneriaceae I. W. Bailey et A. C. Sm.(保留科名)●☆
123587 Degeneriaceae I. W. Bailey et A. C. Sm.(1942)(保留科名);单心木兰科(退化科)●☆
123588 Degenia Hayek(1910);柠檬芥属■☆
123589 Degenia velebitica (Degen) Hayek;柠檬芥■☆
123590 Degranvillea Determann(1985);圭亚那兰属■☆
123591 Degranvillea dermaptera Determann;圭亚那兰■☆
123592 Deguelia Aubl.(废弃属名) = Derris Lour.(保留属名)●
123593 Deguelia grevei Drake = Pongamiopsis pervilleana (Baill.) R. Vig. ●☆
123594 Deguelia robusta (Roxb.) Taub. = Derris robusta (Roxb.) Benth. ●
123595 Deguelia stuhlmannii Taub. = Xeroderris stuhlmannii (Taub.) Mendonça et E. P. Sousa ●☆
123596 Deguelia trifoliata (Lour.) Taub. = Derris trifoliata Lour. ●
123597 Deguelia uliginosa (Willd.) Baill. = Derris trifoliata Lour. ●
123598 Dehaasia Blume(1837);莲桂属(腰果楠属);Dehaasia ●
123599 Dehaasia caesia Blume;蓝灰莲桂●☆
123600 Dehaasia cairocan (Vidal) C. K. Allen;菲律宾莲桂;Philippine Dehaasia ●☆
123601 Dehaasia cairocan (Vidal) C. K. Allen = Dehaasia hainanensis Kosterm. ●
123602 Dehaasia cuneata Blume;楔形莲桂●☆
123603 Dehaasia curtisii Gamble;短柄莲桂●☆
123604 Dehaasia elliptica Ridl.;椭圆叶莲桂●☆
123605 Dehaasia firma Blume;硬莲桂●☆
123606 Dehaasia hainanensis Kosterm.;莲桂;Dehaasia, Hainan Dehaasia ●
123607 Dehaasia incrassata (Jack) Kosterm.;腰果楠(三蕊莲桂);Incrassate Dehaasia ●◇
123608 Dehaasia kwangtungensis Kosterm.;广东莲桂;Guangdong Dehaasia, Kwangtung Dehaasia ●
123609 Dehaasia lanyuensis (C. E. Chang) Kosterm. = Dehaasia incrassata (Jack) Kosterm. ●◇
123610 Dehaasia triandra Merr. = Dehaasia incrassata (Jack) Kosterm. ●◇
123611 Deherainia Decne.(1876);卵花绿果属●☆
123612 Deherainia smaragdina Decne.;卵花绿果●☆
123613 Deianira Cham. et Schltdl.(1826);代亚龙胆属■☆
123614 Deianira cordifolia Malme;心叶代亚龙胆■☆
123615 Deianira erubescens Cham. et Schltdl.;代亚龙胆■☆
123616 Deidamia E. A. Noronha ex Thouars(1805);马岛西番莲属■☆
123617 Deidamia Thouars = Deidamia E. A. Noronha ex Thouars ■☆
123618 Deidamia alata Thouars;翅马岛西番莲■☆
123619 Deidamia bicolor H. Perrier;二色马岛西番莲■☆
123620 Deidamia bipinnata Tul.;双羽马岛西番莲■☆
123621 Deidamia clematoides Harms = Efulensia clematoides C. H. Wright ■☆
123622 Deilanthe N. E. Br. = Aloinopsis Schwantes ■☆
123623 Deilanthe N. E. Br. = Nananthus N. E. Br. ■☆
123624 Deilanthe hilmarii (L. Bolus) H. E. K. Hartmann = Aloinopsis hilmarii (L. Bolus) L. Bolus ■☆
123625 Deilanthe peersii (L. Bolus) N. E. Br. = Aloinopsis peersii (L. Bolus) L. Bolus ■☆
123626 Deilanthe thudichumii (L. Bolus) S. A. Hammer = Aloinopsis thudichumii L. Bolus ■☆
123627 Deilosma Andrz. ex DC. = Hesperis L. ■
123628 Deina Alef.(1866);波兰小麦属■☆
123629 Deina Alef. = Triticum L. ■
123630 Deina polonica (L.) Alef. = Triticum polonicum L. ■
123631 Deina polonica (L.) Alef. = Triticum turgidum L. subsp. polonicum (L.) Thell. ■
123632 Deinacanthon Mez = Bromelia L. ■☆
123633 Deinacanthon Mez(1896);肖凤梨属■☆
123634 Deinacanthon urbanianum Mez;肖凤梨■☆
123635 Deinandra Greene = Hemizonia DC. ■☆
123636 Deinandra Greene(1897);星香菊属■●☆
123637 Deinandra arida (D. D. Keck) B. G. Baldwin;旱生星香菊■☆
123638 Deinandra clementina (Brandegee) B. G. Baldwin;圣克利门蒂星香菊■☆
123639 Deinandra corymbosa (DC.) B. G. Baldwin;伞序星香菊■☆
123640 Deinandra corymbosa subsp. macrocephala (Nutt.) B. G. Baldwin = Deinandra corymbosa (DC.) B. G. Baldwin ■☆
123641 Deinandra fasciculata (DC.) Greene;簇生星香菊■☆
123642 Deinandra floribunda (A. Gray) Davidson et Moxley;繁花星香菊■☆
123643 Deinandra halliana (D. D. Keck) B. G. Baldwin;霍尔星香菊■☆
123644 Deinandra increscens (H. M. Hall ex D. D. Keck) B. G. Baldwin subsp. foliosa (Hoover) B. G. Baldwin = Deinandra paniculata (A. Gray) Davidson et Moxley ■☆
123645 Deinandra increscens (H. M. Hall ex D. D. Keck) B. G. Baldwin subsp. villosa (Tanowitz) B. G. Baldwin;长柔毛星香菊■☆
123646 Deinandra increscens (H. M. Hall ex D. D. Keck) Tanowitz subsp. foliosa (Hoover) Tanowitz = Deinandra paniculata (A. Gray) Davidson et Moxley ■☆
123647 Deinandra kelloggii (Greene) Greene;凯洛格星香菊■☆
123648 Deinandra lobbii (Greene) Greene;洛布星香菊■☆
123649 Deinandra minthornii (Jeps.) B. G. Baldwin;米氏星香菊■☆
123650 Deinandra mohavensis (D. D. Keck) B. G. Baldwin;莫哈维星香菊■☆
123651 Deinandra pallida (D. D. Keck) B. G. Baldwin;苍白星香菊■☆
123652 Deinandra paniculata (A. Gray) Davidson et Moxley;锥形星香菊■☆
123653 Deinandra pentactis (D. D. Keck) B. G. Baldwin;星香菊■☆
123654 Deinanthe Maxim.(1867);叉叶蓝属(银梅草属);Deinanthe ■

123655 Deinanthe bifida Maxim.;二裂叉叶蓝(二裂叶银梅草,银梅草);Bifid Deinanthe ■☆
123656 Deinanthe bifida Maxim. = Deinanthe coerulea Stapf ■
123657 Deinanthe bifida Maxim. f. rotundifolia Satomi;圆二裂叉叶蓝(圆二裂叶银梅草)■☆
123658 Deinanthe bifida Maxim. f. violacea Hayashi;紫罗兰叉叶蓝(紫罗兰色银梅草)■☆
123659 Deinanthe caerulea Stapf;叉叶蓝(银梅草);Deinanthe ■
123660 Deinbollia Schumach. = Deinbollia Schumach. et Thonn. ●☆
123661 Deinbollia Schumach. et Thonn.(1827);邓博木属;Deinbollia ●☆
123662 Deinbollia acuminata Exell;渐尖邓博木●☆
123663 Deinbollia adusta Radlk. = Deinbollia kilimandscharica Taub. var. adusta (Radlk.) Verdc. ●☆
123664 Deinbollia angustifolia D. W. Thomas;窄叶邓博木●☆
123665 Deinbollia boinensis Capuron;马岛邓博木●☆
123666 Deinbollia borbonica Scheff.;邓博木●☆
123667 Deinbollia borbonica Scheff. f. glabrata Radlk.;光邓博木●☆
123668 Deinbollia borbonica Scheff. f. subcordata Verdc.;亚心形邓博木●☆
123669 Deinbollia brachybotrys Gilg;短穗邓博木●☆
123670 Deinbollia calophylla Gilg ex Radlk.;美叶邓博木●☆
123671 Deinbollia cauliflora Hauman;茎花邓博木●☆
123672 Deinbollia crassipes Hauman;粗梗邓博木●☆
123673 Deinbollia cuneifolia Baker;楔叶邓博木●☆
123674 Deinbollia cuspidata Radlk. = Deinbollia laurentii De Wild. var. cuspidata (Radlk.) Hauman ●☆
123675 Deinbollia dahomensis A. Chev. = Deinbollia pinnata (Poir.) Schumach. et Thonn. ●☆
123676 Deinbollia dasybotrys Gilg ex Radlk.;毛穗邓博木●☆
123677 Deinbollia elliotii Gilg = Deinbollia pinnata (Poir.) Schumach. et Thonn. ●☆
123678 Deinbollia evrardii Hauman;埃夫拉尔邓博木●☆
123679 Deinbollia fanshawei Exell;范肖邓博木●☆
123680 Deinbollia fulvo-tomentella Baker f.;褐绒毛邓博木●☆
123681 Deinbollia giorgii De Wild. = Deinbollia pynaertii De Wild. var. giorgii (De Wild.) Hauman ●☆
123682 Deinbollia gossweileri Exell;戈斯邓博木●☆
123683 Deinbollia grandifolia Hook. f.;大叶邓博木●☆
123684 Deinbollia hierniana Gilg;希尔恩邓博木●☆
123685 Deinbollia indeniensis A. Chev. = Deinbollia grandifolia Hook. f. ●☆
123686 Deinbollia insignis Hook. f.;显著邓博木●☆
123687 Deinbollia kilimandscharica Taub.;基利邓博木●☆
123688 Deinbollia kilimandscharica Taub. var. adusta (Radlk.) Verdc.;煤黑邓博木●☆
123689 Deinbollia laurentii De Wild.;洛朗邓博木●☆
123690 Deinbollia laurentii De Wild. var. cuspidata (Radlk.) Hauman;骤尖邓博木●☆
123691 Deinbollia laurentii De Wild. var. cuspidata (Radlk.) Hauman = Deinbollia cuspidata Radlk. ●☆
123692 Deinbollia laurentii De Wild. var. gymnocarpa Hauman;裸果邓博木●☆
123693 Deinbollia laurifolia Baker;月桂叶邓博木●☆
123694 Deinbollia leptophylla Gilg ex Radlk.;细叶邓博木●☆
123695 Deinbollia longiacuminata Hauman;长尖邓博木●☆
123696 Deinbollia macrantha Radlk.;大花邓博木●☆
123697 Deinbollia macroura Gilg ex Radlk.;大尾邓博木●☆
123698 Deinbollia marginata Radlk. = Deinbollia xanthocarpa (Klotzsch) Radlk. ●☆
123699 Deinbollia maxima Gilg;大邓博木●☆
123700 Deinbollia molliuscula Radlk.;易变邓博木●☆
123701 Deinbollia neglecta Radlk.;忽视邓博木●☆
123702 Deinbollia nyasica Exell;尼亚斯邓博木●☆
123703 Deinbollia nyikensis Baker;尼卡邓博木●☆
123704 Deinbollia oblongifolia (E. Mey. ex Arn.) Radlk.;矩圆叶邓博木●☆
123705 Deinbollia obovata Radlk. = Deinbollia laurifolia Baker ●☆
123706 Deinbollia patentinervis Radlk. = Deinbollia laurifolia Baker ●☆
123707 Deinbollia pervillei (Blume) Radlk.;佩尔邓博木●☆
123708 Deinbollia pinnata (Poir.) Schumach. et Thonn.;羽叶邓博木●☆
123709 Deinbollia pinnata Schumach. et Thonn. = Deinbollia pinnata (Poir.) Schumach. et Thonn. ●☆
123710 Deinbollia polypus Stapf;多足邓博木●☆
123711 Deinbollia pycnophylla Gilg ex Radlk.;密叶邓博木●☆
123712 Deinbollia pynaertii De Wild.;皮那邓博木●☆
123713 Deinbollia pynaertii De Wild. var. giorgii (De Wild.) Hauman;乔治邓博木●☆
123714 Deinbollia ramiflora Taub. = Camptolepis ramiflora (Taub.) Radlk. ●☆
123715 Deinbollia reticulata Gilg ex Radlk.;网状邓博木●☆
123716 Deinbollia stenobotrys Gilg = Deinbollia cuneifolia Baker ●☆
123717 Deinbollia unguiculata Gilg;爪状邓博木●☆
123718 Deinbollia unijuga D. W. Thomas;成双邓博木●☆
123719 Deinbollia variabilis De Wild. = Deinbollia molliuscula Radlk. ●☆
123720 Deinbollia voltensis Hutch. ex Burtt Davy et Hoyle;沃尔特邓博木●☆
123721 Deinbollia xanthocarpa (Klotzsch) Radlk.;黄果邓博木●☆
123722 Deinocheilos W. T. Wang(1986);全唇苣苔属;Deinocheilos ■★
123723 Deinocheilos jiangxiense W. T. Wang;江西全唇苣苔;Jiangxi Deinocheilos ■
123724 Deinocheilos sichuanensis W. T. Wang;全唇苣苔;Deinocheilos ■
123725 Deinosmos Raf. = Erigeron L. ■●
123726 Deinosmos Raf. = Pulicaria Gaertn. ■●
123727 Deinostema T. Yamaz.(1953);泽番椒属;Deinostemma ■
123728 Deinostema adenocaulon (Maxim.) T. Yamaz. = Deinostema adenocaulum (Maxim.) T. Yamaz. ■
123729 Deinostema adenocaulum (Maxim.) T. Yamaz.;有腺泽番椒(毛泽蕃椒,泽蕃椒)■
123730 Deinostema violaceum (Maxim.) T. Yamaz.;泽番椒(堇色泽番椒,紫水八角);Violet Deinostemma ■
123731 Deinostigma W. T. Wang et Z. Y. Li(1992);越南苣苔属■☆
123732 Deinostigma poilanei (Pellegr.) W. T. Wang et Z. Y. Li;越南苣苔■☆
123733 Deiregyne Schltr.(1920);颈柱兰属■☆
123734 Deiregyne confusa Garay;颈柱兰■☆
123735 Deiregynopsis Rauschert = Aulosepalum Garay ■☆
123736 Deiregynopsis Rauschert = Gamosepalum Hausskn. ■●
123737 Dekindtia Gilg = Chionanthus L. ●
123738 Dekindtia Gilg = Linociera Sw. ex Schreb.(保留属名)●
123739 Dekindtia africana Gilg = Chionanthus battiscombei (Hutch.) Stearn ■☆
123740 Dekinia M. Martens et Galeotti = Agastache J. Clayton ex

Gronov. ■

123741 Dela Adans. = Libanotis Haller ex Zinn(保留属名)■

123742 Dela Adans. = Libanotis Hill(废弃属名)■

123743 Dela Adans. = Seseli L. ■

123744 Delabechea Lindl. = Sterculia L. ●

123745 Delaetia Backeb. = Neoporteria Britton et Rose ●■

123746 Delairea Lem. (1844);肉藤菊属(德氏藤属);German-ivy ■☆

123747 Delairea Lem. = Senecio L. ■●

123748 Delairea Post et Kuntze = Baphia Afzel. ex Lodd. ●☆

123749 Delairea Post et Kuntze = Delaria Desv. ●☆

123750 Delairea odorata Lem.;肉藤菊(德国常春藤,德氏藤,蔓茎千里光);Cape-Ivy,German Ivy,German-ivy ●☆

123751 Delairea odorata Lem. = Senecio mikanioides Otto ex Walp. ●☆

123752 Delairea scandens Lem. = Senecio mikanioides Otto ex Walp. ●☆

123753 Delairia Lem. = Senecio L. ■●

123754 Delamerea S. Moore(1900);锐苞菊属■☆

123755 Delamerea procumbens S. Moore;锐苞菊■☆

123756 Delaportea Thorel ex Gagnep. = Acacia Mill. (保留属名)●■

123757 Delarbrea Vieill. (1865);德拉五加属●☆

123758 Delarbrea arborea Vieill. ex R. Vig.;乔木德拉五加●☆

123759 Delarbrea collina Vieill.;德拉五加●☆

123760 Delarbrea longicarpa R. Vig.;长果德拉五加●☆

123761 Delarbrea montana R. Vig.;山地德拉五加●☆

123762 Delarbrea paradoxa Vieill.;奇异德拉五加●☆

123763 Delaria Desv. = Baphia Afzel. ex Lodd. ●☆

123764 Delastrea A. DC. = Labramia A. DC. ●☆

123765 Delavaya Franch. (1886);茶条木属(滇木瓜属,黑枪杆属);Delavaya ●

123766 Delavaya toxocarpa Franch.;茶条木(滇木瓜,黑枪杆,米香树);Yunnan Delavaya ●

123767 Delavaya yunnanensis Franch. = Delavaya toxocarpa Franch. ●

123768 Delia Dumort. = Segetella Desv. ■

123769 Delia Dumort. = Spergularia (Pers.) J. Presl et C. Presl(保留属名)■

123770 Delia segetalis Dumort. = Spergularia segetalis (L.) G. Don ■☆

123771 Delila Pfeiff. = Delia Dumort. ■

123772 Delilea Kuntze = Delilia Spreng. ■☆

123773 Delilia Spreng. (1823);圆苞菊属■☆

123774 Delilia Spreng. = Elvira Cass. ■☆

123775 Delilia biflora (L.) Kuntze;双花圆苞菊■☆

123776 Delima L. = Tetracera L. ●

123777 Delima sarmentosa L. = Tetracera sarmentosa (L.) Vahl ●

123778 Delima sarmentosa L. = Tetracera scandens (L.) Merr. ●

123779 Delimaceae Mart. = Dilleniaceae Salisb. (保留科名)●■

123780 Delimopsis Miq. = Delima L. ●

123781 Delissea Gaudich. (1826);基扭桔梗属●☆

123782 Delissea acuminata Gaudich.;渐尖基扭桔梗●☆

123783 Delissea angustifolia C. Presl;窄叶基扭桔梗●☆

123784 Delissea bicolor (St. John) H. St. John;二色基扭桔梗●☆

123785 Delissea floribunda (E. Wimm.) H. St. John;多叶基扭桔梗●☆

123786 Delissea glabrifolia H. St. John;光叶基扭桔梗●☆

123787 Delissea humilis (H. Wawra) H. St. John;矮基扭桔梗●☆

123788 Delissea lanceolata A. Gray;披针叶基扭桔梗●☆

123789 Delissea longicalyx H. St. John;长萼基扭桔梗●☆

123790 Delissea multispicata (H. Lévl.) H. St. John;多穗基扭桔梗●☆

123791 Delissea platyphylla A. Gray;宽叶基扭桔梗●☆

123792 Delissea pulchra (Rock) H. St. John;美丽基扭桔梗●☆

123793 Delissea purpurea H. St. John;紫基扭桔梗●☆

123794 Delissea pycnocarpa (Hillebr.) H. St. John;密果基扭桔梗●☆

123795 Delivaria Miq. = Acanthus L. ●■

123796 Delivaria Miq. = Dilivaria Juss. ●■

123797 Deloderium Cass. = Leontodon L. (保留属名)■☆

123798 Delognaea Cogn. = Ampelosicyos Thouars ●■☆

123799 Delognea Cogn. = Ampelosycyos Thouars ●■☆

123800 Delognea humblotii Cogn. = Ampelosycios humblotii (Cogn.) Jum. et H. Perrier ●☆

123801 Delonix Raf. (1837);凤凰木属;Flamboyant,Flamboyant Tree,Flamboyanttree,Mohur,Poinciana ●

123802 Delonix adansonioides (R. Vig.) Capuron = Delonix floribunda (Baill.) Capuron ●☆

123803 Delonix baccal (Chiov.) Baker f.;巴卡尔凤凰木●☆

123804 Delonix boiviniana (Baill.) Capuron;博伊文凤凰木●☆

123805 Delonix brachycarpa (R. Vig.) Capuron;短果凤凰木●☆

123806 Delonix decaryi (R. Vig.) Capuron;德氏凤凰木●☆

123807 Delonix elata (L.) Gamble;高凤凰木●☆

123808 Delonix floribunda (Baill.) Capuron;多花凤凰木●☆

123809 Delonix leucantha (R. Vig.) Du Puy,Phillipson et R. Rabev.;白花凤凰木●☆

123810 Delonix pumila Du Puy,Phillipson et R. Rabev.;小凤凰木●☆

123811 Delonix regia (Bojer ex Hook.) Raf.;凤凰木(凤凰花,红花楹,火树,金凤树);Flamboyant,Flamboyant Tree,Flamboyanttree,Flame of the Forest,Flame Tree,Flame-of-the-forest,Fleur-de-paradis,Golden Phoenix Tree,Peacock Flower,Peacock-flower,Poinciana,Royal Poinciana ●

123812 Delonix regia (Bojer) Raf. = Delonix regia (Bojer ex Hook.) Raf. ●

123813 Delonix regia (Bojer) Raf. var. flavida Stehlé = Delonix regia (Bojer ex Hook.) Raf. ●

123814 Delonix tomentosa (R. Vig.) Capuron;毛凤凰木●☆

123815 Delonix velutina Capuron;短毛凤凰木●☆

123816 Delopyrum Small = Polygonella Michx. ■☆

123817 Delopyrum articulatum (L.) Small = Polygonella articulata (L.) Meisn. ■☆

123818 Delopyrum basiramia Small = Polygonella basiramia (Small) G. L. Nesom et V. M. Bates ■☆

123819 Delosperma N. E. Br. (1925);露子花属;Delosperma,Hardy Ice Plant ●☆

123820 Delosperma 'Alba' = Delosperma litorale (Kensit) L. Bolus ■☆

123821 Delosperma aberdeenense (L. Bolus) L. Bolus;花飞鸟■☆

123822 Delosperma aberdeenense L. Bolus = Delosperma aberdeenense (L. Bolus) L. Bolus ■☆

123823 Delosperma abyssinicum (Regel) Schwantes;花宇治■☆

123824 Delosperma abyssinicum Schwantes = Delosperma abyssinicum (Regel) Schwantes ■☆

123825 Delosperma acocksii L. Bolus;阿氏露子花■☆

123826 Delosperma acocksii L. Bolus var. luxurians ? = Delosperma acocksii L. Bolus ■☆

123827 Delosperma acuminatum L. Bolus;渐尖露子花■☆

123828 Delosperma adelaidense Lavis;阿德莱德露子花■☆

123829 Delosperma affine Lavis;近缘露子花■☆

123830 Delosperma algoense L. Bolus;阿尔高露子花■☆

123831 Delosperma alpinum (N. E. Br.) S. A. Hammer et A. P. Dold;高山露子花■☆

123832 Delosperma alticola L. Bolus;高原露子花■☆

123833 Delosperma angustifolium L. Bolus = Corpuscularia angustifolia (L. Bolus) H. E. K. Hartmann ●☆
123834 Delosperma angustipetalum Lavis = Corpuscularia angustipetala (Lavis) H. E. K. Hartmann ●☆
123835 Delosperma annulare L. Bolus;一年露子花■☆
123836 Delosperma asperulum (Salm-Dyck) L. Bolus = Drosanthemum asperulum (Salm-Dyck) Schwantes ■☆
123837 Delosperma bosseranum Marais;博瑟露子花■☆
123838 Delosperma brevipetalum L. Bolus;短瓣露子花■☆
123839 Delosperma brevisepalum L. Bolus;短萼露子花■☆
123840 Delosperma brevisepalum L. Bolus var. majus ? = Delosperma brevisepalum L. Bolus ■☆
123841 Delosperma britteniae L. Bolus = Corpuscularia britteniae (L. Bolus) H. E. K. Hartmann ●☆
123842 Delosperma brunnthaleri (A. Berger) Schwantes = Delosperma brunnthaleri (A. Berger) Schwantes ex Jacobsen ■☆
123843 Delosperma brunnthaleri (A. Berger) Schwantes ex Jacobsen;红鳞菊■☆
123844 Delosperma burtoniae L. Bolus;伯顿露子花■☆
123845 Delosperma caespitosum L. Bolus;丛立鳞菊■
123846 Delosperma caespitosum L. Bolus var. roseum ? = Delosperma caespitosum L. Bolus ■
123847 Delosperma calycinum L. Bolus;萼状露子花■☆
123848 Delosperma carolinense N. E. Br.;卡罗来纳露子花■☆
123849 Delosperma carolinense N. E. Br. var. compacta L. Bolus = Delosperma carolinense N. E. Br. ■☆
123850 Delosperma carterae L. Bolus;卡特拉露子花■☆
123851 Delosperma clavipes Lavis;棒梗露子花■☆
123852 Delosperma concavum L. Bolus;凹露子花■☆
123853 Delosperma congestum L. Bolus;密集露子花;Gold Nugget Ice Plant ■☆
123854 Delosperma cooperi (Hook. f.) L. Bolus;花岚山(软叶鳞菊);Cooper's Hardy Ice Plant, Hardy Ice Plant, Hardy Pink Ice Plant, Pink Ice Plant ■☆
123855 Delosperma cooperi (Hook. f.) L. Bolus var. bicolor L. Bolus = Delosperma cooperi (Hook. f.) L. Bolus ■☆
123856 Delosperma cooperi L. Bolus = Delosperma cooperi (Hook. f.) L. Bolus ■☆
123857 Delosperma crassuloides (Haw.) L. Bolus;拟粗露子花■☆
123858 Delosperma crassum L. Bolus;粗露子花■☆
123859 Delosperma davyi N. E. Br.;戴维露子花■☆
123860 Delosperma deilanthoides S. A. Hammer;番杏露子花■☆
123861 Delosperma denticulatum L. Bolus;细齿露子花■☆
123862 Delosperma dolomiticum Van Jaarsv. = Delosperma vandermerwei L. Bolus ■☆
123863 Delosperma dunense L. Bolus;砂丘露子花■☆
123864 Delosperma dyeri L. Bolus;戴尔露子花■☆
123865 Delosperma dyeri L. Bolus var. laxum = Delosperma dyeri L. Bolus ■☆
123866 Delosperma echinatum (Aiton) Schwantes = Delosperma echinatum (Lam.) Schwantes ■☆
123867 Delosperma echinatum (Lam.) Schwantes;小花鳞菊(花笠)■☆
123868 Delosperma ecklonis (Salm-Dyck) Schwantes;埃氏露子花■☆
123869 Delosperma ecklonis (Salm-Dyck) Schwantes var. latifolia L. Bolus = Delosperma ecklonis (Salm-Dyck) Schwantes ■☆
123870 Delosperma edwardsiae L. Bolus = Delosperma rogersii (Schönland et A. Berger) L. Bolus ■☆
123871 Delosperma erectum L. Bolus;直立露子花■☆
123872 Delosperma esterhuyseniae L. Bolus;埃斯特露子花■☆
123873 Delosperma floribundum L. Bolus;繁花露子花■☆
123874 Delosperma framesii L. Bolus;弗雷斯露子花■☆
123875 Delosperma fredericii Lavis;弗雷德里克露子花■☆
123876 Delosperma frutescens L. Bolus;灌木露子花●☆
123877 Delosperma galpinii L. Bolus;盖尔露子花■☆
123878 Delosperma galpinii L. Bolus var. minus ? = Delosperma galpinii L. Bolus ■☆
123879 Delosperma gerstneri L. Bolus;格斯露子花■☆
123880 Delosperma giffenii Lavis;吉芬露子花■☆
123881 Delosperma gracile L. Bolus;纤细露子花■☆
123882 Delosperma gracile L. Bolus = Corpuscularia gracilis H. E. K. Hartmann ●☆
123883 Delosperma gracillimum L. Bolus = Corpuscularia gracilis H. E. K. Hartmann ■☆
123884 Delosperma gramineum L. Bolus;禾状露子花■☆
123885 Delosperma grandiflorum L. Bolus = Drosanthemum longipes (L. Bolus) H. E. K. Hartmann ●☆
123886 Delosperma grantiae L. Bolus;格兰特露子花■☆
123887 Delosperma gratiae L. Bolus;格拉露子花■☆
123888 Delosperma guthriei Lavis;格斯里露子花■☆
123889 Delosperma hallii L. Bolus = Hartmanthus halii (L. Bolus) S. A. Hammer ●☆
123890 Delosperma hirtum (N. E. Br.) Schwantes;多毛露子花■☆
123891 Delosperma hirtum (N. E. Br.) Schwantes var. bicolor L. Bolus = Delosperma hirtum (N. E. Br.) Schwantes ■☆
123892 Delosperma hollandii L. Bolus;霍兰露子花■☆
123893 Delosperma imbricatum L. Bolus;覆瓦露子花■☆
123894 Delosperma inaequale L. Bolus;不等露子花■☆
123895 Delosperma incomptum (Haw.) L. Bolus;装饰露子花■☆
123896 Delosperma incomptum (Haw.) L. Bolus var. ecklonis (Salm-Dyck) H. Jacobsen = Delosperma invalidum (N. E. Br.) H. E. K. Hartmann ■☆
123897 Delosperma inconspicuum L. Bolus;显著露子花■☆
123898 Delosperma intonsum L. Bolus;须毛露子花■☆
123899 Delosperma invalidum (N. E. Br.) H. E. K. Hartmann;粗壮露子花■☆
123900 Delosperma jansei N. E. Br.;简斯露子花■☆
123901 Delosperma karrooicum L. Bolus;卡卢露子花■☆
123902 Delosperma katbergense L. Bolus;卡特贝赫露子花■☆
123903 Delosperma klinghardtianum (Dinter) Schwantes;克林露子花■☆
123904 Delosperma lavisiae L. Bolus;花巨势(拉维斯露子花)■☆
123905 Delosperma lavisiae L. Bolus var. parisepalum ? = Delosperma lavisiae L. Bolus ■☆
123906 Delosperma laxipetalum L. Bolus;宽瓣露子花■☆
123907 Delosperma lebomboense (L. Bolus) Lavis;莱邦博露子花■☆
123908 Delosperma leendertziae N. E. Br.;伦德茨露子花■☆
123909 Delosperma lehmannii (Eckl. et Zeyh.) Schwantes;夕波■☆
123910 Delosperma lehmannii (Eckl. et Zeyh.) Schwantes = Corpuscularia lehmannii (Eckl. et Zeyh.) Schwantes ●☆
123911 Delosperma leightoniae Lavis;莱顿露子花■☆
123912 Delosperma liebenbergii L. Bolus;利本露子花■☆
123913 Delosperma lineare L. Bolus = Delosperma litorale (Kensit) L. Bolus ■☆
123914 Delosperma lineare L. Bolus var. tenuifolium ? = Delosperma lineare L. Bolus ■☆

123915 Delosperma litorale (Kensit) L. Bolus;海滨露子花;Seaside Delosperma, White Trailing Iceplant ■☆
123916 Delosperma luckhoffii L. Bolus;吕克霍夫露子花■☆
123917 Delosperma luteum L. Bolus;黄露子花■☆
123918 Delosperma lydenburgense L. Bolus;莱登堡露子花■☆
123919 Delosperma lydenburgense L. Bolus var. acutipetalum ? = Delosperma lydenburgense L. Bolus ■☆
123920 Delosperma macellum (N. E. Br.) N. E. Br.;花醍醐■☆
123921 Delosperma macellum N. E. Br. = Delosperma macellum (N. E. Br.) N. E. Br. ■☆
123922 Delosperma macrostigma L. Bolus;大柱头露子花■☆
123923 Delosperma mahonii (N. E. Br.) N. E. Br.;马洪露子花■☆
123924 Delosperma mariae L. Bolus;玛利亚露子花■☆
123925 Delosperma minimum Lavis = Corpuscularia taylori (N. E. Br.) Schwantes ●☆
123926 Delosperma monanthemum Lavis;单花露子花■☆
123927 Delosperma muiri L. Bolus;缪里露子花■☆
123928 Delosperma multiflorum L. Bolus;多花露子花■☆
123929 Delosperma nakurense (Engl.) Herre;呐古尔露子花■☆
123930 Delosperma neethlingiae (L. Bolus) Schwantes;尼特凌露子花■☆
123931 Delosperma nelii L. Bolus;尼尔露子花■☆
123932 Delosperma nubigenum (Schltr.) L. Bolus;云雾露子花;Cloud-loving Hardy Ice Plant, Hardy Ice Plant, Hardy Yellow Ice Plant, Yellow Ice Plant ■☆
123933 Delosperma obtusum L. Bolus;钝露子花■☆
123934 Delosperma oehleri (Engl.) Herre;奥勒露子花■☆
123935 Delosperma ornatulum N. E. Br.;饰冠露子花■☆
123936 Delosperma pachyrhizum L. Bolus;粗根露子花■☆
123937 Delosperma pachyrhizum L. Bolus var. pubescens ? = Delosperma pachyrhizum L. Bolus ■☆
123938 Delosperma pageanum (L. Bolus) Schwantes;纸露子花■☆
123939 Delosperma pallidum L. Bolus;苍白露子花■☆
123940 Delosperma papillatum (L. Bolus) L. Bolus = Drosanthemum papillatum L. Bolus ●☆
123941 Delosperma parviflorum L. Bolus;小花露子花■☆
123942 Delosperma patersoniae (L. Bolus) L. Bolus;帕特森露子花■☆
123943 Delosperma peersii Lavis;皮尔斯露子花■☆
123944 Delosperma peglerae L. Bolus;佩格拉露子花■☆
123945 Delosperma pergamentaceum L. Bolus = Hartmanthus pergamentaceus (L. Bolus) S. A. Hammer ●☆
123946 Delosperma pergamentaceum L. Bolus var. roseum Lavis = Hartmanthus pergamentaceus (L. Bolus) S. A. Hammer ●☆
123947 Delosperma pilosulum L. Bolus;疏毛露子花■☆
123948 Delosperma platysepalum L. Bolus;宽萼露子花■☆
123949 Delosperma pondoense L. Bolus;庞多露子花■☆
123950 Delosperma pontii L. Bolus = Delosperma floribundum L. Bolus ■☆
123951 Delosperma pottsii (L. Bolus) L. Bolus;波茨露子花■☆
123952 Delosperma prasinum L. Bolus;草绿露子花■☆
123953 Delosperma pruinosum (Thunb.) J. W. Ingram;刺叶露子花■☆
123954 Delosperma pruinosum (Thunb.) J. W. Ingram = Delosperma echinatum (Lam.) Schwantes ■☆
123955 Delosperma pubipetalum L. Bolus = Drosanthemum papillatum L. Bolus ●☆
123956 Delosperma purpureum H. E. K. Hartmann;紫露子花■☆
123957 Delosperma repens L. Bolus;匍匐露子花■☆
123958 Delosperma reynoldsii Lavis;雷诺兹露子花■☆
123959 Delosperma robustum L. Bolus;花交野■☆
123960 Delosperma rogersii (Schönland et A. Berger) L. Bolus;罗杰斯露子花■☆
123961 Delosperma rogersii (Schönland et A. Berger) L. Bolus var. glabrescens L. Bolus = Delosperma rogersii (Schönland et A. Berger) L. Bolus ■☆
123962 Delosperma roseopurpureum Lavis;粉紫露子花■☆
123963 Delosperma saturatum L. Bolus;花观世■☆
123964 Delosperma saxicola Lavis;岩地露子花■☆
123965 Delosperma scabripes L. Bolus;糙梗露子花■☆
123966 Delosperma schimperi (Engl.) H. E. K. Hartmann et Niesler;欣珀露子花■☆
123967 Delosperma sphalmanthoides S. A. Hammer;凤卵草露子花■☆
123968 Delosperma stenandrum L. Bolus;窄蕊露子花■☆
123969 Delosperma steytlerae L. Bolus;花御所■☆
123970 Delosperma subclavatum L. Bolus;棍棒露子花■☆
123971 Delosperma subincanum (Haw.) Schwantes;灰毛露子花■☆
123972 Delosperma subpetiolatum L. Bolus;亚柄露子花■☆
123973 Delosperma sulcatum L. Bolus;纵沟露子花■☆
123974 Delosperma sutherlandii (Hook. f.) N. E. Br.;沙坐兰■☆
123975 Delosperma sutherlandii N. E. Br. = Delosperma sutherlandii (Hook. f.) N. E. Br. ■☆
123976 Delosperma suttoniae Lavis;萨顿露子花■☆
123977 Delosperma taylori (N. E. Br.) Schwantes = Corpuscularia taylori (N. E. Br.) Schwantes ●☆
123978 Delosperma taylori (N. E. Br.) Schwantes var. albanense L. Bolus = Corpuscularia taylori (N. E. Br.) Schwantes ●☆
123979 Delosperma testaceum (Haw.) Schwantes;淡褐露子花■☆
123980 Delosperma tradescantioides (A. Berger) L. Bolus;水竹草露子花■☆
123981 Delosperma tradescantioides (A. Berger) L. Bolus var. lebomboense L. Bolus = Delosperma lebomboense (L. Bolus) Lavis ■☆
123982 Delosperma truteri Lavis;特鲁特尔露子花■☆
123983 Delosperma uitenhagense L. Bolus;埃滕哈赫露子花■☆
123984 Delosperma uncinatum L. Bolus;具钩露子花■☆
123985 Delosperma uniflorum L. Bolus;独花露子花■☆
123986 Delosperma vandermerwei L. Bolus;范德露子花■☆
123987 Delosperma velutinum L. Bolus;短绒毛露子花■☆
123988 Delosperma verecundum L. Bolus;羞涩露子花■☆
123989 Delosperma versicolor L. Bolus;变色露子花■☆
123990 Delosperma vinaceum (L. Bolus) L. Bolus;葡萄酒色露子花■☆
123991 Delosperma virens L. Bolus;绿露子花■☆
123992 Delosperma waterbergense L. Bolus;沃特露子花■☆
123993 Delosperma wilmaniae Lavis;维尔曼露子花■☆
123994 Delosperma zoeae L. Bolus;佐薇露子花■☆
123995 Delosperma zoutpansbergense L. Bolus;佐特露子花■☆
123996 Delostoma D. Don(1823);张口紫葳属●☆
123997 Delostoma gracile A. H. Gentry;细张口紫葳●☆
123998 Delostoma integrifolium D. Don;全缘张口紫葳●☆
123999 Delostylis Raf. = Trillium L. ■
124000 Delpechia Montrouz. = Psychotria L.(保留属名)●
124001 Delphidium Raf. = Delphinium L. ■
124002 Delphinacanthus Benoist = Pseudodicliptera Benoist ●■☆
124003 Delphinastrum (DC.) Spach = Delphinium L. ■
124004 Delphinastrum Spach = Delphinium L. ■
124005 Delphiniaceae Baum. -Bod.;翠雀花科■

124006 Delphiniaceae Baum. -Bod. = Ranunculaceae Juss.(保留科名)●■

124007 Delphiniastrum Willis = Delphinastrum (DC.) Spach ■

124008 Delphinium L.(1753);翠雀花属(翠雀属,飞燕草属);Delphinium,Larkspur ■

124009 Delphinium Tourn. ex L. = Delphinium L. ■

124010 Delphinium aconitioides F. H. Chen = Delphinium forrestii Diels ■

124011 Delphinium acuminatissimum W. T. Wang = Delphinium kingianum Brühl ex Huth var. acuminatissimum (W. T. Wang) W. T. Wang ■

124012 Delphinium aemulans Nevski;塔城翠雀花;Tacheng Larkspur ■

124013 Delphinium ajacis L. = Consolida ajacis (L.) Schur ■

124014 Delphinium ajacis L. = Consolida ambigua (L.) P. W. Ball et Heywood ■☆

124015 Delphinium aktoense W. T. Wang;阿克陶翠雀花;Aktao Larkspur ■

124016 Delphinium alabamicum Kral;阿拉巴马翠雀花;Alabama Larkspur ■☆

124017 Delphinium albescens Rydb. = Delphinium carolinianum Walter subsp. virescens (Nutt.) R. E. Brooks ■☆

124018 Delphinium albocoeruleum Maxim.;白蓝翠雀花(白蓝翠雀,翠雀);Bluewhite Larkspur ■

124019 Delphinium albocoeruleum Maxim. var. latilobum Y. Z. Zhao;宽裂白蓝翠雀花■

124020 Delphinium albocoeruleum Maxim. var. przewalskii (Huth) W. T. Wang;贺兰山翠雀花(贺兰翠雀花,贺兰山翠雀,乌药);Przewalsk Larkspur ■

124021 Delphinium albocoeruleum Maxim. var. pumilum Huth = Delphinium albocoeruleum Maxim. ■

124022 Delphinium albomarginatum F. H. Chen = Delphinium chenii W. T. Wang ■

124023 Delphinium albomarginatum Simonova;白边翠雀花■

124024 Delphinium alpestre Rydb.;高山翠雀花;Alpine Larkspur ■☆

124025 Delphinium alpinum Waldst. et Kit. = Delphinium elatum L. ■☆

124026 Delphinium altaicum Newski;阿尔泰翠雀花■

124027 Delphinium altissimum Wall.;高茎翠雀花■

124028 Delphinium altissimum Wall. subsp. drepanocentrum (Brühl ex Huth) Brühl = Delphinium drepanocentrum (Brühl ex Huth) Munz ■

124029 Delphinium altissimum Wall. subsp. drepanocentrum (Brühl ex Huth) Brühl = Delphinium umbrosum Hand. -Mazz. var. drepanocentrum (Brühl ex Huth) W. T. Wang ■

124030 Delphinium altissimum Wall. var. drepanocentrum Brühl ex Huth = Delphinium umbrosum Hand. -Mazz. var. drepanocentrum (Brühl ex Huth) W. T. Wang ■

124031 Delphinium altissimum Wall. var. drepanocentrura Brühl = Delphinium umbrosum Hand. -Mazz. var. drepanocentrum (Brühl ex Huth) W. T. Wang ■

124032 Delphinium amabile Chang Y. Yang et B. Wang = Delphinium wangii M. J. Warnock ■

124033 Delphinium andersonii A. Gray;安氏翠雀花;Anderson's Larkspur ■☆

124034 Delphinium andersonii A. Gray subsp. cognatum (Greene) Ewan = Delphinium andersonii A. Gray ■☆

124035 Delphinium andersonii A. Gray var. scaposum (Greene) S. L. Welsh = Delphinium scaposum Greene ■☆

124036 Delphinium angustipaniculatum W. T. Wang;宕昌翠雀花(狭序翠雀花);Narrowspike Larkspur ■

124037 Delphinium angustirhombicum W. T. Wang;狭菱形翠雀花■

124038 Delphinium anhweiensis W. T. Wang;安徽翠雀花■

124039 Delphinium anthriscifolium Hance;还亮草(车子野芫荽,翠雀儿,对叉草,峨山草乌,飞燕草,蝴蝶菊,还魂草,牛疔草,蛇衔草,桃朱术,鱼灯苏);Chervil Larkspur ■

124040 Delphinium anthriscifolium Hance f. latilobulatum W. T. Wang = Delphinium anthriscifolium Hance ■

124041 Delphinium anthriscifolium Hance var. calleryi (Franch.) Finet et Gagnep. = Delphinium anthriscifolium Hance ■

124042 Delphinium anthriscifolium Hance var. majus Pamp.;大花还亮草(绿花草);Bigflower Chervil Larkspur ■

124043 Delphinium anthriscifolium Hance var. savatieri (Franch.) Munz;卵瓣还亮草(对叉草,蝴蝶菊,还魂草,岩黄连);Ovatepetal Chervil Larkspur ■

124044 Delphinium antoninum Eastw.;安东翠雀花;Anthony Peak Larkspur ■☆

124045 Delphinium apetalum Huth = Aconitum apetalum (Huth) B. Fedtsch. ■

124046 Delphinium arcuatum N. Busch;拱形翠雀花;Arcuate Larkspur ■☆

124047 Delphinium armeniacum A. Heller = Delphinium nudicaule Torr. et Gray ■☆

124048 Delphinium autumnale Hand. -Mazz. = Delphinium kamaonense Huth var. autumnale (Hand. -Mazz.) W. T. Wang ■

124049 Delphinium autumnale N. Busch = Delphinium kamaonense Huth var. autumnale (Hand. -Mazz.) W. T. Wang ■

124050 Delphinium bakeri Ewan;贝克翠雀花;Baker's Larkspur ■☆

124051 Delphinium balansae Boiss. et Reut.;巴兰萨翠雀花■☆

124052 Delphinium baoshanense W. T. Wang = Delphinium delavayi Franch. var. baoshanense (W. T. Wang) W. T. Wang ■

124053 Delphinium barbatum Bunge;髯毛翠雀花■☆

124054 Delphinium barbeyi (Huth) Huth;巴比翠雀花;Barbey's Larkspur,Subalpine Larkspur,Tall Larkspur ■☆

124055 Delphinium basalticum M. J. Warnock;哥伦比亚翠雀花;Columbia Gorge Larkspur ■☆

124056 Delphinium batalinii Huth;巴塔林氏翠雀花;Batalin Larkspur ■☆

124057 Delphinium batangense Finet et Gagnep.;巴塘翠雀花;Batang Larkspur ■

124058 Delphinium beesianum W. W. Sm.;宽距翠雀花;Bees Larkspur,Beesian Larkspur ■

124059 Delphinium beesianum W. W. Sm. f. calcicola (W. W. Sm.) W. T. Wang = Delphinium beesianum W. W. Sm. ■

124060 Delphinium beesianum W. W. Sm. var. latisectum W. T. Wang;粗裂宽距翠雀花;Laisect Beesian Larkspur ■

124061 Delphinium beesianum W. W. Sm. var. malacotrichum Hand. -Mazz. = Delphinium caeruleum Jacq. ex Cambess. ■

124062 Delphinium beesianum W. W. Sm. var. malacotrichum Hand. -Mazz. f. radiatifolium Hand. -Mazz. = Delphinium beesianum W. W. Sm. var. radiatifolium (Hand. -Mazz.) W. T. Wang ■

124063 Delphinium beesianum W. W. Sm. var. radiatifolium (Hand. -Mazz.) W. T. Wang;辐裂宽距翠雀花(辐裂翠雀花);Radiateleaf Beesian Larkspur ■

124064 Delphinium beesianum W. W. Sm. var. radiatifolium (Hand. -Mazz.) W. T. Wang f. ramosum W. T. Wang = Delphinium beesianum W. W. Sm. var. radiatifolium (Hand. -Mazz.) W. T. Wang ■

124065 Delphinium beesianum W. W. Sm. var. radiatifolium f. ramosum

W. T. Wang = Delphinium beesianum W. W. Sm. var. radiatifolium (Hand. -Mazz.) W. T. Wang ■

124066 Delphinium bicolor Nutt.;双色翠雀花;Flathead Larkspur, Little Larkspur, Low Larkspur ■☆

124067 Delphinium biternatum Huth;三出翠雀花(黄白翠雀花,黄花翠雀,黄花飞燕草);Triternate Larkspur, Yellowish-white Larkspur ■

124068 Delphinium blochmaniae Greene = Delphinium parryi A. Gray subsp. blochmaniae (Greene) H. F. Lewis et Epling ■☆

124069 Delphinium bonatii H. Lév. = Delphinium grandiflorum L. ■

124070 Delphinium bonvalotii Franch. = Delphinium potaninii Huth var. bonvalotii (Franch.) W. T. Wang ■

124071 Delphinium bonvalotii Franch. var. eriostylum (H. Lév.) W. T. Wang = Delphinium eriostylum H. Lév. ■

124072 Delphinium bonvalotii Franch. var. hispidum W. T. Wang = Delphinium eriostylum H. Lév. var. hispidum (W. T. Wang) W. T. Wang ■

124073 Delphinium brachycentrum Ledeb.;北极翠雀花;Arctic Larkspur ■☆

124074 Delphinium bracteosum Sommier et H. Lév.;多苞片翠雀花■☆

124075 Delphinium brevisepalum W. T. Wang;短萼翠雀花■

124076 Delphinium brunonianum Royle;囊距翠雀花(甲果贝,囊距翠雀,雀沟勃,麝香翠雀);Musk Larkspur, Saccatespur Larkspur ■

124077 Delphinium brunonianum Royle var. densum Maxim. = Delphinium brunonianum Royle ■

124078 Delphinium brunonianum Royle var. densum Maxim. = Delphinium densiflorum Duthie ex Huth ■

124079 Delphinium bucharicum Popov;布哈尔翠雀花;Bucharic Larkspur ■☆

124080 Delphinium bulleyanum Forrest ex Diels;拟螺距翠雀花;Bulley Larkspur, Falsespiralspur Larkspur ■

124081 Delphinium buschianum Grossh.;布什翠雀花■☆

124082 Delphinium caeruleum Jacq. = Delphinium caeruleum Jacq. ex Cambess. ■

124083 Delphinium caeruleum Jacq. ex Cambess.;蓝翠雀花(雀冈);Coeruleus Larkspur, Skyblue Larkspur ■

124084 Delphinium caeruleum Jacq. ex Cambess. f. album W. T. Wang = Delphinium caeruleum Jacq. ex Cambess. ■

124085 Delphinium caeruleum Jacq. ex Cambess. var. crassicalcaratum W. T. Wang et M. J. Marnock;粗距蓝翠雀花■

124086 Delphinium caeruleum Jacq. ex Cambess. var. majus W. T. Wang;大叶蓝翠雀花;Bigleaf Skyblue Larkspur ■

124087 Delphinium caeruleum Jacq. ex Cambess. var. obtusilobum Brühl ex Huth;钝裂蓝翠雀花;Obtuselob Coeruleus Larkspur, Obtuselobed Coeruleus Larkspur ■

124088 Delphinium caeruleum Jacq. ex Cambess. var. tenuicaule Brühl ex Huth = Delphinium nortonii Dunn ■

124089 Delphinium caeruleum Jacq. var. tenuicaule Brühl ex Huth = Delphinium nortonii Dunn ■

124090 Delphinium calcicola W. W. Sm. = Delphinium beesianum W. W. Sm. ■

124091 Delphinium californicum Torr. et A. Gray;加州翠雀花;Coast Larkspur ■☆

124092 Delphinium californicum Torr. et A. Gray subsp. interius (Eastw.) Ewan;间型加州翠雀花;Hospital Canyon Larkspur ■☆

124093 Delphinium californicum Torr. et A. Gray var. interius Eastw. = Delphinium californicum Torr. et A. Gray subsp. interius (Eastw.) Ewan ■☆

124094 Delphinium calleryi Franch. = Delphinium anthriscifolium Hance ■

124095 Delphinium calophyllum W. T. Wang;美叶翠雀花;Beautifulleaf Larkspur ■

124096 Delphinium calthifolium Q. E. Yang et Y. Luo;驴蹄草叶翠雀花■

124097 Delphinium camptocarpum Fisch. et C. A. Mey.;弯果翠雀花■☆

124098 Delphinium campylocentrum Maxim.;弯距翠雀花;Bentspur Larkspur ■

124099 Delphinium candelabrum Ostenf.;奇林翠雀花(单花翠雀花,奇林翠雀,奇林飞燕草);Candelabrum Larkspur, Qilin Larkspur ■

124100 Delphinium candelabrum Ostenf. var. monanthum Ostenf.;单花翠雀花(奇林翠雀);Oneflower Larkspur ■

124101 Delphinium candidum Hemsl. = Delphinium leroyi Franch. ex Huth ■☆

124102 Delphinium caprorum Ewan = Delphinium glareosum Greene ■☆

124103 Delphinium cardinale Hook.;加州猩红翠雀花;Scarlet Larkspur ■☆

124104 Delphinium carolinianum Walter;卡罗来纳翠雀花;Blue Larkspur, Carolina Larkspur, Desert Larkspur, Prairie Larkspur, Wild Blue Larkspur ■☆

124105 Delphinium carolinianum Walter subsp. penardii (Huth) Warnock = Delphinium carolinianum Walter subsp. virescens (Nutt.) R. E. Brooks ■☆

124106 Delphinium carolinianum Walter subsp. virescens (Nutt.) R. E. Brooks;绿卡罗来纳翠雀花;Carolina Larkspur, Penard's Larkspur, Plains Larkspur, Prairie Larkspur, White Larkspur ■☆

124107 Delphinium carolinianum Walter var. crispum L. M. Perry = Delphinium carolinianum Walter ■☆

124108 Delphinium carolinianum Walter var. nortonianum (Mack. et Bush) L. M. Perry = Delphinium carolinianum Walter ■☆

124109 Delphinium cashmerianum Royle;克什米尔翠雀花■☆

124110 Delphinium caucasicum C. A. Mey.;高加索翠雀花;Caucas Larkspur ■☆

124111 Delphinium caucasicum C. A. Mey. var. tangutica Maxim. = Delphinium albocoeruleum Maxim. ■

124112 Delphinium caudatolobum W. T. Wang;尾裂翠雀花;Caudatelobed Larkspur ■

124113 Delphinium cavaleriense H. Lév. = Delphinium anthriscifolium Hance ■

124114 Delphinium cavaleriense H. Lév. et Vaniot = Delphinium anthriscifolium Hance ■

124115 Delphinium ceratophoroides W. T. Wang;拟角萼翠雀花;Corneredcalyxlike Larkspur ■

124116 Delphinium ceratophorum Franch.;角萼翠雀花(虎耳兰,角瓣翠雀花,角花翠雀,虱子药);Corneredcalyx Larkspur ■

124117 Delphinium ceratophorum Franch. var. breviconiculatum W. T. Wang;短角萼翠雀花;Short Corneredcalyx Larkspur ■

124118 Delphinium ceratophorum Franch. var. brevicorniculatum W. T. Wang f. lobatum W. T. Wang = Delphinium ceratophorum Franch. var. breviconiculatum W. T. Wang ■

124119 Delphinium ceratophorum Franch. var. hirsutum W. T. Wang;毛角萼翠雀花;Hirsute Corneredcalyx Larkspur ■

124120 Delphinium ceratophorum Franch. var. robustum W. T. Wang;粗壮角萼翠雀花;Robust Corneredcalyx Larkspur ■

124121 Delphinium cerefolium H. Lév. et Vaniot = Delphinium anthriscifolium Hance ■

124122 Delphinium chamissonis Pritz. ex Walp. = Delphinium brachycentrum Ledeb. ■☆
124123 Delphinium chayuense W. T. Wang;察隅翠雀花■
124124 Delphinium chefoense Franch.;烟台翠雀花(苦莲,山鸦雀儿);Yantai Larkspur ■
124125 Delphinium chefoense Franch. = Delphinium grandiflorum L. var. gilgianum (Pilg. ex Gilg) Finet et Gagnep. ■
124126 Delphinium cheilanthum Fisch. = Delphinium siwanense Franch. ■
124127 Delphinium cheilanthum Fisch. ex DC.;唇花翠雀花(长距飞燕草);Carland Larkspur,Lipflower Larkspur ■
124128 Delphinium cheilanthum Fisch. ex DC. var. pubescens Y. Z. Zhao;展毛唇花翠雀花■
124129 Delphinium chenii W. T. Wang;白缘翠雀花;Chen Larkspur ■
124130 Delphinium chinense Fisch. = Delphinium grandiflorum L. ■
124131 Delphinium chinense Fisch. ex DC. = Delphinium grandiflorum L. ■
124132 Delphinium chrysotrichum Finet et Gagnep.;黄毛翠雀花;Yellowhiry Larkspur ■
124133 Delphinium chrysotrichum Finet et Gagnep. var. pygmaeum Ostenf. = Delphinium tangkulaense W. T. Wang ■
124134 Delphinium chrysotrichum Finet et Gagnep. var. tsarongense (Hand. -Mazz.) W. T. Wang;察瓦龙翠雀花;Chawalong Larkspur,Tsarong Larkspur ■
124135 Delphinium chumulangmaense W. T. Wang;珠峰翠雀花;Chumlangma Larkspur,Jolmi Larkspur ■
124136 Delphinium chungbaense W. T. Wang;仲巴翠雀花;Chungba Larkspur,Zhongba Larkspur ■
124137 Delphinium coelestinum Franch. = Delphinium hirticaule Franch. ■
124138 Delphinium coleopodum Hand. -Mazz.;鞘柄翠雀花(带角飞燕草);Sheathing Larkspur,Sheathstalk Larkspur ■
124139 Delphinium conaense W. T. Wang;错那翠雀花;Cuona Larkspur ■
124140 Delphinium confusum Popov;易混翠雀■☆
124141 Delphinium consolida L.;琉璃翠雀花(琉璃飞燕草);Branching Larkspur,Field Larkspur,Forking Larkspur ■
124142 Delphinium consolida L. = Consolida ajacis (L.) Schur ■
124143 Delphinium consolida L. = Consolida regalis Gray ■☆
124144 Delphinium consolida L. imperiale ?;壮丽琉璃翠雀花■☆
124145 Delphinium consolida L. var. pubescens DC. = Consolida pubescens (DC.) Soó ■☆
124146 Delphinium cordiopetalum ?;心瓣翠雀花■☆
124147 Delphinium corymbosum Regel;伞房花序翠雀花(伞房花翠雀花);Corymbose Larkspur ■☆
124148 Delphinium cossonianum Batt.;科森翠雀花■☆
124149 Delphinium cossonianum Batt. var. laxiflorum Gatt. et Maire = Delphinium cossonianum Batt. ■☆
124150 Delphinium crassifolium Schrad. ex Ledeb. = Delphinium crassifolium Schrad. ex Spreng. ■
124151 Delphinium crassifolium Schrad. ex Spreng.;厚叶翠雀花(根叶飞燕草);Thickleaf Larkspur ■
124152 Delphinium crassifolium Schrad. ex Spreng. var. tangutica Maxim. = Delphinium kansuense W. T. Wang ■
124153 Delphinium crispulum Rupr.;皱波翠雀花;Crisped Larkspur ■☆
124154 Delphinium cryophilum Nevski;幽谷翠雀花■☆
124155 Delphinium cultorum ? = Delphinium hybridum L. ■
124156 Delphinium cuneatum Steven ex DC.;俄罗斯翠雀花(楔形翠雀花);Cuneate Larkspur ■☆
124157 Delphinium cuyamacae Abrams = Delphinium hesperium A. Gray subsp. cuyamacae (Abrams) H. F. Lewis et Epling ■☆
124158 Delphinium cyananthum Nevski;蓝花翠雀■☆
124159 Delphinium cyanoreios Piper = Delphinium depauperatum Nutt. ■☆
124160 Delphinium cyanoreios Piper f. multiplex Ewan = Delphinium multiplex (Ewan) C. L. Hitchc. ■☆
124161 Delphinium dasyanthum Kar. et Kir.;毛花翠雀■☆
124162 Delphinium dasycarpum Steven ex DC.;毛果翠雀花■
124163 Delphinium dasycaulon Fresen. var. minor Cufod. = Delphinium dasycaulon Fresen. ■☆
124164 Delphinium davidii Franch.;谷地翠雀花(四川飞燕草);David Larkspur ■
124165 Delphinium davidii Franch. = Delphinium eriostylum H. Lév. ■
124166 Delphinium davidii Franch. var. saxatile (W. T. Wang) W. T. Wang;岩生翠雀花;Rockliving Larkspur ■
124167 Delphinium davidii Franch. var. saxatile (W. T. Wang) W. T. Wang = Delphinium saxatile W. T. Wang ■
124168 Delphinium decorum Fisch. et C. A. Mey.;沿海翠雀花;Coast Larkspur,Larkspur ■☆
124169 Delphinium decorum Fisch. et C. A. Mey. subsp. tracyi Ewan;特拉西翠雀花;Tracy's Larkspur ■☆
124170 Delphinium decorum Fisch. et C. A. Mey. var. patens (Benth.) A. Gray = Delphinium patens Benth. ■☆
124171 Delphinium delavayi Franch.;滇川翠雀花(滇北翠雀花,鸡足乌草,细草乌,小草乌);Delavay Larkspur ■
124172 Delphinium delavayi Franch. f. aureum W. T. Wang = Delphinium delavayi Franch. ■
124173 Delphinium delavayi Franch. var. acuminatum Franch. = Delphinium delavayi Franch. ■
124174 Delphinium delavayi Franch. var. baoshanense (W. T. Wang) W. T. Wang;保山翠雀花■
124175 Delphinium delavayi Franch. var. lasiandrum W. T. Wang;毛蕊翠雀花■
124176 Delphinium delavayi Franch. var. pogonanthum (Hand. -Mazz.) W. T. Wang;须花翠雀花(白升麻);Beardflower Delavay Larkspur ■
124177 Delphinium densiflorum Duthie ex Huth;密花翠雀花;Denseflower Larkspur ■
124178 Delphinium denudatum Wall.;裸露翠雀花■☆
124179 Delphinium denudatum Wall. var. yunnanense Franch. = Delphinium yunnanense Franch. ■
124180 Delphinium depauperatum Nutt.;矮小翠雀花;Blue Mountain Larkspur,Dwarf Larkspur,Mountain Larkspur ■☆
124181 Delphinium dictyocarpum Steud.;网果翠雀花(网纹果翠雀花);Reticulatefruit Larkspur ■☆
124182 Delphinium distichum Geyer ex Hook.;二裂翠雀花;Meadow Larkspur,Strict Larkspur ■☆
124183 Delphinium divaricaticum Ledeb.;疏枝翠雀花■☆
124184 Delphinium diversifolium Greene = Delphinium depauperatum Nutt. ■☆
124185 Delphinium diversifolium Greene subsp. harneyense Ewan = Delphinium depauperatum Nutt. ■☆
124186 Delphinium dolichocentroides W. T. Wang;拟长距翠雀花;Longicarcarate Larkspur,Longspurlike Larkspur ■
124187 Delphinium dolichocentroides W. T. Wang var. leiogynum W. T.

Wang;基苞翠雀花■
124188 Delphinium dolichocentroides W. T. Wang var. parvidolium W. T. Wang = Delphinium dolichocentroides W. T. Wang ■
124189 Delphinium dolichocentroides W. T. Wang var. parvifolium W. T. Wang;小叶拟长距翠雀花;Smallleaf Longspurlike Larkspur ■
124190 Delphinium dolichocentroides W. T. Wang var. sericeum W. T. Wang;绢毛翠雀花■
124191 Delphinium dolichocentrum W. T. Wang = Delphinium tenii H. Lév. ■
124192 Delphinium drepanocentrum (Brühl ex Huth) Munz = Delphinium umbrosum Hand. -Mazz. var. drepanocentrum (Brühl ex Huth) W. T. Wang ■
124193 Delphinium eglandulosum Chang Y. Yang et B. Wang;无腺翠雀花;Glanduleless Larkspur ■
124194 Delphinium elatum L.;高翠雀花(高翠雀,高飞燕草,穗花翠雀);Alpine Larkspur, Bee Larkspur, Candle Delphinium, Candle Larkspur, Delphinium, High Larkspur, Tall Larkspur ■☆
124195 Delphinium elatum L. var. sericeum W. T. Wang;绢毛高翠雀花;Sericeous Larkspur, Sericeous Tall Larkspur ■
124196 Delphinium elliptico-ovatum W. T. Wang;长卵苞翠雀花;Elliptic-ovate Bract Larkspur, Longovata Larkspur ■
124197 Delphinium emarginatum C. Presl;无边苞翠雀花■☆
124198 Delphinium emarginatum C. Presl subsp. nevadense (Kunze) C. Blanché et Molero;内华达翠雀花■☆
124199 Delphinium emarginatum C. Presl var. africanum Maire = Delphinium emarginatum C. Presl ■☆
124200 Delphinium emarginatum C. Presl var. nevadense (Kunze) Maire = Delphinium emarginatum C. Presl ■☆
124201 Delphinium emeiense W. T. Wang = Delphinium omeiense W. T. Wang ■
124202 Delphinium emeiense W. T. Wang var. pubescens W. T. Wang = Delphinium omeiense W. T. Wang var. pubescens W. T. Wang ■
124203 Delphinium emiliae Greene = Delphinium variegatum Torr. et A. Gray ■☆
124204 Delphinium eriostylum H. Lév.;毛梗翠雀花(峨山草乌,毛梗川黔翠雀花,水川乌,水乌头);Wodystyle Bonvalot Larkspur ■
124205 Delphinium eriostylum H. Lév. var. hispidum (W. T. Wang) W. T. Wang;糙叶毛梗翠雀花(毛柄川黔翠雀花);Hispid Bonvalot Larkspur ■
124206 Delphinium erlangshanicum W. T. Wang;二郎山翠雀花;Erlangshan Larkspur ■
124207 Delphinium esquirolii H. Lév. = Delphinium yunnanense Franch. ■
124208 Delphinium esquirolii H. Lév. et Vaniot = Delphinium yunnanense Franch. ■
124209 Delphinium exaltatum Aiton;高大翠雀花;Tall Larkspur ■☆
124210 Delphinium exaltatum Aiton var. barbeyi Huth = Delphinium barbeyi (Huth) Huth ■☆
124211 Delphinium exiguum Pritz. = Delphinium anthriscifolium Hance ■
124212 Delphinium fangshanense W. T. Wang = Delphinium grandiflorum L. var. fangshanense (W. T. Wang) W. T. Wang ■
124213 Delphinium fargesii Franch. = Delphinium potaninii Huth var. bonvalotii (Franch.) W. T. Wang ■
124214 Delphinium fargesii Franch. = Delphinium potaninii Huth ■
124215 Delphinium favargeri C. Blanché et al.;法瓦尔热翠雀花■☆
124216 Delphinium fengii W. T. Wang = Delphinium forrestii Diels ■
124217 Delphinium flexuosum M. Bieb.;曲折翠雀花;Bent Larkspur ■☆
124218 Delphinium foetidum Lomakin;臭翠雀花;Fetid Larkspur ■☆
124219 Delphinium forrestii Diels;短距翠雀花;Forrest Larkspur ■
124220 Delphinium forrestii Diels var. leiophyllum W. T. Wang = Delphinium leiophyllum (W. T. Wang) W. T. Wang ■
124221 Delphinium forrestii Diels var. viride (W. T. Wang) W. T. Wang;光茎短距翠雀花;Glabrousstem Larkspur, Smoothstem Forrest Larkspur ■
124222 Delphinium freynii Huth;弗雷翠雀花■☆
124223 Delphinium fugorum Hand. -Mazz. = Delphinium micropetalum Finet et Gagnep. ■
124224 Delphinium georgei Comber = Delphinium taliense Franch. ■
124225 Delphinium geraniifolium Rydb.;老鹳草叶翠雀花;Mogollon Larkspur ■☆
124226 Delphinium geyeri Greene;盖格尔翠雀花;Geyer's Larkspur, Poisonweed ■☆
124227 Delphinium gilgianum Pilg. ex Gilg = Delphinium grandiflorum L. var. gilgianum (Pilg. ex Gilg) Finet et Gagnep. ■
124228 Delphinium giraldii Diels;秦岭翠雀花(虎膝,蓝花草,云雾七);Qinling Larkspur ■
124229 Delphinium glabricaule W. T. Wang;光茎翠雀花;Girald Larkspur, Glabrous Larkspur, Glabrousstem Larkspur ■
124230 Delphinium glaciale Hook. f. et Thomson;冰川翠雀花;Glacial Larkspur ■
124231 Delphinium glareosum Greene;石砾翠雀花;Olympic Mountain Larkspur ■☆
124232 Delphinium glaucescens Rydb.;白霜翠雀花;Glaucous Larkspur ■☆
124233 Delphinium glaucum S. Watson;灰白翠雀花;Brown's Larkspur, Duncecap Larkspur, Giant Larkspur, Hooker's Larkspur, Mountain Larkspur, Pale-flowered Brown's Larkspur, Tall Larkspur, Western Larkspur ■☆
124234 Delphinium gonggaense W. T. Wang;贡嘎翠雀花;Gongga Larkspur ■
124235 Delphinium gracile DC.;纤细翠雀花■☆
124236 Delphinium gracilentum Greene;格林翠雀花;Greene's Larkspur ■☆
124237 Delphinium grandiflorum L.;翠雀花(百部草,瓣根草,翠雀,大花翠雀,大花飞燕草,飞燕草,鸽子花,鸡爪连,猫眼花,土黄连,小草乌,摇咀咀花,鹦哥草,鹦哥花,玉珠色洼);Bouquet Larkspur, Largeflower Larkspur, Siberian Larkspur ■
124238 Delphinium grandiflorum L. 'Blue Butterfly';蓝蝴蝶翠雀■☆
124239 Delphinium grandiflorum L. f. album T. J. Feng et J. X. Huang;白花翠雀;Whiteflower Bouquet Larkspur ■
124240 Delphinium grandiflorum L. f. roseolum Y. Z. Zhao;粉花翠雀;Rose Largeflower Larkspur ■
124241 Delphinium grandiflorum L. var. chinense (Fisch.) DC. = Delphinium grandiflorum L. ■
124242 Delphinium grandiflorum L. var. davidii (Franch.) Brühl = Delphinium davidii Franch. ■
124243 Delphinium grandiflorum L. var. deinocarpum W. T. Wang;安泽翠雀花;Anze Bouquet Larkspur ■
124244 Delphinium grandiflorum L. var. fangshanense (W. T. Wang) W. T. Wang;房山翠雀;Fangshan Bouquet Larkspur ■
124245 Delphinium grandiflorum L. var. gilgianum (Pilg. ex Gilg) Finet et Gagnep.;腺毛翠雀花(瓣根草,蓝盏棵,灭虱草,腺毛翠雀,鹦哥花);Glandular Largeflower Larkspur ■
124246 Delphinium grandiflorum L. var. gilgianum (Pilg.) Finet et

Gagnep. = Delphinium chefoense Franch. ■
124247 Delphinium grandiflorum L. var. glandulosum W. T. Wang = Delphinium grandiflorum L. var. gilgianum (Pilg. ex Gilg) Finet et Gagnep. ■
124248 Delphinium grandiflorum L. var. kamaonense (Huth) Brühl = Delphinium kamaonense Huth ■
124249 Delphinium grandiflorum L. var. kunawarense Brühl = Delphinium caeruleum Jacq. ex Cambess. ■
124250 Delphinium grandiflorum L. var. leiocarpum W. T. Wang;光果翠雀;Smoothfruit Bouquet Larkspur, Smoothfruit Largeflower Larkspur ■
124251 Delphinium grandiflorum L. var. majus W. T. Wang = Delphinium majus (W. T. Wang) W. T. Wang ■
124252 Delphinium grandiflorum L. var. mosoynense (Franch.) Huth;裂瓣翠雀(小草乌);Cleftpetal Largeflower Larkspur ■
124253 Delphinium grandiflorum L. var. obtusilobum (Brühl ex Huth) Brühl = Delphinium caeruleum Jacq. ex Cambess. var. obtusilobum Brühl ex Huth ■
124254 Delphinium grandiflorum L. var. obtusilobum Brühl = Delphinium caeruleum Jacq. ex Cambess. var. obtusilobum Brühl ex Huth ■
124255 Delphinium grandiflorum L. var. pilosum Y. Z. Zhao;疏毛翠雀■
124256 Delphinium grandiflorum L. var. potaninii Brühl = Delphinium potaninii Huth ■
124257 Delphinium grandiflorum L. var. robustum W. T. Wang;粗壮翠雀;Robust Bouquet Larkspur, Robust Largeflower Larkspur ■
124258 Delphinium grandiflorum L. var. robustum W. T. Wang = Delphinium grandiflorum L. var. mosoynense (Franch.) Huth ■
124259 Delphinium grandiflorum L. var. robustum W. T. Wang f. glanduliferum W. T. Wang = Delphinium grandiflorum L. var. mosoynense (Franch.) Huth ■
124260 Delphinium grandiflorum L. var. tenuicaule (Brühl ex Huth) Brühl = Delphinium nortonii Dunn ■
124261 Delphinium grandiflorum L. var. tenuicaulis Brühl = Delphinium nortonii Dunn ■
124262 Delphinium grandiflorum L. var. tigridium Kitag. = Delphinium grandiflorum L. ■
124263 Delphinium grandiflorum L. var. tsangense Brühl = Delphinium caeruleum Jacq. ex Cambess. ■
124264 Delphinium grandiflorum L. var. villosum W. T. Wang;长柔毛翠雀;Villose Bouquet Larkspur, Villose Largeflower Larkspur ■
124265 Delphinium grandiflorum L. var. villosum W. T. Wang = Delphinium grandiflorum L. var. mosoynense (Franch.) Huth ■
124266 Delphinium grandilimbum W. T. Wang;硕边翠雀花■
124267 Delphinium gyalanum C. Marquand et Airy Shaw;拉萨翠雀花;Lasa Larkspur ■
124268 Delphinium gypsophilum Ewan;石膏翠雀花;Gypsum-loving Larkspur ■☆
124269 Delphinium gypsophilum Ewan subsp. parviflorum H. F. Lewis et Epling;小花石膏翠雀花;Small-flowered Gypsum-loving Larkspur ■☆
124270 Delphinium halteratum Sm. subsp. verdunense (Balb.) Asch. et Graebn.;砂丘翠雀花■☆
124271 Delphinium hamatum Franch.;钩距翠雀花;Hamate Larkspur, Hookedspur Larkspur ■
124272 Delphinium handelianum W. T. Wang;淡紫翠雀花;Handel Larkspur, Purplish Larkspur ■
124273 Delphinium hansenii (Greene) Greene;汉森翠雀花;Hansen's Larkspur ■☆
124274 Delphinium hansenii (Greene) Greene subsp. arcuatum (Greene) Ewan = Delphinium hansenii (Greene) Greene ■☆
124275 Delphinium hansenii (Greene) Greene subsp. ewanianum M. J. Warnock;埃万翠雀花;Ewan's Larkspur ■☆
124276 Delphinium hansenii (Greene) Greene subsp. kernense (Davidson) Ewan;克恩翠雀花;Kern County Larkspur ■☆
124277 Delphinium hansenii (Greene) Greene var. kernense Davidson = Delphinium hansenii (Greene) Greene subsp. kernense (Davidson) Ewan ■☆
124278 Delphinium henryi Franch.;川陕翠雀花;Henry Larkspur ■
124279 Delphinium henryi Franch. f. concolor W. T. Wang = Delphinium henryi Franch. ■
124280 Delphinium hesperium A. Gray;西部翠雀花;Western Larkspur ■☆
124281 Delphinium hesperium A. Gray f. pallescens Ewan = Delphinium hesperium subsp. pallescens (Ewan) H. F. Lewis et Epling ■☆
124282 Delphinium hesperium A. Gray subsp. cuyamacae (Abrams) H. F. Lewis et Epling;库亚翠雀花;Cuyamaca Larkspur ■☆
124283 Delphinium hesperium A. Gray var. hansenii Greene = Delphinium hansenii (Greene) Greene ■☆
124284 Delphinium hesperium A. Gray var. recurvatum (Greene) K. C. Davis = Delphinium recurvatum Greene ■☆
124285 Delphinium hesperium A. Gray var. seditiosum Jeps. = Delphinium parryi A. Gray ■☆
124286 Delphinium hesperium subsp. pallescens (Ewan) H. F. Lewis et Epling;苍白翠雀花■☆
124287 Delphinium hillcoatiae Munz;毛茛叶翠雀花;Buttercupleaf Larkspur, Ranunculifoliose Larkspur ■
124288 Delphinium hirticaule Franch.;毛茎翠雀花;Hairystem Larkspur ■
124289 Delphinium hirticaule Franch. var. coelestinum (Franch.) Finet et Gagnep. = Delphinium hirticaule Franch. ■
124290 Delphinium hirticaule Franch. var. micranthum Finet et Gagnep. = Delphinium hirticaule Franch. ■
124291 Delphinium hirticaule Franch. var. mollipes W. T. Wang;腺毛茎翠雀花(白花升麻,腺毛翠雀花);Glandhair Hairystem Larkspur ■
124292 Delphinium hirtifolium W. T. Wang;毛叶翠雀花;Hairyleaf Larkspur ■
124293 Delphinium hohenackeri Boiss.;豪氏翠雀花■
124294 Delphinium honanense W. T. Wang;河南翠雀花;Henan Larkspur, Honan Larkspur ■
124295 Delphinium honanense W. T. Wang var. piliferum W. T. Wang;毛梗河南翠雀花(云雾七);Hairypedicel Larkspur ■
124296 Delphinium hsinganense S. H. Li et Z. F. Fang;兴安翠雀花;Hsing'an Larkspur, Xing'an Larkspur ■
124297 Delphinium huangzhongense W. T. Wang;湟中翠雀花;Huangzhong Larkspur ■
124298 Delphinium hui F. H. Chen;贡嘎翠雀花(稻城翠雀花);Gongga Larkspur, Hu Larkspur ■
124299 Delphinium humilius (W. T. Wang) W. T. Wang;矮粗距翠雀花(乡城翠雀花);Dwarf Thickspurred Larkspur ■
124300 Delphinium hutchinsoniae Ewan;哈钦森翠雀花;Hutchinson's Delphinium, Hutchinson's Larkspur ■☆
124301 Delphinium hybridum L.;杂种翠雀花(大飞燕草);Delphinium, Florists' Larkspur, Larkspur, Mongrel Larkspur ■
124302 Delphinium iliense Huth;伊犁翠雀花(飞燕草,伊犁飞燕草);

Ili Larkspur, Yili Larkspur ■

124303 Delphinium incisolobulatum W. T. Wang;缺刻翠雀花;Incised Larkspur ■

124304 Delphinium inconspicuum Serg.;不显翠雀花;Inconspicuous Larkspur ■☆

124305 Delphinium inopinatum Nevski;偶生翠雀花;Occasional Larkspur ■☆

124306 Delphinium inopinum (Jeps.) H. F. Lewis et Epling;意外翠雀花;Unexpected Larkspur ■☆

124307 Delphinium jugorum Hand.-Mazz. = Delphinium micropetalum Finet et Gagnep. ■

124308 Delphinium junceum DC. = Delphinium peregrinum L. ■☆

124309 Delphinium kamaonense Huth;光序翠雀花;Glabrousinflorescence Larkspur, Kamaon Larkspur ■

124310 Delphinium kamaonense Huth var. autumnale (Hand.-Mazz.) W. T. Wang;秋翠雀花;Autumn Kamaon Larkspur, Autumn Larkspur ■

124311 Delphinium kamaonense Huth var. glabrescens (W. T. Wang) W. T. Wang;展毛翠雀花(长毛翠雀花,稀吐);Glabrescent Larkspur ■

124312 Delphinium kansuense W. T. Wang;甘肃翠雀花;Gansu Larkspur, Kansu Larkspur ■

124313 Delphinium kansuense W. T. Wang var. villosiusculum W. T. Wang;黏毛甘肃翠雀花■

124314 Delphinium kantzeense W. T. Wang;甘孜翠雀花;Ganzi Larkspur, Kuntze Larkspur ■

124315 Delphinium karataviense Pavlov;卡拉塔夫翠雀花■☆

124316 Delphinium karateginii Korsh.;喀拉特氏翠雀花;Karateg Larkspur ■☆

124317 Delphinium kaschgaricum Chang Y. Yang et B. Wang;喀什翠雀花■

124318 Delphinium kawaguchii Tamura = Delphinium gyalanum C. Marquand et Airy Shaw ■

124319 Delphinium kingianum Brühl ex Huth;密叶翠雀花;Denseleaf Larkspur, King Larkspur ■

124320 Delphinium kingianum Brühl ex Huth var. acuminatissimum (W. T. Wang) W. T. Wang;尖裂密叶翠雀花;Sharpcleft Denseleaf Larkspur ■

124321 Delphinium kingianum Brühl ex Huth var. eglandulosum W. T. Wang;少腺密叶翠雀花;Glandless Denseleaf Larkspur ■

124322 Delphinium kingianum Brühl ex Huth var. leiocarpum Brühl ex Huth;光果密叶翠雀花(光果翠雀花);Smoothfruit Denseleaf Larkspur ■

124323 Delphinium kinkiense Munz = Delphinium variegatum Torr. et A. Gray subsp. kinkiense (Munz) M. J. Warnock ■☆

124324 Delphinium knorringianum B. Fedtsch.;克诺氏翠雀花■☆

124325 Delphinium korshinskyanum Nevski;东北高翠雀花(东北高翠雀,科氏飞燕草);Korshinsky Larkspur ■

124326 Delphinium kuanii W. T. Wang = Delphinium winklerianum Huth ■

124327 Delphinium kunlunshanicum Chang Y. Yang et B. Wang;昆仑翠雀花;Kunlunshan Larkspur ■

124328 Delphinium kweichowense W. T. Wang = Delphinium anthriscifolium Hance var. savatieri (Franch.) Munz ■

124329 Delphinium labrungense Ulbr. ex Rehder et Kobuski = Delphinium pylzowii Maxim. var. trigynum W. T. Wang ■

124330 Delphinium lacostei Danguy;拉科斯特翠雀花(帕米尔翠雀花);Lacost Larkspur ■

124331 Delphinium lancisepalum Hand.-Mazz. = Delphinium pachycentrum Hemsl. ex Brühl var. lancisepalum (Hand.-Mazz.) W. T. Wang ■

124332 Delphinium lankongense Franch. = Delphinium pycnocentrum Franch. ■

124333 Delphinium lasiantherum W. T. Wang;毛药翠雀花■

124334 Delphinium lasiocarpum Tamura = Delphinium gyalanum C. Marquand et Airy Shaw ■

124335 Delphinium latirhombicum W. T. Wang;宽菱形翠雀花■

124336 Delphinium laxicymosum W. T. Wang;聚伞翠雀花;Cymose Larkspur, Laxymose Larkspur ■

124337 Delphinium laxicymosum W. T. Wang var. pilostachyum W. T. Wang;毛序聚伞翠雀花;Pilose Laxcymose Larkspur ■

124338 Delphinium laxiflorum DC. = Delphinium aemulans Nevski ■

124339 Delphinium laxiusculum (Boiss.) Rouy;疏松翠雀花■☆

124340 Delphinium leiocarpum Brühl ex Huth;光果翠雀花■

124341 Delphinium leiocarpum Brühl ex Huth = Delphinium kingianum Brühl ex Huth var. leiocarpum Brühl ex Huth ■

124342 Delphinium leiophyllum (W. T. Wang) W. T. Wang;光叶翠雀花(光叶短距翠雀花);Brightleaf Forrest Larkspur, Smoothleaf Forrest Larkspur ■

124343 Delphinium leiostachyum W. T. Wang;光轴翠雀花■

124344 Delphinium leptocarpum Nevski;细果翠雀花■☆

124345 Delphinium leptopogon Hand.-Mazz. = Delphinium siwanense Franch. ■

124346 Delphinium leroyi Franch. ex Huth;大萼翠雀花■☆

124347 Delphinium liangshanense W. T. Wang;凉山翠雀花;Liangshan Larkspur ■

124348 Delphinium likiangense Franch.;丽江翠雀花(丽江翠雀,虱子药);Lijiang Larkspur, Likiang Larkspur ■

124349 Delphinium lilacinum Hand.-Mazz. = Delphinium handelianum W. T. Wang ■

124350 Delphinium lineapetalum Ewan;线瓣翠雀花■☆

124351 Delphinium lingbaoense S. Y. Wang et Q. S. Yang;灵宝翠雀花;Lingbao Larkspur ■

124352 Delphinium lipskyi Korsh.;利普斯基翠雀花;Lipsky Larkspur ■☆

124353 Delphinium longiciliatum W. T. Wang = Delphinium iliense Huth ■

124354 Delphinium longipedicellatum W. T. Wang;长梗翠雀花;Longipedicellate Larkspur, Longpedicel Larkspur ■

124355 Delphinium longipedunculatum Regel et Schmalh.;长总梗翠雀花;Longpeduncled Larkspur ■☆

124356 Delphinium longipes Franch. = Delphinium davidii Franch. ■

124357 Delphinium loscosii Costa = Consolida mauritanica (Coss.) Munz ■☆

124358 Delphinium loscosii Costa var. brevirostratum Pau = Consolida mauritanica (Coss.) Munz ■☆

124359 Delphinium luteum A. Heller;黄翠雀花;Yellow Larkspur ■☆

124360 Delphinium maackianum Regel;宽苞翠雀花(宽苞翠雀,马氏飞燕草);Maack Larkspur ■

124361 Delphinium maackianum Regel f. albiflorum S. H. Li et Z. F. Fang = Delphinium maackianum Regel ■

124362 Delphinium maackianum Regel f. lasiocarpum (Regel) Kitag. = Delphinium maackianum Regel ■

124363 Delphinium maackianum Regel var. lasiocarpum Regel = Delphinium maackianum Regel ■

124364 Delphinium macrocentrum Oliv.;大距翠雀花■☆

124365 Delphinium macropetalum DC.;大瓣翠雀花■☆

124366 Delphinium macrophyllum Wooton = Delphinium scopulorum A. Gray ■☆

124367 Delphinium macrosepalum Engl. = Delphinium leroyi Franch. ex Huth ■☆

124368 Delphinium madrense S. Watson;马德雷翠雀花;Edwards' Plateau Larkspur, Sierra Madre Larkspur ■☆

124369 Delphinium mairei Ulbr. = Delphinium forrestii Diels ■

124370 Delphinium mairei Ulbr. var. viride W. T. Wang = Delphinium forrestii Diels var. viride (W. T. Wang) W. T. Wang ■

124371 Delphinium majus (W. T. Wang) W. T. Wang;金沙翠雀花;Big Larkspur, Jinsha Larkspur ■

124372 Delphinium malacophyllum Hand. -Mazz.;软叶翠雀花;Flaccidleaf Larkspur ■

124373 Delphinium maoxianense W. T. Wang;茂县翠雀花;Maoxian Larkspur ■

124374 Delphinium mariae N. Busch;马里亚氏翠雀花;Maria Larkspur ■☆

124375 Delphinium mauritanicum Coss. = Consolida mauritanica (Coss.) Munz ■☆

124376 Delphinium maximowiczii Franch.;多枝翠雀花;Manybranched Larkspur, Maximowicz Larkspur ■

124377 Delphinium medongense W. T. Wang;墨脱翠雀花■

124378 Delphinium megalanthum Nevski;大花翠雀花■☆

124379 Delphinium menziesii DC.;孟席斯翠雀花;Menzies Larkspur, Menzies' Larkspur ■☆

124380 Delphinium menziesii DC. subsp. pallidum M. J. Warnock;苍白孟席斯翠雀花;White-flowered Menzies' Larkspur ■☆

124381 Delphinium menziesii DC. var. ochroleucum Nutt.;黄白孟席斯翠雀花■☆

124382 Delphinium menziesii DC. var. pyramidalis (Ewan) C. L. Hitchc. = Delphinium menziesii DC. ■☆

124383 Delphinium micropetalum Finet et Gagnep.;小瓣翠雀花;Micropetalous Larkspur, Smallpetal Larkspur ■

124384 Delphinium micropetalum Finet et Gagnep. f. album W. T. Wang = Delphinium micropetalum Finet et Gagnep. ■

124385 Delphinium minutum H. Lév. et Vaniot = Delphinium anthriscifolium Hance var. savatieri (Franch.) Munz ■

124386 Delphinium mirabile Serg.;西伯利亚奇异翠雀花;Wonderful Larkspur ■☆

124387 Delphinium mitzugense Ulbr. = Delphinium taliense Franch. ■

124388 Delphinium mollifolium W. T. Wang;新源翠雀花;Soft-leaved Larkspur ■

124389 Delphinium mollipilum W. T. Wang;软毛翠雀花;Softhairy Larkspur ■

124390 Delphinium monanthum Hand. -Mazz. = Delphinium candelabrum Ostenf. var. monanthum Ostenf. ■

124391 Delphinium montanum DC.;山地翠雀花;Mountain Larkspur ■☆

124392 Delphinium mosoynense Franch. = Delphinium grandiflorum L. var. mosoynense (Franch.) Huth ■

124393 Delphinium motingshanicum W. T. Wang et M. J. Warnock;磨顶山翠雀花■

124394 Delphinium muliense W. T. Wang;木里翠雀花;Muli Larkspur ■

124395 Delphinium muliense W. T. Wang var. minutibracteolatum W. T. Wang;小苞木里翠雀花;Smallbract Muli Larkspur ■

124396 Delphinium multiplex (Ewan) C. L. Hitchc.;多折翠雀花■☆

124397 Delphinium nangchienense W. T. Wang;囊谦翠雀花;Nangqian Larkspur, Nangchien Larkspur ■

124398 Delphinium nangziense W. T. Wang;朗孜翠雀花;Langzi Larkspur ■

124399 Delphinium nanum DC.;低矮翠雀花■☆

124400 Delphinium nanum DC. subsp. elongatum (Boiss.) C. Blanché et al. = Delphinium nanum DC. ■☆

124401 Delphinium naviculare W. T. Wang;船苞翠雀花;Boatshaped-bract Larkspur ■

124402 Delphinium naviculare W. T. Wang var. lasiocarpum W. T. Wang;毛果船苞翠雀花;Hairyfruit Boatshapedbract Larkspur ■

124403 Delphinium nevadense Kuntze = Delphinium emarginatum C. Presl subsp. nevadense (Kunze) C. Blanché et Molero ■☆

124404 Delphinium newtonianum D. M. Moore;奥扎克翠雀花;Ozark Larkspur ■☆

124405 Delphinium ninglangshanicum W. T. Wang;宁朗山翠雀花■

124406 Delphinium nordhagenii Wendelbo;迭裂翠雀花;Nordhagen Larkspur ■

124407 Delphinium nordhagenii Wendelbo var. acutidentatum W. T. Wang;尖齿翠雀花;Acutedentate Nordhagen Larkspur ■

124408 Delphinium nortonii Dunn;细茎翠雀花;Norton Larkspur, Thinstem Larkspur ■

124409 Delphinium novomexicanum Wooton;新墨西哥翠雀花;New Mexico Larkspur, White Mountain Larkspur ■☆

124410 Delphinium nudicaule Torr. et A. Gray;裸茎翠雀花(裸茎翠雀);Red Larkspur, Scarlet Larkspur ■☆

124411 Delphinium nuttallianum Pritz. ex Walp.;纳托尔翠雀花;Nuttall's Larkspur ■☆

124412 Delphinium nuttallianum Pritz. ex Walp. var. fulvum C. L. Hitchc. = Delphinium nuttallianum Pritz. ex Walp. ■☆

124413 Delphinium nuttallianum Pritz. ex Walp. var. levicaule C. L. Hitchc. = Delphinium nuttallianum Pritz. ex Walp. ■☆

124414 Delphinium nuttallianum Pritz. var. lineapetalum (Ewan) C. L. Hitchc. = Delphinium lineapetalum Ewan ■☆

124415 Delphinium nuttallianum Pritz. var. pilosum C. L. Hitchc. = Delphinium bicolor Nutt. ■☆

124416 Delphinium nuttallii A. Gray;纳氏翠雀花;Nuttall's Larkspur ■☆

124417 Delphinium nuttallii A. Gray subsp. ochroleucum (Nutt.) M. J. Warnock;亮黄花翠雀花;Light-yellow-flowered Larkspur ■☆

124418 Delphinium obcordatilimbum W. T. Wang;倒心形翠雀花;Obcordatelimb Larkspur, Obcordifolious Larkspur ■

124419 Delphinium obcordatilimbum W. T. Wang var. minus W. T. Wang;光梗翠雀花;Small Obcordatelimb Larkspur ■

124420 Delphinium obcordatilimbum W. T. Wang var. minus W. T. Wang = Delphinium tenii H. Lév. ■

124421 Delphinium obcordatum DC.;倒心叶翠雀花■☆

124422 Delphinium occidentale (S. Watson) S. Watson var. barbeyi (Huth) S. L. Welsh = Delphinium barbeyi (Huth) Huth ■☆

124423 Delphinium occidentale S. Watson ex Coult.;西方翠雀花(西方翠雀)■☆

124424 Delphinium ochotense Nevski;奥霍特翠雀花■☆

124425 Delphinium ochroleucum Huth = Delphinium biternatum Huth ■

124426 Delphinium ochroleucum Steven ex DC.;黄白翠雀花■☆

124427 Delphinium oliganthum Franch. = Delphinium likiangense Franch. ■

124428 Delphinium omeiense W. T. Wang;峨眉翠雀花(峨山草乌);Emei Larkspur ■

124429 Delphinium omeiense W. T. Wang var. micranthum G. F. Tao;小花峨眉翠雀花;Smallflower Emei Larkspur ■
124430 Delphinium omeiense W. T. Wang var. pubescens W. T. Wang;柔毛峨眉翠雀花(毛峨眉翠雀花);Pubescent Emei Larkspur ■
124431 Delphinium oreganum Howell = Delphinium menziesii DC. ■☆
124432 Delphinium oreophilum Huth;喜山翠雀花;Montane Larkspur ■☆
124433 Delphinium orientale J. Gay = Consolida hispanica (Costa) Greuter et Burdet ■☆
124434 Delphinium orientale J. Gay = Consolida orientalis (J. Gay) Schrödinger ■☆
124435 Delphinium orientale J. Gay subsp. hispanicum (Willk.) Batt. et Trab. = Consolida orientalis (J. Gay) Schrödinger ■☆
124436 Delphinium orientale J. Gay var. hispanicum (Willk.) Huth = Consolida orientalis (J. Gay) Schrödinger ■☆
124437 Delphinium ornatum Bouché;装饰飞燕草■
124438 Delphinium orthocentrum Franch.;直距翠雀花;Songpan Larkspur,Straightspur Larkspur ■
124439 Delphinium osseticum N. Busch;骨质翠雀花;Like Bone Larkspur ■☆
124440 Delphinium oxycentrum W. T. Wang;尖距翠雀花;Sharpspur Larkspur,Sharpspurred Larkspur ■
124441 Delphinium pachycentrum Hemsl. ex Brühl;粗距翠雀花;Thickspur Larkspur,Thickspurred Larkspur ■
124442 Delphinium pachycentrum Hemsl. ex Brühl subsp. hemsleyi Brühl = Delphinium pachycentrum Hemsl. ex Brühl ■
124443 Delphinium pachycentrum Hemsl. ex Brühl subsp. tsangense var. dasycarpum Brühl = Delphinium kingianum Brühl ex Huth ■
124444 Delphinium pachycentrum Hemsl. ex Brühl subsp. tsangense var. leiocaropum Brühl = Delphinium kingianum Brühl ex Huth var. leiocarpum Brühl ex Huth ■
124445 Delphinium pachycentrum Hemsl. ex Brühl var. humilius W. T. Wang = Delphinium humilius (W. T. Wang) W. T. Wang ■
124446 Delphinium pachycentrum Hemsl. ex Brühl var. lancisepalum (Hand. -Mazz.) W. T. Wang;狭萼粗距翠雀花;Narrowsepal Thickspur Larkspur,Narrowsepal Thickspurred Larkspur ■
124447 Delphinium pachycentrum Hemsl. ex Brühl var. lobatum W. T. Wang = Delphinium pachycentrum Hemsl. ex Brühl ■
124448 Delphinium pachycentrum Hemsl. ex Brühl var. pseudolancisepalum W. T. Wang = Delphinium pachycentrum Hemsl. ex Brühl var. lancisepalum (Hand. -Mazz.) W. T. Wang ■
124449 Delphinium pachycentrum Hemsl. ex Brühl var. tenuicaule F. H. Chen = Delphinium muliense W. T. Wang ■
124450 Delphinium pachycentrum Hemsl. subsp. tsangense Brühl var. dasycarpum Brühl = Delphinium kingianum Brühl ex Huth ■
124451 Delphinium pachycentrum Hemsl. subsp. tsangense var. leiocarpum (Brühl ex Huth) Brühl = Delphinium kingianum Brühl ex Huth var. leiocarpum Brühl ex Huth ■
124452 Delphinium pachycentrum Hemsl. var. humilius W. T. Wang = Delphinium humilius (W. T. Wang) W. T. Wang ■
124453 Delphinium pachycentrum Hemsl. var. tenuicaule F. H. Chen = Delphinium muliense W. T. Wang ■
124454 Delphinium pallasii Nevski;帕拉氏翠雀花(巴拉斯翠雀花);Pallas Larkspur ■☆
124455 Delphinium paludicola Ulbr. = Delphinium souliei Franch. ■
124456 Delphinium paniculatum Host;穗花翠雀花;Paniculate ■☆
124457 Delphinium paradoxum Bunge;土耳其斯坦翠雀花■☆
124458 Delphinium parishii A. Gray;帕尔什翠雀花;Parish's Larkspur ■☆
124459 Delphinium parishii A. Gray subsp. pallidum (Munz) M. J. Warnock;苍白帕尔什翠雀花;Pale-flowered Parish's Larkspur ■☆
124460 Delphinium parishii A. Gray subsp. purpureum H. F. Lewis et Epling = Delphinium parryi A. Gray subsp. purpureum (H. F. Lewis et Epling) M. J. Warnock ■☆
124461 Delphinium parishii A. Gray subsp. subglobosum (Wiggins) H. F. Lewis et Epling;亚球形帕尔什翠雀花■☆
124462 Delphinium parishii A. Gray var. inopinum Jeps. = Delphinium inopinum (Jeps.) H. F. Lewis et Epling ■☆
124463 Delphinium parishii A. Gray var. pallidum Munz = Delphinium parishii subsp. pallidum (Munz) M. J. Warnock ■☆
124464 Delphinium parryi A. Gray;帕里翠雀花;Parry's Larkspur ■☆
124465 Delphinium parryi A. Gray subsp. blochmaniae (Greene) H. F. Lewis et Epling;布洛赫曼翠雀花;Blochman's Larkspur, Dune Larkspur ■☆
124466 Delphinium parryi A. Gray subsp. eastwoodiae Ewan;伊斯特伍德翠雀花;Eastwood's Larkspur ■☆
124467 Delphinium parryi A. Gray subsp. maritimum (Davidson) M. J. Warnock;滨海帕里翠雀花;Maritime Larkspur ■☆
124468 Delphinium parryi A. Gray subsp. purpureum (H. F. Lewis et Epling) M. J. Warnock;紫色帕里翠雀花;Mount Pinos Larkspur ■☆
124469 Delphinium parryi A. Gray subsp. seditiosum (Jeps.) Ewan = Delphinium parryi A. Gray ■☆
124470 Delphinium parryi A. Gray var. maritimum Davidson = Delphinium parryi A. Gray subsp. maritimum (Davidson) M. J. Warnock ■☆
124471 Delphinium parryi A. Gray var. montanum Munz = Delphinium patens A. Gray subsp. montanum (Munz) Ewan ■☆
124472 Delphinium patens A. Gray subsp. montanum (Munz) Ewan;山地帕里翠雀花■☆
124473 Delphinium patens Benth.;开展翠雀花■☆
124474 Delphinium patens Benth. subsp. greenei (Eastw.) Ewan = Delphinium gracilentum Greene ■☆
124475 Delphinium pauciflorum Nutt. = Delphinium nuttallianum Pritz. ex Walp. ■☆
124476 Delphinium pediforme Comber = Delphinium spirocentrum Hand. -Mazz. ■
124477 Delphinium pellucidum Busch = Delphinium trichophorum Franch. ■
124478 Delphinium penardii Huth = Delphinium carolinianum Walter subsp. virescens (Nutt.) R. E. Brooks ■☆
124479 Delphinium pentagynum Lam.;五蕊翠雀花■☆
124480 Delphinium pentagynum Lam. var. glabratum Maire = Delphinium pentagynum Lam. ■☆
124481 Delphinium pentagynum Lam. var. heterotrichum Maire = Delphinium pentagynum Lam. ■☆
124482 Delphinium pentagynum Lam. var. homotrichum Maire = Delphinium pentagynum Lam. ■☆
124483 Delphinium pentagynum Lam. var. phialotrichum Maire = Delphinium pentagynum Lam. ■☆
124484 Delphinium peregrinum L.;堇色翠雀花;Violet Larkspur ■☆
124485 Delphinium peregrinum L. subsp. gracile (DC.) Sennen et Pau = Delphinium gracile DC. ■☆
124486 Delphinium peregrinum L. subsp. junceum (DC.) Batt. et Trab. = Delphinium peregrinum L. ■☆

124487 Delphinium peregrinum L. subsp. nanum (DC.) Graebn. = Delphinium nanum DC. ■☆

124488 Delphinium peregrinum L. var. elongatum Boiss. = Delphinium gracile DC. ■☆

124489 Delphinium peregrinum L. var. gracile (DC.) Willk. = Delphinium gracile DC. ■☆

124490 Delphinium peregrinum L. var. halteratum (Sibth. et Sm.) Coss. = Delphinium peregrinum L. ■☆

124491 Delphinium peregrinum L. var. junceum (DC.) Bég. et Vacc. = Delphinium peregrinum L. ■☆

124492 Delphinium peregrinum L. var. laxum Gatt. et Maire = Delphinium peregrinum L. ■☆

124493 Delphinium peregrinum L. var. longipes (Moris) Boiss. = Delphinium gracile DC. ■☆

124494 Delphinium peregrinum L. var. macropetalum (DC.) Ball = Delphinium macropetalum DC. ■☆

124495 Delphinium peregrinum L. var. macrosiphon Sennen = Delphinium peregrinum L. ■☆

124496 Delphinium peregrinum L. var. obcordatum (DC.) Huth = Delphinium obcordatum DC. ■☆

124497 Delphinium peregrinum L. var. rifanum Maire et Sennen = Delphinium peregrinum L. ■☆

124498 Delphinium peregrinum L. var. tribracteolatum (DC.) Maire = Delphinium gracile DC. ■☆

124499 Delphinium pergameneum W. T. Wang;纸叶翠雀花■

124500 Delphinium persicum Boiss.;波斯翠雀花■☆

124501 Delphinium pogonanthum Hand.-Mazz. = Delphinium delavayi Franch. var. pogonanthum (Hand.-Mazz.) W. T. Wang ■

124502 Delphinium poltoratzkii Rupr.;泼氏翠雀花;Poltoratzk Larkspur ■☆

124503 Delphinium polyanthum W. T. Wang = Delphinium bulleyanum Forrest ex Diels ■

124504 Delphinium polycladon Eastw.;密枝翠雀花;High Mountain Larkspur ■☆

124505 Delphinium pomeense W. T. Wang;波密翠雀花;Bomi Larkspur,Pome Larkspur ■

124506 Delphinium potaninii Huth;黑水翠雀花(峨山草乌,峨山飞燕草,卷距飞燕草);Potanin Larkspur ■

124507 Delphinium potaninii Huth var. bonvalotii (Franch.) W. T. Wang;螺距黑水翠雀花(川黔翠雀花,川西翠雀花,峨山草乌,环距黑水翠雀花,铁脚草乌);Bonvalot Larkspur, Bonvalot Potanin Larkspur ■

124508 Delphinium potaninii Huth var. jiufengshanense W. J. Zhang et G. H. Chen;彭州黑水翠雀;Jiufengshan Potanin Larkspur ■

124509 Delphinium potaninii Huth var. latibracteolatum W. T. Wang;宽苞黑水翠雀花;Broadbract Potanin Larkspur,Broadbracteole Potanin Larkspur ■

124510 Delphinium pratense Eastw. = Delphinium gracilentum Greene ■☆

124511 Delphinium propinquum Nevski;亲近翠雀花;Near Larkspur ■☆

124512 Delphinium przewalskii Huth = Delphinium albocoeruleum Maxim. var. przewalskii (Huth) W. T. Wang ■

124513 Delphinium pseudoaemulans Chang Y. Yang et B. Wang;赛塔城翠雀花;False Tacheng Larkspur ■

124514 Delphinium pseudoaemulans Chang Y. Yang et B. Wang = Delphinium shawurense W. T. Wang var. pseudoaemulans (Chang Y. Yang et B. Wang) W. T. Wang et M. J. Warnock ■

124515 Delphinium pseudobrescens W. T. Wang = Delphinium kamaonense Huth var. glabrescens (W. T. Wang) W. T. Wang ■

124516 Delphinium pseudocaeruleum W. T. Wang;拟蓝翠雀花;Falsecoeruleus Larkspur,Sham Skyblue Larkspur ■

124517 Delphinium pseudocampylocentrum W. T. Wang;拟弯距翠雀花;Curvicalvarate Larkspur,False Bentspurred Larkspur ■

124518 Delphinium pseudocampylocentrum W. T. Wang var. glabripes W. T. Wang;光序拟弯距翠雀花;Glabrous Curvicalvarate Larkspur ■

124519 Delphinium pseudocandelabrum W. T. Wang;石滩翠雀花■

124520 Delphinium pseudocyananthum Chang Y. Yang et B。Wang;假深蓝翠雀花■

124521 Delphinium pseudoglaciale W. T. Wang;拟冰川翠雀花;Falseglacial Larkspur,Sham Glacial Larkspur ■

124522 Delphinium pseudograndiflorum W. T. Wang = Delphinium kamaonense Huth var. glabrescens (W. T. Wang) W. T. Wang ■

124523 Delphinium pseudograndiflorum W. T. Wang var. glabrescens W. T. Wang = Delphinium kamaonense Huth var. glabrescens (W. T. Wang) W. T. Wang ■

124524 Delphinium pseudograndiflorum W. T. Wang var. lobatum W. T. Wang = Delphinium kamaonense Huth var. glabrescens (W. T. Wang) W. T. Wang ■

124525 Delphinium pseudohamatum W. T. Wang;宁蒗翠雀花■

124526 Delphinium pseudomosoynense W. T. Wang;条裂翠雀花■

124527 Delphinium pseudomosoynense W. T. Wang var. subglabrum W. T. Wang;疏毛条裂翠雀花■

124528 Delphinium pseudopulcherrimum W. T. Wang;宽萼翠雀花;Falsebroadsepal Larkspur,Latisepalous Larkspur ■

124529 Delphinium psudothibeticum W. T. Wang et M. J. Warnock;拟澜沧翠雀花■

124530 Delphinium psudotongolense W. T. Wang;拟川西翠雀花;False Westsichuan Larkspur,Sham Tongol Larkspur ■

124531 Delphinium psudoyunnanense W. T. Wang et M. J. Warnock;拟云南翠雀花■

124532 Delphinium pubescens DC. = Consolida pubescens (DC.) Soó ■☆

124533 Delphinium pubescens DC. var. dissitiflorum Coss. = Consolida pubescens (DC.) Soó ■☆

124534 Delphinium pulanense W. T. Wang;普蓝翠雀花;Pulan Larkspur ■

124535 Delphinium pulcherrimum W. T. Wang = Delphinium batangense Finet et Gagnep. ■

124536 Delphinium pumilum W. T. Wang;矮翠雀花;Dwarf Larkspur, Short Larkspur ■

124537 Delphinium puniceum Pall.;红紫翠雀花;Redpurple Larkspur ■☆

124538 Delphinium purdomii Craib = Delphinium trichophorum Franch. ■

124539 Delphinium purpurascens W. T. Wang;紫苞翠雀花;Purple Bract Larkspur,Purple Larkspur ■

124540 Delphinium purpusii Brandegee;普尔翠雀花;Purpus' Larkspur,Rose-colored Larkspur ■☆

124541 Delphinium pycnocentroides W. T. Wang = Delphinium thibeticum Finet et Gagnep. ■

124542 Delphinium pycnocentroides W. T. Wang var. latisectum W. T. Wang = Delphinium thibeticum Finet et Gagnep. ■

124543 Delphinium pycnocentrum Franch.;密距翠雀花;Densespurred Larkspur,Eryuan Larkspur ■

124544 Delphinium pycnocentrum Franch. var. lankongense (Franch.) Huth = Delphinium pycnocentrum Franch. ■

124545 Delphinium pylzowii Maxim.;大通翠雀花;Datong Larkspur, Tatung Larkspur ■
124546 Delphinium pylzowii Maxim. var. trigynum W. T. Wang;三果大通翠雀花(青甘翠雀花,三果大通翠雀);Threefruit Datong Larkspur,Threefruit Tatung Larkspur ■
124547 Delphinium pyramidale Royle = Delphinium elatum L. ■☆
124548 Delphinium pyramidatum Albov;尖塔翠雀花;Pyramidal Larkspur ■☆
124549 Delphinium qinghaiense W. T. Wang;青海翠雀花;Qinghai Larkspur ■
124550 Delphinium quercetorum Boiss. et Hausskn.;栎叶翠雀花■☆
124551 Delphinium rangtangense W. T. Wang;壤塘翠雀花;Rangtang Larkspur ■
124552 Delphinium ranunculifolium Wall. = Delphinium pyramidale Royle ■☆
124553 Delphinium rectivenium Royle = Delphinium vestitum Wall. ■
124554 Delphinium recurvatum Greene;山谷翠雀花;Recurved Larkspur,Valley Larkspur ■☆
124555 Delphinium retropilosum (Huth) Sambuk;反折毛翠雀花■☆
124556 Delphinium robertianum H. Lév. et Vaniot = Delphinium anthriscifolium Hance var. savatieri (Franch.) Munz ■
124557 Delphinium robustum Rydb.;粗壮翠雀花;Robust Larkspur ■☆
124558 Delphinium rockii Munz = Delphinium albocoeruleum Maxim. ■
124559 Delphinium rugulosum Boiss. = Consolida rugulosa (Boiss.) Schrödinger ■
124560 Delphinium ruprechtii Nevski;鲁布勒氏翠雀花;Ruprecht Larkspur ■☆
124561 Delphinium ruspolianum Engl. = Delphinium leroyi Franch. ex Huth ■☆
124562 Delphinium sapellonis Tidestr.;萨泊罗翠雀花;Sapello Canyon Larkspur ■☆
124563 Delphinium sauricum Schischk.;蜥蜴翠雀花(和丰);Like Saura Larkspur ■
124564 Delphinium savatieri Franch. = Delphinium anthriscifolium Hance var. savatieri (Franch.) Munz ■
124565 Delphinium saxatile W. T. Wang = Delphinium davidii Franch. var. saxatile (W. T. Wang) W. T. Wang ■
124566 Delphinium scaposum Greene;西方花葶翠雀花■☆
124567 Delphinium scaposum W. T. Wang = Delphinium sinoscaposum W. T. Wang ■
124568 Delphinium scopulorum A. Gray;岩翠雀花;Rocky Mountain Larkspur ■☆
124569 Delphinium scopulorum A. Gray var. glaucum (S. Watson) A. Gray = Delphinium glaucum S. Watson ■☆
124570 Delphinium scopulorum A. Gray var. luporum (Greene) Jeps. = Delphinium polycladon Eastw. ■☆
124571 Delphinium sctumalhausenii Albov;施玛翠雀花■☆
124572 Delphinium semibarbatum Bien. ex Boiss.;半毛翠雀(半髯毛翠雀);Halfbearded Larkspur,Zalil ■☆
124573 Delphinium semiclavatum Nevski;半棒状翠雀花;Half-clavate Larkspur ■☆
124574 Delphinium setiferum Franch. = Delphinium pachycentrum Hemsl. ex Brühl ■
124575 Delphinium shawurense W. T. Wang;萨乌尔翠雀花(沙乌尔翠雀花);Sawur Larkspur ■
124576 Delphinium shawurense W. T. Wang var. albiflorum Chang Y. Yang et B. Wang;白花萨乌尔翠雀花(白花沙乌尔翠雀花);Whiteflower Shawur Larkspur ■
124577 Delphinium shawurense W. T. Wang var. pseudoaemulans (Chang Y. Yang et B. Wang) W. T. Wang et M. J. Warnock;毛茎萨乌尔翠雀花■
124578 Delphinium sherriffii Munz;米林翠雀花;Milin Larkspur, Sherriff Larkspur ■
124579 Delphinium shuichengense W. T. Wang;水城翠雀花;Shuicheng Larkspur ■
124580 Delphinium sierrae-blancae Wooton = Delphinium novomexicanum Wooton ■☆
124581 Delphinium sinoelatum Chang Y. Yang et B. Wang;新疆翠雀花(高翠雀花)■
124582 Delphinium sinopentagynum W. T. Wang;五果翠雀花;Chinese Fivefruit Larkspur,Fivefruit Larkspur ■
124583 Delphinium sinoscaposum W. T. Wang;花葶翠雀花;Barestem Larkspur,Chinese Scapose Larkspur,Larkspur,Scapose Larkspur ■
124584 Delphinium sinovitifolium W. T. Wang;葡萄叶翠雀花;Grapeleaf Larkspur,Vitifolious Larkspur ■
124585 Delphinium siwanense Franch.;细须翠雀花(西湾翠雀花,细须翠雀);N. Hebei Larkspur,North Hopeh Larkspur ■
124586 Delphinium siwanense Franch. var. albopuberulum W. T. Wang;冀北翠雀花(英哥草)■
124587 Delphinium siwanense Franch. var. leptopogon (Hand. -Mazz.) W. T. Wang = Delphinium siwanense Franch. ■
124588 Delphinium smithianum Hand. -Mazz.;宝兴翠雀花;Baoxing Larkspur,Paohsing Larkspur ■
124589 Delphinium songoricum (Kar. -Ker.) Nevski;准噶尔翠雀花■☆
124590 Delphinium sonnei Greene = Delphinium nuttallianum Pritz. ex Walp. ■☆
124591 Delphinium soonmingense F. H. Chen = Delphinium tatsienense Franch. ■
124592 Delphinium sordidecaerulescens Ulbr. = Delphinium kamaonense Huth var. glabrescens (W. T. Wang) W. T. Wang ■
124593 Delphinium souliei Franch.;川甘翠雀花;Soulie Larkspur ■
124594 Delphinium sparsiflorum Maxim.;疏花翠雀花;Laxflower Larkspur ■
124595 Delphinium speciosum M. Bieb.;美丽翠雀花;Showy Larkspur ■☆
124596 Delphinium speciosum M. Bieb. subsp. ranunculifolium ? = Delphinium pyramidale Royle ■☆
124597 Delphinium speciosum M. Bieb. var. ranunculifolium ? = Delphinium pyramidale Royle ■☆
124598 Delphinium spirocentrum Hand. -Mazz.;螺距翠雀花;Spiralspurred Larkspur ■
124599 Delphinium spirocentrum Hand. -Mazz. var. grandibracteolatum W. T. Wang = Delphinium spirocentrum Hand. -Mazz. ■
124600 Delphinium spirocentrum Hand. -Mazz. var. hirsutum F. H. Chen = Delphinium spirocentrum Hand. -Mazz. ■
124601 Delphinium spirocentrum Hand. -Mazz. var. pauciflorum F. H. Chen = Delphinium bulleyanum Forrest ex Diels ■
124602 Delphinium spirocentrum Hand. -Mazz. var. pediforme (Comber) W. T. Wang = Delphinium spirocentrum Hand. -Mazz. ■
124603 Delphinium splendens G. N. Jones = Delphinium glaucum S. Watson ■☆
124604 Delphinium staphisagria L.;尖果翠雀花(虱草);Licebane, Lousebane,Louse-herb,Lousewort,Palmated Larkspur,Stavesacre ■☆
124605 Delphinium stocksianum Boiss.;斯道克翠雀花■☆

124606 Delphinium strictum A. Nelson var. distichiflorum (Hook.) H. St. John = Delphinium distichum Geyer ex Hook. ■☆

124607 Delphinium subglobosum Wiggins = Delphinium parishii A. Gray subsp. subglobosum (Wiggins) H. F. Lewis et Epling ■☆

124608 Delphinium subspathulatum W. T. Wang; 匙苞翠雀花; Spoonbract Larkspur ■

124609 Delphinium sungpanense W. T. Wang = Delphinium sutchuenens Franch. ■

124610 Delphinium sutchuenens Franch.; 松潘翠雀花; Songpan Larkspur, Szechwan Larkspur ■

124611 Delphinium sutchuenense Franch. = Delphinium davidii Franch. ■

124612 Delphinium sutherlandii M. J. Warnock; 萨瑟兰翠雀花; Sutherland's Larkspur ■☆

124613 Delphinium sylvaticum Pomel; 林地翠雀花■☆

124614 Delphinium szechuanieum Ulbr. = Delphinium orthocentrum Franch. ■

124615 Delphinium szovitsianum Boiss.; 绍氏翠雀花■☆

124616 Delphinium tabatae Tamura; 吉隆翠雀花; Jilong Larkspur ■

124617 Delphinium taipaicum W. T. Wang; 太白翠雀花; Taibai Larkspur, Taipai Larkspur ■

124618 Delphinium taliense Franch.; 大理翠雀花; Dali Larkspur, Tali Larkspur ■

124619 Delphinium taliense Franch. var. dolichocentrum W. T. Wang; 长距大理翠雀花; Longspur Dali Larkspur, Longspurred Tali Larkspur ■

124620 Delphinium taliense Franch. var. glabrum W. T. Wang = Delphinium taliense Franch. ■

124621 Delphinium taliense Franch. var. hirsutum W. T. Wang; 硬毛大理翠雀花; Hirsute Dali Larkspur, Hirsute Tali Larkspur ■

124622 Delphinium taliense Franch. var. platycentrum W. T. Wang; 粗距大理翠雀花; Thickspur Dali Larkspur, Thickspurred Tali Larkspur ■

124623 Delphinium taliense Franch. var. pubipes W. T. Wang; 毛梗大理翠雀花; Hairypedicel Tali Larkspur ■☆

124624 Delphinium taliense Franch. var. pubipes W. T. Wang = Delphinium lasiantherum W. T. Wang ■

124625 Delphinium tangkulaense W. T. Wang; 唐古拉翠雀花; Tangkula Larkspur, Tangul Larkspur ■

124626 Delphinium tangkulaense W. T. Wang f. xanthanthum W. T. Wang et S. K. Wu; 库赛翠雀花; Kusaihu Dali Larkspur ■

124627 Delphinium tangkulaense W. T. Wang f. xanthanthum W. T. Wang et S. K. Wu = Delphinium tangkulaense W. T. Wang ■

124628 Delphinium tanguticum Huth = Delphinium albocoeruleum Maxim. ■

124629 Delphinium tarbagataicum Chang Y. Yang et B. Wang; 新塔翠雀花■

124630 Delphinium tatsienense Franch.; 康定翠雀花(虎图辣, 鸡爪乌, 康定翠雀, 细草乌, 小草乌); Kangding Larkspur, Kangting Larkspur ■

124631 Delphinium tatsienense Franch. f. sordidecaerulescens (Ulbr.) Hand. -Mazz. = Delphinium kamaonense Huth var. glabrescens (W. T. Wang) W. T. Wang ■

124632 Delphinium tatsienense Franch. var. chinghaiense W. T. Wang; 班玛翠雀花(青海翠雀花); Chinghai Larkspur, Qinghai Kangding Larkspur ■

124633 Delphinium taxkorganense W. T. Wang; 塔什库尔干翠雀花■

124634 Delphinium tenii H. Lév.; 长距翠雀花; Dengchuan Larkspur, Longspurred Larkspur ■

124635 Delphinium tenuisectum Greene subsp. amplibracteatum (Wooton) Ewan = Delphinium geraniifolium Rydb. ■☆

124636 Delphinium tenuissimum Sibth. et Sm. = Consolida tenuissima (Sibth. et Sm.) Soó ■☆

124637 Delphinium ternatum Huth; 三出叶翠雀花; Ternate Larkspur ■☆

124638 Delphinium tetragynum W. T. Wang; 四果翠雀花; Four-fruited Larkspur ■

124639 Delphinium thibeticum Finet et Gagnep.; 澜沧翠雀花; Lancang Larkspur, Tibet Larkspur ■

124640 Delphinium thibeticum Finet et Gagnep. var. laceratilobum W. T. Wang; 锐裂翠雀花; Laceratelobed Larkspur ■

124641 Delphinium thibeticum Finet et Gagnep. var. schigophyllumn Hand. -Mazz. = Delphinium thibeticum Finet et Gagnep. ■

124642 Delphinium thibeticum Finet et Gagnep. var. subintegrum Finet et Gagnep. = Delphinium thibeticum Finet et Gagnep. ■

124643 Delphinium tianshanicum W. T. Wang; 天山翠雀花; Tianshan Larkspur ■

124644 Delphinium tongolense Franch.; 川西翠雀花(草乌头草); Tongol Larkspur ■

124645 Delphinium treleasei Bush ex K. C. Davis; 特氏翠雀花; Trelease's Larkspur ■☆

124646 Delphinium trichophorum Franch.; 毛翠雀花(白狼毒); Trichophore Larkspur ■

124647 Delphinium trichophorum Franch. f. brevungue H. Lév. = Delphinium delavayi Franch. ■

124648 Delphinium trichophorum Franch. var. platycentrum W. T. Wang; 粗距毛翠雀花(宽距毛翠雀); Broadspurred Trichophore Larkspur, Thickspur Trichophore Larkspur ■

124649 Delphinium trichophorum Franch. var. subglabrrimum Hand. -Mazz.; 光果毛翠雀花; Smoothfruit Trichophore Larkspur, Subglabrousfruit Trichophore Larkspur ■

124650 Delphinium tricorne Michx.; 三角翠雀花; Dwarf Larkspur, Spring Larkspur ■☆

124651 Delphinium trifolioletum Finet et Gagnep.; 三小叶翠雀花; Threefoliolate Larkspur ■

124652 Delphinium trisectum W. T. Wang; 全裂翠雀花; Trisect Larkspur ■

124653 Delphinium triste Fisch. ex DC.; 苦翠雀花; Bitter Larkspur ■☆

124654 Delphinium trolliifolium A. Gray; 金莲花叶翠雀花; Cow-poison, Poison Larkspur ■☆

124655 Delphinium tsarongense Hand. -Mazz. = Delphinium chrysotrichum Finet et Gagnep. var. tsarongense (Hand. -Mazz.) W. T. Wang ■

124656 Delphinium tsarongense Hand. -Mazz. var. patentipilum W. T. Wang = Delphinium chrysotrichum Finet et Gagnep. ■

124657 Delphinium tsoongi W. T. Wang = Delphinium caeruleum Jacq. ex Cambess. var. obtusilobum Brühl ex Huth ■

124658 Delphinium turkestanicum Huth = Delphinium iliense Huth ■

124659 Delphinium turkmenum Lipsky; 土库曼翠雀花; Turkman Larkspur ■☆

124660 Delphinium uliginosum Curran; 沼泽翠雀花; Swamp Larkspur ■☆

124661 Delphinium umbraculorum H. F. Lewis et Epling; 伞状翠雀花; Umbrella Larkspur ■☆

124662 Delphinium umbrosum Hand. -Mazz.; 阴地翠雀花; Shady Larkspur ■

124663 Delphinium umbrosum Hand. -Mazz. subsp. drepanocentrum (Brühl ex Huth) N. P. Chowdhury ex Mukerjee = Delphinium

umbrosum Hand. -Mazz. var. drepanocentrum (Brühl ex Huth) W. T. Wang ■
124664 Delphinium umbrosum Hand. -Mazz. subsp. drepanocentrum (Brühl) Chowdhury ex Mukerjee = Delphinium umbrosum Hand. -Mazz. var. drepanocentrum (Brühl ex Huth) W. T. Wang ■
124665 Delphinium umbrosum Hand. -Mazz. var. drepanocentrum (Brühl ex Huth) W. T. Wang;宽苞阴地翠雀花;Broadbract Shady Larkspur ■
124666 Delphinium umbrosum Hand. -Mazz. var. drepanocentrum (Brühl ex Huth) W. T. Wang et M. J. Warnock = Delphinium drepanocentrum (Brühl ex Huth) Munz ■
124667 Delphinium umbrosum Hand. -Mazz. var. hispidum W. T. Wang;展毛阴地翠雀花;Hispid Shady Larkspur ■
124668 Delphinium unifolium Tamura;单叶翠雀花■☆
124669 Delphinium uralense Nevski;乌拉尔翠雀花;Ural Larkspur ■☆
124670 Delphinium variegatum Torr. et A. Gray;高贵翠雀花;Royal Larkspur ■☆
124671 Delphinium variegatum Torr. et A. Gray subsp. apiculatum (Greene) Ewan = Delphinium variegatum Torr. et A. Gray ■☆
124672 Delphinium variegatum Torr. et A. Gray subsp. kinkiense (Munz) M. J. Warnock;近畿翠雀花;San Clemente Island Larkspur ■☆
124673 Delphinium variegatum Torr. et A. Gray subsp. thornei Munz;托纳翠雀花;Thorne's Larkspur, Thorne's Royal Larkspur ■☆
124674 Delphinium variegatum Torr. et A. Gray var. apiculatum (Greene) Greene = Delphinium variegatum Torr. et A. Gray ■☆
124675 Delphinium vestitum Wall.;浅裂翠雀花;Lobate Larkspur, Shallowdivided Larkspur ■
124676 Delphinium villosum Steven ex DC.;长毛翠雀花■☆
124677 Delphinium virescens Nutt.;草原翠雀花;Prairie Larkspur ■☆
124678 Delphinium virescens Nutt. = Delphinium carolinianum Walter subsp. virescens (Nutt.) R. E. Brooks ■☆
124679 Delphinium virescens Nutt. = Delphinium carolinianum Walter ■☆
124680 Delphinium virescens Nutt. subsp. penardii (Huth) Ewan = Delphinium carolinianum Walter subsp. virescens (Nutt.) R. E. Brooks ■☆
124681 Delphinium virescens Nutt. subsp. wootonii (Rydb.) Ewan = Delphinium wootonii Rydb. ■☆
124682 Delphinium virescens Nutt. var. macroceratilis (Rydb.) Cory = Delphinium carolinianum Walter subsp. virescens (Nutt.) R. E. Brooks ■☆
124683 Delphinium virescens Nutt. var. penardii (Huth) L. M. Perry = Delphinium carolinianum Walter subsp. virescens (Nutt.) R. E. Brooks ■☆
124684 Delphinium virescens Nutt. var. penardii (Huth) L. M. Perry = Delphinium carolinianum Walter ■☆
124685 Delphinium viride F. H. Chen = Delphinium forrestii Diels var. viride (W. T. Wang) W. T. Wang ■
124686 Delphinium viridescens Leiberg;韦纳契翠雀花;Wenatchee Larkspur ■☆
124687 Delphinium viscosum Hook. f. et Thomson;黏毛翠雀花;Viscidhairy Larkspur ■☆
124688 Delphinium viscosum Hook. f. et Thomson var. chrysotrichum Huth;黄黏毛翠雀花;Yellow Viscidhairy Larkspur ■
124689 Delphinium viscosum Hook. f. et Thomson var. sordidum ? = Delphinium unifolium Tamura ■☆
124690 Delphinium vitifolium Finet et Gagnep. = Delphinium sinovitifolium W. T. Wang ■
124691 Delphinium wangii M. J. Warnock;秀丽翠雀花;Elegant Larkspur ■
124692 Delphinium wardii C. Marquand et Airy Shaw;堆拉翠雀花;Ward Larkspur ■
124693 Delphinium weiningense W. T. Wang;威宁翠雀花;Weining Larkspur ■
124694 Delphinium wenchuanense W. T. Wang;文川翠雀花;Wenchuan Larkspur ■
124695 Delphinium wentsaii Y. Z. Zhao;文采翠雀花;Wensai Larkspur ■
124696 Delphinium wilsonii Munz = Delphinium hirticaule Franch. ■
124697 Delphinium winklerianum Huth;温泉翠雀花;Hotspring Larkspur ■
124698 Delphinium wootonii Rydb.;伍顿翠雀花;Wooton's Larkspur ■☆
124699 Delphinium wrightii F. H. Chen;狭序翠雀花;Wright Larkspur ■
124700 Delphinium wrightii F. H. Chen var. subtubulosum W. T. Wang;粗距狭序翠雀花;Thickspur Wright Larkspur ■
124701 Delphinium wuqiaense W. T. Wang;乌恰翠雀花;Wuqia Larkspur ■
124702 Delphinium xichangense W. T. Wang;西昌翠雀花;Xichang Larkspur ■
124703 Delphinium yajiangense W. T. Wang;雅江翠雀花;Yajiang Larkspur ■
124704 Delphinium yangii W. T. Wang;竟生翠雀花;Yang's Larkspur ■
124705 Delphinium yanwaense W. T. Wang;岩瓦翠雀花;Yanwa Larkspur ■
124706 Delphinium yechengense Chang Y. Yang et B. Wang;叶城翠雀花;Yecheng Larkspur ■
124707 Delphinium yongningense W. T. Wang;永宁翠雀花;Yongning Larkspur ■
124708 Delphinium yuanum F. H. Chen;中甸翠雀花;Yuan Larkspur, Zhongdian Larkspur ■
124709 Delphinium yuchuanii Y. Z. Zhao;毓泉翠雀花;Yuquan Larkspur ■
124710 Delphinium yulongshanicum W. T. Wang;玉龙山翠雀花;Yulongshan Larkspur ■
124711 Delphinium yunnanense Franch.;云南翠雀花(倒提壶,滇北翠雀花,鸡脚草乌,惊药,细草乌,小草乌,月下参,云南飞燕草);Yunnan Larkspur ■
124712 Delphinium zalil Aitch. et Hemsl.;扎利尔翠雀花;Yellow Larkspur ■
124713 Delphinium zhangii W. T. Wang;镱锂翠雀花;Zhang's Larkspur ■
124714 Delphyodon K. Schum. (1898);齿囊夹竹桃属●☆
124715 Delphyodon oliganthus K. Schum.;齿囊夹竹桃●☆
124716 Delpinoa H. Ross = Agave L. ■
124717 Delpinoella Speg. = Delpinophytum Speg. ■☆
124718 Delpinophytum Speg. (1903);巴西岩园芥属■☆
124719 Delpinophytum patagonicum Speg.;巴西岩园芥■☆
124720 Delpya Pierre = Paranephelium Miq. ●
124721 Delpya Pierre ex Bonati = Pierranthus Bonati ■☆
124722 Delpya Pierre ex Radlk. = Sisyrolepis Radlk. ●☆
124723 Delpydora Pierre(1897);长被山榄属●☆
124724 Delpydora gracilis A. Chev.;纤细长被山榄●☆
124725 Delpydora macrophylla Pierre;大叶长被山榄●☆
124726 Deltacheilos W. T. Wang = Chirita Buch. -Ham. ex D. Don ●■
124727 Deltaria Steenis(1959);痕轴瑞香属●☆
124728 Deltaria brachyblastophora Steenis;痕轴瑞香●☆

124729 Deltocarpus L'Hér. ex DC. = Myagrum L. ■☆

124730 Deltocheilos W. T. Wang = Chirita Buch. -Ham. ex D. Don ●■

124731 Deltocheilos tenuitubum W. T. Wang = Chirita tenuituba (W. T. Wang) W. T. Wang ■

124732 Deltonea Peckolt = Theobroma L. ●

124733 Delueia DC. = Bidens L. ■●

124734 Delwiensia W. A. Weber et R. C. Wittmann = Artemisia L. ●■

124735 Delwiensia W. A. Weber et R. C. Wittmann(2009);科罗拉多蒿属■☆

124736 Dematophyllum Griseb. = Balbisia Cav.(保留属名)●☆

124737 Dematra Raf. = Euphorbia L. ●■

124738 Demavendia Pimenov(1987);代马前胡属■☆

124739 Demavendia pastinacifolia (Boiss. et Hausskn.) Pimenov;代马前胡■☆

124740 Demazeria Dumort. = Desmazeria Dumort. ■☆

124741 Demetria Lag. = Grindelia Willd. ●■☆

124742 Demeusea De Wild. et Durand = Haemanthus L. ■

124743 Demeusea longifolia De Wild. et T. Durand = Scadoxus longifolius (De Wild. et T. Durand) Friis et Nordal ■☆

124744 Demeusia Willis = Demeusea De Wild. et Durand ■

124745 Demidium DC. = Gnaphalium L. ■

124746 Demidofia Dennst. = Carallia Roxb.(保留属名)●

124747 Demidofia J. F. Gmel. = Dichondra J. R. Forst. et G. Forst. ■

124748 Demidovia Hoffm. = Paris L. ■

124749 Demidovia Pall. = Tetragonia L. ●■

124750 Demidovia tetragonioides Pall. = Tetragonia tetragonioides (Pall.) Kuntze ●■

124751 Demnosa Frič = Cleistocactus Lem. ●☆

124752 Democritea DC. = Serissa Comm. ex Juss. ●

124753 Democritea serissoides DC. = Serissa serissoides (DC.) Druce ●

124754 Demorchis D. L. Jones et M. A. Clem. = Gastrodia R. Br. ■

124755 Demosthenesia A. C. Sm. (1936);凌霄莓属●☆

124756 Demosthenesia buxifolia (Fielding et Gardner) A. C. Sm.;黄杨叶凌霄莓●☆

124757 Demosthenesia buxifolia A. C. Sm. = Demosthenesia buxifolia (Fielding et Gardner) A. C. Sm. ●☆

124758 Demosthenesia cordifolia Luteyn;心叶凌霄莓●☆

124759 Demosthenesia microphylla (Hoerold) A. C. Sm.;小叶凌霄莓●☆

124760 Demosthenesia oppositifolia Luteyn;对叶凌霄莓●☆

124761 Demosthenesia spectabilis A. C. Sm.;凌霄莓●☆

124762 Demosthenia Raf. = Sideritis L. ■●

124763 Demosthenia Raf. = Marrubiastrum Moench ■●

124764 Denckea Raf. = Gentiana L. ■

124765 Dendragrostis B. D. Jacks. = Chusquea Kunth ●☆

124766 Dendragrostis Nees = Chusquea Kunth ●☆

124767 Dendranthema (DC.) Des Moul. (1860);菊属;Daisy, Dendranthema, Florist's Chrysanthemum ■

124768 Dendranthema arctica (L.) Tzvelev = Arctanthemum arcticum (L.) Tzvelev ■☆

124769 Dendranthema arcticum (L.) Tzvelev = Chrysanthemum arcticum L. ■☆

124770 Dendranthema arcticum (L.) Tzvelev subsp. maekawanum (Kitam.) H. Koyama = Chrysanthemum yezoense Maek. ■☆

124771 Dendranthema argyrophyllum (Y. Ling) Y. Ling et C. Shih;银背菊;Silverleaf Daisy, Silverleaf Dendranthema ■

124772 Dendranthema arisanense (Hayata) Y. Ling et C. Shih;阿里山菊(阿里山油菊);Alishan Daisy, Taiwan Dendranthema ■

124773 Dendranthema boreale (Makino) Y. Ling = Dendranthema lavandulifolium (Fisch. ex Trautv.) Kitam. ■

124774 Dendranthema boreale (Makino) Y. Ling = Dendranthema lavandulifolium (Fisch. ex Trautv.) Y. Ling et C. Shih var. seticuspe (Maxim.) C. Shih ■

124775 Dendranthema boreale (Makino) Y. Ling ex Kitam.;北方菊■☆

124776 Dendranthema boreale (Makino) Y. Ling var. tomentellum (Hand. -Mazz.) Kitam. = Dendranthema lavandulifolium (Fisch. ex Trautv.) Y. Ling et C. Shih var. tomentellum (Hand. -Mazz.) Y. Ling et C. Shih ■

124777 Dendranthema boreale (Makino) Y. Ling var. tomentellum (Hand. -Mazz.) Kitam. = Dendranthema lavandulifolium (Fisch. ex Trautv.) Kitam. var. tomentellum (Hand. -Mazz.) Y. Ling et C. Shih ■

124778 Dendranthema chanetii (H. Lév.) C. Shih;小红菊(山野菊);Chanet Daisy, Chanet's Dendranthema ■

124779 Dendranthema crassum (Kitam.) Kitam. = Chrysanthemum crassum (Kitam.) Kitam. ■☆

124780 Dendranthema cuneifolium (Kitam.) H. Koyama;楔叶菊■☆

124781 Dendranthema dichrum C. Shih;异色菊(异菊);Bicolored Dendranthema ■

124782 Dendranthema erubescens (Stapf) Tzvelev = Dendranthema chanetii (H. Lév.) C. Shih ■

124783 Dendranthema foliaceum G. F. Peng, C. Shih et S. Q. Zhang;裂苞菊;Dendranthema ■

124784 Dendranthema glabriusculum (W. W. Sm.) C. Shih;拟亚菊;Glabrous Daisy, Glabrous Dendranthema ■

124785 Dendranthema grandiflorum (Ramat.) Kitam.;菊花(白菊花,长生,滁菊,甘菊,贡菊,杭菊,亳菊,黄花,家菊,节花,节华,金精,金蕊,九花,鞠,秋鞠,秋菊,小白菊,药菊,真菊);Florists Chrysanthemum, Florists Daisy, Florists Dendranthema, Garden Chrysanthemum, Mums ■

124786 Dendranthema grandiflorum (Ramat.) Kitam. = Chrysanthemum morifolium Ramat. ■☆

124787 Dendranthema horaimontana (Masam.) S. S. Ying;蓬莱油菊■

124788 Dendranthema hultenii (Á. Löve et D. Löve) Tzvelev;胡尔滕菊■☆

124789 Dendranthema hypargyrum (Diels) Y. Ling et C. Shih;黄花小山菊(黄花小葶菊);Yellowflower Daisy, Yellowflower Dendranthema ■

124790 Dendranthema indicum (L.) Des Moul.;野菊(鬼仔菊,黄菊,黄菊花,黄菊仔,菊花脑,苦薏,苦薏花,路边黄,路边菊,疟疾草,山九月菊,山菊花,野黄菊,野黄菊花,野菊花,野山菊,油菊);Chinese and Japanese Chrysanthemum, Chrysanthemum, India Chrysanthemum, India Daisy, Indian Dendranthema, Mother Chrysanthemum ■

124791 Dendranthema indicum (L.) Des Moul. = Chrysanthemum indicum L. ■

124792 Dendranthema indicum (L.) Des Moul. = Dendrocalamus strictus (Roxb.) Nees ●

124793 Dendranthema indicum (L.) Des Moul. var. acutum Uyeki;尖齿野菊■

124794 Dendranthema indicum (L.) Des Moul. var. aromaticum Q. H. Liu et S. F. Zhang = Dendranthema lavandulifolium (Fisch. ex Trautv.) Y. Ling et C. Shih var. aromaticum (Q. H. Liu et S. F. Zhang) S. J. Zhou et D. X. Zang ■

124795 Dendranthema indicum (L.) Des Moul. var. huludaoensis G. Y. Zhang, L. J. Yu et X. J. Liu;葫芦岛野菊;Hulidao Dendranthema ■

124796 Dendranthema indicum (L.) Des Moul. var. maruyamanum (Kitam.) Kitam. = Chrysanthemum indicum L. var. maruyamanum Kitam. ■☆

124797 Dendranthema integrifolium (Richardson) Tzvelev = Hulteniella integrifolia (Richardson) Tzvelev ■☆

124798 Dendranthema japonense (Nakai) Kitam. = Chrysanthemum japonense (Makino) Nakai ■☆

124799 Dendranthema japonense (Nakai) Kitam. = Dendranthema occidentalijaponense Kitam. ■☆

124800 Dendranthema japonense (Nakai) Kitam. var. ashizuriense (Kitam.) Kitam. = Chrysanthemum japonense (Makino) Nakai var. ashizuriense Kitam. ■☆

124801 Dendranthema japonense (Nakai) Kitam. var. ashizuriense (Kitam.) Kitam. = Dendranthema occidentalijaponense Kitam. var. ashizuriense (Kitam.) H. Koyama ■☆

124802 Dendranthema japonicum (Makino) Kitam. var. wakasaense (Shimot. ex Kitam.) Kitam. = Chrysanthemum wakasaense Shimot. ex Kitam. ■☆

124803 Dendranthema japonicum (Maxim.) Kitam. = Chrysanthemum makinoi Matsum. et Nakai ■☆

124804 Dendranthema kurilense Tzvelev;北海道菊■☆

124805 Dendranthema kurilense Tzvelev = Chrysanthemum arcticum L. subsp. yezoense (Maek.) H. Ohashi et Yonek. ■☆

124806 Dendranthema lavandulifolium (Fisch. ex Trautv.) Kitam.;甘菊(北野菊,少花野菊,细裂香菊,岩香菊,野菊);Lavenderleaf Daisy, Lavenderleaf Dendranthema ■

124807 Dendranthema lavandulifolium (Fisch. ex Trautv.) Kitam. = Chrysanthemum lavandulifolium (Fisch. ex Trautv.) Makino ■

124808 Dendranthema lavandulifolium (Fisch. ex Trautv.) Kitam. var. aromaticum (Q. H. Liu et S. F. Zhang) S. J. Zhou et D. K. Zang = Dendranthema indicum (L.) Des Moul. ■

124809 Dendranthema lavandulifolium (Fisch. ex Trautv.) Kitam. var. discoideum (Hand. -Mazz.) Y. Ling;隐舌甘菊■

124810 Dendranthema lavandulifolium (Fisch. ex Trautv.) Kitam. var. glabriusculum (Y. Ling) Kitam. = Dendranthema lavandulifolium (Fisch. ex Trautv.) Kitam. ■

124811 Dendranthema lavandulifolium (Fisch. ex Trautv.) Kitam. var. seticuspe (Maxim.) C. Shih = Chrysanthemum seticuspe (Maxim.) Hand. -Mazz. f. boreale (Makino) H. Ohashi et Yonek. ■☆

124812 Dendranthema lavandulifolium (Fisch. ex Trautv.) Kitam. var. tomentellum (Hand. -Mazz.) Y. Ling et C. Shih;毛叶甘菊(新竹油菊);Hairy Lavenderleaf Dendranthema ■

124813 Dendranthema lavandulifolium (Fisch. ex Trautv.) Y. Ling et C. Shih var. aromaticum (Q. H. Liu et S. F. Zhang) S. J. Zhou et D. X. Zang;神农香菊;Aromatic Dendranthema ■

124814 Dendranthema lavandulifolium (Fisch. ex Trautv.) Y. Ling et C. Shih var. aromaticum (Q. H. Liu et S. F. Zhang) S. J. Zhou et D. X. Zang = Dendranthema indicum (L.) Des Moul. ■

124815 Dendranthema lavandulifolium (Fisch. ex Trautv.) Y. Ling et C. Shih var. discoideum (Maxim.) C. Shih = Dendranthema lavandulifolium (Fisch. ex Trautv.) Kitam. var. discoideum (Hand. -Mazz.) Y. Ling ■

124816 Dendranthema lavandulifolium (Fisch. ex Trautv.) Y. Ling et C. Shih var. seticuspe (Maxim.) C. Shih = Dendranthema lavandulifolium (Fisch. ex Trautv.) Kitam. ■

124817 Dendranthema lavandulifolium (Fisch. ex Trautv.) Y. Ling et C. Shih var. seticuspe (Maxim.) C. Shih;野甘菊(北野菊,甘野菊);Seta Daisy, Seta Dendranthema ■

124818 Dendranthema lavandulifolium (Fisch. ex Trautv.) Y. Ling et C. Shih var. seticuspe (Maxim.) C. Shih = Chrysanthemum seticuspe (Maxim.) Hand. -Mazz. f. boreale (Makino) H. Ohashi et Yonek. ■☆

124819 Dendranthema lavandulifolium (Fisch. ex Trautv.) Y. Ling et C. Shih var. tomentellum (Hand. -Mazz.) Y. Ling et C. Shih = Dendranthema lavandulifolium (Fisch. ex Trautv.) Kitam. var. tomentellum (Hand. -Mazz.) Y. Ling et C. Shih ■

124820 Dendranthema lavandulifolium (Fisch. ex Trautv.) Y. Ling et C. Shih = Dendranthema lavandulifolium (Fisch. ex Trautv.) Kitam. ■

124821 Dendranthema leucanthum (Makino) H. Koyama;白花菊■☆

124822 Dendranthema littorale (Maek.) Tzvelev;滨海甘菊■☆

124823 Dendranthema longibracteatum C. Shih, G. F. Peng et S. Y. Jin;线苞菊;Longbract Dendranthema ■

124824 Dendranthema maximowiczii (Kom.) Tzvelev;细叶菊;Maximowicz Daisy, Maximowicz Dendranthema ■

124825 Dendranthema mongolicum (Y. Ling) Tzvelev;蒙菊;Mongol Daisy, Mongolian Dendranthema ■

124826 Dendranthema morifolium (Ramat.) Tzvelev;桑叶菊(白菊花,茶菊花,甘菊,甘菊花,杭菊花,黄菊花,家菊,节华,金精,金蕊菊,菊花,馒头菊,容成,甜菊花,药菊,玉英,簪头菊,真菊);Chrysanthemum, Florist Chrysanthemum, Florists' Chrysanthemum, Florists Dendranthema, Florists Dendranthemum, Mum ■

124827 Dendranthema morifolium (Ramat.) Tzvelev = Chrysanthemum morifolium Ramat. ■☆

124828 Dendranthema morifolium (Ramat.) Tzvelev = Dendranthema grandiflorum (Ramat.) Kitam. ■

124829 Dendranthema morii (Hayata) Kitam.;森氏菊(台湾菊)■

124830 Dendranthema morii (Hayata) Kitam. = Chrysanthemum morii Hayata ■

124831 Dendranthema mutellina (Hand. -Mazz.) Kitam.;锡金菊■☆

124832 Dendranthema naktongense (Nakai) Tzvelev;日本楔叶菊;Cuneate Daisy, Cuneateleaf Dendranthema ■

124833 Dendranthema nankingense (Hand. -Mazz.) X. D. Cui = Dendranthema indicum (L.) Des Moul. ■

124834 Dendranthema occidentalijaponense Kitam. = Chrysanthemum japonense (Makino) Nakai ■☆

124835 Dendranthema occidentalijaponense Kitam. var. ashizuriense (Kitam.) H. Koyama = Chrysanthemum japonense (Makino) Nakai var. ashizuriense Kitam. ■☆

124836 Dendranthema ogawae (Kitam.) H. Koyama;小川菊■☆

124837 Dendranthema okiense (Kitam.) Kitam.;大木菊■☆

124838 Dendranthema oreastrum (Hance) Y. Ling;小山菊(毛山菊,小亭菊);Smollscape Daisy, Smollscape Dendranthema ■

124839 Dendranthema ornatum (Hemsl.) Kitam. = Chrysanthemum ornatum Hemsl. ■☆

124840 Dendranthema ornatum (Hemsl.) Kitam. var. tokarense M. Hotta et Hirai = Chrysanthemum ornatum Hemsl. var. tokarense (M. Hotta et Hirai) H. Ohashi et Yonek. ■☆

124841 Dendranthema pacificum (Nakai) Kitam. = Chrysanthemum pacificum Nakai ■☆

124842 Dendranthema pallasianum (Fisch. ex Besser) Vorosch. = Ajania pallasiana (Fisch. ex Besser) Poljakov ■

124843 Dendranthema pallasianum (Fisch. ex Besser) Vorosch. = Chrysanthemum pallasianum (Fisch. ex Besser) Kom. ■

124844 Dendranthema parvifolium (C. C. Chang) C. Shih;小叶菊■

124845 Dendranthema potentilloides (Hand.-Mazz.) C. Shih;委陵菊;Cinquefoli-like Daisy,Potentilla-like Dendranthema ■

124846 Dendranthema rhombifolium Y. Ling et C. Shih;菱叶菊;Rhombicleaf Daisy,Waterchestnulleaf Dendranthema ■

124847 Dendranthema rupestre (Matsum. et Koidz.) Kitam. = Chrysanthemum rupestre Matsum. et Koidz. ■☆

124848 Dendranthema seticuspe (Maxim.) Kitam. = Chrysanthemum seticuspe (Maxim.) Hand.-Mazz. ■☆

124849 Dendranthema seticuspe (Maxim.) Kitam. f. boreale (Makino) Kitam. = Chrysanthemum seticuspe (Maxim.) Hand.-Mazz. f. boreale (Makino) H. Ohashi et Yonek. ■☆

124850 Dendranthema shiwogiku (Kitam.) Kitam. = Chrysanthemum shiwogiku Kitam. ■☆

124851 Dendranthema shiwogiku (Kitam.) Kitam. var. kinokuniense (Shimot. et Kitam.) Kitam. = Chrysanthemum kinokuniense (Shimot. et Kitam.) H. Ohashi et Yonek. ■☆

124852 Dendranthema sichotense Tzvelev = Chrysanthemum zawadskii Herb. var. alpinum (Nakai) Kitam. ■

124853 Dendranthema sichotense Tzvelev = Dendranthema oreastrum (Hance) Y. Ling ■

124854 Dendranthema sinense (Sabine) Des Moul. = Dendranthema grandiflorum (Ramat.) Kitam. ■

124855 Dendranthema sinense (Sabine) Des Moul. = Dendranthema morifolium (Ramat.) Tzvelev ■

124856 Dendranthema sinuatum (Ledeb.) Tzvelev;深波菊■☆

124857 Dendranthema vestitum (Hemsl.) Y. Ling;毛华菊(艾精,野黄菊);Hairy Daisy,Hairy Dendranthema ■

124858 Dendranthema weyrichii (Maxim.) Tzvelev = Chrysanthemum weyrichii (Maxim.) Miyabe et T. Miyake ■☆

124859 Dendranthema weyrichii (Maxim.) Tzvelev = Dendranthema zawadskii (Herb.) Tzvelev ■

124860 Dendranthema yezoense (Maek.) D. J. N. Hind = Chrysanthemum yezoense Maek. ■☆

124861 Dendranthema yoshinaganthum (Makino ex Kitam.) Kitam. = Chrysanthemum yoshinaganthum Makino ex Kitam. ■☆

124862 Dendranthema zawadskii (Herb.) Tzvelev;紫花野菊(山菊);Zawadsk Daisy,Zawadsk's Dendranthema ■

124863 Dendranthema zawadskii (Herb.) Tzvelev = Chrysanthemum zawadskii Herb. ■

124864 Dendranthema zawadskii (Herb.) Tzvelev var. alpinum (Nakai) Kitam.;高山紫花野菊■

124865 Dendranthema zawadskii (Herb.) Tzvelev var. alpinum (Nakai) Kitam. = Chrysanthemum zawadskii Herb. var. alpinum (Nakai) Kitam. ■

124866 Dendranthema zawadskii (Herb.) Tzvelev var. latilobum (Maxim.) Kitam.;宽叶紫花野菊■

124867 Dendranthema zawadskii (Herb.) Tzvelev var. latilobum (Maxim.) Kitam. = Chrysanthemum zawadskii Herb. var. latilobum (Maxim.) Kitam. ■

124868 Dendrema Raf. = Bessia Raf. ●☆

124869 Dendrema Raf. = Intsia Thouars ●☆

124870 Dendriopoterium Svent. = Bencomia Webb et Berthel. ●☆

124871 Dendriopoterium Svent. = Sanguisorba L. ■

124872 Dendriopoterium menendezii Svent. var. virescens ? = Dendriopoterium menendezii Svent. ■☆

124873 Dendrium Desv. = Leiophyllum (Pers.) R. Hedw. ●☆

124874 Dendroarabis (C. A. Mey. et Bunge) D. A. German et Al-Shehbaz = Arabis L. ●■

124875 Dendrobangia Rusby = Dendrobangia Rusby et R. A. Howard ●☆

124876 Dendrobangia Rusby et R. A. Howard(1896);乔茶萸属●☆

124877 Dendrobangia boliviana Rusby;乔茶萸●☆

124878 Dendrobangia multinervia Ducke;密脉乔茶萸●☆

124879 Dendrobates M. A. Clem. et D. L. Jones = Dendrobium Sw.(保留属名)■

124880 Dendrobenthamia Hutch.(1942);四照花属;Dendrobenthamia,Dogwood,Four-involucre ●

124881 Dendrobenthamia Hutch. = Cornus L. ●

124882 Dendrobenthamia angustata (Chun) W. P. Fang;狭叶四照花(湖北四照花,尖叶四照花,野荔枝);Angustifoliate Dogwood,Narrowleaf Dogwood,Sharpleaf Four-involucre ●

124883 Dendrobenthamia angustata (Chun) W. P. Fang = Cornus elliptica (Pojark.) Q. Y. Xiang et Boufford ●

124884 Dendrobenthamia angustata (Chun) W. P. Fang var. molis (Rehder) W. P. Fang;绒毛尖叶四照花;Hairy Narrowleaf Dogwood,Hairy Sharpleaf Four-involucre ●

124885 Dendrobenthamia angustata (Chun) W. P. Fang var. molis (Rehder) W. P. Fang = Dendrobenthamia angustata (Chun) W. P. Fang ●

124886 Dendrobenthamia angustata (Chun) W. P. Fang var. mollis (Rehder) W. P. Fang = Cornus elliptica (Pojark.) Q. Y. Xiang et Boufford ●

124887 Dendrobenthamia angustata (Chun) W. P. Fang var. wuyishanensis (W. P. Fang et Y. T. Hsieh) W. P. Fang et W. K. Hu;武夷四照花;Wuyishan Dendrobenthamia,Wuyishan Four-involucre ●

124888 Dendrobenthamia angustata (Chun) W. P. Fang var. wuyishanensis (W. P. Fang et Y. T. Hsieh) W. P. Fang et W. K. Hu = Cornus elliptica (Pojark.) Q. Y. Xiang et Boufford ●

124889 Dendrobenthamia brevipedunculata W. P. Fang et Y. T. Hsieh = Cornus hongkongensis Hemsl. subsp. tonkinensis (W. P. Fang) Q. Y. Xiang ●

124890 Dendrobenthamia brevipendunculata W. P. Fang et Y. T. Hsieh = Dendrobenthamia tonkinensis W. P. Fang ●

124891 Dendrobenthamia capitata (Wall. ex Roxb.) Hutch.;头状四照花(鸡嗉子,节节树,癞鸡嗉,山覆盆,山荔枝,石蕉子,头状本氏茱萸,羊梅,野荔枝,一支箭,云母树);Bentham's Cornel,Evergreen Dogwood,Evergreen Four-involucre,Himalayan Dogwood ●

124892 Dendrobenthamia capitata (Wall. ex Roxb.) Hutch. = Cornus capitata Wall. ●

124893 Dendrobenthamia capitata (Wall. ex Roxb.) Hutch. var. emeiensis (W. P. Fang et W. K. Hu) W. P. Fang et W. K. Hu;峨眉头状四照花(峨眉四照花);Emei Dogwood,Emei Four-involucre ●◇

124894 Dendrobenthamia capitata (Wall. ex Roxb.) Hutch. var. emeiensis (W. P. Fang et Y. T. Hsieh) W. P. Fang et W. K. Hu = Cornus capitata Wall. ●

124895 Dendrobenthamia capitata (Wall.) Hutch. = Benthamidia capitata (Wall. ex Roxb.) H. Hara ●

124896 Dendrobenthamia capitata (Wall.) Hutch. = Cornus capitata Wall. ex Roxb. ●

124897 Dendrobenthamia capitata (Wall.) Hutch. = Dendrobenthamia capitata (Wall. ex Roxb.) Hutch. ●

124898 Dendrobenthamia capitata (Wall.) Hutch. var. angustata Chun

= Dendrobenthamia angustata (Chun) W. P. Fang ●

124899 Dendrobenthamia capitata (Wall.) Hutch. var. emeiensis (W. P. Fang et W. K. Hu) W. P. Fang et W. K. Hu = Dendrobenthamia capitata (Wall. ex Roxb.) Hutch. var. emeiensis (W. P. Fang et W. K. Hu) W. P. Fang et W. K. Hu ●◇

124900 Dendrobenthamia elegans (Wall. ex Roxb.) Hutch. var. rotundifolia (W. P. Fang et Y. T. Hsieh) W. P. Fang et W. K. Hu = Dendrobenthamia elegans W. P. Fang et Y. T. Hsieh ●

124901 Dendrobenthamia elegans (Wall. ex Roxb.) Hutch. var. rotundifolia (W. P. Fang et Y. T. Hsieh) W. P. Fang et W. K. Hu = Cornus hongkongensis Hemsl. subsp. elegans (W. P. Fang et Y. T. Hsieh) Q. Y. Xiang ●

124902 Dendrobenthamia elegans W. P. Fang et Y. T. Hsieh = Cornus hongkongensis Hemsl. subsp. elegans (W. P. Fang et Y. T. Hsieh) Q. Y. Xiang ●

124903 Dendrobenthamia elegans W. P. Fang et Y. T. Hsieh var. rotundifolia (W. P. Fang et Y. T. Hsieh) W. P. Fang et W. K. Hu = Cornus hongkongensis Hemsl. subsp. elegans (W. P. Fang et Y. T. Hsieh) Q. Y. Xiang ●

124904 Dendrobenthamia emeiensis W. P. Fang et W. K. Hu = Dendrobenthamia capitata (Wall.) Hutch. var. emeiensis (W. P. Fang et W. K. Hu) W. P. Fang et W. K. Hu ●

124905 Dendrobenthamia emeiensis W. P. Fang et Y. T. Hsieh = Cornus capitata Wall. ●

124906 Dendrobenthamia ferruginea (Y. C. Wu) W. P. Fang = Cornus hongkongensis Hemsl. ●

124907 Dendrobenthamia ferruginea (Y. C. Wu) W. P. Fang = Cornus hongkongensis Hemsl. subsp. ferruginea (Y. C. Wu) Q. Y. Xiang ●

124908 Dendrobenthamia ferruginea (Y. C. Wu) W. P. Fang var. jiangxiensis W. P. Fang et Y. T. Hsieh;江西褐毛四照花(江西四照花);Jiangxi Dogwood,Jiangxi Four-involucre ●

124909 Dendrobenthamia ferruginea (Y. C. Wu) W. P. Fang var. jiangxiensis W. P. Fang et Y. T. Hsieh = Cornus hongkongensis Hemsl. subsp. ferruginea (Y. C. Wu) Q. Y. Xiang ●

124910 Dendrobenthamia ferruginea (Y. C. Wu) W. P. Fang var. jiangxiensis W. P. Fang et Y. T. Hsieh. = Cornus hongkongensis Hemsl. ●

124911 Dendrobenthamia ferruginea (Y. C. Wu) W. P. Fang var. jinyunensis (W. P. Fang et W. K. Hu) W. P. Fang et W. K. Hu;缙云褐毛四照花(缙云四照花);Jinyun Dogwood, Jinyun Four-involucre ●

124912 Dendrobenthamia ferruginea (Y. C. Wu) W. P. Fang var. jinyunensis (W. P. Fang et W. K. Hu) W. P. Fang et W. K. Hu = Cornus hongkongensis Hemsl. subsp. melanotricha (Pojark.) Q. Y. Xiang ●

124913 Dendrobenthamia finyunensis Fang et W. K. Hu = Dendrobenthamia ferruginea (Y. C. Wu) W. P. Fang var. jinyunensis (W. P. Fang et W. K. Hu) W. P. Fang et W. K. Hu ●

124914 Dendrobenthamia gigantea (Hand. -Mazz.) W. P. Fang;大型四照花(大叶四照花,大叶香港四照花);Big Dogwood,Giant Four-involucre,Gigantic Dendrobenthamia ●

124915 Dendrobenthamia gigantea (Hand. -Mazz.) W. P. Fang = Cornus hongkongensis Hemsl. subsp. gigantea (Hand. -Mazz.) Q. Y. Xiang ●

124916 Dendrobenthamia gigantea (Hand. -Mazz.) W. P. Fang var. caudata W. P. Fang et W. K. Hu = Dendrobenthamia gigantea (Hand. -Mazz.) W. P. Fang ●

124917 Dendrobenthamia gigantea var. caudata W. P. Fang et W. K. Hu = Cornus hongkongensis Hemsl. subsp. gigantea (Hand. -Mazz.) Q. Y. Xiang ●

124918 Dendrobenthamia gigantea var. caudata W. P. Fang et W. K. Hu = Cornus hongkongensis Hemsl. subsp. melanotricha (Pojark.) Q. Y. Xiang ●

124919 Dendrobenthamia hongkongensis (Hemsl.) Hutch. = Cornus hongkongensis Hemsl ●

124920 Dendrobenthamia hongkongensis (Hemsl.) Hutch. var. gigantea Hand. -Mazz. = Dendrobenthamia gigantea (Hand. -Mazz.) W. P. Fang ●

124921 Dendrobenthamia hupehensis W. P. Fang = Cornus elliptica (Pojark.) Q. Y. Xiang et Boufford ●

124922 Dendrobenthamia hupehensis W. P. Fang = Dendrobenthamia angustata (Chun) W. P. Fang ●

124923 Dendrobenthamia japonica (DC.) W. P. Fang;日本四照花(东瀛四照花,四照花);Chinese Dogwood, Japan Four-involucre, Japanese Dendrobenthamia, Japanese Dogwood, Japanese Flowering Dogwood, Japanese Strawberry, Japanese Strawberry Tree, Korean Dogwood,Kousa,Kousa Dogwood ●

124924 Dendrobenthamia japonica (DC.) W. P. Fang = Benthamidia japonica (Siebold et Zucc.) H. Hara ●

124925 Dendrobenthamia japonica (DC.) W. P. Fang var. chinensis (Osborn) W. P. Fang;四照花(东瀛四照花,凉子,青皮树,日本四照花,山荔枝,石枣,石楂子树,小六角);Chinese Dogwood, Chinese Kousa Dogwood,Four-involucre,Japanese Dogwood,Japanese Strawberry Tree,Korean Dogwood,Kousa,Kousa Dogwood ●

124926 Dendrobenthamia japonica (DC.) W. P. Fang var. chinensis (Osborn) W. P. Fang = Benthamidia japonica (Siebold et Zucc.) H. Hara var. chinensis (Osborn) H. Hara ●

124927 Dendrobenthamia japonica (DC.) W. P. Fang var. chinensis (Osborn) W. P. Fang = Cornus kousa Buerger ex Hance subsp. chinensis (Osborn) Q. Y. Xiang ●

124928 Dendrobenthamia japonica (DC.) W. P. Fang var. huaxiensis W. P. Fang et W. K. Hu;华西四照花;Huaxi Dogwood, Huaxi Four-involucre ●

124929 Dendrobenthamia japonica (DC.) W. P. Fang var. huaxiensis W. P. Fang et W. K. Hu = Cornus kousa Buerger ex Hance subsp. chinensis (Osborn) Q. Y. Xiang ●

124930 Dendrobenthamia japonica (DC.) W. P. Fang var. leucotricha W. P. Fang et W. K. Hu;白毛四照花;Whitehair Dogwood,Whitehair Four-involucre ●

124931 Dendrobenthamia japonica (DC.) W. P. Fang var. leucotricha W. P. Fang et Y. T. Hsieh = Cornus kousa Buerger ex Hance subsp. chinensis (Osborn) Q. Y. Xiang ●

124932 Dendrobenthamia japonica (Siebold et Zucc.) Hutch. = Dendrobenthamia japonica (DC.) W. P. Fang ●

124933 Dendrobenthamia jinyunensis W. P. Fang et W. K. Hu = Cornus hongkongensis Hemsl. subsp. melanotricha (Pojark.) Q. Y. Xiang ●

124934 Dendrobenthamia jinyunensis W. P. Fang et W. K. Hu = Dendrobenthamia ferruginea (Y. C. Wu) W. P. Fang var. jinyunensis (W. P. Fang et W. K. Hu) W. P. Fang et W. K. Hu ●

124935 Dendrobenthamia kousa ? = Dendrobenthamia japonica (DC.) W. P. Fang ●

124936 Dendrobenthamia latibracteata W. P. Fang et Y. T. Hsieh = Cornus hongkongensis Hemsl. ●

124937 Dendrobenthamia latibracteata W. P. Fang et Y. T. Hsieh =

Dendrobenthamia tonkinensis W. P. Fang ●

124938 Dendrobenthamia longipedunculata S. S. Chang et X. Chen;长梗四照花;Longpedicelled Dogwood ●

124939 Dendrobenthamia longipedunculata S. S. Chang et X. Chen = Cornus elliptica (Pojark.) Q. Y. Xiang et Boufford ●

124940 Dendrobenthamia melanotricha (Pojark.) W. P. Fang;光叶四照花(黑毛四照花);Blackhair Dogwood, Blackhair Four-involucre, Melanohairy Dendrobenthamia ●

124941 Dendrobenthamia melanotricha (Pojark.) W. P. Fang = Cornus hongkongensis Hemsl. subsp. melanotricha (Pojark.) Q. Y. Xiang ●

124942 Dendrobenthamia multinervosa (Pojark.) W. P. Fang;巴蜀四照花(多脉四照花);Manyvein Dogwood, Manyvein Four-involucre, Multinerved Dendrobenthamia ●

124943 Dendrobenthamia multinervosa (Pojark.) W. P. Fang = Cornus multinervosa (Pojark.) Q. Y. Xiang ●

124944 Dendrobenthamia pachyphylla W. P. Fang et W. K. Hu = Dendrobenthamia gigantea (Hand. -Mazz.) W. P. Fang ●

124945 Dendrobenthamia pachyphylla W. P. Fang et W. K. Hu. = Cornus hongkongensis Hemsl. subsp. gigantea (Hand. -Mazz.) Q. Y. Xiang ●

124946 Dendrobenthamia qianxinanica S. S. Chang et X. Chen;册亨四照花;Ceheng Dogwood ●

124947 Dendrobenthamia qianxinanica S. S. Chang et X. Chen = Cornus hongkongensis Hemsl. subsp. tonkinensis (W. P. Fang) Q. Y. Xiang ●

124948 Dendrobenthamia rotundifolia W. P. Fang et Y. T. Hsieh = Dendrobenthamia elegans W. P. Fang et Y. T. Hsieh ●

124949 Dendrobenthamia rotundifolia W. P. Fang et Y. T. Hsieh. = Cornus hongkongensis Hemsl. subsp. elegans (W. P. Fang et Y. T. Hsieh) Q. Y. Xiang ●

124950 Dendrobenthamia tonkinensis var. brevipedunculata (W. P. Fang et Y. T. Hsieh) W. P. Fang et W. K. Hu = Cornus capitata Wall. ●

124951 Dendrobenthamia tonkinensis W. P. Fang = Cornus hongkongensis Hemsl. subsp. tonkinensis (W. P. Fang) Q. Y. Xiang ●

124952 Dendrobenthamia tonkinensis W. P. Fang var. brevipenduncculata (W. P. Fang et Y. T. Hsieh) W. P. Fang et W. K. Hu = Dendrobenthamia tonkinensis W. P. Fang ●

124953 Dendrobenthamia wuyishanensis W. P. Fang et Y. T. Hsieh = Dendrobenthamia angustata (Chun) W. P. Fang var. wuyishanensis (W. P. Fang et Y. T. Hsieh) W. P. Fang et W. K. Hu ●

124954 Dendrobenthamia wuyishanica W. P. Fang et Y. T. Hsieh. = Cornus elliptica (Pojark.) Q. Y. Xiang et Boufford ●

124955 Dendrobenthamia xanthocarpa C. Y. Wu = Cornus hongkongensis Hemsl. ●

124956 Dendrobianthe (Schltr.) Mytnik = Polystachya Hook. (保留属名) ■

124957 Dendrobium Sw. (1799)(保留属名);石斛属(石斛兰属);Dendrobium ■

124958 Dendrobium acinaciforme Roxb.; 剑叶石斛; Swordleaf Dendrobium ■

124959 Dendrobium acinaciforme Roxb. var. minus Ts. Tang et F. T. Wang = Dendrobium spatella Rchb. f. ■

124960 Dendrobium acinaciforme Roxb. var. minus Ts. Tang et F. T. Wang = Dendrobium acinaciforme Roxb. ■

124961 Dendrobium acuminatum Rolfe;锐头石斛■☆

124962 Dendrobium aduncum Wall. = Dendrobium aduncum Wall. ex Lindl. ■

124963 Dendrobium aduncum Wall. ex Lindl.;钩状石斛(大黄草,钩石斛,红兰草,黄草石斛,寄生草,藤蓝);Faulhaber Dendrobium, Hooked Dendrobium ■

124964 Dendrobium aduncum Wall. ex Lindl. var. affine W. L. Sha et C. Y. Lo;相似石斛■

124965 Dendrobium aduncum Wall. ex Lindl. var. faulhaberianum (Schltr.) Ts. Tang et F. T. Wang = Dendrobium aduncum Wall. ex Lindl. ■

124966 Dendrobium aemulum R. Br.;澳大利亚石斛■☆

124967 Dendrobium affine (Decne.) Steud.;马刺石斛■☆

124968 Dendrobium affine Steud. = Dendrobium affine (Decne.) Steud. ■☆

124969 Dendrobium aggregatum Roxb.;密茎石斛■

124970 Dendrobium aggregatum Roxb. = Dendrobium lindleyi Steud. ■

124971 Dendrobium aggregatum Roxb. var. jenkinsii (Lindl.) King et Pantl. = Dendrobium jenkinsii Lindl. ■

124972 Dendrobium aggregatum Roxb. var. jenkinsii (Wall. ex Lindl.) King et Pantl. = Dendrobium jenkinsii Lindl. ■

124973 Dendrobium aggregatum Roxb. var. jenkinsii Lindl. = Dendrobium jenkinsii Lindl. ■

124974 Dendrobium albidotomentosum Blume = Dendrolirium lasiopetalum (Willd.) S. C. Chen et J. J. Wood ■

124975 Dendrobium albosanguineum Lindl. = Dendrobium albosanguineum Lindl. et Paxton ■☆

124976 Dendrobium albosanguineum Lindl. et Paxton;浅血红石斛■☆

124977 Dendrobium alboviride Hayata = Dendrobium linawianum Rchb. f. ■

124978 Dendrobium alboviride Hayata var. majuis Rolfe = Dendrobium lindleyi Steud. var. majus (Rolfe) S. Y. Hu ■

124979 Dendrobium alboviride Hayata var. majus Rolfe = Dendrobium lindleyi Steud. ■

124980 Dendrobium aloefolium (Blume) Rchb. f.;芦荟叶石斛(芦荟石斛);Aloeleaf Dendrobium ■☆

124981 Dendrobium alpestre (Sw.) Sw. = Dendrobium monticola P. F. Hunt et Summerh. ■

124982 Dendrobium alpestre Royle = Dendrobium monticola P. F. Hunt et Summerh. ■

124983 Dendrobium amabile (Lour.) O'Brien = Dendrobium densiflorum Lindl. ex Wall. ■

124984 Dendrobium amabile O'Brien = Dendrobium thyrsiflorum Rchb. f. ■

124985 Dendrobium amboinense Hook.; 安倍那石斛; Amboina Dendrobium ■☆

124986 Dendrobium amethystoglossum Rchb. f.;紫水晶石斛■☆

124987 Dendrobium amoenum Wall. = Dendrobium amoenum Wall. ex Lindl. ■☆

124988 Dendrobium amoenum Wall. ex Lindl.; 可爱石斛; Delightful Dendrobium ■☆

124989 Dendrobium amplexicaulis Blume = Thrixspermum amplexicaule (Blume) Rchb. f. ■

124990 Dendrobium amplum Lindl. = Epigeneium amplum (Lindl. ex Wall.) Summerh. ■

124991 Dendrobium anceps Sw.;二棱石斛■☆

124992 Dendrobium andersonianum F. M. Bailey = Dendrobium undulatum R. Br. ■☆

124993 Dendrobium andersonii J. Scott = Dendrobium draconis Rchb. f. ■☆

124994 Dendrobium angustifolium (Blume) Lindl. = Flickingeria angustifolia (Blume) A. D. Hawkes ■

124995 Dendrobium anosmum Lindl.;无香味石斛(艳香石斛,卓花石斛);Odorless Dendrobium ■☆

124996 Dendrobium aphyllum (Roxb.) C. E. C. Fisch.;兜唇石斛(吊兰花,黄草,金耳环,皮氏石斛,无叶石斛);Poketlip Dendrobium ●■

124997 Dendrobium aphyllum Roxb. = Dendrobium amoenum Wall. ex Lindl. ■☆

124998 Dendrobium arachnites Rchb. f.;蜘蛛石斛;Spider Dendrobium ■☆

124999 Dendrobium atractodes Lindl. = Dendrobium heterocarpum Wall. ex Lindl. ■

125000 Dendrobium atractodes Ridl. = Dendrobium heterocarpum Wall. ex Lindl. ■

125001 Dendrobium atroviolaceum Rolfe;紫堇花石斛;Violet Dendrobium ■☆

125002 Dendrobium augustae-victoriae Kraenzl. = Dendrobium lineale Rolfe ■☆

125003 Dendrobium aurantiacum Rchb. f.;线叶石斛(大黄草,黄花棍棒石斛,金草,金草兰,金黄花石斛,木斛);Linearleaf Dendrobium ■

125004 Dendrobium aurantiacum Rchb. f. = Dendrobium chryseum Rolfe ■

125005 Dendrobium aurantiacum Rchb. f. var. denneanum (Kerr) Z. H. Tsi;叠鞘石斛(大黄草,大马鞭草,迭鞘石斛,棍棒石斛,旱马棒,黄草,马棒草,马鞭草,马鞭杆,石斛,土黄草,岩竹,紫斑金兰,紫斑石斛);Dennes Dendrobium ■

125006 Dendrobium aurantiacum Rchb. f. var. zhaojuense (S. C. Sun et L. G. Xu) Z. H. Tsi = Dendrobium chryseum Rolfe ■

125007 Dendrobium aurantiacum Rchb. f. var. zhaojunense (S. C. Sun et L. G. Xu) Z. H. Tsi;双斑叠鞘石斛(昭觉石斛);Zhaojun Dennes Dendrobium ■

125008 Dendrobium aureum Lindl. = Dendrobium heterocarpum Wall. ex Lindl. ■

125009 Dendrobium auriferum Lindl. = Thrixspermum centipeda Lour. ■

125010 Dendrobium bambusifolium E. C. Parish et Rchb. f. = Dendrobium salaccense (Blume) Lindl. ■

125011 Dendrobium banaense Gagnep. = Dendrobium spatella Rchb. f. ■

125012 Dendrobium batanense Ames et Quisumb. = Dendrobium equitans Kraenzl. ■

125013 Dendrobium bellatulum Rolfe;矮石斛(黑节草,小美石斛);Dwarf Dendrobium ■

125014 Dendrobium bensoniae Rchb. f.;本森氏石斛;Benson Dendrobium ■☆

125015 Dendrobium bigibbum Lindl.;澳洲石斛;Cooktown Orchid ■☆

125016 Dendrobium bigibbum Lindl. var. phalaenopsis (Fitzg.) F. M. Bailey = Dendrobium phalaenopsis Fitzg. ■☆

125017 Dendrobium binoculare Rchb. f. = Dendrobium gibsonii Lindl. ■

125018 Dendrobium boxallii Rchb. f. = Dendrobium gratiosissimum Rchb. f. ■

125019 Dendrobium brachycarpum A. Rich. = Aerangis brachycarpa (A. Rich.) T. Durand et Schinz ■☆

125020 Dendrobium bronckartii De Wild.;布氏石斛;Bronckart Dendrobium ■☆

125021 Dendrobium brymerianum Rchb. f.;长苏石斛(缕唇石斛);Longtassel Dendrobium ■

125022 Dendrobium bullenianum Rchb. f.;布伦石斛■☆

125023 Dendrobium bullerianum Bateman = Dendrobium gratiosissimum Rchb. f. ■

125024 Dendrobium bulleyi Rolfe = Dendrobium longicornu Lindl. ■

125025 Dendrobium caespitosum King et Pantl. = Dendrobium prophyrochilum Lindl. ■

125026 Dendrobium calamiforme Lodd. = Dendrobium teretifolium R. Br. ■☆

125027 Dendrobium calamiforme Lodd. ex Lindl. = Dendrobium teretifolium R. Br. ■☆

125028 Dendrobium calceolaria Carey = Dendrobium moschatum (Buch. -Ham.) Sw. ■

125029 Dendrobium calceolaria Carey ex Hook. = Dendrobium moschatum (Buch. -Ham.) Sw. ■

125030 Dendrobium caliniferum Rchb.;背棱石斛■☆

125031 Dendrobium canaliculatum R. Br.;纵沟石斛;Antelope Orchid ■☆

125032 Dendrobium candidum Wall. = Dendrobium candidum Wall. ex Lindl. ■

125033 Dendrobium candidum Wall. ex Lindl.;白花石斛(吊兰,吊兰花,杜兰,黑节草,黄草,金钗花,禁生,林兰,千年润,石斛,石蓫,铁皮兰,铁皮石斛,鲜石斛);White Dendrobium ■

125034 Dendrobium candidum Wall. ex Lindl. = Dendrobium moniliforme (L.) Sw. ■

125035 Dendrobium candidum Wall. ex Lindl. = Dendrobium officinale Kimura et Migo ■

125036 Dendrobium capillipes Rchb. f.;短棒石斛(丝梗石斛);Hairstalk Dendrobium ■

125037 Dendrobium cariniferum Rchb. f.;翅萼石斛(舟状石斛);Carinate Dendrobium, Wingcalyx Dendrobium ■

125038 Dendrobium cariniferum Rchb. f. var. wattii Hook. f. = Dendrobium wattii (Hook. f.) Rchb. f. ■

125039 Dendrobium carnosum (Blume) Rchb. f.;肉质花石斛;Fleshflower Dendrobium ■☆

125040 Dendrobium castum Bateman ex Rchb. f. = Dendrobium moniliforme (L.) Sw. ■

125041 Dendrobium catenatum Lindl. = Dendrobium moniliforme (L.) Sw. ■

125042 Dendrobium catenatum Lindl. = Dendrobium tosaense Makino ■

125043 Dendrobium cathcartii Hook. f. = Dendrobium salaccense (Blume) Lindl. ■

125044 Dendrobium chameleon Ames;长爪石斛(长距石斛,峦大石斛);Longclaw Dendrobium ■

125045 Dendrobium changjiangense S. J. Cheng et C. Z. Tang = Oxystophyllum changjiangense (S. J. Cheng et C. Z. Tang) M. A. Clem. ■

125046 Dendrobium christyanum Rchb. f.;毛鞘石斛■

125047 Dendrobium chrysanthum Wall. = Dendrobium chrysanthum Wall. ex Lindl. ■

125048 Dendrobium chrysanthum Wall. ex Lindl.;束花石斛(大黄草,喉红石斛,黄草石斛,金兰,巨果石斛,马鞭草,石斛,水打棒,水马棒);Goldenflower Dendrobium ■

125049 Dendrobium chrysanthum Wall. ex Lindl. var. anophthalma Rchb. f. = Dendrobium chrysanthum Wall. ex Lindl. ■

125050 Dendrobium chrysanthum Wall. ex Lindl. var. microphthalma Rchb. f. = Dendrobium chrysanthum Wall. ex Lindl. ■

125051 Dendrobium chryseum Rolfe;金黄花石斛(线叶石斛)■

125052 Dendrobium chryseum Rolfe = Dendrobium aurantiacum Rchb. f. ■

125053 Dendrobium chryseum Rolfe = Dendrobium aurantiacum Rchb.

f. var. denneanum (Kerr) Z. H. Tsi ■
125054 Dendrobium chryseum Rolfe var. bulangense G. X. Ma et J. Xu = Dendrobium chryseum Rolfe ■
125055 Dendrobium chryseum Rolfe var. bulongense G. X. Ma et G. J. Xu;布朗山石斛;Bulangshan Dendrobium ■
125056 Dendrobium chryseum Rolfe var. bulongense G. X. Ma et G. J. Xu = Dendrobium aurantiacum Rchb. f. ■
125057 Dendrobium chrysotis Rchb. f. = Dendrobium hildebrandtii Rolfe ■☆
125058 Dendrobium chrysotoxum Lindl.;鼓槌石斛(斑唇石斛,鼓钟石斛,金弓石斛);Yellowarrow Dendrobium, Yellowbow Dendrobium ■
125059 Dendrobium chrysotoxum Lindl. var. delacourii Gagnep. = Dendrobium chrysotoxum Lindl. ■
125060 Dendrobium chrysotoxum Lindl. var. suavissimum (Rchb. f.) Hook. f. ex Veitch = Dendrobium chrysotoxum Lindl. ■
125061 Dendrobium ciliatum Parish;缘毛石斛■☆
125062 Dendrobium clavatum Lindl. = Dendrobium aurantiacum Rchb. f. var. denneanum (Kerr) Z. H. Tsi ■
125063 Dendrobium clavatum Lindl. var. aurantiacum (Rchb. f.) Ts. Tang et F. T. Wang = Dendrobium aurantiacum Rchb. f. ■
125064 Dendrobium clavatum Roxb. = Dendrobium aurantiacum Rchb. f. var. denneanum (Kerr) Z. H. Tsi ■
125065 Dendrobium clavatum Roxb. = Dendrobium denneanum Kerr ■
125066 Dendrobium clavatum Roxb. = Dendrobium densiflorum Lindl. ex Wall. ■
125067 Dendrobium clavatum Wall. ex Lindl. = Dendrobium aurantiacum Rchb. f. var. denneanum (Kerr) Z. H. Tsi ■
125068 Dendrobium clavatum Wall. ex Lindl. var. aurantiacum Ts. Tang et F. T. Wang = Dendrobium chryseum Rolfe ■
125069 Dendrobium coelogyne Rchb. f.;贝母兰状石斛■☆
125070 Dendrobium coelogyne Rchb. f. = Epigeneium amplum (Lindl. ex Wall.) Summerh. ■
125071 Dendrobium coerulescens Wall. = Dendrobium nobile Lindl. ●■
125072 Dendrobium coerulescens Wall. ex Lindl. = Dendrobium nobile Lindl. ●■
125073 Dendrobium cogniauxianum Kraenzl. ex Warb. = Dendrobium lineale Rolfe ■☆
125074 Dendrobium comatum (Blume) Lindl. = Flickingeria comata (Blume) A. D. Hawkes ■
125075 Dendrobium compactum Rolfe ex W. Hackett;草石斛(小密石斛);Herb Dendrobium ■
125076 Dendrobium compressum Lindl. = Dendrobium lamellatum (Blume) Lindl. ■☆
125077 Dendrobium crassinode Benson et Rchb. f.;粗节石斛■☆
125078 Dendrobium crassinode Benson ex Rchb. f. = Dendrobium pendulum Roxb. ■
125079 Dendrobium crassinode Lindl. = Dendrobium crassinode Benson et Rchb. f. ■☆
125080 Dendrobium crassinode Rchb. = Dendrobium pendulum Roxb. ■
125081 Dendrobium crepidatum Lindl. = Dendrobium crepidatum Lindl. et Paxton ■
125082 Dendrobium crepidatum Lindl. et Paxton;玫瑰石斛(大黄草,水打棒,拖鞋石斛);Crepidate Dendrobium, Rose Dendrobium ■
125083 Dendrobium cretaceum Lindl.;白垩色石斛;Cretaceous Dendrobium ■☆
125084 Dendrobium cretaceum Lindl. = Dendrobium primulinum Lindl. ■
125085 Dendrobium crispulum Kimura et Migo = Dendrobium moniliforme (L.) Sw. ■
125086 Dendrobium cruentum Rchb. f.;暗红石斛■☆
125087 Dendrobium crumenatum Sw.;木石斛(袋状石斛,鸽石斛,木斛);Bagshaped Dendrobium, Pigeon Orchid, Wood Dendrobium ■
125088 Dendrobium crumenatum Sw. var. parviflorum Ames et C. Schweinfurth = Dendrobium crumenatum Sw. ■
125089 Dendrobium crystallinum Rchb. f.;晶帽石斛(结晶状石斛);Crystal Dendrobium, Crystalloid Dendrobium ■
125090 Dendrobium crystallinum Rchb. f. var. hainanense S. J. Cheng et C. Z. Tang;海南晶帽石斛■
125091 Dendrobium crystallinum Rchb. f. var. hainanense S. J. Cheng et C. Z. Tang = Dendrobium crystallinum Rchb. f. ■
125092 Dendrobium ctenoglossum Schltr. = Dendrobium strongylanthum Rchb. f. ■
125093 Dendrobium cucullatum R. Br. = Dendrobium pierardii Roxb. ●■
125094 Dendrobium cucullatum R. Br. ex Lindl. = Dendrobium aphyllum (Roxb.) C. E. C. Fisch. ●■
125095 Dendrobium cucumerinum W. MacLeay;瓜石斛■☆
125096 Dendrobium cultriforme Thouars = Polystachya cultriformis (Thouars) Spreng. ■☆
125097 Dendrobium cultriforme Thouars = Polystachya cultriformis Lindl. ex Spreng. ■☆
125098 Dendrobium cumulatum Lindl.;堆积石斛■☆
125099 Dendrobium cupreum Herb. = Dendrobium moschatum (Buch.-Ham.) Sw. ■
125100 Dendrobium cupreum Herb. ex Lindl. = Dendrobium moschatum (Buch.-Ham.) Sw. ■
125101 Dendrobium cupreum Lindl. = Dendrobium moschatum (Buch.-Ham.) Sw. ■
125102 Dendrobium dalhousieanum Wall. = Dendrobium pulchellum Roxb. ex Lindl. ■☆
125103 Dendrobium daoense Gagnep. = Dendrobium henryi Schltr. ■
125104 Dendrobium dearei Rchb. f.;白妙石斛(白花石斛)■☆
125105 Dendrobium decundum Lindl.;偏花石斛■☆
125106 Dendrobium demissum D. Don = Panisea demissa (D. Don) Pfitzer ■
125107 Dendrobium denneanum Kerr = Dendrobium aurantiacum Rchb. f. var. denneanum (Kerr) Z. H. Tsi ■
125108 Dendrobium densiflorum Lindl. = Dendrobium densiflorum Lindl. ex Wall. ■
125109 Dendrobium densiflorum Lindl. ex Wall.;密花石斛(粗黄草,大黄草,吊硬套哑槽榄,黄草,象尾草);Denseflower Dendrobium, Flowery Dendrobium ■
125110 Dendrobium densiflorum Wall. var. alboluteum Hook. f. = Dendrobium thyrsiflorum Rchb. f. ■
125111 Dendrobium devonianum Lindl. ex Wall. var. rhodoneurum Rchb. f. = Dendrobium devonianun Paxton ■
125112 Dendrobium devonianun Paxton;齿瓣石斛(大黄草,旱马棒,水打棒,中黄草);Devon Dendrobium ■
125113 Dendrobium discolor Lindl.;二色石斛;Discolor Dendrobium ■☆
125114 Dendrobium discolor Lindl. = Dendrobium undulatum R. Br. ■☆
125115 Dendrobium dixanthum Rchb. f.;黄花石斛;Yellow Dendrobium ■
125116 Dendrobium donnaiense Gagnep. = Pinalia donnaiensis (Gagnep.) S. C. Chen et J. J. Wood ■
125117 Dendrobium draconis Rchb. f.;龙状石斛■☆
125118 Dendrobium eburneum Rchb. f. = Dendrobium draconis Rchb.

f. ■☆

125119 Dendrobium eburneum Rchb. f. ex Bateman = Dendrobium draconis Rchb. f. ■☆

125120 Dendrobium ellipsophyllum Ts. Tang et F. T. Wang;反瓣石斛(黄毛石斛);Oblongleaf Dendrobium ■

125121 Dendrobium equitans Kraenzl.;燕石斛(套叶石斛,燕子石斛);Swallow Dendrobium ■

125122 Dendrobium eriiflorum Griff. = Dendrobium monticola P. F. Hunt et Summerh. ■

125123 Dendrobium erythroglossum Hayata = Dendrobium falconeri Hook. ■

125124 Dendrobium erythroxanthum Rchb. f. = Dendrobium bullenianum Rchb. f. ■☆

125125 Dendrobium evaginatum Gagnep. = Dendrobium henryi Schltr. ■

125126 Dendrobium exile Schltr.;景洪石斛;Jinghong Dendrobium ■

125127 Dendrobium exsculptum Teijsm. et Binn. = Dendrobium stuposum Lindl. ■

125128 Dendrobium falconeri Hook.;串珠石斛(红鹂石斛,红鹏石斛,水兰,新竹石斛);Falconer Dendrobium ■

125129 Dendrobium fanjingshanense Z. H. Tsi ex X. H. Jin et Y. W. Zhang;梵净山石斛(梵净石斛);Fanjingshan Dendrobium ■

125130 Dendrobium fargesii Finet = Epigeneium fargesii (Finet) Gagnep. ■

125131 Dendrobium farmeri Paxton;法默石斛■☆

125132 Dendrobium faulhaberianum Schltr. = Dendrobium aduncum Wall. ex Lindl. ■

125133 Dendrobium fimbriatolabellum Hayata = Flickingeria comata (Blume) A. D. Hawkes ■

125134 Dendrobium fimbriatum (Blume) Lindl. = Flickingeria fimbriata (Blume) A. D. Hawkes ■

125135 Dendrobium fimbriatum Hook.;流苏石斛(大黄草,大马鞭草,旱马棒,旱马鞭,马鞭草,马鞭杆,马鞭石斛,石斛);Eyeshaped Dendrobium,Tassel Dendrobium ■

125136 Dendrobium fimbriatum Hook. var. bimaculosum Ts. Tang et F. T. Wang = Dendrobium hookerianum Lindl. ■

125137 Dendrobium fimbriatum Hook. var. gibsonii (Lindl.) Finet = Dendrobium gibsonii Lindl. ■

125138 Dendrobium fimbriatum Hook. var. oculatum Hook.;马鞭石斛(吊兰,吊兰花,杜兰,黄草,金钗花,禁生,林兰,流苏石斛,千年润,石蓫)■

125139 Dendrobium fimbriatum Hook. var. oculatum Hook. = Dendrobium fimbriatum Hook. ■

125140 Dendrobium findlayanum Parl. et Rchb. f.;棒节石斛(芬莱石斛);Findley Dendrobium ■

125141 Dendrobium fitzgeraldii F. M. Bailey = Dendrobium superbiens Rchb. f. ■☆

125142 Dendrobium flavescens (Blume) Lindl. = Polystachya concreta (Jacq.) Garay et H. R. Sweet ■

125143 Dendrobium flavidulum Ridl. ex Hook. f. = Dendrobium stuposum Lindl. ■

125144 Dendrobium flaviflorum Hayata = Dendrobium aurantiacum Rchb. f. ■

125145 Dendrobium flaviflorum Hayata = Dendrobium chryseum Rolfe ■

125146 Dendrobium flexicaule Z. H. Tsi, S. C. Sun et L. G. Xu;曲茎石斛;Flexstem Dendrobium ■

125147 Dendrobium flexuosum Griff. = Dendrobium longicornu Lindl. ■

125148 Dendrobium floribundum D. Don = Mycaranthes floribunda (D. Don) S. C. Chen et J. J. Wood ■

125149 Dendrobium forbesii Ridl. = Dendrobium atroviolaceum Rolfe ■☆

125150 Dendrobium formosanum (Rchb. f.) Masam. = Dendrobium nobile Lindl. ●■

125151 Dendrobium formosanum (Rchb. f.) Masam. = Dendrobium nobile Lindl. var. formosanum Rchb. f. ■

125152 Dendrobium formosum Roxb.;美丽石斛;Beautiful Dendrobium ■☆

125153 Dendrobium formosum Roxb. var. giganteum Van Houtte;大花美丽石斛■☆

125154 Dendrobium fugax Schltr. = Dendrobium hendersonii Hawkes et A. Heller ■☆

125155 Dendrobium funiushanense T. B. Chao, Zhi X. Chen et Z. K. Chen = Dendrobium catenatum Lindl. ■

125156 Dendrobium funiushanense T. B. Chao, Zhi X. Chen et Z. K. Chen = Dendrobium huoshanense C. Z. Tang et S. J. Cheng ■

125157 Dendrobium funiushanense T. B. Chao, Zhi X. Chen et Z. K. Chenn;伏牛石斛;Funiushan Dendrobium ■

125158 Dendrobium furcatopedicellatum Hayata;双花石斛;Biflower Dendrobium ■

125159 Dendrobium fuscatum Lindl. = Dendrobium gibsonii Lindl. ■

125160 Dendrobium fuscescens Griff. = Epigeneium fuscescens (Griff.) Summerh. ■

125161 Dendrobium fusiforme Thouars = Polystachya fusiformis (Thouars) Lindl. ■☆

125162 Dendrobium fytchianum Bateman;法氏石斛■☆

125163 Dendrobium galeatum Sw. = Polystachya galeata (Sw.) Rchb. f. ■☆

125164 Dendrobium galliceanum Linden = Dendrobium thyrsiflorum Rchb. f. ■

125165 Dendrobium gibsonii Lindl.;曲轴石斛(伏牛山石斛,曲茎石斛,紫斑石斛);Gibson Dendrobium ■

125166 Dendrobium goldschmidtianum Kraenzl.;哥氏石斛(红花石斛,金将石斛,金匠石斛)■

125167 Dendrobium gouldii Rchb. f.;古尔德石斛■☆

125168 Dendrobium gouldii Rchb. f. = Dendrobium superbiens Rchb. f. ■☆

125169 Dendrobium gratiosissimum Rchb. f.;杯鞘石斛;Muchlovable Dendrobium ■

125170 Dendrobium guangxiense S. J. Cheng et C. Z. Tang;滇桂石斛;Guangxi Dendrobium ■

125171 Dendrobium guangxiense S. J. Cheng et C. Z. Tang = Dendrobium scoriarum W. W. Sm. ■

125172 Dendrobium haemoglossum Thwaites = Dendrobium salaccense (Blume) Lindl. ■

125173 Dendrobium hainanense Rolfe;海南石斛;Hainan Dendrobium ■

125174 Dendrobium hainanense Rolfe = Dendrobium miyakei Schltr. ■

125175 Dendrobium hanburyanum Rchb. f. = Dendrobium lituiflorum Lindl. ■

125176 Dendrobium hancockii Rolfe;细叶石斛(大马鞭草,细黄草,细竹丫草);Hancock Dendrobium ■

125177 Dendrobium harveyanum Rchb. f.;苏瓣石斛;Tasselpetal Dendrobium ■

125178 Dendrobium heishanense Hayata = Dendrobium moniliforme (L.) Sw. ■

125179 Dendrobium henanense J. L. Lu et L. X. Gao;河南石斛;Henan Dendrobium ■

125180 Dendrobium henanense J. L. Lu et L. X. Gao = Dendrobium flexicaule Z. H. Tsi, S. C. Sun et L. G. Xu ■
125181 Dendrobium hendersonii Hawkes et A. Heller;亨德森石斛■☆
125182 Dendrobium henryi Schltr.;疏花石斛;Henry Dendrobium, Laxflower Dendrobium ■
125183 Dendrobium hercoglossum Rchb. f.;重唇石斛(毫猪尖,鸡爪兰,网脉唇石斛,中黄草);Doublelip Dendrobium ■
125184 Dendrobium hercoglossum Rchb. f. = Dendrobium aduncum Wall. ex Lindl. ■
125185 Dendrobium hercoglossum Rchb. f. var. album S. J. Cheng et Z. Z. Tang;白花重唇石斛;Whiteflower Doublelip Dendrobium ■
125186 Dendrobium hercoglossum Rchb. f. var. album S. J. Cheng et Z. Z. Tang = Dendrobium hercoglossum Rchb. f. ■
125187 Dendrobium heterocarpum Wall. = Dendrobium heterocarpum Wall. ex Lindl. ■
125188 Dendrobium heterocarpum Wall. ex Lindl.;尖刀唇石斛(异形果石斛);Acuteknife Dendrobium, Differentfruit Dendrobium ■
125189 Dendrobium hildebrandtii Rolfe;希氏石斛■☆
125190 Dendrobium hirsutum Griff. = Dendrobium longicornu Lindl. ■
125191 Dendrobium hookerianum Lindl.;金耳石斛;Goldear Dendrobium ■
125192 Dendrobium huoshanense C. Z. Tang et S. J. Cheng;霍山石斛;Huoshan Dendrobium ■
125193 Dendrobium huoshanense C. Z. Tang et S. J. Cheng = Dendrobium catenatum Lindl. ■
125194 Dendrobium hybridum Hort.;杂交石斛■☆
125195 Dendrobium imperatrix Kraenzl. = Dendrobium lineale Rolfe ■☆
125196 Dendrobium infundibulum Lindl.;漏斗唇石斛(高山石斛);Alp Dendrobium, Funnelshaped Dendrobium ■
125197 Dendrobium intermedium Teijsm. et Binn. = Dendrobium salaccense (Blume) Lindl. ■
125198 Dendrobium irayense Ames et Quisumb. = Dendrobium goldschmidtianum Kraenzl. ■
125199 Dendrobium irayense Ames et Quisumb. = Dendrobium miyakei Schltr. ■
125200 Dendrobium japonicum (Blume) Lindl. = Dendrobium moniliforme (L.) Sw. ■
125201 Dendrobium javanicum Sw. = Eria javanica (Sw.) Blume ■
125202 Dendrobium jenkinsii Lindl.;小黄花石斛(鸡背石斛,金黄泽,聚石斛,木虾公,上树虾,树上虾,土杵虾公,虾公草,虾青石斛);Jenkins Dendrobium ■
125203 Dendrobium jiajiangense Z. Y. Zhu;夹江石斛■
125204 Dendrobium johannis Lindl.;约翰石斛■☆
125205 Dendrobium johnsoniae F. Muell.;约翰逊石斛■☆
125206 Dendrobium kingianum Bidwill ex Lindl.;金氏石斛;Australian dendrobium, Pink Rock Orchid ■☆
125207 Dendrobium kingianum Lindl. = Dendrobium kingianum Bidwill ex Lindl. ■☆
125208 Dendrobium kosepangii C. L. Tso = Dendrobium moniliforme (L.) Sw. ■
125209 Dendrobium kosepangii C. L. Tso = Dendrobium wilsonii Rolfe ■
125210 Dendrobium kwangtungense C. L. Tso = Dendrobium moniliforme (L.) Sw. ■
125211 Dendrobium kwangtungense C. L. Tso = Dendrobium wilsonii Rolfe ■
125212 Dendrobium kwashotense Hayata = Dendrobium crumenatum Sw. ■
125213 Dendrobium lamellatum (Blume) Lindl.;鳞片石斛■☆
125214 Dendrobium lamellatum Lindl. = Dendrobium lamellatum (Blume) Lindl. ■☆
125215 Dendrobium lawanum Lindl. = Dendrobium crepidatum Lindl. et Paxton ■
125216 Dendrobium leopardinum Wall. = Bulbophyllum leopardinum (Wall.) Lindl. ■
125217 Dendrobium leporinum J. J. Sm.;兔耳状石斛;Pleasant Dendrobium ■☆
125218 Dendrobium leptocladum Hayata;菱唇石斛(禾叶石斛,细茎石斛);Rhombiclip Dendrobium ■
125219 Dendrobium liguiforme Sw.;舌叶石斛;Tongue Orchid ■☆
125220 Dendrobium liguolla Rchb. f.;红蓝草■
125221 Dendrobium linawianum Rchb. f.;矩唇石斛(白花石斛,长爪石斛,矩形石斛,樱石斛);Linaw Dendrobium, Oblonglipped Dendrobium ■
125222 Dendrobium linawianum Rchb. f. = Dendrobium nobile Lindl. ●■
125223 Dendrobium lindleyanum Griff. = Dendrobium nobile Lindl. ●■
125224 Dendrobium lindleyi Steud.;聚石斛(密茎石斛,虾青石斛);Lindley Dendrobium ■
125225 Dendrobium lindleyi Steud. = Dendrobium aggregatum Roxb. ■
125226 Dendrobium lindleyi Steud. var. majus (Rolfe) S. Y. Hu;大叶聚石斛■
125227 Dendrobium lindleyi Steud. var. majus (Rolfe) S. Y. Hu = Dendrobium lindleyi Steud. ■
125228 Dendrobium lineale Rolfe;狭叶石斛■☆
125229 Dendrobium linearifolium Teijsm. et Binn.;细长叶石斛■☆
125230 Dendrobium linguiforme Sm. = Dendrobium linguiforme Sw. ■☆
125231 Dendrobium linguiforme Sw.;舌状石斛■☆
125232 Dendrobium lituiflorum Lindl.;喇叭唇石斛;Bugle Dendrobium, Forkedflower Dendrobium ■
125233 Dendrobium loddigesii Rolfe;美花石斛(大黄草,吊兰,吊兰花,杜兰,粉花石斛,环草石斛,环钗斛,黄草,金钗草,金钗花,禁生,林兰,千年润,石蚌腿,石斛,石蓫,小环草,小黄草,小金钗);Beautyflower Dendrobium, Loddiges Dendrobium ■
125234 Dendrobium loddigesii Rolfe var. album Ts. Tang et F. T. Wang = Dendrobium loddigesii Rolfe ■
125235 Dendrobium lohohense Ts. Tang et F. T. Wang;罗河石斛(出芽草,黄草,黄竹丫,马鞭草,马草,中黄草,竹亚草);Loho Dendrobium, Luohe Dendrobium ■
125236 Dendrobium longicalcaratum Hayata = Dendrobium chameleon Ames ■
125237 Dendrobium longicornu Lindl.;长距石斛(长角石斛);Longspur Dendrobium, Longspurred Dendrobium ■
125238 Dendrobium longicornu Lindl. var. hirsutum (Griff.) Hook. f. = Dendrobium longicornu Lindl. ■
125239 Dendrobium longifolium Kunth = Eulophia alta (L.) Fawc. et Rendle ■☆
125240 Dendrobium luteolum Bateman;淡黄花石斛;Yellowish Flower Dendrobium ■☆
125241 Dendrobium luzonense Lindl.;吕宋石斛■
125242 Dendrobium maccarthiae Pfitzer = Dendrobium maccarthiae Thwaites ■☆
125243 Dendrobium maccarthiae Thwaites;马克石斛■☆
125244 Dendrobium macfarlanei Rchb. f. = Dendrobium johnsoniae F. Muell. ■☆
125245 Dendrobium macrophyllum A. Rich.;大叶石斛;Largeleaf

Dendrobium ■☆

125246 Dendrobium margaritaceum Finet = Dendrobium christyanum Rchb. f. ■

125247 Dendrobium marseillei Gagnep. = Dendrobium jenkinsii Lindl. ■

125248 Dendrobium milliganii F. Muell. = Dendrobium striolatum Rchb. f. ■☆

125249 Dendrobium minahassae Kraenzl. = Dendrobium heterocarpum Wall. ex Lindl. ■

125250 Dendrobium minutiflorum S. C. Chen et Z. H. Tsi = Dendrobium sinominutiflorum S. C. Chen, J. J. Wood et H. P. Wood ■

125251 Dendrobium mirbelianum Grudz.;米贝尔石斛■☆

125252 Dendrobium mitriferum Aver. = Dendrobium scoriarum W. W. Sm. ■

125253 Dendrobium miyakei Schltr.;红花石斛(红石斛);Red Dendrobium ■

125254 Dendrobium miyakei Schltr. = Dendrobium goldschmidtianum Kraenzl. ■

125255 Dendrobium monile (Thunb. ex A. Murray) Kraenzl. = Dendrobium moniliforme (L.) Sw. ■

125256 Dendrobium monile (Thunb.) Kraenzl. = Dendrobium moniliforme (L.) Sw. ■

125257 Dendrobium moniliforme (L.) Sw.;细茎石斛(吊兰,杜兰,耳环草,环草,黄草,霍山石斛,鸡爪兰,金钗草,金钗石斛,禁生,林兰,罗浮山石兰,清水山石斛,石斛,台湾石斛,铁皮石斛,铜皮兰,铜皮石斛,细草,细环草,小环草,小金钗,小石斛);Moniliform Dendrobium,Thinstem Dendrobium ■

125258 Dendrobium moniliforme (L.) Sw. 'Variegata';花叶细茎石斛■☆

125259 Dendrobium moniliforme (L.) Sw. var. lipuense M. F. Li et G. J. Xu;荔埔石斛■

125260 Dendrobium moniliforme (L.) Sw. var. taiwanianum S. S. Ying = Dendrobium moniliforme (L.) Sw. ■

125261 Dendrobium moniliforme Lindl. = Dendrobium linawianum Rchb. f. ■

125262 Dendrobium monodon Kraenzl. = Dendrobium johnsoniae F. Muell. ■☆

125263 Dendrobium monticola P. F. Hunt et Summerh.;藏南石斛;S. Xizang Dendrobium ■

125264 Dendrobium moschatum (Buch.-Ham.) Sw.;杓唇石斛(美丽囊石斛,囊唇石斛,麝香石斛,香石斛);Musk Dendrobium,Musky Dendrobium ■

125265 Dendrobium moschatum (Buch.-Ham.) Sw. var. cupreum (Lindl.) Rchb. f. = Dendrobium moschatum (Buch.-Ham.) Sw. ■

125266 Dendrobium moschatum (Buch.-Ham.) Sw. var. unguipetalum I. Barua = Dendrobium moschatum (Buch.-Ham.) Sw. ■

125267 Dendrobium moulmeinense E. C. Parish ex Hook. f. = Dendrobium devonianun Paxton ■

125268 Dendrobium moulmeinense H. Low ex Warn. et Will. = Dendrobium aggregatum Roxb. ■

125269 Dendrobium moulmeinense H. Low ex Warn. et Will. = Dendrobium infundibulum Lindl. ■

125270 Dendrobium moulmeinense Rchb. f. = Dendrobium dixanthum Rchb. f. ■

125271 Dendrobium moulmeinense Warn. = Dendrobium aggregatum Roxb. ■

125272 Dendrobium moulmeinense Warn. = Dendrobium infundibulum Lindl. ■

125273 Dendrobium muscicola Lindl. = Conchidium muscicola (Lindl.) Rauschert ■

125274 Dendrobium muscicola Lindl. = Eria muscicola (Lindl.) Lindl. ■

125275 Dendrobium mutabile (Blume) Lindl.;易变石斛;Variable Dendrobium ■☆

125276 Dendrobium nakaharae Schltr. = Epigeneium nakaharai (Schltr.) Summerh. ■

125277 Dendrobium nienkui C. L. Tso = Dendrobium moniliforme (L.) Sw. ■

125278 Dendrobium niveum Rolfe = Dendrobium johnsoniae F. Muell. ■☆

125279 Dendrobium nobile Lindl.;石斛(扁草,扁黄草,扁金钗,春石斛,大黄草,吊花,吊兰,吊兰花,金钗石斛,素雅斗,台湾春石斛,台湾石斛,小黄草,雅斗);Noble Dendrobium ●■

125280 Dendrobium nobile Lindl. f. nobilius (Rchb. f.) M. Hiroe = Dendrobium nobile Lindl. ●■

125281 Dendrobium nobile Lindl. var. alboluteum Huyen et Aver. = Dendrobium nobile Lindl. ●■

125282 Dendrobium nobile Lindl. var. formosanum Rchb. f. = Dendrobium nobile Lindl. ●■

125283 Dendrobium nobile Lindl. var. nobilius Rchb. f. = Dendrobium nobile Lindl. ●■

125284 Dendrobium nobile Lindl. var. pallidiflorum Hook. = Dendrobium primulinum Lindl. ■

125285 Dendrobium nutans C. Presl = Geodorum densiflorum (Lam.) Schltr. ■

125286 Dendrobium ochreatum Lindl. = Dendrobium chrysanthum Wall. ex Lindl. ■

125287 Dendrobium odiosum Finet = Dendrobium hancockii Rolfe ■

125288 Dendrobium officinale Kimura et Migo;铁皮石斛(白花石斛,黑节草,云南铁皮);Iron-sheet Dendrobium ■

125289 Dendrobium officinale Kimura et Migo = Dendrobium catenatum Lindl. ■

125290 Dendrobium officinale Kimura et Migo = Dendrobium guangxiense S. J. Cheng et C. Z. Tang ■

125291 Dendrobium okinawense Hatus. et Ida;琉球石斛(冲绳石斛)■

125292 Dendrobium pallens Ridl. = Flickingeria pallens (Ridl.) A. D. Hawkes ■

125293 Dendrobium pallens Ridl. = Flickingeria tairukounia (S. S. Ying) T. P. Lin ■

125294 Dendrobium paniculatum Sw. = Polystachya paniculata (Sw.) Rolfe ■☆

125295 Dendrobium parciflorum Rchb. f. ex Lindl.;少花石斛;Poorlower Dendrobium ■

125296 Dendrobium parishii Rchb. f.;紫瓣石斛(西施石斛);Parish Dendrobium ■

125297 Dendrobium paxtoni Lindl. = Dendrobium chrysanthum Wall. ex Lindl. ■

125298 Dendrobium paxtonii Paxton = Dendrobium fimbriatum Hook. ■

125299 Dendrobium pendulicaule Hayata = Thrixspermum pendulicaule (Hayata) Schltr. ■

125300 Dendrobium pendulicaule Hayata = Thrixspermum pensile Schltr. ■

125301 Dendrobium pendulum Roxb.;肿节石斛;Bignode Dendrobium ■

125302 Dendrobium pere-fauriei Hayata = Dendrobium catenatum Lindl. ■

125303 Dendrobium pere-fauriei Hayata = Dendrobium tosaense Makino ■

125304 Dendrobium phalaenopsis Fitzg.;蝴蝶石斛(假蝴蝶兰);

Butterfly Dendrobium ■☆
125305 Dendrobium pictum Griff. = Dendrobium devonianun Paxton ■
125306 Dendrobium pictum Griff. ex Lindl. = Dendrobium devonianun Paxton ■
125307 Dendrobium pierardii Roxb. = Dendrobium aphyllum (Roxb.) C. E. C. Fisch. ●■
125308 Dendrobium pierardii Roxb. ex Hook. f. = Dendrobium aphyllum (Roxb.) C. E. C. Fisch. ●■
125309 Dendrobium pierardii Roxb. ex Hook. f. = Dendrobium cucullatum R. Br. ex Lindl. ■
125310 Dendrobium pierardii Roxb. ex Hook. f. var. cucullatum (R. Br.) Hook. f. = Dendrobium cucullatum R. Br. ex Lindl. ■
125311 Dendrobium plicatile Lindl.;折叠石斛(有爪石斛)■☆
125312 Dendrobium plicatile Lindl. = Flickingeria fimbriata (Blume) A. D. Hawkes ■
125313 Dendrobium poilanei Guillaumin = Dendrobium hercoglossum Rchb. f. ■
125314 Dendrobium polyphlebium Rchb. f. = Dendrobium parishii Rchb. f. ■
125315 Dendrobium polystachyum Thouars = Polystachya concreta (Jacq.) Garay et H. R. Sweet ■
125316 Dendrobium praecox (Sm.) Sm. = Pleione praecox (Sm.) D. Don ■
125317 Dendrobium primulinum Lindl.;报春石斛(报春黄石斛,细瓣石斛);Primrose Dendrobium ■
125318 Dendrobium prionochilum F. Muell. et Kraenzl. = Dendrobium mirbelianum Grudz. ■☆
125319 Dendrobium pristinum Ames = Dendrobium stuposum Lindl. ■
125320 Dendrobium prophyrochilum Lindl.;单葶草石斛(单葶石斛,紫唇石斛);Singlescape Dendrobium ■
125321 Dendrobium pseudohainanense Masam. = Dendrobium miyakei Schltr. ■
125322 Dendrobium pseudohainanense Matsum. = Dendrobium goldschmidtianum Kraenzl. ■
125323 Dendrobium pseudotenellum Guillaumin;针叶石斛;Needleleaf Dendrobium ■
125324 Dendrobium pubescens Hook. = Eria lasiopetala (Willd.) Ormerod ■
125325 Dendrobium pulchellum Lodd. = Dendrobium loddigesii Rolfe ■
125326 Dendrobium pulchellum Roxb. = Dendrobium pulchellum Roxb. ex Lindl. ■☆
125327 Dendrobium pulchellum Roxb. ex Lindl.;西方美花石斛■☆
125328 Dendrobium pulchellum Roxb. ex Lindl. var. devonianum (Paxton) Rchb. f. = Dendrobium devonianun Paxton ■
125329 Dendrobium pumilum Sw. = Bulbophyllum pumilum (Sw.) Lindl. ■☆
125330 Dendrobium purpureum Lindl.;紫花石斛■☆
125331 Dendrobium pusillum D. Don = Dendrobium monticola P. F. Hunt et Summerh. ■
125332 Dendrobium quadrangulare Parish.;四棱石斛■
125333 Dendrobium randaiense Hayata = Dendrobium chameleon Ames ■
125334 Dendrobium regium Prain;王石斛;Royal Dendrobium ■☆
125335 Dendrobium reptans Franch. et Sav. = Conchidium japonicum (Maxim.) S. C. Chen et J. J. Wood ■
125336 Dendrobium reptans Franch. et Sav. = Eria reptans (Franch. et Sav.) Makino ■
125337 Dendrobium revolutum Lindl. = Dendrobium ellipsophyllum Ts. Tang et F. T. Wang ■
125338 Dendrobium rhodostictum F. Muell. et Kraenzl. = Dendrobium rhodostictum F. Muell. et Kraenzl. ex Kraenzl. ■☆
125339 Dendrobium rhodostictum F. Muell. et Kraenzl. ex Kraenzl.;赤点石斛■☆
125340 Dendrobium rhonboideum Lindl. = Dendrobium heterocarpum Wall. ex Lindl. ■
125341 Dendrobium rivesii Gagnep. = Dendrobium chryseum Rolfe ■
125342 Dendrobium robustum Blume = Eria robusta (Blume) Lindl. ■
125343 Dendrobium rolfei A. D. Hawkes et A. H. Heller = Dendrobium chryseum Rolfe ■
125344 Dendrobium roseum Dalzell = Dendrobium crepidatum Lindl. et Paxton ■
125345 Dendrobium rotundatum (Lindl.) Hook. f. = Epigeneium rotundatum (Lindl.) Summerh. ■
125346 Dendrobium roylei A. D. Hawkes et A. H. Heller = Dendrobium monticola P. F. Hunt et Summerh. ■
125347 Dendrobium rupicola Rchb. f. = Dendrobium ciliatum Parish ■☆
125348 Dendrobium rupicola Rchb. f. ex Kraenzl.;岩生石斛;Rockliving Dendrobium ■☆
125349 Dendrobium salaccense (Blume) Lindl.;竹枝石斛;Bamboobranch Dendrobium, Bambooshoot Dendrobium ■
125350 Dendrobium sandereae Rolfe;桑德氏石斛■☆
125351 Dendrobium sanguinolentum Lindl.;血红石斛;Bloodred Dendrobium ■☆
125352 Dendrobium sanseiense Hayata;三星石斛■
125353 Dendrobium sanseiense Hayata = Epigeneium nakaharai (Schltr.) Summerh. ■
125354 Dendrobium schmidtianum Kraenzl. = Dendrobium crumenatum Sw. ■
125355 Dendrobium schoeninum Lindl. = Dendrobium striolatum Rchb. f. ■☆
125356 Dendrobium schroederi Hort. = Dendrobium densiflorum Lindl. ex Wall. ■
125357 Dendrobium scoriarum W. W. Sm.;广西石斛■
125358 Dendrobium signatum Rchb. f. = Dendrobium bensoniae Rchb. f. ■☆
125359 Dendrobium sinese Ts. Tang et F. T. Wang;华石斛;China Dendrobium, Chinese Dendrobium ■
125360 Dendrobium sinominutiflorum S. C. Chen, J. J. Wood et H. P. Wood;勐海石斛;Menghai Dendrobium ■
125361 Dendrobium somai Hayata;小双花石斛(细茎石斛);Soma Dendrobium ■
125362 Dendrobium spatella Rchb. f.;拟剑叶石斛(剑叶石斛)■
125363 Dendrobium spathaceum Lindl. = Dendrobium moniliforme (L.) Sw. ■
125364 Dendrobium spathaceum Lindl. = Dendrobium officinale Kimura et Migo ■
125365 Dendrobium speciosum Sm.;娇美石斛(大明石斛,丽石斛,司光兰);Dendrobium Orchid, Pretty Dendrobium, Rock Lily, Rock Orchid ■☆
125366 Dendrobium spectabile (Blume) Miq.;大花石斛;Largeflower Dendrobium ■☆
125367 Dendrobium sphegidoglossum Rchb. f. = Dendrobium stuposum Lindl. ■
125368 Dendrobium stenoglossum Schltr. = Dendrobium strongylanthum Rchb. f. ■

125369 Dendrobium striatum Griff. = Bulbophyllum striatum (Griff.) Rchb. f. ■
125370 Dendrobium stricklandianum Rchb. f. = Dendrobium catenatum Lindl. ■
125371 Dendrobium striolatum Rchb. f. ;小沟石斛■☆
125372 Dendrobium strongylanthum Rchb. f. ;疏唇石斛(梳唇石斛,圆花石斛);Laxlip Dendrobium ■
125373 Dendrobium stuposum Lindl. ;叉唇石斛(长柔毛石斛);Forklip Dendrobium ■
125374 Dendrobium suavissimum Rchb. f. = Dendrobium chrysotoxum Lindl. ■
125375 Dendrobium sulcatum Lindl. ;具槽石斛;Sulcate Dendrobium ■
125376 Dendrobium superbiens Rchb. f. ;华丽石斛■☆
125377 Dendrobium superbum Rchb. f. var. anosmum Rchb. f. = Dendrobium anosmum Lindl. ■☆
125378 Dendrobium taiwanianum S. S. Ying;台湾石斛;Taiwan Dendrobium ■
125379 Dendrobium taiwanianum S. S. Ying = Dendrobium moniliforme (L.) Sw. ■
125380 Dendrobium takahashii Carrière;高桥氏石斛■☆
125381 Dendrobium tenuicaule Hayata = Dendrobium leptocladum Hayata ■
125382 Dendrobium teres Roxb. = Papilionanthe teres (Roxb.) Schltr. ■
125383 Dendrobium teretifolium R. Br. ;圆柱叶石斛(棒石斛);Terete-leaf Dendrobium ■☆
125384 Dendrobium terminale Parl. et Rchb. f. ;刀叶石斛;Terminate Dendrobium ■
125385 Dendrobium tetragonum A. Cunn. = Dendrobium tetragonum A. Cunn. ex Lindl. ■☆
125386 Dendrobium tetragonum A. Cunn. ex Lindl. ;四棱茎石斛;Fourangular Dendrobium ■☆
125387 Dendrobium thyrsiflorum Rchb. f. ;球花石斛(聚伞圆锥花序石斛);Ball-flower Dendrobium,Thyrseflower Dendrobium ■
125388 Dendrobium tibeticum Schltr. = Dendrobium aurantiacum Rchb. f. ■
125389 Dendrobium tibeticum Schltr. = Dendrobium chryseum Rolfe ■
125390 Dendrobium tosaense Makino;黄石斛(黄花石斛,霍山石斛,铁皮石斛);Tosa Dendrobium,Yellowflower Dendrobium ■
125391 Dendrobium tosaense Makino = Dendrobium catenatum Lindl. ■
125392 Dendrobium tosaense Makino var. chingshuishanianum S. S. Ying = Dendrobium moniliforme (L.) Sw. ■
125393 Dendrobium tosaense Makino var. chingshuishanianum S. S. Ying = Dendrobium tosaense Makino ■
125394 Dendrobium tosaense Makino var. pere-fauriei (Hayata) Masam. = Dendrobium tosaense Makino ■
125395 Dendrobium tosaense Makino var. pere-fauriei (Hayata) Masam. = Dendrobium catenatum Lindl. ■
125396 Dendrobium trigonopus Rchb. f. ;翅梗石斛;Wingstipe Dendrobium ■
125397 Dendrobium undulatum R. Br. ;钝齿石斛■☆
125398 Dendrobium uniflorum Griff. ;单花石斛;Singleflower Dendrobium ■☆
125399 Dendrobium velutinum Rolfe = Dendrobium trigonopus Rchb. f. ■
125400 Dendrobium ventricosum Kraenzl. ;燕子石斛;Inflated Dendrobium ■☆
125401 Dendrobium veratrifolium Lindl. = Dendrobium lineale Rolfe ■☆
125402 Dendrobium veratrifolium Lindl. = Dendrobium veratrifolium Lindl. ex Hook. ■☆
125403 Dendrobium veratrifolium Lindl. ex Hook. ;藜芦叶石斛■☆
125404 Dendrobium veratrifolium Lindl. ex Hook. = Dendrobium lineale Rolfe ■☆
125405 Dendrobium verlaquii Costantin = Dendrobium terminale Parl. et Rchb. f. ■
125406 Dendrobium vexans Dammer = Dendrobium hercoglossum Rchb. f. ■
125407 Dendrobium victoriae-reginae Loher;维多利亚女王石斛(红花石斛);Victoria Dendrobium ■
125408 Dendrobium victoriae-reginae Loher var. miyakei (Schltr.) Tang S. Liu et H. J. Su = Dendrobium miyakei Schltr. ■
125409 Dendrobium victoria-reginae Loher var. miyakei (Schltr.) Tang S. Liu et H. J. Su = Dendrobium goldschmidtianum Kraenzl. ■
125410 Dendrobium wangii C. L. Tso;王氏石斛■
125411 Dendrobium wangii C. L. Tso = Dendrobium hercoglossum Rchb. f. ■
125412 Dendrobium wardianum Warner;大苞鞘石斛(腾冲石斛);Bigbract Dendrobium ■
125413 Dendrobium wardianum Warner var. album Williams;白花大苞鞘石斛■☆
125414 Dendrobium wattii (Hook. f.) Rchb. f. ;高山石斛■
125415 Dendrobium williamsonii J. Day et Rchb. f. ;黑毛石斛(鸡爪兰,毛石斛);Blackhair Dendrobium ■
125416 Dendrobium wilmsianum Schltr. = Dendrobium compactum Rolfe ex W. Hackett ■
125417 Dendrobium wilsonii Rolfe;广东石斛(白花铜皮石斛,铜皮兰,中黄草);E. H. Wilson Dendrobium,Guangdong Dendrobium ■
125418 Dendrobium wilsonii Rolfe = Dendrobium moniliforme (L.) Sw. ■
125419 Dendrobium xichouense S. J. Cheng et Z. Z. Tang;西畴石斛;Xichou Dendrobium ■
125420 Dendrobium xiurenense M. F. Li et G. H. Xu;修仁石斛;Xiuren Dendrobium ■
125421 Dendrobium yongfuense M. F. Li et G. H. Xu;永福石斛;Yongfu Dendrobium ■
125422 Dendrobium yunnanense Finet = Dendrobium moniliforme (L.) Sw. ■
125423 Dendrobium zhaojuense S. C. Sun et L. G. Xu = Dendrobium aurantiacum Rchb. f. var. zhaojunense (S. C. Sun et L. G. Xu) Z. H. Tsi ■
125424 Dendrobium zhaojuense S. C. Sun et L. G. Xu = Dendrobium chryseum Rolfe ■
125425 Dendrobium zonatum Rolfe = Dendrobium moniliforme (L.) Sw. ■
125426 Dendrobrium Agardh = Dendrobium Sw. (保留属名)■
125427 Dendrobrychis (DC.) Galushko = Onobrychis Mill. ■
125428 Dendrobrychis Galushko = Onobrychis Mill. ■
125429 Dendrobryon Klotzsch ex Pax = Algernonia Baill. + Tetraplandra Baill. ☆
125430 Dendrocacalia (Nakai) Nakai = Dendrocacalia (Nakai) Nakai ex Tuyama ●☆
125431 Dendrocacalia (Nakai) Nakai ex Tuyama(1936);蟹甲木属●☆
125432 Dendrocacalia (Nakai) Tuyama = Dendrocacalia (Nakai) Nakai ex Tuyama ●☆
125433 Dendrocacalia crepidifolia (Nakai) Nakai;蟹甲木●☆
125434 Dendrocalamopsis (L. C. Chia et H. L. Fung) P. C. Keng =

Bambusa Schreb.(保留属名)●

125435 Dendrocalamopsis (L. C. Chia et H. L. Fung) P. C. Keng (1983);绿竹属;Greenbamboo,Dendrocalamopsis ●

125436 Dendrocalamopsis Q. H. Dai et X. L. Tao = Dendrocalamopsis (L. C. Chia et H. L. Fung) P. C. Keng ●

125437 Dendrocalamopsis atrovirens (T. H. Wen) P. C. Keng ex W. T. Lin = Bambusa oldhamii Munro ●

125438 Dendrocalamopsis atrovirens (T. H. Wen) P. C. Keng ex W. T. Lin = Dendrocalamopsis oldhamii (Munro) P. C. Keng ●

125439 Dendrocalamopsis basihirsuta (McClure) P. C. Keng et W. T. Lin;苦绿竹(扁竹);Basihirsute Dendrocalamopsis, Bitter Greenbamboo ●

125440 Dendrocalamopsis basihirsuta (McClure) P. C. Keng et W. T. Lin = Bambusa basihirsuta McClure ●

125441 Dendrocalamopsis beecheyana (Munro) P. C. Keng;吊丝球竹;Beechey Bamboo, Beechey Bambusa, Beechey Dendrocalamopsis, Beechey Greenbamboo,Silk-ball Bamboo,South-China Sinocalamus ●

125442 Dendrocalamopsis beecheyana (Munro) P. C. Keng = Bambusa beecheyana Munro ●

125443 Dendrocalamopsis beecheyana (Munro) P. C. Keng var. pubescens (P. F. Li) P. C. Keng;大头典竹(大头点竹,大头甜竹,大头竹,吊丝球竹,麻竹舅,竹变)●

125444 Dendrocalamopsis beecheyana (Munro) P. C. Keng var. pubescens (P. F. Li) P. C. Keng = Bambusa beecheyana Munro var. pubescens (P. F. Li) W. C. Lin ●

125445 Dendrocalamopsis bicicatricata (W. T. Lin) P. C. Keng = Bambusa bicicatricata (W. T. Lin) L. C. Chia et H. L. Fung ●

125446 Dendrocalamopsis daii P. C. Keng = Bambusa grandis (Q. H. Dai et X. L. Tao) Ohrnb. ●

125447 Dendrocalamopsis edulis (Odash.) P. C. Keng;乌脚绿(鲎脚绿,胡脚绿,胡脚线,胡绿,南洋竹,四季竹,乌脚绿竹);Blackfoot Greenbamboo,Edible Bamboo,Edible Dendrocalamopsis ●

125448 Dendrocalamopsis edulis (Odash.) P. C. Keng = Bambusa odashimae Hatus. ex Ohrnb. ●

125449 Dendrocalamopsis grandis Q. H. Dai et X. L. Tao = Bambusa grandis (Q. H. Dai et X. L. Tao) Ohrnb. ●

125450 Dendrocalamopsis grandis Q. H. Dai et X. L. Tao = Dendrocalamopsis daii P. C. Keng ●

125451 Dendrocalamopsis grandis Q. H. Dai et X. L. Tao ex P. C. Keng = Bambusa grandis (Q. H. Dai et X. L. Tao) Ohrnb. ●

125452 Dendrocalamopsis grandis Q. H. Dai et X. L. Tao ex P. C. Keng = Dendrocalamopsis daii P. C. Keng ●

125453 Dendrocalamopsis oldhamii (Munro) P. C. Keng;绿竹(草鞋底,长枝竹,蝴蝶尖,毛绿竹,坭竹,石竹,甜竹,乌药竹,效脚绿,玉版笋);Clumping Giant Timber Bamboo, Green Bamboo, Greenbamboo, Oldham Bamboo, Oldham Bambusa, Oldham Dendrocalamopsis,Oldham's Bambusa ●

125454 Dendrocalamopsis oldhamii (Munro) P. C. Keng = Bambusa oldhamii Munro ●

125455 Dendrocalamopsis oldhamii (Munro) P. C. Keng f. revoluta (W. T. Lin et J. Y. Lin) W. T. Lin;花头黄●

125456 Dendrocalamopsis prasina (T. H. Wen) P. C. Keng = Bambusa basihirsuta McClure ●

125457 Dendrocalamopsis prasina (T. H. Wen) P. C. Keng = Dendrocalamopsis basihirsuta (McClure) P. C. Keng et W. T. Lin ●

125458 Dendrocalamopsis stenoaurita (W. T. Lin) P. C. Keng ex W. T. Lin = Bambusa stenoaurita (W. T. Lin) T. H. Wen ●

125459 Dendrocalamopsis valida Q. H. Dai;杜绿竹;Strong Greenbamboo ●

125460 Dendrocalamopsis variostriata (W. T. Lin) P. C. Keng = Bambusa variostriata (W. T. Lin) L. C. Chia et H. L. Fung ●

125461 Dendrocalamus Nees(1835);牡竹属(慈竹属,龙竹属,麻竹属,苏麻竹属);Dendrocalamus,Dragonbamboo,Giant Bamboo,Tree-like Bamboo ●

125462 Dendrocalamus affinis Rendle = Bambusa emeiensis L. C. Chia et H. L. Fung ●

125463 Dendrocalamus affinis Rendle = Neosinocalamus affinis (Rendle) P. C. Keng ●

125464 Dendrocalamus albociliatus (Munro) J. L. Sun = Gigantochloa albociliata (Munro) Kurz ●

125465 Dendrocalamus asper (Schult. et Schult. f.) Backer ex K. Heyne;马来甜龙竹(菲律宾巨草竹,光笋竹,毛笋竹);Giant Bamboo,Indonesian Asper,Malay Sweet Dragonbamboo,Phai-tong, Rough Asper, Rough Bamboo, Scabrous Dendrocalamus, Smooth Giantgrass ●

125466 Dendrocalamus bambusoides J. R. Xue et D. Z. Li;椅子竹;Bambusa-like Dendrocalamus,Chair Dragonbamboo ●

125467 Dendrocalamus barbatus J. R. Xue et D. Z. Li;小叶龙竹;Barbaded Dendrocalamus, Barbed Dendrocalamus, Smallleaf Dragonbamboo ●

125468 Dendrocalamus barbatus J. R. Xue et D. Z. Li var. internodiiradicatus J. R. Xue et D. Z. Li;毛脚龙竹;Hairfoot Dragonbamboo ●

125469 Dendrocalamus brandisii (Munro) Kurz;勃氏甜龙竹;Brandis Dendrocalamus,Brandis Dragonbamboo,Velvet Leaf Bamboo ●

125470 Dendrocalamus burmanicus A. Camus;缅甸龙竹;Burma Dendrocalamus,Burma Dragonbamboo ●

125471 Dendrocalamus calostachyus (Kurz) Kurz;美穗龙竹(美穗竹); Beautiful-spiked Dendrocalamus, Beautiful-stachys Dendrocalamus,Finespike Dragonbamboo ●

125472 Dendrocalamus exilis N. H. Xia et L. C. Chia = Dendrocalamus liboensis J. R. Xue et D. Z. Li ●

125473 Dendrocalamus farinosus (Keng et P. C. Keng) L. C. Chia et H. L. Fung;大叶慈竹(大叶慈)●

125474 Dendrocalamus farinosus (Keng et P. C. Keng) L. C. Chia et H. L. Fung f. flavo-striatus T. P. Yi;黄花梁山慈(大叶慈,梁山慈竹);Mealy Dendrocalamus,Yellowflower Farinose Dendrocalamus ●

125475 Dendrocalamus flagellifer Munro = Dendrocalamus asper (Schult. et Schult. f.) Backer ex K. Heyne ●

125476 Dendrocalamus fugongensis J. R. Xue et D. Z. Li;福贡龙竹;Fugong Dendrocalamus,Fugong Dragonbamboo ●

125477 Dendrocalamus giganteus (Wall.) Munro;龙竹(大麻竹,巨竹,甍浓巨竹,美浓巨竹,苏麻竹,印度麻竹);Dragon Bamboo, Giant Bamboo,Giant Dendrocalamus,Giant Dragonbamboo,Gigantic Bamboo ●

125478 Dendrocalamus gui-yangensis N. H. Xia et L. C. Chia = Dendrocalamus liboensis J. R. Xue et D. Z. Li ●

125479 Dendrocalamus hamiltonii Nees et Arn. ex Munro;版纳甜龙竹;Hamilton Dendrocalamus, Hamilton Sweet Dragonbamboo, Tama Bamboo,Tufted Bamboo ●

125480 Dendrocalamus hookeri Munro var. parishii (Munro) Blatt. = Dendrocalamus parishii Munro ●

125481 Dendrocalamus jianshuiensis J. R. Xue et D. Z. Li;建水龙竹(红竹);Jianshui Dendrocalamus,Jianshui Dragonbamboo ●

125482 Dendrocalamus latiflorus Munro;麻竹(大头典竹,大叶乌竹,吊丝竹,甜竹,正坭竹);Big Jute-bamboo, Broadflower Dendrocalamus, Broad-flowered Dragonbamboo, Large-flowered Dendrocalamus, Ma Bamboo, Sweet Bamboo, Taiwan Giant Bamboo, Wideleaf Bamboo ●

125483 Dendrocalamus latiflorus Munro 'Mei-nung';美浓麻竹;Mei-nung Broadflower Dendrocalamus, Mei-nung Dragonbamboo ●

125484 Dendrocalamus latiflorus Munro 'Subconvex';葫芦麻竹;Subconvex Broadflower Dendrocalamus ●

125485 Dendrocalamus latiflorus Munro f. meinung (W. C. Lin) T. P. Yi = Dendrocalamus latiflorus Munro 'Mei-nung' ●

125486 Dendrocalamus latiflorus Munro f. subconvex (W. C. Lin) T. P. Yi = Dendrocalamus latiflorus Munro 'Subconvex' ●

125487 Dendrocalamus latiflorus Munro var. lagenarius W. C. Lin = Dendrocalamus latiflorus Munro ●

125488 Dendrocalamus latiflorus Munro var. lagenarius W. C. Lin = Dendrocalamus latiflorus Munro 'Subconvex' ●

125489 Dendrocalamus liboensis J. R. Xue et D. Z. Li;荔波吊竹;Libo Dendrocalamus, Libo Dragonbamboo ●

125490 Dendrocalamus membranaceus Munro;黄竹(缅甸麻竹);Burman Bamboo, Membranaceous Dendrocalamus, Membranous Dendrocalamus, Waya Bamboo, Yellow Dragonbamboo ●

125491 Dendrocalamus membranaceus Munro f. fimbriligulatus J. R. Xue et D. Z. Li;流苏黄竹;Fimbriligulate Membranaceous Dendrocalamus, Fimbriligulate Yellow Dragonbamboo ●

125492 Dendrocalamus membranaceus Munro f. pilosus J. R. Xue et D. Z. Li;毛竿黄竹;Pilose Membranaceous Dendrocalamus, Pilose Yellow Dragonbamboo ●

125493 Dendrocalamus membranaceus Munro f. striatus J. R. Xue et D. Z. Li;花竿黄竹;Sriated Membranaceous Dendrocalamus, Sriated Yellow Dragonbamboo ●

125494 Dendrocalamus mianningensis Q. Li et X. Jiang = Ampelocalamus mianningensis (Q. Li et X. Jiang) D. Z. Li et Stapleton ●

125495 Dendrocalamus minor (McClure) L. C. Chia et H. L. Fung;吊丝竹(乌药竹);Small Dendrocalamus, Small Dragonbamboo ●

125496 Dendrocalamus minor (McClure) L. C. Chia et H. L. Fung var. amoenus (Q. H. Dai et C. F. Huang) J. R. Xue et D. Z. Li;花吊丝竹●

125497 Dendrocalamus ovatus N. H. Xia et L. C. Chia;船竹;Ovate Dragonbamboo ●

125498 Dendrocalamus ovatus N. H. Xia et L. C. Chia = Dendrocalamus farinosus (Keng et P. C. Keng) L. C. Chia et H. L. Fung ●

125499 Dendrocalamus pachycladus D. Z. Li;江竹;Pachycldous Dendrocalamus ●

125500 Dendrocalamus pachystachys J. R. Xue et D. Z. Li;粗穗龙竹(白竹,蚌竹,粉竹,甜竹);Big-spiked Dendrocalamus, Thick-spike Dendrocalamus, Thick-spike Dragonbamboo ●

125501 Dendrocalamus parishii Munro;巴氏龙竹;Parish Dendrocalamus, Parish Dragonbamboo ●

125502 Dendrocalamus patellaris Gamble;碟环慈竹;Patelliform Dragonbamboo, Small-plate Dendrocalamus ●

125503 Dendrocalamus peculiaris J. R. Xue et D. Z. Li;金平龙竹(青鞘大竹);Jinping Dendrocalamus, Jinping Dragonbamboo ●

125504 Dendrocalamus pulverulentus L. C. Chia et But;粉麻竹;Powder Dragonbamboo, Powdered Dendrocalamus ●

125505 Dendrocalamus ronganensis Q. H. Dai et D. Y. Huang = Dendrocalamus tsiangii (McClure) L. C. Chia et H. L. Fung ●

125506 Dendrocalamus sapidus Q. H. Dai et D. Y. Huang = Dendrocalamus minor (McClure) L. C. Chia et H. L. Fung ●

125507 Dendrocalamus semiscandens J. R. Xue et D. Z. Li;野龙竹;Field Dragonbamboo, Wild-dragon Dendrocalamus ●

125508 Dendrocalamus sikkimensis Gamble ex Oliv.;锡金龙竹;Bhutan Bamboo, Philippines Sweet-shoot Bamboo, Sikkim Dendrocalamus, Sikkim Dragonbamboo ●

125509 Dendrocalamus sinicus L. C. Chia et J. L. Sun;歪脚龙竹;Askewfoot Dragonbamboo, Chinese Dendrocalamus ●

125510 Dendrocalamus strictus (Roxb.) Nees;牡竹(泥竹,印度实竹);Calcutta Bamboo, Male Bamboo, Male Dendrocalamus, Male Dragonbamboo, Solid Bamboo ●

125511 Dendrocalamus textilis N. H. Xia, L. C. Chia et C. Y. Xia;吊篦竹●

125512 Dendrocalamus textilis N. H. Xia, L. C. Chia et C. Y. Xia = Dendrocalamus tsiangii (McClure) L. C. Chia et H. L. Fung ●

125513 Dendrocalamus tibeticus J. R. Xue et T. P. Yi;西藏牡竹;Xizang Dendrocalamus, Xizang Dragonbamboo ●

125514 Dendrocalamus tomentosus J. R. Xue et D. Z. Li;毛龙竹(野龙竹);Hair Dragonbamboo, Tomentose Dendrocalamus ●

125515 Dendrocalamus tsiangii (McClure) L. C. Chia et H. L. Fung;黔竹;Guizhou Dragonbamboo, Tsiang Dendrocalamus ●

125516 Dendrocalamus tsiangii (McClure) L. C. Chia et H. L. Fung 'Viridistriatus';花黔竹;Viridistriatus Tsiang Dendrocalamus ●

125517 Dendrocalamus tsiangii (McClure) L. C. Chia et H. L. Fung = Dendrocalamus textilis N. H. Xia, L. C. Chia et C. Y. Xia ●

125518 Dendrocalamus tsiangii (McClure) L. C. Chia et H. L. Fung f. viridistriatus X. H. Song = Dendrocalamus tsiangii (McClure) L. C. Chia et H. L. Fung 'Viridistriatus' ●

125519 Dendrocalamus yunnanicus J. R. Xue et D. Z. Li;云南龙竹(大桡竹,大竹);Yunnan Dendrocalamus, Yunnan Dragonbamboo ●

125520 Dendrocalla pricei Rolfe = Thrixspermum formosanum (Hayata) Schltr. ■

125521 Dendrocereus Britton et Rose = Acanthocereus (Engelm. ex A. Berger) Britton et Rose ●☆

125522 Dendrocereus Britton et Rose(1920);树木柱属(树状天轮柱属)●☆

125523 Dendrocereus nudiflorus Britton et Rose;树木柱●☆

125524 Dendrocharis Miq. = Ecdysanthera Hook. et Arn. ●

125525 Dendrocharis Miq. = Echites P. Browne ●☆

125526 Dendrochilum Blume(1825);足柱兰属(垂串兰属,穗花一叶兰属);Dendrochilum, Dendrochilum Orchid, Footstyle-orchis ■

125527 Dendrochilum aurantiacum Blume;橙黄足柱兰;Orange Dendrochilum ■☆

125528 Dendrochilum cobbianum Rchb. f.;考氏足柱兰;Cobb Dendrochilum ■☆

125529 Dendrochilum filiforme Lindl.;丝状足柱兰;Filiform Dendrochilum ■☆

125530 Dendrochilum formosanum (Schltr.) Schltr. = Dendrochilum uncatum Rchb. f. ■

125531 Dendrochilum glumaceum Lindl.;颖穗足柱兰(颖状足柱兰);Glumelike Dendrochilum, Silver Chain ■☆

125532 Dendrochilum latifolium Lindl.;宽叶足柱兰;Broadleaf Dendrochilum ■☆

125533 Dendrochilum longifolium Rchb. f.;长叶足柱兰;Longleaf Dendrochilum ■☆

125534 Dendrochilum occultum Lindl. = Bulbophyllum occultum

Thouars ■☆

125535 Dendrochilum uncatum Rchb. f.；足柱兰(黄穗兰，穗花兰)；Common Footstyle-orchis ■

125536 Dendrochilum uncinatum Rchb. f. var. formosanum (Schltr.) T. Hashim. = Dendrochilum uncatum Rchb. f. ■

125537 Dendrochloa C. E. Parkinson = Schizostachyum Nees ●

125538 Dendrochloa C. E. Parkinson(1933)；乔草竹属●☆

125539 Dendrochloa distans C. E. Parkinson；乔草竹●☆

125540 Dendrocnide Miq. (1851)；火麻树属(树火麻属，树头麻属，咬人狗属)；Giant Nettle，Gympie，Stinger，Woodnettle ●

125541 Dendrocnide basirotunda (C. Y. Wu) Chew；圆基叶火麻树(圆基火麻树，圆基叶树火麻)；Basal-round Woodnettle，Roundbase Woodnettle ●

125542 Dendrocnide basirotunda (C. Y. Wu) Chew = Laportea basirotunda C. Y. Wu ●

125543 Dendrocnide chingiana (Hand. -Mazz.) Chew = Dendrocnide urentissima (Gagnep.) Chew ●◇

125544 Dendrocnide kotoensis (Hayata ex Yamam.) B. L. Shih et Yuen P. Yang；红头咬人狗(兰屿咬人狗)●

125545 Dendrocnide meyeniana (Walp.) Chew；咬人狗(恒春咬人狗，兰屿咬人狗，咬人狗艾麻，咬人狗火麻树，咬人猫)；Mayer Woodnettle，Poisonous Woodnettle，Poisonous Wood-nettle ●

125546 Dendrocnide meyeniana (Walp.) Chew f. subglabra (Hayata) Chew；恒春火麻树(巴丹艾麻，巴丹咬人狗，恒春咬人狗，兰屿咬人狗)；Batan Woodnettle，Batan Wood-nettle，Hengchun Mayer Woodnettle ●

125547 Dendrocnide meyeniana (Walp.) Chew f. subglabra (Hayata) Chew = Dendrocnide meyeniana (Walp.) Chew ●

125548 Dendrocnide sinuata (Blume) Chew；全缘火麻树(把手天门，不可摸，毒树，老虎脷，老虎俐，全缘叶树火麻，天下无敌手，圆齿艾麻，圆齿火麻树)；Entire Woodnettle，Sinuate Woodnettle ●

125549 Dendrocnide sinuata (Blume) Chew = Laportea sinuata (Blume) Miq. ●

125550 Dendrocnide stimulans (L. f.) Chew；海南火麻树(狗狂叶，海南艾麻)；Hainan Woodnettle，Malay Nettle Tree，Malay Nettle-tree ●

125551 Dendrocnide stimulans (L. f.) Chew = Laportea stimulans (L. f.) Miq. ●

125552 Dendrocnide subglabra (Hayata) Chew = Dendrocnide meyeniana (Walp.) Chew f. subglabra (Hayata) Chew ●

125553 Dendrocnide urentissima (Gagnep.) Chew；火麻树(电树，憨掌，麻风树，树火麻)；Stinging Woodnettle，Woodnettle ●◇

125554 Dendrocnide urentissima (Gagnep.) Chew = Laportea urentissima Gagnep. ●◇

125555 Dendrocolla Blume = Thrixspermum Lour. + Pteroceras Hasselt ex Hassk. ■

125556 Dendrocolla amplexicaulis Blume = Thrixspermum amplexicaule (Blume) Rchb. f. ■

125557 Dendrocolla appendiculata Blume = Grosourdya appendiculatum (Blume) Rchb. f. ■

125558 Dendrocolla arachnites Blume = Thrixspermum centipeda Lour. ■

125559 Dendrocolla pricei Rolfe = Thrixspermum formosanum (Hayata) Schltr. ■

125560 Dendrocolla subulata Blume = Thrixspermum subulatum Rchb. f. ■

125561 Dendrocoryne (Lindl. et Paxton) Brieger = Dendrobium Sw. (保留属名)■

125562 Dendrocoryne (Lindl.) Brieger = Dendrobium Sw. (保留属名)■

125563 Dendrocousinia Willis = Dendrocousinsia Millsp. ●

125564 Dendrocousinsia Millsp. = Sebastiania Spreng. ●

125565 Dendrodaphne Beurl. = Ocotea Aubl. ●☆

125566 Dendrokingstonia Rauschert(1982)；金斯敦木属●☆

125567 Dendrokingstonia nervosa (Hook. f. et Thomson) Rauschert；金斯敦木●☆

125568 Dendroleandria Arènes = Helmiopsiella Arènes ●☆

125569 Dendrolirium Blume = Eria Lindl. (保留属名)■

125570 Dendrolirium Blume(1825)；绒兰属■

125571 Dendrolirium lasiopetalum (Willd.) S. C. Chen et J. J. Wood；白绵绒兰(白绵毛兰)；Whitelanose Eria，Whitelanose Hairorchis ■

125572 Dendrolirium pubescens Hook. = Dendrolirium lasiopetalum (Willd.) S. C. Chen et J. J. Wood ■

125573 Dendrolirium tomentosum (J. König) S. C. Chen et J. J. Wood；绒兰(海南毛兰，黄绒毛兰)；Hainan Eria，Hainan Heath，Yellowfloss Hairorchis ■

125574 Dendrolobium (Wight et Arn.) Benth. (1852)；假木豆属(木荚豆属，木山蚂蝗属)；Dendrolobium，Fake Woodbean ●

125575 Dendrolobium Benth. = Dendrolobium (Wight et Arn.) Benth. ●

125576 Dendrolobium dispermum (Hayata) Schindl.；二子假木豆(二节假木豆，二种子假木豆，两节假木豆，双节山蚂蝗)；Two-jointed Dendrolobium，Twoseed Dendrolobium，Two-seed Dendrolobium，Twoseed Fake Woodbean ●

125577 Dendrolobium lanceolatum (Dunn) Schindl.；单节假木豆(单节荚假木豆，小叶山木豆)；Lanceolate Dendrolobium，Lanceolate Fake Woodbean ●

125578 Dendrolobium lanceolatum (Dunn) Schindl. var. microcarpum H. Ohashi；小果单节假木豆●

125579 Dendrolobium rugosum (Prain) Schindl.；多皱假木豆；Rugose Dendrolobium ●

125580 Dendrolobium triangulare (Retz.) Schindl.；假木豆(甲由草，假绿豆，千金不藤，野马蝗，野蚂蝗)；Fake Woodbean，Triangular Dendrolobium，Trigonous-branch Dendrolobium ●

125581 Dendrolobium triangulare (Retz.) Schindl. = Desmodium triangulare (Retz.) Merr. ●

125582 Dendrolobium umbellatum (L.) Benth.；伞花假木豆(白古苏花，白木苏花，胡蝇翼)；Umbellate Dendrolobium，Umbellate Fake Woodbean ●

125583 Dendromecon Benth. (1835)；树罂粟属(木罂粟属，罂粟木属)；Tree Poppy ●☆

125584 Dendromecon harfordii Kellogg；哈福德树罂粟；IslandTree-poppy ●☆

125585 Dendromecon rigidum Benth.；树罂粟(罂粟木)；Bush Poppy，Island Bush Poppy，Tree Poppy ●☆

125586 Dendromecon rigidum Benth. subsp. harfordii (Kellogg) P. H. Raven = Dendromecon harfordii Kellogg ●☆

125587 Dendromyza Danser(1940)；干寄生属(米扎树属)●☆

125588 Dendromyza apiculata Danser；干寄生●☆

125589 Dendropanax Decne. et Planch. (1854)；树参属(木五加属，杞李葠属，隐蓑属)；Dendropanax，Treerenshen ●

125590 Dendropanax acuminatissimus Merr. = Dendropanax proteus (Champ. ex Benth.) Benth. ●

125591 Dendropanax angustilobus (Hu) Merr. = Dendropanax proteus (Champ. ex Benth.) Benth. ●

125592 Dendropanax arboreus (L.) Decne. et Planch.；美洲树参(乔木树参)；American Dendropanax，Angelica Tree ●☆

125593 Dendropanax bilocularis C. N. Ho;双室树参(双室木五加);Bilocular Dendropanax,Bilocular Treerenshen ●

125594 Dendropanax brevistylus Y. Ling;短柱树参(短柱杞李参,短柱杞李葠);Shortstyle Dendropanax, Shortstyle Treerenshen, Short-styled Dendropanax ●

125595 Dendropanax brevistylus Y. Ling = Dendropanax proteus (Champ. ex Benth.) Benth. ●

125596 Dendropanax burmaricus Merr.;缅甸树参(云南树参);Burma Dendropanax,Burma Treerenshen,Yunnan Dendropanax ●

125597 Dendropanax caloneurus (Harms) Merr.;榕叶树参;Figleaf Dendropanax,Fig-leaved Dendropanax,Treerenshen ●

125598 Dendropanax chevalieri (R. Vig.) Merr.;大果树参(大果木五加);Bigfruit Dendropanax, Big-fruited Dendropanax, Largefruit Dendropanax,Largefruit Treerenshen ●

125599 Dendropanax chevalieri (R. Vig.) Merr. var. dentiger (Harms) H. L. Li = Dendropanax dentiger (Harms ex Diels) Merr. ●

125600 Dendropanax chevalieri (Vij. Kumar) Merr. = Dendropanax dentiger (Harms ex Diels) Merr. ●

125601 Dendropanax chevalieri (Vij. Kumar) Merr. var. dentigerus (Harms) H. L. Li = Dendropanax dentiger (Harms ex Diels) Merr. ●

125602 Dendropanax confertus H. L. Li;挤果树参(密花木五加);Confertedfruit Dendropanax, Confused Dendropanax, Densefruit Treerenshen ●

125603 Dendropanax crassifolius Y. F. Deng et H. Peng = Dendropanax kwangsiensis H. L. Li ●

125604 Dendropanax dentiger (Harms ex Diels) Merr.;树参(白半枫荷,白荷,白山鸡骨,半边枫,半枫荷,半荷枫,边荷枫,枫荷,枫荷桂,枫荷梨,疯气树,金鸡趾,梨枫桃,梨荷枫,梨胶木,木荷枫,木荷桃,木五加,偏荷枫,杞李参,三叉一支镖,台湾杞李参,台湾树参,五加皮,小荷枫,谢氏杞李葠,鸭脚板,鸭脚荷,鸭脚木,鸭掌柴,阴阳枫);Chavalier Dendropanax, Dentiferous Dendropanax, Taiwan Dendropanax,Taiwan Treerenshen,Treerenshen ●

125605 Dendropanax dentiger (Harms) Merr. = Dendropanax dentiger (Harms ex Diels) Merr. ●

125606 Dendropanax ferrugineus H. L. Li = Brassaiopsis ferruginea (H. L. Li) G. Hoo ●

125607 Dendropanax ferrugineus H. L. Li = Euaraliopsis ferruginea (H. L. Li) G. Hoo et C. J. Tseng ●

125608 Dendropanax ficifolius C. J. Tseng et G. Hoo = Dendropanax caloneurus (Harms) Merr. ●

125609 Dendropanax gracilis C. J. Tseng et G. Hoo;细梗树参(细梗木五加);Slender Dendropanax, Slenderstalk Dendropanax, Thinstype Treerenshen ●

125610 Dendropanax gracilis C. J. Tseng et G. Hoo = Dendropanax proteus (Champ. ex Benth.) Benth. ●

125611 Dendropanax hainanensis (Merr. et Chun) Chun;海南树参(豆腐木,海南木五加,海南杞李参);Hainan Dendropanax, Hainan Treerenshen ●

125612 Dendropanax hoi C. B. Shang = Dendropanax chevalieri (R. Vig.) Merr. ●

125613 Dendropanax inflatus H. L. Li;胀果树参(胀果木五加);Inflated Dendropanax,Inflated Treerenshen ●

125614 Dendropanax inflatus H. L. Li = Dendropanax dentiger (Harms ex Diels) Merr. ●

125615 Dendropanax inflatus H. L. Li f. multiflorus C. J. Tseng et G. Hoo;多花胀果树参(圆锥胀果树参);Manyflower Inflated Dendropanax,Manyflower Inflated Treerenshen ●

125616 Dendropanax inflatus H. L. Li f. multiflorus C. J. Tseng et G. Hoo = Dendropanax dentiger (Harms ex Diels) Merr. ●

125617 Dendropanax inflatus H. L. Li f. paniculatus C. J. Tseng et G. Hoo;锥花胀果树参(多花胀果树参);Paniculate Inflated Treerenshen ●

125618 Dendropanax inflatus H. L. Li f. paniculatus C. J. Tseng et G. Hoo = Dendropanax dentiger (Harms ex Diels) Merr. ●

125619 Dendropanax inflatus H. L. Li f. prominens C. J. Tseng et G. Hoo;显脉胀果树参;Distinctvein Inflated Dendropanax ●

125620 Dendropanax iriomotensis (Masam.) Hatus.;西表树参●☆

125621 Dendropanax japonicus (Jungh.) Seem. = Dendropanax trifidus (Thunb.) Makino ex H. Hara ●

125622 Dendropanax japonicus Seem. = Dendropanax trifidus (Thunb.) Makino ex H. Hara ●

125623 Dendropanax kwangsiensis H. L. Li;广西树参(广西木五加);Guangxi Dendropanax,Guangxi Treerenshen ●

125624 Dendropanax listeri King = Merrilliopanax listeri (King) H. L. Li ●◇

125625 Dendropanax macrocarpus C. N. Ho = Dendropanax chevalieri (R. Vig.) Merr. ●

125626 Dendropanax morbiferus H. Lév.;朝鲜树参●☆

125627 Dendropanax oligodontus Merr. et Chun;保亭树参;Baoting Dendropanax,Baoting Treerenshen,Oligodontous Dendropanax ●

125628 Dendropanax parvifloroides C. N. Ho;两广树参(假小花木五加,拟小花木五加);Parviflorous Dendropanax, Smallflower-like Dendropanax,Smallflower-like Treerenshen ●

125629 Dendropanax parvifloroides C. N. Ho = Dendropanax kwangsiensis H. L. Li ●

125630 Dendropanax parvifloroides C. N. Ho var. chartaceus K. M. Feng et Y. R. Li;坚纸叶树参;Harpaper Treerenshen ●

125631 Dendropanax parvifloroides C. N. Ho var. chartaceus K. M. Feng et Y. R. Li = Dendropanax proteus (Champ. ex Benth.) Benth. ●

125632 Dendropanax parviflorus (Champ. ex Benth.) Benth. = Dendropanax proteus (Champ. ex Benth.) Benth. ●

125633 Dendropanax parviflorus (Champ.) Benth. = Dendropanax proteus (Champ.) Benth. ●

125634 Dendropanax pellucidopunctatus (Hayata) Kaneh.;台湾树参●

125635 Dendropanax pellucidopunctatus (Hayata) Kaneh. = Dendropanax dentiger (Harms ex Diels) Merr. ●

125636 Dendropanax pellucidopunctatus (Hayata) Merr. = Dendropanax dentiger (Harms ex Diels) Merr. ●

125637 Dendropanax petelotii (Harms) Merr. = Dendropanax hainanensis (Merr. et Chun) Chun ●

125638 Dendropanax productus H. L. Li;长萼树参;Longcalyx Dendropanax,Long-calyxed Dendropanax,Longsepaied Treerenshen ●

125639 Dendropanax proteus (Champ. ex Benth.) Benth.;变叶树参(白半枫,白半枫荷,白皮半枫荷,半枫荷,三层楼,铁锹树,窄叶树参);Biformleaf Dendropanax, Biformleaf Treerenshen, Biform-leaved Dendropanax ●

125640 Dendropanax proteus (Champ.) Benth. = Dendropanax proteus (Champ. ex Benth.) Benth. ●

125641 Dendropanax stellatus H. L. Li;星柱树参(星花木五加);Starstyle Treerenshen, Stellatestyle Dendropanax, Stellate-styled Dendropanax ●

125642 Dendropanax trifidus (Thunb. ex Murray) Makino ex H. Hara = Dendropanax trifidus (Thunb.) Makino ex H. Hara ●

125643 Dendropanax trifidus (Thunb. ex Murray) Makino ex H. Hara f.

aureo-variegata Makino = Dendropanax trifidus (Thunb.) Makino ex H. Hara f. aureo-variegata Makino ●☆

125644 Dendropanax trifidus (Thunb.) Makino ex H. Hara;三裂树参(三菱果树参)●

125645 Dendropanax trifidus (Thunb.) Makino ex H. Hara f. aureo-variegata Makino;黄斑三菱果树参●☆

125646 Dendropanax trifidus Makino = Dendropanax trifidus (Thunb.) Makino ex H. Hara ●

125647 Dendropanax yunnanensis C. J. Tseng et G. Hoo;云南树参;Yunnan Inflated ●

125648 Dendropanax yunnanensis C. J. Tseng et G. Hoo = Dendropanax burmaricus Merr. ●

125649 Dendropemon (Blume) Rchb. = Loranthus Jacq.(保留属名)●

125650 Dendropemon (Blume) Rchb. = Phthirusa Mart. ●☆

125651 Dendropemon Blume = Loranthus Jacq.(保留属名)●

125652 Dendrophorbium (Cuatrec.) C. Jeffrey(1992);千里木属■●☆

125653 Dendrophorbium acuminatissimum (Cabrera) D. J. N. Hind;尖千里木●☆

125654 Dendrophorbium americanum (L. f.) C. Jeffrey;美洲千里木●☆

125655 Dendrophorbium floribundum Cuatrec.;多花千里木●☆

125656 Dendrophorbium lucidum (Sw.) C. Jeffrey;亮千里木●☆

125657 Dendrophorbium multinerve (Klatt) C. Jeffrey;多脉千里木●☆

125658 Dendrophthoaceae Tiegh. = Dendrophthoaceae Tiegh. ex Nakai ●

125659 Dendrophthoaceae Tiegh. ex Nakai = Loranthaceae Juss.(保留科名)●

125660 Dendrophthoaceae Tiegh. ex Nakai;五蕊寄生科●

125661 Dendrophthoe Mart. (1830);五蕊寄生属;Dendrophthoe ●

125662 Dendrophthoe elegans (Cham. et Schltdl.) Mart. = Moquiniella rubra (A. Spreng.) Balle ●☆

125663 Dendrophthoe erecta (Engl.) Danser = Oedina erecta (Engl.) Tiegh. ●☆

125664 Dendrophthoe falcata (L. f.) Ettingsh.;镰形五蕊寄生●☆

125665 Dendrophthoe glauca (Thunb.) Mart. = Septulina glauca (Thunb.) Tiegh. ●☆

125666 Dendrophthoe pendens (Engl. et K. Krause) Danser = Oedina pendens (Engl. et K. Krause) Polhill et Wiens ●☆

125667 Dendrophthoe pentandra (L.) Miq.;五蕊寄生(椆树寄生,乌榄寄生);Fivestamen Dendrophthoe, Five-stamen Dendrophthoe, Pentandrous Dendrophthoe ●

125668 Dendrophthoe virescens (N. E. Br.) Danser = Erianthemum virescens (N. E. Br.) Wiens et Polhill ●☆

125669 Dendrophthora Eichler(1868);美洲槲寄生属●☆

125670 Dendrophthora albescens Urb. et Ekman;渐白美洲槲寄生●☆

125671 Dendrophthora basiandra Kuijt;基蕊美洲槲寄生●☆

125672 Dendrophthora densifolia Kuijt;密叶美洲槲寄生●☆

125673 Dendrophthora elliptica Krug et Urb. ex Urb.;椭圆美洲槲寄生●☆

125674 Dendrophthora filiformis Rizzini;线状美洲槲寄生●☆

125675 Dendrophthora gracilipes Kuijt;细梗美洲槲寄生●☆

125676 Dendrophthora gracilis C. Wright;纤细美洲槲寄生●☆

125677 Dendrophthora grandifolia Eichler;大叶美洲槲寄生●☆

125678 Dendrophthora leucocarpa (Pacz.) Trel.;白果美洲槲寄生●☆

125679 Dendrophthora longipes Urb.;长梗美洲槲寄生●☆

125680 Dendrophthora macrostachya Eichler;大穗美洲槲寄生●☆

125681 Dendrophthora mexicana Kuijt;墨西哥美洲槲寄生●☆

125682 Dendrophthora nitidula (Rizzini) Kuijt;光亮美洲槲寄生●☆

125683 Dendrophthora nuda Proctor;裸美洲槲寄生●☆

125684 Dendrophthora oocarpa Krug et Urb.;卵果美洲槲寄生●☆

125685 Dendrophthora parvispicata Rizzini;小穗美洲槲寄生●☆

125686 Dendrophthora paucifolia (Rusby) Kuijt;寡叶美洲槲寄生●☆

125687 Dendrophthora polyantha Kuijt;多花美洲槲寄生●☆

125688 Dendrophthora purpurea Kuijt;紫美洲槲寄生●☆

125689 Dendrophthora tenuiflora (Steyerm. et Maguire) Kuijt;细花美洲槲寄生●☆

125690 Dendrophthora tenuifolia Kuijt;细叶美洲槲寄生●☆

125691 Dendrophthoraceae Dostal = Dendrophthoaceae Tiegh. ex Nakai ●

125692 Dendrophthoraceae Dostal = Loranthaceae Juss.(保留科名)●

125693 Dendrophylax Rchb. f. (1864);附生兰属(抱树兰属)■☆

125694 Dendrophylax lindenii (Lindl.) Benth. ex Rolfe;林登附生兰;Frog Orchid, Ghost Orchid, Linden's Angurek, Palm-polly, White Butterfly Orchid ■☆

125695 Dendrophyllanthus S. Moore = Phyllanthus L. ●■

125696 Dendropogon Raf. = Tillandsia L. ■☆

125697 Dendropogon usneoides (L.) Raf. = Tillandsia usuneoides (L.) L. ■☆

125698 Dendroportulaca Eggli = Deeringia R. Br. ●■

125699 Dendrorchis Thouars = Dendrorkis Thouars(废弃属名)■

125700 Dendrorchis anceps (Ridl.) Kuntze = Polystachya anceps Ridl. ■☆

125701 Dendrorchis appendiculata (Kraenzl.) Kuntze = Polystachya cultriformis Lindl. ex Spreng. ■☆

125702 Dendrorchis estrellensis (Rchb. f.) Kuntze = Polystachya concreta (Jacq.) Garay et H. R. Sweet ■

125703 Dendrorchis jussieuana (Rchb. f.) Kuntze = Polystachya concreta (Jacq.) Garay et H. R. Sweet ■

125704 Dendrorchis minuta (Aubl.) Kuntze = Polystachya concreta (Jacq.) Garay et H. R. Sweet ■

125705 Dendrorchis minutiflora Kuntze = Polystachya fusiformis (Thouars) Lindl. ■☆

125706 Dendrorchis pyramidalis (Lindl.) Kuntze = Polystachya pyramidalis Lindl. ■☆

125707 Dendrorchis rosea (Ridl.) Kuntze = Polystachya rosea Ridl. ■☆

125708 Dendrorchis rosellata (Ridl.) Kuntze = Polystachya rosellata Ridl. ■☆

125709 Dendrorchis rufinula (Rchb. f.) Kuntze = Polystachya concreta (Jacq.) Garay et H. R. Sweet ■

125710 Dendrorchis tessellata (Lindl.) Kuntze = Polystachya concreta (Jacq.) Garay et H. R. Sweet ■

125711 Dendrorchis wightii (Rchb. f.) Kuntze = Polystachya concreta (Jacq.) Garay et H. R. Sweet ■

125712 Dendrorchis zollingeri (Rchb. f.) Kuntze = Polystachya concreta (Jacq.) Garay et H. R. Sweet ■

125713 Dendrorkis Thouars(废弃属名) = Aerides Lour. ■

125714 Dendrorkis Thouars(废弃属名) = Polystachya Hook.(保留属名)■

125715 Dendrorkis purpurea (Wight) Kuntze = Polystachya concreta (Jacq.) Garay et H. R. Sweet ■

125716 Dendrosenecio (Hauman ex Humbert) B. Nord. (1978);木千里光属(莲座千里木属);Giant Groundsel, Tree Groundsel ●☆

125717 Dendrosenecio (Hedberg) B. Nord. = Dendrosenecio (Hauman ex Humbert) B. Nord. ●☆

125718 Dendrosenecio adnivalis (Stapf) E. B. Knox;雪木千里光●☆

125719 Dendrosenecio adnivalis (Stapf) E. B. Knox subsp. friesiorum (Mildbr.) E. B. Knox;弗里斯雪木千里光●☆

125720 Dendrosenecio adnivalis (Stapf) E. B. Knox var. petiolatus (Hedberg) E. B. Knox;柄叶雪木千里光●☆
125721 Dendrosenecio battiscombei (R. E. Fr. et T. C. E. Fr.) E. B. Knox;巴蒂木千里光●☆
125722 Dendrosenecio brassica B. Nord. = Dendrosenecio keniensis (Baker f.) Mabb. ●☆
125723 Dendrosenecio brassiciformis (R. E. Fr. et T. C. E. Fr.) Mabb.;芸苔叶木千里光●☆
125724 Dendrosenecio cheranganiensis (Cotton et Blakelock) E. B. Knox;切兰加尼木千里光●☆
125725 Dendrosenecio cheranganiensis (Cotton et Blakelock) E. B. Knox subsp. dalei Cotton et Blakelock;达勒木千里光●☆
125726 Dendrosenecio elgonensis (T. C. E. Fr.) E. B. Knox;埃尔贡木千里光■☆
125727 Dendrosenecio elgonensis (T. C. E. Fr.) E. B. Knox subsp. barbatipes (Hedberg) E. B. Knox;毛梗木千里光●☆
125728 Dendrosenecio elgonensis (T. C. E. Fr.) E. B. Knox subsp. barbatipes (Hedberg) E. B. Knox = Dendrosenecio johnstonii (Oliv.) B. Nord. subsp. barbatipes (Hedberg) B. Nord. ●☆
125729 Dendrosenecio erici-rosenii (R. E. Fr. et T. C. E. Fr.) E. B. Knox;埃洛木千里光●☆
125730 Dendrosenecio erici-rosenii (R. E. Fr. et T. C. E. Fr.) E. B. Knox subsp. alticola (Mildbr.) E. B. Knox;高原木千里光●☆
125731 Dendrosenecio johnstonii (Oliv.) B. Nord.;约翰逊木千里光●☆
125732 Dendrosenecio johnstonii (Oliv.) B. Nord. subsp. barbatipes (Hedberg) B. Nord. = Dendrosenecio elgonensis (T. C. E. Fr.) E. B. Knox subsp. barbatipes (Hedberg) E. B. Knox ●☆
125733 Dendrosenecio johnstonii (Oliv.) B. Nord. subsp. battiscombei (R. E. Fr. et T. C. E. Fr.) B. Nord. = Dendrosenecio battiscombei (R. E. Fr. et T. C. E. Fr.) E. B. Knox ●☆
125734 Dendrosenecio johnstonii (Oliv.) B. Nord. subsp. cheranganiensis (Cotton et Blakelock) B. Nord. = Dendrosenecio cheranganiensis (Cotton et Blakelock) E. B. Knox ●☆
125735 Dendrosenecio johnstonii (Oliv.) B. Nord. subsp. cottonii (Hutch. et G. Taylor) B. Nord. = Dendrosenecio kilimanjari (Mildbr.) E. B. Knox subsp. cottonii (Hutch. et G. Taylor) E. B. Knox ●☆
125736 Dendrosenecio johnstonii (Oliv.) B. Nord. subsp. dalei (Cotton et Blakelock) B. Nord. = Dendrosenecio cheranganiensis (Cotton et Blakelock) E. B. Knox subsp. dalei Cotton et Blakelock ●☆
125737 Dendrosenecio johnstonii (Oliv.) B. Nord. subsp. elgonensis (T. C. E. Fr.) B. Nord. = Dendrosenecio elgonensis (T. C. E. Fr.) E. B. Knox ■☆
125738 Dendrosenecio johnstonii (Oliv.) B. Nord. subsp. refractisquamatus (De Wild.) B. Nord. = Dendrosenecio adnivalis (Stapf) E. B. Knox ●☆
125739 Dendrosenecio johnstonii (Oliv.) B. Nord. var. friesiorum (Mildbr.) B. Nord. = Dendrosenecio adnivalis (Stapf) E. B. Knox subsp. friesiorum (Mildbr.) E. B. Knox ●☆
125740 Dendrosenecio keniensis (Baker f.) Mabb.;肯尼亚木千里光●☆
125741 Dendrosenecio keniodendron (R. E. Fr. et T. C. E. Fr.) B. Nord.;凯尼木千里光;Kenia Alpine Groundsel ●☆
125742 Dendrosenecio kilimanjari (Mildbr.) E. B. Knox;基利曼木千里光●☆
125743 Dendrosenecio kilimanjari (Mildbr.) E. B. Knox subsp. cottonii (Hutch. et G. Taylor) E. B. Knox;鹅顿木千里光●☆
125744 Dendrosenecio meruensis (Cotton et Blakelock) E. B. Knox;梅鲁木千里光●☆
125745 Dendroseris D. Don(1832);苦苣木属●☆
125746 Dendroseris macrophylla D. Don;苦苣木●☆
125747 Dendrosicus Raf.(废弃属名) = Amphitecna Miers ●☆
125748 Dendrosicus Raf.(废弃属名) = Enallagma (Miers) Baill.(保留属名)●☆
125749 Dendrosicyos Balf. f. (1882);树葫芦属●☆
125750 Dendrosicyos socotrana Balf. f.;树葫芦;Cucumber Tree ●☆
125751 Dendrosida Fryxell(1971);树锦葵属●☆
125752 Dendrosida batesii Fryxell;树锦葵●☆
125753 Dendrosida oxypetala (Triana et Planch.) Fryxell;尖瓣树锦葵●☆
125754 Dendrosida parviflora Fryxell;小花树锦葵●☆
125755 Dendrosipanea Ducke et Steyerm. = Dendrosipanea Ducke ●☆
125756 Dendrosipanea Ducke(1935);树茜属●☆
125757 Dendrosipanea spigelioides Ducke;树茜●☆
125758 Dendrosma Pancher et Sebert = Genista L. ●
125759 Dendrosma R. Br. = Atherosperma Labill. ●☆
125760 Dendrosma R. Br. ex Cromb. = Atherosperma Labill. ●☆
125761 Dendrospartium Spach = Genista L. ●
125762 Dendrospartum Spach = Genista L. ●
125763 Dendrostellera (C. A. Mey.) Tiegh. (1893);树状狼毒属■☆
125764 Dendrostellera (C. A. Mey.) Tiegh. = Diarthron Turcz. ●■
125765 Dendrostellera Tiegh. = Dendrostellera (C. A. Mey.) Tiegh. ■☆
125766 Dendrostellera Tiegh. = Diarthron Turcz. ●■
125767 Dendrostellera arenaria Pobed.;沙地肖草瑞香■☆
125768 Dendrostellera lessertii (Wikstr.) Tiegh. = Stellera lessertii (Wikstr.) C. A. Mey. ●☆
125769 Dendrostellera linearifolia Pobed.;线叶肖草瑞香■☆
125770 Dendrostellera macrorhachis Pobed.;大肖草瑞香■☆
125771 Dendrostellera olgae Pobed.;奥氏肖草瑞香■☆
125772 Dendrostellera stachyoides Tiegh.;穗状肖草瑞香■☆
125773 Dendrostellera turkmenorum Pobed.;土库曼肖草瑞香■☆
125774 Dendrostigma Gleason = Mayna Aubl. ●☆
125775 Dendrostylis H. karst. et Triana = Mayna Aubl. ●☆
125776 Dendrostylis Triana = Mayna Aubl. ●☆
125777 Dendrothrix Esser(1993);树毛大戟属●☆
125778 Dendrothrix multiglandulosa Esser;多腺树毛大戟●☆
125779 Dendrothrix wurdackii Esser;树毛大戟●☆
125780 Dendrotrichum comata Blume = Flickingeria comata (Blume) A. D. Hawkes ■
125781 Dendrotrophe Miq. (1856);寄生藤属;Dendrotrophe, Parasiticvine ●
125782 Dendrotrophe buxifolia (Blume) Miq.;黄杨叶寄生藤;Boxleaf Dendrotrophe, Boxleaf Parasiticvine, Box-leaved Dendrotrophe ●
125783 Dendrotrophe frutescens (Champ. ex Benth.) Danser = Dendrotrophe varians (Blume) Miq. ●
125784 Dendrotrophe frutescens (Champ. ex Benth.) Danser var. subquinquenervia (P. C. Tam) P. C. Tam;叉脉寄生藤●
125785 Dendrotrophe frutescens (Champ. ex Benth.) Danser var. subquinquenervia P. C. Tam = Dendrotrophe varians (Blume) Miq. ●
125786 Dendrotrophe granulata (Hook. f. et Thomson ex A. DC.) A. N. Henry et B. Roy;疣枝寄生藤;Granulate Parasiticvine ●
125787 Dendrotrophe heterantha (Wall. ex DC.) A. N. Henry et B. Roy = Dendrotrophe platyphylla (Spreng.) N. H. Xia et M. G. Gilbert ●
125788 Dendrotrophe platyphylla (Spreng.) N. H. Xia et M. G. Gilbert;异花寄生藤;Diffrantflower Parasiticvine, Diversiflorous

Dendrotrophe ●

125789 Dendrotrophe polyneura (Hu) D. D. Tao;多脉寄生藤;Manyvein Dendrotrophe, Manyvein Parasiticvine, Multiveined Dendrotrophe ●

125790 Dendrotrophe punctata C. Y. Wu et D. D. Tao = Dendrotrophe varians (Blume) Miq. ●

125791 Dendrotrophe punctata C. Y. Wu ex D. D. Tao;点纹寄生藤;Punctate Dendrotrophe, Punctate Parasiticvine, Speckled Dendrotrophe ●

125792 Dendrotrophe umbellata (Blume) Miq.;伞花寄生藤;Umbel Dendrotrophe, Umbellate Dendrotrophe, Umbellate Parasiticvine ●

125793 Dendrotrophe umbellata (Blume) Miq. var. longifolia (Lecomte) P. C. Tam;长叶伞花寄生藤(长叶寄生藤);Longleaf Parasiticvine ●

125794 Dendrotrophe varians (Blume) Miq.;寄生藤(观音藤,鸡骨香,列子,青藤公,入地寄生,藤酸公,藤香,熊胆藤,左扭香);Shrubby Dendrotrophe, Shrubby Parasiticvine ●

125795 Denea O. F. Cook = Howea Becc. ●

125796 Deneckia Sch. Bip. = Denekia Thunb. ■☆

125797 Denekia Thunb. (1801);青绒草属■☆

125798 Denekia capensis Thunb.;青绒草●☆

125799 Denekia capensis Thunb. var. latifolia DC. = Denekia capensis Thunb. ●☆

125800 Denekia capensis Thunb. var. minor DC. = Denekia capensis Thunb. ●☆

125801 Denekia glabrata DC. = Denekia capensis Thunb. ●☆

125802 Denhamia Meisn. (1837)(保留属名);德纳姆卫矛属;Denhamia ●☆

125803 Denhamia Schott(废弃属名) = Culcasia P. Beauv. (保留属名)■☆

125804 Denhamia Schott(废弃属名) = Denhamia Meisn. (保留属名)●☆

125805 Denhamia heterophylla F. Muell.;异叶德纳姆卫矛●☆

125806 Denhamia leucocarpum Steud.;白果德纳姆卫矛●☆

125807 Denhamia obscura (A. Rich.) Walp.;德纳姆卫矛●☆

125808 Denhamia parvifolia L. S. Sm.;小叶德纳姆卫矛●☆

125809 Denhamia xanthosperma F. Muell.;黄籽德纳姆卫矛●☆

125810 Denira Adans. = Iva L. ■☆

125811 Denisaea Neck. = Phryma L. ■

125812 Deniseia Neck. ex Kuntze = Phryma L. ■

125813 Denisia Post et Kuntze = Denisaea Neck. ■

125814 Denisonia F. Muell. = Pityrodia R. Br. ●☆

125815 Denisophytum R. Vig. = Caesalpinia L. ●

125816 Denisophytum madagascariense R. Vig. = Caesalpinia madagascariensis (R. Vig.) Senesse ●☆

125817 Denmoza Britton et Rose (1922);火焰龙属(绯筒球属);Denmoza ●☆

125818 Denmoza erythrocephalus A. Berger;火焰龙(火焰球);Red-tip Spines Denmoza ■☆

125819 Denmoza rhodacantha Britton et Rose;茜球(茜丸)■☆

125820 Denneckia Steud. = Denekia Thunb. ■☆

125821 Dennettia Baker = Dennettia Baker f. ●☆

125822 Dennettia Baker f. (1913);丹尼木属●☆

125823 Dennettia tripetala Baker f.;丹尼木●☆

125824 Dennisonia F. Muell. = Pityrodia R. Br. ●☆

125825 Dens Fabr. = Taraxacum F. H. Wigg. (保留属名)■

125826 Dens-leonis Ség. = Leontodon L. (保留属名)■☆

125827 Denslovia Rydb. = Gymnadeniopsis Rydb. ■☆

125828 Denslovia Rydb. = Habenaria Willd. ■

125829 Denslovia clavellata (Michx.) Rydb. = Platanthera clavellata (Michx.) Luer ■☆

125830 Densophylis Thouars = Bulbophyllum Thouars(保留属名)■

125831 Dentaria L. (1753);石芥花属;Toothwort ■☆

125832 Dentaria L. = Cardamine L. ■

125833 Dentaria alaunica Golitsin = Cardamine trifida (Lam. ex Poir.) B. M. G. Jones ■

125834 Dentaria bipinnata C. A. Mey.;羽状石芥花■☆

125835 Dentaria bodinieri H. Lév. = Cardamine bodinieri (H. Lév.) Lauener ■

125836 Dentaria bulbifera L.;石芥花;Coral Root, Coralroot, Coralroot Bittercress, Coralwort, Coral-wort, Corulwort, Dog Tooth Violet, Dog-toothed Violet, Tonth Violet, Tooth Cress, Tooth Violet, Toothwort, Violet Tooth Cress ■☆

125837 Dentaria bulbifera L. = Cardamine bulbifera (L.) Crantz ■☆

125838 Dentaria concatenata Michx. = Cardamine concatenata (Michx.) O. Schwarz ■☆

125839 Dentaria concatenata Michx. var. coalescens Fernald = Cardamine concatenata (Michx.) O. Schwarz ■☆

125840 Dentaria dasyloba Turcz. = Cardamine leucantha (Tausch) O. E. Schulz ■

125841 Dentaria diphylla Michx.;双叶石芥花;Crinkleroot, Pepper-root, Toothwort, Two-leaved Toothwort ■☆

125842 Dentaria diphylla Michx. = Cardamine diphylla (Michx.) A. W. Wood ■☆

125843 Dentaria gmelinii Tausch. = Cardamine macrophylla Willd. ■

125844 Dentaria heterophylla Nutt.;互叶石芥花;Slender Toothwort ■☆

125845 Dentaria incisa Small = Cardamine diphylla (Michx.) A. W. Wood ■☆

125846 Dentaria laciniata Muhl.;截叶石芥花;Crowfoot, Cut-leaved Toothwort, Pepper-root, Toothwort ■☆

125847 Dentaria laciniata Muhl. ex Willd. = Cardamine concatenata (Michx.) O. Schwarz ■☆

125848 Dentaria laciniata Muhl. ex Willd. var. integra (O. E. Schulz) Fernald = Cardamine concatenata (Michx.) O. Schwarz ■☆

125849 Dentaria leucantha Tausch = Cardamine leucantha (Tausch) O. E. Schulz ■

125850 Dentaria macrophylla (Willd.) Bunge ex Maxim. = Cardamine macrophylla Willd. ■

125851 Dentaria macrophylla (Willd.) Bunge ex Maxim. var. dasyloba (Turcz.) Makino. = Cardamine leucantha (Tausch) O. E. Schulz ■

125852 Dentaria maxima Nutt.;大石芥花;Large Tooth Cress, Large Toothwort ■☆

125853 Dentaria maxima Nutt. = Cardamine maxima (Nutt.) A. W. Wood ■☆

125854 Dentaria microphylla Willd.;小叶石芥花■☆

125855 Dentaria multifida Muhl.;多裂石芥花;Fine-leaved Toothwort ■☆

125856 Dentaria parvifolia Raf.;小花石芥花■☆

125857 Dentaria pinnata Lam. = Cardamine heterophylla T. Y. Cheo et R. C. Fang ■

125858 Dentaria quinquefolia M. Bieb.;五叶石芥花■☆

125859 Dentaria repens Franch. = Cardamine delavayi Franch. ■

125860 Dentaria repens Franch. = Cardamine repens (Franch.) Diels ■

125861 Dentaria sibirica (O. E. Schulz) N. Busch;西伯利亚石芥花■☆

125862 Dentaria sinomanshurica Kitag. = Cardamine macrophylla

Willd. ■
125863 Dentaria tenuifolia Ledeb.;细叶石芥花■☆
125864 Dentaria tenuifolia Ledeb. = Cardamine trifida (Lam. ex Poir.) B. M. G. Jones ■
125865 Dentaria trifida Lam. ex Poir. = Cardamine trifida (Lam. ex Poir.) B. M. G. Jones ■
125866 Dentaria wallichii G. Don = Cardamine macrophylla Willd. ■
125867 Dentaria willdenowii Tausch. = Cardamine macrophylla Willd. ■
125868 Dentella J. R. Forst. et G. Forst. (1775);小牙草属;Dentella ■
125869 Dentella erecta Roth ex Roem. et Schult. = Wahlenbergia erecta (Roth ex Roem. et Schult.) Tuyn ■☆
125870 Dentella matsudai Hayata = Dentella repens (L.) J. R. Forst. et G. Forst. ■
125871 Dentella perotifolia Willd. ex Roem. et Schult. = Wahlenbergia erecta (Roth ex Roem. et Schult.) Tuyn ■☆
125872 Dentella repens (L.) J. R. Forst. et G. Forst.;小牙草;Creeping Dentella ■
125873 Dentella repens (L.) J. R. Forst. et G. Forst. var. grandis Pierre ex Pit.;长花小牙草;Longflower Creeping Dentella ■
125874 Dentella repens (L.) J. R. Forst. et G. Forst. var. grandis Pierre ex Pit. = Dentella repens (L.) J. R. Forst. et G. Forst. ■
125875 Dentidia Lour. = Perilla L. ■
125876 Dentidia nankinensis Lour. = Perilla frutescens (L.) Britton var. crispa (Benth.) W. Deane ex Bailey ■
125877 Dentidia purpurascens Pers. = Perilla frutescens (L.) Britton var. crispa (Benth.) W. Deane ex Bailey ■
125878 Dentidia purpurea Poir. = Perilla frutescens (L.) Britton var. crispa (Benth.) W. Deane ex Bailey ■
125879 Dentillaria Kuntze = Knoxia L. ■
125880 Dentimetula Tiegh. = Agelanthus Tiegh. ●☆
125881 Dentimetula Tiegh. = Tapinanthus (Blume) Rchb.(保留属名)●☆
125882 Dentimetula dodoneifolia (DC.) Tiegh. = Agelanthus dodoneifolius (DC.) Polhill et Wiens ●☆
125883 Dentoceras Small = Polygonella Michx. ■☆
125884 Dentoceras myriophylla Small = Polygonella myriophylla (Small) Horton ●☆
125885 Deonia Pierre ex Pax = Blachia Baill.(保留属名)●
125886 Depacarpus N. E. Br. = Meyerophytum Schwantes ●☆
125887 Depanthus S. Moore(1921);杯花苣苔属●☆
125888 Depanthus glaber S. Moore;杯花苣苔●☆
125889 Depanthus pubescens Guillaumin;毛杯花苣苔●☆
125890 Deplachne Boiss. = Diplachne P. Beauv. ■
125891 Deplanchea Vieill. (1862);德普紫葳属●☆
125892 Deplanchea tetraphylla (R. Br.) F. Muell.;德普紫葳●☆
125893 Deppea Cham. et Schltdl. (1830);德普茜属●☆
125894 Deppea erythrorhiza Cham. et Schltdl.;德普茜●☆
125895 Deppea floribunda Hemsl.;多花德普茜●☆
125896 Deppea longifolia Borhidi;长叶德普茜●☆
125897 Deppea longipes Standl.;长梗德普茜●☆
125898 Deppea macrocarpa Standl.;小果德普茜●☆
125899 Deppea microphylla Greenm.;小叶德普茜●☆
125900 Deppea splendens Breedlove et Lorence;纤细德普茜●☆
125901 Deppea tenuiflora Benth.;细叶德普茜●☆
125902 Deppia Raf. = Lycaste Lindl. ■☆
125903 Deprea Raf.(废弃属名) = Athenaea Sendtn.(保留属名)●☆
125904 Depremesnilia F. Muell. = Pityrodia R. Br. ●☆
125905 Depresmenilia Willis = Depremesnilia F. Muell. ●☆
125906 Derderia Jaub. et Spach = Jurinea Cass. ●■
125907 Derenbergia Schwantes = Conophytum N. E. Br. ■☆
125908 Derenbergiella Schwantes = Mesembryanthemum L.(保留属名)■●
125909 Derenbergiella luisae Schwantes = Mesembryanthemum guerichianum Pax ■☆
125910 Deresiphia Raf. = Podocaelia (Benth.) A. Fern. et R. Fern. ■☆
125911 Deringa Adans.(废弃属名) = Cryptotaenia DC.(保留属名)■
125912 Deringa canadensis (L.) Kuntze = Cryptotaenia canadensis (L.) DC. ■
125913 Deringa vulgaris (Dunn) Koso-Pol. = Pternopetalum vulgare (Dunn) Hand. -Mazz. ■
125914 Deringia Steud. = Deringa Adans.(废弃属名)■
125915 Derlinia Neraud = Gratiola L. ■
125916 Dermasea Haw. = Saxifraga L. ■
125917 Dermatobotrys Bolus(1890);革穗玄参属●☆
125918 Dermatobotrys saundersii Bolus;革穗玄参●☆
125919 Dermatocalyx Oerst. = Schlegelia Miq. ●☆
125920 Dermatophyllum Scheele = Sophora L. ●■
125921 Dermophylla Silva Manso = Cayaponia Silva Manso(保留属名)■☆
125922 Deroemera Rchb. f. = Holothrix Rich. ex Lindl.(保留属名)■☆
125923 Deroemera Rchb. f. = Peristylus Blume(保留属名)■
125924 Deroemeria Willis = Deroemera Rchb. f. ■☆
125925 Deroemeria acuminata Rendle et Schltr. = Holothrix aphylla (Forssk.) Rchb. f. ■☆
125926 Deroemeria culveri (Bolus) Schltr. = Holothrix culveri Bolus ■☆
125927 Deroemeria ledermannii (Kraenzl.) Schltr. = Holothrix aphylla (Forssk.) Rchb. f. ■☆
125928 Deroemeria montigena (Ridl.) Rolfe = Holothrix montigena Ridl. ■☆
125929 Deroemeria pentadactyla Summerh. = Holothrix pentadactyla (Summerh.) Summerh. ■☆
125930 Deroemeria praecox (Rchb. f.) Rendle et Schltr. ex Rolfe = Holothrix praecox Rchb. f. ■☆
125931 Deroemeria squamata (Hochst. ex A. Rich.) Rchb. f. = Holothrix squamata (Hochst. ex A. Rich.) Rchb. f. ■☆
125932 Deroemeria triloba Rolfe = Holothrix triloba (Rolfe) Kraenzl. ■☆
125933 Deroemeria unifolia Rchb. f. = Holothrix unifolia (Rchb. f.) Rchb. f. ■☆
125934 Derosiphia Raf. = Osbeckia L. ●■
125935 Derosiphia tubulosa (Sm.) Raf. = Osbeckia tubulosa Sm. ●☆
125936 Derouetia Boiss. et Balansa = Crepis L. ■
125937 Derris Lour. (1790)(保留属名);鱼藤属(苦楝藤属,苗栗藤属);Fishvine, Flame Tree, Jewel Vine, Jewelvine ●
125938 Derris Miq. = Derris Lour.(保留属名)●
125939 Derris Miq. = Paraderris (Miq.) R. Geesink ●
125940 Derris alborubra Hemsl.;白花鱼藤(红萼白瓣鱼藤);Whiteflower Fishvine, Whiteflower Jewelvine, White-flowered Jewelvine ●
125941 Derris amazonica Killip;亚马孙鱼藤●☆
125942 Derris bonatiana Pamp. = Pueraria peduncularis (Graham ex Benth.) Benth. ■
125943 Derris brachyptera (Benth.) Baker = Leptoderris brachyptera (Benth.) Dunn ■☆
125944 Derris breviramosa F. C. How;短枝鱼藤;Shortbranch Fishvine,

Short-branched Jewelvine, Short-shoots Jewelvine ●

125945 Derris canarensis (Dalzell) Baker = Paraderris canarensis (Dalzell) Adema ●

125946 Derris caudata Benth.;尾状鱼藤●☆

125947 Derris caudatilimba F. C. How;尾叶鱼藤;Caudata-leaflet Jewelvine, Caudata-leafleted Jewelvine, Tailleaf Fishvine ●

125948 Derris cavaleriei Gagnep.;黔桂鱼藤(贵州鱼藤,嘉氏鱼藤,黔贵鱼藤,黔鱼藤);Cavalerie Fishvine, Cavalerie Jewelvine ●

125949 Derris chinensis Benth.;华鱼藤●

125950 Derris congolensis De Wild. = Leptoderris congolensis (De Wild.) Dunn ■☆

125951 Derris cuneifolia Benth.;楔叶鱼藤(楔形叶鱼藤)●☆

125952 Derris cuneifolia Benth. var. malaccensis Benth. = Derris malaccensis (Benth.) Prain ●

125953 Derris cuneifolia Benth. var. malaccensis Benth. = Paraderris malaccensis (Benth.) Adema ●

125954 Derris dalbergioides Baker = Derris microphylla (Miq.) B. D. Jacks. ●☆

125955 Derris dinghuensis P. Y. Chen = Aganope dinghuensis (P. Y. Chen) T. C. Chen et Pedley ●

125956 Derris elegans Benth. = Derris hancei Hemsl. ●

125957 Derris elliptica (Roxb.) Benth. = Paraderris elliptica (Wall.) Adema ●

125958 Derris elliptica (Wall.) Benth.;毛鱼藤(美丽鱼藤,南亚鱼藤,鱼藤);Derris, Elliptic Fishvine, Elliptic Jewelvine, Tuba, Tuba Root, Tuba-root, Tubarroot, Tubarroot Jewelvine, Tubar-rootted Jewelvine ●

125959 Derris eriocarpa F. C. How;毛果鱼藤(鸡血藤,美丽相思子,藤子甘草,土甘草);Eriocarpous Jewelvine, Hairyfruit Hairyfruit, Hairypod Fishvine, Woolly Fruit Jewelvine ●

125960 Derris ferruginea (Roxb.) Benth.;锈毛鱼藤(老荆藤,荔枝藤,山茶藤,锈叶鱼藤);Rustyhair Fishvine, Rustyhair Jewelvine, Rusty-haired Jewelvine ●

125961 Derris ferruginea Benth. = Derris ferruginea (Roxb.) Benth. ●

125962 Derris floribunda Benth.;多花鱼藤●☆

125963 Derris fordii Oliv.;中南鱼藤(霍氏鱼藤,揭阳鱼藤,苦楝藤);Ford Fishvine, Ford Jewelvine ●

125964 Derris fordii Oliv. var. incida F. C. How;亮叶中南鱼藤(老凉藤,亮叶揭阳鱼藤);Shining Leaves Jewelvine, Shinyleaf Ford Fishvine, Shinyleaf Ford Jewelvine ●

125965 Derris giorgii De Wild. = Leptoderris hypargyrea Dunn ■☆

125966 Derris glauca Merr. et Chun = Paraderris glauca (Merr. et Chun) T. C. Chen et Pedley ●

125967 Derris goetzei Harms = Leptoderris goetzei (Harms) Dunn ■☆

125968 Derris hainanensis Hayata = Paraderris hainanensis (Hayata) Adema ●

125969 Derris hancei Hemsl. = Paraderris hancei (Hemsl.) T. C. Chen et Pedley ●

125970 Derris harrowiana (Diels) Z. Wei;大理鱼藤;Dali Fishvine, Dali Jewelvine, Harrow Jewelvine ●

125971 Derris henryi Thoth.;长果鱼藤●

125972 Derris heptaphylla Merr.;七叶鱼藤;Sevenleaf Fishvine ●

125973 Derris indica (Lam.) Bennet = Pongamia pinnata (L.) Pierre ●

125974 Derris indica (Lamkey) Bennet;印度鱼藤;Indian Derris ●☆

125975 Derris indica (Lamkey) Bennet = Pongamia pinnata (L.) Pierre ●

125976 Derris lasiopetala Hayata = Millettia pachyloba Drake ●■

125977 Derris latifolia Prain = Aganope latifolia (Prain) T. C. Chen et Pedley ●

125978 Derris laxiflora Benth.;疏花鱼藤;Laxflower Fishvine, Laxflower Jewelvine, Lax-flowered Jewelvine, Looseflowered Jewelvine ●

125979 Derris leptorhachis Harms = Millettia griffoniana Baill. ●☆

125980 Derris leucobotrya (Dunn) Roberty = Aganope leucobotrya (Dunn) Polhill ■☆

125981 Derris lianoides Elmer;菲律宾鱼藤●☆

125982 Derris lucida Welw. ex Baker = Aganope lucida (Welw. ex Baker) Polhill ■☆

125983 Derris malaccensis (Benth.) Prain;异翅鱼藤(麻六甲鱼藤,马来鱼藤,马六甲鱼藤);Malacca Fishvine ●

125984 Derris malaccensis (Benth.) Prain = Paraderris malaccensis (Benth.) Adema ●

125985 Derris malaccensis Prain = Derris malaccensis (Benth.) Prain ●

125986 Derris marginata (Roxb.) Benth.;边荚鱼藤(老荆藤,纤毛萼鱼藤);Margined Fishvine, Margined Jewelvine ●

125987 Derris microphylla (Miq.) B. D. Jacks.;小叶鱼藤;Smallleaf Jewelvine ●☆

125988 Derris microptera Benth.;小翅鱼藤●☆

125989 Derris mollis (Benth.) N. F. Mattos;柔毛鱼藤●☆

125990 Derris nobilis Welw. ex Baker = Leptoderris nobilis (Welw. ex Baker) Dunn ●☆

125991 Derris oblonga Benth. = Paraderris canarensis (Dalzell) Adema ●

125992 Derris palmifolia Chun et F. C. How;掌叶鱼藤;Palmate Jewelvine, Palmateileaved Jewelvine, Palmleaf Fishvine ●

125993 Derris pinnata Lour. = Dalbergia pinnata (Lour.) Prain ●

125994 Derris robusta (Roxb. ex DC.) Benth.;大鱼藤树(大鱼藤,竖茎鱼藤,硕大鱼藤);Robust Fishvine, Robust Jewelvine ●

125995 Derris robusta (Roxb.) Benth. = Derris robusta (Roxb. ex DC.) Benth. ●

125996 Derris rubromaculata Chun et F. C. How = Derris fordii Oliv. ●

125997 Derris scabricaulis (Franch.) Gagnep. ex F. C. How;毛枝鱼藤(粗茎鱼藤);Roughstem Fishvine, Roughstem Jewelvine, Rough-stemmed Jewelvine ●

125998 Derris scandens (Roxb.) Benth.;攀登鱼藤(澳洲鱼藤,截耳瓣鱼藤);Climbing Fishvine, Climbing Jewelvine ●

125999 Derris scandens Benth. = Derris scandens (Roxb.) Benth. ●

126000 Derris scheffleri Harms = Philenoptera eriocalyx (Harms) Schrire ●☆

126001 Derris stuhlmannii (Taub.) Harms = Xeroderris stuhlmannii (Taub.) Mendonça et E. P. Sousa ●☆

126002 Derris thyrsiflora (Benth.) Benth. = Aganope thyrsiflora (Benth.) Polhill ●

126003 Derris thyrsiflora Benth.;密花鱼藤(长小苞鱼藤,密锥花鱼藤);Dense-flowered Fishvine, Thyrse-flower Fishvine, Thyrse-flower Jewelvine ●

126004 Derris tinghuensis P. Y. Chen = Aganope dinghuensis (P. Y. Chen) T. C. Chen et Pedley ●

126005 Derris tonkinensis Gagnep.;东京鱼藤(黔桂鱼藤,越南鱼藤);Tonkin Fishvine, Tonkin Jewelvine ●

126006 Derris tonkinensis Gagnep. var. compacta Gagnep.;大叶东京鱼藤;Bigleaf Tonkin Fishvine, Bigleaf Tonkin Jewelvine ●

126007 Derris trifoliata Lour.;鱼藤(毒鱼藤,萎藤,篓藤,露藤,三叶鱼藤,水藤,台湾鱼藤);Marshy Jewelvine, Rifoliate Jewel Vine, Three-leaf Jewel-vine, Trifoliate Fishvine, Trifoliate Jewelvine ●

126008 Derris uliginosa (Roxb. ex Willd.) Benth. = Derris trifoliata

Lour. ●

126009 Derris uliginosa (Willd.) Benth. = Derris trifoliata Lour. ●

126010 Derris uliginosa Benth. = Derris trifoliata Lour. ●

126011 Derris violacea (Klotzsch) Harms = Philenoptera nelsii (Schinz) Schrire ●☆

126012 Derris yunnanensis Chun et F. C. How;云南鱼藤(云南臭藤);Yunnan Fishvine, Yunnan Jewelvine ●

126013 Derwentia Raf.(废弃属名)= Veronica L. ■

126014 Desbordesia Pierre ex Tiegh.(1905);西非黏木属●☆

126015 Desbordesia glaucescens (Engl.) Tiegh.;西非黏木●☆

126016 Desbordesia glaucescens (Engl.) Tiegh. = Desbordesia insignis Pierre ex Tiegh. ●☆

126017 Desbordesia insignis Pierre = Desbordesia glaucescens (Engl.) Tiegh. ●☆

126018 Desbordesia insignis Pierre ex Tiegh. = Desbordesia glaucescens (Engl.) Tiegh. ●☆

126019 Desbordesia oblonga (A. Chev.) A. Chev. ex Heitz = Desbordesia glaucescens (Engl.) Tiegh. ●☆

126020 Desbordesia pallida Tiegh. = Desbordesia glaucescens (Engl.) Tiegh. ●☆

126021 Desbordesia pierreana Tiegh. = Desbordesia glaucescens (Engl.) Tiegh. ●☆

126022 Desbordesia soyauxii Tiegh. = Desbordesia glaucescens (Engl.) Tiegh. ●☆

126023 Desbordesia spirei Tiegh. = Desbordesia glaucescens (Engl.) Tiegh. ●☆

126024 Descantaria Schltdl. = Tripogandra Raf. ■☆

126025 Deschampsia P. Beauv.(1812);发草属(米芒属);Bull Faces, Bull Pates, Hair Grass, Hairgrass, Hair-grass ■

126026 Deschampsia alpina (L.) Roem. et Schultz;高山发草;Alpine Hair Grass, Alpine Hair-grass ■☆

126027 Deschampsia angusta Stapf et C. E. Hubb.;狭发草■☆

126028 Deschampsia aralensis Regel = Bromus gracillimus Bunge ■

126029 Deschampsia arctica (Spreng.) Merr.;北极发草;Arctic Hairgrass ■☆

126030 Deschampsia argentea (Lowe) Lowe;银白发草■☆

126031 Deschampsia atropurpurea (Wahrenb.) Scheele;暗紫发草■☆

126032 Deschampsia atropurpurea (Wahrenb.) Scheele subsp. paramushirensis (Kudo) T. Koyama;幌筵岛发草■☆

126033 Deschampsia atropurpurea (Wahrenb.) Scheele var. paramushirensis Kudo = Deschampsia atropurpurea (Wahrenb.) Scheele subsp. paramushirensis (Kudo) T. Koyama ■☆

126034 Deschampsia beringensis Hultén;伯林发草■☆

126035 Deschampsia borealis (Trautv.) Roshev.;北方发草■☆

126036 Deschampsia bottnica (Wahl) Trin. = Deschampsia cespitosa (L.) P. Beauv. ■

126037 Deschampsia caespitosa (L.) P. Beauv. = Deschampsia cespitosa (L.) P. Beauv. ■

126038 Deschampsia caespitosa (L.) P. Beauv. subsp. koelerioides (Regel) Tzvelev = Deschampsia koelerioides Regel ■

126039 Deschampsia cespitosa (L.) P. Beauv.;发草(丛生发草,深山米芒);Bull Face, Bull Pates, Bullpoll, Bullpull, Fescueleaf Hairgrass, Hassock Grass, Rafted Hair Grass, Small-flowered Tickle Grass, Tufted Hair Grass, Tufted Hairgrass, Tufted Hair-grass, Tussac Grass, Tussock Grass, Tussock-grass ■

126040 Deschampsia cespitosa (L.) P. Beauv. subsp. genuina (Rchb.) Volk = Deschampsia cespitosa (L.) P. Beauv. ■

126041 Deschampsia cespitosa (L.) P. Beauv. subsp. glauca (Hartm.) Hartm.;灰绿发草;Tufted Hair Grass ■☆

126042 Deschampsia cespitosa (L.) P. Beauv. subsp. ivanovae (Tzvelev) S. M. Phillips et Z. L. Wu;短枝发草;Ivanov Hairgrass ■

126043 Deschampsia cespitosa (L.) P. Beauv. subsp. koelerioides (Regel) Tzvelev = Deschampsia koelerioides Regel ■

126044 Deschampsia cespitosa (L.) P. Beauv. subsp. macrothyrsa (Tatew. et Ohwi) Tzvelev = Deschampsia cespitosa (L.) P. Beauv. var. macrothyrsa Tatew. et Ohwi ■☆

126045 Deschampsia cespitosa (L.) P. Beauv. subsp. orientalis Hultén;小穗发草;Smallspike Hairgrass, Smallstachys Awnless Hairgrass ■

126046 Deschampsia cespitosa (L.) P. Beauv. subsp. orientalis Hultén = Deschampsia cespitosa (L.) P. Beauv. var. festucifolia Honda ■

126047 Deschampsia cespitosa (L.) P. Beauv. subsp. pamirica (Roshev.) Tzvelev;帕米尔发草;Pamir Hairgrass ■

126048 Deschampsia cespitosa (L.) P. Beauv. subsp. pamirica (Roshev.) Tzvelev = Deschampsia pamirica Roshev. ■

126049 Deschampsia cespitosa (L.) P. Beauv. subsp. subtriflora (Lag.) Bayer et G. López;亚三花发草■☆

126050 Deschampsia cespitosa (L.) P. Beauv. subsp. sukatschewii (Popl.) Chiapella et Prob. = Deschampsia cespitosa (L.) P. Beauv. subsp. orientalis Hultén ■

126051 Deschampsia cespitosa (L.) P. Beauv. var. abbei B. Boivin = Deschampsia cespitosa (L.) P. Beauv. ■

126052 Deschampsia cespitosa (L.) P. Beauv. var. atlantis Maire = Deschampsia cespitosa (L.) P. Beauv. subsp. subtriflora (Lag.) Bayer et G. López ■☆

126053 Deschampsia cespitosa (L.) P. Beauv. var. crassifolia (Font Quer et Maire) Maire = Deschampsia cespitosa (L.) P. Beauv. subsp. subtriflora (Lag.) Bayer et G. López ■☆

126054 Deschampsia cespitosa (L.) P. Beauv. var. exaristata Z. L. Wu;无芒发草;Awnless Hairgrass ■

126055 Deschampsia cespitosa (L.) P. Beauv. var. festucifolia Honda;细叶发草(发草)■

126056 Deschampsia cespitosa (L.) P. Beauv. var. festucifolia Honda = Deschampsia cespitosa (L.) P. Beauv. subsp. orientalis Hultén ■

126057 Deschampsia cespitosa (L.) P. Beauv. var. genuina Gren. et Godr. = Deschampsia cespitosa (L.) P. Beauv. ■

126058 Deschampsia cespitosa (L.) P. Beauv. var. glauca (Hartm.) Lindm. = Deschampsia cespitosa (L.) P. Beauv. subsp. glauca (Hartm.) Hartm. ■☆

126059 Deschampsia cespitosa (L.) P. Beauv. var. intercotidalis B. Boivin = Deschampsia cespitosa (L.) P. Beauv. ■

126060 Deschampsia cespitosa (L.) P. Beauv. var. latifolia (Hochst. ex A. Rich.) Hook. f. = Deschampsia cespitosa (L.) P. Beauv. ■

126061 Deschampsia cespitosa (L.) P. Beauv. var. levis (Takeda) Ohwi;平滑发草■☆

126062 Deschampsia cespitosa (L.) P. Beauv. var. littoralis (Gaudin) K. Richt. = Deschampsia cespitosa (L.) P. Beauv. ■

126063 Deschampsia cespitosa (L.) P. Beauv. var. longiflora Beal = Deschampsia cespitosa (L.) P. Beauv. ■

126064 Deschampsia cespitosa (L.) P. Beauv. var. macrothyrsa Tatew. et Ohwi;大穗发草■☆

126065 Deschampsia cespitosa (L.) P. Beauv. var. maritima Vasey = Deschampsia cespitosa (L.) P. Beauv. ■

126066 Deschampsia cespitosa (L.) P. Beauv. var. microstachya Roshev. = Deschampsia cespitosa (L.) P. Beauv. subsp. orientalis

Hultén ■

126067 Deschampsia cespitosa (L.) P. Beauv. var. oliveri C. E. Hubb. = Deschampsia cespitosa (L.) P. Beauv. ■

126068 Deschampsia cespitosa (L.) P. Beauv. var. parviflora (Thuill.) Coss. et Germ. = Deschampsia cespitosa (L.) P. Beauv. subsp. glauca (Hartm.) Hartm. ■☆

126069 Deschampsia crassifolia Font Quer et Maire = Deschampsia cespitosa (L.) P. Beauv. subsp. subtriflora (Lag.) Bayer et G. López ■☆

126070 Deschampsia danthonioides (Trin.) Munro ex Benth.;扁芒草状发草■☆

126071 Deschampsia discolor Roem. et Schultz;二色发草;Bicolor Hairgrass ■☆

126072 Deschampsia flexuosa (L.) Nees = Deschampsia flexuosa (L.) Trin. ■

126073 Deschampsia flexuosa (L.) Nees f. pallida Hack. ex Honda = Deschampsia flexuosa (L.) Nees ■

126074 Deschampsia flexuosa (L.) Trin.;曲芒发草(波形发草,米芒);Crinkled Hair Grass,Wavy Hair Grass,Wavy Hairgrass,Wavy Hair-grass ■

126075 Deschampsia flexuosa (L.) Trin. 'Aurea';黄曲芒发草;Golden Wavy Hair-grass ■☆

126076 Deschampsia flexuosa (L.) Trin. subsp. iberica Rivas Mart. et al.;伊比利亚曲芒发草■☆

126077 Deschampsia flexuosa (L.) Trin. var. afromontana C. E. Hubb.;非洲山生曲芒发草■☆

126078 Deschampsia flexuosa (L.) Trin. var. brachyphylla Gay;短叶曲芒发草■☆

126079 Deschampsia flexuosa (L.) Trin. var. longiseta (Pau et Font Quer) Maire = Deschampsia flexuosa (L.) Trin. ■

126080 Deschampsia flexuosa (L.) Trin. var. mairei (Sennen) Maire = Deschampsia flexuosa (L.) Trin. subsp. iberica Rivas Mart. et al. ■☆

126081 Deschampsia flexuosa (L.) Trin. var. montana (L.) Gremli = Deschampsia flexuosa (L.) Trin. ■

126082 Deschampsia foliosa Hack. var. maderensis Hack. et Bornm. = Deschampsia maderensis (Hack. et Bornm.) Buschm. ■☆

126083 Deschampsia glauca Hartm. = Deschampsia cespitosa (L.) P. Beauv. subsp. glauca (Hartm.) Hartm. ■☆

126084 Deschampsia ivanovae Tzvelev = Deschampsia cespitosa (L.) P. Beauv. subsp. ivanovae (Tzvelev) S. M. Phillips et Z. L. Wu ■

126085 Deschampsia kawakamii (Hayata) Honda = Deschampsia flexuosa (L.) Trin. ■

126086 Deschampsia koelerioides Regel;穗发草;Spike Hairgrass ■

126087 Deschampsia latifolia Hochst. ex A. Rich. = Deschampsia cespitosa (L.) P. Beauv. ■

126088 Deschampsia littoralis (Gaudin) Reut.;滨发草;Seashore Hairgrass ■

126089 Deschampsia littoralis (Gaudin) Reut. var. ivanovae (Tzvelev) P. C. Kuo et Z. L. Wu = Deschampsia cespitosa (L.) P. Beauv. subsp. ivanovae (Tzvelev) S. M. Phillips et Z. L. Wu ■

126090 Deschampsia macrothyrsa (Tatew. et Ohwi) Kawano = Deschampsia cespitosa (L.) P. Beauv. var. macrothyrsa Tatew. et Ohwi ■☆

126091 Deschampsia maderensis (Hack. et Bornm.) Buschm.;梅德发草■☆

126092 Deschampsia mairei Sennen = Deschampsia flexuosa (L.) Trin. subsp. iberica Rivas Mart. et al. ■☆

126093 Deschampsia media (Gouan) Roem. et Schult.;中间发草;Middle Hairgrass ■

126094 Deschampsia mildbraedii Pilg.;米尔德发草■☆

126095 Deschampsia multiflora P. C. Kuo et Z. L. Wu;多花发草;Flowery Hairgrass,Manyflower Hairgrass ■

126096 Deschampsia multiflora P. C. Kuo et Z. L. Wu = Deschampsia cespitosa (L.) P. Beauv. subsp. ivanovae (Tzvelev) S. M. Phillips et Z. L. Wu ■

126097 Deschampsia obensis Roshev.;鄂毕发草■☆

126098 Deschampsia orientalis (Hultén) B. S. Sun = Deschampsia cespitosa (L.) P. Beauv. subsp. orientalis Hultén ■

126099 Deschampsia pacifica Tatew. et Ohwi = Deschampsia atropurpurea (Wahrenb.) Scheele subsp. paramushirensis (Kudo) T. Koyama ■☆

126100 Deschampsia pamirica Roshev. = Deschampsia cespitosa (L.) P. Beauv. subsp. pamirica (Roshev.) Tzvelev ■

126101 Deschampsia pamirica Roshev. = Deschampsia koelerioides Regel ■

126102 Deschampsia refracta (Lag.) Roem. et Schult.;下弯发草;Refracted Hairgrass ■

126103 Deschampsia ruwensorensis Chiov. = Deschampsia flexuosa (L.) Trin. var. afromontana C. E. Hubb. ■☆

126104 Deschampsia setacea (Huds.) Hack.;刚毛发草;Bog Hair-grass,Bristle Hairgrass ■

126105 Deschampsia stricta Hack. = Deschampsia flexuosa (L.) Trin. ■

126106 Deschampsia stricta Hack. var. longiseta Pau et Font Quer = Deschampsia flexuosa (L.) Trin. ■

126107 Deschampsia sukatschewii (Popl.) Roshev. = Deschampsia cespitosa (L.) P. Beauv. var. microstachya Roshev. ■

126108 Deschampsia sukatschewii (Popl.) Roshev. = Deschampsia cespitosa (L.) P. Beauv. ■

126109 Deschampsia sukatschewii (Popl.) Roshev. = Deschampsia cespitosa (L.) P. Beauv. subsp. orientalis Hultén ■

126110 Deschampsia sukatschewii (Popl.) Roshev. subsp. orientalis (Hultén) Tzvelev. = Deschampsia cespitosa (L.) P. Beauv. subsp. orientalis Hultén ■

126111 Deschampsia takedana Honda = Deschampsia cespitosa (L.) P. Beauv. var. levis (Takeda) Ohwi ■☆

126112 Descurainia Webb et Berthel. (1836)(保留属名);播娘蒿属;Fixweed,Tansy Mustard,Tansymustard,Tansy-mustard ■

126113 Descurainia artemisioides Svent.;蒿状播娘蒿■☆

126114 Descurainia bourgeauana (E. Fourn.) O. E. Schulz;布尔播娘蒿■☆

126115 Descurainia brachycarpa (Richardson) O. E. Schulz = Descurainia pinnata (Walter) Britton subsp. brachycarpa (Rich.) Detling ■☆

126116 Descurainia gilva Svent.;淡黄褐播娘蒿■☆

126117 Descurainia irio (L.) Webb et Berthel. = Sisymbrium irio L. ■

126118 Descurainia kochii (Petri) O. E. Schulz;考氏播娘蒿■☆

126119 Descurainia millefolia (Jacq.) Webb et Berthel.;粟草叶播娘蒿■☆

126120 Descurainia pinnata (Walter) Britton;羽状播娘蒿;Green Tansy Mustard,Pinnate Tansy Mustard,Tansy Mustard,Western Tansy Mustard ■☆

126121 Descurainia pinnata (Walter) Britton subsp. brachycarpa (Rich.) Detling;短果羽状播娘蒿;Green Tansy Mustard,Pinnate Tansy Mustard,Western Tansy Mustard ■☆

126122 Descurainia pinnata (Walter) Britton var. brachycarpa (Richardson) Fernald = Descurainia pinnata (Walter) Britton ■☆

126123 Descurainia pinnata (Walter) Britton var. brachycarpa (Richardson) Fernald = Descurainia pinnata (Walter) Britton subsp. brachycarpa (Rich.) Detling ■☆

126124 Descurainia preauxiana (Webb) O. E. Schulz;普雷播娘蒿■☆

126125 Descurainia preauxiana (Webb) O. E. Schulz var. briquetii (Pit.) O. E. Schulz = Descurainia preauxiana (Webb) O. E. Schulz ■☆

126126 Descurainia sophia (L.) Prantl var. densiflorum Lange = Descurainia sophia (L.) Webb ex Prantl ■

126127 Descurainia sophia (L.) Webb ex Prantl;播娘蒿(大室,大适,葶蒿,丁历,华东葶苈子,麦里蒿,眉毛蒿);Fine-leaved Hedge Mustard, Flixweed, Flix-weed, Flixweed Tansy Mustard, Flixweed Tansy-mustard, Flixwort Flixweed, Fluxweed, Hedge Mustard, Herb Sophia, Herb-sophia, Sophia Sisymbrium, Sophia Tansymustard, Tansy Mustard, Wisdom of Surgeons ■

126128 Descurainia sophia (L.) Webb ex Prantl = Sisymbrium sophia L. ■

126129 Descurainia sophia (L.) Webb ex Prantl var. glabrata N. Busch = Descurainia sophia (L.) Webb ex Prantl ■

126130 Descurainia sophia (L.) Webb ex Schur = Descurainia sophia (L.) Webb ex Prantl ■

126131 Descurainia sophioides (Fisch. ex Hook.) O. E. Schulz;腺毛播娘蒿;Glandhair Fixweed, Sophia-like Tansymustard ■

126132 Desdemona S. Moore = Basistemon Turcz. ●☆

126133 Desfontaena Vell. = Chiropetalum A. Juss. ●☆

126134 Desfontaina Steud. = Chiropetalum A. Juss. ●☆

126135 Desfontaina Steud. = Desfontaena Vell. ●☆

126136 Desfontainea Kunth = Desfontainia Ruiz et Pav. ●☆

126137 Desfontainea Rchb. = Chiropetalum A. Juss. ●☆

126138 Desfontainea Rchb. = Desfontaena Vell. ●☆

126139 Desfontainesia Hoffmanns. = Fontanesia Labill. ●

126140 Desfontainia Ruiz et Pav. (1794);虎刺叶属(德思凤属,迪氏木属,枸骨叶属,美冬青属)●☆

126141 Desfontainia spinosa Ruiz et Pav.;虎刺叶(丛生德思凤,德思凤,多刺迪氏木,美冬青)●☆

126142 Desfontainiaceae Endl. (1873)(保留科名);虎刺叶科(迪氏木科,枸骨叶科,离水花科,美冬青科)●☆

126143 Desfontainiaceae Endl. (保留科名) = Dialypetalanthaceae Rizzini et Occhioni(保留科名)●☆

126144 Desfontainiaceae Endl. (保留科名) = Potaliaceae C. Mart. ●☆

126145 Desforgia Steud. = Deffоrgia Lam. ●☆

126146 Desforgia Steud. = Forgesia Comm. ex Juss. ●☆

126147 Desideria Pamp. (1926);扇叶芥属(合萼芥属)■

126148 Desideria Pamp. = Christolea Cambess. ■

126149 Desideria baiogoinensis (K. C. Kuan et C. H. An) Al-Shehbaz;藏北扇叶芥(藏北高原芥);N. Xizang Plateaucress, North Tibet Christolea ■

126150 Desideria baiogoinensis (K. C. Kuan et C. H. An) Al-Shehbaz = Christolea baiogoinensis K. C. Kuan et C. H. An ■

126151 Desideria flabellata (Regel) Al-Shehbaz;长毛扇叶芥(长毛高原芥);Longhair Christolea, Longhair Plateaucress ■

126152 Desideria flabellata (Regel) Al-Shehbaz = Christolea flabellata (Regel) N. Busch ■

126153 Desideria haranensis Al-Shehbaz = Solms-laubachia haranensis (Al-Shehbaz) J. P. Yue, Al-Shehbaz et H. Sun ■☆

126154 Desideria himalayensis (Cambess.) Al-Shehbaz;须弥扇叶芥(喜马拉雅高原芥);Himalayan Christolea, Himalayas Plateaucress ■

126155 Desideria himalayensis (Cambess.) Al-Shehbaz = Christolea himalayensis (Cambess.) Jafri ■

126156 Desideria linearis (N. Busch) Al-Shehbaz;线果扇叶芥■

126157 Desideria mirabilis Pamp.;扇叶芥■

126158 Desideria mirabilis Pamp. = Christolea mirabilis (Pamp.) Jafri ■

126159 Desideria pamirica Suslova = Desideria mirabilis Pamp. ■

126160 Desideria prolifera (Maxim.) Al-Shehbaz;丛生扇叶芥(丛生高原芥);Tufted Christolea, Tufted Plateaucress ■

126161 Desideria prolifera (Maxim.) Al-Shehbaz = Christolea prolifera (Maxim.) Jafri ■

126162 Desideria pumila (Kurz) Al-Shehbaz;矮扇叶芥(矮高原芥);Dwarf Christolea, Dwarf Plateaucress ■

126163 Desideria pumila (Kurz) Al-Shehbaz = Christolea pumila (Kurz) Jafri ■

126164 Desideria stewartii (T. Anderson) Al-Shehbaz;少花扇叶芥(少花高原芥);Fewflower Christolea, Fewflower Plateaucress ■

126165 Desideria stewartii (T. Anderson) Al-Shehbaz = Christolea stewartii (T. Anderson) Jafri ■

126166 Desmanthodium Benth. (1872);索果菊属■●☆

126167 Desmanthodium ovatum Benth.;索果菊■●☆

126168 Desmanthus Willd. (1806)(保留属名);合欢草属(草合欢属);Bundleflower, Prairie Mimosa ●■

126169 Desmanthus arborescens Bojer ex Benth. = Dichrostachys arborescens (Bojer ex Benth.) Villiers ●☆

126170 Desmanthus bernierianus (Baill.) Drake = Dichrostachys bernieriana Baill. ●☆

126171 Desmanthus brachylobus Benth.;美洲合欢草■☆

126172 Desmanthus brachypus (Baill.) Baill. ex Drake = Mimosa psoralea (DC.) Benth. ●☆

126173 Desmanthus campenonii Drake = Dichrostachys tenuifolia Benth. ●☆

126174 Desmanthus commersonianus Baill. = Gagnebina commersoniana (Baill.) R. Vig. ●☆

126175 Desmanthus greveanus Drake = Dichrostachys tenuifolia Benth. ●☆

126176 Desmanthus illinoensis (Michx.) MacMill. ex B. L. Rob. et Fernald;伊利诺合欢草;Illinois Bundleflower, Illinois Bundle-flower, Prairie Bundle-flower, Prairie Desmanthus, Prairie Mimosa, Prairie-mimosa, Prickle Weed ■☆

126177 Desmanthus leptolobus Torr. et A. Gray;细荚合欢草;Bundle Flower ■☆

126178 Desmanthus leptostachys DC. = Dichrostachys cinerea (L.) Wight et Arn. var. africana Brenan et Brummitt ●☆

126179 Desmanthus nutans (Pers.) DC. = Dichrostachys cinerea (L.) Wight et Arn. var. africana Brenan et Brummitt ●☆

126180 Desmanthus paucifoliolatus Scott-Elliot = Dichrostachys paucifoliolata (Scott-Elliot) Drake ●☆

126181 Desmanthus pernambucanus (L.) Thell.;合欢草●

126182 Desmanthus pervilleanus Baill. = Dichrostachys pervilleana (Baill.) Drake ●☆

126183 Desmanthus scottianus Drake = Dichrostachys scottiana (Drake) Villiers ●☆

126184 Desmanthus stolonifer DC. = Neptunia oleracea Lour. ■☆

126185 Desmanthus tenuifolius (Benth.) Drake = Dichrostachys tenuifolia Benth. ●☆

126186 Desmanthus trichostachys DC. = Dichrostachys cinerea (L.) Wight et Arn. var. africana Brenan et Brummitt ●☆

126187 Desmanthus virgatus (L.) Willd.;多枝合欢草(草合欢,多枝草合欢);Rayado Bundle Bundleflower, Rayado Bundleflower ●■

126188 Desmaria Tiegh. (1895);链寄生属●☆

126189 Desmaria Tiegh. = Loranthus Jacq.(保留属名)●

126190 Desmaria mutabilis Tiegh.;链寄生●☆

126191 Desmazeria Dumort. (1822)('Demazeria');纽禾属■☆

126192 Desmazeria acutiflora (Nees) T. Durand et Schinz = Tribolium acutiflorum (Nees) Renvoize ■☆

126193 Desmazeria composita Hack. = Tribolium obtusifolium (Nees) Renvoize ■☆

126194 Desmazeria compressa Ovcz. et Shibkova;纽禾■☆

126195 Desmazeria loliacea (Huds.) Nyman;黑麦纽禾■☆

126196 Desmazeria loliacea (Huds.) Nyman = Catapodium marinum (L.) C. E. Hubb. ■☆

126197 Desmazeria loliacea (Huds.) Nyman subsp. syrtica Barratte et Murb. = Catapodium marinum (L.) C. E. Hubb. subsp. syrticum (Murb.) H. Scholz ■☆

126198 Desmazeria loliacea (Huds.) Nyman var. syrtica (Barratte et Murb.) Durand et Barratte = Catapodium marinum (L.) C. E. Hubb. subsp. syrticum (Murb.) H. Scholz ■☆

126199 Desmazeria marina (L.) Druce = Catapodium marinum (L.) C. E. Hubb. ■☆

126200 Desmazeria marina (L.) Druce = Desmazeria loliacea (Huds.) Nyman ■☆

126201 Desmazeria rigida (L.) Tutin;硬纽禾;Fern Grass ■☆

126202 Desmazeria rigida (L.) Tutin = Catapodium rigidum (L.) C. E. Hubb. ex Dony ■☆

126203 Desmazeria rigida (L.) Tutin subsp. hemipoa (Spreng.) Stace = Catapodium hemipoa (Spreng.) Lainz ■☆

126204 Desmazeria sicula (Jacq.) Dumort.;西西里纽禾■☆

126205 Desmazeria tuberculosa (Moris) Batt. et Trab. = Castellia tuberculosa (Moris) Bor ■☆

126206 Desmazeria tuberculosa (Moris) Trab. = Castellia tuberculosa (Moris) Bor ■☆

126207 Desmazeria unioloides Defleurs = Halopyrum mucronatum (L.) Stapf ■☆

126208 Desmesia Raf. = Typhonium Schott + Sauromatum Schott ■

126209 Desmia D. Don = Erica L. ●☆

126210 Desmia aequalis D. Don = Erica polifolia Salisb. ex Benth. ●☆

126211 Desmidochus Rchb. = Desmidorchis Ehrenb. ■☆

126212 Desmidorchis Ehrenb. = Boucerosia Wight et Arn. ■☆

126213 Desmidorchis Ehrenb. = Caralluma R. Br. ■

126214 Desmidorchis acutangula Decne. = Caralluma acutangula (Decne.) N. E. Br. ■☆

126215 Desmidorchis foetidus (E. A. Bruce) Plowes = Caralluma foetida E. A. Bruce ■☆

126216 Desmidorchis penicellatus (Deflers) Plowes = Caralluma penicillata (Deflers) N. E. Br. ■☆

126217 Desmidorchis somalicus (N. E. Br.) Plowes = Caralluma somalica N. E. Br. ■☆

126218 Desmidorchis speciosus (N. E. Br.) Plowes = Caralluma speciosa (N. E. Br.) N. E. Br. ■☆

126219 Desmiograstis Börner = Carex L. ■

126220 Desmitus Raf. = Camellia L. ●

126221 Desmitus reticulata (Lindl.) Raf. = Camellia reticulata Lindl. ●◇

126222 Desmocarpus Wall. = Cadaba Forssk. ●☆

126223 Desmocephalum Hook. f. = Elvira Cass. ■☆

126224 Desmochaeta DC. = Pupalia Juss.(保留属名)■☆

126225 Desmochaeta achyranthoides Kunth = Cyathula achyranthoides (Kunth) Moq. ■☆

126226 Desmochaeta alternifolia (L.) DC. = Digera muricata (L.) Mart. ■☆

126227 Desmochaeta atropurpurea (Lam.) DC. = Pupalia lappacea (L.) A. Juss. ■☆

126228 Desmochaeta atropurpurea DC. = Pupalia lappacea (L.) A. Juss. ■☆

126229 Desmochaeta densiflora Kunth = Cyathula achyranthoides (Kunth) Moq. ■☆

126230 Desmochaeta distorta Hiern = Cyathula cylindrica Moq. ■☆

126231 Desmochaeta flavescens DC. = Pupalia lappacea (L.) A. Juss. ■☆

126232 Desmochaeta prostrata (L.) DC. = Cyathula prostrata (L.) Blume ■

126233 Desmochaeta uncinata Roem. et Schult. = Cyathula achyranthoides (Kunth) Moq. ■☆

126234 Desmochaeta uncinulata Hiern = Cyathula uncinulata (Schrad.) Schinz ■☆

126235 Desmocladus Nees = Loxocarya R. Br. ■☆

126236 Desmocladus Nees(1846);链枝帚灯草属■☆

126237 Desmocladus brunonianus Nees;链枝帚灯草■☆

126238 Desmodiastrum (Prain) A. Pramanik et Thoth. = Alysicarpus Desv.(保留属名)■

126239 Desmodiastrum parviflorum (Dalzell) H. Ohashi = Alysicarpus parviflorus Dalzell ■☆

126240 Desmodiocassia Britton et Rose = Cassia L.(保留属名)●■

126241 Desmodiocassia Britton et Rose = Senna Mill. ●■

126242 Desmodium Desv. (1813)(保留属名);山蚂蝗属(山绿豆属,山马蝗属);Mountain Leech, Tick Clover, Tick Trefoil, Tickclover, Tick-clover, Tick-trefoil ●■

126243 Desmodium acrocarpum Hance = Tadehagi triquetra (L.) H. Ohashi ●

126244 Desmodium acuminatum (Michx.) DC. = Desmodium glutinosum (Muhl. ex Willd.) A. W. Wood ●☆

126245 Desmodium acuminatum (Michx.) DC. f. chandonnetii (Lunell) Fassett = Desmodium glutinosum (Muhl. ex Willd.) A. W. Wood ●☆

126246 Desmodium adscendens (Sw.) DC. = Desmodium griffithianum Benth. ●■

126247 Desmodium adscendens (Sw.) DC. var. robustum B. G. Schub.;粗壮山蚂蝗●☆

126248 Desmodium affine Schltdl.;近缘山蚂蝗●☆

126249 Desmodium akoense Hayata = Desmodium scorpiurus (Sw.) Desv. ex DC. ●■

126250 Desmodium amethystinum Dunn;紫水晶山蚂蝗(光果山蚂蝗);Denselutinous Tickclover ●

126251 Desmodium angulatum DC. = Desmodium multiflorum DC. ●

126252 Desmodium aparine Chiov. = Desmodium repandum (Vahl) DC. ●

126253 Desmodium argenteum Wall. = Desmodium elegans (Lour.) Desv. ●

126254 Desmodium asperum (Poir.) Desr.;粗糙山蚂蝗●☆

126255 Desmodium austrojaponense Ohwi = Desmodium laxum DC. ■

126256 Desmodium barbatum (L.) Benth. = Desmodium barbatum (L.) Benth. et Oerst. ●☆

126257 Desmodium barbatum (L.) Benth. et Oerst.;髯毛山蚂蝗●☆

126258 Desmodium barbatum (L.) Benth. subsp. dimorphum (Welw. ex Baker) J. R. Laundon = Desmodium barbatum (L.) Benth. var. dimorphum (Welw. ex Baker) B. G. Schub. ●☆

126259 Desmodium barbatum (L.) Benth. var. argyreum (Welw. ex Baker) B. G. Schub.;银色山蚂蝗●☆

126260 Desmodium barbatum (L.) Benth. var. dimorphum (Welw. ex Baker) B. G. Schub.;二型山蚂蝗●☆

126261 Desmodium barbatum (L.) Benth. var. procumbens B. G. Schub.;平铺髯毛山蚂蝗●☆

126262 Desmodium barbigerum H. Lév. = Desmodium concinnum DC. ●

126263 Desmodium biarticulatum (L.) F. Muell. = Aphyllodium biarticulatum (L.) Gagnep. ●

126264 Desmodium biarticulatum (L.) F. Muell. = Dicerma biarticulatum (L.) DC. ●

126265 Desmodium blandum Meeuwen = Phyllodium elegans (Lour.) Desv. ●

126266 Desmodium bodinieri H. Lév. = Hylodesmum podocarpum (DC.) H. Ohashi et R. R. Mill ■

126267 Desmodium boivinianum Baill. = Desmodium hirtum Guillaumin et Perr. ●☆

126268 Desmodium bonatianum Pamp. = Uraria sinensis (Hemsl.) Franch. ●

126269 Desmodium bracteosum (Michx.) DC. = Desmodium cuspidatum (Muhl. ex Willd.) DC. ex Loudon ●☆

126270 Desmodium bracteosum (Michx.) DC. var. longifolium (Torr. et A. Gray) B. L. Rob. = Desmodium cuspidatum (Muhl. ex Willd.) DC. ex Loudon var. longifolium (Torr. et A. Gray) B. G. Schub. ●☆

126271 Desmodium buergeri Miq. = Desmodium heterocarpon (L.) DC. ●

126272 Desmodium buergeri Miq. = Desmodium heterophyllum (Willd.) DC. ●■

126273 Desmodium caffrum (E. Mey.) Druce = Desmodium dregeanum Benth. ●☆

126274 Desmodium caffrum (E. Mey.) Druce var. schlechteri Schindl. = Desmodium dregeanum Benth. ●☆

126275 Desmodium caffrum Eckl. et Zeyh. = Desmodium repandum (Vahl) DC. ●

126276 Desmodium callianthum Franch.;美花山蚂蝗;Beautifulflower Tickclover, Beautiful-flowered Tickclover, Tropical Ticktrefoil ●

126277 Desmodium callianthum Franch. = Desmodium elegans DC. subsp. callianthum (Franch.) H. Ohashi ●

126278 Desmodium canadense (L.) DC.;加拿大山蚂蝗(加排钱草);Canada Tick Clover, Canada Tickclover, Canadian Tick-trefoil, Giant Tick Clover, Showy Tick Trefoil, Showy Tick-trefoil, Tick-trefoil ●☆

126279 Desmodium canadense (L.) DC. var. longifolium Torr. et A. Gray = Desmodium cuspidatum (Muhl. ex Willd.) DC. ex Loudon var. longifolium (Torr. et A. Gray) B. G. Schub. ●☆

126280 Desmodium canescens (L.) DC.;灰白山蚂蝗(北美舞草);Hoary Tick Clover, Hoary Tick-trefoil ●☆

126281 Desmodium canum (J. F. Gmel.) Schinz et Thell. = Desmodium incanum (G. Mey.) DC. ●☆

126282 Desmodium capitatum (Burm. f.) A. DC.;头花山蚂蝗●

126283 Desmodium capitatum (Burm. f.) A. DC. = Desmodium stenophyllum Pamp. ●

126284 Desmodium capitatum (Burm. f.) A. DC. = Desmodium styracifolium (Osbeck) Merr. ●■

126285 Desmodium carlesii Schindl. = Desmodium rubrum (Lour.) DC. ●

126286 Desmodium caudatum (Thunb.) DC. = Ohwia caudata (Thunb.) H. Ohashi ●

126287 Desmodium cavalerieri H. Lév. = Desmodium gangeticum (L.) DC. ●

126288 Desmodium cephalotes (Roxb.) Wall. ex Wight et Arn. = Desmodium triangulare (Retz.) Merr. ●

126289 Desmodium cephalotes (Roxb.) Wall. ex Wight et Arn. var. congestum Prain = Desmodium triangulare (Retz.) Merr. ●

126290 Desmodium cephalotes Wall.;大头山蚂蝗(假木豆)●☆

126291 Desmodium ciliare (Muhl. ex Willd.) DC.;毛小叶山蚂蝗;Hairy Small-leaved Tick-trefoil, Slender Tick Clover ●☆

126292 Desmodium ciliatum (Thunb.) DC. = Rhynchosia puberula (Eckl. et Zeyh.) Steud. ●☆

126293 Desmodium cinerascens Franch. = Desmodium elegans DC. ●

126294 Desmodium cinerascens Franch. var. longipes Pamp. = Desmodium elegans DC. ●

126295 Desmodium cinerascens Franch. var. microphylla Franch. = Desmodium elegans DC. ●

126296 Desmodium concinnum DC.;凹叶山蚂蝗(美丽山蚂蝗);Neat Mountain Leech, Showy Tickclover ●

126297 Desmodium congestum Benth. = Leptodesmia congesta Benth. ex Baker f. ●☆

126298 Desmodium congestum Wall. ex Wight et Arn. = Desmodium triangulare (Retz.) Merr. ●

126299 Desmodium cordifoliolatum P. C. Li = Desmodium flexuosum Wall. ex Benth. var. cordifoliolatum (P. C. Li) P. H. Huang ●

126300 Desmodium cordifolium (Harms) Schindl.;心叶山蚂蝗●☆

126301 Desmodium cuspidatum (Muhl. ex Willd.) DC. ex Loudon;大苞山蚂蝗;Bracted Tick-trefoil, Large-bracted Tick-trefoil, Longleaf Tick Clover ●☆

126302 Desmodium cuspidatum (Muhl. ex Willd.) DC. ex Loudon var. longifolium (Torr. et A. Gray) B. G. Schub.;长叶大苞山蚂蝗;Hairy Bracted Tick-trefoil, Large-bracted Tick-trefoil ●☆

126303 Desmodium dasylobum Miq. = Desmodium sequax Wall. ●

126304 Desmodium delicatulum A. Rich. = Desmodium hirtum Guillaumin et Perr. f. delicatulum ? ●☆

126305 Desmodium densum (C. Chen et X. J. Cui) H. Ohashi;菱叶山蚂蝗●

126306 Desmodium densum (C. Chen et X. J. Cui) H. Ohashi = Hylodesmum densum (C. Chen et X. J. Cui) H. Ohashi et R. R. Mill ■

126307 Desmodium dichotomum (Willd.) DC.;二歧山蚂蝗(山蚂蝗,五棱茎山蚂蝗);Bifork Tickclover, Dichotomous Tickclover, Fiveangular Desmodium ●

126308 Desmodium diffusum (Roxb.) DC. = Desmodium unibotryosum C. Chen et X. J. Cui ●

126309 Desmodium diffusum (Willd.) DC. = Desmodium dichotomum (Willd.) DC. ●

126310 Desmodium difusum DC. = Desmodium laxiflorum DC. ●

126311 Desmodium dillenii Darl. = Desmodium paniculatum (L.) DC. var. dillenii (Darl.) Isely ●☆

126312 Desmodium dimorphum Welw. ex Baker = Desmodium barbatum (L.) Benth. var. dimorphum (Welw. ex Baker) B. G. Schub. ●☆

126313 Desmodium dimorphum Welw. ex Baker var. argyreum ? = Desmodium barbatum (L.) Benth. var. argyreum (Welw. ex Baker) B. G. Schub. ●☆

126314 Desmodium discolor Vogel;二色山蚂蝗●☆

126315 Desmodium dispermum Hayata = Dendrolobium dispermum (Hayata) Schindl. ●

126316 Desmodium distortum (Aubl.) J. F. Macbr.;旋扭山蚂蝗●☆

126317 Desmodium diversifolium (Poir.) DC. = Desmodium salicifolium (Poir.) DC. ●☆

126318 Desmodium djalonense A. Chev. = Desmodium linearifolium G. Don ●☆

126319 Desmodium dolabriforme Benth.;斧状山蚂蝗●☆

126320 Desmodium dregeanum Benth.;德雷山蚂蝗●☆

126321 Desmodium dubium Lindl. = Desmodium multiflorum DC. ●

126322 Desmodium duclouxii Pamp. = Hylodesmum longipes (Franch.) H. Ohashi et R. R. Mill ■

126323 Desmodium duclouxii Pamp. = Podocarpium duclouxii (Pamp.) Yen C. Yang et P. H. Huang ■

126324 Desmodium duclouxii Pamp. var. henryi (Schindl.) H. Ohashi = Hylodesmum longipes (Franch.) H. Ohashi et R. R. Mill ■

126325 Desmodium duclouxii Pamp. var. henryi (Schindl.) H. Ohashi = Podocarpium duclouxii (Pamp.) Yen C. Yang et P. H. Huang ●

126326 Desmodium dunnii Merr. = Dendrolobium lanceolatum (Dunn) Schindl. ●

126327 Desmodium elegans (Lour.) Benth. = Phyllodium elegans (Lour.) Desv. ●

126328 Desmodium elegans (Lour.) Desv. = Phyllodium elegans (Lour.) Desv. ●

126329 Desmodium elegans DC.;雅致山蚂蝗(叠钱草,黄皮条,灰毛山蚂蝗,鳞狸蕨,毛排钱草,棉筋,棉筋山蚂蝗,排钱草,麒麟片,山蚂蝗,圆锥山蚂蝗,粘人草,总状花山蚂蝗);Elegant Mountain Leech,Elegant Tickclover,Esquirol Tickclover,Spiffyflower Mountain Leech ●

126330 Desmodium elegans DC. f. albiflorum (P. C. Li) H. Ohashi = Desmodium elegans DC. ●

126331 Desmodium elegans DC. subsp. callianthum (Franch.) H. Ohashi = Desmodium callianthum Franch. ●

126332 Desmodium elegans DC. subsp. stenophyllum (Pamp.) H. Ohashi = Desmodium stenophyllum Pamp. ●

126333 Desmodium elegans DC. subsp. wolohoense (Schindl.) H. Ohashi = Desmodium elegans DC. var. wolohoense (Schindl.) H. Ohashi ●

126334 Desmodium elegans DC. var. albiflorum P. C. Li;白花山蚂蝗;White-flower Elegant Tickclover ●

126335 Desmodium elegans DC. var. albiflorum P. C. Li = Desmodium elegans DC. ●

126336 Desmodium elegans DC. var. argenteum (Wall. ex Benth.) H. Ohashi = Desmodium elegans DC. ●

126337 Desmodium elegans DC. var. callianthum (Franch.) P. C. Li = Desmodium callianthum Franch. ●

126338 Desmodium elegans DC. var. handelii (Schindl.) H. Ohashi;盐源山蚂蝗;Yanyuan Tickclover ●

126339 Desmodium elegans DC. var. nutans (Hook.) H. Ohashi;下垂山蚂蝗;Drooping Tickclover ●☆

126340 Desmodium elegans DC. var. wolohoense (Schindl.) H. Ohashi;川南山蚂蝗●

126341 Desmodium esquirolii H. Lév. = Desmodium elegans DC. ●

126342 Desmodium fallax Schindl. = Desmodium podocarpum DC. subsp. fallax (Schindl.) H. Ohashi ●■

126343 Desmodium fallax Schindl. = Hylodesmum podocarpum (DC.) H. Ohashi et R. R. Mill subsp. fallax (C. K. Schindl.) H. Ohashi et R. R. Mill ■

126344 Desmodium fallax Schindl. = Podocarpium fallax (C. K. Schneid.) C. Chen et X. J. Cui ●■

126345 Desmodium fallax Schindl. var. mandschuricum (Maxim.) Nakai = Hylodesmum podocarpum (DC.) H. Ohashi et R. R. Mill subsp. oxyphyllum (DC.) H. Ohashi et R. R. Mill ■

126346 Desmodium fallax Schindl. var. mandshuricum (Maxim.) Nakai = Desmodium podocarpum DC. subsp. oxyphyllum (DC.) H. Ohashi var. mandshuricum Maxim. ●■

126347 Desmodium fallax Schindl. var. mandshuricum (Maxim.) Nakai = Podocarpium podocarpum (DC.) Yen C. Yang et P. H. Huang var. mandshuricum (Maxim.) P. H. Huang ●■

126348 Desmodium flexuosum Wall. ex Benth.;曲枝山蚂蝗;Flexuous Tickclover ●☆

126349 Desmodium flexuosum Wall. ex Benth. var. cordifoliolatum (P. C. Li) P. H. Huang;心叶曲枝山蚂蝗(心叶山蚂蝗);Heart-leaf Tickclover ●

126350 Desmodium floribundum (D. Don) Sweet ex G. Don = Desmodium multiflorum DC. ●

126351 Desmodium floribundum (G. Don) Sweet = Desmodium multiflorum DC. ●

126352 Desmodium floribundum G. Don = Desmodium multiflorum DC. ●

126353 Desmodium formosanum Hayata = Christia campanulata (Benth.) Thoth. ●

126354 Desmodium formosum Vogel = Lespedeza formosa (Vogel) Koehne ●

126355 Desmodium formosum Vogel = Lespedeza thunbergii (DC.) Nakai subsp. formosa (Vogel) H. Ohashi ●

126356 Desmodium forrestii Schindl. = Desmodium elegans DC. ●

126357 Desmodium franchetii Rehder = Desmodium elegans DC. ●

126358 Desmodium fulvescens B. G. Schub.;黄褐山蚂蝗●☆

126359 Desmodium gangeticum (L.) DC.;大叶山蚂蝗(大叶山绿豆,钩毛荚山蚂蝗,恒河山绿豆,红毛鸡草,红母鸡草,粘草,粘人草);Bigleaf Mountain Leech,Bigleaved Desmodium,Hookedhairypod Tickclover,Hook-hairy-podded Tickclover,Large-leaves Tickclover ●

126360 Desmodium gangeticum (L.) DC. var. maculatum (L.) Baker = Desmodium gangeticum (L.) DC. ●

126361 Desmodium gardneri Benth. = Hylodesmum leptopus (A. Gray ex Benth.) H. Ohashi et R. R. Mill ●

126362 Desmodium gardneri Benth. = Podocarpium leptopum (A. Gray ex Benth.) Yen C. Yang et P. H. Huang ●■

126363 Desmodium glabellum (Michx.) DC.;高大山蚂蝗;Dillenius' Tick-trefoil, Panicled Tick-trefoil, Perplexed Tick-trefoil, Tall Tick Clover ●☆

126364 Desmodium glabellum (Michx.) DC. = Desmodium paniculatum (L.) DC. var. dillenii (Darl.) Isely ●☆

126365 Desmodium glaucophyllum Pamp. = Desmodium elegans DC. ●

126366 Desmodium glutinosum (Muhl. ex Willd.) A. W. Wood;密叶山蚂蝗;Cluster-leaf Tick-trefoil, Large-flowered Tick Clover, Pointed Tick-trefoil,Pointed-leaved Tick-trefoil,Sticky Tick Clover ●☆

126367 Desmodium glutinosum (Muhl. ex Willd.) A. W. Wood f. chandonnetii (Lunell) B. G. Schub. = Desmodium glutinosum (Muhl. ex Willd.) A. W. Wood ●☆

126368 Desmodium gracillimum Hemsl.;细叶山蚂蝗(三角叶山蚂蝗);Slenderleaf Mountain Leech,Slenderleaf Tickclover ●

126369 Desmodium grande E. Mey. = Desmodium salicifolium (Poir.) DC. ●☆

126370 Desmodium grande Kurz = Phyllodium kurzianum (Kuntze) H. Ohashi ●

126371 Desmodium grandiflorum DC.;大花山蚂蝗;Large-flowered Tick-trefoil ●☆

126372 Desmodium grandiflorum DC. = Desmodium cuspidatum (Muhl. ex Willd.) DC. ex Loudon ●☆

126373 Desmodium griffithianum Benth.;疏果山蚂蝗(疏果假地豆,外折荚山蚂蝗);Deflexedpod Tickclover,Laxfruit Tickclover ●■

126374 Desmodium griffithianum Benth. var. leiocarpum X. F. Gao et C. Chen;无毛疏果假地豆;Hairless Laxfruit Tickclover ■

126375 Desmodium griffithianum Benth. var. leiocarpum X. F. Gao et C. Chen = Desmodium griffithianum Benth. ●■

126376 Desmodium griffithianum Benth. var. stigosum Meeuwen;粗毛疏果山蚂蝗●■

126377 Desmodium gyrans (L. f.) DC. = Codariocalyx motorius (Houtt.) H. Ohashi ●

126378 Desmodium gyrans (L. f.) DC. = Desmodium motorium (Houtt.) Merr. ●

126379 Desmodium gyrans (L. f.) DC. var. roylei (Wight et Arn.) Baker = Codariocalyx motorius (Houtt.) H. Ohashi ●

126380 Desmodium gyroides (Roxb. ex Link) DC. = Codariocalyx gyroides (Roxb. ex Link) Hassk. ●

126381 Desmodium gyroides (Roxb.) DC. = Codariocalyx gyroides (Roxb. ex Link) Hassk. ●

126382 Desmodium hainanense Isely = Hylodesmum lateral (Schindl.) H. Ohashi et R. R. Mill ■

126383 Desmodium hainanensis Isely = Podocarpium laxum (DC.) Yen C. Yang et P. H. Huang var. laterale (Schindl.) Yen C. Yang et P. H. Huang ■

126384 Desmodium hamulatum Franch. = Desmodium sequax Wall. ●

126385 Desmodium handelii Schindl. = Desmodium elegans DC. var. handelii (Schindl.) H. Ohashi ●

126386 Desmodium helenae Buscal. et Muschl.;海伦娜山蚂蝗●☆

126387 Desmodium henryi Schindl. = Hylodesmum longipes (Franch.) H. Ohashi et R. R. Mill ■

126388 Desmodium henryi Schindl. = Podocarpium duclouxii (Pamp.) Yen C. Yang et P. H. Huang ■

126389 Desmodium heterocarpon (L.) DC.;假地豆(柏氏小槐花,稗豆,大叶青,狗尾花,假花生,木假地豆,山道根,山花生,通乳草,铜钱射,细叶假花生,小槐花,野花生,异果山绿豆,异果山蚂蝗,异叶山绿豆,中蝶草);Asian Ticktrefoil,Differentfruit Tickclover,Heterocarpous Mountain Leech,Heterocarpous Tickclover ●

126390 Desmodium heterocarpon (L.) DC. subsp. angustifolium (Craib) Ohashi;窄叶假地豆;Narrowleaf Differentfruit Tickclover,Narrowleaf Heterocarpus Mountain Leech ●

126391 Desmodium heterocarpon (L.) DC. subsp. angustifolium (Craib) Ohashi = Desmodium reticulatum Champ. ex Benth. ●

126392 Desmodium heterocarpon (L.) DC. subsp. angustifolium H. Ohashi = Desmodium reticulatum Champ. ex Benth. ●

126393 Desmodium heterocarpon (L.) DC. var. buergeri (Miq.) Hosok. = Desmodium heterocarpon (L.) DC. ●

126394 Desmodium heterocarpon (L.) DC. var. patule-pilosum (Ohwi) Ohwi = Desmodium heterocarpon (L.) DC. ●

126395 Desmodium heterocarpon (L.) DC. var. strigosum Meeuwen;糙毛假地豆(付毛假地豆,直毛假地豆);Asian Ticktrefoil,Strigose Differentfruit Tickclover,Strigose Heterocarpus Mountain Leech ●

126396 Desmodium heterophyllum (Willd.) DC.;异叶山蚂蝗(变叶山蚂蝗,假地豆,田胡蜘蛛,铁线草,异叶山绿豆);Different-leaf Tickclover, Heterophyllous Mountain Leech, Heterophyllous Tickclover,Variable-leaf Ticktrefoil,Varieble-leaved Tickclover ●■

126397 Desmodium hirtum Guillaumin et Perr.;毛叶山蚂蝗●☆

126398 Desmodium hirtum Guillaumin et Perr. f. compressum B. G. Schub.;扁山蚂蝗●☆

126399 Desmodium hirtum Guillaumin et Perr. f. delicatulum ?;姣美山蚂蝗●☆

126400 Desmodium hispidum Franch.;粗硬毛山蚂蝗●

126401 Desmodium homblei De Wild. = Desmodium cordifolium (Harms) Schindl. ●☆

126402 Desmodium horridum Steenis = Trifidacanthus unifoliolatus Merr. ●

126403 Desmodium huillensis (Welw. ex Hiern) K. Schum. = Droogmansia megalantha (Taub.) De Wild. var. pilosa ? ■☆

126404 Desmodium huillensis Welw. = Droogmansia megalantha (Taub.) De Wild. var. pilosa ? ■☆

126405 Desmodium humblotianum Baill. = Desmodium hirtum Guillaumin et Perr. ●☆

126406 Desmodium humifusum (Muhl. ex Willd.) Beck;俯卧山蚂蝗●☆

126407 Desmodium illinoense A. Gray;伊利诺山蚂蝗;Illinois Tick Clover,Illinois Tick-trefoil,Prairie Tick-trefoil ●☆

126408 Desmodium incanum (G. Mey.) DC.;灰毛山蚂蝗(灰色山蚂蝗)●☆

126409 Desmodium incanum (Sw.) DC. = Desmodium incanum (G. Mey.) DC. ●☆

126410 Desmodium intortum (DC.) Urb. = Desmodium intortum (Mill.) Urb. ■

126411 Desmodium intortum (Mill.) Urb.;扭曲山蚂蝗(西班牙三叶草,西班牙山蚂蝗);Spanish Clover ■

126412 Desmodium intortum (Mill.) Urb. var. pilosiusculum (DC.) Fosberg;毛叶扭曲山蚂蝗■☆

126413 Desmodium karensium Kurz = Desmodium megaphyllum Zoll. et Moritzi ●

126414 Desmodium kerstenii O. Hoffm. = Pseudarthria hookeri Wight et Arn. ●☆

126415 Desmodium kurzii Craib = Phyllodium kurzianum (Kuntze) H. Ohashi ●

126416 Desmodium laburnifolium (Poir.) DC. = Desmodium caudatum (Thunb.) DC. ●

126417 Desmodium laevigatum (Nutt.) DC.;平滑山蚂蝗;Smooth Tick Clover ●☆

126418 Desmodium lasiocarpum (P. Beauv.) DC. = Desmodium velutinum (Willd.) DC. ●

126419 Desmodium laterale Schindl. = Hylodesmum laterale (C. K. Schindl.) H. Ohashi et R. R. Mill ■

126420 Desmodium laterale Schindl. = Podocarpium laxum (DC.) Yen C. Yang et P. H. Huang var. laterale (Schindl.) Yen C. Yang et P. H. Huang ■

126421 Desmodium latifolium (Ker Gawl.) DC. = Desmodium velutinum (Willd.) DC. ●

126422 Desmodium latifolium (Ker Gawl.) DC. var. plukenetii Wight et Arn. = Desmodium velutinum (Willd.) DC. ●

126423 Desmodium latifolium (Ker Gawl.) DC. var. virgatum Miq. = Desmodium velutinum (Willd.) DC. ●

126424 Desmodium laxiflorum DC.;大叶拿身草(疏花山蚂蝗,羊带归);Bigleaf Tickclover, Laxflower Tickclover, Lax-flowered Tickclover ●

126425 Desmodium laxiflorum DC. subsp. parvifolium H. Ohashi et T. T. Chen = Desmodium diffusum (Roxb.) DC. ●

126426 Desmodium laxiflorum DC. var. formosense H. Ohwi = Desmodium diffusum (Roxb.) DC. ●

126427 Desmodium laxum DC. = Hylodesmum laxum (DC.) H. Ohashi et R. R. Mill ■

126428 Desmodium laxum DC. = Podocarpium laxum (DC.) Yen C. Yang et P. H. Huang ■

126429 Desmodium laxum DC. subsp. falfolium H. Ohashi = Hylodesmum laxum (DC.) H. Ohashi et R. R. Mill subsp. falfolium (H. Ohashi) H. Ohashi et R. R. Mill ■

126430 Desmodium laxum DC. subsp. laterale (Schindl.) H. Ohashi = Desmodium laterale Schindl. ■

126431 Desmodium laxum DC. subsp. laterale (Schindl.) H. Ohashi = Hylodesmum laterale (C. K. Schindl.) H. Ohashi et R. R. Mill ■

126432 Desmodium laxum DC. subsp. lateraxum H. Ohashi = Hylodesmum laxum (DC.) H. Ohashi et R. R. Mill subsp. lateraxum (H. Ohashi) H. Ohashi et R. R. Mill ■

126433 Desmodium laxum DC. subsp. lateraxum H. Ohashi = Hylodesmum podocarpum (DC.) H. Ohashi et R. R. Mill subsp. szechuenense (Craib) H. Ohashi et R. R. Mill ■

126434 Desmodium laxum DC. subsp. leptopum (A. Gray ex Benth.) H. Ohashi = Desmodium leptopus A. Gray ex Benth. ●■

126435 Desmodium laxum DC. subsp. leptopum (A. Gray ex Benth.) H. Ohashi = Podocarpium leptopum (A. Gray ex Benth.) Yen C. Yang et P. H. Huang ●■

126436 Desmodium laxum DC. subsp. leptopus (A. Gray ex Benth.) H. Ohashi = Hylodesmum leptopus (A. Gray ex Benth.) H. Ohashi et R. R. Mill ●

126437 Desmodium laxum DC. var. laterale (Schindl.) Ohashi = Podocarpium laxum (DC.) Yen C. Yang et P. H. Huang var. laterale (Schindl.) Yen C. Yang et P. H. Huang ■

126438 Desmodium leiocarpum (Spreng.) G. Don;光果山蚂蝗●☆

126439 Desmodium leptopus A. Gray ex Benth. = Hylodesmum leptopus (A. Gray ex Benth.) H. Ohashi et R. R. Mill ●

126440 Desmodium leptopus A. Gray ex Benth. = Podocarpium leptopum (A. Gray ex Benth.) Yen C. Yang et P. H. Huang ●■

126441 Desmodium lespedezioides Benth. = Leptodesmia congesta Benth. ex Baker f. ●☆

126442 Desmodium linearifolium G. Don;线叶山蚂蝗●☆

126443 Desmodium lobatum Schindl.;浅裂山蚂蝗●☆

126444 Desmodium longibracteatum Schindl. = Desmodium velutinum (Willd.) DC. subsp. longibracteatum (Schindl.) H. Ohashi ●

126445 Desmodium longibracteatum Schindl. = Desmodium velutinum (Willd.) DC. var. longibracteatum (Schindl.) Meeuwen ●

126446 Desmodium longifolium (Torr. et A. Gray) Smyth = Desmodium cuspidatum (Muhl. ex Willd.) DC. ex Loudon var. longifolium (Torr. et A. Gray) B. G. Schub. ●☆

126447 Desmodium longipes Craib = Phyllodium longipes (Craib) Schindl. ●

126448 Desmodium luteolum H. Ohashi et T. Nemoto = Ohwia luteola (H. Ohashi et T. Nemoto) H. Ohashi ●

126449 Desmodium lutescens A. DC. = Pycnospora lutescens (Poiret) Schindl. ■●

126450 Desmodium macrophyllum Desv. = Desmodium laxiflorum DC. ●

126451 Desmodium mairei Pamp. = Desmodium multiflorum DC. ●

126452 Desmodium mandschuricum (Maxim.) Schindl. = Hylodesmum podocarpum (DC.) H. Ohashi et R. R. Mill subsp. oxyphyllum (DC.) H. Ohashi et R. R. Mill ■

126453 Desmodium mandshuricum (Maxim.) Schindl.;东北山蚂蝗●■

126454 Desmodium mandshuricum (Maxim.) Schindl. = Desmodium podocarpum DC. subsp. oxyphyllum (DC.) H. Ohashi var. mandshuricum Maxim. ●■

126455 Desmodium mandshuricum (Maxim.) Schindl. = Podocarpium podocarpum (DC.) Yen C. Yang et P. H. Huang var. mandshuricum (Maxim.) P. H. Huang ●■

126456 Desmodium mandshuricum Nakai = Podocarpium podocarpum (DC.) Yen C. Yang et P. H. Huang var. fallax (C. K. Schneid.) Yen C. Yang et P. H. Huang ●■

126457 Desmodium manschuricum (Maxim.) Nakai = Desmodium mandshuricum (Maxim.) Schindl. ●■

126458 Desmodium marilandicum (L.) DC.;马里兰德山蚂蝗;Maryland Tick Clover ●☆

126459 Desmodium mauritianum (Willd.) DC. = Desmodium ramosissimum G. Don ●☆

126460 Desmodium megalanthum Taub. = Droogmansia megalantha (Taub.) De Wild. ■☆

126461 Desmodium megalanthum Taub. var. pilosum ? = Droogmansia megalantha (Taub.) De Wild. var. pilosa? ●☆

126462 Desmodium megaphyllum Zoll. et Moritzi;滇南山蚂蝗;Largeleaf Tickclover, Large-leaved Tickclover, Megaphyllous Tickclover, S. Yunnan Mountain Leech ●

126463 Desmodium megaphyllum Zoll. et Moritzi var. glabrescens Prain;无毛滇南山蚂蝗;Glabrous Largeleaf Tickclover ●

126464 Desmodium menglaense (C. Chen et X. J. Cui) H. Ohashi = Hylodesmum menglaense (H. Ohashi) H. Ohashi et R. R. Mill ■

126465 Desmodium microphyllum (Thunb.) DC.;小叶三点金(八字草,斑鸠鼻,斑鸠窝,辫子草,大叶关门草,地盘茶,红梗草,红夜关门,路路星,马龙通,马尾鞭,马尾草,马尾藤,爬地香,漆大伯,散风散,狮子草,碎米柴,太阳草,天小豆,细鞭打,细叶兰,消毒草,消黄散,逍遥草,小木通,小叶三点金草,小叶山菜豆,小叶山绿豆,小叶山蚂蝗,小叶岩黄耆,哮灵草,追风散);Small-leaf Desmodium, Smallleaf Mountain Leech ●■

126466 Desmodium microphyllum (Thunb.) DC. = Codariocalyx microphyllus (Thunb.) H. Ohashi ●☆

126467 Desmodium microphyllum (Thunb.) DC. var. longipilum Ohwi = Desmodium microphyllum (Thunb.) DC. ●■

126468 Desmodium molliculum (Kunth) DC.;柔软山蚂蝗●☆

126469 Desmodium monospermum Baker = Leptodesmia congesta Benth. ex Baker f. ●☆

126470 Desmodium motorium (Houtt.) Merr. = Codariocalyx motorius (Houtt.) H. Ohashi ●

126471 Desmodium multiflorum DC.;饿蚂蝗(大红袍,多花山蚂蝗,红掌草,烂玉树,山豆根,山蚂蝗,胃痛草,野黄豆,粘身草,紫藤小槐花);Hungry Tickclover, Manyflower Tickclover, Multiflorous Tickclover ●

126472 Desmodium nantouensis Y. C. Liu et F. Y. Lu = Desmodium intortum (Mill.) Urb. ■

126473 Desmodium natalitum Sond. = Desmodium gangeticum (L.)

DC. ●

126474 Desmodium neomexicanum A. Gray;新墨西哥山蚂蝗;New Mexico Tick-trefoil ●☆

126475 Desmodium nepalense Ohashi = Desmodium multiflorum DC. ●

126476 Desmodium nervosum Vogel = Desmodium heterocarpon (L.) DC. var. strigosum Meeuwen ●

126477 Desmodium nudiflorum (L.) DC.;裸花山蚂蝗;Bare-stemmed Tick-trefoil, Naked Tick-trefoil, Naked-flowered Tick-trefoil, Naked-stemmed Tick Clover ●☆

126478 Desmodium nudiflorum (L.) DC. f. foliolatum (Farw.) Fassett = Desmodium nudiflorum (L.) DC. ●☆

126479 Desmodium nudiflorum (L.) DC. f. personatum Fassett = Desmodium nudiflorum (L.) DC. ●☆

126480 Desmodium nutans Wall. = Desmodium elegans DC. ●

126481 Desmodium nutans Wall. ex Hook. = Desmodium elegans DC. ●

126482 Desmodium nuttallii (Schindl.) B. G. Schub.;纳托尔山蚂蝗;Tick Clover ●☆

126483 Desmodium oblatum Baker;圆节山蚂蝗●☆

126484 Desmodium oblatum Baker ex Kurz = Desmodium renifolium (L.) Schindl. ●

126485 Desmodium oblongum Wall. ex Benth.;矩叶山蚂蝗(长圆叶山蚂蝗,断节果,锥序山蚂蝗);Oblongleaf Mountain Leech, Paniculate Tickclover ●

126486 Desmodium obtusum (Muhl. ex Willd.) DC.;硬山蚂蝗;Tick Clover ●☆

126487 Desmodium ochroleucum M. A. Curtis;淡黄山蚂蝗;Tick Clover ●☆

126488 Desmodium oldhamii Oliv. = Hylodesmum oldhamii (Oliv.) H. Ohashi et R. R. Mill ●■

126489 Desmodium oldhamii Oliv. = Podocarpium oldhamii (Oliv.) Yen C. Yang et P. H. Huang ●■

126490 Desmodium oojeinse (Roxb.) Ohashi;奥京山蚂蝗●☆

126491 Desmodium ovalifolium (Schumach.) Walp. = Alysicarpus ovalifolius (K. Schum.) J. Léonard ■

126492 Desmodium ovalifolium Guillaumin et Perr. = Desmodium adscendens (Sw.) DC. ●■

126493 Desmodium oxalidifolium H. Lév. = Codariocalyx gyroides (Roxb. ex Link) Hassk. ●

126494 Desmodium oxalidifolium H. Lév. = Desmodium griffithianum Benth. ●■

126495 Desmodium oxybracteum DC. = Desmodium salicifolium (Poir.) DC. ●☆

126496 Desmodium oxyphyllum A. DC. = Hylodesmum podocarpum (DC.) H. Ohashi et R. R. Mill subsp. oxyphyllum (DC.) H. Ohashi et R. R. Mill ■

126497 Desmodium oxyphyllum DC. = Desmodium podocarpum DC. subsp. oxyphyllum (DC.) H. Ohashi ●■

126498 Desmodium oxyphyllum DC. = Podocarpium podocarpum (DC.) Yen C. Yang et P. H. Huang var. oxyphyllum (DC.) Yen C. Yang et P. H. Huang ●■

126499 Desmodium oxyphyllum DC. var. japonicum Matsum. = Podocarpium podocarpum (DC.) Yen C. Yang et P. H. Huang var. japonicum (Matsum.) P. H. Huang ■

126500 Desmodium oxyphyllum DC. var. mandschuricum (Maxim.) H. Ohashi = Hylodesmum podocarpum (DC.) H. Ohashi et R. R. Mill subsp. oxyphyllum (DC.) H. Ohashi et R. R. Mill ■

126501 Desmodium oxyphyllum DC. var. mandshuricum (Maxim.) H. Ohashi;东北长柄山蚂蝗;Northeastern Podocarpium ●■

126502 Desmodium oxyphyllum DC. var. mandshuricum (Maxim.) H. Ohashi = Desmodium podocarpum DC. subsp. oxyphyllum (DC.) H. Ohashi var. mandshuricum Maxim. ●■

126503 Desmodium oxyphyllum DC. var. mandshuricum (Maxim.) H. Ohashi = Podocarpium podocarpum (DC.) Yen C. Yang et P. H. Huang var. fallax (Schindl.) Yen C. Yang et P. H. Huang ●■

126504 Desmodium oxyphyllum DC. var. szechuenense (Craib) H. Ohashi = Hylodesmum podocarpum (DC.) H. Ohashi et R. R. Mill subsp. szechuenense (Craib) H. Ohashi et R. R. Mill ■

126505 Desmodium oxyphyllum DC. var. szechuenense (Craib) H. Ohashi = Podocarpium podocarpum (DC.) Yen C. Yang et P. H. Huang var. szechuenense (Craib) Yen C. Yang et P. H. Huang ■

126506 Desmodium paleaceum Guillaumin et Perr. = Desmodium salicifolium (Poir.) DC. ●☆

126507 Desmodium paniculatum (L.) DC.;圆锥山蚂蝗;Panicled Tick-trefoil, Tall Tick Clover ●☆

126508 Desmodium paniculatum (L.) DC. var. dillenii (Darl.) Isely;迪勒山蚂蝗;Dillenius' Tick-trefoil, Dillens Tick-trefoil, Panicled Tick-trefoil, Perplexed Tick-trefoil ●☆

126509 Desmodium paniculatum (L.) DC. var. dillenii (Darl.) Isely = Desmodium glabellum (Michx.) DC. ●☆

126510 Desmodium paroifolium DC. = Desmodium microphyllum (Thunb.) DC. ●■

126511 Desmodium parvifolium (Spreng.) DC. = Desmodium microphyllum (Thunb.) DC. ●■

126512 Desmodium parvifolium DC. = Desmodium microphyllum (Thunb.) DC. ●■

126513 Desmodium parvifolium DC. f. yunnanense Pamp. = Desmodium microphyllum (Thunb.) DC. ●■

126514 Desmodium pauciflorum (Nutt.) DC.;疏花山蚂蝗;Few-flowered Tick Clover, Small-flowered Tick Clover ●☆

126515 Desmodium pendenticarpum C. Z. Gao et Q. R. Lai = Desmodium strigillosum Schindl. var. pendenticarpum (C. Z. Gao et Q. R. Lai) P. H. Huang ●

126516 Desmodium pendulum Wall. = Desmodium concinnum DC. ●

126517 Desmodium perplexum B. G. Schub.;紊乱山蚂蝗;Tall Tick Clover ■☆

126518 Desmodium perplexum B. G. Schub. = Desmodium paniculatum (L.) DC. var. dillenii (Darl.) Isely ●☆

126519 Desmodium pilosiusculum DC. = Desmodium intortum (Mill.) Urb. var. pilosiusculum (DC.) Fosberg ■☆

126520 Desmodium pilosum (Thunb.) DC. = Lespedeza pilosa (Thunb.) Siebold et Zucc. ●■

126521 Desmodium plukenettii (Wight et Arn.) Merr. = Desmodium velutinum (Willd.) DC. ●

126522 Desmodium plukenettii (Wight et Arn.) Merr. et Chun = Desmodium velutinum (Willd.) DC. ●

126523 Desmodium podocarpum DC. = Hylodesmum podocarpum (DC.) H. Ohashi et R. R. Mill ■

126524 Desmodium podocarpum DC. = Podocarpium podocarpum (DC.) Yen C. Yang et P. H. Huang ●■

126525 Desmodium podocarpum DC. subsp. fallax (Schindl.) H. Ohashi = Hylodesmum podocarpum (DC.) H. Ohashi et R. R. Mill subsp. fallax (C. K. Schindl.) H. Ohashi et R. R. Mill ■

126526 Desmodium podocarpum DC. subsp. fallax (Schindl.) H. Ohashi = Podocarpium podocarpum (DC.) Yen C. Yang et P. H.

Huang var. fallax (Schindl.) Yen C. Yang et P. H. Huang ●■

126527 Desmodium podocarpum DC. subsp. fallax (Schindl.) H. Ohashi f. album (Sugim.) H. Ohashi;白花宽卵叶长柄山蚂蝗●■☆

126528 Desmodium podocarpum DC. subsp. oxyphyllum (DC.) H. Ohashi f. albiflorum (Iwata) H. Ohashi;白花圆菱叶山蚂蝗●■☆

126529 Desmodium podocarpum DC. subsp. oxyphyllum (DC.) H. Ohashi f. decorum Iwata;装饰山蚂蝗●☆

126530 Desmodium podocarpum DC. subsp. oxyphyllum (DC.) H. Ohashi var. japonicum (Miq.) Maxim. = Podocarpium podocarpum (DC.) Yen C. Yang et P. H. Huang var. japonicum (Matsum.) P. H. Huang ■

126531 Desmodium podocarpum DC. subsp. oxyphyllum (DC.) H. Ohashi var. mandschuricum Maxim. = Hylodesmum podocarpum (DC.) H. Ohashi et R. R. Mill subsp. oxyphyllum (DC.) H. Ohashi et R. R. Mill ■

126532 Desmodium podocarpum DC. subsp. oxyphyllum (DC.) H. Ohashi var. mandshuricum Maxim. f. leucanthum Sugim.;白花东北长柄山蚂蝗●■☆

126533 Desmodium podocarpum DC. subsp. oxyphyllum (DC.) H. Ohashi var. mandshuricum Maxim. = Podocarpium podocarpum (DC.) Yen C. Yang et P. H. Huang var. mandshuricum (Maxim.) P. H. Huang ●■

126534 Desmodium podocarpum DC. subsp. oxyphyllum (DC.) Yen C. Yang et P. H. Huang = Hylodesmum podocarpum (DC.) H. Ohashi et R. R. Mill subsp. oxyphyllum (DC.) H. Ohashi et R. R. Mill ■

126535 Desmodium podocarpum DC. subsp. szechuenense (Craib) H. Ohashi = Hylodesmum podocarpum (DC.) H. Ohashi et R. R. Mill subsp. szechuenense (Craib) H. Ohashi et R. R. Mill ■

126536 Desmodium podocarpum DC. var. japonicum Matsum. = Hylodesmum podocarpum (DC.) H. Ohashi et R. R. Mill subsp. oxyphyllum (DC.) H. Ohashi et R. R. Mill ■

126537 Desmodium podocarpum DC. var. laxum (A. DC.) Baker = Hylodesmum laxum (DC.) H. Ohashi et R. R. Mill ■

126538 Desmodium podocarpum DC. var. szechuenense Craib = Hylodesmum podocarpum (DC.) H. Ohashi et R. R. Mill subsp. szechuenense (Craib) H. Ohashi et R. R. Mill ■

126539 Desmodium polycarpon (Poiret) DC. = Desmodium heterocarpon (L.) DC. var. strigosum Meeuwen ●

126540 Desmodium polycarpum DC. = Desmodium griffithianum Benth. var. stigosum Meeuwen ●■

126541 Desmodium polycarpum DC. f. hirsutum Pamp. = Desmodium heterocarpon (L.) DC. var. strigosum Meeuwen ●

126542 Desmodium polycarpum DC. var. angustifolium Carib = Desmodium reticulatum Champ. ex Benth. ●

126543 Desmodium polycarpum DC. var. trichocaulon Baker = Desmodium heterocarpon (L.) DC. var. strigosum Meeuwen ●

126544 Desmodium praestans Forrest = Desmodium yunnanense Franch. ●

126545 Desmodium prainii Schindl. = Desmodium megaphyllum Zoll. et Moritzi ●

126546 Desmodium prainii Schindl. var. glabrescens (Prain) Schindl. = Desmodium megaphyllum Zoll. et Moritzi var. glabrescens Prain ●

126547 Desmodium procumbens (Mill.) Hitchc.;平铺山蚂蝗●☆

126548 Desmodium pseudotriquetrum DC. = Tadehagi pseudotriquetra (DC.) H. Ohashi ●

126549 Desmodium pulchellum (L.) Benth. = Phyllodium pulchellum (L.) Desv. ●

126550 Desmodium pulchellum Blume ex Miq. = Phyllodium pulchellum (L.) Desv. ●

126551 Desmodium purpureum (Gueldenst.) Fawc. et Rendle;紫花山蚂蝗;Purple-flowered Tickclover ●

126552 Desmodium purpureum (Gueldenst.) Fawc. et Rendle = Desmodium tortuosum (Sw.) DC. ●■

126553 Desmodium purpureum (Mill.) Fawc. et Rehdle = Desmodium tortuosum (Sw.) DC. ●■

126554 Desmodium purpureum Fawc. et Rendle = Desmodium tortuosum (Sw.) DC. ●■

126555 Desmodium racemosum (Thunb.) DC. = Desmodium podocarpum DC. ●■

126556 Desmodium racemosum (Thunb.) DC. = Hylodesmum podocarpum (DC.) H. Ohashi et R. R. Mill subsp. oxyphyllum (DC.) H. Ohashi et R. R. Mill ■

126557 Desmodium racemosum (Thunb.) DC. var. dilatatum (Nakai) Ohwi = Desmodium podocarpum DC. subsp. fallax (Schindl.) H. Ohashi ●■

126558 Desmodium racemosum (Thunb.) DC. var. dilatatum (Nakai) Ohwi = Podocarpium podocarpum (DC.) Yen C. Yang et P. H. Huang var. fallax (Schindl.) Yen C. Yang et P. H. Huang ●■

126559 Desmodium racemosum (Thunb.) DC. var. mandshuricum (Maxim.) Ohwi = Desmodium podocarpum DC. subsp. oxyphyllum (DC.) H. Ohashi var. mandshuricum Maxim. ●■

126560 Desmodium racemosum (Thunb.) DC. var. mandshuricum (Maxim.) Ohwi = Podocarpium podocarpum (DC.) Yen C. Yang et P. H. Huang var. fallax (Schindl.) Yen C. Yang et P. H. Huang ●■

126561 Desmodium racemosum (Thunb.) DC. var. mandshuricum (Maxim.) Ohwi = Podocarpium podocarpum (DC.) Yen C. Yang et P. H. Huang var. mandshuricum (Maxim.) P. H. Huang ●■

126562 Desmodium racemosum (Thunb.) DC. var. pubescens F. P. Metcalf = Podocarpium podocarpum (DC.) Yen C. Yang et P. H. Huang var. oxyphyllum (DC.) Yen C. Yang et P. H. Huang ●■

126563 Desmodium racemosum (Thunb.) DC. var. villosum (Matsum.) Ohwi = Desmodium podocarpum DC. ●■

126564 Desmodium racemosum (Thunb.) DC. var. villosum (Matsum.) Ohwi = Podocarpium podocarpum (DC.) Yen C. Yang et P. H. Huang ●■

126565 Desmodium racemosum DC.;小山蚂蝗●■

126566 Desmodium racemosum DC. = Desmodium podocarpum DC. subsp. oxyphyllum (DC.) H. Ohashi ●■

126567 Desmodium racemosum DC. = Podocarpium podocarpum (DC.) Yen C. Yang et P. H. Huang var. oxyphyllum (DC.) Yen C. Yang et P. H. Huang ●■

126568 Desmodium racemosum DC. var. mandschuricum (Maxim.) Ohwi = Hylodesmum podocarpum (DC.) H. Ohashi et R. R. Mill subsp. oxyphyllum (DC.) H. Ohashi et R. R. Mill ■

126569 Desmodium racemosum DC. var. pubescens Metcalf = Hylodesmum podocarpum (DC.) H. Ohashi et R. R. Mill subsp. oxyphyllum (DC.) H. Ohashi et R. R. Mill ■

126570 Desmodium radiatum Baker = Leptodesmia congesta Benth. ex Baker f. ●☆

126571 Desmodium ramosissimum G. Don;多分枝山蚂蝗●☆

126572 Desmodium recurvatum (Roxb.) Graham ex Wight et Arn. = Desmodium laxiflorum DC. ●

126573 Desmodium remotum (Poir.) Drake = Desmodium repandum (Vahl) DC. ●

126574 Desmodium renifolium (L.) Schindl.;肾叶山蚂蝗;Kidneyleaf Mountain Leech,Kidneyleaf Tickclover ●

126575 Desmodium renifolium (L.) Schindl. var. oblatum (Baker ex Kurz) H. Ohashi;长梗肾叶山蚂蝗;Longpedicelled Kidneyleaf Mountain Leech,Longpedicelled Kidneyleaf Tickclover ●

126576 Desmodium renifolium (L.) Schindl. var. oblatum (Baker ex Kurz) H. Ohashi = Desmodium renifolium (L.) Schindl. ●

126577 Desmodium reniforme (L.) DC. = Desmodium renifolium (L.) Schindl. ●

126578 Desmodium repandum (Vahl) A. DC. = Hylodesmum repandum (Vahl) H. Ohashi et R. R. Mill ●

126579 Desmodium repandum (Vahl) DC. = Podocarpium repandum (Vahl) Yen C. Yang et P. H. Huang ●

126580 Desmodium reticulatum Champ. ex Benth.;显脉山绿豆;Netvein Wild Mung Bean,Reticulate Tickclover ●

126581 Desmodium retroflexum (L.) DC. = Desmodium styracifolium (Osbeck) Merr. ●■

126582 Desmodium rhabdocladum Franch. = Desmodium elegans DC. ●

126583 Desmodium rigidum (Elliott) DC. = Desmodium obtusum (Muhl. ex Willd.) DC. ●☆

126584 Desmodium rockii Schindl.;白绒毛山蚂蝗;Rock Tickclover ●

126585 Desmodium rockii Schindl. = Desmodium yunnanense Franch. ●

126586 Desmodium rotundifolium Wall.;圆叶山蚂蝗;Dollarleaf,Low Tick Trefoil,Prostrate Tick-trefoil ●☆

126587 Desmodium rotundifolium Wall. = Desmodium styracifolium (Osbeck) Merr. ●■

126588 Desmodium rubrum (Lour.) DC.;赤山蚂蝗(赤叶山绿豆,单叶假地豆,飞扬草,假地豆,绢毛山蚂蝗);Red Mountain Leech,Silky Tickclover ●

126589 Desmodium rufescens DC. = Uraria rufescens (DC.) Schindl. ●

126590 Desmodium rufihirsutum Craib = Desmodium velutinum (Willd.) DC. var. longibracteatum (Schindl.) Meeuwen ●

126591 Desmodium rugosum Prain = Dendrolobium rugosum (Prain) Schindl. ●

126592 Desmodium salicifolium (Poir.) DC.;柳叶山蚂蝗●☆

126593 Desmodium salicifolium (Poir.) DC. = Phyllodium kurzianum (Kuntze) H. Ohashi ●

126594 Desmodium salicifolium (Poir.) DC. var. densiflorum B. G. Schub.;密花柳叶山蚂蝗●☆

126595 Desmodium sambuense (D. Don) DC. = Desmodium multiflorum DC. ●

126596 Desmodium sandwicense E. Mey.;夏威夷山蚂蝗;Chili Clover,Hawaii Ticktrefoil,Spanish Clover ●☆

126597 Desmodium scalpe DC. = Desmodium repandum (Vahl) DC. ●

126598 Desmodium scalpe DC. = Hylodesmum repandum (Vahl) H. Ohashi et R. R. Mill ●

126599 Desmodium schweinfurthii Schindl.;施韦山蚂蝗●☆

126600 Desmodium scorpiurus (Sw.) Desv. = Desmodium scorpiurus (Sw.) Desv. ex DC. ●■

126601 Desmodium scorpiurus (Sw.) Desv. ex DC.;蝎尾山蚂蝗(阿猴舞草,虾尾山蚂蝗);Scorpionstail Desmodium,Scorpionstail Mountain Leech ●■

126602 Desmodium sennaarense Schweinf. = Desmodium dichotomum (Willd.) DC. ●

126603 Desmodium sequax Wall.;长波叶山蚂蝗(波叶山蚂蝗,波状叶山马蝗,长波状叶山蚂蝗,饿蚂蝗,过路黄,黄粘粑草,黄粘毛草,菱叶山绿豆,牛巴嘴,山蚂蝗,山毛豆花,瓦子草,乌山黄檀草,野豆子,粘人花);Quickly Spreading Tickclover,Sinuate Mountain Leech,Sinuate Tickclover ●

126604 Desmodium sequax Wall. var. sinuatum (Miq.) Hosok. = Desmodium sequax Wall. ●

126605 Desmodium sessilifolium (Torr.) Torr. et A. Gray;无柄山蚂蝗;Tick Clover ●☆

126606 Desmodium setigerum (E. Mey.) Benth. ex Harv.;刚毛山蚂蝗●☆

126607 Desmodium shimadai Hayata = Desmodium zonatum Miq. ●

126608 Desmodium siliquosum (Burm. f.) DC. = Desmodium heterocarpon (L.) DC. var. strigosum Meeuwen ●

126609 Desmodium simplex G. Don = Desmodium adscendens (Sw.) DC. ●■

126610 Desmodium sinoluteolum H. Ohashi et T. Nemoto = Ohwia luteola (H. Ohashi et T. Nemoto) H. Ohashi ●

126611 Desmodium sinuatum (Miq.) Blume ex Baker = Desmodium sequax Wall. ●

126612 Desmodium sinuatum Blume ex Baker = Desmodium sequax Wall. ●

126613 Desmodium spicatum Rehder = Desmodium elegans DC. ●

126614 Desmodium spirale (Sw.) DC. = Desmodium procumbens (Mill.) Hitchc. ●☆

126615 Desmodium squarrosum (Thunb.) DC. var. acuminatum Eckl. et Zeyh. = Eriosema acuminatum (Eckl. et Zeyh.) C. H. Stirt. ■☆

126616 Desmodium stenophyllum Pamp.;狭叶山蚂蝗(窄叶山蚂蝗);Narrowleaf Mountain Leech,Narrowleaf Tickclover,Narrow-leaved Tickclover ●

126617 Desmodium stolzii Schindl.;斯托尔兹山蚂蝗●☆

126618 Desmodium strangulatum Wight et Arn. = Desmodium repandum (Vahl) DC. ●

126619 Desmodium strangulatum Wight et Arn. var. sinuatum Miq. = Desmodium sequax Wall. ●

126620 Desmodium strictum (Pursh) DC.;窄叶山蚂蝗;Narrow-leaved Tick-clover ●☆

126621 Desmodium strigillosum Schindl. var. pendenticarpum (C. Z. Gao et Q. R. Lai) P. H. Huang;垂果山蚂蝗;Pendulous-fruit Tickclover ●

126622 Desmodium stuhlmannii Taub. = Droogmansia pteropus (Baker) De Wild. ■☆

126623 Desmodium styracifolium (Osbeck) Merr.;广东金钱草(广金钱草,假地豆,假花生,金钱草,落地金钱,马蹄草,马蹄香,铜钱草,铜钱沙,铜钱射草,银蹄草);Guangdong Moneygrass,Kwangtung Moneygrass,Snowbell-leaf Tickclover ●■

126624 Desmodium supinum DC.;拟卧山蚂蝗●☆

126625 Desmodium szechuenense (Craib) Schindl. = Hylodesmum podocarpum (DC.) H. Ohashi et R. R. Mill subsp. szechuenense (Craib) H. Ohashi et R. R. Mill ■

126626 Desmodium szechuenense Schindl. = Podocarpium podocarpum (DC.) Yen C. Yang et P. H. Huang var. szechuenense (Craib) Yen C. Yang et P. H. Huang ■

126627 Desmodium tanganyikense Baker;坦噶尼喀山蚂蝗●☆

126628 Desmodium tashiroi Matsum. = Desmodium leptopus A. Gray ex Benth. ●■

126629 Desmodium tashiroi Matsum. = Hylodesmum leptopus (A. Gray ex Benth.) H. Ohashi et R. R. Mill ●

126630 Desmodium tashiroi Matsum. = Podocarpium leptopum (A. Gray ex Benth.) Yen C. Yang et P. H. Huang ●■

126631 Desmodium tenuiflorum Micheli = Macrotyloma tenuiflorum (Micheli) Verdc. ■☆
126632 Desmodium teres Wall.;圆柱拿身草■
126633 Desmodium thunbergii DC. = Lespedeza formosa (Vogel) Koehne ●
126634 Desmodium thunbergii DC. = Lespedeza thunbergii (DC.) Nakai ●■
126635 Desmodium tiliifolium (D. Don) Wall. = Desmodium elegans DC. ●
126636 Desmodium tiliifolium (G. Don) G. Don = Desmodium elegans DC. ●
126637 Desmodium tiliifolium (G. Don) G. Don f. glabrum Schindl. = Desmodium elegans DC. ●
126638 Desmodium tiliifolium (G. Don) G. Don var. potaninii Schindl. = Desmodium elegans DC. ●
126639 Desmodium tiliifolium (G. Don) G. Don var. rhabdocladum (Franch.) Schindl. = Desmodium elegans DC. ●
126640 Desmodium tiliifolium (G. Don) G. Don var. stenophyllum (Pamp.) Schindl. = Desmodium stenophyllum Pamp. ●
126641 Desmodium tiliifolium (G. Don) Wall. = Desmodium elegans DC. ●
126642 Desmodium tiliifolium (G. Don) Wall. var. potaninii Schindl. = Desmodium elegans DC. ●
126643 Desmodium tiliifolium (G. Don) Wall. var. stenophyllum (Pamp.) Schindl. = Desmodium stenophyllum Pamp. ●
126644 Desmodium tiliifolium (G. Don) Wall. var. subtomentosum E. Peter = Desmodium elegans DC. ●
126645 Desmodium tiliifolium Schindl. = Desmodium elegans DC. ●
126646 Desmodium tomentosum (Thunb.) A. DC. = Lespedeza tomentosa (Thunb.) Siebold ex Maxim. ●
126647 Desmodium tomentosum DC. = Lespedeza tomentosa (Thunb.) Siebold ex Maxim. ●
126648 Desmodium tonkinense Schindl. = Phyllodium longipes (Craib) Schindl. ●
126649 Desmodium tortuosum (Sw.) DC.;南美山蚂蝗;Beggar Weed, Cherokec Tickclover, S. America Mountain Leech ●■
126650 Desmodium triangulare (Retz.) Merr. = Dendrolobium triangulare (Retz.) Schindl. ●
126651 Desmodium triflorum (L.) DC.;三点金(八字草,六月雪,品字草,三点金草,三点桃,三脚虎,哮灵草,蝇翅草,蝇翼草);Fly-wing Tickclover, Threeflower Tickclover, Three-flowered Beggarweed ●■
126652 Desmodium triflorum (L.) DC. subsp. pseudotriquetrum (DC.) Prain = Tadehagi pseudotriquetra (DC.) H. Ohashi ●
126653 Desmodium triflorum (L.) DC. var. adpressum Ohwi = Desmodium triflorum (L.) DC. ●■
126654 Desmodium triquetrum (L.) DC. = Tadehagi triquetra (L.) H. Ohashi subsp. pseudotriquetra (DC.) H. Ohashi ●
126655 Desmodium triquetrum (L.) DC. = Tadehagi triquetra (L.) H. Ohashi ●
126656 Desmodium triquetrum (L.) DC. subsp. pseudotriquetrum (A. DC.) Prain = Tadehagi pseudotriquetra (DC.) H. Ohashi ●
126657 Desmodium umbellatum (L.) DC. = Dendrolobium umbellatum (L.) Benth. ●
126658 Desmodium uncinatum (Jacq.) DC.;西班牙三叶草■
126659 Desmodium uncinatum DC. = Desmodium uncinatum (Jacq.) DC. ■
126660 Desmodium unibotryosum C. Chen et X. J. Cui;单序山蚂蝗●
126661 Desmodium unibotryosum C. Chen et X. J. Cui = Desmodium diffusum DC. ●
126662 Desmodium unifoliolatum (Merr.) Steenis = Trifidacanthus unifoliolatus Merr. ●
126663 Desmodium velutinum (Willd.) DC.;绒毛山蚂蝗(短钩毛山蚂蝗,绒毛山绿豆,绒毛叶山蚂蝗);Floss Mountain Leech, Shorthookedhair Tickclover, Shortihook-haired Tickclover, Velutinous Tickclover, Velvety Tickclover ●
126664 Desmodium velutinum (Willd.) DC. subsp. longibracteatum (Schindl.) H. Ohashi;长苞绒毛山蚂蝗●
126665 Desmodium velutinum (Willd.) DC. var. longibracteatum (Schindl.) Meeuwen = Desmodium velutinum (Willd.) DC. subsp. longibracteatum (Schindl.) H. Ohashi ●
126666 Desmodium velutinum (Willd.) DC. var. plukenetii (Wight et Arn.) Schindl. = Desmodium velutinum (Willd.) DC. ●
126667 Desmodium virgatum Prain = Desmodium velutinum (Willd.) DC. ●
126668 Desmodium viride Vogel = Pycnospora lutescens (Poiret) Schindl. ■●
126669 Desmodium viridiflorum (L.) DC.;绿花山蚂蝗;Tick Clover ●☆
126670 Desmodium williamsii H. Ohashi = Hylodesmum williamsii (H. Ohashi) H. Ohashi et R. R. Mill ■
126671 Desmodium williamsii H. Ohashi subsp. magnibracteatum H. Ohashi = Hylodesmum williamsii (H. Ohashi) H. Ohashi et R. R. Mill ■
126672 Desmodium williamsii H. Ohashi var. magnibracteatum (H. Ohashi) P. C. Li = Hylodesmum williamsii (H. Ohashi) H. Ohashi et R. R. Mill ■
126673 Desmodium wittei B. G. Schub.;维特山蚂蝗●☆
126674 Desmodium wolohoense Schindl. = Desmodium elegans DC. var. wolohoense (Schindl.) H. Ohashi ●
126675 Desmodium yunnanense Franch.;云南山蚂蝗;Yunnan Mountain Leech, Yunnan Tickclover ●
126676 Desmodium yunnanense Franch. subsp. praestans (Forrest) H. Ohashi = Desmodium yunnanense Franch. ●
126677 Desmodium yunnanense Franch. var. rockii (Schindl.) Yen C. Yang et P. H. Huang = Desmodium yunnanense Franch. ●
126678 Desmodium zenkeri Schindl.;岑克尔山蚂蝗●☆
126679 Desmodium zonatum Miq.;单叶拿身草(长荚山绿豆,长荚山蚂蝗);Longpod Tickclover, Long-podded Tickclover, Simple-leaf Tickclover, Simple-leaf Trefoil, Simple-leaved Tickclover ●
126680 Desmofischera Holthuis = Desmodium Desv.(保留属名)●■
126681 Desmofischera Holthuis = Monarthrocarpus Merr. ●■
126682 Desmogymnosiphon Guinea.(1946);西非水玉簪属■☆
126683 Desmogymnosiphon Guinea. = Gymnosiphon Blume ■
126684 Desmogymnosiphon chimeicus Guinea;西非水玉簪■☆
126685 Desmogyne King et Prain = Agapetes D. Don ex G. Don ●
126686 Desmogyne neriifolia King et Prain = Agapetes neriifolia (King et Prain) Airy Shaw ●
126687 Desmonchus Desf. = Desmoncus Mart.(保留属名)●☆
126688 Desmoncus Mart.(1824)(保留属名);美洲藤属(大司蒙古属,黑莓棕属,孔带椰子属,南美藤属,束藤属);American Rattan Palm ●☆
126689 Desmoncus orthacanthos Mart.;直刺美洲藤(南美藤);Picmoc ●☆
126690 Desmoncus schippii Burret;希普美洲藤(藤利棕)●☆

126691 Desmonema Miers = Hyalosepalum Troupin ●■

126692 Desmonema Miers = Tinospora Miers(保留属名)●■

126693 Desmonema Raf. = Euphorbia L. ●■

126694 Desmonema caffra Miers = Tinospora caffra (Miers) Troupin ●☆

126695 Desmonema fragosum I. Verd. = Tinospora fragosa (I. Verd.) I. Verd. et Troupin ●☆

126696 Desmonema gossweileri Exell = Tinospora caffra (Miers) Troupin ●☆

126697 Desmonema mossambicense (Engl.) Diels = Tinospora mossambicensis Engl. ●☆

126698 Desmonema mucronulatum Engl. = Tinospora caffra (Miers) Troupin ●☆

126699 Desmonema oblongifolia Engl. = Tinospora oblongifolia (Engl.) Troupin ●☆

126700 Desmonema pallido-aurantiaca Engl. et Gilg = Tinospora caffra (Miers) Troupin ●☆

126701 Desmonema schliebenii Diels = Tinospora caffra (Miers) Troupin ●☆

126702 Desmonema tenerum (Miers) Diels = Tinospora tenera Miers ●☆

126703 Desmophyla Raf. = Ehretia P. Browne ●

126704 Desmophyllum Webb et Berthel. = Ruta L. ●■

126705 Desmophyllum pinnatum (L. f.) Webb et Berthel. = Ruta pinnata L. f. ●☆

126706 Desmopsis Saff. (1916);类鹰爪属●☆

126707 Desmopsis bibracteata Saff.;类鹰爪●☆

126708 Desmopsis brevipes R. E. Fr.;短梗类鹰爪●☆

126709 Desmopsis erythrocarpa Lundell;红果类鹰爪●☆

126710 Desmopsis heteropetala R. E. Fr.;异瓣类鹰爪●☆

126711 Desmopsis lanceolata Lundell;披针叶类鹰爪●☆

126712 Desmopsis mexicana R. E. Fr.;墨西哥类鹰爪●☆

126713 Desmopsis schippii Standl.;希普类鹰爪●☆

126714 Desmopsis stenopetala R. E. Fr.;窄瓣类鹰爪●☆

126715 Desmopsis trunciflora (Schltdl. et Cham.) G. E. Schatz;截花类鹰爪●☆

126716 Desmos Lour. (1790);假鹰爪属(酒饼叶属,山指甲属);Desmos ●

126717 Desmos chinensis Lour.;假鹰爪(半夜兰,宝塔子,串珠,串珠酒饼叶,狗牙花,鸡香草,鸡爪风,鸡爪兰,鸡爪树,鸡爪藤,鸡爪香,鸡爪枝,酒饼藤,酒饼叶,山橘叶,山指甲,五爪龙);China Desmos,Chinese Desmos ●

126718 Desmos chochinchinchensis sensu Merr. = Desmos chinensis Lour. ●

126719 Desmos cochinchinensis Lour. var. grandifolia (Finet et Gagnep.) Ast = Desmos grandifolius (Finet et Gagnep.) C. Y. Wu ex P. T. Li ●

126720 Desmos dumosus (Roxb.) Saff.;毛叶假鹰爪(灯笼木,都蝶,火神,火绳神,云南山指甲);Piloseleaf Desmos, Pilose-leaved Desmos ●

126721 Desmos grandifolius (Finet et Gagnep.) C. Y. Wu ex P. T. Li;大叶假鹰爪(鸡爪风);Bigleaf Desmos,Big-leaved Desmos ●

126722 Desmos hainanensis (Merr.) Merr. = Fissistigma maclurei Merr. ●

126723 Desmos hainanensis (Merr.) Merr. et Chun = Chieniodendron hainanense (Merr.) Tsiang et P. T. Li ●◇

126724 Desmos hainanensis (Merr.) Merr. et Chun = Oncodostigma hainanense (Merr.) Tsiang et P. T. Li ●◇

126725 Desmos macrocalyx (Finet et Gagnep.) P. T. Li;大萼假鹰爪●

126726 Desmos robinsonii (Ast) P. T. Li;钝叶假鹰爪●

126727 Desmos rostratus (Merr. et Chun) P. T. Li;喙果假鹰爪(白面,白叶皂帽花,喙果皂帽花);Rostratefruit Blackhatflower, Rostratefruit Dasymaschalon,Rostrate-fruited Dasymaschalon ●

126728 Desmos saccopetaloides (W. T. Wang) P. T. Li;亮花假鹰爪●

126729 Desmos sootepensis (Craib) J. F. Maxwell;黄花假鹰爪(黄花皂帽花);Soótep Dasymaschalon, Yellow Blackhatflower, Yellowflower Dasymaschalon ●

126730 Desmos yuunanensis (Hu) P. T. Li;云南假鹰爪(云南山指甲);Yunnan Desmos ●◇

126731 Desmoscelis Naudin(1850);索脉野牡丹属●☆

126732 Desmoscelis lychnitoides Naudin;索脉野牡丹●☆

126733 Desmoschoenus Hook. f. (1853);链莎属■☆

126734 Desmoschoenus Hook. f. = Scirpus L. (保留属名)■

126735 Desmoschoenus spiralis (A. Rich.) Hook. f.;链莎■☆

126736 Desmoschoenus spiralis Hook. f. = Desmoschoenus spiralis (A. Rich.) Hook. f. ■☆

126737 Desmostachya (Hook. f.) Stapf = Desmostachya (Stapf) Stapf ■

126738 Desmostachya (Stapf) Stapf(1900);羽穗草属;Desmostachys ■

126739 Desmostachya Stapf = Desmostachya (Stapf) Stapf ■

126740 Desmostachya Stapf = Stapfiola Kuntze ■

126741 Desmostachya bipinnata (L.) Stapf;羽穗草(狗尾画眉草);Bipinnate Desmostachys ■

126742 Desmostachya cynosuroides Stapf ex Massey;洋狗尾草羽穗草■☆

126743 Desmostachys Planch. ex Miers(1852);佛荐草属■☆

126744 Desmostachys brevipes (Engl.) Sleumer;短梗佛荐草■☆

126745 Desmostachys brevipes (Engl.) Sleumer var. oblongifolia (Engl.) Boutique = Desmostachys oblongifolius (Engl.) Villiers ■☆

126746 Desmostachys oblongifolius (Engl.) Villiers;矩圆叶佛荐草■☆

126747 Desmostachys planchonianus Miers;扁佛荐草■☆

126748 Desmostachys preussii Engl. = Desmostachys tenuifolius Oliv. ■☆

126749 Desmostachys tenuifolius Oliv.;细佛荐草■☆

126750 Desmostachys tenuifolius Oliv. var. angustifolius Pellegr. et Villiers;窄叶佛荐草■☆

126751 Desmostachys vogelii (Miers) Stapf;沃格尔佛荐草■☆

126752 Desmostemon Thwaites = Fahrenheitia Rchb. f. et Zoll. ex Müll. Arg. ●☆

126753 Desmotes Kallunki(1992);索芸香属●☆

126754 Desmotes incomparabilis (Riley) Kallunki;索芸香●☆

126755 Desmothamnus Small = Lyonia Nutt. (保留属名)●

126756 Desmotrichum Blume = Ephemerantha P. F. Hunt et Summerh. ■

126757 Desmotrichum Blume = Flickingeria A. D. Hawkes ■

126758 Desmotrichum angustifolium Blume = Flickingeria angustifolia (Blume) A. D. Hawkes ■

126759 Desmotrichum comata Blume = Flickingeria comata (Blume) A. D. Hawkes ■

126760 Desmotrichum comatum Blume = Flickingeria comata (Blume) A. D. Hawkes ■

126761 Desmotrichum fargesii (Finet) Kraenzl. = Epigeneium fargesii (Finet) Gagnep. ■

126762 Desmotrichum fimbriatolabellum (Hayata) Hayata = Flickingeria comata (Blume) A. D. Hawkes ■

126763 Desmotrichum fimbriatolabellum Hayata = Flickingeria comata (Blume) A. D. Hawkes ■

126764 Desmotrichum fimbriatum Blume = Flickingeria fimbriata (Blume) A. D. Hawkes ■

126765 Despeleza Nieuwl. = Lespedeza Michx. ●■

126766 Desplatsia Bocq. (1866);裂托叶椴属●☆
126767 Desplatsia caudata Pierre;尾状裂托叶椴●☆
126768 Desplatsia chrysochlamys (Mildbr. et Burret) Mildbr. et Burret;金被裂托叶椴●☆
126769 Desplatsia dewevrei (De Wild. et T. Durand) Burret;德韦裂托叶椴●☆
126770 Desplatsia floribunda Burret;繁花裂托叶椴●☆
126771 Desplatsia lutea Hutch. et Dalziel = Desplatsia dewevrei (De Wild. et T. Durand) Burret ●☆
126772 Desplatsia mildbraedii Burret;米尔德裂托叶椴●☆
126773 Desplatsia subericarpa Bocq.;木栓果裂托叶椴●☆
126774 Desplatsia trillesiana (Pierre ex De Wild.) Pierre ex A. Chev.;特里列斯裂托叶椴●☆
126775 Despretzia Kunth = Zeugites P. Browne ■☆
126776 Desrousseauxia Tiegh. = Aetanthus (Eichler) Engl. ●☆
126777 Desrousseauxla Tiegh. = Loranthus Jacq. (保留属名)●
126778 Dessenia Adans. = Gnidia L. ●☆
126779 Dessenia Raf. (1838) = Lasiosiphon Fresen. ●☆
126780 Dessenia Raf. (1840) = Struthiola L. (保留属名)●☆
126781 Destrugesia Gaudich. = Capparis L. ●
126782 Destruguezia Benth. et Hook. f. = Destrugesia Gaudich. ●
126783 Desvauxia Benth. et Hook. f. = Centrolepis Labill. ■
126784 Desvauxia Benth. et Hook. f. = Devauxia R. Br. ■
126785 Desvauxia Post et Kuntze = Devauxia R. Br. ■
126786 Desvauxia Post et Kuntze = Glyceria R. Br. (保留属名)■
126787 Desvauxia R. Br. = Centrolepis Labill. ■
126788 Desvauxia R. Br. = Devauxia R. Br. ■
126789 Desvauxia Spreng. = Centrolepis Labill. ■
126790 Desvauxia Spreng. = Devauxia R. Br. ■
126791 Desvauxiaceae Lindl. = Centrolepidaceae Endl. (保留科名)■
126792 Detandra Miers = Sciadotenia Miers ●☆
126793 Detariaceae Burnett = Fabaceae Lindl. (保留科名)●■
126794 Detariaceae Burnett = Leguminosae Juss. (保留科名)●■
126795 Detariaceae J. Hess = Fabaceae Lindl. (保留科名)●■
126796 Detariaceae J. Hess = Leguminosae Juss. (保留科名)●■
126797 Detarium Juss. (1789);荚髓苏木属(德泰豆属)●☆
126798 Detarium beurmannianum Schweinf.;荚髓苏木●☆
126799 Detarium chevalieri Harms = Detarium microcarpum Guillaumin et Perr. ●☆
126800 Detarium heudelotianum Baill.;休德荚髓苏木●☆
126801 Detarium heudelotianum Baill. = Detarium senegalense J. F. Gmel. ●☆
126802 Detarium letestui Pellegr. = Sindoropsis letestui (Pellegr.) J. Léonard ●☆
126803 Detarium macrocarpum Harms;大果荚髓苏木●☆
126804 Detarium microcarpum Guillaumin et Perr.;小果荚髓苏木●☆
126805 Detarium senegalense J. F. Gmel.;塞内加尔荚髓苏木(塞内加尔德泰豆);Dattock,Tallow Tree,Tallow-tree ●☆
126806 Dethardingia Nees et Mart. = Breweria R. Br. ●☆
126807 Dethawia Endl. (1839);细岩芹属●☆
126808 Dethawia tenuifolia (DC.) Godr.;细岩芹■☆
126809 Detridium Nees = Felicia Cass. (保留属名)●■
126810 Detris Adans. (废弃属名) = Felicia Cass. (保留属名)●■
126811 Detris abyssinica (Sch. Bip. ex A. Rich.) Chiov. = Felicia abyssinica Sch. Bip. ex A. Rich. ■☆
126812 Detris barbata (DC.) Schltr. = Felicia ovata (Thunb.) Compton ●☆
126813 Detris dinteri S. Moore = Felicia muricata (Thunb.) Nees subsp. cinerascens Grau ■☆
126814 Detris ericifolia (Forssk.) Hiern = Macowania ericifolia (Forssk.) B. L. Burtt et Grau ●☆
126815 Detris hyssopifolia (P. J. Bergius) Hiern var. straminea Hiern = Felicia mossamedensis (Hiern) Mendonça ■☆
126816 Detris mossamedensis Hiern = Felicia mossamedensis (Hiern) Mendonça ■☆
126817 Detris richardii (Vatke) Chiov. = Felicia dentata (A. Rich.) Dandy ■☆
126818 Detris smaragdina S. Moore = Felicia smaragdina (S. Moore) Merxm. ●☆
126819 Detris smaragdina S. Moore var. versicolor ? = Felicia smaragdina (S. Moore) Merxm. ●☆
126820 Detris tenella (L.) Moore = Felicia tenella (L.) Nees ●☆
126821 Detzneria Schltr. ex Diels(1929);新几内亚婆婆纳属●☆
126822 Detzneria tubata Diels;新几内亚婆婆纳●☆
126823 Deuterocohnia Mez(1894);德氏凤梨属■☆
126824 Deuterocohnia bracteosa W. Till et L. Hromadnik;苞片德氏凤梨■☆
126825 Deuterocohnia brevifolia (Griseb.) M. A. Spencer et L. B. Sm.;短叶德氏凤梨■☆
126826 Deuterocohnia brevispicata Rauh et L. Hrom.;短穗德氏凤梨■☆
126827 Deuterocohnia chrysantha (R. A. Phil.) Mez;金花德氏凤梨■☆
126828 Deuterocohnia glandulosa E. Gross;腺德氏凤梨■☆
126829 Deuterocohnia longipetala Mez;长瓣德氏凤梨■☆
126830 Deuteromallotus Pax et K. Hoffm. (1914);肖野桐属●☆
126831 Deuteromallotus acuminatus Pax et K. Hoffm.;肖野桐●☆
126832 Deutzia Thunb. (1781);溲疏属;Deutzia ●
126833 Deutzia 'Magician';多变溲疏;Magician Deutzia ●☆
126834 Deutzia 'Pink Minor';粉红小溲疏;Deutzia,Pink Minor ●☆
126835 Deutzia 'Rosealind';玫瑰溲疏;Rosealind Deutzia ●☆
126836 Deutzia × candelabrum Rehder;烛台溲疏●☆
126837 Deutzia × magnifica (Lemoine) Rehder;壮观溲疏●☆
126838 Deutzia acuminata Merr. = Deutzia pulchra Vidal ●
126839 Deutzia albida Batalin;白溲疏(甘肃溲疏);Gansu Deutzia, White Deutzia ●
126840 Deutzia albida Batalin = Deutzia discolor Hemsl. ●
126841 Deutzia amurensis (Regel) Airy Shaw;东北溲疏(阿穆尔小花溲疏,东北小花溲疏);Amur Deutzia,Amur Smallflower Deutzia ●
126842 Deutzia amurensis (Regel) Airy Shaw = Deutzia parviflora Bunge var. amurensis Regel ●
126843 Deutzia aspera Rehder;马桑溲疏(糙枝溲疏);Masang Deutzia,Scabrous Deutzia ●
126844 Deutzia aspera Rehder var. fedorovii (Zaik.) S. M. Hwang;镇康溲疏;Zhenkang Deutzia ●
126845 Deutzia aspera Rehder var. fedorovii (Zaik.) S. M. Hwang = Deutzia aspera Rehder ●
126846 Deutzia baroniana Diels;巴氏溲疏(巴氏大花溲疏,钩齿溲疏,楔叶大花溲疏);Baroni Largeflower Deutzia ●
126847 Deutzia baroniana Diels var. insignis Pamp. = Deutzia grandiflora Bunge ●
126848 Deutzia bartlettii Yamam. = Deutzia pulchra Vidal ●
126849 Deutzia bodinieri Rehder = Deutzia setchuenensis Franch. ●
126850 Deutzia bomiensis S. M. Hwang;波密溲疏;Bomi Deutzia ●
126851 Deutzia bomiensis S. M. Hwang var. dinggyensis (H. T. Pan) S. M. Hwang;定结溲疏;Dingjie Bomi Deutzia ●

126852 Deutzia bomiensis S. M. Hwang var. dinggyensis (H. T. Pan) S. M. Hwang = Deutzia bomiensis S. M. Hwang ●

126853 Deutzia breviloba S. M. Hwang;短裂溲疏;Shortlobed Deutzia ●

126854 Deutzia brunoniana Wall. ex G. Don = Deutzia staminea R. Br. ex Wall. var. brunoniana Hook. f. et Thomson ●☆

126855 Deutzia bungoensis Hatus.;丰后溲疏●☆

126856 Deutzia calycosa Rehder;大萼溲疏(宾川溲疏);Bigcalyx Deutzia,Binchuan Deutzia ●

126857 Deutzia calycosa Rehder var. brachytricha Hand.-Mazz. = Deutzia calycosa Rehder ●

126858 Deutzia calycosa Rehder var. longisepala Zaik. = Deutzia calycosa Rehder ●

126859 Deutzia calycosa Rehder var. macropetala Rehder;大瓣溲疏●

126860 Deutzia calycosa Rehder var. xelophyta (Hand.-Mazz.) S. M. Hwang;旱生溲疏●

126861 Deutzia candida Rehder;白色溲疏;White Deutzia ●☆

126862 Deutzia carnea Rehder;三角萼溲疏;Carnose Deutzia ●☆

126863 Deutzia carnea Rehder var. densiflora Rehder;密花三角萼溲疏;Dense-flower Carnose Deutzia ●☆

126864 Deutzia carnea Rehder var. lactea Rehder;乳白三角萼溲疏;Lacteous Carnose Deutzia ●☆

126865 Deutzia carnea Rehder var. stellata Rehder;窄瓣三角萼溲疏;Narrow-petal Carnose Deutzia ●☆

126866 Deutzia chaffanjonii H. Lév. = Deutzia esquirolii (H. Lév.) Rehder ●

126867 Deutzia chanetii H. Lév. = Philadelphus pekinensis Rupr. ●

126868 Deutzia chunii Hu;陈氏溲疏;Chun Deutzia ●

126869 Deutzia chunii Hu = Deutzia ningpoensis Rehder ●

126870 Deutzia cinerascens Rehder;灰叶溲疏;Grey-leaf Deutzia ●

126871 Deutzia compacta Craib;密序溲疏;Compact Deutzia ●

126872 Deutzia compacta Craib var. multiradiata J. T. Pan = Deutzia hookeriana (C. K. Schneid.) Airy Shaw ●

126873 Deutzia cordatula H. L. Li;心基叶溲疏●

126874 Deutzia cordatula H. L. Li = Deutzia taiwanensis (Maxim.) C. K. Schneid. ●

126875 Deutzia coreana H. Lév.;朝鲜溲疏;Korean Deutzia ●

126876 Deutzia coriacea Rehder;革叶溲疏;Lesther-leaf Deutzia ●

126877 Deutzia corymbiflora Lemoine = Deutzia setchuenensis Franch. var. corymbiflora Rehder ●

126878 Deutzia corymbiflora Lemoine ex André = Deutzia setchuenensis Franch. var. corymbiflora Rehder ●

126879 Deutzia corymbosa R. Br. et G. Don;伞房溲疏;Corymb Deutzia,Corymbose Deutzia,Deutzia ●☆

126880 Deutzia corymbosa R. Br. et G. Don var. dinggyensis J. T. Pan = Deutzia bomiensis S. M. Hwang ●

126881 Deutzia corymbosa R. Br. et G. Don var. dinggyensis J. T. Pan = Deutzia bomiensis S. M. Hwang var. dinggyensis (H. T. Pan) S. M. Hwang ●

126882 Deutzia corymbosa R. Br. et G. Don var. hookeriana Schneid. = Deutzia hookeriana (C. K. Schneid.) Airy Shaw ●

126883 Deutzia corymbosa R. Br. et G. Don var. parviflora Schneid. = Deutzia parviflora Bunge ●

126884 Deutzia corymbosa R. Br. et G. Don var. purpurascens Schneid. = Deutzia hookeriana (C. K. Schneid.) Airy Shaw ●

126885 Deutzia corymbosa R. Br. et G. Don var. yunnanensis Franch. ex Rehder = Deutzia hookeriana (C. K. Schneid.) Airy Shaw ●

126886 Deutzia crassidentata S. M. Hwang;粗齿溲疏;Thick-toothed Deutzia ●

126887 Deutzia crassifolia Rehder;厚叶溲疏;Thick-leaf Deutzia,Thick-leaved Deutzia ●

126888 Deutzia crassifolia Rehder var. humilis Rehder = Deutzia crassifolia Rehder ●

126889 Deutzia crassifolia Rehder var. pauciflora (Rehder) S. M. Hwang;少花溲疏;Fewflower Thick-leaf Deutzia ●

126890 Deutzia crassifolia Rehder var. pauciflora (Rehder) S. M. Hwang = Deutzia crassifolia Rehder ●

126891 Deutzia crenata Siebold et Zucc.;齿叶溲疏(日本溲疏,圆齿溲疏);Crenate Deutzia,Crenate Pride-of-Rochester ●

126892 Deutzia crenata Siebold et Zucc. f. candidissima (Benoist) H. Hara;白花齿叶溲疏●☆

126893 Deutzia crenata Siebold et Zucc. f. macrocarpa Nakai;大果齿叶溲疏●☆

126894 Deutzia crenata Siebold et Zucc. f. plena (Maxim.) C. K. Schneid.;重瓣齿叶溲疏●☆

126895 Deutzia crenata Siebold et Zucc. f. pubescens (Makino) H. Hara;毛齿叶溲疏●☆

126896 Deutzia crenata Siebold et Zucc. f. purpurina Honda;紫色齿叶溲疏●☆

126897 Deutzia crenata Siebold et Zucc. var. floribunda (Nakai) H. Ohba = Deutzia floribunda Nakai ●☆

126898 Deutzia crenata Siebold et Zucc. var. heterotricha (Rehder) H. Hara;异毛齿叶溲疏●☆

126899 Deutzia crenata Siebold et Zucc. var. nakaiana (Engl.) H. Hara = Deutzia floribunda Nakai ●☆

126900 Deutzia crenata Siebold et Zucc. var. pubescens (Koidz.) Kitam. = Deutzia floribunda Nakai ●☆

126901 Deutzia cyanocalyx H. Lév. = Deutzia setchuenensis Franch. ●

126902 Deutzia cymuligera S. M. Hwang;小聚花溲疏;Cymligerous Deutzia ●

126903 Deutzia densiflora Rehder = Deutzia discolor Hemsl. ●

126904 Deutzia discolor Hemsl.;异色溲疏(白花溲疏,粗叶溲疏);Discolour Deutzia ●

126905 Deutzia discolor Hemsl. 'Major';红晕异色溲疏●☆

126906 Deutzia discolor Hemsl. = Deutzia purpurascens (Franch.) Rehder ●

126907 Deutzia discolor Hemsl. = Deutzia silvestrii Pamp. ●

126908 Deutzia discolor Hemsl. f. compacta Diels = Deutzia discolor Hemsl. ●

126909 Deutzia discolor Hemsl. var. albida (Batalin) C. K. Schneid. = Deutzia albida Batalin ●

126910 Deutzia discolor Hemsl. var. bicruristyli P. He = Deutzia discolor Hemsl. ●

126911 Deutzia discolor Hemsl. var. gannaensis L. C. Wang et X. G. Sun = Deutzia discolor Hemsl. ●

126912 Deutzia discolor Hemsl. var. gannanensis L. C. Wang et X. G. Sun;甘南异色溲疏;Gannan Deutzia ●

126913 Deutzia discolor Hemsl. var. gannanensis L. C. Wang et X. G. Sun = Deutzia discolor Hemsl. ●

126914 Deutzia discolor Hemsl. var. major Veitch;大花异色溲疏;Big-discolour Deutzia ●

126915 Deutzia discolor Hemsl. var. major Veitch = Deutzia discolor Hemsl. ●

126916 Deutzia discolor Hemsl. var. purpurascens Franch. ex L. Henry = Deutzia purpurascens (Franch.) Rehder ●

126917 Deutzia discolor Hemsl. var. typica Schneid. = Deutzia discolor Hemsl. ●

126918 Deutzia dumicola W. W. Sm. = Deutzia rehderiana C. K. Schneid. ●

126919 Deutzia elegantissima Rehder;美丽溲疏(美秀溲疏,雅致溲疏);Elegant Deutzia ●☆

126920 Deutzia elegantissima Rehder 'Conspicua';拱枝美丽溲疏●☆

126921 Deutzia elegantissima Rehder 'Elegantissima';秀丽溲疏●☆

126922 Deutzia elegantissima Rehder 'Fasciculata';簇生美丽溲疏(束状雅致溲疏)●☆

126923 Deutzia elegantissima Rehder 'Rosealind';罗莎琳德美丽溲疏(粉簇雅致溲疏)●☆

126924 Deutzia esquirolii (H. Lév.) Rehder;狭叶溲疏;Esquirol Deutzia ●

126925 Deutzia esquirolii H. Lév. = Deutzia esquirolii (H. Lév.) Rehder ●

126926 Deutzia faberi Rehder;浙江溲疏(天台溲疏);Faber Deutzia ●

126927 Deutzia fauriei H. Lév. = Deutzia glabrata Kom. ●

126928 Deutzia fedorovii Zaik. = Deutzia aspera Rehder ●

126929 Deutzia floribunda Nakai;日本多花溲疏●☆

126930 Deutzia fragesii Franch. = Deutzia setchuenensis Franch. var. corymbiflora Rehder ●

126931 Deutzia glaberrima Koehne;崂山溲疏(空疏,紫阳花);Laoshan Deutzia ●

126932 Deutzia glaberrima Koehne = Deutzia glabrata Kom. ●

126933 Deutzia glabrata Kom.;光萼溲疏(光滑溲疏,光叶溲疏,崂山溲疏,千层毛,千层皮,无毛溲疏);Glabrous Deutzia ●

126934 Deutzia glabrata Kom. var. sessifolia (Pamp.) Zaik.;无柄溲疏(喇叭树);Sessile Deutzia ●

126935 Deutzia glauca W. C. Cheng;黄山溲疏(冬菊花);Greyblue Deutzia,Grey-blue Deutzia ●

126936 Deutzia glauca W. C. Cheng var. decalvata S. M. Hwang;斑萼溲疏;Decalvate Greyblue Deutzia ●

126937 Deutzia glaucophylla S. M. Hwang;灰绿溲疏;Glaucous Deutzia ●

126938 Deutzia globosa Duthie;球花溲疏(球状溲疏);Globate-flower Deutzia,Glusterflower Deutzia ●

126939 Deutzia globosa Duthie = Deutzia discolor Hemsl. ●

126940 Deutzia glomeruliflora Franch.;团聚叶溲疏(球花溲疏,团花溲疏);Glomerate Deutzia,Glusterflower Deutzia,Gluster-flowered Deutzia ●

126941 Deutzia glomeruliflora Franch. = Deutzia suhulata Hand.-Mazz. ●

126942 Deutzia glomeruliflora Franch. var. forrestiana Zaik. = Deutzia subulata Hand.-Mazz. ●

126943 Deutzia glomeruliflora Franch. var. lichiangensis (Zaik.) S. M. Hwang;丽江溲疏;Lijiang Deutzia ●

126944 Deutzia glomeruliflora Franch. var. lichiangensjs (Zaik.) S. M. Hwang = Deutzia glomeruliflora Franch. ●

126945 Deutzia glomeruliflora Franch. var. xylophyta (Hand.-Mazz.) Zaik. = Deutzia calycosa Rehder var. xelophyta (Hand.-Mazz.) S. M. Hwang ●

126946 Deutzia gracilis Siebold et Zucc.;细梗溲疏(冰生溲疏,细瘦溲疏,小溲疏);Dwarf Deutzia,Japanese Snowflower,Slender Deutzia,Slender Pride of Rochester,Slenderstalk Deutzia,Thin Deutzia ●

126947 Deutzia gracilis Siebold et Zucc. f. latifolia (Nakai) Sugim.;宽叶细梗溲疏●☆

126948 Deutzia gracilis Siebold et Zucc. f. macrantha (Makino) Sugim.;大花细梗溲疏●☆

126949 Deutzia gracilis Siebold et Zucc. f. nagurae (Makino) Sugim.;名仓溲疏●☆

126950 Deutzia gracilis Siebold et Zucc. var. arisanensis Zaik.;阿里山溲疏;Alishan Slenderstalk Deutzia ●

126951 Deutzia gracilis Siebold et Zucc. var. arisanensis Zaik. = Deutzia taiwanensis (Maxim.) C. K. Schneid. ●

126952 Deutzia gracilis Siebold et Zucc. var. aurea Schell;黄叶冰生溲疏;Yellow Slender Deutzia ●☆

126953 Deutzia gracilis Siebold et Zucc. var. ogatae (Koidz.) Ohwi = Deutzia ogatae Koidz. ●☆

126954 Deutzia gracilis Siebold et Zucc. var. pauciflora Sugim.;少花细梗溲疏●☆

126955 Deutzia gracilis Siebold et Zucc. var. zentaroana (Nakai) Hatus. ex Ohwi = Deutzia zentaroana Nakai ●☆

126956 Deutzia grandiflora Bunge;大花溲疏(华北溲疏,教交柴,喇叭枝,密密梢);Big-flowered Deutzia,Early Deutzia,Large Flower Deutzia,Largeflower Deutzia ●

126957 Deutzia grandiflora Bunge var. baroniana (Diels) Rehder = Deutzia baroniana Diels ●

126958 Deutzia grandiflora Bunge var. glabrata Maxim.;疏毛溲疏(步步楷,望镰倒);Glabrous Largeflower Deutzia ●

126959 Deutzia grandiflora Bunge var. glabrata Maxim. = Deutzia grandiflora Bunge ●

126960 Deutzia grandiflora Bunge var. minor Maxim.;小叶大花溲疏;Minor-leaf Largeflower Deutzia ●

126961 Deutzia grandiflora Bunge var. minor Maxim. = Deutzia grandiflora Bunge ●

126962 Deutzia grandiflora Bunge var. typica Schneid. = Deutzia grandiflora Bunge ●

126963 Deutzia hamata Koehne ex Gilg et Loes.;钩齿溲疏(李叶溲疏,杏叶溲疏);Hooked Deutzia,Plum-leaf Deutzia ●

126964 Deutzia hamata Koehne ex Gilg et Loes. = Deutzia baroniana Diels ●

126965 Deutzia hamata Koehne ex Gilg et Loes. var. baroniana (Diels) Zaik. = Deutzia hamata Koehne ex Gilg et Loes. ●

126966 Deutzia hamata Koehne ex Gilg et Loes. var. baroniana (Diels) Zaik. = Deutzia baroniana Diels ●

126967 Deutzia hatusimae H. Ohba,L. M. Niu et Minamitani;初岛溲疏■☆

126968 Deutzia hayatai Nakai = Deutzia pulchra Vidal ●

126969 Deutzia henryi Rehder = Deutzia crassifolia Rehder ●

126970 Deutzia heterophylla S. M. Hwang;异叶溲疏;Diverseleaf Deutzia ●

126971 Deutzia hookeriana (C. K. Schneid.) Airy Shaw;西藏溲疏;Hooker Deutzia ●

126972 Deutzia hookeriana (C. K. Schneid.) Airy Shaw var. macrophylla S. M. Hwang;大叶溲疏;Big-leaf Hooker Deutzia ●

126973 Deutzia hookeriana (C. K. Schneid.) Airy Shaw var. macrophylla S. M. Hwang = Deutzia hookeriana (C. K. Schneid.) Airy Shaw ●

126974 Deutzia hookeriana (C. K. Schneid.) Airy Shaw var. ovatifolia S. M. Hwang;卵叶溲疏;Ovateleaf Hooker Deutzia ●

126975 Deutzia hookeriana (C. K. Schneid.) Airy Shaw var. ovatifolia S. M. Hwang = Deutzia hookeriana (C. K. Schneid.) Airy Shaw ●

126976 Deutzia hypoglauca Rehder;粉背溲疏(粉背叶溲疏,鸡骨头);Hypoglaucous Deutzia,Whiteback Deutzia,White-backed Deutzia ●

126977 Deutzia hypoglauca Rehder = Deutzia hookeriana (C. K. Schneid.) Airy Shaw ●

126978 Deutzia hypoglauca Rehder var. shawana Zaik.; 青城溲疏; Qingcheng Deutzia ●

126979 Deutzia hypoglauca Rehder var. viridis S. M. Hwang; 绿背溲疏; Green-back Hypoglaucous Deutzia ●

126980 Deutzia hypoglauca Rehder var. viridis S. M. Hwang = Deutzia hypoglauca Rehder var. shawana Zaik. ●

126981 Deutzia intermedia ?; 新西兰溲疏; New Zealand Native Blueberry ●☆

126982 Deutzia jinyangensis P. He et L. C. Hu = Deutzia nanchuanensis (W. T. Wang) Rehder ●

126983 Deutzia kalmiaeflora Lemoine; 山月桂果溲疏; Kalmia Deutzia ●☆

126984 Deutzia kelungensis Hayata = Deutzia taiwanensis (Maxim.) C. K. Schneid. ●

126985 Deutzia kiusiana Koidz. ex Nakai = Deutzia sieboldiana Maxim. ●☆

126986 Deutzia lanceifolia Rehder = Deutzia esquirolii (H. Lév.) Rehder ●

126987 Deutzia leiboensis P. He et L. C. Hu; 雷波溲疏; Leibo Deutzia ●

126988 Deutzia leiboensis P. He et L. C. Hu = Deutzia setchuenensis Franch. ●

126989 Deutzia lemoinei Desjardin; 篘莫尼溲疏(莱蒙恩溲疏); Lemoine Deutzia, Lemoine's Deutzia ●☆

126990 Deutzia lemoinei Desjardin 'Compacta'; 密生篘莫尼溲疏; Compact Lemoine Deutzia ●☆

126991 Deutzia longiflora Franch. var. farreri Airy Shaw = Deutzia discolor Hemsl. ●

126992 Deutzia longifolia Franch.; 长叶溲疏(羊不食); Longleaf Deutzia, Long-leaved Deutzia ●

126993 Deutzia longifolia Franch. var. densitomentosa P. He et L. C. Hu = Deutzia squamosa S. M. Hwang ●

126994 Deutzia longifolia Franch. var. elegans Rehder; 紫花长叶溲疏; Purple-flower Longleaf Deutzia ●

126995 Deutzia longifolia Franch. var. elegans Rehder = Deutzia longifolia Franch. ●

126996 Deutzia longifolia Franch. var. grandiflora Franch. ex Diels = Deutzia calycosa Rehder var. macropetala Rehder ●

126997 Deutzia longifolia Franch. var. macropetala (Rehder) Zaik. = Deutzia calycosa Rehder var. macropetala Rehder ●

126998 Deutzia longifolia Franch. var. pingwuensis S. M. Hwang; 平武溲疏; Pingwu Deutzia ●

126999 Deutzia longifolia Franch. var. sikangensis (W. P. Fang) P. He = Deutzia glomeruliflora Franch. ●

127000 Deutzia longifolia Franch. var. veitchii Rehder; 大花长叶溲疏; Bigflower Longleaf Deutzia ●

127001 Deutzia longifolia Franch. var. veitchii Rehder = Deutzia longifolia Franch. ●

127002 Deutzia longifolia Franch. var. xerophyta Hand. -Mazz. = Deutzia calycosa Rehder var. xelophyta (Hand. -Mazz.) S. M. Hwang ●

127003 Deutzia longifolia Franch. var. yunnanensis Franch.; 衣白皮; Yunnan Longleaf Deutzia ●

127004 Deutzia macrantha Hook. f. et Thomson = Deutzia wardiana Zaik. ●

127005 Deutzia magnifica (Lemoine) Rehder; 壮丽溲疏(大溲疏, 华美溲疏); Magnific Deutzia, Tall Deutzia ●

127006 Deutzia magnifica (Lemoine) Rehder var. eburnea Rehder; 稀花大溲疏; Loose Magnific Deutzia ●☆

127007 Deutzia magnifica (Lemoine) Rehder var. erecta Rehder; 直立大溲疏; Erect Magnific Deutzia ●☆

127008 Deutzia magnifica (Lemoine) Rehder var. formosa Rehder; 台湾大溲疏; Taiwan Magnific Deutzia ●

127009 Deutzia magnifica (Lemoine) Rehder var. latiflora Rehder; 大花大溲疏; Largeflower Magnific Deutzia ●☆

127010 Deutzia magnifica (Lemoine) Rehder var. superba Rehder; 钟花大溲疏; Campanulate Magnific Deutzia ●☆

127011 Deutzia maliflora Rehder; 苹果花溲疏; Apple-flower Deutzia ●☆

127012 Deutzia maximowicziana Makino; 马氏溲疏; Maximowicz Deutzia ●☆

127013 Deutzia micrantha Engl.; 多花溲疏(碎花溲疏); Manyflower Deutzia, Multiflorous Deutzia ●

127014 Deutzia micrantha Engl. = Deutzia parviflora Bunge var. micrantha (Engl.) Rehder ●

127015 Deutzia mollis Duthie; 钻丝溲疏(软毛溲疏); Soft-hair Deutzia ●

127016 Deutzia monbeigii W. W. Sm.; 维西溲疏(蒙氏溲疏); Monbeig Deutzia ●

127017 Deutzia monbeigii W. W. Sm. var. lanceolata S. M. Hwang; 披针叶溲疏; Lanceolate Monbeig Deutzia ●

127018 Deutzia muliensis S. M. Hwang; 木里溲疏; Muli Deutzia ●

127019 Deutzia multiradiata W. T. Wang; 多射线溲疏(多辐溲疏, 多辐线光叶溲疏, 多辐线溲疏); Multiradiate Deutzia ●

127020 Deutzia myriantha Desjardin; 密花溲疏; Dense-flower Deutzia ●☆

127021 Deutzia nagurae (Makino) Makino = Deutzia gracilis Siebold et Zucc. f. nagurae (Makino) Sugim. ●☆

127022 Deutzia nanchuanensis (W. T. Wang) Rehder; 南川溲疏; Nanchuan Deutzia ●

127023 Deutzia naseana Nakai; 乐世溲疏●☆

127024 Deutzia naseana Nakai var. amanoi (Hatus.) Hatus. ex H. Ohba; 天农世溲疏●☆

127025 Deutzia naseana Nakai var. macrantha Hatus. ex H. Ohba = Deutzia naseana Nakai ●☆

127026 Deutzia ningpoensis Rehder; 宁波溲疏(观音竹, 空心副常山, 老鼠竹, 细叶空心柴); Ningbo Deutzia, Ningpo Deutzia ●

127027 Deutzia ningpoensis Rehder f. integrifoliis D. T. Liu et J. Han = Deutzia ningpoensis Rehder ●

127028 Deutzia nitidula W. T. Wang; 光叶溲疏; Glabrous Deutzia, Shaining Deutzia ●

127029 Deutzia nitidula W. T. Wang = Deutzia multiradiata W. T. Wang ●

127030 Deutzia obtusilobata S. M. Hwang; 钝裂溲疏; Obtusilobed Deutzia ●

127031 Deutzia ogatae Koidz.; 绪方溲疏●☆

127032 Deutzia parviflora Bunge; 小花溲疏(唐溲疏); Mongolian Deutzia, Mongolian Pride-of-Rochester, Smallflower Deutzia, Small-flowered Deutzia ●

127033 Deutzia parviflora Bunge = Deutzia setchuenensis Franch. var. corymbiflora Rehder ●

127034 Deutzia parviflora Bunge var. amurensis Regel = Deutzia amurensis (Regel) Airy Shaw ●

127035 Deutzia parviflora Bunge var. bungei Franch. = Deutzia parviflora Bunge var. amurensis Regel ●

127036 Deutzia parviflora Bunge var. micrantha (Engl.) Rehder; 碎花溲疏(多花溲疏); Manyflowers Smallflower Deutzia ●

127037 Deutzia parviflora Bunge var. mongolica Franch. = Deutzia parviflora Bunge ●

127038 Deutzia parviflora Bunge var. ovatifolia Rehder;卵圆叶小花溲疏;Ovate-leaf Smallflower Deutzia ●

127039 Deutzia parviflora Bunge var. ovatifolia Rehder = Deutzia parviflora Bunge ●

127040 Deutzia pauciflora (Rehder) Zaik. = Deutzia crassifolia Rehder ●

127041 Deutzia pilosa Rehder;褐毛溲疏(软毛溲疏,通天行);Pilose Deutzia ●

127042 Deutzia pilosa Rehder var. longiflora P. He et L. C. Hu;长萼褐毛溲疏;Longflower Pilose Deutzia ●

127043 Deutzia pilosa Rehder var. longiloba P. He et L. C. Hu = Deutzia pilosa Rehder ●

127044 Deutzia pilosa Rehder var. longiloba W. T. Wang ex S. M. Hwang;峨眉溲疏;Emei Deutzia ●

127045 Deutzia pilosa Rehder var. longiloba W. T. Wang ex S. M. Hwang = Deutzia pilosa Rehder ●

127046 Deutzia pilosa Rehder var. ochrophloeos Rehder = Deutzia setchuenensis Franch. ●

127047 Deutzia prunifolia Rehder = Deutzia baroniana Diels ●

127048 Deutzia prunifolia Rehder = Deutzia hamata Koehne ex Gilg et Loes. ●

127049 Deutzia pulchra Vidal;优雅溲疏(白埔姜,常绿溲疏,常山,大叶溲疏,美丽溲疏);Beautiful Deutzia,Evergreen Deutzia ●

127050 Deutzia pulchra Vidal var. bartlettii (Yamam.) S. S. Ying = Deutzia pulchra Vidal ●

127051 Deutzia pulchra Vidal var. formosana Nakai = Deutzia pulchra Vidal ●

127052 Deutzia pulchra Vidal var. hayatai (Nakai) Zaik. = Deutzia pulchra Vidal ●

127053 Deutzia pulchra Vidal var. typica Nakai = Deutzia pulchra Vidal ●

127054 Deutzia purpurascens (Franch.) Rehder;紫花溲疏;Purple-flower Deutzia,Purple-flowered Deutzia ●

127055 Deutzia purpurascens (Franch.) Rehder var. lichiangensis Zaik. = Deutzia glomeruliflora Franch. ●

127056 Deutzia purpurascens (Franch.) Rehder var. pauciflora Rehder = Deutzia crassifolia Rehder ●

127057 Deutzia reflexa Duthie;卷瓣溲疏;Reflex Deutzia ●

127058 Deutzia reflexa Duthie = Deutzia discolor Hemsl. ●

127059 Deutzia rehderiana C. K. Schneid.;灌丛溲疏(碎米花,下关溲疏);Rehder's Deutzia ●

127060 Deutzia reticulata Koidz.;网脉溲疏●☆

127061 Deutzia rosea Rehder;粉花溲疏(矮溲疏,粉溲疏);Pink Deutzia,Rose Deutzia,Rosepanicle Deutzia ●☆

127062 Deutzia rosea Rehder 'Campanulata' = Deutzia rosea Rehder var. campanulata Rehder ●☆

127063 Deutzia rosea Rehder 'Carminea' = Deutzia rosea Rehder var. carminea Rehder ●☆

127064 Deutzia rosea Rehder var. campanulata Rehder;钟状粉花溲疏(白花矮溲疏);Campanulate Pink Deutzia ●☆

127065 Deutzia rosea Rehder var. carminea Rehder;胭脂粉花溲疏(紫红矮溲疏);Carmine Pink Deutzia,Pink Deutzia ●☆

127066 Deutzia rosea Rehder var. eximia Rehder;浅粉花溲疏;Superior Pink Deutzia ●☆

127067 Deutzia rosea Rehder var. floribunda Rehder;稠密粉花溲疏;Dense Pink Deutzia ●☆

127068 Deutzia rosea Rehder var. grandiflora Rehder;大花粉花溲疏;Big Pink Deutzia ●☆

127069 Deutzia rosea Rehder var. multiflora Rehder;多花粉花溲疏;Multiflorous Pink Deutzia ●☆

127070 Deutzia rosea Rehder var. venustae Rehder;绿萼粉花溲疏;Green-sepal Pink Deutzia ●☆

127071 Deutzia rubens Rehder;粉红溲疏;Red Deutzia ●

127072 Deutzia scabra Thunb.;溲疏(巨骨,空木,空疏,卯花,野茉莉,圆齿溲疏);Fuzzy Deutzia,Fuzzy Pride-of-Rochester,Scabrous Deutzia ●

127073 Deutzia scabra Thunb. 'Candidissima' = Deutzia scabra Thunb. var. candidissima Rehder ●☆

127074 Deutzia scabra Thunb. 'Plena';重瓣溲疏;Double Scabrous Deutzia ●☆

127075 Deutzia scabra Thunb. 'Pride of Rochester';罗切斯溲疏(罗彻斯特的荣耀溲疏);Scabrous Deutzia,White Deutzia ●☆

127076 Deutzia scabra Thunb. f. aurescens (Nakai) Sugim.;浅黄溲疏 ●☆

127077 Deutzia scabra Thunb. var. angustifolia Voss;狭叶粗涩溲疏;Angustifoliate Scabrous Deutzia ●☆

127078 Deutzia scabra Thunb. var. candidissima Rehder;白花重瓣溲疏(白花溲疏);Pure White Scabrous Deutzia ●☆

127079 Deutzia scabra Thunb. var. crenata Maxim. = Deutzia crenata Siebold et Zucc. ●

127080 Deutzia scabra Thunb. var. cumis-paucifloris Forbes et Hemsl. = Deutzia setchuenensis Franch. var. corymbiflora Rehder ●

127081 Deutzia scabra Thunb. var. marmorata Rehder;黄斑叶溲疏;Yellowspotted Scabrous Deutzia ●☆

127082 Deutzia scabra Thunb. var. plena Rehder = Deutzia scabra Thunb. 'Plena' ●☆

127083 Deutzia scabra Thunb. var. punctata Rehder;白斑叶溲疏;White-spotted Scabrous Deutzia ●☆

127084 Deutzia scabra Thunb. var. sieboldiana (Maxim.) H. Hara = Deutzia sieboldiana Maxim. ●☆

127085 Deutzia scabra Thunb. var. sieboldiana (Maxim.) H. Hara f. microcarpa (Nakai) H. Hara;小果西氏溲疏(小果席氏溲疏)●☆

127086 Deutzia scabra Thunb. var. typica Schneid. = Deutzia crenata Siebold et Zucc. ●

127087 Deutzia scabra Thunb. var. watereri Rehder;彩色溲疏;Carmine Scabrous Deutzia ●☆

127088 Deutzia schneideriana Rehder;长江溲疏(毛叶溲疏,旋氏溲疏);Changjiang Deutzia,Schneider Deutzia ●

127089 Deutzia schneideriana Rehder var. laxiflora Rehder;疏花长江溲疏;Loose Schneider Deutzia ●

127090 Deutzia schneideriana Rehder var. laxiflora Rehder = Deutzia glauca W. C. Cheng ●

127091 Deutzia schneideriana Rehder var. laxiflora Rehder = Deutzia schneideriana Rehder ●

127092 Deutzia sessilifolia Pamp. = Deutzia glabrata Kom. var. sessifolia (Pamp.) Zaik. ●

127093 Deutzia setchuenensis Franch.;川溲疏(白茂树,鹅毛通,四川溲疏,夏季石藤);Sichuan Deutzia,Szechwan Deutzia ●

127094 Deutzia setchuenensis Franch. var. corymbiflora Rehder;伞房川溲疏(多花溲疏);Corymb Szechwan Deutzia,Corymbose Sichuan Deutzia ●

127095 Deutzia setchuenensis Franch. var. longidentata Rehder;长齿叶四川溲疏(长齿溲疏);Longtooth-leaf Sichuan Deutzia ●

127096 Deutzia setifera Zaik. = Deutzia crassifolia Rehder ●

127097 Deutzia setosa Zaik.;刚毛溲疏;Setose Deutzia ●

127098 Deutzia shawana Zaik. = Deutzia hypoglauca Rehder var.

shawana Zaik. ●
127099 Deutzia sieboldiana Maxim.;西氏溲疏(萨波得溲疏,席氏溲疏);Siebold Deutzia ●☆
127100 Deutzia sieboldiana Maxim. var. dippeliana C. K. Schneid.;得帕尔溲疏;Dippel Deutzia ●☆
127101 Deutzia sikangensis W. P. Fang;西康溲疏;Xikang Deutzia ●
127102 Deutzia sikangensis W. P. Fang = Deutzia glomeruliflora Franch. ●
127103 Deutzia silvestrii Pamp.;红花溲疏;Red-flower Deutzia ●
127104 Deutzia squamosa S. M. Hwang;鳞毛溲疏;Scalyhair Deutzia ●
127105 Deutzia staminea R. Br. = Deutzia schneideriana Rehder ●
127106 Deutzia staminea R. Br. ex Wall.;长柱溲疏(短萼溲疏,抛筒根,壮溲疏);Staminate Deutzia ●
127107 Deutzia staminea R. Br. ex Wall. = Deutzia subulata Hand.-Mazz. ●
127108 Deutzia staminea R. Br. ex Wall. var. brunoniana Hook. f. et Thomson;布鲁诺溲疏;Brunon Deutzia ●☆
127109 Deutzia staminea R. Br. var. sikkimensis Schneid. = Deutzia staminea R. Br. ex Wall. var. brunoniana Hook. f. et Thomson ●☆
127110 Deutzia staminea R. Br. var. typica Schneid. = Deutzia staminea R. Br. ex Wall. var. brunoniana Hook. f. et Thomson ●☆
127111 Deutzia subsessilis Rehder = Deutzia glomeruliflora Franch. ●
127112 Deutzia subulata Hand.-Mazz.;钻齿溲疏;Suhulate Deutzia ●
127113 Deutzia taibaiensis W. T. Wang ex S. M. Hwang;太白溲疏;Taibai Deutzia ●
127114 Deutzia taiwanensis (Maxim.) C. K. Schneid.;台湾溲疏(白埔姜,台湾心基溲疏);Taiwan Cordate-leafed Deutzia, Taiwan Deutzia ●
127115 Deutzia taiwanensis (Maxim.) C. K. Schneid. = Deutzia pulchra Vidal ●
127116 Deutzia taiwanensis (Maxim.) C. K. Schneid. var. cordatula (H. L. Li) Y. C. Liu = Deutzia cordatula H. L. Li ●
127117 Deutzia taiwanensis Schneid. = Deutzia pulchra Vidal ●
127118 Deutzia uniflora Shirai;单花溲疏●☆
127119 Deutzia veitchii Veitch = Deutzia longifolia Franch. ●
127120 Deutzia vilmorinae Lemoine et Bois = Deutzia discolor Hemsl. ●
127121 Deutzia vilmorinae Lemoine et Bois ex Vilm. et Bois;长梗溲疏(白面花,毛脉溲疏,米花,卫氏溲疏);Long-stalked Deutzia, Vilmorin Deutzia ●
127122 Deutzia vilmorinae Lemoine et Bois ex Vilm. et Bois = Deutzia discolor Hemsl. ●
127123 Deutzia wardiana Zaik.;宽萼溲疏;Ward Deutzia ●
127124 Deutzia wilsonii Duthie;威尔逊溲疏;Wilson Deutzia ●☆
127125 Deutzia yaeyamensis Ohwi;八重山溲疏●☆
127126 Deutzia yunnanensis S. M. Hwang;云南溲疏;Yunnan Deutzia ●
127127 Deutzia zentaroana Nakai;田代善溲疏●☆
127128 Deutzia zhongdianensis S. M. Hwang;中甸溲疏;Zhongdian Deutzia ●
127129 Deutzianthus Gagnep. (1924);东京桐属;Deutzianthus ●
127130 Deutzianthus tonkinensis Gagnep.;东京桐(越南桐);Tonkin Deutzianthus, Viatnam ●◇
127131 Devauxia Kunth = Glyceria R. Br.(保留属名)■
127132 Devauxia P. Beauv. ex Kunth = Glyceria R. Br.(保留属名)■
127133 Devauxia R. Br. = Centrolepis Labill. ■
127134 Devauxia banksii R. Br. = Centrolepis banksii (R. Br.) Roem. et Schult. ■
127135 Devauxiaceae Dumort. = Centrolepidaceae Endl.(保留科名)■
127136 Deverra DC. (1829);德弗草属■☆
127137 Deverra DC. = Pituranthos Viv. ■☆
127138 Deverra aphylla (Cham. et Schltdl.) DC. = Deverra denudata (Viv.) Pfisterer et Podlech subsp. aphylla (Cham. et Schltdl.) Pfisterer et Podlech ■☆
127139 Deverra aphylla (Cham. et Schltdl.) DC. var. burchellii DC. = Deverra burchellii (DC.) Eckl. et Zeyh. ■☆
127140 Deverra battandieri (Maire) Chrtek;巴坦德弗草■☆
127141 Deverra burchellii (DC.) Eckl. et Zeyh.;伯切尔德弗草■☆
127142 Deverra chlorantha Coss. et Durieu = Deverra denudata (Viv.) Pfisterer et Podlech ■☆
127143 Deverra denudata (Viv.) Pfisterer et Podlech;裸露德弗草■☆
127144 Deverra denudata (Viv.) Pfisterer et Podlech subsp. aphylla (Cham. et Schltdl.) Pfisterer et Podlech;无叶裸露德弗草■☆
127145 Deverra fallax Batt. et Trab. = Deverra scoparia Coss. et Durieu ■☆
127146 Deverra intermedia L. Chevall. = Deverra triradiata Boiss. subsp. intermedia (Chevall.) Pfisterer et Podlech ■☆
127147 Deverra juncea Ball;灯芯草状德弗草■☆
127148 Deverra pituranthos DC. = Deverra denudata (Viv.) Pfisterer et Podlech ■☆
127149 Deverra reboudii Coss. et Durieu;雷博德弗草■☆
127150 Deverra rholfsiana Asch.;罗尔德弗草■☆
127151 Deverra scoparia Coss. et Durieu;帚状德弗草■☆
127152 Deverra scoparia Coss. et Durieu subsp. tripolitana (Andr.) Pfisterer et Podlech = Deverra rholfsiana Asch. ■☆
127153 Deverra togasii M. Hiroe = Seseli togasii (M. Hiroe) Pimenov et Kljuykov ■
127154 Deverra tortuosa (Desf.) DC.;扭曲德弗草■☆
127155 Deverra tortuosa (Desf.) DC. var. virgata Coss. et Kralik = Deverra tortuosa (Desf.) DC. ■☆
127156 Deverra triradiata Boiss.;三射线德弗草■☆
127157 Deverra triradiata Boiss. subsp. intermedia (Chevall.) Pfisterer et Podlech;间型三射线德弗草■☆
127158 Deveya Rchb. = Deweya Torr. et A. Gray ■☆
127159 Deveya Rchb. = Tauschia Schltdl.(保留属名)■☆
127160 Devia Goldblatt et J. C. Manning(1990);戴维鸢尾属■☆
127161 Devia xeromorpha Goldblatt et J. C. Manning;戴维鸢尾■☆
127162 Devillea Bert. ex Schult. f. = Guzmania Ruiz et Pav. ■☆
127163 Devillea Bubani = Ligusticum L. ■
127164 Devillea Tul. et Wedd. (1849);德维尔川苔草属■☆
127165 Devillea flagelliformis Tul. et Wedd.;德维尔川苔草■☆
127166 Devogelia Schuit. (2004);马鲁古兰属■☆
127167 Dewevrea Micheli(1898);德瓦豆属■☆
127168 Dewevrea bilabiata Micheli;德瓦豆■☆
127169 Dewevrea gossweileri Baker f.;戈斯德瓦豆■☆
127170 Dewevrella De Wild. (1907);德瓦夹竹桃属●☆
127171 Dewevrella cochliostema De Wild.;德瓦夹竹桃●☆
127172 Dewevrella congensis Wernham = Secamone dewevrei De Wild. ●☆
127173 Deweya Eaton = Nemopanthus Raf.(保留属名)●☆
127174 Deweya Raf. = Carex L. ■
127175 Deweya Torr. et A. Gray = Tauschia Schltdl.(保留属名)■☆
127176 Dewildemania O. Hoffm. (1903);螺叶瘦片菊属■☆
127177 Dewildemania burundiensis Lisowski;布隆迪螺叶瘦片菊■☆
127178 Dewildemania filifolia O. Hoffm.;线螺叶瘦片菊■☆
127179 Dewildemania glandulosa Lisowski;腺螺叶瘦片菊■☆

127180 Dewildemania lancifolia (O. Hoffm.) Kalanda;剑螺叶瘦片菊■☆
127181 Dewildemania platycephala B. L. Burtt;平头螺叶瘦片菊■☆
127182 Dewildemania stenophylla (Baker) B. L. Burtt;窄螺叶瘦片菊■☆
127183 Dewildemania upembensis Lisowski;乌彭贝螺叶瘦片菊■☆
127184 Dewindtia De Wild. = Cryptosepalum Benth. ●☆
127185 Dewindtia katangensis De Wild. = Cryptosepalum katangense (De Wild.) J. Léonard ●☆
127186 Dewinterella D. Müll. -Doblies et U. Müll. -Doblies (1994);拟澳非麻属■☆
127187 Dewinterella mathewsii (W. F. Barker) D. Müll. -Doblies et U. Müll. -Doblies = Hessea mathewsii W. F. Barker ■☆
127188 Dewinterella pulcherrima (D. Müll. -Doblies et U. Müll. -Doblies) D. Müll. -Doblies et U. Müll. -Doblies = Hessea pulcherrima (D. Müll. -Doblies et U. Müll. -Doblies) Snijman ■☆
127189 Dewinteria van Jaarsv. et A. E. van Wyk = Rogeria J. Gay ex Delile ■☆
127190 Dewinteria van Jaarsv. et A. E. van Wyk (2007);澳非麻属■☆
127191 Deyeuxia Clarion = Calamagrostis Adans. ■
127192 Deyeuxia Clarion ex P. Beauv. (1812);野青茅属;Small Reed, Smallreed ■
127193 Deyeuxia Clarion ex P. Beauv. = Calamagrostis Adans. ■
127194 Deyeuxia P. Beauv. = Calamagrostis Adans. ■
127195 Deyeuxia ampla Keng;长序野青茅;Longinflorescence Small Reed, Longspike Smallreed ■
127196 Deyeuxia angustifolia (Kom.) Y. L. Chang;小叶章;Narrowleaf Smallreed, Small Reed ■
127197 Deyeuxia arundinacea (L.) P. Beauv.;野青茅(短毛野青茅,类芦野青茅);Common Small Reed, Short-haired Caladium, Wild Smallreed ■
127198 Deyeuxia arundinacea (L.) P. Beauv. = Calamagrostis arundinacea (L.) Roth ■☆
127199 Deyeuxia arundinacea (L.) P. Beauv. var. borealis (Rendle) P. C. Kuo et S. L. Lu;北方野青茅;Boreal Small Reed, North Wild Smallreed ■
127200 Deyeuxia arundinacea (L.) P. Beauv. var. brachytricha (Steud.) P. C. Kuo et S. L. Lu;短毛野青茅;Shorthair Small Reed, Shorthair Wild Smallreed ■
127201 Deyeuxia arundinacea (L.) P. Beauv. var. brachytricha (Steud.) P. C. Kuo et S. L. Lu = Calamagrostis brachytricha Steud. ■
127202 Deyeuxia arundinacea (L.) P. Beauv. var. ciliata (Honda) P. C. Kuo et S. L. Lu;纤毛野青茅;Ciliate Small Reed, Ciliate Wild Smallreed ■
127203 Deyeuxia arundinacea (L.) P. Beauv. var. ciliata (Honda) P. C. Kuo et S. L. Lu = Calamagrostis brachytricha Steud. var. ciliata (Honda) Ibaragi et H. Ohashi ■
127204 Deyeuxia arundinacea (L.) P. Beauv. var. collina (Franch.) P. C. Kuo et S. L. Lu;丘生野青茅;Hilly Small Reed ■
127205 Deyeuxia arundinacea (L.) P. Beauv. var. hirsuta (Hack.) P. C. Kuo et S. L. Lu;糙毛野青茅;Hirsute Small Reed, Hirsute Wild Smallreed ■
127206 Deyeuxia arundinacea (L.) P. Beauv. var. latifolia (Rendle) P. C. Kuo et S. L. Lu;宽叶野青茅;Broadleaf Small Reed, Broadleaf Wild Smallreed ■
127207 Deyeuxia arundinacea (L.) P. Beauv. var. laxiflora (Rendle) P. C. Kuo et S. L. Lu;疏花野青茅;Distantflower Small Reed, Laxflower Wild Smallreed ■
127208 Deyeuxia arundinacea (L.) P. Beauv. var. ligulata (Rendle) P. C. Kuo et S. L. Lu;长舌野青茅;Ligulate Small Reed, Longtongue Wild Smallreed ■
127209 Deyeuxia arundinacea (L.) P. Beauv. var. robusta (Franch. et Sav.) P. C. Kuo et S. L. Lu;粗壮野青茅;Robust Small Reed, Robust Wild Smallreed ■
127210 Deyeuxia arundinacea (L.) P. Beauv. var. robusta (Franch. et Sav.) P. C. Kuo et S. L. Lu = Calamagrostis brachytricha Steud. ■
127211 Deyeuxia arundinacea (L.) P. Beauv. var. sciuroides (Franch. et Sav.) P. C. Kuo et S. L. Lu;西塔茅;Squirrel Tail Small Reed, Xita Wild Smallreed ■
127212 Deyeuxia arundinacea P. Beauv. = Deyeuxia arundinacea (L.) P. Beauv. ■
127213 Deyeuxia autumnalis (Koidz.) Ohwi = Calamagrostis autumnalis Koidz. ■☆
127214 Deyeuxia autumnalis (Koidz.) Ohwi var. microtis (Ohwi) Ohwi = Calamagrostis autumnalis Koidz. var. microtis Ohwi ■☆
127215 Deyeuxia biflora Keng;两花野青茅;Twoflower Small Reed, Twoflower Smallreed ■
127216 Deyeuxia bolanderi Vasey;加利福尼亚野青茅■☆
127217 Deyeuxia brachytricha (Steud.) Y. L. Chang;类芦野青茅(短舌野青茅,松田野青茅,台湾禾草,台湾野青茅);Matsuda Small Reed, Matsuda Smallreed, Taiwan Small Reed, Taiwan Smallreed ■
127218 Deyeuxia brachytricha (Steud.) Y. L. Chang = Calamagrostis brachytricha Steud. ■
127219 Deyeuxia canadensis (Michx.) Munro ex Hook. f.;加拿大野青茅;Canadian Small Reed ■☆
127220 Deyeuxia collina (Franch.) Pilg. = Deyeuxia arundinacea (L.) P. Beauv. var. collina (Franch.) P. C. Kuo et S. L. Lu ■
127221 Deyeuxia compacta Munro ex Duthie = Deyeuxia holciformis (Jaub. et Spach) Bor ■
127222 Deyeuxia compacta Munro ex Hook. f. = Calamagrostis holciformis Jaub. et Spach ■
127223 Deyeuxia compacta Munro ex Hook. f. = Deyeuxia holciformis (Jaub. et Spach) Bor ■
127224 Deyeuxia conferta Keng;密穗野青茅;Dense Small Reed, Dense Smallreed ■
127225 Deyeuxia conferta Keng var. guoxuniana N. X. Zhao et M. F. Li;国勋野青茅;Guoxun Small Reed, Guoxun Smallreed ■
127226 Deyeuxia continentalis (Hand. -Mazz.) L. Liou = Deyeuxia petelotii (Hitchc.) S. M. Phillips et Wen L. Chen ■
127227 Deyeuxia deschampsioides (Trin.) Scribn. var. hayachinensis (Ohwi) Ohwi = Calamagrostis nana Takeda subsp. hayachinensis (Ohwi) Tateoka ■☆
127228 Deyeuxia diffusa Keng;散穗野青茅;Diffuse Small Reed, Diffuse Smallreed ■
127229 Deyeuxia effusiflora Rendle;疏穗野青茅;Looseflower Small Reed, Looseflower Smallreed ■
127230 Deyeuxia fauriei (Hack.) Ohwi = Calamagrostis fauriei Hack. ■☆
127231 Deyeuxia filiformis (Griseb.) Hook. f. = Calamagrostis scabrescens Griseb. ■
127232 Deyeuxia filiformis (Griseb.) Hook. f. = Deyeuxia scabrescens (Griseb.) Munro ex Duthie ■
127233 Deyeuxia filipes Keng;细柄野青茅(丝状野青茅);Threadstalk Small Reed, Threadstalk Smallreed ■
127234 Deyeuxia flaccida Keng;柔弱野青茅;Flabby Small Reed, Flabby Smallreed ■
127235 Deyeuxia flavens Keng;黄花野青茅;Yellow Smallreed,

Yellowflower Small Reed, Yellowflower Smallreed ■
127236 Deyeuxia formosana (Hayata) C. C. Hsu = Deyeuxia brachytricha (Steud.) Y. L. Chang ■
127237 Deyeuxia gigas (Takeda) Ohwi = Calamagrostis gigas Takeda ■☆
127238 Deyeuxia gigas (Takeda) Ohwi var. aspera (Honda) Ohwi = Calamagrostis gigas Takeda ■☆
127239 Deyeuxia grata Keng;川野青茅(感野青茅);Sichuan Small Reed, Sichuan Smallreed ■
127240 Deyeuxia gyirongensis P. C. Kuo et S. L. Lu;吉隆野青茅;Jilong Small Reed, Jilong Smallreed ■
127241 Deyeuxia gyirongensis P. C. Kuo et S. L. Lu = Deyeuxia pulchella (Griseb.) Hook. f. ■
127242 Deyeuxia hackelii Bor = Calamagrostis decora Hook. f. ■☆
127243 Deyeuxia hakonensis (Franch. et Sav.) Keng;箱根野青茅(箱根拂子茅);Hakone Smallreed, Small Reed ■
127244 Deyeuxia hakonensis (Franch. et Sav.) Keng = Calamagrostis hakonensis Franch. et Sav. ■
127245 Deyeuxia halleriana Vasey;北美野青茅■☆
127246 Deyeuxia henryi Rendle;房县野青茅(亨利野青茅);Fanghsien Small Reed, Fangxian Smallreed ■
127247 Deyeuxia himalaica L. Liou ex Wen L. Chen;喜马拉雅野青茅;Himalayan Small Reed, Himalayas Smallreed ■
127248 Deyeuxia holciformis (Jaub. et Spach) Bor;青藏野青茅;Qinzang Small Reed, Qinzang Smallreed ■
127249 Deyeuxia holciformis (Jaub. et Spach) Bor = Calamagrostis holciformis Jaub. et Spach ■
127250 Deyeuxia hugoniana Rendle = Poa glauca Vahl ■
127251 Deyeuxia hupehensis Rendle;湖北野青茅;Hubei Small Reed, Hubei Smallreed ■
127252 Deyeuxia kashmeriana Hack. ex Bor = Calamagrostis decora Hook. f. ■☆
127253 Deyeuxia kokonorica Keng;青海野青茅;Qinghai Small Reed, Qinghai Smallreed ■
127254 Deyeuxia kokonorica Keng = Calamagrostis kokonorica (Keng) Tzvelev ■
127255 Deyeuxia langsdorffii (Link) Kunth = Calamagrostis purpurea (Trin.) Trin. subsp. langsdorffii (Link) Tzvelev ■
127256 Deyeuxia langsdorffii (Link) Kunth = Calamagrostis purpurea (Trin.) Trin. ■
127257 Deyeuxia lapponica (Wahlenb.) Kunth;欧野青茅;Europe Smallreed, European Small Reed ■
127258 Deyeuxia lapponica (Wahlenb.) Kunth = Calamagrostis lapponica (Wahlenb.) Hartm. ■
127259 Deyeuxia levipes Keng;光柄野青茅;Smoothstalk Small Reed, Smoothstalk Smallreed ■
127260 Deyeuxia longiflora Keng;长花野青茅;Longflower Small Reed, Longflower Smallreed ■
127261 Deyeuxia longiflora Keng = Deyeuxia flavens Keng ■
127262 Deyeuxia longiseta (Hack.) Ohwi = Calamagrostis longiseta Hack. ■☆
127263 Deyeuxia longiseta (Hack.) Ohwi var. longearistata (Takeda) Ohwi = Calamagrostis grandiseta Takeda ■☆
127264 Deyeuxia macilenta (Griseb.) Keng;瘦野青茅;Thin Small Reed, Thin Smallreed ■
127265 Deyeuxia macilenta (Griseb.) Keng = Calamagrostis macilenta (Griseb.) Litv. ■
127266 Deyeuxia mannii Hook. f. = Agrostis mannii (Hook. f.) Stapf ■☆
127267 Deyeuxia masamunae (Honda) Ohwi = Calamagrostis masamunei Honda ■☆
127268 Deyeuxia matsudanae (Honda) Keng;短舌野青茅■
127269 Deyeuxia matsudanae (Honda) Keng = Deyeuxia brachytricha (Steud.) Y. L. Chang ■
127270 Deyeuxia matsumurae (Maxim.) Ohwi = Calamagrostis matsumurae Maxim. ■☆
127271 Deyeuxia megalantha Keng;大花野青茅;Largeflower Small Reed, Largeflower Smallreed ■
127272 Deyeuxia megalantha Keng = Deyeuxia pulchella (Griseb.) Hook. f. ■
127273 Deyeuxia megalantha Keng ex P. C. Keng = Deyeuxia pulchella (Griseb.) Hook. f. ■
127274 Deyeuxia montana P. Beauv.;山野青茅■
127275 Deyeuxia moupinensis (Franch.) Pilg.;宝兴野青茅;Mouping Small Reed, Mouping Smallreed ■
127276 Deyeuxia neglecta (Ehrh.) Kunth;小花野青茅(忽略野青茅);Littleflower Small Reed, Littleflower Smallreed, Slim-stem Reed Grass, Slim-stem Reed-grass ■
127277 Deyeuxia neglecta (Ehrh.) Kunth = Calamagrostis neglecta (Ehrh.) Gaertn., Mey. et Scherb. ■
127278 Deyeuxia neglecta (Ehrh.) Kunth var. aculeolata (Hack.) Ohwi = Calamagrostis neglecta (Ehrh.) Gaertn., B. Mey. et Scherb. subsp. inexpansa (A. Gray) Tzvelev ■☆
127279 Deyeuxia nepalensis Bor;顶芒野青茅(尼泊尔野青茅);Nepal Small Reed, Nepal Smallreed ■
127280 Deyeuxia nivicola Hook. f.;微药野青茅(云生野青茅);Little Anther Small Reed, Minianther Smallreed ■
127281 Deyeuxia nyingchiensis P. C. Kuo et S. L. Lu;林芝野青茅;Linzhi Small Reed, Linzhi Smallreed ■
127282 Deyeuxia petelotii (Hitchc.) S. M. Phillips et Wen L. Chen = Anisachne gracilis Keng ■
127283 Deyeuxia pseudopoa Jansen = Aniselytron treutleri (Kuntze) Soják ■
127284 Deyeuxia pulchella (Griseb.) Hook. f.;小丽茅(小丽草);Pretty Small Reed, Pretty Smallreed ■
127285 Deyeuxia pulchella (Griseb.) Hook. f. var. laxa P. C. Kuo et S. L. Lu;川藏野青茅;Chuanzang Small Reed, Chuanzang Smallreed ■
127286 Deyeuxia pulchella (Griseb.) Hook. f. var. laxa P. C. Kuo et S. L. Lu = Deyeuxia pulchella (Griseb.) Hook. f. ■
127287 Deyeuxia pulchella Hook. f. = Deyeuxia pulchella (Griseb.) Hook. f. ■
127288 Deyeuxia pulchella Hook. f. var. laxa P. C. Kuo et S. L. Lu = Deyeuxia pulchella (Griseb.) Hook. f. ■
127289 Deyeuxia robusta (Franch. et Sav.) Makino = Deyeuxia arundinacea (L.) P. Beauv. var. robusta (Franch. et Sav.) P. C. Kuo et S. L. Lu ■
127290 Deyeuxia rosea Bor;玫红野青茅;Rose-red Small Reed, Rosy Smallreed ■
127291 Deyeuxia sachalinensis (F. Schmidt) Rendle = Calamagrostis sachalinensis F. Schmidt ■☆
127292 Deyeuxia scabrescens (Griseb.) Munro ex Duthie;糙野青茅;Scabrous Small Reed, Scabrous Smallreed ■
127293 Deyeuxia scabrescens (Griseb.) Munro ex Duthie = Calamagrostis scabrescens Griseb. ■
127294 Deyeuxia scabrescens (Griseb.) Munro ex Duthie var. humilis (Griseb.) Hook. f.;小糙野青茅;Lowly Small Reed, Small Scabrous

Smallreed ■
127295 Deyeuxia sikangensis Keng;西康野青茅;Sikang Small Reed, Xikang Smallreed ■
127296 Deyeuxia sinelatior Keng = Deyeuxia sinelatior Keng ex P. C. Kuo ■
127297 Deyeuxia sinelatior Keng ex P. C. Kuo;华高野青茅;China High Smallreed, Chinese High Small Reed ■
127298 Deyeuxia stenophylla (Hand. -Mazz.) P. C. Kuo et S. L. Lu;会理野青茅;Huili Small Reed, Huili Smallreed ■
127299 Deyeuxia suizanensis (Hayata) Ohwi;水山野青茅;Shuishan Small Reed, Shuishan Smallreed ■
127300 Deyeuxia sylvatica (Schrad.) Kunth = Calamagrostis arundinacea (L.) Roth ■☆
127301 Deyeuxia sylvatica (Schrad.) Kunth = Deyeuxia arundinacea (L.) P. Beauv. ■
127302 Deyeuxia sylvatica (Schrad.) Kunth var. adpressiramea (Ohwi) Ohwi = Calamagrostis adpressiramea Ohwi ■☆
127303 Deyeuxia sylvatica (Schrad.) Kunth var. borealis Rendle = Deyeuxia arundinacea (L.) P. Beauv. var. borealis (Rendle) P. C. Kuo et S. L. Lu ■
127304 Deyeuxia sylvatica (Schrad.) Kunth var. brachytricha (Steud.) Rendle = Calamagrostis brachytricha Steud. ■
127305 Deyeuxia sylvatica (Schrad.) Kunth var. brachytricha Steud. = Deyeuxia arundinacea (L.) P. Beauv. var. brachytricha (Steud.) P. C. Kuo et S. L. Lu ■
127306 Deyeuxia sylvatica (Schrad.) Kunth var. ciliata Honda = Deyeuxia arundinacea (L.) P. Beauv. var. ciliata (Honda) P. C. Kuo et S. L. Lu ■
127307 Deyeuxia sylvatica (Schrad.) Kunth var. collina Franch. = Deyeuxia arundinacea (L.) P. Beauv. var. collina (Franch.) P. C. Kuo et S. L. Lu ■
127308 Deyeuxia sylvatica (Schrad.) Kunth var. hirsuta Hack. = Deyeuxia arundinacea (L.) P. Beauv. var. hirsuta (Hack.) P. C. Kuo et S. L. Lu ■
127309 Deyeuxia sylvatica (Schrad.) Kunth var. latifolia Rendle = Deyeuxia arundinacea (L.) P. Beauv. var. laxiflora (Rendle) P. C. Kuo et S. L. Lu ■
127310 Deyeuxia sylvatica (Schrad.) Kunth var. ligulata Rendle = Deyeuxia arundinacea (L.) P. Beauv. var. ligulata (Rendle) P. C. Kuo et S. L. Lu ■
127311 Deyeuxia sylvatica (Schrad.) Kunth var. robusta Franch. et Sav. = Deyeuxia arundinacea (L.) P. Beauv. var. robusta (Franch. et Sav.) P. C. Kuo et S. L. Lu ■
127312 Deyeuxia sylvatica (Schrad.) Kunth var. sciuroides Franch. et Sav. = Deyeuxia arundinacea (L.) P. Beauv. var. sciuroides (Franch. et Sav.) P. C. Kuo et S. L. Lu ■
127313 Deyeuxia tashiroi (Ohwi) Ohwi = Calamagrostis tashiroi Ohwi ■☆
127314 Deyeuxia tianschanica (Rupr.) Bor;天山野青茅;Tianshan Reedbentgrass, Tianshan Small Reed, Tianshan Smallreed ■
127315 Deyeuxia tianschanica (Rupr.) Bor = Calamagrostis tianschanica Rupr. ■
127316 Deyeuxia tibetica Bor;藏野青茅;Tibet Small Reed, Xizang Smallreed ■
127317 Deyeuxia tibetica Bor var. przevalskyi (Tzvelev) P. C. Kuo et S. L. Lu;矮野青茅(普氏野青茅);Przevalsky Small Reed ■
127318 Deyeuxia treutleri (Kuntze) Stapf = Aniselytron treutleri (Kuntze) Soják ■
127319 Deyeuxia treutleri (Kuntze) Stapf = Aulacolepis treutleri (Kuntze) Hack. ■
127320 Deyeuxia tripilifera (Hook. f.) Keng;三刺野青茅;Three-hairs Small Reed ■
127321 Deyeuxia turczaninowii (Litv.) Y. L. Chang;兴安野青茅;Xing'an Small Reed, Xing'an Smallreed ■
127322 Deyeuxia turczaninowii (Litv.) Y. L. Chang var. nenjiangensis S. L. Lu;长毛野青茅;Longhair Small Reed, Nenjiang Xing'an Smallreed ■
127323 Deyeuxia venusta Keng;美丽野青茅;Beatiful Smallreed, Beauty Small Reed ■
127324 Deyeuxia venusta Keng = Deyeuxia flavens Keng ■
127325 Deyeuxia zangxiensis P. C. Kuo et S. L. Lu;藏西野青茅;W. Xizang Smallreed, Zangxi Small Reed ■
127326 Dhofaria A. G. Mill. (1988);星被山柑属●☆
127327 Dhofaria macleishii A. G. Mill.;星被山柑●☆
127328 Diabelia Landrein = Abelia R. Br. ●
127329 Diabelia Landrein = Linnaea L. ●
127330 Diabelia Landrein(2010);双六道木属●☆
127331 Diacaecarpium Endl. = Alangium Lam.(保留属名)●
127332 Diacaecarpium Endl. = Diacicarpium Blume ●
127333 Diacantha Lag. = Chuquiraga Juss. ●☆
127334 Diacantha Less. = Barnadesia Mutis ex L. f. ●☆
127335 Diacarpa Sim = Atalaya Blume ●☆
127336 Diacarpa alata Sim = Atalaya alata (Sim) H. M. L. Forbes ●☆
127337 Diaceearpium Hassk. = Alangium Lam.(保留属名)●
127338 Diaceearpium Hassk. = Diacicarpium Blume ●
127339 Diachroa Nutt. = Leptochloa P. Beauv. ■
127340 Diachroa Nutt. ex Steud. = Glyceria R. Br.(保留属名)■
127341 Diachyrium Griseb. = Sporobolus R. Br. ■
127342 Diacicarpium Blume = Alangium Lam.(保留属名)●
127343 Diacicarpium rotundifolium Hassk. = Alangium chinense (Lour.) Harms ●
127344 Diacicarpium tomentosum Blume = Alangium chinense (Lour.) Harms ●
127345 Diacicarpium tomentosum Blume = Alangium kurzii Craib ●
127346 Diacidia Griseb. (1858);二裂金虎尾属●☆
127347 Diacidia cordata (Maguire) W. R. Anderson;心形二裂金虎尾●☆
127348 Diacidia ferruginea (Maguire et Phelps) W. R. Anderson;锈色二裂金虎尾●☆
127349 Diacidia galphimioides Griseb.;二裂金虎尾●☆
127350 Diacidia glaucifolia (Maguire) W. R. Anderson;灰叶二裂金虎尾●☆
127351 Diacidia parvifolia Cuatrec.;小叶二裂金虎尾●☆
127352 Diacisperma Kuntze = Leptochloa P. Beauv. ■
127353 Diacisperma Post et Kuntze = Disakisperma Steud. ■
127354 Diacisperma Post et Kuntze = Leptochloa P. Beauv. ■
127355 Diacles Salisb. = Haemanthus L. ■
127356 Diacranthera R. M. King et H. Rob. (1972);光果柄泽兰属■●☆
127357 Diacranthera crenata (Schltdl. ex Mart.) R. M. King et H. Rob.;光果柄泽兰●☆
127358 Diacrium (Lindl.) Bentha. = Caularthron Raf. ■☆
127359 Diacrium Benth. = Caularthron Raf. ■☆
127360 Diacrodon Sprague(1928);双齿茜属☆
127361 Diacrodon compressus Sprague;双齿茜☆
127362 Diadenaria Klotzsch et Garcke(1859);双腺戟属■☆

127363 Diadeniopsis Szlach. (2006);拟双腺兰属■☆
127364 Diadeniopsis Szlach. = Diadenium Poepp. et Endl. ■☆
127365 Diadenium Poepp. et Endl. (1836);双腺兰属■☆
127366 Diadenium micranthum Poepp. et Endl.;双腺兰■☆
127367 Diadesma Raf. (1834) = Modiola Moench ■☆
127368 Diadesma Raf. (1836) = Sida L. ●■
127369 Dialanthera Raf. = Cassia L. (保留属名)●■
127370 Dialesta Kunth = Pollalesta Kunth ●☆
127371 Dialion Raf. = Heliotropium L. ●■
127372 Dialiopsis Radlk. = Zanha Hiern ●☆
127373 Dialiopsis africana Radlk. = Zanha africana (Radlk.) Exell ●☆
127374 Dialissa Lindl. = Stelis Sw. (保留属名)■☆
127375 Dialium L. (1767);摘亚苏木属;Dialium ●☆
127376 Dialium acuminatum De Wild. = Dialium zenkeri Harms ●☆
127377 Dialium angolense Welw. ex Oliv.;安哥拉摘亚苏木●☆
127378 Dialium aubrevillei Pellegr.;奥布摘亚苏木●☆
127379 Dialium bipindense Harms;比平迪摘亚苏木(拜平摘亚苏木)●☆
127380 Dialium cochinchinense Pierre;交趾摘亚苏木●☆
127381 Dialium connaroides Harms ex Baker f. = Dialium bipindense Harms ●☆
127382 Dialium connaroides Harms ex De Wild. = Dialium bipindense Harms ●☆
127383 Dialium corbisieri Staner;穆苏摘亚苏木●☆
127384 Dialium coromandelicum Houtt. = Lannea coromandelica (Houtt.) Merr. ●
127385 Dialium densiflorum Harms;密花摘亚苏木●☆
127386 Dialium dinklagei Harms;科特迪瓦摘亚苏木●☆
127387 Dialium englerianum Henriq.;恩格勒摘亚苏木●☆
127388 Dialium eurysepalum Harms;宽萼摘亚苏木●☆
127389 Dialium evrardii Steyaert = Dialium angolense Welw. ex Oliv. ●☆
127390 Dialium excelsum Louis ex Steyaert;大摘亚苏木●☆
127391 Dialium excelsum Steyaert;高大摘亚苏木●☆
127392 Dialium fleuryi Pellegr.;摘亚苏木●☆
127393 Dialium fleuryi Pellegr. = Dialium bipindense Harms ●☆
127394 Dialium gossweileri Baker f.;戈斯摘亚苏木●☆
127395 Dialium graciliflorum Harms;细花摘亚苏木●☆
127396 Dialium guianense (Aubl.) Sandwith;圭亚那摘亚苏木;Velvet Tamarind ●☆
127397 Dialium guineense Willd.;抗盐摘亚苏木;Velvet Tamarind ●☆
127398 Dialium hexasepalum Harms;六萼摘亚苏木●☆
127399 Dialium holtzii Harms;霍氏摘亚苏木●☆
127400 Dialium indum L.;印度摘亚苏木●☆
127401 Dialium kasaiense Louis ex Steyaert;开赛摘亚苏木●☆
127402 Dialium klainei Pierre ex Harms = Dialium dinklagei Harms ●☆
127403 Dialium lacourtianum De Wild. ex Vermoesen = Dialium englerianum Henriq. ●☆
127404 Dialium latifolium Harms;宽叶摘亚苏木●☆
127405 Dialium laurentii De Wild. = Dialium zenkeri Harms ●☆
127406 Dialium macranthum A. Chev.;大花摘亚苏木●☆
127407 Dialium macranthum A. Chev. = Dialium pachyphyllum Harms ●☆
127408 Dialium madagascariense Baill.;马岛摘亚苏木●☆
127409 Dialium madagascariense Baill. subsp. occidentale Capuron = Dialium occidentale (Capuron) Du Puy et R. Rabev. ●☆
127410 Dialium mayumbense Baker f. = Dialium tessmannii Harms ●☆
127411 Dialium mossambicense Steyaert = Dialium holtzii Harms ●☆
127412 Dialium occidentale (Capuron) Du Puy et R. Rabev.;西方摘亚苏木●☆
127413 Dialium orientale Baker f.;东方摘亚苏木●☆
127414 Dialium ovatum Hutch. et Dalziel = Dialium pobeguinii Pellegr. ●☆
127415 Dialium pachyphyllum Harms;厚叶摘亚苏木●☆
127416 Dialium pentandrum Louis ex Steyaert;五雄摘亚苏木●☆
127417 Dialium pierrei De Wild. = Dialium soyauxii Harms ●☆
127418 Dialium platysepalum Baker;宽萼片摘亚苏木●☆
127419 Dialium pobeguinii Pellegr.;波别摘亚苏木●☆
127420 Dialium poggei Harms;波格摘亚苏木●☆
127421 Dialium polyanthum Harms = Dialium pachyphyllum Harms ●☆
127422 Dialium quinquepetalum Pellegr. = Dialium englerianum Henriq. ●☆
127423 Dialium reticulatum Burtt Davy et MacGregor = Dialium orientale Baker f. ●☆
127424 Dialium reygaertii De Wild.;赖氏摘亚苏木●☆
127425 Dialium schlechteri Harms;施莱摘亚苏木●☆
127426 Dialium simii E. Phillips = Dialium englerianum Henriq. ●☆
127427 Dialium soyauxii Harms;索氏摘亚苏木●☆
127428 Dialium staudtii Harms = Dialium dinklagei Harms ●☆
127429 Dialium tessmannii Harms;泰斯曼摘亚苏木●☆
127430 Dialium unifoliolatum Capuron ex Du Puy et R. Rabev.;单叶摘亚苏木●☆
127431 Dialium yambataense Vermoesen = Dialium pachyphyllum Harms ●☆
127432 Dialium zenkeri Harms;岑克尔摘亚苏木●☆
127433 Dialla Lindl. = Dicella Griseb. ●☆
127434 Diallobus Raf. = Cassia L. (保留属名)●■
127435 Diallosperma Raf. = Aspalathus L. ●☆
127436 Diallosteira Raf. = Collinsonia L. ■☆
127437 Dialyanthera Warb. = Otoba (A. DC.) H. karst. ●☆
127438 Dialycarpa Mast. = Brownlowia Roxb. (保留属名)●☆
127439 Dialyceras Capuron(1962);双角木属●☆
127440 Dialyceras coriaceum (Capuron) J. -F. Leroy;双角木●☆
127441 Dialyceras discolor (Capuron) J. -F. Leroy;二色双角木●☆
127442 Dialyceras parvifolium Capuron;小叶双角木●☆
127443 Dialyceras parvifolium Capuron f. discolore Capuron = Dialyceras discolor (Capuron) J. -F. Leroy ●☆
127444 Dialyceras parvifolium Capuron var. coriaceum Capuron = Dialyceras coriaceum (Capuron) J. -F. Leroy ●☆
127445 Dialypetalanthaceae Rizzini et Occhioni(1948)(保留科名);毛枝树科(巴西离瓣花科,拟素馨科,素馨科,枝树科)●☆
127446 Dialypetalanthaceae Rizzini et Occhioni (保留科名) = Rubiaceae Juss. (保留科名)●■
127447 Dialypetalanthus Kuhlm. (1925);毛枝树属●☆
127448 Dialypetalanthus fuscescens Kuhlm.;毛枝树●☆
127449 Dialypetalum Benth. (1873);分瓣桔梗属●■☆
127450 Dialypetalum floribundum Benth.;繁花分瓣桔梗■☆
127451 Dialypetalum montanum E. Wimm.;山地分瓣桔梗■☆
127452 Dialypetalum stenopetalum E. Wimm.;分瓣桔梗■☆
127453 Dialytheca Exell et Mendonça(1935);安哥拉藤属●☆
127454 Dialytheca gossweileri Exell et Mendonça;安哥拉藤●☆
127455 Diamarips Raf. = Salix L. (保留属名)●
127456 Diamena Ravenna = Anthericum L. ■☆
127457 Diamena Ravenna(1987);肖花篱属■☆
127458 Diamena stenantha (Ravenna) Ravenna;肖花篱■☆
127459 Diamonon Raf. = Solanum L. ●■

127460 Diamorpha Nutt. (1818);聚伞景天属■☆
127461 Diamorpha Nutt. = Sedum L. ●■
127462 Diamorpha cymosa (Nutt.) Britton ex Small;聚伞景天■☆
127463 Diamorpha pusilla Nutt. = Diamorpha cymosa (Nutt.) Britton ex Small ■☆
127464 Diamorpha smallii Britton = Sedum smallii (Britton) H. E. Ahles ■☆
127465 Diana Comm. ex Lam. = Dianella Lam. ex Juss. ●■
127466 Diandranthus L. Liou = Miscanthus Andersson ■
127467 Diandranthus L. Liou(1997);双药芒属;Bistamengrass ■
127468 Diandranthus aristatus L. Liou;芒稃双药芒;Awnlemma Bistamengrass ■
127469 Diandranthus brevipilus (Hand.-Mazz.) L. Liou;短毛双药芒(短毛芒);Shorthair Bistamengrass ■
127470 Diandranthus brevipilus (Hand.-Mazz.) L. Liou = Miscanthus nudipes (Griseb.) Hack. ■
127471 Diandranthus corymbosus L. Liou;伞房双药芒;Corymb Bistamengrass ■
127472 Diandranthus corymbosus L. Liou = Miscanthus nudipes (Griseb.) Hack. ■
127473 Diandranthus eulalioides (Keng) L. Liou;类金茅双药芒(类金茅芒);Goldquitchlike Bistamengrass ■
127474 Diandranthus eulalioides (Keng) L. Liou = Miscanthus nudipes (Griseb.) Hack. ■
127475 Diandranthus nepalensis (Trin.) L. Liou;尼泊尔双药芒(尼泊尔芒);Nepal Bistamengrass,Nepal Silvergrass ■
127476 Diandranthus nepalensis (Trin.) L. Liou = Miscanthus nepalensis (Trin.) Hack. ■
127477 Diandranthus nudipes (Griseb.) L. Liou;双药芒(光柄草,光柄芒);Bianthela Silvergrass,Naked-stalk Bistamengrass,Naked-stalk Silvergrass ■
127478 Diandranthus nudipes (Griseb.) L. Liou = Miscanthus nudipes (Griseb.) Hack. ■
127479 Diandranthus ramosus L. Liou;分枝双药芒;Branchy Bistamengrass ■
127480 Diandranthus szechuanensis (Keng ex S. L. Zhong) L. Liou = Miscanthus nudipes (Griseb.) Hack. ■
127481 Diandranthus taylorii (Bor) L. Liou;紫毛双药芒;Purplehair Bistamengrass ■
127482 Diandranthus taylorii (Bor) L. Liou = Miscanthus nudipes (Griseb.) Hack. ■
127483 Diandranthus tibeticus L. Liou;西藏双药芒;Xizang Bistamengrass ■
127484 Diandranthus tibeticus L. Liou = Miscanthus nudipes (Griseb.) Hack. ■
127485 Diandranthus wardii (Bor) L. Liou = Miscanthus nudipes (Griseb.) Hack. ■
127486 Diandranthus yunnanensis (A. Camus) L. Liou;西南双药芒(川芒);Yunnan Bistamengrass ■
127487 Diandranthus yunnanensis (A. Camus) L. Liou = Miscanthus nudipes (Griseb.) Hack. ■
127488 Diandriella Engl. = Homalomena Schott ■
127489 Diandrochloa De Winter = Eragrostis Wolf ■
127490 Diandrochloa diarrhena (Schult.) Henry = Eragrostis japonica (Thunb.) Trin. ■
127491 Diandrochloa diplachnoides (Steud.) A. N. Henry = Eragrostis japonica (Thunb.) Trin. ■
127492 Diandrochloa diplachnoides (Steud.) Henry = Eragrostis japonica (Thunb.) Trin. ■
127493 Diandrochloa japonica (Thunb.) Henry = Eragrostis japonica (Thunb.) Trin. ■
127494 Diandrochloa namaquensis (Nees ex Schrad.) De Winter = Eragrostis namaquensis Nees ex Schrad. ■☆
127495 Diandrochloa pusilla (Hack.) De Winter;微小双药芒■☆
127496 Diandrolyra Stapf(1906);双药禾属■☆
127497 Diandrolyra bicolor Stapf;双药禾■☆
127498 Diandrostachya (C. E. Hubb.) Jacq.-Fél. = Loudetiopsis Conert ■☆
127499 Diandrostachya chrysothrix (Nees) Jacq.-Fél. = Loudetiopsis chrysothrix (Nees) Conert ■☆
127500 Diandrostachya fulva (C. E. Hubb.) Jacq.-Fél. = Loudetiopsis chrysothrix (Nees) Conert ■☆
127501 Diandrostachya glabrinodis (C. E. Hubb.) J. B. Phipps = Loudetiopsis kerstingii (Pilg.) Conert ■☆
127502 Diandrostachya kerstingii (Pilg.) Jacq.-Fél. = Loudetiopsis kerstingii (Pilg.) Conert ■☆
127503 Dianella Lam. = Dianella Lam. ex Juss. ●■
127504 Dianella Lam. ex Juss. (1789);山菅属(桔梗兰属,山菅兰属);Dianella,Flax Lily,Flax-lily ●■
127505 Dianella caerulea Sims;蓝山菅;Cerulean Flaxlily,Flax Lily,New South Wales Dianella ■☆
127506 Dianella ensifolia (L.) DC. = Dianella ensifolia (L.) DC. ex Redoute ■
127507 Dianella ensifolia (L.) DC. ex Redoute;山菅(白花桔梗兰,碟碟草,家鼠草,假射干,铰箭王,较剪草,桔梗兰,老鼠砒,老鼠药,山大箭兰,山菅兰,山交剪,山猫儿,蛇王修,石兰花,水叶兰,天蒜);Cerulean Flax-lily,Swordleaf Dianella,Umbrella Dracaena ■
127508 Dianella ensifolia (L.) DC. ex Redoute f. albiflora Tang S. Liu et S. S. Ying = Dianella ensifolia (L.) DC. ex Redoute ■
127509 Dianella ensifolia (L.) DC. ex Redoute f. racemulifera (Schlitter) Tang S. Liu et S. S. Ying = Dianella ensifolia (L.) DC. ex Redoute ■
127510 Dianella ensifolia Redoute = Dianella ensifolia (L.) DC. ex Redoute ■
127511 Dianella longifolia R. Br.;长叶山菅;Flax Lily ■☆
127512 Dianella mairei H. Lév. = Stemona mairei (H. Lév.) Krause ■
127513 Dianella nemorosa Lam. = Dianella ensifolia (L.) DC. ex Redoute ■
127514 Dianella nemorosa Lam. f. racemulifera Schlitter = Dianella ensifolia (L.) DC. ex Redoute ■
127515 Dianella nigra Colenso;黑山菅;Blue Berry ■☆
127516 Dianella revoluta R. Br.;外卷山菅;Spreading Hax-lily ■☆
127517 Dianella tasmanica Kunth;塔斯马尼亚山菅;Blueberry Flax Lily,Flax Lily ■☆
127518 Dianella triandra Afzel. = Palisota hirsuta (Thunb.) K. Schum. ■☆
127519 Dianellaceae Salisb.;山菅科(山菅兰科)■
127520 Dianellaceae Salisb. = Hemerocallidaceae R. Br. ■
127521 Dianellaceae Salisb. = Phormiaceae J. Agardh ●■
127522 Diania Noronha ex Tul. = Dicoryphe Thouars ●☆
127523 Dianisteris Raf. = Verbesina L.(保留属名)●■☆
127524 Dianthaceae Drude = Caryophyllaceae Juss.(保留科名)■●
127525 Dianthaceae Vest = Caryophyllaceae Juss.(保留科名)■●
127526 Dianthella Clauson ex Pomel = Petrorhagia (Ser. ex DC.) Link ■

127527 Dianthella Clauson ex Pomel = Tunica (Hallier) Scop. ■
127528 Dianthella Clauson ex Pomel(1860);小石竹属■☆
127529 Dianthella compressa (Desf.) Pomel;小石竹■☆
127530 Dianthella compressa (Desf.) Pomel = Petrorhagia illyrica (Ard.) P. W. Ball et Heywood subsp. angustifolia (Poir.) P. W. Ball et Heywood ■☆
127531 Dianthella compressa (Desf.) Pomel var. australis Batt. = Petrorhagia illyrica (Ard.) P. W. Ball et Heywood subsp. angustifolia (Poir.) P. W. Ball et Heywood ■☆
127532 Dianthera Klotzsch = Cleome L. ●■
127533 Dianthera L. (1753);双药爵床属■☆
127534 Dianthera L. = Justicia L. ●■
127535 Dianthera abyssinica Schweinf. = Cleome angustifolia Forssk. ■☆
127536 Dianthera americana L.;双药爵床■☆
127537 Dianthera americana L. = Justicia americana (L.) Vahl ●☆
127538 Dianthera americana L. var. subcoriacea (Fernald) Shinners = Justicia americana (L.) Vahl ●☆
127539 Dianthera ansellianа (Nees) Benth. ex B. D. Jacks. = Justicia anselliana (Nees) T. Anderson ■☆
127540 Dianthera bicalyculata Retz. = Dicliptera paniculata (Forssk.) I. Darbysh. ■☆
127541 Dianthera bicalyculata Retz. = Peristrophe bicalyculata (Retz.) Nees ■
127542 Dianthera bicalyculata Retz. = Peristrophe paniculata (Forssk.) Brummitt ■☆
127543 Dianthera bicolor Pax = Cleome oxyphylla Burch. ■☆
127544 Dianthera burchelliana Klotzsch ex Sond. = Cleome angustifolia Forssk. var. diandra (Burch.) Kers ■☆
127545 Dianthera carnosa Pax = Cleome carnosa (Pax) Gilg et Gilg-Ben. ■☆
127546 Dianthera collina (T. Anderson) C. B. Clarke = Isoglossa collina (T. Anderson) B. Hansen ■
127547 Dianthera debilis Forssk. = Monechma debile (Forssk.) Nees ■☆
127548 Dianthera flava Vahl = Justicia flava (Vahl) Vahl ■☆
127549 Dianthera hochstetteri Eichler = Cleome angustifolia Forssk. ■☆
127550 Dianthera japonica Thunb. = Peristrophe japonica (Thunb.) Bremek. ■
127551 Dianthera leptostachya (Nees) Blatt. = Justicia calyculata Deflers ■☆
127552 Dianthera malabarica L. f. = Dicliptera paniculata (Forssk.) I. Darbysh. ■☆
127553 Dianthera malabarica L. f. = Peristrophe paniculata (Forssk.) Brummitt ■☆
127554 Dianthera odora Forssk. = Justicia odora (Forssk.) Vahl ■☆
127555 Dianthera paniculata Forssk. = Dicliptera paniculata (Forssk.) I. Darbysh. ■☆
127556 Dianthera paniculata Forssk. = Peristrophe paniculata (Forssk.) Brummitt ■☆
127557 Dianthera pectoralis (Jacq.) Murray = Justicia pectoralis Jacq. ■☆
127558 Dianthera petersiana Klotzsch = Cleome angustifolia Forssk. subsp. petersiana (Klotzsch) Kers ■☆
127559 Dianthera punctata Vahl = Isoglossa punctata (Vahl) Brummitt et J. R. I. Wood ■☆
127560 Dianthera semitetrandra (Sond.) Klotzsch et Sond. = Cleome semitetranda Sond. ■☆
127561 Dianthera sinensis W. W. Sm. = Isoglossa collina (T. Anderson) B. Hansen ■
127562 Dianthera sulcata Vahl = Justicia flava (Vahl) Vahl ■☆
127563 Dianthera trisulca Forssk. = Anisotes trisulcus (Forssk.) Nees ●☆
127564 Dianthera verticillata Forssk. = Dicliptera verticillata (Forssk.) C. Chr. ■☆
127565 Dianthera violacea Vahl = Megalochlamys violacea (Vahl) Vollesen ●☆
127566 Dianthoseris Sch. Bip. = Dianthoseris Sch. Bip. ex A. Rich. ■☆
127567 Dianthoseris Sch. Bip. ex A. Rich. (1842);高山苣属■☆
127568 Dianthoseris rueppellii Sch. Bip. ex Oliv. et Hiern = Launaea rueppellii (Sch. Bip. ex Oliv. et Hiern) Amin ex Boulos ■☆
127569 Dianthoseris schimperi A. Rich. = Dianthoseris schimperi Sch. Bip. ex A. Rich. ■☆
127570 Dianthoseris schimperi Sch. Bip. ex A. Rich.;高山苣■☆
127571 Dianthoveus Hammel et Wilder(1989);双花巴拿马草属■☆
127572 Dianthoveus cremnophilus Hammel et Wilder;双花巴拿马草■☆
127573 Dianthus L. (1753);石竹属;Carnation, Dianthus, Gilliflower, Pink, Pinks, Thrift ■
127574 Dianthus × isensis Hirahata et Kitam.;伊势石竹■☆
127575 Dianthus × nigritus Hirahata et Kitam.;黑石竹■☆
127576 Dianthus abyssinicus R. Br.;阿比西尼亚石竹■☆
127577 Dianthus acantholimonoides Schischk.;刺柠檬石竹■☆
127578 Dianthus acicularis Fisch. ex Ledeb.;针叶石竹;Needle Pink ■
127579 Dianthus albens Aiton;微白石竹■☆
127580 Dianthus alpinus Jacq. = Dianthus alpinus L. ■☆
127581 Dianthus alpinus L.;高山石竹;Alpine Pink ■☆
127582 Dianthus alpinus L. var. repens Regel;匍匐高山石竹■☆
127583 Dianthus alpinus L. var. semenovii Regel et Herder = Dianthus semenovii (Regel et Herder) Vierh. ■
127584 Dianthus alpinus Sibth. et Sm. = Dianthus haematocalyx Boiss. et Heldr. ■☆
127585 Dianthus alpinus Sturm ex Steud. = Dianthus glacialis Haenke ■☆
127586 Dianthus alpinus Vill. = Dianthus neglectus Loisel. ■☆
127587 Dianthus amoenus Pomel = Dianthus crinitus Sm. ■☆
127588 Dianthus amurensis Jacq.;东北石竹;Amur Pink ■
127589 Dianthus amurensis Jacq. = Dianthus chinensis L. ■
127590 Dianthus andronakii Woronow = Dianthus andronakii Woronow ex Schischk. ■☆
127591 Dianthus andronakii Woronow ex Schischk.;安德罗石竹■☆
127592 Dianthus andrzejowskianus (Zapal.) Kulcz.;安氏石竹■☆
127593 Dianthus angolensis Hiern ex F. N. Williams;安哥拉石竹■☆
127594 Dianthus angolensis Hiern ex F. N. Williams subsp. orientalis Turrill = Dianthus excelsus S. S. Hooper ■☆
127595 Dianthus anticarius Boiss. et Reut.;内向石竹■☆
127596 Dianthus arboreus L.;树状石竹(树石竹);Finland Pink, Tree Pink ■☆
127597 Dianthus arbusculus Lindl. = Dianthus caryophyllus L. ■
127598 Dianthus arenarius L.;白香石竹;Finland Pink, Sand Pink ■☆
127599 Dianthus aristidis Batt. = Dianthus sylvestris Wulfen subsp. aristidis (Batt.) Greuter et Burdet ■☆
127600 Dianthus armeria L.;亚美尼亚石竹;Deptford Pink, Maiden Pink ■☆
127601 Dianthus atlantica Romo = Dianthus lusitanus Brot. subsp. sidi (Font Quer) Dobignard ■☆
127602 Dianthus atlanticus Pomel = Dianthus vulturius Guss. et Ten. ■☆
127603 Dianthus attenuatus Ball = Dianthus lusitanus Brot. ■☆

127604 Dianthus balbisii Ser. subsp. medius (Rouy) Maire = Dianthus vulturius Guss. et Ten. ■☆

127605 Dianthus balbisii Ser. subsp. vulturius (Guss. et Ten.) Maire = Dianthus vulturius Guss. et Ten. ■☆

127606 Dianthus barati Duval-Jouve = Dianthus tripunctatus Sm. ■☆

127607 Dianthus barbatus L.;须苞石竹(金蝴蝶,美国石竹,美女抚子,美人草,十样锦,五彩石竹,亚美利加瞿麦);Barbate Pink, Bearded Pink, Bloomy Down, Bloomy-down, Colmenier, London Bob, London Bobs, London Pride, London Tuft, Lorelon Pride, Painted Lady, Pink Beauty, Pretty Willie, Sweet John, Sweet Pink, Sweet William, Sweetwilliam, Sweet-william, Tolmeiner, Tolmeneer, Velvet William ■

127608 Dianthus barbatus L. var. asiaticus Thunb.;头石竹;Asia Sweet Pink ■

127609 Dianthus barbatus L. var. shinanensis Yatabe = Dianthus shinanensis (Yatabe) Makino ■☆

127610 Dianthus basuticus Burtt Davy subsp. fourcadei S. S. Hooper;富尔卡德石竹■☆

127611 Dianthus basuticus Burtt Davy var. grandiflorus S. S. Hooper;大花石竹■☆

127612 Dianthus bicolor Adams;双色石竹■☆

127613 Dianthus boissieri Willk. = Dianthus sylvestris Wulfen subsp. boissieri (Willk.) Dobignard ■☆

127614 Dianthus bolivaris Sennen = Dianthus lusitanus Brot. ■☆

127615 Dianthus bolusii Burtt Davy;博卢斯石竹■☆

127616 Dianthus borbasii Vand.;博尔巴什石竹■☆

127617 Dianthus brachyanthus Boiss.;短花石竹■☆

127618 Dianthus brachyanthus Boiss. var. alpinus Willk. = Dianthus brachyanthus Boiss. ■☆

127619 Dianthus brachyanthus Boiss. var. maroccanus Pau et Font Quer = Dianthus brachyanthus Boiss. ■☆

127620 Dianthus brevicaulis Fenzl;短茎石竹;Mt. Taurus Pink ■☆

127621 Dianthus buergeri Miq.;布氏石竹■☆

127622 Dianthus burchellii Ser.;伯切尔石竹■☆

127623 Dianthus caesius Sm.;欧洲蔓丛石竹;Cheddar Pink ■☆

127624 Dianthus caesius Sm. = Dianthus gratianopolitanus Vill. ■☆

127625 Dianthus caespitosus Eckl. et Zeyh. = Dianthus crenatus Thunb. ■☆

127626 Dianthus caespitosus Thunb.;丛生石竹■☆

127627 Dianthus caespitosus Thunb. subsp. pectinatus (E. Mey. ex Sond.) S. S. Hooper;篦状石竹■☆

127628 Dianthus callizonus Schott et Kotschy;美环石竹;Zoned Pink ■☆

127629 Dianthus calocephalus Boiss.;美头石竹■☆

127630 Dianthus campestris M. Bieb.;田野石竹■☆

127631 Dianthus canesceus K. Koch;灰石竹■☆

127632 Dianthus capitatus Balb.;头状石竹■☆

127633 Dianthus carbonatus Klokov;煤色石竹■☆

127634 Dianthus carthusianorum L.;丹麦石竹;Carthusian Pink, Clusterhead, Clusterhead Pink, Cluster-head Pink, German Pink ■☆

127635 Dianthus caryophyllus L.;香石竹(大花石竹,荷兰瞿麦,红茂草,康纳馨,康乃馨,麝香石竹,狮头石竹,香剪绒花,洋石竹);Blunderbuss, Carnadine, Carnation, Clove Gilliflower, Clove July Flower, Clove July-flower, Clove Pink, Coronation, Cottage Pink, Crimson Lady, Gillyflower, Gypsy Pink, Harry Dobs, Incarnation, Indy Pink, Jack, Ley, May Pink, May-pink, Pheasant's Eye, Pheasant's Eyes, Picotee, Sops-in-wine, Wild Gilliflower ■

127636 Dianthus caryophyllus L. 'Juliet';朱丽叶石竹■

127637 Dianthus caryophyllus L. 'Pixie Delight';快乐石竹■

127638 Dianthus caryophyllus L. subsp. aristidis Batt. = Dianthus sylvestris Wulfen subsp. aristidis (Batt.) Greuter et Burdet ■☆

127639 Dianthus caryophyllus L. subsp. coronarius (L.) Maire = Dianthus caryophyllus L. ■

127640 Dianthus caryophyllus L. subsp. longibracteatus Maire = Dianthus sylvestris Wulfen subsp. longibracteatus (Maire) Greuter et Burdet ■☆

127641 Dianthus caryophyllus L. subsp. siculus (C. Presl) Arcang. = Dianthus sylvestris Wulfen subsp. siculus (C. Presl) Tutin ■☆

127642 Dianthus caryophyllus L. var. boissieri (Willk.) Emb. et Maire = Dianthus sylvestris Wulfen subsp. boissieri (Willk.) Dobignard ■☆

127643 Dianthus caryophyllus L. var. lanceolatus Pau = Dianthus sylvestris Wulfen subsp. longibracteatus (Maire) Greuter et Burdet ■☆

127644 Dianthus caryophyllus L. var. longicaulis (Ten.) Briq. = Dianthus sylvestris Wulfen subsp. boissieri (Willk.) Dobignard ■☆

127645 Dianthus caryophyllus L. var. longifolius Maire = Dianthus sylvestris Wulfen ■☆

127646 Dianthus caryophyllus L. var. mauritanicus Ball = Dianthus sylvestris Wulfen ■☆

127647 Dianthus caryophyllus L. var. mogadorensis Maire = Dianthus sylvestris Wulfen ■☆

127648 Dianthus caryophyllus L. var. puberulus Faure et Maire = Dianthus sylvestris Wulfen ■☆

127649 Dianthus caryophyllus L. var. siculus (C. Presl) Bonnet et Barratte = Dianthus sylvestris Wulfen subsp. siculus (C. Presl) Tutin ■☆

127650 Dianthus caryophyllus L. var. tenuicaulis Maire = Dianthus sylvestris Wulfen ■☆

127651 Dianthus caryophyllus L. var. transiens Maire = Dianthus sylvestris Wulfen ■☆

127652 Dianthus charmelii Sennen et Mauricio = Dianthus sylvestris Wulfen subsp. boissieri (Willk.) Dobignard ■☆

127653 Dianthus chimanimaniensis S. S. Hooper;奇马尼马尼石竹■☆

127654 Dianthus chinensis L.;石竹(大兰,剪绒花,巨句麦,巨麦草,瞿麦,洛阳花,南天竺草,青水红,山瞿麦,十样景花,石竹子,石竹子花,石柱花,石柱子花,竹节草);China Pink, Chinese Pink, Dianthus, French Mignonette, French Pink, Indian Pink, Japanese Pink, Pinks, Rainbow Pink ■

127655 Dianthus chinensis L. 'Fire Carpet';火焰花石竹■☆

127656 Dianthus chinensis L. f. albiflorus T. B. Lee;白花石竹■☆

127657 Dianthus chinensis L. f. ignescens (Nakai) Kitag.;火红石竹■

127658 Dianthus chinensis L. f. ignescens (Nakai) Kitag. = Dianthus chinensis L. ■

127659 Dianthus chinensis L. subsp. repens (Willd.) Vorosch. = Dianthus repens Willd. ■

127660 Dianthus chinensis L. subsp. versicolor (Fisch. ex Link) Vorosch. = Dianthus chinensis L. ■

127661 Dianthus chinensis L. var. amurensis (Jacq.) Katag. = Dianthus chinensis L. ■

127662 Dianthus chinensis L. var. dentosus (Fisch. ex Rchb.) Debeaux = Dianthus chinensis L. ■

127663 Dianthus chinensis L. var. heddewiggii ?;印度石竹;Indian Pink, Japanese Pink ■☆

127664 Dianthus chinensis L. var. ignescens Nakai = Dianthus chinensis L. ■

127665 Dianthus chinensis L. var. jingpoensis G. Y. Zhang et X. Y.

Yuan;镜波湖石竹;Jingbohu Pink ■

127666 Dianthus chinensis L. var. jingpoensis G. Y. Zhang et X. Y. Yuan = Dianthus chinensis L. ■

127667 Dianthus chinensis L. var. laciniatus Körn.;洛阳花(大阪抚子,日本剪绒花,日本石竹,萨摩抚子,伊势抚子,羽瓣石竹)■☆

127668 Dianthus chinensis L. var. liaotungensis Y. C. Chu;辽东石竹(长萼石竹);Liaodong Pink ■

127669 Dianthus chinensis L. var. liaotungensis Y. C. Chu = Dianthus chinensis L. ■

127670 Dianthus chinensis L. var. longisquama Nakai et Kitag.;长苞石竹;Longbract Pink ■

127671 Dianthus chinensis L. var. longisquama Nakai et Kitag. = Dianthus chinensis L. ■

127672 Dianthus chinensis L. var. macrosepalus Franch. ex Bailey = Dianthus chinensis L. ■

127673 Dianthus chinensis L. var. morii (Nakai) Y. C. Chu = Dianthus chinensis L. ■

127674 Dianthus chinensis L. var. semperflorens Makino;常夏石竹(常夏)■☆

127675 Dianthus chinensis L. var. shandongensis J. X. Li et F. Q. Zhou;山东石竹;Shandong Pink ■

127676 Dianthus chinensis L. var. subulifolius (Kitag.) Ma;钻叶石竹(蒙古石竹,丝叶石竹);Awlleaf Pink ■

127677 Dianthus chinensis L. var. subulifolius (Kitag.) Ma = Dianthus chinensis L. ■

127678 Dianthus chinensis L. var. sylvaticus Koch;林生石竹;Wood Pink ■

127679 Dianthus chinensis L. var. sylvaticus Koch = Dianthus chinensis L. ■

127680 Dianthus chinensis L. var. trinervis D. Q. Lu;三脉石竹;Trinerve Pink ■

127681 Dianthus chinensis L. var. trinervis D. Q. Lu = Dianthus chinensis L. ■

127682 Dianthus chinensis L. var. versicolor (Fisch. ex Link) Ma;兴安石竹(北石竹,变色石竹);Xing'an Pink ■

127683 Dianthus chinensis L. var. versicolor (Fisch. ex Link) Ma = Dianthus chinensis L. ■

127684 Dianthus cintranus Boiss. et Reut.;嘎迪石竹■☆

127685 Dianthus cintranus Boiss. et Reut. subsp. atrosanguineus (Emb. et Maire) Greuter et Burdet;暗血红石竹■☆

127686 Dianthus cintranus Boiss. et Reut. subsp. jahandiezii (Maire) Greuter et Burdet;贾汉石竹■☆

127687 Dianthus cintranus Boiss. et Reut. subsp. maroccanus (F. N. Williams) Greuter et Burdet;摩洛哥石竹■☆

127688 Dianthus cintranus Boiss. et Reut. subsp. mauritanicus (Pomel) Greuter et Burdet;毛里塔尼亚石竹■☆

127689 Dianthus cintranus Boiss. et Reut. subsp. occidentalis (Quézel) Mathez;西方石竹■☆

127690 Dianthus colensoi F. N. Williams = Dianthus zeyheri Sond. subsp. natalensis S. S. Hooper ■☆

127691 Dianthus crenatus Edwards = Dianthus caespitosus Thunb. subsp. pectinatus (E. Mey. ex Sond.) S. S. Hooper ■☆

127692 Dianthus crenatus Thunb.;圆齿石竹■☆

127693 Dianthus crinitus Sm.;长软毛石竹■☆

127694 Dianthus crinitus Sm. subsp. soongoricus (Schischk.) Kozhevn. = Dianthus soongoricus Schischk. ■

127695 Dianthus crinitus Sm. var. australis Maire = Dianthus crinitus Sm. ■☆

127696 Dianthus crinitus Sm. var. flaviflorus Emb. = Dianthus crinitus Sm. ■☆

127697 Dianthus crossopetalus Fenzl;粗瓣石竹■☆

127698 Dianthus cruentus Griseb.;深红石竹;Blood Pink, Maiden Pink, Meadow Pink, Wild Pink ■☆

127699 Dianthus cyri Fisch. et C. A. Mey.;西尔石竹■☆

127700 Dianthus deltoides L.;一叶石竹(美女石竹,少女石竹,西洋石竹,洋石竹);Maiden Pink, Meadow Pink ■☆

127701 Dianthus deltoides L. 'Leuchtfunk';闪光西洋石竹■☆

127702 Dianthus dentosus Fisch. ex Rchb. = Dianthus chinensis L. ■

127703 Dianthus diminutus L. = Petrorhagia prolifera (L.) P. W. Ball et Heywood ■☆

127704 Dianthus dinteri Schinz = Dianthus namaensis Schinz var. dinteri (Schinz) S. S. Hooper ■☆

127705 Dianthus discolor Sm.;异色石竹■☆

127706 Dianthus dubius Raf. = Petrorhagia dubia (Raf.) G. López et Romo ■☆

127707 Dianthus elatus Ledeb.;高石竹;Tall Pink ■

127708 Dianthus eugeniae Kleopow;欧根石竹■☆

127709 Dianthus excelsus S. S. Hooper;高大石竹■☆

127710 Dianthus fimbriatus M. Bieb. = Dianthus orientalis Adams ■

127711 Dianthus fischeri Spreng. = Dianthus chinensis L. ■

127712 Dianthus floribundus Boiss.;多花石竹■☆

127713 Dianthus fragrans Adams;芳香石竹(香石竹);Fragrant Pink, Sweet-scented Pink ■☆

127714 Dianthus freyni Vandas;弗雷恩冰石竹;Freyns Pink ■☆

127715 Dianthus furcatus Bourg. ex Nyman;彩石竹;Painted Pink ■☆

127716 Dianthus gaditanus Boiss. = Dianthus cintranus Boiss. et Reut. ■☆

127717 Dianthus gaditanus Boiss. subsp. atrosanguineus Emb. et Maire = Dianthus cintranus Boiss. et Reut. subsp. atrosanguineus (Emb. et Maire) Greuter et Burdet ■☆

127718 Dianthus gaditanus Boiss. subsp. jahandiezii Maire = Dianthus cintranus Boiss. et Reut. subsp. jahandiezii (Maire) Greuter et Burdet ■☆

127719 Dianthus gaditanus Boiss. subsp. maroccanus (Williams) Maire = Dianthus cintranus Boiss. et Reut. subsp. maroccanus (F. N. Williams) Greuter et Burdet ■☆

127720 Dianthus gaditanus Boiss. subsp. mauritanicus (Pomel) Maire = Dianthus cintranus Boiss. et Reut. subsp. mauritanicus (Pomel) Greuter et Burdet ■☆

127721 Dianthus gaditanus Boiss. subsp. occidentalis Quézel = Dianthus cintranus Boiss. et Reut. subsp. occidentalis (Quézel) Mathez ■☆

127722 Dianthus gaditanus Boiss. var. riphaeus Pau et Sennen = Dianthus cintranus Boiss. et Reut. subsp. maroccanus (F. N. Williams) Greuter et Burdet ■☆

127723 Dianthus gallicus Pers.;西部石竹;Jersey Pink, Western Pink ■☆

127724 Dianthus giganteus d'Urv.;巨石竹■☆

127725 Dianthus glacialis Haenke;冰石竹;Ice Pink ■☆

127726 Dianthus glaucus Buch.-Ham. = Dianthus gratianopolitanus Vill. ■☆

127727 Dianthus graniticus Jord.;花崗石竹;Granite Pink ■☆

127728 Dianthus gratianopolitanus Vill.;蓝灰石竹;Cheddar Pink, Cleeve Pink, Cliff Pink ■☆

127729 Dianthus gratianopolitanus Vill. = Dianthus caesius Sm. ■☆

127730 Dianthus grossheimii Schischk.;格罗石竹■☆

127731 Dianthus guttatus M. Bieb.;斑点石竹■☆

127732 Dianthus haematocalyx Boiss. et Heldr. ;红萼石竹■☆
127733 Dianthus hoeltzeri Winkl. ;大苞石竹(宽叶石竹);Bigbract Pink ■
127734 Dianthus holopetalus Turcz. ;全瓣石竹■☆
127735 Dianthus humilis Willd. ex Ledeb. ;矮石竹■☆
127736 Dianthus hybridus Schmidt ex Tausch;杂种石竹■☆
127737 Dianthus imereticus (Rupr.) Schischk. ;伊梅里特石竹■☆
127738 Dianthus incurvus Thunb. = Dianthus albens Aiton ■☆
127739 Dianthus japonicus Thunb. ;日本石竹(滨瞿麦,两面青);Japan Pink,Japanese Pink ■
127740 Dianthus japonicus Thunb. f. albiflorus Ohara ex Nakan. ;白花日本石竹■☆
127741 Dianthus junceus Burtt Davy = Dianthus namaensis Schinz var. junceus (Burtt Davy) S. S. Hooper ■☆
127742 Dianthus kamiesbergensis Sond. ;卡米斯贝赫石竹■☆
127743 Dianthus karataviensis Pavlov;卡拉塔夫石竹■☆
127744 Dianthus kirghizicus Schischk. ;基尔吉斯石竹■☆
127745 Dianthus kirkii Burtt Davy = Dianthus moviensis F. N. Williams subsp. kirkii (Burtt Davy) S. S. Hooper ■☆
127746 Dianthus kiusianus (Yatabe) Makino;九州石竹■☆
127747 Dianthus knappii (Pant.) Asch. et Kanitz ex Borbás;纳普石竹;Hardy Garden Pink,Yellow Dianthus ■☆
127748 Dianthus kubanensie Schischk. ;库班石竹■☆
127749 Dianthus kuschakewiczii Regel et Schmalh. ;长萼石竹;Kuschakewicz Pink ■
127750 Dianthus kusnezovii Marcow. ;库兹石竹■☆
127751 Dianthus laciniatus Makino = Dianthus chinensis L. var. laciniatus Körn. ■☆
127752 Dianthus lanceolatus Stev. ex Rchb. ;剑石竹■☆
127753 Dianthus latifolius Willd. ;宽叶石竹;Button Pink ■☆
127754 Dianthus leptoloma Steud. ex A. Rich. ;细边石竹■☆
127755 Dianthus leptopetalus Willd. ;细瓣石竹■☆
127756 Dianthus longicalyx Miq. ;长萼瞿麦(长萼石竹,长筒瞿麦);Longcalyx Pink,Longtube Fringed Pink ■
127757 Dianthus longicalyx Miq. = Dianthus superbus L. var. longicalycinus (Maxim.) F. N. Williams ■
127758 Dianthus longiglumis Delile;长颖石竹■☆
127759 Dianthus lusitanus Brot. ;葡萄牙石竹■☆
127760 Dianthus lusitanus Brot. subsp. sidi (Font Quer) Dobignard;北非石竹■☆
127761 Dianthus lusitanus Brot. var. imberbis Maire = Dianthus lusitanus Brot. ■☆
127762 Dianthus lusitanus Brot. var. latifolius Maire = Dianthus lusitanus Brot. subsp. sidi (Font Quer) Dobignard ■☆
127763 Dianthus macronyx Fenzl;大刺石竹■☆
127764 Dianthus marschallii Schischk. ;玛莎石竹■☆
127765 Dianthus mauritanicus Pomel = Dianthus cintranus Boiss. et Reut. subsp. mauritanicus (Pomel) Greuter et Burdet ■☆
127766 Dianthus mauritii Sennen;毛里特石竹■☆
127767 Dianthus mecistocalyx F. N. Williams = Dianthus zeyheri Sond. ■☆
127768 Dianthus membranaceus Borbás;膜质石竹■☆
127769 Dianthus mesanidum Litard. et Maire = Dianthus serrulatus Desf. subsp. macranthus Maire ■☆
127770 Dianthus microlepis Boiss. ;小鳞片石竹■☆
127771 Dianthus micropetalus Ser. ;小瓣石竹■☆
127772 Dianthus monspessulanus L. ;蒙彼利石竹;Fringed Pink,Montpellier Pink ■☆
127773 Dianthus morii Nakai = Dianthus chinensis L. ■
127774 Dianthus morrisii Hance = Dianthus caryophyllus L. ■
127775 Dianthus moviensis F. N. Williams;南非石竹■☆
127776 Dianthus moviensis F. N. Williams subsp. kirkii (Burtt Davy) S. S. Hooper;吉尔南非石竹■☆
127777 Dianthus moviensis F. N. Williams var. dentatus Burtt Davy;尖齿石竹■☆
127778 Dianthus multicaulls Boiss. et Huet;多茎石竹■☆
127779 Dianthus myrtinervius Griseb. ;多脉石竹;Carnation,Pink ■☆
127780 Dianthus namaensis Schinz;纳马石竹■☆
127781 Dianthus namaensis Schinz var. dinteri (Schinz) S. S. Hooper;丁特石竹■☆
127782 Dianthus namaensis Schinz var. junceus (Burtt Davy) S. S. Hooper;灯芯草石竹■☆
127783 Dianthus nanteuilii Burnat = Petrorhagia nanteulii (Burnat) P. W. Ball et Heywood ■☆
127784 Dianthus neglectus Loisel. = Dianthus pavonius Tausch ■☆
127785 Dianthus nelsonii F. N. Williams = Dianthus moviensis F. N. Williams ■☆
127786 Dianthus oreadum Hance = Dianthus longicalyx Miq. ■
127787 Dianthus orientalis Adams;缕裂石竹(东方石竹);Oriental Pink ■
127788 Dianthus palinensis S. S. Ying;巴陵石竹(八里石竹);Baling Pink ■
127789 Dianthus pallidiflorus Ser. ;苍白花石竹■☆
127790 Dianthus pavonius Tausch;冰山石竹(孔雀石竹);Galcier Pink ■☆
127791 Dianthus pectinatus E. Mey. ex Sond. = Dianthus caespitosus Thunb. subsp. pectinatus (E. Mey. ex Sond.) S. S. Hooper ■☆
127792 Dianthus petraeus Waldst. et Kit. ;岩地石竹;Rocky Pink ■☆
127793 Dianthus plumarius L. ;常花石竹;Burst-bellies, Burst-belly Pink, Busters, Clove Gilliflower, Clove Pink, Cottage Pink, Dianthus Pinks, Feathered Pink, Garden Pink, Grass Pink, Indian Eye, Indian Eyes, Pheasant's Eye Pink, Pink, Scotch Pink, Small Honesty ■☆
127794 Dianthus polymorphus M. Bieb. ;多型石竹■☆
127795 Dianthus pratensis M. Bieb. ;草原石竹■☆
127796 Dianthus preobrashenskii Klokov;普列奥石竹■☆
127797 Dianthus prolifer L. = Petrorhagia prolifera (L.) P. W. Ball et Heywood ■☆
127798 Dianthus prolifer L. subsp. velutinus (Guss.) Batt. = Petrorhagia dubia (Raf.) G. López et Romo ■☆
127799 Dianthus prostratus Jacq. = Dianthus caespitosus Thunb. subsp. pectinatus (E. Mey. ex Sond.) S. S. Hooper ■☆
127800 Dianthus pseudoarmeria M. Bieb. ;拟假海石竹■☆
127801 Dianthus pungens L. subsp. brachyanthus (Boiss.) R. Bernal et al. = Dianthus brachyanthus Boiss. ■☆
127802 Dianthus pygmaeus Hayata;玉山石竹(小石竹);Yushan Pink ■
127803 Dianthus pygmaeus Hayata f. albiflorus (S. S. Ying) S. S. Ying;白花玉山石竹■
127804 Dianthus pygmaeus Hayata f. albiflorus (S. S. Ying) S. S. Ying = Dianthus pygmaeus Hayata ■
127805 Dianthus pygmaeus Hayata var. albiflorus S. S. Ying = Dianthus pygmaeus Hayata f. albiflorus (S. S. Ying) S. S. Ying ■
127806 Dianthus pyrenaicus Bernh. ex Steud. ;比利牛斯石竹;Pyrenean Pink ■☆
127807 Dianthus raddeanus Vierh. ;拉德石竹■☆

127808 Dianthus ramosissimus Pall. ex Poir.;多枝石竹(多分枝石竹);Branchy Pink ■
127809 Dianthus repens Willd.;簇茎石竹(匍匐石竹);Creeping Pink,Northern Pink ■
127810 Dianthus repens Willd. var. scabripilosus Y. Z. Zhao;毛簇茎石竹;Scabripilose Creeping Pink ■
127811 Dianthus rigidus M. Bieb.;坚挺石竹■☆
127812 Dianthus rogowiczii Kleopow;罗高石竹■☆
127813 Dianthus rupicola Biv.;岩生石竹■☆
127814 Dianthus ruprechtii Schischk.;鲁普石竹■☆
127815 Dianthus saxifragus L. = Petrorhagia saxifraga (L.) Link ■
127816 Dianthus scaber Thunb.;野石竹;Wild Pink ■☆
127817 Dianthus scaber Thunb. = Dianthus thunbergii S. S. Hooper ■☆
127818 Dianthus scaber Thunb. var. graminifolius Fenzl ex Szyszyl. = Dianthus transvaalensis Burtt Davy ■☆
127819 Dianthus seguieri Vill.;西高石竹;Ragged Pink ■☆
127820 Dianthus semenovii (Regel et Herder) Vierh.;狭叶石竹;Narrowleaf Pink ■
127821 Dianthus sequieri Chaix = Dianthus chinensis L. ■
127822 Dianthus sequieri Chaix var. dentosus (Fisch. ex Rchb.) Franch. = Dianthus chinensis L. ■
127823 Dianthus seravschanicus Schischk.;塞拉夫石竹■☆
127824 Dianthus serpae Ficalho et Hiern = Dianthus zeyheri Sond. ■☆
127825 Dianthus serratifolius Sm.;齿叶石竹■☆
127826 Dianthus serrulatus Desf.;小齿石竹■☆
127827 Dianthus serrulatus Desf. subsp. cyrenaicus (Pamp.) Maire;昔兰尼石竹■☆
127828 Dianthus serrulatus Desf. subsp. macranthus Maire;大花小齿石竹■☆
127829 Dianthus serrulatus Desf. subsp. mauritanicus (Pomel) Batt. = Dianthus cintranus Boiss. et Reut. ■☆
127830 Dianthus serrulatus Desf. var. mesanidum (Litard. et Maire) Maire = Dianthus serrulatus Desf. ■☆
127831 Dianthus serrulatus Desf. var. strictus Maire = Dianthus serrulatus Desf. ■☆
127832 Dianthus serrulatus Desf. var. subsimplex Williams = Dianthus serrulatus Desf. ■☆
127833 Dianthus shinanensis (Yatabe) Makino;信浓石竹■☆
127834 Dianthus shinanensis (Yatabe) Makino f. alpinus Hid. Takah. ex T. Shimizu;信浓山石竹■☆
127835 Dianthus siculus C. Presl = Dianthus sylvestris Wulfen subsp. siculus (C. Presl) Tutin ■☆
127836 Dianthus siculus C. Presl var. lanceolatus Pau = Dianthus sylvestris Wulfen subsp. siculus (C. Presl) Tutin ■☆
127837 Dianthus sidi-tualii Font Quer = Dianthus lusitanus Brot. subsp. sidi (Font Quer) Dobignard ■☆
127838 Dianthus sinensis Link = Dianthus chinensis L. ■
127839 Dianthus soongoricus Schischk.;准噶尔石竹;Dzungar Pink ■
127840 Dianthus speciosus Rchb. = Dianthus superbus L. subsp. alpestris Kablikova ex Celak. ■
127841 Dianthus stenocalyx (Trautv.) Juz.;窄萼石竹■☆
127842 Dianthus sternbergii Sieber;南欧石竹;Sternberg Pink ■☆
127843 Dianthus strictus Sibth. et Sm.;直立石竹■☆
127844 Dianthus subacaulis Vill. subsp. brachyanthus (Boiss.) P. Fourn. = Dianthus brachyanthus Boiss. ■☆
127845 Dianthus subacaulis Vill. var. maroccanus (Pau et Font Quer) Maire = Dianthus brachyanthus Boiss. ■☆
127846 Dianthus subulifolius Kitag. = Dianthus chinensis L. ■
127847 Dianthus subulifolius Kitag. = Dianthus chinensis L. var. subulifolius (Kitag.) Ma ■
127848 Dianthus subulifolius Kitag. f. leucopetalus Kitag. = Dianthus chinensis L. ■
127849 Dianthus subulosus Conrath et Freyn ex Freyn;砾沙石竹■☆
127850 Dianthus subvaulis Vill.;无茎石竹■☆
127851 Dianthus sundermannii Bornm.;桑德石竹;Sundermann Pink ■☆
127852 Dianthus superbus L.;瞿麦(稠子花,大菊,大菊蘧麦,大兰,地面,杜母草,红花瞿麦,剪刀花,剪绒花,巨句麦,巨麦草,麦句姜,木碟花,南天竺草,山高粱,山瞿麦,十样景,十样景花,石竹子,石竹子花,燕麦蓊,野麦,竹节草);Fringed Pink,Lilac Pink,Superb Pink ■
127853 Dianthus superbus L. f. albiflora Y. N. Lee;白花瞿麦■☆
127854 Dianthus superbus L. f. chionanthus Okuyama;雪花高山瞿麦■☆
127855 Dianthus superbus L. f. latifolius (Nakai) Kitag.;宽叶瞿麦■☆
127856 Dianthus superbus L. f. leucanthus T. Shimizu;白花高山瞿麦■☆
127857 Dianthus superbus L. f. longicalycinus Maxim. = Dianthus superbus L. var. longicalycinus (Maxim.) F. N. Williams ■
127858 Dianthus superbus L. subsp. alpestris Kablikova ex Celak. = Dianthus superbus L. var. speciosus Rchb. ■
127859 Dianthus superbus L. subsp. longicalycinus (Maxim.) Kitam. = Dianthus superbus L. var. longicalycinus (Maxim.) F. N. Williams ■
127860 Dianthus superbus L. subsp. speciosus (Rchb.) Hayek = Dianthus superbus L. subsp. longicalycinus (Maxim.) Kitam. ■
127861 Dianthus superbus L. var. amoenus Nakai;秀丽石竹■☆
127862 Dianthus superbus L. var. hayatae Ohwi;早田氏瞿麦■☆
127863 Dianthus superbus L. var. longicalycinus (Maxim.) F. N. Williams f. albiflorus Honda;白花长萼瞿麦■☆
127864 Dianthus superbus L. var. longicalycinus (Maxim.) F. N. Williams f. tricolor Honda;三色长萼瞿麦■☆
127865 Dianthus superbus L. var. longicalycinus (Maxim.) F. N. Williams = Dianthus longicalyx Miq. ■
127866 Dianthus superbus L. var. monticola Makino = Dianthus superbus L. subsp. longicalycinus (Maxim.) Kitam. ■
127867 Dianthus superbus L. var. monticola Makino = Dianthus superbus L. var. speciosus Rchb. ■
127868 Dianthus superbus L. var. oreadum (Hance) Pamp. = Dianthus longicalyx Miq. ■
127869 Dianthus superbus L. var. pycnophyllus Kitag.;密叶瞿麦■☆
127870 Dianthus superbus L. var. speciosus Rchb.;高山瞿麦;Alpine Fringed Pink ■
127871 Dianthus superbus L. var. speciosus Rchb. = Dianthus superbus L. subsp. longicalycinus (Maxim.) Kitam. ■
127872 Dianthus superbus L. var. taiwanensis (Masam.) Tang S. Liu et S. S. Ying;台湾瞿麦■
127873 Dianthus superbus L. var. taiwanensis (Masam.) Tang S. Liu et S. S. Ying = Dianthus longicalyx Miq. ■
127874 Dianthus sylvestris Wulfen;林地石竹;Wood Pink,Woodland Pink ■☆
127875 Dianthus sylvestris Wulfen subsp. aristidis (Batt.) Greuter et Burdet;三芒草石竹■☆
127876 Dianthus sylvestris Wulfen subsp. boissieri (Willk.) Dobignard;布瓦西耶石竹■☆
127877 Dianthus sylvestris Wulfen subsp. longibracteatus (Maire) Greuter et Burdet;长苞林地石竹■☆
127878 Dianthus sylvestris Wulfen subsp. longicaulis (Ten.) Greuter et

Burdet;长茎林地石竹■☆
127879 Dianthus sylvestris Wulfen subsp. siculus (C. Presl) Tutin;西西里林地石竹■☆
127880 Dianthus szechuensis F. N. Williams;四川石竹;Sichuan Pink ■
127881 Dianthus szechuensis F. N. Williams = Dianthus superbus L. ■
127882 Dianthus taiwanensis Masam. = Dianthus longicalyx Miq. ■
127883 Dianthus taiwanensis Masam. = Dianthus superbus L. var. taiwanensis (Masam.) Tang S. Liu et S. S. Ying ■
127884 Dianthus talyschensis Boiss. et Buhse;塔里什石竹■☆
127885 Dianthus tetralepis Nevski;四鳞石竹■☆
127886 Dianthus thunbergii S. S. Hooper;通贝里石竹■☆
127887 Dianthus thunbergii S. S. Hooper f. maritimus ?;滨海石竹■☆
127888 Dianthus tianschanicus Schischk.;天山石竹■☆
127889 Dianthus transcaucasicus Schischk.;外高加索石竹■☆
127890 Dianthus transvaalensis Burtt Davy;德兰士瓦石竹■☆
127891 Dianthus tripunctatus Sm.;三斑石竹■☆
127892 Dianthus turkestanicus Preobr.;细茎石竹(土耳其斯坦石竹);Turkestanicus Pink ■
127893 Dianthus uralensis Korsh.;乌拉尔石竹■☆
127894 Dianthus velutinus Guss. = Petrorhagia dubia (Raf.) G. López et Romo ■☆
127895 Dianthus versicolor Fisch. ex Link = Dianthus chinensis L. ■
127896 Dianthus versicolor Fisch. ex Link = Dianthus chinensis L. var. versicolor (Fisch. ex Link) Ma ■
127897 Dianthus versicolor Fisch. ex Link f. leucopetalus (Kitag.) Y. C. Chu = Dianthus chinensis L. ■
127898 Dianthus versicolor Fisch. ex Link subsp. turkestanicus (Preobr.) Kozhevn. = Dianthus turkestanicus Preobr. ■
127899 Dianthus versicolor Fisch. ex Link var. subulifolius (Kitag.) Y. C. Chu = Dianthus chinensis L. var. subulifolius (Kitag.) Ma ■
127900 Dianthus versicolor Fisch. ex Link var. subulifolius (Kitag.) Y. C. Chu = Dianthus chinensis L. ■
127901 Dianthus viscidus Bory et Chaub.;黏性石竹;Viscid Pink ■☆
127902 Dianthus vulturius Guss. et Ten.;兀鹰石竹■☆
127903 Dianthus zeyheri Sond.;泽赫石竹■☆
127904 Dianthus zeyheri Sond. subsp. natalensis S. S. Hooper;纳塔尔泽赫石竹■☆
127905 Diapasis Poir. = Diaspasis R. Br. ■☆
127906 Diapedium K. D. König(废弃属名) = Dicliptera Juss.(保留属名)■
127907 Diapedium albicaule S. Moore = Dicliptera albicaulis (S. Moore) S. Moore ■☆
127908 Diapedium clinopodium Kuntze var. minor S. Moore = Dicliptera clinopodia Nees ■☆
127909 Diapensia Hill = Sanicula L. ■
127910 Diapensia L. (1753);岩梅属;Diapensia ●
127911 Diapensia acutifolia Hand. -Mazz. = Diapensia himalaica Hook. f. et Thomson var. acutifolia (Hand. -Mazz.) W. E. Evans ●
127912 Diapensia bullleyana Forrest ex Diels;黄花岩梅;Yellow Diapensia, Yellowflower Diapensia ●
127913 Diapensia himalaica Hook. f. et Thomson;喜马拉雅岩梅;Himalayan Diapensia, Himalayas Diapensia ●
127914 Diapensia himalaica Hook. f. et Thomson = Diapensia purpurea Diels ●
127915 Diapensia himalaica Hook. f. et Thomson var. acutifolia (Hand. -Mazz.) W. E. Evans;渐尖叶岩梅●
127916 Diapensia lapponica L.;岩梅(北美岩梅);Diapensia, Lapland Diapensia ●☆
127917 Diapensia lapponica L. subsp. obovata (F. Schmidt) Hultén = Diapensia lapponica L. var. obovata F. Schmidt ●☆
127918 Diapensia lapponica L. subsp. obovata (F. Schmidt) Hultén f. rosea Honda;粉花倒卵叶岩梅●☆
127919 Diapensia lapponica L. var. obovata F. Schmidt;倒卵叶岩梅●☆
127920 Diapensia obovata (F. Schmidt) Nakai = Diapensia lapponica L. var. obovata F. Schmidt ●☆
127921 Diapensia purpurea Diels;红花岩梅(白奴花,露寒草,石莲,岩菠菜);Pueple Diapensia, Red Diapensia ●
127922 Diapensia purpurea Diels f. albida W. E. Evans;白花岩梅;Whiteflower Diapensia ●
127923 Diapensia purpurea Diels f. bulleyana (Forrest ex Diels) W. E. Evans = Diapensia bullleyana Forrest ex Diels ●
127924 Diapensia purpurea Diels f. rosea W. E. Evans = Diapensia purpurea Diels ●
127925 Diapensia wardii W. E. Evans;西藏岩梅;Ward Diapensia, Xiznag Diapensia ●
127926 Diapensiaceae Lindl. (1836)(保留科名);岩梅科;Diapensia Family ●■
127927 Diaperia Nutt. (1840);兔烟花属;Rabbit-tobacco, Dwarf Cudweed ■☆
127928 Diaperia Nutt. = Evax Gaertn. ■☆
127929 Diaperia Nutt. = Filago L.(保留属名)■
127930 Diaperia candida (Torr. et A. Gray) Benth. et Hook. f.;白兔烟花;Silver rabbit-tobacco ■☆
127931 Diaperia multicaulis Nutt.;兔烟花■☆
127932 Diaperia prolifera (Nutt. ex DC.) Nutt.;大头兔烟花;Big-head Rabbit-tobacco ■☆
127933 Diaperia prolifera (Nutt. ex DC.) Nutt. var. barnebyi Morefield;巴恩比兔烟花;Barneby Rabbit-tobacco ■☆
127934 Diaperia verna (Raf.) Morefield;多茎兔烟花;Many-stem Rabbit-tobacco, Spring Rabbit-tobacco ■☆
127935 Diaperia verna (Raf.) Morefield var. drummondii (Torr. et A. Gray) Morefield;德拉蒙德兔烟花;Gulf Rabbit-tobacco ■☆
127936 Diaphananthe Schltr. (1914);薄花兰属■☆
127937 Diaphananthe acuta (Ridl.) Schltr.;尖薄花兰■☆
127938 Diaphananthe alfredii Geerinck;艾尔薄花兰■☆
127939 Diaphananthe bidens (Sw. ex Pers.) Schltr.;双齿薄花兰■☆
127940 Diaphananthe bilobata (Summerh.) H. Rasm. = Rhipidoglossum bilobatum (Summerh.) Szlach. et Olszewski ■☆
127941 Diaphananthe brevifolia (Summerh.) Summerh.;短叶薄花兰■☆
127942 Diaphananthe burttii Summerh. = Margelliantha burttii (Summerh.) P. J. Cribb ■☆
127943 Diaphananthe caffra (Bolus) H. P. Linder;开菲尔薄花兰■☆
127944 Diaphananthe candida P. J. Cribb;纯白薄花兰■☆
127945 Diaphananthe cuneata Summerh.;楔形薄花兰■☆
127946 Diaphananthe densiflora (Summerh.) Summerh. = Rhipidoglossum densiflorum Summerh. ■☆
127947 Diaphananthe dorotheae (Rendle) Summerh.;多罗特娅石竹■☆
127948 Diaphananthe eggelingii P. J. Cribb;埃格林石竹■☆
127949 Diaphananthe erecto-calcarata (De Wild.) Summerh.;直距薄花兰■☆
127950 Diaphananthe fragrantissima (Rchb. f.) Schltr.;芳香薄花兰■☆
127951 Diaphananthe globulosocalcarata (De Wild.) Summerh. = Rhipidoglossum globulosocalcaratum (De Wild.) Summerh. ■☆
127952 Diaphananthe kamerunensis (Schltr.) Schltr. = Rhipidoglossum

kamerunense (Schltr.) Garay ■☆
127953 Diaphananthe laticalcar J. B. Hall;宽距薄花兰■☆
127954 Diaphananthe laxiflora (Summerh.) Summerh.;疏花薄花兰■☆
127955 Diaphananthe longicalcar (Summerh.) Summerh.;长距薄花兰■☆
127956 Diaphananthe lorifolia Summerh.;纽叶薄花兰■☆
127957 Diaphananthe microphylla (Summerh.) Summerh.;小叶薄花兰■☆
127958 Diaphananthe mildbraedii (Kraenzl.) Schltr. = Rhipidoglossum mildbraedii (Kraenzl.) Garay ■☆
127959 Diaphananthe montana (Piers) P. J. Cribb et J. Stewart;山地薄花兰■☆
127960 Diaphananthe obanensis (Rendle) Summerh. = Rhipidoglossum obanense (Rendle) Summerh. ■☆
127961 Diaphananthe orientalis (Mansf.) H. Rasm. = Rhipidoglossum orientalis (Mansf.) Szlach. et Olszewski ■☆
127962 Diaphananthe ovalis Summerh. = Rhipidoglossum ovale (Summerh.) Garay ■☆
127963 Diaphananthe oxycentron P. J. Cribb;尖距薄花兰■☆
127964 Diaphananthe pachyrhiza P. J. Cribb;粗根薄花兰■☆
127965 Diaphananthe papagayi (Rchb. f.) Schltr.;帕帕加石竹■☆
127966 Diaphananthe pellucida (Lindl.) Schltr.;透明薄花兰■☆
127967 Diaphananthe pellucida (Lindl.) Schltr. var. gigantiflora W. Sanford;巨大薄花兰■☆
127968 Diaphananthe polyantha (Kraenzl.) H. Rasm. = Rhipidoglossum polyanthum (Kraenzl.) Szlach. et Olszewski ■☆
127969 Diaphananthe polydactyla (Kraenzl.) Summerh. = Rhipidoglossum polydactylum (Kraenzl.) Garay ■☆
127970 Diaphananthe producta (Kraenzl.) Schltr. = Diaphananthe bidens (Sw. ex Pers.) Schltr. ■☆
127971 Diaphananthe pulchella Summerh.;美丽薄花兰■☆
127972 Diaphananthe quintasii (Rolfe) Schltr. = Diaphananthe rohrii (Rchb. f.) Summerh. ■☆
127973 Diaphananthe rohrii (Rchb. f.) Summerh.;勒尔薄花兰■☆
127974 Diaphananthe rutila (Rchb. f.) Summerh. = Rhipidoglossum rutilum (Rchb. f.) Schltr. ■☆
127975 Diaphananthe sanfordiana Szlach. et Olszewski;桑福德石竹■☆
127976 Diaphananthe sarcorhynchoides J. B. Hall;肉喙薄花兰■☆
127977 Diaphananthe schimperiana (A. Rich.) Summerh.;欣珀薄花兰■☆
127978 Diaphananthe stellata P. J. Cribb = Rhipidoglossum stellatum (P. J. Cribb) Szlach. et Olszewski ■☆
127979 Diaphananthe stolzii Schltr.;斯托尔兹薄花兰■☆
127980 Diaphananthe subclavata (Rolfe) Schltr. = Diaphananthe acuta (Ridl.) Schltr. ■☆
127981 Diaphananthe subfalcifolia (De Wild.) Schltr. = Diaphananthe bidens (Sw. ex Pers.) Schltr. ■☆
127982 Diaphananthe suborbicularis Summerh.;近圆形薄花兰■☆
127983 Diaphananthe subsimplex Summerh.;简单薄花兰■☆
127984 Diaphananthe tanneri P. J. Cribb;坦纳薄花兰■☆
127985 Diaphananthe tenerrima (Kraenzl.) Summerh. = Rhipidoglossum mildbraedii (Kraenzl.) Garay ■☆
127986 Diaphananthe tenuicalcar Summerh.;细距薄花兰■☆
127987 Diaphananthe trigonopetala Schltr.;三角瓣薄花兰■☆
127988 Diaphananthe ugandensis (Rendle) Summerh.;乌干达薄花兰■☆
127989 Diaphananthe vandaeformis (Kraenzl.) Schltr.;万代兰石竹■☆
127990 Diaphananthe welwitschii (Rchb. f.) Schltr.;韦尔薄花兰■☆
127991 Diaphananthe xanthopollinia (Rchb. f.) Summerh.;黄粉薄花兰■☆
127992 Diaphane Salisb. = Iris L. ■
127993 Diaphanoptera Rech. f. (1940);膜翅花属■☆
127994 Diaphanoptera afghanica Podlech;阿富汗膜翅花■☆
127995 Diaphanoptera khorasanica Rech. f.;膜翅花■☆
127996 Diaphanoptera stenocalycina Rech. f. et Schiman-Czeika;窄萼膜翅花■☆
127997 Diaphora Lour. = Scleria P. J. Bergius ■
127998 Diaphoranthema Beer = Tillandsia L. ■☆
127999 Diaphoranthema recurvata (L.) Beer = Tillandsia recurvata (L.) L. ■☆
128000 Diaphoranthus Anderson ex Hook. f. = Pringlea T. Anderson ex Hook. f. ■☆
128001 Diaphoranthus Meyen = Polyachyrus Lag. ●■☆
128002 Diaphorea Pers. = Diaphora Lour. ■
128003 Diaphorea Pers. = Scleria P. J. Bergius ■
128004 Diaphractanthus Humb. (1923);腺果瘦片菊属■☆
128005 Diaphractanthus homolepis Humb.;腺果瘦片菊■☆
128006 Diaphycarpus Calest. = Bunium L. ■☆
128007 Diaphycarpus Calest. = Carum L. ■
128008 Diaphyllum Hoffm. = Bupleurum L. ●■
128009 Diaphyllum triradiatum (Adams ex Hoffm.) Hoffm. = Bupleurum triradiatum Adams ex Hoffm. ■
128010 Diarina Raf. = Diarrhena P. Beauv. (保留属名) ■
128011 Diarrhena P. Beauv. (1812)(保留属名);龙常草属;Beak Grain, Beakgrain ■
128012 Diarrhena Raf. = Diarrhena P. Beauv. (保留属名) ■
128013 Diarrhena americana P. Beauv.;北美龙常草;American Beak Grain, American Beakgrain, Beak Grass ■☆
128014 Diarrhena americana P. Beauv. var. obovata Gleason = Diarrhena obovata (Gleason) Brandenb. ■☆
128015 Diarrhena fauriei (Hack.) Ohwi;小果龙常草(法利龙常草,富尔氏龙常草);Faurie Beakgrain ■
128016 Diarrhena fauriei (Hack.) Ohwi = Neomolinia fauriei (Hack.) Honda ■
128017 Diarrhena japonica Franch. et Sav.;日本龙常草;Japanese Beakgrain ■
128018 Diarrhena japonica Franch. et Sav. = Neomolinia japonica (Franch. et Sav.) Honda ■
128019 Diarrhena mandshurica Maxim.;龙常草(东北龙常草,粽心草);Manchuri Beakgrain, Manchurian Beakgrain ■
128020 Diarrhena mandshurica Maxim. = Neomolinia mandshurica (Maxim.) Honda ■
128021 Diarrhena obovata (Gleason) Brandenb.;倒卵叶龙常草;American Beakgrain, Obovate Beak Grain ■☆
128022 Diarrhena yabeana Kitag.;朝鲜龙常草■
128023 Diarthron Turcz. (1832);草瑞香属(粟麻属);Diarthron ●■
128024 Diarthron altaica (Thieb.-Bern.) Kit Tan = Stelleropsis altaica (Thieb.-Bern.) Pobed. ■
128025 Diarthron carinatum Jaub. et Spach. = Diarthron vesiculosum C. A. Mey. ■
128026 Diarthron linifolium Turcz.;草瑞香(山胡麻,粟麻);Flaxleaf Diarthron ■
128027 Diarthron tianschanica (Pobed.) Kit Tan = Stelleropsis tianschanica Pobed. ■

128028 Diarthron vesiculosum C. A. Mey.;囊管草瑞香(短叶草瑞香);Shortleaf Diarthron ■
128029 Diascia Link et Otto(1820);双距花属(二距花属);Twinspur ■☆
128030 Diascia alonsooides Benth.;假面花双距花■☆
128031 Diascia austromontana K. E. Steiner;南方山地双距花■☆
128032 Diascia barberae Hook. f.;红双距花;Barber's Diascia, Twinspur ■☆
128033 Diascia barberae Hook. f. 'Ruby Field';宝石红双距花■☆
128034 Diascia batteniana K. E. Steiner;巴滕双距花■☆
128035 Diascia bergiana Benth. = Diascia sacculata Benth. ■☆
128036 Diascia bergiana Eckl. ex Hiern = Diascia elongata Benth. ■☆
128037 Diascia bergiana Link et Otto;贝格双距花■☆
128038 Diascia bicolor K. E. Steiner;二色双距花■☆
128039 Diascia burchellii Benth. = Diascia parviflora Benth. ■☆
128040 Diascia capensis (L.) Britten;好望角双距花■☆
128041 Diascia capsularis Benth.;裂果双距花■☆
128042 Diascia capsularis Benth. var. flagellaris Hiern = Diascia capsularis Benth. ■☆
128043 Diascia cardiosepala Hiern;心萼双距花■☆
128044 Diascia cordata N. E. Br.;心叶双距花■☆
128045 Diascia cuneata E. Mey. ex Benth.;楔形双距花■☆
128046 Diascia decipiens K. E. Steiner;迷惑双距花■☆
128047 Diascia denticulata Benth. = Nemesia denticulata (Benth.) Grant ex Fourc. ■☆
128048 Diascia dielsiana Schltr. ex Hiern;迪尔斯双距花■☆
128049 Diascia diffusa Benth.;松散双距花■☆
128050 Diascia dissecta Hiern = Alonsoa unilabiata (L. f.) Steud. ●☆
128051 Diascia dissimulans Hilliard et B. L. Burtt;不似双距花■☆
128052 Diascia elongata Benth.;伸长双距花■☆
128053 Diascia engleri Diels;恩格勒双距花■☆
128054 Diascia esterhuyseniae K. E. Steiner;埃斯特双距花■☆
128055 Diascia expolita Hiern = Diascia racemulosa Benth. ■☆
128056 Diascia flanaganii Hiern = Diascia stachyoides Schltr. ex Hiern ■☆
128057 Diascia fragrans K. E. Steiner;芳香双距花■☆
128058 Diascia glandulosa E. Phillips;具腺双距花■☆
128059 Diascia glandulosa E. Phillips var. albiflora ?;白花具腺双距花■☆
128060 Diascia gracilis Schltr.;纤细双距花■☆
128061 Diascia heterandra Benth. = Alonsoa unilabiata (L. f.) Steud. ●☆
128062 Diascia humilis K. E. Steiner;低矮双距花■☆
128063 Diascia insignis K. E. Steiner;显著双距花■☆
128064 Diascia integerrima E. Mey. ex Benth.;全缘双距花;Twinspur ■☆
128065 Diascia integrifolia Spreng. ex Eckl. = Tylophora lycioides (E. Mey.) Decne. ■☆
128066 Diascia jonantha Dinter = Diascia engleri Diels ■☆
128067 Diascia lewisiae K. E. Steiner;刘易斯双距花■☆
128068 Diascia lilacina Hilliard et B. L. Burtt;紫丁香色双距花■☆
128069 Diascia longicornis (Thunb.) Druce;长双距花■☆
128070 Diascia macowanii Hiern = Diascia rigescens E. Mey. ex Benth. ■☆
128071 Diascia macrophylla (Thunb.) Spreng.;大叶双距花■☆
128072 Diascia maculata K. E. Steiner;斑点双距花■☆
128073 Diascia minutiflora Hiern;微花双距花■☆
128074 Diascia mollis Hilliard et B. L. Burtt;绢毛双距花■☆
128075 Diascia moltenensis Hiern = Diascia integerrima E. Mey. ex Benth. ■☆
128076 Diascia monasca Hiern = Diascia patens (Thunb.) Grant ex Fourc. ■☆
128077 Diascia montana (L. f.) Spreng. = Hemimeris racemosa (Houtt.) Merr. ■☆
128078 Diascia namaquensis Hiern;纳马夸双距花■☆
128079 Diascia nana Diels;矮双距花■☆
128080 Diascia nemophiloides Benth. = Diascia capensis (L.) Britten ■☆
128081 Diascia nodosa K. E. Steiner;多节双距花■☆
128082 Diascia nutans Diels = Alonsoa unilabiata (L. f.) Steud. ●☆
128083 Diascia pachyceras E. Mey. ex Benth.;粗距双距花■☆
128084 Diascia parviflora Benth.;小花双距花■☆
128085 Diascia patens (Thunb.) Grant ex Fourc.;铺展双距花■☆
128086 Diascia pentheri Schltr.;彭泰尔双距花■☆
128087 Diascia personata Hilliard et B. L. Burtt;张开双距花■☆
128088 Diascia purpurea N. E. Br.;紫双距花■☆
128089 Diascia racemulosa Benth.;小总花双距花■☆
128090 Diascia ramosa Scott-Elliot;多枝双距花■☆
128091 Diascia ramulosa E. Mey. = Diascia racemulosa Benth. ■☆
128092 Diascia rigescens E. Mey. ex Benth.;密序双距花■☆
128093 Diascia rigescens E. Mey. ex Benth. var. angustifolia Benth. = Diascia rigescens E. Mey. ex Benth. ■☆
128094 Diascia rigescens E. Mey. ex Benth. var. bractescens Hiern = Diascia rigescens E. Mey. ex Benth. ■☆
128095 Diascia rigescens E. Mey. ex Benth. var. montana Diels;山地密序双距花■☆
128096 Diascia rotundifolia Hiern = Diclis rotundifolia (Hiern) Hilliard et B. L. Burtt ■☆
128097 Diascia rudolphii Hiern;鲁道夫双距花■☆
128098 Diascia runcinata E. Mey. ex Benth.;倒齿双距花■☆
128099 Diascia sacculata Benth.;小囊双距花■☆
128100 Diascia scullyi Hiern = Hemimeris racemosa (Houtt.) Merr. ■☆
128101 Diascia sinuata (Sm.) Druce = Alonsoa unilabiata (L. f.) Steud. ●☆
128102 Diascia stachyoides Schltr. ex Hiern;穗状双距花■☆
128103 Diascia stricta Hilliard et B. L. Burtt;刚直双距花■☆
128104 Diascia thunbergiana Spreng. = Diascia longicornis (Thunb.) Druce ■☆
128105 Diascia transkeiana Hilliard et B. L. Burtt = Diascia mollis Hilliard et B. L. Burtt ■☆
128106 Diascia tugelensis Hilliard et B. L. Burtt;图盖拉双距花■☆
128107 Diascia tysonii Hiern = Diascia alonsooides Benth. ■☆
128108 Diascia unilabiata (L. f.) Benth. = Alonsoa unilabiata (L. f.) Steud. ●☆
128109 Diascia veronicoides Schltr.;婆婆纳叶双距花■☆
128110 Diascia vigilis Hilliard et B. L. Burtt;匍匐双距花■☆
128111 Diasia DC. = Melasphaerula Ker Gawl. ■☆
128112 Diaspananthus Miq. = Ainsliaea DC. ■
128113 Diaspananthus palmatus Miq. = Ainsliaea uniflora Sch. Bip. ■☆
128114 Diaspananthus uniflorus (Sch. Bip.) Kitam. = Ainsliaea uniflora Sch. Bip. ■☆
128115 Diaspananthus uniflorus (Sch. Bip.) Kitam. f. niveus M. Mizush.;雪白单花兔儿风■☆
128116 Diaspanthus Kitam. = Diaspananthus Miq. ■
128117 Diaspasis R. Br. (1810);无耳草海桐属■☆
128118 Diaspasis filifolia R. Br.;无耳草海桐■☆
128119 Diasperus Kuntze = Glochidion J. R. Forst. et G. Forst. (保留属名)●

128120 Diasperus Kuntze = Phyllanthus L. ●■
128121 Diasperus Kuntze. = Agyneia L.(废弃属名)●
128122 Diasperus emblica (L.) Kuntze = Phyllanthus emblica L. ●
128123 Diasperus muellerianus Kuntze = Phyllanthus muellerianus (Kuntze) Exell ●☆
128124 Diasperus niruri (L.) Kuntze = Phyllanthus niruri L. ●■
128125 Diasperus pulcher (Wall.) Kuntze = Phyllanthus pulcher Wall. ex Müll. Arg. ●
128126 Diasperus verrucosus (Thunb.) Kuntze = Flueggea verrucosa (Thunb.) G. L. Webster ●☆
128127 Diaspis Nied. = Caucanthus Forssk. ●☆
128128 Diaspis Nied. = Triaspis Burch. ●☆
128129 Diastatea Scheidw. (1841);小顶花桔梗属■☆
128130 Diastatea virgata Scheidw.;小顶花桔梗■☆
128131 Diastella Salisb. = Diastella Salisb. ex Knight ●☆
128132 Diastella Salisb. ex Knight(1809);双星山龙眼属●☆
128133 Diastella bryiflora Salisb. ex Knight = Diastella thymelaeoides (P. J. Bergius) Rourke ●☆
128134 Diastella divaricata (P. J. Bergius) Rourke;叉开双星山龙眼●☆
128135 Diastella divaricata (P. J. Bergius) Rourke subsp. montana Rourke;山地双星山龙眼●☆
128136 Diastella ericifolia Salisb. ex Knight = Diastella proteoides (L.) Druce ●☆
128137 Diastella fraterna Rourke;兄弟双星山龙眼●☆
128138 Diastella myrtifolia (Thunb.) Salisb. ex Knight;香桃木叶双星山龙眼●☆
128139 Diastella parilis Salisb. ex Knight;相似双星山龙眼●☆
128140 Diastella proteoides (L.) Druce;海神双星山龙眼●☆
128141 Diastella serpyllifolia Salisb. ex Knight = Diastella divaricata (P. J. Bergius) Rourke subsp. montana Rourke ●☆
128142 Diastella thymelaeoides (P. J. Bergius) Rourke;欧瑞香双星山龙眼●☆
128143 Diastella thymelaeoides (P. J. Bergius) Rourke subsp. meridiana Rourke;南方双星山龙眼●☆
128144 Diastema Benth. (1845);二雄蕊苣苔属;Diastema ■☆
128145 Diastema L. f. ex B. D. Jacks. = Dalbergia L. f.(保留属名)●
128146 Diastema ochroleucum Hook.;黄白色二雄蕊苣苔;Yellowwish-white Diastema ■☆
128147 Diastemanthe Desv. = Stenotaphrum Trin. ■
128148 Diastemanthe Steud. = Stenotaphrum Trin. ■
128149 Diastemation C. Muell. = Diastema Benth. ■☆
128150 Diastemella Oerst. = Diastema Benth. ■☆
128151 Diastemenanthe Desv. = Stenotaphrum Trin. ■
128152 Diastemenanthe Steud. = Stenotaphrum Trin. ■
128153 Diastemma Lindl. = Diastemation C. Muell. ■☆
128154 Diastrophis Fisch. et C. A. Mey. = Aethionema R. Br. ■☆
128155 Diateinacanthus Lindau = Odontonema Nees(保留属名)●■☆
128156 Diatenopteryx Radlk. (1878);南美无患子属●☆
128157 Diatoma Lour. = Carallia Roxb.(保留属名)●
128158 Diatoma brachiata Lour. = Carallia brachiata (Lour.) Merr. ●
128159 Diatonta Walp. = Coreopsis L. ●■
128160 Diatosperma C. Muell. = Ceratogyne Turcz. ■☆
128161 Diatosperma C. Muell. = Diotosperma A. Gray ■☆
128162 Diatrema Raf.(废弃属名) = Diatremis Raf.(废弃属名)●■
128163 Diatrema Raf.(废弃属名) = Pharbitis Choisy(保留属名)■
128164 Diatremis Raf.(废弃属名) = Ipomoea L.(保留属名)●■
128165 Diatremis Raf.(废弃属名) = Pharbitis Choisy(保留属名)■
128166 Diatropa Dumort. = Bupleurum L. ●■
128167 Diaxulon Raf. = Cytisus Desf.(保留属名)●
128168 Diaxylum Post et Kuntze = Diaxulon Raf. ●
128169 Diazeuxis D. Don = Lycoseris Cass. ●☆
128170 Diazia Phil. = Calandrinia Kunth(保留属名)■☆
128171 Diberara Baill. = Nebelia Neck. ex Sweet ●☆
128172 Dibothrospermum Knaf = Matricaria L. ■
128173 Dibrachia Steud. = Dibrachya (Sweet) Eckl. et Zeyh. ●■
128174 Dibrachia Steud. = Pelargonium L'Hér. ex Aiton ●■
128175 Dibrachion Regel = Homalanthus A. Juss.(保留属名)●
128176 Dibrachion Tul. = Diplotropis Benth. ●☆
128177 Dibrachionostylus Bremek. (1952);卡斯纳雪柱属●☆
128178 Dibrachionostylus kaessneri (S. Moore) Bremek.;卡斯纳雪柱●☆
128179 Dibrachium Walp. = Dibrachion Regel ●
128180 Dibrachium Walp. = Homalanthus A. Juss.(保留属名)●
128181 Dibrachya (Sweet) Eckl. et Zeyh. = Pelargonium L'Hér. ex Aiton ●■
128182 Dibracteaceae Dulac = Callitrichaceae Link(保留科名)■
128183 Dicaelosperma E. G. O. Muell. et Pax = Dicoelospermum C. B. Clarke ■☆
128184 Dicaelospermum C. B. Clarke = Dicoelospermum C. B. Clarke ■☆
128185 Dicalix Lour. = Symplocos Jacq. ●
128186 Dicalix adenopus (Hance) Migo = Symplocos adenopus Hance ●
128187 Dicalix anomalus (Brand) Migo = Symplocos anomala Brand ●
128188 Dicalix austrosinensis (Migo) Migo = Symplocos sumuntia Buch. -Ham. ex D. Don ●
128189 Dicalix bodinieri (Brand) Migo = Symplocos cochinchinensis (Lour.) S. Moore var. laurina (Retz.) Raizada ●
128190 Dicalix boninensis (Rehder et E. H. Wilson) Hara = Symplocos boninensis Rehder et E. H. Wilson ●☆
128191 Dicalix botryanthus (Franch.) Migo = Symplocos sumuntia Buch. -Ham. ex D. Don ●
128192 Dicalix chunii (Merr.) Migo = Symplocos poilanei Guillaumin ●
128193 Dicalix cochinchinensis Lour. = Symplocos cochinchinensis (Lour.) S. Moore ●
128194 Dicalix congestus (Benth.) Migo = Symplocos congesta Benth. ●
128195 Dicalix crassifolia (Benth.) Migo = Symplocos lucida (Thunb.) Siebold et Zucc. ●
128196 Dicalix crassilimbus (Merr.) Migo = Symplocos crassilimba Merr. ●
128197 Dicalix decorus (Hance) Migo = Symplocos sumuntia Buch. -Ham. ex D. Don ●
128198 Dicalix delavayi (Brand) Migo = Symplocos dryophila C. B. Clarke ●
128199 Dicalix ernestii (Dunn) Migo = Symplocos lucida (Thunb.) Siebold et Zucc ●
128200 Dicalix forrestii (W. W. Sm.) Migo = Symplocos dryophila C. B. Clarke ●
128201 Dicalix fusonii (Merr.) Migo = Symplocos anomala Brand ●
128202 Dicalix glauca (Thunb.) Migo = Symplocos glauca (Thunb.) Koidz. ●
128203 Dicalix glomeratus (King ex C. B. Clarke) Migo = Symplocos glomerata King ex C. B. Clarke ●
128204 Dicalix groffii (Merr.) Migo = Symplocos groffii Merr. ●
128205 Dicalix heishanensis (Hayata) Migo = Symplocos heishanensis Hayata ●

128206 Dicalix hookeri (C. B. Clarke) Migo = Symplocos hookeri C. B. Clarke ●

128207 Dicalix javanicus Blume = Symplocos cochinchinensis (Lour.) S. Moore ●

128208 Dicalix kawakamii (Hayata) Hara = Symplocos kawakamii Hayata ●

128209 Dicalix kotoensis (Hayata) Hara = Symplocos cochinchinensis (Lour.) S. Moore ●

128210 Dicalix lancifolia (Siebold et Zucc.) Hara = Symplocos lancifolia Siebold et Zucc. ●

128211 Dicalix lancilimbus (Merr.) Migo = Symplocos viridissima Brand ●

128212 Dicalix laurinus (Retz.) Migo = Symplocos cochinchinensis (Lour.) S. Moore var. laurina (Retz.) Raizada ●

128213 Dicalix liukiuensis (Matsum.) Hara = Symplocos liukiuensis Matsum. ●

128214 Dicalix lucida (Thunb.) Hara = Symplocos kuroki Nagam. ●☆

128215 Dicalix lucida (Thunb.) Hara var. nakaharai Hayata = Symplocos nakaharae (Hayata) Masam. ●☆

128216 Dicalix microcalyx (Hayata) Hara = Symplocos microcalyx Hayata ●

128217 Dicalix myrianthus (Rehder) Migo = Symplocos ramosissima Wall. et G. Don ●

128218 Dicalix myrtacea (Siebold et Zucc.) Hara = Symplocos myrtacea Siebold et Zucc. ●☆

128219 Dicalix obana (Masam.) Hara = Symplocos liukiuensis Matsum. ●

128220 Dicalix oligophlebius (Merr.) Migo = Symplocos adenopus Hance ●

128221 Dicalix oligophlebius Migo = Symplocos adenopus Hance ●

128222 Dicalix pergracilis (Nakai) Hara = Symplocos pergracilis (Nakai) T. Yamaz. ●☆

128223 Dicalix poilanei (Guillaumin) Migo = Symplocos poilanei Guillaumin ●

128224 Dicalix propinquus (Hance) Migo = Symplocos racemosa Roxb. ●

128225 Dicalix prunifolia (Siebold et Zucc.) Hara = Symplocos prunifolia Siebold et Zucc. ●

128226 Dicalix prunifolia (Siebold et Zucc.) Hara var. uiae ? = Symplocos prunifolia Siebold et Zucc. ●

128227 Dicalix pseudolancifolius (Hatus.) Migo = Symplocos lancifolia Siebold et Zucc. ●

128228 Dicalix pseudostellaris Migo = Symplocos aenea Hand. -Mazz. ●

128229 Dicalix pseudostellaris Migo = Symplocos stellaris Brand var. aenea (Hand. -Mazz.) Noot. ●

128230 Dicalix psueodlancifolia (Hatus.) Migo = Symplocos lancifolia Siebold et Zucc. ●

128231 Dicalix punctomarginatus (A. Chev. ex Guillaumin) Migo = Symplocos adenophylla Wall. et G. Don ●

128232 Dicalix schaefferae (Merr.) Migo = Symplocos cochinchinensis (Lour.) S. Moore var. laurina (Retz.) Raizada ●

128233 Dicalix setchuensis (Brand) Migo = Symplocos lucida (Thunb.) Siebold et Zucc. ●

128234 Dicalix setchuensis (Brand) Migo = Symplocos setchuensis Brand ●

128235 Dicalix shinodanus Migo = Symplocos lucida (Thunb.) Siebold et Zucc. ●

128236 Dicalix shunningensis Migo = Symplocos dryophila C. B. Clarke ●

128237 Dicalix stellaris (Brand) Migo = Symplocos stellaris Brand ●

128238 Dicalix swinhoeanus (Hance) Migo = Symplocos sumuntia Buch. -Ham. ex D. Don ●

128239 Dicalix tanakae (Matsum.) Hara = Symplocos tanakae Matsum. ●☆

128240 Dicalix terminalis (Brand) Migo = Symplocos cochinchinensis (Lour.) S. Moore var. laurina (Retz.) Raizada ●

128241 Dicalix theophrastifolia (Siebold et Zucc.) Migo = Symplocos cochinchinensis (Lour.) S. Moore var. laurina (Retz.) Raizada ●

128242 Dicalix theophrastifolia (Siebold et Zucc.) Migo = Symplocos theophrastifolia Siebold et Zucc. ●

128243 Dicalix urceolaris (Hance) Migo = Symplocos sumuntia Buch. -Ham. ex D. Don ●

128244 Dicalix wangii Migo = Symplocos glauca (Thunb.) Koidz. ●

128245 Dicalix wiksitroemiifolia (Hayata) Migo = Symplocos wikstroemiifolia Hayata ●

128246 Dicalix yunnanensis (Brand) Migo = Symplocos sulcata Kurz ●

128247 Dicalymma Lem. = Podachaenium Benth. ex Oerst. ●☆

128248 Dicalyx Poir. = Dicalix Lour. ●

128249 Dicalyx Poir. = Symplocos Jacq. ●

128250 Dicardlotis Raf. = Gentiana L. ■

128251 Dicarpaea C. Presl = Limeum L. ■●☆

128252 Dicarpellum (Loes.) A. C. Sm. (1941);双片卫矛属●☆

128253 Dicarpellum (Loes.) A. C. Sm. = Salacia L. (保留属名)●

128254 Dicarpellum pancheri (Baill.) A. C. Sm.;双片卫矛●☆

128255 Dicarpidium F. Muell. (1857);双果梧桐属●☆

128256 Dicarpidium monoicum F. Muell.;双果梧桐●☆

128257 Dicarpophora Speg. (1926);双果萝藦属☆

128258 Dicarpophora mazzuchii Speg.;双果萝藦☆

128259 Dicaryum Willd. = Geissanthus Hook. f. ●☆

128260 Dicaryum Willd. ex Roem. et Schult. (1819);双果马钱属●☆

128261 Dicaryum Willd. ex Roem. et Schult. = Geissanthus Hook. f. ●☆

128262 Dicaryum serrulatum Willd.;双果马钱●☆

128263 Dicaryum serrulatum Willd. = Geissanthus serrulatus Mez ●☆

128264 Dicella Griseb. (1839);二室金虎尾属●☆

128265 Dicella amazonica Pires;亚马孙二室金虎尾●☆

128266 Dicella bracteosa Griseb.;二室金虎尾●☆

128267 Dicella lancifolia A. Juss.;披针叶二室金虎尾●☆

128268 Dicella macroptera A. Juss.;大翅二室金虎尾●☆

128269 Dicella ovatifolia A. Juss.;卵叶二室金虎尾●☆

128270 Dicella tricarpa Nied.;三果二室金虎尾●☆

128271 Dicellandra Hook. f. (1867);二室蕊属●☆

128272 Dicellandra barteri Hook. f.;巴特二室蕊●☆

128273 Dicellandra barteri Hook. f. var. erecta (Mildbr.) Jacq. -Fél.;直立二室蕊●☆

128274 Dicellandra barteri Hook. f. var. escherichii (Gilg) Jacq. -Fél.;埃舍里奇二室蕊●☆

128275 Dicellandra barteri Hook. f. var. magnifica (Mildbr.) Jacq. -Fél.;壮观巴特二室蕊●☆

128276 Dicellandra descoingsii Jacq. -Fél.;德斯二室蕊●☆

128277 Dicellandra erecta Mildbr. = Dicellandra barteri Hook. f. var. erecta (Mildbr.) Jacq. -Fél. ●☆

128278 Dicellandra escherichii Gilg = Dicellandra barteri Hook. f. var. escherichii (Gilg) Jacq. -Fél. ●☆

128279 Dicellandra glanduligera (Pellegr.) Jacq. -Fél.;腺体二室蕊●☆

128280 Dicellandra magnifica Mildbr. = Dicellandra barteri Hook. f. var. magnifica (Mildbr.) Jacq. -Fél. ●☆

128281 Dicellandra scandens Gilg ex Engl. = Dicellandra barteri Hook. f. ●☆

128282 Dicellandra setosa Hook. f. = Ochthocharis setosa (Hook. f.) Hansen et Wickens ●☆

128283 Dicellostyles Benth. (1862);二室柱属●☆

128284 Dicellostyles axillaris Benth.;二室柱●☆

128285 Dicellostyles jujubifolia (Griff.) Benth. = Kydia jujubifolia Griff. ●

128286 Dicellostyles jujubifolia (Griff.) Benth. = Nayariophyton zizyphifolium (Griff.) D. G. Long et A. G. Mill. ●

128287 Dicellostyles zizyphifolia (Griff.) Phuph. = Nayariophyton zizyphifolium (Griff.) D. G. Long et A. G. Mill. ●

128288 Dicentra Bernh. (1833)(保留属名);荷丹属(荷包牡丹属,藤铃儿草属,藤铃儿属,璎珞牡丹属,指叶紫堇属,紫金龙属);Bleeding Heart, Bleedingheart, Bleeding-heart, Colicweed, Dactylicapnos, Dicentra, Dutchman's Breeches ■

128289 Dicentra Borkh. ex Bernh. = Dicentra Bernh. (保留属名)■

128290 Dicentra burmanica K. R. Stern;缅甸紫金龙■

128291 Dicentra canadensis (Goldie) Walp.;加拿大荷包牡丹;American Squirrel Corn, American Squirrel-corn, Squirrel Corn, Squirrel-corn, Squirrel's Corn, Staggerweed, Turkey Corn, Turkey Pea, Turkey-corn ■☆

128292 Dicentra chrysantha (Hook. et Arn.) Walp.;黄花荷包牡丹;Golden Eardrops ■☆

128293 Dicentra chrysantha Walp. = Dicentra chrysantha (Hook. et Arn.) Walp. ■☆

128294 Dicentra cucullaria (L.) Bernh.;兜状荷包牡丹(白兜荷包牡丹);Dutchman's Breeches, Dutchman's-breeches ■☆

128295 Dicentra cucullaria (L.) Bernh. var. occidentalis (Rydb.) M. Peck = Dicentra cucullaria (L.) Bernh. ■☆

128296 Dicentra eximia (Ker Gawl.) Torr.;缕毛荷包牡丹;Eastern Bleeding-heart, Fringed Bleeding Heart, Plume Bleeding Heart, Turkey Corn, Turkey-corn, Wild Bleeding Heart, Wild Bleeding-heart ■☆

128297 Dicentra formosa (Andréws) Walp.;美丽荷包牡丹(北美华鬘草);Bleeding-heart, Pacific Bleeding-heart, Turkey Corn, Turkey Pea, Turkey-corn, Western Bleeding Heart, Western Bleedingheart, Western Bleeding-heart ■☆

128298 Dicentra formosa (Andréws) Walp. subsp. oregana (Eastw.) Munz;俄勒冈荷包牡丹(奥里根荷包牡丹)■☆

128299 Dicentra formosa (Haw.) Walp. subsp. nevadensis (Eastw.) Munz = Dicentra nevadensis Eastw. ■☆

128300 Dicentra formosa (Haw.) Walp. var. breviflora L. F. Hend. = Dicentra formosa (Andréws) Walp. ■☆

128301 Dicentra grandifoliolata (Merr.) K. R. Stern;厚壳紫金龙■

128302 Dicentra lichiangensis Fedde; 丽江紫金龙; Lichiang Dactylicapnos, Lijiang Dactylicapnos ■

128303 Dicentra macrantha Oliv. = Ichtyoselmis macrantha (Oliv.) Lidén ■

128304 Dicentra macrocapnos Prain;薄壳紫金龙■

128305 Dicentra nevadensis Eastw.;内华达荷包牡丹■☆

128306 Dicentra occidentalis (Rydb.) Fedde = Dicentra cucullaria (L.) Bernh. ■☆

128307 Dicentra ochroleuca Engelm.;淡黄荷包牡丹;White Eardrops ■☆

128308 Dicentra oregana Eastw. = Dicentra formosa (Andréws) Walp. subsp. oregana (Eastw.) Munz ■☆

128309 Dicentra paucinervia K. R. Stern = Dicentra grandifoliolata (Merr.) K. R. Stern ■

128310 Dicentra peregrina (Rudolphi) Makino;奇妙荷包牡丹(中南胡麻草)■

128311 Dicentra peregrina (Rudolphi) Makino = Centranthera cochinchinensis (Lour.) Merr. var. lutea (H. Hara) H. Hara ■

128312 Dicentra peregrina (Rudolphi) Makino f. alba (Okada) Takeda;白花奇妙荷包牡丹■☆

128313 Dicentra peregrina (Rudolphi) Makino var. pusilla Makino = Dicentra pusilla Siebold et Zucc. ■

128314 Dicentra pusilla Siebold et Zucc.;小叶荷包牡丹(小荷包牡丹);Smallleaf Colicweed ■

128315 Dicentra pusilla Siebold et Zucc. = Dicentra peregrina (Rudolphi) Makino ■

128316 Dicentra roylei Hook. f. et Thomson;宽果紫金龙(小藤铃儿草);Broadfruit Dactylicapnos ■

128317 Dicentra saccata (Nutt. ex Torr. et A. Gray) Walp. = Dicentra formosa (Andréws) Walp. ■☆

128318 Dicentra scandens (D. Don) Walp.;紫金龙(川山七,穿山七,串枝莲,大麻药,黑牛膝,攀缘紫金龙,藤铃儿草,豌豆跌打,豌豆七);Dactylicapnos, Twine Dactylicapnos ■

128319 Dicentra schneideri Fedde;粗茎紫金龙■

128320 Dicentra schneideri Fedde = Dicentra scandens (D. Don) Walp. ■

128321 Dicentra spectabilis (L.) Lem. = Lamprocapnos spectabilis (L.) Fukuhara ■

128322 Dicentra thalictrifolia (Wall.) Hook. f. et Thomson = Dicentra scandens (D. Don) Walp. ■

128323 Dicentra torulosa Hook. f. et Thomson;扭果紫金龙(大藤铃儿草,大藤紫金龙,独定子,岩连,野落松);Torulosous Dactylicapnos, Twistfruit Dactylicapnos ■

128324 Dicentra torulosa Hook. f. et Thomson var. yunnanensis Fedde = Dactylicapnos torulosa (Hook. f. et Thomson) Hutch. ■

128325 Dicentra torulosa Hook. f. et Thomson var. yunnanensis Fedde = Dicentra torulosa Hook. f. et Thomson ■

128326 Dicentra uniflora Kellogg;单花紫金龙;Steer's-head ■☆

128327 Dicentra wolfdietheri Fedde = Dicentra torulosa Hook. f. et Thomson ■

128328 Dicentranthera T. Anderson = Asystasia Blume ●■

128329 Dicentranthera T. Anderson(1863);双距爵床属●☆

128330 Dicentranthera macrophylla T. Anderson = Asystasia macrophylla (T. Anderson) Lindau ●☆

128331 Dicera Blume = Gironniera Gaudich. ●

128332 Dicera J. R. Forst. et G. Forst. = Elaeocarpus L. ●

128333 Dicera Zipp. ex Blume = Gironniera Gaudich. ●

128334 Dicerandra Benth. (1830);双角雄属●■☆

128335 Dicerandra Benth. = Ceranthera Elliott ●■☆

128336 Dicerandra densiflora Benth.;双角雄●■☆

128337 Diceras Post et Kuntze = Artanema D. Don(保留属名)■☆

128338 Diceras Post et Kuntze = Dicera J. R. Forst. et G. Forst. ●

128339 Diceras Post et Kuntze = Dicera Zipp. ex Blume ●

128340 Diceras Post et Kuntze = Diceros Blume ●

128341 Diceras Post et Kuntze = Diceros Lour. (废弃属名)■

128342 Diceras Post et Kuntze = Diceros Pers. ■☆

128343 Diceras Post et Kuntze = Elaeocarpus L. ●

128344 Diceras Post et Kuntze = Gironniera Gaudich. ●

128345 Diceras Post et Kuntze = Limnophila R. Br. (保留属名)■

128346 Diceras Post et Kuntze = Vandellia L. ■

128347 Diceratella Boiss.(1844);双钝角芥属■☆
128348 Diceratella alata Jonsell et Moggi;翅双钝角芥■☆
128349 Diceratella canescens (Boiss.) Boiss.;灰双钝角芥■☆
128350 Diceratella elliptica (R. Br. ex DC.) Jonsell;椭圆双钝角芥■☆
128351 Diceratella erlangerana Engl. = Diceratella smithii (Baker f.) Jonsell ■☆
128352 Diceratella floccosa (Boiss.) Boiss.;丛毛双钝角芥■☆
128353 Diceratella incana Balf. f.;灰毛双钝角芥■☆
128354 Diceratella inermis Jonsell;无刺双钝角芥■☆
128355 Diceratella psilotrichoides Chiov.;软毛双钝角芥■☆
128356 Diceratella revoilii (Franch.) Jonsell;雷瓦尔双钝角芥■☆
128357 Diceratella ruspoliana Engl. = Diceratella smithii (Baker f.) Jonsell ■☆
128358 Diceratella sahariana Corti = Morettia philaeana (Delile) DC. ■☆
128359 Diceratella sinuata (Franch.) Oliv. ex L. James et Thrupp = Diceratella incana Balf. f. ■☆
128360 Diceratella smithii (Baker f.) Jonsell;史密斯双钝角芥■☆
128361 Diceratella umbrosa Engl. = Diceratella incana Balf. f. ■☆
128362 Diceratium Boiss. = Diceratella Boiss. ■☆
128363 Diceratium Lag. = Notoceras R. Br. ■☆
128364 Diceratium canescens Boiss. = Diceratella canescens (Boiss.) Boiss. ■☆
128365 Diceratium floccosum Boiss. = Diceratella floccosa (Boiss.) Boiss. ■☆
128366 Diceratium prostratum Lag. = Notoceras bicorne (Aiton) Caruel ■☆
128367 Diceratostele Summerh.(1938);双角柱兰属■☆
128368 Diceratostele gabonensis Summerh.;双角柱兰■☆
128369 Dicercoclados C. Jeffrey et Y. L. Chen(1984);歧笔菊属(歧柱蟹甲草属);Dicercoclados ■★
128370 Dicercoclados triplinervis C. Jeffrey et Y. L. Chen;歧笔菊(歧柱蟹甲草);Dicercoclados ■
128371 Dicerma DC.(1825);二节豆属(两节豆属);Dicerma ●
128372 Dicerma biarticulatum (L.) DC.;二节豆(两节豆);Common Dicerma,Dicerma ●
128373 Dicerma biarticulatum (L.) DC. = Aphyllodium biarticulatum (L.) Gagnep. ●
128374 Dicerma elegans DC. = Phyllodium elegans (Lour.) Desv. ●
128375 Dicerma pulchellum (L.) DC. = Phyllodium pulchellum (L.) Desv. ●
128376 Dicerocaryum Bojer(1835);双角胡麻属■☆
128377 Dicerocaryum eriocarpum (Decne.) Abels;毛果双角胡麻■☆
128378 Dicerocaryum forbesii (Decne.) A. E. van Wyk;福布斯双角胡麻■☆
128379 Dicerocaryum senecioides (Klotzsch) Abels;千里光双角胡麻■☆
128380 Dicerocaryum senecioides (Klotzsch) Abels subsp. transvaalense Abels = Dicerocaryum senecioides (Klotzsch) Abels ■☆
128381 Dicerocaryum sinuatum Bojer = Dicerocaryum zanguebarium (Lour.) Merr. ■☆
128382 Dicerocaryum zanguebarium (Lour.) Merr.;赞古双角胡麻■☆
128383 Dicerocaryum zanguebarium (Lour.) Merr. subsp. eriocarpum (Decne.) Ihlenf. = Dicerocaryum eriocarpum (Decne.) Abels ■☆
128384 Dicerocladus C. Jeffrey et Y. L. Chen.(1984);歧柱蟹甲草属■
128385 Dicerolepis Blume = Gymnanthera R. Br. ●
128386 Dicerolepis paludosa Blume = Gymnanthera oblonga (Burm. f.) P. S. Green ●
128387 Diceros Blume = Lindernia All. ■
128388 Diceros Lour.(废弃属名) = Limnophila R. Br.(保留属名)■
128389 Diceros Pers. = Artanema D. Don(保留属名)■☆
128390 Diceros montanus Blume = Lindernia mollis (Benth.) Wettst. ■
128391 Dicerospermum Bakh. f. = Poikilogyne Baker f. ●☆
128392 Dicerostylis Blume = Hylophila Lindl. ■
128393 Dicerostylis Blume.(1859);双臂兰属■☆
128394 Dicerostylis nipponica Fukuy. = Hylophila nipponica (Fukuy.) S. S. Ying ■
128395 Dicersos Lour. = Diceros Lour.(废弃属名)■
128396 Dicersos Lour. = Limnophila R. Br.(保留属名)■
128397 Dichaea Lindl.(1833);蓖叶兰属(迪西亚兰属);Dichaea ■☆
128398 Dichaea glauca (Sw.) Lindl.;粉绿蓖叶兰(粉绿迪西亚兰);Glaucous Dichaea ■☆
128399 Dichaea morrisii Fawc. et Rendle;莫氏蓖叶兰(莫氏迪西亚兰);Morris Dichaea ■☆
128400 Dichaea muricata (Sw.) Lindl.;粗糙蓖叶兰(粗糙迪西亚兰);Rough Dichaea ■☆
128401 Dichaea panamensis Lindl.;巴拿马蓖叶兰(巴拿马迪西亚兰);Panama Dichaea ■☆
128402 Dichaelia Haw. = Brachystelma R. Br.(保留属名)■
128403 Dichaelia barberae (Harv. ex Hook. f.) Bullock = Brachystelma barberae Harv. ex Hook. f. ■☆
128404 Dichaelia brachylepis Schltr. = Brachystelma circinatum E. Mey. ■☆
128405 Dichaelia breviflora Schltr. = Brachystelma pygmaeum (Schltr.) N. E. Br. ■☆
128406 Dichaelia breviflora Schltr. subsp. pygmaea (Schltr.) Schltr. = Brachystelma pygmaeum (Schltr.) N. E. Br. ■☆
128407 Dichaelia cinerea Schltr. = Brachystelma circinatum E. Mey. ■☆
128408 Dichaelia circinata (E. Mey.) Schltr. = Brachystelma circinatum E. Mey. ■☆
128409 Dichaelia elongata Schltr. = Brachystelma elongatum (Schltr.) N. E. Br. ■☆
128410 Dichaelia forcipata Schltr.;钳蓖叶兰■☆
128411 Dichaelia galpinii Schltr. = Brachystelma circinatum E. Mey. ■☆
128412 Dichaelia gerrardii (Harv.) Harv. = Brachystelma gerrardii Harv. ■☆
128413 Dichaelia gracillima (R. A. Dyer) Bullock = Brachystelma gracillimum R. A. Dyer ■☆
128414 Dichaelia hirtella (Weim.) Bullock = Brachystelma hirtellum Weim. ■☆
128415 Dichaelia macra Schltr. = Brachystelma circinatum E. Mey. ■☆
128416 Dichaelia microphylla S. Moore = Brachystelma circinatum E. Mey. ■☆
128417 Dichaelia natalensis Schltr. = Brachystelma sandersonii (Oliv.) N. E. Br. ■☆
128418 Dichaelia pallida Schltr. = Brachystelma circinatum E. Mey. ■☆
128419 Dichaelia pygmaea Schltr. = Brachystelma pygmaeum (Schltr.) N. E. Br. ■☆
128420 Dichaelia undulata Schltr. = Brachystelma circinatum E. Mey. ■☆
128421 Dichaelia villosa Schltr. = Brachystelma villosum (Schltr.) N. E. Br. ■☆
128422 Dichaelia zeyheri Schltr. = Brachystelma circinatum E. Mey. ■☆
128423 Dichaeopsis Pfitzer = Dichaea Lindl. ■☆
128424 Dichaespermum Hassk. = Aneilema R. Br. ■☆
128425 Dichaespermum Wight = Murdannia Royle(保留属名)■
128426 Dichaeta Nutt. = Baeria Fisch. et C. A. Mey. ■☆
128427 Dichaeta Sch. Bip. = Schaetzellia Sch. Bip. ■●☆

128428 Dichaeta fremontii Torr. ex A. Gray = Lasthenia fremontii (Torr. ex A. Gray) Greene ■☆
128429 Dichaetandra Nand. = Ernestia DC. ☆
128430 Dichaetanthera Endl. (1840);二毛药属●☆
128431 Dichaetanthera aculeolata Hook. f. ex Triana = Dichaetanthera articulata Endl. ●☆
128432 Dichaetanthera africana (Hook. f.) Jacq. -Fél.;非洲二毛药●☆
128433 Dichaetanthera altissima Cogn.;高二毛药●☆
128434 Dichaetanthera arborea Baker;树状二毛药●☆
128435 Dichaetanthera articulata Endl.;关节二毛药●☆
128436 Dichaetanthera asperrima Cogn.;粗糙二毛药●☆
128437 Dichaetanthera bifida Jum. et H. Perrier;双裂二毛药●☆
128438 Dichaetanthera brevicauda Jum. et H. Perrier;短尾二毛药●☆
128439 Dichaetanthera calodendron (Gilg et Ledermann ex Engl.) Jacq. -Fél. = Dichaetanthera corymbosa (Cogn.) Jacq. -Fél. ●☆
128440 Dichaetanthera ciliata Jum. et H. Perrier;缘毛二毛药●☆
128441 Dichaetanthera cordifolia Baker;心叶二毛药●☆
128442 Dichaetanthera cornifrons H. Perrier;角叶二毛药●☆
128443 Dichaetanthera corymbosa (Cogn.) Jacq. -Fél.;伞序二毛药●☆
128444 Dichaetanthera crassinodis Baker;粗节二毛药●☆
128445 Dichaetanthera decaryi H. Perrier;德卡里二毛药●☆
128446 Dichaetanthera echinulata (Hook. f.) Jacq. -Fél.;小刺二毛药●☆
128447 Dichaetanthera erici-rosenii (R. E. Fr.) A. Fern. et R. Fern.;欧石南二毛药●☆
128448 Dichaetanthera grandifolia Cogn.;大叶二毛药●☆
128449 Dichaetanthera heteromorpha (Naudin) Triana;异形二毛药●☆
128450 Dichaetanthera heterostemona A. DC. = Dichaetanthera oblongifolia Baker ●☆
128451 Dichaetanthera hirsuta H. Perrier;粗毛二毛药●☆
128452 Dichaetanthera lanceolata Cogn. = Dichaetanthera crassinodis Baker ●☆
128453 Dichaetanthera lancifolia H. Perrier;剑叶二毛药●☆
128454 Dichaetanthera latifolia Cogn. = Dichaetanthera cordifolia Baker ●☆
128455 Dichaetanthera lutescens H. Perrier;淡黄二毛药●☆
128456 Dichaetanthera madagascariensis Triana;马岛二毛药●☆
128457 Dichaetanthera manongarivensis Jum. et H. Perrier = Dichaetanthera bifida Jum. et H. Perrier ●☆
128458 Dichaetanthera matitanensis Jum. et H. Perrier;马蒂坦●☆
128459 Dichaetanthera oblongifolia Baker;矩圆叶二毛药●☆
128460 Dichaetanthera parvifolia Cogn.;小叶二毛药●☆
128461 Dichaetanthera rhodesiensis A. Fern. et R. Fern.;罗得西亚二毛药●☆
128462 Dichaetanthera rutenbergiana Baill. ex Vatke;鲁滕贝格二毛药●☆
128463 Dichaetanthera sambiranensis H. Perrier;桑比朗二毛药●☆
128464 Dichaetanthera scabra Jum. et H. Perrier;糙二毛药●☆
128465 Dichaetanthera squamata H. Perrier;多鳞二毛药●☆
128466 Dichaetanthera strigosa (Cogn.) Jacq. -Fél.;糙伏毛二毛药●☆
128467 Dichaetanthera subrubra Jum. et H. Perrier = Dichaetanthera oblongifolia Baker ●☆
128468 Dichaetanthera trichopoda Jum. et H. Perrier;毛梗二毛药●☆
128469 Dichaetanthera tsaratananensis Jum. et H. Perrier;察拉塔纳纳二毛药●☆
128470 Dichaetanthera verdcourtii A. Fern. et R. Fern.;韦尔德二毛药●☆
128471 Dichaetanthera villosissima H. Perrier;长柔毛二毛药●☆
128472 Dichaetaria Nees ex Steud. (1854);匍匐木根草属■☆
128473 Dichaetaria wightii Nees ex Steud.;匍匐木根草■☆
128474 Dichaetophora A. Gray(1849);田雏菊属■☆
128475 Dichaetophora campestris A. Gray;田雏菊;Plainsdaisy ■☆
128476 Dichanthelium (Hitchc. et Chase) Gould = Panicum L. ■
128477 Dichanthelium (Hitchc. et Chase) Gould (1974);二型花属;Twoformflower ■
128478 Dichanthelium acuminatum (Sw.) Gould et C. A. Clark;渐尖二型花(绵毛稷,绵毛黍);Acuminate Twoformflower, Hairy Panic Grass, Panic Grass, Subvillous Panic-grass, Woolly Panicgrass, Woolly Panic-grass ■
128479 Dichanthelium acuminatum (Sw.) Gould et C. A. Clark = Panicum acuminatum Sw. ■
128480 Dichanthelium acuminatum (Sw.) Gould et C. A. Clark subsp. columbianum (Scribn.) Freckmann et Lelong;哥伦比亚二型花;Hemlock Panic Grass, Hemlock Panic-grass ■☆
128481 Dichanthelium acuminatum (Sw.) Gould et C. A. Clark subsp. fasciculatum (Torr.) Freckmann et Lelong;西部渐尖二型花;Western Panic Grass ■☆
128482 Dichanthelium acuminatum (Sw.) Gould et C. A. Clark subsp. implicatum (Scribn. ex Nash) Freckmann et Lelong;密枝渐尖二型花;Mat Panic Grass ■☆
128483 Dichanthelium acuminatum (Sw.) Gould et C. A. Clark subsp. lindheimeri (Nash) Freckmann et Lelong;林氏渐尖二型花;Lindheimer's Panic Grass ■☆
128484 Dichanthelium acuminatum (Sw.) Gould et C. A. Clark subsp. lindheimeri (Nash) Freckmann et Lelong = Panicum acuminatum Sw. var. lindheimeri (Nash) Beetle ■☆
128485 Dichanthelium acuminatum (Sw.) Gould et C. A. Clark var. fasciculatum (Torr.) Freckmann = Dichanthelium acuminatum (Sw.) Gould et C. A. Clark subsp. fasciculatum (Torr.) Freckmann et Lelong ■☆
128486 Dichanthelium acuminatum (Sw.) Gould et C. A. Clark var. implicatum (Scribn.) Gould et C. A. Clark = Dichanthelium acuminatum (Sw.) Gould et C. A. Clark subsp. implicatum (Scribn. ex Nash) Freckmann et Lelong ■☆
128487 Dichanthelium acuminatum (Sw.) Gould et C. A. Clark var. lindheimeri (Nash) Gould et C. A. Clark = Dichanthelium acuminatum (Sw.) Gould et C. A. Clark subsp. lindheimeri (Nash) Freckmann et Lelong ■☆
128488 Dichanthelium annulum (Ashe) LeBlond = Dichanthelium dichotomum (L.) Gould ■☆
128489 Dichanthelium annulum (Ashe) LeBlond = Panicum dichotomum L. ■☆
128490 Dichanthelium boreale (Nash) Freckmann;北方二型花;Northern Panic Grass ■☆
128491 Dichanthelium boreale (Nash) Freckmann = Panicum boreale Nash ■☆
128492 Dichanthelium boscii (Poir.) Gould et C. A. Clark = Panicum boscii Poir. ■☆
128493 Dichanthelium boscii (Poir.) Gould et C. A. Clark var. molle (Vasey) Mohlenbr. = Panicum boscii Poir. ■☆
128494 Dichanthelium clandestinum (L.) Gould;鹿舌二型花(隐稷);Corn-grass, Deertongue, Deer-tongue Grass, Hidden Panicgrass, Panic Grass ■☆
128495 Dichanthelium clandestinum (L.) Gould = Panicum

clandestinum L. ■☆

128496 Dichanthelium columbianum (Scribn.) Freckmann = Dichanthelium acuminatum (Sw.) Gould et C. A. Clark subsp. columbianum (Scribn.) Freckmann et Lelong ■☆

128497 Dichanthelium columbianum (Scribn.) Freckmann = Panicum portoricense Desv. ex Ham. ■☆

128498 Dichanthelium commonsianum (Ashe) Freckmann = Dichanthelium ovale (Elliott) Gould et C. A. Clark subsp. pseudopubescens (Nash) Freckmann et Lelong ■☆

128499 Dichanthelium commonsianum (Ashe) Freckmann var. euchlamydeum (Shinners) Freckmann = Dichanthelium ovale (Elliott) Gould et C. A. Clark subsp. pseudopubescens (Nash) Freckmann et Lelong ■☆

128500 Dichanthelium commutatum (Schult.) Gould = Panicum commutatum Schult. ■☆

128501 Dichanthelium commutatum (Schult.) Gould var. ashei (G. Pearson ex Ashe) Mohlenbr. = Panicum commutatum Schult. ■☆

128502 Dichanthelium depauperatum (Muhl.) Gould; 贫弱二型花; Panic Grass, Poverty Panic Grass, Starved Panic Grass, Starved Panic-grass ■☆

128503 Dichanthelium depauperatum (Muhl.) Gould = Panicum depauperatum Muhl. ■☆

128504 Dichanthelium depauperatum (Muhl.) Gould var. perlongum (Nash) B. Boivin = Dichanthelium perlongum (Nash) Freckmann ■☆

128505 Dichanthelium dichotomum (L.) Gould; 二型花; Forked Panic Grass, Panic Grass ■☆

128506 Dichanthelium dichotomum (L.) Gould = Panicum dichotomum L. ■☆

128507 Dichanthelium dichotomum (L.) Gould var. nitidum (Lam.) LeBlond = Dichanthelium dichotomum (L.) Gould ■☆

128508 Dichanthelium dichotomum (L.) Gould var. nitidum (Lam.) LeBlond = Panicum dichotomum L. ■☆

128509 Dichanthelium lanuginosum (Elliott) Gould = Dichanthelium acuminatum (Sw.) Gould et C. A. Clark ■

128510 Dichanthelium lanuginosum (Elliott) Gould = Panicum acuminatum Sw. ■

128511 Dichanthelium lanuginosum (Elliott) Gould var. fasciculatum (Torr.) Spellenb. = Dichanthelium acuminatum (Sw.) Gould et C. A. Clark ■

128512 Dichanthelium lanuginosum (Elliott) Gould var. fasciculatum (Torr.) Spellenb. = Dichanthelium acuminatum (Sw.) Gould et C. A. Clark subsp. fasciculatum (Torr.) Freckmann et Lelong ■☆

128513 Dichanthelium lanuginosum (Elliott) Gould var. lindheimeri (Nash) Fernald; 林氏二型花■☆

128514 Dichanthelium latifolium (L.) Gould et C. A. Clark = Panicum latifolium L. ■☆

128515 Dichanthelium latifolium (L.) Harv.; 宽叶二型花; Broad-leaved Panic Grass ■☆

128516 Dichanthelium laxiflorum (Lam.) Gould; 疏花二型花; Panic Grass ■☆

128517 Dichanthelium laxiflorum (Lam.) Gould = Panicum latifolium L. ■☆

128518 Dichanthelium leibergii (Vasey) Freckmann; 雷氏二型花; Leiberg's Panic Grass, Panic Grass, Prairie Panic Grass ■☆

128519 Dichanthelium leibergii (Vasey) Freckmann = Panicum leibergii (Vasey) Scribn. ■☆

128520 Dichanthelium lindheimeri (Nash) Gould = Dichanthelium acuminatum (Sw.) Gould et C. A. Clark subsp. lindheimeri (Nash) Freckmann et Lelong ■☆

128521 Dichanthelium linearifolium (Scribn. ex Nash) Gould = Dichanthelium linearifolium (Scribn.) Gould ■☆

128522 Dichanthelium linearifolium (Scribn. ex Nash) Gould = Panicum linearifolium Scribn. ex Nash ■☆

128523 Dichanthelium linearifolium (Scribn. ex Nash) Gould var. werneri (Scribn.) Mohlenbr. = Panicum linearifolium Scribn. ex Nash ■☆

128524 Dichanthelium linearifolium (Scribn. ex Nash) Gould var. werneri (Scribn.) Mohlenbr. = Dichanthelium linearifolium (Scribn.) Gould ■☆

128525 Dichanthelium linearifolium (Scribn.) Gould; 线叶二型花(小叶稷); Linear-leaved Panic Grass, Narrow-leaved Panic-grass, Panic Grass, Slender-leaved Panic Grass, Slimleaf Panicgrass ■☆

128526 Dichanthelium linearifolium (Scribn.) Gould var. werneri (Scribn.) Mohlenbr. = Dichanthelium linearifolium (Scribn.) Gould ■☆

128527 Dichanthelium malacophyllum (Nash) Gould = Panicum malacophyllum Nash ■☆

128528 Dichanthelium mattamuskeetense (Ashe) Mohlenbr. = Dichanthelium dichotomum (L.) Gould ■☆

128529 Dichanthelium mattamuskeetense (Ashe) Mohlenbr. = Panicum dichotomum L. ■☆

128530 Dichanthelium meridionale (Ashe) Freckmann = Dichanthelium acuminatum (Sw.) Gould et C. A. Clark subsp. implicatum (Scribn. ex Nash) Freckmann et Lelong ■☆

128531 Dichanthelium microcarpon (Muhl. ex Elliott) Mohlenbr. = Dichanthelium dichotomum (L.) Gould ■☆

128532 Dichanthelium microcarpon (Muhl. ex Elliott) Mohlenbr. = Panicum dichotomum L. ■☆

128533 Dichanthelium nitidum (Lam.) Mohlenbr. = Dichanthelium dichotomum (L.) Gould ■☆

128534 Dichanthelium nitidum (Lam.) Mohlenbr. = Panicum dichotomum L. ■☆

128535 Dichanthelium oligosanthes (Schult.) Gould; 少花二型花; Few-flowered Panic Grass, Heller's Rosette Grass, Panic Grass ■☆

128536 Dichanthelium oligosanthes (Schult.) Gould = Panicum oligosanthes Schult. ■☆

128537 Dichanthelium oligosanthes (Schult.) Gould subsp. scribnerianum (Nash) Freckmann et Lelong; 斯氏二型花; Few-flowered Panic Grass, Scribner's Panic Grass, Scribner's Rosette Grass ■☆

128538 Dichanthelium oligosanthes (Schult.) Gould var. helleri (Nash) Mohlenbr. = Dichanthelium oligosanthes (Schult.) Gould ■☆

128539 Dichanthelium oligosanthes (Schult.) Gould var. helleri (Nash) Mohlenbr. = Panicum oligosanthes Schult. ■☆

128540 Dichanthelium oligosanthes (Schult.) Gould var. scribnerianum (Nash) Gould = Dichanthelium oligosanthes (Schult.) Gould ■☆

128541 Dichanthelium oligosanthes (Schult.) Gould var. scribnerianum (Nash) Gould = Panicum oligosanthes Schult. ■☆

128542 Dichanthelium oligosanthes (Schult.) Gould var. wilcoxianum (Vasey) Gould et C. A. Clark = Dichanthelium wilcoxianum (Vasey) Freckmann ■☆

128543 Dichanthelium ovale (Elliott) Gould et C. A. Clark; 卵叶二型花; Stiff-leaved Panic Grass ■☆

128544 Dichanthelium ovale (Elliott) Gould et C. A. Clark subsp.

praecocius (Hitchc. et Chase) Freckmann et Lelong;硬卵叶二型花;Stiff-leaved Panic Grass ■☆
128545 Dichanthelium ovale (Elliott) Gould et C. A. Clark subsp. pseudopubescens (Nash) Freckmann et Lelong;毛卵叶二型花;Stiff-leaved Panic Grass ■☆
128546 Dichanthelium ovale (Elliott) Gould et C. A. Clark var. addisonii (Nash) Gould et C. A. Clark = Dichanthelium ovale (Elliott) Gould et C. A. Clark subsp. pseudopubescens (Nash) Freckmann et Lelong ■☆
128547 Dichanthelium perlongum (Nash) Freckmann;长柄二型花;Long-stalked Panic Grass ■☆
128548 Dichanthelium perlongum (Nash) Freckmann = Dichanthelium linearifolium (Scribn.) Gould ■☆
128549 Dichanthelium perlongum (Nash) Freckmann = Panicum linearifolium Scribn. ex Nash ■☆
128550 Dichanthelium polyanthes (Schult.) Mohlenbr. = Panicum sphaerocarpon Elliott ■☆
128551 Dichanthelium portoricense (Desv. ex Ham.) B. F. Hansen et Wunderlin = Panicum portoricense Desv. ex Ham. ■☆
128552 Dichanthelium ravenelii (Scribn. et Merr.) Gould = Panicum ravenelii Scribn. et Merr. ■☆
128553 Dichanthelium sabulorum (Lam.) Gould et C. A. Clark;钩吻二型花;Hemlock Rosette Grass ■☆
128554 Dichanthelium sabulorum (Lam.) Gould et C. A. Clark var. thinium (Hitchc. et Chase) Gould et C. A. Clark = Dichanthelium acuminatum (Sw.) Gould et C. A. Clark subsp. columbianum (Scribn.) Freckmann et Lelong ■☆
128555 Dichanthelium sabulorum (Lam.) Gould et C. A. Clark var. thinium (Hitchc. et Chase) Gould et C. A. Clark = Panicum portoricense Desv. ex Ham. ■☆
128556 Dichanthelium scoparium (Lam.) Gould = Panicum scoparium Lam. ■☆
128557 Dichanthelium sphaerocarpon (Elliott) Gould = Panicum sphaerocarpon Elliott ■☆
128558 Dichanthelium sphaerocarpon (Elliott) Gould var. isophyllum (Scribn.) Gould et C. A. Clark = Panicum sphaerocarpon Elliott ■☆
128559 Dichanthelium villosissimum (Nash) Freckmann var. praecocius (Hitchc. et Chase) Freckmann = Dichanthelium ovale (Elliott) Gould et C. A. Clark subsp. praecocius (Hitchc. et Chase) Freckmann et Lelong ■☆
128560 Dichanthelium villosissimum (Nash) Freckmann var. pseudopubescens (Nash) Mohlenbr. = Dichanthelium ovale (Elliott) Gould et C. A. Clark subsp. pseudopubescens (Nash) Freckmann et Lelong ■☆
128561 Dichanthelium wilcoxianum (Vasey) Freckmann;威氏二型花;Wilcox's Panic Grass ■☆
128562 Dichanthelium xanthophysum (A. Gray) Freckmann;纤弱二型花;Pale Panic Grass,Slender Rosette Grass ■☆
128563 Dichanthelium yadkinense (Ashe) Mohlenbr. = Dichanthelium dichotomum (L.) Gould ■☆
128564 Dichanthelium yadkinense (Ashe) Mohlenbr. = Panicum dichotomum L. ■☆
128565 Dichanthium Willemet (1796);双花草属;Biflorgrass, Dichanthium ■
128566 Dichanthium affine (R. Br.) A. Camus;近缘双花草■☆
128567 Dichanthium andringitrense A. Camus;安德林吉特拉山双花草■☆
128568 Dichanthium annulatum (Forssk.) Stapf;双花草;Kleberg's Bluestem,Ringed Biflorgrass,Ringed Dichanthium ■
128569 Dichanthium annulatum (Forssk.) Stapf var. decalvatum (Hack.) Maire et Weiller = Dichanthium annulatum (Forssk.) Stapf ■
128570 Dichanthium annulatum (Forssk.) Stapf var. papillosum (Hochst. ex A. Rich.) de Wet et Harlan;乳头双花草■☆
128571 Dichanthium aristatum (Poir.) C. E. Hubb.;毛梗双花草;Angleton Bluestem,Aristate Biflorgrass,Aristate Dichanthium ■
128572 Dichanthium assimile (Steud.) Deshpande = Capillipedium assimile (Steud.) A. Camus ■
128573 Dichanthium bladhii (Retz.) Clayton = Bothriochloa bladhii (Retz.) S. T. Blake ■
128574 Dichanthium caricosum (L.) A. Camus;单穗草;Figlike Biflorgrass,Monospike Biflorgrass,Monospike Dichanthium ■
128575 Dichanthium foveolatum (Delile) Roberty;小孔双花草■☆
128576 Dichanthium insculptum (Hochst. ex A. Rich.) Clayton = Bothriochloa insculpta (Hochst. ex A. Rich.) A. Camus ■☆
128577 Dichanthium intermedium (R. Br.) De Wet et Harlan = Bothriochloa bladhii (Retz.) S. T. Blake ■
128578 Dichanthium ischaemum (L.) Roberty = Bothriochloa ischaemum (L.) Keng ■
128579 Dichanthium nodosum Willemet = Dichanthium annulatum (Forssk.) Stapf ■
128580 Dichanthium nodosum Willemet = Dichanthium aristatum (Poir.) C. E. Hubb. ■
128581 Dichanthium papillosum (Hochst. ex A. Rich.) Stapf = Dichanthium annulatum (Forssk.) Stapf var. papillosum (Hochst. ex A. Rich.) de Wet et Harlan ■☆
128582 Dichanthium parviflorum (R. Br.) de Wet = Capillipedium parviflorum (R. Br.) Stapf ■
128583 Dichanthium parviflorum (R. Br.) de Wet et J. R. Harlan = Capillipedium parviflorum (R. Br.) Stapf ■
128584 Dichanthium pertusum (L.) Clayton = Bothriochloa pertusa (L.) A. Camus ■
128585 Dichanthium radicans (Lehm.) Clayton = Bothriochloa radicans (Lehm.) A. Camus ■☆
128586 Dichanthium sericeum (R. Br.) A. Camus;绢毛双花草;Silky Bluestem ■☆
128587 Dichanthium tenue A. Camus;细双花草;Slender Bluestem ■☆
128588 Dichapetalaceae Baill. (1886)(保留科名);毒鼠子科;Dichapetalum Family,Poisonrat Family ●
128589 Dichapetalum Thouars (1806);毒鼠子属;Dichapetalium, Dichapetalum,Poisonrat ●
128590 Dichapetalum abrupti-acuminatum De Wild. = Dichapetalum madagascariense Poir. ●☆
128591 Dichapetalum acuminatum De Wild.;渐尖毒鼠子●☆
128592 Dichapetalum acutifolium Engl. = Dichapetalum tomentosum Engl. ●☆
128593 Dichapetalum acutisepalum Engl. = Dichapetalum heudelotii (Planch. ex Oliv.) Baill. ●☆
128594 Dichapetalum adolphi-fridericii Engl. = Dichapetalum heudelotii (Planch. ex Oliv.) Baill. ●☆
128595 Dichapetalum affine (Planch. ex Benth.) Breteler;近缘毒鼠子●☆
128596 Dichapetalum albidum A. Chev. ex Pellegr.;白毒鼠子●☆
128597 Dichapetalum altescandens Engl.;攀缘毒鼠子●☆

128598 Dichapetalum angolense Chodat;安哥拉毒鼠子●☆
128599 Dichapetalum angolense Chodat var. glabriusculum Hauman = Dichapetalum angolense Chodat ●☆
128600 Dichapetalum angolense Chodat var. leucanthum Pellegr. = Dichapetalum angolense Chodat ●☆
128601 Dichapetalum angustisquamulosum Engl. et Ruhland = Dichapetalum heudelotii (Planch. ex Oliv.) Baill. ●☆
128602 Dichapetalum arachnoideum Breteler;蛛网毒鼠子●☆
128603 Dichapetalum arenarium Breteler;沙地毒鼠子●☆
128604 Dichapetalum argenteum Engl. = Dichapetalum bangii (Didr.) Engl. ●☆
128605 Dichapetalum aurantiacum Engl. = Dichapetalum heudelotii (Planch. ex Oliv.) Baill. var. longitubulosum (Engl.) Breteler ●☆
128606 Dichapetalum aureonitens Engl. = Dichapetalum mossambicense (Klotzsch) Engl. ●☆
128607 Dichapetalum bakerianum Exell = Dichapetalum madagascariense Poir. ●☆
128608 Dichapetalum bangii (Didr.) Engl.;班氏毒鼠子●☆
128609 Dichapetalum barbatum Breteler;髯毛毒鼠子●☆
128610 Dichapetalum barbosae Torre;巴尔博萨毒鼠子●☆
128611 Dichapetalum barense Engl. = Dichapetalum tomentosum Engl. ●☆
128612 Dichapetalum barteri Engl.;巴特毒鼠子●☆
128613 Dichapetalum batanganum Engl. et Ruhland = Dichapetalum madagascariense Poir. ●☆
128614 Dichapetalum batesii Engl. = Dichapetalum heudelotii (Planch. ex Oliv.) Baill. var. longitubulosum (Engl.) Breteler ●☆
128615 Dichapetalum baturense K. Krause = Dichapetalum glomeratum Engl. ●☆
128616 Dichapetalum beilschmiedioides Breteler;琼楠毒鼠子●☆
128617 Dichapetalum bellum Breteler;雅致毒鼠子●☆
128618 Dichapetalum beniense Engl. = Dichapetalum madagascariense Poir. ●☆
128619 Dichapetalum bojeri (Tul.) Engl.;博耶尔毒鼠子●☆
128620 Dichapetalum brachysepalum Engl. = Dichapetalum crassifolium Chodat ●☆
128621 Dichapetalum braunii Engl. et K. Krause;布劳恩毒鼠子●☆
128622 Dichapetalum brazzae Pellegr. = Dichapetalum librevillense Pellegr. ●☆
128623 Dichapetalum brazzae Pellegr. var. purpurascens Hauman = Dichapetalum librevillense Pellegr. ●☆
128624 Dichapetalum brevitubulosum Engl. = Dichapetalum madagascariense Poir. ●☆
128625 Dichapetalum bussei Engl. = Dichapetalum pallidum (Oliv.) Engl. ●☆
128626 Dichapetalum butayei De Wild. = Dichapetalum librevillense Pellegr. ●☆
128627 Dichapetalum buvumense Baker f. = Dichapetalum madagascariense Poir. ●☆
128628 Dichapetalum chalotii Pellegr.;沙洛毒鼠子●☆
128629 Dichapetalum chlorinum (Tul.) Engl.;绿毒鼠子●☆
128630 Dichapetalum choristilum Engl. var. louisii Breteler;路易斯毒鼠子●☆
128631 Dichapetalum chrysobalanoides Hutch. et Dalziel = Dichapetalum madagascariense Poir. ●☆
128632 Dichapetalum cicinnatum Engl. = Dichapetalum madagascariense Poir. ●☆
128633 Dichapetalum cinereoviride Engl. = Dichapetalum staudtii Engl. ●☆
128634 Dichapetalum cinereum Engl. = Dichapetalum pallidum (Oliv.) Engl. ●☆
128635 Dichapetalum cinnamomeum Hauman = Dichapetalum fructuosum Hiern ●☆
128636 Dichapetalum claessensii De Wild. = Dichapetalum acuminatum De Wild. ●☆
128637 Dichapetalum congoense Engl. et Ruhland;刚果毒鼠子●☆
128638 Dichapetalum conrauanum Engl. et Ruhland = Dichapetalum heudelotii (Planch. ex Oliv.) Baill. ●☆
128639 Dichapetalum contractum Engl. = Dichapetalum staudtii Engl. ●☆
128640 Dichapetalum cordifolium Hutch. et Dalziel = Dichapetalum reticulatum Engl. ●☆
128641 Dichapetalum corradii Chiov. = Tapura fischeri Engl. ●☆
128642 Dichapetalum corrugatum Exell = Dichapetalum unguiculatum Engl. ●☆
128643 Dichapetalum crassifolium Chodat;厚叶毒鼠子●☆
128644 Dichapetalum crassifolium Chodat var. integrum (Pierre) Breteler;全缘厚叶毒鼠子●☆
128645 Dichapetalum cuneifolium Engl. = Dichapetalum heudelotii (Planch. ex Oliv.) Baill. ●☆
128646 Dichapetalum cymosum (Hook.) Engl.;聚伞毒鼠子●☆
128647 Dichapetalum deflexum (Klotzsch) Engl.;外折毒鼠子●☆
128648 Dichapetalum dewevrei De Wild. et T. Durand;德韦毒鼠子●☆
128649 Dichapetalum dewevrei De Wild. et T. Durand var. donisii Hauman = Dichapetalum dewevrei De Wild. et T. Durand ●☆
128650 Dichapetalum dewevrei De Wild. et T. Durand var. klaineanum (Pellegr.) Breteler;克莱恩毒鼠子●☆
128651 Dichapetalum dewildei Breteler;德维尔德毒鼠子●☆
128652 Dichapetalum dewildemanianum Exell = Dichapetalum zenkeri Engl. ●☆
128653 Dichapetalum dictyospermum Breteler;网籽毒鼠子●☆
128654 Dichapetalum divaricatum De Wild. = Dichapetalum librevillense Pellegr. ●☆
128655 Dichapetalum dodoense Engl. = Dichapetalum madagascariense Poir. ●☆
128656 Dichapetalum dummeri Moss = Tapura fischeri Engl. ●☆
128657 Dichapetalum dundusanense De Wild. = Dichapetalum madagascariense Poir. ●☆
128658 Dichapetalum dusenii Engl. = Dichapetalum affine (Planch. ex Benth.) Breteler ●☆
128659 Dichapetalum echinulatum Exell = Dichapetalum staudtii Engl. ●☆
128660 Dichapetalum edule Engl.;可食毒鼠子●☆
128661 Dichapetalum eickii Ruhland;艾克毒鼠子●☆
128662 Dichapetalum ellipticum R. E. Fr. = Dichapetalum bangii (Didr.) Engl. ●☆
128663 Dichapetalum fadenii Breteler;法登毒鼠子●☆
128664 Dichapetalum fallax Ruhland = Dichapetalum affine (Planch. ex Benth.) Breteler ●☆
128665 Dichapetalum ferrugineotomentosum Engl. = Dichapetalum angolense Chodat ●☆
128666 Dichapetalum ferrugineum Engl. = Dichapetalum heudelotii (Planch. ex Oliv.) Baill. ●☆
128667 Dichapetalum filicaule Breteler;丝茎毒鼠子●☆
128668 Dichapetalum flabellatiflorum Hauman = Dichapetalum

madagascariense Poir. ●☆

128669 Dichapetalum flaviflorum Engl. = Dichapetalum madagascariense Poir. ●☆

128670 Dichapetalum floribundum (Planch.) Engl.;多花毒鼠子●☆

128671 Dichapetalum fructuosum Hiern;多果毒鼠子●☆

128672 Dichapetalum fulvialabastrum De Wild. = Dichapetalum madagascariense Poir. ●☆

128673 Dichapetalum fuscescens Engl. = Dichapetalum heudelotii (Planch. ex Oliv.) Baill. ●☆

128674 Dichapetalum gabonense Engl.;加蓬毒鼠子●☆

128675 Dichapetalum gelonioides (Roxb.) Engl.;毒鼠子(滇毒鼠子,琼南毒鼠子);Common Dichapetalum, How's Dichapetalum, Poisonrat ●

128676 Dichapetalum geminostellatum Breteler;星毒鼠子●☆

128677 Dichapetalum germainii Hauman;杰曼毒鼠子●☆

128678 Dichapetalum gillardinii Hauman = Dichapetalum lujae De Wild. et T. Durand var. gillardinii (Hauman) Breteler ●☆

128679 Dichapetalum gilleti De Wild.;吉莱特毒鼠子●☆

128680 Dichapetalum glandulosum De Wild. = Dichapetalum madagascariense Poir. ●☆

128681 Dichapetalum glomeratum Engl.;团集毒鼠子●☆

128682 Dichapetalum gossweileri Engl. = Dichapetalum madagascariense Poir. ●☆

128683 Dichapetalum gracile Exell = Dichapetalum heudelotii (Planch. ex Oliv.) Baill. var. ndongense (Engl.) Breteler ●☆

128684 Dichapetalum grandifolium Ridl.;大叶毒鼠子●☆

128685 Dichapetalum griseisepalum De Wild. = Dichapetalum unguiculatum Engl. ●☆

128686 Dichapetalum griseoviride Ruhland = Dichapetalum pallidum (Oliv.) Engl. ●☆

128687 Dichapetalum guineense (DC.) Keay = Dichapetalum madagascariense Poir. ●☆

128688 Dichapetalum guinense (DC.) Keay;非洲毒鼠子●☆

128689 Dichapetalum hainanense (Hance) Engl. = Dichapetalum longipetalum (Turcz.) Engl. ●

128690 Dichapetalum heudelotii (Planch. ex Oliv.) Baill.;霍德毒鼠子●☆

128691 Dichapetalum heudelotii (Planch. ex Oliv.) Baill. var. hispidum (Oliv.) Breteler;硬毛霍德毒鼠子●☆

128692 Dichapetalum heudelotii (Planch. ex Oliv.) Baill. var. longitubulosum (Engl.) Breteler;长管霍德毒鼠子●☆

128693 Dichapetalum heudelotii (Planch. ex Oliv.) Baill. var. ndongense (Engl.) Breteler;恩东加毒鼠子●☆

128694 Dichapetalum hispidum (Oliv.) Baill. = Dichapetalum heudelotii (Planch. ex Oliv.) Baill. var. hispidum (Oliv.) Breteler ●☆

128695 Dichapetalum holopetalum Ruhland = Dichapetalum crassifolium Chodat ●☆

128696 Dichapetalum holosericeum Engl. = Dichapetalum bangii (Didr.) Engl. ●☆

128697 Dichapetalum howii Merr. et Chun = Dichapetalum gelonioides (Roxb.) Engl. ●

128698 Dichapetalum hypoleucum Hiern = Dichapetalum pallidum (Oliv.) Engl. ●☆

128699 Dichapetalum inaequale Breteler;不等毒鼠子●☆

128700 Dichapetalum insigne Engl.;显著毒鼠子●☆

128701 Dichapetalum integripetalum Engl.;全缘瓣毒鼠子●☆

128702 Dichapetalum integrum Pierre = Dichapetalum crassifolium Chodat var. integrum (Pierre) Breteler ●☆

128703 Dichapetalum jabassense Engl. = Dichapetalum heudelotii (Planch. ex Oliv.) Baill. ●☆

128704 Dichapetalum johnstonii Engl. = Dichapetalum heudelotii (Planch. ex Oliv.) Baill. ●☆

128705 Dichapetalum kamerunense Engl. = Dichapetalum oblongum (Hook. f. ex Benth.) Engl. ●☆

128706 Dichapetalum keniense Hutch. et Bruce = Dichapetalum zenkeri Engl. ●☆

128707 Dichapetalum klainei Pellegr. = Dichapetalum glomeratum Engl. ●☆

128708 Dichapetalum korupinum Breteler;科鲁普毒鼠子●☆

128709 Dichapetalum kribense Engl. = Dichapetalum tomentosum Engl. ●☆

128710 Dichapetalum kumasiense Hoyle = Dichapetalum heudelotii (Planch. ex Oliv.) Baill. ●☆

128711 Dichapetalum lebrunii Hauman = Dichapetalum stuhlmannii Engl. ●☆

128712 Dichapetalum ledermannii Engl. = Dichapetalum rudatisii Engl. ●☆

128713 Dichapetalum lescrauwaeti De Wild. = Dichapetalum heudelotii (Planch. ex Oliv.) Baill. ●☆

128714 Dichapetalum letestui Pellegr. = Dichapetalum lujae De Wild. et T. Durand var. letestui (Pellegr.) Breteler ●☆

128715 Dichapetalum letouzeyi Breteler;勒图毒鼠子●☆

128716 Dichapetalum leucocarpum Breteler;白果毒鼠子●☆

128717 Dichapetalum leucosepalum Ruhland = Dichapetalum lujae De Wild. et T. Durand ●☆

128718 Dichapetalum liberiae Engl. et Dinkl. = Dichapetalum pallidum (Oliv.) Engl. ●☆

128719 Dichapetalum librevillense Pellegr.;利伯维尔毒鼠子●☆

128720 Dichapetalum linderi Hutch. et Dalziel = Dichapetalum heudelotii (Planch. ex Oliv.) Baill. ●☆

128721 Dichapetalum lindicum Breteler;林迪毒鼠子●☆

128722 Dichapetalum lokanduense De Wild. = Dichapetalum heudelotii (Planch. ex Oliv.) Baill. ●☆

128723 Dichapetalum lolo De Wild. et T. Durand = Dichapetalum heudelotii (Planch. ex Oliv.) Baill. ●☆

128724 Dichapetalum longifolium Engl. = Dichapetalum heudelotii (Planch. ex Oliv.) Baill. ●☆

128725 Dichapetalum longipedicellatum De Wild. = Dichapetalum pedicellatum K. Krause ●☆

128726 Dichapetalum longipetalum (Turcz.) Engl.;海南毒鼠子(长瓣毒鼠子);Hainan Dichapetalum, Hainan Poisonrat, Longpetal Dichapetalum ●

128727 Dichapetalum longitubulosum Engl. = Dichapetalum heudelotii (Planch. ex Oliv.) Baill. var. longitubulosum (Engl.) Breteler ●☆

128728 Dichapetalum lujae De Wild. et T. Durand;白萼毒鼠子●☆

128729 Dichapetalum lujae De Wild. et T. Durand var. brevipile Hauman = Dichapetalum lujae De Wild. et T. Durand ●☆

128730 Dichapetalum lujae De Wild. et T. Durand var. gillardinii (Hauman) Breteler;吉拉尔丹毒鼠子●☆

128731 Dichapetalum lujae De Wild. et T. Durand var. letestui (Pellegr.) Breteler;莱泰斯图毒鼠子●☆

128732 Dichapetalum lujae De Wild. et T. Durand var. leucosepalum (Ruhland) Hauman = Dichapetalum lujae De Wild. et T. Durand ●☆

128733 Dichapetalum lukolelaense De Wild. = Dichapetalum heudelotii (Planch. ex Oliv.) Baill. var. ndongense (Engl.) Breteler ●☆

128734 Dichapetalum luteiflorum De Wild. = Dichapetalum staudtii Engl. ●☆

128735 Dichapetalum macrocarpum Engl. ex K. Krause;大果毒鼠子●☆

128736 Dichapetalum macrophyllum (Oliv.) Engl. = Dichapetalum heudelotii (Planch. ex Oliv.) Baill. var. hispidum (Oliv.) Breteler ●☆

128737 Dichapetalum madagascariense Poir.;马岛毒鼠子●☆

128738 Dichapetalum madagascariense Poir. = Dichapetalum beniense Engl. ●☆

128739 Dichapetalum madagascariense Poir. var. beniense (Engl.) Breteler = Dichapetalum madagascariense Poir. ●☆

128740 Dichapetalum madagascariense Poir. var. brevistylum Breteler;短柱马达加斯加毒鼠子●☆

128741 Dichapetalum madagascariense Thouars = Dichapetalum madagascariense Poir. ●☆

128742 Dichapetalum malchairii De Wild. = Dichapetalum glomeratum Engl. ●☆

128743 Dichapetalum malembense Pellegr. = Dichapetalum crassifolium Chodat ●☆

128744 Dichapetalum martineaui Aubrév. = Dichapetalum heudelotii (Planch. ex Oliv.) Baill. var. ndongense (Engl.) Breteler ●☆

128745 Dichapetalum mathisii Breteler;马蒂斯毒鼠子●☆

128746 Dichapetalum mayumbense Exell = Dichapetalum angolense Chodat ●☆

128747 Dichapetalum mekametane Engl. = Dichapetalum congoense Engl. et Ruhland ●☆

128748 Dichapetalum melanocladum Breteler;黑枝毒鼠子●☆

128749 Dichapetalum mendoncae Torre = Dichapetalum deflexum (Klotzsch) Engl. ●☆

128750 Dichapetalum michelsonii Hauman = Dichapetalum stuhlmannii Engl. ●☆

128751 Dichapetalum micranthum Hauman = Dichapetalum dewevrei De Wild. et T. Durand ●☆

128752 Dichapetalum micropetalum Engl. = Dichapetalum gabonense Engl. ●☆

128753 Dichapetalum mildbraedianum Exell = Dichapetalum heudelotii (Planch. ex Oliv.) Baill. var. ndongense (Engl.) Breteler ●☆

128754 Dichapetalum minutiflorum Engl. et Ruhland;微花毒鼠子●☆

128755 Dichapetalum molundense K. Krause = Dichapetalum zenkeri Engl. ●☆

128756 Dichapetalum mombongense De Wild. = Dichapetalum staudtii Engl. ●☆

128757 Dichapetalum montanum Breteler;山地毒鼠子●☆

128758 Dichapetalum mossambicense (Klotzsch) Engl.;莫桑比克毒鼠子●☆

128759 Dichapetalum mossambicense (Klotzsch) Pires de Lima = Dichapetalum mossambicense (Klotzsch) Engl. ●☆

128760 Dichapetalum mucronulatum Engl. = Dichapetalum parvifolium Engl. ●☆

128761 Dichapetalum mundense Engl. var. seretii (De Wild.) Hauman = Dichapetalum mundense Engl. ●☆

128762 Dichapetalum murinum Breteler ex Den Outer = Dichapetalum pallidum (Oliv.) Engl. ●☆

128763 Dichapetalum ndongense Engl. = Dichapetalum heudelotii (Planch. ex Oliv.) Baill. var. ndongense (Engl.) Breteler ●☆

128764 Dichapetalum neglectum Breteler;忽视毒鼠子●☆

128765 Dichapetalum nitidulum Engl. et Ruhland = Dichapetalum gabonense Engl. ●☆

128766 Dichapetalum nyangense Pellegr.;尼扬加毒鼠子●☆

128767 Dichapetalum obanense (Baker f.) Baker f. ex Hutch. et Dalziel;奥班毒鼠子●☆

128768 Dichapetalum obliquifolium Engl. var. klaineana Pellegr. = Dichapetalum dewevrei De Wild. et T. Durand var. klaineanum (Pellegr.) Breteler ●☆

128769 Dichapetalum oblongum (Hook. f. ex Benth.) Engl.;矩圆毒鼠子●☆

128770 Dichapetalum oddonii De Wild. = Dichapetalum fructuosum Hiern ●☆

128771 Dichapetalum oliganthum Breteler;寡花毒鼠子●☆

128772 Dichapetalum ombrophilum K. Krause = Dichapetalum madagascariense Poir. ●☆

128773 Dichapetalum pallidinervum De Wild. = Dichapetalum zenkeri Engl. ●☆

128774 Dichapetalum pallidum (Oliv.) Engl.;苍白毒鼠子●☆

128775 Dichapetalum palustre Louis ex Hauman = Dichapetalum crassifolium Chodat ●☆

128776 Dichapetalum palustre Louis ex Hauman var. polyanthum Hauman = Dichapetalum crassifolium Chodat ●☆

128777 Dichapetalum paniculatum Thonn. ex Schumach. = Dichapetalum madagascariense Poir. ●☆

128778 Dichapetalum parvifolium Engl.;小花毒鼠子●☆

128779 Dichapetalum patenti-hirsutum Ruhland = Dichapetalum bangii (Didr.) Engl. ●☆

128780 Dichapetalum patenti-hirsutum Ruhland var. longibracteatum Hauman = Dichapetalum bangii (Didr.) Engl. ●☆

128781 Dichapetalum pedicellatum K. Krause;梗花毒鼠子●☆

128782 Dichapetalum petaloideum Breteler;花瓣状毒鼠子●☆

128783 Dichapetalum petersianum Dinkl. et Engl. = Dichapetalum angolense Chodat ●☆

128784 Dichapetalum pierrei Pellegr.;皮埃尔毒鼠子●☆

128785 Dichapetalum poggei Engl. = Dichapetalum heudelotii (Planch. ex Oliv.) Baill. ●☆

128786 Dichapetalum potamophilum Breteler;河生毒鼠子●☆

128787 Dichapetalum pulchrum Breteler;美丽毒鼠子●☆

128788 Dichapetalum pynaertii De Wild. = Dichapetalum madagascariense Poir. ●☆

128789 Dichapetalum rabiense Breteler;拉比毒鼠子●☆

128790 Dichapetalum reticulatum Engl.;网状毒鼠子●☆

128791 Dichapetalum retroversum Hiern = Dichapetalum parvifolium Engl. ●☆

128792 Dichapetalum reygaertii De Wild. = Dichapetalum angolense Chodat ●☆

128793 Dichapetalum rhodesicum Sprague et Hutch.;罗得西亚毒鼠子●☆

128794 Dichapetalum rowlandii Hutch. et Dalziel = Dichapetalum madagascariense Poir. ●☆

128795 Dichapetalum rudatisii Engl.;鲁达蒂斯毒鼠子●☆

128796 Dichapetalum ruficeps Breteler;红头毒鼠子●☆

128797 Dichapetalum rufipile (Turcz.) Engl. = Dichapetalum bangii (Didr.) Engl. ●☆

128798 Dichapetalum rufotomentosum Engl. = Dichapetalum angolense Chodat ●☆

128799 Dichapetalum rufum (Tul.) Engl.;浅红毒鼠子●☆
128800 Dichapetalum ruhlandii Engl.;吕兰毒鼠子●☆
128801 Dichapetalum salicifolium Engl. et Ruhland = Dichapetalum heudelotii (Planch. ex Oliv.) Baill. var. hispidum (Oliv.) Breteler ●☆
128802 Dichapetalum sankuruense De Wild. = Dichapetalum heudelotii (Planch. ex Oliv.) Baill. var. ndongense (Engl.) Breteler ●☆
128803 Dichapetalum sapinii De Wild. = Dichapetalum chalotii Pellegr. ●☆
128804 Dichapetalum scabrum Engl. = Dichapetalum heudelotii (Planch. ex Oliv.) Baill. var. longitubulosum (Engl.) Breteler ●☆
128805 Dichapetalum schliebenii Mildbr. = Dichapetalum stuhlmannii Engl. ●☆
128806 Dichapetalum schweinfurthii Engl. = Dichapetalum heudelotii (Planch. ex Oliv.) Baill. ●☆
128807 Dichapetalum schweinfurthii Engl. var. lolo (De Wild. et T. Durand) Hauman = Dichapetalum heudelotii (Planch. ex Oliv.) Baill. ●☆
128808 Dichapetalum seretii De Wild. = Dichapetalum mundense Engl. ●☆
128809 Dichapetalum soyauxii Engl. = Dichapetalum gabonense Engl. ●☆
128810 Dichapetalum spathulatum Engl. = Dichapetalum crassifolium Chodat ●☆
128811 Dichapetalum staudtii Engl.;施陶毒鼠子●☆
128812 Dichapetalum stenophyllum K. Krause = Dichapetalum gilleti De Wild. ●☆
128813 Dichapetalum stuhlmannii Engl.;斯图尔曼毒鼠子●☆
128814 Dichapetalum subauriculatum (Oliv.) Engl. = Dichapetalum heudelotii (Planch. ex Oliv.) Baill. ●☆
128815 Dichapetalum subcordatum (Hook. f. ex Benth.) Engl. = Dichapetalum madagascariense Poir. ●☆
128816 Dichapetalum subcoriaceum Engl. = Dichapetalum madagascariense Poir. ●☆
128817 Dichapetalum subfalcatum Engl. = Dichapetalum gabonense Engl. ●☆
128818 Dichapetalum suboblongum Engl. = Dichapetalum rudatisii Engl. ●☆
128819 Dichapetalum subsessilifolium Chodat = Dichapetalum heudelotii (Planch. ex Oliv.) Baill. var. hispidum (Oliv.) Breteler ●☆
128820 Dichapetalum subtruncatum Engl. = Dichapetalum tomentosum Engl. ●☆
128821 Dichapetalum sulcatum Engl. = Dichapetalum staudtii Engl. ●☆
128822 Dichapetalum sumbense Breteler;松巴毒鼠子●☆
128823 Dichapetalum tetrastachyum Breteler;四穗毒鼠子●☆
128824 Dichapetalum thollonii Pellegr.;托伦毒鼠子●☆
128825 Dichapetalum thomsonii (Oliv.) Engl. = Dichapetalum madagascariense Poir. ●☆
128826 Dichapetalum thomsonii (Oliv.) Engl. var. obanense Baker f. = Dichapetalum obanense (Baker f.) Baker f. ex Hutch. et Dalziel ●☆
128827 Dichapetalum thonneri De Wild. = Dichapetalum bangii (Didr.) Engl. ●☆
128828 Dichapetalum thonneri De Wild. var. ellipticum (R. E. Fr.) Hauman = Dichapetalum bangii (Didr.) Engl. ●☆
128829 Dichapetalum thonneri De Wild. var. longistipulatum Hauman = Dichapetalum bangii (Didr.) Engl. ●☆
128830 Dichapetalum thonneri De Wild. var. polyneuron Hauman = Dichapetalum bangii (Didr.) Engl. ●☆
128831 Dichapetalum thouarsianum Roem. et Schult. = Dichapetalum madagascariense Poir. ●☆
128832 Dichapetalum thouarsianum Roem. et Schult. var. macrophyllum (Tul.) Desc. = Dichapetalum madagascariense Poir. ●☆
128833 Dichapetalum tomentosum Engl.;毛毒鼠子●☆
128834 Dichapetalum tonkinensis Engl. = Dichapetalum longipetalum (Turcz.) Engl. ●
128835 Dichapetalum toxicarium (G. Don) Baill.;西非毒鼠子;West African Ratbane ●☆
128836 Dichapetalum toxicarium (G. Don) Baill. var. elliptica (Oliv.) De Wild. = Dichapetalum toxicarium (G. Don) Baill. ●☆
128837 Dichapetalum trichocephalum Breteler;毛头毒鼠子●☆
128838 Dichapetalum ubangiense De Wild. = Dichapetalum madagascariense Poir. ●☆
128839 Dichapetalum ugandense M. B. Moss;乌干达毒鼠子●☆
128840 Dichapetalum umbellatum Chodat;小伞毒鼠子●☆
128841 Dichapetalum unguiculatum Engl.;爪状毒鼠子●☆
128842 Dichapetalum varians Pellegr. = Dichapetalum heudelotii (Planch. ex Oliv.) Baill. ●☆
128843 Dichapetalum venenatum Engl. et Gilg = Dichapetalum cymosum (Hook.) Engl. ●☆
128844 Dichapetalum verruculosum Engl. = Dichapetalum gabonense Engl. ●☆
128845 Dichapetalum virchowii (O. Hoffm. et Hildebrandt) Engl.;菲而霍毒鼠子●☆
128846 Dichapetalum warneckei Engl. = Dichapetalum pallidum (Oliv.) Engl. ●☆
128847 Dichapetalum whitei Torre = Dichapetalum heudelotii (Planch. ex Oliv.) Baill. ●☆
128848 Dichapetalum wildemanianum Exell = Dichapetalum zenkeri Engl. ●☆
128849 Dichapetalum zambesiacum Torre = Dichapetalum barbosae Torre ●☆
128850 Dichapetalum zenkeri Engl.;岑克尔毒鼠子●☆
128851 Dichasianthus Ovcz. et Yunusov = Neotorularia Hedge et J. Léonard ■
128852 Dichasianthus Ovcz. et Yunusov = Torularia (Coss.) O. E. Schulz ■☆
128853 Dichasianthus Ovcz. et Yunusov(1978);岐序蚓果芥属■☆
128854 Dichasianthus brachycarpus (Vassilcz.) Sojǘk = Neotorularia brachycarpa (Vassilcz.) Hedge et J. Léonard ■
128855 Dichasianthus brevipes (Kar. et Kir.) Sojǘk = Neotorularia brevipes (Kar. et Kir.) Hedge et J. Léonard ■
128856 Dichasianthus humilis (C. A. Mey.) Sojǘk = Neotorularia humilis (C. A. Mey.) Hedge et J. Léonard ■
128857 Dichasianthus korolkowii (Regel et Schmalh.) Sojǘk = Neotorularia korolkowii (Regel et Schmalh.) Hedge et J. Léonard ■
128858 Dichasianthus torulosus (Desf.) Sojǘk = Neotorularia torulosa (Desf.) Hedge et J. Léonard ■
128859 Dichazothece Lindau(1898);巴东爵床属☆
128860 Dichazothece cylindracea Lindau;巴东爵床☆
128861 Dichelachne Endl. (1833);双毛草属;Dichelachne, Plumegrass ■☆
128862 Dichelachne crinita (L. f.) Hook. f.;双毛草;Bearded Dichelachne, Clovenfoot Plumegrass ■☆
128863 Dichelachne micrantha (Cav.) Domin;小花双毛草;

Plumegrass ■☆

128864 Dichelachne montana Endl. ;山地双毛草■☆

128865 Dichelactina Hance = Emblica Gaertn. ●

128866 Dichelactina Hance = Phyllanthus L. ●■

128867 Dichelactina nodicaulis Hance = Phyllanthus emblica L. ●

128868 Dichelostemma Kunth(1843);丽韭属■☆

128869 Dichelostemma californicum (Torr.) A. W. Wood = Dichelostemma volubile (Kellogg) A. Heller ■☆

128870 Dichelostemma capitatum (Benth.) A. W. Wood;头状丽韭(头状布罗地,头状花韭);Blue Dicks, Blue-dicks, Californian Hyacinth, Covenna, Crow Poison, Fool's Onion, Grass Nuts, Indian Hyacinth, Wild Hyacinth ■☆

128871 Dichelostemma capitatum (Benth.) A. W. Wood = Brodiaea capitata Benth. ■☆

128872 Dichelostemma capitatum (Benth.) A. W. Wood subsp. pauciflorum (Torr.) Keator;少花丽韭■☆

128873 Dichelostemma congestum Kunth;缩口丽韭;Bluedicks, Fork-toothed Ookow, Ookow ■☆

128874 Dichelostemma idamaia (A. Wood) Greene = Brevoortia idamaia A. Wood ■☆

128875 Dichelostemma idamaia Greene = Brevoortia idamaia A. Wood ■☆

128876 Dichelostemma idamaia Greene = Dichelostemma idamaia (A. Wood) Greene ■☆

128877 Dichelostemma insulare (Greene) Burnham = Dichelostemma capitatum (Benth.) A. W. Wood ■☆

128878 Dichelostemma lacuna-vernalis L. W. Lenz = Dichelostemma capitatum (Benth.) A. W. Wood ■☆

128879 Dichelostemma multiflorum (Benth.) A. Heller;多花丽韭;Round-toothed Ookow, Wild Hyacinth ■☆

128880 Dichelostemma parviflorum (Torr. et A. Gray) Keator = Dichelostemma multiflorum (Benth.) A. Heller ■☆

128881 Dichelostemma pauciflorum (Torr.) Standl. = Dichelostemma capitatum (Benth.) A. W. Wood subsp. pauciflorum (Torr.) Keator ■☆

128882 Dichelostemma pulchellum (Salisb.) A. Heller var. capitatum (Benth.) Reveal = Dichelostemma capitatum (Benth.) A. W. Wood ■☆

128883 Dichelostemma pulchellum (Salisb.) A. Heller var. pauciflorum (Torr.) Hoover = Dichelostemma capitatum (Benth.) A. W. Wood subsp. pauciflorum (Torr.) Keator ■☆

128884 Dichelostemma pulchellum A. Heller;艳丽韭;Bluedick, Blue-dicks, Californian Hyacinth ■☆

128885 Dichelostemma venustum (Greene) Hoover = Dichelostemma idamaia Greene ■☆

128886 Dichelostemma volubile (Kellogg) A. Heller;缠绕丽韭;Climbing Grass Nut, Snake Lily, Snake-lily, Twining Brodiaea ■☆

128887 Dichelostylis Endl. = Dichostylis P. Beauv. ■

128888 Dichelostylis Endl. = Scirpus L. (保留属名)■

128889 Dicheranthus Webb(1846);无瓣指甲木属●☆

128890 Dicheranthus plocamoides Webb;无瓣指甲木■☆

128891 Dichilanthe Thwaites(1856);二唇茜属☆

128892 Dichilanthe zeylanica Thwaites;二唇茜☆

128893 Dichiloboea Stapf = Trisepalum C. B. Clarke ■

128894 Dichiloboea Stapf(1913);二唇苣苔属■☆

128895 Dichiloboea birmanica (Craib) Stapf = Trisepalum birmanicum (Craib) B. L. Burtt ■

128896 Dichiloboea speciosa Stapf;二唇苣苔■☆

128897 Dichilos Spreng. = Dichilus DC. ■☆

128898 Dichilus DC. (1826);二分豆属■☆

128899 Dichilus candicans E. Mey. = Melolobium candicans (E. Mey.) Eckl. et Zeyh. ■☆

128900 Dichilus crassifolius E. Mey. = Argyrolobium crassifolium (E. Mey.) Eckl. et Zeyh. ●☆

128901 Dichilus dallonianus Maire = Argyrolobium arabicum (Decne.) Jaub. et Spach ●☆

128902 Dichilus gracilis Eckl. et Zeyh. ;纤细二分豆■☆

128903 Dichilus hypotrichum Spreng. ;里毛二分豆■☆

128904 Dichilus lanceolatus E. Mey. = Argyrolobium lunare (L.) Druce subsp. sericeum (Thunb.) T. J. Edwards ●☆

128905 Dichilus lebeckioides DC. ;拟南非针叶豆■☆

128906 Dichilus microphyllus (L. f.) E. Mey. = Melolobium microphyllum (L. f.) Eckl. et Zeyh. ■☆

128907 Dichilus multiflorus Burtt Davy = Dichilus pilosus Conrath ex Schinz ■☆

128908 Dichilus obovatus E. Mey. = Argyrolobium argenteum (Jacq.) Eckl. et Zeyh. ●☆

128909 Dichilus patens E. Mey. = Dichilus gracilis Eckl. et Zeyh. ■☆

128910 Dichilus pilosus Conrath ex Schinz;疏毛二分豆■☆

128911 Dichilus pilosus Conrath ex Schinz var. multiflorus Burtt Davy = Dichilus pilosus Conrath ex Schinz ■☆

128912 Dichilus pilosus Kensit = Dichilus pilosus Conrath ex Schinz ■☆

128913 Dichilus reflexus (N. E. Br.) A. L. Schutte;反折二分豆■☆

128914 Dichilus sericeus E. Mey. = Argyrolobium trifoliatum (Thunb.) Druce ●☆

128915 Dichilus sericeus Spreng. = Argyrolobium sericeum (Spreng.) Eckl. et Zeyh. ●☆

128916 Dichilus spicatus E. Mey. = Melolobium aethiopicum (L.) Druce ■☆

128917 Dichilus strictus E. Mey. ;刚直二分豆■☆

128918 Dichismus Raf. = Scirpus L. (保留属名)■

128919 Dichocarpum W. T. Wang et P. K. Hsiao(1964);人字果属(银白草属);Dichocarpum, Forkfruit ■

128920 Dichocarpum adiantifolium (Hook. f. et Thomson) W. T. Wang et P. K. Hsiao = Dichocarpum auriculatum (Franch.) W. T. Wang et P. K. Hsiao ■

128921 Dichocarpum adiantifolium (Hook. f. et Thomson) W. T. Wang et P. K. Hsiao var. sutchuenense (Franch.) D. Z. Fu = Dichocarpum sutchuense (Franch.) W. T. Wang ■

128922 Dichocarpum arisanense (Hayata) Ohwi = Dichocarpum arisanense (Hayata) W. T. Wang et P. K. Hsiao ■

128923 Dichocarpum arisanense (Hayata) W. T. Wang et P. K. Hsiao;台湾人字果(阿里山白银草,铁线蕨叶人字果,小花人字果);Alishan Forkfruit, Taiwan Dichocarpum ■

128924 Dichocarpum auriculatum (Franch.) W. T. Wang et P. K. Hsiao;耳状人字果(母猪草,山黄连);Auriculate Dichocarpum, Auriculate Forkfruit ■

128925 Dichocarpum auriculatum Franch. = Dichocarpum adiantifolium (Hook. f. et Thomson) W. T. Wang et P. K. Hsiao ■

128926 Dichocarpum auriculatum Franch. var. puberulum D. Z. Fu;毛叶人字果(毛被叶人字果);Hairy Auriculate Dichocarpum ■

128927 Dichocarpum basilare W. T. Wang et P. K. Hsiao;基叶人字果(地五加);Baseleaf Dichocarpum, Basifoliate Forkfruit ■

128928 Dichocarpum carinatum D. Z. Fu;种脊人字果;Carinate Dichocarpum ■

128929 Dichocarpum dalzielii (Drumm. et Hutch.) W. T. Wang et P. K. Hsiao;蕨叶人字果(土黄连,岩黄连,岩节连,野黄连);Fernleaf Dichocarpum,Fernleaf Forkfruit ■

128930 Dichocarpum dicarpon (Miq.) W. T. Wang et P. K. Hsiao;双果人字果;Twofruit Dichocarpum ■☆

128931 Dichocarpum fargesii (Franch.) W. T. Wang et P. K. Hsiao;纵肋人字果(家莫里,扇叶人字果,野黄瓜);Farges Dichocarpum,Farges Forkfruit ■

128932 Dichocarpum francherii (Finet et Gagnep.) W. T. Wang et P. K. Hsiao = Dichocarpum adiantifolium (Hook. f. et Thomson) W. T. Wang et P. K. Hsiao ■

128933 Dichocarpum franchetii (Finet et Gagnep.) W. T. Wang et P. K. Hsiao;小花人字果;Franchet Dichocarpum,Franchet Forkfruit ■

128934 Dichocarpum hakonense (Maek. et Tuyama ex Ohwi) W. T. Wang et P. K. Hsiao;箱根人字果;Hakone Dichocarpum ■☆

128935 Dichocarpum hypoglaucum W. T. Wang et P. K. Hsiao;粉背叶人字果(小乌头);Backglaucousleaf Dichocarpum,Glaucous Forkfruit ■

128936 Dichocarpum malipoense D. D. Tao;麻栗坡人字果;Malipo Dichocarpum ■

128937 Dichocarpum numajirianum (Makino) W. T. Wang et P. K. Hsiao;近畿人字果■☆

128938 Dichocarpum pipponicum (Franch.) W. T. Wang et P. K. Hsiao;日本人字果;Japanese Dichocarpum ■☆

128939 Dichocarpum pterigionocaudatum (Koidz.) Tamura et Lauener;蕨茎人字果;Fernstem Dichocarpum ■☆

128940 Dichocarpum stoloniferum (Maxim.) W. T. Wang et P. K. Hsiao;匍茎人字果■☆

128941 Dichocarpum sutchuense (Franch.) W. T. Wang;人字果(四川人字果);Sichuang Dichocarpum,Sichuang Forkfruit,Szechwan Dichocarpum ■

128942 Dichocarpum trachyspermum (Maxim.) W. T. Wang et P. K. Hsiao;糙籽人字果■☆

128943 Dichocarpum trifoliolatum W. T. Wang et P. K. Hsiao;三小叶人字果(牛尿草,三角蝉,羊不吃);Trifoliolate Dichocarpum,Trifoliolate Forkfruit ■

128944 Dichodon (Bartl. ex Rchb.) Rchb. = Cerastium L. ■

128945 Dichodon (Rchb.) Rchb. = Cerastium L. ■

128946 Dichodon Bartl. ex Rchb. = Cerastium L. ■

128947 Dichodon Rchb. = Cerastium L. ■

128948 Dichodon cerastoides (L.) Rchb. = Cerastium cerastoides (L.) Britton ■

128949 Dichodon maximum (L.) Á. Löve et D. Löve = Cerastium maximum L. ■

128950 Dichodon viscidum (M. Bieb.) Holub = Cerastium dubium (Bastard) Guépin ■☆

128951 Dichoespermum Wight = Aneilema R. Br. ■☆

128952 Dichoespermum blumei Hassk. = Murdannia blumei (Hassk.) Brenan ■☆

128953 Dichoglottis Fisch. et C. A. Mey. = Gypsophila L. ■●

128954 Dicholactina Hance = Phyllanthus L. ●■

128955 Dichondra J. R. Forst. et G. Forst. (1775);马蹄金属;Dichondra,Kidneyweed,Lawnleaf ■

128956 Dichondra argentea Willd.;银色马蹄金;Silver Dichondra ■☆

128957 Dichondra brachypoda Wooton et Standl.;短梗马蹄金■☆

128958 Dichondra brevifolia Buchanan;短叶马蹄金■☆

128959 Dichondra carolinensis Michx.;卡罗来纳马蹄金;Pony-foot ■☆

128960 Dichondra carolinensis Michx. = Dichondra repens J. R. Forst. et G. Forst. ■

128961 Dichondra caroliniana Willd. ex DC. = Dichondra repens J. R. Forst. et G. Forst. ■

128962 Dichondra evolvulacea Britton = Dichondra repens J. R. Forst. et G. Forst. ■

128963 Dichondra macrocalyx Meisn. = Dichondra repens J. R. Forst. et G. Forst. ■

128964 Dichondra micrantha Urb.;马蹄金(肚脐草,过墙风,荷包草,黄胆草,黄疸草,黄眼草,鸡眼草,金马蹄草,金钱草,金锁匙,金挖耳,九连环,酒杯窝,螺丕草,落地金钱,钮子草,匍匐马蹄金,肉混沌草,铜钱草,蚬壳草,小半边钱,小灯盏,小灯盏菜,小蛤蟆碗,小花马蹄金,小金钱,小金钱草,小马蹄草,小马蹄金,小马香,小铜钱草,小碗碗草,小迎风草,小元宝草,鱼脐草,玉混沌,月亮草);Creeping Dichondra,Dichondra,Kidney Weed,Kidneyweed,Lawnleaf,Pony Foot,Silver Falls dichondra ■

128965 Dichondra occidentalis House;西方马蹄金■☆

128966 Dichondra parvifolia Meisn.;小叶马蹄金■☆

128967 Dichondra peruviana Poir. = Dichondra repens J. R. Forst. et G. Forst. ■

128968 Dichondra recurvata Tharp et M. C. Johnst.;反卷马蹄金■☆

128969 Dichondra repens J. Forst. et G. Forst. = Dichondra micrantha Urb. ■

128970 Dichondra repens J. Forst. et G. Forst. var. carolinensis (Michx.) Choisy = Dichondra carolinensis Michx. ■☆

128971 Dichondra repens J. Forst. et G. Forst. var. micrantha (Urb.) Lu = Dichondra micrantha Urb. ■

128972 Dichondra repens J. R. Forst. et G. Forst.;匍匐马蹄金(马蹄金);Dichondra,Lawnleaf,Ponyfoot,Pony's Foot ■

128973 Dichondra repens J. R. Forst. et G. Forst. = Dichondra micrantha Urb. ■

128974 Dichondra sericea Sw. = Dichondra repens J. R. Forst. et G. Forst. ■

128975 Dichondra villosa Parodi;毛马蹄金■☆

128976 Dichondraceae Dumort. (1829)(保留科名);马蹄金科■

128977 Dichondraceae Dumort.(保留科名) = Convolvulaceae Juss.(保留科名)●■

128978 Dichondropsis Brandegee = Dichondra J. R. Forst. et G. Forst. ■

128979 Dichone Lawson ex Salisb. = Tritonia Ker Gawl. ■

128980 Dichone Salisb. = Ixia L.(保留属名)■☆

128981 Dichopetalum F. Muell. = Dichosciadium Domin ■☆

128982 Dichopetalum Post et Kuntze = Dichapetalum Thouars ●

128983 Dichopogon Kunth(1843);双须吊兰属;Chocolate-lily ■☆

128984 Dichopogon humilis Kunth;双须吊兰■☆

128985 Dichopsis Thwaites = Palaquium Blanco ●

128986 Dichopus Blume = Dendrobium Sw.(保留属名)■

128987 Dichorisandra J. C. Mikan(1820)(保留属名);鸳鸯鸭跖草属(敌克里桑草属,蓝姜属,鸳鸯草属);Dichorisandra ■☆

128988 Dichorisandra albomarginata Linden;墨西哥鸳鸯鸭跖草(墨西哥蓝姜);Mexican Flag ■☆

128989 Dichorisandra dubietiana Schult.;奥别特鸳鸯鸭跖草(奥别特敌克里桑草,奥别特蓝姜);Aubit Dichorisandra ■☆

128990 Dichorisandra mosaica Linden;斑驳鸳鸯鸭跖草(斑驳敌克里桑草,斑驳蓝姜);Mosaic Dichorisandra,Seersucker Plant ■☆

128991 Dichorisandra mosaica Linden var. undata (K. Koch et Linden) W. T. Mill. ex L. H. Bailey = Geogenanthus undatus (K. Koch et Linden) Mildbr. et Strauss ■☆

128992 Dichorisandra mosaica Linden var. undata (K. Koch et Linden) W. T. Mill. ex L. H. Bailey = Geogenanthus poeppigii (Miq.) Faden ■☆
128993 Dichorisandra mosaica Linden var. undata W. T. Mill. = Geogenanthus poeppigii (Miq.) Faden ■☆
128994 Dichorisandra mosaica Linden var. undata W. T. Mill. = Geogenanthus undatus (K. Koch et Linden) Mildbr. et Strauss ■☆
128995 Dichorisandra reginae (L. Linden et Rodigas) W. Ludw.;鸳鸯鸭跖草(鸳鸯草)■☆
128996 Dichorisandra reginae (L. Linden et Rodigas) W. Mill. = Dichorisandra reginae (L. Linden et Rodigas) W. Ludw. ■☆
128997 Dichorisandra thyrsiflora J. G. Mikan;聚伞鸳鸯鸭跖草(聚伞圆锥花序敌克里桑草,蓝姜);Blue Ginger,Brazilian Ginger,Thyrse Dichorisandra ■☆
128998 Dichoropetalum Fenzl = Johrenia DC. ■☆
128999 Dichosciadium Domin(1908);双伞芹属■☆
129000 Dichosciadium ranunculaceum (Hook. f.) Domin;双伞芹■☆
129001 Dichosema Benth. = Mirbelia Sm. ●☆
129002 Dichosma DC. ex Loud. = Agathosma Willd.(保留属名)●☆
129003 Dichospermum C. Muell. = Dichoespermum Wight ■☆
129004 Dichospermum C. Müll. = Aneilema R. Br. ■☆
129005 Dichostachys Kranss = Dichrostachys (A. DC.) Wight et Arn.(保留属名)●
129006 Dichostemma Pierre(1896);双冠大戟属☆
129007 Dichostemma amplum Pax = Dichostemma glaucescens Pierre☆
129008 Dichostemma glaucescens Pierre;灰绿双冠大戟☆
129009 Dichostemma zenkeri Pax;岑克尔双冠大戟☆
129010 Dichostylis P. Beauv. (1819);双柱莎草属■
129011 Dichostylis P. Beauv. = Fimbristylis Vahl(保留属名)■
129012 Dichostylis P. Beauv. = Scirpus L.(保留属名)■
129013 Dichostylis P. Beauv. ex Lestib. = Cyperus L. ■
129014 Dichostylis P. Beauv. ex Lestib. = Echinolytrum Desv. ■
129015 Dichostylis Rikli = Scirpus L.(保留属名)■
129016 Dichostylis aristata (Rottb.) Palla = Mariscus aristatus (Rottb.) Ts. Tang et F. T. Wang ■
129017 Dichostylis fluitans (L.) P. Beauv. ex Rchb. = Isolepis fluitans (L.) R. Br. ■☆
129018 Dichostylis hamulosa (M. Bieb.) Nees;具钩双柱莎草■
129019 Dichostylis micheliana (L.) Nees;米氏双柱莎草■☆
129020 Dichostylis micheliana Nees = Cyperus michelianus (L.) Link ■
129021 Dichostylis nipponica (Franch. et Sav.) Palla = Cyperus nipponicus Franch. et Sav. ■
129022 Dichostylis pacifica (Ohwi) Nakai ex Kitag. = Cyperus pacificus (Ohwi) Ohwi ■☆
129023 Dichostylis pygmaea (Rottb.) Nees;矮小双柱莎草■☆
129024 Dichotoma Sch. Bip. = Sclerocarpus Jacq. ■
129025 Dichotomanthes Kurz (1873);牛筋条属;Dichotomanthus,Oxmuscle ●★
129026 Dichotomanthes tristaniaecarpa Kurz;牛筋条(白牛筋,红眼睛,牛筋藤);Common Dichotomanthus,Common Oxmuscle ●◇
129027 Dichotomanthes tristaniaecarpa Kurz var. glabrata Rehder;光叶牛筋条;Glabrous Dichotomanthus,Glabrous Oxmuscle ●
129028 Dichotophyllum Moench = Ceratophyllum L. ■
129029 Dichotrichum S. Moore = Dichrotrichum Reinw. ■☆
129030 Dichroa Lour. (1790);常山属(黄常山属);Dichroa ●
129031 Dichroa cyanea (Wall.) Schltr. = Dichroa febrifuga Lour. ●
129032 Dichroa daimingshanensis Y. C. Wu;大明常山;Daming Dichroa,Damingshan Dichroa ●
129033 Dichroa febrifuga Lour.;常山(白常山,白常山草,摆子药,大金刀,翻胃木,风骨木,恒山,互草,黄常山,鸡粪草,鸡骨常山,鸡骨风,鸡屎草,南常山,七叶,蜀漆,甜茶,土常山,鸭屎草,野兰子);Antifebrile Dichroa,Blue Evergreen Hydrangea ●
129034 Dichroa febrifuga Lour. var. glabra S. Y. Hu = Dichroa febrifuga Lour. ●
129035 Dichroa henryi H. Lév. = Dichroa febrifuga Lour. ●
129036 Dichroa hirsuta Gagnep.;硬毛常山;Hirsute Dichroa ●
129037 Dichroa latifolia Miq. = Dichroa febrifuga Lour. ●
129038 Dichroa mollissima Merr.;海南常山;Hainan Dichroa ●
129039 Dichroa parviflora Schltr.;小花常山;Smallflower Dichroa ●☆
129040 Dichroa pentandra Schltr.;五雄蕊常山●☆
129041 Dichroa philippinensis Schltr.;菲律宾常山;Philippine Dichroa ●☆
129042 Dichroa platyphylla Merr.;宽叶常山;Broadleaf Dichroa ●☆
129043 Dichroa schumanniana Schltr.;鳞毛常山●☆
129044 Dichroa sylvatica (Reinw.) Merr. = Dichroa febrifuga Lour. ●
129045 Dichroa tristyla W. T. Wang = Hydrangea lingii G. Hoo ●
129046 Dichroa tristyla W. T. Wang et M. X. Nie = Hydrangea lingii G. Hoo ●
129047 Dichroa tristyla W. T. Wang et M. X. Nie = Hydrangea vinicolor Chun ●
129048 Dichroa versicolor Hunt = Dichroa febrifuga Lour. ●
129049 Dichroa versicolor Hunt ex Gentil = Dichroa febrifuga Lour. ●
129050 Dichroa yaoshanensis Y. C. Wu;罗蒙常山;Yaoshan Dichroa ●
129051 Dichroa yunnanensis S. M. Hwang;云南常山;Yunnan Dichroa ●
129052 Dichroanthus Webb et Berthel. = Cheiranthus L. ●■
129053 Dichroanthus cinereus Webb et Berthel. = Erysimum bicolor (Hornem.) DC. ■☆
129054 Dichroanthus mutabilis Webb et Berthel. = Erysimum bicolor (Hornem.) DC. ■☆
129055 Dichroanthus mutabilis Webb et Berthel. var. albescens Webb et Berthel. = Erysimum bicolor (Hornem.) DC. ■☆
129056 Dichroanthus mutabilis Webb et Berthel. var. brevifolius ? = Erysimum bicolor (Hornem.) DC. ■☆
129057 Dichroanthus mutabilis Webb et Berthel. var. latifolius ? = Erysimum bicolor (Hornem.) DC. ■☆
129058 Dichroanthus mutabilis Webb et Berthel. var. longifolius ? = Erysimum bicolor (Hornem.) DC. ■☆
129059 Dichroanthus scoparius Webb et Berthel. = Erysimum bicolor (Hornem.) DC. ■☆
129060 Dichroanthus scoparius Webb et Berthel. var. lindleyi ? = Erysimum bicolor (Hornem.) DC. ■☆
129061 Dichrocephala L'Hér. = Dichrocephala L'Hér. ex DC. ■
129062 Dichrocephala L'Hér. ex DC. (1833);鱼眼草属;Dichrocephala,Fisheyeweed ■
129063 Dichrocephala abyssinica Sch. Bip. ex A. Rich. = Dichrocephala chrysanthemifolia (Blume) DC. ■
129064 Dichrocephala alpina R. E. Fr. = Dichrocephala chrysanthemifolia (Blume) DC. var. alpina (R. E. Fr.) Beentje ■☆
129065 Dichrocephala amphiloba H. Lév. = Dichrocephala benthamii C. B. Clarke ■
129066 Dichrocephala amphiloba H. Lév. et Vaniot = Dichrocephala benthamii C. B. Clarke ■
129067 Dichrocephala auriculata (Thunb.) DC. = Dichrocephala integrifolia (L. f.) Kuntze ■

129068 Dichrocephala auriculata (Thunb.) Druce;鱼眼草(白头菜,地胡椒,地苋菜,茯苓菜,鼓丁草,胡椒草,鸡饴草,口疮叶,馒头草,泥鳅菜,蚯疽草,肉桂草,三仙菜,山胡椒菊,土茯苓,星宿草,夜明草,鱼眼菊);Auriculate Dichrocephala,Auriculate Fisheyeweed,Pig's Dichrocephala ■

129069 Dichrocephala auriculata (Thunb.) Druce = Dichrocephala integrifolia (L. f.) Kuntze ■

129070 Dichrocephala benthamii C. B. Clarke;小鱼眼草(地胡椒,地细辛,鼓丁草,鸡眼菊,蛆头草,三仙菜,小馒头草,小鱼眼菊,星宿草,星秀草,翳子草,鱼眼草);Bentham Fisheyeweed,Bentham's Dichrocephala ■

129071 Dichrocephala bicolor (Roth) Schltdl. = Dichrocephala auriculata (Thunb.) Druce ■

129072 Dichrocephala bicolor (Roth) Schltdl. = Dichrocephala integrifolia (L. f.) Kuntze ■

129073 Dichrocephala bodinieri Vaniot = Dichrocephala benthamii C. B. Clarke ■

129074 Dichrocephala capensis (Less.) DC. = Dichrocephala integrifolia (L. f.) Kuntze ■

129075 Dichrocephala chrysanthemifolia (Blume) DC.;菊叶鱼眼草;Daisy Dichrocephala,Daisy Fisheyeweed ■

129076 Dichrocephala chrysanthemifolia (Blume) DC. var. abyssinica Asch. = Dichrocephala chrysanthemifolia (Blume) DC. ■

129077 Dichrocephala chrysanthemifolia (Blume) DC. var. alpina (R. E. Fr.) Beentje;高山菊叶鱼眼草■☆

129078 Dichrocephala chrysanthemifolia (Blume) DC. var. tanacetoides (Sch. Bip.) J. Kost. = Dichrocephala chrysanthemifolia (Blume) DC. ■

129079 Dichrocephala chrysanthemifolia DC. = Dichrocephala chrysanthemifolia (Blume) DC. ■

129080 Dichrocephala integrifolia (L. f.) Kuntze;星秀草(全缘叶鱼眼草,土茯苓,鱼眼草)■

129081 Dichrocephala integrifolia (L. f.) Kuntze = Dichrocephala auriculata (Thunb.) Druce ■

129082 Dichrocephala integrifolia (L. f.) Kuntze var. sonchifolia (M. Bieb.) Kuntze = Dichrocephala integrifolia (L. f.) Kuntze ■

129083 Dichrocephala latifolia (Lam.) DC. = Dichrocephala integrifolia (L. f.) Kuntze ■

129084 Dichrocephala latifolia (Lam.) DC. var. sonchifolia (M. Bieb.) Asch. = Dichrocephala integrifolia (L. f.) Kuntze ■

129085 Dichrocephala latifolia (Pers.) DC. = Dichrocephala auriculata (Thunb.) Druce ■

129086 Dichrocephala latifolia (Pers.) DC. = Dichrocephala integrifolia (L. f.) Kuntze ■

129087 Dichrocephala leveillei Vaniot = Myriactis nepalensis Less. ■

129088 Dichrocephala linearifolia O. Hoffm. = Grauanthus linearifolius (O. Hoffm.) Fayed ■☆

129089 Dichrocephala minutiflora Vaniot = Cyathocline purpurea (Buch. -Ham. ex D. Don) Kuntze ■

129090 Dichrocephala oblonga Hook. f. = Dichrocephala chrysanthemifolia (Blume) DC. ■

129091 Dichrocephala tanacetoides Sch. Bip. = Dichrocephala chrysanthemifolia (Blume) DC. ■

129092 Dichrocephala tibestica Quézel = Dichrocephala chrysanthemifolia (Blume) DC. ■

129093 Dichrolepidaceae Welw. = Eriocaulaceae Martinov(保留科名)■

129094 Dichrolepis Welw. = Eriocaulon L. ■

129095 Dichrolepis pusilla Welw. = Eriocaulon welwitschii Rendle ■☆

129096 Dichroma Cav. = Ourisia Comm. ex Juss. ■☆

129097 Dichroma Pers. = Dichromena Michx. (废弃属名)■☆

129098 Dichroma caespitosa (Muhl.) Spreng. = Bulbostylis stenophylla (Elliott) C. B. Clarke ■☆

129099 Dichroma cespitosum Muhl. = Bulbostylis stenophylla (Elliott) C. B. Clarke ■☆

129100 Dichromanthus Garay(1982);双色花兰属■☆

129101 Dichromanthus cinnabarinus (La Llave et Lex.) Garay;双色花兰■☆

129102 Dichromena Michx. (废弃属名) = Rhynchospora Vahl(保留属名)■

129103 Dichromena alba (L.) J. F. Macbr. = Rhynchospora alba (L.) Vahl ■

129104 Dichromena cephalotes Britton = Rhynchospora colorata (Hitchc.) H. Pfeiff. ■☆

129105 Dichromena colorata (L.) Hitchc. = Rhynchospora colorata (Hitchc.) H. Pfeiff. ■☆

129106 Dichromena colorata Hitchc. = Rhynchospora colorata (Hitchc.) H. Pfeiff. ■☆

129107 Dichromena cymosa (Elliott) J. F. Macbr.;聚伞双月莎■☆

129108 Dichromena diphylla Torr. = Rhynchospora nivea Boeck. ■☆

129109 Dichromena distans (Michx.) J. F. Macbr. = Rhynchospora fascicularis (Michx.) Vahl ■☆

129110 Dichromena floridensis Britton = Rhynchospora floridensis (Britton) H. Pfeiff. ■☆

129111 Dichromena latifolia Baldwin = Rhynchospora latifolia (Baldwin) W. W. Thomas ■☆

129112 Dichromena leucocephala Michx. = Rhynchospora colorata (Hitchc.) H. Pfeiff. ■☆

129113 Dichromena nivea (Boeck.) Britton = Rhynchospora nivea Boeck. ■☆

129114 Dichromna Schltdl. = Paspalum L. ■

129115 Dichromochlamys Dunlop(1980);异色层菀属■☆

129116 Dichromochlamys dentatifolia (F. Muell.) Dunlop;异色层菀■☆

129117 Dichromus Schltdl. = Paspalum L. ■

129118 Dichronema Baker = Dichromena Michx. (废弃属名)■☆

129119 Dichropappus Sch. Bip. ex Krasehen. = Stenachaenium Benth. ■☆

129120 Dichrophyllum Klotzsch et Garcke = Euphorbia L. ●■

129121 Dichrophyllum Klotzsch et Garcke = Lepadena Raf. ■☆

129122 Dichrophyllum Klotzsch et Garcke(1859);双色叶属■☆

129123 Dichrophyllum bicolor Klotzsch et Garcke;双色叶■☆

129124 Dichrophyllum marginatum (Pursh) Klotzsch et Garcke = Euphorbia marginata Pursh ■

129125 Dichrospermum Bremek. = Spermacoce L. ●■

129126 Dichrospermum congense Bremek. = Spermacoce congensis (Bremek.) Verdc. ■☆

129127 Dichrostachys (A. DC.) Wight et Arn. (1834)(保留属名);色穗木属(柏簕树属,代儿茶属,二色穗属,双色花属);Dichrostachys ●

129128 Dichrostachys Wight et Arn. = Dichrostachys (A. DC.) Wight et Arn. (保留属名)●

129129 Dichrostachys arborea N. E. Br. = Dichrostachys cinerea (L.) Wight et Arn. var. africana Brenan et Brummitt ●☆

129130 Dichrostachys arborescens (Bojer ex Benth.) Villiers;乔木色穗木●☆

129131 Dichrostachys benadirensis Chiov. = Dichrostachys kirkii Benth.

●☆

129132 Dichrostachys bernieriana Baill.;伯尼尔色穗木●☆

129133 Dichrostachys brachypus Baill. = Mimosa psoralea (DC.) Benth. ●☆

129134 Dichrostachys brevipes R. Vig. = Alantsilodendron brevipes (R. Vig.) Villiers ●☆

129135 Dichrostachys caffra Meisn. ex Benth. = Dichrostachys cinerea (L.) Wight et Arn. var. africana Brenan et Brummitt ●☆

129136 Dichrostachys cinerea (L.) Wight et Arn.;色穗木(白凿簕,柏簕儿茶,柏簕树,代儿茶);Aroma, Ashy Dichrostachys, Chinese Lanterns, Dichrostachys, Greyish Dichrostachys, Sickle Pod ●

129137 Dichrostachys cinerea (L.) Wight et Arn. subsp. forbesii (Benth.) Brenan et Brummitt;福布斯色穗木●☆

129138 Dichrostachys cinerea (L.) Wight et Arn. subsp. keniensis Brenan et Brummitt;肯尼亚色穗木●☆

129139 Dichrostachys cinerea (L.) Wight et Arn. subsp. nyassana (Taub.) Brenan;尼亚萨色穗木●☆

129140 Dichrostachys cinerea (L.) Wight et Arn. subsp. platycarpa (Welw. ex W. Bull) Brenan et Brummitt;宽果色穗木(宽果二色穗);Broadfruit Dichrostachys ●☆

129141 Dichrostachys cinerea (L.) Wight et Arn. var. africana Brenan et Brummitt;非洲色穗木●☆

129142 Dichrostachys cinerea (L.) Wight et Arn. var. argillicola Brenan et Brummitt;白土色穗木●☆

129143 Dichrostachys cinerea (L.) Wight et Arn. var. hirtipes Brenan et Brummitt = Dichrostachys cinerea (L.) Wight et Arn. var. africana Brenan et Brummitt ●☆

129144 Dichrostachys cinerea (L.) Wight et Arn. var. lugardiae (N. E. Br.) Brenan et Brummitt = Dichrostachys cinerea (L.) Wight et Arn. var. africana Brenan et Brummitt ●☆

129145 Dichrostachys cinerea (L.) Wight et Arn. var. plurijuga Brenan et Brummitt;多结色穗木●☆

129146 Dichrostachys cinerea (L.) Wight et Arn. var. pubescens Brenan et Brummitt;短柔毛色穗木●☆

129147 Dichrostachys cinerea (L.) Wight et Arn. var. setulosa (Welw. ex Oliv.) Brenan et Brummitt;刚毛色穗木●☆

129148 Dichrostachys cinerea (L.) Wight et Arn. var. tanganyikensis Brenan et Brummitt;坦噶尼喀色穗木●☆

129149 Dichrostachys cinerea R. Vig. = Alantsilodendron pilosum Villiers ●☆

129150 Dichrostachys commersonianus (Baill.) Drake = Gagnebina commersoniana (Baill.) R. Vig. ●☆

129151 Dichrostachys decaryana R. Vig. = Alantsilodendron decaryanum (R. Vig.) Villiers ●☆

129152 Dichrostachys dumetaria Villiers;灌丛毛色穗木●☆

129153 Dichrostachys forbesii Benth. = Dichrostachys cinerea (L.) Wight et Arn. subsp. forbesii (Benth.) Brenan et Brummitt ●☆

129154 Dichrostachys glomerata (Forssk.) Chiov. = Dichrostachys cinerea (L.) Wight et Arn. ●

129155 Dichrostachys glomerata (Forssk.) Chiov. subsp. nyassana (Taub.) Brenan = Dichrostachys cinerea (L.) Wight et Arn. subsp. nyassana (Taub.) Brenan ●☆

129156 Dichrostachys grandidieri Baill. = Dichrostachys unijuga Baker ●☆

129157 Dichrostachys humbertii R. Vig. = Alantsilodendron humbertii (R. Vig.) Villiers ●☆

129158 Dichrostachys integrifolia (L. f.) Kuntze;全缘叶色穗木●☆

129159 Dichrostachys kirkii Benth.;柯克色穗木●☆

129160 Dichrostachys lugardiae N. E. Br. = Dichrostachys cinerea (L.) Wight et Arn. var. africana Brenan et Brummitt ●☆

129161 Dichrostachys mahafalensis R. Vig. = Alantsilodendron mahafalense (R. Vig.) Villiers ●☆

129162 Dichrostachys major Sim = Dichrostachys cinerea (L.) Wight et Arn. subsp. nyassana (Taub.) Brenan ●☆

129163 Dichrostachys myriophylla Baker;多叶色穗木●☆

129164 Dichrostachys nutans (Pers.) Benth.;俯垂色穗木(俯垂二色穗);Nodding Dichrostachys ●☆

129165 Dichrostachys nutans (Pers.) Benth. = Dichrostachys cinerea (L.) Wight et Arn. ●

129166 Dichrostachys nutans (Pers.) Benth. = Dichrostachys cinerea (L.) Wight et Arn. var. africana Brenan et Brummitt ●☆

129167 Dichrostachys nutans (Pers.) Benth. var. setulosa Welw. ex Oliv. = Dichrostachys cinerea (L.) Wight et Arn. var. setulosa (Welw. ex Oliv.) Brenan et Brummitt ●☆

129168 Dichrostachys nutans Benth. = Dichrostachys cinerea (L.) Wight et Arn. ●

129169 Dichrostachys nyassana Taub. = Dichrostachys cinerea (L.) Wight et Arn. subsp. nyassana (Taub.) Brenan ●☆

129170 Dichrostachys paucifoliolata (Scott-Elliot) Drake;少叶色穗木●☆

129171 Dichrostachys pervilleana (Baill.) Drake;佩尔色穗木●☆

129172 Dichrostachys platycarpa Welw. ex W. Bull = Dichrostachys cinerea (L.) Wight et Arn. subsp. platycarpa (Welw. ex W. Bull) Brenan et Brummitt ●☆

129173 Dichrostachys richardiana Baill.;理查德色穗木●☆

129174 Dichrostachys scottiana (Drake) Villiers;司科特色穗木●☆

129175 Dichrostachys tenuifolia Benth.;细叶色穗木●☆

129176 Dichrostachys unijuga Baker;双生色穗木●☆

129177 Dichrostachys venosa Villiers;多脉色穗木●☆

129178 Dichrostachys villosa R. Vig. = Alantsilodendron villosum (R. Vig.) Villiers ●☆

129179 Dichrostylis Nakai = Dichostylis P. Beauv. ■

129180 Dichrostylis Nakai = Scirpus L.(保留属名)■

129181 Dichrotrichum Reinw. (1856);二色种毛苣苔属;Dichrotrichum ■☆

129182 Dichrotrichum Reinw. ex de Vriese = Agalmyla Blume ●☆

129183 Dichrotrichum biflorum Elmer;双花二色种毛苣苔■☆

129184 Dichrotrichum brownii (Koord.) B. L. Burtt;二色种毛苣苔■☆

129185 Dichrotrichum griffithii (Wight) C. B. Clarke = Loxostigma griffithii (Wight) C. B. Clarke ■

129186 Dichylium Britton = Euphorbia L. ●■

129187 Dichylium Britton = Poinsettia Graham ●■

129188 Dichynchosia K. Müll. = Caldcluvia D. Don ●☆

129189 Dichynchosia K. Müll. = Dirhynchosia Blume ●☆

129190 Dichynchosia K. Müll. = Spiraeopsis Miq. ●☆

129191 Dickasonia L. O. Williams (1941);迪卡兰属(狄克兰属);Dickasonia ■☆

129192 Dickasonia vernicosa L. O. Williams;迪卡兰;Varnished Dickasonia ■☆

129193 Dickia Scop. = Matourea Aubl. ■☆

129194 Dickia Scop. = Stemodia L.(保留属名)■☆

129195 Dickinsia Franch. (1885);马蹄芹属(大苞芹属);Dickinsia, Hoofcelery ■★

129196 Dickinsia hydrocotyloides Franch.;马蹄芹(大苞芹,大天胡荽,山荷叶,双叉草);Common Dickinsia, Common Hoofcelery ■

129197 Dicladanthera F. Muell. (1882);双枝药属●☆

129198 Dicladanthera forrestii F. Muell.;轮蕊花●☆
129199 Dicladanthera glabra R. M. Barker;光轮蕊花●☆
129200 Diclemia Naudin = Ossaea DC. ●☆
129201 Diclidanthera Mart.(1827);轮蕊花属●■☆
129202 Diclidantheraceae J. Agardh(1858)(保留科名);轮蕊花科●■☆
129203 Diclidantheraceae J. Agardh(保留科名) = Polygalaceae Hoffmanns. et Link(保留科名)■●
129204 Diclidium Schrad. ex Nees = Mariscus Gaertn. ■
129205 Diclidium Schrad. ex Nees = Torulinium Desv. ex Ham. ■
129206 Diclidium aciculare Schrad. ex Nees = Cyperus odoratus L. ■
129207 Diclidium aciculare Schrad. ex Nees = Torulinium odoratum (L.) S. S. Hooper ■
129208 Diclidium odoratum (L.) Schrad. ex Nees = Cyperus odoratus L. ■
129209 Diclidium odoratum (L.) Schrad. ex Nees = Torulinium odoratum (L.) S. S. Hooper ■
129210 Diclidocarpus A. Gray = Trichospermum Blume ●☆
129211 Diclidostigma Kunze = Melothria L. ■
129212 Diclinanona Diels(1927);双腺花属(秘巴番荔枝属)●☆
129213 Diclinanona calycina (Diels) R. E. Fr.;萼状双腺花(萼状秘巴番荔枝)●☆
129214 Diclinothrys Endl. = Diclinotrys Raf. ■☆
129215 Diclinotris Raf. = Diclinotrys Raf. ■☆
129216 Diclinotrys Raf. = Abalon Adans. ■☆
129217 Diclinotrys Raf. = Abalum Adans. ■☆
129218 Diclinotrys Raf. = Chamaelirium Willd. ■☆
129219 Diclinotrys Raf. = Helonias L. ■☆
129220 Dicliptera Juss.(1807)(保留属名);狗肝菜属(华九头狮子草属);Dicliptera,Dogliverweed ■
129221 Dicliptera aculeata C. B. Clarke;皮刺狗肝菜■☆
129222 Dicliptera adusta Lindau;煤黑狗肝菜■☆
129223 Dicliptera albicaulis (S. Moore) S. Moore;白茎狗肝菜■☆
129224 Dicliptera alternans Lindau;互生狗肝菜■☆
129225 Dicliptera angolensis S. Moore;安哥拉狗肝菜■☆
129226 Dicliptera angustifolia Gilli;短叶狗肝菜■☆
129227 Dicliptera arenaria Milne-Redh.;沙地狗肝菜■☆
129228 Dicliptera bagshawei S. Moore;巴格肖狗肝菜■☆
129229 Dicliptera batesii S. Moore;贝茨狗肝菜■☆
129230 Dicliptera betonicoides S. Moore;药水苏狗肝菜■☆
129231 Dicliptera brachiata (Pursh) Spreng.;短狗肝菜;Dicliptera ■☆
129232 Dicliptera brevispicata I. Darbysh.;短穗狗肝菜■☆
129233 Dicliptera buergeriana Miq. = Peristrophe japonica (Thunb.) Bremek. ■
129234 Dicliptera bupleuoides Nees var. roxburghiana Panigrahi et A. K. Dubey = Dicliptera bupleuroides Nees ■
129235 Dicliptera bupleuroides Nees;印度狗肝菜;India Dogliverweed, Indian Dicliptera ■
129236 Dicliptera bupleuroides Nees var. roxburghiana Panigrahi = Dicliptera bupleuroides Nees ■
129237 Dicliptera burmanni Nees = Dicliptera chinensis (L.) Nees ■
129238 Dicliptera capensis Nees;好望角狗肝菜■☆
129239 Dicliptera capitata Milne-Redh.;头状狗肝菜■☆
129240 Dicliptera carvalhoi Lindau;卡瓦略狗肝菜■☆
129241 Dicliptera cephalantha S. Moore;头花狗肝菜■☆
129242 Dicliptera chinensis (L.) Juss. = Dicliptera chinensis (L.) Nees ■
129243 Dicliptera chinensis (L.) Nees;狗肝菜(半支莲,灯台草,华九头狮子草,假红蓝,假米针,金龙棒,金龙柄,梨根青,六角荚,六角英,路边青,麦穗红,青蛇,青蛇仔,土羚羊,乌面草,小青,羊肝菜,野辣椒,野青仔,中华狗肝菜,猪肝菜,紫燕草);China Dogliverweed,Chinese Dicliptera,Chinese Foldwing ■
129244 Dicliptera ciliaris Juss.;缘毛狗肝菜■☆
129245 Dicliptera clinopodia Nees;风轮菜状狗肝菜■☆
129246 Dicliptera colorata C. B. Clarke;具色狗肝菜■☆
129247 Dicliptera crinita (Thunb.) Nees var. floribunda Hemsl. = Peristrophe floribunda (Hemsl.) C. Y. Wu et H. S. Lo ■
129248 Dicliptera crinita Nees = Peristrophe japonica (Thunb.) Bremek. ■
129249 Dicliptera cubangensis S. Moore;库邦戈狗肝菜■☆
129250 Dicliptera cyclostegia Hand. -Mazz. = Calophanoides chinensis (Champ.) C. Y. Wu et H. S. Lo ex Y. C. Tang ●
129251 Dicliptera decaryi Benoist;德卡里狗肝菜■☆
129252 Dicliptera decorticans (K. Balkwill) I. Darbysh.;脱皮狗肝菜■☆
129253 Dicliptera divaricata Compton;叉开狗肝菜■☆
129254 Dicliptera eenii S. Moore;埃恩狗肝菜■☆
129255 Dicliptera elegans W. W. Sm.;优雅狗肝菜(金江狗肝菜,宽叶九头狮子草);Elegant Dicliptera,Elegant Dogliverweed ■
129256 Dicliptera elliotii C. B. Clarke;埃利狗肝菜■☆
129257 Dicliptera foetida (Forssk.) Blatt.;臭狗肝菜■☆
129258 Dicliptera fruticosa K. Balkwill;灌丛狗肝菜■☆
129259 Dicliptera glanduligera Chiov.;腺点狗肝菜■☆
129260 Dicliptera grandiflora Gilli;大花狗肝菜■☆
129261 Dicliptera hastilis Benoist;戟形狗肝菜■☆
129262 Dicliptera hensii Lindau;亨斯狗肝菜■☆
129263 Dicliptera hereroensis Schinz;赫雷罗狗肝菜■☆
129264 Dicliptera heterostegia Nees;莫桑比克狗肝菜■☆
129265 Dicliptera hirta K. Balkwill;粗毛狗肝菜■☆
129266 Dicliptera humbertii Mildbr.;亨伯特狗肝菜■☆
129267 Dicliptera hyalina Nees;透明狗肝菜■☆
129268 Dicliptera induta W. W. Sm.;毛狗肝菜;Clothed Dicliptera ■
129269 Dicliptera insignis Mildbr. = Dicliptera elliotii C. B. Clarke ■☆
129270 Dicliptera insularis Benoist;海岛狗肝菜■☆
129271 Dicliptera japonica (Thunb.) Makino;日本狮子草■
129272 Dicliptera japonica (Thunb.) Makino = Peristrophe japonica (Thunb.) Bremek. ■
129273 Dicliptera japonica (Thunb.) Makino f. albiflora Hiyama;白花日本狮子草■☆
129274 Dicliptera japonica (Thunb.) Makino var. ciliata (Matsuda) Ohwi = Peristrophe japonica (Thunb.) Bremek. ■
129275 Dicliptera japonica (Thunb.) Makino var. subrotunda Matsuda = Peristrophe japonica (Thunb.) Bremek. ■
129276 Dicliptera kamerunensis Lindau = Dicliptera nilotica C. B. Clarke ■☆
129277 Dicliptera katangensis De Wild.;加丹加狗肝菜■☆
129278 Dicliptera laxispica Lindau;疏穗狗肝菜■☆
129279 Dicliptera leandrii Benoist;利安狗肝菜■☆
129280 Dicliptera leistneri K. Balkwill;莱斯特纳狗肝菜■☆
129281 Dicliptera linifolia Lindau = Megalochlamys linifolia (Lindau) Lindau ●☆
129282 Dicliptera longipedunculata Mildbr.;长梗狗肝菜■☆
129283 Dicliptera maculata Nees;斑点狗肝菜■☆
129284 Dicliptera maculata Nees f. albo-lanata Lanza = Dicliptera maculata Nees ■☆
129285 Dicliptera maculata Nees var. senegambica ? = Dicliptera

hyalina Nees ■☆
129286 Dicliptera madagascariensis Nees;马岛狗肝菜■☆
129287 Dicliptera mairei Benoist = Dicliptera elegans W. W. Sm. ■
129288 Dicliptera marlothii Engl. = Megalochlamys marlothii (Engl.) Lindau ●☆
129289 Dicliptera melleri Rolfe;梅勒狗肝菜■☆
129290 Dicliptera micranthes Nees = Dicliptera ocymoides (Lam.) Juss. ■☆
129291 Dicliptera micranthes Nees = Dicliptera verticillata (Forssk.) C. Chr. ■☆
129292 Dicliptera microchlamys S. Moore;小被狗肝菜■☆
129293 Dicliptera minor C. B. Clarke;较小狗肝菜■☆
129294 Dicliptera minutifolia Ensermu;微叶狗肝菜■☆
129295 Dicliptera monroi S. Moore;门罗狗肝菜■☆
129296 Dicliptera mossambicensis Klotzsch = Dicliptera heterostegia Nees ■☆
129297 Dicliptera nemorum Milne-Redh.;丝状狗肝菜■☆
129298 Dicliptera nilotica C. B. Clarke;尼罗河狗肝菜■☆
129299 Dicliptera nobilis S. Moore;名贵狗肝菜■☆
129300 Dicliptera obanensis S. Moore;奥班狗肝菜■☆
129301 Dicliptera ocymoides (Lam.) Juss.;罗勒狗肝菜■☆
129302 Dicliptera olitoria Mildbr.;菜地狗肝菜■☆
129303 Dicliptera ovata (Nees) C. Presl = Isoglossa ovata (Nees) Lindau ■☆
129304 Dicliptera paniculata (Forssk.) I. Darbysh.;圆锥狗肝菜■☆
129305 Dicliptera propinqua Nees = Dicliptera capensis Nees ■☆
129306 Dicliptera pumila (Lindl.) Dandy;偃伏狗肝菜■☆
129307 Dicliptera quintasii Lindau;昆塔斯狗肝菜■☆
129308 Dicliptera resupinata Nutt. ex Nees;紫狗肝菜;Dicliptera, Foldwing, Purple Drop ■☆
129309 Dicliptera riparia Nees;河畔狗肝菜;Riverbank Dicliptera ■
129310 Dicliptera riparia Nees var. yunnanensis Hand.-Mazz.;滇中狗肝菜(云南狗肝菜);Yunnan Dicliptera ■
129311 Dicliptera riparia Nees var. yunnanensis Hand.-Mazz. = Hypoestes triflora (Forssk.) Roem. et Schult. ●■
129312 Dicliptera ripicola Benoist = Dicliptera elliotii C. B. Clarke ■☆
129313 Dicliptera rogersii Turrill;罗杰斯狗肝菜■☆
129314 Dicliptera roxburghiana Nees = Dicliptera bupleuroides Nees ■
129315 Dicliptera roxburghiana Nees = Dicliptera chinensis (L.) Nees ■
129316 Dicliptera roxburghiana Nees var. bupleuroides (Nees) C. B. Clarke = Dicliptera bupleuroides Nees ■
129317 Dicliptera roxburghiana Nees var. riparia (Nees) Benoist = Dicliptera riparia Nees ■
129318 Dicliptera schumanniana Schinz = Megalochlamys marlothii (Engl.) Lindau ●☆
129319 Dicliptera senegambica (Nees) Benoist = Dicliptera hyalina Nees ■☆
129320 Dicliptera serpenticola (K. Balkwill et Campb.-Young) I. Darbysh.;蛇纹岩狗肝菜■☆
129321 Dicliptera silvestris Lindau;林地狗肝菜■☆
129322 Dicliptera silvicola Lindau = Dicliptera elliotii C. B. Clarke ■☆
129323 Dicliptera sinensis W. W. Sm. = Isoglossa collina (T. Anderson) B. Hansen ■
129324 Dicliptera spinulosa Hochst. ex K. Balkwill;细刺狗肝菜■☆
129325 Dicliptera suberecta (André) Bremek.;直立狗肝菜;Dicliptera, Hummingbird Bush, King's Crown ■☆
129326 Dicliptera suffruticosa Wood;亚灌木狗肝菜●☆
129327 Dicliptera swynnertonii S. Moore;斯温纳顿狗肝菜■☆
129328 Dicliptera syringifolia Merxm.;丁香叶狗肝菜■☆
129329 Dicliptera talbotii S. Moore = Dicliptera elliotii C. B. Clarke ■☆
129330 Dicliptera transvaalensis C. B. Clarke;德兰士瓦狗肝菜■☆
129331 Dicliptera umbellata (Vahl) Juss. = Dicliptera ocymoides (Lam.) Juss. ■☆
129332 Dicliptera umbellata (Vahl) Juss. = Dicliptera verticillata (Forssk.) C. Chr. ■☆
129333 Dicliptera uraiensis Hayata = Peristrophe japonica (Thunb.) Bremek. ■
129334 Dicliptera usambarica Lindau = Dicliptera verticillata (Forssk.) C. Chr. ■☆
129335 Dicliptera verticillata (Forssk.) C. Chr.;轮生狗肝菜■☆
129336 Dicliptera villosior Berhaut = Dicliptera hyalina Nees ■☆
129337 Dicliptera welwitschii S. Moore;韦尔狗肝菜■☆
129338 Dicliptera wittei Mildbr.;维特狗肝菜■☆
129339 Dicliptera zeylanica Nees = Dicliptera foetida (Forssk.) Blatt. ■☆
129340 Diclis Benth. (1836);双盖玄参属■☆
129341 Diclis bambuseti R. E. Fr.;邦布塞特双盖玄参■☆
129342 Diclis ovata Benth.;卵双盖玄参■☆
129343 Diclis petiolaris Benth.;柄叶双盖玄参■☆
129344 Diclis reptans Benth.;匍匐双盖玄参■☆
129345 Diclis reptans Benth. var. serrodentata Kuntze = Diclis reptans Benth. ■☆
129346 Diclis reptans Benth. var. subedentata Kuntze = Diclis rotundifolia (Hiern) Hilliard et B. L. Burtt ■☆
129347 Diclis rotundifolia (Hiern) Hilliard et B. L. Burtt;圆叶双盖玄参■☆
129348 Diclis sessilifolia Diels;无柄叶双盖玄参■☆
129349 Diclis stellarioides Hiern;星状双盖玄参■☆
129350 Diclis tenella Hemsl.;柔弱双盖玄参■☆
129351 Diclis tenuissima Pilg.;极细双盖玄参■☆
129352 Diclis viridis Marloth ex Engl. = Diclis petiolaris Benth. ■☆
129353 Diclythra Raf. = Dicentra Bernh.(保留属名)■
129354 Diclytra Borkh.(废弃属名) = Dicentra Bernh.(保留属名)■
129355 Diclytra scandens D. Don = Dicentra scandens (D. Don) Walp. ■
129356 Diclytra spectabilis (L.) DC. = Dicentra spectabilis (L.) Lem. ■
129357 Diclytra spectabilis (L.) DC. = Lamprocapnos spectabilis (L.) Fukuhara ■
129358 Diclytra spectabilis DC. = Dicentra spectabilis (L.) Lem. ■
129359 Diclytra spectabilis DC. = Lamprocapnos spectabilis (L.) Fukuhara ■
129360 Dicneckeria Vell. = Euplassa Salisb. ex Knight ●☆
129361 Dicocca Thouars = Dicoryphe Thouars ●☆
129362 Dicodon Ehrh. = Linnaea L. ●
129363 Dicoelia Benth. (1879);双穴大戟属☆
129364 Dicoelia affinis J. J. Sm.;近缘双穴大戟☆
129365 Dicoelia beccariana Benth.;双穴大戟☆
129366 Dicoelospermum C. B. Clarke(1897);双腔籽属■☆
129367 Dicoelospermum ritchiei C. B. Clarke;双腔籽■☆
129368 Dicolus Phil. = Zephyra D. Don ■☆
129369 Dicoma Cass. (1817);木菊属(鳞苞菊属)●☆
129370 Dicoma aethiopica S. Ortiz et Rodr. Oubina;埃塞俄比亚木菊●☆
129371 Dicoma angustifolia (S. Moore) Wild = Macledium poggei (O. Hoffm.) S. Ortiz ●☆

129372 Dicoma anomala Sond.;异常木菊●☆

129373 Dicoma anomala Sond. f. microcephala (Harv.) Oliv. et Hiern = Dicoma anomala Sond. subsp. gerrardii (Harv. ex F. C. Wilson) S. Ortiz et Rodr. Oubina ●☆

129374 Dicoma anomala Sond. subsp. attenuata (S. Moore) S. Ortiz et Rodr. Oubina;渐狭异常木菊●☆

129375 Dicoma anomala Sond. subsp. cirsioides (Harv.) Wild = Dicoma anomala Sond. ●☆

129376 Dicoma anomala Sond. subsp. gerrardii (Harv. ex F. C. Wilson) S. Ortiz et Rodr. Oubina;杰勒德木菊●☆

129377 Dicoma anomala Sond. var. attenuata S. Moore = Dicoma anomala Sond. subsp. attenuata (S. Moore) S. Ortiz et Rodr. Oubina ●☆

129378 Dicoma anomala Sond. var. cirsioides Harv. = Dicoma anomala Sond. ●☆

129379 Dicoma anomala Sond. var. latifolia O. Hoffm. = Dicoma anomala Sond. ●☆

129380 Dicoma anomala Sond. var. microcephala Harv. = Dicoma anomala Sond. subsp. gerrardii (Harv. ex F. C. Wilson) S. Ortiz et Rodr. Oubina ●☆

129381 Dicoma anomala Sond. var. sonderi Harv. = Dicoma anomala Sond. ●☆

129382 Dicoma antunesii O. Hoffm.;安图内思木菊●☆

129383 Dicoma arenaria Bremek.;沙地木菊●☆

129384 Dicoma arenicola Muschl. ex Dinter = Dicoma schinzii O. Hoffm. ●☆

129385 Dicoma argyrophylla Oliv. = Macledium zeyheri (Sond.) S. Ortiz subsp. argyrophyllum (Oliv.) S. Ortiz ●☆

129386 Dicoma attenuata (S. Moore) G. V. Pope = Dicoma anomala Sond. subsp. attenuata (S. Moore) S. Ortiz et Rodr. Oubina ●☆

129387 Dicoma auriculata Hutch. et B. L. Burtt = Macledium auriculatum (Hutch. et B. L. Burtt) S. Ortiz ●☆

129388 Dicoma burmanii (Cass.) Less. = Macledium spinosum (L.) S. Ortiz ●☆

129389 Dicoma candidissima Desf. = Otanthus maritimus (L.) Hoffmanns. et Link ●☆

129390 Dicoma capensis Less.;好望角木菊●☆

129391 Dicoma carbonaria (S. Moore) Humbert = Cloiselia carbonaria S. Moore ●☆

129392 Dicoma cinerea S. Ortiz et Rodr. Oubina = Dicoma schimperi (DC.) O. Hoffm. subsp. cinerea (S. Ortiz et Rodr. Oubina) S. Ortiz et Rodr. Oubina ●☆

129393 Dicoma cowanii S. Moore = Dicoma incana (Baker) O. Hoffm. ●☆

129394 Dicoma cuneneensis Wild;库内内木菊●☆

129395 Dicoma diacanthoides Less. = Macledium spinosum (L.) S. Ortiz ●☆

129396 Dicoma dinteri S. Moore;丁特致木菊●☆

129397 Dicoma elegans Welw. ex O. Hoffm.;雅致木菊●☆

129398 Dicoma elliptica G. V. Pope = Macledium ellipticum (G. V. Pope) S. Ortiz ●☆

129399 Dicoma flexuoides Muschl. ex Dinter = Dicoma dinteri S. Moore ●☆

129400 Dicoma flexuosa Dinter et Muschl. = Dicoma dinteri S. Moore ●☆

129401 Dicoma foliosa O. Hoffm.;多叶木菊●☆

129402 Dicoma fruticosa Compton;灌丛木菊●☆

129403 Dicoma galpinii F. C. Wilson;盖尔木菊●☆

129404 Dicoma gerrardii Harv. ex F. C. Wilson = Dicoma anomala Sond. subsp. gerrardii (Harv. ex F. C. Wilson) S. Ortiz et Rodr. Oubina ●☆

129405 Dicoma gillettii Rodr. Oubina et S. Ortiz;吉莱特木菊●☆

129406 Dicoma gnaphaloides Mattei = Dicoma tomentosa Cass. ●☆

129407 Dicoma gossweileri S. Moore = Macledium gossweileri (S. Moore) S. Ortiz ●☆

129408 Dicoma grandidieri (Drake) Humbert = Macledium grandidieri (Drake) S. Ortiz ●☆

129409 Dicoma humilis Lawalrée = Macledium humile (Lawalrée) S. Ortiz ●☆

129410 Dicoma incana (Baker) O. Hoffm.;灰毛木菊●☆

129411 Dicoma karaguensis Oliv. = Dicoma anomala Sond. ●☆

129412 Dicoma kirkii Harv. = Macledium kirkii (Harv.) S. Ortiz ●☆

129413 Dicoma kirkii Harv. subsp. vaginata (O. Hoffm.) G. V. Pope = Macledium kirkii (Harv.) S. Ortiz subsp. vaginatum (O. Hoffm.) S. Ortiz ●☆

129414 Dicoma kirkii Harv. var. angustifolia S. Moore = Macledium kirkii (Harv.) S. Ortiz ●☆

129415 Dicoma kirkii Harv. var. microcephala S. Moore = Macledium zeyheri (Sond.) S. Ortiz subsp. thyrsiflorum (Klatt) Netnou ●☆

129416 Dicoma kurumannii S. Ortiz et Netnou;库卢曼木菊●☆

129417 Dicoma lanata Muschl. ex Dinter = Dicoma schinzii O. Hoffm. ●☆

129418 Dicoma latifolia DC.;阔叶木菊●☆

129419 Dicoma latifolia DC. = Macledium latifolium (DC.) S. Ortiz ●☆

129420 Dicoma macrocephala DC.;大头曼木菊●☆

129421 Dicoma marlothiana Muschl. ex Dinter = Dicoma nachtigallii O. Hoffm. ●☆

129422 Dicoma megacephala Baker = Dicoma anomala Sond. ●☆

129423 Dicoma membranacea S. Moore = Macledium sessiliflorum (Harv.) S. Ortiz var. membranaceum (S. Moore) S. Ortiz ●☆

129424 Dicoma montana Schweick.;山地木菊●☆

129425 Dicoma nachtigallii O. Hoffm.;纳赫蒂加尔木菊●☆

129426 Dicoma nana Welw. ex Hiern = Macledium nanum (Welw. ex Hiern) S. Ortiz ●☆

129427 Dicoma nyikensis Baker = Dicoma anomala Sond. ●☆

129428 Dicoma obconica S. Ortiz;倒圆锥木菊●☆

129429 Dicoma oblonga Lawalrée et Mvukiy. = Macledium oblongum (Lawalrée et Mvukiy.) S. Ortiz ●☆

129430 Dicoma oleifolia Humbert = Cloiselia oleifolia (Humbert) S. Ortiz ●☆

129431 Dicoma paivae S. Ortiz et Rodr. Oubina;派瓦木菊●☆

129432 Dicoma picta (Thunb.) Druce;染色木菊●☆

129433 Dicoma plantaginifolia O. Hoffm. = Macledium plantaginifolium (O. Hoffm.) S. Ortiz ●☆

129434 Dicoma poggei O. Hoffm. = Macledium poggei (O. Hoffm.) S. Ortiz ●☆

129435 Dicoma popeana S. Ortiz et Rodr. Oubina;波普木菊●☆

129436 Dicoma pretoriensis C. A. Sm. = Macledium pretoriense (C. A. Sm.) S. Ortiz ●☆

129437 Dicoma prostrata Schweick.;平卧木菊●☆

129438 Dicoma pygmaea Hutch. = Macledium plantaginifolium (O. Hoffm.) S. Ortiz ●☆

129439 Dicoma quinquenervia Baker = Macledium poggei (O. Hoffm.) S. Ortiz ●☆

129440 Dicoma quinquenervia Baker var. angustifolia (S. Moore) S. Moore = Macledium poggei (O. Hoffm.) S. Ortiz ●☆

129441 Dicoma quinquenervia Baker var. latifolia S. Moore = Macledium

poggei (O. Hoffm.) S. Ortiz ●☆
129442 Dicoma radiata Less. = Dicoma picta (Thunb.) Druce ●☆
129443 Dicoma ramosissima Klatt = Pteronia acuminata DC. ●☆
129444 Dicoma relhanioides Less. = Macledium relhanioides (Less.) S. Ortiz ●☆
129445 Dicoma ringoetii De Wild. = Vernonia musofensis S. Moore var. miamensis (S. Moore) G. V. Pope ●☆
129446 Dicoma saligna Lawalrée = Macledium salignum (Lawalrée) S. Ortiz ●☆
129447 Dicoma schimperi (DC.) O. Hoffm.;欣珀木菊●☆
129448 Dicoma schimperi (DC.) O. Hoffm. subsp. cinerea (S. Ortiz et Rodr. Oubina) S. Ortiz et Rodr. Oubina;灰欣兹木菊●☆
129449 Dicoma schinzii O. Hoffm.;欣兹木菊●☆
129450 Dicoma scoparia Rodr. Oubina et S. Ortiz;帚状木菊●☆
129451 Dicoma seitziana Dinter = Pteronia eenii S. Moore ●☆
129452 Dicoma sessiliflora Harv. = Macledium sessiliflorum (Harv.) S. Ortiz ●☆
129453 Dicoma sessiliflora Harv. subsp. kirkii (Harv.) Wild = Macledium kirkii (Harv.) S. Ortiz ●☆
129454 Dicoma sessiliflora Harv. subsp. stenophylla G. V. Pope = Macledium sessiliflorum (Harv.) S. Ortiz subsp. stenophyllum (G. V. Pope) S. Ortiz ●☆
129455 Dicoma sessiliflora Harv. var. membranacea (S. Moore) S. Ortiz et Rodr. Oubina = Macledium sessiliflorum (Harv.) S. Ortiz var. membranaceum (S. Moore) S. Ortiz ●☆
129456 Dicoma somalensis S. Moore;索马里木菊●☆
129457 Dicoma speciosa DC.;美丽木菊●☆
129458 Dicoma speciosa DC. = Macledium speciosum (DC.) S. Ortiz ●☆
129459 Dicoma spinosa (L.) Druce = Macledium spinosum (L.) S. Ortiz ●☆
129460 Dicoma squarrosa Wild;粗鳞木菊●☆
129461 Dicoma superba S. Moore = Macledium poggei (O. Hoffm.) S. Ortiz ●☆
129462 Dicoma superba S. Moore var. angustifolia ? = Macledium poggei (O. Hoffm.) S. Ortiz ●☆
129463 Dicoma swazilandica S. Ortiz, Rodr. Oubina et Pulgar;斯威士兰木菊●☆
129464 Dicoma thuliniana S. Ortiz et Rodr. Oubina et Mesfin;图林木菊●☆
129465 Dicoma thyrsiflora (Klatt) Thell. = Macledium zeyheri (Sond.) S. Ortiz subsp. thyrsiflorum (Klatt) Netnou ●☆
129466 Dicoma tomentosa Cass.;毛木菊●☆
129467 Dicoma vaginata O. Hoffm. = Macledium kirkii (Harv.) S. Ortiz subsp. vaginatum (O. Hoffm.) S. Ortiz ●☆
129468 Dicoma welwitschii O. Hoffm.;韦尔木菊●☆
129469 Dicoma zeyheri Sond.;泽赫木菊●☆
129470 Dicoma zeyheri Sond. subsp. argyrophylla (Oliv.) G. V. Pope = Macledium zeyheri (Sond.) S. Ortiz subsp. argyrophyllum (Oliv.) S. Ortiz ●☆
129471 Dicoma zeyheri Sond. var. thyrsiflora Klatt = Macledium zeyheri (Sond.) S. Ortiz subsp. thyrsiflorum (Klatt) Netnou ●☆
129472 Diconangia Adans. = Itea L. ●
129473 Diconangia Mitch. ex Adans. = Itea L. ●
129474 Dicondra Raf. = Dichondra J. R. Forst. et G. Forst. ■
129475 Dicophe Roem. = Dicoryphe Thouars ●☆
129476 Dicoria Torr. et A. Gray(1859);双虫菊属;Desert Twinbugs ■●☆
129477 Dicoria brandegeei A. Gray = Dicoria canescens A. Gray ■☆
129478 Dicoria canescens A. Gray;双虫菊;Desert Twinbugs ■☆
129479 Dicoria canescens A. Gray subsp. brandegeei (A. Gray) Cronquist = Dicoria canescens A. Gray ■☆
129480 Dicoria canescens A. Gray subsp. clarkiae (P. B. Kenn.) D. D. Keck = Dicoria canescens A. Gray ■☆
129481 Dicoria canescens A. Gray subsp. hispidula (Rydb.) D. D. Keck = Dicoria canescens A. Gray ■☆
129482 Dicoria canescens A. Gray var. brandegeei (A. Gray) Cronquist = Dicoria canescens A. Gray ■☆
129483 Dicoria canescens A. Gray var. hispidula (Rydb.) Cronquist = Dicoria canescens A. Gray ■☆
129484 Dicoria canescens A. Gray var. witherillii (Eastw.) Cronquist = Dicoria canescens A. Gray ■☆
129485 Dicoria clarkiae P. B. Kenn. = Dicoria canescens A. Gray ■☆
129486 Dicoria paniculata Eastw. = Dicoria canescens A. Gray ■☆
129487 Dicorynia Benth. (1840);双柱苏木属●☆
129488 Dicorynia guianensis Amshoff;圭亚那双柱苏木;Angelique, Basralocus ●☆
129489 Dicorynia paraensis Benth.;帕拉州双柱苏木;Para Angelwood Para Angel-wood ●☆
129490 Dicorypha R. Hedw. = Dicoryphe Thouars ●☆
129491 Dicoryphe Thouars(1804);双扇梅属●☆
129492 Dicoryphe angustifolia Tul.;狭叶双扇梅●☆
129493 Dicoryphe gracilis Tul.;纤细双扇梅●☆
129494 Dicoryphe lanceolata Tul.;披针叶双扇梅●☆
129495 Dicoryphe laurifolia Baker;桂叶双扇梅●☆
129496 Dicoryphe macrophylla Baill.;大叶双扇梅●☆
129497 Dicoryphe platyphylla Tul.;宽叶双扇梅●☆
129498 Dicotyodaphne Blume = Endiandra R. Br. ●
129499 Dicraea Tul. = Dicraeia Thouars ■☆
129500 Dicraeanthus Engl. (1905);叉花苔草属■☆
129501 Dicraeanthus africanus Engl.;非洲叉花苔草■☆
129502 Dicraeanthus ramosus H. E. Hess = Dicraeanthus africanus Engl. ■☆
129503 Dicraeanthus taylorii W. J. de Wilde et Guillaumet = Macropodiella taylorii (W. J. de Wilde et Guillaumet) C. Cusset ■☆
129504 Dicraeanthus zehnderi H. E. Hess;策恩德叉花苔草■☆
129505 Dicraeia Thouars = Podostemum Michx. ■☆
129506 Dicraeia Thouars(1806);叉苔草属■☆
129507 Dicraeia erythrolichen Tul. et Wedd. = Ceratolacis erythrolichen (Tul. et Wedd.) Wedd. ■☆
129508 Dicraeia garrettii C. H. Wright = Macropodiella garrettii (C. H. Wright) C. Cusset ■☆
129509 Dicraeia ledermannii Engl. = Ledermanniella ledermannii (Engl.) C. Cusset ■☆
129510 Dicraeia tenax C. H. Wright = Ledermanniella tenax (C. H. Wright) C. Cusset ■☆
129511 Dicraeia violascens Engl. = Sphaerothylax abyssinica (Wedd.) Warm. ■☆
129512 Dicraeia warmingii Engl. = Letestuella tisserantii G. Taylor ■☆
129513 Dicraeopetalum Harms(1902);二叉豆属●☆
129514 Dicraeopetalum capuronianum (M. Pelt.) Yakovlev;凯普伦二叉豆●☆
129515 Dicraeopetalum mahafaliense (M. Pelt.) Yakovlev;马哈法里二叉豆●☆
129516 Dicraeopetalum stipulare Harms;二叉豆●☆
129517 Dicrama Klatt = Dierama K. Koch et C. D. Bouché ■☆

129518 Dicranacanthus Oerst. = Barleria L. ●■
129519 Dicranacanthus buxifolia (L.) Oerst. = Barleria buxifolia L. ●☆
129520 Dicrananthera C. Presl = Anisanthera Raf. ■☆
129521 Dicrananthera C. Presl(1832);叉药野牡丹属■☆
129522 Dicrananthera hedyotidea C. Presl;叉药野牡丹■☆
129523 Dicranilla (Fenzl) Rchb. = Arenaria L. ■
129524 Dicrannsiegia (A. Gray) Pennell = Cordylanthus Nutt. ex Benth.(保留属名)■☆
129525 Dicranocarpus A. Gray(1854);草耙菊属(叉果菊属);Pitchfork ■☆
129526 Dicranocarpus parviflorus A. Gray;草耙菊(小花叉果菊)■☆
129527 Dicranolepis Planch.(1848);叉鳞瑞香属●☆
129528 Dicranolepis angolensis S. Moore;安哥拉叉鳞瑞香●☆
129529 Dicranolepis baertsiana De Wild. et T. Durand subsp. fulva A. Robyns;黄褐叉鳞瑞香●☆
129530 Dicranolepis buchholzii Engl. et Gilg;布赫叉鳞瑞香●☆
129531 Dicranolepis disticha Planch.;二列叉鳞瑞香●☆
129532 Dicranolepis glandulosa H. Pearson;大腺叉鳞瑞香●☆
129533 Dicranolepis grandiflora Engl.;大花叉鳞瑞香●☆
129534 Dicranolepis incisa A. Robyns;锐裂叉鳞瑞香●☆
129535 Dicranolepis laciniata Gilg;撕裂叉鳞瑞香●☆
129536 Dicranolepis mannii Baill. = Dicranolepis buchholzii Engl. et Gilg ●☆
129537 Dicranolepis montana Gilg et Ledermann ex Engl.;山地叉鳞瑞香●☆
129538 Dicranolepis oligantha Gilg = Dicranolepis buchholzii Engl. et Gilg ●☆
129539 Dicranolepis polygaloides Gilg ex H. Pearson;多节叉鳞瑞香●☆
129540 Dicranolepis pubescens H. Pearson;毛叉鳞瑞香●☆
129541 Dicranolepis pulcherrima Gilg;美丽叉鳞瑞香●☆
129542 Dicranolepis pusilla Aymonin;微小叉鳞瑞香●☆
129543 Dicranolepis pyramidalis Gilg;塔形叉鳞瑞香●☆
129544 Dicranolepis soyauxii Engl.;索亚叉鳞瑞香●☆
129545 Dicranolepis stenura Gilg ex Engl. = Dicranolepis vestita Engl. ●☆
129546 Dicranolepis thomensis Engl. et Gilg;爱岛叉鳞瑞香●☆
129547 Dicranolepis usambarica Gilg;乌桑巴拉叉鳞瑞香●☆
129548 Dicranolepis vestita Engl.;包被叉鳞瑞香●☆
129549 Dicranopetalum C. Presl = Toulicia Aubl. ●☆
129550 Dicranopygium Harling(1954);叉臀草属■☆
129551 Dicranopygium amazonicum Harling;亚马孙叉臀草■☆
129552 Dicranopygium atrovirens (H. Wendl.) Harling;墨绿叉臀草■☆
129553 Dicranopygium gracile (Liebm. ex Matuda) Harling;纤细叉臀草■☆
129554 Dicranopygium microcephalum (Hook. f.) Harling;小头叉臀草■☆
129555 Dicranopygium nanum (Gleason) Harling;矮叉臀草■☆
129556 Dicranopygium pachystemon Harling;粗蕊叉臀草■☆
129557 Dicranopygium polycephalum Harling;多头叉臀草■☆
129558 Dicranopygium robustum Harling;粗壮叉臀草■☆
129559 Dicranostachys Trécul = Myrianthus P. Beauv. ●☆
129560 Dicranostachys Trécul(1847);叉穗伞树属●☆
129561 Dicranostachys serrata Trécul = Myrianthus serratus (Trécul) Benth. et Hook. ●☆
129562 Dicranostegia (A. Gray) Pennell = Cordylanthus Nutt. ex Benth.(保留属名)■☆
129563 Dicranostigma Hook. f. et Thomson(1855);秃疮花属;Dicranostigma, Favusheadflower ■
129564 Dicranostigma franchetianum (Prain) Fedde;滇秃疮花(滇川秃疮花,秃疮花);Franchet Dicranostigma, Franchet Favusheadflower ■
129565 Dicranostigma franchetianum (Prain) Fedde = Dicranostigma leptopodum (Maxim.) Fedde ■
129566 Dicranostigma henanensis S. Y. Wang et L. H. Wu;河南秃疮花;Henan Dicranostigma, Henan Favusheadflower ■
129567 Dicranostigma iliense C. Y. Wu et H. Chuang = Glaucium fimbrilligerum (Trautv.) Boiss. ■
129568 Dicranostigma iliensis C. Y. Wu et H. Chuang;伊犁秃疮花;Yili Dicranostigma, Yili Favusheadflower ■
129569 Dicranostigma iliensis C. Y. Wu et H. Chuang = Glaucium fimbrilligerum Boiss. ■
129570 Dicranostigma lactucoides Hook. f. et Thomson;苣叶秃疮花;Lettuceleaf Favusheadflower, Lettuce-like Dicranostigma ■
129571 Dicranostigma leptopodum (Maxim.) Fedde;秃疮花(勒马回,秃子花);Favusheadflower, Slenderstalk Dicranostigma ■
129572 Dicranostigma platycarpum C. Y. Wu et H. Chuang;宽果秃疮花;Broadfruit Dicranostigma, Broadfruit Favusheadflower ■
129573 Dicranostyles Benth.(1846);叉柱旋花属■☆
129574 Dicranostyles boliviensis Ducke;玻利维亚叉柱旋花■☆
129575 Dicranostyles longifolia Ducke;长叶叉柱旋花■☆
129576 Dicranostyles scandens Benth.;叉柱旋花■☆
129577 Dicranotaenia Finer = Microcoelia Lindl. ■☆
129578 Dicraspidia Standl.(1929);叉盾椴属●☆
129579 Dicraspidia donnell-smithii Standl.;叉盾椴●☆
129580 Dicrastyles Benth. et Hook. f. = Dicrastylis Drumm. ex Harv.(保留属名)●☆
129581 Dicrastylidaceae J. Drumm. ex Harv.;离柱花科●☆
129582 Dicrastylidaceae J. Drumm. ex Harv. = Chloanthaceae Hutch. ●■☆
129583 Dicrastylidaceae J. Drumm. ex Harv. = Labiatae Juss.(保留科名)●■
129584 Dicrastylidaceae J. Drumm. ex Harv. = Lamiaceae Martinov(保留科名)●■
129585 Dicrastylidaceae J. Drumm. ex Harv. = Verbenaceae J. St. -Hil.(保留科名)●■
129586 Dicrastylis Drumm. ex Harv.(1855)(保留属名);离柱花属●☆
129587 Dicrastylis fulva J. Drumm. ex Harv.;离柱花●☆
129588 Dicraurus Hook. f. = Iresine P. Browne(保留属名)●■
129589 Dicroacds Raf. = Coreopsis L. ●■
129590 Dicrobotryum Willd. ex Roem. et Schult. = Guettarda L. ●
129591 Dicrocaulon N. E. Br.(1928);银杯玉属■☆
129592 Dicrocaulon brevifolium N. E. Br.;短叶银杯玉■☆
129593 Dicrocaulon grandiflorum Ihlenf.;大花银杯玉■☆
129594 Dicrocaulon humile N. E. Br.;矮小银杯玉■☆
129595 Dicrocaulon microstigma (L. Bolus) Ihlenf.;小柱头银杯玉■☆
129596 Dicrocaulon nodosum (A. Berger) N. E. Br.;多节银杯玉■☆
129597 Dicrocaulon pearsonii N. E. Br. = Dicrocaulon nodosum (A. Berger) N. E. Br. ■☆
129598 Dicrocaulon ramulosum (L. Bolus) Ihlenf.;多枝银杯玉■☆
129599 Dicrocaulon spissum N. E. Br.;密集银杯玉■☆
129600 Dicrocaulon trichotomum (Thunb.) N. E. Br. = Phyllobolus trichotomus (Thunb.) Gerbaulet ●☆
129601 Dicrocephala Royle = Dichrocephala L'Hér. ex DC. ■
129602 Dicrophyla Raf. = Ludisia A. Rich. ■
129603 Dicrophylla Raf. = Ludisia A. Rich. ■
129604 Dicrosperma H. Wendl. et Drude ex W. Watson = Dictyosperma

H. Wendl. et Drude ●☆
129605 Dicrosperma W. Watson = Dictyosperma H. Wendl. et Drude ●☆
129606 Dicrostylis Post et Kuntze = Dicrastylis Drumm. ex Harv.(保留属名)●☆
129607 Dicrus Reinw. = Voacanga Thouars ●
129608 Dicrypta Lindl. = Maxillaria Ruiz et Pav. ■☆
129609 Dicrypta baueri Lindl. = Maxillaria crassifolia (Lindl.) Rchb. f. ■☆
129610 Dicrypta crassifolia (Lindl.) Lindl. ex Loudon = Maxillaria crassifolia (Lindl.) Rchb. f. ■☆
129611 Dictamnaceae Trautv.;白鲜科■
129612 Dictamnaceae Trautv. = Rutaceae Juss.(保留科名)●■
129613 Dictamnaceae Vest = Rutaceae Juss.(保留科名)●■
129614 Dictamnus Hill. = Origanum L. ●■
129615 Dictamnus L.(1753);白鲜属;Burning Bush, Burningbush, Dittany, Fraxinella, Gas Plant ■
129616 Dictamnus Mill. = Amaracus Gled.(保留属名)●■☆
129617 Dictamnus Zinn = Amaracus Gled.(保留属名)●■☆
129618 Dictamnus albus L.;欧白鲜(白鲜,椤白鲜,朝鲜白鲜,南欧白鲜);Bastard Dittany, Burning Bush, Candle Plant, Dittany, False Dittany, Fraxinella, Gas Plant, Gasplant, Gas-plant, Gas-plant Dittany, White Dittany ■☆
129619 Dictamnus albus L. 'Purpureus';紫花欧白鲜(紫花白鲜)■
129620 Dictamnus albus L. = Dictamnus dasycarpus Turcz. ■
129621 Dictamnus albus L. subsp. clasycarpus (Turcz.) Kitag. = Dictamnus dasycarpus Turcz. ■
129622 Dictamnus albus L. subsp. dasycarpus (Turcz.) L. Winter = Dictamnus dasycarpus Turcz. ■
129623 Dictamnus albus L. subsp. dasycarpus (Turcz.) T. N. Liou et Y. H. Chang = Dictamnus dasycarpus Turcz. ■
129624 Dictamnus albus L. subsp. dasycarpus Kitag. = Dictamnus dasycarpus Turcz. ■
129625 Dictamnus albus L. subsp. dasycarpus L. Winter = Dictamnus dasycarpus Turcz. ■
129626 Dictamnus albus L. var. purpureus Hort. = Dictamnus albus L. 'Purpureus' ■
129627 Dictamnus angustifolium G. Don ex Sweet;狭叶白鲜;Narrowleaf Dittany ■☆
129628 Dictamnus caucasicus (Boiss.) Fisch. et C. A. Mey.;高加索白鲜;Caucasia Dittany ■☆
129629 Dictamnus dasycarpus Turcz.;白鲜(八股牛,八圭牛,白膻,白鲜皮,白藓,白羊鲜,臭根皮,臭骨头,臭哄哄,地羊膻,地羊鲜,好汉拔,胡椒,金雀儿椒,千斤拔,山牡丹,羊蹄草,羊鲜草,羊癣草,野花椒);Densefruit Dittany ■
129630 Dictamnus fraxinella Pers. = Dictamnus albus L. ■☆
129631 Dictamnus gymnostylus Stev.;裸柱白鲜■☆
129632 Dictamnus himalayanus Royle = Dictamnus albus L. ■☆
129633 Dictamnus tadshikorum Vved.;塔什克白鲜■☆
129634 Dictilis Raf. = Clinopodium L. ■●
129635 Dictyaloma Walp. = Dictyoloma A. Juss.(保留属名)●☆
129636 Dictyandra Welw. ex Benth. et Hook. f.(1873);网蕊茜属●☆
129637 Dictyandra Welw. ex Hook. f. = Dictyandra Welw. ex Benth. et Hook. f. ●☆
129638 Dictyandra arborescens Welw. ex Hook. f.;树状网蕊茜●☆
129639 Dictyandra congolana Robbr.;刚果网蕊茜●☆
129640 Dictyandra involucrata (Hook. f.) Hiern = Leptactina involucrata Hook. f. ●☆
129641 Dictyanthex Raf. = Aristolochia L. ■●
129642 Dictyanthus Decne. = Matelea Aubl. ●☆
129643 Dictyocalyx Hook. f. = Cacabus Bernh. ■☆
129644 Dictyocalyx Hook. f. = Exodeconus Raf. ■☆
129645 Dictyocarpus Wight = Sida L. ●■
129646 Dictyocarpus Wight(1837);网果锦葵属●☆
129647 Dictyocarpus truncatus Wight;网果锦葵●☆
129648 Dictyocarpus truncatus Wight = Sida schimperiana Hochst. ex A. Rich. ●☆
129649 Dictyocaryum H. Wendl.(1860);网果棕属(金椰属,网实桐属,网实椰子属,网籽椰属);Princess Palm ●☆
129650 Dictyocaryum fuscum H. Wendl.;网果棕●☆
129651 Dictyochloa (Murb.) E. G. Camus = Ammochloa Boiss. ■☆
129652 Dictyochloa E. G. Camus = Ammochloa Boiss. ■☆
129653 Dictyochloa involucrata (Murb.) E. G. Camus = Ammochloa involucrata Murb. ■☆
129654 Dictyodaphne Blume = Endiandra R. Br. ●
129655 Dictyolimon Rech. f.(1974);网脉补血草属■☆
129656 Dictyolimon macrorrhabdos (Boiss.) Rech. f.;网脉补血草■☆
129657 Dictyoloma A. Juss.(1825)(保留属名);网边芸香属●☆
129658 Dictyoloma vandellianum A. Juss.;网边芸香●☆
129659 Dictyoneura Blume(1849);网脉无患子属●☆
129660 Dictyoneura acuminata Blume;尖网脉无患子●☆
129661 Dictyoneura microcarpa Radlk.;小果网脉无患子●☆
129662 Dictyoneura obtusa Blume;钝网脉无患子●☆
129663 Dictyoneura subhirsuta Radlk.;小毛网脉无患子●☆
129664 Dictyopetalum (Fisch. et C. A. Mey.) Baill. = Oenothera L. ●■
129665 Dictyopetalum Fisch. et C. A. Mey. = Oenothera L. ●■
129666 Dictyophleba Pierre(1898);网脉夹竹桃属●☆
129667 Dictyophleba leonensis (Stapf) Pichon;莱昂网脉夹竹桃●☆
129668 Dictyophleba lucida (K. Schum.) Pierre;光亮网脉夹竹桃●☆
129669 Dictyophleba ochracea (K. Schum. ex Hallier f.) Pichon;淡黄褐网脉夹竹桃●☆
129670 Dictyophleba ochracea (K. Schum. ex Hallier f.) Pichon var. breviflora De Wild. = Dictyophleba ochracea (K. Schum. ex Hallier f.) Pichon ●☆
129671 Dictyophleba ochracea (K. Schum. ex Hallier f.) Pichon var. glabrata (Hallier f.) Pichon = Dictyophleba ochracea (K. Schum. ex Hallier f.) Pichon ●☆
129672 Dictyophleba setosa de Hoogh;刚毛网脉夹竹桃●☆
129673 Dictyophleba stipulosa (S. Moore ex Wernham) Pichon;托叶网脉夹竹桃●☆
129674 Dictyophragmus O. E. Schulz(1933);网篱笆属■☆
129675 Dictyophragmus englerianus (Muschl.) O. E. Schulz;网篱笆■☆
129676 Dictyophyllaria Garay(1986);网叶兰属■☆
129677 Dictyophyllaria dietschiana (Edwall) Garay;网叶兰■☆
129678 Dictyopsis Harv. ex Hook. f. = Behnia Didr. ●☆
129679 Dictyosperma H. Wendl. et Drude(1875);网实棕属(环羽椰属,金棕属,飓风椰属,飓风棕属,双籽棕属,网脉种子棕属,网实椰子属,网籽桐属,网子椰子属);Princess Palm ●☆
129680 Dictyosperma Post et Kuntze = Dyctisperma Raf. ●■
129681 Dictyosperma Post et Kuntze = Rubus L. ●■
129682 Dictyosperma Regel = Pirea T. Durand ■
129683 Dictyosperma Regel = Rorippa Scop. ■
129684 Dictyosperma album (Bory) H. Wendl. et Drude;白网实棕(白网脉种子棕,环羽椰);Common Princess Palm, Hurrivane Palm, Princess Palm ●☆

129685 Dictyosperma aureum H. Wendl. et Drude;金黄网实棕(金黄网脉种子棕);Yellow Princess Palm ●☆
129686 Dictyosperma fibrosum C. H. Wright = Dypsis fibrosa (C. H. Wright) Beentje et J. Dransf. ●☆
129687 Dictyosperma olgae Regel et Schmalh. = Nasturtium microphyllum Boenn. ex Rchb. ■☆
129688 Dictyosperma rubrum H. Wendl. et Drude;红网实棕(红色网脉种子棕);Red Princess Palm ●☆
129689 Dictyospermum Wight = Aneilema R. Br. ■☆
129690 Dictyospermum Wight(1853);网籽草属;Netseed, Netseedgrass ■
129691 Dictyospermum conspicuum (Blume) Hassk.;网籽草(白花鸭跖草);Netseed ■
129692 Dictyospermum ovalifolium Wight;卵叶网籽草■
129693 Dictyospermum protensum Wight = Dictyospermum scaberrimum (Blume) J. K. Morton ex H. Hara ■
129694 Dictyospermum protensum Wight = Rhopalephora scaberrima (Blume) Faden ■
129695 Dictyospermum scaberrimum (Blume) J. K. Morton ex D. Y. Hong = Rhopalephora scaberrima (Blume) Faden ■
129696 Dictyospermum scaberrimum (Blume) J. K. Morton ex H. Hara = Rhopalephora scaberrima (Blume) Faden ■
129697 Dictyospermum vaginatum (L.) D. Y. Hong = Murdannia vaginata (L.) Brückn. ■
129698 Dictyospermum wightii Hassk. var. robustum Hassk. = Pollia subumbellata C. B. Clarke ■
129699 Dictyospora Hook. f. = Dyctiospora Reinw. ex Korth. ●■
129700 Dictyospora Hook. f. = Hedyotis L. (保留属名)●■
129701 Dictyostega Miers(1840);网盖水玉簪属■☆
129702 Dictyostega costata Miers;网盖水玉簪■☆
129703 Dictyostega longistyla Benth. = Gymnosiphon longistylus (Benth.) Hutch. et Dalziel ■☆
129704 Dictysperma Raf. = Dyctisperma Raf. ●■
129705 Dictysperma Raf. = Rubus L. ●■
129706 Dicyclophora Boiss. (1844);双环芹属■☆
129707 Dicyclophora persica Boiss.;双环芹■☆
129708 Dicymanthes Danser = Amyema Tiegh. ●☆
129709 Dicymanthes Danser = Loranthus Jacq. (保留属名)●
129710 Dicymbe Spruce ex Benth. = Dicymbe Spruce ex Benth. et Hook. f. ■☆
129711 Dicymbe Spruce ex Benth. et Hook. f. (1865);天篷豆属■☆
129712 Dicymbe Spruce ex Benth. et Hook. f. = Dicymbe Spruce ex Benth. ■☆
129713 Dicymbe amazonica Ducke;亚马孙天篷豆■☆
129714 Dicymbe corymbosa Spruce ex Benth.;天篷豆■☆
129715 Dicymbopsis Ducke = Dicymbe Spruce ex Benth. ■☆
129716 Dicypellium Nees = Dicypellium Nees et Mart. ●☆
129717 Dicypellium Nees et Mart. (1833);丁香桂属(丁香皮树属,香皮桂属)●☆
129718 Dicypellium caryophyllaceum (C. Mart.) Nees et C. Mart.;丁香桂;Clove Bark Tree, Clove-bark Tree, Glove-bark, Pinkwood ●☆
129719 Dicypellium caryophyllatum (Mart.) Nees = Dicypellium caryophyllaceum (C. Mart.) Nees et C. Mart. ●☆
129720 Dicyrta Regel = Achimenes Pers. (保留属名)■☆
129721 Dicyrta Regel(1849);二弯苣苔属;Dicyrta ■☆
129722 Dicyrta candida Hanst. et Klotzsch;白色二弯苣苔草;White Dicyrta ■☆
129723 Dicyrta parviflora Seem.;小花二弯苣苔■☆
129724 Dicyrta warszewicziana Regel;瓦氏二弯苣苔草;Warszewicz Dicyrta ■☆
129725 Didactyle Lindl. = Bulbophyllum Thouars(保留属名)■
129726 Didactylon Moritzi = Dimeria R. Br. ■
129727 Didactylon Zoll. et Moritzi = Dimeria R. Br. ■
129728 Didactylus (Luer) Luer = Pleurothallis R. Br. ■☆
129729 Didactylus Luer = Pleurothallis R. Br. ■☆
129730 Didaste E. Mey. ex Harv. et Sond. = Acrosanthes Eckl. et Zeyh. ■☆
129731 Didelotia Baill. (1865);代德苏木属●☆
129732 Didelotia africana Baill.;热非代德苏木;Bubinga Didelotia ●☆
129733 Didelotia afzelii Taub.;阿芙泽尔代德苏木●☆
129734 Didelotia brevipaniculata J. Léonard;锥花代德苏木●☆
129735 Didelotia engleri Dinkl. et Harms;恩氏代德苏木●☆
129736 Didelotia idae J. Léonard, Oldeman et de Wit;西非代德苏木●☆
129737 Didelotia ledermannii Harms = Didelotia engleri Dinkl. et Harms ●☆
129738 Didelotia letouzeyi Pellegr.;喀麦隆代德苏木●☆
129739 Didelotia minutiflora (A. Chev.) J. Léonard;小花代德苏木●☆
129740 Didelotia morelii Aubrév.;默勒尔代德苏木●☆
129741 Didelotia unifoliolata J. Léonard;单小叶代德苏木●☆
129742 Didelta L'Hér. (1786)(保留属名);离苞菊属■☆
129743 Didelta annua Less. = Cuspidia cernua (L. f.) B. L. Burtt subsp. annua (Less.) Rössler ■☆
129744 Didelta capensis Lam. = Didelta carnosa (L. f.) W. T. Aiton ■☆
129745 Didelta carnosa (L. f.) W. T. Aiton;离苞菊■☆
129746 Didelta carnosa (L. f.) W. T. Aiton var. tomentosa (Less.) Rössler;毛离苞菊■☆
129747 Didelta cernua (L. f.) Less. = Cuspidia cernua (L. f.) B. L. Burtt ■☆
129748 Didelta obtusifolia Cass. = Didelta carnosa (L. f.) W. T. Aiton ■☆
129749 Didelta spinosa (L. f.) W. T. Aiton;具刺离苞菊■☆
129750 Didelta tetragoniifolia L'Hér. = Didelta carnosa (L. f.) W. T. Aiton ■☆
129751 Didelta tomentosa Less. = Didelta carnosa (L. f.) W. T. Aiton var. tomentosa (Less.) Rössler ■☆
129752 Didemia Naudin = Ossaea DC. ●☆
129753 Diderota Comm. ex A. DC. = Ochrosia Juss. ●
129754 Diderotia Baill. = Alchornea Sw. ●
129755 Diderotia Baill. = Lautembergia Baill. ●☆
129756 Didesmandra Stapf(1900);双链蕊属■☆
129757 Didesmandra aspera Stapf;双链蕊■☆
129758 Didesmus Desv. (1815);双索芥属(匕果芥属)■☆
129759 Didesmus aegyptius (L.) Desv.;埃及匕果芥■☆
129760 Didesmus aegyptius (L.) Desv. var. oblongifolius DC. = Didesmus aegyptius (L.) Desv. ■☆
129761 Didesmus aegyptius (L.) Desv. var. pinnatus Jaub. et Spach = Didesmus aegyptius (L.) Desv. ■☆
129762 Didesmus aegyptius (L.) Desv. var. tenuifolius (Sibth. et Sm.) Heldr. = Didesmus aegyptius (L.) Desv. ■☆
129763 Didesmus bipinnatus (Desf.) DC.;羽状匕果芥■☆
129764 Didesmus bipinnatus (Desf.) DC. var. dissectus Maire = Didesmus bipinnatus (Desf.) DC. ■☆
129765 Didesmus bipinnatus (Desf.) DC. var. intermedius Maire et Weiller = Didesmus bipinnatus (Desf.) DC. ■☆
129766 Didesmus bipinnatus (Desf.) DC. var. tenuifolius Halácsy =

Didesmus bipinnatus (Desf.) DC. ■☆
129767 Didiciea King et Prain = Tipularia Nutt. ■
129768 Didiciea King et Prain(1896);迪迪兰属;Didiciea ■
129769 Didiciea cunninghamii King et Prain = Tipularia cunninghamii (King et Prain) S. C. Chen ■
129770 Didiciea japonica H. Hara = Tipularia japonica Matsum. var. harae F. Maek. ■☆
129771 Didiciea neglecta Fukuy.;双生兰■
129772 Didierea Baill. (1880);刺戟木属(刺戟草属,刺戟属,棘针树属,龙树属)●☆
129773 Didierea ascendens Drake = Alluaudia ascendens (Drake) Drake ●☆
129774 Didierea comosa Drake = Alluaudia comosa (Drake) Drake ●☆
129775 Didierea dumosa Drake = Alluaudia dumosa (Drake) Drake ●☆
129776 Didierea madagascariensis Baill.;马岛刺戟木(刺戟草);Didierea,Octopus Tree ●☆
129777 Didierea mirabilis Baill. = Didierea madagascariensis Baill. ●☆
129778 Didierea procera Drake = Alluaudia procera (Drake) Drake ●☆
129779 Didierea trollii Capuron et Rauh = Didierea madagascariensis Baill. ●☆
129780 Didiereaceae Drake = Didiereaceae Radlk. (保留科名)●☆
129781 Didiereaceae Radlk. (1896)(保留科名);刺戟木科(刺戟草科,刺戟科,棘针树科,龙树科)●☆
129782 Didiereaceae Radlk. ex Drake = Didiereaceae Radlk. (保留科名)●☆
129783 Didimeria Lindl. = Corraea Sm. ●☆
129784 Didiplis Raf. = Lythrum L. ●■
129785 Didiplis diandra (Nutt. ex DC.) A. W. Wood = Peplis diandra Nutt. ex DC. ■☆
129786 Didiplis diandra (Nutt. ex DC.) A. W. Wood f. aquatica (Koehne) Fassett = Peplis diandra Nutt. ex DC. ■☆
129787 Didiplis diandra (Nutt. ex DC.) A. W. Wood f. terrestris (Koehne) Fassett = Peplis diandra Nutt. ex DC. ■☆
129788 Didiscus DC. = Trachymene Rudge ■☆
129789 Didiscus DC. ex Hook. = Trachymene Rudge ■☆
129790 Didissandra C. B. Clarke(1883)(保留属名);漏斗苣苔属(卷丝苣苔属,一面锣属);Didissandra ●■
129791 Didissandra agnesiae Forrest = Briggsia agnesiae (Forrest) Craib ■
129792 Didissandra amabilis Diels = Briggsia kurzii (C. B. Clarke) W. E. Evans ■
129793 Didissandra aurea (Dunn) B. L. Burtt = Didissandra macrosiphon (Hance) W. T. Wang ■
129794 Didissandra beauverdiana H. Lév. = Briggsiopsis delavayi (Franch.) K. Y. Pan ■
129795 Didissandra begoniifolia H. Lév.;大苞漏斗苣苔草;Largebract Didissandra ■
129796 Didissandra bullata Craib = Corallodiscus lanuginosus (Wall. ex R. Br.) B. L. Burtt ■
129797 Didissandra cavaleriei H. Lév. et Vaniot = Loxostigma cavaleriei (H. Lév. et Vaniot) B. L. Burtt ■
129798 Didissandra cordatula Craib = Corallodiscus lanuginosus (Wall. ex R. Br.) B. L. Burtt ■
129799 Didissandra delavayi Franch. = Briggsiopsis delavayi (Franch.) K. Y. Pan ■
129800 Didissandra fargesii Franch. = Isometrum fargesii (Franch.) B. L. Burtt ■
129801 Didissandra flabellata Craib = Corallodiscus lanuginosus (Wall. ex R. Br.) B. L. Burtt ■
129802 Didissandra forrestii J. Anthony = Corallodiscus conchifolius Batalin ■
129803 Didissandra fritschii H. Lév. et Vaniot = Briggsia mihieri (Franch.) Craib ■
129804 Didissandra giraldii Diels = Isometrum giraldii (Diels) B. L. Burtt ■
129805 Didissandra glandulosa Batalin = Isometrum glandulosum (Batalin) Craib ■
129806 Didissandra grandis Craib = Corallodiscus kingianus (Craib) B. L. Burtt ■
129807 Didissandra kingiana Craib = Corallodiscus kingianus (Craib) B. L. Burtt ■
129808 Didissandra labordei Craib = Corallodiscus lanuginosus (Wall. ex R. Br.) B. L. Burtt ■
129809 Didissandra lancifolia Franch. = Isometrum lancifolium (Franch.) K. Y. Pan ■
129810 Didissandra lanuginosa (Wall. ex A. DC.) Clarke = Corallodiscus lanuginosus (Wall. ex R. Br.) B. L. Burtt ■
129811 Didissandra lanuginosa (Wall. ex R. Br.) C. B. Clarke = Corallodiscus lanuginosus (Wall. ex R. Br.) B. L. Burtt ■
129812 Didissandra leucantha Diels = Isometrum leucanthum (Diels) B. L. Burtt ■
129813 Didissandra lineata Craib = Corallodiscus lanuginosus (Wall. ex R. Br.) B. L. Burtt ■
129814 Didissandra longipedunculata C. Y. Wu ex H. W. Li;长梗漏斗苣苔草;Longstalk Didissandra ●■
129815 Didissandra longipes Hemsl. ex Oliv. = Briggsia longipes (Hemsl. ex Oliv.) Craib ■
129816 Didissandra lutea Craib = Corallodiscus lanuginosus (Wall. ex R. Br.) B. L. Burtt ■
129817 Didissandra macrosiphon (Hance) W. T. Wang;长筒漏斗苣苔草;Largetube Didissandra ■
129818 Didissandra mengtzeana Craib = Corallodiscus lanuginosus (Wall. ex R. Br.) B. L. Burtt ■
129819 Didissandra mihieri Franch. = Briggsia mihieri (Franch.) Craib ■
129820 Didissandra muscicola Diels = Briggsia muscicola (Diels) Craib ■
129821 Didissandra notochlaena H. Lév. et Vaniot = Ancylostemon notochlaenus (H. Lév. et Vaniot) Craib ■
129822 Didissandra patens Craib = Corallodiscus lanuginosus (Wall. ex R. Br.) B. L. Burtt ■
129823 Didissandra pinfaensis H. Lév. = Briggsia pinfaensis (H. Lév.) Craib ■
129824 Didissandra plicata Franch. = Corallodiscus lanuginosus (Wall. ex R. Br.) B. L. Burtt ■
129825 Didissandra primuliflora Batalin = Isometrum primuliflorum (Batalin) B. L. Burtt ■
129826 Didissandra rosthornii Diels = Briggsia rosthornii (Diels) B. L. Burtt ■
129827 Didissandra rufa King ex Hook. f. = Corallodiscus kingianus (Craib) B. L. Burtt ■
129828 Didissandra saxatilis Hemsl. = Ancylostemon saxatilis (Hemsl.) Craib ■
129829 Didissandra saxatilis Hemsl. var. microcalyx Hemsl. = Ancylostemon humilis W. T. Wang ■
129830 Didissandra sericea Craib = Corallodiscus flabellatus (Craib)

B. L. Burtt var. sericeus (Craib) K. Y. Pan ■
129831 Didissandra sericea Craib = Corallodiscus lanuginosus (Wall. ex R. Br.) B. L. Burtt ■
129832 Didissandra sesquifolia C. B. Clarke;大叶锣(白毛草,大一面绿,大一面锣);Sessile Didissandra ■
129833 Didissandra sinica (Chun) W. T. Wang;无毛漏斗苣苔(接骨草,绵阳岩白菜,岩白菜);China Didissandra, Chinese Didissandra ●■
129834 Didissandra sinoophiorrhizoides W. T. Wang = Anna ophiorrhizoides (Hemsl.) B. L. Burtt et R. A. Davidson ■
129835 Didissandra speciosa Hemsl. = Briggsia speciosa (Hemsl.) Craib ■
129836 Didissandra taliensis Craib = Corallodiscus lanuginosus (Wall. ex R. Br.) B. L. Burtt ■
129837 Didissandra taliensis Craib = Corallodiscus taliensis (Craib) B. L. Burtt ■
129838 Didissandra taliensis Craib f. robusta Craib. = Corallodiscus lanuginosus (Wall. ex R. Br.) B. L. Burtt ■
129839 Didonica Luteyn et Wilbur(1977);羊乳莓属●☆
129840 Didonica crassiflora Luteyn;厚叶羊乳莓●☆
129841 Didonica panamensis Luteyn et Wilbur;羊乳莓●☆
129842 Didonica pendula Luteyn et Wilbur;垂枝羊乳莓●☆
129843 Didothion Raf. = Epidendrum L.(保留属名)■☆
129844 Didymaea Hook. f. (1873);墨西哥茜属■☆
129845 Didymaea australis (Standl.) L. O. Williams;澳洲墨西哥茜■☆
129846 Didymaea crassifolia Borhidi;厚叶墨西哥茜■☆
129847 Didymaea floribunda Rzed.;繁花墨西哥茜■☆
129848 Didymaea linearis Standl.;线形墨西哥茜■☆
129849 Didymaea mexicana Hook. f.;墨西哥茜■☆
129850 Didymaea microphylla L. O. Williams;小叶墨西哥茜■☆
129851 Didymandra Willd. = Lacistema Sw. ●☆
129852 Didymandra Willd. = Synzyganthera Ruiz et Pav. ●☆
129853 Didymanthus Endl. (1839);双花澳藜属■☆
129854 Didymanthus Klotzsch ex Meisn. = Euplassa Salisb. ex Knight ●☆
129855 Didymanthus roei Endl.;双花澳藜●☆
129856 Didymaotus N. E. Br. (1925);灵石属■☆
129857 Didymaotus lapidiformis (Marloth) N. E. Br.;灵石■☆
129858 Didymelaceae Leandri(1937);双蕊花科(球花科,双颊果科)●☆
129859 Didymeles Thouars(1804);双蕊花属(球花属,双颊果属)●☆
129860 Didymeles madagascariensis Willd.;双蕊花●☆
129861 Didymeria Lindl. = Correa Andréws(保留属名)●☆
129862 Didymeria Lindl. = Didimeria Lindl. ●☆
129863 Didymia Phil. = Mariscus Vahl(保留属名)■
129864 Didymiandrum Gilly(1941);双蕊莎草属■☆
129865 Didymiandrum flexifolium Gilly;双蕊莎草■☆
129866 Didymocarpaceae D. Don = Gesneriaceae Rich. et Juss.(保留科名)■●
129867 Didymocarpus Raf. = Didymocarpus Wall.(保留属名)●■
129868 Didymocarpus Wall. (1819)(保留属名);长蒴苣苔属;Didymocarpus ●■
129869 Didymocarpus Wall. ex Buch. -Ham. = Didymocarpus Wall.(保留属名)●■
129870 Didymocarpus acutidentatus W. T. Wang = Chirita forrestii Anthony ■
129871 Didymocarpus adenocalyx W. T. Wang;腺萼长蒴苣苔草;Glandcalyx Didymocarpus ■
129872 Didymocarpus anachoretus (Hance) H. Lév. = Chirita anachoreta Hance ■
129873 Didymocarpus anthonyanus Hand. -Mazz. = Chirita pumila D. Don ■
129874 Didymocarpus aromaticus Wall. = Didymocarpus primulifolius D. Don ■
129875 Didymocarpus aromaticus Wall. ex D. Don;互叶长蒴苣苔草;Arternateleaf Didymocarpus ■
129876 Didymocarpus aureus (Franch.) Diels ex H. Lév. = Ancylostemon aureus (Franch.) B. L. Burtt ●
129877 Didymocarpus auricula S. Moore = Oreocharis auricula (S. Moore) C. B. Clarke ■
129878 Didymocarpus balansae Pellegr. = Chirita swinglei (Merr.) W. T. Wang ■
129879 Didymocarpus bequaertii De Wild. = Schizoboea kamerunensis (Engl.) B. L. Burtt ■☆
129880 Didymocarpus bicornutus (Hayata) S. Y. Hu = Hemiboea bicornuta (Hayata) Ohwi ■
129881 Didymocarpus bicornutus Hayata = Hemiboea bicornuta (Hayata) Ohwi ■
129882 Didymocarpus brevipes (C. B. Clarke) Hand. -Mazz. = Chirita speciosa Kurz ■
129883 Didymocarpus cavaleriei (H. Lév. et Vaniot) H. Lév. = Loxostigma cavaleriei (H. Lév. et Vaniot) B. L. Burtt ■
129884 Didymocarpus cavaleriei H. Lév. = Anna ophiorrhizoides (Hemsl.) B. L. Burtt et R. A. Davidson ■
129885 Didymocarpus clarkei H. Lév. = Didymostigma obtusum (C. B. Clarke) W. T. Wang ■
129886 Didymocarpus cortusifolius (Hance) H. Lév. = Didymocarpus cortusifolius (Hance) W. T. Wang ■
129887 Didymocarpus cortusifolius (Hance) W. T. Wang;温州长蒴苣苔(长蒴苣苔);Wenzhou Didymocarpus ■
129888 Didymocarpus cruciformis Chun = Chirita cruciformis (Chun) W. T. Wang ■
129889 Didymocarpus cyaneus Ridl.;蓝色长蒴苣苔草;Blue Didymocarpus ■☆
129890 Didymocarpus demissus Hance = Chirita demissa (Hance) W. T. Wang ■
129891 Didymocarpus depressus (Hook. f.) Chun = Chirita depressa Hook. f. ■
129892 Didymocarpus dilesii Borza = Chirita dielsii (Borza) B. L. Burtt ■
129893 Didymocarpus eburneus (Hance) H. Lév. = Chirita eburnea Hance ■
129894 Didymocarpus elegantissimus (H. Lév. et Vaniot) H. Lév. = Briggsia elegantissima (H. Lév. et Vaniot) Craib ■
129895 Didymocarpus esquirolii H. Lév. = Lysionotus serratus D. Don ●
129896 Didymocarpus fangii Chun ex W. T. Wang = Chirita fangii W. T. Wang ■
129897 Didymocarpus fauriei (Franch.) H. Lév. = Chirita eburnea Hance ■
129898 Didymocarpus fimbrisepalus (Hand. -Mazz.) Hand. -Mazz. = Chirita fimbrisepala Hand. -Mazz. ■
129899 Didymocarpus floribundus Chun ex W. T. Wang = Gyrocheilos chorisepalum W. T. Wang ■
129900 Didymocarpus fordii Hemsl. = Chirita fordii (Hemsl.) D. Wood ■
129901 Didymocarpus forrestii (Anthony) Hand. -Mazz. = Chirita forrestii Anthony ■

129902 Didymocarpus fritschii (H. Lév. et Vaniot) H. Lév. = Briggsia mihieri (Franch.) Craib ■

129903 Didymocarpus fritschii H. Lév. = Briggsia mihieri (Franch.) Craib ■

129904 Didymocarpus glandulosus (W. W. Sm.) W. T. Wang;腺毛长蒴苣苔草;Glandhair Didymocarpus ■

129905 Didymocarpus glandulosus (W. W. Sm.) W. T. Wang var. lasiantherus (W. W. Sm.) W. T. Wang;毛药长蒴苣苔草;Hairflower Glandhair Didymocarpus,Hairflower Wood Didymocarpus ■

129906 Didymocarpus glandulosus (W. W. Sm.) W. T. Wang var. minor (W. W. Sm.) W. T. Wang;短萼长蒴苣苔草;Shortcalyx Wood Didymocarpus,Small Glandhair Didymocarpus ■

129907 Didymocarpus grandidentatus (W. T. Wang) W. T. Wang;大齿长蒴苣苔草;Largetooth Didymocarpus ■

129908 Didymocarpus grandifolius (A. Dietr.) F. Dietr. = Chirita macrophylla Wall. ■

129909 Didymocarpus griffithii Wight = Loxostigma griffithii (Wight) C. B. Clarke ■

129910 Didymocarpus hancei Hemsl.;东南长蒴苣苔(石茶,石芥菜,石麻婆子草);Hance Didymocarpus,SE. China Didymocarpus ■

129911 Didymocarpus hedyotidens Chun = Chirita hedyotidea (Chun) W. T. Wang ■

129912 Didymocarpus hemsleyanus H. Lév. = Gyrocheilos chorisepalum W. T. Wang var. synsepalum W. T. Wang ■

129913 Didymocarpus heucherifolius Hand. -Mazz.;闽赣长蒴苣苔草;Heucherialeaf Didymocarpus ■

129914 Didymocarpus hwaianus S. Y. Hu = Hemiboea subcapitata C. B. Clarke ■

129915 Didymocarpus juliae (Hance) H. Lév. = Chirita juliae Hance ■

129916 Didymocarpus kamerunensis Engl. = Schizoboea kamerunensis (Engl.) B. L. Burtt ■☆

129917 Didymocarpus kurzii C. B. Clarke = Briggsia kurzii (C. B. Clarke) W. E. Evans ■

129918 Didymocarpus lanuginosus Wall. = Oreocharis maximowiczii C. B. Clarke ■

129919 Didymocarpus lanuginosus Wall. ex A. DC. = Corallodiscus lanuginosus (Wall. ex A. DC.) B. L. Burtt ■

129920 Didymocarpus lanuginosus Wall. ex R. Br. = Corallodiscus lanuginosus (Wall. ex R. Br.) B. L. Burtt ■

129921 Didymocarpus leiboensis Z. P. Soong et W. T. Wang;雷波长蒴苣苔草;Leibo Didymocarpus ■

129922 Didymocarpus mairei H. Lév. = Ancylostemon mairei (H. Lév.) Craib ■

129923 Didymocarpus mannii (C. B. Clarke) Wonisch = Trachystigma mannii C. B. Clarke ■☆

129924 Didymocarpus margaritae W. W. Sm.;短茎长蒴苣苔草;Shortstem Didymocarpus ■

129925 Didymocarpus martinii H. Lév. = Paraboea martinii (H. Lév. et Vaniot) B. L. Burtt ■

129926 Didymocarpus martinii H. Lév. et Vaniot = Paraboea martinii (H. Lév. et Vaniot) B. L. Burtt ■

129927 Didymocarpus medogensis W. T. Wang;墨脱长蒴苣苔草;Motuo Didymocarpus ■

129928 Didymocarpus mengtze W. W. Sm.;蒙自长蒴苣苔草;Mengzi Didymocarpus ■

129929 Didymocarpus mengtze W. W. Sm. = Didymocarpus pseudomengtze W. T. Wang ■

129930 Didymocarpus mengtze W. W. Sm. var. zhenkangensis (W. T. Wang) H. W. Li = Didymocarpus zhenkangensis W. T. Wang ■

129931 Didymocarpus microsiphon (Hance) H. Lév. = Didissandra macrosiphon (Hance) W. T. Wang ■

129932 Didymocarpus mihieri (H. Lév. et Vaniot) H. Lév. = Briggsia mihieri (Franch.) Craib ■

129933 Didymocarpus mihieri H. Lév. = Briggsia mihieri (Franch.) Craib ■

129934 Didymocarpus minutiserrulatus (Hayata) Yamam. = Chirita anachoreta Hance ■

129935 Didymocarpus minutus Hand. -Mazz. = Chirita speluncae (Hand. -Mazz.) D. Wood ■

129936 Didymocarpus mollifolius W. T. Wang;柔毛长蒴苣苔草;Softhair Didymocarpus ■

129937 Didymocarpus nanophyton C. Y. Wu ex H. W. Li;矮生长蒴苣苔草;Dwarf Didymocarpus ■

129938 Didymocarpus neurophyllus Collett et Hemsl. = Paraboea neurophylla (Collett et Hemsl.) B. L. Burtt ■

129939 Didymocarpus niveolanosus D. Fang et W. T. Wang;绵毛长蒴苣苔草;Whitecotton Didymocarpus ■

129940 Didymocarpus notochlaena (H. Lév. et Vaniot) H. Lév. = Ancylostemon notochlaenus (H. Lév. et Vaniot) Craib ■

129941 Didymocarpus oreocharis Hance = Oreocharis benthamii C. B. Clarke ■

129942 Didymocarpus pedicellatus R. Br.;花梗长蒴苣苔■☆

129943 Didymocarpus pinnatifidus Hand. -Mazz. = Chirita pinnatifida (Hand. -Mazz.) Burtt ■

129944 Didymocarpus polycephalus Chun = Chirita polycephala (Chun) W. T. Wang ■

129945 Didymocarpus praeteritus B. L. Burtt et R. A. Davidson;片马长蒴苣苔草;Pianma Didymocarpus ■

129946 Didymocarpus primulifolius D. Don;藏南长蒴苣苔草;S. Xizang Didymocarpus ■

129947 Didymocarpus primulinus W. T. Wang = Didymocarpus sinoprimulinus W. T. Wang ■

129948 Didymocarpus primuloides Maxim. = Oreocharis primuloides H. Lév. ■☆

129949 Didymocarpus pseudomengtze W. T. Wang;凤庆长蒴苣苔草;Fengqin Didymocarpus ■

129950 Didymocarpus pulcher C. B. Clarke;美丽长蒴苣苔草;Pretty Didymocarpus ■

129951 Didymocarpus purpureobracteatus W. W. Sm.;紫苞长蒴苣苔(维奇长蒴苣苔);Purplebract Didymocarpus,Veitch Didymocarpus ■

129952 Didymocarpus purpureobracteatus W. W. Sm. var. veitchianus (W. W. Sm.) H. W. Li = Didymocarpus purpureobracteatus W. W. Sm. ■

129953 Didymocarpus reniformis W. T. Wang;肾叶长蒴苣苔草;Kidneyleaf Didymocarpus ■

129954 Didymocarpus rotundifolius Hemsl. = Chirita rotundifolia (Hemsl.) D. Wood ■

129955 Didymocarpus salviiflorus Chun;迭裂长蒴苣苔草;Sageflower Didymocarpus ■

129956 Didymocarpus saxatilis (Hemsl.) H. Lév. = Ancylostemon saxatilis (Hemsl.) Craib ■

129957 Didymocarpus secundiflorus Chun = Chirita secundiflora (Chun) W. T. Wang ■

129958 Didymocarpus seguini H. Lév. et Vaniot = Paraboea rufescens

(Franch.) B. L. Burtt ■
129959 Didymocarpus sericeus H. Lév. = Oreocharis auricula (S. Moore) C. B. Clarke ■
129960 Didymocarpus sesquifolius (C. B. Clarke) H. Lév. = Didissandra sesquifolia C. B. Clarke ■
129961 Didymocarpus silvarum W. W. Sm.;林生长蒴苣苔草;Forest Didymocarpus, Wood Didymocarpus ■
129962 Didymocarpus silvarum W. W. Sm. var. glandulosus W. W. Sm. = Didymocarpus glandulosus (W. W. Sm.) W. T. Wang ■
129963 Didymocarpus silvarum W. W. Sm. var. lasiantherus W. T. Wang = Didymocarpus glandulosus (W. W. Sm.) W. T. Wang var. lasiantherus (W. W. Sm.) W. T. Wang ■
129964 Didymocarpus silvarum W. W. Sm. var. minor W. T. Wang = Didymocarpus glandulosus (W. W. Sm.) W. T. Wang var. minor (W. W. Sm.) W. T. Wang ■
129965 Didymocarpus sinensis (Lindl.) H. Lév. = Chirita sinensis Lindl. ■
129966 Didymocarpus sinohenryi Chun = Opithandra sinohenryi (Chun) B. L. Burtt ■
129967 Didymocarpus sinoprimulinus W. T. Wang;报春长蒴苣苔草;Primrose Didymocarpus ■
129968 Didymocarpus speciosus (Hemsl.) H. Lév. = Briggsia speciosa (Hemsl.) Craib ■
129969 Didymocarpus speciosus (Kurz) Hand. -Mazz. = Chirita speciosa Kurz ■
129970 Didymocarpus speluncae Hand. -Mazz. = Chirita speluncae (Hand. -Mazz.) D. Wood ■
129971 Didymocarpus stenanthos C. B. Clarke;狭冠长蒴苣苔草;Narrowflower Didymocarpus ■
129972 Didymocarpus stenanthos C. B. Clarke var. pilosellus W. T. Wang;疏毛长蒴苣苔草;Laxhair Narrowflower Didymocarpus ■
129973 Didymocarpus stenocarpus W. T. Wang;细果长蒴苣苔草;Slenderfruit Didymocarpus ■
129974 Didymocarpus stolzii Engl. = Schizoboea kamerunensis (Engl.) B. L. Burtt ■☆
129975 Didymocarpus stolzii Engl. var. minor Mansf. = Schizoboea kamerunensis (Engl.) B. L. Burtt ■☆
129976 Didymocarpus subalternans Wall. = Didymocarpus pulcher C. B. Clarke ■
129977 Didymocarpus subalternans Wall. ex R. Br. = Didymocarpus aromaticus Wall. ex D. Don ■
129978 Didymocarpus subpalmatinervis W. T. Wang;掌脉长蒴苣苔■
129979 Didymocarpus swinglei Merr. = Chirita swinglei (Merr.) W. T. Wang ■
129980 Didymocarpus tengii Chun ex W. T. Wang = Chirita vestita D. Wood ■
129981 Didymocarpus tibeticus (Franch.) Hand. -Mazz. = Chirita tibetica (Franch.) B. L. Burtt ■
129982 Didymocarpus tonkinensis (Kraenzl.) Hand. -Mazz. = Boeica porosa C. B. Clarke ●
129983 Didymocarpus traillianus (Forrest et W. W. Sm.) Hand. -Mazz. = Chirita speciosa Kurz ■
129984 Didymocarpus uniflorus (Franch.) Borza = Chirita dielsii (Borza) B. L. Burtt ■
129985 Didymocarpus urticifolius (Buch. -Ham. ex D. Don) Wonisch = Chirita urticifolia Buch. -Ham. ex D. Don ■
129986 Didymocarpus urticifolius (D. Don) Wonisch = Chirita urticifolia Buch. -Ham. ex D. Don ■
129987 Didymocarpus veitchianus W. W. Sm. = Didymocarpus purpureobracteatus W. W. Sm. ■
129988 Didymocarpus verecundus Chun = Chirita verecunda (Chun) W. T. Wang ■
129989 Didymocarpus villosus D. Don;长毛长蒴苣苔草;Longhair Didymocarpus ■
129990 Didymocarpus yuenlingensis W. T. Wang;沅陵长蒴苣苔草;Yuanling Didymocarpus ■
129991 Didymocarpus yunnanensis (Franch.) C. E. C. Fisch. = Didymocarpus yunnanensis (Franch.) W. W. Sm. ■
129992 Didymocarpus yunnanensis (Franch.) W. W. Sm.;云南长蒴苣苔(象耳朵叶,新香草);Yunnan Didymocarpus ■
129993 Didymocarpus zhenkangensis W. T. Wang;镇康长蒴苣苔草;Zhenkang Didymocarpus ■
129994 Didymocarpus zhufengensis W. T. Wang;珠峰长蒴苣苔草;Jolmolugma Didymocarpus ■
129995 Didymochaeta Steud. = Agrostis L.(保留属名)■
129996 Didymochaeta Steud. = Deyeuxia Clarion ■
129997 Didymocheton Blume = Dysoxylum Blume ●
129998 Didymochiton Spreng. = Didymocheton Blume ●
129999 Didymochlamys Hook. f. (1872);双被茜属■☆
130000 Didymochlamys connellii N. E. Br.;双被茜■☆
130001 Didymocistus Kuhlm. (1938);双果大戟属■☆
130002 Didymocistus chrysadenius Kuhlm.;双果大戟■☆
130003 Didymococcus Blume = Sapindus L.(保留属名)●
130004 Didymocolpus S. C. Chen = Acanthochlamys P. C. Kao ■★
130005 Didymocolpus nanus S. C. Chen = Acanthochlamys bracteata P. C. Kao ■
130006 Didymodoxa E. Mey. ex Wedd. (1856-1857);非洲荨麻属■☆
130007 Didymodoxa E. Mey. ex Wedd. = Australina Gaudich. ■☆
130008 Didymodoxa Wedd. = Didymodoxa E. Mey. ex Wedd. ■☆
130009 Didymodoxa acuminata (Wedd.) Wedd. = Didymodoxa caffra (Thunb.) Friis et Wilmot-Dear ■☆
130010 Didymodoxa caffra (Thunb.) Friis et Wilmot-Dear;非洲荨麻■☆
130011 Didymodoxa capensis (L. f.) Friis et Wilmot-Dear;好望角非洲荨麻■☆
130012 Didymodoxa capensis (L. f.) Friis et Wilmot-Dear var. integrifolia (Wedd.) Friis et Wilmot-Dear;全叶好望角单蕊麻■☆
130013 Didymodoxa cuneata Wedd. = Didymodoxa caffra (Thunb.) Friis et Wilmot-Dear ■☆
130014 Didymodoxa debilis Wedd. = Didymodoxa capensis (L. f.) Friis et Wilmot-Dear var. integrifolia (Wedd.) Friis et Wilmot-Dear ■☆
130015 Didymodoxa integrifolia (Wedd.) Wedd. = Didymodoxa capensis (L. f.) Friis et Wilmot-Dear var. integrifolia (Wedd.) Friis et Wilmot-Dear ■☆
130016 Didymoecium Bremek. (1934);双屋茜属●☆
130017 Didymoecium Bremek. = Rennellia Korth. ●☆
130018 Didymoecium amoenum Bremek.;肖伦内尔茜●☆
130019 Didymogyne Wedd. = Droguetia Gaudich. ■
130020 Didymogyne abyssinica Wedd. = Droguetia iners (Forssk.) Schweinf. ■☆
130021 Didymomeles Spreng. = Didymeles Thouars ●☆
130022 Didymonema C. Presl = Gahnia J. R. Forst. et G. Forst. ■
130023 Didymopanax Decne. et Planch. (1854);对参属(双参属)●☆
130024 Didymopanax Decne. et Planch. = Schefflera J. R. Forst. et G. Forst.(保留属名)●

130025 Didymopanax calvus (Cham.) Decne. et Planch.;裸露对参●☆
130026 Didymopanax longipetiolatus Marchal;长花序柄对参●☆
130027 Didymopanax navarroi A. Samp.;纳氏对参●☆
130028 Didymopelta Regel et Schmalh. = Astragalus L. ●■
130029 Didymophysa Boiss. (1841);双球芥属■☆
130030 Didymophysa aucheri Boiss.;双球芥■☆
130031 Didymophysa fedtschenkoana Regel;中亚双球芥■☆
130032 Didymoplexiella Garay (1954) ('Didimoplexiella');锚柱兰属;Anchorstyleorchis ■
130033 Didymoplexiella hainanensis X. H. Jin et S. C. Chen = Didymoplexiopsis khiriwongensis Seidenf. ■
130034 Didymoplexiella siamensis (Rolfe ex Downie) Seidenf.;锚柱兰;Siam Anchorstyleorchis ■
130035 Didymoplexiopsis Seidenf. (1997);拟锚柱兰属■
130036 Didymoplexiopsis khiriwongensis Seidenf.;拟锚柱兰■
130037 Didymoplexis Griff. (1843);双唇兰属(鬼兰属);Didymoplexis, Diliporchis ■
130038 Didymoplexis africana Summerh.;非洲双唇兰■☆
130039 Didymoplexis brevipes Ohwi = Chamaegastrodia vaginata (Hook. f.) Seidenf. ■
130040 Didymoplexis brevipes Ohwi = Didymoplexis minor J. J. Sm. ■☆
130041 Didymoplexis micradenia (Rchb. f.) Hemsl.;小双唇兰(日本双唇兰)■
130042 Didymoplexis minor J. J. Sm. = Didymoplexis micradenia (Rchb. f.) Hemsl. ■
130043 Didymoplexis pallens Griff.;双唇兰(吊钟鬼兰,鬼兰);Pale Didymoplexis, Pale Diliporchis ■
130044 Didymoplexis subcampanulata Hayata = Chamaegastrodia vaginata (Hook. f.) Seidenf. ■
130045 Didymoplexis subcampanulata Hayata = Didymoplexis pallens Griff. ■
130046 Didymoplexis sylvatica (Blume) Garay = Chamaegastrodia vaginata (Hook. f.) Seidenf. ■
130047 Didymoplexis verrucosa J. L. Stewart et Hennessy;多疣双唇兰■☆
130048 Didymopogon Bremek. (1940);双毛茜属●☆
130049 Didymopogon sumatranum (Ridl.) Bremek.;双毛茜●☆
130050 Didymosalpinx Keay(1958);双角茜属●☆
130051 Didymosalpinx konguensis (Hiern) Keay;孔古双角茜●☆
130052 Didymosalpinx lanciloba (S. Moore) Keay;披针裂双角茜●☆
130053 Didymosalpinx norae (Swynn.) Keay;诺拉双角茜●☆
130054 Didymosalpinx parviflora Keay = Petitiocodon parviflorum (Keay) Robbr. ●☆
130055 Didymosperma H. Wendl. et Drude = Arenga Labill. (保留属名)●
130056 Didymosperma H. Wendl. et Drude = Didymosperma H. Wendl. et Drude ex Benth. et Hook. f. ●
130057 Didymosperma H. Wendl. et Drude ex Benth. et Hook. f. (1883);双籽棕属(阿萨密椰子属,二种子椰子属,双籽藤属);Assam Palm, Dryas Palm, Twoseed Palm ●
130058 Didymosperma H. Wendl. et Drude ex Benth. et Hook. f. = Arenga Labill. (保留属名)●
130059 Didymosperma H. Wendl. et Drude ex Hook. f. = Arenga Labill. (保留属名)●
130060 Didymosperma H. Wendl. et Drude ex Hook. f. = Didymosperma H. Wendl. et Drude ex Benth. et Hook. f. ●
130061 Didymosperma caudatum (Lour.) H. Wendl. et Drude = Arenga caudata (Lour.) H. E. Moore ●
130062 Didymosperma caudatum (Lour.) H. Wendl. et Drude var. tonkinense Becc. = Didymosperma caudatum (Lour.) H. Wendl. et Drude ●
130063 Didymosperma caudatum (Lour.) H. Wendl. et Drude var. tonkinense Becc. = Arenga caudata (Lour.) H. E. Moore ●
130064 Didymosperma engleri (Becc.) Warb. = Arenga engleri Becc. ●
130065 Didymosperma nanum (Griff.) H. Wendl. et Drude;矮双籽棕(矮双籽藤);Dwarf Two-seeded Palm ●
130066 Didymosperma porphyrocarpa H. Wendl. et Drude = Didymosperma porphyrocarpum (Blume ex Mart.) H. Wendl. et Drude ex Hook. f. ●☆
130067 Didymosperma porphyrocarpum (Blume ex Mart.) H. Wendl. et Drude ex Hook. f.;紫果双籽藤●☆
130068 Didymosperma tonkinensis (Becc.) Gagnep. = Arenga caudata (Lour.) H. E. Moore ●
130069 Didymostigma W. T. Wang(1984);双片苣苔属;Didymostigma ■★
130070 Didymostigma leiophyllum D. Fang et Xiao H. Lu;光叶双片苣苔草;Smoothleaf Didymostigma ■
130071 Didymostigma obtusum (C. B. Clarke) W. T. Wang;双片苣苔(唇柱苣苔);Common Chirita, Common Didymostigma ■
130072 Didymotheca Hook. f. = Gyrostemon Desf. ●☆
130073 Didymotoca E. Mey. = Australina Gaudich. ■☆
130074 Didyplosandra Bremek. = Strobilanthes Blume ●■
130075 Didyplosandra Wight ex Bremek. = Strobilanthes Blume ●■
130076 Didyplosandra Wight. = Strobilanthes Blume ●■
130077 Diectomis Kunth = Andropogon L. (保留属名)■
130078 Diectomis P. Beauv. (废弃属名) = Anadelphia Hack. ■☆
130079 Diectomis P. Beauv. (废弃属名) = Diectomis Kunth ■
130080 Diectomis fastigiata (Sw.) Kunth = Andropogon fastigiatus Sw. ■☆
130081 Diectonis Willis = Diectomis Kunth ■
130082 Diedropetala Galushko = Delphinium L. ■
130083 Dieffenbachia Schott (1829);花叶万年青属(黛粉叶属);Dieffenbachia, Dumb Cane, Leopard Lily, Tuftroot ●■
130084 Dieffenbachia 'Camille';白玉黛粉叶■☆
130085 Dieffenbachia 'Compacta';绿玉黛粉叶■☆
130086 Dieffenbachia 'Exotica';美斑黛粉叶■☆
130087 Dieffenbachia 'Marianne';玛丽安万年青■☆
130088 Dieffenbachia 'Maroba';马王万年青■☆
130089 Dieffenbachia 'Mars';马斯万年青■☆
130090 Dieffenbachia 'Pale Yellow';灰白叶黛粉叶■☆
130091 Dieffenbachia 'Wilson's Delight';乳肋万年青■☆
130092 Dieffenbachia × bausei Regel = Dieffenbachia bausei Engl. ■☆
130093 Dieffenbachia amoena Gentil;大王黛粉叶(大王万年青);Dumb Plant, Giant Dumbcane, Mother-in-law Plant, Tuftroot ■☆
130094 Dieffenbachia amoena Gentil 'Tropic Snow';夏雪万年青■☆
130095 Dieffenbachia baraquiniana Verschaff. et Lem. = Dieffenbachia picta (Lodd.) Schott var. barraquiniana (Verschaff. et Lem.) Engl. ■☆
130096 Dieffenbachia bausei Engl.;星点花叶万年青(星点黛粉叶);Bause Dieffenbachia, Bause Tuftroot, Dumbcane ■☆
130097 Dieffenbachia bowmannii Carrière;白斑万年青(鲍氏花叶万年青);Bowmann Tuftroot ●■
130098 Dieffenbachia brasiliensis H. J. Veitch = Dieffenbachia picta (Lodd.) Schott ●■
130099 Dieffenbachia chelsonii Gentil;银道黛粉叶■☆

130100 Dieffenbachia costata Klotzsch = Dieffenbachia costata Klotzsch ex Schott ■☆
130101 Dieffenbachia costata Klotzsch ex Schott;有棱黛粉叶(中脉花叶万年青)■☆
130102 Dieffenbachia humilis Poepp. et Endl.;矮花叶万年青;Dwarf Tuftroot ●☆
130103 Dieffenbachia imperialis Linden et André;壮丽花叶万年青;Imperial Tuftroot ●☆
130104 Dieffenbachia jenmannii Veitch = Dieffenbachia picta (Lodd.) Schott subvar. jenmannii (Veitch ex Regel) Engl. ■☆
130105 Dieffenbachia leoncae Hort.;花斑万年青(花斑黛粉叶)■☆
130106 Dieffenbachia leopoldii Bull. = Dieffenbachia leopoldii Bull. ex Engl. ●■
130107 Dieffenbachia leopoldii Bull. ex Engl.;白肋万年青;Leopold's Tuftroot ●■
130108 Dieffenbachia leopoldii Bull. ex Engl. = Dieffenbachia seguina (L.) Schott var. liturata Engl. ■
130109 Dieffenbachia liturata Schott = Dieffenbachia seguina (L.) Schott var. liturata Engl. ■
130110 Dieffenbachia longispatha Engl. et Krause;中美黛粉叶■☆
130111 Dieffenbachia macrophylla Poepp. et Endl.;大叶花叶万年青(大叶黛粉叶);Largeleaved Tuftroot ●☆
130112 Dieffenbachia maculata (Lodd.) G. Don = Dieffenbachia seguina (Jacq.) Schott ●■
130113 Dieffenbachia maculata (Lodd.) Sweet;黛粉叶(花叶万年青);Spotted Dumbcane ■☆
130114 Dieffenbachia maculata (Lodd.) Sweet 'Rudolph Roehrs';乳斑万年青■☆
130115 Dieffenbachia maculata (Lodd.) Sweet var. angustior (Engl.) Bunting;狭叶黛粉叶■☆
130116 Dieffenbachia maculata (Lodd.) Sweet var. jenmannii (Verschaff.) Bunting = Dieffenbachia picta (Lodd.) Schott subvar. jenmannii (Veitch ex Regel) Engl. ■☆
130117 Dieffenbachia magnifica Linden et Rodigas;粗壮花叶万年青■☆
130118 Dieffenbachia memoria-corsii Fenzl;撒银黛粉叶■☆
130119 Dieffenbachia oerstedtii Schott;里绿花叶万年青■☆
130120 Dieffenbachia parlatorei Linden et André var. marmorea Linden et André;大理石黛粉叶■☆
130121 Dieffenbachia picta (Lodd.) Schott;花叶万年青(白叶万年青);Spotted Tuftroot, Variable Tuftroot, Variegated Tuftroot ●■
130122 Dieffenbachia picta (Lodd.) Schott 'Golden Snow';金雪黛粉叶■
130123 Dieffenbachia picta (Lodd.) Schott 'Lucky Small';小好运黛粉叶■
130124 Dieffenbachia picta (Lodd.) Schott 'Memoria';花斑黛粉叶■
130125 Dieffenbachia picta (Lodd.) Schott 'Rudolon Roehrs';乳斑黛粉叶■
130126 Dieffenbachia picta (Lodd.) Schott 'Superba';白雪黛粉叶■
130127 Dieffenbachia picta (Lodd.) Schott = Dieffenbachia seguina (Jacq.) Schott ●■
130128 Dieffenbachia picta (Lodd.) Schott subvar. jenmannii (Veitch ex Regel) Engl.;白羽黛粉叶(白粉叶万年青)■☆
130129 Dieffenbachia picta (Lodd.) Schott subvar. magnifica (L. Linden et Rodigas) Engl. = Dieffenbachia magnifica Linden et Rodigas ■☆
130130 Dieffenbachia picta (Lodd.) Schott var. barraquiniana (Verschaff. et Lem.) Engl.;白梗花叶万年青■☆
130131 Dieffenbachia picta (Lodd.) Schott var. barraquiniana Engl. = Dieffenbachia picta (Lodd.) Schott var. barraquiniana (Verschaff. et Lem.) Engl. ●■
130132 Dieffenbachia picta Schott = Dieffenbachia picta (Lodd.) Schott ●■
130133 Dieffenbachia pittieri Engl. et K. Krause;麻点黛粉叶■
130134 Dieffenbachia seguina (Jacq.) Schott = Dieffenbachia seguina (L.) Schott ●■
130135 Dieffenbachia seguina (L.) Schott;彩叶万年青(白斑黛粉叶,花茎黛粉叶,花茎万年青,哑蕉);Dumb Cane, Leopard Lily, Mother-in-law Plant ●■
130136 Dieffenbachia seguina (L.) Schott var. barraquiniana (Verschaff. et Lem.) Engl.;白脉黛粉叶(白脉哑蕉)■☆
130137 Dieffenbachia seguina (L.) Schott var. decora Engl. et Prantl;白柄黛粉叶(白柄哑蕉)■☆
130138 Dieffenbachia seguina (L.) Schott var. liturata Engl.;长苞黛粉叶(长苞哑蕉)■
130139 Dieffenbachia seguina (L.) Schott var. nobilis Engl.;白斑黛粉叶(白斑哑蕉)■☆
130140 Dieffenbachia sublime ?;天堂万年青■☆
130141 Dieffenbachia vesuvins ?;卫士万年青■☆
130142 Diegodendraceae Capuron = Bixaceae Kunth(保留科名)●■
130143 Diegodendraceae Capuron = Diervillaceae Pyck ●
130144 Diegodendraceae Capuron = Ochnaceae DC.(保留科名)●■
130145 Diegodendraceae Capuron(1964);基柱木科(岛樟科,地果莲木科)●☆
130146 Diegodendron Capuron(1964);基柱木属(岛樟属)●☆
130147 Diegodendron humbertii Capuron;基柱木●☆
130148 Dielitzia P. S. Short(1989);层苞鼠麹草属■☆
130149 Dielitzia tysonii P. S. Short;层苞鼠麹草■☆
130150 Dielsantha E. Wimm.(1948);迪尔斯花属■☆
130151 Dielsantha galeopsoides (Engl. et Diels) E. Wimm.;迪尔斯花■☆
130152 Dielsia Gilg ex Diels et E. Pritz. = Dielsia Gilg ■☆
130153 Dielsia Gilg(1904);迪尔斯草属■☆
130154 Dielsia Kudo = Isodon (Schrad. ex Benth.) Spach ●■
130155 Dielsia Kudo = Plectranthus L'Hér.(保留属名)●■
130156 Dielsia Kudo = Skapanthus C. Y. Wu et H. W. Li ■★
130157 Dielsia cygnorum Gilg;迪尔斯草■☆
130158 Dielsia oreophila (Diels) Kudo = Skapanthus oreophilus (Diels) C. Y. Wu et H. W. Li ■
130159 Dielsina Kuntze = Polyceratocarpus Engl. et Diels ●☆
130160 Dielsiocharis O. E. Schulz(1924);中亚庭芥属■☆
130161 Dielsiocharis kotschyi O. E. Schulz;中亚庭芥■☆
130162 Dielsiochloa Pilg.(1943);繁花迪氏草属(繁花代尔草属)■☆
130163 Dielsiochloa floribunda (Pilg.) Pilg.;繁花迪氏草■☆
130164 Dielsiothamnus R. E. Fr.(1955);迪氏木属●☆
130165 Dielsiothamnus divaricatus (Diels) R. E. Fr.;迪氏木●☆
130166 Dielytra Cham. et Schltdl. = Dicentra Bernh.(保留属名)■
130167 Dielytra chrysantha Hook. et Arn. = Dicentra chrysantha (Hook. et Arn.) Walp. ■☆
130168 Dielytra scandens (D. Don) G. Don = Dicentra scandens (D. Don) Walp. ■
130169 Dielytra scandens D. Don = Dicentra scandens (D. Don) Walp. ■
130170 Dielytra spectabilis (L.) DC. = Dicentra spectabilis (L.) Lem. ■
130171 Dielytra spectabilis (L.) G. Don = Lamprocapnos spectabilis

(L.) Fukuhara ■

130172 Dielytra thalictrifolia (Wall.) G. Don = Dicentra scandens (D. Don) Walp. ■

130173 Diemenia Korth. = Licania Aubl. ●☆

130174 Diemenia Korth. = Parastemon A. DC. ●☆

130175 Diemisa Raf. = Carex L. ■

130176 Dieneckeria Vell. = Euplassa Salisb. ex Knight ●☆

130177 Dienia Lindl. (1824);无耳沼兰属■

130178 Dienia Lindl. = Malaxis Sol. ex Sw. ■

130179 Dienia congesta Lindl. = Dienia ophrydis (J. König) Ormerod et Seidenf. ■

130180 Dienia congesta Lindl. = Malaxis latifolia Sm. ■

130181 Dienia cylindrostachya Lindl.;筒穗无耳沼兰(圆柱花序沼兰);Cylindric-inflorescence Addermonth Orchid ■

130182 Dienia latifolia (Sm.) M. A. Clem. et D. L. Jones = Dienia ophrydis (J. König) Ormerod et Seidenf. ■

130183 Dienia montana (Blume) M. A. Clem. et D. L. Jones = Dienia ophrydis (J. König) Ormerod et Seidenf. ■

130184 Dienia mucifera Lindl. ex Wall. = Malaxis muscifera (Lindl.) Kuntze ■☆

130185 Dienia ophrydis (J. König) Ormerod et Seidenf.;无耳沼兰(广叶软叶兰,阔叶兰,阔叶沼兰);Broadleaf Addermonth Orchid, Broadleaf Bogorchis ■

130186 Dierama K. Koch = Dierama K. Koch et C. D. Bouché ■☆

130187 Dierama K. Koch et C. D. Bouché(1855);漏斗花属(天使钓竿属,纤枝花属);Angel's Fishing Rod, Angel's Fishingrod, Angel's Fishing-rod, Elfin Wands, Funnel Flower, Wand Flower, Wandflower ■☆

130188 Dierama ambiguum Hilliard;含糊漏斗花■☆

130189 Dierama argyreum L. Bolus;银色漏斗花■☆

130190 Dierama argyreum L. Bolus var. majus N. E. Br. = Dierama argyreum L. Bolus ■☆

130191 Dierama atrum N. E. Br.;黑色漏斗花■☆

130192 Dierama cooperi N. E. Br.;库珀漏斗花■☆

130193 Dierama cupuliflorum Klatt;杯花漏斗花■☆

130194 Dierama cupuliflorum Klatt = Dierama vagum N. E. Br. ■☆

130195 Dierama cupuliflorum Klatt subsp. caudatum Marais;尾状漏斗花■☆

130196 Dierama davyi N. E. Br. = Dierama insigne N. E. Br. ■☆

130197 Dierama densiflorum Marais;密花漏斗花■☆

130198 Dierama dracomontanum Hilliard;龙山漏斗花■☆

130199 Dierama dubium N. E. Br.;可疑漏斗花■☆

130200 Dierama elatum N. E. Br.;高漏斗花■☆

130201 Dierama ensifolium K. Koch et C. D. Bouché = Dierama pendulum (Thunb.) Baker ■☆

130202 Dierama erectum Hilliard;直立漏斗花■☆

130203 Dierama formosum Hilliard;美丽漏斗花■☆

130204 Dierama galpinii N. E. Br.;盖尔漏斗花■☆

130205 Dierama gracile N. E. Br.;纤细漏斗花■☆

130206 Dierama grandiflorum G. J. Lewis;大花漏斗花■☆

130207 Dierama igneum Klatt;火红漏斗花■☆

130208 Dierama insigne N. E. Br.;显著漏斗花■☆

130209 Dierama inyangense Hilliard;伊尼扬加漏斗花■☆

130210 Dierama jucundum Hilliard;愉悦漏斗花■☆

130211 Dierama latifolium N. E. Br.;宽花漏斗花■☆

130212 Dierama longiflorum G. J. Lewis = Dierama pulcherrimum (Hook. f.) Baker ■☆

130213 Dierama longistylum Marais;长柱漏斗花■☆

130214 Dierama luteoalbidum I. Verd.;黄白漏斗花■☆

130215 Dierama medium N. E. Br.;中间漏斗花■☆

130216 Dierama mobile Hilliard;移动漏斗花■☆

130217 Dierama mossii (N. E. Br.) Hilliard;莫西漏斗花■☆

130218 Dierama nebrownii Hilliard;涅布朗漏斗花■☆

130219 Dierama pallidum Hilliard;苍白漏斗花■☆

130220 Dierama palustre N. E. Br. = Dierama latifolium N. E. Br. ■☆

130221 Dierama pansum N. E. Br. = Dierama igneum Klatt ■☆

130222 Dierama parviflorum Marais;小花漏斗花;Smallflower Funnel Flower ■☆

130223 Dierama pauciflorum N. E. Br.;少花漏斗花;Fewflower Funnel Flower ■☆

130224 Dierama pendulum (L. f.) Baker = Dierama pendulum (Thunb.) Baker ■☆

130225 Dierama pendulum (Thunb.) Baker;悬垂漏斗花(垂花鸢尾);Angel's Fishing-rod ■☆

130226 Dierama pendulum Baker = Dierama pendulum (Thunb.) Baker ■☆

130227 Dierama pictum N. E. Br.;着色漏斗花■☆

130228 Dierama plowesii Hilliard;普洛漏斗花■☆

130229 Dierama pulcherrimum (Hook. f.) Baker;红漏斗花(美丽魔杖花,仙钓竿);Angel's Fishing Rod, Angel's Fishingrod, Angel's Fishing Rods, Angel's Tears, Red Funnel Flower, Wand Flower, Wandflower, Wand-flower ■☆

130230 Dierama pulcherrimum (Hook. f.) Baker = Sparaxis pulcherrima Hook. f. ■☆

130231 Dierama pulcherrimum Baker = Dierama pulcherrimum (Hook. f.) Baker ■☆

130232 Dierama pumilum N. E. Br.;矮漏斗花■☆

130233 Dierama reynoldsii I. Verd.;雷诺兹漏斗花■☆

130234 Dierama robustum N. E. Br.;粗壮漏斗花■☆

130235 Dierama rupestre N. E. Br. = Dierama insigne N. E. Br. ■☆

130236 Dierama trichorhizum (Baker) N. E. Br.;毛根漏斗花■☆

130237 Dierama tyrium Hilliard;紫色漏斗花■☆

130238 Dierama tysonii N. E. Br.;泰森漏斗花■☆

130239 Dierama vagum N. E. Br. = Dierama cupuliflorum Klatt ■☆

130240 Dierbachia Spreng. = Dunalia Kunth ●☆

130241 Diervilla L. = Diervilla Mill. ●☆

130242 Diervilla Mill. (1754);黄锦带属;Bush Honeysuckle, Bush-honeysuckle, Weigela ●☆

130243 Diervilla Tourn. ex L. = Diervilla Mill. ●☆

130244 Diervilla diervilla (L.) MacMill. = Diervilla lonicera Mill. ●☆

130245 Diervilla floribunda Siebold et Zucc. = Weigela floribunda (Siebold et Zucc.) K. Koch ●

130246 Diervilla florida (Bunge) Siebold et Zucc. = Weigela florida (Bunge) A. DC. ●

130247 Diervilla grandiflora Siebold et Zucc. = Weigela coraeensis Thunb. ●

130248 Diervilla hortensis Siebold et Zucc. var. alba Siebold et Zucc. = Weigela hortensis (Siebold et Zucc.) K. Koch f. albiflora (Siebold et Zucc.) Rehder ●☆

130249 Diervilla hortensis Siebold et Zucc. var. rubra Siebold et Zucc. = Weigela hortensis (Siebold et Zucc.) K. Koch ●☆

130250 Diervilla japonica (Thunb.) DC var. sinica Rehder = Weigela japonica Thunb. var. sinica (Rehder) Bailey ●

130251 Diervilla japonica (Thunb.) DC. = Weigela japonica Thunb. ●

130252 Diervilla lonicera Mill. ; 黄锦带(忍冬黄锦带); Bush Honeysuckle, Dwarf Bush-honeysuckle, Northern Bush-honeysuckle ●☆

130253 Diervilla lonicera Mill. var. hypomalaca Fernald = Diervilla lonicera Mill. ●☆

130254 Diervilla sessilifolia Buckley;南方黄锦带(无柄黄锦带);Bush Honeysuckle, Southern Bush Honeysuckle, Southern Bush-honeysuckle ●☆

130255 Diervilla suavis Kom. ;芳香黄锦带●☆

130256 Diervilla trifida Moench = Diervilla lonicera Mill. ●☆

130257 Diervillaceae Pyck = Caprifoliaceae Juss. (保留科名)●■

130258 Diervillaceae Pyck;黄锦带科(夷忍冬科)●

130259 Diervillea Bartl. = Diervilla Mill. ●☆

130260 Diesingia Endl. = Psophocarpus Neck. ex DC. (保留属名)■

130261 Diesingia scandens Endl. = Psophocarpus scandens (Endl.) Verdc. ■☆

130262 Dietegocarpus Willis = Carpinus L. ●

130263 Dietegocarpus Willis = Distegocarpus Siebold et Zucc. ●

130264 Dieteria Nutt. (1840);灰菀属■☆

130265 Dieteria Nutt. = Aster L. ●■

130266 Dieteria asteroides Torr. ;美洲灰菀(腺叶灰菀)■☆

130267 Dieteria asteroides Torr. var. glandulosa (B. L. Turner) D. R. Morgan et R. L. Hartm. ;腺点灰菀■☆

130268 Dieteria asteroides Torr. var. lagunensis (D. D. Keck) D. R. Morgan et R. L. Hartm. ;拉古纳灰菀■☆

130269 Dieteria bigelovii (A. Gray) D. R. Morgan et R. L. Hartm. ;毕氏灰菀(宽叶灰菀)■☆

130270 Dieteria bigelovii (A. Gray) D. R. Morgan et R. L. Hartm. var. commixta (Greene) D. R. Morgan et R. L. Hartm. ;混乱灰菀■☆

130271 Dieteria bigelovii (A. Gray) D. R. Morgan et R. L. Hartm. var. mucronata (Greene) D. R. Morgan et R. L. Hartm. ;短尖毕氏灰菀■☆

130272 Dieteria canescens (Pursh) Nutt. ;灰菀;Hoary-aster ■☆

130273 Dieteria canescens (Pursh) Nutt. var. ambigua (B. L. Turner) D. R. Morgan et R. L. Hartm. ;可疑灰菀■☆

130274 Dieteria canescens (Pursh) Nutt. var. aristata (Eastw.) D. R. Morgan et R. L. Hartm. ;具芒灰菀■☆

130275 Dieteria canescens (Pursh) Nutt. var. glabra (A. Gray) D. R. Morgan et R. L. Hartm. ;光滑灰菀■☆

130276 Dieteria canescens (Pursh) Nutt. var. incana (Lindl.) D. R. Morgan et R. L. Hartm. ;灰毛灰菀■☆

130277 Dieteria canescens (Pursh) Nutt. var. leucanthemifolia (Greene) D. R. Morgan et R. L. Hartm. ;滨菊叶灰菀■☆

130278 Dieteria canescens (Pursh) Nutt. var. nebraskana (B. L. Turner) D. R. Morgan et R. L. Hartm. ;内布灰菀■☆

130279 Dieteria canescens (Pursh) Nutt. var. sessiliflora (Nutt.) D. R. Morgan et R. L. Hartm. ;短梗花灰菀■☆

130280 Dieteria canescens (Pursh) Nutt. var. shastensis (A. Gray) D. R. Morgan et R. L. Hartm. ;沙斯塔灰菀■☆

130281 Dieteria canescens (Pursh) Nutt. var. ziegleri (Munz) D. R. Morgan et R. L. Hartm. ;齐格勒灰菀■☆

130282 Dieteria gracilis Nutt. = Xanthisma gracile (Nutt.) D. R. Morgan et R. L. Hartm. ■☆

130283 Dieteria incana (Lindl.) Torr. et A. Gray = Dieteria canescens (Pursh) Nutt. var. incana (Lindl.) D. R. Morgan et R. L. Hartm. ■☆

130284 Dieteria sessiliflora Nutt. = Dieteria canescens (Pursh) Nutt. var. sessiliflora (Nutt.) D. R. Morgan et R. L. Hartm. ■☆

130285 Dieterica Ser. = Caldcluvia D. Don ●☆

130286 Dieterica Ser. ex DC. = Caldcluvia D. Don ●☆

130287 Dieterlea E. J. Lott(1986);拟笑布袋属■☆

130288 Dieterlea fusiformis E. J. Lott;拟笑布袋■☆

130289 Dieterlea maxima (R. Lira et Kearns) McVaugh;大拟笑布袋■☆

130290 Dietes Salisb. = Dietes Salisb. ex Klatt(保留属名)■☆

130291 Dietes Salisb. ex Klatt(1866)(保留属名);离被鸢尾属■☆

130292 Dietes bicolor (Steud.) Sweet ex Klatt;褐斑离被鸢尾;African Iris, Butterfly Iris, Fortnight Lily ■☆

130293 Dietes bicolor Sweet ex G. Don = Dietes bicolor (Steud.) Sweet ex Klatt ■☆

130294 Dietes butcheriana Gerstner;布氏离被鸢尾■☆

130295 Dietes flavida Oberm. ;浅黄离被鸢尾■☆

130296 Dietes grandiflora N. E. Br. ;大花离被鸢尾;Wild Iris ■☆

130297 Dietes huttonii Baker = Moraea huttonii (Baker) Oberm. ■☆

130298 Dietes iridifolia Salisb. ;扇形离被鸢尾;Fortnight Lily ■☆

130299 Dietes iridioides (L.) Sweet ex Klatt;离被鸢尾;African Iris, Fortnight Lily ■☆

130300 Dietes vegeta (L.) N. E. Br. = Dietes iridioides (L.) Sweet ex Klatt ■☆

130301 Dietes vegeta N. E. Br. = Dietes iridioides (L.) Sweet ex Klatt ■☆

130302 Dietilis Raf. = Otostegia Benth. ●☆

130303 Dietrichia Giseke = Zingiber Mill. (保留属名)■

130304 Dietrichia Giseke(1792);迪特姜属■☆

130305 Dietrichia Tratt. = Rochea DC. (保留属名)●■☆

130306 Dietrichia bicolor (Haw.) Eckl. et Zeyh. = Crassula fascicularis Lam. ■☆

130307 Dietrichia coccinia (L.) Tratt = Crassula coccinea L. ●☆

130308 Dietrichia jasminea (Haw. ex Sims) Eckl. et Zeyh. = Crassula obtusa Haw. ■☆

130309 Dietrichia jasminea (Haw. ex Sims) Eckl. et Zeyh. var. uniflora Eckl. et Zeyh. = Crassula obtusa Haw. ■☆

130310 Dietrichia lampuyang Giseke;迪特姜■☆

130311 Dietrichia media (Haw.) Eckl. et Zeyh. = Crassula fascicularis Lam. ■☆

130312 Dietrichia odoratissima (Andréws) Tratt. ex Eckl. et Zeyh. = Crassula fascicularis Lam. ■☆

130313 Dietrichia versicolor (Burch. ex Ker Gawl.) Eckl. et Zeyh. = Crassula coccinea L. ●☆

130314 Dieudonnaea Cogn. = Gurania (Schltdl.) Cogn. ■☆

130315 Diflugossa Bremek. (1944);叉花草属(疏花马蓝属);Diflugossa ■

130316 Diflugossa Bremek. = Strobilanthes Blume ●■

130317 Diflugossa colorata (Nees) Bremek. ;叉花草(腾越金足草)■

130318 Diflugossa divaricata (Nees) Bremek. ;疏花叉花草(叉花草,叉开紫云菜,疏花马蓝)■

130319 Diflugossa muliensis H. P. Tsui;木里叉花草(琴叶马蓝,琴叶紫云菜);Muli Diflugossa ■

130320 Diflugossa muliensis H. P. Tsui = Pteracanthus panduratus (Hand. -Mazz.) C. Y. Wu et C. C. Hu ■

130321 Diflugossa pinetorum (W. W. Sm.) C. Y. Wu et C. C. Hu;松林叉花草(松林紫云菜);Pineland Conehead ■

130322 Diflugossa scoriarum (W. W. Sm.) E. Hossain;瑞丽叉花草(金足草,墨脱马蓝,疏花马蓝,无柄金足草,蝎序金足草,蝎序马兰);Fewflower Diflugossa ■

130323 Diflugossa shweliensis (W. W. Sm.) E. Hossain = Diflugossa scoriarum (W. W. Sm.) E. Hossain ■

130324 Digaster Miq. = Lauro-Cerasus Duhamel ●

130325 Digaster Miq. = Pygeum Gaertn. ●
130326 Digastrium (Hack.) A. Camus = Ischaemum L. ■
130327 Digastrium A. Camus = Ischaemum L. ■
130328 Digera Forssk. (1775);细柱苋属■☆
130329 Digera alternifolia (L.) Asch. = Digera muricata (L.) Mart. ■☆
130330 Digera alternifolia (L.) Asch. var. stenophylla Suess. = Digera muricata (L.) Mart. var. trinervis C. C. Towns. ■☆
130331 Digera angustifolia Suess. = Digera muricata (L.) Mart. var. trinervis C. C. Towns. ■☆
130332 Digera arvensis Forssk. = Digera muricata (L.) Mart. ■☆
130333 Digera muricata (L.) Mart.;细柱苋■☆
130334 Digera muricata (L.) Mart. var. macroptera C. C. Towns.;大翅细柱苋■☆
130335 Digera muricata (L.) Mart. var. patentipilosa C. C. Towns.;毛细柱苋■☆
130336 Digera muricata (L.) Mart. var. trinervis C. C. Towns.;三脉细柱苋■☆
130337 Digitacalia Pippen(1968);指蟹甲属■☆
130338 Digitacalia jatrophoides (Kunth) Pippen;指蟹甲■☆
130339 Digitalidaceae J. Agardh = Plantaginaceae Juss.(保留科名)■
130340 Digitalidaceae J. Agardh = Scrophulariaceae Juss.(保留科名)●■
130341 Digitalidaceae J. Agardh;毛地黄科■
130342 Digitalidaceae Martinov = Plantaginaceae Juss.(保留科名)■
130343 Digitalidaceae Martinov = Scrophulariaceae Juss.(保留科名)●■
130344 Digitalis L. (1753);毛地黄属(洋地黄属);Digitalis, Fox Glove, Foxglove ■
130345 Digitalis × mertonensis Buxton et Darl.;默顿毛地黄■☆
130346 Digitalis abyssinica (A. Rich.) Stapf;阿比西尼亚毛地黄■☆
130347 Digitalis ambigua Murray = Digitalis grandiflora Mill. ■☆
130348 Digitalis ambigua Roem. ex Steud.;大毛地黄;Large Yellow Foxglove ■☆
130349 Digitalis atlantica Pomel;大西洋毛地黄■☆
130350 Digitalis aurea Lindl. = Digitalis ferruginea L. ■☆
130351 Digitalis ballii H. Lindb. = Digitalis subalpina Braun-Blanq. ■☆
130352 Digitalis brachyantha Griseb. = Digitalis ferruginea L. ■☆
130353 Digitalis canariensis L. = Isoplexis canariensis (L.) Loudon ●☆
130354 Digitalis cedretorum Maire = Digitalis subalpina Braun-Blanq. ■☆
130355 Digitalis ciliata Trautv.;紫毛洋地黄(齿毛洋地黄,流苏毛地黄)■☆
130356 Digitalis cochinchinensis Lour. = Centranthera cochinchinensis (Lour.) Merr. ■
130357 Digitalis eriostachya Besser ex Rchb. = Digitalis lutea L. ■
130358 Digitalis ferruginea L.;锈色洋地黄(铁锈色洋地黄);Rusty Foxglove ■☆
130359 Digitalis floribunda ?;多花洋地黄■☆
130360 Digitalis gloxiniiflora Vilm.;苣苔花毛地黄■☆
130361 Digitalis gloxinioides Carrière;苣苔花状洋地黄■☆
130362 Digitalis glutinosa Gaertn. = Rehmannia glutinosa (Gaertn.) Libosch. ex Fisch. et C. A. Mey. ■
130363 Digitalis grandiflora All. = Digitalis grandiflora Mill. ■☆
130364 Digitalis grandiflora Lam. = Digitalis grandiflora Mill. ■☆
130365 Digitalis grandiflora Mill.;大花毛地黄(大花洋地黄);Large Yellow Foxglove, Yellow Foxglove ■☆
130366 Digitalis laciniata Lindl.;撕裂洋地黄■☆
130367 Digitalis laciniata Lindl. subsp. riphaea (Pau et Font Quer) Pau et Font Quer;山地洋地黄■☆
130368 Digitalis laevigata Waldst. et Kit.;平滑洋地黄(无毛洋地黄)■☆
130369 Digitalis lanata Ehrh.;毛花洋地黄(黄花毛地黄,毛叶洋地黄,希腊毛地黄,狭叶毛地黄,狭叶洋地黄);Austrian Digitalis, Grecian Foxglove, Narrowleaf Foxglove, Woolly Foxglove ■
130370 Digitalis leucophaea Sibth. et Sm.;淡灰洋地黄■☆
130371 Digitalis lutea L.;黄花毛地黄(黄花洋地黄);Small Yellow Foxglove, Straw Foxglove ■
130372 Digitalis mauritanica (Humbert et Maire) Ivanina = Digitalis purpurea L. ■
130373 Digitalis mertonensis Buxton et C. D. Darl.;杂种毛地黄■☆
130374 Digitalis micrantha Roth ex Schweigg.;小花毛地黄(小花洋地黄);Smallflower Foxglove ■☆
130375 Digitalis nervosa Steud. et Hochst.;多脉洋地黄■☆
130376 Digitalis obscura L.;柳叶毛地黄;Dusty Foxglove, Willow Leaf Foxglove ■☆
130377 Digitalis obscura L. subsp. laciniata (Lindl.) Maire = Digitalis laciniata Lindl. ■☆
130378 Digitalis obscura L. var. laciniata (Lindl.) Pau = Digitalis laciniata Lindl. ■☆
130379 Digitalis obscura L. var. rhiphaea Pau et Font Quer = Digitalis laciniata Lindl. subsp. riphaea (Pau et Font Quer) Pau et Font Quer ■☆
130380 Digitalis ochroleuca Jacq. = Digitalis grandiflora Mill. ■☆
130381 Digitalis orientalis Lam.;东方洋地黄;Oriental Foxglove ■☆
130382 Digitalis orientalis Mill. = Digitalis grandiflora Lam. ■☆
130383 Digitalis purpurea L.;毛地黄(洋地黄,紫花洋地黄,自由钟);Bee Catcher, Beehive, Blob, Bloody Bells, Bloody Fingers, Bloody Man's Fingers, Bunch-of-grapes, Bunny Rabbits, Bunny Rabbit's Mouth, Clothes Pegs, Common Foxglove, Coral Bells, Cottagers, Cow Flop, Cowflop, Cowslip, Cowslop, Dead Man, Dead Man's Bellows, Dead Man's Bells, Dead Man's Fingers, Dead Man's Thimbles, Dock's Mouth, Dog's Finger, Dog's Fingers, Dog's Lugs, Dragon's Mouth, Duck's Mouth, Eoxflop, Fairy Bells, Fairy Cap, Fairy Caps, Fairy Dresses, Fairy Fingers, Fairy Hat, Fairy Hats, Fairy Petticoats, Fairy Thimbles, Fairy Weed, Fairy's Dresses, Fairy's Gloves, Fairy's Petticoats, Finger Cap, Finger Gloves, Finger Hut, Finger-cap, Finger-flower, Finger-gloves, Finger-hut, Finger-root, Fingers, Fingers-and-thumbs, Finger-tips, Flabby Dock, Flap Dock, Flap-a-dock, Flapdick, Flapdock, Flapper Dock, Flappy Dock, Flobby Dock, Flop, Flop Dock, Flop Docken, Flop Poppy, Flop-a-dock, Flop-dock, Flopdocken, Flop-poppy, Floppy Dock, Flop-top, Flosdocken, Floss Docken, Floss-docken, Flous Docken, Flous-docken, Flowster Docken, Flowster-docken, Flox, Fox Docken, Fox Finger, Fox Fingers, Fox Glove, Fox Ter-leaves, Fox-and-leaves, Foxglove, Fox-leaves, Fox's Glove, Foxter, Foxter-leaves, Foxtree, Foxy, Foxy-leaves, Frap, Gap-mouth, Goblin's Gloves, Goblin's Thimbles, Goose Flop, Granny's Bonnet, Granny's Bonnets, Granny's Gloves, Green Pop, Green Poppy, Green Pops, Harebell, Hedge Poppy, Hill Poppy, Hollyhock, King's Elwand, Lady's Fingers, Lady's Glove, Lady's Gloves, Lady's Slipper, Lady's Thimble, Lady's Thimbles, Lady's-glove, Lion's Mouth, Long Purples, Lusmore, Polters, Pop Bells, Pop Bladder, Pop Dock, Pop Glove, Pop-a-dock, Pop-bells, Pop-bladder, Pop-dock, Pop-gun, Popper, Poppy, Poppy Dock, Pops, Purple Fingers, Purple Foxglove, Rabbit Flower, Rappers, Scabbit Dock, Scabbit-dock, Scotch Mercury, Sheegie Thimbles, Shilly Thim Bles, Shilly Thimbles, Snapdragon, Snapjack, Snaps, Snauper, Snoxums, Soldiers, Thimble-flower, Thimbles, Thor's Mantle, Throatwort, Throttlewort, Tiger's

Mouth, Turtle Doves, Virgin's Fingers, Wild Mercury, Witch Bells, Witches' Gloves, Witches' Thimbles, Wrappers ■
130384 Digitalis purpurea L. 'Campanulata';重瓣自由钟■
130385 Digitalis purpurea L. subsp. mauretanica (Humbert et Maire) Romo = Digitalis purpurea L. ■
130386 Digitalis purpurea L. var. alba ?;白花毛地黄;Purple Foxglove ■☆
130387 Digitalis purpurea L. var. mauretanica Humbert et Maire = Digitalis purpurea L. ■
130388 Digitalis scalarum ? = Digitalis abyssinica (A. Rich.) Stapf ■☆
130389 Digitalis schischkinii Ivanina;希施毛地黄■☆
130390 Digitalis sinensis Lour. = Adenosma glutinosa (L.) Druce ■
130391 Digitalis subalpina Braun-Blanq.;亚高山毛地黄■☆
130392 Digitalis subalpina Braun-Blanq. var. cedretorum (Maire) Ivanina = Digitalis subalpina Braun-Blanq. ■☆
130393 Digitalis subalpina Braun-Blanq. var. mesatlantica (Maire) Ivanina = Digitalis subalpina Braun-Blanq. ■☆
130394 Digitalis subalpina Braun-Blanq. var. transiens (Maire) Ivanina = Digitalis subalpina Braun-Blanq. ■☆
130395 Digitalis thapsi L.;西班牙洋地黄■☆
130396 Digitalis tomentosa Link et Hoffm. = Digitalis purpurea L. ■
130397 Digitalis transiens Maire = Digitalis subalpina Braun-Blanq. ■☆
130398 Digitalis transiens Maire var. dyris ? = Digitalis subalpina Braun-Blanq. ■☆
130399 Digitalis transiens Maire var. mesatlantica ? = Digitalis subalpina Braun-Blanq. ■☆
130400 Digitalis winterli Roth = Digitalis lanata Ehrh. ■
130401 Digitaria Adans. = Digitaria Haller(保留属名)■
130402 Digitaria Adans. = Tripsacum L. ■
130403 Digitaria Fabr. = Paspalum L. ■
130404 Digitaria Haller(1768)(保留属名);马唐属;Crab Grass, Crabgrass, Crab-grass, Finger Grass, Fingergrass, Finger-grass ■
130405 Digitaria Heist. ex Adans. = Digitaria Haller(保留属名)■
130406 Digitaria Heist. ex Fabr.(废弃属名) = Digitaria Haller(保留属名)■
130407 Digitaria abludens (Roem. et Schult.) Veldkamp;粒状马唐;Granular Crabgrass ■
130408 Digitaria abyssinica (Hochst. ex A. Rich.) Stapf;阿比西尼亚马唐(非洲马唐);African Couchgrass ■☆
130409 Digitaria abyssinica (Hochst. ex A. Rich.) Stapf var. micrantha Peter;小花阿比西尼亚马唐■☆
130410 Digitaria abyssinica (Hochst. ex A. Rich.) Stapf var. scalarum (Schweinf.) Stapf = Digitaria scalarum (Schweinf.) Chiov. ■☆
130411 Digitaria abyssinica (Hochst. ex A. Rich.) Stapf var. velutina (Chiov.) Henrard = Digitaria pearsonii Stapf ■☆
130412 Digitaria abyssinica (Hochst.) Stapf = Digitaria abyssinica (Hochst. ex A. Rich.) Stapf ■☆
130413 Digitaria acrotricha (Steud.) Roberty = Eriochloa fatmensis (Hochst. et Steud.) Clayton ■☆
130414 Digitaria acuminatissima Stapf;尖马唐■☆
130415 Digitaria acuminatissima Stapf subsp. inermis Goetgh. = Digitaria acuminatissima Stapf ■☆
130416 Digitaria acuminatissima Stapf subvar. grandiflora Henrard = Digitaria acuminatissima Stapf ■☆
130417 Digitaria acuminatissima Stapf var. conformis Henrard = Digitaria acuminatissima Stapf ■☆
130418 Digitaria adscendens (Kunth) Henrard = Digitaria ciliaris (Retz.) Koeler ■
130419 Digitaria adscendens (Kunth) Henrard subsp. chrysoblephara (Fig. et De Not.) Henrard = Digitaria ciliaris (Retz.) Koeler ■
130420 Digitaria adscendens (Kunth) Henrard subsp. marginata (Link) Henrard = Digitaria ciliaris (Retz.) Koeler ■
130421 Digitaria adscendens (Kunth) Henrard subsp. nubica (Stapf) Henrard = Digitaria ciliaris (Retz.) Koeler ■
130422 Digitaria adscendens (Kunth) Henrard var. criniformis Henrard = Digitaria ciliaris (Retz.) Koeler ■
130423 Digitaria adscendens (Kunth) Henrard var. fimbriata (Link) Cufod. = Digitaria ciliaris (Retz.) Koeler ■
130424 Digitaria adscendens (Kunth) Henrard var. nubica (Stapf) Henrard = Digitaria ciliaris (Retz.) Koeler ■
130425 Digitaria adscendens (Kunth) Henrard var. rhachiseta Henrard = Digitaria ciliaris (Retz.) Koeler ■
130426 Digitaria adscendens (Kunth) Henrard var. sericea (Honda) Henrard = Digitaria ciliaris (Retz.) Koeler ■
130427 Digitaria adscendens Kunth = Digitaria ciliaris (Retz.) Koeler ■
130428 Digitaria alba Mez = Digitaria xanthotricha (Hack.) Stapf ■☆
130429 Digitaria albomarginata Stent = Digitaria diversinervis (Nees) Stapf ■☆
130430 Digitaria andringitrensis A. Camus;安德林吉特拉山马唐■☆
130431 Digitaria angolensis Rendle;安哥拉马唐■☆
130432 Digitaria ankaratrensis A. Camus;安卡拉特拉马唐■☆
130433 Digitaria annua Van der Veken = Digitaria leptorhachis (Pilg.) Stapf ■☆
130434 Digitaria antunesii Mez = Digitaria gazensis Rendle ■☆
130435 Digitaria apiculata Stent = Digitaria maitlandii Stapf et C. E. Hubb. ■☆
130436 Digitaria apiculata Stent var. hirta Goetgh. = Digitaria maitlandii Stapf et C. E. Hubb. ■☆
130437 Digitaria appropinquata Goetgh.;远离马唐■☆
130438 Digitaria argillacea (Hitchc. et Chase) Fernald;陶土马唐■☆
130439 Digitaria argillacea (Hitchc. et Chase) Fernald f. asetosa Scholz;无刚毛陶土马唐■☆
130440 Digitaria argyrotricha (Andersson) Chiov.;银毛马唐■☆
130441 Digitaria aridicola Napper;旱生马唐■☆
130442 Digitaria aristulata (Steud.) Stapf;芒马唐■☆
130443 Digitaria arushae Clayton;阿鲁沙马唐■☆
130444 Digitaria asiatica (Ohwi) Tzvelev = Digitaria ischaemum (Schreb.) Schreb. ex Muhl. ■
130445 Digitaria asiatica Tzvelev = Digitaria ischaemum (Schreb.) Schreb. ex Muhl. ■
130446 Digitaria asiatica Tzvelev = Digitaria violascens Link ■
130447 Digitaria atrofusca (Hack.) A. Camus;暗褐马唐■☆
130448 Digitaria baliensis Ohwi = Digitaria heterantha (Hook. f.) Merr. ■
130449 Digitaria bangweolensis Pilg. = Digitaria debilis (Desf.) Willd. ■☆
130450 Digitaria bantamensis Ohwi = Digitaria heterantha (Hook. f.) Merr. ■
130451 Digitaria barbata Willd. = Digitaria heterantha (Hook. f.) Merr. ■
130452 Digitaria barbulata Desv. = Digitaria ciliaris (Retz.) Koeler ■
130453 Digitaria bechuanica (Stent) Henrard = Digitaria eriantha Steud. ■☆
130454 Digitaria bicornis (Lam.) Roem. et Schult.;异马唐;Doublehorned Crabgrass ■

130455 Digitaria bicornis (Lam.) Roem. et Schult. = Digitaria heterantha (Hook. f.) Merr. ■

130456 Digitaria bicornis (Lam.) Roem. et Schult. subsp. gamblei Henrard = Digitaria ciliaris (Retz.) Koeler ■

130457 Digitaria bicornis Roem. et Schult. = Digitaria heterantha (Hook. f.) Merr. ■

130458 Digitaria bidactyla Van der Veken;双指马唐■☆

130459 Digitaria biformis Wild. = Digitaria ciliaris (Retz.) Koeler ■

130460 Digitaria biformis Wild. subsp. desvauxii Henrard = Digitaria ciliaris (Retz.) Koeler ■

130461 Digitaria biformis Wild. subsp. willdenowii Henrard = Digitaria ciliaris (Retz.) Koeler ■

130462 Digitaria biformis Willd. = Digitaria bicornis (Lam.) Roem. et Schult. ■

130463 Digitaria biformis Willd. = Digitaria ciliaris (Retz.) Koeler ■

130464 Digitaria boivinii Henrard = Digitaria milanjiana (Rendle) Stapf ■☆

130465 Digitaria borbonica Desv. = Digitaria nuda Schumach. ■☆

130466 Digitaria botryostachya Stapf = Digitaria rivae (Chiov.) Stapf ■☆

130467 Digitaria bovonei Chiov. = Digitaria setifolia Stapf ■☆

130468 Digitaria brazzae (Franch.) Stapf;布拉扎马唐■☆

130469 Digitaria bredoensis Robyns et Van der Veken = Digitaria leptorhachis (Pilg.) Stapf ■☆

130470 Digitaria brevifolia Link = Digitaria ciliaris (Retz.) Koeler ■

130471 Digitaria brevipes Mez = Digitaria macroblephara (Hack.) Stapf ■☆

130472 Digitaria buchananii Mez = Digitaria compressa Stapf ■☆

130473 Digitaria bulbosa Peter = Digitaria milanjiana (Rendle) Stapf ■☆

130474 Digitaria calcarata Clayton;距马唐■☆

130475 Digitaria californica (Benth.) Henrard;亚利桑那马唐;Arizona Cottontop ■☆

130476 Digitaria capitipila Stapf = Digitaria compressa Stapf ■☆

130477 Digitaria chevalieri Stapf = Digitaria leptorhachis (Pilg.) Stapf ■☆

130478 Digitaria chinensis (Nees) A. Camus = Digitaria violascens Link ■

130479 Digitaria chinensis Hornem = Digitaria radicosa (J. Presl et C. Presl) Miq. ■

130480 Digitaria chinensis Hornem = Digitaria violascens Link ■

130481 Digitaria chinensis Hornem.;华马唐;Chinese Crabgrass ■

130482 Digitaria chinensis Hornem. = Digitaria ciliaris (Retz.) Koeler ■

130483 Digitaria chinensis Hornem. = Digitaria radicosa (J. Presl) Miq. ■

130484 Digitaria chinensis Hornem. var. hirsuta (Honda) Ohwi = Digitaria radicosa (J. Presl et C. Presl) Miq. ■

130485 Digitaria chinensis Hornem. var. hirsuta Ohwi = Digitaria radicosa (J. Presl et C. Presl) Miq. var. hirsuta (Honda) C. C. Hsu ■

130486 Digitaria chinensis Hornem. var. hirsuta Ohwi = Digitaria radicosa (J. Presl) Miq. ■

130487 Digitaria chrysoblephara Fig. et De Not. = Digitaria ciliaris (Retz.) Koeler var. chrysoblephara (Fig. et De Not.) R. R. Stewart ■

130488 Digitaria chrysoblephara Fig. et De Not. = Digitaria ciliaris (Retz.) Koeler ■

130489 Digitaria ciliaris (Retz.) Koeler;升马唐(拌根草,雌日芝,毛马唐,纤毛马唐);Ascendent Crabgrass, Crab Grass, Eyelash Crabgrass, Henry's Crabgrass, Southern Crab Grass, Tropical Finger-grass ■

130490 Digitaria ciliaris (Retz.) Koeler subsp. chrysoblephara (Fig. et De Not.) S. T. Blake = Digitaria ciliaris (Retz.) Koeler ■

130491 Digitaria ciliaris (Retz.) Koeler subsp. chrysoblephara (Fig. et De Not.) S. T. Blake = Digitaria ciliaris (Retz.) Koeler var. chrysoblephara (Fig. et De Not.) R. R. Stewart ■

130492 Digitaria ciliaris (Retz.) Koeler subsp. chrysoblephara Blake = Digitaria ciliaris (Retz.) Koeler ■

130493 Digitaria ciliaris (Retz.) Koeler subsp. chrysoblephara Blake = Digitaria chrysoblephara Fig. et De Not. ■

130494 Digitaria ciliaris (Retz.) Koeler subsp. nubica (Stapf) S. T. Blake = Digitaria ciliaris (Retz.) Koeler ■

130495 Digitaria ciliaris (Retz.) Koeler var. chrysoblephara (Fig. De Not.) R. R. Stewart = Digitaria ciliaris (Retz.) Koeler ■

130496 Digitaria ciliaris (Retz.) Koeler var. chrysoblephara (Fig. et De Not.) R. R. Stewart;毛马唐(黄缕马唐);Hair Crabgrass ■

130497 Digitaria ciliaris (Retz.) Koeler var. chrysoblephara (Fig. et De Not.) R. R. Stewart = Digitaria chrysoblephara Fig. et De Not. ■

130498 Digitaria ciliaris (Retz.) Koeler var. criniformis (Henrard) R. R. Stewart = Digitaria ciliaris (Retz.) Koeler ■

130499 Digitaria ciliaris Vanderyst = Digitaria abyssinica (Hochst. ex A. Rich.) Stapf ■☆

130500 Digitaria cilliaris (Retz.) Koeler var. chrysoblephara (Fig. et De Not.) Stewart = Digitaria ciliaris (Retz.) Koeler ■

130501 Digitaria cognata (Schult.) Pilg.;近缘马唐;Fall Witch Grass ■☆

130502 Digitaria commutata (Schult.) Schult. subsp. nodosa (Parl.) Maire = Digitaria nodosa Parl. ■☆

130503 Digitaria commutata Schult. = Digitaria ciliaris (Retz.) Koeler ■

130504 Digitaria complanata Goetgh.;扁平马唐■☆

130505 Digitaria compressa Stapf;扁马唐■☆

130506 Digitaria concinna Schrad. ex Steud. = Digitaria stricta Roth ex Roem. et Schult. ■

130507 Digitaria corradii Chiov. = Digitaria longiflora (Retz.) Pers. ■

130508 Digitaria corymbosa Merr. = Digitaria setigera Roth ex Roem. et Schult. ■

130509 Digitaria cruciata (Nees ex Steud.) A. Camus;十字马唐;Cruciform Crabgrass ■

130510 Digitaria debilis (Desf.) Willd.;弱小马唐■☆

130511 Digitaria debilis (Desf.) Willd. var. gigantea Rendle = Digitaria debilis (Desf.) Willd. ■☆

130512 Digitaria debilis (Desf.) Willd. var. reimarioides (Andersson) Henrard = Digitaria debilis (Desf.) Willd. ■☆

130513 Digitaria decipiens Fig. et De Not. = Digitaria debilis (Desf.) Willd. ■☆

130514 Digitaria decumbens Stent;俯卧马唐;Pangola Grass, Pangola-grass ■☆

130515 Digitaria decumbens Stent = Digitaria eriantha Steud. ■☆

130516 Digitaria delicata Goetgh.;姣美马唐■☆

130517 Digitaria denudata Link;露子马唐;Naked Crabgrass ■

130518 Digitaria denudata Link = Digitaria stricta Roth ex Roem. et Schult. ■

130519 Digitaria denudata Link = Digitaria stricta Roth ex Roem. et Schult. var. denudata (Link) Henrard ■

130520 Digitaria diagonalis (Nees) Stapf var. glabrescens (K. Schum.) Peter = Digitaria diagonalis (Nees) Stapf var. uniglumis (Hochst. ex A. Rich.) Pilg. ■☆

130521 Digitaria diagonalis (Nees) Stapf var. hirsuta (De Wild. et T.

Durand) Troupin;粗毛马唐■☆
130522 Digitaria diagonalis (Nees) Stapf var. uniglumis (Hochst. ex A. Rich.) Pilg.;单颖马唐■☆
130523 Digitaria diamesa (Steud.) A. Chev. = Digitaria nuda Schumach. ■☆
130524 Digitaria didactyla Willd.;蓝马唐;Blue Couch ■☆
130525 Digitaria dilatata (Poir.) H. J. Coste = Paspalum dilatatum Poir. ■
130526 Digitaria dispar Henrard = Digitaria heterantha (Hook. f.) Merr. ■
130527 Digitaria distachya (L.) Pers. = Brachiaria distachya (L.) Stapf ■☆
130528 Digitaria divaricata Henrard = Digitaria velutina (Forssk.) P. Beauv. ■☆
130529 Digitaria diversinervis (Nees) Stapf;异脉马唐■☆
130530 Digitaria diversinervis (Nees) Stapf var. woodiana Henrard = Digitaria diversinervis (Nees) Stapf ■☆
130531 Digitaria dunensis Goetgh.;砂丘马唐■☆
130532 Digitaria eichingeri Mez = Digitaria abyssinica (Hochst. ex A. Rich.) Stapf ■☆
130533 Digitaria elegans Stapf = Digitaria nitens Rendle ■☆
130534 Digitaria elegantula Mez = Digitaria gayana (Kunth) A. Chev. ex Stapf ■☆
130535 Digitaria endlichii Mez = Digitaria milanjiana (Rendle) Stapf ■☆
130536 Digitaria endlichii Mez subsp. meziana Henrard = Digitaria milanjiana (Rendle) Stapf ■☆
130537 Digitaria eriantha Steud.;俯仰马唐;Crabgrass, Digitgrass, Pangola Grass ■☆
130538 Digitaria eriantha Steud. subsp. pentzii (Stent) Kok = Digitaria eriantha Steud. ■☆
130539 Digitaria eriantha Steud. subsp. stolonifera (Stapf) Kok = Digitaria eriantha Steud. ■☆
130540 Digitaria eriantha Steud. subsp. transvaalensis Kok = Digitaria eriantha Steud. ■☆
130541 Digitaria eriantha Steud. var. stolonifera Stapf = Digitaria eriantha Steud. ■☆
130542 Digitaria evrardii Van der Veken;埃夫拉尔马唐■☆
130543 Digitaria exasperata Henrard = Digitaria milanjiana (Rendle) Stapf ■☆
130544 Digitaria exilis (Kippist) Stapf;瘦小马唐;Acha, Fonio, Fundi, Hungry Rice ■☆
130545 Digitaria eylesii C. E. Hubb.;艾尔斯马唐■☆
130546 Digitaria eylesii C. E. Hubb. var. hirta Goetgh. = Digitaria eylesii C. E. Hubb. ■☆
130547 Digitaria fauriei Ohwi;佛欧里马唐■
130548 Digitaria fenestrata (A. Rich.) Rendle = Digitaria velutina (Forssk.) P. Beauv. ■☆
130549 Digitaria fibrosa (Hack.) Stapf = Digitaria setifolia Stapf ■☆
130550 Digitaria fibrosa (Hack.) Stapf ex Craib;纤维马唐;Fibrous Crabgrass ■
130551 Digitaria fibrosa (Hack.) Stapf ex Craib var. yunnanensis (Henrard) L. Liou;云南马唐;Yunnan Fibrous Crabgrass ■
130552 Digitaria fibrosa (Hack.) Stapf ex Craib var. yunnanensis (Henrard) L. Liou = Digitaria fibrosa (Hack.) Stapf ex Craib ■
130553 Digitaria filiculmis (Nees ex Miq.) Ohwi = Digitaria longiflora (Retz.) Pers. ■
130554 Digitaria filiformis (L.) Koeler;线马唐;Hairy Finger Grass, Slender Crab Grass ■☆
130555 Digitaria fimbriata Link = Digitaria ciliaris (Retz.) Koeler ■
130556 Digitaria flaccida Stapf;柔软马唐■☆
130557 Digitaria flexilis Henrard = Digitaria longiflora (Retz.) Pers. ■
130558 Digitaria flexuosa Peter = Axonopus flexuosus (Peter) C. E. Hubb. ■☆
130559 Digitaria floribunda Goetgh. = Digitaria perrottetii (Kunth) Stapf ■☆
130560 Digitaria foliosa Stent = Digitaria polyphylla Henrard ■☆
130561 Digitaria formosana Rendle = Digitaria radicosa (J. Presl et C. Presl) Miq. ■
130562 Digitaria formosana Rendle var. hirsuta (Honda) Henrard = Digitaria radicosa (J. Presl et C. Presl) Miq. ■
130563 Digitaria friesii Pilg. = Digitaria longiflora (Retz.) Pers. ■
130564 Digitaria fujianensis (L. Liou) S. M. Phillips et S. L. Chen;福建薄稃草;Fujian Witchgrass ■
130565 Digitaria fulva Bosser;黄褐马唐■☆
130566 Digitaria fusca Chiov. = Digitaria milanjiana (Rendle) Stapf ■☆
130567 Digitaria fuscescens (J. Presl) Henrard;褐马唐;Yellow Crabgrass ■☆
130568 Digitaria gallaensis Chiov. = Digitaria milanjiana (Rendle) Stapf ■☆
130569 Digitaria gayana (Kunth) A. Chev. ex Stapf;盖伊马唐■☆
130570 Digitaria gazensis Rendle = Digitaria herpoclados Pilg. ■
130571 Digitaria geniculata Stent = Digitaria eriantha Steud. ■☆
130572 Digitaria gentilis Henrard;外来马唐■☆
130573 Digitaria glabra (Schrad.) P. Beauv. = Digitaria ischaemum (Schreb.) Schreb. ex Muhl. ■
130574 Digitaria glabra P. Beauv. = Digitaria ischaemum (Schreb.) Schreb. ex Muhl. ■
130575 Digitaria glabrescens (Bor) L. Liou = Digitaria stricta Roth ex Roem. et Schult. var. glabrescens Bor ■
130576 Digitaria glauca A. Camus;灰绿马唐■☆
130577 Digitaria glauca Stent var. bechuanica ? = Digitaria eriantha Steud. ■☆
130578 Digitaria gracilenta Henrard = Digitaria milanjiana (Rendle) Stapf ■☆
130579 Digitaria grantii C. E. Hubb. = Digitaria diagonalis (Nees) Stapf var. uniglumis (Hochst. ex A. Rich.) Pilg. ■☆
130580 Digitaria granularis (Trin. ex Spreng.) Henrard = Digitaria abludens (Roem. et Schult.) Veldkamp ■
130581 Digitaria granularis (Trin.) Henrard = Digitaria abludens (Roem. et Schult.) Veldkamp ■
130582 Digitaria gymnostachys Pilg.;裸穗马唐■☆
130583 Digitaria gymnotheca Clayton;裸马唐■☆
130584 Digitaria hackelii (Pilg.) Stapf = Digitaria abyssinica (Hochst. ex A. Rich.) Stapf ■☆
130585 Digitaria hainanensis Hitchc. ex Keng = Digitaria setigera Roth ex Roem. et Schult. ■
130586 Digitaria hayatae (Honda) Honda ex Ohwi = Digitaria mollicoma (Kunth) Henrard ■
130587 Digitaria hayatae (Honda) Honda ex Ohwi var. magna Honda = Digitaria magna (Honda) Tsuyama ■
130588 Digitaria hayatae Honda ex Ohwi = Digitaria mollicoma (Kunth) Henrard ■
130589 Digitaria hengduanensis L. Liou;横断山马唐;Hengduanshan Crabgrass ■

130590 Digitaria henryi Rendle;亨利马唐;Henry Crabgrass ■
130591 Digitaria herpoclados Pilg. = Digitaria gazensis Rendle ■☆
130592 Digitaria heterantha (Hook. f.) Merr.;二型马唐(粗穗马唐);Dimorphic Crabgrass,Unequal Crabgrass ■
130593 Digitaria heterantha (Hook. f.) Merr. var. hirtella L. C. Chia = Digitaria heterantha (Hook. f.) Merr. ■
130594 Digitaria hiascens Mez = Digitaria eriantha Steud. ■☆
130595 Digitaria hispida (Thunb.) Spreng. = Arthraxon hispidus (Thunb.) Makino ■
130596 Digitaria hispida Spreng. = Arthraxon hispidus (Thunb.) Makino ■
130597 Digitaria homblei Robyns = Digitaria compressa Stapf ■☆
130598 Digitaria horizontalis Willd.;平展马唐■☆
130599 Digitaria horizontalis Willd. var. porrantha (Steud.) Henrard ex C. E. Hubb. et Vaughan = Digitaria horizontalis Willd. ■☆
130600 Digitaria humbertii A. Camus;亨伯特马唐■☆
130601 Digitaria humifusa Pers. = Digitaria ischaemum (Schreb.) Schreb. ex Muhl. ■
130602 Digitaria hyalina Robyns et Van der Veken;无色马唐■☆
130603 Digitaria hydrophila Van der Veken;喜水马唐■☆
130604 Digitaria incisa Van der Veken;锐裂马唐■☆
130605 Digitaria insularis (L.) Mez ex Ekman;酸马唐;Sourgrass ■☆
130606 Digitaria intecta Stapf;隐藏马唐■☆
130607 Digitaria ischaemum (Schreb.) Muhl. var. mississippiensis (Gatt.) Fernald = Digitaria ischaemum (Schreb.) Schreb. ex Muhl. ■
130608 Digitaria ischaemum (Schreb.) Muhlenb. = Digitaria ischaemum (Schreb.) Schreb. ex Muhl. ■
130609 Digitaria ischaemum (Schreb.) Muhlenb. subsp. asiatica (Ohwi) Tzvelev = Digitaria ischaemum (Schreb.) Schreb. ex Muhl. ■
130610 Digitaria ischaemum (Schreb.) Muhlenb. subsp. stewartiana (Bor) Tzvelev = Digitaria stewartiana Bor ■
130611 Digitaria ischaemum (Schreb.) Muhlenb. var. asiatica Ohwi = Digitaria ischaemum (Schreb.) Schreb. ex Muhl. ■
130612 Digitaria ischaemum (Schreb.) Schreb. ex Muhl.;止血马唐(线马唐,抓秧草);Red Millet,Small Crab-grass,Smooth Crab Grass,Smooth Crabgrass,Smooth Crab-grass,Smooth Finger-grass ■
130613 Digitaria ischaemum (Schreb.) Schreb. ex Muhl. subsp. asiatica (Ohwi) Tzvelev = Digitaria violascens Link ■
130614 Digitaria ischaemum (Schreb.) Schreb. ex Muhl. var. asiatica Ohwi = Digitaria violascens Link ■
130615 Digitaria ischaemum (Schreb.) Schreb. ex Muhl. var. mississippiensis (Gatt.) Fernald = Digitaria ischaemum (Schreb.) Schreb. ex Muhl. ■
130616 Digitaria jubata (Griseb.) Henrard;棒毛马唐;Maned Crabgrass ■
130617 Digitaria jubata (Griseb.) Henrardvar. brevigluma H. Peng;短颖棒毛马唐;Shortglume Maned Crabgrass ■
130618 Digitaria kanehira Ohwi = Digitaria heterantha (Hook. f.) Merr. ■
130619 Digitaria kasamaensis Van der Veken = Digitaria parodii Jacq. -Fél. ■☆
130620 Digitaria katangensis Robyns = Digitaria compressa Stapf ■☆
130621 Digitaria katangensis Robyns var. hirta Goetgh. = Digitaria compressa Stapf ■☆
130622 Digitaria keniensis Pilg. = Digitaria maitlandii Stapf et C. E. Hubb. ■☆
130623 Digitaria kilimandscharica Mez = Digitaria milanjiana (Rendle) Stapf ■☆
130624 Digitaria lancifolia Henrard = Digitaria pearsonii Stapf ■☆
130625 Digitaria lasiostachya Peter = Digitaria diagonalis (Nees) Stapf var. uniglumis (Hochst. ex A. Rich.) Pilg. ■☆
130626 Digitaria lecardii (Pilg.) Stapf = Digitaria argillacea (Hitchc. et Chase) Fernald ■☆
130627 Digitaria leptalea Ohwi;丛立马唐■
130628 Digitaria leptalea Ohwi = Digitaria pertenuis Büse ■☆
130629 Digitaria leptalea Ohwi var. reticulmis Ohwi = Digitaria leptalea Ohwi ■
130630 Digitaria leptorhachis (Pilg.) Stapf;细轴马唐■☆
130631 Digitaria linearis (Krock.) Waga ex Rostaf. = Digitaria ischaemum (Schreb.) Schreb. ex Muhl. ■
130632 Digitaria linearis Crép. = Digitaria ischaemum (Schreb.) Schreb. ex Muhl. ■
130633 Digitaria livida Henrard = Digitaria eriantha Steud. ■☆
130634 Digitaria lomanensis Mez = Digitaria brazzae (Franch.) Stapf ■☆
130635 Digitaria longiflora (Retz.) Pers.;长花马唐;Indian Crabgrass,Wire Crabgrass ■
130636 Digitaria longissima Mez = Digitaria heterantha (Hook. f.) Merr. ■
130637 Digitaria macroblephara (Hack.) Stapf;大脉马唐■☆
130638 Digitaria macroglossa Henrard = Digitaria natalensis Stent ■☆
130639 Digitaria macroglossa Henrard var. prostrata (Stent) Henrard = Digitaria natalensis Stent ■☆
130640 Digitaria magna (Honda) Tsuyama;大绒马唐■
130641 Digitaria magna (Honda) Tsuyama = Digitaria mollicoma (Kunth) Henrard ■
130642 Digitaria magna Tsuyama = Digitaria mollicoma (Kunth) Henrard ■
130643 Digitaria maitlandii Stapf et C. E. Hubb.;梅特兰马唐■☆
130644 Digitaria maitlandii Stapf et C. E. Hubb. var. glabra Van der Veken = Digitaria maitlandii Stapf et C. E. Hubb. ■☆
130645 Digitaria major (Van der Veken) Clayton = Digitaria poggeana Mez ■☆
130646 Digitaria manongarivensis A. Camus;马农加马唐■☆
130647 Digitaria marginata Link = Digitaria ciliaris (Retz.) Koeler ■
130648 Digitaria marginata Link var. fimbriata (Link) Stapf = Digitaria ciliaris (Retz.) Koeler ■
130649 Digitaria marginata Link var. linkii Stapf = Digitaria ciliaris (Retz.) Koeler ■
130650 Digitaria marginata Link var. nubica Stapf = Digitaria ciliaris (Retz.) Koeler ■
130651 Digitaria masambaensis Vanderyst = Digitaria atrofusca (Hack.) A. Camus ■☆
130652 Digitaria melanochila Stapf = Digitaria thouarsiana (Flüggé) A. Camus ■☆
130653 Digitaria melanotricha Clayton;黑毛马唐■☆
130654 Digitaria melinioides Mez = Digitaria nitens Rendle ■☆
130655 Digitaria merkeri Mez = Digitaria abyssinica (Hochst. ex A. Rich.) Stapf ■☆
130656 Digitaria microbachne (J. Presl) Henrard;短颖马唐;Shortglume Crabgrass ■
130657 Digitaria microbachne (J. Presl) Henrard = Digitaria setigera Roth ex Roem. et Schult. ■

130658 Digitaria microbachne (J. Presl) Henrard subsp. calliblepharata Henrard = Digitaria setigera Roth ex Roem. et Schult. ■
130659 Digitaria microstachya Henrard = Digitaria microbachne (J. Presl) Henrard ■
130660 Digitaria microstachya Henrard = Digitaria setigera Roth ex Roem. et Schult. ■
130661 Digitaria milanjiana (Rendle) Stapf; 马达加斯加马唐; Madagascar Crabgrass ■☆
130662 Digitaria milanjiana (Rendle) Stapf var. abscondita Henrard = Digitaria milanjiana (Rendle) Stapf ■☆
130663 Digitaria milanjiana (Rendle) Stapf var. eylesiana Henrard = Digitaria milanjiana (Rendle) Stapf ■☆
130664 Digitaria minoriflora Goetgh.; 较小花马唐■☆
130665 Digitaria minutiflora Stapf = Digitaria pseudodiagonalis Chiov. ■☆
130666 Digitaria mollicoma (Kunth) Henrard; 绒马唐(大绒马唐); Papose Crabgrass ■
130667 Digitaria mombasana C. E. Hubb. = Digitaria milanjiana (Rendle) Stapf ■☆
130668 Digitaria moninensis Rendle = Digitaria brazzae (Franch.) Stapf ■☆
130669 Digitaria monobotrys (Van der Veken) Clayton; 单穗马唐■☆
130670 Digitaria monodactyla (Nees) Stapf; 单指马唐■☆
130671 Digitaria monodactyla (Nees) Stapf var. explicata Stapf = Digitaria monodactyla (Nees) Stapf ■☆
130672 Digitaria monopholis Clayton; 单鳞马唐■☆
130673 Digitaria mutica Rendle = Digitaria abyssinica (Hochst. ex A. Rich.) Stapf ■☆
130674 Digitaria myurus Stapf; 鼠尾马唐■☆
130675 Digitaria nardifolia Stapf = Digitaria setifolia Stapf ■☆
130676 Digitaria natalensis Stent; 纳塔尔马唐■☆
130677 Digitaria natalensis Stent subsp. stentiana Henrard = Digitaria natalensis Stent ■☆
130678 Digitaria neghellensis J.-P. Lebrun; 内盖尔马唐■☆
130679 Digitaria nemoralis Henrard = Digitaria eriantha Steud. ■☆
130680 Digitaria nigritiana (Hack.) Stapf = Digitaria leptorhachis (Pilg.) Stapf ■☆
130681 Digitaria nitens Rendle; 光亮马唐■☆
130682 Digitaria nodosa Parl.; 结节马唐■☆
130683 Digitaria nuda Schumach.; 裸露马唐; Naked Crabgrass ■☆
130684 Digitaria nuda Schumach. subsp. schumacheriana Henrard = Digitaria nuda Schumach. ■☆
130685 Digitaria nuda Schumach. subsp. senegalensis Henrard = Digitaria nuda Schumach. ■☆
130686 Digitaria nyassana Mez = Digitaria gazensis Rendle ■☆
130687 Digitaria otaviensis Launert = Tarigidia aequiglumis (Gooss.) Stent ■☆
130688 Digitaria parlatorei (Steud.) Chiov. = Digitaria nodosa Parl. ■☆
130689 Digitaria parlatorei (Steud.) Chiov. var. phaeotricha Chiov. = Digitaria phaeotricha (Chiov.) Robyns ■☆
130690 Digitaria parodii Jacq.-Fél.; 帕罗德马唐■☆
130691 Digitaria paspalodes Michx. = Paspalum distichum L. ■
130692 Digitaria paspalodes Michx. = Paspalum paspalodes (Michx.) Scribn. ■
130693 Digitaria pearsonii Stapf; 皮尔逊马唐■☆
130694 Digitaria pedicellaris (Trin. ex Hook. f.) Prain = Digitaria abludens (Roem. et Schult.) Veldkamp ■
130695 Digitaria pedicellaris (Trin.) Prain. = Digitaria abludens (Roem. et Schult.) Veldkamp ■
130696 Digitaria pellita Stapf; 遮皮马唐■☆
130697 Digitaria pennata (Hochst.) T. Cooke; 羽状马唐■☆
130698 Digitaria pennata (Hochst.) T. Cooke var. pilosa Chiov. = Digitaria pennata (Hochst.) T. Cooke ■☆
130699 Digitaria pentzii Stent = Digitaria eriantha Steud. ■☆
130700 Digitaria pentzii Stent var. stolonifera (Stapf) Henrard = Digitaria eriantha Steud. ■☆
130701 Digitaria perrottetii (Kunth) Stapf; 佩罗马唐■☆
130702 Digitaria perrottetii (Kunth) Stapf var. gondaensis Henrard = Digitaria perrottetii (Kunth) Stapf ■☆
130703 Digitaria pertenuis Büse; 扩展马唐■☆
130704 Digitaria pertenuis Büse = Digitaria violascens Link ■
130705 Digitaria phaeotricha (Chiov.) Robyns; 褐毛马唐■☆
130706 Digitaria phaeotricha (Chiov.) Robyns var. patens Clayton = Digitaria setifolia Stapf ■☆
130707 Digitaria phaeotricha (Chiov.) Robyns var. paucipilosa Ball. = Digitaria phaeotricha (Chiov.) Robyns ■☆
130708 Digitaria pirifera (Chiov.) Chiov. = Digitaria nodosa Parl. ■☆
130709 Digitaria platycarpha (Trin.) Stapf; 宽果马唐■☆
130710 Digitaria plevansii Stent subsp. peterana Henrard = Digitaria milanjiana (Rendle) Stapf ■☆
130711 Digitaria poggeana Mez; 波格马唐■☆
130712 Digitaria polevansii Stent = Digitaria seriata Stapf ■☆
130713 Digitaria polybotrya Stapf = Digitaria leptorhachis (Pilg.) Stapf ■☆
130714 Digitaria polybotryoides Robyns et Van der Veken; 多穗马唐■☆
130715 Digitaria polyphylla Henrard; 多叶马唐■☆
130716 Digitaria preslii (Kunth) Henrard = Digitaria longiflora (Retz.) Pers. ■
130717 Digitaria procurrens Goetgh.; 伸展马唐■☆
130718 Digitaria propinqua (R. Br.) P. Beauv. = Digitaria longiflora (Retz.) Pers. ■
130719 Digitaria propinqua Gaudich. = Digitaria radicosa (J. Presl et C. Presl) Miq. ■
130720 Digitaria proxima Henrard = Digitaria gazensis Rendle ■☆
130721 Digitaria pruriens (Fisch. ex Trin.) Büse; 刺痒马唐■☆
130722 Digitaria pruriens (Fisch. ex Trin.) Büse = Digitaria setigera Roth ex Roem. et Schult. ■
130723 Digitaria pruriens (Trin.) Büse = Digitaria pruriens (Fisch. ex Trin.) Büse ■☆
130724 Digitaria pruriens (Trin.) Büse = Digitaria setigera Roth ex Roem. et Schult. ■
130725 Digitaria pruriens Büse = Digitaria pruriens (Fisch. ex Trin.) Büse ■☆
130726 Digitaria pseudodiagonalis Chiov.; 微花马唐■☆
130727 Digitaria puberula Link = Digitaria stricta Roth ex Roem. et Schult. ■
130728 Digitaria pulchra Van der Veken; 美丽马唐■☆
130729 Digitaria radicosa (C. Presl) Miq. var. hirsuta (Honda) C. C. Hsu = Digitaria radicosa (J. Presl et C. Presl) Miq. var. hirsuta (Honda) C. C. Hsu ■
130730 Digitaria radicosa (J. Presl et C. Presl) Miq.; 红尾翎(根马唐, 毛马唐, 小马唐); Root Crabgrass, Trailing Crabgrass ■
130731 Digitaria radicosa (J. Presl et C. Presl) Miq. var. hirsuta (Honda) C. C. Hsu; 毛红尾翎(毛马唐) ■
130732 Digitaria radicosa (J. Presl et C. Presl) Miq. var. hirsuta (Ohwi)

C. C. Hsu = Digitaria radicosa (J. Presl et C. Presl) Miq. ■
130733 Digitaria radicosa (J. Presl) Miq. = Digitaria radicosa (J. Presl et C. Presl) Miq. ■
130734 Digitaria redheadii (C. E. Hubb.) Clayton;雷德黑德马唐■☆
130735 Digitaria reflexa Schumach. = Digitaria horizontalis Willd. ■☆
130736 Digitaria remotigluma (De Winter) Clayton = Digitariella remotigluma De Winter ■☆
130737 Digitaria richardsonii Mez = Digitaria leptorhachis (Pilg.) Stapf ■☆
130738 Digitaria rigida Stent = Digitaria natalensis Stent ■☆
130739 Digitaria rivae (Chiov.) Stapf;沟马唐■☆
130740 Digitaria roxburghii Spreng. = Digitaria longiflora (Retz.) Pers. ■
130741 Digitaria royleana (Nees ex Hook. f.) Prain = Digitaria stricta Roth ex Roem. et Schult. ■
130742 Digitaria royleana (Nees) Prain = Digitaria stricta Roth ex Roem. et Schult. ■
130743 Digitaria rukwae Clayton;鲁夸马唐■☆
130744 Digitaria sacculata Clayton;小囊马唐■☆
130745 Digitaria sanguinalis (L.) Scop.;马唐(马饭,羊麻,羊粟,止血马唐);Common Crabgrass, Crab Grass, Crab-grass, Hairy Crab Grass, Hairy Crabgrass, Hairy Crab-grass, Hairy Finger-grass, Large Crab-grass, Northern Crabgrass ■
130746 Digitaria sanguinalis (L.) Scop. subsp. aegyptiaca var. frumentacea Henrard = Digitaria sanguinalis (L.) Scop. ■
130747 Digitaria sanguinalis (L.) Scop. subsp. pectiniformis Henrard = Digitaria sanguinalis (L.) Scop. ■
130748 Digitaria sanguinalis (L.) Scop. subsp. vulgaris var. rottleriana Henrard = Digitaria sanguinalis (L.) Scop. ■
130749 Digitaria sanguinalis (L.) Scop. var. aegyptiaca (Retz.) Maire et Weiller = Digitaria sanguinalis (L.) Scop. ■
130750 Digitaria sanguinalis (L.) Scop. var. ciliaris (Retz.) Maire et Weiller = Digitaria ciliaris (Retz.) Koeler ■
130751 Digitaria sanguinalis (L.) Scop. var. ciliaris (Retz.) Parl. = Digitaria ciliaris (Retz.) Koeler ■
130752 Digitaria sanguinalis (L.) Scop. var. pruriens (Fisch. ex Trin.) Prain = Digitaria setigera Roth ex Roem. et Schult. ■
130753 Digitaria sanguinalis (L.) Scop. var. vulgaris (Döll) Maire et Weiller = Digitaria sanguinalis (L.) Scop. ■
130754 Digitaria sasakii (Honda) Tuyama = Digitaria henryi Rendle ■
130755 Digitaria scaettae Robyns = Digitaria thouarsiana (Flüggé) A. Camus ■☆
130756 Digitaria scaettae Robyns var. glabra ? = Digitaria thouarsiana (Flüggé) A. Camus ■☆
130757 Digitaria scalarum (Schweinf.) Chiov.;梯马唐■☆
130758 Digitaria scalarum (Schweinf.) Chiov. var. elgonensis C. E. Hubb. = Digitaria abyssinica (Hochst. ex A. Rich.) Stapf ■☆
130759 Digitaria schmitzii Van der Veken;施密茨马唐■☆
130760 Digitaria seminuda Stapf = Digitaria atrofusca (Hack.) A. Camus ■☆
130761 Digitaria seriata Stapf;列马唐■☆
130762 Digitaria sericea (Honda) Honda = Digitaria ciliaris (Retz.) Koeler ■
130763 Digitaria sericea (Honda) Honda ex Ohwi;绢毛马唐;Sericeous Crabgrass ■
130764 Digitaria sericea (Honda) Honda ex Ohwi = Digitaria ciliaris (Retz.) Koeler ■
130765 Digitaria sericea (Honda) Honda ex Ohwi = Digitaria heterantha (Hook. f.) Merr. ■
130766 Digitaria setifolia Stapf;刚毛马唐■☆
130767 Digitaria setigera Roth = Digitaria setigera Roth ex Roem. et Schult. ■
130768 Digitaria setigera Roth ex Roem. et Schult.;海南马唐(短颖马唐,刚毛马唐);East Indian Crabgrass, Hainan Crabgrass, Hispid Crabgrass, Itchy Crabgrass, Mau'u-Kukaepua'a ■
130769 Digitaria setigera Roth var. calliblepharata (Henrard) Veldkamp = Digitaria setigera Roth ex Roem. et Schult. ■
130770 Digitaria setigera Roth var. calliblepharata ? = Digitaria setigera Roth ex Roem. et Schult. ■
130771 Digitaria setivalva Stent = Digitaria eriantha Steud. ■☆
130772 Digitaria shimadana Ohwi = Digitaria heterantha (Hook. f.) Merr. ■
130773 Digitaria siderograpta Chiov.;铁色马唐■☆
130774 Digitaria smutsii Stent = Digitaria eriantha Steud. ■☆
130775 Digitaria somalensis Chiov. = Digitaria abyssinica (Hochst. ex A. Rich.) Stapf ■☆
130776 Digitaria spectabilis Peter;壮观马唐■☆
130777 Digitaria spirifera Goetgh. = Digitaria parodii Jacq. -Fél. ■☆
130778 Digitaria stapfii Henrard = Digitaria milanjiana (Rendle) Stapf ■☆
130779 Digitaria stentiana Henrard = Digitaria eriantha Steud. ■☆
130780 Digitaria stewartiana Bor;昆仑马唐;Kunlun Crabgrass ■
130781 Digitaria stoloniferissima Vanderyst;匍枝马唐■☆
130782 Digitaria stolzii Mez = Digitaria nitens Rendle ■☆
130783 Digitaria stricta Roth ex Roem. et Schult.;竖毛马唐(短颖马唐);Erecthair Crabgrass ■
130784 Digitaria stricta Roth ex Roem. et Schult. var. denudata (Link) Henrard;露籽马唐■
130785 Digitaria stricta Roth ex Roem. et Schult. var. denudata (Link) Henrard = Digitaria stricta Roth ex Roem. et Schult. ■
130786 Digitaria stricta Roth ex Roem. et Schult. var. glabrescens Bor;秃穗马唐;Nakespike Crabgrass ■
130787 Digitaria stricta Roth ex Roem. et Schult. var. glabrescens Bor = Digitaria stricta Roth ex Roem. et Schult. ■
130788 Digitaria subsulcata Robyns et Van der Veken;纵沟马唐■☆
130789 Digitaria sulcigluma Chiov. = Digitaria brazzae (Franch.) Stapf ■☆
130790 Digitaria swynnertonii Rendle = Digitaria milanjiana (Rendle) Stapf ■☆
130791 Digitaria tangaensis Henrard = Digitaria abyssinica (Hochst. ex A. Rich.) Stapf ■☆
130792 Digitaria tenuiflora (R. Br.) P. Beauv. = Digitaria longiflora (Retz.) Pers. ■
130793 Digitaria tenuifolia Goetgh.;细叶马唐■☆
130794 Digitaria tenuispica Rendle = Digitaria radicosa (J. Presl et C. Presl) Miq. ■
130795 Digitaria ternata (A. Rich.) Stapf = Digitaria ternata (Hochst. ex A. Rich.) Stapf ex Dyer ■
130796 Digitaria ternata (Hochst. ex A. Rich.) Stapf = Digitaria ternata (Hochst. ex A. Rich.) Stapf ex Dyer ■
130797 Digitaria ternata (Hochst. ex A. Rich.) Stapf ex Dyer;三数马唐;Ternate Crabgrass, Triplex Crabgrass ■
130798 Digitaria ternata (Hochst.) Stapf ex Dyer = Digitaria ternata (Hochst. ex A. Rich.) Stapf ex Dyer ■
130799 Digitaria thouarsiana (Flüggé) A. Camus;图氏马唐■☆

130800 Digitaria thwaitesii (Hack.) Henrard;宿根马唐;Perennial Crabgrass ■
130801 Digitaria thwaitesii (Hack.) Henrard var. tonkinensis Henrard = Digitaria violascens Link ■
130802 Digitaria timorensis (Kunth) Balansa = Digitaria radicosa (J. Presl et C. Presl) Miq. ■
130803 Digitaria timorensis (Kunth) Balansa var. hirsuta (Ohwi) Henrard = Digitaria radicosa (J. Presl et C. Presl) Miq. var. hirsuta (Honda) C. C. Hsu ■
130804 Digitaria tisserantii Jacq. -Fél.;蒂斯朗特马唐■☆
130805 Digitaria tricholaenoides Stapf;毛被马唐■☆
130806 Digitaria tricostulata (Hack.) Henrard = Digitaria thouarsiana (Flüggé) A. Camus ■☆
130807 Digitaria trinervis Van der Veken;三脉马唐■☆
130808 Digitaria ulugurensis Pilg. = Digitaria velutina (Forssk.) P. Beauv. ■☆
130809 Digitaria uniglumis (Hochst. ex A. Rich.) Stapf = Digitaria diagonalis (Nees) Stapf ■☆
130810 Digitaria uniglumis (Hochst. ex A. Rich.) Stapf var. hirsuta (De Wild. et T. Durand) Robyns = Digitaria diagonalis (Nees) Stapf var. hirsuta (De Wild. et T. Durand) Troupin ■☆
130811 Digitaria uniglumis (Hochst. ex A. Rich.) Stapf var. major Stapf = Digitaria diagonalis (Nees) Stapf var. hirsuta (De Wild. et T. Durand) Troupin ■☆
130812 Digitaria usambarica Mez = Digitaria gazensis Rendle ■☆
130813 Digitaria vaginata (Sw.) Magnier ex Debeaux = Paspalum vaginatum Sw. ■
130814 Digitaria valida Stent = Digitaria eriantha Steud. ■☆
130815 Digitaria valida Stent var. glauca ? = Digitaria eriantha Steud. ■☆
130816 Digitaria variabilis Fig. et De Not. = Digitaria debilis (Desf.) Willd. ■☆
130817 Digitaria velutina (Forssk.) P. Beauv.;短绒毛马唐;Velvet Crabgrass ■☆
130818 Digitaria velutina (Forssk.) P. Beauv. var. glabrescens Gilli = Digitaria abyssinica (Hochst. ex A. Rich.) Stapf ■☆
130819 Digitaria velutina P. Beauv. = Digitaria velutina (Forssk.) P. Beauv. ■☆
130820 Digitaria verrucosa C. E. Hubb. = Digitaria angolensis Rendle ■☆
130821 Digitaria vestita Fig. et De Not. = Digitaria abyssinica (Hochst. ex A. Rich.) Stapf ■☆
130822 Digitaria vestita Fig. et De Not. var. scalarum (Schweinf.) Henrard = Digitaria abyssinica (Hochst. ex A. Rich.) Stapf ■☆
130823 Digitaria villosa (Walter) Pers. = Digitaria filiformis (L.) Koeler ■☆
130824 Digitaria villosissima Chiov. = Digitaria gazensis Rendle ■☆
130825 Digitaria violascens Link;紫果马唐(五指草,紫马唐);Kukaipua'a-Uka, Smooth Crabgrass, Violet Crabgrass, Violet Fruit Finger-grass ■
130826 Digitaria violascens Link var. lasiophylla (Honda) Tuyama;光叶紫马唐■
130827 Digitaria violascens Link var. villosa Keng = Digitaria violascens Link ■
130828 Digitaria xanthotricha (Hack.) Stapf;黄毛马唐■☆
130829 Digitaria yokoensis Vanderyst = Digitaria angolensis Rendle ■☆
130830 Digitaria yunnanensis Henrard = Digitaria fibrosa (Hack.) Stapf ex Craib ■
130831 Digitaria zeyheri (Nees) Henrard = Digitaria velutina (Forssk.) P. Beauv. ■☆
130832 Digitariella De Winter = Digitaria Haller(保留属名)■
130833 Digitariella De Winter(1961);小马唐属■☆
130834 Digitariella remotigluma De Winter;小马唐■☆
130835 Digitariella remotigluma De Winter = Digitaria remotigluma (De Winter) Clayton ■☆
130836 Digitariopsis C. E. Hubb. = Digitaria Haller(保留属名)■
130837 Digitariopsis major Van der Veken = Digitaria poggeana Mez ■☆
130838 Digitariopsis monobotrys Van der Veken = Digitaria monobotrys (Van der Veken) Clayton ■☆
130839 Digitariopsis redheadii C. E. Hubb. = Digitaria redheadii (C. E. Hubb.) Clayton ■☆
130840 Digitorebutia Frič et Kreuz. = Rebutia K. Schum. ●
130841 Digitorebutia Frič et Kreuz. ex Buining = Rebutia K. Schum. ●
130842 Digitostigma Velazco et Nevárez = Astrophytum Lem. ●
130843 Diglosselis Raf. = Aristolochia L. ■●
130844 Diglosselis Raf. = Howardia Klotzsch ●☆
130845 Diglossophyllum H. Wendl. ex Drude = Serenoa Hook. f. ●☆
130846 Diglossophyllum H. Wendl. ex Salomon = Serenoa Hook. f. ●☆
130847 Diglossus Cass. = Tagetes L. ■●
130848 Diglottis Nees et Mart. = Angostura Roem. et Schult. ●☆
130849 Diglyphis Blume = Diglyphosa Blume ■
130850 Diglyphis latifolia (Blume) Miq. = Diglyphosa latifolia Blume ■
130851 Diglyphosa Blume(1825);密花兰属;Diglyphosa ■
130852 Diglyphosa latifolia Blume;密花兰;Diglyphosa ■
130853 Diglyphosa macrophylla (King et Pantl.) King et Pantl. = Diglyphosa latifolia Blume ■
130854 Dignathe Lindl. = Leochilus Knowles et Westc. ■☆
130855 Dignathia Stapf(1911);合宜草属■☆
130856 Dignathia aristata Cope;合宜草■☆
130857 Dignathia ciliata C. E. Hubb.;缘毛合宜草■☆
130858 Dignathia gracilis Stapf;纤细合宜草■☆
130859 Dignathia hirtella Stapf;多毛合宜草■☆
130860 Dignathia villosa C. E. Hubb.;长柔毛合宜草■☆
130861 Digomphia Benth. (1846);二叉蕊属●☆
130862 Digomphia ceratophora A. H. Gentry;角梗二叉蕊●☆
130863 Digomphia laurifolia Benth.;二叉蕊●☆
130864 Digomphotis Raf. = Habenaria Willd. ■
130865 Digomphotis Raf. = Peristylus Blume(保留属名)■
130866 Digoniopterys Arènes(1946);二节翅属●☆
130867 Digoniopterys microphylla Arènes;小叶二节翅●☆
130868 Digonocarpus Vell. = Cupania L. ●☆
130869 Digraphis Trin. = Phalaris L. ■
130870 Digraphis Trin. = Typhoides Moench ■
130871 Digraphis arundinacea (L.) Trin. = Phalaris arundinacea L. ■
130872 Diheteropogon (Hack.) Stapf(1922);异芒草属■☆
130873 Diheteropogon amplectens (Nees) Clayton;环抱异芒草■☆
130874 Diheteropogon amplectens (Nees) Clayton var. catangensis (Chiov.) Clayton;卡唐异芒草■☆
130875 Diheteropogon filifolius (Nees) Clayton;线叶异芒草■☆
130876 Diheteropogon grandiflorus (Hack.) Stapf;大花异芒草■☆
130877 Diheteropogon hagerupii Hitchc.;哈格吕普异芒草■☆
130878 Diheteropogon maximus C. E. Hubb. = Diheteropogon filifolius (Nees) Clayton ■☆
130879 Diheteropogon microterus Clayton;小异芒草■☆
130880 Dihetetopogon Stapf = Diheteropogon (Hack.) Stapf ■☆
130881 Diholcos Rydb. = Astragalus L. ●■

130882 Dihylikostigma Kraenzl. = Discyphus Schltr. ■☆

130883 Dilanthes Salisb. = Anthericum L. ■☆

130884 Dilanthes revolutum (L.) Salisb. = Trachyandra revoluta (L.) Kunth ■☆

130885 Dilasia Raf. (废弃属名) = Murdannia Royle(保留属名)■

130886 Dilatridaceae M. Roem. = Haemodoraceae R. Br. (保留科名)■☆

130887 Dilatris P. J. Bergius(1767);单珠血草属■☆

130888 Dilatris caroliniana Lam. = Lachnanthes caroliniana (Lam.) Dandy ■☆

130889 Dilatris corymbosa P. J. Bergius;伞序单珠血草■☆

130890 Dilatris ixioides Lam.;鸟娇花单珠血草■☆

130891 Dilatris paniculata L. f. = Dilatris viscosa L. f. ■☆

130892 Dilatris pillansii W. F. Barker;皮朗斯单珠血草■☆

130893 Dilatris viscosa L. f.;黏单珠血草■☆

130894 Dilax Raf. = Smilax L. ●

130895 Dilema Griff. = Dillenia L. ●

130896 Dilepis Suess. et Merxm. = Flaveria Juss. ■●

130897 Dileptium Raf. = Lepidium L. ■

130898 Dilepyrum Michx. (1803);双壳禾属■☆

130899 Dilepyrum Michx. = Muhlenbergia Schreb. ■

130900 Dilepyrum Raf. = Oryzopsis Michx. ■

130901 Dilepyrum erectum (Schreb.) Farw. = Brachyelytrum erectum (Schreb. ex Spreng.) P. Beauv. ■

130902 Dileucaden (Raf.) Steud. = Panicum L. ■

130903 Dileucaden Raf. = Panicum L. ■

130904 Dilivaria Comm. ex Juss. = Acanthus L. ●■

130905 Dilivaria Juss. = Acanthus L. ●■

130906 Dilivaria ilictfolia (L.) Nees = Acanthus ilicifolius L. ●

130907 Dilkea Mast. (1871);迪尔克西番莲属■☆

130908 Dilkea acuminata Mast.;迪尔克西番莲■☆

130909 Dillandia V. A. Funk et H. Rob. (2001);羽脉黄安菊属■☆

130910 Dillandia chachapoyensis (H. Rob.) V. A. Funk et H. Rob.;羽脉黄安菊■☆

130911 Dillenia Fabr. = Sherardia L. ■☆

130912 Dillenia Heist. ex Fabr. = Sherardia L. ■☆

130913 Dillenia L. (1753);五桠果属(第伦桃属);Dillenia ●

130914 Dillenia alata (DC.) Martelli;昆士兰五桠果;Queensiland Red Beech ●☆

130915 Dillenia aurea Sm.;金第伦桃(金黄五桠果)●☆

130916 Dillenia elliptica Thunb. = Dillenia indica L. ●

130917 Dillenia floribunda Hook. f. et Thomson = Dillenia pentagyna Roxb. ●

130918 Dillenia hainanensis Merr. = Dillenia pentagyna Roxb. ●

130919 Dillenia indica L.;五桠果(第伦桃);Chulta, Elephant Apple, Hondapara, Hondapara Tree, India Dillenia, Indian Dillenia ●

130920 Dillenia ovata Wall.;卵形五桠果;Ovate Dillenia ●

130921 Dillenia papuana Martelli;巴布亚五桠果木●☆

130922 Dillenia pentagyna Roxb.;小花五桠果(大叶山枇杷,牛把,牛彭,五蕊第伦桃,五室第伦桃,小花第伦桃);Littleflower Gentian, Pentagynous Dillenia ●

130923 Dillenia philippinensis Rolfe;菲律宾五桠果(菲岛第伦桃,菲律宾第伦桃);Catmon Dillenia, Philippine Dillenia ●☆

130924 Dillenia retusa Thurb.;凹叶第伦桃●☆

130925 Dillenia speciosa Thunb. = Dillenia indica L. ●

130926 Dillenia suffruticosa (Griff.) Martelli;灌木五桠果;Shrubby Dillenia, Shrubby Simpoh ●☆

130927 Dillenia turbinata Finet et Gagnep.;毛五桠果(大花第伦桃,大花五桠果,海南第伦桃,枇杷树,山牛把,山牛彭);Turbinate Dillenia ●

130928 Dilleniaceae Salisb. (1807)(保留科名);五桠果科(第伦桃科,五丫果科,锡叶藤科);Dilienia Family, Tree-fern Family ●■

130929 Dillonia Sacleux = Catha Forssk. ex Scop. (废弃属名)●

130930 Dillonia Sacleux = Gymnosporia (Wight et Arn.) Benth. et Hook. f. (保留属名)●

130931 Dillonia abyssinica Sacleux = Catha edulis (Vahl) Forssk. ex Endl. ●

130932 Dillwinia Poir. = Dillwynia Sm. ●☆

130933 Dillwynia Roth = Rothia Pers. (保留属名)■

130934 Dillwynia Sm. (1805);鹦鹉豆属;Parrot Pea, Parrot-pea ●☆

130935 Dillwynia floribunda Sm.;繁花鹦鹉豆;Yellow Pea ●☆

130936 Dillwynia hispida Lindl.;毛鹦鹉豆●☆

130937 Dillwynia retorta (Wendl.) Druce;鹦鹉豆;Eggs and Bacon Pea, Parrot Pea, Twisted Parrot-pea ●☆

130938 Dillwynia trifoliata Roth = Rothia indica (L.) Druce ■

130939 Dilobeia Thouars(1806);马岛山龙眼属●☆

130940 Dilobeia boiviniana Baill. = Dilobeia thouarsii Roem. et Schult. ●☆

130941 Dilobeia madagascariensis Chanc. = Dilobeia thouarsii Roem. et Schult. ●☆

130942 Dilobeia tenuinervis Bosser et R. Rabev.;细脉马岛山龙眼●☆

130943 Dilobeia thouarsii Roem. et Schult.;马岛山龙眼●☆

130944 Dilochia Blume = Dilochia Lindl. ■☆

130945 Dilochia Lindl. (1830);蔗兰属(迪劳兰属);Dilochia ■☆

130946 Dilochia cantleyi (Hook. f.) Ridl.;蔗兰(迪劳兰);Cantley Dilochia ■☆

130947 Dilochiopsis (Hook. f.) Brieger = Eria Lindl. (保留属名)■

130948 Dilochopsis (Hook. f.) Brieger = Eria Lindl. (保留属名)■

130949 Dilodendron Radlk. (1878);热美无患子属●☆

130950 Dilodendron costaricenss (Radlk.) A. H. Gentry et Steyerm.;热美无患子●☆

130951 Dilomilis Raf. (1838);弱粟兰属■☆

130952 Dilomilis elata (Benth.) Summerh.;高弱粟兰■☆

130953 Dilomilis montana (Sw.) Summerh.;山地弱粟兰■☆

130954 Dilomilis oligophylla (Schltr.) Summerh.;寡叶弱粟兰■☆

130955 Dilomilis serrata Raf.;弱粟兰■☆

130956 Dilophia Thomson(1853);双脊荠属(双脊草属);Dilophia ■

130957 Dilophia dutreulii Franch. = Dilophia salsa Thomson ■

130958 Dilophia ebracteata Maxim.;无苞双脊荠;Bractless Dilophia ■

130959 Dilophia fontana Maxim. = Taphrospermum fontana (Maxim.) Al-Shehbaz et G. Yang ■

130960 Dilophia fontana Maxim. var. trichocarpa W. T. Wang;毛果双脊荠(小籽沟子荠);Hairfruit Dilophia ■

130961 Dilophia fontana Maxim. var. trichocarpa W. T. Wang = Taphrospermum fontana (Maxim.) Al-Shehbaz et G. Yang subsp. microspermum Al-Shehbaz et G. Yang ■

130962 Dilophia hopkinsonii O. E. Schulz = Dilophia ebracteata Maxim. ■

130963 Dilophia kashgarica Rupr. = Dilophia salsa Thomson ■

130964 Dilophia macrisperma O. E. Schulz = Taphrospermum fontana (Maxim.) Al-Shehbaz et G. Yang ■

130965 Dilophia salsa Thomson;盐泽双脊荠(双脊草);Saltmarsh Dilophia ■

130966 Dilophia salsa Thomson var. hirticalyx Pamp. = Dilophia salsa Thomson ■

130967 Dilophotriche (C. E. Hubb.) Jacq. -Fél. (1960);双毛冠草属

(毛状枝草属)■☆
130968 Dilophotriche occidentalis Jacq. -Fél. = Loudetiopsis occidentalis (Jacq. -Fél.) Clayton ■☆
130969 Dilophotriche pobeguinii Jacq. -Fél. = Loudetiopsis pobeguinii (Jacq. -Fél.) Clayton ■☆
130970 Dilophotriche purpurea (C. E. Hubb.) Jacq. -Fél. = Loudetiopsis tristachyoides (Trin.) Conert ■☆
130971 Dilophotriche tristachyoides (Trin.) Jacq. -Fél. = Loudetiopsis tristachyoides (Trin.) Conert ■☆
130972 Dilophotriche tuberculata (Stapf) Jacq. -Fél. = Loudetiopsis tristachyoides (Trin.) Conert ■☆
130973 Dilosma Post et Kuntze = Cheiranthus L. ●■
130974 Dilosma Post et Kuntze = Deilosma Andrz. ex DC. ●■
130975 Dilwinia Jacques = Dillwynia Sm. ●☆
130976 Dilwynia Pers. = Dillwynia Sm. ●☆
130977 Dimacria Lindl. = Pelargonium L'Hér. ex Aiton ●■
130978 Dimacria andrewsii Sweet = Pelargonium longifolium (Burm. f.) Jacq. ■☆
130979 Dimacria aristata Sweet = Pelargonium aristatum (Sweet) G. Don ■☆
130980 Dimacria recurvata Sweet = Pelargonium aristatum (Sweet) G. Don ■☆
130981 Dimacria rumicifolia Sweet = Pelargonium longiflorum Jacq. ■☆
130982 Dimanisa Raf. = Dianthera L. ■☆
130983 Dimanisa Raf. = Justicia L. ●■
130984 Dimeiandra Raf. = Amaranthus L. ■
130985 Dimeianthus Raf. = Bliton Adans. ■
130986 Dimeianthus Raf. = Blitum Fabr. ■
130987 Dimeianthus Raf. = Dimeiandra Raf. ■
130988 Dimeiostemon Raf. = Andropogon L. (保留属名)■
130989 Dimeium Raf. = Zanthoxylum L. ●
130990 Dimejostemon Post et Kuntze = Andropogon L. (保留属名)■
130991 Dimejostemon Post et Kuntze = Dimeiostemon Raf. ■
130992 Dimenops Raf. = Krameria L. ex Loefl. ●■☆
130993 Dimenostemma Steud. = Dimerostemma Cass. ■☆
130994 Dimerandra Schltr. (1922);裂床兰属■☆
130995 Dimerandra Schltr. = Epidendrum L. (保留属名)■☆
130996 Dimerandra elegans (Focke) Siegerist;雅致裂床兰■☆
130997 Dimerandra emarginata (G. Mey.) Hoehne;无边裂床兰■☆
130998 Dimerandra major Schltr.;大裂床兰■☆
130999 Dimerandra rimbachii Schltr.;裂床兰■☆
131000 Dimerandra tenuicaulis (Rchb. f.) Siegerist;细茎裂床兰■☆
131001 Dimeresia A. Gray(1886);对双菊属■☆
131002 Dimeresia howellii A. Gray;对双菊■☆
131003 Dimereza Labill. = Guioa Cav. ●☆
131004 Dimeria Endl. = Dimesia Raf. ■
131005 Dimeria Endl. = Hierochloe R. Br. (保留属名)■
131006 Dimeria R. Br. (1810);鸭嘴茅属(鸭茅属,雁股茅属,雁茅属);Awlquitch,Dimeria ■
131007 Dimeria acinaciformis R. Br.;剑形鸭茅;Scimitar-shaped Dimeria ■
131008 Dimeria falcata Hack.;镰形鸭茅(镰刀鸭茅);Falcate Dimeria,Sickle Awlquitch ■
131009 Dimeria falcata Hack. var. taiwaniana (Ohwi) S. L. Chen et G. Y. Sheng;台湾鸭茅;Taiwan Awlquitch,Taiwan Dimeria ■
131010 Dimeria falcata Hack. var. taiwaniana (Ohwi) S. L. Chen et G. Y. Sheng = Dimeria falcata Hack. ■
131011 Dimeria falcata Hack. var. tenuior Keng et Y. L. Yang;细镰形鸭茅;Thin Awlquitch,Thin Dimeria ■
131012 Dimeria falcata Hack. var. tenuior Keng et Y. L. Yang = Dimeria falcata Hack. ■
131013 Dimeria gracilis Nees ex Steud.;纤细鸭茅(纤细雁茅)■☆
131014 Dimeria guangxiensis S. L. Chen et G. Y. Sheng;广西鸭茅;Guangxi Awlquitch,Guangxi Dimeria ■
131015 Dimeria heterantha S. L. Chen et G. Y. Sheng;异花鸭茅;Diffranther Awlquitch,Diffranther Dimeria ■
131016 Dimeria heterantha S. L. Chen et G. Y. Sheng = Dimeria ornithopoda Trin. subsp. subrobusta (Hack.) S. L. Chen et G. Y. Sheng ■
131017 Dimeria hirtella B. S. Sun = Dimeria ornithopoda Trin. ■
131018 Dimeria ornithopoda Trin.;鸭茅(雁股茅);Birdleg Awlquitch, Birdleg Dimeria ■
131019 Dimeria ornithopoda Trin. subsp. subrobusta (Hack.) S. L. Chen et G. Y. Sheng;具脊鸭茅■
131020 Dimeria ornithopoda Trin. subsp. subrobusta (Hack.) S. L. Chen et G. Y. Sheng var. nana (Keng et Y. L. Yang) S. L. Chen et G. Y. Sheng;矮鸭茅■
131021 Dimeria ornithopoda Trin. subsp. subrobusta (Hack.) S. L. Chen et G. Y. Sheng var. plurinodis (Keng et Y. L. Yang) S. L. Chen et G. Y. Sheng;多节鸭茅■
131022 Dimeria ornithopoda Trin. var. nana Keng et Y. L. Yang = Dimeria ornithopoda Trin. ■
131023 Dimeria ornithopoda Trin. var. parva Keng et Y. L. Yang = Dimeria parva (Keng et Y. L. Yang) S. L. Chen et G. Y. Sheng ■
131024 Dimeria ornithopoda Trin. var. plurinodis Keng et Y. L. Yang = Dimeria ornithopoda Trin. subsp. subrobusta (Hack.) S. L. Chen et G. Y. Sheng var. plurinodis (Keng et Y. L. Yang) S. L. Chen et G. Y. Sheng ■
131025 Dimeria ornithopoda Trin. var. subrobusta Hack. = Dimeria ornithopoda Trin. subsp. subrobusta (Hack.) S. L. Chen et G. Y. Sheng ■
131026 Dimeria ornithopoda Trin. var. tenera (Trin.) Hack.;细矮鸭茅■☆
131027 Dimeria ornithopoda Trin. var. tenera (Trin.) Hack. = Dimeria ornithopoda Trin. ■
131028 Dimeria ornithopoda Trin. var. tenera (Trin.) Hack. f. microchaeta Hack.;小毛鸭茅■☆
131029 Dimeria parva (Keng et Y. L. Yang) S. L. Chen et G. Y. Sheng;小鸭茅;Small Awlquitch,Small Dimeria ■
131030 Dimeria sinensis Rendle;华鸭茅;China Awlquitch, Chinese Dimeria ■
131031 Dimeria solitaria Keng et Y. L. Yang;单生鸭茅;Single Awlquitch,Single Dimeria ■
131032 Dimeria stipaeformis Miq. = Dimeria ornithopoda Trin. ■
131033 Dimeria taiwaniana Ohwi = Dimeria falcata Hack. var. taiwaniana (Ohwi) S. L. Chen et G. Y. Sheng ■
131034 Dimeria tenera Trin. = Dimeria ornithopoda Trin. ■
131035 Dimerocarpus Gagnep. (1921);双果桑属●☆
131036 Dimerocarpus Gagnep. = Streblus Lour. ●
131037 Dimerocarpus balansae (Hutch.) C. Y. Wu et H. L. Li = Streblus macrophyllus Blume ●
131038 Dimerocarpus brenieri Gagnep. = Streblus macrophyllus Blume ●
131039 Dimerocostus Kuntze(1891);二数闭鞘姜属■☆
131040 Dimerocostus guttierezii Kuntze;二数闭鞘姜■☆

131041 Dimerodisus Gagnep. = Ipomoea L.(保留属名)●■
131042 Dimerostemma Cass.(1817);双冠菊属■☆
131043 Dimerostemma annuum (Hassl.) H. Rob.;一年双冠菊■☆
131044 Dimerostemma rotundifolium S. F. Blake;圆叶双冠菊■☆
131045 Dimesia Raf. = Hierochloe R. Br.(保留属名)■
131046 Dimetia (Wight et Arn.) Meisn. = Hedyotis L.(保留属名)●■
131047 Dimetia Meisn. = Hedyotis L.(保留属名)●■
131048 Dimetopia DC. = Trachymene Rudge ■☆
131049 Dimetra Kerr(1938);双囊木犀属●☆
131050 Dimia Raf. = Diospyros L. ●
131051 Dimia Spreng. = Doemia R. Br. ■☆
131052 Dimitopia D. Dietr. = Dimetopia DC. ■☆
131053 Dimitopia D. Dietr. = Trachymene Rudge ■☆
131054 Dimitria Ravenna = Chilocardamum O. E. Schulz ■☆
131055 Dimitria Ravenna = Sisymbrium L. ■
131056 Dimitria onuridifolia Ravenna;双囊木犀●☆
131057 Dimocarpus Lour.(1790);龙眼属;Dimocarpus,Longan ●
131058 Dimocarpus confinis (F. C. How et C. N. Ho) H. S. Lo;龙荔(肖韶子);Confined Dimocarpus,Lycheeshape Longan ●
131059 Dimocarpus fumatus (Blume) Leenh.;肖韶子龙眼;Common Dimocarpus ●
131060 Dimocarpus fumatus (Blume) Leenh. subsp. calcicola C. Y. Wu;灰岩肖韶子;Calcicolous Dimocarpus, Limestone, Limestone Longan ●
131061 Dimocarpus longan Lour.;龙眼(福圆,桂圆,荔奴,龙目,蜜脾,木弹,泰国龙眼,亚荔枝,羊眼果树,益智,圆眼);Dragon's Eye,Longan,Longen,Longyen,Lungan ●◇
131062 Dimocarpus longan Lour. subsp. malesianus Leenh.;马来西亚龙眼;Cat's Eyes,Mata Kuching,Mata Kucin ●☆
131063 Dimocarpus longan Lour. var. obtusa (Pierre) Leenh.;钝叶龙眼●☆
131064 Dimocarpus yunnanensis (W. T. Wang) C. Y. Wu et T. L. Ming;滇龙眼;Yunnan Longan ●
131065 Dimopogon Rydb. = Drimopogon Raf. ●
131066 Dimorpha D. Dietr. = Diamorpha Nutt. ■☆
131067 Dimorpha Schreb. = Eperua Aubl. ●☆
131068 Dimorphandra Schott(1827);异蕊苏木属(二形花属,二型苏木属,二型雄蕊苏木属,二型药属)●☆
131069 Dimorphandra conjugata Sandwith;异蕊苏木(二型雄蕊苏木)●☆
131070 Dimorphandra mollis Benth.;软异蕊苏木(软二形花)●☆
131071 Dimorphandra multiflora Ducke;多花异蕊苏木●☆
131072 Dimorphandra polyandra Benoist;多雄异蕊苏木(多雄二型雄蕊苏木)●☆
131073 Dimorphanthera (Drude) F. Muell. ex J. J. Sm. = Dimorphanthera (F. Muell. ex Drude) F. Muell. ex J. J. Sm. ●☆
131074 Dimorphanthera (Drude) J. J. Sm. = Dimorphanthera (F. Muell. ex Drude) F. Muell. ex J. J. Sm. ●☆
131075 Dimorphanthera (F. Muell. ex Drude) F. Muell. = Dimorphanthera (F. Muell. ex Drude) F. Muell. ex J. J. Sm. ●☆
131076 Dimorphanthera (F. Muell. ex Drude) F. Muell. ex J. J. Sm. (1890);异药莓属(异蕊莓属)●☆
131077 Dimorphanthera F. Muell. = Dimorphanthera (F. Muell. ex Drude) F. Muell. ex J. J. Sm. ●☆
131078 Dimorphanthera alba J. J. Sm.;白异药莓●☆
131079 Dimorphanthera albiflora Schltr.;白花异药莓●☆
131080 Dimorphanthera alpina J. J. Sm.;高山异药莓●☆
131081 Dimorphanthera bracteata P. F. Stevens;苞片异药莓●☆
131082 Dimorphanthera gracilis Sleumer;细异药莓●☆
131083 Dimorphanthera lancifolia Sleumer;披针叶异药莓●☆
131084 Dimorphanthera leucostoma Sleumer;白口异药莓●☆
131085 Dimorphanthera longifolia Kaneh. et Hatus.;长叶异药莓●☆
131086 Dimorphanthera longistyla P. F. Stevens;长柱异药莓●☆
131087 Dimorphanthera nigropunctata Sleumer;黑斑异药莓●☆
131088 Dimorphanthera obovata J. J. Sm.;倒卵异药莓●☆
131089 Dimorphanthera obtusifolia Sleumer;钝叶异药莓●☆
131090 Dimorphanthera ovatifolia Sleumer;卵叶异药莓●☆
131091 Dimorphanthera parvifolia J. J. Sm.;小叶异药莓●☆
131092 Dimorphanthera viridiflora P. F. Stevens;绿花异药莓●☆
131093 Dimorphanthes Cass.(废弃属名) = Conyza Less.(保留属名)■
131094 Dimorphanthes Cass.(废弃属名) = Eschenbachia Moench(废弃属名)■
131095 Dimorphanthes Meisn. = Dimorphanthus Miq. ●■
131096 Dimorphanthus Miq. = Aralia L. ●■
131097 Dimorphanthus edulis Miq. = Aralia cordata Thunb. ■
131098 Dimorphanthus elatus Miq. = Aralia elata (Miq.) Seem. ●
131099 Dimorphanthus mandshuricus (Rupr. et Maxim.) Maxim. = Aralia elata (Miq.) Seem. var. glabrescens (Franch. et Sav.) Pojark. ●
131100 Dimorphanthus mandshuricus Rupr. et Maxim. = Aralia elata (Miq.) Seem. ●
131101 Dimorphocalyx Hook. f. = Dimorphochlamys Hook. f. ■
131102 Dimorphocalyx Hook. f. = Momordica L. ■
131103 Dimorphocalyx Thwaites(1861);异萼木属;Dimorphocalyx ●
131104 Dimorphocalyx poilanei Gagnep.;异萼木;Poliane Dimorphocalyx ●
131105 Dimorphocarpa Rollins(1979);异果荠属■☆
131106 Dimorphocarpa wislizenii (Engelm.) Rollins;异果荠;Spectacle Pod ■☆
131107 Dimorphochlamys Hook. f. = Momordica L. ■
131108 Dimorphochlamys cabrae Cogn. = Momordica cabrae (Cogn.) C. Jeffrey ■☆
131109 Dimorphochlamys crepiniana Cogn. = Momordica cabrae (Cogn.) C. Jeffrey ■☆
131110 Dimorphochlamys glomerata Cogn. = Momordica cabrae (Cogn.) C. Jeffrey ■☆
131111 Dimorphochlamys mannii Hook. f. = Momordica cabrae (Cogn.) C. Jeffrey ■☆
131112 Dimorphochloa S. T. Blake = Cleistochloa C. E. Hubb. ■☆
131113 Dimorphocladium Britton = Phyllanthus L. ●■
131114 Dimorphoclamys Hook. f. = Momordica L. ■
131115 Dimorphocoma F. Muell. et Tate(1883);异冠层菀属■☆
131116 Dimorphocoma minutula F. Muell. et Tate;异冠层菀■☆
131117 Dimorpholepis (G. M. Barroso) R. M. King et H. Rob. = Grazielia R. M. King et H. Rob. ■●☆
131118 Dimorpholepis A. Gray = Helipterum DC. ■☆
131119 Dimorpholepis A. Gray = Triptilodiscus Turcz. ■☆
131120 Dimorpholepis A. Gray(1851);二形鳞菊属■☆
131121 Dimorpholepis australis A. Gray;二形鳞菊■☆
131122 Dimorphopetalum Bertero = Tetilla DC. ■☆
131123 Dimorphorchis Rolfe = Arachnis Blume ■
131124 Dimorphorchis Rolfe(1919);鸳鸯兰属(异花兰属)■☆
131125 Dimorphorchis lowii (Lindl.) Rolfe;鸳鸯兰■☆
131126 Dimorphorchis Rolfe = Arachnis Blume ■

131127 Dimorphorchis rossii Fowlie;罗斯鸳鸯兰■☆
131128 Dimorphosciadium Pimenov(1975);异伞芹属■
131129 Dimorphosciadium gayoides (Regel et Schmalh.) Pimenov;异伞芹■
131130 Dimorphosciadium shenii Pimenov et Kljuykov = Chamaesciadium acaule C. A. Mey. var. simplex R. H. Shan et F. T. Pu ■
131131 Dimorphostachys E. Fourn. = Paspalum L. ■
131132 Dimorphostemon Kitag. (1939);异蕊芥属(二形芥属,异型芥属);Dimorphostemon ■
131133 Dimorphostemon Kitag. = Dontostemon Andrz. ex C. A. Mey. (保留属名)■
131134 Dimorphostemon Kitag. = Sisymbrium L. ■
131135 Dimorphostemon asper (Pall.) Kitag. = Dontostemon pinnatifidus (Willd.) Al-Shehbaz et H. Ohba ■
131136 Dimorphostemon asper Kitag. = Dontostemon pinnatifidus (Willd.) Al-Shehbaz et H. Ohba ■
131137 Dimorphostemon glandulosus (Kar. et Kir.) Golubk.;腺异蕊芥(花旗杆,腺花旗杆,腺毛念珠芥,腺毛异蕊芥,腺质花旗杆);Glandular Dimorphostemon,Glandulose Dimorphostemon ■
131138 Dimorphostemon glandulosus (Kar. et Kir.) Golubk. = Dontostemon glandulosus (Kar. et Kir.) O. E. Schulz ■
131139 Dimorphostemon pectinatus (DC.) Golubk. = Dontostemon pinnatifidus (Willd.) Al-Shehbaz et H. Ohba ■
131140 Dimorphostemon pectinatus (DC.) Golubk. var. humilior (N. Busch) Golubk. = Dontostemon pinnatifidus (Willd.) Al-Shehbaz et H. Ohba ■
131141 Dimorphostemon pinnatus (Pers.) Kitag.;异蕊芥(羽裂花旗杆,栉叶芥);Common Dimorphostemon ■
131142 Dimorphostemon pinnatus (Pers.) Kitag. = Dontostemon pinnatifidus (Willd.) Al-Shehbaz et H. Ohba ■
131143 Dimorphostemon sergievskianus (Polozhij) Ovchinnikova = Dontostemon glandulosus (Kar. et Kir.) O. E. Schulz ■
131144 Dimorphostemon shanxiensis R. L. Guo et T. Y. Cheo;山西异蕊芥;Shanxi Dimorphostemon ■
131145 Dimorphostemon shanxiensis R. L. Guo et T. Y. Cheo = Dontostemon pinnatifidus (Willd.) Al-Shehbaz et H. Ohba ■
131146 Dimorphotheca Moench = Dimorphotheca Vaill. (保留属名)■●☆
131147 Dimorphotheca Vaill. (1754)(保留属名);异果菊属(非洲金盏花属,铜钱花属,雨菊属);African Daisy,Cape Marigold,Star-of-the-veldt ■●☆
131148 Dimorphotheca acutifolia Hutch.;尖叶异果菊■☆
131149 Dimorphotheca annua Less. = Dimorphotheca pluvialis (L.) Moench ■☆
131150 Dimorphotheca aurantiaca DC.;大花异果菊;African Daisy,Cape Marigold,Namaqua Daisy,Namaqualand Daisy,Star of the Veldt,Star-of-the-veldt ■☆
131151 Dimorphotheca aurantiaca DC. = Dimorphotheca tragus (Aiton) B. Nord. ■☆
131152 Dimorphotheca aurantiaca Hort. = Dimorphotheca sinuata DC. ■☆
131153 Dimorphotheca barberae Harv.;巴尔巴拉异果菊■☆
131154 Dimorphotheca calendulacea Harv. = Dimorphotheca sinuata DC. ■☆
131155 Dimorphotheca calendulacea Harv. var. dubia E. Phillips = Dimorphotheca pluvialis (L.) Moench ■☆
131156 Dimorphotheca calendulacea Harv. var. dubia E. Phillips = Dimorphotheca sinuata DC. ■☆
131157 Dimorphotheca caulescens Harv.;具茎异果菊■☆
131158 Dimorphotheca chrysanthemifolia (Vent.) DC.;茼蒿叶异果菊■☆
131159 Dimorphotheca chrysanthemifolia DC. = Dimorphotheca chrysanthemifolia (Vent.) DC. ■☆
131160 Dimorphotheca cuneata (Thunb.) Less.;楔形异果菊■☆
131161 Dimorphotheca dekindtii O. Hoffm. = Dimorphotheca caulescens Harv. ■☆
131162 Dimorphotheca dentata (DC.) Harv. = Dimorphotheca sinuata DC. ■☆
131163 Dimorphotheca dregei DC.;德雷异果菊■☆
131164 Dimorphotheca dregei DC. var. reticulata (Norl.) B. Nord.;网状异果菊■☆
131165 Dimorphotheca ecklonis DC.;埃氏异果菊■☆
131166 Dimorphotheca ecklonis DC. = Osteospermum calendulaceum L. f. ■☆
131167 Dimorphotheca ecklonis Harv. = Dimorphotheca spectabilis Schltr. ■☆
131168 Dimorphotheca flaccida (Vent.) Thell. = Dimorphotheca tragus (Aiton) B. Nord. ■☆
131169 Dimorphotheca fruticosa (L.) Less.;灌丛异果菊■☆
131170 Dimorphotheca graminifolia (L.) DC. = Dimorphotheca nudicaulis (L.) DC. var. graminifolia (L.) Harv. ■☆
131171 Dimorphotheca hybrida (L.) DC. = Dimorphotheca pluvialis (L.) Moench ■☆
131172 Dimorphotheca incrassata Moench = Dimorphotheca pluvialis (L.) Moench ■☆
131173 Dimorphotheca integrifolia (DC.) Harv. = Dimorphotheca sinuata DC. ■☆
131174 Dimorphotheca jucunda E. Phillips;愉悦异果菊■☆
131175 Dimorphotheca leptocarpa DC. = Dimorphotheca pluvialis (L.) Moench ■☆
131176 Dimorphotheca lilacina Regel et Herder = Dimorphotheca barberae Harv. ■☆
131177 Dimorphotheca montana Norl.;山地异果菊■☆
131178 Dimorphotheca montana Norl. var. amoena ? = Dimorphotheca venusta (Norl.) Norl. var. amoena (Norl.) Norl. ■☆
131179 Dimorphotheca montana Norl. var. venusta ? = Dimorphotheca venusta (Norl.) Norl. ■☆
131180 Dimorphotheca multifida DC. = Garuleum bipinnatum (Thunb.) Less. ■☆
131181 Dimorphotheca nervosa Hutch. = Osteospermum striatum Burtt Davy ■☆
131182 Dimorphotheca nudicaulis (L.) DC.;裸茎异果菊■☆
131183 Dimorphotheca nudicaulis (L.) DC. var. graminifolia (L.) Harv.;禾叶裸茎异果菊■☆
131184 Dimorphotheca nudicaulis (L.) DC. var. kraussii Sch. Bip. = Dimorphotheca nudicaulis (L.) DC. ■☆
131185 Dimorphotheca pinnata (Thunb.) Harv. = Osteospermum pinnatum (Thunb.) Norl. ■☆
131186 Dimorphotheca pluvialis (L.) Moench;异果菊(雨纹铜钱花);African Marigold,Cape Marigold,Rain Daisy ■☆
131187 Dimorphotheca pluvialis Moench = Dimorphotheca pluvialis (L.) Moench ■☆
131188 Dimorphotheca polyptera DC.;多翅异果菊■☆
131189 Dimorphotheca pseudo-aurantiaca Schinz et Thell. = Dimorphotheca sinuata DC. ■☆

131190 Dimorphotheca scabra DC. = Dimorphotheca acutifolia Hutch. ■☆

131191 Dimorphotheca sinuata DC.;雨菊(铜钱花);African Daisy, Cape Marigold, Glandular Cape Marigold, Namaqualand Daisy, Star of the Veldt ■☆

131192 Dimorphotheca spectabilis Schltr.;壮观异果菊■☆

131193 Dimorphotheca tragus (Aiton) B. Nord.;山羊异果菊;Cape Marigold ■☆

131194 Dimorphotheca tragus DC. = Dimorphotheca acutifolia Hutch. ■☆

131195 Dimorphotheca venusta (Norl.) Norl.;雅致异果菊■☆

131196 Dimorphotheca venusta (Norl.) Norl. var. amoena (Norl.) Norl.;秀丽异果菊■☆

131197 Dimorphotheca walliana (Norl.) B. Nord.;瓦利异果菊■☆

131198 Dimorphotheca zeyheri Sond.;泽赫异果菊■☆

131199 Dimorphylia Cortés(1917);哥伦比亚茄属☆

131200 Dinacanthon Post et Kuntze = Deinacanthon Mez ■☆

131201 Dinacria Harv. = Crassula L. ●■☆

131202 Dinacria Harv. ex Sond. = Crassula L. ●■☆

131203 Dinacria filiformis (Eckl. et Zeyh.) Harv. = Crassula filiformis (Eckl. et Zeyh.) D. Dietr. ■☆

131204 Dinacria grammanthoides Schönland = Crassula grammanthoides (Schönland) Toelken ●☆

131205 Dinacria sebaeoides (Eckl. et Zeyh.) Schönland = Crassula sebaeoides (Eckl. et Zeyh.) Toelken ■☆

131206 Dinacrusa G. Krebs = Althaea L. ■

131207 Dinacrusa G. Krebs(1994);欧锦葵属■☆

131208 Dinaeba Delile = Botelua Lag. ■

131209 Dinaeba Delile = Dinebra DC. ■

131210 Dinanthe Post et Kuntze = Deinanthe Maxim. ■

131211 Dineba Delile ex P. Beauv. = Dinebra DC. ■

131212 Dineba P. Beauv. = Dinebra DC. ■

131213 Dinebra DC. = Botelua Lag. ■

131214 Dinebra Jacq. (1809);弯穗草属;Bentspikegrass ■

131215 Dinebra americana P. Beauv.;美洲弯穗草■☆

131216 Dinebra arabica Jacq. = Dinebra retroflexa (Forssk. ex Vahl) Panz. ■

131217 Dinebra arabica Jacq. = Dinebra retroflexa (Vahl) Panz. ■

131218 Dinebra brevifolia Steud. = Dinebra retroflexa (Forssk. ex Vahl) Panz. ■

131219 Dinebra curtipendula (Michx.) P. Beauv. = Bouteloua curtipendula (Michx.) Torr. ■

131220 Dinebra guineensis Franch. = Heteranthoecia guineensis (Franch.) Robyns ■☆

131221 Dinebra perrieri (A. Camus) Bosser;佩里耶弯穗草■☆

131222 Dinebra polycarpha S. M. Phillips;多果弯穗草■☆

131223 Dinebra pubescens K. Schum. = Brachypodium flexum Nees ■☆

131224 Dinebra retroflexa (Forssk. ex Vahl) Panz.;普通弯穗草;Common Bentspikegrass, Viper Grass ■

131225 Dinebra retroflexa (Forssk. ex Vahl) Panz. var. brevifolia (Steud.) T. Durand et Schinz = Dinebra retroflexa (Forssk. ex Vahl) Panz. ■

131226 Dinebra retroflexa (Vahl) Panz. = Dinebra retroflexa (Forssk. ex Vahl) Panz. ■

131227 Dinebra retroflexa (Vahl) Panz. var. condensata S. M. Phillips;密集弯穗草■☆

131228 Dinebra tuaensis Vanderyst = Heteranthoecia guineensis (Franch.) Robyns ■☆

131229 Dinema Lindl. = Epidendrum L. (保留属名)■☆

131230 Dinemagonum A. Juss. (1843);双曲蕊属●☆

131231 Dinemagonum gayanum A. Juss.;双曲蕊●☆

131232 Dinemandra A. Juss. = Dinemandra A. Juss. ex Endl. ●☆

131233 Dinemandra A. Juss. ex Endl. (1840);双雄金虎尾属●☆

131234 Dinemandra ericoides A. Juss. ex Endl.;石南状双雄金虎尾●☆

131235 Dineseris Griseb. = Hyaloseris Griseb. ●☆

131236 Dinetopsis Roberty = Dinetus Buch. -Ham. ex Sweet ●■

131237 Dinetopsis Roberty = Porana Burm. f. ●■☆

131238 Dinetopsis Roberty(1953);藏飞蛾藤属■

131239 Dinetopsis grandiflora (Wall.) Roberty;藏飞蛾藤;Tibet Dinetus ■

131240 Dinetopsis grandiflora (Wall.) Roberty = Dinetus grandiflorus (Wall.) Staples ■

131241 Dinetus Buch. -Ham. ex D. Don = Porana Burm. f. ●■☆

131242 Dinetus Buch. -Ham. ex Sweet(1825);飞蛾藤属(羽萼藤属);Dinetus, Porana ●■

131243 Dinetus decorus (W. W. Sm.) Staples;白藤(白飞蛾藤,红薯细辛,绢毛萼飞蛾藤,小萼飞蛾藤);Hairy Maire Porana, Maire Porana, Ornate Dinetus, Ornate Porana, Smallcalyx Porana, White Porana ●

131244 Dinetus decorus (W. W. Sm.) Staples = Porana decora W. W. Sm. ●

131245 Dinetus dinetoides (C. K. Schneid.) Staples;蒙自飞蛾藤(短萼飞蛾藤,冕宁飞蛾藤);Mengzi Dinetus, Mengzi Porana, Mianning Mengzi Porana, Shortsepal Porana ■●

131246 Dinetus dinetoides (C. K. Schneid.) Staples = Porana dinetoides C. K. Schneid. ■

131247 Dinetus duclouxii (Gagnep. et Courchet) Staples;三列飞蛾藤(腺毛飞蛾藤);Ducloux Dinetus, Ducloux Porana, Glandularhair Porana ■●

131248 Dinetus duclouxii (Gagnep. et Courchet) Staples = Porana duclouxii Gagnep. et Courchet ■

131249 Dinetus grandiflorus (Wall.) Staples = Dinetopsis grandiflora (Wall.) Roberty ■

131250 Dinetus grandiflorus (Wall.) Staples = Porana grandiflora Wall. ●

131251 Dinetus paniculata Sweet = Porana paniculata Roxb. ●

131252 Dinetus paniculata Sweet = Poranopsis paniculata (Roxb.) Roberty ●

131253 Dinetus paniculatus (Roxb.) Sweet = Poranopsis paniculata (Roxb.) Roberty ●

131254 Dinetus racemosus Ham. ex Sweet;飞蛾藤(白花藤,打米花,马郎花,毛叶飞蛾藤,莫汝刚,小元宝,翼萼藤,紫花飞蛾藤);Hairy-leaf Porana, Porana, Purple-flower Porana, Racemose Dinetus, Racemose Porana, Snowcreeper Porana ●

131255 Dinetus racemosus Ham. ex Sweet = Porana racemosa Roxb. ●

131256 Dinetus truncatus (Kurz) Staples;毛果飞蛾藤;Hairy-fruit Porana, Hairy-fruited Dinetus ■

131257 Dinetus truncatus (Kurz) Staples = Porana truncata Kurz ●

131258 Dinizia Ducke(1922);亚马孙豆属●☆

131259 Dinizia excelsa Ducke;亚马孙豆●☆

131260 Dinklagea Gilg = Manotes Sol. ex Planch. ●☆

131261 Dinklagea macrantha Gilg = Manotes macrantha (Gilg) G. Schellenb. ●☆

131262 Dinklageanthus Melch. ex Mildbr. = Dinklageodoxa Heine et Sandwith ●☆

131263 Dinklageanthus volubilis Melch. ex Dinkl. et Mildbr. =

Dinklageodoxa scandens Heine et Sandwith ●☆
131264 Dinklageella Mansf. (1934);丁克兰属■☆
131265 Dinklageella liberica Mansf.;离生丁克兰■☆
131266 Dinklageella minor Summerh.;小丁克兰■☆
131267 Dinklageella scandens P. J. Cribb et Stévart;攀缘丁克兰■☆
131268 Dinklageella villiersii Szlach. et Olszewski;维利尔斯丁克兰■☆
131269 Dinklageodoxa Heine et Sandwith(1962);丁克紫葳属●☆
131270 Dinklageodoxa scandens Heine et Sandwith;丁克紫葳●☆
131271 Dinklageodoxa scandens Heine et Sandwith = Dinklageanthus volubilis Melch. ex Dinkl. et Mildbr. ●☆
131272 Dinklageodoxa scandens Heine et Sandwith = Kigelia dinklagei Aubrév. et Pellegr. ●☆
131273 Dinocanthium Bremek. = Pyrostria Comm. ex Juss. ●☆
131274 Dinocanthium affine Robyns = Pyrostria affinis (Robyns) Bridson ●☆
131275 Dinocanthium bequaertii Robyns = Pyrostria affinis (Robyns) Bridson ●☆
131276 Dinocanthium hystrix Bremek. = Pyrostria hystrix (Bremek.) Bridson ●☆
131277 Dinochloa Büse(1854);藤竹属;Dinochloa,Vinebamboo ●
131278 Dinochloa bambusoides Q. H. Dai = Melocalamus arrectus T. P. Yi ●
131279 Dinochloa compacti-flora (Kurz) McClure = Melocalamus compactiflorus (Kurz) Benth. et Hook. f. ●
131280 Dinochloa compactiflora McClure;密花藤竹;Tight-flowered Scrambling Bamboo ●☆
131281 Dinochloa diffusa (Blanco) Merr. = Schizostachyum diffusum (Blanco) Merr. ●
131282 Dinochloa montana Ridl. = Maclurochloa montana (Ridl.) K. M. Wong ●☆
131283 Dinochloa orenuda McClure;无耳藤竹(东方藤竹);Earless Dinochloa,Earless Vinebamboo ●
131284 Dinochloa puberula McClure;毛藤竹(短藤竹);Puberulent Dinochloa,Puberulent Vinebamboo ●
131285 Dinochloa scandens (Blume) Kuntze;西爪哇藤竹;West-Java Climbing Bamboo ●☆
131286 Dinochloa sublaevigata S. Dransf.;沙巴藤竹;Sabah Climbing Bamboo,Sabah Scrambling Bamboo ●☆
131287 Dinochloa trichogona S. Dransf.;婆罗洲藤竹;Borneo Climbing Bamboo,Borneo Scrambling Bamboo ●☆
131288 Dinochloa utilis McClure;藤竹(可用藤竹);Common Scrambling Bamboo,Useful Dinochloa,Useful Vinebamboo ●
131289 Dinophora Benth. (1849);旋梗野牡丹属●☆
131290 Dinophora spenneroides Benth.;旋梗野牡丹●☆
131291 Dinophora spenneroides Benth. subsp. montana Troupin;山地旋梗野牡丹●☆
131292 Dinophora thonneri Cogn. = Ochthocharis dicellandroides (Gilg) Hansen et Wickens ●☆
131293 Dinoseris Griseb. (1879);旋苣属●☆
131294 Dinoseris Griseb. = Hyaloseris Griseb. ●☆
131295 Dinoseris salicifolia Griseb.;旋苣●☆
131296 Dinosma Post et Kuntze = Deinosmos Raf. ■●
131297 Dinosma Post et Kuntze = Pulicaria Gaertn. ■●
131298 Dinosperma T. G. Hartley = Melicope J. R. Forst. et G. Forst. ●
131299 Dinosperma T. G. Hartley(1997);螺籽芸香属●☆
131300 Dintera Stapf(1900);迪恩玄参属■☆
131301 Dintera pterocaulis Stapf;翼茎迪恩玄参■☆
131302 Dinteracanthus C. B. Clarke ex Schinz = Ruellia L. ■●
131303 Dinteracanthus asper Schinz = Ruellia aspera (Schinz) E. Phillips ■☆
131304 Dinteracanthus marlothii (Engl.) Schinz = Ruellia diversifolia S. Moore ■☆
131305 Dinteracanthus velutinus C. B. Clarke = Ruellia diversifolia S. Moore ■☆
131306 Dinteranthus Schwantes(1926);春桃玉属;Cluster Pea ■☆
131307 Dinteranthus microspermus (Dinter et Derenb.) Schwantes;小籽春桃玉■☆
131308 Dinteranthus microspermus (Dinter et Derenb.) Schwantes subsp. impunctatus N. Sauer;无斑小籽春桃玉■☆
131309 Dinteranthus microspermus (Dinter et Derenb.) Schwantes subsp. puberulus (N. E. Br.) N. Sauer = Dinteranthus puberulus N. E. Br. ■☆
131310 Dinteranthus microspermus (Dinter et Derenb.) Schwantes var. acutipetalus L. Bolus = Dinteranthus microspermus (Dinter et Derenb.) Schwantes ■☆
131311 Dinteranthus pole-evansii (N. E. Br.) Schwantes;埃文斯春桃玉■☆
131312 Dinteranthus puberulus N. E. Br.;微柔毛春桃玉■☆
131313 Dinteranthus punctatus L. Bolus = Dinteranthus puberulus N. E. Br. ■☆
131314 Dinteranthus vanzylii (L. Bolus) Schwantes;万齐春桃玉■☆
131315 Dinteranthus vanzylii (L. Bolus) Schwantes var. lineatus H. Jacobsen = Dinteranthus vanzylii (L. Bolus) Schwantes ■☆
131316 Dinteranthus wilmotianus L. Bolus;维尔莫特春桃玉■☆
131317 Dioaceae Doweld = Zamiaceae Rchb. ●☆
131318 Diocirea Chinnock(2001);澳洲苦槛蓝属●☆
131319 Dioclea Kunth(1824);双被豆属(迪奥豆属);Cluster Pea ■☆
131320 Dioclea Spreng. = Arnebia Forssk. ●■
131321 Dioclea Spreng. = Strobila G. Don ●■
131322 Dioclea hispidissima (Lehm.) Spreng. = Lithospermum hispidissimum Sieber ex Lehm. ■☆
131323 Dioclea megacarpa Rolfe;大果双被豆(大果迪奥豆)■☆
131324 Dioclea pauciflora Rusby;少花双被豆(少花迪奥豆)■☆
131325 Dioclea reflexa Hook. f.;反卷双被豆(反卷迪奥豆);Marbles Vine ■☆
131326 Dioclea sericea Kunth;双被豆(迪奥豆)■☆
131327 Dioclea virgata (Rich.) Amshoff;条纹双被豆■☆
131328 Dioclea wilsonii Standl.;威氏双被豆(威氏迪奥豆);Wilson's Clusterpea ■☆
131329 Dioctis Raf. = Polygonum L. (保留属名)■●
131330 Diodeilis Raf. = Clinopodium L. ■●
131331 Diodella (Torr. et A. Gray) Small = Diodia L. ■
131332 Diodella Small = Diodia L. ■
131333 Diodella teres (Walter) Small = Diodia teres Walter ■☆
131334 Diodia Gronov. ex L. = Diodia L. ■
131335 Diodia L. (1753);双角草属(大钮扣草属)■
131336 Diodia angolensis S. Moore = Diodia apiculata (Willd. ex Roem. et Schult.) K. Schum. ■☆
131337 Diodia apiculata (Willd. ex Roem. et Schult.) K. Schum.;细尖双角草■☆
131338 Diodia aulacosperma K. Schum.;沟双角草■☆
131339 Diodia aulacosperma K. Schum. var. angustata Verdc.;狭沟双角草■☆
131340 Diodia benguellensis Hiern = Spermacoce senensis (Klotzsch)

Hiern ■☆
131341 Diodia breviseta Benth. = Diodia sarmentosa Sw. ■☆
131342 Diodia carnosa Hochst. = Phylohydrax carnosa (Hochst.) Puff ■☆
131343 Diodia dasycephala Cham. et Schltdl.;毛双角草;Carretilla ■☆
131344 Diodia flavescens Hiern;浅黄双角草■☆
131345 Diodia kirkii Hiern = Spermacoce kirkii (Hiern) Verdc. ■☆
131346 Diodia maritima Thonn. = Diodia serrulata (P. Beauv.) G. Taylor ■☆
131347 Diodia natalensis (Hochst.) J. G. Garcia = Spermacoce natalensis Hochst. ●☆
131348 Diodia physotricha Chiov. = Diodia aulacosperma K. Schum. ■☆
131349 Diodia physotricha Chiov. var. cyclophylla ? = Diodia aulacosperma K. Schum. ■☆
131350 Diodia pilosa Schumach. et Thonn. = Diodia sarmentosa Sw. ■☆
131351 Diodia radula Cham. et Schltdl.;糙双角草;Rough Buttonweed ■☆
131352 Diodia rigida (Willd. ex Roem. et Schult.) Cham. et Schltdl. = Diodia apiculata (Willd. ex Roem. et Schult.) K. Schum. ■☆
131353 Diodia sarmentosa Sw.;蔓茎双角草■☆
131354 Diodia sarmentosa Sw. var. bisepala Bremek. = Diodia sarmentosa Sw. ■☆
131355 Diodia scabra Schumach. et Thonn. = Spermacoce ruelliae DC. ■☆
131356 Diodia senensis Klotzsch = Spermacoce senensis (Klotzsch) Hiern ■☆
131357 Diodia serrulata (P. Beauv.) G. Taylor;海滨双角草;Seaside Buttonweed ■☆
131358 Diodia teres Walter;柱形双角草;Buttonweed, Button-weed, Poor-joe, Rough Buttonweed ■☆
131359 Diodia teres Walter var. setifera Fernald et Griscom = Diodia teres Walter ■☆
131360 Diodia vaginalis Benth.;鞘双角草■☆
131361 Diodia virginiana L.;双角草(大钮扣草,维州钮扣草);Buttonweed, Large Buttonweed, Virginia Buttonweed ■
131362 Diodioides Loefl. = Spermacoce L. ●■
131363 Diodois Pohl = Psyllocarpus Mart. ■☆
131364 Diodonta Nutt. = Coreopsis L. ●■
131365 Diodontium F. Muell. (1857);双齿菊属■☆
131366 Diodontium F. Muell. = Glossogyne Cass. ■
131367 Diodontium filifolium F. Muell.;双齿菊■☆
131368 Diodontocheilis Raf. = Clinopodium L. ■●
131369 Diodontocheilis Raf. = Diodeilis Raf. ■●
131370 Diodosperma H. Wendl. = Trithrinax Mart. ●☆
131371 Dioecrescis Tirveng. (1983);南亚茜属●☆
131372 Dioecrescis erythroclada (Kurz) Tirveng.;南亚茜●☆
131373 Diogenesia Sleumer(1934);桂叶莓属●☆
131374 Diogenesia alstoniana Sleumer;桂叶莓●☆
131375 Diogenesia boliviana (Britton) Sleumer;玻利维亚桂叶莓●☆
131376 Diogenesia floribunda (A. C. Sm.) Sleumer;多花桂叶莓●☆
131377 Diogenesia oligantha (A. C. Sm.) Sleumer;少花桂叶莓●☆
131378 Diogoa Exell et Mendonça(1951);迪奥戈木属●☆
131379 Diogoa retivenia (S. Moore) Breteler;迪奥戈木●☆
131380 Diogoa zenkeri (Engl.) Exell et Mendonça;非洲迪奥戈木●☆
131381 Dioicodendron Steyerm. (1963);异株茜属●☆
131382 Dioicodendron cuatrecasasii Steyerm.;异株茜●☆
131383 Diolena Naudin = Triolena Naudin ☆
131384 Diolotheca Raf. = Diototheca Raf. ■
131385 Diolotheca Raf. = Phyla Lour. ■
131386 Diomedea Bertol. ex Colla = Helianthus L. ■
131387 Diomedea Cass. = Borrichia Adans. ●■☆
131388 Diomedea Cass. = Diomedella Cass. ●■☆
131389 Diomedella Cass. = Borrichia Adans. ●■☆
131390 Diomedes Haw. = Narcissus L. ■
131391 Diomedia Willis = Borrichia Adans. ●■☆
131392 Diomedia Willis = Diomedea Cass. ●■☆
131393 Diomma Engl. ex Harms = Spathelia L. (保留属名)●☆
131394 Dion Lindl. = Dioon Lindl. (保留属名)●☆
131395 Dionaea Ellis = Dionaea Sol. ex J. Ellis ■☆
131396 Dionaea L. = Dionaea Sol. ex J. Ellis ■☆
131397 Dionaea Sol. ex J. Ellis(1768);捕蝇草属;Venus Fly Trap, Venus Flytrap, Venus Fly-trap, Venus' Flytrap, Venus'-flytrap ■☆
131398 Dionaeaceae Dumort.;捕蝇草科■☆
131399 Dionaeaceae Dumort. = Droseraceae Salisb. (保留科名)■
131400 Dionaeaceae Raf. = Droseraceae Salisb. (保留科名)■
131401 Dioncophyllaceae Airy Shaw(1952)(保留科名);双钩叶科(二瘤叶科,双钩叶木科)●☆
131402 Dioncophyllum Baill. (1890);双钩叶属(双钩叶木属)●☆
131403 Dioncophyllum dawei Hutch. et Dalziel = Habropetalum dawei (Hutch. et Dalziel) Airy Shaw ●☆
131404 Dioncophyllum peltatum Hutch. et Dalziel = Triphyophyllum peltatum (Hutch. et Dalziel) Airy Shaw ●☆
131405 Dioncophyllum thollonii Baill.;双钩叶●☆
131406 Dionea Raf. = Dionaea Ellis ■☆
131407 Dioneiodon Raf. = Diodia L. ■
131408 Dionycha Naudin(1851);双距野牡丹属●☆
131409 Dionycha alba Jum. et H. Perrier = Amphorocalyx albus Jum. et H. Perrier ●☆
131410 Dionycha boinensis H. Perrier;非洲双距野牡丹●☆
131411 Dionycha bojerii Naudin;双距野牡丹●☆
131412 Dionycha gracilis Cogn. = Dionycha bojerii Naudin ●☆
131413 Dionycha triangularis Jum. et H. Perrier;三距野牡丹●☆
131414 Dionychastrum A. Fern. et R. Fern. (1956);小双距野牡丹属●☆
131415 Dionychastrum schliebenii A. Fern. et R. Fern.;小双距野牡丹●☆
131416 Dionychia Benth. et Hook. f. = Dionycha Naudin ●☆
131417 Dionychia Hook. f. = Dionycha Naudin ●☆
131418 Dionysia Fenzl(1843);垫报春属■☆
131419 Dionysia aretioides Boiss.;裂瓣垫报春■☆
131420 Dionysia hissarica Lipsky;希萨尔垫报春■☆
131421 Dionysia involucrata Zaprjag.;总苞垫报春■☆
131422 Dionysia kossinskyi Czerniak.;科辛斯基垫报春■☆
131423 Dionysia lacei (Hemsl. et Watt) Clay;巴基斯坦垫报春■☆
131424 Dionysia microphylla Wendelbo;小叶垫报春■☆
131425 Dionysia tapetodes Bunge;袖珍垫报春■☆
131426 Dionysis Thouars = Diplecthrum Pers. ■
131427 Dioon Lindl. (1843)(保留属名);多脉苏铁属(双卵凤尾蕉属,双子苏铁属,双子铁属);Dioon, Dion ●☆
131428 Dioon aculeatum Lem. = Dioon edule Lindl. ●☆
131429 Dioon edule Lindl.;食用苏铁(食用双子铁,双子苏铁,双子铁);Chestnut Dioon, Mexican Fern Palm, Seed-edible Dioon, Virgin's Palm ●☆
131430 Dioon edule Lindl. var. angustifolium A. DC.;狭裂食用苏铁●☆
131431 Dioon edule Lindl. var. imbricatum A. DC.;覆瓦叶食用苏铁●☆
131432 Dioon edule Lindl. var. lanuginosum J. Schust.;密毛食用苏铁●☆

131433 Dioon edule Lindl. var. lanuginosum Wittm. = Dioon edule Lindl. var. lanuginosum J. Schust. ●☆

131434 Dioon edule Lindl. var. latipinna Dyer;阔裂食用苏铁●☆

131435 Dioon holmgrenii De Luca, Sabato et Vazq. Torres;赫氏双子铁(双子铁)●☆

131436 Dioon mejiae Standl. et L. O. Williams;洪都拉斯双子铁;Cycad ●☆

131437 Dioon merolae De Luca, Sabato et Vazq. Torres;阔叶双子铁●☆

131438 Dioon pectinatum H. Wendl. = Dioon edule Lindl. var. latipinna Dyer ●☆

131439 Dioon purpusii Rose;普氏苏铁●☆

131440 Dioon spinulosum Dyer = Dioon spinulosum Dyer ex Eichler ●☆

131441 Dioon spinulosum Dyer ex Eichler;长茎双子苏铁(芳香多脉苏铁);Cycad ●☆

131442 Dioon strobilaceum A. DC. = Dioon edule Lindl. ●☆

131443 Diopogon Jord. et Fourr. = Jovibarba (DC.) Opiz ■☆

131444 Diopogon Jord. et Fourr. = Sempervivum L. ■☆

131445 Diora Ravenna(1987);毛果吊兰属■☆

131446 Diora cajamarcaensis (Poelln.) Ravenna;毛果吊兰■☆

131447 Diorimasperma Raf. = Cleome L. ●■

131448 Dioryktandra Hassk. = Rinorea Aubl. (保留属名)●

131449 Dioryktandra Hassk. ex Bakh. = Diospyros L. ●

131450 Diosanthos St. -Lag. = Dianthus L. ■

131451 Dioscorea L. (1753)(保留属名);薯蓣属(龟甲龙属,薯芋属);Elephant's-foot, Tortoise Plant, Yam ■

131452 Dioscorea Plum. ex L. = Dioscorea L. (保留属名)■

131453 Dioscorea abyssinica Hochst. ex Kunth;阿比西尼亚薯蓣■☆

131454 Dioscorea acerifolia Uline ex Diels = Dioscorea nipponica Makino ■

131455 Dioscorea acrotheca Uline ex R. Knuth = Dioscorea tenuipes Franch. et Sav. ■

131456 Dioscorea aculeata Antommarchi = Dioscorea cayenensis Lam. ■☆

131457 Dioscorea aculeata L. var. spinosa Roxb. ex Prain et Burkill = Dioscorea esculenta (Lour.) Burkill var. spinosa (Roxb. ex Wall.) R. Knuth ■

131458 Dioscorea alata L.;参薯(大薯,罐薯,红大薯,红毛薯,鸡窝薯,脚板苕,脚板薯,黎洞薯,落子薯,毛薯,山药,薯子,四棱薯,田薯,条薯,云饼山药,紫薯,紫田薯);Guyana Arrowroot, Guyana Arrow-root Greater Yam, Ten-months Yam, Ten-mouths Yam, Water Yam, White Yam, Winged Yam ■

131459 Dioscorea alata L. var. flabella Makino;大薯■

131460 Dioscorea alata L. var. purpurea (Roxb.) A. Pouchet;红大薯(紫田薯,条薯)■

131461 Dioscorea alata L. var. purpurea (Roxb.) A. Pouchet = Dioscorea alata L. ■

131462 Dioscorea althaeoides R. Knuth;蜀葵叶薯蓣(穿地龙,穿山龙,龙骨七,细山药);Hollyhock-like Yam ■

131463 Dioscorea anchiatasi Harms = Dioscorea quartiniana A. Rich. ■☆

131464 Dioscorea andongensis Rendle = Dioscorea preussii Pax ■☆

131465 Dioscorea angolensis R. Knuth = Dioscorea quartiniana A. Rich. ■☆

131466 Dioscorea angusta R. Knuth = Dioscorea cirrhosa Lour. ■

131467 Dioscorea angustiflora Rendle = Dioscorea cayenensis Lam. ■☆

131468 Dioscorea anthropophagorum A. Chev. = Dioscorea bulbifera L. ■

131469 Dioscorea anthropophagorum A. Chev. var. sylvestris ? = Dioscorea hirtiflora Benth. ■☆

131470 Dioscorea apiculata De Wild. = Dioscorea quartiniana A. Rich. ■☆

131471 Dioscorea arachidna Prain et Burkill;三叶薯蓣;Threeleaves Yam, Trifoliate Yam ■

131472 Dioscorea armata De Wild. = Dioscorea minutiflora Engl. ■☆

131473 Dioscorea asclepiadea Prain et Burkill;马利筋薯蓣■☆

131474 Dioscorea aspersa Prain et Burkill;丽叶薯蓣(撒铺薯蓣);Roughhairy Yam, Scattered Yam ■

131475 Dioscorea balkanica Kosanin;巴尔干薯蓣■☆

131476 Dioscorea banzhuana C. P' ei et C. T. Ting;板砖薯蓣(板砖);Banzhuan Yam, Banzuan Yam ■

131477 Dioscorea bartlettii Morton;巴氏薯蓣;Bartlett Yam ■☆

131478 Dioscorea batatas Benth. = Dioscorea fordii Prain et Burkill ■

131479 Dioscorea batatas Decne. = Dioscorea polystachya Turcz. ■

131480 Dioscorea baya De Wild.;巴亚薯蓣■☆

131481 Dioscorea baya De Wild. var. subcordata ? = Dioscorea baya De Wild. ■☆

131482 Dioscorea beccariana Martelli = Dioscorea quartiniana A. Rich. ■☆

131483 Dioscorea belizensis Lundell;伯里兹薯蓣■☆

131484 Dioscorea belophylloides Prain et Burkill = Dioscorea japonica Thunb. ■

131485 Dioscorea benthamii Prain et Burkill;大青薯(山药薯,田菁,小叶薯莨);Bentham Yam ■

131486 Dioscorea bernoulliana Prain et Burkill;伯纳薯蓣■☆

131487 Dioscorea berteroana Kunth = Dioscorea cayenensis Lam. ■☆

131488 Dioscorea bicolor Prain et Burkill;尖头果薯蓣;Bicolor Yam ■

131489 Dioscorea biformifolia C. P' ei et C. T. Ting;异叶薯蓣(野山药);Biformleaf Yam, Diffrantleaf Yam ■

131490 Dioscorea birmanica Prain et Burkill;独龙薯蓣■

131491 Dioscorea biserialis Prain et Burkill = Dioscorea panthaica Prain et Burkill ■

131492 Dioscorea bonatiana Prain et Burkill = Dioscorea kamoonensis Kunth ■

131493 Dioscorea brevipes Burtt Davy = Dioscorea sylvatica Eckl. var. brevipes (Burtt Davy) Burkill ■☆

131494 Dioscorea brevispicata De Wild. = Dioscorea minutiflora Engl. ■☆

131495 Dioscorea brownii Schinz;布朗薯蓣■☆

131496 Dioscorea buchananii Benth.;布坎南薯蓣■☆

131497 Dioscorea buchananii Benth. var. ukamensis R. Knuth = Dioscorea buchananii Benth. ■☆

131498 Dioscorea buchholziana Engl. = Dioscorea dumetorum (Kunth) Pax ■☆

131499 Dioscorea buergeri Ulinie ex R. Knuth = Dioscorea tokoro Makino ex Miyabe ■

131500 Dioscorea buergeri Ulinie ex R. Knuth var. enneaneura Ulnie ex Diels = Dioscorea tokoro Makino ex Miyabe ■

131501 Dioscorea bulbifera L.;黄独(扒毒散,板薯,草蔸苕,草蔸薯,大苦,独黄,狗嗽,狗嗽子,何首乌芋,淮山薯,黄蛋,黄独零余子,黄狗头,黄金山药,黄虾蟆,黄药,黄药脂,黄药子,黄座簕,家种黄独,金锦吊蛤蟆,金丝吊蛋,金线吊蛋,金线吊蛤蟆,金线吊葫芦,金线吊虾蟆,苦卡拉,苦茅薯,苦药子,雷公薯,零余薯,零余子薯蓣,霝,毛荷叶,毛卵陀,毛薯,木药脂,木药子,山慈姑,山慈菇,山芋,薯瓜乳藤,霜降子,蓑衣包,铁秤砣,土首乌,土芋,香芋,薢草川,药子,野生黄独);Acorn, Acorn Yam, Aerial Yam, Air Potato, Air Potato Yam, Air Yam, Air-potato, Airpotato Yam, Carib Potato, Cluster Yam, Common Yam, Karro Otaheite Potato, Otaheite Yam, Potato Yam, Yam ■

131502 Dioscorea bulbifera L. f. domestica (Makino) Makino et Nemoto

= Dioscorea bulbifera L. ■
131503 Dioscorea bulbifera L. f. domestica Makino et Nemoto = Dioscorea bulbifera L. ■
131504 Dioscorea bulbifera L. var. anthropophagorum (A. Chev.) Summerh. = Dioscorea bulbifera L. ■
131505 Dioscorea bulbifera L. var. sativa Prain = Dioscorea bulbifera L. ■
131506 Dioscorea bulbifera L. var. sativa Prain = Dioscorea bulbifera L. f. domestica (Makino) Makino et Nemoto ■
131507 Dioscorea bulbifera L. var. simbha Prain et Burkill;腾冲薯蓣;Tengchong Yam ■
131508 Dioscorea bulbifera L. var. vera Prain et Burkill;茂汶薯蓣;Maowen Yam ■
131509 Dioscorea burchellii Baker;伯切尔薯蓣■☆
131510 Dioscorea burkillii R. Knuth = Dioscorea delavayi Franch. ■
131511 Dioscorea caillei A. Chev. ex De Wild. = Dioscorea togoensis R. Knuth ■☆
131512 Dioscorea camerunensis R. Knuth = Dioscorea cayenensis Lam. ■☆
131513 Dioscorea camposita Hemsl.;墨西哥薯芋■☆
131514 Dioscorea capillaria Hemsl.;发状薯蓣■☆
131515 Dioscorea caucasica Lipsky;高加索薯蓣;Caucasia Yam ■☆
131516 Dioscorea cayenensis Lam.;黄薯蓣;Attoto Yam,Cut-and-come-again,Guinea Yam,Negro Yam,Twelve-month Yam,Yellow Guinea Yam,Yellow Yam ■☆
131517 Dioscorea cayenensis Lam. subsp. rotundata (Poir.) J. Miège;圆叶黄薯蓣■☆
131518 Dioscorea chaipasensis Matuda;猜帕薯蓣■☆
131519 Dioscorea changjiangensis F. W. Xing et Z. X. Li = Dioscorea pentaphylla L. ■
131520 Dioscorea chevalieri De Wild. = Dioscorea preussii Pax ■☆
131521 Dioscorea chingii Prain et Burkill;山葛薯(三百棒);Ching Yam ■
131522 Dioscorea cirrhosa Lour.;薯莨(避血雷,恶边,孩儿血,红孩儿,红薯莨,红药子,鸡血莲,酱头,金花果,里白叶薯榔,染布薯,茹榔,山羊头,山猪薯,薯良,雄黄七,血当归,血葫芦,血母,血三七,血娃,赭魁,朱砂莲,朱砂七,猪番薯);Shuliang Yam ■
131523 Dioscorea cirrhosa Lour. var. cylindrica C. T. Ting et M. C. Chang;异块茎薯莨(异块茎薯蓣);Cylindric Yam ■
131524 Dioscorea claessensii De Wild.;克莱森斯薯蓣■☆
131525 Dioscorea cliffortiana Lam. = Dioscorea villosa L. ■☆
131526 Dioscorea cochleari-apiculata De Wild.;螺尖薯蓣■☆
131527 Dioscorea codonopsidifolia Kamik.;掌叶薯■
131528 Dioscorea codonopsidifolia Kamik. = Dioscorea pentaphylla L. ■
131529 Dioscorea collettii Hook. f.;叉蕊薯蓣(白山药,大山药,饭沙子,黑薯蓣,黑叶薯蓣,华南薯蓣,黄姑娌,黄姜,黄山药,九子不离母,南华薯蓣,蛇头草);Collett Yam ■
131530 Dioscorea collettii Hook. f. var. hypoglauca (Palib.) C. P'ei et C. T. Ting = Dioscorea collettii Hook. f. var. hypoglauca (Palib.) C. T. Ting et al. ■
131531 Dioscorea collettii Hook. f. var. hypoglauca (Palib.) C. T. Ting et al.;粉背薯蓣(白菝葜,百枝,萆薢,赤节,粉萆薢,黄萆薢,黄姜,黄山姜,金刚,麻甲头,山萆薢,山姜黄,土黄连,土薯蓣,土田薯,硬饭团,竹木);Hypoglaucous Yam,Paleback Yam ■
131532 Dioscorea colocasiifolia Pax = Dioscorea alata L. ■
131533 Dioscorea communis (L.) Caddick et Wilkin;普通薯蓣■☆
131534 Dioscorea composita Hemsl.;菊叶薯蓣(复生薯蓣)■☆
131535 Dioscorea convolvulacea Cham. et Schltdl.;旋花薯蓣■☆
131536 Dioscorea crinita Hook. f. = Dioscorea quartiniana A. Rich. ■☆
131537 Dioscorea cryptantha Baker = Dioscorea quartiniana A. Rich. ■☆
131538 Dioscorea cumingii Prain et Burkill;吕宋薯蓣(兰屿田薯)■
131539 Dioscorea cumingii Prain et Burkill var. inaequifolia (Elmer ex Prain et Burkill) Burkill = Dioscorea cumingii Prain et Burkill ■
131540 Dioscorea cumingii Prain et Burkill var. polyphylla (R. Knuth) Burkill = Dioscorea cumingii Prain et Burkill ■
131541 Dioscorea cyphocarpa B. L. Rob.;弯果薯蓣■☆
131542 Dioscorea daemona Hook. = Dioscorea dumetorum (Kunth) Pax ■☆
131543 Dioscorea daemona Roxb. = Dioscorea hispida Dennst. ■
131544 Dioscorea dawei De Wild. = Dioscorea preussii Pax ■☆
131545 Dioscorea decaisneana Carrière = Dioscorea polystachya Turcz. ■
131546 Dioscorea decipiense Hook. f.;多毛叶薯蓣(黄山药,黏山药);Deceiving Yam,Hairy Yam ■
131547 Dioscorea decipiense Hook. f. var. glabrescens C. T. Ting et M. C. Chang;滇薯;Glabrescent Yam,Glabrous Yam ■
131548 Dioscorea delavayi Franch.;高山薯蓣(滇白药子);Henry Yam ■
131549 Dioscorea deltoidea Wall. = Dioscorea deltoidea Wall. ex Kunth ■
131550 Dioscorea deltoidea Wall. ex Kunth;三角叶薯蓣(山药);Deltoid Yam ■
131551 Dioscorea deltoidea Wall. ex Kunth var. orbiculata Prain et Burkill;圆果三角叶薯蓣(风车草,黄山药,母猪藤,圆叶薯蓣);Roundfruit Yam,Roundleaf Yam ■
131552 Dioscorea deltoidea Wall. var. orbiculata Prain et Burkill = Dioscorea deltoidea Wall. ex Kunth var. orbiculata Prain et Burkill ■
131553 Dioscorea demeusei De Wild. et T. Durand = Dioscorea cayenensis Lam. ■☆
131554 Dioscorea digitaria R. Knuth = Dioscorea diversifolia Griseb. ■☆
131555 Dioscorea dinteri Schinz = Dioscorea quartiniana A. Rich. ■☆
131556 Dioscorea discolor Kunth;异色薯蓣;Ornamental Yam,Whiteband Yam ■☆
131557 Dioscorea dissecta R. Knuth = Dioscorea kamoonensis Kunth ■
131558 Dioscorea doryphora Hance;戟叶田薯■
131559 Dioscorea doryphora Hance = Dioscorea polystachya Turcz. ■
131560 Dioscorea dregeana (Kunth) T. Durand et Schinz;德雷薯蓣■☆
131561 Dioscorea dregeana (Kunth) T. Durand et Schinz var. hutchinsonii Burkill = Dioscorea dregeana (Kunth) T. Durand et Schinz ■☆
131562 Dioscorea dugesii B. L. Rob.;杜氏薯蓣■☆
131563 Dioscorea dumetorum (Kunth) Pax;灌丛薯蓣■☆
131564 Dioscorea dumetorum (Kunth) Pax var. glabrescens A. Chev. = Dioscorea dumetorum (Kunth) Pax ■☆
131565 Dioscorea dumetorum (Kunth) Pax var. lanuginosa A. Chev. = Dioscorea dumetorum (Kunth) Pax ■☆
131566 Dioscorea dumetorum (Kunth) Pax var. vespertilio A. Chev. = Dioscorea dumetorum (Kunth) Pax ■☆
131567 Dioscorea dumetorum Pax;阿比西尼亚三叶薯蓣(阿比西尼亚薯蓣);Bitter Yam,Cluster Yam,Three-leaved Yam,Vigongo ■☆
131568 Dioscorea dusenii Uline = Dioscorea hirtiflora Benth. ■☆
131569 Dioscorea ealensis De Wild. = Dioscorea minutiflora Engl. ■☆
131570 Dioscorea echinulata De Wild. = Dioscorea smilacifolia De Wild. ■☆
131571 Dioscorea edulis Lowe;可食薯蓣■☆
131572 Dioscorea ekolo De Wild. = Dioscorea minutiflora Engl. ■☆
131573 Dioscorea elephantipes (L'Hér.) Engl.;南非薯蓣(蔓龟草);Elephant Foot, Elephant's Foot, Elephant's-foot, Hottentot Bread,

Hottentot-bread,Hottentot's Bread,Tortoise Plant,Turtleback Plant ■☆
131574 Dioscorea engbo De Wild. = Dioscorea minutiflora Engl. ■☆
131575 Dioscorea engleriana R. Knuth = Dioscorea delavayi Franch. ■
131576 Dioscorea enneaneura (Uline ex Diels) Prain et Burkill = Dioscorea tokoro Makino ex Miyabe ■
131577 Dioscorea esculenta (Lour.) Burkill;甘薯(刺薯蓣,山薯,甜薯);Chinese Yam,Edible Yam,Lesser Yam,Potato Yam,Potato-yam ■
131578 Dioscorea esculenta (Lour.) Burkill var. fasciculata (Roxb.) R. Knuth = Dioscorea esculenta (Lour.) Burkill ■
131579 Dioscorea esculenta (Lour.) Burkill var. spinosa (Roxb. ex Wall.) R. Knuth;有刺甘薯(刺薯蓣);Spiny Edible Yam ■
131580 Dioscorea esquirolii Prain et Burkill;七叶薯蓣(白参,盘参,七叶薯,七爪金龙,七爪龙,血参);Esquirol Yam,Septifoliate Yam ■
131581 Dioscorea exalata C. T. Ting et M. C. Chang;无翅参薯;Wingless Yam ■
131582 Dioscorea excisa R. Knuth = Dioscorea quartiniana A. Rich. ■☆
131583 Dioscorea fargesii Franch. = Dioscorea kamoonensis Kunth ■
131584 Dioscorea fasciculata Roxb. = Dioscorea esculenta (Lour.) Burkill ■
131585 Dioscorea firma R. Knuth = Dioscorea kamoonensis Kunth ■
131586 Dioscorea flabellifolia Prain et Burkill;兰屿薯蓣;Lanyu Yam ■
131587 Dioscorea flamignii De Wild. = Dioscorea smilacifolia De Wild. ■☆
131588 Dioscorea floribunda M. Martens et Galeotti;聚花薯蓣(多花薯蓣)■☆
131589 Dioscorea floribunda Mart. et Griseb. = Dioscorea floribunda M. Martens et Galeotti ■☆
131590 Dioscorea floridana Bartlett;佛罗里达薯蓣■☆
131591 Dioscorea forbesii Baker = Dioscorea quartiniana A. Rich. ■☆
131592 Dioscorea fordii Prain et Burkill;山薯(秤根薯,山药,土淮山药);Ford Yam,Wild Yam ■
131593 Dioscorea formosana Knuth;台湾薯蓣;Taiwan Yam ■
131594 Dioscorea formosana Knuth = Dioscorea cirrhosa Lour. ■
131595 Dioscorea friedrichsthalii Kunth;中美薯蓣■☆
131596 Dioscorea fulvida Stapf = Dioscorea schimperiana Hochst. ex Kunth ■☆
131597 Dioscorea futschauensis Uline ex R. Knuth;福州薯蓣(福建薯蓣,猴骨草,绵萆薢,土萆薢);Foochow Yam,Fuzhou Yam ■
131598 Dioscorea garrettii Prain et Burkill;宽果薯蓣(加氏薯蓣);Garrett Yam ■☆
131599 Dioscorea gillettii Milne-Redh.;吉莱特薯蓣■☆
131600 Dioscorea giraldii R. Knuth. = Dioscorea nipponica Makino ■
131601 Dioscorea glabra Roxb.;光叶薯蓣(草巴山药,红山药,苦山药,莨菇,盘薯,羊角山药);Glabrous Yam ■
131602 Dioscorea glabra Roxb. var. longifolia Prain et Burkill = Dioscorea glabra Roxb. ■
131603 Dioscorea glabra Roxb. var. vera Kunth;卵叶野山药;Ovateleaf Yam ■☆
131604 Dioscorea glauca Muhl. ex Bartlett = Dioscorea villosa L. ■☆
131605 Dioscorea glauca Rusby;粉绿薯蓣■☆
131606 Dioscorea gossweileri R. Knuth = Dioscorea quartiniana A. Rich. ■☆
131607 Dioscorea gracillima Miq.;纤细薯蓣(白萆薢,白姜,白薯,萆薢,粉萆薢,黄生姜,金线吊蛤蟆,癞蛤蟆);Thinnest Yam ■
131608 Dioscorea gracillima Miq. var. collettii (Hook. f.) Uline ex Yamam. = Dioscorea collettii Hook. f. ■
131609 Dioscorea grandebulbosa R. Knuth = Dioscorea minutiflora Engl. ■☆
131610 Dioscorea grandifolia Mart. ex Griseb.;大叶薯蓣;Bigleaf Yam ■☆
131611 Dioscorea grata Prain et Burkill;异块茎薯蓣;Heterotuberous Yam ■☆
131612 Dioscorea grata Prain et Burkill = Dioscorea cirrhosa Lour. var. cylindrica C. T. Ting et M. C. Chang ■
131613 Dioscorea hainanensis Prain et Burkill = Dioscorea fordii Prain et Burkill ■
131614 Dioscorea haniltonii Hook. f.;哈氏山药;Hanilton Yam ■☆
131615 Dioscorea hemicrypta Burkill;半蔽薯蓣■☆
131616 Dioscorea hemsleyi Prain et Burkill;黏山药(盛末花,黏口薯,粘黏黏,黏薯);Hemsley Yam ■
131617 Dioscorea henryi (Prain et Burkill) C. T. Ting = Dioscorea delavayi Franch. ■
131618 Dioscorea henryi Uline ex Diels = Dioscorea zingiberensis C. H. Wright ■
131619 Dioscorea heptaphylla Sasaki = Dioscorea cumingii Prain et Burkill ■
131620 Dioscorea hexaphylla Raf. = Dioscorea villosa L. ■☆
131621 Dioscorea hirsuta Mart. et Griseb.;毛薯蓣■☆
131622 Dioscorea hirsuticaulis B. L. Rob.;毛茎薯蓣■☆
131623 Dioscorea hirticaulis Bartlett = Dioscorea villosa L. ■☆
131624 Dioscorea hirtiflora Benth.;毛花薯蓣■☆
131625 Dioscorea hirtiflora Benth. subsp. orientalis Milne-Redh.;东方毛花薯蓣■☆
131626 Dioscorea hirtiflora Benth. subsp. pedicellata Milne-Redh.;小梗毛花薯蓣■☆
131627 Dioscorea hirtiflora Benth. var. grahamii Burkill = Dioscorea hirtiflora Benth. subsp. orientalis Milne-Redh. ■☆
131628 Dioscorea hirtiflora Benth. var. polyantha (Rendle) Burkill = Dioscorea hirtiflora Benth. ■☆
131629 Dioscorea hirtiflora Benth. var. trapnellii Burkill = Dioscorea hirtiflora Benth. ■☆
131630 Dioscorea hispida Dennst.;白薯莨(白米茹粮,白薯榔,白薯浪郎,白薯蓣,板薯,榜花薯,榜薯,大苦薯,大力王,独龙,山朴薯,山什薯,山薯,网脉三叶薯,野葛薯,叶板茨,叶板薯);Hispid Yam ■
131631 Dioscorea hispida Dennst. var. daemona (Roxb.) Prain et Burkill = Dioscorea hispida Dennst. ■
131632 Dioscorea hockii De Wild. = Dioscorea schimperiana Hochst. ex Kunth ■☆
131633 Dioscorea hoffa Cordem. = Dioscorea bulbifera L. ■
131634 Dioscorea hofika Jum. et H. Perrier = Dioscorea bulbifera L. ■
131635 Dioscorea holstii Harms = Dioscorea quartiniana A. Rich. ■☆
131636 Dioscorea hongdurensis Kunth;洪都拉斯薯蓣■☆
131637 Dioscorea hongkongense Uline ex R. Knuth = Dioscorea glabra Roxb. ■
131638 Dioscorea hui R. Knuth = Dioscorea collettii Hook. f. ■
131639 Dioscorea hylophila Harms = Dioscorea preussii Pax subsp. hylophila (Harms) Wilkin ■☆
131640 Dioscorea hypoglauca Palib. = Dioscorea collettii Hook. f. var. hypoglauca (Palib.) C. T. Ting et al. ■
131641 Dioscorea hypotricha Uline = Dioscorea semperflorens Uline ■☆
131642 Dioscorea hystrix R. Knuth = Dioscorea minutiflora Engl. ■☆
131643 Dioscorea inaequifolia Elmer ex Prain et Burkill = Dioscorea cumingii Prain et Burkill ■

131644 Dioscorea izuensis Akahori = Dioscorea collettii Hook. f. var. hypoglauca (Palib.) C. T. Ting et al. ■

131645 Dioscorea japonica Thunb.;日本薯蓣(薄叶野山药,大薯,儿草,风车儿,风车子,狂风藤,坭洞薯,千担苕,千斤拔,山蝴蝶,山药,山芋,薯蓣,薯萸,薯语,土淮山,土薯,小粘狗苕,休脆,野白菇,野山药,野薯,玉延,诸薯);Chinese Yam,Japan Yam,Japanese Yam ■

131646 Dioscorea japonica Thunb. var. kelungensis (R. Knuth) Prain et Burkill = Dioscorea japonica Thunb. var. pseudojaponica (Hayata) Yamam. ■

131647 Dioscorea japonica Thunb. var. oldhamii Uline ex R. Knuth;细叶日本薯蓣(细叶野山药,竹高薯);Oldham Yam ■

131648 Dioscorea japonica Thunb. var. pilifera C. T. Ting et M. C. Chang;毛藤日本薯蓣;Hairvine Yam,Hairy Japanese Yam ■

131649 Dioscorea japonica Thunb. var. pseudojaponica (Hayata) Yamam. = Dioscorea japonica Thunb. ■

131650 Dioscorea japonica Thunb. var. pseudojaponica (Hayata) Yamam. = Dioscorea pseudojaponica Hayata ■

131651 Dioscorea japonica Thunb. var. tenuiaxon Prain et Burkill = Dioscorea japonica Thunb. ■

131652 Dioscorea japonica Thunb. var. vera Prain et Burkill;细穗野山药(细叶野山药)■

131653 Dioscorea japonica Thunb. var. vera Prain et Burkill = Dioscorea japonica Thunb. ■

131654 Dioscorea junodii Burtt Davy;朱诺德薯蓣■☆

131655 Dioscorea kamoonensis Kunth;毛芋头薯蓣(白药子,草黄滇白药子,滇白药子,防粉风党参,加莫苕,马蹄细辛,毛芋头);Kamoon Yam ■

131656 Dioscorea kamoonensis Kunth var. brevifolia Prain et Burkill = Dioscorea kamoonensis Kunth ■

131657 Dioscorea kamoonensis Kunth var. delavayi (Franch.) Prain et Burkill = Dioscorea delavayi Franch. ■

131658 Dioscorea kamoonensis Kunth var. delavayi Prain et Burkill = Dioscorea delavayi Franch. ■

131659 Dioscorea kamoonensis Kunth var. engleriana (R. Knuth) Prain et Burkill = Dioscorea delavayi Franch. ■

131660 Dioscorea kamoonensis Kunth var. engleriana Prain et Burkill = Dioscorea kamoonensis Kunth ■

131661 Dioscorea kamoonensis Kunth var. fargesii (Franch.) Prain et Burkill = Dioscorea kamoonensis Kunth ■

131662 Dioscorea kamoonensis Kunth var. fargesii Prain et Burkill = Dioscorea kamoonensis Kunth ■

131663 Dioscorea kamoonensis Kunth var. henryi Prain et Burkill = Dioscorea delavayi Franch. ■

131664 Dioscorea kamoonensis Kunth var. praccox Prain et Burkill = Dioscorea kamoonensis Kunth ■

131665 Dioscorea kamoonensis Kunth var. straminea Prain et Burkill = Dioscorea kamoonensis Kunth ■

131666 Dioscorea kamoonensis Kunth var. vera Prain et Burkill = Dioscorea kamoonensis Kunth ■

131667 Dioscorea kanconensis Kunth;复叶薯蓣■

131668 Dioscorea kaoi Tang S. Liu et T. C. Huang;圆锥花薯蓣;Kao Yam ■

131669 Dioscorea kaoi Tang S. Liu et T. C. Huang = Dioscorea collettii Hook. f. var. hypoglauca (Palib.) C. T. Ting et al. ■

131670 Dioscorea karatana Wilkin;卡拉特薯蓣●☆

131671 Dioscorea kelungensis Hayata = Dioscorea collettii Hook. f. ■

131672 Dioscorea kiangsiensis R. Knuth = Dioscorea japonica Thunb. ■

131673 Dioscorea knuthiana De Wild.;克努特薯蓣■☆

131674 Dioscorea kumaonensis ? = Dioscorea melanophyma Prain et Burkill ■

131675 Dioscorea kuuroo Honda;薯榔■☆

131676 Dioscorea latifolia Benth. = Dioscorea bulbifera L. ■

131677 Dioscorea latifolia Benth. var. anthropophagorum A. Chev. = Dioscorea bulbifera L. ■

131678 Dioscorea lecardii De Wild. = Dioscorea sagittifolia Pax var. lecardii (De Wild.) Nkounkou ■☆

131679 Dioscorea lecardii De Wild. var. vera Burkill = Dioscorea sagittifolia Pax var. lecardii (De Wild.) Nkounkou ■☆

131680 Dioscorea leonensis G. Don;莱昂薯蓣■☆

131681 Dioscorea lilela De Wild. = Dioscorea minutiflora Engl. ■☆

131682 Dioscorea lindiensis R. Knuth = Dioscorea hirtiflora Benth. subsp. orientalis Milne-Redh. ■☆

131683 Dioscorea lineari-cordata Prain et Burkill;柳叶薯蓣;Linear-cordate Yam,Willowleaf Yam ■

131684 Dioscorea litoie De Wild. = Dioscorea minutiflora Engl. ■☆

131685 Dioscorea lloydiana E. H. L. Krause = Dioscorea villosa L. ■☆

131686 Dioscorea lobata Uline;浅裂薯蓣;Lobate Yam ■☆

131687 Dioscorea longespicata De Wild. et T. Durand = Dioscorea preussii Pax ■☆

131688 Dioscorea longicuspis R. Knuth;长尖薯蓣■☆

131689 Dioscorea longifolia Raf. = Dioscorea villosa L. ■☆

131690 Dioscorea longipetiolata Baudon = Dioscorea bulbifera L. ■

131691 Dioscorea luzonensis Schauer;菲律宾薯蓣■☆

131692 Dioscorea macabiha Jum. et H. Perrier = Dioscorea sansibarensis Pax ■☆

131693 Dioscorea macrostachya Benth.;大穗薯蓣;African Elephant's Foot,Bigspike Yam ■☆

131694 Dioscorea macroura Harms = Dioscorea sansibarensis Pax ■☆

131695 Dioscorea mairei H. Lév. = Dioscorea hemsleyi Prain et Burkill ■

131696 Dioscorea mairei R. Knuth = Dioscorea kamoonensis Kunth ■

131697 Dioscorea malchairii De Wild. = Dioscorea preussii Pax ■☆

131698 Dioscorea mangenotiana J. Miège;芒热诺薯蓣■☆

131699 Dioscorea marlothii R. Knuth = Dioscorea sylvatica Eckl. var. multiflora (Marloth) Burkill ■☆

131700 Dioscorea martini Prain et Burkill;柔毛薯蓣;Martin Yam,Softhair Yam ■

131701 Dioscorea matsudae Hayata = Dioscorea cirrhosa Lour. ■

131702 Dioscorea matsudai Hayata;里白叶薯榔;Matsyda Yam ■

131703 Dioscorea maximowiczii Uline ex R. Knuth = Dioscorea tenuipes Franch. et Sav. ■

131704 Dioscorea megaptera Raf. = Dioscorea villosa L. ■☆

131705 Dioscorea melanophyma Burkill et Prain = Dioscorea melanophyma Prain et Burkill ■

131706 Dioscorea melanophyma Prain et Burkill;黑珠芽薯蓣(白药子,粉渣渣,黑弹子,毛狗卵);Blackbulbil Yam ■

131707 Dioscorea menglaensis H. Li;石山薯蓣;Mengla Yam ■

131708 Dioscorea mengtzeana R. Knuth = Dioscorea kamoonensis Kunth ■

131709 Dioscorea menispermoides C. Y. Wu;黄防己状薯蓣■

131710 Dioscorea mexicana Scheidw.;墨西哥薯蓣■☆

131711 Dioscorea microcuspis Baker = Dioscorea retusa Mast. ■☆

131712 Dioscorea mildbraediana R. Knuth = Dioscorea buchananii Benth. ■☆

131713 Dioscorea mildbraedii R. Knuth = Dioscorea sagittifolia Pax ■☆

131714 Dioscorea militaria B. L. Rob. ;甲胄状薯蓣■☆
131715 Dioscorea minima Rob. et Seaton;极小薯蓣■☆
131716 Dioscorea minutiflora Engl. ;小花薯蓣■☆
131717 Dioscorea minutiflora Engl. var. zenkeri Uline ex Harms = Dioscorea smilacifolia De Wild. ■☆
131718 Dioscorea mollissima Blume = Dioscorea hispida Dennst. ■
131719 Dioscorea moma De Wild. = Dioscorea cayenensis Lam. ■☆
131720 Dioscorea montana Eckl. et Zeyh. ex R. Knuth = Dioscorea sylvatica Eckl. ■☆
131721 Dioscorea montana Eckl. et Zeyh. ex R. Knuth var. glauca R. Knuth = Dioscorea sylvatica Eckl. ■☆
131722 Dioscorea montana Eckl. et Zeyh. ex R. Knuth var. lobata Weim. = Dioscorea sylvatica Eckl. ■☆
131723 Dioscorea morsei Prain et Burkill = Dioscorea collettii Hook. f. var. hypoglauca (Palib.) C. T. Ting et al. ■
131724 Dioscorea multicolor Lindl. et André;多色薯蓣■☆
131725 Dioscorea multiflora Engl. ex Pax = Dioscorea minutiflora Engl. ■☆
131726 Dioscorea multiflora Mart. ex Griseb. ;多花薯蓣;Manyflower Yam ■☆
131727 Dioscorea multiloba Kunth = Dioscorea diversifolia Griseb. ■☆
131728 Dioscorea multinervis Benth. ;多脉薯蓣■☆
131729 Dioscorea mundtii Baker;蒙特薯蓣■☆
131730 Dioscorea nanlaensis H. Li;南腊薯蓣■
131731 Dioscorea natalensis Engl. ;纳塔尔薯蓣■
131732 Dioscorea nelsonii Uline;纳尔逊薯蓣■☆
131733 Dioscorea nigrescens R. Knuth = Dioscorea collettii Hook. f. ■
131734 Dioscorea nipponica Makino;穿龙薯蓣(柴黄姜,常山,穿地龙,穿龙骨,穿山骨,穿山龙,穿山薯蓣,串山龙,地龙骨,粉萆薢,狗骨头,狗山药,海龙七,黄鞭,黄姜,火藤,鸡骨头,金刚骨,龙骨七,爬山虎,山常山,山红苕,山花啦,铁根薯,土常山,土山薯,团扇薯蓣,雄黄,野山药,竹根薯);Chuanlong Yam,Throughhill Yam ■
131735 Dioscorea nipponica Makino f. jamesii (Prain et Burkill) Kitag. = Dioscorea nipponica Makino subsp. rosthornii (Prain et Burkill) C. T. Ting ■
131736 Dioscorea nipponica Makino subsp. rosthornii (Prain et Burkill) C. T. Ting;柴姜黄(柴黄姜,穿山龙,光叶穿龙薯蓣,黄姜子);Rosthorn Yam ■
131737 Dioscorea nipponica Makino var. jamesii Prain et Burkill = Dioscorea nipponica Makino subsp. rosthornii (Prain et Burkill) C. T. Ting ■
131738 Dioscorea nipponica Makino var. rosthornii Prain et Burkill = Dioscorea nipponica Makino subsp. rosthornii (Prain et Burkill) C. T. Ting ■
131739 Dioscorea nitens Prain et Burkill;光亮薯蓣■
131740 Dioscorea nummularia Roxb. = Dioscorea glabra Roxb. ■
131741 Dioscorea nutans R. Knuth;俯垂薯蓣;Shining Yam ■
131742 Dioscorea occidentalis R. Knuth = Dioscorea cayenensis Lam. ■☆
131743 Dioscorea oenea Prain et Burkill = Dioscorea collettii Hook. f. ■
131744 Dioscorea officinalis Z. H. Tsi ?;疗萆薢(绵萆薢)■
131745 Dioscorea opposita Thunb. = Dioscorea polystachya Turcz. ■
131746 Dioscorea oppositifolia Benth. = Dioscorea benthamii Prain et Burkill ■
131747 Dioscorea oppositifolia L. ;对叶薯蓣;Chinese Yam,Cinnamon Vine ■
131748 Dioscorea orbicularis A. Chev. ex De Wild. = Dioscorea smilacifolia De Wild. ■☆
131749 Dioscorea owenii Prain et Burkill;真薯■
131750 Dioscorea paecox Prain et Burkill = Dioscorea hemsleyi Prain et Burkill ■
131751 Dioscorea paleata Burkill;稃薯蓣■☆
131752 Dioscorea paniculata Michx. = Dioscorea villosa L. ■☆
131753 Dioscorea paniculata Michx. var. glabrifolia Bartlett = Dioscorea villosa L. ■☆
131754 Dioscorea panthaica Prain et Burkill;黄山药(滇白药子,猴节莲,黄姜,姜黄草,老虎姜,小哨姜黄,知母山药);Yellow Yam ■
131755 Dioscorea parviflora C. T. Ting = Dioscorea sinoparviflora C. T. Ting ■
131756 Dioscorea pendula R. Knuth = Dioscorea minutiflora Engl. ■☆
131757 Dioscorea pentadactyla Welw. = Dioscorea quartiniana A. Rich. ■☆
131758 Dioscorea pentaphylla (Hochst.) A. Rich. = Dioscorea quartiniana A. Rich. ■☆
131759 Dioscorea pentaphylla L. ;五叶薯蓣(苦卡拉,毛薯藤,毛团子,血参);Fiveleaf Yam ■
131760 Dioscorea perrieri R. Knuth = Dioscorea bulbifera L. ■
131761 Dioscorea persimilis Prain et Burkill;褐苞薯蓣(假山药薯,罗社山薯,山薯,山药,薯仔);Brownbract Yam ■
131762 Dioscorea persimilis Prain et Burkill var. pubescens C. T. Ting et M. C. Chang;毛褐苞薯蓣;Pubescent Brownbract Yam,Pubescent Yam ■
131763 Dioscorea peteri R. Knuth = Dioscorea quartiniana A. Rich. ■☆
131764 Dioscorea phaseoloides Pax = Dioscorea quartiniana A. Rich. ■☆
131765 Dioscorea platanifolia Prain et Burkill = Dioscorea althaeoides R. Knuth ■
131766 Dioscorea platycolpota Uline ex Rob. ;宽鞘薯蓣■☆
131767 Dioscorea plumifera B. L. Rob. ;羽毛薯蓣■☆
131768 Dioscorea poilanei Prain et Burkill;吊罗薯蓣(疏花薯蓣);Poilane Yam ■
131769 Dioscorea poligonoides Kunth;蓼状薯蓣■☆
131770 Dioscorea polyantha Rendle = Dioscorea hirtiflora Benth. ■☆
131771 Dioscorea polyphylla R. Knuth = Dioscorea cumingii Prain et Burkill ■
131772 Dioscorea polystachya Turcz. ;薯蓣(白山药,白苕,白药子,百苕,长山药,大薯,多蕊薯蓣,儿草,佛掌薯,光山药,怀山,怀山药,淮山,淮山药,戟叶薯蓣,戟叶田薯,家山药,九黄姜,零余子,毛山药,面山药,坭洞薯,日本薯蓣,山板薯,山板术,山薯,山薯蓣,山薯,山药,山药蛋,山药薯,山芋,山蓣,扇子薯,蛇芋,暑预,暑预子,薯药,薯蓣果,薯蓣子,薯,薯粮,薯薯,薯芋,田薯,铁拐山药,土薯,王芋,修脆,延草,野白薯,野脚板薯,野牛尾苕,野山豆,野山药,玉延,蓣药,诸署);Chinese Potato,Chinese Yam,Cinnamon Vine,Common Yam,Yam ■
131773 Dioscorea polystachya Turcz. f. elongata ?;长蕊薯蓣■
131774 Dioscorea potaninii Prain et Burkill = Dioscorea polystachya Turcz. ■
131775 Dioscorea praecox Prain et Burkill = Dioscorea hemsleyi Prain et Burkill ■
131776 Dioscorea praehensilis Benth. var. minutiflora (Engl.) Baker = Dioscorea minutiflora Engl. ■☆
131777 Dioscorea prazeri Prain et Burkill;巴拉次薯蓣;Prazer Yam ■☆
131778 Dioscorea preussii Pax;普罗伊斯薯蓣■☆
131779 Dioscorea preussii Pax subsp. hylophila (Harms) Wilkin;喜盐薯蓣■☆
131780 Dioscorea pringlei B. L. Rob. ;巴林薯蓣;Pringler Yam ■☆

131781 Dioscorea pruinosa A. Chev. = Dioscorea cayenensis Lam. ■☆
131782 Dioscorea pruinosa Kunth = Dioscorea villosa L. ■☆
131783 Dioscorea pseudo japonica Hayata = Dioscorea japonica Thunb. ■
131784 Dioscorea pseudojaponica Hayata;假日本薯蓣(基隆野山药,山药薯)■
131785 Dioscorea pseudojaponica Hayata = Dioscorea japonica Thunb. ■
131786 Dioscorea pseudojaponica Hayata = Dioscorea japonica Thunb. var. pseudojaponica (Hayata) Yamam. ■
131787 Dioscorea pterocaulon De Wild. et T. Durand = Dioscorea preussii Pax ■☆
131788 Dioscorea pulverea Prain et Burkill = Dioscorea aspersa Prain et Burkill ■
131789 Dioscorea purpurea Roxb. = Dioscorea alata L. ■
131790 Dioscorea purpurea Roxb. = Dioscorea alata L. var. purpurea (Roxb.) A. Pouchet ■
131791 Dioscorea pynaertii De Wild.;皮那薯蓣■☆
131792 Dioscorea pynaertioides De Wild. = Dioscorea minutiflora Engl. ■☆
131793 Dioscorea quartiniana A. Rich.;夸尔薯蓣■☆
131794 Dioscorea quartiniana A. Rich. var. dinteri (Schinz) Burkill = Dioscorea quartiniana A. Rich. ■☆
131795 Dioscorea quartiniana A. Rich. var. latifolia R. Knuth = Dioscorea quartiniana A. Rich. ■☆
131796 Dioscorea quartiniana A. Rich. var. schliebenii (R. Knuth) Burkill = Dioscorea quartiniana A. Rich. ■☆
131797 Dioscorea quartiniana A. Rich. var. stuhlmannii (Harms) Burkill = Dioscorea quartiniana A. Rich. ■☆
131798 Dioscorea quaternata (Walter) J. F. Gmel.;野薯蓣;Wild Yam ■☆
131799 Dioscorea quaternata (Walter) J. F. Gmel. var. glauca (Muhl.) Fernald = Dioscorea quaternata (Walter) J. F. Gmel. ■☆
131800 Dioscorea quaternata J. F. Gmel. = Dioscorea villosa L. ■☆
131801 Dioscorea quaternata J. F. Gmel. var. glauca (Muhl. ex Bartlett) Fernald = Dioscorea villosa L. ■☆
131802 Dioscorea quinata J. F. Gmel. = Dioscorea villosa L. ■☆
131803 Dioscorea quinqueloba Thunb. = Dioscorea quinquelobata Thunb. ■☆
131804 Dioscorea quinquelobata Thunb.;五裂叶薯蓣;Fivelobed Yam ■☆
131805 Dioscorea raishaensis Hayata = Dioscorea persimilis Prain et Burkill ■
131806 Dioscorea ramotiflora Kunth;疏花薯蓣■☆
131807 Dioscorea rehmannii Baker = Dioscorea sylvatica Eckl. var. rehmannii (Baker) Burkill ■☆
131808 Dioscorea repanda Raf. = Dioscorea villosa L. ■☆
131809 Dioscorea retusa Mast.;微凹薯蓣■☆
131810 Dioscorea rhacodes R. Knuth = Dioscorea buchananii Benth. ■☆
131811 Dioscorea rhipogonoides Hayata = Dioscorea japonica Thunb. ■
131812 Dioscorea rhipogonoides Oliv. = Dioscorea cirrhosa Lour. ■
131813 Dioscorea rosthornii Diels = Dioscorea polystachya Turcz. ■
131814 Dioscorea rotundata Poir.;几内亚薯蓣;Guinea Yam, White Guinea Yam, White Yam ■☆
131815 Dioscorea rotundata Poir. = Dioscorea cayenensis Lam. subsp. rotundata (Poir.) J. Miège ■☆
131816 Dioscorea rotundifoliolata Kunth;过沟藤;Roundfoliolate Yam ■
131817 Dioscorea rotundifoliolata R. Knuth = Dioscorea delavayi Franch. ■
131818 Dioscorea rubiginosa Benth. = Dioscorea hirtiflora Benth. ■☆
131819 Dioscorea rupicola Kunth;岩生薯蓣■☆
131820 Dioscorea sagittifolia Pax;箭叶薯蓣■☆
131821 Dioscorea sagittifolia Pax var. lecardii (De Wild.) Nkounkou;莱卡德薯蓣■☆
131822 Dioscorea saidae R. Knuth = Dioscorea tokoro Makino ex Miyabe ■
131823 Dioscorea salvadorica Büttner;萨尔瓦多薯蓣■☆
131824 Dioscorea sansibarensis Pax;桑给巴尔薯蓣;Zanzibar Yam ■☆
131825 Dioscorea sapinii De Wild. = Dioscorea alata L. ■
131826 Dioscorea sativa L. = Dioscorea bulbifera L. ■
131827 Dioscorea sativa Miq. = Dioscorea tokoro Makino ex Miyabe ■
131828 Dioscorea sativa Thunb. = Dioscorea bulbifera L. ■
131829 Dioscorea schimperiana Hochst. ex Kunth;欣珀薯蓣■☆
131830 Dioscorea schimperiana Hochst. ex Kunth var. adamaouense Jacq. -Fél. = Dioscorea schimperiana Hochst. ex Kunth ■☆
131831 Dioscorea schimperiana Hochst. ex Kunth var. nigrescens R. Knuth = Dioscorea longicuspis R. Knuth ■☆
131832 Dioscorea schimperiana Hochst. ex Kunth var. vestita Pax = Dioscorea schimperiana Hochst. ex Kunth ■☆
131833 Dioscorea schlechteri Harms = Dioscorea semperflorens Uline ■☆
131834 Dioscorea schliebenii R. Knuth = Dioscorea quartiniana A. Rich. ■☆
131835 Dioscorea schweinfurthiana Pax = Dioscorea quartiniana A. Rich. ■☆
131836 Dioscorea scortechinii Prain et Burkill;刺薯蓣(斯氏薯蓣);Scortechini Yam ■☆
131837 Dioscorea scortechinii Prain et Burkill var. parviflora Prain et Burkill;小花刺薯蓣;Smallflower Spiny Yam ■
131838 Dioscorea semperflorens Uline;常花薯蓣■☆
131839 Dioscorea seniavinii Prain et Burkill = Dioscorea collettii Hook. f. ■
131840 Dioscorea septemloba Thunb.;绵萆薢(畚箕斗,萆薢,大萆薢,福州薯蓣,狗粪棵,狗骨草,七裂叶薯蓣,三脚灵,山畚箕,小萆薢);Sevenlobed Yam ■
131841 Dioscorea septemloba Thunb. = Dioscorea spongiosa J. Q. Xi, Mizuae et W. L. Zhao ■
131842 Dioscorea septemloba Thunb. var. platyphylla M. Mizush. ex T. Shimizu;宽叶绵萆薢■☆
131843 Dioscorea sikkimensis Prain et Burkill = Dioscorea prazeri Prain et Burkill ■☆
131844 Dioscorea simulans Prain et Burkill;马肠薯蓣(野山薯);Similar Yam ■
131845 Dioscorea sinoparviflora C. T. Ting = Dioscorea sinoparviflora C. T. Ting et al. ■
131846 Dioscorea sinoparviflora C. T. Ting et al.;小花盾叶薯蓣(苦良姜,老虎姜);Smallflower Yam ■
131847 Dioscorea smilacifolia De Wild.;菝葜叶薯蓣■☆
131848 Dioscorea smilacifolia De Wild. var. alternifolia ? = Dioscorea smilacifolia De Wild. ■☆
131849 Dioscorea spiculiflora Hemsl.;小穗花薯蓣;Spiculiflower Yam ■☆
131850 Dioscorea spinosa Roxb. ex Wall. = Dioscorea esculenta (Lour.) Burkill var. spinosa (Roxb. ex Wall.) R. Knuth ■
131851 Dioscorea spongiosa J. Q. Xi, Mizuae et W. L. Zhao;海绵薯蓣(海绵萆薢,绵萆薢,绵草);Spongy Yam ■
131852 Dioscorea stellatopilosa De Wild. = Dioscorea schimperiana Hochst. ex Kunth ■☆

131853 Dioscorea stellatopilosa De Wild. var. cordata ? = Dioscorea hirtiflora Benth. ■☆

131854 Dioscorea stipulosa Uline ex R. Knuth;托叶薯蓣■☆

131855 Dioscorea stolzii R. Knuth = Dioscorea cochleari-apiculata De Wild. ■☆

131856 Dioscorea stuhlmannii Harms = Dioscorea quartiniana A. Rich. ■☆

131857 Dioscorea subcalva Prain et Burkill;毛胶薯蓣(黄山药,近光薯蓣,黏山药,黏粘粘,牛尾参,黏狗苕,黏芋);Subcalvous Yam ■

131858 Dioscorea subcalva Prain et Burkill var. submollis (R. Knuth) C. T. Ting et P. P. Ling;略毛薯蓣;Subpubescent Yam ■

131859 Dioscorea subfusca R. Knuth = Dioscorea kamoonensis Kunth ■

131860 Dioscorea submollis R. Knuth = Dioscorea subcalva Prain et Burkill var. submollis (R. Knuth) C. T. Ting et P. P. Ling ■

131861 Dioscorea swinhoei Rolfe = Dioscorea polystachya Turcz. ■

131862 Dioscorea sylvatica (Kunth) Eckl. subsp. lydenbergensis Blunden, Hardman et Hind = Dioscorea sylvatica Eckl. var. brevipes (Burtt Davy) Burkill ■☆

131863 Dioscorea sylvatica Eckl.;森林薯蓣;Elephant's Foot, Woodland Yam ■☆

131864 Dioscorea sylvatica Eckl. var. brevipes (Burtt Davy) Burkill;短梗森林薯蓣■☆

131865 Dioscorea sylvatica Eckl. var. multiflora (Marloth) Burkill;多花森林薯蓣■☆

131866 Dioscorea sylvatica Eckl. var. paniculata (Dümmer) Burkill;圆锥森林薯蓣■☆

131867 Dioscorea sylvatica Eckl. var. rehmannii (Baker) Burkill;拉赫曼森林薯蓣■☆

131868 Dioscorea sylvestris De Wild. = Dioscorea bulbifera L. ■

131869 Dioscorea tarokoensis Hayata = Dioscorea benthamii Prain et Burkill ■

131870 Dioscorea tashiroi Hayata = Dioscorea collettii Hook. f. ■

131871 Dioscorea tenii R. Knuth = Dioscorea melanophyma Prain et Burkill ■

131872 Dioscorea tentaculigera Prain et Burkill;卷须状薯蓣(触丝薯蓣);Tendrillous Yam, Tentacle-bearing Yam ■

131873 Dioscorea tenuipes Franch. et Sav.;细柄薯蓣(小黄连,野生姜);Thinstiped Yam ■

131874 Dioscorea testudinaria Knuth;龟状薯蓣■☆

131875 Dioscorea thonneri De Wild. et T. Durand = Dioscorea preussii Pax ■☆

131876 Dioscorea togoensis R. Knuth;多哥薯蓣■☆

131877 Dioscorea tokoro Makino ex Miyabe;山萆薢(粉萆薢,土黄姜,土黄连,小萆薢,竹根薯);Mountain Yam ■

131878 Dioscorea toxicaria Bojer = Dioscorea sansibarensis Pax ■☆

131879 Dioscorea transversa R. Br.;横薯蓣■☆

131880 Dioscorea trifida L. f.;三浅裂薯蓣(南美薯蓣);Cush-cush, Cush-cush Yam, Cush-cush Yampee, Indian Yam, Mapney, Yampee, Yampi, Yampi Yam ■☆

131881 Dioscorea triphylla A. Rich. var. abyssinica R. Knuth = Dioscorea dumetorum (Kunth) Pax ■☆

131882 Dioscorea triphylla A. Rich. var. dumetorum (Kunth) R. Knuth = Dioscorea dumetorum (Kunth) Pax ■☆

131883 Dioscorea triphylla A. Rich. var. rotundata R. Knuth = Dioscorea dumetorum (Kunth) Pax ■☆

131884 Dioscorea triphylla A. Rich. var. tomentosa Rendle = Dioscorea dumetorum (Kunth) Pax ■☆

131885 Dioscorea triphylla L. var. reticulata Prain et Burkill = Dioscorea hispida Dennst. ■

131886 Dioscorea triphylla Schimp. ex Kunth = Dioscorea dumetorum (Kunth) Pax ■☆

131887 Dioscorea tysonii Baker = Dioscorea retusa Mast. ■☆

131888 Dioscorea tysonii Schönland;泰森薯蓣■☆

131889 Dioscorea ulinei Greenm.;尤林薯蓣;Uline Yam ■☆

131890 Dioscorea ulugurensis R. Knuth = Dioscorea quartiniana A. Rich. ■☆

131891 Dioscorea undulata R. Knuth = Dioscorea collettii Hook. f. var. hypoglauca (Palib.) C. T. Ting et al. ■

131892 Dioscorea urceolata Uline;壶状薯蓣■☆

131893 Dioscorea velutipes Prain et Burkill;毡毛薯蓣;Velutinousstipe Yam ■

131894 Dioscorea verdickii De Wild. = Dioscorea quartiniana A. Rich. ■☆

131895 Dioscorea vespertilio Benth. = Illigera vespertilio (Benth.) Baker f. ●☆

131896 Dioscorea villosa L.;绒毛薯蓣(北美大西洋沿岸薯蓣,长柔毛薯蓣);Atlantic Yam, Colic Root, Colic-root, Wild Yam, Wild Yamroot, Yam, Yam Root ■☆

131897 Dioscorea villosa L. f. glabrifolia (Bartlett) Fernald = Dioscorea villosa L. ■☆

131898 Dioscorea villosa L. subsp. floridana (Bartlett) R. Knuth = Dioscorea floridana Bartlett ■☆

131899 Dioscorea villosa L. subsp. glabrifolia (Bartlett) S. F. Blake = Dioscorea villosa L. ■☆

131900 Dioscorea villosa L. subsp. glauca (Muhl. ex Bartlett) R. Kunth = Dioscorea villosa L. ■☆

131901 Dioscorea villosa L. subsp. hirticaulis (Bartlett) R. Knuth = Dioscorea villosa L. ■☆

131902 Dioscorea villosa L. subsp. quaternata (J. F. Gmel.) R. Knuth = Dioscorea villosa L. ■☆

131903 Dioscorea villosa L. var. floridana (Bartlett) H. E. Ahles = Dioscorea floridana Bartlett ■☆

131904 Dioscorea villosa L. var. glabrifolia (Bartlett) S. F. Blake = Dioscorea villosa L. ■☆

131905 Dioscorea villosa L. var. hirticaulis (Bartlett) Ahles = Dioscorea villosa L. ■☆

131906 Dioscorea villosa L. var. vera Prain et Burkill = Dioscorea villosa L. ■☆

131907 Dioscorea violacea Baudon = Dioscorea bulbifera L. ■

131908 Dioscorea vittata Bull ex Baker;斑纹薯蓣;Striped Yam ■☆

131909 Dioscorea wallichii Hook. f.;盈江薯蓣;Wallich Yam ■

131910 Dioscorea waltheri Desf. = Dioscorea villosa L. ■☆

131911 Dioscorea welwitschii Rendle = Dioscorea sansibarensis Pax ■☆

131912 Dioscorea wichurae Uline ex R. Knuth = Dioscorea tokoro Makino ex Miyabe ■

131913 Dioscorea xizanensis C. T. Ting;西藏薯蓣(藏刺薯蓣);Xizang Yam ■

131914 Dioscorea yokusai Prain et Burkill = Dioscorea tokoro Makino ex Miyabe ■

131915 Dioscorea yunnanensis Prain et Burkill;云南薯蓣;Yunnan Yam ■

131916 Dioscorea zara Baudon = Dioscorea sagittifolia Pax ■☆

131917 Dioscorea zingiberensis C. H. Wright;盾叶薯蓣(地黄参,地黄姜,黄姜,黄连参,火柴头,火藤根,火头根,苦良姜,水黄姜,野洋姜,枕头根);Peltate Yam ■

131918 Dioscoreaceae R. Br. (1810)(保留科名);薯蓣科;Black Bryony Family, Yam Family ■●

131919 Dioscoreophyllum Engl. (1895);薯蓣叶藤属(薯蓣叶属)●☆
131920 Dioscoreophyllum chirindense Swynn. = Dioscoreophyllum volkensii Engl. ●☆
131921 Dioscoreophyllum cumminsii (Stapf) Diels = Dioscoreophyllum volkensii Engl. ●☆
131922 Dioscoreophyllum cumminsii (Stapf) Diels var. leptotrichos Troupin = Dioscoreophyllum volkensii Engl. ●☆
131923 Dioscoreophyllum cumminsii Diels = Dioscoreophyllum cumminsii (Stapf) Diels ●☆
131924 Dioscoreophyllum fernandense Hutch. et Dalziel = Dioscoreophyllum volkensii Engl. var. fernandense (Hutch. et Dalziel) Troupin ●☆
131925 Dioscoreophyllum gossweileri Exell;戈斯蓣叶藤●☆
131926 Dioscoreophyllum klaineanum Pierre ex Diels = Dioscoreophyllum volkensii Engl. ●☆
131927 Dioscoreophyllum oligotrichum Diels;寡毛薯蓣叶藤●☆
131928 Dioscoreophyllum podandrium Exell = Dioscoreophyllum volkensii Engl. var. fernandense (Hutch. et Dalziel) Troupin ●☆
131929 Dioscoreophyllum strigosum Engl. = Dioscoreophyllum volkensii Engl. ●☆
131930 Dioscoreophyllum tenerum Engl. = Dioscoreophyllum volkensii Engl. ●☆
131931 Dioscoreophyllum tenerum Engl. var. fernandense (Hutch. et Dalziel) Troupin = Dioscoreophyllum volkensii Engl. var. fernandense (Hutch. et Dalziel) Troupin ●☆
131932 Dioscoreophyllum triandrum Troupin = Dioscoreophyllum gossweileri Exell ●☆
131933 Dioscoreophyllum volkensii Engl.;柯氏薯蓣叶藤(柯氏薯蓣叶,奇遇果,应乐果)●☆
131934 Dioscoreophyllum volkensii Engl. var. fernandense (Hutch. et Dalziel) Troupin;费尔南蓣叶藤●☆
131935 Dioscoreopsis Kuntze = Dioscoreophyllum Engl. ●☆
131936 Dioscorida St. -Lag. = Dioscorea L.(保留属名)■
131937 Dioscoridia St. -Lag. = Dioscorea L.(保留属名)■
131938 Diosma L. (1753);逸香木属(布枯属,迪奥斯玛属,地奥属,香叶木属);Diosma ●☆
131939 Diosma acuminata (J. C. Wendl.) C. C. Gmel. = Agathosma imbricata (L.) Willd. ●☆
131940 Diosma alba Thunb. = Coleonema album (Thunb.) Bartl. et H. L. Wendl. ●☆
131941 Diosma ambigua Bartl. et H. L. Wendl. = Diosma hirsuta L. ●☆
131942 Diosma amoena Lodd. = Adenandra villosa (P. J. Bergius) Licht. ex Roem. et Schult. subsp. orbicularis Strid ■☆
131943 Diosma apetala (Dümmer) I. Williams;无瓣逸香木●☆
131944 Diosma arenicola I. Williams;沙生逸香木●☆
131945 Diosma aristata I. Williams;具芒逸香木●☆
131946 Diosma aspalathoides Lam.;芬芳逸香木●☆
131947 Diosma barbigera L. f. = Macrostylis barbigera (L. f.) Bartl. et H. L. Wendl. ●☆
131948 Diosma bifida Jacq. = Agathosma bifida (Jacq.) Bartl. et H. L. Wendl. ●☆
131949 Diosma bisulca Thunb. = Agathosma bisulca (Thunb.) Bartl. et H. L. Wendl. ●☆
131950 Diosma bolusii Glover = Diosma apetala (Dümmer) I. Williams ●☆
131951 Diosma calycina Steud. = Coleonema calycinum (Steud.) I. Williams ●☆
131952 Diosma calycina Tausch = Adenandra villosa (P. J. Bergius) Licht. ex Roem. et Schult. subsp. sonderi (Dümmer) Strid ■☆
131953 Diosma capitata L. = Audouinia capitata (L.) Brongn. ●☆
131954 Diosma cerefolia Vent. = Agathosma cerefolia (Vent.) Bartl. et H. L. Wendl. ●☆
131955 Diosma ciliata L. = Agathosma ciliata (L.) Link ●☆
131956 Diosma corymbosa Montin = Agathosma corymbosa (Montin) G. Don ●☆
131957 Diosma crenulata L. = Agathosma crenulata (L.) Pillans ●☆
131958 Diosma cuspidata Thunb. = Linconia cuspidata (Thunb.) Sw. ●☆
131959 Diosma decussata Lam. = Diosma oppositifolia L. ●☆
131960 Diosma demissa I. Williams;下垂逸香木●☆
131961 Diosma deusta Thunb. = Linconia cuspidata (Thunb.) Sw. ●☆
131962 Diosma dichotoma P. J. Bergius;二歧逸香木●☆
131963 Diosma echinulata I. Williams;小刺逸香木●☆
131964 Diosma eckloniana (Schltdl.) D. Dietr. = Agathosma serpyllacea Licht. ex Roem. et Schult. ●☆
131965 Diosma eriantha Steud. = Agathosma eriantha (Steud.) Steud. ●☆
131966 Diosma ericoides L.;逸香木(白花糖果木);Breath-of-heaven, Breath of Heaven, White Breath of Heaven, White Confetti Bush ●☆
131967 Diosma ericoides L. = Coleonema album Bartl. et H. L. Wendl. ●☆
131968 Diosma fallax I. Williams;迷惑逸香木●☆
131969 Diosma fragrans Sims = Adenandra fragrans (Sims) Roem. et Schult. ■☆
131970 Diosma glandulosa Thunb. = Agathosma glandulosa (Thunb.) Sond. ●☆
131971 Diosma guthriei P. E. Glover;格斯里逸香木●☆
131972 Diosma hirsuta L.;毛逸香木(迪奥斯玛)●☆
131973 Diosma hirta Lam. = Agathosma hirta (Lam.) Bartl. et H. L. Wendl. ●☆
131974 Diosma hispida Thunb. = Agathosma hispida (Thunb.) Bartl. et H. L. Wendl. ●☆
131975 Diosma latifolia L. f. = Agathosma odoratissima (Montin) Pillans ●☆
131976 Diosma linearis Thunb. = Adenandra uniflora (L.) Willd. ■☆
131977 Diosma longifolia J. C. Wendl. = Diosma hirsuta L. ●☆
131978 Diosma meyeriana Spreng.;迈尔逸香木●☆
131979 Diosma oblonga Thunb. = Agathosma ovata (Thunb.) Pillans ●☆
131980 Diosma obtusata Thunb. = Acmadenia obtusata (Thunb.) Bartl. et H. L. Wendl. ●☆
131981 Diosma obtusifolia Sond. = Diosma passerinoides Steud. ●☆
131982 Diosma odoratissima Montin = Agathosma odoratissima (Montin) Pillans ●☆
131983 Diosma oppositifolia L.;对叶逸香木●☆
131984 Diosma orbicularis Thunb. = Agathosma orbicularis (Thunb.) Bartl. et H. L. Wendl. ●☆
131985 Diosma ovata Thunb. = Agathosma ovata (Thunb.) Pillans ●☆
131986 Diosma parviflora Roem. et Schult. = Agathosma virgata (Lam.) Bartl. et H. L. Wendl. ●☆
131987 Diosma parvula I. Williams;较小逸香木●☆
131988 Diosma passerinoides Steud.;雀逸香木●☆
131989 Diosma pectinata Thunb. = Diosma hirsuta L. ●☆
131990 Diosma pedicellata I. Williams;梗花逸香木●☆
131991 Diosma pilosa I. Williams;疏毛逸香木●☆
131992 Diosma puberula Steud. = Agathosma puberula (Steud.)

Fourc. ●☆
131993 Diosma pulchella L. = Agathosma pulchella (L.) Link ●☆
131994 Diosma ramosissima Bartl. et H. L. Wendl.;多分枝逸香木●☆
131995 Diosma recurva Cham.;反折逸香木●☆
131996 Diosma rourkei I. Williams;鲁尔克逸香木●☆
131997 Diosma rubra L. = Diosma hirsuta L. ●☆
131998 Diosma rugosa Thunb. = Agathosma ciliaris (L.) Druce ●☆
131999 Diosma sabulosa I. Williams;砂地逸香木●☆
132000 Diosma serratifolia Sims = Agathosma serratifolia (Curtis) Spreeth ●☆
132001 Diosma speciosa Sims = Adenandra villosa (P. J. Bergius) Licht. ex Roem. et Schult. subsp. umbellata (J. C. Wendl.) Strid ■☆
132002 Diosma squamosa Roem. et Schult. = Agathosma squamosa (Roem. et Schult.) Bartl. et H. L. Wendl. ●☆
132003 Diosma stenopetala Steud. = Agathosma stenopetala (Steud.) Steud. ●☆
132004 Diosma strumosa I. Williams;多疣逸香木●☆
132005 Diosma subulata J. C. Wendl.;钻形逸香木●☆
132006 Diosma succulenta P. J. Bergius;多汁逸香木(卡罗布枯)●☆
132007 Diosma succulenta P. J. Bergius var. bergiana Sond. = Diosma oppositifolia L. ●☆
132008 Diosma tenella I. Williams;柔弱逸香木●☆
132009 Diosma teretifolia Link = Acmadenia teretifolia (Link) E. Phillips ●☆
132010 Diosma tetragona L. f. = Acmadenia tetragona (L. f.) Bartl. et H. L. Wendl. ●☆
132011 Diosma thyrsophora Eckl. et Zeyh.;聚伞逸香木●☆
132012 Diosma umbellata (J. C. Wendl.) C. C. Gmel. = Adenandra villosa (P. J. Bergius) Licht. ex Roem. et Schult. subsp. umbellata (J. C. Wendl.) Strid ■☆
132013 Diosma umbellata Thunb. = Agathosma bifida (Jacq.) Bartl. et H. L. Wendl. ●☆
132014 Diosma unicapsularis L. f. = Empleurum unicapsulare (L. f.) Skeels ■☆
132015 Diosma uniflora L. = Adenandra uniflora (L.) Willd. ■☆
132016 Diosma villosa Thunb. = Macrostylis villosa (Thunb.) Sond. ●☆
132017 Diosma villosa Willd. = Agathosma glabrata Bartl. et H. L. Wendl. ●☆
132018 Diosma virgata G. Mey. ex Bartl. et H. L. Wendl. = Diosma meyeriana Spreng. ●☆
132019 Diosma virgata Lam. = Agathosma virgata (Lam.) Bartl. et H. L. Wendl. ●☆
132020 Diosma vulgaris Schltdl. = Diosma hirsuta L. ●☆
132021 Diosma vulgaris Schltdl. var. hirsuta (L.) Sond. = Diosma hirsuta L. ●☆
132022 Diosma vulgaris Schltdl. var. longifolia (J. C. Wendl.) Sond. = Diosma hirsuta L. ●☆
132023 Diosma vulgaris Schltdl. var. rubra Sond. = Diosma hirsuta L. ●☆
132024 Diosma wittebergensis Compton = Acmadenia wittebergensis (Compton) I. Williams ●☆
132025 Diosmaceae R. Br. = Rutaceae Juss. (保留科名)●■
132026 Diospermum Hook. f. = Oiospermum Less. ■☆
132027 Diosphaera Buser = Campanula L. ■●
132028 Diospyraceae Tiegh. = Ebenaceae Gürke(保留科名)●
132029 Diospyraceae Vest = Ebenaceae Gürke(保留科名)●
132030 Diospyros L. (1753);柿树属(柿属,乌木属);African Ebony, Date Plum, Date-plum, Ebony, Persimmon, Velvet Apple, Zebrawood ●
132031 Diospyros Roxb. = Diospyros L. ●
132032 Diospyros abyssinica (Hiern) F. White;阿比西尼亚柿树●☆
132033 Diospyros abyssinica (Hiern) F. White subsp. attenuata F. White;渐狭柿●☆
132034 Diospyros abyssinica (Hiern) F. White subsp. chapmaniorum F. White;查普曼柿●☆
132035 Diospyros abyssinica (Hiern) F. White subsp. reticulata F. White;网脉阿比西尼亚柿树●☆
132036 Diospyros acocksii (De Winter) De Winter;阿科柿●☆
132037 Diospyros aculeata H. Perrier;皮刺柿●☆
132038 Diospyros affinis Thwaites;近缘柿●☆
132039 Diospyros aggregata Gürke = Diospyros mannii Hiern ●☆
132040 Diospyros albidum Scott-Elliot = Diospyros gracilipes Hiern ●☆
132041 Diospyros alboflavescens (Gürke) F. White;白黄柿●☆
132042 Diospyros amaniensis Gürke;阿马尼柿●☆
132043 Diospyros ampullacea Gürke = Diospyros crassiflora Hiern ●☆
132044 Diospyros analamerensis H. Perrier;阿纳拉柿●☆
132045 Diospyros anisandra S. F. Blake;异蕊柿;Anisosepalous Persimmon, Diversi-calyx Persimmon ●☆
132046 Diospyros anisocalyx C. Y. Wu;异萼柿;Anisosepalous Persimmon, Diversi-calyx Persimmon ●
132047 Diospyros anitae F. White = Diospyros zombensis (B. L. Burtt) F. White ●☆
132048 Diospyros apiculata A. Chev. = Diospyros vignei F. White ●☆
132049 Diospyros argentea Griff.;法国柿●☆
132050 Diospyros argyi H. Lév. = Diospyros kaki Thunb. var. sylvestris Makino ●
132051 Diospyros armata Hemsl.;瓶兰花(瓶子花,宜昌柿,玉瓶兰);Armour Persimmon, Spiny Persimmon ●
132052 Diospyros armata Hemsl. = Diospyros cathayensis Steward ●
132053 Diospyros armata Hemsl. var. pilosa Z. Y. Zhang;毛果刺柿;Pilose Armour Persimmon ●
132054 Diospyros atropurpurea Gürke;暗紫柿●☆
132055 Diospyros atropurpurea Gürke = Diospyros dendo Hiern ●☆
132056 Diospyros atrotricha H. W. Li = Diospyros hasseltii Zoll. ●◇
132057 Diospyros austro-africana De Winter;南非柿●☆
132058 Diospyros austro-africana De Winter var. microphylla (Burch.) De Winter;小叶南非柿●☆
132059 Diospyros austro-africana De Winter var. rubriflora (De Winter) De Winter;红花南非柿●☆
132060 Diospyros austro-africana De Winter var. rugosa (E. Mey. ex A. DC.) De Winter;皱褶南非柿●☆
132061 Diospyros balfouriana Diels;大理柿;Balfour Persimmon, Dali Persimmon ●
132062 Diospyros baroniana H. Perrier;巴龙柿●☆
132063 Diospyros barteri Hiern;巴特柿●☆
132064 Diospyros batocana Hiern;非洲香柿●☆
132065 Diospyros baumii Gürke = Diospyros kirkii Hiern ●☆
132066 Diospyros bemarivensis H. Perrier;贝马里武柿●☆
132067 Diospyros bequaertii De Wild. = Diospyros canaliculata De Wild. ●☆
132068 Diospyros bequaertii De Wild. var. imbimbo = Diospyros canaliculata De Wild. ●☆
132069 Diospyros bernieri Hiern;伯尼尔柿●☆
132070 Diospyros bicolor Klotzsch = Diospyros mespiliformis Hochst. ex A. DC. ●☆
132071 Diospyros bicolor Winkl. = Diospyros cinnabarina (Gürke) F.

White ●☆
132072 Diospyros bipindensis Gürke;比平迪柿●☆
132073 Diospyros blancoi A. DC. = Diospyros discolor Willd. ●
132074 Diospyros bodinieri H. Lév. = Ilex macrocarpa Oliv. ●
132075 Diospyros boivini Hiern;博伊文柿●☆
132076 Diospyros buesgenii Gürke = Diospyros bipindensis Gürke ●☆
132077 Diospyros burmanica Kurz;缅甸柿;Green Ebony ●☆
132078 Diospyros bussei Gürke;布瑟柿●☆
132079 Diospyros buxifolia (Rottb.) Bakh. = Symplocos lucida (Thunb.) Siebold et Zucc. ●
132080 Diospyros buxifolia Hiern;黄杨叶柿●☆
132081 Diospyros buxifolia Thouars = Diospyros thouarsii Hiern ●☆
132082 Diospyros caloneura C. Y. Wu;美脉柿;Beautiful Persimmon, Prettynerved Persimmon ●
132083 Diospyros calophylla Hiern;美叶柿●☆
132084 Diospyros canaliculata De Wild.;沟柿●☆
132085 Diospyros cardiophylla Merr.;无梗柿;Stalkless Persimmon ●☆
132086 Diospyros cardiophylla Merr. = Diospyros strigosa Hemsl. ●
132087 Diospyros carssiflora Hiern;非洲柿;Africa Persimmon, African Persimmon ●☆
132088 Diospyros castaneifolia A. Chev. = Diospyros gabunensis Gürke ●☆
132089 Diospyros cathayensis Steward;乌柿(丁香,丁香柿,黑塔子,山柿,山柿子,野油柿子);China Persimmon, Chinese Persimmon, Everlasting Persimmon ●
132090 Diospyros cathayensis Steward var. foochowensis (F. P. Metcalf et H. Y. Chen) S. K. Lee;福州柿(黑丁香,黑塔子,野油柿,油柿);Fuzhou Persimmon ●
132091 Diospyros cathayensis Steward var. foochowensis (F. P. Metcalf et H. Y. Chen) S. K. Lee = Diospyros cathayensis Steward ●
132092 Diospyros cauliflora De Wild. = Diospyros canaliculata De Wild. ●☆
132093 Diospyros celebica Bakh.;塞雷比柿(苏拉威西乌木);Macassar Ebony ●☆
132094 Diospyros cerasifolia D. Don = Eurya cerasifolia (D. Don) Kobuski ●
132095 Diospyros chaffanjoni H. Lév. = Cotoneaster horizontalis Decne. ●
132096 Diospyros chamaethamnus Dinter ex Mildbr.;矮柿●☆
132097 Diospyros changii R. H. Miao;张氏柿;Chang's Persimmon ●
132098 Diospyros changii R. H. Miao = Diospyros howii Merr. et Chun ●
132099 Diospyros chevalieri De Wild.;舍瓦利耶柿●☆
132100 Diospyros chinensis Blume = Diospyros kaki Thunb. ●
132101 Diospyros chlamydocarpa Mildbr. = Diospyros canaliculata De Wild. ●☆
132102 Diospyros chloroxylon Roxb.;绿乌木;Green Ebony Persimmon ●☆
132103 Diospyros chrysocarpa F. White;金果柿●☆
132104 Diospyros chunii F. P. Metcalf et H. Y. Chen;崖柿;Chun Persimmon ●
132105 Diospyros cinnabarina (Gürke) F. White;朱红柿●☆
132106 Diospyros cinnabarina Gürke ex De Wild. = Diospyros cinnabarina (Gürke) F. White ●☆
132107 Diospyros cinnamomoides H. Perrier;肉桂柿●☆
132108 Diospyros confertiflora Gürke ex J. D. Kenn. = Diospyros suaveolens Gürke ●☆
132109 Diospyros conifera H. Perrier;球果柿●☆
132110 Diospyros conocarpa Gürke et K. Schum.;毛瓣柿●☆
132111 Diospyros cooperi (Hutch. et Dalziel) F. White;库珀柿●☆
132112 Diospyros corallina Chun et H. Y. Chen;五蒂柿;Coralline Persimmon, Fivecalyx Persimmon ●◇
132113 Diospyros cordifolia Roxb. = Diospyros montana Roxb. ●
132114 Diospyros cordifolia Roxb. var. glabrifolia Merr. = Diospyros diversilimba Merr. et Chun ●
132115 Diospyros cornii Chiov. = Diospyros bussei Gürke ●☆
132116 Diospyros coursiana H. Perrier;库尔斯柿●☆
132117 Diospyros crassiflora Hiern;非洲乌木(非洲黑檀,非洲柿,厚瓣乌木);African Ebony, Black Ebony ●☆
132118 Diospyros cupulifera H. Perrier;杯柿●☆
132119 Diospyros danguyana H. Perrier;当吉柿●☆
132120 Diospyros dasypetala Pierre ex A. Chev. = Diospyros conocarpa Gürke et K. Schum. ●☆
132121 Diospyros daucifolia F. P. Metcalf;粉叶柿;Glaucous Persimmon ●
132122 Diospyros dawei (Hutch.) Brenan = Diospyros natalensis (Harv.) Brenan ●☆
132123 Diospyros decandra Lour.;十蕊柿●☆
132124 Diospyros decaryana H. Perrier;德卡里柿●☆
132125 Diospyros decipiens Gürke = Diospyros bussei Gürke ●☆
132126 Diospyros deltoidea F. White;三角柿●☆
132127 Diospyros dendo Hiern = Diospyros dendo Welw. ex Hiern ●☆
132128 Diospyros dendo Welw. ex Hiern;西非柿木;West African Ebony ●☆
132129 Diospyros dichrophylla (Gand.) De Winter;二色叶柿●☆
132130 Diospyros dicorypheoides H. Perrier;双扇梅柿●☆
132131 Diospyros digyna Jacq. = Diospyros digyna Loudon ●☆
132132 Diospyros digyna Loudon;墨西哥黑柿(二花柱柿,黑柿);Black Persimmon, Black Sapote, Black Sapote-tree, Chocolate Pudding Tree, Zapote Negro ●☆
132133 Diospyros discolor Willd. = Diospyros philippensis (Desr.) Gurke ●
132134 Diospyros diversilimba Merr. et Chun;光叶柿(黑烈树,乌力果);Glabrousleaf Persimmon, Glabrous-leaved Persimmon ●
132135 Diospyros dumetorum W. W. Sm.;岩柿(石柿花,小叶山柿,小叶柿,崖柿);Small-leaf Scrub Persimmon, Small-leaved Persimmon ●
132136 Diospyros ebenaster Retz. = Diospyros digyna Loudon ●☆
132137 Diospyros ebenum J. König ex Retz.;乌木(黑檀,文木,乌樠木,乌木柿,乌文木,翳木);Bombay Ebony, Ceylon Ebony, Ceylon Persimmon, Eastindian Persimmon, Ebony, Ebony Persimon, Ebony Tree ●☆
132138 Diospyros ebenum König = Diospyros ebenum J. König ex Retz. ●☆
132139 Diospyros egbert-walkeri Kosterm.;冲绳柿●
132140 Diospyros ehretioides Wall. ex A. DC.;红枝柿;Ehretia-like Persimmon, Red Shoot Persimmon ●
132141 Diospyros elliotii (Hiern) F. White;埃利柿●☆
132142 Diospyros embryopteris Pers.;椑柿(印度马来黑檀)●☆
132143 Diospyros embryopteris Pers. = Diospyros malabarica Kostel. ●☆
132144 Diospyros engleri Gürke;恩格勒柿●☆
132145 Diospyros eriantha Champ. ex Benth.;乌材(米汉,米来,软毛柿,乌材柿,乌材仔,乌杆子,乌眉,乌木,乌蛇,乌柿,小叶乌椿);Erianthous Persimmon, Woolly-flowered Persimmon ●
132146 Diospyros erythrosperma H. Perrier;红籽柿●☆
132147 Diospyros esquirolii H. Lév.;贵阳柿;Esquirol Persimmon, Guiyang Persimmon ●
132148 Diospyros evila Pierre ex A. Chev. = Diospyros crassiflora Hiern

●☆

132149 Diospyros fanjingshanica S. K. Lee;梵净山柿;Fanjingshan Persimmon ●

132150 Diospyros feliciana Letouzey et F. White;费利奇柿●☆

132151 Diospyros fengchangensis S. Y. Lu = Diospyros vaccinioides Lindl. ●

132152 Diospyros fengii C. Y. Wu;老君柿;Feng Persimmon ●

132153 Diospyros ferrea (Willd.) Bakh.;象牙柿(琉球黑檀,乌皮石苓,乌皮石柃,象牙树);Diospyros Ferrea,Iron Persimmon ●

132154 Diospyros ferrea (Willd.) Bakh. = Diospyros egbert-walkeri Kosterm. ●

132155 Diospyros ferrea (Willd.) Bakh. var. buxifolia (Rottb.) Bakh. = Diospyros ferrea (Willd.) Bakh. ●

132156 Diospyros filipes H. Perrier;丝梗柿●☆

132157 Diospyros fischeri Gürke;菲舍尔柿●☆

132158 Diospyros flavescens Gürke = Diospyros dendo Hiern ●☆

132159 Diospyros flavovirens Gürke = Diospyros bipindensis Gürke ●☆

132160 Diospyros flexilis Hiern = Diospyros kirkii Hiern ●☆

132161 Diospyros foochowensis F. P. Metcalf et H. Y. Chen = Diospyros cathayensis Steward ●

132162 Diospyros foochowensis F. P. Metcalf et H. Y. Chen = Diospyros cathayensis Steward var. foochowensis (F. P. Metcalf et H. Y. Chen) S. K. Lee ●

132163 Diospyros forrestii J. Anthony;腾冲柿;Forrest Persimmon ●

132164 Diospyros fragrans Gürke;香柿●☆

132165 Diospyros fuscovelutina Baker;棕毛柿●☆

132166 Diospyros gabunensis Gürke;加蓬柿●☆

132167 Diospyros galpinii (Hiern) De Winter;盖尔柿●☆

132168 Diospyros gavii (Aubrév. et Pellegr.) F. White = Diospyros cooperi (Hutch. et Dalziel) F. White ●☆

132169 Diospyros gilgiana Gürke = Diospyros gabunensis Gürke ●☆

132170 Diospyros gilletii De Wild.;吉勒特柿●☆

132171 Diospyros gilletii De Wild. var. sapinii ? = Diospyros gilletii De Wild. ●☆

132172 Diospyros glabra (L.) De Winter;光滑柿●☆

132173 Diospyros glandulifera De Winter;腺体柿●☆

132174 Diospyros glaucescens Gürke = Diospyros sanza-minika A. Chev. ●☆

132175 Diospyros glaucifolia F. P. Metcalf;浙江柿(粉叶柿,鸡粪柿,毛梨壳);Zhejiang Persimmon ●

132176 Diospyros glaucifolia F. P. Metcalf = Diospyros japonica Siebold et Zucc. ●

132177 Diospyros glaucifolia F. P. Metcalf var. brevipes S. K. Lee;短柄粉叶柿;Short-stalk Zhejiang Persimmon ●

132178 Diospyros glaucifolia F. P. Metcalf var. pubescens Y. Ling = Diospyros japonica Siebold et Zucc. ●

132179 Diospyros gonoclada Baker = Diospyros sphaerosepala Baker ●☆

132180 Diospyros gracilescens Gürke;纤细柿●☆

132181 Diospyros gracilipes Hiern;细梗柿●☆

132182 Diospyros greenwayi F. White;格林韦柿●☆

132183 Diospyros greveana H. Perrier;格雷弗柿●☆

132184 Diospyros guianensis Gürke;圭亚那柿●☆

132185 Diospyros guineensis A. Chev. = Diospyros heudelotii Hiern ●☆

132186 Diospyros hainanensis Merr.;海南柿(参巴,牛金树,牛筋树,细脚巴,硬壳果);Hainan Persimmon ●

132187 Diospyros haplostylis Boivin = Diospyros haplostylis Boivin ex Hiern ●☆

132188 Diospyros haplostylis Boivin ex Hiern;莫桑比克乌木;Madagascar Ebony,Mozambique Ebony ●☆

132189 Diospyros hasseltii Zoll.;黑毛柿;Black-haired Persimmon,Black-hairs Persimmon ●◇

132190 Diospyros hayatae Odash. = Diospyros oldhamii Maxim. ●

132191 Diospyros hayatae Odash. f. ellipsoidea Odash. = Diospyros oldhamii Maxim. ●

132192 Diospyros hazomainty H. Perrier;马达加斯加乌木;Madagascar Ebony Ebony Veine ●☆

132193 Diospyros helferi C. B. Clarke;柬埔寨乌木;Cambodia Persimmon,Helfer Persimmon ●☆

132194 Diospyros heterosepala H. Perrier;异瓣柿●☆

132195 Diospyros heterotricha (B. L. Burtt) F. White;异毛柿●☆

132196 Diospyros heudelotii Hiern;厄德柿●☆

132197 Diospyros hexamera C. Y. Wu;六花柿;Six-flowered Persimmon ●

132198 Diospyros hirsuta L. f.;硬毛柿(粗硬毛乌木,毛柿)●☆

132199 Diospyros hirta Gürke ex Hutch. et Dalziel = Diospyros barteri Hiern ●☆

132200 Diospyros holtzii Gürke = Diospyros mespiliformis Hochst. ex A. DC. ●☆

132201 Diospyros horsefieldii Hiern = Diospyros hasseltii Zoll. ●◇

132202 Diospyros howii Merr. et Chun;琼南柿(黑皮,镜面柿);How Persimmon ●

132203 Diospyros hoyleana F. White;霍伊尔柿●☆

132204 Diospyros hoyleana F. White subsp. angustifolia F. White;窄叶霍伊尔柿(窄叶柿)●☆

132205 Diospyros humbertiana H. Perrier;亨伯特柿●☆

132206 Diospyros hylobia Gürke = Diospyros cinnabarina (Gürke) F. White ●☆

132207 Diospyros ibo Gürke ex De Wild. = Diospyros mespiliformis Hochst. ex A. DC. ●☆

132208 Diospyros implexicalyx H. Perrier;乱萼柿●☆

132209 Diospyros incarnata Gürke = Diospyros crassiflora Hiern ●☆

132210 Diospyros inflata Merr. et Chun;囊萼柿;Bladdery Persimmon ●

132211 Diospyros inhacaensis F. White;伊尼亚卡柿●☆

132212 Diospyros insculpta Hutch. et Dalziel = Diospyros iturensis (Gürke) Letouzey et F. White ●☆

132213 Diospyros intricata H. Perrier;缠结柿●☆

132214 Diospyros iturensis (Gürke) Letouzey et F. White;伊图里柿●☆

132215 Diospyros ivorensis Aubrév. et Pellegr. = Diospyros mannii Hiern ●☆

132216 Diospyros japonica Siebold et Zucc.;山柿(粉叶柿,琉球柿);Japan Persimmon ●

132217 Diospyros japonica Siebold et Zucc. f. pseudolotus Hatus.;假君迁子●☆

132218 Diospyros kaki L. f. = Diospyros kaki Thunb. ●

132219 Diospyros kaki Thunb.;柿(稗柿,红柿,柿树,柿仔,柿子,油柿子,镇头迦);Chinese Persimmon,Date-plum,Japanese Date Palm,Japanese Persimmon,Kaki,Kaki Persimmon,Oriental Persimmon,Persimmon ●

132220 Diospyros kaki Thunb. 'Fuyu';富裕柿●☆

132221 Diospyros kaki Thunb. 'Hachiya';哈奇亚柿●☆

132222 Diospyros kaki Thunb. 'Izu';伊祖柿●☆

132223 Diospyros kaki Thunb. 'Jiro';吉如柿●☆

132224 Diospyros kaki Thunb. 'Tamopan';塔毛潘柿●☆

132225 Diospyros kaki Thunb. 'Tanenashi';塔尼纳斯柿●☆

132226 Diospyros kaki Thunb. 'Wright's Favorite';宠儿柿●☆

132227 Diospyros kaki Thunb. var. costata André;四棱柿●

132228 Diospyros kaki Thunb. var. domestica Makino = Diospyros kaki Thunb. ●

132229 Diospyros kaki Thunb. var. macrantha Hand. -Mazz. ;大花柿●

132230 Diospyros kaki Thunb. var. sylvestris Makino;野柿(马槟榔,毛柿花,山柿,油柿);Sylvestris Persimmon ●

132231 Diospyros kamerunensis Gürke;喀麦隆柿;Cameroon Persimmon ●☆

132232 Diospyros kanurii F. White;卡努里柿●☆

132233 Diospyros katendei Verdc. ;卡滕代柿●☆

132234 Diospyros kekemi Aubrév. et Pellegr. = Diospyros viridicans Hiern ●☆

132235 Diospyros kerrii Craib;傣柿;Kerr Persimmon ●

132236 Diospyros kilimandscharica Gürke = Diospyros mespiliformis Hochst. ex A. DC. ●☆

132237 Diospyros kimba-kimba De Wild. = Diospyros canaliculata De Wild. ●☆

132238 Diospyros kintungensis C. Y. Wu;景东君迁子(小柿花);Jingdong Persimmon ●

132239 Diospyros kirkii Hiern;柯克柿●☆

132240 Diospyros klaineana Pierre ex De Wild. = Diospyros preussii Gürke ●☆

132241 Diospyros kotoensis T. Yamaz. ;兰屿柿;Lanyu Persimmon ●

132242 Diospyros kupensis Gosline;库普柿●☆

132243 Diospyros kuroiwae Nakai = Diospyros japonica Siebold et Zucc. ●

132244 Diospyros kurzii Hiern;斑纹乌木;Andaman Marble, Andaman Marble Wood, Andaman Zebra-wood ●☆

132245 Diospyros kusanoi Hayata; 草野柿(黄心仔); Kusano Persimmon ●☆

132246 Diospyros kusanoi Hayata = Diospyros maritima Blume ●

132247 Diospyros laevis Bojer ex DC. ;平滑柿●☆

132248 Diospyros lancifolia Roxb. ;披针叶柿;Lance-leaf Persimmon ●☆

132249 Diospyros latifolia Gürke = Diospyros kirkii Hiern ●☆

132250 Diospyros latispathulata H. Perrier;宽匙叶柿●☆

132251 Diospyros laui Merr. = Diospyros susarticulata Lecomte ●

132252 Diospyros ledermannii Gürke = Diospyros preussii Gürke ●☆

132253 Diospyros lenticellata Baker;皮孔柿●☆

132254 Diospyros letestui Pellegr. = Diospyros preussii Gürke ●☆

132255 Diospyros leucocalyx Hiern = Diospyros gracilipes Hiern ●☆

132256 Diospyros liberiensis A. Chev. ex Hutch. et Dalziel;利比里亚柿●☆

132257 Diospyros linderi Hutch. et Dalziel ex G. P. Cooper et Record = Diospyros chevalieri De Wild. ●☆

132258 Diospyros liukiuensis Makino = Diospyros maritima Blume ●

132259 Diospyros lobata Lour. = Diospyros kaki Thunb. ●

132260 Diospyros longibracteata Lecomte;长苞柿(白春,定春,海鸵李,乌茶,乌木);Long Bract Persimmon, Long-bracted Persimmon ●

132261 Diospyros longicaudata Gürke ex Hutch. et Dalziel = Diospyros zenkeri (Gürke) F. White ●☆

132262 Diospyros longiflora Letouzey et F. White;长花柿●☆

132263 Diospyros longshengensis S. K. Lee; 龙胜柿; Longsheng Persimmon ●

132264 Diospyros lotus L. ;君迁子(八棱柿,丁香柿,狗柿子,黑枣,红蓝枣,牛奶柿,软枣,瑛枣,小柿,羊屎枣,樗枣,油柿子);Date Plum, Date-plum, Dateplum Persimmon, Date-plum Persimmon, European Date Plum, European Date-plum, Small Date Plum ●

132265 Diospyros lotus L. f. longifolia Z. Ying Zhang;长叶君迁子;Longleaf Date-plum ●

132266 Diospyros lotus L. var. brideliifolia (Elmer) Ng;菲律宾君迁子●☆

132267 Diospyros lotus L. var. glabra Makino;光君迁子(光黑枣);Glabrous Date-plum ●☆

132268 Diospyros lotus L. var. mollissima C. Y. Wu;多毛君迁子(多毛黑枣);Hairy Date-plum ●

132269 Diospyros loureiriana G. Don;洛雷罗柿●☆

132270 Diospyros loureiriana G. Don f. macrocalyx (Klotzsch) Hiern = Diospyros loureiriana G. Don ●☆

132271 Diospyros loureiriana G. Don subsp. rufescens (Caveney) Verdc. ;浅红柿●☆

132272 Diospyros loureiriana G. Don var. heterotricha Welw. ex Hiern = Diospyros heterotricha (B. L. Burtt) F. White ●☆

132273 Diospyros louveli H. Perrier;卢韦尔柿●☆

132274 Diospyros lujae De Wild. = Diospyros gabunensis Gürke ●☆

132275 Diospyros lycioides Desf. ;蓝叶柿;Bloubos, Bluebush, Japanese Date-plum, Monkey Plum ●☆

132276 Diospyros lycioides Desf. subsp. guerkei (Kuntze) De Winter;盖尔克柿●☆

132277 Diospyros lycioides Desf. subsp. nitens (Harv. ex Hiern) De Winter;光亮柿●☆

132278 Diospyros lycioides Desf. subsp. sericea (Bernh.) De Winter;绢毛柿●☆

132279 Diospyros maclurei Merr. ;琼岛柿(大果柿,乌椿,乌木,乌云墨);Maclure Persimmon, Qiongdao Persimmon ●

132280 Diospyros macrocalyx Klotzsch = Diospyros loureiriana G. Don ●☆

132281 Diospyros macrophylla A. Chev. = Diospyros chevalieri De Wild. ●☆

132282 Diospyros madecassa H. Perrier;马德卡萨柿●☆

132283 Diospyros mafiensis F. White;马菲柿●☆

132284 Diospyros mairei H. Lév. = Diospyros dumetorum W. W. Sm. ●

132285 Diospyros major (G. Forst.) Bakh. ;太平洋柿●☆

132286 Diospyros malabarica (Desr.) Kostel. ;马拉巴尔柿(马拉巴柿);Gaub, Gaub Tree ●☆

132287 Diospyros malabarica Kostel. = Diospyros malabarica (Desr.) Kostel. ●☆

132288 Diospyros mamiacensis Gürke = Diospyros gabunensis Gürke ●☆

132289 Diospyros mangorensis H. Perrier;曼古鲁柿●☆

132290 Diospyros mannii Hiern;曼柿●☆

132291 Diospyros maritima Blume;海边柿(黄心柿,黄心仔,琉球柿); Coast Persimmon, Liuqiu Persimmon, Malaysian Persimmon, Sea-side Persimmon ●

132292 Diospyros marmorata R. Parker;安达乌木;Andaman Edony ●☆

132293 Diospyros masoalensis H. Perrier;马苏阿拉柿●☆

132294 Diospyros mayumbensis Exell = Diospyros canaliculata De Wild. ●☆

132295 Diospyros megacarpa Gürke ex De Wild. = Diospyros kamerunensis Gürke ●☆

132296 Diospyros megaphylla Gürke = Diospyros gabunensis Gürke ●☆

132297 Diospyros megasepala Baker;大萼柿●☆

132298 Diospyros melanoxylon Roxb. ;缅甸乌木(黑木柿,印度乌木);Coromandel Ebony, Ebony Tree, Indian Ebony ●☆

132299 Diospyros mespiliformis Hochst. ex A. DC. ;梨形柿(西非乌木); African Ebony, Calabar Ebony, Jakkalbessie, Lagos Ebony, Monkey Guava, Swamp Ebony, West African Ebony, Zanzibar Ebony

●☆

132300 Diospyros metcalfii Chun et H. Y. Chen;圆萼柿(南海柿);F. P. Metcalf Persimmon, Metcalf Persimmon, South Sea Persimmon ●

132301 Diospyros miaoshanica S. K. Lee;苗山柿;Miaoshan Persimmon ●

132302 Diospyros mimfiensis Gürke = Diospyros dendo Hiern ●☆

132303 Diospyros mollifolia Rehder et E. H. Wilson;毛叶柿(涩藿香,乌木,小叶柿,紫藿香);Pubescentleaf Persimmon ●

132304 Diospyros mollifolia Rehder et E. H. Wilson = Diospyros dumetorum W. W. Sm. ●

132305 Diospyros mollis Griff.;软柿(毛柿,柔毛柿,软毛柿);Makua, Makua-plua, Ma-plua, Soft Persimmon, Thailand Ebony ●☆

132306 Diospyros molundensis Mildbr. = Diospyros canaliculata De Wild. ●☆

132307 Diospyros monbuttensis Gürke;手杖柿;Walking-stick Ebony, Yoruba Ebony ●☆

132308 Diospyros montana Roxb.;山地柿(山柿,山柿树);Bombay Ebony, East Indian Ebony, Mountain Persimmon ●

132309 Diospyros morrisiana Hance;罗浮柿(猴鬼子,猴子公,牛古柿,山红柿,山椑树,山柿,乌材柿,乌蛇木,野柿,野柿花,油柿);Morris Persimmon ●

132310 Diospyros mossambicensis ?;莫桑比克柿●☆

132311 Diospyros multilflora Wall. = Diospyros lancifolia Roxb. ●☆

132312 Diospyros mun (A. Chev.) Lecomte;文柿;Indochina Persimmon ●☆

132313 Diospyros mweroensis F. White;姆韦鲁柿●☆

132314 Diospyros myrtifolia H. Perrier;香桃木叶柿●☆

132315 Diospyros natalensis (Harv.) Brenan;纳塔尔柿●☆

132316 Diospyros navillei H. Lév. = Lysimachia navillei (H. Lév.) Hand. -Mazz. ●■

132317 Diospyros nigerica F. White = Diospyros gracilescens Gürke ●☆

132318 Diospyros nigrescens (Dalzell) C. J. Saldanha;变黑柿;Indian Egony, Kanara Ebony ●☆

132319 Diospyros nigrocortex C. Y. Wu;黑皮柿(黑皮树);Black-bark Persimmon, Black-barked Persimmon ●

132320 Diospyros nipponica Nakai;日本柿●☆

132321 Diospyros nipponica Nakai = Diospyros morrisiana Hance ●

132322 Diospyros nitida Merr.;黑柿;Black Persimmon ●

132323 Diospyros nsambensis Gürke = Diospyros sanza-minika A. Chev. ●☆

132324 Diospyros nummularia Brenan;铜钱柿●☆

132325 Diospyros nummularia Brenan = Diospyros natalensis (Harv.) Brenan ●☆

132326 Diospyros nyangensis Pellegr. = Diospyros dendo Hiern ●☆

132327 Diospyros nyasae Brenan = Diospyros natalensis (Harv.) Brenan ●☆

132328 Diospyros obliquifolia (Hiern ex Gürke) F. White;斜叶柿●☆

132329 Diospyros oblongicarpa Gürke;矩圆果柿●☆

132330 Diospyros occlusa H. Perrier;闭合柿●☆

132331 Diospyros occulta F. White;隐蔽柿●☆

132332 Diospyros odashima Hatus. = Diospyros oldhamii Maxim. ●

132333 Diospyros odorata Hiern ex Greves = Diospyros batocana Hiern ●☆

132334 Diospyros odorata Hiern ex Greves var. rhodesiana Rendle = Diospyros batocana Hiern ●☆

132335 Diospyros oldhamii Maxim.;红柿(俄氏柿,台东柿,台湾柿,椭圆红柿);Elliptic Oldham Persimmon, Oldham Persimmon ●

132336 Diospyros oldhamii Maxim. f. ellipsoidea (Odash.) H. L. Li;台湾柿●

132337 Diospyros oldhamii Maxim. f. ellipsoidea (Odash.) H. L. Li = Diospyros oldhami Maxim. ●

132338 Diospyros oldhamii Maxim. var. chartacea Hayata = Diospyros oldhamii Maxim. ●

132339 Diospyros oleifera W. C. Cheng;油柿(洞柿,方柿,绿柿,椑柿,漆柿,青椑,乌椑,油绿柿);Oily Persimmon ●

132340 Diospyros oliviformis R. H. Miao;榄果柿●

132341 Diospyros onivensis H. Perrier;乌尼韦柿●☆

132342 Diospyros oocarpa Thwaites;卵果柿●☆

132343 Diospyros pachyphylla Gürke;厚叶柿●☆

132344 Diospyros packmannii C. B. Clarke;帕克曼柿●☆

132345 Diospyros pallens (Thunb.) F. White;苍白柿●☆

132346 Diospyros pallescens A. Chev. = Diospyros kamerunensis Gürke ●☆

132347 Diospyros parvifolia Hiern;小叶柿●☆

132348 Diospyros perreticulata H. Perrier;网柿●☆

132349 Diospyros perrieri Jum.;佩里耶乌木(黑乌木);Black Ebony ●☆

132350 Diospyros pervillei Hiern;佩尔柿●☆

132351 Diospyros philippensis (Desr.) Gurke;异色柿(菲律宾柿,菲律宾乌木,毛柿,台湾黑檀,台湾柿,台湾乌木);Butter Fruit, Butterfruit, Mobola, Mobola Persimmon, Philippine Ebony, Philippine Persimmon, Taiwan Ebony, Taiwan Persimmon, Velvet Apple ●

132352 Diospyros physocalycina Gürke;膀胱萼柿●☆

132353 Diospyros piscatoria Gürke;加蓬乌木;African Ebony, Gabon Ebony ●☆

132354 Diospyros platanoides Letouzey et F. White;悬铃木柿●☆

132355 Diospyros platycalyx Hiern;宽萼柿●☆

132356 Diospyros platyphylla Welw. ex Hiern = Diospyros kirkii Hiern ●☆

132357 Diospyros polystemon Gürke;多冠柿●☆

132358 Diospyros potamophila Mildbr. = Diospyros gilletii De Wild. ●☆

132359 Diospyros potingensis Merr. et Chun;保亭柿;Baoting Persimmon ●

132360 Diospyros preussii Gürke;普罗柿●☆

132361 Diospyros pruinosa Hiern;白粉柿●☆

132362 Diospyros pseudaggregata Mildbr. = Diospyros mannii Hiern ●☆

132363 Diospyros pseudebenus (E. Mey.) Parm. = Euclea pseudebenus E. Mey. ●☆

132364 Diospyros punctilimba C. Y. Wu;点叶柿;Dotted Persimmon, Punctate Persimmon ●

132365 Diospyros quaesita Thwaites;斯里兰卡柿;Calamander Wood ●☆

132366 Diospyros quiloensis (Hiern) F. White;奎罗柿●☆

132367 Diospyros rabiensis Breteler;拉比柿●☆

132368 Diospyros ramulosa (E. Mey. ex A. DC.) De Winter;多枝柿●☆

132369 Diospyros reticulinervis C. Y. Wu;网脉柿;Net-veined Persimmon, Retinerved Persimmon ●

132370 Diospyros reticulinervis C. Y. Wu var. glabrescens C. Y. Wu;无毛网脉柿;Glabrous Net-veined Persimmon ●

132371 Diospyros rhombifolia Hemsl.;老鸦柿(丁季李,丁香柿,黑柿子,苦李,菱叶柿,牛奶柿,糯米饭刺,拳李,野山柿,月月有,枝柿);Crow Persimmon, Diamondleaf Persimmon, Rhombic-leaved Persimmon ●

132372 Diospyros rivularis Gürke = Diospyros zenkeri (Gürke) F. White ●☆

132373 Diospyros roi Lecomte = Diospyros corallina Chun et H. Y. Chen ●◇

132374 Diospyros rosea Gürke;粉红柿●☆

132375 Diospyros rotundifolia Hiern;圆叶柿●☆

132376 Diospyros rubicunda Gürke = Diospyros barteri Hiern ●☆

132377 Diospyros rubra Lecomte；青茶柿(海南柿，青茶)；Red Persimmon ●

132378 Diospyros rubrolanata H. Perrier；红毛柿●☆

132379 Diospyros rumphii Bakh.；朗氏柿木；Macassar Ebony ●☆

132380 Diospyros sabiensis Hiern = Diospyros mespiliformis Hochst. ex A. DC. ●☆

132381 Diospyros sakalavarum H. Perrier；萨卡拉瓦柿●☆

132382 Diospyros sanza-minika A. Chev.；热非柿；Liberia Ebony，Liberia Persimmon，Sanza Minik ●☆

132383 Diospyros sasakii Hayata；红花柿；Red-flowered Persimmon ●

132384 Diospyros saxicola Hung T. Chang ex R. H. Miao；石生柿；Rocky Persimmon ●

132385 Diospyros scabra (Chiov.) Cufod.；粗糙柿●☆

132386 Diospyros scabrida (Harv. ex Hiern) De Winter；无柄柿；Hard-leafed Monkey Plum ●☆

132387 Diospyros scabrida (Harv. ex Hiern) De Winter var. cordata (E. Mey. ex A. DC.) De Winter；心形微糙柿●☆

132388 Diospyros schitze Bunge = Diospyros kaki Thunb. ●

132389 Diospyros sclerophylla H. Perrier；硬叶柿●☆

132390 Diospyros senegalensis Perr. ex A. DC. = Diospyros mespiliformis Hochst. ex A. DC. ●☆

132391 Diospyros senensis Klotzsch；塞纳柿●☆

132392 Diospyros serrata Buch. -Ham. ex D. Don = Eurya acuminata DC. ●

132393 Diospyros setigera Mildbr. = Diospyros conocarpa Gürke et K. Schum. ●☆

132394 Diospyros shimbaensis F. White；欣巴柿●☆

132395 Diospyros shirensis Hiern = Diospyros senensis Klotzsch ●☆

132396 Diospyros siamensis Hochr.；暹罗柿(暹罗柿木)●☆

132397 Diospyros sichourensis C. Y. Wu；西畴君迁子；Xichou Persimmon ●

132398 Diospyros siderophylla H. L. Li；山榄叶柿(凌扣，枚辣柿，米亚辣，喃咛)；Iron-requiring Persimmon，Jungleplumleaf Persimmon ●

132399 Diospyros simii (Kuntze) De Winter；西姆柿●☆

132400 Diospyros simulans F. White = Diospyros cinnabarina (Gürke) F. White ●☆

132401 Diospyros sinensis Blume ex Naudin = Diospyros cathayensis Steward ●

132402 Diospyros sinensis Blume ex Naudin = Diospyros kaki Thunb. ●

132403 Diospyros sinensis Hemsl. = Diospyros cathayensis Steward ●

132404 Diospyros sinensis Naudin = Diospyros kaki Thunb. ●

132405 Diospyros soubreana F. White；苏布雷柿●☆

132406 Diospyros soyauxii Gürke et K. Schum.；索亚柿●☆

132407 Diospyros sphaerocarpa Pierre ex De Wild. = Diospyros kamerunensis Gürke ●☆

132408 Diospyros sphaerosepala Baker；球萼柿●☆

132409 Diospyros sphaerosepala Baker var. calyculata H. Perrier = Phanerodiscus diospyroidea Capuron ●☆

132410 Diospyros squamosa Bojer ex DC.；多鳞柿●☆

132411 Diospyros squarrosa Klotzsch；粗鳞柿●☆

132412 Diospyros stapfiana F. White = Diospyros obliquifolia (Hiern ex Gürke) F. White ●☆

132413 Diospyros staudtii Gürke = Diospyros conocarpa Gürke et K. Schum. ●☆

132414 Diospyros strigosa Hemsl.；毛柿(乌前)；Bristly Persimmon，Hairy Persimmon ●

132415 Diospyros striicalyx H. Perrier；沟萼柿●☆

132416 Diospyros stuhlmannii Gürke = Diospyros squarrosa Klotzsch ●☆

132417 Diospyros suaveolens Gürke；芳香柿●☆

132418 Diospyros subcanescens Gürke ex De Wild. = Diospyros cinnabarina (Gürke) F. White ●☆

132419 Diospyros subfalciformis H. Perrier；镰柿●☆

132420 Diospyros subsessilifolia H. Perrier；近无柄柿●☆

132421 Diospyros sunyiensis Chun et H. Y. Chen；信宜柿；Xinyi Persimmon ●

132422 Diospyros susarticulata Lecomte；过布柿(五指山柿)；Subarticulate Persimmon，Wuzhishan Persimmon ●

132423 Diospyros sutchuensis Yen C. Yang；川柿；Sichuan Persimmon ●◇

132424 Diospyros taamii Merr. = Diospyros tutcheri Dunn ●

132425 Diospyros taitoensis Odash. = Diospyros oldhamii Maxim. ●

132426 Diospyros talbotii Wernham = Diospyros mannii Hiern ●☆

132427 Diospyros tampinensis H. Perrier；浅边柿●☆

132428 Diospyros temvoensis De Wild. = Diospyros dendo Hiern ●☆

132429 Diospyros tetraceros H. Perrier；四角柿●☆

132430 Diospyros tetrapoda H. Perrier；四足柿●☆

132431 Diospyros texana Scheele；得克萨斯柿；Blacck Persimmon，Chapote，Mexican Persimmon，Texas Persimmon ●☆

132432 Diospyros thomasii Hutch. et Dalziel；托马斯柿●☆

132433 Diospyros thouarsii Hiern；图氏柿●☆

132434 Diospyros tomentosa Roxb.；绒毛乌木(印度乌木)；India Persimmon，Indian Ebony，Indian Persimmon ●

132435 Diospyros toxicaria Hiern；毒柿木；Madagascar Ebony ●☆

132436 Diospyros trichocarpa R. H. Miao；毛果柿；Hairfruit Ebony ●

132437 Diospyros trichocarpa R. H. Miao = Diospyros kaki Thunb. var. sylvestris Makino ●

132438 Diospyros tricolor (Schumach. et Thonn.) Hiern；三色柿●☆

132439 Diospyros tricolor Hiern = Diospyros tricolor (Schumach. et Thonn.) Hiern ●☆

132440 Diospyros troupinii F. White；图平柿●☆

132441 Diospyros truncatifolia Caveney；截叶柿●☆

132442 Diospyros tsiangii Merr.；延平柿；Tsiang Persimmon，Yangping Persimmon ●

132443 Diospyros tuberculosa Gürke = Diospyros verrucosa Hiern ●☆

132444 Diospyros tutcheri Dunn；岭南柿；S. China Persimmon，South China Persimmon ●

132445 Diospyros ubanghensis A. Chev. = Diospyros abyssinica (Hiern) F. White ●☆

132446 Diospyros unisemina G. Y. Wu；单籽柿(单子柿)；One-seed Persimmon，Uniseminal Persimmon ●

132447 Diospyros urschii H. Perrier；乌尔施柿●☆

132448 Diospyros usambarensis A. DC.；乌桑柿●☆

132449 Diospyros usambarensis F. White = Diospyros loureiriana G. Don ●☆

132450 Diospyros usambarensis F. White subsp. rufescens Caveney = Diospyros loureiriana G. Don subsp. rufescens (Caveney) Verdc. ●☆

132451 Diospyros usaramensis Gürke；乌萨拉姆柿●☆

132452 Diospyros utilis Hemsl. = Diospyros philippensis (Desr.) Gurke ●

132453 Diospyros uzungwaensis Frim. -Moll. et Ndangalasi；乌尊季沃柿●☆

132454 Diospyros vaccinioides Lindl.；小果柿(枫港柿)；Blueberrylike Persimmon，Blueberry-like Persimmon ●

132455 Diospyros vaccinioides Lindl. var. oblongata Merr. et Chun；长叶小果柿；Oblongate Blueberrylike Persimmon ●

132456 Diospyros vaccinioides Lindl. var. oblongata Merr. et Chun = Diospyros vaccinioides Lindl. ●
132457 Diospyros vera A. Chev.;交趾乌木;Indo-China Ebony ●☆
132458 Diospyros vermoesenii De Wild.;韦尔蒙森柿●☆
132459 Diospyros verrucosa Hiern;多疣柿●☆
132460 Diospyros vescoi Hiern;小柿●☆
132461 Diospyros vignei F. White;维涅柿●☆
132462 Diospyros viguieriana H. Perrier;维基耶柿●☆
132463 Diospyros villosa (L.) De Winter;长柔毛柿●☆
132464 Diospyros villosa (L.) De Winter var. parvifolia (De Winter) De Winter;小花长柔毛柿●☆
132465 Diospyros virgata (Gürke) Brenan;条纹柿●☆
132466 Diospyros virginiana L.;美洲柿(白柿,白檀,弗吉尼亚州柿,黄油木,美国柿,枣李木);America Persimmon, American Date Plum, American Date-plum, American Persimmon, Bara-bara, Boa Wood, Butter Wood, Common Persimmon, Date Plum, Date-plum, North American Ebony, North American Persimmon, Ossumwood, Persimmon, Possum Wood, Possumwood, Rsimmon Pe, Simmon, Simmonos-tree, Virginian Datelum, White Ebony ●
132467 Diospyros viridicans Hiern;绿柿●☆
132468 Diospyros wagemansii F. White;瓦格曼斯柿●☆
132469 Diospyros wajirensis F. White;瓦吉尔柿●☆
132470 Diospyros welwitschii Hiern = Diospyros abyssinica (Hiern) F. White ●☆
132471 Diospyros whitei Dows. -Lem. et Pannell;瓦特柿●☆
132472 Diospyros whyteana (Hiern) F. White;空果柿;Bladdernut ●☆
132473 Diospyros winkleri Gürke = Diospyros cinnabarina (Gürke) F. White ●☆
132474 Diospyros xanthocarpa Gürke = Diospyros batocana Hiern ●☆
132475 Diospyros xanthochlamys Gürke = Diospyros physocalycina Gürke ●☆
132476 Diospyros xiangguiensis S. K. Lee;湘桂柿;Hunan-Guangxi Persimmon, Xianggui Persimmon ●◇
132477 Diospyros xishuangbannaensis C. Y. Wu et H. Chu;版纳柿;Xishuangbanna Persimmon ●
132478 Diospyros yunnanensis Rehder et E. H. Wilson;云南柿;Yunnan Persimmon ●
132479 Diospyros zenkeri (Gürke) F. White;岑克尔柿●☆
132480 Diospyros zhenfengensis S. K. Lee;贞丰柿;Zhenfeng Persimmon ●
132481 Diospyros zombensis (B. L. Burtt) F. White;宗巴柿●☆
132482 Diostea Miers(1870);双骨草属●☆
132483 Diostea filifolia Miers;线叶双骨草●☆
132484 Diostea infuscata Miers;双骨草●☆
132485 Diotacanthus Benth. = Phlogacanthus Nees ●■
132486 Diothilophis Schltdl. = Dothilophis Raf. ■☆
132487 Diothilophis Schltdl. = Epidendrum L.(保留属名)■☆
132488 Diothonea Lindl. = Epidendrum L.(保留属名)■☆
132489 Dioticarpus Dunn = Hopea Roxb.(保留属名)●
132490 Dioticarpus Dunn(1920);耳果香属■☆
132491 Dioticarpus barryi Dunn;耳果香■☆
132492 Diotis Desf. = Otanthus Hoffmanns. et Link ■☆
132493 Diotis Schreb. = Ceratoides (Tourn.) Gagnebin ■☆
132494 Diotis Schreb. = Eurotia Adans. ■☆
132495 Diotis candidissima Desf. = Achillea maritima (L.) Ehrend. et Y. P. Guo ■☆
132496 Diotis lanata Pursh = Krascheninnikovia lanata (Pursh) A. Meeuse et A. Smit ●☆
132497 Diotis maritima (L.) Sm. = Achillea maritima (L.) Ehrend. et Y. P. Guo ■☆
132498 Diotocarpus Hochst. = Pentanisia Harv. ■☆
132499 Diotocarpus angustifolius Hochst. = Pentanisia angustifolia (Hochst.) Hochst. ■☆
132500 Diotocarpus prunelloides (Eckl. et Zeyh.) Hochst. = Pentanisia prunelloides (Klotzsch ex Eckl. et Zeyh.) Walp. ■☆
132501 Diotocranus Bremek. = Mitrasacmopsis Jovet ■☆
132502 Diotocranus lebrunii Bremek. = Mitrasacmopsis quadrivalvis Jovet ■☆
132503 Diotocranus lebrunii Bremek. var. sparsipilus ? = Mitrasacmopsis quadrivalvis Jovet ■☆
132504 Diotolotus Tausch = Argyrolobium Eckl. et Zeyh.(保留属名)●☆
132505 Diotosperma A. Gray = Ceratogyne Turcz. ■☆
132506 Diotostemon Salm-Dyck = Pachyphytum Link, Klotzsch et Otto ●☆
132507 Diotostephus Cass. = Chrysogonum A. Juss. ●■
132508 Diotostephus Cass. = Leontice L. ●■
132509 Diototheca Raf. = Phyla Lour. ■
132510 Diouratea Tiegh. = Ouratea Aubl.(保留属名)●
132511 Dioxippe M. Roem. = Glycosmis Corrêa(保留属名)●
132512 Dipanax Seem. = Tetraplasandra A. Gray ●☆
132513 Dipascus lushanensis C. Y. Cheng et T. M. Ai = Dipsacus japonicus Miq. ■
132514 Dipascus tianmuensis C. Y. Cheng et Z. T. Yin = Dipsacus japonicus Miq. ■
132515 Dipcadi Medik. (1790);异被风信子属■☆
132516 Dipcadi anthericoides Engl. et Gilg = Dipcadi longifolium (Lindl.) Baker ■☆
132517 Dipcadi arenarium Baker = Ornithogalum viride (L.) J. C. Manning et Goldblatt ■☆
132518 Dipcadi bakerianum Bolus;贝克异被风信子■☆
132519 Dipcadi bakerianum Schinz = Dipcadi marlothii Engl. ■☆
132520 Dipcadi baumii Engl. et Gilg = Dipcadi vaginatum Baker ■☆
132521 Dipcadi bussei Dammer = Ornithogalum viride (L.) J. C. Manning et Goldblatt ■☆
132522 Dipcadi ciliare (Eckl. et Zeyh. ex Harv.) Baker = Ornithogalum cirrhulosum J. C. Manning et Goldblatt ■☆
132523 Dipcadi ciliatum Engl. et K. Krause = Dipcadi platyphyllum Baker ■☆
132524 Dipcadi cinnabarinum Suess. = Ornithogalum viride (L.) J. C. Manning et Goldblatt ■☆
132525 Dipcadi comosum Welw. ex Baker = Ornithogalum viride (L.) J. C. Manning et Goldblatt ■☆
132526 Dipcadi conrathii Baker = Ornithogalum viride (L.) J. C. Manning et Goldblatt ■☆
132527 Dipcadi crispum Baker = Ornithogalum crispum (Baker) J. C. Manning et Goldblatt ■☆
132528 Dipcadi dahomense A. Chev. = Ornithogalum viride (L.) J. C. Manning et Goldblatt ■☆
132529 Dipcadi dekindtianum Engl.;德金异被风信子■☆
132530 Dipcadi dipcadioides Baker = Dipcadi gracillimum Baker ■☆
132531 Dipcadi durandianum Schinz = Dipcadi marlothii Engl. ■☆
132532 Dipcadi elatum Baker = Ornithogalum viride (L.) J. C. Manning et Goldblatt ■☆
132533 Dipcadi erlangeri Dammer = Ornithogalum viride (L.) J. C. Manning et Goldblatt ■☆
132534 Dipcadi ernesti-ruschii Dinter = Dipcadi marlothii Engl. ■☆

132535 Dipcadi erythraeum Webb et Berthel.;浅红异被风信子■☆
132536 Dipcadi filamentosum Medik. = Ornithogalum viride (L.) J. C. Manning et Goldblatt ■☆
132537 Dipcadi filifolium Baker = Ornithogalum viride (L.) J. C. Manning et Goldblatt ■☆
132538 Dipcadi firmifolium Baker = Dipcadi longifolium (Lindl.) Baker ■☆
132539 Dipcadi fulvum (Cav.) Webb et Berthel.;黄褐异被风信子■☆
132540 Dipcadi garuense Engl. et K. Krause;加鲁异被风信子■☆
132541 Dipcadi geniculatum Dinter et Suess. = Ornithogalum viride (L.) J. C. Manning et Goldblatt ■☆
132542 Dipcadi glaucum (Burch. ex Ker Gawl.) Baker;灰绿异被风信子■☆
132543 Dipcadi gourmaense A. Chev. ex Hutch. = Ornithogalum viride (L.) J. C. Manning et Goldblatt ■☆
132544 Dipcadi gracilipes K. Krause = Dipcadi glaucum (Burch. ex Ker Gawl.) Baker ■☆
132545 Dipcadi gracillimum Baker;细长异被风信子■☆
132546 Dipcadi helenae P. Beauv. = Ornithogalum viride (L.) J. C. Manning et Goldblatt ■☆
132547 Dipcadi hockii De Wild. = Ornithogalum viride (L.) J. C. Manning et Goldblatt ■☆
132548 Dipcadi hyacinthoides (Spreng.) Baker = Dipcadi bakerianum Bolus ■☆
132549 Dipcadi kelleri Baker = Ornithogalum viride (L.) J. C. Manning et Goldblatt ■☆
132550 Dipcadi kerstingii Dammer = Dipcadi longifolium (Lindl.) Baker ■☆
132551 Dipcadi lanceolatum Baker = Ornithogalum viride (L.) J. C. Manning et Goldblatt ■☆
132552 Dipcadi lateritium Welw. ex Baker = Ornithogalum viride (L.) J. C. Manning et Goldblatt ■☆
132553 Dipcadi ledermannii Engl. et K. Krause;莱德异被风信子■☆
132554 Dipcadi lividescens Engl. et Gilg = Ornithogalum viride (L.) J. C. Manning et Goldblatt ■☆
132555 Dipcadi longibracteatum Schinz = Dipcadi glaucum (Burch. ex Ker Gawl.) Baker ■☆
132556 Dipcadi longicauda Engl. et K. Krause = Ornithogalum viride (L.) J. C. Manning et Goldblatt ■☆
132557 Dipcadi longifolium (Lindl.) Baker;长叶异被风信子■☆
132558 Dipcadi magnum Baker = Dipcadi glaucum (Burch. ex Ker Gawl.) Baker ■☆
132559 Dipcadi marlothii Engl.;马洛斯异被风信子■☆
132560 Dipcadi mechowii Engl.;梅休异被风信子■☆
132561 Dipcadi minimum (Steud. ex A. Rich.) Webb et Berthel. = Ornithogalum viride (L.) J. C. Manning et Goldblatt ■☆
132562 Dipcadi ndellense A. Chev.;恩代尔异被风信子■☆
132563 Dipcadi nitens K. Krause = Ornithogalum viride (L.) J. C. Manning et Goldblatt ■☆
132564 Dipcadi occidentale Baker = Ornithogalum viride (L.) J. C. Manning et Goldblatt ■☆
132565 Dipcadi oligotrichum Baker = Dipcadi marlothii Engl. ■☆
132566 Dipcadi oxylobum Baker;尖裂片异被风信子■☆
132567 Dipcadi palustre Baker = Ornithogalum viride (L.) J. C. Manning et Goldblatt ■☆
132568 Dipcadi papillatum Oberm.;乳突异被风信子■☆
132569 Dipcadi platyphyllum Baker;宽叶异被风信子■☆
132570 Dipcadi polyphyllum Baker = Dipcadi gracillimum Baker ■☆
132571 Dipcadi rhodesiacum Weim. = Dipcadi rigidifolium Baker ■☆
132572 Dipcadi rigidifolium Baker;硬叶异被风信子■☆
132573 Dipcadi rupicola Chiov. = Ornithogalum viride (L.) J. C. Manning et Goldblatt ■☆
132574 Dipcadi sansibaricum Engl. = Dipcadi longifolium (Lindl.) Baker ■☆
132575 Dipcadi serotinum (L.) Medik.;迟花异被风信子■☆
132576 Dipcadi serotinum (L.) Medik. subsp. fulvum (Cav.) Maire et Weiller = Dipcadi fulvum (Cav.) Webb et Berthel. ■☆
132577 Dipcadi serotinum (L.) Medik. subsp. lividum (Pers.) Maire et Weiller;蓝迟花异被风信子■☆
132578 Dipcadi serotinum (L.) Medik. var. pruinosum Gatt. et Weiller = Dipcadi fulvum (Cav.) Webb et Berthel. ■☆
132579 Dipcadi serotinum Medik. = Dipcadi serotinum (L.) Medik. ■☆
132580 Dipcadi stenophyllum Dinter = Ornithogalum viride (L.) J. C. Manning et Goldblatt ■☆
132581 Dipcadi sulcatum Suess. = Ornithogalum viride (L.) J. C. Manning et Goldblatt ■☆
132582 Dipcadi tacazzeanum (Hochst. ex A. Rich.) Baker = Ornithogalum viride (L.) J. C. Manning et Goldblatt ■☆
132583 Dipcadi tenuifolium A. Chev.;细叶异被风信子■☆
132584 Dipcadi thollonianum Hua = Dipcadi vaginatum Baker ■☆
132585 Dipcadi tortile R. A. Dyer = Ornithogalum crispum (Baker) J. C. Manning et Goldblatt ■☆
132586 Dipcadi umbonatum (Baker) Baker = Ornithogalum viride (L.) J. C. Manning et Goldblatt ■☆
132587 Dipcadi uniflorum Baker = Dipcadi viride (L.) Moench ■☆
132588 Dipcadi unifolium Baker = Ornithogalum viride (L.) J. C. Manning et Goldblatt ■☆
132589 Dipcadi vaginatum Baker;具鞘异被风信子■☆
132590 Dipcadi venenatum Schinz = Dipcadi longifolium (Lindl.) Baker ■☆
132591 Dipcadi viride (L.) Moench = Ornithogalum viride (L.) J. C. Manning et Goldblatt ■☆
132592 Dipcadi welwitschii (Baker) Baker;韦尔异被风信子■☆
132593 Dipcadi wentzelianum Engl. = Ornithogalum viride (L.) J. C. Manning et Goldblatt ■☆
132594 Dipcadi zambesiacum Baker = Dipcadi longifolium (Lindl.) Baker ■☆
132595 Dipcadioides Medik. (1790);拟异被风信子属■☆
132596 Dipcadioides Medik. = Lachenalia J. Jacq. ex Murray ■☆
132597 Dipcadioides maculata Medik.;拟异被风信子■☆
132598 Dipelta Maxim. (1877);双盾木属(双楯属);Dipelta ●★
132599 Dipelta Regel et Schmalh. = Astragalus L. ●■
132600 Dipelta elegans Batalin;优美双盾木;Elegant Dipelta ●◇
132601 Dipelta floribunda Maxim.;双盾木(鸡骨头,满山红,双楯);Rosy Dipelta ●
132602 Dipelta floribunda Maxim. var. parviflora Rehder = Dipelta floribunda Maxim. ●
132603 Dipelta ventricosa Hemsl. = Dipelta yunnanensis Franch. ●
132604 Dipelta wenxianensis Y. F. Wang et Lian = Dipelta floribunda Maxim. ●
132605 Dipelta yunnanensis Franch.;云南双盾木(垂枝双盾,垂枝双盾木,鸡骨菜,云南双楯);Yunnan Dipelta ●
132606 Dipelta yunnanensis Franch. var. brachycalyx Hand.-Mazz.;短萼云南双盾木●

132607 Dipentaplandra Kuntze = Pentadiplandra Baill. ●☆

132608 Dipentodon Dunn(1911);十齿花属(十萼花属);Dipentodon ●

132609 Dipentodon longipedicellatus C. Y. Cheng et J. S. Liu;长梗十齿花;Longpedicel Dipentodon, Longstalk Dipentodon ●

132610 Dipentodon sincicus Dunn = Dipentodon longipedicellatus C. Y. Cheng et J. S. Liu ●

132611 Dipentodon sinicus Dunn;十齿花(青铁果,十萼花);Chinese Dipentodon Dipentodon ●◇

132612 Dipentodontaceae Merr. (1941)(保留科名);十齿花科(十萼花科)●

132613 Dipera Spreng. = Disperis Sw. ■

132614 Dipera tenera Spreng. = Disperis capensis (L. f.) Sw. var. brevicaudata Rolfe ■☆

132615 Diperis Wight = Dipera Spreng. ■

132616 Diperis Wight = Disperis Sw. ■

132617 Diperium Desv. = Mnesithea Kunth ■

132618 Diperium cylindricum Desv. = Mnesithea laevis (Retz.) Kunth ■

132619 Dipetalanthus A. Chev. = Hymenostegia (Benth.) Harms ■☆

132620 Dipetalanthus felicis A. Chev. = Hymenostegia felicis (A. Chev.) J. Léonard ■☆

132621 Dipetalanthus pellegrinii A. Chev. = Hymenostegia pellegrinii (A. Chev.) J. Léonard ■☆

132622 Dipetalia Raf. (废弃属名) = Oligomeris Cambess. (保留属名) ■●

132623 Dipetalia dregeana Kuntze = Oligomeris dregeana (Müll. Arg.) Müll. Arg. ■☆

132624 Dipetalia spathulata (E. Mey. ex Turcz.) Kuntze = Oligomeris dipetala (Aiton) Turcz. var. spathulata (E. Mey. ex Turcz.) Abdallah ■☆

132625 Dipetalon Raf. = Cuphea Adans. ex P. Browne ●■

132626 Dipetalum Dalzell = Toddalia Juss. (保留属名)●

132627 Dipetalum Dalzell = Vepris Comm. ex A. Juss. ●☆

132628 Diphaca Lour. (废弃属名) = Ormocarpum P. Beauv. (保留属名)●

132629 Diphaca bernieriana Baill. = Ormocarpum bernierianum (Baill.) Du Puy et Labat ●☆

132630 Diphaca cochinchinense Lour. = Ormocarpum cochinchinense (Lour.) Merr. ●

132631 Diphaca discolor (Vatke) Chiov. = Ormocarpum kirkii S. Moore ●☆

132632 Diphaca kirkii (S. Moore) Taub. = Ormocarpum kirkii S. Moore ●☆

132633 Diphaca pervilleana Baill. = Pongamiopsis pervilleana (Baill.) R. Vig. ●☆

132634 Diphaca trachycarpa Taub. = Ormocarpum trachycarpum (Taub.) Harms ●☆

132635 Diphaca trichocarpa Taub. = Ormocarpum trichocarpum (Taub.) Engl. ●☆

132636 Diphaca verrucosa (P. Beauv.) Taub. = Ormocarpum verrucosum P. Beauv. ●☆

132637 Diphalangium Schauer(1847);异双列百合属■☆

132638 Diphasia Pierre = Vepris Comm. ex A. Juss. ●☆

132639 Diphasia Pierre(1898);迪法斯木属●☆

132640 Diphasia angolensis (Hiern) I. Verd. = Vepris hiernii Gereau ●☆

132641 Diphasia dainellii Pic. Serm. = Vepris dainellii (Pic. Serm.) Kokwaro ●☆

132642 Diphasia klaineana Pierre = Vepris hiernii Gereau ●☆

132643 Diphasia madagascariensis H. Perrier = Vepris madagascariensis (H. Perrier) Mziray ●☆

132644 Diphasia mildbraedii Engl.;米尔德迪法斯木●☆

132645 Diphasia morogorensis Kokwaro = Vepris morogorensis (Kokwaro) Mziray ●☆

132646 Diphasia morogorensis Kokwaro var. subalata Kokwaro = Vepris morogorensis (Kokwaro) Mziray ●☆

132647 Diphasia noldeae Exell et Mendonça = Vepris noldeae (Exell et Mendonça) Mziray ●☆

132648 Diphasiopsis Mendonça(1961);拟迪法斯木属●☆

132649 Diphasiopsis fadenii Kokwaro;法登拟迪法斯木●☆

132650 Diphasiopsis fadenii Kokwaro = Vepris fadenii (Kokwaro) Mziray ●☆

132651 Diphasiopsis whitei Mendonça;拟迪法斯木●☆

132652 Diphelypaea Nicolson = Phelypaea L. ■☆

132653 Dipherocarpus Llanos = Nephelium L. ●

132654 Dipholis A. DC. (1844)(保留属名);双鳞山榄属(代弗山榄属)●☆

132655 Dipholis A. DC. = Sideroxylon L. ●☆

132656 Dipholis salicifolia A. DC.;柳叶双鳞山榄(柳叶代弗山榄木);Bustic Willow ●☆

132657 Dipholis stevensonii Standl.;斯氏双鳞山榄(伯利兹代弗山榄木)●☆

132658 Diphorea Raf. = Sagittaria L. ■

132659 Diphragmus C. Presl = Spermacoce L. ●■

132660 Diphryllum Raf. (废弃属名) = Listera R. Br. (保留属名)■

132661 Diphryllum convallarioides (Sw.) Kuntze = Listera convallarioides (Sw.) Elliott ■☆

132662 Diphryllum cordatum (L.) Kuntze = Listera cordata (L.) R. Br. ■☆

132663 Diphryllum japonicum (Blume) Kuntze = Neottia japonica (M. Furuse) K. Inoue ■

132664 Diphryllum micranthum (Lindl.) Kuntze = Neottia karoana Szlach. ■

132665 Diphryllum ovatum (L.) Kuntze = Listera ovata R. Br. ■

132666 Diphyes Blume = Bulbophyllum Thouars(保留属名)■

132667 Diphyes ovalifolia Blume = Bulbophyllum ovalifolium (Blume) Lindl. ■

132668 Diphylax Hook. f. (1889);尖药兰属;Diphylax ■

132669 Diphylax contigua (Ts. Tang et F. T. Wang) Ts. Tang, F. T. Wang et K. Y. Lang;长苞尖药兰;Longbract Diphylax ■

132670 Diphylax griffithii (Hook. f.) Kraenzl.;格氏尖药兰■☆

132671 Diphylax uniformis (Ts. Tang et F. T. Wang) Ts. Tang, F. T. Wang et K. Y. Lang;西南尖药兰;SW. China Diphylax ■

132672 Diphylax urceolata (C. B. Clarke) Hook. f.;尖药兰;Diphylax, Urnshaped Diphylax ■

132673 Diphyleia Raf. = Diphylleia Michx. ■●

132674 Diphyllanthus Tiegh. = Ouratea Aubl. (保留属名)●

132675 Diphyllarium Gagnep. (1915);双苞豆属■☆

132676 Diphyllarium melongense Gagnep.;双苞豆■☆

132677 Diphylleia Michx. (1803);山荷叶属(二叶草属);Umbrella Leaf, Umbrellaleaf, Umbrella-leaf ■●

132678 Diphylleia cymosa Michx.;聚伞山荷叶(鬼臼,美洲山荷叶,山荷叶);American Umbrellaleaf, Umbrella Leaf, Umbrella-leaf ■☆

132679 Diphylleia cymosa Michx. subsp. grayi (F. Schmidt) Kitam. = Diphylleia grayi F. Schmidt ■☆

132680 Diphylleia cymosa Michx. subsp. grayi (F. Schmidt) Kitam. var.

incisa (Takeda) T. Shimizu = Diphylleia grayi F. Schmidt ■☆
132681 Diphylleia cymosa Michx. subsp. sinensis (H. L. Li) T. Shimizu = Diphylleia sinensis H. L. Li ■
132682 Diphylleia grayi F. Schmidt;格氏山荷叶(日本山荷叶,山荷叶);Gray Umbrellaleaf,Japan Umbrellaleaf ■☆
132683 Diphylleia grayi F. Schmidt = Diphylleia sinensis H. L. Li ■
132684 Diphylleia grayi F. Schmidt var. incisa Takeda = Diphylleia grayi F. Schmidt ■☆
132685 Diphylleia sinensis H. L. Li;南方山荷叶(阿儿七,旱禾,江边一碗水,金边七,山荷叶,窝儿参,窝儿七,一把伞,一碗水,中华山荷叶);China Umbrellaleaf,Chinese Umbrellaleaf ■
132686 Diphylleiaceae Schultz Sch.;山荷叶科●■
132687 Diphylleiaceae Schultz Sch. = Berberidaceae Juss.(保留科名)●■
132688 Diphylleiaceae Schultz Sch. = Podophyllaceae + Sarraceniaceae Dumort.(保留科名)■☆
132689 Diphyllopodium Tiegh. = Ouratea Aubl.(保留属名)●
132690 Diphyllopodium klainei Tiegh. = Campylospermum klainei (Tiegh.) Farron ●☆
132691 Diphyllopodium zenkeri Engl. ex Tiegh. = Campylospermum zenkeri (Engl. ex Tiegh.) Farron ●☆
132692 Diphyllum Raf. = Diphryllum Raf.(废弃属名)■
132693 Diphyllum Raf. = Listera R. Br.(保留属名)■
132694 Diphysa Jacq.(1760);双泡豆属■☆
132695 Diphysa echinata Rose;刺双泡豆■☆
132696 Diphysa floribunda Peyr.;多花双泡豆■☆
132697 Diphysa macrocarpa Standl.;大果双泡豆■☆
132698 Diphysa macrophylla Lundell;大叶双泡豆■☆
132699 Diphysa microphylla Rydb.;小叶双泡豆■☆
132700 Diphysa occidentalis Rose;西方双泡豆■☆
132701 Diphysa puberulenta Rydb.;毛双泡豆■☆
132702 Diphysa punctata Rydb.;斑点双泡豆■☆
132703 Diphysa robinioides Benth.;槐双泡豆■☆
132704 Diphystema Neck. = Amasonia L. f.(保留属名)●■☆
132705 Dipidax Lawson ex Salisb. = Onixotis Raf. ■☆
132706 Dipidax Salisb. ex Benth. = Onixotis Raf. ■☆
132707 Dipidax ciliata (L. f.) Baker = Onixotis punctata (L.) Mabb. ■☆
132708 Dipidax ciliata (L. f.) Baker var. bergii (Schltdl.) Baker = Onixotis punctata (L.) Mabb. ■☆
132709 Dipidax ciliata (L. f.) Baker var. garnotiana (Kunth) Baker = Onixotis punctata (L.) Mabb. ■☆
132710 Dipidax ciliata (L. f.) Baker var. gracilis (Desv.) Baker = Onixotis punctata (L.) Mabb. ■☆
132711 Dipidax ciliata (L. f.) Baker var. rubicunda (Kunth) Baker = Onixotis punctata (L.) Mabb. ■☆
132712 Dipidax ciliata (L. f.) Baker var. secunda (Desv.) Baker = Onixotis punctata (L.) Mabb. ■☆
132713 Dipidax punctata (L.) Hutch. = Onixotis punctata (L.) Mabb. ■☆
132714 Dipidax punctata (L.) Hutch. var. bergii (Schltdl.) Adamson = Onixotis punctata (L.) Mabb. ■☆
132715 Dipidax punctata (L.) Hutch. var. garnotiana (Kunth) Adamson = Onixotis punctata (L.) Mabb. ■☆
132716 Dipidax rosea Salisb. = Onixotis stricta (Burm. f.) Wijnands ■☆
132717 Dipidax triquetra (L. f.) Baker = Onixotis stricta (Burm. f.) Wijnands ■☆
132718 Diplachna Kuntze et T. Post = Diplachne R. Br. ex Desf. ●☆
132719 Diplachne Desf. = Verticordia DC.(保留属名)●☆
132720 Diplachne P. Beauv.(1812);双稃草属(青茅属);Diplachne ■
132721 Diplachne P. Beauv. = Leptochloa P. Beauv. ■
132722 Diplachne R. Br. = Verticordia DC.(保留属名)●☆
132723 Diplachne R. Br. ex Desf. = Verticordia DC.(保留属名)●☆
132724 Diplachne acuminata Nash = Leptochloa fusca (L.) Kunth subsp. fascicularis (Lam.) N. Snow ■☆
132725 Diplachne acuminata Nash = Leptochloa fusca (L.) Kunth ■
132726 Diplachne alba Hochst. = Leptochloa fusca (L.) Kunth ■
132727 Diplachne amboensis Roiv.;安博双稃草■☆
132728 Diplachne amboensis Roiv. var. plurinodis Roiv.;多节双稃草■☆
132729 Diplachne barbata Hack. = Schenckochloa barbata (Hack.) J. J. Ortíz ■☆
132730 Diplachne biflora Hack. = Bewsia biflora (Hack.) Gooss. ■☆
132731 Diplachne biflora Hack. var. buchananii Stapf = Bewsia biflora (Hack.) Gooss. ■☆
132732 Diplachne burgarica Roshev.;保加利亚青茅■☆
132733 Diplachne capensis (Nees) Nees = Leptochloa fusca (L.) Kunth ■
132734 Diplachne caudata K. Schum.;尾状双稃草■☆
132735 Diplachne cinerea Hack. = Odyssea paucinervis (Nees) Stapf ■☆
132736 Diplachne cuspidata Launert;骤尖双稃草■☆
132737 Diplachne dummeri Stapf et C. E. Hubb. = Diplachne caudata K. Schum. ■☆
132738 Diplachne fascicularis (Lam.) P. Beauv. = Leptochloa fusca (L.) Kunth subsp. fascicularis (Lam.) N. Snow ■☆
132739 Diplachne fascicularis (Lam.) P. Beauv. = Leptochloa fusca (L.) Kunth ■
132740 Diplachne fascicularis P. Beauv. = Leptochloa fusca (L.) Kunth ■
132741 Diplachne festuciformis H. Scholz;羊茅状双稃草■☆
132742 Diplachne fleckii Hack. = Pogonarthria fleckii (Hack.) Hack. ■☆
132743 Diplachne fusca (L.) P. Beauv. = Leptochloa fusca (L.) Kunth ■
132744 Diplachne fusca (L.) P. Beauv. ex Roem. et Schult.;棕色双稃草■☆
132745 Diplachne fusca (L.) P. Beauv. ex Roem. et Schult. = Diplachne malabarica (L.) Merr. ■
132746 Diplachne fusca (L.) P. Beauv. ex Roem. et Schult. = Leptochloa fusca (L.) Kunth ■
132747 Diplachne fusca (L.) Stapf = Leptochloa fusca (L.) Kunth ■
132748 Diplachne fusca (L.) Stapf var. alba (Steud.) Chiov. = Leptochloa fusca (L.) Kunth ■
132749 Diplachne gatacrei Stapf = Kengia gatacrei (Stapf) Cope ■☆
132750 Diplachne gigantea Launert = Leptochloa gigantea (Launert) Cope et N. Snow ■☆
132751 Diplachne grandiglumis (Nees) Hack. = Trichoneura grandiglumis (Nees) Ekman ■☆
132752 Diplachne hackelii Honda = Cleistogenes hackelii (Honda) Honda ■
132753 Diplachne hackelii Honda = Kengia hackelii (Honda) Packer ■
132754 Diplachne halei Nash = Leptochloa panicoides (J. Presl) Hitchc. ■☆
132755 Diplachne jaegeri Pilg. = Psilolemma jaegeri (Pilg.) S. M. Phillips ■☆

132756 Diplachne kokonorica (K. S. Hao) Conert = Orinus kokonorica (K. S. Hao) Tzvelev ■
132757 Diplachne livida Nees = Leptochloa fusca (L.) Kunth ■
132758 Diplachne malabarica (L.) Merr. = Diplachne fusca (L.) P. Beauv. ■
132759 Diplachne malabarica (L.) Merr. = Leptochloa fusca (L.) Kunth ■
132760 Diplachne menyharthii Hack. = Pogonarthria squarrosa (Roem. et Schult.) Pilg. ■☆
132761 Diplachne nana Nees = Triraphis pumilio R. Br. ■☆
132762 Diplachne pallida Hack. = Leptochloa fusca (L.) Kunth ■
132763 Diplachne panicoides J. Presl = Leptochloa panicoides (J. Presl) Hitchc. ■☆
132764 Diplachne paucinervis (Nees) Hack. = Odyssea paucinervis (Nees) Stapf ■☆
132765 Diplachne repatrix (L.) Druce = Diplachne fusca (L.) P. Beauv. ex Roem. et Schult. ■☆
132766 Diplachne serotina (L.) Link var. aristata Hack. = Cleistogenes hackelii (Honda) Honda ■
132767 Diplachne serotina Link = Cleistogenes kitagawai Honda ■
132768 Diplachne serotina Link var. chinensis Maxim. = Cleistogenes chinensis (Maxim.) Keng ■
132769 Diplachne serotina Link var. chinensis Maxim. = Cleistogenes hackelii (Honda) Honda ■
132770 Diplachne sinensis Hance = Cleistogenes hancei Keng ■
132771 Diplachne sinensis Hance = Cleistogenes kitagawai Honda ■
132772 Diplachne songorica Roshev. = Cleistogenes songorica (Roshev.) Ohwi ■
132773 Diplachne squarrosa (Trin.) Maxim. = Cleistogenes squarrosa (Trin.) Keng ■
132774 Diplachne squarrosa (Trin.) Maxim. var. longe-aristata Rendle = Cleistogenes squarrosa (Trin.) Keng ■
132775 Diplachne thoroldii Stapf ex Hemsl. = Orinus thoroldii (Stapf ex Hemsl.) Bor ■
132776 Diplachne uninerva (J. Presl) Parodi = Leptochloa fusca (L.) Kunth subsp. uninerva (J. Presl) N. Snow ■☆
132777 Diplachne vulpiastrum (De Not.) Schweinf. = Leptocarydion vulpiastrum (De Not.) Stapf ■☆
132778 Diplachne wahlbergii Roiv.;瓦尔贝里双稃草■☆
132779 Diplachyrium Nees = Muhlenbergia Schreb. ■
132780 Diplachyrium Nees = Perotis Aiton ■
132781 Diplachyrium rarum (R. Br.) Nees = Perotis rara R. Br. ■
132782 Diplacorchis Schltr. = Brachycorythis Lindl. ■
132783 Diplacorchis angolensis Schltr. = Brachycorythis tenuior Rchb. f. ■☆
132784 Diplacorchis ashantensis Summerh. = Brachycorythis sceptrum Schltr. ■☆
132785 Diplacorchis conica Summerh. = Brachycorythis conica (Summerh.) Summerh. ■☆
132786 Diplacorchis tenuior (Rchb. f.) Schltr. = Brachycorythis tenuior Rchb. f. ■☆
132787 Diplacrum R. Br. (1810);裂颖茅属(小珠茅属);Diplacrum ■
132788 Diplacrum africanum (Benth.) C. B. Clarke;非洲裂颖茅■☆
132789 Diplacrum caricinum R. Br.; 裂颖茅; Common Diplacrum, Diplacrum ■
132790 Diplacrum caricinum R. Br. = Scleria caricina (R. Br.) Benth. ■
132791 Diplacrum longifolium (Griseb.) C. B. Clarke;长叶裂颖茅■☆
132792 Diplacrum pygmaeum Boeck. = Diplacrum africanum (Benth.) C. B. Clarke ■☆
132793 Diplacrum reticulatum Holttum;网果裂颖茅■
132794 Diplactis Raf. = Aster L. ●■
132795 Diplacus Nutt. = Mimulus L. ●■
132796 Diplacus glutinosus Nutt. = Mimulus aurantiacus Curtis ●☆
132797 Dipladenia A. DC. = Mandevilla Lindl. ●
132798 Dipladenia alboviridis Rusby = Mandevilla alboviridis (Rusby) Woodson ●☆
132799 Dipladenia laxa Ruiz et Pav. = Mandevilla laxa (Ruiz et Pav.) Woodson ●
132800 Dipladenia suaveolens Ruiz et Pav. = Mandevilla laxa (Ruiz et Pav.) Woodson ●
132801 Diplandra Bertero(废弃属名) = Diplandra Hook. et Arn. (保留属名)■☆
132802 Diplandra Bertero(废弃属名) = Elodea Michx. ■☆
132803 Diplandra Hook. et Arn. (1838)(保留属名);双蕊柳叶菜属■☆
132804 Diplandra Hook. et Arn. (保留属名) = Lopezia Cav. ■☆
132805 Diplandra Raf. = Ludwigia L. ●■
132806 Diplandrorchis S. C. Chen = Neottia Guett. (保留属名)■
132807 Diplandrorchis S. C. Chen (1979); 双蕊兰属; Twostamen Orchid ■★
132808 Diplandrorchis sinica S. C. Chen;双蕊兰;Twostamen Orchid ■
132809 Diplanoma Raf. = Abelmoschus Medik. ●■
132810 Diplanthemum K. Schum. = Duboscia Bocquet ●☆
132811 Diplanthemum brieyi De Wild. = Duboscia macrocarpa Bocq. ●☆
132812 Diplanthemum viridiflorum K. Schum. = Duboscia macrocarpa Bocq. ●☆
132813 Diplanthera Banks et Sol. = Deplanchea Vieill. ●☆
132814 Diplanthera Banks et Sol. ex R. Br. = Deplanchea Vieill. ●☆
132815 Diplanthera Gled. = Dianthera L. ■☆
132816 Diplanthera Gled. = Justicia L. ●■
132817 Diplanthera Raf. = Platanthera Rich. (保留属名)■
132818 Diplanthera Schrank = Justicia L. ●■
132819 Diplanthera Thouars = Halodule Endl. ■
132820 Diplanthera beaudettei Hartog = Halodule wrightii Asch. ■☆
132821 Diplanthera pinifolia Miki = Halodule pinifolia (Miki) Hartog ■
132822 Diplanthera tridentata Steintheil = Halodule tridentata (Steinh.) Endl. ex Unger ■☆
132823 Diplanthera uninervis (Forssk.) Asch. = Halodule uninervis (Forssk.) Asch. ■
132824 Diplanthera uninervis Forssk. = Halodule uninervis (Forssk.) Asch. ■
132825 Diplanthera wrightii (Asch.) Asch. = Halodule wrightii Asch. ■☆
132826 Diplantheraceae Baum. -Bod. = Cymodoceaceae Vines(保留科名)■
132827 Diplarchaceae Klotzsch = Ericaceae Juss. (保留科名)●
132828 Diplarche Hook. f. et Thomson (1854);杉叶杜鹃属(杉叶杜属);Diplarche ●
132829 Diplarche multiflora Hook. f. et Thomson;杉叶杜鹃(多花杉叶杜,骨痛药,杉叶杜);Diplarche, Manyflower Diplarche, Multiflorous Diplarche ●
132830 Diplarche pauciflora Hook. f. et Thomson;少花杉叶杜鹃(少花杉叶杜);Fewflower Diplarche, Pauciflorous Diplarche ●
132831 Diplaria Raf. ex DC. = Cassandra D. Don ●
132832 Diplarinus Raf. = Scirpus L. (保留属名)■
132833 Diplarpea Triana = Diplarpea Triana ex Benth. ☆

132834 Diplarpea Triana ex Benth. (1867);双镰野牡丹属☆
132835 Diplarpea paleacea Triana;双镰野牡丹☆
132836 Diplarrena Labill. = Diplarrhena Labill. ■☆
132837 Diplarrhena Labill. (1800);澳菖蒲属■☆
132838 Diplarrhena R. Br. = Diplarrhena Labill. ■☆
132839 Diplarrhena latifolia Benth.;宽叶澳菖蒲■☆
132840 Diplarrhena moraea Labill.;澳菖蒲■☆
132841 Diplarrhinus Endl. = Diplarinus Raf. ■
132842 Diplarrhinus Endl. = Scirpus L.(保留属名)■
132843 Diplasanthera Hook. f. = Diplasanthum Desv. ■
132844 Diplasanthum Desv. = Andropogon L.(保留属名)■
132845 Diplasanthum Desv. = Dichanthium Willemet ■
132846 Diplasanthum lanosum Desv. = Dichanthium aristatum (Poir.) C. E. Hubb. ■
132847 Diplasia Pers. = Diplasia Rich. ■☆
132848 Diplasia Rich. (1805);疏黄鞘莎草属■☆
132849 Diplasia karatifolia Rich.;疏黄鞘莎草■☆
132850 Diplasia pycnostachya Benth.;密穗疏黄鞘莎草■☆
132851 Diplaspis Hook. f. (1847);双盾芹属■☆
132852 Diplaspis Hook. f. = Huanaca Cav. ■☆
132853 Diplaspis cordifolia (Hook.) Hook. f.;双盾芹■☆
132854 Diplatia Tiegh. (1894);双阔寄生属●☆
132855 Diplatia alberticii Tiegh.;双阔寄生●☆
132856 Diplatia grandibractea Tiegh.;光苞双阔寄生●☆
132857 Diplatia tenuifolia Tiegh.;细叶双阔寄生●☆
132858 Diplax Benn. = Ehrharta Thunb.(保留属名)■☆
132859 Diplax Sol. ex Benn. = Microlaena R. Br. ■☆
132860 Diplazoptilon Y. Ling(1965);重羽菊属;Diplazoptilon ■★
132861 Diplazoptilon cooperi (J. Anthony) C. Shih;裂叶重羽菊;Cooper Diplazoptilon ■
132862 Diplazoptilon picridifolium (Hand.-Mazz.) Y. Ling;重羽菊(青木香);Common Diplazoptilon ■
132863 Diplecoala G. Don = Diplycosia Blume ●☆
132864 Diplecthrum Pers. = Satyrium Sw.(保留属名)■
132865 Diplecthrum amoenum Thouars = Satyrium amoenum A. Rich. ■☆
132866 Diplecthrum bicallosum (Thunb.) Pers. = Satyrium bicallosum Thunb. ■☆
132867 Diplecthrum bracteatum (L. f.) Pers. = Satyrium bracteatum (L. f.) Thunb. ■☆
132868 Diplecthrum coriifolium (Sw.) Pers. = Satyrium coriifolium Sw. ■☆
132869 Diplecthrum erectum (Sw.) Pers. = Satyrium erectum Sw. ■☆
132870 Diplecthrum parviflorum (Sw.) Pers. = Satyrium parviflorum Sw. ■☆
132871 Diplecthrum pumilum (Thunb.) Pers. = Satyrium pumilum Thunb. ■☆
132872 Diplecthrum striatum (Thunb.) Pers. = Satyrium striatum Thunb. ■☆
132873 Diplectraden Raf. = Habenaria Willd. ■
132874 Diplectria (Blume) Rchb. (1841);藤牡丹属;Diplectria, Vinepeony ●■
132875 Diplectria Blume ex Rchb. = Diplectria (Blume) Rchb. ●■
132876 Diplectria Rchb. = Diplectria (Blume) Rchb. ●■
132877 Diplectria assamica (C. B. Clarke) Kuntze = Medinilla assamica (C. B. Clarke) C. Chen ●
132878 Diplectria barbata (Wall. ex C. B. Clarke) Franken et M. C. Roos;藤牡丹;Barbate Diplectria, Barbed Diplectria, Bearded Diplectria ●■
132879 Diplectrum Pers. = Satyrium Sw.(保留属名)■
132880 Diplectrum Thouars = Diplecthrum Pers. ■
132881 Diplectrum Thouars = Satyrium Sw.(保留属名)■
132882 Diplegnon Post et Kuntze = Diplolegnon Rusby ■☆
132883 Dipleina Raf. = Actaea L. ■
132884 Diplemium Raf. = Erigeron L. ■●
132885 Diplerisma Planch. = Melianthus L. ●☆
132886 Diplesthes Harv. = Salacia L.(保留属名)●
132887 Diplesthes kraussii Harv. = Salacia kraussii (Harv.) Harv. ●☆
132888 Dipliathus Raf. = Licaria Aubl. ●☆
132889 Diplicosia Endl. = Diplycosia Blume ●☆
132890 Diplima Raf. = Salix L.(保留属名)●
132891 Diplisca Raf. = Colubrina Rich. ex Brongn.(保留属名)●
132892 Diploatephion Raf. = Dyssodia Cav. ■☆
132893 Diplobryum C. Cusset(1972);倍苔草属■☆
132894 Diplobryum minutale C. Cusset;倍苔草■☆
132895 Diplocalymma Spreng. = Thunbergia Retz.(保留属名)●■
132896 Diplocalyx A. Rich. (1850);双萼树属●☆
132897 Diplocalyx A. Rich. = Schoepfia Schreb. ●
132898 Diplocalyx C. Presl = Mitraria Cav.(保留属名)●☆
132899 Diplocalyx chrysophylloides A. Rich.;双萼树●☆
132900 Diplocardia Zipp. ex Blume = Pometia J. R. Forst. et G. Forst. ●
132901 Diplocarex Hayata = Carex L. ■
132902 Diplocarex Hayata(1921);倍蕊苔属■
132903 Diplocarex matsudai Hayata;倍蕊苔■☆
132904 Diplocarex matsudai Hayata = Carex dolichostachys Hayata ■
132905 Diplocarex matsudai Hayata = Carex matsudai (Hayata) Hayata ex Makino et Nemoto ■
132906 Diplocaulobium (Rchb. f.) Kraenzl. (1910);褐茎兰属■☆
132907 Diplocaulobium Kraenzl. = Diplocaulobium (Rchb. f.) Kraenzl. ■☆
132908 Diplocauloblum pentanema (Schltr.) Kraenzl.;褐茎兰■☆
132909 Diplocea Raf. (1817) = Salsola L. ●■
132910 Diplocea Raf. (1818) = Triplasis P. Beauv. ■☆
132911 Diplocea Raf. (1818) = Uralepis Nutt. ■☆
132912 Diploceleba Post et Kuntze = Diplokeleba N. E. Br. ●☆
132913 Diplocentrum Lindl. (1832);印度双距兰属■☆
132914 Diploceras Meisn. = Parolinia Webb ■☆
132915 Diplochaete Nees = Rhynchospora Vahl(保留属名)■
132916 Diplochilus Lindl. = Diplomeris D. Don ■
132917 Diplochilus hirsutus Lindl. = Diplomeris hirsuta Lindl. ■
132918 Diplochilus hirsutus Lindl. var. biflorus Pradhan = Diplomeris hirsuta Lindl. ■
132919 Diplochilus longifolius Lindl. = Diplomeris pulchella D. Don ■
132920 Diplochita DC. = Miconia Ruiz et Pav.(保留属名)●☆
132921 Diplochiton Spreng. = Diplochita DC. ●☆
132922 Diplochlaena Spreng. = Diplolaena R. Br. ●☆
132923 Diplochlamys Müll. Arg. = Mallotus Lour. ●
132924 Diplochonium Fenzl = Trianthema L. ■
132925 Diplochonium sesuvioides Fenzl = Sesuvium sesuvioides (Fenzl) Verdc. ■☆
132926 Diploclinium Lindl. = Begonia L. ●■
132927 Diploclinium evansianum Lindl. = Begonia grandis Dryand. ■
132928 Diploclisia Miers(1851);秤钩风属;Diploclisia ●
132929 Diploclisia affinis (Oliv.) Diels;秤钩风(杜藤,过山龙,花防己,华秤钩风,华防己,青冈藤,清风藤,湘防己,中华秤钩风);

Diploclisia, Similar Diploclisia ●
132930 Diploclisia chinensis Merr. = Diploclisia affinis (Oliv.) Diels ●
132931 Diploclisia glaucescens (Blume) Diels;苍白秤钩风(秤钩风,穿墙风,电藤,粉绿秤钩风,过山龙,茎花防己,九层皮,蛇总管,土防己);Glaucescent Diploclisia, Pale Diploclisia ●
132932 Diploclisia kunstleri (King) Diels = Diploclisia glaucescens (Blume) Diels ●
132933 Diploclisia macrocarpa (Wight et Arn.) Miers = Diploclisia glaucescens (Blume) Diels ●
132934 Diplocnema Post et Kuntze = Diploknema Pierre ●
132935 Diplococea Rchb. = Diplocea Raf. (1818) ■☆
132936 Diplococea Rchb. = Triplasis P. Beauv. ■☆
132937 Diplocoea Rchb. = Diplococea Rchb. ■☆
132938 Diplocoma D. Don = Heterotheca Cass. ■☆
132939 Diploconchium Schauer = Agrostophyllum Blume ■
132940 Diploconchium inocephalum Schauer = Agrostophyllum inocephalum (Schauer) Ames ■
132941 Diplocos Bureau = Streblus Lour. ●
132942 Diplocos zeylanica (Thwaites) Bureau = Streblus zeylanicus (Thwaites) Kurz ●
132943 Diplocrater Benth. = Cathedra Miers ●☆
132944 Diplocrater Hook. f. = Tricalysia A. Rich. ex DC. ●
132945 Diplocrater coriacea (Benth.) Hook. f. ex B. D. Jacks. = Tricalysia coriacea (Benth.) Hiern ●☆
132946 Diplocrater reticulata (Benth.) Hook. f. ex B. D. Jacks. = Tricalysia reticulata (Benth.) Hiern ●☆
132947 Diplocyatha N. E. Br. (1878);复杯角属;Diplocyatha ■☆
132948 Diplocyatha N. E. Br. = Orbea Haw. ■☆
132949 Diplocyatha ciliata (Thunb.) N. E. Br.;复杯角;Ciliate Diplocyatha ■☆
132950 Diplocyatha ciliata (Thunb.) N. E. Br. = Orbea ciliata (Thunb.) L. C. Leach ■☆
132951 Diplocyatha ciliata N. E. Br. = Diplocyatha ciliata (Thunb.) N. E. Br. ■☆
132952 Diplocyathium Heinr. Schmidt = Euphorbia L. ●■
132953 Diplocyathus K. Schum. = Diplocyatha N. E. Br. ■☆
132954 Diplocyclos (Endl.) Post et Kuntze (1903);双轮瓜属(毒瓜属);Poisongourd ■
132955 Diplocyclos decipiens (Hook. f.) C. Jeffrey;迷惑毒瓜■☆
132956 Diplocyclos leiocarpus (Gilg) C. Jeffrey = Diplocyclos decipiens (Hook. f.) C. Jeffrey ■☆
132957 Diplocyclos palmatus (L.) C. Jeffrey;毒瓜(花瓜,双轮瓜);Palmate Poisongourd ■
132958 Diplocyclos schliebenii (Harms) C. Jeffrey;施利本毒瓜■☆
132959 Diplocyclos tenuis (Klotzsch) C. Jeffrey;细毒瓜■☆
132960 Diplocyclus (Endl.) Post et Kuntze = Diplocyclos (Endl.) Post et Kuntze ■
132961 Diplocyclus Post et Kuntze = Diplocyclos (Endl.) Post et Kuntze ■
132962 Diplodiscus Turcz. (1858);二重椴属(海南椴属)●
132963 Diplodiscus paniculatus Turcz.;圆锥二重椴;Baroba Nut ●☆
132964 Diplodiscus trichospermus (Merr.) Y. Tang, M. G. Gilbert et Dorr = Hainania trichosperma Merr. ●◇
132965 Diplodium Sw. (废弃属名) = Pterostylis R. Br. (保留属名)■☆
132966 Diplodon DC. = Diplusodon Pohl ●☆
132967 Diplodon Spreng. = Diplusodon Pohl ●☆
132968 Diplodonta H. karst. = Clidemia D. Don ●☆
132969 Diplodonta H. karst. = Heterotrichum DC. ●☆
132970 Diplodontaceae Dulac = Lythraceae J. St. -Hil. (保留科名)■●
132971 Diplofatsia Nakai = Fatsia Decne. et Planch. ●
132972 Diplofatsia Nakai (1924);二重五加属●
132973 Diplofatsia polycarpa (Hayata) Nakai = Fatsia polycarpa Hayata ●◇
132974 Diplofraetum Walp. = Colona Cav. ●
132975 Diplofraetum Walp. = Diplophractum Desf. ●
132976 Diplogama Opiz = Otites Adans. ■
132977 Diplogama Opiz = Silene L. (保留属名)■
132978 Diplogastra Welw. ex Rchb. f. = Platylepis A. Rich. (保留属名)■☆
132979 Diplogastra angolensis Rchb. f. = Platylepis glandulosa (Lindl.) Rchb. f. ■☆
132980 Diplogatha K. Schum. = Diplocyatha N. E. Br. ■☆
132981 Diplogenaea A. Juss. = Diplogenea Lindl. ●
132982 Diplogenea Lindl. = Medinilla Gaudich. ex DC. ●
132983 Diplogenea viscoides Lindl. = Medinilla viscoides (Lindl.) Triana ●☆
132984 Diploglossis Benth. et Hook. f. = Diploglossum Meisn. ●■
132985 Diploglossum Meisn. = Cynanchum L. ●■
132986 Diploglossum auriculatum (Royle ex Wight) Meisn. = Cynanchum auriculatum Royle ex Wight ●■
132987 Diploglossum auriculatum Meisn. = Cynanchum auriculatum Royle ex Wight ●■
132988 Diploglottis Hook. f. (1862);假酸豆属(类酸豆属)●☆
132989 Diploglottis australis (G. Don) Radlk.;澳洲假酸豆木;Australian Tamarind ●☆
132990 Diploglottis australis Radlk. = Diploglottis australis (G. Don) Radlk. ●☆
132991 Diploglottis campbellii Cheel;红果假酸豆;Small-leaf Tamarind ●☆
132992 Diploglottis cunninghamii (Hook.) Benth.;密毛假酸豆;Australian Native Tamarind, Native Tamarind ●☆
132993 Diploglottis cunninghamii (Hook.) Benth. = Diploglottis australis (G. Don) Radlk. ●☆
132994 Diplogon Poir. = Diplopogon R. Br. ■☆
132995 Diplogon Raf. (废弃属名) = Chrysopsis (Nutt.) Elliott (保留属名)■☆
132996 Diplogon hyssopifolia (Nutt.) Kuntze = Chrysopsis gossypina (Michx.) Elliott subsp. hyssopifolia (Nutt.) Semple ■☆
132997 Diplogon mariana (L.) Raf. = Chrysopsis mariana (L.) Elliott ■☆
132998 Diplogon nuttallii Kuntze = Bradburia pilosa (Nutt.) Semple ■☆
132999 Diplogon pilosa (Walter) Kuntze = Chrysopsis gossypina (Michx.) Elliott ■☆
133000 Diplogon pinifolium (Elliott) Kuntze = Pityopsis pinifolia (Elliott) Nutt. ■☆
133001 Diplogon scabrellum (Torr. et Gray) Kuntze = Chrysopsis scabrella Torr. et A. Gray ■☆
133002 Diplogon villosum (Pursh) Kuntze = Heterotheca villosa (Pursh) Shinners ■☆
133003 Diplogon villosum (Pursh) Kuntze var. discoideum (A. Gray) Kuntze = Heterotheca villosa (Pursh) Shinners var. minor (Hook.) Semple ■☆
133004 Diplokeleba N. E. Br. (1894);双杯无患子属●☆
133005 Diplokeleba floribunda N. E. Br.;双杯无患子●☆

133006 Diploknema Pierre(1884);藏榄属;Diploknema,Zangolive ●
133007 Diploknema butyraceum (Roxb.) H. J. Lam.;藏榄(酪状紫荆木,乳酪雾冰藜);Indian Butter-tree, Tibet Diploknema, Xizang Diploknema,Zangolive ●
133008 Diploknema yunnanense D. D. Tao,Z. H. Yang et Q. T. Zhang;云南藏榄;Yunnan Diploknema,Yunnan Zangolive ●
133009 Diplolabellum F. Maek. (1935);朝鲜双唇兰属■☆
133010 Diplolabellum confluens (Hand. -Mazz.) Garay et W. Kittr. = Oreorchis patens (Lindl.) Lindl. ■
133011 Diplolabellum coreanum (Finet) F. Maek.;朝鲜双唇兰■☆
133012 Diplolaena R. Br. (1814);迪普劳属●☆
133013 Diplolaena grandiflora Desf.;大花迪普劳●☆
133014 Diplolaena microcephala Bartl.;小苞迪普劳●☆
133015 Diplolaenaceae J. Agardh = Rutaceae Juss. (保留科名)●■
133016 Diplolegnon Rusby = Corytoplectus Oerst. ■☆
133017 Diplolepis R. Br. (1810);双鳞萝藦属●
133018 Diplolepis longirostrum K. Schum.;长喙双鳞萝藦●☆
133019 Diplolepis ovata Lindl. = Tylophora ovata (Lindl.) Hook. ex Steud. ●
133020 Diplolobium F. Muell. = Swainsona Salisb. ●■☆
133021 Diploloma Schrenk = Craniospermum Lehm. ■
133022 Diploloma echioides Schrenk = Craniospermum echioides (Schrenk) Bunge ■
133023 Diploloma echioides Schrenk = Craniospermum mongolicum I. M. Johnst. ■
133024 Diplolophium Turcz. (1847);双冠芹属■☆
133025 Diplolophium abyssinicum (Hochst. ex A. Rich.) Benth. = Diplolophium africanum Turcz. ■☆
133026 Diplolophium africanum Turcz.;非洲双冠芹■☆
133027 Diplolophium africanum Turcz. f. kankanense Jaeger et Schnell = Diplolophium africanum Turcz. ■☆
133028 Diplolophium boranense Bidgood et Vollesen;博兰双冠芹■☆
133029 Diplolophium buchananii (Benth. ex Oliv.) C. Norman;布坎南双冠芹■☆
133030 Diplolophium buchananii (Benth. ex Oliv.) C. Norman subsp. swynnertonii (Baker f.) Cannon;斯温纳顿双冠芹■☆
133031 Diplolophium diplolophioides (H. Wolff) Jacq. -Fél.;双蛇叶双冠芹■☆
133032 Diplolophium guineense A. Chev. = Pycnocycla ledermannii H. Wolff ■☆
133033 Diplolophium somaliense Verdc.;索马里双冠芹■☆
133034 Diplolophium swynnertonii (Baker f.) C. Norman = Diplolophium buchananii (Benth. ex Oliv.) C. Norman subsp. swynnertonii (Baker f.) Cannon ■☆
133035 Diplolophium tisserantii C. Norman = Diplolophium diplolophioides (H. Wolff) Jacq. -Fél. ■☆
133036 Diplolophium zambesianum Hiern;赞比西双冠芹■☆
133037 Diploma Raf. = Gentiana L. ■
133038 Diplomeris D. Don(1825);合柱兰属;Diplomeris ■
133039 Diplomeris boxallii Rolfe = Diplomeris pulchella D. Don ■
133040 Diplomeris chinensis Rolfe = Amitostigma pingguiculum (Rchb. f. et S. Moore) Schltr. ■
133041 Diplomeris hirsuta Lindl.;硬毛合柱兰■
133042 Diplomeris pulchella D. Don;合柱兰(独肾草,鸡肾草);Common Diplomeris ■
133043 Diplomorpha Griff. = Sauropus Blume ●■
133044 Diplomorpha Griff. = Synostemon F. Muell. ●■
133045 Diplomorpha Meisn. = Wikstroemia Endl. (保留属名)●
133046 Diplomorpha Meisn. ex C. A. Mey. = Wikstroemia Endl. (保留属名)●
133047 Diplomorpha albiflora (Yatabe) Nakai = Wikstroemia albiflora Yatabe ●☆
133048 Diplomorpha canescens (Wall. ex Meisn.) C. A. Mey. = Wikstroemia canescens (Wall.) Meisn. ●
133049 Diplomorpha capitellata H. Hara = Daphnimorpha capitellata (H. Hara) Nakai ●☆
133050 Diplomorpha chamaedaphne (Bunge) C. A. Mey. = Wikstroemia chamaedaphne Meisn ●
133051 Diplomorpha dolichantha (Diels) Hamaya = Wikstroemia dolichantha Diels ●
133052 Diplomorpha dolichantha (Diels) Hamaya var. effusa (Rehder) Hamaya = Wikstroemia dolichantha Diels ●
133053 Diplomorpha dolichantha (Diels) Hamaya var. pilosa (Hamaya) Hamaya = Wikstroemia trichotoma (Thunb.) Makino ●
133054 Diplomorpha dolichantha (Diels) Hamaya var. pilosa (W. C. Cheng) Hamaya = Wikstroemia pilosa W. C. Cheng ●
133055 Diplomorpha dolichantha (Diels) Hamaya var. pubescens (Domke) Hamaya = Wikstroemia dolichantha Diels ●
133056 Diplomorpha ellipsocarpa (Maxim.) Nakai = Diplomorpha trichotoma (Thunb.) Nakai ●
133057 Diplomorpha ganpi (Siebold et Zucc.) Nakai = Wikstroemia ganpi (Siebold et Zucc.) Maxim. ●
133058 Diplomorpha japonica (Siebold et Zucc.) Endl. = Wikstroemia trichotoma (Thunb.) Makino ●
133059 Diplomorpha japonica Endl. = Diplomorpha trichotoma (Thunb.) Nakai ●
133060 Diplomorpha kudoi (Makino) Masam. = Daphnimorpha kudoi (Makino) Nakai ●☆
133061 Diplomorpha ohsumiensis (Hatus.) Hamaya = Wikstroemia ohsumiensis Hatus. ●☆
133062 Diplomorpha pauciflora (Franch. et Sav.) Nakai = Wikstroemia pauciflora (Franch. et Sav.) Franch. et Sav. ex Shirai ●☆
133063 Diplomorpha pauciflora (Franch. et Sav.) Nakai var. yakushimensis (Makino) T. Yamanaka = Wikstroemia pauciflora (Franch. et Sav.) Franch. et Sav. ex Shirai var. yakusimensis Makino ●☆
133064 Diplomorpha phymatoglossa (Koidz.) Nakai = Wikstroemia phymatoglossa Koidz. ●☆
133065 Diplomorpha sikokiana (Franch. et Sav.) Honda = Wikstroemia sikokiana Franch. et Sav. ●☆
133066 Diplomorpha sikokumontana Akasawa = Diplomorpha pauciflora (Franch. et Sav.) Nakai var. yakushimensis (Makino) T. Yamanaka ●☆
133067 Diplomorpha trichotoma (Thunb.) Nakai = Wikstroemia trichotoma (Thunb.) Makino ●
133068 Diplomorpha trichotoma (Thunb.) Nakai f. pilosa Hamaya = Diplomorpha trichotoma (Thunb.) Nakai ●
133069 Diplomorpha trichotoma (Thunb.) Nakai f. pilosa Hamaya = Wikstroemia trichotoma (Thunb.) Makino ●
133070 Diplomorpha yakushimensis (Makino) Masam. = Diplomorpha pauciflora (Franch. et Sav.) Nakai var. yakushimensis (Makino) T. Yamanaka ●☆
133071 Diplonema G. Don = Euclea L. ●☆
133072 Diplonix Raf. = Wisteria Nutt. (保留属名)●

133073 Diplonyx Raf.(废弃属名)= Wisteria Nutt.(保留属名)●
133074 Diploon Cronquist(1946);缺蕊山榄属●☆
133075 Diploon cuspidatum (Hoehne) Cronquist;缺蕊山榄●☆
133076 Diplopanax Hand. -Mazz.(1933);马蹄参属(大果五加参属,大果五加属);Diplopanax,Hoofrenshen ●★
133077 Diplopanax stachyanthus Hand. -Mazz.;马蹄参(大果参,大果五加,山枇杷树,双参,野枇杷);China Hoofrenshen, Chinese Diplopanax ●◇
133078 Diplopappus Cass. = Aster L. ●■
133079 Diplopappus Cass. = Chrysopsis (Nutt.) Elliott(保留属名)■☆
133080 Diplopappus albus (Nutt.) Lindl. ex Hook. = Solidago ptarmicoides (Torr. et A. Gray) B. Boivin ■☆
133081 Diplopappus amygdalinus (Lam.) Hook. = Doellingeria umbellata (Mill.) Nees ■☆
133082 Diplopappus asper Less. = Aster bakerianus Burtt Davy ex C. A. Sm. ■☆
133083 Diplopappus asper Less. var. pleiocephalus Harv. = Aster pleiocephalus (Harv.) Hutch. ■☆
133084 Diplopappus asperrimus DC. = Aster trinervius D. Don ■
133085 Diplopappus asperrimus DC. = Aster trinervius Roxb. ex D. Don ■
133086 Diplopappus baccharoides Benth. = Aster baccharoides (Benth.) Steetz ■
133087 Diplopappus chinensis (L.) Less. = Callistephus chinensis (L.) Nees ■
133088 Diplopappus cornifolius (Muhl. ex Willd.) Less. ex Darl. = Doellingeria infirma (Michx.) Greene ■☆
133089 Diplopappus diplostephioides (DC.) Hook. f. et Thomson ex Hook. f. = Aster diplostephioides Benth. et Hook. f. ■
133090 Diplopappus dysentericus Bluff. et Fingerh. = Pulicaria dysenterica (L.) Gaertn. ■
133091 Diplopappus elegans Hook. f. et Thomson = Aster neo-elegans Grierson ■
133092 Diplopappus elegans Hook. f. et Thomson ex Hook. f. = Aster neoelegans Grierson ■
133093 Diplopappus elongatus DC. = Felicia filifolia (Vent.) Burtt Davy subsp. schlechteri (Compton) Grau ■☆
133094 Diplopappus ericoides Less. = Ericameria ericoides (Less.) Jeps. ●☆
133095 Diplopappus extenuatus (Nees) DC. = Felicia fruticosa (L.) G. Nicholson ●
133096 Diplopappus filifolius (Vent.) DC. = Felicia filifolia (Vent.) Burtt Davy ●☆
133097 Diplopappus filifolius (Vent.) DC. var. elongatus (DC.) Harv. = Felicia filifolia (Vent.) Burtt Davy subsp. schlechteri (Compton) Grau ■☆
133098 Diplopappus filifolius (Vent.) DC. var. teretifolius (Less.) DC. = Felicia filifolia (Vent.) Burtt Davy ●☆
133099 Diplopappus filifolius Hook. = Erigeron filifolius (Hook.) Nutt. ■☆
133100 Diplopappus fruticosus (L.) Levyns = Felicia fruticosa (L.) G. Nicholson ●
133101 Diplopappus fruticulosus (Willd.) Less. = Felicia fruticosa (L.) G. Nicholson ●
133102 Diplopappus hispida Hook. = Heterotheca villosa (Pursh) Shinners var. minor (Hook.) Semple ■☆
133103 Diplopappus incanus Lindl. = Dieteria canescens (Pursh) Nutt. var. incana (Lindl.) D. R. Morgan et R. L. Hartm. ■☆
133104 Diplopappus laevigatus Sond. = Aster laevigatus (Sond.) Kuntze ■☆
133105 Diplopappus lanatus Cass. = Chrysopsis gossypina (Michx.) Elliott ■☆
133106 Diplopappus linearis Hook. = Erigeron linearis (Hook.) Piper ■☆
133107 Diplopappus marianus (L.) Cass. ex Hook. = Chrysopsis mariana (L.) Elliott ■☆
133108 Diplopappus molliusculus Lindl. ex DC. = Aster molliusculus (DC.) C. B. Clarke ■
133109 Diplopappus natalensis Sch. Bip. = Aster bakerianus Burtt Davy ex C. A. Sm. ■☆
133110 Diplopappus obovatus (Nutt.) Torr. et A. Gray = Oclemena reticulata (Pursh) G. L. Nesom ■☆
133111 Diplopappus pinnatifidus Hook. = Xanthisma spinulosum (Pursh) D. R. Morgan et R. L. Hartm. ●■☆
133112 Diplopappus ptarmicoides (Nees) Lindl. = Solidago ptarmicoides (Torr. et A. Gray) B. Boivin ■☆
133113 Diplopappus pulicarius Bluff. et Fingerh. = Pulicaria prostrata (Gilib.) Asch. ■
133114 Diplopappus pulicarius Ledeb. = Pulicaria prostrata (Gilib.) Asch. ■
133115 Diplopappus roylei Lindl. ex DC. = Aster molliusculus (DC.) C. B. Clarke ■
133116 Diplopappus scaber Hook. = Heterotheca subaxillaris (Lam.) Britton et Rusby ■☆
133117 Diplopappus sericeus Hook. = Pityopsis graminifolia (Michx.) Nutt. var. latifolia (Fernald) Semple et F. D. Bowers ■☆
133118 Diplopappus serrulatus Harv. = Aster harveyanus Kuntze ■☆
133119 Diplopappus teretifolius Less. = Felicia filifolia (Vent.) Burtt Davy ●☆
133120 Diplopappus trichophyllos (Nutt.) Hook. = Chrysopsis gossypina (Michx.) Elliott ■☆
133121 Diplopappus umbellatus (Mill.) Hook. = Doellingeria umbellata (Mill.) Nees ■☆
133122 Diplopappus villosus (Pursh) Hook. = Heterotheca villosa (Pursh) Shinners ■☆
133123 Diplopapus Raf. = Diplopappus Cass. ●■
133124 Diplopeltis Endl.(1837);双盾无患子属●☆
133125 Diplopeltis huegelii Endl.;双盾无患子●☆
133126 Diplopenta Alef. = Pavonia Cav.(保留属名)●■☆
133127 Diploperianthium F. Ritter = Calymmanthium F. Ritter ●☆
133128 Diplopetalon Spreng. = Cupania L. ●☆
133129 Diplopetalon Spreng. = Dimereza Labill. ●☆
133130 Diplopetalon Spreng. = Guioa Cav. ●☆
133131 Diplophractum Desf. = Colona Cav. ●
133132 Diplophragma (Wight et Arn.) Meisn. = Hedyotis L.(保留属名)●■
133133 Diplophragma Meisn. = Hedyotis L.(保留属名)●■
133134 Diplophragma tetrangulare Korth. = Hedyotis tetrangularis (Korth.) Walp. ■
133135 Diplophyllum Lehm. = Oligospermum D. Y. Hong ■
133136 Diplophyllum Lehm. = Veronica L. ■
133137 Diplophyllum cardiocarpum Kar. et Kir. = Veronica cardiocarpa (Kar. et Kir.) Walp. ■
133138 Diplopia Raf. = Salix L.(保留属名)●
133139 Diplopilosa Dvořák = Hesperis L. ■
133140 Diplopogon R. Br.(1810);澳双芒草属■☆

133141 Diplopogon setaceus R. Br. ;澳双芒草■☆
133142 Diploprion Viv. = Medicago L. (保留属名)●■
133143 Diploprora Hook. f. (1890);蛇舌兰属(倒吊兰属);Diploprora ■
133144 Diploprora bicaudata (Thwaites) Schltr. = Diploprora championii (Lindl. ex Benth.) Hook. f. ■
133145 Diploprora championii (Lindl. ex Benth.) Hook. f. ;蛇舌兰(船唇兰,倒吊兰,黄吊兰,爬石兰);Champion Diploprora ■
133146 Diploprora championii (Lindl. ex Benth.) Hook. f. var. uraiensis (Hayata) S. S. Ying = Diploprora championii (Lindl. ex Benth.) Hook. f. ■
133147 Diploprora championii (Lindl.) Hook. f. = Diploprora championii (Lindl. ex Benth.) Hook. f. ■
133148 Diploprora championii (Lindl.) Hook. f. var. uraiensis (Hayata) S. S. Ying = Diploprora championii (Lindl. ex Benth.) Hook. f. ■
133149 Diploprora kusukusensis Hayata = Diploprora championii (Lindl. ex Benth.) Hook. f. ■
133150 Diploprora uraiensis Hayata = Diploprora championii (Lindl. ex Benth.) Hook. f. ■
133151 Diploptera C. A. Gardner = Strangea Meisn. ●☆
133152 Diplopterys A. Juss. (1838);双翅金虎尾属●☆
133153 Diplopterys cabrerana (Cuatrec.) Gates;双翅金虎尾●☆
133154 Diplopterys involuta Nied. ;狭萼双翅金虎尾●☆
133155 Diplopteryx Dalla Torre et Harms = Diplopterys A. Juss. ●☆
133156 Diplopyramis Welw. = Oxygonum Burch. ex Campd. ●■☆
133157 Diplorhipia Drude = Mauritia L. f. ●☆
133158 Diplorhynchus Welw. ex Ficalho et Hiern(1881);双喙夹竹桃属(双喙桃属)●☆
133159 Diplorhynchus angolensis Büttner = Diplorhynchus condylocarpon (Müll. Arg.) Pichon ●☆
133160 Diplorhynchus angustifolia Stapf = Diplorhynchus condylocarpon (Müll. Arg.) Pichon ●☆
133161 Diplorhynchus condylocarpon (Müll. Arg.) Pichon;瘤果双喙夹竹桃●☆
133162 Diplorhynchus condylocarpon (Müll. Arg.) Pichon f. angustifolius (Stapf) P. A. Duvign. = Diplorhynchus condylocarpon (Müll. Arg.) Pichon ●☆
133163 Diplorhynchus condylocarpon (Müll. Arg.) Pichon f. microphylla P. A. Duvign. = Diplorhynchus condylocarpon (Müll. Arg.) Pichon ●☆
133164 Diplorhynchus condylocarpon (Müll. Arg.) Pichon subsp. angolensis (Büttner) P. A. Duvign. = Diplorhynchus condylocarpon (Müll. Arg.) Pichon ●☆
133165 Diplorhynchus condylocarpon (Müll. Arg.) Pichon subsp. mossambicensis (Benth. ex Oliv.) P. A. Duvign. = Diplorhynchus condylocarpon (Müll. Arg.) Pichon ●☆
133166 Diplorhynchus condylocarpon (Müll. Arg.) Pichon var. psilopus (Welw. ex Ficalho et Hiern) P. A. Duvign. = Diplorhynchus condylocarpon (Müll. Arg.) Pichon ●☆
133167 Diplorhynchus mossambicensis Benth. ;莫桑比克双喙夹竹桃;Rhodesian Rubber ●☆
133168 Diplorhynchus mossambicensis Benth. ex Oliv. = Diplorhynchus condylocarpon (Müll. Arg.) Pichon ●☆
133169 Diplorhynchus poggei K. Schum. = Diplorhynchus condylocarpon (Müll. Arg.) Pichon ●☆
133170 Diplorhynchus psilopus Welw. ex Ficalho et Hiern = Diplorhynchus condylocarpon (Müll. Arg.) Pichon ●☆
133171 Diplorrhiza Ehrh. = Coeloglossum Hartm. ■
133172 Diplorrhiza Ehrh. = Satyrium Sw. (保留属名)■
133173 Diplosastera Tausch = Coreopsis L. ●■
133174 Diploscyphus Liebm. = Scleria P. J. Bergius ■
133175 Diplosiphon Decne. = Blyxa Noronha ex Thouars ■
133176 Diplosiphon oryzetorum Decne. = Blyxa aubertii Rich. ■
133177 Diplosoma Schwantes(1926);怪奇玉属■☆
133178 Diplosoma leipoldtii L. Bolus = Diplosoma retroversum (Kensit) Schwantes ■☆
133179 Diplosoma luckhoffii (L. Bolus) Schwantes ex Ihlenf. ;吕克霍夫怪奇玉■☆
133180 Diplosoma retroversum (Kensit) Schwantes;怪奇玉■☆
133181 Diplospora DC. (1830);狗骨柴属;Dogbonbavin ●
133182 Diplospora africana Sim = Tricalysia africana (Sim) Robbr. ●☆
133183 Diplospora andamanensis (Thoth.) M. Gangop. et Chakrab. ;安达曼狗骨柴●☆
133184 Diplospora buisanensis Hayata = Diplospora dubia (Lindl.) Masam. ●
133185 Diplospora dubia (Lindl.) Masam. ;狗骨柴(白鸡金,白秋铜盘,狗骨仔,狗骨子,观音茶,青凿树,三萼木);Common Dogbonbavin, Common Tricalysia, False Coffee, Yellowflower Tricalysia ●
133186 Diplospora fruticosa Hemsl. ;毛狗骨柴(狗骨柴,黄鸡脚,小狗骨柴);Hair Dogbonbavin,Hairy Tricalysia ●
133187 Diplospora mollissima Hutch. ;云南狗骨柴(白花苦灯笼,多毛狗骨柴);Soft Haired Tricalysia, Soft Hairy Tricalysia, Yunnan Dogbonbavin ●
133188 Diplospora tanakai Hayata = Diplospora dubia (Lindl.) Masam. ●
133189 Diplospora viridiflora DC. = Diplospora dubia (Lindl.) Masam. ●
133190 Diplosporopsis Wernham = Belonophora Hook. f. ■☆
133191 Diplosporopsis coffeoides Wernham = Belonophora wernhamii Hutch. et Dalziel ■☆
133192 Diplosporopsis talbotii Wernham = Belonophora talbotii (Wernham) Keay ■☆
133193 Diplostegium D. Don = Tibouchina Aubl. ●■☆
133194 Diplostelma A. Gray = Chaetopappa DC. ■☆
133195 Diplostelma Raf. = Chaetanthera Nutt. ■☆
133196 Diplostelma Raf. = Chaetopappa DC. ■☆
133197 Diplostelma bellioides A. Gray = Chaetopappa bellioides (A. Gray) Shinners ■☆
133198 Diplostemma DC. = Amasonia L. f. (保留属名)●■☆
133199 Diplostemma DC. = Diphystema Neck. ●■☆
133200 Diplostemma Steud. et Hochst. ex DC. = Geigeria Griess. ■●☆
133201 Diplostemma alatum (Hochst. et Steud.) DC. = Geigeria alata (Hochst. et Steud.) Oliv. et Hiern ■☆
133202 Diplostemon DC. ex Steud. = Ammannia L. ■
133203 Diplostephium Kunth(1818);双冠菀属(长冠菀属)●☆
133204 Diplostephium amygdalinum (Lam.) Cass. = Doellingeria umbellata (Mill.) Nees ■☆
133205 Diplostephium amygdalinum var. humilius DC. = Doellingeria umbellata (Mill.) Nees ■☆
133206 Diplostephium canum A. Gray = Hazardia cana (A. Gray) Greene ●☆
133207 Diplostephium cornifolium (Muhl. ex Willd.) DC. = Doellingeria infirma (Michx.) Greene ■☆
133208 Diplostephium dichotomum (Elliott) DC. = Oclemena reticulata

(Pursh) G. L. Nesom ■☆
133209 Diplostephium extenuatum Nees = Felicia fruticosa (L.) G. Nicholson ●
133210 Diplostephium filifolium (Vent.) Nees = Felicia filifolia (Vent.) Burtt Davy ●☆
133211 Diplostephium fruticulosum (Willd.) Nees = Felicia fruticosa (L.) G. Nicholson ●
133212 Diplostephium lavandulifolium Kunth;双冠菀●☆
133213 Diplostephium longipes Cass. = Felicia fruticosa (L.) G. Nicholson ●
133214 Diplostephium teretifolium (Less.) Nees = Felicia filifolia (Vent.) Burtt Davy ●☆
133215 Diplostephium umbellatum (Mill.) Cass. = Doellingeria umbellata (Mill.) Nees ■☆
133216 Diplostigma K. Schum. (1895);双柱萝藦属☆
133217 Diplostigma canescens K. Schum.;双柱萝藦☆
133218 Diplostylis H. Karst. et Triana = Rochefortia Sw. ☆
133219 Diplostylis Sond. = Adenocline Turcz. ■☆
133220 Diplosyphon Matsum. = Blyxa Noronha ex Thouars ■
133221 Diplosyphon Matsum. = Diplosiphon Decne. ■
133222 Diplotaenia Boiss. (1844);双带芹属■☆
133223 Diplotaenia Boiss. = Peucedanum L. ■
133224 Diplotaenia cachyridifolia Boiss.;双带芹■☆
133225 Diplotax Raf. = Cassia L.(保留属名)●■
133226 Diplotaxis DC. (1821);二行芥属(二列芥属);Wall Rocket, Wallrocket ■
133227 Diplotaxis Wall. ex Kurz = Chisocheton Blume ●
133228 Diplotaxis acris (Forssk.) Boiss.;锐尖二行芥■☆
133229 Diplotaxis acris (Forssk.) Boiss. subsp. sahariensis Chevassut et Quézel = Diplotaxis duveyrieriana Coss. ■☆
133230 Diplotaxis acris (Forssk.) Boiss. var. duveyrieriana (Coss.) Coss. = Diplotaxis duveyrieriana Coss. ■☆
133231 Diplotaxis acris (Forssk.) Boiss. var. griffithii (Hook. f. et Thomson) Coss. = Diplotaxis griffithii (Hook. f. et Thomson) Boiss ■☆
133232 Diplotaxis acris (Forssk.) Boiss. var. hesperidiflora (DC.) Maire et Weiller = Diplotaxis acris (Forssk.) Boiss. ■☆
133233 Diplotaxis acris (Forssk.) Boiss. var. tibestica Chevassut et Quézel;提贝斯提二行芥■☆
133234 Diplotaxis antoniensis Rustan;安东尼亚二行芥■☆
133235 Diplotaxis assurgens (Delile) Thell.;上升二行芥■☆
133236 Diplotaxis assurgens (Delile) Thell. subsp. tetragona (Maire) Nègre;四角上升二行芥■☆
133237 Diplotaxis assurgens (Delile) Thell. var. dissecta Maire = Diplotaxis assurgens (Delile) Thell. ■☆
133238 Diplotaxis assurgens (Delile) Thell. var. integrifolia Nègre = Diplotaxis assurgens (Delile) Thell. ■☆
133239 Diplotaxis auriculata Durieu = Diplotaxis tenuisiliqua Delile ■☆
133240 Diplotaxis brachycarpa Godr. = Diplotaxis virgata (Cav.) DC. subsp. brachycarpa (Godr.) Nègre ■☆
133241 Diplotaxis brevisiliqua (Coss.) Mart. -Laborde;短荚二行芥■☆
133242 Diplotaxis catholica (L.) DC.;普通二行芥■☆
133243 Diplotaxis catholica (L.) DC. subsp. siifolia (Kuntze) Maire = Diplotaxis siifolia Kuntze ■☆
133244 Diplotaxis catholica (L.) DC. var. bipinnatifida Coss. = Diplotaxis siifolia Kuntze ■☆
133245 Diplotaxis catholica (L.) DC. var. dasycarpa Willk. = Diplotaxis catholica (L.) DC. ■☆
133246 Diplotaxis catholica (L.) DC. var. latirostris (Braun-Blanq.) Maire = Diplotaxis catholica (L.) DC. ■☆
133247 Diplotaxis catholica (L.) DC. var. maritima O. E. Schulz = Diplotaxis catholica (L.) DC. ■☆
133248 Diplotaxis catholica (L.) DC. var. maroccana Pau = Diplotaxis catholica (L.) DC. ■☆
133249 Diplotaxis catholica (L.) DC. var. rivulorum (Braun-Blanq. et Maire) Maire = Diplotaxis catholica (L.) DC. ■☆
133250 Diplotaxis catholica (L.) DC. var. siettiana (Maire) Nègre = Diplotaxis siettiana Maire ■☆
133251 Diplotaxis catholica (L.) DC. var. tenuirostris Maire = Diplotaxis catholica (L.) DC. ■☆
133252 Diplotaxis cinerea (Desf.) Pomel = Ammosperma cinereum (Desf.) Baill. ■☆
133253 Diplotaxis cossoniana (Reut.) O. E. Schulz = Diplotaxis erucoides (L.) DC. subsp. cossoniana (Reut.) Mart. -Laborde ■☆
133254 Diplotaxis cossoniana (Reut.) O. E. Schulz var. sahariensis (Coss.) Maire = Diplotaxis virgata (Cav.) DC. ■☆
133255 Diplotaxis crassifolia (Raf.) DC. = Diplotaxis harra (Forssk.) Boiss. subsp. crassifolia (Raf.) Maire ■☆
133256 Diplotaxis cretacea Kotov;白垩二行芥■☆
133257 Diplotaxis cyrenaica (Durand et Barratte) Maire et Weiller = Diplotaxis virgata (Cav.) DC. subsp. cyrenaica (Durand et Barratte) Nègre ■☆
133258 Diplotaxis decumbens (A. Chev.) Rustan et L. Borgen = Diplotaxis hirta (A. Chev.) Rustan et L. Borgen ■☆
133259 Diplotaxis delagei Pomel ex Batt. = Diplotaxis virgata (Cav.) DC. ■☆
133260 Diplotaxis duveyrieriana Coss.;迪韦里耶二行芥■☆
133261 Diplotaxis erucoides (L.) DC.;芝麻菜二行芥■☆
133262 Diplotaxis erucoides (L.) DC. subsp. cossoniana (Reut.) Mart. -Laborde;科森二行芥■☆
133263 Diplotaxis erucoides (L.) DC. subsp. longisiliqua (Coss.) Gómez-Campo;长荚二行芥■☆
133264 Diplotaxis erucoides (L.) DC. var. cyrenaica E. A. Durand et Barratte = Diplotaxis virgata (Cav.) DC. subsp. cyrenaica (Durand et Barratte) Nègre ■☆
133265 Diplotaxis erucoides (L.) DC. var. dasycarpa O. E. Schulz = Diplotaxis erucoides (L.) DC. ■☆
133266 Diplotaxis erucoides (L.) DC. var. leiocarpa Maire et Weiller = Diplotaxis erucoides (L.) DC. ■☆
133267 Diplotaxis erucoides (L.) DC. var. valentina (Pau) O. E. Schulz;瓦伦特芝麻菜二行芥■☆
133268 Diplotaxis erucoides DC.;白二行芥;White Rocket, White Wall Rocket, White Wallrocket ■☆
133269 Diplotaxis glauca (J. A. Schmidt) O. E. Schulz;灰绿二行芥■☆
133270 Diplotaxis gorgadensis Rustan;戈尔加德二行芥■☆
133271 Diplotaxis gracilis (Webb) O. E. Schulz;纤细二行芥■☆
133272 Diplotaxis griffithii (Hook. f. et Thomson) Boiss;格氏二行芥■☆
133273 Diplotaxis griquensis (N. E. Br.) Sprague = Erucastrum griquense (N. E. Br.) O. E. Schulz ■☆
133274 Diplotaxis harra (Forssk.) Boiss.;哈拉二行芥■☆
133275 Diplotaxis harra (Forssk.) Boiss. subsp. crassifolia (Raf.) Maire;厚叶二行芥■☆
133276 Diplotaxis harra (Forssk.) Boiss. var. fontanesii (Willk.) Maire et Weiller = Diplotaxis harra (Forssk.) Boiss. ■☆
133277 Diplotaxis harra (Forssk.) Boiss. var. hispida (DC.) Nègre =

Diplotaxis harra (Forssk.) Boiss. ■☆

133278 Diplotaxis harra (Forssk.) Boiss. var. intricata (Willk.) O. E. Schulz = Diplotaxis harra (Forssk.) Boiss. ■☆

133279 Diplotaxis harra (Forssk.) Boiss. var. maroccana Nègre = Diplotaxis harra (Forssk.) Boiss. ■☆

133280 Diplotaxis harra (Forssk.) Boiss. var. subglabra (DC.) O. E. Schulz = Diplotaxis harra (Forssk.) Boiss. ■☆

133281 Diplotaxis hirta (A. Chev.) Rustan et L. Borgen;多毛二行芥■☆

133282 Diplotaxis hispida DC. = Diplotaxis harra (Forssk.) Boiss. ■☆

133283 Diplotaxis hispida DC. var. subglabra ? = Diplotaxis harra (Forssk.) Boiss. ■☆

133284 Diplotaxis inopinata Sprague = Erucastrum arabicum Fisch. et C. A. Mey. ■☆

133285 Diplotaxis lagascana DC. = Diplotaxis harra (Forssk.) Boiss. ■☆

133286 Diplotaxis muralis (L.) DC.;二行芥(二列芥,双趋芥);Annual Wallrocket, Annual Wall-rocket, Sand Mustard, Sand Rocket, Stinking Wall Rocket, Stinking Wall-rocket, Stinkweed, Strirking Wallrocket, Wall Cress, Wall Mustard, Wall Rocket, Wild Rocket ■

133287 Diplotaxis muralis (L.) DC. subsp. ceratophylla (Batt.) Mart. -Laborde;角叶二行芥■☆

133288 Diplotaxis muralis (L.) DC. subsp. simplex (Viv.) Jafri = Diplotaxis simplex (Viv.) Spreng. ■☆

133289 Diplotaxis muralis (L.) DC. var. ceratophylla Batt. = Diplotaxis muralis (L.) DC. subsp. ceratophylla (Batt.) Mart. -Laborde ■☆

133290 Diplotaxis muralis (L.) DC. var. pinnatifida Noulet = Diplotaxis muralis (L.) DC. ■

133291 Diplotaxis muralis (L.) DC. var. scaposa (DC.) O. E. Schulz = Diplotaxis scaposa DC. ■☆

133292 Diplotaxis muralis (L.) DC. var. simplex (Viv.) El Naggar = Diplotaxis simplex (Viv.) Spreng. ■☆

133293 Diplotaxis ollivieri Maire;奥里维尔二行芥■☆

133294 Diplotaxis ollivieri Maire var. fallax ? = Diplotaxis ollivieri Maire ■☆

133295 Diplotaxis ollivieri Maire var. fluminea Nègre = Diplotaxis ollivieri Maire ■☆

133296 Diplotaxis ollivieri Maire var. tenuisecta ? = Diplotaxis ollivieri Maire ■☆

133297 Diplotaxis parvula Schrenk = Thellungiella parvula (Schrenk) Al-Shehbaz et O'Kane ■

133298 Diplotaxis pendula (Desf.) DC. = Diplotaxis harra (Forssk.) Boiss. ■☆

133299 Diplotaxis pitardiana Maire;皮塔德二行芥■☆

133300 Diplotaxis platystylis Pomel = Diplotaxis muralis (L.) DC. ■

133301 Diplotaxis rivulorum Braun-Blanq. et Maire = Diplotaxis catholica (L.) DC. ■☆

133302 Diplotaxis scaposa DC.;花茎二行芥■☆

133303 Diplotaxis siettiana Maire;谢特二行芥■☆

133304 Diplotaxis siifolia Kuntze;锥毛二行芥;Cone-hair Wallrocket ■☆

133305 Diplotaxis siifolia Kuntze subsp. bipinnatifida (Coss.) Mart. -Laborde;双羽裂二行芥■☆

133306 Diplotaxis siifolia Kuntze var. bipinnatifida Coss. = Diplotaxis siifolia Kuntze ■☆

133307 Diplotaxis siifolia Kuntze var. hispidissima Emb. = Diplotaxis siifolia Kuntze ■☆

133308 Diplotaxis siifolia Kuntze var. maroccana Pau = Diplotaxis siifolia Kuntze ■☆

133309 Diplotaxis simplex (Viv.) Spreng.;简单二行芥■☆

133310 Diplotaxis simplex (Viv.) Spreng. var. philaenorum Maire et Weiller = Diplotaxis simplex (Viv.) Spreng. ■☆

133311 Diplotaxis simplex (Viv.) Spreng. var. pumila O. E. Schulz = Diplotaxis simplex (Viv.) Spreng. ■☆

133312 Diplotaxis sundingii Rustan;松德林二行芥■☆

133313 Diplotaxis tenuifolia (L.) DC.;细叶二行芥;Perennial Wallrocket, Perennial Wall-rocket, Slim-leaved Wall Rocket, Slim-leaved Wall-rocket, Wall Rocket ■☆

133314 Diplotaxis tenuisiliqua Delile;细荚二行芥■☆

133315 Diplotaxis tenuisiliqua Delile subsp. rupestris (Ball) Mart. -Laborde;岩生二行芥■☆

133316 Diplotaxis tenuisiliqua Delile var. auriculata (Durieu) Maire et Weiller = Diplotaxis tenuisiliqua Delile ■☆

133317 Diplotaxis tenuisiliqua Delile var. dasycarpa O. E. Schulz = Diplotaxis tenuisiliqua Delile subsp. rupestris (Ball) Mart. -Laborde ■☆

133318 Diplotaxis tenuisiliqua Delile var. leiocarpa Nègre = Diplotaxis tenuisiliqua Delile ■☆

133319 Diplotaxis tenuisiliqua Delile var. rupestris Ball = Diplotaxis tenuisiliqua Delile subsp. rupestris (Ball) Mart. -Laborde ■☆

133320 Diplotaxis tenuisiliqua Delile var. subclavata Balansa = Diplotaxis tenuisiliqua Delile ■☆

133321 Diplotaxis tetragona Maire = Diplotaxis assurgens (Delile) Thell. subsp. tetragona (Maire) Nègre ■☆

133322 Diplotaxis valentina Pau = Diplotaxis erucoides (L.) DC. var. valentina (Pau) O. E. Schulz ■☆

133323 Diplotaxis varia Rustan;变异二行芥■☆

133324 Diplotaxis viminea (L.) DC.;葡萄园二行芥;Vineyard Stinkweed ■☆

133325 Diplotaxis viminea (L.) DC. var. balearica Sennen = Diplotaxis viminea (L.) DC. ■☆

133326 Diplotaxis viminea (L.) DC. var. integrifolia Guss. = Diplotaxis viminea (L.) DC. ■☆

133327 Diplotaxis viminea (L.) DC. var. platystylis (Pomel) Batt. = Diplotaxis muralis (L.) DC. ■

133328 Diplotaxis virgata (Cav.) DC.;条纹二行芥■☆

133329 Diplotaxis virgata (Cav.) DC. f. saharensis Coss. = Diplotaxis virgata (Cav.) DC. ■☆

133330 Diplotaxis virgata (Cav.) DC. subsp. brachycarpa (Godr.) Nègre;短果二行芥■☆

133331 Diplotaxis virgata (Cav.) DC. subsp. cavanillesiana Maire et Weiller = Diplotaxis brevisiliqua (Coss.) Mart. -Laborde ■☆

133332 Diplotaxis virgata (Cav.) DC. subsp. cossoniana (Reut.) Maire et Weiller = Diplotaxis erucoides (L.) DC. subsp. cossoniana (Reut.) Mart. -Laborde ■☆

133333 Diplotaxis virgata (Cav.) DC. subsp. cyrenaica (Durand et Barratte) Nègre;昔兰尼二行芥■☆

133334 Diplotaxis virgata (Cav.) DC. subsp. platystylis (Pomel) Maire et Weiller = Diplotaxis muralis (L.) DC. ■

133335 Diplotaxis virgata (Cav.) DC. subsp. syrtica Murb. = Diplotaxis simplex (Viv.) Spreng. ■☆

133336 Diplotaxis virgata (Cav.) DC. var. aissae Hochr. = Diplotaxis virgata (Cav.) DC. ■☆

133337 Diplotaxis virgata (Cav.) DC. var. brevisiliqua Coss. = Diplotaxis brevisiliqua (Coss.) Mart. -Laborde ■☆

133338 Diplotaxis virgata (Cav.) DC. var. cavanillesiana (Maire et Weiller) Nègre = Diplotaxis brevisiliqua (Coss.) Mart. -Laborde ■☆

133339 Diplotaxis virgata (Cav.) DC. var. delagei (Batt.) Maire et Weiller = Diplotaxis virgata (Cav.) DC. subsp. brachycarpa (Godr.) Nègre ■☆
133340 Diplotaxis virgata (Cav.) DC. var. glabrescens Faure et Maire = Diplotaxis erucoides (L.) DC. subsp. cossoniana (Reut.) Mart. - Laborde ■☆
133341 Diplotaxis virgata (Cav.) DC. var. humilis Coss. = Diplotaxis virgata (Cav.) DC. subsp. brachycarpa (Godr.) Nègre ■☆
133342 Diplotaxis virgata (Cav.) DC. var. longisiliqua Coss. = Diplotaxis erucoides (L.) DC. ■☆
133343 Diplotaxis virgata (Cav.) DC. var. platystylos Willk. = Diplotaxis virgata (Cav.) DC. ■☆
133344 Diplotaxis virgata (Cav.) DC. var. pubescens Nègre = Diplotaxis virgata (Cav.) DC. ■☆
133345 Diplotaxis virgata (Cav.) DC. var. tenuirostris Maire = Diplotaxis virgata (Cav.) DC. ■☆
133346 Diplotaxis vogelii (Webb) Cout.;沃格尔二行芥■☆
133347 Diploter Raf. = Tetracera L. ●
133348 Diplotheca Hochst. = Astragalus L. ●■
133349 Diplotheca abyssinica Hochst. = Astragalus atropilosulus (Hochst.) Bunge var. abyssinicus (Hochst.) J. B. Gillett ■☆
133350 Diplotheca atropilosula Hochst. = Astragalus atropilosulus (Hochst.) Bunge ■☆
133351 Diplotheca venosa Hochst. = Astragalus atropilosulus (Hochst.) Bunge var. venosus (Hochst.) J. B. Gillett ■☆
133352 Diplothemium Mart. = Allagoptera Nees ●☆
133353 Diplothorax Gagnep. = Streblus Lour. ●
133354 Diplothorax tonkinenis Gagnep. = Streblus asper Lour. ●
133355 Diplothria Walp. = Diplothrix DC. ●■
133356 Diplothrix DC. = Zinnia L. (保留属名)●■
133357 Diplothrix acerosa DC. = Zinnia acerosa (DC.) A. Gray ●☆
133358 Diplotropis Benth. (1837);双龙瓣豆属;Sucupira ●☆
133359 Diplotropis martiusii Benth.;马氏双龙瓣豆●☆
133360 Diplotropis purpurea (Rich.) Amshoff;紫双龙瓣豆●☆
133361 Diplotropis racemosa (Hoehne) Amshoff;总花双龙瓣豆●☆
133362 Diplousodon Meisn. = Diplusodon Pohl ●☆
133363 Diplukion Raf. (废弃属名) = Iochroma Benth. (保留属名)●☆
133364 Diplusion Raf. (废弃属名) = Salix L. (保留属名)●
133365 Diplusodon Pohl(1827);双齿千屈菜属●☆
133366 Diplusodon alatus T. B. Cavalc.;翅双齿千屈菜●☆
133367 Diplusodon arboreus Poepp. et Endl.;乔木双齿千屈菜●☆
133368 Diplusodon argyrophyllus T. B. Cavalc.;银叶双齿千屈菜●☆
133369 Diplusodon bolivianus T. B. Cavalc.;玻利维亚双齿千屈菜●☆
133370 Diplusodon buxifolius DC.;黄杨叶双齿千屈菜●☆
133371 Diplusodon ciliatiflorus T. B. Cavalc.;毛花双齿千屈菜●☆
133372 Diplusodon cordifolius Lourteig;心叶双齿千屈菜●☆
133373 Diplusodon cryptanthus T. B. Cavalc.;隐花双齿千屈菜●☆
133374 Diplusodon floribundus Pohl;多花双齿千屈菜●☆
133375 Diplusodon gracilis Koehne;细双齿千屈菜●☆
133376 Diplusodon lanceolatus Pohl;披针叶双齿千屈菜●☆
133377 Diplusodon leucocalycinus Lourteig;白萼双齿千屈菜●☆
133378 Diplusodon macrodon Koehne;大齿双齿千屈菜●☆
133379 Diplusodon microphyllus Pohl;小叶双齿千屈菜●☆
133380 Diplusodon mononeuros Pilg.;单脉双齿千屈菜●☆
133381 Diplusodon montanus Casar. ex Koehne;山地双齿千屈菜●☆
133382 Diplusodon nigricans Koehne;黑双齿千屈菜●☆
133383 Diplusodon ovatus Pohl;卵双齿千屈菜●☆
133384 Diplusodon saxatilis Lourteig;岩地双齿千屈菜●☆
133385 Diplycosia Blume(1826);两型萼杜鹃属(簇白珠属)●☆
133386 Diplycosia adenothrix (Miq.) Nakai = Gaultheria adenothrix (Miq.) Maxim. ●☆
133387 Diplycosia adenothrix Nakai = Gaultheria adenothrix (Miq.) Maxim. ●☆
133388 Diplycosia alboglauca Merr. = Gaultheria dumicola W. W. Sm. ●
133389 Diplycosia brachyantha Sleumer;短花两型萼杜鹃●☆
133390 Diplycosia cordifolia Ridl.;心叶两型萼杜鹃●☆
133391 Diplycosia latifolia Blume;宽叶两型萼杜鹃●☆
133392 Diplycosia macrophylla Becc.;大叶两型萼杜鹃●☆
133393 Diplycosia semi-infera C. B. Clarke = Gaultheria semi-infera (C. B. Clarke) Airy Shaw ●
133394 Diplycosia tenuifolia (L.) DC.;细叶两型萼杜鹃;Slimleaf Wallrocket ■☆
133395 Dipodium R. Br. (1810);迪波兰属;Dipodium ■☆
133396 Dipodium paludosum Rchb. f.;沼泽迪波兰;Swampy Dipodium ■☆
133397 Dipodium parviflorum J. J. Sm.;小花迪波兰;Little Flower Dipodium ■☆
133398 Dipodium punctatum (Sm.) R. Br. = Dipodium squamatum (G. Forst.) Sm. ■☆
133399 Dipodium squamatum (G. Forst.) Sm.;斑点迪波兰;Hyacinth Orchid, Punctate Dipodium ■☆
133400 Dipodophyllum Tiegh. (1895);双足叶属●☆
133401 Dipodophyllum Tiegh. = Loranthus L. (废弃属名)●
133402 Dipodophyllum diguetii Tiegh.;双足叶●☆
133403 Dipogon Durand = Diopogon Jord. et Fourr. ■☆
133404 Dipogon Durand = Sempervivum L. ■☆
133405 Dipogon Liebm. (1854);香豌豆藤属■☆
133406 Dipogon Steud. = Sorghastrum Nash ■☆
133407 Dipogon Willd. ex Steud. = Chrysopogon Trin. (保留属名)■
133408 Dipogon lignosus (L.) Verdc.;香豌豆藤■☆
133409 Dipogonia P. Beauv. = Diplopogon R. Br. ■☆
133410 Dipoma Franch. (1886);蛇头荠属(双果属);Dipoma, Snakeheadcress ■★
133411 Dipoma iberideum Franch.;蛇头荠;Snakehead Dipoma, Snakeheadcress ■
133412 Dipoma iberideum Franch. f. pilosius O. E. Schulz = Dipoma iberideum Franch. ■
133413 Dipoma iberideum Franch. var. dacycarpum O. E. Schulz;刚毛蛇头荠;Thickfruit Snakehead Dipoma ■
133414 Dipoma iberideum Franch. var. dasycarpum O. E. Schulz = Dipoma iberideum Franch. ■
133415 Dipoma iberideum Franch. var. pilosius O. E. Schulz;叉毛蛇头荠;Pilose Snakehead Dipoma ■
133416 Dipoma tibeticum O. E. Schulz = Taphrospermum tibeticum (O. E. Schulz) Al-Shehbaz ■
133417 Diporidium H. L. Wendl. = Ochna L. ●
133418 Diporidium H. L. Wendl. ex Bartl. et Wendl. f. = Ochna L. ●
133419 Diporidium acutifolium (Engl.) Tiegh. = Ochna holstii Engl. ●☆
133420 Diporidium arboreum (Burch. ex DC.) H. Wendl. = Ochna arborea Burch. ex DC. ●☆
133421 Diporidium cinnabarinum (Engl. et Gilg) Tiegh. = Ochna cinnabarina Engl. et Gilg ●☆
133422 Diporidium delagoense Eckl. et Zeyh. = Ochna arborea Burch. ex DC. ●☆

133423 Diporidium hoepfneri Tiegh. = Ochna pygmaea Hiern ●☆
133424 Diporidium holstii (Engl.) Tiegh. = Ochna holstii Engl. ●☆
133425 Diporidium inerme (Forssk.) Tiegh. = Ochna inermis (Forssk.) Schweinf. ●☆
133426 Diporidium macrocalyx (Oliv.) Tiegh. = Ochna macrocalyx Oliv. ●☆
133427 Diporidium macrocarpa (Engl.) Tiegh. = Ochna macrocalyx Oliv. ●☆
133428 Diporidium natalitium Meisn. = Ochna natalitia (Meisn.) Walp. ●☆
133429 Diporidium purpureocostatum (Engl.) Tiegh. = Ochna mossambicensis Klotzsch ●☆
133430 Diporidium rovumensis (Gilg) Tiegh. = Ochna rovumensis Gilg ●☆
133431 Diporidium schimperi Tiegh. = Ochna inermis (Forssk.) Schweinf. ●☆
133432 Diporidium schweinfurthianum (F. Hoffm.) Tiegh. = Ochna schweinfurthiana F. Hoffm. ●☆
133433 Diporidium serrulatum Hochst. = Ochna serrulata (Hochst.) Walp. ●☆
133434 Diporochna Tiegh. = Ochna L. ●
133435 Diporochna Tiegh. = Porochna Tiegh. ●
133436 Diporochna brazzae Tiegh. = Ochna latisepala (Tiegh.) Bamps ●☆
133437 Diporochna hiernii Tiegh. = Ochna hiernii (Tiegh.) Exell ●☆
133438 Diporochna latisepala Tiegh. = Ochna latisepala (Tiegh.) Bamps ●☆
133439 Diporochna membranacea (Oliv.) Tiegh. = Ochna membranacea Oliv. ●☆
133440 Diporochna oliveri Tiegh. = Ochna membranacea Oliv. ●☆
133441 Diporochna quintasii Tiegh. = Ochna membranacea Oliv. ●☆
133442 Diposis DC. (1829);双夫草属☆
133443 Diposis bulbocastanum DC.;双夫草☆
133444 Dipsacaceae Juss. (1789)(保留科名);川续断科(山萝卜科,续断科);Teasel Family ■●
133445 Dipsacella Opiz = Dipsacus L. ■
133446 Dipsacella Opiz = Virga Hill ■
133447 Dipsacella Opiz(1838);小川续断属■☆
133448 Dipsacella pilosa (L.) Soják = Dipsacus pilosus L. ■☆
133449 Dipsacella setigera Opiz = Dipsacus pilosus L. ■☆
133450 Dipsacozamia Lehm. ex Lindl. = Ceratozamia Brongn. ●☆
133451 Dipsacus L. (1753);川续断属(山萝菔属);Teasel, Teazel, Teazle ■
133452 Dipsacus acaulis (Steud. ex A. Rich.) Napper = Dipsacus pinnatifidus Steud. ex A. Rich. ■☆
133453 Dipsacus appendiculatus Steud. ex A. Rich. = Dipsacus pinnatifidus Steud. ex A. Rich. ■☆
133454 Dipsacus asper Wall.;川续断(川旦,川断,鼓锤草,和尚头,接骨,接骨草,龙豆,南草,山萝卜,续断,蕒,蕒断,印度续断);Aspel-like Teasel, Himalayan Teasel, India Teasel, Teasel ■
133455 Dipsacus asper Wall. ex C. B. Clarke = Dipsacus asper Wall. ■
133456 Dipsacus asperoides C. Y. Cheng et T. M. Ai = Dipsacus asper Wall. ex C. B. Clarke ■
133457 Dipsacus asperoides C. Y. Cheng et T. M. Ai var. emeiemsis Z. T. Yin;峨眉续断;Emei Teasel, Omei Teasel ■
133458 Dipsacus asperoides C. Y. Cheng et T. M. Ai var. emeiensis Z. T. Yin = Dipsacus asper Wall. ex C. B. Clarke ■
133459 Dipsacus atratus Hook. f. et Thomson ex C. B. Clarke;紫花续断;Purple Teasel ■
133460 Dipsacus atropurpureus C. Y. Cheng et Z. T. Yin;深紫续断(卢汉,陆汗);Darkpurple Teasel, Deeppurple Teasel ■
133461 Dipsacus azureus Schrenk;天蓝续断(北疆头花草,北疆头序花,兰花川续断);Skyblue Teasel ■
133462 Dipsacus bequaertii De Wild. = Dipsacus pinnatifidus Steud. ex A. Rich. ■☆
133463 Dipsacus chinensis Batalin;大头续断(大花续断,华续断,续断,中华续断);China Teasel, Chinese Teasel ■
133464 Dipsacus cyanocapitatus C. Y. Cheng et T. M. Ai;蓝花续断;Blueflower Teasel ■
133465 Dipsacus cyanocapitatus C. Y. Cheng et T. M. Ai = Dipsacus asper Wall. ex C. B. Clarke ■
133466 Dipsacus daliensis T. M. Ai;大理续断;Dali Teasel ■
133467 Dipsacus daliensis T. M. Ai = Dipsacus asper Wall. ex C. B. Clarke ■
133468 Dipsacus daliensis T. M. Ai var. multifidus H. B. Chen;多裂续断;Multifid Dali Teasel ■
133469 Dipsacus daliensis T. M. Ai var. multifidus H. B. Chen = Dipsacus asper Wall. ex C. B. Clarke ■
133470 Dipsacus dipsacoides (Kar. et Kir.) V. I. Bochantsev = Dipsacus azureus Schrenk ■
133471 Dipsacus enshiensis C. Y. Cheng et T. M. Ai;恩施续断;Enshi Teasel ■
133472 Dipsacus enshiensis C. Y. Cheng et T. M. Ai = Dipsacus asper Wall. ex C. B. Clarke ■
133473 Dipsacus eremocephalus Pic. Serm. = Dipsacus pinnatifidus Steud. ex A. Rich. ■☆
133474 Dipsacus ferox Loisel.;多刺续断■☆
133475 Dipsacus fulingensis C. Y. Cheng et T. M. Ai;涪陵续断;Fuling Teasel ■
133476 Dipsacus fulingensis C. Y. Cheng et T. M. Ai = Dipsacus atropurpureus C. Y. Cheng et Z. T. Yin ■
133477 Dipsacus fullonum Huds. = Dipsacus sativus (L.) Garsault ■
133478 Dipsacus fullonum L.;起绒草(拉毛草);Ablamoth, Barber's Brushes, Brush, Brush-and-comb, Brushes-and-combs, Bullrushes, Burler's Teasel, Burtons, Card Teasel, Card Thistle, Church Broom, Cleavers, Clothes Brush, Cockle Dock, Cock's Comb, Comb-and-brush, Common Teasel, Donkey's Thistle, Draper's Teasel, Fairy's Broom, Fairy's Fire, Fuller's Teasel, Fuller's Thistle, Gypsy's Comb, Hairbrush, Johnny Prick-finger, Lady's Brush, Lady's Brush-and-comb, Lady's Brushes, Lady's Brushes-and-combs, Little Brush, Our Lady's Basin, Pincushion, Poor Man's Brush, Prickly Beehive, Prickyback, Shepherd's Rod, Shepherd's Staff, Shepherd's Yard, Sweep's Brush, Teasel, Venus' Basin, Venus' Bath, Venus-cup Teasel, Wild Teasel, Wild Teazle, Wolf's Comb, Wolf's Teasel ■
133479 Dipsacus fullonum L. = Dipsacus sativus (L.) Garsault ■
133480 Dipsacus fullonum L. f. albidus Steyerm. = Dipsacus fullonum L. ■
133481 Dipsacus fullonum L. subsp. sativus L. = Dipsacus sativus (L.) Garsault ■
133482 Dipsacus fullonum L. subsp. sylvestris (Huds.) Clapham;林地起绒草;Common Teasel, Wild Teasel ■☆
133483 Dipsacus fullonum L. subsp. sylvestris (Huds.) Clapham = Dipsacus fullonum L. ■
133484 Dipsacus fullonum L. subsp. sylvestris (Huds.) Clapham =

Dipsacus sylvestris Mill. ■☆
133485 Dipsacus fullonum L. var. sativus (Garsault) Thell. = Dipsacus sativus (L.) Honck. ■
133486 Dipsacus gmelini M. Bieb.;格氏续断;Gmelin Teasel ■☆
133487 Dipsacus inermis Wall.;藏续断(劲直续断);Spineless Teasel, Straight Teasel ■
133488 Dipsacus inermis Wall. var. mitis (D. Don) Y. J. Nasir;滇藏续断(尼泊尔续断,软续断);Gentle Teasel ■
133489 Dipsacus inermis Wall. var. mitis (D. Don) Y. J. Nasir = Dipsacus inermis Wall. ■
133490 Dipsacus japonicus Miq.;日本续断(假续断,小血转,续断);Japan Teasel, Japanese Teasel ■
133491 Dipsacus kangdingensis T. M. Ai et F. X. Feng = Dipsacus asper Wall. ex C. B. Clarke ■
133492 Dipsacus kangdingensis T. M. Ai et X. F. Feng;康定续断;Kangding Teasel ■
133493 Dipsacus kigesiensis Good = Dipsacus pinnatifidus Steud. ex A. Rich. ■☆
133494 Dipsacus laciniatus L.;分割续断(裂叶续断);Cutleaf Teasel, Cut-leaved Teasel ■☆
133495 Dipsacus lijiangensis T. M. Ai et H. B. Chen;丽江续断;Lijiang Teasel ■
133496 Dipsacus lijiangensis T. M. Ai et H. B. Chen = Dipsacus chinensis Batalin ■
133497 Dipsacus lushanensis C. Y. Cheng et T. M. Ai;庐山续断;Lushan Teasel ■
133498 Dipsacus meyeri Chabert;迈尔续断■☆
133499 Dipsacus mitis D. Don = Dipsacus inermis Wall. ■
133500 Dipsacus mitis D. Don = Dipsacus inermis Wall. var. mitis (D. Don) Y. J. Nasir ■
133501 Dipsacus narcisseanus Lawalrée;水仙续断■☆
133502 Dipsacus pilosus L.;疏毛续断;Hairy Teasel, Pilose Teasel, Shepherd's Rod, Small Teasel ■☆
133503 Dipsacus pinnatifidus Steud. ex A. Rich.;羽裂续断■☆
133504 Dipsacus pinnatifidus Steud. ex A. Rich. var. integrifolius Engl. = Dipsacus pinnatifidus Steud. ex A. Rich. ■☆
133505 Dipsacus sativus (L.) Garsault;拉毛果(拉毛草,起绒草);Cuctirating Teasel, Cultivate Teasel, Fuller's Teasel, Indian Teasel, Leazel, Teasel, Teazle ■
133506 Dipsacus sativus (L.) Honck. = Dipsacus sativus (L.) Garsault ■
133507 Dipsacus setosus Hiern = Dipsacus pinnatifidus Steud. ex A. Rich. ■☆
133508 Dipsacus silvester Kern.;林续断;Forest Teasel ■☆
133509 Dipsacus silvester Kern. = Dipsacus fullonum L. ■
133510 Dipsacus simaoensis Y. Y. Qian;思茅续断;Simao Teasel ■
133511 Dipsacus simaoensis Y. Y. Qian = Dipsacus asper Wall. ex C. B. Clarke ■
133512 Dipsacus strictus D. Don = Dipsacus inermis Wall. ■
133513 Dipsacus strigosus Willd. ex Roem. et Schult.;硬毛续断;Yellow-flowered Teasel ■☆
133514 Dipsacus sylvestris Huds. = Dipsacus fullonum L. ■
133515 Dipsacus sylvestris Huds. = Dipsacus fullonum L. subsp. sylvestris (Huds.) Clapham ■☆
133516 Dipsacus sylvestris Huds. f. albidus Steyerm. = Dipsacus fullonum L. ■
133517 Dipsacus sylvestris Mill.;野续断(野起绒草);Common Teasel, Venus Cup Teasel, Venus-cup Teasel, Wild Teasel ■☆
133518 Dipsacus tianmuensis C. Y. Cheng et Z. T. Yin;天目续断;Tianmu Teasel, Tianmushan Teasel ■
133519 Dipsacus xinjiangensis Y. K. Yang, J. K. Wu et T. Abdulla;新疆续断;Xinjiang Teasel ■
133520 Dipsacus xinjiangensis Y. K. Yang, J. K. Wu et T. Abdulla = Dipsacus azureus Schrenk ■
133521 Dipsacus yulongensis T. M. Ai et L. J. Yang;玉龙续断;Yulong Teasel ■
133522 Dipsacus yulongensis T. M. Ai et L. J. Yang = Dipsacus asper Wall. ex C. B. Clarke ■
133523 Dipseudochorion Buchen. = Limnophyton Miq. ■☆
133524 Dipseudochorion sagittifolium (Willd.) Buchenau = Limnophyton obtusifolium (L.) Miq. ■☆
133525 Diptanthera Schrank ex Steud. = Diplanthera Gled. ●■
133526 Diptanthera Schrank ex Steud. = Justicia L. ●■
133527 Diptera Borkh. = Saxifraga L. ■
133528 Diptera mengtzeana (Engl. et Irmsch.) Losinsk. = Saxifraga mengtzeana Engl. et Irmsch. ■
133529 Diptera sarmentosa (L. f.) Losinsk. = Saxifraga stolonifera Curtis ■
133530 Diptera sinensis Losinsk. = Saxifraga rufescens Balf. f. ■
133531 Dipteracanthus Nees = Ruellia L. ■●
133532 Dipteracanthus Nees (1832);双翅爵床属(芦莉草属,楠草属);Dipteracanthus ■
133533 Dipteracanthus calycinus Champ. = Strobilanthes cusia (Nees) Kuntze ●
133534 Dipteracanthus cyaneus Nees = Ruellia cyanea (Nees) T. Anderson ■☆
133535 Dipteracanthus dejectus Nees = Ruellia prostrata Poir. ■☆
133536 Dipteracanthus lanceolatus Nees = Dipteracanthus repens (L.) Hassk. ■
133537 Dipteracanthus matutinus C. Presl = Ruellia patula Jacq. ■☆
133538 Dipteracanthus monanthos Nees = Ruellia monanthos (Nees) Bojer ex T. Anderson ■☆
133539 Dipteracanthus patulus (Jacq.) Nees = Ruellia patula Jacq. ■☆
133540 Dipteracanthus prostrata (Poir.) Nees = Ruellia prostrata Poir. ■☆
133541 Dipteracanthus repens (L.) Hassk.;双翅爵床(芦莉草,芦利草,楠草,匍匐消);Creeping Dipteracanthus ■
133542 Dipteracanthus serpyllifolius Nees = Dyschoriste serpyllifolia (Nees) Benoist ■☆
133543 Dipteracanthus sudanicus Schweinf. = Ruellia sudanica (Schweinf.) Lindau ■☆
133544 Dipteracanthus zeyheri Sond. = Ruellia pilosa L. f. ■☆
133545 Dipteraceae Lindl. = Dipterocarpaceae Blume(保留科名)●
133546 Dipteranthemum F. Muell. = Ptilotus R. Br. ■●☆
133547 Dipteranthus Barb. Rodr. (1882);双翅兰属■☆
133548 Dipteranthus pseudobulbiferus Barb. Rodr.;双翅兰■☆
133549 Dipterix Willd. = Dipteryx Schreb. (保留属名)●☆
133550 Dipterocalyx Cham. = Lippia L. ●■☆
133551 Dipterocarpaceae Blume (1825) (保留科名);龙脑香科;Dipterocarpus Family, Gurjun Family, Gurjun Oil Tree Family, Gurjunoiltree Family, Gurjun-oiltree Family ●
133552 Dipterocarpus C. F. Gaertn. (1805);龙脑香属;Gurjun, Gurjun Balsam, Gurjun Oil, Gurjun Oil Tree, Gurjunoiltree, Gurjun-oiltree,

Keruing ●

133553 Dipterocarpus alatus Roxb. et G. Don;翅龙脑香(高大龙脑香,假龙脑香);Andaman Gurjun,Gurjun Balsam,Siam Gurjun,Siamese Gurjun ●☆

133554 Dipterocarpus apterus Foxw.;无翼假龙脑香●☆

133555 Dipterocarpus austro-yunnanicus Y. K. Yang et J. K. Wu = Dipterocarpus retusus Blume var. macrocarpus (Vesque) P. S. Ashton ●

133556 Dipterocarpus baudii Korth.;东南亚假龙脑香●☆

133557 Dipterocarpus chartaceus Symington;纸质龙脑香;Chartaceous Gurjun ●☆

133558 Dipterocarpus cornutus Dyer;角状龙脑香●☆

133559 Dipterocarpus costatus C. F. Gaertn.;中脉龙脑香(中脉羯布罗香);Dau,Indochina Gurjun,Midrib Gurjun ●☆

133560 Dipterocarpus crinitus Dyer;长毛假龙脑香●☆

133561 Dipterocarpus dyeri Pierre;印支龙脑香(棉兰老假龙脑香,印支羯布罗香);Dyer Gurjun,Indochina Gurjun ●☆

133562 Dipterocarpus gracilis Blume;纤细龙脑香(纤细羯布罗香);Indonesian Gurjun ●

133563 Dipterocarpus grandiflorus (Blanco) Blanco;大花龙脑香;Apitong,Bagac,Gurjun,Large-flower Gurjun ●☆

133564 Dipterocarpus grandiflorus Blanco = Dipterocarpus grandiflorus (Blanco) Blanco ●☆

133565 Dipterocarpus insularis Hance;海岛龙脑香(海岛羯布罗香);Insular Gurjun ●☆

133566 Dipterocarpus intricatus Dyer;缠结龙脑香(缠结羯布罗香);Intricate Gurjun ●☆

133567 Dipterocarpus jourdainii Pierre = Dipterocarpus turbinatus C. F. Gaertn. ●

133568 Dipterocarpus laevis Buch. -Ham. = Dipterocarpus turbinatus C. F. Gaertn. ●

133569 Dipterocarpus luchunensis Y. K. Yang et J. K. Wu = Dipterocarpus retusus Blume var. macrocarpus (Vesque) P. S. Ashton ●

133570 Dipterocarpus macrocarpus Vesque = Dipterocarpus retusus Blume var. macrocarpus (Vesque) P. S. Ashton ●

133571 Dipterocarpus mangachapoi (Blanco) Blanco = Vatica mangachapoi Blanco ●◇

133572 Dipterocarpus mannii King ex Kanjilal et al. = Dipterocarpus retusus Blume ●◇

133573 Dipterocarpus obtusifolius Teijsm. ex Miq.;钝叶龙脑香(钝叶羯布罗香);Hiang,Obtuseleaf Gurjun ●☆

133574 Dipterocarpus occi-dentoyunnanensis Y. K. Yang et J. K. Wu = Dipterocarpus retusus Blume ●◇

133575 Dipterocarpus pilosus Roxb. = Dipterocarpus gracilis Blume ●

133576 Dipterocarpus pubescens Koord. et Valeton = Dipterocarpus retusus Blume ●◇

133577 Dipterocarpus retusus Blume;东京龙脑香(越南龙脑香);Obtuse-leaved Gurjun,Retuse Gurjun,Vietnam Gurjun ●◇

133578 Dipterocarpus retusus Blume subsp. macrocarpus (Vesque) Y. K. Yang et J. K. Wu = Dipterocarpus retusus Blume var. macrocarpus (Vesque) P. S. Ashton ●

133579 Dipterocarpus retusus Blume subsp. tonkinensis (A. Chev.) Y. K. Yang et J. K. Wu = Dipterocarpus retusus Blume ●◇

133580 Dipterocarpus retusus Blume var. macrocarpus (Vesque) P. S. Ashton;多毛东京龙脑香●

133581 Dipterocarpus retusus Blume var. yingjiangensis Y. K. Yang et J. K. Wu = Dipterocarpus retusus Blume ●◇

133582 Dipterocarpus skinneri King = Dipterocarpus gracilis Blume ●

133583 Dipterocarpus spanoghei Blume = Dipterocarpus retusus Blume ●◇

133584 Dipterocarpus stellatus Vesque;星芒假龙脑香●☆

133585 Dipterocarpus tonkinensis A. Chev. = Dipterocarpus retusus Blume ●◇

133586 Dipterocarpus trinervis Blume = Dipterocarpus retusus Blume ●◇

133587 Dipterocarpus tuberculatus Roxb.;小瘤龙脑香(小瘤羯布罗香);Eng,Gurjun-oil Tree,Tuberculate Gurjun ●☆

133588 Dipterocarpus turbinatus C. F. Gaertn.;龙脑香(羯波萝香,羯布罗香,陀螺假龙脑香,陀螺叶龙脑香,陀螺状龙脑香);Common Gurjun,Common Gurjun Oil Tree,Common Gurjunoiltree,Common Gurjun-oiltree,Durgan Mai Yang,Eng Gurjun Oil Tree,Gurjun,Gurjun-oil Tree,Mai Yang,Yang Khao ●

133589 Dipterocarpus turbinatus C. F. Gaertn. var. ramipiliferus Y. K. Yang et J. K. Wu = Dipterocarpus turbinatus C. F. Gaertn. ●

133590 Dipterocarpus zeylanicus Thwaites;锡兰龙脑香(斯里兰卡假龙脑香,锡兰羯布罗香);Ceylon Gurjun,Hola,Hora ●☆

133591 Dipterocome Fisch. et C. A. Mey.(1835);双角菊属■☆

133592 Dipterocome pusilla Fisch. et C. A. Mey.;双角菊■☆

133593 Dipterocypsela S. F. Blake(1945);双翅斑鸠菊属■☆

133594 Dipterocypsela succulenta S. F. Blake;双翅斑鸠菊■☆

133595 Dipterodendron Radlk. = Dilodendron Radlk. ●☆

133596 Dipteronia Oliv.(1889);金钱槭属;Coin Maple,Coinmaple,Dipteronia,False Maple,Money Maple ●★

133597 Dipteronia chinensis Erdtman = Dipteronia sinensis Oliv. ●

133598 Dipteronia dyeriana Henry;云南金钱槭(飞天子,辣子树);Tinctiria Money Maple,Yunnan Coinmaple,Yunnan Dipteronia ●◇

133599 Dipteronia sinensis Oliv.;金钱槭(双轮果);China Coinmaple,Chinese Dipteronia,Chinese Money Maple ●

133600 Dipteronia sinensis Oliv. f. taipaiensis (W. P. Fang et M. Y. Fang) E. Murray = Dipteronia sinensis Oliv. var. taipeiensis W. P. Fang et M. Y. Fang ●

133601 Dipteronia sinensis Oliv. var. taipeiensis W. P. Fang et M. Y. Fang;太白金钱槭;Taibaishan Chinese Dipteronia,Taibaishan Coinmaple ●

133602 Dipteropeltis Hallier f.(1899);双翅盾属☆

133603 Dipteropeltis macrantha Breteler;大花双翅盾☆

133604 Dipteropeltis mayumbensis R. D. Good;马永巴双翅盾☆

133605 Dipteropeltis poranoides Hallier f.;双翅盾☆

133606 Dipteropeltis poranoides Hallier f. var. acutissima R. D. Good = Dipteropeltis poranoides Hallier f. ☆

133607 Dipteropeltis poranoides Hallier f. var. mucronata R. D. Good = Dipteropeltis poranoides Hallier f. ☆

133608 Dipteropeltis poranoides Hallier f. var. velutina De Wild. = Dipteropeltis poranoides Hallier f. ☆

133609 Dipterosiphon Huber = Campylosiphon Benth. ■☆

133610 Dipterosperma Hassk. = Stereospermum Cham. ●

133611 Dipterosperma personatum Hassk. = Stereospermum colais (Buch. -Ham. ex Dillwyn) Mabb. ●

133612 Dipterospermum Griff. = Gordonia J. Ellis(保留属名)●

133613 Dipterostele Schltr. = Stellilabium Schltr. ■☆

133614 Dipterostemon Rydb. = Brodiaea Sm.(保留属名)■☆

133615 Dipterostemon Rydb. = Dichelostemma Kunth ■☆

133616 Dipterostemon capitatus (Benth.) Rydb. = Dichelostemma capitatum (Benth.) A. W. Wood ■☆

133617 Dipterostemon insularis (Greene) Rydb. = Dichelostemma capitatum (Benth.) A. W. Wood ■☆
133618 Dipterostemon pauciflorus (Torr.) Rydb. = Dichelostemma capitatum (Benth.) A. W. Wood subsp. pauciflorum (Torr.) Keator ■☆
133619 Dipterotheca Sch. Bip. ex Hochst. = Aspilia Thouars ■☆
133620 Dipterotheca kotschyi Sch. Bip. = Aspilia kotschyi (Sch. Bip.) Oliv. ■☆
133621 Dipterotheea Sch. Bip. = Aspilia Thouars ■☆
133622 Dipterygia C. Presl = Asteriscium Cham. et Schltdl. ■☆
133623 Dipterygia C. Presl ex DC. = Asteriscium Cham. et Schltdl. ■☆
133624 Dipterygium Decne. (1835);二翅山柑属●☆
133625 Dipterygium glaucum Decne.;灰绿二翅山柑■☆
133626 Dipteryx Schreb. (1791)(保留属名);二翅豆属(零陵香属,香豆属);Tonka Bean,Tonka-bean ●☆
133627 Dipteryx Schreb. = Coumarouna Aubl. + Taralea Aubl. (废弃属名)●☆
133628 Dipteryx micrantha Harms;小花二翅豆●☆
133629 Dipteryx odorata (Aubl.) Willd.;香二翅豆木(香豆);Dutch Tonka Bean, Dutch Tonka-bean, Tonka Bean, Tonka-bean ●☆
133630 Dipteryx odorata Willd. = Dipteryx odorata (Aubl.) Willd. ●☆
133631 Dipteryx oleifera Benth.;油二翅豆(油香豆);Eboe, Ebor ●☆
133632 Dipteryx oppositifolia Willd.;对叶二翅豆(对叶香豆)●☆
133633 Dipteryx puctata (Blake) Amshoff;斑点二翅豆(斑点香豆)●☆
133634 Dipteryx tetraphylla Benth.;四叶二翅豆●☆
133635 Diptychandra Tul. (1843);小黄花苏木属●☆
133636 Diptychandra aurantiaca Tul.;小黄花苏木●☆
133637 Diptychandra glabra Benth.;光小黄花苏木●☆
133638 Diptychocarpus Regel et Schrank = Clausia Korn.-Trotzky ex Hayek ■
133639 Diptychocarpus Regel et Schrank = Dipteryx Schreb. (保留属名)●☆
133640 Diptychocarpus Trautv. (1860);异果芥属(二型果属,双翼果属);Diptychocarpus ■
133641 Diptychocarpus strictus (Fisch. ex M. Bieb.) Trautv.;异果芥;Strict Diptychocarpus ■
133642 Diptychocarpus strictus (Fisch.) Trautv. = Diptychocarpus strictus (Fisch. ex M. Bieb.) Trautv. ■
133643 Diptychum Dulac = Sesleria Scop. ■☆
133644 Dipyrena Hook. (1830);双核草属●☆
133645 Dipyrena glaberrima Gillies et Hook.;双核草●☆
133646 Dirachma Schweinf. ex Balf. f. (1884);八瓣果属(刺木属)●☆
133647 Dirachma socotranum Schweinf. ex Balf. f.;八瓣果(刺木)●☆
133648 Dirachma somalensis D. A. Link;索马里八瓣果●☆
133649 Dirachmaceae Hutch. (1959);八瓣果科(刺木科)●☆
133650 Dirachmaceae Hutch. = Geraniaceae Juss. (保留科名)■●
133651 Diracodes Blume(废弃属名) = Amomum Roxb. (保留属名)■
133652 Diracodes Blume(废弃属名) = Etlingera Roxb. ■
133653 Diracodes Blume(废弃属名) = Nicolaia Horan. (保留属名)■☆
133654 Dirca L. (1753);革木属(糜木属);Leatherwood ●☆
133655 Dirca mexicana G. L. Nesom et Mayfield;墨西哥革木(墨西哥糜木);Mexican Leatherwood ●☆
133656 Dirca palustris L.;沼生革木(糜木);Atlantic Leatherwood, Eastern Leatherwood, Leatherwood, Moosewood, Mousewood, Ropebark, Rope-bark, Wicopy ●☆
133657 Dircaea Decne. = Corytholoma Decne. ■☆
133658 Dircaea Decne. = Megapleilis Raf. (废弃属名)■☆
133659 Dircaea Decne. = Rechsteineria Regel(保留属名)■☆
133660 Dircaea Decne. = Sinningia Nees ●■☆
133661 Dirhacodes Lem. = Amomum Roxb. (保留属名)■
133662 Dirhacodes Lem. = Diracodes Blume(废弃属名)■
133663 Dirhamphis Krapov. (1970);双钩锦葵属●☆
133664 Dirhamphis balansae Krapov.;双钩锦葵●☆
133665 Dirhamphis mexicana Fryxell;墨西哥双钩锦葵●☆
133666 Dirhynchosia Blume = Caldcluvia D. Don ●☆
133667 Dirhynchosia Blume = Spiraeopsis Miq. ●☆
133668 Dirichletia Klotzsch = Carphalea Juss. ■☆
133669 Dirichletia asperula K. Schum. = Carphalea glaucescens (Hiern) Verdc. ■☆
133670 Dirichletia borziana Mattei = Carphalea glaucescens (Hiern) Verdc. ■☆
133671 Dirichletia duemmeri Wernham = Carphalea pubescens (Klotzsch) Verdc. ■☆
133672 Dirichletia ellenbeckii K. Schum. = Carphalea glaucescens (Hiern) Verdc. ■☆
133673 Dirichletia giumbensis Chiov. = Carphalea glaucescens (Hiern) Verdc. ■☆
133674 Dirichletia glabra Klotzsch = Carphalea pubescens (Klotzsch) Verdc. ■☆
133675 Dirichletia glaucescens Hiern = Carphalea glaucescens (Hiern) Verdc. ■☆
133676 Dirichletia leucophlebia Baker = Triainolepis africana Hook. f. subsp. hildebrandtii (Vatke) Verdc. ■☆
133677 Dirichletia paolii Chiov. = Carphalea glaucescens (Hiern) Verdc. ■☆
133678 Dirichletia pubescens Klotzsch = Carphalea pubescens (Klotzsch) Verdc. ■☆
133679 Dirichletia rogersii Wernham = Carphalea pubescens (Klotzsch) Verdc. ■☆
133680 Dirichletia sphaerocephala Baker = Clerodendrum involucratum Vatke ●☆
133681 Dirtea Raf. = Commelina L. ■
133682 Dirynchosia Post et Kuntze = Dirhynchosia Blume ●☆
133683 Dirynchosia Post et Kuntze = Spiraeopsis Miq. ●☆
133684 Disa P. J. Bergius(1767);双距兰属(迪萨兰属,笛撒兰属);Disa, Table Mountain Orchis ■☆
133685 Disa aconitoides Sond.;乌头双距兰■☆
133686 Disa aconitoides Sond. subsp. concinna (N. E. Br.) H. P. Linder;整洁双距兰■☆
133687 Disa aconitoides Sond. subsp. goetzeana (Kraenzl.) H. P. Linder;格兹双距兰■☆
133688 Disa adolphi-fridericii Kraenzl. = Disa ochrostachya Rchb. f. ■☆
133689 Disa aemula Bolus = Disa cornuta (L.) Sw. ■☆
133690 Disa aequiloba Summerh.;等裂双距兰■☆
133691 Disa affinis N. E. Br. = Disa comosa (Rchb. f.) Schltr. ■☆
133692 Disa alpina Hook. f. = Brownleea parviflora Harv. ex Lindl. ■☆
133693 Disa alticola H. P. Linder;高原双距兰■☆
133694 Disa amblyopetala Schltr. = Disa hircicornis Rchb. f. ■☆
133695 Disa amoena H. P. Linder;秀丽双距兰■☆
133696 Disa andringitrana Schltr.;安德林吉特拉山双距兰■☆
133697 Disa aperta N. E. Br.;无盖双距兰■☆
133698 Disa apetala Kraenzl. = Brownleea parviflora Harv. ex Lindl. ■☆
133699 Disa arida Vlok;旱生双距兰■☆
133700 Disa aristata H. P. Linder;具芒双距兰■☆

133701 Disa atricapilla (Harv. ex Lindl.) Bolus;暗毛双距兰■☆
133702 Disa atropurpurea Sond. = Disa spathulata (L. f.) Sw. ■☆
133703 Disa atrorubens Schltr.;暗红双距兰■☆
133704 Disa attenuata Lindl. = Disa sagittalis (L. f.) Sw. ■☆
133705 Disa aurantiaca Rchb. f. = Disa ochrostachya Rchb. f. ■☆
133706 Disa aurata (Bolus) L. Parker et Koop.;金黄双距兰■☆
133707 Disa auriculata Bolus = Disa densiflora (Lindl.) Bolus ■☆
133708 Disa bakeri Rolfe = Disa stairsii Kraenzl. ■☆
133709 Disa barbata (L. f.) Sw.;髯毛双距兰■☆
133710 Disa barelli de Puydt = Disa uniflora P. J. Bergius ■☆
133711 Disa basutorum Kraenzl. = Disa sankeyi Rolfe ■☆
133712 Disa basutorum Schltr.;巴苏托双距兰■☆
133713 Disa baurii Bolus;鲍尔双距兰■☆
133714 Disa bifida (Thunb.) Sw. = Schizodium bifidum (Thunb.) Rchb. f. ■☆
133715 Disa bisetosa Kraenzl. = Disa aconitoides Sond. subsp. concinna (N. E. Br.) H. P. Linder ■☆
133716 Disa bivalvata (L. f.) T. Durand et Schinz;二果片双距兰■☆
133717 Disa bivalvata (L. f.) T. Durand et Schinz var. atricapilla (Harv. ex Lindl.) Schltr. = Disa atricapilla (Harv. ex Lindl.) Bolus ■☆
133718 Disa bodkinii Bolus;博德金双距兰■☆
133719 Disa bolusiana Schltr.;博卢斯双距兰■☆
133720 Disa brachyceras Lindl.;短角双距兰■☆
133721 Disa bracteata Sw.;具苞双距兰■☆
133722 Disa brevicornis (Lindl.) Bolus = Monadenia brevicornis Lindl. ■☆
133723 Disa brevipetala H. P. Linder;短瓣双距兰■☆
133724 Disa breyeri Schltr. = Disa welwitschii Rchb. f. ■☆
133725 Disa buchenaviana Kraenzl.;马岛双距兰■☆
133726 Disa caffra Bolus;非洲双距兰■☆
133727 Disa calophylla Kraenzl. = Disa welwitschii Rchb. f. ■☆
133728 Disa capricornis Rchb. f.;羊角双距兰(羊角迪萨兰);Goathorn Disa ■☆
133729 Disa capricornis Rchb. f. = Disa gladioliflora Burch. ex Lindl. subsp. capricornis (Rchb. f.) H. P. Linder ■☆
133730 Disa cardinalis H. P. Linder;绯红双距兰■☆
133731 Disa carsonii N. E. Br. = Disa erubescens Rendle subsp. carsonii (N. E. Br.) H. P. Linder ■☆
133732 Disa caulescens Lindl.;具茎双距兰■☆
133733 Disa cedarbergensis H. P. Linder;锡达伯格双距兰■☆
133734 Disa cephalotes Rchb. f.;头状双距兰■☆
133735 Disa cephalotes Rchb. f. subsp. frigida (Schltr.) H. P. Linder;耐寒双距兰■☆
133736 Disa cernua (Thunb.) Sw.;俯垂双距兰■☆
133737 Disa charpenteriana Rchb. f. = Disa multifida Lindl. ■☆
133738 Disa chiovendaei Schltr. = Disa aconitoides Sond. subsp. goetzeana (Kraenzl.) H. P. Linder ■☆
133739 Disa chrysostachya Sw.;金穗双距兰■☆
133740 Disa clavicornis H. P. Linder;棒角双距兰■☆
133741 Disa clavigera (Lindl.) Bolus = Schizodium obliquum Lindl. subsp. clavigerum (Lindl.) H. P. Linder ■☆
133742 Disa coccinea Kraenzl. = Disa robusta N. E. Br. ■☆
133743 Disa cochlearis S. D. Johnson et Liltved;螺状双距兰■☆
133744 Disa coerulea (Harv. ex Lindl.) Rchb. f. = Brownleea coerulea Harv. ex Lindl. ■☆
133745 Disa comosa (Rchb. f.) Schltr.;簇毛双距兰■☆
133746 Disa compta Summerh. = Disa caffra Bolus ■☆
133747 Disa concinna N. E. Br. = Disa aconitoides Sond. subsp. concinna (N. E. Br.) H. P. Linder ■☆
133748 Disa concinna N. E. Br. var. dichroa (Summerh.) Geerinck = Disa dichroa Summerh. ■☆
133749 Disa conferta Bolus;密集双距兰■☆
133750 Disa cooperi Rchb. f.;库珀双距兰■☆
133751 Disa cooperi Rchb. f. var. scullyi (Bolus) Schltr. = Disa scullyi Bolus ■☆
133752 Disa cornuta (L.) Sw.;角状双距兰■☆
133753 Disa cornuta (L.) Sw. var. aemula (Bolus) Kraenzl. = Disa cornuta (L.) Sw. ■☆
133754 Disa crassicornis Lindl.;粗角双距兰(粗角迪萨兰);Thickhorn Disa ■☆
133755 Disa culveri Schltr. = Disa hircicornis Rchb. f. ■☆
133756 Disa cylindrica (Thunb.) Sw.;柱形双距兰■☆
133757 Disa cylindrica Sw.;圆柱双距兰(圆柱迪萨兰);Cylindric Disa ■☆
133758 Disa deckenii Rchb. f. = Disa fragrans Schltr. subsp. deckenii (Rchb. f.) H. P. Linder ■☆
133759 Disa densiflora (Lindl.) Bolus;密花双距兰■☆
133760 Disa dichroa Summerh.;二色双距兰■☆
133761 Disa dracomontana Schelpe ex H. P. Linder;德拉科双距兰■☆
133762 Disa draconis (L. f.) Sw.;龙双距兰■☆
133763 Disa eacemosa L. f.;双距兰(迪萨兰)■☆
133764 Disa ecalcarata (G. J. Lewis) H. P. Linder;无距双距兰■☆
133765 Disa elegans Sond. ex Rchb. f.;雅致双距兰■☆
133766 Disa eminii Kraenzl.;埃明双距兰■☆
133767 Disa engleriana Kraenzl.;恩格勒双距兰■☆
133768 Disa equestris Rchb. f.;马双距兰■☆
133769 Disa equestris Rchb. f. var. concinna (N. E. Br.) Kraenzl. = Disa aconitoides Sond. subsp. concinna (N. E. Br.) H. P. Linder ■☆
133770 Disa erubescens Rendle;变红双距兰■☆
133771 Disa erubescens Rendle subsp. carsonii (N. E. Br.) H. P. Linder;卡森双距兰■☆
133772 Disa erubescens Rendle var. carsonii (N. E. Br.) Geerinck = Disa erubescens Rendle subsp. carsonii (N. E. Br.) H. P. Linder ■☆
133773 Disa erubescens Rendle var. katangensis (De Wild.) Geerinck = Disa katangensis De Wild. ■☆
133774 Disa erubescens Rendle var. leucantha Schltr. = Disa erubescens Rendle ■☆
133775 Disa esterhuyseniae Schelpe ex H. P. Linder;埃斯特双距兰■☆
133776 Disa excelsa (Thunb.) Sw. = Disa tripetaloides (L. f.) N. E. Br. ■☆
133777 Disa falcata Schltr. = Disa tripetaloides (L. f.) N. E. Br. ■☆
133778 Disa fallax Kraenzl. = Disa incarnata Lindl. ■☆
133779 Disa fanniniae Harv. ex Rolfe = Disa nervosa Lindl. ■☆
133780 Disa fasciata Lindl.;带状双距兰■☆
133781 Disa ferruginea (Thunb.) Sw.;锈色双距兰■☆
133782 Disa filicornis (L. f.) Thunb.;线角双距兰■☆
133783 Disa filicornis (L. f.) Thunb. var. latipetala Bolus = Disa filicornis (L. f.) Thunb. ■☆
133784 Disa flexuosa (L.) Sw. = Schizodium flexuosum (L.) Lindl. ■☆
133785 Disa forcipata Schltr.;钳双距兰■☆
133786 Disa forficaria Bolus;叉双距兰■☆
133787 Disa fragrans Schltr.;芳香双距兰■☆
133788 Disa fragrans Schltr. subsp. deckenii (Rchb. f.) H. P. Linder;

德肯双距兰■☆

133789 Disa frigida Schltr. = Disa cephalotes Rchb. f. subsp. frigida (Schltr.) H. P. Linder ■☆

133790 Disa galpinii Rolfe;盖尔双距兰■☆

133791 Disa gerrardii Rolfe = Disa patula Sond. var. transvaalensis Summerh. ■☆

133792 Disa glandulosa Burch. ex Lindl.;具腺双距兰■☆

133793 Disa goetzeana (Kraenzl.) Schltr. = Herschelianthe goetzeana (Kraenzl.) Rauschert ■☆

133794 Disa goetzeana Kraenzl. = Disa aconitoides Sond. subsp. goetzeana (Kraenzl.) H. P. Linder ■☆

133795 Disa gracilis Lindl. = Disa chrysostachya Sw. ■☆

133796 Disa graminifolia Ker Gawl. = Herschelia graminifolia (Ker Gawl.) T. Durand et Schinz ■☆

133797 Disa grandiflora L. f. = Disa uniflora P. J. Bergius ■☆

133798 Disa gregoriana Rendle = Disa stairsii Kraenzl. ■☆

133799 Disa gueinzii (Rchb. f.) Bolus = Schizodium obliquum Lindl. subsp. clavigerum (Lindl.) H. P. Linder ■☆

133800 Disa hallackii Rolfe;哈拉克双距兰■☆

133801 Disa hamatopetala Rendle = Disa baurii Bolus ■☆

133802 Disa harveiana Lindl. subsp. longicalcarata S. D. Johnson et H. P. Linder;长距双距兰■☆

133803 Disa hemisphaerophora Rchb. f. = Disa versicolor Rchb. f. ■☆

133804 Disa hians (L. f.) Spreng.;开裂双距兰■☆

133805 Disa hircicornis Rchb. f.;山羊双距兰■☆

133806 Disa huillensis Fritsch = Disa equestris Rchb. f. ■☆

133807 Disa huttonii Rchb. f. = Disa sanguinea Sond. ■☆

133808 Disa hyacinthina Kraenzl. = Disa welwitschii Rchb. f. ■☆

133809 Disa ignea Kraenzl. = Disa welwitschii Rchb. f. ■☆

133810 Disa incarnata Lindl.;肉色双距兰■☆

133811 Disa inflexa Mundt = Schizodium inflexum Lindl. ■☆

133812 Disa intermedia H. P. Linder;间型双距兰■☆

133813 Disa introrsa Kurzweil, Liltved et H. P. Linder;内向双距兰■☆

133814 Disa jacottetiae Kraenzl. = Disa crassicornis Lindl. ■☆

133815 Disa karooica S. D. Johnson et H. P. Linder;卡鲁双距兰■☆

133816 Disa katangensis De Wild.;加丹加双距兰■☆

133817 Disa kilimanjarica Rendle = Disa fragrans Schltr. subsp. deckenii (Rchb. f.) H. P. Linder ■☆

133818 Disa kraussii Rolfe = Disa pulchra Sond. ■☆

133819 Disa lacera Sw. = Disa hians (L. f.) Spreng. ■☆

133820 Disa lacera Sw. var. multifida N. E. Br. = Disa hians (L. f.) Spreng. ■☆

133821 Disa laeta Rchb. f.;可爱双距兰(可爱迪萨兰);Joyful Disa ■☆

133822 Disa laeta Rchb. f. = Disa hircicornis Rchb. f. ■☆

133823 Disa leopoldii Kraenzl. = Disa walleri Rchb. f. ■☆

133824 Disa leptostachys Sond. = Disa tenuis Lindl. ■☆

133825 Disa leucostachys Kraenzl. = Disa fragrans Schltr. ■☆

133826 Disa linderiana Bytebier et E. G. H. Oliv.;林德双距兰■☆

133827 Disa lineata Bolus;条纹双距兰■☆

133828 Disa lisowskii Szlach.;利索双距兰■☆

133829 Disa longicorna L. f.;长角双距兰(长角迪萨兰);Longhorn Disa ■☆

133830 Disa longifolia Lindl.;长叶双距兰■☆

133831 Disa longilabris Schltr. = Herschelianthe longilabris (Schltr.) Rauschert ■☆

133832 Disa longipetala (Lindl.) Bolus = Schizodium longipetalum Lindl. ■☆

133833 Disa lugens Bolus var. nigrescens (H. P. Linder) H. P. Linder;变黑双距兰■☆

133834 Disa lutea H. P. Linder = Disa tenuifolia Sw. ■☆

133835 Disa luxurians Kraenzl. = Disa stairsii Kraenzl. ■☆

133836 Disa macowanii Rchb. f. = Disa versicolor Rchb. f. ■☆

133837 Disa macrantha Sw. = Disa cornuta (L.) Sw. ■☆

133838 Disa macroceras (Sond.) Rchb. f. = Brownleea macroceras Sond. ■☆

133839 Disa macrostachya (Lindl.) Bolus;大穗双距兰■☆

133840 Disa maculata Harv. ex Lindl. = Disa ocellata Bolus ■☆

133841 Disa maculata L. f.;斑点双距兰■☆

133842 Disa marlothii Bolus;马洛斯双距兰■☆

133843 Disa megaceras Hook. f. = Disa crassicornis Lindl. ■☆

133844 Disa melaleuca (Thunb.) Sw. = Disa bivalvata (L. f.) T. Durand et Schinz ■☆

133845 Disa micrantha (Lindl.) Bolus = Disa bracteata Sw. ■☆

133846 Disa micropetala Schltr.;小瓣双距兰■☆

133847 Disa minax Kraenzl. = Disa galpinii Rolfe ■☆

133848 Disa miniata Summerh.;朱红双距兰■☆

133849 Disa minor (Sond.) Rchb. f.;小双距兰■☆

133850 Disa modesta Rchb. f. = Disa vaginata Harv. ex Lindl. ■☆

133851 Disa montana Sond.;山地双距兰■☆

133852 Disa multifida Lindl.;多裂双距兰■☆

133853 Disa multiflora (Sond.) Bolus = Disa densiflora (Lindl.) Bolus ■☆

133854 Disa natalensis Lindl. = Disa polygonoides Lindl. ■☆

133855 Disa neglecta Sond.;忽视双距兰■☆

133856 Disa nervosa Lindl.;具脉双距兰(具脉迪萨兰);Nerved Disa ■☆

133857 Disa newdigateae L. Bolus;纽迪盖特双距兰■☆

133858 Disa nigerica Rolfe;尼日利亚双距兰■☆

133859 Disa nivea H. P. Linder;雪白双距兰■☆

133860 Disa nubigena H. P. Linder;云雾双距兰■☆

133861 Disa nuwebergensis H. P. Linder;纽沃双距兰■☆

133862 Disa nyassana Schltr. = Disa zombica N. E. Br. ■☆

133863 Disa nyikensis H. P. Linder;尼卡双距兰■☆

133864 Disa obliqua (Lindl.) Bolus = Schizodium obliquum Lindl. ■☆

133865 Disa obtusa Lindl.;钝双距兰■☆

133866 Disa obtusa Lindl. subsp. hottentotica H. P. Linder;霍屯督双距兰■☆

133867 Disa obtusa Lindl. subsp. picta (Sond.) H. P. Linder;着色钝双距兰■☆

133868 Disa occultans Schltr. = Disa welwitschii Rchb. f. subsp. occultans (Schltr.) H. P. Linder ■☆

133869 Disa ocellata Bolus;单眼双距兰■☆

133870 Disa ochrostachya Rchb. f.;苍白穗双距兰■☆

133871 Disa ochrostachya Rchb. f. var. latipetala G. Will. = Disa satyriopsis Kraenzl. ■☆

133872 Disa oligantha Rchb. f.;小花双距兰■☆

133873 Disa ophrydea (Lindl.) Bolus;眉兰转状双距兰■☆

133874 Disa oreophila Bolus;喜山双距兰■☆

133875 Disa oreophila Bolus subsp. erecta H. P. Linder;直立喜山双距兰■☆

133876 Disa ornithantha Schltr.;鸟花双距兰■☆

133877 Disa orthostachya Kraenzl. var. latipetala G. Will. = Disa satyriopsis Kraenzl. ■☆

133878 Disa outeniquensis Schltr. = Disa hians (L. f.) Spreng. ■☆

133879 Disa ovalifolia Sond.;椭圆叶双距兰■☆
133880 Disa paludicola J. L. Stewart et J. C. Manning;沼泽双距兰■☆
133881 Disa pappei Rolfe = Disa obtusa Lindl. subsp. picta (Sond.) H. P. Linder ■☆
133882 Disa parviflora (Harv. ex Lindl.) Rchb. f. = Brownleea parviflora Harv. ex Lindl. ■☆
133883 Disa parvilabris Bolus = Disa oligantha Rchb. f. ■☆
133884 Disa patens (L. f.) Thunb.;铺展双距兰■☆
133885 Disa patens Sw. = Disa filicornis (L. f.) Thunb. ■☆
133886 Disa patula Sond.;张开双距兰■☆
133887 Disa patula Sond. var. transvaalensis Summerh.;德兰士瓦双距兰■☆
133888 Disa perplexa H. P. Linder;缠结双距兰■☆
133889 Disa perrieri Schltr. = Disa caffra Bolus ■☆
133890 Disa physodes Sw.;囊状双距兰■☆
133891 Disa picta Sond. = Disa obtusa Lindl. subsp. picta (Sond.) H. P. Linder ■☆
133892 Disa pillansii L. Bolus;皮朗斯双距兰■☆
133893 Disa poikilantha Kraenzl. = Disa montana Sond. ■☆
133894 Disa polygonoides Lindl.;纳塔尔双距兰■☆
133895 Disa porrecta Sw.;外伸双距兰(外伸迪萨兰);Porrect Disa ■☆
133896 Disa praestans Kraenzl. = Disa robusta N. E. Br. ■☆
133897 Disa praetermissa Schltr. = Disa bracteata Sw. ■☆
133898 Disa prasinata Ker Gawl. = Disa cernua (Thunb.) Sw. ■☆
133899 Disa preussii Kraenzl. = Brownleea parviflora Harv. ex Lindl. ■☆
133900 Disa princeae Kraenzl. = Disa walleri Rchb. f. ■☆
133901 Disa procera H. P. Linder;高大双距兰■☆
133902 Disa propinqua Sond. = Disa spathulata (L. f.) Sw. ■☆
133903 Disa propinqua Sond. var. trifida ? = Disa spathulata (L. f.) Sw. ■☆
133904 Disa pulchella Hochst. ex A. Rich.;美丽双距兰■☆
133905 Disa pulchra Sond.;非洲美丽双距兰■☆
133906 Disa pulchra Sond. var. montana (Sond.) Schltr. = Disa montana Sond. ■☆
133907 Disa pumilio (Lindl.) T. Durand et Schinz = Schwartzkopffia pumilio (Lindl.) Schltr. ■☆
133908 Disa purpurascens Bolus;紫双距兰■☆
133909 Disa pygmaea Bolus;矮小双距兰■☆
133910 Disa racemosa L. f.;总状双距兰■☆
133911 Disa racemosa L. f. var. isopetala Bolus = Disa racemosa L. f. ■☆
133912 Disa racemosa L. f. var. venosa (Sw.) Schltr. = Disa venosa Sw. ■☆
133913 Disa recurvata (Sond.) Rchb. f. = Brownleea recurvata Sond. ■☆
133914 Disa reflexa (Lindl.) Rchb. f. = Disa filicornis (L. f.) Thunb. ■☆
133915 Disa renziana Szlach.;伦兹双距兰■☆
133916 Disa reticulata Bolus;网状双距兰■☆
133917 Disa rhodantha Schltr.;粉红花双距兰■☆
133918 Disa rhodesia Summerh. = Disa satyriopsis Kraenzl. ■☆
133919 Disa richardiana Lehm. ex Bolus;理查德双距兰■☆
133920 Disa robusta N. E. Br.;粗壮双距兰■☆
133921 Disa roeperocharoides Kraenzl.;勒珀兰双距兰■☆
133922 Disa rosea Lindl.;粉红双距兰■☆
133923 Disa rufescens (Thunb.) Sw.;浅红双距兰■☆
133924 Disa rungweensis Schltr.;伦圭双距兰■☆
133925 Disa rungweensis Schltr. subsp. rhodesiaca (Summerh.) Summerh. = Disa zimbabweensis H. P. Linder ■☆
133926 Disa rungweensis Schltr. var. rhodesiaca Summerh. = Disa zimbabweensis H. P. Linder ■☆
133927 Disa rutenbergiana Kraenzl. = Disa buchenaviana Kraenzl. ■☆
133928 Disa sabulosa Bolus;砂地双距兰■☆
133929 Disa sagittalis (L. f.) Sw.;箭状双距兰■☆
133930 Disa sagittalis (L. f.) Sw. var. triloba (Lindl.) Schltr. = Disa triloba Lindl. ■☆
133931 Disa salteri G. J. Lewis;索尔特双距兰■☆
133932 Disa sanguinea Sond.;血红双距兰(血红迪萨兰);Bloodred Disa ■☆
133933 Disa sankeyi Rolfe;桑基双距兰■☆
133934 Disa satyriopsis Kraenzl.;鸟足兰状双距兰■☆
133935 Disa saxicola Schltr.;岩地双距兰■☆
133936 Disa schimperi N. E. Br. = Disa scutellifera A. Rich. ■☆
133937 Disa schizodioides Sond.;裂双距兰■☆
133938 Disa schlechteriana Bolus;施莱双距兰■☆
133939 Disa scullyi Bolus;斯卡里双距兰■☆
133940 Disa scutellifera A. Rich.;矩圆盾状双距兰■☆
133941 Disa secunda (Thunb.) Sw. = Disa racemosa L. f. ■☆
133942 Disa similis Summerh.;相似双距兰■☆
133943 Disa spathulata (L. f.) Sw.;匙形双距兰■☆
133944 Disa spathulata (L. f.) Sw. subsp. tripartita (Lindl.) H. P. Linder;三深裂双距兰■☆
133945 Disa spathulata (L. f.) Sw. var. atropurpurea (Sond.) Schltr. = Disa spathulata (L. f.) Sw. ■☆
133946 Disa stachyoides Rchb. f.;穗状双距兰■☆
133947 Disa stairsii Kraenzl.;斯泰双距兰■☆
133948 Disa stenoglossa Bolus = Disa patula Sond. ■☆
133949 Disa stokoei L. Bolus = Disa hallackii Rolfe ■☆
133950 Disa stolonifera Rendle = Disa eminii Kraenzl. ■☆
133951 Disa stolzii Schltr. = Disa erubescens Rendle subsp. carsonii (N. E. Br.) H. P. Linder ■☆
133952 Disa stricta Sond.;刚直双距兰■☆
133953 Disa subaequalis Summerh. = Disa welwitschii Rchb. f. subsp. occultans (Schltr.) H. P. Linder ■☆
133954 Disa subscutellifera Kraenzl. = Disa engleriana Kraenzl. ■☆
133955 Disa subtenuicornis H. P. Linder;亚细角双距兰■☆
133956 Disa tabularis Sond. = Disa obtusa Lindl. ■☆
133957 Disa tanganyikensis Summerh. = Disa welwitschii Rchb. f. subsp. occultans (Schltr.) H. P. Linder ■☆
133958 Disa tenella (L. f.) Sw.;柔弱双距兰■☆
133959 Disa tenella (L. f.) Sw. subsp. pusilla H. P. Linder;微小双距兰■☆
133960 Disa tenella (L. f.) Sw. var. brachyceras (Lindl.) Schltr. = Disa brachyceras Lindl. ■☆
133961 Disa tenuicornis Bolus;细角双距兰■☆
133962 Disa tenuifolia Sw.;细叶双距兰■☆
133963 Disa tenuis Lindl.;细双距兰■☆
133964 Disa thodei Schltr. ex Kraenzl.;索德双距兰■☆
133965 Disa torta (Thunb.) Sw. = Schizodium satyrioides (L.) Garay ■☆
133966 Disa triloba Lindl.;三裂双距兰■☆
133967 Disa tripartita Lindl. = Disa spathulata (L. f.) Sw. subsp. tripartita (Lindl.) H. P. Linder ■☆
133968 Disa tripetaloides (L. f.) N. E. Br.;三瓣双距兰■☆
133969 Disa tripetaloides (L. f.) N. E. Br. subsp. aurata (Bolus) H. P. Linder = Disa aurata (Bolus) L. Parker et Koop. ■☆

133970　Disa tripetaloides (L. f.) N. E. Br. var. aurata Bolus = Disa aurata (Bolus) L. Parker et Koop. ■☆
133971　Disa tysonii Bolus;泰森双距兰■☆
133972　Disa ukingensis Schltr.;热非双距兰■☆
133973　Disa uliginosa Kraenzl. = Disa saxicola Schltr. ■☆
133974　Disa uncinata Bolus;具钩双距兰■☆
133975　Disa uniflora P. J. Bergius;单花双距兰(单花迪萨兰);Singleflower Disa ■☆
133976　Disa vaginata Chiov. = Disa aconitoides Sond. subsp. goetzeana (Kraenzl.) H. P. Linder ■☆
133977　Disa vaginata Harv. ex Lindl.;具鞘双距兰■☆
133978　Disa vasselotii Bolus ex Schltr.;瓦瑟罗双距兰■☆
133979　Disa venosa Lindl. = Disa tripetaloides (L. f.) N. E. Br. ■☆
133980　Disa venosa Sw.;多脉双距兰■☆
133981　Disa venusta Bolus;雅丽双距兰■☆
133982　Disa verdickii De Wild.;韦尔迪双距兰■☆
133983　Disa versicolor Rchb. f.;变色双距兰■☆
133984　Disa virginalis H. P. Linder;洁白双距兰■☆
133985　Disa walleri Rchb. f.;瓦勒双距兰■☆
133986　Disa walteri Schltr. = Herschelianthe goetzeana (Kraenzl.) Rauschert ■☆
133987　Disa welwitschii Rchb. f.;韦尔双距兰■☆
133988　Disa welwitschii Rchb. f. subsp. occultans (Schltr.) H. P. Linder;隐蔽韦尔双距兰■☆
133989　Disa welwitschii Rchb. f. var. buchneri Schltr. = Disa welwitschii Rchb. f. ■☆
133990　Disa welwitschii Rchb. f. var. occultans (Schltr.) Geerinck = Disa welwitschii Rchb. f. subsp. occultans (Schltr.) H. P. Linder ■☆
133991　Disa wissmannii Kraenzl. = Disa stairsii Kraenzl. ■☆
133992　Disa woodii Schltr.;伍得双距兰■☆
133993　Disa zeyheri Sond. = Disa porrecta Sw. ■☆
133994　Disa zimbabweensis H. P. Linder;津巴布韦双距兰■☆
133995　Disa zombaensis Rendle = Disa walleri Rchb. f. ■☆
133996　Disa zombica N. E. Br.;宗巴双距兰■☆
133997　Disa zuluensis Rolfe;祖卢双距兰■☆
133998　Disaccanthus Greene = Streptanthus Nutt. ■☆
133999　Disaccanthus Greene(1906);异花芥属■☆
134000　Disaccanthus luteus Greene;黄异花芥■☆
134001　Disachoena Zoll. et Moritzi = Pimpinella L. ■
134002　Disadena Miq. = Voyria Aubl. ■☆
134003　Disakisperma Steud. = Leptochloa P. Beauv. ■
134004　Disandra L. = Sibthorpia L. ■☆
134005　Disandraceae Dulac = Linaceae DC. ex Perleb(保留科名)■●
134006　Disanthaceae Nakai = Hamamelidaceae R. Br. (保留科名)●
134007　Disanthaceae Nakai;双花木科●
134008　Disantheraceae Dulac = Polygalaceae Hoffmanns. et Link(保留科名)■●
134009　Disanthus Maxim. (1866);双花木属(双花树属,圆叶木属);Disanthus ●
134010　Disanthus cercidifolius Maxim.;双花木(长柄双花木);Common Disanthus, Coupleflower ●
134011　Disanthus cercidifolius Maxim. subsp. longipes (Hung T. Chang) K. Y. Pan = Disanthus cercidifolius Maxim. var. longipes Hung T. Chang ●◇
134012　Disanthus cercidifolius Maxim. var. longipes Hung T. Chang;长柄双花木;Longstipe Disanthus, Long-stiped Disanthus ●◇
134013　Disarrenum Labill. (废弃属名) = Hierochloe R. Br. (保留属名)■
134014　Disarrhenum P. Beauv. = Disarrenum Labill. (废弃属名)■
134015　Disarrhenum P. Beauv. = Hierochloe R. Br. (保留属名)■
134016　Disaster Gilli = Trymalium Fenzl ●☆
134017　Discalyxia Markgr. = Alyxia Banks ex R. Br. (保留属名)●
134018　Discanthera Torr. et A. Gray = Cyclanthera Schrad. ■
134019　Discanthus Spruce = Cyclanthus Poit. ex A. Rich. ■☆
134020　Discaria Hook. (1829);刺鼠李属(棘鼠李属)●☆
134021　Discaria toumatou Raoul;刺鼠李;Wild Irishman ●☆
134022　Dischanthlum Kunth = Andropogon L. (保留属名)■
134023　Dischanthlum Kunth = Dichanthium Willemet ■
134024　Dischema Voigt = Hitchenia Wall. ■☆
134025　Dischidanthus Tsiang (1936);马兰藤属(假瓜子金属);Dischidanthus, Malanvine ●■
134026　Dischidanthus urceolatus (Decne.) Tsiang;马兰藤(假瓜子金,金腰带);Malanvine, Urnshaped Dischidanthus ●■
134027　Dischidia R. Br. (1810);眼树莲属(豆蔓藤属,风不动属,瓜子金属,树眼莲属);Dischidia ●■
134028　Dischidia alboflava Costantin = Dischidia tonkinensis Costantin ■
134029　Dischidia australis Tsiang et P. T. Li;尖叶眼树莲(川甲草,马榴根,南瓜子金,上树瓜,石瓜子);Shrpleaf Dischidia ■
134030　Dischidia balansae Costantin = Dischidia tonkinensis Costantin ■
134031　Dischidia bengalensis Colebr.;孟加拉眼树莲■☆
134032　Dischidia chinensis Champ. ex Benth.;眼树莲(瓜子核,瓜子金,瓜子藤,金瓜核,乳汁藤,上树鳖,上树瓜子,石瓜子,石仙桃,树上瓜子,望水王仙桃,小耳环,奕鱼草,翼鱼草);China Dischidia, Chinese Dischidia ■
134033　Dischidia chinghungensis Tsiang et P. T. Li;云南眼树莲;Yunnan Dischidia ●
134034　Dischidia chinghungensis Tsiang et P. T. Li = Hoya chinghuangensis (Tsiang et P. T. Li) M. G. Gilbert, P. T. Li et W. D. Stevens ●
134035　Dischidia esquirolii (H. Lév.) Tsiang;金瓜核(个兴丁,金瓜核藤,黔越瓜子金);Esquirol Dischidia ■
134036　Dischidia esquirolii (H. Lév.) Tsiang = Dischidia tonkinensis Costantin ■
134037　Dischidia formosana Maxim.;台湾眼树莲(蝙蝠藤,风不动);Taiwan Dischidia ■
134038　Dischidia major (Vahl) Merr.;大眼树莲(瓜子金,拉氏眼树莲)■☆
134039　Dischidia minor (Vahl) Merr. = Dischidia nummularia R. Br. ■
134040　Dischidia nummularia R. Br.;圆叶眼树莲(瓜子金,如意草,上别木,小眼树莲,小叶眼树莲,圆眼树莲);Smallleaf Dischidia ■
134041　Dischidia nummularia R. Br. var. rhombifolia ? = Dischidia formosana Maxim. ■
134042　Dischidia obcordata (N. E. Br.) Maxwell et Donckelaar = Micholitzia obcrodata N. E. Br. ●
134043　Dischidia orbicularis Decne. = Dischidia nummularia R. Br. ■
134044　Dischidia pectinoides Pers.;栉状眼树莲■☆
134045　Dischidia platyphylla Schltdl.;阔叶眼树莲■☆
134046　Dischidia rafflesiana Wall. = Dischidia major (Vahl) Merr. ■☆
134047　Dischidia tonkinensis Costantin;滴锡眼树莲(滴锡藤,金瓜核,金瓜核藤);Whiteyellow Dischidia ■
134048　Dischidia torricellensis (Schltr.) P. I. Forst. = Spathidolepis torricelliensis Schltr. ■☆
134049　Dischidia yunnanensis H. Lév. = Biondia yunnanensis (H. Lév.) Tsiang ●

134050 Dischidiopsis Schltr. (1904);类眼树莲属■☆
134051 Dischidiopsis philippinensis Schltr.;菲律宾类眼树莲■☆
134052 Dischidium (Ging.) Opiz = Viola L. ■●
134053 Dischidium Rchb. = Viola L. ■●
134054 Dischimia Rchb. = Dischisma Choisy ■●☆
134055 Dischisma Choisy(1823);二裂玄参属■●☆
134056 Dischisma affine Schltr. = Dischisma spicatum (Thunb.) Choisy ■☆
134057 Dischisma arenarium E. Mey.;沙地二裂玄参■☆
134058 Dischisma capitatum (Thunb.) Choisy;头状二裂玄参■☆
134059 Dischisma chamaedryfolium (Link ex Jaroscz) Walp. = Dischisma ciliatum (P. J. Bergius) Choisy subsp. erinoides (L. f.) Rössler ■☆
134060 Dischisma ciliatum (P. J. Bergius) Choisy;缘毛二裂玄参■☆
134061 Dischisma ciliatum (P. J. Bergius) Choisy subsp. erinoides (L. f.) Rössler;狐地黄二裂玄参■☆
134062 Dischisma ciliatum (P. J. Bergius) Choisy subsp. flaccum (E. Mey.) Rössler;柔软缘毛二裂玄参■☆
134063 Dischisma ciliatum (P. J. Bergius) Choisy var. crassifolium E. Mey.;厚叶二裂玄参■☆
134064 Dischisma clandestinum E. Mey.;隐匿二裂玄参■☆
134065 Dischisma clandestinum Schltr. = Dischisma capitatum (Thunb.) Choisy ■☆
134066 Dischisma crassum Rolfe;粗二裂玄参■☆
134067 Dischisma erinoides (L. f.) Sweet = Dischisma ciliatum (P. J. Bergius) Choisy subsp. erinoides (L. f.) Rössler ■☆
134068 Dischisma flaccum E. Mey. = Dischisma ciliatum (P. J. Bergius) Choisy subsp. flaccum (E. Mey.) Rössler ■☆
134069 Dischisma fruticosum (L. f.) Rolfe;灌丛二裂玄参■☆
134070 Dischisma hispidum (Lam.) Sweet = Dischisma ciliatum (P. J. Bergius) Choisy ■☆
134071 Dischisma leptostachyum E. Mey.;细穗二裂玄参■☆
134072 Dischisma occludens Schltr. = Dischisma clandestinum E. Mey. ■☆
134073 Dischisma spicatum (Thunb.) Choisy;穗状二裂玄参■☆
134074 Dischisma squarrosum Schltr.;粗鳞二裂玄参■☆
134075 Dischisma struthioloides Killick;花束二裂玄参■☆
134076 Dischisma tomentosum Schltr.;绒毛二裂玄参■☆
134077 Dischistocalyx T. Anderson ex Benth. = Dischistocalyx T. Anderson ex Benth. et Hook. f. ●☆
134078 Dischistocalyx T. Anderson ex Benth. et Hook. f. (1876) ('Distichocalyx');二裂萼属●☆
134079 Dischistocalyx T. Anderson ex Lindau = Pseudostenosiphonium Lindau ●■
134080 Dischistocalyx alternifolius Champl. et Lejoly;互叶二裂萼●☆
134081 Dischistocalyx angustifolius C. B. Clarke = Dischistocalyx hirsutus C. B. Clarke ●☆
134082 Dischistocalyx bignoniiflorus (S. Moore) Lindau = Ruellia bignoniiflora S. Moore ■☆
134083 Dischistocalyx brevifolius C. B. Clarke = Dischistocalyx strobilinus C. B. Clarke ●☆
134084 Dischistocalyx buchholzii Lindau = Dischistocalyx grandifolius C. B. Clarke ●☆
134085 Dischistocalyx capitellatus C. B. Clarke = Dischistocalyx strobilinus C. B. Clarke ●☆
134086 Dischistocalyx confertiflorus Lindau = Acanthopale confertiflora (Lindau) C. B. Clarke ●☆
134087 Dischistocalyx fulvus Lindau ex Bremek. = Dischistocalyx hirsutus C. B. Clarke ●☆
134088 Dischistocalyx grandifolius C. B. Clarke;大叶二裂萼●☆
134089 Dischistocalyx hirsutus C. B. Clarke;毛二裂萼●☆
134090 Dischistocalyx insignis Bremek. = Dischistocalyx grandifolius C. B. Clarke ●☆
134091 Dischistocalyx insignis Bremek. var. parviflorus ? = Dischistocalyx grandifolius C. B. Clarke ●☆
134092 Dischistocalyx klainei Benoist;克莱恩二裂萼●☆
134093 Dischistocalyx laxiflorus Lindau = Acanthopale decempedalis C. B. Clarke ●☆
134094 Dischistocalyx lithicola Champl. et Ngok;岩生二裂萼●☆
134095 Dischistocalyx obanensis S. Moore;奥班二裂萼●☆
134096 Dischistocalyx polyneurus C. B. Clarke = Dischistocalyx hirsutus C. B. Clarke ●☆
134097 Dischistocalyx pubescens Lindau = Acanthopale pubescens (Lindau) C. B. Clarke ■☆
134098 Dischistocalyx rivularis Bremek.;溪边二裂萼●☆
134099 Dischistocalyx ruellioides S. Moore = Dischistocalyx thunbergiiflorus (T. Anderson) Benth. ex C. B. Clarke ●☆
134100 Dischistocalyx ruficaulis Bremek. = Dischistocalyx hirsutus C. B. Clarke ●☆
134101 Dischistocalyx staudtii Bremek.;施陶二裂萼●☆
134102 Dischistocalyx strobilinus C. B. Clarke;球果二裂萼●☆
134103 Dischistocalyx thunbergiiflorus (T. Anderson) Benth. ex C. B. Clarke;通贝里二裂萼●☆
134104 Dischistocalyx togoensis Lindau = Ruellia togoensis (Lindau) Heine ■☆
134105 Dischistocalyx walkeri Benoist = Dischistocalyx hirsutus C. B. Clarke ●☆
134106 Dischizolaena (Baill.) Tiegh. = Tapura Aubl. ●☆
134107 Dischlis Phil. = Distichlis Raf. ■☆
134108 Dischoriste D. Dietr. = Dyschoriste Nees ■●
134109 Disciphania Eichler(1864);盘金藤属●☆
134110 Disciphania cardiophylla Standl.;心叶盘金藤●☆
134111 Disciphania micrantha Diels;小花盘金藤●☆
134112 Discipiper Trel. et Stehle = Piper L. ●■
134113 Discladium Tiegh. = Ochna L. ●
134114 Discladium mossambicensis (Klotzsch) Tiegh. = Ochna mossambicensis Klotzsch ●☆
134115 Discocactus Pfeiff. (1837);圆盘玉属(孔雀花属,圆盘球属)●☆
134116 Discocactus boliviensis Backeb. = Discocactus heptacanthus Britton et Rose ●☆
134117 Discocactus heptacanthus Britton et Rose;七刺圆盘玉●☆
134118 Discocactus heptacanthus Britton et Rose subsp. magnimammus (Buin. et Brederoo) N. P. Taylor et Zappi;扁球圆盘玉●☆
134119 Discocactus horstii Buining et Brederoo;赫氏圆盘玉●☆
134120 Discocactus insignis Pfeiff.;圆盘玉●☆
134121 Discocactus zehntneri Britton et Rose;蔡氏圆盘玉●☆
134122 Discocactus zehntneri Britton et Rose var. boomianus (Buining et Brederoo) P. J. Braun;小球圆盘玉●☆
134123 Discocalyx (A. DC.) Mez(1902);盘萼属●☆
134124 Discocalyx Mez = Discocalyx (A. DC.) Mez ●☆
134125 Discocalyx dissectus Kaneh.;深裂盘萼●☆
134126 Discocapnos Cham. et Schltdl. (1826);翅果烟堇属■☆
134127 Discocapnos dregei Hutch. = Discocapnos mundtii Cham. et Schltdl. ■☆

134128 Discocapnos mundtii Cham. et Schltdl.;翅果烟堇■☆
134129 Discocarpus Klotzsch(1841);盘果大戟属■☆
134130 Discocarpus Liebm. = Discocnide Chew ●☆
134131 Discocarpus essequeboensis Klotzsch;盘果大戟■☆
134132 Discocarpus hirtus (L. f.) Pax et K. Hoffm. = Lachnostylis hirta (L. f.) Müll. Arg. ■●☆
134133 Discocatus Walp. = Disocactus Lindl. ●☆
134134 Discocatus Walp. = Phyllocactus Link ●
134135 Discoclaoxylon (Müll. Arg.) Pax et K. Hoffm. (1914);盘桐树属●☆
134136 Discoclaoxylon Pax et K. Hoffm. = Discoclaoxylon (Müll. Arg.) Pax et K. Hoffm. ●☆
134137 Discoclaoxylon hexandrum (Müll. Arg.) Pax et K. Hoffm.;六蕊盘桐树●☆
134138 Discoclaoxylon hexandrum (Pax) Pax et K. Hoffm. = Discoclaoxylon hexandrum (Müll. Arg.) Pax et K. Hoffm. ●☆
134139 Discoclaoxylon occidentale (Müll. Arg.) Pax et K. Hoffm.;西方盘桐树●☆
134140 Discoclaoxylon occidentale (Müll. Arg.) Pax et K. Hoffm. var. pubescens Pax et K. Hoffm. = Discoclaoxylon pubescens (Pax et K. Hoffm.) Exell ●☆
134141 Discoclaoxylon pedicellare (Müll. Arg.) Pax et K. Hoffm.;梗花盘桐树●☆
134142 Discoclaoxylon pubescens (Pax et K. Hoffm.) Exell;短柔毛盘桐树●☆
134143 Discocleidion (Müll. Arg.) Pax et K. Hoffm. (1914);假奓包叶属;Discocleidion ●
134144 Discocleidion Pax et K. Hoffm. = Discocleidion (Müll. Arg.) Pax et K. Hoffm. ●
134145 Discocleidion glabrum Merr.;光假奓包叶;Bright Dischidia, Glabrous Discocleidion ●
134146 Discocleidion rufescens (Franch.) Pax et K. Hoffm.;假奓包叶(艾桐,老虎麻);Dischidia, Reddish Discocleidion ●
134147 Discocleidion ulmifolium (Müll. Arg.) Pax et K. Hoffm.;大假奓包叶(榆叶棒柄花)●☆
134148 Discocnide Chew(1965);盘果麻属●☆
134149 Discocnide mexicana (Liebm.) Chew;盘果麻■☆
134150 Discocoffea A. Chev. = Coffea L. ●
134151 Discocoffea A. Chev. = Tricalysia A. Rich. ex DC. ●
134152 Discocrania (Harms) M. Kral = Cornus L. ●
134153 Discoglypremna Prain(1911);喀麦隆双蕊苏木属●☆
134154 Discoglypremna caloneura (Pax) Prain;喀麦隆双蕊苏木●☆
134155 Discoglypremna caloneura (Pax) Prain var. rigidifolia Pax = Discoglypremna caloneura (Pax) Prain ●☆
134156 Discoglypremna caloneura Prain = Discoglypremna caloneura (Pax) Prain ●☆
134157 Discogyne Schltr. = Ixonanthes Jack ●
134158 Discolenta Raf. = Polygonum L. (保留属名)■●
134159 Discolobium Benth. (1837);盘豆属■☆
134160 Discolobium pulchellum Benth.;盘豆■☆
134161 Discoluma Baill. = Pouteria Aubl. ●
134162 Discoma O. F. Cook = Chamaedorea Willd. (保留属名)●☆
134163 Discomela Raf. = Helianthus L. ■
134164 Discophis Raf. = Drypetes Vahl ●
134165 Discophora Miers(1852);盘茱萸属●☆
134166 Discophytum Miers = Calycera Cav. (保留属名)■☆
134167 Discopleura DC. = Ptilimnium Raf. ■☆
134168 Discoplis Raf. = Mercurialis L. ■
134169 Discopodium Hochst. (1844);盘足茄属(盘茄属)■☆
134170 Discopodium Steud. = Tricostularia Nees ■
134171 Discopodium grandiflorum Cufod.;大花盘足茄■☆
134172 Discopodium penninervium Hochst.;盘足茄■☆
134173 Discopodium penninervium Hochst. var. holstii (Dammer) Bitter = Discopodium penninervium Hochst. ■☆
134174 Discopodium penninervium Hochst. var. magnifolium Chiov. = Discopodium penninervium Hochst. ■☆
134175 Discopodium penninervium Hochst. var. sparsearaneosum Bitter = Discopodium penninervium Hochst. ■☆
134176 Discorea Miq. = Dioscorea L. (保留属名)■
134177 Discoseris (Endl.) Kuntze = Gochnatia Kunth ●
134178 Discoseris (Endl.) Post et Kuntze = Gochnatia Kunth ●
134179 Discoseris (Endl.) Post et Kuntze = Richterago Kuntze ●☆
134180 Discospermum Dalzell = Diplospora DC. ●
134181 Discostigma Hassk. (1842);盘柱藤黄属;Discostigma ●☆
134182 Discostigma Hassk. = Garcinia L. ●
134183 Discostigma rostratum Hassk.;盘柱藤黄●☆
134184 Discovium Raf. (1819);盘路芥属■☆
134185 Discovium Raf. = Lesquerella S. Watson ■☆
134186 Discovium gracile Raf.;盘路芥■☆
134187 Discretitheca P. D. Cantino(1999);平行弓蕊灌属●☆
134188 Discretitheca nepalensis (Moldenke) P. D. Cantino;平行弓蕊灌●☆
134189 Discurainia Walp. = Descurainia Webb et Berthel. (保留属名)■
134190 Discurea (C. A. Mey. ex Ledeb.) Schur = Descurainia Webb et Berthel. (保留属名)■
134191 Discurea Schur = Discurainia Walp. ■
134192 Discyphus Schltr. (1919);双杯兰属■☆
134193 Discyphus scopulariae (Rchb. f.) Schltr.;双杯兰■☆
134194 Disecocarpus Hassk. = Commelina L. ■
134195 Disella Greene = Sida L. ●■
134196 Disella Greene(1906);小迪萨兰属■☆
134197 Disella hederacea (Douglas) Greene;小迪萨兰■☆
134198 Diselma Hook. f. (1857);对鳞柏属●☆
134199 Diselma archeri Hook. f.;对鳞柏●☆
134200 Diselmaceae A. V. Bobrov et Melikyan;对鳞柏科●☆
134201 Disemma Labill. = Passiflora L. ●■
134202 Disemma horsfieldii Miq. = Passiflora cochinchinensis Spreng. ■
134203 Disemma horsfieldii Miq. var. teysmanniana Miq. = Passiflora cochinchinensis Spreng. ■
134204 Disemma horsfleldii Miq. = Passiflora moluccana Reinw. ex Blume var. teysmanniana (Miq.) Wilde ■
134205 Disemma penangiana (Wall.) Miq. = Adenia penangiana (Wall. ex G. Don) J. J. de Wilde ●
134206 Diseomela Raf. = Helianthus L. ■
134207 Disepalum Hook. f. (1860);九重皮属(双萼木属)●
134208 Disepalum acuminatissimum Boerl. et Koord.;尖九重皮●☆
134209 Disepalum anomalum Hook. f.;九重皮●☆
134210 Disepalum grandiflorum Ridl.;大花九重皮●☆
134211 Disepalum longipes King;长梗九重皮●☆
134212 Disepalum petelotii (Merr.) D. M. Johnson = Polyalthia petelotii Merr. ●
134213 Disepalum plagioneurum (Diels) D. M. Johnson = Polyalthia plagioneura Diels ●
134214 Disepalum platypetalum Merr.;宽瓣九重皮●☆

134215 Disepalum pulchrum (King) J. Sinclair;美丽九重皮●☆
134216 Diseris Wight = Disperis Sw. ■
134217 Diserneston Jaub. et Spach = Dorema D. Don ■☆
134218 Disgrega Hassk. = Tripogandra Raf. ■☆
134219 Disinstylis Raf. = Chlora Adans. ■☆
134220 Disiphon Schltr. = Vaccinium L. ●
134221 Disisocactus Kunze = Disocactus Lindl. ●☆
134222 Disisorhipsalis Doweld = Rhipsalis Gaertn.(保留属名)●
134223 Diskion Raf. = Saracha Ruiz et Pav. ●☆
134224 Diskyphogyne Szlach. et R. González(1837);双凸蕊兰属■☆
134225 Dismophyla Raf. = Drosera L. ■
134226 Disocactus Lindl.(1845);姬孔雀属(双重仙人掌属)●☆
134227 Disocactus ackermannii (Haw.) Barthlott = Disocactus ackermannii (Lindl.) Barthlott ■☆
134228 Disocactus ackermannii (Lindl.) Barthlott;白花姬孔雀■☆
134229 Disocactus eichlamii (Weing.) Britton et Rose;艾氏姬孔雀(艾氏双重仙人掌)■☆
134230 Disocactus flagelliformis (L.) Barthlott;鞭状姬孔雀;Rat Tail, Rat Tail Cactus ■☆
134231 Disocactus macranthus (Alexander) Kimnach et Hutchison;大花姬孔雀(大花双重仙人掌)■☆
134232 Disocactus speciosus (Cav.) Barthlott;优美姬孔雀;Pitaya De Cerro, Santa Marta ■☆
134233 Disocereus Frič et Kreuz. = Disocactus Lindl. ●☆
134234 Disodea Pers. = Lygodisodea Ruiz et Pav. ●■
134235 Disodea Pers. = Paederia L.(保留属名)●■
134236 Disomene A. DC. = Dysemone Sol. ex G. Forst. ■☆
134237 Disomene A. DC. = Gunnera L. ■☆
134238 Disoon A. DC. = Myoporum Banks et Sol. ex G. Forst. ●
134239 Disoxylon Rchb. = Dysoxylum Blume ●
134240 Disoxylum A. Juss. = Dysoxylum Blume ●
134241 Disoxylum Benth. et Hook. f. = Amoora Roxb. ●
134242 Disoxylum Benth. et Hook. f. = Dysoxylum Blume ●
134243 Dispara Raf. = Cristatella Nutt. ■☆
134244 Disparago Gaertn.(1791)(保留属名);多头帚鼠麹属●☆
134245 Disparago anomala Schltr. ex Levyns;异常多头帚鼠麹●☆
134246 Disparago barbata Koekemoer;南非多头帚鼠麹■☆
134247 Disparago ericoides (P. J. Bergius) Gaertn.;石南状多头帚鼠麹●☆
134248 Disparago gongylodes Koekemoer;好望角多头帚鼠麹■☆
134249 Disparago hoffmanniana Schltr. = Disparago ericoides (P. J. Bergius) Gaertn. ●☆
134250 Disparago kraussii Sch. Bip.;克劳斯多头帚鼠麹●☆
134251 Disparago lasiocarpa Cass. = Disparago ericoides (P. J. Bergius) Gaertn. ●☆
134252 Disparago laxifolia DC.;疏叶多头帚鼠麹■☆
134253 Disparago pilosa Koekemoer;毛多头帚鼠麹■☆
134254 Disparago seriphioides DC. = Disparago ericoides (P. J. Bergius) Gaertn. ●☆
134255 Dispeltophorus Lehm. = Menonvillea R. Br. ex DC. ■●☆
134256 Disperanthoceros Mytnik et Szlach.(2007);尼日利亚多穗兰属■☆
134257 Disperanthoceros Mytnik et Szlach. = Polystachya Hook.(保留属名)■
134258 Disperis Sw.(1800);双袋兰属;Disperis ■
134259 Disperis afzelii Schltr.;阿芙泽尔双袋兰■☆
134260 Disperis allisonii Rolfe = Disperis cooperi Harv. ■☆
134261 Disperis ankarensis H. Perrier;安卡拉双袋兰■☆
134262 Disperis anomala Schltr. = Disperis stenoplectron Rchb. f. ■☆
134263 Disperis anthoceros Rchb. f. var. grandiflora Verdc.;大花双袋兰■☆
134264 Disperis aphylla Kraenzl.;无叶双袋兰■☆
134265 Disperis aphylla Kraenzl. subsp. bifolia Verdc.;双叶双袋兰■☆
134266 Disperis aphylla Kraenzl. var. bifolia (Verdc.) Szlach. et Olszewski = Disperis aphylla Kraenzl. subsp. bifolia Verdc. ■☆
134267 Disperis atacorensis A. Chev. = Disperis togoensis Schltr. ■☆
134268 Disperis bathiei Bosser et la Croix;巴西双袋兰■☆
134269 Disperis bicolor Rolfe = Disperis tysonii Bolus ■☆
134270 Disperis bifida P. J. Cribb;双裂双袋兰■☆
134271 Disperis bodkinii Bolus;博德金双袋兰■☆
134272 Disperis bolusiana Schltr.;博卢斯双袋兰■☆
134273 Disperis bolusiana Schltr. subsp. macrocorys (Rolfe) J. C. Manning;大双袋兰■☆
134274 Disperis bosseri la Croix et P. J. Cribb;博瑟双袋兰■☆
134275 Disperis breviloba Verdc.;短双袋兰■☆
134276 Disperis buchananii Rolfe = Disperis cooperi Harv. ■☆
134277 Disperis capensis (L. f.) Sw.;好望角双袋兰■☆
134278 Disperis capensis (L. f.) Sw. var. brevicaudata Rolfe;短尾好望角双袋兰■☆
134279 Disperis capensis (L. f.) Sw. var. tenera (Spreng.) Sond. = Disperis capensis (L. f.) Sw. ■☆
134280 Disperis cardiopetala Summerh. = Disperis togoensis Schltr. ■☆
134281 Disperis centrocorys Schltr. = Disperis nemorosa Rendle ■☆
134282 Disperis ciliata Bosser;缘毛双袋兰■☆
134283 Disperis circumflexa (L.) T. Durand et Schinz subsp. aemula (Schltr.) J. C. Manning;匹敌双袋兰■☆
134284 Disperis circumflexa (L.) T. Durand et Schinz var. aemula (Schltr.) Kraenzl. = Disperis circumflexa (L.) T. Durand et Schinz subsp. aemula (Schltr.) J. C. Manning ■☆
134285 Disperis comorensis Schltr. = Disperis humblotii Rchb. f. ■☆
134286 Disperis concinna Schltr.;整洁双袋兰■☆
134287 Disperis cooperi Harv.;库珀双袋兰■☆
134288 Disperis cordata Summerh. = Disperis togoensis Schltr. ■☆
134289 Disperis crassicaulis Rchb. f.;粗茎双袋兰■☆
134290 Disperis cucullata Sw.;僧帽状双袋兰■☆
134291 Disperis decipiens Verdc.;迷惑双袋兰■☆
134292 Disperis discifera H. Perrier;盘状双袋兰■☆
134293 Disperis egregia Summerh.;优秀双袋兰■☆
134294 Disperis ermelensis Rolfe = Disperis cooperi Harv. ■☆
134295 Disperis falcatipetala P. J. Cribb et la Croix;镰瓣双袋兰■☆
134296 Disperis fanniniae Harv.;范尼双袋兰■☆
134297 Disperis flava Rolfe = Disperis wealei Rchb. f. ■☆
134298 Disperis gracilis Schltr. = Disperis wealei Rchb. f. ■☆
134299 Disperis hildebrandtii Rchb. f.;赫氏双袋兰■☆
134300 Disperis humblotii Rchb. f.;洪布双袋兰■☆
134301 Disperis javanica J. J. Sm. = Disperis neilgherrensis Wight ■
134302 Disperis johnstonii Rchb. f. ex Rolfe;约翰斯顿双袋兰■☆
134303 Disperis kamerunensis Schltr.;喀麦隆双袋兰■☆
134304 Disperis katangensis Summerh.;加丹加双袋兰■☆
134305 Disperis katangensis Summerh. var. minor Verdc.;较小双袋兰■☆
134306 Disperis kermesina Rolfe = Disperis tysonii Bolus ■☆
134307 Disperis kerstenii Rchb. f.;克斯滕双袋兰■☆
134308 Disperis kilimanjarica Rendle;基利曼双袋兰■☆
134309 Disperis lanceana H. Perrier;披针状双袋兰■☆

134310 Disperis lanceolata Bosser et la Croix;披针形双袋兰■☆
134311 Disperis lantauensis S. Y. Hu = Disperis neilgherrensis Wight ■
134312 Disperis latigaleata H. Perrier;宽盔双袋兰■☆
134313 Disperis leuconeura Schltr.;白脉双袋兰■☆
134314 Disperis lindleyana Rchb. f.;林德利双袋兰■☆
134315 Disperis macowanii Bolus;麦克欧文双袋兰■☆
134316 Disperis macrocorys Rolfe = Disperis bolusiana Schltr. subsp. macrocorys (Rolfe) J. C. Manning ■☆
134317 Disperis majungensis Schltr.;马任加双袋兰■☆
134318 Disperis masoalensis P. J. Cribb et la Croix;马苏阿拉双袋兰■☆
134319 Disperis mckenii Harv. = Disperis woodii Bolus ■☆
134320 Disperis micrantha Lindl.;小花双袋兰■☆
134321 Disperis mildbraedii Schltr. ex Summerh.;米尔德双袋兰■☆
134322 Disperis mozambicensis Schltr.;莫桑比克双袋兰■☆
134323 Disperis namaquensis Bolus = Disperis purpurata Rchb. f. ■☆
134324 Disperis nantauensis S. Y. Hu;香港双袋兰;Hongkong Disperis ■
134325 Disperis natalensis Rolfe = Disperis stenoplectron Rchb. f. ■☆
134326 Disperis neilgherrensis Wight;双袋兰(兰屿草兰,远东双袋兰);Siam Disperis,Thailand Disperis ■
134327 Disperis nelsonii Rolfe = Disperis virginalis Schltr. ■☆
134328 Disperis nemorosa Rendle;森林双袋兰■☆
134329 Disperis nitida Summerh.;光亮双袋兰■☆
134330 Disperis oppositifolia Sm.;对叶双袋兰■☆
134331 Disperis orientalis Fukuy. = Disperis neilgherrensis Wight ■
134332 Disperis orientalis Fukuy. = Disperis philippinensis Schltr. ■
134333 Disperis orientalis Fukuy. = Disperis siamensis Rolfe ex Downie ■
134334 Disperis oxyglossa Bolus;尖舌双袋兰■☆
134335 Disperis palawensis (Tuyama) Tuyama = Disperis neilgherrensis Wight ■
134336 Disperis paludosa Harv. ex Lindl.;沼泽双袋兰■☆
134337 Disperis papuana Michol. et Kraenzl. = Disperis neilgherrensis Wight ■
134338 Disperis parvifolia Schltr.;小叶双袋兰■☆
134339 Disperis perrieri Schltr.;佩里耶双袋兰■☆
134340 Disperis philippinensis Schltr. = Disperis neilgherrensis Wight ■
134341 Disperis preussii Rolfe = Disperis kamerunensis Schltr. ■☆
134342 Disperis purpurata Rchb. f.;紫双袋兰■☆
134343 Disperis purpurata Rchb. f. subsp. pallescens Bruyns;变白紫双袋兰■☆
134344 Disperis purpurata Rchb. f. var. parviflora Bolus = Disperis bolusiana Schltr. ■☆
134345 Disperis pusilla Verdc.;微小双袋兰■☆
134346 Disperis reichenbachiana Welw. ex Rchb. f.;赖兴巴赫双袋兰■☆
134347 Disperis renibractea Schltr.;肾苞双袋兰■☆
134348 Disperis rhodoneura Schltr. = Disperis neilgherrensis Wight ■
134349 Disperis secunda (Thunb.) Sw. var. aemula Schltr. = Disperis circumflexa (L.) T. Durand et Schinz subsp. aemula (Schltr.) J. C. Manning ■☆
134350 Disperis siamensis Rolfe ex Downie = Disperis philippinensis Schltr. ■
134351 Disperis similis Schltr.;相似双袋兰■☆
134352 Disperis stenoglossa Schltr. = Disperis woodii Bolus ■☆
134353 Disperis stenoplectron Rchb. f.;纳塔尔双袋兰■☆
134354 Disperis stolzii Schltr. = Disperis johnstonii Rchb. f. ex Rolfe ■☆
134355 Disperis teleplana F. Maek. = Disperis neilgherrensis Wight ■
134356 Disperis teleplana F. Maek. = Disperis siamensis Rolfe ex Downie ■
134357 Disperis thomensis Summerh.;爱岛双袋兰■☆
134358 Disperis thorncroftii Schltr.;托恩双袋兰■☆
134359 Disperis togoensis Schltr.;多哥双袋兰■☆
134360 Disperis trilineata Schltr.;三线双袋兰■☆
134361 Disperis tripetaloidea (Thouars) Lindl.;三瓣双袋兰■☆
134362 Disperis tysonii Bolus;泰森双袋兰■☆
134363 Disperis uzungwae Verdc.;乌尊季沃双袋兰■☆
134364 Disperis villosa (L. f.) Sw.;长柔毛双袋兰■☆
134365 Disperis virginalis Schltr.;弗州双袋兰■☆
134366 Disperis walkerae Rchb. f. = Disperis neilgherrensis Wight ■
134367 Disperis wealei Rchb. f.;威尔双袋兰■☆
134368 Disperis woodii Bolus;伍得双袋兰■☆
134369 Disperis zeylanica Trimen = Disperis neilgherrensis Wight ■
134370 Disperis zeylanica Trimen var. neilgherrensis (Wight) Pradhan = Disperis neilgherrensis Wight ■
134371 Disperma C. B. Clarke = Duosperma Dayton ■☆
134372 Disperma J. F. Gmel. = Duosperma Dayton ■☆
134373 Disperma J. F. Gmel. = Mitchella L. ■
134374 Disperma affine (Lindau) S. Moore = Duosperma quadrangulare (Klotzsch) Brummitt ■☆
134375 Disperma angolense C. B. Clarke = Duosperma quadrangulare (Klotzsch) Brummitt ■☆
134376 Disperma crenatum (Lindau) Milne-Redh. = Duosperma crenatum (Lindau) P. G. Mey. ■☆
134377 Disperma densiflorum C. B. Clarke = Duosperma densiflorum (C. B. Clarke) Brummitt ■☆
134378 Disperma dentatum C. B. Clarke = Duosperma quadrangulare (Klotzsch) Brummitt ■☆
134379 Disperma eremophilum Milne-Redh. = Duosperma longicalyx (Deflers) Vollesen ■☆
134380 Disperma gossweileri S. Moore = Duosperma sessilifolium (Lindau) Brummitt ■☆
134381 Disperma kilimandscharicum (Lindau) C. B. Clarke var. bracteolatum C. B. Clarke = Duosperma kilimandscharicum (C. B. Clarke) Dayton ■☆
134382 Disperma kilimandscharicum C. B. Clarke = Duosperma kilimandscharicum (C. B. Clarke) Dayton ■☆
134383 Disperma nudantherum (C. B. Clarke) Milne-Redh. = Duosperma nudantherum (C. B. Clarke) Brummitt ■☆
134384 Disperma parviflorum (Lindau) C. B. Clarke = Duosperma crenatum (Lindau) P. G. Mey. ■☆
134385 Disperma quadrangulare (Klotzsch) C. B. Clarke = Duosperma quadrangulare (Klotzsch) Brummitt ■☆
134386 Disperma quadrisepalum C. B. Clarke = Duosperma crenatum (Lindau) P. G. Mey. ■☆
134387 Disperma quadrisepalum C. B. Clarke var. grandifolium S. Moore = Duosperma crenatum (Lindau) P. G. Mey. ■☆
134388 Disperma scabridum S. Moore = Duosperma sessilifolium (Lindau) Brummitt ■☆
134389 Disperma sessilifolium (Lindau) S. Moore = Duosperma sessilifolium (Lindau) Brummitt ■☆
134390 Disperma trachyphyllum Bullock = Duosperma trachyphyllum (Bullock) Dayton ■☆
134391 Disperma transvaalense Schinz = Duosperma transvaalense (Schinz) Vollesen ■☆
134392 Disperma viscidissimum S. Moore = Duosperma quadrangulare (Klotzsch) Brummitt ■☆

134393 Dispermotheca P. Beauv. = Parentucellia Viv. ■☆
134394 Disphyma N. E. Br. (1925);圆棒玉属;Dew-plant,Pigface ■☆
134395 Disphyma crassifolium (L.) L. Bolus;圆棒玉;Pigface,Purple Dew-plant ■☆
134396 Disphyma dunsdonii L. Bolus;杜恩圆棒玉■☆
134397 Displaspis Klatt = Diplaspis Hook. f. ■☆
134398 Displaspis Klatt = Huanaca Cav. ■☆
134399 Disporocarpa A. Rich. = Crassula L. ●■☆
134400 Disporopsis Hance(1883);竹根七属(假宝铎花属,假万寿竹属);False Fairybells,Solomon's Seal ■
134401 Disporopsis arisanensis Hayata = Disporopsis fuscopicta Hance var. arisanensis (Hayata) S. S. Ying ■
134402 Disporopsis arisanensis Hayata = Disporopsis pernyi (Hua) Diels ■
134403 Disporopsis aspera (Hua) Engl. ex Krause;散斑竹根七(大玉竹,黄鳝七,马鞭七,散斑假万寿竹,小玉竹,玉竹,竹根七);Asper False Fairybells ■
134404 Disporopsis fuscopicta Hance;竹根七(黄精,假万寿竹,盘龙七,石边七,石竹子,血蜈蚣);Common False Fairybells ■
134405 Disporopsis fuscopicta Hance var. arisanensis (Hayata) S. S. Ying;阿里山假宝铎花(薄叶万寿竹)■
134406 Disporopsis jinfushanensis Z. Y. Liu;金佛山竹根七(金佛竹根七);Jinfoshan False Fairybells ■
134407 Disporopsis kwangsiensis F. T. Wang et Ts. Tang;广西竹根七■
134408 Disporopsis leptophylla Hayata = Disporopsis fuscopicta Hance var. arisanensis (Hayata) S. S. Ying ■
134409 Disporopsis leptophylla Hayata = Disporopsis pernyi (Hua) Diels ■
134410 Disporopsis longifolia Craib;长叶竹根七(长叶假万寿竹,黄精,三子果);Longleaf False Fairybells ■
134411 Disporopsis mairei H. Lév. = Polygonatum punctatum Royle ex Kunth ■
134412 Disporopsis pernyi (Hua) Diels;深裂竹根七(黄脚鸡,假宝铎花,剑叶假万寿竹,十样错,玉竹,竹根假万寿竹,竹根七,竹节参);Perny False Fairybells ■
134413 Disporopsis taiwanensis S. S. Ying;台湾假宝铎花■
134414 Disporopsis taiwanensis S. S. Ying = Disporopsis pernyi (Hua) Diels ■
134415 Disporopsis undulata M. N. Tamura et Ogisu;峨眉竹根七■
134416 Disporum Salisb. = Disporum Salisb. ex D. Don ■
134417 Disporum Salisb. ex D. Don(1812);万寿竹属(宝铎草属,宝铎花属);Fairy Bells,Fairy Lantern,Fairybells ■
134418 Disporum acuminatissimum W. L. Sha;尖被万寿竹;Acuminate Fairybells ■
134419 Disporum austrosinense H. Hara = Disporum trabeculatum Gagnep. ■
134420 Disporum bodinieri (H. Lév. et Vaniot) F. T. Wang et Ts. Tang;短蕊万寿竹(长蕊万寿竹,倒竹散,落得打,牛尾参,万寿竹,小伸筋,玉竹参,竹凌霄);Bodinier Fairybells,Shortstamen Fairybells ■
134421 Disporum brachystemon F. T. Wang et Ts. Tang = Disporum bodinieri (H. Lév. et Vaniot) F. T. Wang et Ts. Tang ■
134422 Disporum cahnae Farw. = Prosartes maculata (Buckley) A. Gray ■☆
134423 Disporum calcaratum D. Don;距花万寿竹(宝铎草,倒竹散,狗尾巴参,狗尾巴草,距花宝铎草);Longspurred Fairybells ■
134424 Disporum calcaratum D. Don var. hamiltonianum (D. Don) Baker = Disporum calcaratum D. Don ■
134425 Disporum cantoniense (Lour.) Merr.;万寿竹(白根药,白龙七,白龙须,白毛七,白子草,草竹叶,倒竹散,广东万寿竹,豪猪七,黑龙须,老虎姜,山竹花,石竹根,小竹根,一线香,迎风不动草,玉竹草,竹根七,竹节参,竹节草,竹叶参,竹叶草,竹叶七);Canton Fairybells,Cantonese Fairy Bells,Fairybells ■
134426 Disporum cantoniense (Lour.) Merr. f. brunneum (C. H. Wright) H. Hara = Disporum cantoniense (Lour.) Merr. ■
134427 Disporum cantoniense (Lour.) Merr. var. brunneum (C. H. Wright) Hand. -Mazz. = Disporum cantoniense (Lour.) Merr. ■
134428 Disporum cantoniense (Lour.) Merr. var. kawakamii (Hayata) H. Hara = Disporum kawakamii Hayata ■
134429 Disporum cavaleriei H. Lév. = Disporum longistylum (H. Lév. et Vaniot) H. Hara ■
134430 Disporum chinense (Ker Gawl.) Kuntze = Disporum cantoniense (Lour.) Merr. ■
134431 Disporum esquirolii H. Lév. = Tricyrtis pilosa Wall. ■
134432 Disporum flavens Kitag. = Disporum uniflorum Baker ex S. Moore ■
134433 Disporum hainanense Merr.;海南万寿竹■
134434 Disporum hamiltonianum D. Don = Disporum calcaratum D. Don ■
134435 Disporum hookeri (Torr.) G. Nicholson = Prosartes hookeri Torr. ■☆
134436 Disporum hookeri (Torr.) G. Nicholson var. oblongifolium (S. Watson) Britton = Prosartes hookeri Torr. ■☆
134437 Disporum hookeri (Torr.) G. Nicholson var. oreganum (S. Watson) Q. Jones = Prosartes hookeri Torr. ■☆
134438 Disporum hookeri (Torr.) G. Nicholson var. trachyandrum (Torr.) Q. Jones = Prosartes hookeri Torr. ■☆
134439 Disporum hookeri Britton;加州万寿竹(胡克宝铎草);Fairy Lantern ■
134440 Disporum jiangchengense Y. Y. Qian = Disporum calcaratum D. Don ■
134441 Disporum kawakamii Hayata;台湾万寿竹(台湾宝铎花);Taiwan Fairybells ■
134442 Disporum lanuginosum (Michx.) G. Nicholson;绵毛万寿竹;Fairy Bells,Yellow Mandarin ■☆
134443 Disporum lanuginosum (Michx.) G. Nicholson = Prosartes lanuginosa (Michx.) D. Don ■☆
134444 Disporum lanuginosum G. Nicholson = Disporum lanuginosum (Michx.) G. Nicholson ■☆
134445 Disporum latipetalum Collett et Hemsl. = Disporum calcaratum D. Don ■
134446 Disporum leschenaultianum D. Don;阔叶宝铎草■☆
134447 Disporum longistylum (H. Lév. et Vaniot) H. Hara;长蕊万寿竹■
134448 Disporum luzoniense Merr. = Disporopsis fuscopicta Hance ■
134449 Disporum maculatum (Buckley) Britton = Prosartes maculata (Buckley) A. Gray ■☆
134450 Disporum maculatum Britton;斑点万寿竹;Nodding Mandarin,Spotted Disporum ■☆
134451 Disporum mairei H. Lév. = Stemona mairei (H. Lév.) Krause ■
134452 Disporum megalanthum F. T. Wang et Ts. Tang;大花万寿竹(白龙须,山竹花);Largeflower Fairybells ■
134453 Disporum nantoense S. S. Ying;南投万寿竹(南投宝铎花);Nantou Fairybells ■
134454 Disporum oreganum (S. Watson) W. T. Mill. = Prosartes hookeri Torr. ■☆

134455 Disporum ovale Ohwi = Streptopus ovalis (Ohwi) F. T. Wang et Y. C. Tang ■
134456 Disporum parvifolium (S. Watson) Britton = Prosartes hookeri Torr. ■☆
134457 Disporum pedunculatum H. Li et J. L. Huang; 总梗万寿竹; Pedunculate Fairybells ■
134458 Disporum pedunculatum H. Li et J. L. Huang = Disporum calcaratum D. Don ■
134459 Disporum pullum Salisb. = Disporum cantoniense (Lour.) Merr. ■
134460 Disporum pullum Salisb. ex Hook. f. = Disporum cantoniense (Lour.) Merr. ■
134461 Disporum pullum Salisb. ex Hook. f. var. brunneum C. H. Wright = Disporum cantoniense (Lour.) Merr. ■
134462 Disporum pullum Salisb. var. brunnea C. H. Wright = Disporum cantoniense (Lour.) Merr. ■
134463 Disporum pullum Salisb. var. ovalifolium H. Lév. = Disporum bodinieri (H. Lév. et Vaniot) F. T. Wang et Ts. Tang ■
134464 Disporum pullum Salisb. var. ovalifolium H. Lév. = Disporum brachystemon F. T. Wang et Ts. Tang ■
134465 Disporum schaffneri Moldenke = Prosartes maculata (Buckley) A. Gray ■☆
134466 Disporum senpomonticola Yamam. = Disporum hainanense Merr. ■
134467 Disporum sessile (Thunb.) D. Don; 宝铎草(白薇,白尾笋,百尾笋,遍地姜,淡竹花,倒竹散,狗尾巴,黄牛尾巴,凉水竹,山丫黄,石竹根,万花梢,万寿竹,小伸筋草,竹林梢,竹林霄,竹凌霄); Common Fairybells ■
134468 Disporum sessile (Thunb.) D. Don 'Variegatum'; 斑叶宝铎草■☆
134469 Disporum sessile (Thunb.) D. Don subsp. flavens (Kitag.) Kitag. = Disporum uniflorum Baker ex S. Moore ■
134470 Disporum sessile (Thunb.) D. Don var. flavens (Kitag.) Y. C. Tang; 黄宝铎草■☆
134471 Disporum sessile (Thunb.) D. Don var. flavens Kitag. = Disporum uniflorum Baker ex S. Moore ■
134472 Disporum sessile (Thunb.) D. Don var. pachyrrhizum Hand.-Mazz.; 粗根宝铎草■
134473 Disporum sessile (Thunb.) D. Don var. pachyrrhizum Hand.-Mazz. = Disporum uniflorum Baker ex S. Moore ■
134474 Disporum sessile (Thunb.) D. Don var. shimadae (Hayata) H. Hara = Disporum shimadai Hayata ■
134475 Disporum sessile (Thunb.) D. Don var. shimadae (Hayata) H. Hara f. intermedium H. Hara = Disporum nantoensis S. S. Ying ■
134476 Disporum sessile (Thunb.) D. Don var. stenophyllum Franch. et Sav. = Disporum uniflorum Baker ex S. Moore ■
134477 Disporum sessile D. Don ex Schult. f. minus (Miq.) H. Hara; 小宝铎草■☆
134478 Disporum sessile D. Don ex Schult. f. stenophyllum (Franch. et Sav.) Hayashi ex H. Hara; 窄叶宝铎草■☆
134479 Disporum sessile D. Don ex Schult. subsp. flavens (Kitag.) Kitag. = Disporum uniflorum Baker ex S. Moore ■
134480 Disporum sessile D. Don ex Schult. var. minus Miq. = Disporum sessile D. Don ex Schult. f. minus (Miq.) H. Hara ■☆
134481 Disporum sessile D. Don ex Schult. var. pachyrrhizum Hand.-Mazz. = Disporum uniflorum Baker ex S. Moore ■
134482 Disporum sessile D. Don ex Schult. var. shimadae (Hayata) H. Hara f. intermedium H. Hara = Disporum nantoensis S. S. Ying ■
134483 Disporum sessile D. Don ex Schult. var. shimadae (Hayata) H. Hara = Disporum shimadae Hayata ■
134484 Disporum shimadae Hayata; 山万寿竹(山宝铎花); Shimada Fairybells ■
134485 Disporum shimadai Hayata = Disporum uniflorum Baker ex S. Moore ■
134486 Disporum smilacinum A. Gray; 山东万寿竹(儿百合); Shandong Fairybells ■
134487 Disporum smilacinum A. Gray var. album Maxim. = Disporum smilacinum A. Gray ■
134488 Disporum smilacinum A. Gray var. ramosum Nakai = Disporum smilacinum A. Gray ■
134489 Disporum smilacinum A. Gray var. rotundatum Satake = Disporum smilacinum A. Gray ■
134490 Disporum smilacinum A. Gray var. variegatum Nakai = Disporum smilacinum A. Gray ■
134491 Disporum smilacinum A. Gray var. viridescens (Maxim.) Maxim. = Disporum viridescens (Maxim.) Nakai ■
134492 Disporum smilacinum C. H. Wright = Disporum viridescens (Maxim.) Nakai ■
134493 Disporum smithii (Hook.) Piper = Prosartes smithii (Hook.) Utech ■☆
134494 Disporum smithii Piper; 长花万寿竹; Fairy Lantern, Fairy Lanterns ■☆
134495 Disporum taipingense M. N. Tamura et Kawano = Disporum nantoensis S. S. Ying ■
134496 Disporum taiwanense S. S. Ying; 红花宝铎花; Redflower Fairybells ■
134497 Disporum taiwanense S. S. Ying = Disporum kawakamii Hayata ■
134498 Disporum trabeculatum Gagnep.; 横脉万寿竹■
134499 Disporum trachyandrum (Torr.) Britton = Prosartes hookeri Torr. ■☆
134500 Disporum trachycarpum (S. Watson) Benth. et Hook. f. = Prosartes trachycarpa S. Watson ■☆
134501 Disporum trachycarpum (S. Watson) Benth. et Hook. f. var. subglabrum E. H. Kelso = Prosartes trachycarpa S. Watson ■☆
134502 Disporum trachycarpum ?; 糙果万寿竹; Wartberry Falrybell ■☆
134503 Disporum uniflorum Baker ex S. Moore; 少花万寿竹(黄花万寿竹)■
134504 Disporum viridescens (Maxim.) Nakai; 宝珠草(绿宝珠草); Virescent Fairybells ■
134505 Disquamia Lem. = Aechmea Ruiz et Pav. (保留属名)■☆
134506 Dissanthelium Trin. (1836); 燥原禾属■☆
134507 Dissanthelium brevifolium Swallen et Tovar; 短叶燥原禾■☆
134508 Dissanthelium densum Swallen et Tovar; 密集燥原禾■☆
134509 Dissanthelium giganteum Tovar; 大燥原禾■☆
134510 Dissanthelium laxifolium Swallen et Tovar; 疏叶燥原禾■☆
134511 Dissanthelium longifolium Tovar; 长叶燥原禾■☆
134512 Dissanthelium minimum Pilg.; 小燥原禾■☆
134513 Dissecocarpus Hassk. = Commelina L. ■
134514 Dissecocarpus Hassk. = Disecocarpus Hassk. ■
134515 Dissiliaria F. Muell. (1867); 澳北大戟属☆
134516 Dissiliaria F. Muell. ex Baill. = Dissiliaria F. Muell. ☆
134517 Dissiliaria laxinervis Airy Shaw; 疏脉澳北大戟☆
134518 Dissocarpus F. Muell. (1858); 聚花澳藜属●☆
134519 Dissocarpus biflorus F. Muell.; 双花聚花澳藜●☆

134520 Dissocarpus latifolius (J. Black) Paul G. Wilson;宽叶聚花澳藜●☆
134521 Dissocarpus paradoxus (R. Br.) Ulbr.;奇异聚花澳藜●☆
134522 Dissochaeta Blume(1831);双毛藤属●☆
134523 Dissochaeta barthei Hance = Barthea barthei (Hance) Krasser ●
134524 Dissochaeta barthei Hance ex Benth. = Barthea barthei (Hance) Krasser ●
134525 Dissochaeta heteromorpha Naudin = Dichaetanthera heteromorpha (Naudin) Triana ●☆
134526 Dissochaeta macrosepala Staf;大瓣双毛藤●☆
134527 Dissochaeta microcarpa Naudin;小果双毛藤●☆
134528 Dissochaeta ovalifolia Naudin;卵叶双毛藤●☆
134529 Dissochaeta ramosii Merr.;拉氏双毛藤(拉莫双毛藤)●☆
134530 Dissochaeta sarcorhiza Baill. = Medinilla sarcorhiza (Baill.) Cogn. ●☆
134531 Dissochondrus (W. F. Hillebr.) Kuntze(1891);双花狗尾草属■☆
134532 Dissochondrus Kuntze = Setaria P. Beauv.(保留属名)■
134533 Dissochondrus Kuntze ex Hack. = Dissochondrus (W. F. Hillebr.) Kuntze ■☆
134534 Dissochondrus bifidus (Hillebr.) Kuntze;双花狗尾草■☆
134535 Dissochondrus bifidus Kuntze = Setaria biflora Hillebr. ■☆
134536 Dissochroma Post et Kuntze = Dyssochroma Miers ●☆
134537 Dissolaena Lour. = Rauvolfia L. ●
134538 Dissolaena verticillata Lour. = Rauvolfia verticillata (Lour.) Baill. ●
134539 Dissolena Lour. = Rauvolfia L. ●
134540 Dissomeria Hook. f. ex Benth. (1849);西非刺篱木属●☆
134541 Dissomeria crenata Hook. f. ex Benth.;西非刺篱木●☆
134542 Dissomeria glanduligera Sleumer;腺点西非刺篱木●☆
134543 Dissopetalum Miers = Cissampelos L. ●
134544 Dissorhynchium Schauer = Habenaria Willd. ■
134545 Dissosperma Soják = Corydalis DC.(保留属名)■
134546 Dissothrix A. Gray(1851);长毛修泽兰属■☆
134547 Dissothrix imbricata (Gardner) B. L. Rob.;长毛修泽兰■☆
134548 Dissotis Benth. (1849)(保留属名);异荣耀木属●☆
134549 Dissotis afzelii Hook. f. = Melastomastrum afzelii (Hook. f.) A. Fern. et R. Fern. ●☆
134550 Dissotis alata A. Fern. et R. Fern.;具翅异荣耀木●☆
134551 Dissotis amplexicaulis Jacq. -Fél. = Heterotis amplexicaulis (Jacq. -Fél.) Aké Assi ●☆
134552 Dissotis angolensis Cogn. = Heterotis angolensis (Cogn.) Jacq. -Fél. ●☆
134553 Dissotis angusii A. Fern. et R. Fern. = Dissotis falcipila Gilg ●☆
134554 Dissotis angustifolia A. Fern. et R. Fern. = Antherotoma angustifolia (A. Fern. et R. Fern.) Jacq. -Fél. ■☆
134555 Dissotis antennina (Sm.) Triana = Heterotis antennina (Sm.) Benth. ●☆
134556 Dissotis aprica Engl.;向阳异荣耀木●☆
134557 Dissotis aquatica De Wild.;水生异荣耀木●☆
134558 Dissotis arborescens A. Fern. et R. Fern.;树状异荣耀木●☆
134559 Dissotis bamendae Brenan et Keay = Dissotis princeps (Kunth) Triana ●☆
134560 Dissotis bangweolensis R. E. Fr. = Antherotoma debilis (Sond.) Jacq. -Fél. ■☆
134561 Dissotis barteri Hook. f.;巴特异荣耀木●☆
134562 Dissotis benguellensis A. Fern. et R. Fern.;本格拉异荣耀木●☆
134563 Dissotis benguellensis A. Fern. et R. Fern. var. parviflora ? = Dissotis benguellensis A. Fern. et R. Fern. ●☆
134564 Dissotis brazzae Cogn.;布拉扎异荣耀木●☆
134565 Dissotis buettneriana (Cogn. ex Büttner) Jacq. -Fél. = Heterotis buettneriana (Cogn. ex Büttner) Jacq. -Fél. ●☆
134566 Dissotis buraevii (Cogn.) A. Fern. et R. Fern.;布拉异荣耀木●☆
134567 Dissotis buraevii (Cogn.) A. Fern. et R. Fern. f. gilletii (De Wild.) A. Fern. et R. Fern.;吉莱异荣耀木●☆
134568 Dissotis buraevii (Cogn.) A. Fern. et R. Fern. var. pauciramosa (Jacq. -Fél.) A. Fern. et R. Fern.;少枝异荣耀木●☆
134569 Dissotis bussei Engl.;布瑟异荣耀木●☆
134570 Dissotis caloneura Engl.;美脉异荣耀木●☆
134571 Dissotis caloneura Engl. var. confertiflora A. Fern. et R. Fern.;密花异荣耀木●☆
134572 Dissotis caloneura Engl. var. pilosa A. Fern. et R. Fern.;疏毛异荣耀木●☆
134573 Dissotis candolleana Cogn. = Dissotis princeps (Kunth) Triana var. candolleana (Cogn.) A. Fern. et R. Fern. ●☆
134574 Dissotis canescens (E. Mey. ex R. A. Graham) Hook. f.;灰白异荣耀木●☆
134575 Dissotis canescens (E. Mey. ex R. A. Graham) Hook. f. = Heterotis canescens (E. Mey. ex R. A. Graham) Jacq. -Fél. ●☆
134576 Dissotis canescens (E. Mey. ex R. A. Graham) Hook. f. var. sudanense Jacq. -Fél. = Heterotis canescens (E. Mey. ex R. A. Graham) Jacq. -Fél. ●☆
134577 Dissotis capitata (Vahl) Hook. f. = Melastomastrum capitatum (Vahl) A. Fern. et R. Fern. ●☆
134578 Dissotis carrissoi A. Fern. et R. Fern.;卡里索异荣耀木●☆
134579 Dissotis castroi A. Fern. et R. Fern.;卡斯特罗异荣耀木●☆
134580 Dissotis chevalieri Gilg ex Engl.;舍瓦利耶异荣耀木●☆
134581 Dissotis cincinnata Gilg = Antherotoma senegambiensis (Guillaumin et Perr.) Jacq. -Fél. var. alpestris Taub. ●☆
134582 Dissotis cinerascens Hutch. = Heterotis cinerascens (Hutch.) Jacq. -Fél. ●☆
134583 Dissotis congolensis (Cogn. ex Büttner) Jacq. -Fél.;刚果异荣耀木●☆
134584 Dissotis conraui Gilg ex Engl.;康氏异荣耀木●☆
134585 Dissotis controversa (A. Chev. et Jacq. -Fél.) Jacq. -Fél. = Melastomastrum theifolium (G. Don) A. Fern. et R. Fern. var. controversum (A. Chev. et Jacq. -Fél.) Jacq. -Fél. ●☆
134586 Dissotis cordata Gilg;心形异荣耀木●☆
134587 Dissotis cordifolia A. Fern. et R. Fern.;心叶异荣耀木●☆
134588 Dissotis cornifolia (Benth.) Hook. f. = Melastomastrum cornifolium (Benth.) Jacq. -Fél. ●☆
134589 Dissotis crenulata Cogn.;细圆齿异荣耀木●☆
134590 Dissotis cryptantha Baker f.;隐花异荣耀木●☆
134591 Dissotis dasytricha Gilg et Ledermann ex Engl.;硬毛异荣耀木●☆
134592 Dissotis debilis (Sond.) Triana;弱小异荣耀木■☆
134593 Dissotis debilis (Sond.) Triana = Antherotoma debilis (Sond.) Jacq. -Fél. ■☆
134594 Dissotis debilis (Sond.) Triana var. lanceolata (Cogn.) A. Fern. et R. Fern. = Antherotoma debilis (Sond.) Jacq. -Fél. ■☆
134595 Dissotis debilis (Sond.) Triana var. pedicellata A. Fern. et R. Fern. = Antherotoma debilis (Sond.) Jacq. -Fél. ■☆
134596 Dissotis debilis (Sond.) Triana var. postpluvialis (Gilg) A.

Fern. et R. Fern. = Antherotoma debilis (Sond.) Jacq. -Fél. ■☆
134597 Dissotis debilis (Sond.) Triana var. prostrata A. Fern. et R. Fern. = Antherotoma debilis (Sond.) Jacq. -Fél. ■☆
134598 Dissotis debilis (Sond.) Triana var. pusilla (R. E. Fr.) A. Fern. et R. Fern. = Antherotoma debilis (Sond.) Jacq. -Fél. ■☆
134599 Dissotis decandra (Sm.) Triana = Osbeckia decandra (Sm.) DC. ●☆
134600 Dissotis decumbens (P. Beauv.) Triana = Heterotis decumbens (P. Beauv.) Jacq. -Fél. ●☆
134601 Dissotis decumbens (P. Beauv.) Triana var. minor Cogn. = Heterotis decumbens (P. Beauv.) Jacq. -Fél. ●☆
134602 Dissotis degasparisiana Buscal. et Muschl. = Dissotis simonis-jamesii Buscal. et Muschl. ●☆
134603 Dissotis deistelii Gilg ex Engl. = Heterotis prostrata (Thonn.) Benth. ●☆
134604 Dissotis densiflora (Gilg) A. Fern. et R. Fern. = Antherotoma densiflora (Gilg) Jacq. -Fél. ■☆
134605 Dissotis denticulata A. Fern. et R. Fern.;细齿异荣耀木●☆
134606 Dissotis dichaetantheroides Wickens;二毛药异荣耀木●☆
134607 Dissotis djalonis A. Chev. = Dissotis grandiflora (Sm.) Benth. ●☆
134608 Dissotis echinata A. Fern. et R. Fern.;具刺异荣耀木●☆
134609 Dissotis elegans (Robyns et Lawalrée) A. Fern. et R. Fern.;雅致异荣耀木●☆
134610 Dissotis elliotii Gilg = Dissotis thollonii Cogn. ex Büttner var. elliotii (Gilg) Jacq. -Fél. ●☆
134611 Dissotis elliotii Gilg var. setosior Keay et Brenan = Dissotis thollonii Cogn. ex Büttner ●☆
134612 Dissotis entii J. B. Hall = Heterotis entii (J. B. Hall) Jacq. -Fél. ●☆
134613 Dissotis erecta (Guillaumin et Perr.) Dandy = Melastomastrum capitatum (Vahl) A. Fern. et R. Fern. ●☆
134614 Dissotis erici-rosenii R. E. Fr. = Dichaetanthera erici-rosenii (R. E. Fr.) A. Fern. et R. Fern. ●☆
134615 Dissotis eximia (Sond.) Hook. f. = Dissotis princeps (Kunth) Triana ●☆
134616 Dissotis falcipila Gilg;镰异荣耀木●☆
134617 Dissotis fenarolii A. Fern. et R. Fern.;费纳罗利异荣耀木●☆
134618 Dissotis floribunda A. Chev. = Dissotis thollonii Cogn. ex Büttner var. elliotii (Gilg) Jacq. -Fél. ●☆
134619 Dissotis formosa A. Fern. et R. Fern.;美丽异荣耀木●☆
134620 Dissotis fruticosa (Brenan) Brenan et Keay;灌丛异荣耀木●☆
134621 Dissotis gilgiana De Wild.;吉尔格异荣耀木●☆
134622 Dissotis gilgiana De Wild. var. petiolata De Wild. ex A. Fern. et R. Fern.;柄叶吉尔格异荣耀木●☆
134623 Dissotis gilgiana De Wild. var. witteana Jacq. -Fél.;维特异荣耀木●☆
134624 Dissotis gilletii De Wild. = Dissotis buraevii (Cogn.) A. Fern. et R. Fern. f. gilletii (De Wild.) A. Fern. et R. Fern. ●☆
134625 Dissotis glaberrima A. Fern. et R. Fern.;光滑异荣耀木●☆
134626 Dissotis glandulicalyx Wickens;腺萼异荣耀木●☆
134627 Dissotis glandulosa A. Fern. et R. Fern.;具腺异荣耀木●☆
134628 Dissotis glauca Keay = Heterotis rupicola (Gilg ex Engl.) Jacq. -Fél. ●☆
134629 Dissotis gossweileri Exell;戈斯异荣耀木●☆
134630 Dissotis gracilis Cogn. = Antherotoma gracilis (Cogn.) Jacq. -Fél. ■☆
134631 Dissotis graminicola Hutch.;草莺异荣耀木●☆
134632 Dissotis grandiflora (Sm.) Benth.;大花异荣耀木●☆
134633 Dissotis grandiflora (Sm.) Benth. var. lambii (Hutch.) Keay;拉氏大花异荣耀木●☆
134634 Dissotis greenwayi A. Fern. et R. Fern. = Dissotis melleri Hook. f. var. greenwayi (A. Fern. et R. Fern.) A. Fern. et R. Fern. ●☆
134635 Dissotis helenae Buscal. et Muschl. = Dissotis speciosa Taub. ●☆
134636 Dissotis hensii Cogn.;亨斯异荣耀木●☆
134637 Dissotis hirsuta Hook. f. = Melastomastrum capitatum (Vahl) A. Fern. et R. Fern. ●☆
134638 Dissotis homblei (De Wild.) A. Fern. et R. Fern.;洪布勒异荣耀木●☆
134639 Dissotis humilis A. Chev. et Jacq. -Fél.;低矮异荣耀木●☆
134640 Dissotis incana (Walp.) Triana = Heterotis canescens (E. Mey. ex R. A. Graham) Jacq. -Fél. ●☆
134641 Dissotis incana (Walp.) Triana var. gilgiana A. Chev. = Heterotis amplexicaulis (Jacq. -Fél.) Aké Assi ●☆
134642 Dissotis irvingiana Hook. f. = Antherotoma irvingiana (Hook. f.) Jacq. -Fél. ■☆
134643 Dissotis irvingiana Hook. f. f. abyssinica (Gilg) A. Fern. et R. Fern. = Antherotoma irvingiana (Hook. f.) Jacq. -Fél. ■☆
134644 Dissotis irvingiana Hook. f. f. alpestris ? = Antherotoma senegambiensis (Guillaumin et Perr.) Jacq. -Fél. var. alpestris Taub. ●☆
134645 Dissotis irvingiana Hook. f. f. osbeckioides A. Fern. et R. Fern. = Antherotoma senegambiensis (Guillaumin et Perr.) Jacq. -Fél. var. alpestris Taub. ●☆
134646 Dissotis jacquesii A. Chev. = Heterotis jacquesii (A. Chev.) Aké Assi ●☆
134647 Dissotis johnstoniana Baker f.;约翰斯顿异荣耀木●☆
134648 Dissotis johnstoniana Baker f. var. strigosa Brenan;糙伏毛异荣耀木●☆
134649 Dissotis kassneri Gilg ex De Wild. = Dissotis peregrina A. Fern. et R. Fern. ●☆
134650 Dissotis kassneriana Kraenzl. = Antherotoma senegambiensis (Guillaumin et Perr.) Jacq. -Fél. ●☆
134651 Dissotis kerstingii Gilg ex Engl.;克斯廷异荣耀木●☆
134652 Dissotis kundelungensis De Wild. = Antherotoma naudinii Hook. f. ■☆
134653 Dissotis laevis (Benth.) Hook. f. = Heterotis decumbens (P. Beauv.) Jacq. -Fél. ●☆
134654 Dissotis lambii Hutch. = Dissotis grandiflora (Sm.) Benth. var. lambii (Hutch.) Keay ●☆
134655 Dissotis lanata A. Fern. et R. Fern.;绵毛异荣耀木●☆
134656 Dissotis lanceolata Cogn. = Antherotoma debilis (Sond.) Jacq. -Fél. ■☆
134657 Dissotis lebrunii (Robyns et Lawalrée) A. Fern. et R. Fern.;勒布伦异荣耀木●☆
134658 Dissotis lecomteana Hutch. et Dalziel = Melastomastrum afzelii (Hook. f.) A. Fern. et R. Fern. var. lecomteanum (Hutch. et Dalziel) Jacq. -Fél. ●☆
134659 Dissotis leonensis Hutch. et Dalziel;莱昂异荣耀木●☆
134660 Dissotis linearis Jacq. -Fél.;线状异荣耀木●☆
134661 Dissotis loandensis Exell;罗安达异荣耀木●☆
134662 Dissotis longicaudata Cogn.;长尾异荣耀木●☆
134663 Dissotis longisepala A. Fern. et R. Fern.;长萼异荣耀木●☆
134664 Dissotis longisetosa Gilg et Ledermann ex Engl.;长刚毛异荣耀

木●☆

134665 Dissotis louisii Jacq. -Fél. ;路易斯异荣耀木●☆

134666 Dissotis luxenii (De Wild.) A. Fern. et R. Fern. ;卢森异荣耀木●☆

134667 Dissotis macrocarpa Gilg = Dissotis speciosa Taub. ●☆

134668 Dissotis mahonii Hook. f. ;马洪异荣耀木●☆

134669 Dissotis melleri Hook. f. ;梅勒异荣耀木●☆

134670 Dissotis melleri Hook. f. var. greenwayi (A. Fern. et R. Fern.) A. Fern. et R. Fern. ;格林威异荣耀木●☆

134671 Dissotis mildbraedii Gilg = Dissotis trothae Gilg ●☆

134672 Dissotis minor Gilg = Melastomastrum segregatum (Benth.) A. Fern. et R. Fern. ●☆

134673 Dissotis mirabilis Bullock = Dissotis speciosa Taub. ●☆

134674 Dissotis modesta Stapf = Heterotis decumbens (P. Beauv.) Jacq. -Fél. ●☆

134675 Dissotis muenzneri Engl. = Dissotis princeps (Kunth) Triana var. candolleana (Cogn.) A. Fern. et R. Fern. ●☆

134676 Dissotis multiflora (Sm.) Triana;多花异荣耀木●☆

134677 Dissotis pachytricha R. E. Fr. ;粗毛异荣耀木●☆

134678 Dissotis pachytricha R. E. Fr. var. grandisquamulosa Wickens;大鳞粗毛异荣耀木●☆

134679 Dissotis pachytricha R. E. Fr. var. orientalis A. Fern. et R. Fern. ;东方粗毛异荣耀木●☆

134680 Dissotis paludosa Gilg = Antherotoma debilis (Sond.) Jacq. -Fél. ■☆

134681 Dissotis paucistellata Stapf = Melastomastrum afzelii (Hook. f.) A. Fern. et R. Fern. var. paucistellatum (Stapf) Jacq. -Fél. ●☆

134682 Dissotis pauwelsii Jacq. -Fél. ;保韦尔斯异荣耀木●☆

134683 Dissotis penicillata Gilg = Antherotoma debilis (Sond.) Jacq. -Fél. ■☆

134684 Dissotis peregrina A. Fern. et R. Fern. ;外来异荣耀木●☆

134685 Dissotis perkinsiae Gilg;珀金斯异荣耀木●☆

134686 Dissotis petiolata Hook. f. = Melastomastrum capitatum (Vahl) A. Fern. et R. Fern. ●☆

134687 Dissotis phaeotricha (Hochst.) Hook. f. = Antherotoma phaeotricha (Hochst.) Jacq. -Fél. ■☆

134688 Dissotis phaeotricha (Hochst.) Hook. f. var. hirsuta (Cogn.) A. Fern. et R. Fern. = Antherotoma phaeotricha (Hochst.) Jacq. -Fél. ■☆

134689 Dissotis phaeotricha (Hochst.) Hook. f. var. villosissima A. Fern. et R. Fern. = Antherotoma phaeotricha (Hochst.) Jacq. -Fél. ■☆

134690 Dissotis plumosa (D. Don) Hook. f. = Heterotis rotundifolia (Sm.) Jacq. -Fél. ●☆

134691 Dissotis pobeguinii Hutch. et Dalziel = Heterotis pobeguinii (Hutch. et Dalziel) Jacq. -Fél. ●☆

134692 Dissotis polyantha Gilg;繁花异荣耀木●☆

134693 Dissotis princeps (Kunth) Triana var. candolleana (Cogn.) A. Fern. et R. Fern. ;康氏大花异荣耀木●☆

134694 Dissotis princeps Triana = Dissotis princeps (Kunth) Triana ●☆

134695 Dissotis procumbens A. Fern. et R. Fern. ;平铺异荣耀木●☆

134696 Dissotis prostrata (Thonn.) Hook. f. = Heterotis prostrata (Thonn.) Benth. ●☆

134697 Dissotis pterocaulos Wickens;翅茎异荣耀木●☆

134698 Dissotis pulcherrima Gilg = Dissotis thollonii Cogn. ex Büttner ●☆

134699 Dissotis pulchra A. Fern. et R. Fern. ;美花异荣耀木●☆

134700 Dissotis pusilla R. E. Fr. = Antherotoma debilis (Sond.) Jacq. -Fél. ■☆

134701 Dissotis pygmaea A. Chev. et Jacq. -Fél. = Heterotis pygmaea (A. Chev. et Jacq. -Fél.) Jacq. -Fél. ●☆

134702 Dissotis quinquenervis De Wild. ;五脉异荣耀木●☆

134703 Dissotis radicans Hook. f. = Melastomastrum capitatum (Vahl) A. Fern. et R. Fern. ●☆

134704 Dissotis rhinanthifolia (Brenan) A. Fern. et R. Fern. var. exellii A. Fern. et R. Fern. ;埃克塞尔异荣耀木●☆

134705 Dissotis riparia Gilg et Ledermann ex Engl. ;河岸异荣耀木●☆

134706 Dissotis romiana De Wild. ;罗米异荣耀木●☆

134707 Dissotis rotundifolia (Sm.) Triana;圆叶异荣耀木;Pinklady ●☆

134708 Dissotis rotundifolia (Sm.) Triana f. buettneriana (Cogn. ex Büttner) A. Fern. et R. Fern. = Heterotis buettneriana (Cogn. ex Büttner) Jacq. -Fél. ●☆

134709 Dissotis rotundifolia (Sm.) Triana var. fruticosa Brenan = Dissotis fruticosa (Brenan) Brenan et Keay ●☆

134710 Dissotis rotundifolia (Sm.) Triana var. prostrata (Thonn.) Jacq. -Fél. = Heterotis prostrata (Thonn.) Benth. ●☆

134711 Dissotis ruandensis Engl. ;卢旺达异荣耀木●☆

134712 Dissotis rupicola Gilg ex Engl. = Heterotis rupicola (Gilg ex Engl.) Jacq. -Fél. ●☆

134713 Dissotis rupicola Hutch. et Dalziel = Heterotis rupicola (Gilg ex Engl.) Jacq. -Fél. ●☆

134714 Dissotis scabra Gilg = Dissotis perkinsiae Gilg ●☆

134715 Dissotis schliebenii Markgr. = Heterotis prostrata (Thonn.) Benth. ●☆

134716 Dissotis schweinfurthii Gilg = Dissotis perkinsiae Gilg ●☆

134717 Dissotis segregata (Benth.) Hook. f. = Melastomastrum segregatum (Benth.) A. Fern. et R. Fern. ●☆

134718 Dissotis senegambiensis (Guillaumin et Perr.) Triana = Antherotoma senegambiensis (Guillaumin et Perr.) Jacq. -Fél. ●☆

134719 Dissotis senegambiensis (Guillaumin et Perr.) Triana var. alpestris (Taub.) A. Fern. et R. Fern. = Antherotoma senegambiensis (Guillaumin et Perr.) Jacq. -Fél. var. alpestris Taub. ●☆

134720 Dissotis senegambiensis Triana = Antherotoma senegambiensis (Guillaumin et Perr.) Jacq. -Fél. ●☆

134721 Dissotis seretii De Wild. = Heterotis seretii (De Wild.) Jacq. -Fél. ●☆

134722 Dissotis seretii De Wild. var. gracilifolia Wickens = Heterotis seretii (De Wild.) Jacq. -Fél. ●☆

134723 Dissotis sessili-cordata Wickens;无柄心形异荣耀木●☆

134724 Dissotis sessilis Hutch. ex Brenan et Keay;无柄异荣耀木●☆

134725 Dissotis simonis-jamesii Buscal. et Muschl. ;西詹异荣耀木●☆

134726 Dissotis sizenandii Cogn. var. brevipilosa A. Fern. et R. Fern. ;短毛异荣耀木●☆

134727 Dissotis sizenandii Cogn. var. longifolia A. Fern. et R. Fern. ;长叶异荣耀木●☆

134728 Dissotis speciosa Taub. ;艳丽异荣耀木●☆

134729 Dissotis spectabilis Gilg = Dissotis cryptantha Baker f. ●☆

134730 Dissotis splendens A. Chev. et Jacq. -Fél. ;光亮异荣耀木●☆

134731 Dissotis swynnertonii (Baker f.) A. Fern. et R. Fern. ;斯温纳顿异荣耀木●☆

134732 Dissotis sylvestris Jacq. -Fél. = Heterotis sylvestris (Jacq. -Fél.) Jacq. -Fél. ●☆

134733 Dissotis talbotii Baker f. = Tristemma oreophilum Gilg ●☆

134734 Dissotis tanganyikae Kraenzl. = Dissotis brazzae Cogn. ●☆

134735 Dissotis tenuis A. Fern. et R. Fern. = Antherotoma tenuis (A.

Fern. et R. Fern.) Jacq. -Fél. ●☆
134736 Dissotis theifolia (G. Don) Hook. f. = Melastomastrum theifolium (G. Don) A. Fern. et R. Fern. ●☆
134737 Dissotis thollonii Cogn. ex Büttner;托伦异荣耀木●☆
134738 Dissotis thollonii Cogn. ex Büttner var. elliotii (Gilg) Jacq. -Fél. ;埃利异荣耀木●☆
134739 Dissotis tisserantii Jacq. -Fél. = Antherotoma tisserantii (Jacq. -Fél.) Jacq. -Fél. ●☆
134740 Dissotis tristemmoides Cogn. = Dissotis multiflora (Sm.) Triana ●☆
134741 Dissotis trothae Gilg;特罗塔异荣耀木●☆
134742 Dissotis tubulosa (Sm.) Triana = Osbeckia tubulosa Sm. ●☆
134743 Dissotis venulosa Hutch. = Dissotis caloneura Engl. ●☆
134744 Dissotis verdickii De Wild. = Dissotis falcipila Gilg ●☆
134745 Dissotis verticillata De Wild. = Dissotis princeps (Kunth) Triana ●☆
134746 Dissotis villosa Hook. f. = Antherotoma phaeotricha (Hochst.) Jacq. -Fél. ■☆
134747 Dissotis welwitschii Cogn. ;韦尔异荣耀木●☆
134748 Dissotis whytei Baker = Dissotis melleri Hook. f. ●☆
134749 Dissotis wildei Jacq. -Fél. ;维尔德异荣耀木●☆
134750 Dissotis wildemaniana Gilg ex Engl. = Dissotis gilgiana De Wild. ●☆
134751 Distandra Link = Disandra L. ■☆
134752 Distandra Link = Sibthorpia L. ■☆
134753 Distasis DC. = Chaetopappa DC. ■☆
134754 Distasis concinna Hook. et Arn. = Erigeron concinnus (Hook. et Arn.) Torr. et A. Gray ■☆
134755 Distasis pumilus Nutt. var. concinnus (Hook. et Arn.) Dorn = Erigeron concinnus (Hook. et Arn.) Torr. et A. Gray ■☆
134756 Disteganthus Lem. (1847);离花凤梨属(菊状花属,卧花凤梨属)■☆
134757 Disteganthus basi-lateralis Lem. ;离花凤梨■☆
134758 Distegia Klatt = Didelta L'Hér. (保留属名)■☆
134759 Distegia Raf. = Lonicera L. ●■
134760 Distegia acida Klatt = Didelta carnosa (L. f.) W. T. Aiton ■☆
134761 Distegia involucrata (Richardson) Cockerell = Lonicera involucrata (Rich.) Banks ex Spreng. ●☆
134762 Distegocarpus Siebold et Zucc. = Carpinus L. ●
134763 Distegocarpus carpinoides Siebold et Zucc. = Carpinus carpinoides Makino ●
134764 Distegocarpus carpinoides Siebold et Zucc. = Carpinus japonica Blume ●☆
134765 Distegocarpus cordata A. DC. = Carpinus cordata Blume ●
134766 Distegocarpus erosa A. DC. = Carpinus cordata Blume ●
134767 Distegocarpus laxiflora Siebold et Zucc. = Carpinus laxiflora (Siebold et Zucc.) Blume ●☆
134768 Disteira Raf. = Martynia L. ■
134769 Distemma Lem. = Disemma Labill. ●■
134770 Distemma Lem. = Passiflora L. ●■
134771 Distemon Bouché = Canna L. ■
134772 Distemon Ehrenb. ex Asch. = Anticharis Endl. ■●☆
134773 Distemon Wedd. = Neodistemon Babu et A. N. Henry ■☆
134774 Distemon campanularis Ehrenb. et Hempr. ex Asch. = Anticharis arabica Endl. ■☆
134775 Distemonanthus Benth. (1865);双蕊苏木属●☆
134776 Distemonanthus benthamianus Baill. ;双蕊苏木(阿胭,两蕊苏木,尼日利亚椴木,尼日利亚双蕊苏木);Anyaran, Ayan, Bongassi, Bonsamdua, Movingue, Nigerian Satin Wood, Nigerian Satinwood ●☆
134777 Distephana (DC.) Juss. = Passiflora L. ●■
134778 Distephana (DC.) Juss. ex M. Roem. = Passiflora L. ●■
134779 Distephana (DC.) M. Roem. = Passiflora L. ●■
134780 Distephana Juss. = Distephana Juss. ex M. Roem. ●■
134781 Distephana Juss. ex M. Roem. = Passiflora L. ●■
134782 Distephana Juss. ex M. Roem. = Tacsonia Juss. ●■
134783 Distephania Gagnep. = Indosinia J. E. Vidal ●☆
134784 Distephania Steud. = Distephana Juss. ex M. Roem. ●■
134785 Distephanus (Cass.) Cass. (1817);黄鸠菊属●■
134786 Distephanus (Cass.) Cass. = Vernonia Schreb. (保留属名)●■
134787 Distephanus Cass. = Distephanus (Cass.) Cass. ●■
134788 Distephanus Cass. = Vernonia Schreb. (保留属名)●■
134789 Distephanus angolensis (O. Hoffm.) H. Rob. et B. Kahn;安哥拉黄鸠菊●☆
134790 Distephanus angulifolius (DC.) H. Rob. et B. Kahn;窄叶黄鸠菊■☆
134791 Distephanus anisochaetoides (Sond.) H. Rob. et B. Kahn;异毛黄鸠菊■☆
134792 Distephanus antandroyi (Humbert) H. Rob. et B. Kahn;安坦德罗黄鸠菊●☆
134793 Distephanus cloiselii (S. Moore) H. Rob. et B. Kahn;克卢塞尔黄鸠菊●☆
134794 Distephanus divaricatus (Steetz) H. Rob. et B. Kahn;叉开黄鸠菊●☆
134795 Distephanus eriophyllus (Drake) H. Rob. et B. Kahn;毛叶黄鸠菊●☆
134796 Distephanus forrestii (Anthony) H. Rob. et B. Kahn;滇西黄鸠菊●☆
134797 Distephanus garnierianus (Klatt) H. Rob. et B. Kahn;嘎尔黄鸠菊●☆
134798 Distephanus glandulicinctus (Humbert) H. Rob. et B. Kahn;腺带黄鸠菊●☆
134799 Distephanus glutinosus (DC.) H. Rob. et B. Kahn;黏性黄鸠菊●☆
134800 Distephanus henryi (Dunn) H. Rob. ;亨利黄鸠菊●☆
134801 Distephanus lastellei (Drake) H. Rob. et B. Kahn;拉斯泰勒黄鸠菊●☆
134802 Distephanus mahafaly (Humbert) H. Rob. et B. Kahn;马哈法里黄鸠菊●☆
134803 Distephanus manambolensis (Humbert) H. Rob. et B. Kahn;马南布卢黄鸠菊●☆
134804 Distephanus mangokensis (Humbert) H. Rob. et B. Kahn;曼戈基黄鸠菊●☆
134805 Distephanus nummulariifolius (Klatt) H. Rob. et B. Kahn;铜钱叶黄鸠菊●☆
134806 Distephanus ochroleucus (Baker) H. Rob. et B. Kahn;白绿黄鸠菊●☆
134807 Distephanus plumosus (O. Hoffm.) Mesfin;羽状黄鸠菊●☆
134808 Distephanus polygalifolius (Less.) H. Rob. et B. Kahn;远志叶黄鸠菊●☆
134809 Distephanus rochonioides (Humbert) H. Rob. et B. Kahn;绒菀木黄鸠菊●☆
134810 Distephanus streptocladus (Baker) H. Rob. et B. Kahn;扭枝黄鸠菊●☆
134811 Distephanus subluteus (Scott-Elliot) H. Rob. et B. Kahn;淡黄

黄鸠菊●☆

134812 Distephanus swinglei (Humbert) H. Rob. et B. Kahn;斯温格尔黄鸠菊●☆

134813 Distephanus trinervis Bojer ex DC.;三脉黄鸠菊●☆

134814 Distephia Salisb. ex DC. = Distephana (DC.) M. Roem. ●■

134815 Distephia Salisb. ex DC. = Passiflora L. ●■

134816 Disterepta Raf. = Cassia L.(保留属名)●■

134817 Disterigma (Klotzsch) Nied. (1889);双柱杜鹃属(拟越橘属)●☆

134818 Disterigma Nied. = Disterigma (Klotzsch) Nied. ●☆

134819 Disterigma Nied. ex Drude = Disterigma (Klotzsch) Nied. ●☆

134820 Disterigma acuminatum Nied.;尖双柱杜鹃●☆

134821 Disterigma micranthum A. C. Sm.;小花双柱杜鹃●☆

134822 Disterigma ovatum (Rusby) S. F. Blake;卵双柱杜鹃●☆

134823 Disterigma pachyphyllum S. F. Blake;厚叶双柱杜鹃●☆

134824 Disterigma pallidum A. C. Sm.;苍白双柱杜鹃●☆

134825 Disterigma pilosum Wilbur;毛双柱杜鹃●☆

134826 Distetraceae Dulac = Thymelaea Mill.(保留属名)●■

134827 Distiacanthus Baker = Karatas Mill. ■☆

134828 Distiacanthus Linden = Karatas Mill. ■☆

134829 Disticheia Ehrh. = Brachypodium P. Beauv. ■

134830 Disticheia Ehrh. = Bromus L.(保留属名)■

134831 Distichella Tiegh. = Dendrophthora Eichler ●☆

134832 Distichia Nees et Meyen(1843);双列灯芯草属■☆

134833 Distichia macrocarpa Wedd. ex Buchenau;大果双列灯芯草■☆

134834 Distichia muscoides Nees et Meyen;双列灯芯草■☆

134835 Distichirhops Haegens(2000);新婆大戟属●☆

134836 Distichis Lindl. = Liparis Rich.(保留属名)■

134837 Distichis Thouars = Liparis Rich.(保留属名)■

134838 Distichis Thouars = Malaxis Sol. ex Sw. ■

134839 Distichis Thouars ex Lindl. = Liparis Rich.(保留属名)■

134840 Distichlis Raf. (1819);盐草属;Alkali Grass ■☆

134841 Distichlis apical (L.) Greene;顶生盐草■☆

134842 Distichlis spicata (L.) Greene;穗花盐草;Alkali Grass, Coastal Salt Grass, Inland Salt Grass, Salt Grass, Spike Grass, Spike-grass ■☆

134843 Distichlis spicata (L.) Greene subsp. stricta (Torr.) Thorne = Distichlis spicata (L.) Greene ■☆

134844 Distichlis spicata (L.) Greene var. borealis (J. Presl) Beetle = Distichlis spicata (L.) Greene ■☆

134845 Distichlis spicata (L.) Greene var. divaricata Beetle = Distichlis spicata (L.) Greene ■☆

134846 Distichlis spicata (L.) Greene var. nana Beetle = Distichlis spicata (L.) Greene ■☆

134847 Distichlis spicata (L.) Greene var. stolonifera Beetle = Distichlis spicata (L.) Greene ■☆

134848 Distichlis spicata (L.) Greene var. stricta (Torr.) Beetle = Distichlis spicata (L.) Greene ■☆

134849 Distichlis stricta (Torr.) Rydb. = Distichlis spicata (L.) Greene ■☆

134850 Distichlis stricta (Torr.) Rydb. var. dentata (Rydb.) C. L. Hitchc. = Distichlis spicata (L.) Greene ■☆

134851 Distichlis sudanensis Beetle = Coelachyrum lagopoides (Burm. f.) Senaratna ■☆

134852 Distichmus Endl. = Distichlis Raf. ■☆

134853 Distichmus Endl. = Scirpus L.(保留属名)■

134854 Distichocalyx Benth. = Dischistocalyx T. Anderson ex Benth. et Hook. f. ●☆

134855 Distichocalyx T. Anderson ex Benth. et Hook. f. = Dischistocalyx T. Anderson ex Benth. et Hook. f. ●☆

134856 Distichochlamys M. F. Newman(1995);二列姜属■☆

134857 Distichochlamys citrea M. F. Newman;二列姜■☆

134858 Distichochlamys orlowii K. Larsen et M. F. Newman;奥尔二列姜■☆

134859 Disticholiparis Marg. et Szlach. (2004);二列羊耳蒜属■☆

134860 Disticholiparis Marg. et Szlach. = Liparis Rich.(保留属名)■

134861 Distichorchis M. A. Clem. et D. L. Jones = Dendrobium Sw.(保留属名)■

134862 Distichorchis ellipsophylla (Ts. Tang et F. T. Wang) M. A. Clem. = Dendrobium ellipsophyllum Ts. Tang et F. T. Wang ■

134863 Distichoselinum Garcia Mart. et Silvestre(1983);二列芹属■☆

134864 Distichoselinum ternifolium (Lag.) Garcia Martin et Silvestre;二列芹■☆

134865 Distichostemon F. Muell. (1857);二列蕊属●☆

134866 Distichostemon phyllopterus F. Muell.;二列蕊●☆

134867 Distictella Kuntze(1903);小红钟藤属●☆

134868 Distictella angustifolia Urb.;窄叶小红钟藤●☆

134869 Distictella cuneifolia (DC.) Sandwith;楔叶小红钟藤●☆

134870 Distictella dasytricha Sandwith;粗毛小红钟藤●☆

134871 Distictella lutescens Freire et A. Samp.;浅黄小红钟藤●☆

134872 Distictella obovata Sandwith;倒卵小红钟藤●☆

134873 Distictella reticulata A. H. Gentry;网脉小红钟藤●☆

134874 Distictis Bureau = Distictella Kuntze ●☆

134875 Distictis Mart. ex DC. = Distictis Mart. ex Meisn. ●☆

134876 Distictis Mart. ex Meisn. (1840);红钟藤属(红钟花属)●☆

134877 Distictis buccinatoria (DC.) A. H. Gentry;红钟藤(红钟花);Mexican Blood Flower, Scarlet Trumpet Vine ●☆

134878 Distictis laxiflora Greenm.;疏花红钟藤;Vanilla Trumpet Vine ●☆

134879 Distigocarpus Sargent = Carpinus L. ●

134880 Distigocarpus Sargent = Distegocarpus Siebold et Zucc. ●

134881 Distimake Raf. = Ipomoea L.(保留属名)●■

134882 Distimum Steud. = Distimus Raf. ■

134883 Distimus Raf. = Cyperus L. ■

134884 Distimus Raf. = Pycreus P. Beauv. ■

134885 Distira Post et Kuntze = Disteira Raf. ■

134886 Distira Post et Kuntze = Martynia L. ■

134887 Distixila Raf. = Myrtus L. ●

134888 Distoecha Phil. = Hypochaeris L. ■

134889 Distomaea Spenn. = Listera R. Br.(保留属名)■

134890 Distomaea Spenn. = Neottia Guett.(保留属名)■

134891 Distomaea cordata (L.) Spenn. = Listera cordata (L.) R. Br. ■☆

134892 Distomaea ovata (L.) Spenn. = Listera ovata R. Br. ■

134893 Distomanthera Turcz. (1862);双口虎耳草属☆

134894 Distomischus Dulac = Festuca L. ■

134895 Distomocarpus O. E. Schulz = Rytidocarpus Coss. ■☆

134896 Distomocarpus maroccanus O. E. Schulz = Rytidocarpus maroccanus (O. E. Schulz) Maire ■☆

134897 Distomomischus Dulac = Vulpia C. C. Gmel. ■

134898 Distrepta Miers = Tecophilaea Bertero ex Colla ■☆

134899 Distreptus Cass. = Elephantopus L. ■

134900 Distreptus Cass. = Pseudelephantopus Rohr(保留属名)■

134901 Distreptus spicatus (Juss. ex Aubl.) Cass. = Pseudelephantopus spicatus (Juss. ex Aubl.) Gleason ■

134902 Distrianthes Danser(1929);二畦花属●☆

134903 Distrianthes lamii Danser;二畦花●☆

134904 Distrianthes molliflora Danser;软叶二畦花●☆

134905 Distyliopsis P. K. Endress = Sycopsis Oliv. ●

134906 Distyliopsis P. K. Endress(1970);假蚊母树属(类蚊母树属,类蚊母属)●

134907 Distyliopsis dunnii (Hemsl.) P. K. Endress;尖叶假蚊母树(达代拟蚊母树,假蚊母,尖水丝梨,尖叶水丝梨);Sharpleaf Fighazel, Sharp-leaved Fighazel ●

134908 Distyliopsis dunnii (Hemsl.) P. K. Endress = Sycopsis dunnii Hemsl. ●

134909 Distyliopsis laurifolia (Hemsl.) P. K. Endress;樟叶假蚊母树(樟叶假蚊母,樟叶水丝梨);Laurelleaf Fighazel, Laurel-leaved Fighazel ●

134910 Distyliopsis laurifolia (Hemsl.) P. K. Endress = Sycopsis laurifolia Hemsl. ●

134911 Distyliopsis salicifolia (H. L. Li) P. K. Endress;柳叶假蚊母树(柳叶水丝梨);Willowleaf Fighazel, Willow-leaved Fighazel ●

134912 Distyliopsis salicifolia (H. L. Li) P. K. Endress = Sycopsis salicifolia H. L. Li ●

134913 Distyliopsis tutcheri (Hemsl.) P. K. Endress;钝叶假蚊母树(钝叶水丝梨);Obtuseleaf Fighazel, Tutcher Fighazel ●

134914 Distyliopsis tutcheri (Hemsl.) P. K. Endress = Sycopsis tutcheri Hemsl. ●

134915 Distyliopsis yunnanensis (Hung T. Chang) C. Y. Wu;滇假蚊母树(滇假蚊母,滇水丝梨,云南水丝梨);Yunnan Fighazel ●

134916 Distyliopsis yunnanensis (Hung T. Chang) C. Y. Wu = Sycopsis yunnanensis Hung T. Chang ●

134917 Distylis Gaudich. = Calogyne R. Br. ■

134918 Distylium Siebold et Zucc. (1841);蚊母树属;Distylium, Mosquitoman ●

134919 Distylium buxifolium (Hance) Merr.;小叶蚊母树(石头棵子);Boxleaf Distylium, Boxleaf Mosquitoman, Box-leaved Distylium ●

134920 Distylium buxifolium (Hance) Merr. var. rotundum C. C. Chang;圆头蚊母树;Round Boxleaf Distylium, Roundhead Mosquitoman ●

134921 Distylium buxifolium (Hance) Merr. var. rotundum Hung T. Chang = Distylium buxifolium (Hance) Merr. ●

134922 Distylium chinense (Franch. ex Hemsl.) Diels;中华蚊母树(川鄂蚊母树,河边蚊母树,石头棵子,水杨柳,蚊母树);China Mosquitoman, Chinese Distylium ●

134923 Distylium chinense (Franch.) Diels = Distylium chinense (Franch. ex Hemsl.) Diels ●

134924 Distylium chinense Rehder = Distylium dunnianum H. Lév. ●

134925 Distylium chinense Rehder et E. H. Wilson = Distylium buxifolium (Hance) Merr. ●

134926 Distylium chingii Chun ex Walker = Eustigma balansae Oliv. ●

134927 Distylium chungii (F. P. Metcalf) W. C. Cheng;闽粤蚊母树;Chung Distylium, Chung Mosquitoman ●

134928 Distylium cuspidatum Hung T. Chang;尖尾蚊母树;Cuspidate Distylium, Cuspidate Mosquitoman ●

134929 Distylium dunnianum H. Lév.;窄叶蚊母树(大叶浆木,狭叶蚊母树);Dunn Distylium, Dunn Mosquitoman ●

134930 Distylium elaeagnoides Hung T. Chang;鳞毛蚊母树;Elaeagnuslike Distylium, Elaeagnus-like Distylium, Elaeagnuslike Mosquitoman ●

134931 Distylium formosanum Kaneh. = Sycopsis sinensis Oliv. ●

134932 Distylium gracile Nakai;台湾蚊母树(细叶蚊母树,小叶蚊母树);Small-leaved Distylium, Taiwan Distylium, Taiwan Mosquitoman ●

134933 Distylium lanceolatum Chun ex W. C. Cheng = Distylium dunnianum H. Lév. ●

134934 Distylium lepidotum Nakai;具鳞蚊母树●☆

134935 Distylium lipoense Y. K. Li et X. M. Wang;荔波蚊母树;Libo Distylium, Libo Mosquitoman ●

134936 Distylium lipoense Y. K. Li et X. M. Wang = Distylium buxifolium (Hance) Merr. ●

134937 Distylium macrophyllum Hung T. Chang;大叶蚊母树;Largeleaf Distylium, Largeleaf Mosquitoman, Large-leaved Distylium ●

134938 Distylium myricoides Hemsl.;杨梅叶蚊母树(大果萍柴,夹心,瓢柴,萍柴,挺香,杨梅蚊母树,野茶);Myrica-like Distylium, Myrica-like Mosquitoman ●

134939 Distylium myricoides Hemsl. var. macrocarpum C. Y. Wu;大果萍柴;Big-fruit Myrica-like Distylium, Big-fruit Myrica-like Mosquitoman ●

134940 Distylium myricoides Hemsl. var. nitidum Hung T. Chang;亮叶蚊母树;Shiningleaf Distylium, Shiningleaf Mosquitoman ●

134941 Distylium myricoides Hemsl. var. nitidum Hung T. Chang = Distylium myricoides Hemsl. ●

134942 Distylium pingpienense (Hu) Walker;屏边蚊母树;Pingbian Distylium, Pingbian Mosquitoman ●

134943 Distylium pingpienense (Hu) Walker var. serratum Walker;锯齿蚊母树(齿叶蚊母树);Serrate Distylium, Serrate Mosquitoman ●

134944 Distylium pingpienense (Hu) Walker var. serratum Walker = Distylium pingpienense (Hu) Walker ●

134945 Distylium racemosum Matsum. et Hayata = Sycopsis sinensis Oliv. ●

134946 Distylium racemosum Siebold et Zucc.;蚊母树(米心树,蚊子树);Flowers Acemes Distylium, Isu Tree, Racemose Distylium, Racemose Mosquitoman ●

134947 Distylium racemosum Siebold et Zucc. f. angustifolium (Masam.) H. Ohba;狭叶蚊母树●☆

134948 Distylium racemosum Siebold et Zucc. f. pendulum (Makino) Okuyama;垂枝蚊母树;Pendulous Racemose Distylium ●☆

134949 Distylium racemosum Siebold et Zucc. var. angustifolium Masam. = Distylium racemosum Siebold et Zucc. f. angustifolium (Masam.) H. Ohba ●☆

134950 Distylium racemosum Siebold et Zucc. var. chinense Franch. ex Hemsl. = Distylium chinense (Franch. ex Hemsl.) Diels ●

134951 Distylium racemosum Siebold et Zucc. var. lepidotum (Nakai) Hatus. = Distylium lepidotum Nakai ●☆

134952 Distylium strictum Hemsl. = Distylium buxifolium (Hance) Merr. ●

134953 Distylium tsiangii Chun ex Walker;黔蚊母树;Tsiang Distylium, Tsiang Mosquitoman ●

134954 Distylium velutinum Hu = Distyliopsis laurifolia (Hemsl.) P. K. Endress ●

134955 Distylium velutinum Hu = Sycopsis laurifolia Hemsl. ●

134956 Distylodon Summerh. (1966);双齿柱兰属■☆

134957 Distylodon comptum Summerh.;双齿柱兰■☆

134958 Disynanthes Rchb. = Disynanthus Raf. ■●

134959 Disynanthus Raf. = Antennaria Gaertn. (保留属名)■●

134960 Disynapheia Steud. = Disynaphia DC. ●☆

134961 Disynaphia DC. = Disynaphia Hook. et Arn. ex DC. ●☆

134962 Disynaphia Hook. et Arn. ex DC. (1838);旋泽兰属●☆

134963 Disynia Raf. = Salix L. (保留属名) ●
134964 Disynoma Raf. = Aethionema R. Br. ■☆
134965 Disynstemon R. Vig. (1951);双合豆属■☆
134966 Disynstemon madagascariense R. Vig. = Disynstemon paullinioides (Baker) M. Pelt. ■☆
134967 Disynstemon paullinioides (Baker) M. Pelt.;双合豆■☆
134968 Disyphonia Griff. = Dysoxylum Blume ●
134969 Ditassa R. Br. (1810);双饰萝藦属●☆
134970 Ditassa acerifolia Lasser et Maguire;尖叶双饰萝藦●☆
134971 Ditassa anderssonii Morillo;安氏双饰萝藦●☆
134972 Ditassa ciliata (Moldenke) Morillo;睫毛双饰萝藦●☆
134973 Ditassa cordata (Turcz.) Fontella;心形双饰萝藦●☆
134974 Ditassa crassinervia Decne.;粗脉双饰萝藦●☆
134975 Ditassa gracilipes Schltr.;细梗双饰萝藦●☆
134976 Ditassa laevis Mart.;平滑双饰萝藦●☆
134977 Ditassa lanceolata Decne.;披针叶双饰萝藦●☆
134978 Ditassa longisepala (Hua) Fontella et E. A. Schwarz;长萼双饰萝藦●☆
134979 Ditassa mexicana Brandegee;墨西哥双饰萝藦●☆
134980 Ditassa montana Schltr.;山地双饰萝藦●☆
134981 Ditassa multinervia (Morillo) Morillo;多脉双饰萝藦●☆
134982 Ditassa nigrescens (E. Fourn.) W. D. Stevens;黑双饰萝藦●☆
134983 Ditassa nitida E. Fourn.;光亮双饰萝藦●☆
134984 Ditassa obovata Morillo;倒卵双饰萝藦●☆
134985 Ditassa rotundifolia (Decne.) K. Schum.;圆叶双饰萝藦●☆
134986 Ditassa verticillata Morillo;轮生双饰萝藦●☆
134987 Ditaxia Endl. = Celsia L. ■☆
134988 Ditaxia Endl. = Ditoxia Raf. ■☆
134989 Ditaxis Vahl ex A. Juss. = Argythamnia P. Browne ●☆
134990 Diteilis Raf. = Liparis Rich. (保留属名) ■
134991 Diteilis sootenzanensis (Fukuy.) M. A. Clem. et D. L. Jones = Liparis sootenzanensis Fukuy. ■
134992 Diteilis wrayi (Hook. f.) M. A. Clem. et D. L. Jones = Liparis barbata Lindl. ■
134993 Ditelesia Raf. = Dilasia Raf. (废弃属名) ■
134994 Ditelesia Raf. = Murdannia Royle (保留属名) ■
134995 Ditepalanthus Fagerl. (1938);马岛菰属■☆
134996 Ditepalanthus afzelii Fagerl. = Ditepalanthus malagasicus (Jum. et H. Perrier) Fagerl. ■☆
134997 Ditepalanthus malagasicus (Jum. et H. Perrier) Fagerl.;马岛菰■☆
134998 Ditereia Raf. = Evolvulus L. ●■
134999 Ditheca Miq. = Ammannia L. ■
135000 Dithecina Tiegh. = Helixanthera Lour. ●
135001 Dithecoluma Baill. = Pouteria Aubl. ●
135002 Dithrichum DC. = Ditrichum Cass. ■☆
135003 Dithrichum DC. = Verbesina L. (保留属名) ●■☆
135004 Dithrix (Hook. f.) Schltr. = Habenaria Willd. ■
135005 Dithrix Schltr. = Habenaria Willd. ■
135006 Dithyraea Endl. = Dithyrea Harv. ■☆
135007 Dithyrea Harv. (1845);奇果芥属■☆
135008 Dithyrea wislizenii Engelm;奇果芥;Spectacle Pod, Spectacle-pod ■☆
135009 Dithyrea wislizenii Engelm. = Dimorphocarpa wislizenii (Engelm.) Rollins ■☆
135010 Dithyria Benth. = Swartzia Schreb. (保留属名) ●☆
135011 Dithyridanthus Garay = Schiedeella Schltr. ■☆
135012 Dithyridanthus Garay (1982);双口兰属■☆
135013 Dithyridanthus densiflorus (C. Schweinf.) Garay;双口兰■☆
135014 Dithyrocarpus Kunth = Floscopa Lour. ■
135015 Dithyrocarpus glabrata Kunth = Floscopa confusa Brenan ■☆
135016 Dithyrostegia A. Gray = Angianthus J. C. Wendl. (保留属名) ■●☆
135017 Dithyrostegia A. Gray (1851);舟苞鼠麹草属■☆
135018 Dithyrostegia amplexicaulis A. Gray;舟苞鼠麹草■☆
135019 Dithyrostegia gracilis P. S. Short;细舟苞鼠麹草■☆
135020 Ditinnia A. Chev. = Remusatia Schott ■
135021 Ditinnia rupicola A. Chev. = Remusatia vivipara (Roxb.) Schott ■
135022 Ditmaria Spreng. = Erisma Rudge ●☆
135023 Ditoca Banks et Sol. ex Gaertn. = Scleranthus L. ■☆
135024 Ditoca Banks ex Gaertn. = Mniarum J. R. Forst. et G. Forst. ■☆
135025 Ditoca Banks ex Gaertn. = Scleranthus L. ■☆
135026 Ditomaga Raf. = Irlbachia Mart. ■☆
135027 Ditomostrophe Turcz. = Guichenotia J. Gay ●☆
135028 Ditoxia Raf. = Celsia L. ■☆
135029 Ditremexa Raf. = Cassia L. (保留属名) ●■
135030 Ditremexa marilandica (L.) Britton et Rose = Senna marilandica (L.) Link ●☆
135031 Ditremexa medsgeri (Shafer) Britton et Rose = Senna marilandica (L.) Link ●☆
135032 Ditrichospermum Bremek. = Strobilanthes Blume ●■
135033 Ditrichum Cass. = Verbesina L. (保留属名) ●■☆
135034 Ditriclita Raf. = Saxifraga L. ■
135035 Ditrisynia Raf. = Ditrysinia Raf. ●
135036 Ditrisynia Raf. = Sebastiania Spreng. ●
135037 Ditritra Raf. = Euphorbia L. ●■
135038 Ditroche E. Mey. ex Moq. = Limeum L. ■●☆
135039 Ditrysinia Raf. = Sebastiania Spreng. ●
135040 Ditta Griseb. (1860);西印度大戟属☆
135041 Ditta myricoides Griseb. et Urb.;西印度大戟☆
135042 Dittelasma Hook. f. = Sapindus L. (保留属名) ●
135043 Dittelasma rarak (DC.) Benth. et Hook. f. = Sapindus rarak DC. ●
135044 Dittelasma rarak (DC.) Hiern = Sapindus rarak DC. ●
135045 Dittoceras Hook. f. (1883);双角萝藦属●☆
135046 Dittoceras andersonii Hook. f.;双角萝藦●☆
135047 Dittoceras maculatum Kerr;斑点双角萝藦●☆
135048 Dittostigma Phil. = Nicotiana L. ●■
135049 Dittrichia Greuter (1973);臭蓬属(迪里菊属);Fleabane ■☆
135050 Dittrichia graveolens (L.) Greuter;臭蓬(香迪里菊);Stinking Fleabane, Stinkwort ■☆
135051 Dittrichia viscosa (L.) Greuter;黏蓬(黏迪里菊,黏性旋覆花);False Yellowhead, Woody Fleabane ■
135052 Dituilis Raf. = Liparis Rich. (保留属名) ■
135053 Ditulima Raf. = Dendrobium Sw. (保留属名) ■
135054 Ditulium Raf. = Diascia Link et Otto ■☆
135055 Diuranthera Hemsl. (1902);鹭鸶兰属(鹭鸶草属);Diuranthera, Egretgrass ■★
135056 Diuranthera Hemsl. = Chlorophytum Ker Gawl. ■
135057 Diuranthera chinglingensis J. Q. Xing et T. C. Cui;秦岭鹭鸶兰(秦岭鹭鸶草) ■
135058 Diuranthera inarticulata F. T. Wang et K. Y. Lang;南川鹭鸶兰(南川鹭鸶草);Nanchuan Diuranthera, Nanchuan Egretgrass ■

135059 Diuranthera major Hemsl.;鹭鸶兰(不死草,长生草,鹭鸶草,山韭菜,书带草,土洋参,野敞冬,野韭菜,野麦冬);Common Diuranthera,Common Egretgrass,Diuranthera ■

135060 Diuranthera minor C. H. Wright ex Hemsl.;小鹭鸶兰(白千针万线草,大兰花参,漏芦,鹭鸶草,鹭鸶兰,山韭菜,蛇咬药,天生草,土漏芦,土洋参,小鹭鸶草);Small Diuranthera, Small Egretgrass ■

135061 Diuratea Post et Kuntze = Diouratea Tiegh. ●

135062 Diuratea Post et Kuntze = Ouratea Aubl.(保留属名)●

135063 Diuris Sm.(1798);簇叶兰属;Diuris,Donkey Orchid ■☆

135064 Diuris alba R. Br.;白花簇叶兰;White Diuris ■☆

135065 Diuris aurea Sm.;黄花簇叶兰;Yellowflower Diuris ■☆

135066 Diuris macula Sm.;小花簇叶兰■☆

135067 Diuroglossum Turcz. = Guazuma Mill. ●☆

135068 Diurospermum Edgew. = Utricularia L. ■

135069 Diurospermum album Edgew. = Utricularia kumaonensis Oliv. ■☆

135070 Diyaminauclea Ridsdale(1979);斯里兰卡茜属●☆

135071 Diyaminauclea zeylanica (Hook. f.) Ridsdale;斯里兰卡茜●☆

135072 Dizonium Willd. ex Schltdl. = Geigeria Griess. ■●☆

135073 Dizygandra Meisn. = Ruellia L. ■●

135074 Dizygostemon (Benth.) Radlk. ex Wettst.(1891);二对蕊属■☆

135075 Dizygostemon Radlk. = Dizygostemon (Benth.) Radlk. ex Wettst. ■☆

135076 Dizygostemon Radlk. ex Wettst. = Dizygostemon (Benth.) Radlk. ex Wettst. ■☆

135077 Dizygostemon angustifolius Giul.;窄叶二对蕊■☆

135078 Dizygostemon floribundus Radlk. ex Wettst.;二对蕊■☆

135079 Dizygotheca N. E. Br.(1892);孔雀木属(假楤木属,假五加属);False Aralia,Threadleaf ●☆

135080 Dizygotheca N. E. Br. = Schefflera J. R. Forst. et G. Forst.(保留属名)●

135081 Dizygotheca elegantissima (Veitch) R. Vig. et Guillaumin;孔雀木(秀丽假五加)●☆

135082 Dizygotheca elegantissima (Veitch) R. Vig. et Guillaumin 'Castor Variegata';斑叶矮孔雀木;Angelica-tree, Aralia, False Aralia ●☆

135083 Dizygotheca veitchii N. Taylor;维奇孔雀木(维奇假五加)●☆

135084 Djaloniella P. Taylor(1963);小九子母属■☆

135085 Djaloniella ypsilostyla P. Taylor;小九子母■☆

135086 Djeratonia Pierre = Landolphia P. Beauv.(保留属名)●☆

135087 Djinga C. Cusset(1987);喀麦隆苔草属■☆

135088 Djinga felicis C. Cusset;喀麦隆苔草■☆

135089 Dobera Juss.(1789);苏丹香属●☆

135090 Dobera glabra (Forssk.) Poir.;光滑苏丹香●☆

135091 Dobera glabra (Forssk.) Poir. var. macalusoi (Mattei) Fiori = Dobera loranthifolia (Warb.) Harms ●☆

135092 Dobera glabra (Forssk.) Poir. var. subcoriacea Engl. et Gilg = Dobera loranthifolia (Warb.) Harms ●☆

135093 Dobera loranthifolia (Warb.) Harms;桑寄生花苏丹香●☆

135094 Dobera macalusoi Mattei = Dobera loranthifolia (Warb.) Harms ●☆

135095 Dobera roxburghii Planch.;苏丹香●☆

135096 Doberia Pfeiff. = Dobera Juss. ●☆

135097 Dobinea Buch. -Ham. ex D. Don(1825);九子母属(九子不离母属);Dobinea ●■

135098 Dobinea delavayi (Baill.) Baill.;九子母(大接骨,大九股牛,多槟榔,九子不离母,绿天麻,羊角天麻);Delavay Dobinea ■

135099 Dobinea vulgaris Buch. -Ham. ex D. Don;贡山九子母(贡山九子不离母);Common Dobinea ●

135100 Dobrowskya C. Presl = Monopsis Salisb. ■☆

135101 Dobrowskya laevicaulis C. Presl = Monopsis unidentata (Aiton) E. Wimm. subsp. laevicaulis (C. Presl) Phillipson ■☆

135102 Dobrowskya scabra (Thunb.) A. DC. = Monopsis scabra (Thunb.) Urb. ■☆

135103 Dobrowskya scabra (Thunb.) A. DC. var. glabrata Sond. = Monopsis unidentata (Aiton) E. Wimm. subsp. intermedia Phillipson ■☆

135104 Dobrowskya serratifolia A. DC. = Monopsis simplex (L.) E. Wimm. ■☆

135105 Dobrowskya stellarioides C. Presl = Monopsis stellarioides (C. Presl) Urb. ■☆

135106 Dobrowskya stricta C. Presl = Monopsis unidentata (Aiton) E. Wimm. ■☆

135107 Dobrowskya tenella (Thunb.) Sond. = Monopsis unidentata (Aiton) E. Wimm. ■☆

135108 Docanthe O. F. Cook = Chamaedorea Willd.(保留属名)●☆

135109 Dochafa Schott = Arisaema Mart. ●■

135110 Dochafa flava (Forssk.) Schott = Arisaema flavum (Forssk.) Schott ■

135111 Dochafa flava Schott = Arisaema flavum (Forssk.) Schott ■

135112 Dockrillia Brieger = Dendrobium Sw.(保留属名)■

135113 Docynia Decne.(1874);移栋属(多胜属,多衣果属,多衣木属,移棟属,移衣海棠属,移衣属);Docynia ●

135114 Docynia delavayi (Franch.) C. K. Schneid.;云南移栋(多衣,酸多李,酸移栋,桃梗,西南移栋,西南移棟,移栋,楂子树);Yunnan Docynia ●

135115 Docynia docynioides (C. K. Schneid.) Rehder = Docynia indica (Wall.) Decne. ●

135116 Docynia docynioides Rehder = Docynia indica (Wall.) Decne. ●

135117 Docynia doumeri (Bois) C. K. Schneid. = Malus doumeri (Bois) A. Chev. ●

135118 Docynia grifithiana Decne. = Docynia indica (Wall.) Decne. ●

135119 Docynia hookeriana Decne. = Docynia indica (Wall.) Decne. ●

135120 Docynia indica (Wall.) Decne.;移栋(红叶移栋,红叶移棟,印度移衣);India Docynia,Indian Docynia ●

135121 Docynia indica (Wall.) Decne. var. doumeri (Bois) A. Chev. = Malus doumeri (Bois) A. Chev. ●

135122 Docynia indica (Wall.) Decne. var. laosensis (Cardot) A. Chev. = Malus doumeri (Bois) A. Chev. ●

135123 Docynia indica (Wall.) Decne. var. laosensis A. Chev. = Malus doumeri (Bois) A. Chev. ●

135124 Docynia rufifolia (H. Lév.) Rehder = Docynia indica (Wall.) Decne. ●

135125 Docyniopsis (C. K. Schneid.) Koidz.(1934);拟移栋属●☆

135126 Docyniopsis (C. K. Schneid.) Koidz. = Malus Mill. ●

135127 Docyniopsis (C. K. Schneid.) Koidz. = Micromeles Decne. ●☆

135128 Docyniopsis prattii (Hemsl.) Koidz. = Malus prattii (Hemsl.) C. K. Schneid. ●

135129 Docyniopsis prattii Koidz. = Malus prattii (Hemsl.) C. K. Schneid. ●

135130 Docyniopsis yunnanensis (Franch.) Koidz. = Malus yunnanensis (Franch.) C. K. Schneid. ●

135131 Docyniopsis yunnanensis Koidz. = Malus yunnanensis (Franch.) C. K. Schneid. ●

135132 Dodartia L.(1753);野胡麻属(斗达草属,多德草属,野胡椒属);Dodartia ■
135133 Dodartia indica L. = Lindenbergia indica (L.) Vatke ■☆
135134 Dodartia orientalis L.;野胡麻(刺儿草,倒打草,倒爪草,道爪草,斗达草,多德草,牛哈水,牛含水,牛汉水,牛汗水,紫花草,紫花秧);Oriental Dodartia ■
135135 Dodecadenia Nees = Litsea Lam.(保留属名)●
135136 Dodecadenia Nees(1831);单花木姜子属(大花檀属);Monoflower,Oneflower ●
135137 Dodecadenia grandiflora Nees;单花木姜子(大花单花木姜子,大花木姜子);Monoflower,Oneflower Litse,Uniflorous Litse ●
135138 Dodecadenia grandiflora Nees var. griffithii (Hook. f.) D. G. Long;无毛单花木姜子●
135139 Dodecadenia griffithii Hook. f. = Dodecadenia grandiflora Nees var. griffithii (Hook. f.) D. G. Long ●
135140 Dodecadenia paniculata Hook. f. = Litsea liyuyingi H. Liu ●
135141 Dodecadenia robusta Zoll. et Moritzi = Litsea liyuyingi H. Liu ●
135142 Dodecadia Lour. = Lauro-Cerasus Duhamel ●
135143 Dodecadia Lour. = Pygeum Gaertn. ●
135144 Dodecahema Reveal et Hardham(1989);加州刺苞蓼属;Spinyherb ■☆
135145 Dodecahema leptoceras (A. Gray) Reveal et Hardham;加州刺苞蓼;Ramona Spineflower ■☆
135146 Dodecas L. f. = Crenea Aubl. ●☆
135147 Dodecasperma Raf. = Bomarea Mirb. ■☆
135148 Dodecaspermum Forst. ex Scop. = Decaspermum J. R. Forst. et G. Forst. ●
135149 Dodecastemon Hassk. = Drypetes Vahl ●
135150 Dodecastigma Ducke(1932);十二戟属☆
135151 Dodecastigma amazonicum Ducke;十二戟☆
135152 Dodecastigma integrifolium (Lanj.) Lanj. et Sandwith;全缘十二戟☆
135153 Dodecatheon L.(1753);流星花属(十二花属,十二神属);America Cow Slip,American Cowslip,Shooting Star,Shooting-star ■☆
135154 Dodecatheon alpinum (A. Gray) Greene;高山流星花;Alpine Shooting Star ■☆
135155 Dodecatheon amethystinum (Fassett) Fassett;紫晶流星花(紫罗兰流星花);Amethyst Shooting Star,Jeweled Shooting-star ■☆
135156 Dodecatheon amethystinum (Fassett) Fassett f. margaritaceum Fassett = Dodecatheon amethystinum (Fassett) Fassett ■☆
135157 Dodecatheon amethystinum Fassett = Dodecatheon amethystinum (Fassett) Fassett ■☆
135158 Dodecatheon clevelandii Greene;克氏流星花;Shooting Star ■☆
135159 Dodecatheon dentatum Hook.;齿叶流星花■☆
135160 Dodecatheon frenchii (Vasey) Rydb.;弗氏流星花;French's Shooting Star,French's Shooting-star ■☆
135161 Dodecatheon frigidum Cham. et Schltdl.;冷地流星花■☆
135162 Dodecatheon hendersonii A. Gray;宽叶流星花;Henderson Shooting Star,Henderson's Shooting Star ■☆
135163 Dodecatheon hugeri Small = Dodecatheon meadia L. ■☆
135164 Dodecatheon latifolium Piper = Dodecatheon hendersonii A. Gray ■☆
135165 Dodecatheon meadia L.;流星花(北美十二花);American Cowslip,Bird's Bill,Cyclamen-primula,Eastern Shooting-star,Prairie Rooster Bill,Pride-of-ohio,Shooting Star,Shooting Stars,Virginian Cowslip,Wild Cyclamen ■☆
135166 Dodecatheon meadia L. f. alba J. F. Macbr.;白流星花■☆
135167 Dodecatheon meadia L. f. sedens Fassett = Dodecatheon meadia L. ■☆
135168 Dodecatheon meadia L. subsp. membranaceum R. Knuth = Dodecatheon frenchii (Vasey) Rydb. ■☆
135169 Dodecatheon meadia L. var. amethystinum Fassett = Dodecatheon amethystinum (Fassett) Fassett ■☆
135170 Dodecatheon meadia L. var. frenchii Vasey = Dodecatheon frenchii (Vasey) Rydb. ■☆
135171 Dodecatheon meadia L. var. genuinum Fassett = Dodecatheon meadia L. ■☆
135172 Dodecatheon meadia L. var. obesum Fassett = Dodecatheon meadia L. ■☆
135173 Dodecatheon meadia L. var. stanfieldii (Small) Fassett = Dodecatheon meadia L. ■☆
135174 Dodecatheon pauciflorum Greene = Dodecatheon meadia L. ■☆
135175 Dodecatheon pauciflorum Greene = Dodecatheon pulchellum (Raf.) Merr. ■☆
135176 Dodecatheon pauciflorum Greene var. salinum (A. Nelson) R. Knuth = Dodecatheon pulchellum (Raf.) Merr. ■☆
135177 Dodecatheon pauciflorum Greene var. watsonii (Tidestr.) C. L. Hitchc. = Dodecatheon pulchellum (Raf.) Merr. ■☆
135178 Dodecatheon pulchellum (Raf.) Merr.;美丽流星花;Few-flowered Shooting Star,Jewelled Shooting-star,Western Shooting-star ■☆
135179 Dodecatheon pulchellum (Raf.) Merr. = Dodecatheon amethystinum (Fassett) Fassett ■☆
135180 Dodecatheon pulchellum (Raf.) Merr. var. watsonii (Tidestr.) B. Boivin = Dodecatheon pulchellum (Raf.) Merr. ■☆
135181 Dodecatheon pulchellum (Raf.) Merr. var. watsonii (Tidestr.) C. L. Hitchc. = Dodecatheon pulchellum (Raf.) Merr. ■☆
135182 Dodecatheon pulchellum (Raf.) Merr. var. watsonii (Tidestr.) Reveal = Dodecatheon pulchellum (Raf.) Merr. ■☆
135183 Dodecatheon pulchellum (Raf.) Merr. var. zionense (Eastw.) S. L. Welsh = Dodecatheon pulchellum (Raf.) Merr. ■☆
135184 Dodecatheon pulchellum Raf. subsp. pauciflorum (Greene) Hultén = Dodecatheon meadia L. ■☆
135185 Dodecatheon radicatum Greene = Dodecatheon pulchellum (Raf.) Merr. ■☆
135186 Dodecatheon radicatum Greene subsp. watsonii (Tidestr.) H. J. Thomps. = Dodecatheon pulchellum (Raf.) Merr. ■☆
135187 Dodecatheon salinum A. Nelson = Dodecatheon pulchellum (Raf.) Merr. ■☆
135188 Dodecatheon stanfieldii Small = Dodecatheon meadia L. ■☆
135189 Dodecatheon zionense Eastw. = Dodecatheon pulchellum (Raf.) Merr. ■☆
135190 Dodonaea Adans. = Comocladia P. Browne ●☆
135191 Dodonaea Böhm. = Ptelea L. ●
135192 Dodonaea Mill.(1754);车桑子属(坡柳属);Aali,Dodonaea,Hop Seed Bush,Hopbush,Hop-bush,Hopseed Bush,Hop-seed Bush,Hopseedbush ●
135193 Dodonaea Plum. ex Adans. = Comocladia P. Browne ●☆
135194 Dodonaea angustifolia L. f.;窄叶车桑子●
135195 Dodonaea angustifolia L. f. = Dodonaea viscosa (L.) Jacq. ●
135196 Dodonaea angustifolia L. f. = Dodonaea viscosa (L.) Jacq. var. angustifolia (L. f.) Benth. ●
135197 Dodonaea angustissima DC.;纤细车桑子;Slender Hop-bush ●☆
135198 Dodonaea attenuata A. Cunn.;渐尖车桑子(渐尖坡柳)●☆

135199 Dodonaea boroniifolia G. Don;蕨叶车桑子;Fern-leaf Hopbush, Hairy Hopbush, Hairy Hop-bush ●☆
135200 Dodonaea burmanniana DC. = Dodonaea viscosa (L.) Jacq. ●
135201 Dodonaea burmanniana DC. = Dodonaea viscosa (L.) Jacq. var. vulgaris Benth. f. burmanniana (DC.) Radlk. ●☆
135202 Dodonaea caffra Eckl. et Zeyh. = Combretum caffrum (Eckl. et Zeyh.) Kuntze ●☆
135203 Dodonaea camfieldii Maiden et Betche;卡氏车桑子(卡氏坡柳);Camfield's Hop Bush ●☆
135204 Dodonaea dioica Roxb. = Dodonaea viscosa (L.) Jacq. ●
135205 Dodonaea lanceolata F. Muell.;披针叶车桑子(披针叶坡柳)●☆
135206 Dodonaea lobulata F. Muell.;红果车桑子;Lobed-leaf Hopbush ●☆
135207 Dodonaea madagascariensis Radlk.;马岛车桑子●☆
135208 Dodonaea multijuga G. Don;丰果车桑子●☆
135209 Dodonaea pinnata Sm.;羽叶车桑子;Pinnate Hopbush ●☆
135210 Dodonaea senegalensis Blume = Dodonaea viscosa (L.) Jacq. ●
135211 Dodonaea thunbergiana Eckl. et Zeyh.;通贝里车桑子(桑柏坡柳,通贝里坡柳)●☆
135212 Dodonaea thunbergiana Radlk. = Dodonaea viscosa (L.) Jacq. var. angustifolia (L. f.) Benth. ●
135213 Dodonaea thunbergiana Radlk. var. linearis Sond. = Dodonaea viscosa (L.) Jacq. var. angustifolia (L. f.) Benth. ●
135214 Dodonaea triquetra J. C. Wendl.;黄果车桑子;Hop Bush, Large-leaf Hop-bush, Native Hopbush ●☆
135215 Dodonaea viscosa (L.) Jacq.;车桑子(白石楝,炒米柴,车桑仔,车闩仔,毛乳,明油脂,明油子,明子柴,坡柳,山相思,铁扫把,溪柳);Akeake, Ake-ake, Clammy Hop Seed Bush, Clammy Hopseed Bush, Clammy Hop-seed bush, Clammy Hopseedbush, Florida Hopbush, Florida Hopseedbush, Hop Bush, Hopbush, Hop-bush, Hopseed Bush, Horse-seed Bush, Native Hops, Pichon, Sticky Hop-bush, Sweetch Sorrel ●
135216 Dodonaea viscosa (L.) Jacq. 'Purpurea';紫叶车桑子(紫色车桑子)●☆
135217 Dodonaea viscosa (L.) Jacq. = Dodonaea angustifolia L. f. ●
135218 Dodonaea viscosa (L.) Jacq. f. angustifolia (Benth.) Sherff = Dodonaea angustifolia L. f. ●
135219 Dodonaea viscosa (L.) Jacq. f. angustifolia (L. f.) Sherff = Dodonaea angustifolia L. f. ●
135220 Dodonaea viscosa (L.) Jacq. f. angustifolia (L. f.) Sherff = Dodonaea viscosa (L.) Jacq. var. angustifolia (L. f.) Benth. ●
135221 Dodonaea viscosa (L.) Jacq. f. burmaniana (DC.) Radlk. = Dodonaea viscosa (L.) Jacq. ●
135222 Dodonaea viscosa (L.) Jacq. f. burmanniana (Schum. et Thonn.) Radlk. = Dodonaea angustifolia L. f. ●
135223 Dodonaea viscosa (L.) Jacq. f. epandar (Schum. et Thonn.) Radlk. = Dodonaea angustifolia L. f. ●
135224 Dodonaea viscosa (L.) Jacq. f. linearis Harv. et Sond. = Dodonaea angustifolia L. f. ●
135225 Dodonaea viscosa (L.) Jacq. f. repanda (Schum. et Thonn.) Radlk.;浅波叶车桑子●
135226 Dodonaea viscosa (L.) Jacq. subsp. angustifolia (L. f.) J. G. West = Dodonaea angustifolia L. f. ●
135227 Dodonaea viscosa (L.) Jacq. subsp. angustifolia (L. f.) J. G. West = Dodonaea viscosa (L.) Jacq. ●
135228 Dodonaea viscosa (L.) Jacq. subsp. angustifolia (L. f.) J. G. West = Dodonaea viscosa (L.) Jacq. var. angustifolia (L. f.) Benth. ●
135229 Dodonaea viscosa (L.) Jacq. subsp. burmanniana (DC.) J. G. West = Dodonaea viscosa (L.) Jacq. ●
135230 Dodonaea viscosa (L.) Jacq. subsp. burmanniana (DC.) J. G. West = Dodonaea burmanniana DC. ●☆
135231 Dodonaea viscosa (L.) Jacq. var. angustifolia (L. f.) Benth. = Dodonaea viscosa (L.) Jacq. ●
135232 Dodonaea viscosa (L.) Jacq. var. angustifolia (L. f.) Benth. = Dodonaea angustifolia L. f. ●
135233 Dodonaea viscosa (L.) Jacq. var. angustifolia Benth. = Dodonaea angustifolia L. f. ●
135234 Dodonaea viscosa (L.) Jacq. var. linearis (Harv. et Sond.) Sherff;线叶车桑子●
135235 Dodonaea viscosa (L.) Jacq. var. linearis (Harv. et Sond.) Sherff f. angustifolia (Benth.) Sherff = Dodonaea angustifolia L. f. ●
135236 Dodonaea viscosa (L.) Jacq. var. linearis (Sond.) Sherff = Dodonaea viscosa (L.) Jacq. var. angustifolia (L. f.) Benth. ●
135237 Dodonaea viscosa (L.) Jacq. var. vulgaris Benth. = Dodonaea viscosa (L.) Jacq. ●
135238 Dodonaea viscosa (L.) Jacq. var. vulgaris Benth. f. burmanniana (DC.) Radlk.;缅甸车桑子(夜闹子)●☆
135239 Dodonaeaceae Kunth ex Small(1903)(保留科名);车桑子科●
135240 Dodonaeaceae Kunth ex Small(保留科名) = Sapindaceae Juss.(保留科名)●■
135241 Dodonaeaceae Link = Dodonaeaceae Kunth ex Small(保留科名)●
135242 Dodonaeaceae Link = Sapindaceae Juss.(保留科名)●■
135243 Dodsonia Ackerman(1979);多德森兰属■☆
135244 Dodsonia falcata Ackerman;镰形多德森兰■☆
135245 Dodsonia saccata (Garay) Ackerman;多德森兰■☆
135246 Doellia Sch. Bip. = Blumea DC.(保留属名)■●
135247 Doellia Sch. Bip. = Doellia Sch. Bip. ex Walp. ■☆
135248 Doellia Sch. Bip. ex Walp.(1843);毛红脂菊属■☆
135249 Doellia bovei (DC.) Anderb. = Blumea bovei (DC.) Vatke ■☆
135250 Doellia cafra (DC.) Anderb.;毛红脂菊■☆
135251 Doellia cafra (DC.) Anderb. = Blumea cafra (DC.) O. Hoffm. ■☆
135252 Doellingeria Nees = Aster L. ●■
135253 Doellingeria Nees (1832);东风菜属;Doellingeria, Tall Flat-topped Aster ■
135254 Doellingeria amygdalina (Lam.) Nees = Doellingeria umbellata (Mill.) Nees ■☆
135255 Doellingeria cornifolia (Muhl. ex Willd.) Nees = Doellingeria infirma (Michx.) Greene ■☆
135256 Doellingeria humilis (Willd.) Britton = Doellingeria infirma (Michx.) Greene ■☆
135257 Doellingeria infirma (Michx.) Greene;柔弱东风菜;Cornel-leaf Whitetop Aster, Cornel-leaved Whitetop Aster ■☆
135258 Doellingeria marchandii (H. Lév.) Y. Ling;短冠东风菜(白花菜,白仙草,菊花暗消,土白前,胃药);Marchand Doellingeria ■
135259 Doellingeria obovata (Nutt.) Nees = Oclemena reticulata (Pursh) G. L. Nesom ■☆
135260 Doellingeria ptarmicoides Nees = Solidago ptarmicoides (Nees) B. Boivin ■☆
135261 Doellingeria pubens (A. Gray) Rydb. = Aster umbellatus Mill. var. pubens A. Gray ■☆
135262 Doellingeria pubens (A. Gray) Rydb. = Doellingeria umbellata

(Mill.) Nees var. pubens (A. Gray) Britton ■☆
135263 Doellingeria reticulata (Pursh) Greene = Oclemena reticulata (Pursh) G. L. Nesom ■☆
135264 Doellingeria scabra (Thunb.) Nees;东风菜(白云草,草三七,疙瘩药,尖叶山苦荬,盘龙草,山白菜,山蛤芦,山田七,天狗,天狗胆,土苍术,土田七,仙白草,小叶青,钻山狗);Scabrous Doellingeria ■
135265 Doellingeria scabra (Thunb.) Nees = Aster scaber Thunb. ■
135266 Doellingeria sericocarpoides Small;南方东风菜;Southern Tall Flat-topped, Southern Whitetop Aster ■☆
135267 Doellingeria sericocarpoides Small = Aster umbellatus Mill. var. pubens A. Gray ■☆
135268 Doellingeria umbellata (Mill.) Nees;伞状东风菜;Parasol Whitetop, Tall Flat-topped White Aster ■☆
135269 Doellingeria umbellata (Mill.) Nees = Aster umbellatus Mill. ■☆
135270 Doellingeria umbellata (Mill.) Nees var. humilis (Willd.) W. Stone = Doellingeria infirma (Michx.) Greene ■☆
135271 Doellingeria umbellata (Mill.) Nees var. latifolia (A. Gray) House = Aster umbellatus Mill. var. pubens A. Gray ■☆
135272 Doellingeria umbellata (Mill.) Nees var. latifolia (A. Gray) House = Doellingeria sericocarpoides Small ■☆
135273 Doellingeria umbellata (Mill.) Nees var. oneidica House = Doellingeria umbellata (Mill.) Nees ■☆
135274 Doellingeria umbellata (Mill.) Nees var. pubens (A. Gray) Britton;毛伞状东风菜;Aster Pubescent ■☆
135275 Doellingeria umbellata (Mill.) Nees var. pubens (A. Gray) Britton = Aster umbellatus Mill. var. pubens A. Gray ■☆
135276 Doellingeria umbellata subsp. pubens (A. Gray) Á. Löve et D. Löve = Doellingeria umbellata (Mill.) Nees var. pubens (A. Gray) Britton ■☆
135277 Doellochloa Kuntze = Gymnopogon P. Beauv. ■☆
135278 Doellochloa Kuntze = Monochaete Döll ■☆
135279 Doemia R. Br. = Pergularia L. ■☆
135280 Doerpfeldia Urb. (1924);古巴鼠李属●☆
135281 Doerpfeldia cubensis Urb.;古巴鼠李●☆
135282 Doerriena Borkh. = Cerastium L. ■
135283 Doerriena Borkh. = Moenchia Ehrh. (保留属名) ■☆
135284 Doerrienia Dennst. = Acronychia J. R. Forst. et G. Forst. (保留属名)●
135285 Doerrienia Rchb. = Genlisea A. St. -Hil. ■☆
135286 Doerriera Steud. = Cerastium L. ■
135287 Doerriera Steud. = Doerriena Borkh. ■
135288 Dofia Adans. = Dirca L. ●☆
135289 Doga (Baill.) Baill. ex Nakai = Storckiella Seem. ●☆
135290 Doga (Baill.) Nakai = Storckiella Seem. ●☆
135291 Doga Baill. = Storckiella Seem. ●☆
135292 Dolabrifolia (Pfitzer) Szlach. et Romowicz = Mystacidium Lindl. ■☆
135293 Dolabrifolia (Pfitzer) Szlach. et Romowicz(2007);斧叶兰属■☆
135294 Dolia Lindl. = Nolana L. ex L. f. ■☆
135295 Dolianthus C. H. Wright = Amaracarpus Blume ●☆
135296 Dolichandra Cham. (1832);长蕊紫葳属●☆
135297 Dolichandra cynanchoides Cham.;长蕊紫葳●☆
135298 Dolichandra obtusifolia Baker = Markhamia obtusifolia (Baker) Sprague ●☆
135299 Dolichandrone (Fenzl) Seem. (1862)(保留属名);栓翅树属(老猫尾木属,马尔汉木属,猫尾木属,猫尾树属);Cat-tail Tree, Dolichandrone ●
135300 Dolichandrone (Fenzl) Seem. (保留属名) = Markhamia Seem. ex Baill. ●
135301 Dolichandrone Fenzl ex Seem. = Dolichandrone (Fenzl) Seem. (保留属名)●
135302 Dolichandrone alba (Sim) Sprague = Spathodea alba Sim ●☆
135303 Dolichandrone alba Sprague = Spathodea alba Sim ●☆
135304 Dolichandrone alternifolia (R. Br.) F. M. Bailey = Spathodea alternifolia R. Br. ●☆
135305 Dolichandrone arcuata C. B. Clarke;拱猫尾木●☆
135306 Dolichandrone atrovirens Schum. = Dolichandrone falcata Seem. ●☆
135307 Dolichandrone brunonis F. Muell. ex Steenis = Dolichandrone alternifolia (R. Br.) F. M. Bailey ●☆
135308 Dolichandrone caudafelina (Hance) Benth. et Hook. f.;猫尾木(猫尾);Cat-tail, Cattail Dolichandrone, Cat-tail Dolichandrone, Cat-tail Tree, Dolichandrone ●
135309 Dolichandrone caudafelina (Hance) Benth. et Hook. f. = Markhamia stipulata (Wall.) Seem. ex K. Schum. var. kerrii Sprague ●
135310 Dolichandrone columnaris Santisuk;东南亚猫尾木●☆
135311 Dolichandrone crispa Seem.;皱波猫尾木●☆
135312 Dolichandrone falcata Seem.;镰状猫尾木●☆
135313 Dolichandrone falcata Seem. = Dolichandrone spathecea (L. f.) Seem. ●
135314 Dolichandrone hildebrandtii Baker;希氏猫尾木●☆
135315 Dolichandrone hildebrandtii Baker = Markhamia lutea (Benth.) K. Schum. ●☆
135316 Dolichandrone hirsuta Baker = Markhamia zanzibarica (Bojer ex DC.) K. Schum. ●☆
135317 Dolichandrone latifolia Baker;宽叶猫尾木●☆
135318 Dolichandrone latifolia Baker = Markhamia zanzibarica (Bojer ex DC.) K. Schum. ●☆
135319 Dolichandrone lawii Seem.;劳氏猫尾木●☆
135320 Dolichandrone lutea Benth. ex Hook. et Jacks. = Markhamia lutea (Benth.) K. Schum. ●☆
135321 Dolichandrone lutea Seem.;黄花猫尾木(黄花马尔汉木,老猫尾木);Golden-yellow Dolichandrone ●☆
135322 Dolichandrone obtusifolia Baker;钝叶猫尾木(钝叶马尔汉木);Obtuseleaf Dolichandrone ●☆
135323 Dolichandrone obtusifolia Baker = Markhamia obtusifolia (Baker) Sprague ●☆
135324 Dolichandrone platycalyx Baker;宽萼老猫尾木●☆
135325 Dolichandrone platycalyx Baker = Markhamia lutea (Benth.) K. Schum. ●☆
135326 Dolichandrone smithii Baker = Stereospermum kunthianum Cham. ●☆
135327 Dolichandrone spathecea (L. f.) Seem.;海滨猫尾木(佛焰猫尾木);Spathose Dolichandrone ●
135328 Dolichandrone stenocarpa (Welw. ex Seem.) Baker = Markhamia zanzibarica (Bojer ex DC.) K. Schum. ●☆
135329 Dolichandrone stenocarpa Seem.;狭果猫尾木●☆
135330 Dolichandrone stipulata (Wall.) Benth. et Hook. f. = Markhamia stipulata (Wall.) Seem. ex K. Schum. ●
135331 Dolichandrone stipulata (Wall.) Benth. et Hook. f. var. kerrii (Sprague) C. Y. Wu et W. C. Yin = Markhamia stipulata (Wall.) Seem. ex K. Schum. var. kerrii Sprague ●
135332 Dolichandrone stipulata (Wall.) Benth. et Hook. f. var. velutina

(Kurz) C. B. Clarke;齿叶猫尾木;Toothleaf SW. China Cattailtree ●
135333 Dolichandrone stipulata (Wall.) Benth. et Hook. f. var. velutina (Kurz) C. B. Clarke = Markhamia stipulata (Wall.) Seem. ex K. Schum. ●
135334 Dolichandrone stipulata (Wall.) C. B. Clarke = Markhamia stipulata (Wall.) Seem. ex K. Schum. ●
135335 Dolichandrone stipulata (Wall.) C. B. Clarke var. kerrii (Sprague) C. Y. Wu et W. C. Qin = Markhamia stipulata (Wall.) Seem. ex K. Schum. var. kerrii Sprague ●
135336 Dolichandrone stipulata (Wall.) C. B. Clarke var. velutina (Kurz) C. B. Clarke = Markhamia stipulata (Wall.) Seem. ex K. Schum. ●
135337 Dolichandrone tomentosa (Benth.) Benth. et Hook. f. = Markhamia tomentosa (Benth.) K. Schum. ex Engl. ●☆
135338 Dolichandrone tomentosa Benth. et Hook. f.;毛猫尾木●☆
135339 Dolichandrone zanzibaricum (Bojer ex DC.) K. Schum.;桑给巴尔猫尾木●☆
135340 Dolichangis Thouars = Angraecum Bory ■
135341 Dolichanthera Schltr. et K. Krause = Morierina Vieill. ☆
135342 Dolichlasium Lag. (1811);阿根廷菊属●☆
135343 Dolichlasium Lag. = Trixis P. Browne ■●☆
135344 Dolichlasium lagascae Gillies ex D. Don;阿根廷菊●☆
135345 Dolichocentrum (Schltr.) Brieger = Dendrobium Sw. (保留属名) ■
135346 Dolichochaete (C. E. Hubb.) J. B. Phipps = Tristachya Nees ■☆
135347 Dolichochaete bequaertii (De Wild.) J. B. Phipps = Tristachya bequaertii De Wild. ■☆
135348 Dolichochaete bicrinita J. B. Phipps = Tristachya bicrinita (J. B. Phipps) Clayton ■☆
135349 Dolichochaete longispiculata (C. E. Hubb.) J. B. Phipps = Tristachya nodiglumis K. Schum. ■☆
135350 Dolichochaete nodiglumis (K. Schum.) J. B. Phipps = Tristachya nodiglumis K. Schum. ■☆
135351 Dolichochaete rehmannii (Hack.) J. B. Phipps = Tristachya rehmannii Hack. ■☆
135352 Dolichochaete rehmannii (Hack.) J. B. Phipps var. helenae (Buscal. et Muschl.) Phipps = Tristachya rehmannii Hack. ■☆
135353 Dolichochaete rehmannii (Hack.) J. B. Phipps var. pilosa (C. E. Hubb.) J. B. Phipps = Tristachya rehmannii Hack. ■☆
135354 Dolichodeira Hanst. = Achimenes Pers. (保留属名) ■☆
135355 Dolichodelphys K. Schum. = Dolichodelphys K. Schum. et K. Krause ☆
135356 Dolichodelphys K. Schum. et K. Krause (1908);长室茜属☆
135357 Dolichodiera Hanst. = Sinningia Nees ●■☆
135358 Dolichoglottis B. Nord. (1978);雪雏菊属■☆
135359 Dolichoglottis scorzoneroides (Hook. f.) B. Nord.;雪雏菊;March-white Snow Marguerite ■☆
135360 Dolichogyne DC. = Nardophyllum (Hook. et Arn.) Hook. et Arn. ●☆
135361 Dolichokentia Becc. = Cyphokentia Brongn. ●☆
135362 Dolicholasium Spreng. = Dolichlasium Lag. ●☆
135363 Dolicholasium Spreng. = Trixis P. Browne ■●☆
135364 Dicholobium A. Gray (1859);长裂茜属☆
135365 Dolicholobium acuminatum Burkill;尖长裂茜☆
135366 Dolicholobium glabrum M. E. Jansen;光长裂茜☆
135367 Dolicholobium graciliflorum Valeton;细花长裂茜☆
135368 Dolicholobium oblongifolium A. Gray;矩圆叶长裂茜☆
135369 Dolicholobium philippinense Trel.;菲律宾长裂茜☆
135370 Dolicholobium rubrum Schltr. ex Valeton;红长裂茜☆
135371 Dolicholoma D. Fang et W. T. Wang (1983);长檐苣苔属;Dolicholoma ■★
135372 Dolicholoma jasminiflorum D. Fang et W. T. Wang;长檐苣苔;Common Dolicholoma ■
135373 Dolicholus Medik. (废弃属名) = Rhynchosia Lour. (保留属名) ●■
135374 Dolicholus ambacensis Hiern = Rhynchosia ambacensis (Hiern) K. Schum. ■☆
135375 Dolicholus huillensis Hiern = Rhynchosia huillensis (Hiern) K. Schum. ■☆
135376 Dolicholus luteolus Hiern = Rhynchosia luteola (Hiern) K. Schum. ■☆
135377 Dolicholus memnonia Hiern var. candida Welw. ex Hiern = Rhynchosia candida (Welw. ex Hiern) Torre ■☆
135378 Dolicholus minimus (L.) Hiern = Rhynchosia minima (L.) DC. ■☆
135379 Dolicholus procurrens Hiern = Rhynchosia procurrens (Hiern) K. Schum. ■☆
135380 Dolicholus totta (Thunb.) Kuntze = Rhynchosia totta (Thunb.) DC. ■☆
135381 Dolicholus venulosus Hiern = Rhynchosia totta (Thunb.) DC. var. fenchelii Schinz ■☆
135382 Dolicholus violaceus Hiern = Rhynchosia viscosa (Roth) DC. ■
135383 Dolicholus violaceus Hiern = Rhynchosia viscosa (Roth) DC. subsp. violacea (Hiern) Verdc. ■☆
135384 Dolichometra K. Schum. (1904);长腹茜属☆
135385 Dolichometra leucantha K. Schum.;长腹茜☆
135386 Dolichonema Nees = Moldenhawera Schrad. ●☆
135387 Dolichopentas Kårehed et B. Bremer = Pentas Benth. ●■
135388 Dolichopentas Kårehed et B. Bremer (2007);长叶五星花属●■☆
135389 Dolichopetalum Tsiang (1973);金凤藤属(金凤藤属);Dolichopetalum, Dracaena ●★
135390 Dolichopetalum kwangsiense Tsiang;金凤藤(广西长梗藤,金凤藤);Guangxi Dolichopetalum, Guangxi Dracaena, Kwangsi Dolichopetalum ●◇
135391 Dolichopsis Hassl. (1907);类镰扁豆属(巴拉圭长豆属)■☆
135392 Dolichopsis paraguariensis Hassl.;类镰扁豆■☆
135393 Dolichopterys Kosterm. = Lophopterys A. Juss. ●☆
135394 Dolichorhynchus Hedge et Kit Tan = Douepea Cambess. ■☆
135395 Dolichorhynchus Hedge et Kit Tan (1987);阿拉伯长嘴芥属■☆
135396 Dolichorhynchus arabicus Hedge et Kit Tan;阿拉伯长嘴芥■☆
135397 Dolichorrhiza (Pojark.) Galushko (1970);长根菊属■☆
135398 Dolichorrhiza caucasica (M. Bieb.) Galushko;高加索长根菊■☆
135399 Dolichos L. (1753) (保留属名);镰扁豆属(扁豆属,大麻药属,鹊豆属);Dolichos, Sicklehairicot ■
135400 Dolichos aciphyllus R. Wilczek;尖叶镰扁豆■☆
135401 Dolichos adenophorus Harms = Adenodolichos punctatus (Micheli) Harms subsp. bussei (Harms) Verdc. ■☆
135402 Dolichos aeschynome-sesban Forssk. = Sesbania sesban (L.) Merr. ●
135403 Dolichos africanus R. Wilczek = Macrotyloma africanum (R. Wilczek) Verdc. ■☆
135404 Dolichos altissimus Jacq. = Mucuna altissima (Jacq.) DC. ■☆
135405 Dolichos anchietae Hiern = Adenodolichos rhomboideus (O. Hoffm.) Harms ■☆

135406 Dolichos andongensis Welw. ex Baker = Dolichos trilobus L. ■
135407 Dolichos angularis Willd. = Vigna angularis (Willd.) Ohwi et H. Ohashi ■
135408 Dolichos angustifolius Eckl. et Zeyh.;窄叶镰扁豆■☆
135409 Dolichos angustifolius Guillaumin et Perr. = Vigna vexillata (L.) A. Rich. var. angustifolia (Schumach. et Thonn.) Baker ■☆
135410 Dolichos angustissimus E. Mey.;狭镰扁豆■☆
135411 Dolichos antunesii Harms;安图内思镰扁豆■☆
135412 Dolichos appendiculatus Hand.-Mazz.;丽江镰扁豆;Lijiang Dolichos,Lijiang Sicklehairicot,Likiang Dolichos ■
135413 Dolichos argenteus Willd. = Pseudovigna argentea (Willd.) Verdc. ■☆
135414 Dolichos argyrophyllus Harms = Dolichos kilimandscharicus Taub. var. argyrophyllus (Harms) Verdc. ■☆
135415 Dolichos argyros R. Wilczek;银色镰扁豆■☆
135416 Dolichos axillaris E. Mey. = Macrotyloma axillare (E. Mey.) Verdc. ■☆
135417 Dolichos axillaris E. Mey. var. glaber ? = Macrotyloma axillare (E. Mey.) Verdc. var. glabrum ? ■☆
135418 Dolichos axillaris E. Mey. var. macranthus Brenan = Macrotyloma axillare (E. Mey.) Verdc. var. macranthum (Brenan) Verdc. ■☆
135419 Dolichos axilliflorus Verdc.;腋花镰扁豆■☆
135420 Dolichos barbatus Wall. = Cajanus goensis Dalzell ●
135421 Dolichos baumannii Harms = Macrotyloma tenuiflorum (Micheli) Verdc. ■☆
135422 Dolichos bellus Harms = Dolichos kilimandscharicus Taub. ■☆
135423 Dolichos benadirianus Chiov. = Macrotyloma uniflorum (Lam.) Verdc. var. benadirianum (Chiov.) Verdc. ■☆
135424 Dolichos bengalensis Jacq. = Lablab purpureus (L.) Sweet subsp. bengalensis (Jacq.) Verdc. ■
135425 Dolichos bequaertii De Wild. = Dolichos kilimandscharicus Taub. ■☆
135426 Dolichos bianoensis R. Wilczek;比亚诺镰扁豆■☆
135427 Dolichos bianoensis R. Wilczek subsp. orientalis Verdc.;东方比亚诺镰扁豆■☆
135428 Dolichos bieensis Torre = Macrotyloma bieense (Torre) Verdc. ■☆
135429 Dolichos biflorus L.;双花镰扁豆(双花扁豆)■☆
135430 Dolichos biflorus L. var. occidentalis Harms = Macrotyloma biflorum (Schumach. et Thonn.) Hepper ■☆
135431 Dolichos bongensis Taub. ex Harms = Dolichos schweinfurthii Harms ■☆
135432 Dolichos brachypterus Harms;短翅镰扁豆■☆
135433 Dolichos brachypus Harms = Neorautanenia brachypus (Harms) C. A. Sm. ■☆
135434 Dolichos brevicaulis Baker = Macrotyloma brevicaule (Baker) Verdc. ■☆
135435 Dolichos brevidentatus Mackinder;短齿镰扁豆■☆
135436 Dolichos buchananii Harms = Dolichos kilimandscharicus Taub. ■☆
135437 Dolichos cardiophyllus Harms;心叶镰扁豆■☆
135438 Dolichos cardiophyllus Harms var. subsessilis Baker f.;近无柄镰扁豆■☆
135439 Dolichos carnosus A. Chev. = Dolichos schweinfurthii Harms ■☆
135440 Dolichos catjang Burm. f. = Vigna unguiculata (L.) Walp. subsp. cylindrica (L.) Verdc. ■
135441 Dolichos chevalieri Harms = Droogmansia chevalieri (Harms) Hutch. et Dalziel ■☆
135442 Dolichos chloryllis Harv. = Dolichos pratensis (E. Mey.) Taub. ■☆
135443 Dolichos chrysanthus A. Chev. = Macrotyloma biflorum (Schumach. et Thonn.) Hepper ■☆
135444 Dolichos chrysanthus A. Chev. var. occidentalis (Harms) R. Wilczek = Macrotyloma biflorum (Schumach. et Thonn.) Hepper ■☆
135445 Dolichos complanatus De Wild.;扁平镰扁豆■☆
135446 Dolichos compressus R. Wilczek;扁镰扁豆■☆
135447 Dolichos corymbosus R. Wilczek;伞序镰扁豆■☆
135448 Dolichos cultratus Thunb.;鹊豆■☆
135449 Dolichos daltonii Webb = Macrotyloma daltonii (Webb) Verdc. ■☆
135450 Dolichos debilis Hochst. ex A. Rich. = Dolichos trilobus L. ■
135451 Dolichos decumbens Thunb.;外倾镰扁豆■☆
135452 Dolichos densiflorus Welw. ex Baker = Macrotyloma densiflorum (Welw. ex Baker) Verdc. ■☆
135453 Dolichos dewildemanianum R. Wilczek = Macrotyloma dewildemanianum (R. Wilczek) Verdc. ■☆
135454 Dolichos dillonii Delile = Vigna oblongifolia A. Rich. ■☆
135455 Dolichos dinklagei Harms;丁克镰扁豆■☆
135456 Dolichos dubius De Wild. = Dolichos kilimandscharicus Taub. ■☆
135457 Dolichos elatus Baker;高镰扁豆●☆
135458 Dolichos ellenbeckii Harms = Neorautanenia mitis (A. Rich.) Verdc. ■☆
135459 Dolichos ellipticus R. E. Fr. = Macrotyloma ellipticum (R. E. Fr.) Verdc. ■☆
135460 Dolichos emarginata P. Beauv. = Vigna luteola (Jacq.) Benth. ■
135461 Dolichos ensiformis L. = Canavalia ensiformis (L.) DC. ■
135462 Dolichos ensiformis Thunb. = Lablab purpureus (L.) Sweet ■
135463 Dolichos erectus Baker f. = Sphenostylis erecta (Baker f.) Hutch. ex Baker f. ■☆
135464 Dolichos erectus De Wild. = Macrotyloma dewildemanianum (R. Wilczek) Verdc. ■☆
135465 Dolichos erectus De Wild. var. brevifolius ? = Macrotyloma dewildemanianum (R. Wilczek) Verdc. ■☆
135466 Dolichos eriocaulus Harms = Macrotyloma ellipticum (R. E. Fr.) Verdc. ■☆
135467 Dolichos erosus L. = Pachyrhizus erosus (L.) Urb. ■
135468 Dolichos esculentus De Wild. = Macrotyloma fimbriatum (Harms) Verdc. ■☆
135469 Dolichos fabaeformis L'Hér. = Cyamopsis tetragonoloba (L.) Taub. ■
135470 Dolichos falcatus Klein ex Willd. = Dolichos trilobus L. ■
135471 Dolichos falcatus Willd. = Dolichos trilobus L. ■
135472 Dolichos falciformis E. Mey.;镰形镰扁豆■☆
135473 Dolichos ficifolius (Benth. ex Harv.) Harms = Neorautanenia ficifolia (Benth. ex Harv.) C. A. Sm. ■☆
135474 Dolichos ficifolius Graham;无花果叶镰扁豆(无花果叶扁豆)■☆
135475 Dolichos filifoliolus Verdc.;丝小叶镰扁豆■☆
135476 Dolichos fimbriatus Harms = Macrotyloma fimbriatum (Harms) Verdc. ■☆
135477 Dolichos fischeri Harms = Macrotyloma stipulosum (Welw. ex Baker) Verdc. ■☆
135478 Dolichos formosoides Harms = Dolichos trilobus L. ■

135479 Dolichos formosus A. Rich. = Dolichos sericeus E. Mey. subsp. formosus (A. Rich.) Verdc. ■☆

135480 Dolichos galpinii Burtt Davy = Decorsea galpinii (Burtt Davy) Verdc. ■☆

135481 Dolichos gardneri Baker f. = Vigna monophylla Taub. ■☆

135482 Dolichos genistiformis Chiov. = Tephrosia interrupta Hochst. et Steud. ex Engl. ●☆

135483 Dolichos gibbosus Thunb. = Dipogon lignosus (L.) Verdc. ■☆

135484 Dolichos giganteus Willd. = Mucuna gigantea (Willd.) DC. ●

135485 Dolichos glabratus R. Wilczek;光滑镰扁豆■☆

135486 Dolichos glabrescens R. Wilczek;渐光镰扁豆■☆

135487 Dolichos gladiatus Jacq. = Canavalia gladiata (Savi) DC. ■

135488 Dolichos goetzei Harms = Dolichos kilimandscharicus Taub. ■☆

135489 Dolichos gracilis Guillaumin et Perr. = Vigna gracilis (Guillaumin et Perr.) Hook. f. ■☆

135490 Dolichos grahamianus (Wight et Arn.) Niyomdham = Wajira grahamiana (Wight et Arn.) Thulin et Lavin ■☆

135491 Dolichos grandifolius Graham ex Wall. = Pueraria lobata (Willd.) Ohwi subsp. thomsonii (Benth.) H. Ohashi et Tateishi ●

135492 Dolichos grandistipulatus Harms;大托叶镰扁豆■☆

135493 Dolichos gululu De Wild.;热非镰扁豆■☆

135494 Dolichos hastatus Lour.;戟形镰扁豆■☆

135495 Dolichos hastifolius Schnizl. = Vigna unguiculata (L.) Walp. var. spontanea (Schweinf.) Pasquet ■☆

135496 Dolichos hastiformis E. Mey.;戟状镰扁豆■☆

135497 Dolichos helicopus (E. Mey.) Steud. = Vigna luteola (Jacq.) Benth. ■

135498 Dolichos hendrickxii De Wild. = Macrotyloma densiflorum (Welw. ex Baker) Verdc. ■☆

135499 Dolichos henryi Harms = Dolichos junghuhnianus Benth. ■

135500 Dolichos hirsutus Thunb. = Pueraria lobata (Willd.) Ohwi ●■

135501 Dolichos hirtus Andréws = Rhynchosia hirta (Andréws) Meikle et Verdc. ■☆

135502 Dolichos hockii De Wild. = Macrotyloma hockii (De Wild.) Verdc. ■☆

135503 Dolichos homblei De Wild.;洪布勒镰扁豆■☆

135504 Dolichos hosei Craib = Vigna hosei (Craib) Backer ex K. Heyne ■

135505 Dolichos huillensis Welw. ex Romariz;威拉镰扁豆■☆

135506 Dolichos japonicus Spreng. = Pueraria lobata (Willd.) Ohwi ●■

135507 Dolichos jumellei R. Vig. = Alistilus jumellei (R. Vig.) Verdc. ●☆

135508 Dolichos junghuhnianus Benth.; 滇 南 镰 扁 豆; S. Yunnan Sicklehairicot, South Yunnan Dolichos ■

135509 Dolichos junodii (Harms) Verdc.;朱诺德镰扁豆■☆

135510 Dolichos karaviaensis R. Wilczek;卡拉维亚镰扁豆■☆

135511 Dolichos kasaiensis R. Wilczek = Macrotyloma rupestre (Welw. ex Baker) Verdc. ■☆

135512 Dolichos katangensis De Wild. = Macrotyloma stipulosum (Welw. ex Baker) Verdc. ■☆

135513 Dolichos kilimandscharicus Taub.;吉利镰扁豆■☆

135514 Dolichos kilimandscharicus Taub. subsp. parviflorus Verdc.;小花吉利镰扁豆■☆

135515 Dolichos kilimandscharicus Taub. var. argyrophyllus (Harms) Verdc.;银叶吉利镰扁豆■☆

135516 Dolichos kosyunensis Hosok. = Dolichos trilobus L. ■

135517 Dolichos lablab L. = Lablab purpureus (L.) Sweet ■

135518 Dolichos lablab L. subsp. bengalensis (Jacq.) Rivals = Lablab purpureus (L.) Sweet ■

135519 Dolichos lablab L. subsp. bengalensis (Jacq.) Rivals = Lablab purpureus (L.) Sweet subsp. bengalensis (Jacq.) Verdc. ■

135520 Dolichos lablab L. var. crenatifructus Rivals = Lablab purpureus (L.) Sweet var. uncinatus Verdc. ■☆

135521 Dolichos lablab L. var. rhomboideus Schinz = Lablab purpureus (L.) Sweet var. rhomboideus (Schinz) Verdc. ■☆

135522 Dolichos lablab L. var. uncinatus (Schweinf.) Chiov. = Lablab purpureus (L.) Sweet var. uncinatus Verdc. ■☆

135523 Dolichos lacteus Raf. = Baptisia alba (L.) Vent. var. macrophylla (Larisey) Isely ■☆

135524 Dolichos lagopus Dunn = Sinodolichos lagopus (Dunn) Verdc. ■

135525 Dolichos lelyi Hutch. = Dolichos schweinfurthii Harms ■☆

135526 Dolichos lignosus L. = Dipogon lignosus (L.) Verdc. ■☆

135527 Dolichos lignosus L. = Dolichos lablab L. ■

135528 Dolichos linearifolius I. M. Johnst.;线叶镰扁豆■☆

135529 Dolichos linearis E. Mey.;线状镰扁豆■☆

135530 Dolichos lineatus Thunb. = Canavalia lineata (Thunb.) DC. ■

135531 Dolichos lobatus Willd. = Pueraria lobata (Willd.) Ohwi ●■

135532 Dolichos longipes Buchw.;长梗镰扁豆■☆

135533 Dolichos longistipellatus Harms = Macrotyloma rupestre (Welw. ex Baker) Verdc. ■☆

135534 Dolichos longistipellatus Harms f. angustifoliola Baker f. = Macrotyloma bieense (Torre) Verdc. ■☆

135535 Dolichos longistipellatus Harms var. gossweileri Baker f. = Macrotyloma rupestre (Welw. ex Baker) Verdc. ■☆

135536 Dolichos lupiniflorus N. E. Br. = Dolichos kilimandscharicus Taub. ■☆

135537 Dolichos lupinoides Baker = Dolichos kilimandscharicus Taub. ■☆

135538 Dolichos luteolus Jacq. = Vigna luteola (Jacq.) Benth. ■

135539 Dolichos luteus Sw. = Vigna marina (Burm.) Merr. ■

135540 Dolichos macrodon Graham = Rhynchosia rothii Benth. ex Aitch. ●■

135541 Dolichos macrothyrsus Harms = Adenodolichos macrothyrsus (Harms) Harms ■☆

135542 Dolichos magnificus Verdc.;华丽镰扁豆●☆

135543 Dolichos maitlandii Baker = Dolichos oliveri Schweinf. ■☆

135544 Dolichos malosanus Baker = Dolichos kilimandscharicus Taub. ■☆

135545 Dolichos maranguensis Taub. = Vigna parkeri Baker subsp. maranguensis (Taub.) Verdc. ■☆

135546 Dolichos maritimus Aubl. = Canavalia maritima (Aubl.) Thouars ■

135547 Dolichos maritimus Aubl. = Canavalia rosea (Sw.) DC. ■

135548 Dolichos mendoncae Torre;门东萨镰扁豆■☆

135549 Dolichos minimus L. = Rhynchosia minima (L.) DC. ■

135550 Dolichos minutiflorus R. Vig.;微花镰扁豆●☆

135551 Dolichos mitis A. Rich. = Neorautanenia mitis (A. Rich.) Verdc. ■☆

135552 Dolichos mnemonium Delile = Rhynchosia minima (L.) DC. var. memnonia (Delile) T. Cooke ■☆

135553 Dolichos monophyllus Taub. = Dolichos xiphophyllus Baker ■☆

135554 Dolichos montana Lour. = Pueraria lobata (Willd.) Ohwi subsp. thomsonii (Benth.) H. Ohashi et Tateishi ●

135555 Dolichos montanus Lour.;山地镰扁豆■☆

135556 Dolichos nanus Harms = Macrotyloma stipulosum (Welw. ex Baker) Verdc. ■☆

135557 Dolichos nervosus Schumach. et Thonn.;多脉镰扁豆■☆
135558 Dolichos niloticus Delile = Vigna luteola (Jacq.) Benth. ■
135559 Dolichos nimbaensis Schnell;尼恩巴镰扁豆■☆
135560 Dolichos obcordatus Roxb. = Canavalia maritima (Aubl.) Thouars ■
135561 Dolichos obliquifolius Schnizl. = Vigna unguiculata (L.) Walp. var. spontanea (Schweinf.) Pasquet ■☆
135562 Dolichos obtusifolius Lam. = Canavalia maritima (Aubl.) Thouars ■
135563 Dolichos obtusifolius Lam. = Canavalia rosea (Sw.) DC. ■
135564 Dolichos oleraceus Schumach. et Thonn. = Vigna adenantha (G. Mey.) Maréchal, Mascherpa et Stainier ■
135565 Dolichos oliganthus Brenan = Macrotyloma oliganthum (Brenan) Verdc. ■☆
135566 Dolichos oliveri Schweinf.;奥里弗镰扁豆■☆
135567 Dolichos oliveri Schweinf. var. kilimandscharicus Taub. ex Baker f. = Dolichos oliveri Schweinf. ■☆
135568 Dolichos orbicularis (Welw. ex Baker) Baker f. = Neorautanenia mitis (A. Rich.) Verdc. ■☆
135569 Dolichos pachyrhizus Harms ex De Wild.;粗根镰扁豆■☆
135570 Dolichos paniculatus Hua = Adenodolichos paniculatus (Hua) Hutch. et Dalziel ■☆
135571 Dolichos pearsonii Hutch. = Lablab purpureus (L.) Sweet var. rhomboideus (Schinz) Verdc. ■☆
135572 Dolichos peglerae L. Bolus;佩格拉镰扁豆■☆
135573 Dolichos petiolatus R. Wilczek;柄叶镰扁豆■☆
135574 Dolichos phaseoloides Roxb. = Pueraria phaseoloides (Roxb.) Benth. ■
135575 Dolichos pilosus Klein ex Willd. = Vigna pilosa (Klein ex Willd.) Baker ex K. Heyne ■
135576 Dolichos platypus Baker = Droogmansia pteropus (Baker) De Wild. var. platypus (Baker) Verdc. ■☆
135577 Dolichos polystachios Forssk. = Canavalia virosa (Roxb.) Wight et Arn. ■
135578 Dolichos praecox R. E. Fr. = Dolichos gululu De Wild. ■☆
135579 Dolichos pratensis (E. Mey.) Taub.;草原镰扁豆■☆
135580 Dolichos pratorum Harms = Dolichos trinervatus Baker ■☆
135581 Dolichos protractus (E. Mey.) Steud. = Vigna unguiculata (L.) Walp. subsp. protracta (E. Mey.) B. J. Pienaar ■☆
135582 Dolichos pruriens L. = Mucuna pruriens (L.) DC. ●■
135583 Dolichos pseudocajanus Baker;假木豆镰扁豆■☆
135584 Dolichos pseudocajanus Baker var. paniculatus R. Wilczek = Dolichos brevidentatus Mackinder ■☆
135585 Dolichos pseudocomplanatus R. Wilczek;假扁平镰扁豆■☆
135586 Dolichos pseudodebilis Harms;假弱小镰扁豆■☆
135587 Dolichos pseudopachyrhizus Harms;假粗根镰扁豆(假粗根扁豆)■☆
135588 Dolichos pseudopachyrhizus Harms = Neorautanenia mitis (A. Rich.) Verdc. ■☆
135589 Dolichos pseudopachyrhizus Harms var. kilimandscharicus ? = Neorautanenia mitis (A. Rich.) Verdc. ■☆
135590 Dolichos pseudopachyrhizus Harms var. subintegrifolius ? = Neorautanenia mitis (A. Rich.) Verdc. ■☆
135591 Dolichos psoraloides Lam. = Cyamopsis tetragonoloba (L.) Taub. ■
135592 Dolichos pteropus Baker = Droogmansia pteropus (Baker) De Wild. ■☆
135593 Dolichos pubescens L. = Macrotyloma uniflorum (Lam.) Verdc. ■
135594 Dolichos purpureus L. = Lablab purpureus (L.) Sweet ■
135595 Dolichos quarrei R. Wilczek;卡雷镰扁豆■☆
135596 Dolichos repens L. = Vigna luteola (Jacq.) Benth. ■
135597 Dolichos reptans Verdc.;匍匐镰扁豆■☆
135598 Dolichos reticulatus Schltr. = Vigna nervosa Markötter ■☆
135599 Dolichos retusus E. Mey. = Vigna marina (Burm.) Merr. ■
135600 Dolichos rhombifolius (Hayata) Hosok.;菱叶镰扁豆(菱叶扁豆);Rhomicleaf Dolichos, Rhomicleaf Sicklehairicot ■
135601 Dolichos rhomboideus O. Hoffm. = Adenodolichos rhomboideus (O. Hoffm.) Harms ■☆
135602 Dolichos ringoetii De Wild. = Macrotyloma densiflorum (Welw. ex Baker) Verdc. ■☆
135603 Dolichos robustus Bolus;粗壮镰扁豆■☆
135604 Dolichos roseus Sw. = Canavalia maritima (Aubl.) Thouars ■
135605 Dolichos roseus Sw. = Canavalia rosea (Sw.) DC. ■
135606 Dolichos rupestris Welw. ex Baker = Macrotyloma rupestre (Welw. ex Baker) Verdc. ■☆
135607 Dolichos rupestris Welw. ex Baker var. gossweileri (Baker f.) Torre = Macrotyloma rupestre (Welw. ex Baker) Verdc. ■☆
135608 Dolichos saponarius De Wild. = Dolichos pseudocajanus Baker ■☆
135609 Dolichos scarabaeoides L. = Cajanus scarabaeoides (L.) Thouars ●
135610 Dolichos schlechteri Harms ex Burtt Davy = Decorsea schlechteri (Harms) Verdc. ■☆
135611 Dolichos schliebenii Harms;施利本镰扁豆■☆
135612 Dolichos schweinfurthii Harms;施韦镰扁豆■☆
135613 Dolichos seineri Harms;巨根镰扁豆■☆
135614 Dolichos seineri Harms = Neorautanenia brachypus (Harms) C. A. Sm. ■☆
135615 Dolichos sericeus E. Mey.;绢毛镰扁豆■☆
135616 Dolichos sericeus E. Mey. subsp. formosus (A. Rich.) Verdc.;美丽绢毛镰扁豆■☆
135617 Dolichos sericeus E. Mey. subsp. glabrescens Verdc.;渐光绢毛镰扁豆■☆
135618 Dolichos sericeus E. Mey. subsp. pseudofalcatus Verdc.;假镰扁豆■☆
135619 Dolichos sericophyllus R. Wilczek;绢毛叶镰扁豆■☆
135620 Dolichos serpens De Wild.;蛇形镰扁豆■☆
135621 Dolichos sesquipedalis L. = Vigna unguiculata (L.) Walp. subsp. sesquipedalis (L.) Verdc. ■
135622 Dolichos shuterioides Baker = Dolichos sericeus E. Mey. ■☆
135623 Dolichos simplicifolius Hook. f.;单叶镰扁豆■☆
135624 Dolichos sinensis L. = Vigna unguiculata (L.) Walp. ■
135625 Dolichos sofa L. = Glycine max (L.) Merr. ■
135626 Dolichos splendens Baker;光亮镰扁豆■☆
135627 Dolichos stenocarpus Hochst. ex A. Rich. = Sphenostylis stenocarpa (Hochst. ex A. Rich.) Harms ■☆
135628 Dolichos stenophyllus Harms = Macrotyloma stenophyllum (Harms) Verdc. ■☆
135629 Dolichos stipulosus Welw. ex Baker = Macrotyloma stipulosum (Welw. ex Baker) Verdc. ■☆
135630 Dolichos stipulosus Welw. ex Baker f. angustifoliolatus Baker f. = Macrotyloma stipulosum (Welw. ex Baker) Verdc. ■☆
135631 Dolichos stipulosus Welw. ex Baker var. randii Baker f. = Macrotyloma stipulosum (Welw. ex Baker) Verdc. ■☆

135632 Dolichos stolzii Harms = Dolichos kilimandscharicus Taub. ■☆
135633 Dolichos subcapitatus R. Wilczek;亚头状镰扁豆■☆
135634 Dolichos subcapitatus R. Wilczek var. angustifolium Mackinder;窄叶亚头状镰扁豆■☆
135635 Dolichos swynnertonii Baker f. = Dolichos kilimandscharicus Taub. ■☆
135636 Dolichos taubertii Baker f. = Macrotyloma maranguense (Taub.) Verdc. ■☆
135637 Dolichos tenuicaulis (Baker) Craib;细茎镰扁豆(大麻药,细茎扁豆)■☆
135638 Dolichos tenuiflorus (Micheli) R. Wilczek = Macrotyloma tenuiflorum (Micheli) Verdc. ■☆
135639 Dolichos tenuis (E. Mey.) Steud. = Vigna unguiculata (L.) Walp. var. tenuis (E. Mey.) Maréchal et Mascherpa et Stainier ■☆
135640 Dolichos tetragonolobus L. = Psophocarpus tetragonolobus (L.) DC. ■
135641 Dolichos thorelii Gagnep.;海南镰扁豆(越南扁豆);Hainan Dolichos, Hainan Sicklehairicot ■
135642 Dolichos tomentosus Roth = Rhynchosia rothii Benth. ex Aitch. ●■
135643 Dolichos tonkouiensis Portères;越南镰扁豆■☆
135644 Dolichos tricostatus Baker = Dolichos trinervatus Baker ■☆
135645 Dolichos trilobatus L. = Vigna trilobata (L.) Verdc. ■
135646 Dolichos trilobus L.;镰扁豆(大豆荚,大九荚,大麻药,豆叶百步还阳,镰藕豆,镰果扁豆,镰叶山扁豆,麻里麻,麻三段,三极方,三裂叶扁豆,山豆根,野饭豆根);Sicklehairicot, Three-lobes Dolichos ■
135647 Dolichos trilobus L. subsp. occidentalis Verdc.;西方镰扁豆■☆
135648 Dolichos trilobus L. subsp. transvaalicus Verdc.;德兰士瓦镰扁豆■☆
135649 Dolichos trilobus L. var. kosyunensis (Hosok.) H. Ohashi et Tateishi;三裂叶扁豆■
135650 Dolichos trilobus L. var. richardsiae Mackinder;理查德镰扁豆■☆
135651 Dolichos trilobus L. var. stenophyllus Verdc.;狭叶镰扁豆■☆
135652 Dolichos trilobus Lour. = Pueraria lobata (Willd.) Ohwi subsp. thomsonii (Benth.) H. Ohashi et Tateishi ●
135653 Dolichos trilobus Thunb. = Vigna unguiculata (L.) Walp. subsp. stenophylla (Harv.) Maréchal et al. ■☆
135654 Dolichos trinervatus Baker;三脉镰扁豆■☆
135655 Dolichos trinervis De Wild. = Dolichos trinervatus Baker ■☆
135656 Dolichos ufiomensis Heering = Dolichos kilimandscharicus Taub. subsp. parviflorus Verdc. ■☆
135657 Dolichos umbellatus Thunb. = Vigna umbellata (Thunb.) Ohwi et H. Ohashi ■
135658 Dolichos uncinatus L. = Teramnus uncinatus (L.) Sw. ■☆
135659 Dolichos uncinatus Schweinf. = Lablab purpureus (L.) Sweet var. uncinatus Verdc. ■☆
135660 Dolichos unguiculatus L. = Vigna unguiculata (L.) Walp. ■
135661 Dolichos uniflorus Lam. = Macrotyloma uniflorum (Lam.) Verdc. ■
135662 Dolichos uniflorus Lam. var. stenocarpus Brenan = Macrotyloma uniflorum (Lam.) Verdc. var. stenocarpum (Brenan) Verdc. ■☆
135663 Dolichos urens L. = Mucuna urens (L.) Medik. ●☆
135664 Dolichos venulosus Hiern = Rhynchosia venulosa (Hiern) K. Schum. ■☆
135665 Dolichos verdickii De Wild. = Dolichos kilimandscharicus Taub. ■☆
135666 Dolichos virgatus Rich. = Dioclea virgata (Rich.) Amshoff ■☆
135667 Dolichos virosus Roxb. = Canavalia virosa (Roxb.) Wight et Arn. ■
135668 Dolichos volkensii Taub. = Dolichos oliveri Schweinf. ■☆
135669 Dolichos xiphophyllus Baker;剑叶镰扁豆■☆
135670 Dolichos zovuanyi R. Wilczek;佐瓦尼镰扁豆■☆
135671 Dolichosiphon Phil. = Jaborosa Juss. ●☆
135672 Dolichostachys Benoist(1962);长穗爵床属■☆
135673 Dolichostachys elongata Benoist;长穗爵床■☆
135674 Dolichostegia Schltr. (1915);长盖萝藦属☆
135675 Dolichostegia boholensis Schltr.;长盖萝藦☆
135676 Dolichostemon Bonati. (1924);长蕊玄参属☆
135677 Dolichostemon verticillatus Bonati;长蕊玄参☆
135678 Dolichostigma Miers = Jaborosa Juss. ●☆
135679 Dolichostylis Cass. = Barnadesia Mutis ex L. f. ●☆
135680 Dolichostylis Cass. = Turpinia Bonpl. (废弃属名)●
135681 Dolichostylis Cass. = Turpinia Vent. (保留属名)●
135682 Dolichostylis Turcz. = Draba L. ■
135683 Dolichostylis Turcz. = Stenonema Hook. ■
135684 Dolichotheca Cass. = Campylotheca Cass. ■●
135685 Dolichothele (K. Schum.) Britton et Rose = Mammillaria Haw. (保留属名)●
135686 Dolichothele (K. Schum.) Britton et Rose(1923);长疣球属(金星属);Dolichothele ■☆
135687 Dolichothele Britton et Rose = Mammillaria Haw. (保留属名)●
135688 Dolichothele aylostera (Werderm.) Backeb.;长疣仙人球■☆
135689 Dolichothele longimamma (DC.) Britton et Rose;金星(长疣八卦掌,大长疣仙人球);Long Mammilated Cactus, Longbreast Dolichothele ■☆
135690 Dolichothele sphaerica (A. Dietr.) Britton et Rose = Mammillaria sphaerica A. Dietr. ■☆
135691 Dolichothele sphaerica (Dietr.) Britton et Rose;八卦掌(小长疣仙人球);Sphaere Dolichothele ■☆
135692 Dolichothele uberiformis (Schum.) Britton et Rose;海王星(乳头长疣球,乳头状仙人球,圆疣仙人球)■☆
135693 Dolichothrix Hilliard et B. L. Burtt(1981);黄花帚鼠麹属●■☆
135694 Dolichothrix ericoides (Lam.) Hilliard et B. L. Burtt;黄花帚鼠麹■☆
135695 Dolichothrtx ericoides Hilliard et B. L. Burtt = Dolichothrix ericoides (Lam.) Hilliard et B. L. Burtt ■☆
135696 Dolichoura Brade(1959);长尾野牡丹属☆
135697 Dolichoura kollmannii R. Goldenb. et R. Tav.;柯尔曼长尾野牡丹☆
135698 Dolichoura spiritusanctensis Brade;长尾野牡丹☆
135699 Dolichovigna Hayata = Vigna Savi(保留属名)■
135700 Dolichovigna Hayata(1920);台豆属■
135701 Dolichovigna formosana Hayata;台豆■
135702 Dolichovigna formosana Hayata = Vigna pilosa (Klein ex Willd.) Baker ex K. Heyne ■
135703 Dolichovigna rhombifolia Hayata = Dolichos rhombifolius (Hayata) Hosok. ■
135704 Dolichus E. Mey. = Dolichos L. (保留属名)■
135705 Dolichus uniflorum Lam. = Macrotyloma uniflorum (Lam.) Verdc. ■
135706 Dolicokentia Becc. = Cyphokentia Brongn. ●☆
135707 Dolicokentia Becc. = Dolichokentia Becc. ●☆
135708 Dolicotheea Benth. et Hook. f. = Campylotheca Cass. ■●

135709 Dolicotheea Benth. et Hook. f. = Dolichotheca Cass. ■●
135710 Doliocarpus Rol. (1756);蕴水藤属(伪果藤属)●☆
135711 Doliocarpus dentatus (Aubl.) Standl.;齿状蕴水藤(齿状伪果藤)●☆
135712 Doliocarpus major J. F. Gmel.;大蕴水藤(大伪果藤)●☆
135713 Dollinea Post et Kuntze = Dollinera Endl. ●
135714 Dollinera Endl. (1840);饿蚂蝗属●
135715 Dollinera Endl. = Desmodium Desv. (保留属名)●■
135716 Dollinera floribunda (G. Don) Endl. = Desmodium floribundum G. Don ●
135717 Dollinera floribunda (G. Don) Endl. = Desmodium multiflorum DC. ●
135718 Dollinera sambuensis (G. Don) Endl. = Desmodium multiflorum DC. ●
135719 Dollinera sequax (Wall.) Schindl. = Desmodium sequax Wall. ●
135720 Dollineria Saut. = Draba L. ■
135721 Dolomiaea DC. (1833);川木香属(藏菊属);Dolomiaea, Vladimiria ■
135722 Dolomiaea berardioides (Franch.) C. Shih;厚叶川木香(厚叶木香,理木香,木香,青木香);Thickleaf Dolomiaea, Thickleaf Vladimiria ■
135723 Dolomiaea berardioides (Franch.) C. Shih = Vladimiria berardioides (Franch.) Y. Ling ■
135724 Dolomiaea calophylla Y. Ling;美叶川木香(多罗菊,美叶藏菊,美叶多罗菊);Calophyllous Dolomiaea ■
135725 Dolomiaea cooperi (Anthony) Y. Ling = Diplazoptilon cooperi (J. Anthony) C. Shih ■
135726 Dolomiaea crispo-undulata (C. C. Chang) Y. Ling;皱叶川木香;Wrinkleleaf Dolomiaea ■
135727 Dolomiaea crispo-undulata (C. C. Chang) Y. Ling var. chienii Y. Ling = Dolomiaea crispo-undulata (C. C. Chang) Y. Ling ■
135728 Dolomiaea denticulata (Y. Ling) C. Shih;越隽川木香(越隽木香,越西木香);Toothed Dolomiaea ■
135729 Dolomiaea denticulata (Y. Ling) C. Shih = Vladimiria denticulata Y. Ling ■
135730 Dolomiaea edulis (Franch.) C. Shih;菜木香(长苞菜木香,大理木香,具茎菜木香,有苞菜木香,有茎菜木香);Edible Vladimiria, Vegetable Dolomiaea ■
135731 Dolomiaea forrestii (Diels) C. Shih;膜缘川木香(膜缘木香);Forrest Dolomiaea, Forrest Vladimiria ■
135732 Dolomiaea georgii (J. Anthony) C. Shih;腺叶川木香;Georg Dolomiaea, Glandleaf Dolomiaea ■
135733 Dolomiaea lateritia C. Shih;红冠川木香;Redcrown Dolomiaea ■
135734 Dolomiaea platylepis (Hand. -Mazz.) C. Shih;平苞川木香;Flatbract Dolomiaea ■
135735 Dolomiaea salwinensis (Hand. -Mazz.) C. Shih;怒江川木香;Nujiang Dolomiaea, Salwin Dolomiaea ■
135736 Dolomiaea saussureoides (Hand. -Mazz.) Y. L. Chen = Bolocephalus saussureoides Hand. -Mazz. ■
135737 Dolomiaea saussureoides (Hand. -Mazz.) Y. L. Chen et C. Shih = Bolocephalus saussureoides Hand. -Mazz. ■
135738 Dolomiaea scabrida (C. Shih et S. Y. Jin) C. Shih;糙羽川木香(糙羽木香);Rugged Dolomiaea, Scabrous Dolomiaea ■
135739 Dolomiaea souliei (Franch.) C. Shih;川木香(越隽木香);Common Vladimiria, Dolomiaea, Soulie Dolomiaea ■
135740 Dolomiaea souliei (Franch.) C. Shih var. mirabilis (J. Anthony) C. Shih;灰毛川木香(灰背川木香,毛川木香,木里木香);Greyhair Dolomiaea ■
135741 Dolomiaea wardii (Hand. -Mazz.) Y. Ling;西藏川木香(南藏菊);S. Xizang Dolomiaea, Ward's Dolomiaea ■
135742 Dolophragma Fenzl = Arenaria L. ■
135743 Dolophragma juniperinum (D. Don) Fenzl = Arenaria densissima Wall. ex Edgew. et Hook. f. ■
135744 Dolophyllum Salisb. (废弃属名) = Thujopsis Siebold et Zucc. ex Endl. (保留属名)●
135745 Dolosanthus Klatt = Vernonia Schreb. (保留属名)●■
135746 Dolosanthus sylvaticus Klatt = Vernonia anthelmintica (L.) Willd. ■
135747 Dolpojestella Farille et Lachard = Chamaesium H. Wolff ■★
135748 Dolpojestella Farille et Lachard(2002);尼泊尔矮芹属■☆
135749 Doma Lam. = Hyphaene Gaertn. ●☆
135750 Doma Poir. = Hyphaene Gaertn. ●☆
135751 Dombeia Raeusch. = Araucaria Juss. ●
135752 Dombeia Raeusch. = Dombeya Cav. (保留属名)●☆
135753 Dombeya Cav. (1786)(保留属名);丹氏梧桐属(丹比亚木属,窦比属,多贝梧桐属,铃铃属);Dombeya ●☆
135754 Dombeya L'Hér. (废弃属名) = Dombeya Cav. (保留属名)●☆
135755 Dombeya L'Hér. (废弃属名) = Tourrettia Foug. (保留属名)■☆
135756 Dombeya Lam. = Araucaria Juss. ●
135757 Dombeya Lam. = Dombeya Cav. (保留属名)●☆
135758 Dombeya acerifolia Baker;尖叶丹氏梧桐●☆
135759 Dombeya acerifolia Baker var. montana Hochr. = Dombeya montana (Hochr.) Arènes ●☆
135760 Dombeya acerifolia Baker var. typica Hochr. = Dombeya acerifolia Baker ●☆
135761 Dombeya acuminatissima Hochr.;尖丹氏梧桐●☆
135762 Dombeya acutangula Cav.;棱角丹氏梧桐●☆
135763 Dombeya aethiopica Gilli;埃塞俄比亚丹氏梧桐●☆
135764 Dombeya alascha K. Schum. = Dombeya quinqueseta (Delile) Exell ●☆
135765 Dombeya albiflora K. Schum. = Dombeya torrida (J. F. Gmel.) Bamps ●☆
135766 Dombeya albisquama Arènes;白鳞丹氏梧桐●☆
135767 Dombeya albotomentosa Arènes;白绒毛丹氏梧桐●☆
135768 Dombeya alleizettei Arènes;阿雷丹氏梧桐●☆
135769 Dombeya amaniensis Engl.;阿马尼丹氏梧桐●☆
135770 Dombeya ambongensis Arènes;安邦丹氏梧桐●☆
135771 Dombeya ambositrensis Arènes;安布西特拉丹氏梧桐●☆
135772 Dombeya amplifolia Arènes;大叶丹氏梧桐●☆
135773 Dombeya andapensis Arènes;安达帕丹氏梧桐●☆
135774 Dombeya angustipetala Arènes;窄瓣丹氏梧桐●☆
135775 Dombeya ankaratrensis Arènes;安卡拉特拉丹氏梧桐●☆
135776 Dombeya ankazobeensis Arènes;阿卡祖贝丹氏梧桐●☆
135777 Dombeya antunesii Exell et Mendonça = Dombeya burgessiae Gerr. -Corn. ex Harv. et Sond. ●☆
135778 Dombeya aquifoliopsis Hochr.;冬青丹氏梧桐●☆
135779 Dombeya atacorensis A. Chev. = Dombeya quinqueseta (Delile) Exell ●☆
135780 Dombeya auriculata K. Schum. = Dombeya burgessiae Gerr. -Corn. ex Harv. et Sond. ●☆
135781 Dombeya australis Scott-Elliot;南部丹氏梧桐●☆
135782 Dombeya autumnalis I. Verd.;秋丹氏梧桐●☆
135783 Dombeya bagshawei Baker f. = Dombeya buettneri K. Schum. ●☆
135784 Dombeya baronii Baker;巴龙丹氏梧桐●☆

135785 Dombeya bathiei Hochr.;巴西丹氏梧桐●☆
135786 Dombeya bemarivensis (Hochr.) Arènes;贝马里武丹氏梧桐●☆
135787 Dombeya bequaertii De Wild. = Dombeya torrida (J. F. Gmel.) Bamps ●☆
135788 Dombeya biumbellata Baker;双小伞丹氏梧桐●☆
135789 Dombeya bogoriensis De Wild. = Dombeya torrida (J. F. Gmel.) Bamps ●☆
135790 Dombeya brachystemma Milne-Redh. = Dombeya wittei De Wild. et Staner ●☆
135791 Dombeya bracteopoda K. Schum. = Dombeya viburniflora Bojer ●☆
135792 Dombeya brevistyla Arènes;短柱丹氏梧桐●☆
135793 Dombeya bruceana A. Rich. = Dombeya torrida (J. F. Gmel.) Bamps ●☆
135794 Dombeya buettneri K. Schum.;比特纳丹氏梧桐●☆
135795 Dombeya burgessiae Gerr. -Corn. ex Harv. et Sond.;粉红丹氏梧桐(多贝梧桐,非洲丹氏梧桐,非洲芙蓉,粉红丹比亚木,婚礼花,铃铃花);Pink Dombeya,Pink Wild Pear ●☆
135796 Dombeya burgessiae Gerrard ex Harv. var. crenulata Szyszyl. = Dombeya burgessiae Gerr. -Corn. ex Harv. et Sond. ●☆
135797 Dombeya burttii Exell = Dombeya burgessiae Gerr. -Corn. ex Harv. et Sond. ●☆
135798 Dombeya cacuminum Hochr.; 渐尖丹氏梧桐; Strawberry Snowball Tree ●☆
135799 Dombeya calantha K. Schum. = Dombeya burgessiae Gerr. -Corn. ex Harv. et Sond. ●☆
135800 Dombeya cannabina (Bojer) Baill.;大麻丹氏梧桐●☆
135801 Dombeya capuroniana Arènes;凯普伦丹氏梧桐●☆
135802 Dombeya capuronii Arènes;凯丹氏梧桐●☆
135803 Dombeya catatii Hochr.;卡他丹氏梧桐●☆
135804 Dombeya cayeuxii L. E. André;杂种丹氏梧桐(红铃花);Hybrid Dombeya,Pink Ball Dombeya,Pink Snowball,Pink Snowball Tree ●☆
135805 Dombeya cerasiflora Exell = Dombeya rotundifolia (Hochst.) Planch. ●☆
135806 Dombeya chapelieri Baill. = Dombeya spectabilis Bojer ●☆
135807 Dombeya cincinnata K. Schum. = Dombeya acutangula Cav. ●☆
135808 Dombeya claessensii De Wild. = Dombeya buettneri K. Schum. ●☆
135809 Dombeya cloiselii Arènes;克卢塞尔丹氏梧桐●☆
135810 Dombeya concinna K. Schum. = Dombeya burgessiae Gerr. -Corn. ex Harv. et Sond. ●☆
135811 Dombeya condensata Hochr.;密集丹氏梧桐●☆
135812 Dombeya condensiflora De Wild. = Dombeya rotundifolia (Hochst.) Planch. ●☆
135813 Dombeya coursii Arènes;库尔斯丹氏梧桐●☆
135814 Dombeya cuanzensis (Hiern) Welw. ex K. Schum. = Dombeya rotundifolia (Hochst.) Planch. ●☆
135815 Dombeya cymosa Harv.;聚伞丹氏梧桐●☆
135816 Dombeya damarana K. Schum. = Dombeya rotundifolia (Hochst.) Planch. ●☆
135817 Dombeya danguyi Hochr.;当吉丹氏梧桐●☆
135818 Dombeya dawei Sprague = Dombeya burgessiae Gerr. -Corn. ex Harv. et Sond. ●☆
135819 Dombeya decaryana Arènes;德卡里梧桐●☆
135820 Dombeya decaryi Hochr.;德氏梧桐●☆
135821 Dombeya delevoyi De Wild. = Dombeya rotundifolia (Hochst.) Planch. ●☆
135822 Dombeya delilei Planch. = Dombeya quinqueseta (Delile) Exell ●☆
135823 Dombeya densiflora Planch. ex Harv. = Dombeya rotundifolia (Hochst.) Planch. ●☆
135824 Dombeya dichotoma Hochr.;二叉丹氏梧桐●☆
135825 Dombeya dichotoma Hochr. f. pubescens Hochr. = Dombeya pubescens (Hochr.) Arènes ●☆
135826 Dombeya dichotomopsis Hochr.;拟二叉丹氏梧桐●☆
135827 Dombeya digyna Hochr.;二雄蕊丹氏梧桐●☆
135828 Dombeya digynopsis Hochr.;拟二雄蕊丹氏梧桐●☆
135829 Dombeya dinteri Schinz = Dombeya rotundifolia (Hochst.) Planch. ●☆
135830 Dombeya divaricata Arènes;叉开丹氏梧桐●☆
135831 Dombeya dregeana Sond. = Dombeya tiliacea (Endl.) Planch. ●☆
135832 Dombeya elegans K. Schum. = Dombeya burgessiae Gerrard ex Harv. ●☆
135833 Dombeya elliotii K. Schum. et Engl. = Dombeya torrida (J. F. Gmel.) Bamps ●☆
135834 Dombeya elliptica Bojer;椭圆丹氏梧桐●☆
135835 Dombeya elskensii De Wild. = Dombeya buettneri K. Schum. ●☆
135836 Dombeya emarginata E. A. Bruce = Dombeya buettneri K. Schum. ●☆
135837 Dombeya endlichii Engl. et K. Krause = Dombeya burgessiae Gerrard ex Harv. ●☆
135838 Dombeya erythroclada Bojer;红枝丹氏梧桐●☆
135839 Dombeya erythroleuca K. Schum. = Dombeya torrida (J. F. Gmel.) Bamps subsp. erythroleuca (K. Schum.) Seyani ●☆
135840 Dombeya faucicola K. Schum. = Dombeya torrida (J. F. Gmel.) Bamps ●☆
135841 Dombeya ficulnea Baill. = Dombeya punctata Cav. ●☆
135842 Dombeya ficulnea var. tsaratananensis Hochr. = Dombeya tsaratananensis (Hochr.) Arènes ●☆
135843 Dombeya flabellifolia Arènes;扇叶丹氏梧桐●☆
135844 Dombeya floribunda Baker;繁花丹氏梧桐●☆
135845 Dombeya gallana K. Schum. et Engl. = Dombeya torrida (J. F. Gmel.) Bamps ●☆
135846 Dombeya gallana K. Schum. et Engl. var. floribunda Fiori = Dombeya aethiopica Gilli ●☆
135847 Dombeya gamwelliae Exell = Dombeya burgessiae Gerrard ex Harv. ●☆
135848 Dombeya gemina Baker = Dombeya biumbellata Baker ●☆
135849 Dombeya gilgiana K. Schum. = Dombeya kirkii Mast. ●☆
135850 Dombeya gilgiana K. Schum. var. scaberula ? = Dombeya kirkii Mast. ●☆
135851 Dombeya gillettii Gilli = Dombeya kirkii Mast. ●☆
135852 Dombeya glaberrima Arènes;光滑丹氏梧桐●☆
135853 Dombeya glabripes Arènes;光梗丹氏梧桐●☆
135854 Dombeya glandulosissima Arènes;多腺丹氏梧桐●☆
135855 Dombeya globiflora Staner = Dombeya burgessiae Gerrard ex Harv. ●☆
135856 Dombeya goetzenii K. Schum.;格氏多贝梧桐●☆
135857 Dombeya goetzenii K. Schum. = Dombeya torrida (J. F. Gmel.) Bamps ●☆
135858 Dombeya gossweileri Exell = Dombeya rotundifolia (Hochst.) Planch. ●☆
135859 Dombeya greenwayi Wild = Dombeya burgessiae Gerrard ex

Harv. ●☆

135860 Dombeya guazumifolia Baill. = Helmiopsis rigida (Baill.) Dorr ●☆

135861 Dombeya heterotricha Mildbr. = Dombeya torrida (J. F. Gmel.) Bamps ●☆

135862 Dombeya hildebrandtii Baill.;希尔德梧桐●☆

135863 Dombeya hilsenbergii Baill.;希尔森梧桐●☆

135864 Dombeya hirsuta (Hochst. ex Schweinf. et Asch.) K. Schum. = Dombeya torrida (J. F. Gmel.) Bamps ●☆

135865 Dombeya hirsuta Bojer;多毛丹氏梧桐●☆

135866 Dombeya hispidicyma Arènes;硬毛丹氏梧桐●☆

135867 Dombeya huillensis (Hiern) K. Schum. = Dombeya rotundifolia (Hochst.) Planch. ●☆

135868 Dombeya humbertiana Arènes;亨伯特梧桐●☆

135869 Dombeya humblotii Baill. = Dombeya spectabilis Bojer ●☆

135870 Dombeya ivohibeensis Arènes;伊武希贝丹氏梧桐●☆

135871 Dombeya johnstonii Baker = Dombeya burgessiae Gerrard ex Harv. ●☆

135872 Dombeya kandoensis De Wild. et Staner = Dombeya buettneri K. Schum. ●☆

135873 Dombeya katangensis De Wild. et T. Durand = Dombeya shupangae K. Schum. ●☆

135874 Dombeya kefaensis Friis et Bidgood;咖法丹氏梧桐●☆

135875 Dombeya kindtiana De Wild. = Dombeya burgessiae Gerr. -Corn. ex Harv. et Sond. ●☆

135876 Dombeya kirkii Mast.;柯克梧桐●☆

135877 Dombeya kituiensis Baill. ex Hochr. = Dombeya kirkii Mast. ●☆

135878 Dombeya laevissima Hochr.;平滑丹氏梧桐●☆

135879 Dombeya lancea Cordem. = Dombeya punctata Cav. ●☆

135880 Dombeya lantziana Baill. = Dombeya spectabilis Bojer ●☆

135881 Dombeya lasiostylis K. Schum. = Dombeya burgessiae Gerr. -Corn. ex Harv. et Sond. ●☆

135882 Dombeya lastii K. Schum.;拉斯特丹氏梧桐●☆

135883 Dombeya latipetala Arènes;阔瓣丹氏梧桐●☆

135884 Dombeya laurifolia (Bojer) Baill.;月桂叶丹氏梧桐●☆

135885 Dombeya laxiflora K. Schum. = Dombeya kirkii Mast. ●☆

135886 Dombeya leachii Wild;利奇丹氏梧桐●☆

135887 Dombeya leandrii Arènes;利安丹氏梧桐●☆

135888 Dombeya lecomtei Hochr.;勒孔特丹氏梧桐●☆

135889 Dombeya lecomteopsis Arènes;拟勒孔特丹氏梧桐●☆

135890 Dombeya ledermannii Engl.;莱德丹氏梧桐●☆

135891 Dombeya leiomacrantha Hochr.;马岛光花丹氏梧桐●☆

135892 Dombeya leucoderma K. Schum. = Dombeya torrida (J. F. Gmel.) Bamps ●☆

135893 Dombeya leucomacrantha Hochr.;白花丹氏梧桐●☆

135894 Dombeya leucorrhoea K. Schum. = Dombeya acutangula Cav. ●☆

135895 Dombeya linearifolia Hochr.;线叶丹氏梧桐●☆

135896 Dombeya lokohensis Arènes;洛克赫丹氏梧桐●☆

135897 Dombeya longebracteolata Seyani;长苞片丹氏梧桐●☆

135898 Dombeya longicuspidata Arènes;小长尖丹氏梧桐●☆

135899 Dombeya longicuspis Baill.;长尖丹氏梧桐●☆

135900 Dombeya longifolia Baill.;长叶丹氏梧桐●☆

135901 Dombeya longipedicellata Hochr.;长梗丹氏梧桐●☆

135902 Dombeya longipes Baill.;长柄丹氏梧桐●☆

135903 Dombeya loucoubensis Baill. = Dombeya mollis Hook. ●☆

135904 Dombeya lucida Baill.;光亮丹氏梧桐●☆

135905 Dombeya lucida var. bemarivensis Hochr. = Dombeya bemarivensis (Hochr.) Arènes ●☆

135906 Dombeya macrantha Baker;大花丹氏梧桐●☆

135907 Dombeya macropoda Hochr.;大足丹氏梧桐●☆

135908 Dombeya macrotis K. Schum. = Dombeya torrida (J. F. Gmel.) Bamps subsp. erythroleuca (K. Schum.) Seyani ●☆

135909 Dombeya magnifolia Arènes;木兰叶丹氏梧桐●☆

135910 Dombeya majungensis Arènes;马任加丹氏梧桐●☆

135911 Dombeya malacoxylon K. Schum. = Dombeya torrida (J. F. Gmel.) Bamps subsp. erythroleuca (K. Schum.) Seyani ●☆

135912 Dombeya mananarensis Arènes;马纳纳拉丹氏梧桐●☆

135913 Dombeya mandrakensis Arènes;曼兹拉基翁丹氏梧桐●☆

135914 Dombeya mangorensis Arènes;曼古鲁丹氏梧桐●☆

135915 Dombeya manongarivensis Hochr.;马农加丹氏梧桐●☆

135916 Dombeya marojejyensis Arènes;马罗丹氏梧桐●☆

135917 Dombeya mastersii Hook. f. = Dombeya burgessiae Gerr. -Corn. ex Harv. et Sond. ●☆

135918 Dombeya megaphylla Baker = Dombeya lucida Baill. ●☆

135919 Dombeya megaphyllopsis Hochr.;拟大叶丹氏梧桐●☆

135920 Dombeya melanostigma K. Schum. et Engl. = Dombeya rotundifolia (Hochst.) Planch. ●☆

135921 Dombeya merika Arènes;默克丹氏梧桐●☆

135922 Dombeya mildbraedii Engl. = Dombeya buettneri K. Schum. ●☆

135923 Dombeya minor (Endl.) Planch. = Dombeya quinqueseta (Delile) Exell ●☆

135924 Dombeya modesta Baker;适度丹氏梧桐●☆

135925 Dombeya modestiformis Arènes;马岛适度丹氏梧桐●☆

135926 Dombeya mollis Hook.;柔软丹氏梧桐●☆

135927 Dombeya montana (Hochr.) Arènes;山地丹氏梧桐●☆

135928 Dombeya monticola K. Schum. = Dombeya torrida (J. F. Gmel.) Bamps subsp. erythroleuca (K. Schum.) Seyani ●☆

135929 Dombeya mukole Sprague = Dombeya kirkii Mast. ●☆

135930 Dombeya multiflora (Endl.) Planch. = Dombeya quinqueseta (Delile) Exell ●☆

135931 Dombeya multiflora (Endl.) Planch. var. senegalensis (Planch.) Aubrév. = Dombeya quinqueseta (Delile) Exell ●☆

135932 Dombeya multiflora (Endl.) Planch. var. vestita K. Schum. = Dombeya quinqueseta (Delile) Exell ●☆

135933 Dombeya muscosa Hochr.;苔地丹氏梧桐●☆

135934 Dombeya myriantha K. Schum. = Dombeya rotundifolia (Hochst.) Planch. ●☆

135935 Dombeya nairobensis Engl. = Dombeya burgessiae Gerr. -Corn. ex Harv. et Sond. ●☆

135936 Dombeya natalensis Sond. = Dombeya tiliacea (Endl.) Planch. ●☆

135937 Dombeya niangaraensis De Wild. = Dombeya torrida (J. F. Gmel.) Bamps ●☆

135938 Dombeya nyasica Exell = Dombeya burgessiae Gerr. -Corn. ex Harv. et Sond. ●☆

135939 Dombeya oblongifolia Arènes;矩圆叶丹氏梧桐●☆

135940 Dombeya oblongipetala Arènes;矩圆瓣丹氏梧桐●☆

135941 Dombeya obovalis Baill.;倒卵叶丹氏梧桐●☆

135942 Dombeya oligantha Arènes;寡花丹氏梧桐●☆

135943 Dombeya palmatisecta Hochr.;掌状全裂丹氏梧桐●☆

135944 Dombeya parkeri Baill. = Dombeya laurifolia (Bojer) Baill. ●☆

135945 Dombeya parviflora Boivin ex Baill. = Dombeya mollis Hook. ●☆

135946 Dombeya parvifolia K. Schum. = Dombeya burgessiae Gerr. -Corn. ex Harv. et Sond. ●☆

135947 Dombeya parvipetala Arènes;小瓣丹氏梧桐●☆
135948 Dombeya pedunculata K. Schum. = Dombeya buettneri K. Schum. ●☆
135949 Dombeya pentagonalis Arènes;五角丹氏梧桐●☆
135950 Dombeya perrieri Arènes;佩里耶梧桐●☆
135951 Dombeya pervillei Baill. = Dombeya punctata Cav. ●☆
135952 Dombeya platanifolia Bojer;悬铃木丹氏梧桐●☆
135953 Dombeya platypoda K. Schum. = Dombeya burgessiae Gerr. -Corn. ex Harv. et Sond. ●☆
135954 Dombeya praetermissa Dunkley = Dombeya taylorii Baker f. ●☆
135955 Dombeya pubescens (Hochr.) Arènes;短柔毛丹氏梧桐●☆
135956 Dombeya pulchra N. E. Br.;美丽丹氏梧桐●☆
135957 Dombeya punctata Cav.;斑点丹氏梧桐;Bourbon Dombeya ●☆
135958 Dombeya punctatopsis Hochr.;拟斑点丹氏梧桐●☆
135959 Dombeya punctatopsis Hochr. var. subfurcata Hochr. = Dombeya biumbellata Baker ●☆
135960 Dombeya quarrei De Wild. = Dombeya buettneri K. Schum. ●☆
135961 Dombeya quinqueseta (Delile) Exell;五毛丹氏梧桐●☆
135962 Dombeya quinqueseta (Delile) Exell var. senegalensis (Planch.) Keay = Dombeya quinqueseta (Delile) Exell ●☆
135963 Dombeya quinqueseta (Delile) Exell var. vestita (K. Schum.) Bamps = Dombeya quinqueseta (Delile) Exell ●☆
135964 Dombeya rariflora Arènes;稀花丹氏梧桐●☆
135965 Dombeya repanda Baker;浅波状丹氏梧桐●☆
135966 Dombeya reticulata Mast. = Dombeya quinqueseta (Delile) Exell ●☆
135967 Dombeya rigida Baill. = Helmiopsis rigida (Baill.) Dorr ●☆
135968 Dombeya ringoetii De Wild. = Dombeya rotundifolia (Hochst.) Planch. ●☆
135969 Dombeya robynsii De Wild. = Dombeya buettneri K. Schum. ●☆
135970 Dombeya rosacea Arènes;蔷薇丹氏梧桐●☆
135971 Dombeya rosea Baker f. = Dombeya burgessiae Gerr. -Corn. ex Harv. et Sond. ●☆
135972 Dombeya roseiflora Arènes;粉红花丹氏梧桐●☆
135973 Dombeya rottleroides Baill.;野桐丹氏梧桐●☆
135974 Dombeya rotunda Arènes;圆形丹氏梧桐●☆
135975 Dombeya rotundifolia (Hochst.) Planch.;圆叶丹氏梧桐(南非丹比亚木,圆叶窦比);South African Wild Pear ●☆
135976 Dombeya rotundifolia (Hochst.) Planch. var. velutina I. Verd.;短绒毛丹氏梧桐●☆
135977 Dombeya rotundifolia Bojer = Dombeya spectabilis Bojer ●☆
135978 Dombeya rotundifolia Planch. = Dombeya rotundifolia (Hochst.) Planch. ●☆
135979 Dombeya rubricuspis Arènes;红尖丹氏梧桐●☆
135980 Dombeya runsoroensis K. Schum. = Dombeya torrida (J. F. Gmel.) Bamps ●☆
135981 Dombeya ruwenzoriensis De Wild. = Dombeya torrida (J. F. Gmel.) Bamps ●☆
135982 Dombeya sahatavyensis Arènes;萨哈塔维丹氏梧桐●☆
135983 Dombeya sakamaliensis Arènes;萨卡马利丹氏梧桐●☆
135984 Dombeya schimperiana A. Rich. = Dombeya torrida (J. F. Gmel.) Bamps ●☆
135985 Dombeya schimperiana A. Rich. var. glabrata K. Schum. = Dombeya torrida (J. F. Gmel.) Bamps ●☆
135986 Dombeya schoenodoter K. Schum. = Dombeya torrida (J. F. Gmel.) Bamps subsp. erythroleuca (K. Schum.) Seyani ●☆
135987 Dombeya senegalensis Planch. = Dombeya quinqueseta (Delile) Exell ●☆
135988 Dombeya seretii De Wild. = Dombeya buettneri K. Schum. ●☆
135989 Dombeya seyrigiana Arènes;塞里格丹氏梧桐●☆
135990 Dombeya seyrigii Arènes;塞里丹氏梧桐●☆
135991 Dombeya shupangae K. Schum.;舒潘噶丹氏梧桐●☆
135992 Dombeya shupangae K. Schum. var. glabrescens Bamps = Dombeya shupangae K. Schum. ●☆
135993 Dombeya shupangae K. Schum. var. katangensis (De Wild. et T. Durand) Bamps = Dombeya shupangae K. Schum. ●☆
135994 Dombeya sisyrocarpa Gilli = Dombeya torrida (J. F. Gmel.) Bamps subsp. erythroleuca (K. Schum.) Seyani ●☆
135995 Dombeya sparmannioides (Hiern) K. Schum. = Dombeya burgessiae Gerr. -Corn. ex Harv. et Sond. ●☆
135996 Dombeya spectabilis Bojer;壮观丹氏梧桐●☆
135997 Dombeya sphaerantha Gilli = Dombeya burgessiae Gerr. -Corn. ex Harv. et Sond. ●☆
135998 Dombeya squarrosa Engl. = Dombeya buettneri K. Schum. ●☆
135999 Dombeya stipulacea Baill.;托叶丹氏梧桐●☆
136000 Dombeya stipulosa Chiov. = Dombeya torrida (J. F. Gmel.) Bamps ●☆
136001 Dombeya stuhlmannii K. Schum. = Dombeya acutangula Cav. ●☆
136002 Dombeya suarezensis Arènes;苏亚雷斯丹氏梧桐●☆
136003 Dombeya subdichotoma De Wild. = Dombeya rotundifolia (Hochst.) Planch. ●☆
136004 Dombeya subsquamosa Hochr.;多鳞丹氏梧桐●☆
136005 Dombeya subviscosa Hochr.;黏丹氏梧桐●☆
136006 Dombeya superba Arènes;华美丹氏梧桐●☆
136007 Dombeya tanganyikensis Baker f. = Dombeya burgessiae Gerr. -Corn. ex Harv. et Sond. ●☆
136008 Dombeya taylorii Baker f.;泰勒丹氏梧桐●☆
136009 Dombeya tiliacea (Endl.) Planch.;椴叶丹氏梧桐(椴叶丹比亚木);Forest Dombeya,Nastal-cherry,Natal Wedding Flower ●☆
136010 Dombeya tiliacea Planch. = Dombeya tiliacea (Endl.) Planch. ●☆
136011 Dombeya tomentosa Cav.;绒毛丹氏梧桐●☆
136012 Dombeya torrida (J. F. Gmel.) Bamps;托尔丹氏梧桐●☆
136013 Dombeya torrida (J. F. Gmel.) Bamps subsp. erythroleuca (K. Schum.) Seyani;红白丹氏梧桐●☆
136014 Dombeya tremula Hochr.;颤丹氏梧桐●☆
136015 Dombeya tremuliformis Arènes;拟颤丹氏梧桐●☆
136016 Dombeya trichoclada Mildbr. = Dombeya burgessiae Gerr. -Corn. ex Harv. et Sond. ●☆
136017 Dombeya trimorphotricha Arènes;三型毛丹氏梧桐●☆
136018 Dombeya tsaratananensis (Hochr.) Arènes;察拉塔纳纳丹氏梧桐●☆
136019 Dombeya tubulosoviscosa Hochr.;黏管丹氏梧桐●☆
136020 Dombeya tulearensis Arènes;图莱亚尔丹氏梧桐●☆
136021 Dombeya umbraculifera K. Schum. = Dombeya kirkii Mast. ●☆
136022 Dombeya urschiana Arènes;乌尔施丹氏梧桐●☆
136023 Dombeya valimpony R. Vig. et Humbert = Dombeya laurifolia (Bojer) Baill. ●☆
136024 Dombeya valimpony R. Vig. et Humbert f. obovalopsis Hochr. = Dombeya laurifolia (Bojer) Baill. ●☆
136025 Dombeya velutina De Wild. et Staner = Dombeya burgessiae Gerr. -Corn. ex Harv. et Sond. ●☆
136026 Dombeya venosa Arènes;多脉丹氏梧桐●☆
136027 Dombeya viburniflora Bojer;荚蒾丹氏梧桐●☆

136028 Dombeya viburniflorpsis Arènes;拟荚蒾丹氏梧桐●☆
136029 Dombeya vonoa Arènes = Dombeya rotunda Arènes ●☆
136030 Dombeya wallichii (Lindl.) K. Schum.;沃利克丹氏梧桐●☆
136031 Dombeya wallichii Benth. et Hook. f.;瓦丹氏梧桐(大叶丹比亚木,大叶丹氏梧桐);Dombeya ●☆
136032 Dombeya warneckei Engl. = Dombeya kirkii Mast. ●☆
136033 Dombeya wittei De Wild. et Staner;维特丹氏梧桐●☆
136034 Dombeya xiphosepala Baker;剑萼丹氏梧桐●☆
136035 Dombeya xiphosepalopsis Arènes;剑瓣丹氏梧桐●☆
136036 Dombeyaceae (DC.) Bartl. = Sterculiaceae Vent.(保留科名)●■
136037 Dombeyaceae Kunth = Sterculiaceae Vent.(保留科名)●■
136038 Dombeyaceae Schultz Sch. = Malvaceae Juss.(保留科名)●■
136039 Dombeyaceae Schultz Sch. = Sterculiaceae Vent.(保留科名)●■
136040 Dombrowskya Endl. = Dobrowskya C. Presl ■☆
136041 Dombrowskya Endl. = Monopsis Salisb. ■☆
136042 Domeykoa Phil.(1860);多梅草属■☆
136043 Domeykoa oppositifolia Phil.;多梅草■☆
136044 Dominella E. Wimm.(1953);多明草属■☆
136045 Dominella crassomarginata E. Wimm.;多明草■☆
136046 Domingoa Schltr.(1913);多明我兰属■☆
136047 Domingoa hymenodes Schltr.;多明戈兰■☆
136048 Dominia Fedde = Uldinia J. M. Black ■☆
136049 Domkeocarpa Markgr. = Tabernaemontana L. ●
136050 Domohinea Léandri(1941);多莫大戟属■☆
136051 Domohinea perrieri Léandri;多莫大戟■☆
136052 Donacium Fr. = Arundo L. ●
136053 Donacium Fr. = Donax P. Beauv. ●
136054 Donacodes Blume = Amomum Roxb.(保留属名)■
136055 Donacopsis Gagnep. = Eulophia R. Br.(保留属名)+ Arundina Blume ■
136056 Donacopsis Gagnep. = Eulophia R. Br.(保留属名)■
136057 Donaldia Klotzsch = Begonia L. ●■
136058 Donaldsonia Baker = Moringa Rheede ex Adans. ●
136059 Donaldsonia Baker f. = Moringa Rheede ex Adans. ●
136060 Donatia Bert. ex J. Rémy = Lastarriaea J. Rémy ■☆
136061 Donatia J. R. Forst. = Donatia J. R. Forst. et G. Forst.(保留属名)●■☆
136062 Donatia J. R. Forst. et G. Forst.(1775)(保留属名);陀螺果属(离瓣花柱草属)●■☆
136063 Donatia Loefl. = Avicennia L. ●
136064 Donatia fascicularis J. R. Forst. et G. Forst.;陀螺果●☆
136065 Donatiaceae B. Chandler(1911)(保留科名);陀螺果科●☆
136066 Donatiaceae Hutch. = Donatiaceae B. Chandler(保留科名)●☆
136067 Donatiaceae Skottsb. = Donatiaceae B. Chandler(保留科名)●☆
136068 Donatiaceae Skottsb. = Stylidiaceae R. Br.(保留科名)●■
136069 Donatophorus Zipp. = Harpullia Roxb. ●
136070 Donax K. Schum. = Clinogyne Salisb. ex Benth. ■☆
136071 Donax K. Schum. = Schumannianthus Gagnep. ■☆
136072 Donax Lour.(1790);竹叶蕉属;Dona ■
136073 Donax P. Beauv. = Arundo L. ●
136074 Donax arillata K. Schum. = Marantochloa purpurea (Ridl.) Milne-Redh. ■☆
136075 Donax arundastrum Lour. = Donax canniformis (G. Forst.) K. Schum. ■
136076 Donax arundinacea Merr. = Maranta arundinacea L. ■
136077 Donax arundinaceus P. Beauv. = Arundo donax L. ●
136078 Donax azurea K. Schum. = Halopegia azurea (K. Schum.) K. Schum. ■☆
136079 Donax bengalensis (Retz.) P. Beauv. = Arundo donax L. ●
136080 Donax borealis Trin. = Scolochloa festucacea (Willd.) Link ■
136081 Donax canniformis (G. Forst.) K. Schum.;竹叶蕉(戈燕,兰屿竹芋);Cannalike Dona ■
136082 Donax canniformis (G. Forst.) Rolfe = Donax canniformis (G. Forst.) K. Schum. ■
136083 Donax congensis K. Schum. = Marantochloa congensis (K. Schum.) J. Léonard et Mullend. ■☆
136084 Donax leucantha K. Schum. = Marantochloa leucantha (K. Schum.) Milne-Redh. ■☆
136085 Donax oligantha K. Schum. = Marantochloa filipes (Benth.) Hutch. ■☆
136086 Donax ugandensis K. Schum. = Marantochloa leucantha (K. Schum.) Milne-Redh. ■☆
136087 Doncklaeria Hort. ex Loudon = Centradenia G. Don ●■☆
136088 Doncklaeria Hort. ex Loudon = Donkelaaria Hort. ex Lem. ●■☆
136089 Dondia Adans. = Lerchia Zinn ●■
136090 Dondia Adans. = Suaeda Forssk. ex J. F. Gmel.(保留属名)●■
136091 Dondia Spreng. = Hacquetia Neck. ex DC. ■☆
136092 Dondia conferta Small = Suaeda conferta (Small) I. M. Johnst. ■☆
136093 Dondia tampicensis Standl. = Suaeda tampicensis (Standl.) Standl. ■☆
136094 Dondia taxifolia Standl. = Suaeda taxifolia (Standl.) Standl. ■☆
136095 Dondisia DC. = Canthium Lam. ●
136096 Dondisia DC. = Plectronia L. ●☆
136097 Dondisia Rchb. = Dondia Spreng. ■☆
136098 Dondisia Rchb. = Hacquetia Neck. ex DC. ■☆
136099 Dondisia Scop. = Raphanus L. ■
136100 Dondodia Luer = Cryptophoranthus Barb. Rodr. ■☆
136101 Dondodia Luer(2006);南美窗兰属■☆
136102 Donella Pierre = Chrysophyllum L. ●
136103 Donella Pierre ex Baill. = Chrysophyllum L. ●
136104 Donella ambrensis Aubrév. = Chrysophyllum ambrense (Aubrév.) G. E. Schatz et L. Gaut. ●☆
136105 Donella analalavensis Aubrév. = Chrysophyllum analalavense (Aubrév.) G. E. Schatz et L. Gaut. ●☆
136106 Donella delphinensis Aubrév. = Chrysophyllum delphinense (Aubrév.) G. E. Schatz et L. Gaut. ●☆
136107 Donella fenerivensis Aubrév. = Chrysophyllum fenerivense (Aubrév.) G. E. Schatz et L. Gaut. ●☆
136108 Donella griffoniana Pierre ex Pellegr. = Chrysophyllum ogoouense A. Chev. ●☆
136109 Donella lanceolata (A. Chev.) A. Chev. ex Pellegr. var. malagassica Aubrév. = Chrysophyllum lanceolatum (Blume) A. DC. ●
136110 Donella lanceolata (A. Chev.) A. Chev. ex Pellegr. var. stellatocarpon (P. Royn) X. Y. Chang = Chrysophyllum lanceolatum (Blume) A. DC. var. stellatocarpon P. Royen ●
136111 Donella letestuana (A. Chev.) A. Chev. ex Pellegr. = Chrysophyllum ubangiense (De Wild.) D. J. Harris ●☆
136112 Donella masoalensis Aubrév. = Chrysophyllum masoalense (Aubrév.) G. E. Schatz et L. Gaut. ●☆
136113 Donella ogoouensis (A. Chev.) Aubrév. et Pellegr. = Chrysophyllum ogoouense A. Chev. ●☆
136114 Donella parvifolia Lecomte = Chrysophyllum pruniforme Pierre

ex Engl. ●☆

136115 Donella pentagonocarpa (Engl. et K. Krause) Aubrév. et Pellegr. = Chrysophyllum ubangiense (De Wild.) D. J. Harris ●☆

136116 Donella perrieri Lecomte = Chrysophyllum perrieri (Lecomte) G. E. Schatz et L. Gaut. ●☆

136117 Donella perrieri Lecomte var. pubescens Aubrév. = Chrysophyllum perrieri (Lecomte) G. E. Schatz et L. Gaut. ●☆

136118 Donella perrieri Lecomte var. sambiranensis Lecomte = Chrysophyllum perrieri (Lecomte) G. E. Schatz et L. Gaut. ●☆

136119 Donella pruniformis (Pierre ex Engl.) Aubrév. et Pellegr. = Chrysophyllum pruniforme Pierre ex Engl. ●☆

136120 Donella sambiranensis (Lecomte) Aubrév. = Chrysophyllum perrieri (Lecomte) G. E. Schatz et L. Gaut. ●☆

136121 Donella ubangiensis (De Wild.) Aubrév. = Chrysophyllum ubangiense (De Wild.) D. J. Harris ●☆

136122 Donella viridifolia (J. M. Wood et Franks) Aubrév. et Pellegr. = Chrysophyllum viridifolium J. M. Wood et Franks ●☆

136123 Donella welwitschii (Engl.) Pierre ex Aubrév. et Pellegr. = Chrysophyllum welwitschii Engl. ●☆

136124 Donepea Airy Shaw = Douepea Cambess. ■☆

136125 Donia G. Don et D. Don = Clianthus Sol. ex Lindl.(保留属名)●

136126 Donia G. Don et D. Don ex G. Don = Clianthus Sol. ex Lindl.(保留属名)●

136127 Donia Nutt. = Aster L. ●■

136128 Donia R. Br. (1813) = Grindelia Willd. ●■☆

136129 Donia R. Br. (1819) = Oxyria Hill ■

136130 Donia Raf. = Donia R. Br. ●■☆

136131 Donia ciliata Nutt. = Grindelia ciliata (Nutt.) Spreng. ■☆

136132 Donia formosa G. Don;丽唐菊■☆

136133 Donia glutinosa (Cav.) R. Br.;唐菊■☆

136134 Donia glutinosa R. Br. = Donia glutinosa (Cav.) R. Br. ■☆

136135 Donia lanceolata Hook. = Pyrrocoma lanceolata (Hook.) Greene ■☆

136136 Donia squarrosa Pursh = Grindelia squarrosa (Pursh) Dunal ■☆

136137 Donia uniflora Hook. = Pyrrocoma uniflora (Hook.) Greene ■☆

136138 Doniana Raf. = Donia R. Br. (1813) ●■☆

136139 Doniana Raf. = Grindelia Willd. ●■☆

136140 Donidsia G. Don = Dondisia DC. ●☆

136141 Donidsia G. Don = Plectronia L. ●☆

136142 Doniophyton Wedd. (1855);羽刺菊属■☆

136143 Doniophyton andicola Wedd.;羽刺菊■☆

136144 Donkelaaria Hort. ex Lem. = Centradenia G. Don ●■☆

136145 Donkelaaria Lem. = Guettarda L. ●

136146 Donnellia C. B. Clarke = Neodonnellia Rose ■☆

136147 Donnellia C. B. Clarke = Tripogandra Raf. ■☆

136148 Donnellia C. B. Clarke ex Donn. Sm. = Neodonnellia Rose ■☆

136149 Donnellia C. B. Clarke ex Donn. Sm. = Tripogandra Raf. ■☆

136150 Donnellsmithia J. M. Coult. et Rose(1890);道斯芹属■☆

136151 Donnellsmithia mexicana (B. L. Rob.) Mathias et Constance;墨西哥道斯芹■☆

136152 Donnellsmithia ovata (J. M. Coult. et Rose) Mathias et Constance;卵形道斯芹■☆

136153 Donningia A. Gray = Downingia Torr.(保留属名)■☆

136154 Dontospermum Neck. ex Sch. Bip. = Asteriscus Mill. ●■☆

136155 Dontospermum Sch. Bip. = Odontospermum Neck. ex Sch. Bip. ■☆

136156 Dontostemon Andrz. ex C. A. Mey. (1831)(保留属名);花旗杆属(花旗竿属);Dontostemon ■

136157 Dontostemon Andrz. ex DC. = Dontostemon Andrz. ex C. A. Mey.(保留属名)■

136158 Dontostemon Andrz. ex Ledeb. = Dontostemon Andrz. ex C. A. Mey.(保留属名)■

136159 Dontostemon asper (Pall.) Schischk. = Dontostemon pinnatifidus (Willd.) Al-Shehbaz et H. Ohba ■

136160 Dontostemon asper Schischk. = Dontostemon pinnatifidus (Willd.) Al-Shehbaz et H. Ohba ■

136161 Dontostemon brevipes Bunge = Malcolmia karelinii Lipsky ■

136162 Dontostemon crassifolius (Bunge) Maxim.;厚叶花旗杆(厚叶花旗竿);Thickleaf Dontostemon ■

136163 Dontostemon dentatus (Bunge) Ledeb.;花旗杆(齿叶花旗杆,花旗竿,苦葶苈,米蒿,腺花旗杆);Dentate Dontostemon, Glandular Dentate Dontostemon ■

136164 Dontostemon dentatus (Bunge) Ledeb. var. glandulosus Maxim. ex Franch. et Sav. = Dontostemon dentatus (Bunge) Ledeb. ■

136165 Dontostemon eglandulosus (DC.) Ledeb. = Dontostemon integrifolius (L.) Ledeb. ■

136166 Dontostemon eglandulosus (DC.) Ledeb. = Dontostemon integrifolius (L.) Ledeb. var. eglandulosus (DC.) Turcz. ■

136167 Dontostemon eglandulosus (DC.) Ledeb. = Dontostemon perennis C. A. Mey. ■

136168 Dontostemon eglandulosus C. A. Mey. = Dontostemon dentatus (Bunge) Ledeb. ■

136169 Dontostemon elegans Maxim.;扭果花旗杆(扭果花旗竿);Elegant Dontostemon ■

136170 Dontostemon elegans Maxim. var. semiamplexicaulis (H. L. Yang) H. L. Yang et M. S. Yan = Dontostemon elegans Maxim. ■

136171 Dontostemon glandulosus (Kar. et Kir.) O. E. Schulz = Dimorphostemon glandulosus (Kar. et Kir.) Golubk. ■

136172 Dontostemon hispidus Maxim.;毛蕊花旗杆(毛花旗杆)■

136173 Dontostemon integrifolius (L.) Ledeb.;全缘叶花旗杆(条叶连蕊芥,无腺花旗杆,无腺花旗竿,线叶花旗杆,线叶花旗竿);Entireleaf Dontostemon, Glanduleless Dontostemon, Linearleaf Dontostemon, Linearleaf Systemon ■

136174 Dontostemon integrifolius (L.) Ledeb. var. eglandulosus (DC.) Turcz. = Dontostemon integrifolius (L.) Ledeb. ■

136175 Dontostemon integrifolius (L.) Ledeb. var. eglandulosus Turcz. = Dontostemon perennis C. A. Mey. ■

136176 Dontostemon integrifolius (L.) Ledeb. var. glandulosus Turcz. = Dontostemon integrifolius (L.) Ledeb. ■

136177 Dontostemon intermedius Vorosch. = Dontostemon dentatus (Bunge) Ledeb. ■

136178 Dontostemon matthioloides Franch. = Oreoloma matthioloides (Franch.) Botsch. ■

136179 Dontostemon micranthus C. A. Mey.;小花花旗杆(小花花旗竿);Littleflower Dontostemon ■

136180 Dontostemon oblongifolia Ledeb. = Dontostemon dentatus (Bunge) Ledeb. ■

136181 Dontostemon pectinatus (DC.) Golubk. = Dontostemon pinnatifidus (Willd.) Al-Shehbaz et H. Ohba ■

136182 Dontostemon pectinatus (DC.) Ledeb. = Dontostemon pinnatifidus (Willd.) Al-Shehbaz et H. Ohba ■

136183 Dontostemon pectinatus (DC.) Ledeb. var. humilior N. Busch = Dontostemon pinnatifidus (Willd.) Al-Shehbaz et H. Ohba ■

136184 Dontostemon perennis C. A. Mey.;多年生花旗杆(多年生花旗竿);Perennial Dontostemon ■

136185 Dontostemon pinnatifidus (Willd.) Al-Shehbaz et H. Ohba;羽裂花旗杆■
136186 Dontostemon pinnatifidus (Willd.) Al-Shehbaz et H. Ohba = Dimorphostemon pinnatus (Pers.) Kitag. ■
136187 Dontostemon pinnatifidus (Willd.) Al-Shehbaz et H. Ohba subsp. linearifolius (Maxim.) Al-Shehbaz et H. Ohba;线叶花旗杆■
136188 Dontostemon pinnatus (Pers.) Kitag. = Dontostemon pinnatifidus (Willd.) Al-Shehbaz et H. Ohba ■
136189 Dontostemon scorpioides Bunge = Malcolmia scorpioides (Bunge) Boiss. ■
136190 Dontostemon semiamplexicaulis H. L. Yang = Dontostemon elegans Maxim. ■
136191 Dontostemon senilis Maxim.;白毛花旗杆(白毛花旗竿);Whitehair Dontostemon ■
136192 Dontostemon sergievskianus (Polozhij) Ovchinnikova = Dontostemon glandulosus (Kar. et Kir.) O. E. Schulz ■
136193 Dontostemon shanxiensis R. L. Guo et T. Y. Cheo = Dontostemon pinnatifidus (Willd.) Al-Shehbaz et H. Ohba ■
136194 Dontostemon tibeticus (Maxim.) Al-Shehbaz et H. Ohba;西藏花旗杆(西藏花旗竿)■
136195 Donzella Lem. = Donzellia Ten. ●
136196 Donzellia Ten. = Flacourtia Comm. ex L'Hér. ●
136197 Doodia Roxb. = Uraria Desv. ●■
136198 Doodia crinita Roxb. = Uraria crinita (L.) Desv. ex DC. ●■
136199 Doodia hamosa Roxb. = Uraria rufescens (DC.) Schindl. ●
136200 Doodia lagopodioides (L.) Roxb. = Uraria lagopodioides (L.) Desv. ex DC. ■
136201 Doodia lagopodioides Roxb. = Uraria lagopodioides (L.) Desv. ex DC. ■
136202 Doodia picta (Jacq.) Roxb. = Uraria picta (Jacq.) Desv. ex DC. ●
136203 Doodia picta Roxb. = Uraria picta (Jacq.) Desv. ex DC. ●
136204 Doona Thwaites(1851)(保留属名);杜纳香属●☆
136205 Doona Thwaites(保留属名) = Shorea Roxb. ex C. F. Gaertn. ●
136206 Doornia de Vriese = Pandanus Parkinson ex Du Roi ●■
136207 Doosera Roxb. ex Wight et Arn. = Mollugo L. ■
136208 Dopatrium Buch.-Ham. = Dopatrium Buch.-Ham. ex Benth. ■
136209 Dopatrium Buch.-Ham. ex Benth. (1835);虻眼草属(虻眼属);Dopatricum ■
136210 Dopatrium angolense V. Naray.;安哥拉虻眼草■☆
136211 Dopatrium baoulense A. Chev.;巴乌莱虻眼草■☆
136212 Dopatrium caespitosum P. Taylor;丛生虻眼草■☆
136213 Dopatrium dawei Hutch. et Dalziel = Dopatrium senegalense Benth. ■☆
136214 Dopatrium junceum (Roxb.) Buch.-Ham. ex Benth.;虻眼草;Horsefly's Eye, Rushlike Dopatricum ■
136215 Dopatrium lobelioides Benth.;半边莲状虻眼草■☆
136216 Dopatrium longidens V. Naray.;长齿虻眼草■☆
136217 Dopatrium luteum Engl. = Dopatrium macranthum Oliv. ■☆
136218 Dopatrium macranthum Oliv.;大花虻眼草■☆
136219 Dopatrium nanum Scott-Elliot = Dopatrium senegalense Benth. ■☆
136220 Dopatrium nudicaule Buch.-Ham. ex Benth.;裸茎虻眼草■☆
136221 Dopatrium peulhorum A. Chev. = Dopatrium senegalense Benth. ■☆
136222 Dopatrium pusillum P. Taylor;微小虻眼草■☆
136223 Dopatrium schweinfurthii Wettst. = Dopatrium macranthum Oliv. ■☆
136224 Dopatrium senegalense Benth.;塞内加尔虻眼草■☆
136225 Dopatrium stachytarphetioides Engl. et Gilg;假马鞭虻眼草■☆
136226 Dopatrium tenerum (Hiern) Eb. Fisch.;柔软虻眼草■☆
136227 Dopatrium tricolor Engl. = Dopatrium macranthum Oliv. ■☆
136228 Dopatrium tricolor Wettst. = Dopatrium macranthum Oliv. ■☆
136229 Doraena Thunb. = Maesa Forssk. ●
136230 Doraena japonica Thunb. = Maesa japonica (Thunb.) Moritzi ex Zoll. ●
136231 Doranthera Steud. = Anticharis Endl. ■●☆
136232 Doranthera Steud. = Doratanthera Benth. ex Endl. ■●☆
136233 Doranxylum Neraud = ? Doratoxylon Thou. ex Benth. et Hook. f. ●☆
136234 Doratanthera Benth. ex Endl. = Anticharis Endl. ■●☆
136235 Doratanthera linearis Benth. = Anticharis senegalensis (Walp.) Bhandari ■☆
136236 Doratanthera senegalensis Walp. = Anticharis senegalensis (Walp.) Bhandari ■☆
136237 Doratium Sol. ex J. St.-Hil. = Zanthoxylum L. ●
136238 Doratolepis (Benth.) Schltdl. = Leptorhynchos Less. ■☆
136239 Doratolepis Schltdl. = Leptorhynchos Less. ■☆
136240 Doratometra Klotzsch = Begonia L. ●■
136241 Doratometra bowringiana (Champ. ex Benth.) Seem. = Begonia palmata D. Don var. bowringiana (Champ. ex Benth.) Golding et Kareg. ■
136242 Doratometra bowringiana Seem. = Begonia palmata D. Don var. bowringiana (Champ. ex Benth.) Golding et Kareg. ■
136243 Doratophora Lem. = Doryphora Endl. ●☆
136244 Doratoxylon Thouars ex Benth. et Hook. f. (1862);矛材属●☆
136245 Doratoxylon Thouars ex Hook. f. = Doratoxylon Thouars ex Benth. et Hook. f. ●☆
136246 Doratoxylon alatum (Radlk.) Capuron;具翅矛材●☆
136247 Doratoxylon apetalum (Poir.) Radlk.;无瓣矛材●☆
136248 Doratoxylon littorale Capuron;滨海矛材●☆
136249 Doratoxylon mauritianum Thouars = Doratoxylon apetalum (Poir.) Radlk. ●☆
136250 Doratoxylon stipulatum Capuron;托叶矛材●☆
136251 Dorcoceras Bunge = Boea Comm. ex Lam. ■
136252 Dorcoceras crassifolium (Hemsl.) Schltr. = Paraboea crassifolia (Hemsl.) B. L. Burtt ■
136253 Dorcoceras hygrometrica Bunge = Boea hygrometrica (Bunge) R. Br. ■
136254 Dorcoceras philippense (C. B. Clarke) Schltr. = Boea philippensis C. B. Clarke ■
136255 Dorcoceras rufescens (Franch.) Schltr. = Paraboea rufescens (Franch.) B. L. Burtt ■
136256 Dorella Bubani = Camelina Crantz ■
136257 Dorema D. Don(1831);氨草属(屈谟属);Sumbul ■☆
136258 Dorema aitchisonii Korovin;埃氏氨草■☆
136259 Dorema ammoniacum D. Don;氨草;Ammoniac Plant, Bombay Sumbul, Sumbul ■☆
136260 Dorema aureum Stocks;黄氨草■☆
136261 Dorema glabrum Fisch. et C. A. Mey.;光氨草■☆
136262 Dorema gummiferum (Jaub. et Spach) K. M. Korol.;胶氨草■☆
136263 Dorema hyrcanum Koso-Pol.;希尔康氨草■☆
136264 Dorema karataviense Korovin;卡拉塔夫氨草■☆
136265 Dorema microcarpum Korovin;小果氨草■☆
136266 Dorema namanganicum K. M. Korol.;纳曼干氨草■☆

136267 Dorema pruinosum K. M. Korol.;粉氨草■☆
136268 Dorema sabulosum Litv.;砂地氨草■☆
136269 Doria Thunb. = Othonna L. ●■☆
136270 Doria Thunb. = Senecio L. ■●
136271 Doria abrotanifolia Harv. = Othonna abrotanifolia (Harv.) Druce ●☆
136272 Doria alata Thunb. = Hertia alata (Thunb.) Kuntze ■☆
136273 Doria arbuscula (Thunb.) DC. = Othonna arbuscula (Thunb.) Sch. Bip. ●☆
136274 Doria bipinnata Thunb. = Senecio pinnatifidus (P. J. Bergius) Less. ■☆
136275 Doria carnosa DC. = Othonna retrofracta Jacq. ●☆
136276 Doria ceradia Harv. = Othonna furcata (Lindl.) Druce ●☆
136277 Doria chromochaeta DC. = Othonna chromochaeta (DC.) Sch. Bip. ■☆
136278 Doria ciliata Harv. = Hertia ciliata (Harv.) Kuntze ■☆
136279 Doria cluytiifolia DC. = Hertia cluytiifolia (DC.) Kuntze ■☆
136280 Doria cneorifolia DC. = Lopholaena cneorifolia (DC.) S. Moore ■☆
136281 Doria concinna (A. Nelson) Lunell = Solidago missouriensis Nutt. ■☆
136282 Doria denticulata Thunb. = Senecio othonniflorus DC. ■☆
136283 Doria digitata (L. f.) Less. = Othonna digitata L. f. ●☆
136284 Doria diversifolia DC. = Othonna diversifolia (DC.) Sch. Bip. ●☆
136285 Doria dumetorum (Lunell) Lunell = Solidago gigantea Aiton ■☆
136286 Doria eriocarpa DC. = Othonna eriocarpa (DC.) Sch. Bip. ●☆
136287 Doria erosa Thunb. = Cineraria erosa (Thunb.) Harv. ■☆
136288 Doria gilvocanescens (Rydb.) Lunell = Solidago altissima L. subsp. gilvocanescens (Rydb.) Semple ■☆
136289 Doria glaberrima (M. Martens) Lunell = Solidago missouriensis Nutt. ■☆
136290 Doria glaberrima (M. Martens) Lunell var. montana (A. Gray) Lunell = Solidago missouriensis Nutt. ■☆
136291 Doria gymnodiscus DC. = Othonna gymnodiscus (DC.) Sch. Bip. ●☆
136292 Doria incana (Torr. et A. Gray) Lunell = Solidago mollis Bartl. ■☆
136293 Doria incisa Thunb. = Senecio erosus L. f. ■☆
136294 Doria kraussii Sch. Bip. = Hertia kraussii (Sch. Bip.) Fourc. ■☆
136295 Doria lasiocarpa DC. = Othonna lasiocarpa (DC.) Sch. Bip. ●☆
136296 Doria linearifolia DC. = Othonna lineariifolia (DC.) Sch. Bip. ●☆
136297 Doria lingua Less. = Othonna lingua (Less.) Sch. Bip. ■☆
136298 Doria longipes Harv. = Lopholaena longipes (Harv.) Thell. ■☆
136299 Doria mollis (Bartl.) Lunell = Solidago mollis Bartl. ■☆
136300 Doria perfoliata (Thunb.) DC. = Othonna perfoliata Thunb. ●☆
136301 Doria pinnatifida Thunb. = Senecio tuberosus (DC.) Harv. ■☆
136302 Doria pitcheri (Nutt.) Lunell = Solidago gigantea Aiton ■☆
136303 Doria pulcherrima (A. Nelson) Lunell = Solidago nemoralis Aiton subsp. decemflora (DC.) Brammall ex Semple ■☆
136304 Doria serrata Thunb. = Senecio glanduloso-pilosus Volkens et Muschl. ■☆
136305 Doria undulata Thunb. = Senecio petiolaris DC. ■☆
136306 Doricera Verdc. (1983);矛角茜属☆
136307 Doricera trilocularis (Balf. f.) Verdc.;矛角茜☆
136308 Doriclea Raf. = Leucas Burm. ex R. Br. ●■
136309 Doriena Endl. = Acronychia J. R. Forst. et G. Forst.(保留属名)●
136310 Dorisia Gillespie = Mastixiodendron Melch. ●☆
136311 Dorisia Gillespie et A. C. Sm. = Mastixiodendron Melch. ●☆
136312 Doritis Lindl. (1833);五唇兰属(朵丽兰属);Doritis ■
136313 Doritis braceana Hook. f. = Kingidium braceanum (Hook. f.) Seidenf. ■
136314 Doritis braceana Hook. f. = Phalaenopsis braceana (Hook. f.) Christenson ■
136315 Doritis deliciosa (Rchb. f.) T. Yukawa et K. Kita = Phalaenopsis deliciosa Rchb. f. ■
136316 Doritis hainanensis (Ts. Tang et F. T. Wang) T. Yukawa et K. Kita = Phalaenopsis hainanensis Ts. Tang et F. T. Wang ■
136317 Doritis honghenensis (F. Y. Liu) T. Yukawa et K. Kita = Phalaenopsis honghenensis F. Y. Liu ■
136318 Doritis latifolia (Thwaites) Benth. et Hook. f. = Phalaenopsis deliciosa Rchb. f. ■
136319 Doritis lobbii (Rchb. f.) T. Yukawa et K. Kita = Phalaenopsis lobbii (Rchb. f.) H. R. Sweet ■
136320 Doritis pulcherrima Lindl.;五唇兰(朵丽兰,美丽五唇兰,擎天蛾兰);Pretty Doritis ■
136321 Doritis pulcherrima Lindl. var. buyssoniana (Rchb. f.) R. Sagarik;大五唇兰(大擎天蛾兰)■☆
136322 Doritis stobartiana (Rchb. f.) T. Yukawa et K. Kita = Phalaenopsis stobariana Rchb. f. ■
136323 Doritis taenialis (Lindl.) Hook. f. = Kingidium taeniale (Lindl.) P. F. Hunt ■
136324 Doritis taenialis (Lindl.) Hook. f. = Phalaenopsis taenialis (Lindl.) Christenson et Pradhan ■
136325 Doritis wightii (Rchb. f.) Benth. et Hook. f. = Phalaenopsis deliciosa Rchb. f. ■
136326 Doritis wilsonii (Rolfe) T. Yukawa et K. Kita = Phalaenopsis wilsonii Rolfe ■
136327 Doritis zhejiangensis (Z. H. Tsi) T. Yukawa et K. Kita = Nothodoritis zhejiangensis Z. H. Tsi ■
136328 Dornera Heuff. ex Schur = Carex L. ■
136329 Dorobaea Cass. (1827);羽莲菊属■☆
136330 Dorobaea Cass. = Senecio L. ■●
136331 Dorobaea pimpinellifolia Cass.;羽莲菊■☆
136332 Doroceras Steud. = Dorcoceras Bunge ■
136333 Doronicum L. (1753);多榔菊属;Doronicum, Leopard's Bane, Leopard's-bane ■
136334 Doronicum Tourn. ex L. = Doronicum L. ■
136335 Doronicum × excelsum (N. E. Br.) Stace 'Harpur Crewe';哈普尔车前状多榔菊;Harpur-Crewe's Leopard's-bane ■☆
136336 Doronicum × excelsum (N. E. Br.) Stace = Doronicum plantagineum L. 'Excelsum' ■☆
136337 Doronicum × willdenowii Rouy;韦氏多榔菊;Wildenow's Leopard's-bane ■☆
136338 Doronicum acaule Walter = Arnica acaulis (Walter) Britton, Sterns et Poggenb. ●☆
136339 Doronicum altaicum Pall.;阿尔泰多榔菊(太白小紫菀,小紫菀);Altai Doronicum ■
136340 Doronicum altaicum Pall. = Doronicum gansuense Y. L. Chen ■
136341 Doronicum asplenifolium Lam. = Gerbera linnaei Cass. ■☆
136342 Doronicum atlanticum Rouy = Doronicum plantagineum L. subsp. atlanticum (Rouy) Greuter ■☆
136343 Doronicum atlanticum Rouy var. atlanticum Chabert = Doronicum plantagineum L. subsp. atlanticum (Rouy) Greuter ■☆

136344 Doronicum austriacum Jacq.; 奥地利多榔菊; Austrian Leopard's-Bane ■☆
136345 Doronicum bargusinense Serg. var. pilosum C. H. An; 毛多头多榔菊■
136346 Doronicum caucasicum M. Bieb.; 高加索多榔菊; Caucasian Leopard's-bane ■☆
136347 Doronicum caucasicum M. Bieb. = Doronicum orientale Hoffm. ■☆
136348 Doronicum clusii Tausch; 克鲁斯多榔菊■☆
136349 Doronicum columnae Ten.; 心叶多榔菊; Eastern Leopard's-bane ■☆
136350 Doronicum conaense Y. L. Chen; 错那多榔菊; Cuona Doronicum ■
136351 Doronicum cordatum K. Koch = Doronicum caucasicum M. Bieb. ■☆
136352 Doronicum cordatum Lam. = Doronicum pardalianches L. ■☆
136353 Doronicum cordatum Sch. Bip. = Doronicum columnae Ten. ■☆
136354 Doronicum cordifolium Sternb. = Doronicum columnae Ten. ■☆
136355 Doronicum gansuense Y. L. Chen; 甘肃多榔菊; Gansu Doronicum, Kansu Doronicum ■
136356 Doronicum grandiflorum Lam.; 大花多榔菊■☆
136357 Doronicum hookeri C. B. Clarke ex Hook. f. = Nannoglottis hookeri (Hook. f.) Kitam. ■
136358 Doronicum latisquamatum C. E. C. Fisher = Aster platylepis Y. L. Chen ■
136359 Doronicum limprichtii Diels = Doronicum thibetanum Cavill. ■
136360 Doronicum longifolium Griseb. et Schenk; 长叶多榔菊■☆
136361 Doronicum macrophyllum Fisch.; 大叶多榔菊■☆
136362 Doronicum oblongifolium DC.; 长圆叶多榔菊(长圆多榔菊); Oblongleaf Doronicum ■☆
136363 Doronicum oblongifolium DC. var. leiocarpum Trautv. = Doronicum turkestanicum Cavill. ■
136364 Doronicum orientale Hoffm.; 东方多榔菊; Oriental Leopard's-bane ■☆
136365 Doronicum pardalianches L.; 多花多榔菊(心叶多榔菊); Crayfish, Crayfish Leopard's Bane, Great False Leopardbane, Great Leopard's Bane, Great Leopard's-bane, Leopard's-bane, Mountain Marygold ■☆
136366 Doronicum plantagineum L.; 车前状多榔菊(车前多榔菊); Leopard's Bane, Plantain False Leopard Bane, Plantain-leaved Leopard's-bane ■☆
136367 Doronicum plantagineum L. 'Excelsum' = Doronicum × excelsum (N. E. Br.) Stace 'Harpur Crewe' ■☆
136368 Doronicum plantagineum L. subsp. atlanticum (Rouy) Greuter; 大西洋多榔菊■☆
136369 Doronicum plantagineum L. var. africanum Barratte = Doronicum plantagineum L. ■☆
136370 Doronicum plantagineum Wild. = Doronicum willdenowii (Rouy) A. W. Hill ■☆
136371 Doronicum ramosum Walter = Erigeron strigosus Muhl. ex Willd. ■☆
136372 Doronicum rotundifolium Desf. = Bellis rotundifolia (Desf.) Boiss. et Reut. ■☆
136373 Doronicum schischkinii Serg.; 希施多榔菊■☆
136374 Doronicum soliei Cavill. = Doronicum stenoglossum Maxim. ■
136375 Doronicum stenoglossum Maxim.; 狭舌多榔菊(多榔菊); Narrowtongued Doronicum ■
136376 Doronicum thibetanum Cavill.; 西藏多榔菊; Xizang Doronicum ■
136377 Doronicum thibetanum Cavill. = Doronicum gansuense Y. L. Chen ■
136378 Doronicum tianshanicum C. H. An; 天山多榔菊; Tianshan Doronicum ■
136379 Doronicum turkestanicum Cavill.; 中亚多榔菊(突厥多榔菊); Centrol Asia Doronicum ■
136380 Doronicum tussilaginis (L'Hér.) Sch. Bip. = Pericallis tussilaginis (L'Hér.) D. Don ■☆
136381 Doronicum webbii Sch. Bip. = Pericallis webbii (Sch. Bip.) Bolle ■☆
136382 Doronicum wightii DC. = Senecio wightii (DC. ex Wight) Benth. ex C. B. Clarke ■
136383 Doronicum wightii DC. ex Wight = Senecio wightii (DC. ex Wight) Benth. ex C. B. Clarke ■
136384 Doronicum willdenowii (Rouy) A. W. Hill; 威尔多榔菊■☆
136385 Doronicum willdenowii Rouy = Doronicum willdenowii (Rouy) A. W. Hill ■☆
136386 Doronicum yunnanense Franch. ex Diels = Doronicum stenoglossum Maxim. ■
136387 Dorothea Wernham = Aulacocalyx Hook. f. ●☆
136388 Dorothea divergens Hutch. et Dalziel = Aulacocalyx divergens (Hutch. et Dalziel) Keay ●☆
136389 Dorothea letestui Pellegr. = Aulacocalyx pallens (Hiern) Bridson et Figueiredo subsp. letestui (Pellegr.) Figueiredo ●☆
136390 Dorothea talbotii Wernham = Aulacocalyx talbotii (Wernham) Keay ●☆
136391 Dorotheanthus Schwantes(1927); 彩虹花属■☆
136392 Dorotheanthus acuminatus L. Bolus = Dorotheanthus bellidiformis (Burm. f.) N. E. Br. ■☆
136393 Dorotheanthus apetalus (L. f.) N. E. Br.; 无瓣彩虹花■☆
136394 Dorotheanthus bellidiformis (Burm. f.) N. E. Br.; 彩虹花(彩虹菊); Ice-plant, Livingstone Daisy ■☆
136395 Dorotheanthus bellidiformis (Burm. f.) N. E. Br. 'Magic Carpet'; 魔毯彩虹花; Livingstone Daisy ■☆
136396 Dorotheanthus bidouwensis L. Bolus = Dorotheanthus bellidiformis (Burm. f.) N. E. Br. ■☆
136397 Dorotheanthus clavatus (Haw.) Struck; 棍棒彩虹花; Fig Marigold ■☆
136398 Dorotheanthus flos-solis (A. Berger) L. Bolus = Dorotheanthus bellidiformis (Burm. f.) N. E. Br. ■☆
136399 Dorotheanthus gramineus (Haw.) Schwantes; 禾状彩虹花■☆
136400 Dorotheanthus gramineus (Haw.) Schwantes = Dorotheanthus apetalus (L. f.) N. E. Br. ■☆
136401 Dorotheanthus gramineus (Haw.) Schwantes f. albus (Haw.) G. D. Rowley = Dorotheanthus clavatus (Haw.) Struck ■☆
136402 Dorotheanthus gramineus (Haw.) Schwantes f. roseus (Haw.) G. D. Rowley = Dorotheanthus clavatus (Haw.) Struck ■☆
136403 Dorotheanthus hallii L. Bolus = Dorotheanthus bellidiformis (Burm. f.) N. E. Br. ■☆
136404 Dorotheanthus littlewoodii L. Bolus = Dorotheanthus bellidiformis (Burm. f.) N. E. Br. ■☆
136405 Dorotheanthus luteus N. E. Br. = Dorotheanthus bellidiformis (Burm. f.) N. E. Br. ■☆
136406 Dorotheanthus martinii L. Bolus = Dorotheanthus bellidiformis (Burm. f.) N. E. Br. ■☆
136407 Dorotheanthus maughanii (N. E. Br.) Ihlenf. et Struck; 莫恩彩虹花■☆

136408 Dorotheanthus muiri L. Bolus = Dorotheanthus bellidiformis (Burm. f.) N. E. Br. ■☆

136409 Dorotheanthus oculatus N. E. Br.;小眼彩虹花■☆

136410 Dorotheanthus oculatus N. E. Br. = Dorotheanthus bellidiformis (Burm. f.) N. E. Br. ■☆

136411 Dorotheanthus oculatus N. E. Br. var. saldhanensis L. Bolus = Dorotheanthus bellidiformis (Burm. f.) N. E. Br. ■☆

136412 Dorotheanthus rourkei L. Bolus;鲁尔克彩虹花■☆

136413 Dorotheanthus stayneri L. Bolus = Dorotheanthus bellidiformis (Burm. f.) N. E. Br. ■☆

136414 Dorotheanthus tricolor (Willd.) L. Bolus = Dorotheanthus clavatus (Haw.) Struck ■☆

136415 Dorrienia Engl. = Doerrienia Rchb. ■☆

136416 Dorrienia Engl. = Genlisea A. St. -Hil. ■☆

136417 Dorstenia L. (1753);多坦草属(多尔斯藤属,琉桑属,墨西哥桑属);Dorstenie,Toms Herb ■●☆

136418 Dorstenia achtenii De Wild. = Dorstenia benguellensis Welw. ■☆

136419 Dorstenia achtenii De Wild. var. laxiflora Hauman = Dorstenia benguellensis Welw. ■☆

136420 Dorstenia africana (Baill.) C. C. Berg;非洲多坦草●☆

136421 Dorstenia afromontana R. E. Fr.;非洲山生多坦草●☆

136422 Dorstenia alta Engl.;翅多坦草(翅墨西哥桑)●☆

136423 Dorstenia alternans Engl. = Dorstenia mannii Hook. f. var. alternans (Engl.) Hijman ■☆

136424 Dorstenia amboniensis De Wild. = Dorstenia tayloriana Rendle ■☆

136425 Dorstenia amoena A. Chev. = Dorstenia kameruniana Engl. ■☆

136426 Dorstenia angusta Engl. = Dorstenia poinsettiifolia Engl. var. angusta (Engl.) Hijman et C. C. Berg ■☆

136427 Dorstenia angusticornis Engl.;窄角多坦草●☆

136428 Dorstenia annua Friis et Vollesen;一年多坦草●☆

136429 Dorstenia arabica Hemsl. = Dorstenia foetida (Forssk.) Schweinf. ■☆

136430 Dorstenia aspera A. Chev. = Dorstenia turbinata Engl. ■☆

136431 Dorstenia asteriscus Engl. = Dorstenia ciliata Engl. ■☆

136432 Dorstenia barnimiana Schweinf.;巴尼姆多坦草■☆

136433 Dorstenia barnimiana Schweinf. var. angustior Engl. = Dorstenia barnimiana Schweinf. ■☆

136434 Dorstenia barnimiana Schweinf. var. ophioglossoides (Hochst. ex Bureau) Engl. = Dorstenia barnimiana Schweinf. ■☆

136435 Dorstenia barnimiana Schweinf. var. telekii (Schweinf.) Engl. = Dorstenia barnimiana Schweinf. ■☆

136436 Dorstenia barnimiana Schweinf. var. tropaeolifolia (Schweinf.) Lye = Dorstenia tropaeolifolia (Schweinf.) Bureau ■☆

136437 Dorstenia barteri Bureau;巴特多坦草●☆

136438 Dorstenia barteri Bureau var. multiradiata (Engl.) Hijman et C. C. Berg;多射线巴特多坦草●☆

136439 Dorstenia barteri Bureau var. paucinervis Hijman et C. C. Berg;少脉巴特多坦草●☆

136440 Dorstenia barteri Bureau var. subtriangularis (Engl.) Hijman et C. C. Berg;三角巴特多坦草●☆

136441 Dorstenia batesii Rendle = Dorstenia lujae De Wild. var. batesii (Rendle) Hijman ■☆

136442 Dorstenia benguellensis Welw.;本格拉多坦草■☆

136443 Dorstenia benguellensis Welw. var. capillaris Hauman = Dorstenia benguellensis Welw. ■☆

136444 Dorstenia bequaertii De Wild. = Dorstenia dinklagei Engl. var. bequaertii (De Wild.) Hijman ■☆

136445 Dorstenia bergiana Hijman;贝格多坦草●☆

136446 Dorstenia bicaudata Peter;双尾多坦草■☆

136447 Dorstenia bicaudata Peter var. quercifolia ? = Dorstenia bicaudata Peter ■☆

136448 Dorstenia bicornis Schweinf. = Dorstenia psilurus Welw. ■☆

136449 Dorstenia binzaensis De Wild. = Dorstenia dinklagei Engl. var. binzaensis (De Wild.) Hijman ■☆

136450 Dorstenia botziana Engl. = Dorstenia poinsettiifolia Engl. var. angusta (Engl.) Hijman et C. C. Berg ■☆

136451 Dorstenia brasiliensis Lam.;巴西多坦草■☆

136452 Dorstenia brasiliensis Lam. f. balansae Chodat = Dorstenia brasiliensis Lam. ■☆

136453 Dorstenia brasiliensis Lam. var. guaranitica Chodat et Vischer = Dorstenia brasiliensis Lam. ■☆

136454 Dorstenia brasiliensis Lam. var. major Chodat et Hassl. = Dorstenia brasiliensis Lam. ■☆

136455 Dorstenia brasiliensis Lam. var. palustris Hassl. = Dorstenia brasiliensis Lam. ■☆

136456 Dorstenia brasiliensis Lam. var. tomentosa (Fisch. et C. A. Mey.) Hassl. = Dorstenia brasiliensis Lam. ■☆

136457 Dorstenia brasiliensis Lam. var. tubicina (Ruiz et Pav.) Chodat et Vischer = Dorstenia brasiliensis Lam. ■☆

136458 Dorstenia braunii Engl. = Dorstenia hildebrandtii Engl. ■☆

136459 Dorstenia brevifolia Peter = Dorstenia hildebrandtii Engl. ■☆

136460 Dorstenia brieyi De Wild. = Dorstenia dinklagei Engl. var. brieyi (De Wild.) Hijman ■☆

136461 Dorstenia brownii Rendle;布朗多坦草■☆

136462 Dorstenia buchananii Engl.;布坎南多坦草■☆

136463 Dorstenia buchananii Engl. var. longepedunculata Rendle;长梗布坎南多坦草■☆

136464 Dorstenia buesgenii Engl. = Dorstenia turbinata Engl. ■☆

136465 Dorstenia campanulata Hauman = Dorstenia benguellensis Welw. ■☆

136466 Dorstenia carnosula De Wild. = Dorstenia hildebrandtii Engl. var. schlechteri (Engl.) Hijman ■☆

136467 Dorstenia caudata Engl. = Dorstenia buchananii Engl. ■☆

136468 Dorstenia caulescens Schweinf. ex Engl. = Dorstenia cuspidata Hochst. ex A. Rich. ■☆

136469 Dorstenia ciliata Engl.;睫毛多坦草■☆

136470 Dorstenia contrajerva L.;墨西哥桑;Contra Hierba, Contra Yerba,Contrayerva,Contrayerva Root,Torus Herb,Tusilla ●☆

136471 Dorstenia convexa De Wild.;弯曲多坦草■☆

136472 Dorstenia convexa De Wild. var. oblonga (Compère) Hijman;矩圆多坦草■☆

136473 Dorstenia crispa Engl. = Dorstenia foetida (Forssk.) Schweinf. ■☆

136474 Dorstenia crispa Engl. var. lancifolia Rendle = Dorstenia foetida (Forssk.) Schweinf. subsp. lancifolia (Rendle) Friis ■☆

136475 Dorstenia crispa Engl. var. pachypoda Chiov. = Dorstenia foetida (Forssk.) Schweinf. subsp. lancifolia (Rendle) Friis ■☆

136476 Dorstenia cuspidata Hochst. ex A. Rich.;骤尖多坦草■☆

136477 Dorstenia cuspidata Hochst. ex A. Rich. var. debilis (Baill.) Léandri = Dorstenia cuspidata Hochst. ex A. Rich. ■☆

136478 Dorstenia cuspidata Hochst. ex A. Rich. var. humblotiana (Baill.) Léandri = Dorstenia cuspidata Hochst. ex A. Rich. ■☆

136479 Dorstenia cuspidata Hochst. ex A. Rich. var. preussii (Schweinf. ex Engl.) Hijman;普罗伊斯多坦草■☆

136480 Dorstenia debeerstii De Wild. et T. Durand = Dorstenia benguellensis Welw. ■☆
136481 Dorstenia debeerstii De Wild. et T. Durand var. multibracteata (R. E. Fr.) Rendle = Dorstenia benguellensis Welw. ■☆
136482 Dorstenia debilis Baill. = Dorstenia cuspidata Hochst. ex A. Rich. ■☆
136483 Dorstenia denticulata Peter = Dorstenia hildebrandtii Engl. var. schlechteri (Engl.) Hijman ■☆
136484 Dorstenia dinklagei Engl.;丁克多坦草■☆
136485 Dorstenia dinklagei Engl. var. bequaertii (De Wild.) Hijman;贝卡尔多坦草■☆
136486 Dorstenia dinklagei Engl. var. binzaensis (De Wild.) Hijman;宾扎多坦草■☆
136487 Dorstenia dinklagei Engl. var. brieyi (De Wild.) Hijman;布里多坦草■☆
136488 Dorstenia dinklagei Engl. var. reducta (De Wild.) Hijman;退缩多坦草■☆
136489 Dorstenia dorstenioides (Engl.) Hijman et C. C. Berg;普通多坦草■☆
136490 Dorstenia drakena L.;德拉多坦草■☆
136491 Dorstenia edeensis Engl. = Dorstenia turbinata Engl. ■☆
136492 Dorstenia ellenbeckiana Engl.;埃伦多坦草■☆
136493 Dorstenia elliptica Bureau;椭圆多坦草■☆
136494 Dorstenia embergeri Mangenot;恩贝格尔多坦草■☆
136495 Dorstenia equatorialis Rendle = Dorstenia dinklagei Engl. var. brieyi (De Wild.) Hijman ■☆
136496 Dorstenia foetida (Forssk.) Schweinf.;臭多坦草■☆
136497 Dorstenia foetida (Forssk.) Schweinf. subsp. lancifolia (Rendle) Friis;剑叶臭多坦草■☆
136498 Dorstenia foetida (Forssk.) Schweinf. var. obovata (A. Rich.) Schweinf.;倒卵叶臭多坦草■☆
136499 Dorstenia frutescens Engl. = Dorstenia elliptica Bureau ■☆
136500 Dorstenia gabunensis Engl. = Dorstenia poinsettiifolia Engl. ■☆
136501 Dorstenia gaussenii Troch. et Koechlin = Dorstenia dinklagei Engl. var. binzaensis (De Wild.) Hijman ■☆
136502 Dorstenia gilletii De Wild. = Dorstenia psilurus Welw. var. scabra Bureau ■☆
136503 Dorstenia goetzei Engl.;格兹多坦草■☆
136504 Dorstenia goetzei Engl. var. angustibracteata ? = Dorstenia goetzei Engl. ■☆
136505 Dorstenia goossensii De Wild. = Dorstenia yambuyaensis De Wild. ■☆
136506 Dorstenia gourmaensis A. Chev. = Dorstenia cuspidata Hochst. ex A. Rich. ■☆
136507 Dorstenia gourmaensis A. Chev. var. floribunda ? = Dorstenia cuspidata Hochst. ex A. Rich. ■☆
136508 Dorstenia gypsophila Lavranos;喜钙多坦草■☆
136509 Dorstenia harmsiana Engl. = Dorstenia ciliata Engl. ■☆
136510 Dorstenia harmsiana Engl. var. batesii (Rendle) Hijman et C. C. Berg = Dorstenia lujae De Wild. var. batesii (Rendle) Hijman ■☆
136511 Dorstenia heringeri Carauta et Valente = Dorstenia brasiliensis Lam. ■☆
136512 Dorstenia hildebrandtii Engl.;希尔德多坦草■☆
136513 Dorstenia hildebrandtii Engl. var. schlechteri (Engl.) Hijman;施莱多坦草■☆
136514 Dorstenia hispida Peter = Dorstenia schliebenii Mildbr. ■☆
136515 Dorstenia hockii De Wild. = Dorstenia benguellensis Welw. ■☆
136516 Dorstenia hockii De Wild. var. multibracteata (R. E. Fr.) Rendle = Dorstenia benguellensis Welw. ■☆
136517 Dorstenia holstii Engl.;霍尔多坦草■☆
136518 Dorstenia holstii Engl. var. grandifolia ? = Dorstenia holstii Engl. ■☆
136519 Dorstenia holstii Engl. var. longestipulata Hijman;长托叶多坦草■☆
136520 Dorstenia holtziana Engl. = Dorstenia zanzibarica Oliv. ■☆
136521 Dorstenia homblei De Wild. = Dorstenia benguellensis Welw. ■☆
136522 Dorstenia humblotiana Baill. = Dorstenia cuspidata Hochst. ex A. Rich. ■☆
136523 Dorstenia infundibuliformis Lodd. = Dorstenia brasiliensis Lam. ■☆
136524 Dorstenia intermedia Engl. = Dorstenia mannii Hook. f. ■☆
136525 Dorstenia involuta Hijman et C. C. Berg;内卷多坦草■☆
136526 Dorstenia jabassensis Engl. = Dorstenia ciliata Engl. ■☆
136527 Dorstenia jabassensis Engl. var. cuneata ? = Dorstenia ciliata Engl. ■☆
136528 Dorstenia kameruniana Engl.;喀麦隆多坦草■☆
136529 Dorstenia katangensis De Wild. = Dorstenia benguellensis Welw. ■☆
136530 Dorstenia katubensis De Wild. = Dorstenia benguellensis Welw. ■☆
136531 Dorstenia klainei Heckel = Dorstenia psilurus Welw. var. scabra Bureau ■☆
136532 Dorstenia kribensis Engl. = Dorstenia mannii Hook. f. var. alternans (Engl.) Hijman ■☆
136533 Dorstenia kyimbilaensis De Wild. = Dorstenia schliebenii Mildbr. ■☆
136534 Dorstenia lactifera De Wild. = Dorstenia benguellensis Welw. ■☆
136535 Dorstenia laikipiensis Rendle = Dorstenia tayloriana Rendle var. laikipiensis (Rendle) Hijman ■☆
136536 Dorstenia latibracteata Engl. = Dorstenia warneckei Engl. ■☆
136537 Dorstenia laurentii De Wild. = Dorstenia zenkeri Engl. ■☆
136538 Dorstenia lavranii T. A. McCoy et Massara;拉夫连多坦草■☆
136539 Dorstenia ledermannii Engl. = Dorstenia turbinata Engl. ■☆
136540 Dorstenia letestui Pellegr.;莱泰斯图多坦草■☆
136541 Dorstenia librevillensis De Wild. = Dorstenia poinsettiifolia Engl. var. librevillensis (De Wild.) Hijman et C. C. Berg ■☆
136542 Dorstenia liebuschiana Engl. = Dorstenia goetzei Engl. ■☆
136543 Dorstenia longicauda Engl. = Dorstenia poinsettiifolia Engl. var. longicauda (Engl.) Hijman et C. C. Berg ■☆
136544 Dorstenia longipedunculata De Wild. = Dorstenia buchananii Engl. var. longepedunculata Rendle ■☆
136545 Dorstenia lujae De Wild.;卢亚多坦草■☆
136546 Dorstenia lujae De Wild. var. batesii (Rendle) Hijman;贝茨多坦草■☆
136547 Dorstenia lukafuensis De Wild. = Dorstenia psilurus Welw. ■☆
136548 Dorstenia mannii Hook. f.;曼氏多坦草■☆
136549 Dorstenia mannii Hook. f. var. alternans (Engl.) Hijman;互生曼氏多坦草■☆
136550 Dorstenia mannii Hook. f. var. humilis (Hijman et C. C. Berg) Hijman;矮小多坦草■☆
136551 Dorstenia mannii Hook. f. var. mungensis (Engl.) Hijman;蒙戈多坦草■☆
136552 Dorstenia mannii Hook. f. var. stipulata (Rendle) Hijman;托叶多坦草■☆

136553 Dorstenia maoungouensis De Wild. = Dorstenia hildebrandtii Engl. var. schlechteri (Engl.) Hijman ■☆

136554 Dorstenia marambensis Peter = Dorstenia hildebrandtii Engl. ■☆

136555 Dorstenia massonii Bureau = Dorstenia psilurus Welw. var. scabra Bureau ■☆

136556 Dorstenia mildbraediana Peter = Dorstenia benguellensis Welw. ■☆

136557 Dorstenia mirabilis R. E. Fr. = Dorstenia benguellensis Welw. ■☆

136558 Dorstenia mogandjensis De Wild. = Dorstenia zenkeri Engl. ■☆

136559 Dorstenia montana Herzog = Dorstenia brasiliensis Lam. ■☆

136560 Dorstenia montevidensis Miq. = Dorstenia brasiliensis Lam. ■☆

136561 Dorstenia multiradiata Engl. = Dorstenia barteri Bureau var. multiradiata (Engl.) Hijman et C. C. Berg ●☆

136562 Dorstenia mundamensis Engl. = Dorstenia mannii Hook. f. var. mungensis (Engl.) Hijman ■☆

136563 Dorstenia mungensis Engl. = Dorstenia mannii Hook. f. var. mungensis (Engl.) Hijman ■☆

136564 Dorstenia mungensis Engl. var. bipindensis ? = Dorstenia mannii Hook. f. var. mungensis (Engl.) Hijman ■☆

136565 Dorstenia nyangensis Pellegr. = Dorstenia dinklagei Engl. var. brieyi (De Wild.) Hijman ■☆

136566 Dorstenia nyungwensis Troupin;尼永圭多坦草■☆

136567 Dorstenia obanensis Hutch. et Dalziel = Dorstenia turbinata Engl. ■☆

136568 Dorstenia oblonga Compère = Dorstenia convexa De Wild. var. oblonga (Compère) Hijman ■☆

136569 Dorstenia obovata Hochst. ex A. Rich. = Dorstenia foetida (Forssk.) Schweinf. var. obovata (A. Rich.) Schweinf. ■☆

136570 Dorstenia obtusibracteata Engl. = Dorstenia tenera Bureau var. obtusibracteata (Engl.) Hijman et C. C. Berg ■☆

136571 Dorstenia oligogyna (Pellegr.) C. C. Berg;寡蕊多坦草■☆

136572 Dorstenia ophiocoma K. Schum. et Engl. = Dorstenia mannii Hook. f. ■☆

136573 Dorstenia ophiocoma K. Schum. et Engl. var. alternans (Engl.) Hijman et C. C. Berg = Dorstenia mannii Hook. f. var. alternans (Engl.) Hijman ■☆

136574 Dorstenia ophiocoma K. Schum. et Engl. var. longipes Engl. = Dorstenia mannii Hook. f. ■☆

136575 Dorstenia ophiocoma K. Schum. et Engl. var. minor Rendle = Dorstenia mannii Hook. f. ■☆

136576 Dorstenia ophiocoma K. Schum. et Engl. var. mungensis (Engl.) Hijman et C. C. Berg = Dorstenia mannii Hook. f. var. mungensis (Engl.) Hijman ■☆

136577 Dorstenia ophiocoma K. Schum. et Engl. var. stipulata (Rendle) Hijman et C. C. Berg = Dorstenia mannii Hook. f. var. stipulata (Rendle) Hijman ■☆

136578 Dorstenia ophiocomoides Engl. = Dorstenia mannii Hook. f. var. mungensis (Engl.) Hijman ■☆

136579 Dorstenia ophioglossoides Hochst. ex Bureau = Dorstenia barnimiana Schweinf. ■☆

136580 Dorstenia orientalis De Wild. = Dorstenia alta Engl. ●☆

136581 Dorstenia palmata Engl. = Dorstenia barnimiana Schweinf. ■☆

136582 Dorstenia palmata Engl. var. integrifolia Chiov. = Dorstenia barnimiana Schweinf. ■☆

136583 Dorstenia papillosa Hauman = Dorstenia benguellensis Welw. ■☆

136584 Dorstenia papillosa Hauman var. plethorica ? = Dorstenia benguellensis Welw. ■☆

136585 Dorstenia paucibracteata De Wild.;少苞多坦草■☆

136586 Dorstenia paucidentata Rendle = Dorstenia tenera Bureau ■☆

136587 Dorstenia pectinata Peter = Dorstenia tayloriana Rendle var. laikipiensis (Rendle) Hijman ■☆

136588 Dorstenia peltata Engl. = Dorstenia tropaeolifolia (Schweinf.) Bureau ■☆

136589 Dorstenia penduliflora Peter = Dorstenia brownii Rendle ■☆

136590 Dorstenia phillipsiae Hook. f. = Dorstenia foetida (Forssk.) Schweinf. ■☆

136591 Dorstenia picta Bureau;着色多坦草■☆

136592 Dorstenia pierrei De Wild. = Dorstenia poinsettiifolia Engl. var. librevillensis (De Wild.) Hijman et C. C. Berg ■☆

136593 Dorstenia piscaria Hutch. et Dalziel = Dorstenia barteri Bureau var. subtriangularis (Engl.) Hijman et C. C. Berg ●☆

136594 Dorstenia piscicelliana Buscal. et Muschl. = Dorstenia benguellensis Welw. ■☆

136595 Dorstenia poggei Engl. = Dorstenia benguellensis Welw. ■☆

136596 Dorstenia poggei Engl. var. meyeri-johannis ? = Dorstenia benguellensis Welw. ■☆

136597 Dorstenia poinsettiifolia Engl.;猩猩木叶多坦草■☆

136598 Dorstenia poinsettiifolia Engl. var. angusta (Engl.) Hijman et C. C. Berg;狭多坦草■☆

136599 Dorstenia poinsettiifolia Engl. var. glabrescens Hijman et C. C. Berg;渐光多坦草■☆

136600 Dorstenia poinsettiifolia Engl. var. grossedentata ? = Dorstenia poinsettiifolia Engl. ■☆

136601 Dorstenia poinsettiifolia Engl. var. librevillensis (De Wild.) Hijman et C. C. Berg;利伯维尔多坦草■☆

136602 Dorstenia poinsettiifolia Engl. var. longicauda (Engl.) Hijman et C. C. Berg;长尾多坦草■☆

136603 Dorstenia poinsettiifolia Engl. var. staudtii (Engl.) Hijman et C. C. Berg;施陶多坦草■☆

136604 Dorstenia poinsettiifolia Engl. var. subdentata ? = Dorstenia poinsettiifolia Engl. ■☆

136605 Dorstenia polyactis Peter = Dorstenia hildebrandtii Engl. var. schlechteri (Engl.) Hijman ■☆

136606 Dorstenia preussii Engl. var. latidentata ? = Dorstenia cuspidata Hochst. ex A. Rich. var. preussii (Schweinf. ex Engl.) Hijman ■☆

136607 Dorstenia preussii Schweinf. = Dorstenia cuspidata Hochst. ex A. Rich. var. preussii (Schweinf. ex Engl.) Hijman ■☆

136608 Dorstenia preussii Schweinf. ex Engl. = Dorstenia cuspidata Hochst. ex A. Rich. var. preussii (Schweinf. ex Engl.) Hijman ■☆

136609 Dorstenia preussii Schweinf. var. latidentata Engl. = Dorstenia cuspidata Hochst. ex A. Rich. var. preussii (Schweinf. ex Engl.) Hijman ■☆

136610 Dorstenia prorepens Engl.;匍匐多坦草■☆

136611 Dorstenia prorepens Engl. var. robustior Rendle = Dorstenia prorepens Engl. ■☆

136612 Dorstenia psiluroides Engl. = Dorstenia psilurus Welw. ■☆

136613 Dorstenia psiluroides Engl. f. subintegra ? = Dorstenia psilurus Welw. ■☆

136614 Dorstenia psilurus Welw.;平滑多坦草■☆

136615 Dorstenia psilurus Welw. var. brevicaudata Rendle = Dorstenia psilurus Welw. ■☆

136616 Dorstenia psilurus Welw. var. compacta De Wild. = Dorstenia psilurus Welw. ■☆

136617 Dorstenia psilurus Welw. var. scabra Bureau;粗糙多坦草■☆

136618 Dorstenia psilurus Welw. var. subintegrifolia De Wild. = Dorstenia psilurus Welw. var. scabra Bureau ■☆
136619 Dorstenia quarrei De Wild. = Dorstenia cuspidata Hochst. ex A. Rich. ■☆
136620 Dorstenia quercifolia R. E. Fr. = Dorstenia hildebrandtii Engl. var. schlechteri (Engl.) Hijman ■☆
136621 Dorstenia radiata Lam. = Dorstenia foetida (Forssk.) Schweinf. ■☆
136622 Dorstenia reducta De Wild. = Dorstenia dinklagei Engl. var. reducta (De Wild.) Hijman ■☆
136623 Dorstenia renneyi Airy Shaw et Taylor = Dorstenia hildebrandtii Engl. var. schlechteri (Engl.) Hijman ■☆
136624 Dorstenia rhodesiana R. E. Fr. = Dorstenia benguellensis Welw. ■☆
136625 Dorstenia rhomboidea Peter = Dorstenia hildebrandtii Engl. ■☆
136626 Dorstenia rosenii R. E. Fr. = Dorstenia benguellensis Welw. ■☆
136627 Dorstenia rosenii R. E. Fr. var. multibracteata ? = Dorstenia benguellensis Welw. ■☆
136628 Dorstenia ruahensis Engl. = Dorstenia buchananii Engl. var. longepedunculata Rendle ■☆
136629 Dorstenia ruahensis Engl. var. appendiculosa Peter = Dorstenia buchananii Engl. var. longepedunculata Rendle ■☆
136630 Dorstenia rugosa Peter = Dorstenia tayloriana Rendle var. laikipiensis (Rendle) Hijman ■☆
136631 Dorstenia ruwenzoriensis De Wild. = Dorstenia brownii Rendle ■☆
136632 Dorstenia sabanensis Cuatrec. = Dorstenia brasiliensis Lam. ■☆
136633 Dorstenia sacleuxii De Wild. = Dorstenia warneckei Engl. ■☆
136634 Dorstenia scabra (Bureau) Engl. = Dorstenia psilurus Welw. var. scabra Bureau ■☆
136635 Dorstenia scabra (Bureau) Engl. var. denticulata Engl. = Dorstenia psilurus Welw. var. scabra Bureau ■☆
136636 Dorstenia scabra (Bureau) Engl. var. longicaudata Engl. = Dorstenia psilurus Welw. var. scabra Bureau ■☆
136637 Dorstenia scabra (Bureau) Engl. var. subintegrifolia (De Wild.) Rendle = Dorstenia psilurus Welw. var. scabra Bureau ■☆
136638 Dorstenia scaphigera Bureau;舟形多坦草■☆
136639 Dorstenia schlechteri Engl. = Dorstenia hildebrandtii Engl. var. schlechteri (Engl.) Hijman ■☆
136640 Dorstenia schliebenii Mildbr.;施利本多坦草■☆
136641 Dorstenia schulzii Carauta, Valente et Dunn de Araujo = Dorstenia brasiliensis Lam. ■☆
136642 Dorstenia seretii De Wild. = Dorstenia dinklagei Engl. var. brieyi (De Wild.) Hijman ■☆
136643 Dorstenia sessilis R. E. Fr.;无柄多坦草■☆
136644 Dorstenia sessilis R. E. Fr. = Dorstenia benguellensis Welw. ■☆
136645 Dorstenia sessilis R. E. Fr. var. undulata Engl.;波状多坦草■☆
136646 Dorstenia smythei Sprague = Dorstenia turbinata Engl. ■☆
136647 Dorstenia solheidii De Wild. = Dorstenia zenkeri Engl. ■☆
136648 Dorstenia spathulibracteata Engl. = Dorstenia turbinata Engl. ■☆
136649 Dorstenia staudtii Engl. = Dorstenia poinsettiifolia Engl. var. staudtii (Engl.) Hijman et C. C. Berg ■☆
136650 Dorstenia stenophylla R. E. Fr. = Dorstenia benguellensis Welw. ■☆
136651 Dorstenia stipitata De Wild. = Dorstenia benguellensis Welw. ■☆
136652 Dorstenia stipulata Rendle = Dorstenia mannii Hook. f. var. stipulata (Rendle) Hijman ■☆
136653 Dorstenia stolzii Engl. = Dorstenia psilurus Welw. ■☆
136654 Dorstenia subrhombiformis Engl. = Dorstenia poinsettiifolia Engl. var. staudtii (Engl.) Hijman et C. C. Berg ■☆
136655 Dorstenia subtriangularis Engl. = Dorstenia barteri Bureau var. subtriangularis (Engl.) Hijman et C. C. Berg ●☆
136656 Dorstenia talbotii Rendle = Dorstenia lujae De Wild. ■☆
136657 Dorstenia tanneriana Peter = Dorstenia hildebrandtii Engl. ■☆
136658 Dorstenia tayloriana Rendle;泰勒多坦草■☆
136659 Dorstenia tayloriana Rendle var. laikipiensis (Rendle) Hijman;卡伊基皮多坦草■☆
136660 Dorstenia telekii Schweinf. = Dorstenia barnimiana Schweinf. ■☆
136661 Dorstenia tenera Bureau;极细多坦草■☆
136662 Dorstenia tenera Bureau var. obtusibracteata (Engl.) Hijman et C. C. Berg;钝苞多坦草■☆
136663 Dorstenia tenuifolia Engl. = Dorstenia psilurus Welw. var. scabra Bureau ■☆
136664 Dorstenia tenuiradiata Mildbr.;细射线多坦草■☆
136665 Dorstenia tessmannii Engl. = Dorstenia poinsettiifolia Engl. ■☆
136666 Dorstenia tetractis Peter = Dorstenia cuspidata Hochst. ex A. Rich. ■☆
136667 Dorstenia thikaensis Hijman;锡卡多坦草■☆
136668 Dorstenia tomentosa Fisch. et C. A. Mey. = Dorstenia brasiliensis Lam. ■☆
136669 Dorstenia triternata Chiov. = Dorstenia barnimiana Schweinf. ■☆
136670 Dorstenia tropaeolifolia (Schweinf.) Bureau;旱金莲叶多坦草■☆
136671 Dorstenia tubicina Ruiz et Pav. = Dorstenia brasiliensis Lam. ■☆
136672 Dorstenia tubicina Ruiz et Pav. f. major (Chodat et Hassl.) Hassl. = Dorstenia brasiliensis Lam. ■☆
136673 Dorstenia tubicina Ruiz et Pav. f. subexcentrica Hassl. = Dorstenia brasiliensis Lam. ■☆
136674 Dorstenia turbinata Engl.;陀螺形多坦草■☆
136675 Dorstenia ulugurensis Engl.;乌卢古尔多坦草■☆
136676 Dorstenia unicaudata Engl. = Dorstenia buchananii Engl. ■☆
136677 Dorstenia unyikae Engl. = Dorstenia cuspidata Hochst. ex A. Rich. ■☆
136678 Dorstenia usambarensis Engl. = Dorstenia holstii Engl. ■☆
136679 Dorstenia variegata Engl. = Dorstenia picta Bureau ■☆
136680 Dorstenia variifolia Engl.;异叶多坦草■☆
136681 Dorstenia verdickii De Wild. et T. Durand = Dorstenia benguellensis Welw. ■☆
136682 Dorstenia verdickii De Wild. et T. Durand var. scaberrima Hauman = Dorstenia benguellensis Welw. ■☆
136683 Dorstenia vermoesenii De Wild. = Dorstenia mannii Hook. f. ■☆
136684 Dorstenia vivipara Welw.;珠芽多坦草■☆
136685 Dorstenia volkensii Engl. = Dorstenia zanzibarica Oliv. ■☆
136686 Dorstenia walleri Hemsl. = Dorstenia cuspidata Hochst. ex A. Rich. ■☆
136687 Dorstenia walleri Hemsl. var. minor Rendle = Dorstenia cuspidata Hochst. ex A. Rich. ■☆
136688 Dorstenia warneckei Engl.;沃内克多坦草■☆
136689 Dorstenia wellmannii Engl. = Dorstenia benguellensis Welw. ■☆
136690 Dorstenia yambuyaensis De Wild.;扬布亚多坦草■☆
136691 Dorstenia yangambiensis J. Léonard;扬甘比多坦草■☆
136692 Dorstenia zambesiaca Hijman;赞比西多坦草■☆
136693 Dorstenia zanzibarica Oliv.;桑给巴尔多坦草■☆
136694 Dorstenia zenkeri Engl.;岑克尔多坦草■☆
136695 Dorsteniaceae Chev. = Hemerocallidaceae R. Br. ■

136696 Dortania A. Chev. = Acidanthera Hochst. ■
136697 Dortania A. Chev. = Gladiolus L. ■
136698 Dortania amoena A. Chev. = Gladiolus chevalieranus Marais ■☆
136699 Dortiguea Bubani = Erinus L. ■☆
136700 Dortmania Neck. = Dortmanna Hill ●■
136701 Dortmanna Hill = Lobelia L. ●■
136702 Dortmanna trigona (Roxb.) Kuntze = Lobelia alsinoides Lam. ■
136703 Dortmannaceae Rupr. = Campanulaceae Juss.(保留科名)■●
136704 Dortmannia Hill = Lobelia L. ●■
136705 Dortmannia Kuntze = Dortmanna Hill ●■
136706 Dortmannia Neck. = Lobelia L. ●■
136707 Dortmannia Steud. = Dortmanna Hill ●■
136708 Dortmannia Steud. = Lobelia L. ●■
136709 Dortmannia alsinoides Kuntze = Lobelia alsinoides Lam. ■
136710 Dortmannia campanuloides Kuntze = Lobelia chinensis Lour. ■
136711 Dortmannia chinensis Kuntze = Lobelia chinensis Lour. ■
136712 Dortmannia colorata Kuntze = Lobelia colorata Wall. ■
136713 Dortmannia erecta Kuntze = Lobelia erectiuscula H. Hara ■
136714 Dortmannia pyramidalis Kuntze = Lobelia pyramidalis Wall. ■
136715 Dortmannia radicans Kuntze = Lobelia chinensis Lour. ■
136716 Dortmannia reinwardtiana Kuntze = Lobelia zeylanica L. ■
136717 Dortmannia succulenta Kuntze = Lobelia zeylanica L. ■
136718 Dortmannia trigona (Roxb.) Kuntze var. terminalis (C. B. Clarke) Kuntze = Lobelia terminalis C. B. Clarke ■
136719 Dortmannia trigona Kuntze = Lobelia alsinoides Lam. ■
136720 Dortmannia trigona Kuntze var. affinis Kuntze = Lobelia zeylanica L. ■
136721 Dortmannia trigona Kuntze var. terminalis Kuntze = Lobelia terminalis C. B. Clarke ■
136722 Dorvalia Comm. ex Lam. = Fuchsia L. ●■
136723 Dorvalia Hoffmanns. = Fuchsia L. ●■
136724 Dorvalla Comm. ex DC. = Dorvalia Comm. ex Lam. ●■
136725 Doryalis E. Mey. = Dovyalis E. Mey. ex Arn. ●
136726 Doryalis E. Mey. ex Arn. = Dovyalis E. Mey. ex Arn. ●
136727 Doryalis Warb. = Dovyalis E. Mey. ex Arn. ●
136728 Doryanthaceae R. Dahlgren et Clifford(1985);矛缨花科(矛花科)■☆
136729 Doryanthes Corrêa(1802);矛缨花属(矛花属);Gymea Lily, Spear-lily ■☆
136730 Doryanthes excelsa Corrêa;矛缨花;Gymea Lily, Gymea-lily ■☆
136731 Doryanthes palmeri W. Hill ex Benth.;棱叶矛缨花■☆
136732 Dorycheile Rchb. = Cephalanthera Rich. ■
136733 Dorychilus Post et Kuntze = Dorycheile Rchb. ■
136734 Dorychnium Brongn. = Dorycnium Mill. ●■☆
136735 Dorychnium Moench = Cullen Medik. ●■
136736 Dorychnium Moench = Psoralea L. ●■
136737 Dorychnium Royen ex Moench = Psoralea L. ●■
136738 Dorycinopsis Lem. = Anthyllis L. ■☆
136739 Dorycinopsis Lem. = Dorycnopsis Boiss. ■☆
136740 Dorycnium Mill.(1754);加那利豆属;Canary-clover ●■☆
136741 Dorycnium Mill. = Lotus L. ■
136742 Dorycnium broussonetii Webb et Berthel.;布鲁索内加那利豆■☆
136743 Dorycnium eriophthalmum (Webb et Berthel.) Webb et Berthel.;红眼加那利豆■☆
136744 Dorycnium gracile Jord. = Dorycnium herbaceum Vill. subsp. gracile (Jord.) Nyman ■☆
136745 Dorycnium graecum Ser. = Dorycnium latifolium Willd. ●☆
136746 Dorycnium herbaceum Vill.;草本加那利豆;Herb Canary-clover ■☆
136747 Dorycnium herbaceum Vill. subsp. gracile (Jord.) Nyman;纤细加那利豆■☆
136748 Dorycnium hirsutum (L.) Ser.;粗毛加那利豆■☆
136749 Dorycnium intermedium Ledeb. = Dorycnium herbaceum Vill. ■☆
136750 Dorycnium latifolium Willd.;宽叶加那利豆●☆
136751 Dorycnium lutescens Steud.;浅黄加那利豆■☆
136752 Dorycnium pentaphyllum Scop.;五叶加那利豆■☆
136753 Dorycnium pentaphyllum Scop. subsp. gracile (Jord.) Rouy = Dorycnium herbaceum Vill. ■☆
136754 Dorycnium pentaphyllum Scop. subsp. jordanianum (Willk.) Batt. = Dorycnium herbaceum Vill. subsp. gracile (Jord.) Nyman ■☆
136755 Dorycnium pentaphyllum Scop. subsp. suffruticosum (Vill.) Rouy = Dorycnium pentaphyllum Scop. ■☆
136756 Dorycnium rectum (L.) Ser.;直加那利豆■☆
136757 Dorycnium rectum (L.) Ser. var. glaber Cavara et Trotter = Dorycnium rectum (L.) Ser. ■☆
136758 Dorycnium rectum (L.) Ser. var. pauciflorum Ball = Dorycnium rectum (L.) Ser. ■☆
136759 Dorycnium spectabile Webb et Berthel.;壮观加那利豆■☆
136760 Dorycnium suffruticosum Vill. = Dorycnium pentaphyllum Scop. ■☆
136761 Dorycnium villosum Blatt. et Hallb. = Lotus makranicus Rech. f. et Esfand. ■☆
136762 Dorycnopsis Boiss. = Anthyllis L. ■☆
136763 Doryctandea Hook. f. et Thomson = Dioryktandra Hassk. ●
136764 Doryctandea Hook. f. et Thomson = Rinorea Aubl.(保留属名)●
136765 Dorydium Salisb. = Asphodeline Rchb. ■☆
136766 Doryphora Endl.(1837);檫木香属(多瑞弗拉属,矛桂属,矛雄香属);Sassafras ●☆
136767 Doryphora aromatica (F. M. Bailey) L. S. Sm.;矛桂(多瑞弗拉)●☆
136768 Doryphora vieillardi Baill.;新南威尔士檫木香;New South Wales Sassafras, Sassafras ●☆
136769 Dorystaechas Boiss. et Heldr. = Dorystaechas Boiss. et Heldr. ex Benth. ●☆
136770 Dorystaechas Boiss. et Heldr. ex Benth.(1848);土耳其山灌属●☆
136771 Dorystaechas hastata Boiss. et Heldr. ex Benth.;土耳其山灌●☆
136772 Dorystephania Warb. = Sarcolobus R. Br. ●☆
136773 Dorystigma Gaudich. = Pandanus Parkinson ex Du Roi ●■
136774 Dorystigma Miers = Jaborosa Juss. ●☆
136775 Dorystigma Miers(1845);矛柱露兜树属●☆
136776 Dorystigma mauritianum Gaudich.;矛柱露兜树●☆
136777 Dorystoechas Boiss. et Heldr. = Dorystaechas Boiss. et Heldr. ex Benth. ●☆
136778 Dorystoechas Boiss. et Heldr. ex Benth. = Dorystaechas Boiss. et Heldr. ex Benth. ●☆
136779 Doryxylon Zoll.(1857);矛材木属●☆
136780 Doryxylon spinosum Zoll.;矛材木●☆
136781 Doschafa Post et Kuntze = Arisaema Mart. ●■
136782 Doschafa Post et Kuntze = Dochafa Schott ●■
136783 Dossifluga Bremek. = Strobilanthes Blume ●■
136784 Dossinia C. Morren(1848);多新兰属(道西兰属);Dossinia ■☆
136785 Dossinia marmorata (Blume) C. Morren;多新兰(道西兰);Marbled Dossinia ■☆

136786 Dossinia obliqua (Blume) Miq. = Hetaeria obliqua Blume ■
136787 Dothieroa Raf. = Phlogacanthus Nees ●■
136788 Dothilis Raf. = Chloraea Lindl. ■☆
136789 Dothilis Raf. = Spiranthes Rich. (保留属名) ■
136790 Dothilis Raf. = Ulantha Hook. ■☆
136791 Dothills Raf. = Chloraea Lindl. ■☆
136792 Dothilophis Raf. = Epidendrum L. (保留属名) ■☆
136793 Douarrea Montrouz. = Psychotria L. (保留属名) ●
136794 Douepea Cambess. (1839);阿拉伯芥属■☆
136795 Douepia Hook. f. et Thomson = Douepea Cambess. ■☆
136796 Douepia tortuosa Cambess.;阿拉伯芥■☆
136797 Douglasdeweya C. Yen, J. L. Yang et B. R. Baum = Agropyron Gaertn. ■
136798 Douglasdeweya C. Yen, J. L. Yang et B. R. Baum (2005);肖冰草属■☆
136799 Douglasia Lindl. (1827) (保留属名);金地梅属■☆
136800 Douglasia Lindl. (保留属名) = Androsace L. ■
136801 Douglasia nivalis Lindl.;金地梅■☆
136802 Douglassia Heist. = Nerine Herb. (保留属名) ■☆
136803 Douglassia Mill. (废弃属名) = Clerodendrum L. ●■
136804 Douglassia Mill. (废弃属名) = Douglasia Lindl. (保留属名) ■☆
136805 Douglassia Mill. (废弃属名) = Volkameria Burm. f. ●
136806 Douglassia Rchb. = Douglasia Lindl. (保留属名) ■☆
136807 Douglassia Schreb. = Aiouea Aubl. ●☆
136808 Douma Poir. = Hyphaene Gaertn. ●☆
136809 Doupea D. Dietr. = Douepea Cambess. ■☆
136810 Douradoa Sleumer (1984);杜拉木属●☆
136811 Douradoa consimilis Sleumer;杜拉木●☆
136812 Dovea Kunth (1841);多夫草属■☆
136813 Dovea aggregata Mast. = Chondropetalum aggregatum (Mast.) Pillans ■☆
136814 Dovea binata Steud. = Hypodiscus laevigatus (Kunth) H. P. Linder ■☆
136815 Dovea bolusii Mast. = Chondropetalum hookerianum (Mast.) Pillans ■☆
136816 Dovea chartacea Pillans = Askidiosperma chartaceum (Pillans) H. P. Linder ■☆
136817 Dovea cylindrostachya Mast. = Chondropetalum tectorum (L. f.) Raf. ■☆
136818 Dovea ebracteata Kunth = Chondropetalum ebracteatum (Kunth) Pillans ■☆
136819 Dovea macrocarpa Kunth;大果多夫草■☆
136820 Dovea marlothii Pillans = Chondropetalum marlothii (Pillans) Pillans ■☆
136821 Dovea microcarpa Kunth = Chondropetalum microcarpum (Kunth) Pillans ■☆
136822 Dovea mucronata (Nees) Mast. = Chondropetalum mucronatum (Nees) Pillans ■☆
136823 Dovea nitida Mast. = Askidiosperma nitidum (Mast.) H. P. Linder ■☆
136824 Dovea nuda (Rottb.) Pillans = Chondropetalum nudum Rottb. ■☆
136825 Dovea paniculata Mast. = Askidiosperma paniculatum (Mast.) H. P. Linder ■☆
136826 Dovea racemosa (Poir.) Mast. = Elegia racemosa (Poir.) Pers. ■☆
136827 Dovea recta Mast. = Chondropetalum rectum (Mast.) Pillans ■☆
136828 Dovea rigens Mast. = Chondropetalum microcarpum (Kunth) Pillans ■☆
136829 Dovea tectorum (L. f.) Mast. = Chondropetalum tectorum (L. f.) Raf. ■☆
136830 Dovea thyrsoidea Mast. = Elegia thyrsoidea (Mast.) Pillans ■☆
136831 Dovyalis E. Mey. = Dovyalis E. Mey. ex Arn. ●
136832 Dovyalis E. Mey. ex Arn. (1841);木莓属(斯里兰卡莓属,西苔栗属);Dovyalis ●
136833 Dovyalis Warb. = Dovyalis E. Mey. ex Arn. ●
136834 Dovyalis abyssinica (A. Rich.) Warb.;阿比西尼亚木莓●☆
136835 Dovyalis acuminata Gilg ex Engl. = Dovyalis zenkeri Gilg ●☆
136836 Dovyalis adolfi-frederici Mildbr. ex Gilg = Dovyalis macrocalyx (Oliv.) Warb. ●☆
136837 Dovyalis afzelii Gilg = Dovyalis zenkeri Gilg ●☆
136838 Dovyalis antunesii Gilg = Dovyalis macrocalyx (Oliv.) Warb. ●☆
136839 Dovyalis caffra (Hook. f. et Harv.) Hook. f.;开木莓;Kav-apple, Kei Apple, Umkokolo ●☆
136840 Dovyalis cameroonensis Cheek et Ngolan;喀麦隆木莓●☆
136841 Dovyalis celastroides Sond. = Dovyalis rotundifolia (Thunb.) Thunb. et Harv. ●☆
136842 Dovyalis chirindensis Engl. = Dovyalis macrocalyx (Oliv.) Warb. ●☆
136843 Dovyalis engleri Gilg = Dovyalis abyssinica (A. Rich.) Warb. ●☆
136844 Dovyalis giorgii De Wild. = Dovyalis zenkeri Gilg ●☆
136845 Dovyalis glandulosissima Gilg = Dovyalis macrocalyx (Oliv.) Warb. ●☆
136846 Dovyalis hebecarpa (Gardner) Warb.;木莓(酸味果,锡兰醋栗,锡兰莓);Ceylon Gooseberry, Kitembilla ●
136847 Dovyalis hispidula Wild;细毛木莓●☆
136848 Dovyalis longispina (Harv.) Warb.;长刺木莓●☆
136849 Dovyalis lucida Sim;亮木莓●☆
136850 Dovyalis luckii R. E. Fr. = Dovyalis macrocalyx (Oliv.) Warb. ●☆
136851 Dovyalis macrocalyx (Oliv.) Warb.;大萼木莓●☆
136852 Dovyalis macrocarpa Bamps;大果木莓●☆
136853 Dovyalis maliformis Gilg = Dovyalis spinosissima Gilg ●☆
136854 Dovyalis mildbraedii Gilg = Dovyalis macrocalyx (Oliv.) Warb. ●☆
136855 Dovyalis mollis (Oliv.) Warb.;绢毛木莓●☆
136856 Dovyalis retusa Robyns et Lawalrée = Dovyalis macrocalyx (Oliv.) Warb. ●☆
136857 Dovyalis revoluta Thom = Dovyalis zeyheri (Sond.) Warb. ●☆
136858 Dovyalis rhamnoides (Burch. ex DC.) Burch. et Harv.;鼠李木莓●☆
136859 Dovyalis rotundifolia (Thunb.) Thunb. et Harv.;圆叶木莓●☆
136860 Dovyalis salicifolia Gilg = Dovyalis macrocalyx (Oliv.) Warb. ●☆
136861 Dovyalis somalensis Gilg = Dovyalis verrucosa (Hochst.) Warb. ●☆
136862 Dovyalis spinosissima Gilg;刺木莓●☆
136863 Dovyalis tenuispina Gilg = Dovyalis zenkeri Gilg ●☆
136864 Dovyalis tristis (Sond.) Warb. = Dovyalis zeyheri (Sond.) Warb. ●☆
136865 Dovyalis verrucosa (Hochst.) Warb.;瘤木莓●☆
136866 Dovyalis xanthocarpa Bullock;黄果木莓●☆
136867 Dovyalis zenkeri Gilg;岑克尔木莓●☆
136868 Dovyalis zenkeri Gilg var. vestita Tisser. et Sillans = Dovyalis zenkeri Gilg ●☆

136869 Dovyalis zeyheri (Sond.) Warb.;蔡氏木莓●☆
136870 Dovyalis zizyphoides E. Mey. ex Sond. = Dovyalis rhamnoides (Burch. ex DC.) Burch. et Harv. ●☆
136871 Dowea Steud. = Dovea Kunth ■☆
136872 Downingia Torr. (1857)(保留属名);唐宁草属;Californian Lobelia ■☆
136873 Downingia elegans Torr.;唐宁草;Californian Lobelia ■☆
136874 Doxantha Miers = Bignonia L.(保留属名)+ Doxanthemum D. R. Hunt ●☆
136875 Doxantha Miers = Bignonia L.(保留属名)●
136876 Doxantha Miers = Macfadyena A. DC. ●
136877 Doxantha Miers(1863);猫爪草属(荣花属)●☆
136878 Doxantha capreolata Miers = Bignonia capreolata L. ●
136879 Doxantha lanceolata Miers;剑叶猫爪草■☆
136880 Doxantha macfadyena ? = Macfadyena unguis-cati (L.) A. H. Gentry ●
136881 Doxantha mexicana Miers;墨西哥猫爪草■☆
136882 Doxantha unguis-cati (L.) Rehder = Macfadyena unguis-cati (L.) A. H. Gentry ●
136883 Doxantha unguis-cati Rehder = Macfadyena unguis-cati (L.) A. H. Gentry ●
136884 Doxanthemum D. R. Hunt(1864);北美紫葳属●☆
136885 Doxanthes Raf. = Phaeomeria Lindl. ex K. Schum. ■☆
136886 Doxema Raf. = Ipomoea L.(保留属名)●■
136887 Doxema Raf. = Quamoclit Moench ●■
136888 Doxomma Miers = Barringtonia J. R. Forst. et G. Forst.(保留属名)●
136889 Doxosma Raf. = Epidendrum L.(保留属名)■☆
136890 Doyerea Grosourdy = Corallocarpus Welw. ex Benth. et Hook. f. ■☆
136891 Doyerea Grosourdy = Doyerea Grosourdy ex Bello ■☆
136892 Doyerea Grosourdy ex Bello(1864);道氏瓜属(道耶瓜属)■☆
136893 Doyerea emetocathartica Grosourdy;吐泻道氏瓜(吐泻道耶瓜)■☆
136894 Doyleanthus Sauquet(2003);多伊尔豆蔻属●☆
136895 Doyleanthus arillata Capuron ex Sauquet;多伊尔豆蔻●☆
136896 Draaksteinia Post et Kuntze = Dalbergia L. f.(保留属名)●
136897 Draaksteinia Post et Kuntze = Drakenstenia Neck. ●
136898 Draba Dill. ex L. = Draba L. ■
136899 Draba L. (1753);葶苈属(山芥属);Draba, Whitlow Grass, Whitlowgrass, Whitlow-grass, Whitlowwort, Whitlow-wort ■
136900 Draba affghanica Pohle = Draba oreades Schrenk ■
136901 Draba aizoides L.;黄刚毛葶苈;Yellow Whitlow Grass, Yellow Whitlow-grass ■☆
136902 Draba aizon Wahlenb.;常绿葶苈■☆
136903 Draba alajica Litv.;阿拉伊葶苈■
136904 Draba alajica Litv. var. leiocarpa Pohle = Draba alajica Litv. ■
136905 Draba albertii Regel et Schmalh.;阿氏葶苈■☆
136906 Draba algida Adams var. brachycarpa Bunge = Draba oreades Schrenk ■
136907 Draba alpicola Klotzsch = Draba oreades Schrenk var. alpicola (Klotzsch) O. E. Schulz ■
136908 Draba alpicola Klotzsch = Draba oreades Schrenk ■
136909 Draba alpina Clairv = Draba fladnizensis Wulfen ■
136910 Draba alpina Hook. f. et Thomson = Draba oreades Schrenk ■
136911 Draba alpina L.;高山葶苈;Rockcress Draba, Rock-cress Draba, Rockcress Whitlowgrass ■
136912 Draba alpina L. var. involucrata W. W. Sm. = Draba involucrata (W. W. Sm.) W. W. Sm. ■
136913 Draba alpina L. var. korshinskyi O. Fedtsch. = Draba korshinskyi (O. Fedtsch.) Pohle ■
136914 Draba alpina L. var. korshinskyi O. Fedtsch. = Draba pamirica (O. Fedtsch.) Pohle ■☆
136915 Draba alpina L. var. leiophylla Franch. = Draba involucrata (W. W. Sm.) W. W. Sm. ■
136916 Draba alpina L. var. pamirica O. Fedtsch. = Draba pamirica (O. Fedtsch.) Pohle ■☆
136917 Draba alpina L. var. rigida Franch. = Draba oreades Schrenk ■
136918 Draba altaica (C. A. Mey.) Bunge;阿尔泰葶苈;Altai Draba, Altai Whitlowgrass ■
136919 Draba altaica (C. A. Mey.) Bunge var. foliosa O. E. Schulz = Draba altaica (C. A. Mey.) Bunge ■
136920 Draba altaica (C. A. Mey.) Bunge var. glabrescens Lipsky = Draba altaica (C. A. Mey.) Bunge ■
136921 Draba altaica (C. A. Mey.) Bunge var. microcarpa O. E. Schulz;小果阿尔泰葶苈;Smallfruit Altai Draba, Smallfruit Altai Whitlowgrass ■
136922 Draba altaica (C. A. Mey.) Bunge var. microcarpa O. E. Schulz = Draba altaica (C. A. Mey.) Bunge ■
136923 Draba altaica (C. A. Mey.) Bunge var. modesta (W. W. Sm.) W. T. Wang;苞叶阿尔泰葶苈;Leafybract Altai Draba, Leafybract Altai Whitlowgrass ■
136924 Draba altaica (C. A. Mey.) Bunge var. modesta (W. W. Sm.) W. T. Wang = Draba altaica (C. A. Mey.) Bunge ■
136925 Draba altaica (C. A. Mey.) Bunge var. pusilla (Kar. et Kir.) Fedtsch.;矮阿尔泰葶苈;Darf Altai Draba, Darf Altai Whitlowgrass ■
136926 Draba altaica (C. A. Mey.) Bunge var. racemosa O. E. Schulz;总序阿尔泰葶苈;Racemose Altai Draba, Racemose Altai Whitlowgrass ■
136927 Draba altaica (C. A. Mey.) Bunge var. racemosa O. E. Schulz = Draba altaica (C. A. Mey.) Bunge ■
136928 Draba alticola Kom. = Draba melanopus Kom. ■
136929 Draba amplexicaulis Franch.;抱茎葶苈;Amplexicaul Draba, Amplexicaul Whitlowgrass ■
136930 Draba amplexicaulis Franch. var. bracteata O. E. Schulz;具苞抱茎葶苈;Bract Amplexicaul Draba, Bract Amplexicaul Whitlowgrass ■
136931 Draba amplexicaulis Franch. var. bracteata O. E. Schulz = Draba surculosa Franch. ■
136932 Draba amplexicaulis Franch. var. dasycarpa O. E. Schulz = Draba calcicola O. E. Schulz ■
136933 Draba amplexicaulis Franch. var. dolichocarpa O. E. Schulz;长果抱茎葶苈;Longfruit Amplexicaul Draba, Longfruit Amplexicaul Whitlowgrass ■
136934 Draba amplexicaulis Franch. var. dolichocarpa O. E. Schulz = Draba amplexicaulis Franch. ■
136935 Draba aprica O. E. Schulz = Draba calcicola O. E. Schulz ■
136936 Draba arabisans Michx.;岩地葶苈;Rock Whitlow-grass ■☆
136937 Draba arabisans Michx. var. superiorensis Butters et Abbe = Draba arabisans Michx. ■☆
136938 Draba araratica Rupr.;亚拉腊葶苈■☆
136939 Draba armata Schott, Nyman et Kotschy = Draba longirostra Schott, Nyman, et Kotschy ■☆
136940 Draba arseniewii (B. Fedtsch.) Gilg ex Tolm.;阿尔葶苈■☆
136941 Draba atlantica Pomel = Draba hispanica Boiss. ■☆

136942 Draba atlantica Pomel var. maroccana O. E. Schulz = Draba hispanica Boiss. ■☆
136943 Draba aucheri Boiss.;奥氏葶苈■☆
136944 Draba aurea Vahl ex Hornem.;金黄葶苈■☆
136945 Draba baicalensis Tolm.;拜卡尔葶苈■☆
136946 Draba balangshanica W. T. Wang;巴郎山葶苈;Balangshan Draba ■
136947 Draba balangshanica W. T. Wang = Draba surculosa Franch. ■
136948 Draba barbata Pohle;髯毛葶苈■☆
136949 Draba behringii Tolm.;拜尔葶苈■☆
136950 Draba bhutanica DC.;不丹葶苈■
136951 Draba borealis DC.;北方葶苈;Nothern Draba, Nothern Whitlowgrass ■
136952 Draba borealis DC. var. kurilensis F. Schmidt;千岛葶苈■☆
136953 Draba borealis DC. var. leiocarpa Pohle = Draba borealis DC. ■
136954 Draba brachycarpa Nutt. ex Torr. et A. Gray;短果葶苈;Shortpod Draba, Whitlow Grass ■☆
136955 Draba breweri S. Watson var. cana (Rydb.) Rollins = Draba cana Rydb. ■☆
136956 Draba bruniifolia Steven;褐叶葶苈■☆
136957 Draba cachemirica Gand.;克什米尔葶苈■
136958 Draba cachemirica Gand. var. koelzii O. E. Schulz = Draba cachemirica Gand. ■
136959 Draba cachemirica Gand. var. stoliczkae O. E. Schulz = Draba cachemirica Gand. ■
136960 Draba cacuminum Ekman;北欧葶苈■☆
136961 Draba calcicola O. E. Schulz = Draba moupingensis Franch. var. calcicola (O. E. Schulz) W. T. Wang ■
136962 Draba cana Rydb. = Draba lanceolata DC. ■☆
136963 Draba cardaminiflora Kom.;碎米荠叶葶苈■☆
136964 Draba caroliniana Walter = Draba reptans (Lam.) Fernald ■☆
136965 Draba caroliniana Walter f. stellifera O. E. Schulz = Draba reptans (Lam.) Fernald ■☆
136966 Draba chamissonis G. Don;哈米葶苈■☆
136967 Draba cholaensis W. W. Sm.;大花葶苈■
136968 Draba cholaensis W. W. Sm. var. leiocarpa H. Hara = Draba cholaensis W. W. Sm. ■
136969 Draba cinerea Adams;北极葶苈;Arctic Draba ■☆
136970 Draba composita O. E. Schulz = Draba senilis O. E. Schulz ■
136971 Draba cossonii O. E. Schulz = Draba hederifolia Coss. subsp. cossoniana (O. E. Schulz) Maire ■☆
136972 Draba cuneifolia Nutt. ex Torr. et A. Gray;肾叶葶苈;Wedgeleaf Draba, Whitlow Grass ■☆
136973 Draba cuspidata M. Bieb.;尖叶葶苈■☆
136974 Draba daochengensis W. T. Wang;稻城葶苈;Daocheng Draba ■
136975 Draba daochengensis W. T. Wang = Draba lichiangensis W. W. Sm. ■
136976 Draba darwasica Lipsky;达尔瓦斯葶苈■☆
136977 Draba dasyastra Gilg et O. E. Schulz;柱形葶苈;Densepulvinate Draba, Densepulvinate Whitlowgrass ■
136978 Draba dasyastra Gilg et O. E. Schulz = Draba winterbottomii (Hook. f. et Thomson) Pohle ■
136979 Draba dasycarpa C. A. Mey. = Draba subamplexicaulis C. A. Mey. ■
136980 Draba diversifolia Boiss.;异叶葶苈■☆
136981 Draba dolichotricha W. T. Wang;长毛葶苈;Longhair Draba, Longhair Whitlowgrass ■
136982 Draba dolichotricha W. T. Wang = Draba oreodoxa W. W. Sm. ■
136983 Draba draboides (Maxim.) Al-Shehbaz = Coelonema draboides Maxim. ■
136984 Draba dubia Suter;柔毛葶苈;Woolly Draba ■☆
136985 Draba dubia Suter subsp. laevipes (DC.) Braun-Blanq.;平滑葶苈■☆
136986 Draba elata Hook. f. et Thomson;高茎葶苈;Tall Draba, Tall Whitlowgrass ■
136987 Draba elata W. W. Sm. et Cave = Draba polyphylla O. E. Schulz ■
136988 Draba elisahethae N. Busch;爱丽萨葶苈■☆
136989 Draba ellipsoidea Hook. f. et Thomson;椭圆果葶苈;Ellipticfruit Draba, Ellipticfruit Whitlowgrass ■
136990 Draba eriopoda Turcz.;毛葶苈;Woollystalk Draba, Woollystalk Whitlowgrass ■
136991 Draba eriopoda Turcz. var. kamensis Pohle = Draba eriopoda Turcz. ■
136992 Draba eriopoda Turcz. var. sinensis Maxim. = Draba eriopoda Turcz. ■
136993 Draba eschscholtzii Pohle ex N. Bosch;爱绍氏葶苈■☆
136994 Draba fedtschenkoi (Pohle) Gilg;范氏葶苈■☆
136995 Draba fladnizensis Wulfen;福地葶苈(福拉尼山葶苈);Arctic Draba, Arctic Whitlowgrass ■
136996 Draba franchetii O. E. Schulz;弗朗谢葶苈■☆
136997 Draba fuhaiensis C. H. An;福海葶苈;Fuhai Draba ■
136998 Draba glabella Pursh;光葶苈;Glabrous Draba, Smooth Whitlow-grass ■☆
136999 Draba glacialis Adams;冰河葶苈(冰川葶苈);Glacier Draba, Glacier Whitlowgrass ■☆
137000 Draba glacialis Adams = Draba cachemirica Gand. ■
137001 Draba glacialis Adams = Draba setosa Royle ■
137002 Draba glacialis Adams f. incompta Regel = Draba incompta Steven ■
137003 Draba globifera Ledeb.;球葶苈■☆
137004 Draba glomerata Royle;球果葶苈(球序葶苈);Clustered Draba, Clustered Whitlowgrass ■
137005 Draba glomerata Royle var. dasycarpa O. E. Schulz;粗球果葶苈;Hairyfruit Clustered Whitlowgrass, Hairyfruit Draba ■
137006 Draba glomerata Royle var. dasycarpa O. E. Schulz = Draba glomerata Royle ■
137007 Draba glomerata Royle var. leiocarpa Pamp. = Draba lasiophylla Royle ■
137008 Draba gmelinii Adams = Draba sibirica (Pall.) Thell. ■
137009 Draba gracillima Hook. f. et Thomson;纤细葶苈;Thinnest Draba, Thinnest Whitlowgrass ■
137010 Draba grandis Langsd. ex DC.;日本大葶苈■☆
137011 Draba granitica Hand.-Mazz.;岩葶苈(花岗岩生葶苈);Granite-living Draba, Granite-living Whitlowgrass ■
137012 Draba granitica Hand.-Mazz. = Draba gracillima Hook. f. et Thomson ■
137013 Draba handelii O. E. Schulz;矮葶苈(贡山葶苈,韩氏葶苈);Dwarf Draba, Dwarf Whitlowgrass ■
137014 Draba hederifolia Coss.;常春藤叶葶苈■☆
137015 Draba hederifolia Coss. subsp. cossoniana (O. E. Schulz) Maire;科森葶苈■☆
137016 Draba hederifolia Coss. var. orientalis Quézel = Draba hederifolia Coss. subsp. cossoniana (O. E. Schulz) Maire ■☆
137017 Draba hicksii Grierson = Draba lichiangensis W. W. Sm. ■

137018 Draba himalayensis Klotzsch = Thlaspi coclearioides Hook. f. et Thomson ■☆

137019 Draba hirta L.；硬毛葶苈（毛绒葶苈）；Hairy Draba, Hairy Whitlowgrass ■

137020 Draba hirta L. var. leiocarpa Maxim. = Draba mongolica Turcz. ■

137021 Draba hirta L. var. leiocarpa Regel f. parviflora Regel = Draba parviflora (Regel) O. E. Schulz ■

137022 Draba hirta L. var. subamplexicaulis (C. A. Mey.) Regel = Draba subamplexicaulis C. A. Mey. ■

137023 Draba hispanica Boiss.；西班牙葶苈■☆

137024 Draba hispanica Boiss. subsp. djurdjurae (Batt.) Greuter；朱尔朱拉山葶苈■☆

137025 Draba hispanica Boiss. var. atlantica (Pomel) Batt. = Draba hispanica Boiss. ■☆

137026 Draba hispanica Boiss. var. cladotricha Maire = Draba hispanica Boiss. subsp. djurdjurae (Batt.) Greuter ■☆

137027 Draba hispanica Boiss. var. djudjurae Batt. = Draba hispanica Boiss. ■☆

137028 Draba hispanica Boiss. var. longistyla Batt. = Draba hispanica Boiss. ■☆

137029 Draba hispanica Boiss. var. macrobotrys Maire = Draba hispanica Boiss. subsp. djurdjurae (Batt.) Greuter ■☆

137030 Draba hispanica Boiss. var. maroccana (O. E. Schulz) Emb. = Draba hispanica Boiss. ■☆

137031 Draba hispanica Boiss. var. rhatica Quézel = Draba hispanica Boiss. ■☆

137032 Draba hispida Willd.；粗毛葶苈■☆

137033 Draba hissarica Lipsky；希萨尔葶苈■☆

137034 Draba huetii Boiss.；中亚葶苈（惠特葶苈）■

137035 Draba humillima Boiss.；微小葶苈■

137036 Draba hystrix Hook. f. et Thomson；豪猪葶苈■☆

137037 Draba igarashii S. Watan. = Draba kitadakensis Koidz. ■☆

137038 Draba imeretica Rupr.；伊梅里特葶苈■☆

137039 Draba incana L.；灰白葶苈（灰葶苈）；Hoary Whitlow-grass, Twisted Draba, Twisted Whitlow Grass, Twisted Whitlowgrass, Twisted-podded Whitlow Grass ■

137040 Draba incana L. f. altera Cham. et Schltdl. = Draba hirta L. ■

137041 Draba incana L. var. borealis Torr. et Gray = Draba borealis DC. ■

137042 Draba incana L. var. flaccida Maxim. = Draba ladyginii Pohle ■

137043 Draba incana L. var. microphylla W. W. Sm. = Draba ladyginii Pohle ■

137044 Draba incana L. var. mongolica (Turcz.) Regel = Draba mongolica Turcz. ■

137045 Draba incana L. var. mongolica Regel = Draba mongolica Turcz. ■

137046 Draba incompta Steven；星毛葶苈；Staellatehair Draba, Staellatehair Whitlowgrass ■

137047 Draba incompta Steven = Draba winterbottomii (Hook. f. et Thomson) Pohle ■

137048 Draba incurvata A. N. Vassilcz. et Golosk.；内折葶苈■☆

137049 Draba involucrata (W. W. Sm.) W. W. Sm.；总苞葶苈（苞花葶苈）；Involucrate Draba, Involucrate Whitlowgrass ■

137050 Draba involucrata (W. W. Sm.) W. W. Sm. var. lasiocarpa W. T. Wang；毛果苞花葶苈；Hairfruit Involucrate Draba, Hairfruit Involucrate Whitlowgrass ■

137051 Draba involucrata (W. W. Sm.) W. W. Sm. var. lasiocarpa W. T. Wang = Draba involucrata (W. W. Sm.) W. W. Sm. ■

137052 Draba japonica Maxim.；日本葶苈；Japanese Draba ■☆

137053 Draba jucunda W. W. Sm.；愉悦葶苈；Agreeable Draba, Agreeable Whitlowgrass ■

137054 Draba kamtschatica N. Busch；勘察加葶苈■☆

137055 Draba kamtschatica N. Busch var. yesoensis Nakai = Draba kitadakensis Koidz. ■☆

137056 Draba kitadakensis Koidz.；信州北岳葶苈■☆

137057 Draba kizylarti (Korsh.) N. Busch = Draba oreades Schrenk ■

137058 Draba kjellmanii Lidén ex Ekman；杰曼氏葶苈；Kjelman Draba ■☆

137059 Draba korshinskyi (O. Fedtsch.) Pohle；科氏葶苈■

137060 Draba korshinskyi (O. Fedtsch.) Pohle var. setosa Pohle = Draba cachemirica Gand. ■

137061 Draba kurilensis (Turcz.) F. Schmidt；库里尔葶苈■☆

137062 Draba kuznetzovii (Turcz.) Hayek；库兹葶苈■☆

137063 Draba lactea Adams；乳白葶苈；Milky Draba, Milky Whitlowgrass ■☆

137064 Draba lactea Adams = Draba fladnizensis Wulfen ■

137065 Draba ladyginii Pohle；苞序葶苈（线果葶苈，穴乌萝卜）；Ladygin Draba, Ladygin Whitlowgrass ■

137066 Draba ladyginii Pohle var. trichocarpa O. E. Schulz；毛果苞序葶苈（毛线果葶苈）；Hairfruit Ladygin Draba, Hairfruit Ladygin Whitlowgrass ■

137067 Draba ladyginii Pohle var. trichocarpa O. E. Schulz = Draba lasiophylla Royle ■

137068 Draba laevipes DC. = Draba dubia Suter subsp. laevipes (DC.) Braun-Blanq. ■☆

137069 Draba lanceolata Royle；锥果葶苈（灰白葶苈）；Hoary Whitlow-grass, Lanceolate Draba, Lanceolate Whitlowgrass ■

137070 Draba lanceolata Royle = Draba cana Rydb. ■☆

137071 Draba lanceolata Royle = Draba gracillima Hook. f. et Thomson ■

137072 Draba lanceolata Royle var. brachycarpa O. E. Schulz；短锥果葶苈；Short Lanceolate Draba, Short Lanceolate Whitlowgrass ■

137073 Draba lanceolata Royle var. brachycarpa O. E. Schulz = Draba lanceolata Royle ■

137074 Draba lanceolata Royle var. chingii O. E. Schulz；紫茎锥果葶苈；Ching Lanceolate Draba, Ching Lanceolate Whitlowgrass ■

137075 Draba lanceolata Royle var. chingii O. E. Schulz = Draba ladyginii Pohle ■

137076 Draba lanceolata Royle var. latifolia O. E. Schulz = Draba ladyginii Pohle ■

137077 Draba lanceolata Royle var. leiocarpa O. E. Schulz；光锥果葶苈；Smooth Lanceolate Draba, Smooth Lanceolate Whitlowgrass ■

137078 Draba lanceolata Royle var. leiocarpa O. E. Schulz = Draba lanceolata Royle ■

137079 Draba lanceolata Royle var. sonamargensis O. E. Schulz = Draba lanceolata Royle ■

137080 Draba lanjarica O. E. Schulz = Christolea lanuginosa (Hook. f. et Thomson) Ovcz. ■

137081 Draba lanjarica O. E. Schulz = Eurycarpus lanuginosus (Hook. f. et Thomson) Botsch. ■

137082 Draba lasiocarpa Adams = Draba aizoides L. ■☆

137083 Draba lasiophylla Royle；毛叶葶苈；Hairyleaf Draba, Hairyleaf Whitlowgrass ■

137084 Draba lasiophylla Royle f. leiocarpa Pamp. = Draba lasiophylla Royle ■

137085 Draba lasiophylla Royle var. leiocarpa (Pamp.) O. E. Schulz；光果毛叶葶苈；Smoothfruit Hairyleaf Draba, Smoothfruit Hairyleaf

Whitlowgrass ■
137086 Draba lasiophylla Royle var. leiocarpa (Pamp.) O. E. Schulz = Draba lasiophylla Royle ■
137087 Draba lasiophylla Royle var. royleana Pamp. = Draba lasiophylla Royle ■
137088 Draba lichiangensis W. W. Sm.;丽江葶苈;Lijiang Draba, Lijiang Whitlowgrass, Likiang Draba ■
137089 Draba lichiangensis W. W. Sm. var. microcarpa O. E. Schulz = Draba lichiangensis W. W. Sm. ■
137090 Draba lichiangensis W. W. Sm. var. trichocarpa O. E. Schulz = Draba lichiangensis W. W. Sm. ■
137091 Draba linearifolia L. L. Lou et T. Y. Cheo;线叶葶苈;Linearleaf Draba, Linearleaf Whitlowgrass ■
137092 Draba linearis Hook. f. et T. Anderson = Draba stenocarpa Hook. f. et Thomson ■
137093 Draba lipskyi Tolm.;利普斯基葶苈■☆
137094 Draba longirostra Schott, Nyman, et Kotschy;长嘴葶苈■☆
137095 Draba longisiliqua Bornm.;长果葶苈■☆
137096 Draba ludingensis W. T. Wang;泸定葶苈;Luding Draba ■
137097 Draba ludingensis W. T. Wang = Draba oreodoxa W. W. Sm. ■
137098 Draba lutea Gilib. = Draba nemorosa L. var. leiocarpa Lindblom ■
137099 Draba lutescens Coss.;淡黄葶苈■☆
137100 Draba macrocarpa Adams;大果葶苈;Bigfruit Draba ■☆
137101 Draba macroloba Turcz. = Draba nemorosa L. ■
137102 Draba magna (N. Busch) Tolm.;大葶苈■☆
137103 Draba mairei H. Lév. = Draba surculosa Franch. ■
137104 Draba matangensis O. E. Schulz;马塘葶苈;Matang Draba, Matang Whitlowgrass ■
137105 Draba media Litv. = Draba stenocarpa Hook. f. et Thomson ■
137106 Draba media Litv. var. leiocarpa Lipsky = Draba stenocarpa Hook. f. et Thomson ■
137107 Draba melanopus Kom.;天山葶苈;Tianshan Draba, Tianshan Whitlowgrass ■
137108 Draba melanopus Kom. f. hebecarpa Pohle;毛果天山葶苈■
137109 Draba micrantha Nutt. = Draba reptans (Lam.) Fernald ■☆
137110 Draba microcarpella A. N. Vassilcz. et Golosk.;小果葶苈■☆
137111 Draba micropetala Hook.;小瓣葶苈;Smallpetal Draba ■☆
137112 Draba micropetala Hook. = Draba oblongata R. Br. ■☆
137113 Draba modesta W. W. Sm. = Draba altaica (C. A. Mey.) Bunge ■
137114 Draba mollissima Stev.;高加索葶苈■☆
137115 Draba mongolica Turcz.;蒙古葶苈;Mongol Whitlowgrass, Mongolian Draba ■
137116 Draba mongolica Turcz. var. chinensis Pohle = Draba mongolica Turcz. ■
137117 Draba mongolica Turcz. var. elongata Pohle = Draba mongolica Turcz. ■
137118 Draba mongolica Turcz. var. trichocarpa O. E. Schulz;毛果蒙古葶苈;Hairyfruit Mongolian Draba ■
137119 Draba mongolica Turcz. var. trichocarpa O. E. Schulz = Draba mongolica Turcz. ■
137120 Draba mongolica Turcz. var. turczaninoviana Pohle = Draba mongolica Turcz. ■
137121 Draba moupinensis Franch. = Draba surculosa Franch. ■
137122 Draba moupinensis Franch. var. calcicola (O. E. Schulz) W. T. Wang = Draba calcicola O. E. Schulz ■
137123 Draba moupinensis Franch. var. dasycarpa O. E. Schulz = Draba calcicola O. E. Schulz ■
137124 Draba moupingensis Franch.;宝兴葶苈;Baoxing Whitlowgrass, Mouping Draba ■
137125 Draba moupingensis Franch. = Draba surculosa Franch. ■
137126 Draba moupingensis Franch. var. calcicola (O. E. Schulz) W. T. Wang;灰岩葶苈;Calcareus Draba, Calcareus Whitlowgrass, Whitlow Grass, Whitlow Wort ■
137127 Draba moupingensis Franch. var. dasycarpa O. E. Schulz;毛果宝兴葶苈;Hairfruit Baoxing Whitlowgrass, Hairfruit Mouping Draba ■
137128 Draba multiceps Kitag. = Stevenia cheiranthoides DC. ■
137129 Draba muralis L.;墙生葶苈;Wall Draba, Wall Whitlow Grass, Wall Whitlow-grass ■☆
137130 Draba nakaiana H. Hara = Draba kitadakensis Koidz. ■☆
137131 Draba nemoralis DC.;林生葶苈■☆
137132 Draba nemoralis Ehrh. = Draba nemorosa L. ■
137133 Draba nemorosa L.;葶苈(大室,大适,丁历,公荠,狗荠,靡草,荠菜);Wood Draba, Woodland Draba, Woodland Whitlow-grass, Woolly Draba ■
137134 Draba nemorosa L. f. acaulis Sommier;短茎葶苈;Shortstem Woolly Draba ■
137135 Draba nemorosa L. f. latifolia M. Bieb.;宽叶葶苈;Broadleaf Woolly Draba ■
137136 Draba nemorosa L. f. leiocarpa (Lindblom) Kitag.;光果宽叶葶苈;Smoothfruit Woolly Draba, Woodland Draba, Woodland Whitlow-grass ■☆
137137 Draba nemorosa L. var. brevisilicula Zapal. = Draba nemorosa L. ■
137138 Draba nemorosa L. var. hebecarpa Lindblom = Draba nemorosa L. ■
137139 Draba nemorosa L. var. leiocarpa Lindblom = Draba nemorosa L. ■
137140 Draba nemorosa L. var. leiocarpa Lindblom = Draba nemorosa L. f. leiocarpa (Lindblom) Kitag. ■☆
137141 Draba nichanaica O. E. Schulz = Draba ladyginii Pohle ■
137142 Draba nichanaica O. E. Schulz = Draba lanceolata Royle ■
137143 Draba nipponica Makino;本州葶苈;Honshu Draba, Honshu Whitlowgrass ■☆
137144 Draba nipponica Makino f. ramosa S. Watan.;分枝本州葶苈■☆
137145 Draba nipponica Makino var. linearis (Satake) Kitam.;线形本州葶苈■☆
137146 Draba nivalis Lilj.;雪线葶苈;Snow Draba, Snow Whitlowgrass ■☆
137147 Draba norvegica Gunnerus;挪威葶苈(岩生葶苈);Cliff Draba, Cliff Whitlowgrass, Norway Draba, Norway Whitlowgrass, Norwegian Draba, Rock Whitlow-grass ■☆
137148 Draba oblongata R. Br.;椭果葶苈;Oblongate Draba ■☆
137149 Draba obscura Dunn = Lignariella obscura (Dunn) Jafri ■☆
137150 Draba ochroleuca Bunge;白绿葶苈■☆
137151 Draba okamotoi Ohwi;岡田葶苈■☆
137152 Draba olgae Regel et Schmalh.;奥尔嘎葶苈■
137153 Draba olgae Regel et Schmalh. var. chitralensis O. E. Schulz = Draba olgae Regel et Schmalh. ■
137154 Draba olympica Sibth. ex DC.;奥林匹克葶苈;Olimpic Draba, Olimpic Whitlowgrass ■☆
137155 Draba oreades Schrenk;喜山葶苈(毛萼葶苈,石菠菜);Mountain-loving Draba, Mountain-loving Whitlowgrass ■
137156 Draba oreades Schrenk var. alpicola (Klotzsch) O. E. Schulz;喜高山葶苈;Alpine-loving Draba, Alpine-loving Whitlowgrass ■

137157 Draba oreades Schrenk var. chinensis O. E. Schulz ex H. Limpr.;中国喜山葶苈;Chinese Mountain-loving Draba, Chinese Mountain-loving Whitlowgrass ■

137158 Draba oreades Schrenk var. ciliolata O. E. Schulz;毛果喜山葶苈;Hairyfruit Mountain-loving Draba, Hairyfruit Mountain-loving Whitlowgrass ■

137159 Draba oreades Schrenk var. ciliolata O. E. Schulz = Draba oreades Schrenk ■

137160 Draba oreades Schrenk var. commutata (Regel) O. E. Schulz;矮喜山葶苈;Dwarf Mountain-loving Draba, Dwarf Mountain-loving Whitlowgrass ■

137161 Draba oreades Schrenk var. commutata (Regel) O. E. Schulz = Draba oreades Schrenk ■

137162 Draba oreades Schrenk var. dasycarpa O. E. Schulz = Draba oreades Schrenk ■

137163 Draba oreades Schrenk var. depauperata O. E. Schulz = Draba oreades Schrenk ■

137164 Draba oreades Schrenk var. estylosa O. E. Schulz = Draba oreades Schrenk ■

137165 Draba oreades Schrenk var. glabra L. L. Lou et T. Y. Cheo;光叶山景葶苈;Smoothleaf Mountain-loving Draba, Smoothleaf Mountain-loving Whitlowgrass ■

137166 Draba oreades Schrenk var. glabrescens O. E. Schulz = Draba oreades Schrenk ■

137167 Draba oreades Schrenk var. leiocarpa L. L. Lou et T. Y. Cheo;光果伊犁葶苈;Smoothfruit Mountain-loving Draba, Smoothfruit Mountain-loving Whitlowgrass ■

137168 Draba oreades Schrenk var. occulata O. E. Schulz = Draba oreades Schrenk ■

137169 Draba oreades Schrenk var. racemosa O. E. Schulz = Draba oreades Schrenk ■

137170 Draba oreades Schrenk var. tafellii O. E. Schulz;长纤毛喜山葶苈;Mountain-loving Whitlowgrass, Tafell Mountain-loving Draba ■

137171 Draba oreades Schrenk var. tafellii O. E. Schulz = Draba oreades Schrenk ■

137172 Draba oreadum Maire;山葶苈■☆

137173 Draba oreadum Maire var. anremerica ? = Draba oreadum Maire ■☆

137174 Draba oreodoxa W. W. Sm.;山景葶苈;Snowrange Draba, Snowrange Whitlowgrass ■

137175 Draba oreodoxa W. W. Sm. var. glabra L. L. Lou et T. Y. Cheo = Coelonema draboides Maxim. ■

137176 Draba ossetica (Rupr.) Sommier et H. Lév.;骨质葶苈■☆

137177 Draba pakistanica Jafri = Draba olgae Regel et Schmalh. ■

137178 Draba pallida A. Heller = Draba lanceolata DC. ■☆

137179 Draba pamirica (O. Fedtsch.) Pohle;帕米尔葶苈■☆

137180 Draba parviflora (Regel) O. E. Schulz;小花葶苈;Draba, Littleflower Parviflorous Whitlowgrass, Littleflower Whitlowgrass ■

137181 Draba parvisiliquosa Tolm.;小荚葶苈■☆

137182 Draba physocarpa Kom.;囊果葶苈■☆

137183 Draba piepuneasis O. E. Schulz;匍匐葶苈(中甸葶苈);Creeping Draba, Creeping Whitlowgrass, Zhongdian Draba ■

137184 Draba piepunensis O. E. Schulz = Draba senilis O. E. Schulz ■

137185 Draba pilosa DC.;疏毛葶苈■☆

137186 Draba pilosa DC. var. commutata Regel = Draba oreades Schrenk ■

137187 Draba pilosa DC. var. oreades (Schrenk) Regel = Draba oreades Schrenk ■

137188 Draba pingwuensis Z. M. Tan et S. C. Zhou = Draba eriopoda Turcz. ■

137189 Draba pohlei Tolm.;波莱葶苈■☆

137190 Draba polyphylla O. E. Schulz;多叶葶苈;Manyleaves Draba, Manyleves Whitlowgrass, Polyphyllous Whitlowgrass ■

137191 Draba polytricha Ledeb.;多毛葶苈■☆

137192 Draba praecox Stev. = Draba verna L. ■☆

137193 Draba praecox Stev. = Erophila verna (L.) Chevall. subsp. praecox (Steven) Walters ■☆

137194 Draba primuloides Turcz.;报春葶苈■☆

137195 Draba prozorovskii Tolm.;普罗葶苈■☆

137196 Draba pseudopilosa Pohle;假毛葶苈■☆

137197 Draba pygmaea Turcz.;矮小葶苈■☆

137198 Draba pyrenaica L. = Petrocallis pyrenaica (L.) R. Br. ■☆

137199 Draba pyriformis Pohle = Draba setosa Royle ■

137200 Draba qinghaiensis L. L. Lou;青海葶苈;Qinghai Draba, Qinghai Whitlowgrass ■

137201 Draba qinghaiensis L. L. Lou = Draba oreades Schrenk ■

137202 Draba ramosissima Desv.;分枝葶苈;Branching Draba, Branching Whitlowgrass ■☆

137203 Draba remotiflora O. E. Schulz;疏花葶苈;Remoteflower Draba, Remoteflower Whitlowgrass ■

137204 Draba repens M. Bieb. = Draba sibirica (Pall.) Thell. ■

137205 Draba reptans (Lam.) Fernald;卡罗来纳葶苈;Carolina Whitlow-grass, Common Whitlow-grass, White Whitlow Wort ■☆

137206 Draba reptans (Lam.) Fernald f. micrantha (Nutt.) C. L. Hitchc. = Draba reptans (Lam.) Fernald ■☆

137207 Draba reptans (Lam.) Fernald subsp. stellifera (O. E. Schulz) Abrams = Draba reptans (Lam.) Fernald ■☆

137208 Draba reptans (Lam.) Fernald var. micrantha (Nutt.) Fernald = Draba reptans (Lam.) Fernald ■☆

137209 Draba reptans (Lam.) Fernald var. stellifera (O. E. Schulz) C. L. Hitchc. = Draba reptans (Lam.) Fernald ■☆

137210 Draba reptans (Lam.) Fernald var. typica C. L. Hitchc. = Draba reptans (Lam.) Fernald ■☆

137211 Draba rhodantha Rech. f. et Edelb. = Draba hystrix Hook. f. et Thomson ■☆

137212 Draba rigida Willd.;坚硬葶苈;Rigid Draba, Rigid Whitlowgrass ■☆

137213 Draba rockii O. E. Schulz;沼泽葶苈;Rock Draba, Rock Whitlowgrass ■

137214 Draba rockii O. E. Schulz = Draba oreades Schrenk ■

137215 Draba rosea Turcz. = Braya rosea (Turcz.) Bunge ■

137216 Draba rostrata Pohle = Draba affghanica Pohle ■

137217 Draba rupestris Bunge = Draba altaica (C. A. Mey.) Bunge ■

137218 Draba rupestris R. Br. var. altaica C. A. Mey. = Draba altaica (C. A. Mey.) Bunge ■

137219 Draba rupestris R. Br. var. pusilla Kar. et Kir. = Draba altaica (C. A. Mey.) Bunge ■

137220 Draba rupestris Willd. ex DC. = Draba norvegica Gunnerus ■☆

137221 Draba rupestris Willd. ex DC. var. pusilla Kar. et Kir. = Draba altaica (C. A. Mey.) Bunge var. pusilla (Kar. et Kir.) Fedtsch. ■

137222 Draba sachalinensis (F. Schmidt) Trautv.;库页葶苈■☆

137223 Draba sachalinensis (F. Schmidt) Trautv. var. shinanomontana (Ohwi) Okuyama;信浓葶苈■☆

137224 Draba sachalinensis F. Schmidt;库页岛葶苈;Sachalin Draba ■

137225 Draba sakuraii Makino;樱井葶苈■☆
137226 Draba sakuraii Makino var. nipponica (Makino) Takeda = Draba nipponica Makino ■☆
137227 Draba scabra C. A. Mey.;粗糙葶苈■☆
137228 Draba sekiyana Ohwi;台湾葶苈(台湾山荠);Taiwan Draba, Taiwan Whitlowgrass ■
137229 Draba senilis O. E. Schulz;衰老葶苈;Senencent Draba, Senencent Whitlowgrass ■
137230 Draba serpens O. E. Schulz;中甸葶苈(蛇行葶苈,蛇状葶苈);Serpent Draba,Serpent Whitlowgrass,Zhongdian Whitlowgrass ■
137231 Draba setosa Royle;刚毛葶苈;Setose Draba, Setose Whitlowgrass ■
137232 Draba setosa Royle var. glabrata O. E. Schulz;变光刚毛葶苈;Glabrous Draba,Glabrous Whitlowgrass ■
137233 Draba setosa Royle var. pyriformis (Pohle) O. E. Schulz = Draba setosa Royle ■
137234 Draba setosa Royle var. pyriformis (Pohle) O. E. Schulz subvar. glabrata O. E. Schulz = Draba setosa Royle ■
137235 Draba shinanomontana Ohwi = Draba sachalinensis (F. Schmidt) Trautv. var. shinanomontana (Ohwi) Okuyama ■☆
137236 Draba shiroumana Makino;白马岳葶苈■☆
137237 Draba sibirica (Pall.) Thell.;西伯利亚葶苈;Creeping Draba, Siberia Whitlowgrass,Siberian Draba,Siberian Whitlowgrass ■
137238 Draba sikkimensis (Hook. f. et Thomson) Pohle;锡金葶苈;Sikkim Draba,Sikkim Whitlowgrass ■
137239 Draba sikkimensis (Hook. f. et Thomson) Pohle f. thoroldii O. E. Schulz = Draba sikkimensis (Hook. f. et Thomson) Pohle ■
137240 Draba sikkimensis (Hook. f. et Thomson) Pohle var. chitralensis O. E. Schulz = Draba tibetica Hook. f. et Thomson ■
137241 Draba siliquosa M. Bieb.;多果葶苈■☆
137242 Draba stenocarpa Hook. f. et Thomson;狭果葶苈;Narrowfruit Draba,Narrowfruit Whitlowgrass ■
137243 Draba stenocarpa Hook. f. et Thomson var. leiocarpa (Lipsky) L. L. Lou;无毛狭果葶苈;Smooth Narrowfruit Draba, Smooth Narrowfruit Whitlowgrass ■
137244 Draba stenocarpa Hook. f. et Thomson var. leiocarpa (Lipsky) L. L. Lou = Draba stenocarpa Hook. f. et Thomson ■
137245 Draba stenocarpa Hook. f. et Thomson var. media (Litv.) O. E. Schulz = Draba stenocarpa Hook. f. et Thomson ■
137246 Draba stenopetala Trautv.;狭瓣葶苈■☆
137247 Draba stepposa L. L. Lou et T. Y. Cheo = Coelonema draboides Maxim. ■
137248 Draba stylaris J. Gay ex E. A. Thomas;伊宁葶苈(森林葶苈);Style-like Draba,Style-like Whitlowgrass,Yining Whitlowgrass ■
137249 Draba stylaris J. Gay ex E. A. Thomas var. leiocarpa L. L. Lou et T. Y. Cheo;光果伊宁葶苈;Smoothfruit Style-like Draba,Smoothfruit Style-like Whitlowgrass ■
137250 Draba stylaris J. Gay ex W. D. J. Koch = Draba cana Rydb. ■☆
137251 Draba stylaris J. Gay. = Draba lanceolata Royle ■
137252 Draba stylaris J. Gay. var. leiocarpa L. L. Lou et T. Y. Cheo = Draba lanceolata Royle ■
137253 Draba subamplexicaulis C. A. Mey.;半抱茎葶苈(土耳其斯坦葶苈);Turkenstan Draba,Turkenstan Whitlowgrass ■
137254 Draba subamplexicaulis C. A. Mey. var. hirsutifolia Pohle = Draba parviflora (Regel) O. E. Schulz ■
137255 Draba subcapitata Simmons;头状葶苈;Subcapitate Draba ■☆
137256 Draba subglabra (Rupr.) Tolm.;近光葶苈■☆
137257 Draba subsecunda Sommier et H. Lév.;单侧葶苈■☆
137258 Draba surculosa Franch.;山菜葶苈;Longsurculose Draba, Longsurculose Whitlowgrass ■
137259 Draba talassica Pohle;塔拉斯葶苈■☆
137260 Draba tenerrima O. E. Schulz = Erophila tenerrima (O. E. Schulz) Jafri ■☆
137261 Draba tenerrima O. E. Schulz var. trichocarpa O. E. Schulz = Erophila tenerrima (O. E. Schulz) Jafri ■☆
137262 Draba thomsonii (Hook. f. et Thomson) Pohle = Draba tibetica Hook. f. et Thomson ■
137263 Draba thomsonii (Hook. f. et Thomson) Pohle var. lasiocarpa (Lipsky) Pohle = Draba tibetica Hook. f. et Thomson ■
137264 Draba thomsonii (Hook. f. et Thomson) Pohle var. leiocarpa (Lipsky) Pohle = Draba tibetica Hook. f. et Thomson ■
137265 Draba tianschanica Pohle = Draba oreades Schrenk ■
137266 Draba tibetica Hook. f. et Thomson;西藏葶苈;Tibet Draba, Tibet Whitlowgrass,Xizang Draba,Xizang Whitlowgrass ■
137267 Draba tibetica Hook. f. et Thomson = Draba lichiangensis W. W. Sm. ■
137268 Draba tibetica Hook. f. et Thomson = Draba sikkimensis (Hook. f. et Thomson) Pohle ■
137269 Draba tibetica Hook. f. et Thomson var. chitralensis (O. E. Schulz) Jafri = Draba tibetica Hook. f. et Thomson ■
137270 Draba tibetica Hook. f. et Thomson var. duthiei O. E. Schulz;光果西藏葶苈;Duthie Tibet Draba,Duthie Tibet Whitlowgrass,Duthie Xizang Whitlowgrass ■
137271 Draba tibetica Hook. f. et Thomson var. duthiei O. E. Schulz = Draba tibetica Hook. f. et Thomson ■
137272 Draba tibetica Hook. f. et Thomson var. sikkimensis Hook. f. et Thomson = Draba sikkimensis (Hook. f. et Thomson) Pohle ■
137273 Draba tibetica Hook. f. et Thomson var. thomsonii Hook. f. et Thomson = Draba tibetica Hook. f. et Thomson ■
137274 Draba tibetica Hook. f. et Thomson var. turkestanica (Regel et Schmalh.) O. E. Schulz = Draba tibetica Hook. f. et Thomson ■
137275 Draba tibetica Hook. f. et Thomson var. turkestanica (Regel et Schmalh.) O. E. Schulz subvar. leiocarpa O. E. Schulz = Draba tibetica Hook. f. et Thomson ■
137276 Draba tibetica Hook. f. et Thomson var. winterbottomii Hook. f. et Thomson = Draba winterbottomii (Hook. f. et Thomson) Pohle ■
137277 Draba tomentosa Quézel = Draba dubia Suter ■☆
137278 Draba torticarpa L. L. Lou et T. Y. Cheo;扭果葶苈;Tortuousfruit Draba,Tortuousfruit Whitlowgrass ■
137279 Draba torticarpa L. L. Lou et T. Y. Cheo = Draba lasiophylla Royle ■
137280 Draba tranzschelii Litv. = Draba tibetica Hook. f. et Thomson ■
137281 Draba trinervis O. E. Schulz;三脉葶苈■☆
137282 Draba turczaninovii Pohle et N. Busch;屠氏葶苈;Turczaninov Draba ■
137283 Draba turkestanica Regel et Schmalh. = Draba tibetica Hook. f. et Thomson ■
137284 Draba turkestanica Regel et Schmalh. var. lasiocarpa Lipsky = Draba tibetica Hook. f. et Thomson ■
137285 Draba turkestanica Regel et Schmalh. var. leiocarpa Lipsky = Draba tibetica Hook. f. et Thomson ■
137286 Draba ussuriensis Pohle;乌苏里葶苈;Ussuri Draba, Ussuri Whitlowgrass ■
137287 Draba valida Pissjauk.;刚直葶苈■☆

137288 Draba verna L. = Erophila verna (L.) Chevall. ■☆
137289 Draba verna L. var. aestivalis Lej. = Draba verna L. ■☆
137290 Draba verna L. var. boerhaavii H. C. Hall = Draba verna L. ■☆
137291 Draba verna L. var. major Stur = Draba verna L. ■☆
137292 Draba verna L. var. majuscula (Jord.) Debeaux = Erophila verna (L.) Chevall. ■☆
137293 Draba violacea DC.;紫花葶苈;Violet Draba, Violet Whitlowgrass ■☆
137294 Draba wardii W. W. Sm. = Draba gracillima Hook. f. et Thomson ■
137295 Draba winterbottomii (Hook. f. et Thomson) Pohle;绵毛葶苈(棉毛葶苈);Winterbottom Draba, Winterbottom Whitlowgrass ■
137296 Draba winterbottomii (Hook. f. et Thomson) Pohle var. stracheyi O. E. Schulz;光果棉毛葶苈;Strachey Winterbottom Draba, Strachey Winterbottom Whitlowgrass ■
137297 Draba winterbottomii (Hook. f. et Thomson) Pohle var. stracheyi O. E. Schulz = Draba alajica Litv. ■
137298 Draba yunnanensis Franch.;云南葶苈;Yunnan Draba, Yunnan Whitlowgrass ■
137299 Draba yunnanensis Franch. var. gracilipes Franch.;细梗云南葶苈;Slenderstalk Yunnan Draba, Slenderstalk Yunnan Whitlowgrass ■
137300 Draba yunnanensis Franch. var. gracilipes Franch. = Draba yunnanensis Franch. ■
137301 Draba yunnanensis Franch. var. latifolia O. E. Schulz;宽叶云南葶苈;Broadleaf Yunnan Draba, Broadleaf Yunnan Whitlowgrass ■
137302 Draba yunnanensis Franch. var. latifolia O. E. Schulz = Draba yunnanensis Franch. ■
137303 Draba yunnanensis Franch. var. microcarpa O. E. Schulz = Draba yunnanensis Franch. ■
137304 Draba yunnanensis Franch. var. nivalis Diels;雪线云南葶苈;Snow Yunnan Draba, Snow Yunnan Whitlowgrass ■
137305 Draba yunnanensis Franch. var. ramosa O. E. Schulz = Draba amplexicaulis Franch. ■
137306 Draba zangbeiensis L. L. Lu;藏北葶苈;N. Xizang Draba, N. Xizang Whitlowgrass ■
137307 Drabaceae Martinov = Brassicaceae Burnett(保留科名)■●
137308 Drabaceae Martinov = Cruciferae Juss.(保留科名)■●
137309 Drabaceae Martinov;葶苈科■
137310 Drabastrum (F. Muell.) O. E. Schulz(1924);亚高山葶苈属■☆
137311 Drabastrum O. E. Schulz = Drabastrum (F. Muell.) O. E. Schulz ■☆
137312 Drabastrum alpestre (F. Muell.) O. E. Schulz;亚高山葶苈■☆
137313 Drabella (DC.) Fourr.(1868);小葶苈属■☆
137314 Drabella (DC.) Fourr. = Draba L. ■
137315 Drabella Fourr. = Draba L. ■
137316 Drabella Fourr. = Drabella (DC.) Fourr. ■☆
137317 Drabella Nábelek = Draba L. ■
137318 Drabella Nábelek = Thylacodraba (Nábelek) O. E. Schulz ■
137319 Drabella muralis Fourr.;小葶苈■☆
137320 Drabopsis C. Koch = Drabopsis K. Koch ■
137321 Drabopsis K. Koch(1841);假葶苈属■
137322 Drabopsis brevisiliqua Naqshi et Javeid = Drabopsis nuda (Bél.) Stapf ■
137323 Drabopsis brevisiliqua Naqshi et Javeid = Drabopsis verna K. Koch ■
137324 Drabopsis nuda (Bél. ex Boiss.) Stapf = Drabopsis verna K. Koch ■
137325 Drabopsis nuda (Bél.) Stapf = Drabopsis verna K. Koch ■
137326 Drabopsis oronotica Stapf = Olimarabidopsis pumila (Stephan) Al-Shehbaz, O'Kane et R. A. Price ■
137327 Drabopsis verna K. Koch;假葶苈■
137328 Drabopsis verna K. Koch = Drabopsis nuda (Bél.) Stapf ■
137329 Dracaena Vand. = Dracaena Vand. ex L. ●■
137330 Dracaena Vand. ex L.(1767);龙血树属(虎斑木属);Dracaena, Dracena, Dragon Tree, Dragonbood ●■
137331 Dracaena acaulis Baker;无茎龙血树■☆
137332 Dracaena acutissima Hua;尖龙血树●☆
137333 Dracaena adamii Hepper;阿达姆龙血树●☆
137334 Dracaena afromontana Mildbr.;非洲山生龙血树●☆
137335 Dracaena afzelii Baker = Dracaena ovata Ker Gawl. ●☆
137336 Dracaena aletriformis (Haw.) Bos;粉龙血树●☆
137337 Dracaena angustifolia Roxb.;长花龙血树(番仔林投,龙血树,狭叶龙血树,竹木参);Narrowleaf Dracaena, Narrowleaf Dragonbood, Narrow-leaved Dracaena ●
137338 Dracaena arborea (Willd.) Link var. baumannii Engl. = Dracaena arborea (Willd.) Link ●☆
137339 Dracaena arborea (Willd.) Link.;大龙血树(乔木龙血树)●☆
137340 Dracaena atropurpurea Roxb. var. gracilis (Baker) Baker = Dracaena elliptica Thunb. ●
137341 Dracaena atropurpurea Roxb. var. gracilis (Baker) Baker = Dracaena gracilis Wall. ex Baker ●
137342 Dracaena aubryana Brongn. ex E. Morren;长柄竹蕉(长柄千年木,大叶富贵竹);Thalia-like Dracena ●☆
137343 Dracaena australis G. Forst. = Cordyline australis (G. Forst.) Hook. f. ●
137344 Dracaena bequaertii De Wild.;贝卡尔龙血树●☆
137345 Dracaena betschleriana K. Koch = Dracaena concinna Kunth ●☆
137346 Dracaena bicolor Hook.;二色龙血树●☆
137347 Dracaena borealis Aiton = Clintonia borealis Raf. ■☆
137348 Dracaena braunii Engl.;布劳恩龙血树●☆
137349 Dracaena buettneri Engl.;比特纳龙血树●☆
137350 Dracaena butayei De Wild. = Dracaena fragrans (L.) Ker Gawl. ●
137351 Dracaena calocephala Bos;美头龙血树●☆
137352 Dracaena cambadiana Gagnep. = Dracaena cochinchinensis (Lour.) S. C. Chen ●◇
137353 Dracaena cambodiana Pierre ex Gagnep.;海南龙血树(海腊,柬埔寨龙血树,剑叶木,木血竭,麒麟血,山海带,山铁树,乌猿蔗,乌猿蔗,小花龙血树,血竭);Cambodia Dracaena, Cambodia Dragonbood ●◇
137354 Dracaena cambodiana Pierre ex Gagnep. = Dracaena cochinchinensis (Lour.) S. C. Chen ●◇
137355 Dracaena camerooniana Baker;喀麦隆龙血树●☆
137356 Dracaena capitulifera De Wild. et T. Durand = Dracaena camerooniana Baker ●☆
137357 Dracaena cerasifera Hua;角龙血树●☆
137358 Dracaena cinnabari Balf. f.;朱红龙血树●☆
137359 Dracaena cochinchinensis (Lour.) S. C. Chen;剑叶龙血树(岩棕);Swordleaf Dracaena, Swordleaf Dragonbood, Sword-leaved Dracaena ●◇
137360 Dracaena concinna Kunth;紫边龙血树(红覆轮千年木)●☆
137361 Dracaena congoensis Hua;刚果龙血树●☆
137362 Dracaena curtisii Ridl.;垂叶龙血树(垂叶竹蕉)●☆
137363 Dracaena cuspidibracteata Engl.;苞片龙血树(苞片竹蕉)●☆
137364 Dracaena cylindrica Hook. f. = Dracaena bicolor Hook. ●☆

137365 Dracaena deisteliana Engl. = Dracaena fragrans (L.) Ker Gawl. ●
137366 Dracaena densifolia Baker;密叶龙血树●☆
137367 Dracaena deremensis Engl.;异味龙血树(德利龙血树,竹蕉);Corn Plant,Striped Dracaena,Unpleasant Dracaena,Unpleasant Dragonbood ●☆
137368 Dracaena deremensis Engl. 'Bausei';大白纹竹蕉(大白纹龙血)●
137369 Dracaena deremensis Engl. 'Boehr's Gold';黄绿龙血树●
137370 Dracaena deremensis Engl. 'Compacta';密叶竹蕉●
137371 Dracaena deremensis Engl. 'Janet Craig';亮绿龙血树(珍妮龙血树)●
137372 Dracaena deremensis Engl. 'Jumbo';巨大竹蕉●
137373 Dracaena deremensis Engl. 'Lance Linear';线叶竹蕉●
137374 Dracaena deremensis Engl. 'Longii';白纹龙血树●
137375 Dracaena deremensis Engl. 'Roehrs Gold';黄绿纹龙血●☆
137376 Dracaena deremensis Engl. 'Snow Queen';雪后龙血树●
137377 Dracaena deremensis Engl. 'Virens Compacta';太阳神●☆
137378 Dracaena deremensis Engl. 'Warneckii Compacta';密叶银线龙血●☆
137379 Dracaena deremensis Engl. 'Warneckii Striata';缟叶竹蕉●
137380 Dracaena deremensis Engl. 'Warneckii';银线龙血树(银纹龙血树)●
137381 Dracaena deremensis Engl. = Dracaena fragrans (L.) Ker Gawl. ●
137382 Dracaena draco (L.) L.;龙血树(渴留,麒麟竭,香龙血树);Dracaena, Dragon, Dragon Blood Tree, Dragon Dracaena, Dragon Tree, Dragon's Blood Tree, Dragonbood, Dragon's-blood, Dragontree, Dragon-tree ●
137383 Dracaena draco L. = Dracaena draco (L.) L. ●
137384 Dracaena dundusanensis De Wild. = Dracaena camerooniana Baker ●☆
137385 Dracaena elegans Hua = Dracaena laxissima Engl. ●☆
137386 Dracaena ellenbeckiana Engl.;埃伦龙血树●☆
137387 Dracaena elliptica Thunb.;细枝龙血树(纹千年木)●
137388 Dracaena elliptica Thunb. var. gracilis Baker = Dracaena elliptica Thunb. ●
137389 Dracaena elliptica Thunb. var. gracilis Baker = Dracaena gracilis Wall. ex Baker ●
137390 Dracaena elliptica Thunb. var. maculata Roxb.;斑叶细枝龙血树●☆
137391 Dracaena ensifolia L. = Dianella ensifolia (L.) DC. ex Redoute ■
137392 Dracaena ensifolia Wall. = Dracaena angustifolia Roxb. ●
137393 Dracaena ensiformis Wall. ex Voigt = Dracaena angustifolia Roxb. ●
137394 Dracaena erecta L. f. = Asparagus striatus (L. f.) Thunb. ■☆
137395 Dracaena ferrea L. = Cordyline fruticosa (L.) A. Chev. ●
137396 Dracaena fischeri Baker;菲舍尔龙血树●☆
137397 Dracaena fragrans (L.) Ker Gawl.;香龙血树(花虎斑木,香千年木);Corn Plant, Fragrant Dracaena, Fragrant Dragonbood, Happy Plant, Variegated Corn Plant ●
137398 Dracaena fragrans (L.) Ker Gawl. 'Lindeniana';黄边香龙血树(黄叶香龙血树)●☆
137399 Dracaena fragrans (L.) Ker Gawl. 'Lindenii' = Dracaena fragrans (L.) Ker Gawl. 'Lindeniana'●☆
137400 Dracaena fragrans (L.) Ker Gawl. 'Massangeana Compacta';中斑密叶香龙血树●☆
137401 Dracaena fragrans (L.) Ker Gawl. 'Massangeana';花叶香龙血树(巴西铁树,花叶千年木,中斑香龙血树);Color Marking Dracaena, Color Marking Dragonbood ●☆
137402 Dracaena fragrans (L.) Ker Gawl. 'Santa Rosa';黄边纹龙血树●☆
137403 Dracaena fragrans (L.) Ker Gawl. 'Victoriae';金边香龙血树(垂叶香龙血树)●☆
137404 Dracaena fragrans Ker Gawl. = Dracaena fragrans (L.) Ker Gawl. ●
137405 Dracaena frommii Engl. et K. Krause = Dracaena camerooniana Baker ●☆
137406 Dracaena gabonica Hua;加蓬龙血树●☆
137407 Dracaena gentilii De Wild. = Dracaena camerooniana Baker ●☆
137408 Dracaena glomerata Baker;团集龙血树●☆
137409 Dracaena godseffiana Baker;星龙血树(撒金千年木,星点木,星虎斑木,星千年木);Gold-dust Dracaena, Gold-dust Dragonbood, Goldic Dracaena ●
137410 Dracaena godseffiana Baker 'Florida Beauty';白星千年木●
137411 Dracaena godseffiana Baker 'Friedmanii';中道星点木●
137412 Dracaena godseffiana Baker 'Maculata';油点木●
137413 Dracaena godseffiana Baker 'Rausei';黄道星点木●
137414 Dracaena godseffiana Sander ex Mast. = Dracaena surculosa Lindl. ●☆
137415 Dracaena goldieana Bull. = Dracaena goldieana Sander ex Mast. ●
137416 Dracaena goldieana Sander ex Mast.;虎斑千年木(虎斑木,花斑龙血树);Goldie Dracaena, Goldie Dracena, Goldie Dragonbood ●
137417 Dracaena gracilis (Baker) Hook. f. = Dracaena elliptica Thunb. ●
137418 Dracaena gracilis (Baker) Hook. f. = Dracaena gracilis Wall. ex Baker ●
137419 Dracaena gracilis Wall. ex Baker;马来细枝龙血树(细枝龙血树);Slender-branch Dracaena, Slender-branch Dragonbood, Slender-branched Gragontree ●
137420 Dracaena graminifolia (L.) L. = Liriope graminifolia (L.) Baker ■
137421 Dracaena graminifolia L. = Liriope graminifolia (L.) Baker ■
137422 Dracaena hanningtoni Baker;汉宁顿龙血树●☆
137423 Dracaena hirsuta Thunb. = Palisota hirsuta (Thunb.) K. Schum. ■☆
137424 Dracaena hokouensis G. Z. Ye;河口龙血树;Hekou Dragonbood ●
137425 Dracaena hookeriana K. Koch;胡克龙血树;Kooker Dragonbood ●☆
137426 Dracaena hookeriana K. Koch = Dracaena aletriformis (Haw.) Bos ●☆
137427 Dracaena humilis Baker = Dracaena aubryana Brongn. ex E. Morren ●☆
137428 Dracaena impressivenia Y. H. Yan et H. J. Guo;深脉龙血树●
137429 Dracaena indivisa Andersen = Cordyline indivisa (G. Forst.) Kunth ●☆
137430 Dracaena indivisa G. Forst. = Cordyline indivisa (G. Forst.) Kunth ●☆
137431 Dracaena indivisa Kunth = Cordyline indivisa (G. Forst.) Kunth ●☆
137432 Dracaena interrupta Baker = Dracaena camerooniana Baker ●☆
137433 Dracaena interrupta Haw. ex Loudon = Dracaena surculosa Lindl. var. maculata Hook. f. ●☆
137434 Dracaena kindtiana De Wild. = Dracaena aubryana Brongn. ex E. Morren ●☆

137435 Dracaena knerkiana K. Koch = Dracaena arborea (Willd.) Link ●☆

137436 Dracaena latifolia Regel = Dracaena aletriformis (Haw.) Bos ●☆

137437 Dracaena laurentii De Wild.;洛朗龙血树●☆

137438 Dracaena laxissima Engl.;极松龙血树●☆

137439 Dracaena lecomtei Hua = Dracaena camerooniana Baker ●☆

137440 Dracaena ledermannii Engl. et K. Krause;莱德龙血树●☆

137441 Dracaena leonensis Lodd. ex Loudon = Dracaena ovata Ker Gawl. ●☆

137442 Dracaena letestui Pellegr.;莱泰斯图龙血树●☆

137443 Dracaena linderi Hort. = Dracaena fragrans (L.) Ker Gawl. ●

137444 Dracaena loureiri Gagnep.;岩棕●☆

137445 Dracaena loureiri Gagnep. = Dracaena cochinchinensis (Lour.) S. C. Chen ●◇

137446 Dracaena mannii Baker;曼氏龙血树;Asparagus Bush ●☆

137447 Dracaena marginata Lam.;缘叶龙血树(红边龙血树,红边铁树,红边朱蕉,红边竹蕉,金丝竹,千年木,细叶竹蕉);Madagascar Dragonbood,Madagascar Dragon-tree,Madagascar Gragon Tree,Red Margins Dracaena ●☆

137448 Dracaena marginata Lam. 'Tricolor';彩虹铁树(三色红边龙血树)●☆

137449 Dracaena mayumbensis Hua = Dracaena camerooniana Baker ●☆

137450 Dracaena medeoloides L. f. = Asparagus asparagoides (L.) Druce ■☆

137451 Dracaena menglaensis G. Z. Ye;勐腊龙血树;Mengla Dracaena, Mengla Dragonbood ●

137452 Dracaena mildbraedii K. Krause;米尔德龙血树●☆

137453 Dracaena monostachya Baker;单穗龙血树●☆

137454 Dracaena monostachya Baker = Dracaena aubryana Brongn. ex E. Morren ●☆

137455 Dracaena monostachya Baker var. angolensis ? = Dracaena aubryana Brongn. ex E. Morren ●☆

137456 Dracaena nemorosa Lam. = Dianella ensifolia (L.) DC. ex Redoute ■

137457 Dracaena nitens Welw. et Baker;光亮龙血树●☆

137458 Dracaena nitens Welw. ex Baker = Dracaena mannii Baker ●☆

137459 Dracaena nyangensis Pellegr.;尼扬加龙血树●☆

137460 Dracaena oculata Linden = Dracaena phrynioides Hook. ●☆

137461 Dracaena oddonii De Wild.;奥顿龙血树●☆

137462 Dracaena odum Engl. et K. Krause = Dracaena camerooniana Baker ●☆

137463 Dracaena ombet Kotschy et Peyr.;非洲龙血树;Abyssiana Dragon Tree,Dragonbood ●☆

137464 Dracaena ovata Ker Gawl.;卵叶龙血树●☆

137465 Dracaena papau Engl. = Dracaena steudneri Engl. ●☆

137466 Dracaena parviflora Baker;小花龙血树●☆

137467 Dracaena perrottetii Baker;佩罗龙血树●☆

137468 Dracaena perrottetii Baker = Dracaena mannii Baker ●☆

137469 Dracaena perrottetii Baker var. minor ? = Dracaena mannii Baker ●☆

137470 Dracaena phanerophlebia Baker;显脉龙血树●☆

137471 Dracaena phrynioides Hook.;柊叶龙血树●☆

137472 Dracaena phrynioides Hook. var. staudtii Engl. = Dracaena phrynioides Hook. ●☆

137473 Dracaena poggei Engl.;波格龙血树●☆

137474 Dracaena praetermissa Bos;疏忽龙血树●☆

137475 Dracaena preussii Engl. = Dracaena bicolor Hook. ●☆

137476 Dracaena prolata C. H. Wright = Dracaena ovata Ker Gawl. ●☆

137477 Dracaena pseudoreflexa Mildbr.;假红果龙血树●☆

137478 Dracaena reflexa Lam.;红果龙血树(百合竹,富贵竹);Long-tuft Leaves Dracaena,Pleomele,Song of India ●☆

137479 Dracaena reflexa Lam. 'Song of India';绿心红果龙血树(绿心富贵竹,印度之歌红果龙血树)●☆

137480 Dracaena reflexa Lam. 'Song of Jamaica';金黄红果龙血树(金黄百合竹)■☆

137481 Dracaena reflexa Lam. 'Variegata';黄边红果龙血树(黄边百合竹,金边百合竹)■☆

137482 Dracaena reflexa Lam. var. nitens (Welw. ex Baker) Baker = Dracaena mannii Baker ●☆

137483 Dracaena rhabdophylla Chiov. = Dracaena ombet Kotschy et Peyr. ●☆

137484 Dracaena rubro-aurantiaca De Wild.;红黄龙血树●☆

137485 Dracaena rumphii (Hook.) Regel = Dracaena aletriformis (Haw.) Bos ●☆

137486 Dracaena rumphii Hook. = Dracaena angustifolia Roxb. ●

137487 Dracaena sanderiana Sander;银叶龙血树(白边富贵竹,喀麦隆龙血树,宽边龙血树,仙达,镶边朱蕉,银边竹蕉,银纹龙血树,银叶虎斑木,银叶千年木);Dragonbood,Ribbon Plant,Sander's Dracaena ●

137488 Dracaena sanderiana Sander 'Borinquensis';银线银叶龙血树(银线富贵竹)■☆

137489 Dracaena sanderiana Sander 'Celica';金边银叶龙血树(金边富贵竹)■☆

137490 Dracaena sanderiana Sander 'Virens';绿银叶龙血树(富贵竹)●☆

137491 Dracaena sanderiana Sander ex Mast. = Dracaena braunii Engl. ●☆

137492 Dracaena scabra Bos;粗糙龙血树●☆

137493 Dracaena schizantha Baker = Dracaena ombet Kotschy et Peyr. ●☆

137494 Dracaena scoparia A. Chev. = Dracaena cerasifera Hua ●☆

137495 Dracaena sessiliflora C. H. Wright = Dracaena ovata Ker Gawl. ●☆

137496 Dracaena silvatica Hua = Dracaena camerooniana Baker ●☆

137497 Dracaena smithii Baker ex Hook. f. = Dracaena fragrans (L.) Ker Gawl. ●

137498 Dracaena soyauxiana Baker;索亚龙血树●☆

137499 Dracaena stenophylla K. Koch;狭叶龙血树●☆

137500 Dracaena steudneri Engl.;斯托德龙血树●☆

137501 Dracaena steudneri Engl. var. kilimandscharica ? = Dracaena fragrans (L.) Ker Gawl. ●

137502 Dracaena striata L. f. = Asparagus striatus (L. f.) Thunb. ■☆

137503 Dracaena stricta Endl. = Cordyline stricta Endl. ●

137504 Dracaena surculosa Lindl.;金斑龙血树(吸枝龙血树)●☆

137505 Dracaena surculosa Lindl. 'Florida Beauty';得州龙血树(得州星点木)●☆

137506 Dracaena surculosa Lindl. var. capitata Hepper = Dracaena surculosa Lindl. var. maculata Hook. f. ●☆

137507 Dracaena surculosa Lindl. var. maculata Hook. f.;斑点龙血树●☆

137508 Dracaena talbotii Rendle;塔尔博特龙血树●☆

137509 Dracaena terminalis (L.) L. = Cordyline fruticosa (L.) A. Chev. ●

137510 Dracaena terminalis (L.) L. = Cordyline terminalis (L.) Kunth ●

137511 Dracaena terminalis L. = Cordyline terminalis (L.) Kunth ●

137512 Dracaena terniflora Roxb.;矮龙血树(大剑叶木);Dwarf

Dracaena, Dwarf Dragonbood, Dwarf Gragontree ●
137513 Dracaena tessmannii Engl. et K. Krause;泰斯曼龙血树●☆
137514 Dracaena thalioides E. Morren;长柄龙血树(长柄千年木,长柄竹蕉,大叶富贵竹);Thalia-like Dracena ●☆
137515 Dracaena thalioides E. Morren = Dracaena aubryana Brongn. ex E. Morren ●☆
137516 Dracaena tholloniana Hua = Dracaena aubryana Brongn. ex E. Morren ●☆
137517 Dracaena thomsoniana H. J. Veitch ex Mast. et T. Moore = Dracaena mannii Baker ●☆
137518 Dracaena transvaalensis Baker;德兰士瓦龙血树●☆
137519 Dracaena ueleensis De Wild. = Dracaena camerooniana Baker ●☆
137520 Dracaena ugandensis Baker = Dracaena fragrans (L.) Ker Gawl. ●
137521 Dracaena undulata L. f. = Asparagus undulatus (L. f.) Thunb. ■☆
137522 Dracaena usambarensis Engl. = Dracaena mannii Baker ●☆
137523 Dracaena vaginata Hutch. = Dracaena mildbraedii K. Krause ●☆
137524 Dracaena vanderystii De Wild.;范德龙血树●☆
137525 Dracaena viridiflora Engl. et K. Krause = Dracaena mildbraedii K. Krause ●☆
137526 Dracaenaceae Salisb. (1866)(保留科名);龙血树科●
137527 Dracaenaceae Salisb. (保留科名) = Droseraceae Salisb. (保留科名)■
137528 Dracaenaceae Salisb. (保留科名) = Ruscaceae M. Roem. (保留科名)●
137529 Dracaenopsis Planch. (1850-1851);类龙血树属●☆
137530 Dracaenopsis Planch. = Cordyline Comm. ex R. Br. (保留属名)●
137531 Dracaenopsis australis (G. Forst.) Planch. = Cordyline australis (G. Forst.) Hook. f. ●
137532 Dracaenopsis calocoma H. Wendl. = Cordyline australis (G. Forst.) Hook. f. ●
137533 Dracaenopsis lineata Rodig. = Cordyline indivisa (G. Forst.) Kunth ●☆
137534 Dracamine Nieuwl. = Cardamine L. ■
137535 Dracena Raf. = Dracaena Vand. ex L. ●■
137536 Draco Crantz = Calamus L. ●
137537 Draco Fabr. = Dracaena Vand. ex L. ●■
137538 Draco hookeriana (K. Koch) Kuntze = Dracaena aletriformis (Haw.) Bos ●☆
137539 Draco humilis (Baker) Kuntze = Dracaena aubryana Brongn. ex E. Morren ●☆
137540 Draco thalioides (E. Morren) Kuntze = Dracaena aubryana Brongn. ex E. Morren ●☆
137541 Dracocactus Y. Ito = Neoporteria Britton et Rose ●■
137542 Dracocactus Y. Ito = Pyrrhocactus (A. Berger) Backeb. et F. M. Knuth ●■
137543 Dracocephalium Hassk. = Dracocephalum L. (保留属名)■●
137544 Dracocephalon All. = Dracocephalum L. (保留属名)■●
137545 Dracocephalum L. (1753)(保留属名);青兰属(龙头花属,枝子花属);Dragon Head, Dragon's Head, Dragonhead, Dragon-head, Dragon's-head, Greenorchid ■●
137546 Dracocephalum acanthoides Edgew. ex Benth. = Dracocephalum heterophyllum Benth. ■
137547 Dracocephalum alpinum Salisb. = Dracocephalum nutans L. ■
137548 Dracocephalum altaiense Laxm. = Dracocephalum grandiflorum L. ■
137549 Dracocephalum altaiense Laxm. = Dracocephalum rupestre Hance ■
137550 Dracocephalum argunense Fisch. = Dracocephalum argunense Fisch. ex Link ■
137551 Dracocephalum argunense Fisch. ex Link;光萼青兰(北青兰);Argun Dragonhead, Argun Greenorchid, Japanese Dragon-head, Ruyschianum Dragonhead ■
137552 Dracocephalum argunense Fisch. ex Link = Dracocephalum ruyschianum L. ■
137553 Dracocephalum argunense Fisch. ex Link f. album T. B. Lee;白光萼青兰■☆
137554 Dracocephalum argunense Fisch. ex Link var. japonicum A. Gray = Dracocephalum argunense Fisch. ex Link ■
137555 Dracocephalum austriacum L.;奥地利青兰;Austrian Dragonhead, Austrian Dragon's-head ■☆
137556 Dracocephalum biondianum Diels = Glechoma biondiana (Diels) C. Y. Wu et C. Chen ■
137557 Dracocephalum bipinnatum Rupr.;羽叶枝子花(羽叶青兰);Bipinnate Dragonhead, Bipinnate Greenorchid ■
137558 Dracocephalum bipinnatum Rupr. var. biflorum C. Y. Wu;二花羽叶枝子花;Twoflower Bipinnate Dragonhead, Twoflower Bipinnate Greenorchid ■
137559 Dracocephalum bipinnatum Rupr. var. biflorum C. Y. Wu = Dracocephalum bipinnatum Rupr. ■
137560 Dracocephalum bipinnatum Rupr. var. brevilobum C. Y. Wu et W. T. Wang;短裂羽叶枝子花;Shotlobed Bipinnate Dragonhead, Shotlobed Bipinnate Greenorchid ■
137561 Dracocephalum bipinnatum Rupr. var. brevilobum C. Y. Wu et W. T. Wang = Dracocephalum bipinnatum Rupr. ■
137562 Dracocephalum botryoides Steven;总状花序青兰■☆
137563 Dracocephalum breviflorum Turrill;短花枝子花;Shortflower Dragonhead, Shortflower Greenorchid ■
137564 Dracocephalum bullatum Forrest ex Diels;皱叶毛建草;Bullate Dragonhead, Wrinkleleaf Greenorchid ■
137565 Dracocephalum bungeanum Schischk. et Serg.;本氏青兰■☆
137566 Dracocephalum calanthum C. Y. Wu = Dracocephalum imbricatum C. Y. Wu et W. T. Wang ■
137567 Dracocephalum calophyllum Hand.-Mazz.;美叶青兰;Beautifulleaf Greenorchid, Prettyleaf Dragonhead ■
137568 Dracocephalum canariense L. = Cedronella canariensis (L.) Webb et Berthel. ●☆
137569 Dracocephalum cavaleriei H. Lév. = Meehania henryi (Hemsl.) Y. Z. Sun ex C. Y. Wu ■
137570 Dracocephalum charkeviczii Prob. = Dracocephalum argunense Fisch. ex Link ■
137571 Dracocephalum cochinchinense Lour. = Nosema cochinchinensis (Lour.) Merr. ■
137572 Dracocephalum coerulescens (Maxim.) Dunn = Nepeta coerulescens Maxim. ■
137573 Dracocephalum coerulescens Dunn = Nepeta coerulescens Maxim. ■
137574 Dracocephalum denticulatum Aiton = Physostegia virginiana (L.) Benth. ■☆
137575 Dracocephalum discolor Bunge;异色青兰;Discolour Dragonhead ■☆
137576 Dracocephalum discolor Bunge = Dracocephalum paulsenii

Briq. ■

137577 Dracocephalum diversifolium Rupr.;异叶青兰(紫花异叶青兰);Diverseleaf Dragonhead,Diverseleaf Greenorchid ■

137578 Dracocephalum esquirolii H. Lév. = Meehania pinfaensis (H. Lév.) Y. Z. Sun ex C. Y. Wu ■

137579 Dracocephalum faberi Hemsl. = Meehania faberi (Hemsl.) C. Y. Wu ■

137580 Dracocephalum fargesii H. Lév. = Meehania fargesii (H. Lév.) C. Y. Wu ■

137581 Dracocephalum foetidum Bunge;烈味青兰;Fetid Dragonhead ■☆

137582 Dracocephalum formosius (Lunell) Rydb. = Physostegia virginiana (L.) Benth. ■☆

137583 Dracocephalum formosum Gontsch.;美丽青兰■☆

137584 Dracocephalum forrestii W. W. Sm.;松叶青兰(傅氏青兰);Forrest Dragonhead,Forrest Greenorchid ■

137585 Dracocephalum forrestii W. W. Sm. var. calyphyllum (Hand.-Mazz.) Kudo = Dracocephalum calophyllum Hand.-Mazz. ■

137586 Dracocephalum fragile Turcz. ex Benth.;易断青兰;Fragile Dragonhead ■☆

137587 Dracocephalum fruticulosum Steph. ex Willd.;灌木青兰(沙地青兰,线叶青兰,小灌木状青兰);Smallshrublike Dragonhead,Smallshrublike Greenorchid ●

137588 Dracocephalum fruticulosum Steph. ex Willd. subsp. psammophilum (C. Y. Wu et W. T Wang) H. C. Fu et S. Chen = Dracocephalum psammophilum C. Y. Wu et W. T. Wang ■

137589 Dracocephalum fruticulosum Steph. subsp. psammophilum (C. Y. Wu et W. T. Wang) H. C. Fu et S. Chen = Dracocephalum psammophilum C. Y. Wu et W. T. Wang ■

137590 Dracocephalum fumosum Gontsch.;烟色青兰(烟灰色青兰);Smoky Dragonhead ■☆

137591 Dracocephalum grandiflorum L.;大花青兰(大花毛建草);Bigflower Greenorchid,Largeflower Dragonhead ■

137592 Dracocephalum grandiflorum L. = Dracocephalum rupestre Hance ■

137593 Dracocephalum grandiflorum L. var. purdomii (W. W. Sm.) Kudo = Dracocephalum purdomii W. W. Sm. ■

137594 Dracocephalum hemsleyanum (Oliv. ex Prain) Prain ex C. Marquand et Airy Shaw = Nepeta hemsleyana Oliv. ex Prain ■

137595 Dracocephalum henryi Hemsl. = Meehania henryi (Hemsl.) Y. Z. Sun ex C. Y. Wu ■

137596 Dracocephalum heterophyllum Benth.;白花枝子花(白花甜蜜蜜,白花夏枯草,白甜蜜蜜,戈壁青兰,蜜罐罐,异叶青兰);Whiteflower Dragonhead,Whiteflower Greenorchid ■

137597 Dracocephalum hoboksarensis G. J. Liu;和布克赛尔青兰(和布克赛青兰);Hebukesair Dragonhead,Hebukesair Greenorchid ■

137598 Dracocephalum hookeri C. B. Clarke;长齿青兰;Hooker Dragonhead,Hooker Greenorchid ■

137599 Dracocephalum imberbe Bunge;无髭毛建草(光青兰,无芒毛建草);Beardless Dragonhead,Beardless Greenorchid ■

137600 Dracocephalum imbricatum C. Y. Wu et W. T. Wang;覆苞毛建草;Imbricate Dragonhead,Imbricate Greenorchid ■

137601 Dracocephalum integrifolium Bunge;全缘叶青兰(甘青青兰,全叶青兰,全缘青兰);Entireleaf Dragonhead,Entireleaf Greenorchid ■

137602 Dracocephalum integrifolium Bunge var. albiflorus G. J. Liou;白花全缘叶青兰(白花全叶青兰);Whiteflower Entireleaf Dragonhead,Whiteflower Entireleaf Greenorchid ■

137603 Dracocephalum isabellae Forrest ex W. W. Sm.;白萼青兰;Whitecalyx Dragonhead,Whitecalyx Greenorchid ■

137604 Dracocephalum kaitcheense H. Lév. = Meehania henryi (Hemsl.) Y. Z. Sun ex C. Y. Wu var. kaitcheensis (H. Lév.) C. Y. Wu ■

137605 Dracocephalum kaschgaricum Rupr. = Dracocephalum heterophyllum Benth. ■

137606 Dracocephalum komarovii Lipsky;科马罗夫青兰;Komarov Dragonhead ■☆

137607 Dracocephalum krylovii Lipsky;柯瑞洛夫青兰;Krylov Dragonhead ■☆

137608 Dracocephalum linearifolium C. H. Hu;线叶青兰;Linearleaf Dragonhead,Linearleaf Greenorchid ●

137609 Dracocephalum linearifolium C. H. Hu var. etokense C. H. Hu;鄂托克青兰;Etuoke Dragonhead,Etuoke Greenorchid ■

137610 Dracocephalum linearifolium C. H. Hu. = Dracocephalum fruticulosum Steph. ex Willd. ●

137611 Dracocephalum microflorum C. Y. Wu et W. T. Wang;小花毛建草;Smallflower Dragonhead,Smallflower Greenorchid ■

137612 Dracocephalum microphyllum Turcz. = Dracocephalum nutans L. ■

137613 Dracocephalum moldavicum L.;香青兰(炒面花,臭蒿,臭兰香,蓝秋花,摩尔达维亚青兰,摩眼子,青兰,山薄荷,山青兰,山香,香花子,小兰花,野青兰,枝子花,紫藻);Fragrant Dragonhead,Fragrant Greenorchid, Moldavian Balm, Moldavian Dragonhead, Moldavian Dragon's-head ■

137614 Dracocephalum multicaule Montbret et Aucher;多茎青兰;Manystem Dragonhead ■☆

137615 Dracocephalum multicolor Kom.;多色青兰;Multicoloured Dragonhead ■☆

137616 Dracocephalum nodulosum Rupr.;多节青兰(分节青兰);Manynode Dragonhead,Nodose Greenorchid ■

137617 Dracocephalum nutans L.;垂花青兰(多节青兰,俯垂青兰);Nodding Dragonhead,Pendentflower Greenorchid ■

137618 Dracocephalum nutans L. var. alpinum Kar. et Kir. = Dracocephalum nutans L. ■

137619 Dracocephalum oblongifolium Regel;椭圆叶青兰;Oblongleaf Dragonhead ■☆

137620 Dracocephalum origanoides Steph. ex Willd.;铺地青兰(牛至青兰);Origanumlike Dragonhead,Origanumlike Greenorchid ■

137621 Dracocephalum palmatoides C. Y. Wu et W. T. Wang;掌叶青兰(掌状青兰);Palmate Dragonhead,Palmleaf Greenorchid ■

137622 Dracocephalum palmatum Stephan;掌状青兰■

137623 Dracocephalum pamiricum Briq. = Dracocephalum heterophyllum Benth. ■

137624 Dracocephalum parviflorum Nutt.;小花青兰;American Dragonhead,American Dragon's-head,Dragonhead ■☆

137625 Dracocephalum paulsenii Briq.;宽齿青兰;Paulsen Dragonhead,Paulsen Greenorchid ■

137626 Dracocephalum pedunculatum Sessé et Moc. = Meehania fargesii (H. Lév.) C. Y. Wu var. pedunculata (Hemsl.) C. Y. Wu ■

137627 Dracocephalum peregrinum L.;刺齿枝子花;Spinetooth Dragonhead,Spinetooth Greenorchid ■

137628 Dracocephalum pinfaense H. Lév. = Meehania pinfaensis (H. Lév.) Y. Z. Sun ex C. Y. Wu ■

137629 Dracocephalum pinnamm L. var. songaricum Lipsky = Dracocephalum origanoides Steph. ex Willd. ■

137630 Dracocephalum pinnatum L.;羽状青兰;Pinnate Dragonhead ■☆

137631 Dracocephalum pinnatum L. = Dracocephalum origanoides Steph. ex Willd. ■

137632 Dracocephalum pinnatum L. var. songaricum Lipsky. = Dracocephalum origanoides Steph. ex Willd. ■

137633 Dracocephalum politovii Gand. = Dracocephalum peregrinum L. ■

137634 Dracocephalum prattii (H. Lév.) Hand.-Mazz. = Nepeta prattii H. Lév. ■

137635 Dracocephalum propinquum W. W. Sm.;多枝青兰;Branchy Greenorchid, Manybranch Dragonhead ■

137636 Dracocephalum prunelliforme Maxim. = Prunella prunelliformis (Maxim.) Makino ■☆

137637 Dracocephalum psammophilum C. Y. Wu et W. T. Wang;沙地青兰;Sandy Dragonhead, Sandy Greenorchid ■

137638 Dracocephalum pulchellum Briq. = Dracocephalum ruyschianum L. ■

137639 Dracocephalum pulchellum Briq. = Fedtschenkiella staminea (Kar. et Kir.) Kudr. ■

137640 Dracocephalum purdomii W. W. Sm.;岷山毛建草;Minshan Dragonhead, Purdom Greenorchid ■

137641 Dracocephalum radicans Vaniot = Meehania fargesii (H. Lév.) C. Y. Wu var. radicans (Vaniot) C. Y. Wu ■

137642 Dracocephalum rigidulum Hand.-Mazz.;微硬毛建草;Rigid Dragonhead, Slightlyhard Greenorchid ■

137643 Dracocephalum robustum Nakai et Kitag. = Nepeta prattii H. Lév. ■

137644 Dracocephalum rockii Diels = Marmoritis complanata (Dunn) A. L. Budantzev ■

137645 Dracocephalum rockii Diels = Phyllophyton complanatum (Dunn) Kudo ■

137646 Dracocephalum royleanum Benth. = Lallemantia royleana (Benth.) Benth. ■

137647 Dracocephalum rupestre Hance;毛建草(白花岩青兰,毛尖,毛尖茶,岩青兰);Rupestrine Dragonhead, Rupestrine Greenorchid ■

137648 Dracocephalum rupestre Hance var. albiflorum Schischk.;白花毛建草(白花岩青兰)■

137649 Dracocephalum ruprechtianum Regel = Dracocephalum bipinnatum Rupr. ■

137650 Dracocephalum ruprechtii Regel = Dracocephalum bipinnatum Rupr. ■

137651 Dracocephalum ruyschianum L.;青兰(路易斯青兰);Common Dragonhead, Common Greenorchid, Dragon's-head, Ruyschianum Dragon's-head ■

137652 Dracocephalum ruyschianum L. = Dracocephalum argunense Fisch. ex Link ■

137653 Dracocephalum ruyschianum L. var. arguneanse Nakai = Dracocephalum argunense Fisch. ex Link ■

137654 Dracocephalum ruyschianum L. var. speciosum Ledeb. = Dracocephalum argunense Fisch. ex Link ■

137655 Dracocephalum scrobiculatum Regel;有孔青兰;Scrobiculate Dragonhead ■☆

137656 Dracocephalum sibiricum (L.) L. = Nepeta sibirica L. ■

137657 Dracocephalum sibiricum L. = Nepeta sibirica L. ■

137658 Dracocephalum simplex Vaniot = Meehania fargesii (H. Lév.) C. Y. Wu var. pinetorum (Hand.-Mazz.) C. Y. Wu ■

137659 Dracocephalum sinense S. Moore = Meehania urticifolia (Miq.) Makino ■

137660 Dracocephalum souliei (H. Lév.) Hand.-Mazz. = Nepeta souliei H. Lév. ■

137661 Dracocephalum speciosum Benth. = Dracocephalum bullatum Forrest ex Diels ■

137662 Dracocephalum speciosum Benth. = Dracocephalum wallichii Sealy ■

137663 Dracocephalum speciosum Ledeb. = Dracocephalum argunense Fisch. ex Link ■

137664 Dracocephalum speciosum Ledeb. ex Gartener = Dracocephalum argunense Fisch. ex Link ■

137665 Dracocephalum spinulosum Popov;小刺青兰;Sping Dragonhead ■☆

137666 Dracocephalum stachydifolium H. Lév. = Meehania henryi (Hemsl.) Y. Z. Sun ex C. Y. Wu var. stachydifolia (H. Lév.) C. Y. Wu ■

137667 Dracocephalum stamineum Kar. et Kir. = Fedtschenkiella staminea (Kar. et Kir.) Kudr. ■

137668 Dracocephalum stellerianum Steud. ex Benth.;星状青兰■☆

137669 Dracocephalum stewartianum (Diels) Dunn = Nepeta stewartiana Diels ■

137670 Dracocephalum subcapitatum (Kuntze) Lipsky;头状青兰;Subcapitate Dragonhead ■☆

137671 Dracocephalum taliense Forrest ex W. W. Sm.;大理青兰;Dali Dragonhead, Dali Greenorchid ■

137672 Dracocephalum tanguticum Maxim.;甘青青兰(甘肃青兰,陇塞青兰,唐古特青兰,杂毕样,在羊古,智洋顾);Tangut Dragonhead, Tangut Greenorchid ■

137673 Dracocephalum tanguticum Maxim. var. cinereum Hand.-Mazz.;灰毛甘青青兰■

137674 Dracocephalum tanguticum Maxim. var. nanum C. Y. Wu et W. T. Wang;矮生甘青青兰;Dwarf Tangut Dragonhead, Dwarf Tangut Greenorchid ■

137675 Dracocephalum tenuiflorum (Diels) Dunn = Nepeta tenuiflora Benth. ■

137676 Dracocephalum tenuiflorum (Diels) Dunn = Schizonepeta tenuifolia (Benth.) Briq. ■

137677 Dracocephalum thymiflorum L.;香花青兰(百里香青兰);Fragrantflower Dragonhead, Fragrantflower Greenorchid, Thyme-flowered Dragonhead, Thyme-flowered Dragon's-head, Thymeleaf Dragonhead ■

137678 Dracocephalum truncatum Y. Z. Sun ex C. Y. Wu;截萼毛建草;Trucatecalyx Dragonhead, Trucatecalyx Greenorchid ■

137679 Dracocephalum turkestanicum Gand. = Dracocephalum grandiflorum L. ■

137680 Dracocephalum urticifolium (Miq.) Makino var. angustifolium (Dunn) Hand.-Mazz. = Meehania pinfaensis (H. Lév.) Y. Z. Sun ex C. Y. Wu ■

137681 Dracocephalum urticifolium Miq. = Meehania urticifolia (Miq.) Makino ■

137682 Dracocephalum urticifolium Miq. var. angustifolia (Dunn) Hand.-Mazz. = Meehania pinfaensis (H. Lév.) Y. Z. Sun ex C. Y. Wu ■

137683 Dracocephalum urticifolium Miq. var. pedunculatum Hemsl. = Meehania fargesii (H. Lév.) C. Y. Wu var. pedunculata (Hemsl.) C. Y. Wu ■

137684 Dracocephalum urticifolium Miq. var. pinetorum Hand.-Mazz. = Meehania fargesii (H. Lév.) C. Y. Wu var. pinetorum (Hand.-

Mazz.) C. Y. Wu ■
137685 Dracocephalum urticifolium Miq. var. typica f. carnosa Dunn = Meehania faberi (Hemsl.) C. Y. Wu ■
137686 Dracocephalum urticifolium Miq. var. typica f. racemosa Dunn = Meehania henryi (Hemsl.) Y. Z. Sun ex C. Y. Wu ■
137687 Dracocephalum veitchii (Duthie) Dunn = Nepeta veitchii Duthie ■
137688 Dracocephalum velutinum C. Y. Wu et W. T. Wang;绒叶毛建草;Greenorchid, Velvety Dragonhead ■
137689 Dracocephalum velutinum C. Y. Wu et W. T. Wang var. intermedium C. Y. Wu et W. T. Wang;圆齿绒叶毛建草;Flossleaf Dragonhead, Flossleaf Greenorchid ■
137690 Dracocephalum virginianum L. = Physostegia virginiana (L.) Benth. ■☆
137691 Dracocephalum wallichii Sealy;美花毛建草(美丽青兰);Specious Dragonhead, Wallich Greenorchid, Wallich Dragonhead ■
137692 Dracocephalum wallichii Sealy var. platyanthum C. Y. Wu et W. T. Wang;宽花美花毛建草;Broadflower Wallich Dragonhead, Broadflower Wallich Greenorchid ■
137693 Dracocephalum wallichii Sealy var. proliferum C. Y. Wu et W. T. Wang;复序美花毛建草;Composite Dragonhead, Composite Greenorchid ■
137694 Dracocephalum wilsonii (Duthie) Dunn = Nepeta wilsonii Duthie ■
137695 Dracocephalus Asch. = Dracocephalum L.(保留属名)■●
137696 Dracomonticola H. P. Linder et Kurzweil = Platanthera Rich.(保留属名)■
137697 Dracomonticola H. P. Linder et Kurzweil(1995);山龙兰属■☆
137698 Dracomonticola virginea (Bolus) H. P. Linder et Kurzweil;纯白毛建草■☆
137699 Draconanthes (Luer) Luer = Pleurothallis R. Br. ■☆
137700 Draconia Fabr. = Artemisia L. ●■
137701 Draconia Heist. ex Fabr. = Artemisia L. ●■
137702 Dracontia (Luer) Luer = Pleurothallis R. Br. ■☆
137703 Dracontiaceae Salisb. = Araceae Juss.(保留科名)■●
137704 Dracontioides Engl.(1911);拟小龙南星属■☆
137705 Dracontioides desciscens Engl.;拟小龙南星■☆
137706 Dracontium Hill = Dracunculus Mill. ■☆
137707 Dracontium L.(1753);小龙南星属■☆
137708 Dracontium foetidum L. = Symplocarpus foetidus (L.) Salisb. ex W. P. C. Barton ■
137709 Dracontium paeoniifolium Dennst. = Amorphophallus paeoniifolius (Dennst.) Nicholson ■
137710 Dracontium polyphyllum L.;多叶小龙南星■☆
137711 Dracontium spinosum L. = Lasia spinosa (L.) Thwaites ■
137712 Dracontocephalium Hassk. = Dracocephalum L.(保留属名)■●
137713 Dracontocephalum St.-Lag. = Dracontocephalium Hassk. ■●
137714 Dracontomelon Blume (1850);人面子属;Dragon Plum, Dragonplum, Dragon-plum, New Guinea Walnut, Pacific Walnut, Papuan Walnut ●
137715 Dracontomelon dao (Blanco) Merr. et Rolfe;道人面子;Dao, Paldao ●
137716 Dracontomelon duperreanum Pierre;人面子(人面树,银莲果,银稔);Indochina Dragon Plum, Indochina Dragonplum, Indochina Dragon-plum ●
137717 Dracontomelon edule Merr.;可食人面子;Edible Dragon Plum, Edible Dragonplum ●☆
137718 Dracontomelon macrocarpum H. L. Li;大果人面子;Bigfruit Dragonplum, Bigfruit Dragon-plum, Largefruit Dragonplum ●◇
137719 Dracontomelon mangiferum Blume = Dracontomelon dao (Blanco) Merr. et Rolfe ●
137720 Dracontomelon mangiferum Blume = Dracontomelon duperreanum Pierre ●
137721 Dracontomelon sinense Stapf = Dracontomelon duperreanum Pierre ●
137722 Dracontopsis Lem. = Dracopis (Cass.) Cass. ■
137723 Dracophilus (Schwantes) Dinter et Schwantes(1927);龙幻属■☆
137724 Dracophilus Dinter et Schwantes = Dracophilus (Schwantes) Dinter et Schwantes ■☆
137725 Dracophilus dealbatus (N. E. Br.) Walgate;龙幻■☆
137726 Dracophilus delaetianus (Dinter) Dinter et Schwantes;纳米比亚龙幻■☆
137727 Dracophilus delaetianus Dinter et Schwantes = Dracophilus delaetianus (Dinter) Dinter et Schwantes ■☆
137728 Dracophilus montis-draconis (Dinter) Dinter et Schwantes = Dracophilus dealbatus (N. E. Br.) Walgate ■☆
137729 Dracophilus proximus (L. Bolus) Walgate = Dracophilus dealbatus (N. E. Br.) Walgate ■☆
137730 Dracophyllum Labill.(1800);龙草树属(龙血石南属,龙叶树属)●☆
137731 Dracophyllum lessonianum A. Rich.;里桑龙叶树●☆
137732 Dracophyllum secundum R. Br.;龙草树;Dragon Heath ●☆
137733 Dracophyllum strictum Hook. f.;尖叶龙草树●☆
137734 Dracophyllum townsonii Cheeseman;粗枝龙草树●☆
137735 Dracophyllum traversii Hook. f.;高山龙草树;Mountain Neinei ●☆
137736 Dracopis (Cass.) Cass. = Rudbeckia L. ■
137737 Dracopis Cass. = Rudbeckia L. ■
137738 Dracopis amplexicaulis (Vahl) Cass. = Rudbeckia amplexicaulis Vahl ■
137739 Dracosciadium Hilliard et B. L. Burtt(1986);龙伞芹属■☆
137740 Dracosciadium italae Hilliard et B. L. Burtt;龙伞芹■☆
137741 Dracula Luer(1978);小龙兰属■☆
137742 Dracula chimaera (Rchb. f.) Luer;小龙兰■☆
137743 Dracunculus Ledeb. = Artemisia L. ●■
137744 Dracunculus Mill.(1754);龙芋属(龙木芋属,紫花海芋属);Arum, Dragon Arum, Stink Dragon ■☆
137745 Dracunculus canariensis Kunth;加那利龙芋■☆
137746 Dracunculus marschallianus Less. = Artemisia marschalliana Spreng. ■
137747 Dracunculus vulgaris Schott;龙芋(龙木芋,普通疆南星);Brook Leek, Brookleek, Common Dracunculus, Common Dragon's Plant, Common Stink Dragon, Dragance, Dragon Arum, Dragon's Arum, Dragons, Dragon's Female, Dragonwort, Edderwort, Faverole, Serpentine, Snake Horemint ■☆
137748 Draikaina Raf. = Dracaena Vand. ex L. ●■
137749 Drakaea Lindl.(1840);西南澳兰属■☆
137750 Drakaea elastica Lindl.;西南澳兰■☆
137751 Drakaea huntiana F. Muell. = Thynninorchis huntiana (F. Muell.) D. L. Jones et M. A. Clem. ■☆
137752 Drakea Endl. = Drakaea Lindl. ■☆
137753 Drakebrockmania A. C. White et B. Sloane = Caralluma R. Br. ■
137754 Drakebrockmania A. C. White et B. Sloane = White-Sloanea Chiov. ■☆
137755 Drake-Brockmania Stapf(1912);德雷草属■☆

137756 Drakebrockmania crassa (N. E. Br.) A. C. White et B. Sloane = White-Sloanea crassa (N. E. Br.) Chiov. ■☆

137757 Drake-Brockmania haareri (Stapf et C. E. Hubb.) S. M. Phillips;哈氏德雷草■☆

137758 Drake-Brockmania somalensis Stapf;德雷草■☆

137759 Drakensteinia DC. = Drakenstenia Neck. ●

137760 Drakenstenia Neck. = Dalbergia L. f. (保留属名)●

137761 Dransfieldia W. J. Baker et Zona = Ptychosperma Labill. ●☆

137762 Draparnalda St. -Lag. = Draparnaudia Montrouz. ●☆

137763 Draparnaudia Montrouz. = Xanthostemon F. Muell. (保留属名) ●☆

137764 Draperia Torr. (1868);德雷珀麻属■☆

137765 Draperia systyla Torr.;德雷珀麻■☆

137766 Drapetes Banks = Drapetes Banks ex Lam. ●☆

137767 Drapetes Banks ex Lam. (1792);细灌瑞香属●☆

137768 Drapetes Lam. = Drapetes Banks ex Lam. ●☆

137769 Drapetes muscosa Lam.;细灌瑞香●☆

137770 Drapiezia Blume = Disporum Salisb. ex D. Don ■

137771 Draytonia A. Gray = Saurauia Willd. (保留属名)●

137772 Drebbelia Zoll. = Olax L. ●

137773 Drebbelia Zoll. et Moritzi = Spatholobus Hassk. ●

137774 Dregea E. Mey. (1838)(保留属名);南山藤属(华他卡藤属,假夜来香属);Dregea ●

137775 Dregea E. Mey. (保留属名) = Peucedanum L. ■

137776 Dregea E. Mey. (保留属名) = Sciothamnus Endl. ●☆

137777 Dregea Eckl. et Zeyh. (废弃属名) = Dregea E. Mey. (保留属名)●

137778 Dregea Eckl. et Zeyh. (废弃属名) = Peucedanum L. ■

137779 Dregea Eckl. et Zeyh. (废弃属名) = Sciothamnus Endl. ●☆

137780 Dregea abyssinica (Hochst.) K. Schum.;阿比西尼亚南山藤●☆

137781 Dregea africana (Decne.) Martelli = Dregea abyssinica (Hochst.) K. Schum. ●☆

137782 Dregea capensis (Thunb.) Eckl. et Zeyh. = Peucedanum capense (Thunb.) Sond. ●■☆

137783 Dregea capensis (Thunb.) Eckl. et Zeyh. var. angustifolia Eckl. et Zeyh. = Peucedanum capense (Thunb.) Sond. ●■☆

137784 Dregea collina Eckl. et Zeyh. = Peucedanum striatum (Thunb.) Sond. ■☆

137785 Dregea corrugata C. K. Schneid. = Dregea sinensis Hemsl. var. corrugata (C. K. Schneid.) Tsiang et P. T. Li ●

137786 Dregea crinita (Oliv.) Bullock;长软毛南山藤●☆

137787 Dregea cuneifolia Tsiang et P. T. Li;楔叶南山藤;Cunealleaf Dregea, Wedge-leaf Dregea, Wedge-leaved Dregea ●

137788 Dregea faulknerae Bullock;福克纳南山藤●☆

137789 Dregea floribunda E. Mey.;多花南山藤●☆

137790 Dregea formosana T. Yamaz. = Dregea volubilis (L. f.) Benth. ex Hook. f. ●

137791 Dregea macrantha Klotzsch;大花南山藤●☆

137792 Dregea montana Eckl. et Zeyh. = Peucedanum dregeanum D. Dietr. ■☆

137793 Dregea rubicunda K. Schum.;稍红南山藤●☆

137794 Dregea schimperi (Decne.) Bullock;欣珀南山藤●☆

137795 Dregea sinensis Hemsl.;苦绳(白浆草,白浆藤,白丝藤,刀疮药,刀愈药,隔山撬,华南山藤,奶浆藤,南山藤,藤木通,通关散,通光散,通炎散,小木通,野泡通,中华南山藤);China Dregea, Chinese Dregea ●

137796 Dregea sinensis Hemsl. var. corrugata (C. K. Schneid.) Tsiang et P. T. Li;贯筋藤(刀口药,奶浆果);Corrugate Chinese Dregea, Corrugate Dregea ●

137797 Dregea stelostigma (K. Schum.) Bullock;柱头南山藤●☆

137798 Dregea virgata (Cham. et Schltdl.) Eckl. et Zeyh. = Peucedanum capense (Thunb.) Sond. var. lanceolatum Sond. ●■☆

137799 Dregea volubilis (L. f.) Benth. ex Hook. f.;南山藤(春筋藤,大果咀彭,隔山消,华他卡藤,假猫豆,假夜来香,苦莱藤,苦凉菜,双根藤,台湾华他卡藤,台湾球兰,牙节);Taiwan Dregea, Twisting Dregea ●

137800 Dregea volubilis (L. f.) Benth. ex Hook. f. = Wattakaka volubilis (L. f.) Stapf ●

137801 Dregea volubilis (L. f.) Hook. f. = Wattakaka volubilis (L. f.) Stapf ●

137802 Dregea yunnanensis (Tsiang) Tsiang et P. T. Li;丽子藤(白血藤,滇假夜来香,奶浆藤);Yunnan Dregea ●

137803 Dregea yunnanensis (Tsiang) Tsiang et P. T. Li var. major (Tsiang) Tsiang et P. T. Li;大丽子藤;Big Yunnan Dregea ●

137804 Dregea yunnanensis (Tsiang) Tsiang et P. T. Li var. major (Tsiang) Tsiang et P. T. Li = Dregea yunnanensis (Tsiang) Tsiang et P. T. Li ●

137805 Dregeochloa Conert(1966);岩地扁芒草属■☆

137806 Dregeochloa calviniensis Conert;岩地扁芒草■☆

137807 Dregeochloa pumila (Nees) Conert;小岩地扁芒草■☆

137808 Drejera Nees(1847);类虾衣花属■☆

137809 Drejera boliviensis Nees;玻利维亚类虾衣花■☆

137810 Drejera parviflora Buckley;小花类虾衣花■☆

137811 Drejera polyantha Rizzini;多花类虾衣花■☆

137812 Drejerella Lindau = Justicia L. ●■

137813 Drejerella Lindau(1900);小虾衣花属(虾衣草属,虾衣花属);Drejerella ■

137814 Drejerella guttata (Brandegee) Bremek. = Calliaspidia guttata (Brandegee) Bremek. ■●

137815 Drepachenia Raf. = Sagittaria L. ■

137816 Drepadenium Raf. = Croton L. ●

137817 Drepanandrum Neck. = Topobea Aubl. ■☆

137818 Drepananthna Maingay ex Hook. f. = Cyathocalyx Champ. ex Hook. f. et Thomson ●

137819 Drepananthus Maingay ex Hook. f. et Thomson(1872);镰花番荔枝属●☆

137820 Drepananthus filiformis (Ast) T. B. Nguyen;线形镰花番荔枝●☆

137821 Drepananthus longiflorus C. B. Rob.;长花镰花番荔枝●☆

137822 Drepananthus philippinensis Merr.;菲律宾镰花番荔枝●☆

137823 Drepania Juss. = Tolpis Adans. ●■☆

137824 Drepania barbata (L.) Desf. = Tolpis barbata (L.) Gaertn. ■☆

137825 Drepanocarpus G. Mey. = Machaerium Pers. (保留属名)●☆

137826 Drepanocarpus africanus G. Don = Machaerium lunatum (L. f.) Ducke ●☆

137827 Drepanocarpus lunatus (L. f.) G. Mey. = Machaerium lunatum (L. f.) Ducke ●☆

137828 Drepanocaryum Pojark. (1954);镰果草属■☆

137829 Drepanocaryum sewerzowii (Regel) Pojark.;镰果草■☆

137830 Drepanolobus Nutt. ex Torr. et A. Gray = Hosackia Douglas ex Benth. ■☆

137831 Drepanometra Hassk. = Begonia L. ●■

137832 Drepanophyllum Wibel = Falcaria Fabr. (保留属名)■

137833 Drepanophyllum Wibel(1799);镰叶草属■☆

137834 Drepanophyllum fulvum Hook.;镰叶草■☆
137835 Drepanophyllum heterophyllum (Greene) Koso-Pol.;异叶镰叶草■☆
137836 Drepanophyllum latifolium Koso-Pol.;宽叶镰叶草■☆
137837 Drepanophyllum luteum Eichw.;黄镰叶草■☆
137838 Drepanospermum Benth. = Campnosperma Thwaites(保留属名)●☆
137839 Drepanospermum Benth. = Cyrtospermum Benth.●☆
137840 Drepanostachyum P. C. Keng = Sinarundinaria Nakai●
137841 Drepanostachyum P. C. Keng (1983);镰序竹属;Drepanostachyum,Sicklebamboo●
137842 Drepanostachyum ampullare (T. P. Yi) Demoly;樟木镰序竹(樟木箭竹);Camphor Arrowbamboo,Zhangmu Fargesia●
137843 Drepanostachyum ampullare (T. P. Yi) Demoly = Fargesia ampullaris T. P. Yi●
137844 Drepanostachyum breviligulatum T. P. Yi;钩竹(钓竹);Breviligulate Sicklebamboo●
137845 Drepanostachyum breviligulatum T. P. Yi = Ampelocalamus breviligulatus (T. P. Yi) Stapleton et D. Z. Li●
137846 Drepanostachyum breviligulatum T. P. Yi f. discrepans T. P. Yi;盐巴竹;Differing Breviligulate Sicklebamboo●
137847 Drepanostachyum calcareus (C. D. Chu et C. S. Chao) J. R. Xue;贵州镰序竹;Guizhou Drepanostachyum●
137848 Drepanostachyum exauritum W. T. Lin;无耳镰序竹;Earless Drepanostachyum●
137849 Drepanostachyum falcatum (Nees) P. C. Keng;镰序竹;Blue Bamboo,Himalayan Bamboo,Nepalese Blue Bamboo,Sicklebamboo●
137850 Drepanostachyum falconeri (Munro) J. J. N. Campb. ex D. C. McClint. = Himalayacalamus falconeri (Munro) P. C. Keng●
137851 Drepanostachyum fractiflexum (T. P. Yi) D. Z. Li;扫把竹;Besom Arrowbamboo,Broom Bamboo,Multiple-fractate Fargesia●
137852 Drepanostachyum fractiflexum (T. P. Yi) D. Z. Li = Fargesia fractiflexa T. P. Yi●
137853 Drepanostachyum hirsutissimum W. D. Li et Y. C. Zhong = Ampelocalamus hirsutissimus (W. D. Li et Y. C. Zhong) Stapleton et D. Z. Li●
137854 Drepanostachyum hookerianum (Munro) P. C. Keng = Himalayacalamus hookerianus (Munro) Stapleton●☆
137855 Drepanostachyum luodianense (T. P. Yi et R. S. Wang) P. C. Keng = Ampelocalamus luodianense T. P. Yi et R. S. Wang●
137856 Drepanostachyum melicoideum P. C. Keng = Ampelocalamus melicoideus (P. C. Keng) D. Z. Li et Stapleton●
137857 Drepanostachyum membranaceum (T. P. Yi) D. Z. Li;膜箨镰序竹(膜箨箭竹);Membrana Arrowbamboo●
137858 Drepanostachyum membranaceum (T. P. Yi) D. Z. Li = Fargesia membranacea T. P. Yi●
137859 Drepanostachyum mianningense (Q. Li et X. Jiang) T. P. Yi = Ampelocalamus mianningensis (Q. Li et X. Jiang) D. Z. Li et Stapleton●
137860 Drepanostachyum microphyllum (J. R. Xue et T. P. Yi) J. R. Xue et T. P. Yi = Ampelocalamus microphyllus (J. R. Xue et T. P. Yi) J. R. Xue et T. P. Yi●
137861 Drepanostachyum microphyllum (J. R. Xue et T. P. Yi) P. C. Keng ex T. P. Yi;坝竹;Littleleaf Sicklebamboo,Small-leaved Drepanostachyum●
137862 Drepanostachyum naibunense (Hayata) P. C. Keng = Ampelocalamus naibunensis (Hayata) T. H. Wen●
137863 Drepanostachyum naibunensoides W. T. Lin et Z. M. Wu;紫斑镰序竹;Naibun-like Drepanostachyum●
137864 Drepanostachyum saxatile (J. R. Xue et T. P. Yi) P. C. Keng ex T. P. Yi = Ampelocalamus saxatilis (J. R. Xue et T. P. Yi) J. R. Xue et T. P. Yi●
137865 Drepanostachyum scandens (J. R. Xue et W. D. Li) P. C. Keng ex T. P. Yi = Ampelocalamus scandens J. R. Xue et W. D. Li●
137866 Drepanostachyum semiorbiculatum (T. P. Yi) Stapleton;圆芽镰序竹(圆芽箭竹);Roundbud Arrowbamboo,Semiorbicular Fargesia●
137867 Drepanostachyum semiorbiculatum (T. P. Yi) Stapleton = Fargesia semiorbiculata T. P. Yi●
137868 Drepanostachyum yongshanense (J. R. Xue et D. Z. Li) P. C. Keng ex T. P. Yi = Ampelocalamus yongshanensis J. R. Xue et D. Z. Li●
137869 Drepanostemma Jum. et H. Perrier(1911);镰冠萝藦属■☆
137870 Drepanostemma luteum Jum. et H. Perrier;镰冠萝藦■☆
137871 Drepanostemma luteum Jum. et H. Perrier = Cynanchum decorsei (Costantin et Gallaud) Liede et Meve●☆
137872 Drepaphyla Raf. = Acacia Mill.(保留属名)●■
137873 Drepilia Raf. = Thermia Nutt.■
137874 Drepilia Raf. = Thermopsis R. Br. ex W. T. Aiton■
137875 Dresslerella Luer(1976);小玉兔兰属■☆
137876 Dresslerella pertusa (Dressler) Luer;小玉兔兰■☆
137877 Dressleria Dodson(1975);玉兔兰属■☆
137878 Dressleria allenii H. G. Hills;阿伦玉兔兰■☆
137879 Dressleria fragrans Dodson;香玉兔兰■☆
137880 Dressleriella Brieger = Jacquiniella Schltr.■☆
137881 Dresslerìopsis Dwyer = Lasianthus Jack(保留属名)●
137882 Dresslerothamnus H. Rob.(1978);红丝菊属●☆
137883 Dresslerothamnus angustiradiatus (T. M. Barkley) H. Rob.;红丝菊●☆
137884 Driessenia Korth.(1844);德里野牡丹属●■☆
137885 Driessenia axantha Korth.;德里野牡丹●☆
137886 Driessenia microthrix Stapf;小毛德里野牡丹●☆
137887 Driessenia minutiflora O. Schwartz;小花德里野牡丹●☆
137888 Driessenia sinensis H. Lév. = Gonostegia hirta (Blume ex Hassk.) Miq.■
137889 Drimeaceae Tiegh. = Winteraceae R. Br. ex Lindl.(保留科名)●
137890 Drimia Jacq. = Drimia Jacq. ex Willd.■☆
137891 Drimia Jacq. ex Willd.(1799);长被片风信子属(锥米属)■☆
137892 Drimia acuminata Lodd. = Ledebouria revoluta (L. f.) Jessop■☆
137893 Drimia albiflora (B. Nord.) J. C. Manning et Goldblatt;白花叶长被片风信子■☆
137894 Drimia alta R. A. Dyer = Drimia elata Jacq.■☆
137895 Drimia altissima (L. f.) Ker Gawl.;高大叶长被片风信子■☆
137896 Drimia altissima Hook. f. = Drimia elata Jacq.■☆
137897 Drimia angustifolia Baker;窄叶长被片风信子■☆
137898 Drimia angustitepala Engl. = Ledebouria kirkii (Baker) Stedje et Thulin■☆
137899 Drimia anomala (Baker) Benth.;异常长被片风信子■☆
137900 Drimia apertiflora Baker = Ledebouria ensifolia (Eckl.) S. Venter et T. J. Edwards■☆
137901 Drimia arenicola (B. Nord.) J. C. Manning et Goldblatt;沙地长被片风信子■☆
137902 Drimia barkerae Oberm. ex J. C. Manning et Goldblatt;巴尔凯拉长被片风信子■☆
137903 Drimia barteri Baker = Drimia altissima (L. f.) Ker Gawl.■☆

137904 Drimia bolusii Baker;澳非长被片风信子■☆
137905 Drimia brachystachys (Baker) Stedje;短穗长被片风信子■☆
137906 Drimia brevifolia Baker = Ledebouria revoluta (L. f.) Jessop ■☆
137907 Drimia burchellii Baker = Drimia elata Jacq. ■☆
137908 Drimia calcarata (Baker) Stedje;距长被片风信子■☆
137909 Drimia capensis (Burm. f.) Wijnands;好望角长被片风信子■☆
137910 Drimia capitata Baker = Drimia sphaerocephala Baker ■☆
137911 Drimia ciliaris Jacq. ex Willd. = Drimia elata Jacq. ■☆
137912 Drimia ciliata (L. f.) J. C. Manning et Goldblatt;缘毛长被片风信子■☆
137913 Drimia coleae Baker = Ledebouria somaliensis (Baker) Stedje et Thulin ■☆
137914 Drimia concolor Baker = Drimia elata Jacq. ■☆
137915 Drimia confertiflora Dammer = Ledebouria somaliensis (Baker) Stedje et Thulin ■☆
137916 Drimia congesta Bullock;密集长被片风信子■☆
137917 Drimia convallarioides (L. f.) J. C. Manning et Goldblatt;环绕长被片风信子■☆
137918 Drimia cooperi (Baker) Benth.;库珀长被片风信子■☆
137919 Drimia cremnophila Van Jaarsv.;悬崖长被片风信子■☆
137920 Drimia cuscutoides (Burch. ex Baker) J. C. Manning et Goldblatt;菟丝子状长被片风信子■☆
137921 Drimia cyanelloides (Baker) J. C. Manning et Goldblatt;蓝色长被片风信子■☆
137922 Drimia delagoensis (Baker) Jessop;迪拉果长被片风信子■☆
137923 Drimia depressa (Baker) Jessop;凹陷长被片风信子■☆
137924 Drimia dregei (Baker) J. C. Manning et Goldblatt;德雷长被片风信子■☆
137925 Drimia duthieae (Adamson) Jessop;达泰长被片风信子■☆
137926 Drimia elata Jacq.;高长被片风信子■☆
137927 Drimia elgonica Bullock = Drimia elata Jacq. ■☆
137928 Drimia ensifolia Eckl. = Ledebouria ensifolia (Eckl.) S. Venter et T. J. Edwards ■☆
137929 Drimia exigua Stedje;埃塞长被片风信子■☆
137930 Drimia exuviata (Jacq.) Jessop;脱落长被片风信子■☆
137931 Drimia fasciata (B. Nord.) J. C. Manning et Goldblatt;带状长被片风信子■☆
137932 Drimia filifolia (Jacq.) J. C. Manning et Goldblatt;线叶长被片风信子■☆
137933 Drimia fischeri Baker = Drimiopsis fischeri (Engl.) Stedje ●☆
137934 Drimia flagellaris T. J. Edwards et D. Styles et N. R. Crouch;鞭状长被片风信子■☆
137935 Drimia forsteri (Baker) Oberm. = Drimia capensis (Burm. f.) Wijnands ■☆
137936 Drimia fragrans (Jacq.) J. C. Manning et Goldblatt;芳香长被片风信子■☆
137937 Drimia fugax (Moris) Stearn = Urginea fugax (Moris) Steinh. ■☆
137938 Drimia gawleri Schrad. = Ledebouria ovalifolia (Schrad.) Jessop ■☆
137939 Drimia glaucescens (Engl. et K. Krause) Scholz;灰绿长被片风信子■☆
137940 Drimia haworthioides Baker;拟霍沃斯长被片风信子■☆
137941 Drimia hesperantha J. C. Manning et Goldblatt;夕花长被片风信子■☆
137942 Drimia hildebrandtii Baker = Ledebouria kirkii (Baker) Stedje et Thulin ■☆
137943 Drimia hyacinthoides Baker;普通长被片风信子■☆
137944 Drimia incerta A. Chev. = Drimia elata Jacq. ■☆
137945 Drimia indica (Roxb.) Jessop;印度长被片风信子■☆
137946 Drimia intricata (Baker) J. C. Manning et Goldblatt;缠结长被片风信子■☆
137947 Drimia involuta (J. C. Manning et Snijman) J. C. Manning et Goldblatt;内卷长被片风信子■☆
137948 Drimia karooica (Oberm.) J. C. Manning et Goldblatt;卡鲁长被片风信子■☆
137949 Drimia kniphofioides (Baker) J. C. Manning et Goldblatt;火炬花长被片风信子■☆
137950 Drimia lanceolata Schrad. = Ledebouria ovalifolia (Schrad.) Jessop ■☆
137951 Drimia laxiflora Baker;疏花长被片风信子■☆
137952 Drimia ledermannii Engl. et K. Krause;莱德长被片风信子■☆
137953 Drimia ludwigii Miq.;路德维格长被片风信子■☆
137954 Drimia macrantha (Baker) Baker;大花长被片风信子■☆
137955 Drimia macrocarpa Stedje;大果长被片风信子■☆
137956 Drimia macrocentra (Baker) Jessop;小刺长被片风信子■☆
137957 Drimia marginata (Thunb.) Jessop;具边长被片风信子■☆
137958 Drimia maritima (L.) Stearn;滨海长被片风信子;Red Squill, Sea Onion, Squill ■☆
137959 Drimia maritima (L.) Stearn = Charybdis maritima (L.) Speta ■☆
137960 Drimia maritima (L.) Stearn var. hesperia (Webb et Berthel.) A. Hansen et Sunding = Charybdis maura (Maire) Speta ■☆
137961 Drimia media Jacq.;中间长被片风信子■☆
137962 Drimia minor (A. V. Duthie) Jessop;小长被片风信子■☆
137963 Drimia modesta (Baker) Jessop = Drimia calcarata (Baker) Stedje ■☆
137964 Drimia montana A. P. Dold et E. Brink;山地被片风信子■☆
137965 Drimia multifolia (G. J. Lewis) Jessop;多叶被片风信子■☆
137966 Drimia multisetosa (Baker) Jessop;多刚毛长被片风信子■☆
137967 Drimia mzimvubuensis Van Jaarsv.;姆津长被片风信子■☆
137968 Drimia namibensis (Oberm.) J. C. Manning et Goldblatt;纳米布长被片风信子■☆
137969 Drimia nana (Snijman) J. C. Manning et Goldblatt;矮小长被片风信子■☆
137970 Drimia neriniformis Baker = Drimia sphaerocephala Baker ■☆
137971 Drimia nervosa (Burch.) Jessop = Scilla nervosa (Burch.) Jessop ■☆
137972 Drimia nitida Eckl. = Ledebouria concolor (Baker) Jessop ■☆
137973 Drimia noctiflora (Batt. et Trab.) Stearn;夜花长被片风信子■☆
137974 Drimia ollivieri (Maire) Stearn = Urginea ollivieri Maire ■☆
137975 Drimia ovalifolia Schrad. = Ledebouria ovalifolia (Schrad.) Jessop ■☆
137976 Drimia paolii Chiov. = Drimia altissima (L. f.) Ker Gawl. ■☆
137977 Drimia pauciflora Baker;少花长被片风信子■☆
137978 Drimia physodes (Jacq.) Jessop;囊状长被片风信子■☆
137979 Drimia platyphylla (B. Nord.) J. C. Manning et Goldblatt;宽叶长被片风信子■☆
137980 Drimia porphyrantha (Bullock) Stedje;紫花长被片风信子■☆
137981 Drimia purpurascens Jacq. = Drimia elata Jacq. ■☆
137982 Drimia pusilla Jacq. ex Willd. = Drimia elata Jacq. ■☆
137983 Drimia revoluta (A. V. Duthie) J. C. Manning et Goldblatt = Drimia hesperantha J. C. Manning et Goldblatt ■☆
137984 Drimia revoluta (L. f.) Kunth = Ledebouria revoluta (L. f.) Jessop ■☆

137985 Drimia rigidifolia Baker = Drimia elata Jacq. ■☆
137986 Drimia robusta Baker = Drimia elata Jacq. ■☆
137987 Drimia rudatisii Schltr. = Drimia elata Jacq. ■☆
137988 Drimia salteri (Compton) J. C. Manning et Goldblatt;索尔特长被片风信子■☆
137989 Drimia sanguinea (Schinz) Jessop;血红长被片风信子■☆
137990 Drimia saniensis (Hilliard et B. L. Burtt) J. C. Manning et Goldblatt;萨尼长被片风信子■☆
137991 Drimia sclerophylla J. C. Manning et Goldblatt;硬叶长被片风信子■☆
137992 Drimia secunda (B. Nord.) J. C. Manning et Goldblatt;单侧长被片风信子■☆
137993 Drimia simensis (Hochst. ex A. Rich.) Stedje;锡米长被片风信子■☆
137994 Drimia sphaerocephala Baker;球头长被片风信子■☆
137995 Drimia sudanica Friis et Vollesen = Ebertia pauciflora (Baker) Speta ●☆
137996 Drimia tazensis (Maire) Stearn = Charybdis tazensis (Maire) Speta ■☆
137997 Drimia uitenhagensis Eckl. = Drimia altissima (L. f.) Ker Gawl. ■☆
137998 Drimia undulata Jacq. = Ledebouria undulata (Jacq.) Jessop ■☆
137999 Drimia undulata Stearn = Charybdis undulata (Desf.) Speta ■☆
138000 Drimia uniflora J. C. Manning et Goldblatt;单花长被片风信子■☆
138001 Drimia uranthera (R. A. Dyer) J. C. Manning et Goldblatt;尾药长被片风信子■☆
138002 Drimia villosa Lindl. = Drimia elata Jacq. ■☆
138003 Drimia virens (Schltr.) J. C. Manning et Goldblatt;绿长被片风信子■☆
138004 Drimia viridiflora Kunze;绿花长被片风信子■☆
138005 Drimia zombensis Baker = Drimia elata Jacq. ■☆
138006 Drimiopsis Lindl. et Paxton(1851);拟辛酸木属■☆
138007 Drimiopsis aroidastrum A. Chev. = Drimiopsis barteri Baker ●☆
138008 Drimiopsis atropurpurea N. E. Br.;暗紫拟辛酸木●☆
138009 Drimiopsis barteri Baker;巴特拟辛酸木●☆
138010 Drimiopsis botryoides Baker;葡萄拟辛酸木●☆
138011 Drimiopsis botryoides Baker subsp. prostrata Stedje;平卧拟辛酸木●☆
138012 Drimiopsis burkei Baker;伯克辛酸木●☆
138013 Drimiopsis burkei Baker subsp. stolonissima U. Müll. -Doblies et D. Müll. -Doblies;匍匐伯克辛酸木●☆
138014 Drimiopsis bussei Dammer = Drimiopsis botryoides Baker ●☆
138015 Drimiopsis comptonii U. Müll. -Doblies et D. Müll. -Doblies;康普顿拟辛酸木●☆
138016 Drimiopsis crenata Van der Merwe = Drimiopsis burkei Baker ●☆
138017 Drimiopsis engleri K. Krause = Ledebouria undulata (Jacq.) Jessop ■☆
138018 Drimiopsis erlangeri Dammer = Drimiopsis botryoides Baker ●☆
138019 Drimiopsis fischeri (Engl.) Stedje;菲舍尔拟辛酸木●☆
138020 Drimiopsis holstii Engl. = Drimiopsis botryoides Baker ●☆
138021 Drimiopsis kirkii Baker;柯克拟辛酸木●☆
138022 Drimiopsis kirkii Baker = Drimiopsis botryoides Baker ●☆
138023 Drimiopsis lachenalioides (Baker) Jessop = Resnova lachenalioides (Baker) Van der Merwe ■☆
138024 Drimiopsis maculata Lindl. et Paxton;斑叶拟辛酸木●☆
138025 Drimiopsis maxima Baker = Resnova humifusa (Baker) U. Müll. -Doblies et D. Müll. -Doblies ■☆
138026 Drimiopsis minor Baker = Drimiopsis maculata Lindl. et Paxton ■☆
138027 Drimiopsis perfoliata Baker;穿叶拟辛酸木●☆
138028 Drimiopsis purpurea Van der Merwe = Drimiopsis atropurpurea N. E. Br. ●☆
138029 Drimiopsis pusilla U. Müll. -Doblies et D. Müll. -Doblies;微小拟辛酸木●☆
138030 Drimiopsis reilleyana U. Müll. -Doblies et D. Müll. -Doblies;赖利拟辛酸木●☆
138031 Drimiopsis rosea A. Chev.;粉红拟辛酸木●☆
138032 Drimiopsis saundersiae Baker = Resnova humifusa (Baker) U. Müll. -Doblies et D. Müll. -Doblies ■☆
138033 Drimiopsis seretii De Wild.;赛雷拟辛酸木●☆
138034 Drimiopsis stuhlmannii Baker = Drimiopsis botryoides Baker ●☆
138035 Drimiopsis volkensii (Engl.) Baker = Drimiopsis botryoides Baker ●☆
138036 Drimiopsis woodii Baker;伍得拟辛酸木●☆
138037 Drimophyllum Nutt. (1842);辛叶樟属●☆
138038 Drimophyllum Nutt. = Umbellularia (Nees) Nutt. (保留属名) ●☆
138039 Drimophyllum pauciflorum Nutt. = Umbellularia californica (Hook. et Arn.) Nutt. ●☆
138040 Drimopogon Raf. = Spiraea L. ●
138041 Drimya Lem. = Drimia Jacq. ex Willd. ■☆
138042 Drimyaceae Tiegh. = Winteraceae R. Br. ex Lindl. (保留科名) ●
138043 Drimycarpus Hook. f. (1862);辛果漆属;Drimycarpus ●
138044 Drimycarpus anacardiifolius C. Y. Wu et T. L. Ming;大果辛果漆; Big-fruited Drimycarpus, Cashew-leaved Drimycarpus, Drimycarpus, Large Drimycarpus ●◇
138045 Drimycarpus racemosus (Roxb.) Hook. f. ex Marchand;辛果漆;Racemose Drimycarpus ●
138046 Drimyidaceae Baill. = Hyacinthaceae Batsch ex Borkh. ■
138047 Drimyphyllum Burch. ex DC. (1936);辛叶菊属●☆
138048 Drimyphyllum Burch. ex DC. = Petrobium R. Br. (保留属名) ●☆
138049 Drimyphyllum helenianum Burch. ex DC.;辛叶菊●☆
138050 Drimyphyllum helenianum Burch. ex DC. = Petrobium arboreum R. Br. ●☆
138051 Drimys J. R. Forst. et G. Forst. (1775)(保留属名);辛酸木属(德米斯属,林仙属,南洋木莲属,辛辣木属,辛酸八角属);Drimys, Mountain Pepper ●☆
138052 Drimys aromatica (DC.) F. Muell.;芳香辛酸木; Mountain Pepper ●☆
138053 Drimys axillaris J. R. Forst. et G. Forst. = Pseudowintera axillaris (J. R. Forst. et G. Forst.) Dandy ●☆
138054 Drimys colorata Raoul = Pseudowintera colorata (Raoul) Dandy ●☆
138055 Drimys glauca ? = Drimys winteri J. R. Forst. et G. Forst. ●☆
138056 Drimys lanceolata Miers;狭叶辛酸木(澳洲林仙);Mountain Pepper ●☆
138057 Drimys latifolia ? = Drimys winteri J. R. Forst. et G. Forst. ●☆
138058 Drimys winteri J. R. Forst. et G. Forst.;辛酸木(林仙,辛辣木,辛酸八角,智利桂皮);Winter's Bark, Winter's Bark Drimys, Winter's Cinnamon ●☆
138059 Drimyspermum Reinw. = Phaleria Jack ●☆
138060 Dripax Noronha ex Thouars = Rinorea Aubl. (保留属名) ●
138061 Drobowskia Brongn. = Dobrowskya C. Presl ■☆

138062 Drobowskia Brongn. = Monopsis Salisb. ■☆
138063 Drobrowskia B. D. Jacks. = Drobowskia Brongn. ■☆
138064 Drobrowskia Brongn. = Drobowskia Brongn. ■☆
138065 Droceloncia J. Léonard(1959);德罗大戟属■☆
138066 Droceloncia rigidifolia (Baill.) J. Léonard;德罗大戟■☆
138067 Drogouetia Steud. = Droguetia Gaudich. ■
138068 Droguetia Gaudich. (1830);单蕊麻属;Droguetia ■
138069 Droguetia ambigua Wedd.;可疑单蕊麻■☆
138070 Droguetia burchellii N. E. Br. = Droguetia iners (Forssk.) Schweinf. subsp. burchellii (N. E. Br.) Friis et Wilmot-Dear ■☆
138071 Droguetia debilis Rendle;弱小单蕊麻■☆
138072 Droguetia diffusa Wedd. = Droguetia iners (Forssk.) Schweinf. ■☆
138073 Droguetia diffusa Wedd. = Droguetia iners (Forssk.) Schweinf. subsp. urticoides (Wight) Friis et Wilmot-Dear ■
138074 Droguetia iners (Forssk.) Schweinf.;迟缓单蕊麻■☆
138075 Droguetia iners (Forssk.) Schweinf. subsp. burchellii (N. E. Br.) Friis et Wilmot-Dear;伯切尔迟缓单蕊麻■☆
138076 Droguetia iners (Forssk.) Schweinf. subsp. pedunculata Friis;梗花迟缓单蕊麻■☆
138077 Droguetia iners (Forssk.) Schweinf. subsp. urticoides (Wight) Friis et Wilmot-dear;单蕊麻;Droguetia,Fewflower Droguetia ■
138078 Droguetia pauciflora (Steud.) Wedd. = Droguetia iners (Forssk.) Schweinf. ■☆
138079 Droguetia thunbergii N. E. Br. = Droguetia iners (Forssk.) Schweinf. ■☆
138080 Droguetia umbricola Engl. = Didymodoxa caffra (Thunb.) Friis et Wilmot-Dear ■☆
138081 Droguetia urticifolia Wedd. = Droguetia ambigua Wedd. ■☆
138082 Droguetia urticoides (Wight) Wedd. = Droguetia iners (Forssk.) Schweinf. subsp. urticoides (Wight) Friis et Wilmot-Dear ■
138083 Droguetia woodii N. E. Br. = Droguetia iners (Forssk.) Schweinf. ■☆
138084 Dromophylla Lindl. = Cayaponia Silva Manso(保留属名)■☆
138085 Dromophylla Lindl. = Dermophylla Silva Manso ■☆
138086 Droogmansia De Wild. (1902);德罗豆属■☆
138087 Droogmansia angolensis Torre;安哥拉德罗豆■☆
138088 Droogmansia chevalieri (Harms) Hutch. et Dalziel;舍瓦利耶德罗豆■☆
138089 Droogmansia elongata B. G. Schub.;伸长德罗豆■☆
138090 Droogmansia friesii Schindl. = Droogmansia pteropus (Baker) De Wild. var. whytei (Schindl.) Verdc. ■☆
138091 Droogmansia giorgii De Wild. = Droogmansia pteropus (Baker) De Wild. var. giorgii (De Wild.) Verdc. ■☆
138092 Droogmansia gossweileri Torre;戈斯德罗豆■☆
138093 Droogmansia hockii De Wild. = Droogmansia platypus (Baker) Schindl. var. hockii (De Wild.) B. G. Schub. ■☆
138094 Droogmansia homblei De Wild. = Droogmansia reducta De Wild. ■☆
138095 Droogmansia huillensis (Welw. ex Hiern) De Wild. = Droogmansia megalantha (Taub.) De Wild. var. pilosa ? ■☆
138096 Droogmansia lancifolia Schindl.;剑叶德罗豆■☆
138097 Droogmansia ledermannii Schindl.;莱德德罗豆■☆
138098 Droogmansia longestipitata De Wild.;长柄德罗豆■☆
138099 Droogmansia longipes R. E. Fr. = Droogmansia pteropus (Baker) De Wild. var. platypus (Baker) Verdc. ■☆
138100 Droogmansia longirhachis B. G. Schub.;长轴德罗豆■☆
138101 Droogmansia megalantha (Taub.) De Wild.;大花德罗豆■☆
138102 Droogmansia megalantha (Taub.) De Wild. var. longipedicellata De Wild.;长梗大花德罗豆■☆
138103 Droogmansia megalantha (Taub.) De Wild. var. pilosa ?;疏毛大花德罗豆■☆
138104 Droogmansia megalantha De Wild. = Droogmansia megalantha (Taub.) De Wild. ■☆
138105 Droogmansia mildbraedii Schindl.;米尔德德罗豆■☆
138106 Droogmansia montana Jacq. -Fél.;山地德罗豆■☆
138107 Droogmansia platypus (Baker) Schindl. = Droogmansia pteropus (Baker) De Wild. var. platypus (Baker) Verdc. ■☆
138108 Droogmansia platypus (Baker) Schindl. var. hockii (De Wild.) B. G. Schub.;霍克德罗豆■☆
138109 Droogmansia pteropus (Baker) De Wild.;翅足德罗豆■☆
138110 Droogmansia pteropus (Baker) De Wild. f. velutina B. G. Schub. = Droogmansia pteropus (Baker) De Wild. ■☆
138111 Droogmansia pteropus (Baker) De Wild. var. angustipetiolata Verdc.;狭瓣德罗豆■☆
138112 Droogmansia pteropus (Baker) De Wild. var. axillaris Verdc.;腋花德罗豆■☆
138113 Droogmansia pteropus (Baker) De Wild. var. giorgii (De Wild.) Verdc.;乔治德罗豆■☆
138114 Droogmansia pteropus (Baker) De Wild. var. platypus (Baker) Verdc.;宽翅足德罗豆■☆
138115 Droogmansia pteropus (Baker) De Wild. var. quarrei (De Wild.) Verdc.;卡雷德罗豆■☆
138116 Droogmansia pteropus (Baker) De Wild. var. whytei (Schindl.) Verdc.;怀特德罗豆■☆
138117 Droogmansia quarrei De Wild. = Droogmansia pteropus (Baker) De Wild. var. quarrei (De Wild.) Verdc. ■☆
138118 Droogmansia reducta De Wild.;退缩德罗豆■☆
138119 Droogmansia stuhlmannii (Taub.) De Wild. = Droogmansia pteropus (Baker) De Wild. ■☆
138120 Droogmansia tenuis B. G. Schub.;细德罗豆■☆
138121 Droogmansia tenuis B. G. Schub. var. laxa ?;疏细德罗豆■☆
138122 Droogmansia tisserantii Sillans;蒂斯朗特德罗豆■☆
138123 Droogmansia vanderystii De Wild.;范德德罗豆■☆
138124 Droogmansia vanmeelii B. G. Schub.;范米尔德罗豆■☆
138125 Droogmansia velutina B. G. Schub.;短绒毛德罗豆■☆
138126 Droogmansia whytei Schindl. = Droogmansia pteropus (Baker) De Wild. var. whytei (Schindl.) Verdc. ■☆
138127 Drosace A. Nelson = Androsace L. ■
138128 Drosanthe Spach = Hypericum L. ■●
138129 Drosanthe scabra (L.) Spach = Hypericum scabrum L. ●■
138130 Drosanthemopsis Rauschert = Drosanthemum Schwantes ●☆
138131 Drosanthemopsis Rauschert(1982);神刀玉属●☆
138132 Drosanthemopsis salaria (L. Bolus) Rauschert;神刀玉■☆
138133 Drosanthemopsis salaria (L. Bolus) Rauschert = Jacobsenia vaginata (L. Bolus) Ihlenf. ●☆
138134 Drosanthemopsis vaginata (L. Bolus) Rauschert = Jacobsenia vaginata (L. Bolus) Ihlenf. ●☆
138135 Drosanthemum Schwantes(1927);泡叶番杏属(泡叶菊属);Pale Dew-plant ●☆
138136 Drosanthemum acuminatum L. Bolus;渐尖泡叶番杏●☆
138137 Drosanthemum acutifolium L. Bolus;尖叶德罗豆■☆
138138 Drosanthemum albens L. Bolus;微白泡叶番杏●☆
138139 Drosanthemum albiflorum (L. Bolus) Schwantes;白花泡叶菊●☆

138140 Drosanthemum ambiguum L. Bolus;可疑泡叶番杏(可疑泡叶菊)●☆
138141 Drosanthemum anomalum L. Bolus;异常泡叶番杏●☆
138142 Drosanthemum archeri L. Bolus;阿谢尔泡叶番杏●☆
138143 Drosanthemum asperulum (Salm-Dyck) Schwantes;粗糙泡叶番杏●☆
138144 Drosanthemum attenuatum (Haw.) Schwantes;渐狭泡叶番杏●☆
138145 Drosanthemum aureopurpureum L. Bolus;黄紫泡叶番杏●☆
138146 Drosanthemum autumnale L. Bolus;秋泡叶番杏●☆
138147 Drosanthemum barkerae L. Bolus;巴尔凯拉泡叶番杏●☆
138148 Drosanthemum barwickii L. Bolus = Drosanthemum subcompressum (Haw.) Schwantes ●☆
138149 Drosanthemum bicolor L. Bolus;花狂想●☆
138150 Drosanthemum bredai L. Bolus = Drosanthemum asperulum (Salm-Dyck) Schwantes ●☆
138151 Drosanthemum breve L. Bolus;短泡叶番杏●☆
138152 Drosanthemum brevifolium (Aiton) Schwantes;短泡叶菊●☆
138153 Drosanthemum calycinum (Haw.) Schwantes;萼状泡叶番杏●☆
138154 Drosanthemum capillare (Thunb.) Schwantes;细毛泡叶番杏●☆
138155 Drosanthemum chrysum L. Bolus;金黄泡叶番杏●☆
138156 Drosanthemum collinum (Sond.) Schwantes;山丘泡叶番杏●☆
138157 Drosanthemum comptonii L. Bolus;康普顿泡叶番杏●☆
138158 Drosanthemum concavum L. Bolus;凹泡叶番杏●☆
138159 Drosanthemum crassum L. Bolus;粗泡叶番杏●☆
138160 Drosanthemum croceum L. Bolus;镉黄泡叶番杏●☆
138161 Drosanthemum curtophyllum L. Bolus;弯叶泡叶番杏●☆
138162 Drosanthemum deciduum H. E. K. Hartmann et Bruckman;脱落泡叶番杏●☆
138163 Drosanthemum delicatulum (L. Bolus) Schwantes;姣美泡叶番杏●☆
138164 Drosanthemum diversifolium L. Bolus;异叶泡叶番杏●☆
138165 Drosanthemum eburneum L. Bolus;象牙白泡叶番杏●☆
138166 Drosanthemum edwardsiae L. Bolus;爱德华兹泡叶番杏●☆
138167 Drosanthemum erigeriflorum (Jacq.) Stearn;毛花泡叶番杏●☆
138168 Drosanthemum filiforme L. Bolus;丝状泡叶番杏●☆
138169 Drosanthemum flammeum L. Bolus;火红泡叶番杏●☆
138170 Drosanthemum flavum (Haw.) Schwantes;黄泡叶番杏●☆
138171 Drosanthemum floribundum (Haw.) Schwantes;美光;Pale Dew-plant,Rosea Ice Plant,Rosea Iceplant,Showy Dewflower ●☆
138172 Drosanthemum fourcadei (L. Bolus) Schwantes;富尔卡德泡叶番杏●☆
138173 Drosanthemum framesii L. Bolus;弗雷斯泡叶番杏●☆
138174 Drosanthemum fulleri L. Bolus;富勒泡叶番杏●☆
138175 Drosanthemum giffenii (L. Bolus) Schwantes;吉芬泡叶番杏●☆
138176 Drosanthemum giffenii (L. Bolus) Schwantes var. intertextum ? = Drosanthemum giffenii (L. Bolus) Schwantes ●☆
138177 Drosanthemum glabrescens L. Bolus;渐光泡叶番杏●☆
138178 Drosanthemum globosum L. Bolus;球形泡叶番杏●☆
138179 Drosanthemum godmaniae L. Bolus;戈德曼泡叶番杏●☆
138180 Drosanthemum gracillimum L. Bolus;细长泡叶番杏●☆
138181 Drosanthemum hallii L. Bolus;霍尔泡叶番杏●☆
138182 Drosanthemum hirtellum (Haw.) Schwantes;多毛泡叶番杏●☆
138183 Drosanthemum hispidum (L.) Schwantes;硬毛泡叶番杏(花弥生,毛泡叶菊);Hairy Dewflower,Ice Plant,Rosea Ice Plant ●☆
138184 Drosanthemum hispidum (L.) Schwantes var. platypetalum (Haw.) Schwantes = Drosanthemum hispidum (L.) Schwantes ●☆
138185 Drosanthemum hispidum Schwantes = Drosanthemum hispidum (L.) Schwantes ●☆
138186 Drosanthemum hispifolium (Haw.) Schwantes;粗糙泡叶番杏●☆
138187 Drosanthemum inornatum (L. Bolus) L. Bolus;无饰泡叶番杏●☆
138188 Drosanthemum intermedium (L. Bolus) L. Bolus;间型泡叶番杏●☆
138189 Drosanthemum jamesii L. Bolus;詹姆斯泡叶番杏●☆
138190 Drosanthemum karrooense L. Bolus;卡卢泡叶番杏●☆
138191 Drosanthemum latipetalum L. Bolus;宽瓣泡叶番杏●☆
138192 Drosanthemum lavisii L. Bolus;拉维斯泡叶番杏●☆
138193 Drosanthemum laxum L. Bolus;疏松泡叶番杏●☆
138194 Drosanthemum leipoldtii L. Bolus;莱波尔德泡叶番杏●☆
138195 Drosanthemum leptum L. Bolus;细泡叶番杏●☆
138196 Drosanthemum lignosum L. Bolus;木质泡叶番杏●☆
138197 Drosanthemum littlewoodii L. Bolus = Drosanthemum albens L. Bolus ●☆
138198 Drosanthemum longipes (L. Bolus) H. E. K. Hartmann;长梗泡叶番杏●☆
138199 Drosanthemum luederitzii (Engl.) Schwantes;吕德里茨泡叶番杏●☆
138200 Drosanthemum macrocalyx L. Bolus;大萼泡叶番杏●☆
138201 Drosanthemum maculatum (Haw.) Schwantes;斑点泡叶番杏●☆
138202 Drosanthemum marinum L. Bolus;海生泡叶番杏●☆
138203 Drosanthemum martinii L. Bolus ex Jacobsen = Dorotheanthus bellidiformis (Burm. f.) N. E. Br. ■☆
138204 Drosanthemum mathewsii L. Bolus;马修斯泡叶番杏●☆
138205 Drosanthemum micans (L.) Schwantes;花嬉游●☆
138206 Drosanthemum micans Schwantes = Drosanthemum micans (L.) Schwantes ●☆
138207 Drosanthemum montaguense L. Bolus = Drosanthemum praecultum (N. E. Br.) Schwantes ●☆
138208 Drosanthemum muiri L. Bolus;缪里泡叶番杏●☆
138209 Drosanthemum nitidum (Haw.) Schwantes = Phyllobolus nitidus (Haw.) Gerbaulet ●☆
138210 Drosanthemum nordenstamii L. Bolus;努登斯坦泡叶番杏●☆
138211 Drosanthemum oculatum L. Bolus;小眼泡叶番杏●☆
138212 Drosanthemum opacum L. Bolus;暗色泡叶番杏●☆
138213 Drosanthemum otzenianum (Dinter) Friedrich = Lampranthus otzenianus (Dinter) Friedrich ■☆
138214 Drosanthemum pallens (Haw.) Schwantes;苍白泡叶番杏●
138215 Drosanthemum papillatum L. Bolus;乳突泡叶番杏●☆
138216 Drosanthemum parvifolium (Haw.) Schwantes;小花泡叶番杏●☆
138217 Drosanthemum pauper (Dinter) Dinter et Schwantes;贫乏泡叶番杏●☆
138218 Drosanthemum paxianum (Schltr. et Diels) Schwantes = Drosanthemum luederitzii (Engl.) Schwantes ●☆
138219 Drosanthemum praecultum (N. E. Br.) Schwantes;澳非泡叶番杏●☆
138220 Drosanthemum pulchellum L. Bolus;花宝生●☆
138221 Drosanthemum pulchrum L. Bolus;美丽泡叶番杏●☆
138222 Drosanthemum pulverulentum (Haw.) Schwantes;粉粒泡叶番杏●☆
138223 Drosanthemum quadratum Klak;四方形泡叶番杏●☆
138224 Drosanthemum ramosissimum (Schltr.) L. Bolus;多枝泡叶番杏●☆
138225 Drosanthemum robustum L. Bolus = Delosperma crassum L. Bolus ■☆

138226 Drosanthemum roridum L. Bolus = Drosanthemum subcompressum (Haw.) Schwantes ●☆

138227 Drosanthemum roseatum (N. E. Br.) L. Bolus = Drosanthemum pulverulentum (Haw.) Schwantes ●☆

138228 Drosanthemum salicola L. Bolus;柳泡叶番杏●☆

138229 Drosanthemum schoenlandianum (Schltr.) L. Bolus;玉辉●☆

138230 Drosanthemum schoenlandianum L. Bolus = Drosanthemum schoenlandianum (Schltr.) L. Bolus ●☆

138231 Drosanthemum semiglobosum L. Bolus;半球泡叶番杏●☆

138232 Drosanthemum sessile (Thunb.) Schwantes = Ruschia sessilis (Thunb.) H. E. K. Hartmann ●☆

138233 Drosanthemum speciosum (Haw.) Schwantes;艳丽泡叶番杏(花猩猩,美丽泡叶菊);Dew Flower Ice Plant, Royal Dewflower ●☆

138234 Drosanthemum speciosum Schwantes = Drosanthemum speciosum (Haw.) Schwantes ●☆

138235 Drosanthemum splendens L. Bolus;光亮泡叶番杏●☆

138236 Drosanthemum stokoei L. Bolus;斯托克泡叶番杏●☆

138237 Drosanthemum striatum (Haw.) Schwantes;花浓染●☆

138238 Drosanthemum striatum (Haw.) Schwantes var. hispifolium (Haw.) G. D. Rowley = Drosanthemum hispifolium (Haw.) Schwantes ●☆

138239 Drosanthemum striatum (Haw.) Schwantes var. pallens (Haw.) G. D. Rowley = Drosanthemum pallens (Haw.) Schwantes ●

138240 Drosanthemum striatum Schwantes = Drosanthemum striatum (Haw.) Schwantes ●☆

138241 Drosanthemum strictifolium L. Bolus;刚直泡叶番杏(刚直泡叶菊)●☆

138242 Drosanthemum subalbum L. Bolus = Drosanthemum diversifolium L. Bolus ●☆

138243 Drosanthemum subcompressum (Haw.) Schwantes;扁泡叶番杏●☆

138244 Drosanthemum subglobosum (Haw.) Schwantes = Drosanthemum capillare (Thunb.) Schwantes ●☆

138245 Drosanthemum subplanum L. Bolus;平泡叶番杏●☆

138246 Drosanthemum subspinosum (Kuntze) H. E. K. Hartmann;具刺泡叶番杏●☆

138247 Drosanthemum tardum L. Bolus;迟缓泡叶番杏●☆

138248 Drosanthemum thudichumii L. Bolus;图迪休姆泡叶番杏●☆

138249 Drosanthemum thudichumii L. Bolus f. aurantiaca ? = Drosanthemum thudichumii L. Bolus ●☆

138250 Drosanthemum thudichumii L. Bolus f. aurea ? = Drosanthemum thudichumii L. Bolus ●☆

138251 Drosanthemum thudichumii L. Bolus f. gracilius ? = Drosanthemum thudichumii L. Bolus ●☆

138252 Drosanthemum torquatum (Haw.) Schwantes = Drosanthemum floribundum (Haw.) Schwantes ●☆

138253 Drosanthemum uniflorum (L. Bolus) Friedrich ex H. Jacobsen = Lampranthus uniflorus (L. Bolus) L. Bolus ■☆

138254 Drosanthemum vaginatum L. Bolus = Jacobsenia vaginata (L. Bolus) Ihlenf. ●☆

138255 Drosanthemum vandermerwei L. Bolus;范德泡叶番杏●☆

138256 Drosanthemum vespertinum L. Bolus;夕泡叶番杏●☆

138257 Drosanthemum vespertinum L. Bolus var. suffusum ? = Drosanthemum vespertinum L. Bolus ●☆

138258 Drosanthemum wittebergense L. Bolus;维特伯格泡叶番杏●☆

138259 Drosanthemum zygophylloides (L. Bolus) L. Bolus;对称泡叶番杏●☆

138260 Drosanthus R. Br. ex Planch. = Byblis Salisb. ●☆

138261 Drosera L. (1753);茅膏菜属;Daily-dew, Dew Plant, Dew-plant, Sun Dew, Sundew ■

138262 Drosera acaulis L. f.;无茎茅膏菜■☆

138263 Drosera acaulis Thunb. = Drosera acaulis L. f. ■☆

138264 Drosera affinis Welw. ex Oliv.;近缘茅膏菜■☆

138265 Drosera alba E. Phillips;白茅膏菜■☆

138266 Drosera aliciae Raym. -Hamet;澳大利亚茅膏菜;Australian Sundew ■☆

138267 Drosera anglica Huds.;英国茅膏菜(线叶茅膏菜,英国裸盆花);English Sundew, Great Sundew, Great-leaved Sundew ■☆

138268 Drosera atrostyla Debbert;暗柱茅膏菜■☆

138269 Drosera bequaertii Taton;贝卡尔茅膏菜■☆

138270 Drosera binata Labill.;二出茅膏菜;Forked-leaved Sundew ■☆

138271 Drosera brevifolia Pursh;短叶茅膏菜;Dwarf Sundew ■☆

138272 Drosera burkeana Planch.;伯克茅膏菜■☆

138273 Drosera burmanii Vahl;锦地罗(钉地金钱,金雀梅,金线草,金线吊芙蓉,落地金钱,丝线串铜钱,文钱红,乌蝇草,五柱毛毡苔,仙人涎,夜落金钱,一朵芙蓉,一朵芙蓉花,怎地罗);Burmann Sundew ■

138274 Drosera burmannii DC. = Drosera spathulata Labill. ●■

138275 Drosera capensis L.;南非茅膏菜;Cape Sundew ■☆

138276 Drosera capillaris Poir.;毛状茅膏菜;Pink Sundew ■☆

138277 Drosera cistiflora L.;囊花茅膏菜■☆

138278 Drosera compacta Exell et J. R. Laundon = Drosera bequaertii Taton ■☆

138279 Drosera congolana Taton = Drosera madagascariensis DC. ■☆

138280 Drosera cuneifolia L. f.;楔叶茅膏菜■☆

138281 Drosera cuneifolia Thunb. = Drosera cuneifolia L. f. ■☆

138282 Drosera curvipes Planch. = Drosera madagascariensis DC. ■☆

138283 Drosera curviscapa Salter = Drosera aliciae Raym. -Hamet ■☆

138284 Drosera curviscapa Salter var. esterhuyseniae ? = Drosera aliciae Raym. -Hamet ■☆

138285 Drosera dichotoma Banks et Sol. = Drosera binata Labill. ■☆

138286 Drosera dichotoma Banks et Sol. ex Sm. = Drosera binata Labill. ■☆

138287 Drosera elongata Exell et J. R. Laundon;伸长茅膏菜■☆

138288 Drosera erythrorrhiza Lindl.;澳洲茅膏菜■☆

138289 Drosera filiformis Raf.;线形茅膏菜;Dew-thread, Thread-leaved Sundew ■☆

138290 Drosera filiformis Raf. var. tracyi (Macfarl.) Diels = Drosera tracyi Macfarl. ■☆

138291 Drosera flexicaulis Welw. ex Oliv. = Drosera affinis Welw. ex Oliv. ■☆

138292 Drosera glabripes (Harv. ex Planch.) Stein;光梗茅膏菜■☆

138293 Drosera grandiflora Bartl. = Drosera pauciflora Banks ex DC. ■☆

138294 Drosera helianthemum Planch. = Drosera cistiflora L. ■☆

138295 Drosera indica L.;长叶茅膏菜(捕蝇草,满露草);India Sundew, Indian Sundew ■

138296 Drosera indica L. f. albiflora Makino;白花长叶茅膏菜;Whiteflower India Sundew ■

138297 Drosera indica L. var. albiflora (Makino) Makino = Drosera indica L. f. albiflora Makino ■

138298 Drosera insolita Taton;异常茅膏菜■☆

138299 Drosera intermedia Hayne;间型茅膏菜(匙叶裸盆花,匙叶茅膏菜);Intermediate Sundew, Narrow-leaved Sundew, Oblong-leaved Sundew, Spatulate-leaved Sundew, Spoon-leaf Sundew ■☆

138300 Drosera intermedia Hayne = Drosera longifolia L. ■☆

138301 Drosera intermedia Hayne f. subcaulescens Melvill = Drosera intermedia Hayne ■☆

138302 Drosera katangensis Taton;加丹加茅膏菜■☆

138303 Drosera leucantha Shinners;白花茅膏菜;Spatulate-leaved Sundew ■☆

138304 Drosera linearis Goldie;线叶茅膏菜(线形茅膏菜,线叶裸盆花);Linear-leaved Sundew,Slender-leaved Sundew ■☆

138305 Drosera liniflora Debbert;亚麻花茅膏菜■☆

138306 Drosera lobbiana Turcz. = Drosera peltata Sm. ex Willd. ■

138307 Drosera longifolia L.;中型茅膏菜(长叶茅膏菜,中间茅膏菜);Long-leaved Sundew ■☆

138308 Drosera longifolia L. = Drosera anglica Huds. ■☆

138309 Drosera longiscapa Debbert;长花茎茅膏菜■☆

138310 Drosera loureiri Hook. et Arn. = Drosera spathulata Labill. ●■

138311 Drosera loureiri Hook. et Arn. = Drosera spathulata Labill. var. loureirii (Hook. et Arn.) Y. Z. Ruan ■

138312 Drosera lunata Buch. -Ham. = Drosera peltata Sm. ex Willd. ■

138313 Drosera lunata Buch. -Ham. ex DC. = Drosera peltata Sm. ex Willd. ■

138314 Drosera lusitanica L. = Drosophyllum lusitanicum (L.) Link ●☆

138315 Drosera madagascariensis DC.;马岛茅膏菜■☆

138316 Drosera madagascariensis DC. var. curvipes (Planch.) Sond. = Drosera madagascariensis DC. ■☆

138317 Drosera madagascariensis DC. var. major Burtt Davy = Drosera madagascariensis DC. ■☆

138318 Drosera makinoi Masam. = Drosera indica L. ■

138319 Drosera natalensis Diels;纳塔尔茅膏菜■☆

138320 Drosera nipponica Masam. = Drosera peltata Sm. ex Willd. var. lunata (Buch. -Ham.) C. B. Clarke ■

138321 Drosera oblanceolata Y. Z. Ruan;长柱茅膏菜;Longstyle Sundew ■

138322 Drosera obovata Mert. et Koch;倒卵叶茅膏菜;Whiteflower India Sundew ■☆

138323 Drosera pauciflora Banks ex DC.;少花茅膏菜■☆

138324 Drosera pauciflora Banks ex DC. var. acaulis (Thunb.) Sond. = Drosera acaulis L. f. ■☆

138325 Drosera pauciflora Banks ex DC. var. minor Sond. = Drosera cistiflora L. ■☆

138326 Drosera pedata Pers.;鸟足茅膏菜■☆

138327 Drosera peltata Sm. = Drosera peltata Sm. ex Willd. ■

138328 Drosera peltata Sm. ex Willd.;茅膏菜(盾叶茅膏菜,龙牙草);Sundew ■

138329 Drosera peltata Sm. ex Willd. var. glabrata Y. Z. Ruan;光萼茅膏菜(捕虫草,捕蝇草,陈伤子,滴水不干,地下明珠,地下珍珠,盾叶茅膏菜,黄金丝,落地珍珠,茅膏菜,泥里珠,肉珠宝,山胡椒草,石龙芽草,铁扭子,土里珍珠,夏无踪,眼泪草,一粒金丹);Peltate Sundew,Sun Dew,Sundew ■

138330 Drosera peltata Sm. ex Willd. var. glabrata Y. Z. Ruan = Drosera peltata Sm. ex Willd. ■

138331 Drosera peltata Sm. ex Willd. var. lunata (Buch. -Ham. ex DC.) C. B. Clarke;新月茅膏菜(白花叶,捕虫草,捕蛇草,苍蝇草,苍蝇网,陈伤子,寸金黄,滴水不干,地胡椒,地下明珠,地下珍珠,地珍珠,胡椒草,黄金丝,落地珍珠,茅膏菜,茅蒿菜,内宝珠,泥里珠,柔鱼草,山地皮,山胡椒,山胡椒草,山君子,山砒霜,石龙牙草,食虫草,铁秤锤,铁钮子,土地子,夏无踪,野高粱,一滴金丹,一粒金丹,珍珠草);Lunate Sundew ■

138332 Drosera peltata Sm. ex Willd. var. lunata (Buch. -Ham. ex DC.) C. B. Clarke = Drosera peltata Sm. ex Willd. ■

138333 Drosera peltata Sm. ex Willd. var. lunata (Buch. -Ham.) C. B. Clarke = Drosera peltata Sm. ex Willd. var. lunata (Buch. -Ham. ex DC.) C. B. Clarke ■

138334 Drosera peltata Sm. ex Willd. var. lunata (Buch. -Ham.) C. B. Clarke = Drosera peltata Sm. ex Willd. ■

138335 Drosera peltata Sm. ex Willd. var. lunata C. B. Clarke = Drosera peltata Sm. ex Willd. ■

138336 Drosera peltata Sm. ex Willd. var. multisepala Y. Z. Ruan;盾叶茅膏菜(苍蝇王,打古子,地下明珠,筋骨草,茅膏菜,牛打架,球子参,山胡椒草,石龙芽草)■

138337 Drosera peltata Sm. ex Willd. var. multisepala Y. Z. Ruan = Drosera peltata Sm. ex Willd. ■

138338 Drosera peltata Sm. ex Willd. var. nipponica (Masam.) Ohwi;日本茅膏菜■☆

138339 Drosera pilosa Exell et J. R. Laundon;疏毛茅膏菜■☆

138340 Drosera pygmaea Lehm.;小茅膏菜■☆

138341 Drosera ramentacea Burch. ex DC.;芽鳞茅膏菜■☆

138342 Drosera ramentacea Burch. ex DC. var. burchelliana Sond. = Drosera ramentacea Burch. ex DC. ■☆

138343 Drosera ramentacea Burch. ex DC. var. curvipes (Planch.) Sond. = Drosera madagascariensis DC. ■☆

138344 Drosera ramentacea Burch. ex DC. var. glabripes Harv. ex Planch. = Drosera glabripes (Harv. ex Planch.) Stein ■☆

138345 Drosera ramentacea Burch. ex Harv. et Sond. = Drosera ramentacea Burch. ex DC. ■☆

138346 Drosera regia Stephens;高贵茅膏菜■☆

138347 Drosera roridula Thunb. = Roridula dentata L. ●☆

138348 Drosera rotundifolia L.;圆叶茅膏菜(毛毡苔,圆叶裸盆花);Catch-trap,Dew Plant,English Flytrap,Flycatcher,Flytrap,Fly-trap,Iles,London Pride,Lustwort,Moor-gloom,Moor-grass,Oilplant,Old Folks' Herb,Red Rot,Roundleaf Sundew,Round-leaved Sundew,Sticky-back,Sundew,Youthwort ■

138349 Drosera rotundifolia L. = Drosera burmannii DC. ●■

138350 Drosera rotundifolia L. var. furcata Y. Z. Ruan;叉梗茅膏菜;Furcate Roundleaf Sundew ■

138351 Drosera rotundifolia L. var. furcata Y. Z. Ruan = Drosera rotundifolia L. ■

138352 Drosera rubrifolia Debbert;红叶茅膏菜■☆

138353 Drosera rubripetala Debbert;红花茅膏菜■☆

138354 Drosera spathulata Labill.;匙叶茅膏菜(地红花,地毡草,金雀花,金雀梅,毛毡苔,天地花,小毛毡苔);Cape Sundew,Spathulate Sundew ●■

138355 Drosera spathulata Labill. f. chionantha K. Nakaj.;雪花匙叶茅膏菜■☆

138356 Drosera spathulata Labill. subsp. tokaiensis Komiya et C. Shibata = Drosera tokaiensis (Komiya et C. Shibata) T. Nakam. et K. Ueda ■☆

138357 Drosera spathulata Labill. var. loureirii (Hook. et Arn.) Y. Z. Ruan;宽苞茅膏菜;Loureir Spathulate Sundew ■

138358 Drosera spathulata Labill. var. loureiroi (Hook. et Arn.) Y. Z. Ruan = Drosera spathulata Labill. ●■

138359 Drosera speciosa C. Presl = Drosera cistiflora L. ■☆

138360 Drosera tenuifolia Willd. = Drosera filiformis Raf. ■☆

138361 Drosera tokaiensis (Komiya et C. Shibata) T. Nakam. et K. Ueda;东海村茅膏菜■☆

138362 Drosera tokaiensis (Komiya et C. Shibata) T. Nakam. et K.

Ueda subsp. hyugaensis Seno;日向茅膏菜■☆
138363 Drosera tracyi Macfarl.;特氏茅膏菜■☆
138364 Drosera trinervia Spreng.;三脉茅膏菜■☆
138365 Drosera umbellata Lour. = Androsace umbellata (Lour.) Merr. ■
138366 Drosera variegata Debbert;杂色茅膏菜■☆
138367 Drosera venusta Debbert;雅致茅膏菜■☆
138368 Drosera zeyheri T. M. Salter = Drosera cistiflora L. ■☆
138369 Droseraceae Salisb. (1808)(保留科名);茅膏菜科;Sundew Family ■
138370 Drosocarpium Fourr. = Hypericum L. ■●
138371 Drosodendron Roem. = Baeckea L. ●
138372 Drosophorus R. Br. ex Planch. = Byblis Salisb. ●☆
138373 Drosophyllaceae Chrtek, Slaviková et Studnicka = Droseraceae Salisb.(保留科名)■
138374 Drosophyllaceae Chrtek, Slaviková et Studnicka(1989);露叶苔科(黏虫草科);Drosophyllum Family ●☆
138375 Drosophyllum Link (1805);露叶苔属(露叶花属);Drosophyllum ●☆
138376 Drosophyllum lusitanicum (L.) Link;露叶苔草;Dewy Pine, Portuguese Sundew ●☆
138377 Drosophyllum lusitanicum Link = Drosophyllum lusitanicum (L.) Link ●☆
138378 Drossera Gled. = Drosera L. ■
138379 Drouguetia Endl. = Droguetia Gaudich. ■
138380 Drozia Cass. = Perezia Lag. ■☆
138381 Drudea Griseb. = Pycnophyllum J. Rémy ■☆
138382 Drudeophytum J. M. Coult. et Rose = Tauschia Schltdl.(保留属名)■☆
138383 Drummondia DC. = Mitella Tourn. ex L. ■
138384 Drummondia DC. = Mitellopsis Meisn. ■
138385 Drummondia DC. = Pectiantia Raf. ■
138386 Drummondita Harv. (1855);德拉蒙德芸香属●☆
138387 Drummondita Harv. = Philotheca Rudge ●☆
138388 Drummondita ericoides Harv.;德拉蒙德芸香●☆
138389 Drummondita fulva A. S. Markey et R. A. Meissn.;黄德拉蒙德芸香●☆
138390 Drummondita longifolia (Paul G. Wilson) Paul G. Wilson;长叶德拉蒙德芸香●☆
138391 Drummondita microphylla Paul G. Wilson;小叶德拉蒙德芸香●☆
138392 Drumondia Raf. = Drummondia DC. ■
138393 Drumondia Raf. = Mitella L. ■
138394 Drupaceae Gray = Rosaceae Juss.(保留科名)●■
138395 Druparia Silva Manso = Cayaponia Silva Manso(保留属名)■☆
138396 Drupatris Lour. = Symplocos Jacq. ●
138397 Drupatris cochinchinensis Lour. = Symplocos cochinchinensis (Lour.) S. Moore var. laurina (Retz.) Raizada ●
138398 Drupifera Raf. = Camellia L. ●
138399 Drupifera oleosa Raf. = Camellia drupifera Lour. ●
138400 Drupina L. = Curanga Juss. ■☆
138401 Drusa DC. (1807);结晶草属■☆
138402 Drusa glandulosa (Poir.) Bornm. = Bowlesia glandulosa (Poir.) Kuntze ■☆
138403 Drusa oppositifolia DC.;结晶草■☆
138404 Drusanthemum Schwantes.;晶花番杏属●☆
138405 Dryadaceae Gray = Rosaceae Juss.(保留科名)●■
138406 Dryadaea Kuntze = Dryadea Raf. ●■
138407 Dryadaea Kuntze = Dryas L. ●■
138408 Dryadanthe Endl. (1840);四蕊山莓草属■
138409 Dryadanthe Endl. = Sibbaldia L. ■
138410 Dryadanthe bungeana Ledeb. = Sibbaldia tetrandra Bunge ■
138411 Dryadanthe tetrandra (Bunge) Juz.;四蕊山莓草(高山山莓草);Fourstamens Sibbaldia, Tetrandrous Wildberry ■
138412 Dryadanthe tetrandra (Bunge) Juz. = Sibbaldia tetrandra Bunge ■
138413 Dryadea Raf. = Dryas L. ●■
138414 Dryadella Luer(1978);树蛹兰属■☆
138415 Dryadella albicans (Luer) Luer;白树蛹兰■☆
138416 Dryadella aurea Luer et Hirtz;黄树蛹兰■☆
138417 Dryadodaphne S. Moore(1923);林桂属●☆
138418 Dryadodaphne celastroides S. Moore;林桂●☆
138419 Dryadorchis Schltr. (1913);德律阿斯兰属■☆
138420 Dryadorchis barbellata Schltr.;德律阿斯兰■☆
138421 Dryadorchis dasystele Schuit. et de Vogel;毛柱德律阿斯兰■☆
138422 Dryadorchis minor Schltr.;小德律阿斯兰■☆
138423 Dryandra R. Br. (1810)(保留属名);丝头花属(蓟序木属);Dryander ●☆
138424 Dryandra Thunb.(废弃属名) = Dryandra R. Br.(保留属名)●☆
138425 Dryandra Thunb.(废弃属名) = Vernicia Lour. ●
138426 Dryandra arborea C. A. Gardner;乔木丝头花●☆
138427 Dryandra cordata Thunb. = Aleurites cordata (Thunb.) R. Br. ex Steud. ●☆
138428 Dryandra falcata R. Br.;镰形丝头花●☆
138429 Dryandra ferruginea Meisn.;锈色丝头花●☆
138430 Dryandra floribunda R. Br.;繁花丝头花;Manyflower Dryander ●☆
138431 Dryandra formosa R. Br.;美丽丝头花(锯叶蓟序木);Pretty Dryander, Showy Dryandra ●☆
138432 Dryandra longifolia R. Br.;长叶丝头花;Longleaf Dryander ●☆
138433 Dryandra montana A. S. George;山地丝头花●☆
138434 Dryandra nivea (Labill.) R. Br.;雪白丝头花●☆
138435 Dryandra nobilis Lindl.;名贵丝头花;Golden Dryandra ●☆
138436 Dryandra obtusa R. Br.;钝丝头花●☆
138437 Dryandra oleifera Thunb. = Aleurites cordata (Thunb.) R. Br. ex Steud. ●☆
138438 Dryandra pallida A. S. George;苍白丝头花●☆
138439 Dryandra polycephala Benth.;多花丝头花;Many-headed Dryandra ●☆
138440 Dryandra praemorsa Meisn.;冬青叶丝头花;Sea-urchin Dryandra, Urchin Dryandra ●☆
138441 Dryandra pulchella Meisn.;艳丽丝头花●☆
138442 Dryandra quercifolia Meisn.;栎叶丝头花;Oak-leaf Dryandra ●☆
138443 Dryandra sessilis Domin;鹦鹉木;Parrot Bush ●☆
138444 Dryandra speciosa Meisn.;窄叶丝头花;Shaggy Dryandra ●☆
138445 Dryas L. (1753);仙女木属(多瓣木属);Dryad, Dryas, Mountain Avens, Mtn. Avens ●■
138446 Dryas ajanensis Juz. = Dryas octopetala L. var. asiatica (Nakai) Nakai ●
138447 Dryas caucasica Juz.;高加索仙女木●☆
138448 Dryas chamissonis Spreng.;哈米逊仙女木●☆
138449 Dryas crenulata Juz.;细圆齿仙女木●☆
138450 Dryas drummondii Rich.;德拉蒙德仙女木(卵萼仙女木);Drummond's Mountain-avens ●■☆
138451 Dryas grandis Juz.;大仙女木●☆
138452 Dryas henricae Juz.;昂里克仙女木●☆
138453 Dryas integrifolia M. Vahl;全缘叶仙女木;Arctic Avens ●☆

138454 Dryas lanata Stein ex Correvon;软毛仙女木●☆
138455 Dryas nervosa Juz. = Dryas octopetala L. var. asiatica (Nakai) Nakai ●
138456 Dryas octopetala L. = Dryas oxyodonta Juz. ●
138457 Dryas octopetala L. f. asiatica Nakai = Dryas octopetala L. var. asiatica (Nakai) Nakai ●
138458 Dryas octopetala L. subsp. nervosa (Juz.) Hultén = Dryas octopetala L. var. asiatica (Nakai) Nakai ●
138459 Dryas octopetala L. subsp. tschonoskii (Juz.) Hultén = Dryas octopetala L. var. asiatica (Nakai) Nakai ●
138460 Dryas octopetala L. var. asiatica (Nakai) Nakai;东亚仙女木(多瓣木,宽叶仙女木,乔诺仙女木,须川氏仙女木);Asia Dryas, Asiatic Eight-petals Dryas ●
138461 Dryas oxyodonta Juz.;仙女木(多瓣木);Eightpetal Mountain-avens, Eight-petals Dryas, Mount Washington Dryad, Mountain Avens, White Mountain Avens, Wild Betony ●
138462 Dryas punctata Juz.;斑仙女木●☆
138463 Dryas suendermanni Kellerer ex Sund.;修德曼仙女木●☆
138464 Dryas tschonoskii Juz. = Dryas octopetala L. subsp. tschonoskii (Juz.) Hultén ●
138465 Dryas tschonoskii Juz. = Dryas octopetala L. var. asiatica (Nakai) Nakai ●
138466 Dryas viscosa Juz.;黏仙女木●☆
138467 Drymaria Schult. = Drymaria Willd. ex Schult. ■
138468 Drymaria Willd. = Drymaria Willd. ex Schult. ■
138469 Drymaria Willd. ex Roem. et Schult. = Drymaria Willd. ex Schult. ■
138470 Drymaria Willd. ex Schult. (1819);荷莲豆草属(荷莲豆属);Drymaria, Drymary, Seccomaria ■
138471 Drymaria arenarioides Humb. et Bonpl. ex Schult.;沙地荷莲豆草■☆
138472 Drymaria chihuahuensis Briq. = Drymaria laxiflora Benth. ■☆
138473 Drymaria cordata (L.) Willd. ex Roem. et Schult.;心叶荷莲豆草■☆
138474 Drymaria cordata (L.) Willd. ex Roem. et Schult. = Drymaria diandra Blume ■
138475 Drymaria cordata (L.) Willd. ex Roem. et Schult. subsp. diandra (Blume) J. A. Duke = Drymaria diandra Blume ■
138476 Drymaria cordata (L.) Willd. ex Roem. et Schult. subsp. diandra (Blume) J. A. Duke ex Hatus. = Drymaria diandra Blume ■
138477 Drymaria cordata (L.) Willd. ex Roem. et Schult. subsp. diandra (Blume) J. A. Duke = Drymaria cordata (L.) Willd. ex Roem. et Schult. ■☆
138478 Drymaria cordata (L.) Willd. ex Roem. et Schult. var. pacifica M. Mizush.;太平洋荷莲豆草■☆
138479 Drymaria cordata (L.) Willd. ex Schult. = Drymaria cordata (L.) Willd. ex Roem. et Schult. ■☆
138480 Drymaria cordifolia Roxb. = Drymaria cordata (L.) Willd. ex Roem. et Schult. subsp. diandra (Blume) J. A. Duke ■
138481 Drymaria depressa Greene;凹陷荷莲豆草■☆
138482 Drymaria diandra Blume;荷莲豆草(除风草,穿线蛇,串钱草,地花生,二蕊荷莲豆,荷连豆,荷莲草,荷莲豆,荷莲豆菜,两面青,龙鳞草,螺蚬草,落水金钱,痞子草,青钱草,青蛇儿,青蛇仔,青蛇子,十二对草,水冰片,水荷豆,水荷兰,水蓝青,水流冰,水青草,水天诛,团叶鹅儿肠,眼镜草,野荷莲豆,野豌豆,野豌豆菜,野豌豆草,野豌豆尖,野雪豆,有米菜,圆叶鹅儿肠,月亮草);Cordate Drymaria, Drymaria, Pipili ■
138483 Drymaria diandra Blume = Drymaria cordata (L.) Willd. ex Roem. et Schult. subsp. diandra (Blume) J. A. Duke ■
138484 Drymaria diandra Blume = Drymaria cordata (L.) Willd. ex Roem. et Schult. ■☆
138485 Drymaria effusa A. Gray var. depressa (Greene) J. A. Duke = Drymaria depressa Greene ■☆
138486 Drymaria fendleri S. Watson = Drymaria glandulosa Bartl. ■☆
138487 Drymaria glandulosa Bartl.;具腺荷莲豆草■☆
138488 Drymaria laxiflora Benth.;疏花荷莲豆草■☆
138489 Drymaria leptophylla (Cham. et Schltdl.) Fenzl ex Rohrb.;细叶荷莲豆草■☆
138490 Drymaria mairei (H. Lév.) H. Lév. = Arenaria iochanensis C. Y. Wu ■
138491 Drymaria mairei H. Lév. = Arenaria iochanensis C. Y. Wu ■
138492 Drymaria mollugineа (Ser.) Didr.;粟米草荷莲豆草■☆
138493 Drymaria pachyphylla Wooton et Standl.;厚叶荷莲豆草;Inkwood, Thickleaf Drymary ■☆
138494 Drymaria sessilifolia Fiori = Drymaria cordata (L.) Willd. ex Roem. et Schult. ■☆
138495 Drymaria sperguloides A. Gray = Drymaria mollugineа (Ser.) Didr. ■☆
138496 Drymaria stylosa Backer = Drymaria villosa Cham. et Schltdl. ■
138497 Drymaria tenella A. Gray = Drymaria leptophylla (Cham. et Schltdl.) Fenzl ex Rohrb. ■☆
138498 Drymaria villosa Cham. et Schltdl.;毛荷莲豆草;Hair Drymaria ■
138499 Drymarta cordata (L.) Roem. et Schult. = Drymaria cordata (L.) Willd. ex Roem. et Schult. ■☆
138500 Drymeia Ehrh. = Carex L. ■
138501 Drymiphila Juss. = Drymophila R. Br. ■☆
138502 Drymis Juss. = Drimys J. R. Forst. et G. Forst. (保留属名)●☆
138503 Drymispermum Rchb. = Drimyspermum Reinw. ●☆
138504 Drymispermum Rchb. = Phaleria Jack ●☆
138505 Drymoanthus Nicholls(1943);丛林兰属■☆
138506 Drymoanthus flavus St. George et Molloy;黄丛林兰■☆
138507 Drymoanthus minutus Nicholls;丛林兰■☆
138508 Drymocallis Fourr. = Potentilla L. ■●
138509 Drymocallis Fourr. ex Rydb. = Potentilla L. ■●
138510 Drymocallis agrimonioides (Pursh) Rydb. = Potentilla arguta Pursh ■☆
138511 Drymocallis arguta (Pursh) Rydb. = Potentilla arguta Pursh ■☆
138512 Drymocallis rupestris (L.) Soják = Potentilla rupestris L. ■
138513 Drymochloa Holub = Festuca L. ■
138514 Drymocodon Fourr. = Campanula L. ■●
138515 Drymoda Lindl. (1838);林地兰属(德里蒙达兰属);Drymoda ■☆
138516 Drymoda gymnopus (Hook. f.) Garay, Hamer et Siegerist = Bulbophyllum gymnopus Hook. f. ■
138517 Drymoda picta Lindl.;林地兰(德里蒙达兰);Painted Drymoda ■☆
138518 Drymoda siamensis Schltr.;泰国林地兰(泰国德里蒙达兰);Siam Drymoda ■☆
138519 Drymonactes Ehrh. = Bromus L. (保留属名)■
138520 Drymonactes Ehrh. = Festuca L. ■
138521 Drymonactes Fourr. = Festuca L. ■
138522 Drymonactes Steud. = Drymonactes Ehrh. ■
138523 Drymonia Mart. (1829);林苣苔属(锥莫尼亚属);Drymonia ●☆
138524 Drymonia macrophylla (Oerst.) H. E. Moore;大叶林地兰(大

叶锥莫尼亚);Bigleaf Drymonia ■☆
138525 Drymonia mollis Oerst.;软林地兰(毛锥莫尼亚);Hairy Drymonia ■☆
138526 Drymonia parviflora Hanst.;小花林地兰(小花锥莫尼亚);Littleflower Drymonia ■☆
138527 Drymonia serrulata Mart.;锯齿林地兰(锯齿锥莫尼亚);Serrate Drymonia ■☆
138528 Drymonia stenophylla (Donn. Sm.) H. E. Moore;狭叶林地兰(狭叶锥莫尼亚);Narrowleaf Drymonia ■☆
138529 Drymonia strigosa (Oerst.) Wiehler;粗毛林地兰(粗毛锥莫尼亚);Strigose Drymonia ■☆
138530 Drymophila R. Br. (1810);林铃兰属■☆
138531 Drymophila cyanocarpa R. Br.;林铃兰■☆
138532 Drymophloeus Zipp. (1829);林皮棕属(阔羽椰属,榄果椰子属,木果椰属,木椆属,木皮棕属,木棕属);Drymophloeus ●☆
138533 Drymophloeus angustifolius Mart.;狭叶林皮棕(狭叶木皮棕)●☆
138534 Drymophloeus beguinii (Burret) H. E. Moore;比棍林皮棕●☆
138535 Drymophloeus communis Miq.;林皮棕(木皮棕)●☆
138536 Drymophloeus olivaeformis Scheff.;卵果林皮棕;Olive Palm ●☆
138537 Drymopogon Fabr. = Aruncus L. ●■
138538 Drymopogon Raf. = Spiraea L. ●
138539 Drymoscias Kaso-Pol. = Notopterygium H. Boissieu ■★
138540 Drymoscias forbesii (H. Boissieu) Koso-Pol. = Notopterygium franchetii H. Boissieu ■
138541 Drymoscias franchetii (H. Boissieu) Koso-Pol. = Notopterygium franchetii H. Boissieu ■
138542 Drymospartum C. Presl = Genista L. ●
138543 Drymosphace (Benth.) Opiz = Salvia L. ●■
138544 Drymosphace Opiz = Salvia L. ●■
138545 Drymyrrhizae Vent. = Zingiberaceae Martinov(保留科名)■
138546 Drymys Vell. = Drimys J. R. Forst. et G. Forst. (保留属名)●☆
138547 Dryobalanops C. F. Gaertn. (1805);婆罗香属(冰片香属,羯布罗香属,龙脑香属);Borneo Camphor, Brunei Teak, Kapur ●☆
138548 Dryobalanops aromatica C. F. Gaertn.;婆罗香(羯布罗香,龙脑香);Bantu Camphor, Baros Camphor, Barus Camphor, Borneo Camphor, Common Borneo Camphor, Kapur, Sumatra Camphor ●☆
138549 Dryobalanops beccarii Dyer;贝氏婆罗香(贝氏冰片香木)●☆
138550 Dryobalanops camphora Colebr.;樟脑婆罗香;Barus Camphor, Camphor of Borneo, Camphor of Malaysia, Kapur ●☆
138551 Dryobalanops fusca Slooten;褐色婆罗香(褐色冰片香木)●☆
138552 Dryobalanops keithii Symington;基氏婆罗香(基氏冰片香木)●☆
138553 Dryobalanops lanceolata Burck;披针叶婆罗香(披针冰片香木)●☆
138554 Dryobalanops oblongifolia Dyer;矩圆叶婆罗香(长椭圆叶冰片香木)●☆
138555 Dryobalanops rappa Becc.;沼泽婆罗香(沼泽冰片香木)●☆
138556 Dryobalanops robusta (C. F. Gaertn.) Oken;粗壮婆罗香(粗壮羯布罗香)●☆
138557 Dryopacia Roeper = Dryopeia Thouars ■
138558 Dryopeia Thouars = Disperis Sw. ■
138559 Dryopeia oppositifolia (Sm.) Thouars = Disperis oppositifolia Sm. ■☆
138560 Dryopeia tripetaloides Thouars = Disperis tripetaloidea (Thouars) Lindl. ■☆
138561 Dryopetalon A. Gray(1853);北美岩芥属■☆
138562 Dryopetalon membranifolium Rollins;膜叶北美岩芥■☆
138563 Dryopetalon runcinatum A. Gray;北美岩芥■☆
138564 Dryopoa Vickery(1963);丰羊茅属■☆
138565 Dryopoa dives (F. Muell.) Vickery;丰羊茅■☆
138566 Dryopria Thouars = Disperis Sw. ■
138567 Dryopria Thouars = Dryopeia Thouars ■
138568 Dryopria Thouars = Dryorkis Thouars ■
138569 Dryopsila Raf. = Erythrobalanus (Oerst.) O. Schwarz ●
138570 Dryopsila Raf. = Quercus L. ●
138571 Dryorchis Thouars = Disperis Sw. ■
138572 Dryorkis Thouars = Disperis Sw. ■
138573 Dryparia Post et Kuntze = Cayaponia Silva Manso(保留属名)■☆
138574 Dryparia Post et Kuntze = Druparia Silva Manso ■☆
138575 Drypetes Vahl(1807);核果木属(核实木属,核实属,环蕊木属,铁色属);Drypetes, Wood Fern, Shield Fern ●
138576 Drypetes afzelii (Pax) Hutch.;阿芙泽尔核果木●☆
138577 Drypetes ambigua Léandri;可疑核果木●☆
138578 Drypetes angustifolia Pax et K. Hoffm.;窄叶核果木●☆
138579 Drypetes aquifolium (Scott-Elliot) Pax et K. Hoffm. = Drypetes madagascariensis (Lam.) Humbert et Léandri ●☆
138580 Drypetes arborescens (Oliv.) Hutch. = Sibangea arborescens Oliv. ●☆
138581 Drypetes arcuatinervia Merr. et Chun;拱网脉核果木(拱网核果木);Arcuatenerve Drypetes, Curved Drypetes ●
138582 Drypetes arcuatinervia Merr. et Chun var. elongata Merr. et Chun = Drypetes arcuatinervia Merr. et Chun ●
138583 Drypetes arguta (Müll. Arg.) Hutch.;亮核果木●☆
138584 Drypetes armoracia Pax et K. Hoffm. = Drypetes gossweileri S. Moore ●☆
138585 Drypetes aubrevillei Léandri;奥布核果木●☆
138586 Drypetes australasica Pax et K. Hoffm.;澳大利亚核果木;Yellow Tulipwood ●☆
138587 Drypetes aylmeri Hutch. et Dalziel;艾梅核果木●☆
138588 Drypetes bathiei Capuron et Léandri;巴西核果木●☆
138589 Drypetes battiscombei Hutch. = Drypetes gerrardii Hutch. ●☆
138590 Drypetes bipindensis (Pax) Hutch.;比平迪核果木●☆
138591 Drypetes brownii Standl.;布氏核果木●☆
138592 Drypetes calvescens Pax et K. Hoffm.;光秃核果木●☆
138593 Drypetes capillipes (Pax) Pax et K. Hoffm.;细毛核果木●☆
138594 Drypetes capuronii Léandri;凯普伦核果木●☆
138595 Drypetes capuronii Léandri var. grandiflora Léandri = Drypetes bathiei Capuron et Léandri ●☆
138596 Drypetes cauliflora Pax et K. Hoffm. = Drypetes stipularis (Müll. Arg.) Hutch. ●☆
138597 Drypetes celastrinea Pax et K. Hoffm.;南蛇藤核果木●☆
138598 Drypetes chevalieri Beille ex Hutch. et Dalziel;舍瓦利耶核果木●☆
138599 Drypetes cinnabarina Pax et K. Hoffm.;朱红核果木●☆
138600 Drypetes confertiflora Merr. et Chun = Drypetes congestiflora Chun et T. Chen ●
138601 Drypetes congestiflora Chun et T. Chen;密花核果木(红枣,密花核实);Dense-flower Drypetes, Dense-flowered Drypetes ●
138602 Drypetes coriifolia Léandri = Drypetes perrieri Léandri ●☆
138603 Drypetes cumingii (Baill.) Pax et K. Hoffm.;青枣核果木(青枣柯);Cuming Drypetes ●
138604 Drypetes dinklagei (Pax) Hutch.;丁克核果木●☆
138605 Drypetes euryodes (Hiern) Hutch.;宽核果木●☆

138606 Drypetes falcata (Merr.) Pax et K. Hoffm. var. yamadae (Kaneh. et Sasaki) Hurus. = Drypetes littoralis (C. B. Rob.) Merr. ●
138607 Drypetes falcata Pax et K. Hoffm. var. yamadai (Kaneh. et Sasaki) Hurus. = Drypetes littoralis (C. B. Rob.) Merr. ●
138608 Drypetes fallax Pax et K. Hoffm.;迷惑核果木●☆
138609 Drypetes fernandopoana Brenan;费尔南核果木●☆
138610 Drypetes floribunda (Müll. Arg.) Hutch.;繁花核果木●☆
138611 Drypetes formosana (Kaneh. et Sasaki ex Shimada) Kaneh.;台湾核果木(台湾假黄杨);Liodendron,Taiwan Drypetes ●
138612 Drypetes gabonensis (Pierre ex Hutch.) Hutch.;加蓬核果木●☆
138613 Drypetes gabonensis Hutch. = Drypetes gabonensis (Pierre ex Hutch.) Hutch. ●☆
138614 Drypetes gerrardii Hutch.;吉氏核果木●☆
138615 Drypetes gerrardii Hutch. var. angustifolia Radcl.-Sm.;窄叶吉氏核果木●☆
138616 Drypetes gerrardii Hutch. var. tomentosa Radcl.-Sm.;毛吉氏核果木●☆
138617 Drypetes gerrardinoides Radcl.-Sm.;拟杰勒德核果木●☆
138618 Drypetes gilgiana (Pax) Pax et K. Hoffm.;吉尔格核果木●☆
138619 Drypetes glabra (Pax) Hutch.;光滑核果木●☆
138620 Drypetes glomerata (Müll. Arg.) Hutch. = Drypetes fernandopoana Brenan ●☆
138621 Drypetes gossweileri S. Moore;高斯核果木(戈伟核果木)●☆
138622 Drypetes gracilis Pax et K. Hoffm.;纤细核果木●☆
138623 Drypetes griffithii (Hook. f.) Pax et K. Hoffm. = Drypetes indica (Müll. Arg.) Pax et K. Hoffm. ●◇
138624 Drypetes hainanensis Merr.;海南核果木(白梨,海南核实,九巴公);Hainan Drypetes ●
138625 Drypetes hainanensis Merr. var. longistipitata P. T. Li;长柄海南核果木(长柄核果木);Long-stalk Hainan Drypetes ●
138626 Drypetes henriquesii (Pax) Hutch.;亨利克斯核果木●☆
138627 Drypetes hieranensis (Hayata) Pax et K. Hoffm.;喜兰山环蕊木(尖叶环蕊木,南仁铁色);Hieransan Drypetes, Xilanshan Drypetes ●
138628 Drypetes hieranensis (Hayata) Pax et K. Hoffm. = Drypetes indica (Müll. Arg.) Pax et K. Hoffm. ●◇
138629 Drypetes hieranensis (Hayata) Pax et K. Hoffm. = Drypetes karapinensis (Hayata) Pax et K. Hoffm. ●
138630 Drypetes hoaensis Gagnep.;勐腊核果木;Mengla Drypetes ●
138631 Drypetes holtzii Pax et K. Hoffm. = Drypetes natalensis (Harv.) Hutch. ●☆
138632 Drypetes hutchinsonii Pax et K. Hoffm. = Drypetes leonensis Pax ●☆
138633 Drypetes inaequalis Hutch.;不等核果木●☆
138634 Drypetes indica (Müll. Arg.) Pax et K. Hoffm.;核果木(长柄核果木,南仁铁色,琼中核果木,校力坪铁色);Indian Drypetes ●◇
138635 Drypetes integerrima (Koidz.) Hosok.;全缘核果木●☆
138636 Drypetes integrifolia Merr. et Chun;全缘叶核果木;Entire-leaved Drypetes,Integriflious Drypetes ●
138637 Drypetes iturensis Pax et K. Hoffm.;伊图里核果木●☆
138638 Drypetes iturensis Pax et K. Hoffm. var. pilosa ?;疏毛核果木●☆
138639 Drypetes ivorensis Hutch. et Dalziel;伊沃里核果木●☆
138640 Drypetes kamerunica Pax et K. Hoffm. = Drypetes leonensis Pax ●☆
138641 Drypetes karapinensis (Hayata) Pax et K. Hoffm.;校力坪环蕊木(阿里山环蕊木,铁色,校力坪铁色);Karaping Drypetes ●
138642 Drypetes karapinensis (Hayata) Pax et K. Hoffm. = Drypetes indica (Müll. Arg.) Pax et K. Hoffm. ●◇
138643 Drypetes karapinensis (Hayata) Pax et K. Hoffm. var. hieranensis (Hayata) Hurus. = Drypetes indica (Müll. Arg.) Pax et K. Hoffm. ●◇
138644 Drypetes karapinensis Pax et K. Hoffm. = Drypetes matsumurae (Koidz.) Kaneh. ●
138645 Drypetes klaineana (Pierre) Breteler;克莱恩核果木●☆
138646 Drypetes laciniata (Pax) Hutch.;撕裂核果木●☆
138647 Drypetes lancifolia (Hook. f.) Pax et K. Hoffm. = Drypetes indica (Müll. Arg.) Pax et K. Hoffm. ●◇
138648 Drypetes leonensis Pax;莱昂核果木●☆
138649 Drypetes littoralis (C. B. Rob.) Merr.;滨海核果木(铁色,铁色树);Philippine Drypetes,Yamada Drypetes ●
138650 Drypetes liukiuensis Hurus. = Drypetes matsumurae (Koidz.) Kaneh. ●
138651 Drypetes longifolia Pax et K. Hoffm.;长叶铁色;Long-leaved Drypetes ●
138652 Drypetes longipes X. H. Song = Drypetes indica (Müll. Arg.) Pax et K. Hoffm. ●◇
138653 Drypetes madagascariensis (Lam.) Humbert et Léandri;马岛核果木●☆
138654 Drypetes madagascariensis (Lam.) Humbert et Léandri subvar. inermis Humbert et Léandri = Drypetes madagascariensis (Lam.) Humbert et Léandri ●☆
138655 Drypetes madagascariensis (Lam.) Humbert et Léandri var. perrieri Humbert et Léandri = Drypetes madagascariensis (Lam.) Humbert et Léandri ●☆
138656 Drypetes magnistipula (Pax) Hutch.;大托叶核果木●☆
138657 Drypetes major (Pax) Hutch. = Drypetes natalensis (Harv.) Hutch. ●☆
138658 Drypetes matsumurae (Koidz.) Kaneh.;毛药核果木;Hairy-anthered Drypetes,Matsumura Drypetes ●
138659 Drypetes matsumurae (Koidz.) Kaneh. = Putranjiva matsumurae Koidz. ●
138660 Drypetes mildbraedii (Pax) Hutch.;米尔德核果木●☆
138661 Drypetes mindanaensis Pax et K. Hoffm.;棉兰老核果木●☆
138662 Drypetes mindorensis (Merr.) Pax et K. Hoffm. = Drypetes littoralis (C. B. Rob.) Merr. ●
138663 Drypetes mossambicensis Hutch.;莫桑比克核果木●☆
138664 Drypetes mottikoro Léandri = Drypetes aylmeri Hutch. et Dalziel ●☆
138665 Drypetes natalensis (Harv.) Hutch.;纳塔尔核果木●☆
138666 Drypetes natalensis (Harv.) Hutch. var. leiogyna Brenan;光蕊纳塔尔核果木●☆
138667 Drypetes natalensis Hutch. = Drypetes natalensis (Harv.) Hutch. ●☆
138668 Drypetes nienhui Merr. et Chun = Drypetes indica (Müll. Arg.) Pax et K. Hoffm. ●◇
138669 Drypetes obanensis S. Moore;奥班核果木●☆
138670 Drypetes obtusa Merr. et Chun;钝叶核果木(广西核实);Blunt Drypetes,Obtuseleaf Drypetes ●
138671 Drypetes occidentalis (Müll. Arg.) Hutch.;西方核果木●☆
138672 Drypetes oppositifolia Léandri;对叶核果木●☆
138673 Drypetes ovata Hutch. = Drypetes floribunda (Müll. Arg.) Hutch. ●☆
138674 Drypetes parvifolia (Müll. Arg.) Pax et K. Hoffm.;小叶核果木●☆

138675 Drypetes paxii Hutch.;帕克斯核果木●☆

138676 Drypetes paxii Hutch. var. aubrevillei (Léandri) J. Léonard = Drypetes aubrevillei Léandri ●☆

138677 Drypetes pellegrinii Léandri;佩尔格兰核果木●☆

138678 Drypetes peltophora S. Moore;盾梗核果木●☆

138679 Drypetes perreticulata Gagnep.;网脉核果木(白梨么);Net-nerved Drypetes, Reticulate Drypetes ●

138680 Drypetes perrieri Léandri;佩里耶核果木●☆

138681 Drypetes pierreana Hutch. = Drypetes klaineana (Pierre) Breteler ●☆

138682 Drypetes polyantha Pax et K. Hoffm.;多花核果木●☆

138683 Drypetes preussii (Pax) Hutch.;普罗伊斯核果木●☆

138684 Drypetes principum (Müll. Arg.) Hutch.;普林西比核果木●☆

138685 Drypetes radamae Léandri = Drypetes perrieri Léandri ●☆

138686 Drypetes reticulata Pax;网状核果木●

138687 Drypetes rowlandii Pax = Drypetes leonensis Pax ●☆

138688 Drypetes roxburghii (Wall.) Hurus.;无盘核果木●

138689 Drypetes rubriflora Pax et K. Hoffm.;红花核果木●☆

138690 Drypetes salicifolia Gagnep.;柳叶核果木;Willow-leaf Drypetes, Willow-leaved Drypetes ●

138691 Drypetes sassandraensis Aubrév. = Keayodendron bridelioides Léandri ●☆

138692 Drypetes sclerophylla Mildbr.;硬叶核果木●☆

138693 Drypetes similis Hutch. = Sibangea similis (Hutch.) Radcl.-Sm. ●☆

138694 Drypetes spinosodentata (Pax) Hutch.;刺齿核果木●☆

138695 Drypetes staudtii (Pax) Hutch.;施陶核果木●☆

138696 Drypetes stipularis (Müll. Arg.) Hutch.;托叶核果木●☆

138697 Drypetes subdentata Mildbr. = Drypetes gerrardii Hutch. ●☆

138698 Drypetes talbotii S. Moore;塔尔博特核果木●☆

138699 Drypetes taylorii S. Moore;泰勒核果木●☆

138700 Drypetes tessmanniana (Pax) Pax et K. Hoffm.;泰斯曼核果木●☆

138701 Drypetes thouarsiana (Baill.) Capuron;图氏核果木●☆

138702 Drypetes thouarsii (Baill.) Léandri;图阿斯核果木●☆

138703 Drypetes ugandensis (Rendle) Hutch.;乌干达核果木●☆

138704 Drypetes urophylla Pax et K. Hoffm. = Drypetes leonensis Pax ●☆

138705 Drypetes usambarica (Pax) Hutch.;乌桑巴拉核果木●☆

138706 Drypetes usambarica (Pax) Hutch. var. rugulosa Radcl.-Sm. = Drypetes usambarica (Pax) Hutch. var. trichogyna Radcl.-Sm. ●☆

138707 Drypetes usambarica (Pax) Hutch. var. stylosa Radcl.-Sm.;花柱核果木●☆

138708 Drypetes usambarica (Pax) Hutch. var. trichogyna Radcl.-Sm.;毛蕊核果木●☆

138709 Drypetes verrucosa Hutch.;多疣核果木●☆

138710 Drypetes vignei Hoyle = Drypetes pellegrinii Léandri ●☆

138711 Drypetes vilhenae Cavaco;维列纳核果木●☆

138712 Drypetes yamadai (Kaneh. et Sasaki) Kaneh. et Sasaki = Drypetes littoralis (C. B. Rob.) Merr. ●

138713 Drypetes yamadai Kaneh. et Sasaki = Drypetes littoralis (C. B. Rob.) Merr. ●

138714 Drypetes zombensis Dunkley = Drypetes natalensis (Harv.) Hutch. ●☆

138715 Drypis L. (1753);刺叶蝇子草属■☆

138716 Drypis Mich. ex L. = Drypis L. ■☆

138717 Drypis spinosa L.;刺叶蝇子草■☆

138718 Drypsis Duchartre = Dypsis Noronha ex Mart. ●☆

138719 Dryptes Kanjilal et al. = Drypetes Vahl ●

138720 Dryptopetalum Arn. = Gynotroches Blume ●☆

138721 Duabanga Buch.-Ham. (1837);八宝树属(杜滨木属);Duabanga ●

138722 Duabanga grandiflora (Roxb. ex DC.) Walp.;八宝树(大平头树,杜滨木,桑管树);Bigflower Duabanga, Big-flowered Duabanga ●

138723 Duabanga grandiflora (Roxb.) Walp. = Duabanga grandiflora (Roxb. ex DC.) Walp. ●

138724 Duabanga moluccana Blume;摩鹿加八宝树(马鲁古八宝木,摩鹿加杜滨木);Molucca Duabanga ●☆

138725 Duabanga sonneratioides Buch.-Ham. = Duabanga grandiflora (Roxb. ex DC.) Walp. ●

138726 Duabanga taylorii Jayaw.;细花八宝树;Taylor Duabanga ●

138727 Duabangaceae Takht. (1986);八宝树科●

138728 Duabangaceae Takht. = Lythraceae J. St.-Hil. (保留科名)■●

138729 Duania Noronha = Homalanthus A. Juss. (保留属名)●

138730 Dubaea Steud. = Diplusodon Pohl ●☆

138731 Dubaea Steud. = Dubyaea DC. ■

138732 Dubanus Kuntze = Pometia J. R. Forst. et G. Forst. ●

138733 Dubardella H. J. Lam = Pyrenaria Blume ●

138734 Dubautia Gaudich. (1830);轮菊属●■☆

138735 Dubautia plantaginea Gaudich.;轮菊●■☆

138736 Duboisia H. Karst. = Dubois-Reymondia H. Karst. ■☆

138737 Duboisia H. Karst. = Myoxanthus Poepp. et Endl. ■☆

138738 Duboisia H. Karst. = Pleurothallis R. Br. ■☆

138739 Duboisia R. Br. (1810);澳茄属;Duboisia ●☆

138740 Duboisia arenitensis Craven, Lepschi et Haegi;北部澳茄●☆

138741 Duboisia campbellii Morrison;坎佛澳茄●☆

138742 Duboisia hopuwoodii (F. Muell.) F. Muell.;皮特里澳茄(杜什茄,霍氏澳茄);Pedgery, Petgery, Pitcheri, Pitchery, Pituri ●☆

138743 Duboisia leichhardtii (F. Muell.) F. Muell.;雷氏澳茄●☆

138744 Duboisia myoporoides R. Br.;澳茄(澳洲茄);Corkwood, Corkwood Duboisia, Duboisia ●☆

138745 Dubois-Refmondia H. karst. = Myoxanthus Poepp. et Endl. ■☆

138746 Dubois-Reymondia H. karst. = Pleurothallis R. Br. ■☆

138747 Duboscia Bocquet(1866);全缘椴属(热带椴属)●☆

138748 Duboscia macrocarpa Bocq.;大果全缘椴(大果热带椴)●☆

138749 Duboscia polyantha Pierre ex A. Chev.;多花全缘椴(多花热带椴)●☆

138750 Duboscia viridiflora (K. Schum.) Mildbr.;绿花全缘椴(热带椴)●☆

138751 Duboscia viridiflora (K. Schum.) Mildbr. = Duboscia macrocarpa Bocq. ●☆

138752 Duboscia viridiflora Mildbr. = Duboscia viridiflora (K. Schum.) Mildbr. ●☆

138753 Dubouzetia Pancher ex Brongn. et Griseb. (1861);迪布木属●☆

138754 Dubouzetia acuminata Sprague;渐尖迪布木●☆

138755 Dubouzetia australiensis Coode;澳大利亚迪布木●☆

138756 Dubouzetia elegans Brongn. et Gris;雅致迪布木●☆

138757 Dubouzetia parviflora Brongn. et Gris;小花迪布木●☆

138758 Dubreuilia Decne. = Dubrueilia Gaudich. ■

138759 Dubrueilia Gaudich. = Pilea Lindl. (保留属名)■

138760 Dubrueilia microphylla Gaudich. = Pilea microphylla (L.) Liebm. ■

138761 Dubrueilia peploides Gaudich. = Pilea peploides (Gaudich.) Hook. et Arn. ■

138762 Dubyaea DC. (1838);厚喙菊属;Dubyaea ■

138763 Dubyaea amoena (Hand. -Mazz.) Stebbins;棕毛厚喙菊■
138764 Dubyaea atropurpurea (Franch.) Stebbins;紫花厚喙菊(紫舌厚喙菊);Darkpurple Dubyaea ■
138765 Dubyaea bhotanica (Hutch.) C. Shih;不丹厚喙菊;Bhotan Dubyaea ■
138766 Dubyaea chimiliensis (W. W. Sm.) Stebbins = Dubyaea tsarongensis (W. W. Sm.) Stebbins ■
138767 Dubyaea cymiformis C. Shih;伞房厚喙菊(聚伞厚喙菊)■
138768 Dubyaea glaucescens Stebbins;光滑厚喙菊■
138769 Dubyaea gombalana (Hand. -Mazz.) Stebbins;矮小厚喙菊■
138770 Dubyaea grandis Hand. -Mazz. = Dubyaea glaucescens Stebbins ■
138771 Dubyaea hispida (D. Don) DC.;厚喙菊;Hispid Dubyaea ■
138772 Dubyaea hispida (D. Don) DC. = Dubyaea pteropoda C. Shih ■
138773 Dubyaea jinyangensis C. Shih;金阳厚喙菊;Jinyang Dubyaea ■
138774 Dubyaea lanceolata C. Shih;披针叶厚喙菊;Lanceolate Dubyaea ■
138775 Dubyaea muliensis C. Shih;木里厚喙菊;Muli Dubyaea ■
138776 Dubyaea omeiensis C. Shih;峨眉厚喙菊;Emei Dubyaea ■
138777 Dubyaea panduriformis C. Shih;琴叶厚喙菊;Fiddleleaf Dubyaea ■
138778 Dubyaea pteropoda C. Shih;翼柄厚喙菊;Wingstipe Dubyaea ■
138779 Dubyaea rubra Stebbins;长柄厚喙菊;Longstipe Dubyaea ■
138780 Dubyaea stebbinii Ludlow;西藏厚喙菊(朗县厚喙菊)■
138781 Dubyaea tsarongensis (W. W. Sm.) Stebbins;察隅厚喙菊;Chayu Dubyaea ■
138782 Dubyaea tsarongensis (W. W. Sm.) Stebbins subsp. chimiliensis (W. W. Sm.) Stebbins = Dubyaea tsarongensis (W. W. Sm.) Stebbins ■
138783 Ducampopinus A. Chev. = Pinus L. ●
138784 Duchartrea Decne. = Pentarhaphia Lindl. ■☆
138785 Duchartrella Kuntze = Holostylis Duch. ■☆
138786 Duchassaingia Walp. = Erythrina L. ●■
138787 Duchekia Kostel. (废弃属名) = Palisota Rchb. ex Endl. (保留属名)■☆
138788 Duchesnea Post et Kuntze = Duchesnia Cass. ■●
138789 Duchesnea Post et Kuntze = Pulicaria Gaertn. ■●
138790 Duchesnea Sm. (1811);蛇莓属;Mock Strawberry, Mockstrawberry, Mock-strawberry, Yellow-flowered Strawberry ●■
138791 Duchesnea Sm. = Potentilla L. ■●
138792 Duchesnea × hara-kurosawae Naruh. et M. Sugim.;黑泽蛇莓■☆
138793 Duchesnea chrysantha (Zoll. et Moritzi) Miq.;皱果蛇莓(地锦,华氏蛇莓,蛇莓,台湾蛇莓);Snake Mockstrawberry, Wrinkledfruit Mockstrawberry, Yellowflower Mockstrawberry ■
138794 Duchesnea chrysantha (Zoll. et Moritzi) Miq. f. leucocephala (Makino) H. Hara;白头皱果蛇莓■☆
138795 Duchesnea formosana Odash. = Duchesnea chrysantha (Zoll. et Moritzi) Miq. ■
138796 Duchesnea indica (Andréws) Focke;蛇莓(宝珠草,蚕莓,地锦,地莓,地杨梅,疔疮药,哈哈果,和尚头草,红顶果,红毛七,鸡冠果,金蝉草,九龙草,老地蜢,老蛇刺占,老蛇藨,老蛇泡,龙吐珠,龙衔珠,龙珠草,落地杨梅,麻蛇果,三点红,三脚虎,三面风,三皮风,三匹方,三匹风,三叶藨,三叶莓,三叶蛇,三叶蛇扭,三爪风,三爪龙,蛇八瓣,蛇藨,蛇波藤,蛇菠,蛇不见,蛇蛋果,蛇果草,蛇果藤,蛇含草,蛇蒿,蛇莓草,蛇缪草,蛇盘草,蛇泡草,蛇婆,蛇葡萄,蛇蓉草,蛇杨梅,蛇枕头,狮子尾,小草莓,雪丁草,血疔草,野杨梅,一点红,印度草莓,紫莓草);False Strawberry, India Mockstrawberry, Indian Mock Strawberry, Indian Mockstrawberry, Indian Mock-strawberry, Indian Strawberry, Mock-strawberry, Yellow-flowered Strawberry ●■
138797 Duchesnea indica (Andréws) Focke f. albocaput Naruh.;白果蛇莓■☆
138798 Duchesnea indica (Andréws) Focke f. leucocephala (Makino) M. Mizush. = Duchesnea chrysantha (Zoll. et Moritzi) Miq. f. leucocephala (Makino) H. Hara ■☆
138799 Duchesnea indica (Andréws) Focke var. leucocephala Makino f. japonica (Kitam.) M. Mizush. = Duchesnea chrysantha (Zoll. et Moritzi) Miq. ■
138800 Duchesnea indica (Andréws) Focke var. major Makino = Duchesnea indica (Andréws) Focke ●■
138801 Duchesnea indica (Andréws) Focke var. microphylla Te T. Yu et L. T. Lu;小叶蛇莓;Smallleaf Mockstrawberry ■
138802 Duchesnea major (Makino) Makino = Duchesnea indica (Andréws) Focke ●■
138803 Duchesnea wallichiana (Ser.) Nakai ex Hara = Duchesnea chrysantha (Zoll. et Moritzi) Miq. ■
138804 Duchesnea wallichiana Nakai ex Hara = Duchesnea chrysantha (Zoll. et Moritzi) Miq. ■
138805 Duchesnia Cass. = Francoeuria Cass. ■●
138806 Duchesnia Cass. = Pulicaria Gaertn. ■●
138807 Duchesnia crispa (Forssk.) Cass. = Pulicaria undulata (L.) C. A. Mey. ■☆
138808 Duchola Adans. = Omphalandria P. Browne(废弃属名)■☆
138809 Duchola Adans. = Omphalea L. (保留属名)■☆
138810 Duckea Maguire(1958);多谢草属■☆
138811 Duckea cyperaceoidea (Ducke) Maguire;多谢草■☆
138812 Duckeanthus R. E. Fr. (1934);多谢花属(达克花属)●☆
138813 Duckeanthus grandiflorus R. E. Fr.;多谢花●☆
138814 Duckeella Porto = Duckeella Porto et Brade ■☆
138815 Duckeella Porto et Brade(1940);多谢兰属■☆
138816 Duckeella adolphii Porto et Brade;多谢兰■☆
138817 Duckeodendraceae Kuhlm. (1950);核果茄科(核果木科)●☆
138818 Duckeodendraceae Kuhlm. = Solanaceae Juss. (保留科名)●■
138819 Duckeodendron Kuhlm. (1925);核果茄属(核果木属)●☆
138820 Duckeodendron cestroides Kuhlm.;核果茄●☆
138821 Duckera F. A. Barkley = Rhus L. ●
138822 Duckesia Cuatrec. (1961);达克木属●☆
138823 Duckesia verrucosa (Ducke) Cuatrec.;达克木●☆
138824 Ducosia Vieill. ex Guillaumin = Dubouzetia Pancher ex Brongn. et Griseb. ●☆
138825 Ducoudraea Bureau = Tecomaria Spech ●
138826 Ducrosia Boiss. (1844);迪克罗草属■☆
138827 Ducrosia anethefolia (DC.) Boiss.;迪克罗草■☆
138828 Ducrosia ovatiloba Dunn et Williams;卵裂片迪克罗草■☆
138829 Dudleya Britton et Rose(1903);仙女杯属(达德利属,粉叶草属);Live Forever ■☆
138830 Dudleya attenuata (S. Watson) Moran;渐尖仙女杯(渐尖达德利)■☆
138831 Dudleya brittonii Johans.;仙女杯(仙女,仙女盃)■☆
138832 Dudleya caespitosa Britton et Rose;海岸仙女杯(海岸达德利);Coast Live-forever ■☆
138833 Dudleya candida Britton ex Britton et Rose;纯白仙女杯(纯白达德利)■☆
138834 Dudleya densiflora (Rose) Moran;密花仙女杯(密花达德利)■☆

138835 Dudleya farinosa Britton et Rose;白粉仙女杯(白粉达德利);Powdery Dudleya,Powdery Live-forever ■☆
138836 Dudleya formosa Moran;优雅仙女杯;Live-forever ■☆
138837 Dudleya nubigena Britton et Rose;云状仙女杯(云状达德利)■☆
138838 Dudleya pulverulenta (Nutt.) Britton et Rose;粉叶草■
138839 Dudleya saxosa (M. E. Jones) Britton et Rose;岩石仙女杯;Rock Live-forever ■☆
138840 Dudleya saxosa (M. E. Jones) Britton et Rose subsp. collomiae (Rose) Moran;黏胶花仙女杯;Rock Live-forever ■☆
138841 Dufourea Bory = Tristicha Thouars ■☆
138842 Dufourea Bory ex Willd. = Tristicha Thouars ■☆
138843 Dufourea Gren. = Arenaria L. ■
138844 Dufourea Kunth = Breweria R. Br. ●☆
138845 Dufourea Kunth = Prevostea Choisy ●☆
138846 Dufourea trifaria Bory ex Willd. = Tristicha trifaria (Bory ex Willd.) Spreng. ■☆
138847 Dufreania DC. = Valerianella Mill. ■
138848 Dufrenoya Chatin = Dendrotrophe Miq. ●
138849 Dufrenoya granulata (Hook. f. et Thomson ex A. DC.) Stauffer = Dendrotrophe granulata (Hook. f. et Thomson ex A. DC.) A. N. Henry et B. Roy ●
138850 Dufrenoya heterantha (Wall. ex DC.) Chatin = Dendrotrophe platyphylla (Spreng.) N. H. Xia et M. G. Gilbert ●
138851 Dufrenoya platyphylla (Spreng.) Stauffer = Dendrotrophe platyphylla (Spreng.) N. H. Xia et M. G. Gilbert ●
138852 Dufresnia DC. = Valerianella Mill. ■
138853 Dufresnia leiocarpa K. Koch = Valerianella leiocarpa (K. Koch) Kuntze ■☆
138854 Dufresnia orientalis DC. = Valerianella leiocarpa (K. Koch) Kuntze ■☆
138855 Dugagelia Juss. ex Gaudich. = ? Piper L. ●■
138856 Dugaldia (Cass.) Cass. = Helenium L. ■
138857 Dugaldia Cass. = Dugaldia (Cass.) Cass. ■
138858 Dugaldia Cass. = Helenium L. ■
138859 Dugaldia Cass. = Hymenoxys Cass. ■☆
138860 Dugaldia hoopesii (A. Gray) Rydb. = Hymenoxys hoopesii (A. Gray) Bierner ■☆
138861 Dugandia Britton et Killip = Acacia Mill.(保留属名)●■
138862 Dugandiodendron Lozano = Magnolia L. ●
138863 Dugandiodendron Lozano(1975);南美盖裂木属●☆
138864 Dugandiodendron argyrothrichum Lozano;银毛南美盖裂木●☆
138865 Dugandiodendron calophyllum Lozano;美叶南美盖裂木●☆
138866 Dugesia A. Gray(1882);绿纹菊属■☆
138867 Dugesia mexicana A. Gray;绿纹菊■☆
138868 Dugezia Montrouz. = Lysimachia L. ●■
138869 Dugezia Montrouz. ex Beauvis. = Lysimachia L. ●■
138870 Duggena Vahl = Gonzalagunia Ruiz et Pav. ●☆
138871 Duglassia Houst. = Clerodendrum L. ●■
138872 Duglassia Houst. = Douglassia Mill.(废弃属名)●
138873 Dugortia Scop. = Parinari Aubl. ●☆
138874 Duguetia A. St. -Hil. (1824)(保留属名);半聚果属(杜盖木属,杜古番荔枝属)●☆
138875 Duguetia barteri (Benth.) Chatrou;巴特半聚果●☆
138876 Duguetia confinis (Engl. et Diels) Chatrou;邻近半聚果●☆
138877 Duguetia lanceolata A. St. -Hil.;窄叶半聚果(窄叶杜盖木,窄叶杜古番荔枝)●☆
138878 Duguetia neglecta Sandwith;略半聚果(略杜盖木)●☆
138879 Duguetia quitarensis Benth.;基塔尔半聚果;Lancewood ●☆
138880 Duguetia spixiana Mart.;半聚果(杜古番荔枝)●☆
138881 Duguetia staudtii (Engl. et Diels) Chatrou;施陶半聚果●☆
138882 Duguetia vallicola J. F. Macbr.;谷地半聚果(谷地杜盖木,谷地杜古番荔枝)●☆
138883 Duguldea Meisn. = Dugaldia Cass. ■
138884 Duguldea Meisn. = Helenium L. ■
138885 Duhaldea DC. (1836);羊耳菊属■●
138886 Duhaldea DC. = Inula L. ●■
138887 Duhaldea cappa (Buch. -Ham. ex D. Don) Anderb. = Duhaldea chinensis DC. ●■
138888 Duhaldea chinensis DC.;羊耳菊(八面风,白背风,白面风,白面将军,白面猫子骨,白牛胆,白牛胆根,白羊耳,白叶菊,冲天白,大刀药,大力黄,大力王,大麻香,大茅香,过山香,黑骨风,华耳木,金边草,蜡毛香,毛茶,毛柴胡,毛将军,毛老虎,毛山肖,毛舌头,绵毛旋覆花,牛耳风,山白芷,山鹿茸,山芷梅,上大黄,天鹅绒,铁杆香,土白芷,土蒙花,小茅香,寻骨风,羊耳草,羊耳茶,羊耳风,叶下白,猪耳风,壮牛浪);Lanuginose Inula,Sheepear Jurinea ●■
138889 Duhaldea chinensis DC. = Inula cappa (Buch. -Ham.) DC. ●■
138890 Duhaldea cuspidata (DC.) Anderb.;凸尖羊耳菊;Cuspidate Inula ●
138891 Duhaldea eupatorioides (DC.) Anderb. = Duhaldea eupatorioides (DC.) Steetz ●
138892 Duhaldea eupatorioides (DC.) Steetz;泽兰羊耳菊;Bogorchid Jurinea,Eupatorium-like Inula ●
138893 Duhaldea eupatorioides (DC.) Steetz = Inula eupatorioides DC. ●
138894 Duhaldea eupatorioides Steetz = Duhaldea eupatorioides (DC.) Steetz ●
138895 Duhaldea forrestii (J. Anthony) Anderb.;拟羊耳菊;Forrest Inula ●
138896 Duhaldea lanuginosa (C. C. Chang) Anderb. = Duhaldea chinensis DC. ●■
138897 Duhaldea nervosa (Wall. ex DC.) Anderb.;显脉旋覆花(草灵仙,草威灵,黑根,黑根药,黑升麻,黑威灵,铁脚威灵,铁脚威灵仙,铜脚葳灵,威灵菊,威灵仙,葳灵仙,乌草根,乌根草,小黑根,小黑药);Veined Inula,Veined Jurinea ■
138898 Duhaldea pterocaulis (Franch.) Anderb.;翼茎羊耳菊(大黑根,大黑洋参,大黑药,石如意,翼茎旋覆花);Wingedstem Inula,Wingedstem Jurinea ■●
138899 Duhaldea rubricaulis (Wall. ex DC.) Anderb.;赤茎羊耳菊;Redcaudex Inula ●
138900 Duhaldea stuhlmannii (O. Hoffm.) Anderb. = Inula stuhlmannii O. Hoffm. ■☆
138901 Duhaldea wissmanniana (Hand. -Mazz.) Anderb.;滇南羊耳菊;Wissmann Inula ●
138902 Duhamela Raf. = Duhamelia Pers. ●
138903 Duhamela Raf. = Hamelia Jacq. ●
138904 Duhamelia Dombey ex Lam. = Myrsine L. ●
138905 Duhamelia Pers. = Hamelia Jacq. ●
138906 Duidaea S. F. Blake(1931);单脉红菊木属●☆
138907 Duidaea pinifolia S. F. Blake;松叶单脉红菊木●☆
138908 Duidaea rubriceps S. F. Blake;红头单脉红菊木●☆
138909 Duidania Standl. (1931);委内瑞拉茜属☆
138910 Duidania montana Standl.;委内瑞拉茜☆
138911 Dukea Dwyer = Raritebe Wernham ●☆

138912 Dulacia Neck. = Acioa Aubl. ●☆
138913 Dulacia Vell. = Acioa Aubl. ●☆
138914 Dulacia Vell. = Liriosma Poepp. et Endl. ■☆
138915 Dulcamara Hill. = Solanum L. ●■
138916 Dulcamara Moench = Solanum L. ●■
138917 Dulia Adans. = Ledum L. ●
138918 Dulichium Pers. (1805);杜里莎草属;Duliche ■☆
138919 Dulichium arundinaceum (L.) Britton;杜里莎草;Dulichium, Pond Sedge, Threeway Sedge, Three-way Sedge ■☆
138920 Dulichium arundinaceum (L.) Britton var. boreale Lepage;北方杜里莎草■☆
138921 Dulichium spathaceum (L.) Rich. ex Pers. = Dulichium arundinaceum (L.) Britton ■☆
138922 Dulongia Kunth = Phyllonoma Willd. ex Schult. ●☆
138923 Dulongia acuminata Kunth = Phyllonoma ruscifolia Willd. ex Schult. ●☆
138924 Dulongiaceae J. Agardh = Phyllonomaceae Rusby ●☆
138925 Dumaniana Yild. et B. Selvi(2006);杜曼草属☆
138926 Dumartroya Gaudich. = Malaisia Blanco ●
138927 Dumartroya Gaudich. = Trophis P. Browne(保留属名)●☆
138928 Dumasia DC. (1825);山黑豆属(黑山豆属,山黑扁豆属,小鸡藤属);Dumasia, Live Forever ■
138929 Dumasia bicolor Hayata;台湾山黑豆(二色叶山黑豆,台湾山黑扁豆);Bicolor Dumasia ■
138930 Dumasia bicolor Hayata var. fulvescens Hayata;黄褐毛山黑扁豆■
138931 Dumasia bracteosa Gagnep. = Dumasia forrestii Diels ■
138932 Dumasia capensis Eckl. et Zeyh. = Dumasia villosa DC. ■
138933 Dumasia cordifolia Benth. ex Baker;心叶山黑豆(铁脚莲);Cordateleaf Dumasia, Heartleaf Dumasia ■
138934 Dumasia forrestii Diels;大苞山黑豆(雀舌豆,小鸡藤);ChickenVine, Forrest Dumasia ■
138935 Dumasia glaucescens Miq. = Dumasia villosa DC. ■
138936 Dumasia hirsuta Craib;硬毛山黑豆;Hardhair Dumasia, Hirsute Dumasia ■
138937 Dumasia miaoliensis Y. C. Liu et F. Y. Lu;苗栗野豇豆;Miaoli Dumasia ■
138938 Dumasia nitida Chun ex Y. T. Wei et S. K. Lee;瑶山山黑豆;Shining Dumasia, Yaoshan Dumasia ■
138939 Dumasia oblongifoliolata F. T. Wang et Ts. Tang ex Y. T. Wei et S. K. Lee;长圆叶山黑豆;Oblongleaf Dumasia ■
138940 Dumasia pubescens DC. = Dumasia villosa DC. ■
138941 Dumasia rotundifolia (Lour.) Merr. = Dunbaria rotundifolia (Lour.) Merr. ■
138942 Dumasia truncata Siebold et Zucc.;山黑豆;Common Dumasia, Truncate Dumasia ■
138943 Dumasia villosa DC.;柔毛山黑豆(八山子,台湾山黑扁豆);Softhair Dumasia, Villous Dumasia ■
138944 Dumasia yunnanensis Y. T. Wei et S. K. Lee;云南山黑豆;Yunnan Dumasia ■
138945 Dumerilia Lag. ex DC. = Jungia L. f. (保留属名)■●☆
138946 Dumerilia Less. = Perezia Lag. ■☆
138947 Dumoria A. Chev. = Tieghemella Pierre ●☆
138948 Dumoria africana (Pierre) Dubard = Tieghemella africana Pierre ●☆
138949 Dumoria heckelii A. Chev. = Tieghemella heckelii (A. Chev.) Pierre ex Dubard ●☆
138950 Dumreichera Hochst. et Steud. = Senra Cav. ●☆
138951 Dumreichera incana Hochst. et Steud. = Senra incana Cav. ●☆
138952 Dumula Lour. ex Gomes = Severinia Ten. ex Endl. ●
138953 Dumula sinensis Lour. ex Gomes = Atalantia buxifolia (Poir.) Oliv. ●
138954 Dunalia Kunth(1818);杜纳尔茄属●☆
138955 Dunalia Montrouz. = Amorphophallus Blume ex Decne. (保留属名)■●
138956 Dunalia R. Br. = Torenia L. ■
138957 Dunalia Spreng. (废弃属名) = Dunalia Kunth ●☆
138958 Dunalia Spreng. (废弃属名) = Lucya DC. (保留属名)■☆
138959 Dunalia angustifolia Dammer;窄叶杜纳尔茄●☆
138960 Dunalia breviflora (Sendtn. ex Mart.) Sleumer;短花杜纳尔茄●☆
138961 Dunalia ferruginea Sodiro et Dammer;锈色杜纳尔茄●☆
138962 Dunalia macrophylla (Benth.) Sleumer;大叶杜纳尔茄●☆
138963 Dunalia obovata (Ruiz et Pav.) Dammer;倒卵杜纳尔茄●☆
138964 Dunantia DC. = Isocarpha R. Br. ■☆
138965 Dunantia achyranthes DC. = Isocarpha oppositifolia (L.) Cass. var. achyranthes (DC.) D. J. Keil et Stuessy ■☆
138966 Dunbaria Wight et Arn. (1834);野扁豆属;Dunbaria, Fieldhairicot ●■
138967 Dunbaria circinalis (Benth.) Baker;卷圈野扁豆;Circinal Dunbaria, Coil Fieldhairicot ●
138968 Dunbaria fusca (Wall.) Kurz;黄毛野扁豆(褐野扁豆);Brownhair Dunbaria, Yellowhair Fieldhairicot ■
138969 Dunbaria harmandii Gagnep. = Dunbaria nivea Miq. ■
138970 Dunbaria henryi Y. C. Wu;鸽仔豆(凹子豆);Henry Dunbaria, Henry Fieldhairicot ■
138971 Dunbaria merrillii Elmer;麦氏野扁豆;Merill's Dunbaria ■
138972 Dunbaria nivea Miq.;白背野扁豆;Glaucous-leaf Dunbaria, Paleback Fieldhairicot ■
138973 Dunbaria parviffolia X. X. Chen;小叶野扁豆;Littleleaf Fieldhairicot, Smallleaf Dunbaria ■
138974 Dunbaria podocarpa Kurz;长柄野扁豆(金钱风,山绿豆,水芽豆,贼老藤);Longstalk Fieldhairicot, Longstipe Dunbaria ■
138975 Dunbaria rotundifolia (Lour.) Merr.;圆叶野扁豆(家豆薯,假绿豆,罗网藤,小黄藤);Roundleaf Dumasia, Roundleaf Dunbaria, Roundleaf Fieldhairicot ■
138976 Dunbaria scortechinii Prain ex King = Dunbaria nivea Miq. ■
138977 Dunbaria subrhombea (Miq.) Hemsl. = Dunbaria villosa (Thunb.) Makino ■
138978 Dunbaria villosa (Thunb.) Makino;野扁豆(毛野扁豆);Longfloss Fieldhairicot, Villous Dunbaria ■
138979 Dunbaria villosa (Thunb.) Makino var. peduncularis Hand.-Mazz.;毛野扁豆(楚雄野扁豆,野扁豆)■
138980 Dunbaria villosa Makino = Dolichos trilobus L. ■
138981 Dunbaria villosa Makino = Dunbaria henryi Y. C. Wu ■
138982 Duncania Rchb. = Toddalia Juss. (保留属名)●
138983 Duncania Rchb. = Vepris Comm. ex A. Juss. ●☆
138984 Dungsia Chiron et V. P. Castro(2002);邓格西兰属■☆
138985 Dunnia Tutcher(1905);绣球茜属(白萼树属,绣球茜草属);Dunnia ●★
138986 Dunnia sinensis Tutcher;绣球茜(白萼树,假黄杨,绣球茜草);China Dunnia, Chinese Dunnia ●◇
138987 Dunniella Rauschert = Aboriella Bennet ■
138988 Dunniella Rauschert = Pilea Lindl. (保留属名)■

138989 Dunniella Rauschert = Smithiella Dunn ■

138990 Dunniella myriantha (Dunn) Rauschert = Aboriella myriantha (Dunn) Bennet ■

138991 Dunniella myriantha (Dunn) Rauschert = Pilea myriantha (Dunn) C. J. Chen ■

138992 Dunstervillea Garay(1972);邓斯兰属■☆

138993 Dunstervillea mirabilis Garay;邓斯兰■☆

138994 Duosperma Dayton(1945);苞爵床属■☆

138995 Duosperma actinotrichum (Chiov.) Vollesen;星毛苞爵床■☆

138996 Duosperma angolense (C. B. Clarke) Brummitt = Duosperma quadrangulare (Klotzsch) Brummitt ■☆

138997 Duosperma clarae Champl.;克拉拉苞爵床■☆

138998 Duosperma crenatum (Lindau) P. G. Mey.;圆齿苞爵床■☆

138999 Duosperma cuprinum Brummitt;铜色苞爵床■☆

139000 Duosperma densiflorum (C. B. Clarke) Brummitt;密花苞爵床■☆

139001 Duosperma dichotomum Vollesen;二歧苞爵床■☆

139002 Duosperma eremophilum (Milne-Redh.) Brummitt = Duosperma longicalyx (Deflers) Vollesen ■☆

139003 Duosperma fanshawei Brummitt;范肖苞爵床■☆

139004 Duosperma fimbriatum Brummitt;流苏苞爵床■☆

139005 Duosperma glabratum Vollesen;光滑苞爵床■☆

139006 Duosperma grandiflorum Vollesen;大花苞爵床■☆

139007 Duosperma kilimandscharicum (C. B. Clarke) Dayton;基利苞爵床■☆

139008 Duosperma latifolium Vollesen;宽叶苞爵床■☆

139009 Duosperma livingstoniense Vollesen;利文苞爵床■☆

139010 Duosperma longebracteolatum Vollesen;长苞片苞爵床■☆

139011 Duosperma longicalyx (Deflers) Vollesen;长萼苞爵床■☆

139012 Duosperma nudantherum (C. B. Clarke) Brummitt;裸药苞爵床■☆

139013 Duosperma parviflora Hedrén;小花苞爵床■☆

139014 Duosperma porotoense Vollesen;波罗托苞爵床■☆

139015 Duosperma quadrangulare (Klotzsch) Brummitt;棱角苞爵床■☆

139016 Duosperma rehmannii (Schinz) Vollesen;拉赫曼苞爵床■☆

139017 Duosperma sessilifolium (Lindau) Brummitt;无柄叶苞爵床■☆

139018 Duosperma stoloniferum Vollesen;匍匐苞爵床■☆

139019 Duosperma subquadrangulare Vollesen;四角苞爵床■☆

139020 Duosperma tanzaniense Vollesen;坦桑尼亚苞爵床■☆

139021 Duosperma trachyphyllum (Bullock) Dayton;糙叶苞爵床■☆

139022 Duosperma transvaalense (Schinz) Vollesen;德兰士瓦苞爵床■☆

139023 Duotriaceae Dulac = Cistaceae Juss.(保留科名)●■

139024 Duparquetia Baill.(1865);西非蔓属■☆

139025 Duparquetia orchidacea Baill.;西非蔓■☆

139026 Dupathya Vell.(废弃属名) = Paepalanthus Kunth(保留属名)■☆

139027 Dupatya Vell. = Paepalanthus Kunth(保留属名)■☆

139028 Dupatya affinis Kuntze = Paepalanthus affinis Kunth ■☆

139029 Duperrea Pierre ex Pit.(1924);长柱山丹属;Duperrea ●

139030 Duperrea pavettifolia (Kurz) Pit.;长柱山丹;Longstyle Duperrea,Long-styled Duperrea ●

139031 Duperreya Gaudich. = Porana Burm. f. ●■☆

139032 Dupineta Raf. = Dissotis Benth.(保留属名)●☆

139033 Dupineta Raf. = Osbeckia L. ●■

139034 Dupinia Scop. = Ternstroemia Mutis ex L. f.(保留属名)●

139035 Duplipetala Thiv = Canscora Lam. ■

139036 Duplipetala Thiv(2003);重瓣龙胆属■☆

139037 Dupontia R. Br.(1823);杜邦草属■☆

139038 Dupontia fischeri R. Br.;菲舍尔杜邦草■☆

139039 Dupratzia Raf. et Wherry = Eustoma Salisb. ■☆

139040 Dupuisia A. Rich. = Sorindeia Thouars ●☆

139041 Dupuisia juglandifolia A. Rich. = Sorindeia juglandifolia (A. Rich.) Planch. ex Oliv. ●☆

139042 Dupuisia longifolia Hook. f. = Trichoscypha longifolia (Hook. f.) Engl. ●☆

139043 Dupuya J. H. Kirkbr.(2005);迪皮豆属●☆

139044 Dupuya haraka (Capuron) J. H. Kirkbr.;迪皮豆●☆

139045 Dupuya madagascariensis (R. Vig.) J. H. Kirkbr.;马岛迪皮豆●☆

139046 Duquetia G. Don = Duguetia A. St. -Hil.(保留属名)●☆

139047 Durandea Delarbre(废弃属名) = Durandea Planch.(保留属名)●☆

139048 Durandea Delarbre(废弃属名) = Raphanus L. ■

139049 Durandea Planch.(1847)(保留属名);杜兰德麻属●☆

139050 Durandea angustifolia Stapf;窄叶杜兰德麻●☆

139051 Durandea latifolia Stapf;宽叶杜兰德麻●☆

139052 Durandea pallida K. Schum.;苍白杜兰德麻●☆

139053 Durandea serrata Planch.;杜兰德麻●☆

139054 Durandeeldea Kuntze = Acidoton Sw.(保留属名)●☆

139055 Durandeeldia Kuntze = Acidoton Sw.(保留属名)●☆

139056 Durandia Boeck. = Scleria P. J. Bergius ■

139057 Durandoa Pomel = Carthamus L. ■

139058 Durandoa arborescens (L.) Pomel = Carthamus arborescens L. ■☆

139059 Durandoa clausonis Pomel = Carthamus arborescens L. ■☆

139060 Duranta L.(1753);假连翘属(金露花属);Skyflower,Sky-flower ●

139061 Duranta ellisia Jacq. = Duranta plumieri Jacq. ●

139062 Duranta erecta L.;假连翘(番仔刺,花墙刺,金露花,蓝花仔,篱笆树,莲荞,桐青,小本苦林盘,洋刺);Creeping Skyflower,Creeping Sky-flower,Golden Bead Tree,Golden Dew Drop Skyflower,Golden Dewdrop,Golden-dewdrop,Pigeon Berry,Pigeonberry,Pigeon-berry,Sky Flower,Skyflower ●

139063 Duranta erecta L.'Variegata';花叶假连翘●☆

139064 Duranta erecta L. = Duranta repens L. ●

139065 Duranta plumieri Jacq.;普氏假连翘(金露花);Golden Dewdrop,Pigeon-berry,Plumier Skyflower,Sky-flower ●

139066 Duranta plumieri Jacq. = Duranta erecta L. ●

139067 Duranta repens L.;金露花(匍匐山黑豆,小本苦林盘)●

139068 Duranta repens L.'Alba';白花假连翘(白花金露花,花叶假连翘);White-flower Creeping Skyflower ●☆

139069 Duranta repens L. = Duranta erecta L. ●

139070 Duranta repens L. = Duranta plumieri Jacq. ●

139071 Duranta repens L. f. alba (Mast.) Matsuda = Duranta repens L.'Alba' ●☆

139072 Duranta repens L. var. alba Domin = Duranta repens L.'Alba' ●☆

139073 Duranta repens L. var. variegata L. H. Bailey = Duranta erecta L.'Variegata' ●☆

139074 Duranta spinosa L. = Duranta plumieri Jacq. ●

139075 Duranta stenostachya Tod.;细花假连翘;Brazilian Sky Flower ●☆

139076 Durantaceae J. Agardh;假连翘科●

139077 Durantaceae J. Agardh = Verbenaceae J. St. -Hil.(保留科名)●■

139078 Durantia Scop. = Duranta L. ●

139079 Duravia (S. Watson) Greene = Polygonum L.(保留属名)■●

139080 Duravia Greene = Polygonum L.(保留属名)■●
139081 Duravia bolanderi (W. H. Brewer) Greene = Polygonum bolanderi W. H. Brewer ■☆
139082 Duravia californica (Meisn.) Greene = Polygonum californicum Meisn. ■☆
139083 Duretia Gaudich. = Boehmeria Jacq. ●
139084 Duria Scop. = Durio Adans. ●
139085 Duriala (R. H. Anderson) Ulbr. = Maireana Moq. ■●☆
139086 Durieua Boiss. et Reut. = Daucus L. ■
139087 Durieura Merat = Lafuentea Lag. ●☆
139088 Durieura Mérat et Diss. = Lafuentea Lag. ●☆
139089 Durio Adans.(1763);榴莲属(韶子属);Durian ●
139090 Durio carinatus Mast.;龙骨榴莲●☆
139091 Durio zibethinus DC. = Durio zibethinus Murray ●
139092 Durio zibethinus Murray;榴莲(毛荔枝,韶子);Civet Durian, Civet-cat Fruit, Durian, Durian-tree ●
139093 Durionaceae Cheek = Malvaceae Juss.(保留科名)●■
139094 Duroia L. f.(1782)(保留属名);杜氏茜属(杜鲁茜属)●☆
139095 Duroia hirsuta K. Schum.;粗毛杜氏茜(粗毛杜鲁茜)●☆
139096 Durringtonia R. J. F. Hend. et Guymer(1985);杜灵茜属☆
139097 Durringtonia paludosa R. J. F. Hend. et Guymer;沼泽杜灵茜☆
139098 Duschekia Opiz = Alnus Mill. ●
139099 Duschekia mandshurica (Callier ex C. K. Schneid.) Pouzar = Alnus mandshurica (Callier ex C. K. Schneid.) Hand. -Mazz. ●
139100 Duschekia maximowiczii (Callier) Pouzar = Alnus crispa (Aiton) Pursh subsp. maximowiczii (Callier) Hultén ●☆
139101 Duschekia viridis (Chaix) Opiz = Alnus viridis (Vill.) DC. subsp. crispa (Aiton) Turrill ●☆
139102 Dusenia O. Hoffm. = Duseniella K. Schum. ■☆
139103 Dusenia O. Hoffm. ex Dusén = Duseniella K. Schum. ■☆
139104 Duseniella K. Schum.(1902);钝菊属■☆
139105 Duseniella patagonica K. Schum.;钝菊■☆
139106 Dusenla O. Hoffm. ex Dusin(1900) = Duseniella K. Schum. ■☆
139107 Dussia Krug et Urb. = Dussia Krug et Urb. ex Taub. ■☆
139108 Dussia Krug et Urb. ex Taub.(1892);杜斯豆属(杜西豆属)■☆
139109 Dussia atropurpurea N. Zamora, R. T. Penn. et C. H. Stirt.;暗紫杜斯豆■☆
139110 Dussia discolor (Benth.) Amshoff;杂色杜斯豆■☆
139111 Dussia grandifrons I. M. Johnst.;大花杜斯豆■☆
139112 Dussia mexicana Harms;墨西哥杜斯豆■☆
139113 Dussia sanguinea Urb. et Ekman;血红杜斯豆■☆
139114 Dutailliopsis T. G. Hartley(1997);拟迪塔芸香属●☆
139115 Dutaillyea Baill.(1872);迪塔芸香属●☆
139116 Dutaillyea longipes Guillaumin;长梗迪塔芸香●☆
139117 Dutaillyea trifoliolata Baill.;迪塔芸香●☆
139118 Duthiastrum M. P. de Vos(1975);假毛蕊草属■☆
139119 Duthiastrum linifolium (E. Phillips) M. P. de Vos;假毛蕊草■☆
139120 Duthiea Hack.(1896);毛蕊草属;Duthiea ■
139121 Duthiea Hack. ex Procop. = Duthiea Hack. ■
139122 Duthiea brachypodia (P. Candargy) Keng et P. C. Keng;毛蕊草;Shortstalk Duthiea, Shortstipe Duthiea ■
139123 Duthiea bromoides Hack.;雀麦状毛蕊草;Brome-like Duthiea ■☆
139124 Duthiea dura (Keng) P. C. Keng = Duthiea brachypodia (P. Candargy) Keng et P. C. Keng ■
139125 Duthiea nepalensis Bor = Duthiea brachypodia (P. Candargy) Keng et P. C. Keng ■
139126 Duthiea oligostachya (Munro ex Aitch.) Stapf;少穗毛蕊草■☆
139127 Duthiella M. P. de Vos = Duthiastrum M. P. de Vos ■☆
139128 Duthiella linifolia (E. Phillips) M. P. de Vos = Duthiastrum linifolium (E. Phillips) M. P. de Vos ■☆
139129 Dutra Bernh. ex Steud. = Datura L. ●■
139130 Duttonia F. Muell.(1852) = Helipterum DC. ex Lindl. ■☆
139131 Duttonta F. Muell.(1856) = Pholidia R. Br. ●☆
139132 Duvalia Bonpl. = Hypocalyptus Thunb. ■☆
139133 Duvalia Haw.(1812);玉牛角属(小花犀角属);Duvalia ■☆
139134 Duvalia angustiloba N. E. Br.;司牛角;Narrowlobe Duvalia ■☆
139135 Duvalia caespitosa (Masson) Haw.;密簇玉牛角■☆
139136 Duvalia caespitosa (Masson) Haw. subsp. pubescens (N. E. Br.) Bruyns;毛密簇玉牛角■☆
139137 Duvalia caespitosa (Masson) Haw. subsp. vestita (Meve) Bruyns;包被玉牛角■☆
139138 Duvalia caespitosa (Masson) Haw. var. compacta (Haw.) Meve = Duvalia caespitosa (Masson) Haw. ■☆
139139 Duvalia compacta Haw. = Duvalia caespitosa (Masson) Haw. ■☆
139140 Duvalia concolor (Salm-Dyck) Schltr.;单色小花犀角■☆
139141 Duvalia corderoyi (Hook. f.) N. E. Br.;柯氏小花犀角■☆
139142 Duvalia corderoyi N. E. Br. = Duvalia corderoyi (Hook. f.) N. E. Br. ■☆
139143 Duvalia dentata N. E. Br. = Duvalia polita N. E. Br. ■☆
139144 Duvalia eilensis Lavranos;埃勒玉牛角■☆
139145 Duvalia elegans (Masson) Haw.;雅致玉牛角■☆
139146 Duvalia elegans (Masson) Haw. f. magnicorona A. C. White et B. Sloane = Duvalia elegans (Masson) Haw. ■☆
139147 Duvalia elegans (Masson) Haw. var. namaquana N. E. Br. = Duvalia caespitosa (Masson) Haw. subsp. pubescens (N. E. Br.) Bruyns ■☆
139148 Duvalia elegans (Masson) Haw. var. seminuda N. E. Br. = Duvalia elegans (Masson) Haw. ■☆
139149 Duvalia elegans Haw.;玉牛角(玉牛掌);Elegant Duvalia ■☆
139150 Duvalia emiliana A. C. White = Duvalia caespitosa (Masson) Haw. ■☆
139151 Duvalia glomerata Haw.;团集玉牛角■☆
139152 Duvalia gracilis Meve = Duvalia modesta N. E. Br. ■☆
139153 Duvalia hirtella (Jacq.) Sweet = Duvalia caespitosa (Masson) Haw. ■☆
139154 Duvalia hirtella (Jacq.) Sweet var. minor N. E. Br. = Duvalia caespitosa (Masson) Haw. ■☆
139155 Duvalia hirtella (Jacq.) Sweet var. obscura N. E. Br. = Duvalia caespitosa (Masson) Haw. ■☆
139156 Duvalia immaculata (C. A. Lückh.) Bayer ex L. C. Leach;无斑玉牛角■☆
139157 Duvalia jacquiniana (Schult.) Sweet = Duvalia elegans (Masson) Haw. ■☆
139158 Duvalia laevigata Haw.;光滑玉牛角■☆
139159 Duvalia maculata N. E. Br.;斑点玉牛角■☆
139160 Duvalia maculata N. E. Br. var. immaculata C. A. Lückh. = Duvalia immaculata (C. A. Lückh.) Bayer ex L. C. Leach ■☆
139161 Duvalia marlothii N. E. Br. = Duvalia caespitosa (Masson) Haw. ■☆
139162 Duvalia mastodes (Jacq.) Sweet = Duvalia caespitosa (Masson) Haw. ■☆
139163 Duvalia minuta Nel = Duvalia maculata N. E. Br. ■☆
139164 Duvalia modesta N. E. Br.;适度玉牛角■☆

139165 Duvalia parviflora N. E. Br. ;小花玉牛角■☆
139166 Duvalia pillansii N. E. Br. ;皮朗斯玉牛角■☆
139167 Duvalia pillansii N. E. Br. var. albanica ? = Duvalia pillansii N. E. Br. ■☆
139168 Duvalia polita N. E. Br. ;光泽玉牛角■☆
139169 Duvalia polita N. E. Br. f. intermedia A. C. White et B. Sloane = Duvalia polita N. E. Br. ■☆
139170 Duvalia polita N. E. Br. var. parviflora (L. Bolus) A. C. White et B. Sloane = Duvalia polita N. E. Br. ■☆
139171 Duvalia polita N. E. Br. var. transvaalensis (Schltr.) A. C. White et B. Sloane = Duvalia polita N. E. Br. ■☆
139172 Duvalia procumbens R. A. Dyer = Huernia procumbens (R. A. Dyer) L. C. Leach ■☆
139173 Duvalia propinqua A. Berger = Duvalia caespitosa (Masson) Haw. ■☆
139174 Duvalia pubescens N. E. Br. = Duvalia caespitosa (Masson) Haw. subsp. pubescens (N. E. Br.) Bruyns ■☆
139175 Duvalia pubescens N. E. Br. var. major ? = Duvalia caespitosa (Masson) Haw. subsp. pubescens (N. E. Br.) Bruyns ■☆
139176 Duvalia radiata (Sims) Haw. ;反曲玉牛角■☆
139177 Duvalia radiata (Sims) Haw. = Duvalia caespitosa (Masson) Haw. ■☆
139178 Duvalia radiata (Sims) Haw. var. hirtella (Jacq.) A. C. White et B. Sloane = Duvalia caespitosa (Masson) Haw. ■☆
139179 Duvalia radiata (Sims) Haw. var. minor (N. E. Br.) A. C. White et B. Sloane = Duvalia caespitosa (Masson) Haw. ■☆
139180 Duvalia radiata (Sims) Haw. var. obscura (N. E. Br.) A. C. White et B. Sloane = Duvalia caespitosa (Masson) Haw. ■☆
139181 Duvalia reclinata (Masson) Haw. = Duvalia caespitosa (Masson) Haw. ■☆
139182 Duvalia reclinata (Masson) Haw. var. angulata N. E. Br. = Duvalia caespitosa (Masson) Haw. ■☆
139183 Duvalia reclinata (Masson) Haw. var. bifida N. E. Br. = Duvalia caespitosa (Masson) Haw. ■☆
139184 Duvalia reclinata Haw. ;大花犀角■☆
139185 Duvalia replicata (Jacq.) Sweet = Duvalia caespitosa (Masson) Haw. ■☆
139186 Duvalia somalensis Lavranos = Duvalia sulcata N. E. Br. subsp. somalensis (Lavranos) Meve ■☆
139187 Duvalia sulcata N. E. Br. ;纵沟玉牛角■☆
139188 Duvalia sulcata N. E. Br. subsp. seminuda (Lavranos) Meve;半裸纵沟玉牛角■☆
139189 Duvalia sulcata N. E. Br. subsp. somalensis (Lavranos) Meve;索马里纵沟玉牛角■☆
139190 Duvalia sulcata N. E. Br. var. seminuda Lavranos = Duvalia sulcata N. E. Br. subsp. seminuda (Lavranos) Meve ■☆
139191 Duvalia tanganyikensis E. A. Bruce et P. R. O. Bally = Huernia tanganyikensis (E. A. Bruce et P. R. O. Bally) L. C. Leach ■☆
139192 Duvalia transvaalensis Schltr. = Duvalia polita N. E. Br. ■☆
139193 Duvalia transvaalensis Schltr. var. parviflora L. Bolus = Duvalia polita N. E. Br. ■☆
139194 Duvalia vestita Meve = Duvalia caespitosa (Masson) Haw. subsp. vestita (Meve) Bruyns ■☆
139195 Duvaliandra M. G. Gilbert(1980);迪瓦尔萝藦属■☆
139196 Duvaliandra dioscoridis (Lavranos) M. G. Gilbert;迪瓦尔萝藦■☆
139197 Duvaliella F. Heim = Dipterocarpus C. F. Gaertn. ●
139198 Duvaliella F. Heim(1892);假玉牛角属●☆
139199 Duvaliella problematica Heim;假玉牛角●☆
139200 Duvaljouvea Palla = Cyperus L. ■
139201 Duval-jouvea Palla = Cyperus L. ■
139202 Duvaua Kunth = Schinus L. ●
139203 Duvaucellia Bowdich(废弃属名) = Kohautia Cham. et Schltdl. ■☆
139204 Duvaucellia tenuis Bowdich = Kohautia tenuis (S. Bowdich) Mabb. ■☆
139205 Duvernaya Desv. ex DC. = Cuphea Adans. ex P. Browne ●■
139206 Duvernoia E. Mey. = Duvernoia E. Mey. ex Nees ●■☆
139207 Duvernoia E. Mey. ex Nees = Justicia L. ●■
139208 Duvernoia E. Mey. ex Nees(1847);枪木属●■☆
139209 Duvernoia Nees = Duvernoia E. Mey. ex Nees ●■☆
139210 Duvernoia aconitiflora A. Meeuse;柠檬枪木;Lemon Pistol Bush ●☆
139211 Duvernoia adhatodoides E. Mey. et Nees;枪木;Pistol Bush, Snake Bush ●☆
139212 Duvernoia andromeda Lindau = Adhatoda andromeda (Lindau) C. B. Clarke ■☆
139213 Duvernoia asystasioides Lindau = Justicia asystasioides (Lindau) M. E. Steiner ■☆
139214 Duvernoia bolomboensis De Wild. = Justicia bolomboensis De Wild. ■☆
139215 Duvernoia brevicaulis (S. Moore) Lindau = Justicia brevicaulis S. Moore ■☆
139216 Duvernoia buchholzii Lindau = Adhatoda buchholzii (Lindau) S. Moore ■☆
139217 Duvernoia chevalieri Lindau;舍瓦利耶枪木●☆
139218 Duvernoia claessensii De Wild. = Justicia claessensii De Wild. ■☆
139219 Duvernoia dewevrei De Wild. et T. Durand = Justicia extensa T. Anderson ■☆
139220 Duvernoia extensa (T. Anderson) Lindau = Justicia extensa T. Anderson ■☆
139221 Duvernoia interrupta Lindau = Justicia interrupta (Lindau) C. B. Clarke ■☆
139222 Duvernoia irumuensis Lindau = Justicia irumuensis (Lindau) Bamps et Champl. ■☆
139223 Duvernoia latibracteata De Wild. = Adhatoda buchholzii (Lindau) S. Moore ■☆
139224 Duvernoia maxima Lindau = Justicia maxima (Lindau) S. Moore ■☆
139225 Duvernoia orbicularis Lindau = Justicia orbicularis (Lindau) V. A. W. Graham ■☆
139226 Duvernoia paniculata (Benth.) Lindau = Justicia laxa T. Anderson ●☆
139227 Duvernoia pumila Lindau = Dicliptera pumila (Lindl.) Dandy ■☆
139228 Duvernoia pynaertii De Wild. ;皮那枪木●☆
139229 Duvernoia pyramidata Lindau = Justicia laxa T. Anderson ●☆
139230 Duvernoia robusta (C. B. Clarke) Lindau = Justicia baronii V. A. W. Graham ■☆
139231 Duvernoia robusta Lindau = Adhatoda camerunensis Heine ■☆
139232 Duvernoia salviiflora Lindau = Justicia salvioides Milne-Redh. ■☆
139233 Duvernoia somalensis Lindau = Justicia grisea C. B. Clarke ■☆
139234 Duvernoia speciosa Rendle = Justicia rendlei C. B. Clarke ■☆
139235 Duvernoia stachytarphetoides Lindau = Justicia stachytarphet-

oides (Lindau) C. B. Clarke ■☆
139236 Duvernoia stuhlmanni Lindau = Justicia extensa T. Anderson ■☆
139237 Duvernoia trichocalyx Lindau = Adhatoda densiflora (Hochst.) J. C. Manning ■☆
139238 Duvernoia verdickii De Wild.;韦尔枪木●☆
139239 Duvernoya E. Mey. = Duvernoia E. Mey. ex Nees ●■☆
139240 Duvernoya E. Mey. et Nees = Duvernoia E. Mey. ex Nees ●■☆
139241 Duvigneaudia J. Léonard(1959);迪维大戟属●☆
139242 Duvigneaudia inopinata (Prain) J. Léonard = Gymnanthes inopinata (Prain) Esser ●☆
139243 Duvigneaudia leonardii-crispi (J. Léonard) Kruijt et Roebers;迪维大戟●☆
139244 Duvoa Hook. et Arn. = Duvaua Kunth ●
139245 Duvoa Hook. et Arn. = Schinus L. ●
139246 Dyakia Christenson(1986);戴克兰属■☆
139247 Dyakia hendersoniana (Rchb. f.) Christenson;戴克兰■☆
139248 Dyanthus P. Browne = Dianthus L. ■
139249 Dybowskia Stapf = Hyparrhenia Andersson ex E. Fourn. ■
139250 Dybowskia dybowskii (Franch.) Dandy = Hyparrhenia dybowskii (Franch.) Roberty ■☆
139251 Dybowskia seretii (De Wild.) Stapf = Hyparrhenia dybowskii (Franch.) Roberty ■☆
139252 Dychotria Raf. = Psychotria L. (保留属名)●
139253 Dyckia Schult. et Schult. f. = Dyckia Schult. f. ■☆
139254 Dyckia Schult. f. (1830);雀舌兰属(狄克凤梨属,狄克氏花属,狄克亚属,狄克属,剑山属,小雀舌兰属,硬叶凤梨属);Dyckia ■☆
139255 Dyckia altissima Lindl.;高茎雀舌兰(高茎小雀舌兰,硬叶菠萝,硬叶雀舌兰);Tall Dyckia ■☆
139256 Dyckia brevifolia Baker;短叶雀舌兰(厚叶菠萝,小雀舌兰);Sawblade,Shortleaf Dyckia ■☆
139257 Dyckia fosteriana L. B. Sm.;花叶雀舌兰■☆
139258 Dyckia frigida Hook.;耐寒雀舌兰■☆
139259 Dyckia leptostachya Baker;细穗雀舌兰■☆
139260 Dyckia minarum Mez;刺缘雀舌兰■☆
139261 Dyckia rariflora Schult.;疏花雀舌兰(疏花小雀舌兰);Looseflower Dyckia ■☆
139262 Dyckia remotiflora Otto et A. Dietr.;细叶剑山■☆
139263 Dyckia sulphurea K. Koch;硫色雀舌兰(多浆凤梨)■☆
139264 Dyctioloma DC. = Dictyoloma A. Juss. (保留属名)●☆
139265 Dyctiospora Reinw. ex Korth. = Hedyotis L. (保留属名)●■
139266 Dyctisperma Raf. = Rubus L. ●■
139267 Dyctosperma H. Wendl. = Dictyosperma H. Wendl. et Drude ●☆
139268 Dydactylon Zoll. = Didactylon Zoll. et Moritzi ■
139269 Dydactylon Zoll. = Dimeria R. Br. ■
139270 Dyera Hook. f. (1882);竹桃木属(夹竹桃木属)●☆
139271 Dyera costulata (Miq.) Hook. f.;小脉竹桃木(胶桐,山丘胶桐,小脉夹竹桃木);Jelutong,Jelutong Bukit,Jelutong Paya ●☆
139272 Dyera costulata Hook. f. = Dyera costulata (Miq.) Hook. f. ●☆
139273 Dyera laxiflora Hook. f.;疏花竹桃木(疏花夹竹桃木)●☆
139274 Dyera lowii Hook. f.;婆罗洲竹桃木(婆罗洲夹竹桃木)●☆
139275 Dyera polyphylla (Miq.) Steenis;多叶竹桃木●☆
139276 Dyerella F. Heim = Vateria L. ●☆
139277 Dyerella F. Heim(1892);小竹桃属●☆
139278 Dyerella scabriuscula (Thwaites) F. Heim;小竹桃●☆
139279 Dyerocyeas Nakai = Cycas L. ●
139280 Dyerophytum Kuntze(1891);膜萼蓝雪花属●☆
139281 Dyerophytum africanum (Lam.) Kuntze;非洲膜萼蓝雪花■☆
139282 Dyllwinia Nees = Dillwynia Sm. ●☆
139283 Dymezewiczia Horan. = Zingiber Mill. (保留属名)■
139284 Dymondia Compton(1953);垫状灰毛菊属■☆
139285 Dymondia margaretae Compton;垫状灰毛菊■☆
139286 Dynamidium Fourr. = Potentilla L. ■●
139287 Dyneba Lag. = Dinebra Jacq. ■
139288 Dyospyros Dumort. = Diospyros L. ●
139289 Dyplecosia G. Don = Diplycosia Blume ●☆
139290 Dyplostylis H. karst. et Triana = Diplostylis H. karst. et Triana☆
139291 Dyplostylis H. karst. et Triana = Rochefortia Sw. ☆
139292 Dyplotaxis DC. = Diplotaxis DC. ■
139293 Dypontia Dietr. ex Steud. = Dupontia R. Br. ■☆
139294 Dypsidium Baill. (1894);假金果椰属(副戴普司桐属)●☆
139295 Dypsidium Baill. = Dypsis Noronha ex Mart. ●☆
139296 Dypsidium Baill. = Neophloga Baill. ●☆
139297 Dypsidium catatianum Baill. = Dypsis catatiana (Baill.) Beentje et J. Dransf. ●☆
139298 Dypsidium emirnense Baill. = Dypsis heterophylla Baker ●☆
139299 Dypsidium vilersianum Baill. = Dypsis heterophylla Baker ●☆
139300 Dypsis Noronha ex Mart. (1837);金果椰属(岱普椰子属,戴普司桐属,荻棕属,马岛椰属,拟散尾葵属,小竹椰子属);Dypsis ●☆
139301 Dypsis Noronha ex Thouars = Dypsis Noronha ex Mart. ●☆
139302 Dypsis acaulis J. Dransf. = Dypsis acaulis J. Dransf. et Beentje ●☆
139303 Dypsis acaulis J. Dransf. et Beentje;无茎金果椰(无茎狄棕)●☆
139304 Dypsis acuminum (Jum.) Beentje et J. Dransf.;渐尖金果椰●☆
139305 Dypsis albofarinosa Hodel et Marcus;白粉金果椰●☆
139306 Dypsis ambanjae Beentje;马达加斯加金果椰●☆
139307 Dypsis ambositrae Beentje;安布西特拉金果椰●☆
139308 Dypsis andapae Beentje;安达帕金果椰●☆
139309 Dypsis angusta Jum.;狭金果椰●☆
139310 Dypsis angustifolia (H. Perrier) Beentje et J. Dransf.;狭叶金果椰(狭叶狄棕)●☆
139311 Dypsis ankaizinensis (Jum.) Beentje et J. Dransf.;安凯济纳金果椰●☆
139312 Dypsis aquatilis Beentje;水生金果椰●☆
139313 Dypsis arenarum (Jum.) Beentje et J. Dransf.;沙地金果椰●☆
139314 Dypsis baronii (Becc.) Beentje et J. Dransf.;巴龙金果椰●☆
139315 Dypsis basilonga (Jum. et H. Perrier) Beentje et J. Dransf.;长基金果椰●☆
139316 Dypsis beentjei J. Dransf.;比特金果椰●☆
139317 Dypsis bernierana (Baill.) Beentje et J. Dransf.;伯尼尔金果椰●☆
139318 Dypsis boiviniana Becc. = Dypsis paludosa J. Dransf. ●☆
139319 Dypsis bosseri J. Dransf.;博瑟金果椰●☆
139320 Dypsis brevicaulis (Guillaumet) Beentje et J. Dransf.;短茎金果椰●☆
139321 Dypsis cabadae (H. E. Moore) Beentje et J. Dransf.;凤尾葵●☆
139322 Dypsis canaliculata (Jum.) Beentje et J. Dransf.;具沟金果椰●☆
139323 Dypsis canescens (Jum. et H. Perrier) Beentje et J. Dransf.;灰白金果椰●☆
139324 Dypsis catatiana (Baill.) Beentje et J. Dransf.;卡他金果椰●☆
139325 Dypsis caudata Beentje;尾状金果椰●☆
139326 Dypsis ceracea (Jum.) Beentje et J. Dransf.;蜡金果椰●☆
139327 Dypsis commersoniana (Baill.) Beentje et J. Dransf.;科梅逊

金果椰●☆
139328 Dypsis concinna Baker;整洁金果椰●☆
139329 Dypsis confusa Beentje;混乱金果椰●☆
139330 Dypsis cookei J. Dransf.;库克金果椰●☆
139331 Dypsis coriacea Beentje;革质金果椰●☆
139332 Dypsis corniculata (Becc.) Beentje et J. Dransf.;圆锥金果椰●☆
139333 Dypsis coursii Beentje;库尔斯金果椰●☆
139334 Dypsis crinita (Jum. et H. Perrier) Beentje et J. Dransf.;长软毛金果椰●☆
139335 Dypsis curtisii Baker;库特斯金果椰●☆
139336 Dypsis decaryi (Jum.) Beentje et J. Dransf.;三角椰;Three-cornered Palm,Triangle Palm ●☆
139337 Dypsis decipiens (Becc.) Beentje et J. Dransf.;美丽凤尾葵(鸳鸯椰子)●☆
139338 Dypsis delicatula Britt et J. Dransf.;姣美金果椰●☆
139339 Dypsis digitata (Becc.) Beentje et J. Dransf.;指裂金果椰●☆
139340 Dypsis dransfieldii Beentje;德兰斯菲尔德金果椰●☆
139341 Dypsis elegans Beentje;雅致金果椰●☆
139342 Dypsis eriostachys J. Dransf.;毛穗金果椰●☆
139343 Dypsis fasciculata Jum.;簇生金果椰●☆
139344 Dypsis fibrosa (C. H. Wright) Beentje et J. Dransf.;纤维质金果椰●☆
139345 Dypsis forficifolia Mart.;叉叶金果椰●☆
139346 Dypsis furcata J. Dransf.;叉分金果椰●☆
139347 Dypsis glabrescens (Becc.) Becc.;渐光金果椰●☆
139348 Dypsis gracilis Bory ex Mart. = Dypsis pinnatifrons Mart. ●☆
139349 Dypsis gracilis var. sambiranensis Jum. et H. Perrier = Dypsis pinnatifrons Mart. ●☆
139350 Dypsis henrici J. Dransf.,Beentje et Govaerts;昂里克金果椰●☆
139351 Dypsis heteromorpha (Jum.) Beentje et J. Dransf.;异形金果椰●☆
139352 Dypsis heterophylla Baker;互叶金果椰●☆
139353 Dypsis hildebrandtii (Baill.) Becc.;希尔德金果椰●☆
139354 Dypsis hirtula Mart. = Dypsis forficifolia Mart. ●☆
139355 Dypsis humbertii (Jum.) Beentje et J. Dransf. = Dypsis henrici J. Dransf.,Beentje et Govaerts ●☆
139356 Dypsis humbertii (Jum.) Beentje et J. Dransf. var. angustifolia H. Perrier = Dypsis angustifolia (H. Perrier) Beentje et J. Dransf. ●☆
139357 Dypsis integra (Jum.) Beentje et J. Dransf.;全缘金果椰●☆
139358 Dypsis intermedia Beentje;中间金果椰●☆
139359 Dypsis interrupta J. Dransf.;间断金果椰●☆
139360 Dypsis jumelleana Beentje et J. Dransf.;朱迈尔金果椰●☆
139361 Dypsis laevis J. Dransf.;平滑金果椰●☆
139362 Dypsis lantzeana Baill.;兰兹金果椰●☆
139363 Dypsis lantzeana Baill. var. simplicifrons Becc. = Dypsis lantzeana Baill. ●☆
139364 Dypsis lanuginosa J. Dransf.;多绵毛金果椰●☆
139365 Dypsis lastelliana (Baill.) Beentje et J. Dransf.;红茎金果椰(红茎凤尾葵)●☆
139366 Dypsis leptocheilos (Hodel) Beentje et J. Dransf.;红脖椰;Redneck Palm,Teddy Bear Palm ●☆
139367 Dypsis ligulata (Jum.) Beentje et J. Dransf.;舌状金果椰●☆
139368 Dypsis linearis (Becc.) Beentje et J. Dransf. = Dypsis procumbens (Jum. et H. Perrier) J. Dransf.,Beentje et Govaerts ●☆
139369 Dypsis littoralis Jum. = Dypsis forficifolia Mart. ●☆
139370 Dypsis lokohoensis J. Dransf.;洛克赫金果椰●☆
139371 Dypsis longipes Jum. = Dypsis procera Jum. ●☆
139372 Dypsis louvelii Jum. et Perrier;洛氏金果椰(洛氏狄棕)●☆
139373 Dypsis lucens (Jum.) Beentje et J. Dransf.;光亮金果椰●☆
139374 Dypsis lutea (Jum.) Beentje et J. Dransf.;黄金果椰●☆
139375 Dypsis lutescens (H. Wendl.) Beentje et J. Dransf. = Chrysalidocarpus lutescens H. Wendl. ●
139376 Dypsis madagascariensis (Becc.) Beentje et J. Dransf.;马达加斯加凤尾葵●☆
139377 Dypsis madageriensis Hort.;拟散尾葵●☆
139378 Dypsis mananarensis Jum. = Dypsis mocquerysiana (Becc.) Becc. ●☆
139379 Dypsis mananjarensis (Jum. et H. Perrier) Beentje et J. Dransf.;马南扎里金果椰●☆
139380 Dypsis mangorensis (Jum.) Beentje et J. Dransf.;曼古鲁金果椰●☆
139381 Dypsis marojejyi Beentje;马罗金果椰●☆
139382 Dypsis masoalensis Jum. = Dypsis forficifolia Mart. ●☆
139383 Dypsis minuta Beentje;微小金果椰●☆
139384 Dypsis mirabilis J. Dransf.;奇异金果椰●☆
139385 Dypsis mocquerysiana (Becc.) Becc.;莫克里斯金果椰●☆
139386 Dypsis monostachya Jum.;单穗金果椰●☆
139387 Dypsis montana (Jum.) Beentje et J. Dransf.;山地金果椰●☆
139388 Dypsis moorei Beentje;摩里金果椰(摩里凤尾葵)●☆
139389 Dypsis nauseosa (Jum. et H. Perrier) Beentje et J. Dransf.;臭金果椰●☆
139390 Dypsis nodifera Mart.;节状金果椰(单叶棕,节状凤尾葵)●☆
139391 Dypsis nossibensis (Becc.) Beentje et J. Dransf.;诺西波金果椰●☆
139392 Dypsis occidentalis (Jum.) Beentje et J. Dransf.;西方金果椰●☆
139393 Dypsis onilahensis (Jum. et H. Perrier) Beentje et J. Dransf.;乌尼拉希金果椰●☆
139394 Dypsis oreophila Beentje;喜山金果椰●☆
139395 Dypsis pachyramea J. Dransf.;粗金果椰●☆
139396 Dypsis paludosa J. Dransf.;沼泽金果椰●☆
139397 Dypsis pembana (H. E. Moore) Beentje et J. Dransf. = Chrysalidocarpus pembanus H. E. Moore ●☆
139398 Dypsis perrieri (Jum.) Beentje et J. Dransf.;佩氏金果椰(佩氏椰)●☆
139399 Dypsis pilulifera (Becc.) Beentje et J. Dransf.;小球状金果椰●☆
139400 Dypsis pinnatifrons Mart.;羽叶金果椰(迷人拟散尾葵)●☆
139401 Dypsis plurisecta Jum.;多裂金果椰●☆
139402 Dypsis poivreana (Baill.) Beentje et J. Dransf.;普瓦夫尔金果椰●☆
139403 Dypsis polystachya Baker = Dypsis nodifera Mart. ●☆
139404 Dypsis procera Jum.;高大金果椰●☆
139405 Dypsis procumbens (Jum. et H. Perrier) J. Dransf.,Beentje et Govaerts;平铺凤尾葵●☆
139406 Dypsis psammophila Beentje;喜沙金果椰●☆
139407 Dypsis pulchella J. Dransf.;美丽金果椰(美丽凤尾葵)●☆
139408 Dypsis pusilla Beentje;瘦小金果椰(微小凤尾葵)●☆
139409 Dypsis ramentacea J. Dransf.;小鳞片金果椰●☆
139410 Dypsis remotiflora J. Dransf.;疏花金果椰●☆
139411 Dypsis rhodotricha Baker = Dypsis heterophylla Baker ●☆
139412 Dypsis rivularis (Jum. et H. Perrier) Beentje et J. Dransf.;湿生金果椰(湿生凤尾葵)●☆

139413 Dypsis robusta Hodel, Marcus et J. Dransf.;粗壮金果椰●☆

139414 Dypsis sahonofensis (Jum. et H. Perrier) Beentje et J. Dransf.;萨霍诺夫金果椰●☆

139415 Dypsis sambiranensis (Jum. et H. Perrier) Jum. = Dypsis pinnatifrons Mart. ●☆

139416 Dypsis scandens J. Dransf.;攀缘金果椰●☆

139417 Dypsis schatzii Beentje;沙茨金果椰●☆

139418 Dypsis scottiana (Becc.) Beentje et J. Dransf.;司科特金果椰●☆

139419 Dypsis serpentina Beentje;蛇形金果椰●☆

139420 Dypsis singularis Beentje;单一金果椰●☆

139421 Dypsis spicata J. Dransf.;穗状金果椰●☆

139422 Dypsis tanalensis (Jum. et H. Perrier) Beentje et J. Dransf.;塔纳尔金果椰●☆

139423 Dypsis tenuissima Beentje;极细金果椰●☆

139424 Dypsis thermarum J. Dransf.;温泉金果椰●☆

139425 Dypsis thouarsiana Baill. = Vonitra thouarsiana (Baill.) Becc. ●☆

139426 Dypsis trapezoidea J. Dransf.;不等四边金果椰●☆

139427 Dypsis tsaratananensis (Jum.) Beentje et J. Dransf.;东风金果椰(东风凤尾葵)●☆

139428 Dypsis turkii J. Dransf.;图尔克金果椰●☆

139429 Dypsis utilis (Jum.) Beentje et J. Dransf.;有用金果椰●☆

139430 Dypsis viridis Jum.;绿金果椰●☆

139431 Dypterygta Gay = Asteriscium Cham. et Schltdl. ■☆

139432 Dypterygta Gay = Dipterygia C. Presl ex DC. ■☆

139433 Dyschoriste Nees(1832);安龙花属;Dyschoriste ■●

139434 Dyschoriste actinotricha Chiov. = Duosperma actinotrichum (Chiov.) Vollesen ■☆

139435 Dyschoriste adscendens (Hochst. ex Nees) Kuntze;上举安龙花■☆

139436 Dyschoriste alba S. Moore;白安龙花■☆

139437 Dyschoriste albiflora Lindau;白色安龙花■☆

139438 Dyschoriste angusta (A. Gray) Small;小安龙花;Dwarf Twinflower, Pineland Snakeherb ■☆

139439 Dyschoriste angustifolia Kuntze;狭叶安龙花■☆

139440 Dyschoriste bayensis Thulin;巴亚安龙花■☆

139441 Dyschoriste capitata Kuntze;头状安龙花■☆

139442 Dyschoriste cernua Nees = Dyschoriste erecta (Burm.) Kuntze ■

139443 Dyschoriste ciliata Kuntze;缘毛安龙花■☆

139444 Dyschoriste clarkei (Vatke) Benoist;克拉克安龙花■☆

139445 Dyschoriste clinopodioides Mildbr.;风轮菜安龙花■☆

139446 Dyschoriste cunenensis C. B. Clarke;库内内安龙花■☆

139447 Dyschoriste decora S. Moore;装饰安龙花■☆

139448 Dyschoriste decumbens E. A. Bruce = Dyschoriste procumbens E. A. Bruce ■☆

139449 Dyschoriste depressa (L.) Nees;凹陷安龙花■

139450 Dyschoriste depressa (Wall.) Nees = Dyschoriste depressa (L.) Nees ■

139451 Dyschoriste erecta (Burm.) Kuntze;直立安龙花■

139452 Dyschoriste fischeri Lindau;菲舍尔安龙花■

139453 Dyschoriste fleckii Schinz = Ruelliopsis damarensis S. Moore ■☆

139454 Dyschoriste fruticulosa (Rolfe) Chiov.;灌木状安龙花●☆

139455 Dyschoriste geniculata Nees = Ruellia geniculata (Nees) Benoist ■☆

139456 Dyschoriste gracilicaulis (Benoist) Benoist;细茎安龙花■

139457 Dyschoriste gracilis (Nees) Kuntze;纤细安龙花■

139458 Dyschoriste grandiflora H. S. Lo;川黔安龙花■

139459 Dyschoriste heudelotiana (Nees) Kuntze;厄德安龙花■☆

139460 Dyschoriste hildebrandtii (S. Moore) Lindau;希尔安龙花■☆

139461 Dyschoriste hildebrandtii (S. Moore) Lindau var. mollis S. Moore = Dyschoriste mollis (S. Moore) C. B. Clarke ■☆

139462 Dyschoriste hildebrandtii (S. Moore) Lindau var. occidentalis Benoist;西方希尔安龙花■☆

139463 Dyschoriste hispidula (Baker) Benoist;细毛安龙花■☆

139464 Dyschoriste hyssopifolia (Nees) Kuntze;神香草叶安龙花■☆

139465 Dyschoriste keniensis Malombe et Mwachala et Vollesen;肯尼亚安龙花■☆

139466 Dyschoriste keniensis Malombe et Mwachala et Vollesen subsp. glandulifera ?;腺体肯尼亚安龙花■☆

139467 Dyschoriste kilimandscharica (C. B. Clarke) Lindau = Duosperma kilimandscharicum (C. B. Clarke) Dayton ■☆

139468 Dyschoriste kyimbalensis (Lindau) S. Moore;基穆巴尔安龙花■☆

139469 Dyschoriste linearis (Torr. et A. Gray) Kuntze;细叶安龙花;Narrow-leaf Snake Herb ■☆

139470 Dyschoriste linifolia (T. Anderson) C. B. Clarke = Strobilanthopsis linifolia (T. Anderson ex C. B. Clarke) Milne-Redh. ●☆

139471 Dyschoriste lycioides Chiov.;枸杞状安龙花■☆

139472 Dyschoriste madagascariensis (Nees) Kuntze;马岛安龙花■☆

139473 Dyschoriste matopensis N. E. Br.;马托安龙花■☆

139474 Dyschoriste microphylla Kuntze;小叶安龙花■☆

139475 Dyschoriste mollis (S. Moore) C. B. Clarke;柔软安龙花■☆

139476 Dyschoriste monroi S. Moore;门罗安龙花■☆

139477 Dyschoriste multicaulis (T. Anderson) Kuntze;多茎安龙花■☆

139478 Dyschoriste mutica (S. Moore) C. B. Clarke;无尖安龙花■☆

139479 Dyschoriste nudanthera C. B. Clarke = Duosperma nudantherum (C. B. Clarke) Brummitt ■☆

139480 Dyschoriste nummulifolia Chiov.;铜钱安龙花■☆

139481 Dyschoriste nyassica Gilli;尼亚萨安龙花■☆

139482 Dyschoriste pedicellata C. B. Clarke = Dyschoriste heudelotiana (Nees) Kuntze ■☆

139483 Dyschoriste perrottetii (Nees) Kuntze;佩罗安龙花■☆

139484 Dyschoriste petalidioides S. Moore = Dyschoriste mutica (S. Moore) C. B. Clarke ■☆

139485 Dyschoriste pilifera Hutch.;纤毛安龙花■

139486 Dyschoriste principis Benoist;基出安龙花■

139487 Dyschoriste procumbens E. A. Bruce;平铺安龙花■☆

139488 Dyschoriste pseuderecta Mildbr.;假直立安龙花■☆

139489 Dyschoriste radicans Nees;辐射安龙花■☆

139490 Dyschoriste rogersii S. Moore;罗杰斯安龙花■☆

139491 Dyschoriste schiedeana (Nees) Kuntze var. decumbens (A. Gray) Henrickson;俯卧安龙花;Trailing Snake Herb ■☆

139492 Dyschoriste serpyllifolia (Nees) Benoist;百里香叶安龙花■☆

139493 Dyschoriste sessilifolia (Lindau) S. Moore;无柄叶安龙花■☆

139494 Dyschoriste sinica H. S. Lo;安龙花;China Dyschoriste, Chinese Dyschoriste ■

139495 Dyschoriste siphonantha (Nees) Kuntze;管花安龙花●☆

139496 Dyschoriste somalensis Rendle = Dyschoriste hildebrandtii (S. Moore) Lindau ■☆

139497 Dyschoriste subquadrangularis (Lindau) C. B. Clarke;四角安龙花●☆

139498 Dyschoriste tanganyikensis C. B. Clarke;坦噶尼喀安龙花■☆

139499 Dyschoriste tenera Lindau = Hygrophila tenera (Lindau) Heine ■☆
139500 Dyschoriste thunbergiiflora (S. Moore) Lindau;通贝里安龙花■☆
139501 Dyschoriste transvaalensis C. B. Clarke;德兰士瓦安龙花■☆
139502 Dyschoriste trichocalyx (Oliv.) Lindau;毛萼安龙花■☆
139503 Dyschoriste tubicalyx C. B. Clarke;管萼安龙花■☆
139504 Dyschoriste verdickii De Wild.;韦尔安龙花■☆
139505 Dyschoriste verticillaris (T. Anderson ex Oliv.) C. B. Clarke;轮生安龙花■☆
139506 Dyschoriste vestita Benoist;包被安龙花●☆
139507 Dyschoriste volkensii (Lindau) C. B. Clarke;福尔安龙花■☆
139508 Dyscritogyne R. M. King et H. Rob. (1971);腺籽修泽兰属(柄腺修泽兰属)■☆
139509 Dyscritogyne R. M. King et H. Rob. = Steviopsis R. M. King et H. Rob. ■☆
139510 Dyscritogyne adenosperma (Sch. Bip.) R. M. King et H. Rob.;腺籽修泽兰■☆
139511 Dyscritothamnus B. L. Rob. (1922);亮光菊属●☆
139512 Dyscritothamnus filifolius B. L. Rob.;亮光菊●☆
139513 Dysemone Sol. ex G. Forst. = Gunnera L. ■☆
139514 Dysinanthus DC. = Antennaria Gaertn. (保留属名)■●
139515 Dysinanthus DC. = Disynanthus Raf. ■●
139516 Dysinanthus Raf. ex DC. = Antennaria Gaertn. (保留属名)■●
139517 Dysinanthus Raf. ex DC. = Disynanthus Raf. ■●
139518 Dysmicodon (Endl.) Nutt. = Triodanis Raf. ■
139519 Dysmicodon Endl. = Triodanis Raf. ■
139520 Dysmicodon Nntt. = Legousia T. Durand ●■☆
139521 Dysmicodon californicum Nutt. = Triodanis biflora (Ruiz et Pav.) Greene ■
139522 Dysmicodon ovatum Nutt. = Triodanis biflora (Ruiz et Pav.) Greene ■
139523 Dysoda Lour. = Serissa Comm. ex Juss. ●
139524 Dysodia DC. = Dyssodia Cav. ■☆
139525 Dysodia Spreng. = Adenophyllum Pers. ■●☆
139526 Dysodia Spreng. = Dyssodia Cav. ■☆
139527 Dysodidendron Gardner = Saprosma Blume ●
139528 Dysodiopsis (A. Gray) Rydb. (1915);犬茴香属■☆
139529 Dysodiopsis A. Gray = Dysodiopsis (A. Gray) Rydb. ■☆
139530 Dysodiopsis Rydb. = Dysodiopsis (A. Gray) Rydb. ■☆
139531 Dysodiopsis tagetoides (Torr. et A. Gray) Rydb.;犬茴香■☆
139532 Dysodium Pers. = Melampodium L. ■●
139533 Dysodium Rich. = Melampodium L. ■●
139534 Dysodium Rich. ex Pers. = Melampodium L. ■●
139535 Dysodium divaricatum Rich. = Melampodium divaricatum (Rich.) DC. ■
139536 Dysolacoedeae Engl. = Olacaceae R. Br. (保留科名)●
139537 Dysolobium (Benth.) Prain(1897);镰瓣豆属(毛豇豆属,台豆属);Sicklelobe ●■
139538 Dysolobium Prain = Dysolobium (Benth.) Prain ●■
139539 Dysolobium grande (Benth.) Prain;镰瓣豆(台豆);Sicklelobe ●
139540 Dysolobium pilosum (Willd.) Maréchal = Vigna pilosa (Klein ex Willd.) Baker ex K. Heyne ■
139541 Dysophylla Blume = Pogostemon Desf. ●■
139542 Dysophylla Blume ex El Gazzar et Watson = Dysophylla El Gazzar et L. Watson ex Airy Shaw ■
139543 Dysophylla El Gazzar et L. Watson ex Airy Shaw = Eusteralis Raf. ●■
139544 Dysophylla El Gazzar et L. Watson ex Airy Shaw(1967);水蜡烛属(水珍珠草属,小珍珠菜属);Dysophylla, Watercandle ■
139545 Dysophylla auricularia (L.) Blume = Pogostemon auricularius (L.) Hassk. ■
139546 Dysophylla benthamiana Hance = Dysophylla stellata (Lour.) Benth. ■
139547 Dysophylla benthamiana Hance var. hainanensis C. Y. Wu et S. J. Hsuan = Dysophylla stellata (Lour.) Benth. ■
139548 Dysophylla benthamiana Hance var. hainanensis C. Y. Wu et S. J. Hsuan = Dysophylla stellata (Lour.) Benth. var. hainanensis (C. Y. Wu et S. J. Hsuan) C. Y. Wu et H. W. Li ■
139549 Dysophylla benthamiana Hance var. intermedia C. Y. Wu et S. J. Hsuan = Dysophylla stellata (Lour.) Benth. ■
139550 Dysophylla benthamiana Hance var. intermedia C. Y. Wu et S. J. Hsuan = Dysophylla stellata (Lour.) Benth. var. intermedia (C. Y. Wu et S. J. Hsuan) C. Y. Wu et H. W. Li ■
139551 Dysophylla communis Collett et Hemsl. = Elsholtzia communis (Collett et Hemsl.) Diels ■
139552 Dysophylla crassicaulis Benth.;粗茎水蜡烛;Thickstem Dysophylla ■☆
139553 Dysophylla crassicaulis Benth. = Dysophylla stellata (Lour.) Benth. ■
139554 Dysophylla crassicaulis Benth. = Eusteralis pumila Raf. ■☆
139555 Dysophylla crassicaulis Benth. var. pumila ? = Eusteralis pumila Raf. ■☆
139556 Dysophylla cruciata Benth.;毛茎水蜡烛;Cruciate Dysophylla, Cruciate Watercandle ■
139557 Dysophylla cruciata Benth. = Dysophylla stellata (Lour.) Benth. var. hainanensis (C. Y. Wu et S. J. Hsuan) C. Y. Wu et H. W. Li ■
139558 Dysophylla erecta Dalzell;直立水蜡烛;Erect Dysophylla ■☆
139559 Dysophylla esquirolii H. Lév. = Dysophylla stellata (Lour.) Benth. ■
139560 Dysophylla falcata C. Y. Wu = Pogostemon falcatus (C. Y. Wu) C. Y. Wu et H. W. Li ■
139561 Dysophylla gracilis Dalzell;细水蜡烛;Slender Dysophylla ■☆
139562 Dysophylla griffithii Hook. f.;格瑞氏水蜡烛;Griffith Dysophylla ■☆
139563 Dysophylla helferi Hook. f.;何佛水蜡烛;Helfer Dysophylla ■☆
139564 Dysophylla ianthina Maxim. ex Kanitz = Elsholtzia densa Benth. ■
139565 Dysophylla ianthina Maxim. ex Kanitz = Elsholtzia densa Benth. var. ianthina (Maxim. ex Kanitz) C. Y. Wu et S. C. Huang ■
139566 Dysophylla japonica Miq. = Dysophylla stellata (Lour.) Benth. ■
139567 Dysophylla japonica Miq. = Pogostemon stellatus (Lour.) Kuntze ■
139568 Dysophylla jatabeana Makino;雅氏水蜡烛■☆
139569 Dysophylla linearis Benth.;线叶水蜡烛;Linearleaf Dysophylla, Linearleaf Watercandle ■
139570 Dysophylla linearis Benth. = Dysophylla sampsonii Hance ■
139571 Dysophylla linearis Benth. var. yatabeana (Makino) Kudo = Dysophylla yatabeana Makino ■
139572 Dysophylla linearis Benth. var. yatabeana Kudo = Dysophylla yatabeana Makino ■
139573 Dysophylla linearis Benth. var. yatabeana Makino = Eusteralis yatabeana (Makino) Panigrahi ■
139574 Dysophylla lythroides Diels = Dysophylla yatabeana Makino ■
139575 Dysophylla mairei H. Lév. = Elsholtzia pilosa (Benth.) Benth. ■

139576 Dysophylla martini Vaniot = Dysophylla yatabeana Makino ■
139577 Dysophylla myosuroides Benth.；鼠尾草状水蜡烛；Myosuruslike Dysophylla ■☆
139578 Dysophylla peguana Prain = Dysophylla stellata (Lour.) Benth. var. hainanensis (C. Y. Wu et S. J. Hsuan) C. Y. Wu et H. W. Li ■
139579 Dysophylla pentagona C. B. Clarke ex Hook. f.；五棱水蜡烛；Pentagonous Dysophylla, Pentagonous Watercandle ■
139580 Dysophylla quadrifolia Benth.；四叶水蜡烛；Fourleaf Dysophylla ■☆
139581 Dysophylla ramosissima Benth. = Dysophylla stellata (Lour.) Benth. ■
139582 Dysophylla rugosa Hook. f.；皱纹水蜡烛；Rugose Dysophylla ■☆
139583 Dysophylla salicifolia Dalzell ex Hook. f.；柳叶水蜡烛；Willowleaf Dysophylla ■☆
139584 Dysophylla sampsonii Hance；齿茎水蜡烛(斑叶水虎尾，齿叶水蜡烛，蒋氏水蜡烛，森氏水珍珠菜，水芙蓉，水龙，野香芹)；Sampson Dysophylla, Sampson Watercandle ■
139585 Dysophylla stellata (Lour.) Benth.；水虎尾(边氏水珍珠菜，轮叶水蜡烛，水箭草，水老虎)；Stellate Dysophylla, Stellate Watercandle, Verticillate Dysophylla ■
139586 Dysophylla stellata (Lour.) Benth. = Eusteralis stellata (Lour.) Panigrahi ■
139587 Dysophylla stellata (Lour.) Benth. = Pogostemon stellatus (Lour.) Kuntze ■
139588 Dysophylla stellata (Lour.) Benth. var. hainanensis (C. Y. Wu et S. J. Hsuan) C. Y. Wu et H. W. Li；海南水虎尾；Hainan Dysophylla ■
139589 Dysophylla stellata (Lour.) Benth. var. hainanensis (C. Y. Wu et S. J. Hsuan) C. Y. Wu et H. W. Li = Dysophylla stellata (Lour.) Benth. ■
139590 Dysophylla stellata (Lour.) Benth. var. intermedia (C. Y. Wu et S. J. Hsuan) C. Y. Wu et H. W. Li；中间水虎尾(三姐妹)；Intermediate Dysophylla ■
139591 Dysophylla stellata (Lour.) Benth. var. intermedia (C. Y. Wu et S. J. Hsuan) C. Y. Wu et H. W. Li = Dysophylla stellata (Lour.) Benth. ■
139592 Dysophylla stocksii Hook. f.；斯托克水蜡烛；Stocks Dysophylla ■☆
139593 Dysophylla szemaoensis C. Y. Wu et S. J. Hsuan；思茅水蜡烛(思茅水珍珠菜)；Simao Dysophylla, Simao Watercandle, Szemao Dysophylla ■
139594 Dysophylla tetraphylla Wight = Dysophylla cruciata Benth. ■
139595 Dysophylla tisserantii Pellegr. = Pogostemon tisserantii (Pellegr.) J. -P. Lebrun et Stork ■☆
139596 Dysophylla tomentosa Dalzell；茸毛水蜡烛；Tomentose Dysophylla ■☆
139597 Dysophylla tsiangii Y. Z. Sun；蒋氏水蜡烛；Tsiang Dysophylla, Tsiang Watercandle ■
139598 Dysophylla tsiangii Y. Z. Sun ex C. H. Hu = Dysophylla sampsonii Hance ■
139599 Dysophylla verticillata (Roxb.) Benth. = Dysophylla stellata (Lour.) Benth. ■
139600 Dysophylla verticillata (Roxb.) Benth. = Eusteralis stellata (Lour.) Panigrahi ■
139601 Dysophylla verticillata (Roxb.) Benth. = Pogostemon stellatus (Lour.) Kuntze ■
139602 Dysophylla yatabeana Makino；水蜡烛；Common Dysophylla, Watercandle ■
139603 Dysophylla yatabeana Makino = Eusteralis yatabeana (Makino) Panigrahi ■
139604 Dysophylla yatabeana Makino = Pogostemon yatabeanus (Makino) Press ■
139605 Dysopsis Baill. (1858)；胡岛大戟属☆
139606 Dysosma Woodson (1928)；八角莲属(鬼臼属)；Dysosma, Manyflowered May-apple ■★
139607 Dysosma aurantiocaulis (Hand. -Mazz.) Hu；云南八角莲(八角莲，秕鳞八角莲，多花八角莲，七角莲)；Yunnan Dysosma, Yunnan Manyflowered May-apple ■
139608 Dysosma chengii (S. S. Chien) M. Hiroe；郑氏八角莲(八角金盘，白八角莲，六角莲，山荷叶)；Cheng Dysosma ■
139609 Dysosma chengii (S. S. Chien) M. Hiroe = Dysosma pleiantha (Hance) Woodson ■
139610 Dysosma delavayi (Franch.) Hu；西南八角莲(川八角莲，鬼臼，华鬼臼)；Delavay Dysosma ■
139611 Dysosma delavayi (Franch.) Hu = Dysosma veitchii (Hemsl. et E. H. Wilson) S. H. Fu ex T. S. Ying ■
139612 Dysosma difformis (Hemsl. et E. H. Wilson) T. H. Wang ex T. S. Ying；小八角莲(包袱莲，包袱七，单叶一枝花，荷花莲，铁骨散，药中王，一块砖)；Dwarf Manyflowered May-apple, Small Dysosma ■
139613 Dysosma emodii (Wall. ex Hook. f. et Thomson) M. Hiroe = Sinopodophyllum hexandrum (Royle) T. S. Ying ■
139614 Dysosma furfuracea S. Y. Bao = Dysosma aurantiocaulis (Hand. -Mazz.) Hu ■
139615 Dysosma guangxiensis Y. S. Wang = Dysosma majorensis (Gagnep.) T. S. Ying ■
139616 Dysosma hispida (K. S. Hao) Chun；毛八角莲(独叶一枝花，毛耳风)；Hispid Dysosma ■
139617 Dysosma hispida (K. S. Hao) M. Hiroe = Dysosma pleiantha (Hance) Woodson ■
139618 Dysosma lichuanensis Z. Zheng et Y. J. Su；乌云散(八角莲)；Lichuan Dysosma ■
139619 Dysosma lichuanensis Z. Zheng et Y. J. Su = Dysosma majoensis (Gagnep.) M. Hiroe ■
139620 Dysosma mairei (Gagnep.) M. Hiroe = Dysosma aurantiocaulis (Hand. -Mazz.) Hu ■
139621 Dysosma majoensis (Gagnep.) M. Hiroe；贵州八角莲(白八角莲，广西八角莲，梨子草，叶下花，叶子花)；Guangxi Dysosma, Guichow Manyflowered May-apple, Guizhou Dysosma ■
139622 Dysosma majoensis (Gagnep.) T. S. Ying = Dysosma majoensis (Gagnep.) M. Hiroe ■
139623 Dysosma pleiantha (Hance) Woodson；六角莲(八角兵盘七，八角金盘，八角莲，八角盘，八角七，川八角莲，独角莲，独脚莲，独叶一枝花，鬼臼，旱八角，金魁莲，马眼莲，山荷叶，叶下花，一把伞，一朵云，一碗水)；East-chinese Manyflowered May-apple, Sixangular Dysosma ■
139624 Dysosma tonkinensis (Gagnep.) M. Hiroe = Dysosma difformis (Hemsl. et E. H. Wilson) T. H. Wang ex T. S. Ying ■
139625 Dysosma tsayuense T. S. Ying；西藏八角莲；Tibetan Manyflowered May-apple, Xizang Dysosma ■
139626 Dysosma veitchii (Hemsl. et E. H. Wilson) L. K. Fu = Dysosma veitchii (Hemsl. et E. H. Wilson) S. H. Fu ex T. S. Ying ■
139627 Dysosma veitchii (Hemsl. et E. H. Wilson) L. K. Fu ex T. S. Ying = Dysosma delavayi (Franch.) Hu ■

139628 Dysosma veitchii (Hemsl. et E. H. Wilson) S. H. Fu ex T. S. Ying;川八角莲(八角金盘,八角乌,独脚莲,红八角莲,金盘,山荷花,五朵云,一把伞,银盘);Szechuan Manyflowered May-apple, Veitch Dysosma ■

139629 Dysosma versipelis (Hance) M. Cheng ex T. S. Ying;八角莲(八角金盘,八角镜,八角盘,八角乌,独荷草,独荷莲,独角莲,独脚莲,独叶一枝花,鬼臼,害母草,旱荷,红八角莲,江边一碗水,解毒,金星八角,九臼,爵犀,六角莲,马目毒公,马目公,千斤锤,山荷叶,术律草,天臼,窝儿七,羞寒花,羞天花,一碗水);Common Dysosma, Dysosma, Manyflowered May-apple ■

139630 Dysosma versipellis (Hance) M. Cheng = Dysosma versipelis (Hance) M. Cheng ex T. S. Ying ■

139631 Dysosmia (Korth.) Miq. = Mephitidia Reinw. ex Blume ●

139632 Dysosmia (Korth.) Miq. = Saprosma Blume ●

139633 Dysosmia Miq. = Saprosma Blume ●

139634 Dysosmia foetida (L.) M. Roem. = Passiflora foetida L. ■

139635 Dysosmon Raf. = Sesamum L. ■●

139636 Dysoxylon Bartl. = Dysoxylum Blume ●

139637 Dysoxylum Blume(1825);樫木属(臭楝属,葱臭木属,柽木属);Dysoxylum, Pencilwood, Pencil-wood ●

139638 Dysoxylum acutangulum Miq.;锐角樫木●☆

139639 Dysoxylum arborescens (Blume) Miq.;小叶樫木(兰屿柽木);Lanyu Pencil-wood ●

139640 Dysoxylum binectariferum (Roxb.) Hook. f. ex Bedd.;红果樫木(二蜜腺臭楝,红果葱臭木,红罗,山罗);Red-fruit Pencilwood, Red-fruited Pencil-wood ●

139641 Dysoxylum cumingianum C. DC.;兰屿柽木(孔氏樫木,兰屿樫木,邱氏樫木,台湾臭楝);Cuming Pencilwood, Cuming Pencil-wood, Lanyu Pencilwood ●

139642 Dysoxylum cupuliforme H. L. Li;杯萼樫木(杯萼葱臭木);Cup-calyx Pencilwood, Cupular Pencil-wood ●◇

139643 Dysoxylum densiflorum (Blume) Miq.;密花樫木;Dense-flower Pencilwood ●

139644 Dysoxylum excelsum Blume;樫木(葱臭木,高樫木);Common Pencil-wood, Pencilwood ●

139645 Dysoxylum filicifolium H. L. Li = Dysoxylum mollissimum Blume ●

139646 Dysoxylum forsteri C. DC. = Dysoxylum mollissimum Blume ●

139647 Dysoxylum fraseranum Benth. = Dysoxylum fraserianum (A. Juss.) Benth. ●☆

139648 Dysoxylum fraserianum (A. Juss.) Benth.;澳洲樫木;Australian Mahogany, Australian Rosewood ●☆

139649 Dysoxylum gobara (Buch.-Ham.) Merr. = Dysoxylum excelsum Blume ●

139650 Dysoxylum grandifolium H. L. Li = Dysoxylum binectariferum (Roxb.) Hook. f. ex Bedd. ●

139651 Dysoxylum hainanense Merr. = Dysoxylum mollissimum Blume ●

139652 Dysoxylum hainanense Merr. var. glaberrimum F. C. How et F. H. Chen = Dysoxylum mollissimum Blume var. glaberrimum (F. C. How et F. H. Chen) P. Y. Chen ●

139653 Dysoxylum hongkongense (Tutcher) Merr.;香港樫木(红果柽木,香港葱臭木,早望);Hongkong Pencilwood, Hongkong Pencil-wood ●

139654 Dysoxylum kanehirai (Sasaki) Kaneh. et Hatus.;今平樫木(金平樫木);Jinping Pencilwood ●

139655 Dysoxylum kusukusens (Hayata) Kaneh. et Hatus.;台湾樫木(大叶楝树,高士佛樫木,红果柽木,台湾溪桫);Big-leaved Dysoxylum, Henchum Pencil-wood, Hengchun Pencilwood, Taiwan Pencilwood ●

139656 Dysoxylum kusukusens (Hayata) Kaneh. et Hatus. = Dysoxylum hongkongense (Tutcher) Merr. ●

139657 Dysoxylum laxiracemosum C. Y. Wu et H. Li;总序樫木(总序葱臭木);Raceme Pencilwood, Racemose Pencilwood, Racemose Pencil-wood ●◇

139658 Dysoxylum lenticellatum C. Y. Wu;皮孔樫木(皮孔葱臭木,双凸樫木);Lenticel Pencilwood, Lenticellate Pencil-wood ●

139659 Dysoxylum leytense Merr. = Dysoxylum parasiticum (Osbeck) Kosterm. ●

139660 Dysoxylum loureiroi Pierre;檀香樫木●☆

139661 Dysoxylum lukii Merr.;多脉樫木(赤木,多脉葱臭木,陆氏樫木);Luk Pencil-wood, Many-veins Pencilwood, Nervose Pencilwood ●◇

139662 Dysoxylum lukii Merr. var. paucinervium F. C. How et F. H. Chen = Dysoxylum lukii Merr. ●◇

139663 Dysoxylum malabaricum Bedd. ex C. DC.;马拉巴尔樫木;Bombay White Cedar, Malabar Pencilwood ●

139664 Dysoxylum medogense C. Y. Wu et H. Li;墨脱樫木;Medog Pencil-wood, Motuo Pencilwood ●

139665 Dysoxylum mollissimum Blume;海南樫木(大蒜果树,大蒜头果);Hainan Pencilwood, Hainan Pencil-wood, Red Bean, Red Bean-tree ●

139666 Dysoxylum mollissimum Blume var. glaberrimum (F. C. How et T. C. Chen) P. Y. Chen;光叶海南樫木(光叶大蒜果);Glabrous Hainan Pencilwood ●

139667 Dysoxylum muelleri Benth. = Dysoxylum mollissimum Blume ●

139668 Dysoxylum oliganthum C. Y. Wu;少花樫木(黄果树,菁麻木,少花葱臭木);Fewflower Pencilwood, Few-flowered Pencil-wood, Poorflower Pencilwood ●

139669 Dysoxylum parasiticum (Osbeck) Kosterm.;大花樫木;Big-flower Pencilwood, Largeflower Pencilwood ●

139670 Dysoxylum procerum (Wall.) Hiern;臭楝;Tall Pencilwood ●

139671 Dysoxylum procerum (Wall.) Hiern = Dysoxylum excelsum Blume ●

139672 Dysoxylum richii (A. Gray) C. DC.;里奇樫木●☆

139673 Dysoxylum spectabile (A. Juss.) Hook. f.;壮观樫木;Kohekohe ●☆

139674 Dysoxylum spicatum H. L. Li;穗序葱臭木;Spicate Pencilwood, Spicate Pencil-wood ●◇

139675 Dysoxylum spicatum H. L. Li = Dysoxylum binectariferum (Roxb.) Hook. f. ex Bedd. ●

139676 Dyspemptemorion Bremek. = Justicia L. ●■

139677 Dysphania R. Br. (1810);刺藜属(澳藜属,土荆芥属,腺毛藜属)■

139678 Dysphania ambrosioides (L.) Mosyakin et Clemants;土荆芥(臭草,臭蒿,臭藜藿,臭苋,臭杏,鹅脚草,钩虫草,狗咬癀,红泽兰,虎骨香,火油根,藜荆芥,杀虫芥,虱子草,天仙草,香藜草);Ambrosia, American Wormseed, Californian Spearmint, Epazote, Indian Wormseed, Jerusalem Tea, Jesuit Tea, Mexican Grape Herb, Mexican Tea, Mexican-tea, Mexico Tea, Spanish Tea, Sweet Pigweed, Worm-seed, Wormseed Goose foot, Wormseed Goosefoot, Wormseed Tea, Wormseed Weed, Wormweed, Wormwood ■

139679 Dysphania ambrosioides (L.) Mosyakin et Clemants = Chenopodium ambrosioides L. ■

139680 Dysphania anthelmintica (L.) Mosyakin et Clemants;驱虫藜;Wormseed ■☆

139681 Dysphania aristata (L.) Mosyakin et Clemants;刺藜(刺穗藜,灯笼草,红小扫帚苗,鸡冠冠草,铁扫帚苗,野鸡冠子花,针尖藜);Aristate Goosefoot,Wormseed ■

139682 Dysphania botrys (L.) Mosyakin et Clemants;香藜(葡萄状澳藜,总状花藜);Ambrosia,Feather Geranium,Feathered Geranium,Feather-geranium,Goosefoot,Jerusalem Oak,Jerusalem Wormseed,Jerusalem-oak,Jerusalem-oak Goosefoot,Oak of Paradise,Oak-leaf Geranium,Sticky Goosefoot ■

139683 Dysphania botrys (L.) Mosyakin et Clemants = Chenopodium botrys L. ■

139684 Dysphania carinata (R. Br.) Mosyakin et Clemants;龙骨刺藜;Keeled Goosefoot ■☆

139685 Dysphania chilensis (Schrad.) Mosyakin et Clemants;智利刺藜■☆

139686 Dysphania cristata (F. Muell.) Mosyakin et Clemants;冠刺藜■☆

139687 Dysphania graveolens (Willd.) Mosyakin et Clemants;烈味刺藜■☆

139688 Dysphania incisum Poir. = Dysphania graveolens (Willd.) Mosyakin et Clemants ■☆

139689 Dysphania multifida (L.) Mosyakin et Clemants;多裂刺藜;Cut-leaf Goosefoot,Small-leaved Wormseed ■☆

139690 Dysphania plantaginella F. Muell.;矮花刺藜;Dwarf-flower ■☆

139691 Dysphania pumilio (R. Br.) Mosyakin et Clemants;侏儒刺藜■☆

139692 Dysphania pumilio (R. Br.) Mosyakin et Clemants = Chenopodium pumilio R. Br. ■☆

139693 Dysphania schraderiana (Roem. et Schult.) Mosyakin et Clemants;菊叶香藜(臭灰菜,蒿子草,菊叶刺藜,土荆芥,总状花藜);Daisyleaf Goosefoot,Foetid Goosefoot ■

139694 Dysphaniaceae Pax(1927)(保留科名);刺藜科(澳藜科)■

139695 Dysphaniaceae Pax(保留科名) = Amaranthaceae Juss.(保留科名)●■

139696 Dysphaniaceae Pax(保留科名) = Chenopodiaceae Vent.(保留科名)●■

139697 Dysphaniaceae Pax(保留科名) = Ebenaceae Gürke(保留科名)●

139698 Dyssapindaceae Radlk. = Sapindaceae Juss.(保留科名)●■

139699 Dyssochroma Miers = Markea Rich. ●☆

139700 Dyssodia Cav.(1801);异味菊属■☆

139701 Dyssodia Willd. = Adenophyllum Pers. ■●☆

139702 Dyssodia acerosa DC. = Thymophylla acerosa (DC.) Strother ●☆

139703 Dyssodia aurea (A. Gray) A. Nelson = Thymophylla aurea (A. Gray) Greene ■☆

139704 Dyssodia aurea (A. Gray) A. Nelson var. polychaeta (A. Gray) M. C. Johnst. = Thymophylla aurea (A. Gray) Greene var. polychaeta (A. Gray) Strother ■☆

139705 Dyssodia belenidium DC. = Thymophylla pentachaeta (DC.) Small var. belenidium (DC.) Strother ■☆

139706 Dyssodia concinna (A. Gray) B. L. Rob. = Thymophylla concinna (A. Gray) Strother ■☆

139707 Dyssodia greggii (A. Gray) B. L. Rob. = Thymophylla setifolia Lag. var. greggii (A. Gray) Strother ■☆

139708 Dyssodia hartwegii (A. Gray) B. L. Rob. = Thymophylla pentachaeta (DC.) Small var. hartwegii (A. Gray) Strother ■☆

139709 Dyssodia micropoides (DC.) Loes. = Thymophylla micropoides (DC.) Strother ■☆

139710 Dyssodia papposa (Vent.) Hitchc.;恶臭异味菊(异味菊);Dogweed,Fetid Marigold,Fetid-marigold,Stinking-marigold ■☆

139711 Dyssodia pentachaeta (DC.) B. L. Rob. = Thymophylla pentachaeta (DC.) Small ●■☆

139712 Dyssodia pentachaeta (DC.) B. L. Rob. var. belenidium (DC.) Strother = Thymophylla pentachaeta (DC.) Small var. belenidium (DC.) Strother ■☆

139713 Dyssodia pentachaeta (DC.) B. L. Rob. var. hartwegii (A. Gray) Strother = Thymophylla pentachaeta (DC.) Small var. hartwegii (A. Gray) Strother ■☆

139714 Dyssodia pentachaeta (DC.) B. L. Rob. var. puberula (Rydb.) Strother = Thymophylla pentachaeta (DC.) Small var. puberula (Rydb.) Strother ■☆

139715 Dyssodia polychaeta (A. Gray) B. L. Rob. = Thymophylla aurea (A. Gray) Greene var. polychaeta (A. Gray) Strother ■☆

139716 Dyssodia porophylla (Cav.) Cav.;孔叶异味菊(孔叶安龙花)■☆

139717 Dyssodia porophylloides A. Gray;孔叶菊状异味菊(孔叶菊状安龙花);San Felipe Dyssodia ■☆

139718 Dyssodia puberula (Rydb.) Standl. = Thymophylla pentachaeta (DC.) Small var. puberula (Rydb.) Strother ■☆

139719 Dyssodia setifolia (Lag.) B. L. Rob. = Thymophylla setifolia Lag. ■☆

139720 Dyssodia setifolia (Lag.) B. L. Rob. var. greggii (A. Gray) M. C. Johnst. = Thymophylla setifolia Lag. var. greggii (A. Gray) Strother ■☆

139721 Dyssodia tagetoides Torr. et A. Gray = Dysodiopsis tagetoides (Torr. et A. Gray) Rydb. ■☆

139722 Dyssodia tenuiloba (DC.) B. L. Rob. = Thymophylla tenuiloba (DC.) Small ■☆

139723 Dyssodia tenuiloba (DC.) B. L. Rob. var. texana (Cory) Strother = Thymophylla tenuiloba (DC.) Small var. texana (Cory) Strother ■☆

139724 Dyssodia tenuiloba (DC.) B. L. Rob. var. treculii (A. Gray) Strother = Thymophylla tenuiloba (DC.) Small var. treculii (A. Gray) Strother ■☆

139725 Dyssodia tenuiloba (DC.) B. L. Rob. var. wrightii (A. Gray) Strother = Thymophylla tenuiloba (DC.) Small var. wrightii (A. Gray) Strother ■☆

139726 Dyssodia tephroleuca S. F. Blake = Thymophylla tephroleuca (S. F. Blake) Strother ■☆

139727 Dyssodia texana Cory = Thymophylla tenuiloba (DC.) Small var. texana (Cory) Strother ■☆

139728 Dyssodia thurberi (A. Gray) B. L. Rob. = Thymophylla pentachaeta (DC.) Small var. belenidium (DC.) Strother ■☆

139729 Dyssodia treculii (A. Gray) B. L. Rob. = Thymophylla tenuiloba (DC.) Small var. treculii (A. Gray) Strother ■☆

139730 Dyssodia wrightii (A. Gray) B. L. Rob. = Thymophylla tenuiloba (DC.) Small var. wrightii (A. Gray) Strother ■☆

139731 Dystaenia Kitag.(1937);肖藁本属■☆

139732 Dystaenia Kitag. = Ligusticum L. ■

139733 Dystaenia ibukiensis (Y. Yabe) Kitag.;伊吹肖藁本■☆

139734 Dystaenia takeshimana (Nakai) Kitag.;竹岛肖藁本■☆

139735 Dystovomita (Engl.) D'Arcy(1979);热美藤黄属●☆

139736 Dystovomita D'Arcy = Dystovomita (Engl.) D'Arcy ●☆

139737 Dystovomita pittieri (Engl.) D'Arcy;热美藤黄●☆

139738 Dzieduszyckia Rehm. = Ruppia L. ■